Contents

ASM Handbook®

Volume 19
Fatigue and Fracture

Prepared under the direction of the
ASM International Handbook Committee

Steven R. Lampman, Technical Editor
Grace M. Davidson, Manager of Handbook Production
Faith Reidenbach, Chief Copy Editor
Randall L. Boring, Production Coordinator
Amy Hammel, Editorial Assistant
Scott D. Henry, Manager of Handbook Development
William W. Scott, Jr., Director of Technical Publications

Editorial Assistance
Nikki DiMatteo
Kathleen S. Dragolich
Kelly Ferjutz
Heather Lampman
Kathleen Mills
Mary Jane Riddlebaugh

ASM
INTERNATIONAL®
**The Materials
Information Society**

First printing, December 1996

This book is a collective effort involving hundreds of technical specialists. It brings together a wealth of information from world-wide sources to help scientists, engineers, and technicians solve current and long-range problems.

Great care is taken in the compilation and production of this Volume, but it should be made clear that NO WARRANTIES, EXPRESS OR IMPLIED, INCLUDING, WITHOUT LIMITA-TION, WARRANTIES OF MERCHANTABILITY OR FITNESS FOR A PARTICULAR PURPOSE, ARE GIVEN IN CONNECTION WITH THIS PUBLICATION. Although this information is believed to be accurate by ASM, ASM cannot guarantee that favorable results will be obtained from the use of this publication alone. This publication is intended for use by persons having technical skill, at their sole discretion and risk. Since the conditions of product or material use are outside of ASM's control, ASM assumes no liability or obligation in connection with any use of this information. No claim of any kind, whether as to products or information in this publication, and whether or not based on negligence, shall be greater in amount than the purchase price of this product or publication in respect of which damages are claimed. THE REMEDY HEREBY PROVIDED SHALL BE THE EXCLUSIVE AND SOLE REMEDY OF BUYER, AND IN NO EVENT SHALL EITHER PARTY BE LIABLE FOR SPECIAL, INDIRECT OR CONSEQUENTIAL DAMAGES WHETHER OR NOT CAUSED BY OR RESULTING FROM THE NEGLIGENCE OF SUCH PARTY. As with any material, evaluation of the material under enduse conditions prior to specification is essential. Therefore, specific testing under actual conditions is recommended.

Nothing contained in this book shall be construed as a grant of any right of manufacture, sale, use, or reproduction, in connection with any method, process, apparatus, product, composition, or system, whether or not covered by letters patent, copyright, or trademark, and nothing contained in this book shall be construed as a defense against any alleged infringement of letters patent, copyright, or trademark, or as a defense against liability for such infringement.

Comments, criticisms, and suggestions are invited, and should be forwarded to ASM International.

Library of Congress Cataloging-in-Publication Data

ASM International

ASM Handbook.
Fatigue and fracture / prepared under the direction of the
ASM International Handbook Committee.
Includes bibliographical references and index.
1. Fracture mechanics—Handbooks, manuals, etc.
2. Materials-Fatigue—Handbooks, manuals, etc.
I. ASM International. Handbook Committee.
II. ASM Handbook
TA409.F35 1996 620.1'126 96-47310
ISBN 0-87170-385-8
SAN 204-7586

ASM International®
Materials Park, OH 44073-0002

Foreword

The publication of this Volume marks the first time that the *ASM Handbook* series has dealt with fatigue and fracture as a distinct topic. Society members and engineers involved in the research, development, application, and analysis of engineering materials have had a long-standing interest and involvement with fatigue and fracture problems, and this reference book is intended to provide practical and comprehensive coverage of all aspects of these subjects.

Publication of *Fatigue and Fracture* also marks over 50 years of continuing progress in the development and application of modern fracture mechanics. Numerous Society members have been actively involved in this progress, which is typified by the seminal work of George Irwin ("Fracture Dynamics," *Fracturing of Metals*, ASM, 1948). Since that time period, fracture mechanics has become a vital engineering discipline that has been integrally involved in helping to prevent the failure of essentially all types of engineered structures.

Likewise, fatigue and crack growth have also become of primary importance to the development and use of advanced structural materials, and this Volume addresses the wide range of fundamental, as well as practical, issues involved with these disciplines.

We believe that our readers will find this Handbook useful, instructive, and informative at all levels. We also are especially grateful to the authors and reviewers who have made this work possible through their generous commitments of time and technical expertise. To these contributors we offer our special thanks.

William E. Quist
President
ASM International

Michael J. DeHaemer
Managing Director
ASM International

Policy on Units of Measure

By a resolution of its Board of Trustees, ASM International has adopted the practice of publishing data in both metric and customary U.S. units of measure. In preparing this Handbook, the editors have attempted to present data in metric units based primarily on Système International d'Unités (SI), with secondary mention of the corresponding values in customary U.S. units. The decision to use SI as the primary system of units was based on the aforementioned resolution of the Board of Trustees and the widespread use of metric units throughout the world.

For the most part, numerical engineering data in the text and in tables are presented in SI-based units with the customary U.S. equivalents in parentheses (text) or adjoining columns (tables). For example, pressure, stress, and strength are shown both in SI units, which are pascals (Pa) with a suitable prefix, and in customary U.S. units, which are pounds per square inch (psi). To save space, large values of psi have been converted to kips per square inch (ksi), where 1 ksi = 1000 psi. The metric tonne (kg × 10³) has sometimes been shown in megagrams (Mg). Some strictly scientific data are presented in SI units only.

To clarify some illustrations, only one set of units is presented on artwork. References in the accompanying text to data in the illustrations are presented in both SI-based and customary U.S. units. On graphs and charts, grids corresponding to SI-based units appear along the left and bottom edges. Where appropriate, corresponding customary U.S. units appear along the top and right edges.

Data pertaining to a specification published by a specification-writing group, may be given in only the units used in that specification or in dual units, depending on the nature of the data. For example, the typical yield strength of steel sheet made to a specification written in customary U.S. units would be presented in dual units, by the sheet thickness specified in that specification might be presented only in inches.

Data obtained according to standardized test methods for which the standard recommends a particular system of units are presented in the units of that system. Wherever feasible, equivalent units are also presented. Some statistical data may also be presented in only the original units used in the analysis.

Conversions and rounding have been done in accordance with ASTM Standard E 380, with attention given to the number of significant digits in the original data. For example, an annealing temperature of 1570 °F contains three significant digits. In this case, the equivalent temperature would be given as 855 °C; the exact conversion to 854.44 °C would not be appropriate. For an invariant physical phenomenon that occurs at a precise temperature (such as the melting of pure silver), it would be appropriate to report the temperature as 961.93 °C or 1763.5 °F. In some instances (especially in tables and data compilations), temperature values in °C and °F are alternatives rather than conversions.

The policy of units of measure in this Handbook contains several exceptions to strict conformance to ASTM E 380; in each instance, the exception has been made in an effort to improve the clarity of the Handbook. The most notable exception is the use of g/cm^3 rather than kg/m^3 as the unit of measure for density (mass per unit volume).

SI practice requires that only one virgule (diagonal) appear in units formed by combination of several basic units. Therefore, all of the units preceding the virgule are in the numerator and all units following the virgule are in the denominator of the expression; no parentheses are required to prevent ambiguity.

Preface

This volume of the *ASM Handbook* series, *Fatigue and Fracture*, marks the first separate Handbook on an important engineering topic of long-standing and continuing interest for both materials and mechanical engineers at many levels. Fatigue and fracture, like other forms of material degradation such as corrosion and wear, are common engineering concerns that often limit the life of engineering materials. This perhaps is illustrated best by the "Directory of Examples of Failure Analysis" contained in Volume 10 of the 8th Edition *Metals Handbook*. Over a third of all examples listed in that directory are fatigue failures, and well over half of all failures are related to fatigue, brittle fracture, or environmentally-assisted crack growth.

The title *Fatigue and Fracture* also represents the decision to include fracture mechanics as an integral part in characterizing and understanding not only ultimate fracture but also "subcritical" crack growth processes such as fatigue. The development and application of fracture mechanics has steadily progressed over the last 50 years and is a field of long-standing interest and involvement by ASM members. This perhaps is best typified by the seminal work of George Irwin in *Fracturing of Metals* (ASM, 1948), which is considered by many as the one of the key beginnings of modern fracture mechanics based from the foundations established by Griffith at the start of this century.

This Handbook has been designed as a resource for basic concepts, alloy property data, and the testing and analysis methods used to characterize the fatigue and fracture behavior of structural materials. The overall intent is to provide coverage for three types of readers: i) metallurgists and materials engineers who need general guidelines on the practical implications of fatigue and fracture in the selection, analysis or application structural materials; ii) mechanical engineers who need information on the relative performance and the mechanistic basis of fatigue and fracture resistance in materials; and iii) experts seeking advanced coverage on the scientific and engineering models of fatigue and fracture.

Major emphasis is placed on providing a multipurpose reference book for both materials and mechanical engineers with varying levels of expertise. For example, several articles address the basic concepts for making estimates of fatigue life, which is often necessary when data are not available for a particular alloy condition, product configuration, or stress conditions. This is further complemented with detailed coverage of fatigue and fracture properties of ferrous, nonferrous, and nonmetallic structural materials. Additional attention also is given to the statistical aspects of fatigue data, the planning and evaluation of fatigue tests, and the characterization of fatigue mechanisms and crack growth.

Fracture mechanics is also thoroughly covered in Section 4, from basic concepts to detailed applications for damage tolerance, life assessment, and failure analysis. The basic principles of fracture mechanics are introduced with a minimum of mathematics, followed by practical introductions on the fracture resistance of structural materials and the current methods and requirements for fracture toughness testing. Three authoritative articles further discuss the use of fracture mechanics in fracture control, damage tolerance analysis, and the determination of residual strength in metallic structures. Emphasis is placed on linear-elastic fracture mechanics, although the significance of elastic-plastic fracture mechanics is adequately addressed in these key articles.

Further coverage is devoted to practical applications and examples of fracture control in weldments, process piping, aircraft systems, failure analysis, and more advanced topics such as high-temperature crack growth and thermo-mechanical fatigue. Extensive fatigue and fracture property data are provided in Sections 5 through 7, and the Appendices include a detailed compilation of fatigue strength parameters and an updated summary of commonly used stress-intensity factors.

Once again, completion of this challenging project under the auspices of the Handbook Committee is made possible by the time and patience of authors who have contributed their work. Their efforts are greatly appreciated along with the guidance from reviewers and the Editorial Review Board.

S. Lampman
Technical Editor

Authors and Contributors

Peter Andresen
General Electric

Bruce Antolovich
Metallurgical Research Consultants, Inc.

Stephen D. Antolovich
Washington State University

S. Becker
NACO Technologies

C. Quinton Bowles
University of Missouri

David Broek
FractuREsearch

Robert Bucci
Alcoa Technical Center

David Cameron

G.F. Carpenter
NACO Technologies

Kwai S. Chan
Southwest Research Institute

Hans-Jürgen Christ
Universtät-GH-Siegen

Yip-Wah Chung
Northwestern University

Jack Crane

Jeff Crompton
Edison Welding Institute

David L. Davidson
Southwest Research Institute

S.D. Dimitrakis
University of Illinois, Urbana

Norman E. Dowling
Virginia Polytechnic Institute

Darle W. Dudley

Anthony G. Evans
Harvard University

Morris Fine
Northwestern University

Randall German
Pennsylvania State University

William A. Glaeser
Battelle

J. Karen Gregory
Technical University of Munich

Todd Gross
University of New Hampshire

Parmeet S. Grover
Georgia Institute of Technology

B. Carter Hamilton
Georgia Institute of Technology

Mark Hayes
The Centre for Spring Technology

David W. Hoeppner
University of Utah

Stephen J. Hudak, Jr.
Southwest Research Institute

R. Scott Hyde
Timken Research Center

R. Johansson
Avesta Sheffield AB

Steve Johnson
Georgia Institute of Technology

Tarsem Jutla
Caterpillar Inc.

Mitchell Kaplan
Willis and Kaplan Inc.

Gerhardus H. Koch
CC Technologies

George Krauss
Colorado School of Mines

John D. Landes
University of Tennessee

Ronald W. Landgraf
Virginia Polytechnic Institute

Fred Lawrence
University of Illinois, Urbana

Brian Leis
Battelle, Columbus

John Lewandowski
Case Western Reserve University

P.K. Liaw
University of Tennessee

John W. Lincoln
Wright Patterson Air Force Base

Alan Liu
Rockwell International Science Center (retired)

Petr Luká
Academy of Science of the Czech Republic

W.W. Maenning

David C. Maxwell
University of Dayton Research Institute

R. Craig McClung
Southwest Research Institute

David L. McDowell
Georgia Institute of Technology

Arthur J. McEvily
University of Connecticut

William J. Mills

M.R. Mitchell
Rockwell International Science Center

Charles Moyer
The Timken Company (retired)

Christopher L. Muhlstein
Georgia Institute of Technology

W.H. Munse
University of Illinois, Urbana

Ted Nicholas
University of Dayton Research Institute

Glenn Nordmark
Alcoa Technical Center (retired)

Richard Norris
Georgia Institue of Technology

Peter S. Pao
Naval Research Laboratory

C.C. "Buddy" Poe
NASA Langley Research Center

Srinivas Rao
Selectron Corporation

John O. Ratka
Brush Wellman

K.S. Ravichandran
University of Utah

H. Reemsnyder
Bethlehem Steel

Ted Reinhart
Boeing Commercial Airplane Group

Alan Rosenfield
Battelle, Columbus (retired)

Ashok Saxena
Georgia Institute of Technology

Jaap Schijve
Delft University of Technology

Huseyin Sehitoglu
University of Illinois, Urbana

Steven Shaffer
Battelle, Columbus

S. Shanmugham
University of Tennessee

E. Starke, Jr.
University of Virginia

Subra Suresh
Massachusetts Institute of Technology

Thomas Swift
Federal Aviation Administration

Robert Swindeman
Oak Ridge National Laboratory

Peter F. Timmins
Risk Based Inspection, Inc.

James Varner
Alfred University

Semyon Vaynman
Northwestern University

Paul S. Veers
Sandia National Laboratory

Lothar Wagner
Technical University Cottbus

Alexander D. Wilson
Lukens Steel

Timothy A. Wolff
Willis & Kaplan, Inc.

Aleksander Zubelewicz
IBM Microelectronics

Reviewers

Editorial Review Board

Reviewers

Contents

Section 1: Introduction

Industrial Significance of Fatigue Problems

David W. Hoeppner, Department of Mechanical Engineering, The University of Utah

THE DISCOVERY of fatigue occurred in the 1800s when several investigators in Europe observed that bridge and railroad components were cracking when subjected to repeated loading. As the century progressed and the use of metals expanded with the increasing use of machines, more and more failures of components subjected to repeated loads were recorded. By the mid 1800s A. Wohler (Ref 1) had proposed a method by which the failure of components from repeated loads could be mitigated, and in some cases eliminated. This method resulted in the stress-life response diagram approach and the component test model approach to fatigue design.

Undoubtedly, earlier failures from repeated loads had resulted in failures of components such as clay pipes, concrete structures, and wood structures, but the requirement for more machines made from metallic components in the late 1800s stimulated the need to develop design procedures that would prevent failures from repeated loads of all types of equipment. This activity was intensive from the mid-1800s and is still underway today. Even though much progress has been made, developing design procedures to prevent failure from the application of repeated loads is still a daunting task. It involves the interplay of several fields of knowledge, namely materials engineering, manufacturing engineering, structural analysis (including loads, stress, strain, and fracture mechanics analysis), nondestructive inspection and evaluation, reliability engineering, testing technology, field repair and maintenance, and holistic design procedures. All of these must be placed in a consistent design activity that may be referred to as a fatigue design policy. Obviously, if other time-related failure modes occur concomitantly with repeated loads and interact synergistically, then the task becomes even more challenging. Inasmuch as humans always desire to use more goods and place more demands on the things we can design and produce, the challenge of fatigue is always going to be with us.

Until the early part of the 1900s, not a great deal was known about the physical basis of fatigue. However, with the advent of an increased understanding of materials, which accelerated in the early 1900s, a great deal of knowledge has been developed about repeated load effects on engineering materials. The procedures that have evolved to deal with repeated loads in design can be reduced to four:

- The stress-life approach
- The strain-life approach
- The fatigue-crack propagation approach (part of a larger design activity that has become known as the damage-tolerant approach)
- The component test model approach

What is Fatigue?

Fatigue is a technical term that elicits a degree of curiosity. When citizens read or hear in their media of another fatigue failure, they wonder whether this has something to do with getting tired or "fatigued" as they know it. Such is not the case.

One way to explain fatigue is to refer to the ASTM standard definitions on fatigue, contained in ASTM E 1150. It is difficult, if not impossible, to carry on intelligent conversations if discussions on fatigue do not use a set of standard definitions such as E 1150. Within E 1150, there are over 75 terms defined, including the term *fatigue*: "fatigue (Note 1): the *process* of *progressive localized permanent structural change* occurring in a material subjected to conditions that produce *fluctuating stresses* and *strains* at some *point or points* and that may culminate in *cracks or complete fracture* after a sufficient number of fluctuations (Note 2). Note 1—In glass technology static tests of considerable duration are called 'static fatigue' tests, a type of test generally designated as stress-rupture. Note 2—Fluctuations may occur both in load and with time (frequency) as in the case of 'random vibration'." (Ref 2).

The words in italics (emphasis added) are viewed as key words in the definition. These words are important perspectives on the phenomenon of fatigue:

- Process
- Progressive
- Localized
- Permanent structural change
- Fluctuating stresses and strains
- Point or points
- Cracks or complete fracture

The idea that fatigue is a *process* is critical to dealing with it in design and to the characterization of materials as part of design. In fact, this idea is so critical that the entire conceptual view of fatigue is affected by it! Another critical idea is the idea of *fluctuating stresses and strains*. The need to have fluctuating (repeated or cyclic) stresses acting under either constant amplitude or variable amplitude is critical to fatigue. When a failure is analyzed and attributed to fatigue, the only thing known at that point is that the loads (the stresses/strains) were fluctuating. Nothing is necessarily known about the *nucleation* of damage that forms the origin of fatigue cracks.

Design for Fatigue Prevention

In design for fatigue and damage tolerance, one of two initial assumptions is often made about the state of the material. Both of these are related to the need to invoke continuum mechanics to make the stress/strain/fracture mechanics analysis tractable:

- The material is an ideal homogeneous, continuous, isotropic continuum that is free of defects or flaws.
- The material is an ideal homogeneous, isotropic continuum but contains an ideal crack-like discontinuity that may or may not be considered a defect or flaw, depending on the entire design approach.

The former assumption leads to either the stress-life or strain-life fatigue design approach.

These approaches are typically used to design for finite life or "infinite life." Under both assumptions, the material is considered to be free of defects, except insofar as the sampling procedure used to select material test specimens may "capture" the probable "defects" when the specimen locations are selected for fatigue tests. This often has proved to be an unreliable approach and has led, at least in part, to the damage-tolerant approach.

Another possible difficulty with these assumptions is that inspectability and detectability are not inherent parts of the original design approach. Rather, past and current experience guide field maintenance and inspection procedures, if and when they are considered.

The damage-tolerant approach is used to deal with the possibility that a crack-like discontinuity (or multiple ones) will escape detection in either the initial product release or field inspection practices. Therefore, it couples directly to nondestructive inspection (NDI) and evaluation (NDE). In addition, the potential for initiation of crack propagation must be considered an integral part of the design process, and the subcritical crack growth characteristics under monotonic, sustained, and cyclic loads must be incorporated in the design. The final instability parameter, such as plane strain fracture toughness (K_{Ic}), also must be incorporated in design. The damage-tolerant approach is based on the ability to track the damage throughout the entire life cycle of the component/system. It therefore requires extensive knowledge of the above issues, and it also requires that fracture (or damage) mechanics models be available to assist in the evaluation of potential behavior. As well, material characterization procedures are needed to ensure that valid evaluation of the required material "property" or response characteristic is made. NDI must be performed to ensure that probability-of-detection determinations are made for the NDI procedure(s) to be used. This approach has proved to be reliable, especially for safety-critical components.

The above approaches often are used in a complementary sense in fatigue design. The details of all three approaches are discussed in this Volume.

The *fatigue process* has proved to be very difficult to study. Nonetheless, extensive progress on understanding the phases of fatigue has been made in the last 100 years or so. It now is generally agreed that four distinct phases of fatigue may occur (Ref 3, 4):

- Nucleation
- Structurally dependent crack propagation (often called the "short crack" or "small crack" phase)
- Crack propagation that is characterizable by either linear elastic fracture mechanics, elastic-plastic fracture mechanics, or fully plastic fracture mechanics
- Final instability

Each of these phases is an extremely complex process (or may involve several processes) in and of itself. For example, the nucleation of "fatigue" cracks is extremely difficult to study, and even "pure fatigue" mechanisms can be very dependent on the intrinsic makeup of the material. Obviously, when one decides to pursue the nucleation of cracks in a material, one has already either assumed that the material is crack-free or has proved it! The assumption is the easier path and the one most often taken. When extraneous influences are involved in nucleation, such as temperature effects (e.g., creep), corrosion of all types, or fretting, the problem of modeling the damage is formidable. In recent years, more research has been done on the latter issues, and models for this phase of life are beginning to emerge.

Industrial Significance

There is little doubt that fatigue plays a significant role in all industrial design applications. Many components are subjected to some form of fluctuating stress/strain, and thus fatigue potentially plays a role in all such cases. However, it is still imperative that all designs consider those aspects of nucleation processes other than fatigue that may act to nucleate cracks that could propagate under the influence of cyclic loads. The intrinsic state of the material and all potential sources of cracks must also be evaluated.

Nonetheless, fatigue is a significant and often a critical factor in the testing, analysis, and design of engineering materials for machines, structures, aircraft, and power plants. An important engineering advance of this century is also the transfer of the multi-stage fatigue process from the field to the laboratory. In order to study, explain, and qualify component designs, or to conduct failure

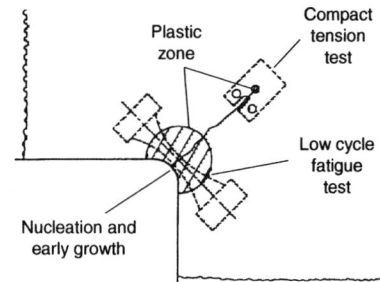

Fig. 1 Laboratory simulation of the multi-stage fatigue process. Source: Ref 5

analyses, a key engineering step is often the simulation of the problem in the laboratory. Any simulation is, of course, a compromise of what is practical to quantify, but the study of the multi-stage fatigue process has been greatly advanced by the combined methods of strain-control testing and the development fracture mechanics of fatigue crack growth rates. This combined approach (Fig. 1) is a key advance that allows better understanding and simulation of both crack nucleation in regions of localized strain and the subsequent crack growth mechanisms outside the plastic zone. This integration of fatigue and fracture mechanics has had important implications in many industrial applications for mechanical and materials engineering.

REFERENCES

1. A. Wohler, *Z. Bauw*, Vol 10, 1860, p 583
2. ASTM E 1150-1987, Standard Definitions of Fatigue, *1995 Annual Book of Standards*, ASTM, 1995, p 753-762
3. D.W. Hoeppner, Estimation of Component Life by Application of Fatigue Crack Growth Threshold Knowledge, *Fatigue, Creep, and Pressure Vessels for Elevated Temperature Service*, MPC-17, ASME, 1981, p 1-85
4. D.W. Hoeppner, Parameters That Input to Application of Damage Tolerant Concepts to Critical Engine Components invited keynote paper, *Damage Tolerance Concepts for Critical Engine Components*, AGARD-CP-393, NATO-AGARD, 1985
5. L.F. Coffin, Fatigue in Machines and Structures, *Fatigue and Microstructure*, American Society for Metals, 1979

Fracture and Structure

C. Quinton Bowles, University of Missouri-Columbia/Kansas City

IT IS DIFFICULT to identify exactly when the problems of failure of structural and mechanical equipment became of critical importance; however, it is clear that failures that cause loss of life have occurred for over 100 years (Ref 1, 2). Throughout the 1800s bridges fell and pressure vessels blew up, and in the late 1800s railroad accidents in the United Kingdom were continually reported as "The most serious railroad accident of the week"! Those in the United States also have heard the hair-raising stories of the Liberty ships built during World War II. Of 4694 ships considered in the final investigation, 24 sustained complete fracture of the strength deck, and 12 ships were either lost or broke in two. In this case, the need for tougher structural steel was even more critical because welded construction was used in shipbuilding instead of riveted plate. In riveted plate construction, a running crack must reinitiate every time it runs out of a plate. In contrast, a continuous path is available for brittle cracking in a welded structure, which is why low notch toughness is a more critical factor for long brittle cracks in welded ships.

Similar long brittle cracks are less likely or rare in riveted ships, which were predominant prior to welded construction. Nonetheless, even riveted ships have provided historical examples of long brittle fracture due, in part, from low toughness. In early 1995, for example, the material world was given the answer to an old question, *"What was the ultimate cause of the sinking of the Titanic?"* True, the ship hit an iceberg, but it now seems clear that because of brittle steel, "high in sulfur content even for its time" (Ref 3), an impact which would clearly have caused damage, perhaps would not have resulted in the ultimate separation of the Titanic in two pieces where it was found in 1985 by oceanographer Bob Ballard. During the undersea survey of the sunken vessel with Soviet Mir submersibles, a small piece of plate was retrieved from 12,612 feet below the ocean's surface. Examination by spectroscopy revealed a high sulfur content, and a Charpy impact test revealed the very brittle nature of the steel (Ref 3). However, there was some concern that the high sulfur content was, in some way, the result of eighty years on the ocean floor at 6,000 psi pressures. Subsequently, the son of a 1911 shipyard worker remembered a rivet hole

plug which his father had saved as a memento of his work on the Titanic. Analysis of the plug revealed the same level of sulfur exibited by the plate from the ocean floor. In the years following the loss of the Titanic metallurgists have become well aware of the detrimental effect of high sulfur content on fracture.

There are numerous other historical examples where material toughness was inadequate for design. The failures of cast iron rail steel for engine loads in the 1800s is one example. A large body of scientific folklore has arisen to explain structural material failures, almost certainly caused by a lack of tools to investigate the failures. The author was recently startled to read an article on the building of the Saint Lawrence seaway that described the effect of temperature on equipment: "The crawler pads of shovels and bulldozers subject to stress cracked and crumbled. Drive chains flew apart, cables snapped and fuel lines iced up...*And anything made of metal, especially cast metal, was liable to crystallize and break into pieces* (Ref 4). It is difficult to realize that there still exists a concept of metal crystallization as a result of deformation that in turn leads to failure. Clearly, the development of fluorescence and diffraction x-ray analysis, transmission and scanning electron microscopes, high-quality optical microscopy, and numerous other analytical instruments in the last 75 years has allowed further development of dislocation theory and clarification of the mechanisms of deformation and fracture at the atomic level.

During the postwar period, predictive models for fracture control also were pursued at the engineering level from the work of Griffith, Orowan, and Irwin. Since the paper of Griffith in 1920 (Ref 5, 6) and the extensions of his basic theory by Irwin (Ref 7) and others, we have come to realize that the design of structures and machines can no longer under all conditions be based on the elastic limit or yield strength. Griffith's basic theory is applicable to all fractures in which the energy required to make the new surfaces can be supplied from the store of energy available as potential energy, in the form of elastic strain energy. The elastic strain energy per unit of volume varies with the square of the stress, and hence increases rapidly with increases in the stress level. One does not need to go to very high stress

levels to store enough energy to drive a crack, even though this crack can be accompanied by considerable plastic deformation, and hence consume considerable energy. Thus, self-sustaining cracks can propagate at fairly low stress levels, a phenomenon that is briefly reviewed in this article along with the microstructural factors that influence toughness.

Fracture Behavior

In most structural failures, final fracture is usually abrupt after some sort of material or design flaw (such as a material defect, improper condition, or poor design detail) that is aggravated by a crack growth process that causes the crack to reach a critical size for final fracture. The cracking process occurs slowly over the service life from various crack growth mechanisms such as fatigue, stress-corrosion cracking, creep, and hydrogen-induced cracking. Each of these cracking mechanisms has certain characteristic features that are used in failure analysis to determine the cause of cracking or crack growth.

In contrast, the final fracture is usually abrupt and occurs from cleavage, rupture, or intergranular fracture (which may involve a combination of rupture and cleavage). Fracture mechanisms also are termed "ductile," although these terms must be defined on either a macroscopic or microscopic level. This distinction is important, because a fracture may be termed "brittle" from an engineering (macroscopic) perspective, while the underlying metallurgical (microscopic) mechanism could be termed either ductile or brittle. For metallurgists, cleavage is often referred to as brittle fracture and dimple rupture is considered ductile fracture. However, these terms must be used with caution, because many service failures occur by dimple rupture, even though most of these failures undergo very little overall (macroscopic) plastic deformation from an engineering point of view.

The majority of structural failures are of the more worrisome type, brittle fracture, and these almost invariably initiate at defects, notches, or discontinuities. Cracks resulting from machining, quenching, fatigue, hydrogen embrittlement, liquid-metal embrittlement, or stress corrosion also

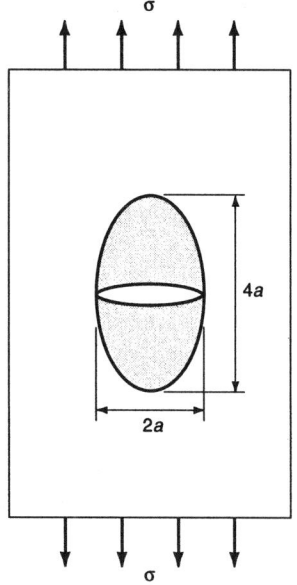

Fig. 1 Schematic illustration of the concept of energy release around a center crack in a loaded plate

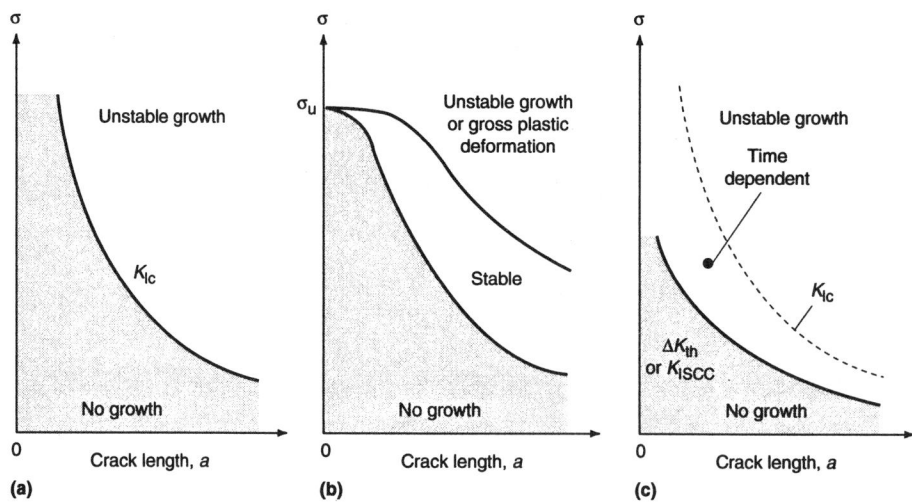

Fig. 2 Relationships between stress and crack length, showing regions and types of crack growth. (a) Linear-elastic. (b) Elastic-plastic. (c) Subcritical

lead to brittle fracture. In fact, the single most prevalent initiator of brittle fracture is the fatigue crack, which conservatively accounts for at least 50% of all brittle fractures in manufactured products by one account (Ref 8).

In contrast, service failure by macroscopic ductile failure is relatively infrequent (although the microscopic mechanisms of ductile fracture can ultimately lead to macroscopic brittle fracture). Typically, macroscopic ductile fracture occurs from overloads as a result of the part having been underdesigned (a term that includes the selection and heat treatment of the materials) for a specific set of service conditions, improperly fabricated, or fabricated from defective materials. Ductile fracture may also be the result of the part having been abused (that is, subjected to conditions of load and environment that exceeded those of the intended use).

This section briefly introduces the macroscopic and microscopic basis of understanding and modeling fracture resistance, while other articles in this Volume expand upon the microscopic and macroscopic basis of fatigue and fracture in engineering research and practice. More detailed information on the mechanisms of ductile and brittle fracture is given in the article "Micromechanisms of Monotonic and Cyclic Crack Growth" in this Volume.

Griffith Theory and the Specific Work of Fracture. The origins of modern fracture mechanics for engineering practice may be traced to Griffith (Ref 5, 6), who established an energy-release-rate criterion for brittle materials. Observations of the fracture strength of glass rods had shown that the longer the rod, the lower the strength. Thus the idea of a distribution of flaw sizes evolved, and it was discovered that the longer the rod, the larger the chance of finding a large natural flaw. This physical insight led to an instability criterion that considered the elastic en-

ergy released in a solid at the time a flaw grew catastrophically under an applied stress.

From the theory of elasticity comes the concept that the strain energy contained in an elastic body per unit volume is simply the area under the stress-strain curve, or:

$$U_0 = \frac{\sigma^2}{2E} \qquad (\text{Eq 1})$$

where σ is the applied stress and E is Young's modulus. However, there is a reduction (that is, a release) of energy in an elastic body containing a flaw or a crack because of the inability of the unloaded crack surfaces to support a load. We shall assume that the volume of material whose energy is released is the area of an elliptical region around the crack (as shown in Fig. 1) times the plate thickness, B; the volume is $\pi(2a) \cdot (a)B$. This is based on the area of an ellipse being $\pi r_a r_b$, where r_a and r_b are the major and minor radii of the ellipse. Then, the total energy released from the body due to the crack is the energy per unit volume times the volume, which is:

$$U = \pi(2a)\,(a)B\,\frac{\sigma^2}{2E} = \frac{\pi\sigma^2 a^2 B}{E} \qquad (\text{Eq 2})$$

In ideally brittle solids, the released energy can be offset only by the surface energy absorbed, which is:

$$W = (2aB)\,(2\gamma_s) = 4aB\gamma_s \qquad (\text{Eq 3})$$

where $2aB$ is the area of the crack and $2\gamma_s$ is twice the surface energy per unit area (because there are two crack surfaces).

Griffith's energy-balance criterion, in the simplest sense, is that crack growth will occur when the amount of energy released due to an increment of crack advance is larger than the amount of energy absorbed:

$$\frac{dU}{da} \geq \frac{dW}{da} \qquad (\text{Eq 4})$$

Performing the derivatives indicated in Eq 4 and rearranging gives the Griffith criterion for crack growth:

$$\sigma\sqrt{\pi a} = \sqrt{2E\gamma_s} \qquad (\text{Eq 5})$$

Fracture theory was built upon this criterion in the early 1940s by considering that the critical strain energy release rate, G_c, required for crack growth was equal to twice an effective surface energy, γ_{eff}:

$$G_c = 2\,\gamma_{\text{eff}} \qquad (\text{Eq 6})$$

This γ_{eff} is predominantly the plastic energy absorption around the crack tip, with only a small part due to the surface energy of the crack surfaces. Then, with the development of complex variable and numerical techniques to define the stress fields near cracks, this energy view was supplemented by stress concepts (i.e., the stress-intensity factor, K, and a critical value of K for crack growth, K_c). Replacing γ_s with γ_{eff} in Eq 5 and noting that the energy and stress concepts are essentially identical (that is, $K = \sqrt{EG}$) gives:

$$K_c = \sqrt{EG_c} = \sigma\sqrt{\pi a} \qquad (\text{Eq 7})$$

which is the crack-growth-criterion equivalent of Eq 1. Thus, K_c is the critical value of K that, when it is exceeded by a combination of applied stress and crack length, will lead to crack growth. For thick-plate plane-strain conditions, this critical value became known as the plane-strain fracture toughness, K_{Ic}, and any combination of applied stress and crack length that exceeds this value could produce unstable crack growth, as indicated schematically in Fig. 2(a) (linear-elastic). This forms the basis for understanding the relation between flaw size and fracture stress, which can be significantly lower than yield strengths, depending on crack length and geometry (Fig. 3).

In work with tougher, lower-strength materials, it was later noted that stable slow crack growth could occur even though accompanied by consid-

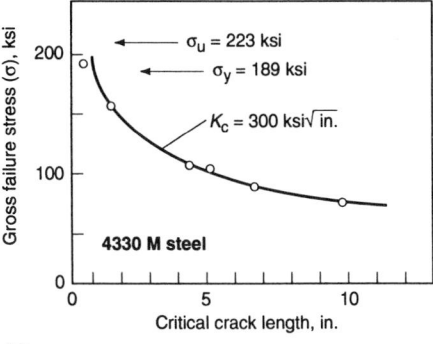

(a)

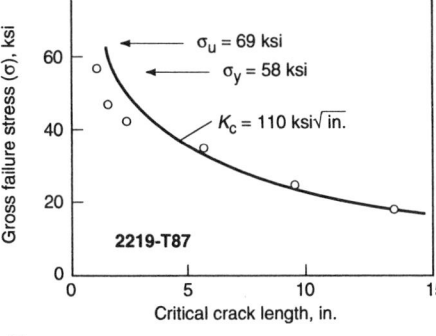

(b)

Fig. 3 Influence of crack length on gross failure stress for center-cracked plate. (a) Steel plate, 36 in. wide, 0.14 in. thick, room temperature, 4330 M steel, longitudinal direction. (b) Aluminum plate, 24 in. wide, 0.1 in. thick, room temperature, 2219-T87 aluminum alloy, longitudinal direction. Source: Ref 9

erable plastic deformation. Such phenomena led to the nonlinear J-integral and R-curve concepts, which can be used to predict the onset of stable slow crack growth and final instability under elastic-plastic conditions, as noted in Fig. 2(b). Finally, the fracture mechanics approach was applied to characterize subcritical crack growth phenomena where time-dependent slow crack growth, da/dt, or cyclic crack growth, da/dN, may be induced by special environments or fatigue loading. For combinations of stress and crack length above some environmental threshold, K_{Iscc}, or fatigue threshold, ΔK_{th}, subcritical growth occurs, as indicated in Fig. 2(c). These concepts form the macroscopic model of fracture for practical engineering use at the component level.

Microscopic Factors in Fracture. Although planar discontinuities (cracks) are the dominant defect in fracture, dislocation theory has been another avenue of research. Quite early in the study of materials and their failure, attempts were made to calculate the theoretical strength of crystals, but of all the possibilities perhaps that of Frenkel (Ref 10) for estimating the theoretical shear strength is most common. Theoretical (or "ideal") shear strength can be related to ductile fracture, because the shearing-off mechanism that is basic to shear lip formation in a tensile test and to the final shearing mode ("internal necking") occurs during void coalescence. However,

for cleavage (brittle fracture, which is by far the most worrisome type of fracture), the corresponding ideal strength is the ideal tensile strength first estimated by Orowan in 1949 and described by Kelly in Ref 11.

The estimate of Frenkel considers two rows of atoms that shear past one another. The spacing between rows is a_r and the spacing between atoms in the slip direction is a_0. The shear stress is σ and is considered to be sinusoidal. The well-known result is:

$$\sigma = b/a(\mu/2\pi)\sin(2\pi x/b) \qquad \text{(Eq 8)}$$

where μ = shear modulus. The maximum value, which is also the point at which the lattice is mechanically unstable and slip occurs, is $\sigma = b/a(\mu/2\pi)$. Because $a \approx b$, the theoretical shear strength is $\sigma_{theo} \approx \mu/2\pi$, which is several orders of magnitude greater than the value usually observed for soft crystals. There have been numerous variations and improvements to Eq 8 in an effort to improve predictions of material strength, but the result remains essentially the same. Unfortunately, the strength of a given material predicted by theoretical calculations is much larger than the observed strength. The question is "why?" Certainly it is important that slip in crystals occurs well below the ultimate stress and that slip occurs by the movement of dislocations, as postulated by Taylor (Ref 12), Orowan (Ref 13), and Polanyi (Ref 14). But these observations do not completely answer the question, and we are led to search for other reasons for weakness.

In looking for points of weakness, we begin by noting that pure metals by definition contain no alloying constituents (and may be single crystals or polycrystalline), while structurally useful materials generally contain alloying constituents for strengthening and may be precipitation hardening, such as many of the aluminum alloys, but may also contain larger second-phase particles. Structural metals may also contain multiple phases, such as the ferrous alloys do, and have grain boundary phases as well as phases within the grain interior. A method that has been used to classify materials as to their mode of failure is that of structure. Shown below are some material properties and their effect on fracture behavior (Ref 15):

Physical property	Increasing tendency for brittle fracture
Electron bond	Metallic → Ionic → Covalent
Crystal structure	Close-packed crystals → Low-symmetry crystals
Degree of order	Random solid → Short-range order → Long-range order solution

For the different classes of materials, crystal structure is of fundamental importance because it influences or determines the competition between flow and fracture. For example, polycrystals of copper are invariably ductile, while magnesium polycrystals are relatively brittle. Magnesium has a close-packed hexagonal crystal structure, with parameters of a = 3.202 Å, c = 5.199 Å, and c/a = 1.624 Å (which is very close to the ratio of 1.633 obtained by piling spheres in

the same arrangement). This structure is basic to much of the physical metallurgy of magnesium and magnesium alloys. At room temperature, slip occurs mainly on (0001) (<11$\bar{2}$0>), with a small amount sometimes seen on pyramidal planes such as (10$\bar{1}$1) <11$\bar{2}$0>. As the temperature is raised, pyramidal slip becomes easier and more prevalent. However, note that the slip directions, whether associated with basal or the pyramidal planes, are coplanar with (0001), a general observation for all observed slip in magnesium and magnesium alloys. Therefore, it is impossible for a polycrystalline piece of magnesium to deform without cracking unless deformation mechanisms other than slip are available. These mechanisms are twinning, banding, and grain-boundary deformation.

At the microstructural level, fracture in engineering alloys can occur by a transgranular (through the grains) or an intergranular (along the grain boundaries) fracture path. However, regardless of the fracture path, there are essentially only four principal fracture modes:

- Ductile fracture from microvoid coalescence
- Brittle fracture from cleavage, intergranular fracture, and crazing (in the case of polymers)
- Fatigue
- Decohesive rupture

These basic fracture modes are discussed in more detail in the article "Micromechanisms of Monotonic and Cyclic Crack Growth" in this Volume (with somewhat more emphasis on cleavage than in this article). Cleavage is perhaps more related to the rapidity of fracturing, as suggested by Irwin's classic paper (Ref 7).

Four major types of failure modes have also been extensively discussed in the literature. A list of classes and the associated modes of failure is shown below (Ref 16):

Dimpled rupture (microvoid coalescence):

- Ductile fracture
- Overload fracture

Ductile striation formation

- Fatigue cracking (subcritical growth)

Cleavage or quasicleavage

- Brittle fracture
- Premature or overload failure
- Quasicleavage from hydrogen embrittlement

Intergranular failure

- Grain boundary embrittlement (by segregation or precipitation)
- Subcritical growth under sustained load (stress-corrosion cracking or hydrogen embrittlement)

A study of these fracture classes normally requires use of the scanning electron microscope or the preparation of replicas that may be examined

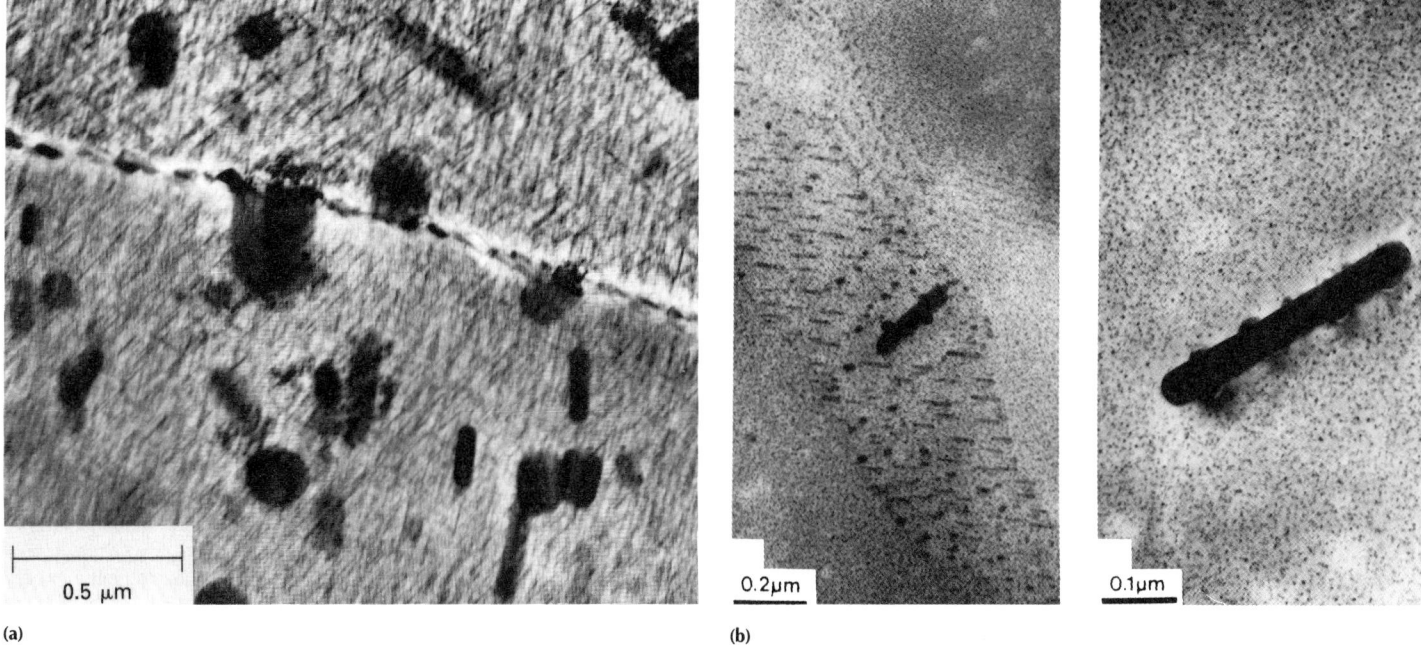

Fig. 4 (a) Platelet formation in a 2*xxx*-series aluminum alloy that was solution heat treated, quenched, cold rolled 6%, and aged 12 h at 190 °C. (b) Spheroidal precipitates in a 7*xxx*-series aluminum alloy. Larger precipitates are seen in the subgrain boundary as well as around the dispersoid particle. Source: *Aluminum: Properties and Physical Metallurgy*, J.E. Hatch, Ed., American Society for Metals, 1984, p 101, 191

in the transmission electron microscope. In some instances it is possible to examine cracked inclusions and second-phase particles using thin foil transmission electron microscopy. An example of this latter behavior can be found in the work of Broek (Ref 17). Optical microscopy can also be of use for examining large inclusion particles.

Precipitation-Hardening Alloys

Precipitation-hardening alloys, such as those of aluminum, can be expected to have dispersed fine precipitates that may range from spherical to platelet, depending on the alloy (Fig. 4a, b). The precipitates may be extremely small and primarily produce lattice strain, such as the case of Guinier-Preston zones, or they may be somewhat larger but still have coherent boundaries with the matrix, as in the case of peak-aged alloys, or be in the overaged condition, which usually results in incoherent boundaries. Precipitates are generally impediments to dislocation motion and therefore tend to raise both the yield strength and the ultimate strength.

Problems begin to arise when laboratory alloys are scaled to commercial production levels. Levels of alloy additions are more difficult to control, and the purity of starting materials can be almost impossible to maintain. As a result, in addition to precipitates there may be larger second-phase particles in the grain interior or the grain boundary (Fig. 4a, b). These particles, which are also called constituent particles, are assumed to be directly related to dimple rupture and are usually

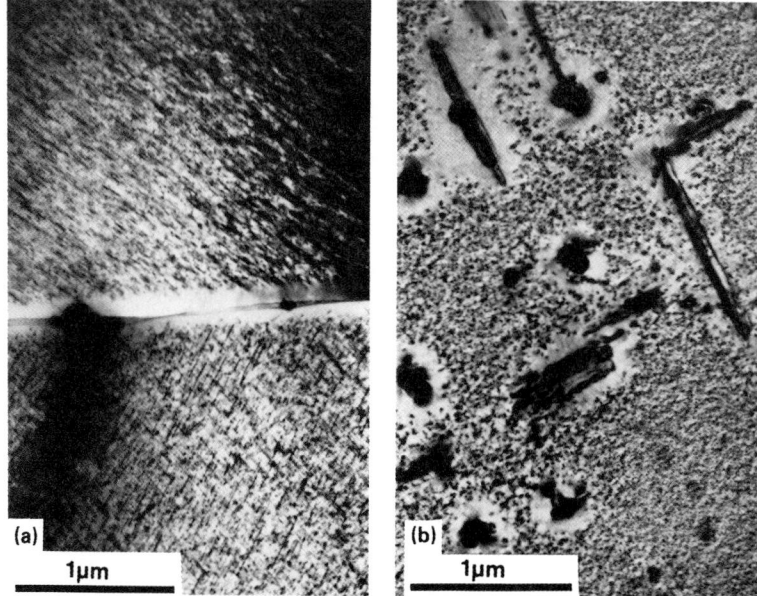

Fig. 5 (a) Precipitate-free zones or denuded zones at a grain boundary in a 6*xxx*-series alloy. (b) Similar denuded regions around dispersoid or constituent particles in a 7*xxx*-series alloy. Source: *Aluminum: Properties and Physical Metallurgy*, J.E. Hatch, Ed., American Society for Metals, 1984, p 102

observed in the bottom of the dimple in fractographs. Finally, there may be denuded zones at grain boundaries that are devoid of precipitates and constituent particles (Fig. 4a, 5a, 5b). These denuded zones may also exist around second-phase particles.

It is certainly possible to consider failure as occurring during tensile overload and general tensile yield. However, we are also interested in failure resulting from an initial fatigue crack that is formed by cyclic loading, followed by crack growth and then final failure. These failures generally result in a more localized plastic deformation behavior that is governed by linear elastic fracture mechanics. With this scenario in mind, Grosskreutz and Shaw (Ref 18), Bowles and Schijve (Ref 19), and McEvily and Boettner (Ref 20), among others, have shown that fatigue cracks generally initiate at larger inclusions that are still larger than the usual second-phase particle. An example of this behavior is shown in Fig. 6. As critical crack lengths are approached, dim-

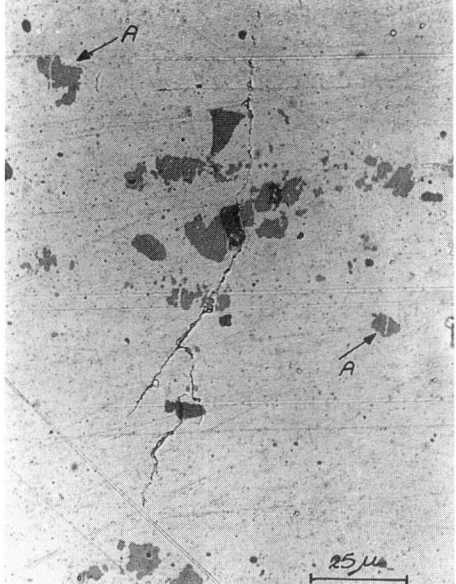

(a)

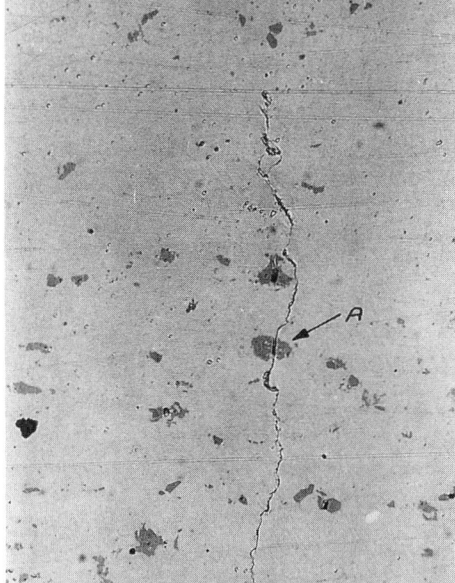

(b)

Fig. 6 Two examples of cracks initiated at inclusions. In figure (a) the crack clearly initiated at a void occurring in a cracked inclusion cluster. In figure (b) the crack appears to have initiated from the side of the inclusion. Cracks were observed after 150,000 cycles. Material was 2024-T3.

ple rupture begins. Numerous examples of dimple rupture have been published, but Broek (Ref 17, 21) was probably the first to demonstrate clearly that void formation begins at the matrix-precipitate or matrix-constituent particle interface and is followed by a linking of other dimples by a mechanism of interface separation leading to final fracture.

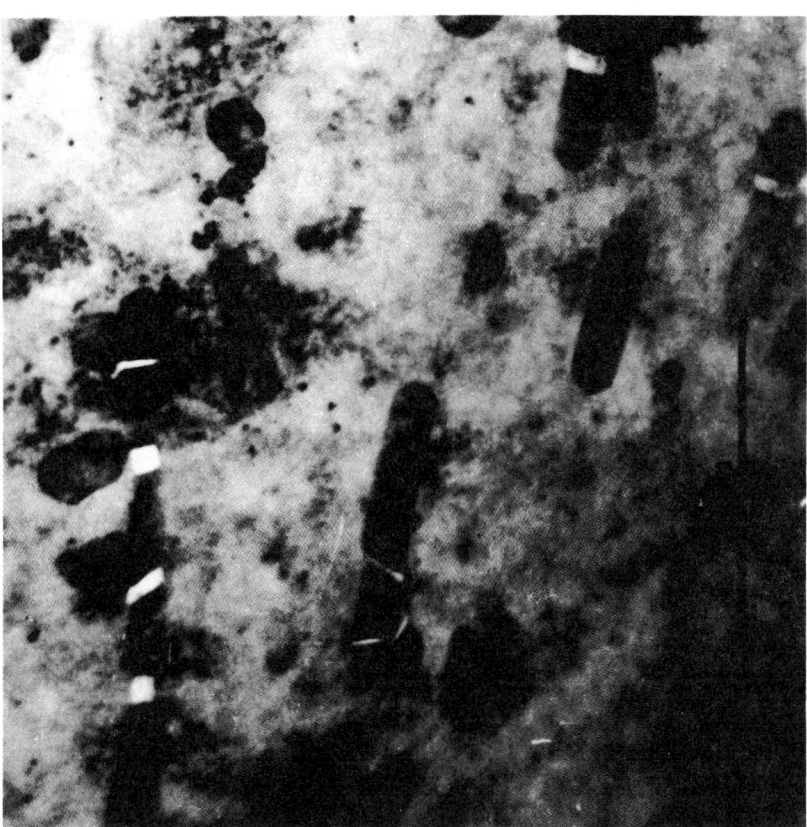

Fig. 7 Transmission electron micrograph of thin foil of an aluminum alloy. Fractured elongated dispersoids can be clearly seen, along with one or two possible interface separations that led to voids. Courtesy of Martinus Nijhoff Publishers. Source: Ref 21

Although fatigue crack initiation in commercial alloys begins at the inclusion-matrix interface, it has been demonstrated that many larger particles are broken during fabrication processes such as forging or plate rolling, or the final stretching that may be part of the heat treatment process. Larger particles can also be broken under tensile loading (Ref 16, 22). Broek (Ref 17) has also observed by means of thin foil electron microscopy that long slender particles probably fracture, whereas smaller, more spherical particles form voids at the particle-matrix interface (Fig. 7). In either case these particles are clearly a potential source of void formation.

Numerous authors have devised schematic diagrams depicting void formation and coalescence. One of the more descriptive schematic diagrams, developed by Broek (Ref 21), is reproduced in Fig. 8. In general the progression is believed to begin with the formation of small voids at the particle-matrix interface, or perhaps the fracture of some particles at low stress levels. As stresses begin to increase, voids grow and ultimately begin to link. The stress distributions shown in Fig. 8 determine the type of dimples that can be expected, and they can be of considerable value to the failure analyst when it is necessary to determine the loading that caused a particular failure. A fractograph of classic dimple fracture in an aluminum alloy, with small particles clearly visible in the bottom of the voids, is shown in Fig. 9.

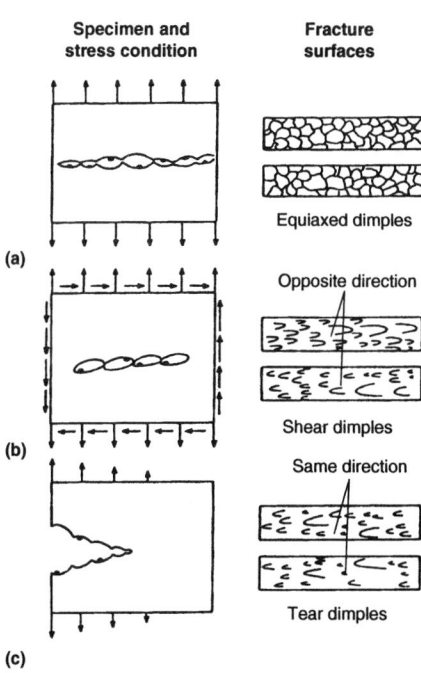

Fig. 8 Different dimple geometries to be expected from three possible loading conditions. The dimple geometry can be valuable to the failure analyst in determining the loading conditions present at the time of a failure. Courtesy of Martinus Nijhoff Publishers. Source: Ref 21

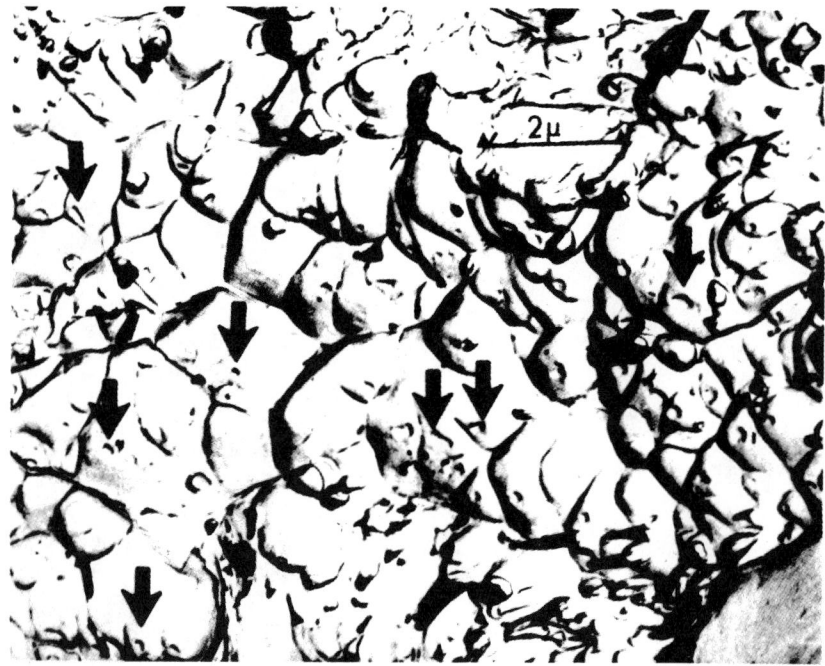

Fig. 9 Fractograph taken from 2024-Al fracture surface replica. Arrows identify small constituent particles at the bottom of dimples that are the origin of the fracture process. Courtesy of Martinus Nijhoff Publishers. Source: Ref 21

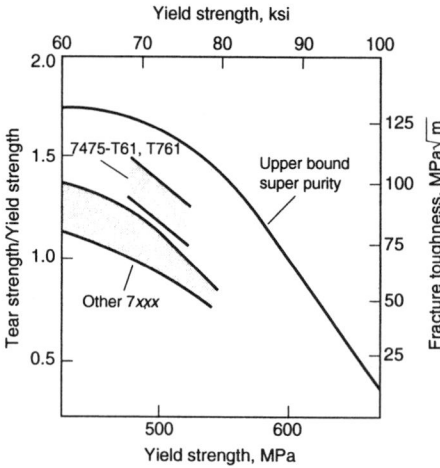

(a)

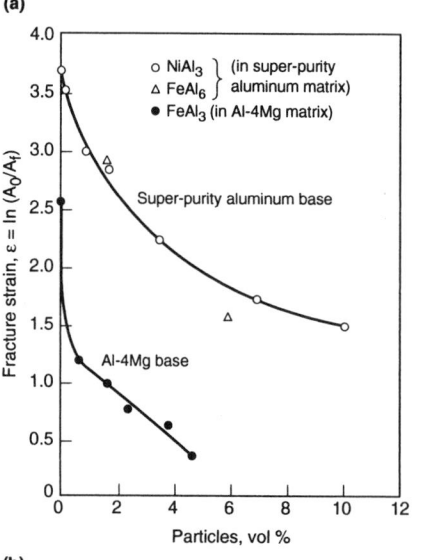

(b)

Fig. 10 (a) The decrease in fracture strain with increase of volume percent of micron-size intermetallic particles for a super-purity aluminum matrix and an Al-4Mg matrix. (b) A comparison of high-purity 7050 aluminum sheet, 7475 sheet, and a 7xxx-series aluminum alloy. Tear strength and fracture toughness are clearly better for the super-purity alloy. Source: *Aluminum: Properties and Physical Metallurgy*, J.E. Hatch, Ed., American Society for Metals, 1984

A study of matching fracture surfaces, also carried out by Broek, showed that the particle is always left in the bottom of one half of the dimple.

The larger constituent particles in aluminum alloys are generally intermetallic compounds and are relatively insoluble $AlCu_2Fe$, Mg_2Si, and $(Fe,Mn)Al_6$. Somewhat more soluble particles, such as $CuAl_2$ and $CuAl_2Mg$, can also be found. However, it is virtually impossible to eliminate these particles by any usual heat treatment once they have formed. A reduced fracture toughness in aluminum alloys can be attributed to the presence of these particles (Ref 23, 24), and a change in toughness of 10 to 15 MPa√m has been observed when efforts have been directed at improving alloy cleanliness by removing copper, chromium, silicon, and iron from commercial alloys. Further evidence of the detrimental role of particles is given in Fig. 10(a), which shows the decrease of fracture strain with increase of volume percent of micron-size intermetallic particles for a super-purity aluminum matrix and an Al-4Mg matrix. By contrast, Fig. 10(b) demonstrates the effect of high-purity 7050 sheet material compared to that of 7475 and other 7xxx-series aluminum alloys. Clearly, tear strength and fracture toughness are improved by increasing purity. Finally, Fig. 11 demonstrates the effect of removing constituent particles on the fracture toughness of 7050 aluminum plate.

Ferrous Alloys

Effect of Second-Phase Particles. Certain fundamental characteristics of fracture that are observed in aluminum alloys are also observed in the fracture of ferrous alloys. For example, the presence of particles such as the sulfide inclusions shown in Fig. 12 results in the typical inclusion-matrix interface failure and the formation of voids, including the possible brittle fracture of the inclusion itself. Either the interface failure or the particle fracture leads to void formation and the linking of voids to give ultimate failure with the usual mechanism of dimple rupture. Still another failure mode prevalent in pearlitic steels is the initial brittle fracture of Fe_3C lamella. These fractures open under continued loading and form voids that can link up to result in larger voids, which in turn further link to give final failure. An example of this type of initial failure, shown in Fig. 13, is taken from the work of Roland (Ref 25), who was examining several possible high-toughness experimental alloys suitable for railroad wheels. Finally, it is not unusual to find fracture surfaces with dimples having small particles at the bottom that have clearly been the sites of initial void formation. In Fig. 14 the resulfurized AISI 4130 had higher strength and lower toughness, while the spheroidized low-sulfur AISI 4130 showed lower strength and higher toughness. Note also that the void geometry of the spheroidized steel is completely different than that of the resulfurized steel. However, ultimate failure was still the result of the linking of voids in both cases.

It is well known that hypoeutectoid steels (those with less than 0.8% C) generally have a proeutectoid grain boundary ferrite that may be continuous or segregated, depending on the composition, and is present in addition to the usual pearlite. Grain boundary ferrite is thought to contribute to crack arrest due to the energy expended in blunting a propagating crack because of the ductile nature of ferrite. Observation of this crack arrest mechanism has been reported by Bouse et

al. (Ref 26) and Fowler and Tetelman (Ref 27). A crack blunting model based on the presence of grain boundary ferrite was developed by Fowler and Tetelman (Ref 27) and is shown in Fig. 15. In contrast to the proeutectoid ferrite found in hypoeutectoid steels, grain boundary carbide resulting from proeutectoid Fe_3C in hypereutectoid steels may also lead to crack arrest. It is very hard and serves as an impediment to crack propagation because of the energy expended in fracturing the hard carbide. However, the iron carbide would also be expected to fail in a brittle manner, and once failed it might lead to lower overall material toughness because of its brittle nature.

In addition to the pearlite colonies and grain boundary ferrite or carbide found in plain carbon steels, there are a variety of second-phase parti-

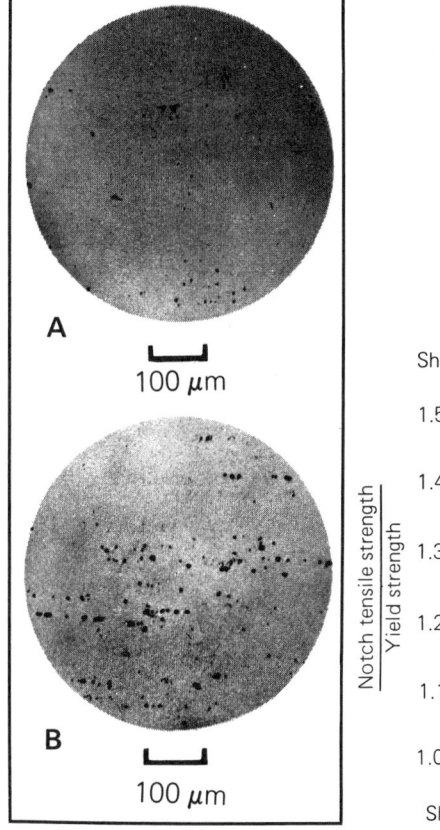

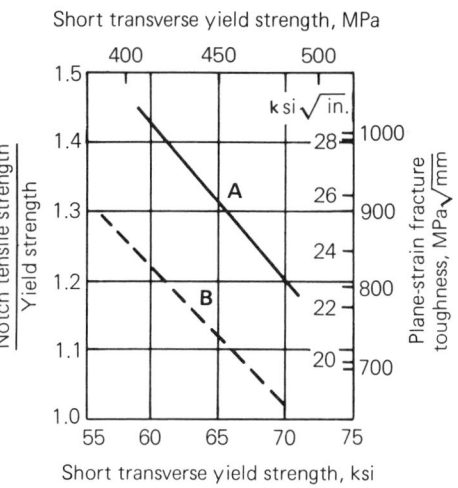

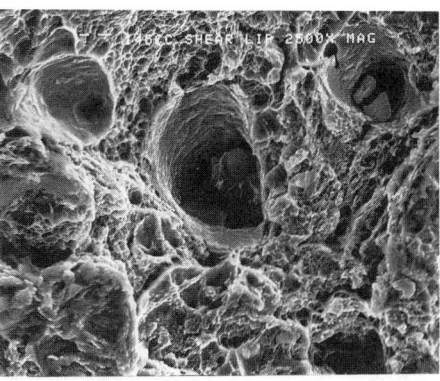

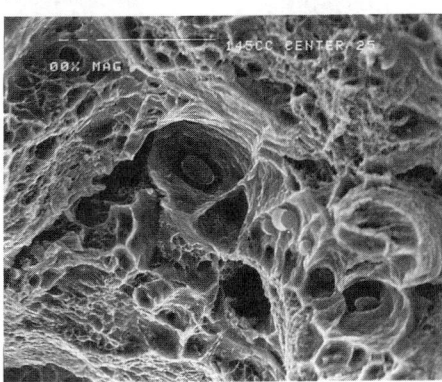

Fig. 11 The effect of decreasing the number of Al_2CuMg constituent particles on the toughness of 7050 is shown in the graph. Both notch tensile strength and plane-strain fracture toughness are improved. Source: *Aluminum: Properties and Physical Metallurgy*, J.E. Hatch, Ed., American Society for Metals, 1984

Fig. 12 Areas from two different fracture surfaces. The fractures are clearly ductile, with varying sizes of dimples and distinct particles in the bottom of the larger dimples. In the upper right-hand corner of (a), the particle is clearly fractured. The material was a low-carbon steel (0.52% C, 0.90% Mn, 0.38% Cr, 0.32% Si) that was being considered for railroad wheels. Source: Ref 19

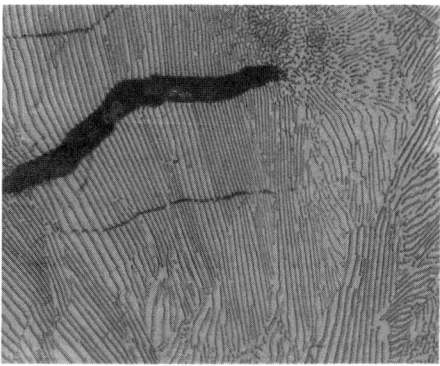

Fig. 13 Optical photograph of polished surface of a highly strained sample of experimental 0.65% C wheel steel. Initial fractures of iron carbide lamella are indicated by the arrow. The fractured lamella result in voids that link up to form a continuous fracture. Source: Ref 19

cles in alloy steels, as well as the usual inclusions that are visible in the optical microscope. A considerable body of literature has examined the effect of particle size and particle distribution on the fracture properties of alloy steels. For example, an empirical relationship relating the effect of particle size to fracture has been given by Priest (Ref 28) for a Ni-Cr-Mo-V steel with 0.45% C:

$$K_{Ic} = 23 \text{ MPa}\sqrt{m} + 7(\sigma^* - \sigma_{ys})(\lambda)^{1/2} \quad \text{(Eq 9)}$$

where $\sigma^* = 2000$ MPa (290 ksi), σ_{ys} is the material yield strength (in MPa) and λ is the average particle spacing between inclusions (in mm). Figure 16 shows that Eq 9 fits the experimental data for large variations in particle spacing as well as for three different test temperatures. In all cases reported, a dimpled rupture surface was the microstructural failure mode.

Schwalbe (Ref 29) has suggested a model as shown in Fig. 17, whereby the crack-tip opening displacement, δ_t, is related to the distance between voids, d. Schwalbe assumes that constancy of volume and plane-strain conditions cause crack advance because of negative strain in the x-direction. The large plastic strains also are responsible for fracture of inclusions (or boundary separation at the inclusion-matrix interface), which leads to void formation. Because the dimple size roughly corresponds to d, one can write:

$$\text{Crack tip opening displacement} = \delta_t \quad \text{(Eq 10)}$$

$$\delta_t = d \quad \text{(Eq 11)}$$

Thus, the more closely spaced the inclusions (and by inference, the larger the density of particles), the smaller the crack-tip opening displacement and the sooner void coalescence with the crack tip begins. Assuming that the onset of instability is related to void coalescence with the crack tip, then an increase in K_{Ic} should be expected with an increase in d. The relationship between volume fraction of inclusions in aluminum and K_{Ic} is shown in Fig. 18(a), and Fig. 16(b), which shows the effect of sulfur content on the fracture properties of 0.45C-Ni-Cr-Mo steels (Ref 32). In Fig. 18(b) the embrittling effect of sulfur results from the dimple formation at sulfides. Finally, note that d in Eq 10 and λ in Eq 9 are essentially equivalent.

Similar relationships are discussed by Hahn (Ref 33), who has examined the relationships between particle size, particle spacing, and the results of tensile tests, Charpy V-notch impact tests (CVN), and fracture toughness (K_{Ic}) results. Hahn's results seem to show that the important variable is the size of the stressed volume. Thus, CVN samples, which have a larger stressed volume in front of a somewhat blunt notch, tend to be more strongly influenced by the particle size in the stressed volume, whereas K_{Ic} samples, which have a much smaller stressed volume at the tip of a fatigue crack, typically give results that are more dependent on the particle spacing in the volume in front of the crack tip. Hahn also advances the interesting hypothesis that the differ-

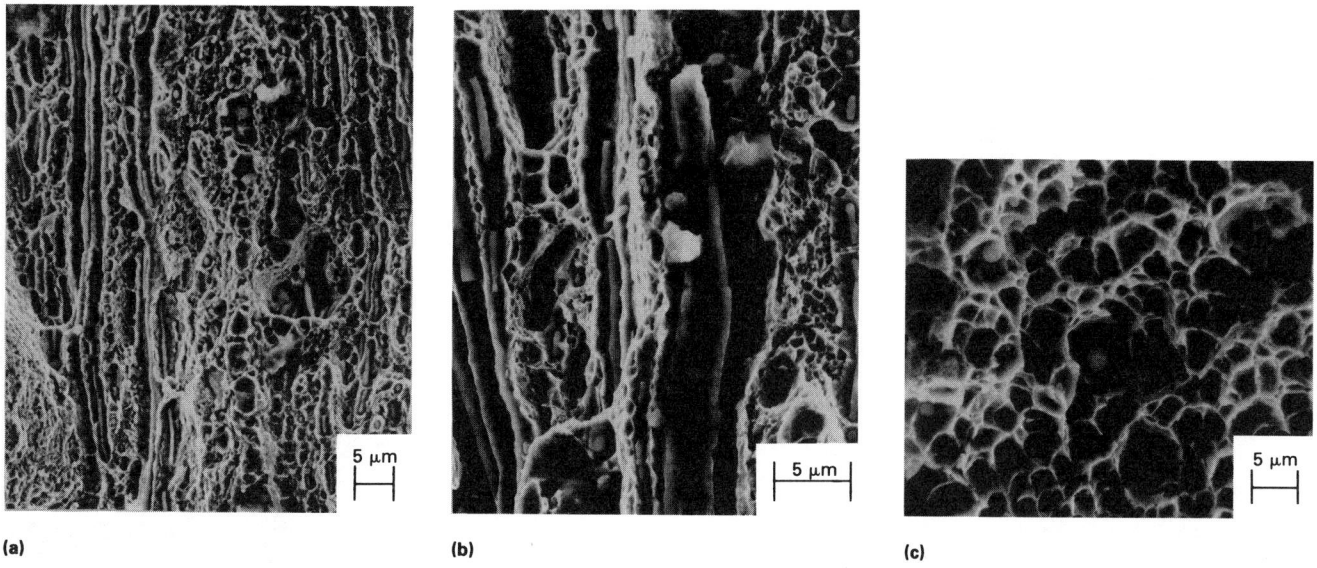

(a) (b) (c)

Fig. 14 Scanning electron micrographs of AISI 4130 steel. (a) and (b) Fractures of resulfurized steel that had been quenched and tempered to 1400 MPa. (c) Low-sulfur AISI 4130 steel that had been spheroidized to 600 MPa. In all three photographs, particles can be found in the dimples. Source: *Metals Handbook*, 9th ed., Vol 8

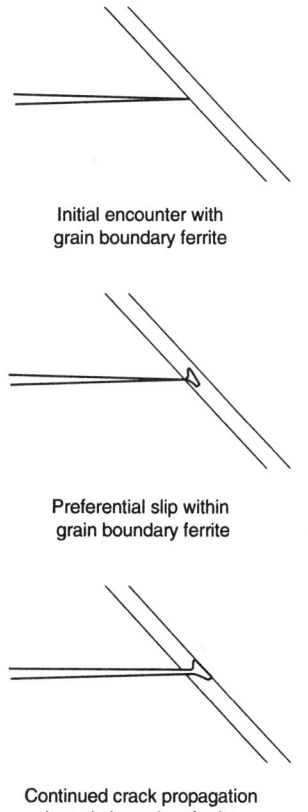

Initial encounter with
grain boundary ferrite

Preferential slip within
grain boundary ferrite

Continued crack propagation
in grain boundary ferrite

Fig. 15 A crack-blunting mechanism resulting from crack propagation into grain boundary ferrite in proeutectoid alloys. Courtesy of American Society for Testing and Materials

ing dependencies of CVN and K_{Ic} tests explain the lack of an all-inclusive, single equation that is able to correlate CVN and K_{Ic} results for all steels. An example of noncorrelating CVN and K_{Ic} toughness measurements is shown in Table 1.

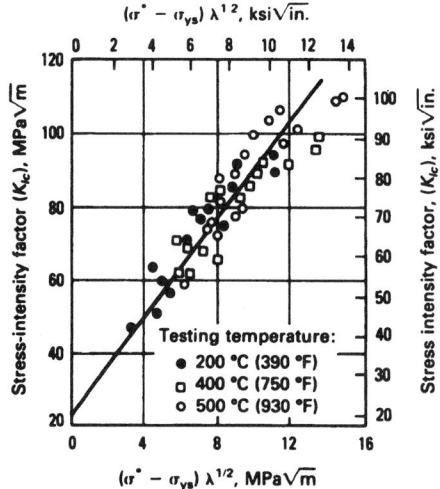

Fig. 16 Plot of Eq 1 from Priest (Ref 28), demonstrating the relationship between constituent particle spacing, material yield stress, and fracture toughness (Ref 15). Experimental data points are for the 0.45C-Ni-Cr-Mo-V steel. Source: Ref 16

Effect of Matrix. Although void formation and the role of second-phase particles and small inclusions are important, it is well known that properties of the matrix may also have an important influence on fracture toughness behavior. For example, Rice (Ref 35) has shown that an increase in matrix strength results in an increase in plastic zone normal stress, such that:

$$\sigma_y = \sigma_Y (1 + \pi/2) \qquad (Eq\ 12)$$

where σ_y is the stress normal to the crack path and σ_Y is the material yield strength. Thus, higher yield strengths result in smaller particles, contributing to dimple formation that in turn results in a smaller average effective inclusion spacing. Similar observations were found by Psioda and Low (Ref 36)

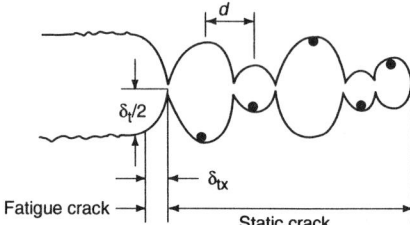

Fig. 17 Model of static crack advance after Schwalbe (Ref 29). The crack-tip opening displacement is equal to the dimple spacing or inclusion spacing.

Table 1 Example of noncorrelating Charpy V-notch (CVN) and K_{Ic} toughness measurements of AISI 4340 steel

Condition(a)	CVN energy, J	K_{Ic}, MPa√m	K_{Id}, MPa√m
A	6.6	70	52
B	9.5	34	33

(a) Condition A—1 h at 1200 °C, salt quench to 870 °C, 1 h at 870 °C, oil quench to room temperature, σ_0 = 1592 MPa. Condition B—1 h at 870 °C, oil quench to room temperature, σ_0 = 1592. Source: Ref 34

during a study of maraging steels. Generally, any change that increases yield strength (such as lower temperature, high deformation rate, or heat treatment) results in a decrease in K_{Ic}. Of course, microstructural changes (such as change in particle size as a result of heat treatment) negates this statement. Pellissier (Ref 37) concludes that a fine, homogeneous distribution of particles of intermetallic compounds results in a high fracture toughness, whereas in martensitic steels the higher carbide content due to high carbon is detrimental to toughness. It should also be noted that Eq 12 is for an elastic perfectly plastic material and that with strain hardening, substantial increases in σ_y can develop.

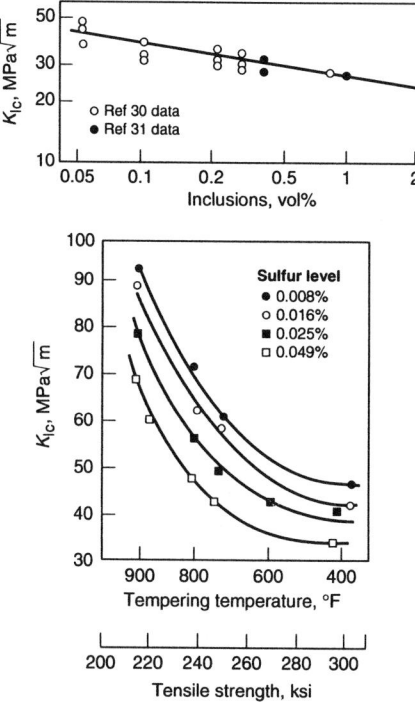

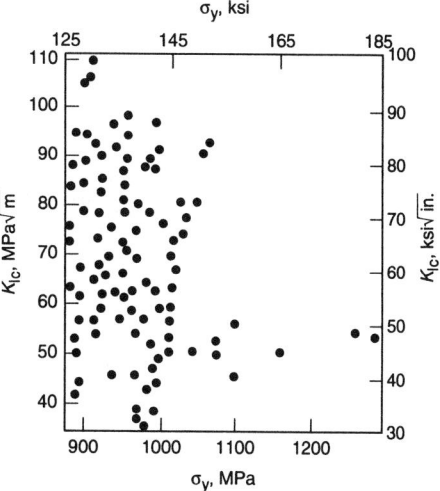

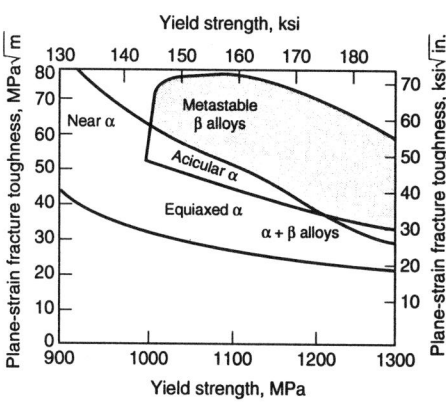

Fig. 18 (a) Fracture toughness of some aluminum alloys vs. volume fraction of inclusions. (b) Fracture toughness of 0.45C-Ni-Cr-Mo steels as a function of sulfur content and tensile strength

Fig. 19 Effect of variations in microstructure on the fracture toughness properties of a Ti-6Al-4V alloy. Source: Ref 40

Fig. 20 Diagram demonstrating the relationship between alloy strength and alloy microstructure for titanium alloys. Source: Ref 42

Table 2 Relation between K_{Ic} and fraction of transformed structure in Ti-6Al-4V

Heat treat temperature		Fraction of transformed structure, %	K_{Ic}	
°C	°F		MPa√m	ksi√in.
1050	1920	100	69.0	63
950	1740	70	61.5	56
850	1560	20	46.5	42
750	1380	10	39.5	36

Numerous workers have examined high-strength steels such as AISI 4340 and AISI 4130. Low tempering temperatures led to a carbide film at the martensite lath boundaries and thus led to low toughness for 4340, according to Wei (Ref 38), whereas Parker (Ref 39) suggests that fracture toughness in the as-quenched condition of AISI 4340 and similar steels is determined by precipitation at prior-austenite grain boundaries.

Titanium Alloys

The titanium alloys are somewhat unique in that they can exist in the alpha (face-centered cubic), beta (body-centered cubic), or alpha + beta condition, depending on the alloy composition and heat treatment. However, in any case, interface weaknesses between the phases tend to lead to failure. For example, Gerberich and Baker (Ref 40) have shown that a change from an equiaxed alpha to a platelet alpha structure gives an increase in K_{Ic} of approximately 25%, with 5% or less change in yield strength and ultimate strength. The authors concluded that the properties changed because of the change in fracture path that resulted from the change in microstructure. The authors also noted that an increase in oxygen content tended to cause embrittlement of the alpha phase, with a subsequent decrease in toughness. In another paper, Gerberich (Ref 16) reiterates the importance of both composition and microstructural effects on the toughness of Ti-6Al-4V. For example, he points out that the alpha platelets in the alpha + beta Widmanstätten matrix may be either detrimental or beneficial, depending on the oxygen content. However, there apparently is no processing route that provides a toughness greater than 55 MPa√m for yield strengths greater than about 1080 MPa.

Finally, an experimental alpha + beta alloy was studied where the strength was held constant in both the equiaxed alpha and transformed microstructural conditions. For equiaxed alpha, toughness increased with beta grain boundary area per unit volume. In the transformed condition, toughness increased with an increase in the percentage of primary alpha. Table 2 gives the relationship between K_{Ic} and the fraction of transformed structure for Ti-6Al-4V.

Harrigan (Ref 41) has examined the effect of microstructures on the fracture properties of titanium alloys and concludes that variations in microstructures can result in large scatter of experimental results. This is suggested by Fig. 19, where no correlation is evident between toughness and yield strength. Still another representation of the relationship between microstructure and toughness for titanium alloys was given by Rosenfield and McEvily (Ref 42), who conclude that toughness depends on the size, shape, and distribution of the phases that are present (Fig. 20). Metastable beta alloys appear to have

the highest toughness, while alpha + beta alloys are generally less tough.

REFERENCES

1. W.D. Biggs, *The Brittle Fracture of Steel*, McDonald and Evans, 1960
2. W.E. Anderson, *An Engineer Reviews Brittle Fracture History*, Boeing, 1969
3. R. Gannon, What Really Sank the Titanic, *Popular Science*, Feb 1995, p 45
4. D.J. McConville, "Seaway to Nowhere," *Am. Heritage Invent. Technol.*, Vol 11 (No. 2), 1995, p 34-44
5. A.A. Griffith, The Phenomena of Rupture and Flow in Solids, *Phil. Trans. Roy. Soc. London*, Series A, Vol 221, 1920, p 163-198
6. A.A. Griffith, The Theory of Rupture, *Proc. First International Congress for Applied Mechanics*, Delft, The Netherlands, 1924, p 55-63
7. G.R. Irwin, Fracture Dynamics, *Trans. ASM*, Vol 40A, 1948, p 147-166
8. G. Vander Voort, Ductile and Brittle Fractures, *Metals Handbook*, 9th ed., Vol 11, 1982, p 85
9. J. Collins, *Failure of Materials in Mechanical Design*, John Wiley, 1993, p 51
10. J. Frenkel, *Zeitshrift der Physik*, Vol 37, 1926, p 572
11. A. Kelly, *Strong Solids*, Oxford University Press, 1973
12. G.I. Taylor, *Proceedings of the Royal Society*, Vol A145, 1934, p 632
13. E. Orowan, *Zeitshrift der Physik*, Vol 89, 1934, p 605
14. M. Polanyi, *Zeitshrift der Physik*, Vol 89, 1934, p 60
15. R. Hertzberg, *Deformation and Fracture Mechanics of Engineering Materials*, 4th ed., John Wiley & Sons, Inc., 1996
16. W.W. Gerberich, Microstructure and Fracture, *Mechanical Testing*, Vol 8, *Metals Handbook*, 9th ed., ASM International, 1985, p 476-491
17. D. Broek, Ph.D. thesis, Delft University of Technology, Delft, The Netherlands, 1971

18. J.C. Grosskreutz and G. Shaw, Critical Mechanisms in the Development of Fatigue Cracks in 2024-T4 Aluminum, *Fracture*, Chapman and Hall, 1969, p 620-629
19. C.Q. Bowles and J. Schijve, The Roll of Inclusions in Fatigue Crack Initiation in an Aluminum Alloy, *Int. J. Fract.*, Vol 9, 1973, p 171-179
20. A.J. McEvily and R.C. Boettner, A Note on Fatigue and Microstructure, *Fracture of Solids*, Interscience Publishers, 1963, p 383-389
21. D. Broek, *Elementary Fracture Mechanics*, 4th ed., Martinus Nijhoff Publishers, 1986, p 51-55
22. D. Broek, The Role of Inclusions in Ductile Fracture and Fracture Toughness, *Eng. Fract. Mech.*, Vol 5, 1973, p 55-66
23. R.H. Van Stone, J.R. Low, Jr., and R.H. Merchant, *Investigation of the Plastic Fracture of High Strength Aluminum Alloys*, ASTM STP 556, ASTM, 1974, p 93-124
24. J.G. Kaufman and J.S. Santner, Fracture Properties of Aluminum Alloys, *Application of Fracture Mechanics for Selection of Metallic Structural Materials*, J.E. Campbell, W.W. Gerberich, and J.A. Underwood, Ed., ASM International, 1982, p 169-211
25. J.R. Roland, "The Fracture Resistance of Experimental Alloy and Class U Carbon Steel Wrought Railroad Wheels," M.S. thesis, University of Missouri-Columbia, Columbia, MO, 1986
26. G.K. Bouse, I.M. Bernstein, and D.H. Stone, Role of Alloying and Microstructure on the Strength and Toughness of Experimental Rail Steels, in STP 644, ASTM, 1978, p 145-161
27. G.J. Fowler and A.S. Tetelman, Effect of Grain Boundary Ferrite on Fatigue Crack Propagation in Pearlitic Rail Steels, in STP 644, ASTM, 1978, p 363-382
28. A.H. Priest, "Effect of Second-Phase Particles on the Mechanical Properties of Steel," The Iron and Steel Institute, London, 1971
29. K.-H. Schwalbe, On the Influence of Microstructure on Crack Propagation Mechanisms and Fracture Toughness of Metallic Materials, *Eng. Fract. Mech.*, Vol 9, 1977, p 795-832
30. J.H. Mulherin and H. Rosenthal, Influence of Nonequilibrium Second-Phase Particles Formed during Solidification upon the Mechanical Behavior of Aluminum Alloys, *Met. Trans.*, Vol 2, 1971, p 427
31. J.R. Low, Jr., R.H. Van Stone, and R.H. Merchant, Technical Report 2, NASA Grant NGR-39-087-003, 1972
32. A.J. Birkle, R.P. Wei, and G.E. Pellissier, Analysis of Plane-Strain Fracture in a Series of 0.45C-Ni-Cr-Mo Steels with Different Sulfur Contents, *Trans. ASM*, Vol 59, 1966, p 981
33. G.T. Hahn, The Influence of Microstructure on Brittle Fracture Toughness, *Met. Trans.*, Vol 15A, 1984, p 947-959
34. R. Ritchie, B. Francis, and W.L. Server, *Met. Trans.*, Vol 7A, 1976, p 831
35. J.R. Rice, Mechanics of Crack Tip Deformation and Extension By Fatigue, STP 415, ASTM, 1967
36. J.A. Psioda and J.R. Low, Jr., "The Effect of Microstructure and Strength on the Fracture Toughness of an 18Ni, 300 Grade Maraging Steel," Technical Report 6, NASA, 1974
37. G.E. Pellissier, Effects of Microstructure on the Fracture Toughness of Ultrahigh-Strength Steels, *Eng. Fract. Mech.*, Vol 1, 1968, p 55
38. R.P. Wei, Fracture Toughness Testing in Alloy Development, in STP 381, ASTM, 1965
39. E.R. Parker and V.F. Zackay, Enhancement of Fracture Toughness in High Strength Steel by Microstructural Control, *Eng. Fract. Mech.*, Vol 5, 1973, p 147
40. W.W. Gerberich and G.S. Baker, Toughness of Two-Phase 6Al-4V Titanium Microstructures, in STP 432, ASTM, 1968, p 80-99
41. M.J. Harrigan, *Met. Eng. Quart.*, May 1974
42. A.R. Rosenfield and A.J. McEvily, Report 610, NATO AGARD, Dec 1973, p 23

Fatigue Properties in Engineering

D.W. Cameron, Allegany, NY, and D.W. Hoeppner, Department of Mechanical Engineering, University of Utah

FATIGUE PROPERTIES are an integral part of materials comparison activities and offer information for structural life estimation in many engineering applications. They are a critical element in the path relating the materials of construction to the components and must take into account as many influences as possible to reflect the actual product situation. In application, fatigue is a detail analysis, trying to assess what will occur at a particular location of a component or assembly under cyclic loading.

The topic of fatigue properties is very broad and is typically based on testing coupons. To be applicable, determined properties must support one of the fatigue design philosophies that may be applied to the part. In this article the three general approaches to fatigue design are stated, with discussion of their respective attributes, and their individual property requirements are described. The intent here is not to present a comprehensive catalog of properties; that would take many volumes this size. Instead, the purpose is to provide the basic insights necessary to examine those properties that can be found, review some of the common presentation formats, and recognize their inherent characteristics. It is important to review information critically for any use, to know when a direct "apples to apples" comparison can be made, and potentially to know how to manipulate some of the data to put it on equal footing with information gathered from diverse sources. The susceptibility of mechanical properties to variation through microstructural manipulation and structural consideration can be substantial.

The importance of testing in property generation is reviewed briefly, and material, property, and structure relations are discussed. Three sections then cover properties specific to each of the major design approaches: stress-life, strain-life, and fracture mechanics. The individual sections offer selected examples of properties that reflect some detail of each approach.

Although life estimation is not the subject of this article, it is obvious that this is one of the main uses of the "properties" development. Basically, data on test coupons are only good at estimating the life of test coupons; other structures may not be as amenable to estimation. A life estimation within a factor of 2 would be excep-

tional, and perhaps one within an order of magnitude would not be considered too outrageous, depending on the quality of information, appropriateness of technique, and "property" data. The substantial amount of scatter in results is one of the contributing features to these difficulties. Certainly verification of life estimations should be considered an important activity to confirm the calculations.

For the sake of brevity, we limit our discussion to constant-amplitude loading. Often, variable-amplitude loading is necessary to correctly replicate structural situations. It is essential to understand that variable-amplitude loading can produce different rankings than constant-amplitude results. Another concession to brevity is that within the fracture mechanics area, only plane-strain considerations are included. Among other critical aspects not covered specifically here are: crack nucleation models and the basic physics of this process and as well as that of crack extension; the extremely important extrinsic factor of environment on both 'initiation' and propagation characteristics; and other phenomenon such as fretting discussed in more detail elsewhere in this Volume.

Fatigue Design Philosophies

To be usable in anything other than a comparative sense, fatigue properties must be consistent with one of three general fatigue design philosophies. Each of these has a concomitant design methodology and one or more means of representing testing data that provide the 'properties' of interest. These are:

Design philosophy	Design methodology	Principal testing data description
Safe-life, infinite-life	Stress-life	S-N
Safe-life, finite-life	Strain-life	ε-N
Damage tolerant	Fracture mechanics	da/dN – ΔK

These "lifing" or assessment techniques correspond to the historical development and evolution of fatigue technology over the past 150 to 200 years.

The safe-life, infinite-life philosophy is the oldest of the approaches to fatigue. Examples of attempts to understanding fatigue by means of

properties, determinations, and representations that relate to this method include August Wöhler's work on railroad axles in Germany in the mid-1800s (Ref 1). The design method is stress-life, and a general property representation would be S-N (stress vs. log number of cycles to failure). Failure in S-N testing is typically defined by total separation of the sample.

General applicability of the stress-life method is restricted to circumstances where continuum, "no cracks" assumptions can be applied. However, some design guidelines for weldments (which inherently contain discontinuities) offer what amount to residual life and runout determinations for a variety of process and joint types that generally follow the safe-life, infinite-life approach (Ref 2). The advantages of this method are simplicity and ease of application, and it can offer some initial perspective on a given situation. It is best applied in or near the elastic range, addressing constant-amplitude loading situations in what has been called the long-life (hence infinite-life) regime.

The stress-life approach seems best applied to components that look like the test samples and are approximately the same size (this satisfies the similitude associated with the use of total separation as a failure criterion). Much of the technology in application of this approach is based on ferrous metals, especially steels. Other materials may not respond in a similar manner. Given the extensive history of the stress-life method, substantial property data are available, but beware of the testing conditions employed in producing older data.

Through the 1940s and 1950s, mechanical designs pushed to further extremes in advanced machinery, resulting in higher loads and stresses and thus moving into the plastic regime of material behavior and a more explicit consideration of finite-lived components. For these conditions, the description of local events in terms of strain made more sense and resulted in the development of assessment techniques that used strain as a determining quantity. The general data (property) presentation is in terms of ε-N (log strain vs. log number of cycles or number of reversals to failure). The failure criterion for samples is usually the detection of a "small" crack in the sample or some equivalent measure related to a substantive change in load-deflection response, although failure may also be defined by separation.

Employment of strain is a consistent extension of the stress-life approach. As with the safe-life, infinite-life approach, the strain-based safe-life, finite-life philosophy relies on the "no cracks" restriction of continuous media. While considerably more complicated, this technique offers advantages: it includes plastic response, addresses finite-lived situations on a sounder technical basis, can be more readily generalized to different geometries, has greater adaptability to variable-amplitude situations, and can account for a variety of other effects. The strain-life method is better suited to handling a greater diversity of materials (e.g., it is independent of assuming steel-like response for modification factors). Because it does not necessarily attempt to relate to total failure (separation) of the part, but can rely on what has become known as "initiation" for defining failure, it has a substantial advantage over the stress-life method. Difficulties in applying the method arise because it is more complex, is more computationally intensive, and has more complicated property descriptions. In addition, because this method does not have as extensive a history, "properties" may not be as readily available.

The ability to generate and model both S-N and ε-N data effectively is clearly very important. Three good sources for increasing the understanding of this are ASTM STP 91A (Ref 3), ASTM STP 588 (Ref 4), and Ref 5. Specifications covering the individual areas are indicated below.

From a design standpoint, there are some circumstances where inspection is not a regularly employed practice, impractical, unfeasible, or occasionally physically impossible. These situations are prime candidates for the application of the safe-life techniques when coupled with the appropriate technologies to demonstrate the likelihood of failure to be sufficiently remote.

The notable connection between the two techniques described above is the necessary assumption of continuity (i.e., "no cracks"). Many components, assemblies, and structures, however, have crack-like discontinuities induced during service or repair or as a result of primary or secondary processing, fabrication, or manufacturing. It is abundantly clear that in many instances, parts containing such discontinuities do continue to bear load and can operate safely for extended periods of time. Developments from the 1960s and before have produced the third design philosophy, damage tolerant. It is intended expressly to address the issue of "cracked" components.

In the case where a crack is present, an alternative controlling quantity is employed. Typically this is the mode I stress-intensity range at the crack tip (ΔK_I), determined as a function of crack location, orientation, and size within the geometry of the part. This fracture mechanics parameter is then related to the potential for crack extension under the imposed cyclic loads for either subcritical growth or the initiation of unstable fracture of the part. It is markedly different from the other two approaches. Property descriptions for the crack extension under cyclic loading are typically da/dN – ΔK_I curves (log crack growth rate vs. log stress-intensity range).

The advantage of the damage tolerant design philosophy is obviously the ability to treat cracked objects in a direct and appropriate fashion. The previous methods only allow for the immediate removal of cracked structure. Use of the stress-intensity values and appropriate data (properties) allows the number of cycles of crack growth over a range of crack sizes to be estimated and fracture to be predicted. The clear tie of crack size, orientation, and geometry to nondestructive evaluation (NDE) is also a plus. Disadvantages are: possibly computationally intensive stress-intensity factor determinations, greater complexity in development and modeling of property data, and the necessity to perform numerical integration to determine crack growth. In addition, the predicted lives are considerably influenced by the initial crack size used in the calculation, requiring quantitative development of probability of detection for each type of NDE technique employed. Related to the initial crack size consideration is the inability of this approach to model effectively that the component was actually suitable for modeling as a continuum, which eliminates the so-called "initiation" portion of the part life.

Considerations in Conducting Tests

Having properties implies testing of materials to make such determinations. Even approximations of properties assume some model of behavior, and initially testing was employed to provide that information. Thus, testing per se warrants some discussion.

The principal question to be asked when considering testing is whether or not the desired information will be produced by the testing. This is reflected directly in the "properties" that will result. Fatigue testing can vary from a few preliminary tests to elaborate, sophisticated programs. In support of deriving the necessary data in the best manner, the principal author identifies three critical aspects of testing programs as the three E's of testing: efficacy, efficiency, and economy. These are certainly not unique to fatigue. They are stated in order of importance and are interpreted as follows:

- *Efficacy:* The testing *must* tell you what you want to know and provide it to the required confidence level. It must be physically capable of generating the desired information, and it must be designed to discriminate to the degree necessary to sort out the details or subtleties of response that is required. (It is sometimes the task of testing to determine whether any difference can be distinguished.) This may call for extensive experimental design up front, and statistical examination of data. Established and consistent test procedures are always a requirement.

- *Efficiency:* The testing should be scheduled, consistent with maintaining efficacy, to generate the greatest amount of usable/desired information as early as possible in the test program.

- *Economy:* Testing should proceed in the most economical manner without compromising efficacy, while meeting the desired information generation levels as well as possible.

Both efficiency and economy are necessarily subservient to efficacy. A substantial amount of work may have to be done before testing begins, to maximize the likelihood of success. If test program manipulations are to be done, they must be done only to balance efficiency/economy issues. If the testing proceeds without the ability to generate the necessary information (e.g., effectively identify subtleties), or if it is altered midstream with the same result, the integrity of the program is breached and there is little or no justification to run or continue the tests (at least according to the original intent of the project). Explicit consideration of the statistical nature of fatigue data should always be part of a testing program.

Assessing Fatigue Characteristics

Supporting Information and What to Look for in Fatigue "Properties" Data. The ability to assess properties information is one of the critical points in deciding if the data found are applicable and usable. Testing should have been done to a stated, set procedure or standard, and all information germane to the testing and resulting data should have been recorded. With the multitude of influential variables, obviously this list can get quite long, but without it the relative value of the information cannot be determined. Dogs and horses have pedigrees, so do data.

For example, in trying to find fatigue properties of rather heavy 7075-T6 aluminum alloy forging, a fatigue curve is found that indicates it is for this alloy and condition. The plot indicates a single line drawn on S_{max} versus log N coordinates, and that's all. What use is it? There is no other description provided. Recommendation: ignore it, or call the originator for clarification. Use of the data would be risky, because there is not sufficient information present to make a defensible assessment. Many necessary pieces of data are simply missing. A partial list might be:

- What were the coupon size and geometry?
- Was there a stress concentration?
- What was the temperature?
- Was an environment other than lab air employed?
- What was the specimen orientation in the original material?
- Does the line represent minimum, mean, or median response?
- How many samples were tested?
- What was the scatter?
- If the plot is based on constant-amplitude data, what were the frequency and waveform?

- Was testing performed using variable-amplitude loading? What spectrum?
- What was the failure criterion?
- If there were runouts, how were they handled and represented?
- If the data found describe a thin sheet response, it is the wrong data.
- If the product form is correct, but the plot represents testing done at $R = 0.3$ and fully reversed data are required, the plot may be helpful, but it is not what is desired.

Material chemistry; product form, condition, and strength level; coupon geometry, size, orientation, and preparation; testing equipment, procedures, parameters, failure criterion, and number of samples; data treatment; and sequence of testing are just some of the contributing and possibly controlling features represented by the single line on the graph. An example of what should be considered important as supporting facts can be found in ASTM E 468-90, "Presentation of Constant Amplitude Fatigue Test Results for Metallic Materials" (Ref 6). It provides guidelines for presenting information other than just final data.

Finding, characterizing, and critical review are clearly extremely challenging parts of attempting to apply materials properties data, with critical designs requiring the most stringent consideration. In some cases this is extremely difficult: necessary data may be sparse, proprietary, and/or poorly documented and very careful use of any information the only choice available. Different criterion, however, apply to the use of property descriptions as examples, representative of *potential* responses for purposes of demonstration and illustration (as they are employed here). While full documentation is still desirable, the use of information in this context only requires that the data be judged an adequately sound depiction of the archetypal behavior.

As in all disciplines, the definition and standardization of both terms and nomenclature are extremely important. A survey of different books, articles, and other literature sources indicates that the fatigue community is no exception to maintaining consistency in this area. The reader is directed to ASTM E 1150-87 (1993), "Standard Definitions of Terms Relating to Fatigue" (Ref 7), for terminology. The authors have attempted to adhere to those definitions.

One point should be made very clear: to establish the nature of any constant-amplitude fatigue data, two dynamic variables *must* be stated, or as a poor second, implied by the nature of the testing. Many dynamic variables apply to constant-amplitude loads: P_{max}, P_{min}, P_m, P_a, and ΔP, which indicate load maximum, minimum, mean, amplitude, and range, respectively. Two load ratio quantities are also frequently encountered: R and A, defined as P_{min}/P_{max} and P_a/P_m, respectively. Note that P, as load, is used in a generic sense here, with other possibilities including S_a, alternating stress; ε_m, mean strain; ΔK_I, stress-intensity range; and so on. These dynamic variables are related such that if any two are known, all the others can be determined, but two *must* be known.

As an example, if a series of tests are conducted at a constant R value (S_{min}/S_{max}), and the alternating stress is used as the other independent dynamic variable, an S-N curve for that situation can be produced and all dynamic variables can be determined. If only one variable is given (e.g., S_a or S_{max}), there is insufficient information to tell what the test conditions were and the data are virtually useless.

Material-Property-Structure Interrelations

It is important to discriminate between fatigue "properties" and structural fatigue response within the context of this article. The term *fatigue properties* is used to describe the response of a test coupon, with all the necessary standardization and other work that this implies. This point is then used to determine the "properties" content of the rest of the article. While test coupons are indeed mini-structures, they are frequently items of geometric convenience, designed for the exigencies of testing, prepared especially for the investigation, and idealized for specific testing or influence determinations. Their relation to any specific structure can be very remote.

A few comments on what ends up as the *material* for a structure should also be made. First is a composition, essentially the basic chemistry of an alloy or the specific components of a composite. Producing the structure may require a few or many steps beyond this chemistry/components combination. Primary processing plays an important role. As examples, an investment cast superalloy blade will have different characteristics depending on whether it is made using an equiaxed, directionally solidified, or single-crystal process; and fiber-reinforced composites clearly have numerous wrap/lay configurations that can influence their response. Subsequent thermal treatment for an alloy, or curing conditions for a composite, also contribute to the end product. Many metallic alloys, far from being the uniform homogenous materials often envisioned, are carefully orchestrated arrangements of microconstituents designed to provide specific property balances from these in situ composites.

Effects of scale in the production of a material can have controlling effects. Examples are graphite size in gray iron, transformation characteristics of steels or titanium alloys in heavy sections, mechanical working in forgings and extrusions, distributions of fibers in chopped-fiber reinforced polymer parts, and phase and discontinuity distribution in ceramics. Further considerations might include machining processes, plating, shot-peening, adhesive bonding, welding, and a myriad of other influences that confound what initially appears to be the desired, rather straightforward association between the material content and structure. This is coupled with the geometric requirements of shape necessary to provide the geometry of the structure. Indeed, it is equally true that the material defines the structure and the structure defines the material.

A small shaft simply loaded in rotating bending may behave quite like specimens tested in a similar manner. On the other hand, a composite wing, built up from multiple parts joined by adhesives and mechanical fasteners, should not be expected to behave in the same manner as a small simple-configuration test coupon of skin material. Attributes of the material, coupon, or structure, along with testing conditions, contribute to the structure-sensitive mechanical behavior identified here as fatigue properties. These have been aptly categorized by Hoeppner (Ref 8) as intrinsic and extrinsic factors, and substantial progress has been made in understanding and controlling both. Design of the materials covers the intrinsic characteristics (e.g., composition, grain size, cleanliness level, layup geometry, and cure cycle). Mechanical design for a specific application addresses the extrinsic influences of the scale, geometry, stress state, loading rates, environment, etc. Both material design and mechanical design play synergistic, substantial, and possibly determining roles in controlling the structural response to cyclic loads.

Does this eliminate the importance of testing and property determinations? Certainly not, but it does increase awareness of the limitations of testing and suggests that they at least be recognized and included in actual structural assessments.

The three following sections provide examples of property determinations from each of the three major groups (S-N, ε-N, and da/dN). Each example demonstrates the general and/or specific aspects of the information within the context of the design philosophy it supports. Where examples of data are offered, the reader should regard the information as indicative only of the specific material/process/product combination involved.

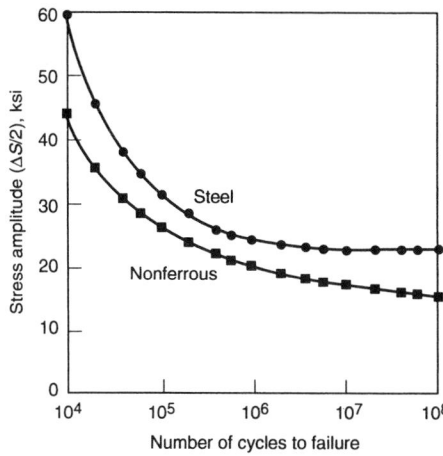

Fig. 1 Schematic S-N representation of materials having fatigue limit behavior (asymptotically leveling off) and those displaying a fatigue strength response (continuously decreasing characteristics)

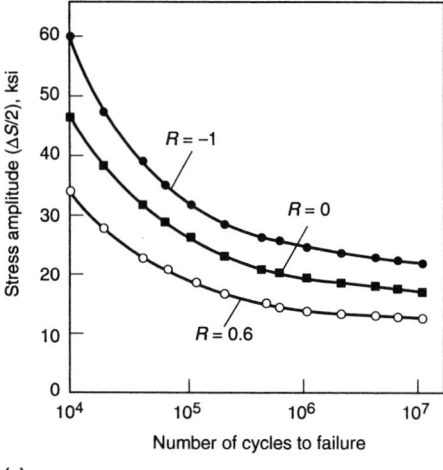

(a)

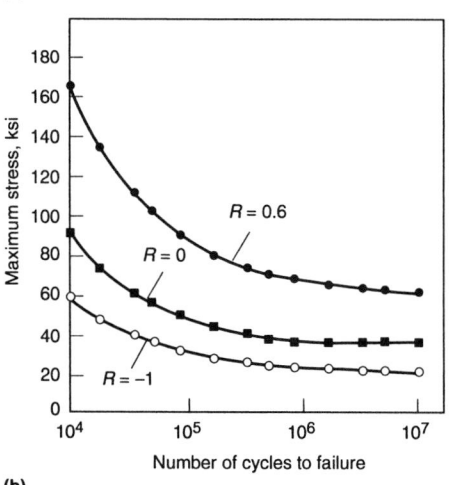

(b)

Fig. 2 The influence of method of S-N data presentation on the perceived effect of R value. (a) Stress amplitude vs. N. (b) Maximum stress vs. N

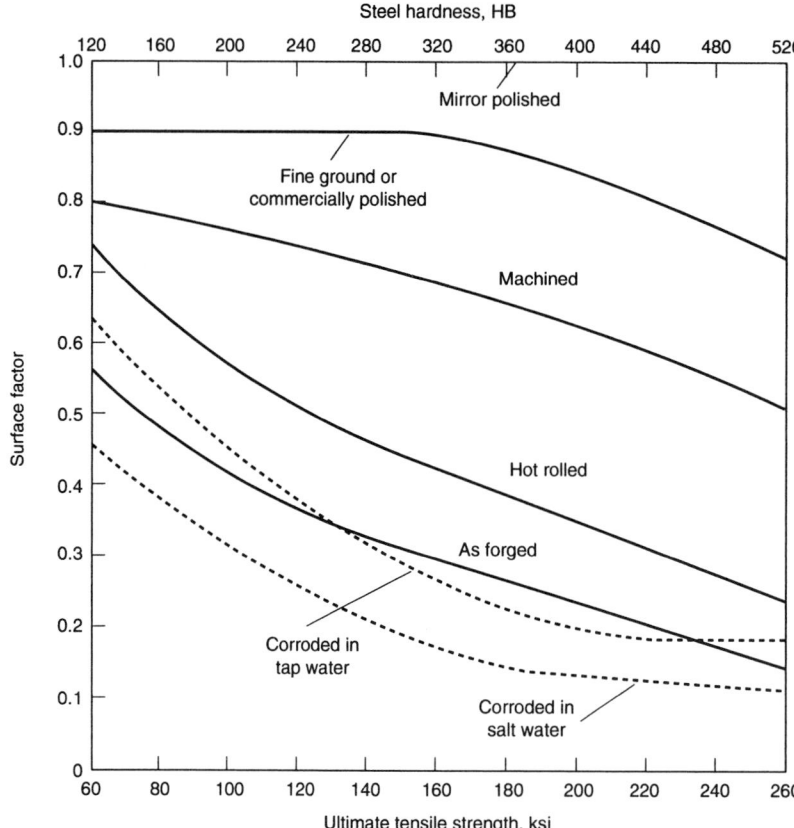

Fig. 3 A plot of reduction factor for use in estimating the effect of surface finish on the S-N fatigue limit of steel parts. Source: Ref 17

Infinite-Life Criterion (S-N Curves)

Safe-life design based on the infinite-life criterion reflects the classic approach to fatigue. It was initially developed through the 1800s and early 1900s because the industrial revolution's increasingly complex machinery produced dynamic loads that created an increasing number of failures. The safe-life, infinite-life design philosophy was the first to address this need.

As stated earlier, the stress-life or S-N approach is principally one of a safe-life, infinite-life regime. It is generally categorized as a "high cycle fatigue" methodology, with most considerations based on maintaining elastic behavior in the sample/components/assemblies examined. The "no cracks" requirement is in place, although all test results inherently include the influence of the discontinuity population present in the samples.

This methodology is one where the influence of steel seems virtually overwhelming, despite the fact that substantial work has been done on other alloys and materials. There are many reasons for this, including the place of steel as the predominant metallic structural material of the

century: in land transportation, in power generation, and in construction. The "infinite-life" aspect of this approach is related to the asymptotic behavior of steels, many of which display a fatigue limit or "endurance" limit at a high number of cycles (typically >10^6) under benign environmental conditions. Most other materials do not exhibit this response, instead displaying a continuously decreasing stress-life response, even at a great number of cycles (10^6 to 10^9), which is more correctly described by a fatigue strength at a given number of cycles. Figure 1 shows a schematic comparison of these two characteristic results. Many machine design texts cover this method to varying degrees (Ref 9–14).

What about the S-N data presentation? Stress is the controlling quantity in this method. The most typical formats for the data are to plot the log number of cycles to failure (sample separation) versus either stress amplitude (S_a), maximum stress (S_{max}), or perhaps stress range (ΔS) (Ref 15). Remember that one other dynamic variable needs to be specified for the data to make sense. Figures 2(a) and (b) provide plots for three constant-R value tests (R is the second dynamic variable). Note the apparent reversal of the effect of R, although the data are identical. Clearly, while the analytical result must be identical regardless of which graphic means is employed, the visual influence in interpretation varies with the method of presentation.

Many applications of this technique require estimations of initial properties and provision for approximating other effects. Overall influences of various conditions (e.g., heat treatment, surface finish, and surface treatment) were determined using substantial empiricism: test and report results. Consequently, much of the challenge was met by testing coupons/components with variations in processing to establish some guidelines for the effect of each such alteration (i.e., see Ref 16). Thus, various correction factors were developed for a variety of conditions, including load type, stress concentration, surface finish, and size. The influences of these intrinsic and extrinsic effects on the properties are typically accounted for by graphics (e.g., Fig. 3), tabular presentations, or mathematical expressions. Reference 18 is an excellent example of this approach, presented in the form of a standard.

Mean stress influences are very important, and each design approach must consider them. According to Bannantine et al. (Ref 13), the archetypal mean (S_m) versus amplitude (S_a) presentation format for displaying mean stress effects in the safe-life, infinite-life regime was originally proposed by Haigh (Ref 19). The Haigh diagram can be a plot of real data, but it requires an enormous amount of information for substantiation. A slightly more involved, but also more useful, means of showing the same information incorporates the Haigh diagram with S_{max} and

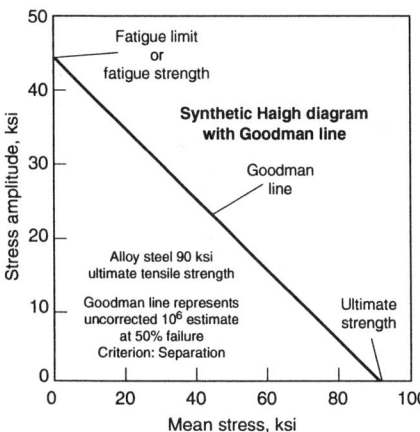

Fig. 4 A synthetically generated Haigh diagram based on typically employed approximations for the axes intercepts and using the Goodman line to establish the acceptable envelope for safe-life, infinite-life combinations

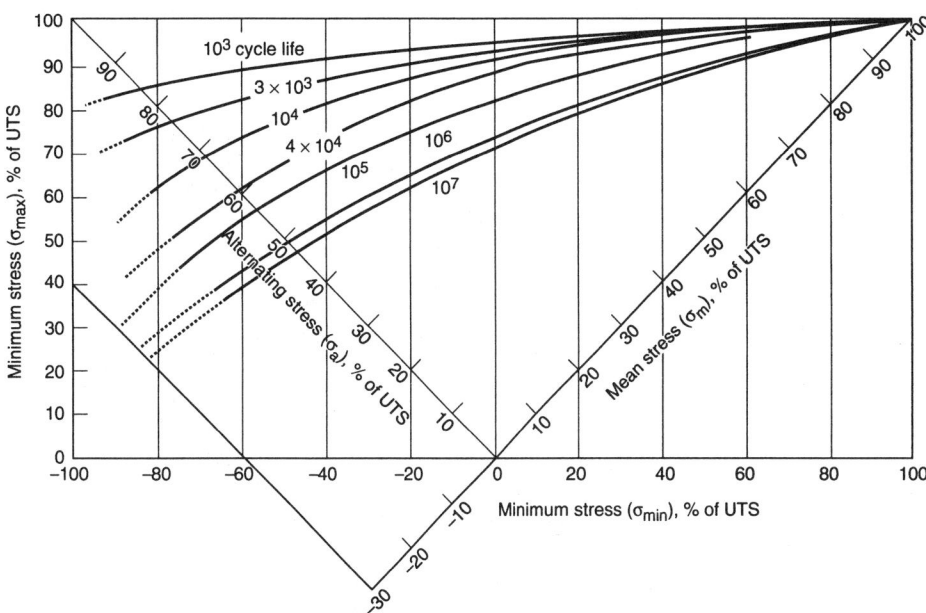

Fig. 5 A constant-life diagram for alloy steels that provides combined axes for more ready interpretation. Note the presence of safe-life, finite-life lines on this plot. This diagram is for average test data for axial loading of polished specimens of AISI 4340 steel (ultimate tensile strength, UTS, 125 to 180 ksi) and is applicable to other steels (e.g., AISI 2330, 4130, 8630). Source: Ref 20

S_{min} axes to produce a constant-life diagram. Examples of these are provided below.

For general consideration of mean stress effects, various models of the mean-amplitude response have been proposed. A commonly encountered representation is the Goodman line, although several other models are possible (e.g., Gerber and Soderberg). The conventional plot associated with this problem is produced using the Haigh diagram, with the Goodman line connecting the ultimate strength on S_m, and the fatigue limit, corrected fatigue limit, or fatigue strength on S_a. This line then defines the boundary of combined mean-amplitude pairs for anticipated safe-life response. The Goodman relation is linear and can be readily adapted to a variety of manipulations.

In many cases Haigh or constant-life diagrams are simply constructs, using the Goodman representation as a means of approximating actual response through the model of the behavior. For materials that do not have a fatigue limit, or for finite-life estimates of materials that do, the fatigue strength at a given number of cycles can be substituted for the intercept on the stress-amplitude axis. Examples of the Haigh and constant-life diagrams are provided in Fig. 4 and 5. Figure 5 is of interest also because of its construction in terms of a percentage of ultimate tensile strength for the strength ranges included.

What are some other examples of metallic response to cyclic loading in this regime? First, consider the behavior of an aluminum alloy 2219-T85 in Fig. 6, consistent with current MIL-HDBK-5 presentations, showing a S_{max} versus log N plot with the supporting data shown. Figure 7 shows the constant-life diagram for Ti-6Al-4V, solution treated and aged, from another MIL-HDBK-5 case: it includes both notched and unnotched behavior, and constant-life lines for various finite-life situations.

Plastics and polymeric composites are interesting materials for the variety of responses they can present under mechanical loading, with dynamic excitation being no exception. The nature of hy-

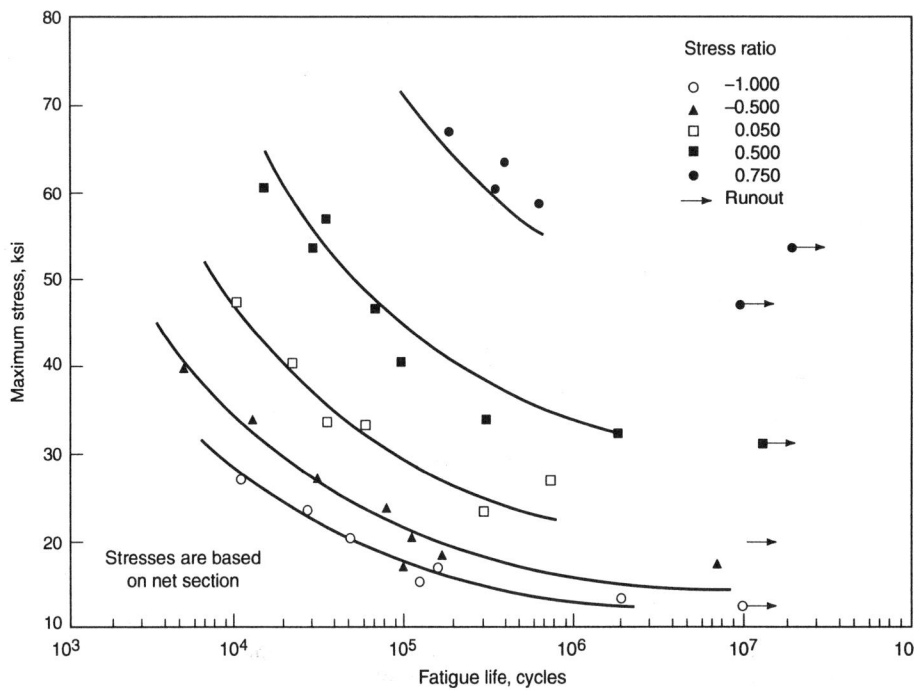

Fig. 6 Best-fit S/N curves for notched, $K_t = 2.0$, 2219-T851 aluminum alloy plate, longitudinal direction. This is a typical S-N diagram from MIL-HDBK-5D showing the fitted curve as the actual data that support the diagram. This is the currently required approach for representing this type of information in that handbook. Source: Ref 21

drocarbon bonding results in substantially more hysteresis losses under cyclic loading and a greater susceptibility to frequency effects. An example of S-N-type results for a variety of materials is provided in Fig. 8 (which is missing one dynamic variable). Also, different specifications are used for fatigue testing of plastics (e.g., Ref 24). The plastics industry also employs tests to

determine a "static" fatigue response, which is a sustained load test similar to a stress-rupture or creep test of metallic materials.

In application, this method is in its simplest form for steels in a benign environment. The task is to compare the S_a determined in the part to a S_a versus N curve at the necessary R value. If the operational S_a is less than the fatigue limit, then

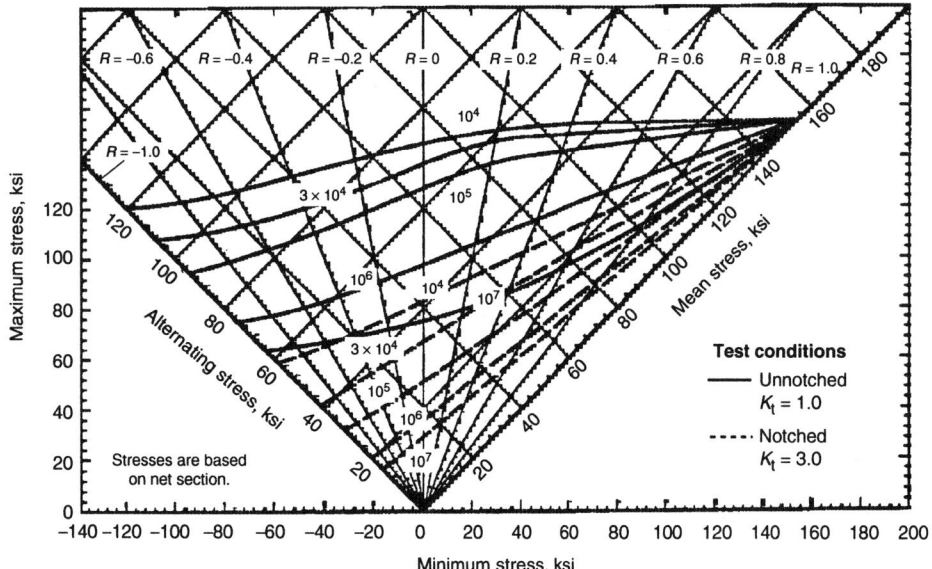

Fig. 7 Typical constant-life diagram for solution-treated and aged Ti-6Al-4V alloy plate at room temperature, longitudinal direction. Notched and smooth behavior are indicated in this constant-life diagram in addition to the finite-life lines. The influence notches is one of the critical effects on the fatigue of component details. Source: Ref 22

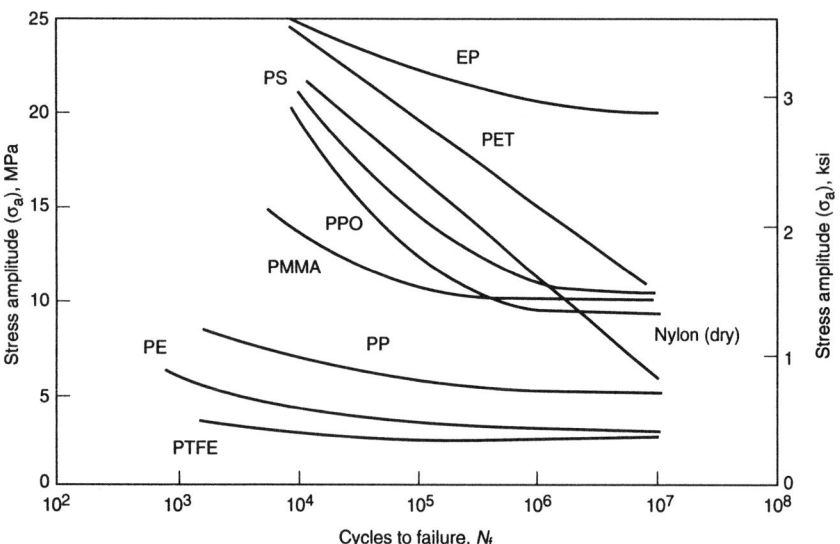

Fig. 8 Typical fatigue-strength curves for several polymers (30 Hz test frequency). Source: Ref 23

an acceptable safe-life, infinite-life situation exists (for whatever reliability was implied). In a slightly more complex scenario, the S_m, S_a pair operating in a component is compared to the appropriately determined Goodman line on a Haigh diagram with two possible results: results on or under the Goodman line indicate an acceptable safe-life, infinite-life situation; or while results above the Goodman line indicate a finite-life situation that can be managed if the general boundary conditions of the method are not heavily abused.

Difficulties occur in multiaxial stress states (discussed in a separate article elsewhere in this Volume) because of the difficulty in identifying an appropriate "stress." The assumption of the failure criterion associated with separation can be problematic in disparate coupon-structure situations. While cumulative damage can be accounted for using this technique, there is no means of including load sequence effects in variable-amplitude loading (which are known to be important).

The stress-life technique offers a variety of advantages. Its extension using strain as a controlling quantity is a natural progression of technology.

Finite-Life Criterion (ε-N Curves)

With more advanced and highly loaded components, it became obvious that stress-based techniques alone would not be sufficient to handle the full range of problems that needed to be addressed using continuum assumptions. The occurrence of plasticity, for example, and the accompanying lack of proportionality between stress and strain in this regime led to the use of strain as a controlling quantity. This was an evolutionary, not revolutionary, change in technology.

Strain-life is the general approach employed for continuum response in the safe-life, finite-life regime. It is primarily intended to address the "low-cycle" fatigue area (e.g., from approximately 10^2 to 10^6 cycles). The basic approaches and modeling, however, also make it amenable to the treatment of the "long-life" regime for materials that do not show a fatigue limit. The use of a consistent quantity, strain, in dealing with both, rather arbitrarily described "high-" and "low-cycle" fatigue ranges, has considerable advantages.

Work in this area was underway in the 1950s (Ref 25, 26). Cyclic thermal cracking problems contributed some of the stimulus for investigation, but the primary driving forces seem to have come from the power generation, gas turbine, and reactor communities. While the general approaches have remained consistent since that time, other outgrowths have offered variations on the theme (Ref 27-29). A simple summary of the strain-life approach can be found in Ref 30.

From a properties standpoint, the representations of strain-life data are similar to those for stress-life data. Rather than S-N, there are now ε-N plots, with a log-log format being most common. The curve represents a series of points, each associated with an individual test result. The vertical axis can have different strain quantities plotted, however. While total strain amplitude seems to be the most common quantity presented, total strain range, plastic strain range, or other determined strain measures can also be found. In ε-N tests the strain can be monitored either axially or diametrally (watch for this possible variable). Again, be aware of the type of presentation, and consider critically what the independent variable is. Also, look for the necessary two dynamic load quantities to define the testing conditions and the specific failure criterion employed.

For data generation to support the ε-N method, there are standards by which testing is conducted (e.g., Ref 31, which includes suggestions for the information to be recorded with the results). According to Ref 31, any of the following may be used as the failure criterion: separation, modulus ratio, microcracking ("initiation"), or percentage of maximum load drop.

Testing for strain-life data is not as straightforward as the simple load-controlled (stress-controlled) S-N testing. Monitoring and controlling using strain requires continuous extensometer capability. In addition, the developments of the technique may make it necessary to determine certain other characteristics associated with either monotonic or cyclic behavior. The combined output of the extensometer and load cell provides the displacement-load trace from which the hysteresis loop is formed. After several to several hundred strain excursions, the hysteresis

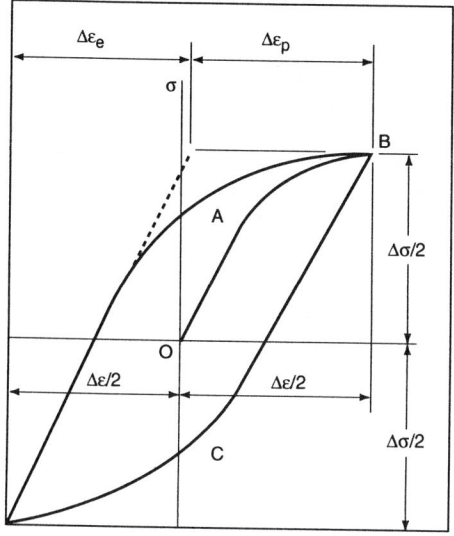

Fig. 9 Stress-strain hysteresis in a constant-amplitude strain-controlled fatigue test. Source: Ref 32

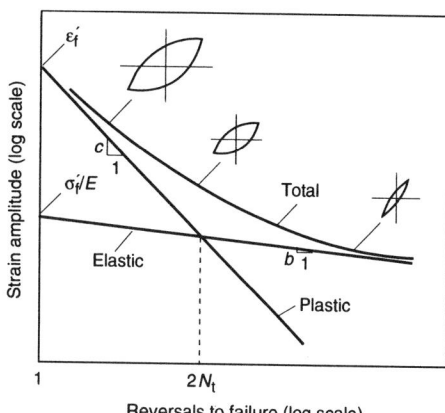

Fig. 10 Representation of total strain amplitude vs. number of reversals to failure, including elastic and plastic portions as well as the combined curve N_t, transition life from plastic (low-cycle) regime to the elastic (high-cycle, regime)

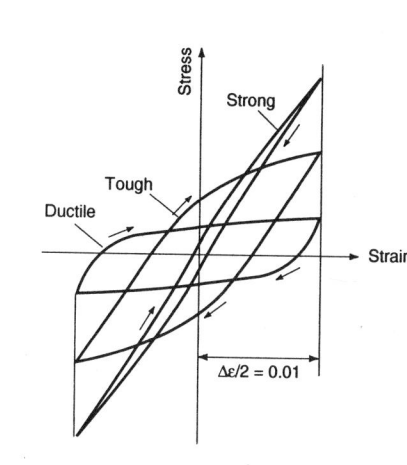

Fig. 11 Schematic representation of the cyclic strain resistance of idealized metals. Response to the strain-controlled testing has resulted in several generalizations of material behavior, which this figure displays in two different formats for a better appreciation of the descriptions. Source: Ref 35

loop typically stabilizes. This stabilized loop is shown in Fig. 9, which indicates the partitioning of the response into elastic and plastic portions. A stabilized loop of this type is formed during every constant-amplitude test and should be recorded as part of test procedures.

Any given stabilized hysteresis loop represents only one of many such loops that would result from conducting the series of tests that are required to develop an ε-N curve. The sequential connection of the vertices of these loops (e.g. point B of Fig. 9) conducted at different strain levels from what is known as the cyclic stress-strain curve. Some of the parameters used in developing the response models for strain-life technology are derived from the cyclic stress-strain curve. Later sections deal with this topic more extensively and additional material on this important subject can be found in the references provided here.

In some cases, strain control is discontinued after loop stabilization and the test proceeds under load control (usually used on long-life samples). If the failure criterion is other than separation or load drop, other monitoring/inspection capabilities may also be required. With one sample per data point and several to many samples to generate an entire curve, replicate tests are important to gage both mean behavior and scatter.

Modeling of the ε-N curve currently employs the separated elastic and plastic strain contributions described above. The total strain amplitude, $\Delta\varepsilon/2$, is considered as follows (note the use of half the range for strain amplitude, instead of ε_a):

$$\Delta\varepsilon/2 = \Delta\varepsilon_e/2 + \Delta\varepsilon_p/2$$

$$= (\sigma_f'/E) \cdot (2N_f)^b + \varepsilon_f' \cdot (2N_f)^c \quad \text{(Eq 1)}$$

where $\Delta\varepsilon/2$ is the total strain amplitude, $\varepsilon_e/2$ is the elastic strain amplitude, $\varepsilon_p/2$ is the plastic strain amplitude, σ_f' is the fatigue strength coefficient, b is

the fatigue strength exponent, ε_f' is the fatigue ductility coefficient, c is the fatigue ductility exponent, and $2N_f$ is the number of reversals to failure (2 reversals = 1 cycle).

A graphical representation of this modeling practice is shown in Fig. 10 (Ref 33). The coefficients and exponents either represent determined cyclic characteristics or can be approximated from monotonic tests. Further appreciation of these terms, means of approximating the necessary coefficients, and the variety of related technology can be gained in either Bannantine (Ref 13) or Conway (Ref 5). The use of approximations can result in synthetic or constructed ε-N plots that contain no real data, similar to the creation of S-N curves or Goodman lines and should be acknowledged as such.

The use of the number of reversals to failure as opposed to the number of cycles to failure seems to be an artifact of early developments in the field. The relationship is simple: a cycle consists of two reversals. There appears to be no argument for its retention in the context of the strain-life expression, but it has become a working part of this technological "package." Note that *a reversal need not imply fully reversed loading* ($R = -1$), but may only indicate a change in direction in load.

As with all methods, there must be a mechanism for treating mean stresses, while mean strain effects are apparently considered negligible (Ref 34). One of the factors that are readily implemented in the strain-life expression is a Morrow-type correction factor in the elastic term of Eq 1:

$$\Delta\varepsilon/2 = \varepsilon_e + \varepsilon_p$$

$$= [(\sigma_f' - \sigma_0)/E] \cdot (2N_f)^b + \varepsilon_f' \cdot (2N_f)^c \quad \text{(Eq 2)}$$

where σ_0 is the mean stress (as determined from the hysteresis loop developed at the detail, not the mean elastic stress). The convenience of the mathematical

representation is readily evident here, and the inclusion of this term generally follows the actual data. Although it requires the mean stress from the hysteresis loop (a supplementary determination or calculation), this is a complete expression. In practice, the application would require the estimation of the strain amplitude and resulting mean stress at the detail, then an iterative solution for the number of reversals to failure, $2N_f$.

The important steps, though, are to review properties and offer examples of the various behaviors. Figure 11 shows generalizations of the response of metallic materials to strain-controlled testing. The terms *strong*, *tough*, and *ductile* are general descriptors of the response.

Because most examples of these data are quite similar, only a selected few are reviewed here.

Figures 12 and 13 offer composite plots of several steels and aluminum alloys. Note that these plots use strain amplitude on the ordinate; there was no second dynamic variable or failure criterion provided. The display of monotonic and cyclic response of the materials produces an interesting plot. It is instructive to reflect on the generalization of Fig. 11 as it is represented in Fig. 12 and 13.

The "low-cycle fatigue" characterization of nickel-base superalloys is an area of considerable interest for various high-temperature applications. Several alloys are shown in Fig. 14, which represents the responses at 850 °C. This plot utilizes total strain range, no R or A value or failure criterion was specified. At elevated temperatures, wave-form, frequency, holdtime, and other effects may be more evident, and occasionally material instabilities may contribute to the response. Creep-fatigue interactions can alter an assumed "simple" fatigue situation to a considerable degree. Plastics and composites also can be approached in this manner (Fig. 15). Orientation effects can dominate this response, and loading must be carefully considered. Two strain-life plots show varying responses in fiber-reinforced composites in Fig. 15.

A distinct advantage of the strain-life method is its ability to deal with variable-amplitude loading through improved cumulative "damage" assessment. Cyclic plasticity responses are accounted for, and load sequence effects are reflected in the analysis and results, one area where the concepts of reversals and the development of closed loops remains important. In addition, advanced methods have been developed to address elevated-temperature situations where creep and fatigue are active simultaneously.

Multiaxial loads as well as in- and out-of-phase loading remain a problem and have not yet been addressed successfully in a general sense. Each situation should be reviewed carefully for possible interactions, and situation-specific testing may be required. More detailed coverage is provided in the article "Multiaxial Fatigue Strength" in this Volume.

Application of the strain-life method in its simplest form is to compare the total strain amplitude ($\Delta\varepsilon/2$) at a detail of the part to a ε-N curve having the necessary mean strain (stress) effects included. The assumption here is that the detail on the part, perhaps in a high-constraint area, will respond identically to a specimen that is inherently a smooth bar in plane stress, albeit at the same strain level. The life, of course, corresponds to the intercept of the strain level and the ε-N curve. In many instances, no actual visual comparison is done; instead, the determination is readily done through calculation using the mathematical model of the ε-N curve. The result is typically a safe-life, finite-life estimate, consistent with the reliability and failure criterion of the model.

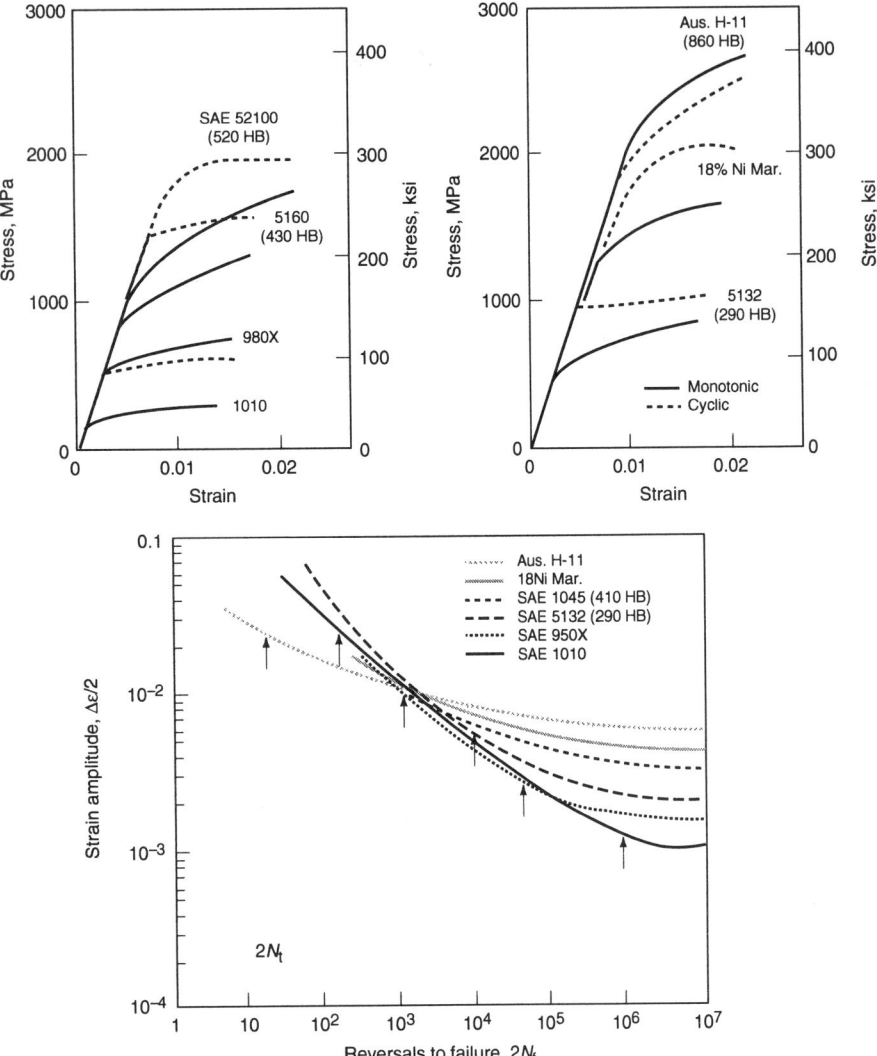

Fig. 12 Examples of the fatigue response of several steels, including their monotonic and cyclic strain-stress curves and their ε-N response. Source: Ref 36

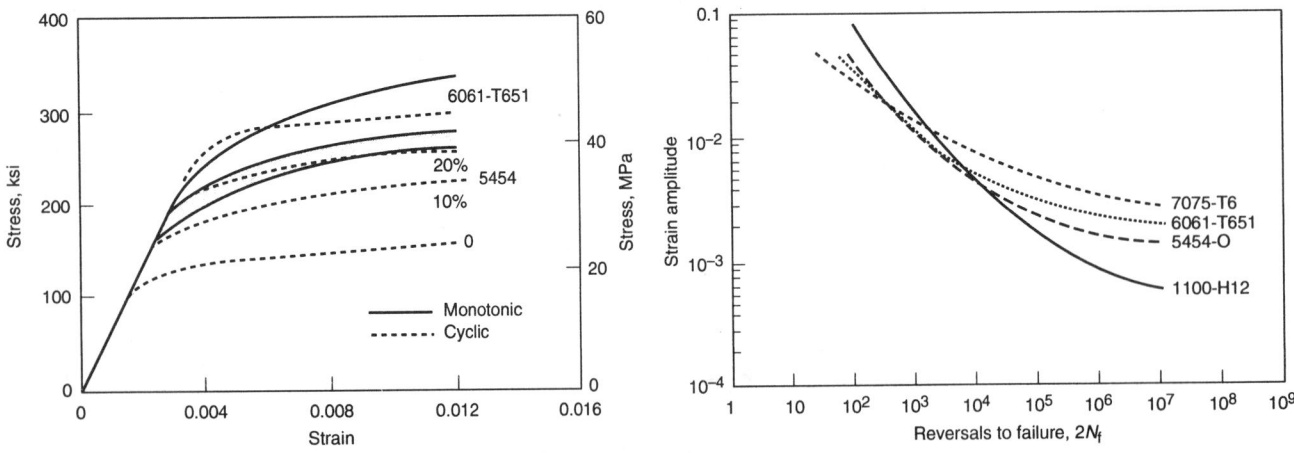

Fig. 13 Fatigue behavior of several aluminum alloys. Aluminum alloys are readily characterized using the strain-controlled methods. The general lack of a fatigue limit in these materials is well represented by the ε-N method. Source: Ref 37

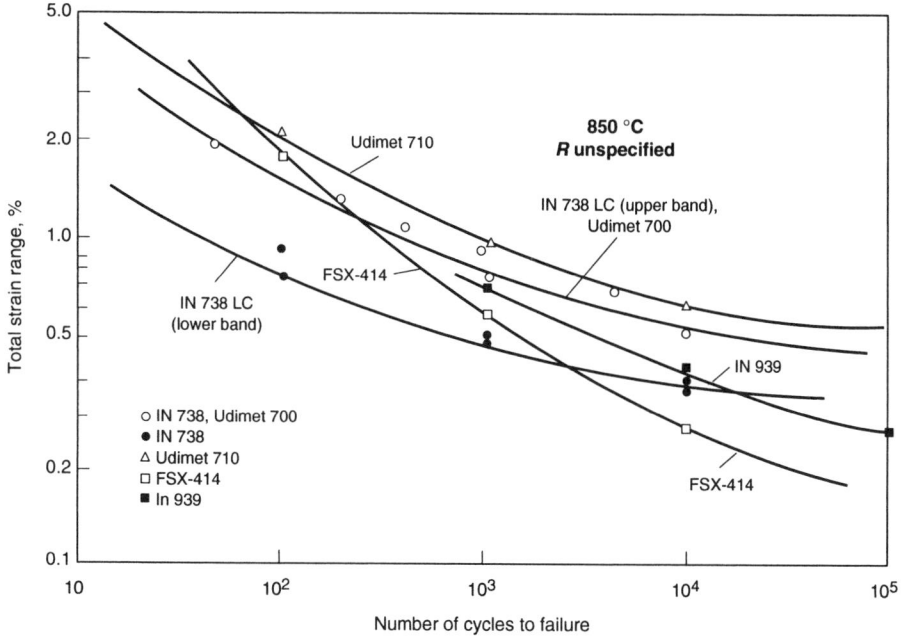

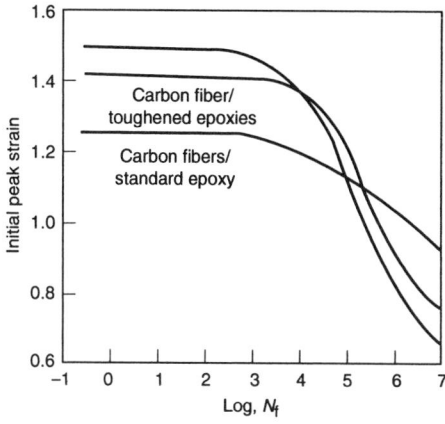

(a)

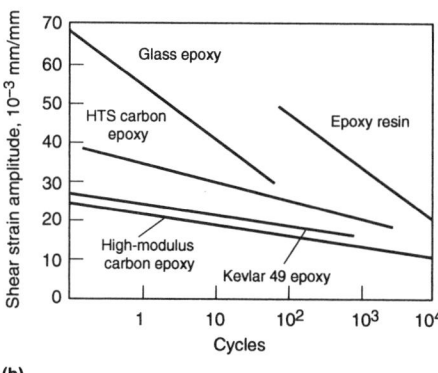

(b)

Fig. 14 Low-cycle fatigue curves for superalloys at 850 °C (1560 °F). Superalloys used under high-load, high-temperature situations are frequently characterized in the safe-life, finite-life regime. This comparison at 850 °C (1560 °F) shows that different alloys can be "better" depending on the specific life desired for the coupon. Source: Ref 38

Fig. 15 Fatigue strain-life data. (a) For unidirectional carbon-fiber composites with the same high-strain in different epoxy matrices. (b) Torsional shear strain-cycle diagram for various 0° fiber-reinforced composites. Source: Ref 39

Damage Tolerant Criterion (*da/dN* vs. Δ*K*)

The *S-N* and ε-*N* techniques are usually appropriate for situations where a component or structure can be considered a continuum (i.e., those meeting the 'no cracks' assumption). In the event of a crack-like discontinuity, however, they offer no support. The mandate is either to "attempt to remove the crack" or "remove the parts." The fact that components with "cracks" may continue to bear load is generally unaddressable using either *S-N* or ε-*N* methods (except through residual life testing).

So what has made these two techniques no longer usable? One point is the inability of the controlling quantities to make sense of the presence of a crack. A brief review of basic elasticity calculations shows that both stress and strain become astronomical at a discontinuity such as a crack, far exceeding any recognized property levels that might offer some sort of limitation. Even invoking plasticity still leaves inordinately large numbers or, conversely, extremely low tolerable loads. An alternative concept and controlling quantity must be used.

That quantity is *stress intensity*, a characterization and quantification of the stress field at the crack tip. It is fundamental to linear elastic fracture mechanics. It recognizes the singularity of stress at the tip and provides a tractable controlling quantity and measurable material property. (Note: The stress intensity as used here *is not* the same as the stress intensity identified with the ASME Boiler and Pressure vessel calculations, which use this term to define the difference between the maximum and minimum principal stresses.)

The development of fracture mechanics has roots in the early 1920s and has developed considerably since the late 1940s and early 1950s. Examples of applicable texts are Ref 40 and 41.

A very basic expression for the stress intensity is its determination for a semi-infinite center-cracked panel having a through-thickness crack of length 2*a* in a uniform stress field that is operating normal to the opening faces of the crack. The resulting stress intensity is as follows (Ref 42):

$$K_I = \sigma \cdot (\pi \alpha)^{0.5} \qquad (Eq\ 3)$$

where σ is the far field stress responsible for opening mode loading (mode I) and *a* is the crack depth in from the edge of the plate. This formula allows an immediate appreciation of the combined influence of stress and crack length common to all stress intensity determinations. Specifically, stress intensity depends directly, but not singularly, on stress, and secondly it depends on crack length.

In a more general format, stress intensities might be expressed as:

$$K_I = \sigma \cdot Y \cdot (\pi a)^{0.5} \qquad (Eq\ 4)$$

where Y is a geometric factor allowing the representation of other geometries. For example, the correction for finite width (*W*) of Eq 3 is (Ref 43):

$$K_I = \sigma \cdot \{sec[\pi(a/W)]\}^{0.5} \cdot (\pi a)^{0.5} \qquad (Eq\ 5)$$

In addition, many geometries (geo), including test specimens, do not readily lend themselves to stress determinations, but the applied loads

(forces) are known, so the stress intensity would take the form:

$$K_I = P \cdot (geo) \cdot f(a/W) \qquad (Eq\ 6)$$

From a philosophical standpoint, a stress can never be applied, but a load can. Stress is always a resultant and determined quantity; it is not measurable. It is a mathematical device that has some very useful characteristics and provides a wealth of interpretations and insights, especially in reflecting an areal rationalized force (load) path through the structure. Structures and materials, however, only experience loads (mechanical, thermal, chemical, etc.) and respond with strains and displacements.

In some cases, however, where complexity precludes a simple "stress" approach, analytical techniques do allow the calculation of stress intensity factors under the imposed loads.

The connection of stress intensity, K_I, as a controlling quantity for fracture is a direct consequence of a physical model for linear elastic fracture under plane-strain conditions. Its limit is K_{Ic}, the critical plane-strain fracture toughness. The use of the stress intensity range, ΔK_I, as a controlling quantity for crack extension under cyclic loading is simply by correlation. The ability of

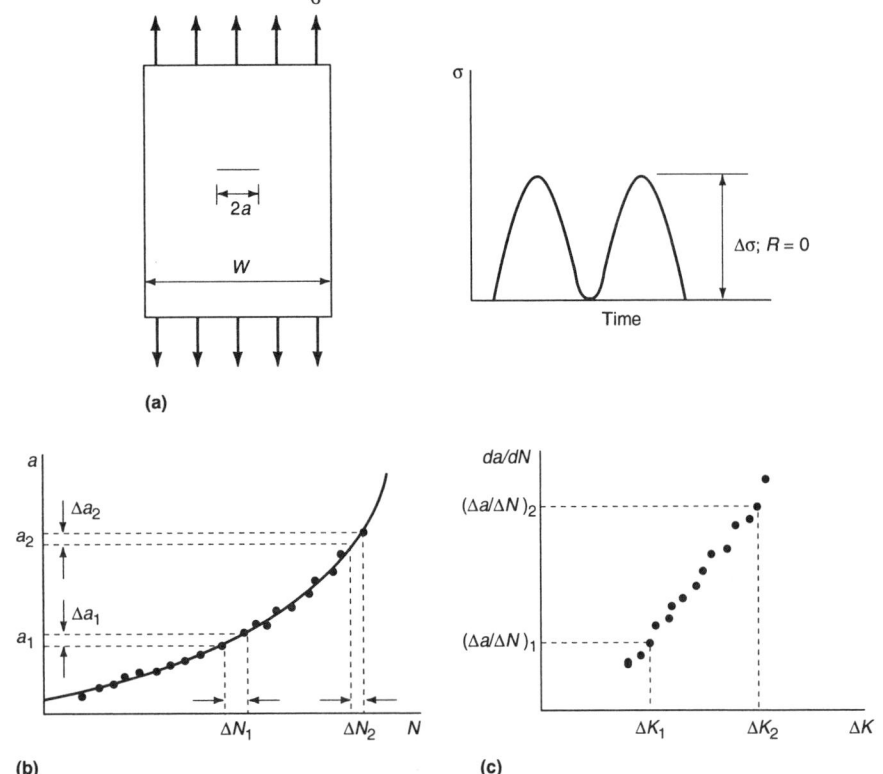

(a)

(b)

(c)

Fig. 16 Schematic representation of the specimen, data, and modeling process for generating fatigue crack growth rate ($da/dN - \Delta K$) data. (a) Specimen and loading. (b) Measured data. (c) Rate data. Source: Ref 44

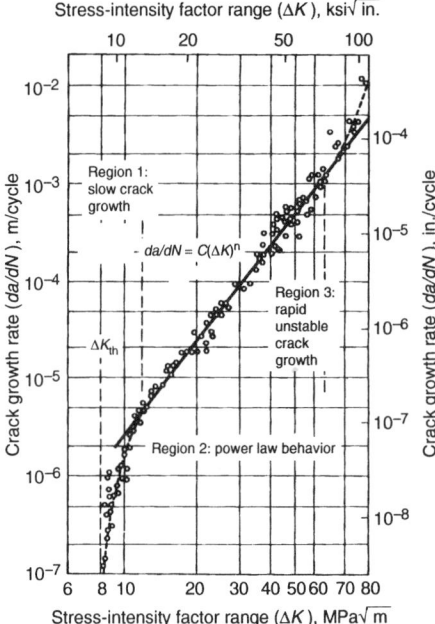

Fig. 17 Entire da/dN vs. ΔK plot for A533 steel showing asymptotic behavior at either end of the curve and a relatively linear portion in the center. Yield strength 470 MPa (70 ksi). Test conditions: $R = 0.10$; ambient room air, 24 °C (75 °F). Source: Ref 46

the stress intensity to reflect crack-tip conditions remains mathematically correct, but the correlation of ΔK_I to crack growth is a successful application by repeated demonstration. By altering Eq 3 using $\Delta\sigma$ instead of σ, ΔK_I results:

$$\Delta K_I = \Delta\sigma \cdot (\pi a)^{0.5} \qquad \text{(Eq 7)}$$

The stress intensity range to a certain extent simply reflects an extension of the stress-based practices. However, the testing to support fracture mechanics-based fatigue data is done differently than in the *S-N* or ε-*N* methods because of the necessity to monitor crack growth. Crack growth testing is performed on samples with established K_I versus a characteristics. Under the controlled load specified using two dynamic variables, the crack length is measured at successive intervals to determine the extension over the last increment of cycles. Crack length measurement can be done visually or by mechanical or electronic interrogation of the sample using established techniques that allow for automation of the process.

The immediate results from the testing then are not da/dN, but a versus N. Subsequent manipulation of the a-N data set using numerical differentiation provides da/dN versus a. Coupling this latter data with a stress intensity expression (K_I as a function of load and crack length) for the specific sample results in the final desired plot of da/dN versus ΔK_I. This process is shown schematically in Fig. 16. Details of this procedure can be found in Ref 45. The da/dN versus ΔK_I curve has a sigmoidal shape, and a full data set covers

crack growth rates that range from threshold to separation. It is important to note that this data represents only "long crack" behavior; that is, the cracks are substantially greater in size than any controlling microstructural unit (e.g., grain size) and typically exceed several millimeters in length. A second important assumption is that of a plane-strain stress state; therefore, a plane-stress descriptor is not required.

A real test of modeled da/dN vs. ΔK_I expressions is whether, under reintegration, the original a-N data will be reproduced. This type of review should be consistently employed to assess the integrity of the modeling process.

The generation of da/dN versus ΔK_I data is obviously considerably more involved than either *S-N* or ε-*N* testing. It does have the advantage, however, of producing multiple data points from a given test.

Figure 17 reflects interesting features at each extreme of the da/dN vs. ΔK_I curve. First, at the upper limit of ΔK_I, it reaches the point of instability and the crack growth rates become extremely large as fracture is approached. The second point of interest is the lower end of the ΔK_I range where crack growth rates essentially decrease to zero; this is identified as the fatigue crack growth threshold, $\Delta K_{I, th}$.

The existence of threshold behavior at low ΔK_I values is analogous, in some senses, to the fatigue limit of some ferrous materials in *S-N* response. If, with the appropriate R ratio, the stress intensity range is below the threshold value, $<\Delta K_{I,th}$, cracks will not extend under the applied load(s).

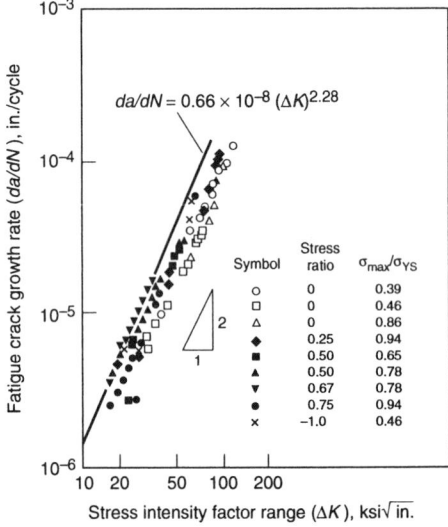

Fig. 18 An example of the use of a Paris relation to model the linear portion of a crack growth rate curve. Effect of stress ratio on the fatigue crack propagation rate in a 140 ksi yield strength martensitic steel. Source: Ref 47

Thus, an assessment of nonpropagating crack(s) can be made under the appropriate circumstances.

A reflection on failure criterion is appropriate here. Much as K_I is a quantity for assessing the point of initiation of propagation of unstable fracture, $\Delta K_{I,th}$ functions as the limit for the initiation of crack propagation (for "long" cracks) under cyclic loading. Above $\Delta K_{I,th}$ and below instabil-

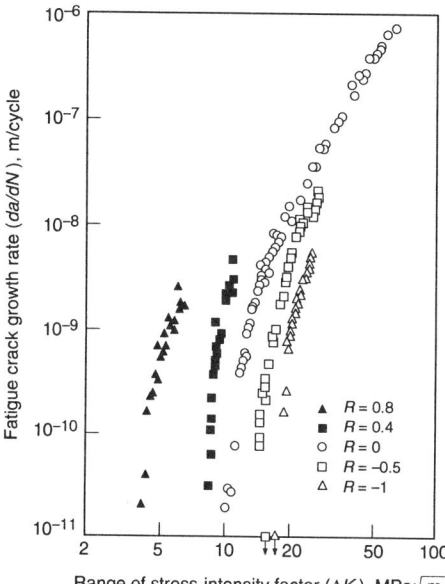

Fig. 19 The influence of R value on the fatigue crack growth characteristics of a steel. Fatigue crack propagation properties of JIS SS41 steel. Source: Ref 48

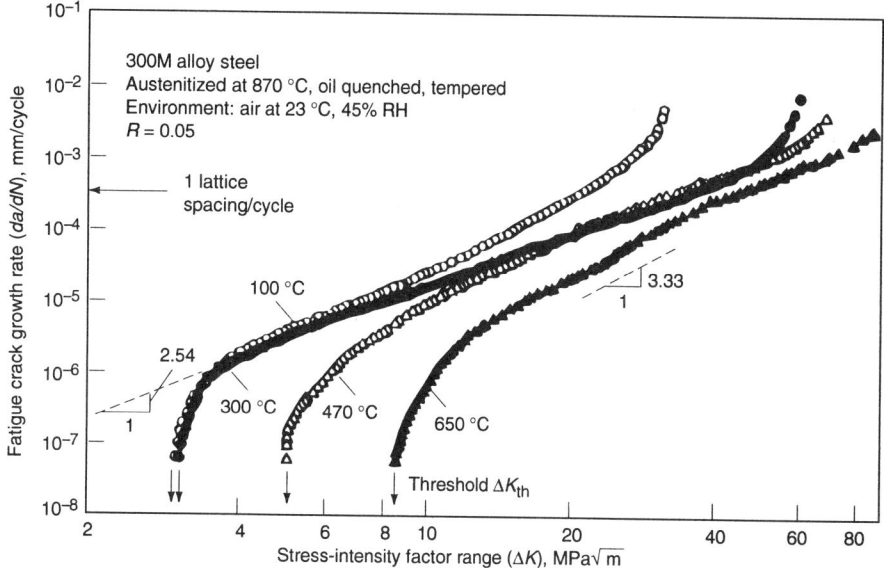

Fig. 20 R value effects on threshold and fracture instability behavior of 300M steel. Source: Ref 49

ity, the criterion for subcritical extension is satisfied and the rate is as determined by the curve.

Modeling of the central portion of the da/dN versus ΔK_I curve is frequently done using the Paris equation:

$$da/dN = C\,(\Delta K_I)^n \qquad \text{(Eq 8)}$$

This is a very simple exponential relation that can readily be curve fit to the desired portion of the data (see Fig. 18, where one of the data sets is modeled as indicated).

Stress or load ratio (mean stress) effects exist in crack propagation data as well. The influence is that increasing R ($R = K_{min}/K_{max}$) decreases both the threshold value at the low end and the instability characteristic at the upper end of the da/dN versus ΔK_I curve. Examples of this are provided in Fig. 19 and 20.

Microstructural features must influence fatigue crack growth characteristics as well as all other properties. An example of this is provided in Fig. 21, which shows the compound influence of both gamma prime and grain sizes on fatigue crack propagation in Waspaloy.

Plastics also can be analyzed using this technique. Figure 22 shows a variety of materials that are displayed in the conventional form. Many polymeric materials exhibit substantial frequency effects, and this should be considered in the generation of data.

In application, use of da/dN versus ΔK_I is completely different than either the S-N or ε-N continuum method. Instead of providing an immediate life estimate in association with a given stress or strain combination and a test coupon's modeled failure criterion, a more complicated determination is required. Using the a versus ΔK_I relation in the part, the applied loads are employed to assess crack extension over incremental changes

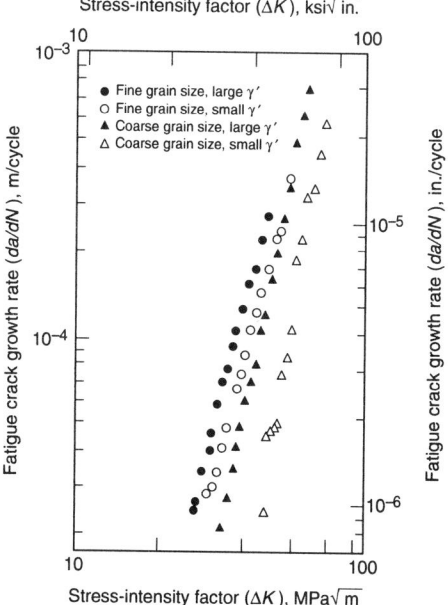

Fig. 21 Effect of grain size on fatigue crack growth of Waspaloy superalloy. Source: Ref 50

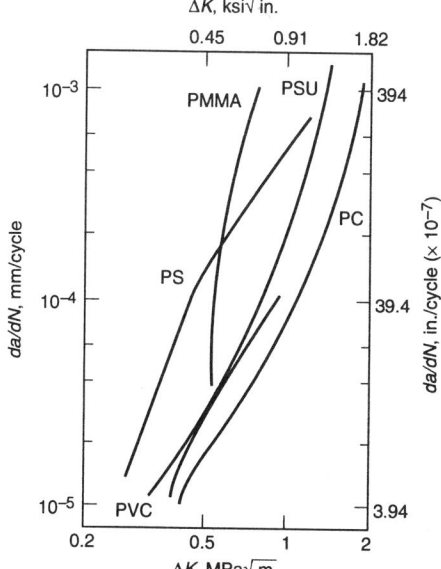

Fig. 22 Fatigue crack propagation behavior of various polymers. Source: Ref 51

in length, and they are continuously summed to reflect the total increase. In essence, this is the regeneration of the a–N curve for the specific part. Crack growth can be assessed until fracture (achieving a critical crack size) or some other intermediate point.

It is common in several industries to use the above technique to determine intervals between directed inspections to ensure structural integrity. This ability to formulate inspection schedules that maximize the likelihood of detecting an extending discontinuity prior to it becoming critical is one of the principal advantages of the damage

tolerant approach. Even using this technique, a crack, once discovered, cannot necessarily be left in place. Removal or structural modification may be the only acceptable alternative (e.g. airframes). In other instances, the predictive aspects of the technique can justify continued operation of equipment under full or derated conditions while waiting for programmed replacement parts or with a stated finite-life limit for the unit (supplemental inspections may be required). Probabilistic methods can be used to produce a quantified risk assessment for these situations.

Extreme caution is advised regarding units, especially for da/dN expressions, where conversions among units can be very confusing because of the mathematical relationships involved. Multiaxiality in this instance is reflected in complex loading modes (e.g., combinations of modes I, II, and III). In a very general sense, the cracks tend to extend in directions that are normal to maximum principal stresses.

REFERENCES

1. A. Wöhler, Versuche über die Festigkeit der Eisenbahnwagenachsen, *Zeitschrift für Bauwesen*, 1860
2. Dynamically Loaded Structures, "*AWS Structural Welding Code*," ANSI/AWS D1.1-92, American Welding Society, 1992, p 185-201
3. *A Guide for Fatigue Testing and the Statistical Analysis of Fatigue Data*, STP 91A, ASTM, 1963
4. *Manual on Statistical Planning and Analysis*, STP 588, ASTM, 1975
5. J.B. Conway and L.H. Sjodahl, *Analysis and Representation of Fatigue Data*, ASM International, 1991
6. ASTM E 468-90, Presentation of Constant Amplitude Fatigue Test Results for Metallic Materials, *Annual Book of ASTM Standards*, Vol 03.01, ASTM, 1995
7. ASTM E 1150-87 (1993), Standard Definitions of Terms Relating to Fatigue, *Annual Book of ASTM Standards*, Vol 03.01, ASTM, 1995
8. D.W. Hoeppner, Estimation of Component Life by Application of Fatigue Crack Growth Threshold Knowledge, *Fatigue, Creep, and Pressure Vessels for Elevated Temperature Service*, MPC-17, ASME, 1981, p 1-84
9. C. Lipson, G.C. Noll, and L.S. Clock, *Stress and Strength of Manufactured Parts*, McGraw-Hill, 1950
10. J.E. Shigley and L.D. Mitchell, *Mechanical Engineering Design*, McGraw-Hill, 4th ed., 1983
11. A.H. Burr, *Mechanical Analysis and Design*, Elsevier, 1981
12. H.O. Fuchs and R.I. Stephens, *Metal Fatigue in Engineering*, John Wiley and Sons, 1980
13. J.A. Bannantine, J.J. Comer, and J.L. Handrock, *Fundamentals of Metal Fatigue Analysis*, Prentice-Hall, 1990
14. *Fatigue Design Handbook*, Society of Automotive Engineers, 2nd ed., 1988
15. ASTM E 468-90, Standard Practice for Presentation of Constant Amplitude Fatigue Test Results for Metallic Materials, *Annual Book of ASTM Standards*, Vol 03.01, ASTM, 1995
16. H.J. Grover, S.A. Gordon, and L.R. Jackson, *Fatigue of Metals and Structures*, NAVAER 00-25-534, Prepared for Bureau of Aeronautics, Department of the Navy, 1954
17. R.C. Juvinall, *Engineering Considerations of Stress, Strain, and Strength*, McGraw-Hill, 1967, p 234
18. "Design of Transmission Shafting," ANSI/ASME B106.1M-1985, American Society of Mechanical Engineers, 1991
19. J.A. Bannantine, J.J. Comer, and J.L. Handrock, *Fundamentals of Metal Fatigue Analysis*, Prentice-Hall, 1990, p 6
20. R.C. Juvinall, *Engineering Considerations of Stress, Strain, and Strength*, McGraw-Hill, 1967, p 274
21. MIL-HDBK-5D, Military Standardization Handbook, *Metallic Materials and Elements for Aerospace Vehicle Structures*, 1983, p 3-164
22. MIL-HDBK-5D, Military Standardization Handbook, *Metallic Materials and Elements for Aerospace Vehicle Structures*, 1983, p 5-87
23. A. Moet and H. Aglan, Fatigue Failure, *Engineering Plastics*, Vol 2, *Engineered Materials Handbook*, ASM International, 1988, p 742
24. ASTM D 671-93, Test Method for Flexural Fatigue of Plastics by Constant-Amplitude-of-Force, *Annual Book of ASTM Standards*, Vol 08.01, ASTM, 1995
25. S.S. Manson, Fatigue: A Complex Subject—Some Simple Approximations, *Experimental Mechanics*, July 1965
26. S.S. Manson, *Thermal Stress and Low-Cycle Fatigue*, McGraw-Hill, 1966
27. J.A. Bannantine, J.J. Comer, and J.L. Handrock, *Fundamentals of Metal Fatigue Analysis*, Prentice-Hall, 1990
28. *Fatigue Design Handbook*, Society of Automotive Engineers, 2nd ed., 1988
29. J.B. Conway and L.H. Sjodahl, *Analysis and Representation of Fatigue Data*, ASM International, 1991
30. "Technical Report on Fatigue Properties," SAE J1099, Society of Automotive Engineers, 1985
31. ASTM E 606-92, Standard Practice for Strain Controlled Fatigue Testing, *Annual Book of ASTM Standards*, Vol 03.01, ASTM, 1995
32. R. Viswanathan, *Damage Mechanisms and Life Assessment of High Temperature Components*, ASM International, 1989, p 119
33. R.W. Langraf, The Resistance of Metals to Cyclic Loading, *Achievement of High Fatigue Resistance in Metals and Alloys*, STP 467, ASTM, 1970, p 24
34. J.A. Bannantine, J.J. Comer, and J.L. Handrock, *Fundamentals of Metal Fatigue Analysis*, Prentice-Hall, 1990, p 67
35. R.W. Langraf, The Resistance of Metals to Cyclic Loading, *Achievement of High Fatigue Resistance in Metals and Alloys*, STP 467, ASTM, 1970, p 27
36. *Fatigue Design Handbook*, Society of Automotive Engineers, 2nd ed., 1988, p 35
37. *Fatigue Design Handbook*, Society of Automotive Engineers, 2nd ed., 1988, p 41
38. R. Viswanathan, *Damage Mechanisms and Life Assessment of High Temperature Components*, ASM International, 1989, p 431
39. B. Jang, Design for Improved Fatigue Resistance of Composites, *Advanced Polymer Composites: Principles and Applications*, ASM International, 1994
40. H.L. Ewalds and R.J.H. Wanhill, *Fracture Mechanics*, Edward Arnold, 1984
41. K. Hellan, *Introduction to Fracture Mechanics*, McGraw-Hill, 1984
42. D. Broek, *The Practical Use of Fracture Mechanics*, Kluwer Academic Publishers, 1989, p 52
43. D. Broek, *The Practical Use of Fracture Mechanics*, Kluwer Academic Publishers, 1989, p 54
44. D. Broek, *The Practical Use of Fracture Mechanics*, Kluwer Academic Publishers, 1989, p 127
45. ASTM E 647-95, Standard Test Method for Measurement of Fatigue Crack Growth Rates, *Annual Book of ASTM Standards*, Vol 03.01, ASTM, 1995
46. J.H. Underwood and W.W. Gerberich, Concepts of Fracture Mechanics, *Application of Fracture Mechanics of Metallic Structural Materials*, American Society for Metals, 1982, p 18
47. J.M. Barsom and S.T. Rolfe, *Fracture and Fatigue Control in Structures*, Prentice-Hall, 1987, p 299
48. *Fatigue Design Handbook*, Society of Automotive Engineers, 2nd ed., 1988, p 42
49. *Fatigue Design Handbook*, Society of Automotive Engineers, 2nd ed., 1988, p 43
50. S.D. Antolovich and J.E. Campbell, Fracture Properties of Superalloys, *Application of Fracture Mechanics for Selection of Metallic Structural Materials*, American Society for Metals, 1982, p 271
51. A. Moet and H. Aglan, Fatigue Failure, *Engineering Plastics*, Vol 2, *Engineered Materials Handbook*, ASM International, 1988, p 747

Alloy Design for Fatigue and Fracture

Stephen D. Antolovich, Washington State University

FRACTURE MECHANICS is a very powerful tool for predicting the loads and crack lengths at which fracture can occur. Broken down to its essential form, it allows an engineer to predict the onset of fracture if the following information is available:

- Load/crack geometry (usually available from NDI)
- A formula for the so-called stress-intensity parameter, K, for the load/crack geometry of interest (the result of sophisticated mathematical analysis but available in handbooks, such as Ref 1)
- The numerical value of the fracture toughness (generally denoted K_{Ic}), which is determined experimentally through well-defined procedures (Ref 2)
- For fatigue crack propagation, knowledge of the crack growth rate as a function of the stress-intensity parameter

With this information and the use of fracture mechanic methods (as briefly described in the next section for monotonic and fatigue fracture), it is possible to compute the life without any consideration of those processes that determine the values of the fracture toughness or the crack growth rates. Such procedures as are outlined below are obviously of great value in carrying out engineering calculations for existing or contemplated components. However, in some instances the properties are insufficient to meet the engineering requirements. In such cases it is necessary to consider alternate materials or, in some instances, to develop alternate heat treatments and compositions that yield properties that allow the requirements to be met. Clearly in these instances it is important to have a clear understanding of the basic processes that affect toughness and fatigue so that effective changes can be made.

Another reason for understanding basic processes is that conditions can change. For example, if an alloy is selected for intended service at 298 K with no impact loading and the temperatures change to 250 K and impact loading occurs, what will be the effect of these changes be on the fracture properties of the material that has been selected? Is there perhaps another material that is more forgiving in terms of temperature and loading changes? These questions must be addressed if

materials are to be put into service (or developed) with confidence. In this context, the goals of this article are:

- To review the basic processes of fracture and fatigue
- To show how these processes occur in materials of engineering interest, such as iron-, aluminum-, titanium-, and nickel-base alloys
- To provide a reference of practical data for engineering alloys and to describe typical applications

Fracture Mechanic Methods

Fracture Mechanics in Monotonic Loading. In many applications there may be a pre-existing flaw that results from processing or from fabrication. Typically such flaws can result from welding, riveting, machining, or, in some cases, inherent material defects that may be considered to be "flaw-like." It is important to know whether the part can operate at the intended stresses given a particular flaw size. Of course, it may be possible in some instances to derate the operating stress to a level that is safe given the flaw that exists. In either case, an analysis can be carried out to determine if the component can be safely used. As an example of the way in which fracture mechanics is used to predict the onset of fracture, consider the following example (Ref 3):

Maximum Stress to Fracture. Suppose that the fracture toughness of a titanium alloy has been determined to be 44 MPa√m and a penny-shaped crack of diameter 1.6 cm (0.016 m) has been located in a thick plate that is to be used in uniaxial tension. If we assume plane-strain conditions and a material yield stress of 900 MPa (130 ksi), then the maximum allowable stress without fracture is calculated as follows.

Solution: The stress-intensity parameter formula for a penny-shaped crack is given by:

$$K = 2\sigma \left(\frac{a}{\pi}\right)^{1/2} \tag{Eq 1}$$

where a is the crack radius and σ is the applied stress. At fracture, the applied stress intensity is equal to the plane-strain fracture toughness: $K = K_{Ic}$.

Rearranging Eq 1 and substituting appropriate values gives:

$$\sigma = K/2[\pi/a]^{1/2} = 44/2[\pi/0.008]^{1/2}$$

$$\sigma_f = 436 \, \text{MPa} \tag{Eq 2}$$

Thus, fracture will occur well below the yield stress of the material. This calculation shows that there is no guarantee that fracture will not occur simply because the nominal applied stresses are below the yield stress.

Fracture Mechanics in Fatigue Loading. In the preceding example it is assumed that there is a pre-existing crack and that the only item of interest is the maximum load that can be applied without failure. While such a situation is certainly important, the more usual situation is that there is a pre-existing crack and cyclically applied loads are present whose magnitude is below that which would cause immediate fracture. In this case, the repeated application of a load (such that $K < K_{Ic}$) causes the crack to grow, slowly at first but more rapidly as the crack increases in length. The question now is how many cycles can be applied before the crack becomes so long that complete separation occurs? In order to determine the number of cycles, the crack growth rate as a function of the stress intensity parameter is required. This is usually available for materials of engineering interest in the form:

$$\frac{da}{dN} = f(\Delta K) \tag{Eq 3}$$

where N is the number of cycles and $\Delta K = K_{max} - K_{min}$. In this equation, ΔK is known as the *stress-intensity parameter range*. The stress-intensity parameter range characterizes the cyclic stresses and strains ahead of the crack tip and uniquely characterizes the crack growth rate through a relationship such as Eq 3. Perhaps one of the simplest, yet most widely used forms of Eq 3 is the Paris equation, which is used to describe crack growth behavior over a fairly broad range of ΔK. Equation 3 then takes on the specific form:

$$\frac{da}{dN} = C(\Delta K)^n \tag{Eq 4}$$

where C and n are material parameters that depend on temperature, frequency, and load ratio.

The cyclic life is computed by integration of the crack growth rate equation or by numerical integration of crack growth rate data. This is illustrated in a straightforward way by integration of the Paris equation, as shown below.

Estimation of Fatigue Life Using Paris Equation. The crack growth rate of 7075-T6 Al is given by:

$$\frac{da}{dN} = 5 \times 10^{-10} (\Delta K)^4$$

where ΔK is given in units of ksi$\sqrt{\text{in.}}$ and da/dN is given in units of in./cycle. Assume that a part contains a center crack that is 0.20 in. long. The stresses vary from 0 to 30 ksi and the fracture toughness is 25 ksi$\sqrt{\text{in.}}$ The life of the part is computed as follows.

Solution: For this geometry, the stress-intensity parameter is given by:

$$K = \sigma\sqrt{\pi a}$$

$$\Delta K = \Delta\sigma\sqrt{\pi a}$$

Using the information given in the problem statement and the above expression, the crack growth rate is given by:

$$\frac{da}{dN} = 5 \times 10^{-10} \times 30^4 \times \pi^2 \times a^2$$

$$= 4 \times 10^{-3} a^2$$

This equation can be integrated from the initial condition of $N = 0$ and $a = a_0 = 0.10$ in. to the final condition of $N = N_f$ and $a = a_f$. The final crack length a_f is the crack length at which fracture occurs, and it corresponds to the condition $K_{max} = K_{Ic}$. Here K_{max} corresponds to the value of K at the maximum stress. Integrating and rearranging terms gives the following expression:

$$N_f = 250 \int_{a_0}^{a_f} \frac{da}{a^2}$$

Integration of the right-hand side of the equation gives:

$$N_f = 250 \left[\frac{1}{a_0} - \frac{1}{a_f} \right]$$

The only remaining problem is to compute the crack length at failure. As mentioned above, a_f depends on the fracture toughness. For the geometry being considered, this is given by:

$$a_f = \frac{1}{\pi}\left(\frac{K_{Ic}}{\sigma}\right)^2 = \left(\frac{25}{30}\right)^2 = 0.22 \text{ in. (5.6 mm)}$$

Substituting into the expression for life, we have:

$$N_f = 250 \left(\frac{1}{0.1} - \frac{1}{0.22}\right) = 1363$$

The part is thus expected to last slightly over 1300 cycles.

Monotonic Fracture

As pointed out in the preceding section, fracture below the yield stress can occur if there are sufficiently large pre-existing flaws. There are different mechanisms by which such fracture can occur, and these mechanisms depend, to some extent, on the material being considered. In this section, a brief overview of the mechanisms and some related models are given for appropriate classes of materials. More detailed information on fracture mechanisms is given in the article "Micromechanisms of Monotonic and Cyclic Crack Growth" in this Volume.

In general, fracture toughness can be viewed as a property that depends on both strength and ductility, modified by the complex stress states that occur within the plastic zone and that trigger certain failure processes. Fracture properties are improved by fine grain sizes, low volume fractions of inclusions that are widely spaced, and special microstructural features (e.g., transformation-induced plasticity in steels, TRIP), and the mechanisms for crack deflection that exist in Al-Li alloys and certain morphologies of titanium alloys. These microstructural factors are briefly reviewed in this section as well.

While for a given system the toughness usually decreases with increasing strength, there are wide variations, and it is frequently possible to manipulate the microstructure and/or chemistry in order to increase the toughness while at the same time maintaining acceptable strength levels. Reducing the grain size of most alloys results in both an increase in toughness and an increase in strength. Thus, grain size control has been a popular mechanism for obtaining desirable combinations of mechanical properties.

Processing Effects on Toughness. Most alloys used in engineering applications must be processed in some way before they can be used. Processing frequently involves mechanical working at intermediate and high temperatures. Hot working can produce highly directional grain structures (e.g., in aluminum alloys) and a significant degree of texturing (e.g., in titanium alloys). The fracture toughness is affected by both of these types of directionality, and it is customary to see fracture toughness values tabulated in terms of the orientation relative to the principal direction of working. Generally, the fracture toughness will be high for orientations in which the grains are elongated normal to the crack plane (e.g., L-S, L-T) and proportionally lower for crack planes parallel to the elongated grain direction (e.g., S-T, S-L).

Of course, other processing steps also have a large effect on the fracture toughness. For example, if a material is produced by a powder metallurgy technique, there is the possibility of prior-particle boundary (PPB) contamination with gaseous species such as oxygen. The absorbed gases can weaken the interfaces of the compacted and sintered material through an effect on the interatomic bonding or through formation of gas bubbles in the material.

Basically, all processing will have an effect on the microstructure, chemistry, or dislocation substructure, which will in turn affect the fracture toughness.

Fabrication Effects on Toughness. Once processed and heat treated, materials must somehow be fabricated to produce a final component. Fabrication frequently involves material removal steps and joining steps, both of which affect the performance of the manufactured component. For example, both machining and welding may be involved.

Machining usually is not a source of reduced fracture toughness (although, as we will see below, it can have a significant effect on fatigue behavior). However, under some circumstances inappropriate machining may damage the surface of a semibrittle material sufficiently that fracture occurs below the net section yield. Materials that are used in applications where wear resistance is of primary concern are susceptible to such a situation, because they are generally very hard in order to improve the wear properties.

Welding often gives rise to fracture problems. In steels, there is always the possibility of developing an untempered martensitic structure or a severe stress pattern that causes microcracks. For example, in welding of ground vehicle frames, weld cracks may occur. The propensity toward crack formation is increased if alloy steels are welded or if thick sections are used. Not only can welding cause microcracking, but it also has the effect of radically altering the microstructure in ways that usually result in lower fracture toughnesses. This can be a result of segregation within the weld solidification structure or the development of coarse grains, to name but two such effects. It is interesting, however, to note that in α/β titanium alloys, welding of fairly thick sections results in rapid cooling and the formation of a transformed microstructure that, as we have seen, is actually beneficial for the fracture toughness. Weighed against this beneficial structure is the fact that interstitials can be absorbed during the weld process; the benefits of a desirable structure may be negated by an increased interstitial content.

However desirable a microstructure may be, processing and fabrication methods affect the fracture toughness and may degrade toughness properties. We will see in the following sections that these considerations are even more important for fatigue properties.

Fracture Mechanisms and Models

It is important to define the term *fracture toughness* and how fracture toughness relates, in a general way, to mechanical and microstructural parameters. Toughness may be thought of as the amount of energy that is absorbed when the crack advances one unit of area. In other words, it is the energy absorbed per unit area of crack advance. As such, it has the same units as the results of a Charpy test (i.e., force $\times$ length/length^2). In fracture mechanics, this value is denoted by the symbol G_{Ic}. This definition of toughness is particu-

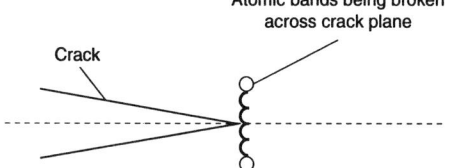

Fig. 1 Cleavage crack formation. Bands are broken across the idealized crack plane.

Fig. 2 Schematic of microvoid coherence

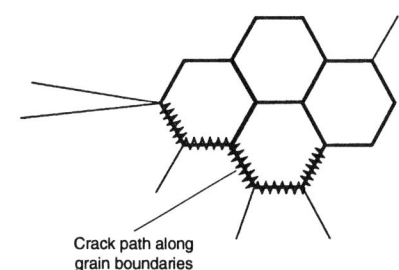

Fig. 3 Schematic of grain boundary cracking

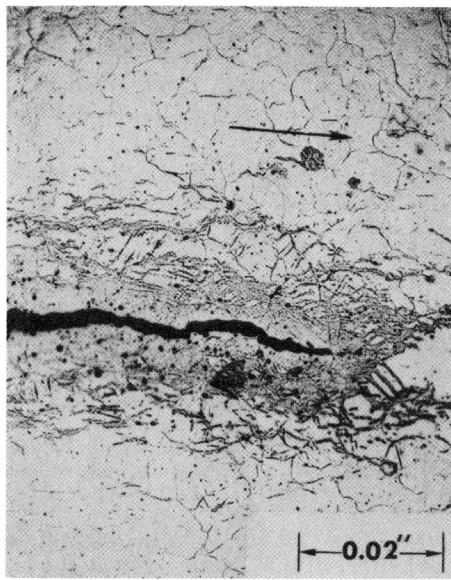

(a)

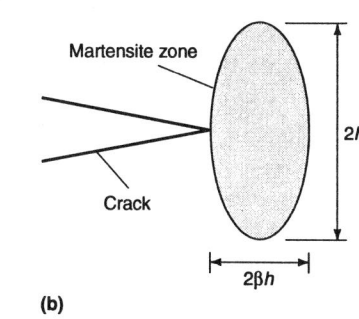

(b)

Fig. 4 (a) Martensite formation in the center of a specimen of steel I that was reduced 45% at 460 °C and fractured at room temperature. The crack has extended approximately 60 mils beyond the fatigue crack. (b) Schematic of the above process

larly attractive to materials developers because it focuses on the crack-tip processes. However, it has become customary to report toughness in terms of the critical value of the stress-intensity parameter at which fracture occurs. As seen above, this is denoted by the symbol K_{Ic} and is called the plane-strain fracture toughness. There is a simple, well-known relationship between these two parameters:

$$G_{Ic} = K_{Ic}^2 / E \qquad \text{(Eq 5)}$$

where E is Young's modulus. This means that it is a simple matter to go from an energy representation (i.e., G) to a stress-intensity representation (i.e., K). In general, the fracture toughness can be computed by calculating all of the energy absorbed in bringing the crack-tip elements to the point of fracture and dividing by the amount of crack extension associated with this critical event. While simple in concept, there are significant problems in implementing such computations, including knowledge of the stress/strain fields ahead of the crack tip in the plastic zone and, of course, the criterion for crack extension. A simple guide for relating the fracture toughness to the mechanical properties based on these concepts is (Ref 4):

$$K_{Ic} = n \, (E\sigma_{ys}e_f)^{1/2} \qquad \text{(Eq 6)}$$

where n is the strain hardening exponent, σ_{ys} is the yield strength, and e_f is the fracture strain.

Equation 6 and variants thereof provide some guidance for understanding the effects of microstructure on the fracture toughness of many engineering materials.

Cleavage is a mechanism whereby in the ideal case, crack extension occurs by breaking of bonds across an atomic plane. In principle, little energy is absorbed in this case and materials tend to behave in a brittle manner. Of course, one must add the caveat that the toughness or brittleness of a material is determined by the total energy that is absorbed, not by the mechanism. We shall see a case in which cleavage is the final failure mechanism but the material is overall very tough. Cleavage occurs frequently in steels and in titanium alloys and is discussed below in more detail. Cleavage cracking is illustrated schematically in Fig. 1. Factors that affect the fracture toughness of materials that fracture by cleavage are the size, strength, and distribution of the cleavage nuclei (i.e., carbide particles in steels) and the cleavage facet size and orientation.

Microvoid Coalescence. In many materials, particles (which may be intentional or unintentional) are found within a ductile matrix. These particles act as initiators of fracture. This initiation can be the result of interface decohesion, fracture of the particle, or a combination. In steels, such fracture occurs above the ductile/brittle transition temperature (DBTT) and may involve both inclusions and small, dispersed carbides. In aluminum alloys, this is the predominant fracture mode at all temperatures and is frequently initiated by large constituent particles, followed by void formation and growth around other particles. A schematic model for microvoid coalescence is shown in Fig. 2. In materials that fail by microvoid coalescence, important factors affecting the fracture toughness are the mechanical properties of the particles, the properties of the particle/matrix interface, the size and distribution of the particles, and the plastic properties of the matrix.

Boundary Decohesion. In many materials, failure along grain boundaries (or prior-grain boundaries) is an important mode of fracture. This fracture mode will occur when the boundary is the "weak link" in the microstructure, which occurs under the following conditions:

- The grain interiors do not deform, but the cleavage strength is higher than the boundary strength (typical of some ceramics).
- The boundary has been weakened by segregation of a damaging species (typical of phosphorus segregation to boundaries in steel).
- The boundary is strong, but brittle particles have nucleated on the boundaries and provide both crack initiators and an easy fracture path (typical of the formation of MnS on prior-austenite boundaries in steel).
- The boundary is strong but is attacked by an external species in the application environment. This situation is similar to the second

condition in this list, but it is time dependent because the damaging species must diffuse into the material ahead of the crack tip (typical of stress-corrosion cracking).

Grain boundary cracking is illustrated in Fig. 3. Clearly it can be influenced by chemistry as well as by heat treatments that control the distribution of boundary particles. For example, a continuous brittle boundary film could be disastrous, whereas a heat treatment that produces discrete boundary particles would increase the toughness.

Phase Changes during Deformation. In some materials (e.g., steels, ceramics, titanium alloys) it is possible to have a phase change ahead of the crack tip during loading such that additional energy is absorbed and the material is effectively toughened. The classical example of this is in TRIP steels. Here a metastable austenite transforms to martensite as the crack propagates. Even though the martensite is significantly more brittle than the austenite that it replaces, it transforms along planes of maximum shear, and the invariant shear associated with the transformation represents an additional deformation mode that absorbs energy. In many cases the TRIP process

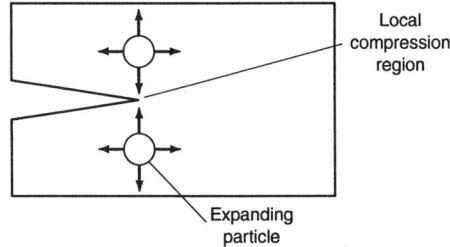

Fig. 5 Toughening mechanism for ceramics: residual compression at the crack tip. Selected regions of a microstructure are induced to expand in the vicinity of the crack tip such that a local state of compression counteracts the externally applied tensile stress.

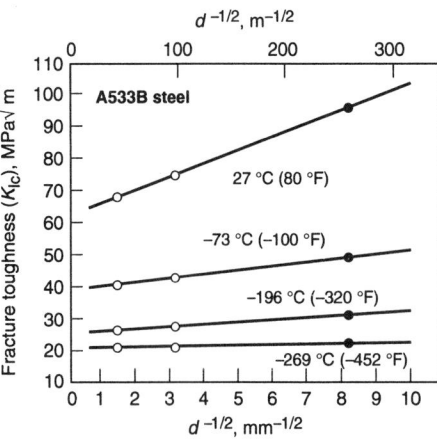

Fig. 6 Relationship of fracture toughness to inverse square root of grain size. Dependence of fracture toughness on prior-austenite grain size at four temperatures. Source: Ref 8

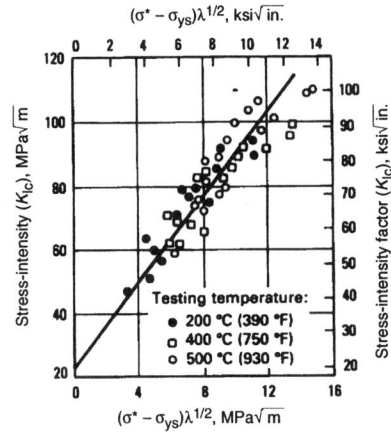

Fig. 7 K_{Ic} for a martensitic 0.45C-Ni-Cr-Mo-V steel as a function of inclusion spacing and yield strength. Source: Ref 9

results in materials that are significantly tougher than those with similar properties but do not transform during crack extension. These ideas have been described quantitatively and verified experimentally (Ref 5). They have also been successfully applied to materials in which there are multiple toughening mechanisms (Ref 6). The process is illustrated in Fig. 4. Factors that influence toughness with such a mechanism are the invariant shear as well as the mechanical properties of the initial and final phases. In ceramics (Ref 3) (and to a lesser extent in metals) there is also a significant volume change (e.g., in zirconia) when the transformation occurs and compressive stress fields are generated ahead of the crack tip, which reduces the driving force for crack extension. This is shown schematically in Fig. 5.

Fracture in Steels

Because steels are used in many microstructural conditions, this section is limited to a few examples. The subject is treated in more detail later in this Volume. This section provides an overview of some of the major microstructural considerations in carbon and alloy steels that affect the fracture toughness, such as the ferrite grain size for low-temperature fracture, the prior-austenite grain size, the size, spacing, and character of inclusions, and phase transformations, if any.

For fracture of a mild steel at low temperatures, it is reasonable to assume that fracture will occur when the conditions at the crack tip are such that the plastic constraint factor (pcf) elevates the yield stress to a level equivalent to the fracture stress. This simple idea may be expressed by the condition:

$$\text{pcf} \cdot \sigma_{ys} = \sigma_f \qquad \text{(Eq 7)}$$

It is well known that both the yield stress and the fracture stress increase with decreasing grain size. It is also known that the pcf increases with K_{Ic} and decreases with σ_{ys}. This can be understood by considering the case in which the fracture toughness is

low and the yield strength is high. In this case the low fracture toughness doesn't allow crack tip stresses to build up to sufficiently high values to exert significant constraint.

These ideas have been combined (Ref 7) to yield an equation of the form:

$$K_{Ic} = A + Bd^{-1/2} \qquad \text{(Eq 8)}$$

The values of A and B would be expected to correspond to the particular steel being considered and whether the correlation is being made with the ferrite grain size or the austenite grain size. Graphs showing the effects of the austenite grain size are shown in Fig. 6 (Ref 8), where it can be seen that for a wide range of temperatures the form of Eq 8 is followed.

Steels frequently contain inclusions, and these inclusions, in conjunction with other precipitates that might be present, have a major influence on the fracture toughness of steels. For example, MnS is frequently present, and for a given size of inclusion, it is known that the toughness increases with decreasing volume fraction of MnS. This is equivalent to saying that the toughness increases with increasing spacing of the inclusions for a given inclusion size. Physically, this is quite understandable, because as the crack is loaded, the crack-tip opening displacement (Eq 6) will not be able to reach its maximum value, due to cracking/decohesion of the inclusion, and link up with the crack tip. In fact K_{Ic} would be expected to vary directly with the square root of the inclusion spacing. It would also be expected that at a given inclusion spacing, the fracture toughness would increase with increasing strength of the inclusion/matrix interface above the yield strength. This variation is shown for a 0.45C-Ni-Cr-Mo-V steel (similar to 4340) in Fig. 7 for a wide range of conditions (Ref 9). The correlation follows an equation of the form:

$$K_{Ic} = A + (\sigma^* - \sigma_{ys}) \cdot \lambda^{1/2} \qquad \text{(Eq 9)}$$

where A is a material-dependent constant, σ^* is a fitting constant related to the strength of the inclusion/matrix interface, and λ is the spacing of the inclusions.

In Eq 9 it should be noted that the relationship is strictly valid only for inclusions of a particular size. If the volume fraction is held constant and the size of the inclusions is increased, it is easier to nucleate voids at the inclusion/matrix interface and the toughness will be lowered at a given spacing. This is accounted for in the term σ^*, which decreases with increasing size. It is clear from this equation and the data of Fig. 7 that it is very important to control the inclusion character (size and volume fraction) in order to increase the fracture toughness. It should also be noted that there should be a "cutoff" to the behavior seen in Fig. 7, because if all inclusions were removed, the fracture toughness would reach some limiting value that is characteristic of the material being considered.

Other inclusions also play a role. For example, in quenched-and-tempered steels the incoherent carbide particles are rather large, and Low (Ref 10) has suggested that the toughness of these steels is limited by the formation of a "void sheet" between the inclusion voids associated with decohesion between the carbides and the matrix. Void sheet nucleation is shown in Fig. 8 (Ref 11). Nucleation of this void sheet will have the effect of limiting the fracture toughness, because the energy absorption process associated with growth of the larger voids to the point of impingement will be interrupted and early linkup will occur. When such mechanisms operate it is usual to see both large and small voids on the fracture surface, such as is seen in Fig. 9 for a 250-grade maraging steel (Ref 12). The nucleation of the void sheet is easier the larger the particles. This mechanism explains the relatively high fracture toughness of maraging steels, which contain extremely fine strengthening precipitates compared to the carbides in common quenched-and-tempered steels.

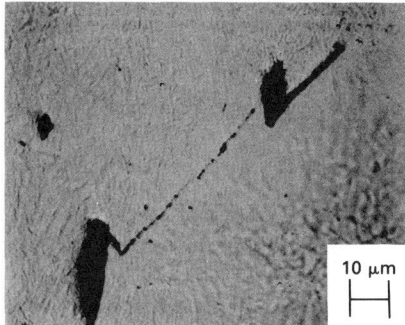

Fig. 8 Effect of particles on fracture toughness. Source: Ref 11

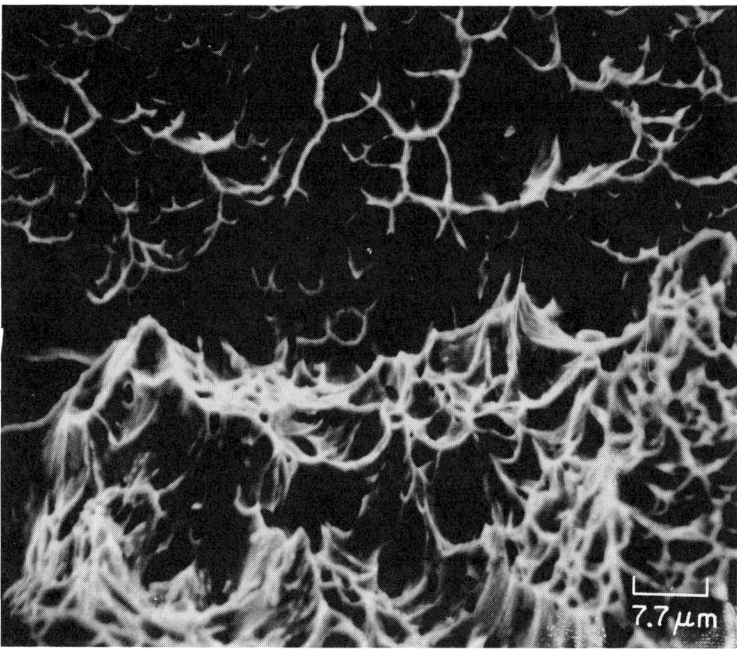

Fig. 9 An SEM view of the surface of the tensile-test fracture in 18% Ni, grade 300 maraging steel, showing a portion of the central zone of the fracture, close to the origin. The surface here is composed of equiaxed dimples of two sizes. The large dimples probably formed at Ti(C,N) particles; all the dimples were caused by particles of some sort. Source: Ref 12

As mentioned above, high toughness levels can be attained when austenite transforms to martensite at the crack tip. When this occurs a significant amount of energy may be absorbed. The energy absorption ahead of the crack tip is given by (Ref 5):

$$\Delta U^{A \to M} = (\sigma_M \cdot \varepsilon_{IS}/\sqrt{3}) \cdot V_{pz} \cdot V_{fM} \qquad \text{(Eq 10)}$$

where σ_M is the stress at which martensite forms, ε_{IS} is the invariant shear associated with martensite formation, V_{pz} is the volume of plastic zone (Ref 13), and V_{fM} is the volume fraction of martensite in the plastic zone. Equation 10 has been used to obtain an expression for the fracture toughness. Without going into the details, the energy absorption is increased for a large plastic zone that contains a large amount of martensite. It is also increased for relatively high stress levels of martensite formation and large shear strains associated with the transformation. Figure 10 illustrates the incremental toughness due to martensite formation (Ref 5).

Aluminum Alloy Fracture Toughness

Aluminum alloys are widely used in the aerospace industry for airframes and in other applications that require a good combination of strength and light weight. They generally consist of a face-centered cubic (fcc) solid solution matrix with various complex precipitate particles. While the physical metallurgy of these alloys is not reviewed here, it is important to note that they generally contain very small, coherent precipitates that are responsible in large measure for the observed strength of these alloys. In addition, other particles are present that usually have a negative effect on the fracture toughness. Some of these particles are rather large and are referred to as "constituent" particles. There are also intermediate-sized particles that cannot be resolutioned on heat treating. Because of the microstructure, a duplex distribution of dimples is usually seen on the fracture surface, and the normal fracture mode is by the formation and coalescence of microvoids, with "void sheet" formation

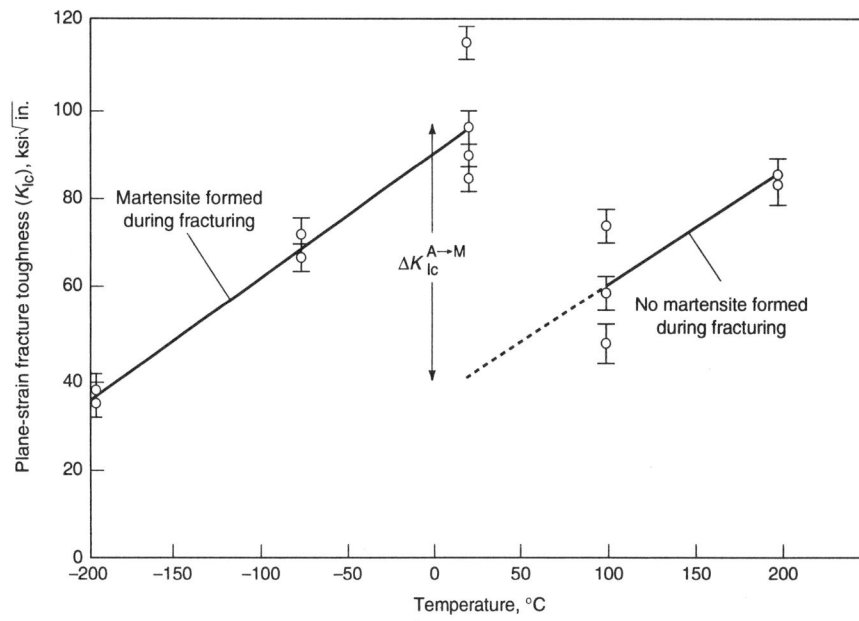

Fig. 10 Toughness vs. temperature for alloys deformed 75% at 460 °C. Source: Ref 5

being an important aspect of fracture in these alloys. This fracture mode is very similar to that discussed above for steels, and many of the conclusions drawn apply also to aluminum alloys.

The fracture toughness of the traditional aluminum alloys shows, as do most materials, a general decrease in toughness with increases in yield strength (Fig. 11) (Ref 14). As the strength increases, the strain-hardening exponent generally decreases and void sheet nucleation occurs more

easily, thus limiting the crack-tip opening displacement. These effects then account for the decreased fracture toughness.

While the strength generally causes a decrease in the toughness, the effects of the large constituent particles should not be overlooked. At any given strength level the toughness of the 2xxx and 7xxx alloys, for example, can be increased through careful control of the insoluble constituent particles that contain iron and/or silicon, such

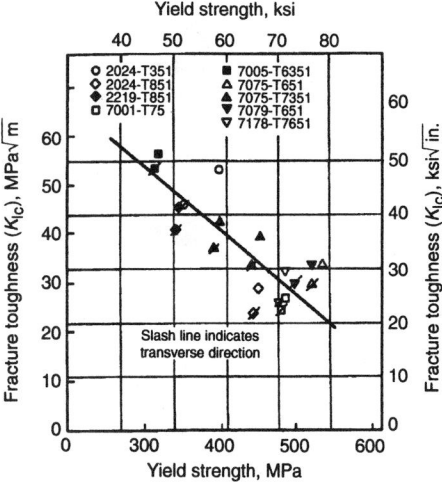

Fig. 11 K_{Ic} measurements on specimens from plate 25 to 38 mm (1.0 to 1.5 in.) thick. Source: Ref 14

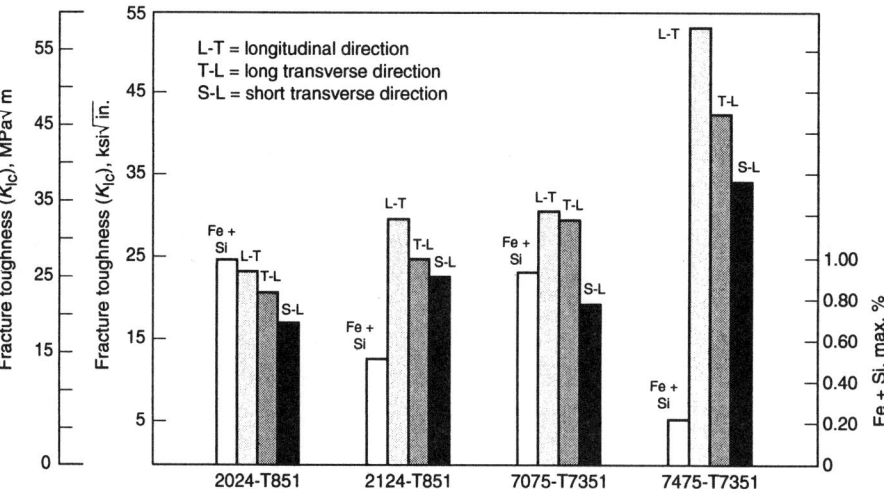

Fig. 12 Fracture toughness comparisons of aluminum alloys 2024, 2124, 7075 and 7475. Source: Ref 15

as Al_7Cu_2Fe, Mg_2Si, and $(Fe,Mn)Al_2$. These particles are quite large (i.e., in excess of 1 μm) and nucleate voids. The toughness can be controlled through control of those elements that effectively increase the spacing of such particles. This results in an increase in the toughness, similar to what was observed for steels (Fig. 8). The efficacy of Fe + Si control is illustrated in Fig. 12 for a series of aluminum alloys of differing orientation (Ref 15). Reducing the Fe + Si content improves the toughness of both the 2xxx and 7xxx alloys.

Another class of alloys is Al-Li alloys, which contain lithium to form coherent Al_3Li precipitates. These precipitates are based on the $L1_2$ structure (i.e., ordered "fcc" with lithium atoms at the cube edges and aluminum atoms at the face centers) and are fully coherent with the matrix. In this regard they are similar to the γ′ precipitates found in nickel-base superalloys. These alloys are strong and light and have a significantly higher modulus than the conventional alloys used in the aerospace industry.

The principal problem with Al-Li alloys is that a significant degree of segregation attends their fabrication, and during mechanical processing (e.g., rolling) the interdendritically segregated regions are rolled out into sheets such that the transverse properties are very low. The corresponding fracture toughness is shown in Fig. 13. Note that for both alloys the fracture toughness depends on the orientation to some extent. In Fig. 13 the first letter refers to the load direction and the second to the expected direction of crack propagation. Thus in both the S-L and S-T orientations, the crack propagates on a plane parallel to that formed by the strong segregation and would be expected to be relatively low. Indeed, other data for which fracture toughness is available shows that the fracture toughness of 8090 in the L-T direction is approximately double that of the S-L and S-T orientations (Ref 16). Based on that result, fracture toughnesses in excess of 30 MPa√m would be expected and indeed have been measured. The reason for the increased toughness

in that orientation for 8090 (and indeed for most other wrought aluminum alloys) is that the precipitate sheets (and associated precipitate free zones PFZ's) promote debonding ahead of the crack tip. This debonding tends to both relax the constraint and blunt the crack, giving rise to an apparent fracture toughness increase. Such boundaries have been studied in maraging steel composites (Ref 17), where phenomenal increases in fracture toughness have been observed. The low short transverse properties of the Al-Li alloys are a significant barrier to wide application of these materials, despite their promise.

Titanium Alloy Fracture Toughness

Titanium is light and strong, with a relatively high melting point and good resistance to both oxidation and attack by chlorides. As a result, it is an attractive alloy for use in aerospace applications such as fan disks. The most widely used alloy is Ti-6Al-4V, which is discussed elsewhere in this Volume.

There are three major groups of titanium, depending on the amount of α (hexagonal close-packed phase) and β (body-centered cubic phase) constituents in the microstructure:

- α alloys such as Ti-6Al-2Sn-4Zr-2Mo
- α + β alloys such as Ti-6Al-4V
- β alloys such as Ti-10V-2Fe-3Al

The α alloys and the α + β alloys have similar structure-property relations, as described below, while β alloys are more distinct and are discussed separately. No single β alloy has the same broad applicability as Ti-6Al-4V, although the near-β alloy Ti-10V-2Fe-3Al is an important alloy for aerospace structures. In general, retained β alloys are used for workability, corrosion resistance, and the ability to heat treat larger section sizes in which β has been retained. Beta alloys also tend to have higher density and lower elastic modulus values than α alloys.

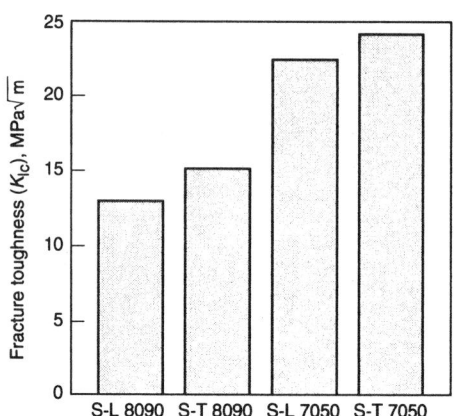

Fig. 13 Graph of fracture toughness test results for 8090-T852 and 7050-T7452 material in the S-L and S-T orientations

Alpha and Alpha-Beta Alloys

Fracture toughness can be varied within a nominal titanium alloy by as much as a multiple of two or three by manipulating alloy chemistry, microstructure, and texture. Some tradeoffs of other desired properties may be necessary to achieve high fracture toughness. There are significant differences among titanium alloys, but there is also appreciable overlap in their properties. Table 1 gives examples of typical plane-strain fracture toughness ranges for α-β titanium alloys. From these data it is apparent that the basic alloy chemistry affects the relationship between strength and toughness. From Table 1 it is also evident that transformed microstructures may greatly enhance toughness while only slightly reducing strength.

Within the permissible range of chemistry for a specific titanium alloy and grade, oxygen is the most important variable insofar as its effect on toughness is concerned. This is readily shown by the data of Ferguson and Berryman (Ref 18), who reported strength and K_{Ic} values for specimens of

Table 1 Typical fracture toughness of high-strength titanium alloys

Alloy	Alpha morphology	Yield strength MPa	Yield strength ksi	Fracture toughness (K_{Ic}) MPa√m	Fracture toughness (K_{Ic}) ksi√m
Ti-6Al-4V	Equiaxed	910	130	44-66	40-60
	Transformed	875	125	88-110	80-100
Ti-6Al-6V-2Sn	Equiaxed	1085	155	33-55	30-50
	Transformed	980	140	55-77	50-70
Ti-6Al-2Sn-4Zr-6Mo	Equiaxed	1155	165	22-23	20-30
	Transformed	1120	160	33-55	30-50

Source: Ref 24

Table 2 Effect of hydrogen content on room-temperature K_{Ic} in alloy Ti-6Al-4V after furnace cooling from 927 °C (1700 °F)

Hydrogen content, ppm	K_{Ic} at room temperature(a) MPa√m	K_{Ic} at room temperature(a) ksi√in.
At 0.16 wt% oxygen		
8	145	132
36	118	107
53	104	95
122	100	91
At 0.05 wt% oxygen		
9	133	121
36	125	114
50	96	87
125	101	92

(a) Specimens were tested in accord with ASTM E 399 but were loaded rapidly (total testing time = 10 s).

α-β processed and recrystallization annealed Ti-6Al-4V. Regression analysis of their data shows that for each 0.01% increase in oxygen, toughness is reduced by about 3.7 MPa√m (3.4 ksi√in.). Whether this is a direct effect or an indirect effect, in the sense that oxygen increases strength and the strength increase reduces K_{Ic}, remains to be determined. Multiple regression analysis of the Ferguson and Berryman data, where both oxygen content and tensile strength are assumed to be independent variables, shows that tensile strength is the dominant variable (the residual effect of oxygen does not reach statistical significance). This implies that, if oxygen affects K_{Ic}, it does so through its strengthening effect. The solid solution strengthening effect of oxygen is further complicated by the fact that oxygen tends to promote the formation of Ti$_3$Al.

As might be expected, hydrogen also has an effect on toughness. The work of Meyn (Ref 19) shows that very low hydrogen contents (less than about 40 ppm) enhance toughness. This effect is particularly dramatic with hydrogen contents below 10 ppm. Table 2 illustrates the essential results for Ti-6Al-4V at two different oxygen levels.

Effects of Microstructure. Improvements in K_{Ic} can be obtained by providing either of two basic types of microstructures: (a) transformed structures, or structures transformed as much as possible, because such structures provide tortu-ous crack paths; and (b) equiaxed structures composed mainly of regrowth α that have both low dislocation densities and low concentrations of aluminum and oxygen (the so-called "recrystallization annealed" structures). It is not yet known (in 1995) whether or not combinations of these two types of structures would further enhance K_{Ic} values.

Transformed structures appear to be tough primarily because fractures in such structures must proceed along tortuous, many-faceted crack paths. According to the work of Hall and Hammond (Ref 20), K_{Ic} is proportional to the fraction of transformed microstructure in alloy Ti-6Al-4V (see Table 3). These authors, however, propose that it is strain-induced transformation of the retained laths of β phase that leads to enhanced fracture toughness. Evidently, their idea is that this TRIP mechanism enhances "ductility" in front of each crack tip. However, in comparing β alloys deformed by either slip or TRIP mechanisms, Wardlaw et al. (Ref 21) could find no advantage in ductility for the TRIP alloys. Curtis and Spurr (Ref 22) suggested that it is primarily the α platelet size and efficient dispersion of the β phase that enhance toughness. In any event, most direct evidence indicates that crack tortuosity is an important variable affecting K_{Ic}.

Effects of Environment. Effects of temperature on toughness are usually less abrupt for titanium than for common low-alloy steel. For example, Tobler (Ref 23) reported a gradual K_{Ic} transition temperature between –196 and –143 °C (–320 and –215 °F) for recrystallization annealed Ti-6Al-4V extra-low interstitial (ELI). For temperatures at and above –143 °C (–215 °F), his K_{Ic} values were typically about 90 MPa√m (82 ksi√in.). At –196 °C (–320 °F), his values were typically 60 to 65 MPa√m (55 to 60 ksi√in.). The

loss is about 30%. The early conclusion by Christian and Hurlich (Ref 24) that Ti-6Al-4V ELI may be used to cryogenic temperatures thus has some justification. The same may not be true of standard-grade Ti-6Al-4V.

Texture Effects. Titanium alloys, like most other practical engineering alloys, are very dependent on the texture. In the hot working operations that generally attend the fabrication of titanium alloys, all of the **c** directions from grain to grain tend to be parallel. Thus, different plastic responses will be elicited, depending on the direction of crack propagation. The effects of direction on toughness are illustrated for a Ti-6Al-2Sn-4Zr-6Mo plate in Table 4 (Ref 25). Similar trends are observed for Ti-6Al-4V alloys.

Beta Alloys

Beta titanium alloys, which include both near-β alloys and β-rich α-β alloys such as Ti-10V-2Fe-3Al, can be heat treated to a wide range of strength levels and can be tailored to strength-toughness combinations for specific applications. That is, moderate strength with high toughness or high strength with moderate toughness can be achieved. This is generally not possible for other types of titanium alloys because they cannot be heat treated over a very wide range. At moderate strength levels, say 965 MPa (140 ksi) and above, the fracture toughness of the β alloys can be processed to achieve higher values than for the other types (α and α-β alloys).

To accomplish these higher toughnesses, however, the processing window is tighter than that normally used for the other alloy types. For the less highly β-stabilized alloys, such as Ti-10V-2Fe-3Al, Ti-17, and β-CEZ, the thermomechanical process is critical to the properties combinations achieved, because this has a strong influence on the final microstructure and the resultant tensile strength and fracture toughnesses that may be achieved.

This is somewhat less important in the more highly β-stabilized alloys such as Ti-3Al-8V-6Cr-4Mo-4Zr (β-C) and Ti-15V-3Cr-3Al-3Sn. In these the final microstructure, precipitated α, is so fine that microstructural manipulation through thermomechanical processing is not as effective. In these cases the aging heat treatments—sequence and temperature—are more critical. The key is to obtain a uniform precipitation. This may be obtained by a low-high aging sequence, or, with residual cold or warm work, possibly a high-

Table 3 Relationship between K_{Ic} and fraction of transformed structure in alloy Ti-6Al-4V

Heat treating temperature(a) °C	Heat treating temperature(a) °F	Fraction of transformed structure, %	K_{Ic} MPa√m(b)	K_{Ic} ksi√in.
1050	1922	100	69.0 (69.9)	64
950	1742	70	61.5 (60.4)	55
850	1562	20	46.5 (44.6)	40
750	1382	10	39.5 (41.5)	38

(a) Heated for 1 h at indicated temperature and then air cooled. (b) Values in parentheses calculated from linear least-squares expression relating percent transformation to K_{Ic}

Table 4 Effect of test direction on mechanical properties of textured Ti-6Al-2Sn-4Zr-6Mo plate

Test direction(a)	Tensile strength, MPa	Yield strength, MPa	Elongation, %	Reduction in area, %	Elastic modulus, GPa	K_{Ic} MPa√m	K_{Ic} ksi √in.	K_{Ic} specimen orientation
L	1027	952	11.5	18.0	107	75	68	L-T
T	1358	1200	11.3	13.5	134	91	83	L-T
S	938	924	6.5	26.0	104	49	45	S-T

(a) High basal pole intensities reported in the transverse direction, 90° from normal, and also intensity nodes in positions 45° from the longitudinal (rolling) direction and about 40° from the plate normal. Source: Ref 25

Table 5 Typical mechanical properties of selected beta alloys

Alloy	Tensile yield strength MPa	Tensile yield strength ksi	Ultimate tensile strength MPa	Ultimate tensile strength ksi	Elongation, %	Reduction in area, %	K_{Ic} MPa√m	K_{Ic} ksi√in.
Beta C(a)	1090	158	1143	165.8	5.5	8.6	72.2	65.6
Beta 21S	1150	164	1057	151	8	...	101(b)	92(b)
Ti-15-3(c)	...	...	1520	220	...	16	59	53.6
Beta CEZ(d)	1200	174	1315	191	10	26	75	69
Ti-10-2-3 (high strength)(e)	1185	172	1250	181	8	18	52	47
Ti-10-2-3 (medium strength)(e)	1080	157	1160	168	16	44	73	66
Ti-10-2-3 (low strength)(e)	940	136	1020	148	22	61	102	93

(a) 457 mm (18 in.) diam × 238 mm (9⅜ in.) I.D. extrusion, air cooled from 815 °C (1500 °F) (above the transus), aged 24 h at 565 °C (1050 °F). (b) K_c for 1.3 mm (0.050 in.) strip, aged 8 h at 595 °C (1100 °F). (c) Solution treated (above the beta transus) at 850 °C (1560 °F), aged for 11 ks at 600 °C (1112 °F), re-aged at 500 °C (932 °F) for 43 ks. (d) Nominal composition, Ti-5Al-2Sn-4Zr-4Mo-2Cr-1Fe; 300 mm diam pancake, processed through the beta transus to 600 °C (1112 °F), reheated to 830 °C (1472 °F), water quench, plus 570 °C (1054 °F) 8 h, air cool. (e) For forgings, Ti-10-2-3 is heat treated 16 to 33 °C (60-100 °F) below the beta transus (744-765 °C, or 1370-1410 °F) and water quenched. Aging temperatures range from 480 to 620 °C (900-1100 °F) depending on desired strength. Source: *Fatigue Databook: Light Structural Alloys*, ASM International, 1995

low aging sequence. When highly alloyed β alloys, such as β-C, are cold worked prior to aging, high strength can be obtained with good ductility because cold work induces finer and more uniform precipitation.

The thermomechanical processing must, however, be controlled to provide a uniform microstructure throughout the cross-section of the material and, in conjunction with the heat treatment, avoid the occurrence of extensive grain boundary α or a precipitate-free zone near the grain boundaries. Some characteristic mechanical properties of β alloys are given in Table 5. Comparison of Table 5 with Table 1 indicates that the β alloys have a higher fracture toughness at higher yield strength than the α-β alloys.

Fatigue Crack Propagation

This section briefly reviews fatigue crack propagation (FCP) behavior in some major structural alloy families, with emphasis on general microstructural factors. Like fracture toughness, FCP behavior is subject to microstructural control. In some instances, however, improvements in the fracture toughness may produce a structure in which the FCP properties are degraded and vice versa. For example, decreasing the grain size generally leads to improvements in the fracture toughness but can degrade both the fatigue threshold, ΔK_{th}, and the FCP rate at high ΔK levels. In other instances, most notably for titanium alloys, a microstructure that leads to improvements in K_{Ic} also produces improved FCP properties. These structure-property relations can be important factors in meeting design-critical properties.

This section begins with a brief review of FCP mechanisms and models. While it is not the goal of this section to make a complete review of models of FCP, it is nonetheless of use to briefly examine the physical basis of FCP and to indicate a few models that have been proposed to understand FCP behavior in engineering alloys. It is also of use to briefly discuss empirical models that have been used to correlate FCP behavior with extrinsic variables. Fatigue crack growth in major structural alloy families is then discussed.

FCP Mechanisms and Models

Damage Accumulation at Crack Tip. An intuitively appealing model of FCP is to imagine that the high stresses and strains at the crack tip produce reversed plastic deformation and damage. As a result of this damage the crack advances a certain distance, which is essentially defined by the microstructure. This distance could be related to the grain size, to the cell size for materials in which cells are formed, or to other microstructural features. Several models based on these ideas have been reviewed in detail elsewhere (Ref 26). One model of this type that has been proposed (Ref 27) predicts a crack growth rate as follows:

$$da/dN = C/(\sigma_{ys}\varepsilon_f E)^{1/\beta} \cdot (1/l^{1/\beta-1}) \cdot \Delta K^{2/\beta} \qquad \text{(Eq 11)}$$

where C is a constant, ε_f is the fracture strain of cyclically deformed material, β is the Coffin-Manson exponent, and l is the process zone size. This model predicts that the FCP rate should decrease with increasing process zone size, which for many materials would be expected to be the grain size. Such a dependence is indeed observed for many materials. It furthermore predicts that the Paris exponent (discussed below) should be simply related to the Coffin-Manson exponent for low-cycle fatigue (LCF). This is also observed. Thus, one mechanism for improving FCP behavior would be to increase the grain size of alloys of interest, especially for alloys that exhibit planar glide and for which the primary obstacles are the grain boundaries. There is extensive experimental evidence that such an approach is indeed very effective for many materials (Ref 26).

Reduced Crack Growth Rate by Crack Tip Interference (Roughness-Induced Closure). Somewhat related to the above discussion is the notion that if the fracture surface is very "rough," then the opposite faces of the crack tip will come into premature contact with one another during

Table 6 Tensile and fracture toughness properties of Inconel 718 at room and subzero temperatures

Testing temperature		Yield strength		Tensile strength		Elongation,	Reduction in area,	Fracture toughness	
°C	°F	MPa	ksi	MPa	ksi	%	%	MPa√m	ksi√in.
22	72	1172	170	1404	204	15.4	18.2	96.3	87.8
−196	−320	1342	197	1649	239	20.6	19.8	103	94
−269	−452	1408	204	1816	263	20.6	20.2	112	102

Note: Heat treatment: 980 °C (1800 °F) ¾ h, air cool; double age 720 °C (1325 °F) 8 h, furnace cool to 620 °C (1150 °F), hold 10 h; air cool. Source: Ref 35

unloading. Thus, during FCP, the nominal stress-intensity range ΔK_{nom} is reduced to ΔK_{eff} and the crack growth rate is correspondingly reduced. The point at which contact of opposite faces occurs is used in computing the minimum K level in the ΔK_{eff} expression. Because ΔK has been reduced, the crack growth rate is also correspondingly reduced. While these concepts make sense physically in certain situations, their applicability depends sensitively on the elastic/plastic properties of the material being considered and the geometry of the fracture surface. Furthermore, these ideas, while appealing, are very difficult to implement in an engineering application, because it is not usually possible to measure the closure load. The utility of this insight appears to be qualitative in nature, predicting that rough fracture surfaces should exhibit low FCP rates. However, rough fracture surfaces are associated with large-grained planar slip materials, and it is difficult to see how the roughness arguments can be separated from those discussed previously. A more detailed discussion of these ideas is found elsewhere (Ref 28-30). Additional information on crack closure is in the article "Micromechanisms of Monotonic and Cyclic Crack Growth" in this Volume.

Phenomenological Models for FCP. The best-known empirical equation for correlating the FCP rate is the Paris equation (Eq 4). While it is now recognized that this equation is an oversimplification of a very complex phenomenon, Paris was the first to recognize that FCP could be correlated with a global parameter that describes the stresses and strains in the vicinity of the crack tip (Ref 31). The Paris equation accurately represents the behavior of many materials in the mid-ΔK range, which is of significant engineering interest. This equation does not recognize the effects of load ratio (P_{min}/P_{max}), the existence of a threshold ΔK_{th}, or very high acceleration of the FCP rate as the stress intensity approaches the fracture toughness of the material.

Fatigue crack propagation has been extensively characterized experimentally. The primary objective of such studies is to develop data that can be used to predict the life of a component given the stress state and the crack geometry. For analytical convenience, models have been developed to correlate the data. Typical of these models is the hyperbolic sine (Ref 32):

$$\log (da/dN) = C_1 \sinh \{C_2 \log \Delta K + C_3\} + C_4 \quad \text{(Eq 12)}$$

where C_1 is a material constant and the other constants are functions of load ratio, temperature, and frequency. The constants can be determined by a regression analysis. This model and other similar models are symmetric about an inflection point and this property is not always in consonance with actual behavior. A model developed for use in the electric power industry, which allows for a more general fit of the data (Ref 33) is given by:

$$\frac{1}{da/dN} = \frac{A_1}{(\Delta K)^{n_1}} + \frac{A_2}{(\Delta K)^{n_2}} + \frac{A_2}{(K_c(1-R))^{n_2}} \quad \text{(Eq 13)}$$

where the various constants and exponents are fitting parameters and R is the load ratio. Temperature and frequency effects are easily accommodated in the fitting parameters. At values of the crack growth rate that are relatively high (e.g., $da/dN > 2.5 \times 10^{-5}$ mm/cycle), data are well represented by the familiar Paris equation.

Another simple equation, known as the Forman equation, has found wide application in correlating FCP data for aluminum alloys (Ref 34):

$$\frac{da}{dN} = \frac{C\Delta K^n}{(1-R) K_c - \Delta K} \quad \text{(Eq 14)}$$

It must be emphasized that the real utility of these representations lies in data correlation and life calculations. It is not possible to use such equations to gain insight into the mechanisms associated with the FCP process or to design new alloys for increased fatigue resistance.

Nickel-Base Alloys

Nickel-base alloys generally consist of coherent γ' precipitates in an fcc matrix. Typically there will also be grain boundary carbides in polycrystalline alloys. These are present, especially in cast forms used in turbine blade applications, to limit grain boundary sliding at high temperatures, and they also usually limit the conventional fracture toughness to rather low values. However, fracture toughness is not a design-critical parameter for most of these applications, and the low fracture toughness levels are not a particular concern as compared to creep and LCF. On the other hand, alloys that are designed for use in turbine disks usually exhibit significantly higher fracture toughness values and do not contain significant amounts of carbides. A very widely used disk material is In 718, which is strengthened primarily by fine, coherent γ'' precipitates in an fcc matrix. The fracture toughness of In 718 depends on factors such as grain size and inclusions, much like some of the systems mentioned previously. The fracture toughness and other mechanical properties of In 718 are shown in Table 6 (Ref 35). We see that the fracture toughness is substantial over a very wide range of temperatures.

The FCP of nickel-base superalloys has been extensively reviewed (Ref 36, 37) and is considered in detail in this book. Thus only some of the major features of FCP in these systems are discussed below. These alloys are used in many fatigue-critical applications in jet engines, and much of their development has been made with the goal of improving fatigue properties.

FCP in Conventionally Processed Alloys. In an extensive study of FCP in commercial alloys it was shown that the FCP rate tended to decrease with increasing strength, as predicted by Eq 11. However, these commercial alloys differed in many ways and were not particularly suitable for consistent comparisons. It was also shown in this study, as well as in other studies (Ref 39-42), that increasing the grain size in a given system also had the effect of decreasing the crack growth rate, again consistent with Eq 11.

One of these studies (Ref 41) demonstrated the effects of slip mode and microstructure using Waspaloy, an alloy that has been widely used in the jet engine industry as a disk material. Heat treatments resulted in coarse and fine-grained specimens (ASTM 3 and 9). For each grain size, further heat treatments resulted in precipitate sizes of either 8 or 90 nm. Based on well-known physical metallurgy principles, the most planar slip would then occur for specimens containing small precipitates and coarse grains, while the least planar slip would occur for the fine-grained specimens containing large particles. The other microstructures would be expected to have deformation modes intermediate between these.

The results of FCP experiments are shown in Fig. 14. These results make it quite clear that microstructures that showed the most planar slip also had the most crystallographic or "rough" surfaces and the lowest crack growth rates. Similar results were also obtained for In 718 (Ref 43). In this study, heat treatments were used to produce four combinations of grain size (250 vs. 25 μm) and precipitate size (150 vs. 20 nm diameter disks). Independent of precipitate size, the FCP rate was lowest for the larger grain sizes. The effect of precipitate size was more complicated and is discussed in more detail in the articles on FCP of nickel-base alloys. Testing was also done at a very high load ratio ($R = 0.75$) to attempt to assess the contribution of roughness-induced closure on lowering the FCP rates. While increasing

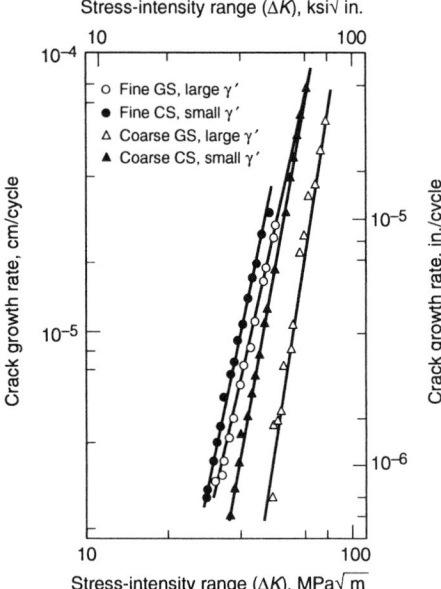

Fig. 14 Composite plot of Waspaloy fatigue crack propagation data. GS, grain size

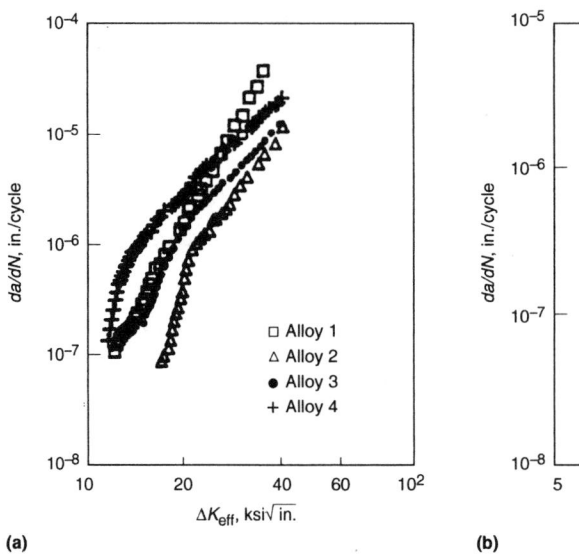

Fig. 15 Fatigue crack propagation response of small γ′ materials. (a) Results plotted vs. ΔK_{eff}, thus accounting for crack closure. (b) Material response at R = 0.8

the R-ratio did increase the FCP rate for all microstructures, the relative ranking of the FCP rates associated with the different microstructures remained unchanged from the tests that were carried out at R = 0.05. These results strongly suggest that differences in closure mechanisms (roughness-induced, oxide-induced) were not responsible for the observed behavior and that intrinsic microstructural features and slip mode parameters play at least as important a role in influencing the FCP behavior as does closure.

This is not to imply that closure may not be important in other systems, and in fact it has been shown in some systems that differences in FCP rates do disappear when testing is done at high R ratios (Ref 44). A detailed study of the effects of slip mode has been carried out on model nickel-base alloys (Ref 45). The results again indicate that the lowest FCP rates occur for materials that exhibit planar reversible glide. The results are shown in Fig. 15.

FCP in Single-Crystal Alloys. At present, single crystalline nickel-base alloys are not used in turbine components that are FCP-critical. However, there is current interest in introducing such requirements into military engines, and some FCP studies have been conducted.

Fatigue crack propagation in the single-crystal alloy designated as N4 has been characterized over a range of temperatures from 25 °C to 1000 °C (Ref 46). The crack propagation direction was [110] and the loading direction was [001]. At room temperature, crack propagation was crystallographic, with cracking occurring primarily on {111} planes with some instances of cracking on {100} planes (Fig. 16). The preferred crack plane was the one having the highest linear combination of resolved shear stress on the slip system (i.e., on the slip plane in the Burgers vector direction) and normal stress across the slip plane. These criteria for multiaxial failure were moti-

Fig. 16 Crystallographic cracking along the {111} and {100} planes of the single-crystal nickel-base alloy designated as N4. Note the less crystallographic nature of the fracture on (001). Source: Ref 46

vated by the fact that the shear stresses cause dislocation motion, cause corresponding damage in the form of dislocation debris, and consequently weaken the slip plane. The normal stress is required for final separation. The FCP rates appeared to separate into two groups, depending on temperature (Fig. 17). Above 925 °C (1700 °F), FCP rates increased dramatically and the fracture planes were macroscopically normal to the loading axis, except for the initial amount of crack extension, which was on a plane inclined to the load axis.

The anisotropic nature of single crystals renders a straightforward application of fracture mechanics problematic. These issues were addressed in a recent study (Ref 47). Using a finite element approach to determine the stresses ahead of the crack tip, and taking into account elastic

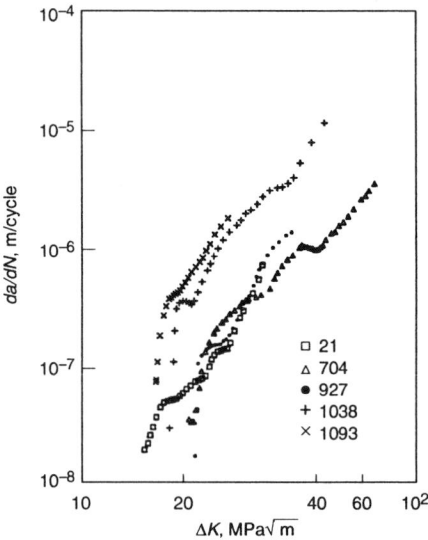

Fig. 17 The effect of temperature on crack growth rates (R = 0.1). At T ≤ 927 °C, the crack growth rates in the Paris regime are similar. At the two higher temperatures, the growth rates are an order of magnitude larger than at the lower temperatures.

anisotropy, it was demonstrated that the features on the fracture surface depended on the critical combinations of normal and shear stress as well as temperature, with environment also playing an important role. At higher temperatures and in areas where the normal stresses were high, cracking was macroscopically on {100} planes. Closer examination revealed that γ′ particles were present on the surface and that cracking was limited to the channels between the γ′ precipitates. This can be understood in terms of the morphology and temperature dependence of the hardness of γ′. It is well known that γ′ become harder with increasing temperature up to about 800-900 °C

(1470-1650 °F), depending on the system. Thus, there is a tendency for dislocations to be confined to the regions between the precipitates, and even though deformation in the matrix between the precipitates probably occurs on the {111} planes, the heavily damaged regions are macroscopically parallel to {100} and final fracture would occur in these areas, assisted, naturally, by environmental interactions. Similar results were obtained for Mar-M002 crystals tested with a [001] load axis at 25, 600, and 850 °C (Ref 48). Crack propagation was in the [100], [210], and [110] directions.

In another study of FCP in nickel-base alloys (Ref 49), Mar-M200 was tested under various loading and crack propagation directions. It was found that cracking occurred on {111} planes, with the preferred plane(s) being the one(s) that had the highest and/or second highest resolved shear stress in the direction of potential Burgers vectors on the slip plane. There was considerable crack branching, and the fracture surfaces were macroscopically "rough," defined by various combinations of {111} planes. When crack branching and surface roughness were taken into account, it was possible to correlate the crack growth rate with the effective stress-intensity parameter. The effective stress-intensity parameter takes into account the fact that there may be modes I, II, and III present in a single crystal that is nominally loaded in mode I. Physically, the effective stress-intensity parameter relates to the elastic strain energy driving the crack and may be defined mathematically as:

$$\Delta K_{\mathrm{eff}} = \left[\Delta K_{\mathrm{I}}^2 + \frac{C_2}{C_1} \Delta K_{\mathrm{II}}^2 + \frac{C_3}{C_1} \Delta K_{\mathrm{III}}^2 \right] \qquad \text{(Eq 15)}$$

where ΔK_{I}, K_{II}, and K_{III} are mode I, II, and III stress-intensity parameters and C_1, C_2, and C_3 are constants related to the elastic anisotropy of the material and loading direction. Acceptable correlations of the data were obtained using ΔK_{eff}. Those orientations for which crack deviation and closure were minimized showed the most rapid crack growth rates.

Environmental Factors. Because nickel-base alloys are used at temperatures as high as 1125 °C (2050 °F) in air environments, the environment becomes a significant factor in influencing FCP behavior. Fatigue crack propagation in nickel-base alloys has been shown to be very sensitive to both oxygen (Ref 36) and hydrogen (Ref 50). The effect of oxygen has been demonstrated to depend on an interplay between two factors: embrittlement due to oxygen diffusion (detrimental) and increased closure due to oxide formation (beneficial). As temperature is increased, the FCP rate first decreases (closure effect) and then increases very rapidly due to the oxygen ingress and embrittlement of the crack-tip region (Ref 36). When an activation energy analysis was carried out on the FCP behavior of René 95, it was shown that an activation energy of about 30 kcal/mol represented the process quite well. This number is in reasonable agreement with the activation energy for diffusion of oxygen in nickel. The environ-

ment also has the effect of changing the fracture surface morphology, which can cloud the effects of the environment. In a study using model nickel-base alloys (Ref 51), testing in air and testing in vacuum at elevated temperatures revealed similar crack growth rates. However, in air the fracture surface exhibited crystallographic facets and was macroscopically rough. In vacuum, the surface was smooth. When tests were carried out on a single specimen, first in air and then in vacuum, the crack growth rate decreased significantly in vacuum, leading to the conclusion that the environment was in fact playing a critical role even at temperatures where obvious oxide films were not formed. These results are shown in Fig. 18.

FCP of Steels

Steels are microstructurally and compositionally complex, so this section is limited to an overview of some of the principal features of FCP behavior of steels.

Near-Threshold Crack Growth. The form of the da/dN versus ΔK curves for steels (indeed for all materials) follows the form predicted by Eq 12. That is, at relatively low values of ΔK, the crack growth rate will go to zero at some value of ΔK, which is customarily defined as ΔK_{th}. In the near-threshold regime, the effects of load ratio are significant. As the load ratio is increased, ΔK_{th} decreases according to an equation of the form:

$$\Delta K_{\mathrm{th}} = C(1 - 0.85R) \qquad \text{(Eq 16)}$$

Thus, increases in R have a pronounced effect on the threshold. The value of the constant C will depend on the composition and heat treatment being considered. However, for many steels the following values for C have been determined (Ref 52):

$$C = 6.4 \text{ for } R \geq 0.1$$

$$C = 5.5 \text{ for } R < 0.1 \qquad \text{(Eq 17)}$$

where K is in ksi $\sqrt{\mathrm{in}}$. Because the threshold regime is where good design and inspection would place the operating conditions, small changes in R would lead to huge changes in the FCP rate and corresponding decreases in the fatigue life. It is in the threshold regime that the most significant effects of microstructure are observed. As was the case for the nickel-base alloys considered previously, increasing the grain size (in this case the ferrite grain size) has the effect of increasing the threshold. This can be understood in terms of basic fracture mechanics. For example, it might be hypothesized that crack propagation could occur only if the plane-strain reversed plastic zone is on the order of the grain size (Ref 7). However, we know that the fatigue zone size is given by the approximate equation:

$$R_{\mathrm{p}} = (\Delta K / \sigma_{\mathrm{ys}})^2 / (24\pi) \qquad \text{(Eq 18)}$$

Substituting the grain size d for the plastic zone size and ΔK_{th} for ΔK gives the following result:

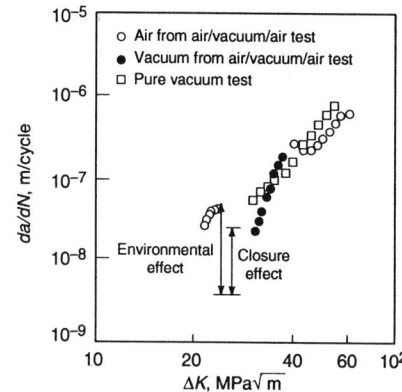

Fig. 18 Vacuum and air/vacuum/air fatigue crack propagation tests. Results for alloy II-S at 427 °C and R = 0.1. Environmental and closure effects can be separated.

$$\Delta K_{\mathrm{th}} = 6.14 \sigma_{\mathrm{ys}} \cdot d^{1/2} \qquad \text{(Eq 19)}$$

where K is in ksi$\sqrt{\mathrm{in}}$. The predictions of Eq 20 are actually followed for low-carbon and low-alloy steels (Ref 7). If Eq 16 and 19 are combined, we can describe the effects of grain size, yield strength, and R-ratio:

$$\Delta K_{\mathrm{th}} = 6.14 \sigma_{\mathrm{ys}} \cdot d^{1/2} \cdot (1 - 0.85R) \qquad \text{(Eq 20)}$$

An extensive compendium of threshold and FCP data for mild steels, alloy steels, stainless steels, aluminum alloys, titanium alloys, and nickel-base alloys is available (Ref 52). All data are evaluated in terms of validity, and structural information (e.g., grain size, heat treatment) is provided where available. The interested reader is referred to this work for further information.

Stress-Ratio Effects at Intermediate Crack Growth Rates. At higher levels of ΔK, the crack growth rate follows Eq 4 very closely with little effect of R-ratio. Thus, it would be unduly conservative to use the Forman equation (Eq 14), which overemphasizes the effect of R-ratio in steels. In the Paris law regime, the upper-bound FCP rates of common steels may be described by the following equations (Ref 53):

$$da/dN = 0.66 \times 10^{-8} (\Delta K)^{2.25} \qquad \text{(Eq 21)}$$

(Martensitic steels)

$$da/dN = 3.6 \times 10^{-10} (\Delta K)^{3.0} \qquad \text{(Eq 22)}$$

(Ferrite/pearlite steels)

As was the case for the nickel-base alloys, the effects of temperature and environment are complex, depending on composition, microstructure, and the aggressive species in the environment. In general, ferritic steels will exhibit higher crack growth rates at temperatures below the onset of severe microcleavage. At higher temperatures, many steels are susceptible to oxidation, which accelerates the FCP rates. Cr-Mo steels are typical of those that are subject to a strong environmental effect. A more detailed description of FCP in steels is provided in the chapter on steels.

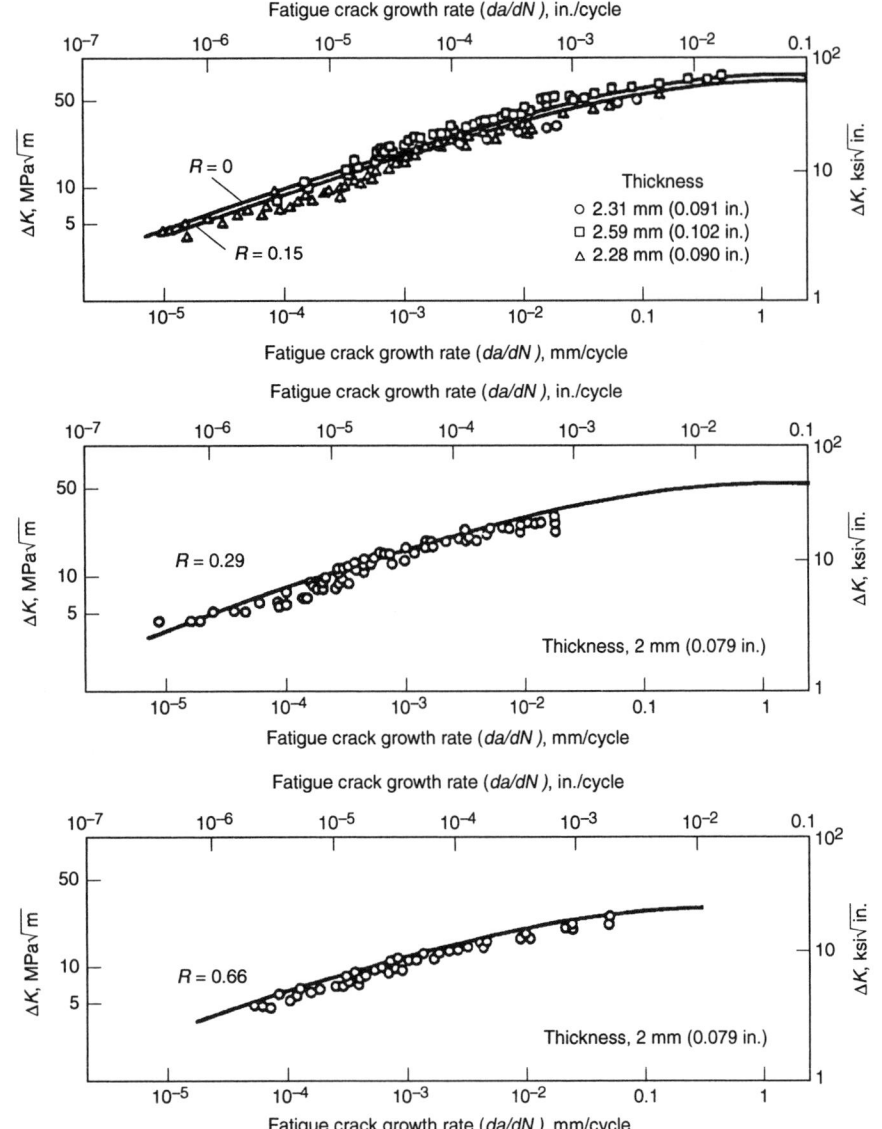

Fig. 19 Effect of *R* on fatigue crack growth rates in aluminum alloy 7075-T6. Source: Ref 54

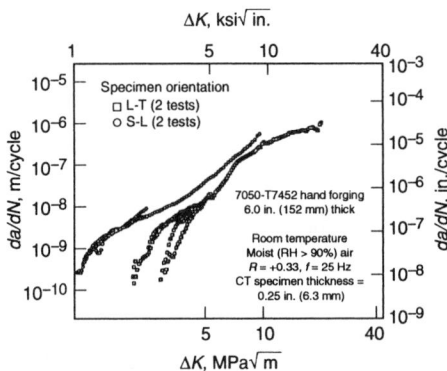

Fig. 20 Fatigue crack propagation data for 7050-T7452 hand forgings. Source: Ref 55

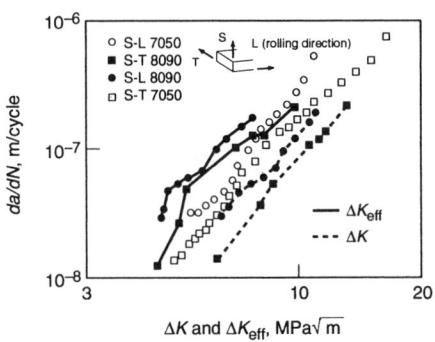

Fig. 21 Effect of crack closure: *da/dN* vs. ΔK_{eff} for 8090-T852 material compared with *da/dN* vs. ΔK_{app} for 8090 and 7050-T7452. Source: Ref 57

Aluminum-Alloy FCP

Aluminum alloys find wide application in fatigue-critical structures, especially in the aerospace industry, where they are used throughout airframes (skins, spars, bulk heads, etc.). As expected, there is considerable detailed knowledge of the FCP behavior of these materials. In addition, due to the continuing push for lighter, stronger structures, new alloys are being introduced, of which the most promising appear to be the Al-Li alloys, which are lighter than conventional alloys and have a higher modulus of elasticity. The key features of most aluminum alloys are grain size, precipitates, dispersoids, and constituent particles.

Aluminum alloys are fairly sensitive to mean stress effects, so Eq 14 can be used with reasonable success for correlating FCP data of most aluminum alloys. However, it should be remembered that the main purpose of such equations is

to put data in a format that can be readily integrated, and equations like Eq 12 and 13 can provide better fits because of their increased number of experimentally determined parameters. Indeed, if life analysis is the sole objective, then numerical integration of the data can be performed directly without fitting it to *any* model. Use of Eq 14 is demonstrated in the data shown in Fig. 19 for 7075-T6 Al in various tempers (Ref 54).

As with most wrought forms, fatigue behavior of aluminum alloys can have significant directionality effects, especially in the threshold regime, as shown in Fig. 20 for 7050-T7452 hand forgings tested at room temperature in moist air (Ref 55). This can be understood quite easily in terms of the grain structure of wrought forms. In the L-T orientation the crack is free to propagate intergranularly in the intermetallic-dominated regions. As a result, the fracture surface would be expected to be quite flat with little or no closure.

This effect means that the crack would "feel" the full stress-intensity range. On the other hand, in the S-T direction the crack could not propagate between the grains very effectively. In essence it would be retarded by the grains and could propagate only slightly in the narrow intergranular regions. Furthermore, the crack surface would be rougher, giving rise to an enhanced closure effect. Thus, at any nominal ΔK level, ΔK_{eff} would be much lower and the driving force for crack propagation would be reduced. As the stress intensity reaches the Paris law regime, the effects of orientation are reduced, as has been pointed out elsewhere (Ref 56). The effects of orientation and closure were investigated in the S-L and S-T orientations for a 7050-T7452 alloy and an 8090 Al-Li alloy. The results are shown in Fig. 21 for $R = 0.1$ (Ref 57). In this study, the 7050 showed minimal closure in these orientations, while the 8090 showed significant closure. It should be noted that in these experiments, extraordinary care was taken to ensure that true closure levels were being measured, as recommended and detailed elsewhere (Ref 58). In this way all artifacts of the closure measurement (which are in fact quite common) were eliminated or reduced to insignificance.

It can be seen that when closure is not accounted for, the FCP rates of the Al-Li alloys appear to be superior. However, when closure is

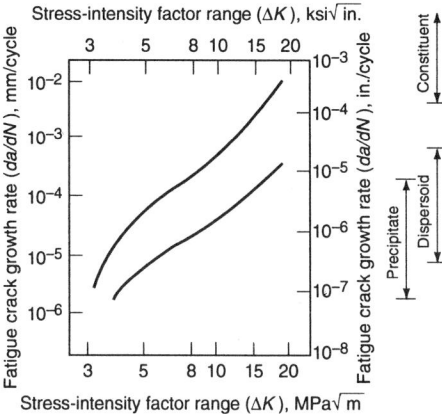

Fig. 22 Comparison of typical particle sizes in aluminum alloys with crack advance per cycle on fatigue loading

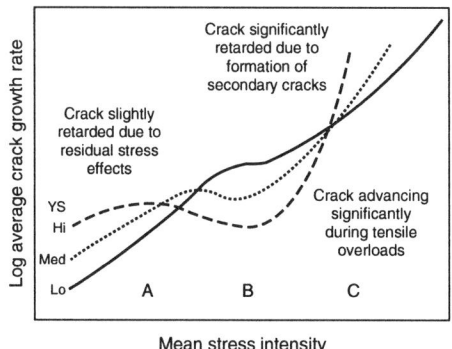

Fig. 23 Schematic effect of yield strength (peak and overaged tempers) on fatigue crack growth rate under spectrum loading. Source: Ref 60

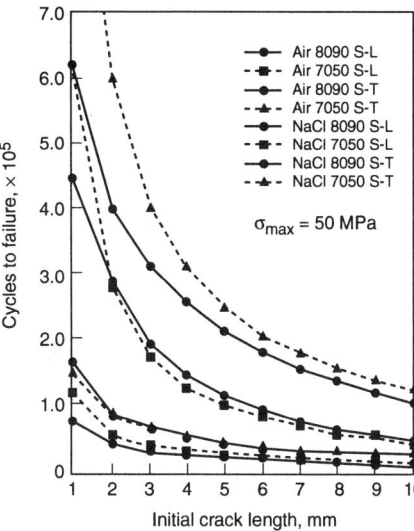

Fig. 24 Comparison of damage tolerance life of 8090-T852 and 7050-T7452 in normal air and saline environments for $R = 0.1$. Source: Ref 57

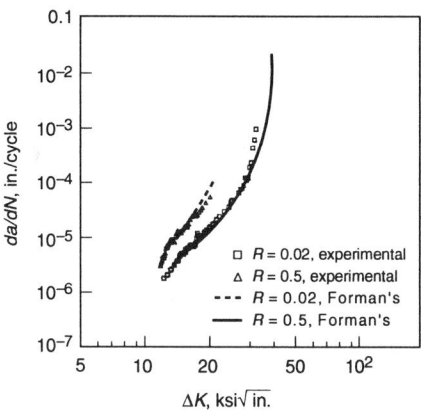

Fig. 25 Comparison of fatigue crack propagation data from Forman's equation and experimental results at room temperature, $R = 0.02$ and $R = 0.5$. Source: Ref 62

taken into account, the effective value of ΔK is decreased at the same level of da/dN. Thus the da/dN versus ΔK curves for the 8090 shift to lie *above* those of the 7050. These results make it very clear that the reason for the observed superior FCP properties of the Al-Li alloys is primarily due to crack closure. It is important to understand this, because at higher R levels, closure effects would be eliminated (or at least significantly reduced) and the FCP properties of the 7050 would be expected to be superior.

The effect of the various types of particles found in conventional aluminum alloys has been characterized (Ref 59). As might be expected, the larger constituent particles begin to play a role at the higher ΔK levels, where normal rupture modes become important. The precipitates influence the FCP behavior at the lowest ΔK levels, while the dispersoid particles, which have the effect of controlling the grain size, have an effect at slightly higher ΔK levels. These ideas are summarized in Fig. 22.

It is interesting to note that the relative ranking of aluminum alloys in constant-amplitude FCP is not preserved in variable-amplitude or peak overload FCP environments. This is related to the effects that overloads have on debonding large particles and the subsequent reduced constraint

during lower-ΔK cycling. This is illustrated in Fig. 23 for three levels of yield strength (Ref 60). At the lower ΔK levels, the higher-yield-strength material exhibits the highest FCP rate, with the lower-strength material showing the lowest FCP rate. As the base ΔK level increases, the overloads are higher and the large constituent particles in the higher-strength material crack locally and reduce the constraint, thereby lowering the FCP rate during subsequent cycling at the base ΔK levels. When the base ΔK level increases to the highest levels, there is significant crack advance per cycle in the high-strength material, corresponding to its lower fracture toughness, and the relative ordering of the alloys goes back to what is seen in constant-amplitude cycling. These ideas would also apply to a given class of alloys with differences in the constituent particle size and density. It can easily be envisioned that alloys with a higher density of large particles will show decreased crack growth rates under tension-dominated spectrum loading.

The effects of the strengthening precipitates would be expected to play an important role in the FCP behavior of aluminum alloys because they influence, in a profound manner, the basic slip characteristics, which as we have seen influence the crack propagation behavior. However, great care must be taken in the interpretation of test results, because the complexity of these alloys means that several potentially competing mechanisms may operate simultaneously. These ideas are discussed more fully in the section on aluminum alloys.

Aluminum alloys are used in environments that can be very aggressive. Saline environments would be typical, for example, of aircraft on carriers or of aircraft operating in warm, coastal areas such as Hawaii or California. The effects of composition, orientation, and environment are illustrated in Fig. 24 (Ref 57). In this figure, FCP curves have been integrated based on data that was generated for air and saline solution environ-

ments. The assumed geometry was a wide edge-cracked plate subjected to a remote stress of 50 MPa. The crack sizes and stresses are representative of what would be expected for military aircraft under normal inspections and operating conditions. The integrated life results show spectacular differences for different environments, discernible differences for different orientations, and modest differences for different alloys.

Titanium-Alloy FCP

Titanium alloys, like the other systems that we have discussed so far, have FCP properties that are sensitive to chemical composition, microstructure, environment, and loading variables. In general, titanium alloys, like most others, exhibit three regimes of behavior: a near-threshold regime (low ΔK values), a Paris law regime (intermediate ΔK levels), and a regime where normal rupture modes dominate as K_{Ic} is approached (high ΔK levels). The effects of these variables on FCP are complex, and only major trends will be discussed here.

The effect of the R ratio on FCP for many alloys is as predicted by Eq 14. The R ratio effect is largest at low and high values of ΔK and somewhat reduced for the intermediate values, as has been pointed out elsewhere (Ref 61). The fit of the data to Forman's equation is well illustrated for the near-α T-6242 alloy tested at two R ratios (Fig. 25) (Ref 62). As predicted by the form of Eq 14, the FCP rate is higher for the higher R ratio.

Interstitial oxygen is known to embrittle the α phase in α-β alloys, and this effect would also be expected in the FCP behavior of these alloys to the extent that the FCP behavior is influenced by those factors which influence K_{Ic}. Figure 26 shows that for a Ti-6Al-4V alloy, the crack growth rate is higher the higher the oxygen content (Ref 63). Of course, the precise composition of titanium alloys has a significant effect on FCP. For example, depending on the composition, some alloys may be more prone to texturing than others. Although this is not a direct compositional effect, texturing does have a significant influence,

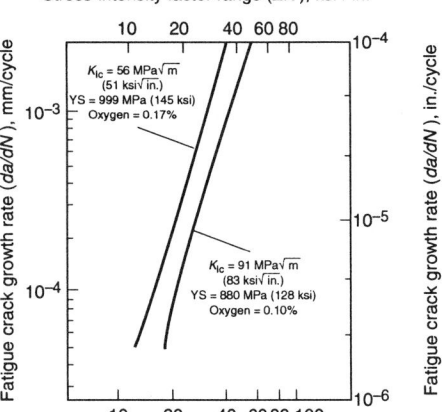

Fig. 26 Variation of fatigue crack propagation rate with yield strength, K_{Ic}, and other variables for annealed Ti-6Al-4V forgings. R = 0.02; 10 Hz; air, argon, and JP4 environments pooled; average of six tests per trend line. Source: Ref 63

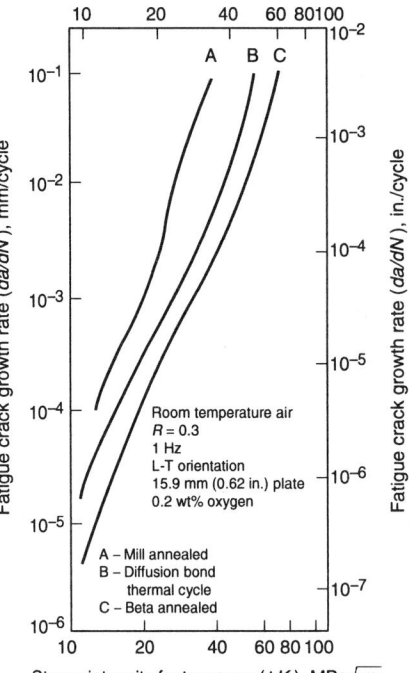

Fig. 27 Effect of heat treatment on fatigue crack growth rate of alloy Ti-6Al-4V. Source: Ref 64

and in this regard it may be considered a composition-related effect.

Microstructural factors are very important in influencing the FCP rates of titanium alloys, and it has been demonstrated that the FCP rate may be changed an order of magnitude depending on the microstructure, especially at low values of ΔK. As has been pointed out elsewhere (Ref 61), FCP properties generally track trends in K_{Ic}. FCP properties are generally quite desirable for transformed structures and for microstructures produced by recrystallization anneals. The effect of microstructure is shown in Fig. 27 (Ref 64). Part of the improvement in the transformed structures is surely related to the meandering nature of the crack and the increased closure levels associated with such a crack surface. In fact, it has been shown that the FCP rate decreases with the size of the Widmanstätten α packet size (Ref 65). Some authors have attributed the reduced crack growth rates to a longer crack path. However, this effect is rather small and can account for at most a factor of 2 compared to the order-of-magnitude changes in FCP rates that occur.

Titanium alloys are fairly resistant to the effects of environment, except at fairly high temperatures. However, it has been established through a very detailed analysis of experimental data that in a $3\frac{1}{2}$% NaCl solution, the rates of crack extension (fatigue and static loading) can be associated with an activation energy very close to that of hydrogen in titanium (Ref 66). This effect is in agreement with the results of other investigators who have found that hydrogen plays a significant role in embrittling titanium alloys and causing higher rates of crack extension.

The fatigue behavior of titanium alloys is very sensitive to the environment and in general corresponds with well-understood principles of fracture mechanics. There is consequently great opportunity for controlling the fatigue behavior through heat treatment and compositional control to achieve desired goals.

Specific Design for Improved Fatigue Resistance

While the details of improving the FCP properties depend on the class of materials being considered, some general principles appear to be valid for the alloy systems considered here. For example, while microstructural effects are observed in all regimes, they are most pronounced in the near-threshold regime. Larger grain sizes generally tend to produce lower FCP rates, either through lowering the average plastic strain at the tip of a propagating crack or (in some sense equivalently) through producing a rougher fracture surface and increased closure levels. In systems with simple microstructures and well-defined slip mechanisms (e.g., nickel-base superalloys), control of the precipitate size also provides a means for controlling FCP. In general, the degree of slip reversibility will depend on the size of the precipitates, and the size of the precipitate can be adjusted to promote increased reversibility. Of course this can also be put in the context of closure, because reversible slip, as we have seen, also produces a fracture surface with considerable surface roughness.

REFERENCES

1. H. Tada, P.C. Paris, and G.R. Irwin, *The Stress Analysis of Cracks Handbook*, 2nd ed., Paris Productions Inc., St. Louis, MO, 1985
2. "Standard Method for Plane-Strain Fracture Toughness Testing of Metallic Materials," ASTM E 399-91, ASTM, Vol 3.01, 1991, p 488-512
3. J.P. Schaffer, A. Saxena, S.D. Antolovich, T. Sanders, and S. Warner, *The Science and Design of Engineering Materials*, Irwin Publishers, 1995
4. G.T. Hahn and A.R. Rosenfield, in STP 432, ASTM, 1968, p 5-32
5. S.D. Antolovich and B. Singh, On the Toughness Increment Associated with the Austenite to Martensite Phase Transformation in TRIP Steels, *Met. Trans.*, Vol 2, 1971, p 2135-2141
6. A. Saxena and S.D. Antolovich, Increased Fracture Toughness in a 300 Grade Maraging Steel as a Result of Thermal Cycling, *Met. Trans.*, Vol 5, 1974, p 623-632
7. W.W. Gerberich, R.H. Van Stone, and A.W. Gunderson, *Application of Fracture Mechanics for the Selection of Metallic Structure Materials*, J. Campbell, W.W. Gerberich, and J.H. Underwood, Ed., American Society for Metals, 1982, p 41-103
8. F.R. Stonsifer and R.W. Armstrong, *Fracture 1977*, Vol 2A, D.M.R. Taplin, Ed., Pergamon Press, 1977, p 1-6
9. A.H. Priest, in *Effect of Second-Phase Particles on the Mechanical Properties of Steel*, The Iron and Steel Institute, London, 1971
10. T.B. Cox and J.R. Low, Jr., An Investigation of the Plastic Fracture of AISI 4340 and 18 Nickel-200 Grade Maraging Steels, *Met. Trans.*, Vol 5, 1974, p 1457-1470
11. K.H. Schwalbe, *Eng. Fract. Mech.*, Vol 9, 1977, p 795-832
12. *Fractography*, Vol 12, *Metals Handbook*, 9th ed., ASM International, 1987, p 384
13. S.D. Antolovich and B. Singh, Observations of Martensite Formation and Fracture in TRIP Steels, *Met. Trans.*, Vol 2, 1970, p 3463-3465
14. J.G. Kaufman, Fracture Toughness of Aluminum Alloy Plate—Tension Tests of Large Center Slotted Panels, *Journal of Materials*, Vol 2 (No. 4), 1967, p 889-914
15. R.J. Bucci, *Eng. Fract. Mech.*, Vol 12, 1979, p 407-441
16. K.T. Rao and R.O. Ritchie, *Internat. Mater. Rev.*, Vol 37, 1992, p 153-185
17. S.D. Antolovich, P.M. Shete, and G.R. Chanani, Fracture Toughness of Duplex Structures, Part I: Tough Fibers in a Brittle Matrix, STP 514, ASTM, 1972, p 114-134
18. R.R. Ferguson and R.G. Berryman, "Fracture Mechanics Evaluation of B-1 Materials," Report AFML-TR-76-137, Vol I, Rockwell International, Los Angeles, 1976
19. D.A. Meyn, Effect of Hydrogen on Fracture and Inert-Environment Sustained Load Cracking Resistance of Alpha-Beta Titanium Alloys, *Met. Trans.*, Vol 5A (No. 11), Nov 1974, p 2405-2414
20. T.W. Hall and C. Hammond, The Relation between Crack Propagation Characteristics and Fracture Toughness in Alpha+Beta Titanium Alloys, *Titanium Science and Technology*, Vol 2, Plenum Press, 1973, p 1365-1376
21. T.L. Wardlaw, H.W. Rosenberg, and W.M. Parris, "Development of Economical Sheet Titanium Alloy, Report AFML-TR-73-296

22. R.E. Curtis and W.F. Spurr, Effect of Microstructure on Fracture Properties of Titanium Alloys in Air and Salt Solution, *ASM Transactions Quarterly*, Vol 61 (No. 1), March 1968, p 115-127

23. R.L. Tobler, Low-Temperature Fracture Behavior of a Ti-6Al-4V Alloy and the Electron-Beam Welds, *Toughness and Fracture Behavior of Titanium*, STP 651, ASTM, 1978, p 267-294

24. J.L. Christian and A. Hurlich, "Physical and Mechanical Properties of Pressure Vessel Materials for Applications in a Cryogenic Environment," Report ASD-TDR-628, Part 2, April 1963

25. M.J. Harrigan et al., The Effect of Rolling Texture on the Fracture Mechanics of Ti-6Al-2Sn-4Zr-6Mo Alloy, *Titanium Science and Technology*, Vol 2, Plenum Press, 1973, p 1297-1317

26. J.P. Bailon and S.D. Antolovich, The Effect of Microstructure on Fatigue Crack Propagation: A Review of Existing Models and Suggestions for Further Research, STP 811, ASTM, 1983, p 313-349

27. A. Saxena and S.D. Antolovich, Low Cycle Fatigue, Fatigue Crack Propagation, and Substructures in a Series of Polycrystalline Cu-Al Alloys, *Met. Trans.*, Vol 6A, 1975, p 1809-1828

28. R. Bowman, S.D. Antolovich, and R.C. Brown, A Demonstration of Problems Associated with Crack Closure Measurement Techniques, *Eng. Fract. Mech.*, Vol 31, 1988, p 703-712

29. W.J. Drury, A.M. Gokhale, and S.D. Antolovich, Effect of Crack Surface Geometry on Fatigue Crack Closure, *Met. Trans.*, Vol 26A, 1995, p 2651-2663

30. H.Y. Jung and S.D. Antolovich, Experimental Characterization of Roughness-Induced Crack Closure in Al-Li 2090 Alloy, *Scripta Met.*, Vol 33, 1995, p 275-281

31. P. Paris and F. Erdogan, *J. Basic Engineering, Trans. ASME*, Series D, Vol 85, 1963, p 528-539

32. R.M. Wallace, C.G. Annis, Jr., and D. Sims, Report AFML-TR-76-176, Vol I, Rockwell International, Los Angeles, 1976

33. A. Saxena, S.J. Hudak, Jr., and G.M. Jouris, *Eng. Fract. Mech.*, Vol 12 (No. 1), 1979, p 103-115

34. A.G. Forman, *J. Basic Engineering, Trans. ASME*, Series D, Vol 89, 1967, p 459-469

35. R.L. Tobler, Low Temperature Effects on the Fracture Behavior of a Nickel Base Superalloy, *Cryogenics*, Vol 16 (No. 11), Nov 1976, p 669-674

36. S.D. Antolovich and J.E. Campbell, *Application of Fracture Mechanics for the Selection of Metallic Structure Materials*, J. Campbell, W.W. Gerberich, and J.H. Underwood, Ed., American Society for Metals, 1982, p 253-310

37. S.D. Antolovich and B. Lerch, Cyclic Deformation and Fatigue in Ni-Base Alloys, *Superalloys, Supercomposites, and Superceramics*, J.K. Tien and T. Caufield, Ed., Academic Press, 1989, p 363-411

38. R. Miner and J. Gayda, *Met. Trans.*, Vol 14A, 1983, p 2301-2308

39. H.F. Merrick and S. Floreen, *Met. Trans.*, Vol 9A (No. 2), 1978, p 231-233

40. J. Bartos and S.D. Antolovich, *Fracture 1977*, D.M.R. Taplin, Ed., Vol 2, University of Waterloo Press, Waterloo, Canada, 1977, p 996-1006

41. B. Lawless, S.D. Antolovich, C. Bathias, and B. Boursier, *Fracture: Interactions of Microstructure, Mechanisms and Mechanics*, J.M. Wells and J.D. Landes, Ed., TMS-AIME, 1985

42. W.J. Mills and L.A. James, Publication 7-WA/PUP-3, ASME, 1979

43. D. Krueger, S.D. Antolovich, and R.H. Van Stone, *Met. Trans.*, Vol 18A, 1987, p 1431-1449

44. K.T. Rao, W. Yu, and R.O. Ritchie, *Fatigue '87*, R.O. Ritchie and E.A. Starke, Ed., EMAS, West Midland, U.K., 1987, p 291-301

45. R. Bowman and S.D. Antolovich, The Effect of Microstructure on the Fatigue Crack Growth Resistance of Nickel Base Superalloys, *Superalloys 1988*, D.N. Duhl, G. Maurer, S. Antolovich, C. Lund, and S. Reichman, Ed., AIME, 1988, p 565-574

46. B.A. Lerch and S.D. Antolovich, Fatigue Crack Propagation Behavior of a Single Crystalline Superalloy, *Met. Trans.*, Vol 21A, 1990, p 2169-2177

47. B.F. Antolovich, A. Saxena, and S.D. Antolovich, Fatigue Crack Propagation in Single Crystal CMXS-2 at Elevated Temperature, *Superalloys 1992*, S.D. Antolovich, et al., Ed., AIME, 1992, p 727-736

48. J.S. Crompton and J.W. Martin, *Met. Trans.*, Vol 15A, 1984, p 1711-1719

49. K.S. Chan and G.R. Leverant, *Met. Trans.*, Vol 18A, 1987, p 593-602

50. P.J. Peters, C.M. Biando, and D.P. DeLuca, Effects of Hydrogen on FCP of Ni-base Alloys, United Technologies Pratt and Whitney Government Engines and Space Propulsion, Report FR-24007, NASA Grant NA58-39050, 1995

51. Arnaud de Bussac, M.S. thesis, Georgia Institute of Technology, 1992

52. D. Taylor, *A Compendium of Fatigue Thresholds and Growth Rates*, EMAS, West Midlands, U.K., 1985

53. S.T. Rolfe and J.M. Barsom, *Fracture and Fatigue Control in Structures—Applications of Fracture Mechanics*, Prentice-Hall, 1977

54. R.G. Forman, R.E. Kearney, and R.M. Engle, *J. Basic Engineering, Trans. ASME*, Vol 89, 1967, p 459-464

55. L.N. Mueller, "ALCOA Aluminum Alloy 7050," Green Letter 220, ALCOA Center, PA, Oct 1985

56. J.G. Kaufmann and J.S. Santner, *Application of Fracture Mechanics for the Selection of Metallic Structure Materials*, J. Campbell, W.W. Gerberich, and J.H. Underwood, Ed., American Society for Metals, 1982, p 169-211

57. Kevin Lemke, Georgia Institute of Technology, personal communication, 1993

58. R. Bowman, S.D. Antolovich, and R.C. Brown: A Demonstration of Problems Associated with Crack Closure Measurement Techniques, *Eng. Fract. Mech.*, Vol 31, 1988, p 703-712

59. J.T. Staley, *Fracture Mechanics*, N. Perrone et al., Ed., University Press of Virginia, 1978, p 671

60. T.H. Sanders, Jr., R.R. Sawtell, J.T. Staley, R.J. Bucci, and A.B. Thakker, "Effect of Microstructure on Fatigue Crack Growth of 7xxx Al Alloys under Constant Amplitude and Spectrum Loading," Naval Air Development Center Contract N00019-76-C-0482, Final Report, 1978

61. H.W. Rosenberg, J.C. Chesnutt, and H. Margolin, *Application of Fracture Mechanics for the Selection of Metallic Structure Materials*, J. Campbell, W.W. Gerberich, and J.H. Underwood, Ed., American Society for Metals, 1982, p 213-252

62. L.S. Vessier and S.D. Antolovich, Fatigue Crack Propagation in Ti-6242 as a Function of Temperature and Waveform, *Eng. Fract. Mech.*, Vol 37, 1990, p 753-775

63. Report MDC-A0913, McDonnell Aircraft Co., 1971

64. R.R. Ferguson and R.G. Berryman, Report AFML-TR-76-137, Vol I, Rockwell International, Los Angeles, 1976

65. G.R. Yoder and D. Eylon, *Met. Trans*, Vol 10A, 1979, p 1808-1810

66. P. Bania and S.D. Antolovich, Activation Energy Dependence on Stress Intensity Level in Stress Corrosion, STP 610, ASTM, 1976, p 157-175

Micromechanisms of Monotonic and Cyclic Crack Growth

Todd S. Gross, University of New Hampshire, and Steven Lampman, ASM International

FRACTURE is the complete separation of a material that occurs when a crack reaches a critical size and impairs strength below the service load. The final fracture is usually abrupt, but it is generally preceded by a cracking process that occurs slowly over the service life from various crack growth mechanisms (e.g., see Fig. 1) such as fatigue, stress-corrosion cracking, creep, and hydrogen-induced cracking. Each of these cracking mechanisms has certain characteristic features that are used in failure analysis to determine the cause of cracking or crack growth.

The purpose of this article is to provide a brief description of the different types of micromechanisms of monotonic and cyclic fracture. General information on what material variables have the most beneficial effect on resistance to failure will be presented. Each micromechanism may have a particular manifestation in a given material system, but there are general features of each micromechanism that are common to all relevant systems. In this article, the micromechanisms are divided into monotonic and cyclic mechanisms. This distinction is somewhat artificial in that one can observe evidence of "monotonic" micromechanisms on fracture surfaces that failed under cyclic loading. Also, the effect of environment on the micromechanism is considered as a variable that may change the micromechanism under a given loading condition rather than a unique type of fracture. Polymer fracture and fatigue crack growth occur by micromechanisms that are not observed for monolithic crystalline materials. Therefore, polymer failure will be discussed separately in each section. The failure of fiber-reinforced composite materials also has unique aspects and is covered in a separate volume, *Engineered Materials Handbook*, Volume 1, *Composites* (ASM International, 1987).

Types of Fracture

The classification of fractures has been approached in many different ways. In general, fracture modes can be classified into four general categories, based on the appearance of fracture surfaces: dimple rupture, cleavage, fatigue, or decohesive rupture (or intergranular fracture from crack growth mechanisms such as creep or stress corrosion) (Ref 1). This classification is viable, because each mode has a characteristic fracture surface appearance and an underlying mechanism for crack growth. However, it is also useful to make a distinction between progressive crack growth mechanisms and abrupt or "instantaneous" fracture, as noted above. This general distinction is shown in Table 1 with distinguishing features of various fracture modes.

Confusion arises when fractures are classified as either ductile or brittle, because correct usage of these terms depends on whether one is referring to a fracture micromechanism or the macroscopic work of fracture. For example, if an unnotched tensile sample of a ductile alloy is pulled to fracture, overall plastic deformation will occur and thus the fracture is ductile. However, if a notched sample is loaded, then plastic deformation can be restricted to the notched area, in which case the overall plastic deformation may be substantially reduced (relative to the percent elongation of the unnotched sample). In this sense, most fractures from an engineering perspective are termed brittle, because plastic deformation is restricted to the notched section or cracked area where fracture failures occur.

Yet even though the fracture can be classified as brittle when plastic deformation is confined to the fracture path, the micromechanism of fracture may be ductile. In fact, many fractures occur by a ductile micromechanism, even though the macroscopic fracture is termed brittle with little or no overall plastic deformation or macroscopic work. In this sense, the terms *brittle* and *ductile fracture* are most clearly defined in a macroscopic context. Most components are designed with stresses in the elastic range, when a crack propagates from a flaw, only a relatively small volume of material plastically deforms and absorbs energy. Because the energy to cause fracture is "small" compared to that expected from a ductile fracture, the failure

is often described as a brittle fracture. This characterization is often applied even if the micromechanism of material separation involves plastic flow and would otherwise be considered a ductile micromechanism.

Ductile fractures are characterized by tearing of metal accompanied by appreciable gross plastic deformation and expenditure of considerable energy. Ductile tensile fractures in most materials have a gray, fibrous appearance and are classified on a macroscopic scale as either flat-face (square) or shear-face (slant-shear) fractures.

Brittle fractures are characterized by rapid crack propagation with less expenditure of energy than in ductile fractures and without appreciable gross plastic deformation. Brittle tensile fractures are of the flat-face type, and are produced under plane-strain conditions with little or no necking (Fig. 2).

Fracture Micromechanisms. The micromechanisms of fracture can be defined as ductile and brittle fracture, provided that the terms are clearly distinguished from the geometric and mechanical constraints that distinguish brittle and ductile fracture on a component or structural level. Nonetheless, ductile fracture is clearly associated with dimpled rupture, while brittle fracture mechanisms produce the well-known fracture features of:

- Transgranular cleavage or quasicleavage
- Intergranular separation
- Features on transgranular facets, such as river marks, herringbone patterns, or tongues
- Crazing (in polymers)

In this context, the micromechanisms of fracture can be appropriately defined as ductile or brittle, depending on whether or not the micromechanisms *require* plastic flow for material separation. A crystalline material can separate by a combination of two micromechanisms: plastic flow and physical separation of atomic planes. While separation of atomic planes generally requires a high stress, the energy absorbed to create the two surfaces is very small from an engineer-

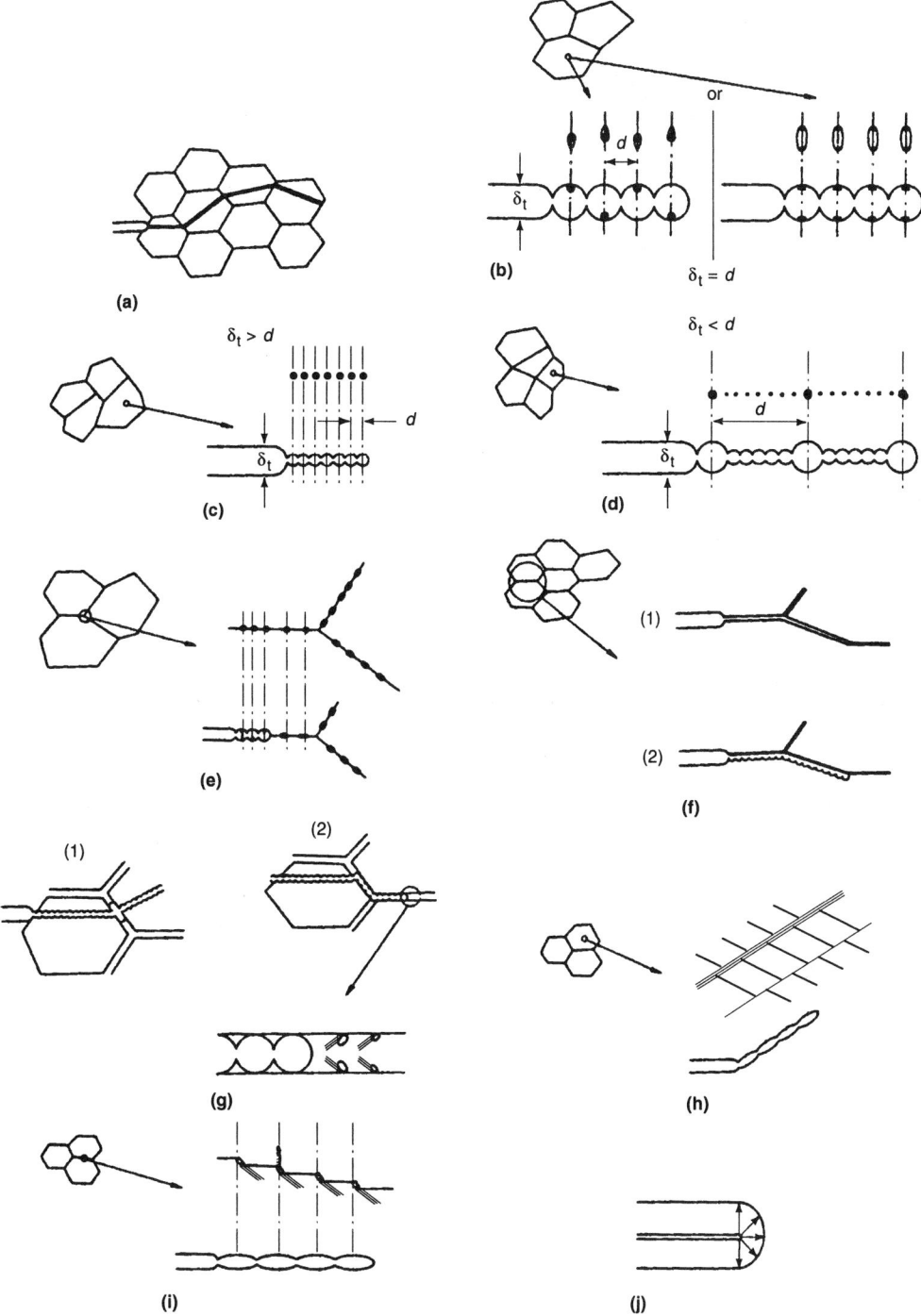

Fig. 1 Crack propagation mechanisms: (a) cleavage crack propagation. (b) Dimple fracture due to coarse particles. (c) Dimple fracture due to fine particles. (d) Dimple fracture due to coarse and fine particles. (e) Intergranular crack propagation due to grain boundary precipitates. (f) Intergranular crack propagation due to a hard phase grain boundary film. (g) Crack propagation mechanisms when a soft phase grain boundary film is present. (h) Crack propagation by slip plane/slip plane intersection. (i) Crack propagation by slip plane/grain boundary intersection. (j) Crack propagation solely by plastic blunting

ing standpoint. Fracture micromechanisms that occur primarily by separation of atomic planes are described as brittle. Fracture by plastic flow takes on the order of 10^6 times as much energy as fracture by separation of atomic planes. The atomic surface energy of pure metals is in the range of 0.1 to 10 J/m^2 (Ref 2, 3), and the J_{Ic}, a measure of surface energy, is in the range of 10^4 to 10^6 J/m^2 for typical engineering materials (Ref

4). Fracture micromechanisms that occur with a significant amount of plastic flow are generally termed ductile.

Fracture Micromechanism Maps. Putting the distinction between brittle and ductile fracture mechanisms in a broader perspective, Ashby et al. (Ref 5) and Ghandi and Ashby (Ref 6) constructed a series of deformation mechanism maps for crystalline materials. Several of these maps

are shown in Fig. 3 (Ref 7). They classified brittle fractures into three distinct types: modes I, II, and III. (These are different from the three modes of crack growth: mode I, opening; mode II, in-plane shear; and mode III, anti-plane shear.) Mode I brittle fractures originate from flaws or cracks in the structure and occur at lower normalized stresses. Little or no plastic deformation occurs. In contrast, microscopic yielding occurs before

Table 1 Fracture mode identification chart

Method	Instantaneous failure mode(a)		Progressive failure mode			
	Ductile overload	Brittle overload	Fatigue	Corrosion	Wear	Creep
Visual, 1 to 50× (fracture surface)	Necking or distortion in direction consistent with applied loads Dull, fibrous fracture Shear lips	Little or no distortion Flat fracture Bright or coarse texture, crystalline, grainy Rays or chevrons point to origin	Flat progressive zone with beach marks Overload zone consistent with applied loading direction Ratchet marks where origins join	General wastage, roughening, pitting, or trenching Stress-corrosion and hydrogen damage may create multiple cracks that appear brittle	Gouging, abrasion, polishing, or erosion Galling or storing in direction of motion Roughened areas with compacted powdered debris (fretting) Smooth gradual transitions in wastage	Multiple brittle-appearing fissures External surface and internal fissures contain reaction-scale coatings Fracture after limited dimensional change
Scanning electron microscopy, 20 to 10,000× (fracture surface)	Microvoids (dimples) elongated in direction of loading Single crack with no branching Surface slip band emergence	Cleavage or intergranular fracture Origin area may contain an imperfection or stress concentrator	Progressive zone: worn appearance, flat, may show striations at magnifications above 500× Overload zone: may be either ductile or brittle	Path of penetration may be irregular, intergranular, or a selective phase attacked EDS(b) may help identify corrodent	Wear debris and/or abrasive can be characterized as to morphology and composition Rolling contact fatigue appears like wear in early stages	Multiple intergranular fissures covered with reaction scale Grain faces may show porosity
Metallographic inspection, 50 to 1000× (cross section)	Grain distortion and flow near fracture Irregular, transgranular fracture	Little distortion evident Intergranular or transgranular May relate to notches at surface or brittle phases internally	Progressive zone: usually transgranular with little apparent distortion Overload zone: may be either ductile or brittle	General or localized surface attack (pitting, cracking) Selective phase attack Thickness and morphology of corrosion scales	May show localized distortion at surface consistent with direction of motion Identify embedded particles	Microstructural change typical of overheating Multiple intergranular cracks Voids formed on grain boundaries or wedge-shaped cracks at grain triple points Reaction scales or internal precipitation Some cold flow in last stages of failure
Contributing factors	Load exceeded the strength of the part Check for proper alloy and processing by hardness check or destructive testing, chemical analysis Loading direction may show failure was secondary Short-term, high-temperature, high-stress rupture has ductile appearance (see creep)	Load exceeded the dynamic strength of the part Check for proper alloy and processing as well as proper toughness, grain size Loading direction may show failure was secondary or impact induced Low temperatures	Cyclic stress exceeded the endurance limit of the material Check for proper strength, surface finish, assembly, and operation Prior damage by mechanical or corrosion modes may have initiated cracking Alignment, vibration, balance High cycle, low stress: large fatigue zone; low cycle, high stress: small fatigue zone	Attack morphology and alloy type must be evaluated Severity of exposure conditions may be excessive; check pH, temperature, flow rate, dissolved oxidants, electrical current, metal coupling, aggressive agents Check bulk composition and contaminants	For gouging or abrasive wear: check source of abrasives Evaluate effectiveness of lubricants Seals or filters may have failed Fretting induced by slight looseness in clamped joints subject to vibration Bearing or materials engineering design may reduce or eliminate problem Water contamination High velocities or uneven flow distribution, cavitation	Mild overheating and/or mild overstressing at elevated temperature Unstable microstructures and small grain size increase creep rates Ruptures occur after long exposure times Verify proper alloy

(a) Fractographers often refer to "overload fracture" as a way to distinguish the failure from fatigue or stress-corrosion crack growth. However, a true overload failure is rare. Indication of overload fracture may just indicate an undetected crack or defect leading to fracture. (b) EDS, energy-dispersive spectroscopy. Source: *Metals Handbook*, 9th ed., Vol 11, p 80

and during mode II brittle fractures, but the plastic strains are not significant. Mode II and mode III brittle fractures occur in materials in which the flaws or inclusions are smaller than the relevant microstructural unit, usually the grain size. During mode III brittle fracture, macroscopic yielding occurs before fracture. As the material deforms, the capacity for further deformation is exhausted and the material fractures by a brittle micromechanism. In all cases, the fracture path can be transgranular (cleavage) or intergranular. The three brittle fracture mechanisms are schematically shown in Fig. 4, along with stress-strain curves for each mode and for ductile fracture. While high-temperature fracture may absorb a significant amount of energy, the inelastic deformation is strongly rate dependent, in contrast to that for low-temperature fracture, and is therefore

classified as a distinct mechanism. Most low-temperature fracture is brittle (Table 2). Constraint (Fig. 2a) can also effect abrupt, brittle-like fracture in plane-strain conditions.

Face-centered cubic (fcc) metals are unique in that they do not exhibit any brittle modes of fracture. (Iridium and rhodium are the exceptions to this statement.) As seen in the fracture mechanism maps, body-centered cubic (bcc) metals and hexagonal close-packed (hcp) metals undergo a transition from mode III brittle fracture to ductile fracture. For bcc metals, this transition is related to the strong temperature and rate dependence of the flow stress. At low temperatures, the flow stress is above the stress required to nucleate voids, which then propagate by cleavage or intergranular fracture (both of which are discussed in the following sections). For hcp metals, the lim-

ited number of slip systems active at low temperatures prevents significant plastic deformation, and the metal fractures in a brittle manner. At higher homologous temperatures, secondary slip systems are activated, enabling the macroscopic plasticity required for ductile fracture.

Alkali halides, refractory oxides, and covalently bound materials exhibit only brittle modes of fracture and high-temperature fracture. For alkali halides (e.g., NaCl), the stress to initiate plastic flow can be less than the stress to initiate a void and cause mode II or mode III brittle fracture, but (as for hcp metals at low temperatures) the limited number of slip systems prevents the distribution of plastic strain necessary for ductile behavior. Refractory oxides (e.g., MgO and Al_2O_3) have stronger bonding (mixed ionic, covalent) than the ionically bonded alkali halides.

Table 2 Characteristics of the various low-temperature fracture modes

Fracture mode	Fracture nucleation	Initial crack propagation	Final crack propagation	Fracture preceded by macroscopic yielding
Mode I brittle fracture	By presence of pre-existing flaws	None	By cleavage or intergranular fracture with limited crack-tip plasticity	No
Mode II brittle fracture	By microscopic yielding at yield strength	To something on the order of the size of a grain	By cleavage or intergranular fracture with limited crack-tip plasticity. For bcc transition metals, immediately following yielding, because $\sigma_y > \sigma_F$.	No
Mode III brittle fracture	By microscopic yielding at yield strength	To something on the order of the size of a grain; because $\sigma_y < \sigma_F$, multiple microcracking is found	By cleavage between microcracks or by intergranular fracture between microcracks. Extensive crack-tip plasticity. Some fibrous tearing at the highest temperatures of mode III fracture.	Limited
Ductile fracture	By inhomogeneous plastic deformation at multiple locations—frequently sites of second-phase and nondeforming particles	Void growth at the locations of fracture nucleation. Growth by plastic deformation processes.	By void linkup via ductile microrupture or shearing	Extensive

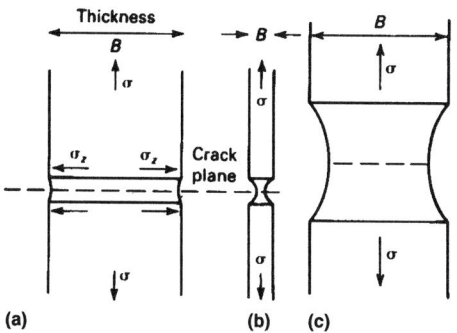

Fig. 2 Typical constraint conditions for plane-strain and plane-stress fracture in a tensile bar. (a) Low-stress, thick-plate, thin-cylinder, no-contraction plane-strain. (b) Low-stress, thin-plate, free-contraction plane-stress. (c) High-stress, thick-plate, thick-cylinder, free-contraction plane-stress. Stresses and strains are high at the crack tip, and the high strains in the x- and y-directions at a crack tip induce a contraction in the z-direction. If the contraction (plastic) zone is long and thin, contraction is prevented by the surrounding material, which thus constrains the contraction and exerts a stress in the z-direction so as to keep $\varepsilon_0 = 0$ such that $\varepsilon_z = (\sigma_z/E) - [v(\sigma_x + \sigma_y)/E] = 0$ or $\sigma_z = v(\sigma_x + \sigma_y)$, where v is Poisson's ratio. If plasticity occurs, the value of v approaches 0.5. The above is for plane-strain. For plane-stress, contraction is unhindered, in which case $\varepsilon_z = -v[(\sigma_x + \sigma_y)/E]$ and $\sigma_z = 0$.

Accordingly, the flow stress is much higher than the stress to initiate voids, and ductile fracture is not observed. The same is true for covalently bound materials (e.g., Si, SiC, Si_3N_4).

Polymeric materials and blends can be classified into modes of brittle fracture similar to those for metals, alkali halides, ceramics, and covalently bound materials, although the time-dependent nature of polymeric deformation suggests classification as a high-temperature fracture. Ductile fracture is also observed for polymers. A stress-strain curve for a polymer that exhibits ductility is schematically represented in Fig. 3. Large, inhomogeneous plastic strain occurs prior to ductile fracture of polymers.

Micromechanisms of Monotonic Fracture

Within the general context of the previous discussions on fracture modes, this section presents generic descriptions of monotonic fracture micromechanisms. For fcc, hcp, and bcc metals, the ductile micromechanism of fracture is microvoid coalescence. The brittle micromechanisms of fracture are cleavage, intergranular fracture, and crazing (for polymeric materials). Some evidence of the first three micromechanisms can be observed on fracture surfaces of many structural metals. However, every attempt is usually made to encourage the occurrence of microvoid coalescence, because it absorbs the most energy during fracture.

Microvoid Coalescence. The only "low-temperature" ductile micromechanism of monotonic fracture (Table 2) is microvoid coalescence. Another term used to describe this mode is dimpled rupture. While it is a ductile micromechanism, the presence of microvoid coalescence on a fracture surface does not guarantee that the fracture absorbed a significant amount of energy. There are a number of situations where dimpled rupture is observed for embrittled materials. Also, dimpled fracture surfaces are observed when voids form on grain boundaries during high-temperature deformation.

Figure 5 is a scanning electron micrograph of a steel fracture surface in which the material failed by microvoid coalescence. Voids initiate at second-phase particles or inclusions either by particle-matrix decohesion, cracking of the particle, or a combination. Upon continued plastic deformation, the voids grow and cause fracture when they impinge on one another. Microvoid coalescence is also observed in polymers. Broek (Ref 8) has shown that the dimple diameter and the inclusion spacing correlate quite well in aluminum alloys, as shown in Fig. 6.

The sequence of fracture is void initiation, void growth, and finally coalescence. Accordingly, the effect of microstructural variables on each step has been studied in detail. Two reviews provide summaries of both experimental studies and analytical models of microvoid coalescence (Ref 9, 10). These two articles summarize two to three decades of the study of fracture and are gateways to the vast literature on the subject. Several textbooks also provide useful and more detailed summaries of the phenomenon of microvoid coalescence (Ref 7, 11, 12). The following paragraphs briefly summarize the important observations.

Void initiation or nucleation occurs by particle-matrix decohesion or particle cracking. Particle-matrix decohesion can be negatively affected by interfacial segregation of trace elements. Smaller particles require a greater plastic strain for void initiation. In some materials, a critical strain defines void initiation; in others, void initiation is more closely correlated with stress level. In some systems, the void-initiating particles are damaged by the initial hot or cold forming process, so void initiation by cracking occurs at low stresses.

For systems containing large volume fractions of particles, the stress fields around the particles interact in such a way as to minimize the effect of particle size on initiation. The void initiation process is affected by the state of stress. Figure 7 compares Al_2O_3 particle-reinforced 6061 aluminum alloy tensile specimens fractured in tension with 300 MPa hydrostatic compression to a specimen fractured at atmospheric pressure. The increased reduction of area in the specimen fractured at high pressure was attributed, in part, to suppression of void initiation (Ref 13). Accordingly, a hydrostatic tensile stress (such as that present at a crack tip) would tend to enhance void initiation.

Void growth, which is strongly affected by microstructure, is thought to absorb most of the work of fracture. Void growth can occur by plastic flow of the matrix or by linkup of smaller voids surrounding the large initial void. Triaxial stresses, such as those occurring at a notch, accelerate void growth and tend to encourage the formation of spherical dimples. Anything that short-circuits the void growth process will degrade the fracture resistance. Conversely, any microstructural feature that inhibits or interferes with void growth will enhance fracture resistance.

Void coalescence occurs by several mechanisms. The voids can grow until the remaining ligament ruptures. Void impingement absorbs the most energy. Voids can also coalesce by shear rupture. Because the volume of material that absorbs energy is more restricted, shear rupture is thought to absorb less energy. Void coalescence can also occur by linkup of smaller voids formed at smaller particles along bands of shear between large voids. This phenomenon, termed void sheet coalescence, degrades the fracture toughness. The classic micrograph, showing void sheet for-

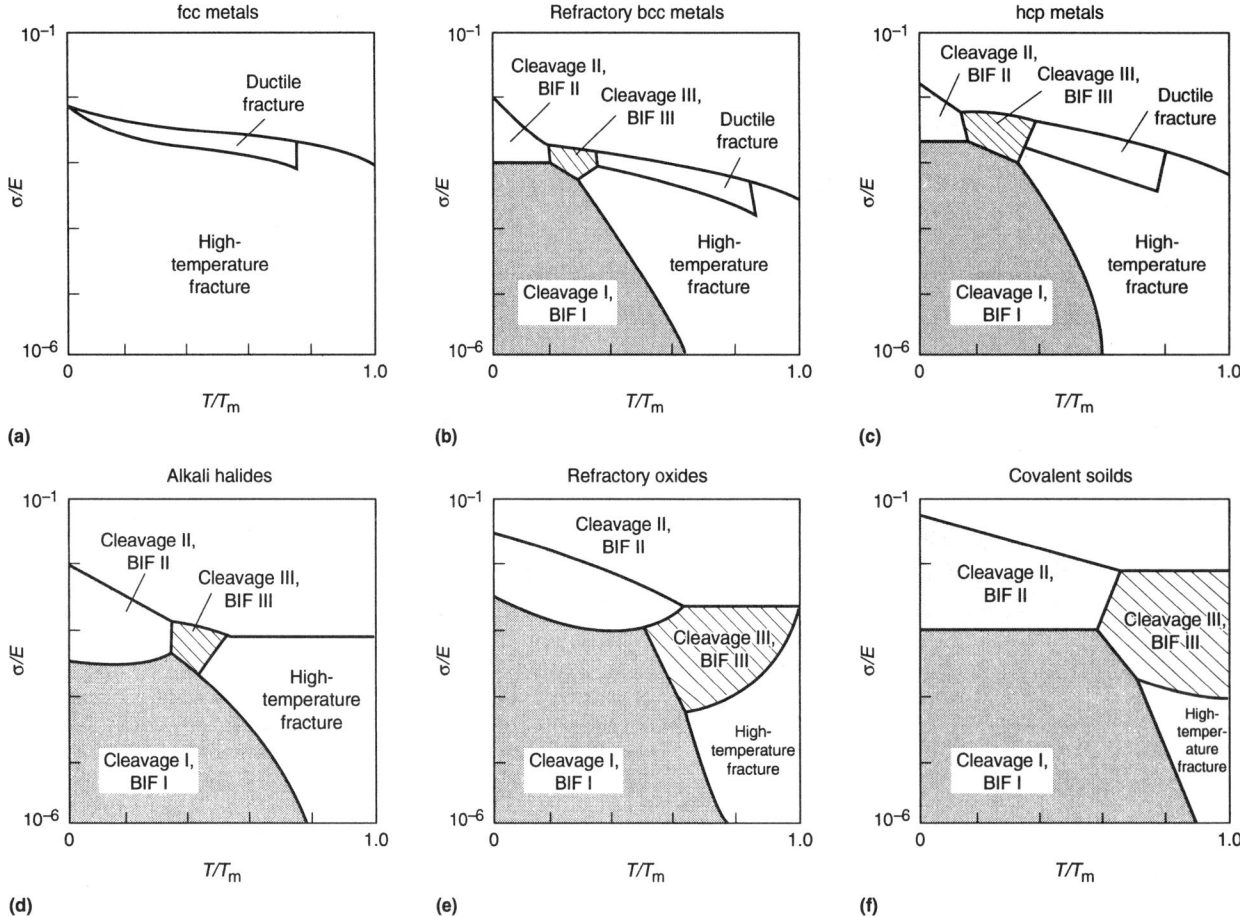

Fig. 3 Fracture mechanism maps for fcc, bcc, and hcp metals, alkali halides, refractory oxides, and covalently bound materials. Note that fcc metals do not exhibit brittle modes of failure, bcc and hcp metals exhibit a limited range of ductile behavior, and alkali halides, refractory oxides, and covalently bound materials do not exhibit ductile modes of failure. BIF, brittle intergranular fracture. Source: Ref 7

mation between two larger voids formed at sulfide inclusions in 4340 steel, is shown in Fig. 8.

The anisotropy of fracture toughness in wrought metals is attributed to the anisotropic distribution and shape of void-initiating particles that result from forming operations. The effect of the shape and distribution of the inclusions is clearly exhibited on the fracture surfaces. Shape control of sulfide inclusions in steel is a well-known method to improve the short-transverse fracture toughness.

Crack growth by microvoid coalescence exhibits a zig-zag pattern in cross section, as shown by Fig. 9 and 10. The reason for this morphology is that the plastic strain at a crack tip is maximum at an angle 45° from the macroscopic plane of growth (Ref 15). Void initiation and growth are thought to be controlled by plastic strain. When the specimen or component is thick with respect to the plastic zone radius, the crack is in a state of plane strain and the crack plane is perpendicular to the maximum principal stress. The dimples are elongated in the same direction on opposing sides of the fracture surface. When the component is thin with respect to the plastic zone radius, the fracture surface is inclined at approximately 45° to the principal stress and is in a state of plane

stress. The inclined fracture surfaces are loaded in shear, and the dimples are elongated in opposite directions on opposing sides of the fracture surface.

Cleavage. Cleavage fracture is a transgranular, low-energy fracture that occurs primarily by separation of atomic bonds on low-index atomic planes. Because cleavage occurs along well-defined crystallographic planes within each grain, a cleavage fracture will change directions when it crosses grain or subgrain boundaries (Fig. 11). Engineering materials contain second-phase constituents; therefore, true featureless cleavage is difficult to obtain, even within a single grain. Transition of the crack from one grain to another and interaction with tilt/twist boundaries generate characteristic river lines (Fig. 12). In hcp metals, a feature called a cleavage tongue is caused when the crystallographic crack interacts with small twins (Ref 16). Although cleavage fracture occurs by separation of atomic bonds, the river lines are tear ridges that form by plastic deformation and absorb energy. The difference between the calculated thermodynamic surface energy, 14 J/m^2, and the experimentally measured energy, 23 J/m^2, to form a cleavage crack in Fe-4Si has been attributed in part to the energy absorbed by

tear ridge formation and crack reinitiation at grain boundaries (Ref 17, 18). The energy to reinitiate the crack on crossing grain boundaries is one reason that toughness increases with a decrease in grain size.

Cleavage fracture occurs in materials that exhibit little or no capacity for plastic deformation (covalently bound materials, alkali halides, refractory oxides) as well as materials that exhibit significant plastic deformation (bcc and hcp metals). In materials with limited capacity for plastic deformation, it is not too surprising that the mechanism of fracture is by separation of atomic bonds. However, the cleavage fracture of bcc and hcp metals has perplexed and fascinated researchers for decades. The focus has been the ductile-to-brittle transition in steels because of its obvious technological significance.

Cleavage fracture in steels has been difficult to explain because the crack tip stress is below the theoretical fracture stress. Due to crack-tip blunting and redistribution of stresses due to yielding, the crack-tip stress is typically no more than 3 to 4 times the yield stress. Considering a range of yield strength from 400 to 1000 MPa, this indicates that crack-tip stresses are 4000 MPa at best. This is sufficiently below the theoretical fracture stress ($E/\pi \sim 70{,}000$ MPa for iron) to cause fracture that one must invoke some other mechanism to explain the brittle fracture. In steels, the most commonly accepted explanation is that carbides fracture ahead of the crack tip, producing sharp, Griffith-type cracks that have crack-tip stresses equal to the theoretical fracture stress (Ref 19). The carbides fracture due to stresses from dislocation pileups. The key to this mechanism is that the crack remains sharp. Cleavage cracks in iron can arrest at grain boundaries if the crack blunts or if the local stresses are below that required for continued growth.

In the oft-quoted Ritchie, Knott and Rice (RKR) model (Ref 20), it is proposed that the crack-tip stress must exceed the stress required to propagate the crack from the carbide over a critical distance before the crack will propagate. It is suggested that this distance is one to two grain diameters. Subsequent researchers have proposed various statistical arguments to explain the grain size dependence, the thickness dependence, and the scatter in fracture strength values (Ref 21-23). While these are all generally based on a weakest link argument, considerable progress has been made in understanding cleavage fracture in steels.

Cleavage fracture is also observed for metals and ceramics in conditions of stress-corrosion cracking and hydrogen-assisted cracking. In fact, fcc austenitic stainless steel, which normally fails by microvoid coalescence, will exhibit stress-corrosion crack growth by transgranular cleavage in certain aqueous solutions (e.g., NaCl solutions). The simplistic explanation of stress-corrosion cleavage crack growth is that the corrosive environment acts to break the crack-tip bonds and thereby lowers the apparent thermodynamic surface energy. The phenomenon is considerably more complicated, and the reader is referred to Ref 24 for recent literature on stress-corrosion cracking.

Intergranular fracture is always associated with low toughness and low ductility and is therefore an undesirable fracture mode. The fracture surface is faceted and is often characterized as being similar to rock candy. The explanation is simply that the grain boundaries are usually stronger than the interior, but when they are not, the grain boundary is the failure path. Some of the causes of intergranular fracture are:

- Intrinsic weakness of the grain boundaries
- Precipitation of a brittle phase or segregation of an embrittling element on the grain boundary
- Embrittlement of the grain boundary by hydrogen or other aggressive environments
- Intergranular corrosion

Thompson and Knott (Ref 17) provide a brief description of current theories of intergranular fracture. The theories either focus on the reason for intrinsic grain boundary weakness or on the effect of segregation of embrittling elements.

Crazing of Polymers. Above the glass temperature, amorphous polymers generally fail by rupture that may be assisted by void initiation.

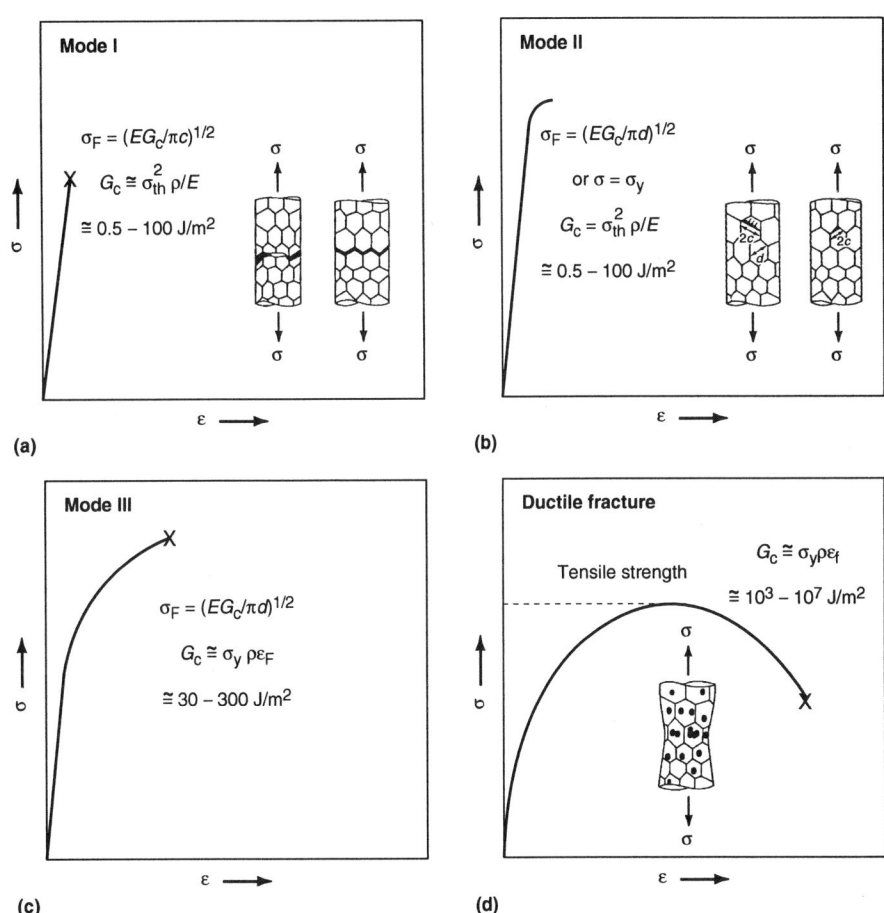

Fig. 4 Schematic of stress-strain curves of brittle and ductile fracture (G_c = critical strain energy release rate). (a) Mode I brittle fracture occurs without plastic deformation, except perhaps to a limited degree in the material adjacent to the crack tip. Mode I fracture may propagate transgranularly or intergranularly brittle intergranular fracture, BIF. (b) Mode II brittle fracture is preceded by microscopic, but not macroscopic, plastic deformation. Indeed, plastic deformation nucleates cracks that may propagate by cleavage or in an intergranular mode (as shown schematically). (c) Macroscopic deformation takes place before fracture by mode III brittle fracture. A limited reduction in area is obtained in a tensile test, but fracture propagation by cleavage or BIF takes place prior to necking. (d) Ductile tensile fracture propagation is preceded by necking. Microscopic voids form throughout the material, and subsequently grow and coalesce by means of plastic deformation processes. Voids are nucleated frequently at inclusions but may also be formed in regions of intense, heterogeneous slip. Void linkup is restricted to the necked region of the tensile sample. Source: Ref 7

The usage of amorphous polymers above the glass temperature is generally not for structural purposes, and therefore fracture is not a significant problem. Nylon is a crystalline polymer used for structural applications and the glass transition temperature is 55 °C. Polymer failure above the glass temperature is quite ductile. Below the glass temperature, both amorphous and crystalline polymers can fail by propagation of crazes. An individual craze is similar in appearance to a crack, except that bundles of polymer chains form fibrils that bridge the boundaries of the craze. The density of the craze is significantly less than that of the parent matrix. The schematic in Fig. 13 shows side and top views of a craze. The craze is a preferred fracture path. Material separation typically occurs at the craze-matrix interface. This can occur in patches or bands (Fig. 14). Fibril failure is sometimes observed.

Failure of amorphous polymers below the glass temperature, T_g, is often incorrectly associated with brittle fracture behavior. This is because failure by propagation of a single craze does not absorb much energy. The brittle tensile failure of polymethylmethacrylate (PMMA, $T_g \sim 105$ °C) at room temperature typically occurs by craze propagation. However, polycarbonate, an amorphous polymer with $T_g \sim 145$ °C, exhibits considerable ductility (one application is "bulletproof" glass). The explanation for this effect is that amorphous polymers can yield by craze formation and/or shear yielding. The yield surfaces for crazing and shear yielding under biaxial stress is depicted in Fig. 15. When a polymer yields by shear band formation, prior to craze formation, fracture is suppressed until the ability of the matrix to form shear bands is exhausted. Then, fracture will occur by propagation of a craze. When craze formation is favored over shear band formation, the polymer initiates a craze that typically propagates to failure.

It is interesting, however, to note that the plane-strain fracture toughness of PMMA and polycarbonate are comparable. For PMMA, $K_c = 0.9$ to 1.4 MPa$\sqrt{m}$; for polycarbonate, $K_c = 1.0$ to 2.6 MPa$\sqrt{m}$ (Ref 25). This is in contrast to the tensile

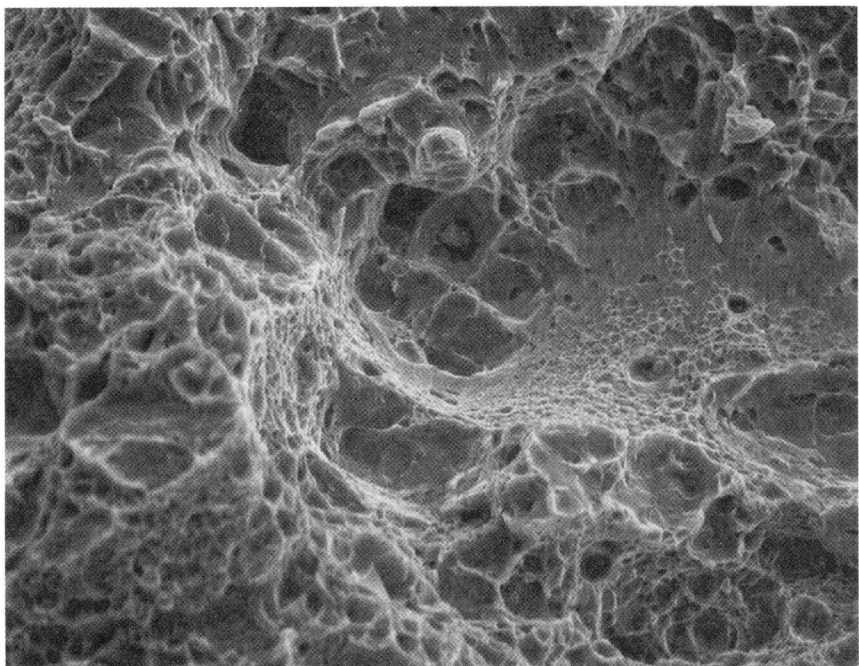

Fig. 5 Microvoid coalescence in quenched-and-tempered (200 °C) 4140 steel. One can see void-initiating particles at the bottom of some dimples.

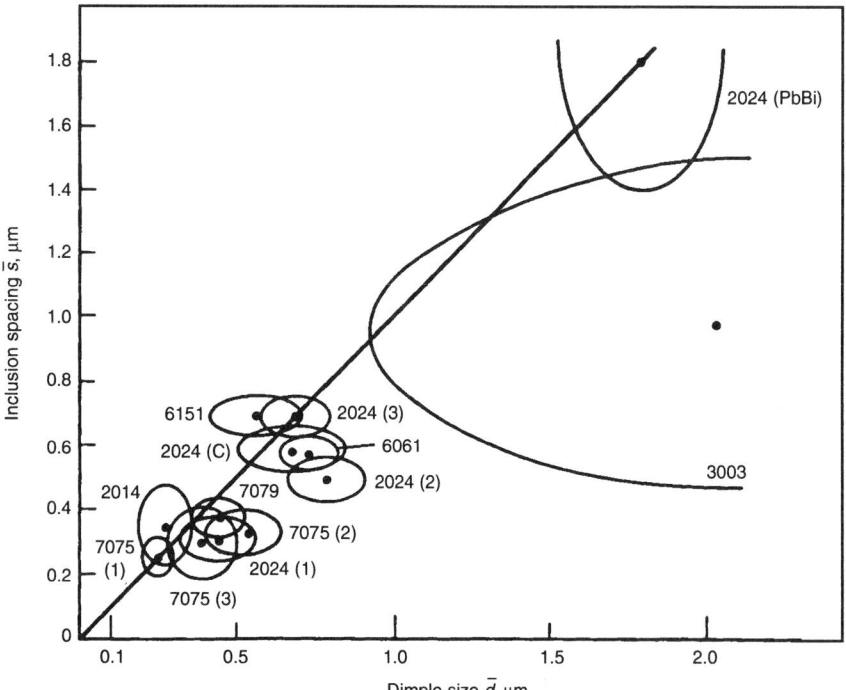

Fig. 6 Correspondence of the dimple size to the inclusion spacing in several aluminum alloys. This provides strong evidence that microvoids are nucleated at inclusions. Numbers represent aluminum alloy designations; ellipses indicate scatter. Source: Ref 8

toughness test favors craze formation, resulting in lower apparent fracture toughness. The blunt notch of an Izod impact test would have a lower triaxial tensile stress and would tend to favor shear band formation in polycarbonate.

Polymers that yield by craze formation can be toughened by distribution of micrometer-sized (or less) rubber particles in the brittle matrix. The rubber particles act as stress concentrators that initiate and ultimately blunt crazes. In addition, because the particles are widely distributed, they encourage the formation of a dense network of crazes that absorbs a considerable amount of energy. Also, the compliance of the crazed material is greater than that of the glassy matrix, so the stress concentration at the crack tip is reduced.

References 28 to 30 provide additional details on polymeric fracture.

Cyclic Crack Growth

Crack growth during uniaxial cyclic deformation occurs by initiation of a microcrack that propagates on a plane close to that for maximum shear. This is termed a stage I fatigue crack. As the crack extends, it rotates from the plane of maximum shear to the plane normal to the principal stress and is then termed a stage II fatigue crack. This transition is shown in Fig. 16. This section focuses first on the factors influencing the initiation of fatigue cracks, then on the micromechanisms and the effects of microstructure and environment on long fatigue crack growth.

Fatigue of uncracked metals is traditionally divided into low-cycle and high-cycle fatigue. The transition is usually associated with about 1000 to 2000 cycles to failure. Low-cycle fatigue involves bulk plasticity, whereas deformation in high-cycle fatigue is primarily elastic. The micromechanisms of crack growth at stresses in the low-cycle fatigue region are quite similar to monotonic crack growth. This discussion is restricted to crack growth rates less than approximately 1 μm per cycle, which corresponds to the high-cycle fatigue region.

Fatigue Crack Initiation

In single-phase, flaw-free, fcc metals, crack initiation is associated with persistent slip bands. As the dislocations exit the surface and possibly try to re-enter, the surface roughens. A protrusion that is composed of intrusions and extrusions forms at the intersection of the persistent slip band and the surface. The surface roughening and protrusion further localize the stresses and cyclic plastic strain until a microcrack initiates in the persistent slip band (Ref 32). If the environment is completely benign, the surfaces of the cracks can reweld thereby retarding the formation of a crack. Rewelding is strongly inhibited by oxidation of the intrusion-extrusion surfaces if they are exposed to an oxidizing environment (Ref 32).

Environment also has an effect on slip morphology and crack initiation in bcc metals (Ref 33). Samples of vacuum-arc remelted iron fatigued in high vacuum exhibit generalized sur-

ductility of 5% for PMMA and 130% for polycarbonate (Ref 26). The Izod impact toughness is 0.5 ft · lb/in. for PMMA and 16 ft · lb/in. for polycarbonate. Clearly, the competition between shear band formation and craze formation has a major

impact on the toughness, but the testing methods may yield conflicting results. The tendency for craze formation is enhanced by a hydrostatic tensile stress (Ref 27). Presumably, the triaxial tensile stress at the tip of a sharp crack in a fracture

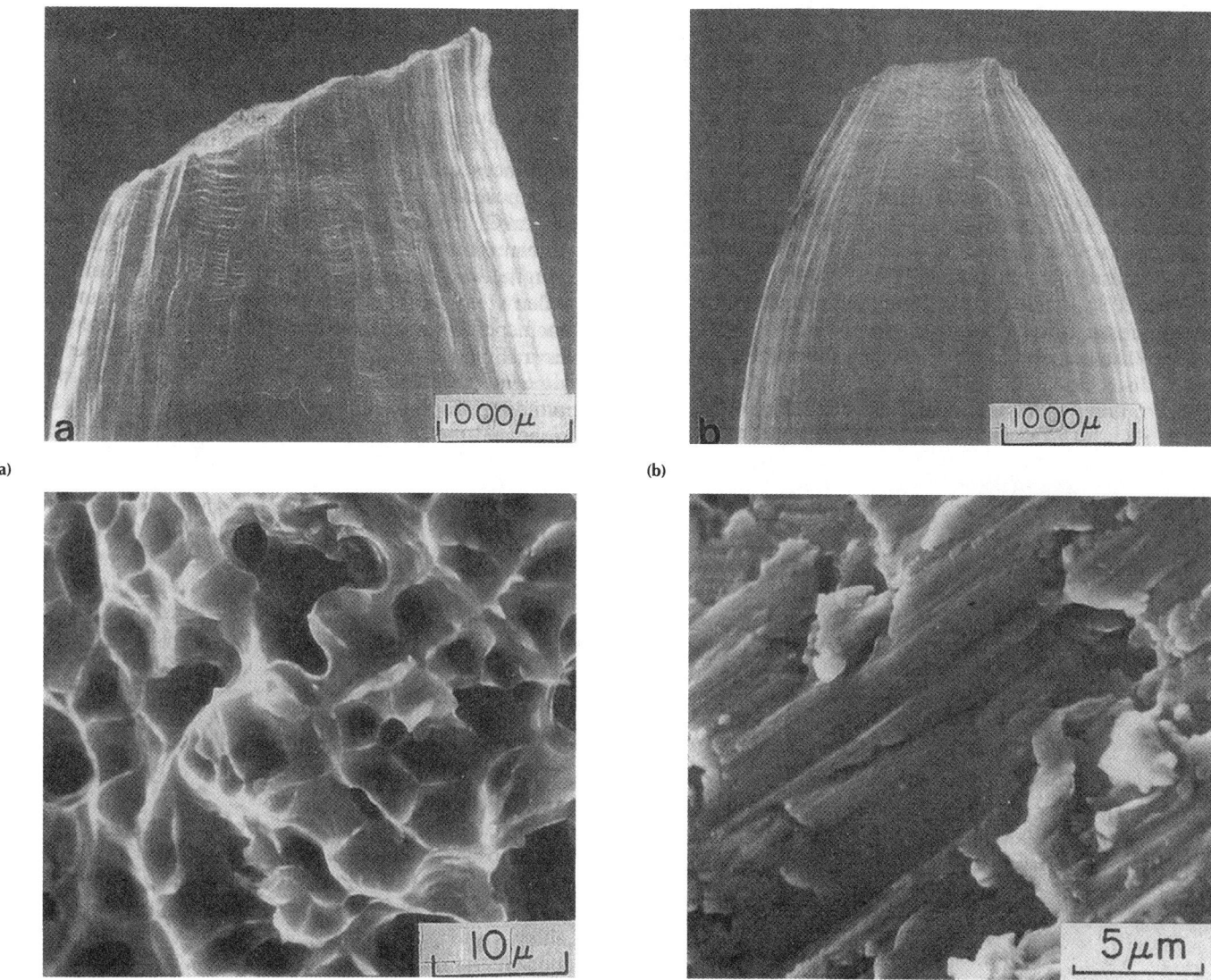

Fig. 7 Effect of atmospheric pressure on fracture of a discontinuously reinforced aluminum metal-matrix composite. Fractured tensile specimens at (a) 0.1 MPa and (b) 300 MPa and respective fracture surfaces. The increased reduction of area in the high-pressure fracture is attributed to suppression of void initiation and growth.

face rumpling, and the cracks initiate exclusively on the grain boundaries. The same samples fatigued in air exhibit well-defined slip lines, and cracks grow along the slip lines (Ref 33). Presumably, the reversibility of surface slip is inhibited in oxygen-containing environments, leading to concentration of slip and crack initiation. The presence of water vapor has been shown to cause grain boundary sliding and microvoid formation on boundaries (Ref 33).

In commercial alloys, fatigue crack initiation occurs at any feature that concentrates the local stress and plastic strain. Some of these features are voids, inclusions, twin boundaries, grain boundaries, stronger second phases, and notches. Crack initiation at voids occurs at the persistent slip bands emanating from the void. Crack initiation at inclusions can occur in a similar manner to that for voids, because a void can be considered an inclusion with an elastic modulus of zero. Sometimes inclusions are already cracked as a result of thermomechanical processing, and crack propagation occurs from the cracked inclusion. Sometimes the inclusion is cracked from the stress concentration resulting from blocking the persistent slip band. The inclusion can also debond from the matrix, and subsequent crack propagation will be similar to that from a void.

Crack initiation under multiaxial loading can be quite complex. Usually, but not always, the crack initiates on the plane of maximum shear (Ref 34). Figure 17 shows how the plane of crack initiation corresponds to the plane of maximum shear for several combinations of axial and torsional loading on a hollow tube. In contrast to axial loading, shear cracks under torsional loading can propagate to significant lengths before rotating to the plane of maximum principal stress range. The higher the stress range, the longer the duration of shear crack growth. The article "Multiaxial Fatigue Strength" in this Volume discusses fatigue under multiaxial loading in greater detail.

The crack size at the transition from a microcrack to a macrocrack is usually associated with some microstructural unit (e.g., grain size or inclusion spacing). More importantly, it is associated with rotation of the crack from the plane of maximum shear strain range to that of maximum tensile stress range. Another descriptor used to describe cracks is short cracks versus long cracks. One of the central assumptions of linear elastic fracture mechanics is that the crack length is "large" with respect to the extent of the process zone at the crack tip and with respect to the extent of the stress concentration. Short cracks do not meet this condition and are discussed in the article "Behavior of Small Fatigue Cracks" in this Volume.

Because a significant portion of the high-cycle fatigue lifetime includes the initiation and growth of microcracks, many investigators have conducted studies of the growth of short cracks. In some of those studies, a long crack is shortened

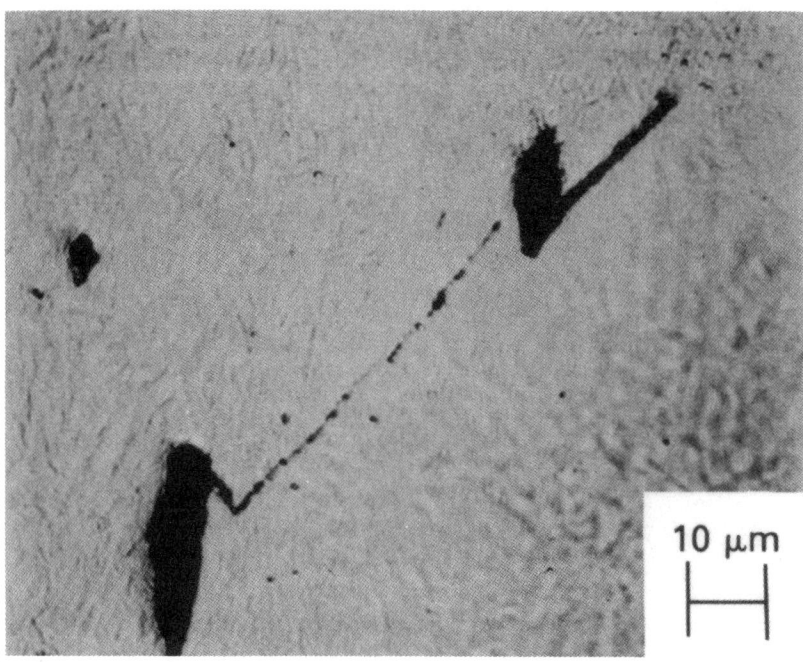

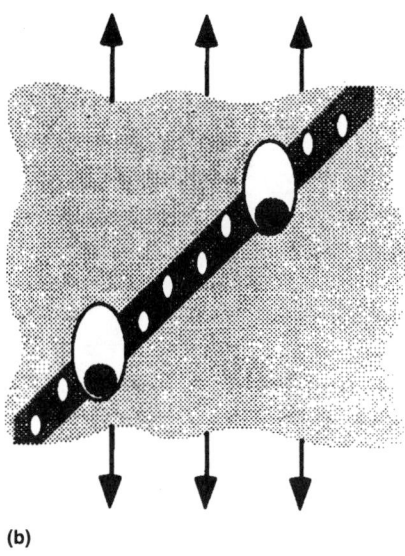

Fig. 8 Large voids in AISI 4340 linked by narrow void sheets consisting of small microvoids. (a) Section through the necked region of a 4340 steel specimen showing the formation of a void sheet between two voids formed at larger inclusions. (b) Schematic of nucleation at smaller particles along the deformation bands

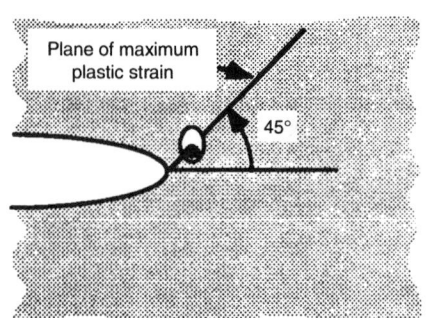

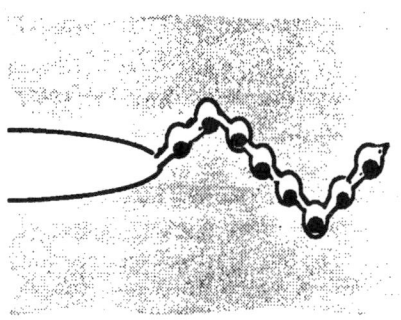

Fig. 9 Ductile crack growth in a 45° zig-zag pattern. The plane of maximum strain is 45° from the macroscopic growth direction, which causes microvoids to grow in a zig-zag pattern.

by removing material up to a short distance from the crack tip. In others, crack initiation from an artificial notch is studied. Yet another approach is to observe crack initiation and growth for a large number of cracks in a fixed area on a specimen. All three techniques have demonstrated that the rate of crack growth is discontinuous and is often greater than that extrapolated from "long" crack data. It is also a common observation that the crack-tip displacements exhibit both shear and opening displacements (Ref 35).

Fatigue Crack Growth Mechanisms

Stage I Growth. Fatigue crack initiation and growth during stage I occurs principally by slip-plane cracking. A typical stage I fatigue fracture is shown in Fig. 19. Stage I fatigue fracture surfaces are faceted, often resemble cleavage, and generally do not exhibit regularly spaced fatigue striations. Stage I fatigue is normally observed on high-cycle, low-stress fractures and is frequently absent in low-cycle, high-stress fatigue.

The tendency toward stage I crack propagation is strongly affected by anything that concentrates slip and by environments that affect slip reversibility (Ref 36, 37). Repeated cutting and disordering of coherent precipitates is suggested to lead to concentrated slip for age hardened aluminum alloys. Figure 18 shows that stage I fatigue crack growth occurs in dry nitrogen at any frequency whereas stage II fatigue crack growth is observed at low frequencies in laboratory air. The oxidation of the fresh surface prevents the slip reversibility required for stage I fatigue crack growth. Many titanium and aluminum alloys are known to exhibit highly localized, concentrated slip and therefore to have a tendency toward stage I crack propagation. Conversely, microstructure

that tends to homogenize slip (e.g., incoherent, closely spaced dispersoids or fine grains) will suppress the tendency for cracks to propagate by stage I, shear crack growth.

In polycrystalline materials, the tendency toward localized strain is affected by the size of the crack-tip cyclic plastic zone relative to the grain or microstructural feature size (Ref 38). If the grain or microstructural packet is much bigger than the cyclic plastic zone, it is possible to have an extended, localized slip band for subsequent crack extension. The crack extends on the most favorably oriented plane in a given grain. Because the grains in a polycrystalline metal will have many different orientations, the macroscopic crack will have a zig-zag appearance. When the cyclic plastic zone is on the order of or bigger than the relevant microstructural unit, the tendency toward slip band cracking is suppressed and the crack becomes nearly planar.

Stage II Crack Growth and Fatigue Striations. The largest portion of a fatigue fracture consists of stage II crack growth, which generally occurs by transgranular fracture and is more influenced by the magnitude of the alternating stress than by the mean stress or microstructure. Fatigue fractures generated during stage II fatigue may exhibit crack marks known as fatigue striations (Fig. 20), which are a visual record of the position of the fatigue crack front during crack propagation through the material.

During stage II fatigue, the crack often propagates on multiple plateaus that are at different elevations with respect to one another (Fig. 21). A plateau that has a concave surface curvature

Fig. 10 Ductile crack growth in a high-strength low-alloy steel (A 710). The zig-zag crack growth results from void initiation and growth on the plane of maximum strain, as illustrated in Fig. 9.

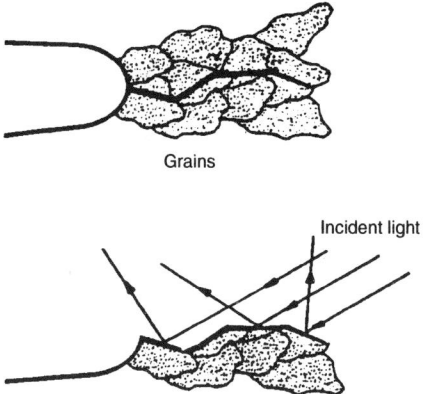

Fig. 11 Schematic of cleavage through grains at a blunt crack tip. Each facet through a grain is flat and so provides a highly reflective surface.

exhibits a convex contour on the mating fracture face (Ref 39). The plateaus are joined either by tear ridges or walls that contain fatigue striations. Fatigue striations often bow out in the direction of crack propagation and generally tend to align perpendicular to the principal (macroscopic) crack propagation direction. However, variations in local stresses and microstructure can change the orientation of the plane of fracture and alter the direction of striation alignment.

Large second-phase particles and inclusions in a metal can change the local crack growth rate and resulting fatigue striation spacing. When a fatigue crack approaches such a particle, it is briefly retarded if the particle remains intact or is accelerated if the particle cleaves. In both cases, however, the crack growth rate is changed only in the immediate vicinity of the particle and therefore does not significantly affect the total crack growth rate. However, for low-cycle, high-stress fatigue, the relatively large plastic zone at the crack tip can cause cleavage and matrix separation at the particles at a significant distance ahead of the advancing fatigue crack. In effect, the cleaved or matrix-separated particles behave as cracks or voids that promote a tear or shear fracture between themselves and the fatigue crack, thus significantly advancing the crack front. Relatively small, individual particles have no significant effect on striation spacing (Fig. 20b).

The distinct, periodic markings sometimes observed on fatigue fracture surfaces are known as "tire tracks," because they often resemble the tracks left by the tread pattern of a tire (Fig. 22). These rows of parallel markings are the result of a particle or a protrusion on one fatigue fracture surface being successively impressed into the surface of the mating half of the fracture during the closing portion of the fatigue cycle (Ref 39, 40). Tire tracks are more common for the tension-compression than the tension-tension type of fatigue loading. The direction of the tire tracks and the change in spacing of the indentations within

the track can indicate the type of displacement that occurred during the fracturing process, such as lateral movement from shear or torsional loading. The presence of tire tracks on a fracture surface that exhibits no fatigue striations may indicate that the fracture occurred by low-cycle, high-stress fatigue.

Fatigue Crack Growth Rates and Striation Spacings. Fatigue striations provide very useful mechanistic information on fatigue crack growth, and a striation spacing and formation mechanisms. The spacing of fatigue striations provides important evidence for understanding the fatigue crack growth process, not because the spacing necessarily reveals much about the process itself, but because striations constitute unambiguous, quantitative evidence of the *average increment* by which a fatigue crack advances.

Typically, one striation forms on each loading cycle, and it is clear from early experiments that one striation often evidences an increment of crack growth and arrest. More recent observations, by numerous investigators, have shown that one striation does not necessarily correspond to one cycle; tens to thousands of cycles may be required to obtain one striation (Ref 41). Data for both results appear to be correct, and many of the differences found have been due to studying different parts of the crack growth rate curve. Early work examined relatively large crack growth rates ($>10^{-6}$ m/cycle), where striation spacing approximately equals the crack growth rate. Later studies have emphasized lower growth rates where striation spacing is larger than crack growth rate.

In the near-threshold (stage I) regime, striation spacing can exceed crack growth rates by orders of magnitude (Ref 41). Based on a large amount of information collected for numerous materials, Grinberg (Ref 42) was the first to recognize the important fact that the minimum striation spacing observed, regardless of material, is approximately 0.1 µm. This has been found to be true for

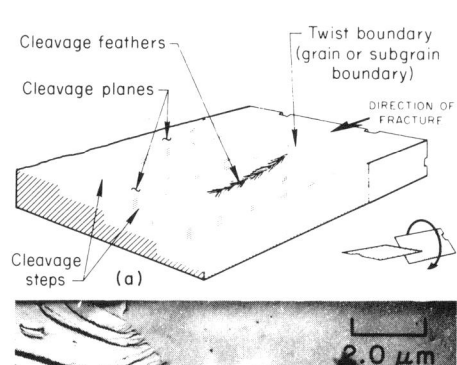

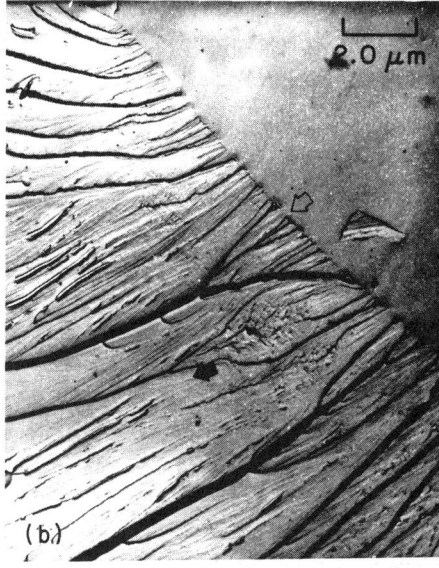

Fig. 12 Effect of twist boundary on cleavage decoration. (a) Fracture models showing a twist boundary and the new cleavage steps that develop as the propagating crack crosses a twist boundary. (b) Cleavage steps initiating at a twist boundary (open arrow) on a fracture surface (TEM, plastic-carbon replica) B-66 (5% Mo, 5% V, 1% Zr). Solid black arrow indicates fracture direction. Feather marks exist between steps.

aluminum and its alloys; magnesium, nickel, and titanium alloys; and many steels. For cases in which the striation spacing increases with ΔK, the minimum spacing occurs near ΔK_{th}, or if data were not obtained at such low stress intensities, values obtained at higher ΔK extrapolate to ~0.1

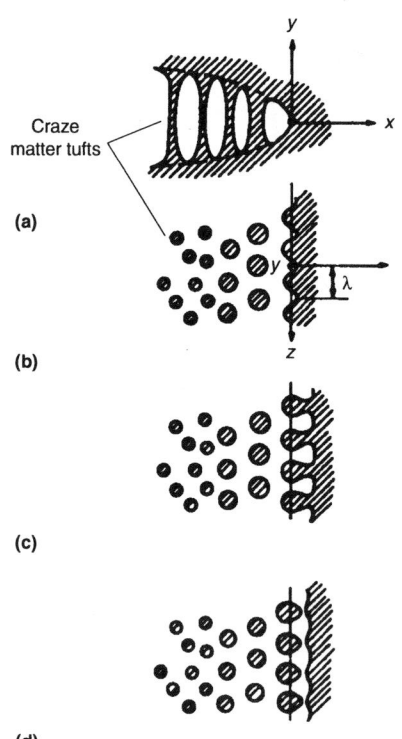

Fig. 13 Schematic of craze formation. (a) Outline of a craze tip. (b) Cross section in the craze plane across craze matter tufts. (c, d) Advance of the craze front by a completed period of interface convolution

μm at the threshold. The fact that striation spacing and crack growth rate diverge with decreasing ΔK, and that the minimum striation spacing is approximately constant for all materials, may have significant mechanistic implications, which are explored in Ref 41.

For a stage II fatigue crack propagating under conditions of reasonably constant cyclic loading frequency and advancing within the nominal range of 10^{-5} to 10^{-3} mm/cycle, the Paris relation of crack growth rate ($da/dN = C(\Delta K)^m$) is applicable. If a fatigue striation is produced on each loading cycle, da/dN represents the striation spacing. The Paris equation does not adequately describe stage I or stage III fatigue crack growth rates; it tends to overestimate stage I and often underestimates stage III growth rates.

Stage III is the terminal propagation phase of a fatigue crack in which the striation-forming mode is progressively displaced by the static fracture modes, such as dimpled rupture or cleavage. The rate of crack growth increases during stage III until the fatigue crack becomes unstable and the part fails. Because the crack propagation is increasingly dominated by the static fracture modes, striation formation per cycle is expected to be intermittent.

The profile of the fatigue fracture can vary, depending on the material and state of stress. Materials that exhibit fairly well-developed striations display a sawtooth-type profile (Fig. 23a) with valley-to-valley or groove-to-groove matching (Ref 40). Low compressive stresses at the crack tip favor the sawtooth profile; however,

high compressive stresses promote the groove-type fatigue profile, as shown in Fig. 23(c) (Ref 43, 44). Jagged, poorly formed, distorted, and unevenly spaced striations (Fig. 23b), sometimes termed quasi-striations (Ref 40), show no symmetrical matching profiles. Even distinct sawtooth and groove-type fatigue surfaces may not show symmetrical matching. The local microscopic plane of a fatigue crack often deviates from the normal to the principal stress. Consequently, one of the fracture surfaces will be deformed more by repetitive cyclic slip than its matching counterpart (Ref 39). Thus, one fracture surface may show well-developed striations while its counterpart exhibits shallow, poorly formed striations.

Under normal conditions, each striation is the result of one load cycle and marks the position of the fatigue crack front at the time the striation was formed. However, when there is a sudden decrease in the applied load, the crack can temporarily stop propagating, and no striations are formed. The crack resumes propagation only after a certain number of cycles are applied at the lower stress (Ref 40, 45). This phenomenon of crack arrest is believed to be due to the presence of a residual compressive stress field within the crack-tip plastic zone produced after the last high-stress fatigue cycle (Ref 40, 45).

Fatigue crack propagation and therefore striation spacing can be affected by a number of variables, such as loading conditions, strength of the material, microstructure, and the environment (e.g., temperature and the presence of corrosive or embrittling gases and fluids). Considering only the loading conditions—which would include the mean stress, the alternating stress, and the cyclic frequency—the magnitude of the alternating stress ($\sigma_{max} - \sigma_{min}$) has the greatest effect on striation spacing. Increasing the magnitude of the alternating stress produces an increase in the striation spacing. While rising, the mean stress can also increase the striation spacing; this increase is not as great as one for a numerically equivalent increase in the alternating stress. Within reasonable limits, the cyclic frequency has the least effect on striation spacing. In some cases, fatigue striation spacing can change significantly over a very short distance. This is due in part to changes in local stress conditions as the crack propagates on an inclined surface.

Mechanisms of Fatigue Striation Formation. Fatigue striations are found frequently on the fracture surfaces of aluminum alloys cycled in moist air, while for other alloys, striations are not always found. It is not clear in many instances why fatigue striations do not form, but it is a subject that has attracted considerable investigation. However, even if striations are not observed, the absence of fatigue striations does not mean that the basic mechanism by which cracks advance has been changed. This fact is illustrated by several examples in Ref 41. A lack of striations on a fatigue fracture surface may be due also to their having been obscured by crack-closure-induced rubbing of the fracture surfaces, as caused by mode II sliding. The debris caused by this

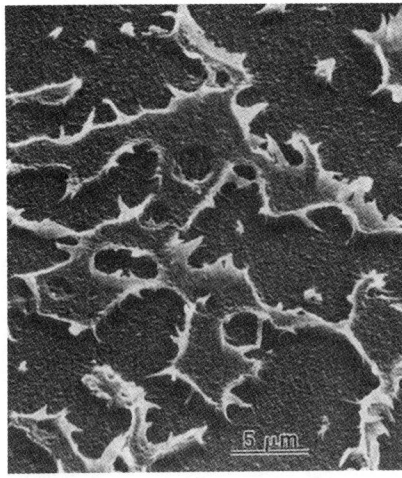

(a)

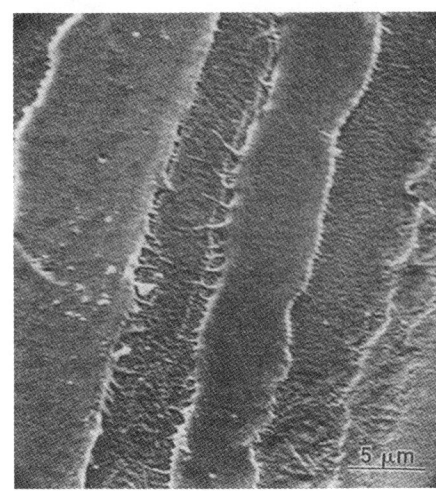

(b)

Fig. 14 Craze fracture surface (a) patch morphology and (b) band morphology

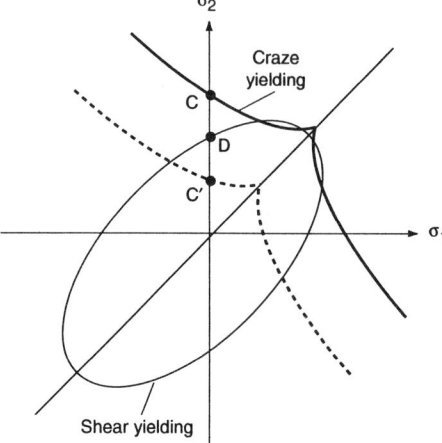

Fig. 15 Shear and yield loci under biaxial stresses. Only shear yielding is observed for biaxial compressive stresses ($\sigma_1, \sigma_2 < 0$). Points C and D are the uniaxial tensile stresses required for crazing and shear yielding, respectively. Application of a hydrostatic stress moves the craze yield locus away from the origin. The dashed line shows the craze yield locus for lower hydrostatic pressure with C′ lower than D. A tensile triaxial stress would move the craze locus closer to the origin.

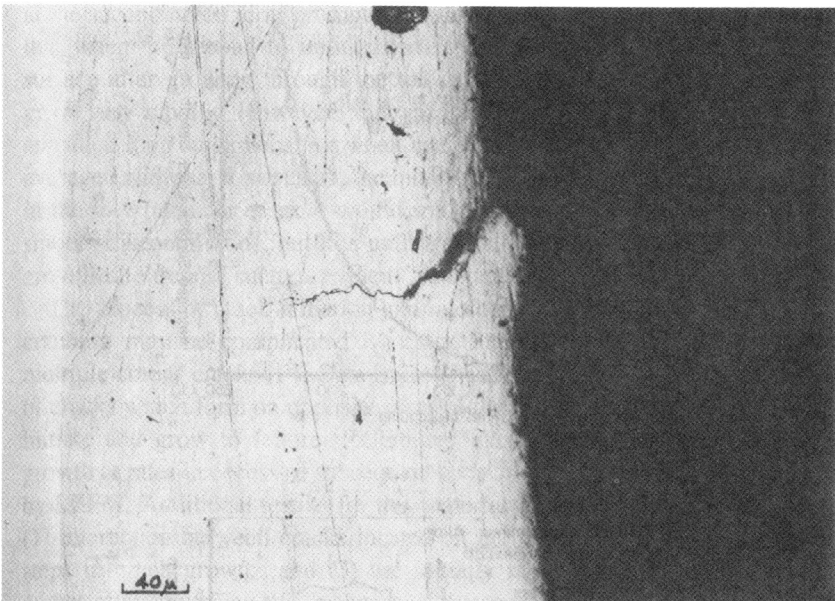

Fig. 16 Cross section through a crack that initiates on the maximum shear plane and grows as a stage I fatigue crack until it rotates normal to the maximum tensile stress range and becomes a stage II fatigue crack. Source: Ref 31

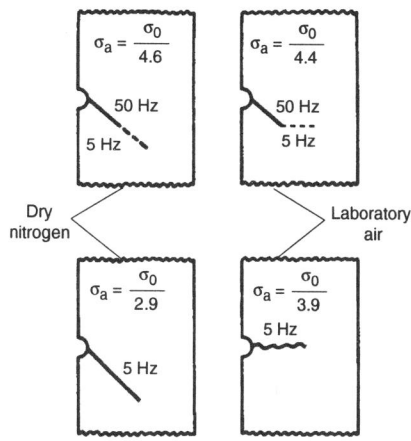

Fig. 18 Schematic comparing crack orientations for testing in dry nitrogen and (humid) laboratory air. Stage I crack growth is enhanced in the inert environment. Source: Ref 36

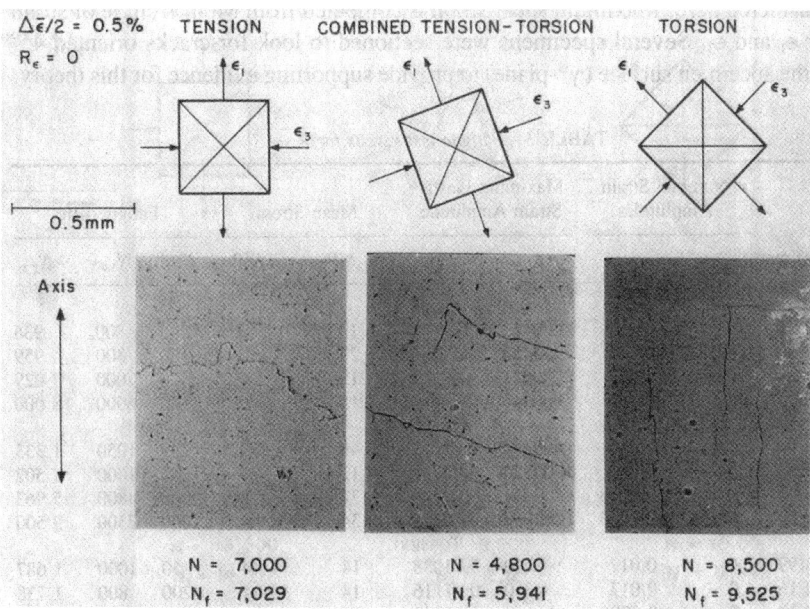

Fig. 17 Crack directions under various biaxial load types. The crack directions correspond to the plane of maximum shear for different combinations of in-phase tension-torsion loading. The material is Inconel 718 ($R = 0$, 0.5% strain amplitude). Source: Ref 34

process, reported by many investigators, is particularly prevalent when thick oxides form on newly exposed fracture surfaces.

The mechanism of striation formation generally is a two-step process as a result of crack-tip blunting on the rising portion of the load, followed by resharpening as the load is released. The model of this mechanism is based on slip at the crack tip, although slip may not occur precisely at the crack tip due to the presence of lattice or microstructural imperfections (Ref 46-48). If stress is concentrated at a fatigue crack, plastic deformation (slip) will be confined to a small region at the tip of the crack while the remainder of the material is subjected to elastic strain. As shown in Fig. 24(a), the crack opens on the rising-tension portion of the load cycle by slip on alternating slip planes. As slip proceeds, the crack tip blunts, but it is resharpened by partial slip reversal during the declining-load portion of the fatigue cycle. This results in a compressive stress at the crack tip due to the relaxation of the residual elastic tensile stresses induced in the uncracked portion of the material during the rising load cycle (Fig. 24b). The closing crack does not reweld, because the new slip surfaces created during

the crack-opening displacement are instantly oxidized, which makes complete slip reversal unlikely. Some fracture surfaces containing widely spaced fatigue striations exhibit slip traces on the leading edges of the striation and relatively smooth trailing edges, as predicted by the model (Fig. 24). Not all fatigue striations, however, exhibit distinct slip traces, as suggested by Fig. 24, which is a simplified representation of the fatigue process.

Striations are more commonly observed in air than in vacuum. The essential absence of striations on fatigue fracture surfaces of metals tested in vacuum tends to support the assumption that oxidation reduces slip reversal during crack closure, which results in the formation of striations. The lack of oxidation in hard vacuum promotes a more complete slip reversal, which results in a smooth and relatively featureless fatigue fracture surface. For fatigue crack growth in a vacuum, evidence points to a two-step process that begins with crack-tip blunting followed by a step evidencing less plasticity (Ref 41). The crack-tip blunting is accompanied first by the generation of numerous dislocations. As the crack advances, many of these dislocations do not return to the crack tip, but remain nearby and are detected in the transmission electron microscopy foils as a tangle of dislocations (Ref 41). The second step in striation formation is related to breakdown along a slip line, which forms during the crack blunting step. Formation and fracture of the slip line evidently leaves few dislocations behind, which might be construed as brittle materials behavior (Ref 41).

In air, striation formation is also a two-step process, although the sequence of effects may depend on type A or type B striation (Fig. 23), as reported in Ref 41 from results of Nix and Flower (Ref 49). Nix and Flower state that a band of deformation accompanies the leading edge of the type A striation. But for the type B striation, the opposite is true, with the first part of the striation being caused by cleavage, followed by crack-tip blunting and dislocation generation. As such the

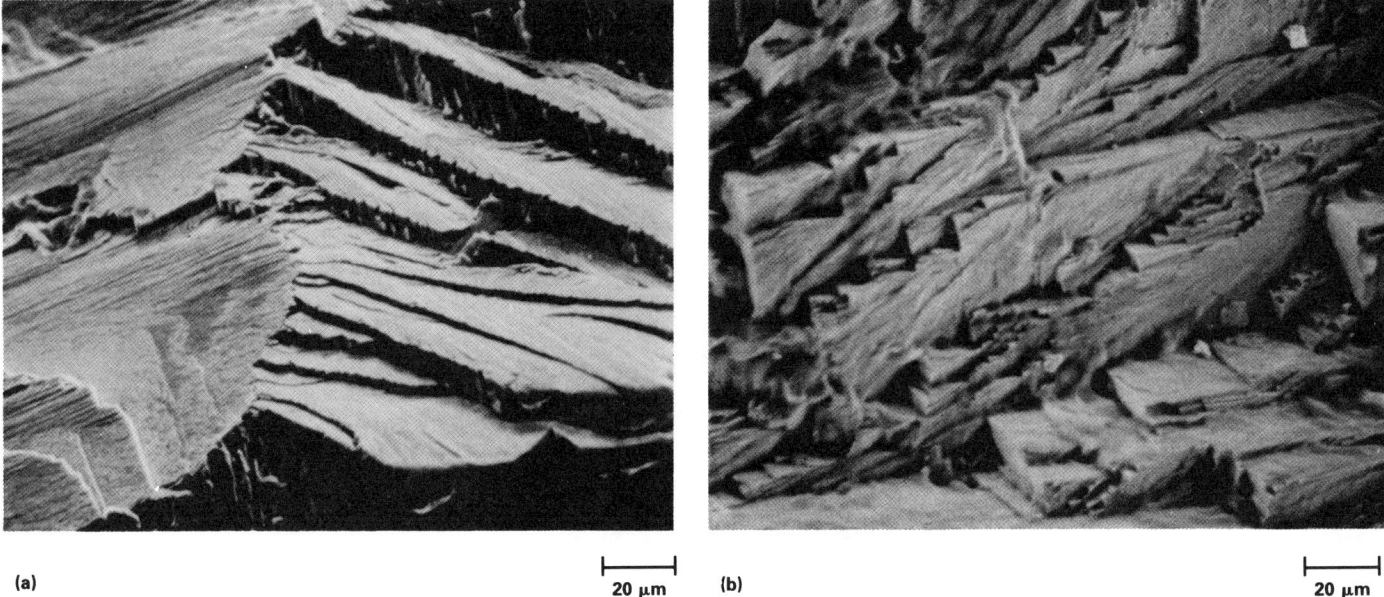

(a)

20 µm

(b)

20 µm

Fig. 19 Cleavagelike, crystallographically oriented stage I fatigue fracture in a cast Ni-14Cr-4.5Mo-1Ti-6Al-1.5Fe-2.0(Nb+Ta) alloy

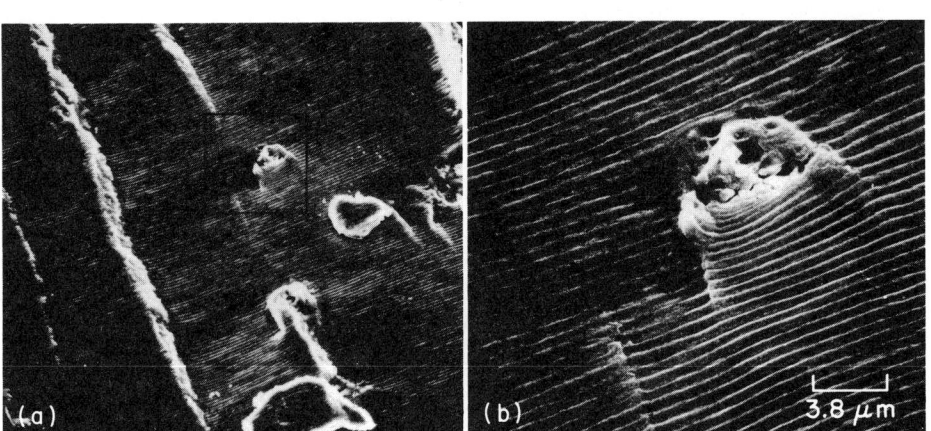

(a)

(b)

3.8 µm

Fig. 20 Uniformly distributed fatigue striations in an aluminum 2024-T3 alloy. (a) Tear ridge and inclusion (outlined by rectangle). (b) Higher-magnification view of the region outlined by the rectangle in (a), showing the continuity of the fracture path through and around the inclusion

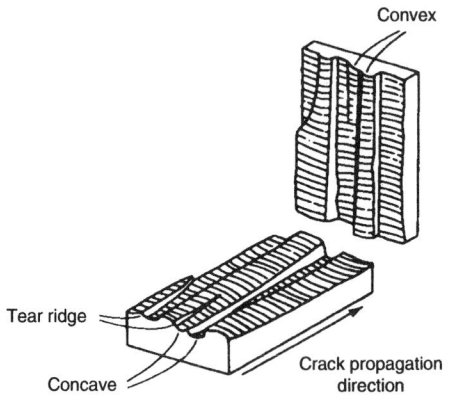

Fig. 21 Schematic illustrating fatigue striations on plateaus

type of striation that forms may depend on the combination of cleavage, dictated by the local tensile stress, followed by shear, controlled by the local shear stress, and the detailed profile developed depends on the orientation of the cleavage plane and the slip systems to the tensile and shear stresses.

Fatigue Crack Growth in Duplex Microstructures. Duplex microstructures (such as in dual-phase steels, α-β titanium alloys, or metal-matrix composites) can have unique fatigue crack growth features, depending on whether the soft phase is continuous or whether the hard phase is continuous. A series of classic experiments were performed (Ref 50, 51) in which these two types of microstructures were obtained with martensite and ferrite. In both structures, crack initiation occurred at the ferrite-martensite interface. In the continuous ferrite microstructure, the crack devi-

ated to avoid the martensite, leading to a more tortuous, rougher fracture surface. The crack path in the continuous martensite structure was flatter than the continuous ferrite, and the crack grew in the martensite, although the crack occasionally deviated by interaction with large ferrite grains. The crack growth rate was slower in the continuous martensite microstructure. While the crack path was rougher in the continuous ferrite structure, the continuous martensite structure exhibited higher closure loads, which is opposite to what one would expect from roughness-induced crack closure. It has been proposed that there is a semicohesive zone behind the crack tip (Fig. 25) in which ligaments of unfractured material limit the driving force for crack extension (Ref 52). Perhaps ductile ferrite ligaments in the continuous martensite structure caused the crack faces to come into contact at a higher load and exhibit a

greater degree of crack closure. The fatigue fracture surfaces often have cleaved and ductile regions in duplex microstructures such as α-β titanium (e.g., Ref 53) or metal-matrix composites.

Cyclic Crack Growth in Polymers. The low thermal conductivity of polymers can cause hysteretic heating during fatigue cycling (Ref 54). This greatly complicates the prediction of fatigue behavior and lifetime. The elastic modulus can decrease one to three orders of magnitude for amorphous polymers as they are heated above the glass temperature. Also, the amount of viscoelastic deformation will become prominent above the glass temperature. Therefore, the fatigue response of a given polymer will depend on frequency, thermal mass of the component, and the proximity of ambient temperature to the glass temperature. The data from S-N curves may or may not accurately represent component behavior, because S-N curves are usually obtained on (thermally) small samples at frequencies between

Fig. 22 Tire tracks on the fatigue fracture surface of a quenched-and-tempered AISI 4140 steel. TEM replica. (Courtesy of J. Le May, Metallurgical Consulting Services Ltd.)

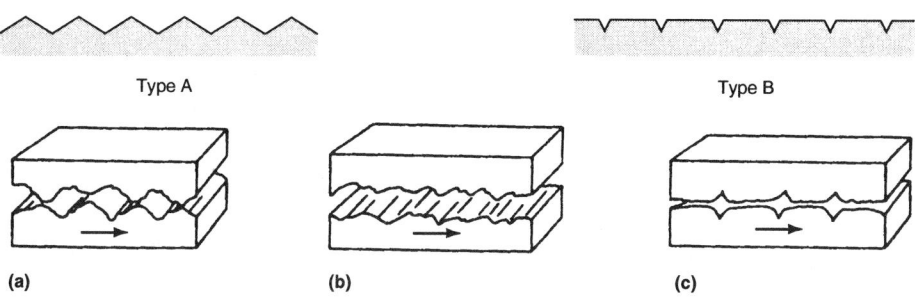

Fig. 23 Sawtooth and groove-type (type B) fatigue fracture profiles of a single fracture (top) and matching fracture surface (bottom). Arrows show crack propagation direction. (a) Distinct sawtooth profile (aluminum alloy). (b) Poorly formed sawtooth profile (steel). (c) Groove-type profile (aluminum alloy). It should be noted that matching surfaces can be either self-similar or of the opposite type. Mughrabi et al. (Ref 43) discuss five different shapes of striations formed in air. Much more irregular and poorly formed periodic markings have been found on fracture surfaces formed by fatigue crack growth in vacuum. Striations formed in a vacuum tend to be rounded but closer to type A than type B. Source: Ref 1, 41

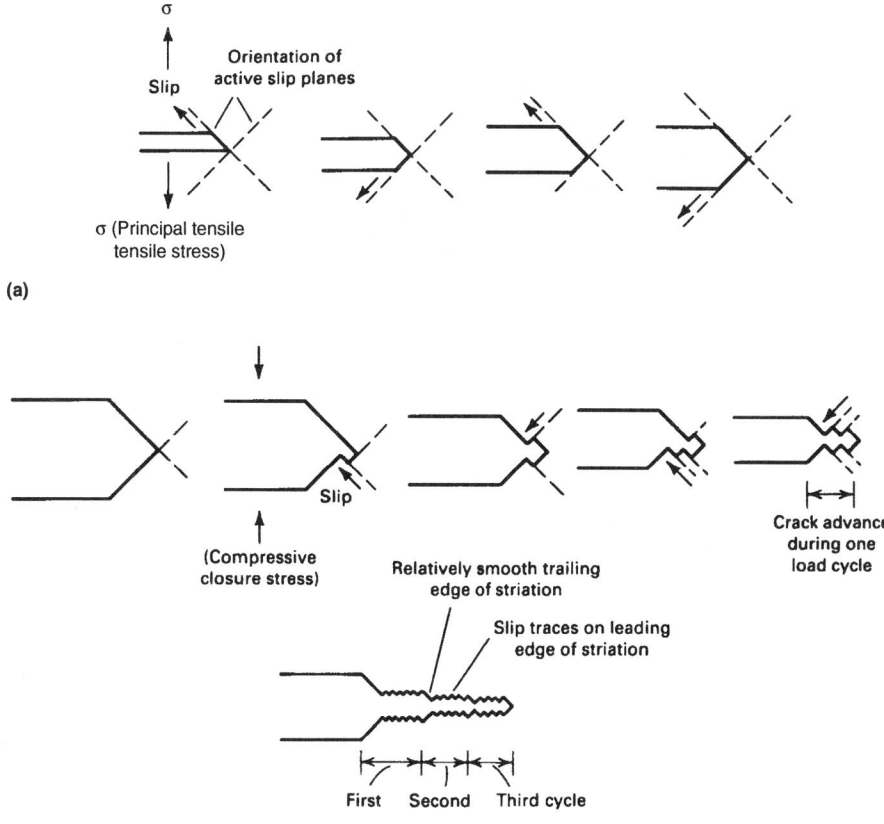

Fig. 24 Mechanism of fatigue crack propagation by alternate slip at the crack tip. Sketches are simplified to clarify the basic concepts. (a) Crack opening and crack-tip blunting by slip on alternate slip planes with increasing tensile stress. (b) Crack closure and crack-tip resharpening by partial slip reversal on alternate slip planes with increasing compressive stress

5 and 100 Hz. A large component may experience a greater temperature rise than the fatigue specimens and a small component may experience less heating. One should use data from fatigue lifetime curves with some caution.

Surprisingly little has been published on the factors influencing fatigue crack initiation in polymers. The primary focus is on fatigue crack growth. Polymer fatigue fracture surfaces often have the classic beach marks or clam shell markings. As with metals, while the macroscopic features represent the positions of the crack front, the spacing of the prominent parallel features more closely corresponds to the spacing between high-load cycles. At higher magnification, fatigue fracture surfaces may be featureless, have patches of striations, have long bands of striations, exhibit a patchy rumpled surface, or have a combination of the above with microvoid coalescence. These features can be related to the mechanisms of crack extension.

Patches of striations can be observed where the striation spacing corresponds to the macroscopic growth rate (Ref 54, 55). This is only observed for crack growth rates in excess of ~0.1 μm per cycle. The striations can have fine features oriented parallel to the crack growth direction, which are tear ridges connecting crack fronts on different levels. The surface can also have fine features parallel to the striations. The mechanism of crack growth is similar to that for metals, where the crack tip blunts by homogeneous inelastic flow on the rising part of the load cycle. The crack then resharpens as the load is released.

Striation-like features have also been observed in which the spacing is many times the macroscopic crack growth rate (Ref 54, 55). These features, termed discontinuous growth bands, are formed by crack extension along a craze that extends from the crack tip, as seen in Fig. 26. The crack grows by progressive breakdown of the craze. The fibrils in the craze are drawn out of the uncrazed matrix and may undergo orientation hardening. At some point, the load-carrying capacity of the fibrils is exceeded and the crack propagates across the craze. Crack growth can be accompanied by audible clicks and pops. At this point, a new craze is initiated and the process repeats itself.

In amorphous polymers, a single craze will emanate from the crack tip, as can be seen in Fig. 26. Because the single craze is nearly planar, a discontinuous growth band can laterally extend all the way across a typical crack growth specimen. The lateral extent of the striation-like feature may be one way to determine whether it is a true striation or a discontinuous growth band. In crystalline polymers, crazes may emanate on several levels, as can be seen in the microtomed sections through a crack in nylon in Fig. 27. A patch-like morphology results as tear ridges form when the different levels link up (Ref 56).

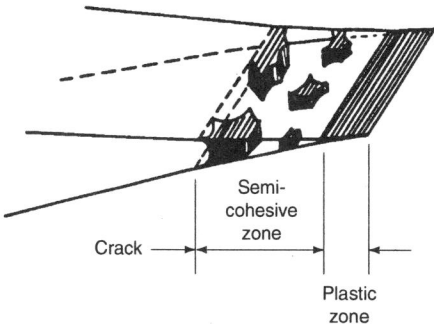

Fig. 25 Schematic of semicohesive zone behind the crack tip. Source: Ref 52

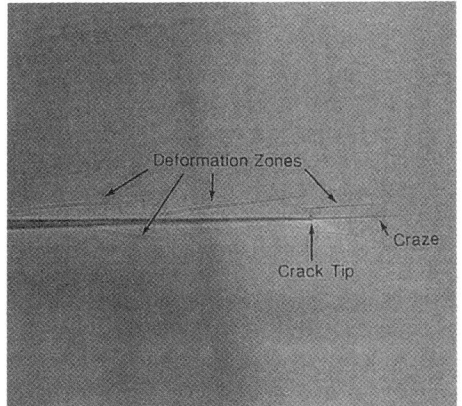

Fig. 26 Craze emanating from a fatigue crack tip in amorphous nylon. A discontinuous growth band will form as the crack propagates through the craze. Under polarized light, 560×. Source: Ref 56

As for monotonic fracture, there is competition between the two mechanisms of yielding, crazing and shear banding. At the tip of a plane-strain crack, the stress field has a strong hydrostatic tensile component that favors formation of crazes. For plane-stress cracks, the large shear component stress field and reduced hydrostatic tensile component has been shown to favor shear banding (Ref 57). Shear banding may also be promoted by the introduction of rubber particles and other second phases. In both cases, anything that promotes shear banding will increase the size of the damage zone and thereby increase the work of fatigue crack growth.

Fatigue Crack Closure

Since 1970 (Ref 58, 59) premature crack closure during unloading has been considered important in the evaluation of the crack tip driving force necessary to quantify fatigue crack growth data. Fatigue crack closure is an important effect because it alters the relationship between the applied stress-intensity factor (calculated from applied stress and crack length) and that actually experienced by the crack tip. In linear elastic fracture mechanics under maximum and minimum loading, the K_I fields ahead of the fatigue crack tip uniquely define the rate of fatigue crack propagation. However, with the discovery of crack closure by Elber (Ref 58, 59), it is now known that crack growth rates are influenced not only by the conditions ahead of the crack tip, but also by the nature of crack face contact behind the crack tip during unloading. Elber suggested that a zone of residual tensile deformation occurs in the wake of a fatigue crack tip, which thus reduces crack opening displacements and causes contact between faces in the wake of the crack tip.

Crack closure reduces the applied stress-intensity amplitude, ΔK_{apl}, by a factor related to the stress intensity at which closure occurs, K_{cl}. The magnitude of K_{cl}, which is generally measured by observing the change in the compliance at the point of closure, depends on material, microstructure, environment, and loading conditions. The effective stress-intensity factor amplitude at the crack tip, ΔK_{eff}, is given by

$$\Delta K_{eff} = K_{max} - K_{cl} < \Delta K_{apl}, \text{ if } K_{cl} > K_{min} = \Delta K_{apl} = K_{max} - K_{min}, \text{ if } K_{cl} \leq K_{min}$$

where K_{max} and K_{min} are the stress intensities at maximum and minimum loads, respectively. For tests involving high R ratios (K_{max}/K_{min}) i.e., for $R > 0.6$ to 0.7, $K_{min} > K_{cl}$, there is little or no crack closure (Ref 60) and the effective driving force is equal to the applied driving force.

Fatigue crack closure is not completely understood, but it is a factor in the retardation of crack growth rates. Under constant-amplitude loading, the applied stress-intensity factor range is decreased by crack closure, which thus has the effect of retarding crack growth rates. In addition, with the suggestion of wake effects by Elber, the importance of prior history is also a factor in the possible retardation or acceleration of fatigue crack growth rates. While crack closure has become a major concern in fatigue crack growth which cannot be easily predicted or measured, its effect is considered significant, particularly for near-threshold crack growth at low R ratios. Crack closure is considered primarily responsible for several fatigue crack growth phenomena, such as the effects of the following:

- R ratio on fatigue crack thresholds, ΔK_{th}, and on fatigue crack growth rates, *da/dN*
- Vacuum, air, humidity, hydrogen, and corrosive environments
- Microstructure that contributes to crack path tortuosity
- Retardation due to overloads, underloads, and thermal history
- Acceleration of short cracks

Mechanisms of Fatigue Crack Closure. Although a considerable amount of research has been devoted to the study of closure for the last two decades, the mechanisms and effects of closure are only partially understood, and in some cases they are controversial. This section only briefly reviews possible mechanisms, while the next section discusses predictive models of closure effects.

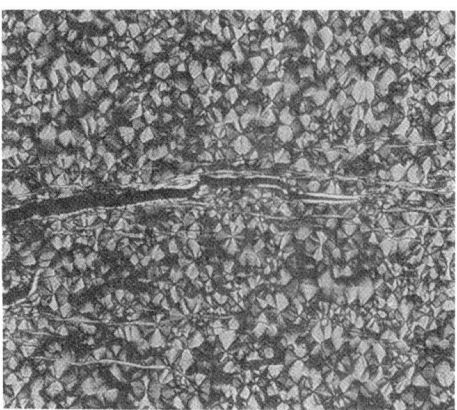

Fig. 27 Microtomed cross section showing multiple crazes emanating from a fatigue crack in crystalline nylon. Source: Ref 56

Fatigue crack closure may be caused by several different phenomena (Fig. 28):

- The plastic deformation of material in the region near to the crack tip (both ahead and behind)
- Crack surface roughness
- The growth of oxides within the crack, usually caused by interaction of the material with the environment in which the crack is being grown

Other closure mechanisms include viscous-fluid induced closure and transformation-induced closure.

Oxide-induced closure is perhaps the best understood and least controversial of these closure mechanism, even though actual observation of a crack volume filled with oxides or a corrosion product has not been documented extensively. Oxide-induced closure has been an effective concept when measuring load-ratio effects on threshold stress-intensity ranges (ΔK_{th}) for fatigue crack growth in corrosive and inert environments. In general, fatigue crack growth rates typically are lower in an inert environment and a vacuum (than in air or other corrosive environment). In the near-threshold (low ΔK) regime, however, crack growth rates at lower R ratios are higher in an inert environment as opposed to air or wet environments (see, for example, Fig. 29a, Ref 61). This effect is rationalized by the concept of crack closure, whereby oxide deposits promote premature contact of the fracture surfaces during unloading, thereby reducing the effective stress intensity at the crack tip ($K_{eff} = K_{max} - K_{cl}$). This concept is further demonstrated by considering the effect of load ratio (R) on the threshold ΔK_{th} in dry and wet environments (Fig. 30). At lower load ratios ($R < 0.5$), the oxides on the fracture surface promote earlier closure, thereby increasing K_{cl} and reducing K_{eff}. As load ratio is increased, the threshold stress-intensity ranges in air and wet environments decrease below that of in dry helium, because premature closure during unloading is less pronounced at higher load ratios. Therefore, load-ratio effects on ΔK_{th} in corrosive environments is a key demonstration of crack-closure concepts, although efforts have been made to explain this phenomena without

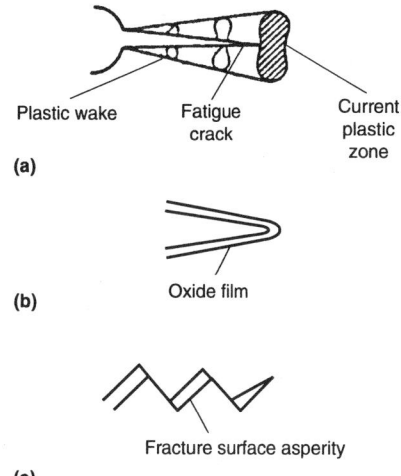

(a)

(b)

(c)

Fig. 28 Schematic illustration of three mechanisms that promote retardation of fatigue crack growth in constant-amplitude fatigue. (a) Plasticity-induced crack closure. (b) Oxide-induced crack closure. (c) Roughness-induced crack closure. Other mechanisms include fluid-induced crack closure, transformation-induced crack closure, crack deflection, crack bridging by fibers, crack bridging (trapping) by particles, crack shielding by microcracks, and crack shielding by dislocations. Source: Ref 64

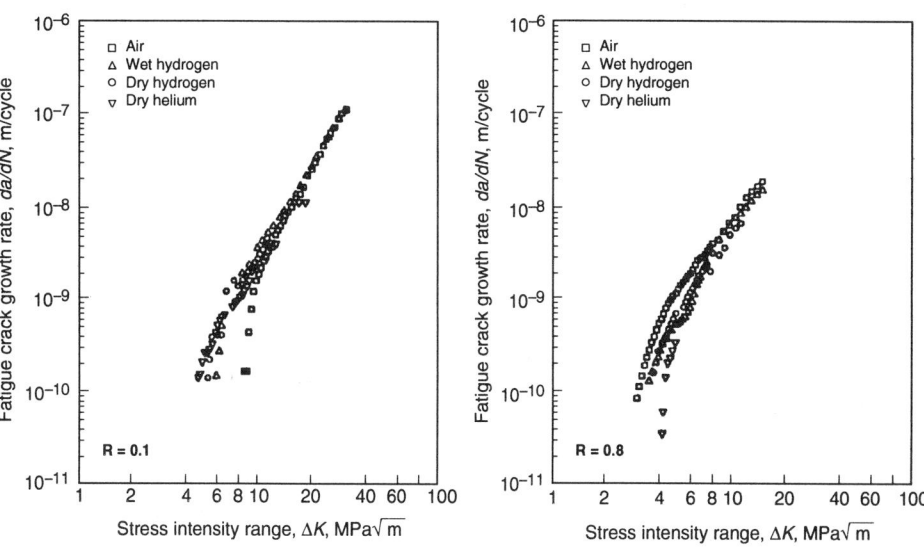

Fig. 29 Effect of load ratio on fatigue cracking threshold of 4340 steel (695 MPa yield strength). (a) $R = 0.1$ (b) $R = 0.8$. Source: Ref 61

invoking closure (see "New Concepts for Fatigue Thresholds" at the end of this article).

Asperity (or Surface-Roughness)-Induced Closure. Fatigue crack closure is often attributed to crack-wake surface roughness, and many studies have addressed this potential closure-effect on fatigue crack growth rates. Several examples of roughness-induced closure effects are described in the article "Fatigue Crack Thresholds" in this Volume. Additional references of other examples are listed in the "Selected References" at the end of this article.

However, direct demonstration of roughness-induced closure is less definitive than oxide-induced closure. *Partial crack closure* can occur *locally* at asperities, but their effects are nominal on the crack-tip stress field (Ref 62). Load ratio effects on ΔK_{th} (Ref 67) also indicate that roughness-induced closure is less significant than oxide-induced closure. Therefore, even if surface roughness is a cause of crack closure, its effect on crack growth may be less critical unless more definitive work is done.

One review (Ref 41) even suggests that roughness is not a factor in closure and that all closure can be described as a plasticity effect. It does seem evident, however, that closure occurs when asperities come into contact behind the crack tip and wedge open the crack at loads above the minimum load, P_{min} (see Gray et al, Horng and Fine, Baker and Mayers, Minakawa and McEvily, and Suresh and Ritchie in "Selected References"). This closure effect is more likely at low stress ratios in the near-threshold regime due to lower crack opening displacements.

Closure from Plastic Deformation. Plastic deformation of material just ahead of the crack tip can cause closure when residual tensile strains are left in the material behind the advancing crack front. Material elements just ahead of the crack tip are elongated by strain concentration at the tip, and this deformation may not be fully reversed when the element breaks as it passes into the wake of the crack. On unloading, the crack surfaces of these "stretched" elements may touch before minimum load is reached, thereby resulting in closure.

In terms of plasticity effects, some success has been obtained by estimating the mode I value of ΔK_{th} using the simple concept that crack growth at threshold will only occur if dislocations are emitted from the crack tip (Ref 63). Using the equations of elasticity, the following expression is easily derived for large cracks:

$$\Delta K_{th} = \sigma_y \sqrt{2\pi r_s} \qquad (Eq\ 1)$$

where σ_y is the yield stress and r_s is the distance dislocations move through the material before being stopped. This equation assumes that a slip line occurs at the crack tip, and that dislocations are emitted and move r_s, with σ_y being the stress at the end of the slip line. The problem with using the above equation to predict ΔK_{th} is that it is difficult to determine r_s. The most success using this concept was obtained for particulate-reinforced composites where r_s was controlled by the size and volume fraction of silicon carbide (Ref 64). This concept seems to hold promise as a method for computing ΔK_{th}, but it has not been proven and remains a hypothesis to be further tested.

Like roughness-induced closure, a definitive consensus has not been reached on the significance (and perhaps existence) of plasticity-induced closure, despite an overwhelming amount of published literature on its measurement and implications (see "Selected References: Plasticity-Induced Closure" at the end of this article for a limited bibliography). Plasticity-induced closure is considered dominant at higher ΔK levels (i.e., in the Paris regime), based on the observation that the Elber closure ratio ($U = \Delta K_{eff}/\Delta K$) decreases rapidly as the threshold is approached.

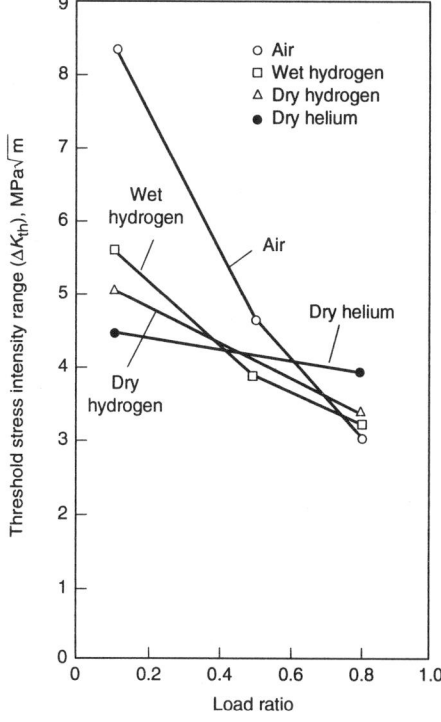

Fig. 30 Effect of load ratio and environments on fatigue threshold of 4340 steel. Source: Ref 61

However, the ability of plasticity-induced closure in accounting for R ratio effects on FCG rates at high ΔK levels has been questioned by some (T. Shih and R. Wei, *Engr. Frac. Mech.,* Vol 6, 1974, p 19-32). In the low-ΔK threshold regime, interpretation of closure effects from possible plasticity effects is complicated by possible roughness- or oxide-induced closure.

Using basic concepts from dislocation theory, Vasudevan et al suggest that crack-tip plasticity cannot contribute to crack closure because crack opening displacements are always greater than

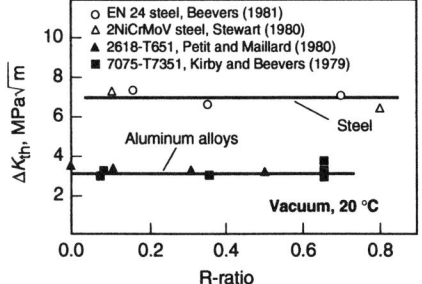

Fig. 31 Apparent lack of closure due to independence of ΔK_{th} versus load ratio in a vacuum. Source: Ref 62

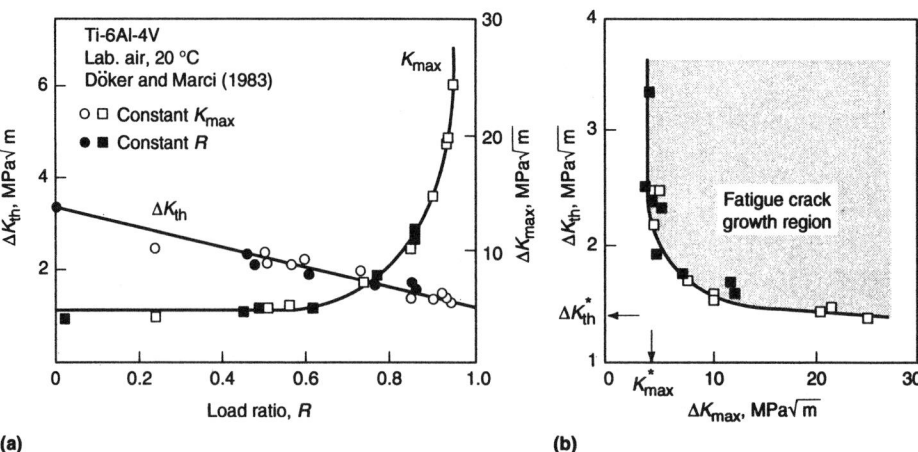

Fig. 32 Near threshold fatigue crack growth data of Ti-6V-4Al alloy in lab air. (a) Variation in ΔK_{th} and K_{max} with load R, and (b) a replot of (a) in terms of ΔK_{th} vs K_{max}, as a fatigue map

crack closing displacements (Ref 62, 65). The experimental basis of this position is based on limited fatigue threshold data, as a function of load ratio R in vacuum (Fig. 31). If there is any plasticity-induced closure, it should manifest more readily in a material tested in vacuum, wherein closure contributions from oxidation or corrosion are absent. Figure 31 shows that ΔK_{th} is essentially independent of load ratio R. Since $K_{max} = \Delta K/(1 - R)$, which increases nonlinearly with R, the contribution from plasticity-induced closure should be particularly apparent at large values of R. The observed independence of ΔK_{th} on R suggests that crack closure due to plasticity is either nonexistent or insignificant (Ref 62).

Oxide-Induced Crack Closure. The mechanism of oxide-induced crack closure evolved as a consequence of attempts to rationalize apparent anomalies in the effects of environment on near-threshold fatigue crack growth in steels and aluminum alloys, as previously noted. Another classic example of this has been demonstrated by Suresh and Ritchie (Ref 66) for a martensitic 2¼Cr-1Mo steel in moist laboratory air and in dehumidified, ultra-high-purity hydrogen and helium (138 kPa) at $R = 0.05$ and 0.75. The near-threshold crack propagation in both the dry gaseous environments are up to two orders of magnitude higher, and the ΔK_{th} is about 50% lower, than in air at the lower R ratio (Ref 66).

The mechanism of this near-threshold behavior can be explained in terms of oxide-induced closure. During the propagation of a fatigue crack, the presence of a moist atmosphere leads to oxidation of the freshly formed fracture surfaces. At low amplitudes of cyclic crack-tip opening displacements (i.e., at near-threshold ΔK levels and low R ratios), the possibility of repeated crack face contact during tensile fatigue is enhanced as a consequence of locally mixed-mode crack opening and closure. In contrast, at high R ratios, the possibility of crack face contact is minimized because of the larger crack-tip opening displacements. Likewise, at high ΔK levels, the rate of crack advance is generally too rapid to allow oxide buildup at any R value.

Models of Fatigue Crack Closure. Fatigue crack closure is not fully understood, but the mechanisms mentioned above have been proposed over the years for various alloys. Other mechanisms that impede constant-amplitude fatigue crack growth are also of interest in composites and nonmetallic materials such as ceramics.

These mechanisms include crack deflection, crack bridging or trapping, and crack shielding due to microcracking, phase transformations, or dislocations. These processes are briefly discussed in the article "Fatigue of Brittle Materials" in this Volume.

Predictive modeling of closure mechanisms is a formidable task. Closure is influenced by several variables, such as microstructure, ΔK magnitude, crack size, environment, variations in load amplitude, R ratio, and temperature. However, Eq 1 is a tentative basis for considering plasticity effects, even though closure probably results from a combination of crack-tip plasticity and wedging open of the crack along its wake. Models that consider both crack wake stretch and plastic zone residual stress include that of Newman (Ref 67), strip yield models (Ref 68, 69) and finite element models (Ref 67, 70).

Oxide-induced closure is also a very complicated process that cannot be predicted in all instances. However, Suresh provides a simple analogy for a qualitative estimate of oxide-induced effects on the near-threshold "driving force" (K) (Ref 60). Consider a rigid wedge of constant thickness d_0 inside a linear elastic fatigue crack. If the wedge extends all along the length of the fatigue crack and its edge terminates at a distance $2l$ behind the crack tip, then closure occurs during unloading of a stress-intensity factor of:

$$K_{cl} = \frac{d_0 E}{4\sqrt{\pi l}\,(1-\nu^2)}$$

where E and ν are the Young's modulus and the Poisson's ratio of the cracking material, respectively. Taking typical values of $l = 0.2$ μm and $d_0 = 0.2$ μm from experiments on lower-strength steels, it is found that $K_{cl} \approx 2.3$ MPa√m. This estimate of the closure K due to wedge contact, although too simplistic to account for all the effects associated with the complex process of oxide-induced closure, provides a justification for the experimentally observed differences in the ΔK_{th} values of 2¼Cr-1Mo steels in dry and moist environments at low load ratios documented in Ref 66.

New Concepts for Fatigue Thresholds. Despite an overwhelming amount of publications on crack closure measurements and its implications, a unanimous view of its significance on crack growth has not been reached based on some critical re-examination of crack closure concepts (Ref 62, 65). This re-examination has lead to a two-parameter analysis of fatigue crack growth based on a critical stress-intensity (ΔK^*_{th}) and critical maximum stress intensity (K^*_{max}). The two loading driving force parameters have to be satisfied simultaneously for a crack to advance, a concept that is independent of the crack-closure phenomena. Such a description gives an intrinsic description of fatigue that can be related to microstructure, slip modes, and environment. The concepts also suggest, that for a complete fatigue description, it is necessary to get a systematic set of data at various load ratios and crack growth rates. Details of the mechanistic descriptions on the fatigue behavior are given in Ref 62, 65, and 71-73. An overall classification scheme of fatigue crack growth rates is also given in Ref 74 based on this two-parameter method. With the two parameters ΔK^*_{th} and K^*_{max}, the entire fatigue data of ΔK_{th} vs R dependence can be fully explained without invoking closure.

For example, this description is shown in Fig. 32(a) for alloy Ti-6Al-4V. At low R, fatigue is K_{max} controlled, meaning that ΔK_{th} increases with decreasing R to meet the K^*_{max}. Similarly, at high R values, fatigue is ΔK_{th} controlled and K_{max} then increases to meet the ΔK^*_{th}. Both these critical parameters can be determined by plotting ΔK_{th} vs K_{max} (as in Fig. 32b). This plot provides the interrelation between the two driving force parameters defining the regimes where ΔK^*_{th} provides the minimum cyclic amplitude required to establish a characteristic fatigue damage, and K^*_{max} provides the critical stress required to break open the crack-tip bonds in the fatigue damaged region.

Figure 32(b) provides a plot showing the interdependence of these two parameters to ensure the conditions necessary for fatigue crack growth. These two critical parameters are dependent on the microstructure, slip mode, and crack tip environment; for example, with an increasing aggressiveness of the crack tip environment (oxygen, hydrogen, water vapor, or NaCl) the role of K^*_{max} becomes increasingly important than ΔK^*_{th}, since the environment affects the crack tip bonds more than the fatigue damage zone ahead of the crack tip. The two parametric description is applicable at all crack growth rates starting from the near threshold region. Details on this topic are given in Ref 71-73.

REFERENCES

1. V. Kerlins and A. Phillips, Modes of Fracture, *Metals Handbook*, 9th ed., Vol 12, 1987, p 12-71

2. H. Jones, The Surface Energy of Solid Metals, *Metal Science Journal*, Vol 5, 1971, p 15

3. D.A. Porter and K.E. Easterling, *Phase Transformations in Metals and Alloys*, Van Nostrand Reinhold, 1981, p 113

4. B. Faucher and W.R. Tyson, in *Fracture Mechanics, 18th Symposium*, STP 945, ASTM, 1985, p 164-178

5. M.F. Ashby, C. Ghandi, and D.M.R. Taplin, *Acta Metallurgica*, Vol 27, 1979, p 699

6. C. Ghandi and M.F. Ashby, *Acta Metallurgica*, Vol 27, 1979, p 1565

7. T.H. Courtney, *Mechanical Behavior of Materials*, McGraw-Hill, 1990

8. D. Broek, *Engineering Fracture Mechanics*, Vol 5, 1973, p 55

9. R.H. van Stone, T.B. Cox, J.R. Low, Jr., and J.A. Psioda, Microstructural Aspects of Fracture by Dimpled Rupture, *International Metals Reviews*, Vol 30, 1985, p 157-180

10. W.M. Garrison, Jr. and N.R. Moody, Ductile Fracture, *Journal of Physics and Chemistry of Solids*, Vol 48, 1987, p 1035-1074

11. T.L. Anderson, *Fracture Mechanics: Fundamentals and Applications*, CRC Press, 2nd ed., 1995

12. R.W. Hertzberg, *Deformation and Fracture Mechanics of Engineering Materials*, Wiley and Sons, 3rd ed., 1989

13. D.S. Liu and J.J. Lewandowski, The Effects of Superimposed Hydrostatic Pressure on Deformation and Fracture, Part I: Monolithic 6061 Aluminum, *Metallurgical Transactions*, Vol 24A, 1993, p 601-608

14. L. Luckyx, J.R. Bell, A. McClean, and A. Korchynsky, *Metallurgical Transactions*, Vol 1, 1970, p 3341

15. R.M. McMeeking and D.M. Parks, On Criteria for J-Dominance of Crack Tip Fields in Large-Scale Yielding, in STP 668, ASTM, 1979, p 175-194

16. J.M. Berry, Cleavage Step Formation in Brittle Fracture, *ASM Transactions*, Vol 51, 1959, p 556-588

17. A.W. Thompson and J.F. Knott, Micromechanisms of Brittle Fracture, *Metallurgical Transactions*, Vol 24A, 1993, p 523-534

18. W.W. Gerberich and E. Kurman, *Scripta Metallurgica*, Vol 19, 1995, p 295-298

19. C.J. McMahon, Jr. and M. Cohen, Initiation of Fracture in Polycrystalline Iron, *Acta Metallurgica*, Vol 13, 1965, p 591-604

20. R.O. Ritchie, J.F. Knott, and J.R. Rice, On the Relationship between Critical Tensile Stress and Fracture Toughness in Mild Steel, *J. Mech. Phys. of Solids*, Vol 21, 1973, p 395-410

21. D.A. Curry and J.F. Knott, Effect of Microstructure on Cleavage Fracture Toughness in Mild Steel, *Metal Science*, Vol 13, 1979, p 341-345

22. A.G. Evans, Statistical Aspects of Cleavage Fracture in Steel, *Metallurgical Transactions*, Vol 14A, 1983, p 1349-1355

23. T.L. Anderson, D.I.A. Steinstra, and R.H. Dodds, Jr., A Theoretical Framework for Addressing Fracture in the Ductile-Brittle Transition Region, *Fracture Mechanics: Symposium*, STP 1207, ASTM, 1995

24. W.F. Flanagan, L. Zhong, and B.D. Lichter, A Mechanism for Transgranular Stress Corrosion Cracking, *Metallurgical Transactions*, Vol 24A, 1993, p 553-559

25. M.F. Ashby and D.R.H. Jones, *Engineering Materials 1: An Introduction to Their Properties and Applications*, Pergamon Press, 1980, p 127

26. D.R. Askeland, *The Science and Engineering of Materials*, PWS Publishing, 3rd ed., 1994, p 498

27. S.S. Sternstein and L. Ongchin, *Polymer Preprints*, Vol 10, 1969, p 1117

28. I.M. Ward and D.W. Hadley, *An Introduction to the Mechanical Properties of Solid Polymers*, J. Wiley and Sons, 1993

29. R.W. Hertzberg and J.A. Manson, *Fatigue of Engineering Plastics*, Academic Press, 1980

30. A.J. Kinloch and R.J. Young, *Fracture Behavior of Polymers*, Applied Science Publishers, 1983

31. B.J. Leis, J. Ahmad, and M.F. Kanninen, Effect of Local Stress State on the Growth of Short Cracks, *Multiaxial Fatigue*, STP 853, ASTM, 1985, p 314-339

32. A. Hunsche and P. Neumann, Crack Nucleation in Persistent Slip Bands, *Basic Questions in Fatigue*, Vol 1, STP 924, ASTM, 1988, p 26-38

33. D. Majumdar and Y-W. Chung, Surface Deformation and Crack Initiation during Fatigue of Vacuum Melted Iron: Environmental Effects, *Metallurgical Transactions*, Vol 14A, 1983, p 1421-1425

34. D.F. Socie, L.A. Waill, and D.F. Dittmer, Biaxial Fatigue of Inconel 718 Including Mean Stress Effects, *Multiaxial Fatigue*, STP 853, ASTM, 1985, p 463-481

35. K. Tanaka, Mechanics and Micromechanics of Fatigue Crack Propagation, *Fracture Mechanics: Perspectives and Directions (20th Symposium)*, STP 1020, ASTM, 1989, p 151-183

36. M. Wilhelm, M. Nagesararao, and R. Meyer, Factors Influencing Stage I Crack Propagation in Age Hardened Alloys, *Fatigue Mechanisms*, STP 675, ASTM, 1979, p 214-233

37. E.A. Starke, Jr. and J.C. Williams, Microstructure and the Fracture Mechanics of Fatigue Crack Propagation, *Fracture Mechanics: Perspectives and Directions (20th Symposium)*, STP 1020, ASTM, 1989, p 184-205

38. G.R. Yoder, L.A. Cooley, and T.W. Crooker, *Fracture Mechanics: 14th Symposium*, STP 791, ASTM, 1983, p 348-365

39. Beachem, *Trans. ASM*, Vol 60, 1967, p 325

40. R. Koterazawa, M. Mori, T. Matsni, and D. Shimo, *J. Eng. Mater. Technol., (Trans. ASME)*, Vol 95 (No. 4), 1973, p 202

41. D.L. Davidson and J. Lankford, Fatigue Crack Growth in Metals and Alloys: Mechanisms and Micromechanics, *International Materials Review*, Vol 37, 1992, p 45-76

42. N. Grinberg, *Int. J. Fatigue*, Vol 6, 1984, p 143-148

43. H. Mughrabi, R. Prass, H.-J. Christ, and D. Puppel, in *Chemistry and Physics of Fracture*, R.M. Latanison and R.H. Jones, Ed., Martinus Nijhoff, Amsterdam, 1987, p 443-448

44. J.C. McMillan and R.M.N. Pelloux, *Eng. Fract. Mech.*, Vol 2, 1970, p 81-84

45. R.W. Hertzberg, Fatigue Fracture Surface Appearance, *Fatigue Crack Propagation*, STP 415, ASTM, 1967, p 205

46. R.M.N. Pelloux, *Trans. ASM*, Vol 62, 1969, p 281-285

47. D. Broek and G.Q. Bowles, *Int. J. Fract. Mech.*, Vol 6, 1970, p 321-322

48. P. Neumann, *Acta Metall.*, Vol 22, 1974, p 1155-1178

49. K.J. Nix and H.M. Flower, *Acta Metall.*, Vol 30, 1982, p 1549-1559

50. K. Minakawa, Y. Matsuo, and A.J. McEvily, The Influence of Duplex Microstructure in Steels on Fatigue Crack Propagation in the Near Threshold Region, *Metallurgical Transactions*, Vol 13A, 1982, p 439-445

51. R.M. Ramage, K.V. Jata, G.J. Shiflet, and E.A. Starke, Jr., The Effect of Phase Continuity on Fatigue and Crack Closure of a Dual Phase Steel, *Metallurgical Transactions*, Vol 18A, 1987, p 1291-1298

52. W.W. Gerberich and N.R. Moody, A Review of Fatigue Fracture Topology Effects on Threshold and Growth Mechanisms, *Fatigue Mechanisms*, STP 675, ASTM, 1979, p 292-341

53. J.C. Chestnutt, C.G. Rhodes, and J.C. Williams, *Fractography—Microscopic Cracking Processes*, STP 600, ASTM, 1976, p 99

54. R.W. Hertzberg, *Fatigue of Engineering Plastics*, Academic Press, 1980

55. R.W. Hertzberg, Fracture Surface Morphology in Engineering Solids, *Fractography of Modern Engineering Materials*, STP 948, ASTM, 1987, p 5-40

56. M.G. Wyzgoski and G.E. Novak, Influence of Thickness and Processing on Fatigue Fracture of Nylon 66, Part II: Crack Tip Morphology, *Polymer Engineering and Science*, Vol 32, 1992, p 1114-1125

57. E.J. Moskala and T.J. Pecorini, Fatigue Crack Propagation Behavior of Cellulose Esters, *Polymer Engineering and Science*, Vol 34, 1994, p 1387

58. W. Elber, Fatigue Crack Closure under Cyclic Tension, *Engineering Fracture Mechanics*, Vol 2, 1970, p 37-45

59. W. Elber, The Significance of Fatigue Crack Closure, *Damage Tolerance in Aircraft Structures*, STP 486, ASTM, 1971, p 230-242

60. S. Suresh, *Fatigue of Materials*, Cambridge University Press, 1991, Chapter 7 and p 239-240

61. P.K. Liaw, T.R. Leax, and J.K. Donald, Gaseous-Environment Fatigue Crack Propagation Behavior of a Low-Alloy Steel, *Fracture Mechanics Perspectives and Directions*, R. Wei and R. Gangloff, Ed., ASTM STP 1020, 1989, p 581-603

62. N. Louat et al, *Met. Trans.*, Vol 24A, 1993, p 2225

63. D.L. Davidson, *Acta Metall.*, Vol 36, 1988, p 2275-2282

64. D. Davidson, *Eng. Fract. Mech.*, Vol 33, 1989, p 965-977

65. A. Vasudevan, K. Sadananda, and N. Louat, *Scripta Metall et Materialia*, Vol 27, 1992, p 1673

66. S. Suresh, G. Zamiski, and R. Ritchie, *Metallurgical Transactions*, Vol 12A, 1982, p 1435-1443

67. J.C. Newman, in *Mechanics of Crack Growth*, STP 590, ASTM, 1976, p 281-301

68. J.C. Newman, in *Methods and Models for Predicting Fatigue Crack Growth Under Random Loading*, STP 748, ASTM, 1981, p 53-84

69. S.R. Daniewicz, J.A. Collins, and D.R. Houser, *Int. Journal of Fatigue*, Vol 16, 1984, p 123-133

70. R.C. McClung and H. Sehitoglu, *Eng. Fracture Mech.*, Vol 33, 1989, p 237-252

71. A. Vasudevan, K. Sadananda, and N. Louat, *Scripta Metall. et Materialia*, Vol 28, 1993, p 65

72. A. Vasudevan, K. Sadananda, and N. Louat, *Mater. Sci. Eng.*, Vol A188, 1994, p 1

73. K. Sadananda and A. Vasudevan, ASTM STP-1220, 1995, p 484

74. A. Vasudevan and K. Sadananda, *Metallurgical and Materials Transactions*, Vol 26A, 1995, p 1221

Selected References: Roughness-Induced Closure

- T.J. Baker and I.C. Mayers, *Fat. Engng. Mater. Struct.*, Vol 4, 1981, p 79

- S. Dhar, Evaluation of Fatigue Crack Closure in the Near-Threshold Level in High Strength Steel, Fatigue 90, Materials and Component Engineering Publications, 1990 p 1261-1266

- H. Jung and S. Antolovich, Experimental Characterization of Roughness-induced Crack Closure in Al-Li, 2090 Alloy, *Scr. Metall. Mater.*, Vol 33, 1995, p 275-281

- G.T. Gray et al, *Met. Trans.*, Vol 14A, 1983, p 421

- J. Horng and M. Fine, in *Fatigue Crack Growth Threshold Concepts*, TMS, 1983, p 115

- K. Li and A. Thompson, Fatigue Threshold and Closure Effect of CP Titanium at 10, 30 Hz in Laboratory Air, *Fatigue 93*, 5th International Conference on Fatigue and Fatigue Thresholds, Vol I, Engineering Materials Advisory Services Ltd., 1993, p 513-518

- S. Li et al, Geometric Model for Fatigue Crack Closure Induced by Fracture Surface Roughness Under Mode I Displacements, *Materials Science and Engineering A*, Vol A150, Feb 1992

- S. Li, L. Sun, and Z. Wang, The Relationship Between Crack Closure Ratio and K_{max} at Near-Threshold Levels, *Scripta Metallurgica et Materialia*, Vol 27, No. 11, 1992, p 1669-1672

- J. Llorca, Roughness-Induced Fatigue Crack Closure: a Numerical Study, *Fatigue and Fracture of Engineering Materials and Structures*, Vol 15, No. 7, 1992, p 655-669

- J. Llorca, Roughness-Induced Closure of Through-Thickness Short Fatigue Cracks, *Fatigue 90*, Materials and Component Engineering Publications, 1990, p 1301-1306

- K. Makhlouf and J.W. Jones, Near-Threshold Fatigue Crack Growth Behaviour of a Ferritic Stainless Steel at Elevated Temperatures, *International Journal of Fatigue*, Vol 14, No. 2, Mar 1992, p 97-104

- K. Minakawa and A. McEvily, *Scripta Met.*, Vol 15, 1981, p 633

- T. Ogawa, K. Tokaji, and K. Ohya, The Effect of Microstructure and Fracture Surface Roughness on Fatigue Crack Propagation in a Ti—6Al—4V Alloy, *Fatigue Fract. Eng. Mater. Struct.*, Vol 16, No. 9, 1993, p 973-982

- R.V. Prakash, M.N. Srinivasan, and K.K. Brahma, Effect of Stress Ratio and Microstructure on Fatigue Threshold of Ductile Iron, *ECF 8 Fracture Behaviour and Design of Materials and Structures*, Vol III, Engineering Materials Advisory Services Ltd., 1990, p 1270-1275

- K.S. Ravichandran, A Theoretical Model for Roughness Induced Crack Closure, *Int. J. Fract.*, Vol 44, No. 2, July 1990, p 97-110 and p R23-R26

- S. Suresh and R. Ritchie, *Met. Trans.*, Vol 13A, 1982, p 937

- Y.S. Zheng, Roughness-Induced Shear Resistance of Mode II Crack Growth, *Acta Metallurgica Sinica (China)*, Vol 29, No. 6, June 1993, p A253-A261

- W. Zhu, Fatigue Crack Growth Process of Steels at the Near-Threshold Region in Air and Vacuum, *Xi'an Jiaotong Daxue Xuebao (Journal Of Xi'an Jiaotong University)*, Vol 25, No. 2, 1991, p 127-134

Selected References: Plasticity-Induced Closure

- A.M. Abdel Mageed, R.K. Pandey, and R. Chinadurai, Effect of Measurement Location and Fatigue-Loading Parameters on Crack Closure Behaviour, *Materials Science and Engineering A*, Vol A150, No. 1, 1992, p 43-50

- G.H. Bray and E.A. Starke Jr., Fatigue Crack Retardation in a Dispersion Strengthened PM Aluminium Alloy, *Fatigue 93*, 5th International Conference on Fatigue and Fatigue Thresholds, Vol III, Engineering Materials Advisory Services Ltd., 1993, p 1587-1593

- C. Bull and R. Hermann, Fatigue Crack Growth and Closure in Aluminum Alloys, *Scr. Metall. Mater.*, Vol 30, No. 10, 1994, p 1337-1342

- P.J. Cotterill and J.F. Knott, Effects of Temperature on Overload Retardation in 9Cr—1Mo Steel, *ECF 7. Failure Analysis—Theory and Practice*, Vol II, Engineering Materials Advisory Services Ltd., 1988, p 1156-1158

- P.J. Cotterill and J.F. Knott, Overload Retardation of Fatigue Crack Growth in a 9% Cr—1% Mo Steel at Elevated Temperatures, *Fatigue and Fracture of Engineering Materials and Structures*, Vol 16, No. 1, 1993, p 53-70

- R. Hermann, Plasticity-Induced Crack Closure Study by the Shadow Optical Method, *Scripta Metallurgica Et Materialia*, Vol 25, No. 1, 1991, p 207-212

- R.M.J. Kemp, R.N. Wilson, and P.J. Gregson, The Role of Crack Closure in Corrosion Fatigue of Aluminium Alloy Plate for Aerospace Structures, *Fracture of Engineering Materials and Structures*, Elsevier Science Publishers Ltd., 1991, p 700-705

- B. Lou et al, Fatigue Crack Growth and Closure Behaviours in Carburized and Hardened Case, *Fatigue 90*, Materials and Component Engineering Publications, 1990, p 1161-1166

- G. Lu, J. Wang, and Y. Tan, Crack Closure Effect on the Threshold Behaviour of Fatigue Crack Growth for a Medium Carbon Steel, *Fatigue 93*, 5th International Conference on Fatigue and Fatigue Thresholds, Vol I, Engineering Materials Advisory Services Ltd., 1993 p 603-608

- A.J. McEvily and Z. Yang, On Crack Closure in Fatigue Crack Growth, *ECF 7, Failure Analysis—Theory and Practice*, Vol II, Engineering Materials Advisory Services Ltd., 1988, p 1231-1248

- K. Ogura, I. Nishikawa, and Y. Miyoshi, Detrimental Effect of Compressive Applied Load in Fatigue Crack Growth Rate at Elevated Temperatures, *Residual Stresses III: Science and Technology*, Vol 1, Elsevier Science Publishers Ltd., 1992, p 531-536

- C.S. Shin and S.H. Hsu, On the Mechanisms and Behaviour of Overload Retardation in AISI 304 Stainless Steel, *International Journal of Fatigue*, Vol 15, No. 3, 1993, p 181-192

- D.M. Shuter and W. Geary, The Influence of Specimen Thickness on Fatigue Crack Growth Retardation Following an Overload, *Int. J. Fatigue*, Vol 17, No. 2, 1995, p 111-119

Section 2: Fatigue Mechanisms, Crack Growth, and Testing

Fatigue Failure in Metals*

Morris E. Fine and Yip-Wah Chung, R.R. McCormick School of Engineering & Applied Science, Department of Materials Science & Engineering, Northwestern University

FATIGUE is the progressive, localized, and permanent structural change that occurs in a material subjected to repeated or fluctuating strains at nominal stresses that have maximum values less than (and often much less than) the static yield strength of the material. Fatigue may culminate into cracks and cause fracture after a sufficient number of fluctuations. Fatigue damage is caused by the simultaneous action of cyclic stress, tensile stress, and plastic strain. If any one of these three is not present, a fatigue crack will not initiate and propagate. The plastic strain resulting from cyclic stress initiates the crack; the tensile stress promotes crack growth (propagation). Although compressive stresses will not cause fatigue, compressive loads may result in local tensile stresses. Microscopic plastic strains also can be present at low levels of stress where the strain might otherwise appear to be totally elastic.

During fatigue failure in a metal free of crack-like flaws, microcracks form, coalesce, or grow to macrocracks that propagate until the fracture toughness of the material is exceeded and final fracture occurs. Under usual loading conditions, fatigue cracks initiate near or at singularities that lie on or just below the surface, such as scratches, sharp changes in cross section, pits, inclusions, or embrittled grain boundaries.

Microcracks may be initially present due to welding, heat treatment, or mechanical forming. Even in a flaw-free metal with a highly polished surface and no stress concentrators, a fatigue crack may form. If the alternating stress amplitude is high enough, plastic deformation (i.e., long-range dislocation motion) takes place, leading to slip steps on the surface. Continued cycling leads to the initiation of one or more fatigue cracks. Alternately, the dislocations may pile up against an obstacle, such as an inclusion or grain boundary, and form a slip band, a cracked particle, decohesion between particle and matrix, or decohesion along the grain boundary.

The initial cracks are very small. Their size is not known well because it is difficult to determine when a slip band or other deformation feature

becomes a crack. Certainly, however, cracks as small as a fraction of a micron can be observed using modern metallographic tools such as the scanning electron microscope or scanning tunneling microscope. The microcracks then grow or link up to form one or more macrocracks, which in turn grow until the fracture toughness is exceeded.

The fatigue failure process thus can be divided into five stages:

1. Cyclic plastic deformation prior to fatigue crack initiation
2. Initiation of one or more microcracks
3. Propagation or coalescence of microcracks to form one or more microcracks
4. Propagation of one or more macrocracks
5. Final failure

These stages in the process of fatigue failure are complicated and are influenced by many factors. This introductory article summarizes some fundamental aspects of the first four stages, prior to the fifth stage of final failure. Further coverage on both crack initiation and crack propagation is given in more detail in subsequent articles in this Volume. In particular, other articles in this section provide detailed information on the fundamental and applied aspects of fatigue crack propagation. Nonetheless, this article briefly considers crack propagation with particular emphasis on the measurement and relation of plastic work with fatigue crack propagation. This article refers primarily to ambient air or an inert environment.

The variability in fatigue behavior in similar specimens or parts is well known. This is true for cycles to failure, cycles to crack initiation, microcrack propagation rate, and macrocrack propagation rate. A fatigue database should include this variability, as well as a clear description of the specimen or part geometry, surface finish, microstructure, and stress state. The design engineer needs to know about such variability in order to predict probability of failure for a given design. Often reported data do not contain sufficient information about variations in fatigue properties to be useful in a database used for design engineers. Also S-N_f or ε_p-N_f curves of smooth (unnotched) specimens are often pre-

sented. Macrofatigue crack propagation rates are also presented as smoothed curves based on averaged data points; even when data points are plotted, they represent a trailing average of three to seven points. A fatigue crack does not propagate at a uniformly increasing rate, but frequently slows down or speeds up due to local conditions along the crack front. Real-world spectrum loading further complicates matters.

As the resolution of inspection instruments has increased, the portion of the fatigue lifetime ascribed to fatigue crack "initiation" has decreased. For practical purposes, it is useful to define the "initiation stage" as that portion of the fatigue lifetime before a crack is detectable by usual nondestructive evaluation techniques. This is typically ½ to 1 mm. Initiation of such an "engineering" crack represents the preponderance of the fatigue lifetime, except in the regime of very low cycles to failure. After formation of such an engineering crack, the remaining fatigue lifetime is usually relatively short. Thus, the design engineer can design with more precision, utilizing cycles to formation of a small crack combined with crack propagation to failure, than with cycles to failure alone or with propagation rates of macrofatigue cracks alone. More data that include statistical variations on formation of small cracks in a form useful to the design engineer are needed.

Cyclic Deformation Prior to Fatigue Crack Initiation

As previously mentioned, even if no stress risers, notches, or inclusions are present, small microcracks may initiate due to a sufficiently high alternating plastic strain amplitude. When a dislocation emerges at the surface, a slip step of one Burgers vector is created. During perfect reversed loading on the same slip plane, this step is canceled; however, slip occurs on many planes, and the reversal is never perfect. Accumulation of slip steps in a local region leads to severe roughening of the surface. Thus, resistance to fatigue failure is greater for alloys that are not subject to severe localization of plastic strain.

*Updated from M.E. Fine, Campbell Lecture, *Metall. Trans. A*, Vol 11A, March 1980, p 368-379

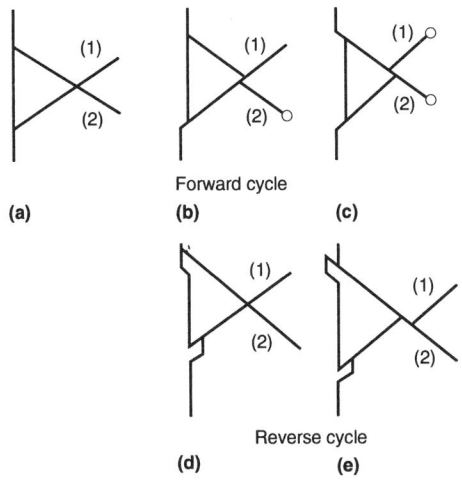

Fig. 1 Cottrell-Hull model for formation of intrusions and extrusions. Operation of two intersecting slip systems is assumed to occur in the sequence shown.

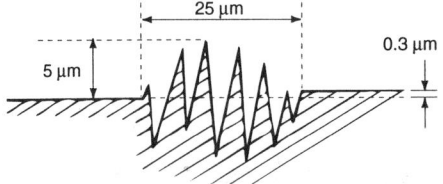

Fig. 2 Approximate profile of surface at a persistent slip band in copper determined from an interferogram. The copper single crystal was cycled over the plastic strain range 0.0025 for 30,000 cycles. Source: Ref 7

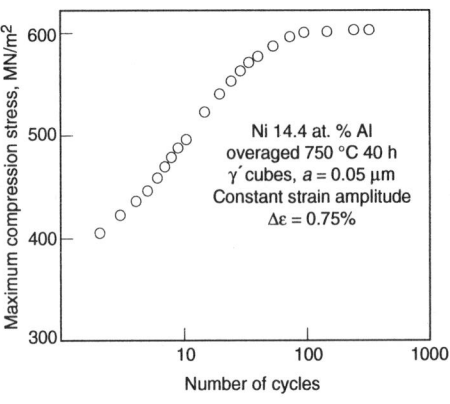

Fig. 3 Maximum compressive stress, $\Delta\sigma/2$, versus cycles at a constant strain amplitude of $\Delta\varepsilon/2 = 0.75\%$ for overaged Ni-14Al (at.%). Source: Ref 12

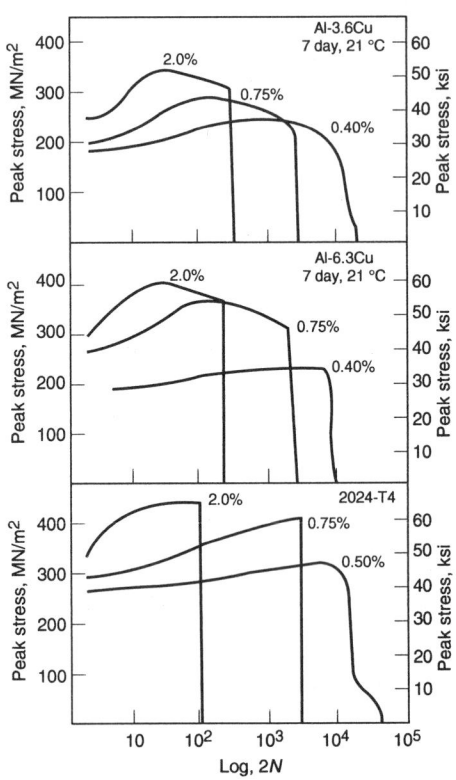

Fig. 4 Stress amplitude, $\Delta\sigma/2$, versus reversals, $2N$, at constant strain amplitude, $\Delta\varepsilon/2$, in binary aluminum-base copper alloys and in 2024 aluminum alloy aged at room temperature. The Al-3.6Cu alloy shows cyclic softening at large N. This is reduced in the Al-6.3Cu alloy at low strain amplitude. This alloy has 5 to 10 µm θ dispersoids to reduce strain localization. In alloy 2024, which has manganese to form MnAl$_6$ dispersoid particles approximately 0.5 µm in diameter, the cyclic softening is absent. The precipitous drop in $\Delta\sigma/2$ is due to cracking. Source: Ref 14

Sometimes features called extrusions and intrusions clearly form on the surface. A simple mechanism, described by Cottrell and Hull (Ref 1), for forming an intrusion-extrusion pair is illustrated in Fig. 1. Sequential slip on two intersecting slip planes is imagined to occur. In the first half-cycle, first one slip system and then the other is imagined to operate, producing indentation (Fig. 1c). Alternately, a protrusion would form if the dislocations have Burgers vectors of opposite sign. During the second half-cycle, the first slip system and then the second is imagined to operate again, giving rise to an intrusion and extrusion pair (Fig. 1e).

It is not believed that intrusions and extrusions form exactly by the Cottrell-Hull mechanism, but this mechanism serves to illustrate the type of process that must be operative. Laird and Duquette (Ref 2) pointed out that intrusion-extrusion pairs have been observed when there is only one operative slip system, and it is not clear whether intrusions and extrusions are always paired (Ref 3). Neumann (Ref 4) postulated that an intrusion or extrusion may form by a dislocation avalanche along parallel neighboring slip planes containing dislocation pileups of opposite sign.

The term "persistent slip band" was introduced by Thompson et al. (Ref 5). They examined polished surfaces of copper and nickel after various amounts of cyclic deformation and observed many slip bands. Although most were removed easily by electropolishing, some required extensive electropolishing for removal; when the samples were retested, slip bands formed again in these places. The authors called such slip bands "persistent." Persistent slip bands in copper have a ladderlike dislocation structure (Ref 3, 6, 7). Ungrouped veins or braids of dislocations occur in the surrounding matrix. The ladderlike structure has a lower flow stress than the surrounding matrix, so very extensive plastic deformation occurs in the persistent slip bands. Lukas and Klesnil (Ref 8) considered that stacking-fault energy

and the attendant ease of cross-slip would play a role in formation of the ladderlike structure, and they thus studied 15 and 31% Zn brasses. The stacking-fault energy is progressively reduced and ease of cross-slip is progressively decreased as zinc is added to copper. The ladderlike structure was found in 15% Zn brass but not in 31% Zn brass. Kuhlmann-Wilsdorf and Laird (Ref 9) have presented an extensive discussion of dislocation models for forming persistent slip bands and attendant surface roughness. Their results are shown in Fig. 2. The persistent slip band protruded slightly from the surface by 0.3 µm, but there were sharp hills and valleys of up to 5 µm in the band.

In polycrystalline specimens of copper, the ladderlike structure and the persistent slip bands are thought to form only near the surface (except in gears and railroad rails, where compressive contact stress dominates). Although the ladderlike structure does not form in all metals, even in 31% Zn brass coarse surface relief is formed from an aggregation of fine slip lines (Ref 8).

The term "persistent slip band" has probably been used somewhat loosely to mean any coarse slip band. Nevertheless, plastic deformation in slip bands produces regions of intense surface roughness, and this is quite general for all metals. Coarse slip bands or plastic strain localization occurs more readily as a result of cyclic deformation than during monotonic deformation. Localized regions of surface roughness occur because certain regions are softer or have become softer than others. For example, the soft regions may be less work hardened, or work softening may have occurred, as from unpinning of dislocations in iron.

During cyclic plastic deformation, cyclic hardening or cyclic softening may occur. Much has been written on this subject (Ref 10, 11). Annealed single-phase alloys usually cyclically harden, while cold-worked alloys cyclically soften. Alloys with coherent precipitates, such as γ' (Ni$_3$Al)-type precipitates in nickel-base alloys, show intense cyclic hardening (Fig. 3) (Ref 12). In binary aluminum-copper containing Guinier-Preston zones, intense hardening is followed by softening (Ref 13). Introducing a dispersed second phase to distribute the plastic deformation more uniformly eliminates the softening (Fig. 4) (Ref 14). Commercial aluminum-base alloys contain dispersoids for grain refinement, but these dispersoids also act to eliminate cyclic softening.

The cyclic behavior of quenched-and-tempered steels is particularly interesting. As-quenched steels cyclically harden, whereas quenched-and-tempered steels soften by as much as 40% (Ref 15). Monotonic and cyclic stress-strain curves for 4140 steel quenched and then

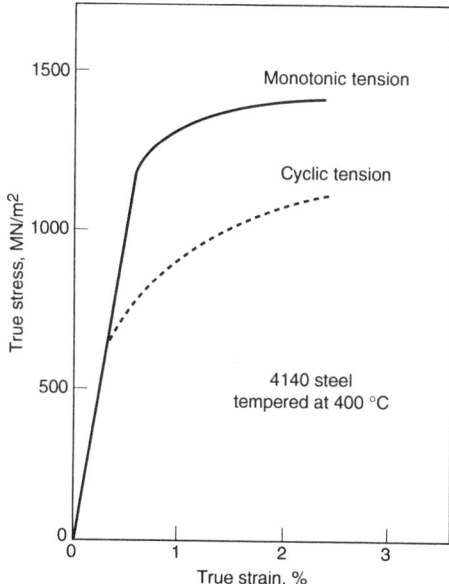

Fig. 5 Monotonic and cyclic stress-strain curves for 4140 steel tempered at 400 °C (750 °F). Large cyclic softening is observed. The cyclic stress-strain curve was obtained by the incremental strain technique. Source: Ref 15

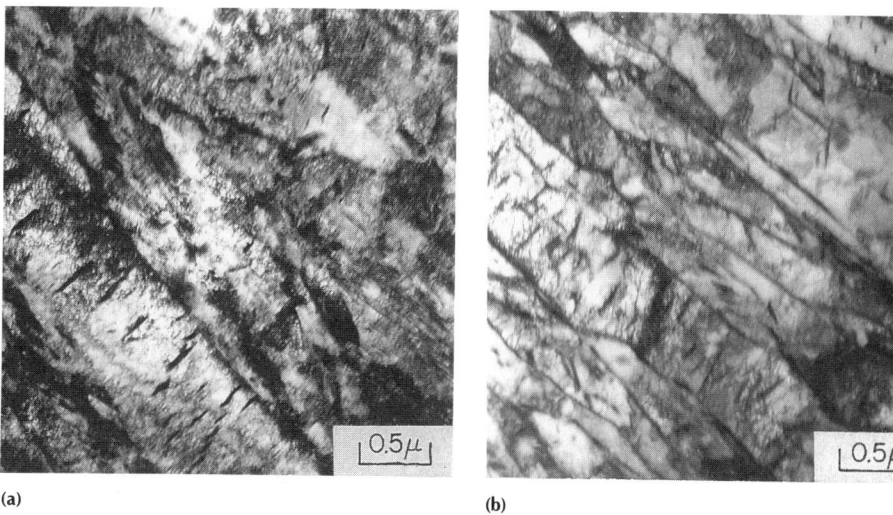

(a)   (b)

Fig. 6 Transmission electron microscopy structures of 4140 steel tempered at 400 °C (750 °F) before (a) and after (b) cycling at $\Delta\varepsilon/2 = 2.5\%$. There has been a large reduction in dislocation density. Source: Ref 15

tempered at 400 °C (750 °F) are shown in Fig. 5 (Ref 15); a large amount of softening is evident. This softening is due largely to a change in dislocation structure with cycling (Ref 15). Even though the steel has been tempered at 400 °C (750 °F), many dislocations remain from the martensite transformation. Cyclic straining reduces the dislocation density, and the remaining dislocations form cell walls. Figure 6 shows a comparison of the structure before and after tempering. Coherent precipitates result in cyclic hardening; therefore, addition of precipitates such as Ni$_3$Al to quenched-and-tempered steels counteracts cyclic softening (Fig. 7) (Ref 16).

Initiation of Microcracks

Initiation of fatigue cracks has been observed to occur along slip bands, in grain boundaries, in second-phase particles, and in inclusion or second-phase interfaces with the matrix phase. The mode of fatigue crack initiation depends on which occurs most easily. If weak, brittle precipitates are present, then they will probably play a dominant role. Slip is discontinuous across grain boundaries, and many slip systems must be active to keep the grains from pulling apart. Therefore, grain boundaries are particularly susceptible to fatigue crack initiation.

In the authors' opinion, the sequence of events and modes of initiation of fatigue cracks occur as follows. During cyclic plastic deformation, dislocations either emerge at the surface of the metal or pile up against obstacles. If the dislocations continuously emerge at the surface rather than pile-up against obstacles, then slip bands that eventually become cracks appear in the central portions of the grains, where the flow stress is

lower. As expected, resistance to slip-band initiation in the central portion of a grain decreases with increasing grain size, following the Hall-Petch relation. A slip-band crack in the central portion of a grain in 2124 aluminum alloy aged at room temperature is shown in Fig. 8 (Ref 17).

Many types of obstacles can cause dislocation pile-ups during cycling, including grain boundaries, inclusions, oxide films, and domain boundaries. Dislocation pile-ups result in an increase in elastic strain energy. When the strain energy density exceeds twice the surface free energy, a condition of instability occurs that energetically favors the initiation of microcracks. This can lead to a slip-band crack in the matrix, decohesion along a grain boundary, or cracking of a second-phase particle that may lie in the matrix or grain boundary. If the grain-boundary regions in a precipitation-hardened alloy are free of precipitates (i.e., precipitate-free zones), then plastic flow at low plastic strains may be concentrated in these regions and initiate fatigue cracks. Even if an isolated second-phase particle cracks, it must spread into the matrix for initiation of a fatigue crack to occur. Thus, not all cracked inclusions initiate fatigue cracks. All of these types of cracks have been observed.

Figure 9 depicts how paired dislocation pile-ups might give rise to an intrusion or extrusion by an avalanche (Ref 18). Here an oxide on the surface is imagined to be the obstacle, but other obstacles would have similar effects. A slip band that is thought to be cracked and that may have formed in this way in a high-strength, low-alloy (HSLA) steel is shown in Fig. 10 (Ref 19). A slip-band crack emanating from an inclusion is shown in Fig. 11 (Ref 17). It appeared very suddenly.

The mode by which inclusions aid in fatigue crack initiation, as already stated, may be for the inclusion to crack first and then extend into the matrix. As shown by Morris (Ref 20), this is the case for 2219-T851 aluminum alloy. Chang et al.

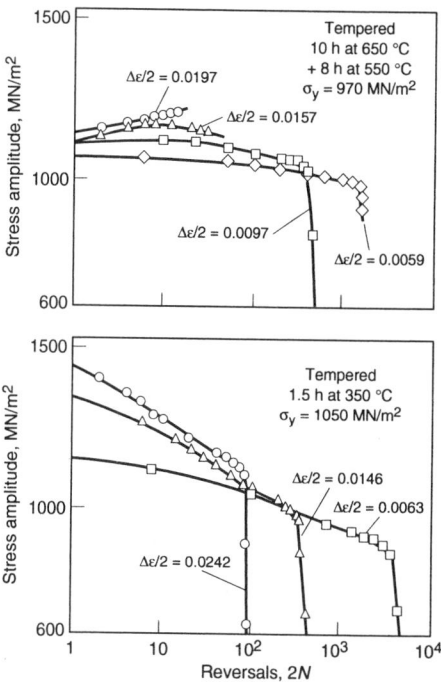

Fig. 7 Stress amplitude, $\Delta\sigma/2$, in Fe-0.3C-4Ni-1Al-1Cu steel versus reversals, $2N$, for various strain amplitudes, $\Delta\varepsilon/2$. Data for two treatments are shown. After tempering 1.5 h at 350 °C (660 °F), only Fe$_3$C is present and much cyclic softening is observed. When tempered 10 h at 650 °C (1200 °F) plus 8 h at 550 °C (1020 °F), coherent NiAl and copper precipitates are present. There is very little cyclic softening for this treatment. The cycles to fatigue crack initiation in a notched specimen were more than ten times greater for the latter treatment at a nominal stress amplitude of 196 MN/m^2. Source: Ref 16

(Ref 21) modeled this process. Particle size determines whether cracking of the particle from a dislocation pile-up or advance of the fatigue crack into the matrix is rate controlling. Advance

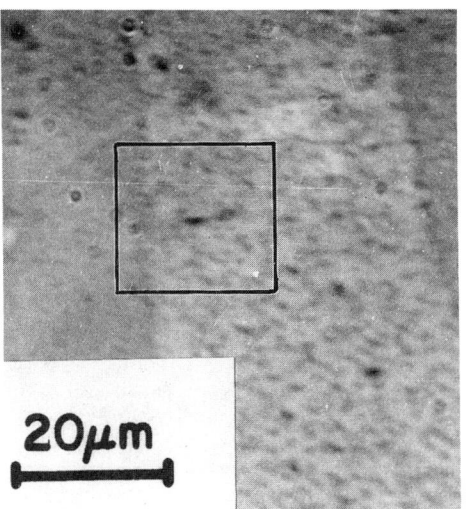

Fig. 8 Slip-band crack in central portion of grain in 2124-T4 aluminum alloy. This occurred after 240 cycles at a nominal stress of 87 MN/m² in a polished notch with an elastic stress-concentration factor of 4.4. The stress-concentration factor, taking plasticity into consideration, was 2.6. Source: Ref 17

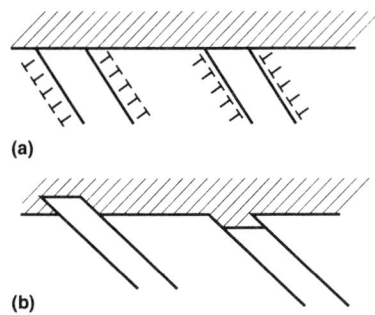

Fig. 9 Paired dislocation pile-ups (from sources not shown) against obstacle on metal surface are imagined to grow with cyclic straining until they reach a critical size. An avalanche then occurs, producing an intrusion or extrusion, depending on the dislocation sign. Source: Ref 18

of the fatigue crack into the matrix was predicted to be rate controlling for large particles.

Eid and Thomason (Ref 22) showed that in a quenched-and-tempered medium-carbon molybdenum alloy steel containing Al_2O_3 inclusions up to 25 μm in diameter, there was debonding between the particles and the matrix as well as cracking of the particles, both initiating fatigue cracks. Their stress field analysis pointed out the importance of the ratio of Young's modulus in the particle to that of the matrix. A ratio greater than 1:1, which is the case for Al_2O_3 in iron, results in large tensile stress concentrations at the polar points of the particles. Neither MnS inclusions nor Fe_3C particles, whose moduli are smaller than that of iron, were sources of fatigue cracks in the same alloy. In a high-strength aluminum alloy prepared by hot working of pressed powder compacts, Al_2O_3 particles, which have approximately ten times the modulus of the matrix, may be the origin of fatigue crack initiation in grain boundaries (Ref 23).

Inclusion size also plays a role, at least in 2024-T4 aluminum alloy. Kung and Fine (Ref 17) de-

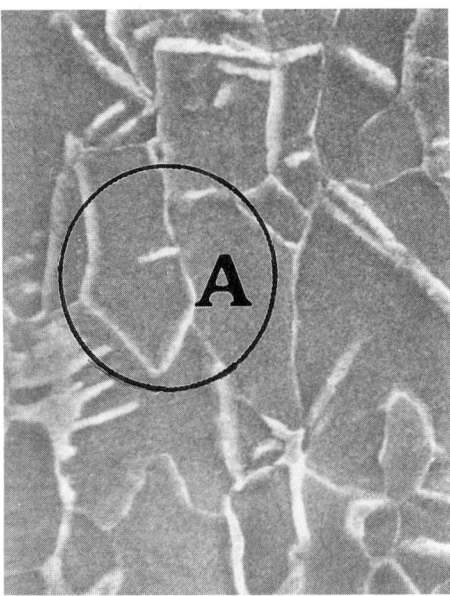

Fig. 10 Scanning electron micrograph of polished notch in Fe-0.03Nb-0.06C steel after 300,000 cycles at $\Delta K/\sqrt{\rho}$ of 850 MN/m², where ρ is the radius of curvature of the notch. Slip band in circle is thought to be already cracked. It appeared suddenly, emanating from a grain boundary. Source: Ref 19

termined that the probability for an Al_2CuMg or Al_7Cu_2Fe constituent particle to initiate a slip-band crack decreases rapidly as its size decreases below 7 μm.

Whether cyclic hardening or softening occurs probably has only a minor effect on the fatigue limit, because the plastic strain amplitude corresponding to the fatigue limit is quite small. The severity of the surface disturbance or the size of the dislocation pile-up produced by cyclic loading depends on the plastic strain amplitude in the localized region where the fatigue cracks form, which in turn depends on the yield stress. It is well known that the fatigue limit is not a definite fraction of the yield stress for all metals and alloys. Two additional factors are important: the degree of strain localization and the critical localized plastic strain to initiate the crack.

The plastic strain amplitude over the gage length, the quantity usually measured in a monotonic or cyclic stress-strain curve, may be made up taking the two extremes of a small number of severe local plastic strains or a large number of small plastic strains. The latter case is much better for fatigue resistance, which benefits from any reduction in strain localization. The chromium present in commercial 7075 aluminum alloy forms $Cr_2Mg_3Al_{18}$ dispersed particles approximately 0.5 μm in size. Hornbogen and Lütjering (Ref 24) showed that these dispersoid particles inhibit the formation of coarse slip bands and improve the fatigue limit by approximately one-third. A similar result was obtained in 2024-type alloys (Ref 25). The precipitate-free zones near grain boundaries, often present in precipitation-hardened alloys aged at elevated temperatures, are also regions of strain localization. Thermomechanical processing that produces jagged

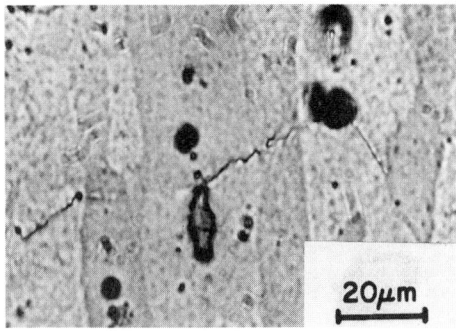

Fig. 11 Slip-band crack in polished notch emanating from Al_7Cu_2Fe inclusion in 2024-T4 aluminum alloy after 12,000 cycles. Crack appeared suddenly. Other cracks are also visible. Nominal stress was 84 MN/m². See Fig. 8 for stress-concentration factor. Source: Ref 17

grain boundaries improves fatigue resistance (Ref 26).

The local critical plastic strain to initiate a crack is related to notch sensitivity. If a material is highly notch sensitive, then only a small surface intrusion or extrusion is needed to initiate fatigue cracking. Notch sensitivity is often considered to increase with yield strength, but this is not a general rule. Other factors are also related to notch sensitivity. For example, removal of large iron- and silicon-containing inclusions in aluminum alloys improves the fatigue limit (Ref 27). It is also well known that removal of inclusions in steels improves the fatigue limit. The local plastic strain necessary for crack initiation has been little studied. This would appear to be a very worthwhile field for research, both to develop fatigue-resistant alloys and to help theoreticians model fatigue crack initiation.

Recently, Mura et al. (Ref 28, 29) adopted a free-energy approach to derive basic equations governing initiation of microcracks. They modeled the fatigue deformation process as one of injecting dislocations along slip bands impinging onto certain obstacles. The buildup of local plastic strain and hence strain energy eventually leads to energy instability that results in the initiation of microcracks. Mura et al. showed that the fatigue limit is directly proportional to the friction stress and that environmental sensitivity is related to the degree of slip reversibility.

Propagation and Coalescence of Microcracks

A fatigue crack must be a certain length before it can be observed; thus, some microcrack growth has always occurred before the measured cycles to "initiation" reported in the literature. The values will depend on the resolving power of the measuring instrument used. Mura et al. (Ref 28, 29) showed that the size of just-initiated cracks is on the order of 0.1 μm or less. Therefore, in order to track the evolution of microcracks from initiation, one must use instruments with resolution better than 0.1 μm. Recently, scanning tunneling microscopy has been used successfully to track the initiation of microcracks during fatigue defor-

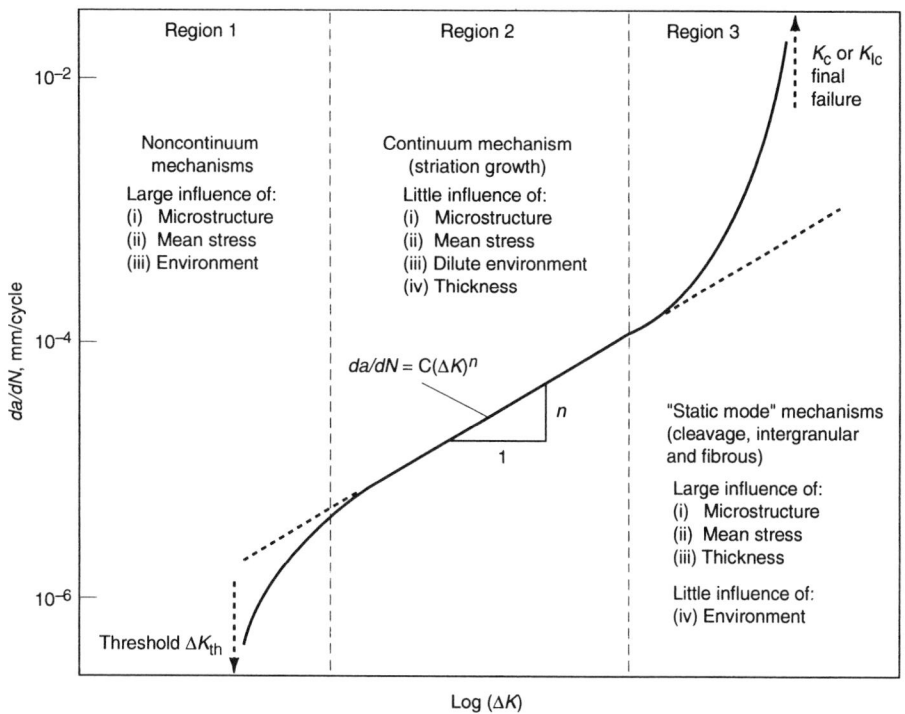

Fig. 12 Schematic plot of fatigue crack propagation rate, da/dN, versus stress-intensity range, ΔK, on a log-log scale. See text for discussion of regions I to III.

rate-controlling parameter under the condition that only small-scale yielding occurs at the tip of the crack. Here ΔK is $\Delta\sigma\sqrt{af(g)}$, where $\Delta\sigma$ is $\sigma_{max} - \sigma_{min}$, the maximum nominal stress minus the minimum nominal stress; a is the crack length; and $f(g)$ is a function of specimen geometry, loading conditions, and the ratio of crack length to specimen width, a/w. Formulas for $f(g)$ have been derived for a wide variety of conditions and are tabulated in a number of texts.

Fatigue crack propagation rates are typically plotted as log-log graphs of da/dN versus ΔK (Fig. 12). Fatigue crack growth behavior is characterized by three regimes (Fig. 12), which may include the region where the scale of yielding at the crack tip is no longer necessarily small. Full discussion of crack growth behavior and analysis is provided elsewhere in this Volume. However, brief descriptions of the three regimes are provided below:

- Region I: Threshold and near-threshold region where da/dN decreases rapidly with decrease in ΔK to a threshold value (ΔK_0 or ΔK_{th})
- Region II: Midregion where the Paris relation (Ref 34) holds:

$$da/dN = C(\Delta K)^n \text{ where } C \text{ and } n \text{ are constants} \quad \text{(Eq 1)}$$

- Region III: High-rate region where the maximum stress intensity, K_{max}, approaches the critical stress intensity for static failure, K_c

Region III will not be discussed. By the time region III has been reached, the fatigue crack is moving so rapidly that for practical purposes a failure has already occurred.

These three regions are modeled mathematically by the McEvily-Forman empirical equation (Ref 35):

$$da/dN = C'\left(\Delta K^2 - \Delta K_0^2\right)\frac{(1 + \Delta K)}{(K_c - K_{max})} \quad \text{(Eq 2)}$$

In this equation, the exponent 2 replaces n in the Paris relation. As shall be discussed later, while an n of 2 is observed in some materials, a value closer to 4 is more common (Ref 36). Equation 2 should be modified to allow for a variable exponent on ΔK.

Threshold and Near-Threshold Growth of Macrocracks. Considering region I first, macrocracks will not grow at all values of ΔK; a threshold stress-intensity range, ΔK_0, is thought to exist below which the crack does not propagate. The rate da/dN increases rapidly as ΔK increases from ΔK_0, merging into the Paris relation region. The values of ΔK_0 and the near-threshold behavior established for cracks a few millimeters or more in length do not hold for shorter cracks. They may propagate at ΔK values smaller than ΔK_0.

Crack closure is another factor that influences near-threshold behavior. The concept of crack closure was first put forth by Elber (Ref 37).

mation of silver single crystals (Ref 30). In order to minimize tip artifacts, the tip must be sharp with a small opening angle. Also, imaging conditions must be chosen to minimize tip-induced surface damage. Atomic force microscopy can also be used, with the same precautions.

As a consequence of stress or plastic strain cycling above the fatigue limit, the initial microcracks grow or coalesce until a macrocrack may be reasonably defined as larger than a number of grain cross sections in an area, with smaller cracks being defined as microcracks. In single crystals, a 500 μm crack diameter may be a reasonable dividing line.

It is not certain to what extent microcracks form in materials cycled below the fatigue limit and then stop growing. This is another subject that requires study. Microcracks that have formed may not grow further for a number of reasons. For example, a fatigue crack may initiate within an inclusion, but could be stopped by the interface because, as suggested by Chang et al. (Ref 21), the stress required to extend the crack into the matrix is larger than that required to initiate the crack in the inclusion. Fatigue cracks that are isolated in inclusions are often observed, for example, in Al_7Cu_2Fe inclusions in aluminum alloys (Ref 17). In two-phase alloys or composites, microcracks may form in a noncontinuous phase but be unable to propagate into the continuous phase. The stress-concentration factor falls off with distance from a notch, hole, scratch, pit, or other flaw. Thus, a crack that forms near or at flaws may stop growing at a nominal stress amplitude just below the fatigue limit. It is now well known that microcrack growth is impeded by grain boundaries (Ref 17, 31).

Based on reviews beginning with Ewing and Humfrey (Ref 32), as well as research at Northwestern University, the number of microcracks that form during fatigue depends on stress or plastic strain amplitude. At high amplitudes, many cracks form, and coalescence across grain boundaries is the dominant mode of microcrack growth. At low stresses, growth of individual microcracks is more important. At the endurance limit, the macrocrack is expected to form most often from a single source. The basic mechanism for micro- or macrofatigue crack propagation in metals is by plastic deformation, except perhaps when there is very extreme grain-boundary embrittlement. A small crack often propagates initially along a crystallographic plane (Forsythe's stage I) (Ref 33). The resistance to microcrack propagation offered by grain boundaries arises because plastic deformation is discontinuous across grain boundaries. In a precipitation-hardened high-strength aluminum alloy made by forging or extruding a pressure-sintered powdered compact, the cracks initiate in the grain boundary, but the propagation mode quickly changes to transgranular (Ref 23).

A significant fraction of the fatigue lifetime is taken up by propagation of microfatigue cracks. Building microcrack stoppers into the microstructure is feasible and should be investigated more fully.

Growth of Macrocracks

Once a macrocrack a few millimeters or more in length has formed, the stress-intensity range, ΔK, of linear fracture mechanics becomes the

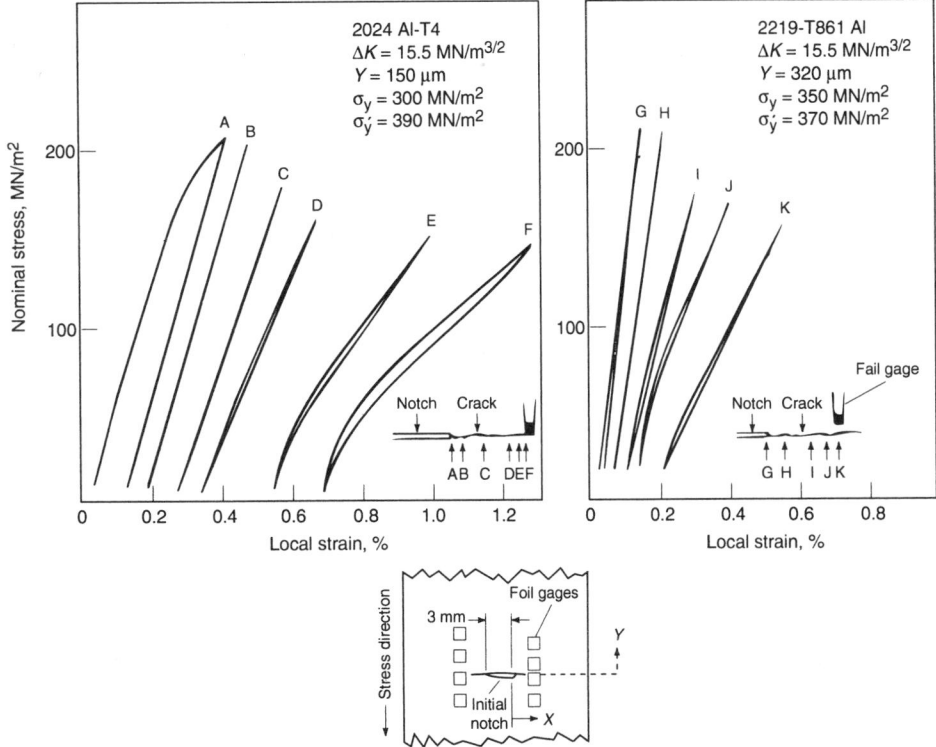

Fig. 13 Typical nominal stress/local strain curves versus distances from crack tip determined using 200 μm foil strain gages. The distances A-F and G-K are approximately 2 mm. Center notch is 3 mm long and 0.2 mm wide. R was 0.05. Source: Ref 36

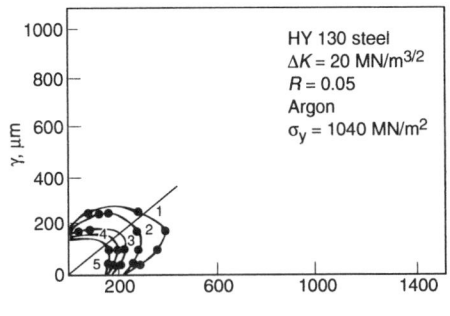

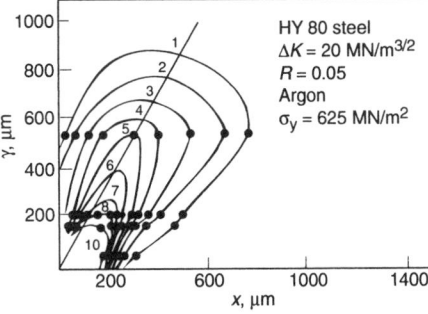

Fig. 14 Contour maps of plastic zones at ΔK of 20 MPa√m in pressure vessel steels HY 80 (0.16% C, 3% Ni, 1.5% Cr, and 0.4% Mo, tempered 1 h at 700 °C, or 1290 °F) and HY 130 (0.10% C, 3% Ni, 0.5% Cr, 0.45% Mo, and 0.10% V, tempered 1 h at 600 °C, or 1110 °F). The contour lines are lines of equal local plastic work, U_{XY}, in units of 10^{11} J/m^4 as indicated by the numbers. Source: Ref 48

When ΔK is relatively large, plastic stretching of the material at the crack tip imposes a compressive stress on the reverse loading, causing the crack to close prior to the tensile stress becoming zero. Elber defined an effective stress-intensity range, $\Delta K_{eff} = \Delta K - \Delta K_{closure}$. Near-threshold plastic stretching is not thought to be important, because the maximum stress intensity is small; however, there are other sources of crack closure, such as oxide bridging of the crack and mixed shear-tensile mode (Ref 38). If the crack propagates partly in a shear mode and partly in a tensile mode, the crack surface is rough and the rough surfaces on each side of the crack face become displaced to each other as the crack propagates. Therefore, the surfaces in the wake touch before the minimum or zero stress is reached. This has been quantitatively related to $\Delta K_{closure}$ (Ref 39, 40). Also, the increase in ΔK_0 with grain size has been shown to be almost completely due to closure effects; the variation of $\Delta K_{0,eff}$ is either very small or nonexistent (Ref 41). Further, the crack closure stress depends on specimen geometry being larger for a side notch specimen than for a center notch specimen, due to the larger bending moment at the crack tip in the former (Ref 42). The bending moment is even larger for a compact tension specimen.

Short cracks (i.e., less than 1 to 2 mm long) are commonly observed to propagate differently than long cracks. This is at least partly due to crack closure increasing with crack length. For 100 to 1000 μm cracks in VAN 80 steel loaded above ΔK_0, a large variation from long crack data was observed. However, when da/dN was plotted versus ΔK_{eff} in the 3 to 10 MPa√m range, the data for many specimens fell into a small scatterband with the long crack data (Ref 43).

Short cracks are also observed to propagate at ΔK values below ΔK_0. Of course, when the cracks are in the microcrack range, the ΔK values at a given stress range are very small, yet microcracks grow sporadically, as already discussed. Thus, fracture-mechanics-based estimates are not valid. From an engineering point of view, the fatigue process can be divided into an "initiation" stage, which continues to first detection of a crack by nondestructive inspection, and a crack propagation stage. The practical detectable crack size limit is usually about 1 mm, and this takes up most of the lifetime. Therefore, the traditional S-N_f curve or the newer $\Delta\varepsilon_p$ versus N_f curve is still often the most practical design basis, provided that the statistics of the variation in N_f (the cycles to failure) as a function of S (the applied stress range) are known. Corrections for mean stress effects are, of course, routinely made.

Paris Relation Region of Fatigue Crack Propagation Rate. Many theoretical studies have resulted in equations of the form of the Paris relation (Eq 1). Based on the concept that plastic work must be provided to advance the crack, the following can be derived (Ref 44):

$$\frac{da}{dN} = A\frac{(\Delta K)^n}{\mu\sigma_y^2 U} \tag{Eq 3}$$

where A and n are dimensionless constants, μ is the shear modulus, σ is an appropriate yield stress, and U is the plastic work required to advance the crack by a unit area.

The Weertman (Ref 45) general theory predicts n values of 2 or 4 as two limiting approximations. For $n < 4$, U was predicted to be a function of ΔK such that:

$$U = \beta(\Delta K)^{4-n} \tag{Eq 4}$$

where β is the proportionality constant. When n is 4, U is independent of ΔK, but when n is 2, U is proportional to ΔK^2.

Experimentally determined values of n range from 2 to 8. Fatigue crack propagation rate data compiled by Frost et al. (Ref 46) for 14 alloys gave an average value of n of 3.94 with a standard deviation of ±0.86. For eight alloys (four aluminum alloys, three steels, and one nickel alloy), Izumi et al. (Ref 36, 47) observed $n = 3.7 \pm 0.5$. There are alloys, however, where n is close to 2 (Ref 35) that were not included in these compilations. Values of n greater than 4 may be due to inclusion of a portion of regions I or III in the data range analyzed.

In order to test Eq 3, Ikeda et al. (Ref 47) devised a method for measuring U. Foil strain gages, 200 by 210 μm, were strategically cemented ahead of the fatigue crack. Holding ΔK constant by reducing $\Delta\sigma$ as the crack lengthened, the nominal stress/local strain curves were measured in these gages as the fatigue crack approached. Figure 13 shows typical results, where

Table 1 Compilation of measured values of the plastic work of fatigue crack propagation, U, for various alloys and calculated values of A in Eq 3

Metal(a)	σ_y (0.2%), MPa	ΔK, MPa$\sqrt{m}$	U, J/m² × 10⁻⁵	da/dN, n/cycle × 10⁻⁸	n	A × 10³	Ref
Steels							
0.05C annealed	170	8.0	53	15	4.6	4.3	49
Nb-HSLA hot rolled	340	12.4	12	0.8	3.5	3.5	47
		15.5	8	1.7	...	2.1	...
		19.5	12	3.7	...	2.7	...
Nb-HSLA 400 °C temper	600	20	30	0.6	3.9	3.2	52
Nb-HSLA 550 °C temper	688	20	7.6	1.5	4.1	2.6	52
HY 80	521	20	7.7	3.5	3.4	3.6	48
HY 130	868	20	2.2	5.0	3.6	4.0	48
1.4 Cu-0.28C, 13 min, 500 °C	710	19.5	3	1.8	3.5	1.6	53
1.4 Cu-0.45C, 200 min, 500 °C	780	19.5	1.3	5.6	3.0	2.3	53
Nickel alloys							
Ni-7.2Al, 2 days, 625 °C	670	15.5	4.8	2.0	4.5	6.3	36
Aluminum alloys							
99.9+ cold rolled	42	2.5	18	0.16	4.0	3.3	54
1100 annealed	49	2.8	12	0.29	4.2	3.5	54
2219 T861	370	7.8	2.4	1.5	4.0	2.5	36
		15.5	1.6	25	...	3.5	...
2219 overaged	260	9.3	2.1	6.8	4.0	3.4	36
		15.5	1.4	32	...	1.4	...
2024-T4	390	7.8	2.6	1.6	3.0	4.4	36
		15.5	3.2	14.0	...	3.1	...
7050-T4	410	12.4	0.5	30	3.5	2.8	47
7050-T76	510	15.5	0.6	30	4.0	2.3	36
6.3Cu-T4	230	9.3	10.5	1.2	...	2.3	36
		10.8	6.1	2.0	4.0	1.2	...
		12.4	5.8	4.7	...	1.6	...
Mean ± standard deviation					3.8±0.5	2.9±1.1	

(a) Compositions given in weight percent

X is defined as the distance from the center of the gage to the crack tip along the crack plane and Y is defined as the distance from the center of the gage to the crack plane. A series of nominal stress/local strain curves of decreasing X for two different Y values are shown. The initial curves A and G do not close; however, the subsequent curves close to a first approximation, forming loops. The nonclosure represents permanent strain, which is negligibly small after the first cycle. The curves have been arbitrarily shifted to the right on the strain axis to set them apart. When the gages are far from the crack tip, the hysteresis loop width is too small to be shown on the scale of the figure (curves B and H). Curve D shows slight hysteresis, which increases with decrease of X (loops E, F, I, and J). When the crack enters the gage, it is broken. When Y is large, the crack bypasses the gage (e.g., loop K).

For determination of U, local stress rather than nominal stress is needed. This was determined from the stress-strain hysteresis loops for unnotched specimens using the strain amplitudes determined with the foil strain gages.

The plastic work U is made up of two parts, hysteretic U_h and permanent U_p. The latter was measured in several alloys and found to be less than 1% of U. Since the major part of U is hysteretic, the areas of the loops are of primary interest.

From the local stress/local strain hysteresis loops, the local plastic work per unit area of crack advance for the coordinates in the plastic zones X and Y, U_{XY}, was determined:

$$U_{XY} = \frac{\int_{\epsilon_1}^{\epsilon_2} \sigma_u d\epsilon - \int_{\epsilon_1}^{\epsilon_2} \sigma_l d\epsilon}{da/dN} \qquad \text{(Eq 5)}$$

where σ_u and σ_l refer to local stresses in the upper and lower curves, respectively, of the hysteresis loop. The experiment is done at constant ΔK, so according to fracture mechanics a constant plastic zone advances with the crack. Thus, a contour map of U_{XY} in the plastic zone can be drawn, as shown in Fig. 14 for HY 80 and HY 130 pressure vessel steels (Ref 48). The plastic zones are butterfly-wing shaped, in keeping with the maximum shear stress being 45° from the loading direction and the crack plane. The plastic zone size increases with decreasing yield stress. The plastic work was obtained by integrating over the plastic zone:

$$U = \iint U_{XY} dX dY \qquad \text{(Eq 6)}$$

In order to obtain U_{XY} closer than 100 μm to the crack tip (Ref 49), U_{XY} for a constant small Y was extrapolated to small X on a log-log scale and, following the solution proposed by Rice (Ref 50) for a mode III crack (the present case is mode I, i.e., tensile stress normal to crack face), the contour lines were assumed to be circles touching the crack tip and symmetrical to the crack tip. Using this method

of extrapolation, U is finite because X approaches zero faster than U_{XY} approaches infinity. The contribution to U from closer than 1 μm to the crack tip is negligible.

Davidson and Lankford (Ref 51) devised a different method for measuring U utilizing electron channeling to measure the cell size in 0.05% C steel versus distance from the crack. The hysteresis loop areas were then obtained from a calibration of cell size versus loop area determined on unnotched specimens. Liaw et al. (Ref 49) compared the contributions to U from closer than 100 μm to the crack tip in the same low-carbon steel as determined by both techniques. The agreement was very good, with the two values in agreement to within 15%.

The results of measurements of U for alloys with n ranging from 3 to 4.6 are given in Table 1 along with other pertinent data and values of A. This set of data is discussed in another paper by Liaw et al. (Ref 55). The mean value of the "constant" A in Eq 3 is 2.9×10^{-3}, with a standard deviation of $\pm 1.1 \times 10^{-3}$. The theoretical treatment of Mura and Villman (Ref 56) obtained 2×10^{-3} for A, which is close to the mean experimental value. In Table 1, yield stress varies from 42 to 868 MPa, shear modulus (μ) varies from 26 to 84 GPa, da/dN varies from 0.16×10^{-8} to 32×10^{-8} m/cycle, ΔK varies from 2.5 to 20 MPa$\sqrt{m}$, and U varies from 0.5 to 53×10^{-5} J/m². When the individual data are inserted into Eq 3, a "constant" A emerges with a relatively small standard deviation. It should, of course, be recognized that data are included for alloys ranging in n from 3 to 4.6, that the 0.2% offset cyclic yield stress may not strictly be the correct yield stress for Eq 3, and that there are errors in measurement of U (Ref 36, 47). The latter two factors lead to errors in A.

Dependence of measured U on ΔK will be considered next. As already mentioned, when n is 4, U is expected to be independent of ΔK. This was investigated in a steel (Ref 47) and several aluminum alloys (Ref 36), as shown in Table 1. The variation observed is considered to be within experimental error. When n is less than 4, ΔK is theoretically expected to be a function of U, with Eq 3 still holding. Noteworthy among n values near 2 reported in the literature is the analysis of martensitic steels by Barsom (Ref 57), where $n = 2.25$ fit a large set of data.

Recently, Kwun and Fine (Ref 48) observed n of 2.3 in MA 87 powder metallurgy aluminum-base alloy containing 6.5 wt% Zn, 2.5 wt% Mg, 1.5 wt% Cu, and 0.4 wt% Co aged 25 h at 120 °C (250 °F) followed by 4 h at 163 °C (325 °F). It was decided to measure U at two different values of ΔK, 10 and 17 MPa$\sqrt{m}$, to check Eq 4. The resulting values of U were 1.7 and 4.3×10^4 J/m^2. Substituting into Eq 3 gave A values of 2.2 and 2.3×10^{-3}, which are close to the mean value of Table 1. Taking U proportional to $(\Delta K)^{1.7}$, the predicted ratio of U for the two ΔK values is 2.5, which is the observed ratio.

It has thus been proved that the plastic work per unit area of fatigue crack propagation, U, is an important material parameter, along with yield stress, for controlling the rate of fatigue crack propagation. At a constant ΔK, da/dN varies inversely as $\mu U(\sigma_y)^2$. For alloys of the same major

component, μ varies little with alloying, and the product $U(\sigma_y)^2$ is controlling.

Many experimental studies have observed that increasing yield strength in an alloy system, such as by changing the tempering temperature in steels, changes da/dN at constant ΔK relatively little. This can be understood by U and σ_y changing in opposite directions. Because U depends in part on the size of the plastic zone, decrease in U with increase in alloy strength is not unexpected, and most of the data in Table 1 bear this out. Of all the alloys studied, annealed 0.05 wt% C steel has the highest U, but $\partial a/\partial N$ (at constant ΔK) is high because σ_y is low. A similar conclusion is reached for 99.99+ aluminum. Fortunately, there are exceptions to the rule that U is inversely related to $(\sigma_y)^2$. The niobium-alloyed steel when quenched and tempered at 400 °C (750 °F) has a very high U for its yield strength and a correspondingly low fatigue crack propagation rate. The binary Al-6.3Cu alloy (aged at room temperature to form Guinier-Preston zones) also has a high value of U for aluminum alloys.

Preventing strain localization is certainly important for increasing U. If the plastic deformation is localized in the plastic zone, then small U is expected. The Al-6.3Cu alloy has excess copper, which forms 5 to 10 μm spherical CuAl$_2$ particles dispersed uniformly in the matrix. These prevent strain localization, which is characteristic of an alloy containing only Guinier-Preston zones.

Even enhancement of the fatigue crack propagation rate by a corrosive environment can be understood by its effect on U. Davidson and Lankford (Ref 51) measured fatigue crack propagation rate and U in dry and moist air. The faster rate in moist air was accompanied by a reduction in U.

Appendix: Scanning Probe Microscopy of Fatigue

Scanning probe microscopy refers to a class of surface diagnostic techniques that operate by scanning a fine probe across a specimen surface. The first such instrument is the topographiner developed by Young in the early 70s (Ref 58), followed by the now well-known scanning tunneling microscope. The scanning tunneling microscope (STM) was invented by Rohrer and Binnig of IBM's Zurich Research Laboratory in Switzerland in 1982 (Ref 59). One of the most interesting aspects of this new microscopy technique is its ability to perform high resolution imaging of surfaces over hundreds of microns, and that such high resolution is achieved in vacuum, air and liquid environments, thus making this a convenient technique for use in various branches of physical and biological sciences, including the study of fatigue-induced surface deformation.

Principle of STM Imaging

Consider a sharp conducting tip brought to within one nm of a specimen surface (see Fig. 15). Typically, a bias of 0.01-1 volt is applied between the tip and the specimen. Under these conditions, the tip-surface spacing (s) is sufficiently small that electrons can tunnel from the tip to the specimen. As a result, a current (i) flows across this gap which can be shown to vary with s as follows:

$$i \propto \exp(-10.25\sqrt{\varphi}\, s)$$

where φ is the effective work function in eV (3-4 eV for most systems) and s is in nm. One can see that if the tip-surface spacing is increased (decreased) by 0.1 nm, the tunneling current will decrease (in-

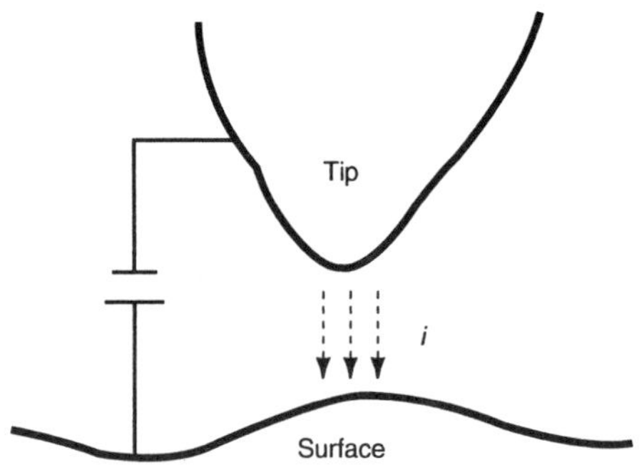

Fig. 15 Electron tunneling from probe tip to specimen surface

crease) by about a factor ten for an effective work function of 4 eV.

One can then exploit this sensitive dependence of the tunneling current i on the tip-surface spacing for topographic imaging as follows. In scanning the tip horizontally across the specimen, any change in the tip-surface spacing results in a large change in the tunneling current i. One can use some feedback mechanism to move the tip up or down to maintain a constant tunneling current. According to the above equation, this implies that one is maintaining a constant tip-surface spacing (assuming constant φ). In other words, the up-and-down motion of the tip traces out the topography of the surface, analogous to the conventional technique of stylus profilometry, except that the tip never touches the surface in STM. This is known as constant current imaging, the most common imaging mode used in scanning tunneling microscopy.

Because of the proximity of the tip to the surface and the nature of tunneling, the tunneling electron beam diameter can be very small. For a tunneling junction with work function of 4 eV, the full-width at half-maximum is approximately equal to $0.2\sqrt{z}$ (Ref 60), where z is the sum of R, the local radius of curvature of the tip and s, the tip-surface spacing in nm. For example, for $R = 0.2$ nm, and $s = 0.5$ nm, the electron beam diameter is on the order of 0.2 nm. This implies that the tunneling current is self-focused into a region with atomic dimensions. Scanning tunneling microscopy has been demonstrated to yield atomic resolution in many cases.

Experimental Aspects

Coarse Motion Control. In order to bring the tip to within tunneling range, one must move the tip over macroscopic distances (hundreds of microns) with precision ~100 nm. Two schemes are commonly used. One involves the use of piezo-electric inchworms. Another one is based on purely mechanical means. For example, consider a conventional 80-pitch screw, i.e., the screw advances by one inch (2.54 cm) after 80 turns. This translates into a motion advance of about 880 nm for one degree of screw rotation. Using a cantilever beam with mechanical advantage of 10, one can achieve a precision of 88 nm for one degree of screw rotation.

Fine Motion Control. The precision required for tip positioning relative to the specimen surface during image acquisition has to be better than 0.1 nm. This is achieved by piezoelectric positioners. Piezoelectric materials expand or contract upon the application of an electric field. Lead zirconium titanate (PZT) is the material of choice in the STM community. Most STMs are designed with response ranging from 1-300 nm/volt. Since voltages can be controlled and monitored in the submillivolt level easily, sub-nanometer control can be readily attained.

Two types of STM scanners are being used. In one design, three-axis scanning is accomplished using three separate pieces of piezoelectric bars held together in an orthogonal arrangement. In another design, a single piezoelectric tube with four separately biased quadrants provides the capability of three-axis scanning (Ref 61). The major advantage of the tube scanner is its improved rigidity.

Tip Preparation. Two tip materials are widely used, viz. tungsten and platinum alloys (e.g., Pt-Ir and Pt-Rh). Tungsten is strong and can be fabricated into a sharp tip easily, but it tends to oxidize rapidly in air. On the other hand, Pt alloys are stable in air, but they may not survive occasional tip crashes on surfaces. Several methods can be used to create sharp tips of these materials. These include electropolishing, cutting and grinding, momentary application of a high bias voltage (a few volts), or simply waiting for a few minutes after achieving tunneling. In order to image rough surfaces with minimal distortion as one normally encounters in fatigue studies, sharp tips with large aspect ratios should be used. For further details, see Ref 62.

Vibration Isolation. Most STMs are supported using damped springs, air tables or stacked stainless steel plates separated by viton dampers. The goal in all these designs is to keep the tip-surface spacing immune to external vibration. The general design is that one should support the STM on a soft platform and design a microscope with high rigidity.

Data Acquisition and Analysis. In a typical experiment, a bias of 0.01-1 volt is applied between the tip and the specimen. The tunneling current so obtained is then compared with a preset value (typically 1-10 nA). The error signal then drives a feedback circuit whose output is used to control a fast high voltage operational amplifier which feeds voltage to the Z electrode of the scanner (which moves the tip perpendicular to the surface). At the same time, raster-scanning is accomplished by using two computer-controlled digital-to-analog converters to control the output of two high voltage operational amplifiers feeding voltages to the X and Y electrodes of the scanner. At each step, the Z voltage required to maintain a constant tunneling current is read by the computer via an analog-to-digital converter (either through AC coupling or potential dividers). This Z voltage, as discussed earlier, corresponds to the surface height at the XY location.

This information can then be displayed in real time as gray level images on a video monitor or stored as two-dimensional integer arrays which can later be retrieved for further processing.

Atomic Force Microscopy

In scanning tunneling microscopy, the tip-surface spacing is sensed by the tunneling current. This sensing technique does not work well for highly resistive materials. An alternative scheme to sense the tip-surface spacing is by measuring the force of interaction between the tip and the specimen surface (Ref 63). This is the basis of atomic force microscopy. The tip is normally part of a small wire or microfabricated cantilever. The force of interaction between the tip and the surface results in a deflection of the cantilever. In most modern designs, the cantilever deflection is sensed either by detecting the reflection of a light beam from the back of the cantilever or by optical interferometry. Other operational aspects of AFM (i.e., motion control, vibration isolation, etc.) are identical to STM. By operating under small loads (~nN), most surfaces can be imaged with high resolution without damage. One important strength of AFM is its ability to obtain images from insulator surfaces. In addition, one can modify an atomic force microscope to study friction and surface mechanical properties on the nanometer scale. For further details, see Ref 63.

Limitations of Scanning Probe Microscopy

One limitation of scanning probe microscopy is the small range of piezoelectric scanners (~100 microns). This limitation is solved in most modern designs by incorporation of an optical microscope, which serves to search for the region of interest. The probe can then be positioned over this region for a close-up view. Another limitation is tip geometry. With the exception of atomically smooth surfaces, images obtained from all other surfaces are due to the convolution of the actual surface topography and the tip geometry, an important consideration in applying scanning probe techniques in fatigue studies (Ref 64). Unfortunately, mathematical deconvolution cannot recover the original topography even if the tip geometry is exactly known. The only practical solution is to image with the sharpest possible tip with large aspect ratio.

REFERENCES

1. A.H. Cottrell and D. Hull, *Proc. R. Soc. (London) A,* Vol A242, 1957, p 211
2. C. Laird and D.J. Duquette, Mechanisms of Fatigue Crack Nucleation, *Corrosion Fatigue,* National Association of Corrosion Engineers, p 88
3. P. Lukas, M. Klesnil, and J. Krejci, *Phys. Status Solidi,* Vol 27, 1968, p 545
4. P. Neumann, *Acta Metall.,* Vol 17, 1969, p 1219
5. N. Thompson, N.J. Wadsworth, and N. Louat, *Philos. Mag.,* Vol 1, 1956, p 113
6. P.J. Woods, *Philos. Mag.,* Vol 28, 1973, p 155
7. J.M. Finney and C. Laird, *Philos. Mag.,* Vol 31, 1975, p 339
8. P. Lukas and M. Klesnil, *Phys. Status Solidi,* Vol 37, 1970, p 833
9. D. Kuhlmann-Wilsdorf and C. Laird, *Mater. Sci. Eng.,* Vol 27, 1977, p 137
10. C. Laird, General Cyclic Stress Strain Response in Al Alloys, STP 637, ASTM, 1977, p 31
11. M.E. Fine and S.P. Bhat, Cyclic Hardening of Alloys with Coherent and Incoherent Dispersed Phases, *Proc. 4th Risø Int. Symp. Metallurgy and Materials Science, Deformation of Multi-Phase and Particle Containing Materials,* Risø National Laboratory, Roskilde, Denmark, 1983, p 151-256
12. D. L. Anton and M.E. Fine, *Metall. Trans. A,* Vol 13A, 1982, p 1187
13. C. Calabrese and C. Laird, *Mater. Sci. Eng.,* Vol 13, 1974, p 141; *Mater. Sci. Eng.,* Vol 13, 1974, p 159; *Metall. Trans.,* Vol 5, 1974, p 1785
14. M.E. Fine and J.S. Santner, *Scr. Metall.,* Vol 9, 1975, p 1239
15. P.N. Thielen, M.E. Fine, and R.A. Fournelle, *Acta Metall.,* Vol 24, 1976, p 1
16. R.A. Fournelle, E.A. Grey, and M.E. Fine, *Metall. Trans. A,* Vol 7A, 1976, p 669
17. C.Y. Kung and M.E. Fine, *Metall. Trans. A,* Vol 10A, 1979, p 603
18. M.E. Fine and R.O. Ritchie, in *Fatigue and Microstructure,* American Society for Metals, 1979, p 24
19. Y.H. Kim and, M.E. Fine, *Metall. Trans. A,* Vol 13A, 1982, p 59
20. W.L. Morris, *Metall. Trans. A,* Vol 11A, 1980, p 1117
21. R. Chang, W.L. Morris, and O. Buck, *Scr. Metall.,* Vol 13, 1979, p 191
22. N.M.A. Eid and P.F. Thomason, *Acta Metall.,* Vol 27, 1979, p 1239
23. S. Hirose and M.E. Fine, *Metall. Trans. A,* Vol 14A, 1983, p 1189
24. E. Hornbogen and G. Lütjering, *Proc. 6th Int. Conf. Light Metals,* Aluminum-Verlag, Düsseldorf, 1975, p 40
25. G. Lütjering, H. Döker, and D. Munz, *Proc. 3rd Int. Conf. Strength of Metals and Alloys,* Cambridge, U.K., 1973, p 427
26. E.A. Starke, Jr. and G. Lütjering, *Fatigue and Microstructure,* American Society for Metals, 1979, p 205
27. W.H. Reimann and A.W. Brisbane, *Eng. Fract. Mech.,* Vol 5, 1973, p 67
28. G. Venkataraman, Y. Nakasone, Y.W. Chung, and T. Mura, *Acta Metall.,* Vol 38, 1990, p 31
29. G. Venkataraman, Y.W. Chung, and T. Mura, *Acta Metall.,* Vol 39, 1991, p 2621; Vol 39, 1991, p 2631
30. T.S. Sriram, C.M. Ke, and Y.W. Chung, *Acta Metall.,* Vol 41, 1993, p 2515
31. W.L. Morris, *Metall. Trans. A,* Vol 10A, 1979, p 5
32. J.A. Ewing and J.C.W. Humfrey, *Philos. Trans. R. Soc. (London) A,* Vol 200, p 241

33. P.J.E. Forsythe, *The Physical Basis of Metal Fatigue*, Blackie, 1969

34. P.C. Paris, in *Proc. 10th Sagamore Army Materials Research Conf.*, Syracuse University Press, 1964, p 107

35. A.J. McEvily, in *The Microstructure and Design of Alloys*, Vol 2, *Proc. 3rd Int. Conf. Strength of Metals and Alloys*, Cambridge, U.K., 1973

36. Y. Izumi and M.E. Fine, *Eng. Fract. Mech.*, Vol 11, 1979, p 791

37. W. Elber, in STP 486, ASTM, 1971, p 230

38. S. Suresh and R.O. Ritchie, *Metall. Trans. A*, Vol 13A, 1982, p 1627

39. D.H. Park and M.E. Fine, *Fatigue Crack Growth Threshold Concepts*, D.L. Davidson and S. Suresh, Ed., TMS-AIME, 1984, p 145

40. J.L. Norng and M.E. Fine, *Scr. Metall.*, Vol 17, 1983, p 1427

41. G.M. Lin and M.E. Fine, *Scr. Metall.*, Vol 16, 1982, p 1249

42. M.E. Fine, J.L. Horng, and D.H. Park, *Proc. 2nd Int. Conf. Fatigue and Fatigue Thresholds*, C.J. Beevers, Ed., University of Birmingham, 1984, p 73

43. F. Heubaum and M.E. Fine, *Scr. Metall.*, Vol 18, 1984, p 1235

44. M.E. Fine and D.L. Davidson, in *Fatigue Mechanisms: Advances in Quantitative Measurement of Physical Damage*, STP 811, J. Lankford, D.L. Davidson, W.L. Morris, and R.P. Wei, Ed., ASTM, 1983, p 350

45. J. Weertman, in *Fatigue and Microstructure*, American Society for Metals, 1979, p 279

46. N.E. Frost, K.J. Marsh, and L.P. Pook, *Metal Fatigue*, Clarendon Press, 1974, p 212

47. S. Ikeda, Y. Izumi, and M.E. Fine, *Eng. Fract. Mech.*, Vol 9, 1977, p 123

48. S.I. Kwun and M.E. Fine, *Scr. Metall.*, Vol 14, 1980, p 155

49. P.K. Liaw, M.E. Fine, and D.L. Davidson, *Fatigue of Engineering Materials and Structures*, Vol 3, 1980, p 59

50. J.R. Rice, in STP 415, ASTM, 1967, p 247

51. D.L. Davidson and J. Lankford, in *Environment Sensitive Fracture of Engineering Materials*, Z.A. Foroulis, Ed., TMS-AIME, 1979, p 581

52. S.I. Kwun and R.A. Fournelle, *Metall. Trans. A*, Vol 11A, 1980, p 1429

53. S. Ideda, T. Sakai, and M.E. Fine, *J. Mater. Sci.*, Vol 12, 1977, p 675

54. P.K. Liaw and M.E. Fine, *Metall. Trans. A*, Vol 12A, 1981, p 1927

55. P.K. Liaw, S.I. Kwun, and M.E. Fine, *Metall. Trans. A*, Vol 12A, 1981, p 49

56. T. Mura and C. Villman, Paper No. 79-WA-APM-26, American Society of Mechanical Engineers, 1979; C. Villman, Ph.D. thesis, Northwestern University, 1979

57. N.M. Newmark and W.J. Hall, Ed., Prentice-Hall, 1977

58. R. Young, J. Ward, and F. Scire, *Rev. Sci. Instrum.*, Vol 43, 1972, p 999

59. G. Binnig, H. Rohrer, Ch. Gerber and E. Weibel, *Phys. Rev. Lett.*, Vol 49, 1982, p 57

60. J. Tersoff and D.R. Hamann, *Phys. Rev.*, Vol B31, 1985, p 2

61. G. Binnig and D.P.E. Smith, *Rev. Sci. Instrum.*, Vol 57, 1986, p 1688

62. A.J. Melmed, *J. Vac. Sci. Technol.*, Vol B9 (No. 2, Part II), 1991, p 601

63. D. Rugar and P. Hansma, *Physics Today*, Vol 43 (No. 10), 1990, p 23

64. T.S. Sriram, C.M. Ke, and Y.W. Chung, *Acta Met.*, Vol 41, 1993, p 2515

Cyclic Stress-Strain Response and Microstructure

Hans-Jürgen Christ, Universität-GH-Siegen

FATIGUE has been known for more than 150 years (Ref 1-3), especially for metallic materials that exhibit a lower strength under cyclic loading conditions compared to static load. For more than a century, common engineering practice has been to characterize the fatigue resistance of materials against cyclic loading by means of Wöhler curves or so-called *S-N* plots. Wöhler (Ref 4), who studied the cyclic strength of steel railroad axes and various other structural parts, was solely interested in the fatigue limit, which is the threshold stress amplitude below which fatigue failure will not occur, even for a very large number of cycles. For many decades it was a mystery as to why the stress level of the fatigue limit often occurs not only below the rupture strength but even below the yield stress.

For many years, and especially since the work of Coffin (Ref 5) and Manson (Ref 6), it has been known and well-accepted that fatigue failure has to be attributed to repeated cyclic plastic straining. The stress amplitudes leading to fatigue failure are in most cases too small to cause "macroyielding," but they are at least large enough to give rise to cyclic "microplastic" strains that are measurable and of the order of 10^{-5} to 10^{-4} at the fatigue limit. Consequently, fatigue fracture has to be considered as a result of repeated plastic straining, where the plastic-strain amplitude rather than the stress amplitude represents the decisive loading parameter. Thus, fundamental studies on the nature of fatigue damage must be based on well-designed cyclic deformation experiments in combination with a detailed evaluation of the microstructural changes that occur during cyclic deformation. The dislocations, their interaction among themselves and with second-phase particles, grain boundaries, and so on, and their behavior in cyclic strain localization play an important role. Even localized events during fatigue, such as crack initiation and crack propagation, which lead to what is commonly referred to as fatigue damage, can be considered a consequence of bulk microstructural changes that normally occur relatively early in fatigue life.

Figure 1 (Ref 7) tries to separate the single processes that finally lead to fatigue failure according to the chronological order of their occurrence. Although this scheme represents a very simplified description of the real situation, it applies to the behavior of many metallic materials under cyclic loading conditions and allows us to define the term *cyclic stress-strain behavior* (or *cyclic deformation behavior*). According to Fig. 1, this expression means the area of fatigue that comprises all aspects that deal with the global mechanical and microstructural response of a material to cyclic loading, without giving much consideration to damage processes, which are mostly localized. Nevertheless, it should be emphasized that fatigue damage usually evolves from microstructural changes as a consequence of cyclic plastic deformation, and that the study of the cyclic stress-strain response and its correlation with microstructure leads to a better understanding of the early stages of the fatigue failure process.

This article, which is restricted to the cyclic stress-strain behavior of metallic materials addresses the microstructural processes that take place during plastic deformation. Cyclic-stress-strain response is discussed on the basis of microstructure, because microstructural processes underlie and influence the examination, interpretation, and understanding of the macroscopic behavior.

From an engineering point of view, it might appear somewhat unsatisfactory and disappointing that no direct answers can be deduced from the cyclic stress-strain behavior to the important question of cyclic life and its assessment. Therefore, it appears necessary to indicate that many methods, especially the more advanced methods, used for cyclic lifetime prediction in engineering practice are based on knowledge of the cyclic deformation behavior of the material considered. Even in simple phenomenological approaches, most damage parameters contain a combination of stress amplitude and strain (or plastic-strain) amplitude (e.g., Ref 8). As an example of more sophisticated concepts for cyclic lifetime assessment, the methods based on the cyclic *J*-integral, ΔJ, can be cited. The transfer of the *J*-integral of elastic-plastic fracture mechanics as a parameter to characterize the situation at the crack tip under

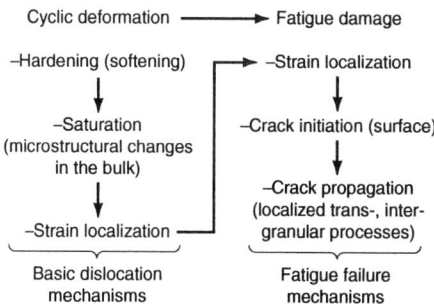

Fig. 1 The sequence of processes during fatigue of metallic materials. Source: Ref 7

monotonic straining (Ref 9, 10) to cyclic loading (Ref 11) has been shown to be successful in numerous studies (e.g., Ref 12-14). The value of ΔJ that actually describes fatigue crack propagation is usually determined by evaluating the stress-strain hysteresis loop, which is the fundamental diagram to represent the cyclic stress-strain response. Moreover, the application of *J* to cyclic loading demands a certain kind of cyclic stress-strain behavior (commonly termed Masing behavior). This behavior, which is considered in more detail below, should normally be checked first, before ΔJ is used.

This article is constructed in such a way that the level of difficulty increases from section to section. The following section gives a plain phenomenological and general description of the cyclic stress-strain response. Then, the microstructural aspects of cyclic deformation are described, taking into account the role of the dislocation slip character and the type of crystal lattice. Examples illustrate the effect of deformation-induced phase transformations on cyclic deformation behavior. The interaction of dislocations and strengthening second-phase particles is subsequently discussed briefly. Most materials used for engineering applications are optimized regarding their mechanical properties by means of a thermal and/or mechanical pretreatment, so the effect of a mechanical history is considered in a separate section. Finally, in the last section, a simple

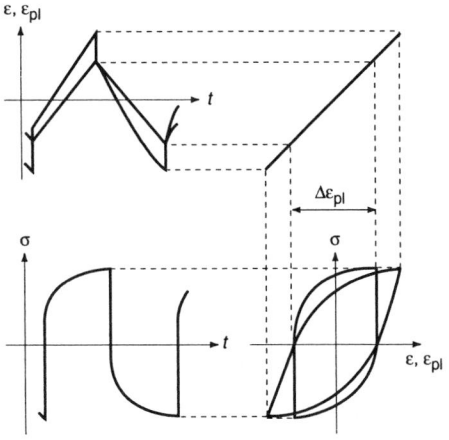

Fig. 2 Mechanical hysteresis loop, constructed from the courses of σ and ε (ε_pl) vs. time

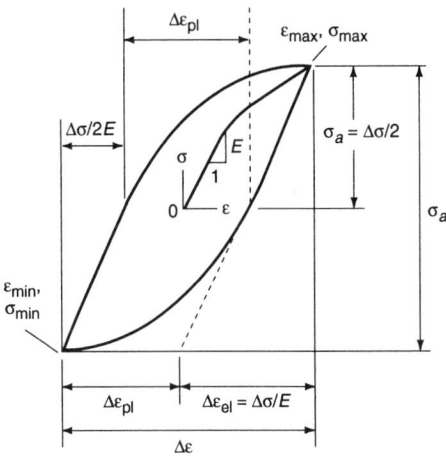

Fig. 3 Primary quantities of the hysteresis loop

method for modeling the cyclic stress-strain response is introduced.

The Mechanical Stress-Strain Response

The Hysteresis Loop. In order to obtain reproducible and unambiguous results, laboratory studies of cyclic stress-strain response are mostly conducted using unnotched samples of high surface quality (i.e., mechanically or electrochemically polished). A sufficient cylindrical gage length allows a precise determination of the uniform axial strain ε by means of a suitable extensometer. Servohydraulic test systems, normally used for cyclic loading experiments, can be run in closed-loop stress and strain control, respectively, as the standard control modes. The fatigue damage and the dislocation motion and rearrangement are determined by the plastic strain amplitude involved, so in some studies the control of the plastic strain is preferred (Ref 15).

The basic information on the cyclic stress-strain behavior of a material is provided in the form of the stress-strain hysteresis loop. In contrast to monotonic loading, cyclic deformation does not lead to a unique relationship between stress σ and strain ε, but rather to a hysteresis loop for each loading cycle. Figure 2 illustrates the way in which the courses of ε versus time t and σ versus t are combined. The example considered represents a fatigue experiment in which the test system forces the sample to follow a triangular signal of the plastic strain ε_{pl}. Total strain consists of plastic strain and elastic strain ε_{el}:

$$\varepsilon = \varepsilon_{pl} + \varepsilon_{el} \qquad (Eq\ 1)$$

The elastic strain can be calculated from stress via Hooke's law, so the plastic strain can be determined easily and continuously if the stress is measured. This allows the use of the plastic strain signal as feedback signal in the closed loop of a testing system:

$$\varepsilon_{pl} = \varepsilon - \frac{\sigma}{E} \qquad (Eq\ 2)$$

where E is Young's modulus.

In addition to the hysteresis loop in the form of σ versus ε, Fig. 2 also contains the loop as a plot of σ versus ε_{pl}. According to Eq 2, the representations are equivalent and can easily be converted one into the other. The representation against ε_{pl} is advantageous if the portion of elastic strain in total strain is high and the hysteresis loop resembles an elastic line in the plot of σ versus ε.

In principle, the hysteresis loop depends not only on the material, but also on the load frequency and control mode (strain or stress control). The hysteresis loop represents the microscopical deformation processes occurring during a load cycle in an integral form.

Some important quantities can be taken directly from the hysteresis loop (Fig. 3). The unloading after the load reversal point in tension, which is described by the coordinates ε_{max} and σ_{max} for maximum strain and stress, occurs with a slope of the tangent that represents the value of Young's modulus. This holds true also for the unloading in compression after having passed the minimum (ε_{min}, σ_{min}). The stress range Δσ follows from:

$$\Delta\sigma = \sigma_{max} - \sigma_{min} \qquad (Eq\ 3)$$

and the strain range Δε can analogously be obtained by:

$$\Delta\varepsilon = \varepsilon_{max} - \varepsilon_{min} \qquad (Eq\ 4)$$

The corresponding amplitudes of stress and strain are determined as the half ranges.

The plastic strain range $\Delta\varepsilon_{pl}$ is equal to the distance between the points of intersection of the hysteresis loop and the strain axis (Fig. 2). Depending on the material, a more or less pronounced back-deformation may occur during unloading, which gives rise to a difference between $\Delta\varepsilon_{pl}$ and $\varepsilon_{max} - \varepsilon_{min}$. This difference is sometimes

attributed to a *"reversible plastic strain,"* an expression that itself seems contradictory.

Figure 3 depicts symmetrical tension-compression loading with a mean stress $\sigma_m = (\sigma_{max} + \sigma_{min})/2 \approx 0$ and a mean strain $\varepsilon_m = (\varepsilon_{max} + \varepsilon_{min})/2 \approx 0$.

Transient cyclic deformation behavior (such as cyclic hardening or softening) refers to a continuous change in the cyclic strength that may occur throughout a test or at least in the first stage of cyclic deformation. Schematic examples of cyclic hardening (Fig. 4a) and cyclic softening (Fig. 4b) (Ref 16) show the stress course and the hysteresis loop shape from a symmetrical total-stress-controlled test performed applying a triangular demand signal with constant amplitude. Cyclic hardening leads to an increase in the stress amplitude, and consequently the hysteresis loop becomes larger. Cyclic softening has the opposite effect: a decrease of Δσ/2 and a reduction of the size of the hysteresis loop. The type of transient behavior is mainly determined by the pretreatment of the material tested. It is plausible that, for instance, heavy cold working prior to cyclic loading could cause subsequent cyclic softening, whereas a recrystallization treatment could give rise to cyclic hardening. Furthermore, deformation-induced microstructural changes may also be the reason for transient deformation behavior.

It is typical of asymmetrical cyclic deformation that transient processes take place that tend to reduce asymmetry. Figure 4(c) depicts the situation in a strain-controlled test in which a mean strain is superimposed. As a consequence, a mean stress arises that slowly diminishes. This process is termed cyclic relaxation. If the test is performed under stress control and a mean stress is applied, the material may show cyclic creep. The mean strain increases continuously, leading to a steady shift of the hysteresis loop to the right.

Cyclic Saturation. Transient effects are often neglected in the description of the cyclic stress-strain behavior due to their very strong dependence on material, testing condition, and mechanical history. A general formalism for a mathematical treatment of the transient cyclic deformation behavior is not known and probably does not exist. Data collections in handbooks usually provide an approximate representation of the cyclic deformation behavior by referring only to the stabilized deformation condition, which is usually termed the cyclic saturation state.

In Fig. 5, two different types of cyclic deformation curves are shown. Figure 5(a) refers to the case where the stress amplitude Δσ/2 is held constant during the test. The cyclic deformation response of the material is therefore plotted in the form of the course of the plastic-strain amplitude $\Delta\varepsilon_{pl}/2$ versus the number of cycles N. In a plastic-strain-controlled test (where $\Delta\varepsilon_{pl}/2$ is constant), the stress amplitude exhibits characteristic changes (Fig. 5b). If tests with considerably different numbers of cycles to fracture N_f are to be compared, instead of N, often the cumulative plastic strain $\varepsilon_{pl,cum}$ is used:

$$\varepsilon_{pl,cum} = 2\,N\,\Delta\varepsilon_{pl} \qquad (Eq\ 5)$$

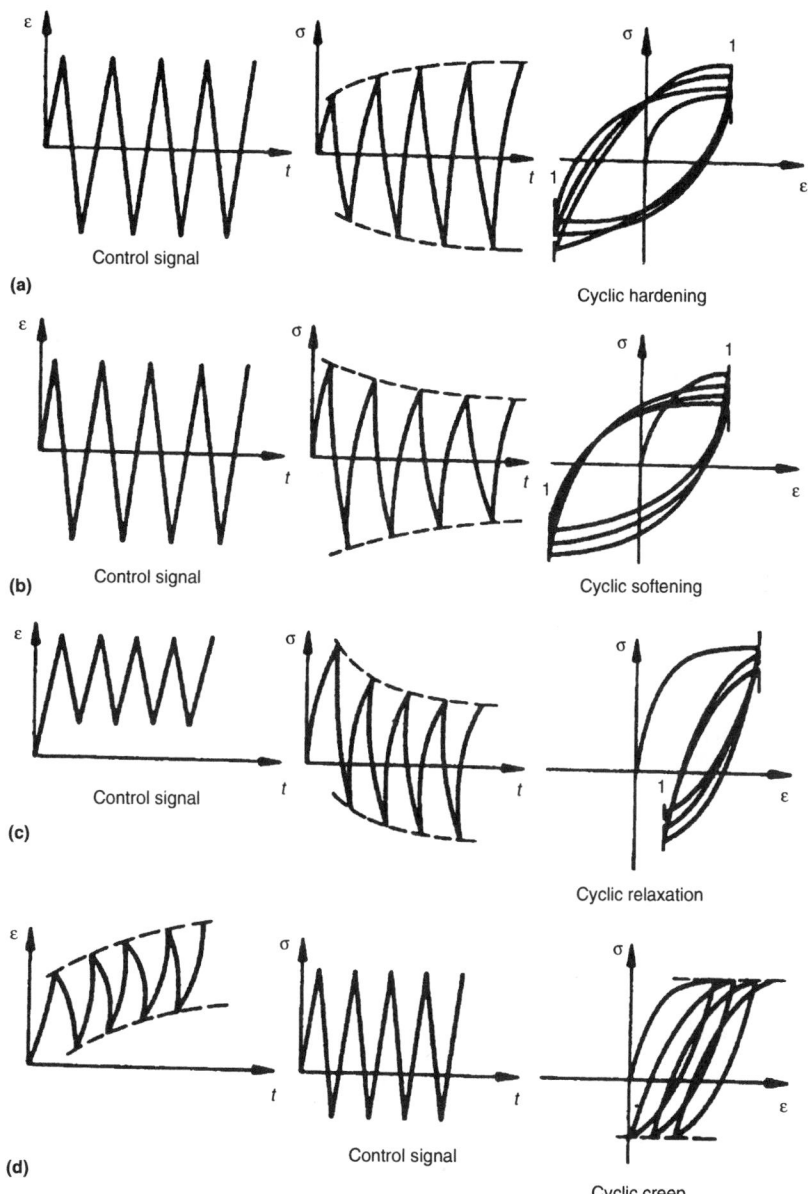

Fig. 4 Schematic representation of transient cyclic deformation processes. Source: Ref 16

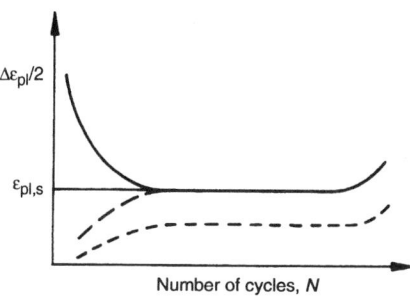

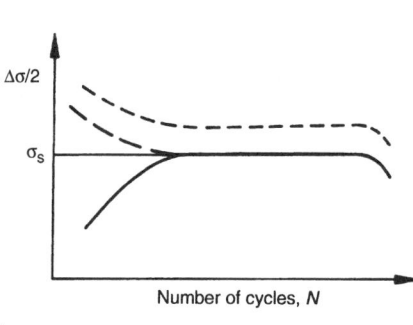

Fig. 5 Schematic representation of cyclic deformation curves for tests where (a) $\Delta\sigma/2$ is constant and (b) $\Delta\varepsilon_{pl}/2$ is constant

The value of $\varepsilon_{pl,cum}$ represents the total plastic strain that the sample undergoes during cyclic deformation. The final value at failure far exceeds the sustainable plastic strains of monotonic loading (typical values: monotonic loading 50%, cyclic loading 10.000%).

Many metallic materials show a cyclic deformation curve of the type shown in Fig. 5, which can be separated into three stages. At the beginning cyclic deformation causes a cyclic hardening or a cyclic softening, as described in the previous section. Cyclic hardening means that the plastic-strain amplitude decreases (solid line in Fig. 5a) and the stress amplitude increases (solid line in Fig. 5b). In the case of cyclic softening, the reverse changes take place (short dashed lines).

After this initial stage, a second region often follows in which $\Delta\varepsilon_{pl}/2$ and $\Delta\sigma/2$ are approximately constant. In this region a quasisteady condition exists. The prefix "quasi" is necessary to indicate that only the amplitude is constant, while the material is still continuously deformed along the hysteresis loop. The values of $\Delta\varepsilon_{pl}/2$ and $\Delta\sigma/2$ in this region of cyclic saturation ($\varepsilon_{pl,s}$ and σ_s, respectively) are of great importance, because frequently the stabilized behavior occupies a major part of fatigue life. For this reason, it is permissible to use the steady-state amplitudes $\varepsilon_{pl,s}$ and σ_s as rough mean values for an estimate of the "average behavior" during fatigue life. The stabilized conditions established in various tests with different loading amplitudes are used to define the cyclic stress-strain (CSS) curve (Ref 17).

In the third region of the cyclic deformation curve, the cyclic stress-strain behavior is affected by the propagation of a fatigue crack. In this region the crack size is already in the same range as the specimen dimensions, so the determined values of stress and strain are no longer of physical validity.

If the material has suffered mechanical deformation, it must not be taken for granted that the same stabilized condition will be established as on a deformation-free sample. This is illustrated in Fig. 5 by the long dashed lines, which define history-dependent behavior. In contrast, if the transient cyclic deformation merges into the plateau of an initially annealed sample (short dashed lines), the cyclic deformation is considered to be history-independent.

Memory of Prior Deformation. A phenomenon that might seem astonishing at first, but that provides some fundamental insights on important microstructural processes, is the "memory" of materials of their prior load history. If, as schematically shown in Fig. 6(a), a material is deformed in tension into the plastic region up to point B, and subsequently unloaded down to point C, a plastic back-deformation will take place if the unloading is sufficiently large. The fact, whether point C is connected with positive (tensile) or negative (compressive) stress, is therefore not the main criterion for plastic back deformation to take place.

Since the work of Bauschinger (Ref 20) more than a century ago, it has been known that the elastic limit after plastic deformation in one direction will be reduced if the loading direction is reversed (Fig. 6b) The Bauschinger effect is a basis for describing the reversibility of deformation, as shown graphically in Fig. 6(b). Its magnitude depends on a variety of factors, including amount of prestrain, deformation mode, and microstructure. Although the Bauschinger effect is an attractive parameter to describe deformation reversibility quantitatively, and thus to describe fatigue resistance, it should be remembered that it is a bulk phenomenon, whereas fatigue is a

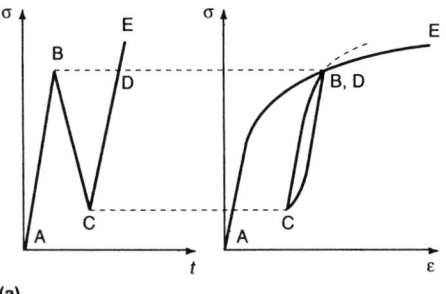

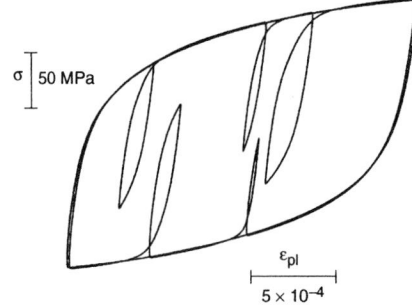

Fig. 7 Experimentally determined stress-strain path measured on polycrystalline copper

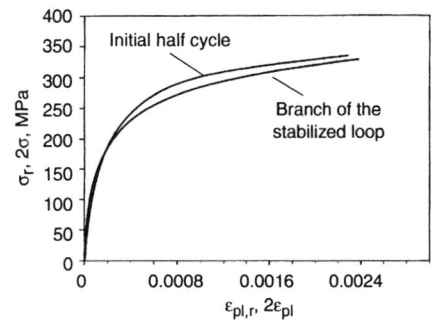

Fig. 8 Comparison of the initial half cycle (multiplied by a factor of 2) after cyclic neutralization with the ascending branch of a stabilized hysteresis loop in relative coordinates, measured on polycrystalline copper

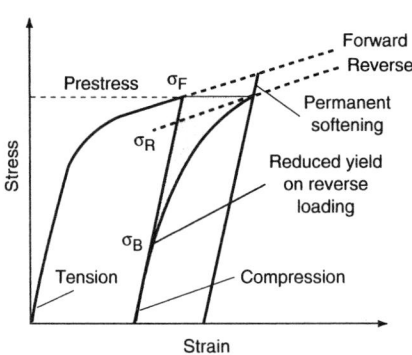

Fig. 6 Schematic of: (a) memory of a prior deformation and (b) Bauschinger effect with softening exaggerated for clarity. The hysteresis loop in (b) shows that on unloading, plastic deformation begins at a lower backward stress, σ_B, than reached in the forward direction, σ_F. The initial rounded portion of the reverse curve is due to short-range effects of weak obstacles, and the downward displacement of the subsequent region is related to the back stress. Source: Ref 18,19

local process. Therefore, although small inhomogeneities in the material may not measurably affect the Bauschinger effect, fatigue life may be drastically lowered.

The memory of mechanical history leads to the observation that a new increase of the stress, which exceeds the first maximum level B, does not give the smooth stress-strain path plotted as the dashed curve in Fig. 6(a) (starting from D). Rather, at D, where the stress level of the former maximum B is reached, the material proceeds along the initial predeformation σ-ε path (the strain path A-E without the interruption BCD). The small hysteresis loop (BCD) is "forgotten." The ability of a material to maintain σ-ε behavior along the original A-E path after BCD cycling is a feature which causes understanding problems if the deformation is considered to be uniformly distributed, as discussed below.

In the case of cyclic deformation with varying amplitude, the memory effect ensures a consistent continuation when stress-strain paths are interrupted by small cycles. The original σ–ε cycle continues as soon as the interrupting cycles are closed. If an interrupting cycle is again interrupted by another cycle, and so on, the memory is termed "second order," "third order," and so on. In Fig. 7, for example, an experimentally determined stress-strain path is represented that has been observed on polycrystalline copper, a relatively soft and ductile metallic material. The specimen had been cycled at $\Delta\varepsilon_{pl}/2 = 1 \cdot 10^{-3}$ for

25,000 cycles into the region of cyclic saturation. Figure 7 exhibits three large hysteresis loops, the second of which has been interrupted by five small unloadings in tension and compression, respectively. It is due to the memory effect that the enclosing loop (No. 2) is almost identical to the stabilized hysteresis loops (No. 1 and 3). The interruptions do not affect the size and the shape of the following loop (No. 3), so the cyclic stabilized state is maintained. However, in contrast to the schematic representation in Fig. 6, the transition from the small hysteresis loop back to the enveloping loop takes place smoothly, without a kink. Transient processes, those described above, can gradually remove the memory. If, for example, in Fig. 7 one of the small plastic strain cycles in tension were constantly applied, a mean stress relaxation would gradually eliminate the influence of the enclosing hysteresis loop. Nevertheless, it is very important to take the memory effect into account if the cyclic stress-strain behavior under variable-amplitude loading is considered. This knowledge has led to the development of appropriate methods for cycle counting (e.g., range pair method or rainflow method, Ref 21).

Memory Effect and the 1:2 Rule. The term *cyclic neutralization* denotes an experimental technique whereby the stress-strain origin in a cyclic deformation test can be found. For this purpose, starting from a (stabilized) hysteresis loop, the stress or strain amplitude is continuously reduced in small increments down to zero. This method is helpful in order to check whether a clip gage extensometer has slipped, because then a strain signal different from zero exists at zero load in the cyclic neutralized state.

If the cyclic neutralization doesn't take too many loading cycles, the microstructure of the material remains almost unchanged in its saturation condition. Therefore, if the original ampli-

tude is again applied, immediately after the initial half cycle the stabilized hysteresis loop is expected to be established. Normally this initial half cycle after cyclic neutralization differs strongly from the monotonic stress-strain curve (first loading curve) because of microstructural changes that have taken place during cyclic loading. Materials that do not undergo such changes are scarce and are called cyclically neutral (in other words, the first and the initial loading curves coincide).

It has been shown frequently (e.g., Ref 22, 23) that the initial half cycle is approximately identical to the ascending branch of the stabilized hysteresis loop if the branch is related to the load reversal point (in compression) and multiplied with a scale factor of $\frac{1}{2}$. An example is given in Fig. 8, which was obtained on polycrystalline copper. The coordinates used are called relative coordinates σ_r, ε_r, and $\varepsilon_{pl,r}$.

$$\sigma_r = \sigma - \sigma_{min} \qquad \text{(Eq 6)}$$

$$\varepsilon_r = \varepsilon - \varepsilon_{min} \qquad \text{(Eq 7)}$$

$$\varepsilon_{pl,r} = \varepsilon_{pl} - \varepsilon_{pl,min} \qquad \text{(Eq 8)}$$

This simple relation between the hysteresis loop branch and the initial half cycle was first reported by Morrow (Ref 17) and is termed the 1:2 rule or Masing hypothesis. It is shown in the last section of this article that this rule can easily be understood on the basis of Masing's simple multicomponent model. Moreover, an identical relation exists between the CSS curve and the hysteresis loop branch, indicating that the CSS curve equals the initial half cycle.

Microstructural Aspects of Cyclic Loading

The underlying microstructural processes for the above-described cyclic mechanical behavior of metallic materials is a highly complex topic covered in an enormous number of publications. Therefore, a reasonable restriction is necessary.

The emphasis in this section is on single-phase materials tested in initially soft, dislocation-poor conditions resulting from a prior heat treatment. The aspects of a mechanical pretreatment and the strengthening effect of a second phase are dealt with in the subsequent sections "Cyclic Deformation in Structural Alloys" and "Variables and Modeling," respectively, in this article.

Dislocation Arrangement of Cyclic Saturation

Factors Determining the Slip Character of a Material. The slip character of a material is a basic parameter that determines the type of dislocation arrangement formed during cyclic loading and therefore also the cyclic stress-strain response. The term slip character describes the tendency of a material to form a three-dimensional dislocation arrangement. The two extreme cases of slip character are represented by materials that show on the one hand a pure planar dislocation slip and on the other hand a pure wavy dislocation slip. Some examples for typical microstructures illustrating the consequences of different slip characters, are summarized below.

When the cyclic stress-strain behavior of a material is estimated, it would be helpful if "slip character" could be quantified from independent materials data. The factors determining the slip character of a material are a subject of continuing research. Gerold and Karnthaler (Ref 24) describe the origin of planar slip in face-centered cubic (fcc) alloys, emphasizing the role of short-range order in slip planarity. Another paper (Ref 25) discusses several factors, including short-range order and a quantitative basis for distinguishing wavy and planar slip. References 24 and 25 are mainly complementary, although Ref 25 does demonstrate that two of the conclusions in Ref 24 are inconsistent with the evidence. The most important factor promoting planarity of slip is a high friction stress, irrespective of its cause. There is good evidence of a role for interstitial solutes in promoting slip planarity, which clearly does not involve short-range order.

The early studies on the influence of loading amplitude on dislocation arrangement were almost exclusively carried out on copper, Cu-Al alloys, and α-brass (Cu-Zn) (e.g., Ref 26-31). The stacking fault energy of copper ($\approx$40 mJ/m^2, e.g., Ref 32) decreases with increasing concentration of alloying element. Because α-brass and Cu-Al alloys of higher aluminum content exhibit planar slip character, the idea seems to be reasonable that the stacking fault energy might be the decisive quantity for the ease of forming a three-dimensional dislocation arrangement. This idea is based on the assumption that the stacking fault energy determines the possibility of cross slip of screw dislocations.

Cross Slip in Pure Metals. As a brief review, if a fcc crystal structure is considered, gliding of a dislocation leads to a shift of a {111} plane in the <110> direction. The atomic arrangement of the fcc lattice can be envisioned to result from simply stacking close-packed {111} planes in a stacking sequence of ABCABC. This is illustrated in Fig. 9, in which the atoms of layer A are drawn and the positions of the atoms of layers B and C are marked. If a dislocation of the Burgers vector $\mathbf{b}_1$

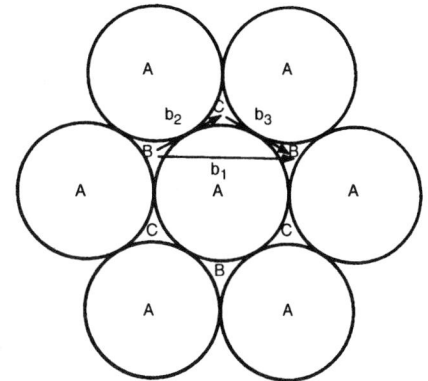

Fig. 9 Part of a {111} plane of the fcc crystal lattice, showing the positions of the atoms of the two next planes stacked above (B and C)

= <110> glides on plane A, the atoms of plane B will reach again B positions. Therefore, the atomic arrangement in front of and behind the gliding dislocation is undisturbed. It can clearly be seen in Fig. 9 that the movement of a B atom along $\mathbf{b}_1$ is unfavorable compared to the path following $\mathbf{b}_2$ to a C position and then $\mathbf{b}_3$ to the final B position. This zigzag course is possible if the dislocation with the Burgers vector $\mathbf{b}_1$ dissociates into two partial dislocations with the Burgers vectors $\mathbf{b}_2$ and $\mathbf{b}_3$ according to the reaction:

$$\mathbf{b}_1 \rightarrow \mathbf{b}_2 + \mathbf{b}_3 \tag{Eq 9}$$

or in the notation of Miller's indices:

$$\frac{1}{2}\langle 110 \rangle \rightarrow \frac{1}{6}\langle 211 \rangle + \frac{1}{6}\langle 12\bar{1} \rangle \tag{Eq 10}$$

As the elastic energy per unit length of a dislocation is proportional to $G\mathbf{b}^2$, where G denotes the shear modulus, this dissociation is connected with an energy reduction ($\mathbf{b}_1^2 < \mathbf{b}_2^2 + \mathbf{b}_3^2$), and hence it is favorable.

The elastic interaction of the partial dislocations leads to a repulsion that is balanced by the energy of the stacking fault existing between the partial dislocations. The stacking fault (e.g., B atoms in C positions) is a disturbance of the perfect lattice, and the energy increase per unit area defines the stacking fault energy γ_{SF}. From a simple consideration of force equilibrium, it follows that the width of splitting up (the distance between corresponding partial dislocations) can be calculated by means of the expression

$$d = \frac{G\mathbf{b}^2}{4\pi\gamma_{SF}} \tag{Eq 11}$$

Therefore, γ_{SF} can be determined if d is measured (e.g., by means of high-resolution transmission electron microscopy, TEM, or by applying the weak-beam technique in TEM, see Ref 32). The value of γ_{SF} depends very sensitively on alloy composition (Ref 33) and increases in almost all fcc metals and alloys with increasing temperature.

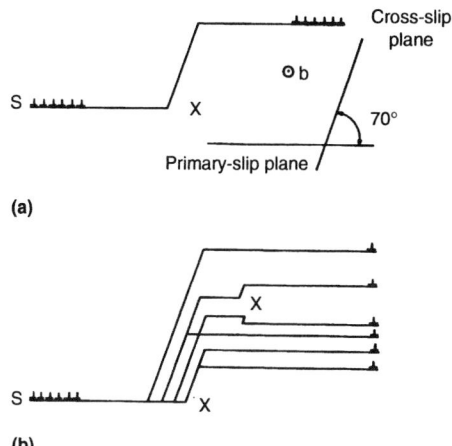

Fig. 10 Schematic representation of a cross slip process of several dislocations at an obstacle (X), if (a) planar slip or (b) wavy slip prevails. Source: Ref 24

The link between the slip character and the stacking fault energy is given by the following reasoning. A three-dimensional arrangement of dislocations demands that dislocations can leave their slip plane. Screw dislocations can do so simply by cross slip. If, however, a screw dislocation has dissociated into partial dislocations, a recombination of these partial dislocations leading back to a complete dislocation must first take place, because the character of partial dislocations does not allow them to cross slip. It is obvious that force is required to push the partial dislocations against each other for recombination and that recombination as a prerequisite for cross slip is easier, if the dissociation width is small, that is, if γ_{SF} is high (see Eq 11). According to this reasoning, wavy slip character is to be expected if the value of the stacking fault energy is high, whereas in materials with low stacking fault energy, planar slip should prevail.

If the type of dislocation arrangement established in reality under cyclic loading conditions is compared with this expectation, it is found that the correlation with γ_{SF} works for pure metals at best, but the agreement is poor for alloys. Therefore, different approaches need to be used to describe the physical origin of the slip character in alloys.

Slip Character in Alloys. As previously noted, Hong and Laird (Ref 25) have treated in a semiquantitative model the role of the dislocation friction stress and the structure of the stacking fault, in order to explain the influence of the content of alloying elements on the slip character. The basic idea is that friction effects impede the recombination of partial dislocations and are the main reason for planar slip. The resultant equation expresses that in addition to a low value of γ_{SF}, a high atomic misfit, a high shear modulus, and a high concentration of solved foreign atoms favor planarity of slip.

According to Gerold and Karnthaler (Ref 24) it is essential for the slip character that a short-range order exists (confirmed by Wolf et al. in Ref 34). The metals and alloys considered in Ref 24 (mostly taken from Ref 35) can be classified into

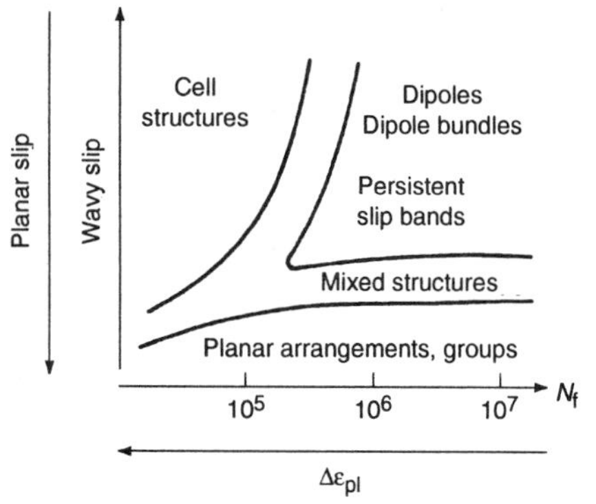

Fig. 11 Dislocation arrangement in cyclically deformed fcc metals as a function of slip character and number of cycles to fracture. Source: Ref 27, 28

two groups. One group contains the materials free of short-range order, which all show wavy slip, whereas the other materials exhibit planar slip and short-range order phenomena are present.

The basic idea is depicted in Fig. 10. If in a slip plane a first dislocation is moving, the short-range order that might exist will be destroyed. Because the short-range order is energetically favorable, a locally higher stress is required to move this first dislocation (e.g., provided by a dislocation pile-up). Once the short-range order has been destroyed, the following dislocations, which might be emitted from a dislocation source S, can move quite easily. Therefore, there is no need to activate new slip planes, and the plastic deformation is confined to individual planes (slip localization).

According to this idea, planar slip character does not mean that cross slip does not take place, or takes place only rarely. Rather, the course of cross slip is different, as shown in Fig. 10. In the case of wavy slip, the dislocations cross slip independently of each other on individual slip planes (e.g., due to the presence of an obstacle) and continue to move in the initial direction on numerous planes (Fig. 10b). This process occurs as a combined movement of a whole group of dislocations in a short-range ordered material. Therefore, the dislocations remain localized in certain planes, giving rise to an overall planar dislocation arrangement.

Loading Amplitude, Slip Character, and Dislocation Arrangement. The results of various studies on fcc metals and alloys are summarized in Fig. 11. The different types of dislocation arrangements are mapped as a function of the number of cycles to fracture (abscissa) and the slip character of the material (ordinate). The value of N_f depends on the loading amplitude, in the sense that, for example, a high plastic-strain range corresponds to a low cycle number to fracture, and vice versa. It is assumed that a state of cyclic saturation is established (i.e., the dislocation arrangement included in Fig. 11 is that of stabilized cyclic deformation behavior).

The representation in Fig. 11 was introduced in 1968 by Feltner and Laird (Ref 27), who used a three-dimensional plot with the homologous temperature T/T_m (T_m denotes the melting temperature) as the third axis. Because the temperature range studied was relatively small ($T/T_m < 0.25$), the influence of temperature can be neglected in this temperature interval. However, for certain materials, such as copper (Ref 36) or aluminum (Ref 37), representations have been developed that show a pronounced effect of temperature, mainly because with increasing temperature, recovery processes gain importance and may lead to an instability of low-temperature dislocation arrangements.

Figure 11, which has been confirmed quantitatively by Lukáš and Klesnil (Ref 28), roughly classifies materials into those showing wavy slip and those showing planar slip. The dislocation arrangement in cyclic saturation of wavy-slip metals depends very strongly on the loading amplitude. At low amplitudes (high values of N_f), arrangements of edge dislocation dipoles are found that form mainly due to single slip. Dislocations agglomerate to so-called bundles or veins, which are separated from each other by regions of low dislocation density (channels). Embedded in this matrix, persistent slip bands (PSB) can form. At higher amplitudes, multiple slip takes place and gives rise to labyrinth and cell structures.

In planar-slip metals and alloys, the dislocation motion is confined to the slip plane. Planar arrangements form, mainly consisting of edge dislocations that occupy a slip plane and line up parallel to each other. Secondary slip contributes to cyclic deformation at high amplitudes, although secondary slip also seems important for PSB formation, even at low amplitudes.

In the following sections the characteristic dislocation arrangements of cyclically deformed metals and alloys are treated in more detail according to slip character (wavy and planar) and crystal lattice structure (fcc and body-centered cubic, bcc).

fcc Metals with Wavy Slip

Most studies on materials showing wavy dislocation slip were performed on copper as a model material. This should not be considered a serious restriction of the general validity of the results, as comparative studies show (e.g., on nickel or austenitic stainless steel). Numerous important results can be obtained on single crystals because of the possibility of adjusting simple and exactly defined conditions regarding slip geometry and resolved shear stress. The transfer of these findings to polycrystals has often proved to be key to developing insight into the processes taking place under complex conditions. Therefore, a brief description of the behavior of single crystals is given first.

Cyclic Stress-Strain of Single Crystals

Dislocation Arrangement in Saturation. The dislocation arrangement established in cyclic saturation of fully reversed fatigue experiments under fixed amplitudes of resolved plastic shear strain, γ_{ap}, can be discussed on the basis of the

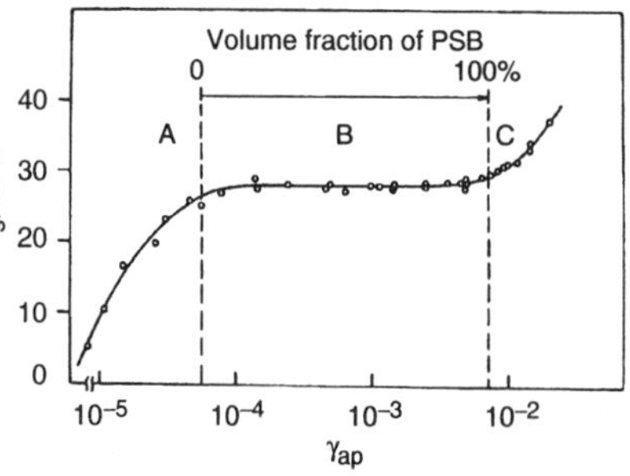

Fig. 12 Cyclic stress-strain curve of monocrystalline copper oriented for single slip. Source: Ref 38

CSS curve of single crystals (Fig. 12), which represents the saturation value of the resolved shear stress, τ_s, versus γ_{ap}. Figure 12 refers to copper, but very similar curves have been found for other wavy-slip fcc metal alloys (see Ref 39 for more details).

In Fig. 12, three distinct regions, marked A, B, and C, can be distinguished. At very low amplitudes of plastic shear ($\gamma_{ap} < 6 \times 10^{-5}$), the shear stress amplitude increases with increasing γ_{ap} (region A). A plateau denoted region B follows, where the saturation stress is independent of the plastic strain (plateau stress). If the plastic shear strain amplitude exceeds 7.5×10^{-3}, region C is reached, where again an increase of τ_s with γ_{ap} occurs.

The shape of the CSS curve is relatively insensitive to the orientation of the single crystal. In most orientations, τ_s values agree with those represented in Fig. 12. In cases where the load axis is oriented in such a way that two or more slip systems are equivalent, a basically different behavior results (e.g., Ref 40-42). The grains in polycrystalline materials sometimes behave like single crystals oriented for single slip, but textures favoring multislip orientation are common and can have large effects.

Region A. At low values of γ_{ap}, the cyclic hardening of well-annealed single crystals is almost entirely due to the accumulation of primary dislocations. The dislocations have mainly edge character and form dipoles that agglomerate to bundles. The saturation condition is characterized by an equilibrium between these dislocation-rich bundles and their relatively dislocation-poor surrounding. Figure 13 tries to give a three-dimensional image of the dislocation arrangement in region A. The top area shows the position of the dislocations in the slip plane (111), whereas the area at the right side represents a ($1\bar{2}1$) plane that is perpendicular to the slip plane and contains the Burgers vector [$\bar{1}01$] of the primary slip system. The portion of screw dislocations in the bundles is small. With increasing saturation stress the volume fraction of the bundles is enhanced up to about 50% at the transition to region B; the bundles become so-called veins. These veins are separated by dislocation-poor channels that are oriented parallel to the veins, (i.e., with their long axis parallel to the dislocation line of primary edge dislocations).

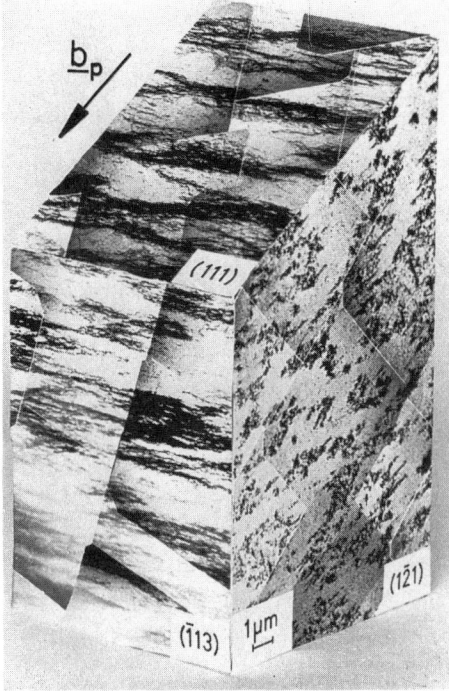

Fig. 13 Three-dimensional image of the saturation dislocation arrangement of monocrystalline copper at $\gamma_{ap} = 2.6 \times 10^{-5}$ (region A). Source: Ref 43

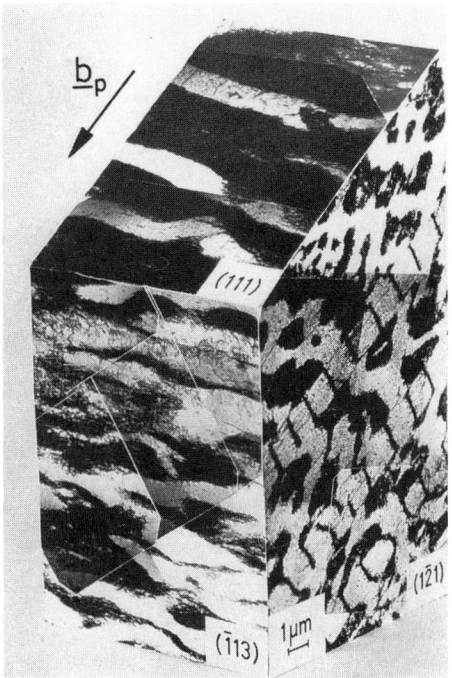

Fig. 14 Three-dimensional image of the dislocation arrangement of cyclic saturation of monocrystalline copper at $\gamma_{ap} = 1.5 \times 10^{-3}$ (region B). Source: Ref 46

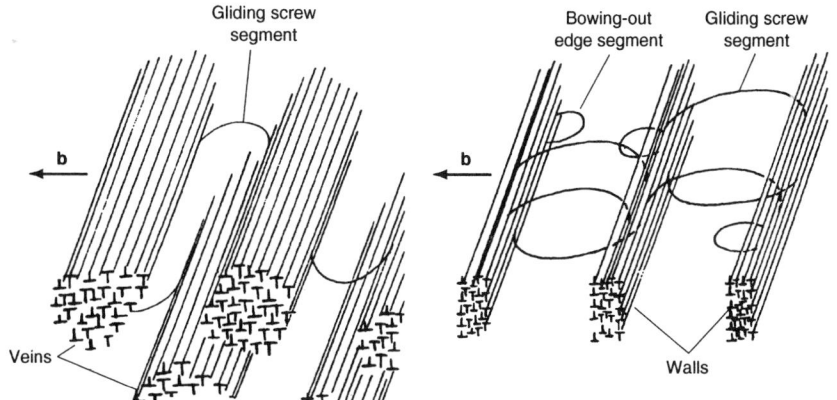

Fig. 15 Schematic representation of the dislocation arrangements in (a) a matrix structure and (b) a persistent slip band. Source: Ref 58

Region B. Under region A conditions, only fine slip markings are observed on the surface of the sample and the number of cycles to fracture is almost infinite. In contrast, a slip concentration process characterizes the behavior in region B and limits fatigue life. As soon as the plastic shear strain amplitude exceeds the limit of $\gamma_{ap} = 6 \times 10^{-5}$, PSBs are formed, regions of localized slip in which plastic deformation is much larger than the overall deformation of the crystal. The volume fraction of PSBs increases in region B with increasing γ_{ap}, starting from 0% at the boundary to region A, up to about 100% at the beginning of region C. The local shear strain amplitude of a PSB and the surrounding matrix, respectively, can be seen as the values of γ_{ap} that limit region

B in Fig. 12. This documents that a PSB deforms about 100 times more than the matrix. The heterogeneous deformation within the plateau of the CSS curve can be described by Winter's rule of mixtures (Ref 44):

$$\gamma_{ap} = f_{PSB} \cdot \gamma_{ap,PSB} + (1 - f_{PSB})\gamma_{ap,M} \qquad (Eq\ 12)$$

where f_{PSB} is the volume fraction of PSBs, $\gamma_{ap,PSB}$ is the local plastic shear strain amplitude acting in a PSB, and $\gamma_{ap,M}$ is the plastic shear strain amplitude of the matrix. $\gamma_{ap,PSB}$ has to be interpreted as an average value, because the deformation amplitude varies from PSB to PSB and also within a PSB, as shown, for instance, by means of interferometric surface observations (Ref 45). Nevertheless, Eq 12

explains reasonably and in accord with the experimental results why τ_s is constant in region B, if one assumes that $\gamma_{ap,PSB}$ and $\gamma_{ap,M}$ are fixed amplitudes and f_{PSB} changes linearly with γ_{ap}.

Figure 14 shows the typical ladder-like structure of PSBs in a ($1\bar{2}1$) plane. The "rungs" of this ladder consist of edge dislocations (more precisely of dislocation dipoles). Between the rungs there are channels with low dislocation density.

PSBs are embedded in a matrix structure whose dislocation arrangement depends on the value of γ_{ap} applied. In the low-amplitude part of region B, the matrix consists of a bundle/vein structure in which single slip prevails (as in the PSBs), whereas secondary slip gains importance at high amplitudes, leading to an arrangement that resembles a labyrinth structure (Ref 47).

The plateau value of τ_s is reported to be 28 MPa for copper single crystals (Ref 38). Similar values of τ_s/G ($\approx 6.5 \times 10^{-4}$) are obtained for other fcc metals (Ref 46). However, τ_s (and hence also the surface and dislocation structure) is affected by the testing mode that is used to produce PSBs (e.g., Ref 48-50). Furthermore, the test temperature plays an important role, in the sense that τ_s depends strongly on temperature (e.g., Ref 51), PSBs cannot exist at high temperatures, and extended wall structures are found instead of PSBs at low temperatures (e.g., Ref 52).

PSBs deform in region B at a constant stress amplitude. An increase of γ_{ap} leads to an increase of the volume fraction of PSBs, which is connected with a brief rise of the shear stress amplitude until the plateau value is established again. According to Blochwitz and Veit (Ref 53), who tried to determine the "true CSS curves" of PSB and matrix, respectively, the plateau value of the plastic shear amplitude of the matrix can be considered as that value at which the corresponding stress amplitude in the matrix equals the nucleation stress of PSBs. Therefore, an increase of γ_{ap} triggers an increase of the PSB volume fraction at constant τ_s, whereas a decrease of γ_{ap} leads to a reduction in τ_s, as a result of a smaller local deformation amplitude in the PSB at a volume fraction that is unchanged (at least for the first moment).

By means of a mathematical estimate of the dislocation multiplication and annihilation processes occurring within a PSB (Ref 54, 55) and by in situ cyclic deformation experiments performed in TEM (Ref 56, 57), it could be shown that both the walls and the dislocation-poor channels take part in plastic deformation. The main mechanism of the macroyielding of the PSB (local shear strain amplitude of about 1%) is that edge dislocations bow out from the walls, traverse the channels, and penetrate partially into the opposite wall. This leads to the existence of screw dislocation segments that glide along the channels (Fig. 15), increasing the length of the edge-dislocation part.

Screw dislocations can annihilate if their distance is smaller than the so-called annihilation distance. The edge dislocations are incorporated into the walls and are partially annihilated. Altogether a highly dynamic process of dislocation

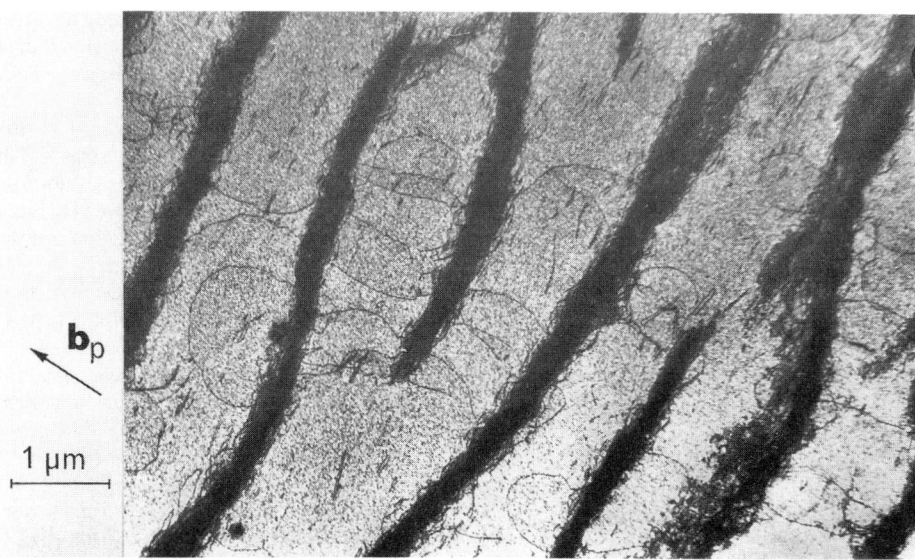

Fig. 16 TEM micrograph of the wall structure of a persistent slip band under load, in the section parallel to the slip plane. Source: Ref 46

Fig. 17 Three-dimensional image of the dislocation arrangement of monocrystalline copper in cyclic saturation at $\gamma_{ap} = 1.45 \times 10^{-2}$. Source: Ref 43

formation and annihilation results, so that after about every 50 cycles all dislocations are renewed, as can be shown by a rough estimate (Ref 59). A micrograph of the screw dislocations spanning the channels (Fig. 16) shows only about 10% of the screw dislocations, which are produced in one cycle and glide between the walls. It should be noted that plastic yielding of the hard walls is a consequence of the requirement of compatible deformation.

In the matrix structure (Fig. 15a), the local plastic deformation is small. Therefore, this deformation can be provided by a quasireversible to-and-fro bowing of the screw dislocations in the channels, essentially by elastic polarization of the hard veins (e.g., Ref 60, 61), and maybe the isolated occurrence of dipole flip-flop (i.e., two edge dislocations forming a dipole change from one stable 45° position to the other).

The importance of PSBs for fatigue is a consequence of their high localized plastic deformation. It has been demonstrated in numerous studies that PSBs form through the bulk of the single crystal and mark their egress at the specimen surface. These slip lines at the surface (which are parallel to the primary glide plane, about 1 to 2 μm high, and show a rough topography) have been given the attribute "persistent" by Thompson et al. (Ref 62), because they reappear at the same sites during continued cycling, even after a thin layer of surface containing these bands has been removed by electropolishing.

PSBs are considered to be responsible for crack initiation. The interrelation of fatigue damage and the effect of PSBs at the surface has been proven by Thompson et al. (Ref 62) in an experiment that has been repeated quite often. The cyclic lifetime of a sample can be extended if a surface layer is removed by electrochemical polishing.

The crack initiation in single crystals takes place at extrusions (e.g., Ref 63-71), which probably form primarily as a consequence of a high

concentration of point defects arising from the dislocation annihilation processes in the PSBs. Polycrystals are prone to surface grain boundary cracking at low strain amplitudes, which is a result of the interaction of PSBs with grain boundaries (e.g., Ref 72-79) or twin boundaries (e.g., Ref 19).

Region C. If the plastic shear amplitude exceeds a critical value (7.5×10^{-3} for copper), secondary slip gains importance, leading to a dislocation cell structure. Figure 17 gives an impression of the resultant dislocation arrangement. Long cells are formed, the geometry of which resembles the wall structure of the PSBs. However, the typical ladder structure of PSBs is not observed. In addition to primary dislocations, secondary ones are present in the walls, so these walls are sometimes termed secondary walls (Ref 80). The long shape of cells in Fig. 17 is only one type of appearance; equiaxed cells are often observed, especially at high amplitudes. The plastic deformation takes place in a homogeneous manner; that is no slip localization exists.

Development of dislocation arrangements in single-crystal fcc metals with wavy slip is discussed below for constant amplitude and for ramp loading.

Behavior at Constant Amplitude. The description given above deals with the situation where the dislocation structure has reached a steady state (cyclic saturation). Detailed TEM observations of the development of the dislocation arrangement with number of loading cycles have been carried out, for example by Hancock and Großkreutz (Ref 81) at an amplitude in region C and by Basinski et al. (Ref 82) at low values of γ_{ap}, leading to basically identical results.

In contrast to monotonic loading, there is no rotation of the slip system with respect to the loading axes. Therefore, the primary slip system remains the most highly stressed. After only a few cycles the dislocations are heterogeneously distributed in the material. Bundles form that are

oriented perpendicular to the primary Burgers vector with their long axis, and parallel to the dislocation lines of edge dislocations. The bundles consist mainly of edge dislocations, indicating that screw dislocations can annihilate by cross slip. The edge dislocations interact with each other, forming dipoles or multipoles over a large part of their length. At high plastic-shear amplitude after only about 100 cycles, a dislocation arrangement is formed that resembles a cell structure with appreciable contribution of secondary dislocations.

From these observations it can be concluded that the formation of bundles is a consequence of mutual trapping of primary edge dislocations. The bundles and veins are obstacles to further plastic deformation because they partially impede dislocation motion on the primary slip system. The dislocation density within the veins and the number of veins per unit volume both increase with cycle number, giving rise to a pronounced cyclic hardening. The dislocation multiplication is considered to occur according to a Frank-Read mechanism, a bowing out of dislocations into the areas adjacent to the veins.

The mean dislocation spacing in the veins is relatively small (≈ 30 nm) and can be related to the so-called trapping distance of the edge dislocations. As a consequence of the approximately equal number of positive and negative edge dislocations, the average Burgers vector of the veins is close to zero. Thus, no long-range internal stresses are produced. This is an important distinction between monotonic and cyclic hardening of fcc single crystals at low amplitudes of imposed strains. It should be noted here that hardening under monotonic loading occurs much faster

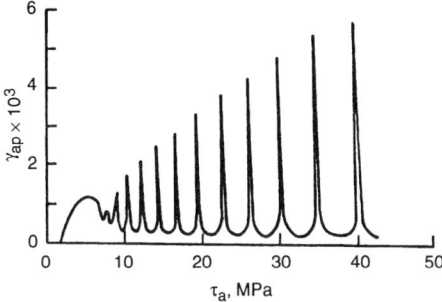

Fig. 18 Strain bursts observed in a copper single crystal subjected to ramp loading at 90K. Source: Ref 39, 94

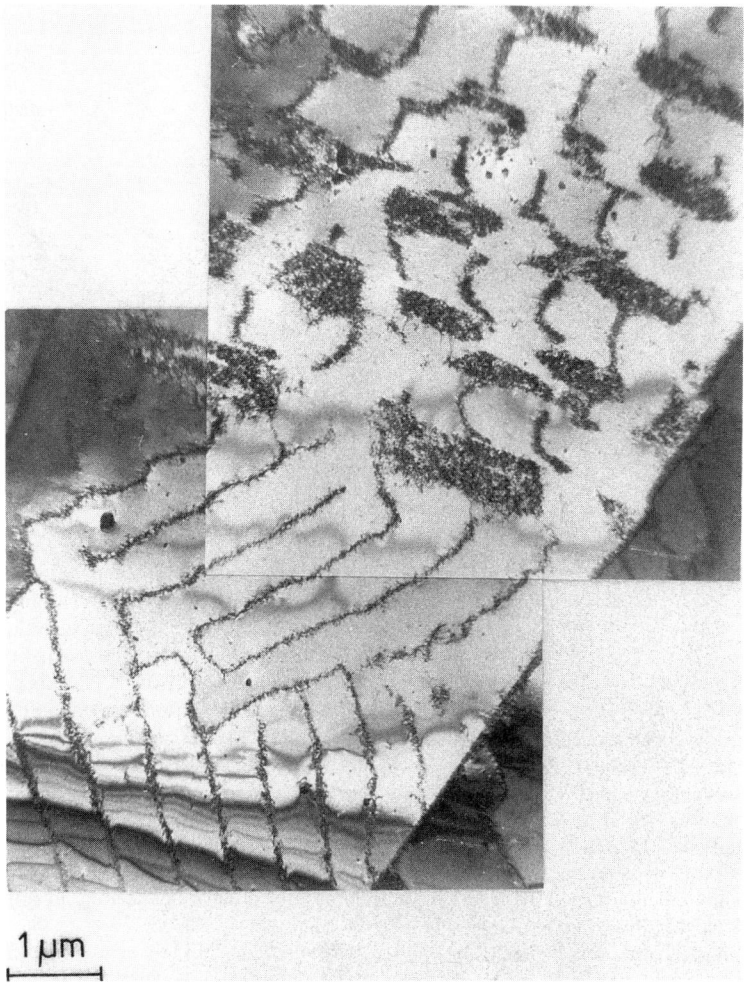

Fig. 19 Different dislocation arrangements within a single grain of copper. $\Delta\varepsilon_{pl}/2 = 5 \times 10^{-4}$, $N \approx 10^5$. Source: Ref 101

than that under cyclic loading, if as a basis of such a comparison the cumulative plastic strain of cyclic deformation is assumed to be analogous to monotonic plastic strain.

As hardening proceeds, the dislocation arrangement develops perpendicular to the primary slip plane, and at high loading amplitudes a labyrinth or cell structure is finally formed as a consequence of an increased contribution of secondary slip.

In spite of numerous models proposed in the literature, there is no fully satisfactory explanation of the establishment of characteristic dislocation arrangements in fatigue. In various works related to this issue, the emerging structures are treated as low-energy dislocation structures (e.g., Ref 43, 83, 84). Attempts have also been made to develop analytical models of continuous concentration fields of dislocations in fatigued crystals (Ref 85, 86). In this approach, the to-and-fro motion of dislocations under cyclic stress is modeled as a diffusion phenomenon, which is treated mathematically by means of a formalism similar to that developed by Cahn and Hillert for spinodal decomposition. A third viewpoint is based on the theory on self-organization of dissipative structures (nonequilibration systems, Ref 87). Taking into account the high dynamics of the dislocation reactions and rearrangements, the last-mentioned approach seems to be most promising.

Even the formation of the geometrically highly regular PSBs, albeit fundamentally significant, has remained a poorly understood phenomenon (e.g., Ref 88). Noting that the transformation to PSBs starts from a vein structure, some investigators believe that the instability of the vein structure relative to the dislocation arrangement found in PSBs is the main point. According to Kuhlmann-Wilsdorf and Laird (Ref 89), a critical value of the dislocation density in the veins needs to be reached in order to trigger the PSB formation.

A quantitative description of the formation of dislocation structures within veins and PSBs can potentially be obtained from calculations of the equilibrium positions of finite populations of dislocations. The basis for most of the models is the Taylor-Nabarro lattice (Ref 90, 91), which consists of a regular arrangement of parallel edge dislocations extending to infinity. It has been

shown by Neumann (Ref 92, 93) that in diamond-shaped sections of the Taylor-Nabarro lattice, the application of a stress results in the emergence of dipolar walls of dislocations from the polarization of the initial regular distribution. The initial configuration begins to decompose into walls only if a certain critical value of the shear stress is exceeded. Moreover, it could be shown that the wall structure is more stable than the vein structure.

Instabilities During Ramp Loading. If single crystals are subjected to a steadily increasing stress amplitude (ramp loading), the resultant strain amplitude shows a somewhat surprising and unusual behavior in the form of perfectly periodic strain maxima (Fig. 18), which were termed strain bursts by Neumann (Ref 94). Strain bursts have been observed in single crystals of copper, aluminum, magnesium, and zinc and of Cu-Al (Ref 94, 95).

The microstructural processes that are responsible for the occurrence of these maxima in strain amplitudes (Ref 92, 96) provide an insight into the basic mechanisms of cyclic hardening, so a brief description seems to be appropriate here. As mentioned above, cyclic loading leads to a hardening that is due to the dislocation multiplication

and mutual trapping of dislocations. As a consequence, one would expect a progressively reducing γ_{ap} if loading takes place at constant shear stress amplitude. This effect continues until the mean free path for dislocation motion is smaller than their mean spacing. Therefore, the plastic strain is accommodated mainly by a polarization of the dipoles in the bundles and dipole flip-flop. If the stress amplitude is slowly raised, dislocations of dipoles of smaller cohesion are separated, leading to an avalanche of free dislocations. This process can be observed macroscopically in the occurrence of a strain burst. After a while, the released dislocations are trapped again, and the cycle repeats.

This mechanism is also expected to occur during cyclic loading at constant amplitude. However, no strain bursts are found, because they do not occur in a "coherent" manner but rather are local events that take place asynchronously and lead at most to a slightly oscillating course of the shear stress amplitude τ_a.

Cyclic Stress-Strain Behavior of Polycrystals (fcc, Wavy Slip)

The next step is whether the basic information on the fatigue mechanisms of single crystals is

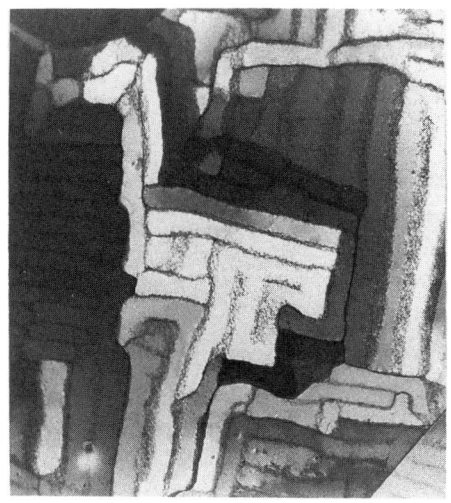

1.5 μm

Fig. 20 Labyrinth/cell structure with marked misorientation. $\Delta\varepsilon_{pl}/2 = 2 \times 10^{-3}$, $N \approx 10^4$. Source: Ref 101

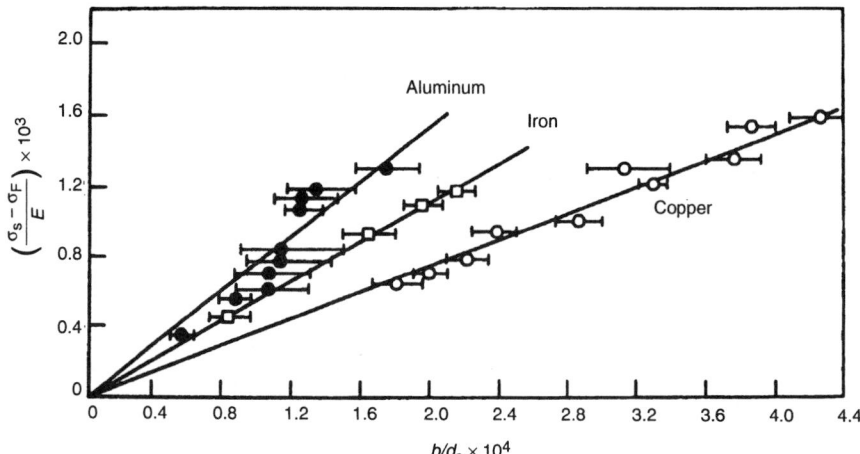

Fig. 21 Relation between cell diameter and normalized saturation stress amplitude. Source: Ref 121

also applicable to polycrystalline materials (e.g., Ref 80, 97-102), which are closer to actual technical alloys for structural use. Because the presence of precipitates, impurities, inclusions, and additional phases gives rise to a CSS response that may significantly deviate from that described for single crystals, single-phase polycrystalline metals and alloys are considered first.

It is obvious that the existence of grains of different orientation that form a solid unit and have to deform in a compatible manner leads to a wide spectrum of the resolved shear stress for the dislocation motion. Therefore, it is not to be expected that only one of the characteristic dislocation arrangements observed in single crystals will become established. When the stabilized microstructure of polycrystalline copper is compared to that of copper single crystals, first of all a distinction between surface grains and grains in the interior is necessary (e.g., Ref 103). PSBs are connected with high local plastic deformation, which is hard to realize in grains located in the interior of a material. It has been documented in various studies (e.g., Ref 97, 99-101, 103-106) that PSBs are also formed in the interior of a material, but that their number is much smaller than in the surface grains (about 10% smaller). As with single crystals, three plastic-strain amplitude regions can be distinguished by means of prevailing dislocation arrangement. At low amplitudes, loose veins are formed; in an intermediate range, veins with PSBs are observed; and at sufficiently high values of $\Delta\varepsilon_{pl}/2$, cells prevail. Comparing the boundaries of these regions with those of the regions A, B, and C of the single-crystal deformation, it becomes evident that as a direct consequence of compatibility constraints, the single-slip dislocation arrangements are restricted to relatively low plastic-strain amplitudes. That means that the cell structure expands its region of dominance drastically in polycrystals ($\Delta\varepsilon_{pl}/2 > 6 \times 10^{-4}$ for copper).

Different dislocation arrangements may exist not only in different grains of a polycrystalline material, but also within a single grain providing the physical basis for the development of a "composite-grain" model (Ref 107). This is illustrated by the TEM micrograph shown in Fig. 19. In the upper part, the ladder-like structure of PSBs embedded in a bundle/vein structure can be seen, whereas the lower part shows a kind of labyrinth structure. If a higher amplitude is applied, mainly cells with long or equiaxed shape form. Beside these cells, labyrinth structures can be found that exhibit misorientations leading to differences in brightness (Fig. 20).

Whether the CSS curves of single crystals and polycrystals are (in certain regions) related to each other depends strongly on the orientation factor M used to convert the loading quantities:

$$\frac{\Delta\sigma}{2} = M \cdot \tau_a \qquad \text{(Eq 13)}$$

$$\frac{\Delta\varepsilon_{pl}}{2} = \frac{\gamma_{ap}}{M} \qquad \text{(Eq 14)}$$

The well-known Taylor factor, $M = 3.06$, is based on the assumption of multiple slip in all grains, whereas the Sachs factor, $M = 2.24$, is reasonable if single slip exists even in the mostly stressed grains. The maximum value of the locally effective shear stress results from the reciprocal Schmid factor, $M = 2$, which demands that both slip plane and slip direction be ideally oriented (i.e., 45° with respect to stress axis).

Whether the CSS curve of polycrystalline copper exhibits a plateau analogous to the behavior of single crystals has been discussed in the literature very inconsistently. Systematic studies on the effect of grain size (e.g., Ref 108-113) have shown that in tests where $\Delta\varepsilon_{pl}/2$ is constant and the grains are small or medium, no plateau exists. Small horizontal regions in the CSS curves of polycrystals (e.g., Ref 42,102,114) either are due to a special testing mode or are a consequence of very large grains. In the case of copper, at large grain sizes a "hard" annealing texture affects the

CSS behavior and leads to a "reversed" grain size effect (i.e., lower saturation stress amplitude for smaller-grained material) (e.g., Ref 111), whereas nickel shows a normal grain size effect (Ref 115).

At low values of $\Delta\varepsilon_{pl}/2$, the CSS curves of copper single crystals and polycrystals coincide if the Sachs orientation factor is applied (Ref 80, 97). This is easily understood, because single slip prevails in the low-amplitude region. However, with increasing values of $\Delta\varepsilon_{pl}/2$, secondary slip starts early in polycrystals. This gives rise to high stress amplitudes, which correspond to region C of single crystals at much higher shear strain amplitudes. Hence, the application of the Taylor factor leads to a reasonable "average" transformation of single-crystal to polycrystal data (Ref 116).

In view of the fatigue cracks that result from the action of PSBs, it is interesting that the lower limit of the existence of PSBs in polycrystals can be estimated by using the Sachs or the Schmid factor to convert single-crystal data (i.e., the transition from region A to region B).

In the case of wavy-slip material, the characteristic dimensions of the dislocation arrangement are in unique relation to the loading amplitude. At sufficiently high amplitudes the mean diameter of the cells, d_c, existing in cyclic saturation serves as a geometric quantity. In 1966, Pratt showed in experiments at room temperature (Ref 117) and at the temperature of liquid nitrogen (Ref 118) that the saturation stress amplitude of polycrystalline copper is inversely proportional to d_c. Since then, this finding has been confirmed for many other metals and alloys (e.g. Ref 119-121).

Figure 21 represents this relation in the form:

$$\frac{\sigma_s - \sigma_F}{E} \propto \frac{\mathbf{b}}{\mathbf{d_c}} \qquad \text{(Eq 15)}$$

where σ_F denotes the stress that must be exceeded to overcome the friction of the dislocations in the crystal lattice (friction stress) and **b** is the length of the Burgers vector. The values of **b** and E are used in Eq 15 to apply this relation to different materials and temperatures.

Figure 21 shows some results of monotonic tests. This indicates that Eq 15 is not restricted to cyclic loading only. However, the establishment of an equiaxed cell structure under monotonic loading normally requires large plastic strains or high temperatures (e.g., Ref 122).

At small values of $\Delta\varepsilon_{pl}/2$, the mean cell diameter cannot be used as the characteristic quantity of the dislocation structure, because cells do not form or exist in a few grains only. Then a relation similar to Eq 15 holds under single-slip conditions, if the width of the channels in the bundle/vein structure and the distance between the walls in the PSBs are used instead of d_c (e.g., Ref 123-125).

Secondary Cyclic Hardening (fcc, Wavy Slip)

Cyclic deformation of a metallic material starting from a dislocation-poor (annealed) condition gives rise to a hardening that merges into cyclic saturation. As mentioned above repeatedly, the dislocation arrangement of saturation condition is considered to be in a quasisteady state; that is, no changes can be observed with cycle number. As a consequence, the CSS response (hysteresis loop shape and dimensions) remains constant.

A more detailed analysis of the deformation behavior indicates that a state of cyclic saturation does not exist in the strict sense. During cyclic saturation, minor microstructural processes take place that lead to a gradual change of the dislocation structure. If this change is observable by an increase of $\Delta\sigma/2$ in the cyclic hardening curve, a "secondary cyclic hardening" occurs. Its existence depends mainly on whether this rise (starting from σ_s) begins earlier than fatigue failure occurs.

Figure 22 shows two cyclic deformation curves of polycrystalline copper measured at $\Delta\varepsilon_{pl}/2 = 5 \times 10^{-4}$. The curves were obtained in air and vacuum, respectively. Because the crack formation occurs normally at the surface and is affected by the surrounding environment, tests in vacuum lead to prolonged cyclic life and may give the opportunity to observe processes that take place at cycle numbers above N_f in air. The lifetime extension in Fig. 22 approximately corresponds

to a factor of 10. The course of the curves is identical up to the number of cycles to fracture in air. Various studies on the development of dislocation arrangement and the degree of surface coverage with PSBs have proved that the environment has no influence, provided that samples are compared that have been cyclically loaded up to the same cycle number in air and in vacuum (e.g., Ref 100, 101, 106, 126-128).

The increased cyclic lifetime in vacuum in Fig. 22 enables the detection of secondary cyclic hardening, which means that after the plateau (corresponding to cyclic saturation) the stress amplitude steadily increases again. At the amplitude considered in Fig. 22, the drop of $\Delta\sigma/2$ in air, which occurs shortly before failure due to crack propagation, coincides with the beginning of secondary hardening (observable in vacuum). At higher amplitude, secondary hardening can also be found in air, because the stress amplitude starts to rise at cycle numbers less than N_f in air.

The microstructural processes responsible for secondary cyclic hardening have been studied best on single crystals (Ref 126, 127). When a single crystal is deformed in region B of the CSS curve (Fig. 12), the primary ladder structure of the PSBs hardens by a gradual increase of the density of secondary dislocations. Contrast experiments in TEM have documented that secondary dislocations are preferentially pinned at the boundary between PSBs and matrix. The local shear strain amplitude is continuously decreased and the PSBs are gradually transformed into a cell structure. The hardening of PSBs forces the formation of new PSBs in the matrix. This leads to increasing coverage of the surface with PSBs. The process continues until the specimen fails or a cell structure exists throughout the whole volume.

Secondary cyclic hardening is responsible for a passivation of PSBs, which can be found if a test is interrupted at certain numbers of cycles and a surface layer is removed each time by polishing. Some formerly active PSBs no longer form slip traces at the surface, but additional new PSBs are initiated. This process may give the appearance of slip homogenization in vacuum (compared to air) if failed samples are studied, but it is simply a consequence of the different cyclic life in air and vacuum (Ref 106, 126-129).

The microstructural processes that take place in polycrystals are basically identical to those discussed thus far, but they are hard to quantify due to the heterogeneity of the dislocation arrangement. It is typical of specimens that have undergone secondary cyclic hardening that a cell structure prevails and that almost no matrix structure exists, whereas in specimens that were cyclically loaded up to a cycle number before the onset of secondary hardening, a high portion of single-slip arrangements can be found. This observation is in accord with studies on alloys (Ref 130, 131) in which secondary hardening was observed only when "transformable" dislocation structures (Ref 131) were completely converted into "nontransformable" arrangements.

The description given above indicates that the term *cyclic saturation* must be used with care. Obviously, even during cyclic deformation in the saturation state, a continuous small deviation from a steady-state deformation process exists, leading under certain circumstances to a hardening by means of a cumulative process. This behavior can be understood if the dynamic equilibrium between dislocation multiplication and annihilation is considered, which is required for a steady-state cyclic deformation. The primary dislocations quickly reach a high density during initial hardening, so that a dynamic equilibrium can be established. In comparison, the formation of secondary dislocations at low plastic-strain amplitudes is a slow process, and the secondary dislocation density is too low to cause steady-state conditions.

fcc Metals With Planar Slip

The idea that planar slip results from the presence of short-range order seems to be at least a promising starting point for a necessary improvement of the classical textbook statement that planar dislocation slip takes place in materials of low stacking-fault energy (due to a reduced ability of screw dislocations to cross slip).

As a rule of thumb, the cyclic deformation behavior of alloys is similar to that of the base metal in a relatively wide range of composition. Only at high concentrations of alloying elements do significant deviations exist. Typical planar-slip materials are Cu-Al alloys, Cu-Zn (α-brass, e.g., Ref 132) and some austenitic stainless steels at low amplitudes (e.g., Ref 133).

Very fundamental studies on the microstructural development in planar-slip materials were conducted by Feltner and Laird (Ref 27, 31), who compared copper with a Cu-Al alloy of 7.5% Al. In a planar-slip alloy, dislocations are localized in discrete bands between which relatively dislocation-poor regions exist. These bands form parallel to the primary slip plane and contain mainly parallel edge dislocations of the primary slip system. If the plastic-strain amplitude is raised, the dislocation density increases and the distance between the bands decreases. Even at very high values of $\Delta\varepsilon_{pl}/2$ (corresponding to $N_f < 10^4$), no cell formation can be observed.

The alloy system Cu-Zn (α-brass) was studied by Lukáš and Klesnil in the early 1970s (Ref 28-30), who applied loading amplitudes to single crystals that corresponded to numbers of cycles to fracture in the range between 4×10^4 and 4×10^6. Whereas α-brass containing 15% Zn showed a behavior very similar to that of pure copper, brass with 31% Zn exhibited planar-slip behavior. TEM micrographs of foils cut parallel to the primary slip plane showed long segments of primary dislocations predominantly of edge character (Fig. 23). In (121) sections (i.e., perpendicular to the slip plane and parallel to the slip direction of the primary slip system), the localization of the dislocations in single active slip planes was visible. In contrast to copper, there were no detect-

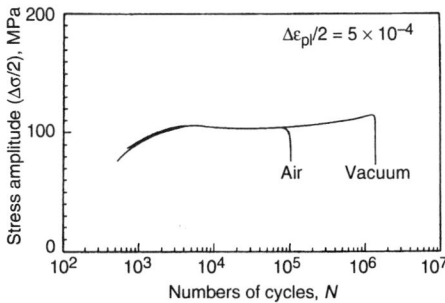

Fig. 22 Cyclic deformation curve of polycrystalline copper (mean grain size 55 μm) in air and in vacuum. Source: Ref 106

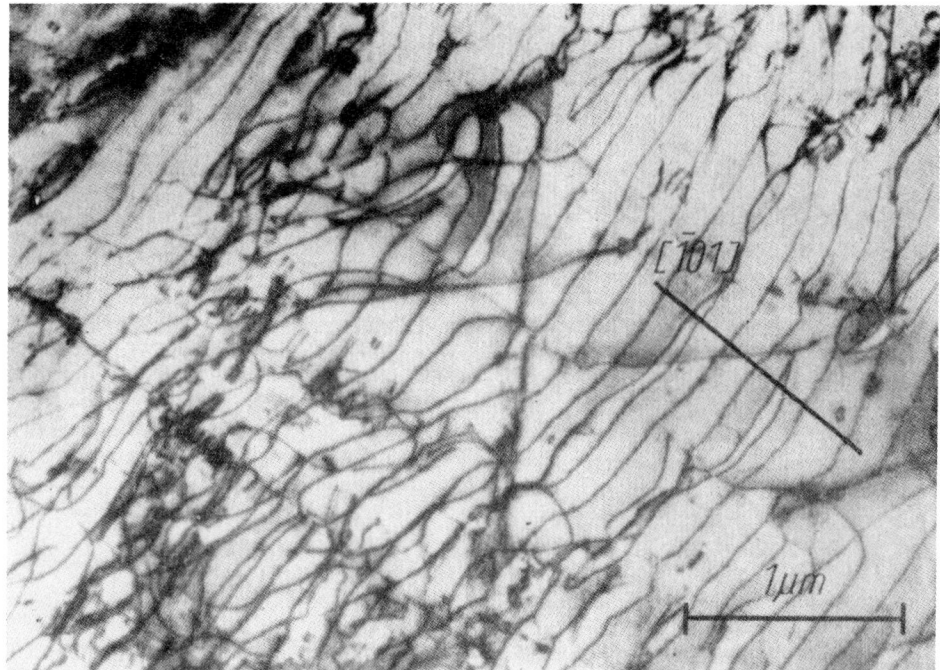

Fig. 23 Dislocation arrangement in a section parallel to (111) of a 69Cu-31Zn alloy (α-brass). Source: Ref 29

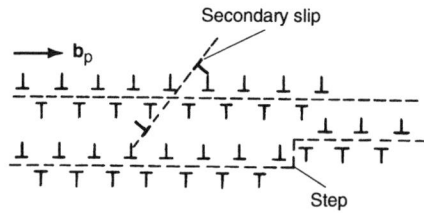

Fig. 24 Schematic representation of dislocation arrangement in a persistent Lüders band. Source: Ref 43

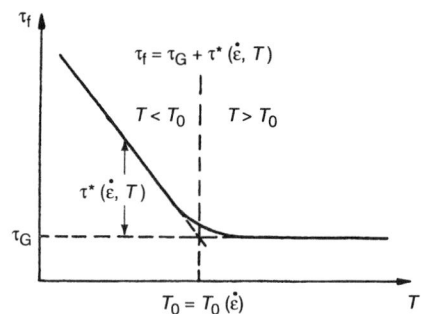

Fig. 25 Dependence of flow stress on temperature and strain rate in the case of bcc crystal structure. Source: Ref 144, 145

able differences between regions near the surface and those in the interior of the material.

Since these first systematic works, the alloy system Cu-Al has been the subject of investigations in many studies (e.g., Ref 134-136). Many efforts have been made to adjust the slip character specifically by means of the aluminum content. Regarding the short-range order as a possible origin of planar-slip behavior, Abel et al. (Ref 134, 136) observed that strain bursts occur in alloys with an aluminum content higher than 7%. These maxima of slip differ from those shown in Fig. 18, in that they are not formed during ramp loading but at constant amplitude and are less pronounced in the cyclic deformation curve. As a possible interpretation, strain bursts can be considered to occur as a result of a local destruction of short-range order by gliding dislocations. This leads to a softening of individual slip planes along which dislocations emitted from an active dislocation source can glide easily. As a consequence, the appearance of new slip bands/lines at the surface is connected with the occurrence of these strain bursts and changes in the shape of the hysteresis loop.

In newer studies on Cu-16Al (Ref 137-139), these fluctuations in the cyclic deformation curve were confirmed in single crystals as well as in polycrystals. The results obtained on single crystals indicate a (small) plateau in the CSS curve, which might result from a strain localization in slip bands (analogous to the behavior of wavy-slip single crystals). However, due to the slow cyclic hardening typical of planar slip, it is questionable whether a stabilized condition can be reached at all and whether therefore only approximately constant values of $\Delta\sigma/2$ at only a few amplitudes of plastic strain can be interpreted as a plateau of cyclic saturation.

The plastic deformation is almost completely localized in small bands, as proven by interferometric surface observations (Ref 137). These bands are termed persistent Lüders bands (PLB). The dislocation arrangement in a PLB is shown schematically in Fig. 24 and differs fundamentally from that in a PSB (Fig. 15b). PLBs are about 3 to 5 μm thick and are oriented parallel to the slip plane. They consist of neighboring planes that are fully packed with dislocations. In addition, secondary dislocations exist which cut the groups of primary dislocations while gliding on their slip planes. The local dislocation density within a PLB varies strongly.

Due to the lack of systematic studies of different planar-slip alloy systems, it is not known whether the formation of PLBs is a specific feature of the Cu-Al system or a general property of planar-slip materials. However, observations in the Cu-Zn (brass) system (Ref 140) document a considerable parallelism.

bcc Metals and Alloys

Influence of Temperature and Strain Rate. The aforementioned fatigue mechanisms seem to have good applicability to hexagonal close-packed (hcp) materials (e.g., Ref 141, 142) and to almost all technical bcc alloys (e.g., Ref 143). Therefore, the following discussion refers to differences and particularities in the cyclic deformation behavior and in the basic dislocation processes of the bcc crystal structure with respect to fcc metals and alloys. The conditions under which a deviant behavior exists need to be considered carefully (more detailed information is given in Ref 46 and 144).

The mechanical properties of pure bcc metals depend very strongly on temperature and strain rate and are sensitive to small amounts of interstitial impurities (such as carbon, nitrogen, and oxygen). In Fig. 25, a schematic representation of the temperature dependence of the flow stress, τ_f, is shown. According to Seeger (Ref 145), τ_f consists of a thermal component, τ^*, and an athermal component, τ_G:

$$\tau_f = \tau_G + \tau^* (\dot{\varepsilon}_{pl}, T) \tag{Eq 16}$$

As indicated in Eq 16, τ^* depends on temperature and plastic strain rate, $\dot{\varepsilon}_{pl}$, and is essentially a consequence of the extended core structure of the screw dislocations, which has a threefold symmetry and is the reason for a large pseudo-Peierls stress. Therefore, τ^* is caused mainly by lattice friction stress for screw dislocations, whereas in the fcc structure τ^* is determined by the dislocation/dislocation interaction. The gliding of the screw dislocations occurs by thermal activation, giving rise to the strong temperature dependence of τ^*.

The athermal stress component τ_G is usually considered to be connected with the dislocation density, ρ:

$$\tau_G = \alpha \, G \, \mathbf{b} \sqrt{\rho} \tag{Eq 17}$$

where α is a geometric constant ($\alpha \approx 0.1 \ldots 0.4$) that depends on the dislocation arrangement. The origin on τ_G is the elastic interaction of dislocations during deformation.

An increase in $\dot{\varepsilon}_{pl}$ has basically the same effect as a decrease in temperature, and vice versa. The transition temperature T_0 (Fig. 25), which defines

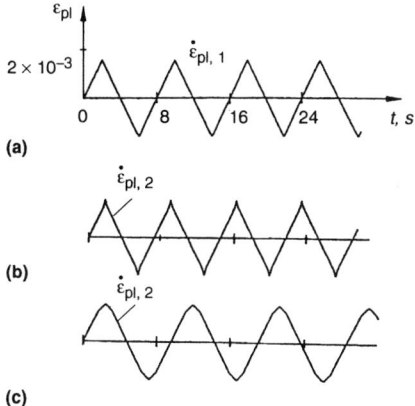

Fig. 26 Changes of plastic strain rate within each cycle. Source: Ref 146

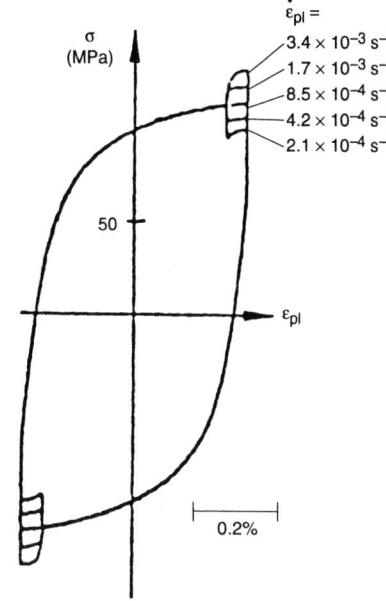

Fig. 27 Hysteresis loops with changes of $\dot{\varepsilon}_{pl}$, obtained on high-purity polycrystalline α-Fe. Source: Ref 15

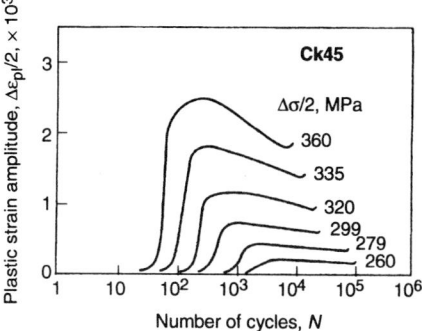

Fig. 28 Cyclic deformation curves of a normalized steel containing 0.45% C at various stress amplitudes. Source: Ref 153

the boundary between the high-temperature and the low-temperature behavior, is shifted to higher temperatures if the deformation rate increases.

A very illustrative example, in which the flow stress of α-Fe is affected by a change of the plastic strain rate, has been given by Mughrabi (Ref 15, 146). Figure 26 shows the course of the (controlled) plastic strain signal used in this study. Based on the triangular course of ε_{pl} in Fig. 26(a), which keeps the (absolute) value of $\dot{\varepsilon}_{pl}$ constant during the test, the strain rate is changed within each cycle. For this purpose the slope of the demand signal is increased or decreased shortly before the reversal points are reached. As a result, one restricts the changed plastic strain rate $\dot{\varepsilon}_{pl, 2}$ to a small time interval around the load reversals, whereas during the main part of the cycle, $\dot{\varepsilon}_{pl, 1}$ continues to exist.

The resultant change in the shape of the hysteresis loop can be seen in Fig. 27. According to Fig. 25, an increase of $\dot{\varepsilon}_{pl}$ leads to an increase of the stress, and vice versa. The shape of the hysteresis loop in the part of the cycle where $\dot{\varepsilon}_{pl}$ remains unchanged is not affected, indicating that the very short intervals with changed values of $\dot{\varepsilon}_{pl}$ do not alter the dislocation arrangement. Therefore, this testing method is an elegant way to determine the dependence of τ^* on $\dot{\varepsilon}_{pl}$ (and T) without changing τ_G at the same time (Ref 147).

The marked temperature dependence of the peak stress in Fig. 27 is due to the fact that cyclic deformation has been carried out in the low-temperature region of Fig. 25. This is a consequence of the rather high strain rate usually applied in a typical fatigue test (e.g., $\dot{\varepsilon} \approx 10^{-2} \times s^{-1}$). Under this condition, the temperature T_0 is shifted to values above room temperature. In other words, the cyclic deformation behavior of pure bcc metals is governed essentially by the low-temperature deformation mode. This mode is characterized by a large difference in the relative mobilities of screw and nonscrew (edge) dislocations. As a consequence of the pseudo-Peierls stress for the screw dislocations, their mobility is very limited.

A sufficiently small plastic-strain amplitude can be accommodated by a quasi-reversible to-and-fro displacement of the more mobile non-

screw segments while the screw segments are only stretched and do not move. In this region of cyclic microyielding, no microstructural changes take place and no cyclic hardening is noticeable. The CSS curve reflects this behavior by exhibiting a small plateau at low plastic-strain amplitudes.

If the material is subjected to an intermediate or high plastic-strain amplitude, the lack of mobile (primary) dislocations causes early secondary (multiple) glide. The more mobile edge dislocations draw out long screw dislocations. If the temperature is not too low, these screw dislocations are at least sufficiently mobile to interact and form cell structures. Consequently, there is a high tendency to form a cell structure at relatively small (compared to fcc metals) plastic-strain amplitudes.

If the screw dislocations are forced to glide in this low-temperature, high-amplitude regime, the threefold symmetry of the core structure leads to an asymmetric slip in tension and compression. In other words, slip occurs on different planes during tension and compression in each fatigue cycle, leading to a change in shape. In the case of a single crystal, an initially circular cross section becomes elliptical (Ref 148-151). The shape changes also play an important role in the fatigue failure of bcc polycrystals. In contrast to the constrained grains in the interior, the surface grains can develop shape changes due to asymmetric slip rather freely. A surface roughness results, with the periodicity of the grain structure leading to incompatibilities and stress concentrations at the grain boundaries. These stress concentrations give rise to crack initiation at these loci, along with a remarkable tendency to the development of intergranular cracks.

The behavior of bcc metals in the high-temperature regime (i.e., high temperature and/or low

strain rate) is quite similar to that of fcc metals. From an engineering point of view, it is important to note that small amounts of interstitial atoms enhance this tendency. Obviously, the effect of interstitials on the mobility of edge dislocations leads to a kind of balancing of the velocities of edge and screw dislocations. Under these circumstances (i.e., high-temperature behavior), PSBs have been observed that appear as dislocation-poor channels embedded in a matrix of veins (Ref 46, 144). PSBs are also known to form in both the surface and interior grains of polycrystalline low-carbon steels (Ref 152).

In summary, the cyclic deformation behavior of technical bcc metals and alloys is mostly very similar to that of wavy-slip fcc materials.

Dynamic Lüders Band Propagation. In any discussion of the CSS behavior of bcc metals and alloys, the phenomenon of dynamic Lüders band propagation must be considered, because it clearly relates the macroscopic deformation behavior to processes occurring on the atomic scale. It is well known that a large number of structural alloys exhibit discontinuous yielding in monotonic straining. This is typical for bcc metals containing interstitial atoms and showing distinguished upper and lower yield stresses in the monotonic stress-strain curve. Normalized low-carbon steels are probably the most important alloy system in this context.

The occurrence of Lüders band propagation under cyclic loading conditions depends very strongly on the testing mode applied. In the case of plastic strain control, the sample is forced to plastically deform up to the amplitude value of the control signal during the first quarter of the first cycle (where a quarter cycle is the same as a tensile test with Lüders band propagation). After the first or first few cycles, a homogeneous plastic deformation of the gage length prevails. The situation is even more definite if in a stress-controlled test a stress amplitude higher than the yield point is applied. Then, in order to reach this stress amplitude, the whole region of Lüders strain must be passed through completely within the first load increase up to the stress amplitude.

From a technical point of view, the case is far more interesting when the stress amplitude is lower than the yield point. In Fig. 28, the cyclic deformation curves obtained in a study on a normalized steel containing 0.45% C (with yield point between 420-440 MPa) are represented in a plot of the plastic-strain amplitude versus the number of cycles for different stress amplitudes. It can be seen that during the first cycles the material behaves almost purely elastically ($\Delta\varepsilon_{pl}/2 \approx 0$). After a certain cycle number, which depends on the stress amplitude applied, plastic deformation begins. The cyclic softening leads to a marked increase of $\Delta\varepsilon_{pl}/2$, which occurs earlier at high stress amplitudes. After deformation passes through a maximum of $\Delta\varepsilon_{pl}/2$, a continuous cyclic hardening follows.

The course of the curves depicted in Fig. 28, which is typical of normalized low-alloy steels under stress control (e.g., Ref 154, 155), can be described quantitatively on the basis of the microstructural processes in reasonable agreement with the experimental observations (Ref 156, 157). Qualitatively, the shape of the curve can be explained as follows. During the first half cycle,

a small number of dislocations are unpinned in individual isolated grains. This process is repeated during the following cycles and leads to an increase of the number of mobile dislocations. Hence, the material is in a plastically inhomogeneous condition, because besides the plastically deforming grains there are those that behave essentially elastically. The steady increase of the volume fraction of regions containing mobile dislocations leads to an increase in $\Delta\varepsilon_{pl}/2$. In these regions, the processes taking place are typical of the cyclic deformation of wavy-slip materials (i.e., dislocation multiplication and the formation of a heterogeneous dislocation arrangement). The resultant local cyclic hardening dominates when its effect is no longer overcompensated by the formation of new plastically deformable regions. The maximum in the cyclic deformation curve corresponds in a reasonable approximation to the condition in which for the first time the whole gage length undergoes plastic deformation.

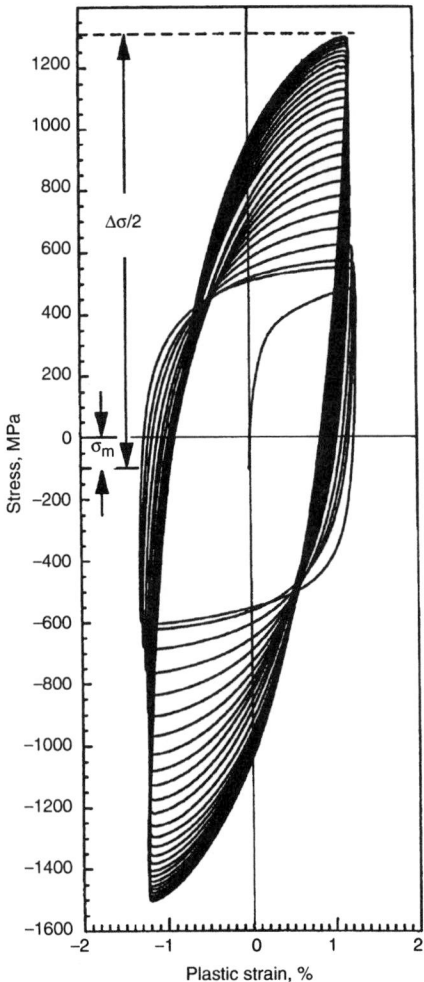

Fig. 29 Stress-strain path of the first 30 cycles at 103 K of a metastable austenitic stainless steel under plastic-strain control ($\Delta\varepsilon_p/2 = 1.26 \times 10^{-2}$). Source: Ref 158

Cyclic Deformation in Structural Alloys

The role of microstructure in the cyclic deformation of structural alloys depends, to a large extent, on the operative strengthening mechanism. For example, ferrite-pearlite structural steels are (at least) two-phase materials because they contain pearlite (α-Fe + Fe$_3$C). Nevertheless, carbon steels can be treated as "quasi-single-phase," because plastic deformation processes are almost completely restricted to the ferritic grains up to a carbon content of about 0.45% (Ref 154). In this case, the second phase does not have a pronounced effect on deformation mechanisms. In contrast, age-hardenable alloys are very dependent on a second phase for increased strength. The effect of strengthening particles on the cyclic deformation is addressed briefly in this section after discussion of deformation-induced phase transformation. Selected examples, which illustrate marked phase transformations caused by cyclic plastic deformation, are described for some metastable alloy systems that are prone to phase transformations. The mechanisms depend very strongly on the particular material, and therefore a comprehensive discussion is far beyond the scope of this article. Nonetheless, the two examples below provide some insight into phase transformation effects on the CSS response.

Deformation-Induced Transformations

Austenitic Strainless Steels. Phase transformations during cycling have been studied particularly carefully in steels where austenite-martensite transformations occur. Cyclic-deformation-induced martensitic transformation is beneficial in high-strength TRIP (transformation-induced plasticity) steels and in lower-

strength metastable austenitic stainless steels, depending on the testing mode and the amplitude. The following example describes deformation-induced martensite formation in an austenitic stainless steel.

At sufficiently low temperatures (below the martensite start temperature, M_S) a spontaneous martensitic transformation takes place. This temperature limit can be raised by means of elastic stresses or plastic deformation, which induce an additional driving force. The threshold temperature for deformation-induced transformation is termed M_d.

Figure 29 shows the stress-strain path of the first 30 cycles of an austenitic stainless steel deformed under plastic-strain control with $\Delta\varepsilon_{pl}/2 = 1.26 \times 10^{-2}$. The test was carried out at 103 K in order to promote martensite formation. The stress amplitude increased strongly, whereas the hardening rate decreased slowly. The maximum stress amplitude was reached after about 40 cycles. The marked cyclic hardening was mainly due to deformation-induced martensitic transformation of austenite, which has been investigated in several steels of the 300 series (Ref 159-167).

The formation of α'-martensite can occur directly from austenite or via the intermediate phase ε-martensite. The transformation starts if a threshold value of the plastic-strain amplitude is exceeded and a certain cumulative "incubation" plastic strain (which increases with decreasing amplitude) is reached. In the case of the alloy corresponding to Fig. 29, it could be shown unambiguously that the martensite forms as a result of cyclic plastic deformation (i.e., deformation-induced and not stress-induced), because quenching (even down to the temperature of liquid helium) and purely elastic cyclic loading did not trigger austenite transformation.

The increase of $\Delta\sigma/2$ shown in Fig. 29 can be attributed directly to an increase in the volume fraction of martensite, as shown by independent measurements of the martensite content. The negative mean stress σ_m that develops during cycling stems from the volume expansion (the martensite formation is connected with a volume expansion of about 2%). It is interesting to note that by means of an optimized low-temperature cyclic predeformation, the residual room-temperature fatigue life can be improved by a factor of almost 2.

Commercial Age-Hardened Aluminum Alloy. Figure 30 shows various cyclic deformation curves obtained on an aluminum alloy in three different conditions (Ref 168). The material studied is a commercial age-hardenable aluminum alloy (corresponding to aluminum alloy 7022) that was heat treated prior to fatigue testing in three different ways: to obtain a precipitate-free condition, a peak-aged precipitation structure, and an overaged condition. Surprisingly enough, in the initially precipitate-free condition the stress amplitude increased drastically during cyclic deformation and exceeded even the corresponding values of the peak-aged condition after suffi-

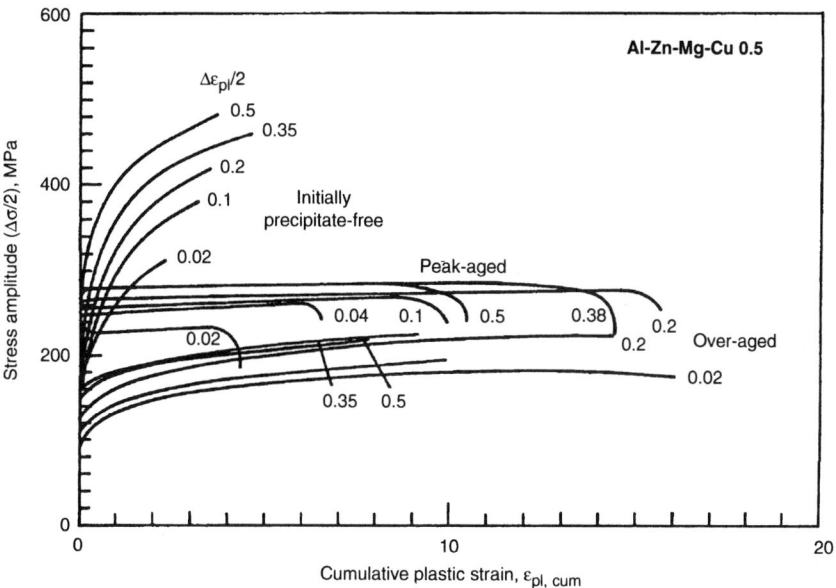

Fig. 30 Cyclic deformation curves of three conditions of an Al-Zn-Mg-0.5Cu alloy at various values of $\Delta\varepsilon_{pl}/2$. Source: Ref 168

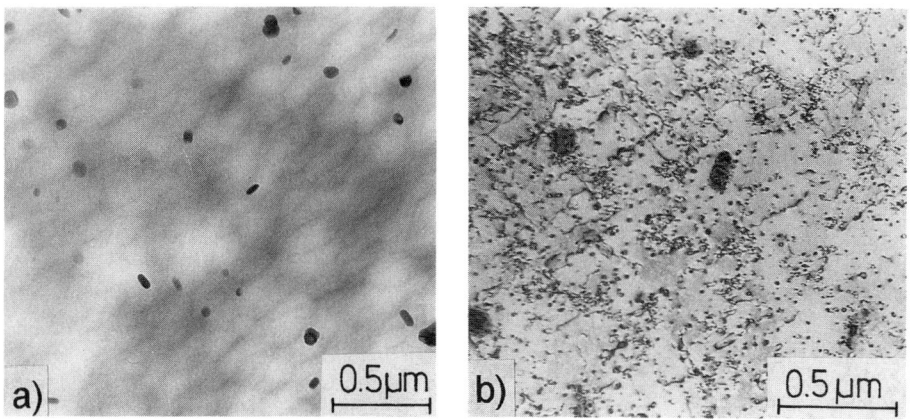

Fig. 31 Effect of cyclic deformation on Al-Zn-Mg-0.5Cu in the initially precipitate-free condition. (a) Before cyclic loading. (b) After cyclic loading at $\Delta\varepsilon_{pl}/2 = 2 \times 10^{-4}$. Source: Ref 168

ciently high cumulative plastic strain. The gain in strength took place at the expense of cyclic lifetime.

An interpretation of the mechanical behavior can be given on the basis of corresponding TEM studies. In Fig. 31(a), the microstructure of the quenched condition is shown as it exists prior to cyclic deformation. Particles of the hardening second phase are not detectable. Only the relatively large dispersoids are present, because they cannot be dissolved during the solution annealing. The reason for the tremendous increase of $\Delta\sigma/2$ during cycling becomes evident in Fig. 31(b). Even at the lowest value of $\Delta\varepsilon_{pl}/2$ used, very small and uniformly distributed precipitates are formed that interact very strongly with the moving dislocations. The TEM micrographs indicate that particle cutting is the most important deformation mechanism. The deformation-induced precipitation is more pronounced with higher values of $\Delta\varepsilon_{pl}/2$. Similar cyclic-induced

precipitation processes have been reported to occur in other aluminum alloys (e.g., Ref 169, 170).

Effect of Second-Phase Particles

Precipitation hardening is an important and commonly used method to raise the strength of alloys. The particles of a second phase act as obstacles to the dislocation motion and markedly influence the dislocation arrangement compared to the precipitate-free situation. The large number of precipitation-hardening alloy systems and the broad range of possible dislocation-particle interactions make a general and uniform summary beyond the scope of this article. Therefore, this section focuses on the underlying mechanisms and tries to illustrate the main effects in selected alloy systems. More detailed information can be found in numerous review articles (e.g., Ref 171-173).

The most important parameters determining particle-dislocation interaction are the size and distance of the precipitates. If the strengthening particles are coherent, small, and close-spaced, they are cut by dislocations as a result of the resolved shear stress. This process is typical of underaged alloys. Overaging leads to a relatively large particle diameter (in combination with a large mean particle distance in the slip plane) at which dislocation motion between the obstacles by the Orowan mechanism (bowing out of the dislocations and finally forming Orowan dislocation loops around the precipitates) is favored. The same is true if the particles are incoherent (e.g., dispersoids).

Single-Crystal Alloys

Shearable Precipitates. Systematic studies of the effect of particle strengthening on fatigue behavior have been carried out, especially in the systems Al-Cu (e.g., Ref 174-187) and Cu-Co (e.g., Ref 188-192). Al-Cu alloys are very well suited to these investigations, because the precipitation behavior, the sequence of the phases forming, and the resultant effects on monotonic properties are well known (e.g., Ref 193). Furthermore, in Cu-Co a condition of solid solution can be established as a reference state that exhibits the dislocation arrangement of copper reasonably well.

If particle-strengthened single crystals are subjected to a cyclic deformation at constant plastic-shear amplitude, first a cyclic hardening up to a maximum of the shear stress amplitude is found, followed by a marked softening. This behavior is observed in a wide range of γ_{ap} values. A state of cyclic saturation is not established in air (but is sometimes found in vacuum as a consequence of an extended lifetime), so that a CSS curve cannot be determined according to its definition. Following a suggestion of Wilhelm (Ref 194), a pseudo-CSS curve can be constructed, taking the maximum values of τ_a instead of the saturation values τ_s for those plastic shear strain amplitudes at which no cyclic saturation is reached.

Figure 32 shows the corresponding representation for Al-3.8Cu. Analogous to the CSS curve of copper (Fig. 12), three regions A, B, and C can be distinguished, and region B defines a marked plateau. Whereas in region A a state of cyclic saturation is reached, in region B the above-mentioned pronounced maximum in the cyclic deformation curve is observed. Therefore, the values of τ_a represented are essentially higher than the true saturation stress amplitudes. The reason for the plateau stress lies in the tremendous change in the dislocation arrangement that takes place during the initial phase of hardening, shortly before the maximum of τ_a is reached. PSBs are formed which, however, differs essentially from those observed in pure metals and single-phase alloys. These PSBs are very thin (<0.1 µm), and they can carry large local plastic strains of 30 to 60% (Ref 181). The volume fraction of PSBs increases with increasing γ_{ap} in region B, but a value of 100% is never reached. In contrast to single-phase wavy-slip materials,

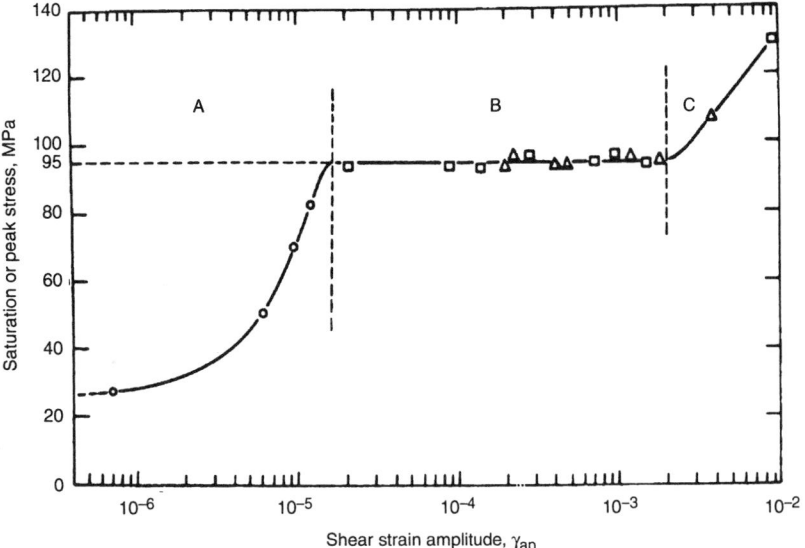

Fig. 32 Pseudo-CSS curve of Al-3.8Cu single crystals containing shearable precipitates. Source: Ref 179

no cell structures are formed in region C. Rather, PSBs exist in different intersecting slip systems.

The pronounced cyclic softening (in region B), which is undesirable in technical applications, is caused by the destruction of the strengthening effect of the particles within the PSBs as a result of particle shearing. If the precipitates are not ordered, as is the case in Cu-Co, the repeated cutting of the particles by the to-and-fro motion of dislocations may ultimately lead to a complete redissolution when the particles become too small to be stable. This random chopping-up process has been reproduced by computer simulation (e.g., Ref 195). The dislocation slip character can be considered wavy. This is also confirmed by the width of the PSBs formed (about 0.1 μm).

Especially the ordered γ′ precipitates in superalloys suffer shearing by planar slip bands that consist of only one or few slip planes (e.g., Ref 196). These planar bands result from a local softening along active slip planes, which itself is a consequence of a reduction in strength of the sheared particles due to the loss of order.

In Al-Cu, Laird et al. (Ref 180, 176) did not find destruction of the particles within the PBSs. The contrast provided by the slip bands in TEM could be attributed to an orientation difference between PSBs and the surrounding matrix. By slightly tilting the sample in the microscope, the precipitates in the PSBs could be detected again. This observation led to the interpretation that the loss of order within the θ″ particles is responsible for softening. However, in other aluminum alloys a redissolution of the ordered precipitates within the PSBs has been proven to occur (e.g., Ref 197).

Cyclic plastic deformation can also strongly affect the kinetics of precipitation and the coarsening processes, due to the high point defect concentrations connected with dislocation reactions. Therefore, processes such as overaging or

Ostwald ripening may occur in an accelerated manner, especially if areas of strain localization exist. In many aluminum alloys, aging processes take place at room temperature, leading to a particle size distribution that is a strong function of time and affects the CSS response.

Non-shearable Particles. With increasing overaging, cyclic hardening with subsequent softening ceases, and a cyclic saturation state is established. In this case, the CSS behavior depends much more strongly on the crystal orientation than in the situation when shearable particles are present (Ref 183). The strain is not localized, because the relatively coarse precipitates reduce the slip length and prevent the formation of PSBs. Therefore, a plateau in the CSS curve is not observed. Furthermore, cyclic softening, which must be considered a consequence of strain localization, does not occur.

At small and moderate strain amplitudes, the dislocation arrangement in Al-Cu consists of dipolar walls that are formed mainly from primary dislocations and are located at the broad sides of the θ′ plates. With increasing amplitude the portion of secondary dislocations in the networks surrounding the θ′ plates increases. The θ′ plates form the basic skeleton of the dislocation arrangement.

Because cutting of particles does not take place, the precipitation structure is quite stable during cyclic loading. As a consequence, changes in test frequency or test interruptions have only minor influence on the CSS behavior (e.g., Ref 184, 185).

Pure Polycrystals

The results obtained on precipitation-hardened single crystals can be transferred to polycrystals provided that these materials are pure (i.e., that they contain only elements necessary to form the hardening second phase). In addition to the orientation differences between the grains of a poly-

crystal, the grain boundaries themselves influence the deformation behavior, because the precipitation conditions in the vicinity of grain boundaries can be different from those in the matrix.

In Fig. 33, cyclic deformation curves are depicted that were measured on Al-4Cu polycrystals by Calabrese and Laird (Ref 176-178) using three different precipitation states. In the case of shearable coherent θ″ particles (Fig. 33a), a course is found that exhibits a maximum and resembles that of the corresponding single crystals. The initial hardening is attributed to an increase of the friction stress due to the coherent stress fields of the precipitates, the formation of anti-phase boundaries during cutting, and mainly the occurring dislocation accumulation. Even before the stress maximum is reached, slip bands are visible on the surface that are formed first in favorably oriented grains. These surface bands correspond to the deformation bands (PSBs) that could be found in the grain interior by means of TEM. The cyclic softening can (in the case of Al-Cu) be traced back to the disordering of the ordered precipitates due to the repeated cutting by dislocations within the PSBs.

The cyclic deformation curves of the conditions with nonshearable particles (Fig. 33b, c) are fundamentally different. At small values of $\Delta\varepsilon_{pl}/2$, cyclic saturation occurs almost immediately; at high values it occurs after few cycles. Monotonic and CSS curves are approximately identical, indicating the stability of the precipitation condition and the dominant role of the precipitates regarding the deformation behavior. This is confirmed by the results of accompanying TEM studies. The particles are the dominant obstacles to dislocation motion as long as the amplitude is not too high. Dislocations are preferentially located at the phase boundary particle/matrix. The plastic strain can be accomplished by the slip of only a few dislocations between the particles, because the "mean cell size" is determined by the small distance between the precipitates.

If a high loading amplitude is applied to a material with relatively large particle distance, a real cell formation can occur. In other words, additional dislocation walls form between the precipitates. The saturation stress amplitude is determined by the same mechanisms that prevail in single-phase alloys. Dislocations bow out from the cell walls, cross the dislocation-poor cell interior, and react with the opposite wall. Therefore, the deformation behavior of the dislocation cell structure gains importance with increasing $\Delta\varepsilon_{pl}/2$.

The question of whether there is a plateau in the CSS curve of polycrystals containing shearable particles, as found in the pseudo-CSS curve of corresponding single crystals (Fig. 32), was checked in a study on Al-Cu with three different grain sizes (0.15, 0.32, and 1 mm) (Ref 182). Only the coarsest grain structure led to a very small plateau. With decreasing grain size, the behavior deviated more and more from that of single crystals, in accordance with the expecta-

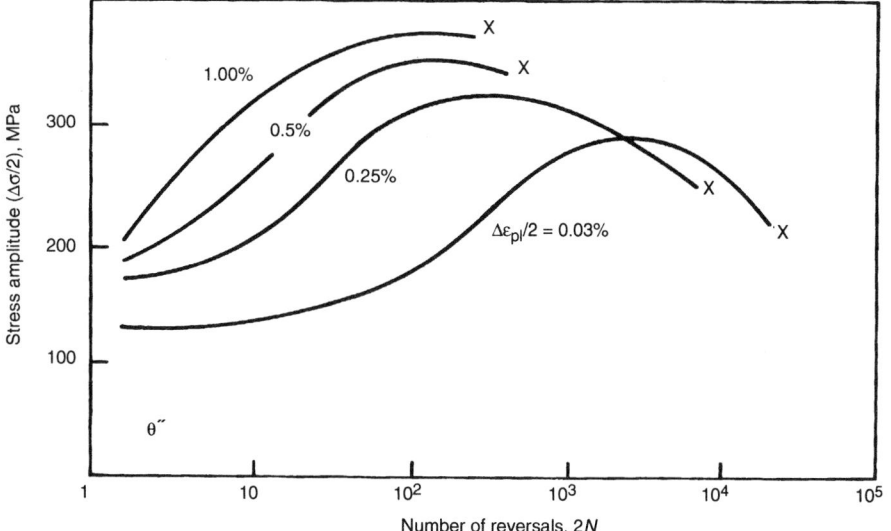

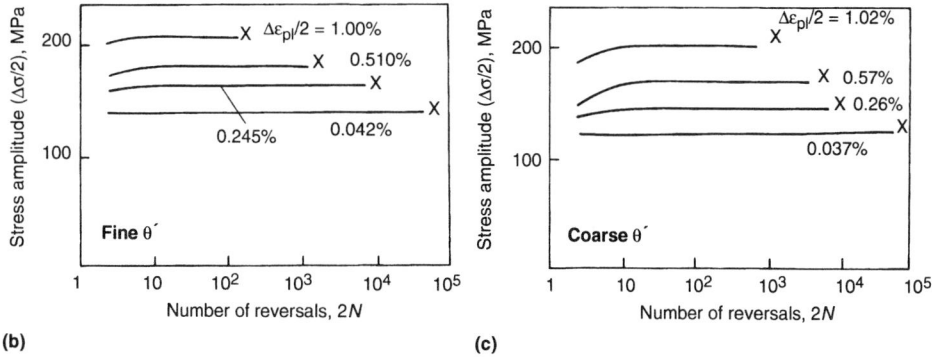

Fig. 33 Cyclic deformation curves of polycrystalline Al-4Cu in different conditions. (a) θ″. (b) Fine θ′. (c) Coarse θ′. Source: Ref 176-178

Table 1 Effect of microstructural modifications on the fatigue resistance of alloys

Modification to microstructure	Shearable precipitates	Precipitate-free zones
Overaging	Improves	No effect
Dispersoids	Improves	No effect
Unrecrystallized structures	Improves	Improves
Reduction of grain size	Improves	Improves
Steps in grain boundaries	No effect	Improves
Alignment of grain boundaries	No effect	Improves

Source: Ref 198

tions based on the observations on single-phase materials.

Grain boundaries can be of significance for the distribution of precipitates, leading to a modification of the particle population. Usually relatively coarse particles precipitate on grain boundaries, whereas in the vicinity of grain boundaries, precipitate-free zones may exist. These areas are mechanically softer than the grain interior, so that localized deformation is promoted. A preferential crack initiation at these locations may result.

Commercial Alloys

In more complex, commercial structural alloys, the fatigue properties (including both resistance against crack initiation and crack propagation) are a function of not only the precipitation condition but also the whole microstructure. For example, the effect of two microstructural features (shearable precipitates and precipitate-free zones) on fatigue resistance are summarized in Table 1. The effect of specific microstructural features on the fatigue properties depends on the degree of plastic-strain localization, which is primarily determined by the slip length and degree of age hardening. Because extensive age hardening and corresponding high yield strength are desirable in commercial alloys, emphasis often is placed on improving the fatigue life by reducing the slip length (reducing grain size, for example). Intermetallic inclusions, however, have mostly a negative influence. They are too large to affect the stress-strain behavior. Rather, they can cause cracks and therefore determine the cyclic lifetime.

The relationship between microstructure and cyclic plastic-deformation behavior is known to be complex, especially for high-strength commercial alloys. It is clear, however, that any microstructural feature that causes an inhomogeneous distribution of plastic strain will have a deleterious effect on fatigue life. In general, a material that cyclically hardens will restrict cyclic plastic deformation, which is a desired effect. However, under strain control, the imposed strain in conjunction with the loss of ductility associated with the higher cyclic strength can lead to a poor low-cycle fatigue life.

The concept of reversible deformation also may not be applicable in complex, commercial alloys. As described above, slip reversibility is a desirable property for fatigue resistance. Planar-slip alloys, with fine, narrow slip bands, are more resistant to fatigue than wavy-slip materials, in which coarse slip bands lead to early crack initiation. Planar slip increases the reversibility of deformation and thus improves the fatigue resistance for some low-stacking-fault-energy materials. However, in commercial alloys strengthened by dislocation substructures, precipitates, and carbides, planar slip can cause softening in the slip bands, resulting in strain localization. Consequently, slip reversibility may have little or no meaning in complex, engineering-type materials (Ref 198).

Structural aluminum alloys contain various alloying elements that are added, for example, to adjust the proper grain size or to improve the corrosion resistance. Typically the following types of precipitated phases can be found (Ref 172):

- The main strengthening second phase is precipitated in the form of finely distributed particles. This phase is in most cases ordered and coherent. The shape of the particles depends on the alloy system; particles may be spherical, plate-shaped, or needle-shaped. Normally the distribution of these particles is uniform, except that precipitate-free zones form at grain boundaries.
- Dispersoids are formed during cooling-down in the solid state, and they can hardly be dissolved during the heat treatment that is necessary to adjust the proper microstructure. The main role of the dispersoids is to keep the grain size small by avoiding recrystallization after forming at high temperatures.
- Intermetallic inclusions grow up to several microns. They are mostly formed due to iron and silicon impurities in aluminum.

Besides the strengthening particles of the second phase, the dispersoids are important for the CSS behavior of technical alloys. Because they cannot be cut by the dislocations, they tend to prevent slip localization. This manifests itself by the lack of a pronounced softening region in the cyclic deformation curve (e.g., Ref 173). Cyclic lifetime is essentially prolonged, compared to that of pure alloys, as a result of the effect of dispersoids. A full discussion of the fatigue and fracture resistance of structural aluminum alloys is contained in the article "Selecting Aluminum Alloys to Resist Failure by Fracture" in this Volume.

Variables and Modeling

Cyclic stress-strain behavior and its relation with microstructure can be complex, because almost every structural alloy undergoes thermomechanical treatment prior to service. In order to classify the reactions of a material to predeformation or thermomechanical treatments, the terms *history dependence* and *history independence* are commonly applied. A history-independent CSS behavior means that after a transition stage, a saturation condition is reached that is exactly the same as the saturation state established in a nonpredeformed (well-annealed) sample. Accordingly, a history-dependent material never reaches the saturation state of the corresponding annealed specimen, even after extended cycling.

Effect of Predeformation

Set-Up and Control Effects. The effect of predeformation on the CSS behavior is complicated by the fact that the application of a particular start-up procedure (e.g., frequency and/or choice of control mode) may affect the saturation state via the dislocation arrangement established (Ref 199, 200). As an example, the plateau stress amplitude of copper single crystals varies between 24 and 33 MPa, depending on how saturation is reached (Ref 48).

This influence on the CSS response occurs because of the different transient deformation behavior under different testing modes. Usually an initially annealed sample first undergoes cyclic hardening before the cyclic saturation state is reached. If plastic-strain control is applied, $\Delta\varepsilon_{pl}/2$ is held constant during the test and cyclic hardening is moderate, because during the first few cycles the maximum (minimum) in plastic strain can be reached quite easily (i.e., at low load). Stress control, however, can force the sample in

the first cycles (soft condition) to plastically deform tremendously in order to accomplish the given stress amplitude. Therefore, it seems reasonable to treat these first cycles as a special type of predeformation. As a consequence, somewhat different microstructural processes (e.g., a higher portion of secondary slip) can occur that give rise to an altered stabilized behavior.

Systematic studies on single-phase materials show (e.g., Ref 201) that the CSS curve, which represents the stabilized condition as a function of amplitude, does not depend on the control mode applied during single-step fatigue testing if the material exhibits wavy slip and the amplitude of testing is not too low. Especially in the case of planar-slip materials, appreciable deviation can be observed in the high-cycle fatigue range (i.e., at low amplitudes).

Effect of Deformation History on Cyclic Deformation. In order to characterize the history dependence of cyclic deformation behavior, mostly either a monotonic predeformation (e.g., Ref 202-206) or a reduction of the cross section by means of swaging (e.g., Ref 26, 31, 207, 208) is applied. These types of predeformation lead to a high flow stress, so that a cyclic softening occurs during the subsequent cyclic deformation.

Differences between wavy-slip and planar-slip materials are again important. Materials showing wavy dislocation slip exhibit a history-independent CSS curve at sufficiently high amplitudes of the plastic strain ($\Delta\varepsilon_{pl}/2 \geq 5 \times 10^{-4}$ in the case of polycrystalline copper). In this region, the dislocation arrangements found after cyclic loading in specimens that were annealed prior to cycling are identical to those of predeformed samples. However, in the low-amplitude region, cold-worked and annealed samples do not show identical stress amplitude values in cyclic saturation established in tests where $\Delta\varepsilon_{pl}/2$ is constant. Here, the dislocation arrangement resulting from

predeformation is not completely changed to the low-amplitude fatigue configuration (Fig. 11), because the tensile deformation (or swaging) leads to a dislocation cell structure that cannot be converted into the single-slip structures typical of wavy-slip materials at low amplitudes.

In materials showing planar dislocation glide, the CSS curve is strongly influenced by predeformation in both the high-amplitude and the low-amplitude regions. In contrast to the mechanical behavior, the microstructure is not markedly affected and shows planar dislocation arrays located on distinct glide planes independent of the prehistory suffered.

The above results have been only partially confirmed by newer and probably more sensitive measurements (Ref 209, 210). Figure 34 shows cyclic deformation curves of α-brass (planar slip) observed after monotonic predeformation in tension up to certain strains and after swaging down to a reduction in diameter of 27%. The behavior of the reference condition (annealed) is represented as a dotted line. In spite of the relatively high plastic strain range of 5×10^{-3} applied during cyclic loading, a history-dependent behavior is present, because the saturation stress amplitude depends strongly on the degree of predeformation. This is in accordance with the observations reported above (e.g., Ref 26, 31).

However, deviations from the "literature behavior" are found in the case of wavy-slip materials. Figure 35 depicts the CSS curves determined on polycrystalline copper samples with and without prestraining. The saturation stress amplitude is raised with increasing degree of predeformation. Swaging, which corresponds to a tensile prestrain of about 87%, gives rise to the highest increase of σ_s. Figure 35 indicates that the influence of predeformation decreases with increasing plastic-strain amplitude, but that it does not cease to exist even at very high $\Delta\varepsilon_{pl}/2$ values.

From these observations it can be concluded that the CSS behavior is generally history-dependent. Therefore, any assumption of history independence must be considered as a specific

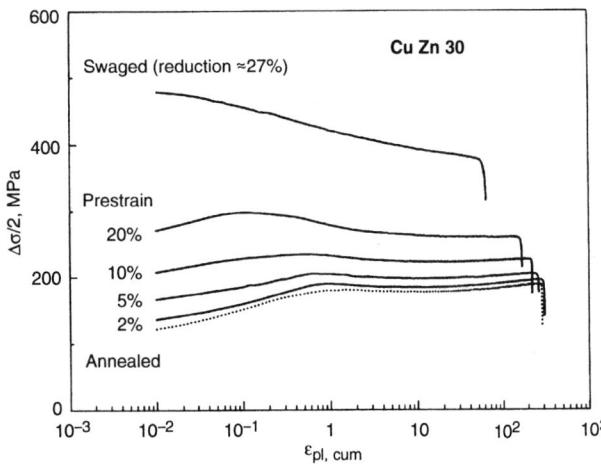

Fig. 34 Cyclic deformation curve of α-brass at $\Delta\varepsilon_{pl} = 5 \times 10^{-3}$ as a function of the degree of monotonic (tensile) predeformation. Source: Ref 209

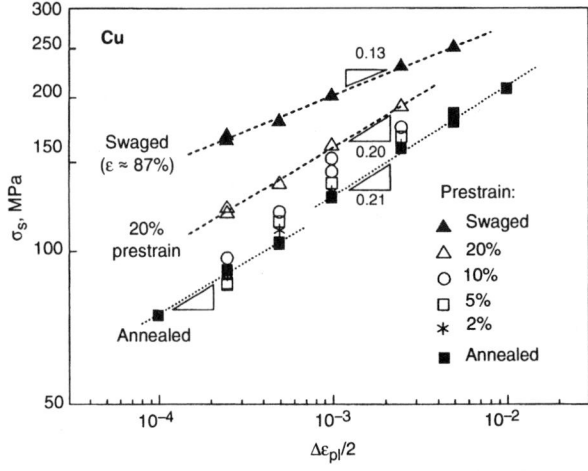

Fig. 35 Cyclic stress-strain curves of annealed and prestrained copper specimens. Source: Ref 210

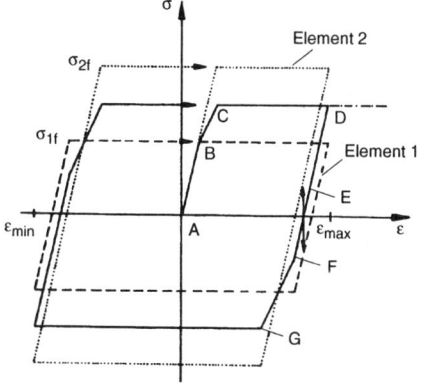

Fig. 36 Stress-strain path of a two-element composite assuming parallel arrangement and ideally elastic-plastic behavior. Source: Ref 220

technical limit, such that the deviation between the saturation state of annealed samples and that of predeformed samples does not exceed a given limit in each situation. Nevertheless, wavy-slip materials generally are less history dependent than planar-slip materials. The behavior of particle-strengthened alloys lies somewhere between these extreme cases. If cyclic deformation is mainly controlled by the dislocation/particle interaction (e.g., in top-aged condition) and particle cutting takes place, a planar-slip-type behavior can be expected. If, on the contrary, dislocation/dislocation interactions prevail due to large particle size, wavy-slip-type behavior may exist.

Effect of Amplitude Changes. As a consequence of the relationship between dislocation arrangements formed during cyclic loading and those formed during monotonic loading (e.g., Fig. 21), cyclic predeformation can be considered a special case of (repeatedly) monotonic prestraining in some respects. Again, the combination of wavy-slip behavior with high plastic-strain amplitude can be understood most easily. Under these conditions, a change from one constant value of $\Delta\epsilon_{pl}/2$ to another leads to a more or

less reversible adaptation of the dislocation cell structure. For example, if $\Delta\epsilon_{pl}/2$ is decreased, a rearrangement of the dislocations is possible in such a way that the dislocation cell size increases to the value that would have been established if the cyclic predeformation had not have been applied (e.g., Ref 26, 202). This rearrangement, which leads to the saturation microstructure of the initially annealed condition, can of course not (completely) take place if a change leaves the high-amplitude range by leading to an amplitude in the low-amplitude range in which single-slip prevails. Then the cell structure is preserved, and normally the stress-strain response differs from that of non-predeformed samples. This agrees with the assumption that cell structures are relatively stable dislocation configurations. However, some indications are given in the literature that under certain conditions a transformation of cells into single-slip arrangements might be possible (e.g., Ref 211).

In the case of planar-slip materials, an effect of the cyclic predeformation always applies, which is still not understood microstructurally. Usually this effect is more pronounced if the loading amplitude is lowered than if the amplitude is increased.

Sometimes multiple-step tests are applied in order to determine the CSS curve of a material in a sample-saving manner. From the discussion so far, it can be deduced that single-step tests are to be preferred for this purpose. If multiple-step tests are unavoidable, the tests should at least be carried out with step-wise increasing amplitude, small steps, and ample cycling at each step.

Incremental Step Test. A technique that has been proposed as a time-saving method to determine the CSS curve with one sample is the incremental step test (IST) (Ref 212, 213). In each block the amplitude is first increased linearly with time up to a maximum amplitude, then decreased to zero (or a minimum amplitude value). This loading sequence (block) is applied repeatedly until a stabilized behavior is established. Then the CSS curve is obtained simply by connecting the load reversal points in the stress-strain

course. From previous discussion, it appears very doubtful that this test procedure gives a CSS curve identical to that determined in single-step tests at different amplitudes. As described above, each amplitude leads to a characteristic dislocation distribution and arrangement that cannot be expected to exist at each single cycle of the IST.

In fact, it can be shown (Ref 214, 215) that a high number of blocks are required to attain a state of saturation or at least approximate saturation. In this state, where no further changes in the mechanical stress-strain response are appreciable, the microstructure is stabilized (i.e., it remains not only unchanged from block to block, but, more important, it is also independent of the immediately acting amplitude). A comparison of this microstructure with those existing in saturation of single-step tests documents that the IST leads to the same dislocation arrangement that is established in a single-step test at an amplitude slightly lower than the maximum amplitude used in the IST. In other words, the microstructural effect of the IST can be described by an "effective" amplitude, and therefore the CSS behavior on the IST differs from that of single-step tests, because it does not take into account the amplitude-dependent dislocation arrangement. However, this behavior at constant microstructure is of interest from an engineering point of view, because under service conditions, if the loading amplitude changes sufficiently fast and often, a similar behavior may exist.

Modeling of the CSS

In concluding this article, a very brief description of a basic concept used to model the CSS response on a physical basis is given. Numerous approaches have been reported, mostly either starting from a description of the individual dislocation behavior or based on plain numerical relations without appreciable physical foundation. However, a more reasonable and promising starting point for a microstructure-related de-

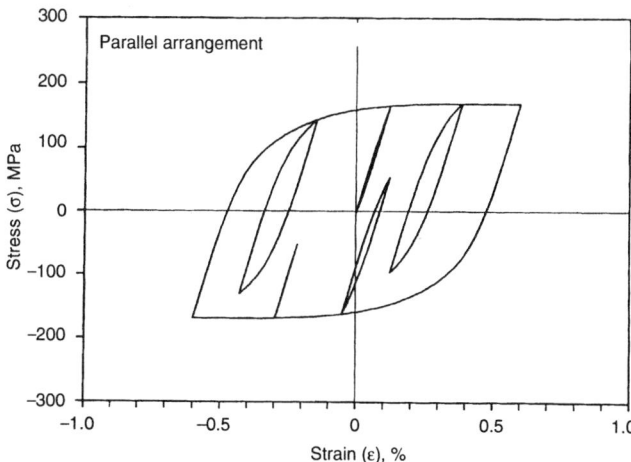

Fig. 37 Calculated stress-strain path for variable-amplitude condition showing the memory of prior load history. Source: Ref 220, 228

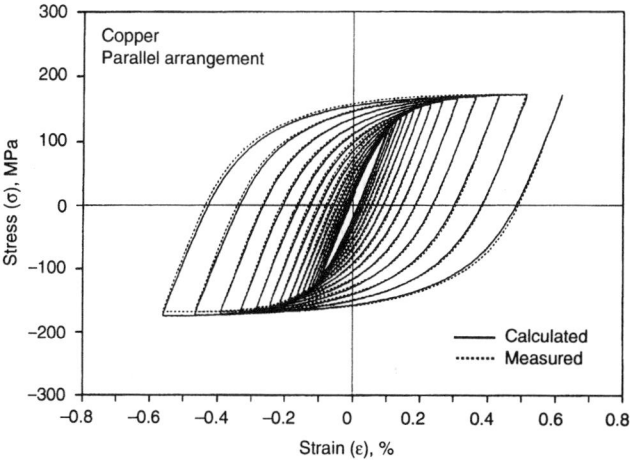

Fig. 38 Comparison of calculated and experimentally determined stress-strain courses of a half block of an incremental step test on copper. Source: Ref 220, 228

scription is provided by so-called multicomponent models.

Cyclic deformation (i.e., the macroscopic behavior) is mostly connected with the formation of a heterogeneous dislocation distribution (Fig. 13, 14, 16, 17, 19, 20). Independent of the details of the dislocation arrangements (the microscopic scale), dislocation-rich and dislocation-poor areas are formed within a few cycles due to bifurcation of the dislocation density (e.g., cell wall and cell interior). It is easily understood that different dislocation densities should lead to different yield stresses. Therefore, on a mesoscopic scale a local variation of the (microscopic) yield stress is to be expected. Consequently, even a single-phase single-crystalline material should be treated like a composite consisting of local areas of different stress-strain behavior (Ref 55, 58, 216, 217). Then, the macroscopic behavior is composed of the behavior of all elements. The relations that determine how the elements are combined result from the assumptions applied to describe the geometric arrangement of the elements.

In order to illustrate the basic idea and the main features of such a multicomponent model, let us consider the parallel arrangement of ideally elastic-plastic elements, which was proposed by G. Masing in 1923 (Ref 218, 219). For the sake of simplicity, only two elements should exist, which have a local yield strength of σ_{1f} and σ_{2f}, respectively, and occupy an area fraction of 50% each. Figure 36 shows the stress-strain path of this two-element model during one cycle, starting from the deformation-free condition. As a consequence of the parallel arrangement, each element undergoes the same strain as the composite. Furthermore, the stress acting on the composite can easily be obtained graphically from the σ-ε courses of the individual elements, because it is the arithmetic mean due to the identical area fractions of the elements.

Starting from the deformation-free condition in point A, a purely elastic deformation occurs until the "weakest" element reaches its yield strength (point B). Further stress increase shows a slope $E/2$, because element 1 no longer contributes to the stress increase. As soon as the yield stress of element 2 is reached (point C), the composite deforms ideally plastically (without hardening). A load reversal (point D) leads first to an elastic back-deformation until, at point F, the plastic deformation of element 1 begins, and the cycle repeats.

In spite of the simplicity of this two-component model, some important consequences that arise from the "composite idea" can be deduced:

- In the macroscopically stress-free condition (point E), internal stresses exist. The weaker element contains a compressive stress, whereas in the stronger element a tensile stress is acting (e.g., Ref 221).
- As a consequence of the compressive internal stress in the weaker element at $\sigma = 0$, plastic deformation after a prior load reversal (point D) begins at a reduced macroscopic yield stress (in absolute values, point F). This behavior is commonly termed the Bauschinger effect (Fig. 6b).
- If the monotonic stress-strain curve is spread by a scale factor of 2, the curve equals the branch of the hysteresis loop (represented in relative coordinates). This 1:2 rule was introduced above as a typical feature of CSS behavior.
- The memory of a prior load history is implicitly contained in this model (the interested reader is invited to "construct" the σ-ε path after a load reversal between point F and point G). A prior load history is stored in the form of the internal stress distribution (memory of the material).

Obviously, a two-element model is not suitable for precise reproduction and prediction of the stress-strain behavior. More advanced models (e.g., Ref 222-228) use a statistical mathematical treatment in which an infinite number of elements is considered and the distribution of the yield stresses is described by a distribution density function. This does not affect the basic consequences of the composite idea as listed above.

Figure 37 shows the result of such a calculation, with small cycles interrupting a large cycle. The calculation is based exclusively on the stress levels of the reversal points and their sequence. The memory of a prior load history is "automatically" taken into account.

In order to give an impression of the quality of prediction, Fig. 38 compares the calculated stress-strain path with that measured on copper in a (half block of an) incremental step test. The only input function for the calculation is the distribution density function of the elementary yield stresses, which has been obtained unambiguously from one ascending branch of the hysteresis loop measured in a single-step test at the effective amplitude. The excellent agreement of measured and calculated data is due to very careful consideration of the microstructural aspects of CSS behavior in terms of the application of a multicomponent model and the correct value of the effective amplitude.

REFERENCES

1. W.A.J. Albert, *Arch. Min. Geo. Berg. Hüttenk.*, Vol 10, 1938, p 215
2. J.V. Poncelet, *Introduction á la Mècanique, Industrielle, Physique our Expèrimentale*, Second ed., Imprimerie de Gauthier-Villars, Paris,1839, p 317-318
3. W.J.M. Rankine, *Proc. Inst. Civ. Engrs.*, Vol 2, 1843, p 195
4. A. Wöhler, *Z. f. Bauwesen*, Vol 8, 1858, p 642-652
5. L.F. Coffin, *Trans. ASME*, Vol 76, 1954, p 931
6. S.S. Manson, NASA TN-2933, 1953
7. H. Mughrabi, in *Conf. Proc. Dislocations and Properties of Real Materials*, No. 323, The Institute of Metals, London, 1985, p 244-261
8. R. Danzer, *Lebensdauerprognose hochfester metallischer Werkstoffe im Bereich hoher Temperaturen*, Gebrüder Bornträger, Stuttgart, 1988
9. J.W. Hutchinson, *J. Mech. Phys. Solids*, Vol 16, 1968, p 13-31
10. J.C. Rice and G.F. Rosengreen, *J. Mech. Phys. Solids*, Vol 16, 1968, p 1-12
11. N.E. Dowling and J.A. Begley, in *Mechanics of Crack Growth*, STP 590, 1976, p 82-103
12. W. Hoffelner, in *Proc. Second Conf. on Low Cycle Fatigue and Elasto-Plastic Behaviour of Materials*, K.-T. Rie, Ed., Elsevier, Amsterdam, 1987, p 622-629
13. J. Bressers, W. Weise, and T. Hollstein, in *Proc. Second Conf. on Low Cycle Fatigue and Elasto-Plastic Behaviour of Materials*, K.-T. Rie, Ed., Elsevier, Amsterdam, 1987, p 655-660
14. H. Heitmann, H. Vehoff, and P. Neumann, in *Proc. Sixth Internat. Conf. on Fracture*, Vol I, T. Rama Rao, S.R. Vallori, and J.F. Knott, Ed., Pergamon Press, Oxford, 1984, p 3599-3606
15. H. Mughrabi, in *Berichtsband DGM-Fachkonferenz: Werkstoffverhalten und Bauteilbemessung*, E. Macherauch, Ed., DGM-Informationsgesellschaft, Oberursel, 1986, p 49-65
16. R. Schubert, *Verformungsverhalten und Rißwachstum bei Low Cycle Fatigue*, Ph.D. thesis, Technische Universität Braunschweig, 1989
17. J. Morrow, in *Internal Friction, Damping and Cyclic Plasticity*, STP 378, American Society of Testing and Materials, 1964, p 45-87
18. H. H. Heitmann, *Betriebsfestigkeit von Stahl: Vorhersage der technischen Anrißlebensdauer unter Berücksichtigung des Verhaltens von Mikrorissen*, Ph.D. thesis, RWTH-Aachen, 1983
19. P. Neumann and A. Tönnessen, in *Fatigue 87*, R. Ritchie and E. Starke, Ed., EMAS, Warley, 1987, p 3-22
20. J. Bauschinger, *Mittheilungen aus dem Mechanisch-Technischen Laboratorium der Königlich Technischen Hochschule in München*, Vol 13, 1886, p 1-116
21. N.E. Dowling, *J. Materials*, Vol 7, 1972, p 71-87
22. J. Burbach, in *Low Cycle Fatigue and Elasto-Plastic Behaviour*, K.-T. Rie and E. Haibach, Ed., Deutscher Veband für Materialforschung (DVM), Berlin, 1979, p 183-197
23. J. Burbach, *Techn. Mitt. Krupp, Forsch.-Ber.*, Vol 28, 1970, p 55-101
24. V. Gerold and H.P. Karnthaler, *Acta Metall.*, Vol 37, 1989, p 2177-2183
25. S.I. Hong and C. Laird, *Acta Metall. Mater.*, Vol 38, 1990, p 1581-1594
26. C.E. Feltner and C. Laird, *Acta Metall.*, Vol 15, 1967, p 1621-1632
27. C.E. Feltner and C. Laird, *TMS-AIME*, Vol 242, 1968, p 1253-1257
28. P. Lukáš and M. Klesnil, in *Proc. 2nd Int. Conf. on Corrosion Fatigue*, O.J. Devereux, A.J. McEvily, and R.W. Staehl, Ed., National Association of Corrosion Engineers NACE, Houston, 1972, p 118-132

29. P. Lukáš and M. Klesnil, *Phys. Stat. Sol.,* Vol 37, 1970, p 833-842

30. P. Lukáš and M. Klesnil, *Phys. Stat. Sol. (A),* Vol 5, 1971, p 247-258

31. C.E. Feltner and C. Laird, *Acta Metall.,* Vol 15, 1967, p 1633-1653

32. C.B. Carter, in *Dislocation 1984, Comptes rendus du Colloque International du C.N.R.S. Dislocations: Structure de Cøeur et Propriétés Physiques,* Éditions du Centre National de la Recherche Scientifique, Paris, 1984, p 227-251

33. R.E. Schramm and R.P. Reed, *Met. Trans. A,* Vol 6A , 1975, p 1345-1351

34. K. Wolf, H.-J. Gudladt, H.A. Calderon, and G. Kostorz, *Acta Metall. Mater.,* Vol 42, 1994, p 3759-3765

35. N. Clément, in *Éditions de Physique,* École d'Hilver, Aussois, France, 1984, p 167

36. H. Shirai and J.R. Weertmann, *Scripta Metall.,* Vol 17, 1983, p 1253-1258

37. P. Charsley, U. Bangert, and L.J. Appleby, *Mater. Sci. Eng. A,* Vol 113, 1989, p 231-236

38. H. Mughrabi, *Mater. Sci. Eng.,* Vol 33, 1978, p 207-223

39. S. Suresh, *Fatigue of Materials,* Cambridge University Press, Cambridge, 1991

40. N.Y. Jin and A.T. Winter, *Acta Metall.,* Vol 32, 1984, p 1173-1176

41. T.K. Lepistö, V.-T. Kuokkala, and P.O. Kettunen, in *Mater. Sci. Eng.,* Vol 81, 1986, p 457-463

42. V.-T. Kuokkala and T.K. Lepistö, in *Conf. Proc. Basic Mechanisms in Fatigue,* P. Lukáš and J. Polák, Ed., Academia, 1988, p 153-160

43. C. Laird, P. Charsley, and H. Mughrabi, *Mater. Sci. Eng.,* Vol 81, 1986, p 433-450

44. A.T. Winter, *Phil. Mag. (Ser. 8),* Vol 30, 1974, p 719-738

45. J.M. Finney and C. Laird, *Phil. Mag.,* Vol 31, 1975, p 339-366

46. H. Mughrabi, F. Ackermann, and K. Herz, in *Fatigue Mechanisms,* STP 675, J.T. Fong, Ed., American Society for Testing and Materials, 1979, p 69-105

47. F. Ackermann, L.P. Kubin, J. Lepinoux and H. Mughrabi, *Acta Metall.,* Vol 32, 1984, p 715-725

48. L. Buchinger and C. Laird, *Mater. Sci. Eng.,* Vol 76, 1985, p 71-76

49. B.-D. Yan, A. Hunsche, P. Neumann, and C. Laird, *Mater. Sci. Eng.,* Vol 79, 1986, p 9-14

50. H.P. Karnthaler, S. Kong, B. Mingler, R. Stickler, and B. Weiss, *Acta Metall. Mater.,* Vol 43, 1995, p 3017-3026

51. U. Holzwarth and U. Essmann, *Mater. Sci. Eng. A,* Vol 164, 1993, p 206-210

52. J. Bretschneider, C. Holste, and W. Kleinert, *Mater. Sci. Eng. A,* Vol 191, 1995, p 61-72

53. C. Blochwitz and U. Veit, *Crystal Res. Technol.,* Vol 17, 1982, p 529-551

54. U. Essmann and H. Mughrabi, *Phil. Mag. A,* Vol 40, 1979, p 731-756

55. H. Mughrabi, in *Continuum Models of Discrete Systems 4,* O. Brulin and R.K.T. Hsieh, Ed., North-Holland Publishing Company, 1981, p 241-257

56. J. Lepinoux and L.P. Kubin, *Phil. Mag. A,* Vol 51, 1985, p 675-696

57. T. Tabata, H. Fujita, M.-A. Hiraoka, and K. Onishi, *Phil. Mag. A,* Vol 47, 1983, p 841-857

58. H. Mughrabi, in *Proc. Fifth Internat. Conf. Strength of Metals and Alloys,* P. Haasen, V. Gerold, and G. Kostorz, Ed., Pergamon Press, Oxford, 1980, p 1615

59. U. Essmann, U. Gösele, and H. Mughrabi, *Phil. Mag. A,* Vol 44, 1981, p 405-426

60. J.C. Grosskreutz and H. Mughrabi, in *Constitutive Equations in Plasticity,* A.S. Argon, Ed., MIT Press, 1975, p 251-326

61. D. Kuhlmann-Wilsdorf, *Mater. Sci. Eng.,* Vol 39, 1979, p 127-139

62. N. Thompson, N.J. Wadsworth, and N. Louat, *Phil. Mag. (Ser. 8),* Vol 1, 1956, p 113

63. Z.S. Basinski, P. Pascual, and S.J. Basinski, *Acta Metall.,* Vol 31, 1983, p 591-602

64. Z.S. Basinski and S.J. Basinski, *Acta Metall.,* Vol 33, 1985, p 1307-1317

65. Z.S. Basinski and S.J. Basinski, *Acta Metall.,* Vol 33, 1985, p 1319-1327

66. B.-T. Ma and C. Laird, *Acta Metall.,* Vol 37, 1989, p 357-368

67. B.-T. Ma and C. Laird, *Acta Metall.,* Vol 37, 1989, p 325-336

68. B.-T. Ma and C. Laird, *Acta Metall.,* Vol 37, 1989, p 337-348

69. B.-T. Ma and C. Laird, *Acta Metall.,* Vol 37, 1989, p 349-355

70. B.-T. Ma and C. Laird, *Acta Metall.,* Vol 37, 1989, p 369-379

71. A. Hunsche, *Untersuchung zur Rißbildung in Ermüdungsgleitbändern,* Ph.D. thesis, RWTH-Aachen, 1982

72. J.C. Figueroa and C. Laird, *Mater. Sci. Eng.,* Vol 60, 1983, p 45-58

73. H. Mughrabi, in *Conf. Proc. Defects, Fracture and Fatigue,* G.C. Sih, and J.W. Provan, Ed., Martinus Nijhoff Publishers, 1983, p 139-146

74. H.-J. Christ, *Mater. Sci. Eng. A,* Vol 117, 1989, p L25-L29

75. L. Cordero, A. Ahmadieh, and P. K. Mazumdar, *Scripta Metall.,* Vol 22, 1988, p 1761-1764

76. B.-T. Ma and C. Laird, *Scripta Metall.,* Vol 23, 1989, p 1029-1032

77. F.L. Liang and C. Laird, *Mat. Sci. Eng. A,* Vol 117, 1989, p 83-93

78. F.L. Liang and C. Laird, *Mat. Sci. Eng. A,* Vol 117, 1989, p 95-102

79. F.L. Liang and C. Laird, *Mat. Sci. Eng. A,* Vol 117, 1989, p 103-113

80. H. Mughrabi and R. Wang, *Proc. Second Risø Intern. Symp. on Metallurgy and Materials Science,* N. Hansen, A. Horsewell, T. Leffers, and H. Lilholt, Ed., Risø National Laboratory, Roskilde, Denmark, 1981, p 87-98

81. J.R. Hancock and J.C. Großkreutz, *Acta Metall.,* Vol 17, 1969, p 77-97

82. S.J. Basinski and Z.S. Basinski, *Phil. Mag.,* Vol 19, 1969, p 899-923

83. P. Charsley and D. Kuhlmann-Wilsdorf, *Phil. Mag. A,* Vol 44, 1981, p 1351-1361

84. D. Kuhlmann-Wilsdorf, *Phys. Stat. Sol. A,* Vol 104, 1987, p 121-144

85. D. Walgraef and E.C. Aifantis, *J. Applied Physics,* Vol 58, 1985, p 688-691

86. E.C. Aifantis, *Internat. J. Plasticity,* Vol 3, 1987, p 211-247

87. G. Nicolis and I. Prigogine, *Self-Organization in Non-equilibrium Systems,* Wiley-Interscience Publisher, 1977

88. K. Differt and U. Essmann, *Mater. Sci. Eng. A,* Vol 164, 1993, p 295-299

89. D. Kuhlmann-Wilsdorf and C. Laird, *Mater. Sci. Eng.,* Vol 46, 1980, p 209-219

90. G.I. Taylor, *Proc. Royal Society A,* Vol 145, 1934, p 362-387

91. F.R.N. Nabarro, *Advances in Physics,* Vol 1, 1952, p 269-395

92. P. Neumann, in *Physical Metallurgy,* R.W. Cahn and P. Haasen, Ed., Elsevier Science, Amsterdam, 1983, p 1554-1593

93. P. Neumann, *Mater. Sci. Eng.,* Vol 81, 1986, p 465-475

94. P. Neumann, *Z. Metallkde,* Vol 59, 1968, p 927-934

95. M.P.E. Desvaux, *Z. Metallkde,* Vol 61, 1970, p 206-213

96. P. Neumann, in *Constitutive Equations in Plasticity,* A.S. Argon, Ed., MIT Press, Cambridge, Mass., 1975, p 449-468

97. R. Wang, *Untersuchungen der mikroskopischen Vorgänge bei der Wechselverformung von Kupferein-und-vielkristallen,* Ph.D. thesis, Universität Stuttgart, 1982

98. H. Mughrabi and R. Wang, in *Conf. Proc. Basic Mechanisms in Fatigue,* P. Lukáš and J. Polák, Ed., Academia, 1988, p 1-14

99. A. T. Winter, O.B. Pedersen, and K.V. Rasmussen, *Acta Metall.,* Vol 29, 1981, p 735-748

100. G. Klein, *Ermüdungsverhalten vielkrsitalliner Kupferproben an Luft und Hochvakuum,* diploma thesis, Universität Erlangen-Nürnberg, 1985

101. C. Wittig-Ling, *Untersuchungen zum Ermüdungsverhalten vielkristalliner Kupferproben,* diploma thesis, Universität Erlangen-Nürnberg, 1986

102. J.C. Figueroa, S.P. Bhat, R. de la Veaux, S. Murzenski, and C. Laird, *Acta Metall.,* Vol 29, 1981, p 1667-1678

103. P. Lukáš and L. Kunz, in *Conf. Proc. Basic Mechanisms in Fatigue,* P. Lukáš and J. Polák, Ed., Academica, 1988, p 161-168

104. G. Hoffmann, *Untersuchung des Ermüdungsverhaltens von vielkristallinem Kupfer in Abhängigkeit von Temperatur und Umgebungsmedium,* diploma thesis, Universität Erlangen-Nürnberg, 1990

105. M. Bayerlein, *Kritische Überprüfung des Incremental-Step-Tests zur Bestimmung der zyklischen Spannungs-Dehnungskurve (auf mikrostruktureller Grundlage),* diploma thesis, Universität Erlangen-Nürnberg, 1986

106. H.-J. Christ, H. Mughrabi, and C. Wittig-Link, in *Conf. Proc. Basic Mechanisms in Fatigue,* P. Lukáš and J. Polák, Ed., Academica, 1988, p 83-92

107. L. Llanes, J.L. Bassani, and C. Laird, *Acta Metall. Mater.,* Vol 42, 1994, p 1279-1288

108. P. Lukáš and L. Kunz, *Mater. Sci. Eng.*, Vol 85, 1987, p 67-75

109. P. Lukáš and L. Kunz, *Mater. Sci. Eng.*, Vol 74, 1985, p L1-L5

110. C.D. Liu, M.N. Bassim, and D. X. You, *Acta Metall. Mater.*, Vol 42, 1994, p 3695-3704

111. L. Llanes, A.D. Rollett, C. Laird, and J.L. Bassani, *Acta Metall. Mater.*, Vol 41, 1993, p 2667-2679

112. L. Llanes, A. D. Rollett, and C. Laird, *Mater. Sci. Eng. A*, Vol 167, 1993, p 37-45

113. D. J. Morrison and V. Chopra, *Mater. Sci. Eng. A*, Vol 177, 1994, p 29-42

114. K.V. Rasmussen and O.B. Pedersen, *Acta Metall.*, Vol 28, 1980, p 1467-1478

115. D.J. Morrison, *Mater. Sci. Eng. A*, Vol 187, 1994, p 11-21

116. P. Lukáš and L. Kunz, *Mater. Sci. Eng. A*, Vol 189, 1994, p 1-7

117. J.E. Pratt, *J. Mater.*, Vol 1, 1966, p 77-88

118. J.E. Pratt, *Acta Metall.*, Vol 15, 1967, p 319-327

119. E.S. Kayali and A. Plumtree, *Met. Trans. A*, Vol 13, 1982, p 1033-1041

120. H. Abdel-Raouf, A. Plumtree, and T. H. Topper, *Met. Trans.*, Vol 5, 1974, p 267-277

121. A. Plumtree, in *Proc. Second Conf. on Low Cycle Fatigue and Elasto-Plastic Behaviour of Materials*, K.-T. Rie, Ed., Elsevier, Amsterdam, 1987, p 19-30

122. G. König and W. Blum, *Acta Metall.*, Vol 28, 1980, p 519-537

123. P. Lukáš and L. Kunz, *Mater. Sci. Eng. A*, Vol 103, 1988, p 233-239

124. Z.S. Basinski, A.S. Korbel, and S.J. Basinski, *Acta Metall.*, Vol 28, 1980, p 191-207

125. H.-J. Christ and H. Mughrabi, in *Proc. Third Internat. Conf. on Low-Cycle Fatigue and Elasto-Plastic Behaviour of Materials*, K.-T. Rie, Ed., Elsevier Applied Science, London, 1992, p 56-69

126. D.E. Witmer, G.C. Farrington, and C. Laird, *Acta Metall.*, Vol 35, 1987, p 1865-1909

127. R. Wang, H. Mughrabi, S. McGovern, and M. Rapp, *Mater. Sci. Eng.*, Vol 65, 1984, p 219-233

128. R. Wang and H. Mughrabi, *Mater. Sci. Eng.*, Vol 63, 1984, p 147-163

129. A. Hunsche and P. Neumann, in *Symp. Fundamental Questions and Critical Experiments in Fatigue*, STP 924, Vol I, J.T. Fong and A.J. Fields, Ed., American Society for Testing and Materials, 1988, p 26-38

130. M. Gerland and P. Voilan, *Mater. Sci. Eng.*, Vol 84, 1986, p 23-33

131. M. Gerland, J. Mendez, P. Violan, and B.A. Saadi, *Mater. Sci. Eng. A*, Vol 118, 1989, p 83-95

132. Z. Wang, *Mater. Sci. Eng. A*, Vol 183, 1994, p L13-L17

133. Y. Li and C. Laird, *Mater. Sci. Eng. A*, Vol 186, 1994, p 65-86

134. A. Abel and V. Gerold, *Mater. Sci. Eng.*, Vol 37, 1979, p 187-200

135. E.S. Kayali and A. Plumtree, *Met. Tans. A*, Vol 13, 1982, p 1033-1041

136. A. Abel and V. Gerold, *Z. Metallkde*, Vol 70, 1979, p 577-581

137. B.-D. Yan, A.S. Cheng, L. Buchinger, S. Stanzl, and C. Laird, *Mater. Sci. Eng.*, Vol 80, 1986, p 129-142

138. C. Laird, S. Stanzl, R. de la Veaux, and L. Buchinger, *Mater. Sci. Eng.*, Vol 80, 1986, p 143-154

139. L. Buchinger, A.S. Cheng, S. Stanzl, and C. Laird, *Mater. Sci. Eng.*, Vol 80, 1986, p 155-167

140. H. Kaneshiro, K. Katagiri, H. Moro, C. Makabe, and T. Yafuso, *Met. Trans. A*, Vol 10, 1988, p 1257-1262

141. R. Kwadjo and L.M. Brown, *Acta Metall.*, Vol 26, 1978, p 1117-1138

142. W.R. Scoble and S. Weissman, *Crystal Lattice Defects*, Vol 4, 1973, p 123-136

143. G. Gonzales and C. Laird, *Met. Trans. A*, Vol 14, 1983, p 2507-2515

144. H. Mughrabi, K. Herz, and X. Stark, *Internat. J. Fracture*, Vol 17, 1981, p 193-220

145. A. Seeger, *Phil. Mag. (Ser. 7)*, Vol 45, 1954, p 771-773

146. H. Mughrabi, in *Proc. ICSMA 7*, J.I. Dickson, J.J. Jones, M.G. Akben, et al., Ed., Pergamon Press, Oxford, 1986, p 1917-1942

147. F. Ackermann, H. Mughrabi, and A. Seeger, *Acta Metall.*, Vol 31, 1983, p1353-1366

148. H.D. Nine, *J. Applied Physics*, Vol 44, 1973, p 4875-4881

149. R. Neumann, *Z. Metallkde*, Vol 66, 1975, p 26-32

150. H. Mughrabi and Ch. Wüthrich, *Phil. Mag. A*, Vol 33, 1976, p 963-984

151. F. Guiu and M. Anglada, *Phil. Mag. A*, Vol 42, 1980, p 271-276

152. K. Pohl, P. Mayr, and E. Macherauch, *Scripta Metall.*, Vol 4, 1980, p 1167-1169

153. G. Pilo, W. Reik, P. Mayr, and E. Macherauch, *Arch. Eisenhüttenwes.*, Vol 48, 1977, p 575-578

154. D. Eifler, in *Ermüdungsverhalten metallischer Werkstoffe*, D. Munz, Ed., DGM-Informationsgesellschaft, Oberursel, 1985, p 73-105

155. D. Eifler and E. Macherauch, *Internat. J. Fatigue*, Vol 12, 1990, p 165-174

156. M. Klesnil and P. Lukáš, *J. Iron Steel Inst.*, Vol 205, 1967, p 746

157. D. Lefebvre and F. Ellyin, *Internat. J. Fatigue*, Vol 6, 1984, p 9-15

158. H.J. Maier, B. Donth, M. Bayerlein, H. Mughrabi, B. Meier, and M. Kesten, *Z. Metallkde*, Vol 84, 1993, p 820-826

159. S.S. Hecker, M.G. Stout, K.P. Staudhammer, and J.L. Smith, *Met. Trans. A*, Vol 13, 1982, p 619

160. R. Lagneborg, *Acta Metall.*, Vol 12, 1964, p 823-843

161. F. Lecroisey and A. Pineau, *Met. Trans.*, Vol 2, 1972, p 387-396

162. D. Hennessy, G. Steckel, and C. Altstetter, *Met. Trans. A*, Vol 7, 1976, p 415-424

163. T. Suzuki, H. Kojima, K. Suzuki, T. Hashimoto, and M. Ichihara, *Acta Metall.*, Vol 25, 1977, p 1151

164. K. Tsuzaki, T. Maki, and I. Tamura, *J. Physique, Colloque C4*, Suppl. 43, 1982, p 423-428

165. J. Singh, *J. Mater. Sci.*, Vol 20, 1985, p 3157-3166

166. M. Bayerlein, H.-J. Christ, and H. Mughrabi, *Mater. Sci. Eng. A*, Vol 114, 1989, p L11-L16

167. M. Bayerlein, H. Mughrabi, M. Kesten, and B. Meier, *Mater. Sci. Eng. A*, Vol 159, 1992, p 35-41

168. H.-J. Christ, K. Lades, L. Völkl, and H. Mughrabi, in *Proc. Third Internat. Conf. on Low-Cycle Fatigue and Elasto-Plastic Behaviour of Materials*, K.-T. Rie, Ed., Elsevier Applied Science, London, 1992, p 106-111

169. M. Sade, R. Rapacioli, and M. Ahlers, *Acta Met.*, Vol 33, 1985, p 487

170. C. Laird, V.J. Langelo, M. Hollrah, N.C. Yang, and R. de la Veaux, *Mater. Sci. Eng.*, Vol 32, 1978, p 137-160

171. H. Mughrabi, in *Conf. Proc. Deformation of Multi-Phase and Particle Containing Materials*, J.B. Sørensen, N. Hansen, A. Horsewell, T. Leffers, and H. Lilholt, Ed., Risø National Laboratory, Roskilde, Denmark, 1983, p 65-82

172. V. Gerold, *Mater. Sci. Forum*, Vol 13/14, 1987, p 175-194

173. C. Laird, in *Cyclic Stress-Strain Response and Plastic Deformation Aspects of Fatigue Crack Growth*, STP 637, American Society for Testing and Materials, 1977, p 3-35

174. S. Horibe and C. Laird, *Acta Metall.*, Vol 35, 1987, p 1919-1927

175. S. Horibe, J.-K. Lee, and C. Laird, *Mater. Sci. Eng.*, Vol 63, 1984, p 257-265

176. C. Calabrese and C. Laird, *Mater. Sci. Eng.*, Vol 13, 1974, p 141-157

177. C. Calabrese and C. Laird, *Mater. Sci. Eng.*, Vol 13, 1974, p 159-174

178. C. Calabrese and C. Laird, *Met. Trans.*, Vol 5, 1974, p 1785-1793

179. J.-K. Lee and C. Laird, *Mater. Sci. Eng.*, Vol 54, 1982, p 39-51

180. J.-K. Lee and C. Laird, *Mater. Sci. Eng.*, Vol 54, 1982, p 53-64

181. J.-K. Lee and C. Laird, *Phil. Mag. A*, Vol 47, 1983, p 579-597

182. S. Horibe, J.-K. Lee, and C. Laird, *Fatigue Engng. Mater. Struct.*, Vol 7, 1984, p 145-154

183. S. Horibe and C. Laird, *Acta Metall.*, Vol 31, 1983, p 1567-1579

184. S. Horibe and C. Laird, *Acta Metall.*, Vol 33, 1985, p 819-825

185. S. Horibe and C. Laird, in *Conf. Proc. Basic Mechanisms in Fatigue*, P. Lukáš and J. Polák, Ed., Academia, 1988, p 197-204

186. S.P. Bhat and C. Laird, *Acta Metall.*, Vol 27, 1979, p 1861-1871

187. S.P. Bhat and C. Laird, *Acta Metall.*, Vol 27, 1979, p 1873-1883

188. V. Gerold, B.A. Lerch, and D. Steiner, *Z. Metallkde*, Vol 75, 1984, p 546-553

189. D. Steiner, R. Beddoe, V. Gerold, G. Kostorz, and R. Schmelzer, *Scripta Metall.*, Vol 17, 1983, 733-736

190. V. Gerold and D. Steiner, *Scripta Metall.*, Vol 16, 1982, p 405-408

191. D. Steiner, W. Müller, and V. Gerold, *Scripta Metall.*, Vol 18, 1984, p 693-698

192. D. Steiner and V. Gerold, *Mater. Sci. Eng.*, Vol 84, 1986, p 77-88
193. J.G. Byrne and M.E. Fine, *Phil. Mag.*, Vol 6, 1961, p 1119-1145
194. M. Wilhelm, *Mater. Sci. Eng.*, Vol 48, 1981, p 91-106
195. K. Differt, U. Essmann, and H. Mughrabi, *Phys. Stat. Sol. A*, Vol 104, 1987, p 95-106
196. M. Clavel and A. Pineau, *Mater. Sci. Eng.*, Vol 55, 1982, p 157-171
197. C.A. Stubbington and P.J.E. Forsyth, *Acta Metall.*, Vol 14, 1966, p 5-12
198. E.A. Starke and G. Lütjering, in *Conf. Proc. Fatigue and Microstructure*, American Society for Metals, 1979, p 205-243
199. H. Mayer and C. Laird, *Mater. Sci. Eng. A*, Vol 187, 1994, p 23-35
200. L. Llanes and C. Laird, *Mater. Sci. Eng. A*, Vol 161, 1993, p 1-12
201. J. Polák, M. Klesnil, and P. Lukáš, *Mater. Sci. Eng.*, Vol 15, 1974, p 231-237
202. P. Lukáš and M. Klesnil, *Mater. Sci. Eng.*, Vol 11, 1973, p 345-356
203. H. Abdel-Raouf, P.P. Benham, and A. Plumtree, *Can. Metall. Q.*, Vol 10, 1971, p 87-95
204. H. Abdel-Raouf and A. Plumtree, *Met. Trans.*, Vol 2, 1971, p 1251-1254
205. H.-F. Chai and C. Laird, *Mater. Sci. Eng.*, Vol 93, 1987, p 159-174
206. A. Plumtreee and L.D. Pawlus, in *Basic Questions in Fatigue*, Vol I, STP 924, J.T. Fong and R.J. Fields, Ed., American Society for Testing and Materials, 1988, p 81-97
207. C.E. Feltner and C. Laird, *TMS-AIME*, Vol 245, 1969, p 1372-1373
208. C. Laird, J.M. Finney, A. Schwartzman, and R. de la Veaux, *J. Testing Eval.*, Vol 3, 1975, p 435-441
209. G. Hoffmann, O. Öttinger, and H.-J. Christ, in *Proc. Third Internat. Conf. on Low-Cycle Fatigue and Elasto-Plastic Behaviour of Materials*, K.-T. Rie, Ed., Elsevier Applied Science, London, 1992, p 100-105
210. H.-J. Christ, G. Hoffmann, and O. Öttinger, *Mater. Sci. Eng. A*, Vol 201, 1995, p 1-12
211. B.-T. Ma and C. Laird, *Mater. Sci. Eng. A*, Vol 102, 1988, p 247-258
212. R. Landgraf, J. Morrow, and T. Endo, *J. Materials*, Vol 4, 1969, p 176-188
213. R.W. Landgraf, in *Achievement of High Fatigue Resistance in Metals and Alloys*, STP 467, American Society for Testing and Materials, 1970, p 3-36
214. M. Bayerlein, H.-J. Christ, and H. Mughrabi, in *Proc. Second Conf. on Low Cycle Fatigue and Elasto-Plastic Behaviour of Materials*, K.-T. Rie, Ed., Elsevier, Amsterdam, 1987, p 149-154
215. H. Mughrabi, M. Bayerlein, and H.-J. Christ, in *Conf. Proc. Constitutive Relations and their Physical Basis*, S.I. Anderson, J.B. Bilde-Sørensen, N. Hansen, et al., Ed., Risø National Laboratory, Roskilde, Denmark, 1987, p 447-452
216. H. Mughrabi, *Acta Metall.*, Vol 31, 1983, p 1367-1379
217. H. Mughrabi, T. Ungár, W. Kienle, and M. Wilkens, *Phil. Mag. A*, Vol 53, 1986, p 793-813
218. G. Masing, *Wissenschaftl. Veröffentl. aus dem Siemens-Konzern*, Vol 3, 1923, p 231-239
219. G. Masing, *Z. Techn. Physik*, Vol 6, 1925, p 569-573
220. H.-J. Christ, *Wechselverformung von Metallen*, Springer, Berlin, 1991
221. T. Ungár, H. Biermann, and H. Mughrabi, *Mater. Sci. Eng. A*, Vol 164, 1993, p 175-179
222. J. Polák and M. Klesnil, *Fatigue Engng. Mater. Struct.*, Vol 5, 1982, p 19-32
223. C. Holste and H.-J. Burmeister, *Phys. Stat. Sol. A*, Vol 57, 1980, p 269-280
224. H.-J. Burmeister and C. Holste, *Phys. Stat. Sol. A*, Vol 64, 1981, p 611-624
225. J. Polák, M. Klesnil, and J. Helesic, *Fatigue Engng. Mater. Struct.*, Vol 5, 1982, p 33-44
226. J. Polák, M. Klesnil, and J. Helesic, *Fatigue Engng. Mater. Struct.*, Vol 5, 1982, p 45-56
227. C. Holste, F. Lange, and H.-J. Burmeister, *Phys. Stat. Sol. A*, Vol 63, 1981, p 213-221
228. H.-J. Christ, in *Proc. Fifth Internat. Conf. on Fatigue and Fatigue Threshold*, J.-P. Bailon and J.I. Dickson, Ed., EMAS, London, 1993, p 115-120

SELECTED REFERENCES

• S. Suresh, *Fatigue of Materials*, Cambridge University Press, Cambridge, 1991
• M. Klesnil and P. Lukáš, *Fatigue of Metallic Materials*, Materials Science Monographs, C. Laird, Ed., Vol 71, Elsevier Science Publisher, Amsterdam, 1992
• H.-J. Christ, *Wechselverformung von Metallen*, WFT 9, B. Ilschner, Ed., Springer, Berlin, 1991
• J. Polák, *Cyclic Plasticity and Low Cycle Fatigue Life of Metals*, Materials Science Monographs, C. Laird, Ed., Vol 63, Elsevier Science Publisher, Amsterdam, 1991

Fatigue Crack Nucleation and Microstructure

Petr Lukáš, Institute of Physics of Materials, Academy of Sciences of the Czech Republic

THE FATIGUE PROCESS can be roughly divided into four stages: cyclic hardening/softening, crack nucleation, crack propagation, and overload (fracture). In flaw-free materials a significant fraction of the total lifetime is spent before the first detectable microcracks appear. At low amplitudes the nucleation stage can occupy the majority of the lifetime. At high amplitudes nucleation is usually accomplished within a small fraction of the fatigue life. Another fraction of the lifetime is needed for propagation of the microstructurally small cracks (cracks small compared to microstructural size scales) to reach the size of the physically small cracks (i.e., cracks of the size ≤0.5-1 mm). Again, this fraction can be quite high at low amplitudes. Propagation of physically small cracks and macrocracks can be quantitatively described by means of fracture mechanics. On the other hand, there is no generally accepted quantitative description of the nucleation process, and there is no widely applicable description of the propagation of microstructurally small cracks. Thus, any numerical analysis of these stages in the fatigue process must be preceded by deeper understanding of the physical mechanisms.

Role of Surface Conditions in Fatigue. It is well established that the fatigue process is very sensitive to surface state and is influenced by the surface finish and surface treatment. The reason is that fatigue cracks in most cases nucleate from free surfaces of cyclically loaded metals. This has been repeatedly demonstrated impressively by the observation that when a specimen is fatigued for a substantial fraction of its fatigue life and its surface is then removed by electropolishing, the specimen in a subsequent test exhibits a fatigue life as long as that of a virgin specimen (Ref 1).

Crack nucleation, as well as the whole fatigue process, is controlled by the cyclic plastic deformation. Therefore, it can be expected that cracks nucleate at positions where cyclic plastic deformation is higher than average. There are basically two reasons why the cyclic plastic deformation is higher just at the surface: concentration of plastic deformation due to higher stresses near the surface; and lower degree of constraint of the near-surface volumes of cyclically loaded material.

In complex engineering components, higher surface stresses result from either notches or bending and twisting, both of which lead to stress gradients with the highest stress on the surface. Even in nominally uniformly stressed parts, a small degree of eccentricity in the axial load is practically unavoidable, again leading to small bending or twisting moments and consequently to higher surface stresses in the surface layer. Besides these macroscopic sources of stress concentration on the surface, there are also microscopic stress concentrators, which are effective even under conditions of ideal uniaxial loading. The stress level at the surface is sensitive to surface topography, and the surface is never perfectly smooth. For example, very fine grinding produces grooves with depths of the order of 0.1 μm, which very locally can increase the stress by about 10%. Fortunately, the surface finish often produces a thin plastically deformed surface layer with compressive residual stress, which may balance or even overbalance the detrimental effect of microscopic stress concentration on the resulting life.

A further type of microscopic stress concentrator is the surface step produced by dislocations leaving the metal during plastic deformation. The cycling itself can produce localized stress concentration at the surface. Second-phase particles, such as inclusions and precipitates, have elastic properties different from those of the matrix and generally also serve as stress concentrators. Argon (Ref 2) showed very instructively that the effect of second-phase particles at the free surface is higher than in the interior. Figure 1, taken from his paper, compares the stress-concentration (K_t) and the stress-intensity (K_I) factors for idealized shapes of particles of zero modulus at the surface and in the interior. The first case is that of a spherical particle. For equal biaxial stress, $K_t =$

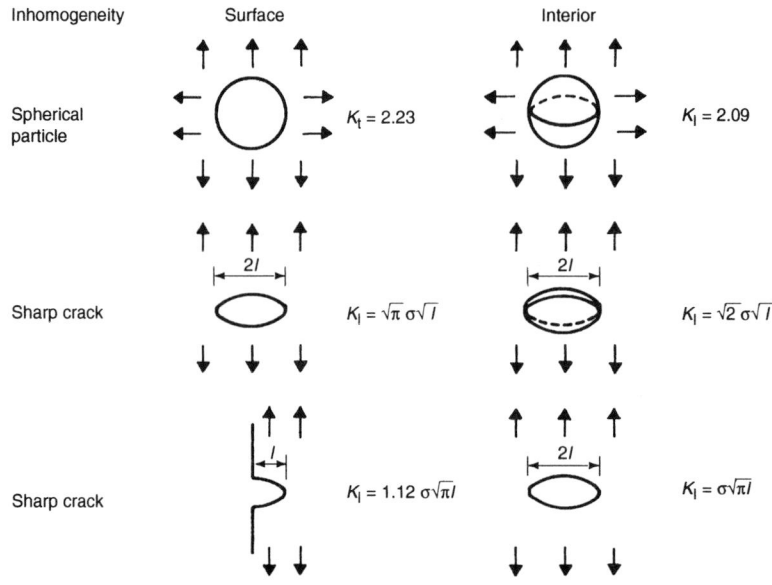

Fig. 1 Comparison of stress-concentration (K_t) and stress-intensity (K_I) factors for inhomogeneities of the same type on the surface and in the interior

2.23 was found for a hemispherical depression in the surface of a semi-infinite body (for Poisson constant $\nu = \frac{1}{4}$). This is about 7% higher than for the corresponding case of a spherical cavity ($K_t = 2.09$ for $\nu = \frac{1}{4}$) (Ref 3). The second example is the sharp, penny-shaped crack in the interior and the sharp crack of the same cross-sectional dimensions in plane-strain. The stress-intensity factor for the plane-strain crack is higher by a factor $\sqrt{\pi/2}$. The third example is the sharp plane-strain crack and the sharp plane-strain half-crack on a surface. The stress-intensity factor for the surface crack is higher by 12%. Although the comparison in Fig. 1 does not correspond perfectly to the real situation (because it assumes zero modulus and idealized shape), the predictions clearly indicate that inclusions at the surface should produce a higher stress concentration than inclusions in the interior.

Higher cyclic plastic deformation at or near the surface is higher, also due to the reduced constraint associated with deformation within the surface layer. This can be best explained in the case of a polycrystalline metal. In the bulk material, each interior grain is under constraints imposed by neighboring grains. Surface grains have a lower number of neighboring grains than interior grains and, consequently, the constraint is relaxed. In less constrained surface grains, single slip is more easily accommodated than in strongly constrained bulk grains, in which multiple slip may be more necessary (Ref 1). Experimentally it has been found by measuring local plasticity in individual surface grains (Ref 4) that the local surface strains can exceed the average strain of the specimen by orders of magnitude (Ref 5).

Sites of Crack Initiation

Significant experimental observations performed by either light or electron microscopy have shown that in homogeneous materials, microcracks always originate at free surfaces (Ref 6). In macroscopically inhomogeneous materials, such as surface hardened steels, cracks often start below the surface where constraint effects induce high triaxial stresses. For the same plastic strain amplitude in the surface layer, a much higher stress is required in comparison to the softer matrix. This is why the microcracks can originate at the interface between the hard layer and the soft matrix. In engineering components, cracks may start from internal defects, where cracks nucleate at internal surfaces or interfaces.

Direct observations of surfaces have shown that there are three types of nucleation sites:

- *Nucleation at fatigue slip bands* is perhaps the most frequent type. Its nature is slip concentration within the grains.
- *Nucleation at grain boundaries* is typical for high-strain fatigue, especially at higher temperatures.
- *Nucleation at surface inclusions* is typical for alloys containing large enough particles.

Common to all three types of nucleation is local plastic strain concentration at or near the surface. Nucleation in the fatigue slip bands is a basic type of nucleation, not only because this is the most frequent case, but mainly because the cyclic slip processes and formation of fatigue slips bands often precede nucleation at grain boundaries (Ref 7, 8) or at surface inclusions. From this point of view, nucleation at inclusions can be understood as cyclic slip localization (Ref 9) due to the stress-concentrating effect of the inclusion. See the section "Easy-Cross-Slip Metals (Low-Amplitude Cycling)" in this article. Nucleation at inclusions can lead either to decohesion of the inclusion-matrix interface or to cracking of the inclusion. Both of these microcracks have been observed experimentally (Ref 10).

There is strong evidence that nucleation at grain boundaries is also conditioned by the cyclic slip processes. Kim and Laird (Ref 11) concluded that the nature of cross-slip within grains, and the compatibility of slip at grain boundaries, are the most important factors in defining crack sites. Mughrabi et al. (Ref 12) showed that nucleation at grain boundaries occurs at sites where persistent slip bands (PSBs) impinge, and they formulated a dislocation model of this kind of crack initiation.

This article emphasizes nucleation at fatigue slip bands. A typical example of the development of fatigue slip bands with an increasing number of loading cycles is shown for polycrystalline copper in Fig. 2. The intensity and number of fatigue slip bands increase with the number of strain cycles, and the surface exhibits a more marked notch-peak topography. The overall appearance of the fatigue slip bands depends mainly on the stress or strain amplitude and on the slip character of the material. Wavy-slip materials are typified by lower numbers of intense bands, sometimes irregular and wavy. Under similar testing conditions, planar-slip materials exhibit higher numbers of less intense, highly regular bands lying exactly along the slip planes.

The initial appearance of fatigue slip bands typically coincides with the end of the macroscopic hardening/softening. The first microcracks are detectable within the bands later during cycling. Figure 3 shows fatigue slip bands with short microcracks in copper. In this replica micrograph, the microcracks cast a shadow. The typical length of microcracks at first detection is of the order of 0.1 to 1.0 μm, depending on the material and the experimental technique.

A large number of microcracks usually form during the first 20 to 40% of the total fatigue life. With continued cyclic loading, some of the nucleated microcracks grow, and practically no new ones are nucleated. This is shown in Fig. 4 which shows the measured crack density versus number of load cycles for an Al-Cu-Mg alloy (Ref 10). In this case, the microcracks emanate from cracked or debonded intermetallic particles, and their shapes are approximately semicircular. After the microcracks nucleate, the density undergoes only minor changes. The slight increase in microcrack density by the end of the lifetime is due to the

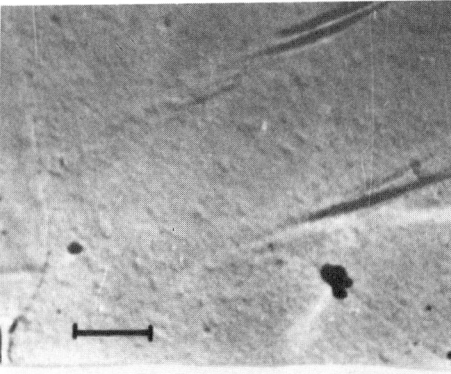
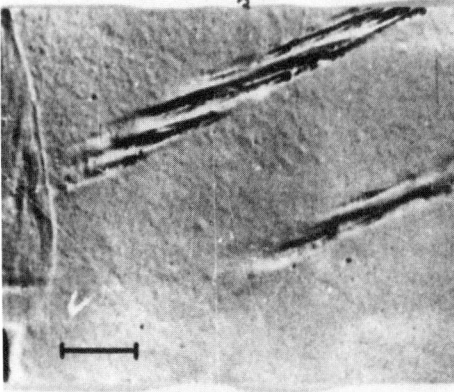
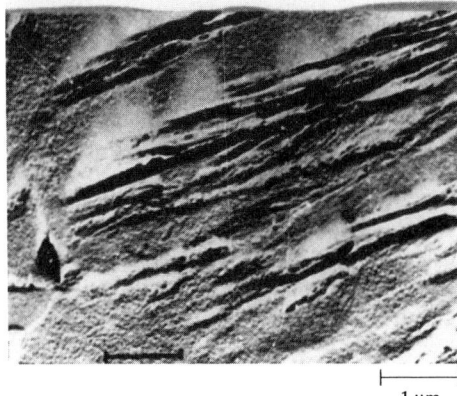

1 μm

Fig. 2 Evolution of fatigue slip bands during cycling of copper. (a) $N = 5 \times 10^3$ (b) $N = 10^4$ (c) $N = 10^5$ cycles. Push-pull symmetrical loading ($R = -1$) with low stress amplitudes. Number of cycles to fracture, $N_f = 1.2 \times 10^6$ cycles

formation of new microcracks in the plastic zone of the propagating macrocracks.

Crack nucleation usually takes place after saturation of the bulk mechanical properties. In the interior of metals cycled at low stress or strain amplitudes, the dislocation processes are reversible at this stage. The intensification of fatigue slip bands on the surface and the formation of microcracks show that the dislocation processes in the surface layer are irreversible.

Relation of Dislocation Structures and Surface Relief

To understand the process of nucleation, it is necessary to understand the dislocation processes

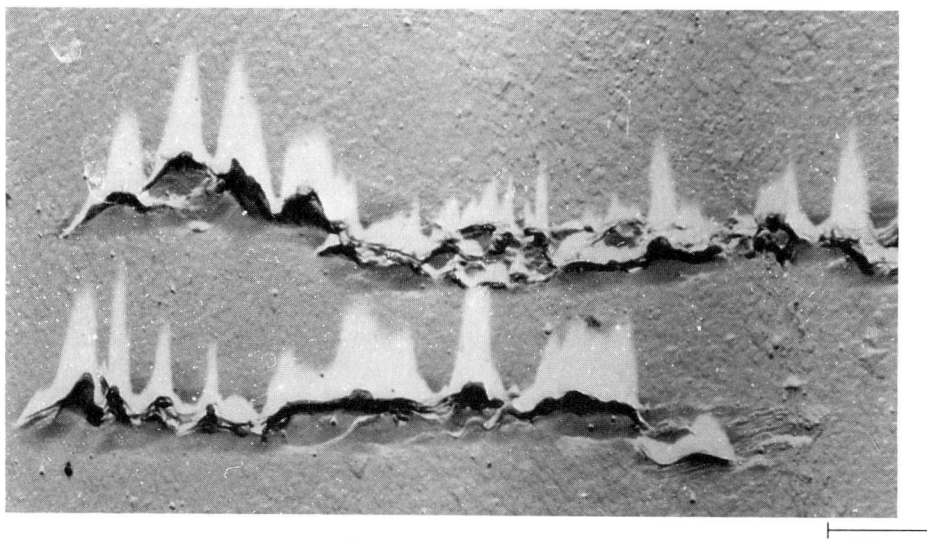

0.5 μm

Fig. 3 Replica micrograph of the first microcracks in fatigue slip bands on the surface of cycled copper. Same loading as in Fig. 2; replica stripped off at $N = 10^5$ cycles

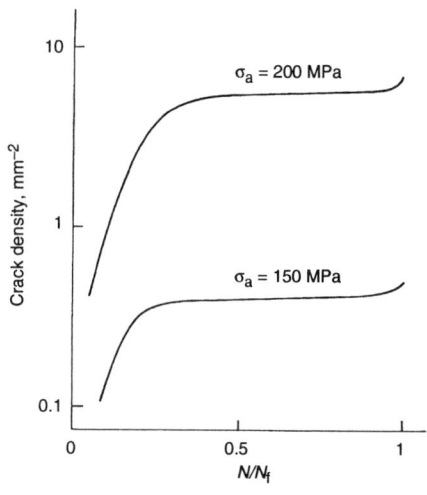

Fig. 4 Density of microcracks for two stress amplitudes (σ_a) in Al-Cu-Mg alloy. N, number of cycles; N_f, number of cycles to fracture. Axial load, $R = -1$

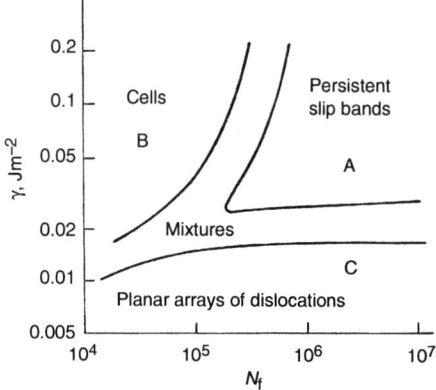

Fig. 5 Summary of near-surface dislocation structures as a function of amplitude (expressed here through number of cycles to fracture, N_f) and stacking-fault energy, γ. This summary covers materials that were well annealed before cycling.

in the surface layer. Perhaps the most substantial part of our current understanding of crack nucleation comes from direct electron microscopy observations of dislocation structures in surface layer after cyclic loading. There have been numerous experimental observations of near-surface dislocation structures in single-phase materials and those behaving like single-phase materials. It is now well established that in single crystals, there are no substantial differences between dislocation structures in the surface layer and in the interior of a crystal.

On the other hand, there are clear differences in polycrystals. This point was a matter of controversy in the literature for many years. The pioneering work of Hirsch et al. from 1959 (Ref 13) showed virtually no difference between near-surface structures and the structure typical of the whole volume, in a polycrystalline austenitic steel with low stacking-fault energy. Segall et al. (Ref 14) in 1961 came to the same conclusion for copper, aluminum, nickel, and gold, which have high stacking-fault energies. A few years later, considerable differences between near-surface and interior structures were unequivocally proved for low-carbon steel (Ref 15) and copper (Ref 16). More specifically, it was found that PSBs in the surface grains exhibit a structure that differs from the structure in the surrounding matrix and interior grains. Near-surface structures were studied in a variety of polycrystalline materials, but the controversy in the literature continued. While the earlier papers (Ref 15, 16) indicated that PSBs are formed only in surface grains, later papers also showed PSBs in interior grains (Ref 17-19). Systematic investigations (Ref 1, 20, 21) proved that this is a difference of quantitative nature: PSBs occur much less frequently in interior grains than in surface grains. The reason might lie in the fact that the cyclic plastic deformation of surface grains is less constrained than the cyclic plastic deformation of interior grains.

In summary, there is no difference between surface and interior grains as to the types of dislocation structures produced during cycling. Besides the difference in the quantity of the PSBs in wavy-slip materials, there is a minor difference between surface and interior dislocation structures in both single crystals and polycrystals: the dislocation density in the volumes near the surface is somewhat lower than the dislocation density in the interior of the cycled specimens (Ref 22-24). The lower dislocation densities near the surface is an effect from the so-called "mirror" or "image" forces at the surface, which attract dislocations to the free surface. This certainly contributes to surface relief formation (Ref 24).

The principal result of all investigations is that the types of near-surface structures depend, exactly in the same way as interior structures depend, on two parameters: the difficulty of cross-slip and the stress or strain amplitude. The difficulty of cross-slip is governed by the stacking-fault energy and to a lower extent by short range order and the yield strength (Ref 25-27). In a simplified sense, the ease of cross-slip can be identified with the stacking-fault energy, which can be expressed (contrary to other parameters affecting the cross-slip) quantitatively. Figure 5 correlates the types of dislocation structures observed in FCC alloys with stacking-fault energy, γ, and loading history, expressed here as number of cycles to fracture. For example, dislocation cells develop in high stacking-fault energy materials when subjected to high stress amplitudes (that is, low number of cycles to failure), while persistent slip bands are observed in the same material at low stress amplitudes (high number of cycles to failure). The following detailed description of the near-surface structures is given according to the regions in Fig. 5 (Ref 28).

Easy-Cross-Slip Metals (Low-Amplitude Cycling). The dislocation structures and corresponding surface relief described below are typical for both fcc metals with high stacking-fault

energies, cycled at any strain rate and at almost any temperature, and body-centered cubic (bcc) metals, cycled at low strain rates and moderate temperatures (Ref 29, 30). Fatigue slip bands, which lie along the intersection lines of the surface with the active slip plane, develop on the surface provided the specimen was cycled far enough in saturation. The dislocation structure just beneath the surface intrusions and extrusions, and in their nearest vicinity, differs considerably from the dislocation structure in the surrounding area and in the interior of the specimen. Examples are presented in Fig. 6 to 9. The crystallographic planes and directions are explained in Fig. 10.

Figure 6 is an electron micrograph obtained by "method-of-one-side" thin foils from the surface grain of a low-carbon steel cycled in push-pull loading at a frequency of 83 Hz (Ref 15). Figure 7 is an electron micrograph taken from the surface replica of the same specimen (but not of the same grain). In Fig. 6, there are well-developed bands with a dislocation structure (very dense dislocation patches indicated by arrow "A")

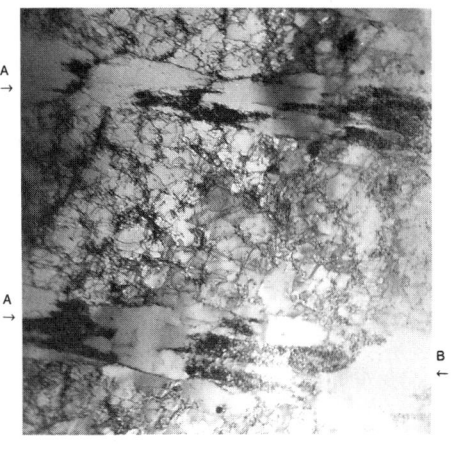

Fig. 6 Transmission electron micrograph showing the dislocation structure of persistent slip bands in the surface grain of low-carbon steel

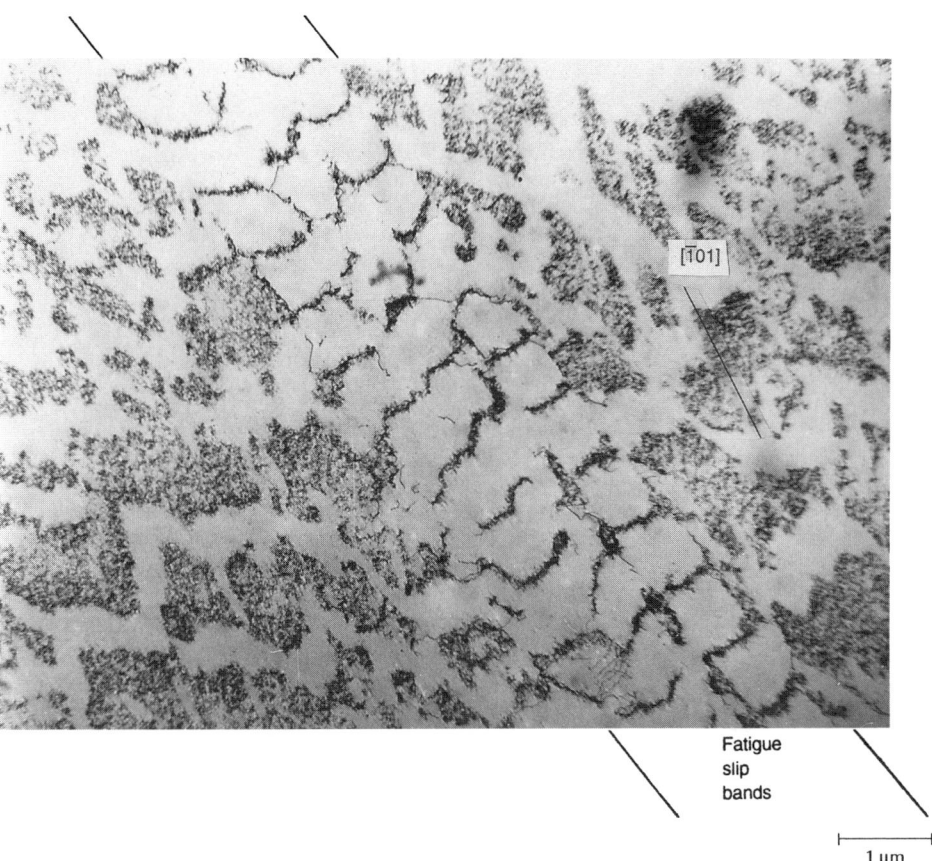

Fatigue slip bands

1 μm

Fig. 8 Structure of persistent band in a section perpendicular to the primary slip plane in copper single crystal. [T01] is the primary slip direction.

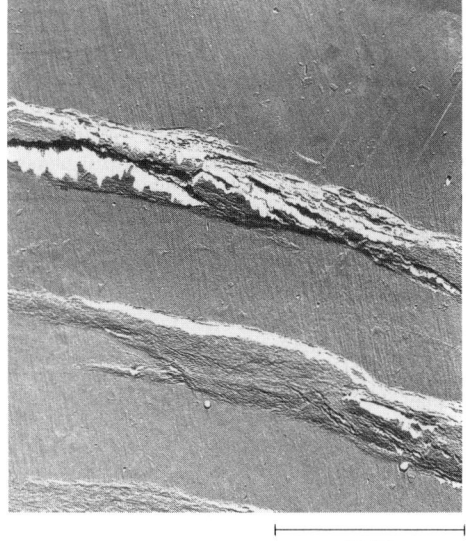

10 μm

Fig. 7 Replica micrograph of persistent slip bands containing microcracks on the surface of low-carbon steel

which is different from the structure of the observed surrounding matrix (inexpressive cell structure, arrow "B"). In Fig. 7, a complicated surface relief can be seen. There is probably a one-to-one relation between the zones of specific dislocation structures beneath the surface (Fig. 6) and the fatigue slip bands on the surface consisting of extrusions and intrusions (Fig. 7).

Figure 8 (Ref 31) presents dislocation structure as observed a few microns below the (1$\bar{2}$1) surface of a cycled single crystal of copper (see the schematic drawing of the specimen in Fig. 10). As mentioned above, there is no substantial difference between near-surface and interior dislocation structures in single crystals. Thus, the same structure can be observed in the center of the crystal. Two distinctly different structures can

be seen here. The structure of the fatigue slip band, lying along the [$\bar{1}$01] crystallographic direction, resembles here three irregular ladders, while the surrounding matrix structure is one of perpendicularly cut veins.

Figure 9 is a transmission electron micrograph obtained in the following way: After cycling, the specimen (not the same one used for Fig. 8) was electroplated. The foils parallel to the (1$\bar{2}$1) plane (see Fig. 10) were then prepared from the interface of the specimen with the electrodeposited layer. In such foils it is possible to see simultaneously the dislocation structure and the corresponding surface relief. Here the broad zone of ladder-like structure lying along the trace of the primary slip plane ends on the specimen surface in a strong extrusion. The structure surrounding this zone is the typical interior structure (perpendicularly cut veins).

The differences in dislocation structures within bands and in the surrounding matrix cause the difference in etching response. Even before electron microscopy methods were applied to the study of structures in fatigued metals, it was established that polishing a few microns from a surface could cover fatigue slip bands, which then can become visible again after etching. That is why these fatigue slip bands were termed PSBs, persistent slip bands.

In the light of current knowledge, the PSB can be defined as a zone that fulfills three conditions:

there is a cyclic strain localization in this zone; its dislocation structure differs from that in the surrounding matrix; and the zone ends on the specimen surface in intrusions and extrusions. The structure shown in Fig. 8 and 9 represents the section (1$\bar{2}$1) perpendicular to the slip plane (111) and containing the slip direction [$\bar{1}$01]; that is, the PSB structure is here seen from the profile. In the section parallel to the primary slip plane, the structure associated with the PSBs can be described as a series of dislocation walls or "hedges" with a [$\bar{1}$01] normal, extended tens of microns in the [1$\bar{2}$1] direction.

The PSB structure is not always of this type; it is often formed by completely closed cells. Thus there are two types of PSBs: ladder-like PSBs consisting of parallel walls perpendicular to the primary slip direction; and cell-like PSBs consisting of fully closed cells. The ladder structure is typical for new PSBs, while the cell structure is typical for old PSBs (Ref 32, 33). This change in dislocation structure within the PSBs during cycling is responsible for the slow secondary hardening after saturation is reached (Ref 33, 34). In other words, the PSBs become hardened by secondary dislocations, and thus they continue to deform, but at a reduced local strain amplitude.

The distance between walls in ladder-like PSBs (the mean distance between the centers of neighboring walls) is highly regular. For example, in copper single crystals cycled at room tempera-

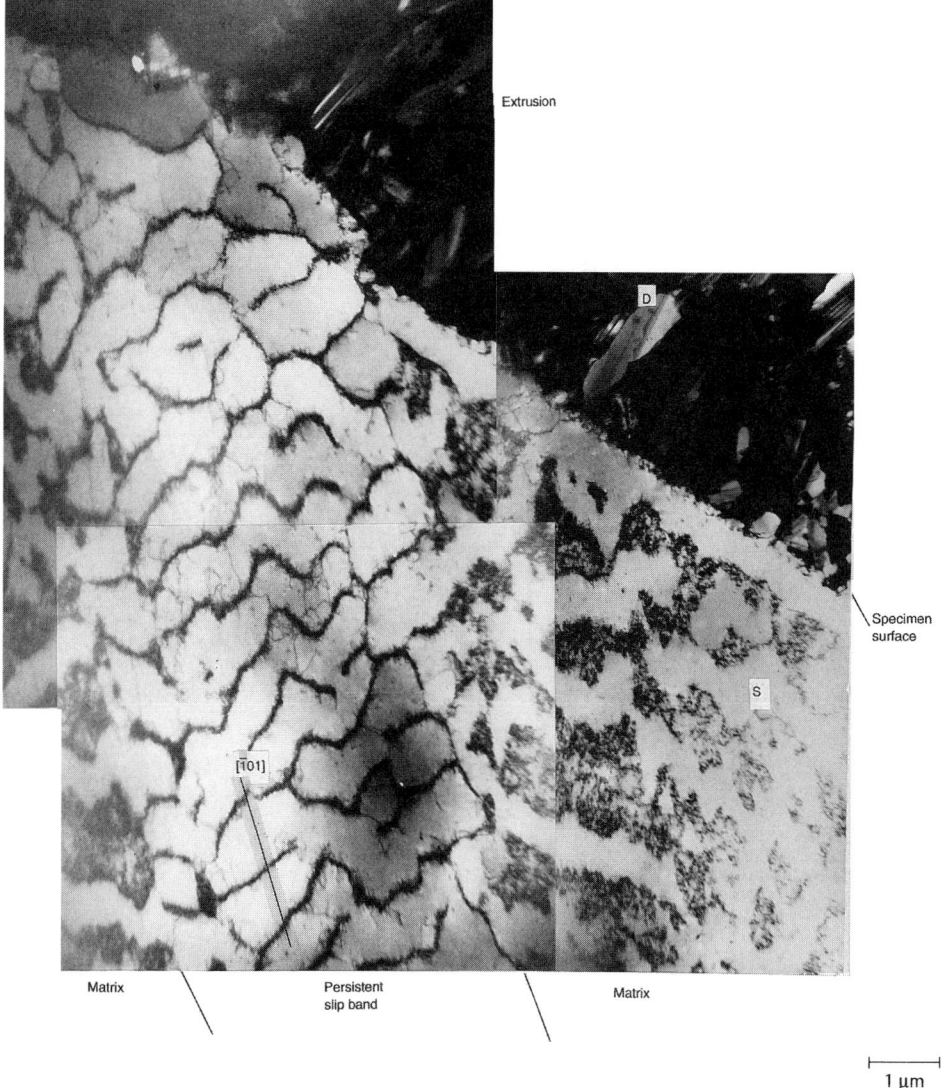

Fig. 9 Surface relief and underlying dislocation structure in a section perpendicular to the specimen surface and the primary slip plane in copper single crystal. D, electrodeposited layer; S, specimen TEM

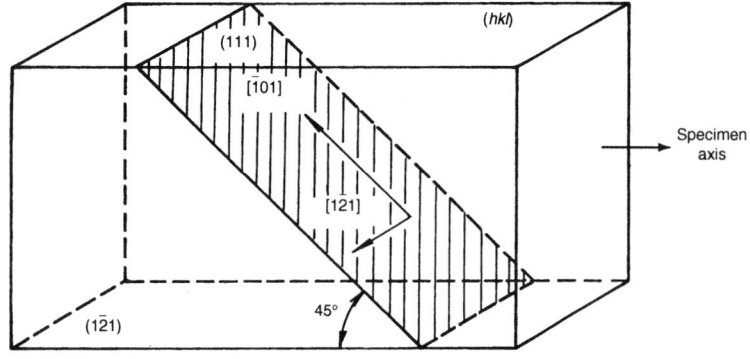

Fig. 10 Notation of crystallographic planes and directions

ture, the distance is 1.4 μm. This ladder spacing depends strongly on temperature, both in single crystals (Ref 35-37) and in polycrystals (Ref 21). The lower the temperature, the smaller the ladder spacing. There is an inverse proportionality between the ladder spacing and the saturation stress amplitude.

The highly regular dislocation structures presented in Figs. 8 and 9 are typical for the "well-behaved" materials like copper. In more compli-

cated materials the degree of dislocation structure regularity is lower. See, for example, the PSB in low-carbon steel presented in Fig. 6. It does not change anything on the above given definition of the PSB.

Despite some minor discrepancies between experimental results concerning the details of PSB structure, there is agreement on the following points (Ref 28, 38, 39):

- PSBs lie along the slip plane, and their structure is different from the surrounding matrix.
- There are certain stress and strain requirements for PSB formation. PSBs have never been observed below a threshold value of the plastic strain amplitude. This threshold is typically of the order of 10^{-5}. For some materials it is slightly higher, but it does not exceed 2×10^{-4}. The corresponding threshold values of the stress amplitude depend very strongly on the material, in contrast to the plastic strain amplitude. These two values (thresholds of the stress and strain) are connected directly through the cyclic stress-strain curve.
- In polycrystalline materials, PSBs are confined mainly to the surface grains. In single crystals, their volume fraction increases with both the number of cycles and the loading amplitude. After a sufficiently high number of loading cycles, the entire crystal is filled with PSBs.
- The plastic strain amplitude in PSBs is higher than in the matrix. For example, from the cyclic stress-strain curve for copper single crystal (Ref 29), it can be deduced that the shear plastic strain amplitude in the PSBs is about 7.5×10^{-3} (end of the plateau), while its matrix value is only about 6×10^{-5} (beginning of the plateau). Thus, the local plastic strain amplitude in the PSBs is higher by a factor of about 100 than the plastic strain amplitude in the surrounding matrix.

In the region of easy-cross-slip metals cycled at low amplitudes, the surface relief is two-stage: (1) a fine slip, with traces observed on the surface between the PSBs (Ref 16, 40, 41); and (2) a coarse slip, that is, PSBs with strong extrusions and intrusions. Fine slip lines are formed during fatigue hardening, and coarse slip markings (i.e., PSBs) appear on the surface at saturation. The PSBs cannot be detected metallographically on the specimen surface right from the onset of cycling, but rather only at the end of hardening/softening process when the interior structure is already formed. Therefore, PSBs are probably formed by transferring the matrix structure into the PSB structure. This process certainly requires cross-slip, and that is probably why no PSBs are formed in very difficult-cross-slip materials.

Some experimental findings show that the interior dislocation structure is on the boundary of stability. Basinski et al. (Ref 41) observed that during the early stages of fatigue of copper single crystals cycled with constant plastic strain amplitude, a very slight increase in strain amplitude resulted in quick formation of intensive slip markings on the surface. Neumann (Ref 42) cycled single crystals with linearly increasing stress

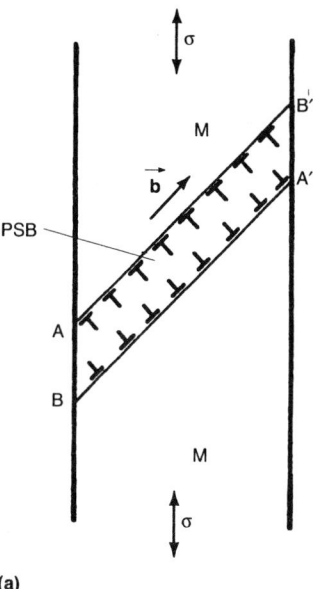

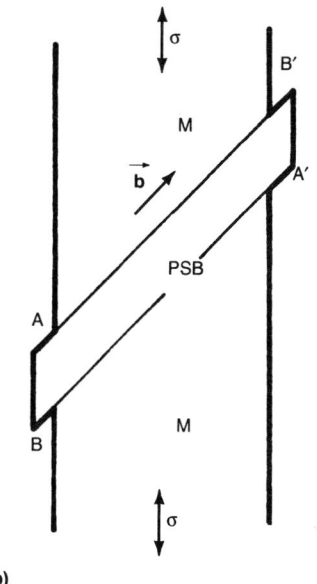

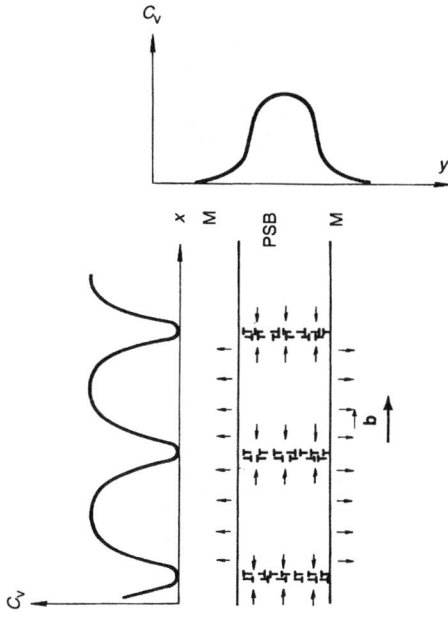

Fig. 11 Model by Essmann et al. (Ref 46) for formation of extrusions. (a) Arrangement of interface dislocations corresponding to an interstitial-type dislocation-dipole layer. (b) Extrusions formed by emergence of interface dislocations. **b,** Burgers vector; M, matrix; PSB, persistent slip band

Fig. 12 Model by Polák (Ref 47) for surface relief formation. A single persistent slip band is shown with vacancy concentrations in the x- and y-directions (vacancy fluxes are denoted by arrows). **b,** Burgers vector; c_v, vacancy concentration; M, matrix; PSB, persistent slip band

amplitude. The plastic strain amplitude showed very strong maxima during testing; these peaks have been called strain bursts since that time. Holzwarth and Essmann (Ref 43) used transmission electron microscopy to prove the transformation of matrix structure into PSB structure on copper single crystals. All these results have shown that matrix structure is not stable with respect to slight variations of stress or strain amplitude.

The formation of PSBs can be approximated as follows: During the early stages of cycling, the interior structure is formed, and simultaneously fine slip lines appear on the specimen surface because dislocations are escaping to the free surface. Near the surface, the stress conditions are a little different from those at greater depths, because of the factors discussed above. Therefore, at or near the surface, the stress in localized places can exceed average. Because of the instability of the interior structure, described above, the dislocation structure in such places must respond to the increased stress. In easy-cross-slip materials, this response can be very quick (in a few loading cycles), and it results in the formation of a new type of structure, the PSB structure. The cyclic plastic strain within these PSBs leads to formation of surface intrusions and extrusions, which in turn give rise to stress concentration and thus enhance the process of PSBs broadening and growing to greater depths. A model for this cyclic plastic strain, based on the idea of low-energy dislocation configurations, has been proposed by Kuhlmann-Wilsdorf and Laird (Ref 44).

Prenucleation Stage Prior to Microcrack Nucleation. To give a concise picture of the prenucleation stage, the outline of the kinetics of PSB formation must be complemented by a more detailed explanation of the processes that lead to formation of the surface relief. Several models

have been proposed to explain the formation of surface extrusions and intrusions due to the processes within the PSBs. These models are based on the slip processes, on the formation and migration of point defects, or on a combination of both.

It seems plausible to correlate the formation of surface intrusions and extrusions with the processes within the whole PSB, not with processes taking place near the surface only (as some earlier models proposed). A qualitative model accounting for these phenomena in terms of slip processes was best formulated by Finney and Laird (Ref 45). In the interior of the specimen, primary-edge dislocation segments bow out from the walls comprising the PSB structure until they become incorporated in the opposite wall. As this is taking place, the stress imbalance causes a segment in the opposite wall to become active, producing a cascading effect that channels numerous dislocations down the PSB. Upon reverse loading, the same or similar segments travel in the opposite direction, and by this means reversibility occurs. However, the segments that emerge at the surface are not completely reversible. Thus, the formation of surface intrusions and extrusions is due to dislocation motion within the whole PSB.

The model by Essmann et al. (Ref 46) is based on the role of vacancies and their agglomerates, produced by nonconservative jog dragging and opposite-edge dislocation annihilation. These vacancies lead to the deposition of edge dislocations at the PSB-matrix interfaces, as shown in Fig. 11(a). These interface dislocations correspond to a multipolar array built up of interstitial-type dipoles. The number of atoms contained in the extra atomic planes is equivalent to the number of vacancies that prevail in the PSBs. Under the action of the applied stress, interface dislocations glide out of the crystal at A and A' during

the tensile phases of cycling and at B and B' during the compressive phases, respectively. This process leads to the formation of extrusions on both sides of the PSBs, as shown in Fig. 11(b). This extrusion does not remain smooth; the random irreversible slip leads to the development of surface roughness.

A model by Polák (Ref 47) specifies the flux of vacancies in the PSBs (Fig. 12). Vacancies are formed both in the walls and in the channel areas between the walls in the PSBs. The densities of dislocations in the walls and in the channels differ considerably. Because edge dislocations serve as sinks for vacancies, the excess vacancy concentration also differs. The excess vacancy concentration and vacancy fluxes are shown in Fig. 12. The flux of vacancies in one direction is equivalent to the flow of atoms in the opposite direction. This results in the accumulation of mass in two directions perpendicular to the direction of flow. Thus, the material is extruded from the crystal along the channels and is intruded into the crystal along the walls.

The models of surface relief formation based on vacancies imply temperature dependence of PSB profile geometry on the mobility of vacancies. Basinski and Basinski (Ref 48, 49) observed formation of expressive surface relief over the PSBs also at temperatures where vacancies are immobile. The existence of immobile vacancies limits the applicability of vacancy-based models to the temperature range where vacancies can migrate. In summary, the formation of the surface relief connected with the PSBs results both from the slip processes and from the production and migration of point defects.

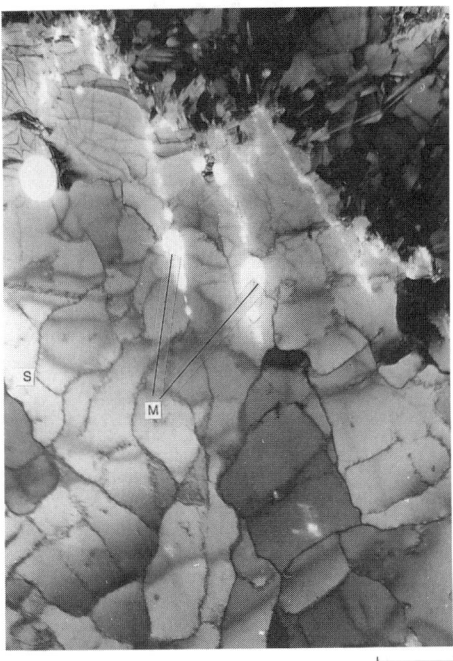

Fig. 13 Surface relief microcracks and dislocation structure in surface layer. Section perpendicular to the specimen surface and the primary slip plane in copper single crystal. D, electrodeposited layer; S, specimen; M, microcracks

The coarse slip markings on the surface are not always related to the zones of specific dislocation structure that differ from the structure of the surrounding matrix. For example, Bretschneider et al. (Ref 50) found no characteristic arrangements of dislocations that could be correlated with the extrusions on the surface of nickel single crystals cycled at low amplitudes at 77 K. Thus, zones with specific dislocation structures are not pre-

requisites for the formation of surface coarse slip bands and for subsequent microcrack nucleation.

Easy-Cross-Slip Metals (High-Amplitude Cycling). High-amplitude cycling of easy-cross-slip materials has been found to form the cell structure both in the interior and the surface layer of specimens (Ref 28). No clear differences have been found between interior and near-surface cell structures. On the other hand, the finer parameters of the cell structure in the surface layer, such as disorientation and the Burgers vectors of dislocations in the walls, have never been measured. Therefore, differences between interior and near-surface cell structure in these finer parameters is a remaining question.

Figure 13 (Ref 28) shows the surface relief and the underlying dislocation structure in a copper single crystal cycled by a high strain amplitude. The foil parallel to the (12̄1) plane (see Fig. 10) was prepared by the method of "electroplated samples" described above (see text to Fig. 9). Very marked surface relief can be seen; the surface extrusions and intrusions can be described as valleys and hills that lie predominantly along the intersection lines of the slip plane with the surface of the specimen. As in the previous case, formation of the notch-peak topography must be related to the collective motion of dislocations within the whole layer of cells lying along the slip plane. Again, the basic mechanism probably lies in bowing-out dislocation segments from cell walls that reach the nearest wall and cause an avalanche along the whole layer of cells. On reverse load, the process is repeated in the opposite direction, but the surface steps produced by dislocations emerging at the surface are not completely reversible, again because of cross slip. Contrary to the low-amplitude case, the interaction of slip activity of different systems must be taken into account in future detailed explanations.

Figure 13 also shows the first microcracks (denoted M) that start from sharp surface intrusions.

In both high-amplitude and low-amplitude cases, the first microcracks are connected with sharp intrusions. The microcracks shown in Fig. 13 follow the trace of the primary slip plane, which proves that they lie along the primary slip plane.

Difficult-Cross-Slip Metals. In the case of materials with very low stacking-fault energies, interior and near-surface structures were similar and described as planar arrays of dislocations. However, near-surface structure differs from interior structure in a minor way: the dislocation density is lower by a factor of 2 to 3 (Ref 22).

An example of the planar array of dislocations is shown in Fig. 14 (Ref 52). The foil was taken from the interior of a Cu-22Zn single crystal cycled at a low strain amplitude; the notation of the crystallographic planes and directions is shown in Fig. 10. The dislocations on the active primary slip planes (111) appear as thin bands in this (12̄1) section. The dislocation density within some of these bands obviously exceeds the dislocation density in the others. Cycling at higher amplitudes results in a higher number of these dense bands (Ref 22). The same type of dislocation structure was found in polycrystalline planar slip alloys after cycling (Ref 53-56). Laird et al. (Ref 23, 57-60) call the slabs of activated primary slip planes (the dense bands in Fig. 14) "persistent Lüder's bands" (PLBs). These bands are assumed to represent zones of localized cyclic slip. Contrary to the PSBs in wavy-slip metals, PLBs are not permanent and do not represent zones of dislocation structure that are distinctly different from surrounding matrix. PLBs are not stable; the localized strain moves around the gage length, and the active life of the slip band is short.

One example of the near-surface planar-slip structure is shown in Fig. 15, a TEM section of a Cu-31Zn single crystal (Ref 28). The sample was electroplated after cycling, and the foil was prepared in such a way that it contains the interface between the specimen and the (hkl) surface (see

Fig. 14 Dislocation structure in Cu-22wt%Zn single crystals

Fig. 15 Surface relief and underlying dislocation structure in a section perpendicular to the specimen surface and the primary slip plane in α-brass single crystal (Cu-31wt%Zn). D, electrodeposited layer; S, specimen

Fig. 10). The roughness profile on the specimen surface can be seen simultaneously with the underlying dislocation structure. The relation between the hill-and-valley surface topography and the slabs of the planar arrays of dislocations is obvious.

The surface morphology in planar-slip alloys is different from that in wavy-slip metals. The surface slip relief in wavy-slip metals is related to the slip activity of PSBs. In many cases a one-to-one correspondence between a ladder-like PSB and a coarse slip band on the surface can be proved (see Fig. 9). In planar-slip crystals, a more or less regular hill-and-valley surface morphology is often found after a sufficiently high number of loading cycles. This, together with the fact that PLBs do not represent zones of distinctly different dislocation structures, leads to the conclusion that surface relief forms because homogeneous cyclic slip takes place simultaneously and with the same intensity over the whole gage length.

A quantitative approach to the homogeneous type of surface roughness formation, based on computer simulation of random slip and on slip irreversibility at the surface (Ref 12), led to reasonable agreement with the experimentally observed surface roughness profile of the type shown in Fig. 15.

Careful experiments performed by Hong and Laird (Ref 23, 59, 61) on Cu-16at.%Al crystals offer a different interpretation of the kinetics of surface morphology formation. Hong and Laird found that the fraction of the gage length occupied by slip bands (in the sense of surface markings) increases with an increasing number of cycles, so that surface relief is not formed simultaneously over the whole gage length. Most important is that the active slip bands repeatedly move around the gage length (i.e., the slip bands can cease their activity and be reactivated again). In fact, the final result of this process is also a regular and rather homogeneous hill-and-valley surface morphology. The cracks are observed after the whole gage length becomes completely filled with slip bands (Ref 60). This confirms again that the prerequisites for microcrack nucleation are a sufficient notch-peak topography, hardening at nucleation sites, and local concentration of cyclic plastic deformation. The way that these necessary conditions are achieved seems to be of less importance.

Damage in Nucleation Stage

Microcrack nucleation is always preceded by cyclic slip localization. The magnitude of the local cyclic plastic strain amplitude is a key parameter governing crack nucleation. The fatigue damage in the prenucleation stage is related to the local cyclic plastic strain, but this fact is of almost no practical use, because the local cyclic plastic strain amplitude at the nucleation sites cannot be measured. Microcracks start from the surface intrusions (i.e., from surface micronotches formed by cyclic plastic deformation). The other, related possibility in polycrystals is grain-boundary nucleation due to the interaction of PSBs with the

grain boundaries, which also leads to the formation of surface micronotches. Crack nucleation on the PSBs and in the grain boundaries often occurs concurrently. The relative density of these two nucleation sites depends on the amplitude and the environment (Ref 67).

As mentioned above, a number of experiments have shown that removal of the surface layer prior to nucleation considerably increases fatigue life. Electropolishing removes the surface roughness produced by cyclic plastic deformation, but it does not change the dislocation structure. Experiments of this type have led to the conclusion that the surface layer as a whole, and the surface stress raisers especially, are the principal sources of fatigue damage. More specifically, the damage can be related primarily to the sharpness and density of the surface notches and, to a lesser extent, to the state of the dislocation structure surrounding them.

The term *fatigue damage* can be defined precisely during the stage of stable crack propagation (i.e., stage II). At this stage, fatigue damage is proportional to crack length. To define fatigue damage during nucleation (stage I), it would be necessary to find a quantitative characteristic of the acuity of the surface relief. Generally, the expressiveness of the surface relief (i.e., the height of extrusions and the depth of intrusions) increases with the number of cycles (Ref 69, 70) and with increasing local plastic strain amplitude in the nucleation sites. It seems plausible that this local plastic strain amplitude increases with increasing overall plastic strain amplitude, and thus that fatigue damage in the prenucleation stage is an increasing function of the plastic strain amplitude and the number of cycles. Irrespective of the mechanism of microcrack initiation, the first microcracks in flaw-free materials nucleate simultaneously in many places on the surface. Thus, the initiation process should be understood as a formation of the whole system of microcracks, not as an isolated event concerning just one nucleation site. The interaction of the particular microcrack with the surrounding system of other microcracks cannot be generally neglected.

For all materials, the population and size distribution of microcracks are strong functions of the stress or strain amplitude and the number of cycles (Ref 10, 62-64). Moreover, these quantities depend strongly on the type of material. We shall discuss here two examples: an Al-Cu-Mg alloy exhibiting a relatively low microcrack density (Fig. 4); and α-iron exhibiting a high density of microcracks (Fig. 16). The degree of mutual interaction among the microcracks varies accordingly.

Figure 4 shows the dependence of microcrack density on the relative number of cycles in an Al-Cu-Mg alloy for two stress amplitudes (Ref 10), both greater than the endurance limit. A high number of microcracks are formed during the first 20 to 30% of total fatigue life. With continued cycling, some of the nucleated microcracks grow, and practically no new ones are nucleated. In addition, the microcracks do not coalesce and the microcrack density does not decrease. Thus

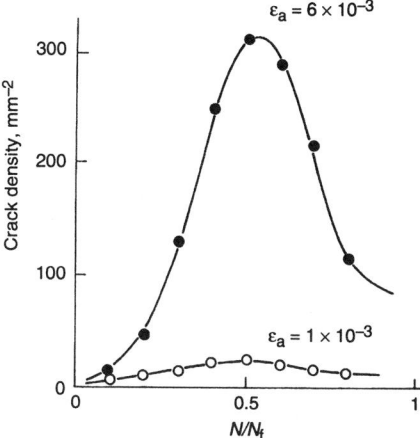

Fig. 16 Microcrack density in dependence on relative number of cycles in α-iron for two strain amplitudes. ε_a, strain amplitude; N, number of cycles; N_f, number of cycles to fracture

the mutual interaction is weak. The microcracks emanate from cracked or debonded intermetallic particles, and their shape is roughly semicircular. After the whole family of microcracks nucleated, their density undergoes only minor change. The same holds for the size of these microcracks. The mean depth of the microcracks is about 6 μm for the lower amplitude and about 9 μm for the higher amplitude. This implies that only a few microcracks can grow after they have reached a critical size, which is obviously strongly stress dependent.

Figure 16 shows the effect of strain amplitude and number of cycles on the density of microcracks for α-iron (Ref 63). There are two clear differences from Al-Cu-Mg alloy: the density of microcracks in α-iron is considerably higher (for comparable fatigue lives); and the density increases with the number of cycles, reaches a maximum, and then decreases. The average crack length on the surface corresponding to the maxima is about 20 μm for the lower amplitude and about 80 μm for the higher amplitude. The curves prior to $N/N_f = 0.5$ represent a continuing nucleation of new microcracks. At the beginning portion of fatigue life (at low values of cycles, N), microcracks nucleate and grow independently. With increased N, additional microcracks nucleate and existing cracks grow. With crack growth eventually there are interactions between microcracks, the local stress is relieved, and microcrack nucleation ceases. For $N/N_f > 0.5$, crack growth and coalescence dominate and the total number of cracks decrease. The subsequent coalescence of the microcracks is the reason for the decrease in microcrack density.

The strain redistribution due to microcrack growth was considered by Ma and Laird (Ref 65) to determine the behavior of the whole system of nucleated microcracks. The strain redistribution leads to the result that shallower cracks stop growing, while a few deeper ones grow faster and become the fatal cracks. This suggests that once the microcrack mean spacing reaches a certain critical value, the fatigue life preceding macrocrack propagation must be close to exhaustion.

Figures 4 and 16 concern the population of freshly nucleated microcracks at stress or strain levels above the fatigue limit. There is now enough experimental evidence to show that microcracks are nucleated and propagated up to the critical size, even at stress or strain levels at or slightly below the fatigue limit. At stress levels at or below the fatigue limit, propagation of nucleated microcracks stops. This is in contrast to stress levels above the fatigue limit, where some microcracks continue to propagate. Thus, the fatigue limit can be understood as the stress (or strain) level that represents the threshold for propagation of critical microcracks. At the fatigue limit, the largest microcrack of the whole population is more important than the mean value of the overall population of microcrack sizes.

The extent of damage due to an array of microcracks depends primarily on the mean spacing between the microcracks. For large mean spacing (i.e., mutual interaction of microcracks is insignificant), the damage can be best defined as the crack length of the dominant crack. This interpretation can probably be applied to all materials subjected to cycling at levels near the fatigue limit. For higher stress or strain levels, this definition can be used only for materials with low microcrack density, such as Al-Cu-Mg (see Fig. 4). Otherwise, in high stress or strain amplitude cycling, the microcrack density should be used as the basis for the definition of damage. For example, Hua and Socie (Ref 66) defined the damage as weighted crack density, taking into account the fact that cracks forming late in the life should not be as damaging as cracks that form earlier.

Mechanisms of Microcrack Nucleation

A large number of models have been proposed for nucleation of microcracks. In light of the experimental results considered above, it is clear that the proposed mechanisms may operate only in nucleation sites (i.e., at the root of intrusions in fatigue slip bands, in the vicinity of an inclusion, or near grain boundaries).

In order to model crack nucleation, the difference between sharp intrusions and the microcracks must be defined. For example, on the basis of experimental data obtained mainly on copper single crystals, Neumann states (Ref 68) that "intrusion formation and crack nucleation are two different processes." Basinski and Basinski conclude (Ref 69), also on the basis of experimental observation of cycled copper single crystals, that intrusions continue to grow with an increasing number of cycles, so that near to failure there is a large population of cracks. The different observations with respect to the transition between intrusions and microcracks leads to different models. Some of the models consider the process of crack nucleation to be indistinguishable from the process of intrusion formation. Other models distinguish intrusions from

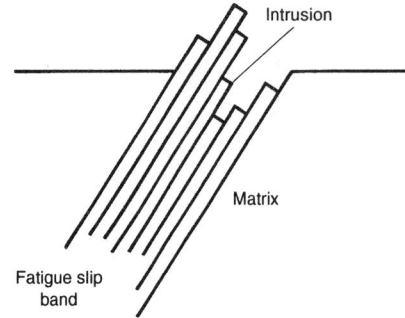

Fig. 17 Model of card slip in fatigue slip band

microcracks, and they require qualitatively different processes for intrusion formation and microcrack formation. None of the proposed models can be experimentally verified to the full extent, so all are more or less hypothetical.

The proposed mechanisms of microcrack nucleation can be roughly divided into five groups, described below. The basis for the division here is the mechanism of the critical event leading to a loss of cohesion, i.e. to the microcrack formation. All of the models have merits and have been justified by experimental findings. None of them has been worked out quantitatively, which would make it possible to express the influence of both external and internal parameters on the rate of nucleation.

Models That Do Not Distinguish between Intrusions and Microcracks. In some models, microcrack formation is identical with continuous growth of intrusions. Growth may occur by repeated slip on one or more slip systems. The basic idea for the one-slip system is the relative motion of parallel "cards" (Fig. 17). Wood (Ref 71) assumes that intrusions act as a stress raiser and promote further slip just in the "notch root." May (Ref 72) showed theoretically on the basis of statistical formalism that with continued cycling, progressively deeper intrusions would result from random slip. Lin et al. (Ref 73, 74) calculated the relative motion of two non-neighboring "cards" or slices. They showed that the local plastic shear strain in both slices (one positive and the other negative) can reach very high values within a relatively short number of cycles. These large plastic strains cause the continuous "squeezing out" or "sucking in" of the layer between the slices. In other words, in the second case we get a continuously deepening intrusion.

The model by Lynch (Ref 75), also using the idea of soft layers being extruded or intruded during cycling, yields a similar result: the fatigue cracks initiate and grow by a mechanism of intrusion that occurs when soft layers are "sucked in." A modern version of the models based on intrusions is represented by computer simulations of random irreversible slip. The model by Rosenbloom and Laird (Ref 76) operates by determining the number of active slip planes during each half cycle, randomly choosing the corresponding number of slip planes, and then randomly distributing the slip among those slip planes after ensur-

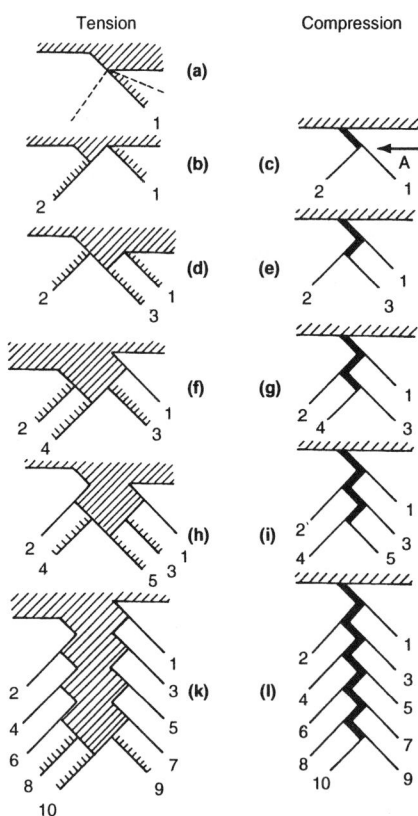

Fig. 18 Neumann's model of crack nucleation. In part (c), A represents a crack nucleus. Source: Ref 42

ing that each chosen slip plane has received at least one increment of slip. Crack nucleation, defined as an intrusion of 4 mm depth, occurred in good agreement with the observed number of cycles.

Neumann (Ref 42) proposed a model for the formation of cracks by coarse slip on alternating parallel slip planes (Fig. 18). In this model the crack develops from coarse slip steps. In tension (Fig. 18a), slip plane 1 is activated; excess dislocations of one sign remain on this slip plane. The slip step that is produced acts as a stress raiser, which also helps to activate slip plane 2 under the same tensile load. This leads to the configuration shown in Fig. 18(b) and to excess dislocations of one sign on plane 2. During the next compression, excess dislocations on slip planes 1 and 2 run back, leading to the configuration shown in Fig. 18(c). It is assumed that at A in Fig. 18(c) the surfaces are not "rewelded" but rather only touch macroscopically. Thus, A in Fig. 18(c) already represents a crack nucleus. Repetition of this process takes place on further glide planes of the same slip systems, leading to continuous increase of microcrack length. Note that in this model, the crack nucleation is on an average plane which is perpendicular to the maximum tensile stress as opposed to parallel to the shear plane as in the case of intrusion formation.

Harvey et al. (Ref 77) proposed a model based on slip band spacing, slip height displacement, and cumulative plastic strain. The quantitative

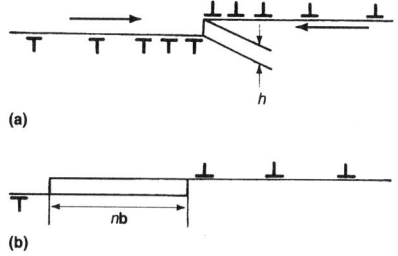

Fig. 19 Fujita's model of crack nucleation. See text for definitions of symbols. Source: Ref 79

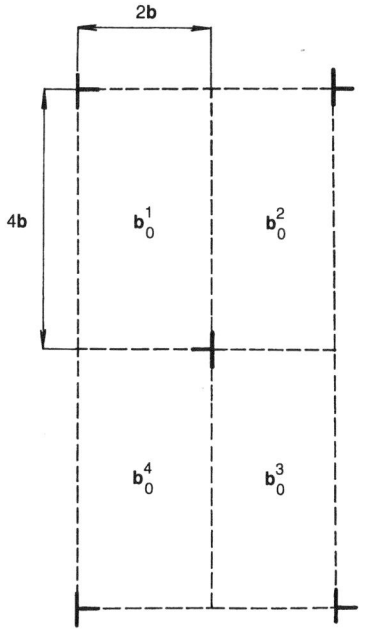

Fig. 20 Oding's model of crack nucleation. **b**, Burgers vector. Source: Ref 80

treatment is based on two assumptions: that surface displacements are crack-like and that crack-tip displacement controls nucleation. The number of cycles to crack initiation is the number of cycles necessary to reach a certain critical displacement. The model predicts that the number of cycles to nucleate a crack increases with a decrease in grain size, an observation which is consistent with experimental results (see, e.g., Ref 78).

Crack Nucleation based on Critical Conditions for Local Brittle Fracture. The concept of local brittle fracture distinguishes clearly between intrusions and cracks. A trivial example is the cracking of a brittle second-phase particle at the site of a stress concentration due to fatigue notch-peak topography.

There is quite a high probability that the critical conditions for local brittle fracture can be reached also in very ductile materials like copper. Figure 13 shows microcracks starting from the surface intrusions. The appearance of the microcracks distinguishes them clearly from intrusions. Thus the mechanism of continuous growth of intrusions does not come into question. Much more acceptable is the explanation that the stress concentration and the stress triaxiality due to the intrusions reach their critical values at or ahead of the tips of the intrusions. The dislocations cannot anymore relax the local stress concentration and the microcracks are formed by local brittle fracture. The role of dislocations in this type of microcrack formation is thus only indirect, through the building up of the surface relief and the hardening of the crystal around the intrusions.

Condensation of Vacancies (Ref 82). Cyclic deformation produces a higher number of vacancies than monotonic loading. This may be due mainly to the to-and-fro motion of dislocations with jogs. As PSBs exhibit higher plastic strain amplitude, it is probable that the vacancies generated could condense to form voids, thus nucleating a crack. This model implicitly requires diffusion of vacancies, which is strongly temperature dependent. There are papers which show that the fatigue processes (crack nucleation and propagation) take place even at cryogenic temperatures (Ref 35), that is, temperatures where practically no diffusion is possible. This limits the applicability of the vacancy model to high temperatures, where creep phenomena and diffusion play critical roles.

Loss of Coherency across a Slip Plane due to Accumulation of Defects. The basic idea of loss-of-coherency models is the formation of dislocation configurations in critical sites, which leads to local increases in stress or energy sufficient to destroy crystal coherency in small regions (of the order of nanometers and less).

Fujita (Ref 79) showed theoretically that a dislocation dipole with a small separation of the two components of the dipole may lead to crack nucleation via annihilation. His model is shown in Fig. 19. On two parallel slip planes, the pileup configurations of opposite signs are formed during cycling, where h is the separation of the planes. The calculation shows that as a function of h, one of two events can happen: for $h > 1$ nm, the two sets of pileups pass each other; or for $h < 1$ nm, the leading dislocations annihilate, even though they do not lie in the same slip plane. By this process a small area with destroyed coherency is formed. If not only the leading dislocations annihilate, but also n dislocations from each pileup (Fig. 19b), then coherency is lost in a region of length nb (where **b** is the Burgers vector) and height h, and a microcrack is formed. This mechanism can operate when each pileup consists of at least a few tens of dislocations.

The model of Oding (Ref 80) is based on the assumption that the multipole dislocation configuration of the type shown in Fig. 20 is built up during cycling. The distances among the particular dislocations continuously decrease with increasing number of cycles. The elastic energy, having its peak values at points b_1 to b_4, increases with decreasing distances among the dislocations. After these distances reach the values shown in Fig. 20, the peak values of the elastic energy are comparable with the latent melting heat. This is considered by Oding to be equivalent

to the destruction of the coherency at the critical points. The area with lost coherency is, in turn, identical with the microcrack. The formation of the dislocation configuration in Fig. 20 again requires high local stress concentration. One of its sources, also in the model of Fujita, is a set of pileup dislocations, but a sharp surface micronotch could, in principle, also produce the required stress.

Dislocation configurations of the type shown in Fig. 19 have never been observed in cycled metals, and therefore the mechanism by Fujita in its original formulation is not applicable. However, its more sophisticated modifications by Mura (Ref 81) seems quite realistic. Mura considers two adjacent planes where positive and negative dislocations are accumulated. The dislocation dipoles are increased by each cycle of loading. Thus, the elastic strain energy increases with the number of cycles. Mura has shown that there exists a critical number of cycles beyond which the dislocation dipole accumulation becomes energetically unstable. The dislocation dipoles are annihilated to form a microvoid (crack).

Nucleation of Cracks in Grain Boundaries. Basically, two kinds of models for nucleation in grain boundaries have been proposed, one based on plastic instability (Ref 83) and one that takes into account the interaction of slip within the grain with the grain boundary (Ref 11, 12).

The first kind of model assumes a very high degree of homogenous cyclic plastic strain across the whole surface layer of surface grains. Because the boundary hinders plastic deformation (the displacement perpendicular to the surface is negligible at the boundary), the plastic instability can occur on a microscale in such a way that the depth of a crease at a grain boundary deepens with an increasing number of cycles, until the strain concentration of the crease becomes so large that it constitutes a microcrack.

From models based on slip band interaction with grain boundaries, the model by Mughrabi et al. (Ref 12) is worked out in a semiquantitative way. This model represents an extension of the model by Essmann et al. (Fig. 11) (Ref 46), proposed for the growth of surface extrusions above PSBs in single crystals. In polycrystals, the interaction between the PSB and the grain boundary leads to a stress concentration that can ultimately cause a decohesion along the grain boundary.

In fcc metals, twin boundaries have often been found to be nucleation sites (Ref 84, 85). Twin boundaries can promote microcrack nucleation in two ways: PSBs form preferentially in highly stressed region near the twin boundary; and in the stress concentrations, twinning dislocations move along the boundary, which is effectively equivalent to a motion of the twin boundary. The region over which the boundary moves undergoes a high cyclic strain that promotes nucleation.

End of the Nucleation Stage

Several interpretations have been used to define the end of the nucleation stage, all of which

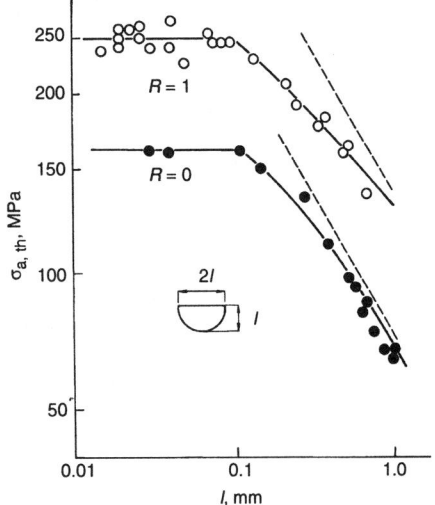

Fig. 21 Kitagawa-Takahashi diagram for natural surface cracks in low-carbon steel at stress ratios of $R = -1$ and $R = 0$

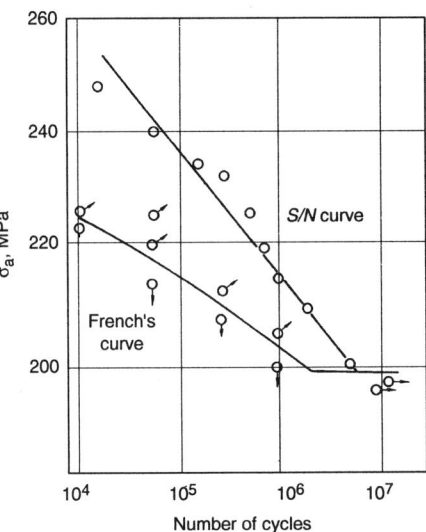

Fig. 22 S/N curve and French's curve for low-carbon steel

are based on a characteristic crack size and spacing. Each interpretation of them has its experimental justification. From the section "Damage in the Nucleation Stage" in this article, it does not seem plausible to relate the end of nucleation with the appearance of the first detectable microcracks. The transition from nucleation to propagation is rather the transition from the system of microcracks governed by cyclic plastic strain to crack propagation governed by fracture mechanics.

In cases in which there is substantial interaction among microcracks, the idea of a critical degree of strain relaxation, discussed above (Ref 65), is a good basis for the definition of the end of the nucleation process. When a critical degree of mean microcrack spacing is reached by crack multiplication, strain relaxation effectively hinders nucleation of new microcracks, and the strain redistribution accelerates stage II crack growth. Nevertheless, it is difficult to formulate this definition quantitatively.

Vašek and Polák (Ref 63) adopted a similar point of view. They assumed every nucleated microcrack leads to strain relaxation in its vicinity. The total area of the surface, at which the strain is relaxed below the value needed for nucleation, increases with an increasing number and size of microcracks. When the number and size of microcracks reach critical values, no new nucleation is possible, and further material degradation is caused by growth of the largest cracks. Vašek and Polák identify the number of cycles that correspond to the maximum of microcrack density in Fig. 16 with the transition from the nucleation stage to the crack propagation stage. The corresponding representative microcrack length strongly depends on strain amplitude, being considerably lower for the low amplitude (20 µm) than for the high amplitude (80 µm). The fraction of cycles spent in the nucleation stage is independent of the strain amplitude, namely about 50% of the total life. This contradicts the generally accepted view that the nucleation process at

high amplitudes is completed within a negligible fraction of total life. Thus the assumption that the end of nucleation is given by the position of the maximum in Fig. 16 is obviously not correct. It is probably another characteristic of the family of nucleated microcracks, which characterizes the end of nucleation stage.

A long period of cycling at stress equal to or slightly lower than the fatigue limit produces nonpropagating microcracks with a size comparable to the grain size (Ref 86, 87). It follows that the fatigue limit is the threshold for small cracks that nucleated (at the same stress level), grew to a critical size, and then ceased to grow (Ref 88). The existence of such a critical crack size implies another possible definition for the end of the nucleation stage: as the number of loading cycles needed to produce a crack of a critical size.

The short crack threshold can be conveniently described by means of the Kitagawa-Takahashi plot (Ref 89), which relates the short threshold stress amplitude with crack size (Fig. 21). The Kitagawa-Takahashi plot introduces a "demarcation line" below, which the cracks cannot propagate. This threshold presentation in terms of the threshold stress amplitude automatically involves the fact that the highest possible short crack threshold stress amplitude is the fatigue limit of smooth specimens. Figure 21 (Ref 90) is an experimentally determined Kitagawa-Takahashi diagram for two R-ratios. Up to a critical size, the cracks are nondamaging. This critical size is about 0.1 mm for both the R-ratios, which corresponds approximately to the prior-austenite grain size. The threshold stress amplitudes at the horizontal parts of the curves (i.e., the threshold stresses for cracks up to the critical size) are identical with the independently determined fatigue limits.

Many years ago, French (Ref 91) proposed the "critical-damage curve." The determination of this damage curve, also called French's curve for a material of known S-N curve, can be performed by the following procedure. A specimen is cycled at a chosen stress level for a chosen number of cycles; then the stress is decreased to the level of fatigue limit and the cycling is continued. If the specimen fractures after a (high) number of further cycles, the original stress level lies above French's curve. If the specimen does not fracture, even after a high number of cycles, the original stress level lies below French's curve. A repetition of this procedure for a number of specimens enables the investigator to locate the position of French's curve quite exactly.

An example of the experimentally determined French's curve is presented in Fig. 22 (Ref 92). In agreement with the definition of French's curve, each specimen was cycled for a chosen number of cycles at a chosen stress level. The stress level was then decreased to the fatigue limit and the cycling was continued. Points marked by upward arrows denote specimens that fractured; points marked by downward arrows denote specimens that did not fracture during the course of 10^7 loading cycles. Experimentally, it was found for low-carbon steel that at French's curve the PSBs

contain cracks extending from grain boundary to grain boundary, in some cases even across two or three grains. This is independent of the stress level. Thus, French's curve represents the curve of constant crack size. It is important that the crack size corresponding to French's curve (Fig. 22) be roughly equal to the critical crack size determined from the Kitagawa-Takahashi diagram (Fig. 21).

In summary, two concepts can be used to define the end of the nucleation stage. One is based on the relaxation of strain around microcracks, and the other is based on the size of the largest crack that cannot propagate below the fatigue limit. The latter definition gives considerably larger cracks at the end of nucleation than the former definition. In a way, the transition from microstructurally small cracks to physically small cracks is well compatible with the latter definition. At present, there is no physically sound basis for a particular choice of definition of the end of the nucleation stage.

Factors That Influence Crack Nucleation

There is no clearcut demarcation between nucleation and early-stage propagation, so it is difficult to define the end of the nucleation stage (see the previous section). For practical purposes, however, such a definition is often necessary. The only possibility is either a convention based on the density of microcracks and their depth and length along the surface, or a convention based on the dimensions of the largest crack. Let us denote the number of loading cycles necessary to complete the nucleation stage as N_0 (for an arbitrarily chosen definition of the end of nucleation) and the number of cycles to fracture as N_f. Then the ratio N_0/N_f is a measure of the length of the nucleation stage in terms of the relative fatigue life.

The relative number of cycles N_0/N_f depends mainly on the amplitude and asymmetry of cycling, the shape of the specimen or engineering component, the material parameters, the environment, the temperature, and the surface layer.

Cycling Amplitude and Asymmetry. The value of N_0/N_f decreases with increasing amplitude. In the low-amplitude region, N_0 can represent a significant percent of the total fatigue life. For very high amplitudes, the nucleation is very quick, N_0 is negligible with respect to N_f, and essentially the whole fatigue life is spent in crack propagation. Nucleation is also strongly influenced by the stress cycle asymmetry. For example, in an extreme case of repeated compression, no cracks at all were found on the surface of cycled single crystals of copper (Ref 93).

Specimen Shape. Notches generally significantly reduce the value of N_0/N_f. For very sharp notches and especially for crack defects, the nucleation stage is almost completely missing and the whole fatigue life is given by the crack propagation stage.

Environment has a strong effect on crack initiation (Ref 94). Ample experimental data show that the fatigue life of all materials tested in vacuum is considerably longer than the fatigue life in any other environment. A part of the increase in fatigue life in vacuum is due to the fact that the growth rate of cracks (especially of small cracks) is smaller in vacuum that in air or other environment. Another substantial part of this increase is due to the inhibited crack initiation. For example, crack initiation in copper single crystals tested in vacuum has been found to be 1 to 2 orders of magnitude slower than that in air (Ref 68). This can be explained by rewelding of newly formed slip steps on slip reversion in vacuum. In air, every slip step is covered by adsorbed atoms or molecules from the environment. After slip reversion, this adsorption layer prevents annihilation of the newly formed surface of the slip step.

Temperature decreases lead to an increase in N_0/N_f for stress-cycled metals exhibiting crack nucleation in fatigue slip bands. For materials in which cracks nucleate at surface inclusions, the decrease in temperature should result in a decrease in N_0/N_f. At higher temperatures, nucleation in slip bands may be by nucleation at grain boundaries.

The surface layer has a very strong effect on fatigue life. The nature of this strong dependence lies mainly in the influence on crack nucleation. Surface treatment of any type leads to one or more of these effects:

Surface Roughness. The surface topography, especially surface scratches, act as stress concentrators and thus shorten the nucleation stage.

Residual Stresses. Macroscopic residual stresses can be detected on the surface after almost all types of surface treatment. Tensile residual stresses are detrimental (they enhance nucleation), whereas compressive residual stresses are beneficial (they inhibit nucleation). The essence of the explanation lies in the superposition of the external stress with the residual stress: the higher the tensile mean stress, the lower the number of cycles necessary for nucleation. This is justified by the above-mentioned experimental result that cracks do not nucleate in the compressive stress cycle.

Phase and Chemical Composition. The effect of phase and chemical composition either is deliberate (as in surface quenching, carbonitriding, coating, ion implantation, laser hardening, etc.) or occurs as a side effect of heat treatment (e.g., decarburization of the surface layer). The phase and chemical composition may influence the nucleation both beneficially and detrimentally, depending on the resistance of the surface layer to cyclic plastic deformation.

Work hardening of the surface layer inevitably occurs as a result of machining and finishing the surface simultaneously, due to the occurrence of residual stress. Cyclic loading removes or reduces work hardening during fatigue softening.

A corrosive environment generally shortens the nucleation stage. The effect of gaseous environments on fatigue crack initiation is a controversial subject. If there is any influence at all, it is probably not strong. However, aqueous environments have been found to significantly shorten the nucleation stage, perhaps without exception. Theories explaining this strong effect can be divided into the following categories:

- *Pitting:* Local etching, either selectively at places of higher slip activity (i.e., at fatigue slip bands) or nonselectively at any place on the surface, produces pits that act as stress raisers. Probably more important is the pitting or preferential dissolution at areas of higher slip activity, where slight differences in electrochemical potential inside and outside slip bands enhance the process of stress raiser formation.
- *Destruction of protective oxide films:* The surface of a metal exposed to an aqueous environment is covered by a thin oxide film that is cathodic with respect to the metal. Slip processes can easily destroy the oxide film locally, especially in places of high slip activity, at the fatigue slip bands. The electrochemical cell (the small anodic region at the site of oxide layer destruction) formed as a result can then very effectively speed up local dissolution at the slip band and thus produce micronotches.
- *Reduction of surface energy by adsorption:* The decrease in surface energy by an adsorbing species in an aqueous environment facilitates the process of surface slip formation. Thus, the formation of fatigue slip bands and, consequently, nucleation are easier.

Common to all of these explanations is the idea that a corrosive environment promotes slip activity in the surface layer of cycled metal. The mechanism of microcrack nucleation is probably the same as in the absence of environment.

Summary

A basic understanding of the fatigue process on a submicroscopial level is important in safe design against fatigue. Fatigue crack nucleation is perhaps the most difficult stage of the fatigue process to study. This is due mainly to the fact that the microcrack nucleation is a highly localized event taking place in a very small part of the total volume. At present, the sites of microcrack nucleation are relatively well known both in the model materials and in the engineering materials. The basic features of the microscopic mechanisms of the nucleation are partly understood on the qualitative level. No quantitative description of the nucleation mechanisms covering explicit expressions for all critical parameters is available. Thus it is not surprising that the present-day research on the mechanisms of fatigue crack nucleation aims to the quantification of the knowledge gathered over years.

The aim of this article is to give an overview on the fatigue crack nucleation from the point of view of the material microstructure and its evolution during cycling. The article describes the sites of microcrack nucleation at the free surfaces, discusses the relation of dislocation structures and surface relief and offers a review of the current mechanisms of crack nucleation. Moreover the meaning of the "damage" of material due to crack nucleation, the extent (in terms of the number of cycles) of the nucleation stage and the factors influencing crack nucleation are covered. The experimental findings discussed in the article concern mainly relatively simple model materials. Further data on crack nucleation in complicated engineering materials can be found in the "Selected References List of Crack Nucleation in Structural Alloys."

REFERENCES

1. H. Mughrabi, *Scripta Metall. Mater.,* Vol 26, 1992, p 1492
2. A.S. Argon, in *Corrosion Fatigue,* O.F. Devereux, A.J. McEvily, and R.W. Staehle, Ed., NACE, 1972, p 176
3. R.E. Peterson, *Stress Concentration Factors,* John Wiley and Sons, 1974
4. M.R. James and W.L. Morris, STP 811, American Society for Testing and Materials, 1983, p 46
5. P. Neumann, *Scripta Metall. Mater.,* Vol 26, 1992, p 1535
6. S. Kocanda, *Fatigue Failure of Metals,* Sihthoff and Noordhoff International Publishers, 1978
7. G.H. Kim, I.B. Kwon, and M.E. Fine, *Mat. Sci. Eng.,* Vol 142A, 1991, p 177
8. P. Šittner, V. Novák, and J. Brádler, *Scripta Metall. Mater.,* Vol 27, 1992, p 705
9. B. Velten, A.K. Vasudevan, and E. Hornbogen, *Z. Metallkde,* Vol 80, 1989, p 21
10. V. Sedláček, M. Ruščák, and J. Cmakal, in *Basic Mechanisms in Fatigue of Metals,* P. Lukáš and J. Polák, Ed., Academia/Elsevier, Prague/Amsterdam, 1988, p 73
11. W.H. Kim and C. Laird, *Acta Met.,* Vol 26, 1978, p 777
12. H. Mughrabi, R. Wang, K. Differt, and U. Essmann, STP 811, American Society for Testing and Materials, 1983, p 5

13. P.B. Hirsch, P.G. Partridge, and R. L. Segal, *Phil. Mag.*, Vol 4, 1959, p 721
14. R.L. Segal, P.G. Partridge, and P.B. Hirsch, *Phil. Mag.*, Vol 6, 1961, p 1493
15. M. Klesnil and P. Lukáš, *J. Iron Steel Inst.*, Vol 203, 1965, p 1043
16. P. Lukáš, M. Klesnil, J. Krejcí, and P. Ryš, *Phys. Stat. Sol.*, Vol 15, 1966, p 71
17. K. Pohl, P. Mayr, and E. Macherauch, *Scripta Metall.*, Vol 14, 1980, p 1167
18. A.T. Winter, O.B. Pedersen, and K.V. Rasmussen, *Acta Metall.*, Vol 29, 1981, p 735
19. J. Polák and M. Klesnil, *Mat. Sci. Eng.*, Vol 63, 1984, p 189
20. G. Gonzales and C. Laird, *Met. Trans.*, Vol 14A, 1983, p 2507
21. P. Lukáš and L. Kunz, *Mat. Sci. Eng.*, Vol 103A, 1988, p 233
22. P. Lukáš, L. Kunz, and J. Krejcí, *Scripta Metall. Mater.*, Vol 26, 1992, p 1511
23. S.I. Hong, C. Laird, H. Margolin, and Z. Wang, *Scripta Metall. Mater.*, Vol 26, 1992, p 1517
24. R.R. Keller, W. Zielinski, and W.W. Gerberich, *Scripta Metall. Mater.*, Vol 26, 1992, p 1523
25. H. Mughrabi and R. Wang, *Proc. of First Int. Symp. on Defects and Fracture*, Tuczno, Poland, 1980, eds. G.C. Sih and H. Zorski, Martinus Nijhoff Publishers, The Hague, 1982, p 15
26. V. Gerold and H.P. Karnthaler, *Acta Metall.*, Vol 37, 1989, p 2177
27. S.I. Hong and C. Laird, *Acta Metall. Mater.*, Vol 38, 1990, p 1581
28. P. Lukáš and M. Klesnil, in *Corrosion Fatigue*, O.F. Devereux, A.J. McEvily, and R.W. Staehle, Ed., NACE, 1972, p 118
29. H. Mughrabi, F. Ackermann, and K. Herz, STP 675, American Society for Testing and Materials, 1979, p 69
30. B. Šesták, V. Novák, and S. Libovický, *Phil. Mag.*, Vol 57A, 1988, p 353
31. P. Lukáš, M. Klesnil, and J. Krejcí, *Phys. Stat. Sol.*, Vol 27, 1968, p 545
32. C. Laird, P. Charsley, and H. Mughrabi, *Mat. Sci. Eng.*, Vol 81, 1986, p 433
33. R. Wang and H. Mughrabi, *Mat. Sci. Eng.*, Vol 63, 1984, p 147
34. D.E. Witmer, G.C. Farrington, and C. Laird, *Acta Metall.*, Vol 35, 1987, p 1895
35. Z.S. Basinski, A.S. Korbel, and S.J. Basinski, *Acta Metall.*, Vol 28, 1980, p 191
36. K. Mecke, J. Bretschneider, C. Holste, and W. Kleinert, *Cryst. Res. Tech.*, Vol 21, 1986, p K135
37. L.L. Lisiecki and J.R. Weertman, *Acta Metall. Mater.*, Vol 38, 1990, p 509
38. C. Laird, in *Plastic Deformation of Materials*, R.J. Arsenault, Ed., Academic Press, 1975, p 101
39. H. Mughrabi, in *Dislocation and Properties of Real Materials*, The Institute of Metals, London, 1985, p 244
40. E.E. Laufer and W.N. Roberts, *Phil. Mag.*, Vol 14, 1966, p 65
41. S.J. Basinski, Z.S. Basinski, and A. Howie, *Phil. Mag.*, Vol 19, 1969, p 899
42. P. Neumann, *Acta Metall.*, Vol 17, 1969, p 1219

43. U. Holzwarth and U. Essmann, *Appl. Phys.*, Vol 57A, 1993, p 131
44. D. Kuhlmann-Wildsdorf and C. Laird, *Mat. Sci. Eng.*, Vol 46, 1980, p 209
45. J.M. Finney and C. Laird, *Phil. Mag.*, Vol 31, 1975, p 339
46. U. Essmann, U. Goesele, and H. Mughrabi, *Phil. Mag.*, Vol 44, 1981, p 405
47. J. Polák, *Mat. Sci. Eng.*, Vol 92, 1987, p 71
48. Z.S. Basinski and S.J. Basinski, *Acta Metall.*, Vol 37, 1989, p 3263
49. Z.S. Basinski and S.J. Basinski, *Scripta Metall. Mater.*, Vol 26, 1992, p 1505
50. J. Bretschneider, C. Holste, and W. Kleinert, *Mat. Sci. Eng.*, Vol 191A, 1995, p 61
51. P. Lukáš and M. Klesnil, *Phys. Stat. Sol.*, Vol 5A, 1971, p 247
52. P. Lukáš, L. Kunz, and J. Krejcí, *Mat. Sci. Eng.*, Vol 158A, 1992, p 177
53. J. Boutin, N. Marchand, J.P. Bailon, and J.I. Dickson, *Mat. Sci. Eng.*, Vol 67, 1984, p L23
54. C. Laird, S. Stanzl, R. de la Veaux, and L. Buchinger, *Mat. Sci. Eng.*, Vol 80, 1986, p 143
55. H. Kaneshiro, K. Katagiri, C. Makabe, T. Yafuso, and H. Kobayashi, *Met. Trans.*, Vol 21A, 1990, p 667
56. H.J. Christ, *Wechselverformung von Metallen*, Springer Verlag, Berlin, 1991, p 229-239
57. C. Laird and L. Buchinger, *Met. Trans.*, Vol 16A, 1985, p 2201
58. L. Buchinger, A.S. Cheng, S. Stanzl, and C. Laird, *Mat. Sci. Eng.*, Vol 80, 1986, p 155
59. S.I. Hong and C. Laird, *Mat. Sci. Eng.*, Vol 128A, 1990, p 55
60. S.I. Hong and C. Laird, *Fatigue Fract. Engng. Mater. Struct.*, Vol 14, 1991, p 143
61. S.I. Hong and C. Laird, *Mat. Sci. Eng.*, Vol 124A, 1990, p 183
62. Bao-Tong Ma and C. Laird, *Acta Metall.*, Vol 37, 1989, p 337
63. A. Vašek and J. Polák, *Kovové Materiály*, Vol 29, 1991, p 113
64. O. Kessler, J. Walla, H. Bomas, and P. Mayr, in *Mechanical Behaviour of Materials*, ICM-6, M. Jono and T. Inoue, Ed., Pergamon Press, 1991, Vol 4, p 501
65. Bao-Tong Ma and C. Laird, *Acta Metall.*, Vol 37, 1989, p 349
66. C.T. Hua and D.F. Socie, *Fatigue Fract. Engng. Mater. Struct.*, Vol 7, 1984, p 165
67. C.V. Cooper and M.E. Fine, *Met. Trans.*, Vol 16A, 1985, p 641
68. P. Neumann, in *Physical Metallurgy*, R.W. Cahn and P. Haasen, Ed., Elsevier, Amsterdam, 1983, p 1554
69. Z.S. Basinski and S.J. Basinski, *Acta Metall.*, Vol 33, 1985, p 1307
70. A. Hunsche and P. Neumann, *Acta Metall.*, Vol 34, 1986, p 207
71. W.A. Wood, in *Fatigue in Aircraft Structures*, A.M. Freudenthal, Ed., Academic Press, 1956, p 1
72. A.N. May, *Nature*, Vol 186, 1960, p 573
73. T.H. Lin and Y.M. Ito, *J. Mech. Phys. Solids*, Vol 17, 1969, p 511
74. T.H. Lin, *Advances in Applied Mechanics*, Vol 29, 1992, p 1

75. S.P. Lynch, *Met. Sci.*, Vol 9, 1975, p 401
76. S.N. Rosenbloom and C. Laird, *Acta Metall. Mater.*, Vol 41, 1993, p 3473
77. S.E. Harvey, P.G. Marsh, and W.W. Gerberich, *Acta Metall. Mater.*, Vol 42, 1994, p 3493
78. K.S. Chan, *Scripta Metall. Mater.*, Vol 32, 1995, p 235
79. F.E. Fujita, *Acta Metall.*, Vol 6, 1958, p 543
80. J.A. Oding, *Reports of Academy of Sciences USSR*, 1960, p 3
81. T. Mura, *Mat. Sci. Eng.*, Vol 176A, 1994, p 61
82. N. Thompson and N.J. Wadsworth, *Adv. in Phys.*, Vol 7, 1958, p 72
83. C. Laird and A.R. Krause, *Int. J. Fract. Mech.*, Vol 4, 1968, p 219
84. P. Neumann and A. Tönnessen, in *Fatigue 87*, Vol 1, R.O. Ritchie and E.A. Starke Jr., Ed., EMAS, Warley, U.K., 1987, p 1
85. L. Llanes and C. Laird, *Mat. Sci. Eng.*, Vol 157A, 1992, p 21
86. M. Hempel, in *Fatigue in Aircraft Structures*, ed. A.M. Freudenthal, Academic Press, New York, 1956, p 83
87. T. Kunio, M. Shimizu, K. Yamada, and M. Tamura, in *Fatigue 84*, ed. C.J. Beevers, EMAS, Warley, 1984, p 817
88. K.J. Miller, *Fatigue Fract. Engng. Mater. Struct.*, Vol 10, 1987, p 93
89. H. Kitagawa and S. Takahashi, in *Proc. Second Int. Conf. on Mechanical Behavior of Materials*, American Society for Metals, 1976, p 627
90. P. Lukáš and L. Kunz, in *Short Fatigue Cracks*, ESIS 13, K.J. Miller and E.R. de los Rios, Ed., Mechanical Engineering Publications, London, 1992, p 265
91. H.J. French, *Trans. Am. Chem. Soc. Steel Treatment*, Vol 21, 1933, p 899
92. P. Lukáš and L. Kunz, *Mat. Sci. Eng.*, Vol 47, 1981, p 93
93. H.I. Kaplan and C. Laird, *Trans. AIME*, Vol 239, 1967, p 1017
94. T.S. Sriram, M.E. Fine, and Y.W. Chung, *Scripta Metall.*, Vol 24, 1990, p 279

Selected References List of Crack Nucleation in Structural Alloys

Aluminum Alloys

- C.A. Stubbington and P.J.E. Forsyth, *Acta Metall.*, Vol 14, 1966, p 5
- C.Q. Bowles and J. Schijve, *Int. J. Fract.*, Vol 9, 1973, p 171
- W.L. Morris, *Met. Trans.*, Vol 9A, 1978, p 1345
- C.Y. Kung and M.E. Fine, *Met. Trans.*, Vol 10A, 1979, p 603
- W.L. Morris and M.R. James, *Met. Trans.*, Vol 11A, 1980, p 850
- D. Sigler, M.C. Montpetit, and W.L. Haworth, *Met. Trans.*, Vol 14A, 1983, p 931
- S. Hirose and M.E. Fine, *Met. Trans.*, Vol 14A, 1983, p 1189
- I. Cerný, V. Sedlácek, and J. Polák, *Kovové Materialy*, Vol 23, 1985, p 715
- W.J. Baxter and T.R. McKinney, *Met. Trams.*, Vol 19A, 1988, p 83

- W.L. Morris, B.N. Cox, and M.R. James, *Acta Metall.,* Vol 37, 1989, p 457
- B. Velten, A.K. Vasudevan, and E. Hornbogen, *Z. Metallkde,* Vol 80, 1989, p 21
- A. Plumtree and B.P.D. O'Connor, *Fatigue Fract. Engng. Mater. Struct.,* Vol 14, 1991, p 171

Titanium Alloys
- C.J. Beevers and M.D. Halliday, *Metal. Sci. J.,* Vol 3, 1969, p 74
- J.J. Lucas and P.P. Konieczny, *Trans. Met.,* Vol 2, 1971, p 911
- D.K. Benson, J.C. Grosskreutz, and G.G. Shaw, *Met. Trans.,* Vol 3, 1972, p 1239
- D.F. Neal and P.A. Blenkinsop, *Acta Mettall.,* Vol 24, 1976, p 59
- A.W. Funkenbusch and L.F. Coffin, *Met. Trans.,* Vol 9A, 1978, p 1159
- Ai Suhua, Wang Zhongguang, and Xia Yuebo, *Scripta Metall.,* Vol 19, 1985, p 1089
- M.A. Däubler, H. Gray, L. Wagner, and G. Lüthering, *Z. Metallkde.,* Vol 78, 1987, p 406
- J.L. Gilbert and H.R. Piehler, *Met. Trans.,* Vol 20A, 1989, p 1715
- O. Umezawa, K. Nagai, and K. Ishikawa, *Mat. Sci. Eng.,* Vol A 129, 1990, p 217

- D.L. Davidson, J.B. Campbell, and R.A. Page, *Met. Trans.,* Vol 22A, 1991, p 377

Superalloys
- J. Gayda, R.V. Miner, *Int. J. Fatigue,* Vol 5, 1983, p 135
- D.L. Anton and M.E. Fine, *Mat. Sci. Eng.,* Vol 58, 1983, p 135
- T.P. Gabb, J. Gayda, and R.V. Miner, *Met. Trans.,* Vol 7A, 1986, p 497
- M.A. Daeubler, A.W. Thompson, and I.M. Bernstein, *Met. Trans.,* Vol 19A, 1988, p 301
- D.M. Elzey and E. Arzt, *Met. Trans.,* Vol 22A, 1991, p 837

Carbon Steels
- Po-We Kao and J.G. Byrne, *Met. Trans.,* Vol 13A, 1982, p 855
- C.M. Suh, R. Yuuki, and H. Kitagawa, *Fatigue Fract. Engng. Mater. Struct.,* Vol 8, 1985, p 193
- J.K. Solberg, *Mat. Sci. Eng.,* Vol 101, 1988, p 39
- L. Yumen, *Mat. Sci. Tech.,* Vol 6, 1990, p 731
- M. Goto, *Fatigue Fract. Engng. Mater. Struct.,* Vol 14, 1991, p 833
- X.J. Wu, *Fatigue Fract. Engng. Mater. Struct.,* Vol 14, 1991, p 369

Fully Pearlitic Steels
- G.T. Gray, III, A.W. Thompson, and J.C. Williamson, *Met. Trans.,* Vol 16A, 1985, p 753
- C.D. Liu, M.N. Bassim, and S. Stlawrence, *Mat. Sci. Eng.,* Vol 167A, 1993, p 107

High Strength Steels
- J. Lankford, *Engng. Fracture Mech.,* Vol 9, 1977, p 617
- T. Kunio, M. Shimizu, K. Yamada, K. Sakura, and T. Yamamoto, *Int. J. Fract.,* Vol 17, 1981, p 111
- Y.H. Kim and M.E. Fine, *Met. Trans.,* Vol 13A, 1982, p 59
- J.H. Beatty, G.J. Shiflet, and K.V. Jata, *Met. Trans.,* Vol 19A, 1988, p 973
- D. Wang, H. Hua, M.E. Fine, and H.S. Cheng, *Mat. Sci. Eng.,* Vol A 118, 1989, p 113

Stainless Steels
- S. Usami, Y. Fukuda, and S. Shida, *J. Pressure Vessel Tech.,* Vol 108, 1986, p 214
- C.M. Suh, J.J. Lee, and Y.G. Kang, *Fatigue Fract. Engng. Mater. Struct.,* Vol 13, 1990, p 487
- A. Heinz and P. Neumann, *Act Metall. Mater.,* Vol 38, 1990, p 1933

Fatigue Crack Growth under Variable-Amplitude Loading

J. Schijve, Delft University of Technology

FATIGUE is a practical problem for all kinds of structures subjected to a spectrum of many load cycles under normal service utilization conditions. A single load application need not be harmful, but a repetition of many load cycles can initiate a fatigue crack. The crack will grow until collapse of the structure, unless it is found by inspection. The variety of practical fatigue problems is large because of the many types of structures, materials, load spectra, and other design variables.

Fatigue failures can have significant consequences in practice, which can be highly undesirable for reasons of economy. Fatigue failures in expensive structures built in small numbers are practically unacceptable. Another important argument is safety. Disastrous fatigue failures occurred in the past with fatalities, serious damage to the environment, and liability problems afterwards. As a consequence, the concept of *designing against fatigue* has attracted much attention from industry, research institutes, universities, and the authorities responsible for safety regulations to protect society against fatal accidents. The economic and social impact of fatigue failures will not be discussed here, but designing against fatigue obviously is a matter of concern. It encompasses various design options, and needless to say, experience and engineering judgment are essential. Fatigue predictions are then necessary to quantify the fatigue problem in terms of fatigue life and crack growth.

A general survey of a fatigue prediction scenario is given in Fig. 1. It illustrates that design includes choosing among various options (first column). The second column includes data on material fatigue properties and calculations, basically stress analysis problems. The third column includes information on the fatigue loads in service, the dynamic response of the structure, and the environment. All aspects of the input information have to be used for predictions on fatigue life and crack growth. A pertinent question then is: Do we have reliable prediction models? If so, do we obtain accurate indications of the fatigue behavior of a structure in service? Is it desirable to verify the predictions by fatigue experiments? If

that is necessary, how are we going to simulate the reality of service conditions in a fatigue test? An evaluation of these questions requires a fundamental understanding of the fatigue mechanisms occurring in structural materials under conditions applicable to the real structure.

This paper summarizes fatigue phenomena in metallic materials, discusses fatigue under variable-amplitude (VA) loading, where the emphasis is on crack growth, and presents prediction models. Its aim is to survey the state of the art. It should be useful for further research, but at the same time, it should indicate possibilities and

limitations of fatigue predictions in a practical engineering environment.

Fatigue Phenomena in Metallic Materials

It is useful to consider the fatigue life as consisting of two periods:

- *The crack initiation period*, including crack nucleation and microcrack growth

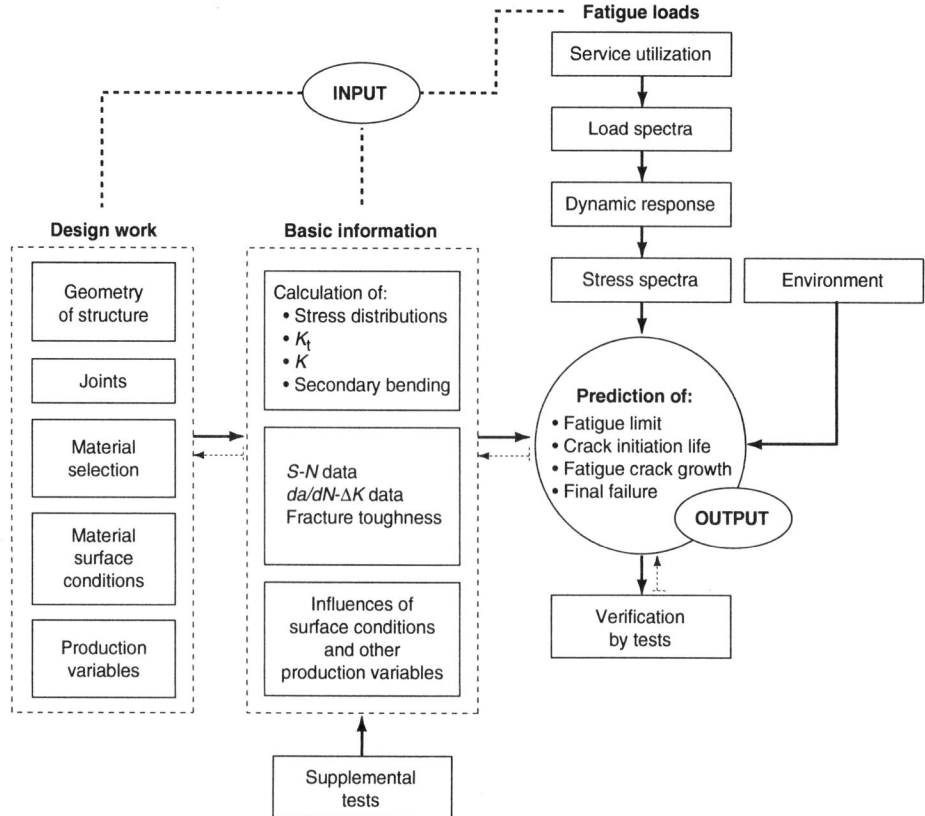

Fig. 1 Diagram of the fatigue prediction problem in practical applications. Dotted arrows indicate feedback.

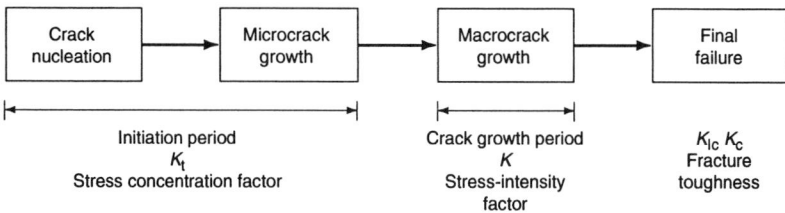

Fig. 2 Different phases of fatigue life and relevant factors

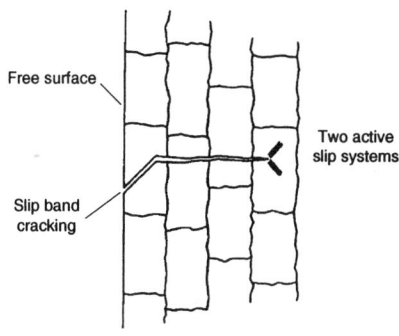

Fig. 4 Cross section of a microcrack

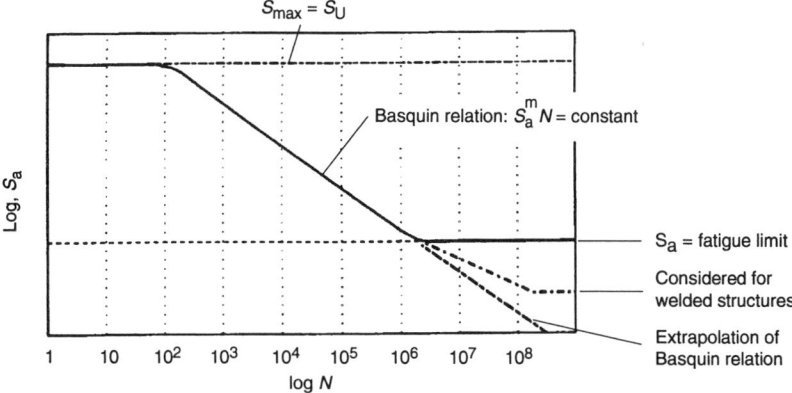

Fig. 3 S-N curve with extrapolations below the fatigue limit

- *The crack growth period*, covering the growth of a visible crack (Fig. 2)

There is an obvious question of defining the transition from the initiation period to the crack growth period, but that will be addressed later.

In a fatigue curve (S-N curve, Wöhler curve), the fatigue life (N) until failure is plotted as a function of the stress amplitude (S_a). Such curves apply to so-called constant-amplitude (CA) loading, that is, cyclic loading with a constant amplitude, but also a constant mean load. Quite often the fatigue curve turns out to be approximately linear in a double logarithmic plot (see Fig. 3, the Basquin relation). However, there are two cutoffs (i.e., two horizontal asymptotes). The upper one is associated with static failure, because the maximum load of the fatigue cycle exceeds the static strength. The lower one is usually referred to as the fatigue limit (S_f). For amplitudes below the fatigue limit, failure no longer occurs, even after a very high number of cycles. The fatigue limit is often defined as the stress amplitude for which the fatigue life becomes infinite, or as the maximum stress amplitude for which failure does not occur. A better definition is that the fatigue limit is the minimum stress amplitude that can still nucleate a crack that grows until failure. It does not imply that a microcrack cannot be initiated below the fatigue limit, but it does not grow into macrocracks. Apparently, the microcrack is arrested at some microstructural barrier.

Cyclic Plasticity, Microcrack Initiation, and Microcrack Growth. Fatigue cracks generally start at the material surface, for practical and fundamental reasons:

- Practical reasons: Higher stress level, K_t always >1; and secondly surface roughness and other small-scale stress concentrations
- Fundamental reasons: Lower restraint on cyclic plasticity, and in addition environmental effects

Of course there are notorious exceptions, such as subsurface crack nucleations associated with inside material defects, inhomogeneous residual stress distributions, and a more fatigue-resistant material structure at the surface (shot peened surface layer, nitriding, etc.). The fundamental reasons are given more attention below, because they are significant for considering threshold problems and the relevance of applications of fracture mechanics to fatigue, also in relation to VA loading.

Grains at the material surface are not supported by other grains at one side (i.e., the side of the environment). As a consequence, cyclic slip can occur more easily than it does inside the material, where slip is more restrained by the surrounding material. Because of the lower restraint on slip in a surface grain, it can occur at a lower stress level. It is one of the reasons why crack nucleation generally starts in surface grains, or slightly subsurface (e.g., at an inclusion). There are different theories on microcrack nucleation, which will not be surveyed here. They explain how cyclic slip in just a few cycles can lead to a physical microcrack at the surface. In several materials the initial microcrack is growing in a slipband. The microshear stress concentration in a slip band depends on the crystal lattice orientations and grain shapes. Due to slip during uploading, the reversed shear stress during unloading will again

be high in the same slip band. Cyclic slip and the initial microcrack growth will thus concentrate in slip bands. Cyclic slip is not a reversible phenomenon (if it were, material fatigue would not be a problem), partly because of strain hardening, but also because of the environmental interaction with slip steps and cracked material. In air it implies strongly adhering oxide monolayers. Aggressive environments promote the initiation of microcracks in cyclic slip bands.

As long as the size of the microcrack is still on the order of a single grain, there is a microcrack in an elastically anisotropic material with a crystalline structure and a number of different slip systems. The microcrack causes an inhomogeneous stress distribution on a microlevel, with a stress concentration at the tip of the microcrack. If that activates more than one slip system, the microcrack growth direction can deviate from the initial slip band orientation. Cracks then tend to grow perpendicular to the loading direction (Fig. 4). Microcrack growth depends on the material structure, crystallography, possible slip systems, the ease of cross-slip (stacking fault energy), the grain lattice orientation (texture), and the grain size. As a result, crack nucleation and the first microcrack growth cannot be expected to be similar phenomena for different materials. As an example, Al alloys usually have small grains, the elastic anisotropy is low, and cross-slip is relatively easy. For Ni alloys, grains can be large, the elastic anisotropy is much larger, and cross-slip is relatively difficult.

In general, a large number of grains at the material surface are nominally loaded to the same cyclic stress level, even if we consider a notched specimen. An obvious question then is why microcracks are not nucleated in all grains. Actually, if the strain amplitude at the surface is large, microscopic investigations have shown that there will be a high number of microcracks in Al alloys (Ref 1-3). Due to the low elastic anisotropy of aluminum, the stress level from grain to grain does not change much, and as a consequence, there will be many grains with a high local stress level. Microcracks can coalesce after some growth and continue to grow as a single crack. Macroscopic fractography usually shows only one or a few dominant fatigue crack nuclei. The number of visible nuclei depends on the fatigue

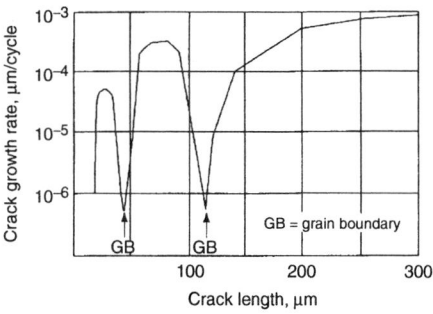

Fig. 5 Grain boundary (GB) effect on microcrack growth in an Al alloy. Source: Ref 4

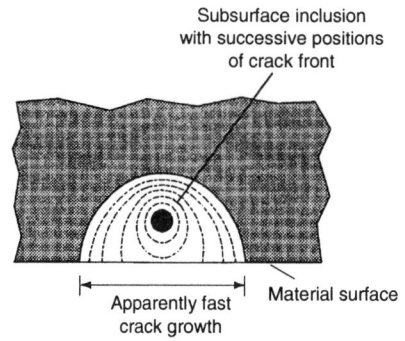

Fig. 7 Subsurface crack nucleation at inclusion, erroneously suggesting initial fast crack growth. Source: Ref 9

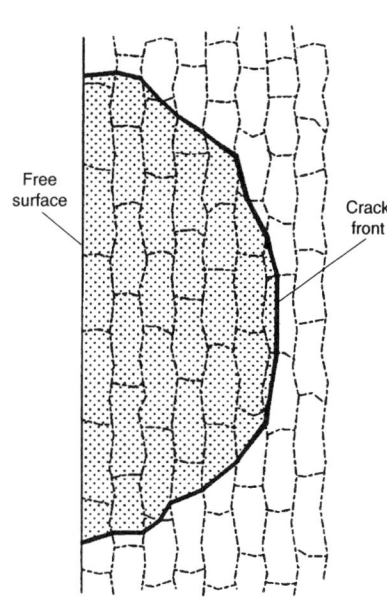

Fig. 8 Top view of crack with crack front through many grains

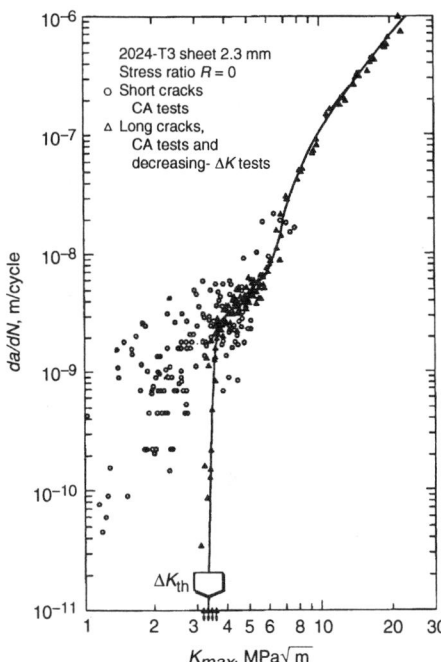

Fig. 6 Growth rates of small and large cracks plotted together as a function of ΔK. CA, constant amplitude. Source: Replotted in Ref 8 from AGARD Report No. 732, 1988

load. At a stress level close to the fatigue limit, only one crack nucleus is observed. This appears to be logical, because the fatigue limit is a threshold stress level. Only one crack nucleus will be successful in growing until final failure. Statisticians call it the *weakest link* in the material.

Because microcrack growth depends on cyclic plasticity, barriers to slip can imply a threshold for crack growth. This has indeed been observed. Illustrative results were published by Blom et al. (Ref 4) for an aluminum alloy (Fig. 5). The crack rate decreases when the crack tip approaches a grain boundary. After passing a third grain boundary, the microcrack continues to grow with a steadily increasing crack growth rate. Reinitiation in a second (subsurface) grain has been shown by fractographic work of Lankford (Ref 5). In low-carbon steel it has been shown that pearlite colonies considerably hamper fast microcrack growth in the ferrite matrix (Ref 6). In

the literature there are several observations on initially inhomogeneous microcrack growth, starting with a relatively high crack rate, which is slowed down or even stopped by material structural barriers. Suresh (Ref 7) introduced the term *microstructurally short cracks* for this behavior.

If the growth rate of microcracks is plotted as a function of ΔK together with results of large cracks, a confusing picture can result (Fig. 6). The apparent paradox is that large macrocracks do not grow if ΔK < ΔK_{th}, whereas microcracks in surface grains can grow in a low-ΔK regime. It appears to be a paradox, but as pointed out above, cyclic slip can occur relatively easily close to the material surface. It allows an initially fast development of a microcrack at the surface. Also, a microcrack initiated at a subsurface inclusion can attain an initially high crack rate during breakthrough to the material surface (Fig. 7). Moreover, it should be realized that ΔK for a microcrack at the material surface is a nominally calculated ΔK. It is not necessarily a meaningful concept for small microcracks. Basically, the stress-intensity factor is meaningful for a crack in a homogeneous material for the stress distribution in close proximity to the crack tip, as long as the plastic zone is very small compared to the crack length. These conditions are simply not satisfied for microcracks with a size of 1 or 2 grain diameters. The literature on small cracks and crack growth at ΔK values below ΔK_{th} is rather extensive (Ref 10, 11).

The crack front of larger cracks passes through a number of grains, as schematically shown in Fig. 8. Because the crack front must remain a coherent crack front, the crack cannot grow in each grain in an arbitrary direction and at any growth rate independent of crack growth in adjacent grains. This coherence prevents significant gradients of the crack growth rate along the crack front. As soon as the number of grains along the crack front becomes sufficiently large, the local crack growth rate can be considered to be well approximated by local averages. Crack growth will occur as a more or less continuous process. The crack front can be approximated by a simple continuous line (e.g., semielliptical curve). How fast the crack will grow depends on the crack-growth resistance of the material, which then is considered to be a bulk property of the material.

(The fatigue crack growth resistance for the long transverse direction can differ from the resistance for the short transverse direction.) The applicability of fracture mechanics may become relevant as soon as the crack extension of a fatigue crack nucleus is controlled by the balance between the crack driving force along the crack front and the material crack growth resistance.

The previous discussion leads to two important conclusions:

- Microcrack initiation is a surface phenomenon
- Continued crack growth is controlled by bulk properties of the material

The microcrack initiation life time primarily depends on the surface conditions of the material. It thus can be sensitive to a large scatter if the surface conditions do not represent a constant surface quality. Continued crack growth occurs away from the material surface; it does not depend on the material surface quality. As a consequence, it does not exhibit the large sensitivity to scatter of crack initiation. The previous discussion also implies that the applicability of fracture mechanics to small microcracks is questionable, but that it can be a useful tool for describing macrocrack growth.

Growth of Macrofatigue Cracks. In this article, fatigue cracks are referred to as macrocracks if crack growth has become a regular growth process along the entire crack front. Macrocracks can still be rather small, and they are not necessarily visible cracks. According to this definition, the transition from the microcrack growth period to the macrocrack growth period depends on the type of material. It can occur in an Al alloy at a short crack length (100 to 200 μm) (Ref 12), whereas in certain Ni alloys the transition may occur at a much longer crack length. In any event, the transition will not be a very sharply defined

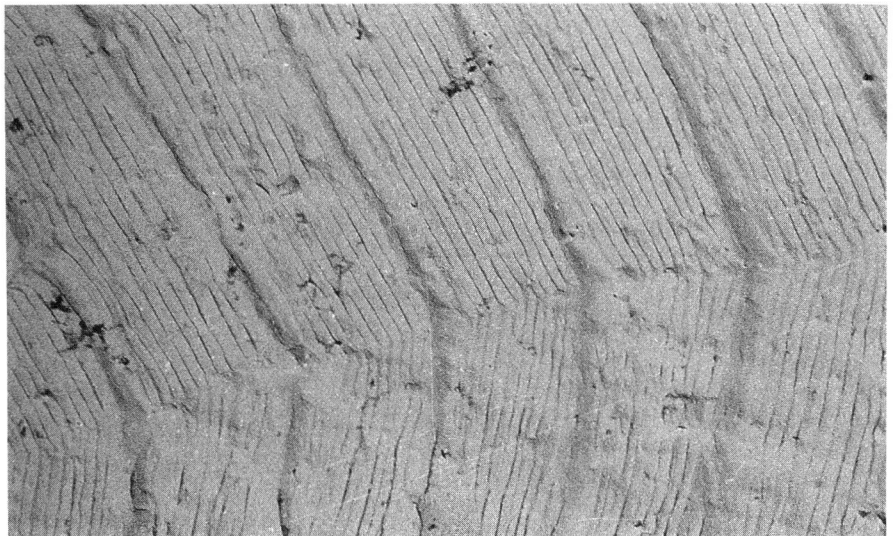

(a)

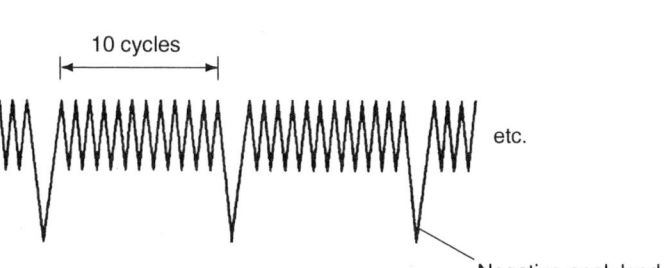

(b)

Fig. 9 Striation pattern corresponding to periodic variable-amplitude load sequence. Fatigue crack in 2024-T3 sheet. Courtesy of the National Aerospace Laboratory NLR, Amsterdam

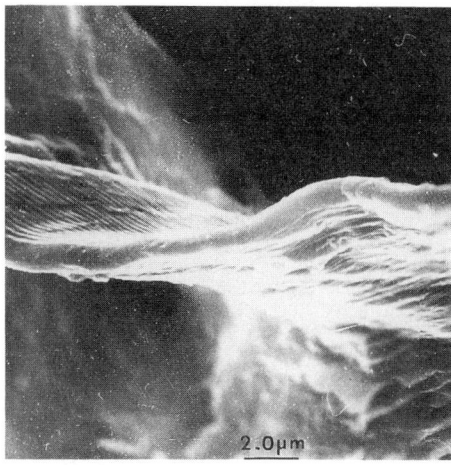

Fig. 10 Plastic casting of fatigue crack in 2024-T3. Note the striations, wavy crack front, and rounded crack tip. Source: Ref 13

crack length. The transition crack length is a function of material structure and structural dimensions.

A characteristic observation on the growth of macrocracks is the occurrence of striations on the fatigue fracture surface (Fig. 9). The correlation between the cyclic load (10 small cycles + 1 larger cycle, repeated) and the striation pattern strongly suggests that crack extension occurs in every cycle. The striations are supposed to be remainders of microplastic deformations, but the mechanism need not be the same for all materials. Moreover, striations are not observed in all materials, at least not equally clearly. The visibility of striations also depends on the severity of the load cycle. Furthermore, microscopic fractography of a macrocrack has shown that the crack front is not a simple straight line and that the crack tip is not necessarily a very sharp crack. New information on the geometry of the crack front in aluminum alloys became available when Bowles (Ref 13, 14) carried out vacuum infiltration experiments. A plastic casting of the crack tip with the crack front was obtained and could be observed in the SEM (Fig. 10). There are interesting observations to be made in this figure:

- The crack front is not a straight line.
- The crack tip is rounded.

- Striations appear on the upper and the lower side of the crack tip casting (i.e., striations from both sides of the fatigue crack appear in one picture).

These observations were made for visible macrocracks. Apparently, the geometry of the macrocrack on a microscopic level does not agree with the classical concept of a crack in elementary fracture mechanics (perfectly flat, straight, or elliptical crack front). However, for these cracks, fracture mechanics applications have been proven to be possible.

The observation of cycle-by-cycle crack extension has stimulated various prediction models on fatigue crack growth. It is a basic concept for models on crack growth under VA loading. Another important concept used in these models is crack closure. Plasticity-induced crack closure was discovered by Elber (Ref 15, 16) in 1968. It implies that fatigue cracks can be fully or partly closed while the material is still under tension. It occurs as a consequence of plastic deformation left in the wake of the crack along the crack flanks. The plastic deformation remains from crack-tip plasticity of previous load cycles. As long as the crack tip is still closed, there is no stress singularity at the physical crack tip. During cycling, the crack opening stress level, S_{op}, can be

between S_{min} and S_{max}. The crack tip is fully open if $S \geq S_{op}$. Elber defined an effective stress range $\Delta S_{eff} = S_{max} - S_{op}$, and similarly an effective ΔK value by:

$$\Delta K_{eff} = \beta \, \Delta S_{eff} \, \sqrt{\pi a} \qquad \text{(Eq 1)}$$

where β is the geometry correction factor. According to Elber:

$$\Delta K_{eff}/\Delta K = U(R) \qquad \text{(Eq 2)}$$

where $U(R)$ is a function of the stress ratio $R = S_{min}/S_{max}$. Several $U(R)$ relations have been proposed in the literature (Ref 17), partly based on fatigue tests results, and for another part supported by finite-element calculations. It turned out that empirical $U(R)$ relations could describe the effect of the R-ratio on crack growth under CA loading by using ΔK_{eff}. Plasticity-induced crack closure has significantly contributed to our understanding of fatigue crack growth under VA loading. Other mechanisms for crack closure have been proposed in the literature, such as roughness-induced crack closure (Ref 18), but they are not considered here for the problem of VA loading.

The literature on VA fatigue investigations has steadily increased through the years and is extensive now. Many test programs were carried out to check the famous Miner rule ($\Sigma n/N = 1$), which Miner published in 1945 (Ref 19). The rule was published earlier by Pålmgren in 1924 (Ref 20). Another noteworthy publication came from Langer in 1937 (Ref 21). He divided the fatigue life into an initiation period and a crack growth period, then postulated that $\Sigma n/N = 1$ is valid for each of the two periods, where N had to be $N_{initiation}$ and $N_{crack\ growth\ life}$ for the two periods, respectively. Langer did not tell how $N_{initiation}$ had to be obtained.

Numerous test series found that the Miner rule was unreliable. $\Sigma n/N$ values much smaller and much larger than one were obtained. In spite of this negative result, a certain understanding of

Table 1 Types of variable-amplitude tests and main variables

Type of test	Main variables
Simple tests	
Constant amplitude with OL	Single OL
	Repeated OLs
	Blocks of OLs
	Magnitude of OLs (including R-effects)
	Sequence in OL cycles
Block tests	2 blocks, Hi-Lo and Lo-Hi sequence
	Repeated blocks
	Magnitude of steps (including R-effects)
Moderate complexity	
Program tests	Sequence of amplitudes
	Size of period of blocks
	Distribution function of amplitudes
Complex tests	
Random load tests	Spectral density function (narrow band or broad band)
	Crest factor (clipping ratio)
	Irregularity factor
Service simulation tests	Variable of service load history to be simulated

OL, overload

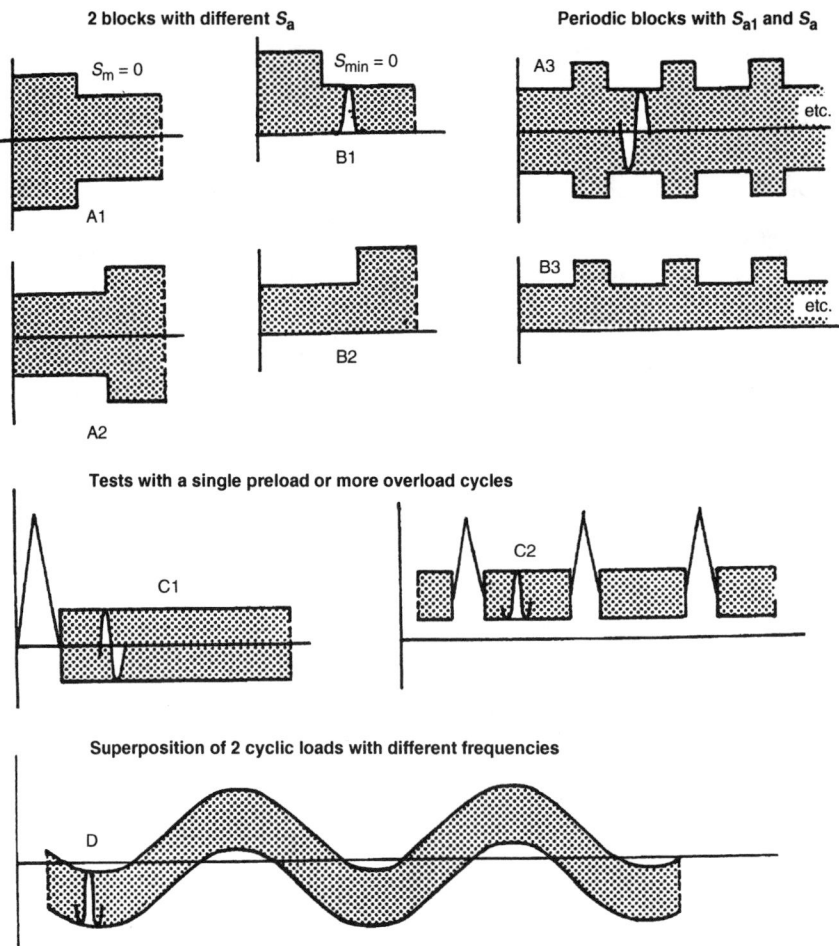

Fig. 11 Simple variable-amplitude load sequences. Source: Ref 22

fatigue damage accumulation emerged. Illustrative results are summarized in this article.

VA Load Sequences

The increased complexity of load histories applied in VA fatigue tests became possible by the development of modern fatigue machines (closed-loop computerized load control). A survey of different types of fatigue tests is given in Table 1, which illustrates the increasing complexity of load histories. It also indicates that the number of variables is large, even for simple tests, as will be shown by the test results. Examples of test load sequences are presented in Fig. 11 and 12 for simple and more complex load sequences, respectively. The most simple but elementary sequences are A1, A2, B1, and B2 (Fig. 11). These sequences are labeled as Hi-Lo and Lo-Hi (Hi-Lo if a high-amplitude block of cycles is followed by a low one, and Lo-Hi for the reversed sequence).

The program test was introduced by Gassner in 1939 (Ref 23) as a first attempt to simulate a VA load spectrum in a test (Fig. 12a). At that time, fatigue machines could not yet simulate more realistic load sequences. The Lo-Hi-Lo sequence of the program test was replaced later by a randomized sequence (Fig. 12b; see, e.g., Ref 24). However, in each block the number of cycles is still large. In general, such a test cannot be considered a realistic simulation of a service load history. Many loads in service have a random character, although there are different types of randomness. A structure with a predominant resonance frequency response is quite often vibrating in a narrow random mode (Fig. 12c). If resonance is less significant, the load history can

be a broad-band random load (Fig. 12d). These sequences can now be applied in fatigue tests.

For aircraft it was recognized in the 1950s that the service load history is a mixture of *random loads* and *deterministic loads* (non-random loads, e.g., ground-air-ground transition loads or maneuvers). Initially both types of loads were applied in fatigue tests, with the random loads reduced to one or two amplitudes (Fig. 12e) for reasons of simplicity. This reduction was done by a Miner calculation, with the aim being that the CA cycles of the test should have the same fatigue damage as the random spectrum. Unfortunately, the Miner rule is fully unreliable for this purpose. In tests on aircraft structures and components, as well as on other types of structures, it is now recognized that a realistic simulation of the service load sequence is essential to obtain a similar fatigue damage accumulation (Fig. 12f). Although it looks quite simple to adopt a simulation of service load histories as a basis for realistic fatigue tests, actually, there are a few inherent problems:

- The service load history must be known. By so-called mission analysis (Ref 25), deterministic fatigue loads may be obtained. Random

loads, however, in the best case are known by statistical distribution functions only.

The sequence of random loads is by nature unknown. Fortunately, techniques for measuring fatigue loads in service have developed considerably. Equipment for that purpose is commercially available, the size is small and it can sample load histories for a long time as standalone equipment (Ref 26). If we wish, we can be well informed about loads in service by relatively easy measurement programs.

- A fatigue test with a service simulation load history is in theory valid only for the load history applied in the test. Load histories adopted in such tests are usually selected to be conservative in order to cover severe service.

- A well-known and easily recognized problem of service simulation fatigue tests is that they must be completed in a limited time period. As a consequence, the service simulation fatigue test is an accelerated fatigue test. If there are time-dependent effects in fatigue, we have a problem. The classical one is fatigue in a corrosive environment. This obviously applies to welded offshore structures in salt water. For

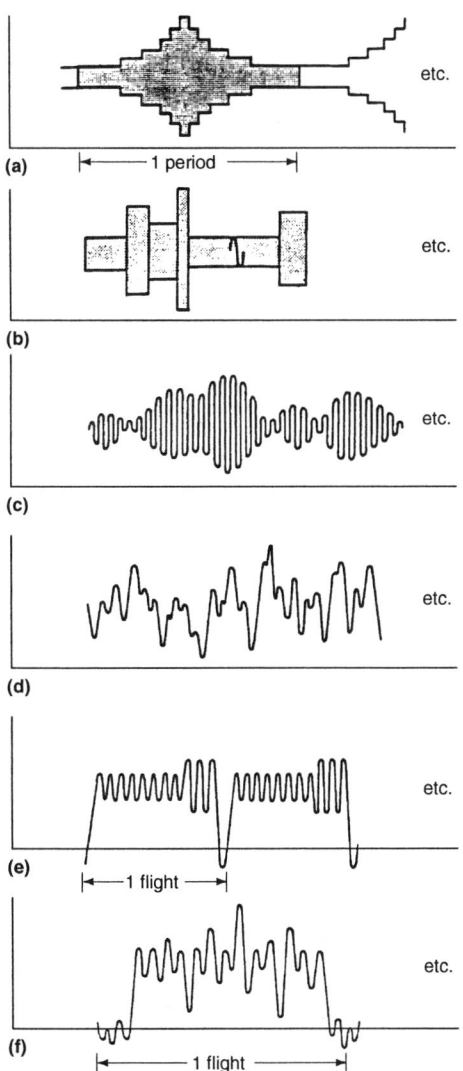

Fig. 12 Examples of more complex variable-amplitude load histories for fatigue tests. (a) Program loading with Lo-Hi-Lo sequence of S_a. (b) Randomized block loading. (c) Narrow-band random loading. (d) Broad-band random loading. (e) Simple flight simulation loading. All flights are equal. Two blocks with different S_a. (f) Complex flight simulation loading. Source: Ref 22

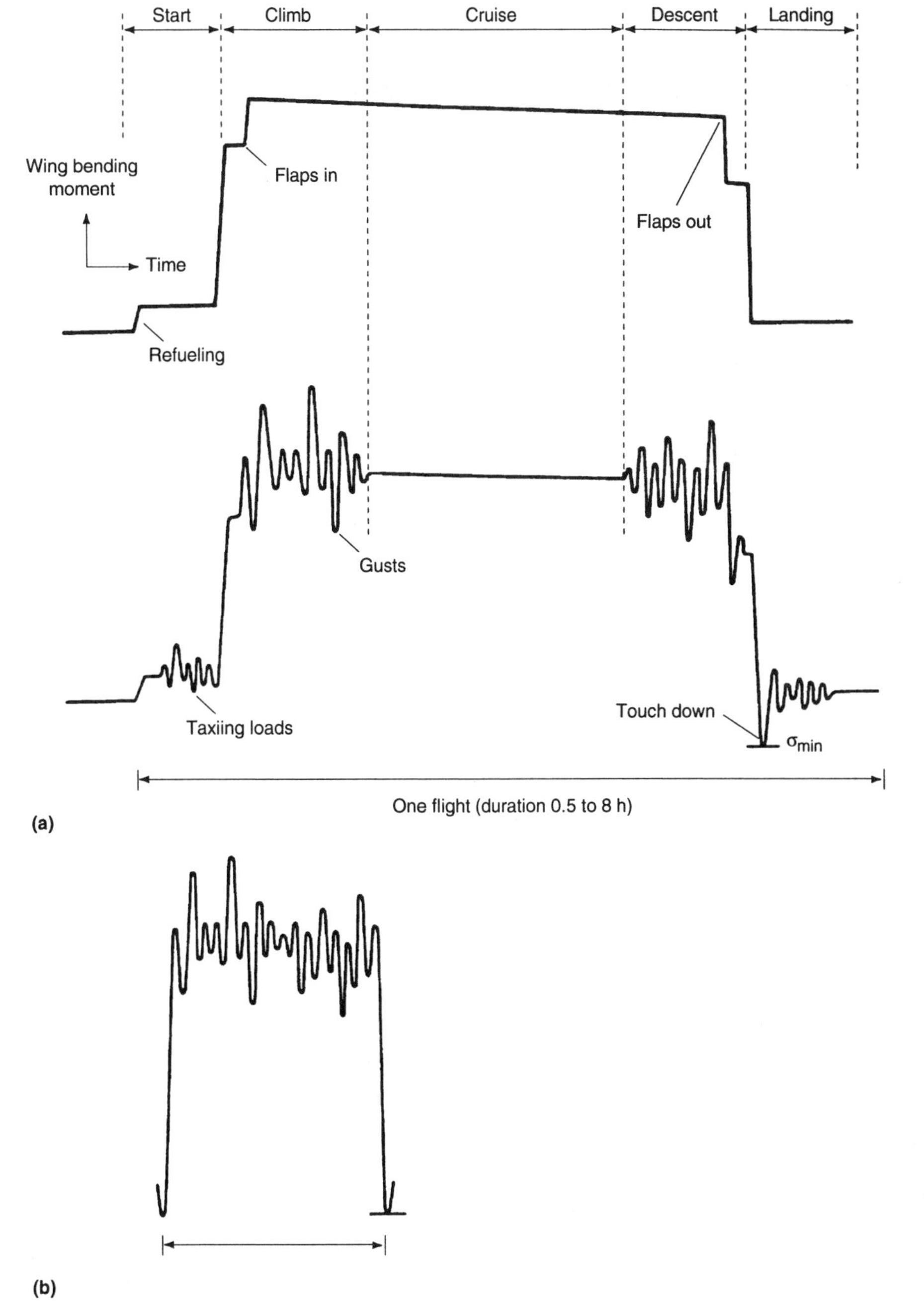

Fig. 13 Flight simulation load history of a single flight. (a) The top curve shows the deterministic load, and the bottom curve shows the superposition of two types of random loads. (b) Time-compressed flight simulation. Same S_{min} as for the bottom curve in part (a). Source: Ref 27

aircraft structures the situation is less dramatic, as briefly reported below.

Figure 13 shows the principle of a flight simulation fatigue test for problems related to the tension skin structure of an aircraft wing. In Fig. 13(a), the top curve shows the deterministic load, which can be obtained by calculation. The bottom curve shows the superposition of two types of random loads, turbulence (gust loads) and taxiing loads. At cruising altitude, gust loads are generally negligible. During landing and takeoff taxiing, loads occur as a result of runway roughness. Turbulence is a matter of weather conditions, so gust severity is different from flight to flight. It is usual to simulate some eight to ten different weather conditions in a flight simulation fatigue test. A sample load record is shown in Fig. 14 (Ref 28). In view of the time scale (flight duration in service in terms of hours), such a load profile cannot be used in a fatigue test. Acceleration occurs by leaving out the time that the load does not vary. Small taxiing loads, if they may be supposed to be nondamaging, are also omitted. Finally, in a fatigue test the load variations are applied at a higher loading rate than in service. As a consequence, a simulated flight in a full-scale test occurs in a few minutes, while it occurs in a laboratory fatigue test on specimens at a rate on the order of 10 flights per minute.

It now can be questioned whether such accelerated tests can still give reliable information. The time scale has been considerably modified. Actu-ally, what is left is the simulation of going from peak load to peak load, from maximum to minimum to maximum, and so on. For the process as related to microplasticity, these load turning points are indeed the decisive events. However, if time-dependent effects (and thus frequency-dependent effects) on fatigue crack extension are significant, the compression of the time scale should have an influence on the test result. For fatigue of Al alloys in air and in other gaseous environments the water vapor content (absolute humidity) has a significant influence on fatigue (Ref 29-31), whereas oxygen is not important. Under normal humidity, cyclic loads with fre-

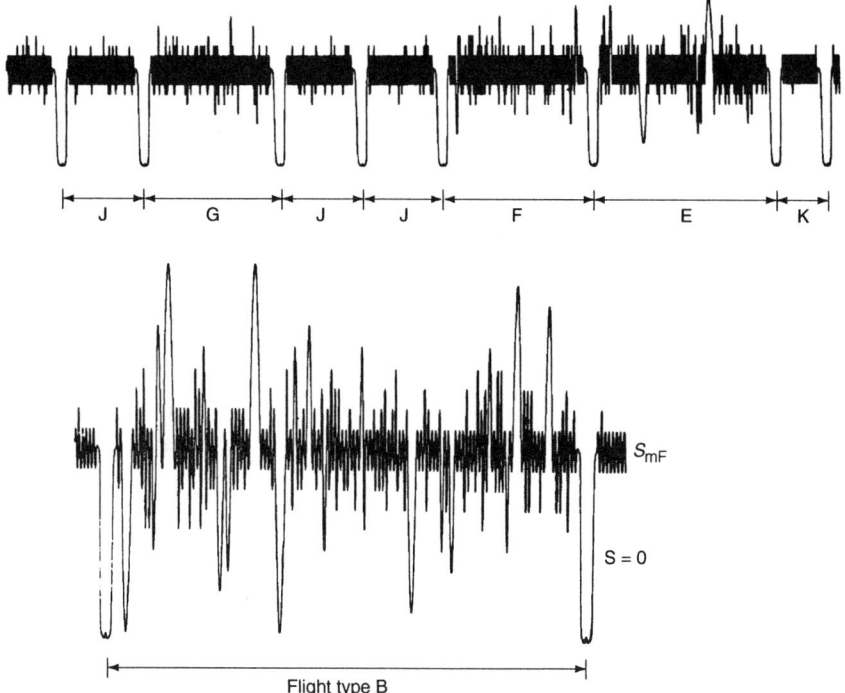

Fig. 14 Samples of a load history applied in flight simulation tests according to the Fokker F-28 wing load spectrum. Source: Ref 28

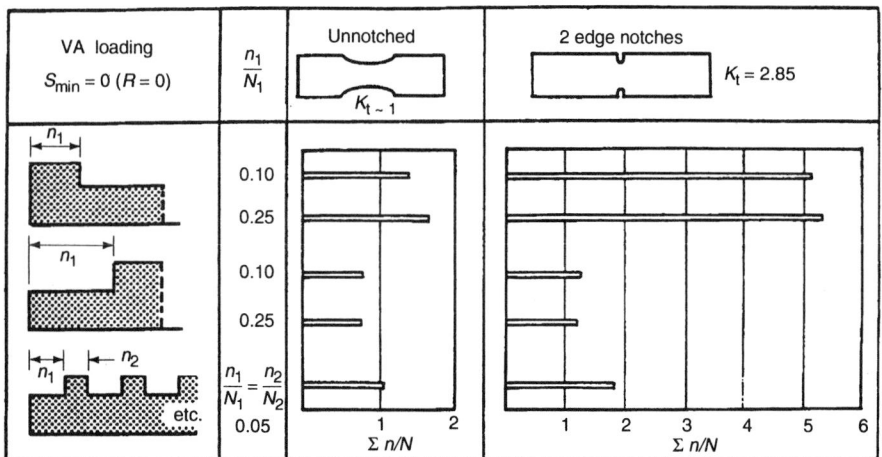

Fig. 15 Sequence effects in unnotched and notched specimens of 2024-T3. Source: Ref 37

Table 2 Survey of standardized service simulation load histories

Year	Name	Load history for:
1973	TWIST	Transport aircraft lower wing skin
1976	FALSTAFF	Fighter aircraft lower wing skin
1977	GAUSSIAN	Random loading
1979	miniTWIST	Shortened TWIST
1983	HELIX/FELIX	Helicopter main rotor blades
1987	ENSTAFF	Tactical aircraft composite wing skin
1987	Cold TURBISTAN	Fighter aircraft engine, cold engine disks
1990	Hot TURBISTAN	Fighter aircraft engine, hot engine disks
1990	WASH	Offshore structures
1990	CARLOS	Car components
19xx	WALZ	Steel mill drive
1991	WISPER/WISPERX	Horizontal axis wind turbine blades

Source: Ref 35, 36

structures mentioned in the table. The sequences of loads in these standardized load histories are fully defined in a numerical format. The load scale can still be selected. A major problem in arriving at some of the standards was the omission of numerous small cycles. If these cycles were included, tests with some standardized sequences could still take a very long time. The main goal of the standardized load histories is the application in general fatigue research programs, where specific variables are studied (usually comparative tests in view of material selection, joint design, surface treatments, etc.).

Results of Simple VA Fatigue Tests

Crack Initiation Life. VA tests results strictly on the crack initiation period are rare. However, numerous VA test series until failure have been carried out on unnotched specimens and simple notched specimens. In such specimens the crack growth period is relatively short, and the total fatigue life thus gives approximate information on the initiation period.

Test results for the VA load sequences B1, B2, and B3 (sequences in Fig. 11) are presented in Fig. 15 (Ref 37). The most noticeable results are obtained for the notched specimens. In the Hi-Lo sequence, $\Sigma n/N$ is much larger than 1. The cycles at the high amplitude in the first block increase the fatigue life at the low amplitude in the second block approximately five times. This large effect is considered to be due to residual compressive stress at the notch root introduced by the first block of cycles.

Another illustrative example, the load sequence A1 (Hi-Lo), is presented in Fig. 16 (Ref 38). It shows results of 2024-T3 Al alloy specimens notched by two holes and tested at zero mean stress. Plastic deformation occurs at the root of the notches. The first block of cycles is followed by a block with a much lower amplitude. However, there is a small but essential difference between the two load programs in Fig. 16(b) and (c). In Fig. 16(b), the transition from the first block to the second block occurs after a positive peak load of the high-load cycles,

quencies of about 10 Hz and lower give the same maximum environmental contribution to fatigue crack growth. However, an experimental proof is not easy. Flight simulation tests have been carried out on 2024-T3 and 7075-T6 sheet specimens with test frequencies of 10 Hz, 1 Hz, and 0.1 Hz (Ref 32). Especially the latter frequency leads to very long testing times. The results have confirmed that the same crack growth rates are found for the three frequencies. This limited experimental verification indicates that time-dependent effects may not be significant, because under both low- and high-frequency load histories, there is sufficient time for the same environmental damage contribution to crack cycles.

The situation can be quite different for other materials and other environments. As an example, for fatigue of steel in salt water, a systematic frequency effect was clearly observed long ago (Ref 33). A detrimental salt water effect has also been found in random-load fatigue tests on steel for off-shore structures tested under a sea wave spectrum (Ref 34). For accurate predictions this is a rather unpleasant problem, which is pragmatically solved by applying empirical life reduction factors.

In the last two decades, several standardized service-simulation load histories have been developed. A survey is given in Table 2. The load spectra are supposed to be characteristic for the

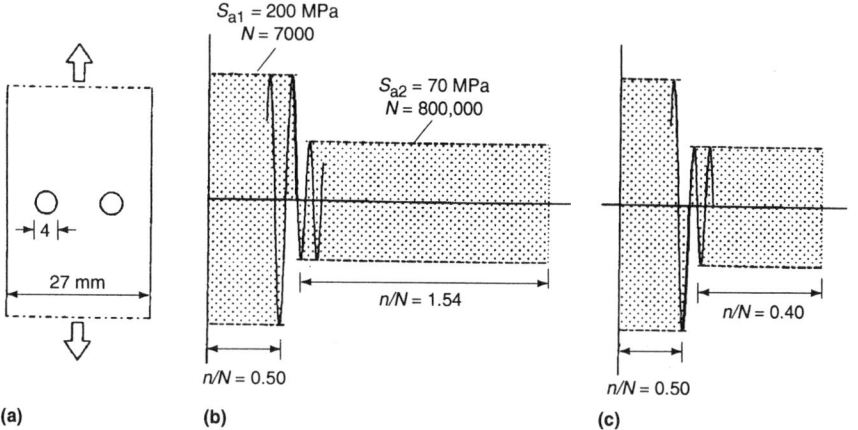

Fig. 16 Hi-Lo tests on notched Al alloy specimens. Note the effects of compressive or tensile residual stress at the notch root. (a) Two-hole specimen. (b) $\Sigma n/N = 2.04$. (c) $\Sigma n/N = 0.90$. Source: Ref 38

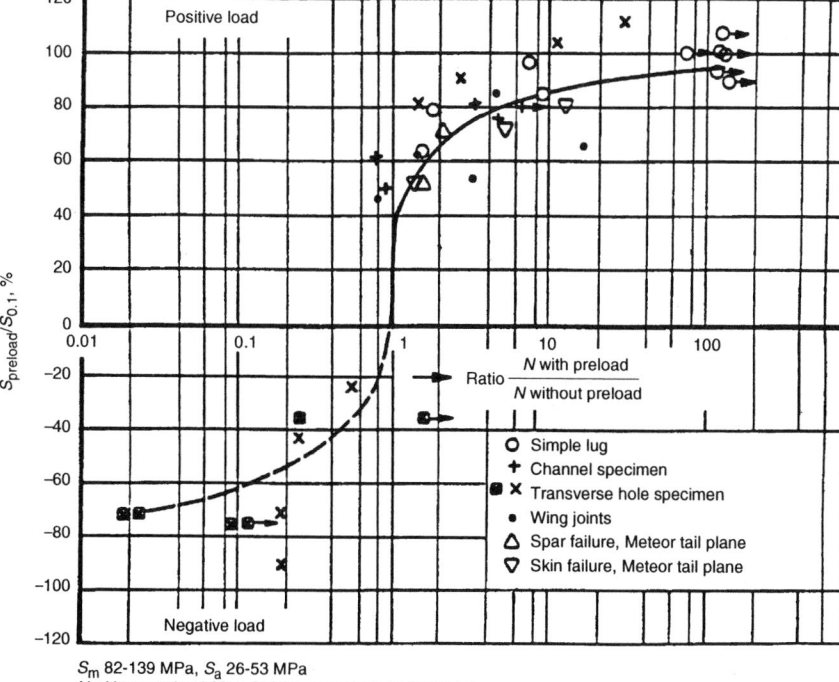

S_m 82-139 MPa, S_a 26-53 MPa
N without preload 80 to 850 kc, except for ⊠ (7700 kc)

Fig. 17 Effect of positive and negative preloads on the fatigue life of notched elements. Source: Ref 39

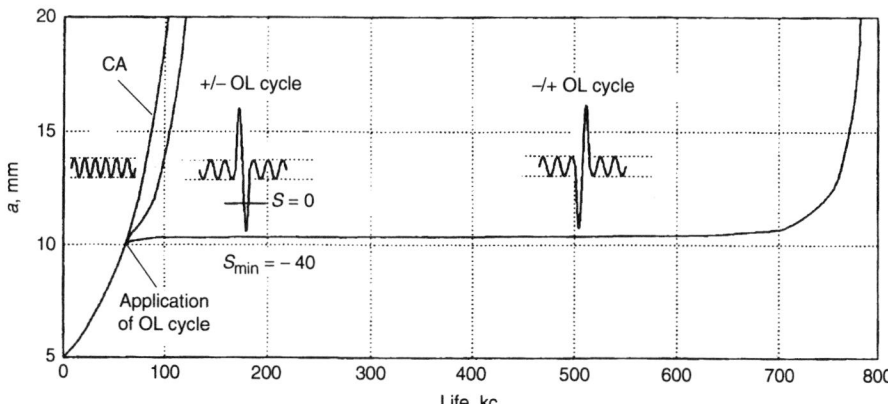

Fig. 18 Effect of two different overload cycles on fatigue crack growth in 2024-T3. Baseline cycle: $S_a = 25$ MPa, $S_m = 80$ MPa. Overload cycle: $S_a = 120$ MPa. CA, constant amplitude. Source: Ref 43

whereas in Fig. 16(c) it occurs after a negative load cycle of the first block. In Fig. 16(b) the last positive peak load leaves a residual compressive stress field at the root of the notch, which is favorable for fatigue in the second block. In Fig. 16(c) the last negative peak load leaves a residual tensile stress field at the notch root, which is unfavorable for fatigue in the second block. As a result, the fatigue life in Fig. 16(b) is significantly longer than predicted by the Miner rule, whereas in Fig. 16(c) it is (slightly) shorter than the Miner prediction. In the latter case there is a kind of damage accelerating effect. After the first block, small cracks must have been present in both types of tests, but the crack length was still much smaller than the hole radius. As a consequence, the plastic deformation was still largely controlled by the geometry of the notch. It does affect the initial growth of a small crack.

Similar indications of the effect of residual stresses at the root of notches were obtained by Heywood (Ref 39) in tests with high preloads (C1 in Fig. 11). Figure 17 shows results obtained for a variety of notched elements. The magnitude of the preload along the vertical axis is presented as the ratio of the preload stress and the 0.1% yield stress. The horizontal axis of the figure gives the life improvement factor (i.e., the ratio of the fatigue life after preloading and the life without preloading). The results clearly demonstrate the large and favorable effect of a positive preload, which induces favorable compressive residual stresses. Fatigue lives were increased up to more than 100×. The smaller number of tests with a negative preload (compression) confirm that the tensile residual stresses do reduce the fatigue life, and this effect can be large.

These tests were carried out after the Comet accidents. Part of the Comet fuselage had been fatigue tested before the accidents, which gave a life until cracking on the order of 15 times the life in service until the accidents. However, that part of the fuselage had been statically tested until the ultimate design load before the fatigue test. Due to this high preload, a highly unconservative test result was unfortunately obtained (Ref 40).

Heywood's results were confirmed by Boissonat in tests on notched specimens and joints (Al alloys, Ti alloy, and low-alloy steel) (Ref 41). Boissonat also observed that a periodical repeating of a high load was much more effective than a single preload.

Crack Growth and Overload (OL) Cycles. Simple load sequences have also been adopted in many test series on macrocrack growth (e.g., Ref 42). Figure 18 shows crack growth curves as recorded under CA loading and under CA loading interrupted by a single OL cycle applied at $a = 10$ mm. The OL cycle starting with the minimum peak, followed by the maximum peak (+/− OL cycle) caused a very large retardation of the fatigue crack growth. The maximum peak load caused a large plastic zone at the crack tip, which left compressive residual stresses in this zone. That will retard subsequent crack growth when a crack grows through this zone. The explanation can also be formulated in terms of the plasticity-

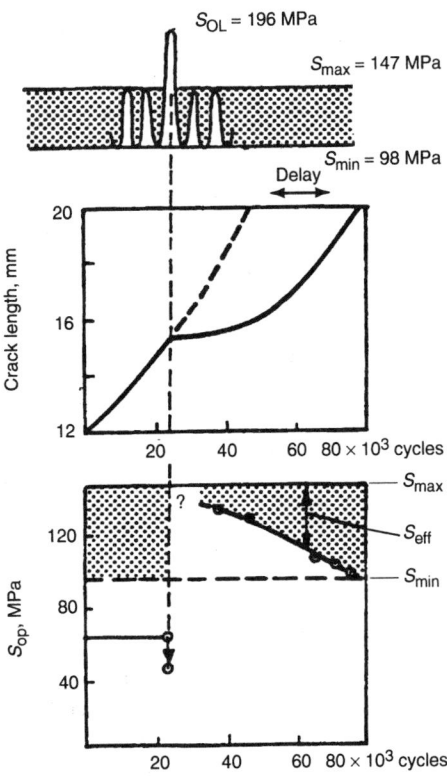

Fig. 19 Crack growth delay after an overload and the influence on S_{op} in 2024-T3 sheet. Source: Ref 44

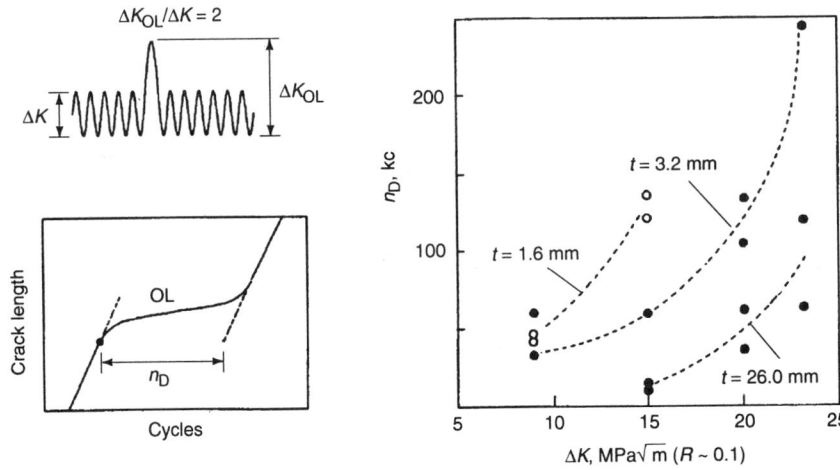

Fig. 20 Effect of material thickness on crack growth delay due to an overload cycle in constant-ΔK tests in 2024-T3. Source: Ref 45

induced crack closure phenomenon (Elber mechanism). Due to the plastic deformation of the OL, more crack closure will occur after the OL has been applied. S_{op} is increased and ΔS_{eff} is reduced.

In the experiments with the reversed OL cycle (+/– OL cycle), a relatively small crack growth delay was observed. The positive peak load again produced a large crack-tip plastic zone. However, the positive peak load was followed by a negative peak load. That will lead to significant reversed plastic deformation, also because the crack was opened and blunted by the preceding positive peak load. The remaining tensile plastic strain was considerably reduced, and the remaining residual stress field was much less intensive. As a consequence, there was less crack closure and a modest crack growth delay was found.

It is easily recognized that macrocracks are closed under a compression load, but due to plasticity in the wake of the crack, that occurs already at a positive load. Because a closed crack is no longer a stress raiser, large negative plastic strains cannot be introduced. This is a fundamental difference with the hole-notched specimen of Fig. 16. If a notched specimen (e.g., with an open hole) is subjected to a high compressive load, there can be significant plastic strains in compression with tensile residual stresses at the root of the notch as a result. That will have a considerable effect on subsequent microcrack growth in that region. The difference between the behavior of notches and cracks has consequences for predic-

tion models on the crack initiation period and the crack growth period under VA loading.

Some elementary tests on crack closure before and after an OL have been carried out (Ref 44). Crack closure measurements were made during a CA test ($R = 0.67$) with an OL as shown in Fig. 19. The delay caused by the OL can easily be observed from the crack growth curve. The crack closure measurements carried out before the application of the OL indicated $S_{op} \sim 62$ MPa. Directly after the OL the S_{op} level was reduced to about 45 MPa. Because the OL opens the crack by crack-tip plasticity, such a trend should be expected. Crack closure measurements made after the OL application indicated S_{op} values above S_{min} of the CA cycles. However, S_{op} decreased later below S_{min}. At the moment that $S_{op} = S_{min}$, the crack growth delay had finished. This should also be expected because crack closure no longer occurred during the CA cycles at $R = 0.67$. Of course, it must be admitted that accurate crack closure measurements are difficult, but the trend of Fig. 19 is considered to be correct.

Crack growth retardation after an OL is generally related to the size of the plastic zone, because crack closure results from the crack-tip plasticity induced by the OL. Unfortunately, the size of the plastic zone is different for plane strain and for plane stress. In a thin sheet the state of stress at the crack tip is predominantly plane stress, whereas in a thick plate it is predominantly plane strain. It then should be expected that the retardation effects are different for fatigue cracks in thin sheets and thick plates. This is very nicely confirmed by results of Mills and Hertzberg (Ref 45) in Fig. 20. They carried out constant-ΔK tests and found a constant crack growth rate, da/dN, as expected. The OL cycle then systematically reduced the crack growth during a delay period, after which the growth rate returned to its original constant value. The delay period (n_D cycles) can then be defined in a simple way (see the inset figure in Fig. 20). Two trends are obvious from the test results: the delay period is larger for thinner materials (larger plastic zone), and the delay period increases at higher stress intensities

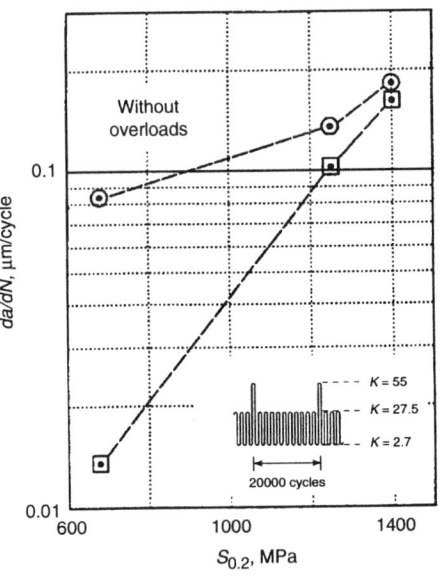

Fig. 21 Effect of material yield stress on crack growth retardation by overload cycles in HP-9Ni-4Co-30C (0.34C-7.5Ni-1.1Cr-1.1Mo-4.5Co). $t = 9$ mm. Heat treated to three different stress levels (675, 1235, and 1400 MPa). Source: Ref 46

(also larger plastic zones). Both trends agree with the effect of the plastic zone size on crack growth delay.

Another instructive example, shown in Fig. 21, has been obtained by Petrak (Ref 46) for an alloy steel. The material was heat treated to three different yield stress levels. Petrak also carried out constant-ΔK tests, but he introduced periodic OL cycles after each 20,000 cycles. In tests without peak loads, the crack growth rate was larger if the steel was heat treated to a higher yield stress. The periodic OL cycles reduced the crack growth rate. The reduction was large for a low-yield-stress material (larger plastic zone) and much smaller for the high-yield-stress material (small plastic zone).

Crack Growth and OL Blocks, Multiple OLs and Delayed Retardation. As discussed

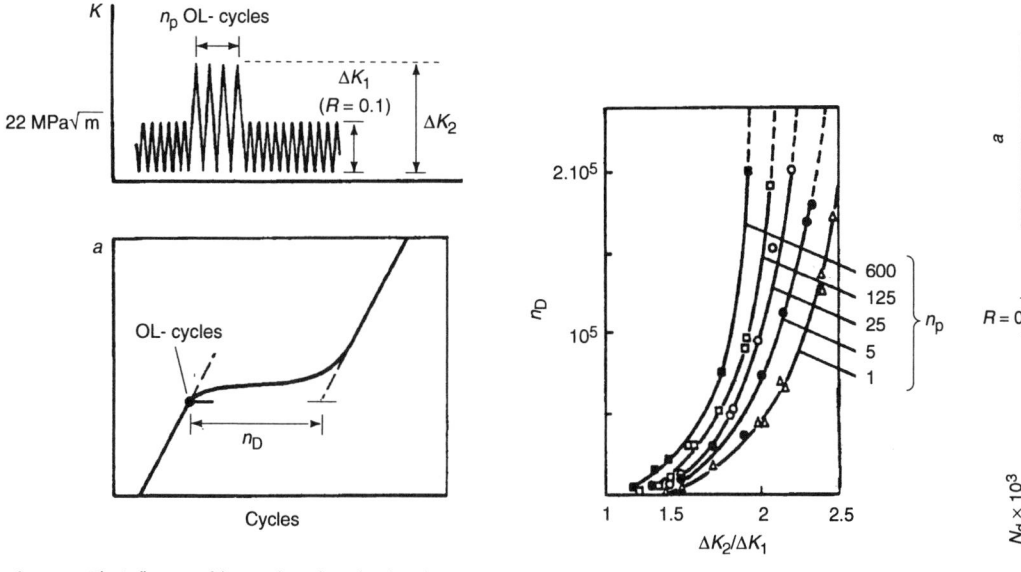

Fig. 22 The influence of the number of overload cycles on the crack growth delay period. Tests on compact-tension specimens in 0.2C steel. Source: Ref 47

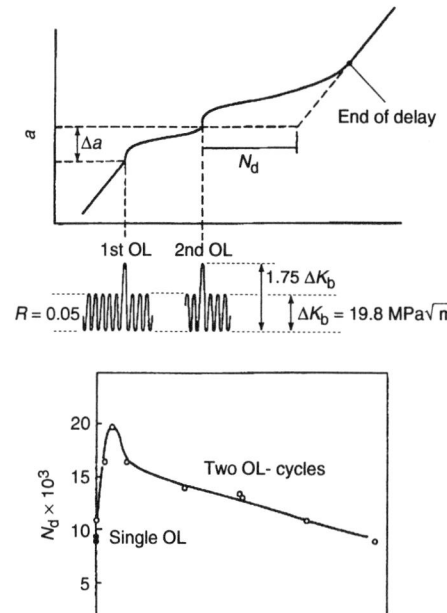

Fig. 23 Crack growth delay after two overload cycles as affected by the number of cycles between the overloads. Source: Ref 48

above, one OL cycle can considerably delay crack growth. However, it has also been observed that more OL cycles give a larger delay. Illustrative results for a carbon steel are presented in Fig. 22. Dahl and Roth (Ref 47) also carried out constant-ΔK tests and adopted the same delay period definition as Mills and Hertzberg. The test results show that the delay period is larger for higher OLs. However, it is noteworthy that larger numbers of OL cycles systematically increased the delay period. The latter trend may be explained by considering that crack extension occurs during the OL cycles. More OL cycles then will leave more plastic deformation in the wake of the crack behind the crack tip. This is a simple explanation based on the Elber crack closure mechanism.

In Fig. 22 the effect of a block of OL cycles is illustrated. A related problem was investigated by Mills and Hertzberg (Ref 48). They considered the effect of two OL cycles in constant-ΔK tests, with a certain number of cycles between the two OLs as a variable (Fig. 23). The second OL cycle can be applied at the moment that the crack growth retardation of the first one is still effective. The results indicate that the delay of the second OL cycle is dependent on the interval between the two OLs (see the lower graph in Fig. 23). According to Mills and Hertzberg, the maximum interaction between the two single OLs is obtained when the crack growth increment between the overloads is about 25% of the plastic zone of the first OL. This multiple OL effect was introduced by de Koning in his CORPUS model, discussed below. The multiple OL effect has recently been confirmed by Tür and Vardar (Ref 49). They applied periodic OLs in CA crack growth tests, with the number of CA cycles (n_{CA}) between the OLs as a variable. Initially the retardation increased for increasing n_{CA}, but for a larger n_{CA} it decreased again.

Delayed retardation has been observed by several research workers. The more reliable indica-

tions should come from observations on striation spacings. Delayed retardation implies that the maximum reduction of the crack growth rate does not occur immediately after the OLs. It requires some crack growth before da/dN has reached its minimum (Fig. 24a). Illustrative results have been obtained by Ling and Schijve (Ref 50) in tests with periodic blocks of overload cycles (type B3 in Fig. 11). In tests with more low-amplitude cycles (100 as compared to 50), delayed crack growth occurred in the same way, but the crack rate could be reduced for a longer time to a lower level (Fig. 24). It apparently requires some crack growth into the plastic zone of the OL

cycles to give the maximum increase of S_{op} (and the minimum ΔK_{eff}). In Fig. 24 that point had not yet been reached.

Incompatible Crack Front Orientation under VA Loading. Shear lips are well known for Al alloys, but they have been observed for several other materials as well (Ref 51-53). When the

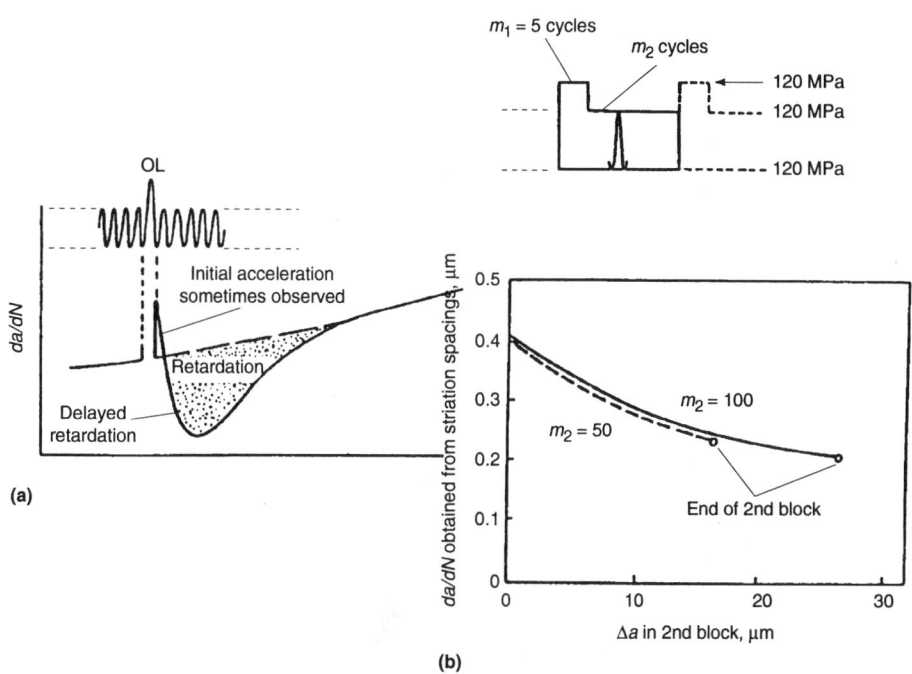

Fig. 24 Delayed retardation after overload and after a block of high-amplitude cycles. (a) Overload effect. (b) Delayed retardation. Source: Ref 50

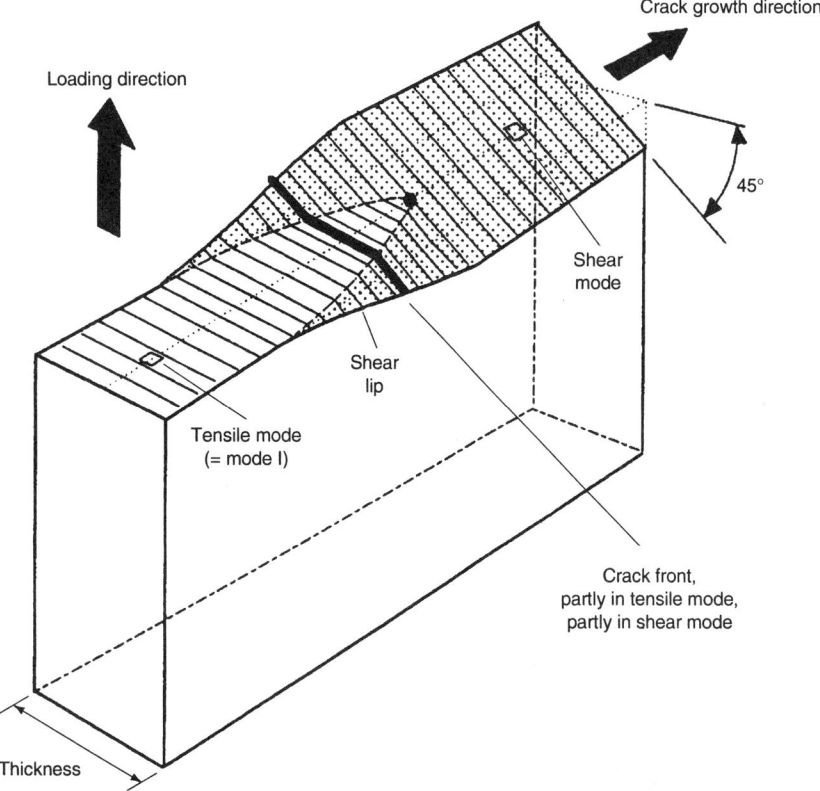

Fig. 25 Fatigue crack growth with shear lips

crack growth rate increases (CA loading assumed), the shear lip width also increases (Fig. 25, 26). It can lead to a full transition from a tensile-mode crack to a shear-mode crack, depending on the material thickness and the stress cycle (Ref 53). Under VA loading the transition can easily imply incompatible crack front orientations, a topic rarely covered in the literature. A simple example is shown on the two fracture surfaces in Fig. 26 (Ref 43). The central cracks in both specimens were already fully in the shear mode under high CA loading when a batch of low-amplitude cycles was introduced. It caused a narrow bright band on the fracture surface (arrows in Fig. 26). The normal fracture mode of the low-amplitude cycles in CA loading at that crack length is the tensile mode (with minute shear lips). This is not compatible with the existing shear mode. There was indeed a tendency to grow again in the tensile mode, which gave the band a stepped appearance. The growth rates in the bands of the two specimens were 2.5 and 8 times lower than observed in normal CA tests at the same crack length. The incompatibility caused a strong retardation effect.

The reverse case is perhaps more relevant, that is, when high-amplitude cycles occur between many low-amplitude cycles. The fracture surface then can be largely in the tensile mode, whereas the failure mode corresponding to the nominal ΔK cycle of the high-amplitude cycle in a CA test may be the shear mode. In elementary tests (Ref 43) such cycles produced dark bands on the fracture surface and a growth rate far in excess of the corresponding CA results. In this case the incompatible crack front caused an accelerated crack growth. It is interesting to note that five high-amplitude cycles produced approximately the same band width as a single high-amplitude cycle. In other words, the major contribution came from the first cycle.

Crack Growth Retardation by Crack Closure and/or Residual Stress in Crack-Tip Plastic Zone. Dahl and Roth (Ref 47) have raised the question whether crack growth delay after an OL is due only to crack closure, or whether there is also an effect of the residual compressive stress in the plastic zone ahead of the crack tip. The question turns up from time to time in discussions. In this respect, interesting experiments were carried out in 1970 by Blazewicz (Ref 54). He made ball impressions on 2024-T3 sheet specimens before the crack growth test was started (Fig. 27). As a result there was a zone between the impressions

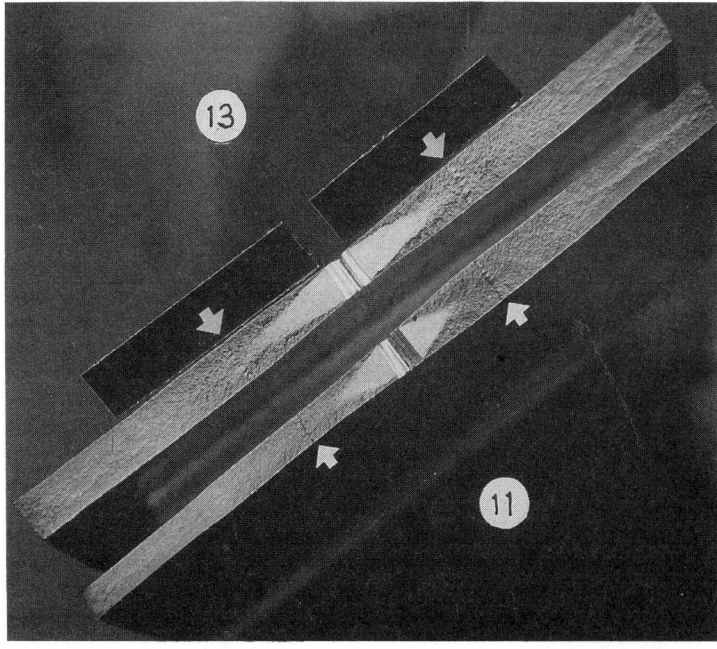

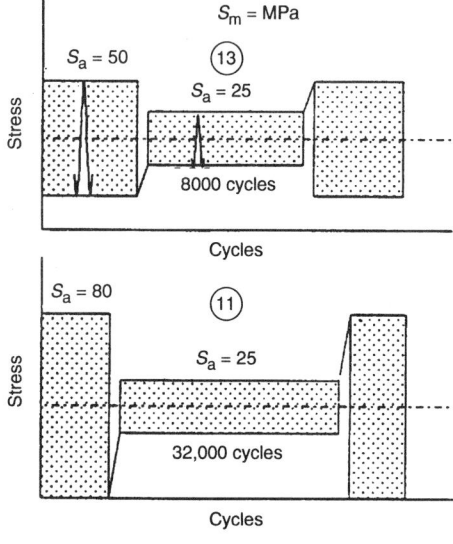

Fig. 26 Incompatible crack front orientation, which occurs if low-amplitude cycles are applied when the crack front is already in the shear mode. Source: Ref 43

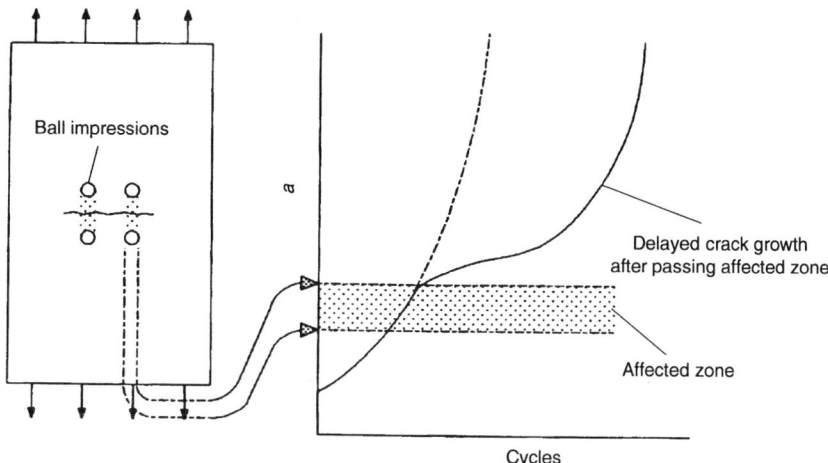

Fig. 27 Crack growth retardation by residual stress in the wake of the crack. Source: Ref 54

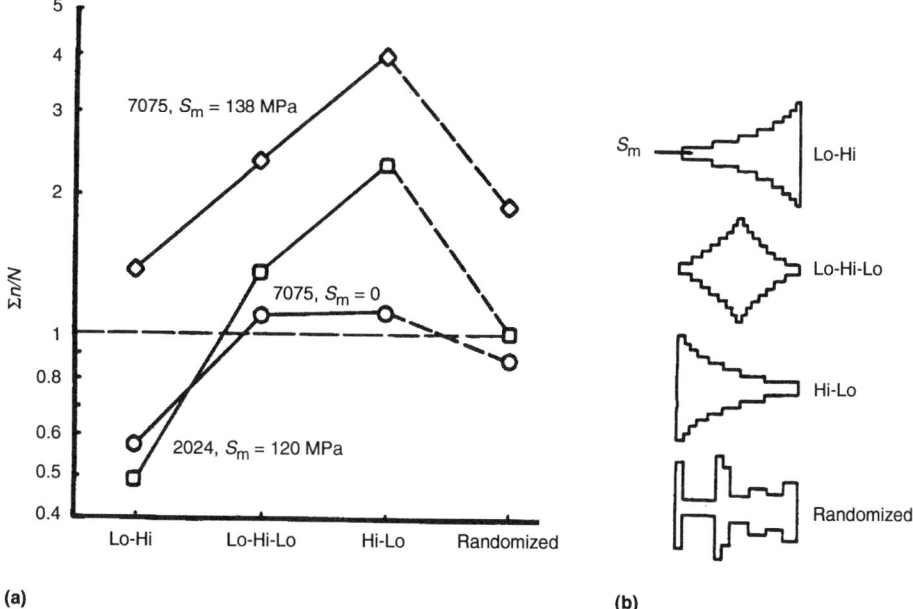

(a)   **(b)**

Fig. 28 The effect of the amplitude block sequence on the fatigue life in program fatigue tests. (a) Tests on edge-notched Al alloy specimens, $K_t = 4$. (b) Block sequence in one period. Source: Ref 24

= 3 for plane strain. It thus should be expected that crack closure will be more significant near the material surface and will occur to a lesser degree at mid-thickness. This is confirmed by finite-element calculations (Ref 57), but there is also experimental confirmation (Ref 58). McEvily (Ref 59) studied crack growth after an OL in Al alloy specimens (6061), which gave a significant crack growth delay. He then reduced the thickness of the specimen immediately after the OL and observed a much smaller crack growth delay. Similarly, Ewalds and Furnee (Ref 60) measured a lower S_{op} after removal of surface layers. In the VA crack growth prediction models discussed below, an averaged S_{op} (averaged over the material thickness) is generally adopted.

Interaction Effects. The observations discussed above are referred to as *interaction effects*. Interaction effects imply that fatigue damage accumulation in a certain load cycle is affected by fatigue in the preceding load cycles of a different magnitude. In other words, a fatigue cycle will affect damage accumulation in subsequent load cycles. As an example, the crack extension in the OL cycle in Fig. 18, although too small to be visible in the graph, was larger than expected without interaction effects. It implies that Δa in the OL cycle was longer than it would have been in a CA test with OL cycles only. The large crack growth retardations induced by OL cycles are a prominent illustration of interaction effects.

Results of More Complex VA Fatigue Tests

Crack Initiation Life. As previously noted, few results on the crack initiation life are available, but data for notched specimens may be representative, assuming that the crack growth period is relatively short. The total life then gives an approximate indication of the crack initiation life. In the past, large numbers of tests were carried out to check the validity of the Miner rule. An enormous scatter of $\Sigma n/N$ at failure was observed, which amply confirmed that the Miner rule is far from accurate. Surveys can be found in Ref 61 and 62. Schütz (Ref 61) reports values of $\Sigma n/N$ in the range of 0.1 to 3.0, which implies that significant interactions must have occurred. In terms of the arguments discussed above, it can be understood that low $\Sigma n/N$ values are to be expected for unnotched specimens and for $S_m = 0$. Large $\Sigma n/N$ values are possible for $S_m > 0$ and notched specimens in view of introducing favorable compressive residual stresses. Some illustrative data are presented below.

Results of a NASA investigation (Ref 24) on edge-notched specimens are presented in Fig. 28. Program tests were carried out with three different sequences, and a randomized sequence was also adopted. The number of cycles in one period was 30,000 to 100,000, while the number of cycles for the eight amplitudes varied from 1 to 82,000 in one block. Two S_m levels were used for 7075-T6 specimens ($S_m = 0$ and $S_m = 138$ MPa). The fatigue lives ($\Sigma n/N$ values) at the positive S_m are about two to four times larger than for $S_m = 0$. It confirms the effect of favorable residual

with residual compressive stresses, which delayed the crack growth. The delay was small during the growth through the zone between the impressions, but it was significant at a later stage. It simply suggests that the crack growth retardation should be explained by crack closure only. In terms of crack growth mechanisms, it appears logical that the crack must be opened before crack extension can start. The efficiency of creating a crack length increment (Δa) depends on the plasticity right at the crack tip (the fracture process zone), not on residual stresses ahead of the crack tip. The residual stress in the crack-tip plastic zone can have an indirect effect on the cyclic plasticity at the crack tip, but opening the crack tip is the decisive mechanism to have crack extension.

Blazewicz also made saw cuts along a fatigue crack and removed part of the plastically de-

formed material in the wake of the crack. That eliminated crack closure, and crack growth retardation was effectively removed. This observation also confirms the significance of the crack closure contribution to crack growth retardation, rather than residual stresses in the crack-tip plastic zone.

Crack closure has also been removed by heat treatments after OLs (Ref 55, 56), and crack growth retardation is thus eliminated.

More Crack Closure at the Material Surface. At the surface of a material the crack tip is loaded under plane-stress conditions. Depending on the material thickness, the state of stress at mid-thickness approaches plane-strain conditions. The plastic zone size under plane stress is significantly larger than under plane strain. Irwin plastic zone size estimates are $r_p = 1/(\alpha\pi)(K/S_{0.2})^2$, with $\alpha = 1$ for plane stress and α

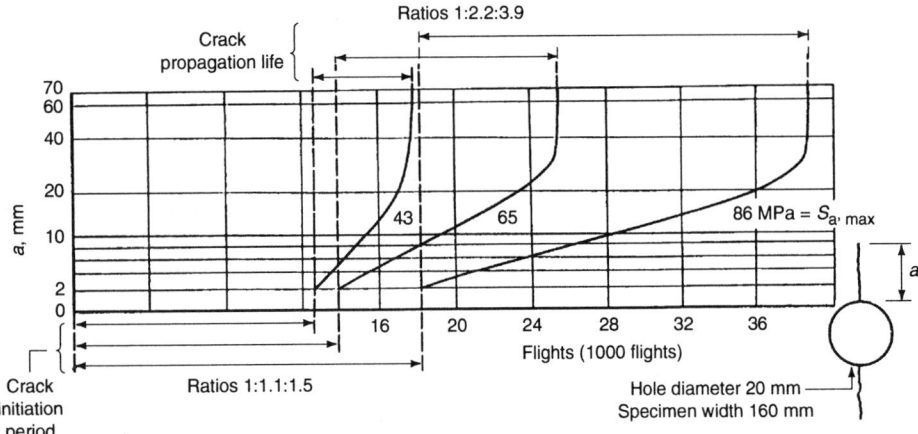

Fig. 29 Effect of the truncation level ($S_{a,max}$) on the crack initiation period (until $a = 2$ mm) and the crack growth period. Results of flight simulation tests on 2024-T3 sheet specimens with a central hole. Source: Ref 63

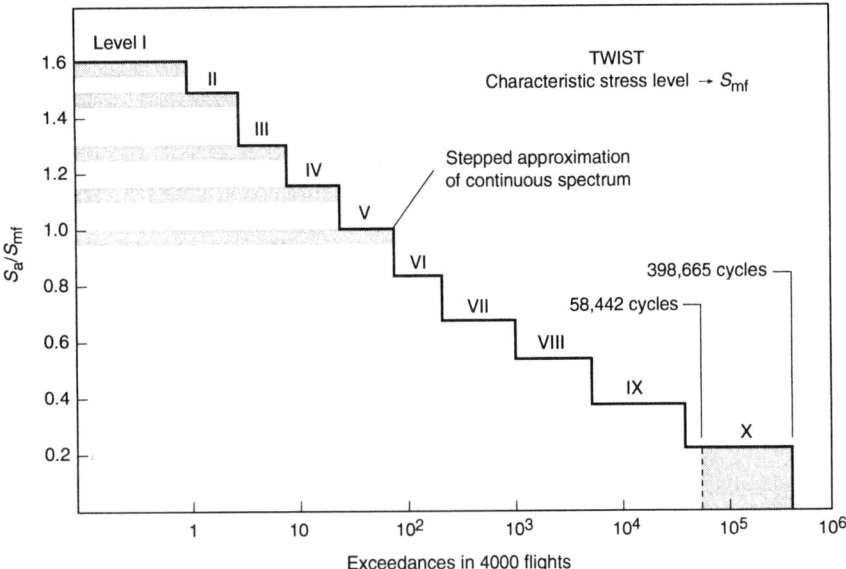

Fig. 30 Standardized gust spectra TWIST and miniTWIST with ten amplitude levels to simulate the continuous spectrum ($S_{ground}/S_{mf} = 0.5$). The levels I to V have been used in experimental investigations.

firmed by comparing the results to the CA reference curve. The reference curve also shows that the crack growth rates during the increasing-amplitude part of a period is faster than expected (acceleration). In terms of crack closure, it implies that S_{op} during the decreasing S_a cycles was higher than in CA tests, whereas it was lower during the increasing part of the sequence. Qualitatively, this appears to be a reasonable explanation for the observed interaction effects.

Another illustrative test program was carried out on fatigue crack growth in 2024-T3 sheet material (Ref 65). The load spectrum used was a gust load spectrum of 40,000 cycles with seven different amplitudes. The same spectrum was used with a random sequence and with programmed sequences (Fig. 32). Two random sequences were applied with full cycles (i.e., each cycle consisted of two half cycles with the same amplitude). However, in one random load history, each cycle started with the positive half cycle, followed by the negative one, while in the other random load history this sequence was reversed. As shown by the results in Fig. 32, the difference between the crack growth lives was rather small.

In the program tests with the full spectrum in one period (40,000 cycles), the results indicate two remarkable trends: (a) There was a systematic sequence effect with the longer crack growth life for the Hi-Lo sequence. This trend agrees with the results of Ryan for D6AC steel. (b) Even more remarkable, and actually more disturbing, the crack growth lives were considerably longer than for the random sequence (i.e., about three times longer for the Lo-Hi-Lo sequence). This is an unconservative result! Fractographic observations indicated that the fracture surfaces of the specimens tested in programmed sequences were rougher than those of specimens tested in random sequences. It then should be concluded that fatigue crack growth is just not the same phenomenon for the two sequences.

In a third group of tests, programmed sequences were used again, but with a much shorter period (average of 40 cycles). In order to accommodate the full 40,000 cycles in these tests, periods with different amplitude contents had to be applied. The crack growth life results were of the same order of magnitude as for the random sequence, and the fracture surface appearance was also similar. The important lesson is that a load spectrum with a random load sequence in service should not be simulated by a programmed sequence with a long period. The programmed sequence is an artificial simulation that cannot be accepted as a simplification.

Crack Growth in Flight Simulation Fatigue Tests. The number of fatigue crack growth experiments under flight simulation load histories is fairly extensive. A survey was published in 1985 (Ref 27). Several more investigations have been published since then, but most have been carried out for comparative purposes and materials evaluations. Several trends on interactions, as observed under the more simple VA load sequences, also apply to the more complex flight-simulation

stresses at the notch root mentioned above. The results in Fig. 28 further show a most significant sequence effect. The effect should be attributed to variations of the residual stress at the notch root, but it is not a simple question to suggest how the variation did occur in detail.

Another example is given in Fig. 29, test results of flight simulation tests. In such tests it is a rather delicate problem to decide whether rarely occurring but very severe fatigue loads with a high amplitude should be included. Such loads can extend the fatigue life considerably, as amply demonstrated in many investigations, surveyed in Ref 27. Unfortunately, the life enhancement of such loads may give unconservative fatigue life results. As a consequence, truncation of high load amplitudes (also called clipping) must be considered (Fig. 30). Test results for three different truncation levels are presented in Fig. 29, both for the crack initiation life (until a 2 mm crack) and for the crack growth life. The maximum S_a level

(truncation level) did systematically affect the crack initiation life (i.e., there were longer fatigue lives if some cycles of the spectrum with a higher S_{max} were introduced). That leads to more favorable residual stresses. A similar trend was found for the crack growth period, but it is noteworthy that the effect on the crack growth period is significantly larger (Fig. 29).

Crack Growth under Program Loading. Ryan (Ref 64) studied fatigue crack growth in a high-strength D6AC steel under a Lo-Hi-Lo program loading at $R = 0$ (Fig. 31a). As a result of the stepwise amplitude changes, fatigue bands could be observed on the fracture surface. It turned out that the band width was smaller in the Hi-Lo part of a period than in the Lo-Hi part. That is also illustrated by the crack growth rates in Fig. 31(b). In view of the $R = 0$ condition, it should be expected that the block of the 20 largest cycles had a favorable effect on crack growth in following cycles with a lower S_{max}. This is con-

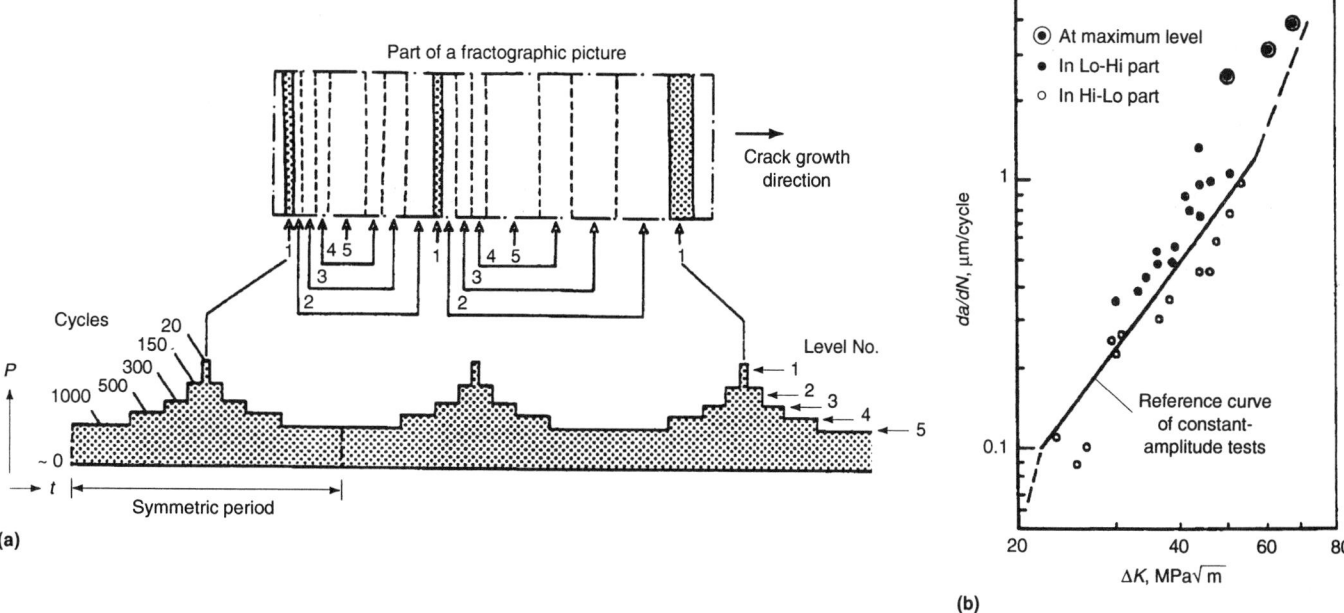

Fig. 31 Sequence effects during fatigue crack growth in D6AC high-strength steel (0.42C-1Cr-0.65Ni-0.1Mo-0.12V) under program fatigue loading. $S_{0.2}$ = 1500 MPa, S_U = 1650 MPa. (a) Program load history and corresponding bands on fatigue fracture. (b) Different da/dN in Lo-Hi and Hi-Lo parts. Source: Ref 64

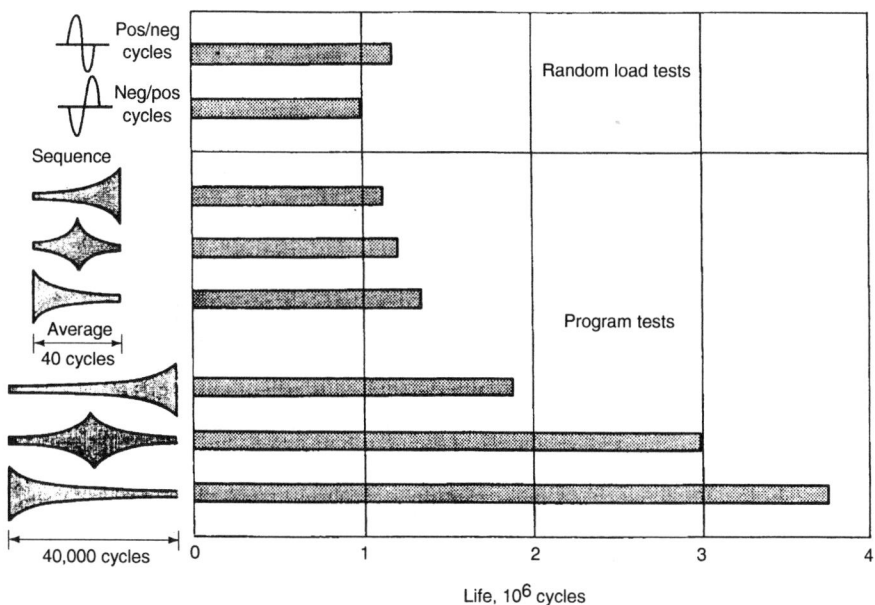

Fig. 32 A comparison between fatigue crack growth under random loading and different types of program loading in 2024-T3. t = 2 mm; a = 12-50 mm. Source: Ref 65

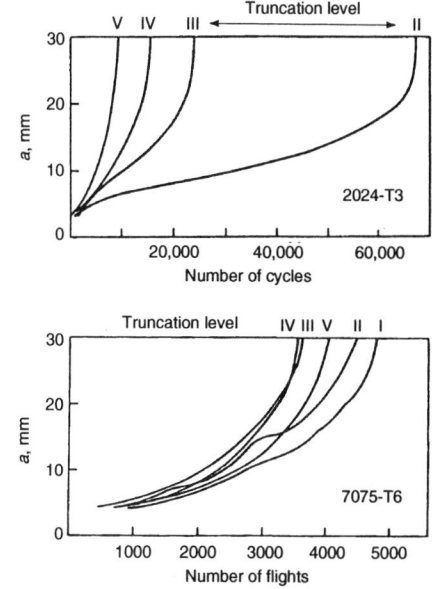

Fig. 33 Effect of the spectrum truncation level (see Fig. 30) on crack growth in two Al alloys under flight simulation loading at S_{mf} = 70 MPa. Source: Ref 67

load histories. Some examples are presented below.

Truncation Effect. The effect of truncating a gust spectrum is shown in Fig. 33. Cycles with the largest amplitudes of the spectrum were truncated to different lower levels (Fig. 30). Level I is the highest gust level, reached once in 4000 flights (TWIST spectrum, Ref 66). Truncation to lower levels (II, III, IV, and V) implies that the maximum stress level is reached 3, 8, 26, or 78 times per 4000 flights, respectively. The effect on the crack growth curves is shown in Fig. 33 for two Al alloys, 2024-T3 and 7075-T6 (Ref 67).

The mean stress in flight was S_{mf} = 70 MPa. The 2024-T3 material has a much better fatigue crack growth resistance than the 7075-T6 alloy. The 2024-T3 alloy has a relatively low yield stress (i.e., $S_{0.2}$ = 377 MPa, as compared to $S_{0.2}$ = 473 MPa for the 7075-T6 material). As a consequence, the crack-tip plastic zone sizes were considerably larger in the 2024-T3 sheet specimens than in the 7075-T6 specimens. That explains the large truncation effect observed in the 2024-T3 experiments. A higher truncation level gives more crack growth retardation and a significantly longer crack growth life. In the 7075-T6 speci-

mens, the plastic zones were smaller and so the retardation effects were less significant. Although the flight simulation test results of 7075-T6 show an effect of the truncation level, it is relatively small, and it is also not fully systematic.

Thickness Effect. Saff and Holloway (Ref 68) carried out flight simulation tests based on a maneuver-dominated load spectrum (F4 aircraft). Crack growth was observed in center-cracked tension specimens of different thicknesses, vary-

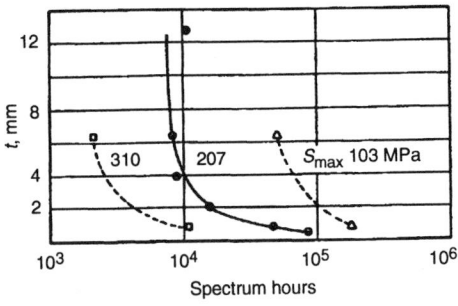

Fig. 34 Thickness effect on crack growth in 7075-T6 under flight simulation loading. Load spectrum F-4; center-cracked tension specimen ($W = 101.6$ mm). Source: Ref 68

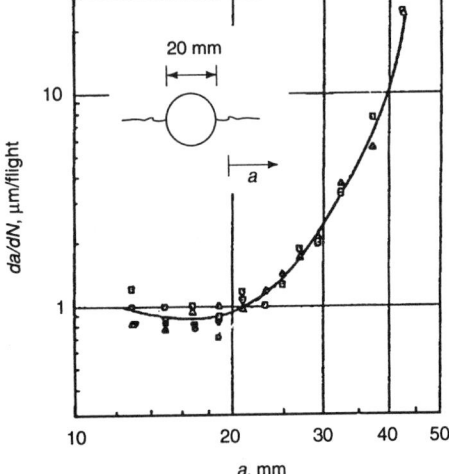

Fig. 35 Initial fast crack growth under flight simulation loading in 2024-T3. Load spectrum F-28; $t = 2$ mm; $S_{mf} = 68.7$ MPa; $S_{a,max}/S_{mf} = 1.26$. Source: Ref 69

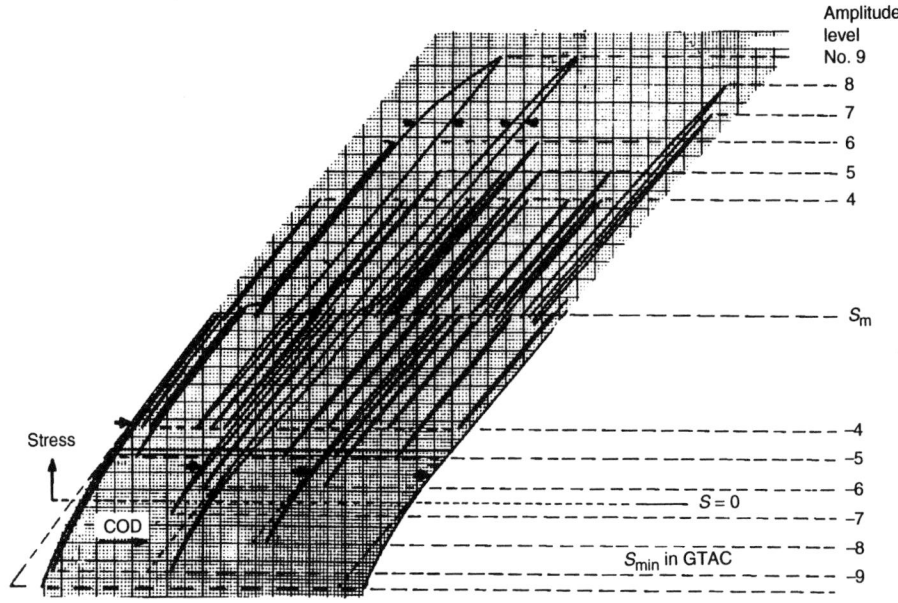

Fig. 36 Crack closure measurements during a flight simulation test and different nonlinearities for two peak loads in 2024-T3. Truncation level 9; $a = 13.5$ mm. Source: Ref 72

ing from 0.5 to 12.7 mm. Three stress levels were used. The results in Fig. 34 clearly show a systematic thickness effect (i.e., lower endurances for thicker material). An increased thickness leads to more plane strain at the crack front, and thus to smaller plastic zones and less crack growth retardation. The crack growth lives are smaller then. The life for the thin sheet material was about 10 times longer than for the thick plate material!

Initial Fast Crack Growth at a Notch. Figure 35 shows the crack growth rate of a crack starting at the edge of an open hole. It is quite remarkable that the crack growth rate between a crack length interval from $l = 2$ mm to $l = 10$ mm slightly decreases, in spite of an increasing stress intensity. In this crack length interval, the load cycle with the maximum stress level ($S_{max} = S_{mf} + S_{a,max}$) has been applied in approximately 30 flights. A similar behavior has frequently been observed in specimens with a central saw cut as a crack starter notch (Ref 70). The initial decrease of the growth rate is attributed to the development of a plastic wake field behind the crack tip. That requires some crack growth, and during the growth the plastic zone sizes of the maximum

loads will increase, while they are still overlapping. The behavior is similar to the multiple OL effect discussed in relation to Fig. 23.

Actually, the behavior of Fig. 35 has also been observed for an edge crack of an open hole under CA loading, as described by Broek (Ref 71). It also requires some crack growth to develop effective crack closure in the wake of the crack. However, under flight simulation loading, the behavior is more pronounced due to the occasional high loads of the load spectrum and some multiple OL effect.

Crack Closure Observed in a Single Severe Flight. About 20 years ago we made crack closure measurements (S-COD) during flight simulation tests on 2024-T3 and 7075-T6 center-cracked sheet specimens, width 160 mm, thickness 2 mm (Ref 72). Figure 36 gives an example of records made during a severe flight. The stress on the specimen is plotted as a function of the crack opening displacement (COD) measured in the center of the specimen. There were 101 cycles in that flight, but records of the smaller cycles were not made because the behavior was linear in these cycles. Moreover, after each half cycle the COD-axis was given a small horizontal shift to separate the records of the various half cycles. Some interesting observations can be made:

- The lower, nonlinear parts of the S-COD records show the typical crack closure behavior. Arrows indicate that S_{op} is reduced after the first high positive load. S_{op} should be associated with plastic deformation at the crack tip, which initially will keep the crack open at a lower stress level.
- During the flight, there were two high positive loads, which showed a hysteresis-type nonlinearity near the maximum stress. This non-

linearity is caused by crack-tip plastic deformation and crack extension. It is noteworthy that the width of the hysteresis is significantly larger during the first high cycle and significantly smaller in the second one.

As part of the CORPUS model, de Koning (Ref 73) has introduced the terms *primary plastic deformation* and *secondary plastic deformation*. Primary plastic deformation occurs at the crack tip if plastic deformation penetrates into elastic material that has not been plastically deformed by previous load cycles. Secondary plastic deformation refers to crack-tip plasticity that remains inside a primary plastic zone. More recently, de Koning and Dougherty (Ref 74) have proposed that crack extension during primary plastic deformation is much more effective than during secondary plastic deformation. In a load cycle, plastic deformation and crack extension will always start with secondary plastic deformation (Fig. 37). If S_{max} is high enough, it will turn over into primary plastic deformation at $K > K^*$. Crack extension, as long as $K < K^*$, will occur in agreement with a Paris-type equation and a related crack growth mechanism. Above K^*, crack extension will occur as a kind of stable crack growth under a quasistatically increasing load.

This appears to be a sound concept, and the data in Fig. 36 support it. The first peak load in the flight caused a good deal of primary plastic deformation. The second one gave a much smaller contribution of primary plastic deformation, due to the primary plastic zone of the first high load. The same explanation also applies to the wide giant striation occurring in the first cycle of a block of OLs, as discussed above. The corresponding plastic zones for this case are schematically indicated in Fig. 38. The exceptional behavior of the first high-S_{max} cycle can be explained

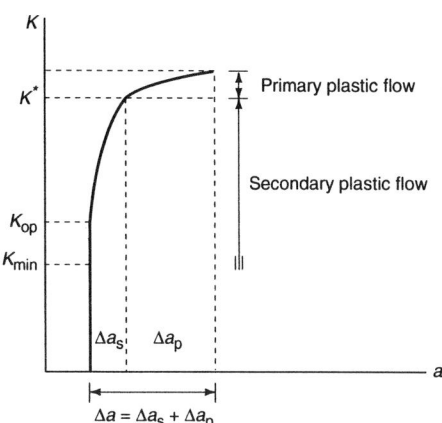

Fig. 37 Different crack increments during secondary and primary plastic deformation. Source: Ref 74

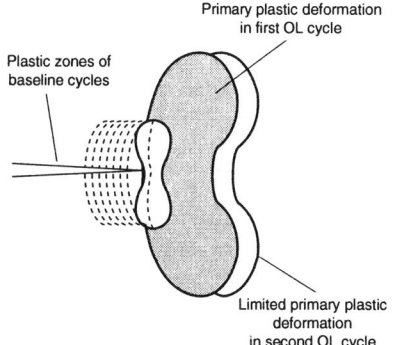

Fig. 38 Plastic zones in crack growth test with blocks of overload cycles. Source: Ref 75

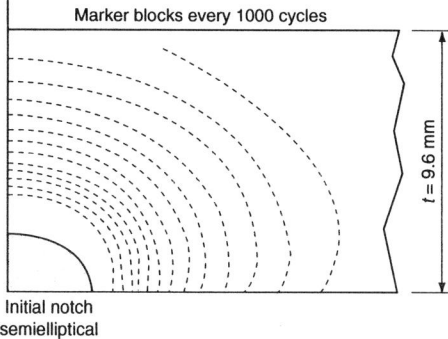

Fig. 39 Trailing crack fronts at material surface. Source: Ref 76

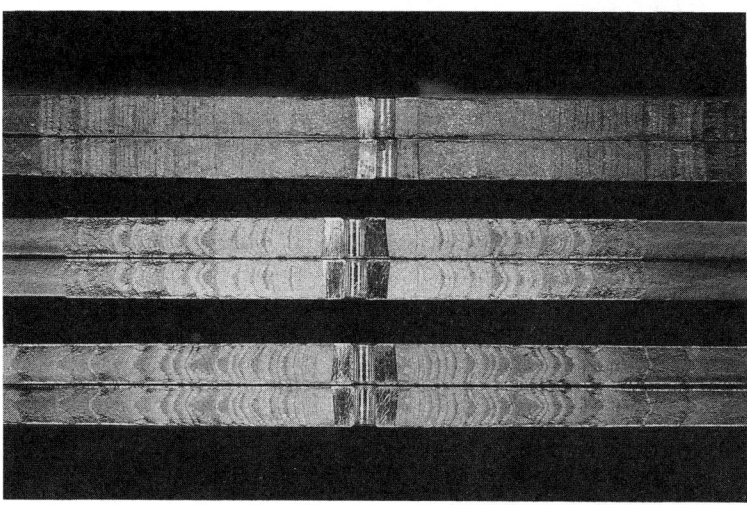

Fig. 40 Fracture surfaces of three center-cracked tension specimens (one 2024-T3 specimen, two 7075-T6 specimens) 2 mm thickness; tested under a flight simulation load history. Relatively flat fractures with limited shear lips; tongue-shaped crack extensions in the two 7075-T6 specimens. Source: Ref 77

by considering the more significant primary plastic deformation during this cycle and the limited primary plastic deformation during subsequent cycles. Similar confirmation was obtained by considering striation spacings in fatigue tests with intermediate heat treatments, as discussed in Ref 75.

Crack Closure of Surface Cracks. So far the above text has been on through-cracks. However, semielliptical surface cracks or quarterelliptical corner cracks can show trailing crack fronts at the material surface (Fig. 39). The specimen was loaded with a simplified flight simulation load sequence and occasional blocks of marker load cycles. The crack fronts in Fig. 39 give the locations of the marker bands. Trailing crack fronts occur at the material surface, where plastic zones are larger, with more crack closure and lower ΔK_{eff} values as a consequence. It is a surface phenomenon, which might be ignored in prediction models (Ref 76). Actually, trailing crack fronts are also observed in CA tests. Some of the strip-yield models, such as FASTRAN, account for more closure at the surface (crack) growth than in the interior, but the nonelliptical shapes are not accounted for.

Incompatible Crack Fronts during Flight Simulation Tests. Relatively flat fracture surfaces have been observed many times for fatigue cracks in sheet specimens, loaded under realistic flight simulation load histories (Ref 28, 63, 69, 72). These cracks have grown more or less in mode I. However, the failure mode for the high-amplitude cycles of the load spectrum is the shear mode in CA tests (or at least significant shear lips). In other words, there is a situation of incompatible crack front orientations, as discussed above for simple VA load sequences (Fig. 25, 26). It is quite possible that crack extension in those high-amplitude cycles is significantly larger than in CA tests at the same nominal ΔK_{eff} value. The mismatch of the crack surface mode can even lead to tongue-shaped crack extensions in mode I (Fig. 40). These tongues do not occur in CA tests. This interaction effect is a complication for prediction models.

A Summary of Observations. We now can list a number of specific observations on fatigue crack growth under VA loading (Table 3). In the following discussion on prediction models it will be worthwhile to consider whether the observations are taken into account by the models.

Table 3 Specific interaction phenomena during crack growth under variable-amplitude loading

Observation	Related aspects
1. Crack growth retardation or acceleration after overloads	Delayed retardation Retardation acceleration
2. Plane-strain/plane-stress transition	Effect of thickness
	Differences between crack closure at the material surface and at mid-thickness
	Trailing crack fronts
3. Multiple overload effect	Initial decreasing growth rate if a crack starts from a notch
4. Different crack growth mechanism for primary and secondary plastic zones	Large Δa during high loads Large striations
5. Incompatible crack front orientation	Tongue-shaped crack extension during high loads
	Accelerated crack growth in mode I

Fatigue Prediction Models for VA Loading

In general, prediction models published in the literature employ basic material fatigue data as a reference. Such data can be fatigue limits, S-N data, fatigue diagrams, crack growth data, and the fracture toughness for final failure. The data are obtained with simple specimens, unnotched specimens for the fatigue data, and simple precracked specimens (center-cracked tension or compact-type specimens) for the fatigue crack growth data. The fatigue load on the specimens should also be of a "fundamental simplicity" (i.e., a cyclic load with a sinusoidal wave shape and a constant S_a and S_m). The data are supposed to be characteristic fatigue properties of a material, characterizing the fatigue resistance or the fatigue crack growth resistance. These properties are used as the material data in predictions on fatigue under VA load histories. They emphasize that fatigue is thought to be primarily a material problem.

The prediction models in principle adopt a *similarity approach* (also called *similitude*): similar stress cycles or similar strain cycles should give the same fatigue damage. Also, similar ΔK_{eff} cycles should give similar crack length increments. This approach implies that fatigue data for the most simple conditions are extrapolated to more realistic engineering conditions. The fatigue model is the frame of the extrapolation procedures, but the extrapolation steps can be quite large. As a consequence, prediction models require empirical verifications. However, to judge the reliability of models, a physical understanding of a model is essential. Because problems involved with crack initiation and crack growth are different, models will be discussed in two categories: models for the crack initiation period and models for fatigue crack growth.

Prediction of Crack Initiation under VA Loading

In the literature, prediction models are rarely presented as models for the crack initiation period. However, several models simply ignore fatigue crack growth. The predicted fatigue life then is the fatigue life until failure. If the macrocrack growth period is relatively short, the total life until failure is mainly covered by the fatigue crack initiation period. Under such conditions, we may consider the perspectives of prediction models for the initiation period under VA loading. The literature covers two approaches: fracture mechanics applied to the initial microcrack growth and models based on stress or strain histories, disregarding microcrack growth.

Fracture mechanics applied to the crack initiation period involves some fundamental problems:

- Crack growth life predictions based on fracture mechanics concepts cannot start from zero crack length, because then there will be no crack growth. They thus must start from some initial crack length. The size of the initial crack length, a_0, may be associated with some initial structural defect, such as an inclusion. Unfortunately, the predicted crack growth life is very sensitive to the size of such an initial defect. Because of the small size, K-values are small and predicted crack growth rates are very low. As a consequence, a large part of the fatigue life is covered by the initial growth of a small crack. Different initial sizes, say 10 μm and 100 μm, can imply a large difference of the predicted life (Ref 78). The choice of a_0 thus has a large effect on the prediction result.
- As discussed before, fracture mechanics concepts have a limited relevance to microstructurally short cracks. Probably Al alloys offer the best conditions for predictions of very short crack lengths. In general, it is still hard to believe that microcrack growth prediction can be made with some reasonable accuracy for VA loading. For many load histories in Table

2, AGARD Reports R-732 and R-767 show overwhelming results in small-crack growth prediction. However, surface effects are very important and must be considered in future applications of "small-crack theory."

- A third problem is a very practical one. It was concluded above that crack initiation is a surface phenomenon. As a consequence, the crack initiation life is sensitive to various surface conditions.

Environmental factors

Crack initiation life is influenced by several factors such as those listed in Table 4. The influence of these factors on fatigue can be large, and several of the effects are not easily accounted for in a strictly rational way. It must then be admitted that fracture mechanics predictions of crack initiation life are quite limited.

The Miner Approach. Miner published his famous rule ($\Sigma n/N = 1$) 50 years ago. Initially many test were carried out to check the validity of the rule, which was rather frustrating in view of the discrepancies between test results and Miner predictions. Some simple arguments can easily prove why the rule cannot be correct:

- If a small fatigue crack is initiated by load cycles with $S_a > S_f$ (where S_f = fatigue limit), load cycles with $S_a < S_f$ can propagate the crack and thus contribute to fatigue damage. According to the Miner rule, that should not be true because $N = \infty$ for $S_a < S_f$.
- In a notched element, plastic deformation at the notch root can be induced by a high S_{max}. It introduces residual stresses that affect the fatigue damage contribution of later cycles with a lower S_{max} (see Fig. 15, 16). This interaction is not recognized by the Miner rule.
- The Miner rule implies that fatigue damage is fully described by a single parameter, $\Sigma n/N$, which can vary between 0 (virgin specimen) and 1 (final failure). Final failure should always stand for the same amount of damage. However, a high S_{max} leads to failure at a small crack length, whereas a low S_{max} requires a much larger crack (i.e., a different amount of damage). The Miner rule presumes that an S-N curve is a line of constant damage, and that is simply not true.

In a certain way, the first objection (damage contributions of cycles with $S_a < S_f$) can be complied with by extrapolating the S-N curve below the fatigue limit (see Fig. 3). Fatigue cycles below the fatigue limit then contribute to fatigue damage. This might appear to be a reasonable approach, but it does not imply that accurate predictions will be obtained. The second objection was related to notch root plasticity and the introduction of favorable or unfavorable residual stresses as a consequence. Quite often, fatigue critical elements carry a positive mean stress. That is one of the reasons why they can become fatigue-critical. The probability of introducing residual stresses by the VA load history depends on the shape of the load spectrum. However, in general, favorable residual stresses are more likely

Table 4 Factors influencing crack initiation life

Material surface and production factors:
Surface roughness
Surface defects
Surface treatments
Material structure at the surface
Residual stress at the surface

Geometrical factors:
Notch effect (K_t)
Size effect (root radius ρ)
Aspects of joints (e.g., fretting in clamped joints, geometrical aspects of weld toe, etc.)

Environmental factors

if the load spectrum is associated with a positive mean stress. Although the Miner rule does not account for residual stress variations, it thus may be thought that ignoring the residual stresses should not necessarily lead to unconservative life predictions. It must be realized, however, that the predictions cannot be accurate. A rough estimate is the best result to be obtained. Another significant argument for this conclusion is that the prediction also depends on the reliability of the S-N curve to be used.

A warning must be made here: the Miner rule is fully unreliable for comparing the severity of different load spectra. As a simple illustration, compare a load spectrum to a modification of that spectrum obtained by adding a small number of high-load cycles. According to the Miner rule the addition should lead to somewhat shorter fatigue lives, whereas in general it leads to significant fatigue life improvements.

The Strain History Prediction Model. Plastic deformation at the root of a notch is not accounted for in the Miner rule. In low-cycle VA fatigue, however, plastic deformation at the root of a notch can occur in every cycle. This has led to fatigue predictions based on the strain history at the notch root (Ref 79, 80). This approach was stimulated by two developments. Low-cycle fatigue experiments under constant-strain amplitudes have indicated an approximately linear relation between log $\Delta\varepsilon$ and log N (the Coffin-Manson relation). Secondly, predictions of the plastic strain at a notch root could be made by adopting an analytical relation of Neuber (Ref 81) between K_σ and K_ε (concentration factors for stress and strain, respectively), that is, $K_\sigma \cdot K_\varepsilon = K_t^2$. The relation was derived for a prismatic notch loaded in shear (mode III), but it was assumed to be valid for notches loaded in tension as well. In order to solve the strain at the notch, the cyclic stress strain curve was adopted as a second equation.

Figure 41 schematically shows the procedures to be used for calculation of the strain history at the notch under VA loading and for the subsequent life prediction. The prediction model has recently been discussed in detail by Dowling (Ref 82). The main steps are mentioned here in order to show the advantages and weaknesses of the approach. Additional information on the strain-life method (including the use of total-strain-life and mean-stress rules commonly used for life

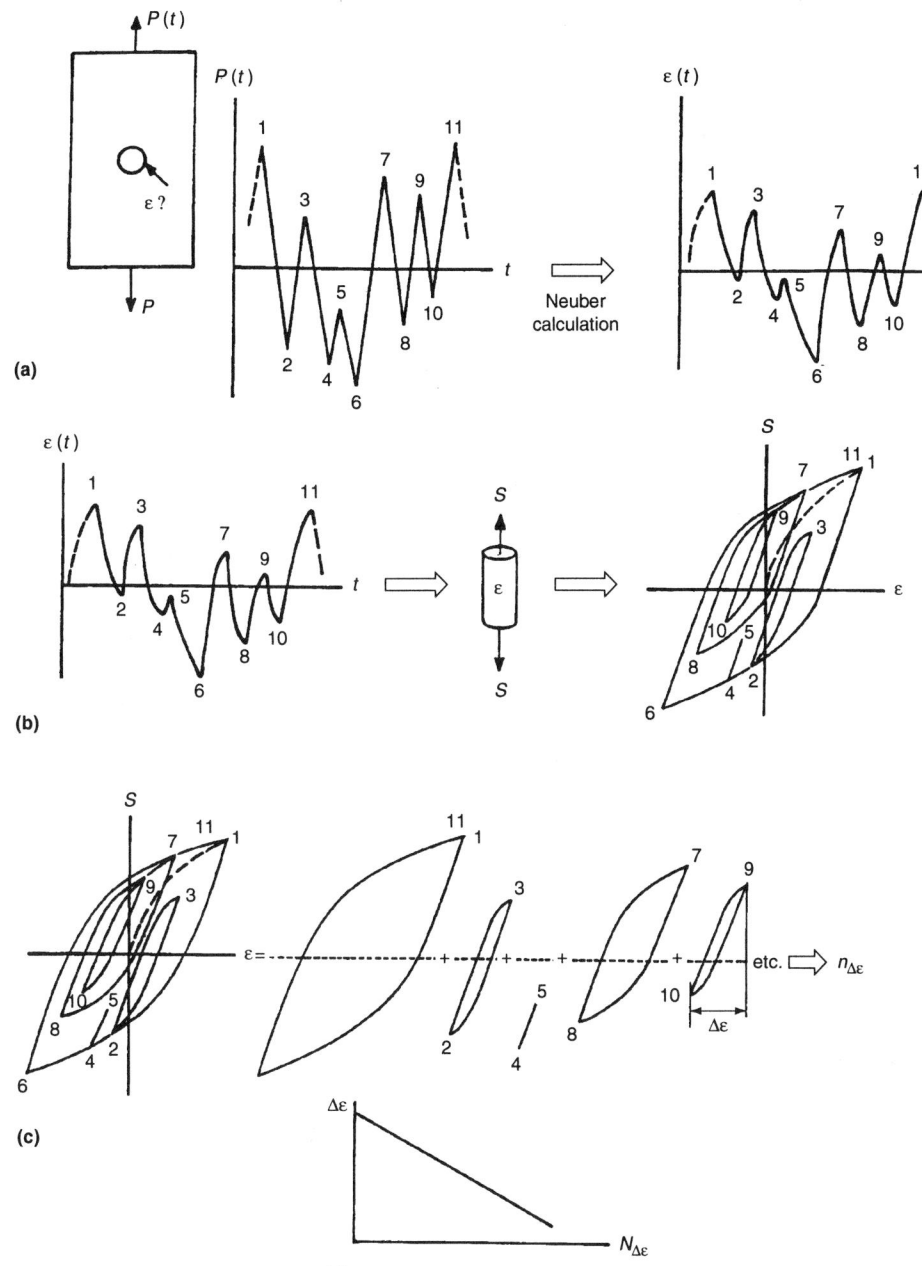

Fig. 41 Principles of the strain-based life prediction model. Failure criterion: $\Sigma\,(n_{\Delta\varepsilon}/N_{\Delta\varepsilon}) = 1$ (see parts c and d). (a) Load history (left graph) and strain history (right graph). (b) Material response. (c) Cycles as closed loops. (d) Material fatigue resistance. Source: Ref 80

cycle fatigue is no longer a surface phenomenon as it is for high-cycle fatigue. At the same time, limitations of the strain-history model are easily recognized. The failure criterion is again the Miner rule, for which physical arguments can hardly be mentioned. Secondly, crack initiation and crack growth are fully ignored. Moreover, the model is restricted to notched elements, for which a theoretical stress concentration factor has a realistic meaning. As a consequence, application to joints is generally impossible. It was emphasized by Dowling (Ref 82) that the merits of the model should be looked for in low-cycle VA problems. Actually verification experiments are still rather limited.

There is a noteworthy comment to be made on the decomposition in Fig. 41(c). The individual cycles obtained are the same as the cycles obtained with the rainflow count method. This implies that this counting method finds some justification in the material memory for previous plastic deformation.

Prediction of Crack Growth under VA Loading

The literature on prediction models for fatigue crack growth under VA loading is extensive. Observations on crack growth retardation after OLs and the occurrence of crack closure have stimulated the development of several prediction models on crack growth under VA loading. Most literature sources on prediction models give verification test data of crack growth in Al alloy sheet and plate material, mainly because VA loading and fatigue crack growth are important for aircraft structures. Acceleration and retardation must also both be considered. Predictive models that do not address acceleration do not appear effective (Chang and Hudson in ASTM STP 748).

Simple Approach to Crack Growth under VA Loading (Noninteraction). The most simple VA load sequence consists of two blocks of load cycles, where the second block is continued until a final crack length $a = a_f$ is reached (Fig. 42a). The sequence may be Hi-Lo (as in Fig. 42a) or Lo-Hi. The simplest prediction model is obtained if all possible interaction effects are ignored. Crack growth then follows the growth curve applicable to the load cycle in the first block (Fig. 42b). After the stress level is changed, crack growth continues along the curve valid for the load cycle of the second block. There is a simple noninteraction transition from one crack growth curve to the other one. The predicted life is $N_p = n_1 + n_2$.

The two curves in Fig. 42(b) can also be presented as a function of n/N. The beginning and the end of the two curves then coincide at $n/N = 0$ and $n/N = 1$, respectively. The fatigue damage, D, represented by the crack length a in Fig. 42(b), is converted to the crack increment $(a - a_0)$ relative to the total crack increment to be covered $(a_f - a_0)$. It is a kind of damage parameter defined by:

$$D = \frac{a - a_0}{a_f - a_0} \qquad \text{(Eq 3)}$$

prediction) is also provided in the articles "Fundamentals of Modern Fatigue Analysis for Design" and "Fatigue Strength in Practical Applications" in this Volume.

In the first step (Fig. 41a), the strain history $\varepsilon(t)$ is derived from the load history $P(t)$ by employing the Neuber postulate and the cyclic stress-strain curve. In the second step (Fig. 41b), the σ-ε response of the material (at the root of the notch) is derived from $\varepsilon(t)$. This derivation presumes a certain plastic hysteresis behavior based on the material memory for previous plastic deformation. In the third step (Fig. 41c), the cyclic hysteresis history is decomposed into closed hysteresis loops. Each loop represents a full strain cycle. In

the last step (Fig. 41d), the $\Delta\varepsilon$-$N_{\Delta\varepsilon}$ curve (adjusted for mean stress with a mean stress rule) is used as the material property characterizing the material resistance against low-cycle fatigue. The Miner rule is then adopted as the failure criterion.

The material properties required for the strain-history model are the cyclic stress-strain curve and the Coffin-Manson relation. Both types of data are considered to be unique for a material. This is an advantage over the stress-based S-N fatigue data, which depend on mean stress and surface quality. The surface quality is much less important for low-cycle fatigue, because the plastic strains are larger and as such depend on the material bulk behavior. It might be said that low-

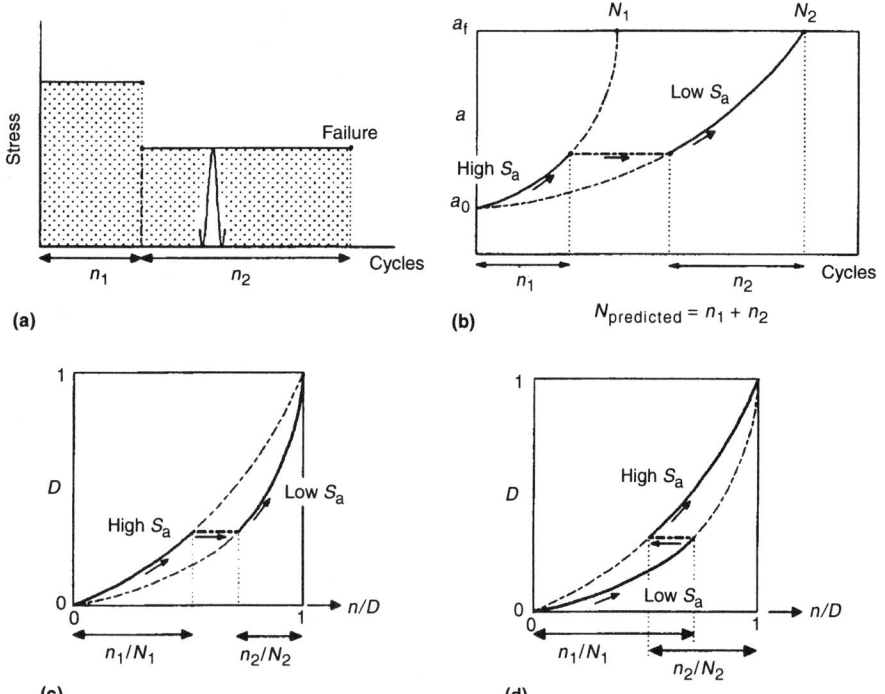

Fig. 42 Noninteraction fatigue crack growth and fatigue damage in Hi-Lo and Lo-Hi tests. $D = (a - a_0) / (a_f - a_0)$. (a) Hi-Lo. (b) Hi-Lo. $N_{predicted} = n_1 + n_2$. (c) Hi-Lo. $\Sigma n/N < 1$. (d) Lo-Hi. $\Sigma n/N > 1$.

Table 5 Three categories of crack growth prediction models

Type of model	Crack closure used?	Crack closure relation
Yield zone models	No	...
Crack closure models	Yes	Empirical
Strip yield models	Yes	Calculated

where D varies from 0 ($a = a_0$) to 1 ($a = a_f$). Considering crack growth along the two curves in Fig. 42(c), it is obvious that it leads to $\Sigma n/N < 1$. For the reversed block sequence (Lo-Hi), it leads to $\Sigma n/N > 1$ (Fig. 42d). This suggests that there is a sequence effect, although interaction effects are disregarded.

An elementary statement can now be made (Ref 83): If fatigue damage is fully characterized by a single damage parameter, interaction effects are impossible. The reverse statement can also be made: If interaction effects do occur, fatigue damage cannot be fully described by one single damage parameter.

There is another interesting observation. If the two curves in Fig. 42(c) and (d) coincide, crack growth leads straightforwardly to $\Sigma n/N = 1$. More generally, if the same damage curves apply to any cyclic stress level, and if interaction effects do not occur, then the Miner rule is valid for any load sequence (Ref 83). In other words, if a damage function can be written as:

$$D = f\left(\frac{n}{N}\right) \qquad (Eq\ 4)$$

which is valid for any cyclic stress level, it leads to the Miner rule, independent of the shape of the

function. (As discussed in Ref 83, the function $f(n/N)$ should be a monotonously increasing function in order that D has a unique value for any n/N.) According to Miner, $f(n/N)$ is a linear function, but nonlinear functions have been proposed in the literature (e.g., Ref 84).

Interaction Models for Prediction of Fatigue Crack Growth under VA Loading. The most well-known prediction models for fatigue crack growth under VA can be characterized by whether crack closure is involved and whether that is done in an empirical way or by calculation. Three categories are listed in Table 5.

The models were developed in the order shown in the table. It was thought that the crack closure models were an improvement of the more primitive yield zone models, and strip yield models were considered superior to the initial crack closure models. As stated above, the models were primarily verified for through-cracks in Al alloy sheet and plate specimens, but experiments on other materials were done. In all models, plastic zone sizes are significant, whereas relaxation of residual stress and plastic shakedown are not included. It appears that the models are considered applicable for high-strength alloys with a limited ductility. Actually, these materials are the most fatigue-critical materials. Due to its special yielding behavior and its high ductility, mild steel is a class of materials of its own. However, fatigue crack growth in low-carbon steel under VA loading is becoming an increasingly relevant problem in welded structures.

Yield Zone Models. The models of Willenborg et al. (Ref 85) and Wheeler (Ref 86) were proposed to explain crack growth delays caused by OLs. The models consider the plastic zone sizes

indicated in Fig. 43, but the concepts are different. In both models it was recognized that new plastic zones are created inside the large plastic zone of the OL. Moreover, the possibility was considered that these new plastic zones could be large enough to grow outside the OL plastic zone.

The Willenborg model starts from a strange assumption that the delay is due to a reduction of K_{max} instead of a reduction of ΔK_{eff}. This is physically incorrect. Crack closure in the model is supposed to occur only if $K_{min} < 0$. From a mechanistic point of view, the Willenborg model does not agree with the present understanding of crack closure. Wheeler introduced a retardation factor β, defined by:

$$\left(\frac{da}{dN}\right)_{VA} = \beta \cdot \left(\frac{da}{dN}\right)_{CA,\ same\ K\ cycle} \qquad (Eq\ 5)$$

The factor β is supposed to be a power function of the ratio r_{pi}/λ_i:

$$\beta = (r_{p,i}/\lambda)^m \qquad (Eq\ 6)$$

The empirical "constant" m is not a material constant, because it depends on the type of the VA load history. Both models can predict crack growth retardation only ($\beta < 1$), not acceleration. After an OL the maximum retardation occurs immediately. Delayed retardation is not predicted. A more extensive summary is given in Ref 87.

Modifications of the two models have been proposed in the literature, which leads to more empirical constants. Crack closure, however, is not included. As a consequence, the models lack a background in sufficient agreement with the present understanding.

Crack Closure Models for Predicting Crack Growth under VA Loading. The crack closure models are based on the phenomenon of plasticity-induced crack closure. The Elber crack closure concept is used (i.e., there is an S_{op} in each cycle and the effective stress range is $\Delta S_{eff} = S_{max} - S_{op}$). A cycle-by-cycle variation of S_{op} has to be predicted (Fig. 44). The cycle-by-cycle calculations then follow apparently simple equations:

$$a = a_0 + \Sigma \Delta a_i \qquad (Eq\ 7)$$

$$\Delta a_i = (da/dN)_i = f(\Delta K_{eff,i}) \qquad (Eq\ 8)$$

$$\Delta K_{eff,i} = C_i (S_{max,i} - S_{op,i})\sqrt{\pi a_i} \qquad (Eq\ 9)$$

The crack extension Δa_i in cycle i is supposed to be a function of ΔK_{eff} in that cycle, while $\Delta K_{eff,i}$ is a function of $S_{max,i}$ and the predicted $S_{op,i}$ for cycle i. The geometry factor C_i depends on the crack size, a_i. The crack opening stress level $S_{op,i}$ depends on the previous load history, but $S_{max,i}$ is part of the imposed load history (i.e., input data). In the models to be discussed below, the Paris relation is used for Eq 8:

$$da/dN = C\ \Delta K_{eff}^m \qquad (Eq\ 10)$$

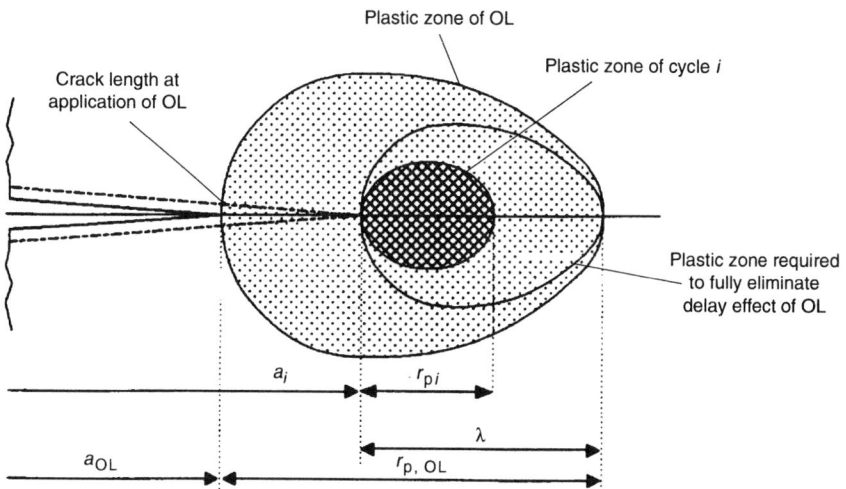

Fig. 43   Plastic zone size concepts in the models of Willenborg (Ref 85) and Wheeler (Ref 86)

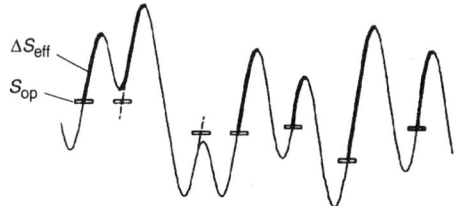

Fig. 44   Variable-amplitude load with cycle-by-cycle variation of S_{op}

Another relation (e.g., interpolation in a table) can also be used.

Four models are briefly discussed below:

- The ONERA model (Ref 88)
- The CORPUS model (Ref 73)
- The modified CORPUS model (Ref 87)
- The PREFFAS model (Ref 89)

The models were developed primarily for applications to flight simulation load histories. They all calculate a variation of S_{op} during the flight simulation load history. The variation depends on the previous load history. It implies that information characteristic of the previous load history must be stored in a memory. The characteristic information is associated with the larger positive and negative peak loads. These loads either have introduced significant plastic zones for the determination of S_{op} or have reduced S_{op}, respectively. There are also significant differences between the models, which will not be discussed here in detail. The PREFFAS model is the simplest; the CORPUS model is the most detailed and also presents the most explicit picture about crack closure between the crack flanks. The differences between the models are associated with the assumptions made for the plane-strain/plane-stress transition during crack growth, the calculation of the plastic zone sizes, the empirical equations for calculating S_{op} (Elber-type relations), the decay of S_{op} during crack growth, the multiple OL effect, and in general the method of deriving S_{op} from the previous load history.

An analysis and comparison of the models has been made by Padmadinata (Ref 87) with extensive verifications, primarily for realistic flight simulation load histories and test results of two Al alloys, 2024-T3 and 7075-T6. However, simplified flight simulation tests were also included. As an example, comparative results for a realistic load spectrum are presented in Fig. 45. The test variables include the stress level, characterized by the mean stress in flight (S_{mf}), the gust spectrum severity, and the downward severity of the ground load during landing. Noninteraction pre-

dictions are also shown in this figure. Unfortunately, this is not always done in model verifications, but differences between noninteraction predictions and predictions of improved models are part of the motivation for the new models. Moreover, these differences indicate whether significant interaction effects have occurred in the test. The results in Fig. 45 clearly show that the noninteraction predictions did systematically underestimate the crack growth life in the tests to a large extent. The test life on the average was 5.3 times longer. The predictions of all models were significantly superior to the noninteraction prediction. Some comments on the results can be made:

- The PREFFAS model does not predict any effect of the ground stress level. That is a

consequence of clipping negative loads in this model to zero.
- The predictions of the CORPUS model and the ONERA model are fairly close to the test results. The test results indicate a significant reduction of the crack growth life for a more severe ground stress level. This trend is not always predicted by the CORPUS model, especially if the gust spectrum is more severe. The maximum downward gust load occurs only once in a large number of flights (2500 flights in Fig. 45). However, the ground load occurs in every flight. In CORPUS its effect is small if the most negative gust is more severe downward. That overrules the ground stress level. This was the reason that the CORPUS model was modified. The modified model is still largely the same as the original, but due to a modified memory effect for downward loads the modified CORPUS model gives a better prediction for the above-mentioned conditions (Ref 87, 90).

It may now be asked if all observations listed in Table 3 are covered to some extent by the basic assumptions on which the crack closure models are based. It then turns out that:

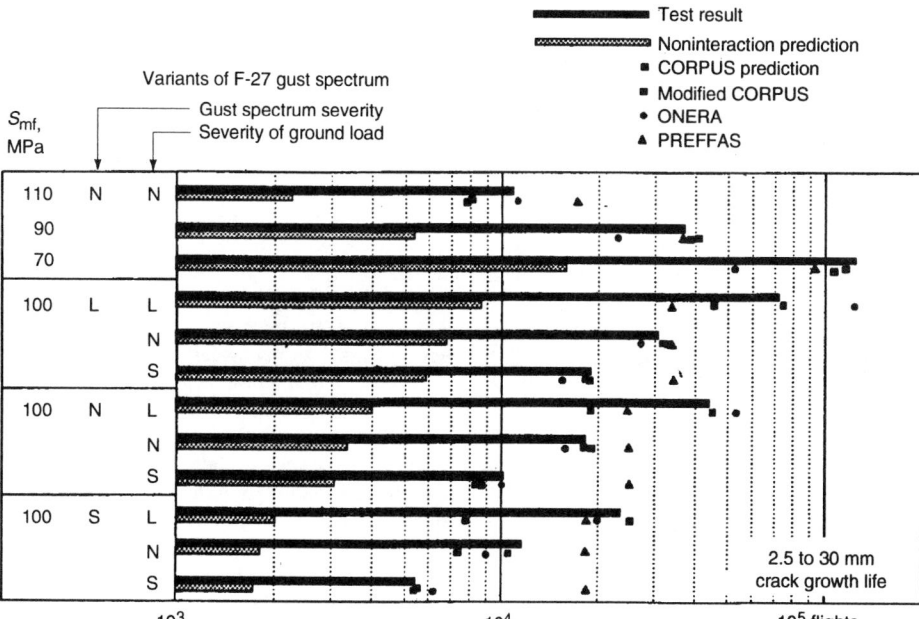

Fig. 45   Comparison between test results and predictions of fatigue crack growth life under flight simulation loading in 2024-T3 (t = 2 mm). S, severe; N, normal; L, light. Source: Ref 87

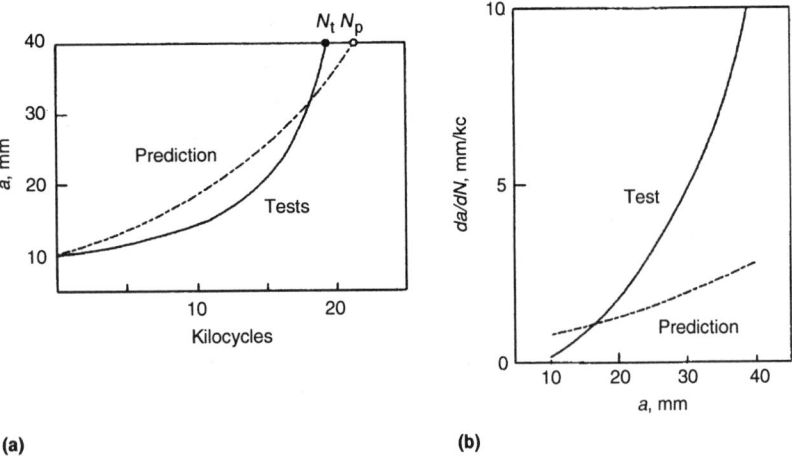

(a)   **(b)**

Fig. 46 Two comparisons between test results and predictions for the same data

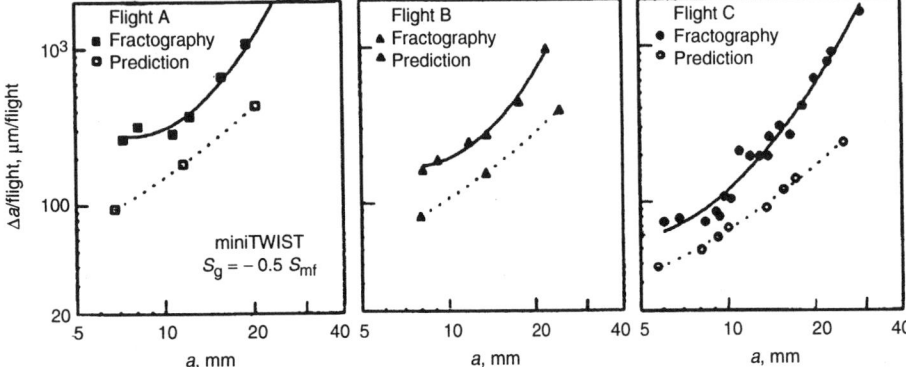

Fig. 47 Δa in the most severe flights of a flight simulation test in 2024-T3 (t = 2 mm). Δa is larger than the crack length increments predicted by the modified CORPUS model. Source: Ref 91

verification of a prediction model also requires a comparison on a microscopic level. Without such observations delayed retardation cannot really be documented.

Strip Yield Models. The empirical crack closure models discussed above are based on the occurrence of crack closure in the wake of the crack. Assumptions are made to account for crack closure under VA loading, but plastic deformation in the wake of the crack is not calculated. This was done in some finite-element modeling studies (Ref 92, 93), which confirmed the occurrence of crack closure and simple interaction effects in qualitative agreement with empirical observations. Such calculations cannot be made for many cycles, so the Dugdale model was adopted and extended to arrive at a crack growth model that leaves plastic deformation in the wake of the crack. This type of work was started by Führing and Seeger (Ref 94, 95). In the Dugdale plastic zone model, plastic deformation occurs in a thin strip with a rigid perfectly plastic material behavior. Because the crack grows into the plastic zone, a plastic wake field is created, which can induce crack closure at positive stress levels.

Quantitative strip yield models have been proposed by Dill et al. (Ref 96, 97), Newman (Ref 98), DeKoning et al. (Ref 99), and Wang and Blom (Ref 100). The models are rather complex, which is a consequence of the nonlinear material behavior and the changing geometry (crack closure and crack opening). Reversed plastic deformation in the wake field can occur when the crack is closed and locally under compression. Iterative solution procedures are to be used, which require significant computer capacity for a cycle-by-cycle calculation. They also require a number of plastic elements in the plastic zone and in the wake of the crack (Fig. 48). Newman has introduced local averages of S_{op} to avoid excessive computer time. Plane-strain/plane-stress transitions are included by changing the yield stress used in the Dugdale model. This has led to a so-called plastic constraint factor α, developed by Newman and defined by him as the ratio of normal stresses in the plastic zone to the flow stress under tension. A separate α factor is defined by Newman for compression. DeKoning's interpretation of Newman's α factor is the ratio between the yield stress in tension and the yield stress in compression. Several predictions are reported for both simple tests with overload/underload cycles and flight simulation tests. In general, good agreement is reported.

The models cannot be discussed here in any detail. However, in comparison to the crack closure models, the improvements appear to be that:

- Empirical equations for crack closure levels are replaced by the calculation of S_{op} as a function of the history of previous plastic deformations. Elber's assumption that $U(R)$ is independent of the crack length is no longer necessary.
- Delayed retardation is predicted (Ref 101).
- In the strip yield model of de Koning, the concept of primary and secondary plastic

- Crack growth retardation after OLs is predicted, but delayed retardation is not. The retardation starts immediately after the overload.
- Plane-strain/plane-stress transition is included in the CORPUS and ONERA models, although not in the same way. It leads to a thickness effect, but variations along the crack front are averaged out. The transition is not included in the PREFFAS model, but the model requires empirical data for the OL effect representative of the thickness considered.
- Multiple OL effects do occur according to the CORPUS and the ONERA models, although they are not modeled in the same way. The CORPUS model predicts an increasing S_{op} during stationary flight simulation loading, which is necessary to predict the initially decreasing crack growth rate (Fig. 35).
- The different crack growth mechanisms for primary and secondary crack-tip plastic zones are not included.
- Incompatible crack front orientations and related phenomena are not covered.

Some comments should be made here on the verification of models. In Fig. 45 a comparison is made between predicted and experimental crack growth lives. This is a primitive comparison, and

if the crack growth lives do agree, a disagreement between crack growth curves is still possible (Fig. 46). In Fig. 46(a) the agreement between $N_{predicted}$ and N_{test} is satisfactory, but that is not true for the crack growth curves or for the crack growth rate development during the crack growth lives. Even if the predicted and the measured crack growth curves do match quite well, that is not necessarily true for the crack rate in small batches of individual cycles. This question has been studied for crack growth under flight simulation loading (Ref 87, 91). The crack extension in the more severe flights has been determined by fractographic analysis. As shown by the results in Fig. 47 (Ref 91), the crack extension in the more severe flights was considerably larger than predicted by the modified CORPUS model, although the agreement for the macroscopic crack growth curve was quite good. This is not strange, because the number of the most severe flights in a flight simulation test is small. Incorrect predictions for the severe flights thus have a minor effect on the general behavior. There are two possible explanations for the discrepancy for the severe flights: the larger Δa for primary plastic deformation (Fig. 37) and incompatible crack front orientation. The discrepancy indicates that the model is not reliable in all details. A real

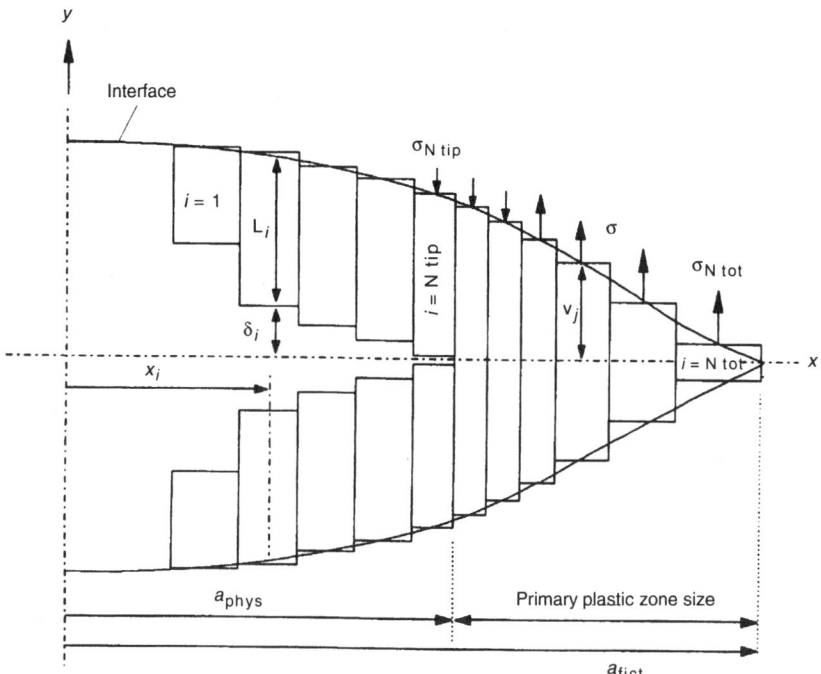

Fig. 48 A strip yield model with discrete plastically stretched parts ahead of the physical crack tip and in the wake of the crack. Source: Ref 101

zones is introduced, which accounts for large Δa values of peak loads (prediction verification in Ref 99).

- Multiple OL effects are predicted by the strip yield model if the modeling is sufficiently refined (see ASTM STP 761, 1982, for example).
- The plane-strain/plane-stress transition is still covered by assumptions.
- Incompatible crack front interactions are not covered.

Strip yield models are superior to the crack closure models because the physical concept has been improved. In general terms, the calculation of the crack driving force, ΔK_{eff}, is based on calculations of the history of the plastic deformations in the crack-tip zone and in the wake of the crack. However, it still may be questioned whether the models are sufficiently realistic from a mechanistic point of view (Table 3) in order to arrive at accurate predictions. There is a lot of verification work to be done.

Engineering Applications

As a result of many experimental investigations, we have obtained a reasonably detailed picture about fatigue in metallic materials under variable amplitude loading. The understanding should lead us to the question whether it is possible to aim at accurate and reliable prediction models for engineering purposes. That is a problem schematically surveyed in Fig. 1 about designing against fatigue. The alternative to making predictions is to carry out experiments for specific fatigue questions when they arise. Unfortunately, testing is not always possible. Moreover,

it is not at all easy to accomplish experimental fatigue conditions that will give a relevant answer to our question. In many cases Miner calculations are made as a first life estimate, but as discussed above, they can lead to conservative estimates if a realistic *S-N* curve is extrapolated below the fatigue limit. Also, a noninteraction fatigue crack growth prediction can certainly lead to a conservative prediction (although even then it is a recommended practice to extrapolate the *da/dN-ΔK* relation below ΔK_{th}), but for macrocracks the noninteraction approach might well lead to *over-conservative* predictions. In other words, there are still good arguments to continue our efforts for improved fatigue prediction methods. An extensive verification of a new model must be recommended to cover all possible conditions associated with engineering applications. The physical understanding to see whether a model is feasible is necessary, but verification of the accuracy is essential in order to be confident that the application can be justified. More comments on practical aspects of Fig. 1 are made in Ref 102.

A final comment should be made on types of material. As stated above, most information on fatigue under VA loading has been obtained in research on aircraft materials. However, mild steel is abundantly used in many welded structures, and fatigue and crack growth are highly relevant issues for this type of material. Mild steel differs from many high-strength structural materials because of its own characteristic plastic yielding behavior. Plastic zone shapes are different for mild steel and high-strength alloys. For mild steel, Dugdale (Ref 103) observed that the shape is a narrow slit in line with the crack. The Dugdale concept for calculating the plastic zone shape for mild steel has been adopted in the VA

strip yield models. Ironically, the plastic zone shape observed in high-strength alloys agrees better with the butterfly shape that is obtained in elastic-plastic finite-element calculations.

REFERENCES

1. W.L. Morris, O. Buck, and H.L. Marcus, Fatigue Crack Initiation and Early Propagation in Al 2219-T851, *Met. Trans. A*, Vol 7, 1976, p 1161-1165
2. C.Y. Kung and M.E. Fine, Fatigue Crack Initiation and Microcrack Growth in 2024-T4 and 2124-T4 Aluminum Alloys, *Met. Trans. A*, Vol 10, 1979, p 603-610
3. D. Sigler, M.C. Montpetit, and W.L. Haworth, Metallography of Fatigue Crack Initiation in an Overaged High-Strength Aluminum Alloy, *Met. Trans. A*, Vol 14, 1983, p 931-938
4. A.E. Blom, A. Hedlund, A.W. Zhao, A. Fathalla, B. Weiss, and R. Stickler, Short Fatigue Crack Growth in Al 2024 and Al 7475, *Behaviour of Short Fatigue Cracks*, K.J. Miller and E.R. de los Rios, Ed., EGF 1, Mechanical Engineering Publications, 1986
5. J. Lankford, The Growth of Small Fatigue Cracks in 7075-T6 Aluminum, *Fatigue Fract. Engng. Mater. Struct.*, Vol 5, 1982, p 233-248
6. D. Craig, F. Ellyin, and D. Kujawski, The Behaviour of Small Corner Cracks in a Ferritic/Pearlitic Steel: Experiments and Analysis, *Fatigue Fract. Engng. Mater. Struct.*, Vol 18, 1995, p 861-873
7. S. Suresh, *Fatigue of Materials*, Cambridge University Press, 1991
8. R.J.H. Wanhill, Durability Analysis Using Short and Long Fatigue Crack Growth Data, *Aircraft Damage Assessment and Repair*, Inst. of Eng. Australia, Barton, Australia, 1991, p 100-104
9. J. Schijve, The Practical and Theoretical Significance of Small Cracks: An Evaluation, *Fatigue 84*, C.J. Beevers, Ed., EMAS, Warley, U.K., 1984, p 751-771
10. K.J. Miller, The Behaviour of Short Fatigue Cracks and Their Initiation, Part I: A Review of Two Recent Books; Part II: A General Survey, *Fatigue Fract. Engng. Mater. Struct.*, Vol 10, 1987, p 57-91, 93-113
11. K.J. Miller, The Two Thresholds of Fatigue Behaviour, *Fatigue Fract. Engng. Mater. Struct.*, Vol 16, 1993, p 931-939
12. J. Schijve, Difference between the Growth of Small and Large Fatigue Cracks: The Relation to Threshold K-Values, *Fatigue Thresholds, Fundamentals and Engineering Applications*, J. Bäcklund et al., Ed., EMAS, Warley, U.K., 1982, p 881-908
13. C.Q. Bowles, "The Role of Environment, Frequency and Wave Shape during Fatigue Crack Growth in Aluminum Alloys," Ph.D. dissertation, Delft University of Technology, 1978
14. C.Q. Bowles and J. Schijve, Crack Tip Geometry for Fatigue Cracks Grown in Air and Vacuum, *Advances in Quantitative Measurement of Physical Damage*, J. Lankford et al., Ed., STP 811, ASTM, 1983, p 400-426

15. W. Elber, "Fatigue Crack Propagation," Ph.D. dissertation, University of New South Wales, Australia, 1968

16. W. Elber, The Significance of Fatigue Crack Closure, *Damage Tolerance in Aircraft Structures*, STP 486, ASTM, 1971, p 230-242

17. R. Kumar, Review on Crack Closure for Constant Amplitude Loading in Fatigue, *Eng. Fract. Mech.*, Vol 42, 1992, p 389-400

18. R.O. Ritchie and S. Suresh, Mechanics and Physics of the Growth of Small Cracks, *Behaviour of Short Cracks in Airframe Components*, AGARD CP-328, 1983

19. M.A. Miner, Cumulative Damage in Fatigue, *J. Appl. Mech.*, Vol 12, 1945, p A159-A164

20. A. Pålmgren, Die Lebensdauer von Kugellagern, *Z. Ver. Deut. Ing.*, Vol 68, 1924, p 339-341

21. B.F. Langer, Fatigue Failure from Stress Cycles of Varying Amplitude, *J. Appl. Mech.*, Vol 4, 1937, p A160-A162

22. J. Schijve, *The Accumulation of Fatigue Damage in Aircraft Materials and Structures*, AGARDograph 157, 1972

23. E. Gassner, Strength Experiments under Cyclic Loading in Aircraft Structures, *Luftwissen*, Vol 6, 1939, p 61-64 (in German)

24. E.C. Naumann, H.R. Hardrath, and E.C. Guthrie, "Axial Load Fatigue Tests of 2024-T3 and 7075-T6 Aluminum Alloy Sheet Specimens under Constant- and Variable-Amplitude Loads," Report TN D-212, NASA, 1959

25. J.B. de Jonge, Assessment of Service Load Experience, *Aeronautical Fatigue in the Electronic Era*, A. Berkovits, Ed., EMAS, Warley, U.K., 1989, p 1-42

26. J.B. de Jonge, National Aerospace Laboratory, Amsterdam, private communication, 1995

27. J. Schijve, The Significance of Flight Simulation Fatigue Tests, *Durability and Damage Tolerance in Aircraft Design*, A. Salvetti and G. Cavallini, Ed., EMAS, Warley, U.K., 1985, p 71-170

28. J. Schijve, F.A. Jacobs, and P.J. Tromp, "Fatigue Crack Growth in Aluminium Alloy Sheet Material under Flight-Simulation Loading: Effect of Design Stress Level and Loading Frequency," Report TR 72018, National Aerospace Laboratory, Amsterdam, 1972

29. A. Hartman, On the Effect of Oxygen and Water Vapour on the Propagation of Fatigue Cracks in 2024-T3 Alclad Sheet, *Int. J. Fract. Mech.*, Vol 1, 1965, p 167

30. F.J. Bradshaw and C. Wheeler, The Effect of Gaseous Environment and Fatigue Frequency on the Growth of Fatigue Cracks in Some Aluminium Alloys, *Int. J. Fract. Mech.*, Vol 6, 1969, p 225

31. D. Broek, "Fatigue Crack Growth and Residual Strength of Aluminium Sheet at Low Temperature," Report TR 72096, National Aerospace Laboratory, Amsterdam, 1972

32. J. Schijve, Effects of Test Frequency on Fatigue Crack Propagation under Flight-Simulation Loading, *Symp. on Random Load Fatigue*, AGARD CP 118, 1972

33. K. Endo and Y. Miyao, Effects of Cyclic Frequency on the Corrosion Fatigue Strength, *Bulletin Japan. Soc. Mech. Engineers*, Vol 1, 1958, p 374-380

34. J.C.P. Kam, Recent Development in the Fast Corrosion Fatigue Analysis of Offshore Structures Subject to Random Wave Loading, *Int. J. Fatigue*, Vol 12, 1990, p 458-468

35. W. Schütz, Standardized Stress-Time Histories: An Overview, *Development of Fatigue Load Spectra*, STP 1006, ASTM, 1989, p 3-16

36. A.A. ten Have, "Wisper and Wisperx: A Summary Paper Describing Their Background, Derivation and Statistics," Report TP 92410, National Aerospace Laboratory, Amsterdam, 1992

37. J. Schijve and F.A. Jacobs, "Fatigue Tests on Notched and Unnotched Clad 24 S-T Sheet Specimens to Verify the Cumulative Damage Hypothesis," Report M, 1982, National Aerospace Laboratory, Amsterdam, 1955

38. G. Wållgren, Review of Some Swedish Investigations on Fatigue during the Period June 1959 to April 1961, Report FFA-TN-HE 879, FFA, Stockholm, 1961

39. R.B. Heywood, The Effect of High Loads on Fatigue, *Colloquium on Fatigue*, W. Weibull and F.K.G. Odquist, Ed., Springer Verlag, 1956, p 92-102

40. C.M. Sharp, *D.H.—An Outline of De Havilland History*, Faber and Faber, London, 1960

41. J. Boissonat, Experimental Research on the Effects of a Static Preloading on the Fatigue Life of Structural Components, *Fatigue of Aircraft Structures*, W. Barrois and E. Ripley, Ed., Pergamon Press, 1963, p 97-113

42. *Fatigue Crack Growth under Spectrum Loads*, STP 595, ASTM, 1976

43. J. Schijve, Fatigue Damage Accumulation and Incompatible Crack Front Orientation, *Eng. Fract. Mech.*, Vol 6, 1974, p 245-252

44. J. Schijve, Observations on the Prediction of Fatigue Crack Propagation under Variable-Amplitude Loading, *Fatigue Crack Growth under Spectrum Loads*, STP 595, ASTM, 1976, p 3-23

45. W.J. Mills and R.W. Hertzberg, The Effect of Sheet Thickness on Fatigue Crack Retardation in 2024-T3 Aluminum Alloy, *Eng. Fract. Mech.*, Vol 7, 1975, p 705-711

46. G.S. Petrak, Strength Level Effects on Fatigue Crack Growth and Retardation, *Eng. Fract. Mech.*, Vol 6, 1974, p 725-733

47. W. Dahl and G. Roth, "On the Influence of Overloads on Fatigue Crack Propagation in Structural Steels," Tech. Un. Aachen, 1979

48. W.J. Mills and R.W. Hertzberg, Load Interaction Effects on Fatigue Crack Propagation in 2024-T3 Aluminum Alloy, *Eng. Fract. Mech.*, Vol 8, 1976, p 657-667

49. Y.K. Tür and Ö. Vardar, Periodic Tensile Overloads in 2024-T3 Al-Alloy, *Eng. Fract. Mech.*, Vol 53, 1996, p 69-77

50. M.R. Ling and J. Schijve, Fractographic Analysis of Crack Growth and Shear Lip Development under Simple Variable-Amplitude Loading, *Fatigue Fract. Engng. Mater. Struct.*, Vol 13, 1990, p 443-456

51. T.C. Lindley and C.E. Richards, The Relevance of Crack Closure to Fatigue Crack Propagation, *Mater. Sci. Eng.*, Vol 14, 1974, p 281-293

52. F.A. Veer, "The Effect of Shear Lips, Loading Transitions and Test Frequency on Constant ΔK and Constant Load Amplitude Fatigue Tests," Ph.D. dissertation, Delft University of Technology, 1993

53. J. Zuidema, "Square and Slant Fatigue Crack Growth," Ph.D. dissertation, Delft University of Technology, 1995

54. J. Schijve, Four Lectures on Fatigue Crack Growth, *Eng. Fract. Mech.*, Vol 11, 1979, p 176-221

55. M.R. Ling and J. Schijve, The Effect of Intermediate Heat Treatments and Overload Induced Retardations during Fatigue Crack Growth in an Al-Alloy, *Fatigue Fract. Engng. Mater. Struct.*, Vol 15, 1992, p 421-430

56. P.J. Bernard, T.C. Lindley, and C.E. Richards, Mechanisms of Overload Retardation during Fatigue Crack Propagation, *Fatigue Crack Growth under Spectrum Loading*, STP 595, ASTM, 1976, p 78-96

57. R.G. Chermahini, K.N. Shivakumar, and J.C. Newman, Jr., Three Dimensional Finite-Element Simulation of Fatigue-Crack Growth and Closure, *Mechanics of Fatigue Crack Closure*, J.C. Newman, Jr. and W. Elber, Ed., STP 982, ASTM, 1988, p 398-413

58. A.F. Grandt, "Three-Dimensional Measurements of Fatigue Crack Closure," Report CR-175366, NASA, 1984

59. A.J. McEvily, "Current Aspects of Fatigue: Appendix—Overload Experiments," paper presented at Fatigue 1977, University of Cambridge

60. H.L. Ewalds and R.T. Furnee, Crack Closure Measurements along the Crack Front in Center Cracked Specimens, *Int. J. Fract.*, Vol 14, 1978, p R53

61. W. Schütz, The Prediction of Fatigue Life in the Crack Initiation and Propagation Stages: A State of the Art Survey, *Eng. Fract. Mech.*, Vol 11, 1979, p 405-421

62. P. Heuler and W. Schütz, Fatigue Life Prediction in the Crack Initiation and Crack Propagation Stages, *Durability and Damage Tolerance in Aircraft Design*, A. Salvetti and G. Cavallini, Ed., EMAS, Warley, U.K., 1985, p 33-69

63. J. Schijve, Cumulative Damage Problems in Aircraft Structures and Materials, *Aero. J.*, Vol 74, 1970, p 517-532

64. N.E. Ryan, "The Influence of Stress Intensity History on Fatigue-Crack Growth," Report ARL/Met. 92, Aerospace Research Laboratory, Melbourne, Australia, 1973

65. J. Schijve, Effect of Load Sequences on Crack Propagation under Random and Program Loading, *Eng. Fract. Mech.*, Vol 5, 1973, p 269-280

66. J.B. de Jonge, D. Schütz, H. Lowak, and J. Schijve, "A Standardized Load Sequence for Flight Simulation Tests on Transport Aircraft

Wing Structures," Report TR-73029, National Aerospace Laboratory, Amsterdam, 1973

67. J. Schijve, A.M. Vlutters, S.P. Ichsan, and J.C. ProvoKluit, Crack Growth in Aluminium Alloy Sheet Material under Flight-Simulation Loading, *Int. J. Fatigue*, Vol 7, 1985, p 127-136

68. C.R. Saff and D.R. Holloway, Evaluation of Crack Growth Gages for Service Life Tracking, *Fract. Mech.*, R. Roberts, Ed., STP 743, ASTM, 1981, p 623-640

69. J. Schijve, F.A. Jacobs, and P.J. Tromp, "Crack Propagation in Aluminium Alloy Sheet Materials under Flight-Simulation Loading," Report TR 68117, National Aerospace Laboratory, Amsterdam, 1968

70. R.J.H. Wanhill, "The Influence of Starter Notches on Flight Simulation Fatigue Crack Growth," Report MP 95127, National Aerospace Laboratory, Amsterdam, 1995

71. D. Broek, *Elementary Engineering Fracture Mechanics*, 4th ed., Martinus Nijhoff Publishers, The Hague, 1985

72. J. Schijve, F.A. Jacobs, and P.J. Tromp, "Flight-Simulation Tests on Notched Elements," Report TR 74033, National Aerospace Laboratory, Amsterdam, 1974

73. A.U. de Koning, A Simple Crack Closure Model for Prediction of Fatigue Crack Growth Rates under Variable-Amplitude Loading, *Fracture Mechanics*, R. Roberts, Ed., STP 743, ASTM, 1981, p 63

74. A.U. de Koning and D.J. Dougherty, Prediction of Low and High Crack Growth Rates under Constant and Variable Amplitude Loading, *Fatigue Crack Growth under Variable Amplitude Loading*, J. Petit et al., Ed., Elsevier, 1989, p 208-217

75. J. Schijve, Fundamental Aspects of Predictions on Fatigue Crack Growth under Variable-Amplitude Loading, *Theoretical Concepts and Numerical Analysis of Fatigue*, A.F. Blom and C.J. Beevers, Ed., EMAS, Warley, U.K., 1992, p 111-130

76. S.P. Ichsan, "Fatigue Crack Growth Predictions of Surface Cracks under Constant-Amplitude and Variable-Amplitude Loading," Ph.D. dissertation, Delft University of Technology, 1994

77. J. Schijve and P. de Rijk, "The Crack Propagation in Two Aluminium Alloys in an Indoor and Outdoor Environment under Random and Programmed Load Sequences," Report TR M.2156, National Aerospace Laboratory, Amsterdam, 1965

78. E.P. Phillips and J.C. Newman, Jr., Impact of Small-Crack Effects on Design-Life Calculations, *Exp. Mech.*, June 1989, p 221-224

79. J.F. Martin, T.H. Topper, and G.M. Sinclair, Computer Based Simulation of Cyclic Stress-Strain Behavior with Applications to Fatigue, *Mat. Res. Stand.*, Vol 11, 1971, p 23-28, 50

80. R.M. Wetzel, Ed., *Fatigue under Complex Loading: Analysis and Experiments*, Vol 7, *Advances in Engineering*, SAE, 1977

81. H. Neuber, Theory of Stress Concentration for Shear Strained Prismatical Bodies with Arbitrary Nonlinear Stress-Strain Law, *J. Applied Mech.*, Vol 28, 1961, p 544-550

82. N.E. Dowling, *Mechanical Behavior of Materials: Engineering Methods for Deformation, Fracture, and Fatigue*, Prentice-Hall, 1993

83. J. Schijve, Some Remarks on the Cumulative Damage, *Minutes Fourth ICAF Conf.*, 1956

84. F.R. Shanley, A Proposed Mechanism of Fatigue Failure, *Colloquium on Fatigue*, W. Weibull and F.K.G. Odquist, Ed., Springer Verlag, 1956, p 251-259

85. J. Willenborg, R.M. Engle, and H.A. Wood, "A Crack Growth Retardation Model Using an Effective Stress Concept," Report TR71-1, Air Force Flight Dynamic Laboratory, Wright-Patterson Air Force Base, 1971

86. O.E. Wheeler, Spectrum Loading and Crack Growth, *J. Basic Eng.*, Vol 94, 1972, p 181-186

87. U.H. Padmadinata, "Investigation of Crack-Closure Prediction Models for Fatigue in Aluminum Sheet under Flight-Simulation Loading," Ph.D. dissertation, Delft University of Technology, 1990

88. G. Baudin and M. Robert, Crack Growth Life Time Prediction under Aeronautical Type Loading, *Proc. Fifth European Conf. on Fracture*, 1984, p 779

89. D. Aliaga, A. Davy, and H. Schaff, A Simple Crack Closure Model for Predicting Fatigue Crack Growth under Flight Simulation Loading, *Durability and Damage Tolerance in Aircraft Design*, A. Salvetti and G. Cavallini, Ed., EMAS, Warley, U.K., 1985, p 605-630

90. U.H. Padmadinata and J. Schijve, Prediction of Fatigue Crack Growth under Flight-Simulation Loading with the Modified CORPUS Model, *Advanced Structural Integrity Methods for Airframe Durability and Damage Tolerance*, C.E. Harris, Ed., Conf. Publication 3274, NASA, 1994, p 547-562

91. J. Siegl, J. Schijve, and U.H. Padmadinata, Fractographic Observations and Predictions on Fatigue Crack Growth in an Aluminium Alloy

under miniTWIST Flight-Simulation Loading, *Int. J. Fatigue*, Vol 13, 1991, p 139-147

92. J.C. Newman and H. Armen, Elastic-Plastic Analysis of a Propagating Crack under Cyclic Loading, *AIAA J.*, Vol 13, 1975, p 1017-1023

93. K. Ohji, K. Ogura, and Y. Ohkubo, Cyclic Analysis of a Propagating Crack and its Correlation with Fatigue Crack Growth, *Eng. Fract. Mech.*, Vol 7, 1975, p 457-463

94. H. Führing and T. Seeger, Structural Memory of Cracked Components under Irregular Loading, *Fracture Mechanics*, C.W. Smith, Ed., STP 677, ASTM, 1979, p 1144-1167

95. H. Führing and T. Seeger, Dugdale Crack Closure Analysis of Fatigue Cracks under Constant Amplitude Loading, *Eng. Fract. Mech.*, Vol 11, 1979, p 99-122

96. H.D. Dill and C.R. Saff, Spectrum Crack Growth Prediction Method Based on Crack Surface Displacement and Contact Analysis, *Fatigue Crack Growth under Spectrum Loads*, STP 595, ASTM, 1976, p 306-319

97. H.D. Dill, C.R. Saff, and J.M. Potter, Effects of Fighter Attack Spectrum and Crack Growth, *Effects of Load Spectrum Variables on Fatigue Crack Initiation and Propagation*, D.F. Bryan and J.M. Potter, Ed., STP 714, ASTM, 1980, p 205-217

98. J.C. Newman, Jr., A Crack-Closure Model for Predicting Fatigue Crack Growth under Aircraft Spectrum Loading, *Methods and Models for Predicting Fatigue Crack Growth under Random Loading*, J.B. Chang and C.M. Hudson, Ed., STP 748, ASTM, 1981, p 53-84

99. D.J. Dougherty, A.U. de Koning, and B.M. Hillberry, Modelling High Crack Growth Rates under Variable Amplitude Loading, *Advances in Fatigue Lifetime Predictive Techniques*, STP 1122, ASTM, 1992, p 214-233

100. G.S. Wang and A.F. Blom, A Strip Model for Fatigue Crack Growth Predictions under General Load Conditions, *Eng. Fract. Mech.*, Vol 40, 1991, p 507-533

101. A.U. de Koning and G. Liefting, Analysis of Crack Opening Behavior by Application of a Discretized Strip Yield Model, *Mechanics of Fatigue Crack Closure*, J.C. Newman, Jr. and W. Elber, Ed., STP 982, ASTM, 1988, p 437-458

102. J. Schijve, Predictions on Fatigue Life and Crack Growth as an Engineering Problem: A State of the Art Survey, *Fatigue 96*, Elsevier, to be published

103. D.S. Dugdale, Yielding of Steel Sheets Containing Slits, *J. Mech. Phys. Solids*, Vol 8, 1960, p 100-104

Fatigue Crack Thresholds

A.J. McEvily, University of Connecticut

THE FATIGUE CRACK THRESHOLD is a function of a number of variables, including the material, the test conditions, the R-ratio, and the environment. ASTM E 647 defines the fatigue crack growth threshold, ΔK_{th}, as that asymptotic value of ΔK at which da/dN approaches zero. For most materials an operational, although arbitrary, definition of ΔK_{th} is given as that ΔK which corresponds to a fatigue crack growth rate of 10^{-10} m/cycle.

Figure 1 (Ref 1) depicts the form of the da/dN versus ΔK plot, where a is the crack length, N is the number of cycles, and ΔK is the range of the stress-intensity factor in a loading cycle. The curve shown is bounded by two limits, the upper limit being the fracture toughness of the material and the lower limit being the threshold.

It is appropriate to review some background information before dealing specifically with the topic of thresholds. For example, although the subject of fatigue has been investigated since the mid-nineteenth century, attention was not focused on the fatigue crack growth aspect of the fatigue process until the 1950s. Several events occurred in this latter period that led to increased interest in concern about fatigue crack growth. One of these was the investigation of the crashes of the Comet jet-aircraft, which raised awareness of the importance of fatigue crack growth. A second development was the emergence of the field of fracture mechanics, which permitted the quantitative analysis of fatigue crack growth. The third major development was the advent of the transmission electron microscope and a bit later

the scanning electron microscope, which permitted detailed analysis of the fractographic features associated with fatigue crack growth.

Some of the early studies that relate to what we now refer to as the threshold were carried out by Frost and Dugdale (Ref 2). They observed that under certain loading conditions, nonpropagating cracks formed at notch roots, and Fig. 2 is an example of the type of plot they developed. The plot shows three regions. In one region, the stress amplitude was sufficient to result in complete fracture. In a second region, where the stress amplitude was smaller than the endurance limit divided by K_T, the theoretical stress concentration factor, no cracks formed. In a third region, nonpropagating cracks formed. These cracks exhibited a decreasing rate of growth with increase in crack length before reaching a growth rate of zero, over many millions of cycles, at a crack length of the order of a millimeter.

Frost (Ref 3) also determined an empirical relationship between crack length and the stress necessary for crack growth. Notched test specimens were cycled to introduce fatigue cracks, then reprofiled to remove the notches. The specimens were then stress relieved to minimize any residual stresses introduced during the precracking. The fatigue-cracked specimens were then subjected to fully reversed cycling, and the subsequent fatigue crack growth behavior was noted. Tests in which a crack did not grow were continued for at least 50×10^6 cycles. The fatigue limit

for cracked specimens is shown in Fig. 3 as a function of crack length for copper plates. Frost observed that the equation $\sigma_a^3 a = C$, where σ_a is the stress amplitude, a is the crack length, and C is a constant, described the fatigue limit for such precracked specimens. At the shortest fatigue crack lengths, the plain fatigue strength is plotted as an upper limit to the stress amplitude, an indication that Frost realized that there was a transition from crack length control of the fatigue limit to material control in the very short crack length region. Further mention of this transition is made below. Fatigue-limit data of this nature were later analyzed by Frost et al. (Ref 4) in terms of the fracture mechanics parameter ΔK, with ΔK taken to be equal to $1.1 \times \sigma_a (\pi a)^{1/2}$. Table 1 gives both the constant C as well as the corresponding ΔK_{th} values. The ΔK_{th} values decrease with decrease in crack length, and this may be associated with crack closure, as discussed below.

Frost and Greenan (Ref 5) also determined the critical stress for growth of a fatigue crack as a function of R, where R is the ratio of the minimum to maximum stress in a loading cycle. Table 2 gives their results. As R increases, the value of ΔK_{th} decreases.

Paris et al. (Ref 6) were the first to determine the threshold value, ΔK_{th}, at crack growth rates of the order of 2.5×10^{-11} m/cycle. Paris was concerned about the situation where existing material flaws are small and lightly stressed but are subjected to a large number of cycles over a lifetime.

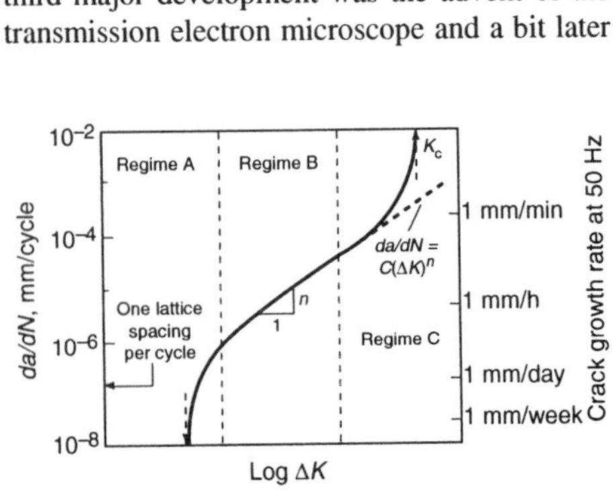

Fig. 1 Schematic illustration of the different regimes of stable fatigue crack propagation. Source: Ref 1

Table 1 Fatigue crack thresholds compared with the constant $C = (\sigma_a^3) \cdot a$

Material	Stress relieved 1 h at:	Tensile strength, MPa	Plain fatigue strength, 50×10^6 cycles, MPa	C (MPa)$^3 \cdot$ m	ΔK_{th} (crack length 0.5-5 mm), MPa√m	ΔK_{th} (crack length 0.025-0.25 mm), MPa√m
Inconel	600 °C in vacuum	655	±220	750	6.4	...
Nickel	500 °C in vacuum	455	±140	700	5.9	...
18/8 austenitic steel	600 °C in vacuum	685	±360	540	6.0	...
Low-alloy steel	570 °C in vacuum	835	±460	510	6.3	...
Mild steel	650 °C in vacuum	530	±200	510	6.4	4.2
Nickel-chromium alloy steel	570 °C in vacuum	925	±500	510	6.4	3.3
Monel	500 °C in vacuum	525	±240	360	5.6	...
Phosphor bronze	500 °C in air	325	±130	160	3.7	...
60/40 brass	550 °C in air	330	±105	94	3.1	...
Copper	600 °C in vacuum	225	±62	56	2.7	1.6
4.5Cu-Al	...	450	±140	19	2.1	1.2
Aluminum	320 °C in vacuum	77	±27	4	1.02	...

Table 2 Critical stress required to cause a crack to grow

Material, Stress-relieved 1 h at:	Tensile strength, MPa	Stress ratio, R	C, MPa · m	ΔK_{th} (crack length 0.5-5 mm), MPa√m
Mild steel 650 °C in vacuum	430	0.13	56	6.6
		0.35	37	5.2
		0.49	17	4.3
		0.64	14	3.2
		0.75	15	3.8
18/8 austenitic steel 4 h at 500 °C in vacuum	665	0	65	6.0
		0.33	37	5.9
		0.62	22	4.6
		0.74	18	4.1
Aluminum 320 °C in vacuum	77	0	0.93	1.7
		0.33	0.75	1.4
		0.53	0.56	1.2
4.5Cu-Al (BS L65)	495	0	2.8	2.1
		0.33	1.9	1.7
		0.5	1.2	1.5
		0.67	0.6	1.2
Copper 600 °C in vacuum	215	0	4.7	2.5
		0.33	1.9	1.8
		0.56	1.4	1.5
		0.69	1.1	1.4
		0.80	1.1	1.3
Commercially pure titanium 700 °C in vacuum	540	0.60	3.3	2.2
Nickel 2 h at 850 °C in vacuum	430	0	93	7.9
		0.33	65	6.5
		0.57	28	5.2
		0.71	14	3.6
Low-alloy steel 650 °C in vacuum	680	0	79	6.6
		0.33	32	5.1
		0.50	19	4.4
		0.64	9	3.3
Monel 2 h at 800 °C in vacuum	525	0	61	7.0
		0.33	39	6.5
		0.50	24	5.2
		0.67	12	3.6
Maraging steel(a)	2000	0.67	4.7	2.7
Phosphor bronze 550 °C in air	370	0.33	14	4.1
		0.50	11	3.2
		0.74	4	2.4
60/40 brass 550 °C in air	325	0	14	3.5
		0.33	11	3.1
		0.51	4	2.6
		0.72	3	2.6
Inconel 2 h at 800 °C in vacuum	650	0	93	7.1
		0.57	28	4.7
		0.71	14	4.0

(a) Heat treated after fatigue cracking: 1 h at 820 °C, air cooled, 3 h at 480 °C

It was therefore of interest to examine slow rates of growth of fatigue cracks as well as the secondary variables that may affect these rates, such as mean stress, environment, and temperature. In these studies, compact tension specimens and a load shedding technique were used to approach threshold. Figure 4 shows the results for an A533 B-1 steel tested at $R = 0.1$.

Interest in fatigue testing at low fatigue crack growth rates grew during the 1970s to the extent that an international symposium on fatigue thresholds was held in Stockholm in 1981. This was followed in 1983 by a second symposium on fatigue crack growth threshold concepts. Since that time, research has increased considerably.

Two other developments have played an important role in explaining fatigue crack behavior at the threshold level. One of these is crack closure, a phenomenon discovered by Elber (Ref 7). The other development is modelling of short crack behavior.

Crack closure can occur during the unloading portion of a fatigue cycle and is defined as the contacting of the opposing surfaces of a crack before the minimum of the loading cycle is reached. The crack opening load is that particular load level during the loading portion of a cycle at which the crack surfaces become fully separated. Usually the crack opening process is viewed in a continuum sense as an unzippering process, in which contact between the opposing crack surfaces is first lost at some distance behind the crack tip, then progressively closer to the crack tip until all contact is lost at the opening level. Only that portion of a loading cycle above the opening level is considered to be effective in propagating the fatigue crack (i.e., $\Delta K_{eff} = K_{max} - K_{op}$). The rate of fatigue crack growth then becomes a function of ΔK_{eff}, provided that the nature of the cracking process at the crack tip is similar (i.e., we are not comparing branched cracks with nonbranched cracks). ASTM STP

982 provides a comprehensive review of this subject.

To illustrate the importance of crack closure on the propagation or nonpropagation of fatigue cracks, consider the results obtained by Pippan with Armco iron (Ref 8). He prepared specimens that contained slightly blunted, closure-free long cracks. These specimens were then cyclically loaded at various R values that included compression-compression cycling. Pippan observed that the initial fatigue crack growth rate was independent of the R value and was a function only of ΔK, which initially was identical to ΔK_{eff} because of the closure-free starting condition. As roughness-induced closure developed, the rate of crack growth decreased with decrease in ΔK_{eff} and became sensitive to the R value, with crack arrest occurring under compression-compression loading. The conditions of loading, as well as K_{op}, are shown in Fig. 5. For each test the corresponding R and ΔK values are indicated in parentheses. Cracks became nonpropagating if the K_{max} value did not exceed ΔK_{th}. Further, the fact that the initial rate was independent of R implies that the cracks were initially closing only at K_{min}, even for compression-compression loading. This means that even if K_{min} is in compression, its value is initially significant in defining ΔK_{eff}. The observed behavior can be analyzed with the aid of the following constitutive relation:

$$\frac{da}{dN} = A(\Delta K_{eff} - \Delta K_{eff(th)})^2 \qquad \text{(Eq 1)}$$

To allow for the development of crack closure with crack advance, Eq 1 becomes:

$$\frac{da}{dN} = A[(K_{max} - K_{min})$$
$$- \left(1 - e^{-k\lambda}\right)(K_{opmax} - K_{min}) - \Delta K_{eff(th)}]^2 \quad \text{(Eq 2)}$$

For Armco iron, the value of A is 1.8×10^{-10} $(MPa)^{-2}$, λ is the length of the crack that develops from the blunted crack, and k is a parameter that relates to the rate of crack closure development in the wake of the new crack. For Armco iron it has a value of 2 mm $^{-1}$ with λ measured in millimeters. Figure 6 compares the experimental results with the predicted results based upon Eq 2, and the agreement is quite good, illustrating the importance of crack closure in interpreting some fatigue crack growth phenomena. Other constitutive relations involving the threshold level have been proposed. For example, Ohta et al. (Ref 9) have used the nonlinear equation $da/dN = C[(\Delta K)^m - (\Delta K_{th})^m]$ to fit da/dN versus ΔK data by a regression method to evaluate the 99% confidence intervals. Experimental results on fatigue crack propagation properties of welded joints in several low-alloy steels (SM50B, HT80, SB42, and SPV50) were compared by using these confidence intervals.

The types of closure mechanisms include plasticity-induced, roughness-induced, oxide-induced, and fretting-debris-induced. The type of closure mechanism can vary with test condition and material. For example, if an overload is ap-

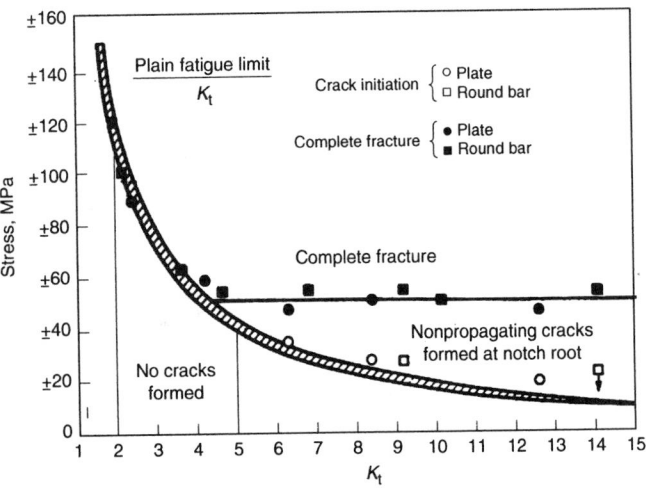

Fig. 2 Frost and Dugdale plot of nominal alternating stress versus K_t for reversed direct stress mild steel specimens having notches 5 mm deep. Source: Ref 2

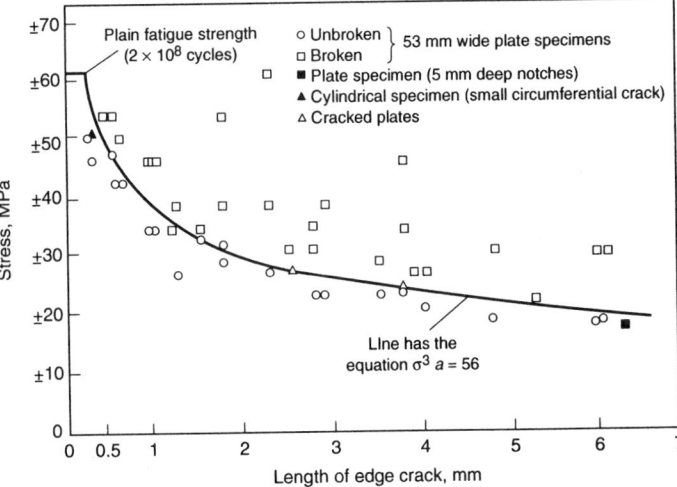

Fig. 3 Fatigue limit of cracked copper plate. Source: Ref 3

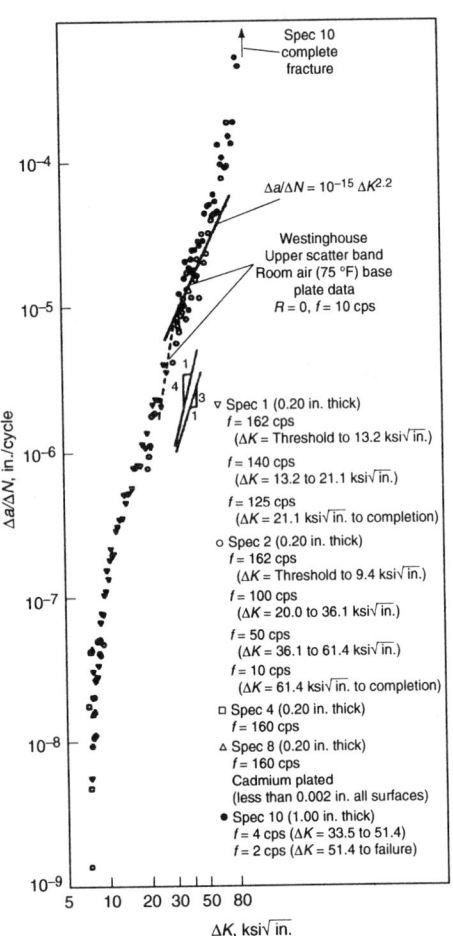

Fig. 4 Paris data for fatigue crack propagation of ASTM A533 B-1 steel, $R = 0.10$, ambient room air, 75 °F. Source: Ref 6

plied near threshold, plasticity-induced closure in the plane-stress, surface regions of a specimen may be important (Ref 10). More often, roughness-induced closure accompanied by differing degrees of wear in various materials is the impor-

tant type of closure at threshold. Increasing fracture surface roughness tends to correlate with lower fatigue crack growth rates, and this has been related to variations in the extent of crack closure (Ref 11).

There is also an influence of the R level on the extent of roughness. For example, at $R = -1$, lower thresholds are found than at small positive R values. This is due to the development of smoother crack surfaces due to the compressive loads and consequently less roughness-induced crack closure (Ref 12). It has been noted by Ohta et al. (Ref 13) that the fracture surface appearance can differ significantly at a given growth rate as a function of the R value. Blom (Ref 12) found, in studies of near-threshold fatigue crack growth and crack closure in 17-4 PH steel and 2024-T3 aluminum alloy, evidence for oxide-induced closure at room temperature. It was definitely a contributing factor to the closure level of steels at elevated temperature (Ref 14). Kobayashi et al. (Ref 15) found that crack closure resulting from fretting oxide debris is of particular importance to the near-threshold characteristics of A508-3 steel.

The fracture surface appearance near threshold differs from that in the mid-range of crack growth rates where mode I growth dominates and where in ductile materials the fatigue striations typical of a fatigue crack can be found. In the near-threshold range, mode II growth is often found to dominate (Ref 16), and Fig. 7 (Ref 17) emphasizes this point.

An analysis of fatigue crack closure caused by asperities using a modified Dugdale model developed by Newman (Ref 18) has been presented by Nakamura and Kobayashi (Ref 19). This analysis involved the rigidity of the asperities, the asperity length, the asperity thickness, and the distance from the crack tip. However, despite the overwhelming amount of data relating to crack closure, a unanimous view as to its significance has not as yet been reached (see, e.g., Ref 20).

Short-Fatigue-Crack Behavior. In 1973, Pearson (Fig. 8) drew attention to the fact that

short fatigue cracks could grow at stress intensity levels below the threshold level for macroscopic cracks, a process referred to as anomalous fatigue crack growth behavior. Such behavior is now better understood. For example, it has been shown with respect to Fig. 2 that cracks that are initially closure free can propagate below ΔK_{th}. Such behavior clearly demonstrates that use of the macroscopic threshold level as a design criterion to guard against the growth of fatigue cracks is not applicable in the realm of short fatigue cracks. In fact, as a crack under consideration is made smaller and smaller, there is a transition from linear elastic fracture mechanics (LEFM) treatment of long cracks at threshold to endurance-limit-dominated behavior of short cracks, as shown by Kitagawa and Takahashi for a steel of 725 MPa yield strength tested under $R = 0$ conditions. When the surface crack length, $2a$, is larger than 0.5 mm, then a simple conventional fracture mechanics law can be applied to calculate the threshold condition, (i.e., ΔK_{th} is constant). However, below a surface crack length of 0.5 mm, the threshold stress range departs gradually from its macroscopic value, and as $2a$ decreases, $\Delta\sigma_{th}$ asymptotically approaches a constant stress range level that is approximately equal to the fatigue limit of unnotched smooth specimens of this material. The type of diagram depicting this situation, shown in Fig. 9, is referred to as a Kitagawa diagram.

Morris and James (Ref 21), in an investigation of the growth threshold for short cracks, observed that the stochastic growth rate variations found experimentally were attributable to crack closure and to the reduced stress intensity that accompanied irregularities in the crack path. More information on this topic is in the next article "Behavior of Small Fatigue Cracks" in this Volume.

A comprehensive review of threshold data has been provided by Taylor (Ref 22). Liaw (Ref 23) reviewed the effects of microstructure, environment, loading condition, and crack size on near-threshold fatigue crack growth rate and concluded that the crack closure concept led to the

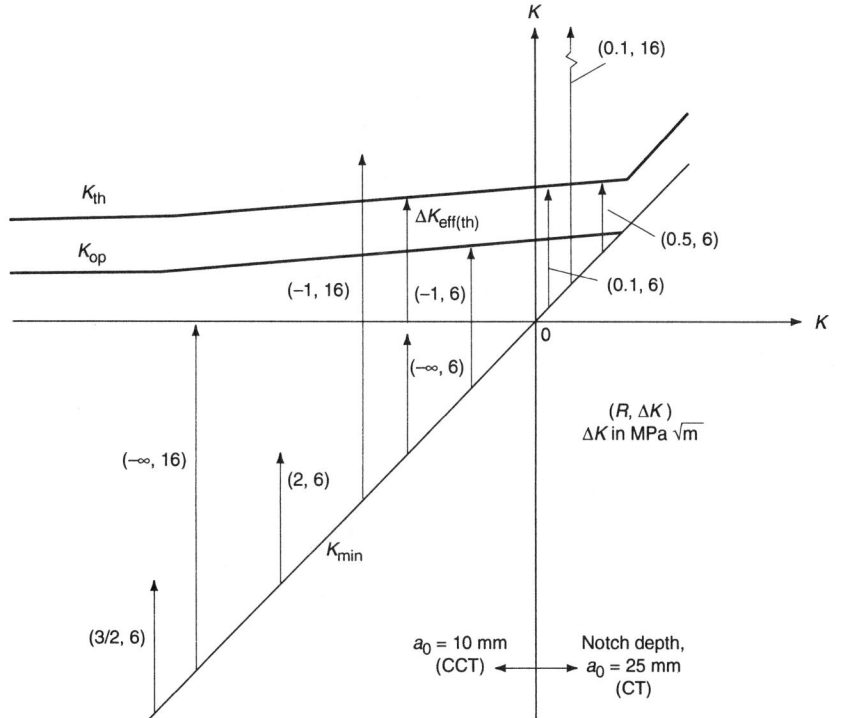

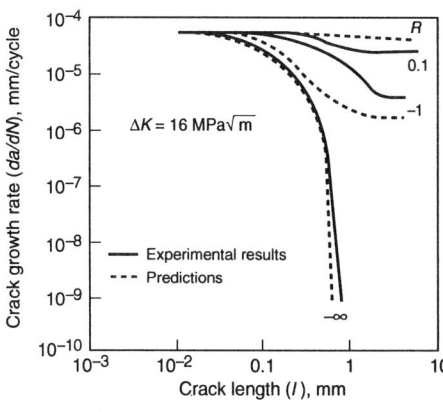

(a)

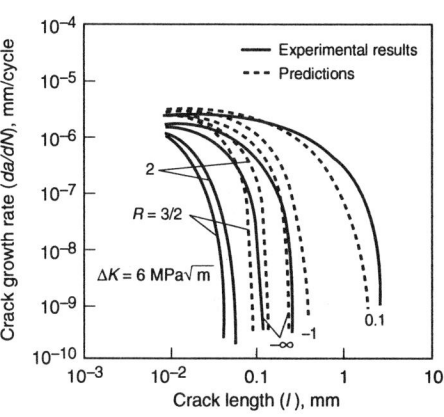

(b)

Fig. 5 Pippan loading conditions with the variation of K_{th} and K_{op} as a function of K_{min} in Armco iron. The vertical lines indicate the range of ΔK used by Pippan (Ref 8) at each R value.

Fig. 6 Pippan data for fatigue crack growth rate as a function of crack length and R in Armco iron. Solid lines represent the experimental findings of Pippan (Ref 8). Dashed lines are predicted values (A. McEvily and Z. Yang, *Met. Trans.*, Vol 22A, 1991, p 1079). (a) $\Delta K = 16$ MPa$\sqrt{m}$. (b) $\Delta K = 6$ MPa$\sqrt{m}$.

correlation of much fatigue crack growth data. Beevers et al. (Ref 24) have discussed crack closure in relation to ΔK_{th}, and Beevers and Carlson (Ref 25) have considered the significant factors controlling fatigue thresholds. In many instances the fatigue cracks are initiated at small defects that originate from microstructural or fabrication flaws. The development of these small defects involves, in many instances, stress intensities near the threshold regime and fatigue crack growth rates in the range of 10^{-8} to 10^{-13} m/cycle. Because most of the "lifetime" is spent in this low growth rate regime, variables such as microstructure, stress state, and environment have an appreciable influence on ΔK_{th}.

Test Techniques

In order to establish a valid threshold value experimentally, it is necessary to reduce gradually the applied stress-intensity factor range. ASTM 647 recommends that the rate of load shedding with increasing crack length should be gradual enough to: (a) preclude anomalous data resulting from reductions in the stress-intensity factor and concomitant transient growth rates; and (b) allow the establishment of about five da/dN, ΔK data points of approximately equal spacing per decade of crack growth rate. These requirements can be met by limiting the normalized K-gradient, $C = (1/K)(dK/da)$, to a value equal to or greater than -0.08 mm^{-1}. The ASTM procedure further recommends that the load ratio, R, and C be maintained constant during K-decreasing testing. A procedure for standardizing

crack closure levels has been proposed by Donald (Ref 26).

The following procedure, given in ASTM 647, provides an operational definition of the threshold stress-intensity factor range for fatigue crack growth, ΔK_{th}, that is consistent with the general definition. Determine the best-fit straight line from a linear regression plot of log da/dN versus

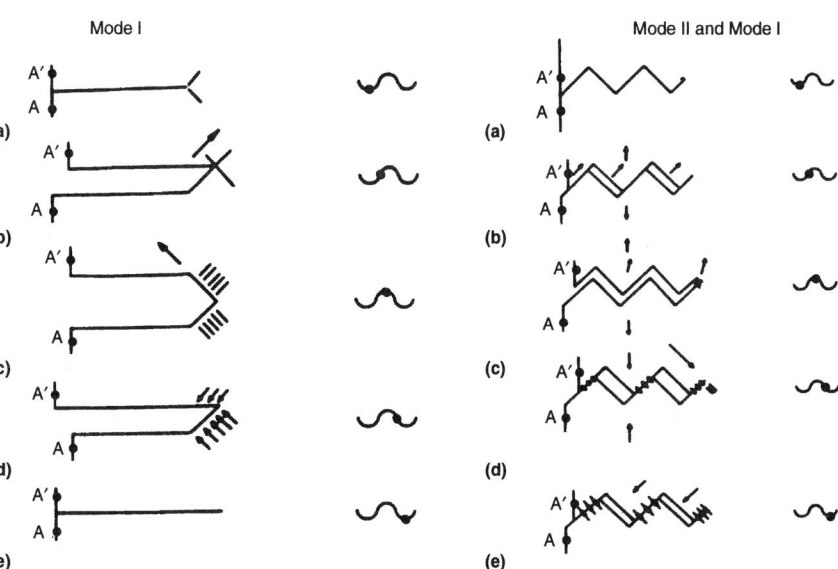

Fig. 7 Schematic illustration of mode I and II fatigue crack growth processes

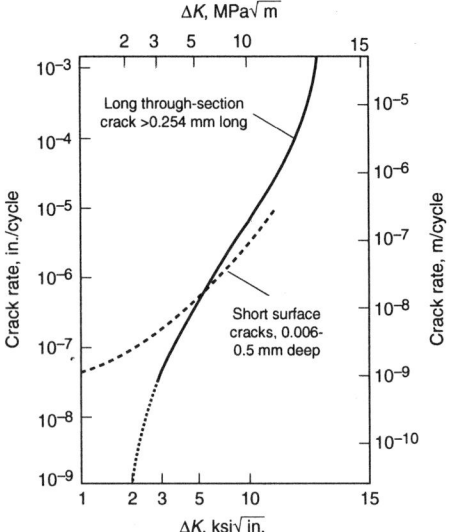

Fig. 8 Pearson plot of crack growth rate as a function of K for short surface cracks and through-cracks. Source: *Engr. Fracture Mech.*, Vol 7, 1975, p 235-247

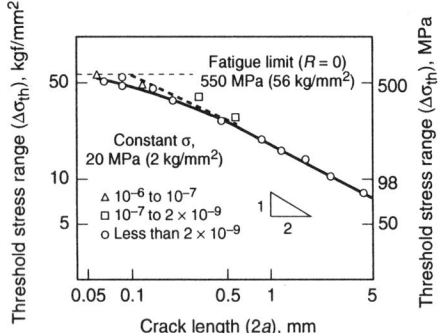

Fig. 9 Kitagawa plot of the effect of crack length on the threshold stress range for fatigue crack growth. Source: *2nd Intl. on Mech. Behavior of Materials*, ASM, 1976, p 627-631

log ΔK using a minimum of five da/dN, ΔK data points of approximately equal spacing between 10^{-9} and 10^{-10} m/cycle. Calculate the ΔK value that corresponds to a growth rate of 10^{-10} m/cycle using the fitted line. This value of ΔK is defined as ΔK_{th} according to the operational definition of this test method. The requirements for obtaining economic fatigue crack growth data in the threshold regime in inert, gaseous, elevated-temperature, and aqueous environments can be facilitated by the development of remote crack growth monitoring techniques (Ref 27).

Some investigators have checked on the effect of the rate of load shedding on the threshold level. In one case it was reported that for type 316 stainless steel in air at 24 °C, load shedding rates greater than the maximum rate recommended in the ASTM test procedure were found to have no substantial effect on the threshold behavior. At very low ΔK levels, crack growth rates were apparently dependent on environmental effects and the degree of plastic constraint (Ref 28). However, it has also been noted that a rapid load

reduction can increase the threshold by 10 to 100% in the aluminum alloy 2024, and that after a rapid decrease in vacuum the crack may begin to propagate again after 5×10^7 cycles (Ref 29).

A number of investigators have developed other modes of load shedding, in the attempt to avoid crack closure during the load shedding process, by testing at effectively high R values where closure is not a factor, or by maintaining K_{max} constant and gradually decreasing ΔK so that the R value continually shifts to a higher value as load is shed. The threshold determined by such procedures, being free of closure, is designated ΔK_{eff}, a conservative lower bound to long crack threshold values. One such method, designated "P_{max} constant, ΔK decreasing" was introduced to avoid the closure effect in fatigue crack growth testing near the threshold region. It was useful in investigating the effect of the environment, because it allowed a direct evaluation of da/dN versus ΔK_{eff} relations (Ref 30). The K_{max} constant, decreasing ΔK method has been used to determine the threshold level in liquid helium (Ref 31). ΔK-decreasing threshold fatigue crack propagation data under conditions of constant maximum stress intensity (K_{max}) has been generated by Herman et al. (Ref 32), by Ohta et al. (Ref 33), and by Matsuoka et al. (Ref 34). The influence of test variables on ΔK_{th} has been discussed by Priddle (Ref 35), and Ref 36 provides a comparison of test methods for the determination of ΔK_{th} in titanium at elevated temperature.

In addition, crack initiation and growth under cyclic compression has been demonstrated to be a useful method for quickly obtaining estimates of fatigue crack growth thresholds while minimizing some of the uncertainties inherent in the conventional (load shedding) procedures (Ref 37).

Even under closure-free conditions, some uncertainties remain. For example, a hysteresis effect on the value of ΔK_{th} has been observed for near-threshold fatigue crack propagation behavior of a high-strength steel investigated in laboratory air under closure-free conditions ($R = 0.7$). Also, the ΔK curves obtained on the same specimen during the ΔK-decreasing and the ΔK-increasing tests may not be identical in the threshold regime (Ref 38).

Other test procedures relating to threshold behavior have also been used. For example, under narrow-band random loading, the threshold for fatigue crack growth may be lower than that observed under sinusoidal loading. It has been suggested that small, regular overloads under random loading help to keep the crack faces apart, and thereby prevent closure and assist in crack growth (Ref 39). On the other hand, in tests carried out in salt water environments, multiple overloads produced a much larger increase in ΔK_{th} than a single overload (Ref 40). The effect of underload cycles on the reduction of the threshold level of a structural steel has also been investigated (Ref 41). Debris in salt water solutions has been shown to significantly affect the near-threshold growth through its influence on crack closure and the transport processes occur-

ring at the crack tip (Ref 42). Also, under fretting conditions the threshold for S45C steel fell to the critical threshold value of the order of 1 MPa$\sqrt{m}$ (Ref 43).

Specialized techniques relating to the threshold studies have also been employed. For example, by using thermometrical techniques, the crack opening load and the stress distribution during crack growth near the threshold value can be determined (Ref 44). Ultrasonics has been used to investigate the degree of contact of asperities during crack closure, and the existence of a threshold has been related to crack closure (Ref 45).

Ultrasonic fatigue testing involves cyclic stressing of material at frequencies typically in the range of 15 to 25 kHz. The major advantage of using ultrasonic fatigue is its ability to provide near-threshold data within a reasonable length of time. High-frequency testing also provides rapid evaluation of the high-cycle fatigue limit of engineering materials as described in the article "Ultrasonic Fatigue Testing" in Volume 8, *Metals Handbook*, 9th edition.

Some of the values of ΔK_{th} and the corresponding minimum crack growth rate are presented in Table 3 for several pure metal and alloy systems, from threshold testing at ultrasonic frequencies. The minimum crack growth rates obtained at ultrasonic frequency are decades below the value of one lattice parameter per cycle. At these low crack growth rates and ΔK values, the crack tip plastic size is extremely small.

Acoustic emission also has potential for determining threshold levels (Ref 46). Testing at 20 KHz has been employed to reduce total test time, and crack growth rates between $10^{-12} < da/dN < 10^{-9}$ m/cycle have been measured (Ref 47).

Aluminum Alloy Crack Growth Thresholds

Figure 10(a) shows the ΔK_{th} values for three aluminum alloys as a function of the load ratio, R, and environment: laboratory air (50% relative humidity) versus vacuum (10^{-5} torr). Figure 10(b) shows the same data after correction for crack closure (i.e., $\Delta K_{eff(th)}$). Figure 10 shows that the dependence of the threshold value on R is about the same in air as in vacuum, and that $\Delta K_{eff(th)}$ is within experimental error independent of R. Threshold data obtained in air in other investigations generally show a similar R dependency, whereas the threshold data obtained in vacuum may not. For example, Lafarie-Frenot and Gasc (Ref 48) found little effect of R in vacuum on ΔK_{th} or $\Delta K_{eff(th)}$, which means that the extent of closure above K_{min} had to be the same at $R = 0.5$ as at $R = 0.1$, which is unexpected. Beevers (Ref 49) has found the threshold level determined in vacuum to be independent of R for high-strength aluminum alloys and En24 steel; however, no closure data were obtained. It is usually found that alloys exhibit a higher $\Delta K_{eff(th)}$ in vacuum as compared to air, but in at least one case the reverse has been reported. Jono (Ref 50)

Table 3 Threshold stress intensity, ΔK_{th}, determined by ultrasonic resonance test methods

Material	Loading mode	R	Test conditions Environment	Temperature	Young's Modulus MPa	psi × 10⁶	ΔK_{th} at da/dN MPa √m	ksi √in.	m/cycle	ft/cycle	Remarks
Al	Axial	−1	Air	20 °C (68 °F)	69,700-72,000	10.1-10.4	1.33	1.21	1×10^{-13}	3.28×10^{-13}	Grain size and work effects
Cu	Axial	−1	Air	20 °C (68 °F)	122,000-126,000	17.7-18.2	1.4-2.0	1.3-1.8	1×10^{-13}	3.28×10^{-13}	Grain size effect
					126,000-126,500	17.6-18.3	1.4-2.6	1.3-2.3	1×10^{-15}	3.28×10^{-13}	Cold work effect
					138,000-184,000	20.0-26.7	1.6-2.3	1.6-2.1	1×10^{-13}	3.28×10^{-13}	Single crystals
Low-carbon steel	Axial	−1	Oil	293 K	...		3.8	3.4	6×10^{-14}	1.97×10^{-13}	Comparison with NaCl solutions(a)
AISI 304	Axial	−1	Oil	293 K	...		7.0	6.4	5×10^{-13}	1.64×10^{-12}	Comparison with NaCl solutions
X10Cr13	Axial	−1	Oil	23 °C (73 °F)	219,600	31.8	6.7	6.1	6×10^{-14}	1.97×10^{-13}	
34CrMo4	Axial	−1	Air	20 °C (68 °F)	196,200	28.4	2.45	2.2	1×10^{-12}	3.28×10^{-12}	
P/M Mo	Axial	−1	Air	20 °C (68 °F)	322,000	46.7	4.8-5.2	4.4-4.7	1×10^{-13}	3.28×10^{-13}	Grain size effect
A 286	Axial	−1	Air	20 °C (68 °F)	...		13.0	11.8	3×10^{-12}	0.98×10^{-11}	
IN-738	Axial	−1	Air	20 °C (68 °F)	200,000	29.0	3.13	2.84	1×10^{-12}	3.28×10^{-12}	
IN-792	Axial	−1	Air	20 °C (68 °F)	206,600	30.0	4.48	4.07	1×10^{-12}	3.28×10^{-12}	
IN-600	Transverse	0.3	Air	20 °C (68 °F)	...		approx 5	4.5	1×10^{-11}	3.28×10^{-11}	

(a) Including comparison with low-frequency test data. Source: R. Stickler and B. Weiss, in *Ultrasonic Fatigue*, TMS, 1982, p 135

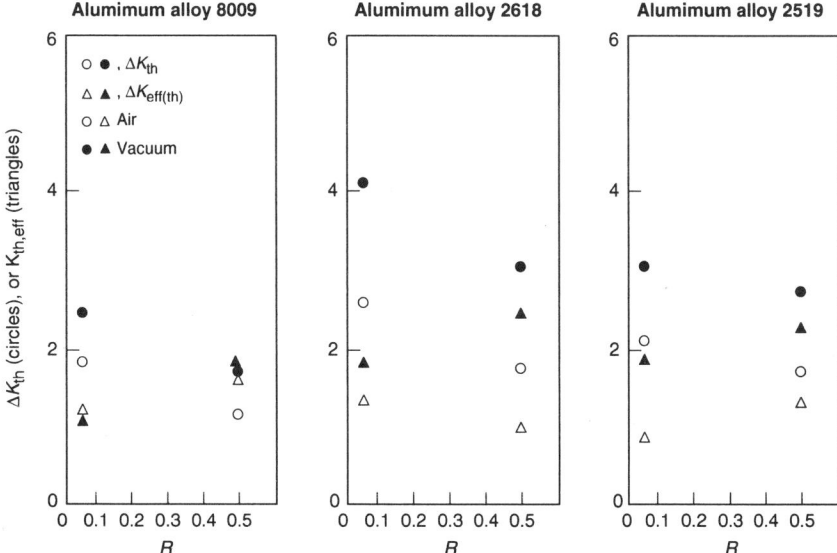

Fig. 10 ΔK_{th} and $\Delta K_{eff(th)}$ data on three aluminum alloys. Source: M. Renauld, University of Connecticut

observed that fatigue cracks in aluminum alloys grew in vacuum below the $\Delta K_{eff(th)}$ for crack growth in air. He attributed this circumstance to the ease of deformation of aluminum in vacuum. A review of the near-threshold fatigue crack growth behavior of 7xxx and 2xxx alloys has been given by Vasudevan and Bretz (Ref 51).

Bretz et al. (Ref 52) determined the effects of grain size and stress ratio on fatigue crack growth in wrought P/M 7091 aluminum alloy and found that increasing the grain size by thermomechanical processing can significantly increase near-threshold fatigue crack growth resistance. A factor of 2 increase in ΔK_{th} with grain size was measured at both R ratios. Bretz et al. concluded that crack closure alone was not responsible for grain size effects on fatigue behavior at a particular R ratio, but that crack tip deviation was an important mechanism by which near-threshold fatigue crack growth rates were reduced as grain size increased.

The influences of load ratio, R (at values of −2, −1, 0, 0.33, 0.5, and 0.7), and crack closure on fatigue crack growth thresholds in 2024-T3 aluminum alloy have been investigated by Phillips (Ref 53). He found that values of ΔK_{th} varied significantly with R, whereas values of $\Delta K_{eff(th)}$ did not. The influence of load ratio on fatigue crack growth in 7090-T6 and IN9021-T4 P/M aluminum alloys was examined by Minakawa et al. (Ref 54). A principal difference between these two alloys was grain size, which was <1 μm for the alloy IN9021 and >5 μm for the alloy 7090. Crack closure was observed in the 7090 alloy, and there was also a dependency of the threshold level on R (Fig. 11). No closure was found in the fine-grained IN 9021 alloy, nor was there any dependency of ΔK_{th} on R (Fig. 12). Such results support a $\Delta K_{eff(th)}$ interpretation of the effect of R on the threshold level.

Park and Fine (Ref 51) studied the near-threshold fracture characteristics of an Al-3%Mg alloy

(σ_y = 52 MPa) and found that the shear-mode areal fraction of the fracture surface increased in dry argon as the threshold was approached, with the increase greater at R = 0.05 than at R = 0.5. The mechanism for crack closure at low load ratio appeared to be surface roughness coupled with shear mode displacements. (In Al-Zn-Mg single crystals, crystallographic cracks propagated in vacuum near threshold, even as K_{eff} approached zero and mode II was dominant (Ref 55). Park and Fine also observed that the roughness of the fracture surface decreased as the threshold was approached, and that the roughness also decreased with decrease in R. Such results suggest that crack-surface wear was responsible for the reduction in roughness. This wear, due to rubbing of the mating fracture surfaces, increased with closure level near the threshold, where an increasingly large number of cycles is required to advance the crack a given increment.

Wanhill (Ref 56) compared the low-stress-intensity fatigue crack growth of 2024 aluminum alloy in the naturally aged T3 and T351 conditions. Particular attention was paid to crack growth curve transitions in the near-threshold regime. These transitions corresponded to monotonic or cyclic plane-strain plastic zone dimensions becoming equal to characteristic microstructural dimensions, and changes in fracture surface topography were also associated with the transitions.

Harrison and Martin (Ref 57) studied the effect of dispersoids on near-threshold fatigue crack propagation in an Al-Zn-Mg alloy and found that manganese-bearing dispersoids lowered ΔK_{th}. They proposed that manganese-bearing dispersoids homogenized the dislocation distribution, which reduced the tendency for slip reversibility. Zinc-bearing dispersoids did not homogenize slip, and their effect was to raise ΔK_{th} without changing the predominantly intergranular fracture mode. It was suggested that the action of zinc-bearing dispersoids was to reduce hydrogen embrittlement. Near-threshold fatigue crack propagation and crack closure in Al-Mg-Si alloys with varying manganese concentrations have also

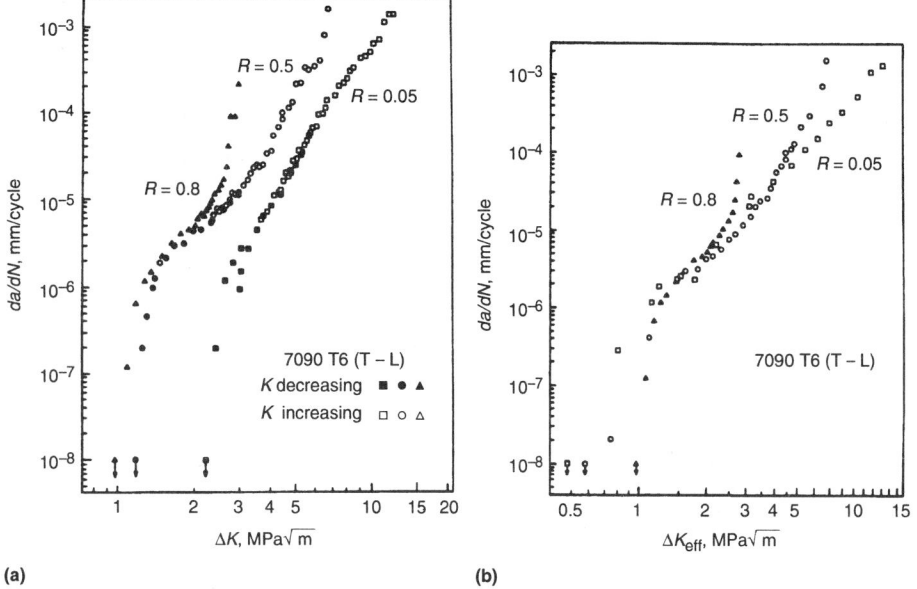

(a)   (b)

Fig. 11 Fatigue crack growth rate at various R ratios. (a) As a function of ΔK for the P/M aluminum alloy 7090-T6. Grain size 1-20 μm. (b) Plotted in terms of ΔK_{eff}. Source: Ref 54

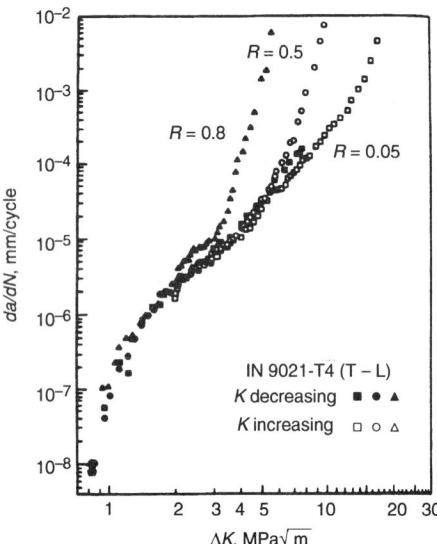

Fig. 12 Fatigue crack growth rate at various R ratios as a function of ΔK for the P/M aluminum alloy IN9021-T4. Grain size 0.1-1.0 μm. Source: Ref 54

been investigated by Scheffel and Detert (Ref 58).

Manganese dispersoids were also found to have a deleterious effect on near-threshold fatigue crack growth in 2134 type alloys by Jata et al. (Ref 59). They tested the alloy in the under- and overaged conditions as a function of manganese additions ranging from 0 to 1.02 wt%. The additions of manganese resulted in a continuous decrease of the nominal threshold in both conditions, with the effect more pronounced in the overaged condition. Crack deflections, closure, and fractography suggested that roughness-induced crack closure was dominant in all alloys. The fractographic evidence also suggested that

large manganese particles contributed to local microcrack acceleration, resulting in an intrinsic lowering of the fatigue thresholds and faster crack propagation rates as compared to a similarly overaged 2124 alloy.

Venkateswara et al. (Ref 60) have observed that overaging the Al-Li alloy 2090 led to a decrease in strength and toughness, principally through the formation of platelike copper-rich grain boundary precipitates and associated copper-depleted and δ' precipitate-free zones. This overaging also was found to result in increased fatigue crack growth rates, except near-threshold rates. Such behavior was related to a diminished role of crack-tip shielding during crack extension in overaged microstructures, resulting from less crack deflection and lower roughness-induced crack closure levels because of the more linear crack paths.

Yoder et al. (Ref 61), in a study of the Al-Li alloy 2090 at $R = 0.1$ in ambient air, observed that the fracture surface exhibited an extraordinary tortuosity, with considerable oxide debris attributable to fretting giving rise to a macroscopically blackish appearance. Associated with this tortuosity, the fracture surface exhibited asperities of unusual height, comprised of adjacent pairs of slip-band facets. This height was a consequence of an extraordinary textural intensity and an uncommon propensity for a planar slip mode in Al-Li alloys. Thus, individual, well-defined slip-band facets were formed that could traverse multiple grains at a time to give asperities of unusual height, which gave rise to high closure levels at stress-intensity ranges much above near-threshold values. Moreover, it was shown that the characteristic included angle between an adjacent pair of slip-band facets that comprise an individual asperity was a consequence of the texture.

Welch and Picard (Ref 62) observed an effect of texture on fatigue crack propagation in alumi-

num alloy 7075. Their results showed a small but distinct variation in crack propagation rates for the three orientations, with the effect somewhat more marked near the fatigue threshold.

In the Al-Li alloy Lital-A, Anandan et al. (Ref 63) observed a rough sawtooth-type fracture in both plate and sheet material. The material exhibited a higher ΔK_{th} and lower fracture toughness than 2024, and the stress ratio effect was pronounced over the entire growth rate range.

Fatigue crack propagation behavior has been examined in a commercial 12.7 mm thick plate of Al-Cu-Li-Zr alloy 2090 by Yu and Ritchie (Ref 64) with specific emphasis on the effect of single compression overload cycles. Based on low-R-value experiments on cracks arrested at ΔK_{th}, it was found that crack growth at ΔK_{th} could be promoted through the application of periodic compression cycles of magnitude two times the peak tensile load. Similar to 2124 and 7150 aluminum alloys, such compression-induced crack growth at the threshold decelerated progressively until the crack rearrested, consistent with the reduction and subsequent regeneration of crack closure. The compressive loads required to cause such behavior, however, are far smaller in the 2090 alloy. Such diminished resistance of Al-Li alloys to compression cycles was discussed in terms of their enhanced "extrinsic" crack growth resistance from crack path deflection and resultant crack closure, and the reduction in the closure from the compaction of fracture surface asperities by moderate compressive stresses.

Venkateswara et al. (Ref 65) found that artificial aging of commercial Al-Li alloys to peak strength had a mixed influence on the long crack resistance. Although behavior at higher growth rates was relatively unaffected, in 2091, the nominal ΔK_{th} values were increased by 17%, whereas in 8090 and 8091 they were decreased by 16 to 17%. Aging to peak strength also resulted in a decrease in $\Delta K_{eff(th)}$. For three Al-Cu-Li-Mg-Ag alloys (Weldalite 049, X2095, and MD 345) (Ref 66), the threshold level increased with increasing strength.

Tintillier et al. (Ref 67) showed that for the 8090 alloy, alloying with lithium produced a significant improvement of the near-threshold crack growth resistance as compared to that of 2024-T351 and 7075-T651. This improvement was considered to be a consequence of the planar slip mechanism observed in the δ' hardened matrix, which resulted in substantial roughness-induced closure effects. The influences of load ratio and texture were shown to be mainly related to crack closure and the propagation behavior was rationalized in terms of the effective stress-intensity factor range ΔK_{eff} for given environmental and aging conditions. The influence of environment was discussed in terms of water vapor embrittlement.

Good crack growth resistance in Al-Li, Al-Li-Zr, and 8090 alloys has been observed by Xiao and Bompard (Ref 68), with ΔK_{th} higher than 9 MPa√m. High closure levels were due to crack propagation into persistent slip bands or grain boundaries. Expressed in terms of ΔK_{eff}, the re-

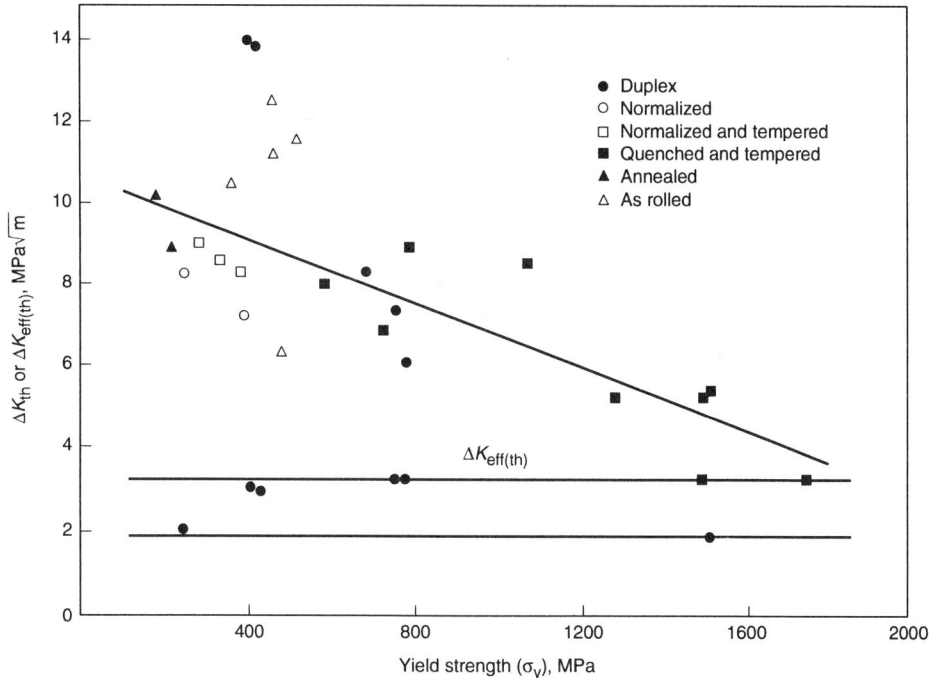

Fig. 13 Threshold as a function of yield strength ΔK_{th} at $R = 0\text{-}0.05$ vs. yield strength for various types of steels. Solid symbols indicate ΔK_{eff} at threshold. The line is obtained by the least square method ($\Delta K_{th} = -4.0 \times 10^{-3} \sigma_y + 10.7$). Source: K. Minakawa and A. McEvily, in *Fatigue Thresholds*, Vol 1, EMAS, 1982, p 373

level of closure was independent of ΔK. The presence of oxygen in the ambient environment (50% relative humidity) increased the overall closure level by as much as 50% in the case of 4135 and 2.25Cr-1Mo steels, but it did not contribute to closure in the case of the 9Cr-1Mo steel (Fig. 14). The level of roughness-induced closure increased with increase in tempering temperature, but $\Delta K_{eff(th)}$ was fairly independent of tempering temperature (Fig. 15), being about 4 MPa√m in vacuum and 2.8 MPa√m in air. Clearly the environment as well as closure played a role in reducing $\Delta K_{eff(th)}$. It was also concluded that rewelding was not responsible for the high threshold levels found in vacuum, because the opening levels in vacuum were never higher than those observed in air. Oxide film rupture as well as hydrogen embrittlement were deemed to be responsible for the increased rate of fatigue crack growth in air over that observed in vacuum.

A study of the effect of tempering temperature on near-threshold fatigue crack behavior in quenched and tempered 4140 steel (Ref 71) indicated that as the yield strength increased (lower tempering temperature), the crack growth rate increased at a given ΔK, and ΔK_{th} decreased from 9.5 MPa√m (700 °C temper) to 2.8 MPa√m (200 °C temper) (Ref 72). Another study of the influence of carbon content and tempering temperature on ΔK_{th} of a low-alloy steel showed that while a tempering treatment increased ΔK_{th}, increasing the carbon content from 0.13 to 0.8% significantly decreased the ΔK_{th} value by more than 100%. The threshold stress-intensity level could be expressed as $\Delta K_{th} = 8.74 - 3.42 \times 10^{-3} (\sigma_y)$ MPa√m.

Yu et al. (Ref 73) determined ΔK_{th} and crack opening levels for a 1010 steel. As the stress ratio and the magnitude of the compressive peak stress were increased, ΔK_{th} decreased linearly. ΔK_{th} also decreased linearly as the yield strength was increased by cold rolling, and severe cold rolling decreased the opening stress in the near-threshold region to near zero. Crack opening measurements

sistance of the alloys was comparable to that of other aluminum alloys.

At 150 °C, the threshold for 8090 has been found to be lower than at room temperature (Ref 69). Increases in slip homogenization, coarsening, and precipitate-free zone formation were associated with decrease.

Steel Crack Growth Thresholds

Figure 13 is a plot of ΔK_{th} and $\Delta K_{eff(th)}$ as a function of yield strength for a number of steels.

ΔK_{th} decreases with increase in yield strength, indicating that the extent of roughness decreases with refinement in grain size and microstructure as strength increases. On the other hand, the value of $\Delta K_{eff(th)}$ remains fairly constant.

A comparison of the crack opening and fatigue crack growth characteristics of three tempered martensitic steels (a modified 4135, 2.25Cr-1Mo, and a modified 9Cr-1Mo) in ambient air and in vacuum (3×10^{-5} torr) at $R = 0.05$ was made by Zhu et al. (Ref 70). It was found that in vacuum, roughness-induced closure was responsible for closure in the near-threshold region and that the

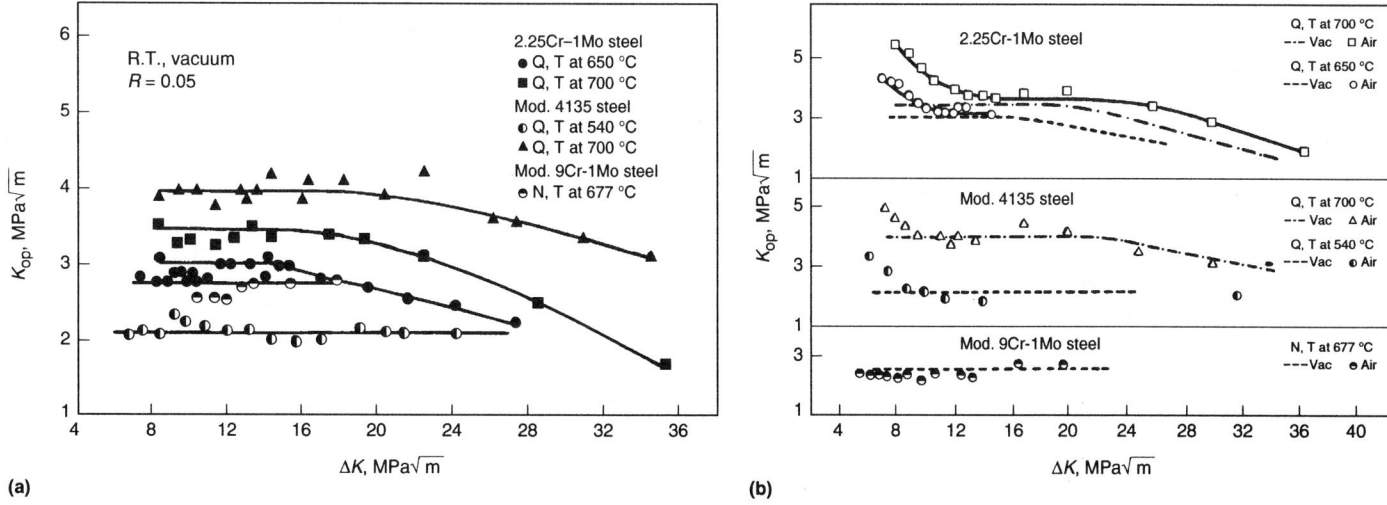

Fig. 14 Crack closure in air and in vacuum. (a) K_{op} level as a function of ΔK in vacuum for 4135 and 2.25Cr-1Mo steels. (b) Comparison of K_{op} levels determined in air with those determined in vacuum. Q, T = quenched and tempered; N, T = normalized and tempered. Source: Ref 70

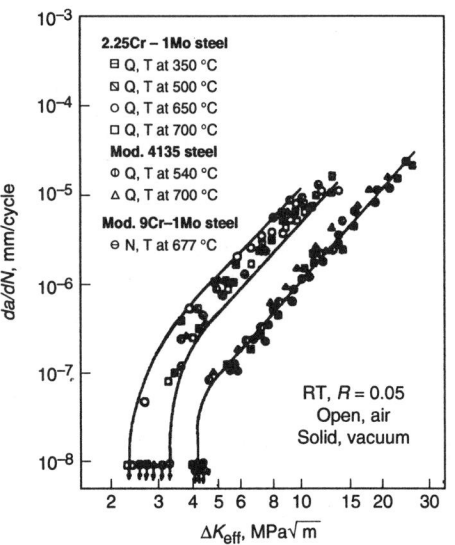

Fig. 15 Effect of tempering on fatigue crack growth rate as a function of the range of the effective stress-intensity factor, ΔK_{eff}. Source: Ref 70

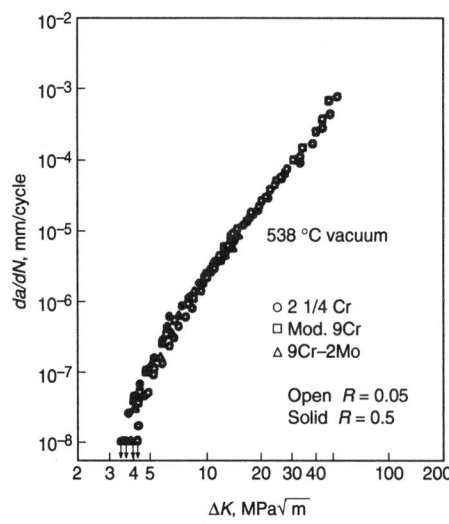

Fig. 16 Fatigue crack growth rate as a function of ΔK for mod. 9Cr-1Mo, 9Cr-2Mo, 2Mo, and 2.25Cr-1Mo steels in vacuum at 538 °C. Source: Ref 90

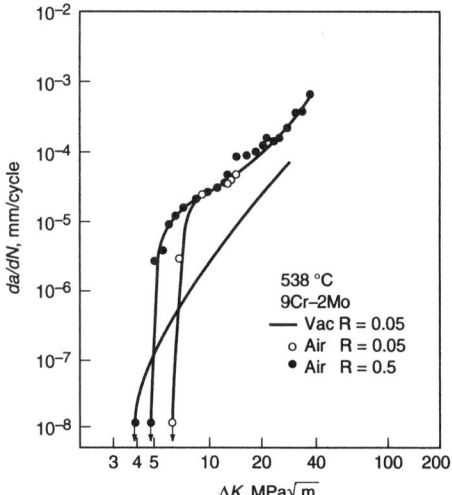

Fig. 17 Fatigue crack growth rate as a function of ΔK for 9Cr-2Mo steel in air and vacuum at 20 °C, and in vacuum at 538 °C. Source: Ref 90

showed that the measured threshold was composed of two parts: an intrinsic threshold stress intensity range, $\Delta K_{\text{eff(th)}}$, and an opening stress intensity, K_{op}. Whereas K_{op} decreased with increasing magnitude of the compressive peak stress, $\Delta K_{\text{eff(th)}}$ was not significantly affected. In a constant-amplitude, load-controlled test, the ratio of K_{op} to K_{max} decreased as the maximum stress intensity (crack length) increased, and when the net stress approached the yield strength of the material, no crack closure could be observed. A similar study was carried out by Yu and Topper with 1045 steel (Ref 74). The effect of crack closure on the near-threshold behavior of structural steels has also been studied (Ref 75).

An investigation into the micromechanics of fatigue crack growth in the near-threshold region of a high-strength steel (Fe-0.32C-1.2Si-1.1Mn-0.97Cr-0.22Ti) under three different temper levels has been made (Ref 76). Tempering at high temperatures resulted in a strong dependence of ΔK_{th} on the prior austenitic grain size by the virtue of strong interaction between the crack-tip plastic zone and prior austenitic grain boundaries. Tempering at a low temperature resulted in a high-strength, high-strain-hardening microstructure and a weak dependence of ΔK_{th} on the prior austenitic grain size.

Bulloch (Ref 77) observed for granular bainitic microstructures of differing carbon contents that the ΔK_{th} values markedly decreased with increasing area fraction martensite, and that a 0.3% C steel at area fraction martensite values approaching 0.4 exhibited ΔK_{th} values that were below those for a 0.13% C steel. The influence of R ratio and microstructure on the threshold fatigue crack growth characteristics of spheroidal graphite cast irons has been investigated by Bulloch and Bulloch (Ref 78). A study by Bulloch (Ref 79) of the effect of material segregation on the near-threshold fatigue crack propagation characteristics of a low-alloy pressure vessel steel in various environments showed that with the exception of high

R ratio air results, segregation effects had little effect on the fatigue crack growth characteristics in air, argon, or vacuum environments.

Fatigue thresholds of isothermally transformed cast steel and nodular cast iron were determined by Zhou et al. (Ref 80). Cu-Mo nodular cast iron exhibited the lowest ΔK_{th}, and silicon cast steel and plain nodular iron and superior ΔK_{th} values together with adequate mechanical properties. Another study (Ref 81) showed that the fracture surface roughness was greater in as-cast material than in heat-treated material.

Reference 82 shows that for 300-series stainless steels, with the exception of 310S and 304HN, the influence of specimen thickness on ΔK_{th} was large at 300 K. On the other hand, the influence was relatively small on $\Delta K_{\text{eff(th)}}$ at 300 K or on ΔK_{th} and $\Delta K_{\text{eff(th)}}$ at 4 K. The influence of load ratio was greater on ΔK_{th} than on $\Delta K_{\text{eff(th)}}$ at both 300 and 4 K. The influence of yield strength on ΔK_{th} at 4 K was relatively small. ΔK_{th} and $\Delta K_{\text{eff(th)}}$ values at 4 K for 310S, a stable stainless steel, was 1.7 to 1.8 times greater than for 300-series metastable stainless steel, excluding 304HN. This phenomenon is believed to be mainly due to the nonexistence of the α' martensitic transformation.

The influence of R ratio and orientation on ΔK_{th} in a low-alloy free-machining (0.31% S) steel was studied by Cadman et al. (Ref 83). They found that the effects of R were dependent on the orientation of the crack with respect to the rolling direction. For both the orientations considered, an increase in the R ratio not only decreased the threshold values but also led to marked changes in the fracture appearance.

Residual stresses can affect the threshold level, as indicated by a study of laser surface hardening on fatigue crack growth rate in AISI-4130 steel (Ref 84). Residual compressive stresses retarded the crack growth rate near ΔK_{th}, but this beneficial effect disappeared as the ΔK value increased.

The near-threshold fatigue crack growth behavior of a ferrite/martensite dual-phase steel with different volume fractions of martensite was investigated in laboratory air at R values of 0 and 0.5 (Ref 85). The volume fraction of martensite had a significant effect on the fatigue threshold. The threshold value of the dual-phase steel first increased and then decreased as the martensite content increased, with a maximum at a volume fraction of approximately 35% martensite.

A study of the influence of prestrain and aging on near-threshold fatigue crack propagation in as-rolled and heat-treated dual-phase steels (Ref 86) revealed that ΔK_{th} increased with increasing grain size and decreasing yield stress. A combination of 10% prestraining with aging at 175 °C for 30 min showed almost no effect on the threshold level of as-rolled dual-phase steel but decreased that of heat-treated dual-phase steels more than 37%. This difference in behavior was suggested to result from the differences in grain size and volume fraction of martensite in these two kinds of dual-phase steels.

The role of crack-tip shielding in retarding fatigue crack growth has been examined (Ref 87) in ferritic-martensitic duplex microstructures, with the objective of achieving maximum resistance to fatigue through crack deflection and resultant crack closure. ΔK_{th} values were 100% higher than in normalized structures, (i.e., greater than 20 MPa$\sqrt{\text{m}}$), the highest thresholds reported for a metallic alloy at that time. Duplex as well as ferritic and austenitic single-phase materials have been tested (Ref 88). Ferritic specimens exhibited the highest ΔK_{th} levels, while austenite had the lowest. Prestraining by 8% led to a significant drop in the threshold level for all materials, while the crack closure level decreased solely in the single-phase austenitic and ferritic materials.

The near-threshold properties of a ferritic/austenitic stainless steel were studied (Ref 89). Cold rolling of the originally hot-rolled, banded microstructure increased ΔK_{th} by 25%. A further increase in ΔK_{th} of the same order was caused by annealing the cold-rolled structure for 120 h at 475 °C, resulting in the spinodal decomposition of the ferritic phase. ΔK_{th} was also raised by high-temperature annealing, which broke up the banded structure. These improvements were caused by an increase in the closure level, K_{cl}, and also in $\Delta K_{eff(th)}$. The results were interpreted in terms of changes in the fracture surface topography and the flow properties.

The near-threshold fatigue crack growth behavior at elevated temperatures is a matter of interest. A study by Nakamura et al. (Ref 90) involved 2.25Cr-1Mo, 9Cr-1Mo, and 9Cr-2Mo steels as influenced by both temperature and environment. At 538 °C, crack closure in vacuum in these alloys was not detected, and the da/dN results for all three alloys at R values of 0.05 and 0.5 fell along a single curve (Fig. 16). When the change in modulus with temperature was accounted for, the in-vacuum results for the 9Cr steels at both room temperature and 538 °C fell along a single line. However, in air at 538 °C, the ΔK_{th} level was above that in vacuum (Fig. 17), due to the effects of oxidation-induced closure, and a sharp break appeared in the da/dN plot just above threshold due to oxide rupture at the crack tip.

The effect of crack surface oxidation on near-threshold fatigue crack growth characteristics and crack closure has been investigated at elevated temperatures (80 to 350 °C) in an A508-3 steel (Ref 91). ΔK_{th} decreased with increasing temperature up to 100 °C and increased thereafter. Oxidation of the crack surfaces had an important role on the near-threshold characteristics. Below 150 °C, a thin oxide layer formed that prevented the formation of fretting oxide debris during crack growth. At 288 and 350 °C, a thick oxide layer formed that induced crack closure.

The near-threshold fatigue crack growth properties at elevated temperature for 1Cr-1Mo-0.25V steel and 12Cr stainless steel were investigated by Matsuoka et al. (Ref 92). Fatigue tests were conducted at 0.5, 5, and 50 Hz, in a manner designed to avoid crack closure. The effective value of threshold stress-intensity range increased with increasing temperature and with decreasing frequency for the Cr-Mo-V steel, whereas the effective threshold stress-intensity range was independent of temperature and frequency in the case of the more oxidation-resistant SUS403 steel. The observed threshold levels and crack growth behavior were closely related to the oxidation process of the bare surface formed at the crack tip during each loading cycle.

Nishikawa et al. (Ref 93) investigated the near-threshold fatigue crack growth and crack closure behavior in SS41, SM41A, and SUS304 steels and A2218-T6 aluminum alloy at temperatures up to 500 °C. The fatigue threshold increased with increasing test temperature in all the steels tested, while it decreased in 2218-T6 alloy. Oxide-induced crack closure played an important role in the increase of ΔK_{th} at elevated temperatures in SM41A but played a less important role in SS41 and SUS304. It was concluded that oxide

products on the fracture surface enhance crack closure only when the crack-tip opening displacement at threshold remains small at elevated temperatures. ΔK_{th} for type 304 stainless steel at 650 °C was lower than that at 550 °C (Ref 94).

Near-threshold fatigue crack propagation (FCP) behavior was studied in an 18Cr-Nb stabilized ferritic stainless steel (Ref 95) as a function of elevated temperature. Crack closure measurements were obtained from room temperature to 700 °C. At a stress ratio of 0.1, increasing the test temperature from room temperature to 500 °C resulted in an increase of the growth rates in the midrange growth regime and a sharply defined threshold at a ΔK level higher than the room-temperature threshold, giving rise to a crossover type of behavior of a type similar to that shown in Fig. 14. A constant-K_{max} increasing R-ratio (CKIR) test procedure was utilized at room temperature and at 500 °C in an attempt to identify near-threshold FCP data in the absence of crack closure. The type of crossover behavior identified with constant R ratio tests at room temperature and 500 °C was also observed in the CKIR tests, an indication that at low ΔK levels, even at high R ratios, oxidation-induced closure may still be effective as a shielding mechanism.

The effect of R ratio on near-threshold fatigue crack growth in a stainless steel and a metallic glass has been studied by Alpas et al. (Ref 96).

Titanium Alloy Crack Growth Thresholds

There have been a number of investigations of the effect of microstructure and load ratio on ΔK_{th} in titanium alloys, for example, Ref 97. Fatigue threshold levels have been determined for $\alpha_2 + \beta$ forged Ti-24Al-11Nb (Ref 98). Both roughness-induced and phase-transformation-induced types of crack closure were studied in the metastable β alloy Ti-10V-2Fe-3Al (Ref 99). Such crack-tip shielding mechanisms can provide a means of increasing the fatigue threshold. The near-threshold behavior of P/M Ti-6Al-4V has been determined using a high-frequency test method (20 KHz) at room temperature for crack growth rates between 10^{-12} and 10^{-9} m/cycle (Ref 47). It has been reported for Ti-6Al-4V with a Widmanstatten colony microstructure that two transitions during fatigue crack growth can occur at ΔK values where the cyclic and monotonic plastic zones become equal to the α lath size. Data were obtained on near-threshold fatigue crack growth behavior and crack closure for this microstructure (Ref 100).

Near-threshold fatigue crack growth behavior of Ti-6Al-4V alloy was investigated as a function of Widmanstatten microstructure with emphasis on the effect of colony size on ΔK_{th} and $\Delta K_{eff(th)}$ (Ref 101). It was found that crack growth rates were strongly affected by microstructural sizes such as colony size and α lath size. The microstructural units controlling crack growth in fast-cooled aligned microstructures were colonies, whereas they were α laths in relatively slow-

cooled ones. This distinction was brought about by the thick continuous interplatelet β phase present in slowly cooled structures. In rapidly cooled structures, thin discontinuous β phase appears to be ineffective in arresting cracks. The crack growth rates and the magnitudes of ΔK_{th} and $\Delta K_{eff(th)}$ were correlated with the controlling microstructural units, with crack closure levels being dependent on colony size. It has also been determined that for high ΔK_{th} values at low R in Ti-6Al-4V, the best microstructural condition is a coarse lamellar structure with high (0.2%) oxygen content, age-hardened at 500 °C (Ref 102).

The fatigue crack growth rates of small surface cracks have received much attention in the last several years, because it has been observed that small cracks can propagate not only much faster than long cracks under nominally identical ΔK values, but also well below the ΔK_{th} values of long cracks (Ref 103). One study was made to determine the effect of microstructural parameters on propagation of small surface cracks in two representative titanium alloys (Ti-8.6Al and Ti-6Al-4V). It was concluded that aside from the absence of significant crack closure in the early stages of growth of small surface cracks, there must be other contributing factors, because the threshold value of long cracks were significantly higher than that of small cracks, even after the long crack data was corrected for closure. It was also found that the ranking of different microstructures with respect to the resistance to crack growth of small surface cracks could be the reverse of that of long through-cracks. Smaller grains and finer phase dimensions led to lower growth rates of small surface cracks, while for long through-cracks these parameters exhibited an opposite effect.

The influence of crack closure and load history on near-threshold crack growth behavior in surface flaws has been studied in Ti-6Al-6Mo-4Zr-2Sn (Ref 104). Four types of loading histories were used to reach a threshold condition. Results from all four test types indicated that a single value of $\Delta K_{eff(th)}$ was obtained that was independent of stress ratio, R, or load history. Crack growth rate data in the near-threshold regime, on the other hand, appeared to have a dependence on R, even when ΔK_{eff} was used as a correlating parameter. In another study of the propagation of small surface cracks in titanium alloys in vacuum and in laboratory air, it was also found that small semielliptical surface cracks propagated faster and below the near-threshold stress-intensity factors of long through-cracks (Ref 105).

A study of the effect of an in situ phase transformation on ΔK_{th} in Ti-Ni shape-memory alloys has shown that the value of ΔK_{th} can vary from 5.4 MPa$\sqrt{m}$ in a stable austenitic microstructure, to 1.6 MPa$\sqrt{m}$ in an unstable (reversible) austenitic microstructure (Ref 106).

A number of papers have dealt with near-threshold fatigue crack growth phenomena at elevated temperature in titanium alloys (see, e.g., Ref 107). Crack closure data at moderately elevated temperatures have been obtained for several structural alloys, including Ti-6Al-4V alloy,

Inconel 600, and A2218-T6 Al alloy, and the effects of crack closure as well as negative R values on crack propagation have been discussed (Ref 108). In a study of fatigue crack growth behavior of Ti-6Al-4V at 300 °C in high vacuum, it was observed that near the threshold, a crystallographic stage I type of crack propagation occurred, and that the transition from the stage I to stage II type of propagation was sensitive to loading conditions and temperature (Ref 109).

The effect of stress ratio on the near-threshold fatigue crack growth behavior of Ti-8Al-1Mo-1V has been studied at 24 and 26 °C in laboratory air (Ref 110). The effects of stress ratio at a constant temperature could be explained in terms of crack closure and ΔK_{eff}. However, crack closure did not account for the effects of temperature at a fixed stress ratio of 0.1, because higher near-threshold crack growth rates were observed at 260 °C than at 24 °C when the data were plotted as a function of ΔK_{eff}. This difference in crack growth rate was believed to be attributable to significant crack front branching and secondary cracking.

Nickel-Base Alloys

In a study of the near-threshold behavior of nickel-base superalloys (Ref 111), the crack closure level was a function of ΔK. The existence of ΔK-dependent closure in nickel-base superalloys resulted in microstructurally sensitive crack growth rates, even at high R ratios. This behavior is in contrast to that of steels and titanium alloys, for which crack closure levels are often found to be ΔK independent, and for which an increase in the R ratio has a larger influence on near-threshold crack growth than on region II crack growth.

Single-Crystal Superalloy. In single crystals of the nickel-base superalloy Udimet 720, Reed and King found that stage I growth occurred along slip planes of maximum resolved shear stress giving rise to faceted fatigue fracture surfaces. Short crack growth behavior was observed in that the crack growth rates were higher than for short cracks in polycrystals, an indication that grain boundaries in the polycrystals retarded fatigue crack growth. The threshold level for the single crystals at $R = 0.5$ was about 3 MPa$\sqrt{m}$ as compared to about 6 MPa$\sqrt{m}$ for polycrystalline material. In polycrystals an R-effect was observed which was attributed to surface roughness associated with small-scale faceted growth. Additional results implied that crack closure played a role in single crack growth crystals as well.

Plasma-Sprayed Alloy. In a porous, plasma-sprayed 80Ni-20Cr alloy it was found that there was little or no effect of the R ratio on ΔK_{th}. This finding was attributed to the material's fine grain size of the material as well as to regions of porosity (Ref 111).

Metal Matrix Composites/Intermetallics

This section briefly reviews results for fiber-reinforced, whisker-reinforced, and particulate-reinforced metal-matrix alloys. In addition, reference will be made to ARALL, a composite of aluminum sheets reinforced with layers of aramid fibers.

Results for a number of aluminum alloys reinforced with SiC particulate (SiC_p) have shown the threshold for $R = 0.1$ to be in the range 2.5 to 4.7 MPa$\sqrt{m}$ (Ref 112). In 6061 with 15 vol% SiC_p there was a 50% increase in the threshold level as a result of increased closure and crack deflection. In 2024 reinforced with SiC_p, an increase in the crack closure level led to a higher threshold (Ref 113). The addition of SiC to 6061 led to a decrease in the threshold range due to a decrease in grain size associated with the reinforcement (Ref 114). Davidson (Ref 115) found that reinforcing an Al-4Mg alloy with SiC led to an increase in ΔK_{th} but also resulted in higher growth rates above the threshold level. On the other hand, a 6061/SiC_p composite was found to be superior to the 6061 matrix alloy over the whole range of ΔK values studied, including ΔK_{th}. Because the grain size and microhardness were identical in the two materials, it was concluded that the superiority of the composite was solely due to the SiC particles. Detailed crack profile analysis showed that the crack was deflected by the particles, leading to higher crack closure, which resulted in slower crack growth rates (Ref 116).

In 2024, both SiC_p and SiC whiskers (SiC_w) raised the threshold level (Ref 117). Some of this improvement resulted from roughness-induced closure, which develops as the crack meanders to avoid particles, as in a cast aluminum alloy composite reinforced with 15% alumina (Ref 118). It has also been reported that coarser distributions of SiC were more effective in raising the threshold level in SiC/Al alloys (Ref 119). However, Shang and Ritchie (Ref 120) found that whereas a coarse-particle distribution resulted in higher ΔK_{th} values at low R ratios, fine particles gave higher threshold values at high R ratios. Such behavior was analyzed in terms of the interaction of SiC_p with the crack path, both in terms of the promotion of roughness-induced crack closure at low R ratios and the trapping of the crack by particles at high R ratios. Consideration of the latter mechanism yielded the limiting requirement for the intrinsic threshold condition in these materials that the maximum plastic zone size must exceed the effective mean particle size. This implies that for near-threshold crack advance, the tensile stress in the matrix must exceed the yield strength of the material beyond the particle. It was also noted that as ΔK levels increased from near-threshold levels, there was a gradual transition of fracture mode, with a high incidence of reinforcement particle/matrix decohesion at low ΔK and a predominance of particle cracking at higher ΔK levels (Ref 121).

High-strength 2025 aluminum alloy reinforced with SiC_w also showed an improvement in ΔK_{th} (Ref 122). Stretching after quenching to relieve residual stresses in 8090 aluminum alloy reinforced with SiC_w resulted in a decrease in threshold as well as a decrease in closure due to the elimination of compressive stress on stretching (Ref 123). In a SiC_w-reinforced aluminum alloy, short fatigue cracks were observed to grow at ΔK levels below the threshold for long cracks. The fatigue crack growth characteristics (with emphasis on ΔK_{th}, considered to be one of the most important fracture mechanical properties for ensuring the composite structural integrity) were investigated for a whisker-reinforced high-strength aluminum alloy, continuous fiber-reinforced aluminum, and composites with titanium alloy matrices (Ref 124). In a study of the cyclic crack growth behavior of extrusions of TiC_p-reinforced P/M Ti-6Al-4V metal-matrix composites, ΔK_{th} was typically below 10 MPa$\sqrt{m}$ (Ref 125).

For long fatigue cracks, both roughness and crack deflection have been observed to reduce the driving force (Ref 126). The presence of alumina fibers in squeeze cast 6061 aluminum alloy resulted in higher closure levels and a significant increase in ΔK_{th} (Ref 127). The near-threshold transverse fatigue crack growth characteristics of unidirectionally continuous-fiber-reinforced metals have been discussed by Hirano (Ref 128).

Fatigue cracks have been grown in five-layer aluminum alloy 2024-T8-aramid fiber laminate composite ARALL-4 over the range of cyclic stress-intensity factors (ΔK) from 3.5 to 91 MPa$\sqrt{m}$. ΔK_{th} was about the same as for unreinforced aluminum alloys, and the extent of crack closure depended on the crack length, with fiber bridging influencing the results (Ref 129). In ARALL the threshold level has also been found to increase with crack length, and this behavior has been attributed to fiber bridging (Ref 130).

Fatigue-crack propagation along ceramic/metal interfaces at 10^{-9} m/cycle has also been investigated (Ref 131).

Crack Growth Thresholds in Gamma Titanium Aluminide. Davidson and Campbell have found that ΔK_{th} for fatigue crack growth through the $\gamma + \alpha_2$ lamellar microstructure of an alloy based on TiAl was lower at 25 °C than at 800 °C, and the lamellar microstructure was found to have a strong influence on crack tip behavior (Ref 132).

Effect of Environment on ΔK_{th}

At room temperature, the ambient environment generally has a deleterious effect on ΔK_{th}. For example, in high-frequency (20 kHz) testing of 2024-T3 aluminum alloys, ΔK_{th} determined at 10^{-13} m/cycle was found to be 2.1 MPa$\sqrt{m}$ in moist air, whereas in vacuum it was 3.3 MPa$\sqrt{m}$. The decrease was attributed to hydrogen embrittlement (Ref 133). A large number of factors are involved in dealing with the effects of the environment. For example, Vosikovsky et al. (Ref 134) considered the influence of sea water temperature on corrosion fatigue crack growth in structural steels. Often hydrogen embrittlement is considered to be a factor, and it has been found (Ref 135) that both external and internal hydrogen can play similar roles in the degradation of ΔK_{th} in high-strength steels.

Pao et al. have studied the influences of yield strength and microstructure on environmentally assisted fatigue crack growth in 7075 and 7050 high-strength aluminum alloys. The influences were analyzed on the basis of a model for transport-controlled fatigue crack growth that incorporated metallurgical, mechanical, and environmental variables. The model was based on the assumption that when the crack driving force is below that of the stress-corrosion cracking threshold, the rate of fatigue crack growth in a deleterious environment is the sum of the rate of fatigue crack growth in an inert environment plus a corrosion fatigue component (Ref 136).

Piascik and Gangloff (Ref 137) have investigated the effect of gaseous environments on fatigue crack propagation in the Al-Li-Cu alloy 2090 in a peak-aged condition. For the moderate ΔK/low R regime as well as the low ΔK/high R regime, crack growth rates decreased and ΔK_{th} increased when the environment was changed from purified water vapor to moist air, helium, or oxygen. The gaseous environmental effects were pronounced near threshold and were not closure dominated. The deleterious effect of low levels of H_2O (ppm) supports a hydrogen embrittlement mechanism and suggests that molecular-transport-controlled cracking, established for high ΔK/low R, is modified near threshold. Localized crack-tip reaction sites or high R crack opening shape may enable the strong environmental effect at low levels of ΔK. The similarity of crack growth in helium and oxygen ruled out the contribution of surface films to fatigue damage in alloy 2090. In a comparison of 2090 and 7075, both alloys exhibited similar environmental trends, but the Al-Li-Cu alloy was more resistant to intrinsic corrosion fatigue crack growth. Another study found that 2090 exhibited the lowest threshold in salt water, a somewhat higher threshold in air, and the highest in vacuum (Ref 138).

It has also been found (Ref 139) that Al-Li alloys exhibit environmental fatigue crack growth characteristics similar to those of the conventional 2000-series alloys and are more resistant to environmental fatigue than 7000-series alloys. The superior fatigue crack growth behavior of Al-Li alloys 2090, 2091, 8090, and 8091 was related to crack closure caused by a tortuous crack path morphology and crack surface corrosion products. At high R and reduced closure, the chemical environmental effects were pronounced, resulting in accelerated near-threshold da/dN values. The "chemically small crack" effect observed in other alloy systems was not pronounced in Al-Li alloys. Modeling of environmental fatigue in Al-Li-Cu alloys related accelerated fatigue crack growth in moist air and salt water to hydrogen embrittlement.

For steam turbine rotor steels (Ref 140), pure water at 160 °C reduced the fatigue strength by about 25% compared to the value in air at room temperature. Fatigue crack propagation rates in water at 100 °C were higher than at 160 °C and were about three times higher than in air at room temperature. A deaerated water environment reduced ΔK_{th} by approximately 20%.

An increase in ΔK_{th} in sea water was observed by Todd et al. (Ref 141) for ASTM A710 steel cathodically protected at an applied potential of −1.0 V. This increase was not attributable to calcareous deposit formation, but rather appeared to be a result of hydrogen embrittlement. Such embrittlement led to the development of metal wedges in the crack wake that contributed to a new mechanism of crack closure.

The effect of laboratory air, dry hydrogen, and dry helium gaseous environments on the fatigue crack propagation behavior of low-alloy 4340 steel has been investigated (Ref 142). Below an R value of 0.5, ΔK_{th} in the air environment was larger than in the dry environments. ΔK_{th} in wet hydrogen was between the values in air and dry environments. At a high load ratio of 0.8, however, ΔK_{th} was insensitive to test environment. It was concluded that oxide-induced crack closure governed the kinetics of gaseous-environment, near-threshold crack propagation behavior. However, thick oxide deposits in wet hydrogen did not cause high levels of crack closure.

Near-threshold fatigue crack growth of HY80 (Q1N) alloy steel was investigated in air and in a vacuum by James and Knott (Ref 143). The applied stress ratio affected crack growth rates in air but had little effect on rates in the vacuum environment. Additional studies have been carried out by Kendall and Knott (Ref 144).

The effects of moisture on the fatigue crack growth behavior of a low-alloy 2Ni-Cr-Mo-V rotor steel near threshold were investigated by Smith (Ref 145). At $R = 0.14$, the growth rates in moist air were much lower than in dry air. This difference was associated with the formation of oxides on the fracture surface, with moisture modifying the type and extent of oxidation observed. Observations of the transient crack growth following environmental changes suggested that fracture surface oxides within approximately 0.3 mm of the crack tip exerted a strong retarding influence on crack growth, although oxides up to at least 3 mm from the tip may also have had some retarding effect.

In a study at high frequency of the near-threshold behavior of stage I corrosion fatigue of an austenitic stainless steel (316L), Fong and Tromans (Ref 146) found that at high anodic potentials with good mixing between the crack solution and bulk solution, crack retardation and arrest effects due to surface-roughness-induced closure were minimized by electrochemical erosion.

In a study of the environmental influence on the near-threshold behavior of a high-strength steel, it was concluded (Ref 147) that fatigue crack growth rates measured in ambient air depend on three processes: intrinsic fatigue crack propagation as observed in vacuum; adsorption of water vapor molecules on freshly created rupture surfaces, which enhances crack propagation; and a subsequent step of hydrogen-assisted cracking. A reduction of sea water temperature from room temperature to 0 °C decreased the fatigue crack growth rates at free corrosion potential by a factor of almost 2. At −1.04 V, the plugging of cracks by calcareous deposits re-

duced the effective stress-intensity range and increased the apparent ΔK_{th} level.

Esaklul and Gerberich (Ref 148) observed that the presence of internal hydrogen through cathodic charging had a substantial influence on the near-threshold fatigue behavior of a high-strength, low-alloy steel with an as-received yield strength of 365 MPa. The results of fatigue crack propagation tests indicated higher crack propagation rates and lower threshold stress intensities in the presence of internal hydrogen. These effects were dependent on strength, R ratio, and test temperature. The enhancement in the crack propagation process was more severe at higher strength levels and at higher mean stresses.

Under freely corroding conditions, $\Delta K_{eff(th)}$ values in 3% sodium chloride (NaCl) aqueous solution were found by Matsuoka et al. (Ref 149) to be lower than in air for all of the structural-grade steels investigated. In particular, $\Delta K_{eff(th)}$ values for carbon and high-strength steels were almost equal to a theoretical $\Delta K_{eff(th)}$ value of approximately 1 MPa√m, calculated on the basis of the dislocation emission from the crack tip.

The effects of free corrosion and cathodic protection on fatigue crack growth in structural steel (BS4360-50D) in synthetic sea water were studied by Bardal (Ref 150). At an R value of 0.5, free corrosion in sea water led to higher crack growth rates and a lower threshold value than in air, while cathodic protection had the opposite effect.

Komai (Ref 151) has found that due to a corrosion-product-induced wedge effect, crack growth rates in HT55 steel (tensile strength 580 MPa) were significantly reduced in NaCl solution, with ΔK_{th} greater than in air. In corrosion-resistant stainless steel SUS304, the corrosion-product-induced wedge effect diminished. In the absence of closure, two factors were considered to increase the growth rate: a suppression of reversed slip by water molecule adsorption, and in the case of SUS304, the hydrogen embrittlement of the stress-induced martensite formed at the crack tip.

For Ti-10V-2Fe-3Al tested in vacuum as well as in 3.5% NaCl solution, it was found that the corrosive environment led to near-ΔK_{th} values of 2 to 3 MPa√m, compared to 4 to 5 MPa√m in vacuum (Ref 152).

Environment/mechanical interaction processes and hydrogen embrittlement of titanium alloys (IMI115, IMI130, IMI155) have been investigated (Ref 153). While interstitial hydrogen was found to have little effect, a significant increase in the resistance to fatigue crack propagation was observed with increasing interstitial oxygen content. In contrast, when hydrogen was present in the form of hydride precipitates, crack growth rate was significantly increased, particularly in the threshold and high-ΔK regions of the da/dN versus ΔK curve. The results showed that crack propagation in the matrix containing hydrides occurred mainly through the hydride/matrix interface without any significant hydride cracking.

In a determination of ΔK_{th} at cryogenic temperatures for aluminum alloys (2024, 2124, Al-3Mg), copper, steels (304, Fe-4.0Si, Fe-0.1C-9Ni, Fe-0.15C-4Mn), nickel alloy (Inconel 706), and titanium alloy (Ti-30Mo), it was observed that resistance to near-threshold fatigue crack propagation generally improved with decreasing temperature. Although crack closure could account for the influence of load ratio on low-temperature near-threshold crack propagation behavior, it alone could not account for the temperature effect (Ref 154). It is possible that the absence of any deleterious environmental effect also played a role.

Additional aspects of the interaction of microstructure and environment in the near-threshold range have been discussed by Petit (Ref 155) and Bailon et al. (Ref 156). There is also information available on the effect of overloads on the corrosion fatigue crack growth behavior of low-alloy steel in the threshold region in 3.5% NaCl solution (Ref 157).

Welds

Investigations of the effect on ΔK_{th} of the tensile residual stresses in steels resulting from welding have shown that such stresses lower ΔK_{th} (Ref 158, 159). When these tensile residual stresses were high, ΔK_{th} became equal to $\Delta K_{eff(th)}$ as a lower limit and was no longer a function of R. In a study of near-threshold fatigue crack propagation in welded joints under random loading, Ohta et al. (Ref 160) found that for specimens of HT80 steel in which tensile residual stresses were present at the crack tips, da/dN could be estimated from constant-amplitude tests, assuming a linear cumulative damage law. In Ref 161 it was found that MIG welding yielded higher values of ΔK_{th} than shielded metal arc welding, and that ΔK_{th} was highest when the fatigue crack propagated through the weldment.

Modes II and III. Otsuka et al. (Ref 162) have shown that ΔK_{IIth} values for mode II growth in the heat-affected zone of the aluminum alloy 7N01-T4 and in 2017-T3 and -T4 base metal were quite low, about 1 MPa√m. They noted that for other materials, ΔK_{IIth} values fell between 6 and 10 MPa√m.

A comparison of fatigue crack propagation in modes I and III has been provided by Ritchie (Ref 163), and Ref 164 discusses near-threshold fatigue crack growth in steels under mixed mode II and III loading. In a study of the fatigue crack direction and threshold behavior of a medium-strength structural steel under mixed mode I and III loading, the experimental fatigue crack growth threshold data were close to a lower-bound failure envelope, based on the premise that the event controlling failure is the propagation of mode I branch cracks (Ref 165). It has also been observed (Ref 166) that mode I thresholds shifted toward higher values when mode III superimposed loads were increased, with this increase more pronounced for $R = -1$ than for $R = 0$ and 0.5. Roughness-induced crack closure was assumed to be the main closure mechanism in explaining this result. It has also been concluded that a crack cannot grow by mode III shear without the presence of a mode II component (Ref

167), and Ref 163 considers whether or not there is a fatigue threshold for mode III crack growth.

Modeling Threshold Behavior

One of the fundamental considerations regarding near-threshold fatigue crack growth is whether or not a fatigue crack, in the absence of environmental effects, can advance an increment per cycle somewhere along the crack front. Opinions on this point differ. For example, it has been observed (Ref 168) for a ferritic steel that for over four orders of magnitude in near-threshold region fatigue crack propagation rates, the striation spacing was independent of the ΔK, which might imply that the number of cycles necessary to form one striation was greater than one. On the other hand, the observed striation spacing happened to be equal to the dislocation cell size, and an apparent striation may have been created where a crack, growing at a spacing per cycle too small to be resolved, crossed the cell boundary. No such striation-like markings were observed in aluminum alloy in which no cell substructure formed (Ref 169).

The threshold itself can be considered the dividing line between the propagation and nonpropagation of a fatigue crack, and a number of proposals for the threshold condition have been put forth. Among these are the emission of dislocations from the crack tip (Ref 170) and the blockage of slip from the crack tip by some barrier such as a grain boundary or more resistant phase. Radon and Guerra-Rosa (Ref 171) have developed a model for the threshold based on the tensile and cyclic properties of the material. McClintock in 1963 proposed that crack growth could occur when the local strain or accumulated damage at the crack tip reached a critical value. Such proposals are purely mechanical in nature, whereas in tests in air the effects of the environment are always superimposed. In some mechanical models, propagation is a go no-go situation, whereas when environmental effects are present, corrosion may continuously reduce the threshold, much as it reduces the long-life portion of the S/N curve. Another aspect of the environment is that it may introduce a discontinuity into the growth process, particularly near threshold, if the development of a critical extent of corrosion takes time rather than cycles to be accomplished. Also, because there is a transition from the LEFM ΔK_{th} value to the endurance limit with decreasing crack size, it is to be expected that some of the factors that govern the endurance limit also affect the threshold level. In fact, the existence of a lower limit for fatigue crack growth was postulated by McEvily and Illg in 1956. They proposed that $K_N S_{net} = EL$, where K_N is the stress-concentration factor for a fatigue crack, computed according to Neuber's procedures for a crack tip of effective radius of the order of 0.05 mm; S_{net} is the net section stress; and EL is the endurance limit. The term $K_N S_{net}$ is directly related to the stress-intensity factor, and the equation can provide a smooth transition of the type observed by Kitagawa and Takahashi in the small crack re-

gime between the endurance limit and the macroscopic threshold level (Fig. 9).

Use of a strain-intensity factor (K/E) for crack growth near threshold resulted in a narrow scatter band in high-R tests where closure was absent (Ref 172).

In considering the models for the threshold condition, it is clear that Young's modulus is an important parameter (Ref 172). Other mechanical properties play a much smaller role on the relationship between the rate of crack growth and ΔK_{eff}. On the other hand, grain size does affect ΔK_{th}, because it influences the degree of roughness and hence the level of closure at low R values. The effect of grain size on ΔK_{th} is principally due to a larger degree of crack deflection in coarse-grained structures and the accompanying high levels of crack closure as a consequence of zig-zag crack growth (Ref 173). A theoretical model (Ref 174) for the effects of grain size on the magnitude of roughness-induced crack closure at ΔK_{th} considered a crack propagating incrementally along planar slip bands and being deflected at grain boundaries to create an idealized zig-zag crack path. The effective slip band length was taken to be equal to the grain size. It was assumed that the dislocations emitted from the crack tip upon loading to form the pile-up were completely irreversible to produce a combined mode I and II displacement at the crack tip. The magnitude of ΔK_{clth} can then be expressed in terms of slip length or grain size, macroscopic yield stress, critical resolved shear stress, and the angle between slip plane and crack plane (Ref 175).

Taira et al. (Ref 176) proposed a micromechanistic model for the fatigue limit to relate a Petch-type dependence of the fatigue limit on grain size. A model for the threshold condition was developed that involved a microscopic stress-intensity factor at the tip of a crack blocked at a grain boundary. Fatigue crack propagation depended on whether or not a slipband near the crack tip propagated into an adjacent grain. Tanaka and Nakai (Ref 177) extended this model to include the development of crack closure with crack length in considering the mechanics of the threshold for the growth of small cracks. Fatigue crack growth at near-threshold rates has also been modeled using microstructurally-controlled micromechanical crack tip parameters (Ref 178). The model is based on the concept of crack opening by means of local slip lines whose length and dislocation density are controlled by the alloy microstructure.

Gerberich et al. (Ref 179) considered dislocation cell networks important features in cyclic strain hardening and crack-tip advance in Fe-4Si. Near-threshold fatigue crack behavior was considered, together with evidence of crack-tip interactions with dislocation cells, and a computer simulation of slip band pile-ups interacting with an idealized cell network was developed. A threshold model was derived that included the flow stress, cell size, test frequency, and strain rate sensitivity.

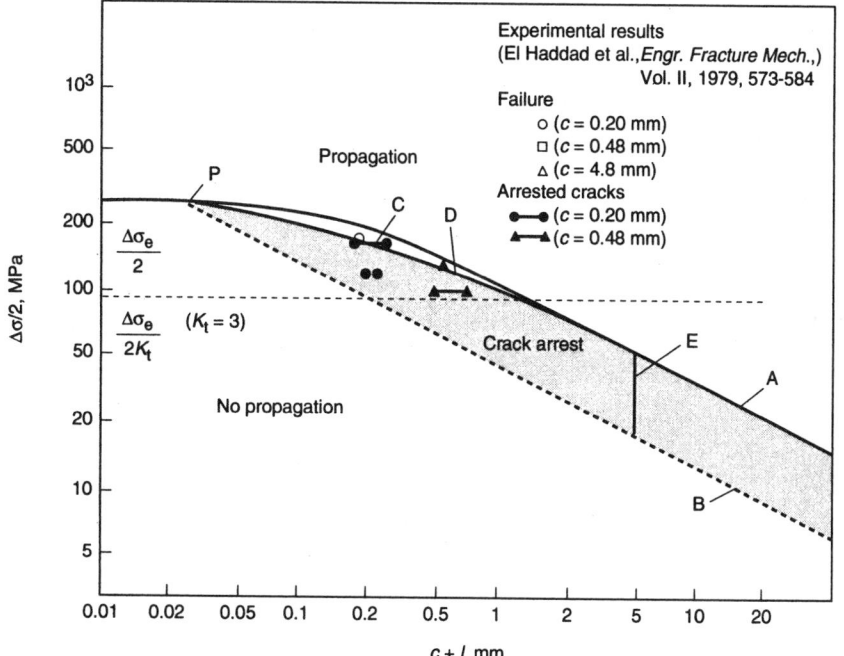

Fig. 18 Modified Kitagawa plot for the influence of crack closure on the stress required to propagate fatigue cracks as a function of notch or flaw size. Source: *Fracture* (Wells and Landes, Ed.), AIME, 1984, p 215-234

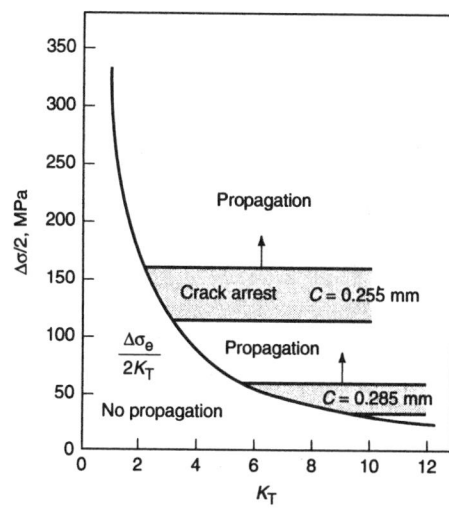

Fig. 19 Modified Frost plot of arrest conditions as a function of initial notch size and K_T. Shaded areas indicate regions in which fatigue cracks will form and then become nonpropagating. For a given initial notch size, cracks will not form below the level of the corresponding shaded area. Source: *Scripta Met.*, Vol 18, 1984, p 71

In pearlitic steels it has been shown and related to a theoretical model (Ref 173) that while the interlamellar spacing explicitly controls the yield strength, a similar effect on ΔK_{th} cannot be expected. On the other hand, the pearlitic colony size was shown to strongly influence ΔK_{th} and K_{clth} through the deflection and retardation of cracks at colony boundaries. An increase in ΔK_{th} and K_{clth} with colony size was found. Further, $\Delta K_{eff(th)}$ was found to be insensitive to colony size and interlamellar spacing.

Mura and Weertman (Ref 180) reviewed the dislocation models that have been applied to the near-threshold stress intensity factor region. They concluded that because of the sparseness of existing theory, this region of the fatigue crack growth curve is as yet not well understood.

Thresholds in Design

Figure 18 is an example of a modified Kitagawa diagram, where c is the length of the crack-initiating notch and l is the crack length measured from the notch. The modification consists of plotting both the $\Delta K_{eff(th)}$ (line B) and the ΔK_{th} (line A) conditions for long cracks. It is clear that with respect to the initiation and growth of fatigue cracks from flaws or notches, $\Delta K_{eff(th)}$ is a much more significant parameter than ΔK_{th}. Lines C, D, and E indicate the stress amplitude required to maintain a fatigue crack growth rate of 10^{-11} m/cycle as crack closure develops in the wake of a newly formed fatigue crack. If cracks are initiated at notches at stress amplitudes between the dashed horizontal line and the maximum value of curves C or D, nonpropagating cracks will develop, as has been observed by El Haddad et al. (Ref 181). The shaded area indicates the region

on this diagram where crack arrest due to the development of crack closure is predicted to occur. Below an initial notch depth of the order of 10 μm, the material is insensitive to the presence of cracks and the endurance limit is the dominant parameter. If one wanted to design a notched component so as to avoid any fatigue crack growth, then depending on the initial notch or flaw size, the allowable stress amplitude would have to fall within the indicated area of "no propagation." However, a number of factors can shrink this area in service: corrosion, surface damage, the endurance limit, and the threshold value. The selection of a material of higher strength to improve the endurance limit would most likely result in a decrease in ΔK_{th} but not in $\Delta K_{eff(th)}$, so that there should be some expansion of the no-propagation region. An increase in R value over that shown here for $R = -1$ conditions should lead to a decrease in the endurance limit and hence a decrease in the no-propagation region. At high R values, $\Delta K_{eff(th)}$ and ΔK_{th} would merge, and the nonpropagation of fatigue cracks should not be observed because closure would be absent.

It can also be noted from Fig. 18 that fatigue notch sensitivity is related to crack closure, in that higher stress amplitudes are required for crack propagation from small notches than from large notches of the same geometrical shape.

Figure 19, based on Fig. 18, shows the crack and no-propagation regions as a function of the initial stress-concentration factor. This figure is of the same type as that originally developed by Frost and Dugdale (Fig. 2) and provides a rationale, based on crack closure, for their observations.

In a meaningful analysis of near-threshold fatigue crack growth behavior in service, a number

of complicating factors may also have to be considered. For example, Koterazawa (Ref 182) has observed that crack propagation rates were accelerated by more than 100 times in some cases by understressing below the threshold, with this effect more pronounced in low-strength materials. It has also been observed (Ref 183) that a very small number of cycles of overstress, applied intermittently during a very large number of cycles of understress below threshold, caused significant acceleration in crack growth rate as compared to steady cyclic stress in moist air, dry air, and nitrogen. It has also been observed (Ref 184) that prolonged in-service exposure of a rotor steel at elevated temperature led to a decrease in ΔK_{th} due to the precipitation of carbides in the material. Geary and King (Ref 185) have demonstrated that residual stresses can also exert a strong influence on near-threshold fatigue crack behavior.

As indicated above, in design the effect of notches may have to be considered. Lukáš et al. (Ref 186) have examined the limiting case of nondamaging notches in fatigue, and Ogura et al. (Ref 187) have dealt with the threshold behavior of small fatigue cracks at notches in type 304 stainless steel, with nonpropagation occurring when the value of ΔK_{eff} reached its threshold level.

The practical significance of the fatigue crack growth threshold condition has also been discussed in relation to engineering design considerations by, for example, Austen and Walker (Ref 188), who considered corrosion fatigue crack growth and lifetime predictions for offshore environments. Harrison has written on damage-tolerant design (Ref 189), and Brook (Ref 190) has assessed the significance of the threshold as a design parameter. In a study of the influence of orientation on the fatigue strength of Ni-Cr-Mo-V steels, Nix and Lindley (Ref 191) used LEFM to calculate ΔK_{th} values for fatigue crack growth from inclusions.

REFERENCES

1. S. Suresh, *Fatigue of Materials,* Cambridge University Press, 1991
2. N.E. Frost and D.S. Dugdale, *J. Mech. Phys. Solids,* Vol 5, 1957, p 182
3. N.E. Frost, *Proc. Instn. Mech. Engrs.,* Vol 173, 1959, p 811
4. N.E. Frost, L.P. Pook, and K. Denton, *Eng. Fracture Mech.,* Vol 3, 1971, p 109
5. N.E. Frost and A.F. Greenan, *J. Mech. Eng. Sci.,* Vol 12, 1970, p 159
6. P.C. Paris et al, Extensive Study of Low Fatigue Crack Growth Rates in A533 and A508 Steels, ASTM STP 513, 1972, p 141-176
7. W. Elber, *Eng. Fract. Mech.,* Vol 2, 1970, p 37-45
8. R. Pippan, *Fatigue Fract. Eng. Mater. Struct.,* Vol 9, 1987, p 319-328
9. A. Ohta, I. Soya, S. Nishijima, and M. Kosuge, Statistical Evaluation of Fatigue Crack Propagation Properties Including Threshold Stress Intensity Factor, *Eng. Fract. Mech.,* Vol 24 (No. 6), 1986, p 789-802
10. A.J. McEvily and Z. Yang, The Nature of the Two Opening Loads Following an Overload in Fatigue Crack Growth, *Met. Trans.,* Vol 21A, 1990, p 2717-2727
11. G.T. Gray III, A.W. Thompson, and J.C. Williams, The Effect of Microstructure on Fatigue Crack Path and Crack Propagation Rate, *Fatigue Crack Growth Threshold Concepts,* The Metallurgical Society/AIME, 1984, p 131-143
12. A.F. Blom, Near-Threshold Fatigue Crack Growth and Crack Closure in 17-4 PH Steel and 2024-T3 Aluminum Alloy, *Fatigue Crack Growth Threshold Concepts,* The Metallurgical Society/AIME, 1984, p 263-279
13. N. Suzuki, T. Mawari, and A. Ohta, Minor Role of Fractographic Features in Basic Fatigue Crack Propagation Properties, *Int. J. Fracture,* Vol 54 (No. 2), 1992, p 131-138
14. H. Kobayashi, T. Ogawa, H. Nakamura, and H. Nakazawa, Oxide Induced Fatigue Crack Closure and Near-Threshold Characteristics in A508-3 Steel, *Advances in Fracture Research (Fracture 84),* Vol 4, Pergamon Press Ltd., 1984, p 2481-2488
15. H. Kobayashi, T. Ogawa, H. Nakamura, and H. Nakazawa, Oxide Induced Fatigue Crack Closure and Near-Threshold Characteristics in A508-3 Steel (Retroactive Coverage), *ICF International Symposium on Fracture Mechanics—Proceedings,* VNU Science Press, 1984, p 718-723
16. A. Otsuka, K. Mori, and T. Miyata, *Eng. Fract. Mech.,* 1975, Vol 7, p 429
17. K. Minakawa and A.J. McEvily, On Crack Closure in the Near-Threshold Region, *Scripta Metall.,* Vol 15, 1981, p 633-636
18. J.C. Newman, Jr., in *Mechanics of Crack Growth,* ASTM STP 590, 1976, p 281-301
19. H. Nakamura and H. Kobayashi, Analysis of Fatigue Crack Closure Caused by Asperities

20. A.K. Vasudevan, K. Sandananda, and N. Louat, Critical Evaluation of Crack Closure and Related Phenomena, *Fatigue '93,* Vol 1, J.-P. Bailon and J.I. Dickson, Ed., Engineering Materials Advisory Services Ltd., Warley, U.K., 1993, p 565-582
21. W.L. Morris and M.R. James, Investigation of the Growth Threshold for Short Cracks, *Fatigue Crack Growth Threshold Concepts,* The Metallurgical Society/AIME, 1984, p 479-495
22. D. Taylor, *Fatigue Thresholds,* Butterworths, London, 1989
23. P.K. Liaw, Overview of Crack Closure at Near-Threshold Fatigue Crack Growth Levels, *Mechanics of Fatigue Crack Closure,* ASTM, 1988, p 62-92
24. C.J. Beevers, K. Bell, and R.L. Carlson, Fatigue Crack Closure and the Fatigue Threshold, *Fatigue Crack Growth Threshold Concepts,* The Metallurgical Society/AIME, 1984, p 327-340
25. C.J. Beevers and R.L. Carlson, A Consideration of the Significant Factors Controlling Fatigue Thresholds, *Fatigue Crack Growth: 30 Years of Progress,* 20 Sept 1984, Pergamon Press Ltd., Cambridge, U.K., 1986, p 89-101
26. K.V. Jata, J.A. Walsh, and E.A. Starke, Jr., Effects of Manganese Dispersoids on Near Threshold Fatigue Crack Growth in 2134 Type Alloys, *Fatigue '87,* Vol I, Engineering Materials Advisory Services Ltd., Warley, U.K., 1987, p 517-526
27. P.M. Sooley and D.W. Hoeppner, A Low-Cost Microprocessor-Based Data Acquisition and Control System for Fatigue Crack Growth Testing, *Automated Test Methods for Fracture and Fatigue Crack Growth,* ASTM, 1985, p 101-117
28. W.J. Mills and L.A. James, Near-Threshold Fatigue Crack Growth Behavior for 316 Stainless Steel, *ASTM J. Test. Eval.,* Vol 15 (No. 6), Nov 1987, p 325-332
29. H.R. Mayer, S.E. Stanzl, and E.K. Tschegg, Fatigue Crack Propagation in the Threshold Regime after Rapid Load Reduction, *Eng. Fract. Mech.,* Vol 40 (No. 6), 1991, p 1035-1043
30. S. Nishijima, S. Matsuoka, and E. Takeuchi, Environmentally Affected Fatigue Crack Growth (Retroactive Coverage), *Fatigue '90,* Materials and Component Engineering Publications, 1990, p 1761-1770
31. R.L. Tobler, J.R. Berger, and A. Bussiba, Long-Crack Fatigue Thresholds and Short Crack Simulation at Liquid Helium Temperature, *Advances in Cryogenic Engineering,* Vol 38A, Plenum Publishing Corp., 1992, p 159-166
32. W.A. Herman, R.W. Hertzberg, and R. Jaccard, Prediction and Simulation of Fatigue Crack Growth under Conditions of Low Crack Closure, *ICF 7: Advances in Fracture Research,* Vol 2, Pergamon Press Ltd., 1989, p 1417-1426

Using the Modified Dugdale Model, *Mechanics of Fatigue Crack Closure,* ASTM, 1988, p 459-474

33. A. Ohta, M. Kosuge, and S. Nishijima, Conservative Data for Fatigue Crack Propagation Analysis, *Int. J. Pressure Vessels Piping,* Vol 33 (No. 4), 1988, p 251-268
34. S. Matsuoka, E. Takeuchi, and M. Kosuge, A Method for Determining Conservative Fatigue Threshold while Avoiding Crack Closure, *ASTM J. Test. Eval.,* Vol 14 (No. 6), Nov 1986, p 312-317
35. E.K. Priddle, The Influence of Test Variables on the Fatigue Crack Growth Threshold, *Fatigue Fract. Eng. Mater. Struct.,* Vol 12 (No. 4), 1989, p 333-345
36. G.C. Salivar and F.K. Haake, *Engr. Fracture Mech.,* Vol 37, 1990, p 505-517
37. S. Suresh, T. Christman, and C. Bull, Crack Initiation and Growth under Far-Field Cyclic Compression: Theory, Experiments and Applications, *Small Fatigue Cracks,* The Metallurgical Society/AIME, 1986, p 513-540
38. W.V. Vaidya, Fatigue Threshold Regime of a Low Alloy Ferritic Steel under Closure-Free Testing Conditions, Part II: Hysteresis in Near-Threshold Fatigue Crack Propagation—An Experimental Assessment, *ASTM J. Test. Eval.,* Vol 20 (No. 3), May 1992, p 168-179
39. R.S. Gates, Fatigue Crack Growth in C-Mn Steel Plate under Narrow Band Random Loading at Near-Threshold Vibration Levels, *Mater. Sci. Eng.,* Vol 80 (No. 1), June 1986, p 15-24
40. T. Ogawa, K. Tokaji, S. Ochi, and H. Kobayashi, The Effects of Loading History on Fatigue Crack Growth Threshold, *Fatigue '87,* Vol II, Engineering Materials Advisory Services Ltd., Warley, U.K., 1987, p 869-878
41. D. Damri, The Effect of Underload Cycling on the Fatigue Threshold in a Structural Steel, *Scripta Metall. Mater.,* Vol 25 (No. 2), Feb 1991, p 283-288
42. W.O. Soboyejo and J.F. Knott, An Investigation of Environmental Effects on Fatigue Crack Growth in Q1N(HY80) Steel, *Met. Trans.,* Vol 21A (No. 11), Nov 1990, p 2977-2983
43. K. Tanaka and Y. Mutoh, Fretting Fatigue under Variable Amplitude Loading, *Fatigue Crack Growth under Variable Amplitude Loading,* Elsevier Applied Science Publishers, 1988, p 64-75
44. K. Muller and H. Harig, Thermometrical Investigations on the Near Threshold Fatigue Crack Propagation Behavior, *Fatigue '87,* Vol II, Engineering Materials Advisory Services Ltd., Warley, U.K., 1987, p 809-818
45. R.B. Thompson, O. Buck, and D.K. Rehbein, Ultrasonic Characterization of Fatigue Crack Closure, *Fracture Mechanics: Twenty-Third Symposium,* ASTM, 1993, p 619-632
46. M.D. Banov, E.A. Konyaev, V.P. Pavelko, and A.I. Urbakh, Determination of the Threshold Stress Intensity Factor by the Method of Acoustic Emission, *Strength of Materials (USSR),* Vol 23 (No. 4), April 1991, p 439-443
47. B. Weiss and R. Stickler, High-Cycle Fatigue Properties of PM-TiAl6V4 Specimens, *Horizons of Powder Metallurgy: Part I,* Verlag Schmid, Dusseldorf, FRG, 1986, p 511-514

48. "Lafarie-Frenot and Gasc (Fat. of Eng Mats and Struct. Vol 6, 1983, p 329)"
49. Fatigue Thresholds, Backlund, Blom, and Beevers, Ed., Engineering Advisory Service, 1982
50. M. Jono, Fatigue Crack Growth Resistance of Structural Materials in Vacuum, *Advanced Materials for Severe Service Applications*, Elsevier Applied Science Publishers, 1987, p 303-324
51. A.K. Vasudevan and P.E. Bretz, Near-Threshold Fatigue Crack Growth Behavior of 7XXX and 2XXX Alloys: A Brief Review, *Fatigue Crack Growth Threshold Concepts*, The Metallurgical Society/AIME, 1984, p 25-42
52. P.E. Bretz, J.I. Petit, and A.K. Vasudevan, The Effects of Grain Size and Stress Ratio on Fatigue Crack Growth in 7091 Aluminum Alloy, *Fatigue Crack Growth Threshold Concepts*, The Metallurgical Society/AIME, 1984, p 163-183
53. E.P. Phillips, The Influence of Crack Closure on Fatigue Crack Growth Thresholds in 2024-T3 Aluminum Alloy, *Mechanics of Fatigue Crack Closure*, ASTM, 1988, p 505-515
54. K. Minakawa, G. Levan, and A.J. McEvily, The Influence of Load Ratio on Fatigue Crack Growth in 7090-T6 and 1N9021-T4 P/M Aluminum Alloys, *Met. Trans.*, Vol 17A (No. 10), Oct 1986, p 1787-1795
55. H.-J. Gudladt, K. Kosche, and J. Petit, Microstructural Aspects of Fatigue Crack Propagation in Al-Zn-Mg Alloys, *Zeitschrift fur Metallkunde*, Vol 84 (No. 5), May 1993, p 301-306
56. R.J.H. Wanhill, "Low Stress Intensity Fatigue Crack Growth in 2024-T3 and T351," Report PB89-146013/XAB, National Aerospace Laboratory (Netherlands), 30 March 1987
57. M. Harrison and J.W. Martin, The Effect of Dispersoids on Fatigue Crack Propagation in Al-Zn-Mg Alloy, *ECF6—Fracture Control of Engineering Structures*, Vol III, Engineering Materials Advisory Services Ltd., Warley, U.K., 1986, p 1503-1510
58. R. Scheffel and K. Detert, Near Threshold Fatigue Crack Propagation and Crack Closure in Al-Mg-Si Alloys with Varying Manganese Concentrations, *ECF6—Fracture Control of Engineering Structures*, Vol III, Engineering Materials Advisory Services Ltd., Warley, U.K., 1986, p 1511-1521
59. J.A. Walsh, K.V. Jata, and E.A. Starke, The Influence of Mn Dispersoid Content and Stress State on Ductile Fracture in 2134 Type Aluminum Alloys, *Acta Metall.*, Vol 37, 1989, p 2861-2871
60. K.T.V. Rao and R.O. Ritchie, Effect of Prolonged High-Temperature Exposure on the Fatigue and Fracture Behavior of Aluminum-Lithium Alloy 2090, *Mater. Sci. Eng.*, Vol 100 (No. 1-2), April 1988, p 23-30
61. G.R. Yoder, P.S. Pao, M.A. Imam, and L.A. Cooley, Unusual Fracture Mode in the Fatigue of an Al-Li Alloy, *ICF 7: Advances in Fracture Research*, Vol 2, Pergamon Press Ltd., 1989, p 919-927

62. P.I. Welch and A.C. Pickard, The Effect of Texture on Fatigue Crack Propagation in Aluminium Alloy 7075, *Aluminium*, Vol 61 (No. 5), May 1985, p 332-335
63. K. Anandan, K.K. Bramha, and K.N. Raju, Fatigue Crack Propagation in Al-Li Alloys, *Science and Technology of Aluminium-Lithium Alloys*, The Aluminium Association of India, 1989, p 187-197
64. W. Yu and R.O. Ritchie, Fatigue Crack Propagation in 2090 Aluminum-Lithium Alloy: Effect of Compression Overload Cycles, *J. Eng. Mater. Technol. (Trans. ASME)*, Vol 109 (No. 1), Jan 1987, p 81-85
65. K.T. Venkateswara, V. Rao, and R.O. Ritchie, Mechanical Properties of Al-Li Alloys, Part II: Fatigue Crack Propagation, *Mater. Sci. Technol.*, Vol 5 (No. 9), Sept 1989, p 896-907
66. C.S. Lee, D.L. Jacobson, and K.S. Shin, Fatigue Crack Growth Behavior of Al-Cu-Li-Mg-Ag Alloys, *Light Materials for Transportation Systems*, Pohang University of Science and Technology, 1993, p 161-169
67. R. Tintillier, H.S. Yang, N. Ranganathan, and J. Petit, Near Threshold Fatigue Crack Growth in an 8090 Lithium-Containing Aluminium Alloy (Retroactive Coverage), *J. Phys. (France)*, Vol 48 (Supp. C3), Sept 1987, p 777-784
68. Y. Xiao and P. Bompard, Low Cycle Fatigue and Fatigue Crack Growth in Al-Li, Al-Li-Zr and 8090 Alloys (Retroactive Coverage), *J. Phys. (France)*, Vol 48 (Supp. C3), Sept 1987, p 737-743
69. H.D. Dudgeon and J.W. Martin, Near-Threshold Fatigue Crack Growth at Room Temperature and an Elevated Temperature in Al-Li Alloy 8090, *Mater. Sci. Eng.*, Vol A150 (No. 2), 29 Feb 1992, p 195-207
70. W. Zhu, K. Minakawa, and A.M. McEvily, On the Influence of the Ambient Environment on the Fatigue Crack Growth Process in Steels, *Eng. Fract. Mech.*, Vol 25 (No. 3), 1986, p 361-375
71. B. London, D.V. Nelson, and J.C. Shyne, The Effect of Tempering Temperature on Near-Threshold Fatigue Crack Behavior in Quenched and Tempered 4140 Steel, *Met. Trans.*, Vol 19A (No. 10), Oct 1988, p 2497-2502
72. J.H. Bulloch and D.J. Bulloch, Influence of Carbon Content and Tempering Temperature on Fatigue Threshold Characteristics of a Low Alloy Steel, *Int. J. Pressure Vessels Piping*, Vol 47 (No. 3), Sept 1991, p 333-354
73. M.T. Yu, T.H. Topper, D.L. DuQuesnay, and M.A. Pompetzki, Fatigue Crack Growth Threshold and Crack Opening of a Mild Steel, *ASTM J. Test. Eval.*, Vol 14 (No. 3), May 1986, p 145-151
74. M.T. Yu and T.H. Topper, The Effects of Material Strength, Stress Ratio, and Compressive Overload on the Threshold Behavior of a SAE 1045 Steel, *J. Eng. Mater. Technol. (Trans. ASME)*, Vol 107 (No. 1), Jan 1985, p 19-25
75. O.N. Romaniv, A.N. Tkach, and Y.N. Lenets, Effect of Fatigue Crack Closure on Near-

Threshold Crack Resistance of Structural Steels, *Fatigue Fract. Eng. Mater. Struct.*, Vol 10 (No. 3), 1987, p 203-212
76. K.S. Ravichandran and D.S. Dwarakadasa, Micromechanics of Fatigue Crack Growth at Low Stress Intensities in a High Strength Steel, *Trans. Indian Institute of Metals*, Vol 44 (No. 5), Oct 1991, p 375-396
77. J.H. Bulloch, Fatigue Crack Growth Threshold Behaviour of Granular Bainitic Microstructures of Differing Carbon Content, *Res. Mech.*, Vol 25 (No. 1), 1988, p 51-69
78. D.J. Bulloch and J.H. Bulloch, The Influence of R-Ratio and Microstructure on the Threshold Fatigue Crack Growth Characteristics of Spheroidal Graphite Cast Irons, *Int. J. Pressure Vessels Piping*, Vol 36 (No. 4), 1989, p 289-314
79. J.H. Bulloch, The Effect of Material Segregation on the Near Threshold Fatigue Crack Propagation Characteristics of a Low Alloy Pressure Vessel in Various Environments, *Int. J. Pressure Vessels Piping*, Vol 33 (No. 3), 1988, p 197-218
80. H.J. Zhou, J. Zeng, H. Gu, and D.Z. Guo, Fatigue Thresholds of Isothermally Transformed Cast Steel and Nodular Cast Iron, *Strength of Metals and Alloys (ICSMA 7)*, Vol 3, Pergamon Press Ltd., 1985, p 2123-2128
81. R.-I. Murakami, Y.H. Kim, and W.G. Ferguson, The Effect of Microstructure and Fracture Surface Roughness on Near Threshold Fatigue Crack Propagation Characteristics of a Two-Phase Cast Stainless Steel, *Fatigue Fract. Eng. Mater. Struct.*, Vol 14 (No. 7), July 1991, p 741-748
82. K. Suzuki, J. Fukakura, and H. Kashiwaya, Near-Threshold Fatigue Crack Growth of Austenitic Stainless Steels at Liquid Helium Temperature, *Advances in Cryogenic Engineering*, Vol 38A, Plenum Publishing Corp., 1992, p 149-158
83. A.J. Cadman, C.E. Nicholson, and R. Brook, Influence of R Ratio and Orientation on the Fatigue Crack Threshold ΔK_{th}, and Subsequent Crack Growth of a Low-Alloy Steel, *Fatigue Crack Growth Threshold Concepts*, The Metallurgical Society/AIME, 1984, p 281-288
84. J.-L. Doong and T.-J. Chen, Effect of Laser Surface Hardening on Fatigue Crack Growth Rate in AISI-4130 Steel, *The Laser vs. the Electron Beam in Welding, Cutting and Surface Treatment: State of the Art*, Bakish Materials Corp., 1987, p 129-143
85. Z.G. Wang, D.L. Chen, X.X. Jiang, C.H. Shih, B. Weiss, and R. Stickler, The Effect of Martensite Content on the Strength and Fatigue Threshold of Dual-Phase Steel (Retroactive Coverage), *Fatigue '90*, Materials and Component Engineering Publications, 1990, p 1363-1368
86. Y. Zheng, Z. Wang, and S. Ai, The Influence of Prestrain and Aging on Near-Threshold Fatigue-Crack Propagation in As-Rolled and Heat-Treated Dual-Phase Steels, *Steel Research*, Vol 62 (No. 5), May 1991, p 223-227

87. J.K. Shang and R.O. Ritchie, On the Development of Unusually High Fatigue Crack Propagation Resistance in Steels: Role of Crack Tip Shielding in Duplex Microstructures, *Mechanical Behaviour of Materials V*, Vol 1, Pergamon Press Ltd., 1988, p 511-519

88. M. Nystrom, B. Karlsson, and J. Wasen, Fatigue Crack Growth of Duplex Stainless Steels, *Duplex Stainless Steels '91*, Vol 2, Les Editions de Physique, 1992, p 795-802

89. J. Wasen, B. Karlsson, and M. Nystrom, Fatigue Crack Growth Properties of SAF 2205, *Nordic Symposium on Mechanical Properties of Stainless Steels*, Institute for Metallforskning, 1990, p 122-135

90. H. Nakamura, K. Murali, K. Minakawa, and A.J. McEvily, Fatigue Crack Growth in Ferritic Steels as Influenced by Elevated Temperature and Environment, *Proc. Int. Conf. on Microstructure and Mechanical Behavior of Materials*, Engineering Materials Advisory Services Ltd., Warley, U.K., 1986, p 43-57

91. H. Kobayashi, H. Tsuji, and K.D. Park, Effect of Crack Surface Oxidation on Near-Threshold Fatigue Crack Growth Characteristics in A508-3 Steel at Elevated Temperature, *Fracture and Strength '90*, Trans Tech Publications, 1991, p 355-360

92. S. Matsuoka, E. Takeuchi, S. Nishijima, and A.J. McEvily, Near-Threshold Fatigue Crack Growth Properties at Elevated Temperature for 1Cr-1Mo-0.25V Steel and 12Cr Stainless Steel, *Met. Trans.*, Vol 20A (No. 4), April 1989, p 741-749

93. I. Nishikawa, T. Gotoh, Y. Miyoshi, and K. Ogura, The Role of Crack Closure on Fatigue Threshold at Elevated Temperatures, *JSME Int. J. I*, Vol 31 (No. 1), Jan 1988, p 92-99

94. K. Ohji, S. Kubo, and Y. Nakai, Near-Threshold Fatigue Crack Growth Behavior at High Temperatures, *Creep: Characterization, Damage and Life Assessments*, ASM International, 1992, p 379-388

95. K. Makhlouf and J.W. Jones, Near-Threshold Fatigue Crack Growth Behaviour of a Ferritic Stainless Steel at Elevated Temperatures, *Int. J. Fatigue*, Vol 14 (No. 2), March 1992, p 97-104

96. A.T. Alpas, L. Edwards, and C.N. Reid, The Effect of R-Ratio on Near Threshold Fatigue Crack Growth in a Metallic Glass and a Stainless Steel, *Eng. Fract. Mech.*, Vol 36 (No. 1), 1990, p 77-92

97. J.C. Chesnutt and J.A. Wert, Effect of Microstructure and Load Ratio on ΔK_{th} in Titanium Alloys, *Fatigue Crack Growth Threshold Concepts*, The Metallurgical Society/AIME, 1984, p 83-97

98. W.O. Soboyejo, An Investigation of the Effects of Microstructure on the Fatigue and Fracture Behavior of $\alpha_2 + \beta$ Forged Ti-24Al-11Nb, *Met. Trans.*, Vol 23A (No. 6), June 1992, p 1737-1750

99. G. Haicheng and S. Shujuan, Microstructural Effect on Fatigue Thresholds in β Titanium Alloy Ti-10V-2Fe-3Al (Retroactive Coverage), *Fatigue '90*, Materials and Component Engineering Publications, 1990, p 1929-1934

100. K.S. Ravichandran and E.S. Dwarakadasa, Fatigue Crack Growth Transitions in Ti-6Al-4V Alloy, *Scripta Metall.*, Vol 23 (No. 10), Oct 1989, p 1685-1690

101. K.S. Ravichandran, Fatigue Crack Growth Behavior Near Threshold in Ti-6Al-4V Alloy: Microstructural Aspects (Retroactive Coverage), *Fatigue '90*, Materials and Component Engineering Publications, 1990, p 1345-1350

102. G. Lutjering, A. Gysler, and L. Wagner, Fatigue and Fracture of Titanium Alloys, *Light Metals: Advanced Materials Research and Developments for Transport 1985*, Les Editions de Physique, 1986, p 309-321

103. L. Wagner and G. Lutjering, Propagation of Small Fatigue Cracks in Titanium Alloys, *Sixth World Conference on Titanium I*, Les Editions de Physique, 1988, p 345-350

104. J.R. Jira, T. Nicholas, and D.A. Nagy, Influences of Crack Closure and Load History on Near-Threshold Crack Growth Behavior in Surface Flaws, *Surface-Crack Growth: Models, Experiments and Structures*, ASTM, 1990, p 303-314

105. C. Gerdes, A. Gysler, and G. Lutjering, Propagation of Small Surface Cracks in Titanium Alloys, *Fatigue Crack Growth Threshold Concepts*, The Metallurgical Society/AIME, 1984, p 465-478

106. R.H. Dauskardt, T.W. Duerig, and R.O. Ritchie, Effects of in Situ Phase Transformation on Fatigue-Crack Propagation in Titanium-Nickel Shape-Memory Alloys, *Shape Memory Materials*, Vol 9, *Proc. MRS International Meeting on Advanced Materials*, Materials Research Society, 1989, p 243-249

107. J.E. Allison and J.C. Williams, Near-Threshold Fatigue Crack Growth Phenomena at Elevated Temperature in Titanium Alloys, *Scripta Metall.*, Vol 19 (No. 6), June 1985, p 773-778

108. K. Ogura and I. Nishikawa, Fatigue Threshold and Closure at Moderately Elevated Temperatures (Retroactive Coverage), *Fatigue '90*, Materials and Component Engineering Publications, 1990, p 1413-1418

109. J. Petit, W. Berata, and B. Boucher, Fatigue Crack Growth Behavior of Ti-6Al-4V at Elevated Temperature in High Vacuum, *Scripta Metall. Mater.*, Vol 26 (No. 12), 15 June 1992, p 1889-1894

110. G.C. Salivar, J.E. Heine, and F.K. Haake, The Effect of Stress Ratio on the Near-Threshold Fatigue Crack Growth Behavior of Ti-8Al-1Mo-1V at Elevated Temperature, *Eng. Fract. Mech.*, Vol 32 (No. 5), 1989, p 807-817

111. J.H. Bulloch and I. Schwartz, Fatigue Crack Extension Behavior in Porous Plasma Spray 80Ni-20Cr Material: The Influence of R-Ratio, *Theoret. Appl. Fract. Mech.*, Vol 15 (No. 2), July 1991, p 143-154

112. D.M. Knowles and J.E. King, Fatigue Crack Propagation Testing of Particulate MMCs, *Test Techniques for Metal Matrix Composites*, IOP Publishing Ltd., 1991, p 98-109

113. K. Tanaka, M. Kinefuchi, and Y. Akiniwa, Fatigue Crack Propagation in SiC Whisker Reinforced Aluminum Alloy, *Fatigue '90*, Mate-

114. D.M. Knowles, T.J. Downes, and J.E. King, Crack Closure and Residual Stress Effects in Fatigue of a Particle-Reinforced Metal Matrix Composite, *Acta Metall. Mater.*, Vol 41 (No. 4), April 1993, p 1189-1196

115. D.L. Davidson, Fracture Characteristics of Al-4%Mg Mechanically Alloyed with SiC, *Met. Trans.*, Vol 18A (No. 12), Dec 1987, p 2115-2128

116. M. Levin, B. Karlsson, and J. Wasen, The Fatigue Crack Growth Characteristics and Their Relation to the Quantitative Fractographic Appearance in a Particulate Al 6061/SiC Composite Material, *Fundamental Relationships between Microstructures and Mechanical Properties of Metal Matrix Composites*, The Minerals, Metals and Materials Society, 1990, p 421-439

117. C. Masuda, Y. Tanaka, Y. Yamamoto, and M. Fukazawa, Fatigue Crack Propagation Properties and Its Mechanism for SiC Whiskers or SiC Particulates Reinforced Aluminum Alloys Matrix Composites, *Structural Composites: Design and Processing Technologies*, ASM International, 1990, p 565-573

118. G. Liu, D. Yao, and J.K. Shang, Fatigue Crack Growth Behaviour of a Cast Particulate Reinforced Aluminum-Alloy Composite, *Advances in Production and Fabrication of Light Metals and Metal Matrix Composite*, Canadian Institute of Mining, Metallurgy and Petroleum, 1992, p 665-671

119. J.K. Shang and R.O. Ritchie, Fatigue of Discontinuously Reinforced Metal Matrix Composites, *Metal Matrix Composites: Mechanisms and Properties*, Academic Press Inc., 1991, p 255-285

120. J.K. Shang and R.O. Ritchie, On the Particle-Size Dependence of Fatigue-Crack Propagation Thresholds in SiC-Particulate-Reinforced Aluminum-Alloy Composites: Role of Crack Closure and Crack Trapping, *Acta Metall.*, Vol 37 (No. 8), Aug 1989, p 2267-2278

121. C.P. You and J.E. Allison, Fatigue Crack Growth and Closure in a SiCp-Reinforced Aluminum Composite, *ICF 7: Advances in Fracture Research*, Vol 4, 1989, p 3005-3012

122. K. Hirano and H. Takizawa, Evaluation of Fatigue Crack Growth Characteristics of Whisker-Reinforced Aluminium Alloy Matrix Composite, *JSME Int. J.*, Series I, Vol 34 (No. 2), April 1991, p 221-227

123. M. Levin and B. Karlsson, Influence of SiC Particle Distribution and Prestraining on Fatigue Crack Growth Rates in Aluminum AA 6061/SiC Composite Material, *Mater. Sci. Technol.*, Vol 7 (No. 7), July 1991, p 596-607

124. K. Hirano, Fatigue Crack Growth Characteristics of Metal Matrix Composites, *Mechanical Behaviour of Materials VI*, Vol 3, Pergamon Press Ltd., 1992, p 93-100

125. J.-K. Shang and R.O. Ritchie, Monotonic and Cyclic Crack Growth in a TiC-Particulate-Reinforced Ti-6Al-4V Metal-Matrix Composite,

Scripta Metall. Mater., Vol 24 (No. 9), Sept 1990, p 1691-1694

126. H. Toda and T. Kobayashi, Fatigue Crack Initiation and Growth Characteristics of SiC Whisker Reinforced Aluminum Alloy Composites, *Mechanisms and Mechanics of Composites Fracture*, ASM International, p 55-63

127. M. Levin and B. Karlsson, Fatigue Behavior of a Saffil-Reinforced Aluminium Alloy, *Composites*, Vol 24 (No. 3), 1993, p 288-295

128. K. Hirano, Near-Threshold Transverse Fatigue Crack Growth Characteristics of Unidirectionally Continuous Fiber Reinforced Metals, *Proceedings of the Fourth Japan-U.S. Conference on Composite Materials*, Technomic Publishing Co., Inc., 1989, p 633-642

129. D.L. Davidson and L.K. Austin, Fatigue Crack Growth through ARALL-4 at Ambient Temperature, *Fatigue Fract. Eng. Mater. Struct.*, Vol 14 (No. 10), 1991, p 939-951

130. S.E. Stanzl-Tschegg, M. Papakyriacou, H.R. Mayer, J. Schijve, and E.K. Tschegg, High-Cycle Fatigue Crack Growth Properties of Aramid-Reinforced Aluminum Laminates, *Composite Materials: Fatigue and Fracture*, Vol 4, ASTM, 1993, p 637-652

131. J.-K. Chang and R. Ritchie, *Scripta Metall. Mater.*, Vol 24, 1990, p 1691-1694

132. D.L. Davidson and J.B. Campbell, Fatigue Crack Growth through the Lamellar Microstructure of an Alloy Based on TiAl at 25 and 800 °C, *Met. Trans.*, Vol 24A (No. 7), July 1993, p 1555-1574

133. S.F. Stanzl, H.R. Mayer, and E.K. Tschegg, The Influence of Air Humidity on Near-Threshold Fatigue Crack Growth of 2024-T3 Aluminum Alloy, *Mater. Sci. Eng.*, Vol A147 (No. 1), 30 Oct 1991, p 45-54

134. O. Vosikovsky, W.R. Neill, D.A. Carlyle, and A. Rivard, The Effect of Sea Water Temperature on Corrosion Fatigue-Crack Growth in Structural Steels, *Can. Metall. Q.*, Vol 26 (No. 3), July-Sept 1987, p 251-257

135. W.W. Gerberich, Fatigue and Hydrogen Diffusion (Retroactive Coverage), *Hydrogen Degradation of Ferrous Alloys*, Noyes Publications, 1985, p 366-413

136. P.S. Pao, M. Gao, and R.P. Wei, Environmentally Assisted Fatigue-Crack Growth in 7075 and 7050 Aluminum Alloys, *Scripta Metall.*, Vol 19 (No. 3), March 1985, p 265-270

137. R.S. Piascik and R.P. Gangloff, Intrinsic Fatigue Crack Propagation in Aluminum-Lithium Alloys: The Effect of Gaseous Environments, *ICF 7: Advances in Fracture Research*, Vol 2, Pergamon Press Ltd., 1989, p 907-918

138. K.S. Shin and S.S. Kim, Environmental Effects on Fatigue Crack Propagation of a 2090 Al-Li Alloy, *Hydrogen Effects on Material Behavior*, The Minerals, Metals and Materials Society, 1990, p 919-928

139. R.S. Piascik, "Environmental Fatigue in Aluminum-Lithium Alloys," Report N92-32423/5/XAB, NASA Langley Research Center, 1992

140. R.B. Scarlin, C. Maggi, and J. Denk, Corrosion Fatigue Failure Mechanisms of Steam Turbine Rotor Materials (Retroactive Coverage), *Fatigue '90*, Materials and Component Engineering Publications, 1990, p 1857-1862

141. J.A. Todd, P. Li, G. Liu, and V. Raman, A New Mechanism of Crack Closure in Cathodically Protected ASTM A710 Steel, *Scripta Metall.*, Vol 22 (No. 6), June 1988, p 745-750

142. P.K. Liaw, T.R. Leax, and J.K. Donald, Gaseous-Environment Fatigue Crack Propagation Behavior of a Low-Alloy Steel, *Fracture Mechanics: Perspectives and Directions—20th Symposium*, ASTM, 1989, p 581-604

143. M.N. James and J.F. Knott, Near-Threshold Fatigue Crack Closure and Growth in Air and Vacuum, *Scripta Metall.*, Vol 19 (No. 2), Feb 1985, p 189-194

144. J.M. Kendall and J.F. Knott, Near-Threshold Fatigue Crack Growth in Air and Vacuum, *Basic Questions in Fatigue*, Vol II, ASTM, 1988, p 103-114

145. P. Smith, The Effects of Moisture on the Fatigue Crack Growth Behaviour of a Low Alloy Steel Near Threshold, *Fatigue Fract. Eng. Mater. Struct.*, Vol 10 (No. 4), 1987, p 291-304

146. C. Fong and D. Troman, High Frequency Stage I Corrosion Fatigue of Austenitic Stainless Steel (316L), *Met. Trans.*, Vol 19A (No. 11), Nov 1988, p 2753-2764

147. G. Henaff, J. Petit, and B. Bouchet, Environmental Influence on the Near-Threshold Fatigue Crack Propagation Behaviour of a High-Strength Steel, *Int. J. Fatigue*, Vol 14 (No. 4), July 1992, p 211-218

148. K.A. Esaklul and W.W. Gerberich, Internal Hydrogen Degradation of Fatigue Thresholds in HSLA Steel, *Fracture Mechanics: 16th Symposium*, ASTM, 1985, p 131-148

149. S. Matsuoka, H. Masuda, and M. Shimodaira, Fatigue Threshold and Low-Rate Crack Propagation Properties for Structural Steels in 3% Sodium Chrloride Aqueous Solution, *Met. Trans.*, Vol 21A (No. 8), Aug 1990, p 2189-2199

150. E. Bardal, Effects of Free Corrosion and Cathodic Protection on Fatigue Crack Growth in Structural Steel in Seawater, *Eurocorr '87*, Dechema, 1987, p 451-457

151. K. Komai, Corrosion-Fatigue Crack Growth Retardation and Enhancement in Structural Steels, *Current Research on Fatigue Cracks*, Elsevier Applied Science Publishers, 1987, p 267-289

152. B. Dogan, G. Terlinde, and K.-H. Schwalbe, Effect of Yield Stress and Environment on Fatigue Crack Propagation of Aged Ti-10V-2Fe-3Al, *Sixth World Conference on Titanium I*, Les Editions de Physique, 1988, p 181-186

153. P.K. Datta, K.N. Strafford, and A.L. Dowson, Environment/Mechanical Interaction Processes and Hydrogen Embrittlement of Titanium, *Mech. Corros. Prop. A*, No. 8, 1985, p 203-216

154. P.K. Liaw and W.A. Logsdon, Fatigue Crack Growth Threshold at Cryogenic Temperatures: A Review, *Eng. Fract. Mech.*, Vol 22 (No. 4), 1985, p 585-594

155. J. Petit, Some Aspects of Near-Threshold Crack Growth: Microstructural and Environmental Effects, *Fatigue Crack Growth Threshold Concepts*, The Metallurgical Society/AIME, 1984, p 3-24

156. J.P. Bailon, M. El Boujani, and J.I. Dickson, Environmental Effects on Threshold Stress Intensity Factor in 70-30 Alpha Brass and 2024-T351 Aluminum Alloy, *Fatigue Crack Growth Threshold Concepts*, D. Davidson and S. Suresh, Ed., The Metallurgical Society of AIME, 1984, p 63-82

157. A. Sengupta, A. Spis, and S.K. Putatunda, The Effect of Overload on Corrosion Fatigue Crack Growth Behavior of a Low Alloy Steel in Threshold Region, *J. Mater. Eng.*, Vol 13 (No. 3), Sept 1991, p 229-236

158. K. Horikawa, A. Sakakibara, and T. Mori, The Effect of Welding Tensile Residual Stresses on Fatigue Crack Propagation in Low Propagation Rate Region, *Trans. Jpn. Weld. Res. Inst.*, Vol 18 (No. 2), 1989, p 125-132

159. A. Ohta, E. Sasaki, M. Kosuge, and S. Nishijima, Fatigue Crack Growth and Threshold Stress Intensity Factor for Welded Joints, *Current Research on Fatigue Cracks*, Elsevier Applied Science Publishers, 1987, p 181-200

160. A. Ohta, Y. Maeda, S. Machida, and H. Yoshinari, Near-Threshold Fatigue Crack Propagation in Welded Joints under Random Loadings, *Trans. Jpn. Weld. Soc.*, Vol 19 (No. 2), Oct 1988, p 148-153

161. L. Bartosiewicz, A.R. Krause, A. Sengupta, and S.K. Putatunda, Application of a New Model for Fatigue Threshold in a Structural Steel Weldment, *Eng. Fract. Mech.*, Vol 45 (No. 4), July 1993, p 463-477

162. A. Otsuka, K. Mori, and K. Tohgo, Mode II Fatigue Crack Growth in Aluminum Alloys, *Current Research on Fatigue Cracks*, Elsevier Applied Science Publishers, 1987, p 149-180

163. R.O. Ritchie, A Comparison of Fatigue Crack Propagation in Modes I and III, *Fracture Mechanics: 18th Symposium*, ASTM, 1988, p 821-842

164. A.K. Hellier and D.J.H. Corderoy, Near Threshold Fatigue Crack Growth in Steels under Mode II/Mode III Loading, *Australian Fracture Group 1990 Symposium*, 1990, p 164-175

165. L.P. Pook and D.G. Crawford, The Fatigue Crack Direction and Threshold Behaviour of a Medium Strength Structural Steel under Mixed Mode I and III Loading (Retroactive Coverage), *Fatigue under Biaxial and Multiaxial Loading*, Mechanical Engineerings Publications Ltd., 1991, p 199-211

166. E.K. Tschegg, M. Czegley, H.R. Mayer, and S.E. Stanzl, Influence of a Constant Mode III Load on Mode I Fatigue Crack Growth Thresholds (Retroactive Coverage), *Fatigue under Biaxial and Multiaxial Loading*, Mechanical Engineering Publications Ltd., 1991, p 213-222

167. A.K. Hellier, D.J.H. Corderoy, and M.B. McGirr, Some Observations on Mode III Fatigue Thresholds, *Int. J. Fract.*, Vol 29 (No. 4), Dec 1985, p R45-R48

168. H.J. Roven and E. Nes, Cyclic Deformation of Ferritic Steel, Part II: Stage II Crack Propagation, *Acta Metall. Mater.*, Vol 39 (No. 8), Aug 1991, p 1735-1754

169. H. Cai, University of Connecticut, unpublished research

170. R. Pippan, Dislocation Emission and Fatigue Crack Growth Threshold, *Acta Metall. Mater.*, Vol 39 (No. 3), March 1991, p 255-262

171. J.C. Radon and L. Guerra-Rosa, A Model for Ultra-Low Fatigue Crack Growth, *Fatigue '87*, Vol II, Engineering Materials Advisory Services Ltd., Warley, U.K., 1987, p 851-859

172. A. Ohta, N. Suzuki, and T. Mawari, Effect of Young's Modulus on Basic Crack Propagation Properties Near the Fatigue Threshold, *Int. J. Fatigue*, Vol 14 (No. 4), July 1992, p 224-226

173. K.S. Ravichandran, A Rationalisation of Fatigue Thresholds in Pearlitic Steels Using a Theoretical Model, *Acta Metall. Mater.*, Vol 39 (No. 6), June 1991, p 1331-1341

174. K.S. Ravichandran, A Theoretical Model for Roughness Induced Crack Closure, *Int. J. Fract.*, Vol 44 (No. 2), 15 July 1990, p 97-110

175. K.S. Ravichandran and E.S. Dwarakadasa, Theoretical Modeling of the Effects of Grain Size on the Threshold for Fatigue Crack Growth, *Acta Metall. Mater.*, Vol 39 (No. 6), June 1991, p 1343-1357

176. S. Taira, K. Tanaka, and M. Hoshina, Grain Size Effects on Crack Nucleation and Growth in Long-Life Fatigue of Carbon Steel, in ASTM STP 675, 1979, p 135-161

177. K. Tanaka and Y. Nakai, Mechanics of Growth Threshold of Small Fatigue Cracks, *Fatigue Crack Growth Threshold Concepts*, The Metallurgical Society/AIME, 1984, p 497-516

178. J. Lankford, G.R. Leverant, D.L. Davidson, and K.S. Chan, "Study of the Influence of Metallurgical Factors on Fatigue and Fracture of Aerospace Structural Materials," Report AD-A170 218/2/WMS, Southwest Research Institute, Feb 1986

179. W.W. Gerberich, E. Kurman, and W. Yu, Dislocation Substructure and Fatigue Crack Growth, *The Mechanics of Dislocations*, American Society for Metals, 1985, p 169-179

180. T. Mura and J.R. Weertman, Dislocation Models for Threshold Fatigue Crack Growth, *Fatigue Crack Growth Threshold Concepts*, The Metallurgical Society/AIME, 1984, p 531-549

181. M.H. Haddad, T.H. Topper, and K.N. Smith, Fatigue Crack Propagation of Short Cracks, *Eng. Fract. Mech.*, Vol 11, 1979, p 573

182. R. Koterazawa, Acceleration of Fatigue and Creep Crack Propagation under Variable Stresses, *Fatigue Life: Analysis and Prediction*, American Society for Metals, 1986, p 187-196

183. R. Koterazawa and T. Nosho, Acceleration of Crack Growth under Intermittent Overstressing in Different Environments, *Fatigue Fract. Eng. Mater. Struct.*, Vol 15 (No. 1), Jan 1992, p 103-113

184. S.K. Putatunda, I. Singh, and J. Schaefer, Influence of Prolonged Exposure in Service on Fatigue Threshold and Fracture Toughness of a Rotor Steel, *Metallographic Characterization of Metals after Welding, Processing, and Service*, ASM International, 1993, p 441-453

185. W. Geary and J.E. King, Residual Stress Effects during Near-Threshold Fatigue Crack Growth, *Int. J. Fatigue*, Vol 9 (No. 1), Jan 1987, p 11-16

186. P. Lukas, L. Kunz, B. Weiss, and R. Stickler, Non-Damaging Notches in Fatigue, *Fatigue Fract. Eng. Mater. Struct.*, Vol 9 (No. 3), 1986, p 195-204

187. K. Ogura, Y. Miyoshi, and I. Nishikawa, Threshold Behavior of Small Fatigue Crack at Notch Root in Type 304 Stainless Steel, *Eng. Fract. Mech.*, Vol 25 (No. 1), 1986, p 31-46

188. I.M. Austen and E.F. Walker, Corrosion Fatigue Crack Growth Rate Information for Offshore Life Prediction, *SIMS '87 Steel in Marine Structures*, Elsevier Applied Science Publishers, 1987, p 859-870

189. J.D. Harrison, Damage Tolerant Design, *Fatigue Crack Growth: 30 Years of Progress*, Pergamon Press Ltd., 1986, p 117-131

190. R. Brook, An Assessment of the Fatigue Threshold as a Design Parameter, *Fatigue Crack Growth Threshold Concepts*, The Metallurgical Society/AIME, 1984, p 417-429

191. K.J. Nix and T.C. Lindley, The Influence of Orientation on the Fatigue Strength of Ni-Cr-Mo-V Rotor Steels, *Fatigue of Engineering Materials and Structures*, Vol II, Mechanical Engineering Publications, 1986, p 429-436

192. P.A. Reed and J.E. King, Comparision of Long and Short Crack Growth in Polycrystalline and Single Crystals of Udimet 720, in *Short Fatigue Cracks*, Mech Eng Pub, London, 1992 153-168

Behavior of Small Fatigue Cracks

R. Craig McClung, Kwai S. Chan, Stephen J. Hudak, Jr., and David L. Davidson, Southwest Research Institute

FATIGUE CRACKS are small for a significant fraction of the total life of some engineering components and structures. The growth behavior of these small cracks is sometimes significantly different from what would be expected based on conventional (i.e., large-crack) fatigue crack growth (FCG) rate test data and standard FCG design and analysis techniques discussed elsewhere in this Volume. Small fatigue cracks are sometimes observed to grow faster than corresponding large cracks at the same nominal value of the cyclic crack driving force, ΔK. Small cracks have also been observed to grow at nonnegligible rates when the nominal applied ΔK is less than the threshold value, ΔK_{th}, determined from traditional large-crack test methods. Therefore, a structural life assessment based on large-crack analysis methods can be nonconservative if the life is dominated by small-crack growth. In contrast to large-crack growth rates, which generally increase with increasing ΔK, small-crack growth rates are sometimes observed to increase, decrease, or remain constant with increasing ΔK. A variety of typical small-crack growth rate behaviors are illustrated schematically in Fig. 1.

The fundamental reason for this disagreement between measured large-crack and small-crack growth rate data is often a lack of similitude. Although nominal calculated ΔK values for large and small cracks may be the same, the actual driving force for crack growth may be different due to the effects of localized plasticity, crack closure, microstructural influences on crack-tip strain, or localized crack-tip chemistry. In some cases, the basic continuum mechanics assumptions of material homogeneity and small-scale yielding may be violated for small-crack analysis.

Small-crack behavior is a complex subject, due to the variety of factors that may affect small cracks and the variety of microstructures used in engineering structures. Many different researchers have published small-crack data and offered various explanations and models to rationalize these data, and apparent disagreements are not uncommon in the literature. This article is a general introduction to the subject of small cracks that attempts to provide an organizational framework for published data and to summarize the most current understandings of the phenomena. The serious student should consult more extensive review articles (Ref 1-3) and collections of small-crack papers (Ref 4-7) for further details and references.

In this article, different types of small cracks are carefully defined, and different factors that influence small-crack behavior are identified. Appropriate analysis techniques, including both rigorous scientific and practical engineering treatments, are briefly described. Important materials data issues are addressed, including increased scatter in small-crack data and recommended small-crack test methods. Applications where small cracks may be particularly important are highlighted.

Types of Small Cracks

All small cracks are not the same. Different mechanisms are responsible for different types of "small-crack" effects in different settings. Criteria that properly characterize small-crack behavior in one situation may be entirely inappropriate in another situation. It is critical, therefore, to understand the different types of small cracks before selecting suitable analytical treatments. This article considers three types of small cracks: microstructurally small, mechanically small, and chemically small.

One note on nomenclature is needed at the outset. The terms "small crack" and "short crack" both appear in the literature, and sometimes the two appear to be used interchangeably. In recent years, however, the two terms have acquired distinct meanings among many researchers. In the U.S. research community, the currently accepted definition for a "small" crack requires that all physical dimensions (in particular, both the length and depth of a surface crack) are small in comparison to the relevant length scale. The relevant length scale, and hence the specific physical dimensions, vary with the particular material, geometry, and loading of interest. In contrast, a crack is defined as being "short" when only one physical dimension (typically, the length of a through-crack) is small in comparison to the length scale. These definitions are illustrated in Fig. 2. However, it should be noted that this distinction has not always been observed in the literature, and that some current authors (especially in Europe) employ the terms with nearly reverse meanings. Whatever the usage, the reader should carefully observe which type of "little" crack is the subject of a given application. Some of the different implications of short versus small cracks are discussed later in the article.

Microstructurally Small Cracks

A crack is generally considered to be "microstructurally small" when all crack dimensions are small in comparison to characteristic microstruc-

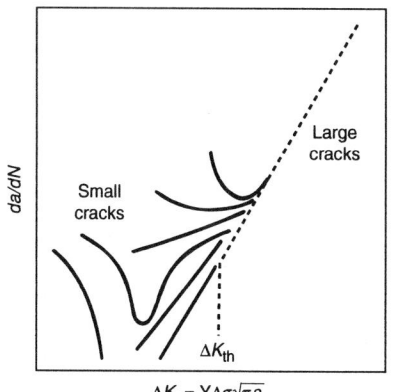

Fig. 1 Typical small-crack growth rate behaviors, in comparison to typical large-crack behavior

Axis labels: da/dN (vertical), $\Delta K = Y\Delta\sigma\sqrt{\pi a}$ (horizontal), Large cracks, Small cracks, ΔK_{th}

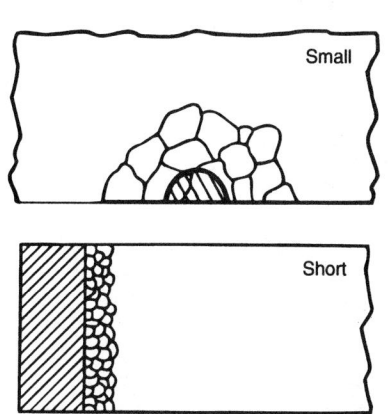

Fig. 2 Schematic of "small" and "short" cracks, including relationship to microstructure

Labels: Small, Short

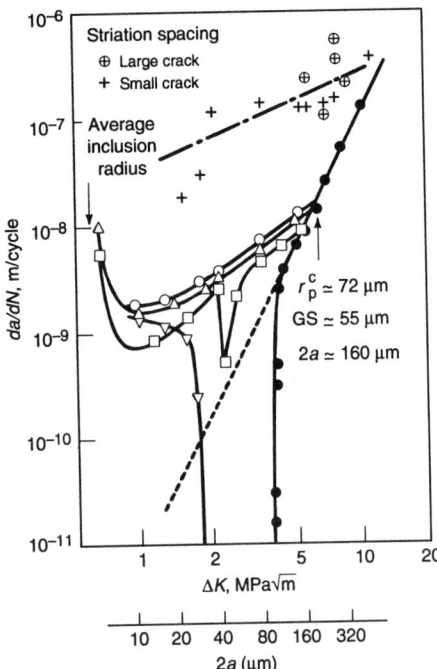

Fig. 3 Typical fatigue crack growth data for microstructurally small cracks and large cracks. Source: Ref 8

tural dimensions. The relevant microstructural feature that defines this scaling may change from material to material, but the most common microstructural scale is the grain size. The small crack and its crack-tip plastic zone may be embedded completely within a single grain, or the crack size may be on the order of a few grain diameters.

Typical crack growth data for microstructurally small cracks are shown for a 7075 aluminum alloy in Fig. 3, along with traditional large-crack data for the same material (Ref 8). Note that small-crack growth can occur at nominal ΔK values below the large-crack threshold. Small-crack growth rates are often faster than would be predicted by the extrapolated large-crack Paris equation (the dashed line in Fig. 3), and the apparent Paris slope for the small-crack data can be smaller than for the large-crack data. Crack arrest (momentary or permanent) can occur at these low ΔK values, and this arrest is often observed to occur when the crack size, a, is on the order of the grain size (GS) (i.e., when the crack tip encounters a grain boundary). However, not all small cracks arrest or even slow down at these microstructural barriers. As the crack continues to grow, the small-crack da/dN data often merge with large-crack data.

Why do microstructurally small cracks behave this way? Several factors are involved, all related to the loss of microstructural and mechanical similitude (Ref 9, 10). When the crack-tip cyclic plastic zone size, r_p^c, (and sometimes the crack itself) is embedded within the predominant microstructural unit (e.g., a single grain), the crack-tip plastic strain range is determined by the properties of individual grains and not by the

continuum aggregate. The growth rate acceleration of small cracks embedded within a single surface grain is primarily due to enhancement of the local plastic strain range that results from a lower yield stress for optimum slip in the surface grains. This microplastic behavior also causes (and, in turn, is affected by) changes in crack closure behavior.

As a small crack approaches a grain boundary, the fatigue crack may accelerate, decelerate, or even arrest, depending on whether or not slip propagates into the contiguous grain. The transmission of slip across a grain boundary in turn depends on the grain orientation, the activities of secondary and cross slip, and the planarity of slip. The transition of the small crack from one grain to another may require a change in the crack path, which may also influence crack closure. The resulting crack growth behavior is therefore very sensitive to the crystallographic orientation and properties of individual grains located within the cyclic plastic zone. As the crack grows, the number of grains interrogated by the crack-tip plastic zone increases, and the statistically averaged material properties become smoother.

However, it is important to note that the fundamental mechanism of crack growth is often the same for small and large cracks in the near-threshold regime. In both cases, FCG occurs as an intermittent process involving strain range accumulation and incremental crack extension, followed by a waiting period during which plastic strain range reaccumulates at the crack tip. Fatigue striations of equivalent spacing have been observed on the fracture surfaces of both large and small fatigue cracks tested under equivalent nominal ΔK ranges, as shown in Fig. 3 for 7075 aluminum alloy. The essential difference between large and small cracks is that the number of fatigue cycles per striation is less for small cracks, due to differences in the local crack driving force.

Other factors may also influence microstructurally small cracks. In some cases, small cracks are stage I shear cracks oriented along preferred crystallographic directions, which exhibit different resistance to crack advance. Microstructurally induced changes in crack path can influence the development of crack closure due to crack surface roughness. In addition, many microstructurally small cracks grow under relatively large applied stresses, which further magnifies near-tip plasticity effects.

How can the behavior of microstructurally small cracks be modeled or predicted analytically? Many different approaches have been developed, ranging from detailed scientific models to simplified engineering treatments. However, no single approach has demonstrated widespread applicability. The fundamental problem is that the customary linear elastic fracture mechanics (LEFM) parameter ΔK is, strictly speaking, an invalid representation of the crack driving force in the presence of enhanced near-tip plasticity and microstructural inhomogeneity. Unfortunately, no obvious alternative to ΔK has been widely accepted as a correlating parameter for

microstructural small-flaw growth. In view of the widespread use of ΔK for large-crack analysis, many researchers and engineers have attempted to describe microstructurally small crack growth in terms of some modified ΔK.

At one extreme, complex micromechanical models attempt to address directly the changes in the local crack driving force and the local microstructure. For example, some models are based on a modified Dugdale crack in an idealized microstructure with microplastic grains and grain boundaries (e.g., Ref 9). The nominal ΔK may be modified by influence functions that explicitly describe the effects of microplastic/macroplastic yield strength, large-scale yielding at the crack tip, and crack closure. More general phenomenological models motivated by detailed experimental measurements of near-tip strains and displacements employ an "equivalent" ΔK incorporating a plastic component and a closure-modified elastic component.

Simpler mechanical treatments have also been proposed to address FCG behavior in the microstructurally small crack regime. The attractive simplicity of these models is that they avoid dealing directly with complex microstructural issues. Small-crack acceleration effects are incorporated through simple modifications to mechanical parameters in the expression for the crack driving force. El Haddad (Ref 11), for example, replaced the actual crack length a by an effective length ($a + a_0$) to calculate ΔK, which enhances the predicted crack growth when a is very small. A more sophisticated approach has been developed by Newman (Ref 12). The Newman model is based on computed changes in plasticity-induced crack closure for small cracks growing out of initiation sites simulated as micronotches. Newman has shown reasonably good success in predicting small-crack growth rates and total fatigue lives for several different materials, but it should be remembered that the simple mechanical treatments do not address the most fundamental causes of the microstructurally small crack effect. Hence, the generality of the models cannot be ensured.

Simpler, more empirical engineering approaches may be useful for some practical applications in which it is not possible or practical to address changes in the driving force explicitly. Stochastic treatments that acknowledge the inherent uncertainties associated with microstructurally small crack growth address this uncertainty through appropriate statistical techniques. Formulation and calibration of these techniques may require extensive analysis of statistical-quality small-crack data, which is a limitation. Variability of small-crack data is discussed further below. Conservative bounding approaches that simply draw some upper bound to the crack growth data in the defined small-crack regime, or fitting approaches that perform regression on small-crack data to generate a new set of Paris equation constants, are also possible. These engineering treatments may be a useful means of avoiding detailed analysis, especially when small-crack data are available for materials and

Fig. 4 Schematic of relationship between mechanically small cracks and plastic zones

load histories representative of service conditions.

Based on these observations and models, several practical suggestions can be offered to predict growth rates for microstructurally small cracks. In general, it appears that the large-crack Paris equation can be extrapolated downward at least to some microstructural limit, neglecting the large-crack threshold. Some treatment of nominal plasticity and crack closure effects on the crack driving force (discussed at more length in the next section) may be useful to improve agreement with large-crack data. However, it must be emphasized that some nonconservatism may remain if the true local microstructural effects have not been addressed. Guidance for addressing these effects can be obtained from various scientific approaches, although practical considerations may dictate the use of more general engineering approaches.

Mechanically Small Cracks

A crack is generally considered to be "mechanically small" when all crack dimensions are small compared to characteristic mechanical dimensions. The relevant mechanical feature is typically a zone of plastic deformation, such as the crack-tip plastic zone or a region of plasticity at some mechanical discontinuity (e.g., a notch). The crack may be fully embedded in the plastic zone, or the plastic zone size may simply be a large fraction of the crack size, as illustrated by Fig. 4. As discussed below, many microstructurally small cracks are also mechanically small, but our focus in this section is on mechanically small cracks that are microstructurally large. The "short" crack, as defined above, also behaves in the same manner as the mechanically small crack. The crack front of a short crack interrogates many different grains and hence is not subject to strong microstructural effects.

Typical crack growth data for mechanically small cracks in unnotched configurations are shown in Fig. 5 for a high-strength, low-alloy (HSLA) steel (Ref 13). Note again that small-crack growth can occur below the large-crack threshold. The slope of the Paris equation often appears to be roughly the same for small- and large-crack data, but the small-crack data sometimes fall above the large-crack trend line when expressed in terms of nominal ΔK.

Small or short cracks growing in notch fields can exhibit a characteristic "fish-hook" growth behavior, as illustrated in Fig. 6 (Ref 14). Here small-crack growth rates are much faster than for comparable large cracks when the cracks are extremely small in comparison to the notch dimen-

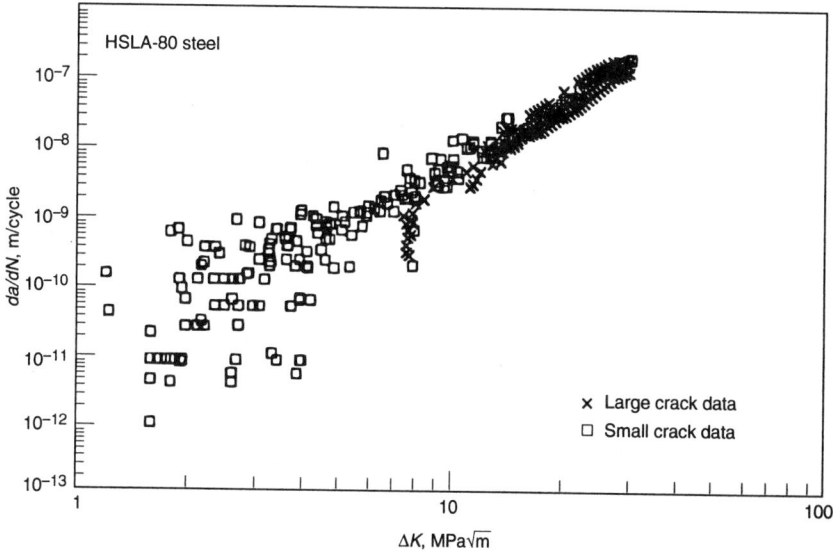

Fig. 5 Typical fatigue crack growth data for mechanically small cracks and large cracks. Source: Ref 13

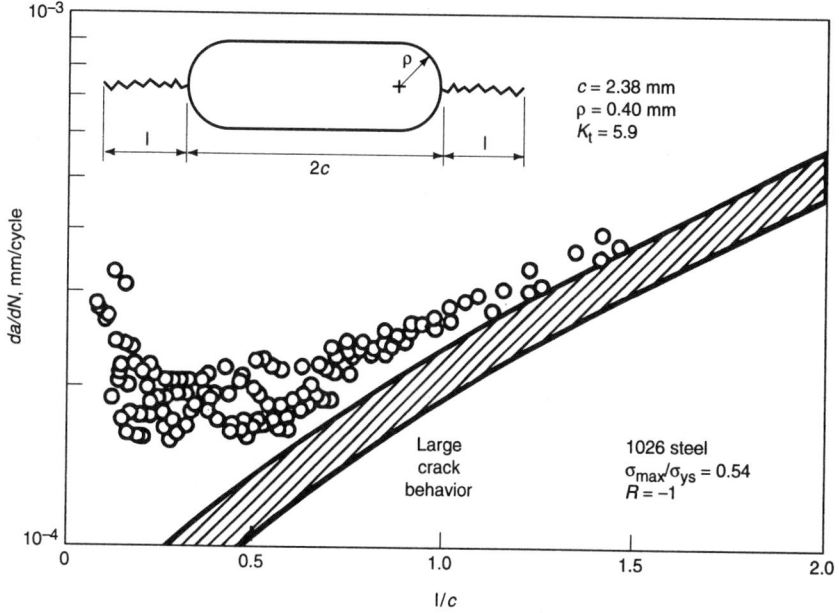

Fig. 6 Typical fatigue crack growth data for short cracks at notches. Source: Ref 14

sions. These small-crack growth rates can actually decrease with increasing crack growth and then eventually merge with large-crack data.

Why do mechanically small cracks grow in this manner? The primary motivation appears to be that local stresses are significantly larger than those encountered under typical small-scale yielding (SSY) conditions, especially at near-threshold values of ΔK. These local stresses may have been elevated by the presence of a stress concentration, or they may simply be large nominal stresses in uniform geometries. These large local stresses significantly enhance crack-tip plasticity, which in turn enhances the crack driving force, either directly through violations of K-dominance, indirectly through changes in plas-

ticity-induced crack closure, or both. The appropriate analytical treatment of the mechanically small crack, then, primarily involves appropriate treatments of the elastic-plastic crack driving force and crack closure.

The nominal elastic formulation of ΔK gradually becomes less accurate as a measure of the crack driving force as the applied stresses become a larger fraction of the yield stress. Under intermediate-scale yielding (ISY), when σ_{max}/σ_{ys} exceeds about 0.7, a first-order plastic correction to ΔK may be useful (Ref 15). This correction may be based on the complete Dugdale formulation for the J-integral, expressed in terms of K, or it may be based on an effective crack size, defined as the sum of the actual crack size and the plastic zone radius. However, in most cases this first-order correction will change the magnitude of ΔK by no more than 10 to 20%. In the large-scale yielding (LSY) regime, when the nominal plastic strain range becomes non-negligible (typically, when the total stress range approaches twice the cyclic yield strength), it will generally be necessary to replace ΔK entirely with some alternative elastic-plastic fracture mechanics (EPFM) parameter (Ref 16), such as a complete ΔJ formulation.

Plasticity-induced crack closure also becomes increasingly significant outside the small-scale yielding regime. Crack opening stresses are a function of the ratio of maximum stress to yield stress, the ratio of minimum to maximum stress (R), and the stress state. Changes in closure behavior are most pronounced for large maximum stresses, low R, and plane stress (typical conditions for mechanically small cracks). Simple closed-form equations based on modified-Dugdale closure models are available to predict normalized crack opening stress as a function of maximum stress, stress ratio, and a constraint factor (Ref 17). Changes in closure behavior are also significant for crack growth at notches, and simple models are available to predict these changes.

If appropriate revisions to the crack driving force based on plasticity and crack closure considerations are carried out, the growth rates of mechanically small cracks can often be predicted successfully by extrapolating the large-crack Paris equation and neglecting the large-crack threshold. This implies that if plastic corrections to ΔK are relatively minor, and if the closure behavior of the small crack does not differ significantly from that of the large cracks used to derive the Paris equation, the small-crack growth rates may be essentially the same as for the large cracks at the same nominal ΔK. It is not entirely clear under what conditions the large crack threshold will be observed by the small cracks, and in the absence of contradicting data, it is probably prudent to neglect the threshold for all mechanically small cracks. If a complete crack closure analysis is not possible or practical, it may be sufficient to predict the growth rates of mechanically small cracks using closure-free (high-stress-ratio) large-crack data (Ref 18).

As noted above, the regimes of mechanically small and microstructurally small cracks can overlap. A more complete organizational scheme for large and small cracks from both microstructural and mechanical perspectives is given in Table 1 (Ref 19). The "microstructurally small" crack discussed earlier in this article is often both microstructurally and mechanically small, although it is also possible to have a crack that is microstructurally small and mechanically large. This can be true of cracks in very large-grained materials, or cracks in single crystals, although single crystals do not exhibit all aspects of small-crack behavior due, in part, to the homogeneity of the microstructure. The traditional "mechanically small" (or "short") crack discussed in this article is typically microstructurally large. Traditional

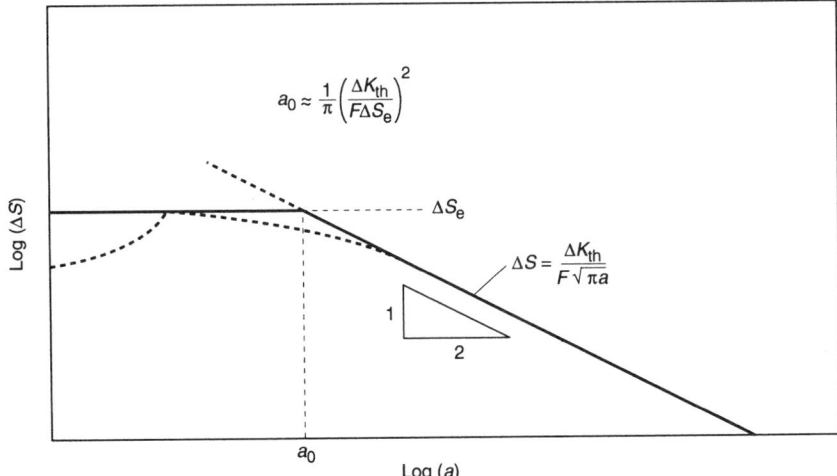

Fig. 7 Diagram for estimating a_0

large cracks are both microstructurally and mechanically large.

Size Criteria for Small Cracks

Table 1 also includes some suggestions for approximate size criteria based on comparisons of the crack dimensions with either the crack-tip plastic zone size, r_p, or the microstructural unit size, M. Cracks are generally considered to be microstructurally small when their size is less than 5 to 10 times the microstructural unit size (typically the grain size). Alternatively, a crack may be microstructurally small when the plastic zone size is roughly less than or equal to the microstructural unit size. A crack often behaves in a mechanically small manner when the ratio of crack size to crack-tip plastic zone size is less than 4 to 20. Here the lower limit corresponds roughly to an applied maximum stress that is about 70% of the yield strength. It should be recognized, however, that these operational definitions of the transition crack size are rough approximations. The actual transition will likely be more gradual than distinct, and identification of the proper criterion is less clear when cracks are both microstructurally and mechanically small.

Another approach to identification of the small-crack regime is based on the relationship between the crack growth threshold and the fatigue limit shown in Fig. 7. From an initiation perspective, failure of a specimen without a preexisting crack should occur only if the applied stress range is greater than the fatigue limit, ΔS_e

(although it should be noted that microstructurally small crack growth can sometimes occur at applied stresses below the fatigue limit). From a fracture mechanics perspective, crack growth should occur only if the applied stress-intensity factor range, $\Delta K = F\,\Delta\sigma\sqrt{\pi a}$, is greater than the threshold value, ΔK_{th}, which is the region above the sloping line. Therefore, the utility of ΔK_{th} as a "material property" appears to be limited to cracks of lengths greater than that given by the intersection of the two lines (a_0). For many materials, a_0 appears to give a rough approximation of the crack size below which microstructural small-crack effects become potentially significant. However, a_0 may underestimate the importance of small-crack effects when crack closure or localized chemistry effects are dominant. Note that the construction of Fig. 7 also indicates that the effective threshold decreases with crack size for cracks smaller than a_0.

Chemically Small Cracks

Experiments on a variety of ferritic and martensitic steels in aqueous chloride environments have shown that under corrosion-fatigue conditions, small cracks can grow significantly faster than large cracks at comparable ΔK values (Ref 20). This phenomenon is believed to result from the influence of crack size on the occluded chemistry that develops at the tip of fatigue cracks. The specific mechanism responsible for this "chemical crack size effect" is believed to be the enhanced production of embrittling hydrogen

Table 1 Classification of crack size according to mechanical and microstructural influences

Microstructural size	Mechanical size	
	Large a/r_p > 4-20 (SSY)	Small a/r_p < 4-20 (ISY and LSY)
Large: (a/M > 5-10) and (r_p/M >> 1)	Mechanically and microstructurally large (LEFM valid)	Mechanically small/microstructurally large (may need EPFM)
Small: (a/M < 5-10) and (r_p/M ~ 1)	Mechanically large/microstructurally small	Mechanically and microstructurally small (inelastic, anisotropic, stochastic)

a, crack size; r_p, crack-tip plastic zone size; M, microstructural unit size; SSY, small-scale yielding; ISY, intermediate-scale yielding; LSY, large-scale yielding; LEFM, linear elastic fracture mechanics; EPFM, elastic-plastic fracture mechanics

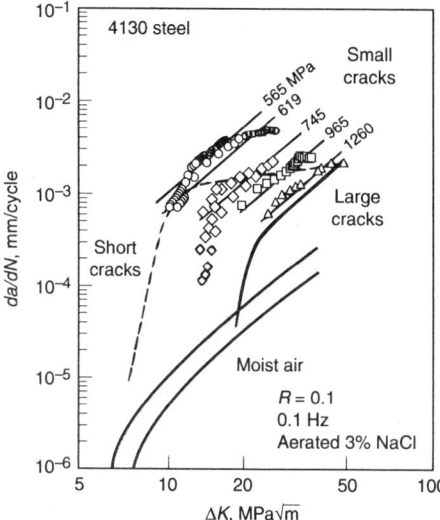

Fig. 8 Typical corrosion-fatigue crack growth data for chemically small cracks and large cracks. Source: Ref 20

within small cracks, resulting from a crack size dependence of one or more factors that control the evolution of the crack-tip environment: convective mixing, ionic diffusion, or surface electrochemical reactions (Ref 21). This mechanism is distinctly different from that responsible for the enhanced rate of crack growth in microstructurally or mechanically small fatigue cracks. However, the enhanced crack-tip plasticity associated with microstructurally or mechanically small cracks could further stimulate the electrochemical reactions through the creation of additional fresh and highly reactive surfaces at the crack tip.

The chemical crack size effect is clearly illustrated by the data of Gangloff (Ref 20) for 4130 steel in an aqueous NaCl environment (see Fig. 8). Note that corrosion-fatigue crack growth rates from small surface cracks (0.1 to 1 mm deep), as well as short through-thickness edge cracks (0.1 to 3 mm), are appreciably faster than corrosion-fatigue crack growth rates from large through-thickness cracks (25 to 40 mm) in standard compact tension specimens. It is also interesting to note that the corrosion-fatigue crack growth rates for small surface cracks decrease with increasing applied stress (at a given ΔK). This trend is opposite to the dependence of applied stress on crack growth rates in mechanically small fatigue cracks. Moreover, all of the corrosion-fatigue crack growth rates in NaCl are enhanced compared to those in a moist laboratory air environment, even though the latter were generated with both small and large cracks. Thus, in relation to the fatigue small-crack effect, the chemical small-crack effect is of potentially greater importance, because it can occur over a much larger range of crack sizes (up to 3 mm).

Not all materials exhibit a chemically small crack effect, and the complexity of the important electrochemical mechanisms makes it difficult, if not impossible, to predict a priori the existence or quantitative extent of this effect in a given application. Changes in alloy and solution chemistry,

electrode potential, oxygen concentration, applied stress and stress ratio, and the specific rate-controlling process in the electrochemical reaction can all influence crack growth rates. In general, experimental data for specific material-geometry-load-chemistry combinations are needed to characterize chemically small crack effects.

Small-Crack Test Methods

Analytical treatments of small-crack growth rate behavior often attempt to derive predictions of small-crack growth rates from large-crack data, which is more commonly available. In some applications, however, this approach will clearly not be adequate (e.g., for some microstructurally small cracks), and it may be necessary to obtain direct experimental evidence for small-crack behavior. Unfortunately, small-crack growth rates cannot usually be measured with the standard test procedures developed for large cracks. Small-crack tests usually require different specimen geometries and different specimen preparation techniques, different crack length measurement techniques and equipment, and different data analysis techniques.

Guidelines for small-crack test methods are now available in appendix X3 to ASTM E 647-95 (Ref 22). This appendix does not prescribe complete, detailed test procedures. Instead, it provides general guidance on the selection of appropriate experimental and analytical techniques and identifies aspects of the testing process that are of particular importance when fatigue cracks are small. A brief summary of these recommendations is provided here for completeness.

Several well-established experimental techniques are available for measuring the size of small fatigue cracks, and hence deducing their growth rates. These techniques include replication, photomicroscopy, potential difference, ultrasonic, laser interferometry, and scanning electron microscopy. Some of these techniques, such as replication and photomicroscopy, are amenable to routine use, while others require significant expertise and expenditures. Each technique has unique strengths and limitations, and different techniques are optimum for different circumstances. All are useful for measuring the growth of fatigue cracks on the order of 50 μm and greater, and some are applicable to even smaller cracks. Detailed descriptions of each technique are collected in Ref 7.

The study of small cracks requires detection of crack initiation and growth while physical crack sizes are extremely small, and this requirement influences specimen design. Today the preferred and most widely used technique is to promote the initiation of naturally small surface or corner cracks in rectangular or cylindrical specimens, rather than growing a large crack and then machining away material in the crack wake to leave a small crack. Early crack detection can be facilitated by using specimens with extremely small artificial flaws or very mild stress concentrations, but the completely natural initiation of a small

crack at a location chosen entirely by the crack itself is sometimes preferred. Near-surface residual stresses and surface roughness induced by specimen fabrication can artificially influence small-crack growth behavior and should be eliminated or minimized prior to testing. However, the growth rates of small surface cracks in engineering components can be influenced by residual stress fields arising from fabrication of the component, so residual stresses should be considered when the laboratory data are applied.

Many small surface cracks develop shapes that are approximately semielliptical, and the standard K solutions for these geometries can be applied during data analysis. However, variations in the crack shape can be a source of scatter in growth rate data, especially for microstructurally small cracks, and some confirmation of crack shape is desirable. Interactions between closely spaced multiple cracks that affect growth rates are more likely to occur in the small-crack regime and must be addressed. Special attention must be given to the minimum interval between successive crack length measurements, Δa. Closely spaced measurements are often needed to capture key crack-microstructure interactions, but measurement error can significantly influence variations in da/dN for extremely small Δa values.

Scatter in Small-Crack Growth Rate Data

Small-crack data often exhibit much more scatter in da/dN than large-crack data, sometimes several orders of magnitude at a single ΔK value (see Fig. 3, 5). Of course, this leads to greater uncertainty in life calculations, especially when the small-crack regime dominates the total life. Analytical approaches based on simple upper bounds to the small-crack regime may be unacceptably overconservative.

This apparent variability can arise from several different sources. Some true variability is due to stochastic microstructural effects. Local resistance to crack growth will vary with local differences in grain orientation, microplastic yield strength, and grain boundary effects, and these can be especially significant when the crack driving force is small (on the same order as the material resistance to crack growth). A small crack embedded in a preferentially oriented microstructure may grow very rapidly, while a similar crack in a contrasting microstructure might arrest completely. Larger cracks simultaneously interrogate many grains and microstructural features along the crack front, and hence there is a smoother average resistance to crack advance.

On the other hand, some apparent variability in da/dN is more artificial and hence will not have a significant impact on variability in total life. Measurement errors become significant when the crack growth increment becomes small relative to the measurement resolution. Other apparent variability can be attributed to mathematical averaging effects. The normal point-to-point variability in growth rates due to local microstructural variations is effectively averaged out for most large cracks, because the crack travels a relatively long distance (through many different microstructural features) during the measurement interval. But because the small crack usually travels only a short distance during the measurement interval, this normal variability has a more dramatic impact on calculated da/dN. Large cracks could exhibit a similar increase in apparent variability if they, too, were measured at much shorter Δa intervals.

The appropriate treatment for small-crack scatter depends, at least in part, on the origin of the scatter. Some scatter that is only apparent can be effectively reduced with improvements in the measurement precision or in the analytical schemes used to process the raw crack growth data, including data filtering and modified incremental polynomial techniques (Ref 22). However, other forms of scatter may require a formal stochastic treatment of the data. Many stochastic FCG models are available in the literature. Unfortunately, many of these models require extensive data of high statistical quality, which is often difficult (expensive) to obtain for small cracks. Other stochastic FCG models designed for practical engineering applications, such as the lognormal random variable model, require fewer data and simpler calculations. However, these models are often not able to address the effects of crack size on scatter.

Applications Where Small Cracks Are Important

Small-crack behavior is not an important issue for applications in which initial defects are large and fatigue cracks of interest are also large, such as welded civil engineering structures. In addition, small cracks are generally not significant for many traditional mechanical and aeronautical engineering design/analysis applications based on damage tolerance concepts, because the initial flaw size (based on conventional nondestructive evaluation inspection limits) is usually beyond the small-crack regime. However, damage tolerance methods are sometimes applied to more highly stressed structures where tolerable flaw sizes are much smaller and nondestructive evaluation requirements are stricter. Small-crack behavior can be very important in these applications, which historically have been treated with safe-life methods based on bulk damage strain-life or stress-life analyses. Note that the total life in many strain-life applications is often dominated by the growth of small cracks, especially in the low-cycle fatigue (LCF) regime where crack formation occurs very early in life and final crack sizes are still relatively small. Therefore, the damage growth process in LCF, which is often treated as an "initiation" problem, is often actually a small-crack growth process. Small-crack analysis techniques may provide valuable new insights

into some difficult LCF lifting problems. Small-crack phenomena, especially small-crack arrest, are thought by some to be the key to high-cycle fatigue (HCF) behavior, including the fatigue limit, but a practical treatment of HCF based on small cracks is not yet available. The relative contributions of crack nucleation and small crack growth for total HCF life are not yet well understood. Small cracks can also be important for fracture mechanics-based durability assessments in which an equivalent initial flaw size (EIFS) is back-calculated from some economic total life. This EIFS is often well within the small-flaw regime.

Discussion/Summary

Small-crack behavior was first documented in the mid-1970s, extensively investigated in the 1980s, and remains an active research topic. The problem is now well enough understood to facilitate some standardization of concepts, test methods, and analysis techniques, but small-crack technology is not yet routinely applied in industrial practice. At this writing, no general-purpose computer codes for fatigue crack growth (FCG) analysis are available that explicitly address small-crack behavior. Furthermore, several important problems remain unresolved. For example, some small-crack effects appear to be accentuated under variable-amplitude loading, but load history effects have not been adequately characterized. In addition, as noted earlier, it is not yet clear if small cracks exhibit a well-defined threshold or nonpropagation condition; if so, how this might be related to the large-crack threshold; or when small cracks observe the large-crack threshold. Nevertheless, the current understandings about when and why small-crack effects occur, how to characterize them experimentally, and how to treat them analytically are adequate to provide significant improvements in the quality of structural integrity assessments.

ACKNOWLEDGMENTS

The substantial support of research on small cracks and related topics at Southwest Research Institute over the past fifteen years by AFOSR, AFWAL, NASA, ARO, ONR, and others is gratefully acknowledged.

REFERENCES

1. S. Suresh and R.O. Ritchie, *Int. Metals Rev.,* Vol 29, 1984, p 445-476
2. S.J. Hudak, Jr., *ASME J. Engng. Mater. Technol.,* Vol 103, 1981, p 26-35
3. K.J. Miller, *Mater. Sci. Technol.,* Vol 9, 1993, p 453-462
4. R.O. Ritchie and J. Lankford, Ed., *Small Fatigue Cracks,* The Metallurgical Society, 1986

5. K.J. Miller and E.R. de los Rios, Ed., *The Behaviour of Short Fatigue Cracks,* EGF 1, Mechanical Engineering Publications, London, 1986
6. K.J. Miller and E.R. de los Rios, Ed., *Short Fatigue Cracks,* ESIS 13, Mechanical Engineering Publications, London, 1986
7. J.M. Larsen and J.E. Allison, Ed., *Small-Crack Test Methods,* STP 1149, American Society for Testing and Materials, 1992
8. J. Lankford, *Fatigue Engng. Mater. Struct.,* Vol 5, 1982, p 233-248
9. K.S. Chan and J. Lankford, *Acta Metall.,* Vol 36, 1988, p 193-206
10. J. Lankford and D.L. Davidson, The Role of Metallurgical Factors in Controlling the Growth of Small Fatigue Cracks, *Small Fatigue Cracks,* R.O. Ritchie and J. Lankford, Ed., The Metallurgical Society, 1986, p 51-71
11. M.H. El Haddad, K.N. Smith, and T.H. Topper, *ASME J. Engng. Mater. Technol.,* Vol 101, 1979, p 42-46
12. J.C. Newman, Jr., *Fatigue Fract. Engng. Mater. Struct.,* Vol 17, 1994, p 429-439
13. D.L. Davidson, K.S. Chan, and R.C. McClung *Metall. and Mater. Trans.,* Vol 27A, 1996, p 2540-2556
14. R.C. McClung and H. Sehitoglu, *ASME J. Engng. Mater. Technol.,* Vol 114, 1992, p 1-7
15. J.C. Newman, Jr., Fracture Mechanics Parameters for Small Fatigue Cracks, *Small-Crack Test Methods,* STP 1149, J.M. Larsen and J.E. Allison, Ed., American Society for Testing and Materials, 1992, p 6-33
16. R.C. McClung and H. Sehitoglu, *ASME J. Engng. Mater. Technol.,* Vol 113, 1991, p 15-22
17. J.C. Newman, Jr., *Int. J. Fract.,* Vol 24, 1984, p R131-R135
18. R. Hertzberg, W.A. Herman, T. Clark, and R. Jaccard, Simulation of Short Crack and Other Low Closure Loading Conditions Utilizing Constant K_{max} ΔK-Decreasing Fatigue Crack Growth Procedures, *Small-Crack Test Methods,* STP 1149, J.M. Larsen and J.E. Allison, Ed., American Society for Testing and Materials, 1992, p 197-220
19. S.J. Hudak, Jr. and K.S. Chan, In Search of a Driving Force to Characterize the Kinetics of Small Crack Growth, *Small Fatigue Cracks,* R.O. Ritchie and J. Lankford, Ed., The Metallurgical Society, 1986, p 379-405
20. R.P. Gangloff, *Metall. Trans. A,* Vol 16, 1985, p 953-969
21. R.P. Gangloff and R.P. Wei, Small Crack-Environment Interactions: The Hydrogen Embrittlement Perspective, *Small Fatigue Cracks,* R.O. Ritchie and J. Lankford, Ed., The Metallurgical Society, 1986, p 239-264
22. "Standard Test Method for Measurement of Fatigue Crack Growth Rates," ASTM E 647-95, *Annual Book of ASTM Standards,* Vol 03.01, 1996

Effect of Crack Shape on Crack Growth

K.S. Ravichandran, The University of Utah

FRACTURES in engineering applications (Ref 1) occur mostly from surface or internal three-dimensional cracks, which generally propagate in all directions and often have irregular shapes. Such shapes may not strictly have an elliptical or circular geometry, although such an approximation is often practiced in research investigations and engineering analyses. This may introduce errors in growth data and the estimated fatigue life, but it also raises several parallel questions. First, what are the factors that make the three-dimensional cracks grow with irregular shapes? Second, how can we describe the growth behavior of regular and irregular cracks exhibiting a continuous change in shape from initiation to failure? Third, how can we predict the growth of these cracks in fatigue leading to unstable fracture?

Recent studies on the effects of crack shape on the behavior of surface and embedded cracks have resolved these issues to some extent. These studies have also clarified several important factors that influence the three-dimensional crack growth behavior, including, for example, loading mode, residual stress, microstructure, and material anisotropy. Additionally, methods have been developed to calculate stress-intensity factors (SIFs) of arbitrarily shaped flaws and to predict failure from these cracks. This article summarizes the aspects of crack shape and irregularity that are relevant to fatigue and fracture of surface cracks. The issues covered are the basic nature of regular surface cracks; variables that influence the shape of surface cracks, such as grain size, residual stresses, texture, loading mode, environment, and crack coalescence; techniques for monitoring crack shape development; methods for calculating SIFs for arbitrarily shaped flaws; and simple approaches to predicting failure or threshold for crack growth from arbitrarily shaped flaws and notches.

Nature of Three-Dimensional Surface Cracks

Two terms pertain to the three-dimensional aspects of surface cracks: the crack shape (semicircular, semielliptical, square, triangular, etc.) and

the crack aspect ratio (a/c, the ratio of half surface length to the distance of maximum depth point in crack front, from surface) (Fig. 1). The two terms are somewhat related: the former provides a qualitative description of crack geometry, whereas the latter is a quantitative measure of depth in relation to the length at surface, irrespective of geometry.

The straight crack fronts of through cracks, as in standard fracture mechanics specimens, allow the characterization of fatigue fracture in terms of two-dimensional fracture mechanics formulations. On the other hand, an elliptical surface crack requires two parameters to describe the fracture process. The surface crack length ($2c$) and the depth (a) are required to adequately describe the stress-intensity factor, K, along the crack front (Fig. 1).

Irwin (Ref 2) formulated the SIF of a semi-elliptical surface crack in an infinite plate, subjected to an applied stress, σ, as:

$$K = \frac{\sigma\sqrt{\pi a}}{\varphi}\left[\left(\frac{a}{c}\right)^2 \cos^2\phi + \sin^2\phi\right] \quad \text{(Eq 1)}$$

where the elliptic integral φ is given by:

$$\varphi = \int_0^{\pi/2}\left[\sin^2\phi + \left(\frac{a}{c}\right)^2 \cos^2\phi\right]^{1/2}$$

In Eq 1, ϕ is the parametric angle (Fig. 1) of the point of interest on the crack front. The ratio a/c is referred to as the "aspect ratio" of the crack and is often used to describe the semicircular ($a/c = 1$), shallow ($a/c < 1$), and deep ($a/c > 1$) crack morphologies found in engineering components. Equation 1 is useful only for surface cracks in infinite bodies. However, employing detailed finite element analysis, Newman and Raju (Ref 3) modified surface crack formulae for wider practical use by incorporating the specimen thickness (t) and width ($2w$). The Newman-Raju formula for surface cracks is given by:

$$K = \sigma\sqrt{a}\, F\, g\, f_\phi f_w \quad \text{(Eq 2)}$$

where:

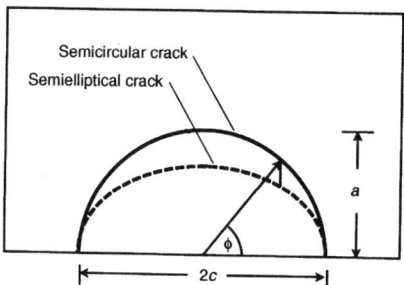

Fig. 1 The geometry of surface cracks

$$F = \left[M_1 + M_2\left(\frac{a}{t}\right)^2 + M_3\left(\frac{a}{t}\right)^4\right]\sqrt{\frac{\pi}{Q}}$$

$$f_w = \sqrt{\sec\left(\frac{\pi c}{2w}\sqrt{\frac{a}{t}}\right)}$$

In the above equations, for $a/c \leq 1$:

$$M_1 = 1.13 - 0.09\left(\frac{a}{c}\right)$$

$$M_2 = -0.54 + \frac{0.89}{0.2 + \left(\frac{a}{c}\right)}$$

$$M_3 = 0.5 - \frac{1}{0.65 + \left(\frac{a}{c}\right)} + 14\left(1 - \frac{a}{c}\right)^{24}$$

$$g = 1 + \left[0.1 + 0.35\left(\frac{a}{t}\right)^2\right](1 - \sin\phi)^2$$

$$f_\phi = \left[\left(\frac{a}{c}\right)^2\cos^2\phi + \sin^2\phi\right]^{1/4}$$

$$Q = 1 + 1.464\left(\frac{a}{c}\right)^{1.65}$$

and for $a/c > 1$:

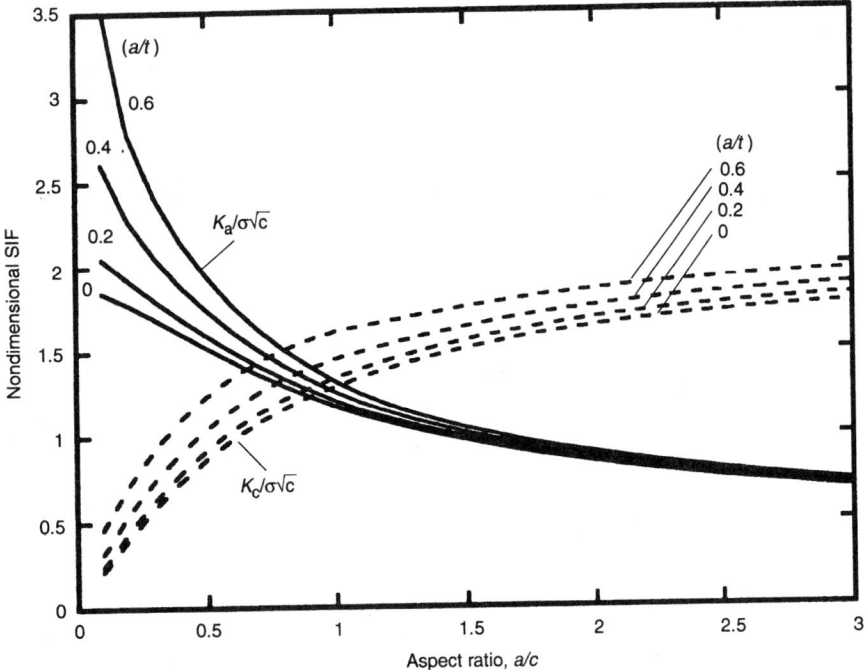

Fig. 2 The variation of stress-intensity factor (SIF) at the surface (K_c) and at the depth (K_a) with the aspect ratio of the surface crack

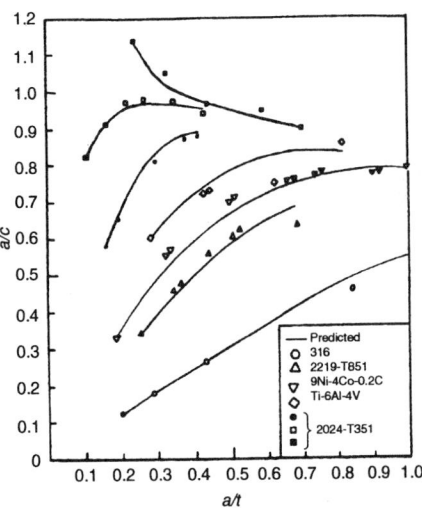

(a)

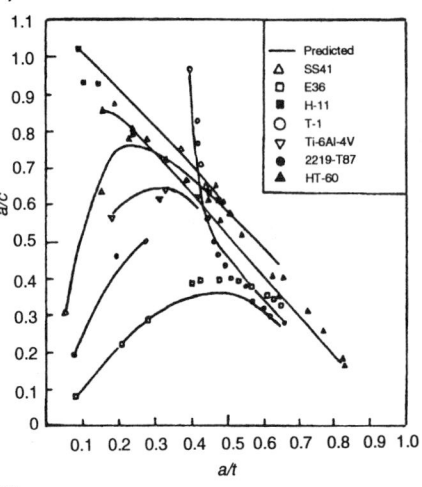

(b)

Fig. 3 The nature of crack aspect ratio variation in (a) tension and (b) bending for various starting crack shapes, in the absence of residual stresses and microstructural influences.

$$M_1 = \sqrt{\frac{c}{a}}\left(1 + 0.04\,\frac{c}{a}\right)$$

$$M_2 = 0.2\left(\frac{c}{a}\right)^4$$

$$M_3 = -0.11\left(\frac{c}{a}\right)^4$$

$$g = 1 + \left[0.1 + 0.35\left(\frac{c}{a}\right)\left(\frac{a}{t}\right)^2\right](1 - \sin\phi)^2$$

$$f_\phi = \left[\left(\frac{c}{a}\right)^2\sin^2\phi + \cos^2\phi\right]^{1/4}$$

$$Q = 1 + 1.464\left(\frac{c}{a}\right)^{1.65}$$

It is noted that Eq 2 is only for pure tensile loading with an aspect ratio (a/c) between 0.2 and 2.0 and for a/t <0.8. The SIF equations for loading and crack conditions are in Ref 3. The SIF in Eq 2 varies with location along the crack front as a function of surface length and crack depth, for a given specimen size. The SIF values at different points along the crack front are nearly the same for a crack of $a/c = 0.85$, and this shape is often termed *equilibrium crack shape* or *semielliptical crack*. Surface cracks in relatively brittle materials grow with this shape in the presence of uniform stress and homogeneous material characteristics along the crack front. In relatively ductile materials, plastic deformation occurs at tips at the surface, due to a lower degree of constraint. This retards the growth of surface tips, causing the equilibrium shape to be nearly semicircular (a/c = 1), even though the K at the surface tip (K_c) is

about 8% higher than that at the depth position (K_a). Hereafter, the term *equilibrium crack shape* means either semielliptical or semicircular, depending on the material.

The SIF distribution changes significantly when there is a deviation from the equilibrium crack shape, depending on whether the crack is shallow ($a/c < 1$) or deep ($a/c > 1$). The distribution becomes very complex if the crack shape is irregular, deviating far from the elliptical geometry. The nature of SIF variations in an elliptical crack can be visualized from Fig. 2, in which the SIFs at the surface ($\phi = 0$), K_c, and at the maximum point at depth ($\phi = 90$), K_a, are plotted as a function of a/c. When the crack is shallow, the SIF at depth is higher than that at the surface tip ($K_a > K_c$). The situation is reversed for the deep crack ($K_c > K_a$). Hence, for example, initially shallow and initially deep cracks grow at depth and surface positions, respectively, in order to make the SIF uniform all around the crack front. Similarly, in irregularly shaped cracks, irrespective of their geometric shape, crack growth at locations of high K occur to move the shape to equilibrium. Although crack growth is dictated by the requirement to maintain equilibrium shape, in practice, surface cracks often maintain irregular shapes due to nonuniformity in structural stress distribution as well as material and microstructural inhomogeneities. These factors are discussed in the following sections.

Variables That Influence Crack Shape

Mechanical Variables. The principal factors that affect the variation in crack shape or aspect ratio are the nature of stress distribution in the

crack plane and residual stresses induced by surface damage, machining, shot peening, and coating.

In the absence of residual stresses, the variation of crack aspect ratio as the crack grows through a plate of rectangular cross-section depends on whether the remote loading is tension or bending in nature. Additionally, the initial crack aspect ratio influences the aspect ratio during growth. For purely tensile loading, the a/c will tend to reach a value of 0.85 after sufficient crack growth from a crack with arbitrary initial aspect ratio. At large crack sizes, when the crack front at the depth approaches the specimen back surface, there is a tendency for the cracks to become shallow. This is due to the fact that even in nominally tensile loading, the bending component becomes significant at small net section sizes, due to specimen rotation with respect to the loading axis. The variation in a/c is illustrated in Fig. 3(a) for different materials with varying initial aspect ratios. The aspect ratio variation can be described by (Ref 4):

$$\frac{da}{dc} = \left\{ \sqrt{\frac{a}{c}} \left[1 + 0.3182 \left(\frac{a}{t}\right)^2 \right] \right\}^{-n} \quad \text{(Eq 3)}$$

where n is the exponent in the Paris law for stage II fatigue crack growth:

$$\frac{da}{dN} = C \, \Delta K^n \quad \text{(Eq 4)}$$

where C is a material constant. From Eq 3, a/c can be determined at any stage during fatigue crack growth using Runge-Kutta numerical technique if the initial aspect ratio, a_0/c_0, and a_0/t are known.

On the other hand, in bending, a/c changes continuously, even for the crack starting with $a/c = 1$, due to the variation of stress in the through-the-thickness direction. As the crack grows, a shallow shape is preferred, because the tensile stress at the depth point is lower than that at the surface, leading to different local K values at the tips in these locations. Cracks with arbitrary initial shapes also follow this trend eventually, after some growth. The variation in aspect ratio in bending is illustrated in Fig. 3(b) for different materials with varying initial aspect ratios. The rate of change of a/c in this case is given by:

$$\frac{da}{dc} = \left\{ \frac{\sqrt{\dfrac{a}{c}} \left[1 + 0.3182 \left(\dfrac{a}{t}\right)^2 \right] H_1}{H_2} \right\}^{-n} \quad \text{(Eq 5)}$$

where:

$$H_1 = 1 - 0.34 \left(\frac{a}{t}\right) - 0.11 \left(\frac{a}{c}\right)\left(\frac{a}{t}\right)$$

$$H_2 = 1 - \left[1.22 + 0.12 \left(\frac{a}{c}\right) \right] \left(\frac{a}{t}\right)$$

$$+ \left[0.55 - 1.05 \left(\frac{a}{c}\right)^{0.75} + 0.47 \left(\frac{a}{c}\right)^{1.5} \right] \left(\frac{a}{t}\right)^2$$

In Fig. 3(a) and (b), the solid lines are the predictions from Eq 3 and 5, respectively, and they are often referred to as preferred propagation paths (PPP). In practice, the initial crack shape depends on the geometry of discontinuities, including notches introduced during component fabrication, cracks forming from inclusions, and so on. Hence, knowing the initial aspect ratio, a_0/c_0, of these defects, the aspect ratio at any stage in fatigue life can be determined numerically from Eq 3 and 5. This is of considerable use in predictions of fatigue failure.

The nature of development of crack shape or aspect ratio is also influenced by stress states other than that due to applied loading. One example is the residual stress introduced by surface modification processes, such as shot peening, surface hardening, and coating. The variation in the shape of surface cracks during fatigue after

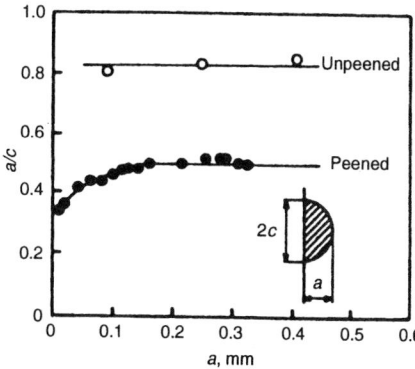

Fig. 4 The trends in aspect ratios of surface cracks in 7010 aluminum alloy, before and after shot peening

shot peening (Ref 5) is shown in Fig. 4, along with the data for unpeened material, for 7010 high-strength aluminum alloy. While the unpeened alloy maintained nearly the equilibrium shape ($a/c = 0.85$), the shot-peened alloy showed shallow crack shape ($a/c = 0.5$) during growth. After shot peening, the growth rate of the surface crack tips was higher than that at the depth. Shot peening of the alloy produced a highly deformed layer at the surface. Due to extensive plastic deformation in the direction normal to the shot-peened surface, the width of grains in this direction was smaller than in other directions. The grains had a layered microstructure. As a result, propagation was difficult normal to these grain layers into the specimen, relative to that in the surface direction. The shallow crack shape also reflects this difficulty, indicating that it is easier for the crack to grow in the surface direction than in the depth direction.

Other factors influence the development of crack shape in surface cracks. Unique circumstances such as crack coalescence can cause a sudden change in aspect ratio (Ref 6). As shown in Fig. 5(a), when two semicircular cracks propagating in the same plane touch each other, the combined crack has a lower aspect ratio ($a/c < 1$), and the crack propagation in surface temporarily ceases. In conventionally manufactured components, residual stresses are invariably present. In autofretagged gun tubes (Ref 7), small cracks, initially pinned from growing in the surface direction, coalesce to larger ones with shallow shape, often resulting in a far more complex shape (Fig. 5b). These shapes are not generally semielliptical, so some inaccuracy is expected when using the known SIF formulas for elliptical cracks. Therefore, alternate methods are required to estimate the SIFs for these cracks.

Changes in load spectrum, either due to change in mean stress or stress ratio, R, or a change in loading mode, can cause a change in the shape of the crack (Ref 8). Figure 5(c) shows how the change in R alters the crack front for a surface crack in a plate under tensile loading. It was suggested that the constraint loss in surface changed the size of the plastic zone through which the crack should enter the body. This in turn led to unusual crack shapes, as shown in the

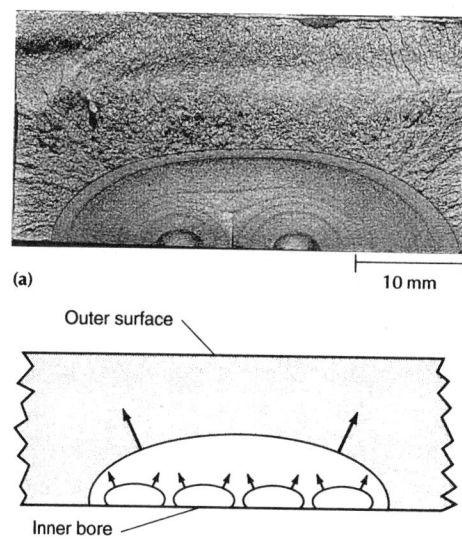

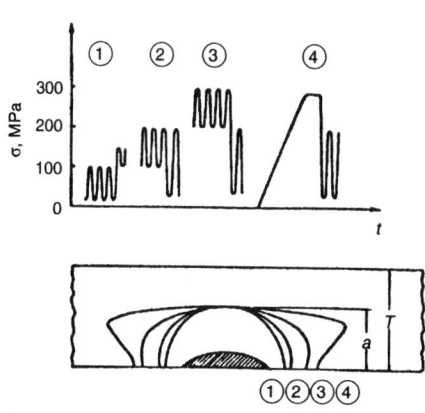

Fig. 5 The nature of crack shape development during (a) the coalescence of two semicircular cracks, (b) the coalescence of multiple cracks in autofrettaged gun tubes, and (c) the application of varying mean stress fatigue loading

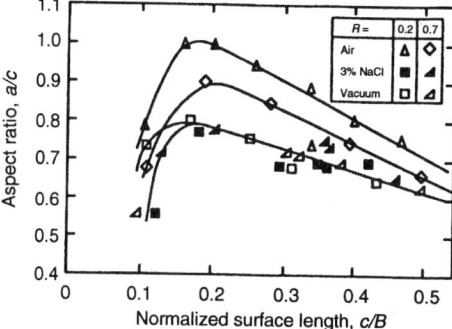

Fig. 6 The effect of environment on crack shape development

figure. Additionally, the resulting stress redistribution influenced the crack shape.

The variations in surface crack aspect ratio, or the PPP, also depend on the environment (Ref 9). Figure 6 shows the effect of environment and stress ratio on the change in a/c of surface cracks in HY80 steel. For crack growth in air, the aspect

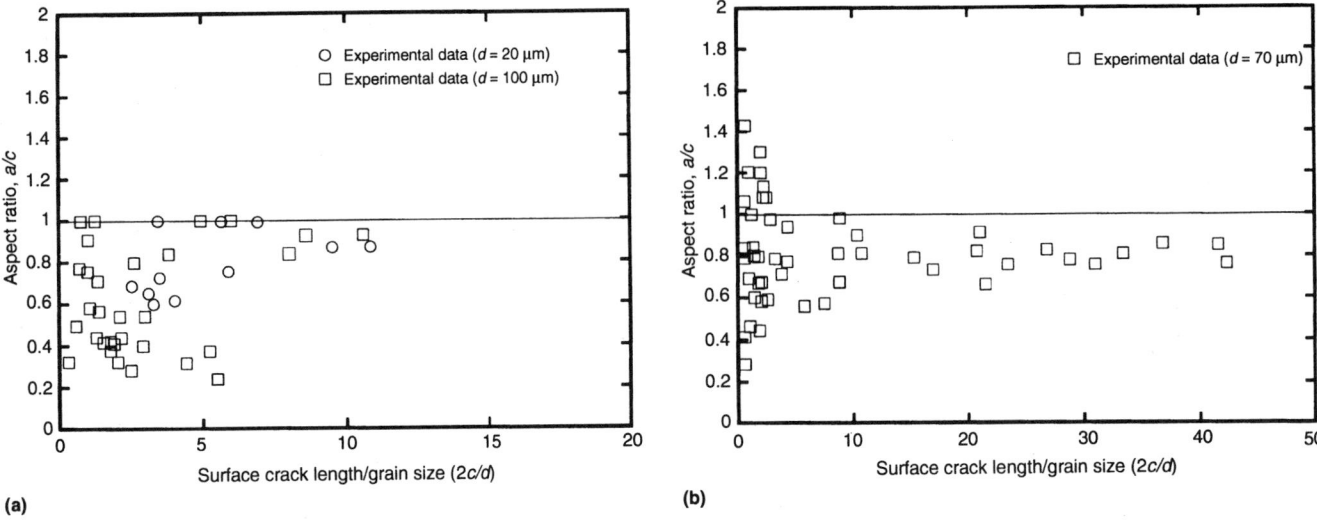

Fig. 7 Microstructure (grain size) induced crack aspect ratio variations in (a) Ti-8Al alloy and (b) stainless steel

ratios were generally lower at $R = 0.7$ than $R = 0.2$, similar to the effect of load spectrum on crack shape, indicated above. However, the effect was not seen for vacuum or saltwater environments. Additionally, the crack aspect ratios in these environments are similar. The reason for this different behavior is not well understood, but the data clearly suggest that the effect of environment on crack shape development can be significant and should be considered in surface crack analyses.

Microstructural Variables. Many microstructural parameters, including grain boundaries, crystallographic orientation/texture, and inclusions, influence the development of crack shape in surface cracks. Despite its importance to microstructure control and fatigue life prediction in general, this subject has received little attention in research, let alone in fatigue life prediction. One of the principal reasons is that microstructure-induced effects occur at very small sizes, making measurement and interpretation difficult. Until recently, nondestructive evaluation procedures in engineering emphasized a lower detectable crack size limit of about 1 mm. At this size, the microstructure-induced effects on crack shape are generally small. However, in high-performance applications, high-strength and inherently anisotropic materials are being used increasingly often, leading to a decrease in the limiting crack size at which unstable failure may occur. Therefore, consideration should be given to microstructural effects on small cracks, in the size range in which microstructural effects are significant.

The shape of inclusions, for example in steel, influences the crack shape at very early stages of fatigue, affecting the fatigue life (Ref 10, 11). This influence is significant on the initial values of crack aspect ratio. As the crack grows away from the inclusion, equilibrium semielliptical shape is reached, provided that factors such as residual stress and microstructural effects are absent.

The most significant microstructural factors that have a documented influence on crack shape

development are grain size and texture. It is to be noted that microstructural effects on crack shape are yet to be fully understood. However, certain experimental data are included here for the reader, to caution the use of surface crack equations at small crack sizes, as well as to appreciate the relevance of microstructure in crack shape variations. The effect of grain size on crack shape is dominant at small crack sizes of the order of a few grain diameters. Figures 7(a) and (b) illustrate a/c values determined by serial electropolishing (Ref 12) and by heat tinting (Ref 13), respectively. Both data sets consist of measurements from cracks grown to different sizes in several specimens. The fluctuations in crack aspect ratio are significant, especially at small crack sizes of the order of a few grain diameters. At large crack sizes, the aspect ratios converge to nearly equilibrium crack shape ($a/c = 0.85$).

The reasons for the grain-induced aspect ratio variations are beginning to be understood (Ref 14). At small crack sizes, the crystallographic orientations of grains influence the crack extension. Perturbations in the crack front occur at locations where the grains ahead of the crack are favorably oriented for cleavage or slip. The crack front is arrested at locations having grains not so oriented. The isolated crack front perturbations are significant at crack sizes of the order of a few times the grain size. These perturbations significantly alter the shape as well as the distribution of SIF along the crack front. As a result, there are wide variations in crack shape and SIF distribution. When the crack grows to a larger size, of the order of several times the grain diameter, the same perturbations become less significant in relation to the size of the crack. Hence, the changes in overall shape and the K distribution along the crack front are less severe. With continued crack growth, the crack approaches the equilibrium crack shape. Therefore, microstructure-induced crack shape variations are limited to few grain diameters, typically of the order of ten times the grain size (Ref 14). This limit can change with a

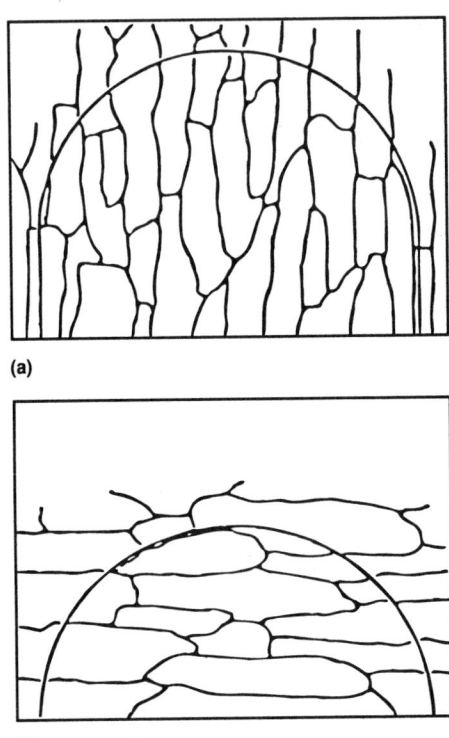

(a)

(b)

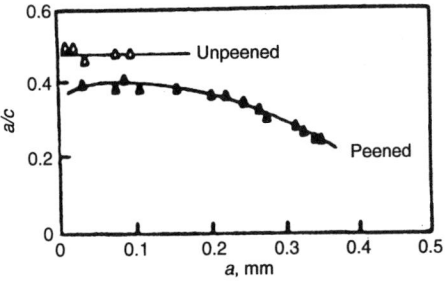

(c)

Fig. 8 (a, b) Schematics of crack shapes observed in different orientations of fatigue tests in 7010 aluminum alloy. (c) The variation of aspect ratio with crack growth in 8090 Al-Li alloy, before and after shot peening

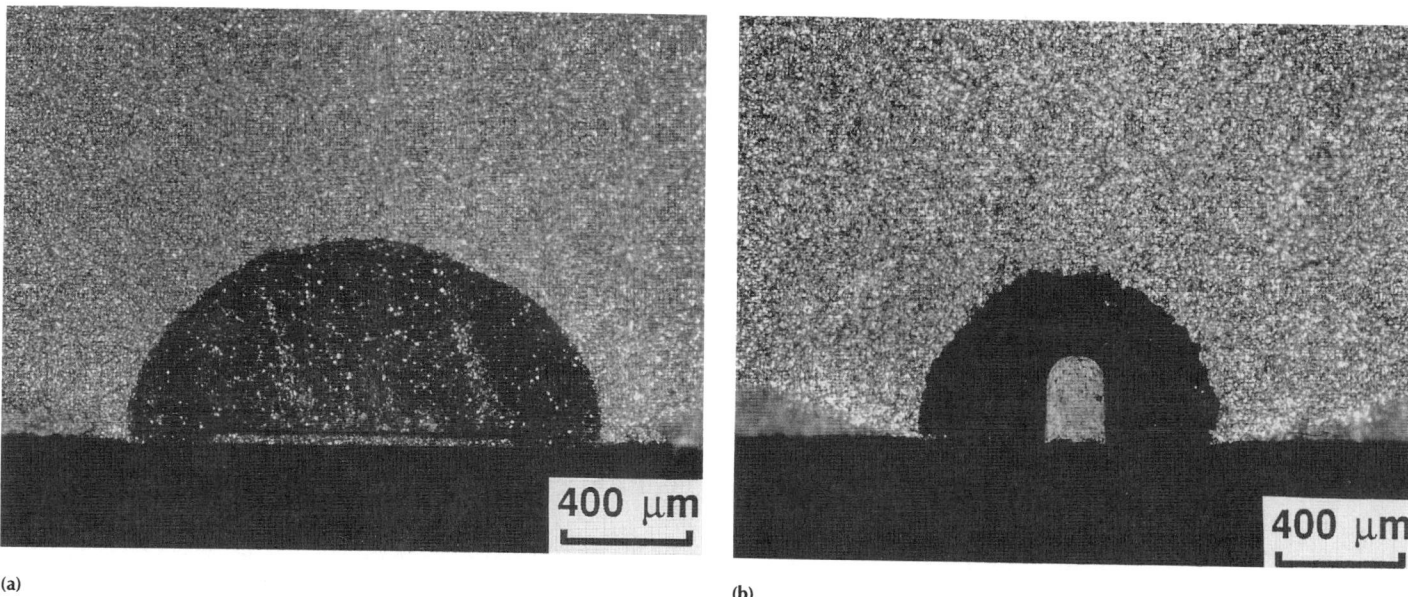

Fig. 9 Shapes of surface cracks, revealed by heat tinting before specimen fracture. The cracks were grown from (a) shallow and (b) deep notches.

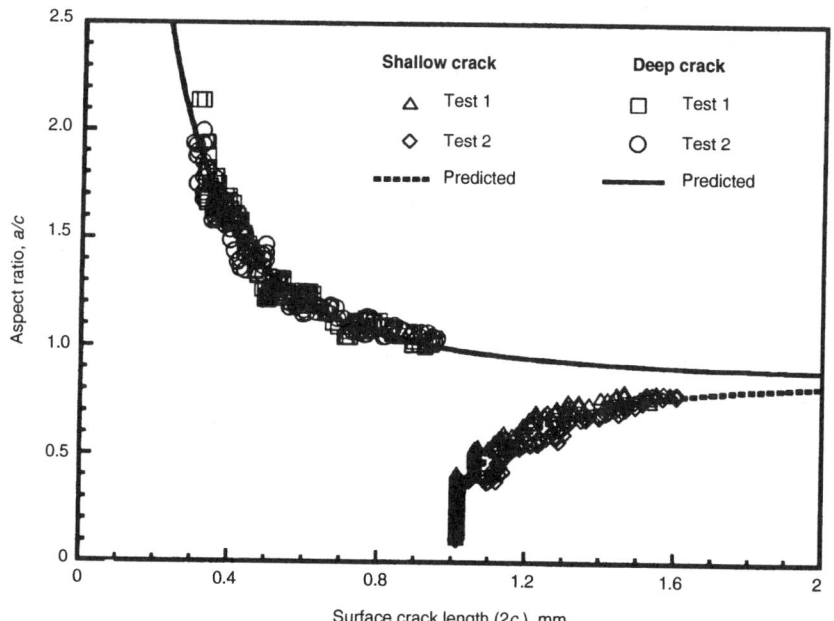

Fig. 10 Comparison of the experimentally measured aspect ratios with the predicted trend during fatigue crack growth from initially shallow and deep notches

material, but it appears to be linked to the ratio of crack size to grain size. Grain-induced crack shape variations are also significant in materials such as beta- processed titanium alloy, in which cracks of the order of 1 mm are known (Ref 15) to exhibit irregular crack shapes due to the coarse colony microstructure.

Texture or grain shape can influence the shape of the surface crack during fatigue crack growth, especially in rolled and extruded materials such as aluminum alloys. This is because the resistance to crack growth is different in the longitudinal, transverse, and short-transverse directions of a rolled plate, leading to different rates of crack

front advance in different directions under the same applied stress range. Hence, nonequilibrium crack shapes occur, either shallow or deep configurations (Ref 16), depending on the relative crack growth resistance at the surface direction compared to that at the depth direction. Figures 8(a) and (b) illustrate the crack shapes as observed on fracture surfaces in different orientations of a 7010 aluminum alloy. The effect of texture or orientation is more significant in the case of Al-Li alloys (Ref 5), in which shallow crack configuration is seen even in the absence of shot peening (Fig. 8c).

Measurement and Analysis

Measurement of crack shapes or aspect ratios during fatigue crack growth can be performed by a number of techniques. Most common are application of high mean stress and low-ΔK loading periodically during the regular cyclic loading to mark the crack front, heating the specimen with the crack in air at 300 to 700 °C (for most steels and titanium alloys) for 1 or 2 h to color the crack surfaces by oxidation (heat tinting), and using dye penetrants or inks to mark the crack front. However, these techniques provide useful information only after specimen fracture. In many instances, a knowledge of the shape of the crack before fracture is required in order to assess the criticality of the structure.

To this end, a method has recently been developed (Ref 17) to continuously track the changes in shape or aspect ratio of the crack during fatigue crack growth, using advanced measurement techniques. The method relies on the measurements of instantaneous crack compliance and surface crack length. A laser interferometric displacement measurement system is used to accurately measure the crack compliance. A photographic camera or replication is used to continuously record the surface crack length at the same time as the compliance measurement. The compliance of a surface crack is a function of surface length and its depth (alternatively, aspect ratio), so the aspect ratio can be estimated if the compliance and the surface length are known. The relationship between the compliance ($2U/\sigma$, where $2U$ is the crack-mouth opening displacement due to a stress, σ, on the specimen), the surface crack length, and the aspect ratio is given by:

$$\frac{2U}{\sigma} = 1.6 \frac{\sqrt{8}}{\sqrt{\pi E}} c \, (1 - \nu^2) \, \lambda \, F f_w \, g \qquad \text{(Eq 6)}$$

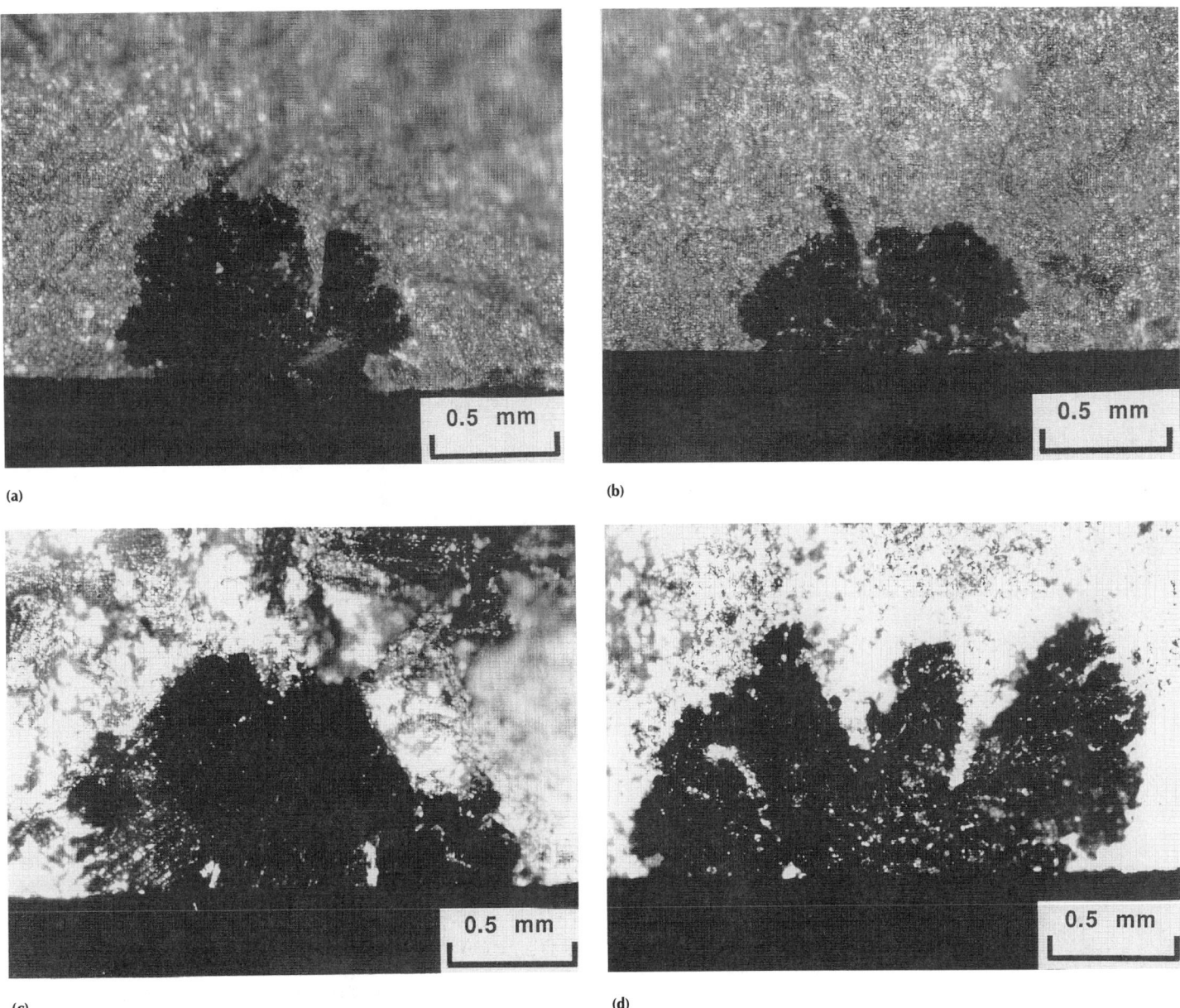

(a)

(b)

(c)

(d)

Fig. 11 Crack shapes observed in a titanium aluminide alloy, revealed by heat tinting

where the parameters F and f_w are the same as in Eq 2. For $a/c < 1$, M_1, M_2, and M_3 are the same as in Eq 2 for $a/c < 1$, and:

$$\lambda = \frac{a}{c}$$

$$g = 1 + \left[0.1 + 0.35 \left(\frac{a}{t} \right)^2 \right]$$

For $a/c > 1$, M_1, M_2, and M_3 are the same as in Eq 2 for $a/c > 1$, and:

$$\lambda = \sqrt{\frac{a}{c}}$$

$$g = 1 + \left[0.1 + 0.35 \left(\frac{c}{a} \right) \left(\frac{a}{t} \right)^2 \right]$$

In Eq 6, E is the tensile modulus, and ν is Poisson's ratio. The validity of Eq 6 is restricted to $0.2 < a/c < 2.0$.

The method presented above was evaluated for the growth of surface cracks initially having shapes different from the equilibrium shape. A shallow notch ($a/c = 0.1$) and a deep notch ($a/c = 2.5$) were introduced in tensile specimens made out of a near α-titanium alloy. Surface cracks are known to exhibit semicircular ($a/c = 1$) shapes in this material (Ref 18). Therefore, cracks initiating from these starter notches are expected to grow with continuous changes in crack aspect ratio, eventually converging to a semicircular crack at crack lengths that are large compared to notch dimensions. Figure 9(a) and (b) illustrate the fracture surfaces, heat tinted before fracture to reveal the final crack shape. The initial notch geometries are also visible. The crack aspect ratios estimated

by the present technique are given in Fig. 10, along with the changes in aspect ratio predicted using the SIF equations. The good agreement between the measured and predicted data suggests that this approach is accurate and reliable.

The difficulties associated with this approach are the cost of instrumentation, set-up time, and the experimental care required. At present, this technique is limited to laboratory investigations. However, extension of this technique to complicated geometries or actual components in service is possible by replacing the laser interferometric system with simpler techniques, such as using a strain gage or miniature linear variable differential transformer to measure crack opening displacements. In this approach, it is also implied that the surface cracks have elliptical geometry, because the compliance relationship (Eq 6) was deduced from the Newman-Raju formula for el-

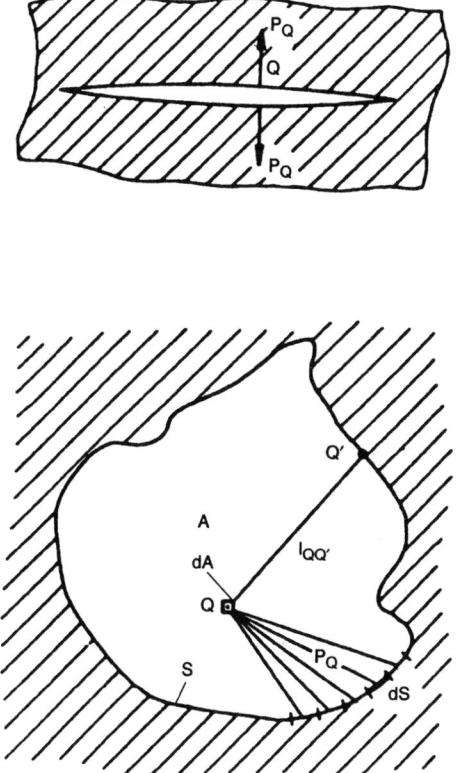

Fig. 12 An irregular crack embedded in an infinite solid subjected to a point force

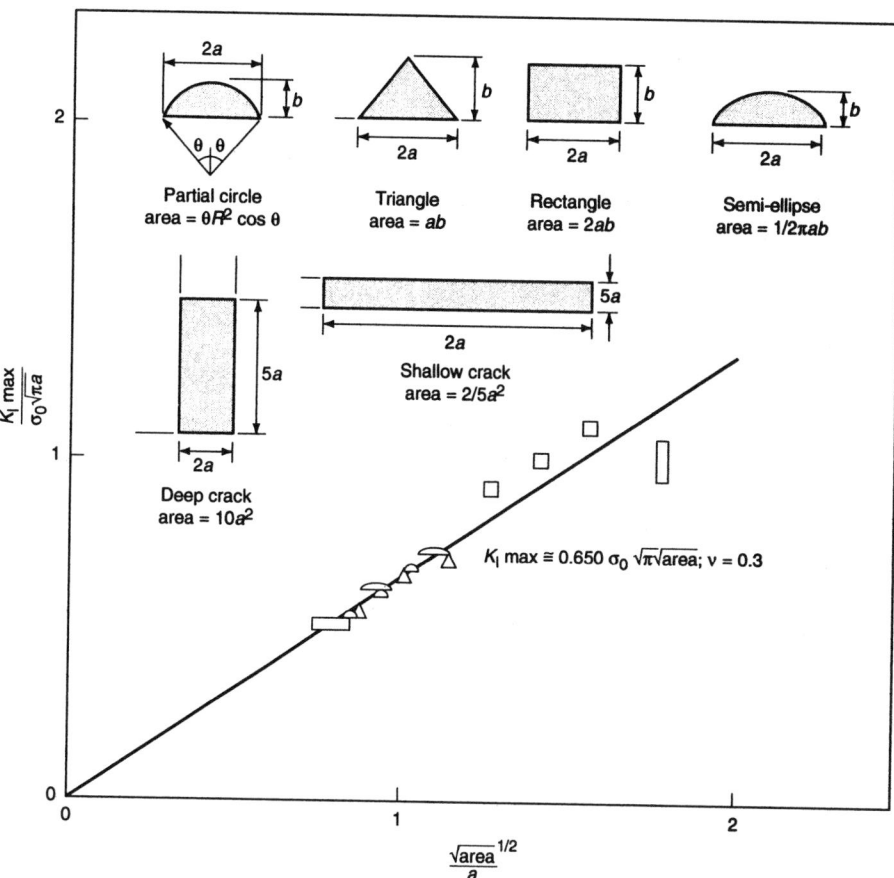

Fig. 13 The uniqueness in the variation of maximum stress-intensity factor of irregular cracks with the square root of the area, for various crack geometries

liptical cracks. However, reasonably accurate measurements of average crack aspect ratio have been made (Ref 19, 20) by approximating irregular cracks to elliptical shapes in a titanium aluminide alloy. Figure 11 illustrates some of the crack shapes at the end of fatigue tests in Ti-24Al-11Nb alloy. The measured aspect ratios reasonably agreed with those measured from fracture surface after heat tinting (Ref 20). Hence, the described technique can provide good estimates of aspect ratios of regular surface cracks, as well as those having limited irregularity in shape, continuously during their growth in fatigue.

Estimation of SIF for Arbitrarily Shaped Cracks. Analytical solutions for straight, circular, and elliptical cracks are readily available, owing to their simplicity of geometry. On the other hand, irregular cracks seldom have simple solutions due to their complex geometry. Often, the finite element method must be applied in order to determine the distribution of SIF (or stress concentration factor, in the case of a pore/cavity). Because of crack irregularity, fatigue crack growth often occurs at points of maximum stress intensity, leading to continuous change in the irregularity of the crack front. Under these circumstances, the finite element method calculations must be repeated to trace the change in crack shape and/or to allow for the loading spectrum. However, this is not practical, due to the cost and time involved.

Based on weight function technique in fracture mechanics, Oore and Burns (Ref 21) developed a simple procedure to calculate the mode I stress-intensity factor at any point along the front of an irregular flat crack embedded in an infinite solid and subjected to an arbitrary normal stress field. The stress-intensity factor, $K_{Q'}$, at any point Q' on the crack front (Fig. 12) is given by:

$$K_{Q'} = \iint_A W_{QQ'} q_Q dA_Q \qquad \text{(Eq 7)}$$

where q_Q is the opening force intensity (pressure) acting at point Q over the area dA_Q and $W_{QQ'}$ is the weight function. If $W_{QQ'}$ is known for each point on the crack surface, $K_{Q'}$ can be calculated for any distribution of pressure on the crack surface. For a circular crack the weight function is readily available (Ref 22). From this, Oore and Burns recognized that the form of weight function depends on the inverse of the square of the distance ($l_{QQ'}$) from the load point (Q) to the point of interest (Q'), the geometry of the crack front, and the location of the load point Q in crack geometry. They arrived at the weight function for an irregular crack as:

$$W_{QQ'} = \cfrac{\sqrt{2}}{\left[\pi\, l_{QQ'}^2 \sqrt{\displaystyle\int_s \frac{dS}{\rho_Q^2}} \right]} \qquad \text{(Eq 8)}$$

where the integral over the crack front, S, captures the irregularity of the crack front and its effect on SIF at different locations, and the variable ρ_Q is the distance from the point load at Q to each infinitesimal portion, dS, of the crack front. Using Eq 7, SIFs can be calculated by simple numerical methods. Equation 7 is a general expression for SIF and is suitable for any shape of the embedded crack. Its application to the prediction of crack front advance of irregular cracks during fatigue crack growth yielded consistent results (Ref 21). Although Eq 7 is applicable to embedded cracks, a modification to surface cracks appears to be possible (Ref 23) by incorporating a magnification factor for the specimen geometry. Hence, this approach can be of significant use in analyzing the growth behavior of arbitrarily shaped flaws, such as those found in weldments, during fatigue.

Methods of Failure Prediction for Arbitrarily Shaped Flaws. Application of fracture mechanics methods to the prediction of failure from through cracks is simple and straightforward, because only crack length in one direction is involved. On the other hand, for surface and em-

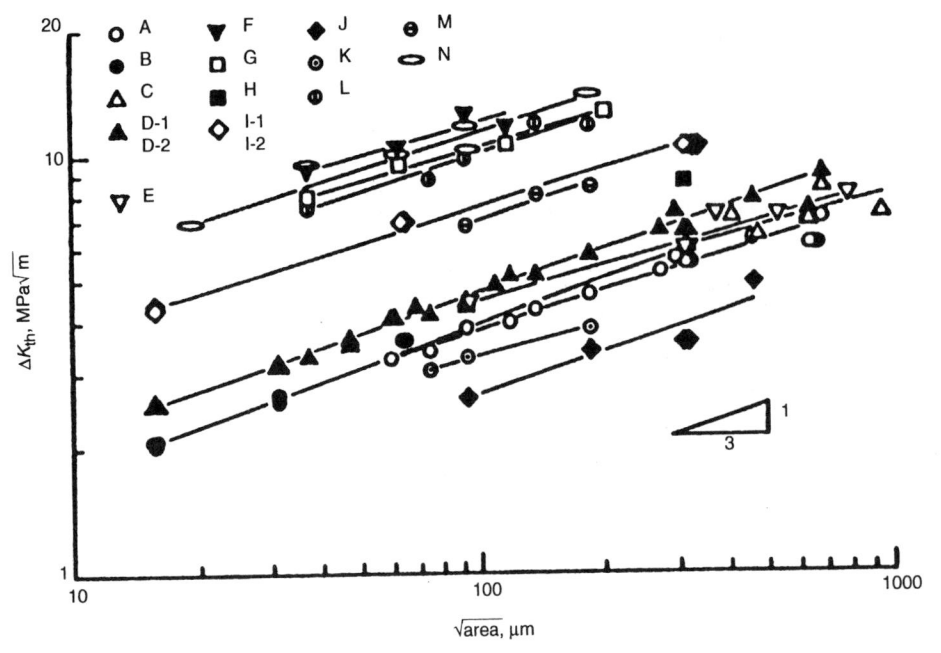

Fig. 14 Relationship between ΔK_{th} and the square root of the area, for various defects and cracks. Letters corresponding to the materials are given in Table 1.

Table 1   Materials in Fig. 14

Material	HV	Defect
A: S10C (annealed)	120	Notch
		Hole
B: S30C (annealed)	153	Notch
C: S35C (annealed)	160	Notch
		Hole
D-1: S45C (annealed)	180	Notch
D-2: S45C (annealed)	170	Hole
E: S50C (annealed)	177	Notch
		Crack
F: S45C (quenched)	650	Hole
G: S45C (quenched and tempered)	520	Hole
H: S50C (quenched and tempered)	319	Notch
I-1: S50C (quenched and tempered)	378	Notch
I-2: S50C (quenched and tempered)	375	Notch
J: 70/30 brass	70	Notch
		Hole
K: aluminum alloy (2017-T4)	114	Hole
L: Stainless steel (SUS 603)	355	Hole
M: Stainless steel (YUS 170)	244	Hole
N: Maraging steel	720	Vickers hardness indentation, hole and notch

bedded cracks with arbitrary shapes, both the size and shape are important. This is because changes in length in any direction can change the projected area of the defect on the plane perpendicular to the principal loading direction, thereby altering the load-bearing cross-sectional area. This would naturally affect the SIF or stress concentration in the vicinity of the crack or cavity, respectively. An appropriate methodology is therefore required to take into account the irregularity of the crack/defect in predictions of failure.

Cracks and cavities encountered in most applications are irregular, far from the circular or elliptical geometries assumed in standard fracture mechanics solutions. Murakami et al. (Ref 24, 25) have developed simple approaches to extend fracture mechanics to irregularly shaped cracks as well as defects of varying geometry. The key to this approach is the observation that the maximum SIF along the crack front is proportional to the square root of crack area. In the case of notches or cavities, the area projected onto the plane normal to the loading direction is considered the crack area. This approach brings cracks and defects of varying geometries to a common base, since the square root of the area and the square root of the projected area, which are dimensionally equivalent to length, are considered for cracks and notches, respectively. Figure 13 shows the normalized maximum SIF as a function of the crack area for several crack geometries having aspect ratios (ratio of major axis to minor axis) restricted to ≤5 (Ref 24).

On this basis, the threshold for the nonpropagation of defects of various geometries can be represented (Ref 25) as a function of the square root of the area, as shown in Fig. 14. The increase in ΔK_{th} with the square root of the area, is due to the increase in threshold with crack size in the short-crack regime of through cracks, an effect arising from crack closure. For several metals, including carbon steels, aluminum alloys, brass, and stainless steel, it has been found that:

$$\Delta K_{th} = 0.0033 (HV + 120) \left(\sqrt{area} \right)^{1/3} \quad \text{(Eq 9)}$$

where HV is Vickers hardness. It has been found that Eq 9 is within 10% of the experimentally observed threshold data of the materials studied and is applicable to cracks of varying size and geometry. The only restriction is the defect or crack aspect ratio (a/b, where a and b are the major and minor dimensions of the projected area of the crack or defect) should not exceed 5, since the approximation of the square root of the area becomes inaccurate for $a/b > 5$. It is evident that this equation is a simple and useful tool to predict the failure of components in engineering practice.

Cracks and cavities with irregular shapes initiate cracks from the location of maximum stress concentration. These cracks propagate to the extent that the projected area of the crack, onto the plane perpendicular to stress, becomes close to a circle. This is the condition of uniform stress intensity or concentration around the crack or cavity. Such crack propagation behavior was observed (Ref 24) in rotating bending fatigue tests of steel specimens having starter notches of various geometries, as well as in steels containing irregularly shaped inclusions. It was then deduced that it is the nonpropagation condition of cracks, not the initiation of cracks at the point of maximum stress concentration, that determines the fatigue limit. On this basis, Murakami et al. correlated the fatigue limit of specimens containing variously shaped notches to the square root of

the area, projected onto the plane normal to applied stress:

$$\sigma_a^n \sqrt{area} = C \quad \text{(Eq 10)}$$

in which σ_a is the fatigue limit at $R = -1$ of a specimen containing a defect of the square root of the area, irrespective of its shape, and n and C are material constants. This approach is generally limited to notches with $a/b < 5$. For several metals, including aluminum alloy, brass, stainless steel, and quenched and tempered martensitic steels, a practically useful correlation has been produced (Ref 25) from a large set of experimental data:

$$\sigma_a = 1.43 \frac{(HV + 120)}{\left(\sqrt{area} \right)^{1/6}} \quad \text{(Eq 11)}$$

It has been found that Eq 11 is within 10% of the experimentally observed fatigue limits of specimens having cracks and notches of varying geometry. Hence, this relationship is useful to predict the effect of defects and notches on the fatigue limit of components in service.

The problem of irregularity of crack front is of foremost importance in common metallurgical situations such as weldments, carburized and surface-hardened materials, and components with notches. Further work is clearly needed to advance the understanding generated to date and to apply it more widely in engineering practice.

REFERENCES

1. *Fractography, Metals Handbook*, Vol 12, ASM International, 1987
2. G.R. Irwin, Crack Extension Force for a Part-through Crack in a Plate, *J. Appl. Mech., Trans. ASME*, Vol 29 (No. 4), 1962, p 651-654
3. J.C. Newman, Jr. and I.S. Raju, Stress Intensity Factor Equations for Cracks in Three-Dimensional Finite Bodies, *Fracture Mechanics: 14th*

Symposium, Vol I, STP 791, ASTM, 1983, p I-238 to I-265

4. Wu Shang-Xian, Shape Change of Surface Crack during Fatigue Crack Growth, *Eng. Fract. Mech.*, Vol 22 (No. 5), 1985, p 897-913

5. Y. Mutoh, G.H. Fair, B. Noble, and R.B. Waterhouse, The Effect of Residual Stresses Induced by Shot-Peening on Fatigue Crack Propagation in Two High Strength Aluminum Alloys, *Fatigue Fract. Engng. Mater. Struct.*, Vol 10 (No. 4), 1987, p 261-272

6. W.O. Soboyejo, K. Kishimoto, R.A. Smith, and J.F. Knott, A Study of the Interaction and Coalescence of Two Coplanar Fatigue Cracks in Bending, *Fatigue Fract. Engng. Mater. Struct.*, Vol 12 (No. 3), 1989, p 167-174

7. J.H. Underwood and D.P. Kendall, Fracture Analysis of Thick Wall Cylindrical Pressure Vessels, *J. Theoret. Appl. Fract. Mech.*, Vol 2 (No. 2), 1984, p 47-58

8. L. Hodulak, H. Kordisch, H. Kunzelmann, and E. Sommer, Influence of the Load Level on the Development of Part Through Cracks, *Int. J. Fract.*, Vol 14, 1984, p R35-R38

9. W.O. Soboyejo and J.F. Knott, An Investigation of Environmental Effects on Fatigue Crack Growth in Q1N (HY80) Steel, *Metall. Trans.*, Vol 21A (No. 11), 1990, p 2977-2983

10. Y. Murakami, S. Kodama, and S. Konuma, Quantitative Evaluation of Effects of Non-Metallic Inclusions on Fatigue Strength of High Strength Steels, Part I: Basic Fatigue Mechanism and Evaluation of Correlation between the Fatigue Fracture Stress and the Size and Location of Non-Metallic Inclusions, *Int. J. Fatigue*, Vol 11 (No. 5), 1989, p 291-298

11. Y. Murakami and H. Usuki, Quantitative Evaluation of Effects of Non-Metallic Inclusions on Fatigue Strength of High Strength Steels, Part II: Fatigue Limit Evaluation based on Statistics for Extreme Values of Inclusion Size, *Int. J. Fatigue*, Vol 11 (No. 5), 1989, p 299-307

12. L. Wagner, J.K. Gregory, A. Gysler, and G. Lutjering, Propagation Behavior of Short Cracks in a Ti-8.6Al Alloy, *Small Fatigue Cracks: Proc. of International Workshop on Small Fatigue Cracks*, TMS-AIME, 1986, p 117-124

13. M. Okazaki, T. Endoh, and T. Koizumi, "Surface Small Crack Growth Behavior on Type 304 Stainless Steel in Low-Cycle Fatigue at Elevated Temperature," *J. Eng. Mater. Tech., Trans. ASME*, Vol 110, 1988, p 9-16

14. K.S. Ravichandran, "Fatigue Crack Growth Behavior of Small and Large Cracks in Titanium Alloys and Intermetallics," WL-TR-94-4030, Wright Patterson, 1994

15. P.J. Hastings, "The Behavior of Short Fatigue Cracks in a Beta Processed Titanium Alloy," Ph.D. thesis, University of Nottingham, 1989

16. R.K. Bolingbroke, "The Growth of Short Fatigue Cracks in Titanium and Aluminum Alloys," Ph.D. thesis, University of Nottingham, 1988

17. K.S. Ravichandran and J.M. Larsen, "An Approach to Measure the Shapes of Three-Dimensional Surface Cracks during Fatigue Crack Growth," *Fatigue Fract. Engng. Mater. Struct.*, Vol 16 (No. 8), 1993, p 909-930

18. W.N. Sharpe, Jr., J.R. Jira, and J.M. Larsen, Real-Time Measurement of Small-Crack Opening Behavior Using an Interferometric Strain/Displacement Gage, *Small-Crack Test Methods*, STP 1149, ASTM, 1992, p 92-115

19. K.S. Ravichandran and J.M. Larsen, Behavior of Small and Large Fatigue Cracks in Ti-24Al-11Nb: Effects of Crack Shape, Microstructure, and Closure, *Fracture Mechanics: 22nd Symposium*, STP 1130, Vol 1, ASTM, 1992, p 727-748

20. K.S. Ravichandran and J.M. Larsen, Microstructure and Crack Shape Effects on the Growth of Small Cracks in Ti-24Al-11Nb, *Mat. Sci. Eng.*, Vol A152, 1992, p 499

21. M. Oore and D.J. Burns, Estimation of Stress Intensity Factors for Embedded Irregular Cracks Subjected to Arbitrary Normal Stress Fields, *J. Press. Vess. Tech., Trans. ASME*, Vol 102 (No. 6), 1980, p 202-211

22. H. Tada, P.C. Paris, and G.R. Irwin, *The Stress Analysis of Cracks Handbook*, Paris Productions Inc., St. Louis, MO, 1985

23. J.L. Desjardins, D.J. Burns, and J.C. Thompson, A Weight Function Technique for Estimating Stress Intensity Factors for Cracks in High Pressure Vessels, *J. Press. Vess. Tech., Trans. ASME*, Vol 113 (No. 2), 1991, p 10-21

24. Y. Murakami and M. Endo, Quantitative Evaluation of Fatigue Strength of Metals Containing Various Small Defects or Cracks, *Eng. Fract. Mech.*, Vol 17 (No. 1), 1983, p 1-15

25. Y. Murakami and M. Endo, Prediction Equation for ΔK_{th} of Various Metals Containing Small Defects in Terms of Vickers Hardness (H_V) and the Square Root of the Projected Area of Defects, *Fracture Mechanics*, Vol 8, *Current Japanese Materials Research*, H. Okamura and K. Ogura, Ed., Elsevier Applied Science Pub., 1990, p 105-124

Fatigue Crack Growth Testing

Ashok Saxena and Christopher L. Muhlstein, Georgia Institute of Technology

FATIGUE is generally understood to be a process dominated by cyclic plastic deformation, such that fatigue damage can occur at stresses below the monotonic yield strength. The process of fatigue cracking generally begins from locations where there are discontinuities or where plastic strain accumulates preferentially in the form of slip bands. In most situations, fatigue failures initiate in regions of stress concentration such as sharp notches, nonmetallic inclusions, or at preexisting crack-like defects. Where failures occur at sharp notches or other stress raisers, cracks first initiate and then propagate to critical size, at which time sudden failure occurs. The fatigue life consists of crack initiation as well as crack propagation. On the other hand, when fatigue failures are caused by large inclusions or pre-existing crack-like defects, the entire life consists of crack propagation. Such situations are commonly encountered in service failures. A typical example of such a failure in a railroad track is shown in Fig. 1. The light area in the photograph is the region of fatigue crack growth, and the surrounding darker area is the region of fast fracture. The dark spot within the light area is the origin of the failure, which is a pre-existing defect due to a hydrogen flake.

Testing of smooth or notched specimens generally characterizes the overall fatigue life of a specimen material. This type of testing, however, does not distinguish between fatigue crack initiation life and fatigue crack propagation life. With this approach, preexisting flaws or crack-like defects, which would reduce or eliminate the crack initiation portion of the fatigue life, cannot be adequately addressed. Therefore, testing and characterization of fatigue crack growth is used extensively to predict the rate at which subcritical cracks grow due to fatigue loading. For components that are subjected to cyclic loading, this capability is essential for life prediction, for recommending a definite accept/reject criterion during nondestructive inspection, and for calculating in-service inspection intervals for continued safe operation.

Fracture Mechanics in Fatigue

Linear elastic fracture mechanics is an analytical procedure that relates the magnitude and distribution of stress in the vicinity of a crack tip to the nominal stress applied to the structure; to the size, shape, and orientation of the crack or crack-like imperfection; and to the crack growth and fracture resistance of the material. The procedure is based on the analysis of stress-field equations, which show that the elastic stress field in the region of a crack tip can be described by a single parameter, K, called the stress-intensity factor. This same procedure is also used to characterize fatigue crack growth rates (da/dN) in terms of the cyclic stress-intensity range parameter (ΔK).

When a component or a specimen containing a crack is subjected to cyclic loading, the crack length (a) increases with the number of fatigue cycles, N, if the load amplitude (ΔP), load ratio (R), and cyclic frequency (v), are held constant. The crack growth rate, da/dN, increases as the crack length increases during a given test. The da/dN is also higher at any given crack length for tests conducted at higher load amplitudes. Thus, the following functional relationship can be derived from these observations:

$$\left(\frac{da}{dN}\right)_{R,v} = f(\Delta P, a) \qquad \text{(Eq 1)}$$

where the function f is dependent on the geometry of the specimen, the crack length, the loading configuration, and the cyclic load range. This general relation is simplified with the use of the ΔK parameter as summarized below.

Correlation between da/dN and ΔK. In 1963, Paris and Erdogan (Ref 1) published an analysis consisting of considerable fatigue crack

Fig. 1 Fatigue failure of a railroad track

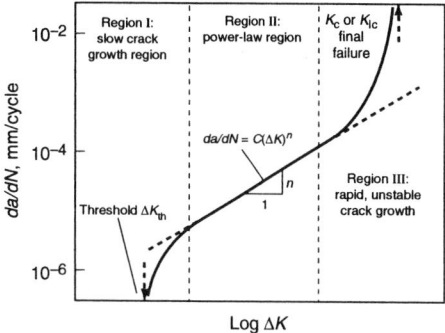

Fig. 2 Fatigue crack growth regimes versus ΔK

growth rate (FCGR) data and demonstrated that a correlation exists between da/dN and the cyclic stress intensity parameter, ΔK. They argued that ΔK characterizes the magnitude of the fatigue stresses in the crack-tip region; hence, it should characterize the crack growth rate. Such a proposition is in obvious agreement with the functional relationships of Eq 1. The parameter ΔK accounts for the magnitude of the load range (ΔP) as well as the crack length and geometry. A number of later studies (Ref 2) have confirmed the findings of Paris and Erdogan. The data for intermediate FCGR values can be represented by the following simple mathematical relationship, commonly known as the Paris equation:

$$\frac{da}{dN} = C(\Delta K)^n \tag{Eq 2}$$

where C and n are constants that can be obtained from the intercept and slope, respectively, of the linear log da/dN versus log ΔK plot. This representation of FCGR is a useful model for midrange FCGR values (Fig. 2).

It has been shown that specimen thickness has no significant effect on the FCGR behavior (Ref 3), although that is not always the case. The ability of ΔK to account for so many variables has tremendous significance in the application of the data. Thus, the FCGR behavior expressed as da/dN versus ΔK can be regarded as a fundamental material property analogous to the yield and ultimate tensile strength, plane strain fracture toughness, K_{Ic}, etc. From the knowledge of this property, prediction of the crack length versus cycles behavior of any component using that material and containing a preexisting crack or cracklike defect can be obtained, as long as the fatigue stresses in the component are known and a K expression for the crack/load configuration is available.

Crack-Tip Plasticity during Fatigue. The cyclic stress-intensity parameter, ΔK, is based on linear elastic fracture mechanics, and characterizes only the elastic stress field beyond the plastic zone. However, fatigue is a process dominated by cyclic plastic deformation. Even when fatigue damage occurs at stresses below the monotonic yield strength, the process of fatigue cracking begins from locations where there are discontinuities, such as nonmetallic inclusions, or

from surfaces where plastic strain accumulates preferentially in the form of slip bands (Ref 4). Therefore, a brief explanation is given why ΔK can characterize fatigue crack growth behavior.

When a cracked body is subjected to cyclic loading, a monotonic plastic zone develops at the crack tip during the first loading cycle. If predominantly linear elastic conditions are maintained during loading, as are necessary for ΔK to be a valid crack-tip parameter, compressive stress develops within this plastic zone during unloading because the elastic forces in the overall body tend to restore its original shape (Ref 2). The magnitude of the maximum compressive stress increases as the crack tip is approached. In a small region within the monotonic plastic zone, the maximum compressive stress exceeds the yield strength, resulting in plastic flow in compression. This small region of reversing plastic flow is called the cyclic plastic zone. A simple estimate of the size of this zone was made by Paris (Ref 2) and Rice (Ref 5) for nonhardening materials by substituting $2\sigma_{ys}$ in place of σ_{ys} in the expression for monotonic plastic zone size and by replacing K with ΔK:

$$r_{cp} = \frac{1}{\pi}\left(\frac{\Delta K}{2\sigma_{ys}}\right)^2 \tag{Eq 3}$$

where r_{cp} is the cyclic plastic zone size under plane-stress conditions. For materials that undergo cyclic hardening or softening, a first-order estimate of the fatigue plastic zone size can be obtained by replacing σ_{ys} with the cyclic yield strength (σ_{cys}) in Eq 3.

General Crack Growth Behavior. When crack growth rates over six to seven decades are plotted against ΔK, the behavior is no longer a straight line on a log-log plot. Results of FCGR tests for nearly all metallic structural materials have shown that the da/dN versus ΔK curves have three distinct regions. The behavior in region I (Fig. 2) exhibits a fatigue crack growth threshold, ΔK_{th}, which corresponds to the stress-intensity factor range below which cracks do not propagate. Equation 2 is applicable in the midrange of da/dN values for FCGR (region II in Fig. 2). Typically, the validity of Eq 2 is limited over a range of two to four decades for midrange crack growth rates. Testing and material factors that affect crack growth behavior in regions I, II, and III of Fig. 2 are discussed in more detail in the article "Fatigue Failures in Metals" in this Volume.

At high ΔK values, region III, the K_{max} approaches the critical K for instability, K_c, and the crack growth rate accelerates. In some cases K_c may be equal to K_{Ic}, but this cannot be generalized because the FCGR specimens or even actual components may not always satisfy size requirements for valid linear elastic plane-strain conditions. In some materials there is also an effect of prior fatiguing on the K value at which instability occurs (Ref 6). In such cases, K_c will not be equal to the K_{Ic} of the material.

At low ΔK values (region I in Fig. 2), the crack growth rate decreases rapidly with decreasing

ΔK, and ultimately ΔK approaches a threshold value, ΔK_{th}, when the crack growth rate approaches zero. In high-cycle fatigue applications, ΔK_{th} is an important design parameter. The above definition of ΔK_{th} is an idealized definition; for practical usage it is important to define its value unambiguously. An operational value of ΔK_{th} is frequently defined as the ΔK value at a da/dN of 10^{-10} m/cycle (Ref 7).

FCGR under Elastic-Plastic Conditions. There are applications when fatigue crack growth occurs under conditions of gross plastic deformation, or at least under conditions for which dominant linear elasticity cannot be ensured. As a crack tip parameter, ΔK breaks down under these conditions and can no longer be expected to uniquely characterize FCGR behavior. Dowling and Begley have defined a cyclic J-integral, ΔJ, which is determined utilizing the loading portion of the load-displacement diagram during cyclic loading (Ref 8, 9).

Metals and alloys can be assumed to deform according to the cyclic stress-strain law given by:

$$\Delta\varepsilon = \frac{\Delta\sigma}{E} + D'\left(\frac{\Delta\sigma}{2\sigma_{cys}}\right)^{m'} \tag{Eq 4}$$

where $\Delta\varepsilon$ is the cyclic strain range, $\Delta\sigma$ is the cyclic stress range, E is the elastic modulus, and D' and m' are empirically determined material constants. The value of ΔJ for such materials can be defined by (Ref 10):

$$\Delta J = \int_{\Gamma}(\Delta W)dy - \Delta T_i\left(\frac{\partial \Delta u_i}{\partial x}\right)ds \tag{Eq 5}$$

The term ΔJ in Eq 5 is a path-independent integral along any given path Γ which originates at the lower crack surface and ends on the upper crack surface traversing along the contour in a counterclockwise direction. The definition of ΔJ is written as a direct analogy to Rice's J-integral (Ref 11) used extensively in characterizing fracture under monotonic loading conditions. The term ΔW in Eq 5 is as follows:

$$\Delta W = \int_0^{\Delta\varepsilon_{ij}}\Delta\sigma_{ij}\,d(\Delta\varepsilon_{ij}) \tag{Eq 6}$$

Other terms in Eqs 5 and 6 are:

- ΔT_i is the range of the traction vector
- Δu_i is the range of displacement
- $\Delta\sigma_{ij}$ and $\Delta\varepsilon_{ij}$ are the ranges of the stress and strain, respectively
- ds is an element along the contour Γ

All range quantities are calculated by subtracting the values at minimum load from the corresponding values at maximum load. When ΔJ is defined in the above manner, its value characterizes the crack-tip stress and strain ranges according to the Hutchinson (Ref 12) and Rice and Rosengren (Ref 13) relationships. It must also be noted that for linear elastic conditions, Eq 5 will yield the following relationship:

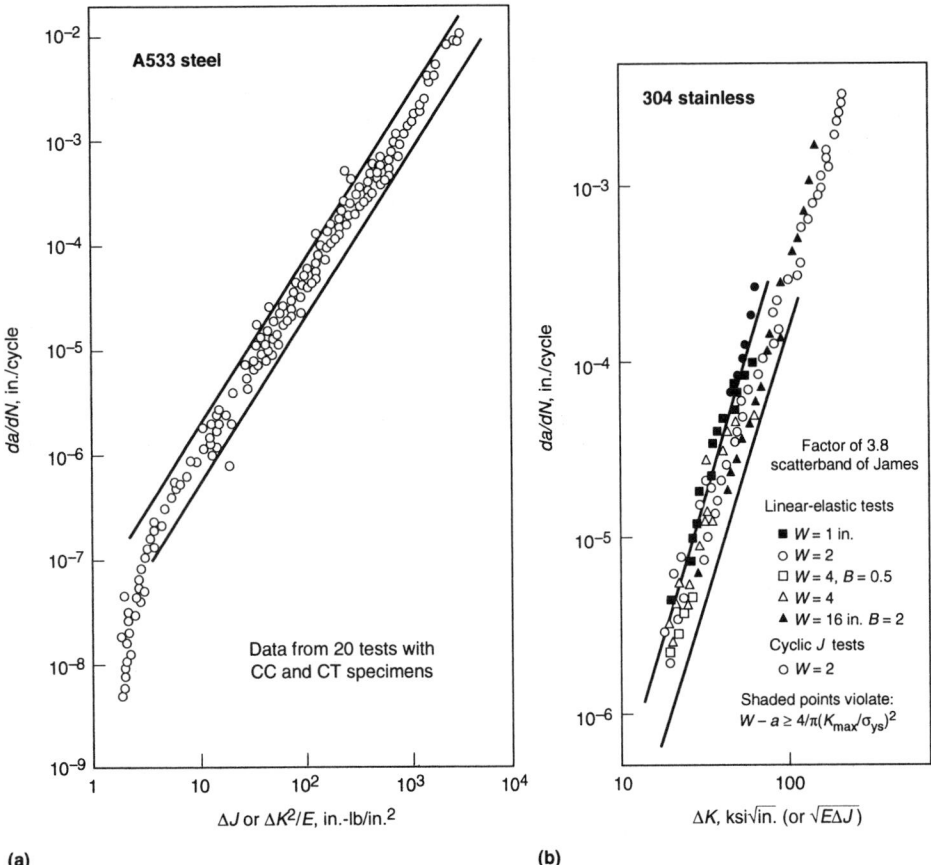

Fig. 3 Fatigue crack growth rate obtained under linear elastic and elastic-plastic conditions in A533 steel (a) and 304 stainless steel (b). CC, center-cracked; CT, compact-type. Source: Ref 9, 14

$$\Delta J = \frac{\Delta K^2}{E} \qquad \text{(Eq 7)}$$

From the above relationship, the data from linear elastic tests and elastic-plastic or fully plastic tests can be combined into a single plot of da/dN with ΔK or $\sqrt{E\Delta J}$. Similarly, the data can be correlated with $\Delta K^2/E$ or ΔJ. Figure 3 shows the FCGR data for A533 and for 304 stainless steel in this manner (Ref 9, 14). These data were developed on specimens of two geometries and more notably on specimens with varying sizes within those geometries. Thus, small specimens exhibited considerable plasticity, and the large specimens were under dominantly elastic conditions. Despite the enormous differences in the scales of plasticity among the various tests, the FCGR data lay in a single scatter band.

Crack Closure. The concept of crack closure was first introduced by Elber (Ref 15, 16) as an effect from a zone of residual deformation that is left in the wake of a growing fatigue crack. According to this concept, crack surfaces at the crack tip might stay closed during a portion of the fatigue cycle due to compressive residual stress acting at the crack tip. Elber further postulated that this portion of the loading cycle is ineffective in growing the fatigue crack and that thus the corresponding load should be subtracted from the

applied ΔP to determine the effective value of ΔK.

Figure 4 shows a series of schematic sketches that show the stress and strain distributions at the crack tip at maximum and minimum load. At the maximum load, A, all the load is borne by the uncracked ligament because cracks are unable to transmit the load. At the minimum load, B, there are compressive stresses to the left of the crack tip because of the contact between opposing crack surfaces within the zone of residual plastic deformation. This causes the effective stiffness of the cracked body to change, which manifests itself in the load-displacement diagram. Thus, the crack closure load can be defined as the load at which this change in stiffness occurs.

Figure 5(a) shows a schematic load-deflection diagram and the crack closure point. Figure 5(b) plots only the deviation between the total deflection and the linearly predicted deflection, thus highlighting the crack closure point.

The importance of crack closure varies with the crack growth regime, crack tip material-microstructure interactions, and the extent of plasticity. Crack closure is more significant in the near-threshold regime (region I) than in region II. Materials in which the crack path is such that rougher crack surfaces are produced usually exhibit enhanced crack closure levels. The crack closure levels can also increase with plasticity.

For example, during fatigue crack growth in the elastic-plastic regime, crack closure levels take on added significance (Ref 8, 9).

Test Methods and Procedures

American Society for Testing and Materials Standard E647 (Ref 7) is the accepted guideline for fatigue crack growth testing and is applicable to a wide variety of materials and growth rates.

FCGR testing consists of several steps, beginning with selecting the specimen size, geometry, and crack length measurement technique. When planning the tests, the investigator must have an understanding of the application of FCGR data. Testing is often performed in laboratory air at room temperature; however, any gaseous or liquid environment and temperature of interest may be used to determine the effect of temperature, corrosion, or other chemical reaction on cyclic loading (see the article "Corrosion Fatigue Testing" in this Volume and the appendix "High-Temperature Fatigue Crack Growth Testing" at the end of this article). Cyclic loading also may involve various waveforms for constant-amplitude loading, spectrum loading, or random loading.

In addition, many of the conventions used in plane-strain fracture toughness testing (ASTM E-399, Ref 17) are also used in FCGR testing. For tension-tension fatigue loading, the K_{Ic} loading fixtures frequently can be used. For this type of loading, both the maximum and minimum loads are tensile, and the load ratio, $R = P_{min}/P_{max}$, is in the range $0 < R < 1$. A ratio of $R = 0.1$ is commonly used for developing data for comparative purposes.

Cyclic Crack Growth Rate Testing in the Threshold Regime. Cyclic crack growth rate testing in the low-growth regime (region I in Fig. 2) complicates acquisition of valid and consistent data, because the crack growth behavior becomes more sensitive to the material, environment, and testing procedures in this regime. Within this regime, the fatigue mechanisms of the material that slow the crack growth rates are more significant (see the article "Fatigue Crack Threshold Behavior and Analysis" in this Volume).

It is extremely expensive to obtain a true definition of ΔK_{th}, and in some materials a true threshold may be nonexistent. Generally, designers are more interested in the fatigue crack growth rates in the near-threshold regime, such as the ΔK that corresponds to a fatigue crack growth rate of 10^{-8} to 10^{-10} m/cycle (3.9×10^{-7} to 10^{-9} in./cycle). Because the duration of the tests increases greatly for each additional decade of near-threshold data (10^{-8} to 10^{-9} to 10^{-10}, etc., m/cycle), the precise design requirements should be determined in advance of the test. Although the methods of conducting fatigue crack threshold testing may differ, ASTM Standard E-647 addresses these requirements.

In all areas of crack growth rate testing, the resolution capability of the crack measuring technique should be known; however, this becomes considerably more important in the threshold re-

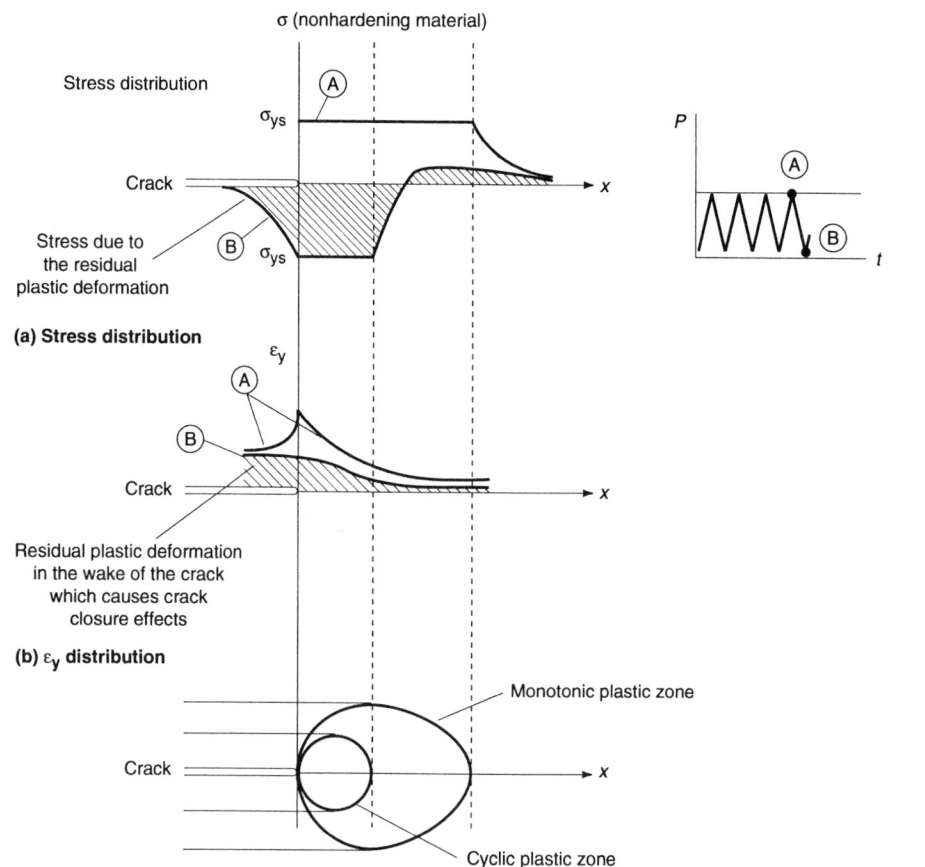

Fig. 4 Schematic representation of the crack-tip conditions during crack closure

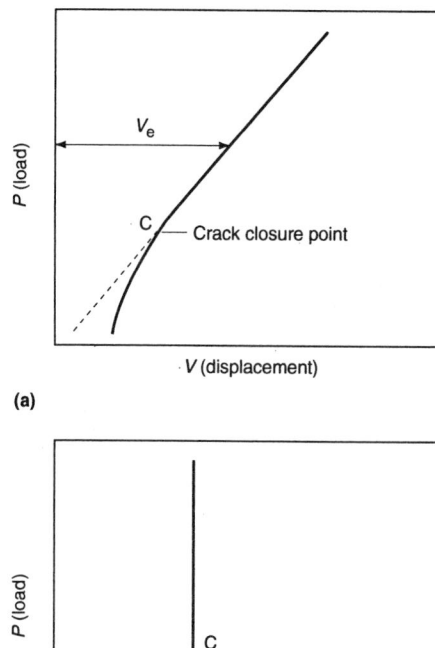

Fig. 5 Load versus displacement diagram showing a change in stiffness at the crack closure point. (b) A plot of total deflection minus the elasticity calculated deflection amplified to highlight crack closure. v_e, elastic displacement

gime. The smallest amount of crack length resolution as possible is desired, because the rate of decreasing applied loads (load shedding) is dependent on how easily the crack length can be measured. The minimum amount of change in crack growth that is measured should be ten times the crack length measurement precision. It is also recommended that for noncontinuous load shedding testing, where $[(P_{max_1} - P_{max_2}) / P_{max_1}] > 0.02$, the reduction in the maximum load should not exceed 10% of the previous maximum load, and the minimum crack extension between load sheds should be at least 0.50 mm (0.02 in.).

In selecting a specimen, the resolution capability of the crack measuring device and the K-gradient (the rate at which K is increased or decreased) in the specimen should be known to ensure that the test can be conducted appropriately. If the measuring device is not sufficient, the threshold crack growth rate may not be achieved before the specimen is separated in two. To avoid such problems, a plot of the control of the stress intensity (K versus a) should be generated before selection of the specimen.

When a new crack-length measuring device is introduced, a new type of material is used, or any other factor is different from that used in previous testing, the K-decreasing portion of the test should be followed with a constant load amplitude (K-increasing) to provide a comparison be-

tween the two methods. Once a consistency is demonstrated, constant-load amplitude testing in the low crack growth rate regime is not necessary under similar conditions.

Specimen Selection and Preparation

The two most widely used types of specimens are the middle-crack tension, M(T), and the compact-type, C(T), specimen (see Fig. 6 and 7). However, any specimen configuration with a known stress-intensity factor solution can be used in fatigue crack growth testing, assuming that the appropriate equipment is available for controlling the test and measuring the crack dimensions.

Specimens used in FCGR testing may be grouped into three categories: pin-loaded (Fig. 6, 7), bend-loaded (Fig. 8a) and wedge-gripped specimens (Fig. 8b, c, d). Precisely machined specimens are essential, and ASTM E 647 specifies the recommended tolerances and K-calibrations for compact-type C(T) and middle-tension M(T) geometries. Single-edge bend SE(B), arc-shaped A(T), and disk-shaped compact DC(T) specimen geometries and their K-calibrations are discussed in ASTM E 399. Comparable tolerances should be specified for "nonstandard" specimens. The selection of an appropriate geometry requires consideration of material avail-

ability and raw form, desired loading condition, and equipment limitations.

Crack Length and Specimen Size. The applicable range of the stress-intensity solution of a specimen configuration is very important. Many stress-intensity expressions are valid only over a range of the ratio of crack length to specimen width (a/W). For example, the expression given in Fig. 6 for the compact-type specimen is valid for $a/W > 0.2$; the expression for the center-cracked tension specimen (Fig. 7) is valid for $2a/W < 0.95$. The use of stress-intensity expressions outside their applicable crack-length region can produce significant errors in data.

The size of the specimen must also be appropriate. To follow the rules of linear elastic fracture mechanics, the specimen must be predominantly elastic. However, unlike the requirements for plane-strain fracture toughness testing, the stresses at the crack tip do not have to be maintained in a plane-strain state. The stress state is considered to be a controlled test variable. The material characteristics, specimen size, crack length, and applied load will dictate whether the specimen is predominantly elastic. Because the loading modes of different specimens vary significantly, each specimen geometry must be considered separately.

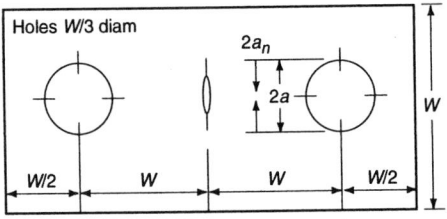

Center-cracked tension specimens ΔK:

$$\Delta K = \frac{\Delta P}{B} \sqrt{\frac{\pi \alpha}{2W}} \ \sec \frac{\pi \alpha}{2}$$

where $\alpha = \dfrac{2a}{W}$; expression valid for $\dfrac{2a}{W} < 0.95$

Fig. 6 Standard center-cracked tension (middle-tension) specimen and ΔK solution. Specimen width (W) ≤ 75 mm (3 in.). $2a_n$, machined notch; a, crack length; B, specimen thickness

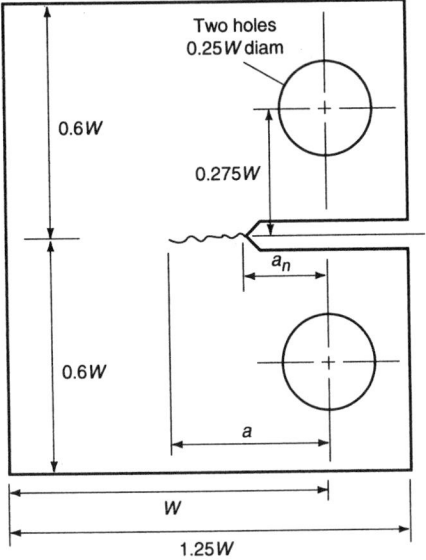

Compact-type specimens ΔK value:

$$\Delta K = \frac{\Delta P (2 + \alpha)}{B \sqrt{W}(1 - \alpha)^{3/2}} \ (0.886 + $$

$$4.64\alpha - 13.32\alpha^2 + 14.72\alpha^3 - 5.6\alpha^4), \text{ where}$$

$$\alpha = \frac{a}{W} \ ; \text{ expression valid for } \frac{a}{W} \geq 0.2$$

Fig. 7 Standard compact-type specimen and ΔK value (per ASTM E 647). Allowable thickness: $W/20 \leq B \leq W/4$. Minimum dimensions. W = 25 mm (1.0 in.) and machined notch size (a_n) = 0.20W

For the center-cracked tension specimen, the following is required:

$$W - 2a \geq \frac{1.25 P_{max}}{B\sigma_{ys}} \qquad \text{(Eq 8)}$$

where $W - 2a$ is the uncracked ligament of the specimen (see Fig. 6) and σ_{ys} is the 0.2% offset yield

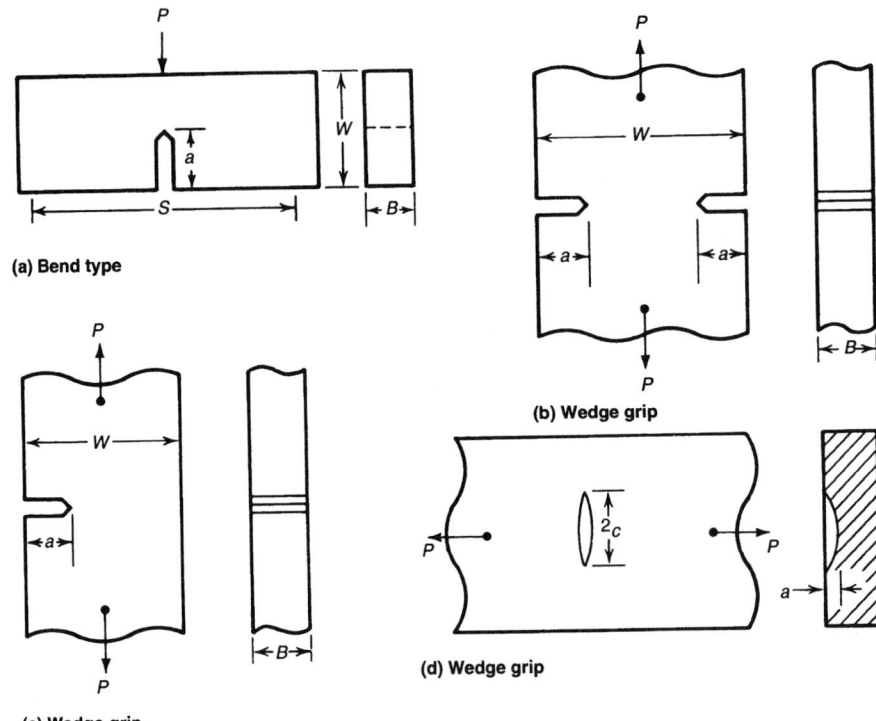

(a) Bend type

(b) Wedge grip

(c) Wedge grip

(d) Wedge grip

Fig. 8 Alternative crack growth specimen geometries. (a) Single-edge-crack bending specimen. (b) Double-edge-crack tension specimen. (c) Single-edge-crack tension specimen. (d) Surface-crack tension specimen

strength at the temperature corresponding to the FCGR data. For the compact-type specimen, the following is required:

$$W - a \geq \frac{4}{\pi} \left(\frac{K_{max}}{\sigma_{ys}} \right)^2 \qquad \text{(Eq 9)}$$

where $W - a$ is the uncracked ligament (see Fig. 7). For the compact-type specimen, the size requirement in Eq 9 limits the monotonic plastic zone in a plane-stress state to approximately 25% of the uncracked ligament. For both Eq 8 and 9, ASTM E 647 recommends the use of the monotonic yield strength. The size requirements in Eq 8 and 9 are appropriate for low-strain hardening materials ($\sigma_u/\sigma_{ys} \leq 1.3$), where σ_u is the ultimate tensile strength of the material. For higher-strain-hardening materials, Eq 8 and 9 may be too restrictive. In such cases, the criteria may be relaxed by replacing the yield strength, σ_{ys}, with the effective yield strength, σ_F:

$$\sigma_F = \frac{(\sigma_{ys} + \sigma_u)}{2} \qquad \text{(Eq 10)}$$

Specimen Thickness. While fatigue crack growth rates have been shown to be relatively insensitive to stress state (i.e., plane-stress versus plane-strain, Ref 3), there are some practical limitations on specimen thickness. ASTM E 647 recommends that generally compact-tension specimen thickness (B) range between 5 and 25% of width ($W/20 \leq B \leq W/4$). Middle-tension speci-

mens may have thicknesses up to 12% of width (≤W/8). For center-cracked tension specimens, thickness should not exceed 25% of width. When other specimen geometries are used, similar ranges for the thicknesses should be employed.

Although specimen thickness can vary significantly, the amount of crack curvature in the specimen will increase as the thickness increases. Because stress-intensity solutions are based on a straight through-crack, a significant amount of curvature, if not properly accounted for, can lead to an error in the data. Crack-curvature correction calculations are detailed in ASTM E 647. The minimum allowable thickness depends on the gripping method used; however, the bending strains should not exceed 5% of the nominal strain in the specimen.

Material Form and Microstructure Considerations. The material and its microstructure play an important role in the selection of an appropriate specimen geometry. Materials with anisotropic microstructures due to processing such as rolling or forging may show large variations in fatigue crack growth rates in different directions (Ref 18). If the experimental crack growth rate data are to be used for life estimates, the orientation of the specimen should be selected to represent loading orientations expected in service.

In order to eliminate grain size effects, it is usually recommended that the specimen thickness (B) be greater than 30 grain diameters (Ref 19, 20). In some cases, such as in large-grain (~3 mm) lamellar γ-α₂ Ti-Al intermetallic or α-β titanium alloys, the required specimen sizes would be prohibitively expensive, test loads would be

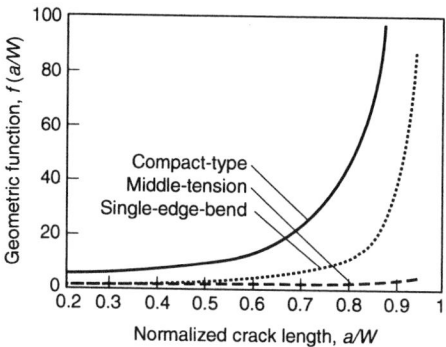

Fig. 9 K-gradients for a number of fatigue crack growth specimens. Source: Ref 7, 17

very high, and the component dimensions would probably be less than 30 times the grain size. In such cases, testing should be performed on thickness representative of the component. Curvature of the crack front and side-to-side variation in crack length due to excessive thickness can also be a problem in thick specimens, as discussed below.

Loading Considerations. The desired loading conditions play an important role in the specimen geometry and size selection process. Loading considerations include load ratio, R, residual stresses, K-gradients, and maintaining small-scale yielding (SSY). All specimen geometries are well suited for tension-tension ($R > 0$) testing. However, tests that call for negative R (i.e., those with minimum loads of less than 0) are restricted to symmetric, wedge-grip loaded specimens such as the middle-tension specimens. This is due to questions about the crack-tip stress field under compressive loads (Ref 7) and difficulties moving through zero load with pin-loaded specimens.

Residual stresses in the material also have a marked effect on FCGR. Depending on the orientation of the residual stresses, specimen dimensions or geometries should be altered. Residual stresses through the thickness of the specimen (i.e., perpendicular to the direction of crack growth) may accelerate or retard crack growth. When these stresses are not uniform, the ASTM E 647 recommends a reduction of the thickness-to-width ratio (B/W).

The rate at which K increases as the crack extends at a constant-load amplitude is given by the geometry function $f(a/W)$ and may be a consideration when selecting the most appropriate specimen geometry. Figure 9 shows the effect of geometry on the K-gradient through a variety of specimen geometries. Specimens with shallower K-gradients are preferable for brittle materials, while the opposite is true for ductile materials.

Equipment Considerations. Specimen size and geometry can also be influenced by laboratory equipment such as the loadframe, loadcell, existing loading fixtures, testing environment, and even the crack length measurement apparatus. To minimize cost, specimen sizes and geometries should be selected to use existing clevises, pins, and other hardware.

Most modern mechanical testing laboratories exclusively use electroservohydraulic loadframes for FCGR investigations. Current controls and data acquisition technology have hydraulic loadframes more versatile than the electromechanical systems used in previous years. When selecting a specimen geometry and size, one must be aware of the load capacity of the actuator and loadframe. Loads that are too high cannot be applied, and those that are too low cannot be controlled with the required accuracy (±2%). In addition, the load cell to be used during testing must be able to measure the maximum applied load and resolve the lowest expected amplitudes, as specified in ASTM E 4.

When testing in environments, specimens fit inside ovens, furnaces, or other chambers with ample space left for clevises, cantilever beam clip gages, and other hardware. Special notch geometries or knife edge attachment locations are often necessary for attaching clipgages or other types of extensometers for nonvisual crack length measurements using compliance techniques.

Notch and Specimen Preparation. The method by which a notch is machined depends on the specimen material and the desired notch root radius (ρ). Sawcutting is the easiest method but is generally acceptable only for aluminum alloys. For a notch root radius of $\rho \leq 0.25$ mm (0.010 in.) in aluminum alloys, milling or broaching is required. A similar notch root radius in low- and medium-strength steels can be produced by grinding. For high-strength steel alloys, nickel-base superalloys, and titanium alloys, electrical discharge machining may be necessary to produce a notch root radius of $\rho \leq 0.25$ mm (0.010 in.).

The specimen is polished to allow measurement of the crack during the precracking and testing phases of the experiment. Many specimens can be polished using standard metallography practices. In some cases, etching of the polished surface may provide better contrast for viewing of the crack. If the specimen is too large or small to be handled, then hand grinders, finishing sanders, or handheld drills can be used with pieces of polishing cloth to locally apply the abrasive and create a satisfactory viewing surface. These techniques are quick and easy to apply, and they are often used when visual measurements are made only during precracking and subsequent measurements are made by automated techniques such as electric potential or compliance.

Precracking. The K-calibration functions found in ASTM E 647 and E 399 are valid for sharp cracks within the range of crack length specified. Consequently, before testing begins a sharp fatigue crack that is long enough to avoid the effects of the machined notch must be present in the specimen (0.1B, or 0.1H, or 1 mm [0.040 in.], whichever is greatest). The process that generates this crack is termed precracking. In general, loads for precracking should be selected such that the K_{max} at the end of precracking does not exceed levels expected at the start of a test.

For most metals, precracking is a relatively simple process that can be performed under load or displacement control conditions. Moderate growth rates (1×10^{-5} m/cycle) can be selected by estimating the necessary ΔK from growth curves in the literature. Precracking of a specimen prior to testing is conducted at stress intensities sufficient to cause a crack to initiate from the starter notch and propagate to a length that will eliminate the effect of the notch. To decrease the amount of time needed for precracking to occur, common practice is to initiate the precracking at a load above that which will be used during testing and to subsequently reduce the load.

Load generally is reduced uniformly to avoid transient (load-sequence) effects. Crack growth can be arrested above the threshold stress-intensity value due to formation of the increased plastic zone ahead of the tip of the advancing crack. Therefore, the step size of the load during precracking should be minimized. Under these circumstances, the loads should be shed no faster than 20% (per increment of crack extension, as discussed below) from the previous load increment (Ref 7). This will eliminate load-sequence effects on growth rates. As the crack approaches the final desired size, this percentage can be decreased.

The amount of crack extension between each load decrease must also be controlled. If the step is too small, the influence of the plastic zone ahead of the crack may still be present. To avoid transient (load-sequence) effects in the test data, as discussed above, the load range in each step should be applied over a crack-length increment of at least $(3\pi) (K'_{max}/\sigma_{ys})^2$, where K'_{max} is the terminal value of K_{max} from the previous load step. This requirement ensures that the crack extension between load sheds is at least three plastic zone diameters.

The influence of the machined starter notch must be eliminated so that the crack tip conditions are stable. For compact-type and center-cracked tension specimens, this requires that the final precrack be at least 10% of the thickness of the specimen or equivalent to the height of the starter notch, whichever is greater (Ref 7).

Two additional considerations regarding crack shape are the amount of crack variation from the front and back sides of the specimen and the amount of out-of-plane cracking. Due to microstructural changes through the specimen thickness, residual stresses (particularly in weldments), or misalignment of the specimen in the grips, the crack may grow unevenly on the two surfaces. If any two crack length measurements vary by more than 0.025W or by more than 0.25B (whichever is less), the precracking operation was not suitable and test results will not be valid. If a fatigue precrack departs more than ±5° from the plane of symmetry, the specimen is not suitable for subsequent testing.

Precracking of Brittle Materials. Brittle materials such as intermetallics and ceramics can be very difficult to precrack. It is not uncommon to initiate a flaw that immediately propagates to failure. This is due, in part, to the increasing

K-gradient found in FCGR specimens and the relatively narrow range of ΔK for stable crack growth.

To improve the chances of successful precracking of brittle materials, chevron notches are advised. Chevron-notched specimens (Fig. 10) are used for determining the fracture toughness of brittle materials that are difficult to fatigue precrack. Chevron notches generate decreasing *K*-gradients at the start of precracking and may be machined as part of the specimen, or they may be added just prior to testing using a thin diamond wafering blade. The maximum slope of the chevron notch should be 45°. Precracking of brittle materials should be performed under displacement control conditions, so that as the crack extends, the load and the applied *K* decrease. Lastly, the loads should be increased slowly from low levels due to the stochastic nature of crack initiation in these materials. If initiation is especially difficult, compressive overloads may assist the process. It is also helpful to monitor the initiation process with a method other than optical observation. Electric potential techniques (bulk and foil) and back face strain compliance techniques are very effective.

Once precracking has been completed, an accurate optical measurement of the initial crack length, a_0, must be made on both sides of the specimen to within 0.10 mm (0.004 in.) or $0.002W$ (whichever is greatest), or to within 0.25 mm (0.01 in.) for specimens where $W > 127$ mm (5 in.). If the crack lengths on the two surfaces differ by more than $0.25B$, then the test will not be valid, because *K*-calibration functions presume the existence of a straight crack front. Middle-tension specimens further require that both halves of the precrack be the same length to within $0.025W$. In addition, ASTM E 647 requires that cracks lie on the centerline such that the crack is no more than ±20° from a centerline over a distance $0.1W$. Once the precrack has been measured and side-to-side variation and distance from centerline have been established, testing may begin. Additional information on fatigue testing of brittle materials is in "Fatigue of Brittle Materials" in this Volume.

Gripping of the specimen must be done in a manner that does not violate the stress-intensity solution requirements. For example, in a single-edge notched specimen, it is possible to produce a grip that permits rotation in the loading of the specimen, or it is possible to produce a rigid grip. Each of these requires a different stress-intensity solution. In grips that are permitted to rotate, such as the compact-type specimen grip, the pin and hole clearances must be designed to minimize friction. It is also advisable to consider lateral movement above and below the grips.

When appropriate, the use of a lubricant is recommended to reduce friction. In thick samples, the amount of bending in the pins should be minimized. Finally, the alignment of the system should be checked carefully to avoid undesirable bending stresses, which generally cause uneven cracking. Alignment can be easily checked using a strain gage specimen of a geometry similar to

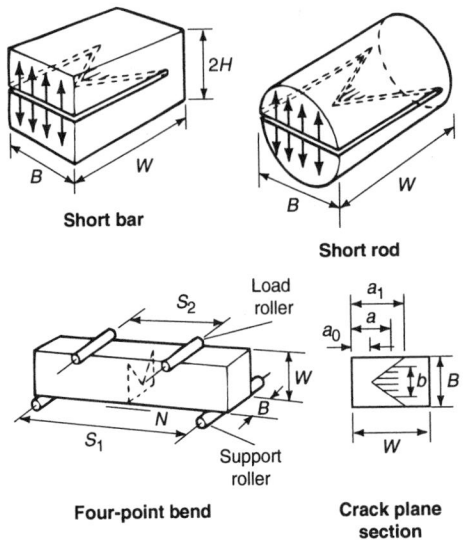

Fig. 10 Schematic of chevron notches in fracture mechanics specimens. The shaded area denoted "b" is the crack area.

that used in the test program. Generally, bending strains should not exceed 5% of the nominal strain to be used in the test program.

Gripping arrangements for compact-type and center-cracked tension specimens are described in ASTM E 64 (Ref 7). For a center-cracked tension specimen less than 75 mm (3 in.) in width, a single pin grip is generally suitable. Wider specimens generally require additional pins, friction gripping, or some other method to provide sufficient strength in the specimen and grip to prohibit failure at undesirable locations, such as in the grips.

Crack Length Measurement

Precise measurements of fatigue crack extension are crucial for the determination of reliable crack growth rates. ASTM E 647 requires a minimum resolution of 0.1 mm (0.004 in.) in crack length measurement. Crack extension measurements are recommended at intervals that are 10 times the minimum required resolution.

Various crack measurement techniques have been applied, including optical (visual and photographic), ultrasonic, acoustic emission, electrical (eddy current and resistance), and compliance (displacement and back face strain gages) methods. Optical, compliance, and electric potential difference are the most common laboratory techniques, and their merits and limitations are reviewed in detail in the following sections. Other references are listed in "Selected References" at the end of this article and in the article "Detection and Monitoring of Fatigue Cracks" in this Volume.

Optical Crack Measurement

Monitoring of fatigue crack length as a function of cycles is most commonly conducted visu-

ally by observing the crack at the specimen surfaces with a traveling low-power microscope at a magnification of 20 to 50×. Crack-length measurements are made at intervals such that a nearly even distribution of *da/dN* versus ΔK is achieved. The minimum amount of extension between readings is commonly about 0.25 mm (0.010 in.).

For planar specimens, the crack length is measured on one or both surfaces, depending on the section thickness. For example ASTM E 647 (Ref 7) specifies a *B/W* value of 0.15 as the limit; measurements on only one side are sufficient if $B/W < 0.15$.

Through-thickness variations in crack length must be considered and corrected for if too severe. Typical behavior is for the crack length to lead at the midplane (crack tunneling). Because this cannot be observed in situ by visual monitoring, post-test observations must be made. Rough alignment of the traveling microscope can be easily achieved by shining a pen light through the eyepiece on the crack-tip region. To ensure accurate crack measurements, obliquely incident light on a well-polished specimen surface is an effective means of highlighting fine cracks. High-intensity strobe lights with adjustable function generators are used to allow "motion free" viewing of cracks during high-frequency tests. The development of extra-long focal length optics has added new functionality to optical techniques. These microscopes allow the in situ observation and image analysis of crack-tip processes while keeping the instruments a reasonable distance (>381 mm, or 15 in.) from the specimen and other testing hardware.

To account for through-thickness crack-length variation, ASTM E 647 recommends measuring the crack length at five points along the crack front contour and averaging the five readings. If the average of the five points exceeds the surface length by more than 5%, the average length is used in computing the growth rate and *K*.

The optical technique is straightforward and, if the specimen is carefully polished and does not oxidize during the test, produces accurate results. However, the process is time consuming, subjective, and can be automated only with complicated and expensive video-digitizing equipment. In addition, many fatigue crack growth rate tests are conducted in simulated-service environments that obscure direct observation of the crack. The trend toward laboratory automation has resulted in the development of indirect methods of determining crack extension, such as specimen compliance and electric potential monitoring.

Compliance Method

Under linear elastic conditions for a given crack size, the displacement, *v*, across the load points or at any other locations across the crack surfaces is directly proportional to the applied load (*P*). The compliance, *C*, of the specimen is defined as

$$C = \frac{v}{P} \qquad \text{(Eq 11)}$$

The relationship between dimensionless compliance, BEC, where B is the thickness and E is the elastic modulus, and the dimensionless crack size, a/W, where W is the specimen width, is unique for a given specimen geometry (Ref 21). Thus:

$$BEC = f\left(\frac{a}{W}\right) \qquad \text{(Eq 12)}$$

The inverse relationship (Ref 21) between crack size and compliance can be written as $a/W = q(u)$ where $u = [1 + BEC]^{-0.5}$. This relationship may be determined numerically using finite element techniques or by experiment. ASTM E 647 also specifies these relationships for compact-type and middle-tension specimens.

The compliance of an elastically strained specimen (expressed as the quotient of the displacement, v, and the tensile load, P, per Eq 11) is determined by measuring the displacement along, or parallel to, the load line. Figure 11 illustrates that the more deeply a specimen is cracked, the greater the amount of v measured for a specific value of tensile load. Additional information on the calculation of compliance and the method can be found in the "Selected References" listed at the end of this article.

Instrumentation. The displacement usually is measured across the crack mouth opening using cantilever beam clip gages, optical (laser and white light) extensometry, or back face strain gages. Linear variable differential transducers have been used, but hysteresis in their response can sometimes be a problem. Each of these techniques has its own advantages and may be used to continuously monitor crack length. An additional benefit of compliance techniques is that the same signal can be used for determining crack closure, as discussed below.

Cantilever beam clip gages based on resistive and capacitance strain gage technology are well suited for elevated (<370 °C) and high-temperature tests (up to 1200 °C), respectively. Deflection of the arms is measured by the output of the strain gages mounted on the clip gage arms. Extensometer and transducer design theory is well established in the literature (Ref 17, 22). Attachment of clip gages to the specimen is achieved through integral, machined knife edges or by knife edge blocks bolted to the front face of the specimen across the crack plane.

Optical extensometry techniques include those based on fiber optics and laser technology. There are two main types of optical extensometry used in compliance measurement. The first are advanced laser systems that track the motion of spots projected on the specimen. In this case the transmitter and receiver are located on the same side of the specimen. A second group of extensometers measure the width of the notch using a transmitter and receiver on opposite sides of the specimen. Laser and white light systems based on these principles are available commercially. Optical systems may have restrictions on their frequency response. Like optical crack length measurement techniques, most optical extensometry techniques are difficult to use when testing at

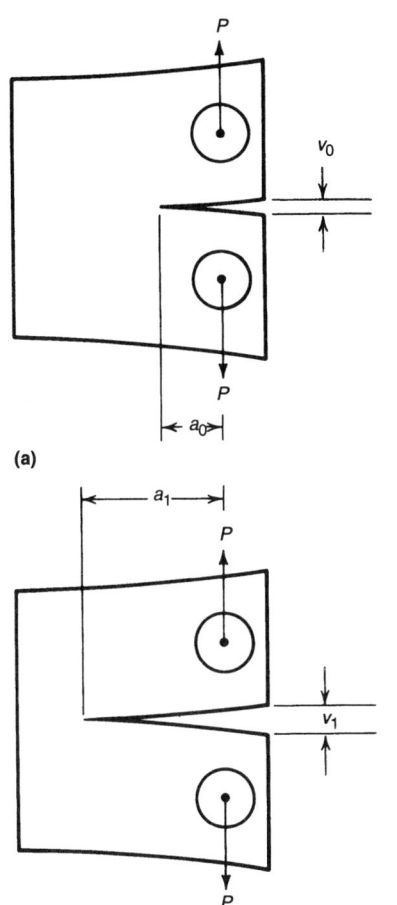

(a)

(b)

Fig. 11 Schematic of the relationship between compliance and crack length. (a) $C(a_0) = v_0/P$. (b) $C(a_1) = v_1/P$

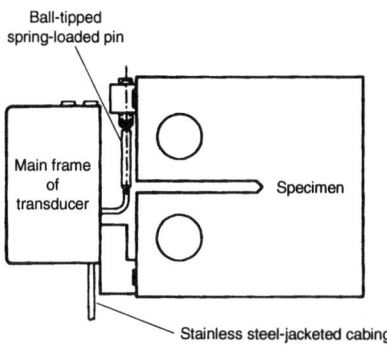

Fig. 12 Bolt-on attachment of crack-opening displacement transducer to fatigue test specimen

elevated temperatures or in environmental chambers.

An electrical resistance strain gage mounted opposite the notch on the back face of the fatigue specimen is termed a back face strain (BFS) gage. Just as with the clip gage, the load-strain signal from the BFS gage may be used to determine the crack length in the specimen. While conventional strain gages are limited to elevated temperatures (<370 °C) in gaseous or aqueous environments, they have the advantage of directly measuring strain without the application of a force. The direct measurement of strain eliminates frequency limitations associated with clip gages at the expense of having no geometric amplification of the strains or the sensitivity benefits of a four active leg Wheatstone bridge (Ref 22). BFS is especially useful for nonmetallic materials where integral knife edges and tapped holes for knife edge blocks are difficult to machine.

The required sensitivity of the systems depends on specimen geometry and sizes. In general, noise-free, amplified output on the order of 1 V direct current (dc) per 1 mm (0.04 in.) of deflection is satisfactory. Similarly, for the load range applied to the specimen, an approximately 1 V dc

change in signal from the load cell is required for accurate calculation of the compliance.

Attachment of Displacement Measurement Hardware. One of the most important factors affecting the accuracy of crack-opening displacement measurements is the manner in which the displacement transducer is attached to the test sample. Transducers for measuring the crack-opening displacement commonly consist of cantilevered arms affixed across the crack. When the crack is opened, deflections either in the arms, or in a flexure attached to the arms, induces measurable strains, which are ultimately converted to displacements.

To prevent slipping of the gage during testing, the gage must have an adequate, well-documented clamping force (~2500 g). This force must be added to the mean tensile load applied to the specimen by the gage during data analysis. For thin or small specimens, this gage-induced mean load may be high enough to preclude testing at the desired load levels. Even with high clamping forces, there will be a limitation on the maximum testing frequency for the gage due to the excitation of resonant modes in the gage or inadequate clamping force. If necessary, higher frequencies can be achieved by bolting the clip gage to the specimen.

The transducer can be bolted across the crack opening at the point of testing, or it can be attached to the specimen through hardened knife-edge pivots that are mechanically or adhesively affixed to the specimen. The transducer can also be affixed via knife-edge contacts that are machined into the test sample. For elevated-temperature testing, feed-rod systems are frequently used.

The bolt-on system of attaching the transducer to the test specimen (Fig. 12) is capable of reacting to high acceleration loads. These result from higher-frequency dynamic testing when the rocking moment generated by the mass of the transducer is carried to the bolt-on attachment through the transducer frame. This attachment system is preferred when a transducer has high mass or an effective mass center that is located a great distance from the specimen contact pads.

The bolt-on system also provides accurate crack-opening displacement measurements on specimens tested under environments that are not

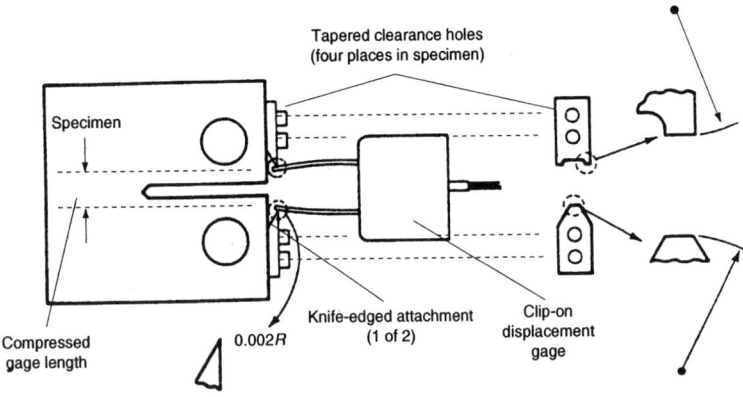

Fig. 13 Bolt-on hardened knife-edge attachment of crack-opening displacement transducer to fatigue test specimen

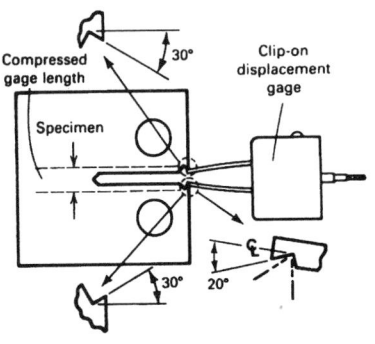

Fig. 14 Attachment of crack-opening displacement transducer to specimen by machined knife-edge contacts

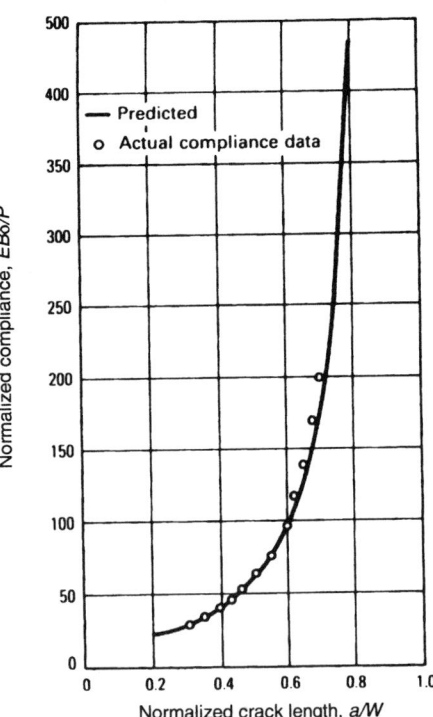

Fig. 15 Comparison of predicted and experimental compliance for a compact-type fatigue specimen

conducive to the use of knife edges, such as elevated-temperature or corrosive environments. In addition, the bolt-on attachment system allows the use of stiffer cabling without disturbing the measurement. For example, a displacement transducer with relatively rigid stainless steel-jacketed cabling can be used to make measurements in pressurized high-temperature water/steam environments.

Hardened knife-edge pivot contacts (Fig. 13) provide a measurement system with minimal sliding action; the knife edge rocks in a hardened seat in the transducer arm. This allows measurements to be made with very low hysteresis levels. Contact and seat ramp angles can be designed for optimal tradeoffs between static and dynamic measurement accuracy, dynamic stability, and contact durability. Male knife-edge contact replacements are relatively low in cost, and various configurations are available, such as three-point contact, line contact, large radius, and small radius.

Knife-edge contacts that are machined into the test sample (Fig. 14) eliminate the possibility of knife-edge screws loosening, which results in slippage and hysteresis. The compressed initial gage length can be machined to the required tolerance.

Computing Normalized Compliance. When measuring the compliance of a fatigue specimen, the usual practice is to compute a normalized compliance (EBv/P). This normalized compliance is plotted against a normalized crack length, a/W. For standard geometries, such as a compact-type specimen, this relationship has the form shown in Fig. 15. Thus, from the measured compliance, a crack length can be obtained from the known analytical relationships as shown in Fig. 15 (Ref 21). Note that when the crack is short ($a/W \sim 0.2$ to 0.4), the compliance is less sensitive to changes in crack length than when the crack length is long ($a/W > 0.5$). Thus, the sensitivity of the compliance method is significantly improved for the longer crack lengths, both because of this relationship and because the amount of crack mouth opening and the resulting displacement gage signal are larger. The amount of displacement or crack mouth opening that is measured is a strong function of the location of the line of measurement of the gage with respect to the load line, which is the reference point for crack extension. The farther away from the crack tip the measurement can be made, the more displacement that will be incurred, and the sensitivity of the method will be improved proportionately.

Data Acquisition and Processing. The signals from the load cell and displacement gage must be obtained simultaneously in order for this method to work to its best advantage. In the most direct case, the two signals can be fed to an *x-y* recorder, with the load applied to the *y*-axis and displacement to the *x*-axis. At various intervals during the test, a trace of the two signals can be made. If the test is being conducted at a reasonably high frequency (>1 Hz), then the frequency will have to be diminished so that the slow rate of the recorder can keep up with the changing voltage. This is not a problem if a transient recorder is used and the results from the two channels (load and displacement) are co-plotted. The slopes of the recorder traces can be measured, multiplied by suitable calibration factors, and used in the compliance to crack length relationship.

A more sophisticated method is to use a computerized data acquisition system to obtain load displacement data. These systems are usually faster and thus can accept data from rather high-frequency waveforms. In addition, software can be developed to perform the calculations involved in processing the compliance data to crack length. Software to perform fatigue crack growth rate measurements is generally available from manufacturers, but most researchers write their own data acquisition packages, perhaps using some of the manufacturer-supplied subroutines that are specific to the hardware involved.

Additionally, data should be taken between about 10 and 90% of the load range. Eliminating the top and bottom fractions of the load range avoids problems of crack closure (at loads approaching zero) or incipient plasticity (near the load maximum, at longer crack lengths). The sets of load-displacement pairs are fitted to a straight line, the slope of which is used in the compliance expression.

Electric Potential Difference Method

The electrical potential, or potential drop, technique has gained increasingly wide acceptance in fracture research as one of the most accurate and efficient methods for monitoring the initiation and propagation of cracks. This method relies on the fact that there will be a disturbance in the electrical potential field about any discontinuity in a current-carrying body, the magnitude of the disturbance depending on the size and shape of the discontinuity.

For the application of crack growth monitoring, the electric potential method entails passing a constant current (maintained constant by external means) through a cracked test specimen and measuring the change in electrical potential

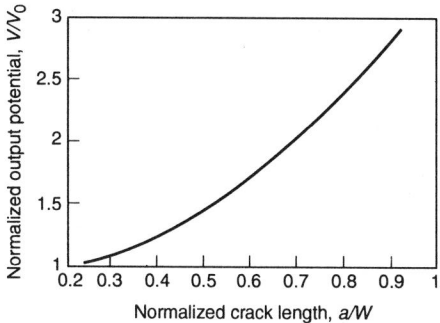

Fig. 16 Potential response for a compact-type specimen

across the crack as it propagates. With increasing crack length, the uncracked cross-sectional area of the test piece decreases, its electrical resistance increases, and thus the potential difference between two points spanning the crack rises. By monitoring this potential increase, V_a, and comparing it with some referencing potential, V_0, the ratio of crack length to width, a/W, can be determined through the use of the relevant calibration curve for the particular test piece geometry concerned. The crack length is expressed as a function of the normalized potential (V/V_0) and the initial crack length (a_0) (Fig. 16).

Accuracy of electrical potential measurements of crack length may be limited by a number of factors, including the electrical stability and resolution of the potential measurement system, electrical contact between crack surfaces where the fracture morphology is rough or where significant crack closure effects are present, and changes in electrical resistivity with plastic deformation. Another key factor is the determination of calibration curves relating changes in potential across the crack (V_a) to crack length (a). In most instances, experimental calibration curves have been obtained by measuring the electrical potential difference: across the machined slots of increasing length in a single test piece; across a growing fatigue crack, where the length of the crack at each point of measurement is marked on the fracture surface by a single overhead cycle or by a change in mean stress; across a growing fatigue crack in thin specimens where the length of the crack is measured by surface observation.

Other experimental calibrations have been achieved using an electrical analog of the test piece, where the specimen design is duplicated, usually with increased dimensions for better accuracy, using graphitized analog paper or thin aluminum foil, and where the crack length can be increased simply by cutting with a razor blade. Such calibration procedures, however, are relatively inaccurate, particularly at short crack lengths, and are tedious to perform. Furthermore, where measurements of crack initiation and early growth are required ahead of short cracks or notches of varying acuity, such procedures demand a new experimental calibration to be obtained for each notch geometry.

Electric potential response may be determined empirically (Ref 23-25) or using numerical methods such as finite element or conformal mapping techniques (Ref 26-30). Johnson's analytical solution of the middle-tension geometry is widely used in experimental work due to its flexibility (Ref 28):

$$a = \frac{W}{\pi} \cos^{-1} \left[\frac{\cosh\left(\frac{\pi}{W} \times Y_0\right)}{\cosh\left[\frac{V}{V_0} \times \cosh^{-1}\left[\frac{\cosh\left(\frac{\pi}{W} \times Y_0\right)}{\cos\left(\frac{\pi}{W} \times a_r\right)}\right]\right]} \right]$$

$$\text{for } 0 \leq 2\frac{a}{W} \leq 1 \qquad \text{(Eq 13)}$$

where a is the crack size, a_r is the reference crack size from other method, W is the specimen width, V is the measured electric potential difference, V_r is the measured voltage corresponding to a_r, and Y_0 is the voltage measurement lead spacing from crack plane. With minor modifications, Eq 13 can be applied to edge-cracked geometries by treating them as half of a middle-tension geometry. Third or higher-order polynomial expressions with coefficients obtained from regression analysis can be used to describe the potential response of the specimens when simplified expressions are required or Eq 13 does not apply.

The electric potential technique may be used with alternating current (ac) or dc power supplies. Alternating current systems have lower power requirements and do not suffer from the thermally induced potentials that plague dc systems. On the other hand, dc systems are widely used because of their relative simplicity. Consequently, this discussion of typical experimental setups is restricted to dc systems.

The main components of the dc electric potential system are shown in Fig. 12. The operating parameters for such a system are applied currents from 5 to 50 A and output voltages from 0.1 to 50 mV. Power supplies must be stable to 1 part in 10^4 or better, and nano- or microvoltmeters with a resolution of 0.05 to 0.5 μV are used (Ref 7). It is crucial that all dc potential measurement equipment (e.g., power supplies, voltage meters, etc.) and the loadframe itself be properly grounded. Before a power supply or nearby electromagnetic field (EMF) source (e.g., induction heater) is faulted for poor performance of the electric potential technique, researchers are reminded to

Table 1 Typical EPD voltages as measured on a standard compact-type specimen

Material	Approximate EPD, mV	Approximate change in crack length for 1 ì V change in EPD, ì m
Aluminum	0.1	300
Steel	0.6	50
Titanium	3.5	9

Based on $a/W = 0.22$, $B = 7.7$ mm, and $W = 50$ mm. Lead geometry per Ref 7 and direct current of 10 A

check that all equipment is properly grounded. In some cases, EMF shielding may be required.

High-resolution, stable, properly grounded equipment does not guarantee reliable performance and high resolution for the dc potential difference technique. Proper selection and use of current and potential leads are essential. High-current (welding) cable is ideal for current input leads, which are usually bolted to the specimen. To reduce noise, the potential leads should be firmly attached to the specimen, shielded, and twisted together. To ensure that current will pass through the specimen, the ratio of the loadtrain resistance to that of the specimen must be on the order of 10^4. If this cannot be achieved, the specimen must be electrically isolated using nonconducting (e.g., alumina) pins and washers or sleeves. The current applied to the specimen should be large enough to produce a measurable potential. Table 1 lists typical current and output voltages for compact-type (CT) specimens of steel, aluminum, and titanium. Excessive current (>10 A) can cause heating of the specimen and should be avoided. Potential leads should be made from fine wire of the same material as the specimen to reduce thermally induced EMF. Potential measurement leads and equipment should be kept away from EMF sources such as transformers to further reduce noise.

Crack tip processes such as fatigue crack closure (see the section "Fatigue Crack Closure" in this article) can reduce the potential of the specimen as the crack faces come together, effectively shortening the crack. This is especially a problem when testing materials that do not form protective, nonconducting oxide layers in the environment of interest. The solution to this problem is to measure the potential output at the peak load. In addition to crack closure, crack-tip plasticity and distributed damage such as microcracking must be considered. Large plastic zones such as those encountered under elastic-plastic conditions disturb the equipotential lines much like the crack (Ref 31). Distributed damage processes can also complicate measurements by making it difficult to define a continuous crack. Hence, optical measurements of the crack should be made to ensure that the electric potential difference technique provides a realistic representation of crack length. Changes in the electrical properties of the material can also limit the effectiveness of dc potential systems.

Changes in conductivity can complicate electric potential measurements. When high-conductivity materials such as aluminum are tested, temperature fluctuations of ±1 °C will cause a change in potential on the order of a few μV due to the temperature dependence of conductivity, and this change may vary with time. This can limit the crack extension resolution. Environmental chambers are useful with high-conductivity materials, even when testing at room temperature.

The primary difficulty with the dc electric potential technique is the junction potentials created at points of current and potential lead attachment. When dissimilar materials are in contact, a potential is generated due to the thermocouple effect, and it may be of the same order of magnitude as the potential generated by the specimen. This thermally induced potential, also known as the thermal voltage, may not be constant. Consequently, care must be taken to separate changes in potential due to fluctuations in thermal voltage from changes due to crack extension. This is especially important when measuring the slow growth rates found in the near-threshold regime.

There are three common approaches to accounting for the thermal voltage. The first method is to periodically turn off the power supply, note the value of thermal voltage, and subtract it from the output of the specimen with the current applied. This approach is acceptable for manually run tests, but it is not very useful when a continuous signal is required for computer-controlled tests. One alternative to manual measurement of the thermal voltage is to apply a current to an uncracked specimen with no applied load in the same environment as the test specimen in the "reference potential" technique. The tendency of the thermal voltage to drift should be the same in both the cracked and uncracked specimens. The drift can then be monitored, and the thermal voltage simply becomes an offset. Attempts have been made to apply the reference potential technique to a single specimen by measuring potentials in areas of the specimen that are "insensitive" to crack extension. The development of high-current-capacity solid-state switches has made the use of fully reversed electric potential drop systems a third method for dealing with thermal voltages. If the direction of current flow is periodically reversed, the thermal voltage, which has a fixed polarity, will shift the maximum and minimum output potentials but will not influence the range or amplitude of the signal. Thus, the amplitude of the output potential can be used to determine the length of the crack.

The electric potential technique may also be applied to nonconducting specimens with the use of conducting thin foils. The foils are applied prior to testing, and they crack with the underlying specimen. Current is applied to the foil instead of to the specimen, and the calibrated response of the foil may be used to monitor the growth of the crack. This technique may be used for room- and elevated-temperature tests, provided that the foil accurately reflects the growth of the crack. Polymer-backed gages sold under the trade name KrakGage require special hardware for mounting and use and may be used with conducting or nonconducting specimens. It is also possible to vapor deposit gages directly to nonconducting specimens or to nonconducting oxide films on conducting or nonconducting (e.g., SiC) materials. The drawback of electric potential foils is the tendency for cracks with small opening displacements to "tunnel" under the gage. This crack extension without breaking the foil will lead to inaccurate growth rates.

Optimization Parameters. In any specimen geometry, there are numerous locations for both the current input leads and the potential measurement probes. Optimization of the technique involves finding the best locations, considering accuracy, sensitivity, reproducibility, and magnitude of output (measurability).

In practice, the accuracy of the electrical potential technique may be limited by several factors, such as the electrical stability and resolution of the potential measurement system, crack front curvature, electrical contact between crack surfaces where the fracture morphology is particularly rough or where significant crack closure effects are present, and changes in electrical resistivity with plastic deformation, temperature variations, or both.

Reproducibility refers to inaccuracies produced by small errors in positioning the potential measurement leads. Such leads are generally fine wires that are spot welded or screwed to the specimen, and accurate positioning is typically no better than to within 0.5 mm (0.02 in.). To maximize reproducibility, these leads should be placed in an area where the calibration curve is relatively insensitive to small changes in position—that is, where dV/dx and dV/dy are small, where x and y are position coordinates— with the origin at the midpoint of the specimens. This consideration is often at variance with sensitivity considerations for measuring small changes in crack length.

To optimize measurability (i.e., signal-to-noise ratios), current input and potential measurement lead locations are chosen to maximize the absolute magnitude of the output voltage signal V_a. As output voltages are generally at the microvolt level and because of the high electrical conductivity of metals, a practical means of achieving measurability is simply to increase the input current. However, there is a limit to this increase, because when the current is too large (typically exceeding 30 A in a 12.7 mm, or 0.5 in., thick $1T$ steel compact-type specimen), appreciable specimen heating can result from contact resistance at current input positions.

Studies have shown that there must be a compromise between the sensitivity, reproducibility, and magnitude of the output signal when using electric potential techniques. In the case of compact-type specimens, it has been shown that potential leads are best placed on the notched side of the specimen, as close to the mouth as possible, as recommended by the ASTM E 647. When using nonstandard geometries, the reader is encouraged to use the above references to ensure a sound basis for lead placement.

Loading Methods

The goal of a fatigue crack growth rate test is to generate a record of crack length (a) versus number of cycles (N) under specified loading conditions. This information can be generated by applying cyclic varying loads of specified amplitude and frequency.

The frequency of the test should, when possible, be kept constant. However, it may be necessary to reduce the frequency of a test in order to make crack length measurements. Frequency effects are usually not observed in metals in laboratory air at room temperature over the range of typical testing frequencies (1 to 100 Hz). Although higher-frequency tests finish more quickly, specimen and loadtrain stiffness, as well as load range, impose a practical limit on the maximum testing frequency. Steel specimens that are 50 mm wide can be run on a typical 90 kN (20 kilo pounds) capacity loadframe at 25 to 50 Hz. If compliance methods are being used to control the test or monitor crack extensions, the frequency response of the clip gage and recording instruments may limit the maximum frequency for testing.

The waveform to be used during a test is usually a sine or sawtooth (ramp) shape. Both waveforms will generate similar data at room temperature in benign environments. However, sine waveforms are easier for servohydraulic systems to control. Ramp waveforms should be used when elevated-temperature FCGR and creep-fatigue interaction are of interest (see the section "High-Temperature Fatigue Crack Growth Testing" in this article) or when testing in aqueous environments (Ref 32).

Five types of FCGR tests are used in laboratories today. How the specimen is loaded defines the type of growth rate test. Different types of tests are often conducted in series to confirm growth rates and to use as much of the specimen as possible. To avoid load sequence effects, tests conducted in series should adhere to the same guidelines specified for precracking.

The simplest test type is one in which the load amplitude is kept constant and the applied ΔK increases as the crack extends. The simplicity of the test is its advantage. However, this test is essentially impractical for crack growth rates below 1×10^{-8} (m/cycle). In a second type of test, loads are shed manually at increments of 10% or less. Although cumbersome because they require constant attention, these tests allow the generation of data for slower crack growth in a more time-efficient manner than the constant-load-amplitude test. The prevalence of personal computers and modern controls technology in today's laboratories has popularized the remaining three types of so-called "continuous loadshedding" or "K-controlled" experiments.

Continuous loadshedding tests are those in which loads are shed at steps of 2% or less for a predetermined increment of crack extension. During these tests the crack length is continuously monitored by electric potential, compliance, or another suitable technique. Loads are shed or increased according to the following relation proposed by Saxena et al. (Ref 33):

$$\Delta K = \Delta K_0 \exp\left[c(a - a_0)\right] \qquad \text{(Eq 14)}$$

where ΔK is the applied range of ΔK, ΔK_0 is the initial range of ΔK, a is the current crack length, a_0 is the crack length at the beginning of the test, and c is the normalized K-gradient. The normalized K-gradient is defined as:

$$c = \frac{1}{K}\left(\frac{dK}{da}\right)[L^{-1}] \qquad \text{(Eq 15)}$$

Table 2 Comparison of electromechanical fatigue systems

Parameter	Forced displacement	Forced vibration	Rotational bending	Resonance	Servomechanical
Tension	Yes	Yes	No	Yes	Yes
Compression	Yes	Yes	No	Yes	Yes
Reverse stress	Yes	Yes	Yes	Yes	Yes
Bending	Yes	Yes	Yes	Yes	Yes
Frequency range	Fixed	Fixed, 1800 rpm	0-10,000 rpm	40-300 Hz	0-1 Hz
Load range	Typically <450 N (<100 lbf)	Up to 220 kN (50,000 lbf)	...	UP to 180 kN (40,000 lbf)	Up to 90 kN (20,000 lbf)
Type					
Control	Open-loop	Open-loop	Open-loop	Closed-loop	Closed-loop
Mode	Displacement	Load	Rotation/bending	Load	Load, displacement, strain
Maximum deflection	...	25.4 mm (1.00 in.)	...	1.0 mm (0.040 in.)	100 mm (4 in.)
Advantages	Simple, straightforward	Versatile, efficient, durable	Efficient, durable, simple	Fully closed-loop, extremely efficient	Fully closed-loop, high precision
Disadvantages	No load control, very limited applications (soft samples)	Fixed frequency, limited control (open-loop)	Rotational bending only, limited applications	Operating frequency directly proportional to sample stiffness	Low frequency only

The use of Eq 14 for changing fatigue loads is ideally suited for personal computers, and it allows testing under K-controlled conditions. If the normalized K-gradient is less than zero, the applied ΔK will be decreased as the crack extends. These are termed K-decreasing tests. Conversely, $c \geq 0$ will lead to increasing ΔK as the crack extends.

The appropriate value of c for a decreasing ΔK test is that which avoids the anomalous growth rates caused by shedding loads too quickly. Investigators have determined that $c = 0.08$ mm^{-1} (-2 in.$^{-1}$) is an appropriate value for decreasing ΔK tests on most metals (Ref 34). This value of c was derived to eliminate load-interaction effects caused by crack-tip plasticity in metals. The same value of c for a K-decreasing test in intermetallics and ceramics is recommended because it ensures that sufficient data can be obtained over the narrow range of stable crack growth, even though plastic zones are considerably smaller or nonexistent in these materials (Ref 35, 36).

Increasing ΔK tests (i.e., $c > 0$) are usually conducted after a decreasing test to confirm the growth rates measured during the previous K-decreasing portion of the test. Increasing ΔK tests may, if necessary, be conducted with larger normalized K-gradients. It is important to note that during an increasing ΔK test, loads may have to be decreased as the crack extends, which could lead to difficulties with control. Hence, it is preferable to use the simple constant amplitude instead of a controlled increasing ΔK test.

The first type of continuous loadshedding test is where the load ratio (R) is held constant. Constant-R tests generate the same type of information as constant-amplitude tests. Low- and high-R ($R = 0.1$ and 0.5, respectively) tests are usually conducted for comparison purposes.

Another type of K-controlled test is a constant-K_{max} test, which is essentially a variable-R test. When K_{max} is held constant as the crack extends, R will vary as shown schematically in Fig. 17. Once again, the value of ΔK to be applied to the specimens is dictated by Eq 14. The advantage of this test is that it quickly establishes the role of R on crack growth rate. For decreasing ΔK in constant-K_{max} tests with negative c, the lower crack growth rates are at very high values of R. The behavior of threshold cracks under these condi-

tions has been used as a measure of "closure free" fatigue crack growth, reflecting the "intrinsic resistance" of the material to fatigue (Ref 37).

The last type of continuous loadshedding fatigue test is a constant-K_{mean} test. Much like the constant-K_{max} test, a constant-K_{mean} test can be used as a comparison with constant K_{max} to help establish the role of K_{mean} versus K_{max} on fatigue crack growth rates. Constant-K_{max} and K_{mean} tests have been popular in the testing of brittle materials where mechanisms for crack advance have yet to be established.

Once testing is complete, the final crack in the specimen should be measured optically on both sides of the specimen. This will be compared to the terminal crack length predicted by other measurement techniques in the analysis of the investigation.

Electromechanical Fatigue Testing Systems. The primary function of electromechanical fatigue testers is to apply millions of cycles to a test piece at oscillating loads up to 220 kN (50,000 lbf) to investigate fatigue life, or the number of cycles to failure under controlled cyclic loading conditions. Variables associated with fatigue-life tests are frequency of loading and unloading amplitude of loading (maximum and minimum loads), and control capabilities. The fundamental data output requirement is the number of cycles to failure, as defined by the application.

A variety of electromechanical fatigue testers have been developed for different applications. Forced-displacement, forced-vibration, rotational-bending, resonance, and servomechanical systems are discussed in this article and are compared in Table 2. Other specialized electromechanical systems are available to perform specific tasks.

Servohydraulic testing machines are particularly well suited for providing the control capabilities required for fatigue testing. Extreme demands for sensitivity, resolution, stability, and reliability are imposed by fatigue evaluations. Displacements may have to be controlled (often for many days) to within a few microns, and forces can range from 100 kN to just a few newtons. This wide range of performance can be obtained with servomechanisms in general and, in particular, with the modular concept of servohydraulic systems.

Usually, the problem of selecting the appropriate system is simply a matter of optimizing the various components to form a system best suited to the given testing application. With any type of control system, the objective is to obtain an output that relates as closely as possible to the programmed input. In a fatigue testing system, it may be desired to vary the force on a specimen in a sinusoidal manner, at a frequency of 1 Hz over a force range of 0 to 100 kN (0 to 22,000 lbf). The only practical means to accomplish this with precision is through the use of a negative-feedback closed-loop system.

Analysis of Crack Growth Data

The two major aspects of FCGR test analysis are to ensure suitability of the test data and to calculate growth rates from the data. In addition to growth rate calculations, analysis also may require the calculation of fatigue crack closure levels and the characterization of fracture surface and metallographic features as discussed in this section. Combining the results from all of these areas is necessary to develop an understanding of the FCGR behavior of the material. Finally, the reliability of "real time" analysis by personal computers must be verified, and the sophisticated

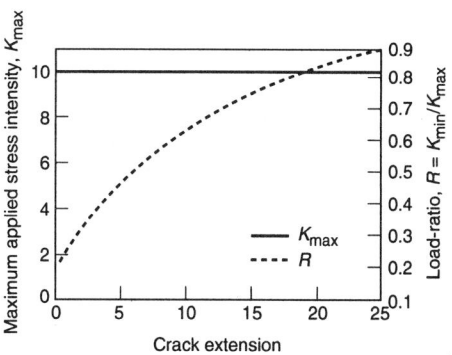

Fig. 17 Constant K_{max} test load ratio

Table 3 Methods for calculating crack growth rates

Incremental Polynomial Method

A least-squares, second-order polynomial is obtained for successive sets of $(2k + 1)$ data points:

$$\hat{a}_i = b_0 + b_1\left(\frac{N_i - C_1}{C_2}\right)$$

$$+ b_2\left(\frac{N_i - C_1}{C_2}\right)^2$$

where $C_1 = (N_{i+k} + N_{i-k})/2$ and $C_2 = (N_{i+k} - N_{i-k})/2$ are centering and scaling constants, respectively, that are introduced to prevent numerical problems in obtaining the least squares fit. The crack growth rate at $\hat{a}_i$, the predicted central crack length at N_i, is given by the derivative:

$$\left.\frac{da}{dN}\right|_{\hat{a}_i} = \frac{b_1}{C_2} + 2b_2\frac{(N_i - C_1)^2}{C_2}$$

Typical values of k are 1, 2, or 3, resulting in the second-order polynomial being estimated on the basis of 3, 5, or 7 successive data points. The estimated crack growth rate function loses k data points at each end. Less apparent scatter is obtained for larger k values.

Secant Method

In the secant method for differentiating the a versus N data, the average crack extension per cycle is calculated for each pair of data points, and ΔK is calculated at the midpoint of the crack lengths:

$$\left.\frac{da}{dN}\right|_{\hat{a}_i} = \frac{a_{i+1} - a_i}{N_{i+1} - N_i}$$

$$\hat{a}_i = \frac{a_{i+1} + a_i}{2}$$

This method is simple but exhibits the most scatter in da/dN values.

The Modified Difference Methods

These methods are finite difference techniques for estimating the derivative at the midpoint of a data set. These methods use numerical derivatives. The formula for estimating the derivative at the midpoint a_i, of three successive data points is given by:

$$\left.\frac{da}{dN}\right|_{a_i} = \frac{a_i - a_{i-1}}{N_i - N_{i-1}} + \left[\frac{N_i - N_{i-1}}{N_{i+1} - N_{i-1}}\right]$$

$$\left[\frac{a_{i+1} - a_i}{N_{i+1} - N_i} - \frac{a_i - a_{i-1}}{N_i - N_{i-1}}\right]$$

---

tests performed in today's laboratories must undergo some degree of analysis. Hence, it is essential that computer-controlled tests generate records of a versus N as well as FCGR.

Validity of the Test Data. The first step is to ensure the validity of the test data and make corrections to the crack length, if necessary. Crack measurement intervals are recommended in ASTM E 647 according to specimen type. For compact-type specimens:

$$\Delta a \leq 0.04W \text{ for } 0.25 \leq \frac{a}{W} \leq 0.40$$

$$\Delta a \leq 0.02W \text{ for } 0.40 \leq \frac{a}{W} \leq 0.60$$

$$\Delta a \leq 0.01W \text{ for } \frac{a}{W} \geq 0.60$$

For center-cracked tension specimens:

$$\Delta a \leq 0.03W \text{ for } \frac{a}{W} < 0.60$$

$$\Delta a \leq 0.03W \text{ for } \frac{a}{W} > 0.60$$

At the end of the test, the final crack length is measured on both sides of the specimen. A comparison between optical measurements and the final predicted crack length by any non-optical techniques used should be made. Differences between the measured crack lengths should be corrected using a linear relationship. In other words, the error between the final measured and predicted crack size is linearly distributed over the crack extension range. If periodic optical measurements were made during the test, other more appropriate correction procedures can be used.

Thicker specimens should be fractured after testing to determine the degree of crack front curvature. Cooling the specimen to liquid nitrogen temperatures allows the brittle fracture of most metallic materials and reveals a clear demarcation between the fatigued and fractured portions of the specimen. Five evenly spaced measurements of crack length should be made across the crack front. The average length should then be

used as the final crack length, and the corrections should be applied using this value instead of the surface measurements.

Crack Growth Rate Calculation. A number of different numerical techniques have been used to calculate crack growth rates from the set of (a_i, N_i) data points of a given crack growth rate test (Table 3). The secant and incremental polynomial methods are the most widely used. When the data are processed to a final smoothed $da/dN = f(\Delta K)$ format, these methods provide approximately equivalent results. However, the scatter of individual da/dN values about the average depends greatly on the data reduction method.

The secant method fits a line between adjacent data points. The slope of the line is the crack growth rate, da/dN. The load at the average crack length of the interval is used to calculate the corresponding ΔK. It is not uncommon to collect data points more closely spaced than the 0.25 mm (0.010 in.) minimum crack extension for growth rate calculations. When more data are available, a straight line through the multiple data points can be fitted by regression analysis to calculate crack growth rates. A 50% overlap between data sets used for calculating successive crack growth rates can considerably reduce scatter in the processed data. This method is termed the *modified secant method*.

The polynomial method uses the derivative of a second-order polynomial fitted to a fixed number of data points (often five or seven). The growth rate and ΔK level are calculated for the average value of crack length in the interval. This method tends to provide data with less scatter than the secant method. In practice, there are no systematic differences in the da/dN versus ΔK trends if the same data are processed by these different techniques (Ref 38). In fact, little difference is observed between the data processed by the modified secant method and the polynomial method.

Variability. Although the apparent variability in the resulting da/dN value depends on the calculation method, none of the methods introduces

a significant bias to an overall mean trend curve. Figure 18(b) plots da/dN versus ΔK, as calculated from the data of Fig. 18(a) using the secant, incremental polynomial, and five-point modified difference methods. Because methods of analyzing and interpreting the scatter in da/dN are not currently available, the simpler techniques—i.e., the secant and incremental polynomial methods— are often chosen.

Crack Closure Analysis. The determination of the fatigue crack closure load is a subject of vigorous debate. Visual inspection of compliance curves can be used to estimate closure levels. However, the method is subjected and not well suited for large amounts of data. ASTM task groups have explored the use of the compliance offset and correlation method for closure analysis (Ref 39). In the compliance offset method, crack closure levels are determined by finding the point that deviates from the linear portion (i.e., is offset) by a set amount (usually 1, 2, 4, or 8%). The correlation method determines the closure level by mathematically representing the "strength" of the linear relationship between load and displacement/strain along the curve. To confirm that the methods are applied correctly, one is encouraged to test algorithms with hypothetical, bilinear compliance curves, for which both methods should yield identical results.

The compliance offset method is generally applied using a computer, because the calculations do not lend themselves to the use of spreadsheets. The procedure is as follows:

1. Collect digitized strain/displacement and load data for a complete load cycle. The data sampling rate should be high enough to ensure that at least one data pair (displacement and load) is taken in every 2% interval of the cyclic load range.
2. Starting with the first data sample below maximum load on the unloading curve, fit a least-squares straight line to a segment of the curve spanning approximately the uppermost 25% of the cyclic load range. The slope of

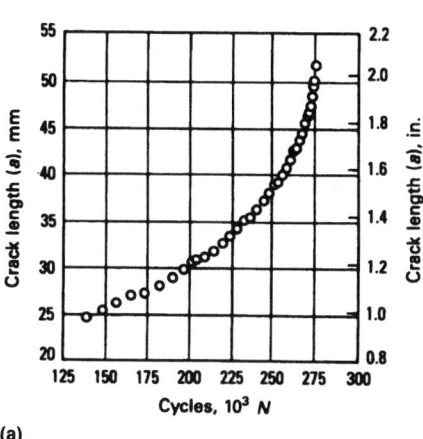

Fig. 18 Comparison of *da/dN* calculation methods. (a) Crack length test data. (b) Plot of calculated *da/dN* rates from the data in (a). Source: *Metals Handbook,* 9th ed., Vol 8, p 679-680

this line is the compliance value that corresponds to the fully open crack configuration.

3. Starting with the first data sample below maximum load on the loading curve, fit least-squares straight lines to segments of the curve that span approximately 10% of the cyclic load range and that overlap each other by approximately 5% of the range. Store the compliance (slope) and the corresponding mean load for each segment in a vertical array with the highest load location at the top.

4. Replace the compliance stored in each location in the array with the corresponding compliance offset, which is computed as a percentage of the "open crack" compliance and is given by:

Compliance offset

$$= \frac{(\text{"Open crack" compliance} - \text{compliance})(100)}{\text{"Open crack" compliance}}$$

$$(Eq 16)$$

5. Identify the highest load location in the array which has a compliance offset greater than the selected offset criterion, and for which all array locations below it have compliance offsets greater than the offset criterion.

6. Starting at the array location identified in step 5, identify the nearest, higher load location which has a compliance offset less than the selected offset criterion, and for which all array locations above it have compliance offsets less than the offset criterion.

7. Determine the opening load corresponding to the selected offset criterion by linear interpolation between the two (compliance offset, load) points identified in steps 5 and 6 (Ref 39).

An alternative to the offset method is the correlation coefficient method. The correlation coefficient, *r*, is defined as (Ref 39):

$$r = \frac{n \Sigma (x_i y_i) - \Sigma x_i \Sigma y_i}{[(n \Sigma x_i^2 - (\Sigma x_i)^2)(n \Sigma y_i^2 - (\Sigma y_i)^2)]^{1/2}} \quad (Eq 17)$$

where x_i are individual load data samples, y_i are individual displacement data samples, n is the number of data pairs, and Σ denotes the summation from

$i = 1$ to n. The closure level is defined as the load at which the correlation coefficient has the highest value. In contrast to the offset method, in which different levels of offset can be selected, there is only one criterion in this method. Both methods generate similar closure levels for the same data. However, issues of data quality are currently the biggest problem for consistent experimental closure analysis. Methods to characterize data quality are presently being explored.

Fracture Surface Characterizations. Cracked specimens are also useful for establishing modes of crack advance and other interactions of the crack with the microstructure. Scanning electron microscopy is used to characterize the fracture surface. Care should be taken to establish the level of ΔK associated with the image of the fracture surface. This is especially important when working with brittle materials where inspection of the surface does not reveal an obvious difference in fracture surface morphology between the fatigued material and post-test fracture.

Portions are often removed from the interior of the specimen and mounted for metallographic preparation. "Crack profiles" can be useful for illustrating the interaction of the crack with the microstructure. There can be marked differences in how cracks interact with microstructures, especially in cases where crack closure and distributed damage are important.

Appendix: High-Temperature Fatigue Crack Growth Testing

David C. Maxwell, University of Dayton, and Theodore Nicholas, Wright-Patterson Air Force Base

The concepts and procedures for fatigue crack growth testing at room or ambient temperature, described in this article, are generally applicable to elevated-temperature conditions. In this appendix, some of the features that are unique to testing and measurement at high temperatures are reviewed. In particular, methods for heating and controlling temperature are discussed briefly. Other aspects related to instrumentation, measurement, and test techniques are also highlighted. While there are no well-established standards for elevated-temperature crack growth testing, one standard testing procedure has been recommended recently based on a cooperative study using 304 stainless steel (Ref 40). While documentation on methods employed in elevated-temperature crack growth testing are scattered widely in the open literature, conferences devoted to this particular subject contain many papers that provide details of individual investigations (see, e.g., Ref 41).

Specimen Design. In general, the specimens used in room-temperature testing are equally applicable for high-temperature testing. In fact, in making comparisons of crack growth rates as a function of temperature, it is considered good practice to retain the same test specimen geometry at all temperatures tested.

An exception to this is the specialized case of testing under thermomechanical fatigue (TMF) crack growth conditions. Here, the temperature as well as the load is cycled. In order to control the cyclic temperature on the specimen, effective means of both heating and cooling the specimen must be employed. This usually requires either a thin planar specimen or a thin-walled tube, both of which minimize the thermal mass and allow for more rapid heating or cooling and better control of temperature gradients. Information on some of the techniques that have been employed for TMF can be found in conference proceedings such as those of ASTM (Ref 42, 43), ASME (Ref

44), or the International Conferences on Fatigue and Fatigue Thresholds (Ref 45, 46).

Another aspect of TMF crack growth that is almost unique to this type of test is the use of strain control as the mode of load application, similar to that used in low-cycle fatigue. A brief discussion of the problems and limitations of the use of strain control in elevated-temperature or TMF crack growth testing can be found in Ref 47.

The final consideration in TMF crack growth testing is the ability to heat and cool the specimen, or the access of heat and cooling air to the specimen. For example, clevises used on compact-type specimens tend to shield the specimen from the heating or cooling source and make it difficult to control the temperature cycle. Therefore, other geometries, such as middle-tension geometries, are generally used for this type of test.

Specimen Gripping. There are two main considerations in gripping specimens at high temperature. The first is the possible introduction of friction at any type of pin joint. Whereas friction can be minimized at room temperature through lubrication or careful surface finishing, these methods are less effective at elevated temperatures. Special high-temperature lubricants or oxidation-resistant materials for pins and clevises are methods for minimizing the development of friction in pin joints. Rigid fixtures and grips, on the other hand, do not require such considerations.

The second aspect of gripping at high temperatures is the possible loss of friction due to mismatches in the coefficient of thermal expansion of grips and specimens. Calculations or experimental evaluations should be made to ensure that the load-carrying capacity of the gripping system does not degrade due to differential thermal expansions so that slippage might occur.

Further, it should be demonstrated that excessive clamping stresses do not develop such that the specimen is crushed or the grips fail. This aspect of high-temperature testing is only applicable to cases where "hot" grips are used, such as when the grips and specimens are both contained within a furnace. Cold grips, on the other hand, produce high-temperature gradients along the length of the specimen and make it more difficult to maintain a constant temperature over the gage length or the crack growth region of the specimen.

Heating Methods. There are several methods available for heating specimens and maintaining a constant and uniform temperature during elevated-temperature crack growth testing. One of the more common methods is to use a commercially available or homemade furnace and power controllers that produce one or more zones of uniform, controlled temperature. Temperature control is achieved from thermocouple feedback, either from within the chamber, where the air temperature is being controlled, or from thermocouples directly attached to the specimen at one or more locations. In either case, temperature control with commercial units is generally accu-

rate and reliable and can be considered to be a mature, state-of-the-art technology.

A second method of heating specimens is through the use of induction heaters, where the specimen is heated by alternating currents produced by an electromagnetic field (EMF) generated by the inductance coil surrounding the test specimen. Shielding of the specimen by grips may cause nonuniform temperature distribution. Therefore, single-edge-tension specimen geometries are generally preferred. The number and spacing of the coils, though calculated from formulas provided with such apparatus, is determined most often from trial and error and usually requires a certain amount of experience for optimum performance. Feedback is provided from one or more thermocouples attached to the specimen, and uniformity of temperature must be checked with some type of temperature mapping system such as multiple thermocouples on a dummy specimen.

Another method of heating used in several laboratories is the use of radiant energy from quartz lamps mounted in reflective and cooled housings. Each lamp focuses its energy over a limited portion of the test specimen, so multiple lamps, multiple thermocouples, and good thermal conductivity across the specimen all lead to more uniform temperature fields. A description of the technique of quartz lamp heating can be found in Hartman (Ref 48).

The last common method of specimen heating is direct resistance heating. Here, a current is passed directly through the specimen from one end to another. The test apparatus must be electrically insulated from the input and output leads. Uniform gage length specimens are better adapted to this method of heating than highly nonuniform ones; that is, a middle-tension specimen would be a much better candidate than a compact-type specimen. Because of the high current passing through the specimen, no conducting leads can be attached to the specimen that would provide an alternative path for the current. Thus, temperature measurement from thermocouples attached directly to the specimen and electric potential crack growth measurements cannot be made with this type of heating. Similarly, extensometry that attaches directly to the specimen must use nonconducting elements.

Temperature Measurement. The science of the measurement of temperature, known as pyrometry, dates back to before World War II. Books on the subject of pyrometry in general (Ref 49), or optical pyrometry in particular (Ref 50) were published in 1941 and provide detailed descriptions of the theory and the methods used in that era. The basic principles have not changed. The most common method for temperature measurement is through the use of thermocouples, or thermoelectric pyrometer, directly attached to the specimen. A thermocouple is made by welding two dissimilar wires together at one end. A change in temperature will generate an EMF which can be recorded on an instrument attached to the other end of the wires. These provide a real-time, continuous record of temperatures at a

given point on a specimen, and they are commonly used for temperature control feedback as well as direct temperature measurement.

Another commonly used method of measuring temperature is through the use of commercially available infrared detectors, which sense the radiation emitted from a sample and convert the frequency of the radiation to temperature after appropriate calibration. The theory and methods of optical pyrometry are documented in numerous places (see, e.g., Ref 51 and 52). The emissivity of the test material whose temperature is being measured is the quantity that is used as the basis of the measurement. Emissivity is the ratio between total radiant energy per square centimeter per second between the specimen being measured and a black body at the same temperature. Thus, as the emissivity of a material decreases from 1, the apparent temperature as measured by an optical pyrometer will deviate from the actual temperature by a greater amount.

The emissivity of a heated specimen will always be less than 1, and the apparent temperature will therefore be less than the desired test temperature. Emissivity can be influenced by surface roughness, the spectral transmission of any windows between the test specimen and the measuring instrument, and the chemical changes of the specimen surface due to oxidation or other environmental degradation. All these issues should be addressed when using optical pyrometry. Commercial units are widely available for this type of measurement and usually come with detailed instructions.

Crack Length Measurements. Electric potential drop is a common method of crack length determination, both at room temperature and at elevated temperatures. The major consideration for elevated-temperature testing is to ensure that temperature fields are uniform and constant between the potential drop leads. Changes in temperature can result in a false indication of crack length changes. The error is due to the resistivity change in the material due to a change in temperature, and this change, in turn, is dependent on the resistivity characteristics of the material. For the direct current potential drop (DCPD) method, a 3 °C (5.5 °F) change in temperature in an aluminum alloy can result in a 1% change in the DCPD signal, whereas for the same change in resistivity, Inconel 718 would require a 100 °C (180 °F) change in temperature (Ref 53). The electric potential drop method of crack length measurement, used commonly in both room- and elevated-temperature isothermal crack growth testing, has also been applied to TMF testing (Ref 54). As with any other technique, modifications and improvements are continually being developed and documented in the literature, as in the work of Shin et al. (Ref 55), where modification to the dc potential drop system is reported.

Compliance methods are equally valid at elevated temperatures as at room temperature, provided that some simple considerations are addressed. Nonconstant thermal gradients must be avoided, because any change in the dimensions of any part of a mechanical extensometer can result

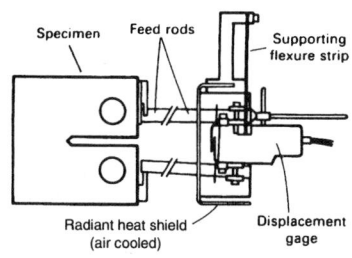

Fig. 19 Feed-rod attachment for high-temperature tests. Feed rods are made of quartz (up to 1000 °C) or ceramics for higher temperatures.

in a change in electrical output. The gage portion, which is normally attached to the specimen through special high-temperature extension arms or feed rods, is usually shielded from the specimen (Fig. 19) and cooled with either blowing air or circulating water. Whatever the setup, the gage should reach thermal equilibrium and stability before measurements are made, in order to avoid false indications due to thermal transients.

Optical methods are less widely used today than years ago because of the availability of automated methods for crack length measurement. Optical methods, such as use of a traveling microscope, are useful for reference measurements and are equally valid at room or elevated temperature. At elevated temperature, however, the resolution of crack length may be reduced because of deterioration of the surface of the specimen through oxidation. When specialized optical methods such as laser interferometry are used, the deterioration of the specimen surface over time should be considered so that the optical information is still available with sufficient resolution.

Special Methods. Other specialized techniques see occasional application in high-temperature crack growth testing. For example, crack growth over prolonged periods of time has been measured using real-time holographic interferometry (Ref 56). Compact-tension specimens were tested under sustained load at 120 °C (250 °F) for 860 h, and crack-opening displacement (COD) was monitored. The COD measurements had to be converted to crack lengths using an elastic-plastic finite-element analysis code and clip gage COD measurements on a reference sample.

Validity of *K* Solutions. The reduction of crack length versus *N* data to a *da/dN-ΔK* curve depends on having a *K* solution for the specimen geometry being used. Stress-intensity solutions are widely available, or can be easily generated, for almost any specimen geometry used in laboratory testing. These solutions, however, are based on the assumption of a homogeneous material and do not take into account any thermal gradients or thermal stresses that may arise in elevated-temperature crack growth testing. For cases where the specimen and grips are at a uniformly high temperature, such as within a furnace, there is no problem. For cases where the temperature is nonuniform, such as with inductance or quartz lamp heating with cooled grips, temperature gradients are developed along the axis of the specimen. These gradients, in turn,

may result in thermal stresses that produce thermal *K*-values that alter the isothermal *K* solution for a given specimen. These thermal *K*-values, therefore, must be taken into account in the *K* analysis of the particular crack geometry.

Ohta et al. (Ref 57) found that crack growth rate data at 300 °C (570 °F) in a middle-tension specimen matched room-temperature data in a low-alloy ferritic steel (SB46) only when the parabolic temperature gradient that existed along the specimen was taken into account using finite element analysis, which was confirmed with experimental measurements. Coker et al. (Ref 58) conducted a finite element analysis of a single-edge crack specimen with thermal gradients between the plane of the crack and the room-temperature grips. For their specific geometry and temperature profile, stress-intensity factors due to thermal stresses were below $2\ MPa\sqrt{m}$ for $a/W \leq 0.8$. While these values are small, near-threshold tests involving very small values of ΔK could be substantially influenced by thermal stresses. It is important, therefore, to have calculations or estimates of thermal-stress-induced values that have nonuniform temperature profiles. Unfortunately, there are very few papers in the literature where this particular problem is addressed. Nonetheless, this can be an important consideration in high-temperature testing, particularly when testing in the near-threshold regime or when thermal gradients are severe.

Thermal gradients and effective values of *K* resulting from direct resistance heating are discussed by Cunningham and Griffin (Ref 59). Their analysis shows when direct resistance heating becomes important during the cyclic conditions of thermal fatigue. The results indicate that stress intensities resulting from thermal gradients due to direct resistance heating are generally small except when thermal cycling frequencies are high. The thermal gradients and resultant *K*-values resulting from thermal cycling are reported in a subsequent publication (Ref 60). There, it is shown that for high thermal frequencies, significant magnitudes of *K* can be developed at the crack tip due to the thermal transients. For those involved in TMF crack growth testing, this information should be considered when determining the stress intensity at the crack tip.

REFERENCES

1. P.C. Paris and F. Erdogan, *J. Basic Eng. (Trans. ASME)*, Series D, Vol 85, 1963, p 528-534
2. P.C. Paris, *Proc. 10th Sagamore Conf.*, Syracuse University Press, 1965, p 107-132
3. J.R. Griffiths and C.E. Richards, *Mater. Sci. Eng.*, Vol 11, 1973, p 305-315
4. J.C. Grosskruetz, Strengthening in Fracture and Fatigue, *Metall. Trans.*, Vol 3, 1972, p 1255-1262
5. J.R .Rice, in *Fatigue Crack Propagation*, STP 415, ASTM, 1967, p 247-311
6. N.E. Dowling, in *Flaw Growth and Fracture*, STP 631, ASTM, 1977, p 139-158
7. "Standard Test Method for Measurement of Fatigue Crack Growth Rates," E 647-91, *Annual Book of ASTM Standards*, Vol 03.01, 1992, ASTM, p 674-701
8. N.E. Dowling and J.A. Begley, in *Mechanics of Crack Growth*, STP 590, ASTM, 1976, p 82-103
9. N.E. Dowling, in *Cracks and Fracture*, STP 601, ASTM, 1977, p 131-158
10. H.S. Lamba, The J-Integral Applied to Cyclic Loading, *Eng. Fract. Mech.*, Vol 7, 1975, p 693-696
11. J.R. Rice, *J. Appl. Mech. (Trans. ASME)*, Vol 35, 1968, p 379-386
12. J.W. Hutchinson, *J. Mech. Phys. Solids*, Vol 16, 1968, p 337-347
13. J.R. Rice and G.F. Rosengren, *J. Mech. Phys. Solids*, Vol 16, 1968, p 1-12
14. W.R. Brose and N.E. Dowling, in *Elastic-Plastic Fracture*, STP 668, ASTM, 1979, p 720-735
15. W. Elber, Fatigue Crack Closure under Cyclic Tension, *Eng. Fract. Mech.*, Vol 2, 1970, p 37-45
16. W. Elber, *The Significance of Fatigue Crack Closure*, STP 486, ASTM, 1971, p 230-242
17. "Standard Method for Plane-Strain Fracture Toughness of Metallic Materials," E 399-90, *Annual Book of ASTM Standards*, Vol 3.01, 1992, ASTM, p 569-596
18. K.T. Venkateswara Rao, W. Yu, and R.O. Ritchie, *Metall. Trans. A*, Vol 19A (No. 3), March 1988, p 549-561
19. A.W. Thompson and R.J. Bucci, *Metall. Trans.*, Vol 4, April 1973, p 1173-1175
20. G.R. Yoder and D. Eylon, *Metall. Trans.*, Vol 10A, Nov 1979, p 1808-1810
21. A. Saxena and S.J. Hudak, Review and Extension of Compliance Information for Common Crack Growth Specimens, *Int. J. Fract.*, Vol 14 (No. 5), 1978, p 453-468
22. J.W. Dally and W.F. Riley, *Experimental Stress Analysis*, 3rd ed., McGraw-Hill, 1991
23. R.O. Ritchie, G.C. Garrett, and J.F. Knott, *Int. J. Fract. Mech.*, Vol 7, 1971, p 462-467
24. C.Y. Li and R.P. Wei, *Mater. Res. Stand.*, Vol 6, 1966, p 392-445
25. R.O. Ritchie and J.F. Knott, *Acta Metall.*, Vol 21, 1973, p 639-648
26. R.O. Ritchie and K.J. Bathe, *Int. J. Fract.*, Vol 15, 1979, p 47-55
27. G. Clark and J.F. Knott, *J. Mech. Phys. Solids*, Vol 23, 1975, p 265-276
28. H.H. Johnson, *Mater. Res. Stand.*, Vol 5, 1965, p 442-445
29. G.H. Aronson and R.O. Ritchie, *J. Test. Eval.*, Vol 7, 1979, p 208-215
30. M.A. Ritter and R.O. Ritchie, *Fat. Eng. Mater. Struct.*, Vol 5, 1982, p 91-99
31. G.M. Wilkowski and W.A. Maxey, *Fracture Mechanics: 14th Symposium—Vol II: Testing and Applications*, STP 791, J.C. Lewis and G. Sines, Ed., ASTM, 1983, p II-266 to II-294
32. R.J. Selines and R.M. Pelloux, Effect of Cyclic Stress Waveform on Corrosion Fatigue Crack Propagation in Al-Zn-Mn Alloys, *Metall. Trans.*, Vol 13, 1972, p 2525-2531
33. A. Saxena, S.J. Hudak, Jr., J.K. Donald, and D.W. Schmidt, *J. Test. Eval.*, Vol 6 (No. 3), May 1978, p 167-174

34. R.J. Bucci, Development of a Proposed ASTM Standard Test Method for Near-Threshold Fatigue Crack Growth Rate Measurement, *Fatigue Crack Growth Measurement and Data Analysis*, S.J. Hudak, Jr. and R.J. Bucci, Ed., STP 738, ASTM, 1981, p 5-28

35. K.T. Venkateswara Rao, Y.W. Kim, C.L. Muhlstein, and R.O. Ritchie, *Mater. Sci. Eng. A*, Vol 192/193, 1995, p 474-487

36. R.J. Dauskardt, W. Yu, and R.O. Ritchie, *J. Am. Ceram. Soc.*, Vol 70 (No. 10), Oct 1987, p C248-C252

37. R.W. Hertzberg, W.A. Herman, and R.O. Ritchie, *Scr. Metall.*, Vol 21, 1987, p 1541

38. W.G. Clark and S.J. Hudak, The Analysis of Fatigue Crack Rate Data, *Application of Fracture Mechanics to Design*, J.J. Burke and V. Weiss, Ed., Vol 22, Plenum, 1979, p 67-81

39. E.P. Phillips, "Results of the Second Round Robin on Opening Load Measurement Conducted by ASTM Task Group E24.04.04 on Crack Closure Measurement and Analysis," Technical Memorandum 109032, National Aeronautics and Space Administration, Nov 1993

40. Y. Asada, M. Kitagawa, N. Shimakawa, T. Kodaira, T. Asayama, and Y. Wada, Standardization of Procedures for the High-Temperature Crack Growth Testing for FBR Materials in Japan, *Nucl. Eng. Des.*, Vol 133, 1992, p 465-473

41. S. Mall and T. Nicholas, Ed., *Elevated Temperature Crack Growth*, MD-Vol 18, ASME, 1990

42. H. Sehitoglu, Ed., *Thermomechanical Fatigue Behavior of Materials*, STP 1186, ASTM, 1993

43. M.J. Verrilli and M.G. Castelli, Ed., *Thermo-Mechanical Fatigue Behavior of Materials*, STP 1263, ASTM, 1995

44. W.J. Jones, Ed., *Thermomechanical Behavior of Advanced Structural Materials*, AD-Vol 34/AMD-Vol 173, ASME, 1993

45. H. Kitagawa and T. Tanaka, Ed., *Fatigue 90*, Materials and Components Engineering Publications, Ltd., Birmingham, U.K., 1990

46. J.-P. Bailon and J.I. Dickson, *Fatigue 93*, Engineering Materials Advisory Services, Ltd., West Midlands, U.K., 1993

47. T. Nicholas, M.L. Heil, and G.K. Haritos, Predicting Crack Growth under Thermo-Mechanical Cycling, *Int. J. Fract.*, Vol 41, 1989, p 157-176

48. G.A. Hartman, A Thermal Control System for Thermal/Mechanical Cycling, *J. Test. Eval.*, Vol 13, 1985, p 363-366

49. W.P. Wood and J.M. Cork, *Pyrometry*, McGraw-Hill, 1941

50. W.E. Forsythe, *Optical Pyrometry: Temperature—Its Measurement and Control in Science and Industry*, Reinhold Publishing, 1941

51. H.J. Kostkowski and R.D. Lee, "Theory and Methods of Optical Pyrometry," Monograph 41, National Bureau of Standards, 1962

52. J.C. Richmond and D.P. DeWitt, Ed., *Applications of Radiation Thermometry*, STP 895, ASTM, 1985

53. J.K. Donald and J. Ruschau, Direct Current Potential Difference Fatigue Crack Measurement Techniques, *Fatigue Crack Measurement: Techniques and Applications*, K.J. Marsh, R.A. Smith, and R.O. Ritchie, Ed., EMAS, West Midlands, U.K., 1991, p 11-37

54. G.A. Hartman and D.A Johnson, DC Electric Potential Method Applied to Thermal/Mechanical Fatigue Crack Growth Testing, *Exp. Mech.*, Vol 11, 1987, p 106-112

55. C.-S. Shin, W.H. Huang, and H.-Y. Chen, An Improved DC Potential Drop System for Crack Length Measurement, *J. Chinese Inst. Eng.*, Vol 16, 1993, p 29-40

56. T.R. Hsu, R. Lewak, and B.J.S. Wilkins, Measurements of Crack Growth in a Solid at Elevated Temperature by Holographic Interferometry, *Exp. Mech.*, Vol 18, 1978, p 297-302

57. A. Ohta, M. Kosuge, S. Matsuoka, E. Takeuchi, Y. Muramatsu, and S. Nishijima, Significant Effect of Thermal Stresses on Fatigue Crack Propagation Properties, *Int. J. Fract.*, Vol 38, 1988, p 207-216

58. D. Coker, B.K. Parida, and N.E. Ashbaugh, "Thermal Stresses due to a Temperature Gradient on a Single Edge Notch Specimen," Report UDR-TM-93-07, University of Dayton Research Institute, 1993

59. S.E. Cunningham and J.H. Griffin, On the Importance of Direct Resistance Heating in Thermo-Mechanical Fatigue, *Int. J. Fract.*, Vol 46, 1990, p 257-270

60. S.E. Cunningham and J.H. Griffin, Estimating the Importance of Cyclic Thermal Loads in Thermo-Mechanical Fatigue, *Int. J. Fract.*, Vol 47, 1991, p 161-180

SELECTED REFERENCES

FCGR Testing

- D. Blatt, R. John, and D. Coker, Single Edge Notched Specimens with Clamped Ends for Automated Crack Growth Testing, *Proc. SEM Spring Conference on Experimental Mechanics*, Society for Experimental Mechanics, 1993

- D. Blatt, R. John, and D. Coker, Stress Intensity Factor and Compliance Solutions for a Single Edge Notched Specimen with Clamped Ends, *Eng. Fract. Mech.*, Vol 47 (No. 4), March 1994

- D.C. Freeman and M.J. Strum, Near Threshold Fatigue Testing, Report DE93009069/XAB, *Gov. Res. Announc. Index*, 1993

- J.C.P. Kam and W.D. Dover, The Recent Development of Advanced Fatigue Crack Growth Testing Technology for Large Scale Offshore Tubular Joints, *Conf. Proc. Engineering Integrity through Testing*, Engineering Materials Advisory Services Ltd., West Midlands, U.K., 1990

- R.K. Nanstad et al., A Computer-Controlled Automated Test System for Fatigue and Fracture Testing, *Conf. Proc. Applications of Automation Technology to Fatigue and Fracture Testing*, STP 1092, ASTM, 1990

- S. Nishijima, S. Matsuoka, and E. Takeuchi, Environmentally Affected Fatigue Crack Growth, *Fatigue 90*, Materials and Component Engineering Publications, Birmingham, U.K., 1990

- H.-H. Over and B. Buchmayr, Collection and Evaluation of Fatigue and Fracture Mechanics Data According to the European High Temperature Materials Databank (Petten) Standard, *Conf. Proc. Applications of Automation Technology to Fatigue and Fracture Testing*, STP 1092, ASTM, 1990

- S. Saki, T. Asakawa, and H. Okamura, Automatic Data Acquisition System for Fatigue Crack Growth in the Range from SSY to LGSY Region, *Conf. Proc. Asian Pacific Conference on Fracture and Strength '93*, Japan Society of Mechanical Engineers, 1993

- J. Solin and J. Hayrynen, Simulation of Mechanical and Environmental Conditions in Fatigue Crack Growth Testing, *Conf. Proc. Applications of Automation Technology to Fatigue and Fracture Testing*, STP 1092, ASTM, 1990

Crack Measurement

- C.J. Beevers, *Advances in Crack Length Measurement*, Engineering Materials Advisory Services Ltd., West Midlands, U.K., 1983

- C.J. Beevers, *The Measurement of Crack Length and Shape during Fracture and Fatigue*, Engineering Materials Advisory Services Ltd., West Midlands, U.K., 1980

- M.S. Domack, Evaluation of K_{ISCC} and da/dt Measurements for Aluminum Alloys Using Precracked Specimens, *Conf. Proc. Environmentally Assisted Cracking: Science and Engineering*, STP 1049, ASTM, 1990

- R.L. Hewitt, Accuracy and Precision of Crack Length Measurement Using a Compliance Technique, *J. Test. Eval.*, Vol 11, 1983, p 150-155

- H.H. Johnson, Calibrating the Electrical Potential Method for Studying Slow Crack Growth, *Mat. Res. Stand.*, Vol 5, 1965, p 442-445

- J.A. Kapp, G.S. Leger, and B. Gross, "Wide Range Displacement Expressions for Standard Fracture Mechanics Specimens," Report ARLCB-TR-84025, Army Armament Research and Development Center, 1984

- C.Y. Li and R.P. Wei, Calibrating the Electrical Potential Method for Studying Slow Crack Growth, *Mat. Res. Stand.*, Vol 6, 1966, p 392-394

- D.C. Maxwell, J.P. Gallagher, and N.E. Ashbaugh, "Evaluation of COD Compliance Determined Crack Growth Rates," Report AFWAL-TR-84-4062, Wright-Patterson Air Force Base

- A.C. Pickard, R.A. Venables, and S.I. Vukelich, Crack Detection and Crack Length Measurement in the Gas Turbine Industry, *Fatigue Crack Measurement: Techniques and Applications*, Engineering Materials Advisory Services Ltd., West Midlands, U.K., 1991

Mechanisms of Corrosion Fatigue

P.S. Pao, Materials Science and Technology Division, Naval Research Laboratory

CORROSION FATIGUE refers to the phenomenon of cracking in materials under the combined actions of fatigue (or cyclic) loading and a corrosive (or deleterious) environment (gaseous or aqueous). This phenomenon is known to occur in many engineering alloys over a broad range of environments and has been recognized as an important cause for failure of engineering structures. Characterization and understanding of corrosion fatigue kinetics and mechanisms are essential to service life prediction, fracture control, and the development of fatigue-resistant alloys. The primary characteristics of corrosion fatigue crack growth are that the crack growth rates can be substantially higher in the corrosive environment than those obtained in a benign environment (such as in vacuum) and that crack growth rates can be dependent upon a large number of chemical and electrochemical variables not present in a benign environment.

Several corrosion fatigue mechanisms have been proposed to explain the enhanced crack growth rates with varying degrees of success. The generalized corrosion fatigue cracking mechanism involves the single or mutual occurrence of hydrogen-induced cracking and/or anodic dissolution at the crack tip. There appears currently to be at least four possible mechanisms of anodic dissolution:

- Slip-dissolution
- Brittle film-rupture
- Corrosion-tunneling
- Selective-dissolution (dealloying)

Hydrogen embrittlement appears to be divided into five types of mechanisms:

- Decohesion
- Pressure
- Adsorption
- Deformation
- Brittle hydride

The balance between various phenomena is very difficult to sort out. Many researchers think that dissolution and hydrogen-induced cracking processes are competitive, so that only one of them makes the major contribution to cracking and the lesser process can be ignored. However, both processes (dissolution and hydrogen evolution) can oc-

cur simultaneously at the crack tips over ranges of potential that have been measured or that are suspected to occur. If the processes were independent and were really competitive, it would only be necessary to determine which process is faster. The anodic and cathodic processes are, however, interdependent, and they may operate simultaneously or sequentially. Another general type of mechanism is the surface energy reduction mechanism, and it will be described later in this article.

In addition to the various mechanisms of corrosion fatigue, there also are many mechanical, metallurgical, and environmental variables that affect fatigue crack growth rates. These variables are discussed in more detail following a brief review of the corrosion fatigue mechanisms associated with hydrogen-assisted cracking (hydrogen embrittlement) and anodic dissolution. Another common environment-assisted cracking phenomenon is stress-corrosion cracking where susceptible materials fail under the conjointed action of corrosion environment and a sustained load. In many engineering alloy/environment systems, because the fatigue process can efficiently rupture the protective surface oxide films and facilitate access to corrosive environments and produce new metal surfaces, threshold stress

intensities for cracking to occur are often lower under corrosion fatigue conditions that under a static stress-corrosion cracking condition. Depending on the stress level and the cyclic frequency, the crack growth rates may be high under both stress-corrosion and corrosion fatigue. However the underlying mechanisms for both corrosion fatigue and stress-corrosion cracking may very well be the same, as the only difference between these two phenomena is the mode of loading (cyclic vs. sustained). Stress-corrosion cracking is covered elsewhere in this Volume.

Hydrogen-Assisted Cracking

In corrosion fatigue, hydrogen (either atomic or protonic) is generated by the reaction of environmental species such as gaseous hydrogen, water vapor, water, and so forth, with the newly cracked material at the crack tip. This hydrogen is absorbed at the metal surface and then transported by diffusion or dislocation sweeping mechanisms into the highly stressed region (plastic zone) at the crack tip, where it causes localized damage and increases the fatigue crack growth rate. Figure 1 illustrates schematically various

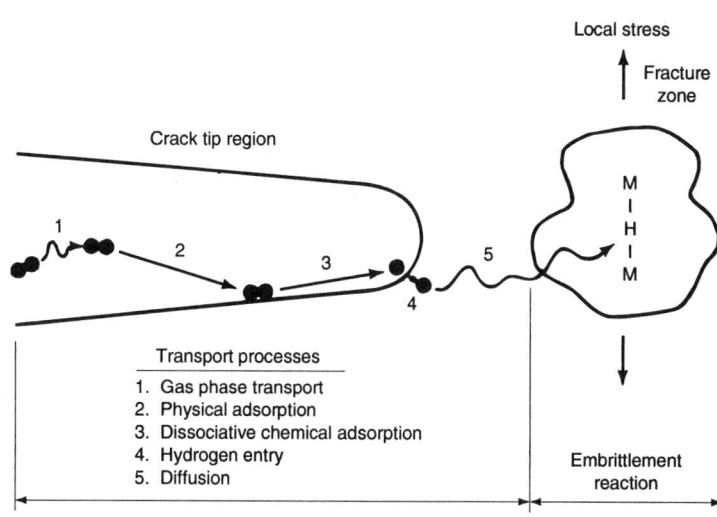

Fig. 1 Various sequential processes involved in corrosion fatigue crack growth in alloys exposed to aggressive environments. Source: Ref 1

processes that might be involved in corrosion fatigue crack growth of ferrous alloys by a hydrogen-assisted cracking mechanism. The processes or steps shown in Fig. 1 are sequential, and the rate of corrosion fatigue crack growth is controlled by the slowest process in this sequence. Additionally, in some cases, hydrogen can enter the material during fabrication without an aggressive external environment. Several hydrogen-assisted cracking mechanisms have been proposed to explain how hydrogen enhances corrosion fatigue crack growth rates. Because of the complex nature of corrosion fatigue phenomenon, it is not likely that a unified hydrogen-assisted cracking mechanism can be assumed to account for all alloy/environment systems. Rather, the actual hydrogen-assisted cracking mechanism may depend on the specific alloy/environment combination. Several principal hydrogen-assisted cracking mechanisms are briefly discussed below.

Pressure Mechanism. Many of the earlier observations of hydrogen embrittlement were associated with the formation of high-pressure hydrogen gas bubbles in internal voids and microcracks. These hydrogen bubbles may form when the alloy is exposed to a high-hydrogen-fugacity environment such as high-pressure hydrogen gas or an extreme cathodic charging condition. The atomic hydrogen derived from metal-environment surface reactions diffuses to crack-tip microstructural heterogeneities such as voids and microcracks, recombines to the molecular form, and builds up a very large internal pressure. For example, the hydrogen pressure under cathodic charging conditions can be as large as 10^5 atm. This high pressure obviously will exert an additional stress and thus will increase the crack-tip driving force and result in a higher crack propagation rate.

Except in the high-hydrogen-fugacity environment, the pressure mechanism is not generally adequate for explaining the phenomenon of alloys cracking in less severe environments (most corrosion fatigue of engineering alloys). For instance, high-strength ferrous alloys exhibit significantly higher corrosion fatigue crack growth rates in gaseous hydrogen environments with pressure considerably less than 1 atm when compared with those obtained in vacuum. These observations suggest that, from a thermodynamic consideration, the hydrogen pressure developed in internal voids in the crack-tip region should not exceed the external hydrogen pressure of 1 atm. Thus, it would be difficult to reconcile these results with the classic pressure mechanism.

The lattice decohesion mechanism postulates that hydrogen as a solute decreases the cohesive bonding forces between metal atoms. Crack growth occurs when the local tensile elastic stress in the crack-tip region exceeds the hydrogen-weakened interatomic cohesive strength. It is also suggested that hydrogen migrates to regions of maximum triaxial stresses under stress-assisted diffusion, and the magnitude of cohesive force reduction depends on the local hydrogen concentration. An accumulation of hy-

drogen is also preferred along grain or second-phase boundaries. This mechanism, though attractive because of its simplicity, is difficult to prove on the atomic scale.

The surface adsorption mechanism proposes that strongly adsorbed hydrogen at the surface serves to lower the surface energy of the metal, and if the classic theory of the Griffith criteria for crack propagation is adopted, facilitates crack extension and increases the crack growth rate. The surface adsorption mechanism and the lattice decohesion mechanism are closely linked. While the adsorption mechanism suggests that adsorbed hydrogen reduces the surface energy needed for crack extension, the lattice decohesion mechanism proposes that hydrogen lowers the atomic bonding strength at the crack tip. The end results of these two mechanisms are the same in that the critical crack-tip drive force required for advancing the crack is reduced by the presence of hydrogen.

One of the major deficiencies in the surface adsorption mechanism is centered on the generally large amount of plastic deformation energy that accompanies crack growth. The plastic deformation energy is usually much larger than the relatively small surface energy (about 1000 to 1). Thus, even a large reduction in surface energy due to hydrogen adsorption should not markedly affect the fracture stress. Another discrepancy is that other environmental species, such as oxygen and nitrogen, also strongly adsorb to the clean metal surfaces and have the potential to reduce surface energy to a greater extent than hydrogen. Yet, neither oxygen nor nitrogen accelerates crack growth rates like hydrogen does.

Hydride Mechanism. Many metals that are susceptible to environmentally assisted cracking are also known to form stable or metastable hydrides. Hydrides have been reported to form in titanium, nickel, niobium, vanadium, zirconium, and so forth. These hydrides are often brittle and in alloy/environment systems where hydride precipitation is feasible, crack propagation is assisted by either cracking through the hydrides or along the hydride-matrix interfaces. Because of the volume expansion that occurs upon formation of metal hydrides, high tensile stress at the crack-tip region would promote the formation of stress-assisted hydrides. In alloys that form stable hydrides, hydride formation is a plausible mechanism for corrosion fatigue cracking. In alloys that form either metastable hydrides, such as AlH, or unstable hydrides, such as FeH, the evidence that links hydrides directly to fracture are lacking and the applicability of the hydride mechanism to these alloys is thus debatable.

Hydrogen-Enhanced Plasticity Mechanism. In contrast to previous hydrogen-assisted cracking mechanisms that suggest hydrogen "embrittles" the material and thus reduces the driving force required for crack extension, the hydrogen-enhanced plasticity mechanism proposes that hydrogen assists the processes of plastic flow by making dislocations move at reduced stresses or by easing the generation of dislocations at the crack tip. By recognizing that the three major

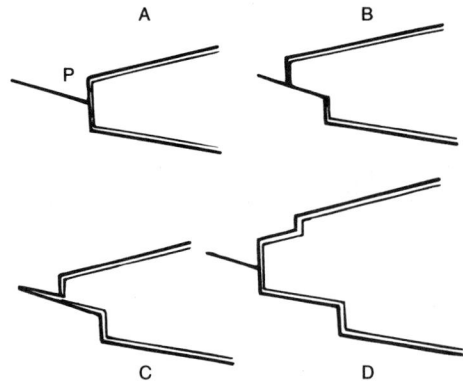

Fig. 2 Sequence of events occurring at the crack tip. Source: Ref 2

categories of fracture paths, namely microvoid coalescence, cleavage, and intergranular cracking, which are observed in benign environments, are also produced in environmentally assisted cracking, it is argued that hydrogen in the lattice merely assists these fracture processes. This mechanism has received some experimental support recently in that, using environmental cell transmission electron microscopy, hydrogen is seen to enhance the dislocation mobility in nickel by reducing the activation enthalpy for dislocation motion.

Anodic Dissolution

Commonly referred to as anodic dissolution, active path dissolution, slip dissolution, strain/stress-enhanced dissolution, and surface film rupture/metal dissolution, this mechanism maintains that crack growth rates are enhanced by anodic dissolution along susceptible paths such as grain boundaries or crack-tip strained metal that are anodic relative to the surrounding matrix. In corrosion fatigue cracking, the anodic dissolution mechanism depends on the rupture of the protective film at the crack tip by the fatigue process and the subsequent repassivation of the newly exposed fresh metal surface. Corrosion fatigue crack growth rates will be controlled by the bare surface anodic dissolution rate, the rate of repassivation, the rate of oxide film rupture, the mass transport rate of reactant to the dissolving surface, and the flux of solvated metal cations away from the surface. The repassivation rate is critical in that it must be rapid enough to avoid extensive and widespread dissolution, which leads to crack-tip blunting and pit formation rather than sharp and directional crack advance, but slow enough to allow significant dissolution at the crack tip.

The dynamic nature of the anodic dissolution and repassivation processes is schematically illustrated in Fig. 2. Under fatigue loading, a slip step forms at the crack tip and ruptures the protective surface film (Fig. 2B). The freshly created surface reacts with the environment and partly dissolves until the crack-tip region is completely repassivated and the protective surface film is repaired (Fig. 2C and 2D). These processes repeat themselves when a slip step again ruptures the protective film and exposes more bare surface.

Anodic dissolution can be reduced or controlled by cathodic protection techniques, which are commonly used to increase the cracking resistance of alloys in marine environments. This phenomenon is consistent with the anodic dissolution mechanism in that it indicates an anodic process controlling the cracking. Cathodic protection will reduce the corrosion fatigue crack growth rate by promoting repassivation, reducing the dissolution rate, and facilitating film repair.

While the anodic dissolution mechanism is often used to explain corrosion fatigue cracking of alloys exposed in deleterious aqueous environments, it is difficult to apply this mechanism to corrosion fatigue of alloys in gaseous environments (such as water vapor, hydrogen, or hydrogen sulfide) or in deaerated distilled water where the electrochemical reaction necessary for dissolution at the crack tip is unattainable. Also, the phenomenon of partial reversibility of preexposure by heat treatment is inconsistent with the anodic dissolution mechanism.

Anodic dissolution also fails to account for several phenomenological observations, especially the fractographic features of steels. The brittlelike appearance of fatigue fracture surfaces, including fan-shaped features emanating from manganese sulfide inclusions in steels and brittle striations on the fans, are difficult to reconcile with the lateral dissolution velocity concept that is an integral part of the anodic dissolution mechanism description (Ref 3). The hydrogen-assisted cracking mechanism has the advantage that it can account for the fractographic features in steels, but it is difficult to quantify the hydrogen effect.

Another concern is that the dissolution mechanism cannot account for the observed threshold behavior and frequency dependence of fatigue crack growth rates in steels without invoking a strong competitive process. For low load ratios ($R \sim 0.1$ to 0.2), it has been demonstrated that very long cyclic periods produce little or no environmental assistance (Ref 4), while for very high load ratios ($R \sim 0.8$ to 0.9), the maximum in growth rate enhancement comes at high frequencies, up to about 10 or 20 Hz (Ref 5). On its own, anodic dissolution would predict a continuing increase in growth rates for increasing cyclic periods. Advocates of the dissolution mechanism usually invoke creep and crack blunting arguments to help handle the low frequency problem.

Surface Energy Reduction Mechanism

This mechanism is similar to the hydrogen surface adsorption mechanism discussed previously in that the species (other than hydrogen) that are

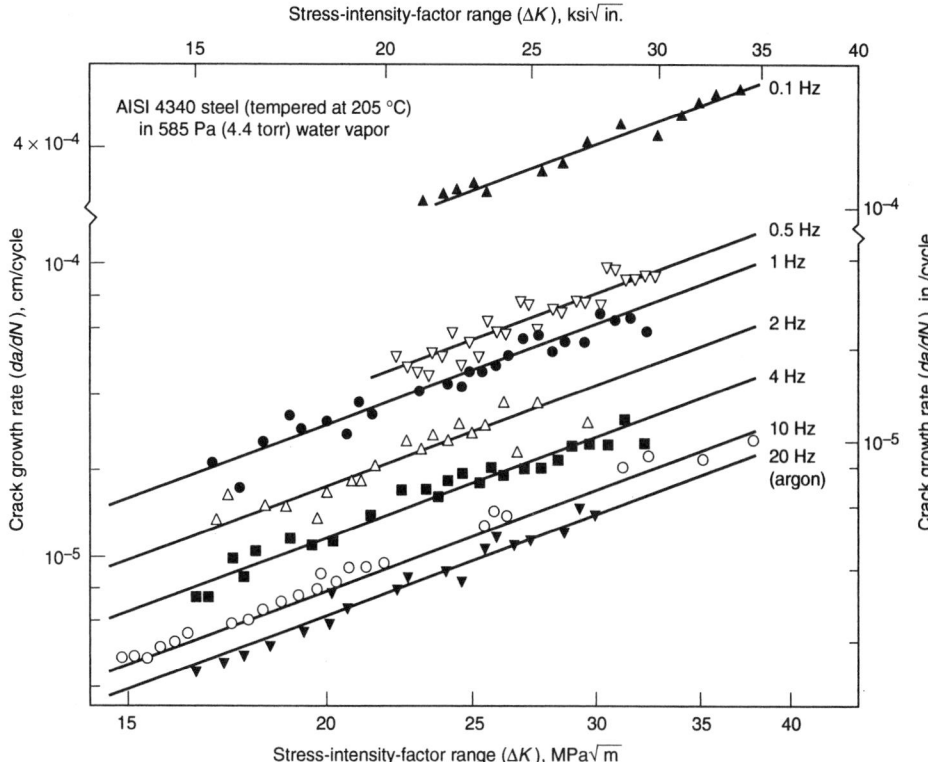

Fig. 3 Room-temperature fatigue crack growth kinetics of AISI 4340 steel in dehumidified argon and in water vapor (585 Pa) at $R = 0.1$. Source: Ref 6

strongly adsorbed at the surfaces serve to lower the surface energy required for the formation of a crack and thus increase corrosion fatigue crack growth rates. The surface energy reduction mechanism is also difficult to reconcile with the problems that hinder the acceptance of the hydrogen adsorption mechanism.

Variables Affecting Corrosion Fatigue

In corrosion fatigue crack growth analysis, the mechanical driving force is normally characterized in terms of the fracture mechanics parameters such as the crack-tip stress-intensity factor, K, or stress-intensity-factor range, ΔK. The assumptions, utility, and restrictions of this approach are discussed in detail elsewhere in this Handbook. Two of the following three interrelated loading variables are commonly used for characterizing corrosion fatigue crack growth: maximum stress-intensity factor, K_{max}; stress-intensity-factor range, ΔK ($\Delta K = K_{max} - K_{min}$); and stress ratio, or load ratio, R ($R = K_{min}/K_{max}$), where K_{min} is the minimum stress-intensity factor in a load cycle.

Many variables can influence corrosion fatigue crack growth. Many of the significant variables have been examined, and the results are available in a number of review papers. The influences of some of the variables on the corrosion fatigue

crack growth are described briefly in the following sections. Some of these variables include, along with the aforementioned loading variables:

Mechanical Variables

- Fatigue load frequency
- Fatigue load ratio
- Fatigue load waveform
- Maximum stress-intensity factor and stress-intensity-factor range
- Load interactions in variable amplitude loading (over/under/spectrum load)
- Residual stress

Geometrical Variables

- Crack size
- Crack geometry
- Specimen thickness (plane strain versus plane stress)

Metallurgical Variables

- Alloy composition
- Microstructure and crystal structure
- Heat treatment
- Grain boundary structure
- Grain shape and size
- Texture
- Distribution of alloy elements and impurities
- Deformation mode (slip character, twining, cleavage)
- Mechanical properties (strength, toughness, etc.)

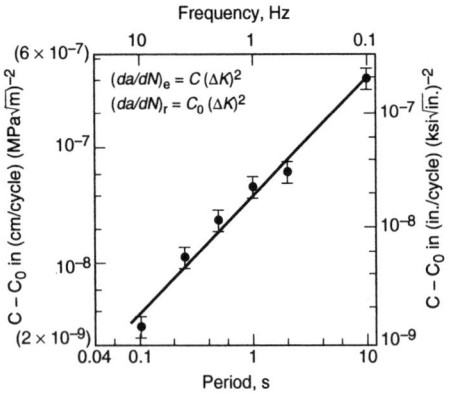

Fig. 4 Environment-dependent component of fatigue crack growth parameter as a function of cyclic load period for AISI 4340 steel in 585 Pa water vapor at room temperature. Source: Ref 6

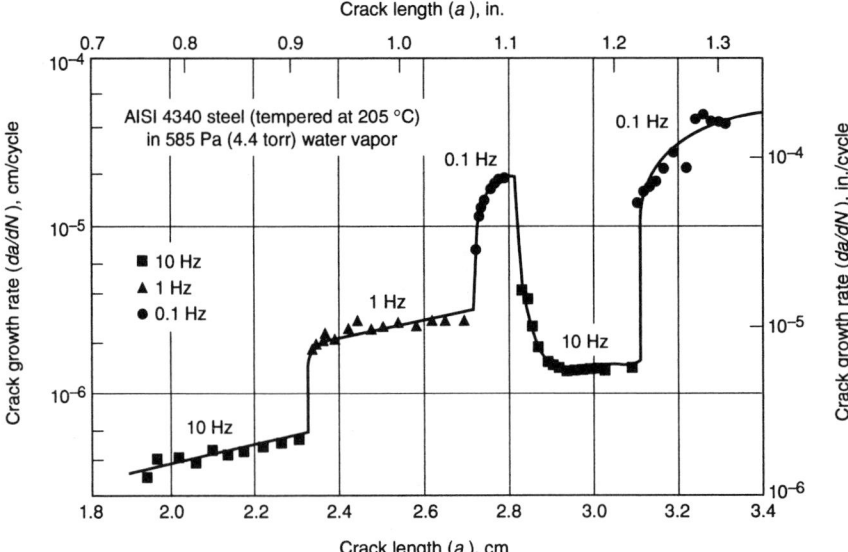

Fig. 5 Crack growth rate response resulting from changes in cyclic load frequency. Source: Ref 6

Environmental Variables

- Type of environments (gaseous or liquid)
- Partial pressure of damaging species in gaseous environments
- Concentration of damaging species in aqueous or other liquid environments
- Temperature
- pH
- Electrochemical potential
- Viscosity of the environment

Effect of Fatigue Load Frequency. An example of the frequency effect is shown in Fig. 3 (Ref 6) for an AISI 4340 steel in a water vapor environment (585 Pa). Data for tests in argon are also included in Fig. 3 for comparison. These data cover ΔK from 15 to 40 MPa$\sqrt{m}$ at a load ratio, R, of 0.1 and the corresponding K_{max} values are below the apparent K_{Iscc} (approximately 55 MPa$\sqrt{m}$) for this water vapor pressure. Figure 3 clearly shows that the environmental effect is much more pronounced at lower cyclic frequencies than that at higher frequencies as the fatigue crack growth rates at 0.1 Hz are more than an order of magnitude higher than those obtained at 10 Hz. As shown in Fig. 4, the environment-induced increase, which is represented by the value $(C - C_0)$, is found to be linearly proportional to the "time-at-load" or period (the inverse of cyclic-load frequency). Fractographic examination of the fracture surfaces also indicated that, at lower frequencies, where the environmental effect was significant, the fracture path in water vapor was primarily intergranular along the prior austenite grain boundaries, similar to those observed in hydrogen and in water under sustained load conditions. At higher test frequencies, the fracture path was primarily transgranular with respect to the prior austenite grains and resembled paths in specimens tested in inert argon environment.

Fatigue crack growth kinetics shown in Fig. 3 are steady-state rates corresponding to a particular cyclic load frequency. If the cyclic load frequency is abruptly changed during fatigue test, a region of transient growth may develop. Figure 5

shows fatigue data obtained at 0.1, 1, and 10 Hz under the same maximum and minimum loads. A region of transient growth is evident following each change in cyclic load frequency, before steady-state growth rate related to the new cyclic load frequency is established. For example, decreasing the test frequency from 10 to 0.1 Hz did not produce an immediate change in crack growth rate to its expected steady-state value. Instead, it resulted in a gradual change that extended over crack growth increments of the order of 0.1 cm. Similarly, an increase in frequency produced a gradual decrease in growth rate to its steady-state value. The extent of crack growth required to establish steady-state growth and the size of the transition zone appeared to depend on the magnitude of the frequency change and on the crack length (or ΔK, for constant load-amplitude fatigue). Because there is no influence of frequency on fatigue crack growth in argon over this

Fig. 6 Influence of frequency on fatigue crack growth in a Ti-6Al-4V alloy exposed to 0.6 M NaCl solution at room temperature and $R = 0.1$. Source: Ref 7

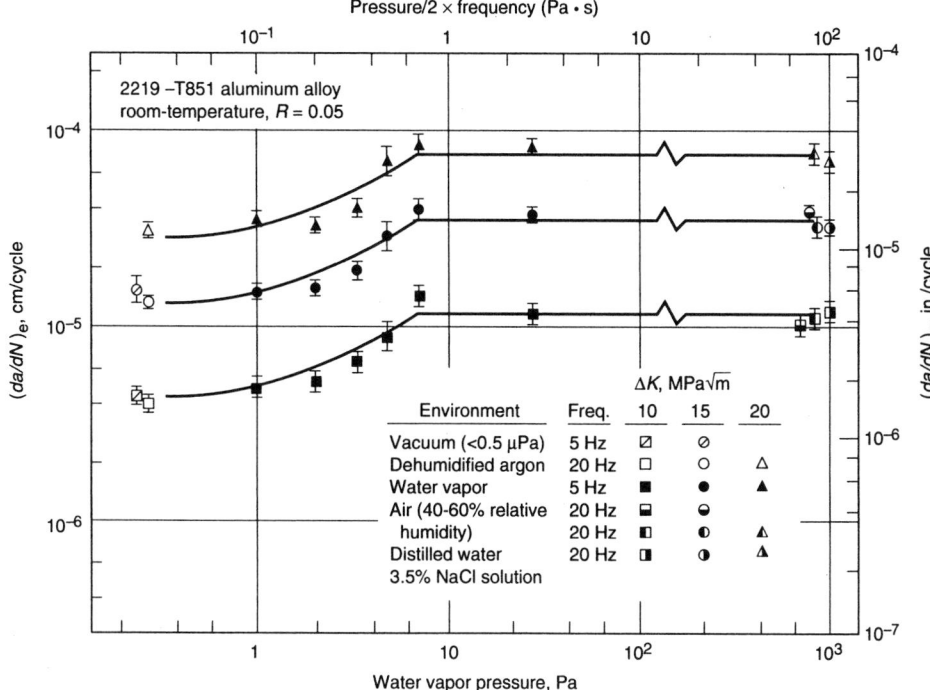

Fig. 7 Influence of water vapor pressure on fatigue crack growth rates in 2219-T851 aluminum alloy at room temperature. Source: Ref 8

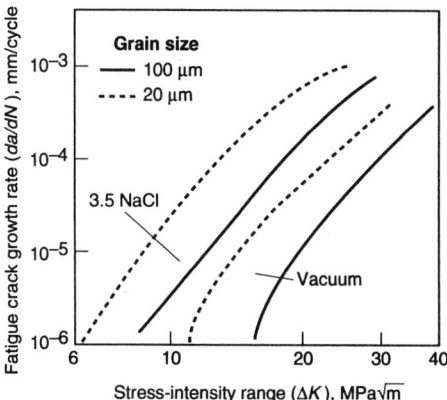

Fig. 8 Effect of grain size on the fatigue crack growth rates of Ti-8.6Al in vacuum and 3.5% NaCl solution. Source: Ref 9

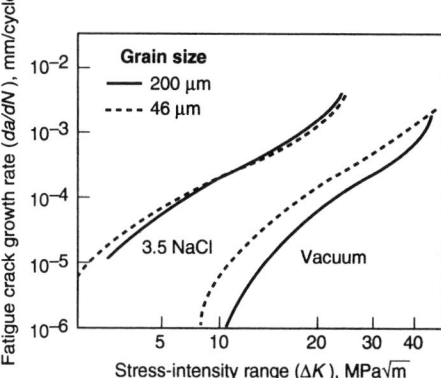

Fig. 9 Effect of grain size on the fatigue crack growth rates of Al-5.7Zn-2.5Mg-1.5Cu in vacuum and 3.5% NaCl solution. Source: Ref 9

range of frequencies, the observed transient phenomenon must result from interactions with the environment.

The above examples of the frequency effect are consistent with the hydrogen-assisted cracking mechanisms. Because condensation is not expected at such low water vapor pressure, it would be difficult to apply the anodic dissolution mechanism. Also, the anodic dissolution mechanism cannot explain adequately the observed transient responses during frequency cycling.

Titanium alloys exhibit an unusual frequency dependency of corrosion fatigue crack growth in salt-water environments. This frequency dependence is illustrated in Fig. 6 by data on a Ti-6Al-4V alloy tested in 0.6 M NaCl solution. The crack growth rates increased with decreasing frequency and reached a maximum that depended on ΔK, and then decreased to rates that are comparable to those observed in vacuum or other inert environments. Further analyses also suggested that, at higher frequencies, corrosion fatigue crack growth rates were inversely proportional to the square root of frequency. This dependence, coupled with the surface reactivity of titanium, is consistent with hydrogen-diffusion-controlled crack growth and suggests a hydride formation mechanism. The abrupt decrease in crack growth rates at the lower frequencies is attributed to the difficulty of hydride formation at the substantially lower strain rates associated with the lower cyclic frequencies.

Effect of Environment (Water Vapor Pressure). Figure 7 shows the effect of water vapor pressure on the fatigue crack growth kinetics of 2219-T851 aluminum alloy at three ΔK levels, along with data from a reference dehumidified

argon environment. As shown in Fig. 7, at the cyclic frequency of 5 Hz, the rate of fatigue crack growth is unaffected by water vapor until a threshold pressure is reached. The rates then increase and reach a maximum within an order of magnitude increase in vapor pressure from this threshold. The maximum fatigue crack growth rate at each ΔK is equal to that obtained in air, distilled water, and 3.5% NaCl solution. This observed crack growth response as a function of water vapor pressure is similar to the water vapor/aluminum alloy surface reaction kinetics and is consistent with the transport limited model. These results strongly suggest that, instead of an anodic dissolution mechanism, the hydrogen produced by the surface reactions is responsible for the enhancement in fatigue crack growth.

Effect of Grain Size. The influence of grain size on the corrosion-fatigue crack growth behavior of an alloy depends on the fracture mode of that particular alloy in the environment. In a benign environment, such as in a vacuum, the fatigue crack propagates generally along transgranular slip bands and the crack growth rates usually decrease with increasing grain size. Examples of this grain size effect are shown in Fig. 8 and Ti-8.6Al and in Fig. 9 for Al-5.7Zn-2.5Mg-1.5Cu. In both cases, the fatigue crack growth rates of coarser-grain alloys are substantially lower than those for finer-grain alloys. This grain size effect in the benign environment can be explained by the concept of reversed slip of dislocations within the plastic zone ahead of the crack tip or by increased closure induced by interference between irregular mating crack surfaces during unloading.

In aggressive environments, the grain size effect will either still be present or greatly diminish, depending on how the fracture mode of the alloy is affected by the environment. For the Ti-8.6Al alloy, the fracture mode in salt-water environments for both small- and larger-grain alloys is primarily transgranular slip-band cracking, similar to that in vacuum environments. As shown in Fig. 8, the corrosion fatigue crack growth rates of large-grain alloys are significantly lower than those of small-grain alloys. The slip reversal concept discussed previously can be applied to explain the observed grain size dependency of the corrosion fatigue crack growth behavior in Ti-8.6Al.

However, for Al-5.7Zn-2.5Mg-1.5Cu in a salt-water environment, the grain size dependence of corrosion fatigue crack growth rates almost disappears, as shown in Fig. 9. Fractographic analyses indicated that the fracture path in salt water changed to intergranular at low ΔK. At higher ΔK, the transgranular fracture surfaces for both small- and large-grain alloys are relatively flat as several slip systems are activated and cross slips become possible. These changes in the fracture mode, particularly the change to intergranular separa-

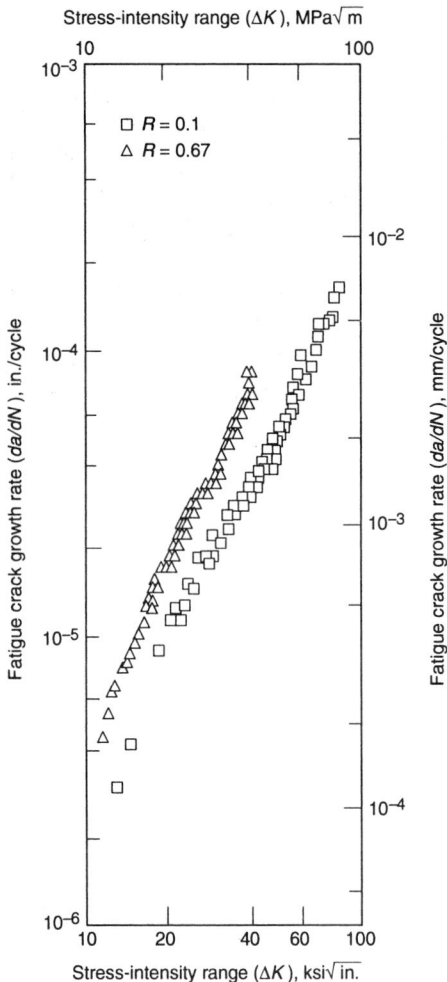

Fig. 10 Effect of load ratio on the corrosion fatigue crack growth rates of MF-80 HSLA steel in 3.5% NaCl solution. Source: Ref 10

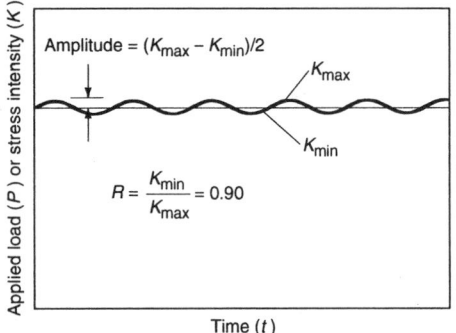

Fig. 11 Ripple load profile

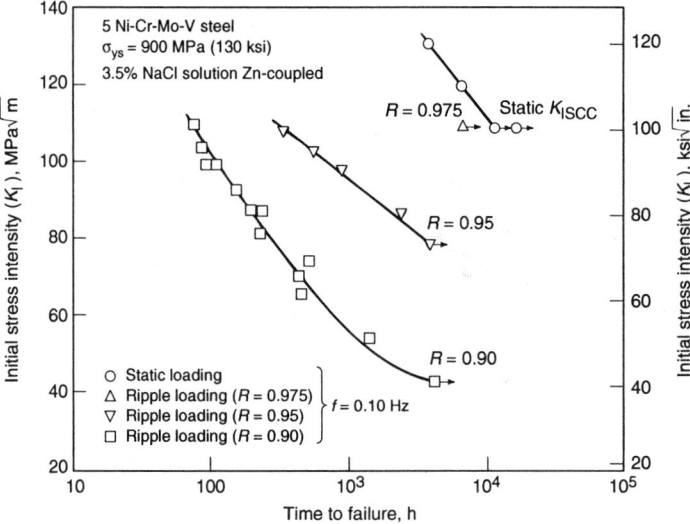

Fig. 12 Ripple-load cracking of 5Ni-Cr-Mo-V steel in a 3.5% NaCl solution. Source: Ref 11

is shown in Fig. 10 for MF-80 HSLA steel in a salt-water environment. The corrosion fatigue crack growth rates of MF-80 HSLA steel in 3.5% NaCl solution at $R = 0.67$ are about two times higher than the rates obtained at $R = 0.10$ at similar stress-intensity levels.

A special case of corrosion fatigue cracking is "ripple load" cracking. "Ripple loading" refers to a loading profile where relatively small amplitude cyclic loads are superimposed on a large sustained load. A typical ripple load profile is schematically shown in Fig. 11 where the stress ratio often exceeds $R = 0.90$. The significance of ripple load cracking is its relation to stress-corrosion cracking (SCC), where the applied load is assumed to be constant. However, in the real world, structures rarely experience a constant load condition, but are far more apt to see a combination of very small amplitude cyclic loads and a large constant load. A typical example is offshore platform structures. Under ripple loading conditions, if a structure is designed solely based on the SCC threshold without considering the possibility of ripple loads, cracking or fracture may occur prematurely.

Ripple load cracking has been approached successfully as high stress ratio corrosion fatigue. An example of the ripple load effect is shown in Fig. 12 for 5Ni-Cr-Mo-V steel in a 3.5% NaCl solution. As shown in Fig. 12, at $R = 0.90$ (representing ripple loads with a cyclic amplitude equal to 5% of the sustained load), cracking can occur by a corrosion fatigue mechanism under ripple loading at stress intensities as much as 60% lower than the static SCC threshold (K_{Iscc}) and the ripple load cracking threshold (K_{IRLC}, which is defined as $\Delta K_{th}/(1 - R)$) is less than half of the K_{Iscc}.

It is significant to note that the ripple effect depends strongly on the size or amplitude of the ripple loads. As the fatigue crack growth threshold, ΔK_{th}, varies with the stress ratio, R, the ripple load cracking threshold correspondingly varies with R. Thus, while a material may exhibit substantial ripple load degradation with larger ampli-

tude ripple loading, the same material may show less or even no ripple load degradation if the existing ripples are smaller. Such ripple size effect is shown in Fig. 12 for 5Ni-Cr-Mo-V steel under SCC and ripple load conditions in a 3.5% aqueous NaCl solution. For the 5Ni-Cr-Mo-V steel, a static K_{Iscc} was established at 110 MPa√m. As shown in Fig. 12, a 1.25% ripple loading (corresponding to $R = 0.975$) did not cause any degradation and the apparent cracking threshold equaled the static K_{Iscc}. However, as the amplitude of the ripples was increased to 2.5% (corresponding to $R = 0.95$) and 5% (corresponding to $R = 0.90$) of the sustained load, ripple loading significantly reduced the values of time to failure and the apparent cracking threshold. For instance, at the initial stress intensity of 90 MPa√m, which is well below the K_{Iscc} of 110 MPa√m, SCC failure will not occur under static loading. However, when small ripple loads with amplitude equal to 2.5% of the static load were added, the test specimen failed in about 1600 h. For a 5% ripple loading, the failure time was further reduced to only 180 h.

Not all materials, however, appear susceptible to the ripple load effect. While materials that exhibit greater SCC resistance under static load conditions, such as 5Ni-Cr-Mo-V steel, are susceptible to ripple load degradation, materials that are less SCC resistant may exhibit little or no ripple load degradation. An example is shown in Fig. 13 for a SCC-susceptible AISI 4340 steel in a 3.5% NaCl solution. As shown in Fig. 13, the SCC threshold, determined from long-term static tests, was established at 33 MPa√m. The two 5% ripple loading tests (corresponding to $R = 0.90$) produced results that fell directly on the static SCC time-to-failure curve. That is, even under a fairly large 5% ripple loading, AISI 4340 steel does not exhibit any ripple load effect. It is also safe to predict, based on the experience with 5Ni-Cr-Mo-V steel, that AISI 4340 steel is not likely to demonstrate any ripple load degradation

tion, would explain the lack of grain size dependence for corrosion fatigue crack growth rates.

The effect of stress ratio on corrosion fatigue crack growth rates depends on the alloy/environment system. In general, higher stress ratios will result in higher corrosion fatigue crack growth rates and lower corrosion fatigue crack growth thresholds. An example of the stress-ratio effect

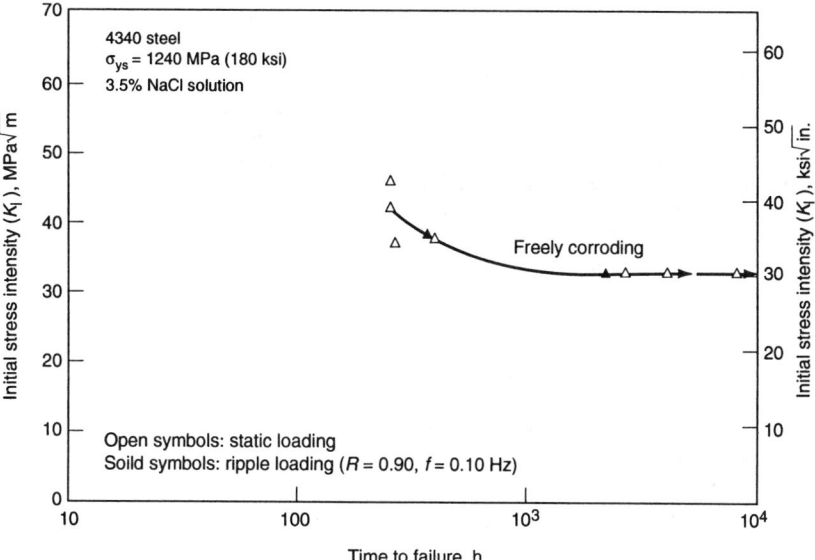

Fig. 13 Ripple-load cracking of AISI 4340 steel in a 3.5% NaCl solution. Source: Ref 11

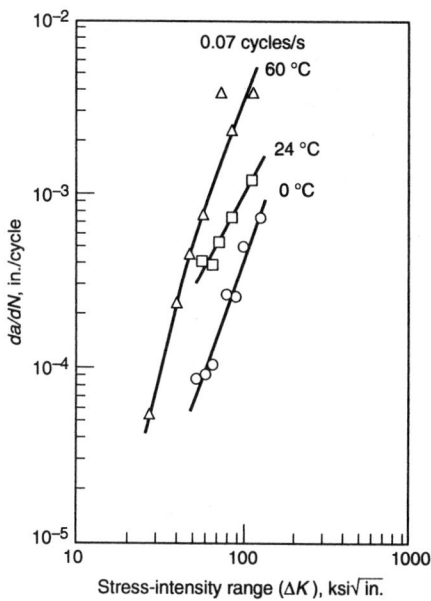

Fig. 16 Effect of temperature on the corrosion fatigue crack growth rate of a metastable austenitic steel in distilled water. Source: Ref 13

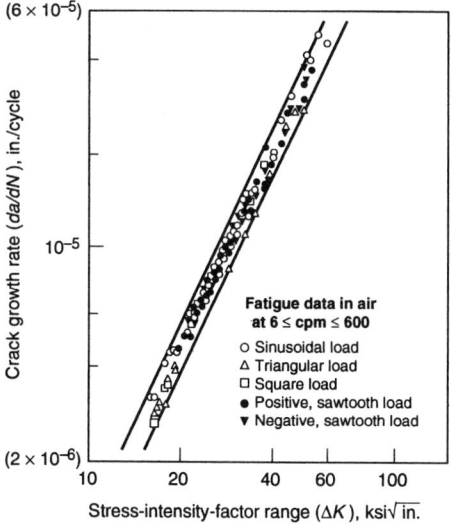

Fig. 14 Effect of cyclic load waveform on the fatigue crack growth rates of 15Ni-5Cr-3Mo steel in ambient air. Source: Ref 12

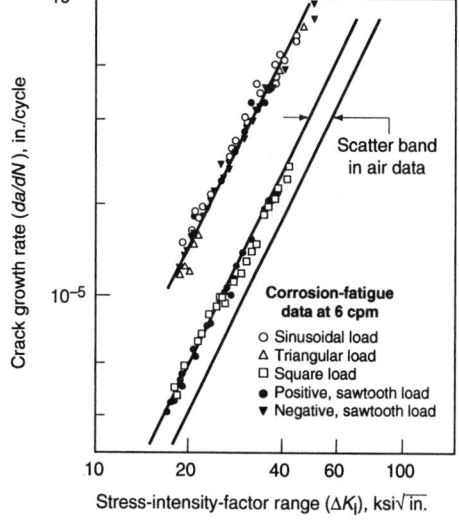

Fig. 15 Effect of cyclic load waveform on the corrosion fatigue crack growth rates of 15Ni-5Cr-3Mo steel in 3% NaCl solution. Source: Ref 12

if the size of the ripple is smaller than 5% of the sustained load.

Effect of Waveform. While cyclic load waveform has little effect on the fatigue crack growth rates in benign environments, available data indicate that corrosion fatigue crack growth rates in aggressive environments may be highly dependent on the shape of the cyclic load waveform. This waveform effect is illustrated in Fig. 14 and 15 for 15Ni-5Cr-3Mo steel in ambient air and in a 3% NaCl solution, respectively. As shown in Fig. 14, fatigue crack growth rates in ambient air are identical and the applied cyclic load waveform has no effect on the fatigue crack growth rates.

The nature of the cyclic load waveform can significantly affect the corrosion fatigue crack

growth rates in aggressive environments. As shown in Fig. 15, corrosion fatigue crack growth rates in 3% NaCl solution under both square and negative sawtooth waveforms, which have very short rising load periods, are identical and are statistically equal to the fatigue crack growth rates in ambient air. That is, no environment-enhanced effect is observed with the square and negative sawtooth waveforms.

On the other hand, environment-enhanced effects on fatigue crack growth rates can be substantial if the cyclic waveform consists of a significant period of rising load. As shown in Fig. 15, the corrosion fatigue crack growth rates of 15Ni-5Cr-3Mo steel in 3% NaCl solution under sinusoidal, triangular, and positive sawtooth waveforms, which all have long rising load peri-

ods, are identical and are about three times higher than the fatigue crack growth rates under square and negative sawtooth waveforms in the same 3% NaCl environment and in air. It has been suggested that the environmental enhancement of fatigue crack growth is caused primarily by the interaction between the environment and the steel during the rising load portion of each loading cycle. The longer the load rise time during each load cycle, the greater the influence of an aggressive environment. In the case of the square cyclic waveform that has a very short rising load period, the aggressive environment has little effect on the fatigue crack growth rates.

Effect of Temperature. Because temperature may influence the environment/metal surface reactions as well as the many transport processes listed in Fig. 1, temperature can be expected to affect corrosion fatigue crack growth rates. In many environment/alloy systems, corrosion fatigue crack growth rates increase with increasing temperature. That is, the corrosion fatigue cracking process is thermally activated. If the activation energy for corrosion fatigue cracking can be measured, then the rate-limiting process and possibly the mechanism of cracking may be determined.

An example of the temperature effect is shown in Fig. 16 for a metastable austenitic steel (composition in wt%: 0.27C-0.43Mn-0.09Si-11.95Cr-1.90Mo-7.96Ni) in distilled water. As shown in Fig. 16, as the temperature increases, the corrosion fatigue crack growth rates also increase. The activation energy for corrosion fatigue crack growth has an averaged value of 35.2 kJ/mol. This activation energy is very close to the activation energy (36 ± 14 kJ/mol) of water vapor reaction with AISI 4340 steel surface (Ref 14). The good agreement between the activation energy for corrosion fatigue crack growth and the activation energy for water vapor/steel surface reaction kinetics suggests that the surface reaction is the rate limiting process for crack growth and the hydrogen produced during the surface reaction is responsible for the embrittlement of this austenitic steel fatigued in distilled water.

Corrosion fatigue is a function of the chemistry of the environment/material system, the microstructure, and the applied loading conditions. Because environmental susceptibility is a complex phenomenon occurring at the atomic/molecular level, it is difficult to predict or verify using specific models or tests. For the present, it is necessary to characterize materials using appropriate real time loading profiles in the service environment to ensure adequate service lives. Obviously, given the widely varying nature of growth behaviors under such loading/environment conditions, variations in the environmental and loading parameters must be considered to ensure that the worst case conditions are captured in the life verification tests.

REFERENCES

1. R.P. Gangloff and R.P. Wei, *Metall. Trans. A*, Vol 8A, 1977, p 1043
2. H.L. Logan, *J. Res. Natl. Bur. Stand.*, Vol 48, 1952, p 99
3. W. Cullen, G. Gabetta, and H. Hanninen, "A Review of the Models and Mechanisms for the Environmentally-Assisted Crack Growth of Pressure Vessel and Piping Steels in PWR Environments," NUREG/CR-4422 MEA-2078, U.S. Nuclear Regulatory Commission, Dec 1985
4. W. Van Der Sluys and R. Emanuelson, Overview of Data Trends in Cyclic Crack Growth Results in LWR Environments, *Proc. 2nd IAEA Specialists' Meeting on Subcritical Crack Growth* (Sendai, Japan), U.S. Nuclear Regulatory Commission, 15-17 May, 1985
5. P.M. Scott and A.E. Truswell, Corrosion Fatigue Crack Growth in Reactor Pressure Vessel Steels in PWR Primary Water, *J. Pressure Vessel Technol.*, Vol 105, 1983, p 245-254
6. P.S. Pao, W. Wei, and R.P. Wei, *Environment-Sensitive Fracture of Engineering Materials*, TMS-TIME, 1979, p 565
7. S. Chiou and R.P. Wei, "Corrosion-Fatigue Cracking Response of Beta Annealed Ti-6Al-4V Alloy in 3.5% NaCl Solution," NADC-83126-60, U.S. Naval Air Development Center, 1984
8. R.P. Wei, P.S. Pao, R.G. Hart, T.W. Weir, and G.W. Simmons, *Metall. Trans. A*, Vol 11A, 1980, p 151
9. J. Lindigkeit, G. Terlinde, A. Gysler, and G. Lütjering, *Acta Metall.*, Vol 27, 1979, p 1717
10. S.J. Gill and T.W. Crooker, *Marine Technol.*, Vol 27, 1990, p 221
11. P.S. Pao, R.A. Bayles, and G.R. Yoder, *J. Eng. Mater. Technol. (Trans. ASME)*, Vol 113, 1991, p 125
12. S.T. Rolfe and J.M. Barsom, *Fracture and Fatigue Control in Structures*, Prentice-Hall, 1977
13. W.W. Gerberich, J.P. Birat, and V.F. Zacky, *Corrosion Fatigue: Chemistry, Mechanics, and Microstructure*, National Association of Corrosion Engineers, 1971, p 396
14. G.W. Simmons, P.S. Pao, and R.P. Wei, *Metall. Trans. A*, Vol 9A, 1978, p 1147

SELECTED REFERENCES

- O.F. Devereux, A.J. McEvily, and R.W. Staehle, Ed., *Corrosion Fatigue: Chemistry, Mechanics, and Microstructure*, National Association of Corrosion Engineers, 1971
- P.R. Swann, F.P. Ford, and A.R.C. Westwood, Ed., *Mechanisms of Environment Sensitive Cracking of Materials*, The Metals Society, 1977
- J.R. Fong, Ed., *Fatigue Mechanisms*, ASTM STP 675, 1978
- T.W. Crooker and B.N. Leis, Ed., *Corrosion Fatigue: Mechanics, Metallurgy, Electrochemistry, and Engineering*, ASTM STP 801, 1983
- C.E. Jaske, J.H. Payer, and V.S. Balint, *Corrosion Fatigue of Metals in Marine Environments*, Metals and Ceramic Information Center, Springer-Verlag, 1981
- R.N. Parkins and Y.M. Kolotyrkin, Ed., *Corrosion Fatigue*, The Metals Society, 1983
- Z.A. Foroulis, Ed., *Environment-Sensitive Fracture of Engineering Materials*, TMS-AIME, 1979

Corrosion Fatigue Testing

Peter L. Andresen, GE Corporate Research & Development

ENVIRONMENTAL EFFECTS play a dominant role in fatigue behavior, as illustrated in fatigue crack growth of laboratory test specimens (Fig. 1) and the general fatigue life behavior of engineering component applications (Fig. 2, 3). Many environments can produce a profound increase in crack growth rates, including seemingly innocuous environments such as high-purity water, laboratory air, and very low partial pressures of oxygen, hydrogen, or water vapor. A general review of these underlying mechanisms of environmental effects on fatigue crack growth is provided in the article "Mechanisms of Corrosion Fatigue" in this Volume.

The objective of this article is to introduce fundamental aspects of environmental crack advance in general, and corrosion fatigue in particular, and then discuss critical experimental issues associated with corrosion fatigue testing. Insufficient emphasis is often given to environmental effects, despite their influence on fatigue. Attention to testing detail is required because of the greater number of critical variables and their interactions. Whereas fatigue crack propagation studies in inert environments may evaluate only a few primary test variables, and require control to be maintained over less than a dozen significant parameters (alloy composition, heat treatment, etc.), the number of variables and control parameters in environmentally assisted cracking investigations is typically multiplied by 3 to 10 times (Fig. 4).

In this article, the primary emphasis is on corrosion fatigue testing of steel in high-temperature water. However, the appendix at the end of this article provides details for corrosion fatigue testing in other aggressive environments. A brief collection of corrosion fatigue data for selected nonferrous alloys is also included in the appendix. Further information on corrosion fatigue of specific alloys is covered in separate articles in this Volume.

Environmentally Assisted Cracking

Because corrosion fatigue is just one part of the complex phenomenon known as environmentally assisted cracking (Fig. 4), a brief review of some fundamentals is useful. Not all aspects of envi-

ronmentally assisted crack advance are identical in all cracking systems, but many interactions between mechanical loading and environmental exposure are common across the spectrum of loading conditions and materials.

In broad terms, the phenomenon of environmentally assisted cracking can encompass relevant crack propagation rates of more than 10^1 mm/s to less than 10^{-10} mm/s depending on:

- *Environment*, including low- and high-temperature aqueous solutions, low- and high-temperature gas, steam, organics, liquid metals, and so on.
- *Load condition*, such as static and monotonically increasing loading (*stress-corrosion cracking*) and cycling loading (*corrosion fatigue*). In most systems, loading variations represent a complete continuum in response, which is not surprising given that distinctions are necessarily blurry when considering ripple (i.e., low-amplitude) loading, long hold time behavior, and so on.

Thus, the choice between the terms *stress-corrosion cracking* and *corrosion fatigue* is often arbitrary. They can refer to the same underlying degradation phenomenon.

- *Material*, ranging from metals and alloys to glasses, plastics, composites, ceramics, and so on. Materials such as very pure metals, which were once considered unsusceptible to envi-

ronmentally assisted cracking, have usually, on more detailed investigation, been shown to be susceptible.

More examples of these variables are given in Table 1.

Because of the large contribution of environment to crack advance, attention to experimental detail is greatly rewarded. Unless the multidisciplinary nature of these studies is recognized however, most efforts will produce seriously misleading results or outright confusion, which is perhaps most responsible for the slow progress in quantifying and understanding environmental cracking. Familiarity with chemical and physical metallurgy, mechanics, chemistry, electrochemistry, and corrosion also is generally considered essential.

Environmental Factors during Crack Advancement. In simple terms and from a thermodynamic perspective, almost all engineering materials are highly reactive in most environments, consistent with the observation that, with rare exceptions, metals are found in nature in their oxidized state. Thus, the use of most engineering materials relies on the presence of a kinetic surface barrier (passivity) to reduce the oxidation rate to manageable proportions. When passivity is disturbed (e.g., by local strains in the underlying material), the reaction (oxidation) rate of the exposed metal is generally very high (>1 A/cm^2, Fig. 5) as the protective film is reformed (as repassivation occurs).

Table 1 Material, mechanical, and environmental variables of environmentally assisted cracking

Metallurgical variables	Environmental variables	Mechanical variables
Alloy composition	Temperature	Maximum stress or stress-intensity factor, σ_{max} or K_{max}
Distribution of alloying elements and impurities	Types of environments: gaseous, liquid, liquid metal, etc.	Cyclic stress or stress-intensity range, $\Delta\sigma$ or ΔK
Microstructure and crystal structure	Partial pressure of damaging species in gaseous environments	Stress ratio (R)
Heat treatment	Concentration of damaging species in aqueous or other liquid environments	Cyclic loading frequency
Mechanical working	Electrical potential	Cyclic load waveform (constant-amplitude loading)
Preferred orientation of grains and grain boundaries (texture)	pH	Load interactions in variable-amplitude loading
Mechanical properties (strength, fracture toughness, etc.)	Viscosity of the environment	State of stress
	Coatings, inhibitors, etc.	Residual stress
		Crack size and shape, and their relation to component size and geometry

In aqueous environments, oxidation reactions can have several repercussions, including the formation of atomic hydrogen from chemical reactions (e.g., $2Fe + 3H_2O \rightarrow Fe_2O_3 + 6H$) or electrochemical reactions (e.g., $2H^+ + 2e^- \rightarrow 2H^0 \rightarrow H_2$), and shifts in pH from localization of specific reactions (e.g., differential aeration crevice cells). Thus, many environmental effects on crack advance are associated with the direct loss (oxidation/dissolution) of metal (known as the *slip oxidation mechanism*); the formation, absorption, and transport of hydrogen (*hydrogen embrittlement*); and other corrosion- or repassivation-related phenomena. An example of the latter is *film-induced cleavage*, whereby cracks that form in a brittle surface layer propagate in the ductile underlying metal. Such brittle layers can form by selective corrosion and dealloying, or by the formation of a brittle, coherent film (Ref 7).

The role of deformation is often of paramount, fundamental importance, because it defines the periodicity of rupture of the protective oxide. For a growing crack, the *crack-tip strain rate* can be associated with reversed slip during cyclic loading, thermal or irradiation creep, or any "crack advance" process (e.g., corrosion) that redistributes the stress/strain fields in front of a crack and produces dislocation motion.

In some alloy environment systems (e.g., carbon steels in carbonate/bicarbonate solutions, stainless steel in high-temperature water), the maximum environmental contribution can be predicted by measuring the dissolution rate on a bare (e.g., scratched or rapidly strained) metal surface (Fig. 6). A large environmental component of crack advance thus may occur with increasing cyclic amplitude or frequency, although this can be misleading because:

- The inert fatigue crack growth rate also increases, and eventually becomes dominant. Thus the environmental enhancement (i.e., the percentage increase above the inert baseline rate) also decreases, which can lead, for example, to transitions from intergranular cracking to the transgranular morphology characteristic of inert fatigue.
- With decreasing cyclic amplitude/frequency, the environmental contribution to cracking can begin to dominate, and the traditional Paris law dependence (e.g., ΔK^4) can shift substantially (e.g., to ΔK^2), resulting in a greatly reduced sensitivity to cyclic amplitude or frequency (Fig. 7, 8).

The interactions that give rise to corrosion fatigue can be quite complex and sometimes nonintuitive. For example, in low-alloy and carbon steels in high-temperature water (Fig. 8), a very large environmental enhancement occurs at intermediate ΔK or frequency conditions. At higher cyclic amplitudes or frequencies, the total crack growth rate becomes dominated by the inert fatigue crack growth rate. As cyclic amplitude or frequency is decreased, the environmental component is more pronounced, and the crack growth rate response follows a shallower ΔK/frequency dependency. With further decreases in cyclic amplitude or frequency, a dramatic decrease in crack growth rate is observed, and the subsequent response parallels the inert fatigue response. This complex behavior is associated with the ability to sustain an aggressive crack chemistry; in these steels, MnS inclusions dissolve and create a sulfur-rich crack chemistry. As the cyclic amplitude or frequency is decreased, eventually the rate of intersection of new MnS inclusions slows to the point that insufficient MnS dissolution occurs and a critical chemistry (for a large environ-

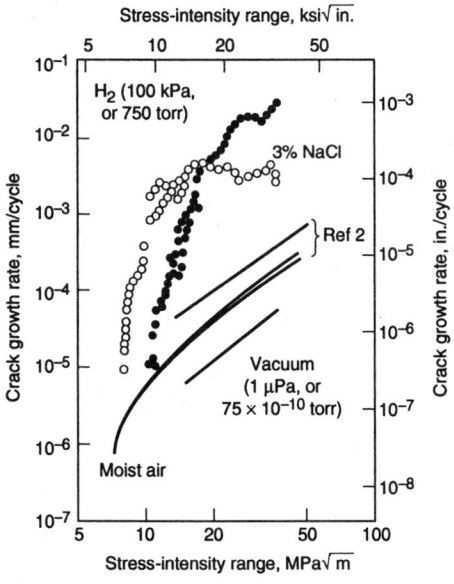

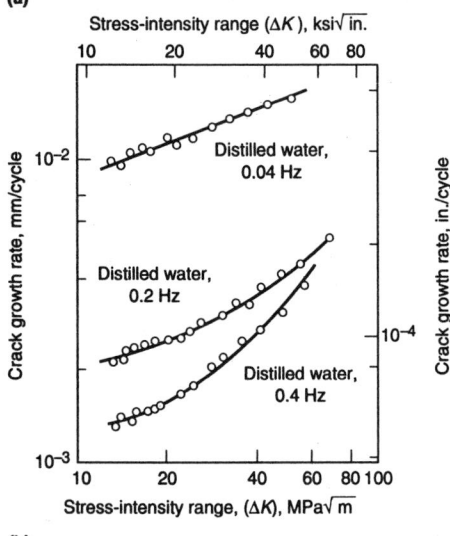

Fig. 1 Room-temperature corrosion fatigue crack growth rates. (a) Effect of environment on fatigue crack propagation in 4130 steel with a yield strength of 1330 MPa (195 ksi). The band of data about the moist air line represents cracking in 13 steels with varying microstructures and yield strength ranging from 300 to 2100 MPa (45 to 3054 ksi). Temperature 23 °C (75 °F), frequency 0.1 Hz, load ratio 0.1. Source: Ref 1, 2. (b) Effect of stress-intensity amplitude and loading frequency on corrosion fatigue crack growth in an ultrahigh-strength 4340 steel exposed to distilled water. Temperature 23 °C. Source: Ref 3

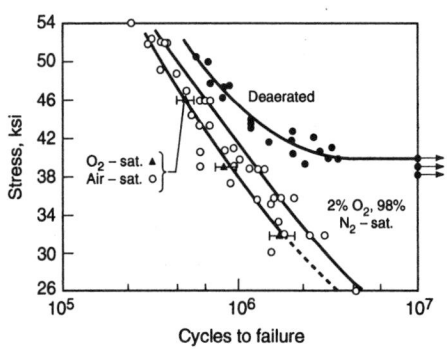

Fig. 2 Stress amplitude vs. cycles-to-failure for corrosion fatigue of 0.18% C steel in 3% NaCl at 25 °C, showing the strong effect of dissolved oxygen in accelerating cracking and eliminating the stress threshold. Source: Ref 4

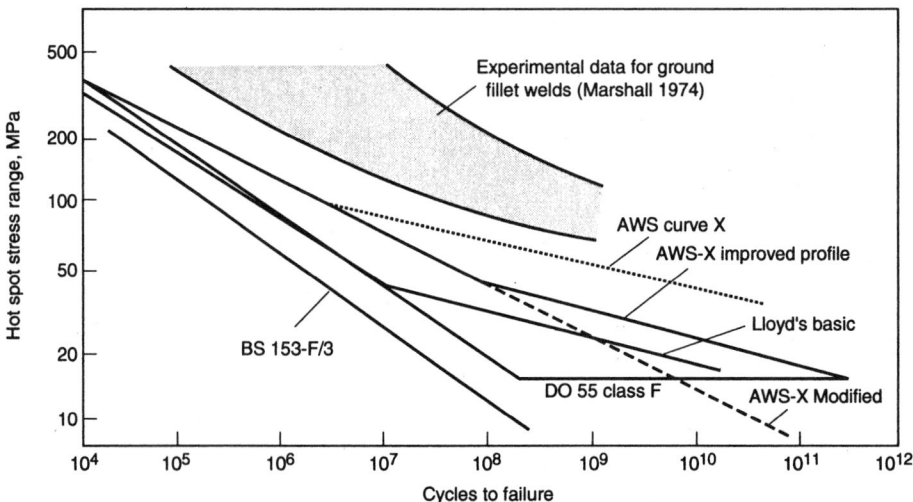

Fig. 3 Empirically derived design codes for corrosion fatigue of offshore welded tube structures, illustrating their invalidity under specific test conditions and their constantly changing formulation. Hot spot stress range refers to local peak stress amplitudes at specific locations of the structure. Source: Ref 5

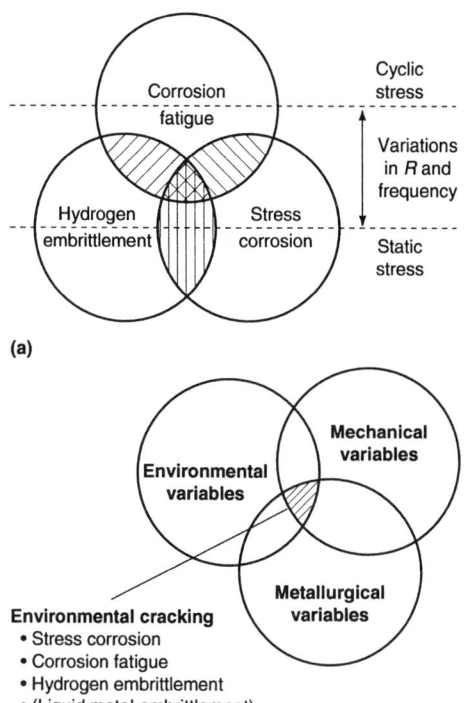

(a)

(b)

Fig. 4 General interrelationships associated with environmentally assisted cracking, which can encompass crack growth rates of more than 10 mm/s to less than 10^{-10} mm/s. (a) Venn diagram of mechanisms. (b) Venn diagram of variables listed in Table 1

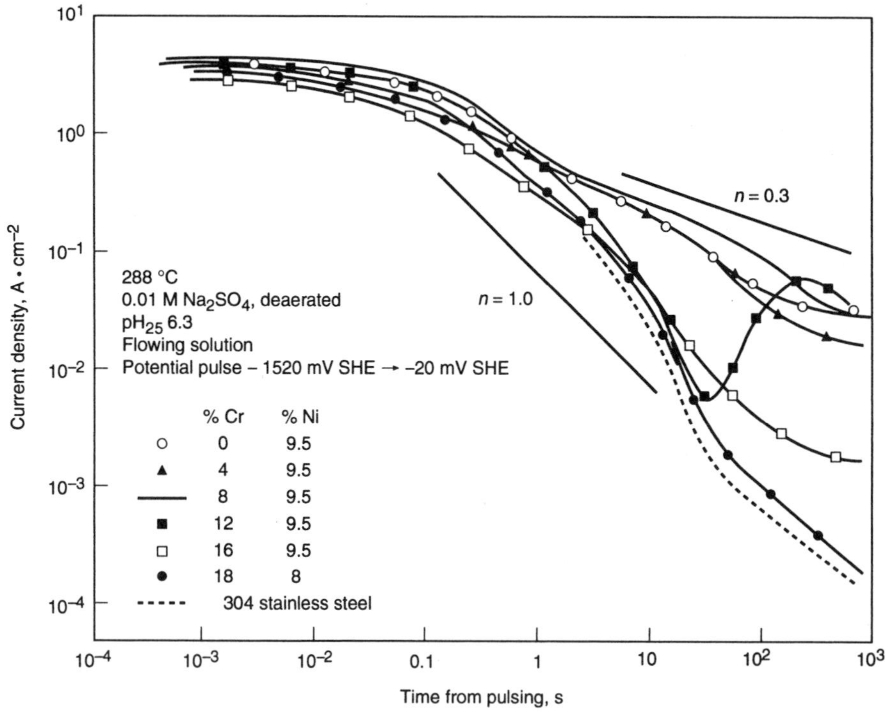

Fig. 5 Oxidation current density vs. time following rupture of the protective oxide on a stainless steel wire by rapid straining in hot water. A high-peak (bare-surface) current density (generally ≥1 A/cm², corresponding to rapid metal dissolution) is followed by repassivation as the protective film re-forms. Source: Ref 6

mental enhancement) is not sustained. Complicating factors include the very high sensitivity to relatively low dissolved-sulfur concentrations ("high-sulfur" behavior is observed above about 1 ppm sulfur) and the large role of dissolved oxygen/corrosion potential (which concentrates sulfur anions in the crack), flow rate (which can flush out the aggressive crack chemistry), and nonuniform distribution of MnS inclusions (Ref 8, 11).

Concepts and Role of Environmentally Assisted Crack Growth. Historically, the progression in the understanding of metal failure has shifted from addressing simple overload and buckling, to addressing repeated application of lesser loads (fatigue), then to addressing environmental effects on cracking. While the role of environment has long been acknowledged, the design and life evaluation codes (e.g., the ASME Boiler and Pressure Vessel Codes) are based largely on a mechanical engineering perspective and may ignore or average, rather than distinguish and quantify, the following:

- Time dependent (vs. cycle dependent) environmental crack growth
- The continuum in loading from corrosion fatigue to stress-corrosion cracking
- The spectrum of environmental conditions
- Variations in material microstructure (e.g., sensitization)

This historical emphasis on mechanical quantification to the exclusion of environmental and material effects can provide either overly conservative predictions, which penalize systems designed for and/or operated under improved conditions, or non-conservative predictions (i.e., creating risk of failure by environmental cracking). Other factors that are unaccounted for include "anomalies" associated with chemically short cracks, metallurgically small cracks, crack-tip shielding (closure), transient crack growth, and net section stresses above the yield strength. These factors are either proscribed in the design, left to the discretion of the engineer, or incorporated (by conservatism) into the code.

Another aspect of corrosion fatigue, of importance both to laboratory studies and component life evaluation, is the difficulty in conceptually or experimentally distinguishing crack initiation from propagation. Many scientists and engineers consider crack initiation to be phenomenologically distinct from crack advance. However, in most instances there is little evidence to support this view, and any "initiation phase" is merely associated with crack detection capability. Thus "initiation" can span the range from complete failure, to 25% load drop in strain-controlled tests, to readily observable in visual examination, to subgrain size. All of these definitions of crack initiation are arbitrary. In many cases the processes that control propagation of long cracks also control very short crack behavior, although there are also instances where crack propagation occurs from localized corrosion (e.g., pitting or grain boundary attack), fracture of inclusions,

and so on, where the processes controlling initiation differ from those that control propagation. In some systems, very small environmentally assisted cracks (e.g., ≤20 μm) have been shown to behave exactly the same as long cracks (e.g., >1 mm) (Ref 12). Clearly the traditional distinctions and delineations associated with "crack initiation" need to be carefully examined.

Similar concerns exist for the traditional concepts of an environmental fatigue threshold (ΔK_{th}) and a threshold stress intensity for (constant-load) stress-corrosion cracking (K_{Iscc}). It has long been known that the nominal (applied) cyclic amplitude (ΔK) can be attenuated at the crack tip by crack closure (crack-tip shielding) (Ref 13), which can be attributed to a variety of phenomena that promote premature contact of the crack flanks, including fracture surface roughness, oxide growth on the walls of the crack, plasticity or phase transformation in the crack wake, and high-viscosity fluids. While usually affecting the behavior primarily at low load ratios (P_{max}/P_{min}), under conditions where copious oxide forms in the crack (e.g., high-temperature water), closure can also occur at high load ratio (Ref 14). The significance of closure is very large, making it difficult to determine whether a "real" (intrinsic) threshold crack tip ΔK_{th}^{ct} exists; closure can shift the observed threshold from less than 2 MPa$\sqrt{m}$ to more than 15 MPa$\sqrt{m}$ (ASTM STP 982, *Mechanics of Fatigue Crack Closure*, 1988).

The role of environment can add substantial complexity by increasing crack growth rates, for example, while perhaps also increasing closure

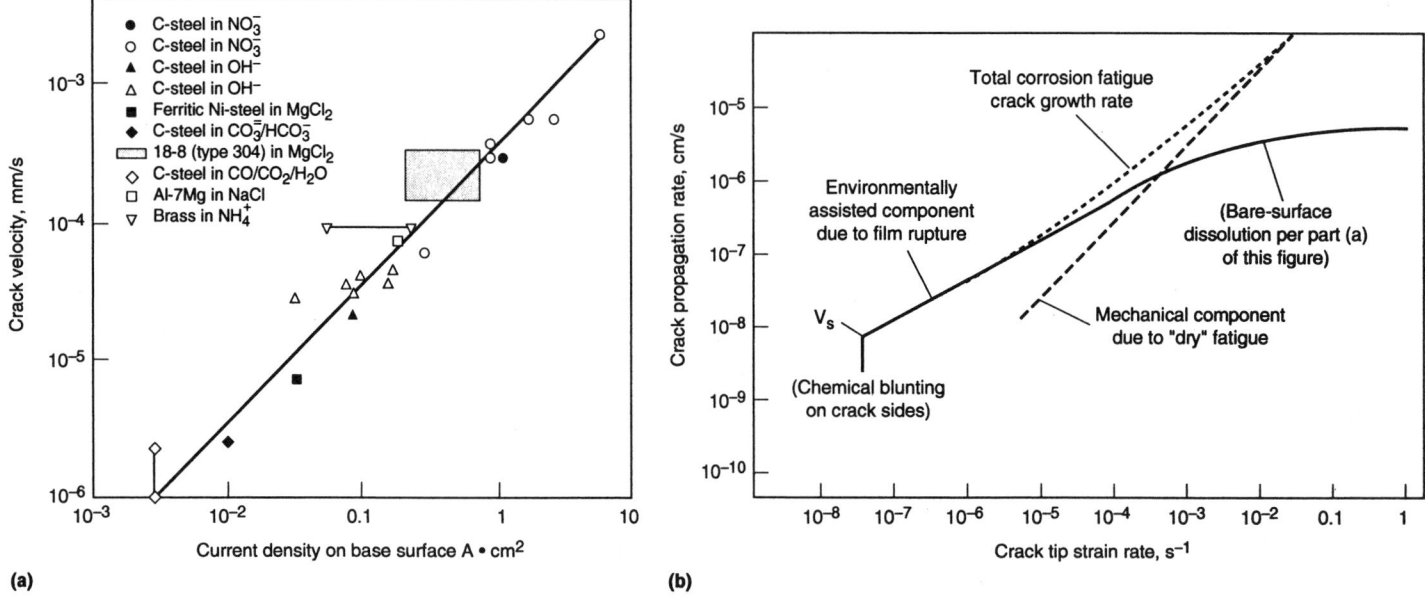

Fig. 6 The relationship between (a) the bare-surface current density and (b) the maximum environmentally assisted crack growth rate for various material/environment systems. This demonstrates the Faradaic agreement in the film rupture/slip oxidation between dissolution kinetics and environmental cracking kinetics. Source: Ref 6

effects, which increases ΔK_{th}. However, the observed ΔK_{th} cannot be considered a constant, given the large effect of environment on oxide formation, oxide solubility (in aqueous systems) (Ref 14), and calcareous deposits (e.g., from sea water) (Ref 15).

These concerns become greater when considering threshold stress intensity for stress-corrosion cracking (K_{Iscc}). No such closure mechanism can be invoked because closure limits the effective *cyclic* amplitude at the crack tip, precisely by maintaining a closure-induced "tare" load at the crack tip. Thus, the maximum load cannot be decreased, and in some instances it may be increased from oxide wedging forces. The origin and significance of K_{Iscc} is controversial, with broad agreement that it is very dependent on test technique (and most test variables) and is rarely, if ever, thermodynamic in origin. In many engineering systems, cracks are observed to grow at stress intensities dramatically lower than the observed K_{Iscc} in laboratory data. There is also increasing awareness that there is a subtle, delicate, and complex interdependence between sustained dynamic strain at the crack tip (which locally disrupts passivity) and the crack advance process itself. Once crack advance stalls, corrosion or small environmental, thermal, or loading fluctuations may be required to reinitiate crack advance.

Fully integrated approaches to mechanistic understanding and life prediction of environmentally assisted cracking are being developed for a variety of systems (Ref 16, 17). Some of these (Ref 18) are specifically designed to address the shortcomings of the traditional codes that address only cyclic-based crack growth (not time-dependent crack growth) and fail to address the continuum in the environmental and material responses, crack initiation, the fundamental role of passivity in most alloy/environment systems, and so on.

General Test Methods

Laboratory fatigue tests can be classified as crack initiation or crack propagation. In crack initiation testing, specimens or parts are subjected to the number of stress (or strain-controlled) cycles required for a fatigue crack to initiate and to subsequently grow large enough to produce failure. In crack propagation testing, fracture mechanics methods are used to determine the crack growth rates of preexisting cracks under cyclic loading. Both methods can be used in a benign environment, or by the combined effects of cyclic stresses and an aggressive environment (corrosion fatigue), as described below. A general review of corrosion fatigue testing is also provided in Ref 19 for these two general methods.

Fatigue Life (Crack Initiation) Testing. In general, fatigue life testing is stress controlled (SN) or strain controlled (ε-N). The test specimens (Fig. 9) are described primarily by the mode of loading, such as:

- Direct (axial) stress
- Plane bending
- Rotating beam
- Alternating torsion
- Combined stress

Testing machines are defined by several classifications: (a) the controlled test parameter (load, deflection, strain, twist, torque, etc.); (b) the design characteristics of the machine (direct stress, plane bending, rotating beam, etc.) used to conduct the specimen test; or (c) the operating characteristics of the machine (electromechanical, servohydraulic, electromagnetic, etc.). Machines range from simple devices that consist of a cam run against a plane cantilever beam specimen in constant-deflection bending to complex servohy-

draulic machines that conduct computer-controlled spectrum load tests.

High-cycle corrosion fatigue tests (performed in the range of 10^5 to 10^9 cycles to failure) are typically done at a relatively high frequency of 25 to 100 Hz to conserve time. Multiple, inexpensive rotating-bend machines are often dedicated to these experiments. Low-cycle corrosion fatigue tests (in the regime where plastic strain, ε_p, dominates) follow from the ASTM standard for low-cycle fatigue testing in air (ASTM E 606) with further technical information provided in Ref 19 and 20.

For aqueous media, the typical cell for corrosion fatigue life testing includes an environmental chamber of glass or plastic that contains the electrolyte. The specimen is gripped outside of the test solution to preclude galvanic effects. The chamber is sealed to the specimen, and solution can be circulated through the environmental cell. The setup should include reference electrodes and counter electrodes to enable specimen (working electrode) polarization with standard potentiostatic procedures. Care should be taken to uniformly polarize the specimen, to account for voltage drop effects, and to isolate counter electrode reaction products. If potential is controlled, control of the oxygen content of the solution may not be necessary (Ref 19), although hightly deaerated solutions are considered prudent.

Environmental containment for high-cycle and low-cycle corrosion fatigue life testing is similar, but the overall setup for low-cycle (strain-controlled) testing is more complicated because gage displacement must be measured. For strain-controlled fatigue life testing in simple aqueous environments, diametral or axial displacement is measured by a contacting but galvanically insulated extensometer, perhaps employing pointed glass or ceramic arms extending from an exten-

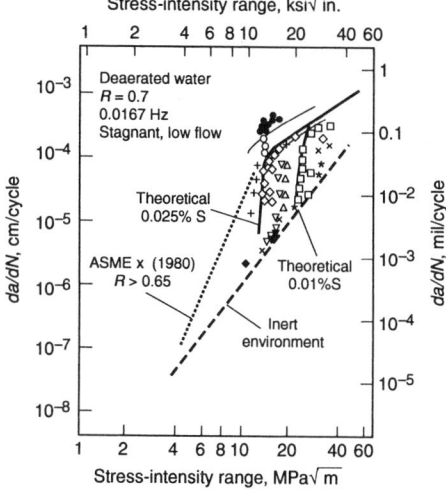

(a)

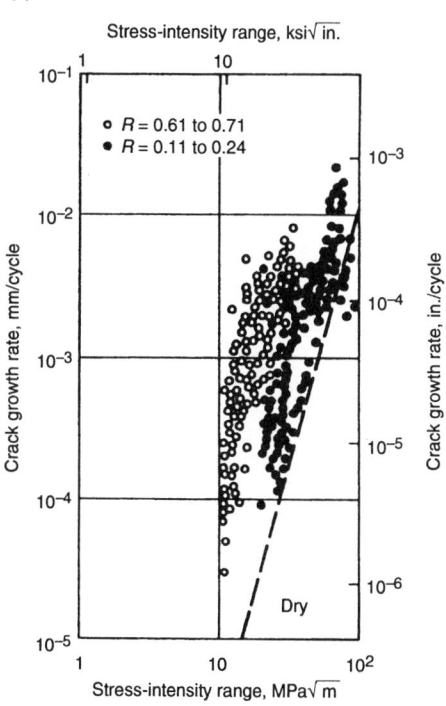

(b)

Fig. 7 High-temperature corrosion fatigue of structural steels in water. (a) Fatigue crack growth of A533B and A508 low-alloy steels with 0.01-0.025% S tested in deaerated high-temperature water at 0.0167 Hz and $R = 0.7$. The nominal ΔK^4 dependency of crack growth rate in inert environments can vary substantially in the environment. Source: Ref 8, 9. (b) Effect of stress ratio on corrosion fatigue crack propagation in A533B and A508 carbon steels exposed to pressurized high-purity water. Temperature 288 °C (550 °F), frequency 0.017 Hz. Average behavior in air is represented by the dashed line labeled "Dry." Source: Ref 10

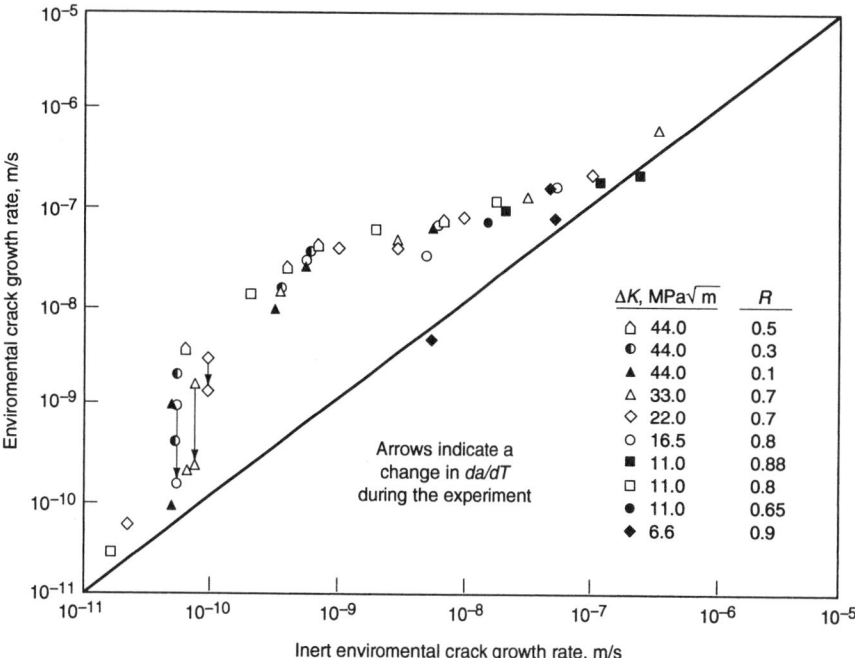

Fig. 8 Corrosion fatigue crack growth rates for A533B low-alloy steel (ASME grade SA533B-1, 0.025% S) in 288 °C pure water. Environmental enhancement is not uniform but reaches a maximum under intermediate ΔK, R, and frequency conditions. Source: Ref 8, 9

someter body located outside of the solution. Hermetically sealed extensometers or linear-variable-differential transducers can be submerged in many electrolytes over a range of temperatures and pressures. Alternately, the specimen can be gripped in a horizontally mounted test machine and be half-submerged in the electrolyte with the extensometer contacting the dry side of the gauge

(Ref 20). For simple and aggressive environments, grip displacement can be measured external to the cell-contained solution, such as for high-temperature water in a pressurized autoclave (Ref 21, 22). It is necessary to conduct low-cycle fatigue tests in air (at temperature), with an extensometer mounted directly on the specimen gauge, to relate grip displacement and specimen strain (Ref 19).

Fracture Mechanics (da/dN) versus ΔK Approach to Corrosion Fatigue. While there is still a strong reliance on smooth-specimen, low- and high-cycle fatigue testing, which is designed to characterize stress or strain amplitude vs. cycles to failure, there is an increasing emphasis on characterizing crack propagation using a fracture mechanics approach. This results from the ambiguities associated with defining or identifying crack "initiation" (addressed above), as well as increasingly successful efforts to unify the two approaches by predicting "initiation" and short crack behavior from a thorough understanding of crack propagation. The advantage of this approach is that corrosion fatigue crack growth (da/dN vs. ΔK) data from laboratory testing is in many cases (though not all, as described below) useable in stress-intensity solutions for practical prediction of component life. For example, Fig. 10 illustrates the predicted 85-year life of a welded pipe based on week-long laboratory measurements of da/dN versus ΔK for steel in an oil environment.

Fracture mechanics is based on the concept of similitude, wherein the stress-intensity factor (K) defines the near-tip driving forces for crack growth and thus is able to characterize crack growth for different geometries and loads. Crack growth rate data also are important to fundamen-

tal studies of corrosion fatigue mechanisms. The fracture mechanics approach isolates crack propagation from initiation and in terms of a precise near-tip mechanical driving force, ΔK. Crack growth rates are related directly to the kinetics of mass transport and chemical reaction that constitute embrittlement. As shown in Fig. 11, prediction of the effect of loading frequency on crack growth rate in salt water (normalized to vacuum) identifies important rate-limiting crack tip electrochemical reactions. Modeling and measurements in Fig. 11 provide a sound basis for extrapolating short-term laboratory data to predict long-term component cracking.

However, the fracture mechanics approach to corrosion fatigue can be compromised by various factors. In addition to the complications arising from crack-tip plasticity (which may affect the assumption of linear, elastic conditions for K) and crack closure effects (which can be accounted for if ΔK_{eff} is known), environmental effects can complicate the requirement of similitude. This is not surprising, because stress intensity is designed to provide only a mechanical description of similitude, which cannot be expected to account for the interaction of chemical and mechanical contributions. Examples of loss of similitude from environmental effects would include any case where a different crack chemistry (or, more generally, chemical contribution to crack advance) develops in small versus deep cracks (where mass transport can vary substantially), or three-sided open cracks (e.g., compact-type specimens) versus 1-side open (thumbnail cracks) (where convection can have a dramatically different effect) (Ref 25).

Another disadvantage of the fracture mechanics approach is that it may not provide a meaningful description of crack "nucleation," especially in cases where cracks are observed to nucleate by processes (e.g., pitting, and corrosion or cracking at inclusions) that are unrelated to crack advance.

Importance of Environmental Definition and Control. The nature and variations of the environment are dominating factors in environmental cracking, and all environments must be considered damaging compared to vacuum or "laboratory air" until proven otherwise. Figure 1(a) shows that, compared to vacuum, the crack propagation rate of a high-strength steel is 4 times higher in moist air, 100 times higher in sodium chloride solutions, and 1000 times higher in gaseous hydrogen. Environmental cracking kinetics tend to be controlled by chemical reaction and transport rates, and much less so by metallurgical variables. For example, the moist air data vary by less than 3 times for a wide range of yield strengths (300 to 2100 MPa) and microstructures (pearlitic, martensitic, and bainitic). The large differences in crack growth rate at constant ΔK correlate with a shift from ductile (reversed slip) transgranular fatigue cracking in vacuum, to brittle intergranular and transgranular cleavage micromechanisms in aggressive environments.

Another example of environmental effects is shown in Fig. 7, 8, and 12 for a low-alloy steel of medium sulfur content tested in high-temperature water, where a very large environmental enhancement in crack growth rate is observed under specific conditions. Figures 7 and 8 highlight the important observation that the environment enhancement is not uniform, for example across the entire range of loading conditions. Indeed, the environmental enhancement tends to decrease at very high loading rates (e.g., at high frequency and ΔK values), and it may also decrease at very low loading rates. Figure 12 shows the impor-

tance of the specific test conditions. Tests at high flow rates on three-side-open compact-type specimens caused the aggressive crack chemistry to be flushed out, resulting in lower crack growth rates.

Key Test Variables

The specific types and influence rankings of experimental variables in corrosion fatigue can vary markedly with specific alloy/environment systems. However, the following factors are crucial in most investigations of corrosion fatigue:

- Stress intensity amplitude (ΔK) or stress amplitude ($\Delta\sigma$)
- Loading frequency (ν)
- Load ratio ($R = P_{min}/P_{max}$) or K_{min}/K_{max})
- Chemical concentration and contaminants (e.g., for aqueous environments: ionic species; pH; and dissolved species/gases, such as oxygen, hydrogen, and copper ion, that influence the corrosion potential)
- Alloy microstructure; yield strength; and often inhomogeneities, such as MnS and other inclusions and second phases, grain boundary enrichment or depletion, etc.

Other variables, such as load waveform, load history, and test temperature may also contribute, but they vary substantially in importance from system to system. Electrode potential should be monitored and, if appropriate, maintained constant during corrosion fatigue experimentation. Often, apparent effects of variables such as solution dissolved oxygen content, flow rate, ion concentration, and alloy composition on corrosion fatigue are traceable to changing electrode potential.

Stress Intensity Amplitude (ΔK). While environmental crack growth rates increase with increasing ΔK, the specific dependency varies

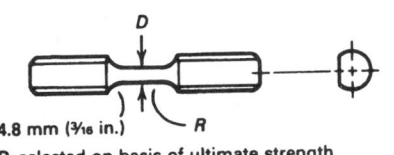

4.8 mm (³/₁₆ in.)   R

D, selected on basis of ultimate strength of material R, 12.7 mm (0.50 in.)

(a)

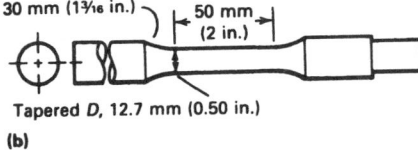

30 mm (1³/₁₆ in.)   50 mm (2 in.)

Tapered D, 12.7 mm (0.50 in.)

(b)

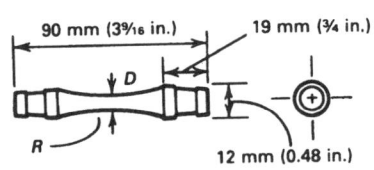

90 mm (3⁹/₁₆ in.)   19 mm (³/₄ in.)
D
R   12 mm (0.48 in.)

D, 5 to 10 mm (0.20 to 0.40 in.) selected on basis of ultimate strength of material R, 90 to 250 mm (3.5 to 10 in.)

(c)

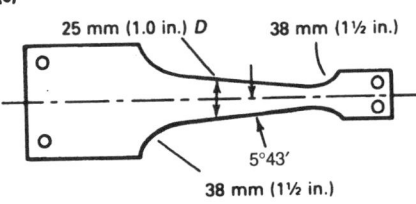

25 mm (1.0 in.) D   38 mm (1½ in.)
5°43'
38 mm (1½ in.)

(d)

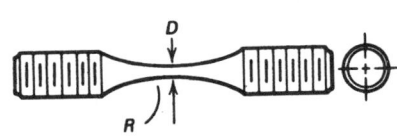

D
R

D, selected on basis of ultimate strength of material R, 75 to 250 mm (3 to 10 in.)

(e)

Fig. 9 Typical fatigue life test specimens. (a) Torsional specimen. (b) Rotating cantilever beam specimen. (c) Rotating beam specimen. (d) Plate specimen for cantilever reverse bending. (e) Axial loading specimen. The design and type of specimen used depend on the fatigue testing machine used and the objective of the fatigue study. The test section in the specimen is reduced in cross section to prevent failure in the grip ends and should be proportioned to use the upper ranges of the load capacity of the fatigue machine (i.e., avoiding very low load amplitudes where sensitivity and response of the system are decreased).

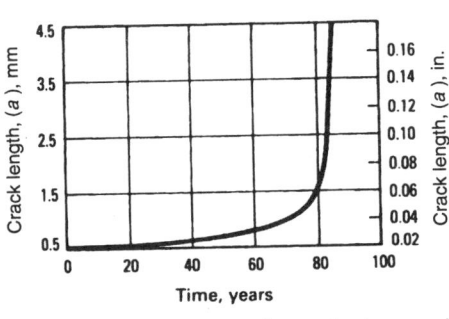

Fig. 10 Predicted fatigue crack extension from a weld toe crack in an API 5LX52 carbon steel pipeline carrying hydrogen-sulfide-contaminated oil. Temperature 23 °C (73 °F). Source: Ref 23

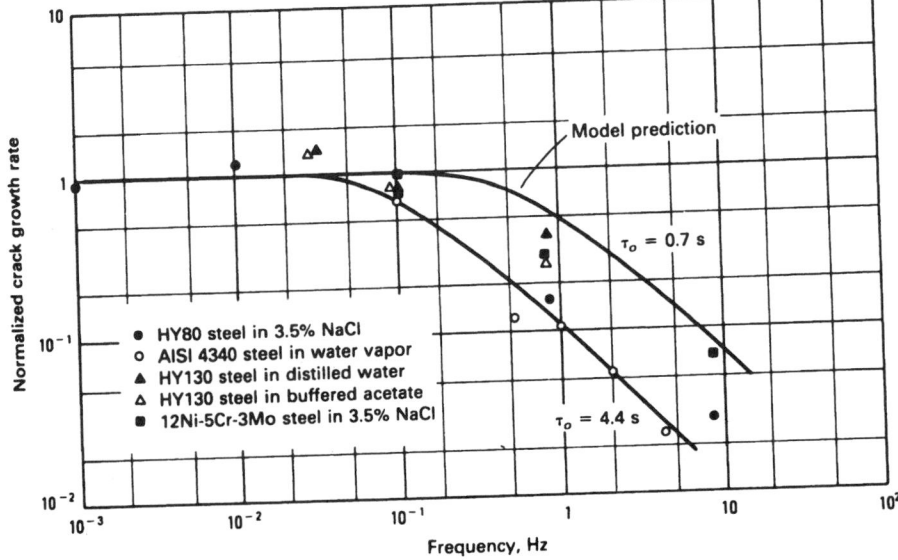

Fig. 11 Modeled effect of loading frequency on corrosion fatigue crack growth in alloy steels in an aqueous chloride solution. The determination of the normalized crack growth rate and the time constants, τ_o, from the model can be found in Ref 24.

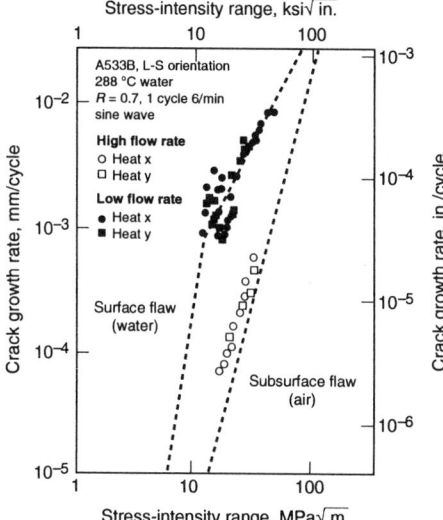

Fig. 12 The effect of solution flow rate on the corrosion fatigue crack growth rate of a medium-sulfur, low-alloy steel tested in deaerated 288 °C (550 °F) water. Tests at high flow rate on the 3-side-open compact-type specimens permit the aggressive crack chemistry to be flushed out, reducing the crack growth rates. Source: Ref 8, 9

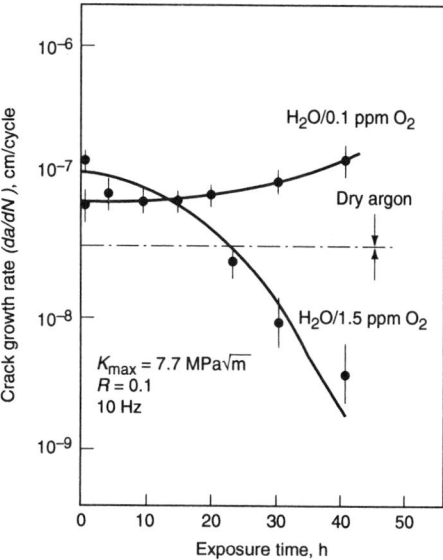

Fig. 13 Cycle-based corrosion fatigue crack growth rates vs. time for an SA333-grade 6 ASME carbon steel tested in 97 °C water. At 0.1 ppm dissolved oxygen, the corrosion rate is low, the crack tip remains sharp, and cracking is sustained. At 1.5 ppm dissolved oxygen, considerable corrosion occurs, the crack tip becomes blunted, and the crack growth rate decays. Source: Ref 26

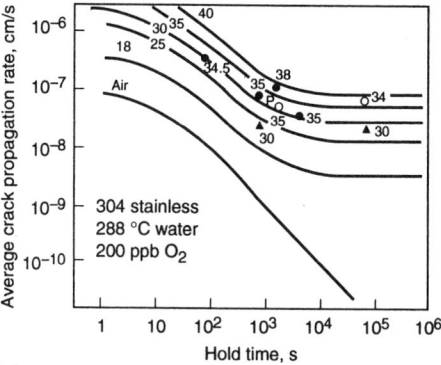

(a)

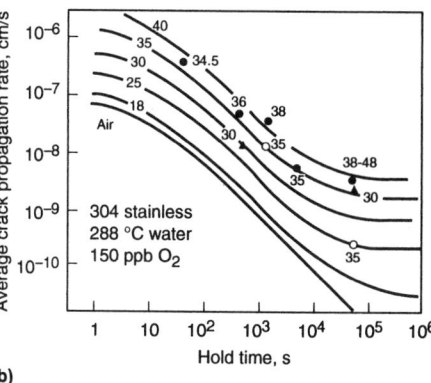

(b)

Fig. 14 Comparison of observed (data) and predicted (curves) crack growth rates for sensitized type 304 stainless steel in 288 °C water. Thumbnail crack specimens were loaded using trapezoidal loading patterns of varying hold time at the maximum stress intensity with net section stresses above yield. Theoretical relations for various net section stresses (in ksi) as noted by numbers. K(max) = 16.5 MPa$\sqrt{m}$ (15 ksi$\sqrt{in.}$); R = 0.1; loading rise and fall times of 5s; E(corr) = +125 mV(SHE); and 0.1 µS/cm. (a) In 200 ppb dissolved oxygen. (b) In 150 ppb dissolved hydrogen. Source: Ref 6

greatly. In some environments, the effect of environment is merely to offset the observed crack growth rate by some fixed factor above the inert rate (e.g., Fig. 1(a) for moist air vs. vacuum; Fig. 7(a), 12 for low-alloy steel in high-temperature water). However, there is often a profound shift in the dependence of ΔK, typically producing a reduced ΔK dependence in aggressive environments, at least in the intermediate region where power law behavior is observed. It is always important to examine the entire relevant ΔK regime, not assuming the observed enhancement at a specific ΔK.

Environments do not always enhance the crack growth rate. The most common origins of crack retardation are associated with increased crack closure and crack blunting. Crack closure is most often increased by thicker oxides and perhaps the rougher (i.e., intergranular, with secondary cracks) fracture surface (Ref 13, 14). Crack blunting results from aggressive environments that result in inadequate passivity. If the flanks of the crack are not adequately passive, then the crack tip will not remain sharp. This has been observed in low-alloy and carbon steels in hot water (Fig. 13) and in other systems.

Shifts in ΔK, K_{max}, or load ratio during testing should be made very gradually, preferably continuously (e.g., under computer control). Changes in K should be limited to less than 10%, preferably much less. Any large change in growth rate should be confirmed using increments of <1%. Data may differ for rising K versus K-shedding conditions. Crack increments should be sufficient to provide statistically significant crack growth rates (e.g., >10 times above the crack length resolution) and should account for effects of plastic zone size under prior conditions during K-shedding. Shifts in frequency and hold time are

not as restrictive, although changes greater than 3 to 10 times can lead to anomalous results.

The presence of an environment can also shift the dependence on stress amplitude ($\Delta\sigma$) or plastic strain amplitude ($\Delta\varepsilon$), not only by decreasing the stress at which a certain cyclic life can be attained, but also by eliminating the stress amplitude threshold altogether (Fig. 2). This, and increased scatter in the data, can lead to differences in estimating environmental effects at different stress amplitudes (Fig. 3). Note also that there is a consistent trend versus time in which the "bounding" curves are periodically shifted lower and to the left in Fig. 3.

Loading Frequency (ν). Because the environment induces a significant time-dependent response, environment enhancement can vary markedly with loading frequency. At high frequency it is common for the environmental enhancement to be substantially eliminated because of inadequate time available for associated chemical reaction and mass transport kinetics.

Transitions in significant environmental enhancement are often apparent when plotting crack growth rate versus frequency or hold time. For example, in Fig. 11 at frequencies below about 0.1 Hz, time-dependent processes completely dominate and there is no effect of loading frequency (i.e., crack growth is not controlled by cycling and would be high at constant load). In contrast, above about 0.1 Hz there is little time dependency, and growth rates are proportional to frequency.

Strong frequency effects are observed in most corrosion fatigue systems. In high-temperature water (Fig. 14), behavior similar to that in Fig. 11 exists, although it is plotted versus loading period or hold time (Ref 6) rather than frequency. Pre-

dictive modeling has been quite successful in accounting for the transition between cycle- and time-dependent behavior as a function of corrosion potential, water purity, and degree of sensitization of the stainless steel (Ref 6, 18).

Load Ratio (R). At higher load ratios (P_{min}/P_{max}), corrosion fatigue crack growth rates are usually higher than in inert environments. This can be viewed as a mean stress effect, and the greater environmental enhancement can be considered to result from the expected increase in contribution of time-dependent crack advance that would occur even under static load conditions. Figure 15 shows the effect of load ratio in low-alloy steel tested in high-temperature water at 0.017 Hz. The increased K_{max} associated with testing at R = 0.7 (e.g., K_{max} = 66.7 MPa$\sqrt{m}$ for ΔK = 20 MPa$\sqrt{m}$) compared to 0.2 (K_{max} = 25 MPa$\sqrt{m}$) is substantial, and it is consistent with an increase in crack-tip strain rate and thereby an increase in the frequency of rupture of the protective oxide film and high growth rates. As expected, the effect of load ratio is frequency dependent. Also, if plotted versus K_{max} rather than

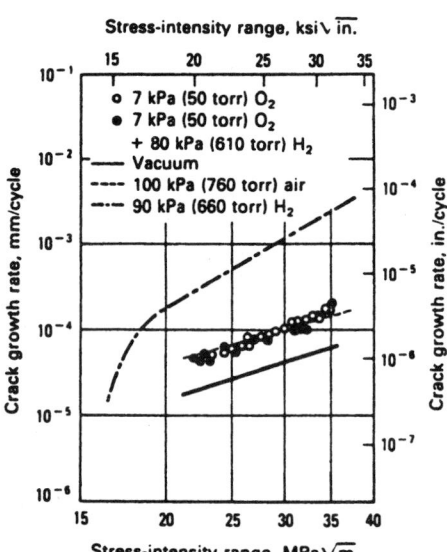

Fig. 15 Corrosion fatigue crack growth rates plotted for medium-sulfur A533B and A508-2 low-alloy steels and weldments in 288 °C deaerated (pressurized water reactor primary) water. Data show a stronger environmental effect at $R = 0.7$ than at $R = 0.2$. Source: Ref 8, 9

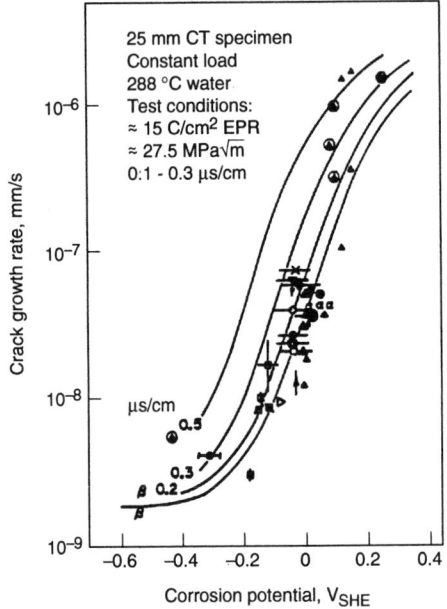

Fig. 17 The effect of dissolved oxygen on the corrosion potential of type 304 stainless steel in 274 °C high-purity water. Important effects on corrosion potential and crack growth rate (Fig. 18) occur at ppb levels of dissolved oxygen, a small fraction of the oxygen-saturated value of ≈42 ppm at standard temperature and pressure. Source: Ref 6

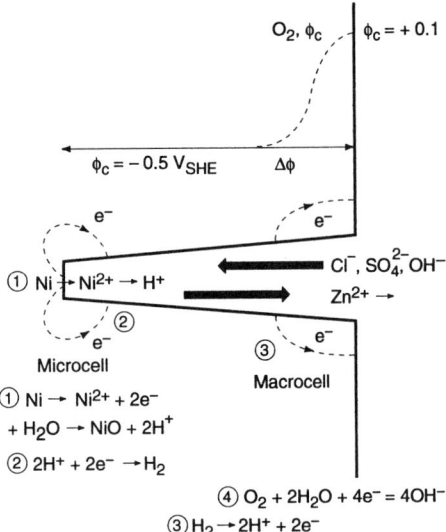

$①$ $Ni \rightarrow Ni^{2+} + 2e^-$
$+ H_2O \rightarrow NiO + 2H^+$

$②$ $2H^+ + 2e^- \rightarrow H_2$

$④$ $O_2 + 2H_2O + 4e^- = 4OH^-$

$③$ $H_2 \rightarrow 2H^+ + 2e^-$

$$J_A = - D_A \Delta C_A - z \mu C_A F \Delta \phi + C_A V$$

Flux = diffusion + ϕ -driven + convection

Fig. 19 Schematic of crack showing the differential aeration macrocell that establishes the crack-tip chemistry and the local microcell that is associated with metal dissolution and crack advance. Because the differential aeration macrocell is not essential to elevated crack growth rates, some coupling of the currents associated with these two cells may occur, but this is unnecessary. Source: Ref 25

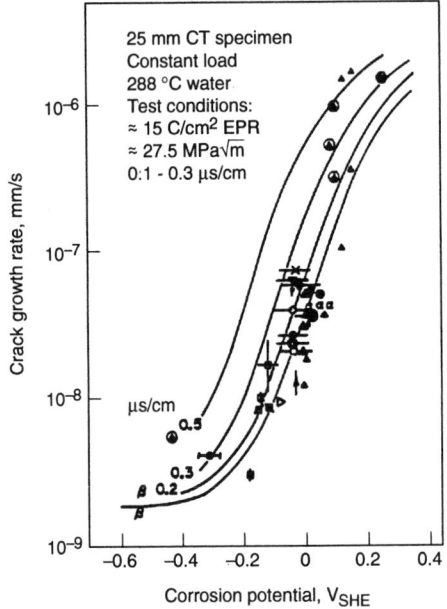

Fig. 16 Effect of oxygen (O_2) contamination on gaseous hydrogen embrittlement of a low-strength AISI/SAE 1020 carbon steel. Frequency 1 Hz. Source: Ref 27

Fig. 18 Comparison of observed (data) and predicted (curves) crack growth rate response of sensitized type 304 stainless steel in 288 °C water. Circled data points represent measured corrosion potentials and crack growth rates under irradiated water chemistry conditions. Source: Ref 6, 18

ΔK, higher crack growth rates should always result with decreasing load ratio.

Test Environment and Chemical Contaminants. Besides the obvious concern of primary species (such as NaCl concentration for salt water) in corrosion fatigue, small amounts of contaminants are also a key variable. A striking example (Ref 27) of an environmental-purity effect is illustrated in Fig. 16 for gaseous hydrogen embrittlement of a low-strength carbon steel. Relative to vacuum, crack growth is accelerated by factors of 3 and 25 for moist air and highly purified low-pressure hydrogen gas, respectively. Small additions of oxygen to the hydrogen environment essentially eliminate the brittle corrosion fatigue component to crack growth, consistent with a trend first reported by Johnson (Ref

28). Similar effects have been reported for carbon monoxide and unsaturated hydrocarbon contamination of otherwise pure hydrogen environments. In aqueous environments, the effects of bulk ionic concentration and pH are often quite pronounced (especially in unbuffered systems), although dissolved oxidants are often of greater consequence (e.g., dissolved oxygen, hydrogen peroxide, and copper and iron ions), as are contaminants (e.g., dissolved sulfur, chloride, lead, mercury).

The primary role of oxidizing and reducing species, especially dissolved oxygen and hydrogen, is in shifting the corrosion potential. Some species, such as nitrate, may directly influ-

ence crack chemistry and, if reduced to ammonia, can be directly responsible for environmental enhancement (e.g., of brasses). In many cracking systems, the role of oxidants (elevated corrosion potential) is an indirect one, because inside the crack the oxidants are generally fully consumed and the corrosion potential is low (Ref 25). In such systems, the role of oxidants is to create a potential gradient, usually near the crack mouth, that causes anions (e.g., Cl^-) to concentrate in the crack and causes the pH to shift.

Oxidants increase the corrosion potential in aqueous environments, which can have very pronounced effects on environmental enhancement. This can occur at exceedingly low concentrations; in high-temperature water, crack growth rates can increase by orders of magnitude merely from the presence of parts-per-billion levels of dissolved oxygen in water (Fig. 17, 18); this is also evident in Fig. 14. Similar enhancements are observed for small concentrations of aqueous impurities (e.g., <10 ppb of sulfate or chloride) or MnS inclusions in low-alloy and carbon steels, which dissolve within the crack to form sulfides. This usually is associated with the formation of a differential aeration cell by complete oxygen consumption within the crack (Fig. 19). Thus, even very small cracks usually advance under deaerated conditions, and the gradient in corrosion potential that is formed from crack mouth to crack tip causes an increase in anion concentration and a shift in pH in the crack. The shift is often acidic, but not necessarily so, because it requires the presence of non-OH^- anions to balance the acidity (H^+). Thus, if only OH^- is present (e.g., from NaOH), the pH shift can only be in the alkaline direction.

Potentiostats can be used to control the specimen potential, although their use (which rarely directly simulates the real situation) can provide misleading data. Primary concerns are associated with:

- *Voltage drop in solution:* The reference electrode reading may be biased by the potential distribution in the solution associated with passage of ionic current.
- *Failure to polarize the crack tip:* Even in highly conductive solutions, the crack-tip potential is rarely significantly affected by external polarization, and therefore crack advance does not occur at the potential that is controlled on the specimen surface.
- *A reversal of surface reactions and shifts in local pH:* In solutions containing oxygen, the reaction on the external surface is cathodic, and inside the crack, anodic reactions occur. With a potentiostat, as the specimen is polarized to more positive potentials, it becomes more anodic and causing oxidation reactions to become predominantly on the metal surface, which can alter the local pH. Cathodic reactions occur predominantly on the relatively remote counter electrode.

To accurately measure potentials, commercial or custom-built reference electrodes are used. Both to measure potentials accurately and to prevent galvanic coupling, it is desirable to electrically isolate the specimen from the linkage and surrounding metal surfaces. If the environment is not very conducting or the potentials of surrounding metal surfaces are not too different from the specimen, electrical isolation may not be critical. However, it is then necessary to place the reference electrode *much closer* to the test specimen than to other (electrically connected) metal surfaces.

Other concerns for the environment include:

- Specimen and grip design (e.g., to avoid failure at crevices by minimizing stress in creviced regions)
- Proper design of environmental cells (e.g., to avoid contamination from leachants from or diffusion of oxygen through plastics)
- Maintenance of proper chemistry, which often requires refreshed/flowing systems for controlling the chemistry in gases or liquids
- Proper stability and measurement of temperature (near the specimen)
- Proper and thorough monitoring/recording of all relevant chemical and electrochemical parameters

Metallurgical Variables. Microstructure and alloy strength influence fatigue crack propagation in embrittling gases and liquids. In general, brittle corrosion fatigue cracking is accentuated by:

- Impurity (e.g., phosphorus or sulfur) segregation at grain boundaries

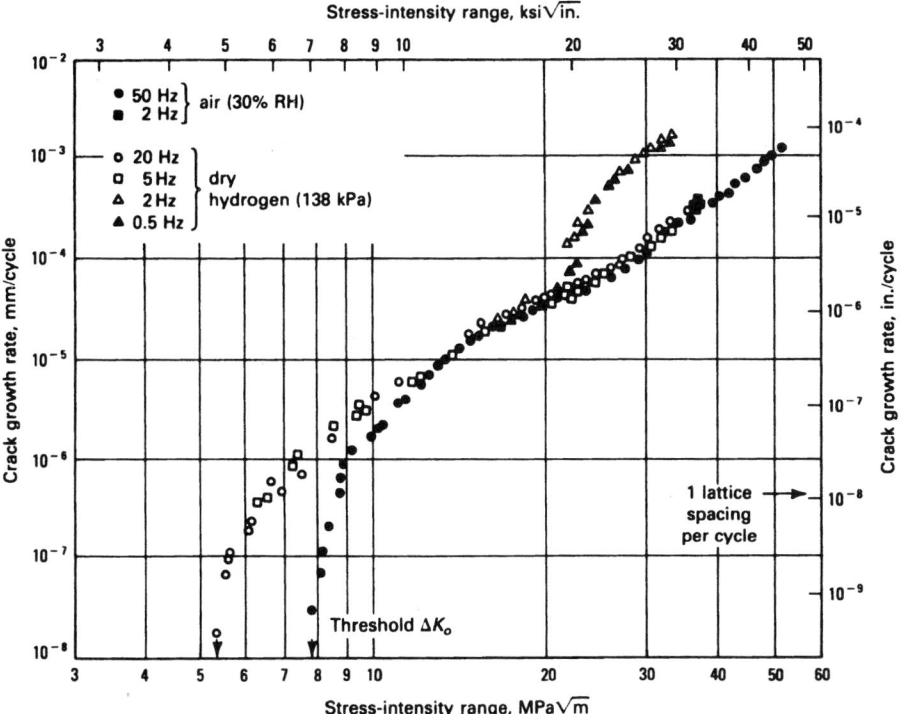

Fig. 20 Corrosion fatigue in 2.25Cr-1Mo pressure-vessel steel in hydrogen due to hydrogen embrittlement at high ΔK and to reduced oxide-induced closure at low ΔK. Stress ratio = 0.05.

- Solute depletion or sensitization (e.g., chromium) about grain boundaries
- Planar deformation associated with ordering or peak aged coherent precipitates
- Increased yield strength or hardness
- Large inclusions (e.g., MnS)

The effects of alloy composition, grain size, and microstructure (e.g., bainitic vs. martensitic steel) vary with environment and brittle cracking mechanism. Laboratory experiments are necessary to establish specific trends. For example, sulfide (MnS) inclusions have already been discussed in connection with accelerated environmental cracking of low-alloy and carbon steels in hot water (Fig. 7, 8, 12, 15).

Yield strength plays a large role in environmental cracking, which has been attributed both to enhanced crack-tip strain rate as well as to complete shifts in the crack advance mechanism. The importance of hydrogen embrittlement in higher-strength materials has been confirmed by experiments in gaseous hydrogen. Its direct role in environmental crack advance is considered to be limited to about 150 °C in iron- and nickel-base alloys, although it may have an indirect role at higher temperatures.

Similar effects have been observed for bulk or surface cold work, which raises the yield strength. Thus, machining and surface treatments such as shot peening can significantly affect cracking. Shot peening and related treatments that produce surface compressive stresses can be very beneficial, provided that cracks do not exist (or form) and that tensile stresses do not exceed them. If sufficient strain occurs, transgranular

cracks often nucleate in the surface-hardened region.

Other important microstructure factors include gamma prime or delta phases on grain boundaries of nickel alloys, martensite formation in steels, carbide formation (sensitization) in stainless steel and nickel alloys, and inhomogeneities (e.g., MnS and nonmetallic inclusions). These often have an even larger role in corrosion fatigue than under inert conditions, and a uniform microstructure or distribution of inhomogeneities can rarely be assumed.

Crack Closure Effects. Premature crack surface contact during unloading, or "crack closure," can greatly reduce rates of fatigue crack propagation. The true (or effective) crack-tip driving force is reduced below the applied ΔK because of the reduced crack-tip displacement range. Closure phenomena are produced by a variety of mechanisms and are particularly relevant to fatigue crack propagation in the near-threshold regime, after large load excursions, or for corrosive environments.

Two mechanisms of crack closure are relevant to corrosion fatigue. Rough intergranular crack surfaces (typical of environmental embrittlement) promote crack closure, because uniaxially loaded cracks open in a complex three-dimensional mode, thus allowing for surface interactions and load transfer. Roughness-induced closure is most relevant to corrosion fatigue at low ΔK and at stress-ratio levels where absolute crack opening displacements (0.5 to 3 μm) are less than fractured grain heights (5 to 50 μm).

Alternately, crack closing is impeded by dense corrosion products within the fatigue crack. For mildly oxidizing environments, such as moist air, this closure mechanism is relevant at low stress-intensity levels and contributes to the formation of a "threshold," as described by Suresh and Ritchie (*Inter. Metals Review*, Vol 29, 1984, p 445-476).

For corrosive bulk environments or localized crack solutions, cracking at high ΔK values may be retarded below the growth rates observed for air or vacuum due to corrosion product formation within the crack. The engineering significance of beneficial crack closure influences depends on the stability of the corrosion product during complex tension-compression loading and fluid conditions.

The complexity of predicting environmentally enhanced cracking is emphasized by data contained in Fig. 20 for a 2.25Cr-1Mo steel (ASTM A542, class 3) stressed in hydrogen at 23 °C (73 °F). At high ΔK levels and low loading frequencies, cracking is accelerated in hydrogen compared to moist air, due to classic hydrogen embrittlement. For low stress intensities, cracking is also enhanced by hydrogen exposure; however, the effect is not due to chemical embrittlement because of the rapid loading frequencies. Oxides form on crack surfaces through a fretting mechanism during cycling in air. Crack growth rates are reduced by oxide-induced closure. Thus crack growth at low frequencies may increase by embrittlement but decrease from increased oxide formation and closure.

Corrosion Fatigue Crack Growth Test Methods

Standard methods of fatigue crack growth (as defined in ASTM E 647 and described in the article "Fatigue Crack Growth Testing" in this Volume) are generally applicable to corrosion fatigue crack growth tests. ASTM E 647 (1991) also contains an appendix specific to crack growth in marine environments. Procedures for other corrosion fatigue environments are not standardized, but various methods have evolved. Some general aspects of corrosion fatigue crack growth are described below, and additional background is provided in Ref 19.

Three problem areas are relevant to corrosion fatigue experimentation. The environment must be contained about the cracked specimen without affecting loading, crack monitoring, or specimen-environment composition. Parameters such as environmental purity, composition, temperature, and electrode potential must be monitored and controlled frequently.

Secondly, the deleterious effect of low cyclic frequency dictates that crack growth rates must be measured at low (often <0.2 Hz) frequencies, which lead to long test times, often from several days to weeks. Load-control and crack-monitoring electronics and environment composition must be stable throughout long-term testing.

Thirdly, crack length must also be measured for calculations of stress intensity and crack growth rate. Optical methods are often precluded by the

environment and test chamber. Indirect methods, based on specimen compliance or electrical potential difference, have been applied successfully to monitor crack growth in a wide variety of hostile environments and are described in more detail below. Experimental and analytical requirements, however, are complex for indirect crack monitoring.

Test Specimens. Finally, specimen geometry and size requirements for ΔK-based crack propagation data, which are scaleable to components through similitude, have not been established completely for subcritical crack growth. In-plane yielding must be limited to the crack tip by guaranteeing that the net section stress is below yield and that the maximum plastic zone size, defined as $\sim 0.2~(K_{max}/\sigma_{ys})^2$, is much less (e.g., 10- to 50-fold) than the uncracked ligament. Specimen thickness, as it influences the degree of plane-strain constraint, and crack size, as it influences the chemical driving force, may affect corrosion fatigue crack speeds. Currently, such effects are unpredictable; specimen thickness and crack geometry must be treated as variables.

Corrosion fatigue testing often is performed at low cyclic frequencies. Consequently, multiple test stations are desirable. For this reason and for general economy, compact tension specimens are frequently used. Such specimens minimize the applied load required to achieve a given crack tip stress intensity, thus permitting the use of low load capacity and less expensive test machines.

In corrosion fatigue, the electrochemistry within the crack is mass transport dependent and can vary with crack depth, and possibly also with specimen geometry and with accessibility of solution in the through-thickness direction via the crack sides. These factors can influence crack growth rates despite the constancy of the range of the stress-intensity factor. The application of fracture mechanics to corrosion fatigue must therefore be considered carefully, because in some circumstances the basic concepts may be invalid. In reports of test data, information regarding crack depth should be quoted or be deducible from the data. An alternative is to determine crack depth and ΔK as independent variables to verify that ΔK is an adequate characterizing parameter for the rate of crack growth.

In applying load to specimens in a test cell, cell friction must not affect load in sealed systems. This is generally not a significant factor in most ambient-temperature applications, however. Insulation between specimens and grips, pin assemblies, and so forth is essential to avoid galvanic effects, but greases should not be used.

Electrode Potential

Monitoring and reporting the electrode potential during corrosion fatigue experiments is important. The potential should be measured using a reference electrode located in the bulk solution adjacent to the specimen. When impressed currents are applied to the specimen, measurement should be made adjacent to the surface using a Luggin capillary to minimize the potential drop between the reference electrode and the metal

surface, the magnitude of which will depend on the solution conductivity and flow of current.

Selection of a reference electrode depends on the particular application, but those most commonly used in laboratory room-temperature tests are the saturated calomel electrode and the silver/silver chloride electrode. For some solutions in which contamination with chloride is undesirable, use of a mercury/mercurous sulfate reference electrode is an option. Contamination can be reduced by using commercially available double-junction electrodes, in which the outer jacket is filled with test solution.

In quoting measured potentials, the potential should be referred to a standard scale such as the standard hydrogen electrode (SHE) or the saturated calomel electrode (SCE) at 25 °C (75 °F). In tests remote from 25 °C (75 °F), allowance must be made for the fact that the half cell potential of the reference electrode varies with temperature. A high-impedance meter ($>10^{12}~\Omega$), such as an electrometer or a pH meter, should be used for monitoring potential, although periodic (short-term) measurements can usually be successfully performed using digital voltmeters whose input impedance is $\geq 10^9$ ohms (usually limited $\leq 2V$ full scale dc ranges).

Near room temperature, it generally is possible to use commercial reference electrodes such as calomel and silver chloride electrodes; some electrode designs permit use near boiling. Designs that place the reference electrode in a separate chamber at a different temperature than the test solution are complicated by formation of a thermal junction potential in the electrolyte, the magnitude of which may be large (over 0.1 V).

At temperatures over boiling, a custom reference electrode generally is necessary. Most investigators use internal or external silver/silver chloride reference electrodes. For internal electrodes, the silver chloride reaction occurs at the test temperature. For external electrodes, the silver chloride reaction occurs at room temperature, but system pressure is applied (so no streaming potentials form), with a temperature gradient occurring in the potassium chloride electrolyte as it enters the autoclave. A porous junction in the autoclave isolates the potassium chloride electrolyte from the autoclave solution. This thermal junction potential has been well characterized over a range of temperatures and potassium chloride concentrations (*Electrochem. Society*, Vol 126, 1979, p 908).

Potentials should be reported on the standard hydrogen electrode (SHE) scale, particularly for elevated-temperature tests, for which the conversion factors to $V_{(SHE)}$ are not widely known. However, when comparing results as a function of temperature, it may be helpful to eliminate the contribution of the standard hydrogen cell, because like other reactions, it has a potential that varies with temperature. It is by convention that the standard hydrogen cell is 0 V at any temperature; relative to the standard hydrogen reaction at 25 °C (75 °F), the potential of the standard hydrogen reaction is about 0.021 V at 50 °C (120 °F), 0.057 V at 100 °C (212 °F), 0.086 V at 150 °C

(300 °F), and about 0.105 V between 210 and 300 °C (410 and 570 °F).

Application of imposed potential using a potentiostat requires electrical isolation of the specimen. In some cases, it may be difficult to insulate the specimen from the loading linkage; instead, the linkage must be insulated from the autoclave, and the measured current flow cannot be attributed only to reactions at the specimen. Ground loops present perpetual problems, because most potentiostats are designed to hold the specimen at ground (or virtual ground). With necessary mechanical and plumbing connections, the autoclave is usually connected to ground; thus, the specimen is effectively connected to the autoclave. The problem is compounded if the autoclave is used as the counter electrode, because the ground loop shorts out the potentiostat. Options include thorough electrical isolation of the autoclave from ground and use of a fully floating potentiostat.

Attachment of a lead to the specimen to permit measurement or application of potential can be a challenge in aggressive environments. Recommendations include use of wire that is either identical to the specimen or a very noble metal, such as platinum. Attachment using a weld bead (e.g., by gas-tungsten arc welding) usually is superior to spot welding. Covering the lead wire with heat-shrink Teflon and, at low test temperatures, covering the weld with an organic "stopoff" coating helps maintain a good connection and minimizes the effects of the wire (via galvanic coupling or its contribution to the measured current). Another technique involves the use of a commercially available plasma-sprayed insulating coating, which can also be used in high-temperature water.

Errors in the potential applied by a potentiostat can occur in solutions of low conductivity. These iR drops, which result from current flow between the counter electrode and working electrode in the high-resistance solution, are detected by the reference electrode and summed with the electrode potential of the specimen. Electronic compensation is possible, but not straightforward in most high-resistivity media. Partial compensation is possible by placing the reference electrode near the specimen, although for small specimens the measured potential becomes very sensitive to electrode positioning. A rough estimate of the possible error can be made by multiplying the resistivity of the solution (preferably determined by measuring the alternating current (ac) flowing between the counter electrode and working electrode when a known 1000 Hz ac voltage is applied) by the potentiostat current that flows during a test.

Monitoring Crack Length

The electrical potential technique and the compliance method (discussed elsewhere in this Volume) are frequently used to monitor fatigue crack growth in solution and in gaseous environments. Visual methods generally are not practical; often, the crack and the test specimen are obscured by the test chamber, or a microscope with a long focal length is needed.

The electrical potential technique is preferred over the compliance method for use inside an environmental test chamber, because the compliance gage may outgas and is a potential source of test environment contamination. Its use in a corrosive environment is also unsuitable. The electric potential technique, however, is noncontaminating and can be used in most environments.

Use of the compliance method is generally limited to compact tension (CT, or compact type) and wedge-opening load specimens. It is not used for center-cracked tension specimens because of limitations in sensitivity and accuracy. The electrical potential technique can be readily applied to all three specimen types. The principal drawback of the electrical potential technique is that the specimen must be electrically conductive; thus, it cannot be applied directly to specimens made of nonelectrically conducting materials, such as polymer-based composites and ceramics. In addition, electrical shorting across the crack surfaces may affect its measurement accuracy, particularly for tests in vacuum. Both the electrical potential and compliance techniques can be readily interfaced with a computer for real-time control of the experiment and for online data acquisition and reduction.

Potential Drop Method. The most common and sensitive in situ crack monitoring technique is reversing direct current (dc) potential drop, which typically applies a constant current to a specimen and measures the changes in potential across the specimen as the crack grows. The current is reversed between readings to eliminate thermoelectric effects, amplifier offsets, and so on. A digital voltmeter is recommended, preferably with an integration time of more than one power line cycle. Hundreds or thousands of readings are often averaged to reduce noise and improve the sensitivity. Some investigators use a "reference probe" pair, whose potential is as insensitive as possible to crack growth, to normalize any changes in current or temperature (which changes the resistivity of metals). Usually, solving the problem at its origin (i.e., using very stable constant current sources; using high-quality, well-tuned digital temperature controllers; and ensuring stable room temperature) works better.

In the environment, there are several special considerations. Solution conductivity can be a major issue; an extreme example is the inability to use potential drop in liquid metal environments. Some deviations in crack length versus measured potential response can also occur in highly conducting environments (e.g., aqueous solutions), and it must be recognized that the crack chemistry can be substantially more conductive and at different pH than the bulk solution. However, despite the small distance between the upper and lower crack flanks, the role of ionic (e.g., aqueous) conductivity is not large compared to that of metal conductivity, because aqueous conductivities are typically measured in 10^{-1} to 10^{-6} S/cm (S, or Siemen, is Ω^{-1}), whereas metal conductivities are typically between 10^5 and 10^6 S/cm. Thus, errors associated with aqueous environments are relatively small, although not always ignorable.

Another concern relates to inaccuracies in indicated crack length because of a nonuniform crack front or because of metal contact along the crack flank during the fatigue cycle. In both cases, an abnormal fraction of the dc current "shorts" through the uncracked metal ligament in the wake of the nominal crack front, and the measured potential and indicated crack length is strongly affected. For example, if the crack front moves forward in a 25 mm compact-type specimen by 3 mm in all locations except along one narrow, rectangular ligament that is only 1 mm wide, the indicated crack advance by potential drop can be very small (i.e., dramatically less than the area average of crack advance). Nonuniform crack fronts are much more common when the environmental contribution to crack advance is high, and static loading (stress-corrosion cracking) is generally much worse than dynamic loading (e.g., corrosion fatigue). Certain microstructures, such as weld metal, can be quite susceptible to accelerated or retarded crack advance in localized regions (i.e., along certain weld dendrites). The "unzipping" of the final metal ligament can lead to anomalously high "apparent" crack growth rates over certain testing periods.

In many instances, the importance of metal contact from the upper to lower crack flank surface during the fatigue cycle is low because of the formation of surface oxides of very limited electrical conductivity. However, in vacuum, hydrogen, or sufficiently reducing aqueous conditions (e.g., nickel-base alloys in hot water containing sufficient dissolved hydrogen), little or no oxide forms, and contact and shorting of dc current occurs during the unloading cycle. In these cases, either a different crack monitoring technique should be used, or potential drop measurements should only be made at maximum load. Even so, making environmental conditions more reducing (e.g., increasing the hydrogen fugacity in high-temperature water testing of nickel-base alloys) can lead to an indicated decrease in crack length by potential drop because of the change in the insulating characteristic of the oxides.

Other concerns for dc potential drop include electrochemical effects, particularly polarization. If a well-designed, ground isolated power supply is used, then all of the dc current that leaves the "+" terminal must return on the "−" terminal, and direct polarization of the specimen is not possible. In most cases, there is little basis for concern for the electrochemical effects of using dc potential drop, although, for example, the small potential difference between the crack flanks could have some influence in tight cracks in conductive solutions. This potential difference is very small near the crack tip, so it is more likely to influence, for example, dissolution of MnS inclusions at some distance toward the crack mouth, where the potential difference across the crack flanks is higher. While the potential difference between the upper and lower surfaces of the crack is small (typically 100 µV in many potential drop implementations), the gradient can be relatively large because of the small separation of the crack faces. The importance of this issue can be quantified by establishing a steady-state crack growth rate and disconnecting the potential drop system for a period of time, then reconnecting it to evaluate its effect (or by comparing a duplicate experiment using an extensometer to monitor the crack growth rate).

Electrochemical effects can also result from improperly insulated dc current leads. Because significant current is passed through leads that are often relatively small, the potential drop in the current leads can be large (e.g., >1 V). If the current leads are not continuously insulated through the entire solution right up to the location where they are spot welded onto the specimen, there is an opportunity for crosstalk with closely adjacent potential leads (where the signal is typically 100 µV). Additionally, biasing of the specimen can occur if the current leads are not continuously insulated through the system seals. Any ionic communication in the tight-fitting seal area permits leakage to the metal (e.g., autoclave), and a circuit is established. The current leads act like a 1 V battery that is shared across two resistors, one representing the water resistivity in the seal and one representing the water resistivity between the specimen and the autoclave. This can cause some polarization of the specimen in conductive solutions, or voltage (iR) drop in low-conductivity solutions. In the latter case, even though no substantial polarization occurs, reference electrodes that are located between the specimen and the autoclave "see" the voltage drop, and the apparent (measured) corrosion potential can be observed to fluctuate as the direction of the dc current is reversed. This represents a good check of the integrity of the dc potential drop system and wire insulation.

Finally, there is a potential concern for self-heating of the specimen by the applied dc current. While this is not a problem in aqueous environments or at common current densities, there have been cases where high current densities coupled with air or vacuum exposure resulted in significant self-heating.

High-quality implementations of dc potential drop are consistently able to achieve a crack length resolution on 1T compact-type specimens of about 1 µm, and an overall accuracy of <5% on the overall increment in crack advance. Current and potential leads can be insulated using Teflon tubing for test temperatures up to 300 °C; above 300 °C, zirconia is generally used. The relationship between measured potential and crack length varies with specimen geometry and placement of the current and potential leads (see the article "Fatigue Crack Growth Testing" in this Volume). Because of this sensitivity to lead placement, it is best to avoid detaching and reattaching leads on specimens (e.g., specimens that are only periodically measured for crack length such as those for in-reactor exposures).

Compliance Method and Other Cracking Monitoring Methods. The next most common

crack following technique is mechanical compliance, which relies on the relationship between crack mouth opening displacement and load during an unload/reload cycle. Resolution is typically limited by the strain gage or proximity sensors (e.g., eddy current or capacitance) that must monitor crack opening displacement in the (high-temperature water) environment. See the article "Fatigue Crack Growth Testing" in this Volume.

Another method is the ac potential drop technique, which relies on the "skin" (surface) effect of high-frequency current in metals. The advantage of the skin effect for detecting crack nucleation is generally more than offset by the higher noise (poorer noise rejection) of the ac measurement, even with sophisticated lock-in amplifiers, although improved instrumentation is closing the gap. Other crack following techniques include burst detection by monitoring pressurized tubes, periodic ultrasonic or eddy current scans to detect small cracks, and periodic interruption and inspection.

In aqueous environments, electrochemical noise can be used as a semiquantitative crack monitoring technique. This technique measures the small variations in corrosion potential and/or corrosion current as cracks (or other corrosion phenomena, such as pitting) nucleate and grow. This technique is good at discriminating the early stages of crack initiation. However, the correlation between crack depth (or number of cracks) and the electrochemical noise signal is at best semiquantitative, because: (a) the noise signal intensity decreases with increasing crack depth, increasing distance between sensors, and the location of cracking on the specimen surface (especially in low-conductivity solutions); and (b) noise from multiple small cracks cannot be distinguished from noise from longer cracks.

Post-Test Analysis

Parametric Measurement, Computer Automation, and Data Analysis. Most aspects of experimental procedure are basic to all types of fatigue testing (see the article "Fatigue Crack Growth Testing" in this Volume). In environments, there are numerous additional parameters to control and measure (e.g., those related to partial pressures of gases, ionic species, pH, corrosion potential, temperature, etc.). These are ideally measured on a "continuous" basis and included as part of the computer data record, and they generally substantially expand the size and complexity of the data acquisition requirements.

Additionally, data analysis should recognize the time-dependent character of environmentally assisted cracking (or high-temperature creep-fatigue interactions). Fatigue processes are historically viewed as (purely) cycle-dependent processes, and this approach must be broadened to avoid missing (or misinterpreting) environmental enhancement. It is particularly important to factor this into experimental design, so that experiments include an evaluation over a wide range of loading frequencies and/or hold times. By contrast, the assumption that the greatest environmental

enhancement occurs at very low frequency/long hold time can also be erroneous, because of the role of dissolving species in the crack (e.g., MnS dissolution in low-alloy and carbon steels, which can require continuing exposure of new MnS inclusions to maintain the sulfur-rich crack-tip chemistry). Thus, corrosion fatigue testing requires recognition that:

- It is important to achieve a steady-state crack growth rate at each test condition, which requires achieving a steady-state surface condition and crack chemistry. The potentially large role of nonuniform distribution of inhomogeneities must also be recognized.
- Time dependency is very important, and therefore the role of mean/maximum stress and frequency/hold time can be very large.
- Unexpected increases in crack growth rate can occur at specific loading conditions (e.g., associated with achieving critical crack chemistry). Some systems also exhibit hysteresis (e.g., in frequency, whereby high crack growth rates can be sustained to low frequency, but once a low growth rate is obtained, the frequency may have to be increased to high values before a return to "high" growth rates is observed). This behavior is often characteristic of any system where there is a strong interaction between the crack chemistry and growth rate.

Calculation of Crack Growth Rate. Crack growth rate is calculated from crack length versus cycle number (da/dN) data. The data for crack growth rate determination should be chosen to eliminate transient effects, which may be caused by changes in stress-intensity range level, test frequency or waveform, or gas partial pressure or temperature. The crack length increment for transient behavior can be estimated from the plastic zone size, R_p:

$$R_p = \frac{1}{3\pi}\left(\frac{\Delta K}{\sigma_{ys}}\right)^2$$

where ΔK is the stress-intensity factor range and σ_{ys} is the yield strength.

Crack length data should not be used for crack growth rate determination until the crack has propagated a distance of at least twice the plastic zone size. Crack growth rate can be calculated by the secant and incremental methods, which are described in the article "Fatigue Crack Growth Testing" in this Volume and in ASTM E 647.

Analysis of the fracture surface is crucial to ensure accuracy of the crack monitoring technique, to identify branching and out-of-plane cracking, and to determine crack morphology. Because pits can act as nucleation sites for cracking in solutions containing chloride, possible interaction between pitting and fatigue behavior should also be examined. Accurate determination of crack growth on a cycle or time basis requires an understanding of the resolution of the monitoring technique under the actual test conditions.

Fracture surfaces of fatigue-fractured test specimens usually are examined by scanning electron microscopy to determine the fracture path and the fracture mode of the test material in relation to its microstructure. Such information may be valuable in identifying the fracture mechanism in certain environment and material combinations and may be used to assess the severity of the deleterious environment and to aid in analyzing service failures. Fatigue in inert environments (including vacuum) generally pro-

duces different fracture modes or fracture paths than does fatigue in deleterious environments. Good qualitative agreement between the observed fracture modes and fatigue crack growth kinetics can be expected. In some cases, excellent quantitative correlation between fracture surface morphology and fatigue crack growth kinetics has been reported.

Finally, if the environment consists of mixed gases, the gas at the lowest partial pressure should be admitted first. If premixed gases are used, they must be thoroughly mixed in the supply reservoir to minimize stratification.

If test conditions such as gas pressure, test frequency, or applied load are changed during fatigue testing, a transient period may occur before the material assumes the steady-state fatigue crack growth rate that corresponds to the new test condition. The duration of this transient period depends on several variables, including the type of material, the test environment, and the magnitude of the change in test conditions.

Appendix: Crack Growth Test Methods for Specific Environments*

The preceding article provides an overview of corrosion fatigue testing, with particular emphasis on corrosion fatigue crack growth of steels in high-temperature water. This appendix briefly reviews test methods for other environments and gives a brief summary of environmental effects on fatigue of selected nonferrous alloys. Environmentally assisted fracture of specific alloys is covered elsewhere in this Volume.

Vacuum and Gases at Room Temperature

One of the most critical considerations for fatigue tests in vacuum and gaseous environments is the maintenance of the purity of (and the reduction and measurement of the impurity level in) the test environment. As mentioned above, small amounts of contaminants (impurities) in the test environment can lead to fatigue crack growth rates that are not representative of the resistance of the material to fatigue crack growth in that environment.

A clean environmental test chamber that provides a very low background pressure and quantifiable impurity levels (below 10^{-7} to 10^{-6} Pa, or 7.5×10^{-10} to 7.5×10^{-9} torr) is essential, even if the tests are to be carried out in gaseous environments at relatively high pressures (i.e., above the background).

Environment Containment. An all-metal environmental test chamber with mechanical-force feedthroughs is preferred for the study of environmentally assisted fatigue crack growth in vacuum and gaseous environments. Stainless steels are suitable materials for the environmental test chamber, with copper used as the gasketing material. The test chamber usually is equipped with a glass viewport that enables the operator to visually monitor the progress of the experiment.

With adequate pumping, the background pressure in the clean test chamber is usually below 10^{-6} Pa (7.5×10^{-9} torr). Maintaining an ultraclean test system is important, because a small amount of impurities can either significantly re-

duce or accelerate the fatigue crack growth rate, depending on the material and the types of impurities.

To achieve a low background pressure, the test chamber frequently is baked out (with the test specimen in place) at a temperature above ambient (60 to 400 °C, or 140 to 750 °F) to remove adsorbed and absorbed gases on the chamber wall. The bakeout temperature should be considerably below the tempering or aging temperature of the test material to ensure that the microstructure and the mechanical properties of the test material are not altered by the bakeout process. For example, the first-step artificial aging temperature for high-strength 7050-T7451 aluminum alloy is 121 °C (250 °F). The bakeout temperature for the test chamber is thus normally kept below 80 °C (175 °F).

Environment. Only high-purity, laboratory-grade gases should be used. Additional purification and dehumidification of the gas is recommended by passing it through a molecular-sieve purifier and a cold trap (–196 °C, or –321 °F) before allowing the gas to enter the test chamber. Gas pressure in the environmental test chamber is usually controlled by admitting the gas through a variable-leak valve.

If the test environment contains a toxic gas (such as hydrogen sulfide) or a combustible gas (such as hydrogen or methane), a protective hood with negative suction pressure should be used to enclose the test chamber or the entire test system. The test chamber should be purged thoroughly with an inert gas, such as argon or nitrogen, before it is reopened to the atmosphere.

If water vapor is used as the test environment, it can be drawn through the variable-leak valve from a high-purity reservoir that is attached to the test chamber. Deionized distilled water in the reservoir should be purified further by subjecting it to repeated freezing/pumping/thawing cycles to remove residual dissolved gases in the water (*Surface Science*, Vol 64, 1977, p 617).

Certain gases can decompose or react with containment vessels over time. For example, hydrogen sulfide can react with a stainless steel container to produce hydrogen. Provision must be made to remove the product gases before the test gas is admitted into the test chamber.

High-Temperature Vacuum and Oxidizing Gases

Fatigue testing in elevated-temperature vacuum and oxidizing environments requires a carefully designed vacuum test chamber. The chamber must keep the test specimen in a vacuum or oxidizing gas environment, allowing forces to be applied to the specimen, a means to measure crack length, and a method of applying and controlling the specimen temperature.

Variables that can affect fatigue crack growth rate at high temperature are time and rate dependent or structure dependent. Examples of time- and rate-dependent variables are oxidation and creep (Fig. 21). Structure-dependent variables include phase transformations, nucleation and growth of new and existing phases, and grain growth. When fatigue crack growth rate test data are reported for these environments, test temperature, vacuum pressure, partial pressure of oxidizing gas, waveform type, waveform frequency, and stress ratio must be reported. Additional information on high-temperature fatigue crack growth testing is given in the article "Fatigue Crack Growth Testing" in this Volume.

Environment Chambers. Materials used in the test chamber should be selected to minimize outgassing in vacuum. For example, many plastic materials contain plasticizers, which slowly outgas in vacuum. These types of materials limit the ultimate vacuum obtainable. Stainless steel is suitable for the manufacture of the main test chamber. Components in the chamber should be designed for fast outgassing. When threaded components are used in the test chamber, channels should be machined in the threads to allow paths for fast outgassing.

For vacuum levels of 6.5×10^{-5} Pa (5×10^{-7} torr), O-rings provide sufficient sealing; for higher vacuum levels, copper gaskets should be used. Electricity, water, radiofrequency, and the thermocouple can be input into the chamber using standard vacuum feedthroughs.

Specimen Heating and Temperature Control. Induction heating is the only suitable method to heat test specimens in vacuum and oxidizing environments. Radiofrequency generators with frequencies of 200 to 500 kHz are used for induction heating of test specimens. The in-

*Adapted and updated from *Mechanical Testing*, Vol 8, *Metals Handbook*, 9th ed., American Society for Metals, 1985, p 410-429

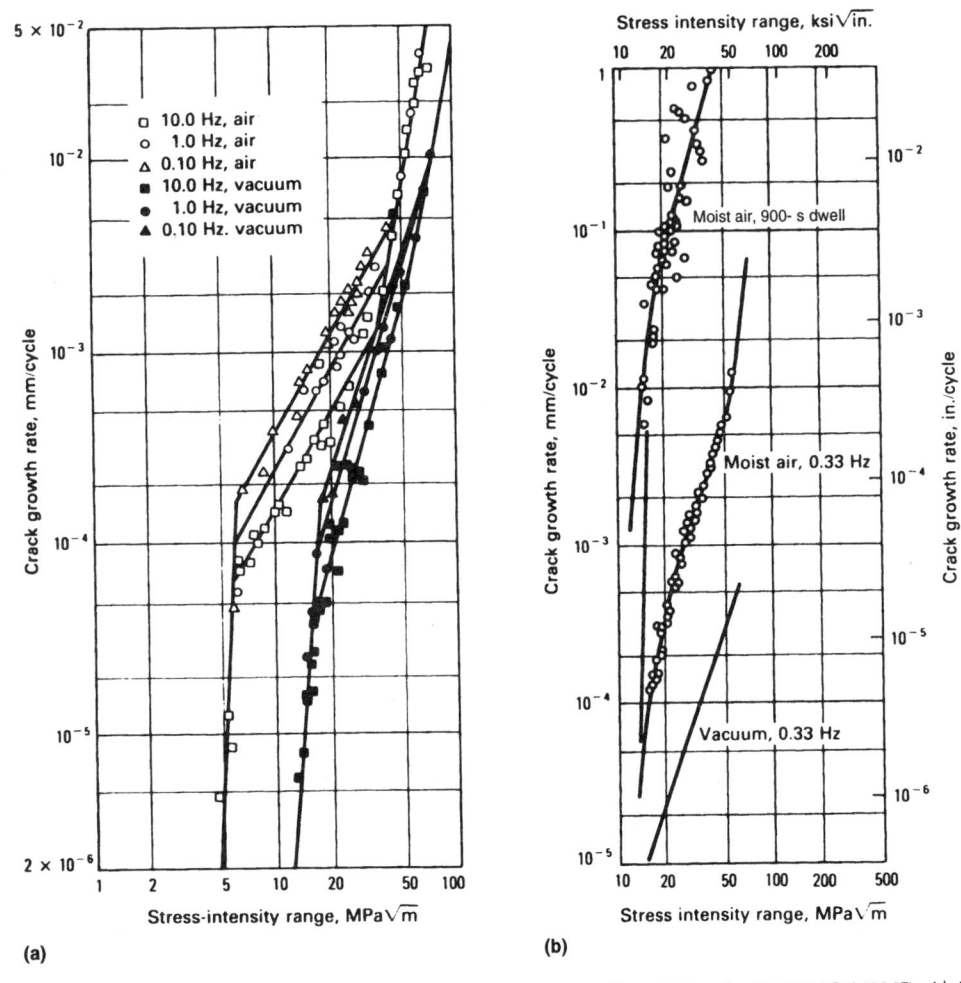

(a)

(b)

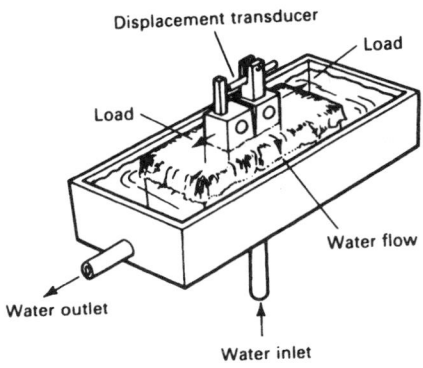

Fig. 22 Typical corrosion fatigue test cell. Maintenance of the equilibrium oxygen concentration is ensured by cascading the solution in the circulation rig.

Fig. 21 High-temperature fatigue crack growth of two nickel-base superalloys. (a) Hastelloy X at 760 °C (1400 °F) with R = 0.05. (b) NASA 11B-7 at 650 °C (1200 °F) with R = 0.05

duction coils should be made of copper and have no insulating coating. When oxidizing gases are introduced into the test chamber, a certain pressure range exists at which the gases will be ionized between the specimen and induction coils. In this pressure range, it is impossible to heat the specimen, because the radio frequency field arcs and shuts off the radiofrequency generator. To continue testing, the gas pressure must be either increased or decreased.

Two types of temperature controllers that are suitable for induction heating are thermocouple and infrared controllers. Each controller type has advantages and disadvantages. Infrared temperature controllers measure and control temperature from the spectral energy density emitted from the test specimen over a certain wavelength range. These measurements are noncontacting but require a clear optical path from the sensor head to the test specimen. Infrared temperature controllers have a minimum temperature measurement capability of approximately 350 °C (660 °F). Two-color infrared controllers eliminate errors due to transmission loss and emissivity changes, but they have a minimum temperature measuring capability of 700 °C (1290 °F).

Thermocouple temperature controllers are also used in vacuum test chambers. A variety of ther-

mocouple types can be used, depending on the temperature range and the required durability of the thermocouple. For example, American National Standards Institute type S and type K thermocouple temperature ranges overlap, but for long-term tests of more than one week, type S thermocouples are preferred because they are more oxidation resistant. This would not be a consideration in a high-vacuum environment. Thin thermocouple wire less than 0.25 mm (0.01 in.) in diameter must be used to eliminate inductive heating of the thermocouple wire. With some temperature controllers, it is necessary to filter out radiofrequency noise in the thermocouple with a passive inductor/capacitor-type filter.

Test Specimens. Because high-temperature vacuum requires specimen heating by induction, many of the standard fracture mechanics test specimens cannot be used. Center-cracked tension and single-edge notched specimens are commonly used, because it is relatively easy to maintain the specimen gage section at uniform temperature with induction heating. When tests are conducted at high vacuum levels or low oxidizing gas partial pressures, specimen thickness may affect crack growth rate, because transport of the oxidizing gas to the crack tip may be the rate-limiting factor.

Aqueous Solutions at Ambient Temperature

Fatigue studies in aqueous solutions at ambient temperatures present fewer problems experimentally than many of the other environments considered in this article. Nevertheless, it is often the case that the most frequent problem in determining the validity of corrosion fatigue data lies with the control and monitoring of the bulk water chemistry and the monitoring and recording of the electrochemical potential.

Environment Containment. Glass and plastics are suitable materials for environmental test chambers and ancillary pipework for aqueous solutions at ambient temperatures. At elevated temperatures (>60 °C, or 140 °F), however, dissolution of silicates from glassware can inhibit corrosion. Dissolution of plasticizers from certain plastics (e.g., polypropylene) is also a concern. Flexible plastics, such as twin-pack casting silicone rubber, have proved to be useful in the vicinity of the fatigue specimen.

A corrosion fatigue test cell that avoids the need for a water-tight seal at the specimen is shown in Fig. 22. Normal specimen movement and any sudden fracture event can be accommodated without catastrophic consequences. Highly effective seals between plastic and metal surfaces can be made with silicone rubber caulking compounds, if necessary, although sufficient time must be allowed for escape of the acetic acid solvent base.

Fatigue specimens of passive metals such as aluminum, titanium, and stainless steel may be subject to crevice corrosion under the caulking compound unless a primer and epoxy paint coat are applied initially to the metal surface. Gasket seals using O-rings, for example, can also form a satisfactory seal, but generally are more expensive to engineer and can also be subject to crevice corrosion in some configurations. The decision to circulate the environment depends on the application and the extent of any problems in controlling water chemistry.

Water Chemistry. The prevailing water chemistry and the electrode potential of the material in its environment in the field are essential factors in any simulation experiment. Accelerated fatigue cracking can occur in a number of environments, including seawater, salt water/salt spray, and body fluids. These must be reproduced as closely as possible in the laboratory, although limitations are necessarily imposed in simulating aspects of complex environments, such as the biological activity of seawater.

The importance of reproducing the service environment as closely as possible is illustrated by comparing the behavior of metals in sodium chloride and in seawater. The buffering action of seawater associated with dissolved bicarbonate/carbonate can result in the formation of calcareous scale under cathodic protection, which can precipitate in cracks and influence the cyclic crack opening and closing, thus affecting crack growth rates. Substitute ocean water, as described in ASTM D 1141, usually is a satisfactory substitute for seawater, but some differences have been observed in relation to the rate of calcareous scale formation and the rate of corrosion fatigue growth.

Laboratory solutions should be prepared using the purest chemicals available in distilled or deionized water. Concentrations at the level of parts per million can have profound effects on electrochemistry and corrosion.

Several variables must be measured and controlled when simulating an aqueous environment: solution purity, composition, temperature, pH, dissolved oxygen content, and the flow (circulation) rate of the solution.

Acidified Chloride

Investigations performed in acidified chloride, particularly at high temperature, pose unique problems. These include not only experimental barriers, such as suitable containment and seal materials and sensitivity to low-level oxidizing species, but also interpretational complexities, such as the effects of pitting and crevice processes on enhancement or retardation (by blunting) of crack initiation and growth. Care must be exercised in designing and conducting experiments to ensure personnel and equipment safety and to ensure proper simulation, control, and monitoring of environmental parameters.

Below 100 °C (212 °F). Materials and techniques for solution containment depend on the test temperature regime. Below the boiling point in solutions containing dissolved oxygen, a primary design concern is to prevent leaks that can damage equipment. A horizontal loading frame helps ensure that sensitive components are not readily damaged by leaks. Additionally, some specimen configurations (such as compact tension) permit the loading linkage to be placed above the solution, simplifying the choice of materials and seal designs.

Testing in deaerated solutions may require careful selection of materials, depending on the sensitivity of the test to low oxygen concentration. For example, the clear, flexible tubing often used in laboratories is very permeable to oxygen. Additionally, some plastics degrade in acidic environments.

Above 100 °C (212 °F), the propensity for pitting and crevice attack increases, the internal pressure rises, the design strength of some materials (e.g., titanium) begins to decrease, and good seal design (particularly for sliding seals) is crucial. Pitting and crevice potential studies show that the resistance of iron- and nickel-base alloys in environments containing chloride decrease from room temperature to about 200 °C (390 °F).

The best approach for selecting pressure boundary materials is to combine published data with recommendations from autoclave manufacturers and metals producers. No assumptions should be made regarding the performance of materials with varying environment. For example, commercial-purity titanium, which is often used in neutral and acidified chloride environments, performs very poorly in acidified chloride under reducing conditions, in acidified environments containing sulfate, and in caustic environments at high temperature. Addition of a small amount (0.2%) of palladium (grade 7) greatly improves resistance in acidified environments that contain sulfate.

Above 200 °C (390 °F), materials selection is particularly difficult. In general, for acidified chlorides, commercial-purity titanium is favored under oxidizing conditions (containing oxygen, iron ion, or copper ion), while zirconium (for example, UNS R60702) is favored for reducing environments. Zirconium alloys are highly intolerant of fluoride. In some cases, high-strength materials, such as Ti-6Al-4V or the Hastelloy C series alloys, are required, although there is generally a loss in corrosion resistance. Liners of Teflon or tantalum are options in some instances.

Because of its effect on the autoclave and test results, control of the oxidizing nature of the environment is often critical. In addition to oxidizing species, such as oxygen, iron ions, and copper ions, care in the use of externally applied potential is required. The autoclave may be polarized into a harmful regime if ground loops exist, or if it is used as the counterelectrode. A similar result can occur if the autoclave contacts a dissimilar metal.

Because of the rate and extent of expansion on leakage, hot pressurized water poses a serious safety hazard. Each autoclave must have a pressure-relief device attached to it, preferably in a fashion that does not permit bypassing or isolation. Selection of the pressure-relief device must account for the pressure, environment (often gold-coated elements are used in rupture disks), and temperature at which the device actually operates. Additionally, autoclaves, particularly when used in aggressive environments, must be examined regularly for damage resulting from pitting, crevice attack, general corrosion, hydriding, and so forth.

Pressure testing coupled with dimensional checks must also be performed. Manufacturers offer this service and will usually provide the test details. Test pressure and dimensional tolerances are a function of autoclave design, material, and temperature of use. Leaks may also occur in tubing and in valves, which are often difficult to

inspect or test. Leaks almost always develop slowly. Nevertheless, a relatively rapid, controlled method for depressurizing the system should be included in the system design.

For some applications, inexpensive miniature autoclaves can be custom fabricated. The small internal volume of these devices is an advantage if a leak occurs in the system.

Liquid Metal Environments

Liquid metals (sodium, potassium, and lithium, for example) are frequently used in heat-transport applications at elevated temperatures. Such applications include liquid-metal-cooled nuclear reactors, first-wall coolant for fusion devices, and heat-transport systems in solar collectors. These applications often involve cyclic temperature and/or pressure fluctuations, as well as other sources of cyclic stresses. For this reason, knowledge of the fatigue crack propagation behavior of structural alloys in the liquid metal environments is sometimes necessary.

Generally, liquid metals react (in some cases, quite violently) with air and/or water vapor; therefore, testing systems must be designed to exclude both air and water. Three basic designs have been developed to expose the specimen (or crack region of a specimen) to the liquid metal environment, while excluding air, water, and other contaminants.

The simplest method uses a sealed environmental chamber attached to the specimen that completely surrounds the notch and crack extension plane in a compact-type specimen. The small environmental chamber contains the liquid metal but does not extend to the region of the loading holes; hence, the loading pins, clevis grips, and remainder of the load train are not subjected to the liquid metal environment.

Relative motion across the notch and crack area is accommodated by bellows. This type of system has the advantages of simplicity and low cost. The main disadvantage is that the liquid metal is static; hence, the characteristics of large heat transport systems (e.g., mass transport due to nonisothermal operation) cannot be studied.

The second type of system, a circulating loop, is much more costly to build and operate, but it can be used to study potential effects on fatigue crack propagation such as mass transport, which occurs during carburizing, decarburizing, and dissolution of alloying elements. A third type of system consists of an open crucible (containing the test specimen immersed in static liquid metal) that is located within an inert gas cell or glovebox. This type of system is relatively inexpensive to build and operate, but it has the greatest potential for exposure to air and other contaminants.

Austenitic stainless steels generally have been used in the construction of current systems, and their use has been satisfactory. System designers should consider, however, that under some conditions mechanical properties (tensile, stress rupture, etc.) can be influenced by long-term exposure to liquid metals.

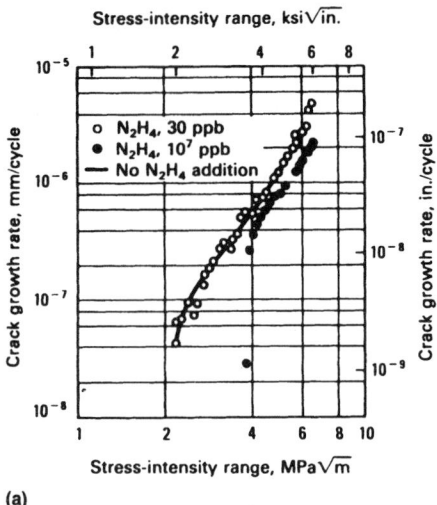

(a)

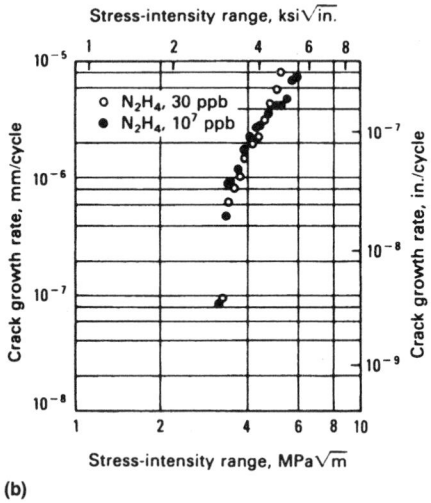

(b)

Fig. 23 Effect of hydrazine on fatigue crack growth rates of (a) 403 stainless and (b) Ti-6Al-4V. Environment: 0.1 g NaCl + 0.1 g Na_2SO_4 (g/100 mL H_2O) in boiling water (100 °C, or 212 °F). Stress ratio = 0.8.

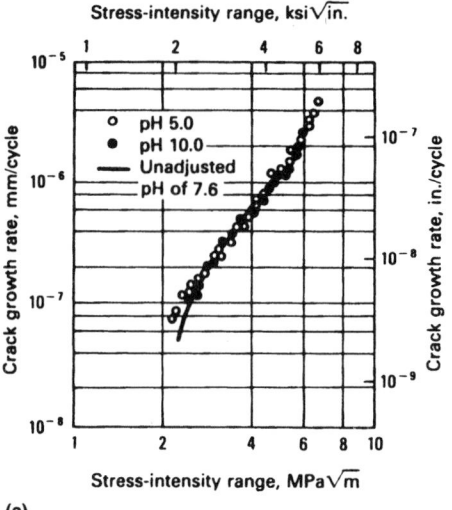

(a)

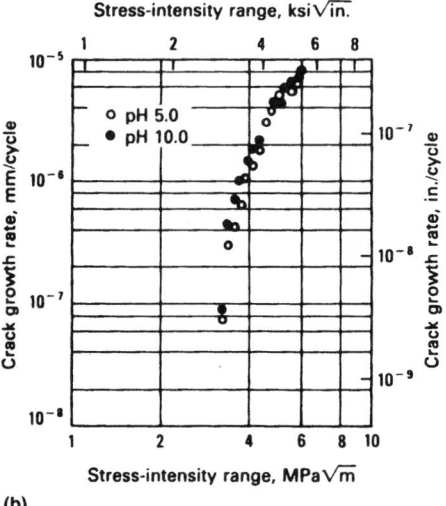

(b)

Fig. 24 Effect of pH on near-threshold fatigue crack growth rates of (a) type 403 stainless and (b) Ti-6Al-4V. Environment: 0.1 g NaCl + 0.1 g Na_2SO_4 (g/100 mL H_2O) in boiling water (100 °C, or 212 °F). Stress ratio = 0.8.

Typical results for fatigue crack propagation behavior of austenitic stainless steels in a liquid sodium environment are documented in Ref 29 and 30. In most cases, fatigue crack propagation rates are lower in sodium environments than in elevated-temperature air environments. The relatively benign nature of sodium environments also leaves the fracture faces in excellent condition for viewing with optical microscopes, scanning electron microscopes, or transmission electron microscopes.

Steam or Boiling Water with Contaminants

Corrosive environments, such as steam or boiling water with contaminants, come in contact with many structural components. To assess the structural integrity of machine hardware, testing in the environments of concern is essential. Fatigue crack growth testing in corrosive environments requires special care because of the presence of corrosive mediums and testing complexity.

Environment Containment. Special designs are required to accommodate fatigue crack growth testing in steam or boiling water with contaminants. If the environmental pressure and temperature are moderate, for example at a pressure of 500 kPa (72.5 psi) and a temperature of 100 °C (212 °F), simple stainless steel O-ring sealed chambers can be clamped to each side of the specimen in which cracking will occur. If necessary, the test environment can be circulated through the chamber at a controlled flow rate.

If the environmental pressure and temperature are high, for example in steam at a pressure of 7.2 MPa (1040 psi) and a temperature of 288 °C (550 °F), a chamber that encloses the test specimens must be constructed. Composition of the test environment must be carefully analyzed before and after the experiment, given the variety of possible chemical effects on crack growth rates. (See Fig. 23 and 24 as examples for selected alloys.)

Dissolved Oxygen. Control and measurement of dissolved oxygen levels in the steam environment are of prime importance, because oxygen can affect fatigue crack propagation rate properties. Oxygen content can be controlled by bubbling argon or nitrogen through the water reservoir, or by maintaining a hydrogen overpressure. Oxygen content can be measured by using a colorimetric technique or by using oxygen analyzers that can continuously monitor oxygen in the parts per billion range.

REFERENCES

1. R.P. Gangloff, Exxon Research and Engineering Co., unpublished research, 1984
2. J.M. Barsom, E.J. Imhoff, and S.T. Rolfe, Fatigue Crack Propagation in High Yield Strength Steels, *Eng. Fract. Mech.*, Vol 2, 1971, p 301-324
3. C.S. Kortovich, Corrosion Fatigue of 4340 and D6AC Steels Below K_{Iscc}, *Proc. 1974 Triservice Conf. on Corrosion of Military Equipment*, AFML-TR-75-43, Air Force Materials Lab, Wright-Patterson Air Force Base, 1975
4. D.J. Duquette and H.H. Uhlig, *Trans. Am. Soc. Metals*, Vol 61, 1968, p 449
5. P.L. Andresen, R.P. Gangloff, L.F. Coffin, and F.P. Ford, Overview—Applications of Fatigue Analysis: Energy Systems, *Proc. Fatigue/87*, EMACS, 1987
6. F.P. Ford, D.F. Taylor, P.L. Andresen, and R.G. Ballinger, "Corrosion Assisted Cracking of Stainless and Low Alloy Steels in LWR Environments," Final Report NP-5064-S, EPRI, 1987
7. *Proc. First International Conf. on Environment Induced Cracking of Metals*, National Association of Corrosion Engineers, 1988
8. F.P. Ford, Status of Research on Environmentally Assisted Cracking in LWR Pressure Vessel Steels, *Trans. ASME, J. Pressure Vessel Technology*, Vol 110, 1988, p 113-128
9. F.P. Ford and P.L. Andresen, Corrosion Fatigue of A533B/A508 Pressure Vessel Steels in Water at 288 °C, *Proc. Third International Atomic Energy Agency Specialists Mtg. on Subcritical Crack Growth*, NUREG/CP-0112 (ANL-90/22), Vol 1, U.S. Nuclear Regulatory Commission, 1990, p 105-124
10. B. Tompkins and P.M. Scott, Environment Sensitive Fracture: Design Considerations, *Met. Tech.*, Vol 9, 1982, p 240-248
11. P.L. Andresen and L.M. Young, Crack Tip Microsampling and Growth Rate Measurements in Low Alloy Steel in High Temperature Water, *Corrosion Journal*, Vol 51, 1995, p 223-233
12. P.L. Andresen, I.P. Vasatis, and F.P. Ford, "Behavior of Short Cracks in Stainless Steel at 188 °C," Paper 495, *Corrosion/90*, National Association of Corrosion Engineers, 1990
13. J.C. Newman, Jr. and W. Elber, Ed., *Mechanics of Fatigue Crack Closure*, STP 982, ASTM, 1988

14. P.L. Andresen and P.G. Campbell, The Effects of Crack Closure in High Temperature Water and Its Role in Influencing Crack Growth Data, *Proc. Fourth International Symp. on Environmental Degradation of Materials in Nuclear Power Systems—Water Reactors*, National Association of Corrosion Engineers, 1990, p 4-86 to 4-110

15. P.M. Scott, Effects of Environment on Crack Propagation, *Developments in Fracture Mechanics—II*, G.G. Shell, Ed., Applied Science Publishers, London, 1979, p 221-257

16. *Proc. Life Prediction of Structures Subject to Environmental Degradation*, Corrosion/96, National Association of Corrosion Engineers, 1996

17. *Proc. Int. Symp. on Plant Aging and Life Prediction of Corrodible Structures*, National Association of Corrosion Engineers, 1995

18. P.L. Andresen and F.P. Ford, Use of Fundamental Modeling of Environmental Cracking for Improved Design and Lifetime Evaluation, *Trans. ASME, J. Pressure Vessel Technology*, Vol 115 (No. 4), 1993, p 353-358

19. R. Gangloff, Corrosion Fatigue, *Corrosion Tests and Standards: Application and Interpretation*, R. Baboian, Ed., ASTM, 1995

20. B. Yan, G.C. Farrington, and C. Laird, *Acta Metall.*, Vol 33, 1985, p 1533-1545

21. T. Magnin and L. Coudreuse, *Matls. Sci. Engr.*, Vol 72, 1985, p 125-134

22. H.M. Chung et al., *Environmentally Assisted Cracking in Light Water Reactors*, Report NUREG/CR-4667 (ANL-93/27), Vol 16, U.S. Nuclear Regulatory Commission, 1993

23. O. Vosikovsky and R.J. Cooke, An Analysis of Crack Extension by Corrosion Fatigue in a Crude Oil Pipeline, *Int. J. Pressure Vessel Piping*, Vol 6, 1978, p 113-129

24. R.P. Wei and G. Shim, Fracture Mechanics and Corrosion Fatigue, *Corrosion Fatigue: Mechanics, Metallurgy, Electrochemistry and Engineering*, STP 801, T.W. Crooker and B.N. Leis, Ed., ASTM, 1984, p 5-25

25. P.L. Andresen and L.M. Young, Characterization of the Roles of Electrochemistry, Convection and Crack Chemistry in Stress Corrosion Cracking, *Proc. Seventh International Symposium on Environmental Degradation of Materials in Nuclear Power Systems—Water Reactors*, National Association of Corrosion Engineers, 1995, p 579-596

26. F.P. Ford, "Mechanisms of Environmental Cracking in Systems Peculiar to the Power Generation Industry," Final Report NP-2589, EPRI, 1982

27. H.G. Nelson, Hydrogen Induced Slow Crack Growth of a Plain Carbon Pipeline Steel under Conditions of Cyclic Loading, *Effect of Hydrogen on the Behavior of Materials*, A.W. Thompson and I.M. Bernstein, Ed., The Metals Society—American Institute of Mining, Metallurgical, and Petroleum Engineers, 1976, p 602-611

28. H.H. Johnson, Hydrogen Brittleness in Hydrogen and Hydrogen-Oxygen Gas Mixtures, *Stress Corrosion Cracking and Hydrogen Embrittlement of Iron Based Alloys*, J. Hochmann, J. Slater, R.D. McCright, and R.W. Staehle, Ed., National Association of Corrosion Engineers, 1976, p 382-389

29. L.A. James and R.L. Knecht, Fatigue-Crack Propagation Behavior of Type 304 Stainless Steel in a Liquid Sodium Environment, *Met. Trans. A*, Vol 6 (No. 1), 1975, p 109-116

30. J.L. Yuen and J.F. Copeland, Fatigue Crack Growth Behavior of Stainless Steel Type 316 Plate and 16-8-2 Weldments in Air and High-Carbon Liquid Sodium, *J. Eng. Mat. Technol.*, Vol 101 (No. 3), 1979, p 214-223

Detection and Monitoring of Fatigue Cracks

S. Shanmugham and P.K. Liaw, Department of Materials Science and Engineering, University of Tennessee

MEASUREMENT OR DETECTION of fatigue cracks and damage can, in general terms, be classified into the following two application areas: laboratory methods and field service assessment methods. Specific techniques for these two areas of application are summarized in Table 1. Several techniques are available to detect crack initiation and measure crack size for laboratory and field applications.

This article describes and compares the test techniques listed in Table 1. An attempt is made to include methods that are available for monitoring crack initiation and crack growth. Some methods (such as x-ray diffraction) for obtaining information on fatigue damage in test specimens are also included. The fatigue damage can be considered as the progressive development of a crack from the submicroscopic phases of cyclic slip and crack initiation, followed by the macroscopic crack propagation stage, to final fracture. These three stages are important in determining the fatigue life of structural components. In many cases, crack initiation can, however, be the dominant event for life analyses and design considerations, such as the applications of S (applied stress) versus N (fatigue-life cycle) curves. Furthermore, crack initiation is the precursor of fatigue failure. If the early stage of crack initiation can be detected and the mechanisms of crack initiation can be better understood, fatigue failure may be prevented.

Each method in Table 1 is summarized in the following sections along with a brief discussion of principles underlying each method. When selecting a method for fatigue crack detection or monitoring, oftentimes the sensitivity or crack size resolution plays a dominant role in the selection. The resolution of crack detection methods can range from 0.1 μm to 0.5 mm as summarized in Table 2. The resolution depends on the specific technique, component geometry, surface condition, physical accessibility, and phenomenon responsible for crack initiation. While selecting a technique for crack detection, the sensitivity or crack size resolution plays a dominant role. For a higher crack size resolution requirement, the choice should be a method having greater sensitivity.

Techniques listed in Table 1 can be used for either lab or field use, with some suitable for both. For example, the eddy current technique is used as an inspection tool and as a laboratory tool. Generally, one technique may not satisfy all requirements, and hence, a combination of two or more techniques may be utilized. For example, one may utilize the compliance technique for measuring the crack initiation and propagation behavior, and the mechanisms involved in the fatigue process could be examined using the scanning electron microscope.

Table 3 is a collection of sample testing and material parameters from several investigations (Ref 1-7) including loading type, specimen type, material, environment, crack initiation site, crack detection method, and sensitivity. For example, loading condition could be bending, axial, reverse bending, tension, and mode II loadings. Specimen types could be plate, welded plate, cylindrical bar, compact-type (CT) specimen, blunt-notched specimen, and three-point bend bar. Test environments have been air, water, vacuum, hydrogen, helium, and oxygen.

Other reviews on techniques for detecting fatigue crack initiation and propagation are provided by Allen et al. (Ref 8) and Liaw et al. (Ref 9). More detailed information on the probability of detecting cracks is addressed in the article

Table 1   Applications of the methods available for detecting fatigue cracks

Method	Application
Optical	Detecting fatigue cracks in the laboratory
Compliance	Detecting fatigue cracks in the laboratory
Electric potential/Krak gage	Detecting fatigue cracks in the laboratory and during service
Gel electrode imaging	Detecting fatigue cracks in the laboratory
Liquid penetrant	Inspecting structural components in the laboratory and during service
Magnetic property	Detecting fatigue damage in the laboratory and inspecting structural components during service
Positron annihilation	Residual life estimation and fatigue damage in the laboratory
Acoustic emission	Laboratory and in-field testing
Ultrasonics	Laboratory and in-field testing
Eddy current	Laboratory and in-field testing
Infrared	Laboratory and in-field testing
Exoelectrons	Residual life estimation in the laboratory
Gamma radiography	Laboratory and in-field testing
Scanning electron microscope	Basic understanding of the crack initiation and growth mechanisms
Transmission electron microscope	Basic understanding of the crack initiation and growth mechanisms
Scanning tunneling microscope	Understanding of the crack nucleation phenomena
Atomic force microscope	Detecting fatigue crack initiation
Scanning acoustic microscope	Detecting fatigue crack initiation
X-ray diffraction	Detecting fatigue damage and residual stresses in the laboratory

Table 2   Summary of the crack detection sensitivity of the methods available for detecting fatigue cracks

Method	Crack detection sensitivity, mm
Gamma radiography	2% of the component thickness
Magnetic particle	0.5
Krak Gage	0.25
Acoustic emission	0.1
Eddy current	0.1
Optical microscope	0.1-0.5
Electric potential	0.1-0.5
Magnetic property	0.076
Ultrasonics	0.050
Gel electrode imaging	0.030
Liquid penetrant	0.025-0.25
Compliance	0.01
Scanning electron microscope	0.001
Transmission electron microscope	0.0001
Scanning tunneling microscope	0.0001

Table 3 Testing parameters adopted by some fatigue researchers

Loading type	Specimen type	Material	Environment	Crack detection method	Sensitivity, mm	Ref
Reverse bending	Plate	Alpha iron	Air and vacuum	Scanning electron microscopy replica	0.001	1
Axial	Plate	Alpha iron	Air	Transmission electron microscopy replica	0.0001	2
Tension	Compact type with blunt notch	316 stainless steel	Air and coal process solvent	Optical microscope and Krak Gage	0.25	3
Bending	Plate	HT-80 steel weldment	Sea water	Computer image	...	4
Mode II	Notched plate	4340 steel	Air, water, and hydrogen	Optical microscope	0.1	5
Axial	Plate	Silver	Helium and oxygen	Scanning tunneling microscope	0.0001	6
Axial	Cylindrical bar	4340 steel	Air	Acoustic emission	0.1	7

"NDE Reliability Data Analysis" in Volume 17 of the *ASM Handbook, Nondestructive Evaluation and Quality Control* (1989, p 689-701).

Crack Measurement for Specimen Testing

Laboratory methods for developing fatigue life (*S-N* or *ε-N*) and crack growth (*da/dN* versus *ΔK*) are described elsewhere in this Volume and in Ref 10. General aspects of *S-N* or *ε-N* testing are discussed in this Volume in the article "Corrosion Fatigue Testing," while crack growth testing and crack monitoring techniques are described in detail in the article "Fatigue Crack Growth Testing." Nonetheless, this section briefly summarizes the key methods for detecting and monitoring fatigue cracks in laboratory specimen testing as reference information prior to discussions of methods suitable for field or service life assessment.

Optical methods are often used to characterize fatigue crack growth, and numerous investigators have utilized this technique. Monitoring crack length is usually done by a traveling microscope, and the crack on the specimen surface is observed usually at a magnification of 20 to 50×. The crack length is measured as a function of cycles at intervals so as to obtain an even distribution of *da/dN* versus *ΔK*. Traveling microscopes usually have a repeatability of 0.01 mm, and the interval between measurements is typically about 0.25 mm. To aid in crack-length measurements, scribe marks are often applied on specimens.

Surface characteristics of a metal object at two different times in its fatigue life can be correlated when coherent optical techniques are employed as shown by Marom and Mueller (Ref 11). They reported that the degree of correlation prevalent between these two states may be used for detecting the onset of fatigue failure and the subsequent formation of fatigue cracks. A stroboscopic light source arrangement to observe specimens during fatigue testing has been reported (Ref 10). Cracks can be detected with good sensitivity provided the light is triggered at the time of the maximum tensile stress, and the specimen observation is conducted at moderate magnification.

The optical technique is simple and inexpensive, and calibration is not required (Ref 12). Accurate measurements can be performed provided corrosion or oxidation products are not formed during testing. Crack length is usually underestimated with this method. This technique has the following limitations:

- It is time consuming.
- Automation is expensive.
- The specimen must be accessible during the testing.

The compliance method is based on the principle that when a specimen is loaded, a change in the strain and displacement of the specimen will occur. These strains and displacements are altered by the length of the initiated crack. Crack length can be estimated from remote strain and displacement measurements. However, each specimen/crack geometry requires separate calibration that can be either experimental or theoretical. The methods used to measure changes in compliance include crack-opening displacement (COD), back-face strain, and crack-tip strain measurements (Ref 12 and 13).

The compliance method typically has a crack-length detection sensitivity of 10 μm, and more detailed discussions on the use of the method in specimen testing is contained in the article "Fatigue Crack Growth Testing" in this Volume. Duggan and Proctor (Ref 14) also have provided a good review of crack-length measurements from specimen compliance changes. Compliance-crack-length relationships has been given for most of the common fatigue crack growth specimen configurations (Ref 15, 16). The compliance method enables crack growth measurements with accuracies similar to optical and electrical methods in the case of long cracks and high crack growth rates (Ref 13). The strain gage method is more suitable than the crack-opening displacement measurement in high-frequency fatigue tests. The unloading elastic compliance method is applicable for both short and long crack measurements

Each compliance method has its own merits and demerits. For example, the COD method is less expensive, the specimen need not be visually accessible, and it provides an average crack-length figure. However, separate calibration tests are required in some cases. Richards (Ref 12) also has summarized the advantages and limitations of the various compliance techniques, such as COD, back-face strain, and crack-tip strain measurements as described below.

The COD method has the following advantages: it can be used from nonaggressive to aggressive environments and for various geometry configurations that behave in an elastic manner; its costs range from low in room-temperature air tests to moderately expensive in high-temperature aggressive environments; it can be used as a remote method and is easily automated; and it produces an average crack-length figure where crack-front curvature occurs. The COD technique, however, has its limitations: separate calibration tests are warranted in some instances, and it is used for specimens where time-dependent, time-independent, and reversed-plasticity effects are small.

The back-face strain method has the following advantages (Ref 12): cost ranges from low in room-temperature tests to moderately expensive in high-temperature tests, remote method, easy automation, and crack length increases of 10 μm can be resolved. However, this technique could be used only for specimens where time-dependent, time-independent, and reversed plasticity effects are small.

The crack tip strain measurement is applicable to various specimen geometries and detects crack initiation even in a large-scale plasticity condition. However, it cannot be used for large specimens where the surface behavior is not identical with the crack growth in the interior.

Electric Potential Measurement. The existence of a crack or defect in an electrical field can introduce a perturbation which, if measured, can be interpreted in terms of crack size and shape. The electric field can be produced by means of direct potential or alternating potential. In this method, a constant current is passed through a cracked test specimen, and the change in the electric potential across the crack, as the crack propagates, is monitored and measured. When the crack length increases, the uncracked cross-sectional area of the specimen decreases and the electrical resistance increases. This is reflected as an increase in the potential difference between two points across the crack. The calibration curves are established by monitoring this potential increase against a reference potential and plotting it as a function of crack length to specimen width ratio. The electrical potential crack monitoring technique is discussed in detail in the article "Fatigue Crack Growth Testing" in this Volume, but a brief description of the direct potential (with a Krak Gage technique) and alternating current (ac) methods are summarized below.

In general, electric potential methods can be used for detecting crack initiation as well as for measuring the propagation rate in the laboratory. If proper calibration is established, this method can be used for predicting residual life as well.

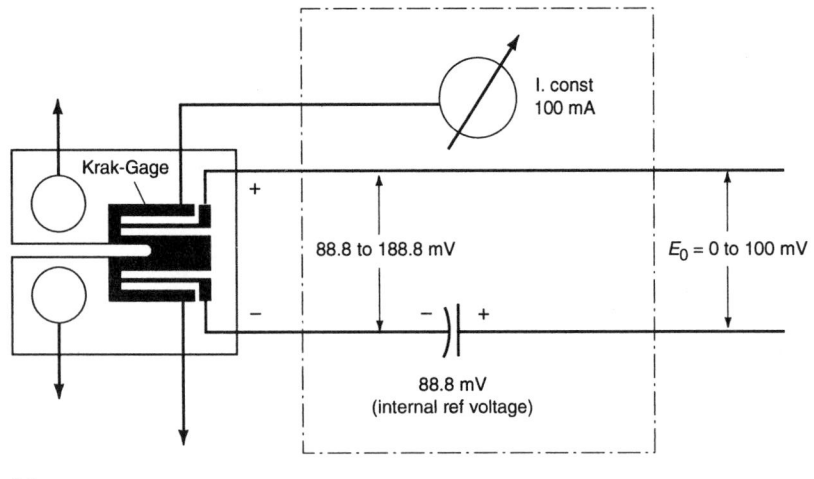

(a)

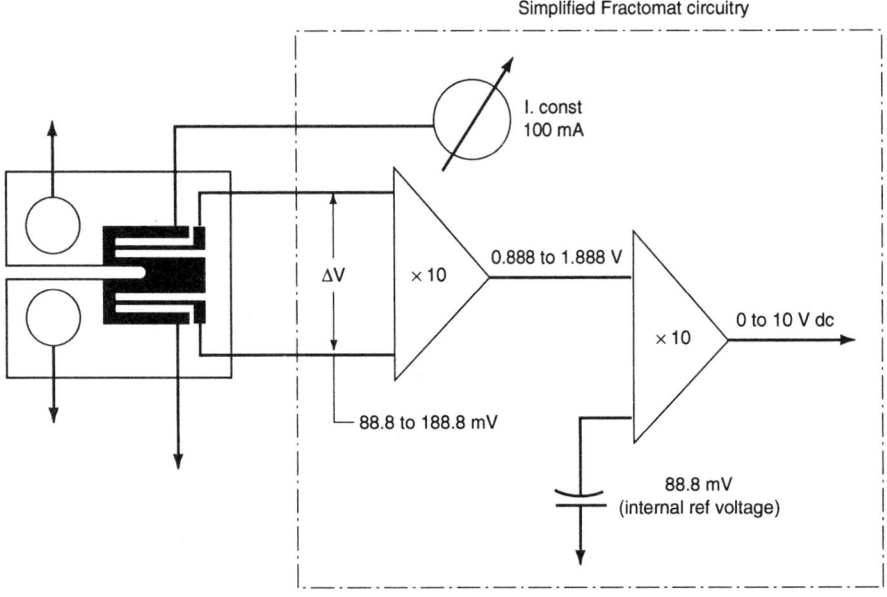

(b)

Fig. 1 Schematic of the Krak Gage technique. (a) Constant current circuitry. (b) Output voltage amplification circuitry. Source: Ref 29

This can be used for room-temperature applications as well as high-temperature applications. Typically, the crack-detection sensitivity of this method ranges from 0.1 to 0.5 mm. For the Krak Gage technique, the crack detection sensitivity is around 0.25 mm. Richards (Ref 12) and Watt (Ref 17) have summarized the relative advantages and disadvantages of dc (direct potential) and ac potential difference methods, as described below.

The direct potential (DP) method uses the changing potential distribution around a growing crack when a constant direct current is passed through the specimen. This is usually monitored by measuring the potential difference between two probes, which are placed on either side of the crack. The technique relies on the relationship between the crack length and the measured potential, which can be determined either by empirical or theoretical means. The basic equipment for

the DP method consists of a source of constant dc current and a means of measuring the potential differences that are produced across the crack plane.

The direct potential technique is simple, robust, and of relatively low cost. It is amenable to automation and for long-term high-temperature testing but is well established for only certain specimen geometries. Theoretical relationships are limited, and hence the potential difference and crack-length relationship needs to be established through calibration tests. Furthermore, the method has the limitation of not distinguishing between the crack extension and external dimensional changes of the specimen that would typically occur during general yielding and is not suitable for large specimens. The possible interference of electrochemical conditions near the crack tip cause some uncertainty in corrosion fatigue and stress-corrosion studies. For the DP

method, the sensitivity level has been reported from 0.1 to 0.5 mm based on a review of the measurements documented in Ref 18 to 28.

The Krak Gage technique utilizes an indirect dc potential measurement method, and Liaw et al. have utilized this gage for fatigue studies (Ref 29-31). Krak Gage is a registered tradename of Hartrun Corperation (Chaska, MN) and is a bondable, thin, electrically insulated metal foil of certain dimensions photoetched from a constantan alloy. The gage backing is made of a flexible epoxy-phenolic matrix that provides the desired insulation and bonding surface area similar to the technology of foil-type strain gages. Conventional and well-established foil strain-gage installation methods can be applied to the bonding and installation of such a gage to test samples. The gage is bonded to the specimen under investigation such that when a crack is initiated in the material, it will also propagate in the bonded Krak Gage.

A constant current source of the order of 100 mA is used to excite the low-resistance gage, as shown in Fig. 1. A propagating crack produces a large change in the resistance of the gage and yields a sufficient dc output, proportional to crack length, of 0 to 100 mV for the full-scale rating of the gage. The output voltage of the gage is further amplified to a 10 V dc full scale and is shown in Fig. 1(b). The precision in the geometry of the gage determines the accuracy and the linear relationship between the output voltage and the crack length. A typical length of the gage equals 20 mm and yields a crack-detection sensitivity of 0.25 mm. The potential generated is further amplified and displayed on a digital voltmeter. Furthermore, analog outputs are provided to readily interface with all conventional recording instrumentation, data acquisition systems, and computers for fully automated crack detection.

The ac electric potential method involves an ac source connected to the specimen such that the current flows perpendicularly to the crack. The ac field is typically limited to the thin skin at the metal surface and hence is effective in measuring crack dimensions at or near the specimen surface compared to the dc method. For the ac technique, the following crack-depth equation is applicable:

$$d = (v/v_o - 1)D/2 \qquad \text{(Eq 1)}$$

where d is the crack depth, v is the measured electric potential, v_o is the initial electric potential, and D is the separation distance of the output leads for measuring the potential.

The ac electric potential method is applicable to all test geometries, involves simple calibration procedures, and has no specimen-size dependence. This technique can be easily automated and has high sensitivity suitable for large-specimen testing and for surface-crack detection in specimens and structures. Similar to the dc method, the ac method produces average crack length values and accommodates relaxation from linear elastic behavior. However, the ac method is relatively expensive, connection wires need to be carefully placed, electrical insulation of specimens are re-

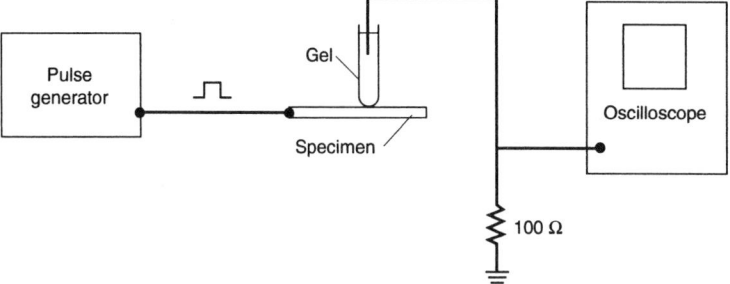

Fig. 2 Calibration curve for ac and dc potential systems. Source: Ref 32

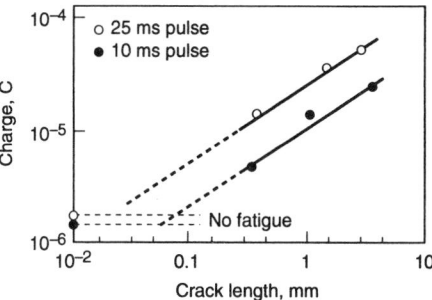

Fig. 3 Schematic of experimental arrangement for developing image and recording current flow. Source: Ref 35

Fig. 4 Effect of crack length on the charge flow during the formation of an image by a 10 V pulse. Source: Ref 35

quired, and long-term stability is difficult to achieve. Erroneous crack-length measurements can occur due to bridging of crack surfaces by corrosion products in both dc and ac methods.

Wei and Brazill (Ref 32) utilized an ac potential method for monitoring fatigue crack growth rates in an ASTM A 542 steel and reported that fatigue crack growth rates could be determined within ±20%. Their setup comprised an excitation circuit that supplied a constant ac current to the system and a measurement circuit that detected the ac potential drop across the system.

For the CT specimen geometry, Wei and Brazill (Ref 32) established the relationship between the normalized ac potential and the crack length through a calibration test in which data pairs of crack length and potential were recorded during fatigue. The calibration crack length was taken as the five-point average of posttest measurements on the fracture surface obtained at the specimen side surface, quarter points, and midpoint. They normalized the potential measurements with respect to the initial potential of the uncracked specimen and obtained a calibration curve of crack length versus normalized ac potential for three specimens (Fig. 2). Also shown in Fig. 2 is the calibration data obtained for two specimens utilizing a dc system, and the curves were fitted using a third-order polynomial. The accuracy of the ac crack-length measurement method was reported to be better than 1% for crack lengths from 20 to 45 mm. The ac system had a resolution better than 0.01 mm for a 20 nV resolution in the electric potential at an operating current of 1 A.

Gel Electrode Imaging Methods. Gel electrode imaging is capable of detecting fatigue crack initiation. It is simple and possesses good sensitivity. Typically, the crack detection sensitivity of this method is of the order of 30 μm. This technique uses a hand-held probe for detecting and imaging short fatigue cracks in metallic components subjected to cyclic loading (Ref 33). The only precondition is that the metal surface be coated with a thin anodic film before fatigue testing. It can be used to follow the fatigue dam-

age process without the need for dismantling the test fixture. Fatigue cracks as small as 0.01 mm can be easily imaged and provides discrimination of features, such as machining marks, scratches, or notches.

The gel electrode imaging method is based upon a redox printing technique developed by Klein (Ref 34). Klein soaked a filter paper in an electrolyte containing potassium iodide, starch, and agar gel and squeezed it between the specimen and a metal cathode. On application of an electric potential, the potassium iodide is anodically oxidized to release iodine ions that react with the starch to form a black adsorption complex. This usually occurs at conductive flaws in the surface oxide film on the metal at the interface between the electrolyte and the positively polarized specimen. Klein mapped the distribution of high conductivity defective areas in anodic oxide films on several valve metals.

Baxter (Ref 35) modified Klein's method and imaged fatigue cracks in 6061-T6 aluminum. He used a liquid drop electrode with a surface skin of dehydrated gel and pressed it against the specimen. During fatigue damage imaging, the current flows preferentially to thinner surface oxide regions that form during fatigue of the underlying metal. A thick layer of the surface oxide on the specimen is grown prior to the fatigue test, and during fatigue loading the thick oxide film develops microcracks exposing fresh metal surfaces. These regions rapidly reoxidize but only to a very thin layer, thus providing sites of high conductivity during subsequent imaging.

Baxter printed the image on the gel tip and photographed it immediately in order to prevent the deterioration of the image that occurs at room temperature within a few hours. The current during imaging was recorded on a Nicolet digital oscilloscope and was then displayed on a recorder. The total charge flow was obtained by measuring the area under the curve. Figure 3 shows the experimental setup for developing image and recording current flow.

The sensitivity and spatial resolution attainable by this method is determined by the amount of charge flow, which depends on the duration of the voltage pulse. The data obtained during imaging of virgin cracks with 10 and 25 ms pulses are shown in Fig. 4. The charge flow in the absence of a fatigue crack is indicated in the figure, and the data extrapolation indicate that fatigue cracks as small as 60 μm can be detected with a 10 ms

pulse. The result corresponded well with the microscopic examination.

Liquid Penetrant Method

The liquid penetrant method involves the penetration of liquid to flow into the minute surface openings through capillary action (Ref 36). The test surface is covered with a penetrating liquid, then the excess liquid is removed after a particular period of time and a developing agent is then applied to the surface. A powder film is formed after drying of the developing agent, and it draws the liquid to the surface from the crack.

The liquid penetrant method provides only crack length information, and cracks of the order of 1 μm can be detected. Typically, the crack detection sensitivity ranges from 0.025 to 0.25 mm. However, the crack depth information cannot be obtained by this technique. The sensitivity level of the method depends on the surface condition, crack morphology, and physical access to the components. The sensitivity spectrum of this method ranges from fine, tight cracks to broad, shallow, and open cracks. Using this method, only crack-length information can be obtained and cracks of the order of 1 μm in width can be detected. For field applications, the crack detection sensitivity ranges from 0.025 to 0.25 mm provided that the surface is clean, polished, and an appropriate penetrant is selected.

The liquid penetrant technique is simple, applicable to nonmagnetic and magnetic materials, and possesses higher sensitivity than the magnetic particle method. However, only surface imperfections can be detected, and in components having high surface roughness or porosity, this method cannot be successfully employed.

The liquid penetrant method is classified into four methods: water washable, postemulsifiable lipophilic, solvent removable, and postemulsifiable hydrophilic. The latter terms indicate the type of media that are required to remove the excess penetrant from the surface. For example, solvent removable requires a solvent, while water washable mandates a water spray. These methods are discussed in detail in *ASM Handbook*, Volume 17, *Nondestructive Evaluation and Quality Control*.

Magnetic Techniques

Magnetic techniques are primarily used as inspection techniques for detecting fatigue cracks in structural components, and in particular, the magnetic particle method is widely used. However, these methods can be employed only for magnetic materials. Magnetic methods provide ways of following how the magnetic properties of materials change as a function of various factors, such as microstructure, heat treatment, chemical composition, and mechanical condition. The crack detection sensitivity of the magnetic method is of the order of 0.076 mm.

Structure-sensitive magnetic properties, such as coercivity, remanence, permeability, susceptibility, and hysteresis loss, are intimately related to the microscale of domain sizes and orientations, and hence their measurements can be used to infer the microstructural state of the steel (Ref 8). Fatigue damage is associated with changes in dislocation density and dislocation structure, and thus could be measured by magnetic techniques. The different magnetic methods could be used for the measurement of fatigue damage and they are described below.

Magnetic Barkhausen Effect. The Barkhausen effect (Ref 37) consists of discontinuous changes in the flux density known as Barkhausen jumps. These jumps are due to sudden irreversible motion of magnetic domain walls when they break away from pinning sites because of changes in the magnetic field H. By placing a search coil in the vicinity of the specimen undergoing a change in magnetization, a series of transient pulses of electromotive force will be induced across it and could be measured individually by counting and amplitude sorting or as a root mean square (rms) signal, as a function of magnetic field or as a scalar rms value (Ref 8).

Karjalainen and Moilanen (Ref 38-39) investigated the effects of plastic deformation and fatigue on the magnetic Barkhausen effect. They utilized a surface coil placed between two magnetizing pole pieces operating at 50 Hz and measured the root-mean-square Barkhausen signal with respect to an applied stress axis from the mild steel tensile sample in both parallel, B_p, and perpendicular, B_t, directions. The authors observed drastic changes in the Barkhausen signals occurring only after 5% of the fatigue life and suggested that the life of components could be determined using this technique.

Magnetoacoustic Emission. Magnetoacoustic emission (MAE) is caused by microscopic changes in strain due to magnetostriction when the discontinuous irreversible domain wall motion of the non-180° domain wall occurs (Ref 40). It arises when ferromagnetic steels are subjected to a time-dependent field. A piezoelectric transducer bonded to the specimen could measure acoustic emissions, and the amplitude of MAE depends on the magnetostriction coefficient, frequency, and amplitude of the driving field. Because the stress alters the magnetocrystalline anisotropy, MAE should also change with the applied stress.

The MAE technique is of recent origin and not well developed but is sensitive to fatigue damage. Ono and Shibata (Ref 41) investigated several carbon steels, A533-B steel, and pure iron, using MAE. The magnetic field was alternated at 60 Hz, and the maximum field was 25.5 kA/m rms. They used two acoustic emission transducers of different resonant frequencies and measured rms voltages at two frequency ranges. Also, the maximum applied stress level was 350 MPa in tension. They observed that the 1020 steel showed the highest acoustic emission response among the materials tested. They also reported that residual stress levels can be determined by monitoring the ratio of the outputs of the two acoustic emission transducers for a given material condition.

Magnetic Particle Method. When a ferromagnetic material is magnetized, magnetic discontinuities that lie in a direction generally transverse to the magnetic field will set up leakage fields. The presence of this leakage field, and hence the crack or discontinuity can be detected by the application of finely divided magnetic particles over the surface, which tend to gather and are held by the leakage field. Thus, an outline of the discontinuity and its location, size, shape, and extent could be obtained from the magnetically held collection of particles. More details about this method are given in *ASM Handbook*, Volume 17, *Nondestructive Evaluation and Quality Control*.

Magnetic particles are available in a variety of highly visible colors or as a fluorescent substance visible under a black light. Crack detection resolution depends on the type of magnetic particles applied. The magnetic particle accumulation could be used to ascertain crack length, but no useful information about crack depth is generated. It is used as an inspection technique for detecting cracks in structural components during service, and cracks with a major dimension of 0.5 mm can be detected (Ref 36).

Magnetic Flux Leakage. When a ferromagnetic material is magnetized, magnetic discontinuities, such as microcracks, voids, inclusions, and local stresses, give rise to magnetic flux leakage. The magnetic flux leakage could be measured utilizing a magnetometer, and the field components could be measured in three directions (perpendicular and parallel to the flaw and normal to the surface).

Barton (Ref 42) monitored fatigue damage during stress cycling of SAE 4140 steel specimens using a high-frequency vibrating magnetic probe (60 kHz). He used various tensile and compressive stress levels and reported that fatigue damage signals were detected in the steel tubes well before gross crack development. Fatigue cracks were easily detected using this method, and he reported that the cracks could be detected with an accuracy of ±0.25 mm. Barton established a functional relationship between signal buildup and fatigue damage so that fatigue life could be predicted with good accuracy. The crack detection sensitivity of this method can be of the order of 0.076 mm (Ref 9).

Magnescope. Jiles et al. (Ref 43) reported a portable inspection device that could be used for nondestructive evaluation of the mechanical condition of steel structures and components outside the laboratory. They showed the dependence of magnetic properties of four identical samples of rail steel as a function of number of fatigue cycles. Jiles et al. followed the changes in remanence and coercivity of the rail steel samples with expended fatigue life. He observed that the coercivity and remanence reduced drastically as the material approached failure. Thus, by measuring coercivity and remanence, one would be able to predict the remaining fatigue life.

Positron Annihilation

Positron annihilation is a nondestructive method that may be utilized for predicting fatigue life. It involves injecting of positrons from a radioactive source and measuring positron lifetime as a function of fatigue cycles. It can provide a basic understanding of what stage the material is subjected to in terms of its total life.

The positron-annihilation method obtains information about the state of imperfection of the solid based on the principle that when a positron is injected into a material, it annihilates with an electron within a few hundred picoseconds (Ref 8). During this process, it emits annihilation radiation in the form of two gamma rays that travel in opposite directions. The average behavior of a large number of positrons are usually determined, and the precision of the measurement has a direct dependence on the square root of the measurement duration.

By measuring the positron lifetime in a solid, the state of imperfection of the solid can be deduced (Ref 44). The defects in crystals, such as dislocation or vacancy, serve as trapping sites for positrons. Hence, trapped positrons survive on an average longer than an untrapped positron. Untrapped positrons, on the other hand, annihilate with an electron in a more perfect region of the lattice. Thus, by measuring the average positron lifetime, the state of crystalline perfection can be ascertained with high sensitivity.

In a similar manner, the state of imperfection of the solid can be deduced by measuring the Doppler broadening of the energies of gamma rays emitted during the annihilation events (Ref 44). Defect trap sites are deprived of higher energy core electrons, and hence a trapped positron has a higher probability of annihilating with a lower

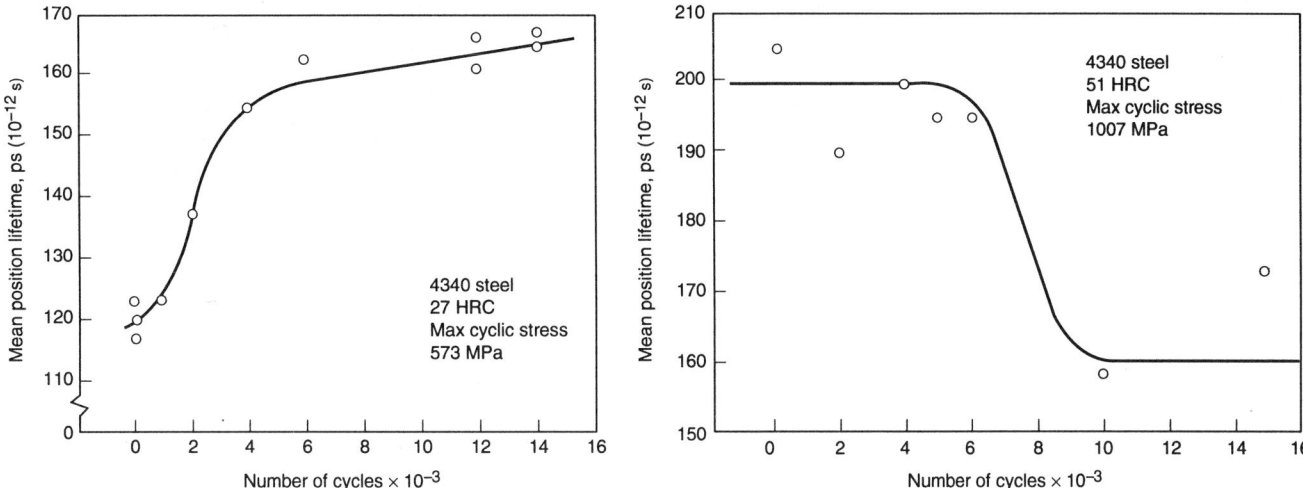

Fig. 5 Mean positron lifetime in picoseconds versus number of fatigue cycles for 4340 steel of initial hardnesses, 27 and 51 HRC. Source: Ref 45

energy conduction electron. This trend is reflected as a narrowing of the energy distribution about the value of 511 keV. If the electron-positron center of the mass was stationary, then 511 keV would be the gamma ray energy. Thus, the Doppler energy shift from the 511 keV can be considered to be due to the energy of the electron involved in the annihilation process.

Fatigue damage increases defect concentration, such as dislocation and vacancies, in the test specimen and hence can be estimated from positron annihilation measurements. The fraction of positrons trapped at the defect sites increases with increase in fatigue damage, and hence saturation can occur at high damage levels. In order for positrons to be useful for actual applications, there has to be a reasonable balance between the trapping and annihilation processes (Ref 8). The existence of such balance and whether fatigue life could be monitored successfully using positron lifetime measurements can be determined only by empirical methods.

In high-cycle fatigue testing, it typically takes several fatigue cycles before fatigue damage can be observed in a test specimen (Ref 8). With the inception of fatigue damage, a rapid buildup of damage occurs with sustained fatigue cycling. This damage can be easily followed using positron lifetime measurements since the positron response increases with increase in fatigue damage. This process continues and beyond a particular damage level, the positron response either flattens or increases very slowly.

The positron mean lifetime measurements are conducted by sandwiching the positron source between two flat-faced portions of the test specimen with the two scintillator detectors positioned on the opposite sides of the sandwich (Ref 8). A ^{22}Na source emits a 1.37 MeV marker gamma ray along with the positron at the same time. By measuring the time lag between the arrival of the marker gamma ray and of one of the annihilation gamma rays in the scintillators, the individual positron lifetime in the sample is established.

The Doppler broadening measurements are typically conducted using a Ge(Li) detector, multichannel analyzer, and digital stabilizer (Ref 44) and has a resolution of 1.24 keV full width half maximum at the total count rate of 14 kHz. The changes in the spectrum of the annihilation photon energies are described using a shape factor. The shape factor represents the sum of counts in a peak region divided by the total counts in two wing regions.

Lynn and Byrne (Ref 45) investigated AISI 4340 steels of Rockwell hardness levels (27 and 51 HRC) using cantilever bending fatigue cycles with a maximum stress of two-thirds of their corresponding yield stresses. Figure 5 summarizes their measurements. The mean positron lifetime decreased during fatigue for 51-HRC steel, and this was due to cyclic fatigue softening. However, fatigue hardening of the soft 27-HRC samples resulted in increasing the mean positron lifetime. The positron lifetime was 119 ps initially, and increased to 165 ps at fracture, thus indicating an increase in the number of defects. Also, the decrease in slope occurred at about 20 % of the total fatigue life.

Alexopoulos and Byrne (Ref 46) conducted x-ray line broadening and positron lifetime measurements on 4340 steel with hardness of 30 HRC. In order to provide a better explanation and understanding for the increase in the positron lifetime with cyclic fatigue of soft steels, they made measurements at much smaller fatigue intervals. They observed that the mean positron lifetime increased to a maximum in the vicinity of 10^4 cycles, and subsequently, instead of failure, there was an interesting undulation in the positron mean lifetime. This undulation persisted till fracture at about 73,000 cycles, and the more frequent interruptions and reapplications of fatigue cycling seem to have considerably increased the fatigue life by a factor of about 7. They called this process "coaxing." The x-ray measurements did not give any indications of corresponding changes in particle size. This trend indicates that the positrons did respond to structural changes

that do not influence the x-ray particle size. Byrne (Ref 44) did an excellent review paper on positron studies of the annealing of the cold-worked state of different materials.

Duffin and Byrne (Ref 47) utilized positron Doppler broadening measurements to detect changes in trapping mechanisms in steels. They cycled 1020 steel at an alternating stress of ±606.7 MPa (much below the yield stress of 1110 MPa) in cantilever bending in a thermomechanically produced condition arrived at by: 75% cold rolling, up-quenching to 751.5 °C for 1 min followed by a brine quench. The Doppler peak to wings parameter was plotted as a function of fatigue cycles, and the variation reflected an increasing degree of damage during cycling. An excellent review of the application of positron annihilation techniques for defect characterization was done by Granatelli and Lynn (Ref 48).

Acoustic Emission Techniques

Acoustic emissions allow the capability of determining fatigue crack initiation and following the crack propagation as the crack generates elastic waves in the material. Acoustic emissions occur as a result of the release of elastic strain energy that accompanies crack extension and other processes involving atomic rearrangement in materials (Ref 49). These elastic waves can be detected by the use of sensitive transducers that are located at the surface of the sample. These piezoelectric transducers normally operate in the range of 20 kHz to 1 MHz.

Acoustic emissions can produce transducer outputs that can vary over many orders of magnitude from less than 10 μV to more than 1 V. Usually, most emissions produce outputs toward the lower end of this range, and hence processing equipment is used. The initial preamplification of the acoustic emission signals involves a gain of 20, 40, or 60 dB. Band-pass filtration is then used, commonly over the range of 100 to 300 kHz, for

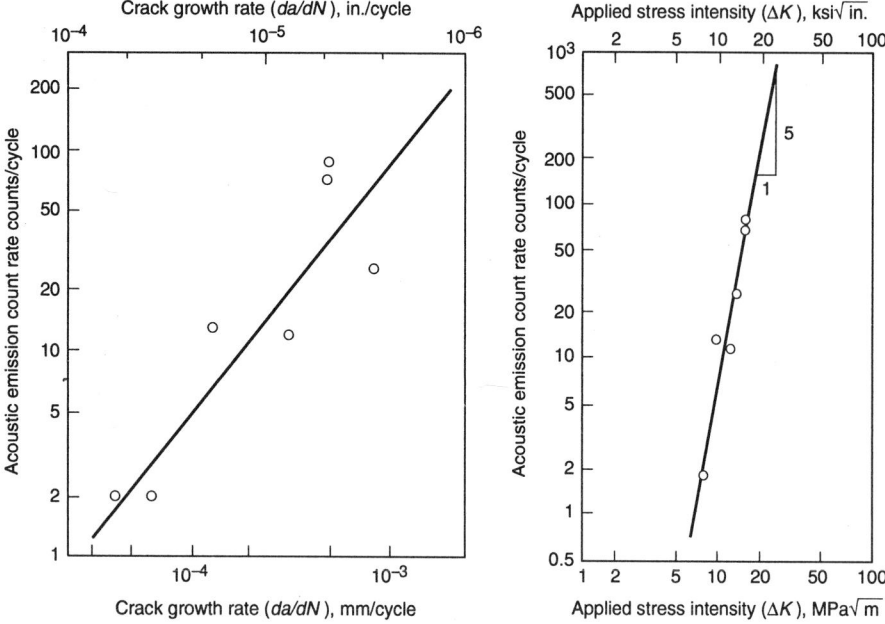

Fig. 6 Crack growth rate and ΔK versus acoustic emission count rate for 2024-T851 aluminum alloy. Source: Ref 51

removing much of the mechanical and electrical background noise before final amplification. This is followed by main amplifiers with gain levels. There are various methods by which the amplified acoustic emission signals can be analyzed. The different methods yield different information about the source responsible for the emissions. They include: ring-down counting, event counting, energy measurements, amplitude measurements, and frequency analyses. A good summary of the analyzing methods, and a review of the literature is provided by Lindley and McIntyre (Ref 50).

Acoustic emission monitoring has been used in the laboratory to study various crack propagation mechanisms including fatigue, corrosion fatigue, stress corrosion, hydrogen embrittlement, and ductile tearing. It could also be useful for predicting the residual fatigue life in specimens, if properly calibrated. The crack detection sensitivity of this method is of the order of 0.1 mm. Examples are given below.

Morton and coworkers (Ref 51, 52) studied the high-cycle fatigue behavior of 2024-T851 aluminum and correlated the peak load acoustic emission rate, N', with the crack growth rate, da/dN, and the applied stress-intensity factor range, ΔK (Fig. 6).

Houssyn-Emam and Bassim (Ref 7) utilized an acoustic emission technique to monitor the onset of crack initiation and to follow the fatigue damage process in low-cycle fatigue of AISI 4340 steel. They plotted total counts against the number of cycles and divided it into three regimes. The first stage is the initial softening that results in a high acoustic emission activity. The second stage corresponds to a quasi-stable stage during which there is relatively little activity. This is followed by a further increase in the acoustic

activity that accompanies the onset of crack initiation and crack propagation to failure.

Ultrasonic Methods

Ultrasonic techniques involve transmitting pulses of elastic waves into the specimen from an ultrasonic probe held on the surfaces of the specimen (Ref 53). It is used for following crack propagation as an in-field or a laboratory technique. The crack detection sensitivity of this technique is around 50 μm.

Ultrasonic techniques are widely used for the detection and sizing of fatigue cracks and monitoring the crack growth both in the laboratory and field. Ultrasonic methods fall into one of the following groups depending on the way the crack size is determined (Ref 53):

- Those methods that calibrate the ultrasonic signal amplitude directly in terms of crack size
- Those techniques in which transmitting and/or receiving probes are displaced over the specimen surface to locate the crack tip at a particular position within the ultrasonic beam
- Those methods that measure crack size by the time of flight of pulses from the transmitter to receiver via the crack tip, irrespective of the pulse amplitude

Ultrasonic measurements comprise two stages. The first stage involves obtaining the signal from the crack, and the second stage involves the interpretation of this signal to estimate crack size and shape. In order to obtain a signal, the most commonly used equipment is the piezoelectric probe and commercial flaw detector, and it is shown in Fig. 7. The wavepackets of ultrasound are transmitted into the specimen, and the scattered pulses

are then received at the probe. These pulses are then reconverted to electric signals and are displayed on an oscilloscope screen as a function of time of flight.

Ultrasonic Amplitude Calibration Methods. In this method, a fixed transmitting probe is used to beam pulses onto a crack, and a fixed receiver is used to receive the signal (Ref 53). The receiver can be located either in the shadow of the crack or positioned so as to receive the specular reflection from the crack face. The amplitude calibration methods use specimens containing known cracks to calibrate either the drop in directly transmitted signals or the amplitude of specular echoes against crack size. If the ultrasonic coupling of the probes and the morphology and orientation of the cracks are reproducible, then accurate results could be obtained.

Lumb et al. (Ref 54) utilized a compression beam to monitor through-thickness growth of fatigue and ductile cracks initiating at shallow surface notches or natural cracks at the toes of the welds. They established the ultrasonic signal versus crack depth calibration curve using milled slots and checked against part-through fatigue cracks from interrupted tests. They reported that growth increments of 0.025 mm can be easily detected and larger amounts of growth measured to ±0.25 mm.

Defebvre and Pouliquen (Ref 55) monitored fatigue tests using surface waves. They observed a sudden increase in the attenuation at about 60,000 cycles of a steel sample and related it to the onset of microcracking. They monitored a total of 170,000 cycles and the crack had propagated to 30% of the width of the specimen during this time.

Recently, Resch and Karpur (Ref 56) utilized a surface acoustic wave technique to detect the initiation of surface microcracks in highly stressed regions of hourglass-shaped 2024-T6 alloy aluminum specimens during fatigue cycling. They used contacting wave transducers to excite the incident waves and to detect the reflected wave signals. They demonstrated the effectiveness of a split spectrum processing algorithm to separate specular reflections of isolated cracks from nonspecular reflections of microstructural features.

Joshi (Ref 57) utilized an ultrasonic attenuation technique to monitor continuously precrack damage and crack propagation in polycrystalline aluminum and steel specimens subjected to cyclic loading. He reported that the measurement of change in ultrasonic attenuation prior to the onset of the stage II crack propagation proved useful in explaining the rate of crack propagation. Also, the specimens that undergo higher precrack damage showed shorter postcrack percent lives.

Probe Displacement Method on Compact Specimens. Clark (Ref 58) developed an equipment that utilized the specularly reflected signal from the crack for use in a wedge-opening load (WOL) fracture-toughness specimen. They used a fixed 10 mm diam, 10 MHz normal compression probe in pulse-echo to observe the increase in echo as the fatigue crack grows. They moved the probe along the surfaces of specimens containing long fatigue cracks to establish the calibration curve of growth against echo amplitude. The accuracy of this method was found to be about ±0.1 mm using beach-marked cracks in steel and aluminum. However, the maximum amount of crack that can be monitored without transducer movement was only 2.5 mm because of the saturation characteristics of the associated instrumentation.

Subsequently, Clark and Ceschini introduced a motor drive to increment the probe's position along the specimen (Ref 59). They used a conventional ultrasonic flaw detector in conjunction with a reflectoscope. Using this method, the position of the transducer on the specimen surface can be related to the extent of crack growth by transducer movement such that a constant flaw signal is maintained from the tip of the propagating crack. This arrangement permitted the crack tip always to be kept near the center of the beam. Their setup is shown schematically in Fig. 8.

First, a 25 mm sweep to peak second back reflection signal was generated through the uncracked portion of the specimen by adjusting the ultrasonic instrumentation. Then the transducer was positioned on the specimen so as to obtain a 5 mm sweep to peak signal from the fatigue precrack tip (Fig. 8, position A). The position A corresponds to the zero crack growth transducer location. This serves as a reference for subsequent crack growth measurements. With an increase in crack length, the flaw signal amplitude increases (Fig. 8, position B) due to the increase in the reflecting area of the crack within the scanning beam. The transducer is then moved to position C in the direction of crack growth till the flaw signal is similar to that of position A. Thus, the transducer movement distance is equivalent to the crack growth increment. By recording the transducer location versus time or cycles, one could deduce the crack growth rate. Using this method, a crack-length measurement sensitivity of ±0.25 mm was reported.

Time of Flight Measuring Techniques. These methods detect and measure the flight time of the ultrasonic pulse diffracted from the crack tip (Ref 60). If the path taken from the transmitter to the receiver via the crack tip is known, one can calculate the position of the tip and hence the crack length.

The probe arrangement for measuring surface cracks along with the electronics is shown in Fig. 9 (Ref 53). Most of the beam that is incident on the crack is either reflected or passed directly on, but a small portion is diffracted. This diffracted signal reaches the receiver. If the transmitted rays emerge effectively from a point on the specimen surface, and the diffracted rays are received at another point, then the time of the flight, t, to the crack depth, a, is related by the equation

$$a = [(ct/2)^2 - h^2]^{1/2} \qquad \text{(Eq 2)}$$

where c is the velocity of sound, and h is the horizontal distance of the receiver from the crack.

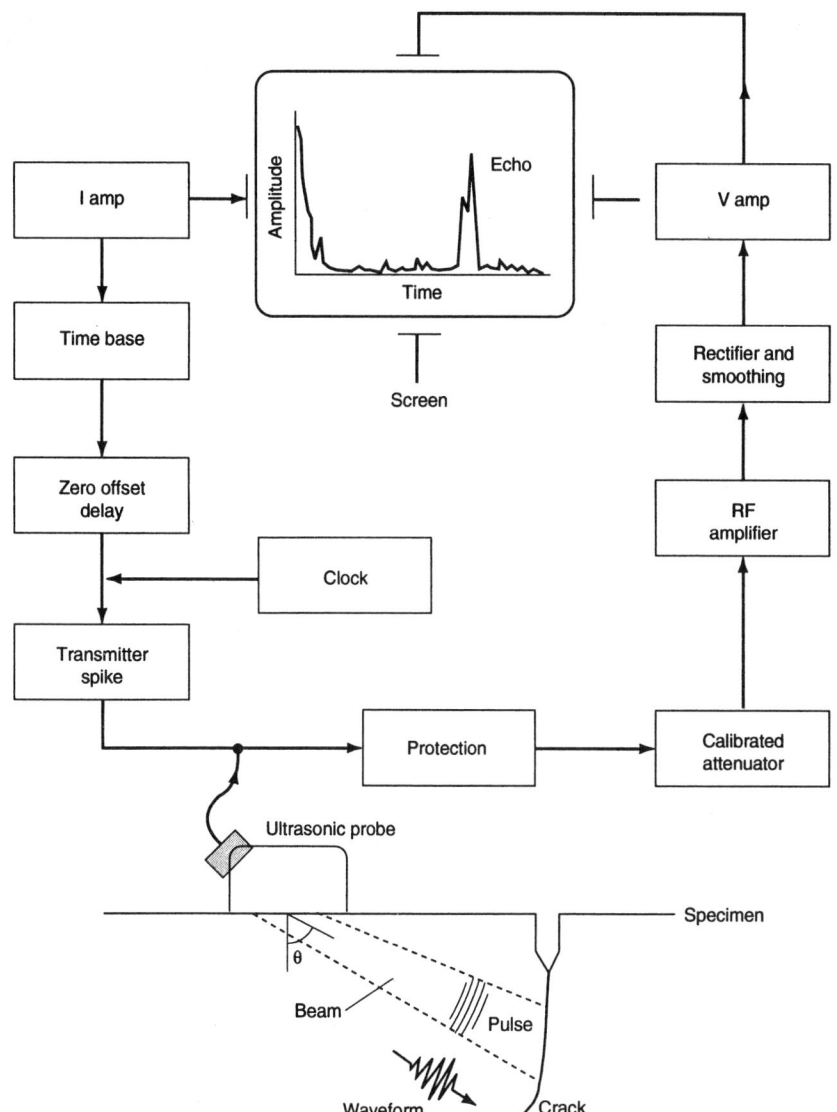

Fig. 7 Block diagram of an ultrasonic probe and flaw detector. Source: Ref 53

The surface crack measurements by timing the diffracted pulses is accurate since the time of flight can be measured precisely to nanoseconds level. Mudge and Whitaker (Ref 61) have measured fatigue precracks in wide plate tests and the onset of ductile tearing in crack-opening displacement specimens. They reported errors in measuring fatigue crack depth within ±0.2 mm.

Silk (Ref 60) pointed out that both surface and subsurface defects can be evaluated using the ultrasonic method. Richards (Ref 12) summarized the relative merits and demerits of the ultrasonic methods for fatigue crack growth monitoring. The merits of this method include: Embedded cracks and crack profiles can be easily measured, it can be easily automated, both metals and nonmetals can be studied, and it accommodates relaxation from linear-elastic behavior. However, the ultrasonic methods have the following limitations: They are neither suited for small specimens nor are they well developed for high-temperature studies, and they are expensive.

Eddy Current Techniques

The eddy current method is essentially a combination of a local resistance measuring technique and a magnetic method. Typically, the crack-length as well as the crack-depth information can be easily obtained using this method. It is widely used in the areospace industry and also for laboratory applications. The crack detection sensitivity of this technique is around 0.1 mm.

The eddy current method is essentially a variation of the alternating current electric potential method. The connection between the specimen and the measuring system is done by electromagnetic induction instead of connecting wires (Ref 9, 62). In this method, an alternating current is passed through a coil adjacent to the sample surface, which contains crack initiation sites or cracks. An alternating magnetic field is created, and this induces eddy current in the sample. The eddy currents result in a secondary current, which adds vectorially to the exciting field, and the

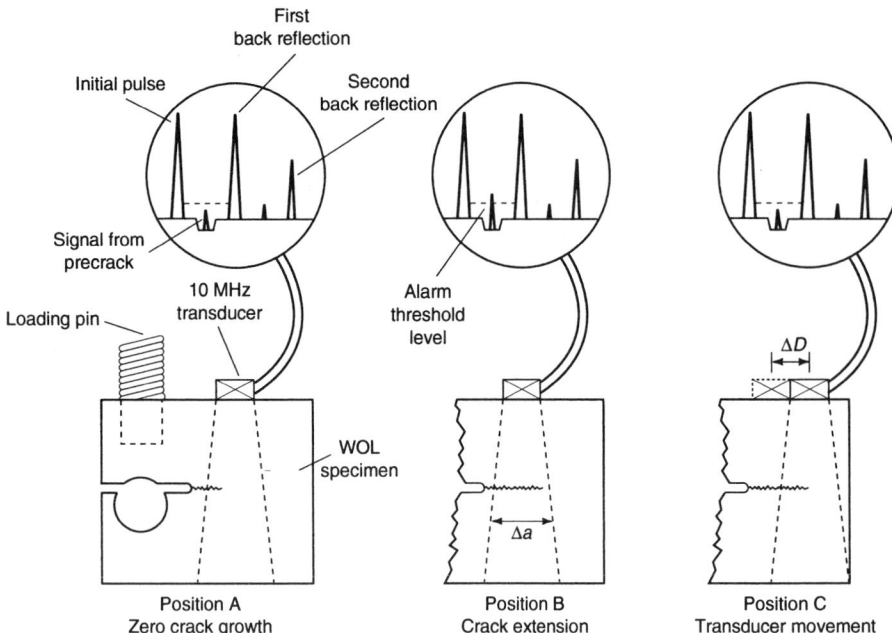

Fig. 8 Clark and Ceschini's ultrasonic setup. Source: Ref 59

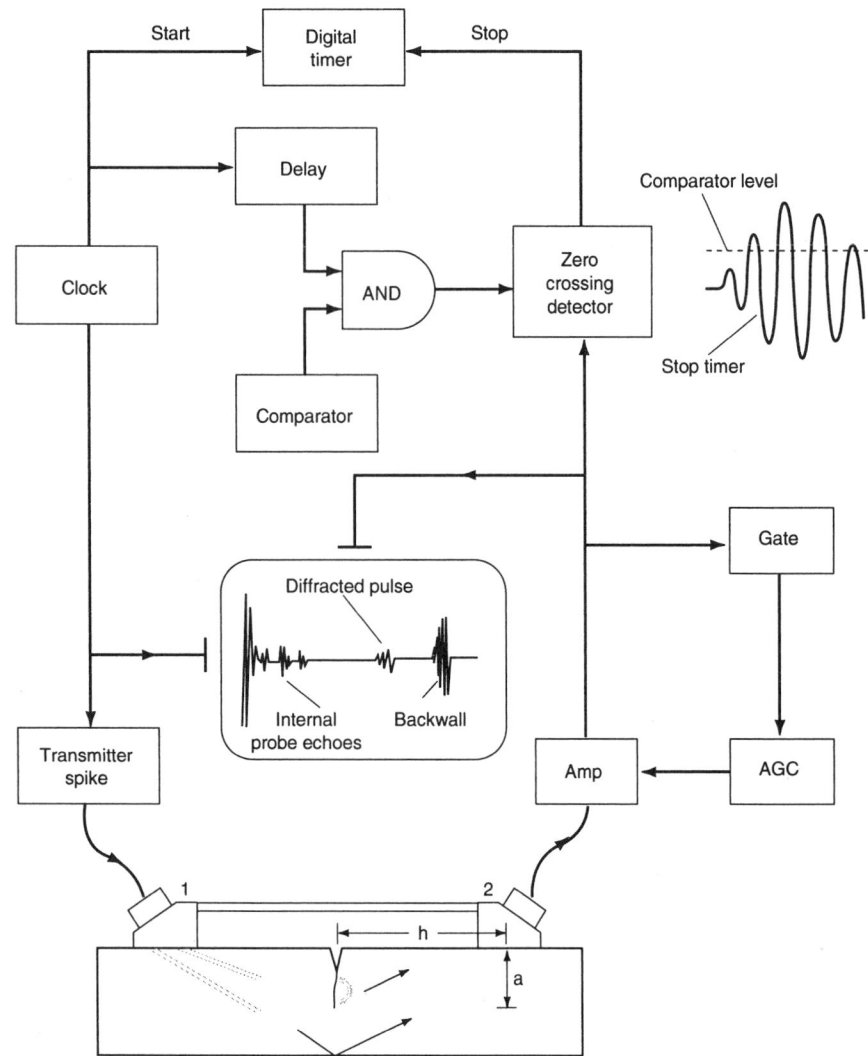

Fig. 9 Schematic measuring the time of flight of a diffracted wave. Source: Ref 53

combined field can then be detected by a secondary coil. However, for crack detection, the variation in the complex impedance of the driving coil is usually determined.

The eddy current method uses a range of frequency from several hundred Hz to several MHz depending upon the type of application. Eddy current excitation is usually on a small, local scale, and hence a traveling probe is commonly employed to scan the whole surface of the test specimens and to detect small defects. Because there are no direct electrical connections to the specimen, the specimen insulation from the test machine is usually not required.

Portable eddy current instruments are available, and they exhibit phase and/or amplitude changes in the eddy currents induced in the presence of a crack. The amplitude or phase variation can then provide estimates of crack length or depth, respectively, and typically for a short crack, crack length or depth is obtained from amplitude or phase variation. The crack length or crack depth is usually several times the eddy current skin depth, S, and is given by the equation

$$S = (\mu\,\mu_0\,\sigma\,\pi f)^{-0.5} \qquad \text{(Eq 3)}$$

where μ is the permeability of the material, μ_0 is that of the free space permeability, σ is the metal conductivity, and f is the frequency. In eddy current measurements, there is always a compromise between high sensitivity at high frequencies and the ability to monitor deeper cracks at lower frequencies.

The eddy current method depends on the change in the inductance of a search coil in the vicinity of a conducting test specimen caused by the generation of electrical currents in the test specimen when it is subjected to a time-varying magnetic field. It can be used for the crack detection because the defects interrupt the flow of the eddy currents generated in the material. This is reflected in a different complex impedance of the eddy current pick-up coil when it is positioned over the flaw in comparison to the signal generated over an undamaged region of the material.

An eddy current system employed for continuous crack monitoring has been reported (Ref 10). In this system, the probe is enclosed in a nylon sheath and is positioned at a fixed distance (0.25 mm) from the sheet surface in order to prevent any damage of the probe when the specimen fractures into two pieces. On the occurrence of a crack, the eddy current off-null signal is used to drive the linear servoactuator horizontally to the right. The probe is then moved physically to the right, and when it reaches the crack tip the off-null signal drops to zero and the servoactuator movement is stopped. Thus, the high-response actuator system is locked onto the tip of the crack. This system is capable of measuring increments in crack growth of less than 0.25 mm. The eddy current method is simple and amenable to automation. However, it is expensive and produces only surface measurements (Ref 12).

Infrared Techniques

Infrared techniques allow detection of fatigue damage from remote locations. It also can be used for predicting the residual fatigue life of components in service. Infrared techniques have been investigated for their potential to detect fatigue damage since the mid-1970s. They can be classified into passive, mechanically activated, or radiation activated (Ref 8). In the passive technique, the heat produced by spontaneous strain release is monitored, and in the case of the mechanically activated method, the rise in temperature around stress concentrations is monitored when the material is subjected to cyclic loading. In the radiation-activated technique, heat is applied to the material and the subsequent heat flow is observed over a period of time. The infrared technique has the following features: Functions in real-time, is nondestructive, and can be used for remote measurements.

Huang et al. (Ref 63) used an infrared-sensing method for monitoring fatigue processes in stainless steels and superalloys during a revolving-bending fatigue test. They used an infrared radiometer to record the temperature changes of the center part of the specimen. They reported an exponential relationship between the temperature rise and stress increment of the fatigue fracture. The rate of increase in the initial temperature for materials with high ductility during high-stress fatigue testing could be related to the life of the fatigue fracture. On the basis of their experiments, they concluded that the infrared technique could be used for monitoring the sudden fracture due to overloading as well as for predicting fatigue life.

Exoelectrons

Exoelectron methods can be used to predict residual fatigue life as well as to follow crack propagation. However, it has limited sensitivity for detecting fatigue cracks. The photoelectron emission from a metal may be enhanced by plastic deformation of the surface (Ref 64). This effect is commonly known as exoelectron emission. Exoelectrons can be produced by unidirectional tensile deformation from slip steps. As a slip step emerges from a brittle natural surface oxide, cracks open to reveal the fresh metal surface of the slip step. These surfaces have a lower photoelectric work function than the surrounding oxide-coated surface and result in enhanced emission.

Baxter investigated the fatigue behavior of a 1018 steel sheet stock in a reverse-bending constant-amplitude mode using the exoelectron approach (Ref 64). The specimen was mounted in a vacuum chamber, and a small spot (~70 μm diam) of ultraviolet radiation was used to scan along its gage length. The light source utilized was a 1 kW mercury arch lamp with a Corning 9-54 filter. The ultraviolet spectral range of interest stability was monitored by diverting the beam through an interference filter, and the transmitted radiation was measured using an RCA 1P28

photomultiplier. The electrons emitted from the sample were accelerated up to 500 eV and were detected by an electron multiplier. Baxter recorded the emission rate as a function of the position of the light spot. Five parallel paths separated by ~300 μm were scanned to provide a more complete and representative picture of the exoelectron emission generated during fatigue.

Baxter demonstrated that the exoelectrons emitted are associated with the accumulation of fatigue damage but are also influenced by pressure. For example, exposure to higher pressures of air results in decreased exoelectron emission. He interrupted the fatigue cycling at 800 cycles and exposed the sample to air at atmospheric pressure for 1 h, thereby eliminating the three emission peaks. On resumption of the fatigue cycling, the emission peaks reappeared, grew rapidly, and followed an apparent extension of the original growth curve which clearly shows the significance of the surface oxide (Fig. 10). With the accumulation of fatigue deformation, the brittle surface oxide cracks open and reveal a fresh metal surface of a lower work function ($\Delta\varphi \sim 1$ eV) that emits exoelectrons. The location of final failure always corresponded to the largest exoelectron peak.

Samples after being fatigued at different strain amplitudes were compared to produce a range of fatigue lives from 27,400 to 942,000 cycles. The normalized exoelectron emission intensity (at 2%) was plotted against the number of fatigue cycles normalized with respect to the number of cycles of failure. The parallel growth curves revealed that the increase of localized exoelectron emission is a very systematic, reproducible, and continuous process, particularly in the range of ~0.7% to 7% of life. Based on his results, Baxter concluded that the intensity of the localized exoelectron emission is a measure of the localized accumulation of fatigue damage. Also, the growth of the emission is not only a function of the number of fatigue cycles at a given strain level but is related to the total accumulated fraction of life. In order to facilitate the extraction of the number of fatigue cycles remaining before failure from the exoelectron emission measurement, he developed a procedure for normalizing the emission intensity. Based on this new procedure, he showed that when the maximum intensity of localized exoelectron emission is 10 times the initial background intensity, the sample is between 0.8 to 3% of its ultimate fatigue life.

Gamma Radiography

The gamma radiography technique is an in-field technique, and the crack detection sensitivity is typically around 2% of the component thickness. It uses penetrating radiation emitted by an isotope source, such as ^{60}Co or ^{192}Ir, on a structural component (Ref 9). This penetrating radiation is either transmitted or attenuated by the component under investigation. Fatigue cracks having major dimensions parallel to the radiation represents regions lacking attenuative material,

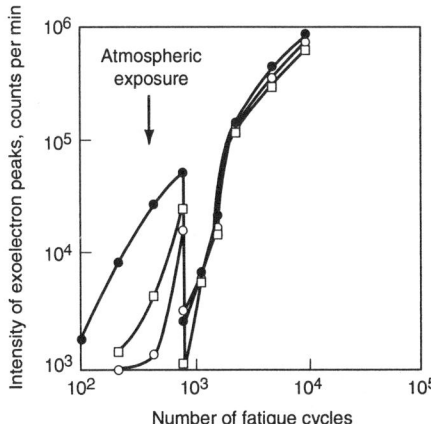

Fig. 10 Growth of three exoelectron peaks with continued fatigue cycling. Test interrupted at 800 cycles and specimen exposed to air at atmospheric pressure for 1 h. Fatigue cycling then resumed under vacuum. Source: Ref 64

and the difference can be easily imaged on a radiographic film.

Gamma radiography is typically an in-field application technique, and its use is restricted to dense or thick metallic materials. Crack detection sensitivity of this method is typically 2% of the thickness of the component. Crack length also can be measured with a sensitivity level that depends on component geometry, crack morphology, and accessibility to the component (Ref 36). Another method similar to gamma radiography is x-ray radiography, which is primarily used for laboratory applications.

Microscopy Methods

Microscopic techniques allow us the capability of understanding the mechanisms involved in fatigue crack initiation and propagation. It is the most widely used technique for characterizing the fatigue damage. It has very high sensitivity for crack detection and can be used for following crack propagation. This feature provides us insights not only in microstructural changes, but also in compositional changes. Crack detection sensitivity of the scanning electron microscopy (SEM) and the transmission electron microscopy (TEM) methods are 1 and 0.1 μm, respectively. Optical techniques serve the same purpose as that of the microscopic techniques but with a lower sensitivity. Atomic force microscopy (AFM), scanning tunneling microscopy (STM), and scanning acoustic microscopy (SAM) are relatively new techniques. They can provide much better insights into the crack nucleation process than any of the other techniques. However, they have to be nurtured and involve elaborate specimen preparation.

Electron Microscopy. Scanning electron microscopy (SEM) and transmission electron microscopy (TEM) are often used to follow fatigue crack initiation and growth behavior to identify the mechanisms involved. The fatigue process can be divided into four stages based on the structural changes that take place when a metal is subjected to cyclic stress as described in the article "Fatigue Failure in Metals" in this Volume and Ref 65:

- *Crack initiation*: This represents the early development of fatigue damage.
- *Slip-band crack growth*: During this stage, the deepening of the initial crack on planes of high shear stress takes place and is referred to as stage I crack growth.
- *Crack growth on planes of high tensile stress*: The crack grows in a direction normal to the maximum tensile stress and is referred to as stage II crack growth.
- *Final ductile fracture*: The crack reaches a length at which the remaining cross section does not have the ability to support the applied load.

For steel alloys subjected to fatigue testing in air or inert environments, the crack can initiate from persistent slip bands, extrusions/intrusions, grain boundaries, inclusions, and porosity (Ref 9). During fatigue testing, the dislocations can move to the specimen surface and form fine lines. These lines are called persistent slip bands. The stage I crack propagates initially along these slip bands, and the fracture surface of stage I fractures are typically featureless. In contrast, the stage II crack propagation fracture surface is marked by a pattern of ripples or fatigue fracture striations.

Three approaches of following fatigue damage using SEM are replication, direct, and in situ techniques. In the replication method using SEM, cellulose acetate films softened with acetone are typically used to replicate the specimen surface for detecting crack initiation (Ref 9). Usually, the replicas are taken at a predetermined number of fatigue cycles to detect crack initiation processes. The replicas are placed or rolled onto the fatigued specimens. The specimen should be under tensile loading as replication proceeds. The above procedure enables the opening of the cracks and the better penetration of the replication material into the potential cracking area. Following acetone evaporation, the acetate films are removed from the specimens to develop replicas.

The replicas are typically coated with gold in a vacuum evaporator. Replicas represent a negative image of the actual surface as the replicas are typically based on the one-stage technique. Hence, small fatigue cracks on the specimen surface appear as protrusions whereas extrusions on the surface appear as valleys in the micrographs. The crack-length detection sensitivity is typically 1 μm. A two-stage replica technique is also available (Ref 66).

In the two-stage replication method, a layer of solder approximately 0.3 μm is vapor deposited on the cellulose acetate, and subsequently a mount of epoxy adhesive containing a set screw is applied to enable the peeling of the solder film. Thus, in this technique, positive impressions of the actual test specimens are obtained. The resolution of crack size detection is typically 0.1 μm.

The replication techniques enable the microstructural evolution of fatigue cracks to be examined easily at the same site on the specimen as a function of the number of fatigue cycles and provides detailed and direct information of small fatigue cracks.

In the direct method, the fatigued specimens are periodically removed from the test machine and inspected for evaluating crack initiation process using SEM. In some instances, a multiple specimen technique is used for studying crack initiation. Each specimen can be fatigued for a given number of cycles and removed from the test machine for SEM examination. This method is expensive and time consuming compared with the replication technique. In the in situ technique, the fatigue machine is installed in the SEM, and the test specimens are fatigued as well as inspected in the SEM (Ref 67-70). This method is effective and convenient for investigating crack initiation, and the detection sensitivity is typically 1 μm.

Two-stage replicas are prepared for TEM examination (Ref 9). Replicas are taken from the fatigued specimens and coated with a thin layer of metal, such as gold. Then, the replicas are generally coated with amorphous carbon to develop the two-stage replicas. Typical crack detection sensitivity is 0.1 μm.

Davidson and Lankford (Ref 71) have provided a comprehensive review of fatigue crack growth in metals and alloys and discuss in detail the origin of striations and crack growth. The spacing of fatigue striations provides important evidence for understanding the fatigue crack growth process. This is because striations provide unambiguous, quantitative evidence of the increment by which a fatigue crack advances. Grinberg (Ref 72) examined the fatigue behavior of annealed iron in moist air. Figure 11 illustrates the fatigue crack growth behavior compared with the average number of cycles required for single striation formation, and the striation spacing was found to be much greater than da/dN (Ref 72).

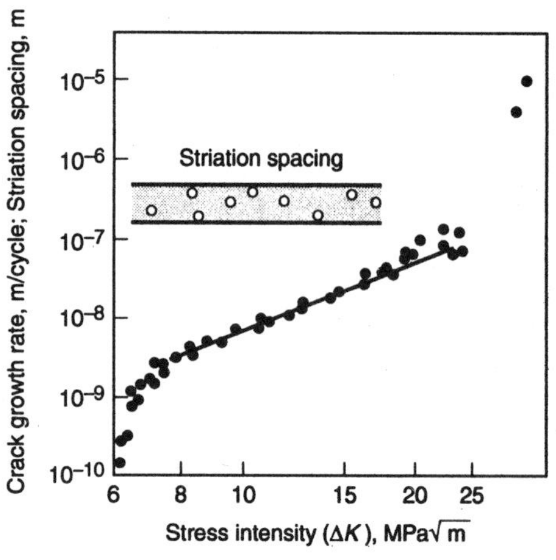

Scanning Tunneling Microscope (STM). The STM is a recent innovation and is capable of resolving surface features down to the atomic level. The STM works on the principle of development of tunneling current (Ref 73). A tunneling current is developed when an electrode is placed close to the specimen surface at a distance of 0.5 to 1.0 nm away from the surface. By maintaining a constant tunneling current, as the probe moves across the specimen, it pops up when there is a protrusion on the surface. It moves down when it comes across a cavity, and the up-and-down motions are recorded by the computer. The topographical data thus gathered provide a sensitive image of the specimen surface.

The STM has a sharp conducting tip that traces the surface contours with atomic resolution, and the tip is moved in three dimensions by means of an x, y, z piezoelectric translator (Ref 74). With the piezoelectric element calibrated to move 1 nm for a 1 V application, the tip will move over approximately three atoms for an incremental potential of 1 V. The voltage applied to the z-piezo element governs the distance between the

Fig. 11 Crack growth rate and striation spacing for an annealed iron tested in moist air. Source: Ref 72

surface of the specimen and the tip. The voltage is determined by a feedback circuit that also measures and controls a small electric current. This current is due to the electrons tunneling between the tip and the sample and is affected by the bias voltage applied to the tip. The tunneling current is maintained constant by the feedback circuit, which modulates the voltage to the z-piezo, as the x-piezo moves the tip across the specimen surface.

The amplitude of the tunneling current is very sensitive to the gap distance between the tip and the specimen surface. For example, as the distance between the tip and the surface changes by 0.1 nm, the tunneling current value changes by a multiple of 2 or greater. This tunneling current sensitivity enables divulging of height differences along the contours to be better than 0.01 of an atomic diameter. However, the lateral resolution along the contours is governed by the radius of curvature of the tip. In a single scan, the voltage applied to the z-piezo is recorded as a function of the voltage applied to the x-piezo. Thus, a complete image is an assembly of multiple scans, with each displaced from the preceding scan by a small shift in the y direction, to form a raster pattern. By virtue of computer-aided image processing, the data can be presented as images that provide topographical information either as a gray level, illuminated filled surfaces, or multicolored elevation maps.

Recently, Venkataraman et al. (Ref 6) used STM to study fatigue crack initiation of silver single crystals oriented for a single slip. They reported that the slip bands could easily be captured using STM, and the fatigue process has a definite crack nucleation stage. An STM image of a just-nucleated crack found within a slip band of a specimen fatigued to crack initiation at 180 K in He-15%O_2 was also captured. Subsequently, Sriram et al. (Ref 75) demonstrated the effect of oxygen partial pressure on fatigue crack initiation in silver single crystals and captured the nucleation process using STM.

Sriram et al. (Ref 76) investigated the role of surface chemistry in the initiation of fatigue cracks for silver single crystals. They conducted fatigue tests in an oxygen environment up to crack initiation on pure silver specimens. The STM can be easily utilized for observing shallow cracks with the lateral resolution restricted by the geometry of the tip. However, the STM can be used only for conducting surfaces. By using STM, the cracks were identified that satisfied the following criteria:

- They were invariably associated with slip bands.
- They were pit- or arrowhead-shaped and usually 1 μm in length along the slip bands and 0.1 μm deep.
- The cracks appeared only after a certain number of cycles in comparison with intrusions or extrusions.

Atomic Force Microscope (AFM). The AFM is a recent invention that produces images that are much closer to simple topographs and can image nonconducting surfaces (Ref 74, 77). The AFM is a combination of the principles of the scanning tunneling microscope and the stylus profilometer.

The AFM operates by measuring the forces between the specimen and the probe. These forces are determined by the nature of the sample, the operating distance between the probe and the sample, the geometry of the probe, and the contaminants present on the specimen surface. The two important properties of the AFM cantilever are spring constant and resonant frequency. The spring constant governs the force between the probe and the specimen when they are close to each other; the spring constant is defined by the material used to build the cantilever. If the cantilever is moved from its equilibrium position and released, it will vibrate at a resonant frequency. This frequency is determined by the cantilever material, dimensions of the cantilever, and the forces acting on the probe.

The AFM records interatomic forces between the apex of a tip and atoms in a sample as the tip is moved over the surface of the sample (Ref 74). During this process, it senses the repulsive forces between the tip and the sample with the tip actually touching the sample. The tip is very sharp, and the tracking force used is small, and the tip traces over individual atoms without damaging the surface of the sample. In this mode of operation, the AFM cantilever is weak with a very low spring constant.

Another mode in which the AFM can be operated involves being sensitive to the attractive forces between the tip and the sample. A feedback system is used in order to prevent the tip from touching and damaging the sample. Also, the resolution attained in this mode of operation is at the expense of decreased lateral resolution.

Recently, Gerberich (Ref 78) used AFM to study fatigued titanium samples. He reported that it can be used to capture images of the surface where fatigue cracks normally initiate, and the slip upset can be directly measured to angstrom accuracy.

Scanning Acoustic Microscope (SAM). The SAM is based on the principle that an acoustic lens having good focusing properties on axis can be used to focus acoustic waves onto a spot on a specimen and receive the acoustic energy from the spot (see *ASM Handbook*, Volume 17, *Nondestructive Evaluation and Quality Control*). By scanning the lens over the specimen systematically, and by sending the intensity of the reflected signal to a synchronous display, a scanned image is built up. Fatigue crack images of an Al-20%Si plain bearing alloy that failed in fatigue have been recorded using SAM (Ref 9).

X-Ray Diffraction

X-ray diffraction can be used for determining the compositional changes, strain changes, and residual stress evaluation during fatigue process. Hence, by utilizing this technique the processes occurring during fatigue damage can be understood. The macroscopic and microscopic properties of materials subjected to fatigue cycling have been studied using XRD techniques by measuring the position and shape of diffraction profiles (Ref 8). The XRD method is widely used for the qualitative and quantitative analysis of samples, precise determination of lattice constants, crystallite size, and lattice strains from line broadening, investigation of preferred orientation and texture, stress measurements, and radial distribution studies of noncrystalline materials.

The phenomenon of XRD by crystals is due to the scattering process in which x-rays are scattered by the electrons of the atoms without change in wavelength (Ref 79). Monochromatic x-rays are usually obtained by the electron bombardment of targets of metallic elements, such as chromium, iron, cobalt, or copper. A diffracted beam will be produced by such scattering only when certain geometrical conditions are satisfied. These geometrical conditions are provided by Bragg's law or the Laue equations. Thus, the resultant diffraction pattern of a crystal that contains both the positions and intensities of the diffraction pattern is a physical property of the substance. Diffraction patterns obtained this way can be recorded by a Debye-Scherrer method, parafocusing technique (powder diffractometer), or a monochromatic pinhole approach. Among the three available methods, the powder diffractometer is the most sensitive.

Diffraction theory predicts that the lines of the powder pattern obtained from a polycrystalline specimen will be exceedingly sharp, if the specimen consists of sufficiently large and strain-free crystallites. Hence, the profile analysis method could be used for assessing fatigue damage, and the broadness of the diffraction line is related to the microscopic structure of polycrystalline materials (Ref 8). The shape and breadth of the profile are determined both by the mean crystallite size or distribution of sizes, and the particular imperfections prevailing in the crystal lattice. Precise diffraction profiles of the material under investigation are usually obtained from the powder diffractometer. Then, by utilizing either Fourier transformation or the iterative method of successive foldings, the line broadening is sepa-

rated into two components related to microstrain and particle size, from which the dislocation density can be calculated.

The residual stress determined by XRD is a macroscopic parameter, and it represents the mean value of microscopic lattice distortions in a surface layer, which is few square millimeters in area and of thickness equal to the depth of penetration of the x-rays (Ref 8). When a polycrystalline piece of metal is deformed elastically such that the strain is uniform over relatively large distances, the lattice plane spacings in the constituent grains change from their stress-free value to some new value. The new lattice plane spacing value corresponds to the magnitude of the applied stress. This uniform macrostrain results in a shift of the diffraction lines to new 2θ positions. This stress is calculated from precise measurements of the peak shifts of diffraction profiles caused by changes in the interplanar spacing from the equilibrium value. The lattice strain is calculated by employing the double-exposure method, which measures the changes in lattice dimensions in two or more directions in the surface layer. Once the strain is determined, the stress can be determined by a calculation involving the mechanically measured elastic constants of the material or by a calibration procedure involving the measurement of the strains produced by known stresses.

Alexoupoulus and Byrne (Ref 80) investigated the fatigue behavior of hard and soft copper using x-ray line broadening and positron annihilation lifetime measurements. The cold-rolled copper was annealed for 1 h at 93.3 °C and had a yield stress of 186.3 MN/m^2. The fatigue testing was conducted at a maximum cyclic stress of 1.5 times the yield stress. The mean positron lifetime and x-ray particle size variation with cycles were determined. They observed that the mean positron lifetime decreased after about 55,000 cycles, and then increased after about 80,000 cycles. In the same fatigue range, the x-ray particle size first increased and then decreased. They explained that the increase in the x-ray particle size is expected because of the occurrence of cyclic softening. In the same study, they did mean positron lifetime and x-ray particle size measurements on cold-rolled copper annealed at 399 °C for 1 h at a maximum cyclic stress of 1.3 times the yield stress. The mean positron lifetime initially increased, then decreased, and again increased prior to fracture. The x-ray particle size measurements followed exactly an opposite behavior. This behavior results from the fact that the present sample initially fatigue hardened and then fatigue softened.

In-Field Application

The following techniques are capable of inspecting components in service for fatigue cracks: magnetic methods, liquid penetrant, eddy current, electric potential, acoustic emission, ultrasonics, radiography, and infrared. Most of the above techniques are also utilized for nondestructive evaluation of components during fabrication as well as manufacturing to detect cracks (not necessarily fatigue cracks). For example, weld defects can be detected by radiography, ultrasonics, magnetic particle, or liquid penetrant method (Ref 81). Hence, in the following paragraphs, in-field applications of each of the above methods for crack detection are summarized.

The magnetic particle method is applicable only to magnetic materials. It is used for inspecting cracks in steel tubular products, pressure vessels, weldments, castings, and forgings. The field applications of other magnetic methods, such as magnetic Barkhausen, magnetoacoustic emissions, and magnetic flux leakage, are well detailed in the article "Magnetic Field Testing" in *ASM Handbook*, Volume 17, *Nondestructive Evaluation and Quality Control*. These methods have been used to detect cracks or flaws in ferromagnetic tubular products (such as gas pipelines, down hole casing, and other steel piping), helicopter rotor blade D-spars, gear teeth, artillery projectiles, drill pipe, collars, steel ropes, and cables, and steel reinforcement in concrete beams. The liquid penetrant method is applicable for both magnetic and nonmagnetic materials. It is used for inspecting cracks in nonmagnetic ferrous tubular products, boilers, pressure vessels, weldments, brazed assemblies, castings, and forgings.

The eddy current method has been used for detecting surface cracks in aircraft structures and engines since the late 1950's (Ref 82). Reference 82 provides a historical development of eddy current testing in aircraft maintenance. The article "Eddy Current Inspection" in *ASM Handbook*, Volume 17, *Nondestructive Evaluation and Quality Control* includes several examples of inspections of aircraft structural and engine components using the eddy current technique. The eddy current method is also used for inspecting tubular products, bars, billets, castings, boilers, pressure vessels, weldments, and forgings. The electric potential method is used for monitoring the crack initiation and propagation behavior in steam turbine components and pipes.

The acoustic emission method is widely used for structural testing of aircraft, spacecraft, bridges, bucket trucks, buildings, dams, military vehicles, pressure vessels, tubular products, rotating machinery, weldments, storage tanks, and other structures. The article "Acoustic Emission Inspection" in *ASM Handbook*, Volume 17, *Nondestructive Evaluation and Quality Control* gives an example of fatigue crack detection in jumbo tube trailers, which transport large volumes of industrial gases at a pressure of about 18.2 MPa. Ultrasonic methods are used for detecting defects in tubular products, bars, boilers, pressure vessels, machine components, weldments, forgings, and castings. The detection of in-service fatigue cracks in machine components has been reported in the article "Ultrasonic Inspection" in *ASM Handbook*, Volume 17.

Radiography methods are used to detect flaws in weldments, pressure vessels, and boilers. In-service radiographic inspection of boilers and pressure vessels is outlined in the article "Boilers and Pressure Vessels" in *ASM Handbook*, Volume 17, *Nondestructive Evaluation and Quality Control*. This article also compares the merits and demerits of the techniques discussed above for nondestructive evaluation of pressure vessels and boilers.

Infra-red techniques are utilized for detecting fatigue cracks in the metallic skin of aircraft and missile structures, and the details are presented in Ref 83.

REFERENCES

1. C.S. Kim, Ph.D. thesis, Northwestern University, 1987
2. C.V. Cooper, Ph.D. thesis, Northwestern University, 1983
3. V.K. Mathews and T.S. Gross, *Trans. ASME*, Vol 110, 1988, p 240
4. K. Komai, K. Minoshima, and G. Kim, *J. Soc. Mater. Sci. Jpn.*, Vol 36, 1981, p 141
5. W.Y. Chu, C.M. Hsiao, and Y.S. Zhao, *Metall. Trans.*, Vol 19A, 1988, p 1067
6. G. Venkataraman, T.S. Sriram, M.E. Fine, and Y.W. Chung, *Scripta Met.*, Vol 24, 1990, p 273
7. M. Houssyn-Eman and M.N. Bassim, *Mater. Sci. Eng.*, Vol 61, 1983, p 79
8. A.J. Allen, D.J. Buttle, C.F. Coleman, F.A. Smith, and R.L. Smith, "In Microstructural Examination of Fatigue Accumulation in Critical LWR Components," EPRI Final Report, NP-5590, Electric Power Research Institute, Jan 1988
9. P.K. Liaw, C.Y. Yang, S.S. Palusamy, and R.D. Rishel, Scientific Paper 92-2TE1-PVRCP-P1, Westinghouse Science and Technology Center, 1992
10. *Handbook of Fatigue Testing*, STP 566, ASTM, 1974
11. E. Marom and R.K. Mueller, *Int. J. Nondestructive Test.*, Vol 3 (No. 2), 1971, p 171
12. C.E. Richards, *The Measurement of Crack Length and Shape during Fracture and Fatigue*, C.E. Beevers, Ed., Engineering Materials Advisory Services, Warley, U.K., 1980, p 461
13. W.F. Deans and C.E. Richards, *J. Test. Eval.*, Vol 7, 1979, p 147
14. T.V. Duggan and M.W. Proctor, *The Measurement of Crack Length and Shape during Fracture and Fatigue*, C.E. Beevers, Ed., Egineering Materials Advisory Services, Warley, U.K., 1980, p 1
15. A. Saxena and S.J. Hudak, *Int. J. Fract.*, Vol 14, 1978, p 453
16. C.E. Richards and W.F. Deans, *The Measurement of Crack Length and Shape during Fracture and Fatigue*, C.E. Beevers, Ed., Engineering Materials Advisory Services, Warley, U.K., 1980, p 28
17. K.R. Watt, *The Measurement of Crack Length and Shape during Fracture and Fatigue*, C.J. Beevers, Ed., Engineering Materials Advisory Services, Warley, U.K., 1980, p 202
18. F.D.W. Charlesworth and W.D. Dover, *Advances in Crack Length Measurement*, C.J. Beevers, Ed., Chamelon Press Ltd., London, 1982, p 253
19. H.H. Johnson, *Mater. Res. Stand.*, 1965, p 442
20. A. Saxena, *Eng. Fract. Mech.*, Vol 13, 1980, p 741
21. W.A. Logsdon, P.K. Liaw, A. Saxena, and V.E. Hulina, *Eng. Fract. Mech.*, Vol 25, 1986, p 259
22. P.K. Liaw, A. Saxena, and J. Schaefer, *Eng. Fract. Mech.*, Vol 32, 1989, p 675
23. P.K. Liaw, G.V. Rao, and M.G. Burke, *Mater. Sci. Eng.*, Vol A131, 1991, p 187
24. R.P. Wei and R.L. Brazill, STP 738, ASTM, 1981, p 103
25. R.P. Gangloff, *Advances in Crack Length Measurement*, C.J. Beevers, Ed., 1982, p 175
26. T.A. Prater and L.F. Coffin, *J. of Pressure Vessel Technology*, Vol 109, 1987, p 124
27. O. Vosikovsky, R. Bell, D.J. Burns, and U.H. Mohaupt, *Steel in Marine Structures*, C. Nordhoek and J. de Back, Ed., 1987
28. C.Y. Li and R.P. Wei, *Mater. Res. Stand.*, Vol 6, 1966, p 392
29. P.K. Liaw, H.R. Hartmann, and E.J. Helm, *Eng. Fract. Mech.*, Vol 18, 1983, p 121
30. P.K. Liaw, W.A. Logsdon, L.D. Roth, and H.R. Hartmann, STP 877, ASTM, 1985, p 177
31. P.K. Liaw, H.R. Hartmann, and W.A. Logsdon, *Eng. Fract. Mech.*, Vol 18, 1983, p 202
32. R.P. Wei and R.L. Brazill, *The Measurement of Crack Length and Shape during Fracture and Fatigue*, C.J. Beevers, Ed., Engineering Materials Advisory Services, Warley, U.K., 1980, p 190
33. W.J. Baxter, *J. Test. Eval.*, Vol 18, 1990, p 430
34. G.P. Klein, *J. Electrochem. Soc.*, Vol 113, 1966, p 345
35. W.J. Baxter, *Metall. Trans.*, Vol 13A, 1982, p 1413
36. A. Vary, "Non Destructive Evaluation Guide," SP-3079, National Aeronautics and Space Administration, 1973
37. D.C. Jiles, *Non Destr. Test. Int.*, Vol 21, 1988, p 311
38. L.P. Karjalainen and M. Moilanen, *NDT Int.*, Vol 12, 1979, p 51
39. L.P. Karjalainen and M. Moilanen, *IEEE Trans. Magnetics*, Vol 3, 1980, p 514
40. D.J. Buttle, G.A.D. Briggs, J.P. Jakubovics, E.A. Little, and C.B. Scruby, *Philos. Trans. R. Soc.*, Vol A320, 1986, p 363
41. K. Ono and M. Shibata. *Advances in Acoustic Emission*, Proc. Int. Conf., H.L. Dunegan and W.F. Hartman, Ed., Dunhart Publishing, 1981, p 154
42. J.R. Barton, *Proc. 5th Annual Symposium on Nondestructive Evaluation of Aerospace and Weapons Systems Components and Materials* (San Antonio, TX), 1965, p 253
43. D.C. Jiles, S. Hariharan, and M.K. Devine, *IEEE Trans. Magnetics*, Vol 26, Sept 1990, p 2577
44. J.G. Byrne, *Metall. Trans.*, Vol 10A, 1979, p 791
45. K.G. Lynn and J.G. Byrne, *Metall. Trans.*, Vol 7A, 1976, p 604
46. P. Alexopoulos and J.G. Byrne, *Metall. Trans.*, Vol 9A, 1978, p 1344
47. R. Duffin and J.G. Byrne, *Mater. Res. Bull.*, Vol 15, 1980, p 635

48. L. Granatelli and K.G. Lynn, Proc. Symposium *Non-destructive Evaluation: Microstructural Characterization and Reliability Strategies* (Pittsburgh), Oct 1980, O. Buck and S.M. Wolf, Ed., Metallurgical Society of AIME, 1981, p 169

49. H.N.G. Wadley, C.B. Scrubby, and J.H. Speake, *Int. Met. Rev.*, Vol 2, 1980, p 41

50. T.C. Lindley and P. McIntyre, *The Measurement of Crack Length and Shape during Fracture and Fatigue*, C.J. Beevers, Ed., Engineering Materials Advisory Services, Warley, U.K., 1980, p 285

51. T.M. Morton, R.M. Harrington, and J.C. Bjeletich, *Eng. Fract. Mech.*, Vol 5, 1973, p 691

52. T.M. Morton, S. Smith, and R.M. Harrington, *Exp. Mech.*, Vol 14, 1974, p 208

53. J.M. Coffey, *The Measurement of Crack Length and Shape during Fracture and Fatigue*, C.J. Beevers, Ed., Engineering Materials Advisory Services, Warley, U.K., 1980, p 345

54. R.F. Lumb, R.J. Hudgell, and P. Winship, "Monitoring Slow Crack Growth by Ultrasonic Methods," Proc. 7th Int. Conf. on NDT, Warsaw, 1973, p 4

55. A. Defebvre and J. Pouliquen, *Ultrasonics Int.*, Vol 79, 1979, p 398

56. M.T. Resch and P. Karpur, *Cyclic Deformation, Fracture and Nondestructive Evaluation of Advanced Materials*, M.R. Mitchell and O. Buck, Ed., STP 1157, 1992, p 323

57. N.R. Joshi, Materials Science Seminar on Fatigue and Microstructure (St. Louis), American Society for Metals, Oct 1978

58. W.G. Clark, *Mater. Eval.*, Vol 25, 1967, p 185

59. W.G. Clark and L.J. Ceschini, *Mater. Eval.*, Vol 27, 1969, p 180

60. M.G. Silk, *Research Techniques in Non Destructive Testing*, R.S. Sharpe, Ed., Academic Press, 1977, p 3

61. P.J. Mudge and J.S. Whitaker, *Weld. Res. Bull.*, Vol 20, 1979, p 6

62. R.D. Shaffer, *Mater. Eval.*, Vol 1, 1992, p 76

63. Y. Huang, S.X. Li, S.E. Lin, and C.H. Shih, *Mater. Eval.*, Vol 42, 1984, p 1020

64. W.J. Baxter, *Metall. Trans.*, Vol 6A, 1975, p 749

65. W.J. Plumbridge and D.A. Ryder, *Metall. Rev.*, Vol 14, 1969, p 136

66. C.W. Brown and G.C. Smith, *Advances in Crack Length Measurement*, C.J. Beevers, Ed., Chamelon Press Ltd., London, 1982, p 41

67. D.L. Davidson and J. Lankford, *Fat. Eng. Mater. Struct.*, Vol 6, 1983, p 241

68. D.R. Williams, D.L. Davidson, and J. Lankford, *Exp. Mech.*, Vol 20, 1980, p 134

69. D.L. Davidson, *M.E. Fine Symposium*, P.K. Liaw, J.R. Weertman, H.L. Marcus, and J.S. Santner, Ed., TMS-AIME, 1991, p 355

70. P.K. Liaw, M.E. Fine, and D.L. Davidson, *Fat. Eng. Mater. Struct.*, Vol 3, 1980, p 59

71. D.L. Davidson and J. Lankford, *Int. Mater. Rev.*, Vol 37, 1992, p 45

72. N.M. Grinberg, *Int. J. Fat.*, Vol 3, 1981, p 143

73. R. Young, J. Ward, and F. Scire, *Rev. Sci. Instrum.*, Vol 43, 1972, p 999

74. P.K. Hansma, V.B. Elings, O. Marti, and C.E. Bracker, *Science*, Vol 242, 1988, p 157

75. T.S. Sriram, M.E. Fine, and Y.W. Chung, *Scr. Metall.*, Vol 24, 1990, p 279

76. T.S. Sriram, M.E. Fine, and Y.W. Chung, *Acta Metall. Mater.*, Vol 40 (No.10), 1992, p 2769

77. G. Binning, C.F. Quate, and Ch. Gerber, *Phys. Rev. Lett.*, Vol 56, 1986, p 930

78. S.E. Harvey, P.G. Marsh, and W.W. Gerberich, *Acta Metall. Mater.*, Vol 42, No. 10, 1994, p 3493

79. B.D. Cullity, *Elements of X-Ray Diffraction*, Addison Wesley, 1978

80. P. Alexopoulos and J.G. Byrne, *Metall. Trans.*, Vol 9A, 1978, p 1829

81. A. De Sterke, Proc. 5th Intl. Conf. on Nondestructive Testing, D.A. Shenstone, Ed., The Queens Printer, Ottawa, Canada, 1969, p 460

82. R.D. Shaffer, Mater. Eval., January 1992, p 76

83. E.J. Kubiak, B.A. Johnson, and R.C. Taylor, Proc. 5th Intl. Conf. on Nondestructive Testing, D.A. Shenstone, Ed., The Queens Printer, Ottawa, Canada, 1969, p 69

Section 3: Fatigue Strength Prediction and Analysis

Fundamentals of Modern Fatigue Analysis for Design*

M.R. Mitchell, Rockwell Science Center

FATIGUE CRACK INITIATION is an important aspect of materials performance in design, and this introductory article summarizes some fundamental concepts and procedures for fatigue life prediction of relatively homogeneous, wrought metals when a major portion of total life is exhausted in crack initiation. Life prediction based on fatigue crack growth involves the concepts of fracture mechanics and is discussed elsewhere in this Volume. Cast and composite materials also are discussed elsewhere in this Volume.

The basic concepts and methods discussed in this article include:

- Cyclic stress-strain mechanical behavior
- Strain-life behavior
- Effects of mean stress and geometric notches
- Local stress-strain and cumulative fatigue damage analysis

Several examples are also given as a way to illustrate the use of strain-based fatigue analysis in the early design stages of components. These methods can reduce costly design alterations (particularly in materials selection) and prototype testing, but by no means imply the elimination of component testing (particularly in the case of "critical" components). The techniques and concepts described in this article are best suited to material selection for specific strain-time histories and comparison of design "A" to design "B" on a relative life improvement basis. They should be employed as early in the design stage as possible in order to circumvent costly prototype development and testing programs.

The strain-life approach is effective in characterizing the fatigue behavior of materials because it accounts for plastic strain, which is a fundamental cause of fatigue crack initiation. Constitutive equations between strain and life are therefore useful because materials are metastable under cyclic loads. Understanding of cyclic strain-strain behavior is necessary for fatigue design.

To predict the crack-initiation life of actual components, the following techniques (with an understanding of strain-life behavior) need to be considered:

1. Mean stress effects need to be accounted for by modification of the strain-life equation
2. Size effects of geometric notches need to be considered
3. Procedures need to relate remotely measured stresses and strains to the stresses and strains at a notch root where plasticity dominates

By combining the above "analytical tools" with an adequate cycle-counting technique that accrues closed hysteresis loops (for example, rainflow or range pair), a means is available to predict fatigue-initiation life of real components or parts.

Explanation of these topics is aimed primarily as a primer on the basic concepts and methods for predicting fatigue crack initiation lifetimes. It should be noted, however, that the techniques outlined in this article are not "the only" or "the best" way to approach an engineering solution to materials selection or the lifetime prediction of materials in design. Other techniques and more complex materials such as composites are therefore covered in a multitude of books, journal articles, and conference proceedings. The major driving forces in development and dissemination of fatigue analysis techniques are the Fatigue Design and Evaluation Committee of the Society of Automotive Engineers (SAE FD&E) and the E 08 Committee on Fatigue and Fracture Mechanics of the American Society for Testing and Materials (ASTM). The SAE FD&E has published its second *Fatigue Design Handbook*, AE10, 1988, and has furthered these general principles to include multiaxial fatigue with the publication of *Multiaxial Fatigue*, AE 14, 1989. The ASTM has numerous Special Technical Publications (STP's) germane to this topic but the most directly applicable are *Advances in Fatigue Lifetime Predictive Techniques*, Vol 1, STP 1122, 1992 and Vol 2, STP 1211, 1993, *Low Cycle Fatigue*, STP 942, 1988 and *Low-Cycle Fatigue and Life Predictions*, STP 770, 1982.

Historical Development

The failure of a metal because of repeated loads was first documented by Albert (circa 1838) (Ref 1). Since that time, considerable attention has been paid to the deformation behavior of metals under a variety of loading conditions. Initially, possibly because of Wöhler (circa 1860) (Ref 2), the fatigue resistance of metals was investigated by conducting rotating-bending experiments; the results were reported as the now familiar S-log N (stress-log cycles to failure) curve, from which the concept of an "endurance limit" (a stress limit below which failure of metal *should* never occur) finds its origin.

We now know that fatigue of metal is the result of to-and-fro slip, or plastic deformation, particularly at a local level. In earlier attempts to describe the fatigue resistance of metals, the rotating-bending stress (S) was calculated by the familiar elasticity relationship:

$$S = Mc/I \qquad \text{(Eq 1)}$$

where M is the applied bending moment to the specimen; c is the distance from neutral axis to the surface of the specimen; and I is the cross-sectional moment of inertia or second moment of area of the specimen. It would seem that these earlier attempts are, at least, questionable because there was no account for plastic deformation.

This article contains an overview of the strain-based, as opposed to stress-based, criterion of material behavior and fatigue analysis. Attention is focused on failure of metals caused by repeated or cyclic loading. *Cyclic stress-strain* behavior of metals is described to illustrate the inadequacy of the monotonic or tensile stress-strain curve in accounting for material instabilities caused by cyclic deformations. The concept of the strain-life curve, that does account for plastic deformation, is also illustrated. Next, the local stress-

* Adapted from the article "Fundamentals of Modern Fatigue Analysis for Design" in *Fatigue and Microstructure*, ASM, 1979.

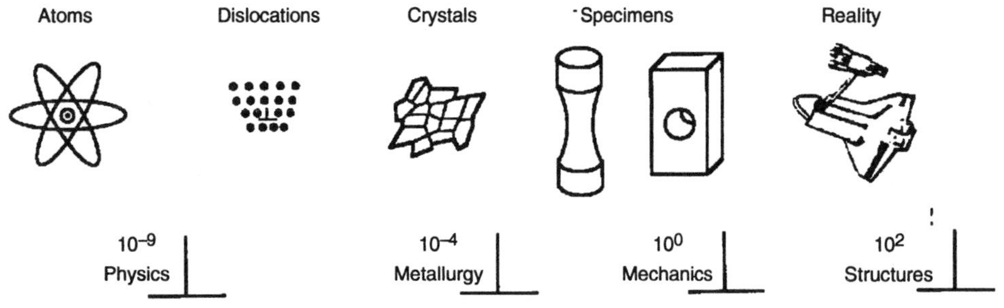

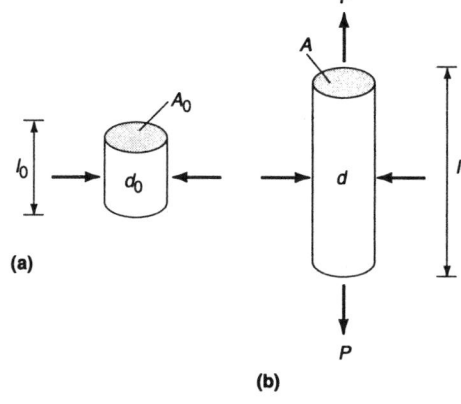

Fig. 1 Relative observation levels for the fatigue process

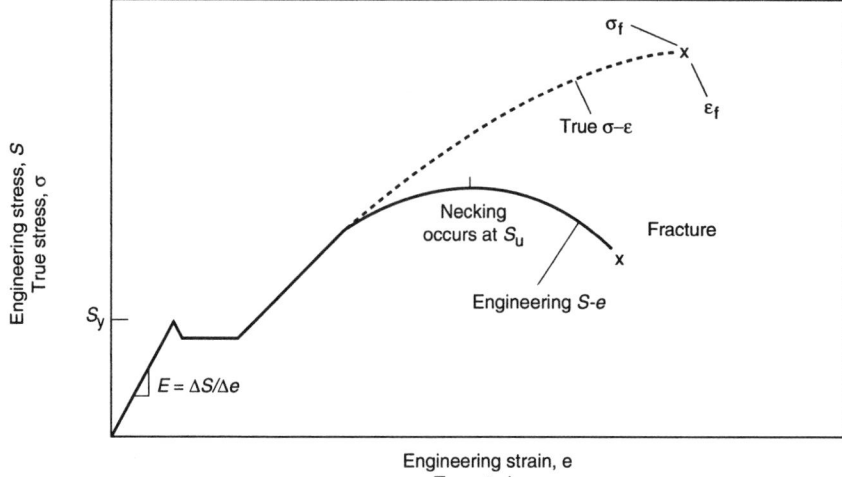

Fig. 2 Original (a) and instantaneous (b) cylindrical section of tension-test specimen

strain approach to fatigue analysis is explained—an approach in which attention is focused on critical locations in a structure where failure is most likely to occur. Finally, a cyclic-plasticity analysis is described for a strain-time history such as that expected in an actual component history. All these concepts are then combined in an attempt to predict the design life of engineering structures, and several examples are used for illustration.

Failure of metals because of repeated loads became a recognized engineering problem with the advent of rotating or reciprocating machinery during the Industrial Revolution of the early 1800s. Metals that were known to be ductile were observed to fail in what appeared on their fracture surfaces to be a "brittle" manner—at what were considered to be "safe" load levels. Since that time, the fatigue problem has plagued engineers. Today it accounts for the vast majority of service failures in ground, air and sea vehicles as well as in many electronic components.

Considerable effort has been expended to determine the nature of the fatigue-damage problem and to find relatively simple methods for coping with it in design. This problem has been investigated from a number of differing viewpoints, or observation levels, as illustrated in Fig. 1. Studies have ranged from dislocation mechanism to phenomenological material behavior to full-scale structural analyses. Many investigators have made pioneering contributions to our present understanding of the fatigue process. For example:

- 1838—Albert in Germany: failure because of repeated loads first documented
- 1839—Poncelet in France: introduces term *fatigue*
- 1849—Institute of Mechanical Engineers in England: "crystallization" theory of metal fatigue debated
- 1860—Wöhler: first systematic investigation of fatigue behavior of railroad axles; rotating-bending test; *S-N* curve; concept of "endurance limit"
- 1864—Fairbairn: first experiments of effects of repeated loads
- 1886—Bauschinger: notes change in "elastic limit" caused by reversed loading or cycling; stress-strain hysteresis loop
- 1903—Ewing and Humfrey: microscopic study disproves old "crystallization" theory;

failure deformation takes place by slip similar to monotonic deformation

- 1910—Bairstow: investigates changes in stress-strain response during cycling; hysteresis loop measured; multiple-step tests; concepts of cyclic hardening and softening
- 1955—Coffin and Manson (working independently): thermal cycling, low-cycle fatigue, plastic-strain considerations
- 1965—Morrow: cyclic plasticity, local stress-strain approach, cumulative damage, life prediction techniques

Stress-Strain Behavior of Materials

The engineering stress-strain behavior of materials is usually determined from a monotonic tension test on smooth specimens with a cylindrical gage section (shown schematically in Fig. 2). "Specimens" as used throughout this paper are axially loaded cylindrical samples with a gage length-to-diameter ratio of approximately two ($l_0/d_0 \cong 2$). Many designs are shown in Ref 3 and ASTM E606-92. For such a test specimen:

$$S = \text{engineering stress} = P/A_0 \qquad \text{(Eq 2)}$$

$$e = \text{engineering strain} = \frac{l - l_0}{l_0} = \frac{\Delta l}{l_0} \qquad \text{(Eq 3)}$$

where P is the applied load: A_0 is the original area; l_0 is the original length; and l is the instantaneous length.

However, because of changes in cross-sectional area during deformation, the true stress, σ, which is greater than the engineering stress in tension (conversely, less in compression), is defined as:

$$\sigma = true \text{ stress} = P/A \qquad \text{(Eq 4)}$$

where A is the instantaneous area.

Similarly, in tension, true strain, ε, is less than engineering strain (up to necking). Ludwik (circa 1909) defined true or natural strain, based on the instantaneous gage length, l, as:

$$\varepsilon = true \text{ strain} = \int_{l_0}^{l} \left(\frac{dl}{l} \right) = \ln \frac{l}{l_0} \qquad \text{(Eq 5)}$$

The use of true stress and true strain merely changes the appearance of the monotonic tension stress-strain curve, as illustrated for a typical low-carbon steel in Fig. 3, and provides an advantage

Fig. 3 Engineering and true stress-strain curves

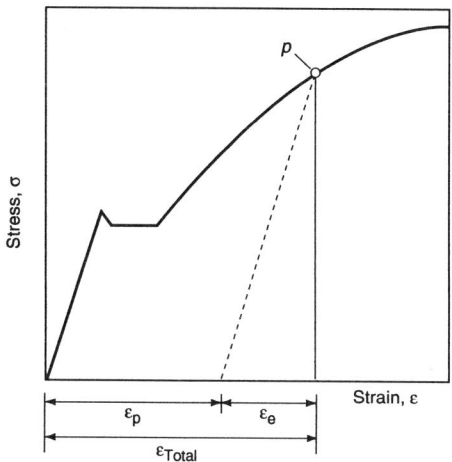

Fig. 4 Illustration of total strain components

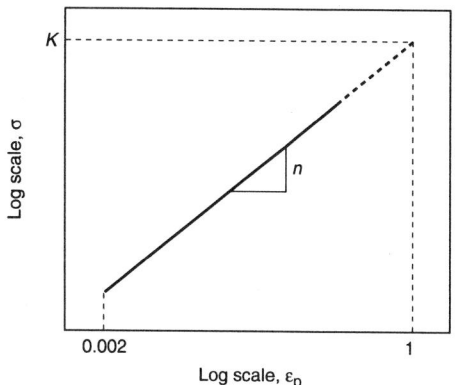

Fig. 5 True stress versus plastic strain (log-log coordinates)

in that it lends itself readily to mathematic description as will be shown subsequently.

The engineering stress and strain may be related to true stress and strain from the following equation for strain:

$$l = l_o + \Delta l \tag{Eq 6}$$

Combining Eq 5 and 6:

$$\varepsilon = \ln\left[\frac{l_o + \Delta l}{l_o}\right] = \ln\left[1 + \frac{\Delta l}{l_o}\right] \tag{Eq 7}$$

From Eq 3, then:

$$\varepsilon = \ln(1 + e) \tag{Eq 8}$$

Note that this relationship is *valid only up to the point of necking* of the specimen during the tension test (that is, when the strain is uniform throughout the gage length of the specimen). It should be noted also that the deviation between the engineering and true strain becomes significant at an engineering strain of approximately 10% [that is, $\varepsilon = \ln(1 + 0.1 = 0.0953)$]. Since the volume of the metal changes by less than 1/1000 during large plastic strains, it is convenient to assume constant volume. Therefore:

$$A_o l_o = A l = \text{Constant} \tag{Eq 9}$$

or

$$\frac{A_o}{A} = \frac{l}{l_o} \tag{Eq 10}$$

So that:

$$\varepsilon = \ln\left(\frac{l}{l_o}\right) = \ln\left(\frac{A_o}{A}\right) = 2\ln\left(\frac{d_o}{d}\right) \tag{Eq 11}$$

where d_o is the original diameter and d is the instantaneous diameter. To relate true stress, σ, to engineering stress, S, from Eq 2, we have $P = SA_o$; and from Eq 3, $P = SA$. Therefore:

$$\sigma = S\left(\frac{A_o}{A}\right) \tag{Eq 12}$$

Up to the inception of necking in the specimen, by combining Eq 8 and 11:

$$\varepsilon = \ln(1 + e) = \ln\left(\frac{A_o}{A}\right) \tag{Eq 13}$$

or:

$$\left(\frac{A_o}{A}\right) = (1 + e) \tag{Eq 14}$$

Thus:

$$\sigma = S(1 + e) \tag{Eq 15}$$

Again, note that this relationship is valid *only up to the point of necking* in the specimens during a monotonic tension test.

The total true strain in a tension test may be separated conveniently into two components, as illustrated in Fig. 4: (1) the linear elastic, or that portion of strain that is recovered upon unloading, ε_e; and (2) the nonlinear plastic strain, that cannot be recovered on unloading, ε_p. Mathematically, this concept is expressed by the equation:

$$\varepsilon = \varepsilon_e + \varepsilon_p \tag{Eq 16}$$

at any point, P, on the true stress-strain curve.

For most metals, a logarithmic plot of true stress versus true plastic strain is a straight line, as shown in Fig. 5. It may be expressed by the power law equation:

$$\sigma = K(\varepsilon_p)^n \tag{Eq 17}$$

or:

$$\varepsilon_p = \left(\frac{\sigma}{K}\right)^{1/n} \tag{Eq 18}$$

where K is the strength coefficient (intercept on a log σ vs. log ε_p plot at $\varepsilon_p = 1$) and n is the strain-hardening exponent (slope).

At the point of fracture, two other quantities, true fracture strength and ductility (shown in Fig. 3), are also quite important. True fracture strength is the true stress at final fracture:

$$\sigma_f = \frac{P_f}{A_f} \tag{Eq 19}$$

where A_f is the area at fracture generally determined from measurements of the averaged minimum diameter on the failed halves of the specimen with an optical comparator at several positions on the necked ligaments. If the material has "sufficient" ductility, a Bridgeman correction factor should be employed to augment the stress due to the triaxiality in the necked section (Ref 4). Likewise, true fracture ductility is the true strain at final fracture:

$$\varepsilon_f = \ln\left(\frac{A_o}{A_f}\right) = 2\ln\left(\frac{d_o}{d_f}\right) = \ln\left(\frac{1}{1 - RA}\right) \tag{Eq 20}$$

where the reduction in area $RA = (A_o - A_f)/A_o$.

Substituting σ_f and ε_f into Eq 17:

$$\sigma_f = K(\varepsilon_f)^n \tag{Eq 21}$$

or $K = \sigma_f / \varepsilon_f^n$. Combining Eq 21 and 17:

$$\varepsilon_p = \left(\frac{\sigma/\sigma_f}{\varepsilon_f^n}\right)^{1/n} = \left(\frac{\sigma \varepsilon_f^n}{\sigma_f}\right)^{1/n} = \varepsilon_f\left(\frac{\sigma}{\sigma_f}\right)^{1/n} \tag{Eq 22}$$

Since the elastic strain is defined by:

$$\varepsilon_e = \sigma/E \tag{Eq 23}$$

we may now express the total strain ($\varepsilon = \varepsilon_e + \varepsilon_p$) as:

$$\varepsilon = \frac{\sigma}{E} + \varepsilon_f\left(\frac{\sigma}{\sigma_f}\right)^{1/n} \tag{Eq 24}$$

Summary of Monotonic Stress-Strain Relationships:

Engineering stress: $S = P/A_o$ (Eq 2)

Engineering strain: $e = \Delta l/l_o$ (Eq 3)

True stress: $\sigma = P/A$ (Eq 4)

True strain: $\varepsilon = \ln(l/l_o) = \ln(A_o/A) = 2\ln(d_o/d)$ (Eq 5)

$$\left.\begin{array}{l} \sigma = S(1 + e) \\ \varepsilon = \ln(1 + e) \end{array}\right\} \textit{Valid only up to necking.} \quad \begin{array}{l}(\text{Eq }15) \\ (\text{Eq }8)\end{array}$$

Strain-hardening exponent, n = slope of log σ versus log ε_p plot

or $n \cong \ln(1 + e_{\text{at necking}})$ (Ref 4)

Table 1 Ranges in monotonic stress-strain properties of metals

Monotonic property	Typical range of engineering metals
Modulus of elasticity, E, ksi	10 to 80×10^3
Tensile yield strength, $S_{0.2\%y}$, ksi	1 to 3×10^2
Ultimate tensile strength, S_u, ksi	10 to 400
Percent reduction in area, % RA	Zero to 90%
True fracture strength, σ_f, ksi	0.5 to 5×10^2
True fracture ductility, ε_f	Zero to 2
Strain-hardening exponent, n	Zero to 0.5

Strength coefficient: $K = \sigma_f / \varepsilon_f^n$ \hfill (Eq 21)

True fracture strength: $\sigma_f = P_f / A_f$ \hfill (Eq 19)

Again, note that the formation of a "neck" in a tensile specimen introduces a complex, triaxial stress state in that region. As such, in ductile metals the quantity, σ_f must be corrected using a Bridgeman correction factor as a function of true strain at fracture (see Ref 4, p 252).

True fracture ductility: $\varepsilon_f = \ln(A_o/A_f) = 2\ln(d_o/d_f)$ \hfill (Eq 20)

$\varepsilon_f = \ln[1/(1-RA)]$

Percent reduction in area: $\%RA = 100[(A_o - A_f)/A_o]$
Total strain = elastic strain + plastic strain:

$\varepsilon = \varepsilon_e + \varepsilon_p = \sigma/E + \varepsilon_f(\sigma/\sigma_f)^{1/n} = \sigma/E + (\sigma/K)^{1/n}$ \hfill (Eq 24)

Cyclic Stress-Strain Behavior of Metals

Table 1 gives typical ranges in many monotonic stress-strain properties of metals, and Table 2 gives specific examples for fairly common steels and aluminum alloys along with their cyclic properties described in this section.

Metals are metastable under application of cyclic loads, and their stress-strain response can be drastically altered when subjected to repeated plastic strains. This is evident by corresponding monotonic and cyclic properties shown in Tables 2 and 3. Depending on the initial state (quenched and tempered, normalized, annealed, cold worked, solution treated and aged, overaged, etc.) and its test condition, a metal may (a) cyclically harden; (b) cyclically soften; (c) be cyclically stable; or (d) have mixed behavior (soften at small strains then harden at greater strains).

In this section, equations similar to those describing the monotonic stress-strain behavior are developed for fatigue analysis. These equations define properties more appropriate to fatigue analyses and are called *fatigue properties*. The reader is also referred to "Recommended Practice for Strain Controlled Fatigue Testing," ASTM E606-92, for the methodology involved in performing such tests.

Determination of constant-amplitude fatigue lives of specimens is customarily performed un-

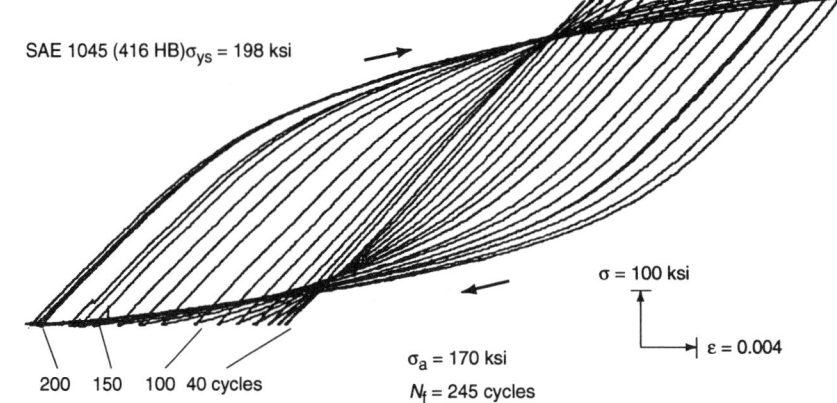

Fig. 6 Cyclic softening of a steel under controlled-stress cycling. Source: Ref 5

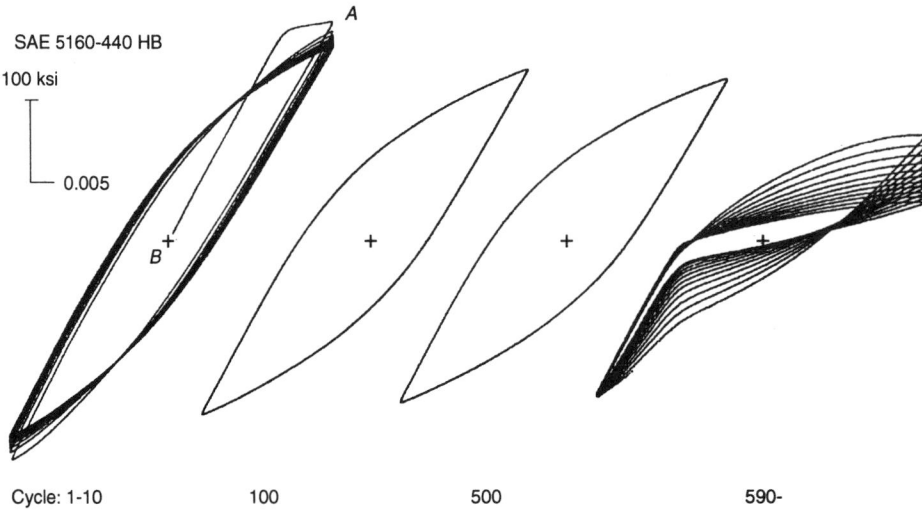

Fig. 7 Cyclic softening of a steel under controlled-strain cycling. Source: Ref 6

der conditions of controlled stress (as in the rotating-bending or cantilever-bending type of test) or controlled strain. As a justification for the use of controlled strain while observing the stress response, the ramifications of controlling stress are illustrated in Fig. 6 (Ref 5). As shown, the applied stress amplitude is less than the initial or monotonic yield strength of the steel (as noted by the "linear elastic" strain response during the first 40 cycles). However, because plastic deformation occurs at a microscopic level, the macrolevel response of the steel is the accrual of ever-increasing amounts of plastic strain. As stress cycling proceeds beyond 40 cycles (in this instance), a "runaway" process occurs as the steel undergoes cyclic softening.

Compare the above response to a steel of similar hardness (as shown in Fig. 7) under conditions of controlled strain. Although the stress limits decrease with increased cycles, no instability is observed, as happened under controlled stress. As Landgraf (Ref 6) points out, these test conditions represent extremes of completely unconstrained or stress-cycling conditions and completely constrained or strain-cycling conditions.

In actual engineering structures, stress-strain gradients do exist, and there is usually a certain degree of structural constraint of the material at critical locations. Such a condition is most reminiscent of strain control. Therefore, it is more advantageous to characterize material response under strain-controlled conditions than under stress-controlled. Also, when an engineering structure is evaluated in a component test arrangement, strain gages are usually affixed to the structure at locations indicated by the most densely cracked locations in a brittle lacquer coating. When used in an analysis, these strains are converted to stress using the modulus of elasticity.

Why not use the strains directly? Consider the cases illustrated in Fig. 8 and 9, in which total strain is controlled and the stress response is observed. As illustrated in Fig. 8, if the stress required to enforce the strain increases on subsequent *reversals*, the metal undergoes cyclic hardening. (*Reversals* are twice the number of cycles in a completely reversed test, $R = -1$. Reversals are preferred to cycles because in pseudo-random spectra, it is impossible to con-

Table 2 Monotonic and cyclic stress-strain properties of selected steels

Alloy	Condition	Monotonic properties								Cyclic properties						
		E, 10^6 psi	S_y, ksi	S_u, ksi	K, ksi	n	%RA	σ_f, ksi	ε_f	S_y', ksi	K', ksi	n'	σ_f, ksi	b	ε_f'	c
A136	As-rec'd	30	46.5	80.6	144	0.21	67	143.6	1.06	47.9	148.8	0.18	115.9	–0.09	0.22	–0.46
A136	150 HB	30	46.0	81.9	...	0.21	69	145.0	1.19	48.9	167.0	0.20	122.7	–0.08	0.20	–0.42
SAE950X	As-rec'd 137 HB	30	62.6	75.8	94.9	0.11	54	...	...	51.2	138.8	0.16	112.0	–0.08	0.34	–0.52
SAE950X	As-rec'd 146 HB	30	56.7	74.0	116.0	0.15	74	141.8	1.34	59.3	136.2	0.13	119.5	–0.08	0.42	–0.57
SAE980X	Prestrained 225 HB	28	83.5	100.8	143.9	0.13	68	176.8	1.15	82.5	385.5	0.25	171.8	–0.10	0.09	–0.48
1006	Hot rolled 85 HB	30	36.0	46.1	60.0	0.14	73	...	...	34.2	196.0	0.28	116.3	–0.12	0.48	–0.52
1020	Annealed 108 HB	27	36.8	56.9	57.9	0.07	64	95.9	1.02	33.8	174.9	0.26	123.3	–0.12	0.44	–0.51
1045	225 HB	29	74.8	108.9	151.8	0.12	44	144.7	...	58.3	170.8	0.17	139.2	–0.08	0.50	–0.52
1045	Q&T 390 HB	29	184.8	194.8	...	0.04	59	269.8	0.89	122.1	216.4	0.09	204.2	–0.07	1.51	–0.85
1045	Q&T 500 HB	29	250.6	283.7	341.0	0.04	38	334.4	...	189.0	672.1	0.20	418.9	–0.09	0.23	–0.56
1045	Q&T 705 HB	29	264.7	299.8	...	0.19	2	309.6	0.02	327.0	618.4	0.10	350.4	–0.07	0.002	–0.47
10B21	Q&T 320 HB	29	144.9	152.0	187.7	0.05	67	217.4	1.13	100.2	143.6	0.06	150.3	–0.10	4.33	–0.85
1080	Q&T 421 HB	30	141.8	195.6	323.0	0.15	32	238.6	...	126.2	460.8	0.21	342.9	–0.10	0.51	–0.59
4340	Q&T 350 HB	29	170.8	179.8	229.2	0.07	57	239.7	0.84	115.6	270.2	0.14	282.0	–0.10	1.22	–0.73
4340	Q&T 410 HB	30	198.8	212.8	...	...	38	225.8	0.48	127.0	282.8	0.13	275.3	–0.09	0.67	–0.64
5160	Q&T 440 HB	30	215.7	230.0	281.4	0.05	39	280.0	0.51	155.2	352.7	0.13	300.0	–0.08	9.56	–1.05
8630	Q&T 254 HB	30	102.8	113.9	153.9	0.08	16	121.8	0.17	87.5	139.4	0.08	152.1	–0.11	0.21	–0.86

Q&T, quenched and tempered. Source: L.E. Tucker, Deere & Co.

Table 3 Monotonic and cyclic stress-strain properties of selected aluminum alloys

Alloy	Condition	Monotonic properties								Cyclic properties						
		E, 10^6 psi	S_y, ksi	S_u, ksi	K, ksi	n	%RA	σ_f, ksi	ε_f	S_y', ksi	K', ksi	n'	σ_f, ksi	b	ε_f'	c
1100	As rec'd	10	14	16	...	...	88	...	2.1	8	23	0.17	28	–0.106	1.8	–0.69
2014	T6	10.6	67	74	...	...	35	91	0.42	65	102	0.073	114	–0.081	0.85	–0.86
2014	T6	10.8	70	78	...	...	...	...	...	73	107	0.062	129	–0.092	0.37	–0.74
2024	T351	10.2	44	69	117	0.20	35	92	0.38	65	114	0.09	147	–0.11	0.21	–0.52
2024	T4	10.6	T/C 55/44	68	T/C 66/92	T/C 0.32/0.17	25	...	0.43	62	95	0.065	160	–0.124	0.22	–0.59
2219	T851	10.3	52	68	...	...	...	81	0.28	48	115	0.14	121	–0.11	1.33	–0.079
5086	F	10.1	30	45	...	...	...	...	0.36	43	87	0.11	83	–0.092	0.69	–0.75
5182	0	10.5	L/T 16/19	L/T 44/49	...	...	L/T 37/44	57	L/0.46 T/0.58	43	68	0.075	122	–0.137	1.76	–0.92
5454	0	10	20	36	...	...	44	53	0.58	34	58	0.084	82	–0.116	1.78	–0.85
5454	10% CR	10	...	...	...	...	...	...	...	34	62	0.098	82	–0.108	0.48	–0.67
5454	20% CR	10	...	...	...	...	35	...	...	37	59	0.081	82	–0.103	1.75	–0.80
5456	H311	10	34	58	...	...	35	76	0.42	51	87	0.086	105	–0.11	0.46	–0.67
6061	T651	10	42	45	53	0.042	58	68	0.86	43	78	0.096	92	–0.099	0.92	–0.78
7075	T6	10.3	68	84	120	0.113	33	108	0.41	75	140	0.10	191	–0.126	0.19	–0.52
7075	T73	10.4	60	70	86	0.054	23	84	0.26	58	74	0.032	116	–0.098	–0.26	–0.73

Source: R.W. Landgraf, Virginia Polytechnic Institute

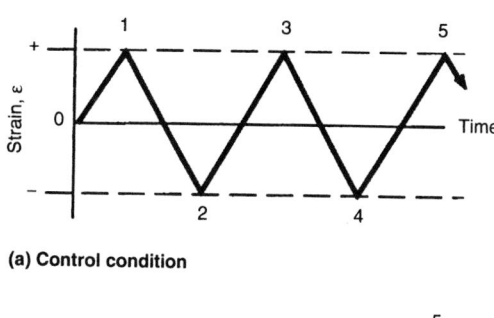

(a) Control condition

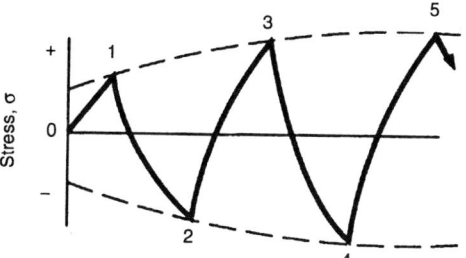

(b) Response variable

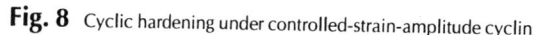

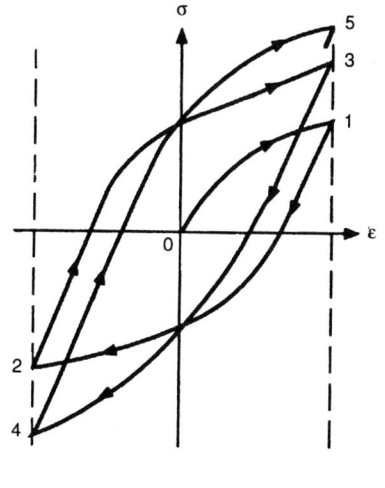

(c)

Fig. 8 Cyclic hardening under controlled-strain-amplitude cycling

veniently define a cycle whereas a reversal is simply a change in sign of a given excursion.) The hardness, yield, and ultimate strength increase. Such behavior is characteristic of annealed pure metals (for example, copper), many aluminum alloys, and as-quenched (untempered) steels.

As illustrated in Fig. 9, the strain amplitude is controlled, but the stress required to enforce the strain decreases with subsequent reversals. This phenomenon is called cyclic softening. It is characteristic of cold worked pure metals and many steels at small strain amplitudes. During cyclic softening, the flow properties (for example, hardness, yield strength, and ultimate strength) decrease.

By plotting the stress amplitude versus reversals from controlled-strain test results, one can observe cyclic strain hardening and softening, as illustrated in Fig. 10. Thus, through cyclic hardening and softening, some intermediate strength level is attained that represents a steady-state condition (in which case the stress required to en-

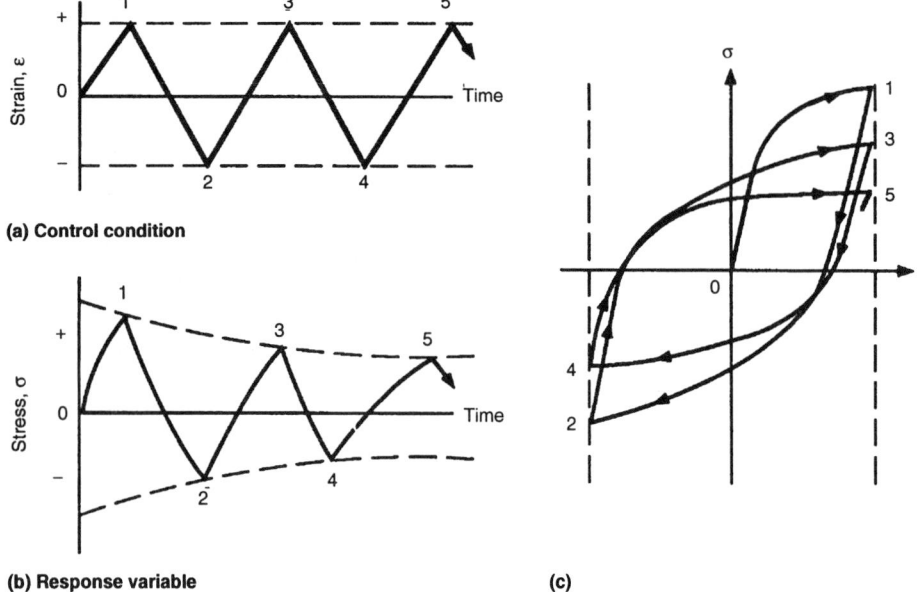

(a) Control condition

(b) Response variable

(c)

Fig. 9 Cyclic softening under controlled-strain-amplitude cycling

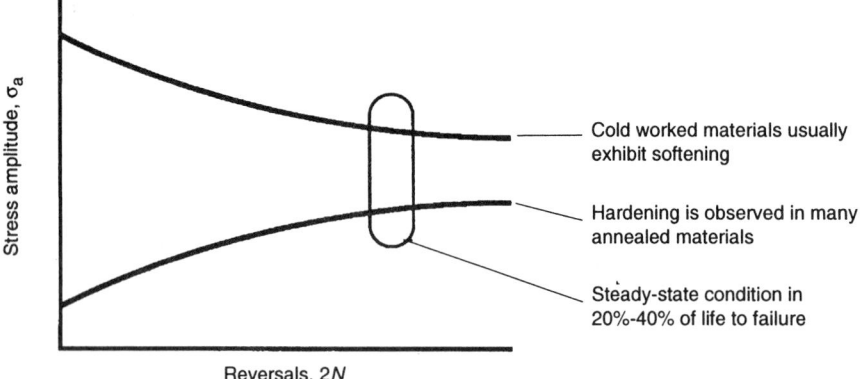

Cold worked materials usually exhibit softening

Hardening is observed in many annealed materials

Steady-state condition in 20%-40% of life to failure

Fig. 10 Steady-state stress response for strain-controlled cycling

force the controlled strain does not vary significantly).

Some metals are cyclically stable, in which case their monotonic stress-strain behavior adequately describes their cyclic response. The steady-state condition is usually achieved in about 20 to 40% of the total fatigue life in either hardening or softening materials. The cyclic behavior of metals is best described in terms of a stress-strain hysteresis loop, as illustrated in Fig. 11.

For completely reversed, $R = -1$, strain-controlled conditions with zero mean strain, the total width of the loop is $\Delta\varepsilon$, or total strain range. (The symbol Δ is used throughout this paper to signify range.)

$$\Delta\varepsilon = 2\varepsilon_a \ (\varepsilon_a = \text{strain amplitude}) \quad \text{(Eq 25)}$$

The total height of the loop is $\Delta\sigma$, or the total stress range:

$$\Delta\sigma = 2\sigma_a \ (\sigma_a = \text{stress amplitude}) \quad \text{(Eq 26)}$$

The difference between the total and elastic strain amplitudes is the plastic-strain amplitude. Since:

$$\frac{\Delta\varepsilon}{2} = \frac{\Delta\varepsilon_e}{2} + \frac{\Delta\varepsilon_p}{2} \quad \text{(Eq 27)}$$

then:

$$\frac{\Delta\varepsilon_p}{2} = \frac{\Delta\varepsilon}{2} - \frac{\Delta\varepsilon_e}{2} = \frac{\Delta\varepsilon}{2} - \frac{\Delta\sigma}{2E} \quad \text{(Eq 28)}$$

Changes in stress response of a metal occur relatively rapidly during the first several percent of the total reversals to failure. The metal, under controlled strain amplitude, will eventually attain a steady-state stress response.

Now, to construct a cyclic stress-strain curve, one simply connects the locus of the points that

represent the tips of the stabilized hysteresis loops from comparison specimen tests at several controlled strain amplitudes (see Fig. 12).

In the particular example shown in Fig. 12, it was presumed that three companion specimens were tested to failure, at three different controlled strain amplitudes. Failure of a specimen is defined, typically, as complete separation into two distinct pieces. Generally, the diameter of specimens are approximately 0.25 to 0.375 inches. In actuality, there is a "propagation" period included in this definition of failure. Other definitions of failure appear in ASTM E606-92.

The steady-state stress response, measured at approximately 50% of the life to failure, is thereby obtained. These stress values are then plotted at the appropriate strain levels to obtain the cyclic stress-strain curve. In actuality one would typically test approximately ten or more companion specimens. The cyclic stress-strain curve can be compared directly to the monotonic or tensile stress-strain curve to quantitatively assess cyclically induced changes in mechanical behavior. This is illustrated in Fig. 13. Note that 50% may not always be the life fraction where steady-state response is attained. Often it is left to the discretion of the interpreter as to where the steady-state cyclic stress-strain occurs. In any event, it should be noted on the cyclic stress-strain curve for the material being tested (i.e., cyclic curve @ 50% life to failure).

In Fig. 13(a), when a material cyclically softens, the cyclic yield strength is considerably lower than the monotonic yield strength. Using monotonic properties in a cyclic application can result in predicting fully elastic strains, when in fact considerable plastic strains are present. In T-1 steels or an equivalent HSLA steel, for example, the cyclic yield strength is only about 50% of the monotonic yield strength.

Whereas the steady-state process consumes 20% to 40% of total life in constant-amplitude testing, a single large overload in actual service-type histories can produce an immediate change from the monotonic curve to the cyclic. Assembly or even driving the completed machine "out the door" can cause an instantaneous loss of 50% of the monotonic yield strength in some materials.

Figure 14 illustrates representative behaviors for aluminum alloys and low-strength steels. Such materials may harden or soften or, depending on the strain amplitude, soften and then harden. The latter phenomenon, known as mixed behavior, is illustrated in Fig. 14(c). Such behavior is common in many HSLA and low-carbon, low-hardness steels (Ref 7, 8).

If we use the same approach as with the monotonic stress-strain curve, a plot of true stress versus true strain from constant-strain-amplitude test data of companion specimens on log-log paper results in a straight line (see Fig. 15). Again, a power-law function between true stress and plastic strain may be represented as:

$$\sigma_a = K' \, (\varepsilon_p)^{n'} \quad \text{(Eq 29)}$$

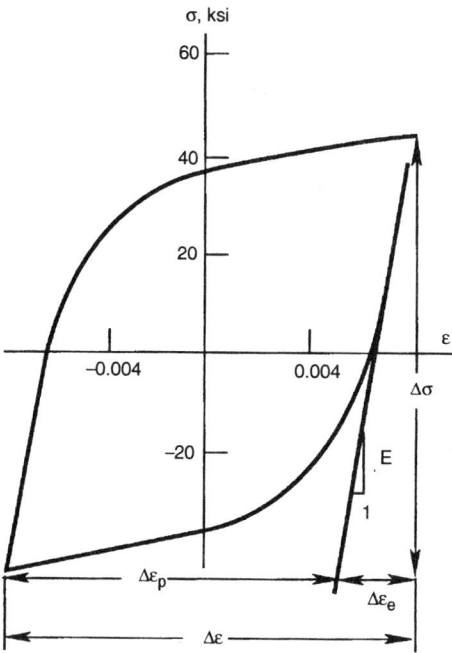

Fig. 11 Steady-state stress-strain hysteresis loop

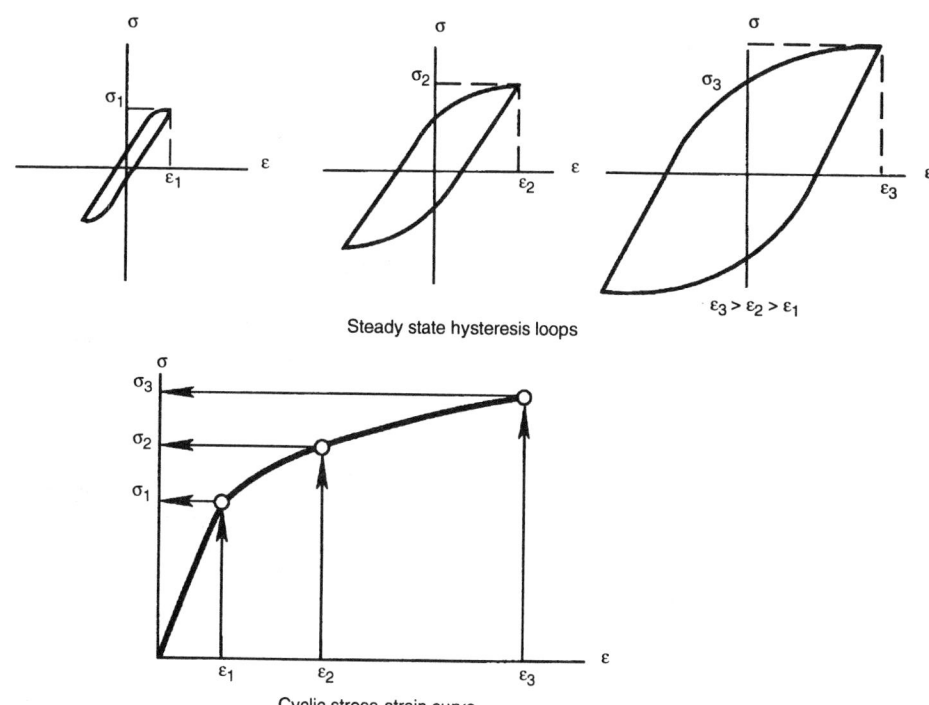

Steady state hysteresis loops

Cyclic stress-strain curve

Fig. 12 Construction of cyclic stress-strain curve by joining tips of stabilized hysteresis loops

where σ_a is the steady-state stress amplitude (measured at 50% of life to failure), ε_p is the plastic-strain amplitude, K' is the cyclic-strength coefficient, and n' is the cyclic-strain-hardening exponent.

Cyclic stress-strain response of a material is characterized by the following relationship:

$$\varepsilon = \frac{\sigma}{E} + \left(\frac{\sigma}{K'}\right)^{1/n'} \qquad \text{(Eq 30)}$$

The value of n' varies between 0.10 and 0.20, with an average value very close to 0.15. In general, if n, the monotonic strain hardening exponent, is initially high it will tend to decrease, or the metal will harden. If n is initially low it will tend to increase, or the metal will soften.

Another method of determining what a metal will do cyclically was proposed by Smith, et al. (Ref 9) and is expressed as:

$$\frac{S_u}{S_{0.2\%y}} > 1.4 \text{ (hardening expected)} \qquad \text{(Eq 31a)}$$

$$\frac{S_u}{S_{0.2\%y}} < 1.2 \text{ softening expected} \qquad \text{(Eq 31b)}$$

where S_u is the monotonic ultimate strength and $S_{0.2\%y}$ is 0.2% offset yield strength. Between the values 1.2 and 1.4, a metal is generally stable but may harden or soften.

Fatigue-Life Behavior

Ever since Wöhler's work on railroad axles subjected to rotating-bending stresses, fatigue data have been presented in the form of an S_a-log N_f curve, where S_a is the stress amplitude and N_f is cycles to failure. This is shown in Fig. 16(a).

Although an "endurance limit" is generally observed for many steels under constant-stress-amplitude testing, such a limit does *not* exist for high-strength steels or such nonferrous metals as aluminum alloys. As a matter of fact, as mentioned previously, a single large overload that is common in many air, sea, ground-vehicle and electronic applications, will unpin dislocations, thereby causing the "endurance limit" to be *eradicated!* This has been shown conclusively by Brose et al. (Ref 10).

Around 1900, Basquin showed that the S_a-log N_f plot could be linearized with full log coordinates [see Fig. 16(b)] and thereby established the exponential law of fatigue. In axial tests using engineering stress, the curve "bends over" at short lives and extrapolates to the ultimate tensile strength (S_u) at 1/4 cycle. Further, in comparing axial test results to rotating-bending test results, we observe that rotating bending gives significantly longer lives, particularly in the low-cycle region (see Fig. 17). The reason for the deviation is the method of calculation of the fiber stress in a bending type of test from Eq 1. This is an elasticity equation, whereas fatigue is caused by plastic deformation (to-and-fro slip). Thus, the assumption of "elastic response" in a cyclic environment can be and is often erroneous. This fact is certainly true in the presence of a notch or other geometric or metallurgical discontinuity. Such possibilities do exist in most common engineering materials.

If true stress amplitudes are used instead of engineering stress, the entire stress-life plot may be linearized, as illustrated in Fig. 18. Thus, stress amplitude can be related to life by another power-law relationship:

$$\sigma_a = \sigma_f' (2N_f)^b \qquad \text{(Eq 32)}$$

$\frac{\Delta\sigma}{2} = \sigma_a$ in zero-mean constant amplitude test,

σ_a = true stress amplitude.

$2N_f$ = reversals to failure (1 cycle = 2 reversals).

σ_f' = fatigue-strength coefficient.

b = fatigue-strength exponent (Basquin's exponent).

The parameters σ_f' and b are fatigue properties of the metal. The fatigue strength coefficient, σ_f', is approximately equal to σ_f for many metals. The fatigue strength exponent, b, varies between approximately −0.05 and −0.12.

Around 1955, Coffin and Manson, who were working independently on the thermal-fatigue problem, established that plastic strain-life data could also be linearized with log-log coordinates (see Fig. 19). As with the true stress-life data the plastic strain-life data can be related by the power-law function:

$$\frac{\Delta\varepsilon_p}{2} = \varepsilon_f' (2N_f)^c \qquad \text{(Eq 33)}$$

where $\frac{\Delta\varepsilon_p}{2}$ = plastic-strain amplitude; ε_f' = fatigue-ductility coefficient; and c = fatigue-ductility exponent. The parameters ε_f' and c are also fatigue properties where ε_f' is approximately equal to ε_f for many metals, and c varies between approximately –0.5 and –0.7 for many metals.

It was mentioned previously that total strain has two components: elastic and plastic, or:

$$\varepsilon = \varepsilon_e + \varepsilon_p \qquad \text{(Eq 16)}$$

Expressed as the strain amplitudes from a constant-amplitude, zero-mean-strain controlled test:

$$\frac{\Delta\varepsilon}{2} = \frac{\Delta\varepsilon_e}{2} + \frac{\Delta\varepsilon_p}{2} \qquad \text{(Eq 27)}$$

Since:

$$\sigma_a = \sigma_f' (2N_f)^b \qquad \text{(Eq 32)}$$

and:

$$\frac{\Delta\varepsilon_e}{2} = \frac{\sigma_a}{E} \qquad \text{(Eq 34)}$$

one can divide Eq 32 by E, the modulus of elasticity, to obtain:

$$\frac{\Delta\varepsilon_e}{2} = \frac{\sigma_f'}{E} (2N_f)^b \qquad \text{(Eq 35)}$$

Combining Eq 27, 33, and 35:

$$\frac{\Delta\varepsilon}{2} = \underbrace{\frac{\sigma_f'}{E} (2N_f)^b}_{\text{Elastic}} + \underbrace{\varepsilon_f' (2N_f)^c}_{\text{Plastic}} \qquad \text{(Eq 36)}$$

Equation 36 is the foundation for the strain-based approach to fatigue and is called the strain-life relationship. Further, the two straight lines, one for the elastic strain, and one for the plastic strain, can be plotted as has been done in Fig. 20.

Several conclusions may be drawn from the total-strain-life curve in Fig. 20. At short lives, less than $2N_t$ (the transition fatigue life where $\Delta\varepsilon_p/2 = \Delta\varepsilon_e/2$), plastic strain predominates and the metal's ductility will control performance. At longer lives, greater than $2N_t$, the elastic strain is more dominant than the plastic, and strength will control performance. An "ideal material" would be one with both high ductility and high strength. Unfortunately, strength and ductility are usually a tradeoff; the optimum compromise must be tailored to the expected load or strain environment being considered in a real history for a fatigue analysis.

By equating the elastic and plastic components of total strain, we can calculate the transition fatigue life as:

$$2N_t = \left(\frac{\varepsilon_f' E}{\sigma_f'}\right)^{1/(b-c)} \qquad \text{(Eq 37)}$$

This is the point on the plot of strain-life where the elastic and plastic strain-life lines intersect

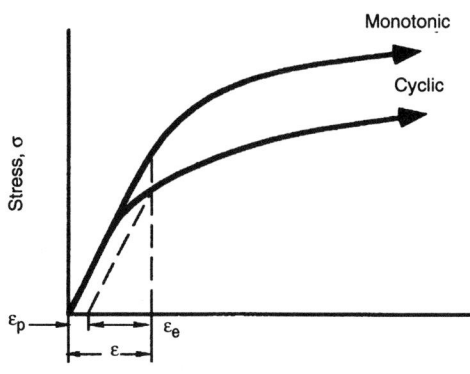

(a) Cyclic softening

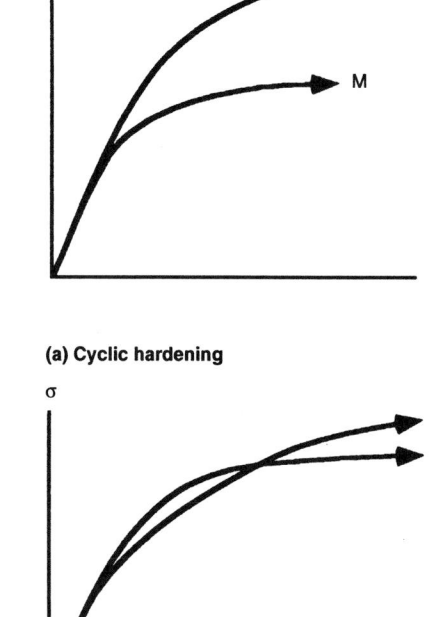

(a) Cyclic hardening

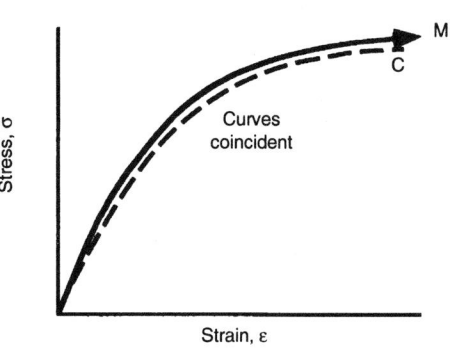

(c) Cyclically stable

(d) Mixed behavior

Fig. 13 Examples of various types of cyclic stress-strain curves

and will prove useful in several calculations shown later in this paper.

Summary of Cyclic Stress-Strain and Strain-Life Relationships. Four fatigue properties have been introduced:

- σ_f', fatigue-strength coefficient
- ε_f', fatigue-ductility coefficient
- b, fatigue-strength exponent
- c, fatigue-ductility exponent

A functional relationship between strain and life has been introduced. A means of accounting for plastic strain, that causes fatigue, is therefore available (Eq 36):

$$\frac{\Delta\varepsilon}{2} = \frac{\sigma_f'}{E} (2N_f)^b + \varepsilon_f'(2N_f)^c$$

These relationships apply to *wrought metals only*. When internal defects govern life (as is the case with cast metals, higher-hardness wrought steels, weldments, composite materials and so forth), these principles are not directly applicable, and appropriate modifications to account for "internal micronotches" may be made (Ref 11).

Cyclic stress-strain material properties may be related in the following manner:

$$K' = \frac{\sigma_f'}{\left(\varepsilon_f'^{n'}\right)} \qquad \text{(Eq 38)}$$

Through energy arguments, Morrow (Ref 12) has shown that:

$$b = -n' / (1 + 5n') \qquad \text{(Eq 39)}$$

and:

$$c = 1 / (1 + 5n') \qquad \text{(Eq 40)}$$

Thus:

$$n' = b / c \qquad \text{(Eq 41)}$$

which allows a relationship between fatigue properties and cyclic stress-strain properties. If average values of b and c (–0.09 and –0.6, respectively) are inserted into Eq 41, $n' \cong 0.15$ results. This is in agreement with the observation that, in general, the average value of n' for most metals is close to 0.15.

As an addendum and caveat to the above, it must be pointed out that the "log-log linear, two straight lines, elastic-plastic approach" doesn't always describe the results of strain-life testing. As early as 1969, Endo and Morrow (Ref 13) showed that several alloys, including SAE 4340, 2024-T4Al, 7075-T6Al, and Ti-8Al-1Mo-1V, did not exhibit a linear relationship for either elastic or plastic strain-life. Sanders and Starke (Ref 14) show that heterogeneous deformation in aluminum alloys also caused deviation from a singular straight line description for elastic and

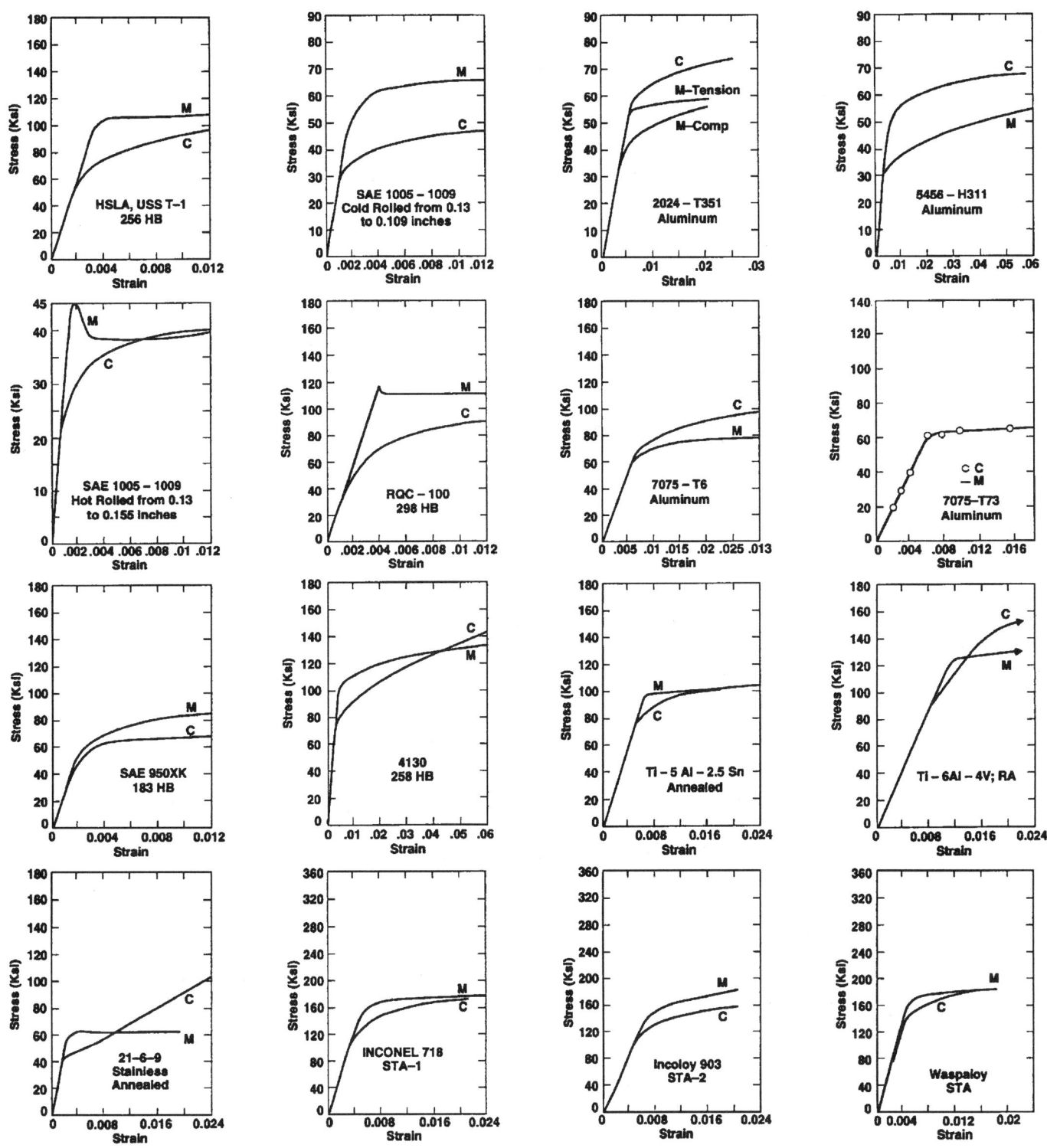

Fig. 14 Cyclic stress-strain response compared with monotonic behavior, for aluminum alloys and low-strength steels

plastic strain-life lines. Also, Radhakrishnan (Ref 15) has demonstrated recently that there is a bilinear Coffin-Manson low cycle fatigue relationship for aluminum-lithium alloys and dual phase steels. But, what is typically employed in a cumulative damage analysis is the total strain-life relationship and the curve may be "approximated" adequately with two straight, log-log, lines.

Approximation of Fatigue Properties from Monotonic Properties. In the absence of adequate data on constant-strain-amplitude, it is often necessary to approximate the strain-life curve from monotonic tensile properties. The Appendix "Estimating Fatigue Curves From Monotonic Properties" in this Volume describes some of the common approximation methods. The following example is a general approach for estimating fatigue behavior of hardened steels. It is an example intended for illustration only.

Example 1: Estimating Fatigue of Hardened Steel

Fatigue-Strength Limit (S_{fl}). For many steels with hardnesses less than approximately 500 HB,

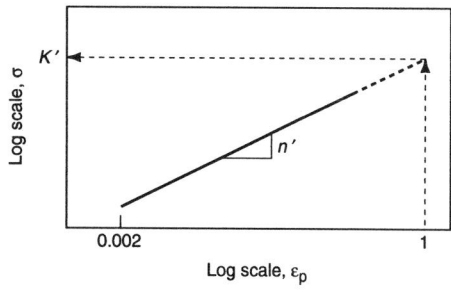

Fig. 15 True stress versus plastic strain for cyclic response (log-log coordinates)

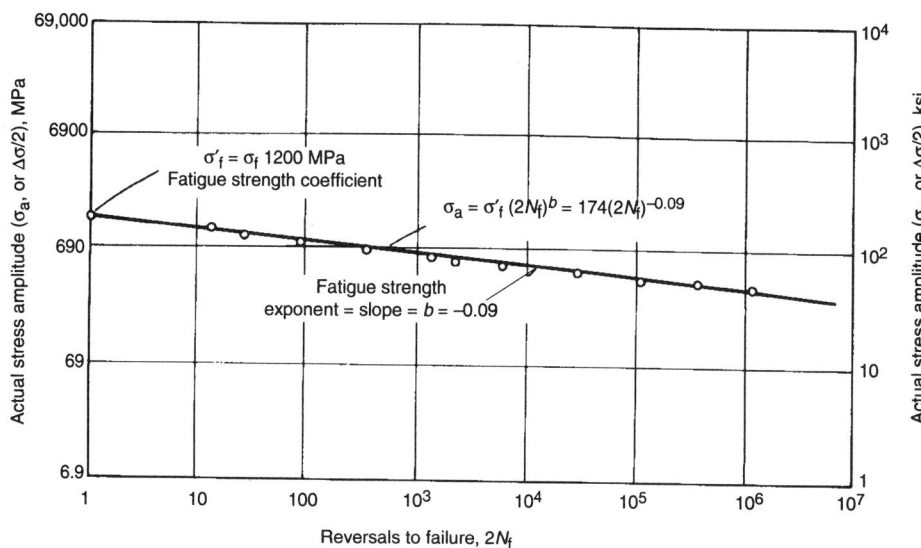

Fig. 18 Log true stress versus log reversals to failure of 4340 steel. Source: *Fatigue Design Handbook,* SAE

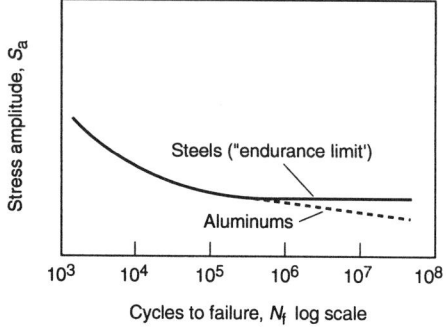

Fig. 16 (a) Stress versus log-cycles-to-failure curve. (b) Log stress versus log-cycles-to-failure curve

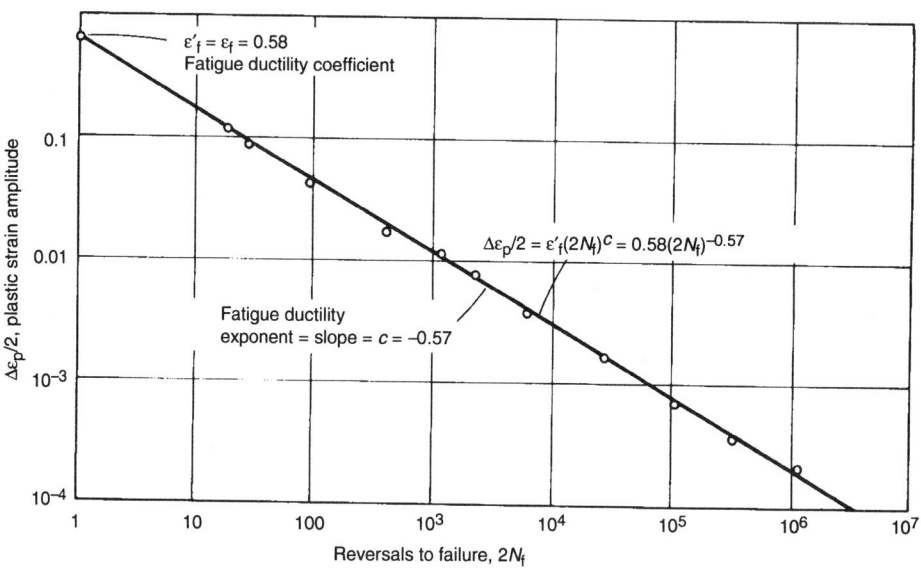

Fig. 19 Log plastic strain versus log reversals to failure of 4340 steel. Source: *Fatigue Design Handbook,* SAE

the fatigue limit S_{fl} at 2×10^6 reversals is approximated by:

$$S_{fl} \text{(ksi)} \cong \frac{S_u}{2} \cong 0.25 \times \text{(HB)} \qquad \text{(Eq 42)}$$

where HB is the Brinell hardness number. For example, for a steel of 200 HB:

$$\frac{S_u}{2} \approx \text{HB} / 2 \cong 100 \text{ ksi} \qquad \text{(Eq 43a)}$$

$$S_{fl} \cong 50 \text{ ksi} \qquad \text{(Eq 43b)}$$

Often the 0.1% offset yield stress from the cyclic stress-strain curve may be used to approximate S_{fl}. For high-strength steels and nonferrous

metals, it is more appropriate to use $1/3 \, S_u \cong S_{fl}$ at 10^8 cycles. In general, however, it is probably more conservative to use $1/2 \, S_u$ at 10^6 cycles for all metals.

Fatigue-Strength Coefficient (σ'_f). A reasonably good approximation for the fatigue-strength coefficient is:

$$\sigma'_f \cong \sigma_f \text{ (corrected for necking)} \qquad \text{(Eq 44)}$$

or for steels to about 500 HB:

$$\sigma_f \text{(ksi)} \cong (S_u + 50) \qquad \text{(Eq 45)}$$

For example, a steel of 200 HB:

$$\sigma'_f \cong \sigma_f \cong 150 \text{ ksi} \qquad \text{(Eq 46)}$$

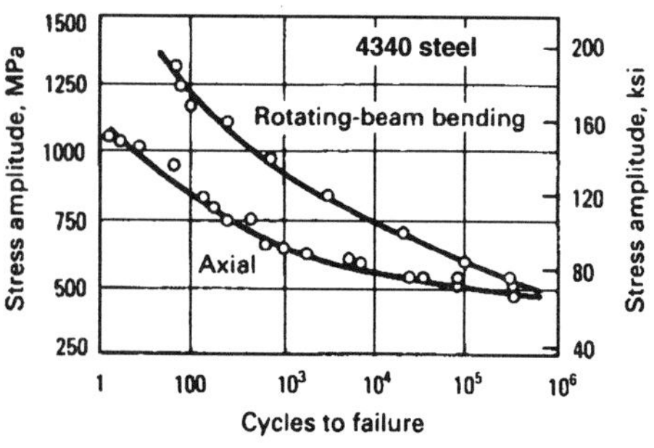

Fig. 17 Stress versus log-cycles-to-failure curves for bending and axial-loading tests of 4340 steel

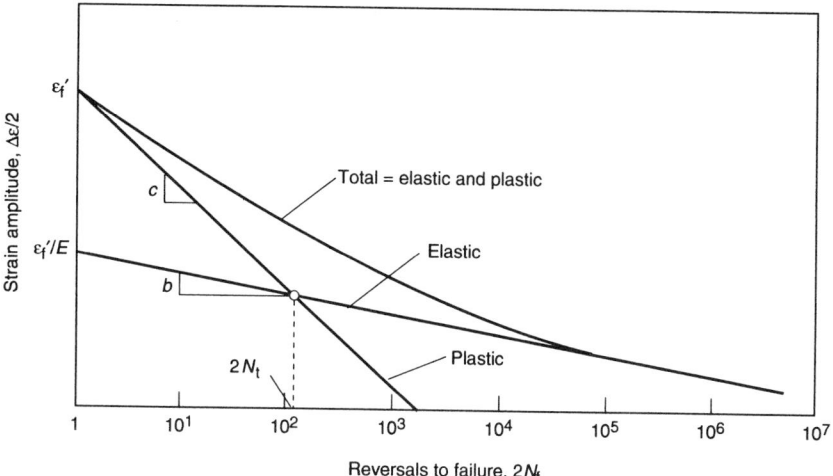

Fig. 20 Log strain versus log reversals to failure

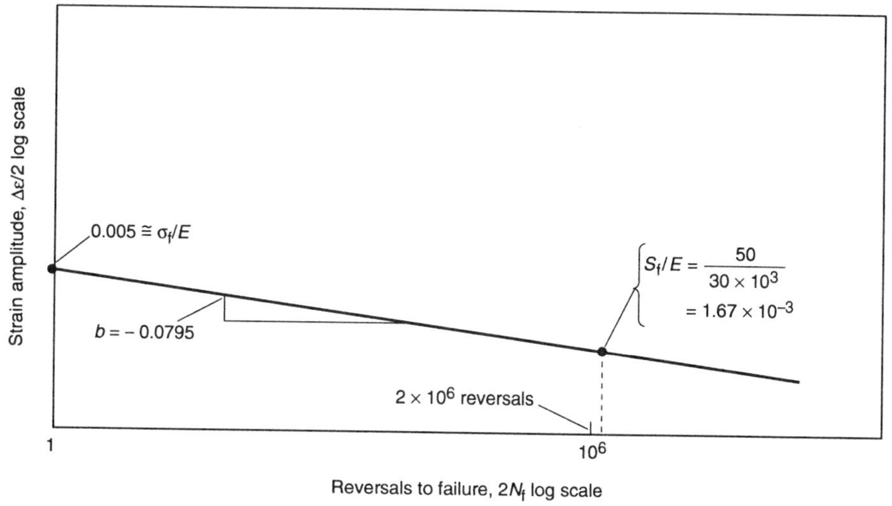

Fig. 21 Log elastic strain versus log reversals to failure

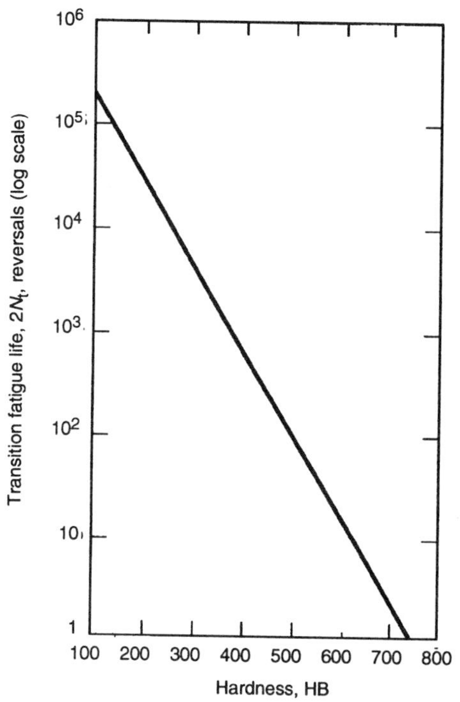

Fig. 22 Log transition failure life versus Brinell hardness for steels. Source: Ref 18

Thus, the intercept at one reversal of the elastic strain-life line is:

$$\frac{\Delta\varepsilon_e}{2} \cong \frac{\sigma_f'}{E} \cong \frac{\sigma_f}{30 \times 10^3} = 0.005 \quad \text{(Eq 47)}$$

Fatigue-Strength Exponent (b). As mentioned previously, b varies from −0.05 to −0.12 and for most metals has an average of −0.085. In approximating the fatigue strength at 2×10^6 reversals with $1/2\ S_u$, it may be shown that:

$$b \cong -1/6 \log\left(\frac{300}{100}\right) = -1/6\ (0.477) = -0.0795 \quad \text{(Eq 48)}$$

One may now construct the elastic-strain-life line as illustrated in Fig. 21, by either the slope and intercept or the intercept and the fatigue limit at 2×10^6 reversals.

Fatigue-Ductility Coefficient (ε_f'). It is a common approximation to set the fatigue-ductil-ity coefficient equal to the true fracture ductility ($\varepsilon_f' \approx \varepsilon_f$).

For the 200-HB steel, that is very ductile, the percent reduction in area is approximately 65% = %RA. Therefore:

$$\varepsilon_f \approx \ln\left(\frac{1}{1-RA}\right) \cong \ln\left(\frac{1}{1-0.65}\right) \cong 1 \quad \text{(Eq 49)}$$

Fatigue-Ductility Exponent (c). The fatigue-ductility exponent, c, is not as well-defined as are the other fatigue properties. According to Coffin (Ref 16), c is approximately −0.5, whereas according to Manson (Ref 17), c is approximately −0.6. Morrow (Ref 12) has shown that for many metals c varies between −0.5 and −0.7, or an average of −0.6.

Plotting of Strain Life Curve. Instead of using a slope, c, to construct the plastic strain-life line, it is advantageous to note the empirical representation of the hardness and transition fatigue life shown in Fig. 22 (Ref 18). For the 200-HB steel in this example, the transition fatigue life is

$2N_t \cong 6 \times 10^4$ reversals. By connecting the intercept of $\varepsilon_f' \cong \varepsilon_f = 1$ and the point on the elastic strain-life line at the value of $2N_t$, we construct the plastic strain-life line. One may now plot the plastic strain-life line, and algebraically add to it the elastic strain-life line to obtain the total strain-life curve, as illustrated in Fig. 23.

It should be clear after the examples given that the manner in which metals resist cyclic straining is dependent on both strength and ductility. An idealized situation is depicted in Fig. 24. Consider the steel at 600 HB (a strong metal that resists strain "elasticity" on the basis of its high strength) compared to the steel at 300 HB (a ductile metal that resists strain "plastically" on the basis of its superior ductility). The "tough" steel at 400 HB resists strain by a combination of both its strength and ductility. This does not, however, mean that the 400-HB steel is the best material for a specific duty cycle that must be resisted in actual design application. The "best" material must be tailored to the application. This hypothesis will be further expounded in a later section.

The strain-life curves in Fig. 24 all intersect at a strain of 0.01 with life to failure of approximately 2×10^3 reversals (1000 cycles). Figure 25 illustrates the real trend for a variety of steels of varying hardnesses and microstructures (Ref 6). Note that the SAE 1010 (a low-carbon, low-hardness steel used in many ground-vehicle components) has a transition fatigue life of approximately 10^5 reversals. Therefore, even at 10^6 reversals, there will be a certain portion of plastic strain present that would not be accounted for in the stress-based approach to fatigue.

The advent of modern, closed-loop electrohydraulic testing machines has made the strain-based test procedure and data presentation fairly commonplace. Interested readers are referred to Ref 19-21 for a compilation of cyclic stress-strain properties and strain-life curves for a variety of materials and conditions.

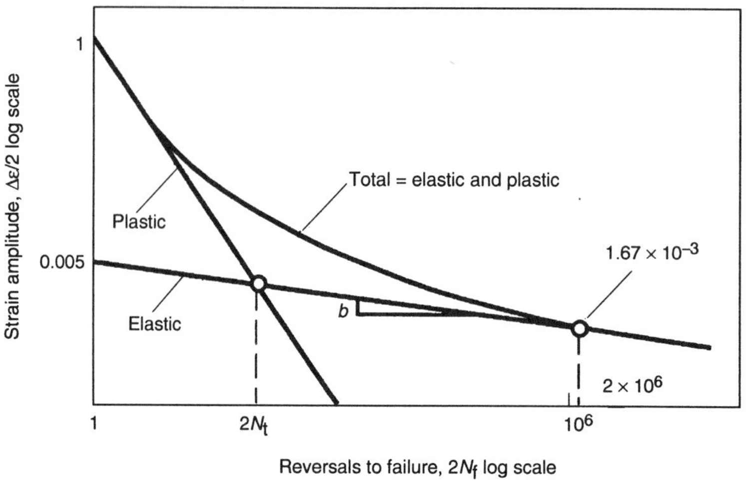

Mean-Stress Effects

To predict the crack-initiation life of actual components, the following need to be considered:

- Mean stress effects
- Size effects of geometric notches
- Relation between remotely measured stresses and strains to stresses and strains at a notch root where plasticity dominates

The methods used to analyze these factors, when combined with an adequate cycle-counting technique that accrues closed hysteresis loops (for example, rainflow or range pair), fatigue-initiation life of real components or parts can be predicted.

The preceding sections have outlined a contemporary presentation for the strain-based description of the fatigue properties of materials. This section considers the effect of mean stress on fatigue life, that would later be factored into a cumulative fatigue-damage analysis.

As illustrated in Fig. 26, the following nomenclature will be used in accounting for mean stresses:

$$\text{Alternating stress amplitude } (\sigma_a) = \frac{\sigma_{max} - \sigma_{min}}{2} \quad \text{(Eq 50)}$$

$$\text{Mean stress } (\sigma_o) = \frac{\sigma_{max} + \sigma_{min}}{2} \quad \text{(Eq 51)}$$

As an illustrative example, let σ_{max} be 15 ksi and σ_{min} be -5 ksi. Then:

$$\sigma_a = \frac{15 + 5}{2} = 10 \text{ ksi}, \ \sigma_o = \frac{15 - 5}{2} = 5 \text{ ksi} \quad \text{(Eq 52)}$$

Mean-stress data are generally presented in terms of constant-life diagrams that are plots of all combinations of alternating and mean stresses resulting in the same finite life to failure. These are illustrated in Fig. 27.

The equations for the lines shown in Fig. 27 are the following:

Line a (Soderberg):

$$\frac{S_a}{S_{cr}} + \frac{S_o}{S_y} = 1 \ \text{ or } \ S_a = S_{cr}\left(1 - \frac{S_o}{S_y}\right) \quad \text{(Eq 53)}$$

Line b (Goodman):

$$\frac{S_a}{S_{cr}} + \frac{S_o}{S_u} = 1 \ \text{ or } \ S_a = S_{cr}\left(1 - \frac{S_o}{S_u}\right) \quad \text{(Eq 54)}$$

Line c (Gerber):

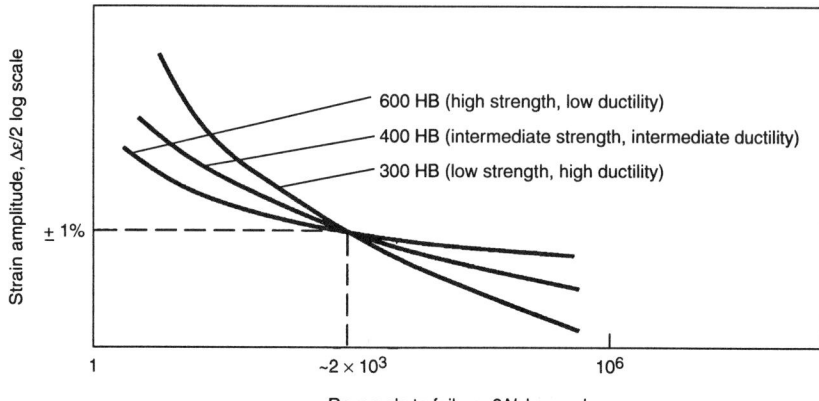

Fig. 23 Estimated curve of log strain versus log reversals to failure for a steel (200 HB)

Fig. 24 Strain-life curves for a steel at three different hardness levels (approximation)

$$\frac{S_a}{S_{cr}} + \left(\frac{S_o}{S_u}\right)^2 = 1 \ \text{ or } \ S_a = S_{cr}\left[1 - \left(\frac{S_o}{S_u}\right)^2\right] \quad \text{(Eq 55)}$$

where S_a is the alternating-stress amplitude; S_{cr} is the completely reversed stress amplitude for a given life (i.e., 10^6, 10^5, etc.); S_u is the ultimate tensile strength of the material; S_y is the yield strength; and S_o is the mean stress. For the case of tensile mean stresses, as a rule-of-thumb:

1. Soderberg's relation is very conservative for most cases.
2. Goodman's relation is good for brittle metals but conservative for ductile metals.
3. Gerber's relation is good for ductile metals.

The above statements apply only to tensile mean stress. Moreover, there are other ways of accounting for mean stresses, and those cited are used only as typical examples.

As an alternative approach, consider that a mean stress alters the value of the fatigue strength coefficient, σ_f, in the stress-life relationship. That is, tensile mean stress would reduce the fatigue strength, whereas a compressive mean stress would increase the fatigue strength. Thus, we have:

$$\sigma_a = \left(\sigma_f' - \sigma_0\right)(2N_f)^b \quad \text{(Eq 56)}$$

In this equation, tensile mean stresses are positive, and compressive ones are negative. Hence for a tensile mean stress, the new intercept constant (σ_f') is decreased relative to σ_f for zero mean stress, and the intercept is increased for a compressive mean stress::

$$(\sigma_f' - \sigma_0) < \sigma_f' \text{ (tensile or + mean)}$$

$$(\sigma_f' - \sigma_0) > \sigma_f' \text{ (compressive or } - \text{mean)} \quad \text{(Eq 57)}$$

In terms of the strain-life relationship:

$$\frac{\Delta \varepsilon}{2} = \left(\frac{\sigma_f' - \sigma_0}{E}\right)(2N_f)^b + \varepsilon_f'(2N_f)^c \quad \text{(Eq 58)}$$

where negative σ_0 is for tensile mean stress, positive σ_0 is for compressive mean stress. Figure 28 illustrates the effect of a tensile mean stress in modifying the strain-life curve. Consistent with expected be-

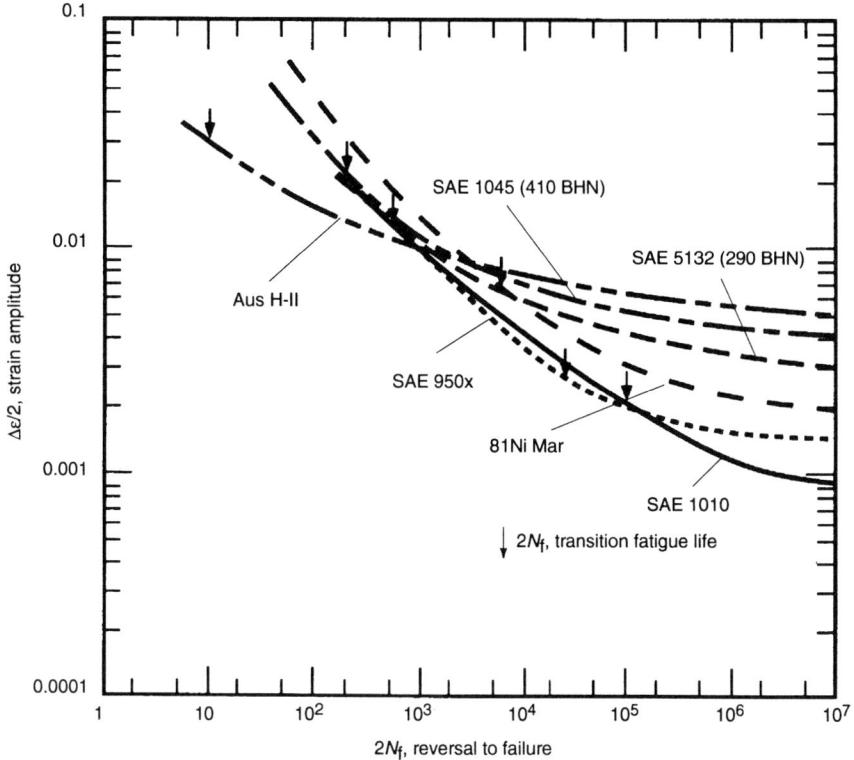

Fig. 25 Strain-life curves for steels with varying microstructures and hardnesses. Source: Ref 6

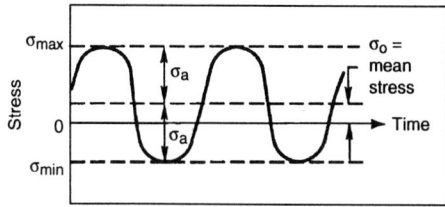

Fig. 26 Stress versus time for nonzero-mean-stress cycling

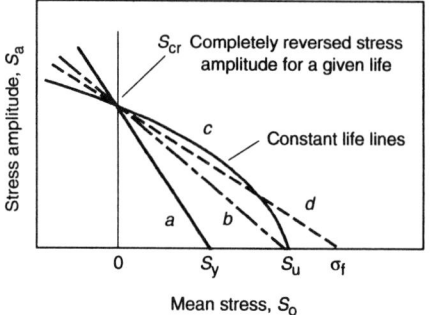

Fig. 27 Various forms of presenting mean-stress data

havior, the effect is most significant in the long-term fatigue region.

As can be seen, there is little or no effect of mean stresses at lives less than approximately the transition fatigue life (the low cycle fatigue region). In this life region, the large amounts of plastic deformation will eradicate any beneficial or detrimental effect of a mean stress, because it will not be sustained. Relaxation of mean stresses (Ref 22) and cyclic-dependent creep (Ref 23) are not covered in detail in this paper, because these phenomena are special instances of material response—particularly when considered in a cumulative-damage analysis (Ref 24, 25).

Cumulative Fatigue Damage

Analysis of cumulative fatigue damage is a method that addresses the following:

- The strain-life behavior of metal fatigue resistance
- The effect of mean stress
- The effect of geometric notches
- The typical strain-time histories of components

To ascertain structural life under other constant-amplitude conditions, one must apply cumulative damage criteria to conditions of varying stress or strain amplitudes.

The simple example of a bilevel loading sequence (Fig. 29) can help illustrate the common

Palmgren-Miner linear cumulative damage rule, which may be mathematically stated as:

$$d_i = \frac{2N_i}{2N_{f_i}} = \frac{\text{(Reversals applied at } \sigma_{ai})}{\left(\text{Reversals failure at } \sigma_{ai}\right)} \quad \text{(Eq 59)}$$

where d = damage.

Accordingly, failure is defined as occurring when:

$$D = \sum_i d_i = \sum_i \frac{2N_i}{2N_{f_i}} = 1 \quad \text{(Eq 60)}$$

In the example shown in Fig. 29, assume that 50 reversals (25 cycles) are applied at σ_{a1}. How many can be applied at σ_{a2} until failure occurs? Using Eq 60, we find that:

$$\sum_i \frac{2N_i}{2N_{f_i}} = \frac{50}{100} + \frac{x}{10,000} = 1 \quad \text{(Eq 61)}$$

Thus, $x = 5,000$ reversals may be applied at σ_{a2} until failure occurs. However, the problem is not quite this simple. Such things as sequence effects, overstressing and understressing also need to be taken into account.

Effect of Overstressing and Understressing. As a simple example of overstressing and understressing phenomena, consider a cyclically softening material subject to the strains corresponding to the steady-state stresses σ_{a1} and σ_{a2} as

shown in Fig. 30. Should the lower strain corresponding to the stress, σ_{a2}, be applied first, the materials response will be "linear elastic" and will follow the curve. If "enough" cycles are applied to the metal, the loop will eventually stabilize to the cyclic response and include some plasticity. This phenomenon is called cyclic-dependent yielding, and the hysteresis loop is depicted in Fig. 31. However, if the larger strain is applied after a few of the lower cycles, we will obtain the same hysteresis loop that we would obtain if the lower strain had not been applied (see Fig. 32). Now imagine that the larger strain had been applied first. The large hysteresis loop would have developed as it did in Fig. 32. As a result, the stress-strain curve would stabilize at the cyclic pattern, and the subsequent application of the lower strain corresponding to the stress σ_{a2} would immediately produce the loop shown in Fig. 31. This is very different from the "fully elastic" loop in that a considerable plastic strain is immediately evident. Thus, the high-low sequence would result in a shorter life than the low-high sequence because of the cyclic-dependent yielding phenomenon.

Sequence Effects. Figure 33(a), shows the importance of accounting for sequence effects in a loading history. Presume the larger strain amplitude, ε_{a1}, is imposed first and after several reversals is transferred to the smaller strain amplitude, ε_{a2}, from the compressive peak (No. 4). Note from the stress response that a self-imposed tensile mean stress, σ_o, develops, as illustrated in Fig. 33(c). Instead of transferring from the large strain to the small strain from the compression peak, reverse the situation and transfer from the tensile peak; then, as Fig. 34(c) shows, a self-im-

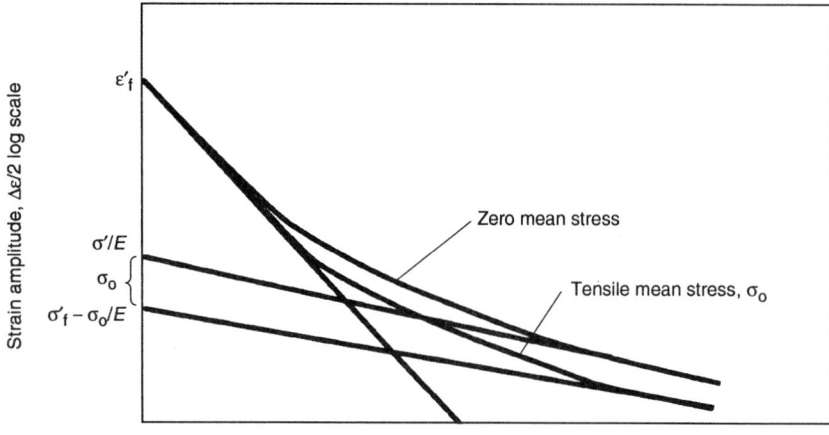

Fig. 28 Mean-stress modification to strain-life curve

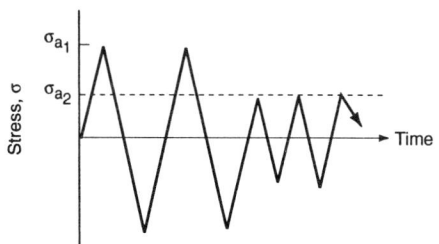

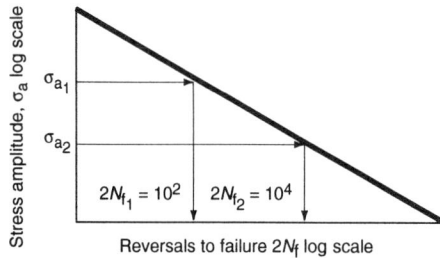

Fig. 29 Example of bilevel loading sequence

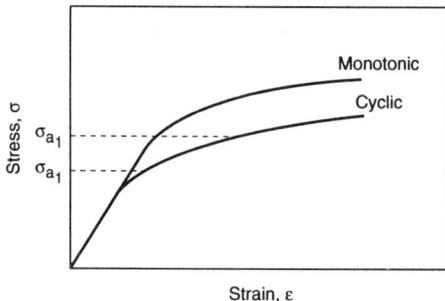

Fig. 30 Monotonic and cyclic stress-strain curves used in bilevel loading example

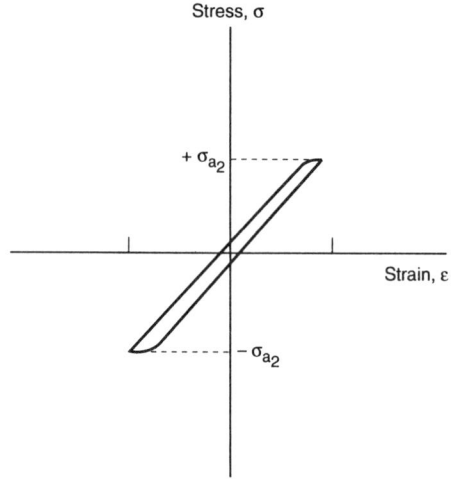

Fig. 31 Hysteresis loop illustrating development of plastic strain from initial "elastic" response

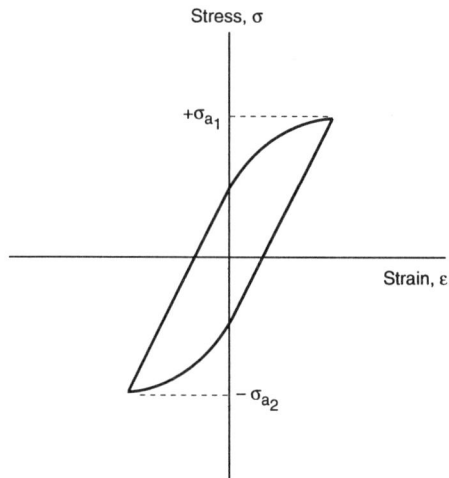

Fig. 32 Hysteresis loop illustrating development of nonelastic response

posed compressive mean stress, σ_o, results because of this particular transfer sequence.

Another example of sequence effects is shown in Fig. 35. Load history A (Fig. 35b) and load history B (Fig. 35c) have similar-appearing strain histories with totally different stress-strain response and fatigue life (Fig. 35a) from slightly different initial transients. Load history A has a tensile leading edge as an initial transient, while load history B has a compressive leading edge and a markedly higher fatigue strength (Fig. 35a). This illustrates the difficulty of applying data to new designs without complete and accurate characterization of anticipated and, occasionally, unanticipated load histories.

Cyclic-Dependent Stress Relaxation. An analogous situation to the cyclic-dependent creep under biased stress-cycling conditions mentioned earlier in this paper is the deformation response in biased strain control known as cyclic-dependent stress relaxation. The idealized situation illustrated in Fig. 33 and 34 (in which the compressive and tensile self-imposed mean stresses result from the transfer sequences) is not precisely accurate. As Fig. 36 shows, there is a relaxation of the mean stress under the biased strain conditions. Relaxation rates depend on the material hardness and imposed strain amplitude. As Fig. 37 shows, the harder the metal, the lesser the relaxation rate of the mean stress. Also, the greater the strain amplitude, the greater the relaxation rate. Both responses depend on the

amount of sustained cyclic plastic deformation that occurs. A softer steel (for example, SAE 1045 at 280 HB) will display greater amounts of plastic deformation, and thus the relaxation rate of the mean stress will be greater than it will for a harder steel (for example, the SAE 1045 at 560 HB). Similarly, the greater the total strain, the greater the plastic strain in proportion and thus the greater relaxation rate for the mean stress. However, such specialized responses are not generally included in cumulative-fatigue-damage analyses unless the component strain-time history would be heavily biased in either tension or compression.

Cycle Counting

The preceding section gave some simple examples of the importance of the sequence of events in a variable-loading history. To assess the fatigue damage for complex histories, one must reduce them to a series of discrete events by employing some type of cycle-counting technique. For purposes of illustration, consider the strain history shown in Fig. 38 (Ref 26).

The stress-time history is quite different from the corresponding strain-time history, and no clear functional relationship exists between them because of the nonlinear (plasticity) material response. Events C-D and E-D have identical mean strains and strain ranges but quite different mean stresses and stress ranges. (Note that a positive strain is indicated at point E in the strain-time history but that the stress response is compressive.) Following the elastic unloading (B-C), the material exhibits a discontinuous accumulation

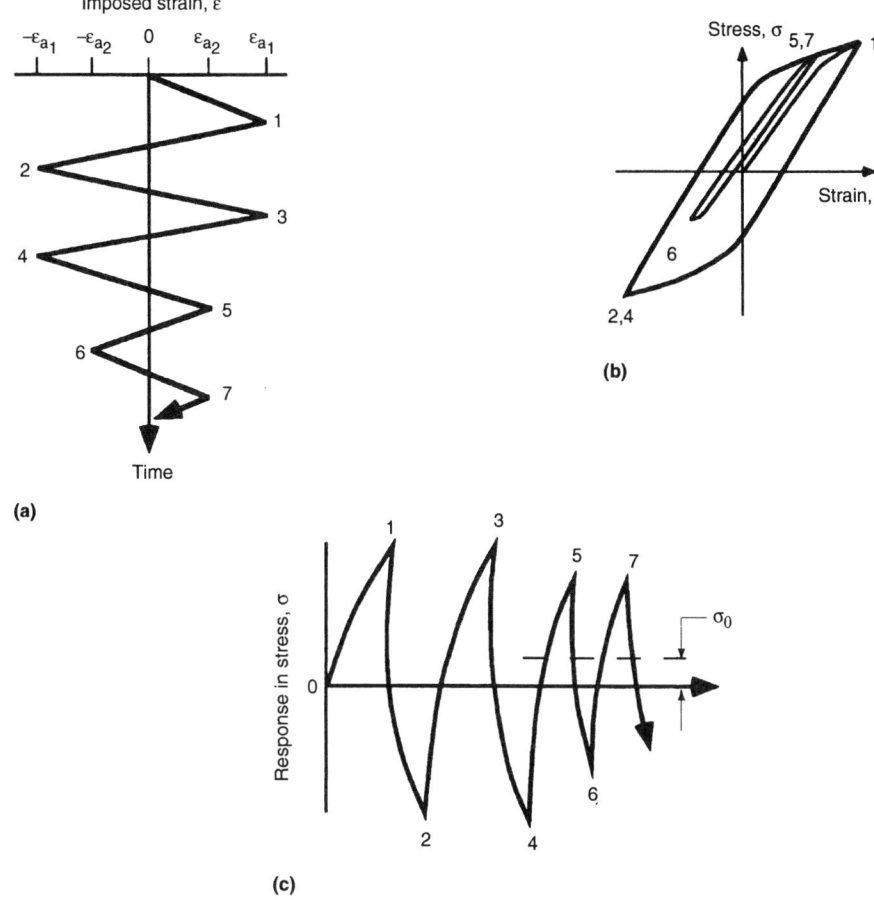

Fig. 33 Development of tensile mean stress because of sequence effect

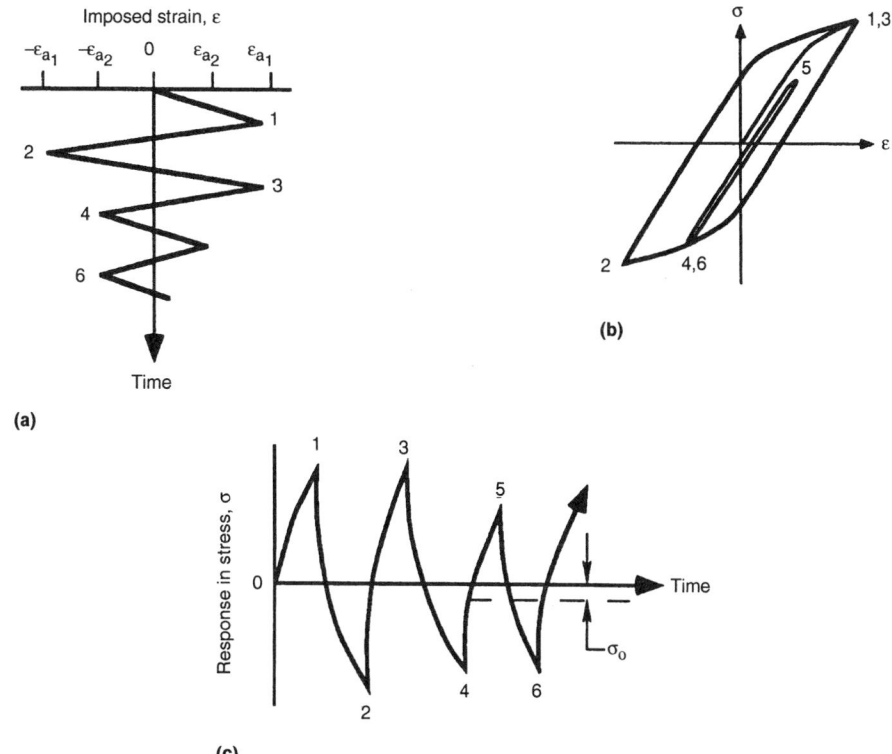

Fig. 34 Development of compressive mean stress because of sequence effect

of plastic strain upon deforming from C to D. When point B is reached, the material "remembers" its prior deformation (A-B), and deforms along path A-D as though event B-C had never occurred.

In this simple sequence, there are four events that resemble constant-amplitude cycling. These events (which are closed hysteresis loops) are: A-D-A, B-C-B, D-E-D and F-G-F. Each event is associated with a strain range and a mean stress. Of the various counting techniques in use (rainflow, range pair, level crossing, and peak counting), rainflow (or its equivalent, range pair) has been shown to produce superior fatigue-life estimates (Ref 27). The apparent reason for the superiority of rainflow counting is that it combines load reversals in a manner that defines cycles by closed hysteresis loops (see Fig. 39).

To implement the rainflow counting technique, plot the strain-time history with the time axis vertically downward and imagine the lines connecting strain peaks to be a series of "pagoda roofs." Several rules are imposed on rain "dripping down" from these roofs so that closed hysteresis loops are defined. The following rules govern the manner in which rain flows:

1. Plot the history so that the largest strain magnitude occurs as the first and last peaks or valleys. This eliminates half-cycles when counting.
2. "Rainflow" is initiated at each peak and is allowed to drip down and continue—except that if it initiates at a maximum (points A, B, D, G) it must stop when it comes opposite a more positive peak than the maximum from which it started. Rainflow dripping from B must stop opposite D because D is more positive than B. The converse rules are also necessary for rainflow initiated at a minimum (points A, C, E, F).
3. Finally, rainflow must stop if it encounters rain from the roof above, as in the event from C to D.

Events A-D and D-A are paired to form on full cycle. Event B-C is paired with the partial cycle formed from C-D. Cycles are also formed from E-D and F-G. Obviously, rainflow counting requires a great deal of bookkeeping and is ideally suited to a digital computer. Several algorithms have been published to reduce computation time (Ref 28) and there is now an ASTM standard, E-1049, "Standard Practice for Cycle Counting in Fatigue Analysis," dedicated to this technique.

Stress Concentrations

Besides material cyclic response and cycle counting in fatigue, changes in geometry act as stress and strain concentrations and therefore affect fatigue. Consideration of notch effects are considered in this section with the following symbols for key variables:

- E = modulus of elasticity

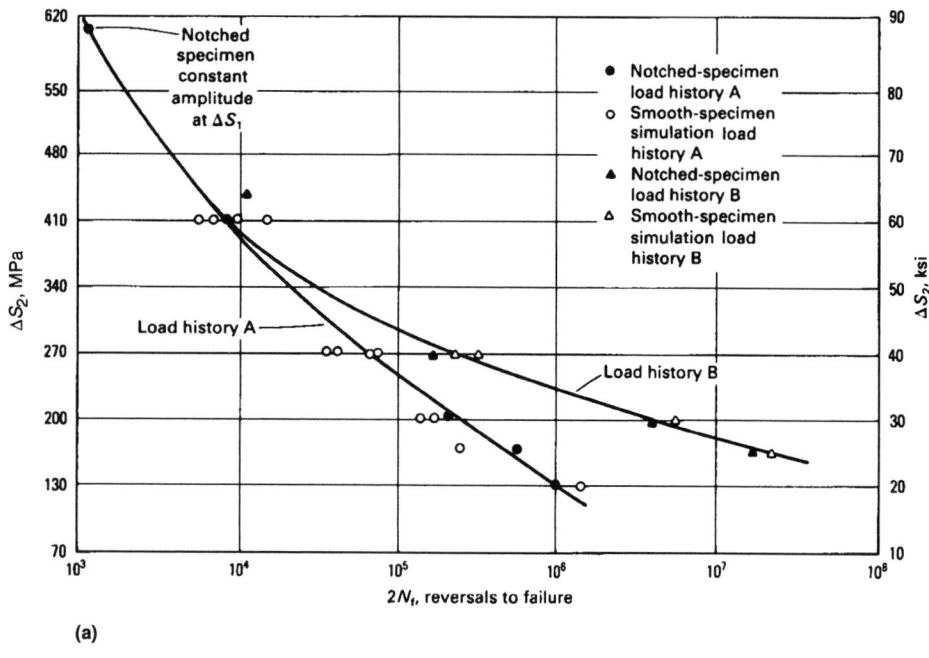

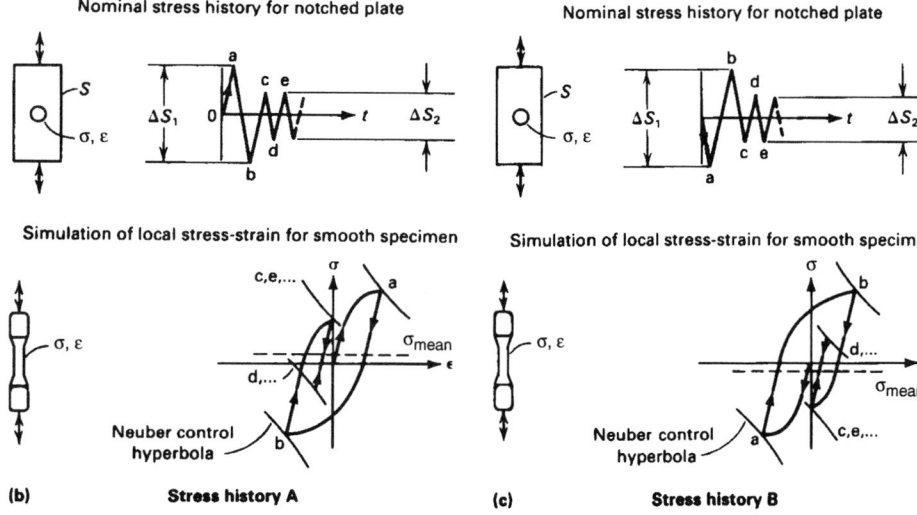

Fig. 35 Fatigue data (a) showing sequence effects for notched-specimen and smooth-specimen simulations (2024-T4 aluminum, $K_f = 2.0$). Load histories A and B have a similar cyclic load pattern (ΔS_2) but have slightly different initial transients (ΔS_1) with either (b) a tensile loading edge (first stress peak at $+\Delta S_1/2$) or (c) a compressive leading edge (first stress peak at $-\Delta S_1/2$). The sequence effect on fatigue life (a) becomes more pronounced as ΔS_2 becomes smaller. Source: D.F. Socie, "Fatigue Life Estimation Techniques," Technical Report 145, Electro General Corporation

- S = nominal stress on a notched member measured remotely from the stress concentration; for example, in an axial test, the axial load divided by the net area
- e = nominal strain (equal to S/E only when the nominal strain is elastic) measured remotely from the stress concentration
- s = actual or local stress at the stress concentration
- e = actual or local strain at the stress concentration
- ΔS, Δe, $\Delta \sigma$, $\Delta \varepsilon$ = peak-to-peak change in the above quantities during one reversal or half-

cycle (Δ represents range, as opposed to amplitude)
- K_t = theoretical (elastic) stress-concentration factor = σ_{max}/S, where σ_{max} is the maximum local stress
- K_σ = stress concentration factor = $\Delta\sigma/\Delta S$
- K_ε = strain concentration factor = $\Delta\varepsilon/\Delta e$
- K_f = fatigue notch factor
- a = material constant with dimensions of length

Fatigue failures nearly always initiate at a geometric discontinuity in wrought products, exclud-

ing inclusions in high-hardness steels, that are considered microdiscontinuities. (An example is explained in a later section of this paper.) Associated with every notch is a theoretical stress-concentration factor, K_t, that is dependent only on geometry and loading mode. In fatigue, notches may be less effective than predicted by K_t. Therefore, a fatigue-notch factor, K_f, is frequently employed. It is often determined by the ratio of unnotched fatigue strength to notched fatigue strength at a given life level:

$$K_f = \frac{\sigma_{unnotched}}{\sigma_{notched}} \text{ at a finite life}$$

$$(\text{for example, } 10^7 \text{ cycles}) \qquad (\text{Eq 62})$$

Often, a notch-sensitivity index is defined as:

$$q = \frac{K_f - 1}{K_t - 1} \qquad (\text{Eq 63})$$

and varies from 0 (no notch effect) to 1 (full theoretical effect).

The value of q is dependent on the material and the radius of the notch root, as illustrated by a plot of the relationship shown in Fig. 40. It should be apparent that small notches are less effective than large notches, and soft metals are less affected than hard metals by geometric discontinuities that reduce the fatigue resistance.

Many attempts have been made to determine values of K_f analytically. One of the more successful is attributed to Peterson (Ref 29) and is expressed as:

$$K_f = 1 + \frac{K_t - 1}{1 + \frac{a}{r}} \qquad (\text{Eq 64})$$

where "a" is a material constant dependent on strength and ductility, and is determined from long-life test data for notched and unnotched specimens of known K_t and tip radius, r.

Fortunately, "a" can be approximated for ferrous-based wrought metals by the following empirical relationship:

$$a \cong \left(\frac{300}{S_u \text{ (ksi)}}\right)^{1.8} \times 10^{-3} \text{ in.} \qquad (\text{Eq 65a})$$

$$a \cong \left(\frac{300}{0.5 \text{ HB}}\right)^{1.8} \times 10^{-3} \cong \left(\frac{600}{\text{HB}}\right)^{1.8} \times 10^{-3} \text{ in.} \qquad (\text{Eq 65b})$$

where S_u (ksi) $\cong 0.5$ HB (Brinell hardness). As a rule-of-thumb, "a" for normalized or annealed steels $\cong 0.01$; for highly hardened steels $\cong 0.001$; and for quenched-and-tempered steels $\cong 0.025$ in.

Figure 41 illustrates the effect on K_f from changing r for a hard-and-soft metal. When r is approximately equal to "a," the effect of changing r and/or a is most apparent (that is, at the inflection point in the sigmoidal curve). When r is greater than $10a$ or less than $a/10$, very little

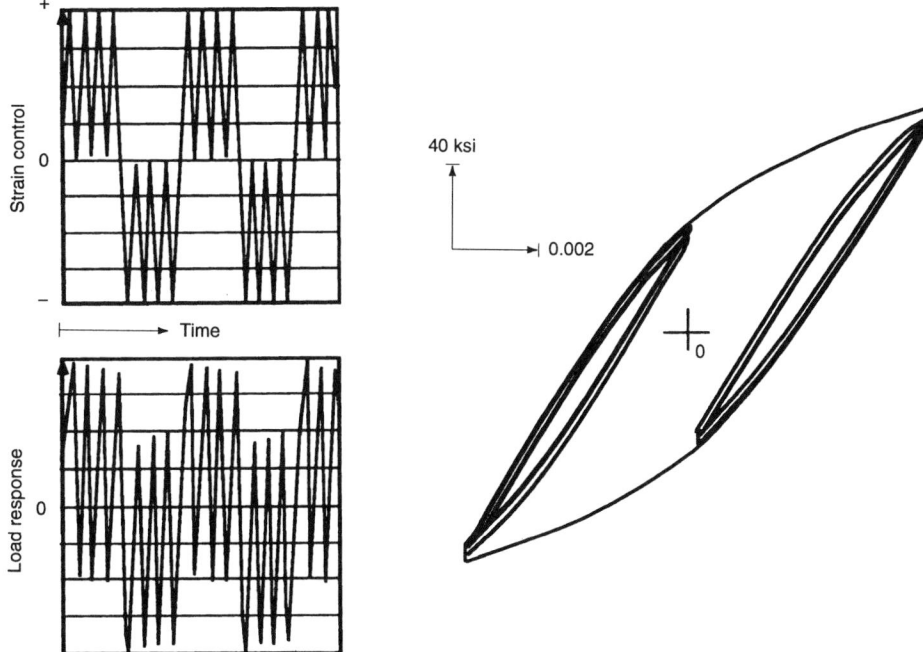

Fig. 36 Relaxation of mean stresses under biased straining of an SAE 1045 steel. Source: Ref 6

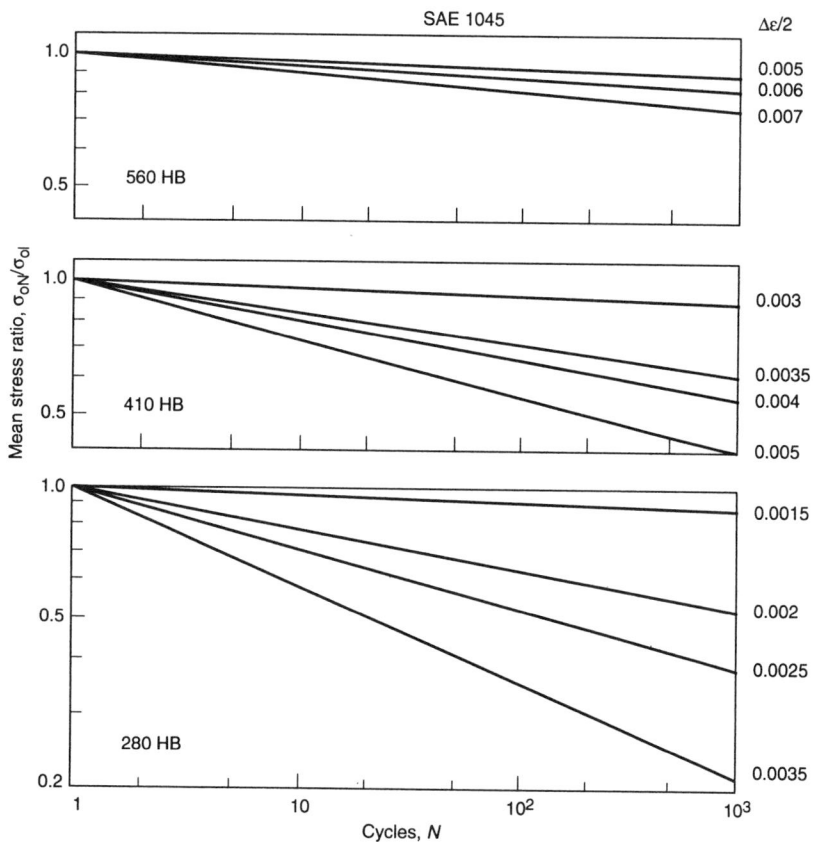

Fig. 37 Effect of strain amplitude and hardness on relaxation rate of mean stress. Source: Ref 6

In the low- and intermediate-life region where yielding can occur at a notch, strain concentration as well as a stress concentration must be considered. When yielding occurs, K_σ and K_ε are no longer equal (see Fig. 42). After yielding, K_ε increases but K_σ decreases. To solve this plasticity problem, employ Neuber's rule (Ref 30), in which the theoretical stress-concentration factor, K_t, is equated to the geometric mean of the stress-concentration factor, K_σ, and the strain-concentration factor, K_ε:

$$K_t = (K_\sigma K_\varepsilon)^{1/2} \tag{Eq 66}$$

For fatigue, K_f is often substituted for K_t (Ref 31), so that Eq 66 may be expressed as:

$$K_f = (K_\sigma K_\varepsilon)^{1/2} \tag{Eq 67}$$

Through the definition of the stress-concentration factor, $K_\sigma = \Delta\sigma/\Delta S$, and the strain-concentration factor, $K_\varepsilon = \Delta\varepsilon/\Delta e$, we may substitute to Eq 67 and obtain

$$K_f = \left(\frac{\Delta\sigma\Delta\varepsilon E}{\Delta S\Delta e E}\right)^{1/2} \tag{Eq 68}$$

where E has been inserted to present the equation in terms of stress units. Therefore:

$$K_f (\Delta S\Delta e E)^{1/2} = (\Delta\sigma\Delta\varepsilon E)^{1/2} \tag{Eq 69}$$

Illustrated schematically, the quantities of interest are shown in Fig. 43.

If the response if nominally elastic, which is often the case in vehicle design, $\Delta S = E\Delta e$, and Eq 69 may be written as:

$$K_f\Delta S = (\Delta\sigma\Delta\varepsilon E)^{1/2} \tag{Eq 70}$$

This approach is convenient because:
1. The relationships relate remotely measured stresses and strains to local response at the critical location of the notch root.
2. They allow the simulation of notch fatigue behavior with smooth specimens.
3. They allow the prediction of notch behavior with smooth-specimen data.

To illustrate the use of this equation, it is convenient (although not necessary) to use the more simplified case of nominally elastic stressing and to rearrange the terms of Eq 70 to the form:

$$\Delta\sigma\Delta\varepsilon = \frac{(K_f\Delta S)^2}{E} \tag{Eq 71}$$

Equation 71 is the relation for a rectangular hyperbola (xy = constant). If a nominal stress, applied to a notched sample starting at zero stress, is increased to some arbitrary "elastic" stress, S_1 (Fig. 44a), it is a relatively simple task to compute the value on the right of Eq 71, if K_f is known.

There is a family of values of the product of local stress range, $\Delta\sigma$, and strain range, $\Delta\varepsilon$, that is

change in K_f will accompany changes in r and/or "a."

The previous discussion is an attempt to account for "size effect" of notches in fatigue. Although a functional relationship such as given in Eq 65 is valid for steels, no clear relationship of this type exists for aluminum alloys. Thus, it is mandatory to conduct notched and unnotched fatigue tests on the aluminum alloy of interest to functionally define "a."

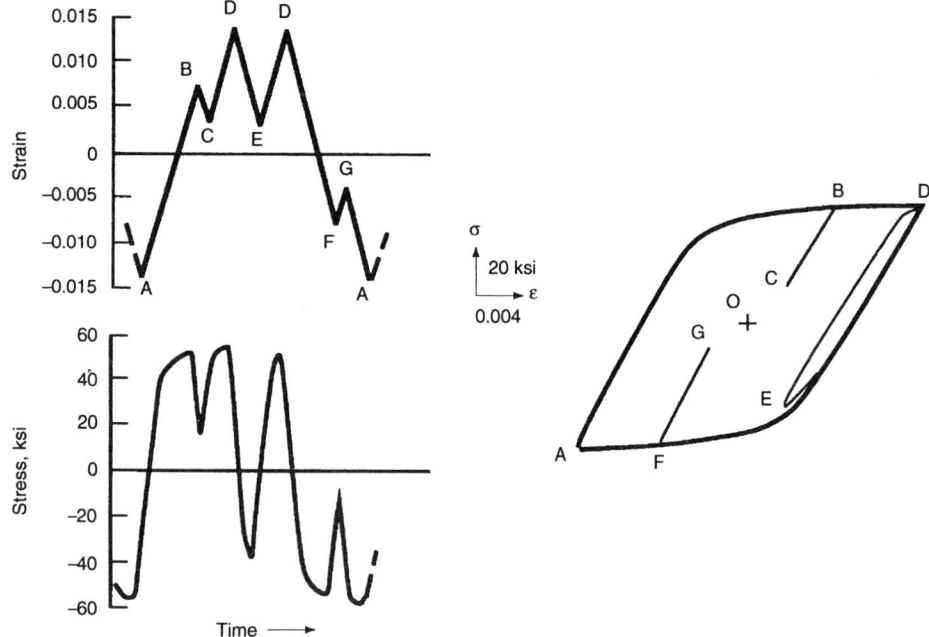

Fig. 38 Imposed-variable-strain history and stress response. Source: Ref 26

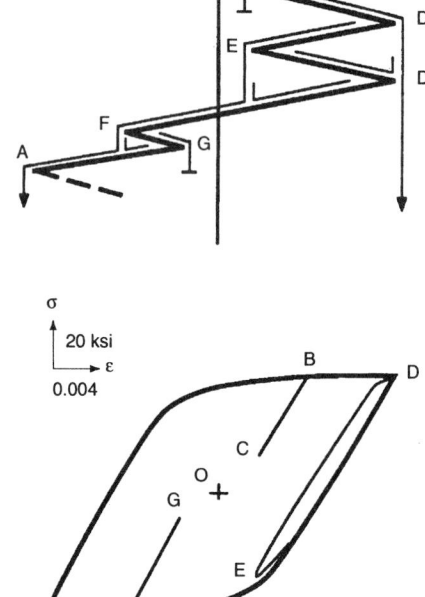

Fig. 39 Schematic of rain flow counting technique. Source: Ref 26

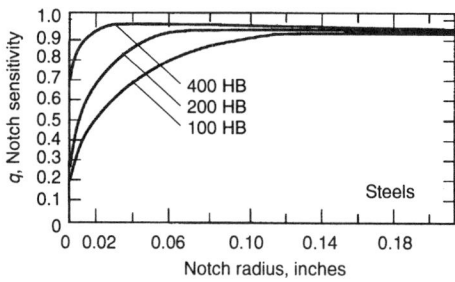

Fig. 40 Notch sensitivity versus notch radius as a function of hardness for steels

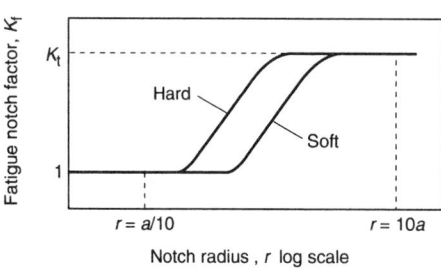

Fig. 41 Fatigue-notch factor versus notch radius as a function of relative hardness

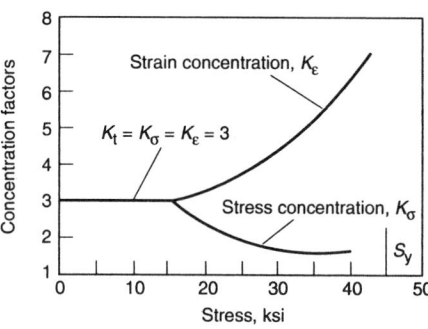

Fig. 42 Schematic of change in stress-concentration and strain-concentration factors as yielding occurs at notch root

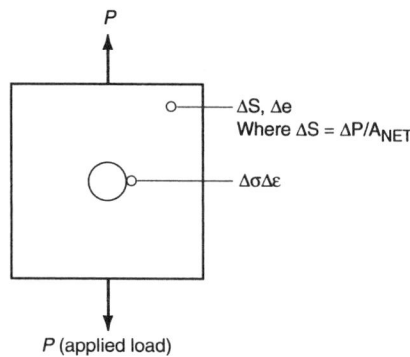

Fig. 43 Quantities of interest in a notch analysis

equal to the constant, $(K_f \Delta S)^2/E$. However, if the cyclic stress-strain curve for the material of interest is traced on rectangular coordinates (Fig. 44b), there is a unique combination of stress and strain ranges that satisfies the equation. This

unique value occurs at the intersection of the cyclic stress-strain curve with the rectangular hyperbola.

If there is a reversal in nominal stress at S_1, the above procedure is repeated; but the origin of the rectangular coordinate system used for the next step in the sequence is located at point P in Fig. 44(b). On unloading or for any subsequent events not starting at zero stress and strain, the cyclic stress-strain curve is magnified by a factor of two in order to trace the hysteresis loop.

As a continuation of our example, assume the nominal stress-time sequence to be analyzed is as shown in Fig. 45, with $S_2 = 0$. By following the same procedure as above, but with the local stress-strain origin fixed at point P, a trace of the second event would be as shown in Fig. 46. Of course, such a point-by-point analysis is tedious; in real life, situations must be computerized. This is often accomplished by employing the equation for the cyclic stress-strain curve in the form:

$$\Delta \varepsilon = \frac{\Delta \sigma}{E} + \left(\frac{\Delta \sigma}{K''}\right)^{1/n'} \tag{Eq 72}$$

and taking the product:

$$\Delta \sigma \Delta \varepsilon = \frac{(\Delta \sigma)^2}{E} + \Delta \sigma \left(\frac{\Delta \sigma}{K'}\right)^{1/n'} \tag{Eq 73}$$

By equating Eq 73 to the constant in Eq 71 we have:

$$\frac{(\Delta \sigma)^2}{E} + \Delta \sigma \left(\frac{\Delta \sigma}{K'}\right)^{1/n'} = \frac{(K_f \Delta S)^2}{2} \tag{Eq 74}$$

This equation is solved relatively easily using the Newton-Raphson iteration technique and standard numerical methods.

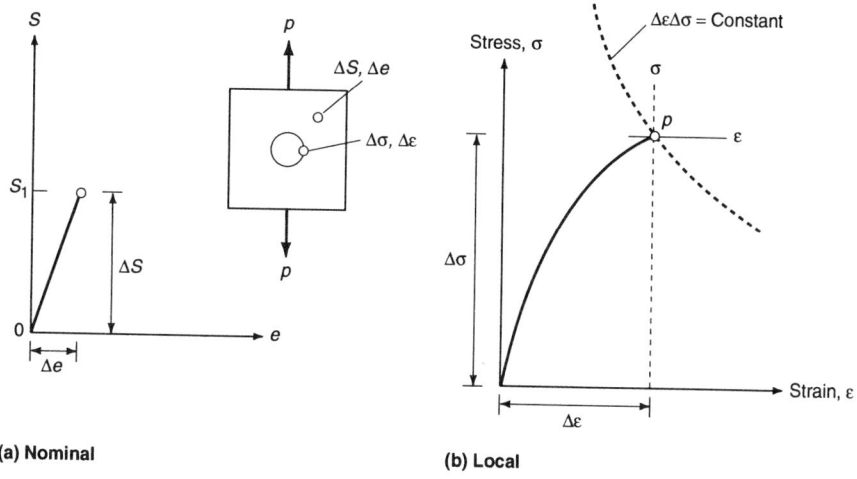

(a) Nominal

(b) Local

Fig. 44 Nominally imposed elastic stress and strain and local changes in stress and strain at a notch root

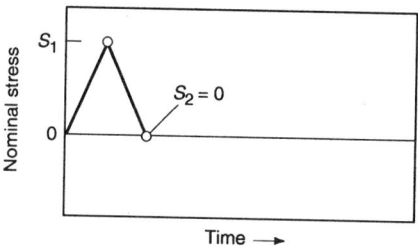

Fig. 45 Nominal elastic unloading to zero stress

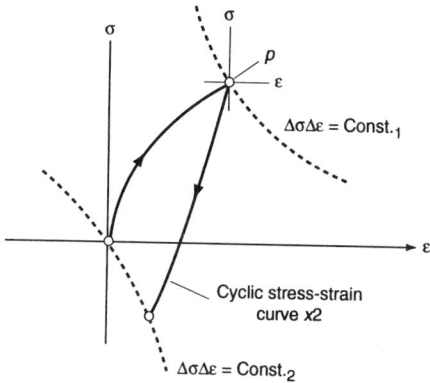

Fig. 46 Changes in local stress and strain on nominal elastic unloading

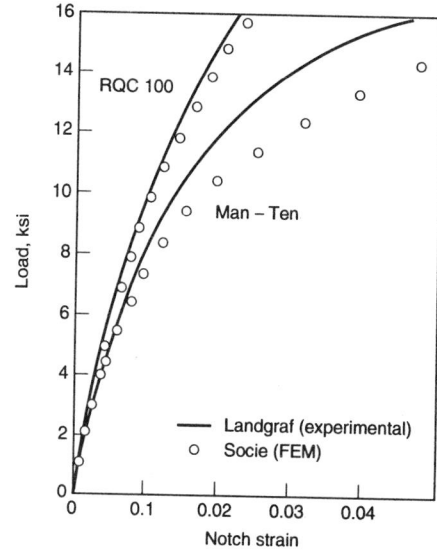

Fig. 47 Component-calibration curve. Source: Ref 33

Summary

The strain-life approach to characterizing the fatigue behavior of materials has been presented. An effective means of accounting for plastic strain, that is the cause of fatigue failures, has been given; and a constitutive equation between strain and life was developed. Because materials are metastable under cyclic loads and because a simple tensile stress-strain curve was shown inadequate for fatigue design, a cyclic strain-strain curve was introduced.

To predict the crack-initiation life of actual components:

1. Mean stresses were taken into account by a simple modification of the strain-life equation.
2. A technique to account for size effect of geometric notches was introduced.
3. Two procedures were given relating remotely measured stresses and strains to stresses and strains at a notch root where plasticity dominates.

By combining the above "analytical tools" with an adequate cycle-counting technique that accrues closed hysteresis loops (for example, rainflow or range pair), a means was developed to analyze pseudo-random load histories of real components or parts to predict fatigue-initiation life. Some examples of application techniques are described below.

Example 2: Component-Calibration Techniques

In many practical problems, engineers and designers are required to evaluate the fatigue resistance of prototype components while they are at the "drawing board" stage of development. One method for performing such an analysis is the component-calibration technique, which requires a relationship between applied load and local strains, such as the one shown in Fig. 47 (Ref 32). Such information may be obtained analytically by using finite element models, or experimentally by testing the component. The component would normally be tested by mounting strain gauges at the critical locations, applying one load-unload cycle and measuring the load-strain response. However, this type of test may produce erroneous data because of the cyclic hardening or softening characteristics of the material. For this reason, an incremental-step strain-type test (Ref 33) should be used to obtain load-strain calibration curves from a single component. In this way, the material is cyclically stabilized. Similarly, cyclically stable material properties should be used in subsequent analytical calculations.

The conversion of applied load to strain is accomplished in the same manner as the conversion of strain into stress. The load-strain response has all the features normally associated with stress-strain response (hysteresis effects, memory, cyclic hardening and softening). Transient material response is normally neglected, so that the load-strain response model accounts only for hysteresis and memory effects. From a computational viewpoint, this technique is the same as those described in the section for stress-strain response. In fact, the load-strain and stress-strain response models can be combined, so that the applied load can be converted to both the local stress and strain with one simple computer algorithm.

Many of the aforementioned methods for analysis of cumulative fatigue damage have been combined with various design philosophies. Several commercially available programs are now readily available to accomplish the necessary calculations for materials selection, design analysis and cumulative fatigue damage. For example, Somat Corporation in Champaign, Illinois has a LifeEst® program that is very user-friendly, and is presently being incorporated into the academic curriculum of several major universities as part of their mechanical design courses. But, the reader is cautioned that a thorough understanding of the basic philosophy outlined in this paper and the introductory literature for these programs should be mastered before attempting their implementation.

Example 3: Variable Histories: Different Steels

Procedures discussed in the preceding sections were used to evaluate the results of an early SAE FD&E Cumulative Fatigue Damage Test Program. Three different load-time histories (see Fig. 48) were applied to the test specimen shown in Fig. 49. Note that the specimen, when loaded,

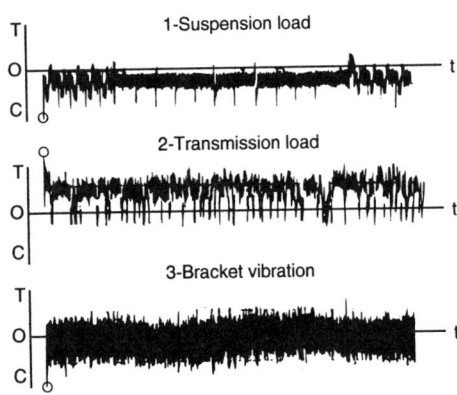

Fig. 48 Three different load-time histories. Source: Ref 28

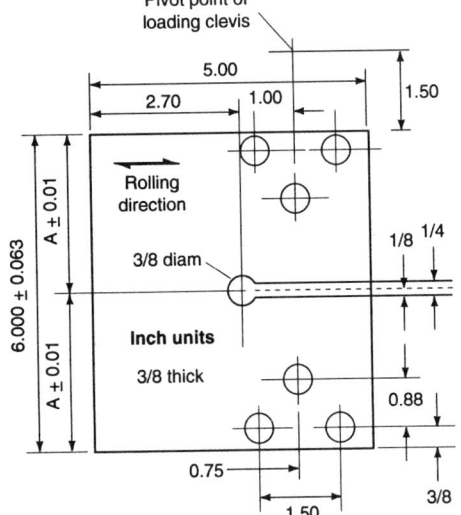

Fig. 49 Specimen design for test program. Source: Ref 28

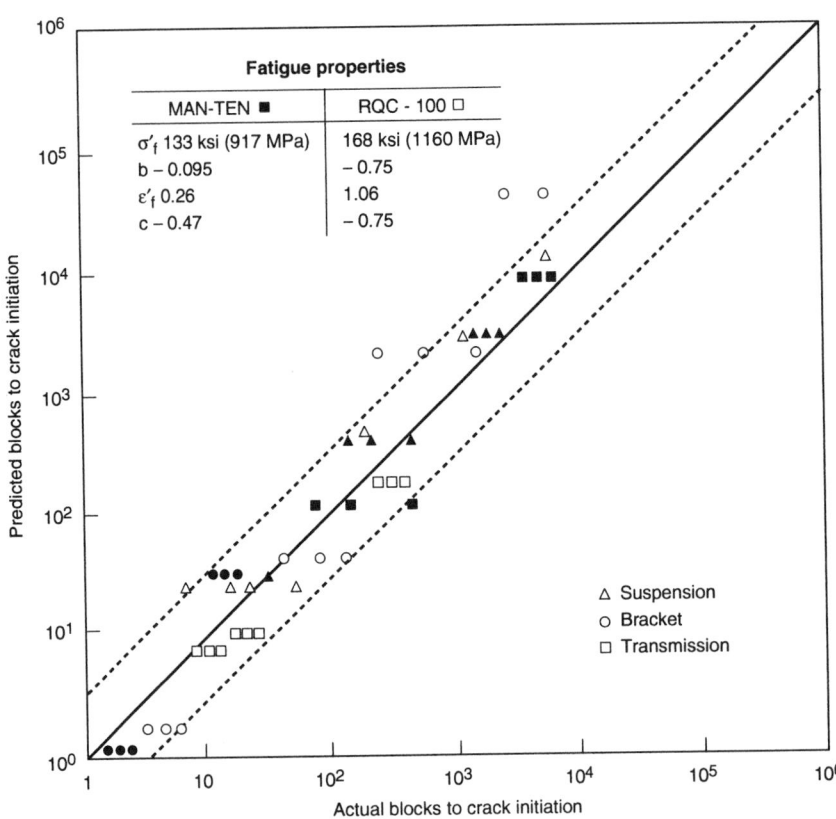

Fig. 50 Predicted versus actual blocks to crack initiation. Source: Ref 28

provides both axial and bending components of stress and strain at the notch root. Two steels were used: U.S. Steel's MAN-TEN and Bethlehem's RQC-100. Tests were conducted at several load levels for each spectrum. Fatigue lives ranged from 10^4 to 10^9 reversals. A complete description of the test program is given in Ref 19.

A summary of predicted and actual crack-initiation lives is shown in Fig. 50. These predictions were made using the load-strain curves shown in Fig. 47. For "perfect" correlation, all the data points should lie along the 45° solid line. All but four of the predicted lives are within the factor-of-three scatterband indicated by the dashed lines. This agreement is good, considering that there are two steels, three types of load-time histories, and at least three different load levels.

The first example cited was included in this paper as an example of an "early" attempt to predict fatigue lifetime behavior with "state-of-the-art" technology at that time (1978). As has been mentioned repeatedly, the techniques employed today are almost a routine part of compo-

nents designed to survive fatigue environments (or, at least, they should be).

Of the many examples available in the open literature, the reader is referred to a recent publication, *Case Histories in Fatigue Design*, ASTM STP 1250, R.I. Stephens, Ed., ASTM, 1994. Also, excellent examples appear in *Fatigue Design Handbook*, SAE AE10, Second Edition, 1988, including wheels made of high strength sheet steel, suspension system components, forged connecting rods and axle shafts. Several examples of fatigue lifetime predictions for cast metals are also given that are quite adequate for the purpose intended.

Example 4: Cast Metals

The basic techniques for describing the cyclic stress-strain resistance of cast metals is not as straightforward as are those for a wrought metal. Cast ferrous-based products (gray and nodular iron and cast steels) are internally defected structures. As such, the stress-strain resistance of the bulk material, which contains second-phase discontinuities in the form of graphite flakes, nodules, and/or gas porosities, is not an adequate representation of the capacity of the material to resist stresses and strains.

By considering cast metals as a homogeneous steel matrix with "micronotches," we can extend the previously described notch analysis to predict the fatigue-life behavior of these products.

Since fatigue-crack initiation generally occurs in regions where the stress-concentrating effect

of the micronotches is greatest, it is justifiable to assume that the fatigue resistance of cast ferrous-based metals is governed by the largest surface discontinuity. For example, in the case of gray iron, the flake type-A (ASTM A247) graphite colonies extend in three-dimensional space to the extent of the eutectic cell walls. Metallographic examination of critical areas in a component can be employed to reveal "the largest" diameter eutectic cell, t. Approximating the graphite flakes as surface slits, the theoretical stress-concentration factor is given by:

$$K_t = 1 + 2\sqrt{\frac{t}{r}} \qquad \text{(Eq 75)}$$

where r is the tip radius of the most notch-effective graphite flake.

The question now is: In a three-dimensional graphite colony, which flake has the most effective tip radius? Mattos and Lawrence (Ref 34) have observed in their treatment of weld flaws that the fatigue-notch factor, K_f, has a maximum value for the special case of an ellipse with fixed major axis but variable tip radius. Using a similar approach, we may substitute the value of K_t into Eq 64:

$$K_f = 1 + \frac{2\sqrt{tr}}{r+a} \qquad \text{(Eq 76)}$$

The maximum value of K_f occurs when:

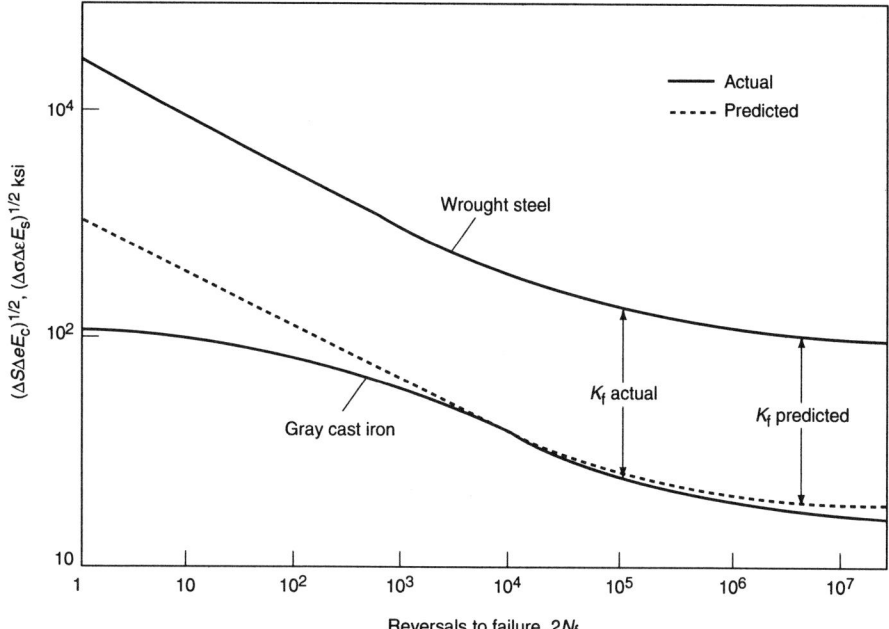

Fig. 51 Neuber-Topper parameter versus reversals to failure for wrought steel (260 HB) and 0.08-in. eutectic cell

$$\frac{dK_f}{dr} = 0 \qquad \text{(Eq 77)}$$

Thus:

$$K_{f_{max}} = 1 + \sqrt{\frac{l}{a}} \qquad \text{(Eq 78)}$$

Peterson's "a" in Eq 78 is defined by Eq 65, but the value of hardness (HB) employed must be that of the matrix metal. (HB of the matrix of gray iron, which is approximately an SAE 9200-series steel, is converted from microhardness readings, for example, Vickers or Knoop.) Upon appropriate substitution, it can be shown that

$$K_{f_{max}} = 1 + 0.1 l^{0.5} HB^{0.9} \qquad \text{(Eq 79)}$$

and the "size effect" of graphite in gray iron (viewed as surface slits) is taken into account.

Next in our example, the strain-life behavior of the matrix steel must be determined. This may be accomplished by performing a constant-amplitude, controlled-strain type of test (outlined previously) on a wrought steel matched in hardness, composition and structure to the matrix of the gray iron of interest. Or, in the absence of strain-controlled test data, the approximations for strain-life behavior shown earlier in this paper may be employed—but using the matrix hardness of the iron.

An example of results from such an analysis for a gray iron with fully pearlitic matrix (260 HB) and type A graphite with a 0.08 in. eutectic-cell diameter is shown in Fig. 51, which is presented in a slightly different form than used previously. The vertical axis is the geometric mean of the product of stress and strain that results from a notch analysis. Note also that the E_c shown in Fig. 51 is the modulus of elasticity of the cast metal that has been employed to account for the limiting case of nominal elastic response.

Similar analyses have also been performed for nodular irons, cast steels and high-hardness wrought steels in which inclusions govern behavior (Ref 11).

If, in retrospect, one examines closely the concept of "crack initiation" in a gray cast iron, it is obvious that there is only a very brief period of "initiation" in these heavily defected materials where life can be dominated by crack growth. This is also true for other cast metals, such as aluminum-silicon alloys where the free silicon is in the form of lenticles, and for many composite materials, such as metal matrix composites and ceramic matrix composites. A much better means of attacking such life predictions (in the author's opinion) is to use a continuum damage mechanics approach as first employed by Downing (Ref 35) for gray cast iron. Many of the procedures outlined for life prediction using the strain-based approach are similar and the baseline materials data collection procedure is similar. Additional collection is, however, made of the rate of change of, for example, modulus or compliance, peak stresses, crack length, etc., with cycles. The "damage" is then viewed as a rate phenomena (i.e., as that fraction of time spent at a given rate corresponding to a specified strain amplitude to the total time to failure at that strain amplitude).

Example 5: Effect of Environment

It is a well-established fact that the fatigue life of materials is in many instances drastically al-

tered by the environment in which the materials must perform. Perhaps one of the most significant instances is exhibited by aluminum alloys in saline environments.

As another example of the versatility of the strain-based approach to fatigue-damage analysis, consider the problem of predicting the stress-life behavior of 7075-T73 aluminum alloy containing a geometric notch ($K_t = 2.52$) in a 3.5 wt% NaCl environment. Obviously, this is a somewhat pedagogical example; nonetheless, it will illustrate the basic concept of how to proceed to a more complex damage analysis under component service histories.

Monotonic and cyclic stress-strain curves for 7075-T73 aluminum alloy tested in laboratory air (20 to 50% relative humidity) are shown in Fig. 52. Note that the material is cyclically stable. Data points for the cyclic stress-strain curve were obtained from companion-specimen results controlled-strain-amplitude tests performed at a constant total strain rate of $\dot{\varepsilon} = 2.4 \times 10^{-3} \sec^{-1}$ ($\dot{\varepsilon} = f \times \varepsilon$ = frequency $\times$ strain amplitude). A saline environment (or, for that matter, even relative humidity) has a more pronounced effect on the long-life fatigue behavior of aluminum than on short lives. Thus, the saline environment can be considered to degrade basic material properties (to alter the slope, b, of the elastic strain-life line). Figure 53 shows the strain-life results of smooth specimens tested in a 3.5 wt% NaCl environment compared with the strain-life curve for the specimens tested in laboratory air. The values of the slope, b, of the respective elastic strain-life lines are –0.15 (3.5 NaCl) and –0.11 (air). (The value of the slope has been modified for periodic overstraining by decreasing the life an order of magnitude of a nonoverstrained value corresponding to 10^7 reversals.) Other material properties of interest for subsequent life predictions are given in Table 4.

Having defined the material properties for 7075-T73 alloy in the 3.5 wt% NaCl environment, the next step in the analysis is the determination of the fatigue-notch factor, K_f, for the geometric notch with $K_t = 2.52$. As mentioned previously, there is no clear functional relationship between K_f and K_t through Peterson's equation because each aluminum alloy, depending on thermomechanical processing, has a different value of the length parameter, "a." It was therefore necessary to conduct long-life fatigue tests (10^7 reversals) of notched specimens of 7075-T73. By the quotient of the fatigue strength at 10^7 reversals of unnotched specimens to the fatigue strength of notched specimens $K_f = \sigma_{unnotched}/\sigma_{notched}$, a value of the fatigue-notch factor was determined to be 2.2.

Next, a cumulative-fatigue-damage-analysis program was developed, similar to that of Landgraf et al. (Ref 26), in which the cyclic stress-strain curve was "modeled" by a series of straight-line segments. This particular program must be initialized at the absolute maxima or minima of a digitized input history that are nominal stresses, but must be "elastic." To relate nominal stresses, ΔS, to notch root stresses and strains,

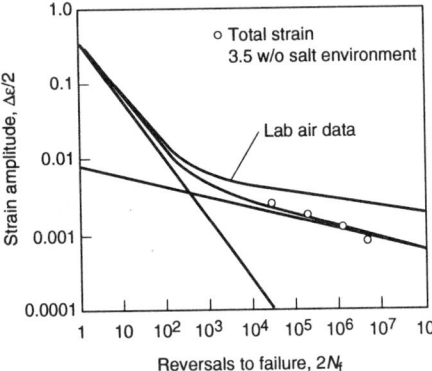

Fig. 52 Monotonic and cyclic stress-strain curves for 7075-T73 aluminum alloy

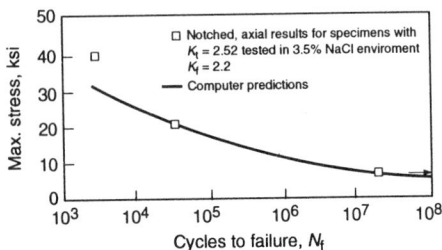

Fig. 53 Strain versus life curves for 7075-T73 aluminum alloy tested in laboratory air and 3.5 wt% NaCl. Source: Ref 36

Fig. 54 Test results and computer predictions for 0-max stressing of 7075-T73 aluminum alloy in a 3.5 wt% NaCl environment. Source: Ref 36

$\Delta\sigma$ and $\Delta\varepsilon$, Neuber's rule was employed in the form of Eq 71. By proper manipulation of the strain-life equation, it can be shown that:

$$\text{Damage/reversal} = \frac{1}{2N_f} = \left[\frac{\sigma'_f}{\varepsilon'_f E}\frac{\Delta\varepsilon_p}{\Delta\varepsilon_e}\right]^{1/(b-c)} \quad \text{(Eq 80)}$$

and mean stresses (σ_o) accounted for by modification of the above equation to:

Table 4 Material properties for 7075-T73 aluminum alloy

Property	Laboratory air	3.5 wt% NaCl environments
Modulus of elasticity, E, ksi	10×10^3	10×10^3
Fatigue-strength coeff., σ'_f, ksi	89.0	89.0
Fatigue-ductility, coeff., ε'_f	0.387	0.387
Fatigue-strength exponent, b	−0.11	−0.15
Fatigue-ductility exponent, c	−0.8	−0.8

$$\text{Damage/reversal} = \left[\frac{\sigma'_f}{\varepsilon'_f E}\frac{\Delta\varepsilon_p}{\Delta\varepsilon_e}\frac{\sigma'_f}{\sigma'_f - \sigma_o}\right]^{1/(b-c)} \quad \text{(Eq 81)}$$

where $\Delta\varepsilon_p$ is the local plastic-strain range and $\Delta\varepsilon_e$ is the local elastic-strain range. In this example, a conditional was employed for input of material-property data: "Cyclic stresses and strains were defined by the laboratory air cyclic stress-strain curve and the effect of environment was to modify only the long-life fatigue resistance of the aluminum alloy (i.e., b increases in absolute value as the environmental severity increases)."

Figure 54 compares zero to maximum stress fatigue results of notched specimens tested in a 3.5 wt% NaCl environment, using the techniques described, to the predicted behavior. The agreement between the prediction and test results appears favorable (Ref 36).

The approach described above is not the only means of accounting for environmental effects on the fatigue life of materials. Interested readers are guided to Ref 37 for an excellent review of the effects of environment, frequency, strain rate, metallurgical variables, wave shape, and thermal cycling on the fatigue behavior of metals. The author employs what have been called "frequency modified" relationships.

ACKNOWLEDGMENTS

This paper has drawn heavily from the course notes of Professor (Emeritus) Jo Dean Morrow, Department of Theoretical and Applied Mechanics, University of Illinois; and Dr. R.W. Landgraf, Virginia Polytechnic Institute. The former was my thesis advisor, the latter a colleague while I worked for Ford Motor Co., Scientific Research Staff. Both are long time friends and colleagues and to both I owe an eternal debt of gratitude. Much of the information given herein was originally presented as an introductory seminar to sponsors of the Fracture Control Program, College of Engineering, University of Illinois at Urbana-Champaign and was published as FCP Report No. 26 in 1976. A later version of similar notes published under the same program at the University of Illinois expanding these concepts and including fatigue crack propagation methodologies for life predictions was written by J.A. Bannantine, J.J. Comer and J.L. Handrock, and titled *Fundamentals of Metal Fatigue Analysis*, 1987.

REFERENCES

1. W.A.J. Albert, "Uber Treibseile am Harz," Archive fur Mineralogie, Geognosie, Bergbau und Huttenkunde, Vol 10, 1838, p 215-234 (in German)
2. A. Wöhler, "Versuche uber die Festigkeit der Eisenbahnwagenachsen," Zeitschrift fur Bauwesen, Vol 10, 1860 (in German), with English summary in *Engineering*, Vol 4, 1867, p 160-161
3. Manual on Low Cycle Fatigue Testing, STP 465, ASTM, Dec 1969
4. G.E. Dieter, *Mechanical Metallurgy*, McGraw-Hill, 1961
5. J. Morrow, G.R. Halford, and J.F. Millan, Optimum Hardness for Maximum Fatigue Strength of Steels, *Proceedings of the First International Conference on Fracture*, (Sendai, Japan), Vol 3, 1965, p 1611-1635
6. R.W. Landgraf, Cycle Deformation Behavior of Engineering Alloys, *Proceedings of Fatigue-Fundamental and Applied Aspects Seminar*, 15-18 August 1977 (Remforsa, Sweden)
7. R.W. Landgraf, M.R. Mitchell, and N.R. LaPointe, "Monotonic and Cyclic Properties of Engineering Materials," Ford Motor Co., June 1972 (Also F. Conle, R. Landgraf, F. Richards, 1990)
8. SAE Handbook, Section J-1099, Society of Automotive Engineers, 1992
9. R.W. Smith, M.H. Hirschberg, and S.S. Manson, "Fatigue Behavior of Materials Under Strain Cycling in Low and Intermediate Life Range," NASA TN D-1574, NASA, April 1963
10. W. Brose, N.E. Dowling, and J. Morrow, "Effect of Periodic Large Strain Cycles on the Fatigue Behavior of Steels," SAE Paper No. 740221, SAE, Automotive Engineering Congress, 25 Feb-1 March 1974 (Detroit, MI)
11. M.R. Mitchell, *A Unified Predictive Technique for the Fatigue Resistance of Cast Ferrous-Based Metals and High Hardness Wrought Steels*, SAE SP 442, Society of Automotive Engineers, 1979
12. J. Morrow, "Cyclic Plastic Strain Energy and Fatigue of Metals," *International Friction Damping and Cyclic Plasticity*, STP 378, ASTM, 1965, p 45-87
13. T. Endo and J. Morrow, Cyclic Stress-Strain and Fatigue Behavior of Representative Aircraft Alloys, *Journal of Materials*, Vol 4, 1969, p 159-175
14. T.H. Sanders, Jr. and E.A. Starke, Jr., The Relationship of Microstructure to Monotonic and Cyclic Straining of Two Age Hardening Aluminum Alloys, *Met. Trans. A*, Vol 7A, Sept 1976, p 1407-1418
15. V.M. Radhakrishnan, On the Bilinearity of the Coffin-Manson Low-Cycle Fatigue Relationship, *Int. Journal Fatigue*, Vol 14 (No. 5), 1992, p 305-311
16. L.F. Coffin, Jr. and J.F. Tavernelli, The Cyclic Straining and Fatigue of Metals, *Trans. Metallurgical Society*, AIME, Vol 215, Oct 1959, p 794-806

17. S.S. Manson, Fatigue: A Complex Subject—Some Simple Approximations, *Experimental Mechanics*, July 1975, p 1-35

18. Y. Higashida and F.V. Lawrence, "Strain Controlled Fatigue Behavior of Weld Metal and Heat-Affected Base Metal in A36 and A514 Steel Welds," Fracture Control Program Report No. 22, University of Illinois, College of Engineering, Aug 1976

19. C.H.R. Boller and T. Seeger, Materials Science Monographs, 42A, Part A: Unalloyed Steels; 42B, Part B: Low-Alloy Steels; 42C, Part C: High-Alloy Steels; 42D, Part D: Aluminum and Titanium Alloys; 42E, Part E: Cast and Weldment Metals, *Materials Data for Cyclic Loading*, Elsevier, 1987

20. A. Baumel, Jr. and T. Seeger, Supplement 1, *Materials Data for Cyclic Loading*, Elsevier, 1990

21. F.A. Conle, R.W. Landgraf, and F.D. Richards, *Materials Data Book—Monotonic and Cyclic Properties of Engineering Materials*, Ford Motor Company, Dearborn, MI, 1988

22. J. Morrow and G.M. Sinclair, *Symposium on Basic Mechanisms of Fatigue*, STP 237, ASTM, 1958, p 83-101

23. R.W. Landgraf, The Resistance of Metals to Cyclic Deformation, *Achievement of High Fatigue Resistance in Metals and Alloys*, STP 467, ASTM, 1970, p 3-36

24. D.A. Woodford and J.R. Whitehead, Ed., *Advances in Life Prediction Methods*, ASME, 1983

25. S.S. Manson and G.R. Halford, Re-Examination of Cumulative Fatigue Damage Analysis—An Engineering Perspective, *Mechanics of Damage and Fatigue*, S.R. Bodner and Z. Hashin, Ed., Pergamon Press, 1986, p 539-571

26. R.W. Landgraf and N.R. LaPointe, "Cyclic Stress-Strain Concepts Applied to Component Fatigue Life Prediction," SAE Paper No. 740280, SAE, Automotive Engineering Congress, 25 Feb-1 March, 1974 (Detroit, MI)

27. N.E. Dowling, Fatigue Life and Inelastic Strain Response under Complex Histories for an Alloy Steel, *Journal of Testing and Evaluation*, Vol 1 (No. 4), 1973, p 271-287

28. Fatigue Under Complex Loading, Advances in Engineering Series, Vol 6, SAE, 1977

29. R.E. Peterson, *Stress Concentration Factors*, John Wiley & Sons, 1974

30. H. Neuber, Theory of Stress Concentration for Shear-Strained Prismatical Bodies with Arbitrary Nonlinear Stress-Strain Law, *Trans. ASME, Journal of Applied Mechanics*, Dec 1961, 544-550

31. T.H. Topper, R.M. Wetzel, and J. Morrow, Neuber's Rule Applied to Fatigue of Notched Specimens, *Journal of Materials*, Vol 4 (No. 1), March 1969, p 200-209

32. D.F. Socie, Fatigue Life Prediction Using Local Stress-Strain Concepts, *Experimental Mechanics*, Vol 17 (No. 2), 1977

33. R.W. Landgraf, J. Morrow, and T. Endo, Determination of the Cyclic Stress-Strain Curve, *Journal of Materials*, Vol 4 (No. 1), March 1969, p 176-188

34. R.J. Mattos and F.V. Lawrence, "Estimation of the Fatigue Crack Initiation Life in Welds Using Low Cycle Fatigue Concepts," Fracture Control Program, Report No. 19, College of Engineering, University of Illinois, Oct 1975

35. S.D. Downing, "Modeling Cyclic Deformation and Fatigue Behavior of Cast Iron Under Uniaxial Loading," UILU-ENG-84-3601, Materials Engineering—Mechanical Behavior, College of Engineering, University of Illinois at Urbana—Champaign, Jan 1984

36. M.R. Mitchell, M.E. Meyer, and N.Q. Nguyen, Fatigue Considerations in Use of Aluminum Alloys, *Proceedings of the SAE Fatigue Conference*, (Dearborn, MI), SAE P-109, 1982, p 249-272

37. L.F. Coffin, Fatigue at High Temperature—Prediction and Interpretation, *Proc. Institution of Mechanical Engineers*, Vol 188, Sept 1974, p 109-127

Estimating Fatigue Life

Norman E. Dowling, Virginia Polytechnic Institute and State University

FATIGUE LIFE ESTIMATES are often needed in engineering design, specifically in analyzing trial designs to ensure resistance to cracking. A similar need exists in the troubleshooting of cracking problems that appear in prototypes or service models of machines, vehicles, and structures.

Three major approaches are in current use: (1) the stress-based (*S-N* curve) approach, (2) the strain-based approach, and (3) the fracture mechanics approach. Both the stress- and strain-based approaches are considered in this article from the viewpoint of their use as engineering methods. Analogous treatment of the fracture mechanics or damage tolerant approach, which is based on following crack growth, is not included here, but is given in several other articles in this Volume (see the Section "Fracture Mechanics, Damage Tolerance, and Life Assessment").

Much of what follows is adapted from selected portions of the book *Mechanical Behavior of Materials: Engineering Methods for Deformation, Fracture, and Fatigue* (Ref 1). The previous article in this Handbook, "Fundamentals of Modern Fatigue Analysis for Design," also contains considerable information of relevance to this article and is frequently referenced.

Stress-Based (*S-N* Curve) Method

Since the well-known work of Wöhler in Germany starting in the 1850s, engineers have employed curves of stress versus cycles to fatigue failure, which are often called *S-N* curves. Although now supplemented and sometimes replaced by more sophisticated approaches, the stress-based approach continues to serve as a useful tool.

Component *S-N* Curves. It is sometimes useful to conduct fatigue tests on an engineering component, such as a machine or vehicle part, or a structural joint. Subassemblies, such as a vehicle suspension system, may also be tested as may a portion of a structure or even an entire machine, vehicle, or structure. The Bailey Bridge panel made from structural steel (Fig. 1) is an example. This is one panel of a modular truss for military

and temporary civilian bridges used by the British in World War II. Bailey Bridges were still being manufactured long after the end of the war, and some were used in situations and for lengths of time (10 years or more) that were not envisioned by the original designers. Hence, a fatigue testing program was undertaken, as reported in a 1968 paper by Webber (Ref 2), to provide information on permissible length and severity of bridge usage.

A constant amplitude *S-N* curve from this work is shown in Fig. 2. This was obtained by applying cyclic loads to an assembly of panels, with these loads oriented in a plane corresponding to vertical loads on a bridge, which is the vertical direction of Fig. 1. Cracks generally started at a weld near the slot for sway brace shown in Fig. 1 and were visibly growing for at least half of the life. The stresses shown in Fig. 2 were calculated by treating the entire panel as a beam, with the bracing averaged as a web, and with the location of the critical slot giving the distance from the neutral axis of this beam. All tests used the same minimum load corresponding to the dead load of a bridge.

Such a curve is useful in assessing the life expected for Bailey Bridges under various vehicle weights and histories of usage. The curve lacks generality in that it is applicable only to this particular component cycled with the particular minimum stress used. However, it automatically includes the effects of details such as complex geometry, surface finish, residual stresses from fabrication, and the complex metallurgy at welds. Such factors are difficult to evaluate by any means other than a structural test.

Mean Stress Effects. *S-N* curves are usually plotted as stress amplitude, S_a, or stress range, $\Delta S = 2S_a$, versus life as cycles to failure, N_f. For a given stress amplitude, the level of mean stress, S_m, affects the life, with tensile values shortening life compared to tests at $S_m = 0$, and compressive values having the opposite effect. Where various mean stresses may occur, component test results can be obtained to generate a family of S_a versus N_f curves, one for each of several S_m values. An alternative means of considering the mean stress effect is to conduct tests at various values of the ratio $R = S_{min}/S_{max}$ and plot S_{max} versus N_f curves for various R-values.

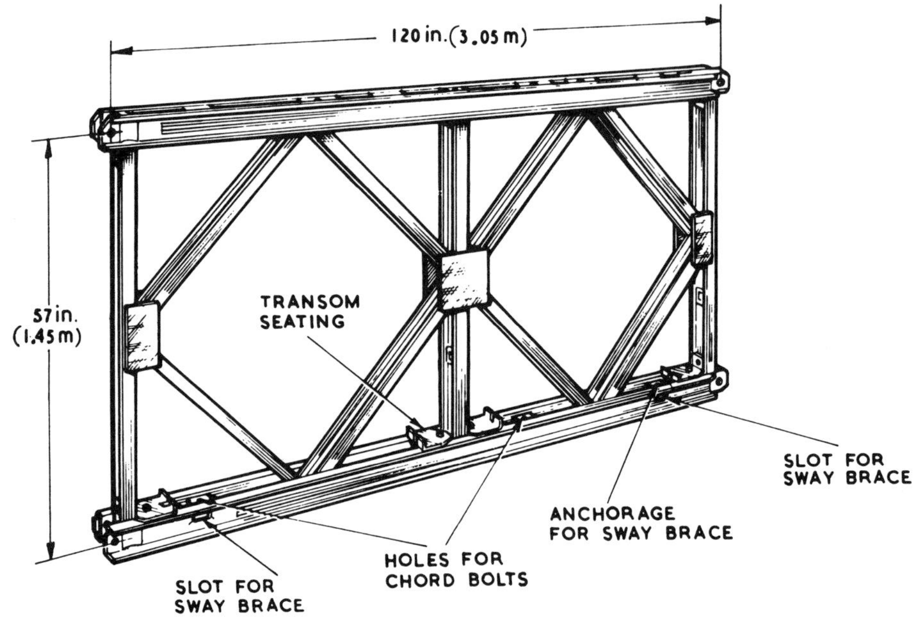

Fig. 1 Bailey Bridge panel. Reprinted with permission of American Society of Testing and Materials. Source: Ref 2

However, component test results are expensive to obtain, especially if the extra variable of mean stress is included. Hence, it may be desirable to estimate mean stress effects. It then becomes necessary to have only the *S-N* curve for one S_m or R value. Usually, the case of completely reversed loading, that is, $S_m = 0$ or $R = -1$, is the one chosen for experimental determination. For nonzero S_m, this curve may be entered with values of an equivalent completely reversed stress amplitude, S_{ar}, to obtain the life. Various mathematical expressions are used to estimate S_{ar}; the most common is based on the modified Goodman diagram equation:

$$S_{ar} = \frac{S_a}{1 - \dfrac{S_m}{\sigma_u}} \qquad \text{(Eq 1)}$$

where σ_u is the ultimate tensile strength of the material. For the above equation and the remainder of this article, mean stresses that are tensile are considered to be positive, and compressive mean stresses are considered negative. Alternative expressions and additional discussion are provided in the previous article in this Volume and in Ref 1 and 3 to 5.

An alternative approach is to choose as the single *S-N* curve one for zero-to-tension loading, that is, $R = 0$. For cases other than $R = 0$, this curve is entered with values of an equivalent zero-to-tension stress S^*. The expression provided by Walker (Ref 6) is often used in this context:

$$S^* = S_{max}(1 - R)^{\gamma} \qquad \text{(Eq 2)}$$

where γ is a material parameter obtained from correlating limited test data of nonzero R. For ductile metals, $\gamma = 0.5$ is a reasonable estimate in the absence of test data, in which case Eq 2 reduces to $S^* = \sqrt{S_{max}\Delta S}$.

Definition of Nominal Stress, S. When working with *S-N* curves for engineering components, or simulated components such as notched members, it is customary to define a nominal or average stress, *S*. For example, such a definition was described above for the Bailey Bridge panel. Some care is needed in defining and using *S* as illustrated in Fig. 3.

For simple axial loading of an unnotched member, as in Fig. 3(a), load *P* is of course divided by area *A* to obtain $S = P/A$. This is a reasonable approximation to the actual stress σ in the member, which is at least approximately uniform. For bending, as in (b), the elementary bending stress formula, $S = Mc/I$, is used to define *S* as the stress at the edge of the member with a cross sectional area moment of inertia, *I*. However, this simple analysis does not give the actual stress if yielding occurs, as a result of the formula being based on the assumption of linear-elastic material behavior. In particular, the actual stress at the edge is lower than *S* as shown by a solid line on the diagram to the right in (b). As a result, if *S-N* curves for bending and axial loading are compared, they do not agree, as they would if the

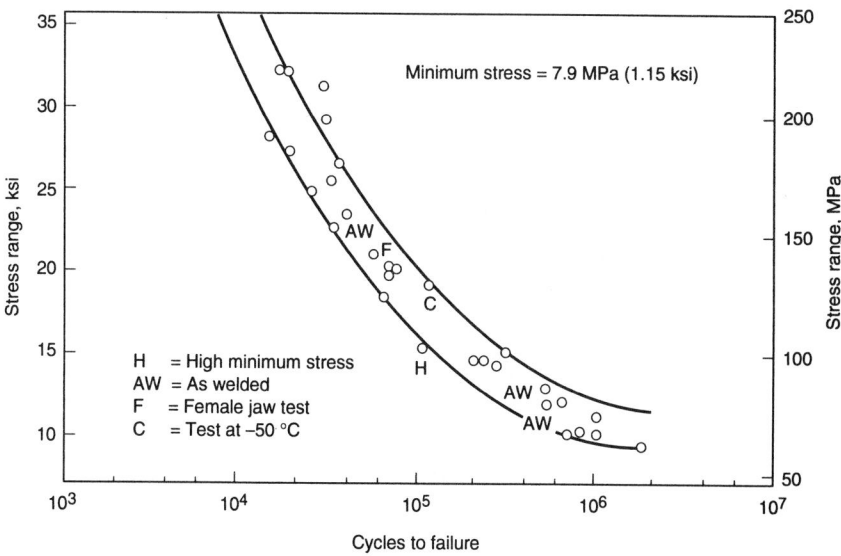

Fig. 2 Fatigue life curve for Bailey Bridge panels. The vertical axis gives maximum stress range, $\Delta S = S_{max} - S_{min}$. Source: Ref 2

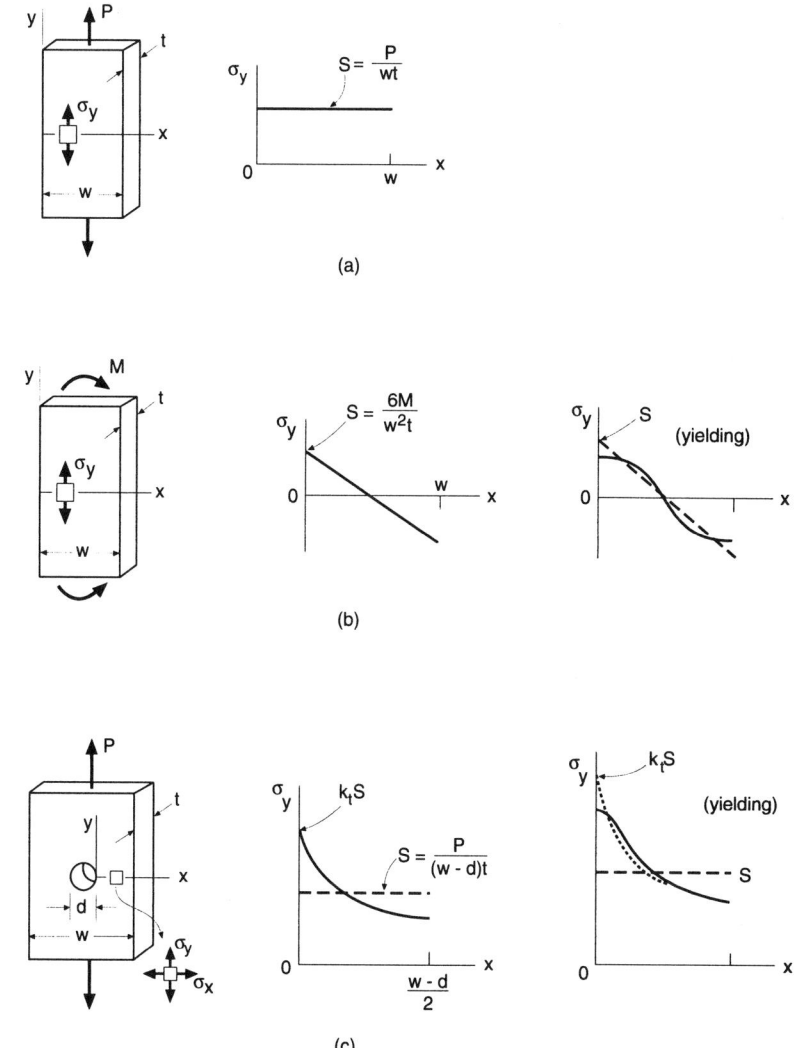

Fig. 3 Actual and nominal stresses for simple tension (a), bending (b), and a notched member (c). Actual stress distributions σ_y versus *x* are shown as solid lines, and hypothetical distributions associated with nominal stresses *S* as dashed lines. In (c), the stress distribution that would occur if there were no yielding is shown as a dotted line. Source: Ref 1 (p 344)

actual stress σ were plotted. (See Fig. 17 in the previous article in this Volume.)

Consider cases where a stress raiser, such as a notch, groove, hole, or fillet, occurs. (For brevity, any such stress raiser will be generically called a notch.) Nominal stress S is conventionally defined in such cases as an axial, elastic bending, or elastic torsional stress, or a combination of these. The cross section used for the area A and the area moment of inertia I is the net area remaining after removal of material to form the notch. For linear-elastic stress-strain behavior, such an S is related to the actual stress at the notch by $\sigma = k_t S$. The quantity k_t is an elastic stress concentration factor, defined to be consistent with the (actually arbitrary) definition of S. Values of k_t are available from a variety of sources, such as Ref 7.

However, as for the unnotched bending case, the linear-elastic material behavior assumed in obtaining k_t does not apply beyond yielding. The actual stress σ now becomes less than $k_t S$ as shown by the solid line on the right in Fig. 3(c). Hence, S-N curves for notched members plotted as either S or $k_t S$ versus life will not agree with curves from simple axial loading. An example is provided by Fig. 4.

Values of S and $k_t S$ as conventionally calculated are always proportional to the applied load, such as axial load P, bending moment M, or torque T. For example, for Fig. 3(c),

$$S = \frac{P}{A}$$

$$k_t S = \frac{k_t P}{A} \qquad \text{(Eq 3)}$$

where A and k_t are noted to be constants. On this basis, it is best to view nominal stress S, and also the elastically calculated notch stress $k_t S$, as being merely the applied load scaled in a convenient manner. Neither S nor $k_t S$ is in general equal to the actual stress σ. This will be important to remember at several points later in this article.

Estimated S-N Curves. Mechanical engineering design books, such as Juvinall (Ref 4) and Shigley (Ref 5), generally give a procedure for estimating component S-N curves (see Fig. 5). First, a life N_e is specified, such as 10^6 cycles for steels, beyond which the S-N curve is assumed to be horizontal. Hence, a fatigue limit, or safe stress below which no fatigue failure is expected, *is assumed* to exist.

This fatigue limit stress σ_{er} is first estimated for unnotched material:

$$\sigma_{er} = m\sigma_u$$

$$m = m_e m_t m_d m_s m_o \qquad \text{(Eq 4)}$$

where m is a reduction factor applied to the ultimate tensile strength. The quantity m is the product of individual reduction factors for several situations that affect S-N curves. In particular, m_e depends on material, m_t on type of loading, m_d on size, m_s on surface finish, and m_o on any other effects judged to be relevant. Values for all of these factors are based on empirical data from fatigue tests.

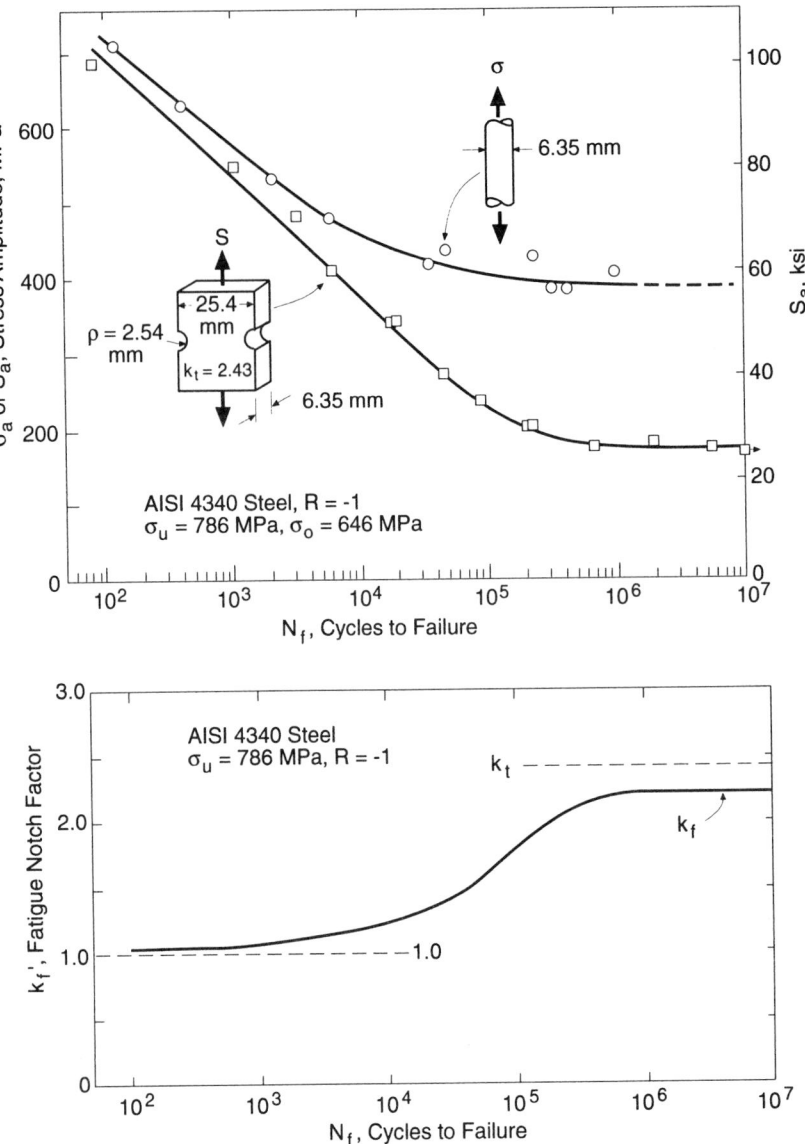

Fig. 4 Test data for a ductile metal illustrating variation of the fatigue notch factor with life. The S-N data in (a) are used to obtain $k_f' = \sigma_a/S_a$ in (b). The notches are half-circular cutouts. Source: Ref 1 (p 409)

The material-specific factor m_e gives an estimate of the fatigue limit in bending for small polished test specimens. A factor of $m_e = 0.5$ is generally applied for steels, and lower values are used for most other metals, such as Juvinall's use of $m_e = 0.4$ for cast irons and 0.35 for magnesium alloys. A value of m_t is assigned that depends on the type of loading, such as $m_t = 1.0$ for bending, and 0.58 for torsion, in both Juvinall and Shigley. The size factor given by m_d reflects statistical effects that cause lower fatigue strengths to be observed in larger size members. For example, Juvinall recommends $m_d = 1$ for diameters less than 10 mm, $m_d = 0.9$ for diameters 10 to 50 mm, $m_d = 0.8$ for diameters 50 to 100 mm, and so forth. Surface finishes other than a fine polish are assigned a factor $m_s < 1$ according to various curves or equations based on empirical data.

For the engineering component itself, a fatigue notch factor, k_f, is needed for the stress raiser (notch) where fatigue resistance is being evaluated. As described in Eq 62 to 64 in the previous article in this Volume and in Ref 1, 3-5, and 7), the value of k_f is obtained by modifying k_t, the elastic stress concentration factor. The various empirical equations employed for this purpose all depend on the notch tip radius and a material constant that is affected by the ultimate tensile strength. The fatigue limit stress S_{er} for the notched component is then estimated to be:

$$S_{er} = \frac{m\sigma_u}{k_f} \qquad \text{(Eq 5)}$$

where the symbol S denotes a nominal stress defined consistently with the k_t value used in evaluating k_f.

The S-N curve is estimated for lives less than N_e by establishing a second point, with both Juvinall and Shigley doing so at $N_f = 10^3$ cycles.

The stress at this point for the notched component is:

$$S'_{ar} = \frac{m' \sigma'_u}{k'_f} \qquad \text{(Eq 6)}$$

where $m' = 0.9$ for bending or torsion according to both Juvinall and Shigley. For axial loading, Juvinall uses $m' = 0.75$, whereas Shigley uses 0.90. The quantity σ'_u is the ultimate strength in tension, σ_u, except that a shear ultimate τ_u is used for torsion. Juvinall applies a notch effect at $N_f = 10^3$ by employing $k'_f = k_f$, whereas Shigley differs dramatically by applying $k'_f = 1$. Although neither Juvinall nor Shigley provide an estimate for lives shorter than 10^3 cycles, it would be reasonable to assume that the curve must pass through the ultimate tensile strength σ_u, or other estimate of component static strength, at $N_f = 1$.

Finally, the curve is drawn by connecting the above-described stress values at $N_f = 1$, 10^3, and N_e cycles with straight lines on a log-log plot as shown in Fig. 5. Hence, any straight-line segment has an equation of the form:

$$S_{ar} = AN_f^B \qquad \text{(Eq 7)}$$

where A and B have one set of values for the interval $1 \le N_f \le 10^3$, and another set for $10^3 \le N_f \le N_e$, and where the curve is horizontal at S_{er} for $N_f \ge N_e$.

Summary on the S-N Method. Estimated S-N curves are difficult to employ for combined loading cases, such as bending plus torsion on a notched shaft. However, even where the estimate is straightforward, the curve should be regarded as providing nothing more than a very crude estimate that is generally expected to be conservative. Comparison of estimates as recommended by different design books (e.g., Ref 3-5) reveal major differences, as do comparisons with test data.

Note that actual component fatigue data, as in Fig. 2, automatically include such effects as size, surface finish, geometric detail, and material condition as altered in component manufacture. Because estimates of such effects may be quite inaccurate, it is clear that any actual fatigue data that are available should be used to the maximum extent possible to improve or replace estimated S-N curves.

Variable Amplitude Loading

Cyclic loading histories that occur in the actual service of machines, vehicles, and structures often involve irregular variations of load with time. Life estimates for such situations may be made by employing the Palmgren-Miner rule along with a cycle counting procedure. Cycle counting permits an irregular time history to be broken down into individual events that may be evaluated from a constant amplitude S-N curve.

The time history and the S-N curve can employ a common variable, which may be actual stress σ, nominal stress S, load P, or strain ε, or else the time history must be transformed to the same

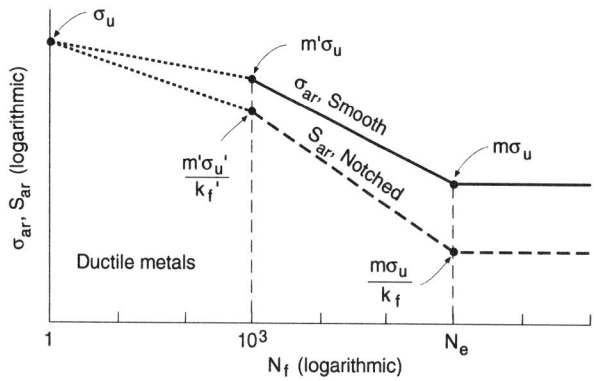

Fig. 5 Estimating completely reversed S-N curves for smooth and notched members according to procedures suggested by Juvinall or Shigley. Source: Ref 1 (p 423)

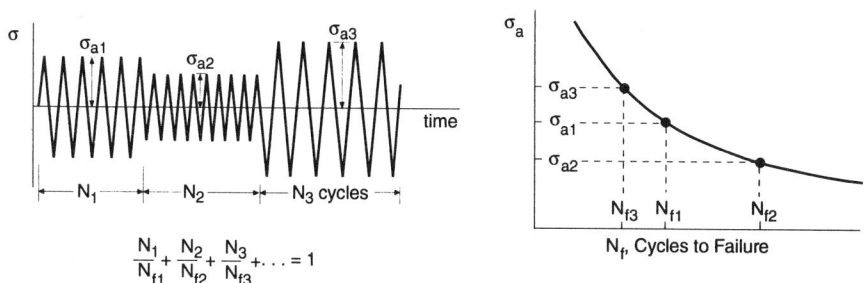

Fig. 6 Use of the Palmgren-Miner rule for life prediction for variable amplitude loading that is completely reversed. Source: Ref 1 (p 383)

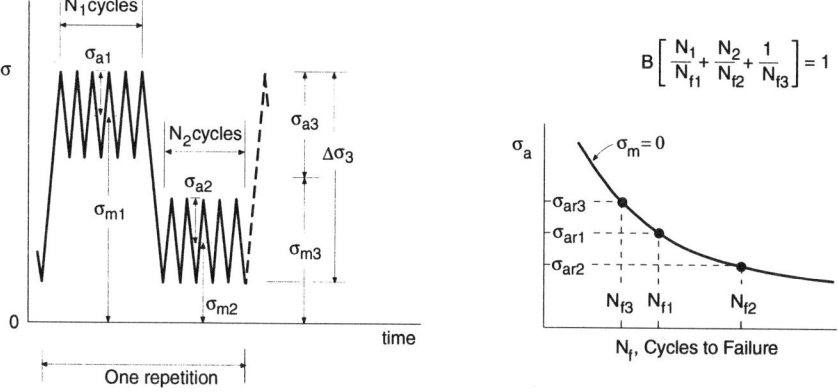

Fig. 7 Life prediction for a repeating stress history with mean level shifts. Source: Ref 1 (p 384)

variable as the S-N curve. The cycle counting procedure is the same for time histories of any of these variables, and stress, σ, is used in this section as a generic variable representing any choice. However, the handling of mean stress effects requires special care as discussed near the end of this section.

Palmgren-Miner Rule. Consider the relatively simple case where the stress amplitude changes one or more times during cyclic loading (see Fig. 6). Let N_1 cycles be applied at the first stress level σ_{a1}. If the S-N curve is entered, the number of cycles to failure at this same stress level, N_{f1}, can be determined. The interpretation

can then be made that a life fraction of N_1/N_{f1} has been exhausted. It is logical to assume that the sum of such life fractions for each stress level will reach unity when fatigue failure occurs:

$$\Sigma \frac{N_j}{N_{fj}} = 1 \qquad \text{(Eq 8)}$$

This simple rule was first proposed for use on ball and roller bearings by A. Palmgren of Sweden in the 1920s, but it was not widely applied until after the publication of a paper by M.A. Miner in 1945. Hence, it is called the Palmgren-Miner (P-M) rule,

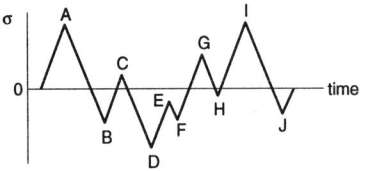

Peaks: A, C
Valleys: B, D
Simple ranges: A-B, B-C
Overall ranges: A-D, D-G

Fig. 8 Definitions for irregular loading. Source: Ref 1 (p 386)

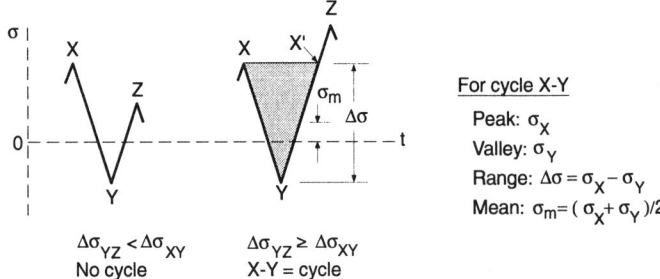

For cycle X-Y

Peak: σ_X
Valley: σ_Y
Range: $\Delta\sigma = \sigma_X - \sigma_Y$
Mean: $\sigma_m = (\sigma_X + \sigma_Y)/2$

$\Delta\sigma_{YZ} < \Delta\sigma_{XY}$
No cycle

$\Delta\sigma_{YZ} \geq \Delta\sigma_{XY}$
X-Y = cycle

Fig. 9 Condition for counting a cycle using the rainflow method. Source: Ref 1 (p 386)

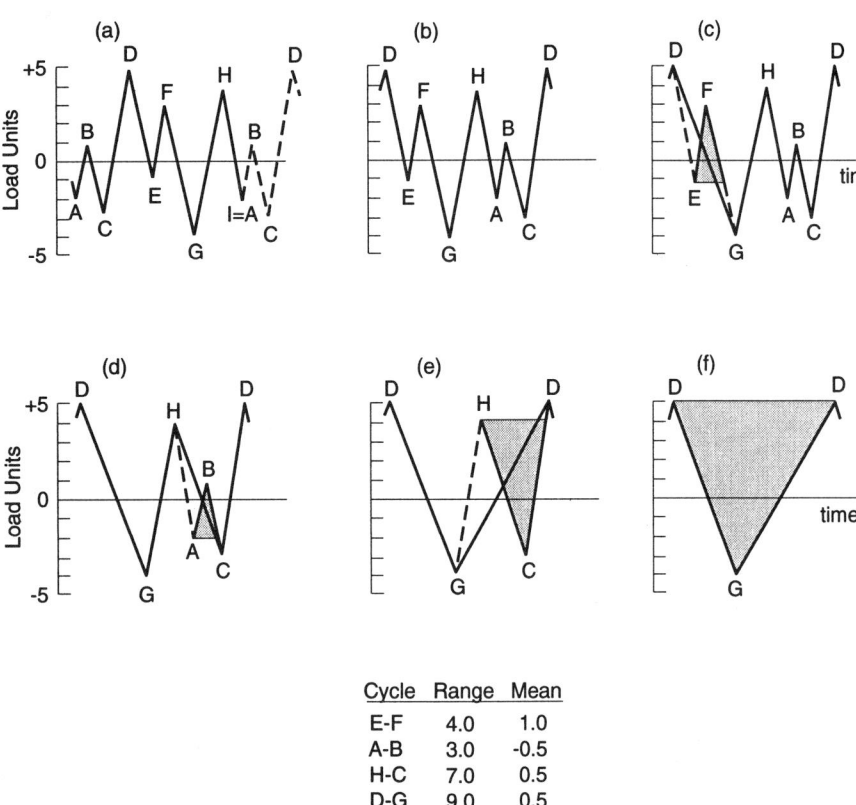

Cycle	Range	Mean
E-F	4.0	1.0
A-B	3.0	-0.5
H-C	7.0	0.5
D-G	9.0	0.5

Fig. 10 Example of rainflow cycle counting. Source: Adapted from Ref 8

although a 1937 paper by B.F. Langer also employed the same approach. (These early papers are cited in Ref 2.)

If typical variable loading is known for one aircraft flight, one machine operating cycle, or other time interval, Eq 8 can be applied for one repetition of this interval:

$$B_f\left[\Sigma\frac{N_i}{N_{fi}}\right]_{\text{one rep}} = 1 \qquad \text{(Eq 9)}$$

where B_f is the number of repetitions to failure. For example, consider the loading of Fig. 7, which is assumed to be repeatedly applied. There are N_1 cycles applied at a particular combination of mean stress and stress amplitude, σ_{a1} and σ_{m1}, and then the mean stress changes, following which N_2 cycles

are applied at a different combination, σ_{a2} and σ_{m2}. However, even if these two amplitudes were so small as to be nondamaging, fatigue failure could eventually occur due to the once-per-repetition application of the single large cycle identified as $\Delta\sigma_3$, having amplitude σ_{a3} and mean σ_{m3}. For the three levels of cycling, equivalent completely reversed stresses, σ_{ar1}, σ_{ar2}, and σ_{ar3}, as from Eq 1, must be computed and these values used with the S-N curve for $\sigma_m = 0$ to obtain N_{f1}, N_{f2}, and N_{f3}. Application of Eq 9 then allows the unknown number of repetitions to failure, B_f, to be calculated. Numerical solutions for two problems of this general type are given in Ref 1 (pp 384-385, 444-445).

Cycle Counting. If the time variation is irregular, as in Fig. 8, it is not obvious how one should identify the cycles for use of the P-M rule. Before proceeding, note that Fig. 8 gives definitions of

some useful terms. The irregular load, stress, or strain history consists of a series of peaks and valleys. A simple range is measured between a peak and the next valley, or between a valley and the next peak. An overall range is measured between a peak and a valley, but the valley occurs later and is more extreme than the one that follows immediately. Similarly, an overall range may be measured between a valley and a later peak.

Although a number of different procedures have been employed for identifying cycles, a consensus appears to have been reached that the preferable method is the rainflow method, or the essentially equivalent range pair method (Ref 8). When performing rainflow cycle counting, a cycle is identified or counted if it meets the criterion illustrated in Fig. 9. A peak-valley-peak or valley-peak-valley sequence X-Y-Z is counted as a cycle if the second range (Y-Z) exceeds the first (X-Y). In particular, the cycle has a stress range equal to that of the first range, $\Delta\sigma_{XY} = \sigma_X - \sigma_Y$, and a mean stress $\sigma_m = (\sigma_X + \sigma_Y)/2$.

Now consider an entire stress history, using the short history of Fig. 10 as an example. First, the history is assumed to be a repeating one that can be assumed to start and stop at any peak or valley. This permits the convenience of assuming that the history begins and ends at the peak or valley having the highest absolute value of stress. Peaks and valleys occurring prior to this extreme event are then moved to the end of the history as shown in Fig. 10(a) and (b).

Then proceed with counting as follows: Start by considering the first three peak or valley events as X-Y-Z of Fig. 9. If a cycle is counted, note its range and mean and remove it from the history to be employed for purposes of further counting. If none is counted, move ahead by one peak or valley and check for a cycle there. Continue counting cycles and moving ahead until the entire history is exhausted. When the process is complete, each peak or valley is noted to have participated in one and only one cycle. Some cycles counted correspond to simple ranges in the original history, but others to overall ranges. The final and largest cycle that is counted always involves the highest peak and lowest valley.

Computer programs for performing the cycle counting are available (Ref 9, 10). For lengthy histories, the range and mean values are often rounded off to discrete values in a range-mean matrix as shown in Fig. 11. A numerical example

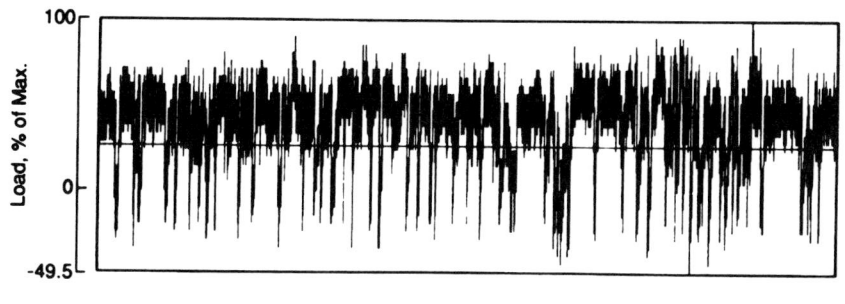

Range	−15	−10	−5	0	5	10	15	20	25	30	35	40	45	50	55	60	65	70	75	All
20	4	1	5	2	2	5	—	—	3	6	15	27	29	32	22	12	6	2	—	173
25	2	4	3	9	8	10	4	6	2	7	17	37	36	43	33	13	7	1	2	244
30	1	1	5	3	1	1	4	3	—	4	13	20	20	23	20	8	6	1	—	134
35	1	1	4	2	3	2	—	1	3	2	8	17	16	11	11	7	2	—	—	91
40	—	1	1	1	2	1	1	—	—	4	7	15	16	9	8	2	—	—	—	68
45	—	1	—	4	3	—	—	—	—	2	1	9	7	2	3	1	—	—	—	33
50	—	—	2	2	2	1	—	—	—	2	2	3	3	1	1	1	1	—	—	21
55	—	—	1	1	—	—	—	—	—	2	2	4	4	2	—	1	—	1	—	18
60	—	1	1	—	—	—	—	—	—	1	1	3	2	1	—	—	—	—	—	10
65	—	—	—	—	—	—	—	—	—	—	2	1	—	—	—	—	—	—	—	3
70	—	—	—	—	—	—	—	—	—	—	2	—	1	—	—	—	—	—	—	3
75	—	—	—	—	—	—	—	1	—	—	1	2	—	—	—	—	—	—	—	4
80	—	—	—	—	—	—	—	—	—	—	—	—	—	—	—	—	—	—	—	—
85	—	—	—	—	1	—	1	3	3	—	—	—	—	—	—	—	—	—	—	8
90	—	—	—	—	—	—	—	4	—	—	—	—	—	—	—	—	—	—	—	4
95	—	—	—	—	1	—	1	4	1	—	—	—	—	—	—	—	—	—	—	7
100	—	—	—	—	—	5	3	1	—	—	—	—	—	—	—	—	—	—	—	9
105	—	—	—	—	—	3	3	3	—	—	—	—	—	—	—	—	—	—	—	9
110	—	—	—	—	—	—	2	3	—	—	—	—	—	—	—	—	—	—	—	5
115	—	—	—	—	—	—	3	—	—	—	—	—	—	—	—	—	—	—	—	3
120	—	—	—	—	1	—	1	1	—	—	—	—	—	—	—	—	—	—	—	3
125	—	—	—	—	—	—	2	—	—	—	—	—	—	—	—	—	—	—	—	2
130	—	—	—	—	—	—	—	—	—	—	—	—	—	—	—	—	—	—	—	—
135	—	—	—	—	—	1	—	—	—	—	—	—	—	—	—	—	—	—	—	1
140	—	—	—	—	—	—	—	—	—	—	—	—	—	—	—	—	—	—	—	—
145	—	—	—	—	—	—	—	—	—	—	—	—	—	—	—	—	—	—	—	—
150	—	—	—	—	—	1	—	—	—	—	—	—	—	—	—	—	—	—	—	1

Fig. 11 An irregular load versus time history from a ground vehicle transmission and a matrix giving numbers of rainflow cycles at various combinations of range and mean. The range and mean values are percentages of the peak load, and these were rounded to the discrete values shown in constructing the matrix. Reprinted with permission from AE-6 *Fatigue under Complex Loading: Analysis and Experiments,* 1977 (Ref 11), Society of Automotive Engineers, Inc.

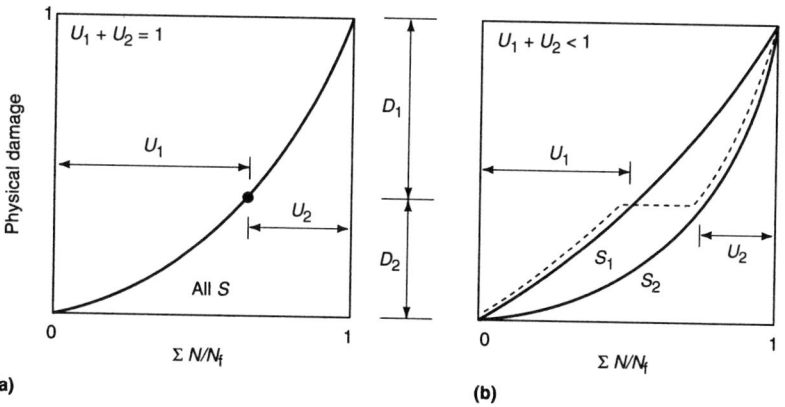

Fig. 12 Physical damage versus life fraction, where the relationship is unique (a) and nonunique (b). Source: Ref 12

of a life calculation that involves cycle counting is given in Ref 1 (p 387-389).

Sequence Effects. For the Palmgren-Miner rule to be valid, the physical damage in the material, D, which could be crack length, crack density, modulus or compliance change, or other relevant parameter, must be uniquely related to the life fraction, $U = N/N_f$. The relationship between D and U need not be linear, as long as there is a single monotonically increasing curve for all stress values. This is illustrated in Fig. 12(a).

However, if the U versus D curve varies with stress level (see Fig. 12b), a sequence effect can occur, such that the summation of cycle ratios differs from unity. Such effects do indeed occur. For example, assume that a few severe loading cycles that cause plastic deformation are applied at the beginning of a fatigue test. These cycles may advance the damage process sufficiently that subsequent cycles at a low level can proceed to propagate this damage, which would ordinarily take much of the life at the lower level to initiate. Some test data illustrating this effect are given in Fig. 13. A few cycles at high strain lower the strain-life curve in the long-life region. The effect on life increases for lower stress levels and is as large as a factor of 10.

The situation of Fig. 13 corresponds to Fig. 12(b), where damage at the beginning of cycling proceeds more rapidly at a higher stress, S_1, than it would have at a lower stress, S_2. Starting at S_1 and changing later to S_2, a high-low stress sequence, causes $\Sigma U < 1$. Conversely, starting at S_2 and changing to S_1, a low-high sequence, causes $\Sigma U > 1$.

Periodic overstrains have an effect similar to a high-low sequence. In steels with a distinct fatigue limit, periodic overstrains have the special effect of eliminating this fatigue limit. Data showing this are given in Fig. 14.

A summation of cycle ratios less than unity, $\Sigma U < 1$, is of course a problem as it corresponds to the P-M rule giving a nonconservative life estimate. Component S-N curves could be adjusted based on overstrain data for the material as in Fig. 13 and 14. For example, component S-N curves are sometimes extrapolated as straight lines on log-log plots, that is, using Eq 7, thus eliminating any distinct fatigue limit that might be observed in constant amplitude data. Initial or periodic overloads could also be applied during the component S-N tests. However, this must be done by a special procedure so that residual stresses affecting the life are not introduced by the overloads.

Local Mean Stress Effects. In addition to the material-damage-related sequence effects just discussed, an additional cause of sequence effects is related to the local mean stresses at notches affecting life. In particular, local mean stresses are altered by overloads that cause local yielding at the notch. This is illustrated schematically in Fig. 15, where two types of overload cycle are shown, along with the resulting local stress-strain (σ-ε) behavior at a notch. The tension-compression overload (Fig. 15a) results in a tensile mean stress at the notch during subsequent cycling, and thus a shorter life, than for the compression-ten-

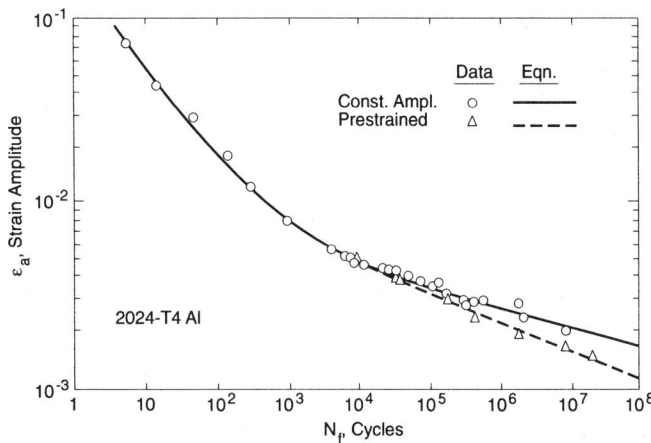

Fig. 13 Effect of initial overstrain (10 cycles at $\varepsilon_a = 0.02$) on the strain-life curve of an aluminum alloy. Adapted from Ref 13 as based on data from Ref 14

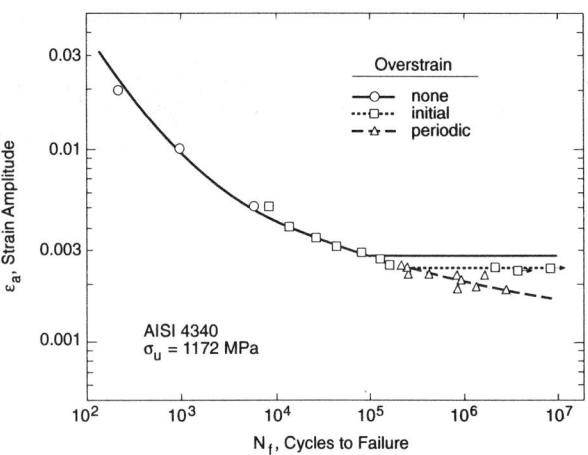

Fig. 14 Effects of both initial and periodic overstrain on the strain-life curve for an alloy steel. The fatigue limit for the no overstrain case is estimated from test data on similar material. From Ref 1 (p 666) as based on data from Ref 15

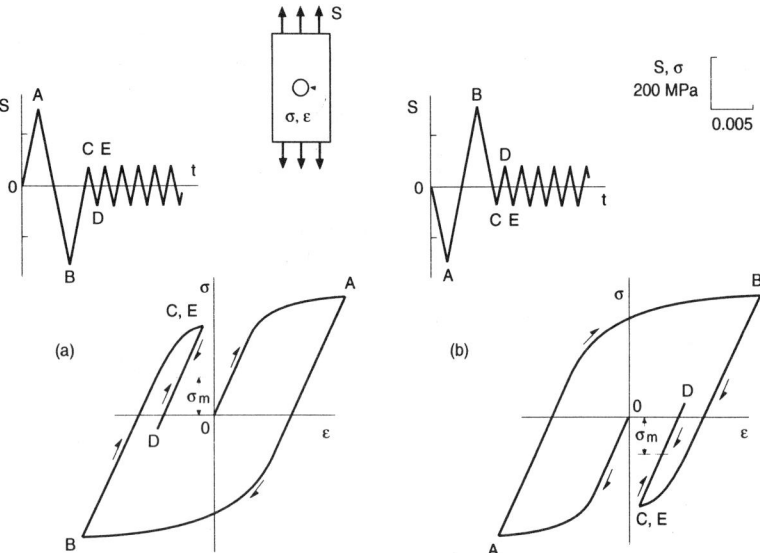

Fig. 15 Two load histories applied to a notched member ($k_t = 2.4$) and the estimated notch stress-strain responses for 2024-T4 Al. The high-low overload in (a) produces a tensile mean stress, and the low-high overload in (b) produces the opposite. Adapted from Ref 16

$$\varepsilon_a = \frac{\sigma_f'}{E}\,(2N_f)^b + \varepsilon_f'\,(2N_f)^c, \text{ (with } \sigma_m = 0) \qquad \text{(Eq 10)}$$

where ε_a is strain amplitude, N_f is cycles to failure for completely reversed cycling, E is the elastic modulus, and σ_f', b, ε_f', and c are material fatigue constants. A special cyclic stress-strain curve, as described in the previous article, is also needed:

$$\varepsilon_a = \frac{\sigma_a}{E} + \left(\frac{\sigma_a}{H'}\right)^{1/n'} \qquad \text{(Eq 11)}$$

where σ_a is stress amplitude, and n' and H' are material constants (and where $H' = K'$ in Eq 29 and 30 of the previous article in this Volume).

Mean Stress Effects. A generalized strain-life curve can address mean stress effects in a relationship proposed by Morrow (Ref 18) that is analogous to Eq 1 as follows:

$$\sigma_{ar} = \frac{\sigma_a}{1 - \dfrac{\sigma_m}{\sigma_f'}} \qquad \text{(Eq 12)}$$

where σ_{ar} is the equivalent completely reversed local stress amplitude and σ_a and σ_m are also local stresses at the notch. Note that σ_u is replaced by σ_f', the constant from Eq 10. The quantity σ_f' is approximately equal to the true fracture strength from a tension test, so that it is larger than σ_u, except for low-ductility metals, where it has a value close to σ_u.

Equations 10 to 12 can be combined to generalize the strain-life curve to include mean stress effects:

$$\varepsilon_a = \frac{\sigma_f'}{E}\,(2N*)^b + \varepsilon_f'\,(2N*)^c$$

$$N_f = \frac{N*}{\left(1 - \dfrac{\sigma_m}{\sigma_f'}\right)^{1/b}} \qquad \text{(Eq 13)}$$

sion overload (Fig. 15b), which produces a compressive mean stress. Note that, without these overloads, the local mean stress σ_m would be zero during the low-level cycling at $S_m = 0$.

Sequence effects related to the local mean stress σ_m represent a fundamental difficulty for a stress-based approach employing nominal stresses, S. The nominal mean stress S_m is simply not the controlling variable, rather, it is the local mean stress, σ_m. This should not be surprising in view of the discussion above and Fig. 3, where it is noted that nominal stresses S are in general not actual stresses; they are essentially conveniently scaled applied loads.

For the situation of Fig. 15, note that $S_m = 0$ for all cycles, so that direct application of Eq 1 would predict no mean stress effect at all, and no difference in life between cases (a) and (b). However, actual test data on notched members show a large

effect (see Ref 17 in this article and Fig. 35 in the previous article in this Volume).

Analysis of local mean stresses requires considering the elasto-plastic stress-strain behavior of the material at the notch to obtain values of σ_m, which can then be used in evaluating fatigue life. Analysis of this type is a key feature of the strain-based approach, which is described in the next section of this article.

Strain-Based Approach

In this approach, local stresses and strains at notches, σ and ε, as in Fig. 15, are estimated and used as the basis of life predictions. The S-N curve used is a strain-life curve, often represented as described in the previous article in this Volume by the following equation:

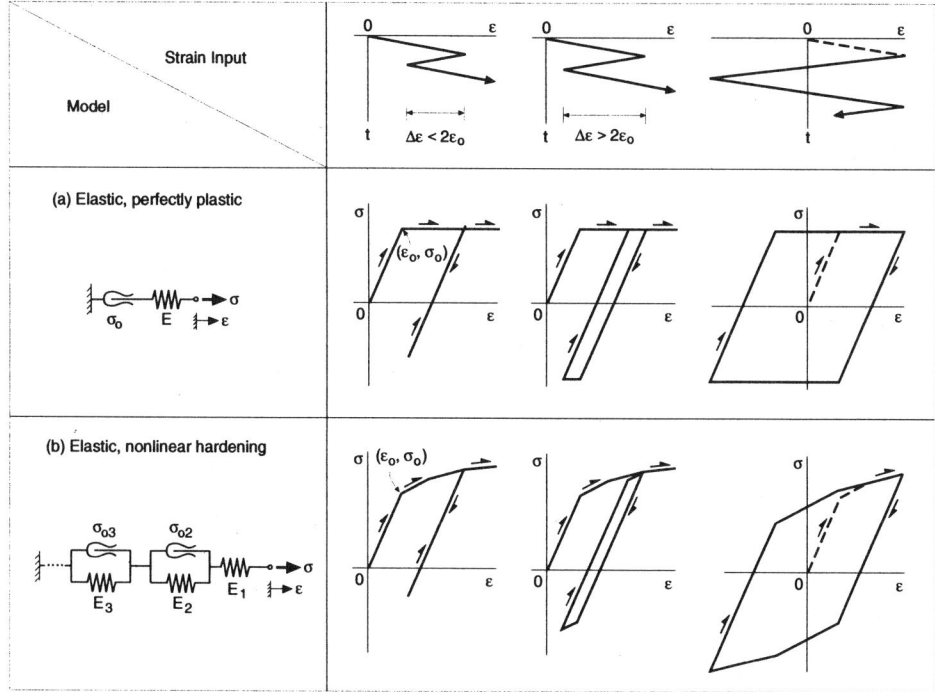

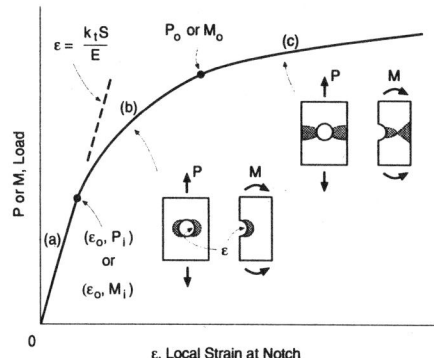

Fig. 18 Load versus local strain behavior of a notched member showing three regions of behavior: no yielding (a), local yielding (b), and fully plastic yielding (c). Source: Ref 1 (p 594)

Fig. 16 Unloading and reloading behavior for two rheological models. The first strain history causes only elastic deformation during unloading, but the second one is sufficiently large to cause compressive yielding. The third history is completely reversed and causes a hysteresis loop that is symmetrical about the origin. Source: Ref 1 (p 545)

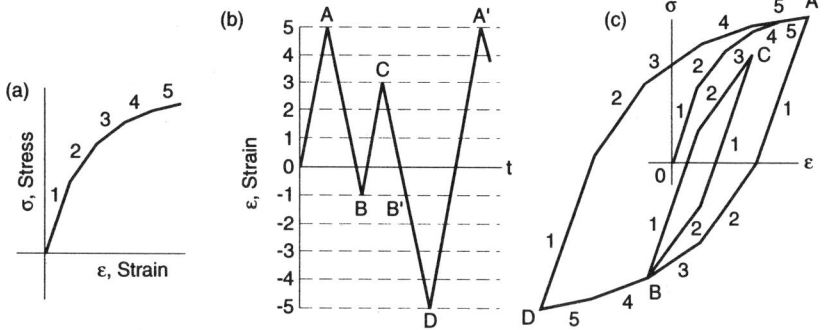

Fig. 17 Behavior of a multistage spring-slider rheological model for an irregular strain history. A model having the monotonic stress-strain curve (a) is subjected to strain history (b), resulting in stress-strain response (c). Adapted from Ref 21

where N^* is the life from the strain-life equation for zero mean stress, and N_f is the actual life as adjusted to include the mean stress effect. A modification of this equation is also used:

$$\varepsilon_a = \frac{\sigma_f'}{E}\left(1 - \frac{\sigma_m}{\sigma_f'}\right)(2N_f)^b + \varepsilon_f'\ (2N_f)^c \qquad \text{(Eq 14)}$$

A graphical or iterative numerical solution is required to obtain life N_f from any of Eq 10, 13, or 14.

An alternative approach to evaluating mean stresses is that of Smith, Watson, and Topper (Ref 19):

$$\sigma_{max}\varepsilon_a = \frac{(\sigma_f')^2}{E}\ (2N_f)^{2b} + \sigma_f'\varepsilon_f'\ (2N_f)^{b+c} \qquad \text{(Eq 15)}$$

where $\sigma_{max} = \sigma_a + \sigma_m$ is the local maximum stress, and the material constants have the same values as in Eq 10. For known values of σ_{max} and ε_a, Eq 15 is solved numerically for N_f. Alternatively, the quantity $\sigma_{max}\varepsilon_a$ can be plotted versus life for particular values of the material constants, and N_f can then be obtained graphically.

Elasto-Plastic Stress-Strain Behavior. To use the strain-based approach, it is necessary to model the elasto-plastic stress-strain behavior that occurs at a notch as in Fig. 15. An analogy with spring and frictional slider rheological models is useful (Ref 20). Consider Fig. 16, model (a). This model corresponds to an elastic, perfectly plastic material. A linear spring of stiffness E gives an initial elastic response, and the frictional slider moves at a yield stress σ_0. Unloading and reloading may cause only elastic deformation, or

the frictional slider may move if the strain excursion is sufficiently large.

In model (b), there are one or more spring and slider parallel combinations, which provides a strain-hardening behavior. If the strain excursion on unloading and reloading is sufficiently large, a stress-strain hysteresis loop is formed as for engineering metals. For completely reversed strain cycling, a loop that is symmetrical about the origin is obtained. [Compare (b) to Fig. 7 of the previous article in this Volume.]

The model (b) parameters (E_i, σ_{0i}) can be adjusted to fit the cyclic stress-strain curve of a particular material, and the model will then provide a reasonable representation of the stable cyclic stress-strain behavior. The transient cyclic hardening or softening stage is not modeled, only the stable behavior after this is complete, nor is the cycle dependent relaxation of mean stress. Such details can be added if desired by making the model parameters act as variables, or by employing a more general plasticity theory. However, the basic nontransient model is sufficient for most applications and will be employed here.

Consider a spring and slider model that fits the cyclic stress-strain curve as in Fig. 17(a). Let this curve be denoted $\varepsilon = f(\sigma)$, with Eq 11 being the specific form that is usually employed. If an irregular strain history is imposed on the model, a set of simple rules is seen to describe its behavior. First, after the model reaches the largest absolute value of strain, as at A in Fig. 17(b), stress-strain paths follow a unique curve that is related to the cyclic stress-strain curve, $\varepsilon = f(\sigma)$, by being expanded with a scale factor of two:

$$\frac{\Delta\varepsilon}{2} = f\left(\frac{\Delta\sigma}{2}\right) \qquad \text{(Eq 16)}$$

The quantities $\Delta\sigma$ and $\Delta\varepsilon$ are stress and strain changes measured relative to each point where the direction of loading changes, with coordinate axes positive in the direction of loading, as at A, B, C, and D in Fig. 17.

The second rule is an exception to the first: When the strain next reaches a value where the loading direction was changed, the stress has the same value as before, and the Eq 16 stress-strain

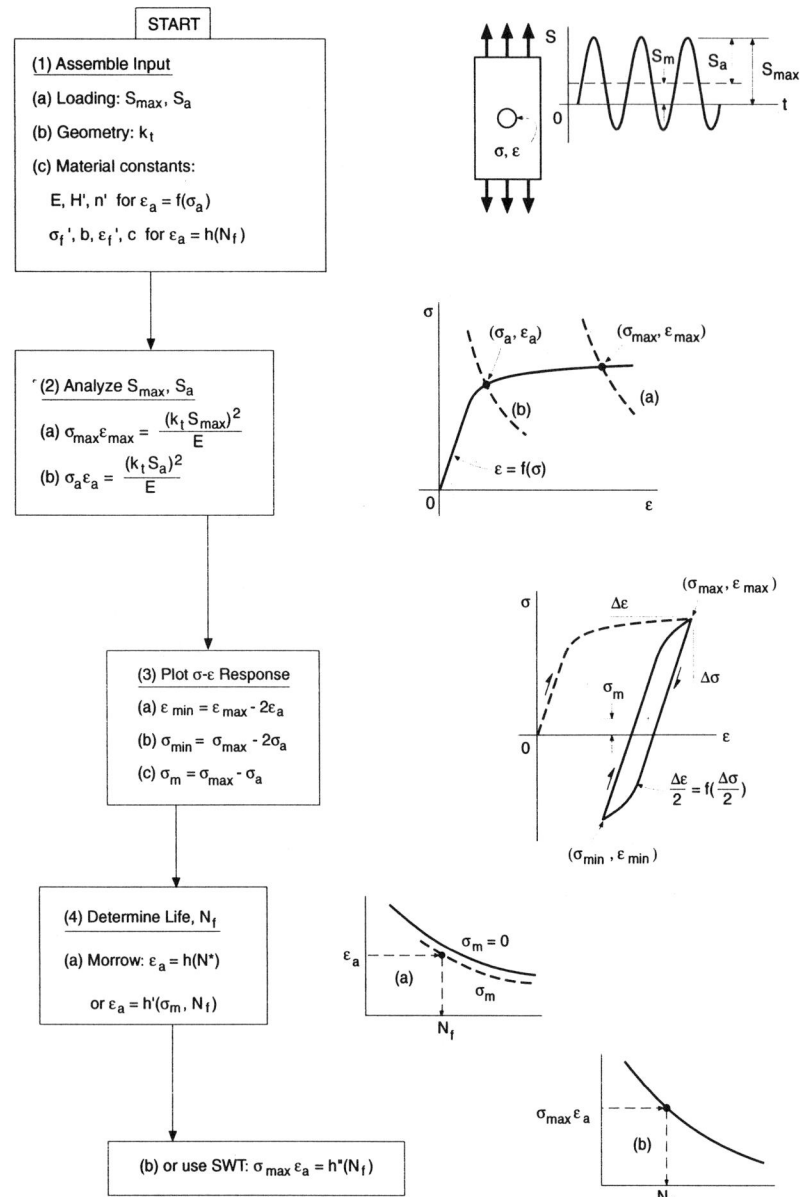

Fig. 19 Steps required in strain-based life prediction for a notched member under constant amplitude loading. Source: Ref 1 (p 648)

extend into the fully plastic region, Neuber's rule predicts that the following relationship applies:

$$\sigma\varepsilon = \frac{(k_t S)^2}{E} \qquad \text{(Eq 17)}$$

When combined with a stress-strain curve, $\varepsilon = f(\sigma)$, values for local stress and strain, σ and ε, can be obtained for any desired value of nominal stress S. In plots of Eq 17 as a hyperbola on σ-ε axes, the intersection with the stress-strain curve provides the desired σ and ε values (see Fig. 44 in the previous article, for example).

As an alternative to Neuber's rule, the strain energy density method has been proposed by Glinka (Ref 23). Its application is similar to that of Neuber's rule, and it can be used in place of Neuber's rule in the descriptions that follow in this article.

Life Estimates for Constant Amplitude Loading. Figure 19 is an illustrated flow chart for the entire procedure for making a life estimate with constant amplitude loading applied to a notched member. The input information required consists of the applied loading expressed in terms of nominal stress, S, the geometry, hence the k_t value, and the cyclic stress-strain and strain-life curves for the material. The latter are denoted as $\varepsilon_a = f(\sigma_a)$ and $\varepsilon_a = h(N_f)$, respectively, with the forms of Eq 10 and 11 often being employed.

The material is assumed to have stable behavior with its initial monotonic stress-strain curve being the same as the cyclic stress-strain curve, Eq 11. Neuber's rule in the form of Eq 17 is then applied to both the maximum and amplitude values of nominal stress, S_{max} and S_a. Hence, the following equations are solved to obtain the local maximum stress and strain, σ_{max} and ε_{max}:

$$\varepsilon_{max} = \frac{\sigma_{max}}{E} + \left(\frac{\sigma_{max}}{H'}\right)^{1/n'}$$

$$\frac{(k_t S_{max})^2}{E} = \sigma_{max}\varepsilon_{max} \qquad \text{(Eq 18)}$$

Similarly, the amplitudes σ_a and ε_a are obtained from

$$\varepsilon_a = \frac{\sigma_a}{E} + \left(\frac{\sigma_a}{H'}\right)^{1/n'}$$

$$\frac{(k_t S_a)^2}{E} = \sigma_a\varepsilon_a \qquad \text{(Eq 19)}$$

Solutions of the above pairs of equations can be thought of graphically as shown in step 2 of Fig. 19, where a hyperbola is intersected with the cyclic stress-strain curve.

Once these stresses and strains are known, the local stress-strain response can be plotted using the factor-of-two expansion of the cyclic stress-strain curve, Eq 16. This is shown as step 3 in Fig. 19. Because relaxation of mean stress is assumed to be small, the mean stress found for the first cycle is assumed to apply throughout the fatigue life.

path returns to the one that was underway prior to the direction change. This occurs in Fig. 17 at point B', beyond which the behavior is the same as if even B-C-B' had not occurred. This behavior of returning to a previously established stress-strain path is called the memory effect.

At points where the memory effect acts, a closed stress-strain hysteresis loop is completed, such as loop B-C-B' in Fig. 17. Also, for histories reordered to start at the most extreme peak or valley as in Fig. 10, the closed stress-strain loops correspond to the cycles from rainflow counting of the strain history. For Fig. 17, loops B-C-B' and A-D-A' correspond to the rainflow cycles for this short history.

The device of a rheological model is actually unnecessary. It is necessary only to use the factor-of-two (Eq 16) scaling of a smooth continuous cyclic stress-strain curve along with the memory

effect to form closed stress-strain hysteresis loops.

Analysis of Notched Members. Consider the local strain ε at a notch and the variation of this with applied load as shown in Fig. 18. An elasto-plastic stress-strain analysis, as by finite elements, could be used to determine the local stresses and strains at the notch. At low loads, only elastic behavior occurs, so that $\sigma = k_t S$ and $\varepsilon = \sigma/E$ applies. Once the yield stress is exceeded, yielding occurs in a small region at the notch, and strains are larger and stresses smaller, than would be the case for simple elastic behavior. Yielding spreads with increasing load, and when the entire cross section becomes involved, fully plastic behavior is said to occur.

The approximate procedure called Neuber's rule is often used to estimate local notch stresses and strains (Ref 22). For loading that does not

$$\sigma_m = \sigma_{max} - \sigma_a \qquad \text{(Eq 20)}$$

Finally, having evaluated ε_a and σ_m, the number of cycles to failure N_f may be calculated from a strain-life curve that includes the mean stress effect, such as Eq 13 or 14. Or the Smith, Watson, and Topper approach can be employed by substituting σ_{max} and ε_a into Eq 15 and solving for N_f. A numerical example is provided by combining Examples 13.3 and 14.3 of Ref 1 (pp 610-611 and 649-650).

Life Estimates for Variable Amplitude Loading. The life estimation procedure just described for constant amplitude loading of notched members can be extended to variable amplitude cases by including cycle counting and the memory effect. An illustrated flow chart of the procedure is shown in Fig. 20.

For this analysis, it is useful to solve Neuber's rule with the cyclic stress-strain curve, or perform other analogous mechanics analysis, to obtain a load-strain curve as in Fig. 18. For Neuber's rule and no loading into the fully plastic region, the implicit relationship between S_a and ε_a of Eq 19 applies. Let this relationship be denoted $\varepsilon = g(S)$. Also, reorder the load history so that it starts and ends with the peak or valley having the largest absolute value. (This reordering step can be avoided, but more general cycle counting and stress-strain modeling procedures become necessary.)

The analysis begins by simultaneously following both the load-strain and stress-strain curves, $\varepsilon = g(S)$ and $\varepsilon = f(\sigma)$, to the first (most extreme) load peak or valley. For example, for point A in Fig. 20, the known S_A and $\varepsilon = g(S)$ gives ε_A, and this ε_A and $\varepsilon = f(\sigma)$ gives σ_A. See step 2 and the corresponding diagram in Fig. 20. Then proceed to each subsequent peak or valley while applying the relationships:

$$\frac{\Delta\varepsilon}{2} = g\left(\frac{\Delta S}{2}\right)$$

$$\frac{\Delta\varepsilon}{2} = f\left(\frac{\Delta\sigma}{2}\right) \qquad \text{(Eq 21)}$$

For example, for range $\Delta S_{AB} = S_A - S_B$, the first of these gives $\Delta\varepsilon_{AB}$, and then the second gives $\Delta\sigma_{AB}$. Because point A was previously located, these fix point B for both the S-ε and σ-ε responses. In addition to the end points, Eq 21 can also be used to plot the entire load-strain and stress-strain paths as shown in Fig. 20, step 3.

However, it is also necessary to apply rainflow cycle counting while proceeding through the history. Whenever a rainflow cycle is completed, the memory effect acts, and closed loops are formed in both the S-ε and σ-ε responses. When a cycle is completed, the initial points of the ranges ΔS, $\Delta\varepsilon$, and $\Delta\sigma$ revert back to the peak or valley that applied prior to the beginning of the cycle. For example, when cycle B-C-B′ is completed in Fig. 20, the range ΔS_{AD} is used with Eq 21 to obtain $\Delta\varepsilon_{AD}$ and $\Delta\sigma_{AD}$, so that point D is located for both the S-ε and σ-ε responses.

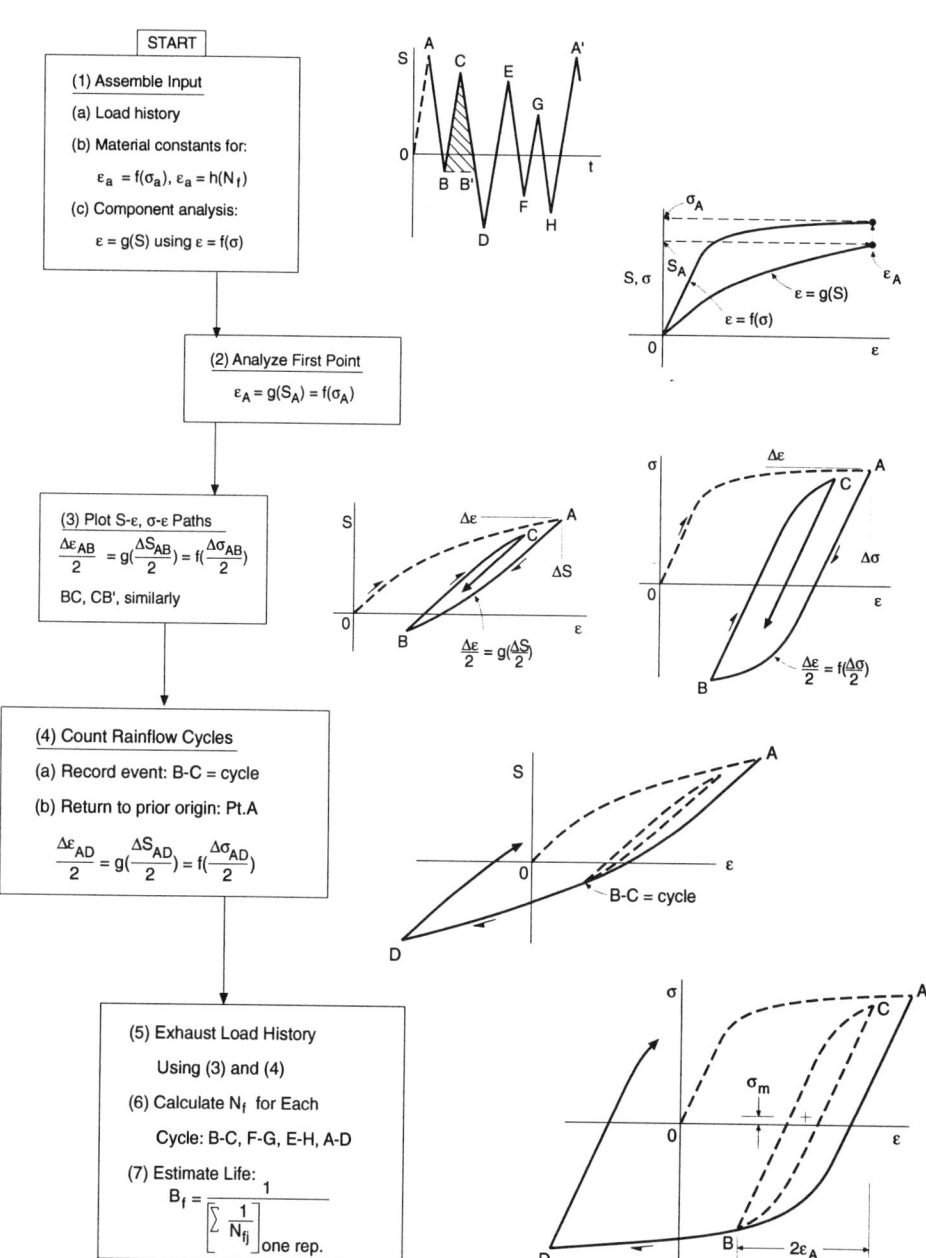

Fig. 20 Life prediction for an irregular load versus time history using the stepwise procedure described in the text. Source: Ref 1 (p 653)

The estimated stress-strain response is completely determined as this analysis proceeds. For the example of Fig. 20, the result is shown in Fig. 21. The stress-strain response consists of a set of closed stress-strain hysteresis loops, each of which corresponds to a rainflow cycle. For this example, the loops (cycles) correspond to load excursions B-C-B′, F-G-F′, E-H-E′, and A-D-A′. Because stresses and strains are known for each peak and valley in the load history, the strain amplitude ε_a and mean stress σ_m are available for each cycle. The corresponding number of cycles to failure N_f for each cycle is then available from either Eq 13 or 14. Alternatively, ε_a and σ_{max} with

Eq 15 also gives an N_f value. These N_f values and the P-M rule then give a life estimate. A numerical example of this type is given in Ref 1 (p 655-658).

Simplified Approach. The procedure just described assumes that the load history is known as a list of peaks and valleys in order. However, in some cases the only information available is the result of rainflow-cycle counting of the load history in the form of a range-mean matrix, as in Fig. 11. Some of the detailed knowledge of the load history has thus been lost. In such a situation, it is possible to perform a simplified strain-based analysis that determines upper and lower bounds on life for all possible sequences of loading giving a particular rainflow matrix.

This is done by noting that the load-strain and stress-strain loops for all cycles cannot lie outside of the loops for the largest cycle. For example, in Fig. 21, S-ε and σ-ε loops F-G must lie inside loops A-D. This is shown in Fig. 22. Also, a similar limitation applies to S-σ loops as also shown. This situation and the known values of S from cycle counting place bounds on the mean stress of each cycle, such as σ_{mQ} and σ_{mP} for loop F-G in Fig. 22. If the worst case (most tensile) mean stresses for all cycles are employed in a life calculation, a lower bound on life is obtained. Similarly, the best case mean stresses give an upper bound on life.

The example load history of Fig. 11 was analyzed in this manner for a particular steel and notched member as shown in Fig. 23. Various scale factors were applied to the load history to generate an entire S-N curve. The indicated comparison with test data is reasonable. Also, the bounds are quite close at all load levels, which is generally the case for irregular load histories. (The special situation of Fig. 15 would, however, give a wide separation between the upper and lower bounds.) A more detailed description of the procedure for making such a bounded analysis is given in Ref 1, and further details and test data are given in Ref 13.

Comparison of Methods

The stress-based and strain-based approaches are discussed and compared below, with some comments on their manner of use and limitations.

Discussion of Stress-Based Approach. The stress-based approach is most applicable when S-N curves are available for actual components, or for component-like test members. Component S-N curves have the major advantage of automatically including effects such as surface finish, residual stresses, weld geometry and metallurgy, and frictional surface contact in joints. The effects of such factors are otherwise generally difficult to include in fatigue life estimates. A major disadvantage is that mean stress effects based on nominal stress may be in error due to sequence effects related to the local notch mean stress actually being the controlling variable (see Fig. 15). Hence, caution is needed for load histories that might cause harmful local notch mean stresses that are not properly analyzed by this approach.

Now consider use of a stress-based approach where no fatigue data are available for the component or for component-like geometries, so that an estimated S-N curve must be relied upon. The estimate of the fatigue limit and the various related reduction factors, and also k_f, are based on empirical data. However, the data are fragmentary, and some of the mechanical design books use empirical factors that were developed many years ago. Furthermore, the data used to originally develop these factors are in some areas limited to steels, and where nonferrous metals are included, the data are generally less extensive. An additional concern is that the multiplicative com-

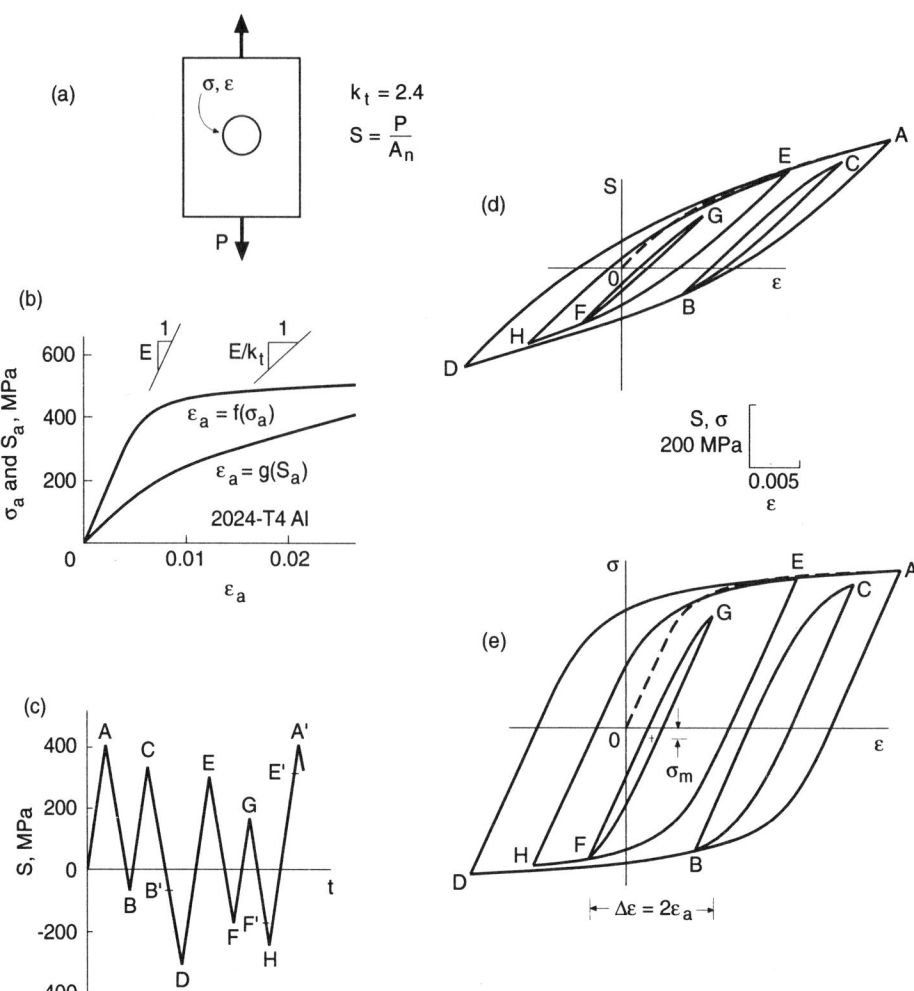

Fig. 21 Analysis of a notched member subjected to an irregular load versus time history. Notched member (a), having cyclic stress-strain and load-strain curves as in (b), is subjected to load history (c). The resulting load versus notch strain response is shown in (d), and the local stress-strain response at the notch in (e). Adapted from Ref 13

bination of the various reduction factors, as in Eq 4, is basically an assumption that has never been adequately verified. The overall effect of this situation is that estimates of the fatigue limit stress should be considered to be rough estimates only, especially for nonferrous metals. Also, analogous procedures for nonmetals are simply not available.

Consider the estimation of S-N curves in the intermediate and low-cycle region, as described previously based on Fig. 5. Here, the estimates are extraordinarily crude, as evidenced by large inconsistencies among various mechanical design books. For example, for axial loading, the use at 10^3 cycles of $k_f' = k_f$ and $m' = 0.75$ by Juvinall (Ref 4) is generally excessively conservative for ductile metals. This differs drastically from the use of $k_f' = 1$ and $m' = 0.9$ by Shigley (Ref 5), which may sometimes produce a nonconservative estimate.

Given the tenuous nature of estimated S-N curves, their use should either be abandoned, or recent and new fatigue data need to be employed to fill in gaps and to refine and extend the estimates.

Comparison of the Stress- and Strain-Based Approaches. The strain-based approach has the major advantage compared with any form of stress-based approach of rationally accounting for mean stresses based on local notch stresses. However, in fairness to the Juvinall book (Ref 4), it will be noted that local mean stresses are indeed estimated there based on an elastic, perfectly plastic stress-strain curve using the monotonic yield strength. In Juvinall, this approach is applied only to constant amplitude loading, but it could be logically extended to the variable amplitude case. The altered yield strength caused by cyclic hardening or softening, as reflected in the cyclic stress-strain curve, would still not be included, however.

Comparing the strain-based approach with estimated S-N curves, it is significant that the crude factors used at intermediate and short life, such as k_f' and m' at 10^3 cycles, are entirely unnecessary in a strain-based approach. These arise primarily from plasticity and notch effects, and the interaction of these, which is handled in a fairly rigorous manner in the strain-based approach through the use of a load-strain curve, $\varepsilon = g(S)$. Although

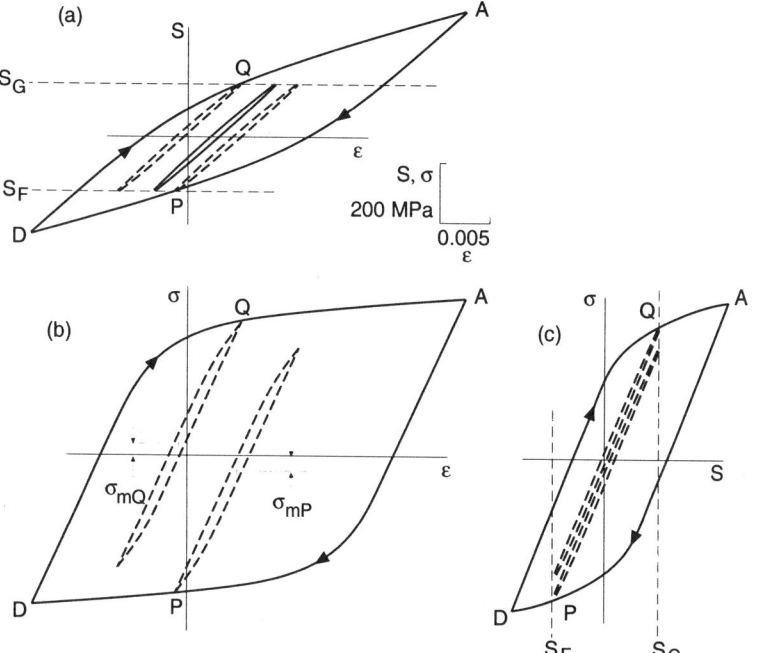

Fig. 22 Simplified procedure that places bounds on the mean stress effect. For cycle F-G of Fig. 21, the mean stress must lie between the values σ_{mP} and σ_{mQ}. Adapted from Ref 13

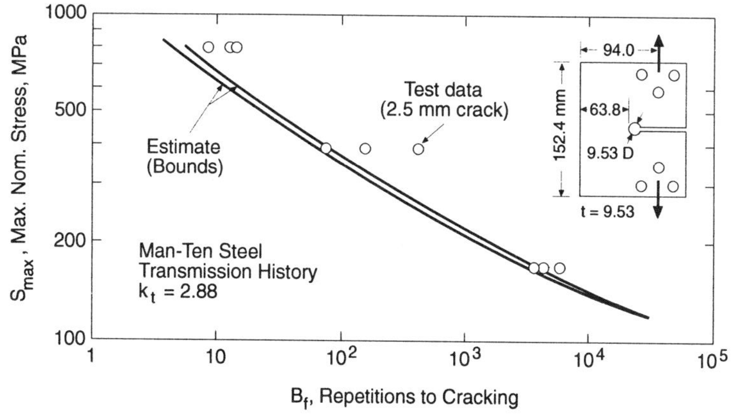

Fig. 23 Maximum nominal stress versus the number of repetitions to cracking, for repeated application of the SAE transmission history of Fig. 11 to the notched member and material indicated. Adapted from Ref 12 with data from Ref 11

S-N curves. However, such factors can also be applied to adjust strain-life curves. For example, because surface finish effects act primarily at long life, the exponent *b* of Eq 10 can be altered based on a surface effect factor m_s to lower the strain-life curve in the long life region. (See Section 14.2.4 of Ref 1 for more detail.) A size-effect correction could also be similarly applied, but it is less clear that the adjustment should be confined to only the exponent *b*. As already mentioned, recent data need to be analyzed and new data obtained, to improve existing methods of empirically adjusting fatigue life curves.

The strain-based approach as described here is similar to current industrial practice and achieves relative simplicity by making compromises in some areas where greater sophistication is possible. Some of these areas are: 1) limitation to local yielding, 2) neglecting mean stress relaxation, and 3) lack of applicability to multiaxial nonproportional loading cases. These areas and some related work are discussed to an extent in Ref 1 (pp 560-563, 600-601, and 644-645, respectively). Concerning mean stress relaxation, it should be noted that the major effect of local notch yielding in altering the mean stress that would exist if there were no yielding is specifically analyzed by the strain-based approach, as illustrated by Fig. 19. What is neglected is the minor effect of subsequent adjustment of the mean stress after a number of cycles has elapsed.

The area of complex multiaxial loading cases, as in shafts under out-of-phase bending and torsion, is of considerable practical importance and represents the most significant limitation of the strain-based approach as described in this section. Research and trial industrial applications are currently underway toward developing a more general approach that addresses this area; see the next article in this Volume for detailed treatment of multiaxial loading.

General Discussion on the Palmgren-Miner Rule. In the preceding description, the simple P-M rule is retained, with specific actions taken as follows to avoid its shortcomings: First, use of rainflow-cycle counting or a similar method is necessary. Otherwise, difficulties in life prediction will be encountered that may appear to be due to the P-M rule. Second, sequence effects can be caused by local yielding at notches altering the local mean stress and thus affecting life. Such effects should be properly analyzed by a strain-based approach. Third, initial or occasional overloads may cause material-damage-related sequence effects. These should be included in life estimates by including them in the stress or strain versus life curve, as in Fig. 13 and 14.

In Ref 26, an approach termed the relative Miner rule is described. This consists essentially of adjusting the P-M rule (Eq 8) so that the sum of cycle ratios is a value other than unity. The adjusted value is obtained from limited data using a load history, stress level, and component geometry as close as possible to the actual application. This provides an empirical adjustment that can account for various uncertainties in life estimates, such as: (1) failure of a stress-based ap-

approximate methods such as Neuber's rule are often used for notched members to obtain $\varepsilon = g(S)$, this can be more precisely determined from elasto-plastic finite element analysis or strain measurements. Also, cyclic yielding of unnotched members in bending or torsion can be analyzed, using the cyclic stress-strain curve to obtain $\varepsilon = g(S)$, so that such cases are also included in the strain-based approach. (See Sections 13.2 and 13.4 in Ref 1.)

Additional Discussion of the Strain-Based Approach. In the descriptions earlier in this article, notch strain estimates are described that employ Neuber's rule used with the elastic stress concentration factor, k_t. However, this is often replaced by k_f, the fatigue notch factor, as in the previous article in this Volume. Such an additional empirical adjustment may improve accuracy in some cases. As discussed in Ref 1, 24, and

25, the need for a k_t to k_f adjustment is thought to be primarily caused by crack growth effects. On this basis, it is the author's opinion that it is preferable to use k_t in estimating the crack initiation life.

The k_t to k_f adjustment will be significant primarily for cases of sharp notches, where cracks are likely to start early, so that crack growth dominates the life. Hence, an alternative to using k_f is to use k_t to estimate the crack initiation life, and then fracture mechanics to estimate the crack growth life, so that the total life is obtained. See Ref 25 and also Ref 1 (pp 665-666) for selecting initial crack length for the fracture mechanics part of this analysis.

Recall that estimated *S-N* curves use adjustments as in Eq 4 for such factors as surface finish and size effect. It might appear at first that this represents an advantage of stress-based estimated

proach to properly handle sequence effects related to local mean stress, (2) material-damage-related sequence effects not otherwise addressed, (3) surface finish, residual stresses, and other fabrication-related details not otherwise accounted for, and (4) inaccuracies in strain-based analysis. The latter category might include the approximate nature of Neuber's rule, stress-relaxation effects, and inaccuracies in mean stress adjustments, as shown by Eq 13 to 15. The relative Miner rule thus has considerable merit. See Ref 26 and other work by the same authors for more details.

Welded members comprise a category that merits special comment. Life prediction is complicated by the geometric and metallurgical complexity and variability involved, by the usual presence of complex residual stress fields, and by the frequent presence of initial cracklike flaws. As a result, component *S-N* data and a stress-based approach are often used. The strain-based approach is difficult to apply except where welds have well-defined geometry and are of very high quality, that is, relatively free of cracklike flaws. An alternative is to use a fracture mechanics approach based on growth of the weld flaws as cracks. Some success in dealing with the complexities involved through a fracture mechanics approach is demonstrated in Ref 27. Weldment fatigue is also addressed specifically in several articles in this Handbook.

REFERENCES

1. N.E. Dowling, *Mechanical Behavior of Materials: Engineering Methods for Deformation, Fracture, and Fatigue*, Prentice Hall, 1993
2. D. Webber, Constant Amplitude and Cumulative Damage Fatigue Tests on Bailey Bridges, *Effects of Environment and Complex Load History on Fatigue Life*, STP 462, ASTM, 1970, p 15-39
3. R.C. Juvinall, *Stress, Strain, and Strength*, McGraw-Hill, 1967
4. R.C. Juvinall and K.M. Marshek, *Fundamentals of Machine Component Design*, 2nd ed., John Wiley & Sons, 1991
5. J.E. Shigley and C.R. Mischke, *Mechanical Engineering Design*, 5th ed., McGraw-Hill, 1989
6. K. Walker, The Effect of Stress Ratio during Crack Propagation and Fatigue for 2024-T3 and 7075-T6 Aluminum, *Effects of Environment and Complex Load History on Fatigue Life*, STP 462, 1970, p 1-14
7. R.E. Peterson, *Stress Concentration Factors*, John Wiley & Sons, 1974
8. *Cycle Counting in Fatigue Analysis*, Vol 03.01 (No. 1049), *1994 Annual Book of ASTM Standards*, ASTM, 1994
9. S.D. Downing and D.F. Socie, Simplified Rainflow Counting Algorithms, *Int. J. Fatigue*, Vol 4 (No. 1), Jan 1982, p 31-40
10. R.C. Rice, Ed., *Fatigue Design Handbook*, 2nd ed., No. AE-10, Society of Automotive Engineers, 1988
11. R.M. Wetzel, Ed., *Fatigue Under Complex Loading: Analyses and Experiments*, No. AE-6, Society of Automotive Engineers, 1977
12. N.E. Dowling, A Review of Fatigue Life Prediction Methods, Paper No. 871966, *Durability by Design*, No. SP-730, Society of Automotive Engineers, 1987
13. N.E. Dowling and A.K. Khosrovaneh, Simplified Analysis of Helicopter Fatigue Loading Spectra, *Development of Fatigue Loading Spectra*, STP 1006, J.M. Potter and R.T. Watanabe, Ed., ASTM, 1989, p 150-171
14. T.H. Topper and B.I. Sandor, Effects of Mean Stress and Prestrain on Fatigue Damage Summation, *Effects of Environment and Complex Load History on Fatigue Life*, STP 462, ASTM, 1970, p 93-104
15. N.E. Dowling, Fatigue Life and Inelastic Strain Response under Complex Histories for an Alloy Steel, *J. Test. Eval.*, Vol 1 (No. 4), July 1973, p 271-287
16. N.E. Dowling, Fatigue Failure Predictions for Complex Load versus Time Histories, Section 7.4, *Pressure Vessels and Piping: Design Technology—1982—A Decade of Progress*, S.Y. Zamrik and D. Dietrich, Ed., Book No. G00213, American Society of Mechanical Engineers, 1982. Also in *J. Eng. Mater. Technol. (Trans. ASME)*, Vol 105, July 1983, p 206-214, with Erratum, Oct 1983, p 321
17. S.J. Stadnick and J. Morrow, Techniques for Smooth Specimen Simulation of the Fatigue Behavior of Notched Members, *Testing for Prediction of Material Performance in Structures and Components*, STP 515, ASTM, 1972, p 229-252
18. J. Morrow, Fatigue Properties of Metals, Section 3.2, *Fatigue Design Handbook*, Society of Automotive Engineers, 1968. (Section 3.2 is a summary of a paper presented at a meeting of Division 4 of the SAE Iron and Steel Technical Committee, 4 Nov 1964.)
19. K.N. Smith, P. Watson, and T.H. Topper, A Stress-Strain Function for the Fatigue of Metals, *J. Mater.*, Vol 5 (No. 4), Dec 1970, p 767-778
20. J.F. Martin, T.H. Topper, and G.M. Sinclair, Computer Based Simulation of Cyclic Stress-Strain Behavior with Applications to Fatigue, *Mater. Res. Stand.*, Vol 11 (No. 2), Feb 1971, p 23-29
21. N.E. Dowling and W.K. Wilson, Analysis of Notch Strain for Cyclic Loading, *Fifth Int. Conf. Structural Mechanics in Reactor Technology*, Vol L, Paper L13/4, North-Holland Publishing, 1979
22. J. Morrow, R.M. Wetzel, and T.H. Topper, Laboratory Simulation of Structural Fatigue Behavior, *Effects of Environment and Complex Load History on Fatigue Life*, STP 462, ASTM, 1970, p 74-91
23. G. Glinka, Energy Density Approach to Calculation of Inelastic Stress-Strain near Notches and Cracks, *Eng. Fract. Mech.*, Vol 22 (No. 3), 1985, p 485-508. See also Vol 22 (No. 5), p 839-854
24. N.E. Dowling, Fatigue at Notches and the Local Strain and Fracture Mechanics Approaches, *Fracture Mechanics*, STP 677, ASTM, 1979, p 247-273
25. N.E. Dowling, Notched Member Fatigue Life Predictions Combining Crack Initiation and Propagation, *Fat. Eng. Mater. Struct.*, Vol 2 (No. 2), 1979, p 129-138
26. A. Buch, T. Seeger, and M. Vormwald, Improvement of Fatigue Life Prediction Accuracy for Various Realistic Loading Spectra by Use of Correction Factors, *Int. J. Fatigue*, Oct 1986, p 175-185
27. S.J. Hudak, Jr., O.H. Burnside, and K.S. Chan, Analysis of Corrosion Fatigue Crack Growth in Welded Tubular Joints, Paper No. OTC-4771, *16th Annual Offshore Technology Conference* (Houston, TX), May 1984

Multiaxial Fatigue Strength

David L. McDowell, Georgia Institute of Technology

MOST ENGINEERING DESIGNS and/or failure analyses involve three-dimensional combinations of stress and strain (multiaxiality) in the vicinity of surfaces and notches, which can be limiting in fatigue applications. This article briefly reviews the state-of-the-art of fatigue correlations for such combined stress states. Basic definitions of multiaxial effective stresses and strains and differences between proportional and nonproportional loading are first introduced to facilitate discussion of various correlating parameters. Some basic correlations for multiaxial fatigue also are presented.

Fatigue crack "initiation" parameters are reviewed, ranging from simple effective stress and strain concepts to more recent critical plane theories. This approach is considered as distinct from fracture mechanics approaches in view of the difficulties in applying the latter to small cracks in rigorous fashion. Typical experimental observations of formation and propagation of small fatigue cracks are considered under various stress states, and the relation to long crack fracture mixed-mode fracture mechanics is explored. Differences between low-cycle fatigue (LCF) and high-cycle fatigue (HCF) behaviors are discussed. State I crystallographic and stage II normal stress-dominated growth of microcracks are discussed, along with some observations regarding the influence of combined stress state on the propagation of small cracks.

Finally, several other features of multiaxial fatigue are discussed, including mean stress effects, sequences of stress/strain amplitude or stress state, nonproportional loading and cycle counting, and HCF fatigue limits.

This article also covers the formation and propagation of cracks on the order of several grain sizes in diameter, typically less than 1 mm in length, in initially isotropic, ductile structural alloys. The propagation of mechanically long cracks is not considered.

Basic Definitions

The stress tensor, σ_{ij}, can be decomposed into hydrostatic and deviatoric components:

- *Hydrostatic stress* $\equiv \sigma_h = \sigma_{kk}/3 = (\sigma_{11} + \sigma_{22} + \sigma_{33})/3 = (\sigma_1 + \sigma_2 + \sigma_3)/3$
- *Deviatoric stress tensor* $\equiv \sigma_{ij}' = \sigma_{ij} - \sigma_h \delta_{ij}$ where $\delta_{ij} = 1$ if $i = j$, 0 if $i \neq j$

Here, σ_1, σ_2, and σ_3 are the three principal stresses. The octahedral shear stress is defined as the resolved shear stress on the Π-plane, the plane making equal angles with the three principal stress directions:

$$\tau_{oct} = \frac{1}{3}\sqrt{(\sigma_1 - \sigma_2)^2 + (\sigma_1 - \sigma_3)^2 + (\sigma_2 - \sigma_3)^2}$$

(Eq 1)

Strain Tensor. Likewise, the strain tensor, ε_{ij}, is decomposed into a hydrostatic (dilatation) component and a deviatoric part such that:

- *Dilatation* $\equiv \varepsilon_{kk} = \varepsilon_{11} + \varepsilon_{22} + \varepsilon_{33} = \varepsilon_1 + \varepsilon_2 + \varepsilon_3 = \Delta V/V_0$ for small strain, where ε_1, ε_2, and ε_3 are the three principal strains, ΔV is the volume change, and V_0 is the initial, reference volume.
- *Deviatoric strain tensor* $\equiv \varepsilon_{ij}' = \varepsilon_{ij} - (\varepsilon_{kk}/3)\delta_{ij}$

The octahedral shear strain may be written as:

$$\gamma_{oct} = \frac{2}{3}\sqrt{(\varepsilon_1 - \varepsilon_2)^2 + (\varepsilon_1 - \varepsilon_3)^2 + (\varepsilon_2 - \varepsilon_3)^2}$$

(Eq 2)

The Scalar Quantities τ_{oct} and γ_{oct} are considered as equivalent shear quantities for a multiaxial stress/strain state. Alternatively, we may consider the maximum shear stress and strain quantities acting on at most three mutually orthogonal sets of planes that intersect the principal stress/strain axes at 45°:

$$\tau_{max} = \frac{max}{i,j}\left|\frac{\sigma_i - \sigma_j}{2}\right|$$

(Eq 3)

$$\frac{\gamma_{max}}{2} = \frac{max}{i,j}\left|\frac{\varepsilon_i - \varepsilon_j}{2}\right|, \quad \frac{\gamma^p_{max}}{2} = \frac{max}{i,j}\left|\frac{\varepsilon^p_i - \varepsilon^p_j}{2}\right|$$

(Eq 4)

Both the maximum shear and the octahedral shear quantities are defined for planes with specific orientation with respect to the applied stress/strain state. For proportional loading, all principal stresses change in proportion, so the plane of maximum shear stress remains fixed in orientation. In this case, the expression in Eq 3 holds for the amplitude for maximum shear stress when the amplitudes of principal stresses are substituted. Similarly, for proportional straining, Eq 4 holds when amplitudes are substituted.

In classical theories of yielding of initially isotropic, ductile metals, it is common practice to consider τ_{oct} and γ_{oct} or τ_{max} and γ_{max} as conjugate scalar pairs that reflect the intensity of the combined stress/strain state. Yielding is assumed to occur when τ_{max} (Tresca theory) or τ_{oct} (Von Mises or distortion energy theory) reaches some critical value. The Rankine failure criterion $\sigma_1 = \sigma_{critical}$ is often applied as a failure criterion for brittle materials.

Uniaxial test data are usually available. In the case of the octahedral shear parameters, the uniaxial equivalent stress and strain quantities are defined as:

$$\bar{\sigma} = \frac{1}{\sqrt{2}}\sqrt{(\sigma_1 - \sigma_2)^2 + (\sigma_1 - \sigma_3)^2 + (\sigma_2 - \sigma_3)^2}$$

(Eq 5)

Likewise, if $\sigma_1 \geq \sigma_2 \geq \sigma_3$ and $\varepsilon_1 \geq \varepsilon_2 \geq \varepsilon_3$, the uniaxial effective stress and strain quantities based on the maximum shear parameters are defined as:

$$\bar{\varepsilon} = \frac{\sqrt{(\varepsilon_1 - \varepsilon_2)^2 + (\varepsilon_1 - \varepsilon_3)^2 + (\varepsilon_2 - \varepsilon_3)^2}}{\sqrt{2}(1 + \nu)}$$

(Eq 6)

$$\bar{\varepsilon}^p = \frac{1}{\sqrt{2}}\gamma^p_{oct} = \sqrt{\frac{2}{3}\left[(\varepsilon^p_1)^2 + (\varepsilon^p_2)^2 + (\varepsilon^p_3)^2\right]}$$

$$= \sqrt{\frac{2}{3}\varepsilon^p_{ij}\varepsilon^p_{ij}}$$

(Eq 7)

$$\sigma_{eff} = 2\tau_{max} = \sigma_1 - \sigma_3$$

(Eq 8)

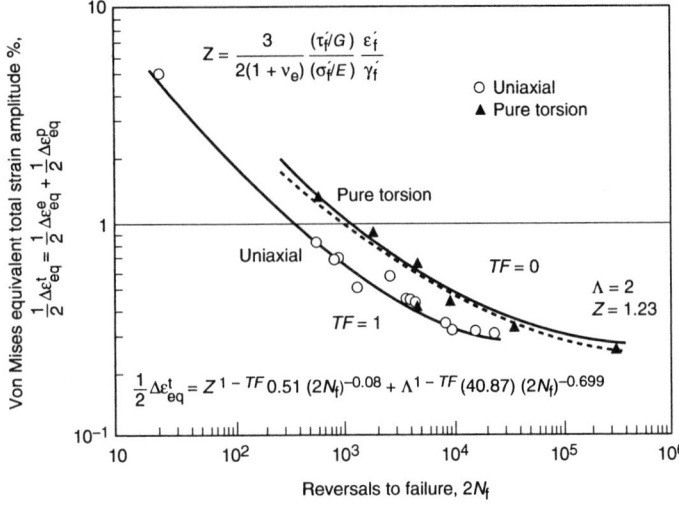

Fig. 1 Correlation of effective strain amplitude versus N_f for axial and torsional fatigue of Haynes 188 cobalt-base alloy at 760 °C. Note that the torsional data are shifted to the right. Source: Ref 19.

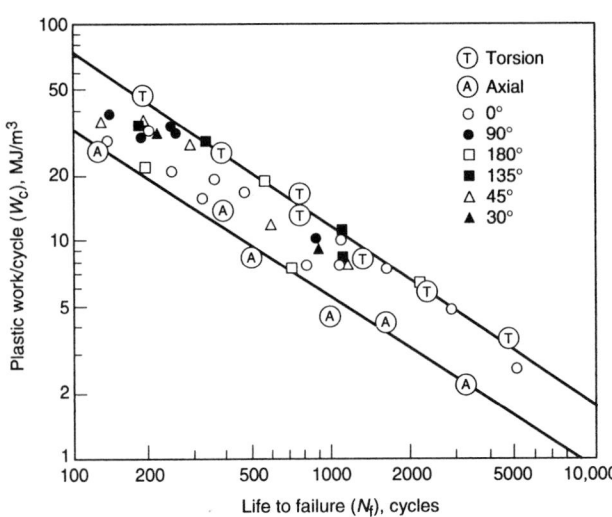

Fig. 2 Correlation of plastic hysteresis energy versus N_f for 1% Cr-Mo-V at 20 °C in the LCF range. Torsional data are shifted to the right. Source: Ref 21

$$\varepsilon_{eff} = \frac{\gamma_{max}}{(1+\nu)} = \frac{\varepsilon_1 - \varepsilon_3}{(1+\nu)}$$

$$\varepsilon_{eff}^p = \frac{2}{3}\gamma_{max}^p = \frac{2}{3}(\varepsilon_1^p - \varepsilon_3^p) \qquad \text{(Eq 9)}$$

The effective Poisson's ratio, ν, ranges from approximately 0.3 under fully elastic conditions to 0.5 for the fully plastic case.

Proportional and Nonproportional Loading. An important consideration for cyclic deformation and fatigue is whether the axes of principal axes of strain (stress) are fixed with respect to the material. If this is the case, the straining (stressing) is considered as proportional, and the components of the stress or strain tensors increase or decrease in constant proportion. Consequently, the octahedral shear plane and planes of maximum shear remain fixed in orientation as well. In terms of cyclic deformation, proportional loading is often considered as equivalent to uniaxial loading on the basis of effective stress and strain. But there are important differences in the formation and propagation of small fatigue cracks among different stress states, even for proportional loading. In uniaxial straining there are an infinite number of octahedral and maximum shear planes making equal angles with the axis of tension-compression loading. In contrast, pure torsion may be identified with a single set of octahedral or maximum shear planes. Moreover, the normal stress and strain amplitudes to these planes differ between uniaxial and shear cases, and among other states of stress as well.

Estimation of the elastic-plastic deformation under nonproportional loading requires the use of incremental cyclic plasticity theory, in general (Ref 1-3). The principal axes of stress or strain may remain fixed in direction with the components varying nonproportionally, or more generally the principal axes may rotate.

Correlating Parameters for Multiaxial Fatigue

The subject of multiaxial fatigue has developed over many decades. More complete reviews of the historical development may be found in several recent reviews (Ref 4-6). Here it is understood that the fatigue crack initiation life, N_f, corresponds to a crack length on the order of 500 to 1000 μm.

Static Yield Criteria. Initial approaches to modeling multiaxial fatigue behavior were based on static yield criteria developed a century ago, as discussed in the previous section. These approaches have been widely used in multiaxial fatigue design, as evidenced by various present-day standards, codes, and design textbooks. Due to their use in plasticity theory, τ_{oct} and γ_{oct} or τ_{max} and γ_{max} have been historically employed in fatigue correlations for small cyclic plastic strains of ductile metals. It is assumed that results for different combined stress states should collapse onto one *universal* curve, provided the loading is completely reversed and proportional in nature. Unfortunately, effective stress and strain approaches do not generally correlate fatigue behavior under various stress states such as uniaxial, torsion, and equibiaxial in-plane loading. A typical example appears in Fig. 1, which demonstrates differences between the correlation of uniaxial fatigue data and torsional fatigue data for smooth specimens based on effective strain range. Torsional loading typically exhibits a much longer life than uniaxial loading for a surface crack length on the order of 1 mm for a given effective strain amplitude. Detailed studies by Socie and colleagues (Ref 6-9) have clearly demonstrated that significant differences exist in the growth of small cracks among various stress states.

Haigh (Ref 10) recognized that effective stress was inadequate to correlate multiaxial HCF.

Gough et al. (Ref 11-12) showed that effective stress amplitude was insufficient to correlate HCF under combined bending and torsion, and they introduced the ellipse quadrant and ellipse arc concepts for ductile and brittle materials, respectively. An historically common form that has been used to distinguish fatigue crack initiation behavior among stress states is given by:

$$\bar{\sigma}_a + g(\sigma_h) = C \qquad \text{(Eq 10)}$$

where C is a constant for a given fatigue life, subscript "a" denotes amplitude, and σ_h denotes either the amplitude or mean value of hydrostatic stress over a cycle. Sines (Ref 13) proposed a form in which g is linear in the mean value of σ_h over a cycle. Fuchs (Ref 14) generalized this approach for nonproportional loading. The function $g(\sigma_h)$ may be interpreted as introducing the effect of mean normal stress on the formation of fatigue cracks. Equation 10 is recognized to be related to a Mohr theory of rupture that augments frictional failure processes with the dependence on normal stress.

Correlation Based on Triaxiality Factor. Libertiny (Ref 15) introduced the dependence of LCF on σ_h as well. The triaxiality factor ($TF = \sigma_{kk}/\bar{\sigma}$, based on either the amplitude or mean value of these quantities, has been introduced by Davis and Connelly (Ref 16) for ductility reduction due to triaxiality. It was later employed by Manjoine (Ref 17) and others to describe constraint effects in fracture that are not fully correlated by the amplitude of the crack tip singularity. This parameter has been employed by Manson and Halford (Ref 18), Zamrik et al. (Ref 19), and others to reflect the dependence of fatigue crack initiation on combined stress state for a wide range of HCF and LCF conditions. Note that $TF = 0$ for torsion, 1 for uniaxial loading, and 2 for in-phase, equibiaxial loading. Typical behavior is exhibited by Haynes 188 cobalt-base alloy at 760 °C (Ref 19), as shown in Fig. 1. Use of the triaxiality factor offers somewhat improved correlation of uniaxial and torsional fatigue data (Ref 19) in terms of the following strain-life relation:

$$\frac{\Delta\bar{\varepsilon}^t}{2} = \frac{\Delta\bar{\varepsilon}^e}{2} + \frac{\Delta\bar{\varepsilon}^p}{2}$$

$$= Z^{1-TF}\frac{\sigma_f'}{E}(2N_f)^b + \Lambda^{1-TF}\varepsilon_f'(2N_f)^c \quad \text{(Eq 11)}$$

where

$$Z = \frac{3}{2(1+\nu_e)}\frac{\tau_f'}{G}\frac{E}{\sigma_f'}\frac{\varepsilon_f'}{\gamma_f'} \quad \text{(Eq 12)}$$

and Λ is a ductility parameter ($\Lambda \approx 2$). The elastic Poisson's ratio is given by ν_e, while E and G are the Young's modulus and shear modulus, respectively, and τ_f' and γ_f' are the coefficients in a pure torsion strain-life relation analogous to the right-hand side of the uniaxial equivalent form in Eq 11. Both LCF and HCF (finite life) regimes are addressed.

Correlation Based on Cyclic Hysteresis Energy. Another method correlates cyclic hysteresis energy with the number of cycles to crack initiation (Ref 20-24). For example, Garud (Ref 21) applied this approach in conjunction with incremental plasticity theory to predict the fatigue crack initiation life under complex nonproportional multiaxial loading conditions. As shown in Fig. 2 for 1% Cr-Mo-V steel, the approach does not typically correlate both the uniaxial and torsional fatigue cases, even for completely reversed loading (no mean stress). Garud suggested differential weighting of the contribution of shear components to the hysteresis energy relative to the normal components in order to account for these differences. To effectively collapse uniaxial and torsional data, Ellyin and Kujawski (Ref 24) introduced explicit dependence on mean stress and stress state in the hysteresis energy parameter

$$\frac{\Delta W_d h(\sigma_m)}{f(\bar{\rho})} = P(2N_f) \quad \text{(Eq 13)}$$

where ΔW_d is the area under the effective stress/strain hysteresis loop (both elastic and plastic parts), σ_m is the mean value of σ_{kk} over the cycle, and $\bar{\rho}$ is a constraint factor, defined by $(1 + \nu)\hat{\varepsilon}_{max} / \hat{\gamma}_{max}$, where $\hat{\varepsilon}_{max}$ is the maximum principal strain in the surface plane and $\hat{\gamma}_{max}$ is the maximum shear strain on a plane that intersects the free surface at 45°. Similar to effective stress or strain approaches, hysteresis energy approaches do not infer specific orientations or planes of microcracking.

Critical Plane Theories. Another class of approaches, called "critical plane theories," devote specific attention to the orientation of small cracks in multiaxial fatigue. These theories assert that the most critically damaged plane is one of maximum shear stress or strain amplitude that experiences the maximum normal strain and/or normal stress. These critical plane theories were preceded by some 20 to 30 years by the HCF theories of Stulen and Cummings (Ref 25), Guest

(Ref 26) and Findley (Ref 27), which augmented the maximum shear stress amplitude with an additive term involving the normal stress to the plane of maximum shear. Their approaches may be summarized as:

$$\frac{(\sigma_1 - \sigma_3)}{2} + F\left(\frac{(\sigma_1 + \sigma_3)}{2}\right) = G \quad \text{(Eq 14)}$$

where F and G are constants for a given life and σ_1 (σ_3) is the value of the largest (smallest) peak principal stress. Equation 14 achieved satisfactory correlation of HCF strength under various stress states, predominantly verified under combined bending and torsion.

For proportional loading, $(\sigma_1 + \sigma_3)/2$ is the amplitude of stress normal to the plane of maximum shear stress amplitude. It is commonly observed that small cracks in ductile polycrystals nucleate and grow early in life on crystallographic planes that are favorably aligned with the maximum shear stress or strain, defined as stage I fatigue crack propagation by Forsyth (Ref 28). Typically, small cracks propagate in this manner until reaching a length on the order of 3 to 10 grain diameters (Ref 29-30), and then follow a macroscopic mode I path normal to the range of maximum principal stress. Hence, the second term in Eq 14 incorporates the assistance of tensile stress normal to the crack plane in opening the crack during shear-dominated growth early in life. For brittle materials that are more sensitive to normal stress to the maximum shear plane, F is relatively larger than for ductile materials. Of course, these global strain or stress parameters pertain to the polycrystalline average response and are somewhat loosely related to local driving forces at the crack tip. Nonetheless, they have demonstrated quantitative agreement with experimentally observed behaviors for a wide range of stress states.

The HCF approach in Eq 14 is a predecessor of similar strain-based relations for stage I microcrack propagation along maximum shear strain amplitude planes under LCF conditions. Brown and Miller (Ref 31) introduced the so-called Γ plane approach, wherein the orientation of the maximum shear strain amplitude planes with respect to the free surface distinguishes two very different types of fatigue crack propagation behaviors, termed cases A and B. They defined a general relationship between the maximum shear strain amplitude, $\Delta\gamma_{max}/2$, and the normal strain amplitude, $\Delta\varepsilon_n/2$, to the plane of maximum shear strain amplitude:

$$U_1\left(\frac{\Delta\gamma_{max}}{2}\right) = U_2\left(\frac{\Delta\varepsilon_n}{2}\right) \quad \text{(Eq 15)}$$

for a given fatigue crack initiation life, where U_1 and U_2 are nonlinear functions of their arguments. Equation 15 assumes different forms for cases A and B, which are defined by the orientation of maximum shear strain range planes relative to the surface, as shown in Fig. 3(a). The case for which vectors normal to the maximum shear strain amplitude planes lie within the specimen surface is termed case

A. Case B is defined by the intersection of the maximum shear strain range planes with the surface. Typical experimental data are plotted in the so-called Γ plane in Fig. 3(b). In some cases, the case B contours are approximately described by $U_2 = 0$, although this is not a general relation. In case A, the functions U_1 and U_2 are approximately quadratic in their arguments for ductile metals. This approach has successfully correlated tension-torsion and tension-tension experiments for completely reversed proportional loading. Lohr and Ellison (Ref 32) introduced a slight variation of this approach that considered case B planes to be always more damaging, even if they are not the planes of maximum shear strain range.

Experiments on thin-walled tubular specimens involving combined axial loading and cyclic internal/external pressure may be used to apply a range of shear strain combined with static mean normal stresses during a cycle. Critical experiments reported by Socie (Ref 6) have shown for several ductile alloys that: (a) augmentation of a maximum shear strain parameter by only hydrostatic stress is insufficient to describe orientations of fatigue cracking that are consistently observed in multiaxial experiments; and (b) even under LCF conditions, the normal stress to the plane of maximum shear strain range plays an important role in delineating the failure plane and correlating mean stress effects. On the basis of observations of the formation and growth of small cracks, Socie (Ref 6, 8) distinguishes between materials that exhibit prolonged stage I propagation behavior along maximum shear planes ("shear dominated") and those that transition at very small crack lengths to stage II ("normal stress-dominated") propagation. Socie has argued that additional effects of the mean normal strain and stress to the plane of maximum shear strain range may be used to augment maximum shear strain amplitude to correlate shear-dominated fatigue behavior.

Fatemi and Socie (Ref 33) and Fatemi and Kurath (Ref 34) have demonstrated robust correlation of fatigue under various stress states for both case A and case B histories with and without mean stress, based on the assumption that peak normal stress to the plane of maximum range of shear strain directly affects the stage I shear-dominated propagation of small cracks. They proposed the correlative parameter

$$\frac{\Delta\gamma_{max}}{2}\left[1 + K\frac{\sigma_n^{max}}{\sigma_y}\right] = f_f(N_f) \quad \text{(Eq 16)}$$

where K is a constant, σ_n^{max} is the maximum normal stress to the plane of maximum shear strain range, and σ_y is the yield stress. Figure 4 presents multiaxial fatigue correlations for both Inconel 718 and 1045 steel with this parameter (Ref 35), typically within a factor of 2 in fatigue life. Furthermore, Socie (Ref 6) has shown that the orientation of microcracking follows the plane(s) of the maximum value of this parameter. Most of the correlations obtained to date pertain to LCF or transition fatigue, rather than HCF. However, the analogy to the HCF relation in Eq 14 suggests more general applicabil-

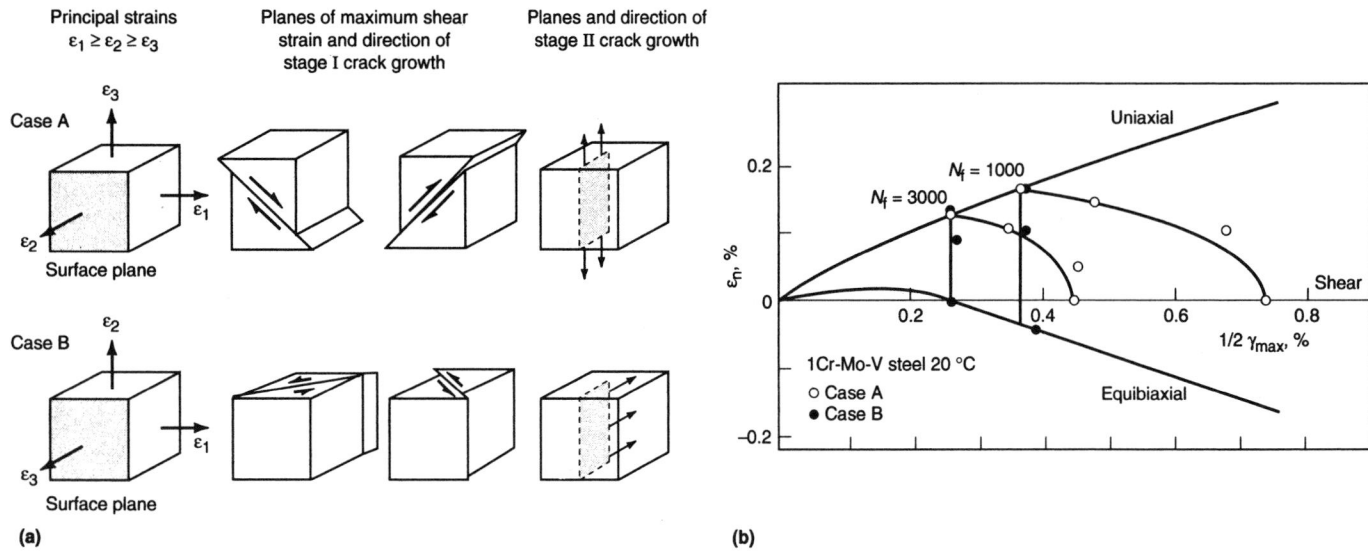

Fig. 3 (a) Distinction between case A and B cyclic strain states. (b) Typical contours of completely reversed fatigue crack initiation data for 1% Cr-Mo-V at 20° C in the Γ plane, where γ_{max} and ε_n represent amplitudes of maximum shear strain and normal strain to this plane for each case. Source: Ref 5

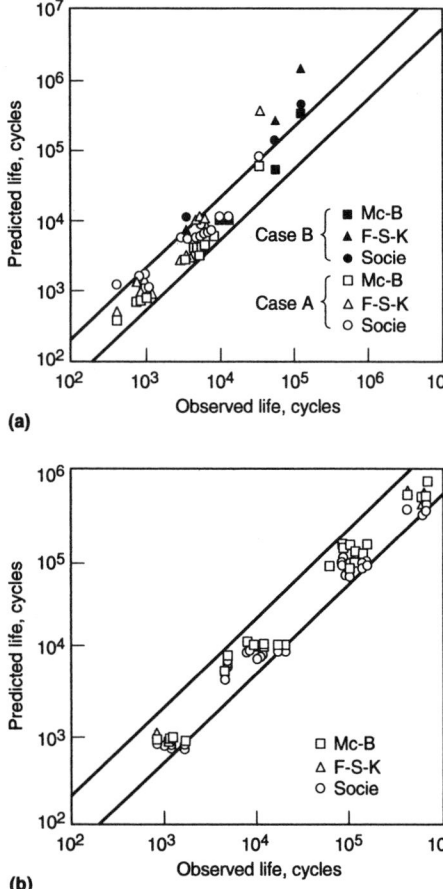

Fig. 4 Correlation of case A and case B completely reversed fatigue data for (a) Inconel 718 and (b) 1045 steel. The Fatemi-Socie-Kurath (F-S-K) and McDowell-Berard (Mc-B) correlations are included. Source: Ref 35

ity, at least in the absence of a fatigue limit. Socie (Ref 8) has proposed a Smith-Watson-Topper (Ref 36) generalization for normal-stress dominated ma-

terials, which may be associated with an early transition to stage II fatigue crack propagation.

Small Crack Growth in Multiaxial Fatigue

Crack nucleation processes, as discussed elsewhere in this Volume, are associated with the generation and coalescence of excess vacancies along persistent slip bands (Ref 37-39) in ductile single crystals or coarse grain polycrystals. In polycrystals, cracks may nucleate via fracture during processing (Ref 37) at intersecting slip bands (or twin) or by blockage of a slip band (or twin) by second-phase particles. A second type of microcracking in polycrystals occurs along grain boundaries due to impurity embrittlement or the presence of voids. Sometimes microcracking in polycrystals occurs at strong grain boundaries due to heterogeneous plastic deformation, governed by the degree of misorientation at the grain boundary, usually associated with a mixed mode of intercrystalline-transcrystalline fracture. These nucleation processes become increasingly dominant at very long lives in materials with minimal processing defects. We focus here on the growth of small cracks in fatigue rather than the nucleation problem. Clearly, propagation of small cracks is an important aspect of the "initiation" of a fatigue crack on the order of 1 mm.

Characteristics of Small Fatigue Cracks. Mixed-mode fatigue crack propagation studies have largely focused on the behavior of mechanically long cracks. The problem of the growth of small cracks in fatigue (from lengths on the order of 1 to 500-1000 μm) has received increased attention. Cracks are considered to be small when all pertinent dimensions are small compared to some characteristic length scale. In the case of microstructurally small cracks, the length scale is on the order of the dimensions of microstructural

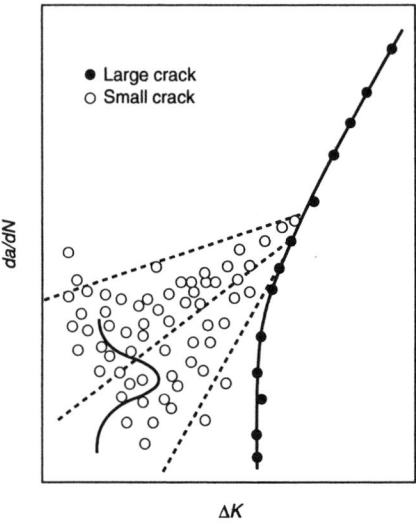

Fig. 5 Typical propagation behavior of small cracks. Note that da/dN is higher for a given ΔK than for long cracks, and the apparent scatter in da/dN is significant. The bottom dashed line is a linear extension of Paris Regime. Source: Ref 40

periodicity (e.g., grain diameter). For physically or mechanically small cracks, it is typically on the order of 5 to 10 times the microstructural scale. Attempting to develop a correlation between da/dN and ΔK, as in the case of mechanically long cracks, the so-called anomalous behavior of microstructurally small cracks has been widely demonstrated. In particular, the cyclic crack growth rate of small cracks may significantly exceed that of long cracks at the same level of ΔK, as shown in Fig. 5. Considerable scatter of the fatigue crack growth rate of small cracks at a given ΔK level is apparent. At low stress amplitudes (HCF), deceleration of crack growth is often observed, associated with a dip in the da/dN versus ΔK behavior. Subsequently, crack growth may accelerate prior to merging with the long crack data. At sufficiently low amplitudes, small cracks may become arrested. As small cracks propagate, their *da/dN* versus Δ*K* responses are typically observed to merge with the long crack response, as shown in Fig. 5.

Experimental observations indicate that the propagation behavior of microstructurally small and physically small cracks depends significantly on both the *R*-ratio and stress amplitude, in addition to stress state. Small crack behavior is subject to more scatter due to greater dependence on microstructure.

There are several prevalent explanations for the nonconformity of small/short crack behavior with that of mechanically long cracks:

- Differences in plasticity-induced closure transients relative to long cracks
- Microstructural roughness-induced closure/bridging
- Interaction with microstructural features, three-dimensional nonplanar growth, and pinning effects
- Violation of validity limits of linear elastic fracture mechanics (LEFM) or elastic-plastic fracture mechanics (EPFM) due to lack of self-similarity of growth and cyclic plastic zone/process zone size on the order of crack length
- Intensification of local driving forces relative to nominal applied stresses and strains due to heterogeneity and anisotropy of cyclic slip in the vicinity of the small crack(s), in addition to reduced constraint due to proximity of the free surface
- Local mixed-mode growth for small cracks, even for remote mode I loading

Some of these factors are more influential at high stress amplitudes and others at low stress amplitudes, for a given *R*-ratio. As pointed out by Suresh (Ref 41), low-strain amplitudes (HCF) promote predominantly mode II crystallographic growth and a higher degree of microstructural roughness along the crack faces, leading to enhanced crack-tip shielding effects. Likewise, predominantly remote shear loading may promote quite different roughness-induced crack face interference, and so on. Relatively few of these aspects have been considered in detail for small cracks. For example, the treatment of plasticity-induced closure (e.g., Ref 42) typically assumes validity of LEFM or EPFM concepts, even for microstructurally small cracks, while neglecting microstructural roughness-induced closure/bridging or interaction with microstructural features. Even with this simplification, the application of plasticity-induced closure models requires considerable idealization. On the other hand, models that consider interaction with periodic microstructural barriers (e.g., Ref 43-45) typically do not consider closure or bridging effects, although they may recognize the lack of applicability of LEFM or EPFM for small cracks. Models for the growth of small cracks have largely been confined to simple uniaxial (mode I) loading conditions; formal treatment of mul-

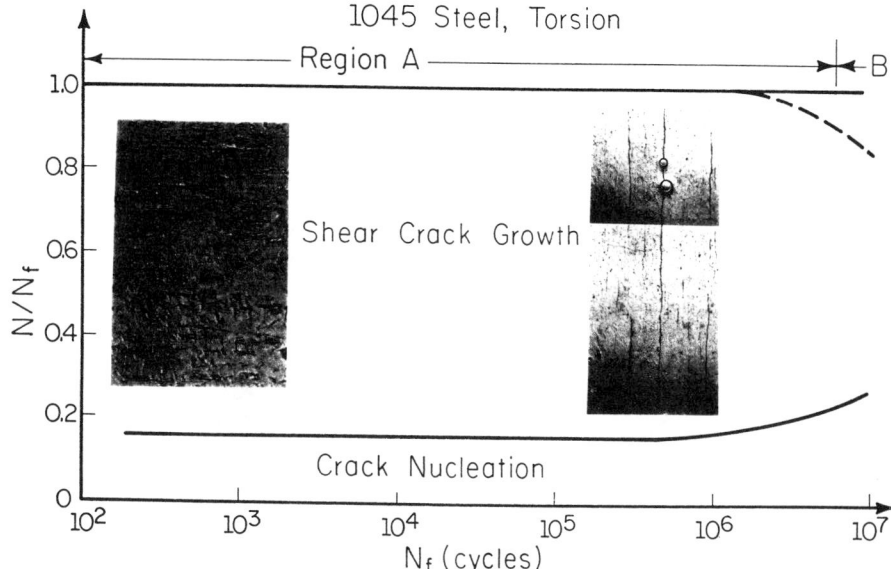

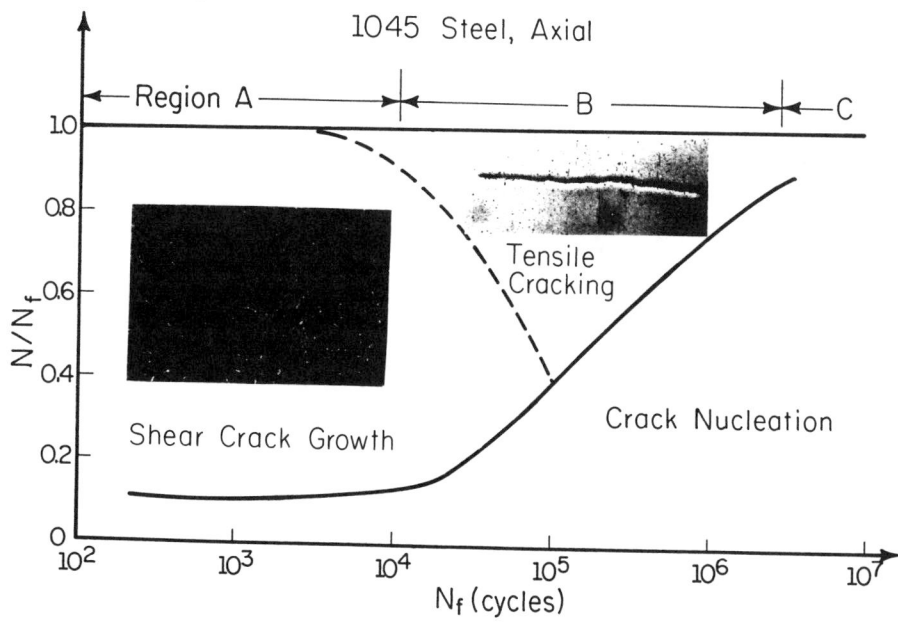

(b)

Fig. 6 Data of Socie on (a and b) 1045 steel and (c and d) IN 718 for life to 0.1 mm and 1 mm cracks ($N/N_f = 1$) for torsional and uniaxial loading. Source: Ref 6

tiaxial loading conditions within the fracture mechanics methodology is challenging in view of the plethora of mechanisms and "local mixity" (a term that represents the combination of different opening and sliding displacements at the crack tip, distinct from the remote loading history). This so-called "local mixity" arises from the nature of crystallographic propagation of stage I cracks. The range of validity of LEFM or EPFM concepts diminishes even further under multiaxial loading conditions.

The data in Fig. 6 clearly illustrate some important aspects of multiaxial fatigue crack growth for constant-amplitude loading of two ductile alloys in tension-compression and in torsion. The

curved contours represent the locus of normalized cycles, N/N_f, to growth to a 0.1 mm surface crack, with N_f corresponding to the number of cycles of growth to a 1 mm surface crack. Regimes of shear-dominated growth (stage I) along maximum shear strain range planes and normal stress-dominated growth (stage II) normal to the range of maximum principal stress are shown. The curve representing the fraction of life to a 0.1 mm crack is termed "crack nucleation" in Fig. 6, but it actually reflects microcrack propagation to this length.

The fraction of 1 mm crack life required for growth to a 0.1 mm crack is approximately 10% at high strain amplitudes (e.g., LCF) for both

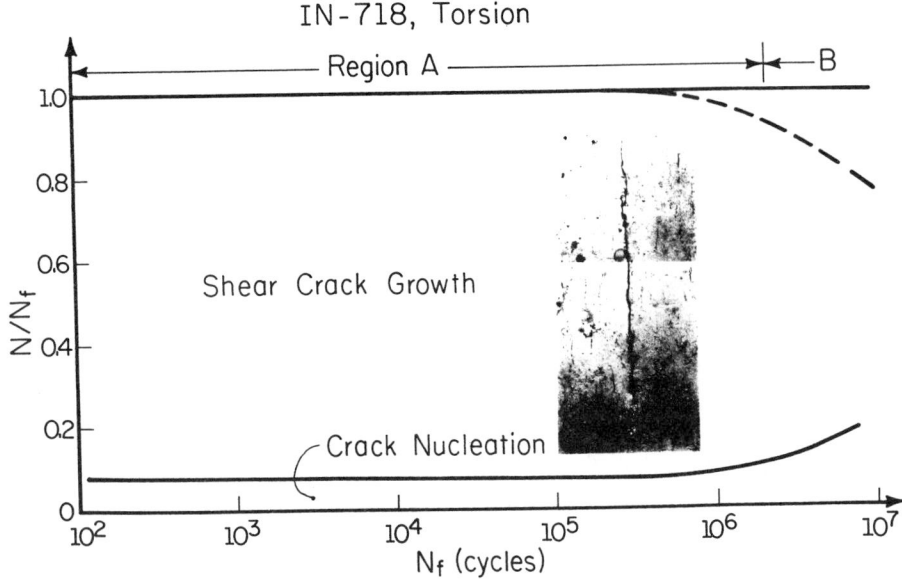

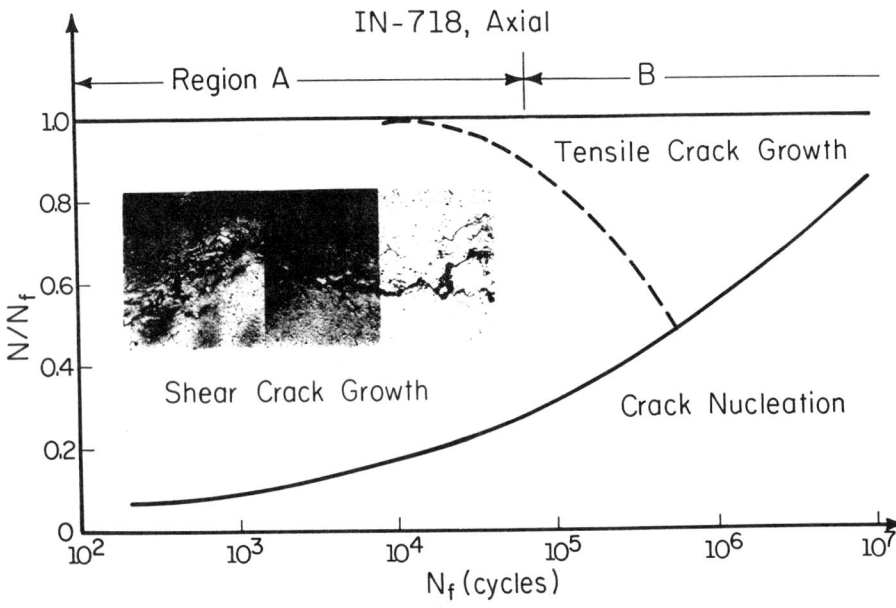

Fig. 6 (continued) Data of Socie on (a and b) 1045 steel and (c and d) IN 718 for life to 0.1 mm and 1 mm cracks ($N/N_i = 1$) for torsional and uniaxial loading. Source: Ref 6

point during or somewhat after this transition. This transition crack length may also depend on stress state and stress amplitude; these issues are not yet fully resolved. It may be related to the balance of competing mode I and mode II growth mechanisms (Ref 8, 46). Some modeling efforts have been devoted to the role of grain boundary blockage and transmission of slip to adjacent grains (e.g., Ref 45) in defining this transition.

J-Integral Correlations of Small Fatigue Cracks. Long crack solutions based on the ΔJ-integral (Ref 47-49) have been employed to correlate the propagation of small/short cracks in fatigue. Although some correlations have been obtained under predominantly uniaxial LCF conditions (Ref 50-52), such treatments ignore the limits of applicability of long crack solutions that assume homogeneity, isotropy, self-similarity, and a small ratio of cyclic plastic zone to crack length. It is essential to recognize the role of local mode mixity on crack growth. Although all three modes are operative (Ref 53), microstructurally sensitive small crack growth has often been idealized as mixed mode I-II as a reasonable approximation. Mode II is primary in stage I, whereas mode I dominates in stage II (Ref 53). There are presently no well-accepted criteria for mixed-mode stage I growth, and the data in Fig. 6 provide some insight into the complexities involved.

Hoshide and Socie (Ref 54) extended the elastic and plastic forms for the standard long crack J-integral of EPFM (Ref 55) to correlate combined mode I-II axial-torsional fatigue:

$$J = \pi \frac{(\sigma^2 + \tau^2)}{E} a_{eff} + \tilde{J}(n, \lambda_\sigma, \xi_\sigma) \overline{\sigma}\, \overline{\varepsilon}^p a = J_e + J_p \quad \text{(Eq 17)}$$

where a and a_{eff} are actual and effective crack lengths, and the stress biaxiality ratios are given by $\lambda_\sigma = \tau/\sigma_{yy}$ and $\xi_\sigma = \sigma_{xx}/\sigma_{yy}$, where τ and σ_{yy} are the far field shear and normal stresses, respectively, and σ_{xx} is the direct stress parallel to the crack. In general, J depends on the biaxiality ratios λ_σ and ξ_σ and on the strain hardening exponent, n. Self-similar crack extension is assumed. The growth law for mixed-mode proportional loading was assumed to follow

$$\frac{da}{dN} = \phi_I(\lambda_\sigma, \xi_\sigma)\Delta J^{M_I} + \phi_{II}(\lambda_\sigma, \xi_\sigma)\Delta J^{M_{II}} \quad \text{(Eq 18)}$$

where ΔJ is generalized from Eq 17 by considering the range of stress and strain as in Ref 47 to 49. Exponents M_I and M_{II} are not equal, in general. Hoshide and Socie used an analogous formulation with $M_I = M_{II}$ to correlate growth of fatigue cracks of length less than 1 mm in Inconel 718. They correlated the data with a growth law of the form

$$\frac{da}{dN} = C_J(\Delta J)^{M_J} \quad \text{(Eq 19)}$$

where C_J and M_J depend on the biaxiality ratios. Exponent M_J varied from 1.31 to 1.45.

Critical Plane Methods. Socie et al. (Ref 7) and Berard et al. (Ref 56, 57) have shown that the

uniaxial and torsional fatigue. Assuming an initial crack size on the order of 10 μm, these data suggest that crack propagation is only weakly dependent on crack length for high strain amplitudes. At increasing lives, the fraction of life spent in growing cracks less than 0.1 mm in length increases, to a much greater extent in uniaxial fatigue than in torsional fatigue. The fact that torsional fatigue exhibits a considerably lower ratio for a given N_f indicates that the differences reside in the crack propagation behavior. The crack growth behavior is quite nonlinear with respect to crack length for cracks shorter than 0.1mm under HCF conditions, particularly for uniaxial fatigue. This has important conse-

quences in terms of the nonlinear growth behavior of small cracks and in terms of both amplitude and stress state sequence effects. Also, the point of departure from stage I shear-dominated crack growth to state II normal stress-dominated growth occurs at higher strain amplitudes for uniaxial fatigue. Torsional fatigue appears to promote extended stage I behavior, perhaps associated with low symmetry slip (e.g., single slip) at the local level. Observations under uniaxial straining reveal that small cracks transition from transgranular stage I growth to stage II growth when the ratio of crack length to grain size is in the range of 3 to 10 (Ref 29-30). The influence of microstructure is also observed to wane at some simple bulk stress and strain range parameters used in critical-plane fatigue crack initiation laws serve to correlate the propagation rate of small cracks in multiaxial LCF. Some studies (Ref 30, 50, 51, 58) have shown that the growth of small cracks does not correlate with a crack length dependence of the ΔJ-integral of conventional EPFM. For HCF, this dependence differs significantly from that of LEFM (Ref 50). Departure from rigorous applicability of fracture mechanics approaches might be expected, particularly for nonplanar cracks with length on the order of microstructure.

McDowell and Berard (Ref 35) introduced an analogue of the ΔJ-integral approach to address the growth of small cracks along critical planes in multiaxial fatigue, addressing both case A and case B cracking. For multiaxial LCF, they proposed the law

$$\frac{da}{dN} = D_{am} C_p (1 + \mu \rho)^m (\beta_p(R_n) R_n + 1)^m \left(\frac{\Delta \tau_n}{2} \frac{\Delta \gamma_{max}^p}{2} \right)^m a^m$$

(Eq 20)

where the constraint parameter is defined by:

$$\rho = \frac{\Delta \sigma_{kk}/2}{2 \Delta \tau_n/2} - R_n$$

(Eq 21)

and $R_n = (\Delta \sigma_n/2)/(\Delta \tau_n/2)$. Here, σ_n and τ_n are the normal and shear stress, respectively, on the plane of maximum range of plastic shear strain. Parameter R_n varies from zero for completely reversed torsional fatigue to unity for uniaxial or biaxial loading conditions. Parameter β_p introduces dependence of the crack-tip fields and/or crack-tip opening and sliding displacements on biaxiality. An additional dependence of the microcrack propagation rate on triaxiality is introduced via constraint parameter ρ, analogous to the TF factor in Eq 11. Inspection of Eq 20 reveals that this form is similar in nature to that of Eq 16, but the plastic hysteresis energy term $\Delta \tau_n \Delta \gamma_{max}^p$ (rather than $\Delta \gamma_{max}^p$) is weighted by the relative effect of the normal stress amplitude acting on the plane of maximum cyclic shear (R_n). Constants μ and m control the influence of constraint and nonlinearity of crack growth, and:

$$C_p = (K_0' \gamma_f'^{(1+n_o')})^{-m}$$

(Eq 22)

recovers the independent LCF Coffin-Manson and cyclic stress-plastic strain laws for completely reversed loading in torsional and uniaxial fatigue, respectively, given by:

$$\frac{\Delta \gamma^p}{2} = \gamma_f' (2N_f)^{c_0}, \frac{\Delta \varepsilon^p}{2} = \varepsilon_f' (2N_f)^c$$

(Eq 23)

$$\frac{\Delta \tau}{2} = K_0' \left(\frac{\Delta \gamma^p}{2} \right)^{n_0'}, \frac{\Delta \sigma}{2} = K' \left(\frac{\Delta \varepsilon^p}{2} \right)^{n'}$$

(Eq 24)

with the additional prescriptions

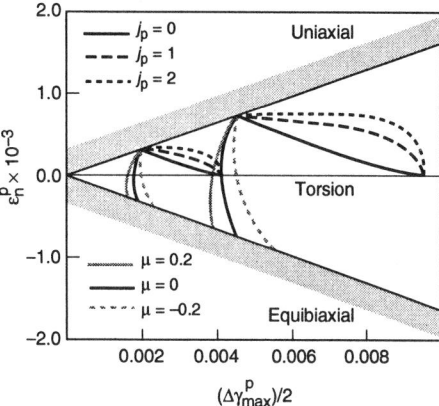

(a)

(b)

Fig. 7 Γ_p plane contour plots for two different low-cycle fatigue lives predicted by the (a) McDowell-Berard model in the Γ_p plane for $n' = 0.15$ and an approximate ratio of plastic work to failure in a torsion test to a tension test of two, and (b) the Fatemi-Socie-Kurath model contours in the Γ_p plane for $n' = 0.2$ and $K'/\sigma_y = 3.5$. Source: Ref 35

$$\beta_p(R_n) = \left[\frac{4}{3} \frac{K_o' (\gamma_f')^{(1+n_o')}}{K'(\varepsilon_f')^{(1+n')}} - 1 \right] (R_n)^{j_p}$$

(Eq 25)

$$m = \frac{-1}{c(R_n)(1 + n'(R_n))}$$

(Eq 26)

$$c(R_n) = c_0 + R_n(c - c_0),$$

$$n'(R_n) = n_0' + R_n(n' - n_0')$$

(Eq 27)

where j_p is a constant. Coefficient D_{am} is determined by integrating the expression for constant-amplitude loading conditions between given initial and final crack lengths:

$$D_{am} = 2 \int_{a_i}^{a_f} a^{-m} da$$

(Eq 28)

Constant-life plots in the Γ plane for completely reversed LCF (plastic strain range much greater than elastic) loading conditions are shown in Fig. 7(a), based on Eq 20, for 1045 steel (Ref 35). Case A contours for several values of j_p are

presented, along with several case B contours, which depend on the value of μ. The overall shape of these case A and B contours is generally in agreement with the form of LCF experimental data (e.g., Fig. 3b). A similar plot for the Fatemi-Socie-Kurath parameter in Eq 16 appears in Fig. 7(b), also in qualitative agreement with data, albeit with less flexible treatment of the shape of the case A and B contours. Such plots form a convenient basis for quickly evaluating the potential of a proposed parameter to correlate both case A and B data, as outlined in Ref 35.

A similar propagation law, consistent with Basquin's Law for uniaxial and torsional loading, was introduced by McDowell and Berard (Ref 35) for predominantly HCF conditions (finite life). The exponent on crack length differed significantly from the LCF case. By superimposing the resulting shear strain-life relations obtained by independent integration of the LCF and HCF relations, a maximum shear strain-life relation was developed to correlate the fatigue life to a crack of length 1 mm. As seen in Fig. 4, the McDowell-Berard method compares well with the Fatemi-Socie-Kurath approach in Eq 16 for completely reversed fatigue of the shear-dominated materials 1045 steel and Inconel 718 under various stress states, ranging from torsion to uniaxial to internal/external pressure. McDowell and Poindexter (Ref 58) extended this approach to address the dependence of the crack propagation rate on stress state and on crack length normalized by transition crack length, which demarcates microstructurally small and physically small crack behavior. Figure 8 shows the key differences between uniaxial and torsional crack propagation as a function of the number of cycles to a crack of length 1 mm for 1045 steel, as described by the McDowell-Poindexter model, based on fitting the data of Socie in Fig. 6.

Some common themes are evident in these critical plane theories. First, the effect of maximum cyclic shear strain is moderated by an additional influence of the normal stress or strain to this plane. This modification is based on the premise that the normal stress assists mode II propagation by opening the crack, thereby reducing crack face asperity or wake plasticity interactions. The notion of constraint may additionally be introduced, as in Eq 20, to reflect the effect of the hydrostatic stress on the local cyclic slip and damage processes ahead of the crack tip, analogous to constraint effects on damage evolution ahead of long cracks. Second, the product of stress and strain ranges in Eq 16 and 20 (and Eq 13, for that matter) is similar to that of EPFM ΔJ-integral for long cracks, which relates to crack-tip opening displacements.

It is clear from the foregoing discussion that the correlation of small crack growth in multiaxial fatigue for finite life may be framed in terms of a selected parameter for a driving force (Ref 7), provided that key elements are present.

In general, forms such as

$$\frac{da}{dN} = L_1 \left(\Delta \gamma_{max}, \sigma_n^{max} \text{ or } \varepsilon_n^{max}, \rho, \frac{a}{\beta} \right)$$

(Eq 29)

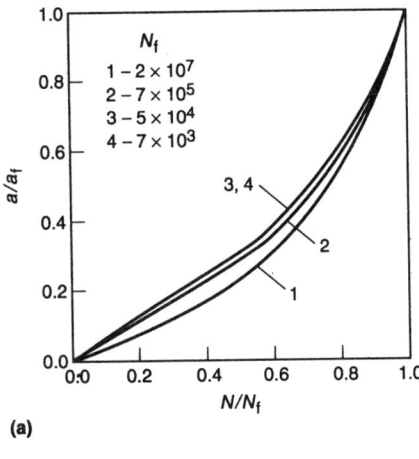

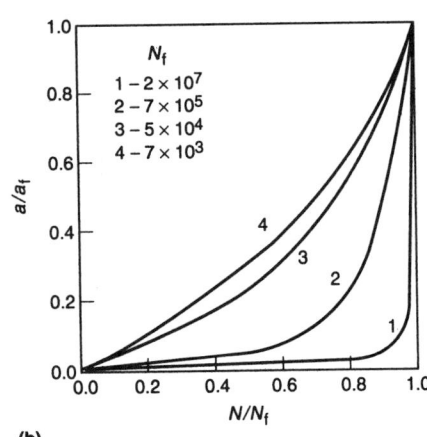

Fig. 8 Predicted nonlinear growth of microcracks for 1045 steel for four different constant-amplitude fatigue lives. Completely reversed (a) torsional fatigue and (b) uniaxial fatigue. Source: Ref 58

for stage I growth, or

$$\frac{da}{dN} = L_{II}\left(\sigma_n^{max}\frac{\Delta\varepsilon_n}{2}, \rho, \frac{a}{\beta}\right) \qquad (Eq\ 30)$$

for stage II growth appears to adhere to these requirements in the microstructurally sensitive regime, where β represents microstructure barrier length scales. For example, Ogata et al. (Ref 59) employed the relation

$$\frac{da}{dN} = C\left[(\Delta\gamma_{max}/2)^2 + B(\Delta\varepsilon_n)^2\right]a \qquad (Eq\ 31)$$

to correlate the propagation behavior of cracks from 100 to 1000 μm in length for in-phase and out-of-phase LCF of austenitic stainless steel at high temperature, where C and B are constants. This may be recognized as the incorporation of the Brown and Miller parameter in Eq 15 as the driving force for propagation, with a weak dependence on crack length, consistent with the data in Fig. 6 in the LCF regime. Fewer detailed studies of small crack propagation exist for the HCF case. A micromechanical or "first-principles" construction of specific forms of Eq 29 and 30 remains an open issue in multiaxial fatigue.

Fig. 9 Orientation and magnitude of stress normal to one of the two planes of maximum shear for (a) uniaxial and (b) torsional cases

Additional Considerations for Multiaxial Fatigue Life Prediction

Mean Stress Effects. Particularly under HCF conditions, mean stresses play a key role in fatigue. Even for proportional loading, the correlation of multiaxial mean stress effects is challenging. It is generally observed that torsional mean stresses do not significantly affect fatigue crack "initiation" life, whereas mean normal stresses have a potentially strong effect (Ref 6). Consequently, the mean value of the equivalent stress $(\bar{\sigma})$ is not a very useful quantity for correlation of mean stress effects, even under proportional loading conditions. Likewise, the mean value of hydrostatic stress has been used prominently in HCF parameters (e.g., Eq 10), but is does not isolate the effects of mean stress normal to the plane of stage I or II cracks, as discussed above.

Within the context of the critical plane theory, one can readily interpret common observations regarding mean stress effects under uniaxial and torsional loading conditions. Mean shear stress in torsional fatigue does not result in mean normal stress on the plane(s) of stage I crack propagation (maximum shear). Figure 9 shows the shear plane orientation of stage I microcracks for cases of completely reversed uniaxial and torsional loading. From a macroscopic viewpoint, the stress amplitude normal to the plane of the stage I microcrack in torsional loading is zero, whereas the stress amplitude normal to the shear plane in the uniaxial case is $\Delta\sigma/4$. For the same range of shear stress driving the mode II growth of the microcrack, the tensile normal stress in the uniaxial

case promotes opening behavior of the stage I crack. This results in a significantly lower life for a given maximum shear stress or effective stress amplitude in completely reversed uniaxial loading as compared to shear, in agreement with experiments such as those in Fig. 1. The effects of tensile mean stress across the crack plane may therefore be understood in terms of an enhanced contribution of mode I opening, as well as an intensification of mode II due to reduction of crack face interference effects induced by local plasticity, crack surface roughness, or a combination of these. It is interesting to note that the fracture mechanics treatment of the fatigue crack propagation of long cracks has long recognized the important role of plasticity-induced closure (Ref 60, 61), as well as that of various other shielding mechanisms that affect the crack-tip driving forces. In contrast, classical crack initiation approaches such as that of Basquin's law for HCF have been modified in somewhat ad hoc fashion to reflect the dependencies on mean stress. In the presence of multiaxial stress states, these ad hoc modifications have adopted many forms, with general recognition of the importance of mean normal stresses in contrast to mean shear stresses.

To account for mean stress effects, critical plane approaches for stage I microcrack propagation may employ the mean normal stress across the plane of maximum alternating shear strain (Ref 35) or peak normal stress to this plane, as in Eq 16 or Eq 29. This form of mean stress dependence correlates complex mean stress experiments rather well (Ref 6, 35, 62, 63). McDowell and Berard (Ref 35) suggested a form that reduces to

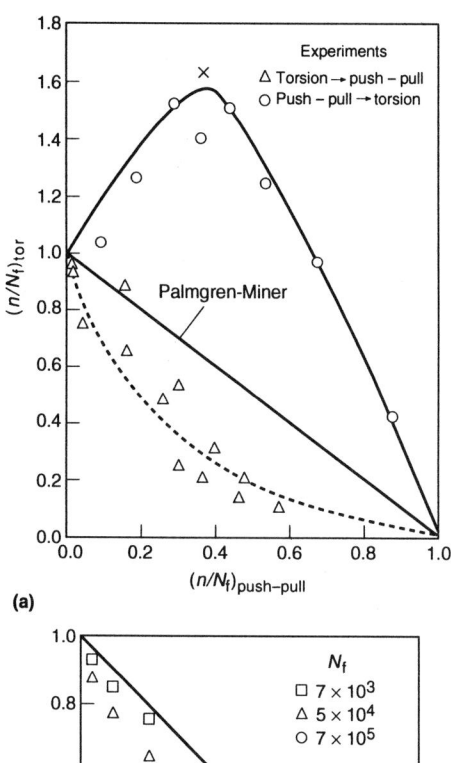

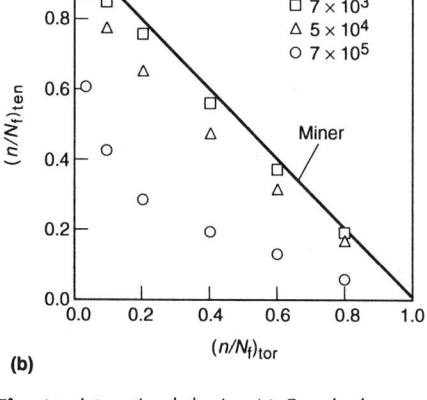

Fig. 10 Interaction behavior. (a) Completely reversed torsion followed by uniaxial push-pull and vice-versa loading sequences. Source: Ref 29. (b) Completely reversed torsion followed by uniaxial push-pull for three different constant-amplitude fatigue lives for 1045 steel, based on the propagation curves shown in Fig. 8, where N_f is the same for the torsional and uniaxial stress amplitudes of each sequence. Source: Ref 58

conventional mean stress approaches under pure uniaxial and pure torsional loading. It is somewhat more difficult to incorporate mean stress effects in plastic work approaches in a physically meaningful manner (Ref 24, 64). This is also the case for effective stress- or strain-based theories of multiaxial fatigue.

The driving force may be modified to include effects of plasticity-induced closure, in analogy to the concepts of ΔK_{eff} or ΔJ_{eff} used for correlation of crack growth rate of long cracks or for short cracks in stage I in the presence of very fine microstructure. However, available models and experimental data (e.g., Ref 61) indicate that small/short cracks are open over nearly the entire stress range under very high cyclic tensile strain (LCF) conditions normal to the microcrack. The HCF torsional mean stress case is not as well understood or characterized for stage I small crack growth, because the interference of crack

faces plays an increasingly strong role in mode II and III dominated growth.

Stress Amplitude Sequence Effects. Fatigue life prediction under variable loading histories is of great practical importance. The growth of microstructurally and physically small cracks with cycles differs between uniaxial and torsional loading, as discussed in reference to Fig. 6. For long fatigue lives, the crack length versus N relation may be extremely nonlinear, particularly for uniaxial fatigue, whereas it can be nearly linear under LCF conditions. Figure 8 shows the differences in the nature of propagation as a function of cycles for 1045 steel, as inferred from the data in Fig. 6 (Ref 58). This leads to strong amplitude sequence effects, particularly in uniaxial fatigue. In contrast, the crack length versus N relation in torsional fatigue is more nearly linear, and amplitude sequence effects are less pronounced. These phenomena are likely largely related to differences in crack face interference effects between uniaxial HCF and shear-dominated stage I growth, with little driving force for crack opening (e.g., torsional fatigue). These interference or shielding effects apparently scale quite differently with crack length and effective strain amplitude in uniaxial and torsional fatigue.

Sequences of Stress State. Sequences of stress state generate potentially strong history effects. A relevant example is that of combined tension and torsion sequences of thin-walled tubular specimens, as discussed by Miller (Ref 29). A sequence of torsional cycling followed by axial cycling results in a lower lifetime than would be anticipated on the basis of a linear damage rule such as Miner's rule, as shown in Fig. 10. In contrast, uniaxial push-pull followed by torsion results in a significant extension of life relative to the linear rule. Stage I microcracks formed in the torsional cycling effectively propagate as stage II cracks during subsequent uniaxial loading, resulting in a shorter life than continued cycling in torsion. On the other hand, uniaxial loading forms stage I microcracks roughly along 45° planes to the surface, and subsequent torsion is less effective in driving these cracks in stage I or stage II regimes.

Two conclusions are as follows. First, the orientation of crack systems formed under a specific loading condition depends on the applied stress state, and it is relevant to the prediction of fatigue life. Second, standard fatigue crack initiation approaches would be unsuitable, in general, for such sequences because they do not specify an orientation for microcrack propagation. Both of these observations point to the applicability of concepts involving propagation of small cracks along critical planes.

A more general type of nonproportional loading history involves rotation of the principal stresses (strains) or nonproportional variation of components of the stress (strain) *during* each cycle (Ref 33, 65, 66). Under LCF conditions, out-of-phase sinusoidal axial-torsional cycling of tubular specimens may lead to a decreased fatigue crack initiation life relative to in-phase (proportional) loading for ductile metals (Ref 33). For

HCF, though, the opposite may be true. In such cases, it is particularly important to employ a suitable incremental cyclic plasticity model (Ref 1-3) to estimate the ranges of shear strain and normal stress in the material on various planes. As discussed by Chu et al. (Ref 67), the critical plane can then be selected, typically associated with the maximum value of the damage parameter (e.g., Eq 15 or 16) over the cycle.

There are complexities associated with defining and counting cycles under conditions of general nonproportional loading, because the normal stress to the plane of maximum shear strain range may vary independently of the shear strain (Ref 66, 67). Further work is necessary to clarify a life estimation methodology for such cases.

Fatigue Limit in Multiaxial HCF. If subgrain-scale small cracks cannot bypass strong barriers at the microstructural scale such as grain boundaries, then a fatigue limit results (Ref 29, 30, 43, 44) at long lives (order of 10^6 to 10^7 cycles and beyond). Likewise, elastic shakedown or cessation of cyclic microplastic flow may occur due to the heterogeneity of yielding among grains, and this also leads to a fatigue limit (Ref 68). Naturally, this fatigue limit will depend on stress state as well, because the heterogeneity of microslip processes depends on constraint. The dependence of the fatigue limit on combined stress state has been studied extensively (e.g. Ref 11-13). It has long been known that the fatigue limit (threshold) in bending, for example, cannot be related to that in torsion using deviatoric plasticity arguments. At present, no general theory exists to relate the fatigue limits among different stress states, and empiricism is employed. The same comment applies to threshold stress intensity factors in modes I, II, and III, a related problem. Recent work by Dang-Van (Ref 68) offers some promise in predicting stress state dependence of the HCF fatigue limit by evaluating a local critical slip plane failure criterion of Mohr-type analogous to Eq 10 within grains embedded in a polycrystalline orientation distribution of grains. Such microstructure/stress state couplings are essential to understand the phenomenon. Indeed, it is clear that the fatigue limit is dependent on stress state, including the mix of shear and normal stress, level of triaxiality, and so on.

It is of paramount importance to recognize the role of heterogeneity of microstructure and free surface proximity effects on the propagation of crack-like defects in polycrystals. Fatigue limit(s) for nonpropagating cracks should be consistent with the notion of threshold(s) for small fatigue crack propagation. It is commonly observed that the threshold for microstructurally small cracks is less than that of long cracks. It should be emphasized that such small crack thresholds and fatigue limits may in general be eradicated by overloads that drive the crack past barriers.

REFERENCES

1. J.L. Chaboche, Constitutive Equations for Cyclic Plasticity and Cyclic Viscoplasticity, *Inter-*

national *Journal of Plasticity,* Vol 5 (No. 3), 1989, p 247

2. N. Ohno, Recent Topics in Constitutive Modeling of Cyclic Plasticity and Viscoplasticity, *Appl. Mech. Rev.,* Vol 43 (No. 11), 1990, p 283-295

3. D.L. McDowell, Multiaxial Effects in Metallic Materials, ASME AD, Vol 43, *Durability and Damage Tolerance,* A.K. Noor and K.L. Reifsnider, Ed., 1994, p 213-267

4. E. Krempl, The Influence of State of Stress on Low Cycle Fatigue of Structural Materials: A Literature Survey and Interpretive Report, STP 549, ASTM, 1974

5. M. Brown and K.J. Miller, Two Decades of Progress in the Assessment of Multiaxial Low-Cycle Fatigue Life, *Low Cycle Fatigue and Life Prediction,* STP 770, C. Amzallag, B. Leis, and P. Rabbe, Ed., ASTM, 1982, p 482-499

6. D.F. Socie, Critical Plane Approaches for Multiaxial Fatigue Damage Assessment, *Advances in Multiaxial Fatigue,* STP 1191, D.L. McDowell and R. Ellis, Ed., ASTM, 1993, p 7-36

7. D.F. Socie, C.T. Hau, and D.W. Worthem, Mixed Mode Small Crack Growth, *Fatigue Fract. Engng. Mater. Struct.,* Vol 10 (No. 1), 1987, p 1-16

8. D. Socie, Multiaxial Fatigue Damage Models, *ASME J. Engng. Mater. Techn.,* Vol 109, 1987, p 293-298

9. J. Bannantine and D. Socie, Observations of Cracking Behavior in Tension and Torsion Low Cycle Fatigue, *Low Cycle Fatigue,* STP 942, H.D. Solomon, G.R. Halford, L.R. Kaisand, and B.N. Leis, Ed., 1988, p 899-921

10. B.P. Haigh, *Reports of the British Association for the Advancement of Science,* 1923, p 358-368

11. H.J. Gough and H.V. Pollard, The Strength of Metals under Combined Alternating Stresses, *Proc. Inst. Mech. Engr.,* Vol 131 (No. 3). 1935, p 3-54

12. H.J. Gough, H.V. Pollard, and W.J. Clenshaw, Some Experiments on the Resistance of Metals under Combined Stress, *Aeronautical Research Council Reports and Memoranda No. 2522,* Ministry of Supply, HMSO, London, 1951

13. G. Sines, Failure of Materials under Combined Repeated Stresses with Superimposed Static Stresses, *NACA Technical Note 3495,* NACA, 1955

14. H.O. Fuchs, *Fatigue Engng. Mater. Struct.,* Vol 2, 1979, p 207-215

15. G. Libertiny, Short Life Fatigue under Combined Stresses, *J. Strain Anal.,* Vol 2 (No. 1), 1967, p 91-95

16. E.A. Davis and F.M. Connelly, Stress Distribution and Plastic Deformation in Rotating Cylinders of Strain-Hardening Materials, *ASME J. Appl. Mech.,* 1959, p 25-30

17. M. Manjoine, Damage and Failure at Elevated Temperature, *ASME J. Press. Ves. Techn.,* Vol 105, 1983, p 58-62

18. S.S. Manson and G.R. Halford, Multiaxial Low-Cycle Fatigue of Type 304 Stainless Steel, *ASME J. Engng. Mater. Techn.,* 1977, p 283-285

19. S.Y. Zamrik, M. Mirdamadi, and D.C. Davis, A Proposed Model for Biaxial Fatigue Analysis Using the Triaxiality Factor Concept, *Advances in Multiaxial Fatigue,* STP 1191, D. L. McDowell and R. Ellis, Ed., ASTM, 1993, p 85-106

20. G.R. Halford and J. Morrow, *Proc. ASTM,* Vol 62, 1962, p 695-709

21. Y.S. Garud, A New Approach to the Evaluation of Fatigue under Multiaxial Loading, *Proc. Symp. on Methods for Predicting Material Life in Fatigue,* W.J. Ostergren and J.R. Whitehead, Ed., ASME, 1979, p 247-264

22. F. Ellyin, A Criterion for Fatigue under Multiaxial States of Stress, *Mechanics Research Communications,* Vol 1, 1974, p 219-224

23. F. Ellyin and K. Golos, Multiaxial Fatigue Damage Criterion, *ASME J. Engng. Mater. Techn.,* Vol 110, 1988, p 63-68

24. F. Ellyin and D. Kujawski, A Multiaxial Fatigue Criterion Including Mean Stress Effect, *Advances in Multiaxial Fatigue,* STP 1191, D.L McDowell and R. Ellis, Ed., ASTM, 1993, p 55-66

25. F.B. Stulen and H.N. Cummings, A Failure Criterion for Multiaxial Fatigue Stresses, *Proc. ASTM,* Vol 54, 1954, p 822-835

26. J.J. Guest, *Proc Instn. Automobile Engrs.,* Vol 35, 1940, p 33-72

27. W.N. Findley, A Theory for the Effect of Mean Stress on Fatigue of Metals under Combined Torsion and Axial Load or Bending, *J. Engng. Industry,* 1959, p 301-306

28. P.J.E. Forsyth, A Two-Stage Process of Fatigue Crack Growth, *Proc. Symp. on Crack Propagatoin,* Cranfield, 1971, p 76-94

29. K.J. Miller, Metal Fatigue—Past, Current and Future, *Proc. Inst. Mech. Engrs.,* Vol 205, 1991, p 1-14

30. K.J. Miller, Materials Science Perspective of Metal Fatigue Resistance, *Mater. Sci. Techn.,* Vol 9, 1993, p 453-462

31. M. Brown and K.J. Miller A Theory for Fatigue Failure under Multiaxial Stress-Strain Conditions, *Proc. Inst. Mech. Engr.,* Vol 187 (No. 65), 1973 p 745-755

32. R. Lohr and E. Ellison, A Simple Theory for Low Cycle Multiaxial Fatigue, *Fatigue Engng. Mater. Struct.,* Vol 3, 1980, p 1-17

33. A. Fatemi and D. Socie, A Critical Plane Approach to Multiaxial Fatigue Damage Including Out of Phase Loading, *Fatigue Fract. Engng. Mater. Struct.,* Vol 11 (No. 3), 1988, p 145-165

34. A. Fatemi and P. Kurath, Multiaxial Fatigue Life Predictions under the Influence of Mean Stress, *ASME J. Engng. Mater. Tech.,* Vol 110, 1988, p 380-388

35. D.L. McDowell and J.-Y. Berard, A ΔJ-Based Approach to Biaxial Fatigue, *Fatigue Fract. Engng. Mater. Struct.,* Vol 15 (No. 8), 1992, p 719-741

36. R.N. Smith, P. Watson, and T.H. Topper, A Stress-Strain Parameter for Fatigue of Metals, *J. Mater.,* Vol 5 (No. 4), 1970, p 767-778

37. M. Sarfarazi and S. Ghosh, Microfracture in Polycrystalline Solids, *Engng. Fracture Mech.,* Vol 27 (No. 3), 1987, p 257-267

38. G. Venkataraman, T. Chung, Y. Nakasone, and T. Mura, Free-Energy Formulation of Fatigue Crack Initiation along Persistent Slip Bands: Calculation of S-N Curves and Crack Depths, *Acta Met. Mater.,* Vol 38 (No. 1), 1990, p 31-40

39. G. Venkataraman, Y. Chung, and T. Mura, Application of Minimum Energy Formalism in a Multiple Slip Band Model for Fatigue, Parts I and II, *Acta Met. Mater.,* Vol 39 (No. 11), 1991, p 2621-2638

40. R.C. McClung, K.S. Chan, S.J. Hudak, Jr., and D.L. Davidson, Analysis of Small Crack Behavior for Airframe Applications, *FAA/NASA Int. Symp. on Advanced Structural Integrity Methods for Airframe Durability and Damage Tolerance,* NASA CP 3274, Part 1, 1994, p 463-479

41. S. Suresh, *Fatigue of Materials,* Cambridge Solid State Science Series, Cambridge University Press, 1991

42. J.C. Newman, Jr., A Review of Modelling Small-Crack Behavior and Fatigue-Life Predictions for Aluminum Alloys, *Fatigue Fract. Engng. Mater. Struct.,* Vol 17 (No. 4), 1994, p 429-439

43. K. Tanaka, Y. Akiniwa, Y. Nakai, and R.P. Wei, Modelling of Small Fatigue Crack Growth Interacting with Grain Boundary, *Engng. Fracture Mech.,* Vol 24 (No. 6), 1986, p 803-819

44. K. Tanaka, Short-Crack Fracture Mechanics in Fatigue Conditions, *Current Research on Fatigue Cracks,* T. Tanaka, M. Jono, and K. Komai, Ed., *Current Japanese Materials Research,* Vol 1, Elsevier, 1987, p 93-117

45. A. Navarro and E.R. De Los Rios, A Model for Short Fatigue Crack Propagation with an Interpretation of the Short-Long Crack Transition, *Fatigue Fract. Engng. Mater. Struct.,* Vol 10 (No. 2), 1987, p 169-186

46. M.W. Brown, K.J. Miller, U.S. Fernando, J.R. Yates, and D.K. Suker, Aspects of Multiaxial Fatigue Crack Propagation, *Proc. Fourth Int. Conf. on Biaxial/Multiaxial Fatigue,* Vol I, SF2M/ESIS, May 31-June 3 1994, p 3-16

47. H.S. Lamba, The J-Integral Applied to Cyclic Loading, *Engng. Fracture Mech.,* Vol 7, 1975, p 693

48. N.E. Dowling and J.A. Begley, *Mechanics of Crack Growth,* STP 590, ASTM, 1976, p 82-103

49. K. Tanaka, The Cyclic J-Integral as a Criterion for Fatigue Crack Growth, *Int. J. Fract.,* Vol 22, 1983, p 91-104

50. H. Nisitani, Behavior of Small Cracks in Fatigue and Relating Phenomena, *Current Research on Fatigue Cracks,* T. Tanaka, M. Jono, and K. Komai, Ed., *Current Japanese Materials Research,* Vol 1, Elsevier, 1987, p 1-26

51. S. Harada, Y. Murakami, Y. Fukushima, and T. Endo, Reconsideration of Macroscopic Low Cycle Fatigue Laws through Observation of Microscopic Fatigue Process on a Medium Carbon Steel, *Low Cycle Fatigue,* STP 942, H.D. Solomon et al., Ed., ASTM, 1988, p 1181-1198

52. T. Hoshide, M. Miyahara, and T. Inoue, Elastic-Plastic Behavior of Short Fatigue Cracks in

Smooth Specimens, *Basic Questions in Fatigue: Volume I,* STP 924, J.T. Fong and R.J. Fields, Ed., ASTM, 1988, p 312-322

53. G. Hau, N. Alagok, M.W. Brown, and K.J. Miller, Growth of Fatigue Cracks under Combined Mode I and Mode II Loads, *Multiaxial Fatigue,* STP 853, K.J. Miller and M.W. Brown, Ed., ASTM, 1985, p 184-202

54. T. Hoshide and D. Socie, Mechanics of Mixed Mode Small Fatigue Crack Growth, *Engng. Fract. Mech.,* Vol 26 (No. 6), 1987, p 842-850

55. C.F. Shih and J.W. Hutchinson, Fully Plastic Solution and Large Scale Yielding Estimates for Plane Stress Crack Problems, *ASME J. Engng. Mater. Techn.,* Vol 98, 1976, p 289-295

56. J.-Y. Berard and D.L. McDowell, A Δ*J* Based Approach to Biaxial Low-Cycle Fatigue of Shear Damaged Materials, *Fatigue under Biaxial and Multiaxial Loading,* ESIS10, K. Kussmaul, D. McDiarmid and D. Socie, Ed., Mech. Engng. Publ., London, 1991, p 413-431

57. J.-Y. Berard, D.L. McDowell, and S.D. Antolovich, Damage Observations of a. Low Carbon Steel under Tension-Torsion Low-Cycle Fatigue, *Advances in Multiaxial Fatigue,* STP 1191, D.L. McDowell and R. Ellis, Ed., ASTM, 1993, p 326-344

58. D.L. McDowell and V. Poindexter, Multiaxial Fatigue Modelling Based on Microcrack Propagation: Stress State and Amplitude Effects., *Proc. Fourth. Int. Conf. on Biaxial/Multiaxial Fatigue,* Vol I, SF2M/ESIS, 1994, p 115-130

59. T. Ogata, A. Nitta, and J.J. Blass, Propagation Behavior of Small Cracks in 304 Stainless Steel under Biaxial Low-Cycle Fatigue at Elevated Temperature, *Advances in Multiaxial Fatigue,* STP 1191, D.L. McDowell and R. Ellis, Ed., ASTM, 1993, p 313-325

60. R.C. McClung and H. Sehitoglu, Closure Behavior of Small Cracks under High Strain Fatigue Histories, STP 982, ASTM, 1988, p 279-299

61. R.C. McClung and H. Sehitoglu, Closure and Growth of Fatigue Cracks at Notches, *ASME J. Engng. Mater. Techn.,* Vol 114, 1992, p 1-7

62. D. Socie, L. Waill, and D. Dittmer, Biaxial Fatigue of IN718 Including Mean Stress Effects, *Multiaxial Fatigue,* STP 853, K.J. Miller and M.W. Brown, Ed., ASTM, 1985, p 463-481

63. C.H. Wang and K.J. Miller, The Effects of Mean and Alternating Shear Stresses on Short Fatigue Crack Growth Rates, *Fatigue Fract. Engng. Mater. Struct.,* Vol 15 (No. 12), 1992, p 1223-1236

64. B. Leis, An Energy-Based Fatigue and Creep-Fatigue Damage Parameter, *ASME J. Press. Ves. Techn.,* Vol 99 (No. 4), 1977, p 524-533

65. J.L. Koch, Proportional and Nonproportional Fatigue of Inconel 718, *Mater. Engng.-Mech. Behavior,* Report 121, University of Illinois, 1985

66. E. Jordan, M. Brown, and K.J. Miller, Fatigue under Severe Nonproportional Loading, *Multiaxial Fatigue,* STP 853, K.J. Miller and M.W. Brown, Ed., ASTM, 1985, p 569-585

67. C.-C. Chu, F.A. Conle, and J.J.F. Bonnen, Multiaxial Stress-Strain Modeling and Life Prediction of SAE Axle Shafts, *Advances in Multiaxial Fatigue,* STP 1191, D.L. McDowell and R. Ellis, Ed., ASTM, 1993, p 37-54

68. K. Dang-Van, Macro-Micro Approach in High Cycle Multiaxial Fatigue, *Advances in Multiaxial Fatigue,* STP 1191, D.L. McDowell and R. Ellis, Ed., ASTM, 1993, p 120-130

Factors Influencing Weldment Fatigue

F.V. Lawrence, S.D. Dimitrakis, and W.H. Munse, University of Illinois at Urbana-Champaign

THERE IS general agreement that the main factors influencing the fatigue life of a weldment are:

- Applied stress amplitude
- Mean and residual stresses
- Material properties
- Geometrical stress concentration effects
- Size and location of welding discontinuities

but there is often disagreement as to the relative importance of each.

This article is intended to help engineers understand why the fatigue behavior of weldments can be such a confusing and seemingly contradictory topic, and hopefully to clarify this complex subject. The endless variations in weldment geometry are a major source of difficulty and an alternative classification system will be suggested. For a given weld geometry, it will be concluded that the behavior of structural weldments depends to a rather large extent on the nature of the industrial application, that is, upon the size of the weldment and upon the quality of the welding and the post-welding procedures employed.

Scope and Sections in this Article. In what follows, the factors influencing the fatigue behavior of an individual weldment will be reexamined

using extensive experimental data and a computer model which simulates the fatigue resistance of weldments. In the next section, the process of fatigue in weldments will be discussed in general terms and the service conditions which favor long crack growth and the conditions which favor crack nucleation will be contrasted. Next, experimental data will be used to show the effect of weldment geometry on fatigue resistance. Several useful geometry classification systems will be compared. In the last section, a computer model will be employed to investigate the behavior of two hypothetical weldments: a discontinuity-containing ("Nominal") weldment and a discontinuity-free ("Ideal") weldment. As will be seen, these two weldments exhibit radically different fatigue behavior and thus are useful paradigms which will help engineers decide how their weldments will behave and how the fatigue resistance of their weldments might be improved.

Metallic Fatigue in Weldments

As with any notched metal component, the process of fatigue in weldments can be divided into three periods: crack nucleation, the develop-

ment and growth of a short crack (stage I), and the growth of a dominant (long) crack to a length at which it either arrests or causes fracture (stage II in Fig. 1).

The boundaries between these periods are ill-defined. Nonetheless, it is useful to think of the total fatigue life of a notched metal component or a weldment (N_T) as the sum of three life periods: fatigue crack nucleation (N_N), short (or stage I) crack growth (N_{P1}), and long (or stage II) crack growth (N_{P2}):

$$N_T = N_N + N_{P1} + N_{P2} \qquad \text{(Eq 1)}$$

The relative contribution of each of these three periods to the fatigue life of weldments is controversial and appears to vary with the geometry of the weld and weldment, the size of the weldment, the nature of the residual stresses present, and the severity of the weld discontinuities existing in the weldment. In this article, it will be useful to imagine that there are two extreme kinds of weldments: "Nominal" weldments, which contain substantial (~0.1 in. depth) weld discontinuities; and "Ideal" weldments, which have blended weld toes and no substantial weld discontinuities (Fig. 2). As will be seen, the fatigue behavior of the

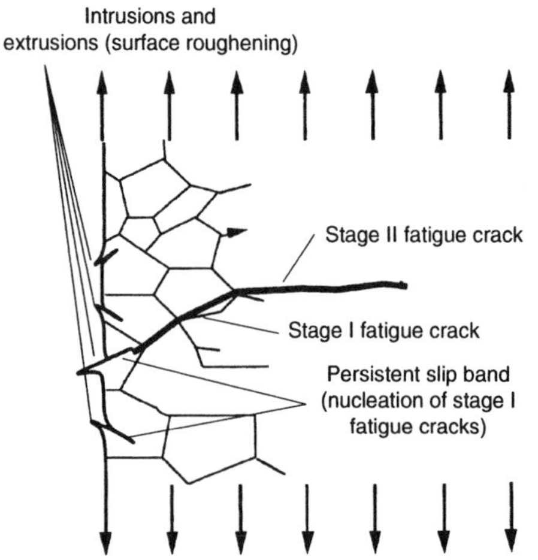

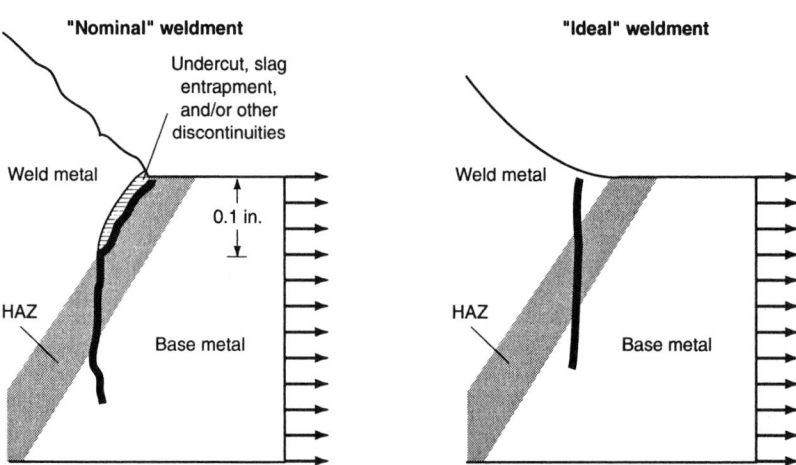

Fig. 1 Metallic fatigue. The stages of fatigue include cyclic slip (crack nucleation) and stage I and stage II crack growth.

Fig. 2 Conceptual drawing of fatigue crack initiation and growth at the toe of (left) a "Nominal" groove welded butt joint having a substantial (≈ 0.1 in. depth) weld discontinuity (slag entrapment) at the root of the critical notch (weld toe) and (right) an "Ideal" weldment with good wetting and no substantial discontinuity at the root of the critical notch. Only the right halves of these weldments are illustrated. In the case of the "Nominal" weldment the fatigue crack initiates at the tip of the preexisting discontinuity, that is, at a depth of ≈ 0.1 in. along the line of fusion; whereas in the case of the "Ideal" weldment, the fatigue crack is presumed to initiate at the weld toe, possibly in weld metal.

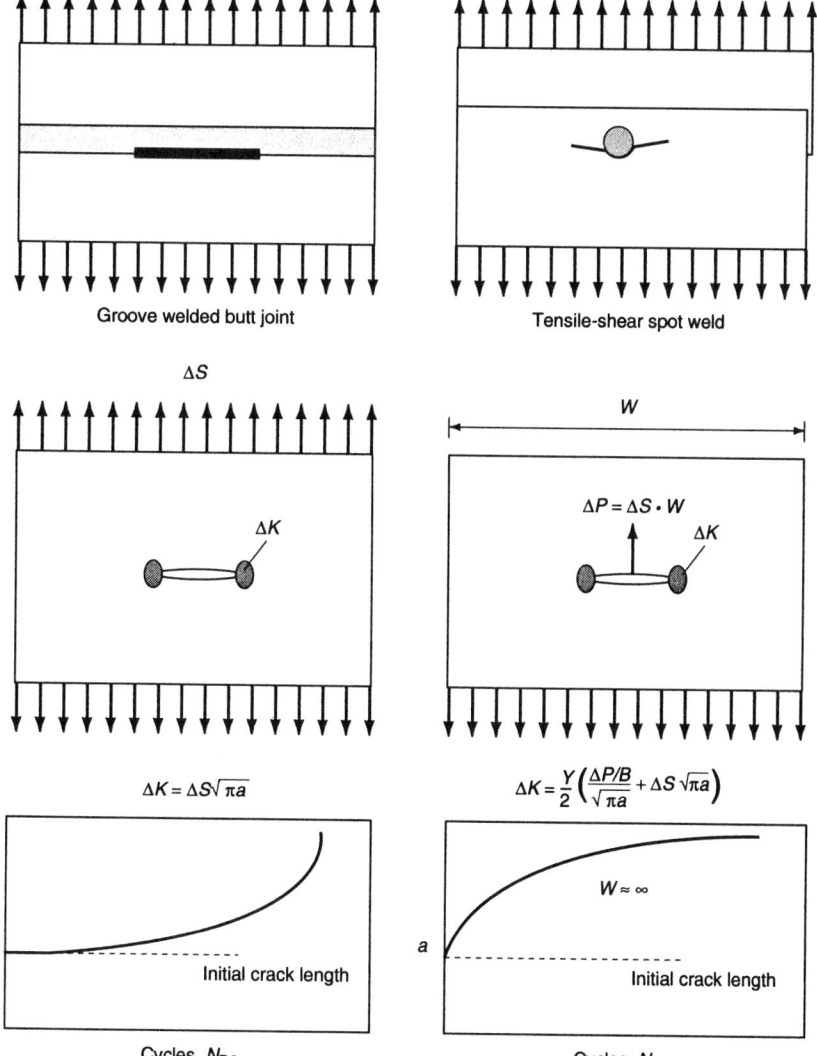

Fig. 3 Two radically dissimilar patterns of Stage II crack growth in weldments. The crack geometry and load path in the groove welded butt joint (top left) is similar to the center cracked panel for which the stress intensity factor increases with crack growth; whereas, the crack geometry and load path in the tensile shear spot weld (top right) is similar to the loading pattern for a bolt or rivet for which the stress intensity factor may decrease with crack growth. The difference between the two weldments favors the acceleration of fatigue crack growth with increasing crack length in the case of the groove welded butt joint and the possible development of nonpropagating cracks in the case of the tensile shear spot weld.

"Nominal" and "Ideal" weldments differs greatly.

Conditions Leading to the Dominance of Long Crack Growth (N_{P2}). For many reasons, stage II crack growth generally dominates the fatigue life of a weldment, while the periods devoted to crack nucleation (N_N) and early crack growth (N_{P1}) are generally relatively short. Engineers for whom a single failure would be catastrophic and who are forced to use low-quality welding procedures must by necessity adopt a very pessimistic view regarding the fatigue life of weldments and make the rather conservative assumption that:

$$N_T \approx N_{P2} \qquad \text{(Eq 2)}$$

The basic geometry and/or loading of some weldments leads to a very desirable phenomenon

in which a stage II crack slows down rather than accelerates as the crack lengthens. Whenever this occurs, the growth [by the Paris power law $da/dN = C(\Delta K)^n$] of long cracks (N_{P2}) can be a major fraction of their fatigue life. Such weldments may never fail but rather may develop long, slow-growing fatigue cracks (Fig. 3). Most weldments have several sites of stress concentration.

Corrosion fatigue is another phenomenon that diminishes the relative importance of crack nucleation (N_N) and small crack growth (N_{P1}) in weldments. Finally, variable load histories containing many large, damaging events may greatly shorten the fatigue life devoted to N_N and N_{P1}.

Conditions Favoring Crack Nucleation and Early Crack Growth. While the deleterious effects of weld discontinuities, corrosion fatigue, and some variable load histories can diminish the importance of N_N and N_{P1} in weldments, one can also adopt an opposite, more optimistic view of

the fatigue life of weldments in which N_N and N_{P1} can be a major part of the fatigue life of a weldment and in which the fatigue life of such an "Ideal" weldment can be greater than N_{P2}. "Flux-less" fusion welding processes such as gas-metal arc welding (GMAW) or gas-tungsten arc welding (GTAW) are capable of producing large weldments in which weld discontinuities at the root of the critical notch are small or even nonexistent. It should be noted that for a weld discontinuity to control the fatigue resistance of a weldment, it must be located at the root of the critical notch so that the worst case can occur, in which the stress concentrations of both the critical notch and the weld discontinuity interact. The fact that fatigue invariably begins at the root of the critical notch reduces the likelihood of randomly distributed weld discontinuities participating in fatigue crack nucleation and early crack growth, which are constrained to the root of the critical notch (i.e., the ripple, the toe, or the root of a weld-ment). It is also possible that the weld reinforcement may be sufficiently irregular that the worst notch can be located in the weld metal; however, this situation can be avoided by proper welding.

All welding processes can produce either "Ideal" welds free of discontinuities or "Nominal" welds with a 0.1 in. crack-like discontinuity. For example, welding processes such as resistance spot welding produce weldments in which large discontinuities are not found. Thus, high-quality structural welds and welds such as resistance spot welds may not contain large discontinuities, and their behavior may approach that of an "Ideal" weldment. In some applications, however, highly stressed welds in critical locations are less likely (compared to the majority of the population of the welded components) to be considered discontinuity free and like the "Ideal" weldment. Many situations may not involve constant-amplitude or pseudo-constant-amplitude loading, and thus the concern about variable load histories may be a factor. However, welding procedures and postweld treatments can substantially improve the fatigue life of a weldment through increases in any or all of the life periods N_N, N_{P1}, and N_{P2}, and many applications may allow the assumption of "Ideal" welds and constant-amplitude conditions. In this circumstance, it is reasonable to think of the fatigue life of a weldment as approaching that of the "Ideal" weldment, as depending on N_N and N_{P1}, and unlike the "Nominal" weldment susceptible to large improvement.

Effects of Weldment Geometry

The Fatigue Behavior of 53 Structural Details. Some of the common structural details encountered in bridge, ship, and ground-vehicle construction have been catalogued by Munse et al. (Ref 1). The shapes of 53 structural details and variations of these details are shown in Fig. 4. The abbreviations used are given in Table 1, and further information regarding the 53 joints is given in Table 2. This catalogue begins with what would seem to be the simplest shapes and pro-

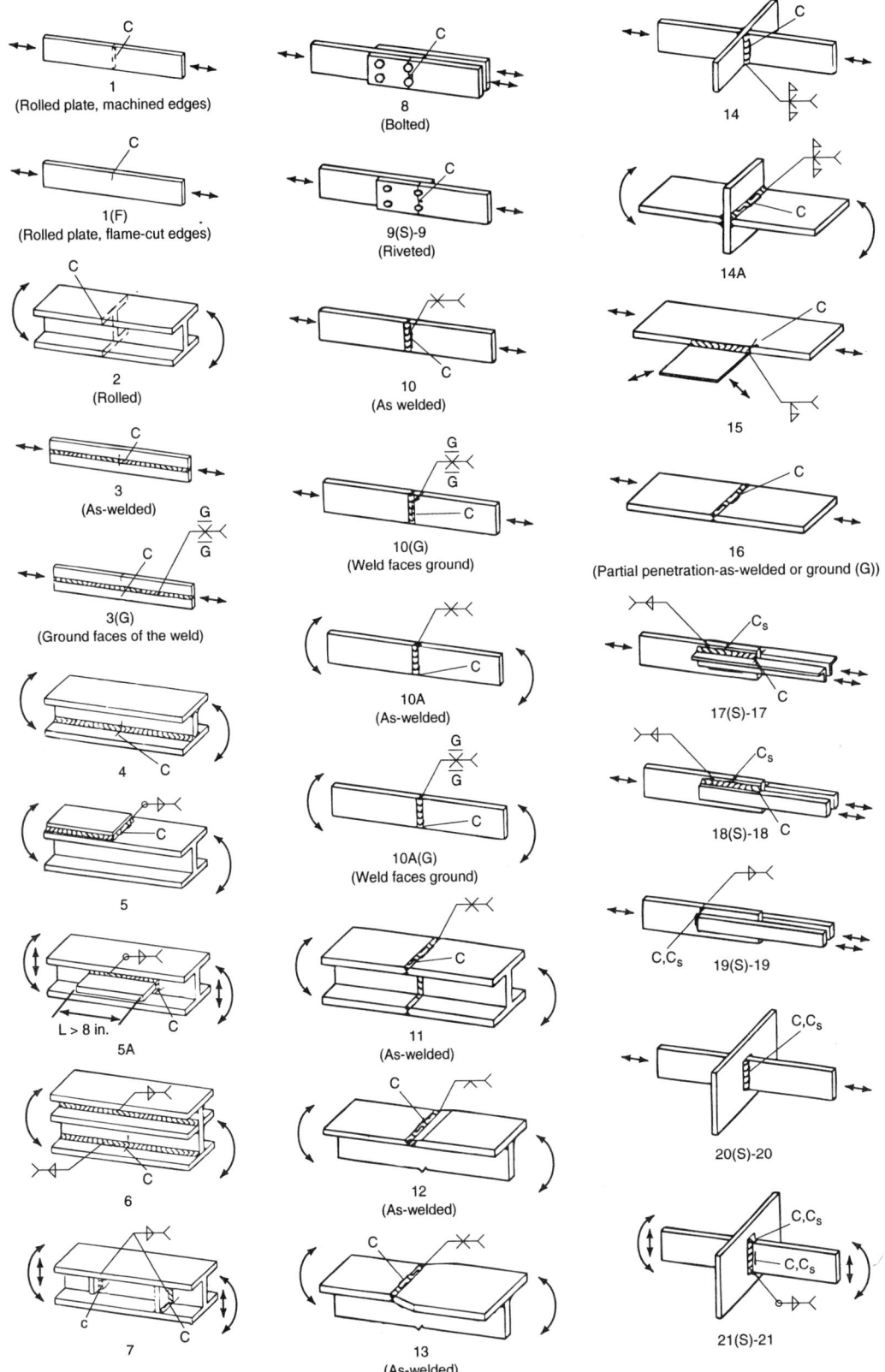

Fig. 4 Selected illustration of 28 details from the 53 structural weld details catalogued by Munse et al. in Ref 1. A complete list of the 53 details is given in Table 2.

ceeds toward the more complex. Some of the final geometries (e.g., #39) are complex weldments and should really be considered structures. Note that the classification system includes bolted and riveted joints (#8 and #9) and plug and spot welds (#27), which as discussed in the previous section behave in a fundamentally different way. Several details (#28 and #29) are not connections at all, simply notched components. For

Table 1 Abbreviations for weld details in Table 2

(F)	Flame cut edges
(G)	Weld ground
(B)	Bending stresses
(M)	Machined surfaces
(P)	Principal stresses
(S)	Shear stresses
A, B, C, ...	Additional description within the same detail number
$C\rightarrow$	Crack initiation site due to tensile stresses
$C_S\rightarrow$	Crack initiation site due to shear stresses
L	Length of intermittent weld
P	Pitch between two intermittent welds
R	Radius
t	Thickness of plate

this reason, this article at first refers to the items in the catalogue as structural details, but later focuses on the welded details.

The Influence of Structural Detail Geometry on Fatigue Strength. The mean strength data in Table 3 suggest that, after the applied stress range, detail geometry is the most important variable affecting a structural detail's fatigue life. The role of geometry can be better assessed if fatigue databank information is edited to suppress the effects of other variables, such as R-ratio and material strength. (*Note:* It is customary to group together fatigue data for all thicknesses, strengths, and R-ratios. This practice is inadvisable and leads to an unnecessarily large scatter in fatigue data information. All databanks should be restricted to a standard strength, R-ratio, and thickness.) In Table 3, the fatigue databank information for many of the structural details listed in Table 2 is reanalyzed and restricted to data for $R = 0$ tests and data for steels having yield strength less than 50 ksi (345 MPa). In several instances, the AISC classification of the joint was altered by this procedure.

Scatter of Structural Detail Fatigue Data Resulting from Classification Systems. The design stresses that an engineer must adopt are as much controlled by the scatter in the fatigue data as by the mean value of strength for a certain design life. Thus, the uncertainty (scatter) in weldment fatigue life is as or more important than the mean value. This scatter has two basic sources: "real" scatter, which results from the random nature of the fatigue variables controlling the fatigue resistance of a detail; and the contribution to the "apparent" scatter, which is an artifact of the classification system imposed. The simplest classification scheme is suggested by Fig. 4 and Table 2. Each detail shape is placed in a class by itself. However, as mentioned in the previous section, grouping together data for tests having different experimental conditions leads to artificially large values in standard deviation of the log of strength(s). Furthermore, as discussed below, the practice producing the greatest amount of apparent scatter is the use of broad classification systems in which details having only roughly similar fatigue resistances are grouped together. Munse and Ang (Ref 2) suggested that the effects of scatter on the desired or required reliability of a particular structural detail could be incorporated into the design procedure by calculating a

Table 2 List of weld details catalogued by Munse et al. in Ref 1

Detail number(a)	Detail description	Loading condition	Fatigue crack initiation site
1	Plain plate, machined edges	Axial	Corners
1(F)	Plain plate, flame-cut edges	Axial	Edges
2	Rolled I-beam	Bending	Corners
2A	Riveted I-beam	Bending	Holes
3	Longitudinally welded plate, as-welded	Axial	Ripple
3(G)	Longitudinally welded plate, weld ground	Axial	Corners or discontinuity
4	Welded I-beam, continuous weld	Bending	Ripple
4A	Welded I-beam, intermittent weld	Bending	End of weld
4B	Welded box, continuous weld	Bending	Ripple
4C	Welded box, intermittent weld	Bending	End of weld
5	I-beam with welded cover plate	Bending	Weld toe
5A	I-beam with welded plate to web	Bending, shear	Weld toe
6	Welded I-beam with longitudinal stiffeners welded to web	Bending	Ripple
7(B)	I-beam with welded stiffeners	Bending	Weld toe
7(P)	I-beam with welded stiffeners	Bending, shear	Weld toe
8	Double shear bolted lap joint	Axial	Holes
8A	Double shear riveted lap joint	Axial	Holes
9	Single shear riveted lap joint	Axial	Holes
10	Transverse butt joint, as-welded	Axial	Weld toe
10(G)	Transverse butt joint, weld ground	Axial	Weld
10A	Transverse butt joint, as-welded	In-plane bending	Weld toe
10A(G)	Transverse butt joint, weld ground	In-plane bending	Weld
11	Transverse butt welded I-beam, as-welded	Bending	Weld toe
11(G)	Transverse butt welded I-beam, weld ground	Bending	Weld
12	Flange splice (unequal thickness), as-welded	Bending	Weld toe
12(G)	Flange splice (unequal thickness), weld ground	Bending	Weld
13	Flange splice (unequal width), as-welded	Bending	Weld toe
13(G)	Flange splice (unequal width), weld ground	Bending	Weld
14	Cruciform joint	Axial	Weld corner
14A	Cruciform joint	Bending	Weld toe
15	Lateral attachment to plate edge	Axial	End of weld
16	Partial penetration butt weld, as-welded	Axial	Weld toe or weld
16(G)	Partial penetration butt weld, weld ground	Axial	Weld metal
17	Angle welded to plate, longitudinal weld only	Axial	End of weld
17(S)	Angle welded to plate, longitudinal weld only	Axial	Weld
17A(S)	Channel welded to plate, longitudinal weld only	Axial	Weld
17A	Channel welded to plate, longitudinal weld only	Axial	End of weld
18	Flat bars welded to plate, longitudinal weld only	Axial	End of weld
18(S)	Flat bars welded to plate, longitudinal weld only	Axial	Weld
19	Flat bars welded to plate, lateral welds only	Axial	Weld
19(S)	Flat bars welded to plate, lateral welds only	Axial	Weld
20	Cruciform joint	Axial	Weld toe
20(S)	Cruciform joint	Axial	Weld
21	Cruciform joint, 1/4 in. weld	In-plane bending	Weld toe
	Cruciform joint, 3/8 in. weld	Shear	Weld toe
21(S)	Cruciform joint, 1/4 in. weld	In-plane bending	Weld
	Cruciform joint, 3/8 in. weld	Shear	Weld toe
22	Attachment of stud to flange	Bending	Weld toe
23	Attachment of channel to flange	Bending	Weld toe
24 (2L < 4 in.)	Attachment of bar to flange	Bending	Weld toe
24A (L ≤ 2 in.)	Attachment of bar to flange	Bending	Weld toe
24B (4 in. < L < 8 in.)	Attachment of bar to flange	Bending	Weld toe
25	Lateral attachments to plate	Axial	Weld toe
25A	Lateral attachment to plate	Axial	Weld toe
25B	Lateral attachment to plate with stiffener	Axial	Weld toe or end of weld
26	Doubler plate welded to plate	Axial	Weld toe
27	Slot or plug welded double lap joint	Axial	End of weld nugget
27(S)	Slot or plug welded double lap joint	Axial	Weld nugget
27A	Spot welded single lap joint	Axial	End of weld nugget
27A(S)	Spot welded single lap joint	Axial	Weld nugget
28	Plain plate with drilled hole	Axial	Edge of hole
28(F)	Plain plate with flame-cut circular hole	Axial	Edge of hole
29	Plain plate with machined rectangular hole (R ≤ 1/4 in.)	Axial	Corner of hole
29R1	Plain plate with machined rectangular hole (1/4 in. < R ≤ 1/2 in.)	Axial	Corner of hole
29R2	Plain plate with machined rectangular hole (1/2 in. < R ≤ 1 in.)	Axial	Corner of hole
29(F)	Plate with flame-cut rectangular hole (R ≤ 1/4 in.)	Axial	Corner of hole
29(F) R1	Plain plate with flame-cut rectangular hole (1/4 in. < R ≤ 1/2 in.)	Axial	Corner of hole
29(F) R2	Plain plate with flame-cut rectangular hole (1/2 in. < R ≤ 1 in.)	Axial	Corner of hole
30	Longitudinal attachments to plate	Axial	Plate at end of weld
30A	Longitudinal attachments to plate	Bending	Plate at end of weld
31	Attachments of plate to edge of flange	Bending	Flange at end of weld
31A	Lateral attachment of plate to flange	Bending	Flange at weld toe
32	Groove welded attachment of radiused plate to edge of flange	Bending	Flange at end of weld
32A	Groove welded attachment of plate to edge of flange	Bending	Flange at end of weld
32B	Butt welded flange (unequal width)	Bending	Weld toe
32C	Butt welded flange (unequal width, radiused transition)	Bending	Weld toe

(continued)

reliability factor (R_F) that shifted the mean curve of a detail's *S-N* diagram downward by an amount (Fig. 5) that would guarantee a desired level of safety (or probability of failure):

$$\Delta S_{design} = \Delta S_{mean} (R_F) \qquad \text{(Eq 3)}$$

A relation between the reliability factor (R_F) shown in Fig. 5 and the coefficient of variation of the mean fatigue strength (Ω_S) is given in Eq 4. As suggested in the last section of this article, a typical value of R_F for a weldment is 0.7:

$$R_F = \exp\left[-2\sqrt{\ln(1 + \Omega_S^2)}\right] \qquad \text{(Eq 4)}$$

Table 3 shows that standardizing the databank information frequently alters the mean fatigue strength (ΔS at 10^6 cycles) and usually reduces the standard deviation in the log of fatigue strength for most structural details. The effects of standardizing the fatigue databank information on the scatter in the fatigue data for a given detail are plotted in the histograms of Fig. 6 and 7. Standardizing databanks greatly reduces the scatter in fatigue data for a given detail and consequently increases the allowable design stresses. (*Note:* The design stress range (ΔS_{design}) at a certain life (10^6 cycles) can be estimated from the mean fatigue strength (ΔS_{weld}) of a weldment at a given life by: log ΔS_{design} = log ΔS_{weld} − 2*s*, where *s* is the standard deviation of log ΔS.

The scatter in fatigue information is increased by grouping structural details into a small number of broad categories of decreasing fatigue resistance. The AISC weld category fatigue design method (Ref 3) and other similar approaches group the data for all strengths of steel, all *R*-ratios, and all "similar" structural detail geometries together into a single databank for each category. This practice of placing individual structural details into such broad classifications greatly increases the apparent scatter in fatigue data, leads to lower design stresses (Fig. 5), and obscures the effects of many variables that influence the fatigue life of weldments. The data in AISC categories A through F exhibit large scatter and force design stresses 40% or more lower than the mean

fatigue strength; that is, this practice results in a reliability factor (R_F) of around 0.4 rather than values of 0.9 to 0.6, which reflect the essential nature of weldments (Fig. 8) (Ref 4).

Table 2 (continued)

Detail number(a)	Detail description	Loading condition	Fatigue crack initiation site
33	Flat bars welded to plate, lateral and longitudinal welds	Axial	End of weld
33(S)	Flat bars welded to plate, lateral and longitudinal welds	Axial	Weld
34	Flat bars welded to plate, lateral and longitudinal welds	In-plane bending	End of weld
34(S)	Flat bars welded to plate, lateral and longitudinal welds	In-plane bending	Weld
35	Butt joint with backing bar	Axial	Weld toe
36	Welded beam with intermittent welds and cope hole in the web	Bending	End of weld or cope hole
36A	Welded beam with staggered intermittent welds	Bending	End of weld
37	Beam connection with sloping flanges	Bending	Weld toe or end of weld at cope hole
37(S)	Beam connection with sloping flanges	Shear	Weld
38	Beam connection with horizontal flanges	Bending	Weld toe
38(S)	Beam connection with horizontal flanges	Shear	Weld
39	Beam bracket without cope hole	Bending	Weld toe
39A	Beam bracket with round cope hole in web	Bending	Weld toe or end of weld at cope hole
39B	Beam bracket with straight cope hole in web	Bending	Weld toe or end of weld at cope hole
40	Interconnecting beams	Bending in perpendicular directions	Weld toe
41	Beam bracket	Axial	Weld toe
42	Lateral attachment of plate to plate with weld beads on both sides	Lateral (reversal)	Weld toe
42A	Lateral attachment of plate to plate with weld beads on both sides	Lateral one direction	Weld toe
42B	Lateral attachment of plate to plate with weld bead on one side	Lateral (reversal)	Weld root
42C	Lateral attachment of plate to plate with weld bead on one side	Lateral one direction toward the weld	Weld root
42D	Lateral attachment of plate to plate with weld bead on one side	Lateral one direction away from the weld	Weld toe away from the weld
42E	Lateral attachment of plate to plate with weld beads on both sides	Axial in attachment	Weld toe
43	Partial penetration butt weld, as-welded	In-plane bending	Weld corner or weld
43A	Partial penetration butt weld, with edges notched at weld	In-plane bending	Weld corner or weld
44	Tube welded to plate	Bending, shear	Weld toe
45	Tube welded to flange plate	Bending, shear	Weld toe
46	Triangular gusset attachments to plate	Axial	End of weld
47	Penetrating tube welded to plate	Axial in plate	Weld toe
47A	Attachment of tube to plate	Axial in plate	Weld toe
48 (R = 2t)	Penetrating rectangular tube welded to plate	Axial in plate	Weld toe
48R (R > 2t)	Penetrating rectangular tube welded to plate	Axial in plate	Weld toe
49	Clearance cut-out	Bending, shear	Weld toe or end of weld
50	Clearance cut-out	Bending, shear	Weld toe or end of weld
51	Clearance cut-out	Bending, shear	Weld toe or end of weld
52	Clearance cut-out	Bending, shear	Weld toe or end of weld
53	Reinforced deck cut-out	Axial	Weld ripple

(a) The numbering system of Munse et al. is related to that of the American Institute of Steel Construction (AISC, Ref 3), but the AISC classification of weldments contains only 27 shapes. Source: Ref 1

Classifying Weldment Geometry on the Basis of the Site of Fatigue Crack Initiation. As mentioned above, eliminating the influence of secondary variables such as *R*-ratio and strength

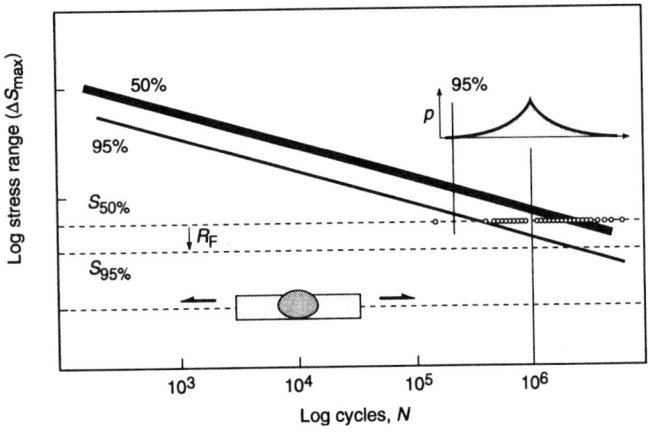

Fig. 5 The reliability factor (R_F) is calculated for the desired level of safety given the scatter in the fatigue data for the detail.

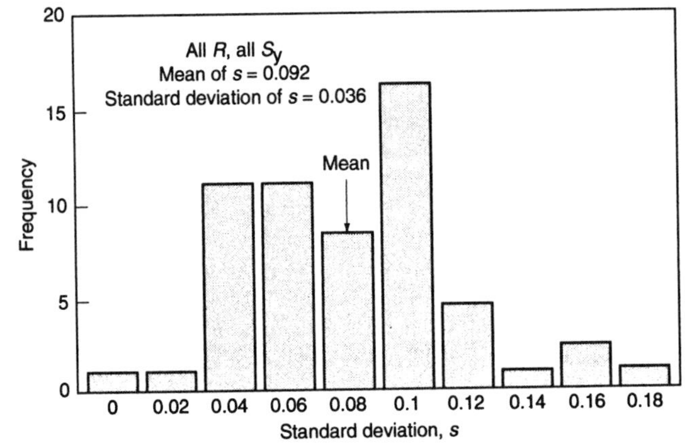

Fig. 6 Frequency vs. the log of the standard deviation in fatigue strength in ksi

Table 3 Comparison of fatigue data for structural details

Details	Mean fatigue strength ΔS at 10^6 cycles, ksi			Standard deviation of log ΔS, ksi		New AISC classification(a)	Fatigue crack initiation sites(b)
	All R All S_y	$R = 0$	$R = 0$ $S_y < 50$ ksi	$R = 0$	$R = 0$ $S_y < 50$ ksi		
1Q	51.8	51	...	0.074	...	A	...
1H	48.2	45.6	39.3	0.06	0.04	A	...
1.A11	44.9	42.1	38.2	0.104	0.042	A	...
1M	37.1	36.2	36.2	0.04	0.04	A	...
8	39.8	39.1	35.4	0.094	0.079	A	...
2	42.1	41	35	0.076	0.017	A	...
10(G)	35.2	32.8	31.6	0.136	0.127	A	Ripple
10Q	31.5	32.7	...	0.114	...	B	Toe
3(G)	31.2	31	31	0.084	0.081	B	Ripple
1(F)	38.4	38.4	30.5	0.117	0.057	B (−1)	...
10A	31.1	28.8	29.7	0.115	0.066	B	Toe
25A	35.8	29.3	29.6	0.109	0.12	B (−1)	Toe
3	29	29.1	29.2	0.049	0.044	B	Ripple
13	27.8	27.3	28.5	0.055	0.057	B (+1)	Toe
28	29.8	28.4	28.1	0.097	0.045	B	...
12(G)	27.2	27.2	27.2	0.072	0.072	C	Ripple
10H	35.2	33.1	25.8	0.102	0.101	C (−1)	Toe
4	27.3	26.8	25.7	0.092	0.095	C	Ripple
6	27.3	26.8	25.7	0.092	0.095	C	Ripple
9	25.7	25.8	25.5	0.079	0.085	C	...
10M	26.4	24.5	24.5	0.093	0.093	C	Toe
16(G)	22.7	24.5	24.5	0.215	0.215	C (+1)	Root
25	24.1	23.9	24.5	0.09	0.08	C (+1)	Toe
7(B)	23.8	23.8	24.4	0.083	0.11	C (+1)	Toe and CT
19	23.2	23.1	...	0.157	...	E?	Toe
30A	23	23	23	0.014	0.014	D	Toe and DT
26	17.4	23	23	0.054	0.054	D (+1)	Toe
14	25.9	22.9	22.9	0.115	0.109	D (−1)	Toe
11	22.7	22.7	22.1	0.078	0.08	D	Toe
21	21.8	21.8	21.8	0.117	0.117	D	Toe
7(P)	21.5	21.5	...	0.075	...	D	Toe and CT
36	20	20	20	0.062	0.062	D	Toe and DT
25B	20	20	20	0.062	0.062	D	Toe or Toe and DT
12	19.7	19.7	19.7	0.055	0.055	D	Toe
16	19.6	19.6	19.6	0.104	0.104	D	Toe or Root
22	19.1	19.5	19.4	0.045	0.044	D	Toe
21(3/8 in.)	17.9	17.9	17.9	0.037	0.037	E	Toe
20	17.5	17.5	17.5	0.099	0.099	E (+1)	Toe
23	18.3	...	...	...	...	E	Toe
24	18.3	...	...	...	...	E	Toe
30	16.7	16.7	16.7	0.051	0.051	E	DT
38	16	16	16	0.058	0.058	F	Toe
17A	16.2	15.8	15.8	0.051	0.051	F	DT
17	14.6	14.6	14.6	0.046	0.046	F	DT
18	12.2	12.8	14.5	0.107	0.148	F (+1)	DT
32A	14.1	14.1	14.1	0.055	0.055	F	DT
27	12.8	13.5	13.5	0.101	0.101	G	
33	11.6	12.9	12.9	0.055	0.055	G	Toe at CT or DT
31A	15.6	15.8	...	0.12	...	F	Toe
46	11.9	...	...	...	...	G	DT
40	11.2	...	...	...	...	G	Toe and DT
32B	11.2	...	...	...	...	G	Toe and DT

(a) The shift in AISC category resulting from restricting database information to $R = 0$ and $S_y < 50$ ksi test results is indicated by +1 or −1, depending on whether the weldment was increased or decreased by one category. (b) CT, continuous termination (wraparound weld); DT, discontinuous termination (simple start or stop)

effects gives a sharper picture of the true effects of structural detail geometry. A simple way of quantifying the severity of the critical notch in a structural detail is to introduce the concept of the fatigue notch factor (K_f), a nondimensional, scalar quantity that is defined as:

$$K_f = \frac{\Delta S_{\text{smooth specimen}}}{\Delta S_{\text{weldment}}} \approx 1.43 \left(\frac{\Delta S_{\text{plain plate}}}{\Delta S_{\text{weldment}}} \right) \qquad \text{(Eq 5)}$$

In the instance of mild steel, K_f can be determined using plain plate data, assuming that the fatigue notch factor for plain plate is $K_f = 1.43$.

The experimental definition of the fatigue notch factor (K_f) and the use of the mean and standard deviation in design are illustrated in Fig. 9.

The fatigue behavior of only the welded structural details of Tables 2 and 3 are reproduced in Table 4. The site at which the fatigue failure initiates in these weldments is inevitably one of four locations: weld ripple, weld toe, weld root, or a weld termination (Fig. 10). As seen from the comments in Table 4, several weldments are not pure cases of fatigue initiation and growth from either the ripple, toe, root, or termination. These unusual weldments will (for the most part) be eliminated from further consideration and termed

"mavericks." For instance, all partial penetration welds must be considered mavericks because their fatigue resistance depends entirely on the size of the incomplete joint penetration (IJP), the magnitude of which is generally unknown. If the "mavericks" are disregarded, it is evident in Table 4 that weldments initiating fatigue cracks at weld ripple and weld toes have the lowest values of K_f and are the welded details having the higher fatigue strengths. All weldments failing from terminations are among the worst welded details and have the largest values of K_f and the least fatigue strength. This weldment has been much studied and is often used as the paradigm for the behavior of all weldments.

Each of the welded details of Table 4 categorized as a "pure" case of fatigue crack initiation from either the weld ripple, weld toe, or weld termination was given a designation:

- *Ripple (R):* Failure initiating from the ripple in a weld.
- *Toe (G):* Failure initiating from the toe of a groove weld.
- *Toe (F):* Failure initiating from the toe of either a full-penetration, load-carrying fillet weld or any non-load-carrying fillet weld.
- *Toe (F′):* Failure initiating from the toe of a partial-penetration, load-carrying fillet weld. This case is actually a "maverick," but it is so important that it is included in the comparisons below.
- *Termination (T):* Failure initiating from the "start" or "stop" of a fillet or groove weld.

The essential distinctions between the four fundamentally different initiation sites (ripple, toe, root, or weld termination) are summarized in Table 5 from a metallurgy and mechanics perspective. Figures 11 to 13 give schematic diagrams of the weld categories based on sites of fatigue crack initiation and growth.

The effect of this system of categorizing weldments is shown in Fig. 14 and 15. In Fig. 14 the standard deviation of the log of strength (ksi units) is plotted versus K_f for each of the details listed in Table 2. Several interesting observations can be made: First, the scatter in fatigue strength is inversely related to K_f. The welds having a higher fatigue resistance exhibit more scatter, presumably because fatigue crack initiation plays a larger role. On the other hand, welds having the lowest fatigue resistance have less scatter in their fatigue data, presumably because their behavior is governed entirely by fatigue crack growth. Second, it is obvious that weldments for which fatigue cracks initiate at the ripple or toe have relatively small values of K_f (1.8 to 2.5); whereas terminations have very high values of K_f (3.0 to 4.5). The "maverick" load-carrying cruciform (Fig. 11) is seen to range in behavior from as bad as the terminations to as good as a non-load-carrying cruciform weldment.

These observations are reinforced by the S-N diagrams of Fig. 15. It is interesting to note that the slopes of the S-N diagrams for the terminations are different from the slopes for the ripple and toe categories. The slope of the S-N diagrams

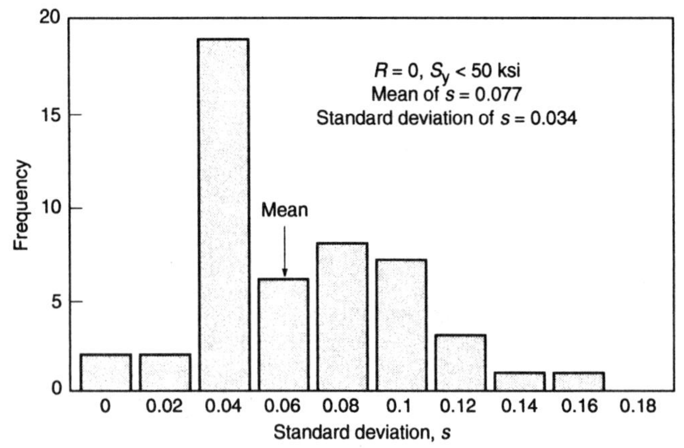

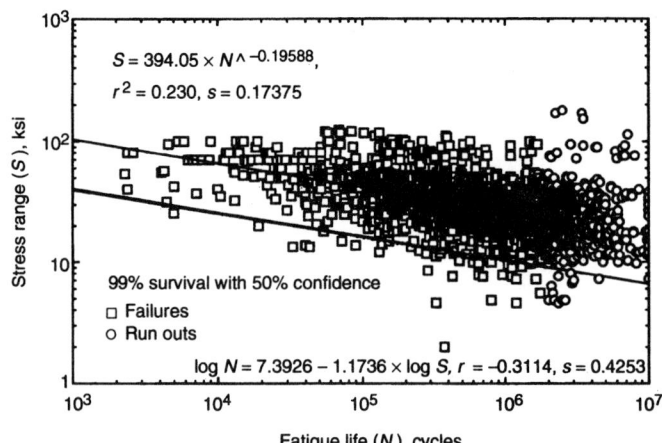

Fig. 7 Frequency vs. the log of the standard deviation in fatigue strength in ksi

Fig. 8 Typical data for AISC category C weld details. Data taken from the University of Illinois at Urbana-Champaign weldment fatigue databank. Data for details 7, 10-14, 19, and 22-25 are in Table 2. Tests that were discontinued before failure are termed "run outs."

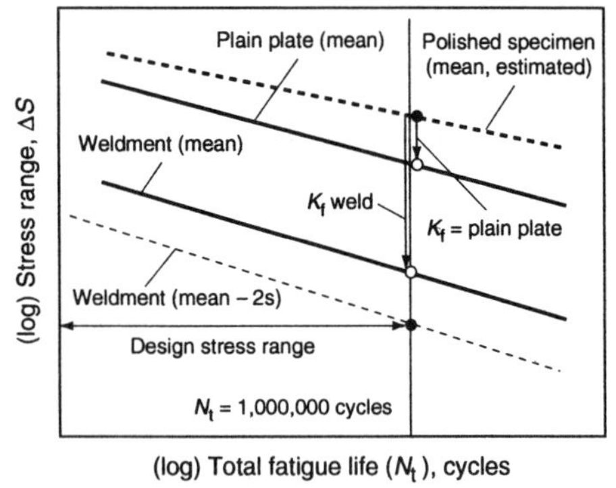

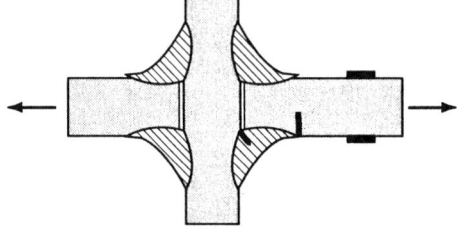

Fig. 9 Definition of K_f and the role of the weldment mean strength ($\Delta S_{weldment}$) and the standard deviation (s) in log of fatigue strength in determining the design stress range permitted for a given service life

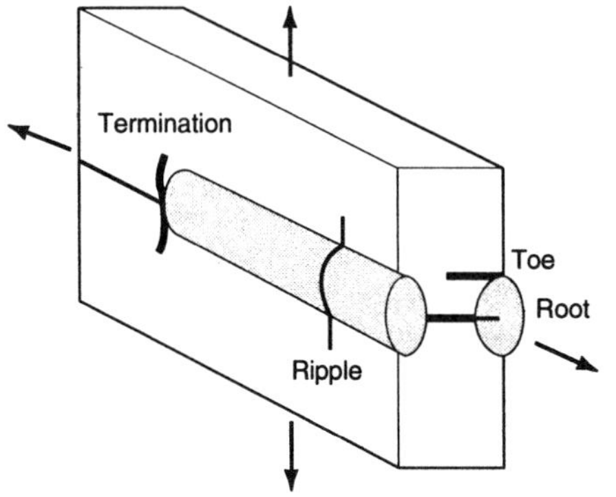

Fig. 10 Failure locations in weldments: ripple, toe, root, or weld termination. The distinction between a wraparound (continuous) termination and a simple termination (stop or start) is not made in this drawing. The termination and ripple are sites of fatigue crack initiation only when the load applied to the weldment is longitudinal. Likewise, the root and toe become fatigue crack initiation sites under transverse loading.

Fig. 11 Example of "maverick" joint. The load-carrying fillet weld is an important case. Failure may occur at either the weld toe or weld root. Applied axial stresses favor root failures. Applied bending stresses favor toe failures. The size of incomplete joint preparation (IJP) controls the notch severity (K_f) of both the root and the toe, with the result that this weld can be as good as a "good" weld or as bad as a termination, depending entirely on the size of the IJP.

for the terminations portrays a situation in which there is very little, if any, contribution from crack nucleation (N_N) and early crack growth (N_{PI}) or at least no crack closure. In such a case, the slope of the S-N diagram is $-1/n$ or $-1/3$ for mild (ferritic-pearlitic) steel. The more nearly horizontal slope for the ripple and toe category welds indicates a substantial crack nucleation (N_N) and early crack growth (N_{PI}) contribution to their total life. The slopes of S-N diagrams are an incontrovertible indication of the importance or unimportance of crack nucleation and early crack growth.

Summary. Joint geometry has a large influence on the fatigue resistance of weldments. While it is an appealing idea to assemble the fatigue data for weldments into a comprehensive "encyclopedia" organized to reflect their fatigue behavior, such efforts may be hopeless because there are just too many different joint geometries. However, collecting weldment fatigue data into a limited number of broad weld "categories" (which may be an appealingly simple concept for designers) increases the apparent scatter in weldment fatigue data and reduces the allowable de-

sign stresses for a required level of safety. The scatter in both the encyclopedia approach and the weld category approach inevitably obscures the effects of the secondary but nonetheless important fatigue variables.

If weldments are classified by the site of fatigue crack initiation, it would seem that "good" weldments (for which fatigue cracks initiate at the weld toe or weld ripple) can be distinguished from "bad" weldments (which are substantially worse than "good" weldments for a variety of reasons) or "mavericks" (for which the fatigue resistance depends largely on the undefined size of a discontinuity or is complicated by ambiguity as to the definition of nominal stress).

Variables of Weldment Fatigue

In this section, the variables influencing the fatigue resistance of an individual joint geometry (a non-load-carrying cruciform weldment) are in-

vestigated with the aid of a computer simulation of weldment fatigue behavior. This weld geometry is used as an example of fatigue behavior of weldments where fatigue failure initiates at a weld toe (i.e., "good" weldments). The behavior of "Nominal" and "Ideal" weldments, that is, non-load-carrying cruciform weldments with and without a 0.1 in. weld discontinuity at the weld toe, is compared and contrasted.

The variables influencing the fatigue life of a weld, such as a non-load-carrying cruciform weldment (Fig. 16), are:

- *Applied stress amplitude:* The remote axial and bending stresses (ΔS_A and ΔS_B) at the weld toe. The bending stresses may be applied or residual stresses resulting from weld fabrication distortions. Welding distortions may not induce secondary bending stresses when the applied load is pure bending.
- *Mean and residual stresses:* Remote mean stresses resulting from the applied load (S_m), welding residual stress at the weld toe (σ_r), and fabrication residual stresses resulting from

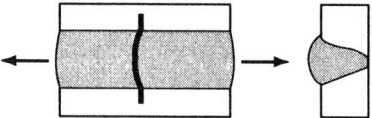

Ripple: Longitudinally loaded groove weld

Toe (G): Single-V groove welded butt joint

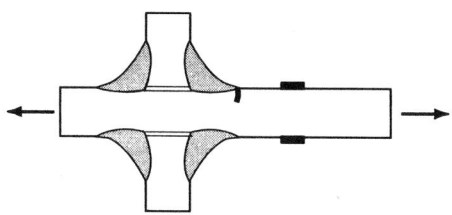

Toe (F): Non-load-carrying cruciform joint

Fig. 12 Examples of "good" welds addressed in this article. Initiation sites at a weld ripple or a weld toe are illustrated.

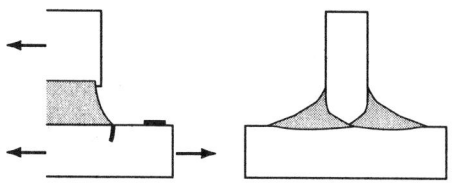

Termination (T): Fillet weld termination

Fig. 13 "Bad" weld with initiation site at the end of a fillet weld

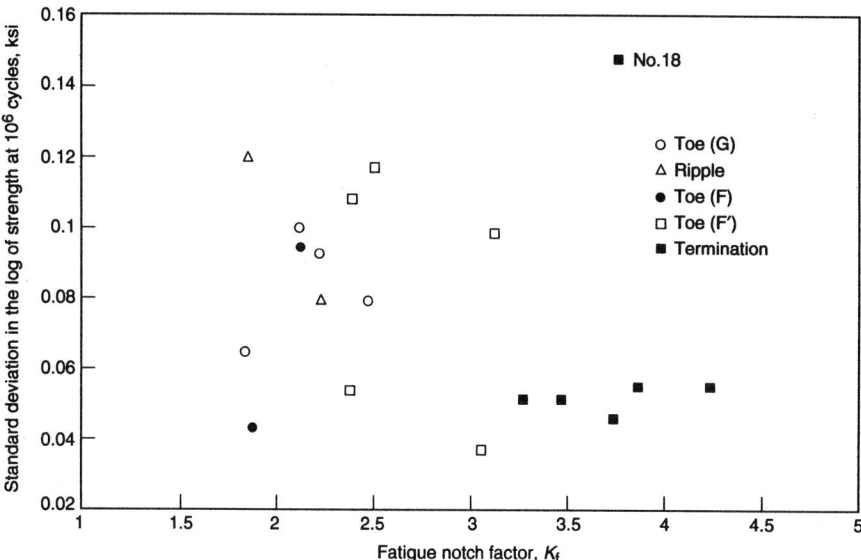

Fig. 14 Variation in the log of the standard deviation in fatigue strength in ksi with fatigue notch factor (K_f). The uncertainty in the fatigue strength of terminations would seem to be generally less than that of the toe and ripple.

Table 4 Welded details and "standardized" fatigue strengths

Details(a)	Loading(b)	Mean fatigue strength (ΔS) at 10^6 cycles, ksi(c)	Standard deviation of log ΔS, ksi	K_f	New values of ΔS design(d)	Fatigue crack initiation sites	Comment
10A	AB	29.7	0.066	1.84	21.9	Toe (G)	...
25A	A	29.6	0.12	1.85	17.0	Toe (F)	...
3	A	29.2	0.044	1.87	23.8	Ripple	...
13*	AB	28.5	0.057	1.92	21.9	Toe	Change in flange width
10H	A	25.8	0.101	2.12	16.2	Toe (G)	...
4	AB	25.7	0.095	2.13	16.6	Ripple	...
6	AB	25.7	0.095	2.13	16.6	Ripple	...
10M	A	24.5	0.093	2.23	16.0	Toe (G)	...
25	A	24.5	0.08	2.23	16.9	Toe (F)	...
07(B)*	AB	24.4	0.11	2.24	14.7	Toe and CT	Toe or termination failure
26	A	23	0.054	2.38	17.9	Toe (F)	...
30A*	B	23	0.014	2.38	21.6	Toe and DT	Pure bending
14	A	22.9	0.109	2.39	13.9	Toe	...
11	AB	22.1	0.08	2.47	15.3	Toe	...
21	AB	21.8	0.117	2.51	12.7	Toe	...
25B*	A	20	0.062	2.73	15.0	Toe or Toe and DT	Toe or termination failure
36*	AB	20	0.062	2.73	15.0	Toe and DT	Toe or termination failure
12*	AB	19.7	0.055	2.77	15.3	Toe	Change in flange slope
16*	A	19.6	0.104	2.79	12.1	Toe or Root	Partial penetration
22*	AB	19.4	0.044	2.82	15.8	Toe	Attachment or cruciform
21(3/8 in.)	AB	17.9	0.037	3.05	15.1	Toe (F′)	...
20	A	17.5	0.099	3.12	11.1	Toe (F′)	...
30	A	16.7	0.051	3.27	13.2	Termination	...
38*	AB	16	0.058	3.41	12.2	Toe	High restraint
17A	A	15.8	0.051	3.46	12.5	Termination	...
17	A	14.6	0.046	3.74	11.8	Termination	...
18	A	14.5	0.148	3.77	7.3	Termination	...
32A	AB	14.1	0.055	3.87	10.9	Termination	...
33	A	12.9	0.055	4.23	10.0	Termination	...

(a) Details listed with an asterisk were labeled "mavericks." (b) A, axial; B, bending; AB, deep section loaded under bending but stress at hot-spot pseudoaxial. (c) $R = 0$, $S_y < 50$. (d) Resulting from standardizing the databank

subsequent remote welding (S_{fab}), which add to the remote mean stresses.

- *Material properties:* Strain-controlled fatigue properties (ε'_f, σ'_f, b, c) determine the resistance to crack nucleation and early crack growth, while the crack growth properties (C, n) control the growth of fatigue cracks. The residual stresses are limited (often controlled by) the metal's yield strength (S_y), so yield strength of the weldment's constituent materials is of great importance in non-stress-relieved weldments.

- *Geometrical stress concentration effects:* The concentration of stress and strain at a notch such as a weld toe magnify the effects of the applied stress, the remote mean stress, and the fabrication stresses. Thus notches reduce the fatigue life, particularly N_N and N_{P1}. The effects of the notch are captured by the fatigue notch factor (K_f), which influences N_N and N_{P1}, and by M_k, which is the elevation of the range in stress-intensity factor at the weld toe.

- *Size and location of welding discontinuities:* The weld discontinuities, both at the notch root and elsewhere, magnify the stress-concentrating effects of the critical notch and can greatly

reduce N_N, N_{P1}, and N_{P2}. The presence of a 0.1 in. planar discontinuity at the weld toe of the non-load-carrying cruciform weldment considered here is the condition that distinguishes the "Nominal" from the "Ideal" weldment.

The Role of Analytical Models. The fatigue of weldments is a complicated topic. No two weldments are identical, and weldment fatigue resistance may depend on many variables in a complex, nonlinear way. The effects of the major variables such as stress range and weld geometry are certainly understood, and one can usually predict what will happen if one of these major variables is changed; however, it is difficult to predict what will happen if these and several

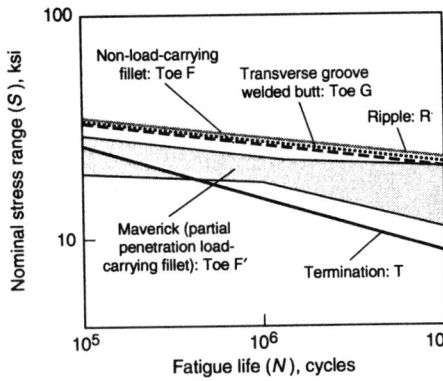

Fig. 15 Average S-N diagrams for the welded details in Table 2. The average S-N curves for ripple (R), toe (G), and toe (F) are similar. The fatigue behavior of the "maverick" toe (F') (partial penetration load-carrying fillet) ranges from being as bad as the terminations to as good as the toe (G) and toe (F) data.

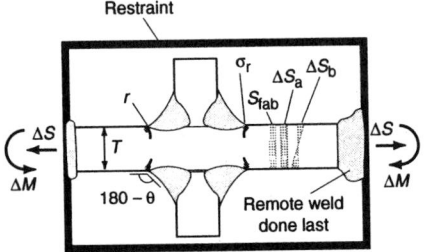

Fig. 16 Schematic diagram of a non-load-carrying cruciform weldment subjected to axial and bending loads as well as to global residual (mean) stresses generated by subsequent welding fabrication (S_{fab}). Welding residual stresses (σ_r) are considered to exist only in a small volume at the weld toe.

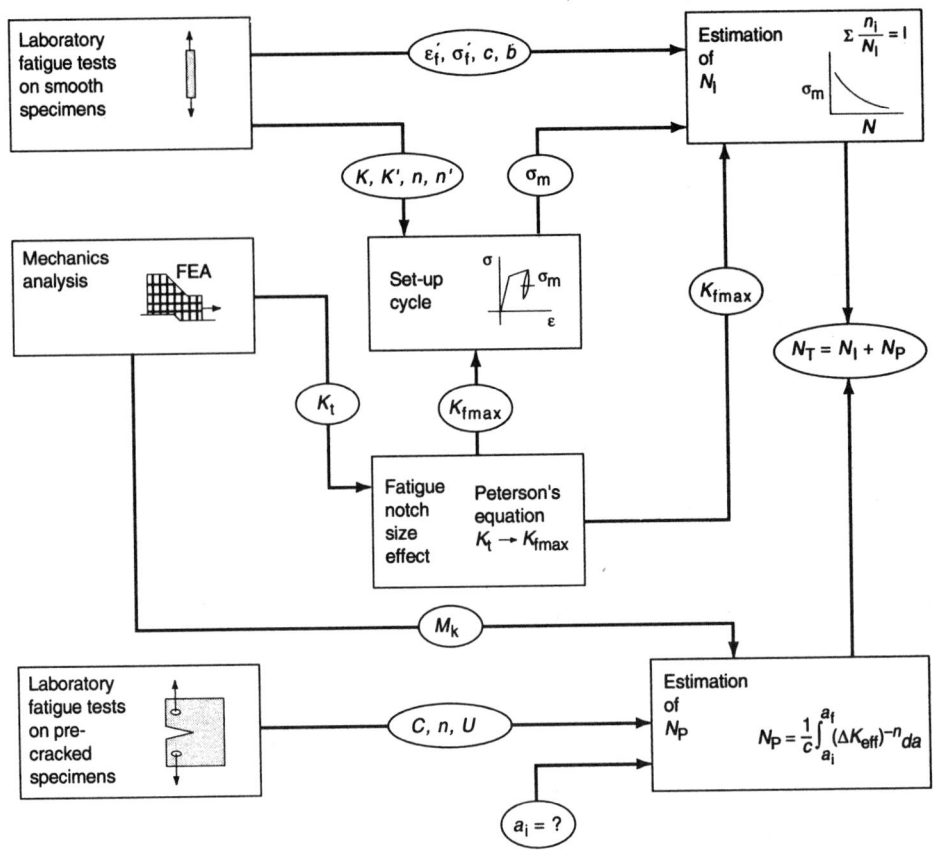

Fig. 17 A schematic diagram showing the information required by and the organization of the initiation-propagation model

Table 5 Essential differences between fatigue crack initiation sites

Fatigue crack initiation site	Relevant material condition	Notch	Residual stresses
Ripple	WM	Weld ripple: a periodic array of small notches on the surface of the weld bead	$+S_y$ WM: No larger than the yield strength of weld metal
Toe	HAZ	Weld toe: a surface notch having no defined depth and variable notch-root radius	$+S_y$ BM: No larger than the yield strength of base metal
Root	(Tempered) WM	Weld root: a sharp notch having an unknown and variable notch-root radius	Unknown: Probably near zero if the fit-up is not tight
Termination	HAZ	Weld toe: As above, except it is possible or even probable that starts will involve a lack of fusion and that stops may involve crater cracks, pipes, or hot cracks	$+S_y$ WM: Possibly as high as the yield strength of weld metal

WM, weld metal; HAZ, heat-affected zone; BM, base metal

secondary variables are changed at once. In such a circumstance, the outcome may be counterintuitive.

Computer models can simulate the behavior of complex weldment fatigue. Fracture mechanics crack growth models for N_{P2} provide the lower-bound estimates of N_{P2} for the "Nominal" weldment, while the initiation-propagation (I-P) model described below, which combines the linear elastic fracture mechanics (LEFM) estimates of N_{P2} (the crack propagation life or "P") with estimates of N_N and N_{P1} (crack initiation life or "I"), can provide estimates of the upper-bound behavior of the "Ideal" weldment (Ref 5).

$$N_T = [N_N + N_{P1}] + N_{P2} = N_I + N_{P2} \quad \text{(Eq 6)}$$

The I-P model is shown schematically in Fig. 17. Relevant material properties for two steels are listed in Table 6. The I-P model predicts the total fatigue life of a weldment (N_T) by making separate estimates of the fatigue crack initiation life (N_I) and the fatigue crack propagation life (N_{P2}) and summing them.

The fatigue crack initiation life (N_I) is thought of as the life period spent in crack nucleation and the growth of small cracks through (roughly) the first 50 to 100 μm of the metal, that is, N_N and

N_{P1}. This life period is captured in the fatigue behavior of smooth specimens, and thus, strain-controlled fatigue life concepts are used to estimate this life period. The severity of the notch presented by the weld toe is quantified using the $K_{f,max}$ hypothesis, a concept for determining the pessimum value of fatigue notch factor K_f using Peterson's equation.

A second noteworthy feature of the N_I part of the I-P model is the use of the "set-up cycle" analysis to determine the notch-root mean stress remaining after the first few cycles. This analysis approximates the effects of notch-root plasticity during the first few applications of load through

the use of Neuber's rule and models the difference in behavior between monotonic (first reversal) and cyclic (subsequent reversals) behavior of the material at the notch root (the grain-coarsened heat-affected zone). This phenomenon is called the Bauschinger effect. The first reversal includes the effects of the initial loading from 0 to S_{max}, the weld toe (welding) residual stresses (σ_r), and the residual fabrication stresses (S_{fab}). In the "set-up cycle" simulation, the notch-root (welding) residual stresses are treated as an equivalent remote stress by dividing the notch-root residual stresses by $K_{f,max}$. The "set-up cycle" analysis provides the initial value of notch-root mean

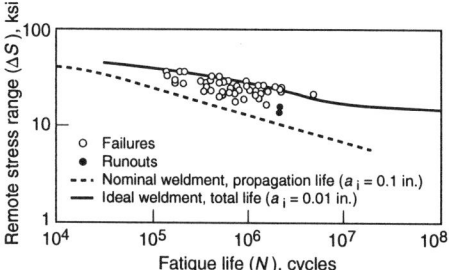

Fig. 18 Comparison of the predictions of the initiation-propagation model with data in the University of Illinois at Urbana-Champaign weldment fatigue databank for a mild steel, non-load-carrying cruciform weldment, $R = 0$

stress for the linear cumulative damage calculation, which considers the exponential decay of the notch-root mean stress during subsequent cycling. Thus, N_I is calculated considering the notch-root mean stresses established during the first few cycles of load application and their relaxation during the fatigue crack initiation period.

The calculation of N_{P2} is based on ΔK_{eff} and values for the effective stress-intensity ratio (U). The M_k value in Ref 6 was used. The R-ratio is redefined for the estimation of the N_P. The notch-root mean stresses are not considered because they exist only in the small volume of material at the notch root; furthermore, in most cases, the I-P model predicts that these notch-root mean stresses substantially diminish during the crack initiation period. Therefore, only the applied mean stresses and the mean stresses resulting from subsequent fabrication (S_{fab}) are assumed to influence crack growth. Crack shape development (Ref 7) is included. For "Ideal" weldments, the initial crack size is arbitrarily taken as $a_i = 0.01$ in. The final crack size (a_f) was determined using LEFM and K_{Ic}.

For the "Nominal" weldment, N_I was neglected and N_{P2} was calculated assuming an initial flaw size $a_i = 0.1$ in. (2.5 mm).

Predicted Fatigue Life and Data Comparison. Figures 18 to 21 show experimental data from the University of Illinois at Urbana-Cham-

paign (UIUC) fatigue databank for a butt joint (#10) and non-load-carrying cruciform weldments (#25) for $R = 0$ and $R = -1$ test conditions. In each figure, the predictions of the I-P model for the "Ideal" and "Nominal" weldments are seen to bound the experimental data. When propagation dominates, the slope of the S-N curve is $1/n$ or $1/3$. When initiation dominates, the slope of the S-N curve is $1/b$ or around $1/8$ to $1/10$. Thus, the slope of theoretical and experimental S-N curves reflect the relative importance of "I" and "P."

The Effect of Residual Stresses. Residual stresses greatly influence the fatigue life of both the "Nominal" and "Ideal" weldments, as shown in Fig. 22. In Fig. 22, the welding residual stresses (σ_r) and the fabrication stresses (S_{fab}) were presumed to be either 0 or their largest possible value, the yield strength of base metal (36 ksi). Both N_I and N_P are much affected by fabrication stresses. As can be seen, there is a very large difference between the total lives of both "Nominal" and "Ideal" weldments with and without fabrication stresses. This effect is probably a major source of the reported large difference between fatigue tests using small, simple testpieces and full-scale fatigue tests on complex welded structures in which large fabrication stresses (S_{fab}) may exist.

The effect of weldment size is a subject of continuing controversy. The predicted effect of size on the fatigue strength at 10^7 cycles is shown for both the "Nominal" and "Ideal" weldments in Fig. 23. The predicted behavior of the "Ideal" weldment is similar to the currently anticipated size effect and has a slope of $\approx -1/3$. Note that "Ideal" weldments with high fabrication stresses may have a slope greater than $-1/3$. The 0.1-in. discontinuity in the "Nominal" weldment leads to essentially no size effect for weldments having $T > 1.0$ in. (25 mm) and a reversal in the size effect when $T < 0.7$ in. (18 mm) due to the large size of the initial flaw (0.1-in.) assumed relative to the smaller plate thicknesses considered.

The Effect of Base Metal Strength. It is generally believed that the fatigue resistance of non-stress-relieved weldments is independent of its base metal tensile strength. Steels with higher

Table 6 Material properties used in estimating weldment fatigue life

Property, symbol (units)	A36 HAZ	A514 HAZ
Tensile properties		
Ultimate strength, S_u (ksi)	97	204
Yield strength, S_y (ksi)	77	171
Base metal yield strength, S_{yBM} (ksi)	35	100
Young's modulus, E (ksi)	2.74×10^4	3.03×10^4
Peterson's constant, a_p (in.)	0.01	0.005
Monotonic strength coefficient, K (ksi)	142	306.0
Strain-controlled fatigue properties		
Cyclic strength coefficient, K' (ksi)	216	256.0
Monotonic strength exponent, n	0.102	0.092
Cyclic strength exponent, n'	0.215	0.103
Cyclic ductility coefficient, ε_f'	0.218	0.783
Cyclic strength coefficient, σ_f' (ksi)	105	290
Cyclic strength exponent, b	-0.066	-0.087
Cyclic ductility exponent, c	-0.492	-0.713
Crack growth properties		
Paris equation, C, (in./cycle)	3.6×10^{-10}	6.6×10^{-9}
Paris equation, C', (in./cycle)	1.21×10^{-9}	1.64×10^{-8}
Paris equation, exponent	3.0	2.25
Fracture toughness, K_{Ic} (ksi$\sqrt{\text{in.}}$)	100	150

HAZ, heat-affected zone

UTS do exhibit a greater fatigue resistance in the absence of (tensile) mean or residual stresses. Unfortunately, higher UTS weldments also have higher yield strengths and thus can sustain much more damaging welding and fabrication residual stresses. So any improvement in intrinsic fatigue resistance resulting from increasing the strength of a base metal is usually more than offfset by the damaging effects of tensile residual stresses, which develop during fabrication. If one can induce compressive residual stresses or reduce the size of the as-welded tensile residual stresses, the fatigue strength of higher strength "Ideal" weldments of high-strength materials can be much improved.

The fatigue strength of a mild steel ($S_y = 36$ ksi) and a quenched-and-tempered steel ($S_y = 100$ ksi) are compared in Fig. 24 below. The higher strength Q&T steels are predicted to perform better when the residual stresses are small or

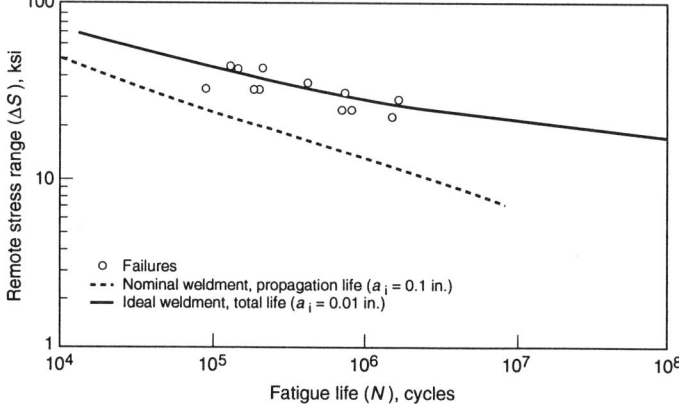

Fig. 19 Comparison of the predictions of the initiation-propagation model with data in the University of Illinois at Urbana-Champaign weldment fatigue databank for a mild steel, non-load-carrying cruciform weldment, $R = -1$

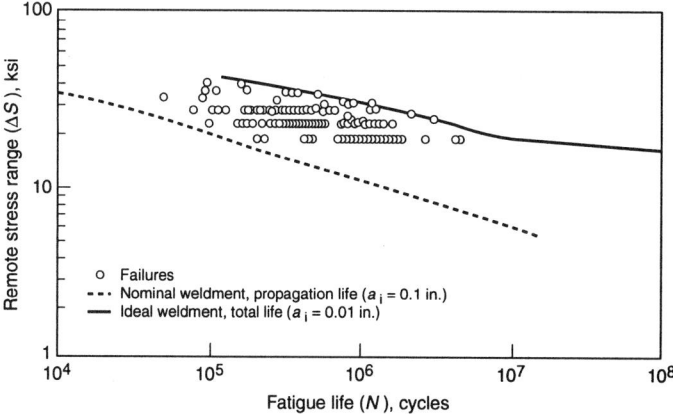

Fig. 20 Comparison of the predictions of the initiation-propagation model with data in the University of Illinois at Urbana-Champaign weldment fatigue databank for a mild steel, double-V butt weldment, $R = 0$

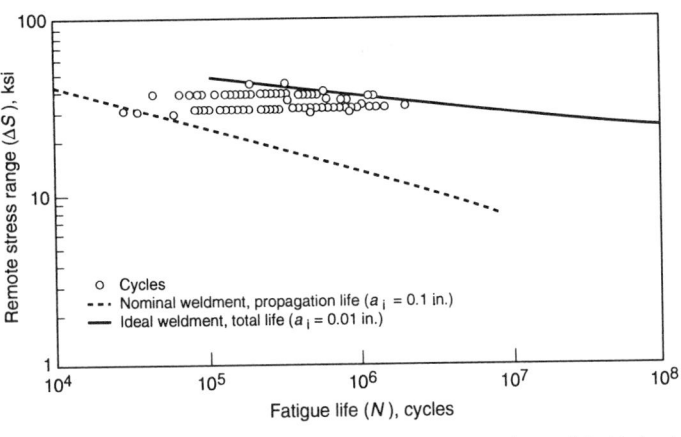

Fig. 21 Comparison of the predictions of the initiation-propagation model with data in the University of Illinois at Urbana-Champaign weldment fatigue databank for a mild steel, double-V butt weldment, $R = -1$

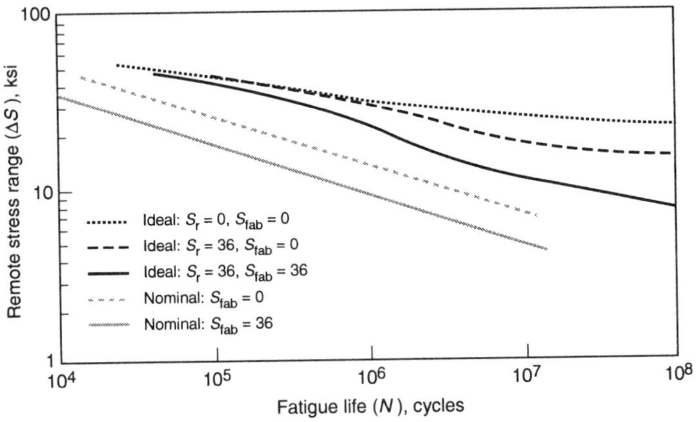

Fig. 22 Effect of residual stresses on the fatigue behavior of "nominal" and "ideal" 1.0 in. plate thickness, mild steel, non-load-carrying cruciform weldments

Table 7 Estimated sources of uncertainty in weldment fatigue strength data

Data and references(a)	$\Omega_{\hat{P}}^2$	$\Omega_{\hat{G}}^2$	$\Omega_{\hat{B}}^2$	$\Omega_{\hat{RS}}^2$	$\Omega_{\hat{MS}}^2$	$\Omega_{\hat{S}}^2$	R_F
LCC1-Load-carrying cruciform (12.7 mm) (Ref 10, 11)	0.000946	0.006410	0.003870	0	0.04507	0.05630	0.626
LCC2-Load-carrying cruciform (6.35 mm) (Ref 10, 11)	0.000947	0.010677	0.004328	0	0.02492	0.02492	0.731
NLCC1-Non-load-carrying cruciform (32 mm) (Ref 9, 12)	0.001200	0.003890	0.006250	0	0	0.01134	0.808
NLCC2-Non-load-carrying cruciform (25 mm) (Ref 13)	0.000883	0.005270	0.000521	0	0	0.01136	0.808
B1-Butt weldment (6 mm) (Ref 14-16)	0.000814	0.000661	0.000811	0.001394	0	0.00365	0.886
B2-Butt weldment (20 mm) (Ref 14-16)	0.000862	0.001280	0.000941	0.024916	0	0.02800	0.717
Data Bank-Butt weldment: Single-V (Ref 17)	0.004060	0.009750	0.081110	0.001790	0	0.09671	0.545

(a) Reported by various investigators in Ref 4

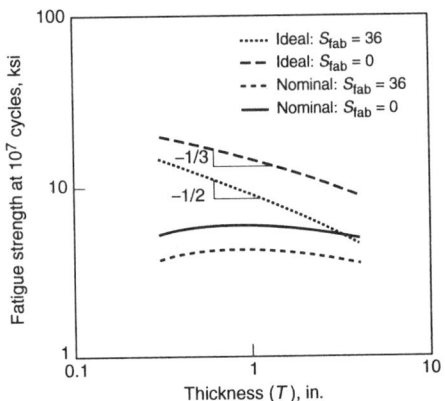

Fig. 23 The predicted effect of weldment size for both a "nominal" and "ideal" mild steel, non-load-carrying cruciform weldment

compressive and when crack growth is relatively unimportant, that is, as in the case of small-thickness, "Ideal" weldments.

The Effect of Post-Weld-Processing Procedures. The predicted effect of various post-weld-processing procedures on the fatigue strength of the hypothetical "Ideal" and "Nominal" mild steel nonload carrying cruciform weldments is shown in Fig. 25. The fatigue strength of "Ideal" weldments can be much improved; whereas, that of "Nominal" weldments cannot because most improvement techniques alter the material properties or residual stresses in the near-surface (notch root) material and thus do not much affect crack growth behavior at a depth of 0.1 in. (2.5 mm). Only the fabrication stresses and applied mean stresses are predicted to influence much crack growth behavior at a depth of 0.1 in.

In general, there are essentially two main strategies for improving weldment fatigue strength: alter the residual and mean stresses or improve the stress-concentrating geometry of the critical notch (weld toe) or a combination of both. As noted above, using higher-strength materials is not effective for nonstress-relieved weldments.

The Combined Effect of Weldment Size and Fabrication Stresses on "Nominal" and "Ideal" Weldments. Weldments or welding applications can be categorized according to weldment size and weld quality as shown in Fig. 26.

- *Light industry applications* in which the weldment size is about 0.5 in. or less. For such weldments it is presumed that the weldments are "simple" and do not therefore engender high fabrication stresses, that is, $S_{fab} \approx 0$.
- *Heavy industry applications* in which the weldment size is about 2.0 in. It is presumed that the weldments are "complex" and therefore do engender high fabrication stresses, that is, $S_{fab} \approx +S_{yBM}$ (base material yield strength).
- *High-quality welding processes* such as gas-tungsten arc welding and gas-metal arc welding in which the weld perfection may approach that of the ideal weldment.
- *Low-quality welding processes* in which the weld perfection is low and a substantial initial weld discontinuity must be assumed to be present. Such weldments may approach the behavior of the "Nominal" weldment ($a_i \approx 0.1$ in., or 2.5 mm).

Modeling the Uncertainty in Weldment Fatigue Strength. In the section "Metallic Fatigue in Weldments" in this article, the scatter or uncer-

tainty in weldment fatigue data overshadowed the effect of most of the fatigue variables discussed in this section. While these variables may have only a moderate effect on the mean fatigue life, they are the source of scatter in the fatigue strength data for a given weldment. This scatter forces low design stresses to avoid frequent failures (as seen in the section "Metallic Fatigue in Weldments" in this article).

A powerful application of the analytical models employed here to estimate the mean fatigue life or strength of a given weldment is to use the model as the basis of a stochastic analysis of weldment fatigue life and to imagine that some or all of the variables considered by the model are stochastic in nature. In this way, the uncertainty in the fatigue resistance of a weldment can be estimated, and the contribution of each fatigue variable to the total uncertainty in fatigue strength of a weldment can be assessed.

The uncertainty in the fatigue life of the "Nominal" weldment has been studied by Engesvik and Moan (Ref 9). For weldments with large welding discontinuities, the major sources of

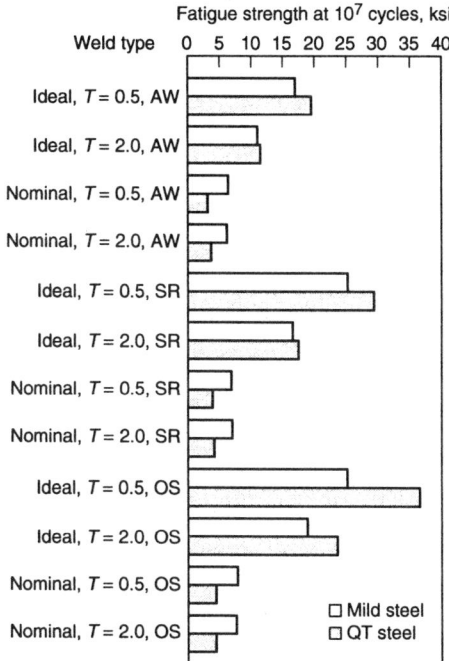

Fig. 24 Predicted fatigue strength of a cruciform weld model (Fig. 16) for mild steel (S_y = 36 ksi, 250 MPa) and quenched-and-tempered (QT) steel (S_y = 100 ksi, 690 MPa). R = 0; T given in inches. Quenched-and-tempered steels show no advantage for all nominal types in as-welded (AW), stress-relieved (SR), as-welded (AW), stress-relieved (SR), and tensile overstressed (OS) conditions.

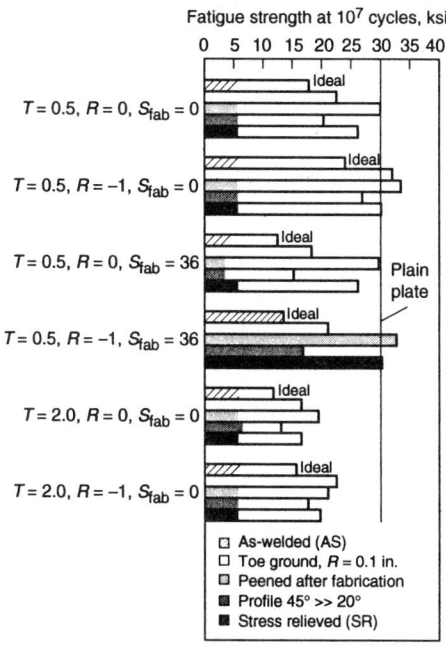

Fig. 25 The predicted effect of various fatigue strength improvement treatments on the fatigue strength of a mild steel, non-load-carrying cruciform weldment. The "nominal" joint fatigue (shaded at the lower strength levels) did not benefit from the indicated treatments as much as the "ideal" joints for a similar treatment (unshaded bars). Some improvement techniques can cause the weldment to equal the fatigue strength of plain plate (30 ksi). Because of its rolled-in surface discontinuities, plain plate has a $K_f \approx 1.43$. Source: Ref 8

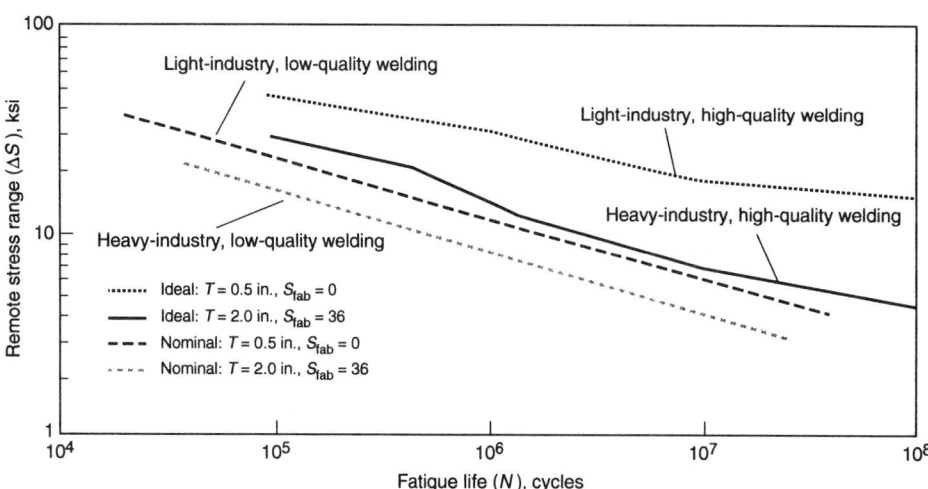

Fig. 26 Comparison of four common industrial situations: Light Industry-High Quality Welding, Light Industry-Low Quality Welding, Heavy Industry-High Quality Welding, Heavy Industry-Low Quality Welding. Mild steel cruciform weldments R = 0. The small-size, high-welding quality weldments typical of the ground vehicle industry may perform substantially better than larger and more complex, or lesser quality weldments.

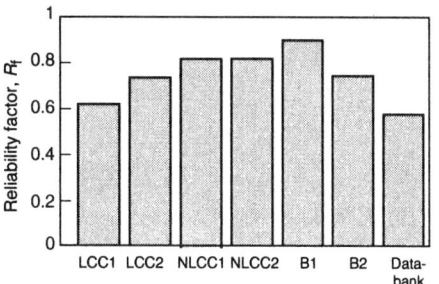

Fig. 27 The reliability factor (R_F) for each of the data sources in Table 7

scatter in fatigue data are variation in the size of the weld discontinuity and in the magnitude of the fabrication stresses.

In contrast, the uncertainty in the fatigue life of the "Ideal" weldment depends on a large number of variables. Lawrence and Chang (Ref 8) suggested a simple, approximate expression based on the Basquin-Morrow equation for the fatigue strength of a weldment at long lives, which assumes that at long lives $N_T \approx N_N + N_{P1}$, that is $\approx N_I$:

$$S_a^T = \frac{(\sigma_f' - \sigma_r)(2N_I)^b}{K_{f,max}^{eff}(1 + \frac{1+R}{1-R}(2N_I)^b)} \qquad (Eq\,7)$$

Equation 7 was factored to isolate five important attributes of weldments that determine their fatigue strength: the notch severity of the discontinuity (G); the mechanical properties of the material in which fatigue crack initiation and short crack growth takes place (P); the applied mean stresses effects (MS); the residual stresses resulting from fabrication and subsequent use of the weldments (RS); and the self-induced stresses caused by the welding distortions (B). Equation 7

was rewritten assuming that the weldments are axially loaded and that the bending stress components are induced by the welding distortions (Ref 4):

$$S_a^A = P \cdot G \cdot B \cdot RS \cdot MS \cdot (2N_I)^b \qquad (Eq\,8)$$

where the factors for the effects mentioned above are as follows:

$$P = \sigma_f' \qquad \text{(material properties)}$$

$$G = \frac{1}{K_{f,max}^A} \qquad \text{(notch severity)}$$

$$B = \frac{1}{1 + x\left(\frac{K_{f,max}^B}{K_{f,max}^A} - 1\right)} \qquad \text{(distortion)}$$

$$RS = 1 - \frac{\sigma_r}{\sigma_f'} \qquad \text{(residual stresses)}$$

$$MS = \frac{1}{(1 + (2N_I)^b \frac{(1+R)}{(1-R)})} \qquad \text{(mean stress)}$$

If the variables P, G, B, RS, and MS can be considered to be normally distributed variates, the coefficient of variation (COV) of the fatigue strength of a weldment (Ω_S) can be approximated (Ref 2) as:

$$\Omega_S^2 \approx \Omega_P^2 + \Omega_G^2 + \Omega_B^2 + \Omega_{RS}^2 + \Omega_{MS}^2 + \Omega_f^2 \qquad (Eq\,9)$$

where the subscripts P through MS designate the COV of the random variable representing the effects of material properties, notch severity of geometry, welding-distortion-induced stresses, notch-root residual stress, and applied mean stress, respectively.

The estimated sources of uncertainty in weldment fatigue strength data reported by various investigators (Ref 4) are tabulated in Table 7 together with the reliability factor (R_F, see Eq 4).

As can be seen, the sources of the scatter in weldment fatigue data depend on the nature of the joint and how it is loaded. The butt joints (B1 and B2) have very little scatter associated with their reported fatigue data. The load-carrying cruciform weldments (LCC1 and LCC2) have enormous scatter, largely due to welding fabrication distortions that induce secondary stresses during gripping and subsequent axial loading. However, the estimated scatter in the fatigue databank data is much larger than that associated with even the "worst" weldment for the reasons discussed in the section "Metallic Fatigue in Weldments" in this article.

These observations are reflected in the plot (Fig. 27) of reliability factor (R_F). The values of R_F were obtained using Eq 4. The value of R_F for an individual weldment may be as high as 0.9 and as low as 0.7. Average values of R_F would seem to be around 0.7. The R_F implied by the use of a fatigue databank entry for a particular weld geometry (e.g., detail 10 in Fig. 4) is estimated at 0.55. The use of weld categories was earlier argued to lead to values of R_F of around 0.4.

The form of Eq 9 suggests some interesting but perhaps obvious strategies for reducing the uncertainty in the fatigue strength of an individual weldment:

- *If there is only one large source of uncertainty,* the uncertainty in weldment fatigue strength can only be improved by its reduction, but the uncertainty in weldment fatigue strength can increase if any of the lesser sources is permitted to grow.
- *If there is no dominant source of uncertainty,* the uncertainty in weldment fatigue strength can only be improved by reducing all sources uniformly, but, as above, the uncertainty in weldment fatigue strength can increase if any one of the sources is permitted to grow.

Summary of Weld Fatigue Variables. The main variables affecting weld fatigue strength are weldment geometry, weld-distortion-induced bending stresses, and residual stresses. These variables are also the main contributors to the uncertainty in fatigue life. Analytical models can be used to estimate the uncertainty in fatigue strength and to identify the contribution of each

source to the overall uncertainty.

ACKNOWLEDGMENTS

This article draws on the work of many people and many studies carried out over a period of years. The authors would like to acknowledge the advice and help of our colleagues at the UIUC in the Departments of Theoretical and Applied Mechanics and Mechanical Engineering. The article is based on several studies that were sponsored at various times by the UIUC Fracture Control Program, The Edison Welding Institute, the U.S. Coast Guard, and the Ship Structures Committee. The line drawings of Fig. 4 were drawn by Dr. Gregorz Banas.

REFERENCES

1. W.H. Munse, T.W. Wilbur, M.L. Tellalian, K. Nicoll, and K. Wilson, "Fatigue Characterization of Fabricated Ship Details for Design," Ship Structure Committee, SSC-318, 1983
2. A.-H.-S. Ang and W.H. Munse, "Practical Reliability Basis for Structural Fatigue," Preprint 2459 for ASCE National Structural Engineering Convention, 1975
3. Manual of Steel Construction, 8th ed., American Institute of Steel Construction, 1980. (A ninth edition, 1995, is also available.)
4. S.K. Park and F.V. Lawrence, Jr., Sources of Uncertainty in Weldment Fatigue Strength, *Proc. Ninth International Conference of Offshore Mechanics and Arctic Engineering*, Vol II, ASME, p 205-214
5. F.V. Lawrence and S.D. Dimitrakis, "I-P Model Simulation of the Factors Influencing Weldment Fatigue Life," North American Welding Research Conference, EWI, 1995
6. S.J. Maddox and R.M. Andrews, Stress Intensity Factors for Weld Toe Cracks, *Proc. Conf. Computer Aided Assessment and Controls of Localized Damage*, Springer Verlag, Berlin, 1990
7. R. Bell and O. Vosikovsky, Fatigue Life Prediction of Welded Joints for Offshore Structures under Variable Amplitude Loading, *Offshore Mechanics and Arctic Engineering, Materials Engineering*, Vol III-B, p 385-393
8. S.-T. Chang and F.V. Lawrence, "Improvement of Weld Fatigue Resistance," FCP Report 46, College of Engineering, University of Illinois at Urbana-Champaign, 1986
9. K.M. Engesvik and T. Moan, Probabilistic Analysis of the Uncertainty in the Fatigue Capacity of Welded Joints, *Engineering Fracture Mechanics*, Vol 18 (No. 4), 1983, p 743-762
10. S.-K. Park and F.V. Lawrence, "A Long-Life Regime Probability-Based Fatigue Design Method for Weldment," FCP Report 142, College of Engineering, University of Illinois at Urbana-Champaign, June 1988
11. S.K. Park and F.V. Lawrence, Monte Carlo Simulation of Weldment Fatigue Strength, *J. Construct. Steel Research*, Vol 12, 1989, p 279-299
12. K.M. Engesvik, "Analysis of Uncertainties in the Fatigue Capacity of Welded Joints," Report UR-82-17, Department of Marine Technology, The University of Trondheim, Norway, 1982
13. T. Lassen and O.I. Eide, "Data for Fracture Mechanics Derivation of S-N Curves," The Ship Research Institute of Norway, 1984
14. M. Nihei, E. Sasaki, M. Kanao, and M. Inagaki, Statistical Analysis of Fatigue Strength of Arc Welded Joints Using Covered Electrodes under Various Welding Conditions with Particular Attention to Toe Shape, *Transactions of the National Research Institute for Metals*, Vol 23 (No. 1), 1981
15. A. Ohta, M. Kamakura, Y. Nihei, M. Inagaki, and E. Sasaki, An Automatic Detection of Fatigue Initiation Life of Welded Joints, *Transactions of the National Research Institute for Metals*, Vol 22 (No. 3), 1980
16. M. Nihei, M. Yohda, and E. Sasaki, Fatigue Properties for Butt Welded Joints of SM50A High Tensile Strength Steel Plate, *Transactions of the National Research Institute for Metals*, Vol 20 (No. 4), 1978
17. J.B. Radziminski, J.B. Srinivasan, R. Moore, C. Thrasher, and W.H. Munse, "Fatigue Data Bank and Data Analysis Investigation," Structural Research Series 405, Civil Engineering Studies, University of Illinois at Urbana-Champaign, 1973

Fatigue of Mechanically Fastened Joints

Harold Reemsnyder, Bethlehem Steel Corporation

MECHANICALLY FASTENED structural joints (such as riveted joints and high-strength bolted joints) are used in many fatigue critical applications, and a considerable amount of fatigue testing of mechanically fastened structural joints has been performed to support the development of fatigue provisions in structural design specifications. In general, the important factors affecting the fatigue strength of riveted (hot-driven) and friction bolted joints are clamping force, grip length, type of fastener, fastener pattern, tension-shear ratio, tension-bearing ratio, and type of steel in fastener or main material. These factors are discussed in various publications such as Ref 1 and 2.

Typically the fatigue resistance of high-strength friction bolted joints is superior to that of riveted joints. For example, the fatigue strength of carbon steel joints fastened with A-325 bolts approaches the yield strength of the connected material (Ref 3 and 4) (Fig. 1). On the other hand, the fatigue resistance of carbon steel joints fastened with hot-driven rivets is practically identical to the fatigue resistance of carbon steel plates perforated with circular holes (Ref 5).

Joint design and loading are probably the most critical factors, static strength of the material may not always be a factor. In some codes, fatigue resistance of bolt joints may be considered independent of static strength (Ref 6). For example, in a Swedish design recommendation for bolted connections (Ref 7), the fatigue strength is actually reduced for higher strength bolts, from a stress range of 100-120 MPa (14.5-17.4 ksi) at $N = 10^7$ cycles for Grade 8.8 bolts, to 80 MPa (11.6 ksi) for Grade 12.9 and higher strength bolts. In the European Convention for Constructional Steelworks' recommendation (Ref 8) a single S-N curve is given for all steel bolt materials up to Grade 10.9 (max UTS – 1040 MPa), where the stress range at $N = 0.5 \times 10^6$ cycles is only 26.5 MPa (3.8 ksi). However, static strength may have a significant effect for rolled threads (Fig. 2).

A summary of many tests on riveted plain carbon and alloy steels (Ref 9) also demonstrates a joint fatigue strength at about 180 MPa (26 ksi) for 2×10^6 cycles, despite a range of static tensile strengths of 415 to 690 MPa (60 to 100 ksi). This illustrates the importance of joint design and loading over that of material strength. For exam-

ple, the tension/shear/bearing ratio is very important in the fatigue life of riveted joints (Fig. 3). In particular, the stress concentrations at the hole surface degrades allowable bearing strength,

which thus promotes the use of high-strength bolts instead of rivets. In a bolted joint of proper design it is possible to carry the shear load by friction.

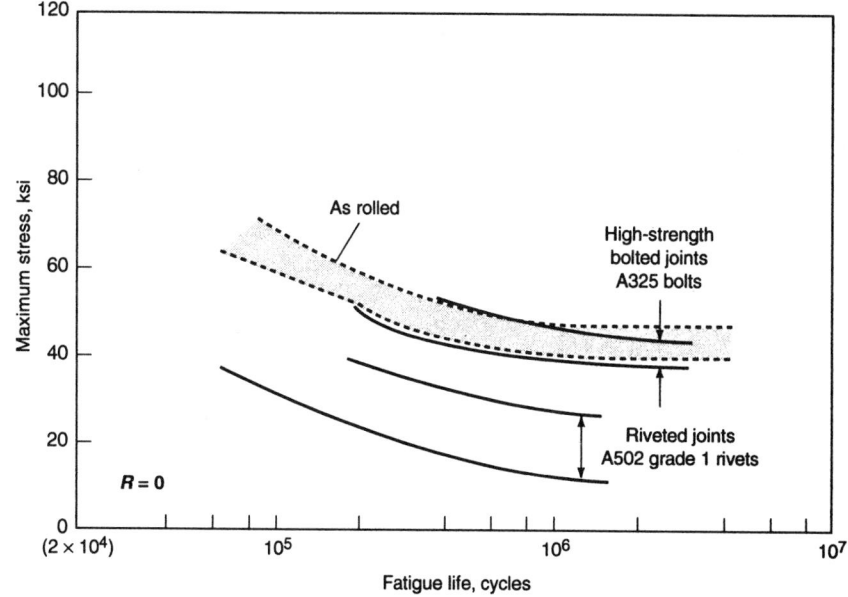

Fig. 1 Fatigue strength of carbon steel structural joints. Source: *Structural Steel Design*, Ronald Press, 1974, p 519-551

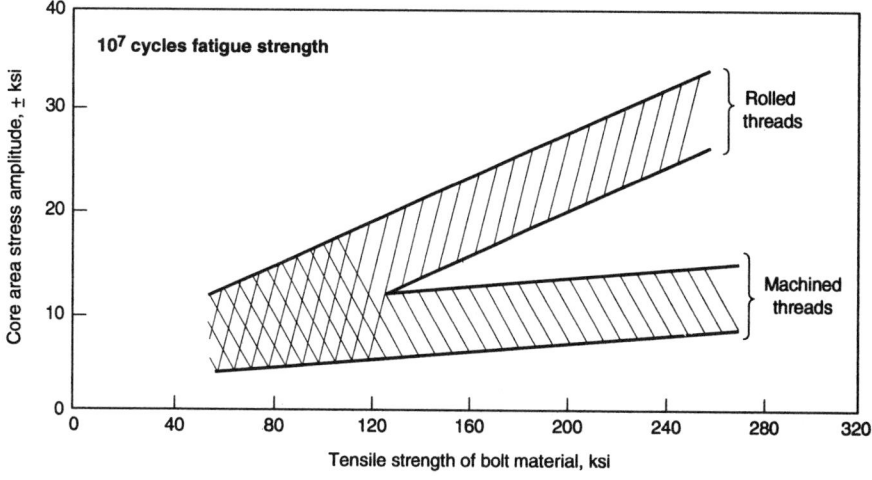

Fig. 2 Axial fatigue strength at 10^7 cycles of bolt-nut assemblies with rolled threads and machined threads ($R = -1$).

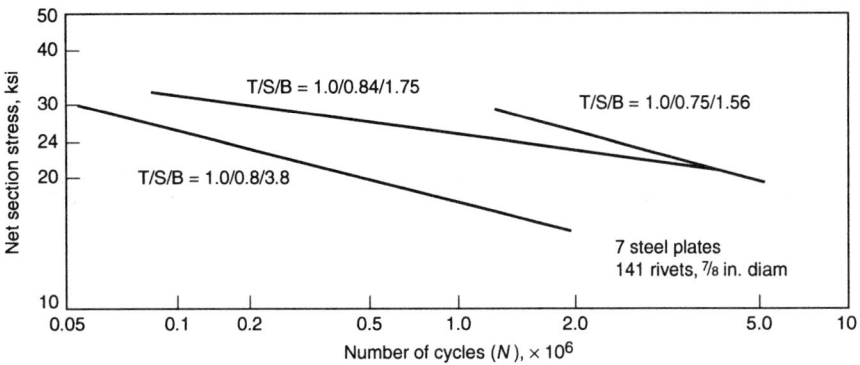

Fig. 3 Fatigue life of riveted joints with different tension-shear-bearing ratios. Source: Ref 9

Table 1 Effect of thread form on fatigue strength of bolt steel

Steel	Tensile strength, ksi	Thread form	Fatigue limit(a), ksi	Notch factor (K_f)(b)	Stress concentration factor (K_t)
0.3% C	57.4	None	37 ksi	...	...
		Whitworth	21	1.76	3.86
		American Std.	13	2.84	5.62
SAE2320	109	None	73	...	...
		Whitworth	22	3.32	3.86
		American Std.	19	3.85	5.82

(a) Repeated tension (R = 0). (b) K_f = ratio of unnotched fatigue/notched fatigue limit. Source: Ref 10

Table 2 Fatigue strength of bolt steels with unified threads

Diameter, in.	Threads per in.	Tensile strength, ksi	Unified thread	Fatigue limit at 10^7 cycles, stress range, ksi
3.8	20	134	Ground	27.5
			Rolled	48.0
		87.5	Lath-cut	17.0
			Rolled	28.5
3/4	10	141	Ground	29.5 (28.0)(a)
			Rolled	62.5 (50.0)
		87.5	Lath-cut	18.0 (16.8)
			Ground	16.5
			Rolled	34.5 (28.0)
2-1/2	10	76.0	Ground	12.5
	16		Ground	11.2

(a) Whitworth thread in parentheses. Source: Ref 11

Threaded Fasteners in Tension

Thread design and clamping forces are two key factors affecting the fatigue resistance of bolted joints. However, fastener strength can be an important variable when threads are rolled (Fig. 2). This is in direct contrast with machined threads, which demonstrate little or no static-strength effect on fatigue endurance (Fig. 2). This demonstrates the importance of surface condition on fatigue crack initiation. Rolled threads demonstrate a superior fatigue resistance due to a combination of compressive residual stresses and smoother surface at the thread root.

The form of the threads, plus any mechanical or metallurgical surface condition is much more important than nominal steel composition in determining the fatigue strength of a particular lot of bolts. Surface composition is, however, a factor. Surface decarburization can cause a significant reduction in the fatigue resistance of steel bolts. Fatigue tests on $\frac{1}{4}$-in. diam. steel bolts with a tensile strength of 917 MPa (133 ksi) showed:

Decarburized layer, in.	Fatigue limit at 10^7 cycles, ksi
None	40
0.001 in.	27
0.002	25

It is recommended that heat treatment be performed after threading to reduce the deleterious effects of decarburization. Thread-rolling also eliminates effects of decarburization.

The weakest point of an axially load standard bolt and nut combination in fatigue is normally in the bolt threads at about one turn in from the loaded, or bearing, face of the nut where the load transfer from nut to bolt is at maximum value. This area of stress concentration occurs because the bolt elongates as the nut is tightened, thus producing increased loads on the threads nearest the bearing face of the nut, which add to normal service stresses. This condition is alleviated to some extent by using nuts of a softer material that will yield and distribute the load more uniformly over the engaged threads.

Fatigue failures occur with much less frequency under the bolt head or at the thread run-out, and are usually machining errors. For example, failures beneath the bolt head can be caused by poor forging practice, grinding burns, etc. Head failures have also been caused by improper upsetting procedures that result in broken flow lines after machining in the critical head-to-shank fillet region. The stress concentration at the runout of the thread can be reduced by reducing the bolt shank diameter to less than the thread root diameter, or by a stress-relieving groove. The nut should never be run up to the end of the thread unless the shank is undercut below the thread.

Shape and size of the head-to-shank fillet are important, as is a generous radius from the thread runout to the shank. In general, the radius of this fillet should be as large as possible while at the same time permitting adequate head-bearing area. This requires a design trade-off between the head-to-shank radius and the head-bearing area to achieve optimum results. Cold working of the head fillet is another common method of preventing fatigue failure because it induces a residual compressive stress and increases the material strength.

Effect of Thread Design. The principal design fracture of a bolt is the threaded section, which establishes a notch pattern in design. Any measures that decrease stress concentration can lead to improved fatigue life. Typical examples of such measures are the use of UNJ increased root radius threads (see MIL-S-8879A) and the use of internal thread designs that distribute the load uniformly over a large number of bolt threads.

The effect of thread form on the fatigue resistance of an axially loaded bolt is shown in Table 1. The Whitworth or British Standard thread (with rounded thread root and crest) is superior in fatigue when compared to the American Standard thread form with flat crest and root (Table 1) (Ref 10). Axial load fatigue tests on "unified" screw threads for use in the USA, Canada, and Britain show a somewhat higher fatigue strength than the British Standard Whitworth (Table 2) (Ref 11). (The unified thread differs from the American Standard thread in that the former has a radius at the thread root while the latter has a flat thread root.)

As previously noted, the use of softer nuts can reduce the stress concentration at the bearing face of the nut-bolt, where maximum stress concentration occurs. For example, the stress concentration factor at the bearing face of the nut can be reduced from 3.4 to 2.5 by using an aluminum, rather than a steel, nut on a steel bolt (Ref 12).

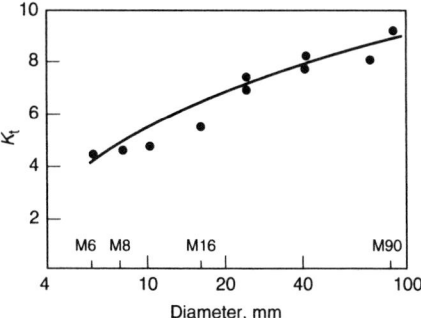

Fig. 4 Stress concentration factor (K_t) versus bolt diameter for bolts with standard metric threads. Source: Ref 14

[Note: A galvanic action may occur in corrosive environments with such a combination of two different metals.] Various nut design modifications have also improved fatigue resistance by improving the load distribution (Ref 2).

A general estimate of the notch effect on fatigue strength can be approximated using Peterson's well-known relation for the fatigue notch factor (K_f = unnotched/notched fatigue strength):

$$K_f = 1 + \frac{K_t - 1}{1 + a/r} \qquad (Eq\ 1)$$

where r is the notch root radius, and a is a material constant which is related to the ultimate material strength such that $a \approx 110/UTS^2$ (a in m, UTS in MPa units). The stress concentration factor (K_t) depends on geometry. For each groove in a threaded bolt the adjacent threads act as relief notches so that the stress concentration factor is much lower than that of a single groove of the same geometry. Otaki (Ref 13) estimated the stress concentration factor (K_t) of a single notch of thread shape to be 4.3 while for similarly shaped parallel grooves as in a threaded bolt was only 2.52. The effect of bolt diameter on K_t is shown in Fig. 4 (Ref 14). This method allows an estimate of fastener fatigue limits from fatigue strength of smooth specimens. Rough estimates of a fatigue limit may agree with the fatigue limit given in some codes, while there are large discrepancies with other codes and rules. These differences in fatigue properties are examined in Ref 6 for some European codes.

Effect of Preload. Tightening, or preloading, an axially loaded bolt-nut combination increases the fatigue strength significantly. A rule of thumb is "pretension to twice the maximum fatigue load" (Ref 2). Higher clamping forces make more rigid joints and therefore reduce the rate of fatigue crack propagation from flexing. In addition, bolts are most likely to fail by fatigue if the assemblies involve soft gaskets or flanges, or if the bolts are not properly aligned and tightened. In many assemblies, a certain minimum clamping force is required to ensure both proper alignment of the bolt in relation to other components of the assembly and proper preload on the bolt. The former ensures that the bolt will not be subjected to undue eccentric loading, and the latter that the

correct mean stress is established for the application.

Preloading, in effect, increases the area carrying the external load from the net area of the bolt to the sum of the clamped faying surfaces plus the bolt net area. The relationship between the bolt load P_B, the bolt pretension Q, and the externally applied load P is expressed by (Ref 2):

$$P_B = Q + \frac{P}{1 + \frac{K_p}{K_b}} \qquad (Eq\ 2)$$

where K_b and K_p are, respectively, the stiffnesses of the bolt and clamped parts. The cyclic load in a pretensioned bolt is a function of the external cyclic load and the stiffness ratio K_p/K_b. The magnitude of the pretension Q establishes the maximum working load, i.e., the load at which the faying surfaces separate. The load carried by the bolt P_B may be decreased by decreasing the bolt stiffness K_b, thus increasing the fatigue resistance of the bolt. For maximum fatigue resistance, soft or flexible inserts between the clamped parts should be avoided. The clamped parts should be flat and tight (i.e., K_p should be high relative to K_b). Methods for computing K_p and K_b are described in Ref 15.

Pretension is gradually reduced during cyclic loading. It appears that loss of pretension is more rapid with machine-cut threads than with rolled threads and also when there are several plys being joined. The interaction of elevated temperature and cyclic loading on bolt preload is also complicated. It appears that fatigue resistance may increase with temperature. However, at long lives, creep, rather than fatigue, may be the controlling failure mode. Bolt preloading is discussed in detail in Ref 1 and 2.

Tightening of Bolts. The determination of the actual preload in a given bolt-nut combination is difficult. Tightening with a torque wrench is the simplest but is likely to be inaccurate unless calibrated. Various devices exist that allow the clamping force-torque relation to be established for each bolt-combination. Bolts with built-in indicators of elongation, crushable washers, and bolts with break-off driving elements are also used to control preload.

It is important that this calibration be made for each lot (size, length, thread combination) of bolt-nut pairs used because the force-torque relation is quite sensitive to thread condition, lubrication, and cleanliness. For example, in the case of new, well-lubricated threads, 90% of the torque is used to overcome friction (friction under nut, 50%, thread friction 40%) and only 10% of the torque is available to develop the axial load in the bolt.

For *estimating* purposes, torque may be expressed approximately by (Ref 16):

$$T = K \cdot D \cdot P_B \qquad (Eq\ 3)$$

where
T = torque
K = torque coefficient

D = nominal bolt size
P_B = bolt tension

The unitless torque coefficient, K, may be taken as 0.2 but is, in reality, a function of thread conditions, geometry, friction between thread contact surfaces, friction between bearing face of turning element (nut or bolt head), etc. A more exact torque-bolt tension relation (Ref 17) for estimating purposes is shown in Fig. 5.

Effect of Mean Stress. Axially loaded mild steel bolts are insensitive to mean stress (Ref 2). A similar insensitivity to mean stress has been shown (Ref 18) for axially loaded high-strength structural bolts. However, in the later case, it was shown that the variability in observed lives increased with decreasing mean stress.

Studs. The fatigue resistance of studs is normally greater than that of bolts because the loading at the first turn of thread engagement is less severe. The fatigue resistances of various preloaded stud configurations are compared in Ref 2.

Bolts and Rivets in Bearing and Shear

Many fasteners transfer load through shear by bearing on the sides of holes in the joined plies. Although proportioned in design by assuming all fasteners carry an equal share of the transferred load, the fasteners in the outermost rows carry more load than those in the inner rows.

Bolted Joints. In bolted shear joints, fatigue failures are initiated at:

- Bolt holes in the outermost row where shear transfer is maximum,
- A region of fretting, or
- Some geometric stress concentration other than a bolt hole.

Failures at the first two sites are more likely than at some stress concentration other than the bolt hole. Fretting caused by relative movement of the plies is more common at long lives and relatively low stresses. In aircraft design, elaborate joints (e.g., scarfed joints with tapered plies), produce a more uniform shear transfer and, therefore, fretting predominates.

The influence of mean stress, bolt patterns and joint dimensional parameters are similar to those of riveted joints.

Cold-Driven Rivet Joints in Sheet. At short lives (less than 10^5 cycles), rivets may fail in shear while, at longer lives, sheet failure occurs through the outermost line of rivets with cracks initiated at the rivet holes. Fatigue life is insensitive to mean stress and is, primarily, a function of stress range.

In single-lap joints, the fatigue resistance increases with the ratio of rivet diameter d to sheet thickness t, with the ratio of pitch p to d, and with the number of rows of rivets. Recommended proportions (Ref 2) of riveted, single-lap joints for high fatigue resistance are listed in Table 3.

Double-lap joints are superior in fatigue resistance to single-lap joints due to offset loading in

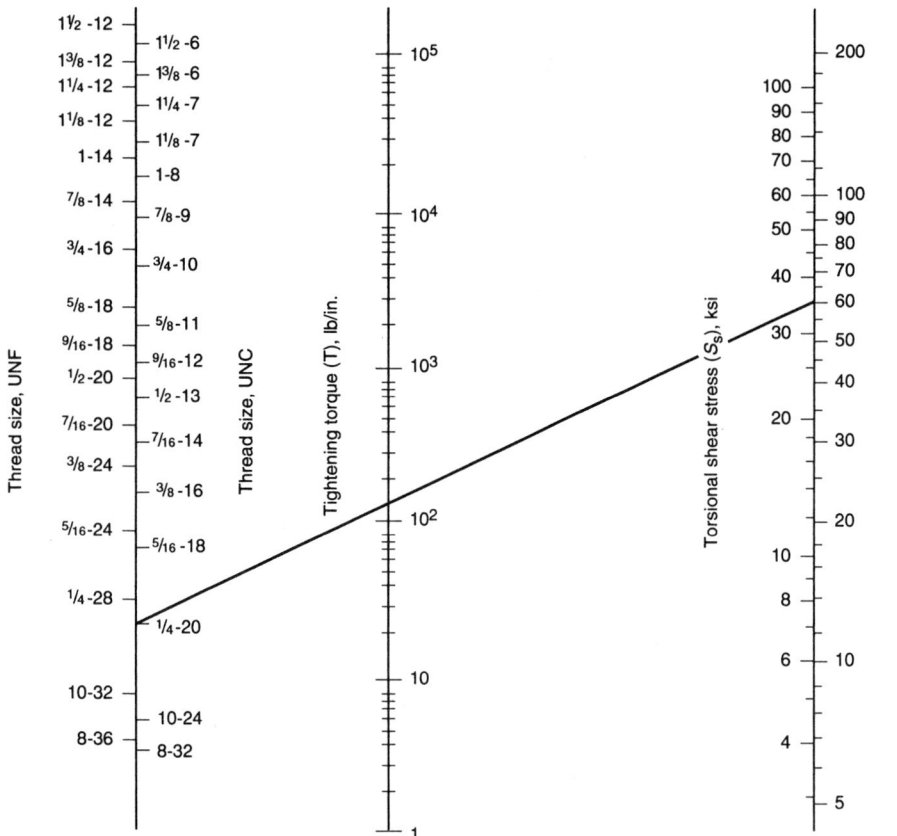

Fig. 5 Nomograph for torque on bolts. Source: Ref 17

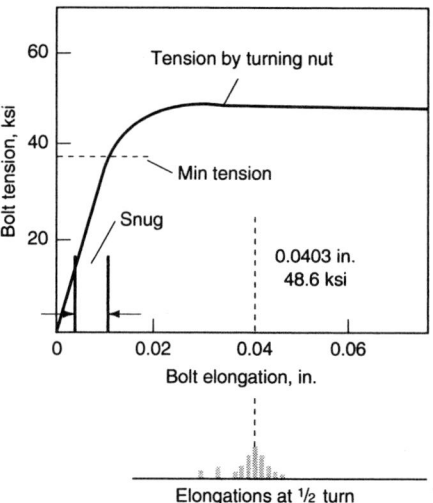

Fig. 6 Measurement of bolt elongation in a typical test joint

$$P_S = K_S \cdot m \cdot n \cdot P_B \qquad (Eq\ 4)$$

where

K_S = coefficient of slip (friction)

m = number of slip planes (number of plies –1)

n = number of bolts

P_B = bolt preload or clamping force.

Coefficients of slip for various structural steels are listed below (dry mill scale):

Steel	K_S
A36 (carbon steel)	0.35 (Ref 19)
A440 (high-strength low-alloy)	0.18 (Ref 20)
A514 (quenched and tempered)	0.20 (Ref 21)

Table 3 Riveted single-lap joint proportions for high fatigue resistance

Joint	d/t	g/d	P/d	Edge distance, inches From row	From line
Single row	3	3	...	2	1-1/2 to 2
Double row:					
In-line	4.5	4	3-1/2	3	3
Staggered	4.5	4	2-7/8 (min)		

d = rivet diameter, g = gage, t = sheet thickness, p = pitch. Source: Ref 2

the latter. The number of rows of rivets in a double-lap joint appear to have less influence on fatigue resistance than in a single-lap joint, but, again, a high d/t ratio is desirable. Also, the outer plys should be at least 0.6 times the thickness of the inner ply. Optimum proportions for double-lap joints are similar to those listed in Table 3.

Joints with flush-head rivets generally have poorer fatigue strength than those with protruding heads (Ref 2). Rivet hole fabrication also has a significant influence on fatigue resistance. In order of decreasing fatigue strength, fabrication techniques are: coin-dimpling, spin-dimpling, drilling, and machine-countersinking.

Hole Sizing and Interference Fits. The stress concentration factor of a circular hole loaded in bearing with a round pin is significantly higher than an open hole in a remotely loaded tensile strip (see section "Pin Joints" in this article). However, these high local stresses can be reduced

by "sizing" or drifting." The hole is plastically expanded to produce compressive residual stresses by forcing an oversize, smooth ball or pin through the hole.

Tapered bolts and self-sizing or interference bolts (e.g., the Huckbolt) also introduce compressive residual stresses in the hole surface through plastic deformation. Interference bolts can be used in multi-bolt joints but, to be effective, the hole patterns must be accurately matched in all plies. This is accomplished by drilling and reaming the plies held together in a jig assembly.

Friction Joints. The most effective means of increasing the fatigue strength of joints is by tensioning the fasteners to clamp the plies together. The load is then transferred from one ply to another by friction of the faying surfaces rather than by shear in the bolts bearing on the plies. The maximum, or slip, load that can be transferred through friction is

An increase in fatigue strength with clamped joints is only achieved if relative slipping is eliminated—otherwise fretting failures are likely. Slipping is best prevented by careful positioning of the bolts over the entire contact area, using smallest possible gage and pitch (allowing for wrench clearance), tightening to a high bolt clamping force, and reducing the contact, or faying, surfaces to a minimum. Unnecessary overhang of plies beyond the first bolt row must be eliminated to prevent fretting.

Clamping causes the plies to behave integrally as if welded together and to more uniformly distribute the total load over the fasteners. The stress concentrations at the holes are reduced by the compressive clamping forces and fatigue failure is generally initiated away from the holes. The fatigue resistance of bolted friction joints is discussed in the next section.

It is essential in assembling bolted friction joints that a consistent, high bolt preload be maintained. This is accomplished through the calibration of torque or pneumatic impact cut-off wrenches. The bolt-tension calibrator is a hydraulic instrument which measures the bolt tension developed by tightening the bolt or nut. The test is made on at least three bolts of each lot (size-

Table 4 Minimum bolt breaking strengths

Bolt diameter, in.	Breaking strength, ksi	
	A325	A490
1/2	17.0	21.3
3/4	40.1	50.1
1	72.7	90.9
1-1/4	102	145
1-1/2	147	211

Source: Ref 22

grip combination) and the torque or cut-off to develop the required tension is averaged (Ref 22).

An alternate criterion for tightening bolts is the "turn-of-nut" method, Fig. 6, developed in the railroad and automotive industries. Bolts are installed to a specified rotation from the "snug-tight" condition. "Snug-tight" condition is achieved when the bolt tension or preload equals 70% of the specified minimum breaking strength of the bolt (Table 4). Nut rotations when both faces are normal to the bolt axis are (Ref 22):

Bolt length, 1	Nut rotation from "snug-tight"
1 < 4 diameters	1/3 turn
4d < 1 < 8d	1/2 turn
8d < 1 < 12d	2/3 turn

The button-head, interference-body bolt, called, variously, interrupted-rib, structural-rib, or Dardelet-rivet bolt, combines the clamping action of a high-strength fastener with that of an interference fastener. The body fit of this fastener is an advantage in the assembly of galvanized materials with low interface friction. The body-bound fit—metal-to-metal contact—also ensures better electrical grounding of structures.

Hot-Driven Riveted Joints of Structural Plate. Important factors affecting the fatigue strength of riveted (hot-driven) and friction bolted joints are clamping force, grip length, type of fastener, fastener pattern, tension-shear ratio, tension-bearing ratio, and type of steel in fastener or main material. Failure through the net section occurs in both riveted and bolted joints, but an increase in clamping force of the bolts can cause failure in the gross section (Ref 3). In a riveted joint, fatigue cracks are initiated at the rivet holes while, in friction bolted joints, they are initiated at the edges of the clamped plies. It has been shown that the fatigue resistance of high-strength friction bolted joints is superior to that of riveted joints and that the fatigue strength of carbon steel joints fastened with A-325 bolts approaches the yield strength of the connected material (Fig. 1).

Some research indicates that no steel shows a superiority over carbon steel as plate material in riveted joints subjected to cyclic loading. It has been suggested that the effect of variation in clamping force within a joint or from one joint to another subordinates the relative superiority of one steel over another and produces the great experimental scatter. In addition the fatigue resistance of riveted joints is very sensitive to the tension-shear-bearing ratio (Fig. 2), which masks variations in static strength. As previously noted, high-strength bolted joints transfer the loads through friction and, therefore, avoid the above sensitivity.

Field Repair and Rehabilitation of Riveted Structures. The generally accepted method of repair or rehabilitation of elderly riveted structures to extend their service life has been the replacement of entire members that either contain fatigue damage or are suspected to be overstressed in cyclic loading. One program (Ref 23) studied the effectiveness of rehabilitation by replacement of rivets with high-strength structural bolts in critical or fatigue-damaged locations as a more economical alternative to replacement of entire members. Constant amplitude fatigue tests and variable amplitude service simulation tests of full-scale model ore bridge joints and constant amplitude tests of joints removed from an ore bridge showed that rehabilitation of fatigue-damaged members by fastener replacement increases fatigue life 2 to 6 times. The effect on fatigue life extension of increasing bolt tension above the minimum value accepted in present specifications and/or the effect of crack length (up to 1 inch long) at rehabilitation were found to be of secondary importance.

Drilling holes at crack tips is not always effective in arresting fatigue cracks in structural members. Tests on welded crane runway girders have shown that, should the drilled hole miss the crack tip, crack growth could be reinitiated (Ref 24). It is often very difficult to ensure that the hole includes the crack tip. It was concluded, therefore, that this technique is neither sufficiently reliable nor applicable to be recommended as a universal repair for cracks. Instead, a drilled hole with a properly torqued high-strength bolt (70% or more of bolt breaking strength), could be considered as an expedient crack arrestor until a more reliable repair can be implemented. The combination of drilled hole and high-strength bolt has also been used as a temporary crack arrestor on railway cars.

Pin Joints

The most extensive review of the stress concentrations and fatigue resistance of pin joints has been made by Heywood (Ref 2). Stress concentration factors for pin joints have also been presented by Lipson and Juvinall (Ref 25) and Peterson (Ref 26).

The fatigue strength of a single pin joint is extremely low in comparison to that of the base material. The fatigue strength is influenced by the stress concentration factor K_t which, in turn, is a function of many parameters—geometric and material. The relationship between mean stress and stress range are further complicated by fretting.

Stress Concentrations in Pin Joints. The stress concentration factor in the lug of a pin joint is defined as

$$K_t = \frac{\text{maximum stress at hole}}{\text{net-section stress}} \qquad (\text{Eq 5})$$

where the net section (on the transverse diameter of the hole) is $(D-d)/t$. D, d, and t are, respectively, lug width, hole diameter, and lug thickness.

As d/D approaches unity, the lug tends to act as a flexible strap and K_t goes to unity. K_t is also a function of the edge distance H and for $H/D = 1$ (Ref 2):

$$K_t = 0.6 + 0.95 \, (D/d) \qquad (\text{Eq 6})$$

and for $H/D = 0.5$,

$$K_t = 0.85 + 0.95 \, (D/d) \qquad (\text{Eq 7})$$

for values of D/d from 0.2 to 0.8. Note: Heywood (Ref 2) uses a stress concentration factor C_t based on the gross area where

$$C_t = K_t/(1 - d/D)$$

In addition to the hole size relative to lug size affecting K_t, the stress concentration factor is also influenced by:

- Lug head shape
- Lug "waisting"
- Clearance and interference of pin
- Pin material and lubrication
- Pin bending

A square-shaped lug head has a lower K_t than a rounded lug head. However, head shape has little effect at large values of D/d. Reducing the lug width on its loaded side (i.e., "waisting") increases K_t, as documented in Ref 2. As the clearance between the pin and hole increases, K_t also increases. Interference fit reduces K_t but this reduction varies inversely with load (Ref 2).

Heywood (Ref 2) suggests that $K_t = 3$ is near-optimum for fatigue strength design. The proportions for this K_t would be $D/d = 2$, $H/d = 1$, and $t/d = 1$. Heywood also discusses in detail the beneficial effects of interference, compressive residual stresses, and preloaded lugs.

Bibliography of Stress-Intensity Factors

The description of fatigue crack growth in mechanically fastened joints by stress intensity factor concepts is complicated by:

- The statically indeterminate nature of the stress distribution in the plies along, and across, the hole array
- The change in stress distribution due to ply stiffness changes with crack growth
- The influence of hole proximity on stress intensity factor, i.e., notch shadow effect

The following sections list some references that address some of these factors for mechanically fastened joints.

Crack Growth from Holes. The use of fracture mechanics in aircraft joints with open, cold-worked, and pin-loaded holes, and with interference fit fasteners is described in Ref 27 and 28. Fatigue crack propagation from rivet holes is studied in Ref 29 and 30. Radial cracks at pin-loaded holes and corner cracks are studied in, respectively, Ref 31 and 32.

Dealing with cracks at holes in engineering structures is discussed in Ref 33. Stress-intensity factor solutions for cracks emanating from both open and loaded holes in finite width plates, lugs, and multi-fastener joints are tabulated and plotted in Ref 34. A weight function method is used in Ref 35 to estimate stress-intensity factors for surface and corner-cracked fastener holes.

Life Extension. Crack growth in riveted and bolted friction structural joints is discussed in Ref 23. The use of high-strength bolts to retard or arrest crack growth in structures is described in Ref 23 and 24.

Axially Loaded Round Bars. The application of fracture mechanics to the fasteners themselves requires stress-intensity-factor solutions for cracks in round bars. Solutions for circumferentially cracked, axially loaded, unnotched, round bars are summarized in Ref 36-38, while the circumferentially notched case is studied in Ref 39. Solutions are available for axially loaded round bars containing:

- Straight crack fronts (Ref 40-42)
- Small curved crack fronts (Ref 43, 44)
- Semi-elliptical crack fronts (Ref 45, 46)
- Sickle-shaped cracks (Ref 47)

Bending in round bars is treated for:

- Straight crack fronts (Ref 40, 48-50)
- Semi-elliptical fronts (Ref 45, 46, 51)
- Edge-cracked bars under tension and bending (Ref 52, 53)
- Round bars in tension and in bending containing straight cracks (straight and semicircular cracks, respectively, Ref 54, 55)

Stress-intensity factors for the surface and the deepest point of cracks varying from nearly straight to semicircular are presented for cylindrical bars in tension and bending in Ref 56. Stress-intensity factor solutions for both straight- and curved-fronted cracks in a bar subjected to either axial load or bending are summarized and compared in Ref 57 and 58.

The stress-intensity factor solutions for cracked, round bars, both with and without threads, subjected to either axial load or bending, are reviewed, compared, and synthesized into a form suitable for the analysis of bolts and studs in Ref 59 (with German and Japanese bibliography).

Compliance solutions are presented for the straight-front edge-cracked solid and hollow round bars in tension (Ref 42) and in bending (Ref 49). The compliance solution of Ref 49 for solid bars in bending has been extended to deeper cracks (Ref 50).

The growth and shape of surface fatigue cracks has been studied and stress-intensity factor solutions have been developed experimentally for tension loaded rods (Ref 60), solid and hollow cylinders (Ref 61), and high-strength bolts (Ref 62). Fracture mechanics predictions of crack shape, size, and life have been compared to observations from tests on shafts in bending and containing semi-elliptical cracks (Ref 51), and surface-cracked shafts in bending with shoulder fillets (Ref 63).

Fasteners. The growth of a fatigue crack in a bolt of a bolt-nut combination has been modeled using finite element analysis and fracture mechanics (Ref 64). Stress-intensity factor solutions for axially-loaded, threaded round bars are presented in Ref 53 (bending as well as tension) and 65-68. It should be noted that the analyses of Ref 53, 64, 65 modeled thread-like circumferential projections (two or more) in round bars while Ref 66-68 modeled thread-like single circumferential grooves in round bars. However, the projections and grooves were not helical. Instead, the planes of the projections and grooves were perpendicular to the axis of the round bar.

More information on the use of fracture mechanics for fasteners is covered in Ref 69-78. Stress-intensity factors are also summarized in the Appendix Section of this Volume.

REFERENCES

1. Carl Osgood, *Fatigue Design*, 2nd ed., Pergamon Press, 1982, p 247-298
2. R.B. Heywood, *Designing Against Fatigue in Metals*, Reinhold, 1962, p 168-285
3. F. Baron and E.W. Larson, Jr., Comparative Behavior of Bolted and Riveted Joints, *Trans. ASCE*, Vol 120, 1955, p 1322-1352
4. F. Baron and E.W. Larson, Jr., *AREA Bulletin*, No. 54 (503), Sept-Oct 1952, p 175-190
5. H.S. Reemsnyder, Fatigue of Riveted and Bolted Joints, Bethlehem Steel Corp., Bethlehem, Pa., 24 July 1969
6. P. Haagensen and T. Slind, Fatigue Design Data for Threaded Fasteners, *Fatigue 90*, Engr. Material Advisory Services, 1990, p 2353-2359
7. "Bolted Connections—Design and Installation," Institutet för Verkstads-teknisk Forskning, Mekanförbundet, IVF-Resultat 82611, 1983 (in Swedish)
8. "European Recommendations for Bolted Connections in Structural Steelwork," Publication No. 38, 4th ed., March 1985
9. C.W. Lewitt, E. Chesson, Jr., and W.H. Munse, *The Effect of Rivet Bearing on the Fatigue Strength of Riveted Joints*, A Progress Report by University of Illinois, Eng. Exp. Station for Illinois Div. of Highways and Dept. of Commerce-Bureau of Public Roads, Urbana, Illinois, Jan 1959
10. R.C.A. Thurston, The Fatigue Strength of Threaded Connections, *Trans. ASME*, Vol 73, Nov 1951, p 1085-1092
11. P.G. Forrest, Fatigue of Metals, Pergamon Press, New York, 1962, p 262-280
12. D.G. Sopwith, The Distribution of Load in Screw Threads, *Proc. Inst. Mech. Engrs.*, Vol 159, 1948, p 373-383
13. H. Otaki, Konstruktion, Vol 31 (No. 3), 1979, p 121-126 (in German)
14. P. Haagensen et al., Size Effects in Machine Components and Welded Joints, OMAE 88, 1988, p 386
15. G. Meyer and D. Strelow, Simple Diagrams Aid in Analyzing Forces in Bolted Joints, *Assembly Engineering*, Jan-March, 1972
16. *Fastener Standards*, The Industrial Standards Institute, Cleveland, 1965, p 343-345
17. Nomogram for Torque on Bolts, *Design News*, Feb 1973, p 76-77
18. J.A. Dunsby, W. Pitman, and A.C. Walker, The Fatigue Strength of High Strength Structural Bolts, National Aeronautical Establishment, National Research Council of Canada, 1971
19. J.L. Rumpf, Riveted and Bolted Connections, Chapter 18 in Structural Steel Design, L. Tall, Ed., Ronald Press, New York, 1974, 2nd ed., p 592-639
20. P.C. Birkemoe and R. Srinivasan, Fatigue of Bolted High Strength Structural Steel, *Proc. ASCE*, Vol 97 (No. ST3), March 1971, p 935-950
21. P.C. Birkemoe, D.F. Meinheit, and W.H. Munse, Fatigue of A514 Steel in Bolted Connections, *Proc. ASCE*, Vol 95 (No. ST10), Oct 1969, p 2011-2030
22. Specification for Structural Joints Using ASTM A-325 or A-490 Bolts, Research Council on Riveted and Bolted Structural Joints, New York, 8 May 1974
23. H.S. Reemsnyder, Fatigue Life Extension of Riveted Connections, *Proc. ASCE*, Vol 101 (No. ST12), Dec 1975, p 2591-2608
24. H.S. Reemsnyder and D.A. Demo, Fatigue Cracking in Welded Crane Runway Girders: Causes and Repair Procedures, *Iron and Steel Engineer*, Vol 5, April 1978, p 52-56
25. C. Lipson and R.C. Juvinall, Handbook of Stress and Strength, McMillan, New York, 1963
26. R.E. Peterson, Stress Concentration Factors, John Wiley & Sons, 2nd ed., New York, 1974
27. A.F. Grandt, Jr. and J.P. Gallagher, "Proposed Fracture Mechanics Criteria to Select Mechanical Fasteners for Long Service Lives, in *Fracture Toughness and Slow-Stable Cracking*, STP 559, American Society for Testing and Materials, Philadelphia, 1974, p 283-297
28. G.J. Petrak and R.P. Stewart, "Retardation of Cracks Emanating from Fastener Holes," *Eng. Frac. Mech.*, Vol 6, 1974, p 275-282
29. I.E. Figge and J.C. Newman, Jr., "Fatigue Crack Propagation in Structures with Simulated Rivet Forces," in *Fatigue Crack Propagation*, STP 415, American Society for Testing and Materials, Philadelphia, 1967, p 71-93
30. D. Brock and H. Vlieger, "Cracks Emanating from Holes in Plane Stress," *Int. J. of Frac. Mech.*, Vol 8, 1972, p 353-356
31. D.J. Cartwright and G.A. Ratcliffe, "Strain Energy Release Rate for Radial Cracks Emanating

from a Pin Loaded Hole," *Int. J. of Frac. Mech.*, Vol 8, 1972, p 175-181

32. A.F. Liu, "Stress Intensity Factor for a Corner Flaw," *Eng. Frac. Mech.*, Vol 4, 1972, p 175-179

33. D. Brock, "Cracks at Structural Holes," MCIC-75-25, Metals and Ceramics Information Center, Columbus, Ohio, March 1975

34. D.L. Ball, "The Development of Mode I, Linear-Elastic Stress Intensity Factor Solutions for Cracks in Mechanically Fastened Joints," *Eng. Frac. Mech.*, Vol 27 (No. 7), 1987, p 653-681

35. W. Zhao and S.N. Atluri, "Stress Intensity Factor for Surface and Corner Cracked Fastener Holes by the Weight Function Method," presented at the *Symposium on Structural Integrity of Fasteners*, American Society for Testing and Materials, Miami, 1992

36. G.C. Sih, "Handbook of Stress Intensity Factors," Institute of Fracture and Solid Mechanics, Lehigh University, Bethlehem, Pa., 1973

37. H. Tada, P.C. Paris, and G.R. Irwin, "The Stress Analysis of Cracks Handbook," 2nd ed., Paris Productions Inc., St. Louis, Mo., 1985

38. Y. Murakami, Editor-in-Chief, "Stress Intensity Factors Handbook," Pergamon Press, New York, 1987, 2 volumes

39. P. Lefort, "Stress Intensity Factors for a Circumferential Crack Emanating from a Notch in a Round Tensile Bar," *Eng. Frac. Mech.*, Vol 10, 1978, p 897-904

40. W.S. Blackburn, "Calculation of Stress Intensity Factors for Straight Cracks in Grooved and Ungrooved Shafts," *Eng. Frac. Mech.*, Vol 8, 1976, p 731-736

41. O.E.K. Daoud, D.J. Cartwright, and M. Carney, "Strain-Energy Release Rate for a Single-Edge-Cracked Circular Bar in Tension," *J. of Strain Analysis*, Vol 13 (No. 2), 1978, p 83-89

42. A.J. Bush, "Stress Intensity Factors for Single-Edge-Crack Solid and Hollow Round Bars Loaded in Tension," *J. of Testing and Evaluation*, Vol 9 (No. 4), July 1981, p 216-223

43. R.P. Gangloff, "Quantitative Measurements of the Growth Kinetics of Small Fatigue Cracks in 10Ni Steel," in *Fatigue Crack Growth Measurement and Data Analysis*, STP 738, American Society for Testing and Materials, Philadelphia, 1981, p 120-138

44. G.G. Trantina, H.G. deLorenzi, and W.W. Wilkening, "Three-Dimensional Elastic-Plastic Finite Element Analysis of Small Surface Cracks," *Eng. Frac. Mech.*, Vol 18 (No. 5), 1983, p 925-938

45. A. Athanassiadis, J.M. Boissenot, P. Brevet, D. Francois, and A. Raharinaivo, "Linear Elastic Fracture Mechanics Computations of Cracked Cylindrical Tensioned Bodies," *Int. J. of Frac.*, Vol 17 (No. 6), Dec 1981, p 553-566

46. I.S. Raju and J.C. Newman, Jr., "Stress Intensity Factors for Circumferential Surface Cracks in Pipes and Rods," in *Fracture Mechanics, Seventeenth Volume*, STP 905, American Society for Testing and Materials, Philadelphia, 1986, p 789-805

47. C. Mattheck, P. Morawictz, and D. Munz, "Stress Intensity Factors of Sickle-Shaped Cracks in Cylindrical Bars," *Int. J. of Fatigue*, Vol 7, Jan 1985, p 45-47

48. O.E.K. Daoud and D.J. Cartwright, "Strain Energy Release Rates for a Straight-Fronted Edge Crack in a Circular Bar Subject to Bending," *Eng. Frac. Mech.*, Vol 19 (No. 4), 1984, p 701-707

49. A.J. Bush, "Experimentally Determined Stress-Intensity Factors for Single-Edge-Crack Round Bars Loaded in Bending," *Experimental Mechanics*, Vol 16 (No. 6), July 1976, p 249-280

50. F. Ouchterlony, "Extension of the Compliance and Stress Intensity Formulas for the Single Edge Crack Round Bar in Bending," in *Fracture Mechanics Methods for Ceramics, Rocks, and Concrete*, STP 905, American Society for Testing and Materials, Philadelphia, 1981, p 237-256

51. T. Lorentzen, N.E. Kjaer, and T.K. Henriksen, "The Application of Fracture Mechanics to Surface Cracks in Shafts," *Eng. Frac. Mech.*, Vol 23 (No. 6), 1986, p 1005-1014

52. O.E.K. Daoud and D.J. Cartwright, "Strain-Energy Release Rate for a Circular-Arc Edge Crack in a Bar under Tension or Bending," *J. of Strain Analysis*, Vol 20 (No. 1), 1985, p 53-58

53. K.J. Nord and T.J. Chung, "Fracture and Surface Flaws in Smooth and Threaded Round Bars," *Int. J. of Frac.*, Vol 30 (No. 1), 1986, p 47-55

54. A. Carpinteri, "Stress Intensity Factors for Straight-Fronted Edge Cracks in Round Bars," *Eng. Frac. Mech.*, Vol 42 (No. 6), 1992, p 1035-1040

55. A.S. Salah el din and J.M. Lovegrove, "Stress Intensity Factors for Fatigue Cracking of Round Bars," *Int. J. of Fatigue*, Vol 3, July 1981, p 117-123

56. M. Caspers and C. Mattheck, "Weighted Average Stress Intensity Factors of Circular-Fronted Cracks in Cylindrical Bars," *Fatigue Frac. Engng. Mater. Struct.*, Vol 9 (No. 5), 1987, p 329-341

57. O.E.K. Daoud and D.J. Cartwright, "Stress Intensity Factors and Strain-Energy Release Rates for Edge-Cracked Circular Bars," in *The Mechanism of Fracture*, V.S. Goel, Ed., American Society for Metals, 1986, p 223-229

58. C.K. Ng and D.N. Fenner, "Stress Intensity Factors for an Edge Cracked Circular Bar in Tension and Bending, Method," *Int. J. of Frac.*, Vol 36, 1988, p 291-303

59. L.A. James and W.J. Mills, "Review and Synthesis of Stress Intensity Factor Solutions Applicable to Cracks in Bolts," *Eng. Frac. Mech.*, Vol 30 (No. 5), 1988, p 641-654

60. D. Wilhem, J. FitzGerald, J. Carter, and D. Dittmer, "An Empirical Approach to Determining K for Surface Cracks," in *Advances in Fracture Research*, Pergamon Press, New York, 1980, p 11-21

61. R.G. Forman and V. Shivakumar, "Growth Behavior of Surface Cracks in the Circumferential Plane of Solid and Hollow Cylinders," in *Fracture Mechanics, Seventeenth Volume*, STP 905, American Society for Testing and Materials, Philadelphia, 1986, p 59-74

62. T.L. MacKay and B.J. Alperin, "Stress Intensity Factors for Fatigue Cracking in High-Strength Bolts," *Eng. Frac. Mech.*, Vol 21 (No. 2), 1985, p 391-397

63. E. Hojfeldt and C.B. Ostervig, "Fatigue Crack Propagation in Shafts with Shoulder Fillets," *Eng. Frac. Mech.*, Vol 25 (No. 4), 1986, p 421-427

64. D.F. Fischer, E.T. Till, and F.G. Rammerstorfer, "Fatigue Cracks in Bolt Threads," in *Fatigue of Steel and Concrete Structures*, International Association of Bridge and Structural Engineers, Eth-Honggerberg ch 8093, Zurich, 1982, p 725-732

65. G. Reibaldi and M. Eiden, "SpaceLab Bolt Fracture Mechanics Analysis," *European Space Agency Working Paper No. 1274*, March 1981

66. A.A. Popov and A.V. Ovchinnikov, "Stress Intensity Factors for Circular Cracks in Threaded Joints," *Strength of Materials*, Vol 15 (No. 11), 1983, p 1586-1589

67. R.C. Cipolla, "Stress Intensity Factor Approximations for Semi-Elliptical Cracks at the Thread Root of Fasteners," in *Improved Technology for Critical Bolting Applications*, MPC-Vol 26, ASME, 1986, p 49-58

68. C.W. Springfield and H.Y. Jung, "Investigation of Stress Concentration Factor—Stress Intensity Factor Interaction for Flaws in Filleted Rods," *Eng. Frac. Mech.*, Vol 31 (No. 1), 1988, p 135-144

69. J. Toribio, "Stress Intensity Factor Solutions for a Cracked Bolt Loaded by a Nut," *Int. J. of Frac.*, Vol 53, 1992, p 367-385

70. M. Makhutov, V. Zatsarinny and V. Kagan, "Initiation and Propagation Mechanics of Low Cycle Fatigue Cracks in Bolts," in *Advances in Fracture Research (Fracture 81)*, D. Francois, Ed., Pergamon Press, Oxford, UK, 1982, Vol 2, p 605-612

71. A.F. Liu, "Evaluation of Current Analytical Methods for Crack Growth in a Bolt," in *Proceedings of the 17th ICAF Symposium*, Stockholm, May 1993

72. W.D. Dover, "Crack Shape Evolution Studies in Threaded Connections Using A.C.F.M.," *Fatigue of Engineering Materials and Structures*, Vol 5 (No. 4), 1982, p 349-353

73. D.H. Michael, R. Collins, and W.D. Dover, "Detection and Measurement of Cracks in Threaded Bolts with an A.C. Potential Difference Method," *Proc. Royal Society*, Vol A 385, 1983, p 145-168

74. H. Nakajima, T. Shoji, M. Kikuchi, H. Niitsuma, and M. Shindo, "Detecting Acoustic Emission during Cyclic Crack Growth in Simulated BWR Environment," in *Fatigue Crack Growth Measurement and Data Analysis*, STP 738, American Society for Testing and Materials, Philadelphia, 1981, p 139-160

75. A.F. Liu, "Behavior of Fatigue Cracks in a Tension Bolt," presented at the *Symposium on Structural Integrity of Fasteners*, American Society for Testing and Materials, Miami, 1992

76. J. Toribo, V. Sanchez-Galvez, and MA. Astiz, "Stress Intensification in Cracked Shank of

Tightened Bolt," *Theo. and Applied Frac. Mech.*, Vol 15, 1991, p 85-97

77. R.C. Cipolla, "Stress Intensity Factor Approximations for Cracks Located at Thread Root Region of Fasteners," presented at the *Symposium on Structural Integrity of Fasteners*, American Society for Testing and Materials, Miami, 1992

78. A. Tafreshi and W. Dover, "Stress Analysis of Drillstring Threaded Connections Using the Finite Element Method," *Int. J. Fatigue*, Vol 15 (No. 5), 1993, p 429-438

Statistical Considerations in Fatigue

Paul S. Veers, Sandia National Laboratories

STATISTICAL METHODS and probability are useful tools when uncertainty influences important outcomes, which is why many fundamental mathematical aspects of statistical theory began with the casinos of Europe (e.g., Monte Carlo). Knowledge of probability and statistics permits the "house" to put large quantities of capital at risk (on the gaming tables) by creating games with odds only slightly in their favor. Although gamblers win almost half the time, on average the house will win. One of the paradigms of stochastic analysis is known as the "Gambler's ruin problem" (Ref 1), where the time to "ruin" (loss of all capital) is estimated given the slight disadvantage to the player. Notice that it is not a matter of *if* but *when*.

The design of structures subjected to fatigue loadings has produced a substantial interest in statistical aspects, for similar reasons. There is substantial uncertainty as to the outcome, that is, the lifetime of each finished product. In fact, even under the controlled conditions of the laboratory with carefully prepared specimens and loadings, the time to failure in identical tests will vary widely. In addition, in the design of structures, the capital at risk is often high due to the high cost of a single installation, the large numbers of units to be produced, or life-critical applications.

In the early years of fatigue design, attention was mainly focused on the phenomenon of endurance limit in steels. The statistical problem is then reduced to simply comparing a single load level with a single strength number, the endurance limit. Figure 1 schematically illustrates the problem of determining the acceptable distance between the average loading and the average endurance limit. The goal is to produce an acceptably low probability that a combination of higher than average loading and lower than average strength will result in failure. This is exactly analogous to the standard approach to static strength design.

The endurance limit approach, as illustrated in Shigley and Mitchell (Ref 2), assumes a carefully prepared test specimen to establish an upper bound on the endurance limit and then applies modifying factors (also called knockdown factors) to reduce the strength to actual conditions. These include factors for surface finish, size effects, reliability, temperature, stress concentra-

tion, and, ominously, miscellaneous effects, which include such diverse possibilities as corrosion, electrolytic plating, metal spraying, strain rates, and fretting. Only the factor called "reliability" is related to the sample-to-sample variation in the endurance limit. The importance of the endurance limit is illustrated by the extensive guidance provided for determining the median endurance limit from laboratory data (Ref 3).

Unfortunately, designers these days rarely have the luxury of keeping the maximum stresses below the endurance limit. Such designs are too heavy and costly for the high efficiency demands driven by weight and cost constraints. In reality, loadings can almost never be described by a single stress level, but contain a distribution of stress amplitudes and means. In addition, under variable-amplitude loading the very existence of an endurance limit depends on *all* the stresses staying below it (Ref 4 and 5). The simple schematic of Fig. 1 is therefore a poor model for the complexity of a variable-amplitude fatigue analysis, which is discussed in more detail in the articles "Fundamentals of Modern Fatigue Analysis for Design" and "Estimating Fatigue Strength" in this Volume.

In general, uncertainty can be divided into random (or aleatory) and statistical (or epistemic) types. Quantities whose uncertainties are random have values that are inherently variable and can be described by distribution functions, exceedance diagrams, or histograms. Random uncertainty cannot be reduced; it is inherent in the quantity. A good example is material strength, which will always vary from sample to sample. The good thing about these uncertainties is that they can be well defined with additional data collection, and acceptable levels of reliability can be selected. However, the variability cannot be reduced. Statistical uncertainties are in some ways harder to deal with because they are usually due to a lack of knowledge of some sort. There may be a single correct value for a parameter, but one doesn't know exactly what the value is. Therefore, this uncertainty can be reduced with additional data collection or analysis, but the distribution of possible values will always be a best guess. Typical examples include a stress concentration that is not known exactly, or uncertain usage and loading, or errors and biases built into

the analysis models. Both statistical and random uncertainties need to be considered and accounted for in producing reliable designs (i.e., designs that have an acceptable probability of failure before the target lifetime). Both are typically described by a distribution of possible values for the uncertain parameters using a probability density function (defined below).

There are therefore two parts to dealing with uncertainty in fatigue design: determine the distributions of possible values for all uncertain inputs; and calculate the probability of failure due to all the uncertain inputs. This article reviews each of these parts in turn. First, the sources of uncertainty in a fatigue analysis are discussed. They are divided into four principal areas:

- Material properties
- Distribution of applied stress levels within a given environment
- Environments or loading intensities
- Modeling or prediction

Then a probabilistic approach for analyzing the uncertainties and determining a level of reliability (probability of failure) is presented.

This article cannot hope to include all the approaches to dealing with the statistical aspects of fatigue. Several references are included for additional reading. Statistical analysis of fatigue test data can be found in ASTM STP 744 (Ref 6). Wirsching (Ref 7) and Wirsching et al. (Ref 8) give a thorough explanation of the basics of linking probability to fatigue analysis. Thoft-Chris-

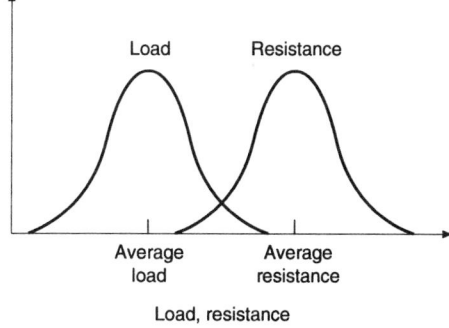

Fig. 1 Schematic of probabilistic definition of load and resistance

tiansen (Ref 9) and Madsen (Ref 10) cover the probabilistic analysis. Tangjitham and Landgraf (Ref 11) and Veers et al. (Ref 12) provide a couple of specific application examples. Scatter and variability in fatigue life tests are also covered in the article "Fatigue Data Analysis" in Volume 8 of the *ASM Handbook*.

Basic Descriptors of Variability and Uncertainty. One descriptor for variability or uncertainty that is used repeatedly in this article and deserves early definition is the coefficient of variation (COV). It is the ratio of the standard deviation and mean of a random quantity. It can be written in terms of samples x_i of a random variable X:

$$COV_X = \frac{\sigma_X}{m_X}; \ m_X = \frac{1}{N}\sum_{i=1}^{N} x_i;$$

$$\sigma_X^2 = \frac{1}{N}\sum_{i=1}^{N}(x_i - m_X)^2 \qquad (Eq\ 1)$$

(For $N < 30$, N is replaced by $N - 1$ in the estimate of the mean (m) and by $N - 2$ in the estimate of the standard deviation, σ, to produce unbiased estimates.) The COV is nondimensional and can be applied to any quantity that varies. The interpretation of the COV may differ with application, however.

How a quantity varies can be described empirically, and most intuitively, by plotting the relative number of times the quantity falls into various ranges in a histogram. If the histogram cells are integrated, upward or downward, it is possible to determine the relative percentage of time the quantity is below some level, or conversely, will exceed the level. In standard probability theory, these histograms are taken to the limit as cell width goes to zero, become continuous functions, and are known as the probability density function, cumulative density function, and inverse cdf, respectively. A good overview of probabilistic concepts treated in an engineering vein can be found in Ang and Tang (Ref 13).

Quantities with near-zero mean values cannot use the COV as a descriptive parameter; the standard deviation with its dimensional requirement is needed. Fortunately, most of the quantities of interest in fatigue analysis are positive valued and the COV is quite descriptive. If restricted to positive outcomes, only low-COV quantities can be normally distributed (Gaussian distribution), while high-COV quantities are often modeled by Weibull and other one-sided distributions, as discussed below.

Uncertainty in S-N Data

As mentioned above, dividing a problem into constituent (albeit sometimes arbitrary) parts can be useful in modeling a large problem. The area that usually comes to mind in fatigue statistics is the scatter in S-N data. It is not the only area of statistical importance, as the following sections are intended to show, but it is definitely an area

where variability is large. The statistics of the fatigue response of the material cannot be ignored. Schijve (Ref 14) provides a very good discussion of this topic. An ASCE series on fatigue reliability suggests that, in most large structure applications, a COV of 0.6 or higher is appropriate for variation of S-N data (Ref 15). Because the uncertainty in an S-N curve can (and usually does) represent over half of the total uncertainty in fatigue life estimation, the modelling deserves special attention.

The S-N curve is usually established based on the results of numerous constant-amplitude tests to failure at several loading amplitudes and mean stresses. To condense everything to a single S-N plot and increase the number of samples in the analysis, the mean and amplitude are usually condensed into an effective stress amplitude using a mean stress correction, such as the Goodman or Gerber models (Ref 16).

The goal of the statistical analysis is generally to find an S-N curve at an acceptable probability level, which represents the relationship between effective stress amplitude and lifetime for which less than some small percentage of samples would fail. This curve can then be applied to a deterministic lifetime calculation. If all the other inputs to the calculation are known perfectly (are not random), the outcome is a prediction of a lifetime at which less than the specified percentage of parts will have failed.

It is rare that all the nonmaterial inputs (loads, stress concentrations, etc.) are known perfectly. Additional safety factors need to be applied. It is therefore useful in probabilistic calculations to have a model for the distribution of possible values for the S-N curve. Then the uncertainty in S-N properties can be combined with the other uncertainties in a unified probabilistic analysis to estimate total probability of failure. Probabilistic analysis is covered below in the section on dealing with uncertainty.

There are numerous and varied approaches to carrying out the statistical analysis of S-N data. The ASTM standard for establishing median S-N curves at various confidence levels is given in Ref 3. A compilation of papers on statistical analysis of fatigue data can be found in Ref 6. Some other examples can be found in Ref 10 and 17. There is also an American Society for Testing and Materials standard for statistical analysis of S-N data (Ref 18).

Two levels of random variable models for S-N uncertainty are described in this article. The first assumes that all the variation can be captured in a single random variable; the second uses two random variables. These are presented as the simplest approaches that are defensible, statistically speaking, but are not necessarily the absolute best in every situation. The emphasis is on simplicity while producing a random variable model for the distribution of possible fatigue lives at a given stress level.

Single-Random-Variable Model. The S-N curve is often assumed to follow the relationship shown in Fig. 2, given by the equation

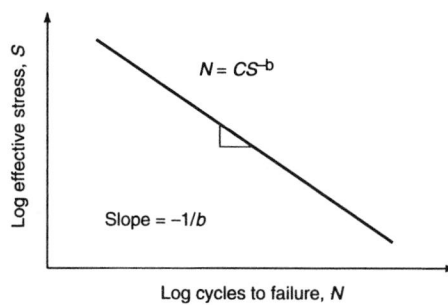

Fig. 2 Schematic of S-N curve

$$N = CS^{-b} \qquad (Eq\ 2)$$

This assumes that the curve will be a straight line on a log-log plot. C is called the S-N coefficient and b is the exponent, which is one over the slope of the curve. The median line can be determined by a linear regression on the logarithms of the data:

$$\ln(N) = \ln(C_0) + (-b_0)\ln(S) \qquad (Eq\ 3)$$

solving for the slope, $-b_0$, and intercept, $\ln(C_0)$. The result is a good estimate of the *median* curve, because a best fit will, on average, result in half of the data on each side of the regression line. The *mean* will not be the same as the median, because the distribution (of the original data) will be skewed. Remember, even though the S-N curve is traditionally plotted with the stress on the ordinate, stress is actually the independent variable and should be treated as such in the regression analysis. To treat cycles as the independent variable can lead to errors in the regression (Ref 19).

With the median value of life determined for all effective stress amplitudes, it is left to determine how much variation about that line is present. If there are sufficient replications, the variation about the median can be established at several stress levels. Without copious replications, the data can again be condensed by estimating the variance of the ratios of the data points divided by the median regression line:

$$X_i = \frac{N_i}{C_0 S_i^{-b_0}} \qquad (Eq\ 4)$$

The overall COV_X is calculated from all the ratios, X_i, using Eq 1.

A distribution of X can be established by plotting the X_i on probability paper of various types and searching for the best fit. Many people have their favorite distribution type: some like log-normal, which results from normally distributed logarithms of the data, and some prefer Weibull, which is a natural form for fatigue data because it was originally derived as a model of "weakest link" behavior for fatigue applications. Both are characterized by two parameters (Table 1), which can be derived from the median and COV. With less than about 50 samples it may be difficult to find a significant difference between the fits to the two distribution types. Figure 3 shows the log-normal and Weibull distributions plotted on

Table 1 Normal, log normal, and Weibull distributions

Statistical distribution	Probability density function (pdf)	Variance	Mean
Normal	$f(x) = \dfrac{1}{\sqrt{2\pi}\sigma} \exp\left[-\frac{1}{2}\left(\dfrac{x-\mu}{\sigma}\right)^2\right]$	σ^2	μ
Log normal	$f(x) = \dfrac{1}{\sqrt{2\pi}\sigma x} \exp\left[-\frac{1}{2}\left(\dfrac{\ln(x)-\mu}{\sigma}\right)^2\right], x > 0$	$(\exp(\sigma^2)-1)\exp(2\mu+\sigma^2)$	$\exp\left(\mu + \dfrac{\sigma^2}{2}\right)$
Weibull	$f(x) = \dfrac{\beta}{\alpha}\left(\dfrac{x}{\alpha}\right)^{\beta-1} \exp\left[-\left(\dfrac{x}{\alpha}\right)^\beta\right], x > 0$	$\sigma^2\left[\Gamma\left(1+\dfrac{2}{\beta}\right) - \Gamma^2\left(1+\dfrac{1}{\beta}\right)\right]$	$\alpha\Gamma\left(1+\dfrac{1}{\beta}\right)$

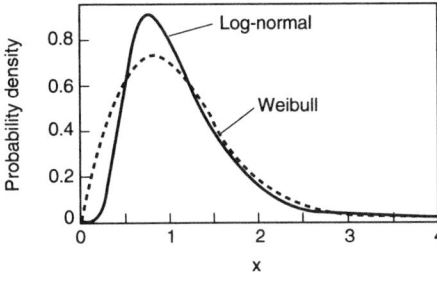

(a)

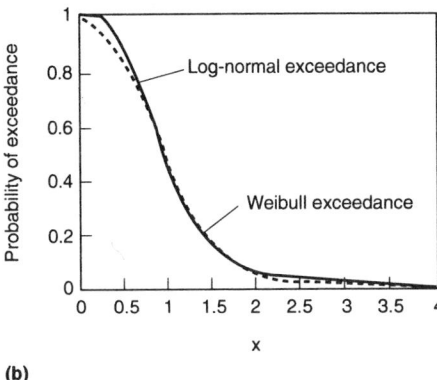

(b)

Fig. 3 Comparison of log-normal and Weibull functions with median = 1.0 and COV = 0.6. (a) Probability density functions. (b) Exceedance diagrams

top of each other for a median of 1 and a COV of 0.6, a typical number for material S-N data (Ref 15). The largest difference occurs on the lower end of the distributions; the log-normal has a fatter tail on the right and the Weibull has a fatter tail on the left. The two are quite similar in the medium to high range. The difference on the low end can lead to tremendous differences in estimated probability of failure, including a factor of 10 in one example (Ref 20). The Weibull distribution is more conservative, providing more probability of significantly lower fatigue lives. When neither can be clearly established as the best fit, it may be practical to be conservative and assume Weibull, or to use both and compare the results.

The probabilistic model derived from this data fitting approach is to let the S-N coefficient, $C = C_0X$, be the only random variable. Because $C = C_0X$, the distribution shape of C is the same as for

X. The S-N exponent is assumed to be a deterministic constant, $b = b_0$. All realizations of the S-N curve then exist as parallel lines of identical slope and variable intercept on a log-log plot, as shown in Fig. 4. (The resulting COV of lifetime is the same at all effective stress levels. Schijve, for example, shows in Ref 14 that this is not necessarily the case. Therefore, this method has a significant limitation of application, but it is simple and often used. See Ref 15 and 21.) The S-N curve representing a given probability of survival, say the 95% line often used in design, is one of these parallels. The Weibull 95% line for Fig. 3 would lie at 0.25 times the number of cycles at the median line, while the log-normal is 0.41 times the median.

Two-Random-Variable Model. Another approach is to create a two-random-variable fit to the data. One can employ standard statistical analysis tools to calculate the linear regression fit to Eq 3 and get estimates of the uncertainty and correlation between the estimated slope and intercept. Most data analysis software packages will do this operation, but it may be difficult to output the actual variance of, and correlation between, the slope and intercept. Some packages output confidence limits only at user-selected levels; more statistically oriented packages provide more detailed information. The results are usually based on the assumption that b and $\ln(C)$ are normally distributed. The correct interpretation of the statistical analysis is therefore to assume that b is normal and that C is log-normal.

It is important, however, to transform the data before beginning the regression analysis, subtracting the average logarithm of lifetime from all the logarithms, $Y = \ln(N_i) - m_{\ln(N)}$, forcing the stress intercept to be at $Y = 0$. Without this shift, there will be high correlation between the slope and intercept due to deterministic coupling (i.e., any change in slope would have to be accompanied by a change in intercept to keep the line running through the data). For example, Ronold (Ref 25) fits the two-random-variable model without a shift and finds a correlation of –0.996. When data do not fit a linear curve, the method described by Yu et al. ("Fatigue Data Pooling and Probabilistic Design," ASTM STP 1106, p 197) may be useful. When both C and b are modeled as random variables, uneven variation at different stress levels will result. The spread is greater

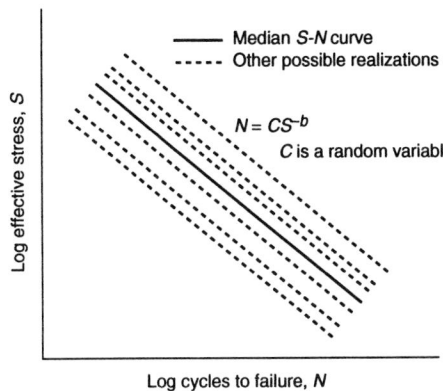

Fig. 4 Single-random-variable model of S-N variability

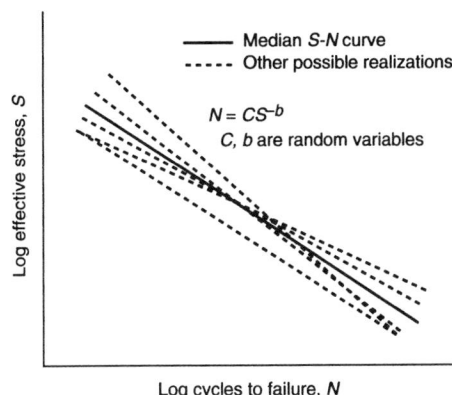

Fig. 5 Two-random-variable model of S-N variability

toward the edges of the test data at both high and low stress levels, as illustrated in Fig. 5.

Applied Stress Levels

There are two levels of statistical description required to define the distribution of applied stresses and the level of certainty for that distribution. Except in the rare case where the cyclic stresses are absolutely fixed in amplitude for all possible situations and over all time, there will be some distribution of loading amplitudes which can and should be described using the tools of probability and statistics. The second kind, or perhaps level, of statistical application is that the exact distribution of cycles to be experienced over the design life is not known.

The fact that stress amplitudes are irregular does not by itself add significantly to the uncertainty in fatigue life estimation. Over most lifetimes, a stable and representative sample of all possible stress amplitudes will occur, one that does not vary significantly in its net damaging effect. The key statistic is the average damage done by the overall distribution of cycles.

The second kind of statistic, one that describes the uncertainty in the actual distribution, does not average out and has a direct impact on the fatigue life uncertainty. The impact can be substantial because the stress amplitudes are raised to the b_{th} power in the lifetime calculation, where b is the exponent of the S-N curve. Because b can range from 3 to >10, depending on the application, the amplification of uncertainties in overall level of stress response can be large. For example, a 5% error in stress amplitude results in a 16% error in damage when $b = 3$, but produces a 63% error when $b = 10$.

Distributions of Stress Amplitudes. Stress amplitudes are traditionally described by an exceedance diagram, which is the equivalent of a probability distribution function, with some important differences. Figure 6 shows some exceedance diagrams from Wirsching and Chen (Ref 22). Figure 7 shows equivalent distributions of cycles as probability density functions (pdf's). The pdf is a continuous and normalized form of a histogram where the area under each segment of the curve represents the percentage of all cycle amplitudes within that range. The integral of the pdf is the cumulative distribution function (cdf), which starts at zero and goes to one with increasing amplitude. The complementary cdf, one minus the cdf, is the equivalent of the exceedance diagram.

The exceedance diagram is usually normalized by the maximum stress level, which drives most statisticians to distraction. There are few statistics less dependable than the maximum value of the samples of a random process. If measurements are made of an irregular loading, the only thing that can generally be guaranteed is that the maximum value or range will be significantly different each time a sample measurement is taken. More stable statistics include the mean value (first moment), the variance (second moment), and sometimes higher moments. These are the types of statistics that can be used to describe standard pdf's.

Figures 6 and 7 show Weibull distributions with different shape factors. Because stress amplitudes are always positive, the distributions that describe them are necessarily one sided. The family of Weibull distributions provides a good basis for matching most stress amplitude data found in nature. The Weibull distribution is a two-parameter distribution described by its mean (or median) and shape factor (which is approximately equal to one over the COV). There is a good description of the use of Weibull distributions for simple fatigue calculations in Ref 8. A third parameter can be added by truncating the data on the low end and shifting the distribution. The use of three-parameter Weibulls is described in Ref 23. Winterstein and Lange (Ref 24) illustrate how a quadratic distortion of a Weibull (or other pdf) can improve the fit to wind-driven structural response. With the fourth parameter (the quadratic distortion of the standard, shifted Weibull), the fit to almost any data set can be very good.

The reason for using standard distribution types to describe loading measurements, rather than applying the empirical data directly to the fatigue analysis, is threefold. First, the statistics on which they are based, means and variances,

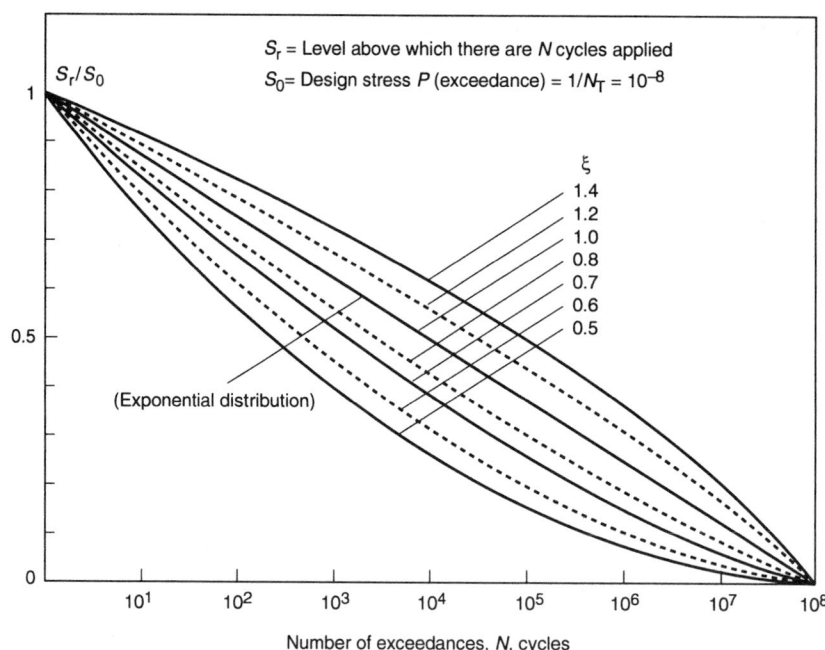

Fig. 6 Exceedance diagrams for Weibull distributions of various shape factors. Source: Ref 22

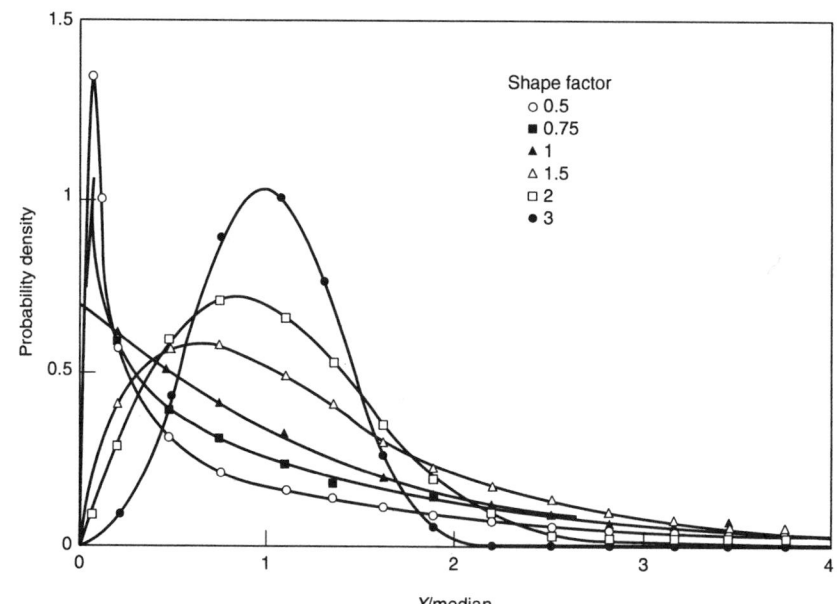

Fig. 7 Weibull probability density functions of various shape factors normalized by the median

are good descriptors of the process as a whole and can be confidently estimated with smaller data samples. The distributions based on these measures should be less sensitive to sample-to-sample variations in measurements of the loading process. Second, the fatigue life calculations can be simplified greatly by using a known analytical form for the distribution of cycle amplitudes, rather than a set of empirical measurements such as histogram cell frequencies. Third, the uncertainty in the parameters of the fitted distribution can be estimated and used in the probabilistic analysis (Ref 25, 26).

There are reasons to believe that Weibull is a good fit to vibration response data. The Weibull with a shape factor of 2 is known as a Rayleigh distribution and is the natural result of narrow-band random vibration of a linear system (Ref 27). A squaring phenomenon, such as might result from fluid drag forces, will distort the Rayleigh into an exponential distribution (Weibull with shape factor 1.0). Likewise, any nonlinearities in the vibration will distort the distribution of stress amplitudes, but the result is often a Weibull distribution with another shape factor. For systems that are linear, but not perfectly narrow band, estimates of correction factors have been published (Ref 28-30).

Many loadings are the result not of vibration but of quasistatic response to variable external forces. When the loadings are driven by natural forces, such as winds or waves, the responses are still often modeled well by analytical distributions. When the loadings are driven by human intervention (e.g., maneuvers of vehicles or aircraft), it may be more difficult to fit the responses to simple analytical forms, although it is often still possible (Ref 30). Empirical distributions based on measurement can always be used. However, because the extreme values in the measurements are highly uncertain, and the higher end of the distribution is usually important in fatigue life estimates, it may be better to fit an analytical form over the entire range of measurements. This permits the estimate of high-amplitude cycles to be somewhat influenced by the body of the measurements, not dictated by the extreme values that happen to be captured in a limited experiment.

Uncertainty in Stress Distribution. One of the greatest benefits of using an analytical form to model the distribution of stress amplitudes is that the uncertainty in the parameters of the analytical form can be estimated. Of the standard statistical techniques that apply, "bootstrapping" may be the best (Ref 31, 32). Bootstrapping involves resampling the data that has already been collected (with replacement) to create pseudoreplicates of the sample. Distributions are fit to the pseudoreplicates, and statistics can be derived for the parameters of the fitted distributions. The result is an analytical form for the distribution of stress cycles with the parameters of the distribution (e.g., mean, COV, or shape factor) treated as random variables.

Environments and Loading Intensities

Many environmental factors that affect fatigue durability are not directly connected to loading. Shigley and Mitchell (Ref 2) list several, including temperature, stress concentration, corrosion, electrolytic plating, strain rates, and fretting. There are a host of others, including ultraviolet or ionizing radiation, moisture, and salt spray. These are usually dealt with using knockdown factors on the S-N curve. Each one will be uncertain and should be described as such. General guidance on these varied sources of uncertainty is not available; each one needs to be dealt with individually.

The description of a particular loading distribution will depend on the intensity of the loading environment. Loading distributions are usually conditional on some environmental parameter. For offshore applications it is "sea state" (Ref 33); for wind loading it is average wind speed (Ref 34); for automobiles, it may be "driver profile" (Ref 35) or road roughness. The total distribution of stresses is the integral over all the environments that the structure (or vehicle) will experience over its lifetime. The distribution of loading intensities can be described similarly to the way that stress amplitude distributions are defined above. The uncertainty in the loading intensities can also be described by making the parameters of the distribution functions random variables, just as was done for stress amplitudes.

Because there is such a wide diversity of applications, it is impossible to make global statements about how to model the variations in environmental intensities. The point to be made is that they should be modeled. It is better not to throw all the stress amplitudes from all the possible loading intensities together into a single pot for fatigue analysis. Rather, stress amplitudes should be defined as a function of loading intensity. Then fatigue analysis can be adjusted to account for applications where the environmental intensities may differ and the fatigue demands may differ as well. Guidance on the cost of operating in different environments, as well as customized inspections or maintenance schedules (Ref 36), can be established if the analysis depends on the intensity of the environment.

Modeling or Prediction Errors

Material properties and applied loads must be combined in a fatigue life prediction using some damage accumulation or crack growth model. However, there may be considerable uncertainty in how accurately those models predict lifetime. The model accuracy is heavily obscured by the tremendous scatter in fatigue lives, which is apparent under constant-amplitude loading where there is no need for a model.

The accuracy of the Miner's rule model has been investigated repeatedly. Miner's rule can be stated simply as the assumption that the "damage state," Δ, is estimated by:

$$\Delta = \sum_i \frac{n_i}{N(S_i)} \qquad \text{(Eq 5)}$$

where n_i is the number of applied cycles at effective stress amplitude S_i, and $N(S_i)$ is the number of cycles to failure at effective stress amplitude S_i. Failure is assumed to occur when $\Delta = 1$. It is a simple model and widely used, especially in high-cycle applications. In experience, the value of Δ at failure is often quite different from 1.0.

Wirshing and Wu (Ref 17) gathered data from 12 publications (Ref 37-48) aimed at addressing the variability in Δ from as far back as 1945 and involving as few as one set of 11 tests to as many as 37 separate sets of variable-amplitude tests. The results are summarized in Table 1. The means range from a low of 0.836 to a high of 1.72, and the COVs range from 0.16 (for a small number of tests) to 0.98 (for a large number of tests). The overall mean is calculated to be about 1.3, and the COV associated with that mean is about 0.8. In two publications, the exact number of tests was not reported, but the mean and COV move up or down by at most 20% when the assumed number of tests in the two incomplete references are doubled and halved.

A COV of 0.8 is a relatively large uncertainty. The Table 2 numbers include the variability inherent in material response, which has been placed at about 0.6 (Ref 15). Even if this is assumed to be 0.6, the remaining COV is about 0.5, also quite high. Inaccuracies in predictions in some of the (perhaps older) references due to misapplication of Miner's rule or other currently discredited damage accumulation models may be a source of additional variability. It had been the practice, in materials with fatigue endurance limits, to assign no damage for cycles with amplitudes less than the endurance limit. We now know that an endurance limit derived from constant-amplitude testing only exists when all the stresses are below it (Ref 4, 5). Cycles below the limit will be damaging in irregular loading where the limit is often exceeded. These variations from report to report due to different modeling approaches would tend to inflate the COV somewhat. It may be that some of the bias and uncertainty in modeling can be reduced by tuning the analysis technique for a specific application. Wirsching (Ref 21), aware of the above results, selected a median value of 1.0 and a COV of 0.3 in offshore appli-

Table 2 Variation in Δ

Reference	Number of tests	Δ, mean	Δ, median	Standard deviation	Coefficient of variation
37	266	1.53	1.3	0.962	0.627
38	~200(a)	1.72	1.23	1.68	0.98
39 (axial)	31	1.27	1.23	0.341	0.269
39 (rotating beam)	29	1.47	1.39	0.519	0.353
40	18	1.15	1.14	0.186	0.161
41	83	0.863	0.823	0.271	0.314
42	11	0.863	0.809	0.219	0.262
43	11	0.98	0.949	0.251	0.256
44	34	1.28	1.15	0.62	0.48
SAE	54	1.46	1.09	1.3	0.889
45	47	1.23	1.06	0.49	0.4
46	~400(a)	1.08	0.9	0.726	0.67
47	43	0.882	0.849	0.249	0.292
48	35	0.792	0.778	0.151	0.191
Combined	~1262	1.29	(b)	1.00	0.78

(a) Estimated (total number of tests not explicitly stated in the reference). (b) Cannot be calculated from the reported data. Source: Ref 17

cations. It appears that a slightly higher COV might be warranted. As in the case of *S-N* properties, log-normal and Weibull distributions are good candidates, with the Weibull distribution being more conservative.

Dealing with Uncertainty

The emphasis of this article to this point has been on modeling the sources of uncertainty and inherent randomness as random variables. This section attempts to point out how these descriptions can be combined to produce a probability of failure result. Attention is centered on the inputs and outputs of the analytical approaches, which are now available as commercial software packages that are easy to use. The intent here is to make the probabilistic modeling capability known and lead the interested reader to a more thorough treatment in the references.

Applying Probability. Let me begin this section with a story. I had just finished a presentation on applying structural reliability methods to dealing with uncertainty in the fatigue analysis of a particular class of problems. The approach (discussed below) requires that distributions (pdf's) be assigned to all uncertain inputs. The probability of failure before a specified lifetime is then calculated. The first audience response was from a distinguished and senior researcher from a highly esteemed university, a man whose knowledge and opinion I hold in high regard. He gave a brief lecture on the danger involved in attempting to design in the absence of a complete set of material test data. To guess at the correct range of possible values without sufficient laboratory-generated data is to invite disaster, he said. The second comment came from the director of engineering for a major corporation. He agreed with the researcher up to a point, but said that, unfortunately, he and his staff are rarely given the luxury of getting all the answers before a decision needs to be made. The constraints of the marketplace (and sometimes the corporation) do not allow sufficient time and money to completely define the critical material or structural response properties. Designers are forced to do the best job they can with the data available and use good "engineering judgment."

The idea that a probabilistic tool will tempt designers away from thorough testing before making design choices is off target; decisions are routinely made without complete data! The feeling that a probabilistic analysis is an attempt to get away without doing your homework in the test laboratory may be widespread, but probabilistic methods should never be used that way. While it is true that any tool may be abused, it is a perversion of probabilistic methods to assume more certainty than you have. The primary issue is establishing exactly what is or is not known. Creating distribution models that describe the relative likelihood of different values of important inputs is a good exercise. If the value of an important input is completely unknown, the distribution of possible values should reflect reality by being very wide. In the best case, prob-

abilistic analysis forces designers to be completely honest about their level of knowledge. It makes them document the distribution of possible values that are assumed for all the inputs, instead of trying to account for all the uncertainties with one or a few safety factors applied at the end. In fact, one very useful application of probabilistic analysis is in establishing safety factors and design criteria that will produce an agreed-on level of reliability, that is, a predetermined probability of failure or, in the fatigue case, of premature failure (Ref 10, 22, 25, 26, 33).

Every designer of fatigue-sensitive structures would like to know the lifetime of the design with perfect accuracy. The design could then be fine tuned to eliminate needless costs while maintaining acceptable durability. Unfortunately, designers are often disappointed with fatigue life predictions. Not only are the techniques difficult to apply, requiring a daunting level of detail of the machine and its environment, but the results are highly sensitive to changes in the inputs (Ref 34). Ranges of plausible answers from two months to ten years erode the value of the results and make the process frustrating. The knowledge that this sensitivity is inherent to the fatigue problem is of little comfort. This sensitivity suggests that an appropriate range of uncertainty be reported, reflecting both natural variability and professional ignorance of precise structural behavior, mechanical fatigue laws, and so forth. A good designer will therefore put appropriate safety factors on all the uncertain quantities that affect fatigue life. It would be beneficial, however, to provide a more quantitative measure of the design conservatism. This in turn suggests that the proper question may be not "What is the actual fatigue life of this component?" but rather, "With what probability will the component meet its target design life?" Such questions are naturally addressed by the theory of structural reliability.

Analytical Methods. The ability to calculate the probability of failure for truly complex (i.e., realistic) problems has developed rapidly and fairly recently. The first probabilistic analysis capability successfully applied to structural reliability was the awkwardly named mean-value first-order second-moment method (Ref 49). To oversimplify greatly, it assumes that everything in the world is Gaussian, uses the nice property that the sum of Gaussian distributed variables is Gaussian, and provides probabilities of failure for the standard normal (or Gaussian) distribution of the difference between loading and resistance. This approach worked fairly well for static loading problems where there are often additive loadings. It suffers from the fact that the world is not Gaussian and not all uncertain factors are additive. Hassofer and Lind (Ref 50) created a generalized safety index that solved many of the problems of the simpler index but was still highly limited in application.

Fatigue life prediction is often composed of many multiplicative, rather than additive, factors. The next logical step was to assume that the logarithm of the world is Gaussian. Then the sum of the logarithms (product of physical variables)

is also Gaussian and many of its nice features can again be exploited. This was done by Wirsching (Ref 21) for fatigue of offshore structures. This approach also has limited application because of the restriction of distribution type and problem formulation.

Rackwitz and Fiessler (Ref 51) greatly expanded the applicability of probabilistic analysis when they created a numerical procedure that transforms all the random variables into uncorrelated Gaussians and then numerically estimates the probability of failure in the transformed space. They first approximated the failure region linearly and called the analysis FORM (first order reliability method). An extension to approximate the failure region with a second-order fit is called SORM (second order reliability method). Both are well described in Ref 9 and 10. With FORM/SORM there is no longer any restriction on the assumed pdf of the random variables. Also, a virtually unlimited number of random variables can be included in the analysis. Other approaches in the same vein, but using novel numerical techniques, have been developed by Wu, including the Wu/FPI algorithm (Ref 52) and the advanced mean value method (Ref 53).

The analyst using these approaches does not need to do the probabilistic analysis; that is accomplished by the computer algorithm, just as rainflow counting is no longer done by drawing drips off pagoda roofs, but is left to software subroutines. The required inputs are the distribution functions and correlations between all the random variables describing the uncertainties. The form of the problem does not matter as long as one can calculate (numerically) the state of the response (failed or not) given values for all the random variables. The numerical analysis searches the space of possible values for the region that contains combinations most likely to occur and that result in premature failure. It then solves for the probability that those combinations will occur. Some codes that perform this probabilistic analysis are marketed under the names RELAX and FPI (USA), STRUREL (Germany), and PROBAN (Norway).

Analyses programmed in the failure evaluation subroutines have been enormously varied, from simple closed-form functions to large finite-element analyses. Good examples can be found for an automotive application (Ref 11, 35, 54), for offshore structures (Ref 10, 36), and for wind turbines (Ref 12, 55, 56).

These probabilistic tools can be applied to fatigue-life and crack-growth problems in the following ways. For fatigue-life problems, one starts with a deterministic analysis capability for the time to failure. An objective function is created by defining a time to failure less than a specified target lifetime as "unsafe." The probabilistic analysis package, given the deterministic capability and the objective function, will select a set of values for all the random inputs (usually starting with the mean values) and calculate the time to failure. The directional gradient of the difference between the calculated time and the target is also determined in the space of the random variables.

Various optimization methods are then used to search for the set of values of the random variables that results in a time to failure equal to the target value. The space occupied by all possible values resulting in "unsafe" time to failure is integrated to determine the total probability of the unsafe condition.

In crack-growth applications the "unsafe" condition can be defined to reflect the possibility of growing to a critical crack length between inspections (or within the component lifetime). In this case the deterministic analysis calculates the crack growth from some initial size (the distribution of which may be determined by the probability of detection characteristics of the inspection technique) until the next scheduled inspection. Probabilistic analysis is done as in the fatigue-life case. The probability that the crack has reached a critical size before the next inspection is estimated. A probability sufficiently large to cause concern means that the inspection must be scheduled earlier. Conversely, if the probability is sufficiently small, the inspection interval can be increased.

FORM/SORM and advanced mean value techniques provide more information than just the probability of failure. Importance factors are also calculated as a natural part of the analysis. Importance factors indicate the relative magnitude of the effect of each random variable on the probability of failure. Variables that could change in value throughout the assigned input distribution without significantly altering the overall probability of failure show low importance. The controlling random variables (uncertainties) are thus flagged for special attention. It is also a fairly simple task to conduct sensitivity studies, because the calculations are usually very fast. The advanced mean value method is particularly good at mapping out the probability of failure versus some parameter, such as probability of failure versus time, although any of the above-mentioned analytical approaches can be used to achieve this result.

Simulation. Another way to calculate the probability of premature failure is to calculate the fatigue lifetime repeatedly, using different values for the uncertain quantities in each calculation. If the values are selected at random from the distribution of possible values, the approach is called "Monte Carlo analysis" (Ref 57, 58). The probability of failure is equal to the fraction of outcomes that have shorter than desired lifetimes. Monte Carlo methods have become quite popular due to the ready availability and ease of repeated numerical analyses. However, if the probability of failure is low, which is usually the desired situation, the number of simulations required to accurately estimate the probability is very high. Depending on the desired accuracy, between 100 and 1000 failure outcomes are necessary. For low fractions of failure outcomes this translates into millions of repeated calculations. Some standard statistical software packages do offer capabilities for Monte Carlo analysis, so the investment in time can be relatively short, and with the ready availability of fast computing, simulation may be the fastest way to answer for the probability of failure. Simulation will not provide importance factors, however, unlike structural reliability methods such as FORM/SORM or the advanced mean value method. Sensitivity studies using simulation may also be so time consuming as to be prohibitive. More advanced simulation techniques have also been developed for probabilistic analysis; a good overview can be found in Bjerager (Ref 59).

REFERENCES

1. S.M. Ross, *Stochastic Processes*, John Wiley & Sons, 1983, p 114-115
2. J.E. Shigley and L.D. Mitchell, *Mechanical Engineering Design*, 4th ed., McGraw-Hill, 1983, p 270-356
3. R.E. Little, *Tables for Estimating Median Fatigue Limits*, STP 731, American Society for Testing and Materials, 1981
4. M.R. Mitchell, Fundamentals of Modern Fatigue Analysis for Design, *Fatigue and Microstructure*, American Society for Metals, 1979
5. N.E. Dowling, Estimation and Correlation of Fatigue Lives for Random Loading, *International Journal of Fatigue*, Vol 10 (No. 3), 1988, p 179-185
6. R.E. Little and J.C. Ekvall, Ed., *Statistical Analysis of Fatigue Data*, STP 744, American Society for Testing and Materials, 1981
7. P.H. Wirsching, Probabilistic Fatigue Analysis, *Probabilistic Structural Mechanics Handbook*, C. Sundarajan, Ed., Chapman & Hall, 1995, p 146-165
8. P.H. Wirsching, T.L. Paez, and K. Ortiz, *Random Vibrations Theory and Practice*, John Wiley & Sons, 1995, p 266-295
9. P. Thoft-Christiansen and M.J. Baker, *Structural Reliability Theory and Its Applications*, Springer-Verlag, New York, 1982
10. H.O. Madsen, S. Krenk, and N.C. Lind, *Methods of Structural Safety*, Prentice Hall, 1986
11. S. Tangjitham and R.W. Landgraf, Probability-Based Methods for Fatigue Analysis, *Fatigue Research and Applications*, SP-1009, Society of Automotive Engineers, 1993, p 225-235
12. P.S. Veers, C.H. Lange, and S.R. Winterstein, FAROW: A Tool for Fatigue and Reliability of Wind Turbines, *Proc. Windpower '93*, American Wind Energy Association, 1993, p 342-349
13. A.H.-S. Ang and W.H. Tang, *Basic Principles*, Vol 1, *Probability Concepts in Engineering Planning and Design*, John Wiley & Sons, 1975
14. J. Schijve, Fatigue Predictions and Scatter, *Fatigue Frac. Engng. Mater. Struct.*, Vol 17 (No. 4), 1994, p 381-396
15. The Committee on Fatigue and Fracture Reliability of the Committee on Structural Safety and Reliability of the Structural Division, Fatigue Reliability: Introduction, *Journal of the Structural Division, American Society of Civil Engineers*, Vol 108 (No. ST1), 1982
16. H.O. Fuchs and R.I. Stephens, *Metal Fatigue in Engineering*, John Wiley & Sons, 1980
17. P.H. Wirsching and Y.-T. Wu, Probabilistic and Statistical Methods of Fatigue Analysis and Design, *Pressure Vessel and Piping Technology—1985: A Decade of Progress*, American Society of Mechanical Engineers, 1985, p 793-819
18. "Standard Practice for Statistical Analysis of Linear or Linearized Stress-Life (S-N) and Strain Life (ε-N) Fatigue Data," E 739-80, American Society for Testing and Materials, 1980
19. J.H. Wilson, Statistical Comparison of Fatigue Data, *Journal of Material Science Letters*, Vol 7, 1988, p 307-308
20. P.S. Veers, H.J. Sutherland, and T.D. Ashwill, Fatigue Life Variability and Reliability of a Wind Turbine Blade, *Proc. Sixth Specialty Conference on Probabilistic Mechanics and Structural and Geotechnical Reliability*, Y.K. Lin, Ed., American Society of Civil Engineers, 1992
21. P.H. Wirsching, Fatigue Reliability of Offshore Structures, *Journal of the Structural Division, Engineering, American Society of Civil Engineers*, Vol 110 (No. 10), 1984
22. P.H. Wirsching and Y.-N. Chen, Considerations of Probability-Based Fatigue Design for Marine Structures, *Marine Structures*, Vol I, 1988, p 23-45
23. P.H. Wirsching, K. Ortiz, and Y.-N. Chen, Fracture Mechanics Fatigue Model in a Reliability Format, *Proc. Sixth International Symp. on Offshore Mechanics and Arctic Engineering (OMAE)*, 1987
24. S.R. Winterstein and C.H. Lange, Load Models for Fatigue Reliability from Limited Data, *Wind Energy 1995*, W.D. Musial, S.M. Hock, and D.E. Berg, Ed., American Society of Mechanical Engineers, 1995, p 73-82
25. K.O. Ronold, J. Wedel-Heinen, C.J. Christensen, and E. Jorgensen, Reliability-Based Calibration of Partial Safety Factors for Design of Wind-Turbine Rotor Blades Against Fatigue, *Proc. Fifth European Wind Energy Conf., II*, 1995, p 106-111
26. C.H. Lange and S.R. Winterstein, Fatigue Design of Wind Turbine Blades: Load and Resistance Factors from Limited Data, *Wind Energy 1996*, ASME, 1996
27. S.H. Crandall and W.D. Mark, *Random Vibrations in Engineering Systems*, Academic Press, 1963
28. P.H. Wirsching and M.C. Light, Fatigue under Wide-Band Random Stress, *Journal of the Structural Division, American Society of Civil Engineers*, Vol 106 (No. ST7), July 1980
29. L.D. Lutes and C.E. Larsen, Improved Spectral Method for Variable Amplitude Fatigue Prediction, *Journal of the Structural Division*, Vol 116 (No. 4), April 1990, p 1149-1164
30. P.S. Veers, S.R. Winterstein, D.V. Nelson, and C.A. Cornell, Variable Amplitude Load Models for Fatigue Damage and Crack Growth, *Development of Fatigue Loading Spectra*, STP 1006, J.R. Potter and R.T. Watanabe, Ed., American Society for Testing and Materials, 1989, p 172-197

31. B. Efron and R.J. Tibshirani, An Introduction to the Bootstrap, *Monographs on Statistics and Applied Probability*, No. 57, Chapman and Hall, 1993

32. B. Efron and R.J. Tibshirani, Bootstrap Methods for Standard Errors, Confidence Intervals, and Other Measures of Statistical Accuracy, *Statistical Science*, Vol 1 (No. 1), 1986, p 54-77

33. *Recommended Practice for Planning, Designing and Constructing Fixed Off-shore Platforms—Load and Resistance Factor Design*, RP2A-LRFD, American Petroleum Institute, 1993

34. H.J. Sutherland, P.S. Veers, and T.D. Ashwill, Fatigue Life Prediction for Wind Turbines: A Case Study in Loading Spectra and Parameter Sensitivity, *Case Studies for Fatigue Education*, STP 1250, R.I. Stephens, Ed., American Society for Testing and Materials, 1994, p 174-207

35. R.W. Landgraf, S. Tangjitham, and R.L. Rider, Automotive Wheel Assembly: A Case Study in Durability Design, *Case Studies in Fatigue Education*, STP 1250, R.I. Stephens, Ed., American Society for Testing and Materials, 1994, p 5-22

36. R. Skjong and R. Torhaug, Rational Methods for Fatigue Design and Inspection Planning of Offshore Structures, *Marine Structures*, Vol 4, 1991, p 381-406

37. J. Crichlow et al., "An Engineering Evaluation of Methods for the Prediction of Fatigue Life in Airframe Structures," Report ASD-TR-61-434, Wright Patterson Air Force Base, 1962

38. J. Schijve, Estimation of Fatigue Performance of Aircraft Structures, STP 338, American Society for Testing and Materials, 1962

39. F.E. Richart and N.M. Newmark, A Hypothesis for the Determination of Cumulative Damage in Fatigue, *Proceedings of the ASTM*, Vol 48, 1948, p 767-799

40. T.H. Topper, B.I. Sandor, and J.D. Morrow, Cumulative Fatigue Damage under Cyclic Strain Control, *Journal of Materials*, Vol 4 (No. 1), March 1969, p 189-200

41. N.E. Dowling, Fatigue Failure Predictions for Complicated Stress-Strain Histories, *Journal of Materials*, Vol 7 (No. 1), March 1972, p 71-87

42. T.H. Topper and B.I. Sandor, Effects of Mean Stress and Pre-Strain on Fatigue Damage Summation, *Effects of Environment and Complex Load History on Fatigue Life*, STP 462, American Society for Testing and Materials, 1970, p 93-104

43. M.A. Miner, Cumulative Damage in Fatigue, *Transactions of the ASME*, Vol 67, 1945, p A159-A164

44. C.G. Schilling et al., Fatigue of Welded Steel Bridge Members under Variable-Amplitude Loadings, *Research Results Digest*, No. 60, Highway Research Board, April 1974

45. S. Berge and O.I. Eide, Residual Stress and Stress Interaction in Fatigue Testing of Welded Joints, Norwegian Institute of Technology, Trondheim, 1981

46. Y.S. Shin and R.W. Lukens, Probability Based High Cycle Fatigue Life Predictions, *Random Fatigue Life Prediction*, American Society of Mechanical Engineers, 1983

47. T.R. Gurney, "Fatigue Tests under Variable Amplitude Loading," Report 220/83, The Welding Institute, Cambridge, U.K., July 1983

48. O.I. Eide and S. Berge, Cumulative Damage of Longitudinal Non-Load Carrying Fillet Welds, *Fatigue 84: Papers Presented at the Second International Conference on Fatigue and Fatigue Thresholds*, Engineering Materials Advisory Services Ltd., Birmingham, England, 1984, p 1039-1055

49. C.A. Cornell, A Probability-Based Structural Code, *Journal of the American Concrete Institute*, Vol 66, p 974-985

50. A.M. Hasofer and N.C. Lind, Exact and Invariant Second-Moment Code Format, *Journal of the Engineering Mechanics Division, American Society of Civil Engineers*, Vol 100, p 111-121

51. R. Rackwitz and B. Fiessler, Structural Reliability under Combined Load Sequences, *Computers and Structures*, Vol 9, 1978, p 489-494

52. Y.-T. Wu and P.H. Wirsching, "A New Algorithm for Structural Reliability Estimation," *Journal of the Engineering Mechanics Division, American Society of Civil Engineers*, Vol 113, 1987, p 1319-1336

53. Y.-T. Wu, H.R. Millwater, and T.A. Cruse, An Advanced Probabilistic Structural Analysis Method for Implicit Performance Functions, *AIAA Journal*, Vol 28, 1990, p 1663-1669

54. R.L. Rider, R.W. Landgraf, and S. Tangjitham, Reliability Analysis of an Automotive Wheel Assembly, *Fatigue Research and Applications*, SP-1009, Society of Automotive Engineers, 1993, p 81-85

55. P.S. Veers, S.R. Winterstein, C.H. Lange, and T.A. Wilson, "User's Manual for FAROW: Fatigue and Reliability of Wind Turbine Components," SAND94-2460, Sandia National Laboratories, Nov 1994

56. R.Y. Rubenstein, *Simulation and the Monte Carlo Method*, John Wiley & Sons, 1981

57. R.E. Melchers, *Structural Reliability Analysis and Prediction*, Ellis Horwood Ltd., 1987, p 88-103

58. C.H. Lange, "Probabilistic Fatigue Methodology and Wind Turbine Reliability," Ph D. dissertation, Stanford University, Stanford, California, April 1996

59. P. Bjerager, "Probability Computation Methods in Structural and Mechanical Reliability," Chapter 2 in *Computational Mechanics of Probabilistic and Reliability Analysis*, Ed. W.K. Liu and T. Belytschki, Ed., Elmepress International, Lausanne, Switzerland, 1989

Planning and Evaluation of Fatigue Tests

Wolfgang-Werner Maennig, Department of Mechanical Engineering, Division of Material Sciences, Technische Fachhochschule Berlin (retired), Germany

PLANNING AND EVALUATION describes the scientific approach to obtaining valid measures of fatigue performance from S/N-type fatigue data. The need for this approach arises from fatigue scatter. If a whole S/N curve is based on only ten specimens, it appears well-defined and may display an obvious fatigue limit. However, as more specimens are tested, failures are discovered below the first-appearing fatigue limit. So what should the real (safe) limit be? When components are tested, fatigue performance and scatter can be even more problematic. For example, the so-called "fatigue limit" of screws is only 5 to 15% of the ultimate tensile strength with a scatter in fatigue strength at a comparable or even a bigger size. Thus errors of 100% and more are possible. For cast metals and welded structures the situation is usually worse. Small changes in the composition of materials and the heat treatment, construction, and processing of parts also can have considerable influence on fatigue strength and scatter.

This extraordinary scatter is explained by the repetition of relatively small loads, which enables weak properties of the materials to become dominant. Such properties include the local fluctuations of the composition; the different sizes and orientations of grains; the distribution and sizes of slag inclusions and of segregations; the distribution and local concentrations of solute atoms and pre-existing dislocations; tiny scratches and

corrosion points at the surface; cavities, blisters, and blow holes; and microstructural defects in general. The superposition of so many weakening factors leads to the evolution of microcracks as well as macrocracks. As a result the cycles to fracture for several parts or specimens will be very different. To overcome this problem, one has either to test an unacceptably large number of specimens or to apply a scientific approach to the planning and evaluation of the tests. This article describes the latter approach based on a determination of the probabilities for detected fracture positions within the observed range of scatter. This method leads to the evaluation of special ranges for stresses (or strains) and the cycles to fracture with well-defined probabilities of fracture suitable for design or engineering use.

The Fatigue Diagrams

Figure 1(a) is a schematic of constant-amplitude ($\pm S_a$) stress cycling about a possible mean value (S_m) that leads to fracture of a single part or specimen after N cycles. This result is condensed to Fig. 1(b), where the tiny vertical dash on the line for the level S_a represents the life result of the one fractured specimen. Testing of $n = 10$ specimens leads to the schematic of Fig. 1(c), which provides some idea of typical scatter. In the case of the so-called "service-load" (in which the

stress amplitude is not constant), a frequently used approach is to attribute the fracture position to the highest existing amplitude S_a^* instead of to the magnitude of that amplitude from the test life or fracture point of the specimen (Fig. 2a and b). The test life of the fractured specimen might be expected at a much higher number of cycles for fracture than in the case of constant amplitude. This is a result of the many cycles with amplitudes that are smaller than S_a^*. As shown in Fig. 2(c), the scatter of $n = 10$ specimens also will become much bigger, because the variation of the random application of the amplitudes gives an additional strong influence on life behavior. Service load programs are extremely different. Therefore it is necessary to give additional information, for example by a name that describes the program. Altering of S_a^* will be done with the other amplitudes in the same relation. The principles of planning and evaluation of fatigue tests are the same for constant amplitudes and service (variable-amplitude) load, although special load history methods are used to quantify variable-amplitude fatigue life (see the article "Estimating Fatigue Life" in this Volume).

Constant amplitudes and service load might be applied also in the form of controlled deformations instead of controlled loads, for example in the form of elongation, torsion, and bending. In strain-controlled fatigue testing, the loads are not controlled and very often will decrease. Under

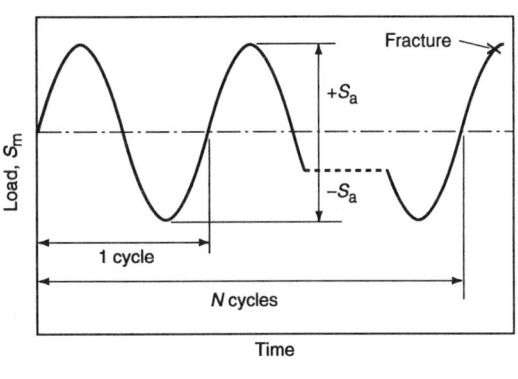

(a)

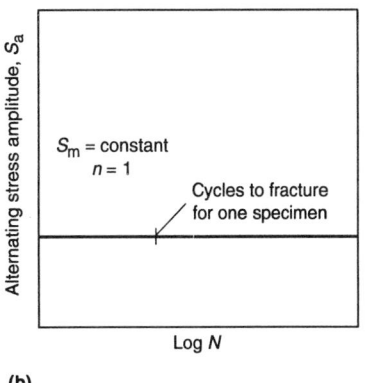

(b)

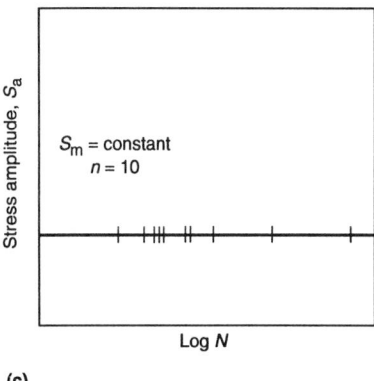

(c)

Fig. 1 Schematic of constant-amplitude (S_a) stress cycling about a mean load S_m. (a) The alternating load plotted vs. time. (b) Fracture for one specimen plotted vs. the number of cycles, N. (c) Fractures for ten specimens at one constant level of alternating load S_a plotted vs. N

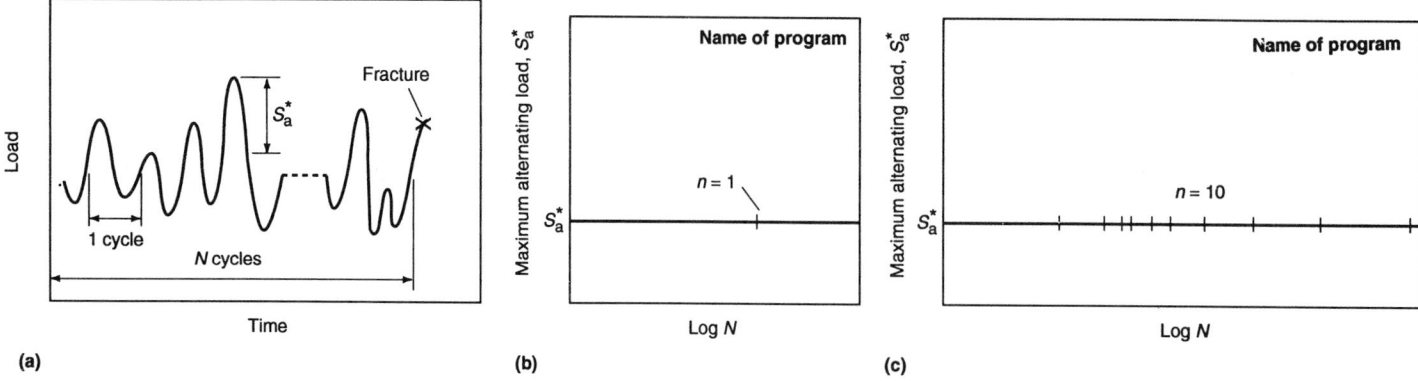

Fig. 2 Variable-amplitude loading and schematic of data presentation. (a) Alternating load varying with time. (b) Fracture of one specimen plotted by maximum alternating load (S_a^*) vs. number of cycles, N. (c) Fracture of ten specimens plotted vs. S_a^* and cycles to failure

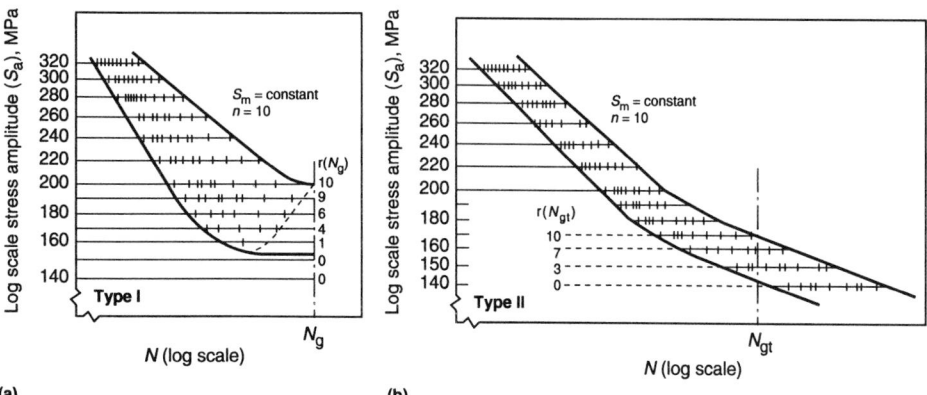

Fig. 3 Fatigue experiments for types I and II of the fatigue diagram. The dotted line in (a) demonstrates the separation of the fractures from the ordinate in N_g. See text for details. Source: Ref 1, 2

constant strain, the stress intensity remains constant as the crack grows; under constant load, it increases as the crack grows and therefore the rate of growth increases also. This explains the difference in life and the higher numbers of cycles for fracture in strain-controlled tests as compared to stress-controlled tests. The scatter of results, however, is the same in principle, although perhaps scatter is larger in strain-controlled fatigue tests. The principles of planning and evaluation are also the same for stress- or strain-controlled cycling.

Testing of specimens in 13 different levels of load typically leads to two types of diagrams (Fig. 3a, b) (Ref 1, 2). In the type I diagram there is no fracture below the level with the alternating load $S_a = 160$ MPa (23 ksi), and the number of broken specimens will decrease below $S_a = 200$ MPa (29 ksi) (Fig. 3a). The fractures appear in the range bordered by the dotted line, well separated from the greatest number of cycles (N_g). No fractures are expected outside this dotted range; the non-fractured specimens are real survivors. N_g is a significant value for this type of diagram. In case of constant load-amplitude it is 10^7, and with acceptable small errors it is 2×10^6 for probabilities of fracture ≤10% (Ref 1-3).

In the case of the type II diagram, all specimens will fracture below 200 MPa (29 ksi) if we cycle

long enough (Fig. 3b). In this case, the performance requirements and the needed number of cycles for a technical design form a basis to choose a "technological greatest number of cycles (N_{gt})" as a "cut off," where all fracture positions above N_{gt} are declared (by definition only) to be survivors. With this definition, there is no gap between N_{gt} and the detected fracture positions below 200 MPa. By this principal difference it is often possible to distinguish between Fig. 3(a) and (b) and thus the diagrams of type I and II. This distinction might become less noticeable if the lower part of the range of fractures in Fig. 3(b) has a very small slope.

A more generalized form of type I is in Fig. 4(a). The ordinate in N_g and its intersection with the shell lines of the fracture positions allows a tripartition. In the range of finite endurance all specimens will fail before N_g, in the range of infinite endurance none will fail, and in the range of transition about 50% will fail. The horizontal border between the transition region and the infinite-endurance region is the famous fatigue endurance limit (noted here as S_{FL}). Below and to the left of the lower shell line, there are no fractures; above and to the right of the upper shell line, all specimens are fractured. The passage from no fractures to all fractures happens between the shell lines, and thus probabilities of

fracture (P_F) of 0% (1%) and 100% (99%) can be assigned to these shell lines,[*] which is the object of this article. *Note:* The probability of survival (P_S) could also be used, but it is not advisable for two reasons. First, it demands inverse thinking in case of comparison of results. Further, $P_S + P_F = 1$ will not hold. There remains a difference, its size depending from the number n of specimens (Ref 4).

The type II fatigue diagram also has the ranges as type I (Fig. 4b), but N_g may approach 10^{12} to 10^{13} cycles, and a meaningful fatigue limit may be small (e.g., unnotched S_{FL} at 20 MPa for aluminum and its alloys and at 40 MPa for copper and its alloys). Such fatigue limit values have poor experimental basis because N_g is too high. Practically, tests are stopped always before N_g and thus a fatigue limit will not be found for the metals with type II diagrams although it may exist (Fig. 4b).

In the type II diagram (Fig. 4b), there is a "small-slope" part (before N_g) that does not exist in the diagram of type I. It begins at about 2×10^6 or 5×10^6 cycles, ending at N_g. This is quite an interesting range for many purposes, for example cars and aircraft. According to the technical needs of the component, an N_{gt} value will be set. This again allows the tripartition of the diagram, but now into "technological (or engineering-based) ranges" of the finite endurance, infinite-endurance, and transition. Thus it becomes possible to evaluate a *technological fatigue limit* (S_{FLt}), which is strongly dependent on the chosen N_{gt}. Therefore it is necessary to write, for example, $S_{FLt} (10^8) = 120$ MPa (17.4 ksi) if $N_{gt} = 10^8$.

The general methods for estimating fracture probabilities are principally the same for both types of the fatigue diagram. From what is known today, the type I diagram belongs with the body-centered cubic metals, such as unalloyed and low-alloy steels, if not hardened; molybdenum and its alloys; probably (though not conclusively) high-alloy ferritic steels; and perhaps Al-Mg and

---

[*] The tested number of specimens is always too small. Therefore, it is technically incorrect to give values at 0% and 100% probability of fracture, even if it is possible to determine them mathematically. Other probability limits (such as 1% and 99%) are preferred.

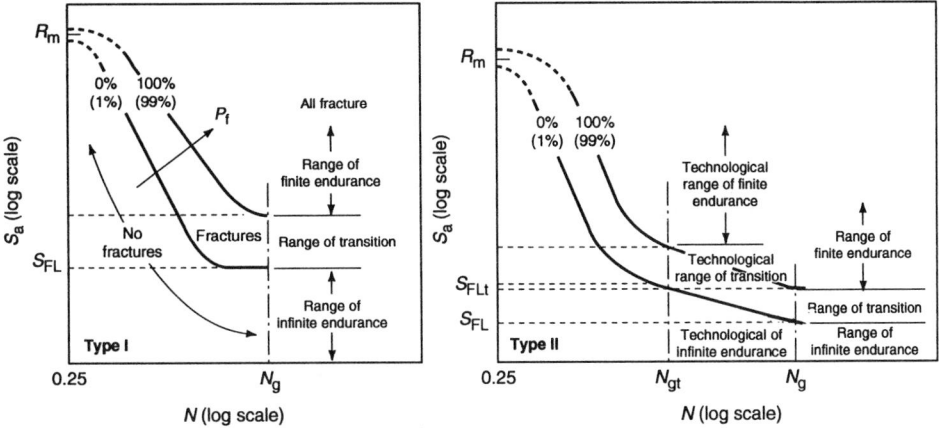

Fig. 4 Results of fatigue diagrams in condensed form and the tripartition of the two types. Shell lines at 0 and 100% have parenthetical values noted at (1%) and (99%), which are preferred because usually the number of test specimens is small and therefore it is difficult to have meaningful 0 and 100% values. (a) Type I diagram. (b) Type II diagram

Table 1 Calculation of the probabilities of fracture in the range of transition for the type I fatigue diagram according to Fig. 3(a) and Eq 1

Alternating load, MPa	r	$(3r-1)/(3n+1)$	P, %
200	10	29/31	93.55
190	9	26/31	83.9
180	6	17/31	54.8
170	4	11/31	35.5
160	1	2/31	6.45
150	0	1/31(a)	3.23

(a) $P(r=0) = 0.5P(r=1)$

Table 2 Calculation of the probabilities of fracture in the technological range of transition for Type II diagram according to Fig. 3(b) and Eq 1

Alternating load, MPa	r	$(3r-1)/(3n+1)$	P, %
170	10	29/31	93.55
160	7	20/31	64.5
150	3	8/31	25.8
140	0	1/31(a)	3.23

(a) $P(r=0) = 0.5P(r=1)$

Cu-Sn alloys in the solution-treated (not age-hardened) condition, though they belong to the face-centered cubic metals (Ref 1, 5). The type II diagram generally occurs for all face-centered cubic metals, such as aluminum and copper and its alloys (exceptions noted above); probably austenitic steels and martensitic steels; and perhaps also the hexagonal close-packed metals such as titanium alloys. The uncertainty of some of the above statements is due to poor planning and evaluation of fatigue tests in the past.

Evaluating the Ranges of Fatigue Diagrams

There are two principal methods for evaluating fatigue curves. First, the (technological) ranges of infinite-endurance and of transition are determined separately. If a whole fatigue diagram is needed, both ranges can be joined afterward, and this is preferable on a scientific basis. More often and nearly always in engineering practice, only one range or even only one value out of one region will be needed and evaluated. The methods to do this are described in the numbered sections below.

Secondly, both ranges can be determined and evaluated all at once by using a homogenous mathematical method called the Wolfsburger Model, developed by R. Müller for Volkswagen (Ref 6). It is a very difficult mathematical procedure, and it needs, for the same level of reliability, no fewer specimens. Perhaps this is one reason why no independent data evaluation and comparison has become known. Yet it is the only model and mathematical treatment of fatigue data that allows one to calculate not only fatigue values but also first confidence intervals in a satisfactory way. Furthermore it has been used down to probabilities of fracture of only 10^{-6} ($10^{-4}\%$). No one other than R. Müller dared to extrapolate so far. One has to consider also that R. Müller had 500 specimens and thus gave experimental proof of the suitability of his method only down to about 0.2% probability of fracture.

Evaluating the Range of Transition

The Classical Way. The classical method of evaluating the (technological) range of transition (i.e., just above the infinite-endurance range in Fig. 4) can be illustrated with the data in Fig. 3(a), where $n = 10$ specimens were tested in seven levels of alternating load in the range of transition and below. In level $S_a = 200$ MPa (29 ksi) $r = 10$ specimens are broken, thus $r = n = 10$. This means that the level is just at the border and belongs also to the range of finite endurance. In the two lower levels $r = 0$ had been found; we use only level $S_a = 150$ MPa (21 ksi) and its result. The levels $S_a = 150$ MPa and $S_a = 200$ MPa are the most important because their evaluations indicate the smallest errors (Ref 7). In Fig. 3(a) and (b) obviously the probability of fracture (P_F) rises with higher stress levels. There has been some discussion about how to calculate P_F. Theoretical studies and some comparative investigations (Ref 4) lead to:

$$P_F = 100 (3r-1)/(3n+1)$$

$$P_F(r=0) = 0.5P_F(r=1) = 100/(3n+1) \qquad \text{(Eq 1)}$$

where P_F is given as a percentage.

The values of $P_F = P_F(r)$ according to Eq 1 are evaluated for each level and plotted in Fig. 5(a) arithmetically on the ordinate over the logarithmic alternating loads S_a of the levels. (This is done here only for demonstration of the logic.) The result is a curve in S-form which is usually symmetrical to the point of deflection. By use of a transformation function (T), the S-curve is transformed into a straight line, which is convenient for extrapolation (Fig. 5b). Substantial research has been done to select the right transformation function (e.g., Ref 1, 2, 6, 8). Only two functions have been left over by these works: the Gaussian cumulative probability, which is available in the forms of Probit tables and of "probability paper" (normality paper), and the $\sqrt[3]{\ln(P)}$ transformation. Both transformations give similar results with some minor differences. The Gaussian function has been tested often. Probability paper is easily available much more

in special shops for technical drawings and papers, yet its applications by computer is more difficult. One has to either store large Probit tables or do more difficult programming. The $\sqrt[3]{\ln(P)}$ function is not only simple for the computer but can also be easily applied on hardcopy tables and graphs.

A sample calculation using the classical method in the case of a Type I diagram is shown in Table 1, where results are implemented into Fig. 5(a) and (b). A straight line allows the evaluation of alternating loads with defined probabilities. In the example of Fig. 5(b), $S_{FL1} = 147.2$ MPa and $S_{FL99} = 211.3$ MPa for probabilities of fracture at 1 and 99%. $S_{FL10} = 159.1$ MPa and $S_{FL90} = 195.3$ MPa for probabilities of fracture at 10 and 90%. An additional point is evaluated at $S_{FL0.1} = 138.5$ MPa.

Which of these values should be used for engineering? The size of scatter in the range of transition can be described as a line or by two or more points. Thus a value such as S_{FLx} alone is not useful. Two materials might have the same value for S_{FLx} and might behave very differently (see "Presentation and Comparison of Results" in this article). Theoretically each combination of two points of the straight line is sufficient. It is only convention to pair S_{FL10} with S_{FL90} and S_{FL1} with S_{FL99}.

The S_{FL10} value is sometimes requested in industrial standards, and the ratio S_{FL90}/S_{FL10} is in this case a good description for the scatter of the distribution. The S_{FL1} value usually is preferred as the lower limit for the range of transition and thus for a good approximation of S_{FL} itself. The ratio S_{FL99}/S_{FL1} is also a good description of not only the scatter of the distribution but also the

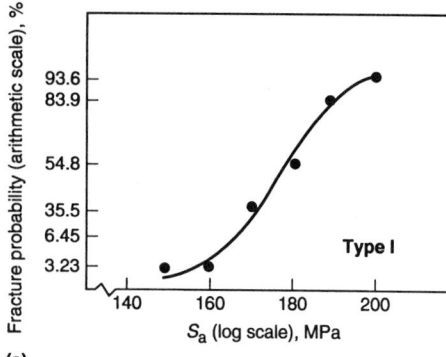

(a)

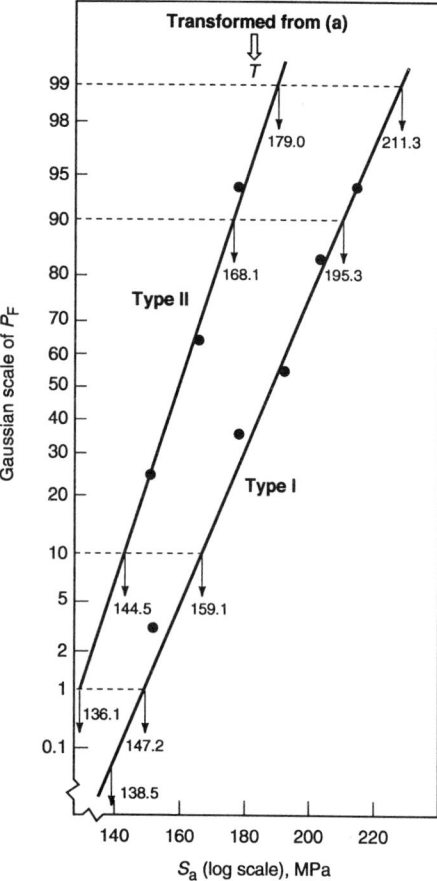

(b)

Fig. 5 The way of evaluating the (technological) range of the transition region. (a) Arithmetic probabilities vs. alternating load for the type I diagram in Fig. 3(a). (b) Transformation of the data in (a) by the transformation T converts the curved line in (a) into a straight line on a Gaussian probability scale for the types I and II diagrams in Fig. 3(a) and (b).

size (or scatter range) of the transition region. This can be seen easily from the experimental results of Fig. 3(a).

The probability paper (of Fig. 5(b)) also allows determinations of values much closer to 0 and 100% probability of fracture (e.g., $S_{FL0.1}$ and $S_{FL99.9}$ or even $S_{FL0.0001}$ and $S_{FL99.9999}$) because the paper is not limited in both directions. The Gaussian normal distribution and (in this case) the Gaussian cumulative normal distribution obviously is only a very good technical approxima-

Table 3 Values d and m for use with the Boundary Technique

Factor	Notched specimen	Smooth specimen	Simple parts	Parts like bolts	Complicated parts, the form:	
					Well defined	Poorly defined
d	0.05-0.15	0.1-0.3	0.2-0.4	0.4-1.2	0.4-1-?	0.6-?
m	1-1.2	1-1.7	1.4-2	2-3.2-?	2.2-3.2-?	2.2-?

Note: d gives an estimation of the size of the (technological) range of the transition region; m is a multiplicating factor to apply on 10 and gives n, the number of specimens needed in one level ($m = 2$, $n = 20$).

tion to the practical distribution in the range of transition. Theoretically it cannot be valid because this would mean $S_{FL0} = 0$ and $S_{FL100} = \infty$. Both results are unacceptable on the ground of experience. In fact, the Gaussian distribution is almost never valid strictly in theory but is often good practically. The function $\sqrt[3]{\ln(P)}$ is limited at $P = 100\%$ and unlimited for $P \to 0$.

It is also necessary to discuss whether $S_{FL0.1}$ is perhaps a better approximation of $S_{FL} = S_{FL0}$ than S_{FL1}. No scientific investigation of this point is known. Indeed, it is not known how the results of Fig. 3(a) and 5(b) would be altered if hundreds of specimens were tested in each level instead of only ten. It is only customary to use S_{FL1}. In Fig. 5(b), it gives a satisfying value also if one compares with Fig. 3(a). The remaining 1% probability of fracture gives a basis for protection against the risks from employing too small a number of specimens and the resulting uncertainty. Such values are used in the calculation of ball bearing life, for example.

The evaluation of the range of transition in a type II diagram (Fig. 3b) is shown in Table 2 and in Fig. 5(b) for the chosen N_{gt} according to Fig. 3(b).

In the classical way of evaluating the (technological) range of transition, a minimum of ten specimens should be tested in each level of alternating load. Basic statistical laws and a bit of experience (Ref 1, 2, 9) suggest that the minimum number of specimens should be increased if the range of the transition region or the ratio S_{FL99}/S_{FL1} becomes bigger. The m values in Table 3 give a first-order estimate of the needed numbers. The number of levels should be about 4. This allows some correction of the statistical influence of the small number of specimens, as has been done in Fig. 5(b) for type I. Preferably there should be an upper level with r near to n and a lower level with r near to zero. This gives the best results because the possible errors become smallest (Ref 4, 7).

Quick Solutions. There are problems in arriving at a firm evaluation using the classical method. First and mainly, the above-described recommendations for the number of levels and specimens in the levels are high, too high in many practical cases. Too much testing time would be needed. Second, the classical way does not give a method for positioning the needed levels. For example, the tested level 140 MPa in Fig. 3(a) could not be evaluated and thus would become a waste of time and money. Three methods try to give some help on this problem.

The two-point strategy (Ref 10) tries to establish two levels close to 50% probability of frac-

ture and to reach a better estimation of S_{FL50}. The first and main objection is that one single point in the range of transition cannot give its description. Second, levels close to 50% probability will have the biggest possible errors out of statistical laws. This can be appreciated in every table of binomial or hypergeometrical distributions (e.g., Ref 7). Thus advantage from establishing two levels near 50% probability is even more doubtful. Third, nobody needs values with 50% probability of fracture really. Finally, this strategy contains no rule to find the two levels.

The staircase method (Ref 11-13) won a lot of attention when it was introduced because it includes a way to find the wanted levels. However, it concentrates itself around the 50% level of probability of fracture. In addition, in the description based on Ref 11 to 13, only half of the tested specimens are used for evaluation, an unbearable waste. In Ref 14 the staircase method was studied comparatively. It was proved that the results depend very strongly on the number of specimens and the properly chosen step-width. Without pretests, 20 specimens were found to be necessary for the evaluation of a fracture probability of 50%, 40 specimens for the probabilities 10 and 90%, and 50 specimens for the probabilities 1 and 99%. This means that the staircase method is less suitable than other known methods. An attempted improvement (Ref 15) brought little or no advantage. Finally, it is difficult to choose the correct step-width.

The Boundary Method (Ref 4, 7, 9, 16-18) uses only two levels and tries to place them near the lower and upper borders of the (technological) range of the transition. According to Fig. 5(b) two such levels are sufficient theoretically. Practical comparative studies (Ref 4) and theoretical studies (Ref 7) have proved the suitability of the method. It leads to conservative values. For example, the values S_{FL1} and S_{FL10} will be preferably too low if there are deviations. If the values happen to be too big, the deviations will be relatively smaller for theoretical reasons. Likewise, values such as S_{FL90} and S_{FL99} will become more preferably too high and therefore safer. Smaller deviations are combined with values too low.

The boundary technique has been tested with success in several laboratories (e.g., Ref 9, 18). Today it is in use worldwide. Even the classical evaluation of the (technological) scatter range of the transition region is often begun with this method. Additional levels of alternating load testing can be well placed afterward. In Ref 9 the Boundary Technique proved best regarding the needed number of specimens, although by an

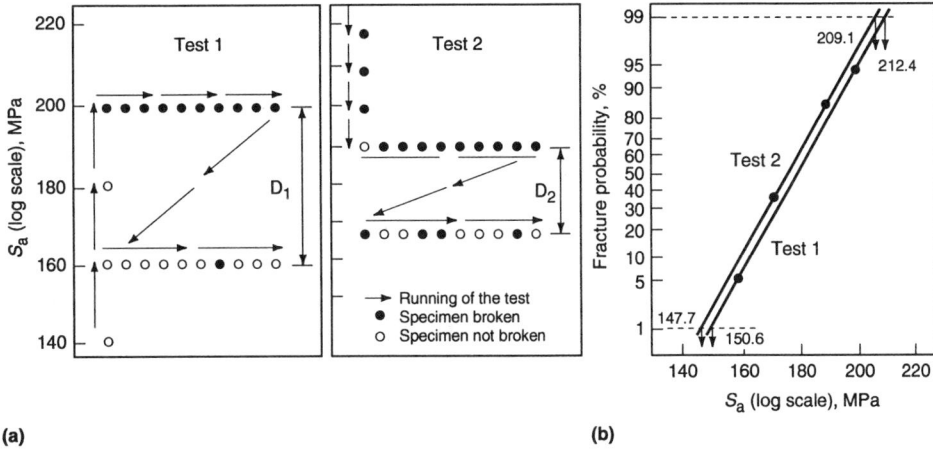

(a)   **(b)**

Fig. 6 The way of (a) running and (b) evaluating two tests using the Boundary Technique

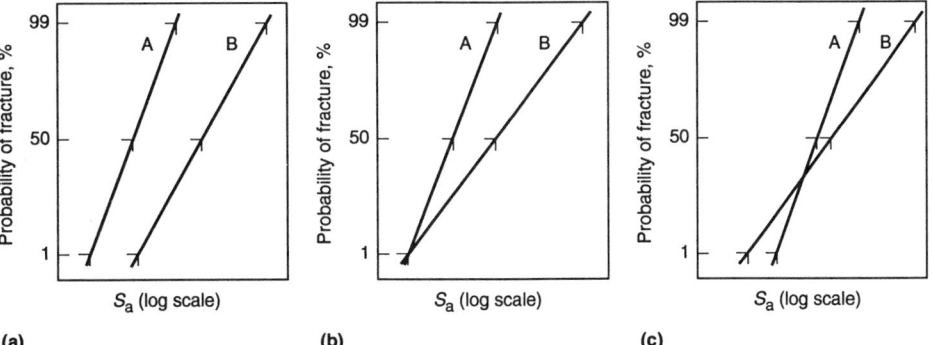

(a)   **(b)**   **(c)**

Fig. 7 Principal possibilities for the comparison of two conditions, A and B, in the (technological) range of the transition. (a) Condition B is better than condition A. (b and c) Condition A is better than condition B. See text for details.

error the most worthy levels of alternating load had not been used for that method.

The boundary technique starts by testing one randomly chosen specimen at any level of alternating load (see Fig. 6). In the case of a type I fatigue diagram it is a private recommendation to use a load near to $R_{P0.001}$, the unidirectional tensile stress that yields a plastic deformation of 0.001%. Tests will then be done until $N = N_g$ or N_{gt}. If no fracture happens before this limit, the next specimens will be tested at higher levels raised stepwise until the opposite event happens, that is, fracture before N_g or N_{gt}. The stress level with the first opposite event becomes the first fully tested level. This means that nine additional specimens have to be tested. If the first tested specimen fractures before N_g or N_{gt}, the next specimens will be tested stepwise on lower levels of alternating load, until the opposite event happens: no fracture!

For the second level of alternating load, the distance D (see Fig. 6) has to be estimated. It is given by:

$$D = (1 - r/n) \cdot d \cdot S_a \text{ if } r \leq 0.5n$$

$$D = (-r/n) \cdot d \cdot S_a \text{ if } r \geq 0.5n \quad \text{(Eq 2)}$$

where S_a is the alternating load of the first tested level and d is the estimated size of the (techno-

logical) range of the transition region according to Table 3.

In Fig. 6, two tests are presented that refer to Fig. 3(a) and its results. Test 1 was started too low. Raising in big steps led to the first opposite event: fracture in $S_a = 200$ MPa. A total of $n = 10$ specimens were tested there, and all fractured. Obviously the full range of transition had been passed during the tracing. Equation 2 with $d = 0.2$ out of Table 3 led to the second level, $S_a = 160$ MPa, with one fractured specimen out of ten. Is it possible that a level of 200 MPa is already too high, deep in the region of finite endurance? No, it is not. According to Fig. 3(a), the highest fracture position is close to $N_g = 10^7$. Further the next lower level had in the minimum the one tested specimen with "no fracture," and the level at 160 MPa had one fracture in ten specimens. As mentioned above, the distribution in the range of transition is symmetric. One has to conclude that the level 200 MPa is just the counterpart of the level 160 MPa, each of them close to the borders of the range, providing a good comprehension of the range of the transition region. Evaluation of test 1 in Fig. 6 under the use of Eq 1 gives values for S_{FL1} and S_{FL99} that are close to those in Fig. 5(b).

Test 2 was started too high, and the alternating load had to be lowered in the next levels. The first opposite event, no fracture, happened at $S_a = 190$

MPa, which thus became the first tested level. Yet no other specimen survived, $r = 9$. This time $d = 0.1$ was taken and according to Eq 2 the second and lower level became $S_a = 170$ MPa with $r = 4$.

Figure 6 shows excellent results compared with Fig. 5(b). Yet normally Fig. 5(b) is not known, because it is the intention of the Boundary Technique to avoid such expensive investigations. Therefore, it is not advisable to trust the results of test 2 because $r = 4$ or $P_F = 35.5\%$ in the lower level, and the possible errors out of principal statistical laws are too big (Ref 7). A third level at $S_a = 150$ MPa should be tested if specimens are still available. Generally, levels with 26 to 74% probability of fracture should be mistrusted. The possible adverse effect of testing such levels is bigger as the adjacent level becomes more distant from the wanted value S_{FLx} (Ref 4, 7).

If in tests 1 and 2 the value $d = 0.3$ had been taken for estimation of D, the second and lower level would have been 140 MPa with $r = 0$, according to Fig. 3(a). Use of that and evaluation would have given a fatigue limit or S_{FL1} too low and therefore too safe. This explains why practitioners tend to take bigger values for d instead of values that are too small if the second level is the lower one. The situation is different if the second level is the upper one. A big value of d might result in placing the upper level too high, outside the (technological) range of transition, already clearly in the (technological) range of finite endurance. This would influence the evaluation of the S_{FLx} values, although not strongly. In Ref 19 different fatigue diagrams were studied for this problem. It was proven that significant deviations will be obtained only if the biggest number of cycles in that upper level is smaller than 10^6.

Tests 1 and 2 needed two and three tracing specimens, which could not be included in the evaluation. This is a good average; in practice one to five will be necessary. Thus a minimum of 25 specimens should be provided for use with the boundary technique. Ten further specimens are advisable for a case that demands a third level, such as test 2.

If the (technological) range of transition is very large, more specimens have to be prepared for each level of alternating load, as suggested by general statistical laws and some experience (e.g., Ref 9). A first approximation is given with the values m in Table 3 from Ref 1. Whether a test is started too high or too low has no influence on the results of the evaluation. Yet starting too high usually results in time savings, because all tracing specimens will fracture before N_g or N_{gt}. That might mean a difference of one or more days.

Presentation and Comparison of Results. In the rare scientific cases, the evaluation results for the (technological) range of transition can be transferred into graphs such as Fig. 4(a) and (b) with numeric scales. This gives the best information to those who have no special training in fatigue. Discussion and comparison of results by using only one value, for example S_{FL1}, should be avoided. For S_{FL50} it should be forbidden. Experienced engineers use diagrams that give full information, such as Fig. 7.

In the case of Fig. 7(a), obviously condition B is better because all its values are to the right of those of condition A. The slightly bigger size of the range of transition will not alter this result. In the case of Fig. 7(b), the values S_{FL99} and S_{FL50} of condition B are bigger than those of condition A. The values S_{FL1} are the same. Nevertheless, an experienced engineer will prefer condition A. First, there is no intention to sell parts, engines, or aircraft if they have probabilities of fracture bigger than 1%. Therefore the bigger values S_{FL50} and S_{FL99} of condition B are not preferable; all values below S_{FL1} or below 1% of condition B are worse than those of condition A. Second, and even more important, the reliability of the lines has to be considered. If both had been evaluated with the same number of specimens and even with similarly well-placed levels of alternating load, condition B is less reliable because in this condition the size of the range of transition is just twice as big as in condition A (which might be expressed by calculating for both conditions the ratios S_{FL99}/S_{FL1}). In order to obtain the same reliability one has to increase the number of tested specimens in condition B with a multiplicating factor: (size range B/size range A)$^{0.5}$ = $2^{0.5}$ = 1.4 from general statistical laws. If the number of specimens was $n = 10$ in the levels of condition A, those for condition B have to be raised to $n = 14$ afterward. Having done that, a new comparison might be started.

In the case of Fig. 7(c), the lines for A and B have an intersection at about 35% probability of fracture. This means that condition B will lead to fractures at lower alternating loads than condition A below 35%. This is unbearable; thus, condition B is worse. Additionally, in condition B the range of transition is again twofold bigger.

Neglecting these simple rules has already brought serious consequences. Often comparisons have been done by using only the values S_{FL50}, or perhaps only poor approximations to that, because no adequate planning had been done. On the basis of S_{FL50} values, condition B always would be better in Fig. 7. One famous example for such a consequence is the so-called *geometrical size effect*. It means that bigger (longer and/or larger-diameter) parts have smaller fatigue values than smaller parts, even if the same material has been used. Figure 7(b) gives an exact description of the situation that will occur if we take condition A for bigger parts and condition B for the smaller ones. Of course, if one concentrates the material of several small parts into a big part, one also concentrates its faults (e.g., inclusions). Thus, it becomes very unlikely that bigger parts will have so small and few faults that fractures will happen at big alternating loads. The line for condition A becomes steeper and the size of the range of transition becomes smaller. Yet there is still the largest defect, which has the largest consequence of fatigue life and has to be in one of the many smaller parts as well as in one of the few bigger parts. Thus, the lower border of the range of transition has to be the same for both conditions A and B. According to Ref 20 the two conditions meet at

about 1%.* Or, there is no geometrical size effect if one does qualified investigations and evaluations.

Evaluation of Finite-Endurance Range

Evaluation of the range (Fig. 4a) or the technological range (Fig. 4b) of finite endurance is similar to that for the transition region. However, each level of alternating load is evaluated separately for the finite-endurance region. There are also some additional problems. Usually the distribution is not symmetrical about the point of inflection, and therefore it is problematical to apply symmetrical T functions for transformation. This means that the experimental values usually fit less well on a straight line for regression. Further, it is not good practice to use functions for estimating the probabilities that are associated with the transformation functions theoretically. The possible influence of the probability functions should be used to correct the suitability of the transformation functions.

Many investigations have been made of this point (e.g., Ref 1, 2, 4, 8, 16, 17). For example, 56 different data groups in Ref 8 were used to test three transformation functions and 10 probability functions in all possible combinations. According to Ref 8, the $\sqrt[3]{\ln(P)}$ transformation function and the Gaussian cumulative normality distributions (Probit function, probability paper) are suitable. For the estimation of the probabilities, the functions $P_F = (i - 0.3)/(n + 1)$ and $P_F = (i - 0.417)/(n + 0.166)$ were best. In this case, i is the order number of the fractured specimens. The functions $(i - 0.535)/(n - 0.07)$; $(i - 0.375)/(n + 0.25)$; $(2i - 1)/2n = (i - 0.5)/n$; and $(3i - 1)/(3n + 1)$ were nearly as good, the first of them being the best in that group.

General Rules: The needed number of specimens in one level is given by (Ref 1):

$$n = 10 \text{ if } N_{max}/N_{min} \leq 10{:}1$$

$$n = 15 \text{ if } 10{:}1 \leq N_{max}/N_{min} \leq 30{:}1$$

$$n = 20 \text{ if } N_{max}/N_{min} \geq 30{:}1 \qquad \text{(Eq 3)}$$

The number of needed levels of alternating load is determined by the rule that the median P_{50} with 50% probability of fracture should not shift from one level to the other more than:

$$\varphi = \sqrt[2]{10} = 3.16 \text{ for engineering decisions,}$$

or $\varphi = \sqrt[3]{10} = 2.16$ for high reliability or small confidence intervals $\qquad$ (Eq 4)

Yet increasing the number of specimens in the levels increases the reliability more than lowering the shift (Ref 2). The positioning of a second level after having evaluated a first one is made easy by using the sufficiently exact equations:

$$N_{50} = C \cdot S_a^{-K}$$
$$(N_{50/2}/N_{50/1}) = (S_{a1}/S_{a2})^K$$

and with Eq 4:

$$S_{a2}/S_{a1} = [^K\sqrt{\varphi}]^{-1} \qquad \text{(Eq 5)}$$

where K is a value for the slope of the range of finite endurance (see Fig. 11). With some experience with K and Table 4 it is possible to choose S_{a2}/S_{a1} at once. If there is no experience at all, one might start with $6 \leq K \geq 9$, preferably with the higher values. The latter will lead to smaller distances of levels. After having the results of two levels one is able to calculate a more correct value using:

$$K = (\log N_{50/1} - \log N_{50/2}) / (\log S_{a2} - \log S_{a1}) \qquad \text{(Eq 6)}$$

In practical work, three different tasks have to be done. First, the (technological) range of finite endurance has to be determined for a given field. Second, the alternating load has to be determined for a given number of cycles and to a given probability. Finally, at a given level of alternating load, the range of fractures has to be determined with caution in terms of statistical safety.

Evaluation of the Range of Finite Endurance, Comparison of Results, and Conjunction with the Range of the Transition Region. With only three additional levels of alternating load, the range of finite endurance will be evaluated with a shift of 1/2 decade for N_{50}. It is described above that the level 200 MPa was determined for the onset of the range of the transition region. All specimens had fractured before $N_g = 10^7$. The observed numbers of cycles were (in 10^3): 723, 1403, 200, 7002, 560, 1963, 408, 4211, 1570, 880. The biggest and smallest numbers are 7002×10^3 and 200×10^3 cycles; therefore N_{max}/N_{min} are about $35{:}1 \geq 30{:}1$. This means that ten additional specimens had to be tested in the same level according to Eq 3. This is very risky, because one or two specimens might not fracture if we regard Fig. 5(b) as having established $S_{FL99} = 211.3$ MPa. Thus, additional specimens in the level 200 MPa might become a waste in our task to determine the range of finite endur-

Table 4 Values of S_{a2}/S_{a1} varying with K and φ

Decade	φ	K								
		4	5	6	7	8	9	10	11	12
1/2	3.16	0.750	0.796	0.825	0.848	0.866	0.880	0.892	0.902	0.908
1/3	2.16	0.826	0.858	0.880	0.896	0.909	0.918	0.926	0.933	0.938

Note: S_{a2} is the lower level.

---

* This experience became one of the experimental reasons to accept S_{FL1} as a good approximation for S_{FL}.

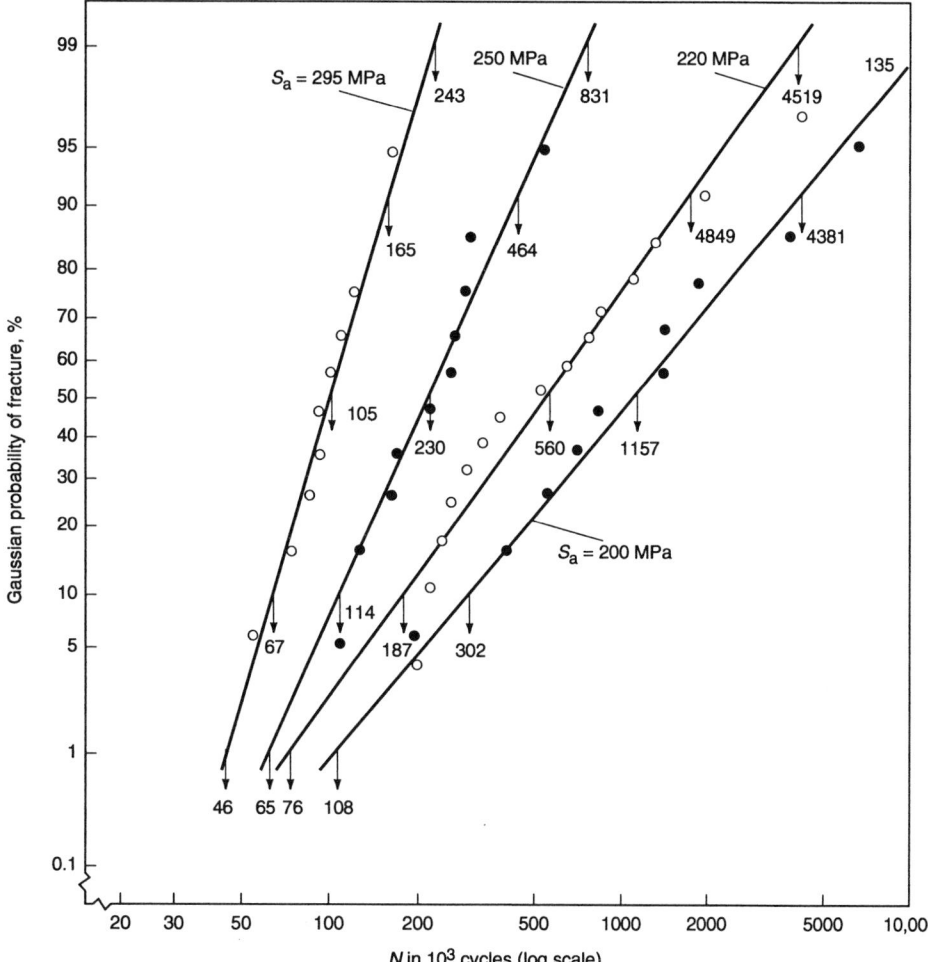

i	N	P, %
1	196×10^3	3.8
2	227×10^3	10.4
3	250×10^3	17
4	271×10^3	23.6
5	308×10^3	30.2
6	347×10^3	36.8
7	393×10^3	43.4
8	548×10^3	50
9	669×10^3	56.6
10	799×10^3	63.2
11	879×10^3	69.8
12	1154×10^3	76.4
13	1388×10^3	83
14	2073×10^3	89.6
15	4257×10^3	96.2

Drawing of the regression line gives $N_1 = 76 \times 10^3$, $N_{10} = 187 \times 10^3$, $N_{50} = 560 \times 10^3$, $N_{90} = 1819 \times 10^3$ and $N_{99} = 4519 \times 10^3$ cycles (Fig. 8). With Eq 6 we control K:

$$K = (\log 1157 - \log 560)/(\log 220 - \log 200) = 0.3151/0.0414 = 7.6$$

Thus, the first estimation is seen to be wrong; therefore, we take $K = 8$. By means of Table 4 and Eq 5 this leads to the next level of alternating load, namely 254 MPa; 250 MPa is taken with results as follows ($S_a = 250$ MPa):

i	N	P, %
1	115×10^3	5.7
2	129×10^3	15.6
3	169×10^3	25.4
4	178×10^3	35.2
5	230×10^3	45.1
6	271×10^3	54.9
7	280×10^3	64.8
8	305×10^3	74.6
9	326×10^3	84.4
10	568×10^3	94.3

In this case, $N_{max}/N_{min} < 10/1$ and we need no further specimens. Evaluation by a regression line gives $N_1 = 65 \times 10^3$, $N_{10} = 114 \times 10^3$, $N_{50} = 230 \times 10^3$, $N_{90} = 464 \times 10^3$, and $N_{99} = 831 \times 10^3$, and the values are plotted in Fig. 8. They call for renewed control of K:

$$K = (\log 560 - \log 230)/(\log 250 - \log 220) = 0.3865/0.0555 = 7.0$$

The next higher level, according to Eq 5 and Table 4, is $250/0.848 = 295$ MPa. At this level of alternating load the results were:

i	N	P, %
1	59×10^3	5.7
2	80×10^3	15.6
3	90×10^3	25.4
4	98×10^3	35.2
5	100×10^3	45.1
6	107×10^3	54.9
7	117×10^3	64.8
8	128×10^3	74.6
9	158×10^3	84.4
10	177×10^3	94.3

The evaluation by a regression line gives the values $N_1 = 46 \times 10^3$, $N_{10} = 67 \times 10^3$, $N_{50} = 105 \times 10^3$, $N_{90} = 165 \times 10^3$, and $N_{99} = 243 \times 10^3$ cycles. Altogether with 35 specimens the range for the finite endurance

Fig. 8 Evaluation of the levels of alternating load 220, 250, and 295 MPa in the range of the finite endurance. The level 200 MPa belongs as well to the range of the finite endurance as to the range of the transition. It was evaluated preferably for determination of N_{50} and estimation of the position of the next level 220 MPa, although not enough specimens were tested.

ance. Yet we will use the existing results to estimate the position of the next level. We arrange the results according to the number of cycles, assigned order numbers i, and calculate:

$$P_F = 100(i - 0.417)/(n + 0.166)$$

where P_F is given as a percentage. For $S_a = 200$ MPa:

i	N	P, %
1	200×10^3	5.73
2	408×10^3	15.6
3	560×10^3	25.4
4	723×10^3	35.2
5	880×10^3	45.1
6	1403×10^3	54.9
7	1570×10^3	64.8
8	1963×10^3	74.6
9	4211×10^3	84.4
10	7002×10^3	94.3

The results are plotted in Fig. 8. By a regression line the cycles N_1, N_{10}, N_{50}, N_{90}, and N_{99} with the probabilities of fracture 1, 10, 50, 90, and 99% are evaluated as follows: $N_1 = 108 \times 10^3$, $N_{10} =$

302×10^3, $N_{50} = 1157 \times 10^3$, $N_{90} = 4381 \times 10^3$, and $N_{99} = 13,570 \times 10^3$ (over 10^7!). We want a shift of a half decade and estimate $K = 11$. According to Table 4, the upper level is thus $S_{a1} = S_{a2}/0.902 = 200/0.902 = 221.7$ MPa and 220 MPa will be used. The tests in this level with ten specimens yielded the following results for $S_a = 220$ MPa (N in 10^3): $N = 4257$ (specimen 1), $N = 879$, 799, 1388, 271, 308, 2073, 227, 347, and $N = 669$ (specimen 10). For this case, $N_{max}/N_{min} = 4257/227 = 19/1$, and we have to test five other specimens according to Eq 3. Results for the additional five specimen tests for $S_a = 220$ MPa (N in 10^3) are:

Specimen	N
11	1154×10^3
12	393×10^3
13	250×10^3
14	196×10^3
15	548×10^3

Indeed, the range became a bit bigger because $N_{min} = 196$ now. For the experimental ranking order, the probabilities P turn out as follows (with $S_a = 220$ MPa):

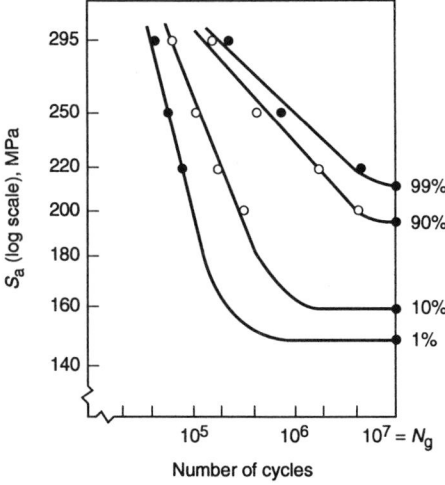

Fig. 9 Evaluation of the range of finite endurance in Fig. 8 and conjunction with the range of transition in Fig. 5(b) to the type I fatigue diagram

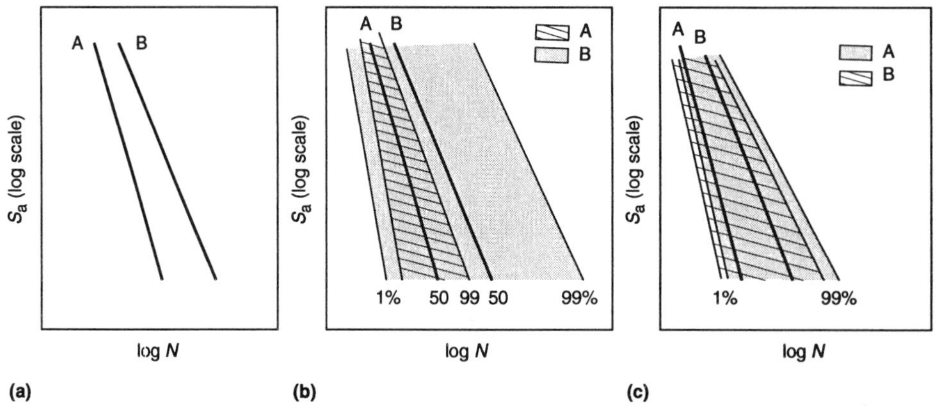

Fig. 10 Comparison of results in the range of the finite endurance. The position of the thick lines is constant in the three subfigures. (a) Condition B is better than condition A. (b) The thick lines represent 50% of probability. Condition A becomes better because the range of condition B is much bigger and its 1% line is left of that of condition A. (c) The thick lines had no defined probabilities of fracture because no qualified evaluation had been done; both are situated a bit extremely in their evaluated ranges, which have about the same width. Condition A is better because its 1% line is right of that of condition B.

has been determined between 46×10^3 and 4519×10^3 cycles.

The four regression lines in Fig. 8 have different slopes. That happens often and explains why the estimation of the expected shift in Eq 5 is related to the median value, N_{50}.

The presentation of the ranges of finite endurance is done by drawings such as Fig. 3(a) and (b) and usually in conjunction with the range of transition also, if it has been determined. This is done in Fig. 9. Obviously the 1, 10, 90, and 99% points in the range of finite endurance are not very good on regression lines. This is quite typical of that range and results from the small number of specimens at each level, which is not big enough to compensate for the strong statistical influence. The regression lines for 1, 10, 90, and 99% give some correction toward the population. No scientific study has yet allowed correction for whole ranges of finite endurance. It exists only for single levels (see below). Out of personal experience the following recommendations are offered: The number of specimens according to Eq 3 is already a minimum, good enough for clean metals such as aluminum- or copper-base alloys or clean steels, but it may be scanty for lower-quality carbon steel (as in this case), and quite probably it is not sufficient for cast metals, welded structures, and service-load conditions.

The joining of the ranges of the finite-endurance and the transition regions are not problematic in the case of Fig. 9. Only the point N_{99} of the level 200 MPa is not in harmony. This will not be a problem, because that level did not have enough specimens tested in the range of finite endurance. Its evaluation there was done only for the estimation of K, using N_{50}. Yet sometimes the conjunction becomes difficult because both ranges have been determined independently in different ways and with too few specimens. Perhaps corrections are necessary. For example, S_{FL99} might be shifted on the ordinate up or down. In this case, some other S_{FLx} may have to be shifted. Accord-

ing to Fig. 5 they have to be corrected by a new straight line through the shifted S_{FL99}.

It is mentioned above that N_g might be lowered to 2×10^6 if only probabilities below 10% are important. This is visible in a typical way in Fig. 9. By that reduction S_{FL1} would be influenced not at all, S_{FL10} a little bit, and S_{FL90} and S_{FL99} heavily. As a fact, 2×10^6 is already partially in the range of the finite-endurance region. If one reduces N_g in order to save time, one has to consider the consequences. For example, the conjunction of the ranges of finite endurance and of transition will become difficult. In that case, S_{FL1} will be fully acceptable and S_{FL10} will be nearly acceptable, but all values above will be too big to a remarkable degree.

The comparison of the ranges of finite endurance has to be done by regarding the *whole* ranges over the full width of probabilities of fracture. In Fig. 10(a), conditions A and B are represented by two lines only. Both conditions were determined with the same number of specimens. Condition B clearly seems to be better than condition A. However, in Fig. 10(b), the thick lines for conditions A and B represent 50%, and the ranges are given fully. The range of condition B is much wider, and its 1% line is left of that of the range of condition A. Thus, condition A is better. First, the values for 1% probability of fracture are at higher cycles; second, the values of condition A are much more reliable, because a much smaller range had been determined with the same number of specimens. For example, a slight change of slag inclusions or composition of high-strength screws might produce the change from condition A to condition B.

In Fig. 10(c), the two lines for conditions A and B were positioned without any planning or evaluation. Thus, the lines do not represent a defined probability of fracture. Full determination of the ranges might show up that the thick line of condition A was just at the left side of its range and that the one for condition B was more at the right side. Again, condition A is better, although both ranges have the same width about. The 1% line of

condition A has the higher values and lies to the right of condition B.

Evaluation of the Alternating Load Belonging to a Given Number of Cycles and a Given Probability of Fracture. Some standards demand the evaluation of the alternating load $S_{ax}(N_d)$ with a given probability of fracture, x, which belongs to a predetermined number of cycles for fracture, N_d. A typical demand is $S_{a10}(10^5)$. This value is determined from previous examples (Fig. 9). A value for the probability 1% might be determined as well.

From Fig. 9 it is easy to see that the S_{a10} value is expected to become 262 MPa. The way (Ref 21) to determine it in an accurate manner and with a minimum of specimens is demonstrated in Fig. 11. In Fig. 11(a) only the ordinate in N_d is known. This value is used instead of N_g or N_{gt} and the Boundary Technique is applied. Figure 11(b) shows results in which the start level happened to be too high. Specimens 1 to 3 all fractured before N_d as S_a was readied at stepwise lowered levels. The first opposite event (i.e., fracture beyond N_d) happened with specimen 4, marked by a thick dash. At this stress level, nine additional specimens will be tested in the beginning. Equation 3 has to be considered, and the number of specimens may have to be raised after a first evaluation. In the case of Fig. 11(b), ten specimens are considered sufficient. Points I and II, with 10 and 90% probability of fracture, respectively, are determined in level 1 by evaluation as in Fig. 8. Point I is not far from the wanted point, WP, which might relate to level 1 as shown in Fig. 11(c). This might lead to the only theoretically correct conclusion: a level 2 would be fine for determination of points III and IV and, in consequence, determination of WP. However, there are severe objections:

- The statistical influence on the scatter of the positions of fracture might be so strong that point F will be found instead of point III, and point Z in the zenith will be evaluated instead of the theoretically correct value WP. In prac-

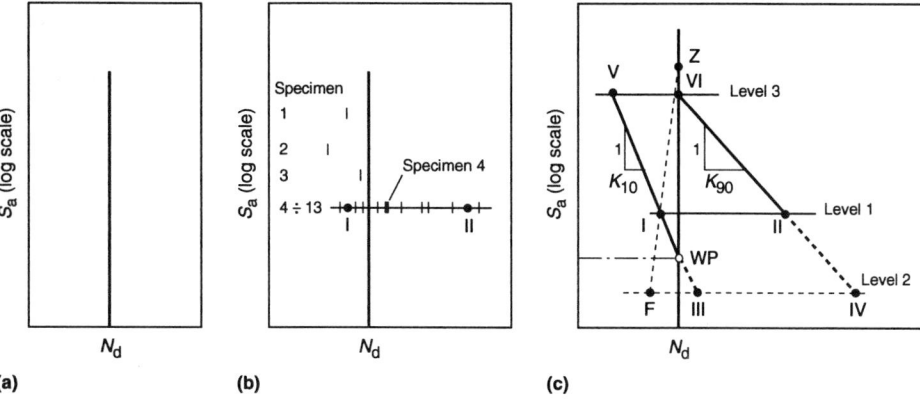

Fig. 11 Evaluation of the alternating load $S_{ax}(N_d)$, which belongs to a predetermined value of probability x and cycles to fracture N_d, where $x = 10\%$. (a) Only N_d is known. (b) By means of the Boundary Technique and three tracing specimens, a level is found where the first opposite event (fracture right of N_d) will be found. Altogether ten specimens are tested there and the points I and II with 10 and 90% probability of fracture are evaluated. (d) The wanted point, WP, is determined after finding and evaluating level 3. See text for details. Source: Ref 21

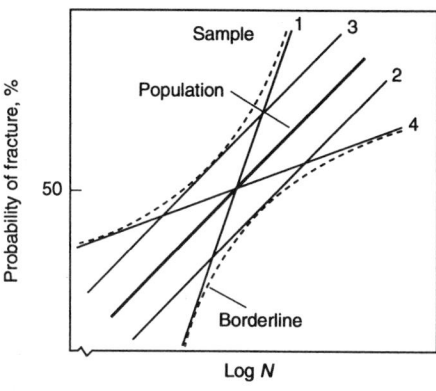

Fig. 12 The regression lines of the population of specimens and of four of many possible samples from the population within its dotted borderline for a given sample size

tical work, evaluation of both points I and III might be slightly inaccurate, such that point I is too high and point III is too low. The effect might be the same. This example of a heavy fault is not very extreme, thus a situation as given by levels 1 and 2 is not reliable enough and should be avoided.

- Level 2 might need more specimens than level 3. Nevertheless, the results of level 2 might be less reliable statistically.
- The position of level 2 might already be in the range of transition, and if so some specimens will not fracture. Thus, level 2 would not belong to the range of the finite-endurance region. This case should be avoided. (*Note:* Nevertheless it will be given sometimes and no further specimens might be available. An evaluation is possible if the number of unfractured specimens is small relatively to the number of tested specimens (<20%). For this purpose the unfractured specimens will get their order numbers too, the highest ones of course. That makes it possible to give the other specimens the correct order numbers and probabilities of fracture. During evaluation (as in Fig. 8) in such a level the unfractured specimens will all get the same position N_g or N_{gt}, yet different probabilities. The evaluation by drawing a straight line will use the positions of the fractured specimens preferably. The reliability of the results is lower.)
- Level 2 not only might need more specimens, but without any doubt it will need much more testing time than level 3. Thus testing at level 2 will be more expensive.

In order to find the upper level 3, a line has to be drawn from point II in the lower level to point VI, the intersection with the ordinate in N_d, giving the position and alternating load of the upper level. Mathematically this can be done by:

$$\log S_{upper} = \log S_{lower} + (\log N_{90lower} - \log N_d)/k \quad \text{(Eq 7)}$$

where $\log N_{90lower}$ means point II.

The Boundary Technique might make the first level the upper one. In this case a line has to be drawn from point V in the upper level to point I. Mathematically this will be done by:

$$\log S_{lower} = \log S_{upper} + (\log N_{10upper} - \log N_d)/k \quad \text{(Eq 8)}$$

where $\log N_{10upper}$ refers to point V. The problem is the estimation of the right value for k. It becomes more difficult if a value of $x = 1\%$ for $S_{ax}(N_d)$ is wanted. No scientific study is known for k, and there have been only a few experiences. The one reported here belongs to 50% values or lines, and those for 1, 10, 90, and 99% are different, as might be seen in Fig. 9 and 11(c), where the values K_{10} and K_{90} are defined. Every laboratory has to develop its own experience. Using k_{50} instead for the beginning will give a higher position of the upper level, because the line from point II to point VI becomes steeper. This is not bad. Yet the use of k_{50} for finding the lower level will result in too low a lower level, and this might become a problem, as explained above.

Alternatively, the first level might be in the midst. This will be seen by the fact that N_d is in the center of the range of fractures in that level, close to the median N_{50}. In such a case it might be dared to use the first choice for the upper level and to go downward afterward. This is given if we take 295 MPa in Fig. 9 as a first and upper level and evaluate by use of Eq 8 to 220 MPa for the lower level. Connection of the crossed 10% points in those two levels by a straight line would lead to practically the same result, $S_{a10}(10^5) = 262$ MPa. Using the other tested levels would give about the same result. Thus all appears to be very fine, but the approach is not practical. In the example of Fig. 9, the same results were always used, and there the points lay on well-defined straight lines. This is not guaranteed in practice. In all likelihood, repetition with the next batch of specimens out of the same population (tested at the same levels) would give different results. No known scientific study has faced that problem and determined safety factors for correction. Safety factors that might be used are explained below, based on the evaluation of one level only.

The evaluated $S_{ax}(N_d)$ should be divided by those factors and thus be lowered in the interest of safety.

The method illustrated in Fig. 11 might also be used for searching a restricted area in the range of the finite-endurance region with a minimum of specimens, as a preliminary measure before more intensive experiments are started. In Ref 21 a more sophisticated method is used to evaluate point WP in Fig. 11(c) directly. Unpublished comparative studies indicate that this method is not safe enough. The fault previously described for point F happened too often.

Safe Evaluation of Fatigue Data in a Predetermined Level of Alternating Load. Consider that an unlimited number of specimens in one level could represent a population. From general statistical laws, we know that for each population there are two borderlines within which the possible results of samples will be found (Fig. 12). This means that we have to expect that some results out of a limited number of samples will differ from that one of the unlimited population. For example, the very steep sample 1 will give much too high a value for the 1% probability of fracture, and sample 4 will give too low a value for N_1. Nevertheless, samples 1 and 4 have the same median as the general population. For the samples 2 and 3 the medians are different, yet the errors for a 1% value are smaller. An enlargement of the sample size will cause the distribution to approach the theoretical borderlines and enlarge their radius. Thus the possible errors will be reduced. To a limited degree, the possible errors are dependent on the steepness of the sample lines, in other words on their ranges or, as a very practical tool, on the relations $R_1 = N_{99}/N_1$ or $R_{10} = N_{90}/N_{10}$.

In Ref 8 this problem was studied on 56 data groups that were taken for approximations to populations and were evaluated for N_1 and N_{10}. Afterward, samples were drawn out of the populations, and their values N_1, N_{10}, R_1, and R_{10} were evaluated. Equation 3 was considered. The comparison with the N_1 and N_{10} values of the populations gave safety factors SF_1 and SF_{10}, which mean the correction of the samples toward the population. The plotting of the safety factors over the values R_1 and R_{10} was done in diagrams such as Fig. 13(a) and (b). In the case of homogeneous populations, all possible values for the safety factors grouped themselves in a restricted area (Fig. 13a). The dotted line of this restricted area might represent a confidence of 95 or 99%. This means that 5 to 1% of all safety factors might be outside of the dotted lines. The restricted area is symmetrical about the value $SF = 1$. Its position and width is typical for the investigated materials (for the scatter of results conforming to Fig. 12). Half of the points are below and above the horizontal line through $SF = 1$. Theoretically this would allow correction of every point by application of the appropriate safety factor. Points below $SF = 1$ mean that the values N_1 or N_{10} of the samples are too safe (too low in regard to those of the population). Yet there is no indication whether the sample is on the safe or unsafe side of the population. For the sake of safety, one has to use only the values of $SF > 1$ and to calculate N_1 and N_{10}:

$$N_{x\,corrected} = N_{x\,sample}/SF_x \qquad (Eq\ 9)$$

In the case of inhomogeneous populations, distributions as in Fig. 13(b) were found.

In practical work, one has only a sample. After evaluation as in Fig. 8, values such as N_1, N_{10}, R_1, and R_{10} are available. For example, one goes with R_1 into Fig. 13(a), finds SF_1, and corrects N_1 by use of Eq 9. Of course, the problem is the source of Fig. 13(a). Interested laboratories have to excavate their stored data and determine the diagrams for safety factors themselves for their special needs.

For the situation of a laboratory with no appropriate population, one option is to cautiously consider whether or not their sample can be represented by the 55 data groups in Ref 8 (excepting the data group for cast iron, for which Eq 3 was not sufficient). The 55 data groups contain alloys such as AlMg, AlCuMg, cheap steels, steels low alloyed with Cr and Mo or with Cr and Ni, one steel with 0.02% C, 18% Ni, 8% Co, and 5% Mo. The alternating loads were deflection with triangular form of momentum; rotating bending with trapezoid form of momentum; prismatic specimens with one-sided notch at the surface; with a thin specimen as used in fracture mechanics in mode I. Possibly, Fig. 14 (Ref 8) might be used as a crutch. The price will be to find bigger risk factors, as in diagrams for one special population. In any case, Fig. 14 should not be used for samples out of cast alloys or welded structures, or to investigate the effects of service load, combined alternating loads, corrosion, or humidity.

Extrapolation from the Range of Finite Endurance into the Range of Transition

It is a common desire to extrapolate from evaluations in the range of the finite endurance into the range of the transition. The needed number of specimens and the testing time, even with the Boundary Technique, is believed to be too high to be feasible in some cases. In Ref 2 it is proven that such an extrapolation is not possible by the use of simple lines (as in Fig. 10a).

Another way is to find a diagram by a basic study, as in Fig. 9, and to test new incoming batches of parts with theoretically the same features at only one level of alternating load, for example 220 MPa. A rejection of the incoming charge of parts will be planned for a defined change in the considered level. For example, an important change might be seen if the range of fracture positions becomes much broader or has a shift to the left in such a way that N_1 will be at 2×10^4 cycles in level 220 MPa. In this case one might conclude that S_{FL1} is much lower than 147 MPa.

This way is full of risks. First, although the upper border of the range of transition approaches the range of the finite-endurance region, there is no reason to believe that the lower border of the range of transition (S_{FL1}) has any logical connection with the range of the finite-endurance region. No scientific proof or indication is known

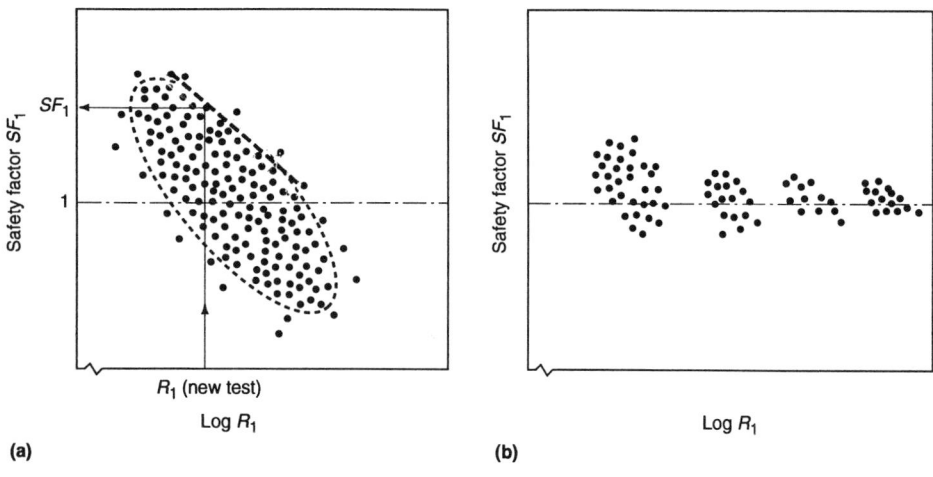

Fig. 13 (a) Scheme of the relation between range R_1 and safety factor SF_1 in the case of homogenous populations. (b) One possible scheme for an inhomogeneous population

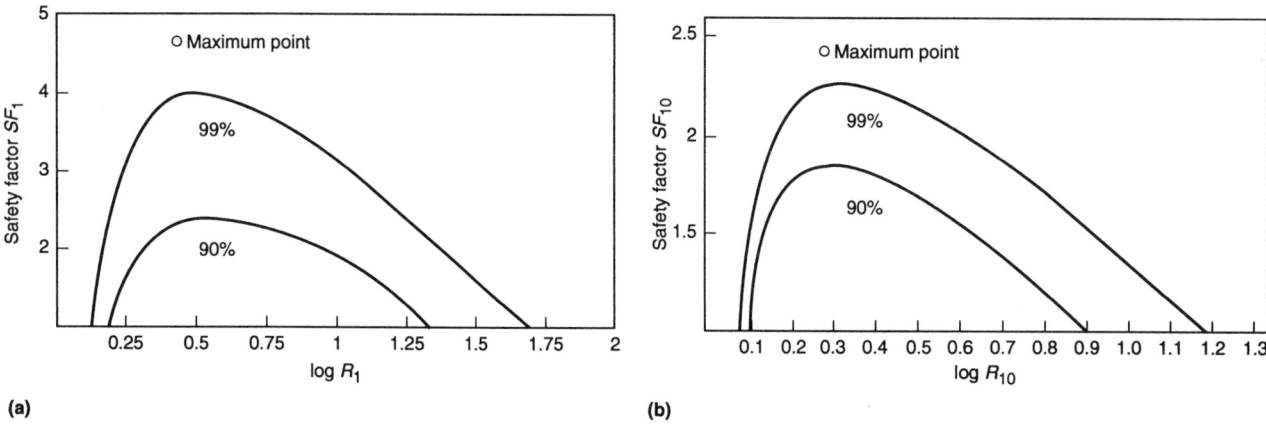

Fig. 14 Safety factor SF_1 varying with range R_1 for 55 data groups combined in one diagram for 1% probability of fracture. Eight random starts; probability function $P = (i - 0.417)/(n + 0.166)$; transformation function $\sqrt[3]{\ln(P)}$ or Gaussian cumulative normality (Probit); confidence lines for 90 and 99% which represent only 90 and 99% of all determined values, not of the population. (b) As in (a), except SF_{10} for 10% probability of fracture and range R_{10}. Source: Ref 8

for that. On the contrary, it is trivial that different phenomena of physical metallurgy play the main role in both ranges. Thus it is conceivable that any change in the properties of the material might enlarge the range of transition without influencing much the range of the finite endurance. In regard to Fig. 9 this might mean that the results in level 220 MPa remain untouched, even though the 1% line goes much lower and S_{FL1} will therefore become smaller.

Second, consider a professional who believes in a logical connection between a level in the range of the finite endurance and the lower border (S_{FL1}) of the range of transition. If the incoming batch of parts has a much broader range N_{99}/N_1 in the level 220 MPa, it will need more specimens for reliable evaluation. Usually those will not be prepared, and the consequence might be that randomly the chosen sample will give values for acceptance nevertheless. This risk becomes much stronger if the planned number of specimens is chosen too small from the very beginning, in this example 10 instead of 15. Another batch of incoming parts might have a range N_{99}/N_1 which is only a bit smaller than usual, say N_{90}/N_{10} as in Fig. 9. Perhaps the charge will not be rejected, although it might belong to a fatigue diagram that has a lower range of finite endurance and thus also a lower range of transition and S_{FL1}. In still another case, the range N_{99}/N_1 of the new charge of parts might be excellent and no rejection will be possible. Also, this time the range of finite-endurance might be much lower and the S_{FL1} value lower as a consequence.

A much more reliable solution would be to choose level 140 MPa in the case of Fig. 9 and to settle the demand for no fractures in ten specimens up to 2×10^6 cycles. This would mean 2×10^7 cycles altogether. A resonance fatigue testing engine might be trimmed to 150 cycles per second, or 540,000 cycles per hour. (Resonance testing machines are relatively cheap and do not need much energy. They are sufficient for the range of transition and for the lower part of the finite-endurance region in case of constant amplitudes, as in Fig. 1.) Thus the ten specimens could be tested within 37 h running time, or easily in 48 h, including machine setup by trained operators. If that is still too much time, one has to use several testing machines, gage them against one another, and run the sample out of the incoming batch of parts on all machines at the same time. (The gaging should be done statically as well as by running fatigue tests with big samples of equal specimens that are from the same population without any doubt. A careful planning of the tests is also necessary.) There are practically no risks involved in choosing the level if it is well established. The risks of using several machines will depend on the quality of the gaging. In any event, the risks should be smaller than those of testing a level in the range of the finite endurance in order to estimate the lower border of another range.

In our example, at the level $S_a = 140$ MPa one has to expect a result of "no fracture" within the 10 specimens. If there happens to be 1 fracture, it is possible and perhaps even likely that this sample has a lower S_{FL1} value. And if there is more than 1 fracture it should be taken as being proven that the S_{FL1} value is lower than in the population. In other words, the new delivery does not belong to the population and has to be refused. An agreement might be made to take and test a second sample from the new delivery if there is 1 fracture in the first sample tested. The delivery might be accepted if there is no fracture in the second sample.

REFERENCES

1. W.-W. Maennig, Planning and Evaluation of Fatigue Tests in Regard to Ultrasonic Frequency Tests, *Ultrasonic Fatigue*, The Met. Soc. of AIME, 1982, p 611-645
2. W.-W. Maennig, Untersuchungen zur Planung und Auswertung von Dauerschwingversuchen an Stahl in den Bereichen der Zeit- und Dauerfestigkeit, doctoral thesis, Technische Universität Berlin, 1966
3. D. Dengel and H. Harig, Zur Frage der Grenzlastspielzahl und deren Einfluß auf den Schätzwert der Dauerfestigkeit, *Materialprüfung*, Vol 16, April 1974, p 88-94
4. W.-W. Maennig, Bemerkungen zur Beurteilung des Dauerschwingverhaltens von Stahl und einige Untersuchungen zur Bestimmung des Dauerfestigkeitsbereiches, *Materialprüfung*, Vol 12, April 1970, p 124-131
5. W.-W. Maennig and H.-J. Taferner, Ursachen der Ausbildung einer Dauerschwingfestigkeitsgrenze bei kubischraumzentrierten: kubischflächenzentrierten und hexagonal dichtest gepackten Metallen, *VDI-Forschungsheft* 611, Verein Deutscher Ingenieure, 1982, Düsseldorf, Germany
6. R. Müller, Das Wolfsburger Modell der Schwingfestigkeit, *VDI-Z* 122, Verein Deutscher Ingenieure, 1980, p 761-768, 841-847
7. W.-W. Maennig, Das Abgrenzungsverfahren, eine kostensparende Methode zur Ermittlung von Schwingfestigkeitswerten—Theorie, Praxis und Erfahrungen, *Materialprüfung*, Vol 19, Aug 1977, p 280-289
8. J. Jahn and W.-W. Maennig, Safe Evaluation of Fatigue-Data in the Range of Finite Endurance, submitted to *Int. J. Fatigue*, 1996
9. W. Thomala, Beitrag zur Dauerhaltbarkeit von Schraubenverbindungen, doctoral thesis, Technical University Darmstadt, Germany, 1978
10. R.E. Little, Estimating the Median Fatigue Limit for Very Small Up-and-Down Quantal Response Tests and for S-N Data with Runouts, in STP 511, ASTM, 1971
11. W.-J. Dixon and A.M. Mood, A Method for Obtaining and Analyzing Sensitivity Data, *J. Am. Statistical Assoc.*, Vol 43, 1948, p 108-126
12. J.T. Ransom and R.F. Mehl, Symposium on Fatigue with Emphasis on Statistical Approach, Part II, in STP 137, ASTM, 1952, p 3-24
13. H. Bühler and W. Schreiber, Lösung einiger Aufgaben der Dauerschwingfestigkeit mit dem Treppenstufenverfahren, *Archiv Eisenhüttenwesen*, Vol 28 (No. 3), 1957, p 153-156
14. W.-W. Maennig, Vergleichende Untersuchung über die Eignung der Treppenstufen-Methode zur Berechnung der Dauerschwingfestigkeit, *Materialprüfung*, Vol 13, 1971, p 6-11
15. M. Hück, Verbesserung der statistischen Aussagefähigkeit von Dauerfestigkeitsversuchen (Treppenstufenversuchen), *IABG-Bericht TF-651*, IABG, 1977, Einsteinstr., D 85521, Ottobrunn, Germany
16. W.-W. Maennig, Statistical Planning and Evaluation of Fatigue Tests: A Survey of Recent Results, *Int. J. Fracture*, Vol 11, 1975, p 123-129
17. W.-W. Maennig and M. Pfender, Zur Problematik der Dauerschwingfestigkeit metallischer Werkstoffe, *Progress in Fatigue and Fracture*, Pergamon Press, 1976
18. C. Kübler, "Darstellung und Vergleich statistischer Verfahren zur Abschätzung der Schwingfestigkeit," Publ. 754598, Materialprüfungsamt Stuttgart, Germany
19. R. Eckert, H.-J. Eng, and W.-W. Maennig, Einfluß von Laststufen im Zeitfestigkeitsgebiet auf die Ermittlung des Übergangsgebietes im Wöhlerdiagram bei Anwendung des Abgrenzungsver fahrens, *Materialprüfung*, Vol 27, 1985, p 162-166
20. K. Heckel and J. Köhler, Experimentelle Untersuchung des statistischen Größeneinflusses im Dauerschwingversuch an ungekerbten Proben, *Z. F. Werkstofftechnik*, Vol 6, 1975, p 52-54
21. W.-W. Maennig, Das Abgrenzungsverfahren im Zeitfestigkeitsgebiet des Schwingfestigkeitsdiagrammes, *Materialprüfung*, Vol 27, 1985, p 216-222

Effect of Surface Condition and Processing on Fatigue Performance

Brian Leis, Battelle-Columbus

SURFACE CONDITION and manufacturing-related surface alterations strongly influence fatigue resistance. Data that characterize their effects provide the means to extend life prediction methods (which are based on the behavior of idealized test specimens) to applications more typically found in industrial design applications. The effect of a surface treatment or modification is case specific. That is, the effect depends on how a process alters local composition, alters or orients local microstructure, introduces long- or short-range self-stresses (residual stresses) due to constraint, and/or alters the surface finish, which depends on the specific process parameters and the material that the process is applied to. The extent to which these surface changes improve fatigue life are measured by comparison to reference data developed using typical smooth laboratory test specimens.

Surface treatments or modifications are effective because many materials are subject to service conditions that degrade material performance by contact of the surface with some external factor, such as an aggressive environment. Surface modifications are particularly effective in applications where the serviceability is limited by fatigue (or a fatigue-related mechanism such as corrosion fatigue) because the fatigue process, which involves reversed slip in homogeneous materials, occurs most readily at a surface. (Slip at a surface is enhanced in the absence of grain-to-grain compatibility and the freedom for slip along slip planes with a component normal to the surface.) Finally, surface modifications are popular because the modification is localized at the surface and so the modification can be made cost-effectively.

Surface modifications have been used to improve fatigue performance for decades. Early applications include improved resistance to environmental effects, as for example the use of clad aluminum, introduced in the 1950s to improve corrosion-fatigue resistance. Early applications also used case heat treatment to harden a thin shell of material to achieve a higher steel hardness, with the anticipated increase in fatigue endurance or wear resistance as in hardened steel shafts and gears. Surface processing has historically also made use of treatments that created local compression residual stresses. This was done either by external working, as with shot peening, or by mechanical loading prior to service, which created compression residual stresses at the surface through local yielding in components with strain gradients. In principle, the same concepts employed in some cases since the 1950s are still being used today. However, the recent schemes involve more sophisticated technologies, such as localized laser processing, ion-beam bombardment, and so on.

This article presents an approach to characterizing the effects of surface treatments to enhance fatigue properties, with particular concern for wear, corrosion, and thermal effects. The effect of processing on performance in this context could be illustrated by experimental data sets representing specific materials, typical test geometries, and a range of different processing methods used to enhance resistance as compared to results for laboratory tests. Such a presentation format would catalog available data, but it would not, nor could it, cover all current applications. More importantly, it could not provide insight useful in assessing the possible effects of new processes or new materials, nor could one identify the best way(s) to improve fatigue resistance in a given application.

For these reasons, the approach taken here is to present and illustrate trends in available data with respect to influences on microstructure, orientation, residual stress effects, and surface effects. In this way the available data provide guidance for both current and future applications. Following this generic discussion, data specific to the more common processes are presented to document the extent of the effect on fatigue resistance and deformation behavior.

The general outline of this article follows some possible procedural steps in evaluating processing effects on fatigue performance (Fig. 1). The article begins with general information on the role of residual stress and the baseline established by typical polished smooth specimen testing for comparison of the effects of surface treatments and modifications. With this backdrop, considerations in selecting fabrication or subsequent surface processing procedures to improve fatigue resistance are discussed in terms of their respective effects on fatigue performance. This article then considers the development of test data to quantify the effect of a procedure on a given application. Thereafter, the less satisfactory alternative is addressed, adapting previously generated data to estimate trends in the improvement from a given processing effect. In this case the extent of the improvement is estimated by adjusting baseline properties to account for the expected improvement, based on patterns established from the literature for a similar material and process.

This article is neither material nor application specific. It reflects the behavior of materials such as steels, cast irons, and aluminum alloys and considers to some extent both crack initiation and propagation to the degree that each is affected by surface treatment and modifications. Fabrication and subsequent processing procedures relying on thermal, thermal-mechanical, and mechanical processes applied to the base material are addressed, and other methods that add to or modify the surface are discussed in terms of surface effects, residual stress effects, and chemistry and microstructure effects. Two examples and details about specific processing and fabrication effects on crack initiation for a range of materials also are included to serve as a guide for applications encountered in design.

Residual Stresses and Processing

It has long been recognized that residual stress and surface condition influence fatigue. Smooth test specimens are typically polished and enhance fatigue, whereas precracked specimen geometries may by virtue of their preparation contain longer-range residual stresses. Actual component fabrication may also involve processes that modify the surface or relieve the residual stresses, altering the fatigue resistance from that anticipated. The stress gradient in a component almost

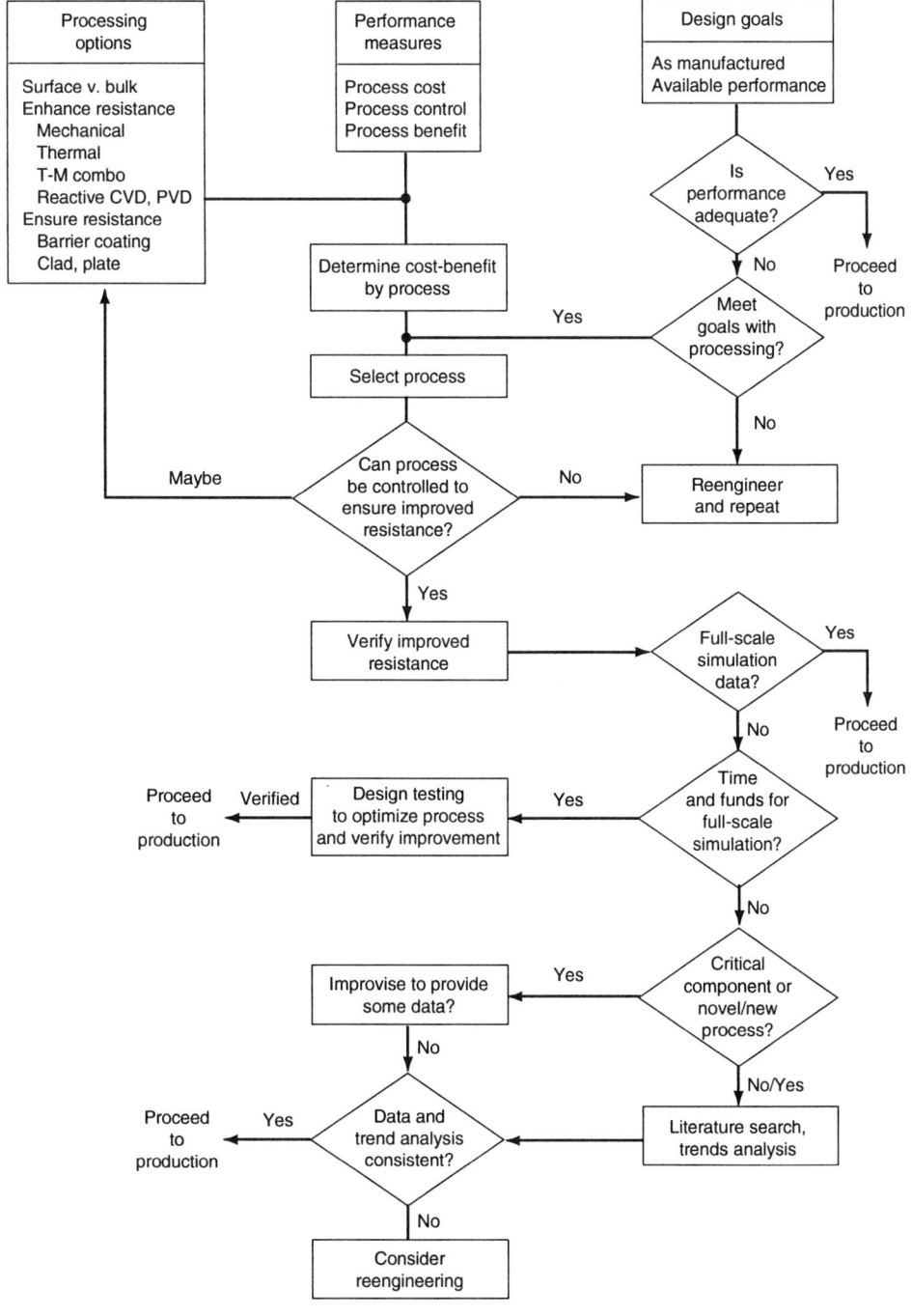

Fig. 1 Conceptual procedure for evaluating processing effects on fatigue performance

polishing, and lapping. To a large extent, the surface residual stresses resulting from these different finishing methods are key. For example, rough grinding typically produces high tensile residual stresses; "fine" grinding, lower tensile stresses; polishing and sanding, moderate compression; lapping and polishing, high (but shallow) compressive stresses, and so on. It is now realized that variations in residual stress produced in the surface by the finishing methods often are the main influence on fatigue life, not the surface roughness itself (Ref 1).

From a systematic study described in Ref 1, general surface integrity and fatigue strength are consistently related by the surface residual stress with some general observations:

- In many cases, surface residual stress is the single most important factor influencing fatigue behavior after the inherent strength of the material itself is considered.
- The residual stress effect is most pronounced in the endurance limit regime (10^7 cycles), although significant effects of residual stress can be seen in the higher stress ranges (low-cycle fatigue region), which yield lives of 10^3 to 10^5 cycles.
- The effect of residual surface stress on fatigue tends to diminish as test temperatures are increased significantly, presumably related to the relaxation of residual stress due to the thermal exposure.
- The effect of surface finish or roughness within reasonable ranges (up to 100 to 200 μin., arithmetic average) on fatigue strength is much less than has been traditionally accepted by the engineering community. Residual stress can be a much more potent factor. Gross surface discontinuities are, however, clearly detrimental.
- Surface residual stress also affects the stress corrosion resistance of those materials/conditions sensitive to this phenomenon. Tensile residual stresses in the same direction as applied external stress can significantly reduce the threshold of crack initiation in sensitive materials/environmental conditions.
- Complex surface conditions, such as those produced by carburizing and subsequent case hardening, exhibit the same trend, although the increase in fatigue strength observed (in this case) is probably attributable to both compressive residual stresses related to the carburizing/hardening plus concomitant phase changes, as well as to the increase in the inherent strength of the higher carbon, higher hardness of the surface.
- Shot peening is particularly complex in its analysis and can be discussed in overview only in general terms. Peening generally produces residual surface compression; it also typically causes work hardening. Both tend to increase fatigue life. Peening, however, also can cause microcracking, microtearing, the formation of small laps and seams, and significant surface roughening in the "overpeened" condition. The net result is commonly an increase in fa-

certainly will be different from that of a test specimen. The correlation, or lack of it, between simple specimen fatigue data and fatigue lives obtained with real components is a well-known concern that is influenced by several possible sources of material differences between the specimen and an application component.

In particular, several generations of mechanical and metallurgical engineers have dealt in some way with the interrelation between the surface condition of a material and the resulting fatigue strength. From the early days of analyzing the failures of rotating and reciprocating machinery,

the detriment of sharp corners, steps in shaft diameters, and tooling marks has been well recognized. In addition, the surface roughness measurement is very significant in relation to exhibited fatigue life; there is a direct quantitative correlation between surface roughness and fatigue.

However, the apparent relation between surface roughness and fatigue is in reality much more complex. Typical plots of fatigue strength versus surface roughness in a given material are often produced experimentally by different surface finishing methods, such as grinding, turning with different tool radii, rough polishing, finish

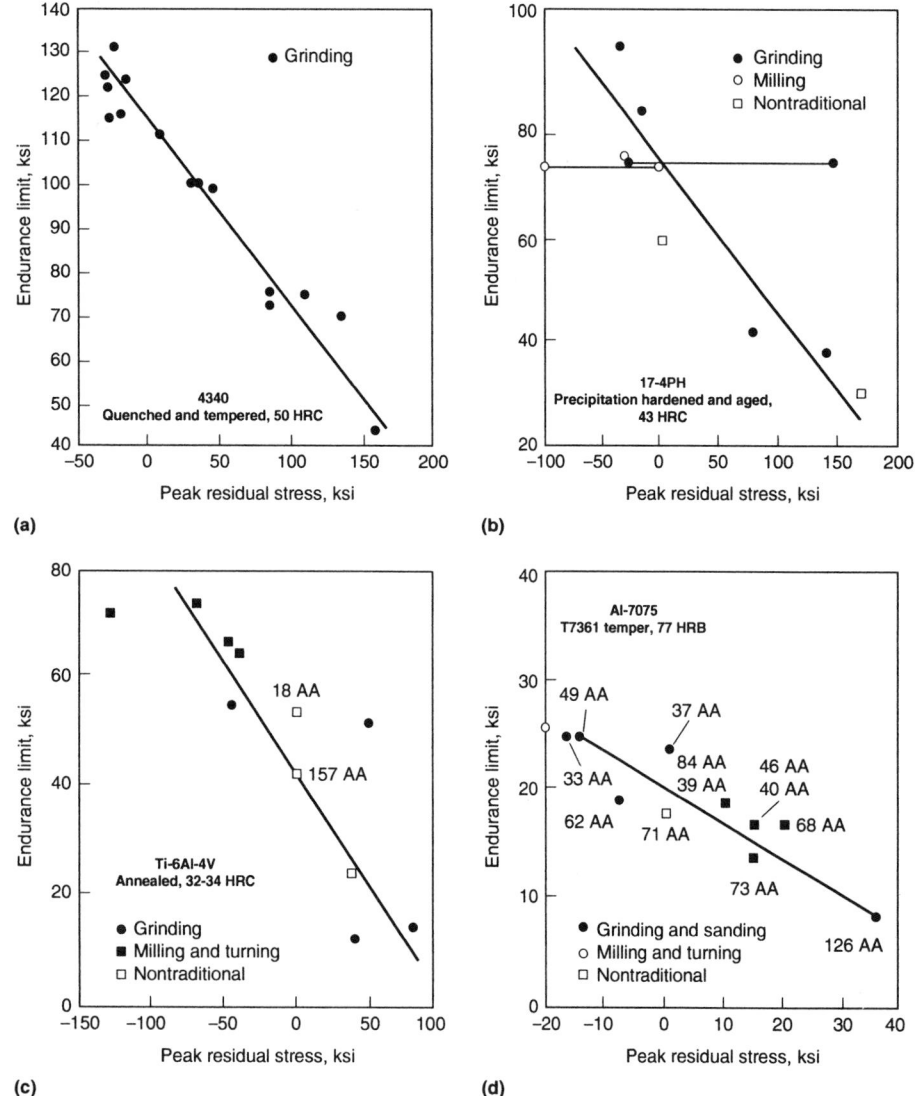

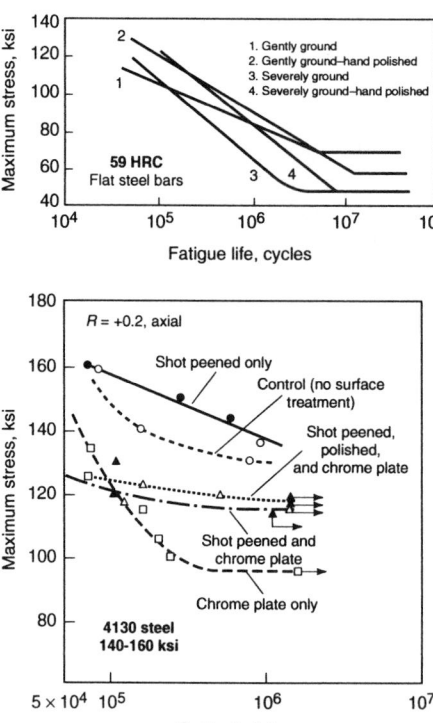

Fig. 3 Influence of surface roughness and finish on fatigue resistance of smooth specimens. Roughness and finish effects include some residual stress effects, as noted in text. Some surface treatments, such as hard chrome plating, can produce surface cracks, which result in a lower life. Source: Ref 2

Fig. 2 Correlation between peak residual surface stress and 10^7 cycles fatigue limit of various structural alloys for various finishing methods. (a) 4340 steel. (b) 17-4PH austenitic stainless steel, with double peaks of residual stress indicated with lines between data points. (c) Ti-6Al-4V. (d) 7075 aluminum alloy. Source: Ref 1

tigue strength with moderate peening but a degradation of properties if multiple or redundant peening cycles are applied. Undoubtedly, the enhancement due to the development of compressive residual stresses is a factor in the behavior of peened surfaces, but is only one of the factors that are active.

The observed relationship between residual stress in the surface zone and fatigue strength as measured by the 10^7 cycle endurance limit is shown in Fig. 2 for 4340 steel; 17-4PH, a typical austenitic steel; Ti-6Al-4V; and aluminum alloy 7075, typical of high-strength aluminum alloys. While Fig. 2(a) involves only grinding, Fig. 2(b) to (d) show data from surface finishing methods including grinding, sanding, milling, and turning, and nontraditional methods such as electrical discharge machining, electrochemical machining, electropolishing, and chem-milling. In Fig. 2(b), two of the 17-4PH specimens exhibited distinct double peaks in the residual stress profile, as

indicated in this figure. In the case of the titanium alloy (Fig. 2c), two different chemical milling procedures (nontraditional) were used that resulted in widely varying surface finishes; both exhibited virtually no residual stress. The aluminum alloy tended to exhibit smaller differences in fatigue strength levels, due in part to the lower strength level in comparison to that of the other materials.

Standard Specimen Fatigue Data as a Benchmark. Fatigue properties are normally determined using standard specimens, which often do not reflect the conditions that exist in real components or reflect the type of surface treatment or modifications that might be used to improve fatigue resistance to meet design objectives. For example, the standard smooth specimen is lathe turned and often has a ground or polished surface.

In contrast, in practical situations fatigue cracks often initiate on a mill-scale, sheared, or plated surface, or at a surface that is roughened

due to plastic flow associated with significant stretching due to forming. In turn, these surfaces may be subjected to some subsequent reworking and processing to offset the reduction in fatigue resistance as compared to the polished smooth specimen. Because fatigue crack initiation in homogeneous materials occurs at the surface, surface conditions that diminish stress raisers (finish) enhance fatigue properties or protect against degradation (wear, corrosion, etc., via functionally gradient materials). A rougher surface thus shortens the life in the high-cycle fatigue regime, although surface treatments can improve fatigue life (Fig. 3). In these cases, it is also important to consider the inherent residual stresses, as noted above.

In most cases, standard specimens are designed to minimize residual stresses. ASTM E 606 (Ref 3) discusses smooth specimen preparation with a view to minimizing initial stresses in the specimens. Likewise, ASTM E 647 (Ref 4) warns of the effect of residual stresses and provides guidance to assess their potential influence on crack propagation data. In contrast, actual hardware may contain residual stresses and related gradients due to processing such as surface hardening methods or shot peening. Mean stresses influence fatigue life, and the effect of residual stress can also be accounted for as a mean stress effect. This influence of residual stress is illustrated in Fig. 4 for precracked (fatigue crack growth) specimen test results (Ref 5).

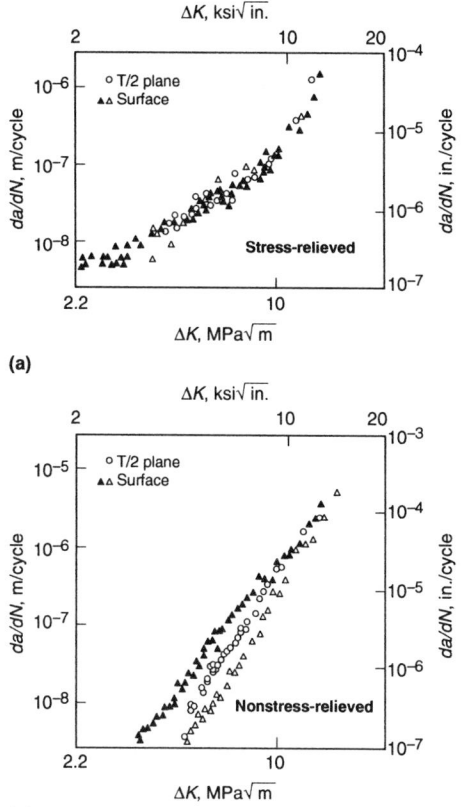

(a)

(b)

Fig. 4 Influence of residual stresses on precracked specimens. Compact tension specimen, $R = 1/3$, frequency 20 Hz, dry air (relative humidity < 10%). (a) Stress relieved. (b) Parent slab. Source: Ref 5

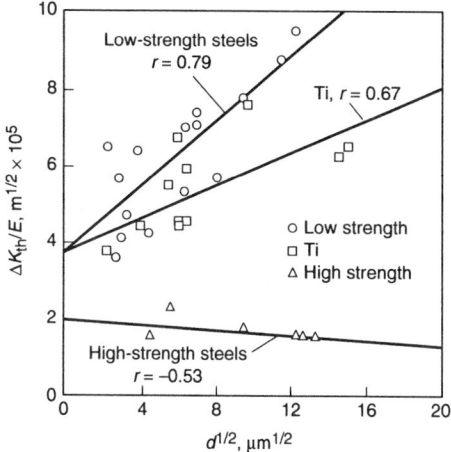

Fig. 5 Influence of grain size on normalized fatigue thresholds in three materials. Source: Ref 6

Table 1 Analysis of the effect of fabrication on fatigue performance of components

Procedure	Physical effect(a)	Relative influence	Typical effect on fatigue resistance
Form or deep draw	Cold work (M, SS)	Significant	Variable
	Orientation (M)	Significant	
	Roughness (S)	Minor to significant	
Forge	Hot work (M)	Minor	Decrease
Extrude	Orientation (M)	Minor	(can increase in
	End grain (S)	Significant	direction of flow)
	Laps (S)	Significant	
Machine	Roughness (S)	Significant	Variable, but often
Punching	Cold work (SS)		decrease
Cutting			
Turn/Mill			
Grinding			
Cast	Orientation (M)	Significant	Decrease with respect
Die	Size (M)		to wrought
	Chemistry (M)		
	Residuals (SS)		
	Voids (S)		
	Cold shuts (S)		
Sand	As for die		
	Roughness (S)		
Weld	Chemistry (M)	Significant	Decrease
Linear or spot	Roughness (S)		
	Residuals (SS)		
	Geometry		
Dressed, contoured	Geometry	Significant	Increase with respect to as-welded
Peened	Residuals (SS)	Significant	
Powder compact	Porosity (S)	Significant	Decrease

Note: Data are as compared to laboratory fatigue test results on polished, residual-stress-free samples. They apply only for intermediate to long-life crack initiation conditions and near-threshold to intermediate crack growth rate conditions. (a) M, material/microstructure change; SS, surface residual stress effect; S, surface alteration

In addition, it may be difficult to represent the microstructural and composition gradients that exist at critical areas and along crack paths using standard specimen sizes and geometries. It also should be emphasized that the "best" material under standard conditions may not be the best under practical conditions. This confusing circumstance develops in situations where the benefit of one material (say higher endurance stresses at long fatigue lives, which arise because of increased hardness in polished smooth specimens) is offset by the rough surface in the application, which reduces the life through notch sensitivity (that has been increased by the increased hardness). Increased notch sensitivity, formation of quench cracks, and a range of other negative factors may develop when uncontrolled changes are introduced in order to enhance fatigue performance.

The final point of emphasis relates to components for which there is a significant fraction of the total life involved in both crack initiation and propagation. Attempts to improve long-life fatigue resistance by using more refined microstructures may lead to decreases in crack growth thresholds (Fig. 5). Refined grain sizes tend to increase the long-life fatigue strength for smooth specimens, but they also tend to decrease the threshold for cracking in precracked specimens, at least for some materials. In such cases, the choice of the "best" material is not straightforward. To complicate matters further, processing may produce a variation in microstructure throughout a component. Figure 5 might suggest that a refined surface structure, coupled with a coarser structure in the midthickness, would represent an optimum processing procedure for some lower-strength steels (in some applications). However, for high-strength steels an optimum processing procedure could involve maintenance of a refined surface structure throughout the thickness.

It follows from the above discussions that standard specimens represent factors that affect fatigue in ways that may differ significantly from the conditions in real components. This does not mean that the standard specimens and related test standards are ill-conceived. Standard specimen designs and test conditions are typically defined so that a common basis for comparison of properties exists between materials. It is this common or standard basis for comparison that permits the engineer to assess which is the "best" material for the application. It remains the responsibility of the engineer to account for differences between standard conditions and service applications. Likewise, it remains the responsibility of the engineer to ensure that the apparent benefits of surface treatments and modifications made to improve fatigue resistance will have the intended effect in practice. Selected full-scale component testing may be necessary to accomplish this. In general, standard specimens and test conditions tend to maximize fatigue resistance or minimize fatigue damage, as compared to practical conditions.

Effect of Processing and Fabrication

Figures 2 to 5 suggest that the design analysis cycle can move forward only if data on the effects of fabrication and surface processing are avail-

Table 2 Analysis of the effect of thermal processing on fatigue performance of components

Procedure	Physical effect(a)	Relative influence	Typical effect on fatigue resistance
Ferrous			
Full-section quench	Structure (M)	Significant	Increase (generally)
Anneal, temper, normalize	Structure (M)	Significant	Decrease
Surface alloying	Composition (M)	Significant	Increase (generally)
	Structure (M)		
	Residuals (SS)		
Aluminum			
Full-section age	Structure (M)	Significant	Increase (generally)
Surface laser	Structure (M)	Significant	Increase (generally)
	Residuals (SS)		

Note: Data are as compared to laboratory fatigue test results on polished, residual-stress-free samples. They apply only for intermediate to long-life crack initiation conditions and near-threshold to intermediate crack growth rate conditions. (a) M, material/microstructure change; SS, surface residual stress effect; S, surface alteration

Table 3 Analysis of the effect of mechanical processing on fatigue performance of components

Procedure	Physical effect(a)	Relative influence	Typical effect on fatigue resistance
Peening	Residuals (SS)	Significant	Increase (generally)
	Cold work (M)	Significant	
	Structure (M)	Minor	
	Roughness (S)	Minor	
Rolling	Residuals (SS)	Significant	Increase
	Cold work (M)		
	Structure (M)		
Nominal overload or strain	Residuals (SS)	Significant	Increase
	Cold work (M)		
	Roughness (S)	Minor	Decrease
Local prestress	Residuals (SS)	Significant	Increase
(coin, expand, etc.)	Cold work (M)		
	Structure (M)		

Note: Data are as compared to laboratory fatigue test results on polished, residual-stress-free samples. They apply only for intermediate to long-life crack initiation conditions and near-threshold to intermediate crack growth rate conditions. (a) M, material/microstructure change; SS, surface residual stress contributor; S, surface alteration

able for the range of viable materials, covering the scope of fabrication and processing methods that are both practical and economically attractive. Unfortunately this is seldom the case. Necessary data simply are not available to cover this spectrum of materials and processes. Designers are often obliged to move forward in the presence of very limited data. In some cases they can develop data using standard tests to quantify the improvement that can be expected from a given surface process for a given set of processing parameters. However, seldom are designers afforded the luxury of data that cover all fabrication and processing situations. Accordingly, this section discusses fabrication and processing effects from the standpoint of why/how the effect develops and how one can assess its significance using either standard specimen data and/or application-specific experiments. Trends in surface processing and related fabrication effects are presented for a range of processing procedures and materials at the end of this article. Throughout it is assumed that the reader is familiar with the various fabrication and processing methods and is acquainted with the key parameters for controlling these methods. Further background on fabri-

cation and processing methods may be found in companion volumes of the *ASM Handbook* series.

The effects of various processing or fabrication procedures are outlined briefly in Tables 1 to 4. The effects are analyzed first as physical effects, and they are described in terms of their tendency to alter the material, the stress field, and/or the surface, as compared to that of the standard smooth or precracked specimens.

Material effects include cold work (increased hardness), microstructure (type, orientation, and size), composition (chemistry), hot work, and electrochemical alteration. When cold work (plastic flow) is confined or occurs locally in a gradient (notch root, bent beam), the cold work also influences the stress field by creating compressive (or tensile) residual stresses that are concentrated at the surface. Surface residual stresses are reacted by lower, nominal stresses in the bulk elastic field. Material and cold-work-induced hardness and residual stresses are both key factors in fabrication, as can be seen from Table 1. Increased hardness and residual stresses due to cold work are also key factors in mechanical processing, as is evident in Table 3.

Residual stresses concentrated at the surface may also be created by thermal gradients during the quench cycle of thermal processing, mechanical gradients due to phase transformation on the quench cycle, or plastic deformation of surface grains by shot peening or rolling. Both processing techniques may cause the surface to yield in tension. When the processing transients are completed, the surface yielding may lead to a steep gradient of compressive residual stress at the surface that is reacted by much lower elastic tensile stresses in the core. Because of this gradient, fatigue cracks may initiate subsurface. A complete fatigue life analysis for such a case probably should consider the behavior at the surface as well as at several sites below the surface. Quenching from high temperatures also produces a change in the material, usually most evident as a refined structure with increased hardness (see Table 2).

It is noteworthy that residual (self or internal) stresses are always balanced within a part. Compressive stresses are normally sought at critical areas to reduce tensile mean stress effects. In contrast, preloading into compression at critical areas may reduce the range of the local stress cycle that will enhance fatigue performance. In both cases, equilibrium requires that the compressive stresses be balanced by tensile stresses. Wise design of a component and its processing/fabrication procedure will ensure that the tensile stresses are located in noncritical areas.

Surface alteration is important for both fabrication and surface processing. The surface that develops in practice is much rougher than that of the polished specimen, so that surface invariably shows up as a negative factor in Tables 1 to 4. In contrast, the tables show that increased hardness (strength), local compressive residual stresses (thermal or mechanical), and changes in material structure (that increase strength or tend to refine/clean up the microstructure) are beneficial.

As noted above, surface roughness may be difficult to separate from the more dominant effect of residual stress in a specific procedure. However, surface condition can be critical, because crack-like features or blunt notches on the surface can quickly sharpen into cracks. Care must be taken to ensure that quality control procedures are implemented to preclude surface defects or roughness. Otherwise, the fatigue life can be orders of magnitude less than standard smooth specimen data. In cases where process variations lead to roughened surfaces, controls on the process should be implemented to preclude such detrimental effects in practice.

Effects on Fatigue Performance

Tables 1 to 4 are useful in selecting fatigue performance data to characterize fabrication and processing effects. For example, consider a part that is most economically produced by forging. The factors in Table 1 suggest some guidelines for designing forging dies and processing proce-

Table 4 Analysis of the effect of surface treatment and modification on fatigue performance of components

Procedure	Physical effect(a)	Relative influence	Typical effect on fatigue resistance
Overlay coatings	Residuals (SS)	Significant	Increase (generally) by avoiding surface
	Structure (M)	Minor	degradation if process parameters are
	Protection		correct
Reactive coatings/topical	Residuals (SS)	Significant	Increase if process parameters are
modification	Structure (M)	Significant	correct
Surface	Roughness (S)	Modest/significant	Increase or decrease, depending on
preparation	Electrochemical (S, M)	Significant	process parameters and material

Note: Data are as compared to laboratory fatigue test results on polished, residual-stress-free samples. They apply only for intermediate to long-life crack initiation conditions and near-threshold to intermediate crack growth rate conditions. (a) M, material/microstructure change; SS, surface residual stress effect; S, surface alteration

dures. For wrought forms such as forgings, it is well known that fatigue resistance is direction-dependent when working creates a strongly oriented structure. An order-of-magnitude decrease in fatigue life can occur when cracks are allowed to initiate and grow along a strongly oriented structure. Obviously, cutting across this elongated structure to expose end grains would cause a major life penalty. It follows that forging dies and subsequent machining should be chosen to avoid either exposing end grains or creating macroscopic grain patterns that could be exposed accidentally in service or through subsequent reworking. Shot peening could be considered as one approach to offset the reduced resistance caused by end grain. As such, Table 1 indicates that care should be taken to account for orientation, laps, and end grains, with secondary concern for hot working.

The factors listed in Table 1 are also useful in assessing how to design specimens to be cut from actual hardware to characterize fatigue performance and in assessing the merits of various surface processing. First, critical locations and probable cracking planes should be identified. Care should be taken in this step to account for stress redistribution and changes in load transfer that may activate other critical locations or redirect the cracking path. The next step makes use of Tables 1 to 4 to identify which areas of the component should be sampled to characterize properties. For the forging example, key areas to be characterized would include crack paths parallel to strongly oriented microstructure, for precracked samples, and areas of exposed (intersected) end grain, for smooth specimens. The next step is to design test specimens that place the cracking direction perpendicular to the test load and provide for adequate gripping. Plausible methods to improve fatigue resistance at these locations can then be identified and samples prepared for comparative fatigue testing. It will often be impossible to obtain standard specimens, so some creativity may be necessary. The final step is to develop data, but care should be taken to avoid aspects of the standard test methods that alter the surface from service conditions, such as polishing.

When the opportunity does not exist to develop application-specific material/fabrication and pro-

cessing fatigue data, the designer can make use of the factors in Tables 1 to 4 and literature data to estimate fatigue performance data and benefits from surface processing. Estimating performance data parallels the steps taken in developing actual test results. As outlined in the preceding paragraph, locations that represent critical areas and crack paths have to be identified and the related fatigue data must be estimated. If the factors in Tables 1 to 4 have been used in selecting the fabrication and processing methods and procedures, and if these factors have then been implemented successfully, there is a reasonable chance that the related change in fatigue life as compared to the smooth specimen data will be realized. The analyst should still estimate the effect of fabrication and processing for use in design, erring on the conservative side unless full-scale testing is done to support the estimation procedure.

The key question in estimating the effect of a fabrication or processing method for a given application is: How does one judge the significance of the effect relative to standard specimen data? Related questions are: How well can fabrication or processing be controlled? How sensitive will the fatigue life be to variations in fabrication or processing? How well do data for similar materials (or perhaps distinctly different materials chosen as a reference in estimating fatigue behavior) represent the material at hand for the procedure being considered?

General answers to these key questions are not possible, but by categorizing fabrication and processing methods according to their physical effects, similarities between methods are apparent. Within a given generic material there are classes of material with comparable structures and strengths, so microstructural and cold work effects may be comparable. Finally, given that surface effects range from the influence of a notch to that of a crack, fracture mechanics for blunt cracks (Ref 7) could be used to estimate the role of the surface effects based on the related toughness. In all cases the estimate should be benchmarked with results from some comparable circumstances to ensure that the estimate is reasonable and conservative. In the absence of benchmark tests, an adequate factor of safety should be used. Here "adequate" is judged by the

criticality of the component to continued safe operation. Care should also be taken to ensure that the factor of safety reflects the degree of uncertainty in the estimation process. When the factor of safety becomes too large because of large uncertainty or high criticality, consideration should be given to developing application-specific data supported by benchmark component tests.

Assessing Fabrication and Surface Processing Effects on Fatigue Resistance. Estimates of fatigue performance dealing with initiation performance embrace all factors indicated in Tables 1 to 3. However, because surface is not a factor in crack propagation, the propagation performance literature focuses more on material effects and to a much lesser extent on residual stress effects. It should be noted that when compressive residual stresses are higher than the imposed tensile stress, the crack tip remains closed with minimal crack growth.

The focus here is on fatigue performance, but it should be remembered that stress-strain response, which is used in some crack initiation life analyses, may also depend on fabrication and processing. In particular, stress-strain response is sensitive to microstructural changes and accumulated strain. For steels the effect of microstructural changes on stress-strain response is reasonably estimated by data for other steels with a similar microstructure at a comparable hardness level. The effect of accumulated strain on stress-strain behavior for ferrous metals varies. Increased hardness generally leads to elastic response at higher strain levels, but cyclic softening might negate this effect. Thus, cyclic strains approaching or exceeding the initial mechanical strain that caused the hardening often generate a stress-strain response similar to the stable curve of the "virgin" material. The behavior of aluminum alloys is less patterned. Testing should be done if there is a significant change from virgin material due to fabrication and processing.

Fatigue performance is tabulated and graphed for a wide variety of steels and aluminum alloys in a range of handbooks. The purpose here is to illustrate how to use such performance data and to cite the most relevant data sources (e.g., Ref 8-11). These data can also be used to estimate the influence of a given procedure for a given material. The following two examples illustrate how to estimate the effect of processing as compared to standard specimens.

Example 1: Effect of Finishing and Surface Modification. The problem is to estimate the effect of fabrication on the fatigue crack initiation performance of a 1045 steel component (235 HB average) subject to fully reversed axial loading.

Solution: Fatigue performance in this case is assessed in terms of allowable stress to survive for a given number of cycles. Specific data representing the fatigue strength at 10^7 cycles for this situation can be found in Ref 9 in the form of a modified Goodman diagram. Taking the polished specimen as the reference ($S/2 = 49$ ksi at 10^7 cycles), the results are as follows:

As forged	~0.38 × fatigue strength of polished specimen
Hot rolled	~0.51
Turned	~0.71
Ground	~0.88
Ion nitriding	~1.5 on a polished surface

This table shows that a very strong penalty on allowable stress is paid if the component incurs stresses near the fatigue limit and the surface is left in an as-forged or an as-hot-rolled condition, as compared to some final machining operation applied in critical areas. The beneficial effects of surface treatment, such as ion nitriding, could also have a role. The same steel after ion nitriding of a polished surface has a hardness at the surface on the order of about 350 HB, with a corresponding increase in long-life fatigue strength of about 1.5. Initiation in this context is subsurface, so that the detrimental effects of finish noted above are not a factor, except that the benefit of the surface modification will be diminished on a lower-quality surface. In this context, the benefits of surface modification follow only after the surface is prepared for modification, and the detrimental effects of surface roughness do not intervene to shorten the time for crack initiation.

Finite-life fatigue behavior can be estimated if the fatigue strength data are used in conjunction with data for polished specimens for this material condition as shown for each of the fabrication methods in Fig. 6. It is common practice (for many steels) to assume that a fatigue limit exists beyond about 2×10^6 cycles (4×10^6 reversals), as indicated in the figure.

Differences between the fatigue strength produced by these fabrication procedures and the smooth specimen response are due primarily to surface condition, because microstructural and hardness differences are normalized by comparing results at a given hardness. Because surface does not significantly influence a "strength" coefficient, it is reasonable to assume that the fatigue strength coefficient for polished specimens represents that for the other fabrication conditions. Therefore, finite-life elastic-strain-life be-

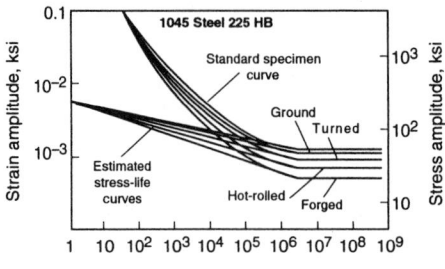

Fig. 6 Example of estimated fatigue life curves using derating factors (see text) from a 49 ksi endurance limit for a 1045 steel standard (smooth) specimen

havior can be conservatively estimated by connecting the fatigue strength coefficient for the standard specimen with the corresponding point at 2×10^6 cycles for the fabrication process of interest. The stress-life behavior can be estimated from the elastic-strain-life behavior through the modulus. Finally, inelastic cycle strain quickly creates surface slip steps that cause significant surface roughness. The influence of fabrication-induced surface roughness on the plastic strain life curve can therefore be assumed to be negligible. Thus, the total strain life response can be estimated by adding the plastic response for the standard smooth specimen to the appropriate elastic response, as shown in Fig. 6.

Example 2: Effect of Fabrication on Fatigue Crack Growth. The problem is to estimate the effect of fabrication on fatigue crack propagation performance on an aluminum alloy (7075-T6) component subject to axial fatigue cycling at $R = -0.33$.

Solution: Fatigue performance in this case is assessed in terms of differences in growth rate at the same effective stress-intensity factor range. For simplicity it is assumed that $\Delta K = \Delta K_{eff}$. Data useful in assessing the influence of fabrication address the dependence of growth rate on microstructural orientation. Such data are tabulated in handbooks (e.g., Ref 10) for a variety of materials and conditions. The longitudinal (L) orientation

is typical of most standard specimen orientations, whereas the transverse (T) orientation is typical of cracking parallel to highly oriented microstructures such as those that may develop in forging, extrusion, drawing, forming, casting, and some weldments. The effect of fabrication as compared to typical bar and plate products is evident in the difference in rates between L and T orientations, with the extent depending on the difference in orientation.

REFERENCES

1. W. Koster, Effects of Residual Stress on Fatigue of Structural Alloys, *Practical Applications of Residual Stress Technology*, ASM International, 1991, p 1-10
2. A.F. Madayag, Ed., *Metal Fatigue*, Wiley, 1969
3. "Standard Recommended Practice for Constant-Amplitude Low-Cycle Fatigue Testing," ASTM Standard E 606
4. "Standard Test Method for Constant-Load-Amplitude Fatigue Crack Growth Rates Above $1 \times 10E-08$ Meters/Cycle," ASTM Standard E 647
5. R.J. Bucci, Effect of Residual Stress on Fatigue Crack Growth Rate Measurement, in *STP 743*, ASTM, 1981, p 28-47
6. W.W. Gerberich and N.R. Moody, Review of Fatigue Fracture Topology Effects on Threshold and Growth Mechanisms, in *STP 675*, ASTM, p 292-341
7. M. Creager and P.C. Paris, Elastic Field Equations for Blunt Cracks with Reference to Stress Corrosion Cracking, *Int. J. Fract. Mech.*, Vol 3, 1967, p 247-252
8. *Aerospace Structural Handbook*, U.S. Navy Publications and Forms Center
9. C. Lipson and R.C. Juvinall, *Handbook of Stress and Strength: Design and Materials Applications*, Macmillan Co.
10. *Damage Tolerant Design Handbook*, MCIC-HB-01, Metals and Ceramics Information Center, Battelle Columbus Div., 1983
11. *Structural Alloys Handbooks*, CINDAS

Fretting Fatigue

S.J. Shaffer and W.A. Glaeser, Battelle Memorial Institute

FRETTING is a special wear process that occurs at the contact area between two materials under load and subject to slight relative movement by vibration or some other force. Damage begins with local adhesion between mating surfaces and progresses when adhered particles are removed from a surface. When adhered particles are removed from the surface, they may react with air or other corrosive environments. Affected surfaces show pits or grooves with surrounding corrosion products. On ferrous metals, corrosion product is usually a very fine, reddish iron oxide; on aluminum, it is usually black. The debris from fretting of noble metals does not oxidize.

Under fretting conditions, fatigue strength or endurance limits can be reduced by as much as 50 to 70% during fatigue testing (e.g., see Fig. 1a). During fretting fatigue, cracks can initiate at very low stresses, well below the fatigue limit of non-fretted specimens. In fatigue without fretting, the initiation of small cracks can represent 90% of the total component life. The wear mode known as fretting can cause surface microcrack initiation within the first several thousand cycles, significantly reducing the component life. Additionally, cracks due to fretting are usually hidden by the contacting components and are not easily detected. If conditions are favorable for continued propagation of cracks initiated by fretting, catastrophic failure can occur (Fig. 1b). As such, prevention of fretting fatigue is essential in the design process by eliminating or reducing slip between mated surfaces.

The initiation of fatigue cracks in fretted regions depends mainly on the state of stress in the surface, particularly stresses caused by high friction. The direction of growth of the fatigue cracks is associated with the direction of contact stresses and takes place in a direction perpendicular to the maximum principal stress in the fretting area. After formation due to fretting, cracks propagate initially under shear (mode II) conditions under the influence of the near-surface shear stress field due to friction of fretting. Beyond that, tensile (mode I) crack propagation under bulk cyclic stresses controls further propagation.

Because mode I bulk crack propagation is covered in detail throughout this Volume, this article focuses on what, if any, measures the design engineer has at his or her command to avoid or minimize crack initiation and fretting fatigue. The topics covered in this article are:

- Mechanisms of fretting and fretting fatigue
- Typical occurrences of fretting fatigue
- Fretting fatigue testing
- Prevention methods

Many investigators have contributed to the theoretical and practical research in the field of fretting and fretting fatigue, and the information in this section is derived from their work. Several general texts are available (Ref 1-4). Reference 5 is another key source for information and illustrations of fretting fatigue failures. In addition, more current discussions and background are also provided in the book *Fretting Fatigue* (R.B. Waterhouse and T.C. Lindley, Ed.) ESIS, Mechanical Engineering Publications, 1994. The article "On Fretting Maps" by O. Vingsbo and S. Söderberg (*Wear*, Vol 126, 1988, p 131-147) is another useful general reference on fretting.

As yet, general techniques or models permitting prediction of crack initiation due to fretting are limited. However, an understanding of the factors contributing to fretting fatigue can help minimize the risk and extent of damage. The examples presented in this article from case studies, theoretical work, and laboratory investigations are intended to assist the reader in recognizing the potential for fretting fatigue in design and materials selection. General principles and practical methods for the abatement or elimination of fretting fatigue are summarized in Table 1.

Fretting and Fretting Fatigue Mechanisms

In general, fretting occurs between two tight-fitting surfaces that are subjected to a cyclic, relative motion of extremely small amplitude. Although certain aspects of the mechanism of fretting are still not thoroughly understood, the

fretting process is generally divided into the following three parts: initial conditions of surface adhesion, oscillation accompanied by the generation of debris, and fatigue and wear in the region of contact.

Fretting wear occurs from repeated shear stresses that are generated by friction during small amplitude oscillatory motion or sliding between two surfaces pressed together in intimate contact. Surface cracks initiate in the fretting wear region. The relative slip amplitude is typically less than 50 μm (0.002 in.), and displacements as small as 10^{-4} μm have produced fret-

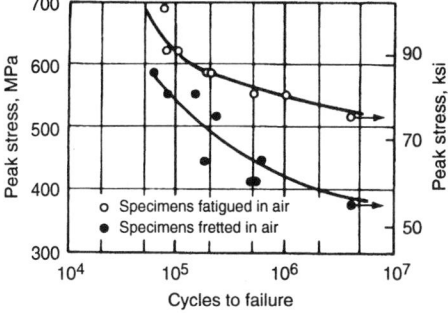

(a)

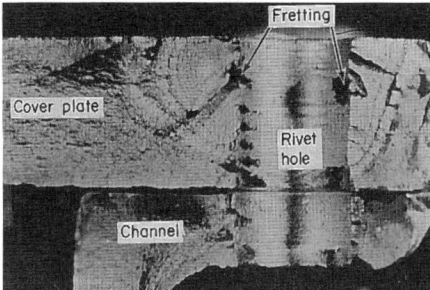

(b)

Fig. 1 (a) Comparison of fatigue life for 4130 steel under fretting and nonfretting conditions. Specimens were water quenched from 900 °C (1650 °F), tempered 1 h at 450 °C (840 °F), and tested in tension-tension fatigue. Normal stress was 48.3 MPa (7 ksi); slip amplitude was 30 to 40 μm. (b) Example of catastrophic failure due to fretting of a flanged joint

Table 1 Reduction or elimination of fretting fatigue

Principle of abatement or mitigation	Practical method
Reduction in surface shear forces	Reduction in surface normal forces
	Reduction in coefficient of friction with coating or lubricants
Reduction/elimination of stress concentrations	Large radii
	Material removal (grooving)
	Compliant spacers
Introduction of surface compressive stress	Shot or bead blasting
	Interference fit
	Nitriding/heat treatment
Elimination of relative motion	Increase in surface normal load
	Increase in coefficient of friction
Separation of surfaces	Rigid spacers
	Coatings
	Compliant spacers
Elimination of fretting condition	Drive oscillatory bearing
	Remove material from fretting contact (pin joints)
	Separation of surfaces (compliant spacers)
Improved wear resistance	Surface hardening
	Ion implantation
	Soft coatings
	Slippery coatings
Reduction of corrosion	Anaerobic sealants
	Soft or anodic coatings

ting. Generation of fine wear debris that usually oxidizes is an indication of fretting wear (Fig. 2). The following factors are known to influence the severity of fretting:

- *Contact Load.* As long as fretting amplitude is not reduced, fretting wear will increase linearly with increasing load.
- *Amplitude.* There appears to be no measurable amplitude below which fretting does not occur. However if the contact conditions are such that deflection is only elastic, it is not likely that fretting damage will occur. Fretting wear loss increases with amplitude. The effect of amplitude can be linear, or there can be a threshold amplitude above which a rapid increase in wear occurs (Ref 4). The transition is not well

established and probably depends on the geometry of the contact.

- *Frequency.* When fretting is measured in volume of material removed per unit sliding distance, there does not appear to be a frequency effect.
- *Number of Cycles.* An incubation period occurs during which fretting wear is negligible. After the incubation period, a steady-state wear rate is observed, and a more general surface roughening occurs as fretting continues.
- *Relative Humidity.* For materials that rust in air, fretting wear is higher in dry air than in saturated air.
- *Temperature.* The effect of elevated temperature on fretting depends on the oxidation characteristics of the material.

In terms of fatigue, the following three primary variables contribute to shear stresses at the surface and hence are important for crack initiation and initial propagation of fretting fatigue cracks:

- Normal load (e.g., contact pressure)
- Relative displacement (slip amplitude)
- Coefficient of friction

The other primary variable is the bulk tensile stresses that control crack propagation beyond the limit of the surface-induced stress field. Secondary factors, including surface roughness, surface contaminants, contact size, debris accumulation, and environment affect fretting fatigue through their influence on the primary variables. Effective lubrication will reduce friction stresses and wear particle accumulation.

Fretting Modes and Contact Conditions. The oscillatory motion responsible for fretting can be induced by system vibrations or by cyclic loading of one of the components. The relative displacement can be either amplitude controlled or load controlled, or a combination of both. Methods to control fretting fatigue depend on which of these two modes dominates the contact conditions.

Stress Conditions. The nominal macroscopic normal stress between the two surfaces is defined by the normal force divided by the nominal area of contact. Subsurface stress distributions can be computed using Hertzian calculations and the macroscopic contact geometry. The normal stress is also influenced by geometric stress concentrations. The real area of contact is limited to the contacting tips of the microscopic asperities on each of the surfaces, and the local (microscopic) normal stress is dictated by the yield strength of the softer of the two materials. Superimposed on the local normal stresses are shear stresses resulting from the friction of relative displacement of the two contacting members. The magnitude of the shear stresses induced by asperity contact depends on the coefficient of friction (due to adhesive forces between, and interpenetration of, asperities), the local load, asperity geometry, the elastic moduli of the two surfaces, and the amplitude of relative displacement.

Strain Conditions. If the amplitude of oscillation is small, the shear strains are elastic, and the contact condition is one of sticking or *no-slip*. Even under microelastic displacements fatigue cracks can form due to reverse bending at the bases of the contacting asperities. If the amplitude of oscillation is large, depending on the strength and ductility of the asperities in contact, and on the adhesive forces acting between them, the asperities will be forced to pass over one another and *slip* occurs[*]. With slip, the possibility of wear exists from either adhesion, abrasion, or delamination. All of these material removal mechanisms lead to a roughening of the surface, the creation of sites for crack initiation, and the generation of wear debris. Cracks can also be initiated by pitting that occurs during fretting.

---

[*] In fretting, the term "slip" is used to denote small amplitude surface displacements, in contrast to "sliding," which denotes macroscopic displacements. Additionally, in this article slip does not refer to the mechanism of fatigue resulting as a consequence of dislocation motion.

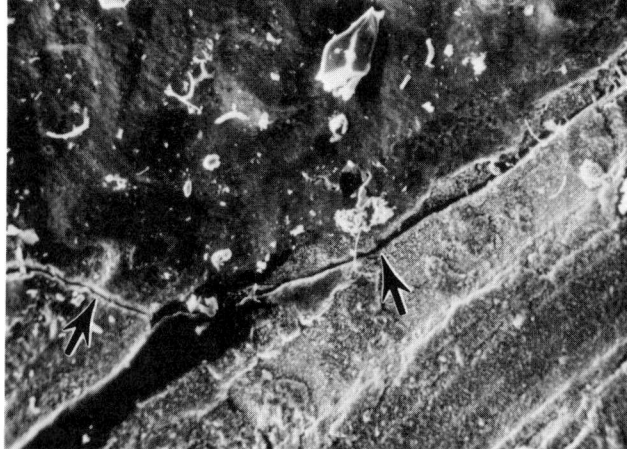

(a)

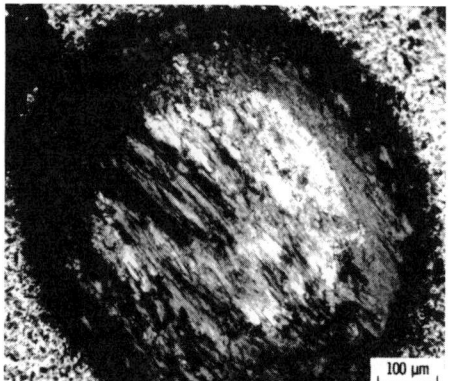

100 μm

(b)

Fig. 2 Fretting wear scars. (a) On steel (arrows indicate fatigue crack). Courtesy of R.B. Waterhouse, University of Nottingham. (b) On high-purity nickel. Courtesy of R.C. Bill, NASA Lewis Research Center

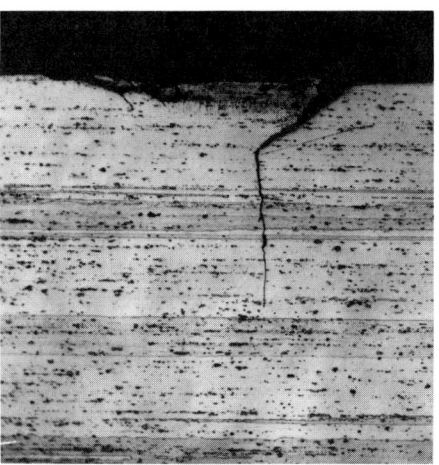

Fig. 3 Stress distribution for hemispherical contact pressed into flat plate. Source: Ref 6

Conditions for Slip.

The local contact conditions may be predominantly displacement controlled or load controlled. In displacement-controlled fretting contacts undergoing full slip, the amplitude of motion is controlled by the external displacements, an example being the relative displacement imposed on adjoining strands of a wire rope passing over a pulley. For force-controlled fretting contacts, the displacement will depend on the macroscopic shear force, the normal force, and the coefficient of friction; for example, the mating forces of a bolted flange or a hub/shaft press fit interface. No slip occurs until the shear stress exceeds the product of the normal force and local coefficient of friction. The condition for slip is met when

$$\tau > \mu\sigma_n \qquad \text{(Eq 1)}$$

where τ is the local shear stress, μ is the coefficient of friction between contacting surfaces, and σ_n is the local normal stress.

Regions of both slip and no-slip can occur at an interface of contacting solids. This is most easily seen for convex contacts using the elastic stress analysis of a sphere pressed into a plate by Mindlin (Ref 6) (Fig. 3). The normal (Hertzian) stress field is an elliptical distribution with the maximum stress occurring under the center of the contact. The shear stresses, which promote relative slip, are a maximum at the edges of the contact (limited by yielding) and a minimum in the center. The forces resisting sliding due to the shear stress are given by μN. The condition pictured is known as *partial slip*. As μN is increased, the region of sticking expands, and vice versa. When the shear forces of fretting are superimposed on this stress field, the result is a smaller "stick" area. This analysis can also apply to a cylindrical contact or can be adapted to microscopic asperity contacts.

For contact between nominally flat surfaces, the stress state is different, though the slip condition is still defined by Eq 1. The macroscopic stress concentrations for well-defined geometries can be computed using finite-element modeling (FEM) analysis, although wear will change the assumed contact geometry. A general treatment of the subject can be found in Ref 3.

Fatigue Crack Nucleation from Fretting. Crack nucleation due to fretting must involve a

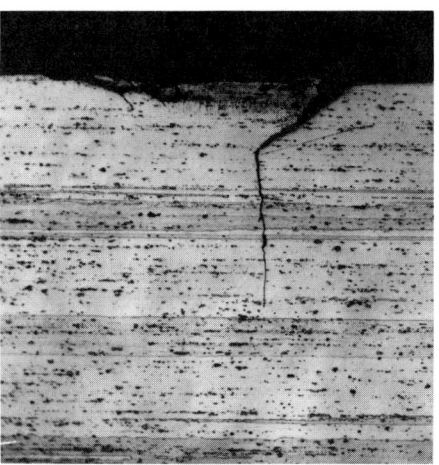

Fig. 4 Example of fretting fatigue crack viewed in cross section. Courtesy of R.B. Waterhouse, University of Nottingham

stress concentration or discontinuity. At the microscopic level, examples include: microcracks formed at the base of asperities due to reverse bending fatigue of the asperities, stress concentrations in the pits left by sheared adhering "cold-welded" asperity junctions, corrosion pits that form due to removal of protective oxides by fretting, grooves due to abrasion, or delamination of a thin surface region whose work-hardening capacity has been exhausted. At the macroscopic level, cracks are proposed to form solely as a result of geometric stress concentrations, usually at the edges of the fretting contact region, where shear stresses are predicted to be highest, at some microscopic inhomogeneity. The two views are not significantly different. The second view is more amenable to modeling and FEM analysis.

The location of crack nucleation depends on the contact conditions. Under *full slip*, in the absence of stress concentrations at the edge of the contact, cracks can nucleate anywhere in the contact region. The number of asperity interactions per cycle depends on the asperity distribution (surface roughness) and the amplitude of relative motion. Several cracks may be formed. Their stress fields can interact and lead to a decrease in the stress field associated with a single crack. This may explain why multiple nonpropagating cracks are often found in association with fretting. Under *partial-slip conditions*, the cracks always form at the border between the slip and the no-slip regions. In this case, multiple cracks are proposed to result from the movement of the slip/no-slip boundary due to the generation of debris (Ref 7). Though less likely, cracks can also form in the region of *no slip* (full sticking) due to reciprocating subsurface shear stresses associated with reversing elastic deformation of the contacting asperities and stress concentrations at their bases leading to local microplastic deformation and fatigue.

Fatigue Crack Propagation during Fretting Fatigue. Crack propagation is initially driven by

the stress state dominated by the surface shearing. As such, when viewed in cross section, the crack direction initially appears at an angle to the surface of between 35 and 55°. Mode II crack propagation dominates this region. The mode II propagation may be dependent on material parameters such as grain size, texture, and phase morphology. Because the surface shear stresses fall off rapidly with depth, the crack will either arrest, or, if static or alternating tensile stresses exist in the bulk material, will change direction and run perpendicular to the surface as the driving forces come under the control of the bulk tensile forces (Fig. 4). The depth at which this occurs depends on the magnitude of the surface shear stresses, which depend on the coefficient of friction and normal contact stress. For Hertzian stresses of convex contacts, this depth is on the order of the half-width of the contact area. Beyond this depth, mode I crack propagation analysis can be used to predict the growth rate under the bulk stress state.

A phenomenon peculiar to fretting is that some of the fatigue cracks do not propagate because the effect of contact stress extends only to a very shallow depth below the fretted surface. At this point, favorable compressive residual stresses retard or completely halt crack propagation. Under full-slip conditions, the wear rate caused by fretting occasionally outpaces the growth rate of surface-initiated fatigue cracks. In this situation, fretting wear preempts fretting fatigue.

Typical Systems and Specific Remedies

Fretting generally occurs at contacting surfaces that are intended to be fixed in relation to each other, but that actually undergo minute alternating relative motion that is usually produced by vibration. Fretting generally does not occur on contacting surfaces in continuous motion, such as ball or sleeve bearings. There are exceptions, however, such as contact between balls and raceways in bearings and between mating surfaces in oscillating bearings and flexible couplings. Common sites for fretting are in joints that are bolted, keyed, pinned, press fitted, or riveted; in oscillating bearings, splines, couplings, clutches, spindles, and seals; in press fits on shafts; and in universal joints, baseplates, shackles, and orthopedic implants.

Three general geometries and loading conditions for fretting fatigue are considered in this section:

- Parallel surfaces clamped together with some type of fastener, such as a bolted flange or riveted lap joint
- Parallel surfaces loaded by means of a press or interference fit, such as a gear or wheel on a shaft
- Convex contacts, as found beneath a convex washer, between crossed cylinders such as wire rope strands, or a sphere or a cylinder in a bearing race

Specific remedies to reduce fretting are given for these common examples. When frettting occurs, it often cannot be eliminated but can be reduced in severity.

Parallel Contact with External Loading (Fastened Joints). Bolted flanges in pipe systems are common locations for fretting fatigue. Cracks can occur in the plate either under a bolt head or washer (load controlled), on the inside diameter of the bolt through-hole (displacement controlled), or on the surface of one plate at the point of contact with the end of the other plate (load controlled) (Fig. 5a). Lap joints are found in both heavy plates and thin sheets such as aircraft skins. Fretting can occur in the joint or under the head of a countersunk screw, bolt, or rivet (Fig. 5b).

For both these geometries, the reduction or abatement of fretting severity depends on whether the motion is load controlled or displacement controlled. If it is displacement controlled, then reducing the contact stress and minimizing coefficient of friction at the interface is recommended. For lap joints however, a reduction in coefficient of friction may result in insufficient load transmitted by the interface, transferring the load to the fasteners and leading to their failure. For load-controlled motions, it may be possible to increase either the clamping force and coefficient of friction to completely eliminate relative motion between the two contacting members. While adhesives can be used to eliminate the relative motion, their use complicates future disassembly. If motion cannot be completely eliminated, then minimizing the coefficient of friction may help, although this will likely lead to an increased slip amplitude. Alternatively, a thin compliant layer, such as rubber or other polymer, may be able to absorb the deflection and prevent contact between the two members.

For pin joints, fretting can occur on diagonally opposite sides of the pin at the points of contact with the hole due to vibrations or reversing loads. In these cases, White (Ref 8) showed that an increase in fatigue strength can be achieved by machining flats on the sides of the pins to prevent contact at the position of maximum stress, thus removing the region where fretting occurs (Fig. 6).

Parallel Surfaces without External Loading. Hubs, flywheels, gears, and other types of press-fit wheels, pulleys, or disks on shafts are subject to fretting fatigue caused by reverse bending strains compounded by the stress concentration where the shaft meets the disk (Fig. 7a). The introduction of lubricant in the interface can make matters worse by increasing the relative slip. In this case, it is best to attempt a strong interference fit. This can be achieved through cooling the shaft and heating the bore of the hole during assembly in order to produce sufficiently high normal stresses to completely eliminate slip within the interface. After assembly, both surfaces will also be in a state of compressive stress, providing further resistance to fatigue crack propagation. Finally, if possible, a stress-reliev-

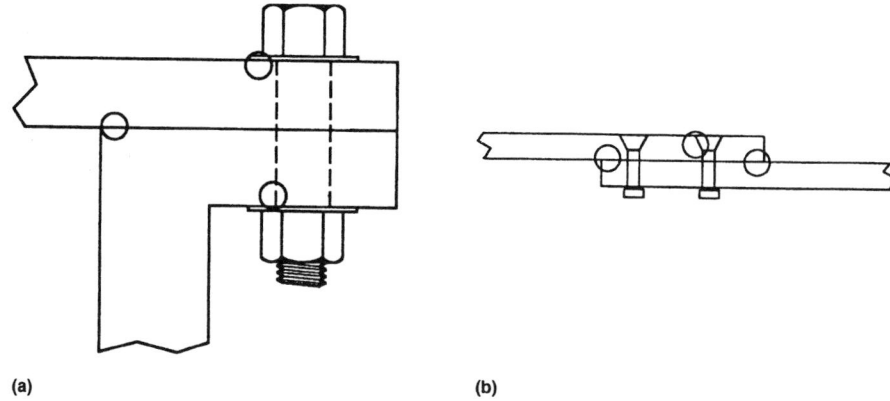

Fig. 5 Typical location of fretting fatigue cracks in (a) a bolted flange, and (b) a lap joint

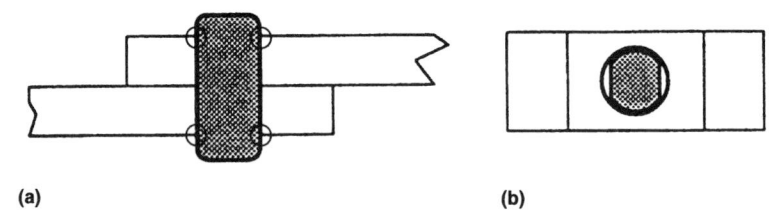

Fig. 6 (a) Fretting locations in pin joint. (b) Material removal to eliminate region of highest stress

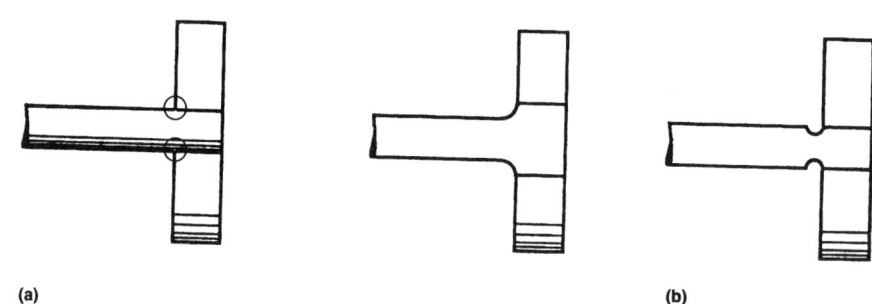

Fig. 7 (a) Location of fretting fatigue cracks in wheel on shaft. (b) Stress reduction grooves

ing groove or large radius on the shaft (Fig. 7b) should be incorporated into the design.

Gas turbine rotor blade roots and other dovetail joints are potential locations of fretting fatigue failures (Fig. 8a). In this case, the loading conditions are variable and depend on the rotational speed. For these situations, stress-relieving grooves can be incorporated into the design (Fig. 8b). Coatings to reduce the coefficient of friction (and hence the surface shear forces) can also help. Experiments by Ruiz and Chen on simulated blade/disk dovetail joints at 600 °C indicated that shot peening followed by electroplating with a 10 μm thick Co/C surface layer was effective (Ref 9). Another example is provided in the article by Johnson in Ref 5 (paper 5).

Convex Surfaces. Fretting fatigue in control cables or wire ropes is caused by small relative displacements between the individual strands as the cable flexes in passing over pulleys, or by varying stresses from wind or water currents (Fig.

9). Stainless steel control cables are particularly susceptible because of the high friction and galling propensity between the strands. Fatigue fractures of the inner strands of the cables make detection by visual inspection virtually impossible until the ends of the fractured strands pop out through the outer strands. Wire rope fretting fatigue in control cables is an example of displacement-controlled contact. As such, large pulley diameters as a function of cable cross section can be specified in the design in order to decrease the displacement and minimize fretting fatigue. Incorporation of lubricant in the cable will reduce strand-to-strand shear forces, but only as long as the lubricant is contained within the rope interior by the outer strands. Takeuchi and Waterhouse report that electrodeposited zinc coatings helped prevent fretting fatigue of wire rope in sea water (Ref 10). The zinc provides both a reduction in the effects of corrosion and a low shear strength surface film that reduces friction.

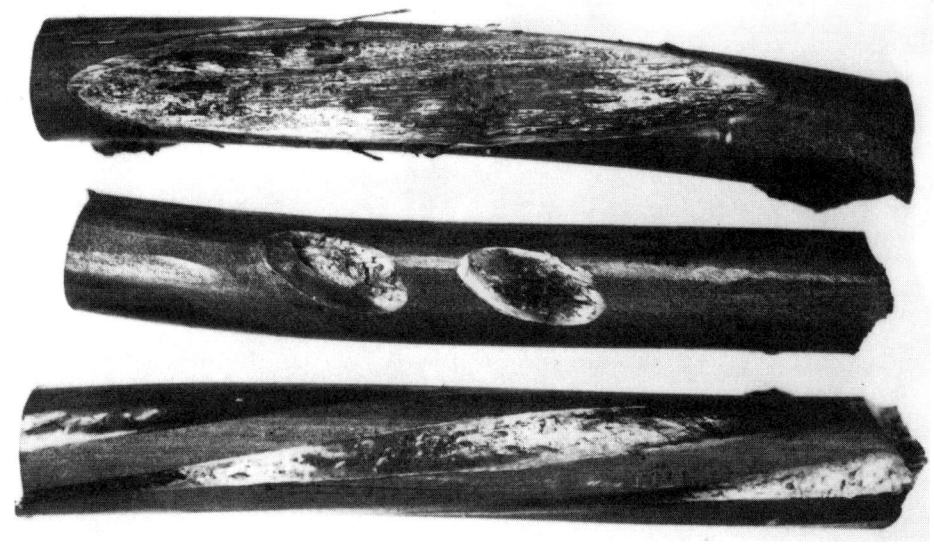

Fig. 8 (a) Location of fretting fatigue cracks in dovetail joint. (b) Stress reduction grooves

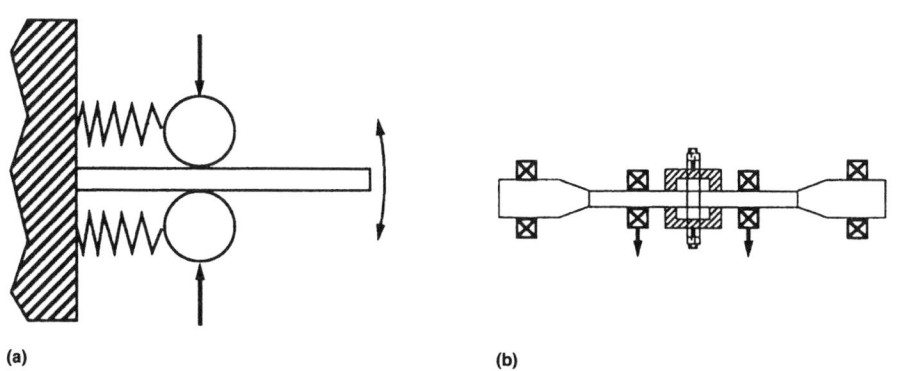

5 mm

Fig. 9 Examples of fretting on inner strands of drag line wire rope

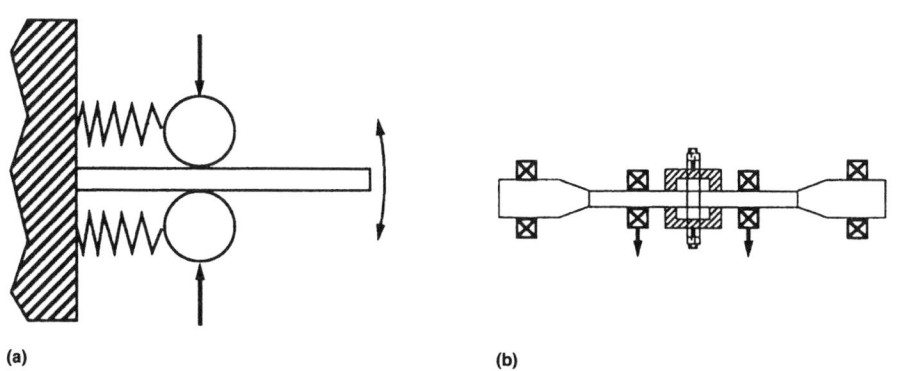

Fig. 10 Examples of fretting fatigue test configurations. (a) Cantilever beam reverse bending with single pads. (b) Rotating fully reversing bending with double foot-pad bridges and proving ring

In rolling-element bearings, high Hertzian normal stresses occur beneath the contact of bearing balls in their races. Control system or oscillatory pivot bearings are often subjected to a low amplitude, but high frequency, dithering motion leading to fretting or false brinelling between the balls and the race. (The difference between false brinelling and fretting is discussed in the article "Fretting Wear," Vol 18 of the *ASM Handbook*.) Continuous rotation of one of the races can help prevent fretting under these circumstances (Ref 11).

For convex washers and other convex contacts, a coating, nonmetallic shim, or lubricating film can help reduce the surface shear forces.

Testing, Modeling, and Analysis

When fretting fatigue is encountered or anticipated, laboratory tests are usually required to find a solution. Test results can also be used to develop an empirical model and are required for validation if a model is to be used for design changes. This section presents types of fretting fatigue tests, the forms of results found in the literature, and the effect of variables on fretting fatigue from different research test programs.

Types of Fretting Fatigue Tests. Fretting fatigue tests are designed to accomplish one of three goals. The first is a test to predict or duplicate field failures and evaluate the effect of design changes or treatments based on replication or simulation of service components and conditions. This is most useful if all the service conditions are known and can be replicated or appropriately scaled, but the results will have limited applicability. A second type of test uses a simple geometry and setup. The contact conditions may not be well defined or directly applicable to a specific application, but they are assumed to be the same for every specimen set and many tests can be conducted at a reasonable cost for screening new material combinations. In the third type of fretting fatigue test, well-defined geometries and controlled and/or monitored loads and displacements are used. These more fundamental tests are intended to evaluate the effect of specific variables such as amplitude, clamping loads, reciprocating stresses, environments, or palliative methods, and to develop and validate fretting fatigue models. In order to apply published test results to a specific component, the test parameters must be well defined and understood for each engineering application. At present, work is under way to standardize test methods in order to assist with this endeavor (Ref 2).

For most fretting fatigue testing, fretting pads are positioned on opposite faces of the sample and can be either single or double footprints of flat or convex contact geometry. The relative displacement between the pads and the sample can be driven independently or can be controlled by the loads and motion of the system (Fig. 10). Of the four methods for cyclic loading of fretting fatigue specimens, general trends indicate that torsional specimens had the smallest drop in fatigue strength, while the largest drop was for tests

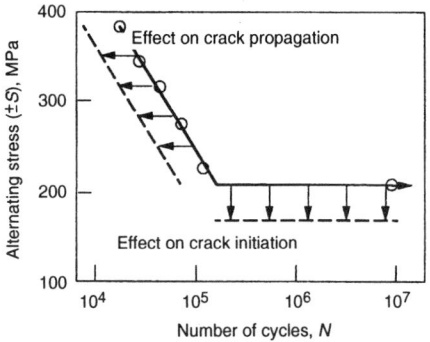

Fig. 11 Effect of fretting on fatigue strength reduction (SRF) through crack initiation. SRF is equal to difference between solid and dashed lines.

carried out under rotating-bending or plane-reverse-bending conditions, with plane push-pull testing falling in between (Ref 7).

The influence of fretting on fatigue strength can be determined by two basic methods. In one method, a sample is subjected to a certain number of cycles of fretting, followed by standard fatigue testing to failure without fretting. Plots of fretting cycles versus total cycles are recorded. This method has been used to determine the number of cycles for fatigue crack detection under the given geometry, material, normal stress, applied reversing stress and relative slip amplitude (Ref 12).

Influence of Fretting on S-N Plot. In the second method, the sample is subjected to fretting for the entire test. The fretting fatigue life or strength is determined by plotting the number of cycles to failure on an S-log N curve where S is the alternating stress. The strength-reduction factor (SRF) is defined by the ratio of the plain fatigue strength to the fretting fatigue strength as shown in Fig. 11 and is attributed to a decrease in crack nucleation time. A shift in cycles to failure at a given strength level is attributed to an increase in crack propagation rate and is often observed under corrosive environments.

The number of cycles to failure can be defined either as full specimen rupture or by initiation of a propagating (into the bulk stress region) crack. Crack length can be determined either from cross-sectional metallography or by specimen compliance methods.

Stress Analysis, Modeling, and Prediction of Fretting Fatigue. Testing, stress analysis, and modeling are complementary techniques required for understanding and predicting fretting fatigue behavior. For well-defined conditions, test results provide input to models. These models aim to predict crack initiation location, time, propagation rate, and the effect of changes in variables on these factors. At present, prediction of fretting fatigue is less developed than for plain (unnotched) fatigue. The main limitation is that continuum mechanics approaches do not consider microstructural inhomogeneities, and crack nucleation is controlled by such factors as well as "short" fatigue crack propagation.

Most of the early work in stress field modeling for fretting fatigue uses, as a starting point, analy-

sis similar to that by Mindlin (Ref 6) of a sphere pressed into a half plane and expand this to consider other geometries and imposed shear loads. If stress fields are computed using FEM analysis, an assumed contact geometry and coefficient of friction are used and loads are imposed at various mesh locations in order to compute stresses and subsequent strains. Alternatively, displacements can be imposed and strains and stresses computed. To facilitate modeling, the stress singularity associated with an abrupt contact geometry change, such as at the edge of a bolted flange or a hub/shaft interface, is accommodated by plastic deformation and a limiting stress is assumed. Current models are limited in that changes in contact geometry due to wear and variations in coefficient of friction due to lubrication or debris accumulation are difficult to take into account.

Experiments have been undertaken and models have been proposed for both the full- and partial-slip regimes and are based on empirical observations. Full-slip and partial-slip conditions can be achieved by varying the test configurations.

In addition, while most fretting contacts are some combination of load and displacement controlled, laboratory experiments can be designed either to drive the fretting pads independently (displacement controlled), or to allow them to move as a consequence of the clamping force and displacement of the "beam" sample (load controlled).

For fretting under conditions of full slip, two early models predict the SRF due to fretting. Nishioka and Hirakawa (Ref 13) derived the following equation to describe the fretting fatigue strength limit determined using their displacement-controlled experiment setup with full-slip conditions under the fretting pads (Fig. 10).

$$\sigma_{fwl} = \sigma_{w1} - \mu p_o \{ 1 - e^{(-S/K)} \} \qquad (Eq\ 2)$$

where σ_{fwl} is the fretting fatigue strength, σ_{w1} is the plain fatigue strength, μ is the coefficient of friction at the fretting interface, p_o is the clamping pressure, S is the slip amplitude (in mm), and K is a constant dependent on the material and surface condition (on the order of 3.4×10^{-3} mm in Ref 13).

In later work of Wharton et al. (Ref 14), a similar form was developed in which notch sensitivity of the base material was taken into account. The reduction in fatigue strength due to fretting was then proposed to also be proportional to the shear stress resulting from the contact pressure of the cylindrical fretting pads, inversely proportional to the contact width, and given by the equation:

$$\sigma_{wf} = \sigma_{wo} - q(8\mu P/\pi b) \qquad (Eq\ 3)$$

where σ_{wf} is the fatigue strength with fretting, σ_{wo} is the fatigue strength without fretting, q is the notch sensitivity factor, μ is the coefficient of friction at the fretting interface, P is the load per unit length, and b is the contact width under the fretting pads (mm). Note that both these predictions show that SRF is worse as μP or μP_o is increased.

For probable location of fretting fatigue crack nucleation in the partial slip regime, the approach of Ruiz and Chen (Ref 9) can be used. In their analysis of a dovetail interface, a fretting parameter representing the energy available for causing fretting damage, and given by the product of the slip amplitude (δ) and shear stress (τ) at points under the interface, was computed. Next, the product $\sigma\tau\delta$, called the fretting fatigue parameter, is computed, where σ is the maximum surface tensile stress (resulting from the bulk cyclic loading). A fretting fatigue crack is predicted to occur where the local value of $\sigma\tau\delta$ in the interface exceeds an empirically determined critical value, or, fretting occurs when:

$$\sigma\tau\delta \geq \sigma\tau\delta_{crit} \qquad (Eq\ 4)$$

If $\sigma\tau\delta$ and $\sigma\tau\delta_{crit}$ can be experimentally determined, then the designer can use this value as a design guide. Note that both the slip amplitude and the shear stress depend on the coefficient of friction (with opposite responses) and the imposed loading. Analysis by Nowells and Hill (Ref 15) of this work provided a theoretical justification and a possible method for predicting "initiation" (or nucleation) time based on the total accumulated incremental strain. With further effort, it appears that the composite parameter approach can be applied to fretting fatigue in the full-slip regime and can be expanded to include plasticity.

Other models may be used to determine whether the conditions at the interface will be of full or partial slip and to predict the location of the partial slip. These are generally FEM studies and make use of assumed macroscopic contact conditions and bulk material properties. The text by Hills and Nowell covers this area; yet it is claimed that no current models exist that can predict crack initiation times strictly from the knowledge of the states of stress, strain, and displacement on a macroscopic scale (Ref 16). Though experiments are required to determine $\sigma\tau\delta_{crit}$, the most probable location of cracking may be predicted using FEM analysis and the criteria of Ruiz and Chen (Ref 9). If conditions are sufficiently well defined, the slip characteristics of the interfaces may also be predicted. Whether the interface is in full slip or the extent of partial slip will help guide the choice of palliative. This article presents some palliatives that have been applied successfully in the past for fretting abatement. However, the designer is advised to apply discretion in applying techniques listed, because results depend on the particular condition. In their review, Gordelier and Chivers (Ref 17) attribute contradictory effects of similar treatments on fretting fatigue to the differing effects on the base materials and to the different contact conditions.

Variables Investigated during Fretting Fatigue Tests

Of the dozens of variables that can potentially affect fretting (Ref 18), the three primary variables that control fatigue crack initiation are surface contact stress, slip amplitude, and coefficient of friction. All other variables are secondary and affect fretting and fretting fatigue through their influence on the primary variables. Unfortunately, much contradictory data appear in the literature (Ref 17). The contradictions concern whether the fretting conditions were displacement controlled or load controlled and the interactions of the secondary variables between each other and the primary variables. This section presents the results of various investigations into the effect of variables on fretting fatigue.

Contact Stress and Alternating Stress. Contact stresses include both normal and shear stresses imposed at the sample surface. The cyclic shear stresses at the surface are the cause of crack nucleation (where the propensity to cracking from the stress state is also affected by pits, corrosion, and other forms of surface degradation). The magnitude of shear stresses depend on the imposed forces and displacements, the coefficient of friction, macroscopic stress concentrations, and local asperity geometries and distribution. Nishioka and Hirakawa (Ref 19) reported that fretting fatigue strength based on fatigue crack "initiation" (or nucleation) was found to decrease linearly with increasing normal force. For fretting fatigue strength based on fracture, they found a critical contact pressure, above which no further degradation occurred. Although their experiments did not reach this level, a sufficiently high normal force can sometimes result in the closure of cracks initiated by fretting through the superposition of a compressive stress.

Alternating stress is the stress imposed on the bulk sample, characterized by an amplitude and a mean stress. For fully reversing bending (which occurs on a rotating shaft), the mean stress is zero. Nishioka and Hirakawa performed experiments in reverse bending on annealed and induction-hardened medium-carbon steel under displacement-controlled conditions (full slip). They found that the mean stress did not affect the range of alternating stress amplitude for fatigue crack initiation due to fretting, but that it did affect crack propagation (Ref 19).

Wharton et al. also found that the percentage of fatigue life reduction due to fretting, defined by crack propagation to failure, was independent of applied alternating stress level for 70/30 brass and for a 0.7% carbon steel (Ref 20). These experiments were performed in rotating bending using flat fretting pad contacts (Ref 21, 22). Ruiz and Chen found that peak contact stress was important at 600 °C, but at room temperature the fretting parameter ($\tau\delta$) dominated (Ref 9). This was attributed to the nature of the oxide and the influence of the wear debris.

Displacement (Slip Amplitude) and Direction. For conditions of full slip, there is general agreement in the literature that the effect is most severe for slip amplitudes of 20 to 25 μm. In work on 4130 steel Gaul and Duquette (Ref 23) found that, for clamping pressures of 20 to 41 MPa (3 to 6 ksi), a minimum in the fatigue life at a given alternating stress occurred at a slip amplitude of 20 μm. At amplitudes higher than this, the wear

rate due to fretting exceeded the rate of crack growth rate just after initiation and the fretting fatigue strength increased. A similar critical amplitude of relative slip, on the order of 15 to 20 μm, was found by Nishioka and Hirakawa (Ref 13) for both induction-hardened and quenched-and-tempered medium-carbon steel samples. Clamping pressures of 120 MPa (17 ksi) were used in their work. They also concluded that the fatigue SRF due to fretting can be minimized if the fretting amplitude (relative slip) can be kept below 5 μm.

An effect related to slip amplitude is that of contact width. Experiments showed that larger-diameter cylinders had a greater detrimental effect on fretting fatigue life for the same line contact stress than small-diameter cylinders (Ref 24). This may be caused by increased partial slip or more asperity contacts per stress cycle.

Nowell and Hills (Ref 15) looked at the effect of slip amplitude through elastic modeling of fretting fatigue in the partial-slip regime. In low-amplitude fretting experiments using cylindrical radii fretting pads pressed against an in-plane tension/compression loaded Al/4Cu alloy sample, they found a critical contact width for fretting fatigue damage. They also found that a transition between long and short fatigue lives occurred for microslip amplitudes of 0.9 to 1.2 μm in the slip zones. This is considerably smaller than the maximum damage displacement of 20 to 30 μm found by Nishioka and Hirakawa and others in their full-slip experiments on steel samples.

The results of work by Collins and Tovey (Ref 25) indicated that fretting motion in the same direction as the cyclic stress had a greater effect than fretting in the perpendicular direction. They used this result to conclude that cracks are nucleated by adhesive wear rather than by abrasive plowing via the expected orientation from each mode.

Coefficient of Friction. The coefficient of friction probably has the greatest influence on fretting fatigue. It influences both slip amplitude and shear stress, though in opposite ways. The influence of different material combinations on fretting fatigue life has been reported to be due to the coefficient of friction (Ref 14). For load-controlled fretting, an increase in coefficient of friction can prevent slip over the whole contact region and reduce or eliminate fretting fatigue. For amplitude-controlled fretting, the opposite effect can be expected and a reduction in the coefficient of friction is desired since the surface shear stresses are reduced. Reduction in coefficient of friction for clamped (bolted or riveted) joints has been shown to be detrimental in some cases because the lower coefficient of friction between the overlapping plates increased the load-carrying requirements of the bolts or rivets leading to failures initiating at the hole edges (Ref 26).

High Temperature. Fretting fatigue strength decreases with increasing temperature for titanium alloys (Ref 27), but was found to increase for the nickel-base alloy Inconel 718 (Ref 28) due to the formation of a protective oxide glaze. Such glazes typically lower the coefficient of friction.

For iron-base alloys, both hard and soft flame-sprayed coatings based on molybdenum have shown success in improving the fretting fatigue strength at 300 °C. Overs et al. (Ref 29) attribute the improvement to MoO_2 glaze formation.

Environment and Corrosive Media. The effect of environment on fretting fatigue depends on the material and its corrodibility. Fretting action readily destroys passivating films on materials that are normally corrosion resistant. Poon and Hoeppner found that when fretting and corrosion occur simultaneously, the effect of corrosion on fatigue is dominant (Ref 30). As such, many palliatives or remedies for fretting fatigue can be viewed in terms of their effectiveness on corrosion fatigue.

In mildly corrosive aqueous environments, such as weak sodium chloride solutions representative of human body fluids, the fretting fatigue strength of materials not resistant to corrosion is reduced compared with fretting fatigue in air. Corrosion-resistant materials such as austenitic stainless steel and titanium have reduced fretting fatigue strengths in both environments due to the disruption by fretting of the otherwise protective oxide (Ref 31).

Endo found that ductile carbon steels are not affected by water vapor in fretting fatigue (Ref 32). Nishioka and Hirakawa found corrosion to be a secondary factor in fretting fatigue in their work on medium-carbon steels by comparing test data from argon and air experiments (Ref 33). Somewhat in contrast, Endo found that the fretting fatigue strength of carbon steel was higher in argon than in air, explaining that while the crack initiation rate is almost the same, the crack propagation rate is lower in argon (Ref 32).

Aluminum alloys are known to be very sensitive to water corrosion under dynamic conditions. Endo found that both crack initiation and propagation are accelerated by traces of water vapor due to corrosive attack, but not by oxygen. In the case of argon versus air experiments, the removal of oxygen was found to decrease the tangential stress due to soft aluminum wear debris accumulating between the mating surfaces (Ref 32). The alumina formed on aluminum would, if broken, be expected to be more abrasive and give rise to higher shear stresses than soft metallic wear debris.

Compared with tests in air, Ti-6Al-4V alloy was found to be adversely affected by corrosive atmospheres of humid argon and 1% NaCl at alternating stress levels above 120 MPa (17 ksi), but improved fatigue life was observed below 90 MPa (13 ksi) (Ref 34). The nature of the corrosion product was proposed to play a major role, in some cases forming a compacted layer that shielded the metal against crack initiation. Hoeppner reports that steel does not exhibit such a strong dependence on corrosion product (Ref 35).

Corrosion or oxidation are not required in the fretting process. Metallic fretting debris will form with gold or platinum contacts, or other nonoxidizable materials (Ref 36).

Microstructure and Material. If the design requirements will permit their use, annealed materials were found to be less susceptible to fretting fatigue than in the work-hardened state (Ref 20). Similarly, cast structures were less susceptible than forged structures (Ref 37). These results infer that if fatigue crack nucleation is through a wear mode involving exhaustion of work hardening, then prior-worked materials have a significantly shorter nucleation period.

Reeves and Hoeppner (Ref 38) found that carbon steel in the martensitic condition was more resistant to fretting fatigue than in the normalized (ferrite-plus-pearlite) steel due to the higher hardness and wear resistance of the martensitic steel. Nishioka and Hirakawa also found the fatigue limit to be higher on induction-hardened versus annealed steel (Ref 19). These results infer that more wear-resistant materials have better fretting fatigue properties.

Copper-Base Alloys. Wharton et al. (Ref 20) found that 70/30 brass did not show a strength limit in either plain fatigue or fretting fatigue. However, they found that the fretting fatigue strength at a given number of cycles was reduced by a fixed proportion over the plain fatigue strength for two microstructural conditions, independent of applied stress. The reduction was 61 and 74%, for annealed and work-hardened brass, respectively.

Ferritic Alloys. Endo and Goto (Ref 39) found that fatigue cracks generally initiated in ferrite grains, and propagated perpendicular to the sliding direction through a pearlitic region, irrespective of the orientation of the pearlite plates. Their experiments were performed under reverse bending. They also reported that, for two-stage tests on a medium-carbon steel (ferrite-plus-pearlite microstructure), no further reduction in fatigue life occurred if the fretting was continued for the entire test or was stopped after one-quarter of the total life. They concluded that the saturation point in the fretting cycles versus total fatigue cycles curve correspond to the point at which cracks that were initiated during fretting had grown to a depth where they propagated solely due to the macroscopic repeated stress (Ref 39).

Titanium Alloys. For three alpha + beta titanium microstructures, a fine or acicular microstructure was found to be more resistant to damage, defined by the number of propagating cracks found after a given number of cycles at a fixed stress, than a coarser-annealed structure (Ref 40). The finer alloy structure also had a lower SRF.

Prevention or Improvement

Fretting can be minimized or eliminated in many cases by one or more of the methods outlined in Table 1. Additional techniques in the design stages to diminish the effect of fretting on fatigue are discussed in more detail below in terms of:

- *Modification of the surface stress state* by introduction of compressive stresses, reduction of surface shear stresses, or elimination or reduction of relative motion
- *Application of principles for wear and fracture toughness* in selection of materials, coatings, and lubricants

Surface Stress Modification

Design for Introduction of Residual Compressive Stress. The only treatment or remedy that has been shown to be universally effective in improving fretting fatigue is the introduction of residual compressive stresses. The compressive stress will reduce the driving force for both crack initiation and propagation. Residual compressive stresses can be imparted through high and uniform clamping forces or interference fits, plastic deformation of the surfaces, phase changes, and precipitation or diffusion/thermal treatments. If sufficiently deep, superposition of the compressive stress can also decrease the tensile stress field on the propagating crack in the base material.

Plastic deformation by means of shot peening, surface rolling, or ballizing the inside diameter of through-holes has the secondary effect of work hardening of the surface. In some cases, this work hardening leads to higher resistance to fretting wear. Waterhouse and Saunders (Ref 41) have attributed the increase in fretting fatigue strength of austenitic stainless steels by shot peening to an increase in surface hardness to 400 HV, compared to 150 HV for the bulk material, resulting in an increase in the fretting fatigue strength to that of the plain fatigue strength without fretting. In contrast, Leadbeater et al. (Ref 42) found that the improved fretting fatigue life in an aluminum 2014A alloy by shot peening was most likely due only to residual compressive stresses. They determined that while increasing surface roughness had a small beneficial effect, work hardening did not influence fretting fatigue properties.

Cold-working methods are not effective in applications where temperatures during service, or those generated locally due to fretting, would lead to annealing of the previously work-hardened surface.

Nitriding is a very effective palliative for steels. It introduces residual compressive stresses in the surface, and local hardening through solid solution strengthening can decrease the areas of real contact of self-mated ferrous material couples. Both chemical and ion implantation nitriding methods can be used. The effectiveness of carburizing will vary. Although the strength and hardness of the surface will be increased, the heat treatment required to create the martensitic transformation, carburizing can produce either compressive or tensile residual stresses depending on the section size and shape. Tensile stresses of course would be detrimental.

Avoidance of Stress Raisers. Reducing geometric stress raisers in the vicinity of the contact will help prevent fretting fatigue crack nucleation. Examples were presented in the section "Parallel Surfaces without External Loading" and in Fig. 7 and 8. Local stress raisers due to pitting by wear or pits caused by corrosion can be prevented by corrosion-resistant coatings such as zinc or by cathodic protection.

Spacers or Shims. If the amplitude of relative displacement is small, it may be possible to absorb all of the transmitted shear stresses through the elasticity of a thin, flexible layer such as rubber. The effectiveness of this method depends on the modulus of the layer, its thickness, the severity of the normal load, and whether the relative motion is displacement controlled. For aluminum alloys used for aircraft skins at temperatures up to 150 °C, Taylor reports in work by Harris (Ref 43) that several investigations have shown the success against fretting fatigue of a joint bonded with isocyanate epoxy resin loaded with MoS_2. It is likely that the solid lubricant lowered the transmitted shear stresses while the resin prevented metal-to-metal contact. It has also been reported that both pure aluminum and copper are effective shim materials to be used against steel for reducing the transmitted shear stresses (Ref 24).

Lubricants. By using lubricants to lower the coefficient of friction, shear stresses resulting from the normal load will be decreased. Again, the amplitude of displacement may be increased. Oils and greases tend to be forced from the interface. Rough surfaces, such as those left by a shot-peening operation, can help retain liquid lubricants. Surface finishes that deliver oil to the fretting site are beneficial. Liquid lubricants can be effective on wire rope, where there is some containment of the lubricant due to the outer strands. Solid lubricants tend to be worn away over time and therefore have limited effectiveness.

Relative Motion. If the relative motion of the two members can be eliminated, the fatigue life can then be computed using standard methods found elsewhere in this Volume. The design concern then shifts to minimizing the stress concentration at the edges of the contact. Methods for eliminating motion include increasing the normal load and/or increasing the coefficient of friction of the interface to expand the region of no-slip to the entire contact.

Surface Roughness. Surprisingly, a deliberately rough surface finish may be the best for minimizing fretting fatigue damage. Waterhouse suggests machining grooves in the surface of one of the two contacting members, preferably the one that is not subjected to the major cyclic stresses (Ref 44). It was suggested that the benefit found in this work on aluminum was due to a minimization of the extent of any one contact area. An alternative explanation is that the debris generated during fretting may be more readily trapped in the grooves, thus promoting both a lubricating effect and a reduction in the local stresses due to support by the compacted debris.

Wear and Cracking Resistance

Material Selection for Fretting Fatigue Resistance. Two guides for materials selection choices should be used. The first is materials and treatments selected for avoidance of fretting dam-

Table 2 Relative fretting resistance of various material combinations

Combination	Fretting resistance
Aluminum on cast iron	Poor
Aluminum on stainless steel	Poor
Bakelite on cast iron	Poor
Cast iron on cast iron, with shellac coating	Poor
Cast iron on chromium plating	Poor
Cast iron on tin plating	Poor
Chromium plating on chromium plating	Poor
Hard tool steel on stainless steel	Poor
Laminated plastic on cast iron	Poor
Magnesium on cast iron	Poor
Brass on cast iron	Average
Cast iron on amalgamated copper plate	Average
Cast iron on cast iron	Average
Cast iron on cast iron, rough surface	Average
Cast iron on copper plating	Average
Cast iron on silver plating	Average
Copper on cast iron	Average
Magnesium on copper plating	Average
Zinc on cast iron	Average
Zirconium on zirconium	Average
Cast iron on cast iron with coating of rubber cement	Good
Cast iron on cast iron with Molykote lubricant	Good
Cast iron on cast iron with phosphate conversion coating	Good
Cast iron on cast iron with rubber gasket	Good
Cast iron on cast iron with tungsten sulfide coating	Good
Cast iron on stainless steel with Molykote lubricant	Good
Cold-rolled steel on cold-rolled steel	Good
Hard tool steel on tool steel	Good
Laminated plastic on gold plating	Good

Source: J.R. McDowell, in *Symposium on Fretting Corrosion*, STP 144, ASTM, 1952

age, hence minimizing crack nucleation (see Table 2). Materials with low propensity for adhesion are described by Rabinowicz (Ref 45) in terms of the inverse of metallurgical compatibility. Typically, dissimilar couples are preferred. Self-lubricating components, such as porous metal washers impregnated with lubricant, can be used. Materials with high work-hardening capacity or dynamic recrystallization characteristics can also minimize fatigue crack initiation.

The second materials selection guide is the use of alloy compositions and thermomechanical treatments for existing alloys that improve their fatigue crack propagation resistance, as described elsewhere in this Volume and in a variety of mechanical properties handbooks (Ref 46-48).

Soft coatings provide a "sacrificial," low shear strength layer that reduces the magnitude of the oscillatory shear stresses transmitted into the substrate. Both a lower coefficient of friction and a lower shear strength contribute. A coating that dynamically recrystallizes under service conditions would be expected to be effective. Situations where creep of the soft coating results must be avoided, as the introduction of excessive clearance or a decrease in preload on bolts or washers can occur. In addition, low friction of soft coatings can lead to larger relative displacements for load-controlled fretting.

Diffusion treatments such as sulfidized coatings on steel were shown to be effective in delaying the initiation of a propagating fatigue crack in carbon steels in the laboratory when impregnated

with a suitable oil-in-water emulsion (Ref 49), and subsequently applied successfully to compressor blade roots (Ref 44).

Hard coatings minimize the areas of real contact and penetration by opposing asperities and are less prone to adhesive wear. However, hard coatings have high friction coefficients. As such, high shear stresses are transmitted into the coating, which can limit their effectiveness. Care must be taken in design of coating thickness to avoid high shear stresses at the depth of the coating/substrate interface, which is typically a plane of weakness.

Additionally, hard coatings such as chromium platings are often filled with cracks, and sprayed molybdenum coatings usually contain pores. These defects in the coatings can serve as initial sites for fatigue cracks to develop and propagate into the substrate. Post-coating shot peening is recommended. Improvement in both electrodeposited chromium and nickel coatings was achieved by thermal diffusion processing followed by shot peening (Ref 44).

While hard coatings can reduce the overall fatigue strength compared with that of samples tested without prior fretting, under fretting fatigue conditions they typically show an improvement over uncoated materials. A pretreatment, such as shot peening or vapor blasting, may be required to compensate for the reduction in normal fatigue strength due to the hard coating. In the case of carbon steel, sprayed molybdenum coatings were found to increase the fretting fatigue strength from 33 to 72% of the normal fatigue strength compared with uncoated samples (Ref 50).

Nonmetallic coatings offer improvement in fretting fatigue resistance through the lowering of the coefficient of friction and the prevention of metal-to-metal welding at asperity contacts. The same types of substances that are considered solid "boundary" lubricants for wear control can be used. Conversion coatings produced by phosphating and anodizing, polymer sheets of nylon or polytetrafluoroethylene, and polymerized epoxy resins can all be used. In the case of porous coatings, impregnation with oils or greases can provide effective means of reducing or eliminating the occurrence of fretting fatigue when operating in the full-slip regime, over the supply life of the lubricant.

REFERENCES

1. R.B. Waterhouse, Ed., *Fretting Fatigue*, Applied Science, 1981
2. M.H. Attia and R.B. Waterhouse, Ed., *Standardization of Fretting Fatigue Test Methods and Equipment*, STP 1159, ASTM, 1992
3. D.A. Hills and D. Nowell, *Mechanics of Fretting Fatigue*, Kluwer Academic Publishers, 1994
4. R.B. Waterhouse, *Fretting Corrosion*, Fretting Fatigue, Pergamon Press, 1972
5. Proc. Specialists Meeting on Fretting in Aircraft Systems, AGARD-CP-161, Advisory Group for Aerospace Research and Development, 1974
6. R.D. Mindlin, Compliance of Elastic Bodies in Contact, *J. Appl. Mech.*, Vol 16, 1949, p 259-268
7. R.B. Waterhouse, Theories of Fretting Processes, *Fretting Fatigue*, Applied Science, 1981, p 203-220
8. D.J. White, *Proc. Inst. Mech. Eng.*, Vol 185, 1970-1971, p 709-716
9. C. Ruiz and K.C. Chen, Life Assessment of Dovetail Joints between Blades and Disks in Aero-Engines, paper C241/86, *Proc. Conf. Fatigue of Engineering Materials and Structures*, Institute of Mechanical Engineers, 1986, p 187-194
10. M. Takeuchi and R.B. Waterhouse, Fretting-Corrosion-Fatigue of High Strength Steel Roping Wire and Some Protective Measures, *Proc. Int. Conf. Evaluation of Materials Performance in Severe Environments*, EVALMAT 89, Iron and Steel Institute of Japan, 1989, p 453-460
11. J.A. Collins, Fretting, Fretting Fatigue, and Fretting Wear, *Failure of Materials in Mechanical Design—Analysis, Prediction, Prevention*, 2nd ed., John Wiley & Sons, 1992, p 504-524
12. K. Nishioka and K. Hirakawa, Fundamental Investigations of Fretting Fatigue, Part 3, Some Phenomena and Mechanisms of Surface Cracks, *Bull. Jpn. Soc. Mech. Eng.*, Vol 12 (No. 51), 1969, p 397-407
13. K. Nishioka and K. Hirakawa, Fundamental Investigations of Fretting Fatigue, Part 5, The Effect of Relative Slip Amplitude, *Bull. Jpn. Soc. Mech. Eng.*, Vol 12 (No. 52), 1969, p 692-697
14. M.H. Wharton, R.B. Waterhouse, K. Hirakawa, and K. Nishioka, The Effect of Different Contact Materials on the Fretting Fatigue Strength of an Aluminum Alloy, *Wear*, Vol 26, 1973, p 253-260
15. D. Nowell and D.A. Hills, Crack Initiation Criteria in Fretting Fatigue, *Wear*, Vol 136, 1990, p 329-343
16. D.A. Hills and D. Nowell, *Mechanics of Fretting Fatigue*, Kluwer Academic Publishers, 1994, p 210
17. S.C. Gordelier and T.C. Chivers, A Literature Review of the Palliatives for Fretting Fatigue, *Wear*, Vol 56, 1979, p 177-190
18. J.A. Collins and S.M. Marco, The Effect of Stress Direction during Fretting on Subsequent Fatigue Life, *Proc. ASTM*, Vol 64, 1964, p 547-560
19. K. Nishioka and K. Hirakawa, Fundamental Investigations of Fretting Fatigue, Part 4, The Effect of Mean Stress, *Bull. Jpn. Soc. Mech. Eng.*, Vol 12 (No. 51), 1969, p 408-414
20. M.H. Wharton, D.E. Taylor, and R.B. Waterhouse, Metallurgical Factors in the Fretting Fatigue Behaviour of 70/30 Brass and 0.7% Carbon Steel, *Wear*, Vol 23, 1973, p 251-260
21. R.B. Waterhouse and D.E. Taylor, The Initiation of Fatigue Cracks in a 0.7% Carbon Steel by Fretting, *Wear*, Vol 17, 1971, p 139-147
22. R.B. Waterhouse and M. Allery, The Effect of Powders in Petrolatum on the Adhesion be-

tween Fretted Surfaces, *Trans. ASLE*, Vol 9, 1966, p 179

23. D.J. Gaul and D.J. Duquette, The Effect of Fretting and Environment on Fatigue Crack Initiation and Early Propagation in a Quenched and Tempered 4130 Steel, *Metall. Trans. A*, Vol 11A, Sept 1980, p 1555-1561

24. R.B. Waterhouse, Fretting Fatigue, *Int. Mater. Rev.*, Vol 37 (No. 2), 1992, p 77-97

25. J.A. Collins and F.M. Tovey, Fretting-Fatigue Mechanisms and the Effect of Motion on Fatigue Strength, *J. Mater.*, Vol 7 (No. 4), 1972, p 460-464

26. Citations 31 and 32 in Ref 17

27. M.M. Hamdy and R.B. Waterhouse, The Fretting Fatigue Behavior of Ti-6Al-4V at Elevated Temperatures, *Wear*, Vol 56, 1979, p 1-8

28. M.M. Hamdy and R.B. Waterhouse, The Fretting Fatigue Behavior of a Nickel-Based Alloy (Inconel 718) at Elevated Temperatures, *Wear of Materials*, American Society of Mechanical Engineers, 1979, p 351-355

29. M.P. Overs, S.J. Harris, and R.B. Waterhouse, in *Wear of Materials*, American Society of Mechanical Engineers, 1979, p 379-387

30. C. Poon and D.W. Hoeppner, The Effect of Environment on the Mechanism of Fretting Fatigue, *Wear*, Vol 52, 1979, p 175-191

31. R.B. Waterhouse, Fretting Fatigue in Aqueous Electrolytes, *Fretting Fatigue*, R.B. Waterhouse, Ed., Applied Science, 1981, p 159-177

32. K. Endo, Practical Observations of Initiation and Propagation of Fretting Fatigue Cracks, *Fretting Fatigue*, R.B. Waterhouse, Ed., Applied Science, 1981, p 127-141

33. K. Nishioka and K. Hirakawa, Some Further Experiments on the Fatigue Strength of Me-dium Carbon Steel, *Proc. Mech. Behav. Mater.*, Vol 3, 1971, p 308-318

34. M.H. Wharton and R.B. Waterhouse, Environmental Effects in the Fretting Fatigue of Ti-6Al-4V, *Wear*, Vol 62, 1980, p 287-297

35. D.W. Hoeppner, Environmental Effects in Fretting Fatigue, *Fretting Fatigue*, R.B. Waterhouse, Ed., Applied Science, 1981, p 143-158

36. D. Godfrey and J. Bailey, Early Stages of Fretting of Copper, Iron and Steels, *Lub. Eng.*, Vol 10, 1954, p 155-159

37. G. Sachs and P. Stefan, Chafing Fatigue Strength of Some Metals and Alloys, *Trans. ASM*, Vol 29, 1941, p 373-401

38. R.K. Reeves and D.W. Hoeppner, Microstructural and Environmental Effects on Fretting Fatigue, *Wear*, Vol 47, 1978, p 221-229

39. K. Endo and H. Goto, Initiation and Propagation of Fretting Fatigue Cracks, *Wear*, Vol 38, 1976, p 311-324

40. In Ref 7, p 212-215. Cited as H. Goto and R.B. Waterhouse, Proc. 4th Int. Conf. on Titanium (Kyoto), May 1980

41. R.B. Waterhouse and D.A. Saunders, The Effect of Shot-Peening on the Fretting Fatigue Behaviour of an Austenitic Stainless Steel and a Mild Steel, *Wear*, Vol 53, 1979, p 381-386

42. G. Leadbeater, B. Noble, and R.B. Waterhouse, The Fatigue of an Aluminum Alloy Produced by Fretting on a Shot Peened Surface, *Advances in Fracture Research* (Fracture 84), Vol 3, Pergamon Press, 4-10 Dec 1984

43. D.E. Taylor, Fretting Fatigue in High Temperature Oxidising Gases, *Fretting Fatigue*, R.B. Waterhouse, Ed., Applied Science, 1981, p 177-202

44. R.B. Waterhouse, Avoidance of Fretting Fa-tigue, *Fretting Fatigue*, R.B. Waterhouse, Ed., Applied Science, 1981, p 221-240

45. E. Rabinowicz, Wear Coefficients—Metals, *Wear Control Handbook*, American Society of Mechanical Engineers, 1980, p 475-501

46. *Aerospace Structural Metals Handbook*, AFML-TR-68-115, Materials and Ceramics Information Center, Battelle Columbus Laboratories, 1991

47. *Damage Tolerant Design Handbook*, MCIC-HB-OIR, Materials and Ceramics Information Center, Battelle Columbus Laboratories, 1983

48. *Military Handbook for Metallic Materials and Elements for Aerospace Vehicle Structures*, MIL-HDBK-5F, U.S. Dept. of Defense, 1990

49. R.B. Waterhouse and M. Allery, The Effect of Non-Metallic Coatings on the Fretting Corrosion of Mild Steel, *Wear*, Vol 8, 1965, p 112-120

50. D.E. Taylor and R.B. Waterhouse, Sprayed Molybdenum Coatings as a Protection Against Fretting Fatigue, *Wear*, Vol 20, 1972, p 401-407

SELECTED REFERENCES

- R.B. Waterhouse, Fretting Wear, in *ASM Handbook*, Vol 18, *Friction Lubrication and Wear Technology*, ASM International, 1992, p 242-256
- R.B. Waterhouse and T.C. Lindley, *Fretting Fatigue*, ESIS 18, Mechanical Engineering Publications, 1994
- O. Vingsbo and S. Söderberg, On Fretting Maps, *Wear*, Vol 126, 1988, p 131-147

Contact Fatigue

W.A. Glaeser and S.J. Shaffer, Battelle Laboratories

CONTACT FATIGUE is a surface-pitting-type failure commonly found in ball or roller bearings. This type of failure can also be found in gears, cams, valves, rails, and gear couplings. Contact fatigue has been identified in metal alloys (both ferrous and nonferrous) and in ceramics and cermets.

Contact fatigue differs from classic structural fatigue (bending or torsional) in that it results from a contact or Hertzian stress state. This localized stress state results when curved surfaces are in contact under a normal load. Generally, one surface moves over the other in a rolling motion as in a ball rolling over a race in a ball bearing. The contact geometry and the motion of the rolling elements produces an alternating subsurface shear stress. Subsurface plastic strain builds up with increasing cycles until a crack is generated. The crack then propagates until a pit is formed. Once surface pitting has initiated, the bearing becomes noisy and rough running. If allowed to continue, fracture of the rolling element and catastrophic failure occurs. Fractured races can result from fatigue spalling and high hoop stresses.

Rolling contact components have a fatigue life (number of cycles to develop a noticeable fatigue spall). However, unlike structural fatigue, contact fatigue has no endurance limit. If one compares the fatigue lives of cyclic torsion with rolling contact, the latter are seven orders of magnitude greater (Ref 1). Rolling contact life involves ten to hundreds of millions of cycles.

Examples of Contact Fatigue

Contact fatigue produces a surface damage that is unique and well recognized. Familiar examples are found in fatigue of ball and roller bearings. A typical failure in a roller bearing is shown in Fig. 1. Although this spall is small, it would grow in size until roller fracture would occur, as bearing operation continues.

One classic shape of a fatigue spall in a ball bearing is a delta shape, as shown in Fig. 2 with a diagram of the pit (Ref 2). The apex of the pit is the initiation point, usually the location of a surface defect like a dent. The pit grows in a fan shape, becoming wider and deeper as it grows in the direction of ball travel. Not all spalls in ball-bearing races are of the shape shown in Fig. 2.

Figure 3 shows a fatigue spall near the race shoulder of a deep-groove ball bearing. The spall appears to have been formed by the joining of several pits. The fact that the spall occurred close to the race shoulder may have distorted the contact state of stress, causing a multiple origin.

Fatigue in roller bearings may differ from ball-bearing contact fatigue. Quite often the pitting occurs in the inner race at the contact zone of the roller ends. In some cases, contact stress peaks at the roller ends and pitting originates in these locations. Roller-end pitting can be a sign of misalignment.

Cams and Gears. Valve lifter cams and rollers are subject to contact fatigue. An example is shown in Fig. 4 (Ref 3). The character of the damage is very similar to that found in rolling contact bearings. The example shown in Fig. 4 was found in both cam nose and lifters during automobile engine tests (Ref 3). Lifters were nodular iron, and cams were flake graphite cast iron. Fatigue cracks were associated with cracked carbides, graphite flakes, and hard inclusions.

Contact fatigue occurs in gears along the pitch line. The geometry of tooth mesh is such that rolling occurs at the pitch line while sliding occurs at the addendum as the gears come out of mesh. An example of pitch line contact fatigue is shown in Fig. 5 (Ref 4). The pits seen on the teeth will grow in size and depth, ultimately resulting in tooth fracture.

Another form of contact fatigue, known as micropitting, occurs in bearings. An example is shown in Fig. 6. This feature can show up over the entire raceway surface. It is often the result of too thin a lubricant film or excessive surface roughness and sometimes heavy loading.

In gears, micropitting is termed frosting and in the present ANSI/AGNA standard it is considered a form of contact fatigue. For bearings,

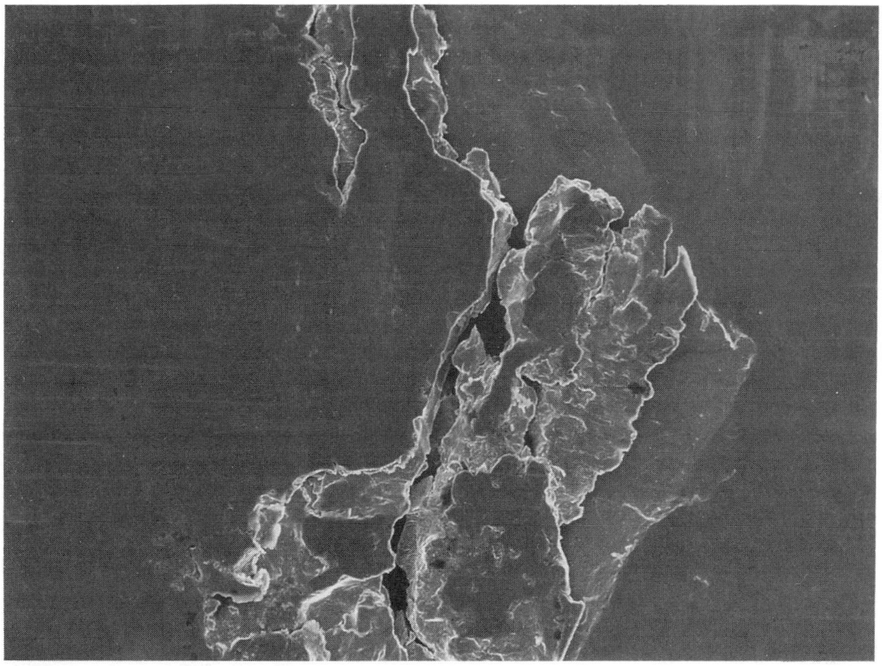

Fig. 1 Scanning electron micrograph of a fatigue spall on a roller from a roller bearing after 630,000 cycles. Roller is AISI 1060 steel, hardened to 600 HV. Spall is 400 μm wide by 700 μm long.

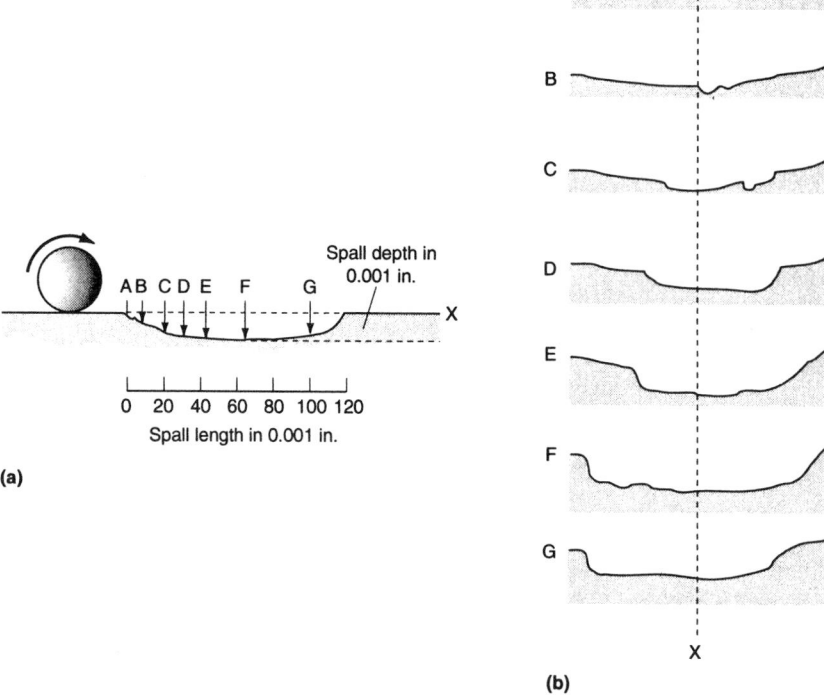

(a)

(b)

Fig. 2 Anatomy of a race spall in a ball bearing. (a) Typical delta shape with the apex at the origin. (b) Profiles of the spall. Source: Ref 2

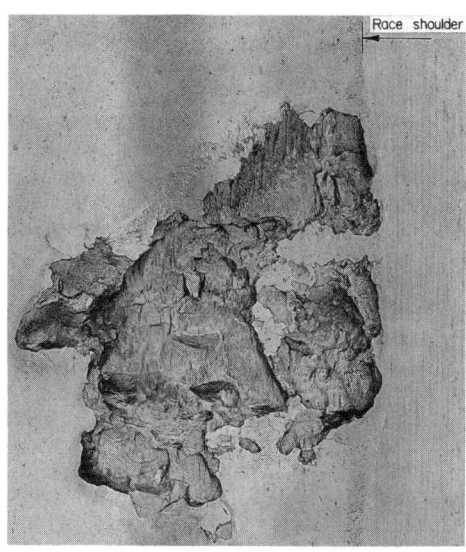

Fig. 3 Multiple spall near a race shoulder

gears, and any contact, micropitting can be reduced by improved surface finish, reduced temperature or loads, and providing sufficient elastohydrodynamic film.

Rails. Spalling and "shelly"-type failures occur on track rails from wheel-track rolling contacts. An example of shelly failure is shown in Fig. 7 from Kilburn (Ref 5). The name comes from the morphology of the fracture surface in the bottom of the spall. Shelly failures are serious because they lead to rail fracture and derailments. Rail spalling has been reduced in recent years by the use of higher carbon steels for rails.

Analysis of the subsurface stress state indicates that a maximum shear stress exists at a given depth below the surface. The stress distribution is shown in Fig. 8. The maximum shear stress is shown increasing with depth below the surface as discussed by Kloos and Schmidt (Ref 6).

The curves shown are based on two different mathematical approaches to the estimation of contact state of stress. Both approaches produce a shear stress distribution quite close to each other. The *z* axis scale is *Z/B*, where *B* is the minor axis length of the contact ellipse and the usual direction of rolling motion. Note in Fig. 8 that the

maximum shear stress depth is about the same as the dimension of *B*. However, with increased traction or tangential force, the maximum shear stress moves closer to the surface. In hardened martensitic steels used in ball and roller bearings, the subsurface shear stresses produce plastic deformation in the martensitic structure. The residual strain increases with increase in rolling cycles. This has been shown by x-ray measurements during rolling contact experiments (Ref 7).

Many researchers have studied the microstructural changes that occur as a result of the buildup of subsurface strain (Ref 8-10). In AISI 52100 steel, a common rolling-contact-bearing material, the accumulation of strain initially is associated with the formation of a dark etching zone below the surface. Further strain causes the formation of light etching bands caused by the formation of a new ferrite phase. Then carbides in the high stress region begin to show decay and break up. Other microstructural features include "butterflies" or

Mechanisms of Contact Fatigue

The state of stress produced by rolling contact is concentrated in a small volume of material and produces intense plastic strain. The strain accumulates as the same volume is stressed with each rolling cycle until a crack is initiated and forms a spall. In the real world of contact fatigue, the mechanisms involved can be quite complex. Most models assume a condition of ideal geometric surfaces and little input by heat generation, environmental conditions, and inhomogeneities of materials. Hertz stress analysis assumes a circular, elliptical, or line contact surface area between curved surfaces (depending on the geometry of the contacts) and a parabolic pressure distribution with the maximum pressure at the center of the contact.

The subsurface state of stress involves a hydrostatic component that inhibits tensile fracture.

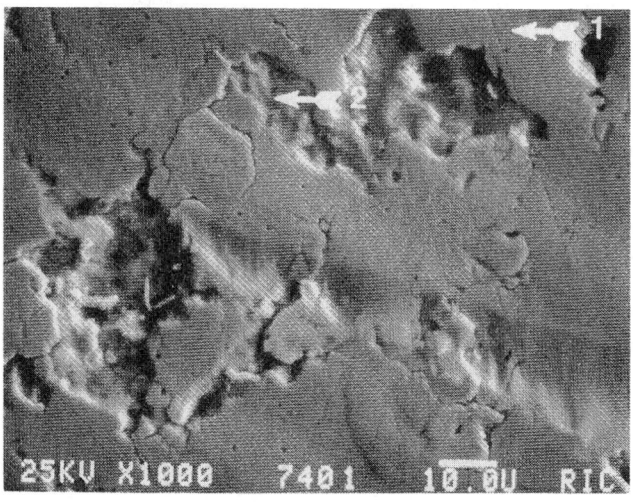

Fig. 4 Contact fatigue spalling of cam lifter surface. Source: Ref 3

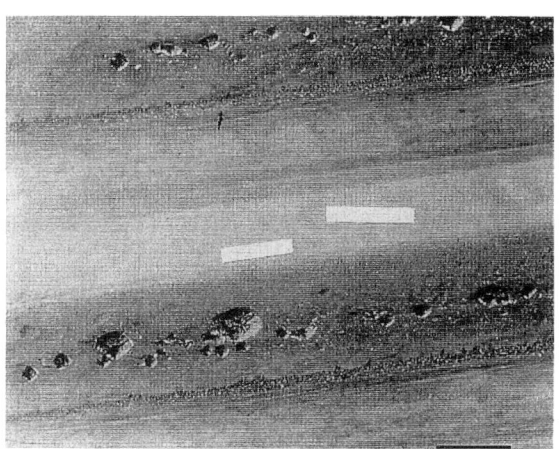

Fig. 5 Pitch line spalling of medium-hardened gears. Source: Ref 4

(a)

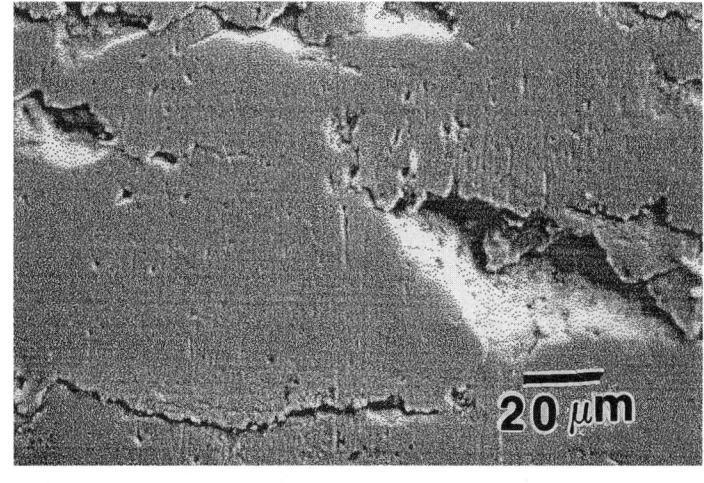

(b)

Fig. 6 Micropitting of roller bearing outer race. Scanning electron micrograph, (a) 57× and (b) at higher magnification

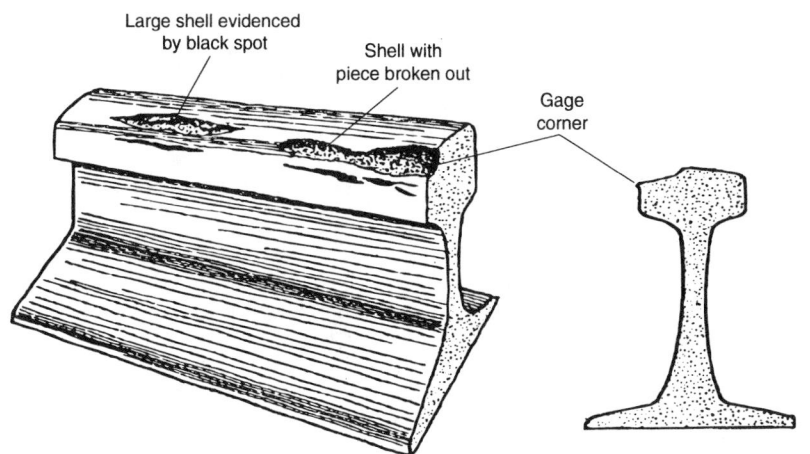

Fig. 7 Shelly rail spall from wheel-rail contact fatigue. Source: Ref 5

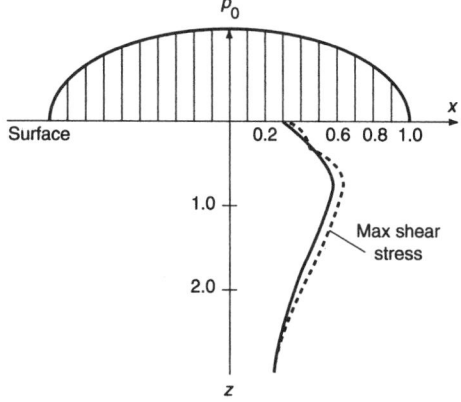

3D contact pressure distribution for curved contacts

Fig. 8 Stress distributions at a contact subsurface

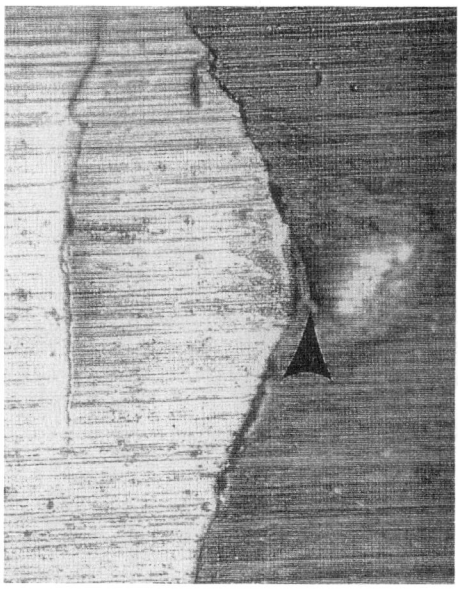

Fig. 9 Developing spall. (a) Top view of developing spall at race surface dent. (b) Section through developing spall showing subsurface cracking Source: Ref 4

white etching wings radiating out from large hard inclusions. Some features depend on the level of contact stress. For example, butterflies are often associated with high contact stress (2000 to 4000 MPa), as discussed in more detail in the article "Contact Fatigue of Hardened Steel" in this Volume.

In a detailed study of butterfly formations in AISI 52100 steel ball bearings by Becker (Ref 11), both through-hardened AISI 52100 steel and carburized SAE 8620 steel bearing races were used. Contact stress was 3280 MPa (480 ksi). Butterflies were found in sectioned posttest races. They were always oriented at about 40° to the surface and oriented in the rolling direction. The "wings" of the butterflies were found to be composed of a mix of heavily strained, ultrafine-grained ferrite and fine carbide particles. Hardness was measured as close to 1000 HV—harder than the martensitic matrix surrounding the butterflies. Fine cracks were also found on the edges of the butterfly wings. The same structures were found in the carburized case in the 8620 steel. Becker says in Ref 11: "The breakdown of the matrix microstructure to ferrite and carbide is caused by very high stress concentration either at hard inclusions or at pre-existing cracks."

Contact fatigue is also surface generated. In fact, surface-originating spalls are more prevalent than subsurface-generated cracks. Proving subsurface fracture origin is difficult because a metallographic section only shows a profile of the crack which, in three dimensions, may have a surface origin. The higher the tangential force or traction, the more likely will be surface-generated

contact fatigue. With shear stresses higher and closer to the surface, surface defects (dents, scratches, etc.) all contribute to higher incidence of surface-originating fatigue. Figure 2 shows a race spall that started at a dent in the race. This produced a delta-shaped spall as the cracking progressed from the origin. Sections through the spall show it to be shallow at the origin and deeper at the other end. Photomicrographs of a developing spall (Fig. 9, Ref 4) caused by a dent shows a ridge between the dent and the crack. This is typical and causes disruption in the oil film. The arrow shows the direction of movement of the balls over the race. The section through the developing spall shows the subsurface crack propagating down into the race at an angle to the surface.

As the developing spall matures, a surface layer loosens and eventually breaks out, leaving a pit. While the pit develops, the loose layer batters the fracture surface, obliterating the surface features. Fractographic analysis is not a likely option for investigating contact fatigue.

Rolling Contact Bearing Life

Ball and roller bearings have been subject to the most extensive life testing of all contact fatigue components. Bearing catalog lives are based on fatigue failure considerations. It is assumed that no ball or roller bearing gives unlimited service. Owing to the special stress state experienced by rolling contact bearings, bending or push-pull tensile fatigue results cannot be applied to their life calculations. There is significant scatter in life tests for rolling contact bearings. The Weibull distribution is used in statistical analysis of bearing-life tests. A typical bearing-life Weibull plot is shown in Fig. 10 (Ref 12).

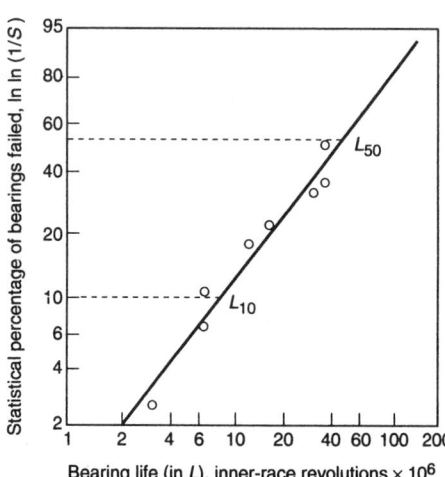

Fig. 10 Weibull plot of ball-bearing lives distribution. Source: Ref 12

Two life values in the distribution are shown. The L_{10} life, or the life at which 10% of all the bearings have failed is used for bearing selection. Lundberg and Palmgren (Ref 13) developed a relationship that can be used to predict bearing life for any load, using the life for standard load in the relation:

$$L = (C/P)^p$$

where L is the fatigue life in revolutions $\times 10^6$; C is the standard load (C is defined as the load that gives an L_{10} life of one million revolutions); P is the selected load; and p is 3 for ball bearings and 10/3 for roller bearings.

The predicted life from the above relationship is, of course, based on bearing tests, analyzed statistically. It does not take into account other

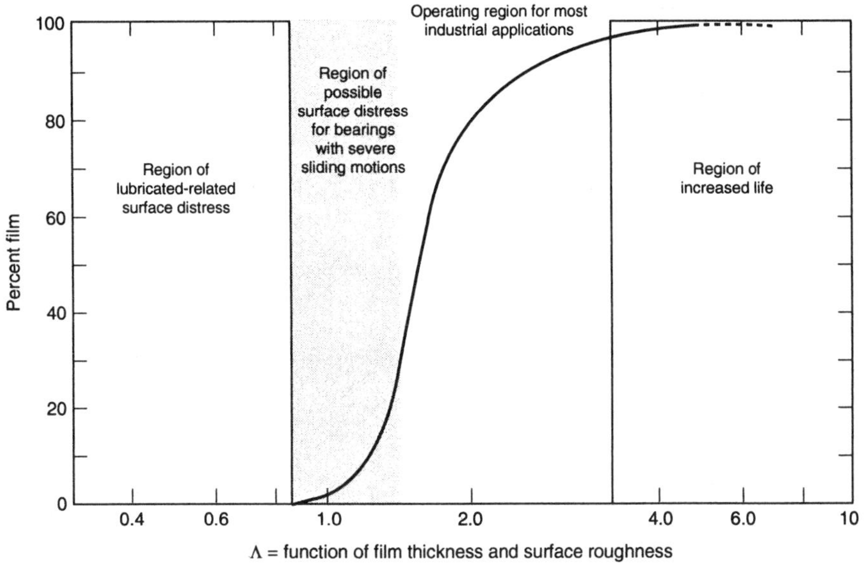

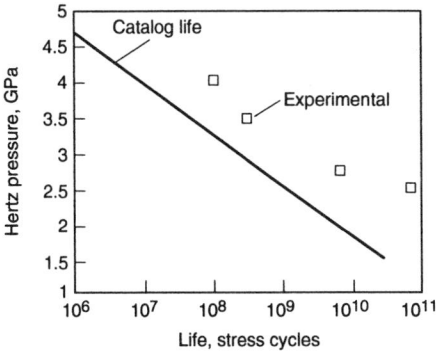

Fig. 11 Ball-bearing performance map. Source: Ref 2

Fig. 12 Contact stress-life plots for lives based on the inverse power load-life law and bearing tests with ideal operating conditions. Source: Ref 14

factors that impact on bearing life. Lubrication is a powerful factor in bearing life. Since the discovery of thin-film lubrication, elastohydrodynamic (EHD) lubrication of rolling contact bearings, the effect of film thickness on bearing life, has received considerable attention. Tests have shown that the lubricant film thickness is influenced by bearing speed and lubricant viscosity and less by load. A bearing performance map was developed by Harris (Ref 2). The performance map is shown in Fig. 11. It has been in general use for a number of years. The lubricant film coefficient, Λ, determined by dividing EHD film thickness by a surface roughness factor, relates to the present film or percentage time the surfaces are totally separated by a lubricant film. If Λ is less than 1, the bearing is likely to not attain the L_{10} life predicted by the Weibull distribution. If Λ exceeds 4, then one might expect longer life than predicted.

Steel microstructure also has a significant effect on bearing life. Of greatest importance is the cleanliness of the steel. Because hard inclusions have been found to enhance the fatigue crack process, steelmaking methods have been modified to eliminate or substantially reduce the production of hard inclusions. Consumable electrode vacuum melting has produced bearing-grade steels that have dramatically improved bearing reliability. In many cases, bearing failure is now related to wear rather than to contact fatigue. Good surface finish is necessary for long bearing life. As was noted, contact fatigue is initiated by surface defects like dents and deep scratches. Surface defects not only cause asperity contact in thin-film lubrication, but dents have been shown to disturb the EHD film and cause local film breakdown.

The possibility of a fatigue limit for rolling contact (deviation from the inverse load power law) has been investigated. Tallian (Ref 14) has analyzed test data from bearing tests run at high Λ values under conditions free of contaminants and debris and found deviations from the theoretical life suggesting a fatigue limit. This is shown in the plot in Fig. 12. Further information on bearing life is described in the article "Fatigue and Life Prediction of Bearings" in this Volume.

Minimizing Contact Fatigue

The study of rolling contact behavior has indicated new approaches that might further improve the contact fatigue resistance of these systems. One important problem in the application of rolling contact systems is the wide scatter in failure lives. Zaretsky (Ref 15) indicates that in a group of 30 ball bearings the ratio of the longest to the shortest life may be as much as 20 times. Bearing-catalog ratings are based on L_{10} lives or the time in which 10% of the bearings have failed. The same scatter can be expected in other rolling contact components such as gears and cams. Bearing-fatigue life is sensitive to bearing load. Generally, it is assumed that life is inversely pro-

portional to maximum Hertz stress to the 9th power. The exponent can be as high as 12 for roller bearings. Lubricant composition, microstructure, and geometry of rolling contacts can influence these exponents of life. Zaretsky, Poplawski, and Peters (Ref 15) give a good summary of life exponents from various tests for a number of bearing configurations. This also includes case-carburized consumable electrode vacuum arc remelted AISI 9310 spur gears. The L_{10} lives varied inversely with the stress to the 8.4 power for the spur gears.

Reducing the scatter in bearing lives so that the distribution would be compressed toward the longer lives would intrinsically improve bearing lives. Improving rolling-element precision, surface finish, and homogeneity of microstructure should reduce scatter somewhat. Lubrication also is effective. Ensuring that a rolling contact component is operating within satisfactory EHD conditions relative to surface finish is essential. Cleanliness of the operating environment and reasonable protection from corrosion are also important.

Because of the sensitivity of contact fatigue life to contact stress, reduction of contact stress can significantly improve bearing life. Of course, accurate estimation of the actual operating contact stress is important. Contact stress can be reduced by spreading out the area contact with a soft solid thin film applied to the surfaces (bearing races, for instance). Conversely, hard coatings have been used to improve fatigue life of bearing steels (Ref 16).

High-speed ball bearings have an increased ball contact stress owing to centrifugal forces. Such increased stress levels are sufficient to cause significant reduction in fatigue life even in very clean precision ball bearings. Reduction in the ball mass can reduce this effect and increase life to reasonable levels. Significant advances have been made by the use of silicon-nitride balls for high-speed bearings. Because of the lower density of silicon nitride, centrifugal forces in the bearing are reduced. Hybrid ball bearings with silicon-nitride balls have surpassed bearing grade steel in rolling contact performance (Ref 17-19). These bearings are finding use in gas turbines and high-speed machine tools.

Race fracture in high-speed ball bearings can be avoided by using a carburizing grade steel with increased fracture toughness (M50 NiL) (Ref 20) instead of through-hardening steels like AISI 52100. Carburizing to a depth below the estimated maximum shear depth will provide the required resistance to contact fatigue. Cleanliness of the steel will still be an important factor in bearing life.

The residual stress state in the near surface of rolling contact elements resulting from heat treatment and machining have an influence on contact fatigue life. By imposing compressive residual stresses, gear life can be improved. This can be accomplished by shot peening and burnishing. Nitriding gear steel will produce the desired compressive residual stresses to inhibit crack propagation.

As power systems become lighter and more compact, bearings, gears, and other rolling elements will have to operate at higher speeds. Although even at this time not all is understood about the mechanisms of contact fatigue, advances in improved reliability and component life are being made. Research and testing continue to try to narrow the life scatter and increase the predicted life of rolling contact parts.

REFERENCES

1. V. Bhargava, G.T. Hahn, and C.A. Rubin, Rolling Contact Deformation and Microstructural Changes in High Strength Bearing Steel, *Wear*, Vol 133, 1989, p 69
2. T. Harris, "The Endurance of Modern Rolling Bearings," AGMA paper 269.01, Oct 1964, *Rolling Bearing Analysis*, John Wiley, 1966
3. S.H. Roby, Investigation of Sequence IIIE Valve Train Wear Mechanisms, *Lubr. Eng.*, Vol 47 (No. 5), p 413-430
4. T.E. Tallian, *Failure Atlas for Hertz Contacts*, ASME, 1992
5. K.R. Kilburn, An Introduction to Rail Wear and Rail Lubrication Problems, *Wear*, Vol 7, 1964, p 255-269
6. K.H. Kloos and F. Schmidt, Surface Fatigue and Wear, *Metallurgical Aspects of Wear*, K.H. Zum Gahr, Ed., Deutsche Gesellschaft fur Metallkunde, 1981, p 163-182
7. K.H. Kloos and F. Schmidt, Surface Fatigue and Wear, *Metallurgical Aspects of Wear*, K.H. Zum Gahr, Ed., Deutsche Gesellschaft fur Metallkunde, 1981, p 163-182
8. J.A. Martin, S.F. Borgese, and A.D. Eberhardt, *Trans. ASME*, Vol 59, 1966, p 555
9. H. Swahn, P.C. Becker, and O. Vingsbo, *Met. Sci.*, Jan 1976, p 35
10. W.D. Syniuta and C.J. Corrow, *Wear*, Vol 15, 1970, p 187
11. P.C. Becker, Microstructural Changes Around Non-metallic Inclusions Caused by Rolling-Contact Fatigue of Ball-Bearing Steels, *Met. Technol.*, June 1981, p 234-243
12. R.J. Boness, W.R. Crecelius, W.A. Ironside, C.A. Moyer, E.E. Pfaffenberger, and J.V. Poplawski, Current Practice, *Life Factors for Rolling Bearings*, E.V. Zaretsky, Ed., Society of Tribologists and Lubrication Engineers, 1992, p 5-7
13. G. Lundberg and A. Palmgren, Dynamic Capacity of Rolling Bearings, *Acta Polytechnia*, Mechanical Engineering Series 1, R.S.A.E.E., No. 3, 7, 1947
14. T.E. Tallian, Unified Rolling Contact Life Model with Fatigue Limit, *Wear*, Vol 107, 1986, p 13-36
15. E.V. Zaretsky, J.V. Popiawski, and S.M. Peters, "Comparison of Life Theories for Rolling-Element Bearings," Preprint 95-AM-3F-3, Society of Tribologists and Lubrication Engineers, May 1995
16. A. Erdemir, Rolling Contact Fatigue and Wear Resistance of Hard Coatings on Bearing Steel Substrates, *Surf. Coat. Technol.*, Vol 54-55 (No. 1-3), 1992, p 482-489
17. R.J. Parker and E.V. Zaretsky, Fatigue Life of High-Speed Ball Bearings with Silicon Nitride Balls, *J. Lubr. Technol. (Trans. ASME)*, 1975, p 350-357
18. F.J. Ebert, Performance of Silicon Nitride Components in Aerospace Bearing Applications, *Proc. Gas Turbine and Aeroengine Congress*, 11-14 June 1990, American Society of Mechanical Engineers
19. M. Hadfield, S. Tobe, and T.A. Stolarski, Sub-surface Crack Investigation of Delaminated Ceramic Elements, *Tribol. Int.*, Vol 27 (No. 4), 1994, p 359-367
20. C.A. Moyer and E.V. Zaretsky, Failure Modes Related to Bearing Life, *Life Factors for Rolling Bearings*, E.V. Zaretsky, Ed., Society of Tribologists and Lubrication Engineers, 1992, p 67

Fatigue and Fracture Control for Powder Metallurgy Components

Randall M. German and Richard A. Queeney, The Pennsylvania State University

POWDER METALLURGY (P/M) is one of the most diverse approaches to metalworking. The main attraction of P/M technology is the ability to fabricate high-quality, complex parts to close tolerances in an economical manner. In essence, P/M converts a metal powder from a semifluid state into a strong, precise, high-performance shape. Key steps include the shaping or compaction of the powder and the subsequent thermal bonding of the particles by sintering. These two steps can be combined into a single operation, for example in hot powder forging or hot isostatic pressing. All P/M processes are fairly automated with relatively low energy consumption, high material utilization, and low capital costs. These characteristics align P/M with current concerns over manufacturing productivity. Consequently, the field is experiencing growth and progressively replacing traditional metalforming operations over a wide range of applications and materials. As illustrations, P/M is used in the fabrication of lamp filaments (tungsten), dental restorations (precious metals), self-lubricating bearings (bronze), automotive transmission gears (steels), armor piercing projectiles (tungsten alloys), welding electrodes (copper), nuclear power fuel elements (uranium dioxide), orthopedic implants (cobalt and titanium alloys), high-temperature filters (stainless steels), aircraft brake pads (iron-copper-tin-carbon), rechargeable batteries (nickel), and jet engine components (superalloys).

There are three basic approaches to powder metallurgy processing (Ref 1). The most common method, termed pressing and sintering, is to fill a die cavity with loose powder and apply a uniaxial compaction pressure to the powder. This pressure deforms and densifies the powder to approximately 85 to 90% of theoretical density. Subsequently, the pressed powder is heated to a temperature where atomic diffusion gives rise to interparticle bonding, but with little densification. Accordingly, the final product is porous. Such a technique is in widespread use for forming moderately complex shapes for mechanical systems using ferrous powders. The open continuous pore structures that exist in these P/M products dominate fracture and fatigue behavior. Additionally, many filters, electrodes, capacitors, batteries, and other porous structures are formed in a similar manner using low compaction pressures. Such high-porosity structures should not be employed in a fatigue-sensitive environment.

Alternatively, small particles are shaped into useful components at low pressures with the assistance of an organic binder, such as wax, by injection molding, tape casting, or extrusion. The particle-packing density is relatively low, typically only 60% of theoretical. After shaping, the binder is removed by either heat or solvent extraction, and the powder is densified by sintering at a high temperature. These approaches give a final density that is usually between 94 and 100% of theoretical. They are slower and more costly than die compaction, but they deliver greater shape complexity and improved mechanical response measures. Thus, techniques such as powder injection molding are in widespread use for computer, biomedical, and firearm applications, especially using stainless steels. Because the pores are small, closed, and spherical, they have less detrimental effect on fracture and fatigue properties.

Finally, a powder can be subjected to a combination of heat and stress simultaneously. This allows full densification and is widely employed in the fabrication of structural metals, composites, and high-temperature alloys. Variations include forging, hot pressing, hot isostatic pressing, extrusion, and roll forming. Alloys fabricated this way are usually based on aluminum, titanium, steel, nickel, or refractory metal systems, but include composites and intermetallics. Because there is no residual porosity, fracture and fatigue properties are totally dependent on the microstructure, especially any inhomogeneities or contaminants. When properly performed, these processes result in full-density P/M products that have mechanical responses superior to those of their wrought equivalents, largely because of the microstructure homogeneity.

P/M Materials

Many metals are available via P/M techniques. Aluminum and its alloys are highly compressible as powders; green densities of 90% of theoretical are common. They can be sintered or hot consolidated using extrusion, forging, hot pressing, and hot isostatic pressing. As summarized in Table 1, typical strengths for the press and sinter approach are in the 200 MPa (29 ksi) range with 2% elongation to fracture. Higher strengths are available by dispersion strengthening and deformation processing, including hot isostatic pressing and extrusion. In the best cases, fatigue endurance limits (or fatigue strength at about 10^7 cycles) approach about 200 MPa (29 ksi). Some of the high-performance rapidly solidified P/M products provide excellent strength retention to high temperatures.

Table 1 Properties attainable in aluminum P/M alloys

Composition, wt % (a)	Fabrication(b)	Density, g/cm³	Yield strength, MPa	Tensile strength, MPa	Elongation, %
4Mg-0.80Si-1.1C	MA + forged	...	550	570	2
4Cu-1.5Mg-0.80Si-1.1C	MA + forged	...	580	600	11
0.4Si-0.6Mg	Cold forged	2.66	90	180	11
4.4Cu-0.8Si-0.5Mg	P+S	2.64	200	250	3
0.4Cu-1.0Mg-0.6Si	P+S	2.45	176	183	1
0.4Cu-1.0Mg-0.6Si	P+S	2.58	230	238	2
4Ti	MA + HIP	2.74	325	380	11
8Fe-2Mo	HIP	2.89	470	490	7

(a) Balance Al. (b) MA, mechanically alloyed; HIP, hot isostatically pressed; P+S, pressed and sintered

Table 2 Common ferrous P/M alloy classes

Designation	Composition
Pure iron (steel)	max 1% C
Copper steel	1-22% Cu, max 1% C
Iron-nickel	1-3% Ni, max 2.5% Cu, max 0.3% C
Nickel steel	1-8% Ni, max 2.5% Cu, max 1% C
Low-alloy steel	0.3-2% Ni, 0.5-1% Mo, 0.4-0.8% C
Infiltrated steel	8-25% Cu, max 1% C
Phosphorus steel	0.4-0.8% P, low C
Sinter-hardened steel	1-3% Cr, 1-2% Mn, 2% Ni, 0.4-0.8% C

Table 3 Sample mechanical properties for Fe-2Cu-0.8C P/M alloys

Pressed and sintered, 1120 °C, ½ h, N_2-H_2 atmosphere

Density, g/cm³	6.65	6.85	7.15
Porosity, %	14.2	11.8	7.9
Hardness, HRB	70	75	85
Yield strength, MPa	365	400	415
Tensile strength, MPa	425	495	620
Elongation, %	1.3	1.8	2.5
Transverse rupture strength, MPa	890	1025	1325
Fatigue strength, MPa	168	198	266

Table 4 Density and heat treatment effects on the properties of an Fe-10Cu-0.3C P/M alloy

Density, g/cm³	6.4	6.4	7.1	7.1
Thermal condition	As-sintered	Heat treated	As-sintered	Heat treated
Hardness (scale)	50 (HRB)	25 (HRC)	80 (HRB)	40 (HRC)
Yield strength, MPa	280	...	395	655
Tensile strength, MPa	310	380	550	690
Elongation, %	0.5	0.5	1.5	0.5
Fatigue strength, MPa	115	145	210	260
Impact energy, J	4	...	11	...
Elastic modulus, GPa	90	90	130	130

Table 5 Mechanical property comparison for Ti-6Al-4V processed by various P/M techniques

Process	Porosity, %	Yield strength, MPa	Tensile strength, MPa	Elongation, %	Reduction in area, %
Blended elemental P+S	2	786	875	8	14
Blended elemental HIP	<1	805	875	9	17
Prealloy HIP	0	880	975	14	26

P+S, pressed and sintered; HIP, hot isostatically pressed

Table 6 Examples of the mechanical properties of tungsten heavy alloys sintered at 1500 °C to 100% density

Composition, wt%	Density, g/cm³	Hardness, HRA	Yield strength, MPa	Tensile strength, MPa	Elongation, %
97W-2Ni-1Fe	18.6	65	610	900	19
93W-5Ni-2Fe	17.7	64	590	930	30
90W-7Ni-3Fe	17.1	63	530	920	30
86W-4Mo-7Ni-3Fe	16.6	64	625	980	24
82W-8Mo-8Ni-2Fe	16.2	66	690	980	24
74W-16Mo-8Ni-2Fe	15.3	69	850	1150	10

Copper, brass, and bronze are sintered from particles, where the typical applications are not fatigue sensitive. Cemented carbides, such as WC-Co and TiC-Ni, are sintered using a liquid phase to deliver a full-density structure, often by the application of a high pressure at the end of the sintering cycle. The elimination of residual pores has considerable impact on fracture resistance, giving a fracture strength in the 1700 to 3000 MPa (245 to 435 ksi) range. Unfortunately, the basic materials are brittle, so fracture toughness is usually in the range of 10 to 20 MPa√m, depending on cobalt (or other matrix phase) content and carbide grain size.

Stainless steel P/M products are usually selected for their corrosion resistance. However, they are capable of highly variable final response measures, depending on the composition, density, and microstructure. For precipitation-hardenable alloys such as 17-4 PH, yield strengths of 1100 MPa (160 ksi) with 12% elongation are possible with fatigue endurance limits in the 500 MPa range (72 ksi). Alternatively, for austenitic stainless steels such as 316L, a sintered yield strength of 250 MPa (36 ksi) and considerable ductility (30% or more) are common.

By far the largest segment of the P/M applications rely on iron-base alloys. Generally, a powder is pressed in uniaxial tooling to near-final dimensions, but not to full density. Dimensional control during sintering is very important, and usually size can be held to within ±0.025 mm of specification, with concentricity to 0.1 mm, squareness to 0.05 mm, and density to 0.1 g/cm³. Strength typically exhibits a small scatter of ±35 MPa (±5 ksi) and elongation exhibits a scatter of ±2%. In most sintered structural steel components, over 90% of the composition is iron. Table 2 gives examples of common P/M alloy compositions. In all cases, the mechanical properties increase with the final density.

Iron-copper-carbon compositions are the most common in production, because copper forms a liquid phase during sintering that greatly aids particle bonding. This system illustrates the properties possible with P/M. Copper and graphite (carbon) are mixed with iron, and during sintering the copper forms a liquid phase. Wrought materials of equivalent compositions are not possible due to extensive segregation in the molten state. The mechanical properties are degraded by whatever pores remain after sintering. Tables 3 and 4 provide examples of the property degradation by listing the hardness, strength, ductility, and impact energy versus density for two Fe-Cu-C alloys. Table 3 shows a density effect for an Fe-2Cu-0.8C alloy, while Table 4 includes both density and heat treatment effects for an Fe-10Cu-0.3C alloy. In these tables the fatigue life was measured at 10^7 fully reversed cycles ($R = -1$). Note that, for example, hardness and strength actually change less with increases in density than does the fatigue endurance strength. This reflects the greater sensitivity of the dynamic properties to pore structure as compared with the quasistatic tensile properties.

Nickel is another common addition to ferrous P/M alloys for improved strength. In low concentrations, phosphorus is used due to its potent hardening of iron and formation of a liquid phase at temperatures above 1050 °C (1920 °F). The liquid aids sintering, pore spheroidization, and alloy hardening, but usually these additives are selected for magnetic properties, not mechanical properties, with a popular composition containing 0.45% P. Most of these alloys are formed by mixing powders that are alloyed as part of the sintering cycle, because of the higher compressibility of the elemental powders as compared with that of prealloyed powders.

There are several other widely employed P/M materials. Tool steels are usually fabricated to full density by liquid-phase sintering or hot isostatic pressing. Cobalt-base alloys, titanium alloys, and superalloys are fabricated to full density by hot isostatic pressing. Table 5 compares the mechanical properties of Ti-6Al-4V alloys fabricated by three processing routes. A very useful group of P/M alloys are the tungsten heavy alloys. These are based on W-Ni-Fe mixtures that are densified by liquid-phase sintering. Table 6 gives the typical mechanical properties of sintered tungsten heavy alloys. These are full-density products, but despite the high sintered density they lack good fatigue properties due to the two-phase microstructure. Like tungsten, most of the other refractory metals (molybdenum, tantalum, titanium, chromium, niobium, and rhenium) are fabricated from powders.

Low-cost composites are fabricated using P/M techniques. Particle reinforcement requires mixing of the constituents and consolidation to full density. A popular combination is Al-SiC. Sample mechanical properties of this composite as fabricated by vacuum hot pressing are given in

Table 7 Mechanical properties of Al-SiC P/M composites

6061 alloy matrix, densified by vacuum hot pressing

SiC, vol%	Elastic modulus, GPa	Density, g/cm³	Yield strength, MPa	Elongation, %
0	69	2.71	430	20
15	97	2.77	435	6
20	103	2.80	450	5
25	114	2.83	475	4
30	121	2.85	510	0
40	138	2.91	379	<1

Table 7, where the SiC content ranges from 0 to 40 vol%. Again, the fatigue properties of these composites are not very attractive because of the differential elastic modulus between the reinforcement and matrix.

Volume 7 of the *ASM Handbook* provides a large compilation of mechanical properties of P/M materials.

Porosity Effects

An inherent physical characteristic of P/M materials is the presence of pores. The role of porosity in determining fatigue endurance in powdered metals is akin to that of porosity that is induced through metal solidification in casting or welding. However, the porosity that is characteristic of sintered powdered metals, and of those materials subsequently deformation processed, may differ in character and influence from solidification porosity. In all cases, porosity catches the attention of the design engineer because it immediately conjures up images of classical stress concentrators. In addition, porosity featuring sharp re-entrant corners, a possibility for marginally equilibrium-sintered powder particle boundaries, may be more accurately viewed as crack precursors. Pore geometry can be altered by modifications to the sintering cycle, such as a longer hold time or higher temperature, wherein the smoother pores improve strength, fatigue life, and fracture resistance.

Ductility is sensitive to the pore structure, but impact, fracture, and fatigue behavior have the greatest sensitivities. In general, dynamic strength responses are the most sensitive. Even in those materials possessing full density, inferior properties can occur due to microstructural defects. Recent applications have pushed P/M into very demanding applications, a good example being the automotive connecting rod. Such rods are formed by hot forging a porous P/M preform using an alloy of Fe-2Cu-0.8C. They weigh nearly 650 g. The ultimate tensile strength is 825 MPa (120 ksi), with a yield strength of 550 MPa (80 ksi) and fatigue endurance limit of 255 MPa (37 ksi). With expansion of P/M fabrication technology into dynamically loaded components, there arise performance limitations associated with fracture and fatigue. For static tensile properties, there is a good basis for predicting the effect of residual porosity on strength (Ref 2-5).

Recent research has had an emphasis on fatigue and fracture behavior to fill the database void.

Porosity is especially important to the high-cycle fatigue life (Ref 6-9). Pores play a role in both crack initiation and propagation, typically increasing the threshold stress intensity for crack initiation but lowering the resistance to crack propagation (Ref 10). For an alloy of Fe-2Ni-0.8C the fatigue endurance limit at 10^7 reverse cycles is between 200 and 250 MPa at a density of 7.1 g/cm³ (approximately 10% porosity). That is approximately 35% of the tensile strength, and many porous injection-molded materials exhibit similar ratios of fatigue strength to tensile strength. Closed pores are less detrimental than those that are interconnected and open to the external surface. Surface pores act as the preferred site of fatigue crack initiation by acting as stress concentrators (Ref 11). The simplest view of a pore would be that of a spherical hole embedded in a continuous matrix whose response parameters are identical with that of a fully dense form of exactly the same metallurgical state. Since the maximum, most positive, principal stress is the controlling load parameter for high-cycle fatigue endurance, the embedded pore in a tensile field is a reasonably approximate model for a stressed porous metal without interacting stress fields (one of less than 10% porosity). One can expect the local pore-dominated stress field to control local events such as crack initiation and threshold stress intensity values (ΔK_{th}). For an embedded pore, the local stress fields are related to the far-field (applied) design stresses (Ref 12):

$$\sigma_{local} = \sigma_{design} \left[\frac{27 - 15v}{2(7 - 5v)} \right] \qquad (Eq 1)$$

Note that, for the average steel, $v = 0.30$, and the local stresses are magnified by a factor of about 2. The concentrated stress field does not persist far beyond the embedded pore, being reduced to 105% of the design value at a distance of $2a$ (where a is the pore radius) from the pore center, thus possibly exerting little influence on far field parameters.

If the pore features sharp re-entrant corners, the elastic concentrated stresses are more accurately predicted from an elliptical hole example. However, these same pores are unlikely to exhibit smooth ellipsoidal morphologies, and their concentration effects can be more usefully predicted by the proportionality (Ref 13):

$$\sigma_{local} \approx 2\sigma_{design} \sqrt{\frac{a}{\rho}} \qquad (Eq 2)$$

Here, a is the major pore length normal to the maximum principal stress direction, and ρ is the radius of curvature at the sharp corner. Suffice to say that the calculated stress concentration value can be large for a nonspheroidized pore.

Due to the stress-raising properties of pores, and the fact that most fatigue failures originate at free surfaces, treatments aimed at surface densification and serendipitous surface strengthening (e.g., coining, shot peening, ausrolling) raise the fatigue endurance limit (Ref 14-16). As a conse-

quence, the current models for fatigue response in porous sintered materials have a major dependence on the pore microstructure: the models address the total porosity, alloying homogeneity in near-pore regions, pore size, pore shape, and interpore separation distance (Ref 6, 17-24). Round pores provide improved resistance to crack propagation. Pores act as linkage sites through which cracks can propagate. Microstructure-based fatigue models for ferrous alloys have had to address seemingly contradictory porosity effects: round pores retard stable fatigue crack propagation but increase crack extension growth rates by contributing linkage sites. While extant theories successfully explain some porosity effects on crack propagation, no total predictive model has been created that embraces microstructure effects (Ref 18, 20, 22). The presence of pores in the reversed plastic zone that is the site of propagating crack damage does not lend itself to facile analysis.

Pore structure changes are obtainable through processing and material variations: powder variables (particle size distribution, particle shape); compaction variables (type of lubricant, amount of lubricant, tool motions, maximum pressure); sintering variables (hold time, maximum temperature, atmosphere); and postsintering treatments. Figure 1 compares the pore structure in two sintered stainless steels to emphasize this point. Smaller particles result in faster sintering and higher strengths and toughness. Associated with the smaller particle sizes are smaller final pores. At lower densities (around 6.6 g/cm³ or 18% porosity in steels), a high relative content of small particles is beneficial to fatigue resistance, while at higher densities (over 7.1 g/cm³ or less than 10% porosity), larger particles prove beneficial. The difference relates to the ligament size between pores, which is the determinant of fatigue. There are three pore microstructure parameters relevant to the fatigue resistance of porous P/M materials: pore size, pore curvature, and pore spacing. These largely reflect the role of stress concentration with respect to the advancing fatigue crack, as noted above (Ref 6). Thus, lower porosity contents, smoother (rounder) pores, and wider interpore separations increase the fatigue endurance strength.

Fatigue cracks have been successfully analyzed with regard to their propagation response, and the cyclic growth of a crack can be predicted by the modified Paris growth law:

$$\frac{da}{dN} = A (\Delta K_{eff})^n \qquad (Eq 3)$$

where the material parameters A and n must be experimentally determined for any given material microstate, including different distributions of porosity. The stress analytical variable ΔK_{eff}, the effective stress intensity range, factors out that portion of the total stress range that relieves the stresses holding the crack flanks closed, the opening stress range $\Delta\sigma_{op}$, leaving only the stress range component that displaces the crack faces relative to each other. In the presence of appreciable levels of mean stress, the

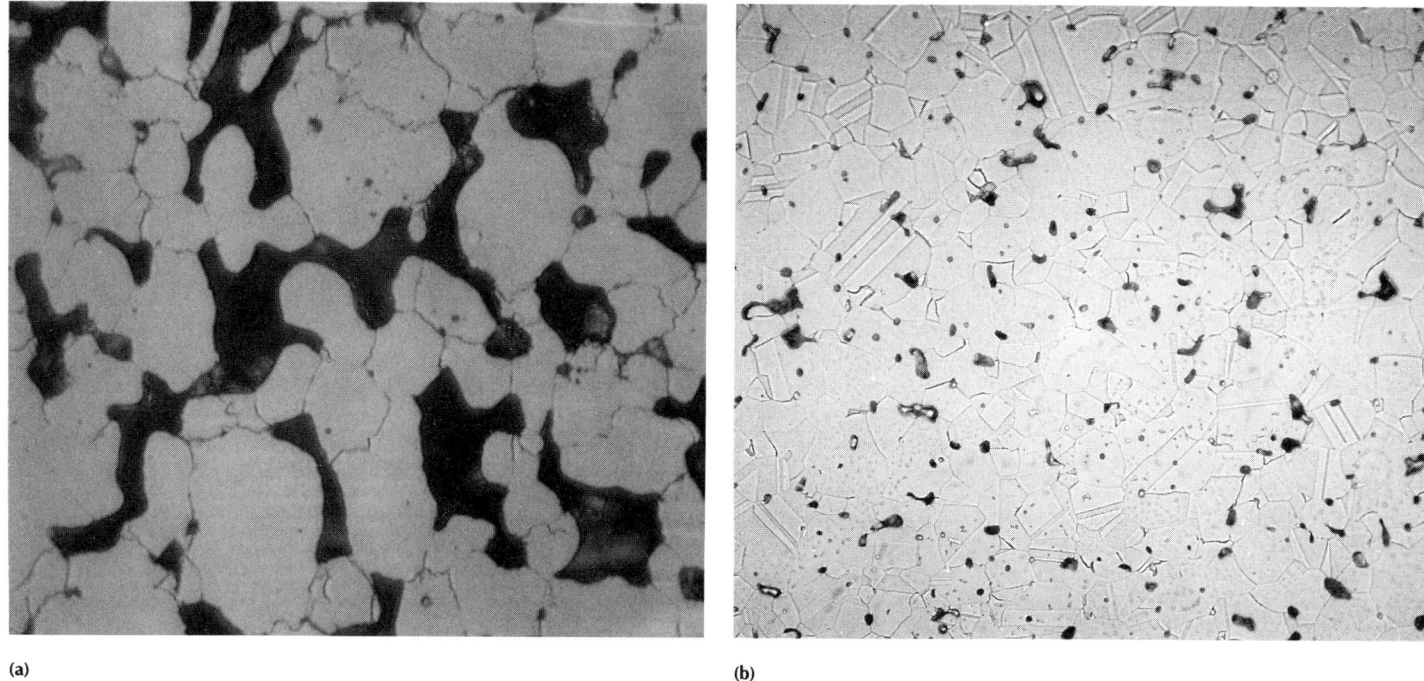

(a)   (b)

Fig. 1 Two stainless steels fabricated by P/M, as demonstrations of the microstructure variations possible by tailoring the powder, compaction, and sintering variables. (a) A high-porosity microstructure useful for filtration, formed by press and sinter. 1000×. (b) A closed-porosity, high-density microstructure useful for mechanical components, formed by injection molding and high-temperature sintering. 200×

relation between load design parameters and fatigue crack propagation rates is given by:

$$\frac{da}{dN} = \frac{C\Delta K^n}{(1-R)K_c - \Delta K} \qquad \text{(Eq 4)}$$

Here, the total stress-intensity factor range ΔK is the load variable, but the fatigue ratio $R = \sigma_{min}/\sigma_{max}$ and the fracture toughness K_c, or K_{Ic} for a low-ductility sintered member, enter into the fatigue crack propagation response, as do the material constants C and n.

Regardless of which of the three fatigue crack propagation relations are relevant in a particular service context, their collective utility lies in predicting member lifetimes, or precise segments of that lifetime. Thus, in the case of the Paris law:

$$N = \int_0^{N_f} dN = \int_{a_0}^{a_c} \frac{da}{A(\Delta K)^n} \qquad \text{(Eq 5)}$$

The calculated endurance cycles to failure, N, can be from an initial flaw size a_0 that may be the minimum detectable to final fracture at $a = a_c$, where:

$$K_{Ic} = \sigma_{max}\sqrt{\pi a_c} \qquad \text{(Eq 6)}$$

Again, the fracture toughness of the material plays a role in determining structural endurance. The fracture resistance K_{Ic}, or K_R in the case of more ductile sintered materials (Ref 25), is known to be porosity sensitive (Ref 26). As a first estimate, the sensitivity is about a 100 MPa√m gain in toughness per percentage point of porosity reduction in quenched and tempered steels. Copper-infiltrated steels have

toughnesses that run from 40 to 50% those of wrought steels (Ref 27, 28). However, their static strengths are equivalent, reflecting the inability of the steel skeleton to absorb the same level of strain energy release as a fully dense body of the same material.

The material parameters in the modified Paris equation (Eq 3) are sensitive to the porosity state, a not unexpected result since the coefficient and exponent are related to the plastic zone size r_p ahead of the advancing crack, the only region of irreversible deformation. The coefficient A in Eq 3 increases with increasing porosity fraction (Ref 29), resulting in faster growth rates for comparable ΔK_{eff} values. With higher coefficient measures but constant exponent n values, the threshold stress intensity range ΔK_{th}, below which no crack growth is thought to occur, also decreases with porosity increase. However, ΔK_{th} values for fully dense materials are sufficiently low that it is not at all clear in what way they could be successfully employed in design practice if service stresses are to be set at appreciable fractions of the yield or tensile strength. The reversed plastic zone size is given by (Ref 30):

$$r_p = \frac{K_I^2}{6\pi\sigma_y^2} \qquad \text{(Eq 7)}$$

When the plastic zone size is of the same size as the average pore diameter, it is effectively enlarged by the high strain field in the vicinity of the pore. The net result of the enlarged effective plastic zone is reflected in higher values of the exponent n (Ref 11), and the enlarged zone is even more pronounced in the response of short cracks driven by the locally

raised stress/strain fields associated with design stress concentrators (Ref 15). Copper-infiltrated steels (Ref 27, 28) are as fatigue and fracture resistant as fully dense wrought medium-strength steels. Although the performance standards are not up to those of fully dense martensitic steels, copper infiltration represents a considerable cost saving over forging to full density, while maintaining the cost advantage inherent in press and sinter P/M.

Typical fatigue endurance limits (or fatigue strengths at about 10^7 cycles) are collected in Table 8 for several P/M materials. This compilation includes several ferrous alloys, reflecting the high interest in P/M fatigue for automotive applications. There are variations in density, alloying, and sintering cycles to show the relative effects on sintered properties. Systematic testing of various alloys has shown that slight changes in the particle size distribution or alloying homogeneity can affect these properties. For example, in the Fe-2Ni-0.8C alloy system, shifts in just the iron powder source lead to ±33 MPa (±4.8 ksi) variations in the fatigue endurance strength. Accordingly, the values in Table 8 are for relative ranking purposes only and cannot be used as an accurate basis for design of fatigue-sensitive components.

Fracture studies of P/M alloys are usually restricted to impact testing, and often this is performed in the unnotched condition because of the low toughness of porous materials. Fracture toughness measurements on P/M materials are relatively rare. Tables 9 and 10 summarize some prior findings. Table 9 demonstrates the density effect on tensile properties and K_{Ic}, for an Fe-Ni-Mo-C steel. Note that as the porosity decreases,

Table 8 Representative P/M materials, processing cycles, and fatigue endurance limit

Composition, wt %	Processing	Density, g/cm³	Testing	Endurance limit, MPa
Al-5Cu-0.5Mg-0.8Si	P+S, 600 °C, 1 h	2.6	...	53
Fe	P+S, 1120 °C, ½ h	6.0	Rotating R=−1	39
Fe	P+S, 1120 °C, ½ h	6.7	Bending R=0	67
Fe	P+S, 1120 °C, ⅔ h	6.9	Rotating R=−1	102
Fe	P+S, 1150 °C, 1 h	7.2	Axial R=−1	65
Fe	P+S, 1120 °C, ½ h	7.3	Rotating R=−1	145
Fe	P+S, 1250 °C, 2 h	7.6	Rotating R=−1	181
Fe	HIP, 1100 °C, 200 MPa	7.86	Axial R=−1	230
Fe-17Cr-4Cu-4Ni (17-4 PH)	PIM, 1350 °C, 2 h, HT	7.5	Rotating R=−1	517
Fe-1.5Cu-0.6C	P+S, 1120 °C, ½ h, HT	7.0	Bending R=0	390
Fe-2Cu-0.5C	P+S, 1120 °C, 30 min	7.1	Rotating R=−1	125
Fe-2Cu-0.8C	P+S, 1120 °C, ½ h	6.7	Rotating R=−1	165
Fe-2Cu-0.8C	P+S, 1120 °C, ½ h	7.0	Rotating R=−1	234
Fe-2Cu-0.8C	P+S, 1120 °C, ½ h	7.15	Rotating R=−1	241
Fe-2Cu-0.8C	P+S, 1330 °C, 1 h	7.1	Rotating R=−1	270
Fe-2Cu-2Ni-0.8C	P+S, 1120 °C, ½ h, HT	7.0	Rotating R=−1	240
Fe-2Cu-2Ni-0.8C	P+S, 1120 °C, ½ h, HT, SP	7.0	Rotating R=−1	282
Fe-2Ni-0.5C	PIM, 1250 °C, 4 h, HT	7.7	Rotating R=−1	239
Fe-2Ni-0.8C	P+S, 1120 °C, ½ h, HT	7.12	Rotating R=−1	159
Fe-2Ni-0.8C	P+S, 1175 °C, ½ h	6.9	Rotating R=−1	192
Fe-2Ni-0.5Mo-0.5C	P+S, 1120 °C, ½ h, HT	7.0	Rotating R=−1	350
Fe-2Ni-0.5Mo-0.5C	DP + DS, 1260 °C, 30 min, HT	7.4	Rotating R=−1	425
Fe-2Ni-0.5Mo-0.5C	P+S, 1120 °C, 30 min, HT	6.8	Rotating R=−1	345
Fe-2Ni-0.5Mo-0.4C	PF, 1150 °C	7.9	Rotating R=−1	780
Fe-2Ni-1Mo-0.9C	P+S, 1275 °C, 1 h, HT	...	Rotating R=−1	390
Fe-2Ni-1.5Cu-0.5Mo-0.5C	HIP, 1160 °C, 3 h, 105 MPa, HT	7.86	Rotating R=−1	480
Fe-4Ni-1.5Cu-0.5Mo-0.6C	P+S, 1120 °C, ½ h	7.1	Rotating R=−1	129
			Bending R=0	148
Fe-4Ni-1.5Cu-0.5Mo-0.6C	P+S, 1120 °C, 2 h	7.1	Rotating R=−1	135
Fe-4Ni-1.5Cu-0.5Mo-0.6C	P+S, 1250 °C, ½ h	7.1	Rotating R=−1	147
Fe-4Ni-1.5Cu-0.5Mo-0.6C	P+S, 1250 °C, 2 h	7.1	Rotating R=−1	195
			Bending R=0	266
Fe-7Ni	PIM, 1250 °C, 1 h	7.71	Rotating R=−1	236
Tool steel (Fe-8Co-6.3W-5Mo-3V-4Cr-1.3C	HIP, 1150 °C, hot roll, HT	8.0	Axial R=−1	950
Ni	P+S, 1000 °C, 1 h	7.8	Rotating R=−1	70
Ni	DP + DS, 1300 °C, 3 h	8.5	Rotating R=−1	121
Ni₃Si	Hot extrude, HT	...	Axial R=−1	579
Ti-6Al-4V	HIP, 925 °C, 3 h, 200 MPa	4.46	Axial R=−1	475

P+S, pressed and sintered; PIM, powder injection molded and sintered; HIP, hot isostatically pressed; HT, heat treated; SP, shot peened; DP + DS, double pressed and double sintered; PF, powder forged

the strength essentially doubles, ductility increases substantially (a nearly tenfold gain), the impact energy goes up by a factor of 4, and fracture toughness increases threefold. Table 10 collects several examples of the mechanical properties of ferrous alloys and one titanium alloy as representative values obtainable via P/M. Clearly, porosity negatively affects the fracture toughness in most materials. The fracture toughness of P/M steels is essentially a linear function of density (Ref 18), with sensitivities of about 100 MPa√m gain in toughness per percentage point of porosity reduction. In low-density P/M materials, the fracture crack propagation is rapid because the pore structure amplifies the stress and provides an easy path. At low porosity levels, an advancing fracture crack can be blunted by the pores, effectively forming microcracks that improve toughness (Ref 31).

Other Factors Determining Fatigue and Fracture Resistance

It is well established that porosity is the major detriment to fatigue life for P/M materials. Beyond porosity, the sintered microstructure is a factor. Even in full-density materials fabricated by hot isostatic pressing, microstructure has a role. Weak links in the microstructure become evident during fracture. For porous structures, these weak links prove to be microstructural inhomogeneities, typically resulting from incomplete diffusional homogenization. Often powders (such as iron, nickel, and graphite) are mixed in the compaction stage. During heating the intent is for the mixed powders to homogenize to form a uniform microstructure, but this often is inhibited by too short a hold time at the peak temperature. Consequently, the alloying elements are poorly distributed and give point-to-point composition and microstructure changes, which are especially evident in postsintering heat treatment response. Accordingly, during fatigue or fracture, the weak links become the preferred failure paths. Most notable are the negative effects from inadequate homogenization of carbon to ensure uniform strength (Ref 18).

Little is known about the sensitivity of fatigue and fracture to loading conditions for P/M material. Table 11 compares the 2×10^6 fatigue endurance strengths for Fe-1.5Cu-0.6C at 7.1 g/cm³ density using bending and axial fatigue tests. The table also includes a comparison with two loading stress ratios (half-cycle and fully reversed, $R = -1$) and two notch conditions (unnotched and notched) (Ref 18). In these cases the unnotched loading shows little sensitivity to axial versus bending fatigue, but a large sensitivity is evident in the presence of notches. The notch sensitivity factor is reported to range between 0.32 and 0.43 for many of the common pressed and sintered P/M alloys (Ref 32).

Full-density materials also suffer from residual microstructure artifacts that degrade the microstructure. In hot isostatically compacted powders, the achievement of 100% density is still insuffi-

Table 9 Mechanical properties of pressed and sintered Fe-1.8Ni-0.5Mo-0.5C P/M compacts

Double pressed and double sintered, 1120 °C, ½ h, tested as-sintered

Porosity, %	Density, g/cm^3	Elastic modulus, GPa	Yield strength, MPa	Ultimate strength, MPa	Elongation, %	Notched impact energy, J	Fracture toughness, MPa√m
16	6.6	110	280	350	2	3	19
10	7.1	145	370	460	3	4	28
5	7.4	180	425	610	5	4	38
0	7.9	190	590	800	19	12	65

Table 10 Representative P/M materials, processing cycles, and fracture toughness

Composition, wt%	Processing	Density, g/cm^3	Fracture toughness, MPa√m
Fe-4.4Cr-9.2Co-7.2V-3.7Mo-9.2W-2.7C	P+S, 1150 °C, 1 h	8.1	13
Fe-1.5Cu-2Ni-0.8C	P+S, 1120 °C, ½ h	6.8	40
Fe-1.8Ni-0.5Mo-1.5C	P+S, 1120 °C, ½ h	6.6	15
Fe-1.8Ni-0.5Mo-1.5C	P+S, 1150 °C, ½ h	6.8	26
Fe-1.8Ni-0.5Mo-1.5C	P+S, 1120 °C, ½ h	7.1	24
Fe-1.8Ni-0.5Mo-1.5C	DP+DS, 1100 °C, ½ h	7.5	21-38
Fe-1.8Ni-0.5Mo-1.5C	HF, 1100 °C	7.85	64
Fe-0.8P-0.3C	P+S, 1120 °C, ½ h	7.0	22
Fe-0.8P-0.3C	DP+DS, 1120 °C, ½ h	7.8	20
Ti-6Al-4V	HIP, 925 °C, 3 h, 200 MPa	4.46	65

P+S, pressed and sintered; HIP, hot isostatically pressed; DP + DS, double pressed and double sintered; HF, hot forged

Table 11 Fatigue properties of Fe-1.5Cu-0.6C

Notch factor, K	Stress ratio, R	Endurance strength, MPa at 2 × 10^6 cycles
Axial		
1.0	−1	165
	0	130
2.8	−1	84
	0	64
Bending		
1.0	−1	160
	0	127
2.8	−1	137
	0	102

Note: Sintered to 7.2 g/cm^3 at 1120 °C for 30 min. Elastic modulus, 153 GPa; yield strength, 418 MPa; tensile strength, 483 MPa. Source: Ref 15

cient to guarantee competitive fracture toughness, fatigue life, or even impact toughness. Thermally induced porosity is a subtle problem in many full-density P/M products. After consolidation the material is pore free, but it may contain small quantities of adsorbed gas. Once the product is put into high-temperature heat treatment or service, this residual gas precipitates to form pores if there is no compressive stress. In hot isostatically pressed titanium alloys, gas precipitation reportedly gives a 10 to 20% decrement in fatigue endurance strength (Ref 33).

Another difficulty rests in slight contaminants located on the interfaces that were previously particle surfaces, a feature termed prior particle boundary decorations. Figure 2 shows such decorations in a fully densified P/M steel. Improper powder handling or cleaning prior to consolidation are the primary detriments. These contaminants remain on the powder interfaces, even though the structure is fully densified. Consequently, a small contamination film runs throughout the structure, providing an easy fracture path that is often traced to a trivial impurity level. The fracture path is along the prior particle boundaries and has a characteristic morphology, as shown in Fig. 3. In hot isostatically pressed Ti-6Al-4V there is substantial fatigue life improvement due to removal of the contaminant, with a change from 450 MPa endurance strength to 600 MPa due to powder cleaning prior to consolidation (Ref 34).

One option for limiting the detrimental effects from prior particle boundary decorations is to forge the structure after hot isostatic pressing, a process often used in producing aerospace structures to ensure ultimate reliability. The forging operation upsets the microstructure and breaks apart the continuous films of contamination. The alternative is to resort to clean handling and processing, where the powder is produced by rapid solidification and kept under inert conditions during handling. These steps, which minimize segregation and contamination, are employed in the production of aerospace components, microelectronic structures, and high-performance filters.

Steps to Improve Fatigue and Fracture Resistance

Surface pores are particularly detrimental to sintered materials with respect to fatigue life (Ref 35). Accordingly, carbonitriding and other surface strengthening and sealing treatments are most useful. Common treatments include shot peening, case hardening, repressing and resintering, coining, sizing, surface ausrolling, and postsintering heat treatments. For example, in pressed and sintered ferrous alloys, the endurance limit can be increased on small cross sections (6 by 6 mm) by at least 20% through shot peening. Carbonitriding is even more effective and can double the fatigue endurance limit. A typical carbonitride cycle involves heating to 940 °C in a mixture of ammonia and carbon dioxide for 4 h to form a 0.5 mm deep carbon-rich layer. Surface grinding is another approach to improved fatigue strength. The larger the cross section of the material, the less benefit possible from surface treatments, because bulk material states will dominate mechanical response.

The double press and double sinter approach was largely the only viable option for improving density and strength in traditional press and sinter P/M. This is more costly and involves extra tooling. A newly employed technique for improved fatigue life and fracture toughness in pressed and sintered ferrous alloys is to sinter at higher temperatures. The typical sintering temperature for steels is about 1120 °C, largely because of conveyor belt limitations in the furnaces. New materials of construction (ceramic belts) and new conveyor mechanisms (pusher plate and walking beam designs) allow higher-temperature processing regimes. Additionally, vacuum sintering usually is not limited in temperature, so it is viable for high-performance components. There is more sintering densification at the higher temperatures, so the density gain alone improves properties. Induced changes in the pore shape and size also improve fracture and fatigue properties. Several examples of the property gains are evident in Table 8. In a comparison of density gains versus sintering temperature effects, it is usually concluded that a change from 1120 to 1280 °C is equivalent to a density gain from 7.1 to 7.4 g/cm^3 in terms of both fracture toughness and fatigue endurance strength.

For small components the surface treatments are most useful, because the compressive forces extend through a major portion of the microstructure. However, for the porous materials with large cross sections, the need is to sinter at higher temperatures to improve fatigue and fracture. Further, designs that minimize density gradients will assist in minimizing fatigue failure. The high-density regions have a higher fatigue strength, and the difference in strength with density often results in failure at the interface between high- and low-density regions. For fatigue-sensitive components, the tolerable range of densities is less than 0.05 g/cm^3 within the structure. As with all fatigue-sensitive components, consideration must be given to surface finishing and processing optimization. The keys to improved performance are reduction in the total porosity, elimination of segregation and contamination, and manipulation of the pore microstructure.

There are some unique design opportunities where the microstructure of P/M materials offers

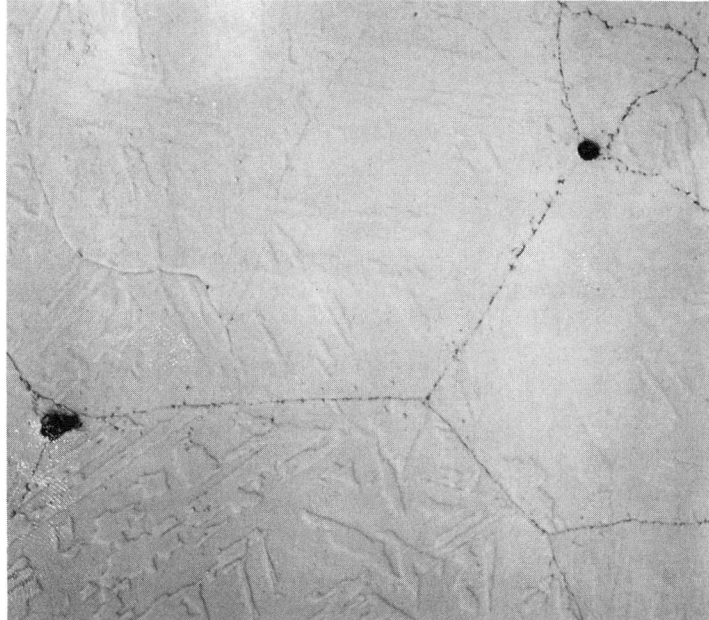

Fig. 2 Prior particle boundary precipitates formed on a hot isostatically pressed steel as the result of contamination during powder fabrication. 500×

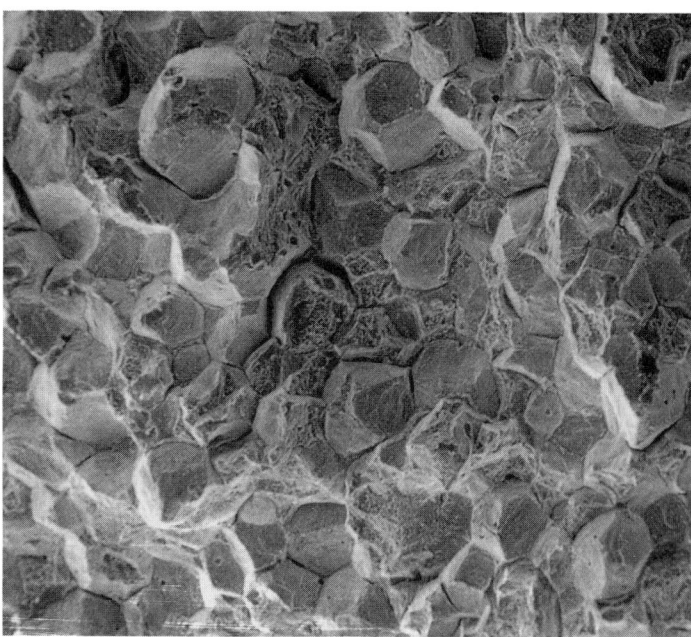

Fig. 3 A fracture surface showing preferential failure along prior particle boundaries. 150×

a fatigue advantage in spite of the porosity. Because many alloys are formed from mixed powders, there is often an enrichment of the alloying addition near pores. The resulting higher strength aids local strength and fatigue life. Surprisingly, when properly exploited this effect gives higher fatigue and fracture strength to porous structures formed from mixed elemental powders, compared to those formed from prealloyed powders (Ref 36). Unfortunately, pores reduce the elastic modulus, strength, ductility, hardness, and other mechanical properties, so a high-density structure usually proves most successful in fatigue-sensitive applications. Indeed, the high sensitivity to porosity mandates that porosity be tightly controlled (density held within 0.05 g/cm³) in regions of high stress concentrations. At a given strength level the P/M steels exhibit less notch sensitivity than wrought steels, because pores inherently act to blunt cracks and redistribute the load, especially under complex loading. Open pores, which dominate the sintered microstructure for densities below 92% of theoretical, retard fatigue crack propagation. Surface densification of P/M steels from shot peening, sizing, coining, surface rolling, or carbonitriding all prove beneficial in improving fatigue strength because of pore closure and surface compressive stresses.

Safety Factors for P/M Materials

Several factors contribute to scatter in the fracture and fatigue properties of P/M materials. There is the obvious error in testing, where the range of highest fatigue strength to lowest fatigue strength for a single test condition may be 4%. Further, there is typically a notch sensitivity and test error. Consequently, a safety factor of 1.4 is

often cited as appropriate for sintered P/M alloys (Ref 18). This means that a peak cyclic stress of about 70% of the fatigue endurance strength is the maximum recommended in service.

REFERENCES

1. R.M. German, *Powder Metallurgy Science*, 2nd ed., Metal Powder Industries Federation, 1994
2. B. Karlsson and I. Bertilsson, Mechanical Properties of Sintered Steels, *Scand. J. Metall.*, Vol 11, 1982, p 267-275
3. G.F. Bocchini, The Influences of Porosity on the Characteristics of Sintered Materials, *Rev. Powder Met. Phys. Ceram.*, Vol 2, 1985, p 313-359
4. R. Haynes, The Mechanical Behavior of Sintered Metals, *Rev. Deform. Behav. Mater.*, Vol 3, 1981, p 1-101
5. S.H. Danninger, G. Jangg, B. Weiss, and R. Stickler, Microstructure and Mechanical Properties of Sintered Iron, Part 1: Basic Considerations and Review of Literature, *Powder Met. Inter.*, Vol 25, 1993, p 111-117
6. B. Weiss, R. Stickler, and H. Sychra, High-Cycle Fatigue Behaviour of Iron-Base PM Materials, *Metal Powder Report*, Vol 45, 1990, p 187-192
7. R. Haynes, Fatigue Behaviour of Sintered Metals and Alloys, *Powder Met.*, Vol 13, 1970, p 465-510
8. S. Oki, T. Akiyama, and K. Shoji, Fatigue Fracture Behavior of Sintered Carbon Steels, *J. Japan Soc. Powder Met.*, Vol 30, 1983, p 229-234
9. W.B. James and R.C. O'Brien, High Performance Ferrous P/M Materials: The Effect of Alloying Method on Dynamic Properties, *Progress in Powder Metallurgy*, Vol 42, Metal Powder Industries Federation, 1986, p 353-372
10. H. Danninger, G. Jangg, B. Weiss, and R. Stickler, The Influence of Porosity on Static and Dynamic Properties of P/M Iron, *PM into the 1990's*, Vol 1, *Proceedings of the World Conference on Powder Metallurgy*, Institute of Materials, London, 1990, p 433-439
11. J. Holmes and R.A. Queeney, Fatigue Crack Initiation in a Porous Steel, *Powder Met.*, Vol 28, 1985, p 231-235
12. S. Timoshenko and J.N. Goodier, *Theory of Elasticity*, McGraw-Hill, 1951, p 359-362
13. F.A. McClintock and A.S. Argon, *Mechanical Behavior of Materials*, Addison Wesley, 1966, p 412
14. C.M. Sonsino, F. Muller, V. Arnhold, and G. Schlieper, Influence of Mechanical Surface Treatments on the Fatigue Properties of Sintered Steels under Constant and Variable Stress Loading, *Modern Developments in Powder Metallurgy*, Vol 21, Metal Powder Industries Federation, 1988, p 55-66
15. C.M. Sonsino, G. Schlieper, and W.J. Huppmann, How to Improve the Fatigue Properties of Sintered Steels by Combined Mechanical and Thermal Treatments, *Modern Developments in Powder Metallurgy*, Vol 16, Metal Powder Industries Federation, 1985, p 33-48
16. J.H. Lange, M.F. Amateau, N. Sonti, and R.A. Queeney, Rolling Contact Fatigue in Ausrolled 1%C 9310 Steel, *Inter. J. Fatigue*, Vol 16, 1994, p 281-286
17. H. Kuroki and Y. Tokunaga, Effect of Density and Pore Shape on Impact Properties of Sintered Iron, *Inter. J. Powder Met. Powder Tech.*, Vol 21, 1985, p 131-137
18. F.J. Esper and C.M. Sonsino, *Fatigue Design for PM Components*, European Powder Metallurgy Association, Shrewsbury, UK, 1994

19. K.D. Christian and R.M. German, Relation between Pore Structure and Fatigue Behavior in Sintered Iron-Copper-Carbon, *Inter. J. Powder Met.*, Vol 31, 1995, p 51-61

20. I. Bertilsson, B. Karlsson, and J. Wasen, Fatigue Properties of Sintered Steels, *Modern Developments in Powder Metallurgy*, Vol 16, Metal Powder Industries Federation, 1985, p 19-32

21. R.C. O'Brien, Impact and Fatigue Characterization of Selected Ferrous P/M Materials, *Progress in Powder Metallurgy*, Vol 43, Metal Powder Industries Federation, 1987, p 749-775

22. P.S. Dasgupta and R.A. Queeney, Fatigue Crack Growth Rates in a Porous Metal, *Inter. J. Fatigue*, Vol 3, 1980, p 113-117

23. T. Prucher, Fatigue Life as a Function of the Mean Free Path between Inclusions, *Modern Developments in Powder Metallurgy*, Vol 18, Metal Powder Industries Federation, 1988, p 143-154

24. K.D. Christian, R.M. German, and A.S. Paulson, Statistical Analysis of Density and Particle Size Influences on Microstructural and Fatigue Properties of a Ferrous Alloy, *Modern Developments in Powder Metallurgy*, Vol 21, Metal Powder Industries Federation, 1988, p 23-39

25. I.J. Mellanby and J.R. Moon, The Fatigue Properties of Heat-Treatable Low Alloy Powder Metallurgy Steels, *Modern Developments in Powder Metallurgy*, Vol 18, Metal Powder Industries Federation, 1988, p 183-195

26. J.T. Barnby, D.C. Ghosh, and K. Dinsdale, Fracture Resistance of a Range of Steels, *Powder Met.*, Vol 16, 1973, pp 55-71

27. E. Klar, D.F. Berry, P.K. Samal, J.J. Lewandowski, and J.D. Rigney, Fracture Toughness and Fatigue Crack Growth Response of Copper Infiltrated Steels, *Inter. J. Powder Met.*, Vol 31, 1995, p 316-324

28. R.A. Queeney, Fatigue and Fracture Response of Metal-Infiltrated Sintered Powder Metals, *Proceedings of ICM3*, Vol 3, Pergamon Press, Oxford, 1979, p 373-381

29. D.A. Gerard and D.A. Koss, The Influence of Porosity on Short Fatigue Crack Growth at Large Strain Amplitudes, *Inter. J. Fatigue*, Vol 13, 1991, p 345-352

30. P.C. Paris, *Fatigue—An Interdisciplinary Approach*, Syracuse University Press, 1964, p 107-117

31. W. Pompe, G. Leitner, K. Wetzig, G. Zies, and W. Grabner, Crack Propagation and Processes Near Crack Tip of Metallic Sintered Materials, *Powder Met.*, Vol 27, 1984, p 45-51

32. A.F. Kravic, The Fatigue Properties of Sintered Iron and Steel, *Inter. J. Powder Met.*, Vol 3 (No. 2), 1967, p 7-13

33. R.L. Dreshfield and R.V. Miner, Effects of Thermally Induced Porosity on an as-HIP Powder Metallurgy Superalloy, *Powder Met. Inter.*, Vol 12, 1980, p 83-87

34. F.H. Froes and C. Suryanarayana, Powder Processing of Titanium Alloys, *Reviews in Particulate Materials*, Vol 1, A. Bose, R.M. German, and A. Lawley, Ed., Metal Powder Industries Federation, 1993, p 223-276

35. J.M. Wheatley and G.C. Smith, The Fatigue Strength of Sintered Iron, *Powder Met.*, Vol 6, 1963, p 141-153

36. U. Engstrom, C. Lindberg, and I. Tengzelius, Powders and Processes for High Performance PM Steels, *Powder Met.*, Vol 35, 1992, p 67-72

Fatigue and Life Prediction of Gears

Darle W. Dudley

GEARS can fail in many different ways, and except for an increase in noise level and vibration, there is often no indication of difficulty until total failure occurs. In general, each type of failure leaves characteristic clues on gear teeth, and detailed examination often yields enough information to establish the cause of failure. The general types of failure modes (in decreasing order of frequency) include fatigue, impact fracture, wear, and stress rupture (Table 1). The leading causes of failure appear to be tooth-bending fatigue, tooth-bending impact, and abrasive tooth wear.

This article summarizes the various kinds of gear wear and failure and how gear life in service is estimated. Hopefully, gears are properly designed, made of good material, and accurately machined so that the life of the gear is adequate for the service intended. The manner in which gears wear in service and the various kinds of failure that may occur determine how gear life in service is estimated. In addition, the kinds of flaws in material that may lead to premature gear failure are discussed. Gear life is influenced primarily by geometric accuracy, gear-tooth contact conditions, and material condition or flaws. Metallurgical quality is just as important as gear-tooth geometric accuracy, and both must be under good control by those designing, making, or inspecting gears.

Gear Tooth Contact

The way in which tooth surfaces of properly aligned gears make contact with each other is responsible for the heavy loads that gears are able to carry. In theory, gear teeth make contact along lines or at points; in service, however, because of elastic deformation of the surfaces of loaded gear teeth, contact occurs along narrow bands or in small areas. The radius of curvature of the tooth profile has an effect on the amount of deformation and on the width of the resulting contact bands. Depending on gear size and loading, the width of the contact bands varies from about 0.38 mm (0.015 in.) for small, lightly loaded gears to about 5 mm (0.2 in.) for large, heavily loaded gears.

Gear tooth surfaces are not continuously active. Each part of the tooth surface is in action for only short periods of time. This continual shifting of the load to new areas of cool metal and cool oil makes it possible to load gear surfaces to stresses approaching the critical limit of the gear metal without failure of the lubricating film.

The maximum load that can be carried by gear teeth also depends on the velocity of sliding between the surfaces, because the heat generated varies with rate of sliding as well as with pressure. If both pressure and sliding speeds are excessive, the frictional heat developed can cause destruction of tooth surfaces. This pressure-velocity factor, therefore, has a critical influence on the probability of galling and scoring of gear teeth. The permissible value of this critical factor is influenced by gear metal, gear design, character of lubricant, and method of lubricant application.

Lubrication is accomplished on gear teeth by the formation of two types of oil films. The reaction film, also known as the boundary lubricant, is produced by physical adsorption and/or chemical reaction to form a desired film that is soft and easily sheared but difficult to penetrate or remove from the surface. The elastohydrodynamic film forms dynamically on the gear tooth surface as a function of the surface speed. This secondary film is very thin, has a very high shear strength, and is only slightly affected by compressive loads as long as constant temperature is maintained.

Certain rules about a lubricant should be remembered in designing gearing and analyzing failures of gears:

- Load is transferred from a gear tooth to its mating tooth through a pressurized oil film. If not, metal-to-metal contact may be detrimental.
- Increasing oil viscosity results in a thicker oil film (keeping load, speed, and temperature constant).
- Heat generation cannot be controlled above a certain maximum viscosity (for a given oil).
- Breakdown of the oil film will occur when the gear tooth surface-equilibrium temperature has reached a specific value.

The scuffing load limit of mating tooth surfaces is speed dependent. With increasing speed, the load required to be supported by the reaction film decreases, while the load that can be supported by the increasing elastohydrodynamic film increases. The result is a decreasing scuffing load limit to a certain speed as the reaction film decreases; then, as the speed picks up to where the elastohydrodynamic film increases, the scuffing load limit increases. This allows an increase in overall load-carrying capacity (assuming no change in temperature that would change viscosity).

- At constant speed, surface-equilibrium temperature increases as load increases, which lowers the scuffing load limit of the reaction film (surface-equilibrium temperature is attained when the heat dissipated from the oil is equal to the heat extracted by the oil).

Damage to and failures of gears can and do occur as a direct or indirect result of lubrication problems.

Spur and Bevel Gears. Spur gear teeth are cut straight across the face of the gear blank, and the mating teeth theoretically meet at a line of contact (Fig. 1a) parallel to the shaft. Straight teeth of bevel gears also make contact along a line (Fig. 1b) that, if extended, would pass through the point of intersection of the two shaft axes. As teeth on either spur or bevel gears pass through mesh, the line of contact sweeps across the face

Table 1 Failure modes of gears

Failure mode	Type of failure
Fatigue	Tooth bending, surface contact (pitting or spalling), rolling contact, thermal fatigue
Impact	Tooth bending, tooth shear, tooth chipping, case crushing, torsional shear
Wear	Abrasive, adhesive
Stress rupture	Internal, external

In an analysis of more than 1500 studies, the three most common failure modes, which together account for more than half the failures studied, are tooth-bending fatigue, tooth-bending impact, and abrasive tooth wear. Source: *Metals Handbook*, 9th ed., Vol 11, *Failure Analysis and Prevention*, ASM International, 1986

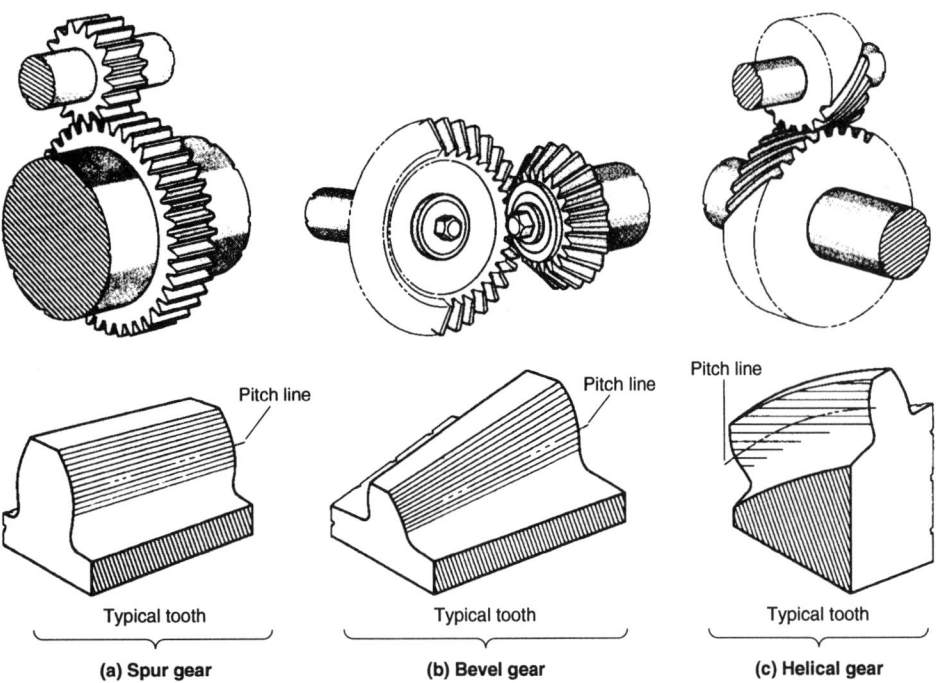

Fig. 1 Tooth contact lines on a spur gear (a), a bevel gear (b), and a low-angle helical gear (c). Lines on tooth faces of typical teeth are lines of contact.

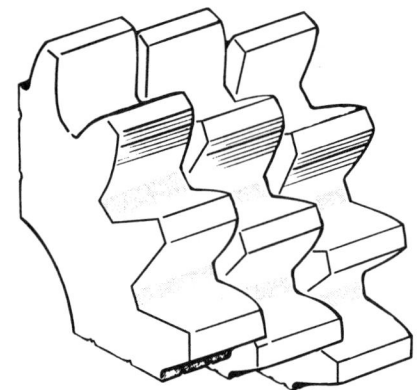

Fig. 2 Lines of contact on a stepped spur gear. The heavy line on a tooth face of each gear section represents the instantaneous line of contact for that section. This offset-contact pattern is typical for helical, spiral bevel, and hypoid gears. Lines on tooth faces are lines of contact.

of each tooth. On the driving tooth, it starts at the bottom and finishes at the tip. On the driven tooth, the line of contact starts at the tip and finishes at the bottom.

Helical, Spiral Bevel, and Hypoid Gears. Gear tooth contact on helical, spiral bevel, and hypoid gears is similar to that developed on a stepped spur gear (Fig. 2). Each section, or lamination, of the spur gear makes contact with its mating gear along a straight line; each line, because of the offset between sections, is slightly in advance of its adjacent predecessor. When innumerable laminations are combined into a smoothly twisted tooth, the short individual lines of contact blend into a smoothly slanted line (Fig. 1c) that extends from one side of the tooth face to the other and sweeps either upward or downward

as the tooth passes through mesh. This slanted-line contact occurs between the teeth of helical gears on parallel shafts, spiral bevel gears, and hypoid gears.

The load pattern is a line contact extending at a bias across the tooth profile, moving from one end of the contact area to the other end. Under load, this line assumes an elliptical shape and thus distributes the stress over a larger area. Also, the purpose of spiral bevel gearing is to relieve stress concentrations by having more than one tooth enmeshed at all times.

The greater the angle of the helix or spiral, the greater the number of teeth that mesh simultaneously and share the load. With increased angularity, the length of the slanted contact line on each tooth is shortened, and shorter but more steeply slanted lines of contact sweep across the faces of several teeth simultaneously. The total length of these lines of contact is greater than the length of the single line of contact between straight spur-gear teeth of the same width. Consequently, the load on these gears is distributed not only over more than one tooth but also over a greater total length of line of contact. On the other hand, the increased angularity of the teeth increases the axial thrust load and thus increases the loading on each tooth. These two factors counterbalance each other; therefore, if the power transmitted is the same, the average unit loading remains about the same.

Helical gears on crossed shafts make tooth contact only at a point. As the teeth pass through mesh, this point of contact advances from below the pitchline of the driving tooth diagonally across the face of the tooth to its top, and from the top of the driven tooth diagonally across its face to a point below the pitchline. Even with several

teeth in mesh simultaneously, this point contact does not provide sufficient area to carry an appreciable load. For this reason, helical gears at angles are usually used to transmit motion where very little power is involved.

Worm Gears. In a single-enveloping worm gear set, in which the worm is cylindrical in shape, several teeth may be in mesh at the same time, but only one tooth at a time is fully engaged. The point (or points) of contact in this type of gear set constitutes too small an area to carry an appreciable load without destruction of the metal surface. As a result, single-enveloping worm gear sets are used in applications similar to those for helical gears on crossed shafts: to transmit motion where little power is involved.

Considerable power must be transmitted by commercial worm gear sets; therefore, the gears of these sets are throated to provide a greatly increased area of contact surface. The gear tooth theoretically makes contact with the worm thread along a line curved diagonally across the gear tooth. The exact curve and slant depend on tooth design and on the number of threads on the worm relative to the number of teeth on the gear. Usually, two or more threads of the worm are in mesh at the same time, and there is a separate line of contact on each meshing tooth. As meshing proceeds, these lines of contact move inward on the gear teeth and outward on the worm threads. To secure smooth operation from a gear of this type, the teeth of the gear and sometimes the threads of the worm are usually altered from theoretically correct standard tooth forms. These alterations result in slightly wider bands of contact, thus increasing the load-carrying capacity of the unit. Load-carrying capacity also depends on the number of teeth in simultaneous contact. Exact tooth design varies from one manufacturer to another, and for this reason, the patterns of contact also vary.

In a double-enveloping worm gear set, the worm is constructed so that it resembles an hourglass in profile. Such a worm partly envelopes the gear, and its threads engage the teeth of the gear throughout the entire length of the worm. The teeth of both the worm and the gear have straight-sided profiles like those of rack teeth, and in the central plane of the gear, they mesh fully along the entire length of the worm.

The exact pattern of contact in double-enveloping (or double-throated) worm gear sets is somewhat controversial and seems to vary with gear design and with method of gear manufacture. It is generally agreed, however, that contact is entirely by sliding with no rolling and that radial contact occurs simultaneously over the full depth of all the worm teeth.

Operating Loads. Gears and gear drives cover the range of power transmission from fractional-horsepower applications, such as hand tools and kitchen utensils, to applications involving thousands of horsepower, such as heavy machinery and marine drives. However, neither the horsepower rating nor the size of a gear is necessarily indicative of the severity of the loading it can withstand. For example, the severity of tooth

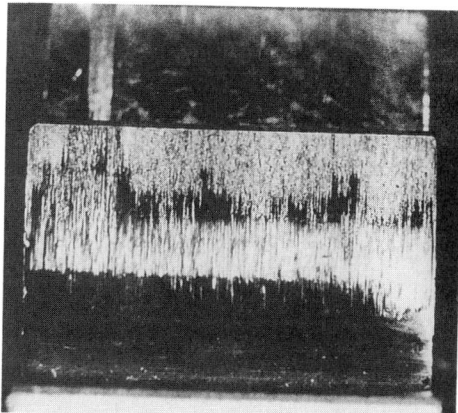

Fig. 3 An example of gear tooth scuffing. Note radial scratch lines.

Fig. 4 Pitted gear teeth. Note micropitting at the pitch line, scattered macropits, and one area of gross pitting near the left end.

loading in the gear train of a 186 W (¼ hp) hand drill may exceed that of the loading in a 15 MW (20,000 hp) marine drive. Factors other than horsepower rating and severity of loading can affect gear strength and durability, particularly duration of loading, operating speed, transient loading, and environmental factors such as temperature and lubrication.

Gear Tooth Surface Durability and Breakage

Metal gears are normally lubricated with an oil or a grease. Some small, nonmetallic gears have limited capability to run without lubrication, due to the self-lubricating properties of the material.

When metal gears are adequately lubricated with a clean lubricant, there is very little abrasive wear of the contacting tooth surfaces. However, the tooth surface may be quickly damaged by scuffing if (in combination with poor surface fin-

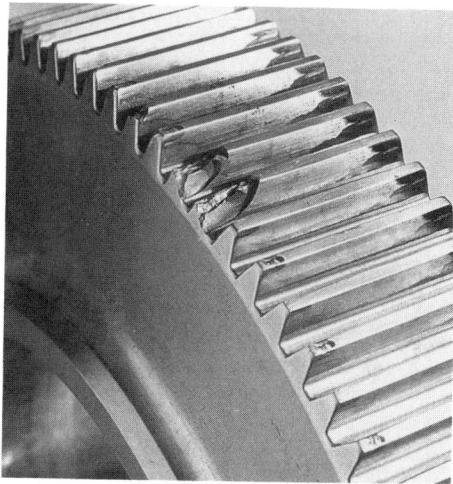

Fig. 5 Misaligned gear has tooth breakage at ends of teeth.

ish and low-viscosity lubricant) the load intensity, temperature, and rubbing speed are too high. Scuffing consists of small radial tears in the tooth surface. It is most apt to happen where the tip of one tooth is contacting the lower flank of a mating tooth (Fig. 3).

Pitting may occur after some millions or billions of tooth contacts. In general, pitting almost never occurs before 10,000 contacts. Pitting is a fatigue failure where small cracks form in the tooth surface and then grow to the point where small, round bits of metal break out of the tooth surface (Fig. 4).

Tooth breakage (such as in Fig. 5) is normally the situation where a crack starts in the root fillet, below the contacting surface, and then the crack grows such that a whole tooth breaks off in a cantilever-beam-type of failure. In wide-face gears, one end of the tooth may break off, leaving the tooth intact for the rest of the face width.

Traditionally, the gear designer first determines a pitch diameter and a face width for the pinion that are large enough for the pinion to last for the required service life with a probability of failure of no more than 1 in 100. This determination is based on a possible pitting fatigue failure. It is assumed in the beginning that the surface finish, the tooth accuracy, the lubrication, and the needed profile and helix modifications will all be carried out well enough to avoid any serious risk of scuffing.

Normally, the pinion is more apt to fail in pitting than the gear, so the sizing of the pinion tends to determine the needed size of the gear. (The gear pitch diameter equals the pinion pitch diameter multiplied by the ratio. If the gear has four times as many teeth as the mating pinion, then the pitch diameter of the gear is four times greater than the pinion pitch diameter.)

After the pitch diameter of the pinion has been determined, the size of the teeth are determined by calculations regarding a possible failure in tooth breakage. A broken tooth tends to be catastrophic to a gear unit, so the designer usually makes the teeth large enough so that they are

definitely less apt to fail in breakage mode than in a pitting mode. This makes the design life of a gear unit primarily dependent on its surface fatigue capacity (pitting resistance) rather than on its cantilever beam capacity (capacity to resist tooth breakage).

Life Determined by Contact Stress

The surface contact stress, S_c, between gear teeth (discussed further in Ref 1 with in-depth presentations of gear rating calculations and concepts) may be calculated by a simplified formula:

$$S_c \text{ (in psi)} = C_k (KC_d)^{0.5} \qquad \text{(Eq 1)}$$

where C_k is the geometry factor for durability (see Table 2), K is the index of tooth loading severity for pitting (see Eq 2), and C_d is the overall derating factor for durability (see Table 3).

The index of tooth loading for pitting is called the "K-factor." For spur or helical gears, the K-factor is:

$$K = \frac{W_t}{Fd}\left(\frac{m_G + 1}{m_G}\right) \qquad \text{(Eq 2)}$$

Table 2 Some typical geometry factors (C_k) for durability

No. of pinion teeth	No. of gear teeth	Standard addendum, $a_p = 1.000, a_G = 1.000$, whole depth 2.35 in.	25% long addendum, $a_p = 1.250, a_G = 0.750$, whole depth 2.35 in.
Spur gears, 20° pressure angle			
25	35	5913	(a)
25	50	5985	5756
25	85	6057	5768
35	35	5756	(a)
35	50	5810	5722
35	85	5877	5726
35	275	5925	5732
Spur gears, 25° pressure angle			
20	35	5564	5371
20	50	5660	5419
20	85	5744	5479
25	35	5419	(a)
25	50	5503	5335
25	85	5575	5380
35	35	5262	(a)
35	50	5329	5269
35	85	5407	5298
35	275	5479	5323
Helical gears, 15° helix angle, 20° normal pressure angle			
25	35	4576	(a)
25	50	4540	4564
25	85	4479	4533
35	35	4516	(a)
35	50	4492	4499
35	85	4419	4469
Helical gears, 15° helix angle, 25° normal pressure angle			
25	35	4444	(a)
25	50	4419	(a)
25	85	4383	(a)
35	35	4395	(a)
35	50	4371	(a)
35	85	4335	(a)

(a) 25% long addendum design not used here. a_p, pinion addendum; a_G, gear addendum

where W_t is the tangential driving force (in pounds), F is the net face width in contact (in inches), d is the pinion pitch diameter (in inches), and m_G is the ratio of gear teeth to pinion teeth.

The tangential driving force is obtained from the horsepower being transmitted by the combination of pinion and gear, the speed of the pinion in revolutions per minute, and the pitch diameter of the pinion in inches. Thus:

$$W_t = \frac{P \times 126{,}050}{n_p \times d} \text{ lb} \qquad \text{(Eq 3)}$$

where P is the horsepower transmitted and n_p is the speed of the pinion (in rpm).

Example 1: Surface Durability of Gear. To show how the above equations work, consider a practical problem of an electric motor (running at a design speed of 1800 rpm) that drives a fan at 450 rpm and transmits power at a constant 30 hp. If the pinion driven by the motor has 22 teeth at a 15° helix angle and the normal diametral pitch of the teeth is 10, what are the calculated K-factor and contact stress when the mating gear has a face width 1.0 times the pinion pitch diameter? The normal pressure angle is 20°. The tooth design is 25% extra addendum of the pinion and 25% shorter addendum for the gear.

The first step is to obtain the pinion pitch diameter:

$$d = \frac{N_p}{p_n \cos \psi} \text{ in.} \qquad \text{(Eq 4)}$$

where N_p is the number of pinion teeth, ψ is the helix angle, and P_n is the normal diametral pitch. For the problem described above, then:

$$d = \frac{22}{10 \cos (15°)} = 2.2776 \text{ in.}$$

$$F = 1.0 \times 2.2776 = 2.28 \text{ in.}$$

$$W_t = \frac{30 \times 126{,}050}{1800 \times 2.2776} = 922.4 \text{ lb}$$

$$K = \frac{922.4}{2.2776 \times 2.28} \frac{(4.0 + 1)}{(4.0)} = 222.0$$

$$S_c = 4558 (222.0 \times 2.5)^{0.5}$$

$$= 107{,}380 \text{ psi}$$

Table 2 does not have 22/88 tooth numbers, so in calculating S_c, C_k was determined from Ref 2.

After the contact stress has been calculated, a rating curve sheet needs to be used to determine what life can be expected. Figure 6 shows typical contact stress values plotted against numbers of tooth contacts. This curve sheet is drawn for gearing having these qualifications:

- Material quality: Normal industry quality (Grade 1 per Ref 2)
- Probability of failure: Not over 1 in 100

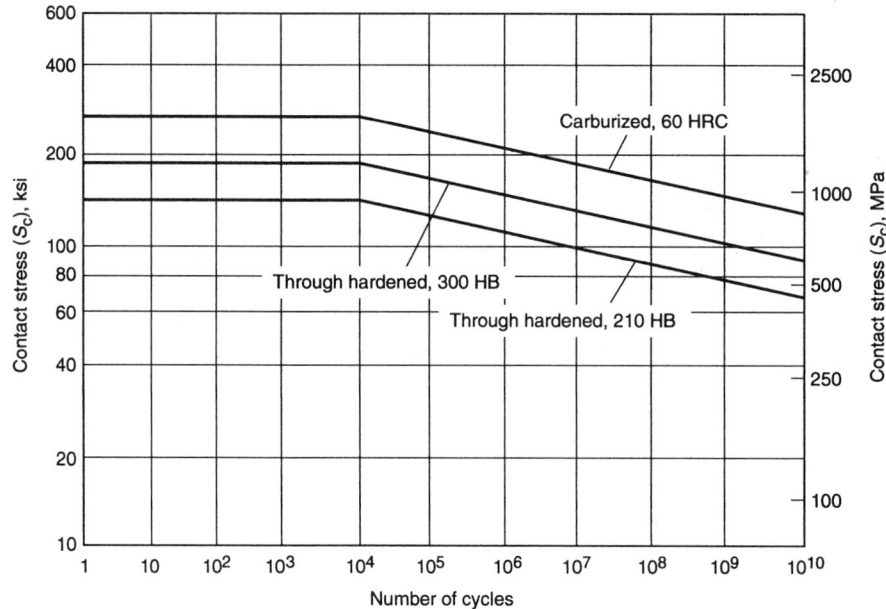

Fig. 6 Surface durability curve for gear life rating (contact stress vs. cycles) for normal industry quality material (Grade 1 per Ref 2)

Table 3 Some typical overall derating factors (C_d)

Aspect ratio, d/F	Accuracy level by AGMA			
	12 or 13	10 or 11	8 or 9	6 or 7
For designs where pinion pitch diameter is about 2 in. (50 mm)				
0.25 or less	1.50	1.65	1.80	2.00
0.50 or less	1.65	1.75	2.00	2.25
0.75 or less	1.75	2.00	2.25	2.50
1.0 or less	2.00	2.25	2.50	2.75
For designs where pinion pitch diameter is about 5.0 in. (125 mm)				
0.25 or less	1.45	1.55	1.65	1.80
0.50 or less	1.55	1.65	1.80	2.00
0.75 or less	1.65	1.80	2.00	2.25
1.0 or less	1.80	2.00	2.25	2.50
For designs where pinion pitch diameter is about 12 in. (300 mm)				
0.25 or less	1.40	1.50	1.60	1.70
0.50 or less	1.50	1.60	1.70	1.80
0.75 or less	1.60	1.70	1.80	2.00
1.0 or less	1.70	1.80	2.00	2.25

Note: These derating factors should be considered illustrative only. For a prime mover that is rough running and a driven device that is rough running, the derating needs to be more than shown above. Likewise, for a very smooth-running prime mover and a low-inertia, smooth-running driven device, lower derating factors may be justified. Field experience in vehicles, aircraft, and ships tends to be the best guide for choosing the right derating factor.

- Lubrication: Full oil film, which is regime III
- Hardness: Both pinion and mating gear at least as hard as value shown on surface durability figure

Determination of Gear Life. To explain how gear life is determined, the discussion can be based on the same sample problem. The calculated contact stress was 107,380 psi. From Fig. 6 the following life values are obtained for steel gears (through hardened):

- 210 HB: 1.1×10^6 cycles
- 300 HB: 2.0×10^8 cycles
- 60 HRC: over 1.0×10^{10} cycles

The next step is to calculate the number of pinion cycles for some different lengths of time at full rated load, where:

$$\text{Cycles} = (\text{rpm}) \times (60 \text{ min/h}) \times (\text{no. of hours})$$
$$= 1800 \times 60 \times \text{no. of hours}$$

Taking some different numbers of hours, the following results are obtained:

Time at rated power, h	No. of pinion cycles
500	54×10^6
2000	216×10^6
20,000	2.16×10^9

Table 4 Situations and concepts of pitting failures in gears

Description	Recommended action
Situation 1	
Some micropitting and a few macropits on the pinion. No gear pitting.	Scrap the pinion, even though it is still quite runnable. It may not last until the next scheduled overhaul.
Situation 2	
Micropitting and appreciable macropitting (e.g., 3-10% of the tooth surface of the pinion). Almost no pitting on gear.	Scrap the pinion, because metal debris in the oil system is a serious hazard to the bearings. Premature bearing failure may occur at any time.
Situation 3	
Micropitting and considerable macropitting (e.g., 15-40% of the surface). One or more gross pits. Damage to both pinion and gear.	Scrap the gears. Damage to the tooth surface may lead to tooth breakage, starting in a badly pitted location.
Situation 4	
Macropitting over 50-100% of the pinion tooth surface. Gross pitting present. Removal of metal thins the teeth and disrupts load sharing between teeth. Gear unit has greatly increased noise and vibration.	Scrap the gears. Complete failure is imminent. Foundation bolts may break. Bearings are probably in the process of failure.
Situation 5	
Macropitting all over the teeth. Also considerable gross pitting. Teeth are thinned so much by wear that the tips are becoming sharp like a knife.	Scrap the gears. A thinned tooth may roll over and jam the drive—if it doesn't break off first.

Determination of adequate gear life requires some reasoning. Inexpensive equipment that is used infrequently and is seldom used at full rated power might be satisfactory if it had only a 500-hour capability at full rating. The design under consideration would be quite inadequate for 500 hours if the teeth were at low hardness, such as 210 HB. However, if the pinion and gear were at 300 HB hardness, they would have enough capacity for 500 hours and almost enough for 2000 hours.

At full hardness of 60 HRC, the situation changes considerably. The calculated life capacity is well over 10 billion cycles (on the pinion). Only a little over 2 billion cycles are needed for a life of 20,000 hours. The gear has only one-quarter of the pinion cycles, which is just a little over 0.5 billion cycles. If the gear hardness is the same as that of the pinion, the gear certainly has adequate capacity.

Derating Factors. In the above calculation, the derating factor (C_d) of 2.50 may seem high but can be explained. The values given in Table 3 are based on an assumption of *average* conditions and a failure from pitting conditions of situation 1, 2, or 3 (see Table 4 and later discussions). A derating factor lower than 2.0 may even be acceptable for a 2 in. (50 mm) pinion if the driving and driven units are very light (low mass) and run *very* smoothly. When a low derating factor such as 1.5 is used, there is usually fairly extensive pitting. If the concept of a "failure" is situation 4 or 5 (Table 4), field experience may show that the gears can survive even though the derating is too low for situations such as situations 1, 2, and 3 in Table 4.

If a derating factor of 1.50 had been used, the calculated stress at 10 billion cycles would be reduced by the square root of the change in the derating factor. This works out to $(1.50/2.50)^{0.50}$ = 0.775. This makes the calculated contact stress drop to 83,220 psi. At this value, a through-hard-

ened steel at 300 HB has a life of over 10 billion cycles. This is enough for a life of 20,000 hours.

Now consider the variables for the 2.50 derating factor. The first of three major factors is a derating factor of 1.4 for typical experience with industrial fan drives. These run somewhat rough, and gear people evaluate this as a 1.4 application factor.

With a face width equal to the pitch diameter, there will normally be some misfit across the wide face. In addition, the accuracy of tooth spacing will not be close to perfect. For an industrial application, the tooth spacing accuracy is probably AGMA quality level 9. This requires a derating factor of about 1.4 for small gearing such as a 2 in. (50 mm) pinion.

It is also necessary to consider dynamic overload due to the masses in the drive system and the tooth spacing accuracy. In this sample problem, the 2.125 pitch diameter pinion running at 1800 rpm has a pitch line velocity of 1073 ft/min (327 m/min). This kind of drive at this speed could be expected to have a dynamic overload factor of 1.00/0.80 = 1.25.

The overall derating factor is obtained by multiplying these three factors such that 1.4 × 1.4 × 1.25 = 2.45 (this is rounded off to 2.50).

For the values shown in Fig. 6, the geometry factor (C_k) may look quite conservative when compared to values used in bearing design. Part of this is due to the relatively high component of sliding in gear tooth contacts. For spur and helical gearing, pure rolling occurs only at the pitch line. Above and below the pitch line there is appreciable sliding. Also, the geometric shape of bearings is simpler than that of toothed gears. This helps in both quality of fit and quality control of material.

Another factor in comparing rolling element bearings to gears is probability of failure. Bearings are considered somewhat expendable and are generally designed with a probability of failure of 10 in 100 rather than the 1 in 100 that is

normal for gearing. In some cases of expendable gearing, a factor of 10 failures in 100 is used for the rating curve. If Fig. 6 was redrawn for the 10 to 100 probability, all curves shown would allow about 17% more stress. The approximate life for a stress of 107,380 psi (740 MPa) works out as follows:

Hardness, HB	Approximate life at probability of failure of:	
	1 out of 100	10 out of 100
210	1.1 million cycles	20 million cycles
300	180 million cycles	3.5 billion cycles

Pitting Failure. The degree of risk to be assumed in a gear design is a most important consideration. The above results show more than a 15 times longer life by going from a very low risk of failure to a still somewhat low risk of failure. If a 10 out of 100 degree of risk was tolerable, then the gear set being considered would have adequate pitting life for 20,000 hours at 300 HB.

The concept and definition for contact stress of pitting failures is another very significant consideration. Pitting failure is not easy to define, and a gear designer needs to use judgment based on experience. In general, a *pitted* gear tooth will still run, whereas broken-off gear teeth will so damage a gear that it is either impossible or impractical to continue to operate the gear unit.

Table 4 shows five different levels of tooth damage due to pitting that govern whether or not a pinion and a mating gear have failed by pitting. In aircraft or space vehicle gearing, situation 1 is cause for failure. In long-life turbine gearing, situation 2 is quite apt to be considered a failure. According to the calculated life given earlier in this article, the gear failed when it reached situation 3.

In some factory and farm applications, gears that have reached situation 4 are still kept in service. The cost of replacement and a forgiving environment where noise and vibration can be tolerated may prompt the gear user to keep running the gears. Situation 5 is the ultimate. If the pinion won't turn under power or the pinion does turn and the gear does not turn, there is just no question about the gears having failed!

From what has been discussed so far, the method used in the gear trade to calculate gear life should be quite clear. It should be evident, though, that much experience and judgment are needed to properly decide when a pitted and worn gear should be taken out of service and scrapped.

Although there are good guidelines for evaluating the things that derate gears, it still takes much data and experience to set numerical values that are just right for a particular product application. For instance, long-life turbine gearing is very expensive, and downtime in rebuilding a broken main gear drive may be much more expensive than replacing a gear unit. This means that the designer of turbine gearing is very concerned about having an adequate overall derating factor and close control over gear material quality. Grade 2 material is often specified rather than grade 1 material. Grade 2 steel is rated to carry about 25% more contact stress than grade 1 when carburized but only about 10% more stress when through hardened.

In the vehicle field (trucks, tractors, and automobiles), gearing is not highly expensive and a replacement gear unit can often be installed within a few hours. The designer of the gears may use a relatively low overall derating. The designer learns by thousands of similar units in service how to set the derating factor and how to control geometric quality and material quality so that adequate gear life is obtained.

The standards of the American Gear Manufacturers Association (AGMA) give recommendations for how to rate many kinds of gears and how to specify geometric and material quality. AGMA standards define established gearing practices for the gear trade. In the industrial field, contracts for geared machinery generally specify that the gearing supplied must be in accordance with applicable AGMA standards.

The allowable contact stress for gears made of grade 1 material (per Ref 2) is drawn for the lubrication condition of regime III (full oil film). Many gears that are small and run at relatively slow pitch line speeds are in regime II (partial oil film) or regime I (wet with oil but with almost no elastohydrodynamic (EHD) oil film thickness).

As of December 1995, there is no national standard for how to rate gears for allowable contact stress when they are in regimes II or I. There is much field experience, though, with gears running with an inadequate EHD oil film thickness. Work is underway to add material to AGMA standards to cover the problem of an inadequate oil film thickness.

Figure 7 shows how the allowable contact stress may change for grade 1 carburized steel gears at 60 HRC. Note the dashed curves for regimes II and I.

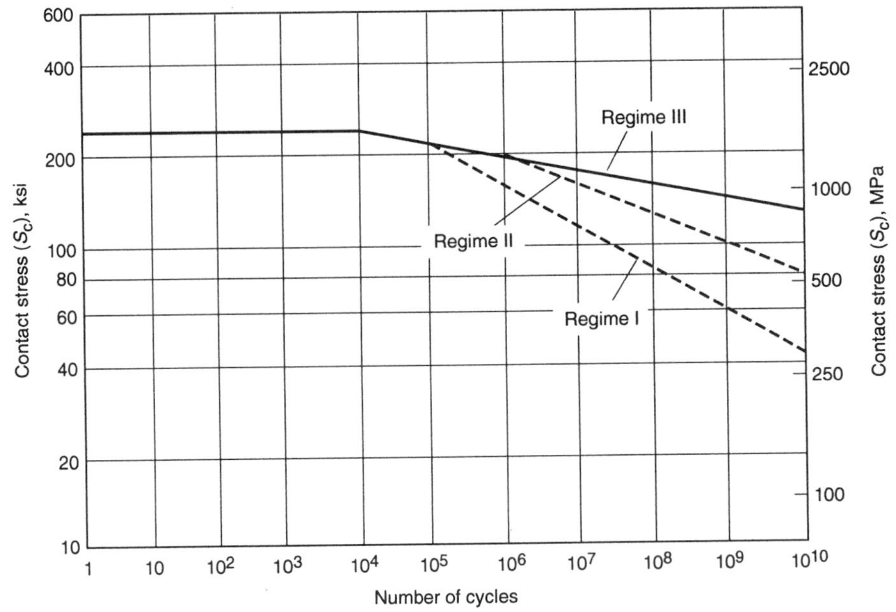

Life Determined by Bending Stress

The bending stress of a pinion or a mating gear tooth (Ref 1, 2) may be estimated by a simplified formula:

$$S_t = K_t U_l K_d \qquad \text{(Eq 5)}$$

where K_t is a geometry factor (see Table 5), U_l is the index of tooth loading severity for breakage (see Eq 6 below), and K_d is the overall derating factor for bending stress (see footnote in Table 3). The bending stress derating factor (K_d) is approximately the same as the overall derating factor (C_d) unless there is a wide face width. When the aspect ratio is around 0.25, the C_d for contact stress and the K_d for bending stress are about equal. At an aspect ratio of 1.0, K_d may be about 80% of C_d.

The index of tooth loading for breakage is often called "unit load." Its value for spur or helical teeth is:

$$U_l = \frac{W_t P_n}{F} \qquad \text{(Eq 6)}$$

Fig. 7 Regimes of lubrication for carburized gears of normal industry quality material (Grade 1 material, per Ref 2)

Table 5 Some typical geometry factors (K_t) for strength

No. of pinion teeth	No. of gear teeth	Standard addendum $a_p = 1.000$, whole depth 2.35	$a_G = 1.000$,	25% long addendum $a_p = 1.250$, whole depth 2.35	$a_G = 0.750$,
Spur gears, 20° pressure angle					
25	35	2.72	2.55	2.37	3.00
25	50	2.68	2.42	2.33	2.74
25	85	2.62	2.22	2.30	2.54
35	35	2.50	2.50	2.21	2.90
35	50	2.42	2.30	2.18	2.70
35	85	2.36	2.15	2.15	2.54
35	275	2.20	2.00	2.10	2.45
Spur gears, 25° pressure angle					
20	35	2.42	2.16	2.08	2.42
20	50	2.38	2.01	2.05	2.22
20	85	2.35	1.90	2.03	2.09
25	35	2.26	2.10	2.00	2.40
25	50	2.23	1.97	1.95	2.20
25	85	2.20	1.86	1.91	2.05
35	35	2.06	2.06	1.88	2.34
35	50	2.02	1.92	1.84	2.15
35	85	1.99	1.80	1.80	2.00
35	275	1.90	1.70	1.75	1.82
Helical gears, 15° helix angle, 20° normal pressure angle					
25	35	1.93	1.83	1.74	2.05
25	50	1.90	1.76	1.72	1.88
25	85	1.87	1.67	1.71	1.75
35	35	1.82	1.82	1.68	1.97
35	50	1.78	1.73	1.65	1.82
35	85	1.75	1.64	1.62	1.73
Helical gears, 15° helix angle, 25° normal pressure angle					
25	35	1.60	1.52	...	...
25	50	1.57	1.47	...	...
25	85	1.54	1.40	...	...
35	35	1.50	1.50	...	...
35	50	1.42	1.42	...	...
35	85	1.35	1.34	...	...

where W_t is the tangential driving force (same value as in Eq 3), F is the net face width in contact, and P_n is the normal diametral pitch (1/in.).

P_n is the inverse of the tooth size in a plane normal to a helical tooth. The diametral pitch (P_d) is the inverse of the tooth size in a plane normal

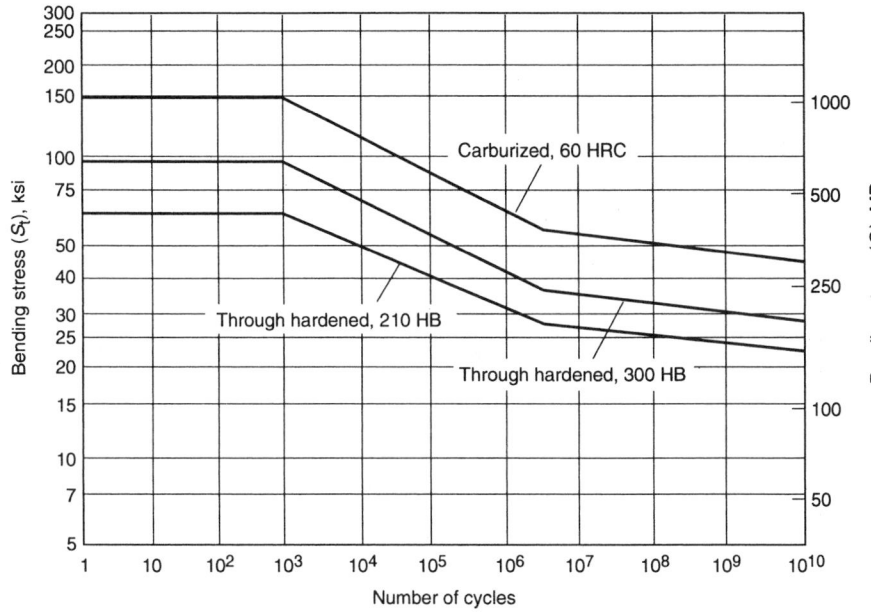

Fig. 8 Bending strength curve for gear life rating of normal industry quality material (Grade 1 per Ref 2)

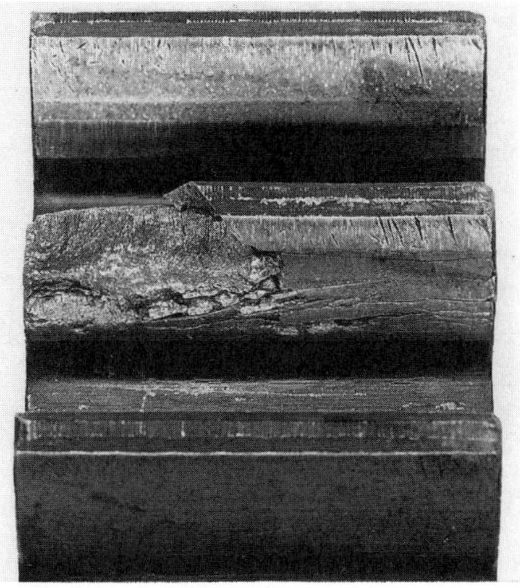

Fig. 9 End of tooth broken because of gross pitting. Note cracks in remaining part of tooth.

to the axis of gear. The relation between the two is:

$$P_d = P_n \cos (\text{helix angle}, \psi) \qquad (\text{Eq 7})$$

The tangential load for tooth strength is the same quantity as the tangential load used for contact stress calculations (Eq 3).

The derating factor for strength is sometimes taken as the same value as is used for surface durability. In some cases of gears with favorable contact ratios, there may be justification to use a somewhat lower derating factor for strength. The points of high beam stress are some distance away from the points of maximum contact stress. This tends to allow the beam stress to spread out more than the contact stress. (In this brief article, it is not practical to show how to reduce the derating factor for the tooth strength calculations.) For the sample problem below it is reasonable to use $K_d = 2.0$. (The pitting derating, C_d, was 2.5.)

Example 2: Rating Gear Life with Bending Stress. In Example 1 of this article, the numerical values are $W_t = 922.4$ lb, $P_n = 10$, and $F = 2.28$. For the data just given, the unit load index is:

$$U_l = \frac{922.4 \times 10}{2.28} = 4045.4 \text{ psi (27.9 MPa)}$$

The geometry factor is taken as 1.75 (see Table 5). The calculated bending stress becomes:

$$S_t = 1.75 \times 4045.4 \times 2.0 = 14{,}159 \text{ psi (97.6 MPa)}$$

The tooth bending stress can now be used to determine allowable cycles from a rating curve, such as Fig. 8. Figure 8 shows typical bending stresses plotted against tooth contacts. This curve sheet is drawn for gearing having these qualifications:

- Material quality: Grade 1
- Probability of failure: Not over 1 in 100
- Lubrication: Full oil film, which is regime III
- Hardness: Both pinion and mating gear at least as hard as value shown on bending strength figure (Fig. 8)

The expected life of both the pinion and the gear, from a tooth breakage standpoint, can now be determined for the calculated stress of 14,159 psi (97.6 MPa) for the pinion and a calculated stress of 14,321 psi (98.7 MPa) for the gear. (The gear has the same unit load as the pinion and the same derating factor. However, the geometry factor for the gear from Ref 1 is 1.77, compared to 1.75 for the pinion.)

Based on Fig. 8, the pinion stress of 14,159 psi (97.6 MPa) is well below all the curves. If a pinion life of 20,000 hours was desired, the bending stress allowable could be about 23,000 psi (158.5 MPa) even for the through-hardened, 210 HB steel. This means that with 22 pinion teeth, the "life" of the pinion is determined by pitting and not by tooth breakage.

Suppose the pinion had the same diameter but had 40 teeth instead of 22 teeth. The bending stress would increase, because the normal diametral pitch would be about 18. The derating would not change for an increased number of teeth, and the geometry factor would tend to decrease a small amount. This means that a 40-tooth pinion meshing with a 160-tooth gear would have a calculated bending stress of about 22,300 psi (153.8 MPa).

If the normal diametral pitch was changed to 18.18182, the pitch diameters of the pinion and gear would not change with 40 pinion teeth and a 15° helix. The K-factor for pitting would not change, and the change in the geometry factor for durability would be a small decrease. This means

that it is possible to change numbers of teeth to adjust bending stress and not have much change in the contact stress.

In this sample problem, "life" determined by contact stress is very short at 210 HB, but it is almost enough for 2000 hours at 300 HB. The pinion could have 40 teeth at 300 HB because the allowable bending stress is up to 32,000 psi (220 MPa) at 2000 hours (216×10^6 million cycles). With the bending stress at 24,274 psi (167.4 MPa) there is a modest amount of extra margin in bending strength, but not nearly as much as at 22 teeth.

From a practical standpoint, it is quite desirable to control the life of a gear set and have an appreciable extra margin of bending strength. Having lots of small teeth instead of fewer and larger teeth is not desirable from two other standpoints: in general, parts with fewer teeth cost less to make; and larger teeth can stand somewhat more wear.

In summary, this example shows that 22 teeth on a pinion is a reasonable design. There is no good reason to use 40 teeth, even though the bending stress would be allowable.

Gear Tooth Failure by Breakage after Pitting

The preceding section covered tooth breakage by the tensile stress in the root fillet. Another kind of common tooth breakage in gearing starts somewhat high on the tooth profile rather than in the root fillet. When a gear is damaged by pitting, a crack can start at a pit, spread lengthwise, and go all the way through the tooth. On wider-face-width gears, a triangular piece of tooth often breaks out. This piece may be as wide as one-third or one-half of the face width. The break may progress as deep as the root fillet at one end and run out at the outside diameter at the other end. The "eye" where the crack started may be at the

Fig. 10 Helical gear with serious ledge wear due to micro- and macropitting

pitch line, or it may be in the lower flank of the tooth.

If a pinion or gear continues in service after an appreciable part of the tooth is gone, the now-overloaded remainder of the tooth is likely to break off. Sometimes, even though several teeth have portions broken off, the gears continue in service because the increased noise and vibration are not enough to attract attention. Then one or more teeth are completely removed by successive chunks breaking off (see Fig. 9).

In narrow-face-width gears, such as those where the face width is about equal to the tooth height, a break starting at a pit tends to remove the tooth across the whole face width.

The normal calculation of bending stresses is not useful in designing gears to avoid tooth breakage starting at pits. The logic of this kind of trouble must be considered *an incident of tooth breakage caused by pitting.*

Gears that are case hardened by carburizing are the most susceptible to this trouble. Gears with a nitrided case are not quite as susceptible because the nitrided case depths tend to be thinner. A pit in a thin case is not quite as great a stress riser as a pit in a deep case. To prevent pitting altogether, though, deep carburized case depths may be needed.

Through-hardened gears in the 300 to 400 HB range are also quite critical. In general, gears are not used that are around 500 HB. At 450 HB and higher, a through-hardened gear is usually so brittle that it will fracture with even a small amount of surface distress.

The concepts of pitting failure given in Table 4 are quite useful in regard to the risk of tooth breakage stemming from pitting. In situation 1, the pitting is relatively minor, so there is very little danger of tooth breakage starting at a pit.

In situation 2, there is a moderate risk of some area on the tooth surface having enough macropits to start a tooth breakage crack when the pinion or gear has a carburized case. Through-hardened parts at 300 HB are not so notch sensitive and will probably not develop cracks from pits. Through-hardened parts at 210 HB are not notch sensitive. There is not much risk of cracks from pits at this level of hardness.

In situation 3, the much heavier pitting is quite risky for carburized gear teeth and moderately risky for through-hardened gears at 300 HB. However, through-hardened gears at 210 HB do not have much risk.

In some special cases where the pinion is quite hard (e.g., 360 HB) and the gear is much lower in hardness (e.g., 300 HB), there may be such serious *ledge wear* on the gear due to micro- and macropitting that a whole layer of metal is worn away from the gear tooth. This can result in the addendum of the gear being forced to carry all the load. This tends to double the stress in the gear root fillet, and gear teeth then break off due to the greatly increased stress.

Ledge wear that occurs in helical gears over a long period of time is not too dangerous if both the pinion and gear have about the same amount of wear. The big danger of tooth breakage comes when one pitted gear has little or no ledge wear and the mating gear has serious ledge wear. Figure 10 shows an example of a gear tooth with ledge wear and in danger of breaking. (The pinion that mated with this gear had some pitting but essentially no ledge wear.)

In situations 4 and 5, hard gears cannot survive without tooth breakage. Low-hardness gears that run at relatively slow pitch line speeds often survive for years without tooth breakage, even though the pitting and wear are severe.

In summary, tooth breakage due to pitting is a high risk with hard gears and may not be much of a risk if the gears are of low hardness and not running fast.

Flaws and Gear Life

Metallurgical flaws may seriously reduce the life of gears. The calculations for the life of gear teeth assume that the material in the gear is not flawed by a mistake in processing the part. If the material is grade 1 (Ref 2), the quality is not as good as for grades 2 or 3, but it is assumed that the quality is normal for the grade specified. For instance, the microstructure for grade 1 is not specified to a limit as close as that for grade 2. Also, the cleanliness of the steel in grade 1 does not have to be as good as the cleanliness for grades 2 or 3.

Some of the more common flaws that affect the life of gear teeth are:

- Case depth for carburized, nitrided, or induction-hardened teeth too thin or too thick
- Grinding burns on case-hardened gears
- Nonhomogenous structure, voids, hard spots, soft spots, etc.
- Alloy segregation
- Core hardness too low near the root diameter or at the middle of the face width
- Forging flow lines pronounced and in the wrong direction
- Quenching cracks
- Composition of steel not within specification limits

When gears fail prematurely, the normal first step of an investigator is to make some calculations of gear life to see if the gear unit size was large enough for the rated life. If these calculations show that a gear unit rated for something like 10 years of life was indeed large enough and had the right materials specified, then it becomes a matter of concern to investigate factors such as lubrication system, behavior of driving and driven equipment, operating environment, and material quality (perhaps some serious metallurgical flaw (or flaws) exist).

Quite often the case depth is not right. The gear drawing should specify the right case depth for the midpoint of the tooth flank. Besides the middle of the tooth, the case depth needs to be right at the lower flank of the tooth. The case depth at the tip of the tooth is also critical.

The case depth may be right at midheight but too thin at the lower flank and root fillet. This is often caused by inadequate carbon gas penetration between teeth. In large gears, heat soak back after quenching may reduce the hardness considerably. Too much case depth at the tooth tip may cause the tooth tips to be so brittle that they break off in service. (This has happened off and on in gears used in mining.)

If large, through-hardened gears are quenched too drastically, there may be hidden cracks in the gear body. On the other hand, an inadequate quench may cause the hardness of the gear teeth to be below specifications in the middle of the face width or near the root diameter of the teeth.

Case-hardened gears can be ground by several different methods. There is always a risk of grinding too fast or having inadequate cooling at the grinding location. Grinding burns and grinding cracks must be avoided to get the potential life out of the gears.

When scrap material is melted to make steel, there is a hazard that something in the melt will produce a local hard spot or a local soft spot. There may also be stringers of inclusions. There is a well-developed art of producing steel ingots that can be made into steel billets of uniform good material for a toothed gear. Good gears require good steel.

When gear parts are forged, the flow lines in the forging should not be in the wrong direction at critical stress points, such as the root fillet of a gear tooth or where a thin web meets a thin gear rim. Rolled bar stock that is not forged may have flow lines in the wrong direction for a critical gear part. Normal gear rating practice is based on the assumption that the forging or rolling of the stock to make the gear blank is done right for the configuration of the gear part.

Another flaw to be concerned about is the chemical composition of the steel in the gear. Sometimes mistakes are made and the steel for a gear is taken out of the wrong bin. Through-hardening steels have different compositions than case-hardening steels.

Sometimes an improper substitution is made. For example, it quite often happens in worldwide operations that a readily available local steel is substituted for the specified steel that was just right for the rating and method of manufacturing used in another country. It is then quite a shock to have gears failing in a year or two that should have lasted for 10 or more years.

In summary, quite a number of metallurgical flaws can seriously reduce gear life. Metallurgical quality in gears is just as important as geometric accuracy in gears. For further data in regard to metallurgy and heat treating, see ANSI/AGMA 2004-B89, *Gear Materials and Heat Treatment Manual*. Appropriate references are made to a substantial number of ASTM, SAE, and military standards. Detailed specifications that define metallurgical requirements for grade 1, 2, and 3 gear materials are given in ANSI/AGMA 2001-C95, *Fundamental Rating Factors and Calculation Methods for Involute Spur and Helical Gear Teeth* (Ref 2). Grade 1 is considered normal industrial quality, grade 2 is extra high quality, and grade 3 is super quality. Grade 2 is used mostly for high speed, long life gearing. Grade 3 is often used for critical aerospace gearing.

Bores and Shafts

Besides proper gear design, geometric accuracy, and good gear material, associated components in a gear train can also be important. This section briefly reviews some components in the design and structure of each gear and/or gear train that must be considered in conjunction with the teeth.

A round bore, closely toleranced and ground, may rotate freely around a ground shaft diameter, rotate tightly against a ground diameter by having a press fit, or be the outer race of a needle bearing that gives freedom of rotation. Each application has unique problems. A ground bore is always subject to tempering, burning, and checking during the grinding operation. A freely rotating bore requires a good lubrication film, or seizing and galling may result. The bore used as a bearing outer race must be as hard as any standard bearing surface and is subject to all conditions of rolling contact, such as fatigue, pitting, spalling, and galling. The bore that is press fit onto a shaft is subject to a definite amount of initial tensile stress. Also, any tendency for the bore to slip under the applied rotational forces will set up a

unique type of wear between the two surfaces that leads to a recognizable condition of fretting damage.

A spur of a helical round bore gear acting as an idler, or reversal, gear between the input member and the output member of a gear train has an extremely complicated pattern of stresses. The most common application is the planet pinion group in wheel-reduction assemblies or in planetary-type speed reducers. Typical stress patterns are as follows:

- The tensile and compressive stresses in the bore are due to bending stresses of the gear being loaded as a ring.
- The maximum tensile and compressive stresses in the bore increase as the load on the teeth increases.
- The maximum tensile and compressive stresses in the bore increase as the clearance between the bore and the shaft is increased.
- The maximum tensile and compressive stresses in the bore increase as the ratio of the size of the bore to the root diameter of the teeth is increased.
- During one revolution of the gear under load, the teeth go through one cycle of complete reversal of stresses, whereas each element of the bore experiences two cycles of reversals.

Three modifications of the round bore alter stress patterns considerably:

- *Oil holes* that extend into the bore are intended to lubricate the rotating surfaces. Each hole may be a stress raiser that could be the source of a fatigue crack.
- *Tapered bores* are usually expected to be shrink fitted onto a shaft. This sets up a very high concentration of stresses, not only along the ends of the bore but also at the juncture on the shaft.
- *A keyway* in the bore also creates a stress-concentration area. A keyway is also required to withstand a very high load that is continuous. In fact, the applied load to the side of the keyway is directly proportional to the ratio of tooth pitchline radius to the radius of the keyway position. Fatigue failure at this point is common.

Splined Bores. The loads applied to the splined-bore area are also directly proportional to the ratio of pitchline radius of the teeth to the pitchline radius of the splines. However, the load is distributed equally onto each spline; therefore, the stress per spline is usually not excessive. It is possible for an out-of-round condition to exist that would concentrate the load on slightly more than two splines. Also, a tapered condition would place all loadings at one end of the splines, which would be detrimental both to the gear splines and to the shaft.

Heat treating of the splined area must be monitored closely for quench cracks. Grinding of the face against the end of the splines can also cause grinding checks to radiate from the corners of the root fillets.

Shafts. The shafts within a gear train, as well as the shank of a pinion that constitutes a shaft in function, are very important to load-carrying capacity and load distribution. They are continually exposed to torsional loads, both unidirectional and reversing. The less obvious stressed condition that is equally important is bending. Bending stresses can be identified as unidirectional, bidirectional, or rotational. When the type of stress is identified, one can explore the causes for such a stress.

There can be a number of stresses applied to a shaft that are imposed by parts riding on it. For example, helical gears will transmit a bending stress, as will straight and spiral bevel gears; round bores may be tight enough to cause scoring and galling; a splined bore may cause high stress concentration at the end face; runout in gears may cause repeated deflections in bending of the shaft; and loose bearings may cause excessive end play and more bending.

REFERENCES

1. D.W. Dudley, *The Handbook of Practical Gear Design*, Technomics Publishing Co., Inc., Lancaster, PA, 1994
2. *Fundamental Rating Factors and Calculation Methods for Involute and Helical Gear Teeth*, ANSI/AGMA 2001-C95, American Gear Manufacturers Association

SELECTED REFERENCES

Fatigue of Gears

- B. Abersek, J. Flasker, and J. Balic, Theoretical Model for Fatigue Crack Propagation on Gears, *Conf. Proc. Localized Damage III: Computer-Aided Assessment and Control*, Computational Mechanics Publications, Billerica, MA, 1994, p 63-70
- A.Y. Attia, Fatigue Failure of Gears of Circular-Arc Tooth-Profile, *J. Tribol.*, Vol 112 (No. 3), 1990, p 453-459
- G.P. Cavallaro et al., Bending Fatigue and Contact Fatigue Characteristics of Carburized Gears, *Surf. Coat. Technol.*, Vol 71 (No. 2), 1995, p 182-192
- G. Dalpiaz and U. Meneghetti, Monitoring Fatigue Cracks in Gears, *Nondestr. Test. Eval. Int.*, Vol 24 (No. 6), 1991, p 303-306
- J. Flasker and B. Abersek, Determination of Plastification of Crack Tip of Short and Long Fatigue Crack on Gears, *Conf. Proc. Computational Plasticity: Fundamentals and Applications*, Pineridge Press Ltd., Mumbles, Swansea, U.K., 1992, p 1585-1595
- Y. Jia, Analysis and Calculation of Fatigue Loading of Machine Tool Gears, *Int. J. Fatigue*, Vol 13 (No. 6), 1991, p 483-487
- B. Jiang and E.A. Shao, Rapid Method for Measurement of Rolling Contact Fatigue Limit of Case-Hardened Gear Materials, *ASTM J. Test. Eval.*, Vol 18 (No. 5), 1990, p 328-331

- B. Jiang, X. Zheng, and M. Wang, Calculation for Rolling Contact Fatigue Life and Strength of Case-Hardened Gear Materials by Computer, *ASTM J. Test. Eval.,* Vol 21 (No. 1), 1993, p 9-13
- B. Jiang, X. Zheng, and L. Wu, Estimate by Fracture Mechanics Theory for Rolling Contact Fatigue Life and Strength of Case-Hardened Gear Materials with Computer, *Eng. Fract. Mech.,* Vol 39 (No. 5), 1991, p 867-874
- G. Mesnard, Super Puma MK II: Rotor and Gearbox Fatigue, *Conf. Proc. ICAF '91: Aeronautical Fatigue—Key to Safety and Structural Integrity,* Ryoin Co., Tokyo, 1991, p 195-215
- M. Nakamura et al., Effects of Alloying Elements and Shot Peening on Fatigue Strength of Gears, *Kobelco Technol. Rev.,* No. 16, 1993, p 14-19
- Y. Obata et al., Evaluation of Fatigue Crack Growth Rate of Carburized Gear by Acoustic Emission Technique, *Acoustic Emission: Current Practice and Future Directions,* STP 1077, ASTM, 1989, p 261-270
- Y. Okada et al., Fatigue Strength Analysis of Carburized Transmission Gears, *JSAE Rev.,* Vol 11 (No. 3), 1990, p 82-84
- T.L. Panontin, Fatigue Fracture of a C130 Aircraft Main Landing Gear Wheel Flange, *Handbook of Case Histories in Failure Analysis,* Vol 1, ASM International, 1992, p 25-29
- W. Pizzichil, Testing Gear Teeth for Low-Cycle Fatigue and Static Bending Strength, *Heat Treating,* Vol 24 (No. 2), 1992, p 14-20
- H.J. Rose, Vibration Signature and Fatigue Crack Growth Analysis of a Gear Tooth Bending Fatigue Failure, *Conf. Proc. Current Practices and Trends in Mechanical Failure Prevention,* Vibration Institute, Willowbrook, IL, p 235-245
- M. Roth and K. Sander, Fatigue Failure of a Carburized Steel Gear from a Helicopter Transmission, *Handbook of Case Histories in Failure Analysis,* Vol 1, ASM International, 1992, p 228-230
- G.M. Tanner and J.R. Harty, Contact Fatigue Failure of a Bull Gear, *Handbook of Case Histories in Failure Analysis,* Vol 2, ASM International, 1993, p 39-44
- D.P. Townsend, "Improvement in Surface Fatigue Life of Hardened Gears by High-Intensity Shot Peening," TM-105678, National Aeronautics and Space Administration, 1992
- D.P. Townsend and J. Shimski, "Evaluation of Advanced Lubricants for Aircraft Applications Using Gear Surface Fatigue Tests," TM-104336, National Aeronautics and Space Administration, 1991
- V.I. Tsypak, Fretting Corrosion and Fatigue of Gears, *Mater. Sci. (Russia),* Vol 29 (No. 6), 1993, p 680-681
- J.-G. Wang, Z. Feng, and J.-C. Zhou, Effect of the Morphology of Carbides in the Carburizing Case on the Wear Resistance and the Contact Fatigue of Gears, *J. Mater. Eng. (China),* Dec 1989, p 35-40
- T.P. Wilks et al., Conditions Prevailing in the Carburising Process and Their Effect on the Fatigue Properties of Carburised Gears, *J. Mater. Process. Technol.,* Vol 40 (No. 1-2), 1994, p 111-125
- J.J. Zakrajsek, D.P. Townsend, and H.J. Decker, An Analysis of Gear Fault Detection Methods as Applied to Pitting Fatigue Failure Data, *Conf. Proc. The Systems Engineering Approach to Mechanical Failure Prevention,* Vibration Institute, Willowbrook, IL, p 199-208

Gear Durability and Failure

- J.W. Blake and H.S. Cheng, A Surface Pitting Life Model for Spur Gears, Part II: Failure Probability Prediction, *J. Tribol.,* Vol 113 (No. 4), 1991, p 719-724
- R. Gnanamoorthy and K. Gopinath, Surface Durability Studies on As-Sintered Low Alloy Steel Gears, *Non-Ferrous Materials,* Vol 6, *Advances in Powder Metallurgy & Particulate Materials,* Metal Powder Industries Federation, 1992, p 237-250
- G.M. Goodrich, Failure of Four Cast Steel Gear Segments, *Handbook of Case Histories in Failure Analysis,* Vol 2, ASM International, 1993, p 45-52
- C. Madhavan and K. Gopinath, Surface Durability of Through-Hardened P/M Gears, *Powder Met. Sci. Technol.,* Vol 2 (No. 4), 1991, p 30-44
- T. Nakatsuji, A. Mori, and Y. Shimotsuma, Pitting Durability of Electrolytically Polished Medium Carbon Steel Gears, *Tribol. Trans.,* Vol 38 (No. 2), 1995, p 223-232
- M. Roth, M. Yanishevsky, and P. Beaudet, Failure Analysis of Aircraft Landing Gear Components, *Conf. Proc. Failure Analysis: Techniques and Applications,* ASM International, 1992, p 109-118
- H.V. Somasundar and R. Krishnamurthy, Surface Durability of HSLA Steel Gears, *Conf. Proc. Ninth National Conference on Industrial Tribology,* Central Machine Tool Institute, Bangalore, India, 1991, p M87-M95
- Y.L. Su, J.S. Lin, and S.K. Hsieh, The Tribological Failure Diagnosis of Spur Gear by an Expert System, *Wear,* Vol 166 (No. 2), 1993, p 187-196
- A. Verma and R.K. Sidhu, Failure Investigation of Gears, *Tool and Alloy Steels,* Vol 27 (No. 6), 1993, p 157-160
- K. Wozniak, Worm Gears of Increased Durability, *Archiwum Budowy Maszyn,* Vol 38 (No. 1), 1991, p 5-13

Fatigue and Life Prediction of Bearings

Charles Moyer, The Timken Company (retired)

ROLLING ELEMENT FATIGUE has been recognized since the turn of the century, with fatigue life testing beginning in the early 1900s by rolling bearing manufacturers. Because of the high stresses imposed on bearings, steel with a hard matrix was required. However, the hard matrix, along with other factors to be discussed, meant that the matrix was very sensitive to loading and other external conditions that could lead to incipient material cracking and thus to contact fatigue. In the early tests, large scatter in fatigue life was experienced. This persists and indicates that close correlation of the incipient cracks and running time should not be expected (Ref 1).

Internal stress raisers, such as nonmetallic inclusions or carbides, are the beginning sites for incipient cracking. Because of the varying severity, the magnitude of the stress raisers is difficult to calculate, so that rolling fatigue, basically a problem of stress concentrations, displays the scatter mentioned or shows a clear statistical nature.

The high stresses involved come about from the geometry of bearings, which produces a concentration of load in the contacts in a small volume of the material within a bearing (Ref 2). Stresses were first computed based on the basic bearing internal geometries, and the stress distribution within and beneath the contact area was determined using the basic equations proposed by Hertz (Ref 3) in 1881. From such fundamental work, along with considerable fatigue test results that established various fatigue and material "constants," the first bearing ratings and fundamental life theories were proposed.

Following the work of Grubin in 1949 (Ref 4), Dowson and Higginson in 1966 (Ref 5), and Dowson and Toyoda in 1978 (Ref 6), to name just a few, the important role of lubricant in the contacts of bearings was next understood. It was recognized that the type of lubricant and the thickness of the lubricant film impacted bearing fatigue life and even influenced modes of failure. Considerable work has been done to relate failure modes, including fatigue, to bearing life.

As knowledge in all these areas grew, it was possible to understand the requirements for a bearing to develop longer life and produce acceptable life in a variety of adverse environments, such as low speed, high speed, elevated temperatures, misalignment, and contamination. This work, drawing heavily on much experimental data, is the basis for the present level of bearing life prediction.

Bearing Materials

As stated, it was determined early in the 1900s that special material was required to give sufficiently long fatigue life for bearings. One percent carbon-chromium through-hardening steels were used for ball and cylindrical roller bearings and other types, perhaps influenced in Europe by familiarity with chromium-type tool steels (Ref 2, 7). U.S. manufacturers included low-carbon (0.20%) case-carburizing steels in bearing manufacture, primarily in the tapered roller bearing industry. Both steel types are still used, and examples of each are listed in Tables 1 and 2. Bearing steel types, microstructure, heat treatment, and specialty bearing steels are discussed in the article "Bearing Steels" in Volume 1 of the *ASM Handbook* (Ref 7).

High-quality steel, free from nonmetallic inclusions or other stress raisers, is a primary requirement to produce bearings that achieve extended fatigue life. Over the years, steel manufacturers have strived to produce cleaner bearing steel. The continual increases in bearing ratings, especially over the last 20 years, can be attributed to these improvements. After World War II, the requirements for significantly improved steels for bearings used in gas turbine engines demonstrated the performance improvements that vastly cleaner steels could provide (Ref 8). Besides clean steel, aerospace bearing applications needed high-speed capabilities (up to and then exceeding 2.4 million DN, where DN is bearing bore in milli-

Table 1 Nominal compositions of high-carbon bearing steels

Grade	Composition, %					
	C	Mn	Si	Cr	Ni	Mo
AISI 52100	1.04	0.35	0.25	1.45	...	...
ASTM A 485-1	0.97	1.10	0.60	1.05	...	...
ASTM A 485-3	1.02	0.78	0.22	1.30	...	0.25
TBS-9	0.95	0.65	0.22	0.50	0.25 max	0.12
SUJ 1(a)	1.02	<0.50	0.25	1.05	<0.25	<0.08
105 Cr6(b)	0.97	0.32	0.25	1.52	...	...
SHKH15-SHD(c)	1.00	0.40	0.28	1.48	<0.30	...

(a) Japanese grade. (b) German grade. (c) Russian grade. Source: Ref 7

Table 2 Nominal compositions of carburizing bearing steels

Grade	Composition, %					
	C	Mn	Si	Cr	Ni	Mo
4118	0.20	0.80	0.22	0.50	...	0.11
5120	0.20	0.80	0.22	0.80	...	...
8620	0.20	0.80	0.22	0.50	0.55	0.20
4620	0.20	0.55	0.22	...	1.82	0.25
4320	0.20	0.55	0.22	0.50	1.82	0.25
3310	0.10	0.52	0.22	1.57	3.50	...
SCM420	0.20	0.72	0.25	1.05	...	0.22
20MnCr5	0.20	1.25	0.27	1.15	...	...

Source: Ref 7

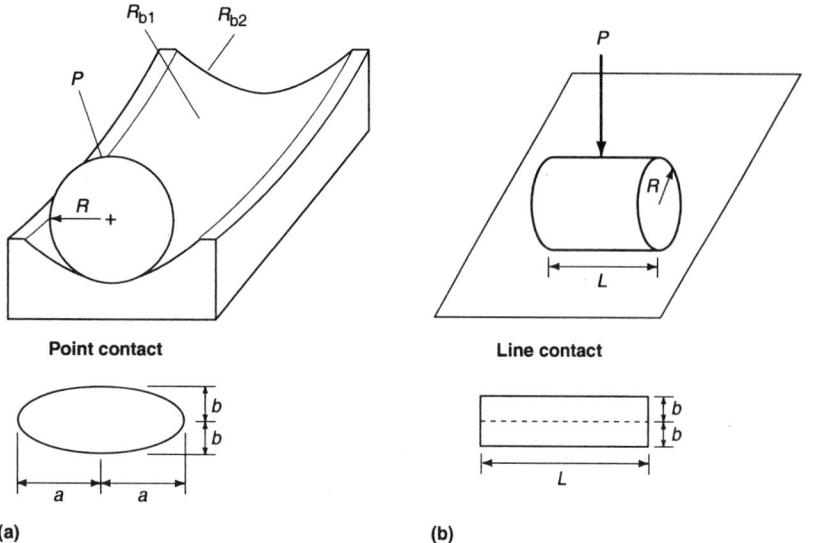

Fig. 1 Geometry for (a) point contact and (b) line contact, showing the general contact shapes produced for each geometry

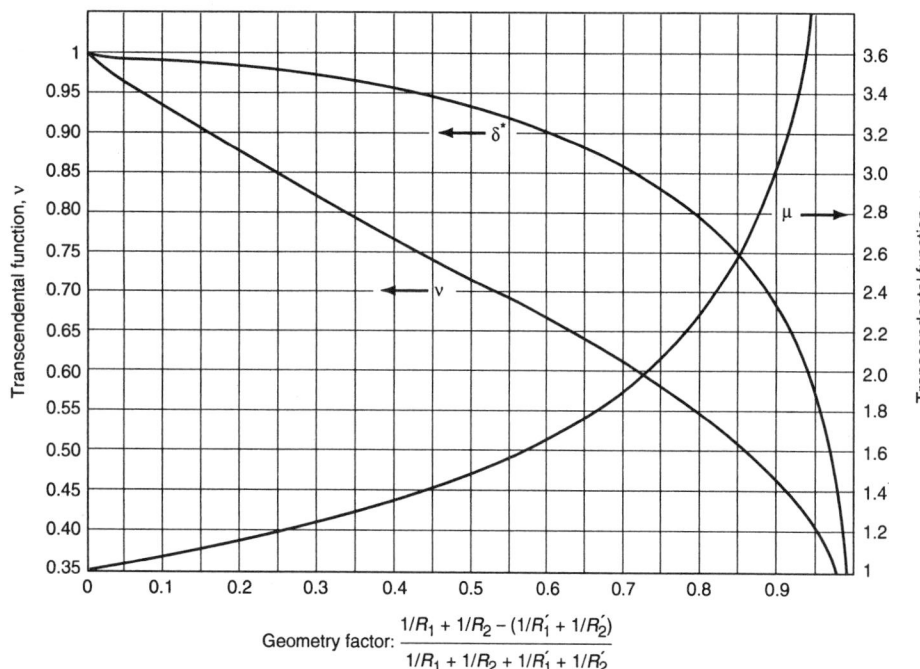

Fig. 2 Plot of geometry factor vs. transcendental functions μ and v in Eq 1 and 2 and related to the semimajor/semiminor axes a/b. δ^* is the dimensionless contact deformation or approach factor for the two bodies in contact.

from tight, interference fits on shafts. Very often, carburizing steels (as given in Table 2) are used for inner races and for bearings that require greater surface ductility, greater resistance to debris particles, and a higher level of core toughness, along with the compressive residual surface stresses that help reduce fatigue crack propagation through the bearing cross section (Ref 7).

Through-hardening steels can carry the somewhat higher contact stresses that are prevalent in ball bearings. Through-hardening steels may also give better dimensional stability over temperature extremes, because of lower retained austenite, and these advantages may dictate their use in some inner ring applications. Care should be taken, however, to minimize tensile residual stresses in through-hardened steel races. It is known that control of surface grinding for bearing steels (avoiding abusive grinding forces and high-temperature gradients) can minimize tensile residual stresses (Ref 14). With soft wheels, low wheel speeds, shallow cuts, and proper grinding fluid, El-Helieby and Rowe produced compressive residual stresses on EN31 steel (chemistry related to AISI 52100 through-hardening steel) following these practices. Although such practices may be too time-consuming for industrial use, eliminating surface damage can lead to improved fatigue life with any type of bearing steel.

Bearing Fundamentals

Surface Contact and Stresses. Bearings are described by their rolling elements as ball or roller bearings or by their contact footprints as point or line (Fig. 1a, b). A point contact forms an ellipse, where the area is defined by the semi-axes a and b (Fig. 1a). For line contact, a rectangle is formed, where the area is defined by L times $2b$ (Fig. 1b). These small, concentrated areas explain the high stresses developed on and below the immediate bearing contact surfaces.

Maximum compressive stress (q_0) and half-width contact (b) for point contact (Fig. 1a) are calculated from the following equations:

$$q_0 = \frac{3P}{2\pi \cdot ab} = 23600 \cdot \frac{P^{1/3}}{\mu v}(S_c\, 1/R)^{2/3} \qquad (Eq\ 1)$$

$$b = 0.004498\, v\, (P/S_c\, 1/R)^{1/3} \qquad (Eq\ 2)$$

where the constants in the above equations assume bearing steel with elastic modulus of 207 MPa (30 $\times 10^6$ psi) and Poisson's ratio of 0.3. Constants for other materials can be found in Ref 8. Approximate values of μ and v can be taken from Fig. 2. S_c is the summation of curvatures in Fig. 1(a) as follows: $S_c\, 1/R = 1/R_{b1} + 1/R_{b2} + 1/R + 1/R$.

Maximum compressive stress (q_0) and half-width contact (b) for line contact (Fig. 1b) is as follows:

$$q_0 = 2290\, (P/L)^{1/2} \cdot (1/R_1 + 1/R_2)^{1/2} \qquad (Eq\ 3)$$

meters times bearing speed in rpm) and materials capable of up to and above 200 °C (400 °F) (Ref 9).

The experience gained in developing and testing such materials provided the bearing industry valuable insight into refinements that could be achieved with industrial-grade bearings by better control of tolerances, better manufacturing methods, and better finishing techniques. Thus, over the past 20 years as cleaner steels improved bearing life, it was recognized that rather than materials limiting fatigue, processing and surface-related shortcomings (e.g., rough surface finish,

surface defects, deep grinding grooves, handling nicks or defects from the ingress of hard debris) were now limiting life (Ref 10-12). The most recent efforts have been to try to lower surface roughness and make material processing changes that will better resist the incipient damage of contaminants, including debris and other life-limiting factors.

The role of residual stresses in bearings after manufacture has also been recognized (Ref 13). Heat treatments that minimize or avoid tensile stresses are recommended, especially for inner raceway rings that develop hoop tensile stresses

$$b = 0.0002779 \, (P)^{1/2}/[L(1/R_1 + 1/R_2)]^{1/2} \qquad \text{(Eq 4)}$$

where again the constants in these equations assume bearing steel with elastic modulus of 207 MPa (30 × 10^6 psi) and Poisson's ratio of 0.3. Constants for other materials can be found in Ref 8.

Fatigue Life of Bearings. Considerable fatigue testing over the years developed the relationships between stress and life and then to bearing load, as evidenced by the work of Clinedinst (Ref 15) and Lundberg and Palmgren (Ref 16). While Clinedinst based his early load ratings on the maximum contact pressure, q_0, Lundberg and Palmgren recognized that fatigue cracks usually start below the contact surface near the depth of a maximum shear stress, such as T_0 (the range of orthogonal shear). They perceived a change in the material near inclusions below the surface, microcrack development, and then propagation of the cracking to the surface with formation of a pit or spall. This mode of subsurface spall development has become the classical mode of fatigue failure for bearings running under good operating conditions. An adequate lubricant film, reasonable temperatures, and reasonable environmental conditions are the conditions for which bearing rating life or higher should be achieved.

The Lundberg-Palmgren power equation (Eq 5), based on a measure of the statistical life scatter of a group of identical bearings (Ref 17), relates the strength dependency (life) with the magnitude (volume) of the stressed material within the bearings (Ref 18). Life dispersion is represented by the Weibull dispersion parameter (slope) "e." The volume of material stressed (V) is related to a bearing by considering the product of:

- a, as the half major axis of the pressure ellipse
- l, as the circumference of the critical raceway
- z_0, as the depth from contact surface to the location of T_0, the maximum shear stress amplitude

The power equation is:

$$\ln(1/S) \sim \frac{T_0^c \cdot N^e \cdot V}{z_0^h} \qquad \text{(Eq 5)}$$

where S is the probability of survival and z_0 is the depth to where the amplitude of the shear stress T_0 is greatest. The depth z_0 is also used to determine volume (V). In Ref 16, it is considered that the depth below the surface where microscopic material changes become the origin site of fatigue also influences fatigue propagation, so z_0 is included twice in Eq 5. Life (N) is expressed in millions of stress cycles. The three exponents "e," "c," and "h" are considered life dispersion factors or material constants, and they are not independent. Equation 5 then becomes:

$$\ln(1/S) \sim \frac{T_0^c \cdot N^e \cdot a \cdot l}{z_0^{(h-1)}} \qquad \text{(Eq 6)}$$

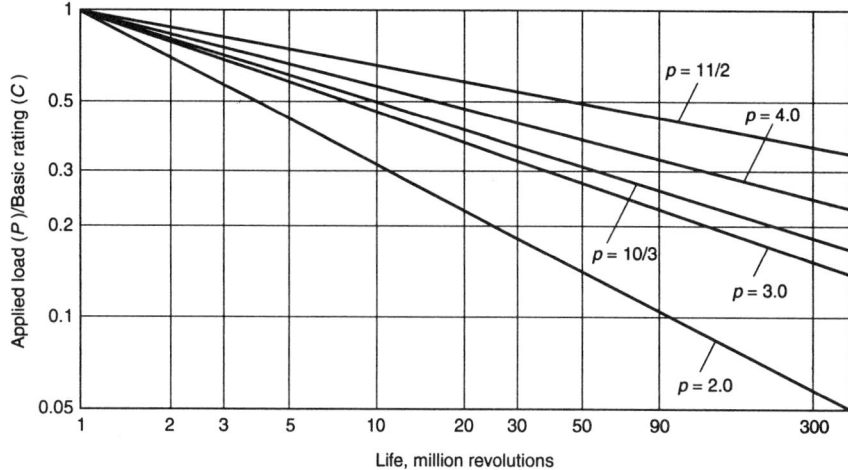

Fig. 3 Effect of load-life exponent with one million revolutions reference point

By considering actual bearing contact geometry, the contact stresses (Eq 1 and 3), the number of rolling elements, their diameter relative to the inner and outer race diameters, and the appropriate material constants, it is possible to develop a relationship of bearing load to life in millions of revolutions. The load at a specific life is defined as the basic dynamic load capacity, "C." Lundberg and Palmgren did this in their original works (Ref 16, 19), but although much of their basic approach still is used, considerable change has taken place. Current load capacity equations are given in the ISO Standard and the two American National Standards (Ref 20-22). There are 12 rating equations, covering radial and thrust ball and roller bearings. Two are cited here, simply to illustrate the source of the present exponents as given in the radial equations, because the exponent values affect bearing life prediction. The ball bearing radial rating equation is:

$$C_r = F_{cm} \, (i \cdot \cos \alpha)^{0.7} \cdot Z^{2/3} \cdot D_w^{1.8} \qquad \text{(Eq 7)}$$

where F_{cm} is the factor based on bearing component geometry, accuracy, manufacturing quality, and normal bearing steel; D_w is the ball diameter in millimeters or inches; i is the number of rows of rolling elements in a bearing; Z is the number of rolling elements per row; and α is the nominal contact angle of the bearing in degrees. When D_w is above 25.4 mm (1 in.), the exponent becomes 1.4 and the rating equation changes (see Ref 21).

The roller bearing radial rating equation is:

$$C_r = F_{cm} \, (i \cdot L_{we} \cdot \cos \alpha)^{7/9} \cdot Z^{3/4} \cdot D_{we}^{29/27} \qquad \text{(Eq 8)}$$

where D_{we} is the roller diameter applicable in the calculation of load ratings, in millimeters or inches; and L_{we} is the roller length used in the calculation of roller bearing load ratings, in millimeters or inches.

Values of F_{cm} are given in standards. Based on bearing geometries that can be measured, it is possible to calculate generic ratings for any rolling element bearing. Because these ratings are developed by consensus within standards working committees, they represent neither minimum nor maximum values. Machine designers should review the most current handbooks of bearing manufacturers for specific ratings and more accurate life estimates.

Besides the significant experimentally determined material and manufacturing constant within F_{cm}, an experimental relationship between load and fatigue life is in all the standards. The relation of rating capacity C and actual bearing load P to L-10 life is:

$$L\text{-}10 = (C/P)^p \times (\text{One million revolutions}) \qquad \text{(Eq 9)}$$

where L-10 is the basic rating life in millions of revolutions with 90% probability of survival, and p = 3 for ball bearings, or 10/3 for roller bearings.

When $P = C$, the L-10 life is exactly one million revolutions. At 500 rpm, L-10 is only 33⅓ hours. Depending on the bearing design, this L-10 is generally above the elastic limit for a bearing. Bearing ratings (C load) are in reality reference loads, used only to calculate L-10 life for bearings with Eq 9. Actually, it is advisable to stay below about 50% of C for most applications. Figure 3 shows how the load-fatigue life changes with five values of p. The three inside exponents, 3, 10/3, and 4, are used in the standards.

Sources of Rating/Life Exponents. Past experimental determination of the load-life exponent p for ball bearings (Ref 16, 23) are the basis of the exponent 3 in the standards. The actual values of the exponent in Ref 23 were 2.87 ± 0.35 at L-10 and 2.80 ± 0.31 at L-50 (median life), so that considering the confidence bands, an exponent of 3.0 was acceptable based on a database of almost 5000 deep-groove ball bearings.

For roller bearings, an exponent of 4.0 (assumed at L-50 life) was determined in Ref 19 from 6 test series of 30 spherical roller bearings each, with loads ranging from about 0.15 to 0.4 P/C. Exponent values of 4.22 at L-10 life and 4.08 at L-50 (median) life were obtained in Ref 24 based on rig testing at 1840 to 2910 MPa (267 to 422 ksi) from four test series with a cylindrical

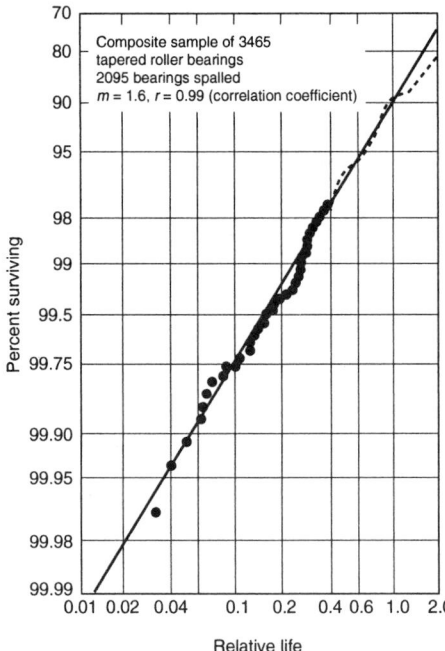

Fig. 4 Example of a group of tested bearings, ranging from 12 to 85 mm bore, all run with SAE 20 oil and 38 °C inlet oil temperature and rationalized so each sample L-10 was equal to 1.0. The fatigued bearings, ordered by increasing life, plot close to the Weibull line (slope) of 1.6.

roller bearing inner raceway against a loading ring. The test specimens in Ref 19 and 24 were not modified to avoid edge loading or end-of-contact stress concentrations. However, with rollers and/or raceways crowned for elliptical contact or profiled for "optimized" contact, a load-life exponent of 10/3 was considered acceptable (Ref 19) and confirmed with recent tests (Ref 25). The exponent 10/3 has been used in Eq 9 for roller bearings in the bearing standards for many years. The ANSI/AFBMA Standard 11-90 (Ref 22) recognizes that depending on the load, whether with crowned or profiled contact in a roller bearing, truncation and edge stresses can occur. Roller bearing contact lengths and internal or end-of-contact modifications are not standardized, so how specific bearings are influenced by truncation must be obtained from the manufacturer.

Besides the material constant in F_{cm} and the load-life exponent (p), the life dispersion parameter (e, or the Weibull slope) is experimentally determined. Figure 4 shows a large group of tested roller bearings with failures plotted in order of increasing life. The plot is on Weibull probability paper, and the data plot is close to the line for the Weibull slope (e), which equals 1.6 for this sample. In Ref 16 and 19, similar plots were made and it was concluded that a slope of 1.11 (= 10/9) could be used for ball bearings and 1.125 (= 9/8) could be used for roller bearings. Lundberg and Palmgren (Ref 16) also ran tests on bearings with a range of roll diameters (D_a sizes) of 1.5 to 25 mm. The resultant exponent on the roll diameter relative to capacity was 1.8. Considering all these experimental factors, it was possi-

ble to solve for the other "material constants" c and h (see Eq 5 and 6). Using the following from Ref 16 (p 15, Eq 51):

$$D_a = 1.8 = (2c + h - 5)/(c - h + 2) \qquad \text{(Eq 10)}$$

From Ref 16 (p 15, using the transformed Eq 54 for point contact):

$$c - h = 3(p)(e) - 2 = 3(3)(10/9) - 2 = 8 \qquad \text{(Eq 11)}$$

From Ref 16 (p 16, using the transformed Eq 55 for line contact):

$$c - h = 2(p)(e) - 1 = 2(4)(9/8) - 1 = 8 \qquad \text{(Eq 12)}$$

Combining Eq 10 and 11 or Eq 10 and 12, the values of $c = 31/3$ and $h = 7/3$ are determined. From these, the experimental basis of the exponents in the present rating equations (Eq 7, 8) can be seen.

Rating/life exponents for ball bearings are:

$$\text{Exponent for } D_w (D_a) = 1.8$$

$$\text{Exponent for } Z = 1 - 1/(p) = 1 - 1/(3) = 2/3$$

$$\text{Exponent for } (i \cos \alpha) = 1 - 1/(p)(e)$$
$$= 1 - 1/(3)(10/9) = 7/10$$

Rating/life exponents for roller bearings are:

$$\text{Exponent for } D_{wc} = (c + h - 3)/(c - h + 1)$$
$$= (31/3 + 7/3 - 3)/(31/3 - 7/3 + 1) = 29/27$$

$$\text{Exponent for } Z = 1 - 1/p = 1 - 1/(4) = 3/4$$

$$\text{Exponent for } (i \cos \alpha) = 1 - 1/pe = 1 - 1/(4)(9/8) = 7/9$$

This background information about the source of the exponents in the rating equations for ball and roller bearings and in Eq 9 illustrates the experimental input in the system of calculating rolling bearing life prediction as contained in the present bearing standards.

The generic forms of the basic dynamic load rating equations in the standards (Ref 20-22) are good reference ratings based on reasonable consensus through the Engineering Committees of the American Bearing Manufacturers Association (ABMA), but as the background discussion illustrates, many of the experimental factors still in the standards may need to be updated. References 26 to 31 attempt to cover some of the changes proposed for better predicting bearing fatigue life. Lorösch (Ref 26) gives the first indication that rolling bearings, run in a clean environment under lower load that causes no microstructural alterations in the stressed contact region, can have significantly extended life. Ioannides and Harris (Ref 27) essentially generalize the Lundberg-Palmgren model (Eq 6) and include a limiting stress below which fatigue failure may not occur. In essence this means that the load-life exponent p may increase at lower loads.

If rolling contact fatigue then follows an S-N curve, one might expect life scatter to increase at lower stress as with other fatigue (Ref 32). There

are limited data to suggest that life scatter does increase (i.e., Weibull slope decreases) for both bearings and gears under lower stress (Ref 33), so that at low loads, near an endurance limit, extended life scatter would make it more difficult to estimate failures. Lubrecht et al. (Ref 28) review the new life theory (Ref 27) and consider a possible range of values for the fatigue limit stress and how surface irregularities can be handled.

Dominik (Ref 29) gives specific factors (constants c, h, p and e), experimentally determined for tapered roller bearings, then proposes a specific rating equation for these bearings. Zhou and Nixon (Ref 30) present another specific bearing rating model that includes the equation from Ref 29, but also includes the film thickness and surface roughness ratio in a micro-macro contact model aimed at predicting bearing life for different surface textures. Tallian (Ref 31) reviews some dozen existing engineering and research models and proposes a new model based on all major life-influencing factors. It is clear that considerable activity leading to improved bearing life prediction is ongoing. The ideal operating conditions that give extended L-10 rating life are fairly limited in real applications, so most recent bearing manufacturer catalogs/handbooks have updated means to calculate L_{na}, which is an L-10 life rating adjusted by several life adjustment factors first introduced in 1972.

Bearing Life Prediction

The bearing standards in 1972 (Ref 34, 35) first included the concept of life adjustment factors. The three factors considered were reliability, material, and application. In equation form this became:

$$L_{na} = a_1 a_2 a_3 (C/P)^p \times (\text{One million revolutions}) \qquad \text{(Eq 13)}$$

or

$$L_{na} = a_1 a_2 a_3 (L\text{-}10) \qquad \text{(Eq 14)}$$

where L_{na} is calculated from the L-10 life rating adjusted by the following combination of factors: a_1, the reliability factor for other than 90%; a_2, the material factor (1972), now called the special bearing properties factor (1990); and a_3, the application or environmental conditions factor, which primarily recognizes the lubricant condition in a bearing but which also recognizes alignment and internal load distribution and contamination, liquid or solid.

Reliability Factor, a_1. For a single bearing, the probability that it will reach the specified life is its reliability. For a group of identical bearings run under identical conditions, the reliability is the percent that reach the specified life. For L-10 rating life there is 90% probability that a bearing will achieve this life. The reliability factor, a_1, given in the standards and many bearing handbooks, is based on a Weibull slope "e" equal to

1.5. The factor a_1 values from 90 to 99% reliability are:

Reliability, %	Life factor, a_1
90	1.00
95	0.62
96	0.53
97	0.44
98	0.33
99	0.21

These life factors indicate that the standards and others (Ref 25, 36) consider that bearings deviate to higher lives for failure probabilities less than 10% (i.e., higher reliabilities). If data are at hand that show that other Weibull slope values could be used to determine the life factor, then the following equation can be used to calculate a_1:

$$a_1 = [9.4912 \ln(100/R)]^{1/e} \qquad \text{(Eq 15)}$$

where a_1 is the reliability life factor ratio, R is the reliability expressed as percent survival, and e is the Weibull slope or life dispersion parameter.

Special Bearing Properties Factor, a_2. The bearing standards have never supplied values for a_2, simply saying in 1972 that a_2 could be used for bearings that incorporated improved materials and processing. By 1990, the a_2 factor was for special bearing properties described as follows: "A bearing may acquire special properties, as regards life, by the use of a special type and quality of material and/or special manufacturing processes and/or special design" (Ref 20, 21).

Significantly cleaner steels are now available. Examples are steels produced by vacuum-arc remelting (VAR) and electroslag melting or remelting (ESR). Other melting practices, such as vacuum induction melting vacuum-arc remelting (VIMVAR), were developed for use in critical applications such as the aerospace industry. Many applications include high-temperature steels for bearing operations at temperatures of 150 to 475 °C (300 to 890 °F), M-50, M-50NL, CBS-600* or CBS-1000M*. Air melt steels for conventional applications have shown significant improvements since 1972 and this is recognized by revised ratings in recent bearing catalogs.

Bearing users need to contact the bearing manufacturer to determine specific ratings and obtain new a_2 factors for current special steels. Values of a_2 range from 0.6 to over 6 for different chemistries, melting practices, cleanness levels, and metal working, heat treatment, and surface modification practices. Information on a_2 is included in Chapter 3 of Ref 37.

Special manufacturing and special design changes have also been significant, leading to improved internal profiles and reduced contact finishes so that under a wider range of operating conditions the contact surfaces are completely separated. By inputting specifics of these designs into a computer program, along with application information and lubricant conditions, it is possi-

*Trademark of the Timken Company

ble to quantify specific a_2 values for bearing users (Ref 28, 30). However, because determining a_2 this way requires fairly extensive computer programs, the details are proprietary and not generally available.

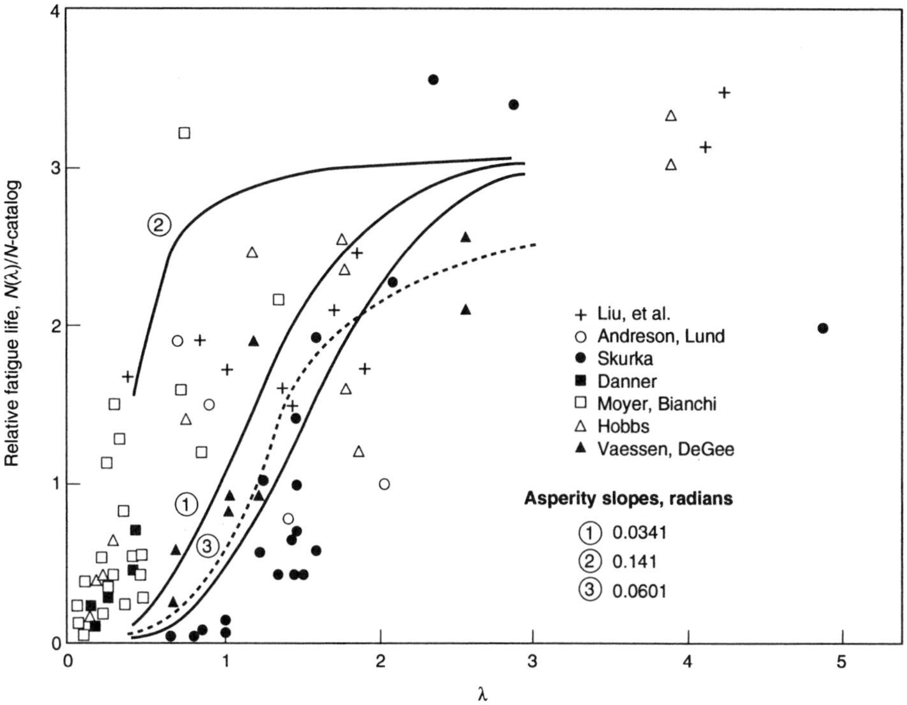

Operating Conditions Factor, a_3. Beyond load, which is important but already accounted for, adequacy of lubrication is central to a_3. Basic rating life assumes that normal lubrication provides a lubricant film that is equal or slightly greater than the composite roughness of the contact surfaces. The film thickness (EHL) divided by composite roughness is termed λ. References 6 and 38 are examples of elastohydrodynamic lubrication (EHL) film calculations for line contact and point contact conditions. Lubricant properties (absolute viscosity and pressure coefficient of viscosity at operating temperature), radii of the contacting surfaces, mean surface velocity of the surfaces, load, and the elastic moduli of the contacting bodies are needed for the lubricant film calculation.

The composite roughness is usually measured in terms of root mean square (RMS) roughness in the rolling direction with a cutoff length of 0.8 mm. There is some indication that the contact width in the rolling direction acts as a functional filter and that measurement cutoff length closer to the contact width might be more meaningful for λ (Ref 39). However, the important point is that the cutoff be known for the actual measurements and for comparisons of several λ values, so that consistent film and composite roughness determinations can be made. The calculation of composite roughness σ_q is:

Fig. 5 Plot of relative lives vs. λ for three asperity slope curves. Source for dashed curve: Ref 37. Source for other curves: Ref 42

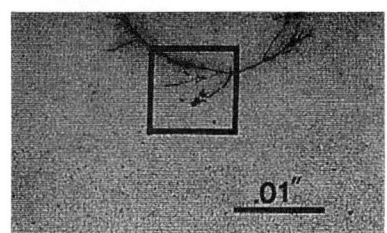

Fig. 6 Near-surface fatigue initiation site from inclusion origin fatigue. (a) Incipient cracking from subsurface inclusion spall. (b) Crack development below the raceway surface

(a) ⊢————⊣ 100 μm

Fig. 7 Appearance of microspalling (peeling or frosting appearance) from thin lubricant film. (a) Magnified shallow pitting along asperity ridges. (b) Transverse section shows shallow cracking from microspalling.

$$\sigma_q = \left(R_{q_1}^2 + R_{q_2}^2 \right)^{1/2} \qquad \text{(Eq 16)}$$

where R_{q1} and R_{q2} are the RMS roughness measurements of the two surfaces.

A considerable amount of published data has shown strong correlation of λ versus bearing fatigue life. The correlation is especially close for tests over a range of λ done with one specific type of bearing (Ref 40, 41). Tallian (Ref 42) collected such data from seven sources, including Ref 40 and 41. The plot of relative fatigue life [N-λ divided by N-catalog versus λ] that he generated is shown in Fig. 5. Some 83 bearing test groups are included and indicate life scatter but also a definite trend of λ versus fatigue life. Considering the asperity slope along with possible traction coefficients for mixed to boundary lubrication, Tallian was able to explain some of the scatter. Bearing type and material differences and all the different testing machines and test lubricants were other contributing factors to the scatter.

The three solid curves in Fig. 5 are based on the model described in Ref 42 for three levels of asperity slope. Curve 1 is for a 0.0341 radian (Rad) (1 degree, 57.2 minutes) RMS composite asperity slope, curve 2 is for 0.141 rad (48.5 minutes), and curve 3 is for 0.0601 rad (3 degrees, 26.6 minutes). The dashed curve, taken from Ref 37 (p 200), shows a similar trend for λ values under 2.0. More recent data indicate that this ASME-STLE curve is probably conservative for the higher λ values. It can be seen that fatigue life

is extended at high values and that surface roughness and asperity slope have little influence, because the contact surfaces are almost completely separated by the lubricant film. As λ is reduced, it is clear that fatigue life decreases, and for λ below 1.0, fatigue occurs below catalog or rating expected life.

Bearing Damage Modes. Bearing fatigue mode is also influenced by λ. For high λ, fatigue is primarily material related, at subsurface or near-surface fatigue initiation sites as outlined in Table 3. Figure 6(a) shows a subsurface inclusion spall, and Fig. 6(b) is a transverse section that shows the crack development below the surface. As a bearing operating condition can be described by lower λ, even going below 1.0, surface-related fatigue is more common. General roughness and higher asperity slopes can lead to surface distress or microspalling. Perhaps because of a thinner film and higher surface traction, the maximum shearing stress approaches the surface, and microhertzian stress at asperities can exceed the macrohertzian stresses that define the classical stressed volume. Very often microspalling is more apparent on the smoother surface in contact, but the rougher surface may have micropits forming at asperity peaks. Surface fatigue can also begin at localized stress raisers such as grinding grooves, surface inclusions, edge conditions, nicks and dents from handling, or hard debris ingested into the bearing contact.

The surface appearance of microspalling (Fig. 7a, b) shows how shallow the cracks formed are. Localized, point surface origin (PSO) fatigue starting at a debris bruise is shown in Fig. 8. Point-surface origin fatigue can propagate at larger λ values, based on the severity of the stress raiser. Under thicker films the propagation may not be as rapid and the arrowhead shape may not be very pronounced. An example of edge loading from geometric stress concentration (GSC) failure is given in Fig. 9. This usually happens in line contact bearings, but it can occur with any type of exaggerated misalignment.

Other Fatigue Failure Considerations. Fatigue below the expected rating life can also occur if a bearing is starved for lubrication, either from lack of supply or by blockage such as entrapped debris before the contact region. Seeing that a full film of lubricant reaches the bearing is a basic requirement.

Improper, excessive fitting practices can lead to fatigue cracks that produce inner raceway section fracture. Sometimes excessive fretting contributes to this fracture or initiates cracks on a

(b) ⊢—⊣ 10 μm

(a)

(b)

Fig. 8 Point surface origin fatigue. (a) Spall initiated at debris bruise on roller (4.55×). (b) Enlarged view of dent (100×)

raceway outside diameter. Because of severe fretting, nonround outer raceway seats, or hard particles left on the shaft or seat, bending fatigue (perhaps aggravated by corrosion) can lead to bearing ring fracture. The separator or cage for the rolling elements of a bearing may be mishandled so that wear takes place or eccentric motion occurs, leading to bridge or ring segment fatigue fracture. Care in mounting bearings includes avoiding any cage damage.

Fig. 9 Geometric stress concentration, edge loading on a bearing outer raceway, misaligned so that only half the race path carried the load

Table 3 Relating contact fatigue damage mode to λ

λ	Contact fatigue mode (initiation)	Material influence	Surface roughness influence	Geometry influence
Ratio > 3.0	Subsurface fatigue, inclusion(a)	Important	Minor	Important
Ratio 3–1.0	Subsurface/near surface, mostly inclusion	Important	"Sharp"/high asperities important	Important
Ratio 1–0.3	Some inclusion, some surface related	Somewhat important	Important for surface and near surface fatigue	Somewhat important
Ratio < 0.3	Surface related(b)	Minor	Important	Less important
Any ratio	Localized stress risers	Mixed	Mixed	Mixed
	PSO—dents, grooves, and surface inclusions	Minor	Somewhat important	Minor
	GSC—edge fatigue	Somewhat important	Minor	Important

PSO, point surface origin, fan-shaped spall propagation starting on the surface. GSC, geometric stress concentration starting at end of line contact. (a) Fatigue originates at nonmetallic inclusion in the maximum shear zone below the surface. (b) Called peeling or micropitting for bearings

Bearing life is also shortened by operating conditions that, beyond promoting contact fatigue, promote wear. It may be adhesive wear that is mild (limited to run-in), or the wear may lead to irreversible scuffing, smearing, or even final seizure. Abrasive wear can also occur, very often from particles causing grooving or denting with ridges that cause serious surface stress raisers, often leading to surface-related fatigue. Corrosive wear from water, acidic contaminants, or aggressive additive breakdown in the lubricant can also shorten bearing life. Fretting is a microscale version of adhesive, abrasive wear, often with corrosion that can again limit bearing performance.

There can also be plastic flow (Ref 43) caused by loading over the material yield point, yielding aggravated by high temperature, or brinelling that usually occurs under impact or high load, nonrotating conditions. Bearings can suffer electrical damage by grounding of electrical current through the bearing contacts. Under stationary conditions or slow rotation, the current causes localized melting and vaporization of individual asperities, while the surrounding material is rehardened and tempered. Under continuous current passage and bearing rotation, the electrical damage has a washboard appearance, called fluting, that is severe and leads to noise and fatigue of the damaged surfaces.

Finally, one of the most common failures for bearings is burnup, caused by inadequate lubricant supply, plugged lubrication lines, inadequate seals, improper/infrequent maintenance, or excessive preload or speed. The result is heat generation at a rate that exceeds heat transfer from the bearing so that material flows plastically and bearing geometry is destroyed. All these failure modes can be avoided if bearings have effective lubrication, so that adequate lubricant films are prevalent within the bearing, temperatures are within limits, and loads and speeds do not extend beyond the operating design parameters established for any specific system.

REFERENCES

1. C.A. Moyer and E.V. Zaretsky, Failure Modes Related to Bearing Life, *STLE Life Factors for Rolling Bearings*, E.V. Zaretsky, Ed., STLE SP-34, 1992, p 47-69
2. E.S. Rowland, Resistance of Materials to Rolling Loads, Chap 6, *Handbook of Mechanical Wear*, C. Lipson and L.V. Colwell, Ed., University of Michigan Press, 1961, p 108-130
3. H. Hertz, *Gesammelte Werke*, Vol 1, *Leibzig*, 1895, English translation, *Miscellaneous Papers*, Macmillan Co., London, 1896
4. A.N. Grubin and I.E. Vinogradova, *Investigation of the Contact of Machine Components*, Book 30, Tsniitmash, Moscow (DSIR, London, Translation 337), 1949
5. D. Dowson and G.R. Higginson, *Elastohydrodynamic Lubrication*, Pergamon Press, 1966
6. D. Dowson and S. Toyoda, A Central Film Thickness Formula for Elastohydrodynamic Line Contacts, *Proc. Fifth Leeds-Lyon Symposium*, 1978, p 60-65
7. H.I. Burrier, Jr., Bearing Steels, *Properties and Selection*, Vol 1, *ASM Handbook*, ASM International, 1990, p 380-388
8. W.J. Anderson, E.N. Bamberger, W.E. Poole, R.L. Thom, and E.V. Zaretsky, Materials and Processings, *STLE Life Factors for Rolling Bearings*, E.V. Zaretsky, Ed., STLE SP-34, 1992, p 71-128
9. R.J. Boness, W.J. Crecelius, W.R. Ironside, C.A. Moyer, E.E. Pfaffenberger, and J.V. Poplawski, Current Practice, *STLE Life Factors for Rolling Bearings*, E.V. Zaretsky, Ed., 1992, p 1-45
10. M.N. Webster, E. Ioannides, and R.S. Sayles, The Effect of Topographical Defects of the Contact Stress and Fatigue Life in Rolling Element Bearings, *Mechanisms of Surface Distress*, Butterworth, 1986, p 207-224
11. R.S. Sayles, Debris and Roughness in Machine Element Contacts: Some Current and Future Engineering Implications, *J. Eng. Tribol.*, Vol 209 (No. J3), 1995, p 149-172
12. T.E. Tallian, Prediction of Rolling Contact Fatigue Life in Contaminated Lubricant, Part II: Experimental, *Trans. ASME, JOLT*, Vol 98 (No. 3), July 1976, p 384-392
13. W.J. Derner and E.E. Pfaffenberger, Rolling Element Bearings, *Handbook of Lubrication*, Vol 2, *Theory and Practice of Tribology*, E.R. Booser, Ed., 1983, p 495-537
14. S.O.A. El-Helieby and G.W. Rowe, A Quantitative Comparison between Residual Stresses and Fatigue Properties of Surface-ground Bearing Steel (En 31), *Wear*, Vol 58, 1980, p 155-172
15. W.O. Clinedinst, Fatigue Life of Tapered Roller Bearings, *ASME Journal of Applied Mechanics*, Vol 4 (No. 4), Dec 1937, p A-143 to A-150
16. G. Lundberg and A. Palmgren, Dynamic Capacity of Rolling Bearings, *Acta Polytech*, Mech. Eng. Series, Vol 1 (No. 3), 1947
17. W. Weibull, A Statistical Theory of the Strength of Materials, *IVA Handlingar* (Royal Swedish Acad. of Eng. Sciences, Proceedings), No. 151, 1939

18. W. Weibull, The Phenomenon of Rupture in Solids, *IVA Handlingar* (Royal Swedish Acad. of Eng. Sciences, Proceedings), No. 153, 1939

19. G. Lundberg and A. Palmgren, Dynamic Capacity of Roller Bearings, *Acta Polytech*, Mech. Eng. Series, Vol 2 (No. 4), 1952

20. "Rolling Bearings—Dynamic Load Ratings and Rating Life," International Standard ISO 281, 1990-12-01

21. "Load Ratings and Fatigue Life for Ball Bearings," American National Standard, ANSI/AFBMA Std 9-1990, 17 July 1990

22. "Load Ratings and Fatigue Life for Roller Bearings," American National Standard, ANSI/AFBMA Std 11-1990, 17 July 1990

23. J. Lieblein and M. Zelen, Statistical Investigation of the Fatigue Life of Deep-Groove Ball Bearings, *J. Res. Nat. Bur. Standards*, Vol 57 (No. 5), 1956, p 273-316

24. G. Lohmann and H.H. Schreiber, Determination of the Rating Life Exponent for Ball and Roller Bearings, *Werksatt u Betreib*, Vol 92 (No. 4), April 1959, p 188-192

25. C.A. Moyer, The Status and Future of Roller Bearing Life Prediction, *Advances in Engineering*, Y. Chung and H.S. Cheng, Ed., STLE SP-31, April 1991, p 89-99

26. H.-K. Lorösch, Influence of Load on the Magnitude of the Life Exponent for Rolling Bearings, *Fatigue Testing of Bearing Steels*, J.J.C. Hoo, Ed., STP 771, ASTM, 1982, p 275-292

27. E. Ioannides and T.A. Harris, A New Fatigue Life Model for Rolling Bearings, *Trans. ASME, JOT*, Vol 107, July 1985, p 367-378

28. A.A. Lubrecht, B.O. Jacobson, and E. Ioannides, *Lundberg-Palmgren Revisited Bearings—Towards the 21st Century*, Mech. Eng. Pub. Ltd., 1990, p 17-20

29. W.K. Dominik, "Rating and Life Formulas for Tapered Roller Bearings," Technical Paper 841121, SAE, 1984, p 1-15

30. R.-S. Zhou and H.P. Nixon, "A Contact Stress Model for Predicting Rolling Contact Fatigue," Technical Paper 921720, SAE, 1992, p 1-8

31. T.E. Tallian, Simplified Contact Fatigue Life Prediction Model, Part II: New Model, *Trans. ASME, JOT*, Vol 114, April 1992, p 214-222

32. R.C. Rice, Fatigue Data Analysis, Statistics and Data Analysis, *Mechanical Testing*, Vol 8, *ASM Handbook*, ASM International, 1989, p 675

33. C.A. Moyer, "The Role of Reliability for Bearings and Gears," Technical Paper FTM8, AGMA, Oct 1992

34. "Load Ratings and Fatigue Life for Ball Bearings," Standard 9, Anti-Friction Bearing Manufacturers' Association, June 1972

35. "Load Ratings and Fatigue Life for Roller Bearings," Standard 11, Anti-Friction Bearing Manufacturers' Association, June 1972

36. T.E. Tallian, Weibull Distribution of Rolling Contact Fatigue Life and Deviations Therefrom, *ASLE Trans.*, Vol 5, 1962, p 183-196

37. E.V. Zaretsky, Ed., *STLE Life Factors for Rolling Bearings*, STLE SP-34, 1992

38. B.J. Hamrock and D. Dowson, Isothermal Elastohydrodynamic Lubrication of Point Contacts, Part III: Fully Flooded Results, *ASME Trans., JOLT*, Vol 99 (No. 3), 1977, p 264-276

39. C.A. Moyer and L.L. Bahney, Modifying the Lambda Ratio to Functional Line Contacts, *STLE Trib. Trans.*, Vol 33 (No. 4), 1990, p 535-542

40. C.A. Danner, Fatigue Life of Tapered Roller Bearings under Minimal Lubricant Films, *ASLE Trans.*, Vol 13 (No. 4), 1970, p 241-251

41. S. Andreason and T. Lund, Ball Bearing Endurance Testing Considering Elastohydrodynamic Lubrication, Paper C36, *Proc. Symp. IMechE*, EHD, 1972, p 138-141

42. T.E. Tallian, Rolling Bearing Life Modifying Factors for Film Thickness, Surface and Friction, *ASME Trans, JOLT*, Vol 103, Oct 1981, p 509-520

43. R.L. Widner and W.E. Littmann, Bearing Damage Analysis, SP-423, Nat. Bur. of Standards, April 1976

Fatigue of Springs

Mark Hayes, Spring Research and Manufacturers' Association

Over half of all metallic springs that fail in testing or service do so by a fatigue mechanism. Sometimes the fatigue crack is initiated by preexisting defects or cracks, corrosion, wear, or mechanical damage, but in the majority of cases there is no significant "fault" identified that has initiated failure. Once initiated in consequence of repeatedly applied stress, a fatigue crack will grow some distance through the spring section until the effective spring section is reduced and can no longer support the maximum applied stress, at which time overload failure will occur. Fatigue is a likely failure mechanism for all types of springs (compression, extension, torsion, leaf, presswork, spiral, constant force, disc, etc.) as well as for all spring sizes (fatigue occurs in springs made from materials 0.1 mm thick to 80 mm diameter). Springs are invariably manufactured from high-strength materials, the most important of which are carbon steels, low-alloy steel, stainless steels, and nickel, cobalt, copper, and titanium alloys, and all are susceptible to fatigue failure. For more information about spring material composition and properties see *Metals Handbook*, Vol 1 (9th edition), p 283-313.

Fatigue Mechanisms and Performance Factors

Short Crack Incubation. More than 95% of all spring fatigue failures initiate at the surface, and high-integrity surfaces are better for resisting fatigue crack initiation. The importance of crack initiation is clear when the following observations are considered regarding fatigue testing of springs:

- Springs mostly fail at one position, even if a long length of material is exposed to the maximum stress in cyclic loading.
- It is very unusual to find any evidence of fatigue cracking away from the fracture itself when failed springs are examined metallographically.
- When a large number of springs are tested over a given stress range, the scatter in cycles to failure is very large (Fig. 1).

Generally the initiation crack is too small to see during metallographic examination, but once propagation starts, it occurs relatively quickly and usually before other cracks in the process of initiating have started to propagate. The variability and importance of initiation for the first small crack results in scatter, which is why spring fatigue data are treated in a statistical manner. In addition, failure of a metallic spring by a fatigue mechanism is almost always sudden and catastrophic: there is no warning of impending failure. The spring is working completely satisfactorily one minute and is in two pieces the next. There also are no nondestructive or other methods available for predicting imminent failure.

The mechanism of crack initiation is typically investigated by examination of the resultant fracture surfaces, and this article provides several examples. However, for most springs with high-integrity surface quality and a clearly defined position of maximum stress, failure is generally initiated at the position of maximum stress (e.g., on the inside surface of an active coil in a parallel sided compression spring, or at the outside surface of a torsion spring at the position where an external tangential leg meets the coiled body of the spring).

Stresses and Fracture Appearances. Most springs are generally subjected to torsion or bending stresses. Helical compression, extension, and volute springs and torsion bars are the most common spring forms that are stressed in torsion. Consequently, the fatigue fracture surface of these springs is frequently at 45° to the wire, strip, or bar axis, and the final overload failure takes up a characteristic helicoidal shape. Torsion springs (so named after their load output, torque); leaf springs; clock, spiral, and constant-force springs; most spring pressings; disc springs or belleville washers; and the hooks of extension springs are all stressed in bending. The fatigue fracture surface of these springs is initially at 90° to the bending stress.

Having established that there are only two main stress regimes for typical spring loading, it is necessary to emphasize that the calculation of stress values in springs is not always straightforward, because there tend to be small bending stresses additional to the main torsional stresses in compression springs, and likewise a small ele-

ment of torsion in springs for which the principal stress mode is bending. Furthermore, there are many possibilities for stress raisers to be present in springs (surface defects, effectively notches, in springs made from wire, bar, or strip material, burrs on the edge of strip materials, holes stamped in flat springs, etc.). In addition, it is not usually possible to obtain an accurate measure of stress by strain gaging, because springs are too small or too sharply curved. Finite element analyses can be helpful for establishing stress levels in some circumstances, but at high deflections and stresses, close to the elastic limit, finite element methods are not so accurate. Hence, whenever stress is referred to in this chapter, it has been calculated by classical mechanics, generally using the formulas given by Wahl (Ref 1), and no allowance has been made for stress raisers or residual stresses.

Residual Stress and Surface Roughness. For a given spring material with a certain strength and surface quality, residual stress is a key variable that influences fatigue resistance. Residual stresses are certain to be generated during spring manufacture, and so the stress threshold for fatigue crack propagation will be reached at much lower applied stresses when residual tensile stresses are present. When compressive residual

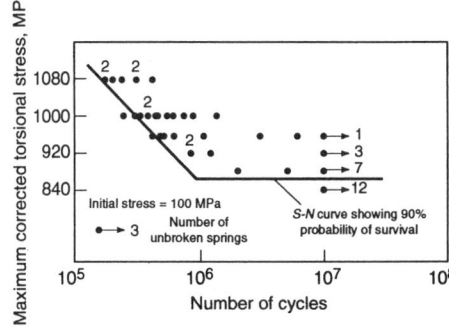

Fig. 1 Scatter in the fatigue life for shot-peened carbon steel spring wire (ASTM A 227, Class II steel). It is not uncommon for some springs to survive 10 million cycles when other springs from the same batch, tested at the same time, have failed at 400,000 cycles. The S-N curve shown here gives the results of 55 fatigue tests carried out on shot-peened springs.

stresses are present, the applied stress necessary to reach the stress threshold for fatigue crack growth will have to be high to negate the residual compressive stress. In addition, fatigue resistance is affected by surface roughness (which is why shot peening rather than shot blasting is widely employed in spring manufacturing).

Residual stress and surface roughness are interrelated phenomena that are correlated, to some degree, by the surface and the process. For example, rough grinding typically produces high tensile residual stresses; "fine" grinding, lower tensile stresses; polishing and sanding, moderate compression; lapping and polishing, high (but shallow) compressive stresses, and so on. In reality, residual stress produced in the surface by the finishing methods is the key effect on fatigue life, not the surface roughness itself.

When the independent effects of residual stress and surface roughness are considered, the effect of residual stress on fatigue resistance appears to be a more potent factor than surface roughness. The effect of surface roughness (within reasonable ranges, say up to 2.5 or 5.0 μm, arithmetic average) is much less than traditionally considered by the engineering community, according to a broad study of various finishing processes (e.g., polishing and peening) on the fatigue life of several alloys (see W.P. Koster, Effect of Residual Stress on Fatigue of Structural Alloys, *Practical Applications of Residual Stress Technology*, ASM International, 1991, p 1-10). Thus, surface residual stress is a predominant factor affecting fatigue life, provided that gross surface discontinuities are avoided.

If the surface residual compressive stress is increased by controlled surface engineering, it is possible that the maximum applied stress occurs below the surface. When these circumstances arise (as they should in a well shot-peened spring), fatigue fracture initiation may not be at the surface of the spring, but rather may occur at between 150 and 400 μm below the wire, bar, or strip surface. However, in order to initiate a fatigue crack from below the surface, it is necessary that a significant stress raiser exist at this position, and almost invariable this stress raiser is a nonmetallic inclusion. In addition, this inclusion has to be greater than 15 μm, based on observations of many investigators (Ref 2-4) and the observations of investigators at the Spring Research and Manufacturers' Association (Ref 5), who have examined hundreds of spring fatigue failures by scanning electron microscopy. Inclusions influence fatigue fracture appearance only when the springs have been shot peened or surface engineered to produce a residual compressive stress, and when the inclusion is larger than 15 μm.

However, it is not only observations of fractures that lead to the conclusion that the inclusion must be 15 μm. A short-crack theory of fatigue mechanisms in high-strength materials (Ref 6) also predicts that short cracks need to "incubate" to 10 to 15 μm before fatigue cracks propagate. Although cracks of 15 μm could be observed metallographically, it is not surprising that such small cracks at the surface, or subsurface, would

be missed when metallographic examinations are undertaken of springs that have failed by fatigue. This is consistent with the common observation of fatigue failure only at one position, with no additional cracking observed when springs are examined metallographically.

Critical Factors Affecting Spring Fatigue Performance. The above discussions have briefly summarized the factors of surface integrity, applied stress, residual stress, and subsurface crack initiation. However, it is also worth listing practical factors that have a direct bearing on spring performance, and the circumstances in which fatigue could occur.

Fatigue failure can occur after as few as 4000 cycles when a spring is subject to very high bending stresses close to the elastic limit of the material, but usually 10,000 to 50,000 cycles are required to cause fatigue failure. The probability of fatigue failure is related to fundamental material factors (such as residual stress), but also to each of the following application factors:

- Spring material type and strength
- Stress conditions such as maximum applied stress, applied stress range, stress regime (bending, torsion, or a combination), and number of cycles at maximum stress range
- Surface quality
- Spring manufacturing processes, residual stress, and surface roughness
- Environment, corrosion, and temperature
- Rate of application of load
- Wear and fretting
- Embrittlement and cracking

There are two notable absentees from this list. First, fatigue crack growth rate is of relatively little concern, because the key to preventing fatigue failure in springs is the avoidance of crack initiation. Second, impact failure seldom occurs in springs (even at very low temperatures at which spring materials are very brittle), because springs deflect elastically as a consequence of impact loading. (However, under high load rates, impact conditions may become important if the spring could become subject to sufficiently high deflections that the coils could impact with one another, the end coil, or adjacent components.)

In addition, even though springs generally fail at only one position under laboratory conditions, in actual service multiple cracking is more frequently observed. The reasons for this difference are believed to include the following:

- In service, compression springs often fail in such a way that the coils interlock and the spring can and does continue to operate after first failure. However, stress levels increase after the first fatigue failure, so a second fatigue crack grows rapidly, causing a second break.
- In service, environmental influences (however slight) are likely to play a role, and a characteristic of springs that fail by corrosion fatigue, or fatigue initiated by small corrosion pits, is that they fail into three or more pieces.

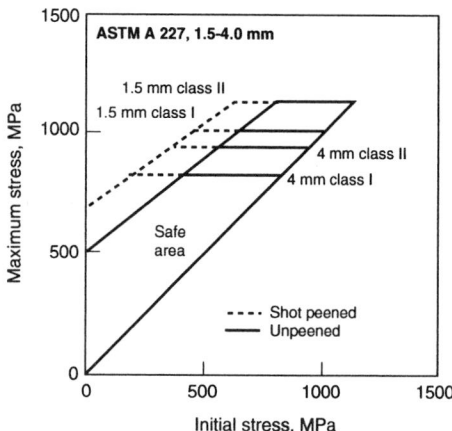

Fig. 2 Goodman diagram for 10 million cycle fatigue life and 95% probability of survival for cold-drawn carbon steel wire spring (ASTM A 227, 1.5-4.0 mm wire sizes)

- In springs that fail in service, apparently by a fatigue mechanism, fatigue crack initiation may be due to the presence of material defects, cracks, burrs, or other preexisting faults.

Material Type and Strength. Typically, springs are only manufactured from high-strength (or high-hardness) materials. The maximum strength utilized is often determined by considerations of formability (in small springs) or from concern over the risk of brittle failure.

Stress conditions,, such as maximum stress and mean stress, are obvious fatigue factors. The fatigue performance of springs is usually illustrated by modified Goodman diagrams, like Fig. 2, which covers actual fatigue data for carbon steel compression springs made from both the lower-strength (class I) and the higher-strength (class II) ASTM A 227 wire. The Goodman diagram does not apply if the spring is subject to periodic or occasional overloads. Cumulative damage due to such overloading is a more difficult question that is addressed elsewhere in this Volume. Moreover, Fig. 1 and 2 only illustrate the case for compression springs loaded between two applied torsional stresses. These diagrams would not be applicable if the stress regime were in bending, or if a combination of torsional and bending stresses were present. Figures 1 and 2 also are limited to springs loaded from only one side of their unloaded position. Some springs can operate from either side of their unloaded position, in which circumstances the benefits to be gained from prestressing are not applicable.

Surface Quality. After a spring material is selected based on strength requirements and environments, the next most important factor is surface quality. Typical surface defects found in spring materials are seams, laps, and decarburization. The maximum depth of such defects can be controlled by raw material producers, and it is limited in the better international spring material specifications (e.g., no defects permitted that exceed a depth of 1% of the wire diameter or 40 μm, whichever is smaller). Forty μm is the smallest

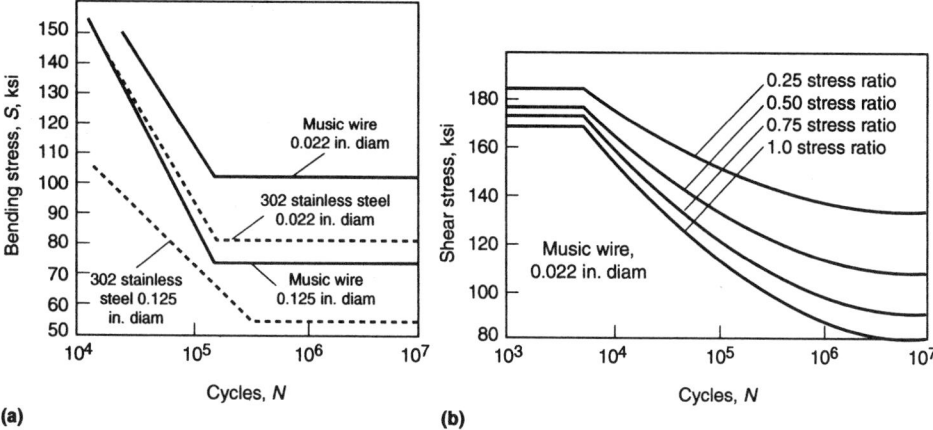

Fig. 3 Bending and shear fatigue strength. (a) Bending fatigue of various spring wire. (b) Shear fatigue curves for music wire (0.022 in. diam) at various stress ratios. Source: Ref 8

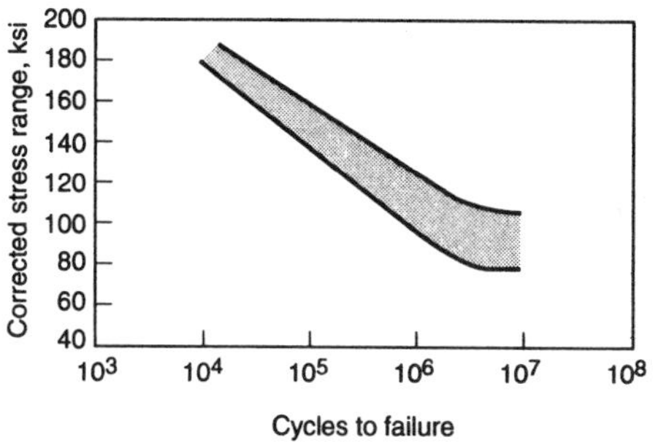

Fig. 4 Typical scatter band in fatigue tests of music wire helical springs, stress range zero to maximum. Wire size 0.022-0.048 in. Source: Ref 9

surface defect that can reliably be detected by eddy-current methods.

Spring Manufacturing Processes. The most common spring manufacturing processes may each have some influence on residual stress or surface roughness and will thereby affect spring fatigue performance. The most important principles involved are that:

- Forming a spring will generally involve bending wire or strip to the required shape. In this process the spring material will be overbent and will then spring back. The residual stresses are those consequent upon the springback.
- Stress relieving, or low-temperature heat treatment, after spring forming will stabilize the spring shape and relieve some, but not all, of the residual stresses introduced during forming. If springs are hardened and tempered or austempered after forming, the residual stresses will be very close to zero.
- Prestressing is a process widely used in the spring industry in which the spring is deflected beyond its normal working position, deliberately causing plastic deformation of the material in the loading direction and thereby inducing a higher elastic limit and beneficial residual stresses.
- Shot peening is the single most important process for fatigue resistance in most dynamically loaded springs.
- Other surface engineering processes that are capable of enhancing residual compressive stresses and/or surface roughness of springs include ion implantation, nitrocarburizing, laser shock processing, carefully controlled physical vapor deposition, and electropolishing, but as yet these processes are of limited commercial importance.
- Corrosion protection processes do not generally have a large influence on fatigue resistance, but they may impart a small residual tensile stress in a spring surface, and they are consequently detrimental. Examples of such processes are the use of pregalvanized wire and electroplating springs with zinc or other metals.

Rate of Application of Load. If a load is applied (or removed) quickly from a spring, the spring may not react as fast as the load is applied. A spring has a natural frequency that determines its maximum speed of operation. Also, under conditions of dynamic loading a spring, or spring mass system, may be prone to resonance. In either of these circumstances, the actual stress range for the spring may be significantly higher than that calculated, and consequently the fatigue performance may be much worse than expected.

Wear and fretting sometimes occur when a spring is in contact with nearby components or itself. The wear resistance of spring materials is quite good, but the wear region is obviously more susceptible to fatigue failure. Fretting fatigue occurs when mated parts undergo small oscillatory displacement relative to each other, as in the case of leaf springs, for example (Ref 7).

Embrittlement or Cracking. Hydrogen or liquid metal embrittlement may cause springs to fail if these embrittlement conditions arise when the spring is under load, without the need for any fatigue mechanism. However, if the embrittlement or cracking occurred as a consequence of residual tensile stresses, then the precracked spring can fail very early as a consequence of dynamic loading. Examples of such fatigue failures have occurred in springs with small quench cracks, stress-corrosion cracks, coiling cracks (through failure to stress relieve silicon chromium steel immediately after coiling), and liquid metal embrittlement by solder or other lead, tin, zinc, or cadmium alloys.

Fatigue of Spring Steels

As previously noted, the selection of spring material is usually based on factors besides strength, such as formability, surface condition, cost, or the risk of embrittlement. Steels for springs include carbon steels in the mid- to high-carbon range (AISI 1050-1095) for cold-formed or wound springs, and low-alloy steels such as AISI 4160, 6150H, and 8660H for hot-wound forms. For small springs, music wire (ASTM A

228) is the highest-quality carbon steel (with a surface quality almost comparable to that of valve spring wire). Hard-drawn wire (ASTM 227) is the least expensive and has the lowest surface quality. Valve spring wires (such as ASTM A 230, A 232, and A 401) are more expensive, and have higher surface quality and are suitable for slightly higher temperature use. For corrosion and heat resistance, cold-drawn stainless steel 302 wire (ASTM A 313) has moderate surface quality and is available in several tempers. It is more expensive than carbon steel for designs requiring diameters over about 0.3 mm (0.012 in.) but less costly than music wire under 0.3 mm (0.012 in.) because of the lack of need for plating. Stainless steel type 316 is good for spring wire because it has better corrosion resistance than 302, particularly in salt water.

The conventional *S-N* diagrams (such as in Fig. 3) are a useful start for general comparisons of fatigue strength. However, *S-N* curves are not available for the wide variety of loading conditions that may occur in spring design. Scatter in fatigue life (Fig. 4) is another key variable that requires statistical analysis of some type.

In addition, shot peening is a key factor for fatigue resistance in most dynamically loaded springs. For leaf springs this process is sometimes carried out in conjunction with prestressing, the so-called strain peening process, in order to gain exceptionally high levels of residual compressive stress. Optimizing the shot peening process has been the subject of much research, but it is almost always beneficial (Ref 10), even when carried out under non-optimum conditions. An early example of shot peening effects on fatigue is shown in Fig. 5.

Scatter and Statistical Analysis of *S-N* Data. The considerable scatter in fatigue strength data is attributable to a variety of factors, including defect frequency and variations in strength with changes in diameter. Two methods of statistical analysis are briefly summarized below.

Confidence Limits. Figure 1 is a typical *S-N* curve for shot-peened compression springs made from hard-drawn carbon steel. In this instance the wire size was 2.5 mm diameter, at which size the tensile strength of this grade of wire was about 1800 MPa. Superimposed upon the actual data points given in Fig. 1, a line has been drawn that represents the 90% probability of survival. For

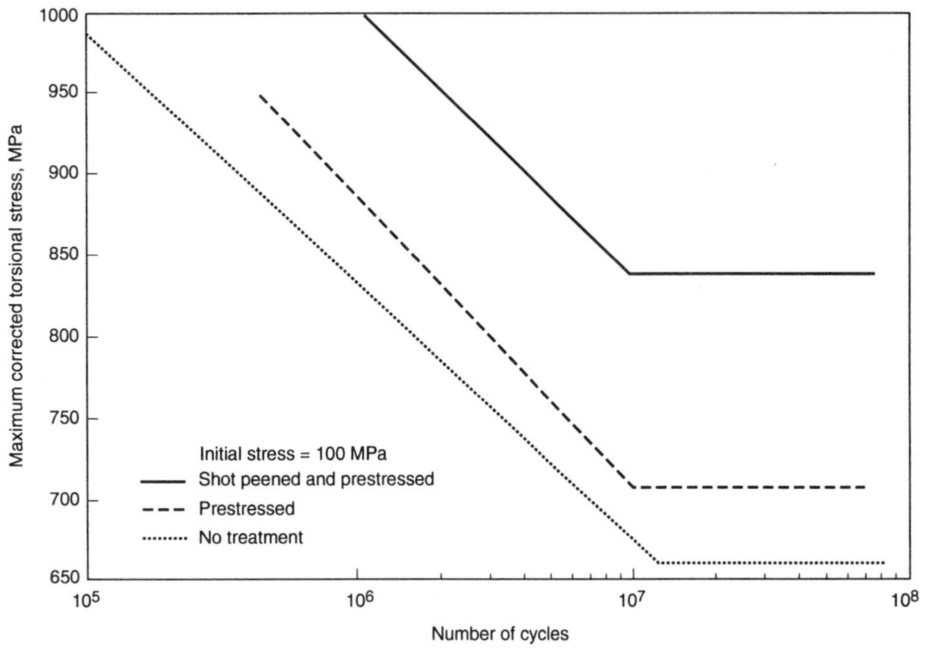

Fig. 5 Effect of prestressing and shot peening on fatigue curves for typical compression springs made of chrome-vanadium wire (ASTM A231, 1.5-4.0 mm wire sizes)

Table 1 Spring fatigue data from Fig. 1

Maximum stress, MPa	Number tested	Number broken	Ideal percentage of failure for probit	Actual percentage
1080	7	7		
1000	12	12	95	100
960	9	8	80-90	89
920	6	3	25-75	50
880	9	2	10-20	22
840	12	0	5	0

each maximum applied stress value, it is a fairly simple mathematical exercise, using Weibull statistics (Ref 11) or other methods, to estimate the number of cycles at which 90% of the springs would have been expected to break. Having calculated the 90% probability of survival at each stress value, then it is relatively easy to fit the 90% curve to the data, using the characteristic shape of line given in Fig. 1.

Probit Method. No one designs for even a 10% chance of failure, and hence a more sophisticated statistical treatment of spring fatigue data is necessary for spring designers to know the stress value at which no springs will be susceptible to fatigue failure in service.

Table 1 shows a pattern to the number of fatigue tests at each maximum stress value in Fig. 1. The testing was deliberately arranged so that *x* springs were tested at the stress at which 25 to 75% would fail, 1.5*x* springs for the stress at which 80 to 90% would fail or survive, and 2*x* springs for the stress at which 95% would fail or survive. This pattern of testing is the type required for the simplest analysis according to the probit method (Ref 12, 13). From probit analysis, it is possible to predict confidence limits for stress values at which no more than one in ten, one in a

thousand, or one in a million springs would be expected to fail. The probit method of statistical analysis is the most widely used method in Europe and Japan for analyzing spring fatigue data so that safe design stresses can be calculated that will lead to very low risks of failure by fatigue in service (Ref 14). In practice, confidence limits vary rather widely unless a very large number of fatigue tests have been undertaken.

A probit analysis is useful only if it has been carried out for precisely the same wire type and size as the spring designer wishes to use, and if the fatigue testing was carried out under appropriate stress conditions (including residual stresses from surface engineering). Hence, probit data is of limited applicability, but it can be more useful if probit analyses are carried out at two or, preferably, three initial stress values for a given spring material type (and surface condition), because this will enable the construction of a Goodman diagram (Fig. 2). In order to draw Fig. 2 it was necessary to carry out at least two probit analyses on shot-peened springs and two probit analyses on springs that had not been shot peened. Experience has shown that the effect of wire size on the shape and size of Goodman diagrams is relatively small, so Fig. 2 is valid for all wire sizes between 1.5 and 4.0 mm, despite the differences in the tensile strength of these wires. Similarly, the diagram is applicable to class I and class II (higher tensile strength) wires, the only modification to the diagram being the higher maximum stress allowable in the class II wire.

The Goodman diagram in Fig. 2 is for springs that are required to operate for 10 million cycles. The diagram would have a larger area, and hence allow larger stress ranges, if it were drawn up for springs that will never operate for more than one

million cycles. For ASTM A 227 material, the diagram would not need to be smaller in area if the springs were required to operate for more than 10 million cycles. It is generally accepted that in this grade, a spring that survives 10 million cycles will survive forever. This is not true for springs made from shot-peened oil-tempered silicon chromium wire (e.g., ASTM A 401) for which failures between ten and fifty million cycles have frequently been recorded, but nonetheless silicon chromium is generally regarded as the best material available today for dynamic performance springs.

Goodman diagrams are the most useful source of fatigue data for designers of springs, but using published Goodman diagrams requires understanding of the applicable design circumstances. If in doubt, fatigue testing of samples is advised. Testing is advised whenever a spring design is close to the design limit given on a Goodman diagram (some of which are drawn up on the basis of only 95% probability of survival), or if the actual spring differs from the published data in terms of surface quality, prestressing, surface engineering, operating environment, larger wire size than that for which the Goodman diagram is applicable, or other factors. With these important qualifications in mind, typical Goodman diagrams are shown in Fig. 2, 6, and 7. Design curves and allowable stresses for dynamic loading of springs should also be considered from various design publications, such as those listed in "Selected References: Spring Design," at the end of this article.

Examination of Failed Springs

Springs may fail as a result of corrosion, various types of embrittlement, relaxation, overload, or fatigue (see, e.g., *Metals Handbook*, Vol II (9th edition), p 550-562). Sometimes these causes act in combination, but the most common is fatigue, which may occur after as little as a few thousand operations. The characteristics of fatigue failure in springs are readily identified by experienced investigators, but these characteristics are briefly summarized below.

When examining spring fatigue failures, it is almost always prudent to check the structure and hardness adjacent to the fracture surface. This is most readily accomplished by taking metallographic transverse and longitudinal sections from behind the fracture initially. These sections should be mounted in plastic, polished to a 1 μm finish and then hardness tested. If the material hardness is approximately correct, the sections should be examined metallographically for surface defects or gross internal faults to check that the material is to specification, particularly with respect to surface quality. Etching will reveal the structure of the material, which should always be uniform throughout, and any surface decarburization present in carbon or low-alloy steels should be within the prescribed limits of the specification. It is self-evident that surface de-

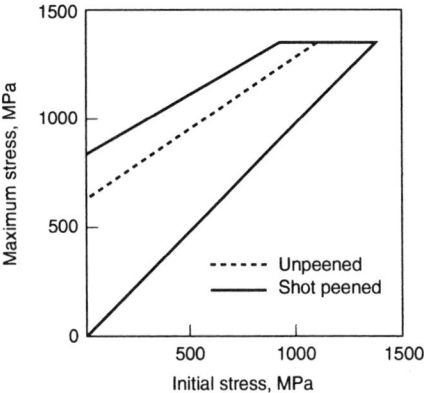

Fig. 6 Goodman diagram of 10^7 cycle fatigue limit for 4.2 mm Si-Cr (ASTM A401) wire compression spring

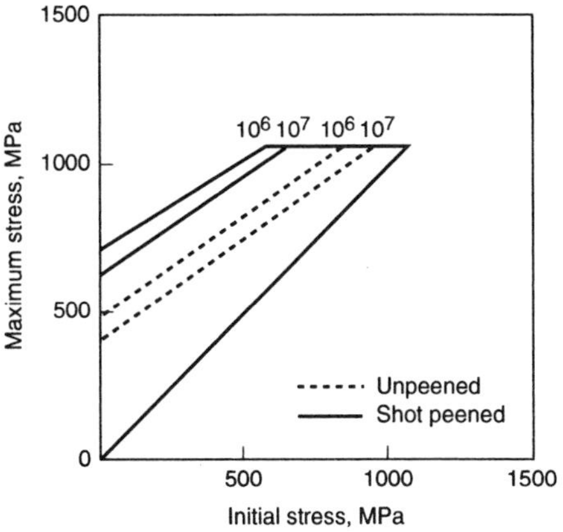

Fig. 7 Good diagram of 10^6 and 10^7 cycle fatigue life and 95% probability of survival for unpeened and shot peened 302 stainless steel springwire (ASTM A313, 1.5-4.0 mm wire)

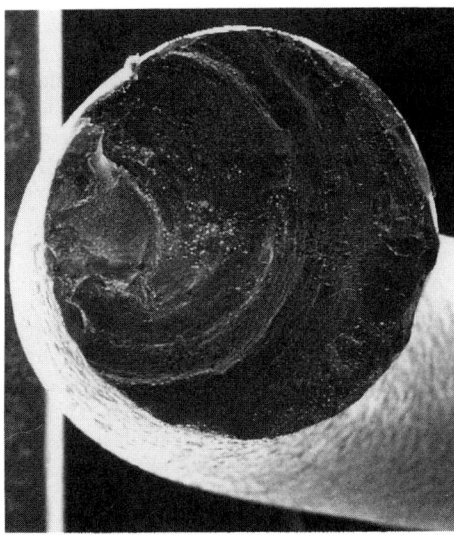

Fig. 8 SEM photomicrograph showing the fracture surface of a shot-peened 302 stainless steel spring that had fatigue initiation at an inclusion that was 300 μm below the wire surface. 21×

fects or decarburization represent sites of weakness from which fatigue cracks are likely to initiate.

Springs may suffer fatigue failures because of some fault in the raw material, non-optimum spring manufacture, or an environmental influence in service. However, springs may also fail because the stress range and number of cycles exceed the Goodman diagram limits for that spring. Hence, the applied stress should be calculated whenever possible, so that an estimation of the likely fatigue performance may be made by comparing the calculated stresses with the best available published data. In estimating stresses it is necessary to consider dynamic aspects as well as the magnitude of stresses at the working positions, because the actual stresses could be considerably higher than is immediately apparent, perhaps due to resonance or impact loading effects. The possibility of stress raisers, wear due to contact with adjacent components, or residual tensile stresses from manufacture also needs to be considered. Stress calculations are best accomplished by use of commercial computer-assisted

design programs, or from first principles of mechanics using texts such as Wahl (Ref 1). Finite element methods in association with load testing may also be used, but experience has shown that these methods may be insufficiently accurate at the large deflections at which springs are usually used.

Beach marks or striations, which are the identifying feature of fatigue in many components, are not always present in spring fatigue fracture surfaces. The absence of beach marks has led some investigators to misdiagnose the reason for spring failure. Beach marks are more likely to be present if the spring material is comparatively soft (40 to 45 HRC), but they are seldom present when higher-strength spring materials fail by fatigue.

General Fracture Appearance. Some fracture features are always present in spring fatigue. For example, the fracture surface from spring fatigue invariably has a region that is flat in topography, often with clear "river markings" showing the position or positions at which the fatigue crack started. This flat region never extends over the whole fracture surface, because the fatigue crack, which caused the flat region, will reduce the effective section from which the spring is manufactured. When the fatigue crack has grown far enough, the remaining material will no longer be able to support the load on the spring and so will fail by an overload mechanism.

The direction of propagation of the fatigue crack will generally be at 90° to the principal axis of the spring section if the spring is stressed in bending, or at 45° if the spring is stressed in torsion. However, if the spring is stressed in torsion, the stage I fatigue crack may initiate in the resolved shear stress direction before switching to the stage II tensile direction. The final overload failure is often referred to as stage III of a fatigue failure. When fatigue cracks initiate below the

surface in compression springs (because of surface compressive stresses from shot peening), the cracks propagate in the 45° direction. A typical example is Fig. 8.

These characteristics can usually be identified by the simple expediency of examining the fracture surface on a binocular microscope. A scanning electron microscope (SEM) is generally not needed to recognize a fatigue failure, but it is necessary to use an SEM to confirm fatigue as the failure mechanism when springs are manufactured from wire or strip that is less than about 1 mm in diameter or thickness. In addition, the SEM is often necessary to reveal details of the position at which a fatigue crack initiated. For instance, a small corrosion pit or some mechanical damage may be present at the fatigue origin, which could easily be irresolvable or overlooked on examination with a binocular microscope alone.

Springs stressed in bending have their fatigue fracture origin at the tensile surface of the bend. Typically, this position will be at the inside of a hook on an extension spring, or where the moving tangential leg meets the body of a torsion spring, in which case the position of maximum stress will be at the outside surface if the torsion spring is loaded in the wind-up direction, as is invariably recommended. In strip springs the position of maximum stress is often at the outside surface of a bend, particularly if the spring "hinges" about this bend. In clock, constant-force, or spiral-type springs there is a very long length of the outside surface of the strip where the stress is maximum and is equal across the width of the strip. In these springs there are often numerous fatigue cracks formed before complete failure occurs.

Example 1: Corrosion Fatigue Failure of 302 Stainless Steel Spring. After only two months of use in a printing operation, type 302 stainless steel springs were breaking into several pieces. The springs were operating over a very small deflection and were regulating the flow of ink, in which they were constantly immersed.

Visual examination revealed clear evidence of fatigue fractures on every piece of the spring, and each of the fractures was oriented at 45° to the wire axis. Figure 9 shows crack arrest marks and fatigue striations (beach marks), which are not always expected in the fracture surface of spring fatigue failures, as noted above. Clear evidence of pitting corrosion at the fatigue fracture origin can also be seen. However, the portions of spring showed no evidence of red rust and had a bright (greenish from the ink) surface, indicating that the corrosion pitting was local in nature.

Analysis of the ink in which the spring was operating revealed that free chloride ions were present. The extent of the pitting damage was surprising, but once initiated the corrosion was rapid because the surface of the stainless steel could not repair itself, because there was no oxygen present in the alcohol-based ink. The solution was to use an alternative ink that contained no free chloride ions, and three years later the springs are working satisfactorily. All the compa-

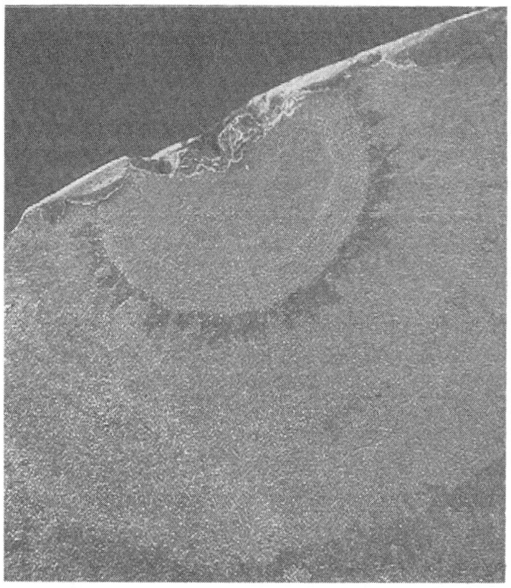

Fig. 9 Optical photograph of the fatigue region of a stainless steel compression spring that failed due to corrosion fatigue. Note beach marks on this fracture. As noted in text, beach marks are often not present in fracture surface of spring fatigue failures, which makes this case somewhat unusual. 40×

nies involved in this problem were surprised that corrosion fatigue was the problem, because no rust could be seen and the operating stresses were insufficient to initiate fatigue.

Example 2: Aqueous Corrosion Fatigue of Type 302 and 304 Stainless Steel Springs. After six months of operating a new chicken house, a farmer noticed that the majority of the water feeders had stopped working. The water feeders operated on the principle that when the chickens pecked a plastic bowl, a compressed spring released a squirt of water. If the spring broke, no water came out.

The small compression springs were made from 0.8 mm diameter type 302 stainless steel, and the operating stresses were safely within the design limits given by the Goodman diagram. These springs were so small that it was necessary to use a SEM to identify why they had failed. Some were in three or four pieces, but most had only one fracture and did not show any evidence of mechanical or corrosion damage. On the SEM it was observed that the springs contained numerous cracks only on their inside surface, and these cracks were all at 45° to the wire axis. The cross patterns made by these cracks were recognized as the classic indication of corrosion fatigue.

The solution was to select a grade of spring steel that would be more corrosion resistant than 302 stainless. Glass bead peening would probably have been beneficial too, but the springs were too small for this process. Type 304 stainless steel springs were tried, but these also failed, albeit after two or three times longer service than 302. Inconel 600 was eventually used, and no

complaints about thirsty chickens have been received since.

Example 3: Effect of Spring Unwinding on Fatigue. This example involves an agricultural tine, which is a relatively large double torsion spring with outer legs that are used to sweep through hay or other crops and turn them over. The tine machinery manufacturer submitted several examples of failed tines made from hard-drawn carbon steel, wanting to learn whether the life could be significantly improved.

Visual examination revealed that the wear pattern on the outside of the long legs of the tines was consistent with them having been used in the wind-up direction, as recommended. The mechanical damage on the broken tines was not thought to be excessive, and the leg angle was close to that specified on the drawing, indicating that it had not been significantly abused in service. Abuse (such as use in very stony ground) would have caused plastic deformation. Inevitably there was some rust on the failed tines, but no evidence of pitting could be seen.

Visual examination of the fracture surface showed that bending fatigue was almost certainly the cause of failure, but despite the clear evidence that the tine had been used in the wind-up direction, the fatigue fracture origin was on the inside surface of the legs, at the point where they joined the coiled body of the spring. No evidence of corrosion pits at the origin of the fatigue crack could be seen.

Inquiries were made concerning the process route used to manufacture the tines. The spring-maker reported that the hard-drawn carbon steel had been coiled, stress relieved at 200 °C, and painted, explaining that the stress relieving temperature had been kept as low as possible in order to maximize the retention of beneficial residual stresses from coiling. This is a good practice for dynamically loaded torsion springs.

In this instance, however, it appeared that the tines were being wound up by loading them with hay. When the load was released, the tines were springing back through the neutral unloaded position and into the unwind direction. This movement into the unwind direction was happening often enough to initiate fatigue. The solution was to increase the stress relieving temperature to reduce the residual stresses from coiling and hence improve fatigue performance. This may give rise to earlier fatigue failure in the wind-up direction, of course, but the net effect is to prolong the tine life significantly.

ACKNOWLEDGMENT

The help of my colleagues at the Spring Research and Manufacturers' Association in preparing and checking this article is gratefully acknowledged.

REFERENCES

1. A.M. Wahl, *Mechanical Springs*, 2nd ed., McGraw-Hill/SMI, 1982
2. P. Hora and V. Leidenroth, *Quality of Coil Springs*, Dr. Riederer Verlag, 1987
3. T. Hagiwara et al., Super-clean Steel for Valve Spring Quality, *Wire Journal International*, April 1991, p 29-34
4. T. Shibata et al., Valve Spring with High Fatigue Resistance for Automotive Engine, Technical Paper 880417, SAE, Feb 1988
5. M. Hayes, Physical Appearance of Spring Fatigue Failures, *The Metallurgist and Materials Technologist*, Nov 1984, p 572-573
6. K.J. Miller, Metal Fatigue—Past, Current and Future, *Proc. Institute of Mechanical Engineers*, Vol 205 (No. C5), 1991
7. Y. Mutoh, K. Tanaka, and K. Takeda, Fretting Fatigue of Automotive Leaf Spring, *Fatigue of Engineering Materials and Structures*, Vol I, Mechanical Engineering Publications, 1986, p 203-209
8. *Design Curves for Neg'ator Springs*, Ametek Corp., 1967
9. C. Osgood, *Fatigue Design*, 2nd ed., Pergamon Press, 1982
10. M. Hayes, On the Factors Affecting the Fatigue Limit of Shot-Peened Springs, *Metal Treatments against Wear, Corrosion, Fretting and Fatigue*, Pergamon Press, 1988, p 41-49
11. L.D. Johnson, *The Statistical Treatment of Fatigue Experiments*, Elsevier Publishing Co., 1963
12. D.J. Finney, *Probit Analysis*, Cambridge University Press, 1952
13. E.S. Pearson and H.O. Hartley, *Biometrika Tables*, Cambridge University Press, 1962
14. M.P. Hayes and L.F. Reynolds, "Analysis of Spring Fatigue Data," Report 431, Spring Research and Manufacturers' Association, 1988
15. C. Osgood, *Fatigue Design*, 2nd ed., Pergamon, 1982, p 393-404

SELECTED REFERENCES: SPRING DESIGN

- H. Carlson, *Spring Designer's Handbook*, Marcel Dekker Inc., 1978
- *Design Handbook*, Associated Spring, 1987
- Deutsche Industrie Norms, DIN2088 (Nov 1992), DIN2089 Part 1 (Dec 1984), DIN 2089 Part 2 (Nov 1992), DIN2091 (June 1981), and DIN2093 (Jan 1992)
- *Helical Springs*, Engineering Design Guide 08, Oxford University Press, 1974
- Japan Industrial Standards and JIS B2709, 1987
- "Manual on Design and Application of Helical and Spiral Springs," SAE HS795, rev. April 1990
- "Manual on Design and Manufacture of Torsion Bar Springs," SAE HS796, rev. July 1990

Section 4: Fracture Mechanics, Damage Tolerance, and Life Assessment

An Introduction to Fracture Mechanics*

Stephen D. Antolovich, FASM, Professor and Director, School of Mechanical and Materials Engineering, Washington State University, and Bruce F. Antolovich, President and Consulting Engineer, Metallurgical Research Consultants Inc.

THE CONCEPTS of fracture mechanics are basic ideas for developing methods of predicting the load-carrying capabilities of structures and components containing cracks. Though virtually all design and standard specifications require the definition of tensile properties for a material, these data are only partly indicative of inherent mechanical resistance to failure in service. Except for those situations where gross yielding or highly ductile fracture represents limiting failure conditions, tensile strength and yield strength are often insufficient requirements for design of failure-resistant structures. Brittle fracture can also occur if toughness, resistance to corrosion, stress corrosion, or fatigue resistance are reduced too much in achieving high strength.

The concepts of fracture mechanics are concerned with the basic methods for predicting the load-carrying capabilities of structures and components containing cracks. The fracture mechanic approach is based on a mathematical description of the characteristic stress field that surrounds any crack in a loaded body. When the region of plastic deformation around a crack is small compared to the size of the crack (as is often true for large structures and high-strength materials), the magnitude of the stress field around the crack is related to the stress-intensity factor, K, with:

$$K = \sigma (\sqrt{a}) Y (a/W)$$

where σ = remotely applied stress, a = characteristic flaw size dimension, Y = geometry factor that depends on the ratio of the crack length, a, to the width, W, determined from linear elastic stress analysis. The stress-intensity factor, K, thus represents a single parameter that includes both the effect of the stress applied to a sample and the effect of a crack of a given size in a sample. The stress-intensity factor can have a simple relation to applied stress and crack length, or the relation can involve complex geometry factors for complex loading, various configurations of real structural components, or variations in crack shapes. In this way, linear elastic analysis of small-scale yielding can be used to define a unique factor, K, that is proportional to the local crack tip stress field outside the small crack tip plastic zone.

These concepts provide a basis for defining a critical stress-intensity factor (K_c) for the onset of crack growth, as a material property independent of specimen size and geometry for many conditions of loading and environment. In general, when the specimen thickness and the in-plane dimensions near the crack are large enough relative to the size of the plastic zone, then the value of K at which growth begins is a constant and generally minimum value called the plane-strain fracture toughness factor, K_{Ic}, of the material. The parameter K_{Ic} is a true material property in the same sense as is the yield strength of a material. The value of K_{Ic} determined for a given material is unaffected by specimen dimensions or type of loading, provided that the specimen dimensions are large enough relative to the plastic zone to ensure plane-strain conditions around the crack tip (strain is zero in the through-thickness or z-direction). Therefore, plane-strain fracture toughness, K_{Ic}, is particularly pertinent in materials selection because, unlike other measures of toughness, it is independent of specimen configuration.

In the plane-strain state, a material is at its lowest point of resistance to unstable fracture. The onset of fracture is abrupt and is most clearly observed in thick sections of low-ductility (high-strength) alloys, as discussed in more detail in this article. Originally, the technology of fracture mechanics was limited to relatively high-strength materials that could be tested in sizes that met certain requirements for linear-elastic displacement during slow testing of specimens of certain configurations. More recent advancements in the state of the art, such as R-curve and J-integral tests, have extended the use of fracture mechanics to elastic-plastic conditions which are associated with lower-strength materials and smaller section sizes. However, when stresses approach or exceed yield values, the elastic stress field surrounding the crack departs from that of plane strain (from the development of an enlarged crack tip plastic zone which generally enhances fracture toughness). With increasing load, slow stable crack extension (tearing) may accompany the increasing plastic zone size. Onset of rapid fracture occurs when increase in crack tip stress field, measured by K (increase in K due to increased nominal stress and crack length), equals or exceeds resistance to crack extension (due to an increase in plastic zone size, crack tip blunting, and change from flat to slant fracture). This behavior is most clearly seen in fracture of relatively tough thin plate and sheet alloys. Unstable fracture under these conditions cannot be described as a material property since events leading to rapid fracture are specimen configuration and size dependent.

Finally, the technology of fracture mechanics has been applied to fatigue crack growth rate assessment under various conditions, environmental and stress-corrosion problems, dynamic fracturing, and determinations of the effects of elevated and cryogenic testing temperatures. These developments, which have occurred in the past 35 years, have led to broad use of fracture mechanics and to greater confidence in the design of fracture-critical structures.

General Fracture Control Concepts

Examples and Sequence of Events Leading to Brittle Fracture. Brittle fracture is defined as fracture that takes place at stresses below the net section yield with very little observable plastic deformation and minimal absorption of energy. Such fracture occurs very abruptly with little or no warning and can take place in all classes of materials. It is a major goal of structural engineering practice to develop methodologies (analytical and experimental) to avoid such fractures as they are associated with massive economic impacts and frequently involve loss of life. One notable example of brittle fracture occurred with the World War II Liberty ships which experienced numerous failures of this type. The most spectacular example of this problem was the U.S.S.

*Sections of this article have been abstracted from Chapter 9 of *The Science and Design of Engineering Materials*, J.P. Schaffer, A. Saxena, S.D. Antolovich, T. Sanders, and S. Warner, Irwin Publishers, 1995

Fig. 1 Fracture of the USS Schenectady at Pier in San Diego

Schnectady whose hull completely fractured while it was docked in San Diego. The fractured ship is shown in Fig. 1. This fracture was in part related to the welding methods that were used to construct the ships. When riveting was introduced to replace some of the welded structures, the incidence of fracture was markedly reduced.

Brittle fracture has also plagued the aviation industry. In the 1950's several Comets, the first commercial jet aircraft, produced in Britain, mysteriously exploded while in level flight. The cause was eventually traced to a design defect in which high stresses around the windows (caused by sharp corners) caused small cracks to initiate from which the fractures initiated. In the late 1960's and early 1970's the U.S. fighter aircraft, the F-111, experienced catastrophic failure of the wing through box (the structure at which the wings join to the fuselage). Failures of the F-111 were related to choice of a very brittle material (D6AC—a high-strength tool steel) and a heat-treating procedure that produced non-uniform microstructures. In 1988, the upper fuselage of a Boeing 737 operated by Aloha Air fractured without warning during level flight over the Pacific ocean! The reasons for this were related to corrosion of the Al alloy that is used as a skin material.

In addition to the above examples can be added numerous bridges, train wheels, heavy equipment, etc. In virtually every case, the reasons for brittle fracture can be found in inappropriate choice of materials, manufacturing defects, faulty design, and a lack of understanding of the effects of loading and environmental conditions. In all of the cases cited, there was severe economic loss and/or loss of life. For these reasons, it is an important engineering and ethical undertaking to reduce such accidents caused by brittle fracture to an absolute minimum.

In the above examples there are some common factors. Brittle fracture generally occurs in high-strength alloys (D6AC steel for the F-111 wing box, high-strength Al alloys for the Comets and 737), welded structures (Liberty ships, bridges) or cast structures (train wheels). It is significant that all failures started at small flaws which had escaped detection during prior inspection (in some cases, e.g., F-111, many previous inspections). Subsequent analysis showed that in most instances, small flaws slowly grew as a result of repeated loads or a corrosive environment (or both) until they reached a critical size. After reaching critical size, rapid, catastrophic failure took place. The following sequence of events is usually associated with brittle fracture:

1. A small flaw forms either during fabrication (e.g., welding, riveting) or during operation (fatigue, corrosion).
2. The flaw then propagates in a stable mode due to repeated loads, corrosive environments or both. The initial growth rate is slow and undetectable by all but the most sophisticated techniques. The crack growth rate accelerates with time but the crack remains "stable."
3. Sudden fracture occurs when the crack reaches a critical size for the prevailing load conditions. Final fracture proceeds at almost the velocity of sound!

Some Experimental Approaches to Fracture Control. Initial approaches to understanding fracture involved recognition that the quantity of toughness was associated with a material's ability to absorb energy. The area under the stress-strain diagram, U, is a measure of the energy required to rupture one unit volume of the material in question. This quantity is one measure of the toughness of the material.

$$U = \int_0^{\varepsilon_f} \sigma\,(\varepsilon)\, d\varepsilon \qquad \text{(Eq 1)}$$

Thus those materials which have a good combination of strength and ductility should, in this view, exhibit good toughness. Thus it is clear that diamond (which is very strong but which has virtually no ductility) will be brittle while steel (which has a good combination of strength and ductility) will be tough.

For materials which do not contain flaws and which are loaded in uniaxial tension, this approach is sufficient. However, in many engineering applications a notch is present and impact loads are applied. The notch has the effect of introducing a three-dimensional state of stress. Such stress states reduce the level of the maximum shear stress. For example, in the extreme case of hydrostatic tension (or compression) the shear stress in the body is zero. However, in metallic materials of interest, plastic deformation occurs as a result of shear stresses. Thus the triaxial stress state reduces the ability of the material to deform plastically and the toughness is correspondingly reduced. Similarly, impact loading reduces the amount of thermal energy available for plastic deformation and the toughness is further reduced.

The Charpy test accounts for both of these effects and is widely used to compare the relative behavior of materials under impact loads in the presence of a notch as a function of temperature. This test is very simple, inexpensive and widely used (as discussed in detail elsewhere in the *ASM Handbook* series). Some typical fractured Charpy specimens at various levels of ductility are shown in Fig. 2. The brittle fractures are flat in appearance with no shear lips (ridges at the sides of the tested specimens, Fig. 2). With increased ductility, the area of the specimen covered by shear lips increases progressively. The macroscopic appearance of the fracture surface (i.e., flat vs. shear lips) is related to the state-of-stress under which fracture took place and will be an important consideration when discussing fracture mechanics below.

Fatigue SN Curves. Traditionally, fatigue behavior has been characterized by the rotating beam test in which the number of cycles to failure of an unnotched specimen is plotted as a function of the applied cyclic stress. This results in the so-called *S/N* curve. When a corrosive medium is present, stressed immersion tests are used to evaluate the stress-corrosion cracking characteristics. The results of these tests (as well as tensile and Charpy tests) are of value for comparative purposes. However they do not account for the fact that cracks may already be present nor do they provide guidance for making valid, quantitative predictions. In the following sections, approaches to fracture that allow quantitative predictions to be made are developed.

Both 400 ×

Temperature, °C (°F)	25 (75)	65 (150)	95 (200)
Energy, J (ft · lb)	34 (25)	134 (99)	152 (112)
Lateral expansion, mm (in.)	0.81 (0.032)	1.85 (0.073)	1.85 (0.073)
% fibrous	65	95	100

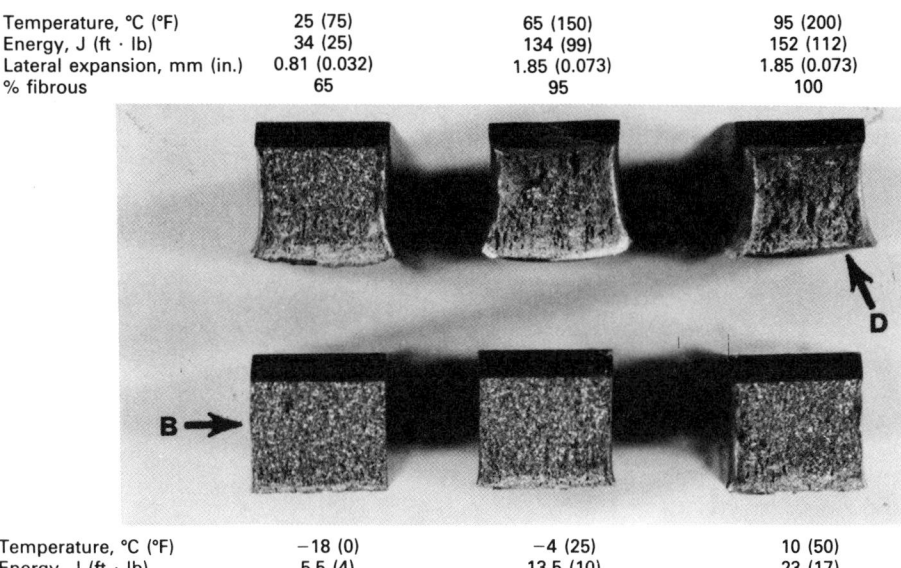

Temperature, °C (°F)	−18 (0)	−4 (25)	10 (50)
Energy, J (ft · lb)	5.5 (4)	13.5 (10)	23 (17)
Lateral expansion, mm (in.)	0.15 (0.006)	0.35 (0.014)	0.53 (0.021)
% fibrous	15	20	40

Fig. 2 Photographs of fractured Charpy specimens with ductile (D) and brittle (B) fractures

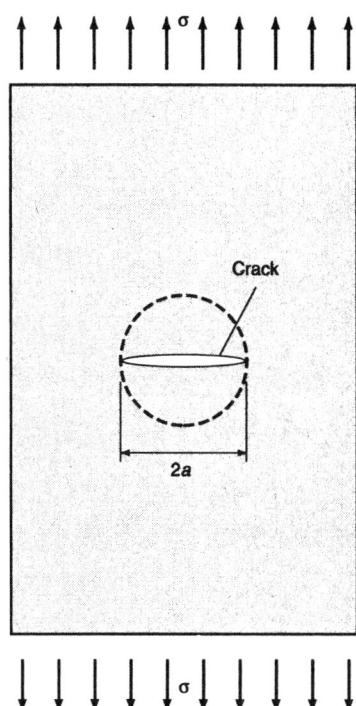

Fig. 3 Classical Griffith crack. The Griffith equation (i.e., Eq 3) represents a minimum condition that must be met for fracture to take place. The dotted circle represents a cylinder of material which carries little or no load.

Griffith Theory: An Analytic Approach to Brittle Fracture.

All of the preceding approaches to studying fracture suffer from the common limitation that they cannot be used to predict fracture loads of components.

In 1924, the British physicist A.A. Griffith (Ref 1) developed an approach to put fracture prediction on an analytic basis. He started by noting that the strength of a brittle material that exhibited linear elastic behavior is given by:

$$\sigma_{th} = \left(\frac{2E\gamma_s}{a_o}\right)^{1/2} \quad \text{(Eq 2)}$$

where:

E = Young's Modulus
γ_s = specific surface energy
a_o = lattice parameter
σ_{th} = theoretical cohesive strength

Equation 2 gives the theoretical strength of a defect-free, brittle material. If reasonable numbers are substituted into Eq 2, strengths are predicted that are orders of magnitude higher than those that are observed. This discrepancy was attributed to small *pre-existing* flaws which could greatly reduce the fracture strength. Using this concept, Griffith developed an analytic approach to predict the conditions under which the flaws would propagate unstably.

He considered a panel of thickness t containing a crack of length $2a$ subjected to a remote stress σ (Fig. 3). By carrying out a sophisticated mathematical analysis, he was able to determine the conditions under which the energy of the system would be reduced should the crack extend. Such conditions favor continued crack extension since systems tend naturally to minimize their energy. The results are expressed in the following equation:

$$\sigma_G = \left[\frac{2E\gamma_s}{\pi a}\right]^{1/2} \quad \text{(Eq 3)}$$

where:

σ_G = stress to fracture
$2a$ = crack size
γ_s = specific surface energy

Equation 3 is known as the Griffith equation. Since it was developed on the basis of minimizing the energy of the system, it can be thought of as an energetically necessary condition for fracture. However, it may not always be sufficient for fracture as explained below.

Equation 3 applies only to brittle materials which do not deform plastically. Examples of such materials are glasses and most ceramic materials such as Al_2O_3 and SiO_2. However most structural components are fabricated from metals which do undergo plastic deformation. The effects of plastic deformation were accounted for by Orowan (Ref 2) by simply noting that the effective surface energy (which includes the work of plastic deformation around the fracture surface) can be substituted for the true surface energy in the Griffith equation. In practice, this effective surface energy was simply an adjustable parameter used to force agreement with observed fracture loads and crack lengths. Conceptually, it is given by:

$$\gamma_e = \gamma_s + \gamma_p \quad \text{(Eq 4)}$$

where:

γ_p = work of plastic deformation
γ_s = true surface energy
γ_e = effective surface energy

For metals and polymers, γ_p is much greater than γ_s and to a good approximation:

$$\gamma_p \approx \gamma_e \quad \text{(Eq 5)}$$

It should be noted that the Griffith equation and the corresponding Orowan modification apply only to the limited geometry shown in Fig. 3. To be useful in an engineering sense, a methodology must be developed which is applicable for a broad range of geometries. In the next section, the basis of fracture mechanics is reviewed.

Linear Elastic Fracture Mechanics

The methods considered thus far are geometry-limited approaches to understanding fracture. Ideally, a material constant is sought as an index of the material's toughness, which is independent of geometry and which along with a good stress analysis can be used to predict fracture loads and critical crack sizes. In other words, this toughness value could be determined using a simple laboratory test and also be used to predict the flaw size at which fracture will occur in flawed components of arbitrary geometry. Conversely, given the flaw size it should be possible to predict the maximum safe operating stress. This is precisely what "fracture mechanics" is all about: specification of appropriate measures of toughness, determination of such measures by experimental techniques and incorporation of numbers into design to predict the conditions under which fracture will occur.

Crack Tip Stresses: The Field Equations. Consider a structural component containing a

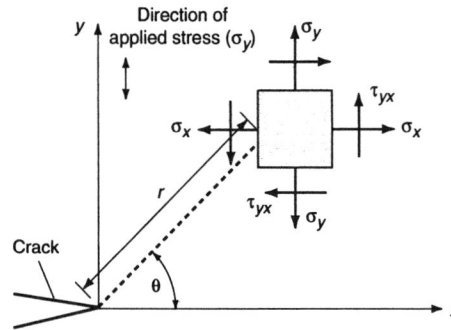

Fig. 4 Mode I crack showing the coordinate system and stress components. A mode I crack opens such that all points on the crack surfaces are displaced parallel to the y-axis.

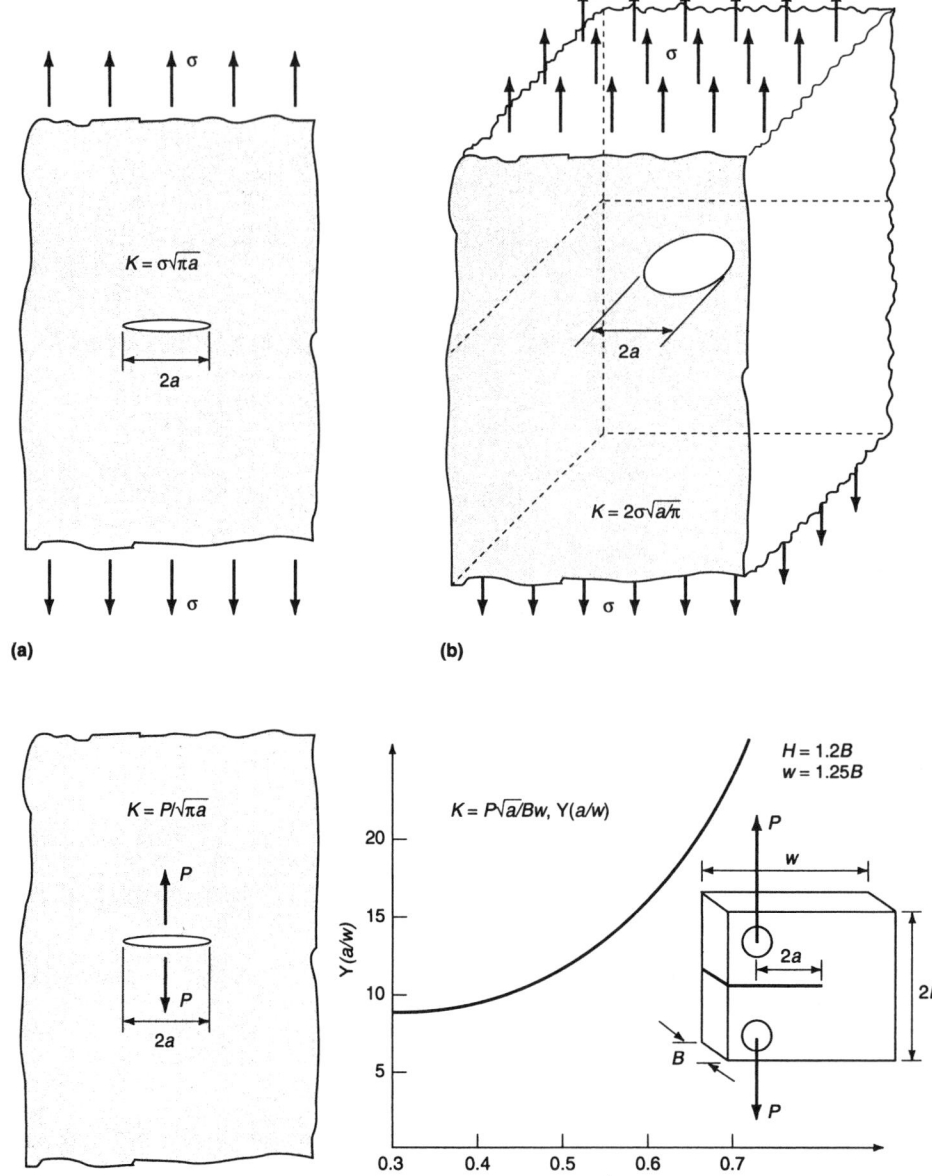

Fig. 5 Some typical load/crack geometries and their corresponding stress-intensity parameters. (a) Tunnel crack, (b) penny crack, (c) wedge opened crack, (d) eccentrically loaded crack

sharp crack. Assume that it is loaded so that the opposite faces of the crack are displaced vertically without horizontal offset. This is called Mode I or opening mode and is shown schematically in Fig. 4. For materials which are homogeneous, isotropic, and linearly elastic (i.e., obey Hooke's law) any stress component in the vicinity of the crack tip is given by the following equation:

$$\sigma_{ij} = \frac{K}{\sqrt{2\pi r}} f_{ij}(\theta) \qquad \text{(Eq 6)}$$

where:

σ_{ij} = the stress component of interest
r, θ = polar coordinates with the origin at the crack tip
$f_{ij}(\theta)$ = function of θ that depends on the stress component being considered

In this notation, the "*i*" subscript refers to the plane on which the stress component acts and the "*j*" component refers to the direction. For example if $i, j = x$, then the stress component is σ_{xx} which is the tensile stress acting in the x direction on the plane normal to the x axis. When $i \neq j$, shear stresses are defined.

The set of equations represented by Eq 6 is called the field equations. Regardless of the method of loading, they always have the same form in r and θ for Mode I. The parameter K is a measure of the magnitude of the stress field in the crack tip region and is called the stress-intensity parameter. It is given in Si units of MPa√m or in English of ksi√in. From the field equations, we see that independent of geometry, when the values of K are the same, the crack tip stress fields are identical.

Specific formulas for K depend on the load/crack geometry and numerous solutions for K can be found in references such as Ref 3 and 4. Some simple load-crack geometries and their corresponding stress intensities are shown in Fig. 5. In Fig. 5 (a-c), the dimensions of the body are assumed to be very large compared to the dimension of the crack. When the crack size is not negligible compared to the dimensions of the component, the stress-intensity parameter increases compared to what it would be for a crack

of the same size and loading conditions in a larger body.

Part of this increase may be understood intuitively by noting that there is less area ahead of the crack through which to transmit force. This means that the stresses in the crack tip region must be higher than for an equivalent crack in a larger body. In this case, the stress-intensity parameter is obtained by applying a multiplier to the corresponding expression for K in a semi-infinite body. This factor depends on the relative dimensions of the crack (as measured by a) and the specimen width (as measured by W) as well as on the load/crack geometry. For example, the multiplier for a center cracked specimen (Fig. 5) is (Ref 4):

$$\alpha = \left[\frac{W}{\pi a} \tan \pi a / W \right]^{1/2} \qquad \text{(Eq 7)}$$

A Fracture Mechanics Based Fracture Criterion. Equation 6 predicts that $\sigma \to \infty$ as $r \to 0$. However, this situation is unrealistic; no material is capable of withstanding infinite stress. Instead, the material will tend to deform plastically near the crack tip and absorb energy in the process. The preceding analysis must be modified to include a small zone of plasticity at the crack tip. The modified stress distribution is shown schematically in Fig. 6. The small plastic enclave is called a plastic zone and its extent directly ahead of the crack is denoted by R_p. This plastic enclave

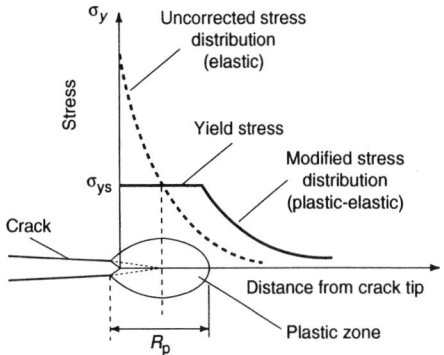

Fig. 6 Stress distribution ahead of crack in which small-scale plasticity is included

is very important in metals and polymers since it is the region in which energy is absorbed and is responsible for the relatively high fracture toughness of metals compared to all other materials.

The plastic zone size depends on both K and σ_{ys} and is given by:

$$R_p = C \left(\frac{K}{\sigma_{ys}} \right)^2 \qquad \text{(Eq 8)}$$

The constant, C, depends on various factors including thickness and deformation characteristics of the material. It usually has a value between $1/\pi$ and $1/3\pi$. Since Eqs 6-8 are valid for linear elastic bodies, they only apply when R_p is much smaller (i.e., R_p <10% of the crack size) than the crack size and other physical dimensions of the body. As K increases (e.g., by increasing the load) R_p also increases and the material in the plastic enclave becomes more severely strained. As a result of this direct relationship between K and the strains at the crack tip, it can be hypothesized that:

$$\text{At fracture}, K \rightarrow K(\text{critical}) \qquad \text{(Eq 9)}$$

In other words, there is a critical value of the stress-intensity parameter at which fracture will take place. This is the fundamental assumption of fracture mechanics and is the basis of computing fracture stresses and critical crack lengths.

Fracture Mechanics and State-of-Stress under Plane Strain (Thick Sections and High-Strength Materials). The value of K(critical) depends not only on the material being considered but also on the thickness of the material in which the crack is found. This can be understood by considering the impediments to plastic deformation that exist in the vicinity of the crack tip and how these impediments vary with thickness. As previously noted, toughness is manifested primarily through plastic deformation at the crack tip; anything that impedes crack tip plastic deformation will reduce the toughness. The high stresses in the crack tip region are such that the material ahead of the crack tip tends to plastically deform normal to the crack plane. Simultaneously, the material tends to contract parallel to the crack plane. If the section size is sufficiently

thick, the contraction is opposed by the bulk of lowly stressed material and an additional stress component develops parallel to the crack plane.

This state-of-stress is called plane strain* and develops when plastic deformation at the crack tip is severely limited. When plasticity is limited, the fracture toughness is also limited since plastic deformation is the most important source of toughness for most structural materials. The value of the stress-intensity parameter at which fracture occurs in thick sections is called the plane-strain fracture toughness and is denoted by the symbol K_{Ic}. K_{Ic} represents the toughness of the material above some critical thickness and is a material constant. It is a realistically conservative measure of the material's toughness and is widely used for engineering calculations, as described in subsequent sections. Plastic restraint at the crack tip is also promoted by high strength and limited ductility. Examples of high-strength materials of limited ductility are tool steels, ceramics, and glass. Mild steels, stainless steels, and many polymers are examples of low-strength, high-toughness materials.

Plane Stress (Thin Sections and Low-Strength Materials). In contrast to the preceding section, low-strength, ductile materials or very thin sections of high-strength materials develop much less constraint to plastic deformation parallel to the crack surface. In the limit, there is no stress opposing through-the-thickness deformation and the only stresses that are present are in the plane of the specimen. Such conditions are termed plane stress** and the value of the stress-intensity parameter at which fracture occurs for under plane stress is called the plane-stress fracture toughness, K_c. The toughness under plane-stress conditions is higher than under plane-strain conditions for a given material under most practical conditions. The effect of thickness on toughness is illustrated in Fig. 7. Numerous experimental results of fracture toughness measurements have shown that for most engineering alloys, plane strain develops when the thickness reaches the following value:

$$B = 2.5 \left(\frac{K_{Ic}}{\sigma_{ys}} \right)^2 \qquad \text{(Eq 10)}$$

Thus for a high-strength aluminum alloy such as 7050 for which K_{Ic} is about 28 MPa$\sqrt{m}$ (25 ksi$\sqrt{in}$.) and the yield strength is about 520 MPa (75 ksi), the thickness at which plane strain occurs is about 0.72 cm (0.28 in.). For 300 grade maraging steel, the critical thickness is 0.25 cm (0.1 in.).

Fracture mechanics is most useful when designing with high-strength materials of limited ductility. For such materials, the plane-strain fracture toughness K_{Ic} is the most important practical measure of toughness. The reason for this is

---

*Plane strain may be thought of as a stress state in which there is no through-the-thickness deformation. Mathematically it is equivalent to saying that $\varepsilon_{zz}, \gamma_{zy}, \gamma_{zx} = 0$; $\varepsilon_{xx}, \varepsilon_{yy}, \gamma_{xy} \neq 0$.
**Plane stress is defined mathematically as that stress state for which $\sigma_{zz}, \tau_{zw}, \tau_{zy} = 0$; $\sigma_{xx}, \sigma_{yy}, \tau_{xy} \neq 0$.

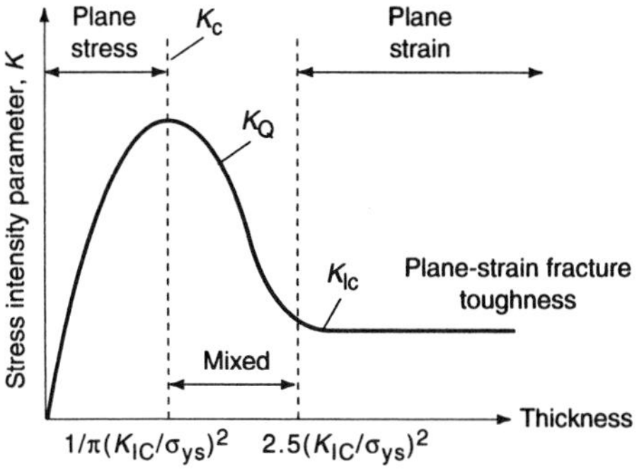

Fig. 7 Effect of thickness on state-of stress and fracture toughness at the crack tip

that plane-strain conditions can develop even for relatively thin section sizes as shown above. The plane-strain fracture toughness, K_{Ic}, represents a practical minimum and is the most widely used measure of fracture toughness. For these reasons, we will limit our discussion to plane-strain conditions. This is consistent with current engineering practice.

The Morphology of Fracture Surfaces. The macroscopic appearance of the fracture surface depends on the amount of plastic deformation which accompanies fracture. This is equivalent to saying that the appearance depends on the state-of-stress. The fracture surface can then be used to gain insight into the conditions prevailing at the time of fracture. For fracture under plane-stress conditions, the fracture surface will be inclined at 45° to the sides of the specimen. For plane strain the fracture surface will be very flat. Schematic crack surface morphologies are shown in Fig. 8.

The fracture surface also can reveal the point of initiation. It frequently occurs that there are "chevron" marks on the fracture surface and these point back towards the origin of fracture as seen in Fig. 9. Fracture surfaces contain other information on a much finer scale about the nature of the fracture process and these are discussed later in this article.

Example: Maximum Stress to Fracture. Suppose that the fracture toughness of an aluminum alloy has been determined to be 30 MPa$\sqrt{m}$ (27 ksi$\sqrt{in}$.) and a penny shaped crack of diameter 1.6 cm (0.63 in.) has been located in a thick plate that is to be used in uniaxial tension. Calculate the maximum allowable stress that can be imposed with fracture. Assume plane-strain conditions prevail. The yield stress of the material is 500 MPa (72.5 ksi).

Solution. The stress-intensity parameter formula for a penny-shaped crack is given by:

$$K = 2\sigma \left(\frac{a}{\pi} \right)^{1/2} \qquad \text{(Eq 11)}$$

where a = crack radius, and σ = applied stress.

At fracture, the applied stress intensity is equal to the plane-strain fracture toughness. In equation form, $K = K_{Ic}$. Rearranging Eq 11 and substituting appropriate values gives:

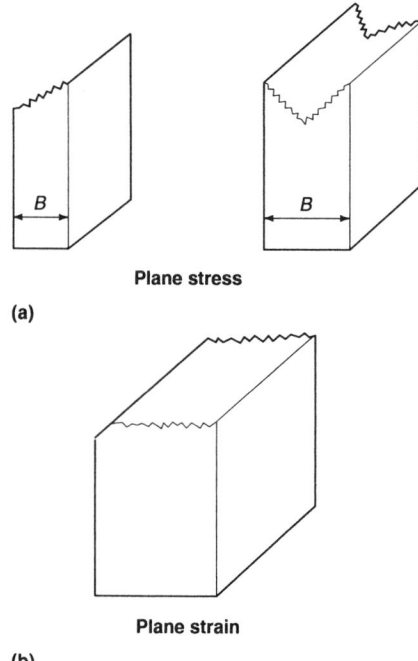

Plane stress

(a)

Plane strain

(b)

Fig. 8 Schematic crack surface morphologies for (a) plane stress and (b) plane strain. The crack direction is normal to the plane of the paper.

$$\sigma = K/2\,[\pi/a]^{1/2} = 30/2\,[\pi/0.008]^{1/2}$$

$$\sigma_f = 297\ \text{MPa}$$

Thus fracture will occur well below the material's yield stress. This calculation shows that there is no guarantee that fracture will not occur simply because the nominal applied stresses are below the yield stress.

Example: Calculation of the Maximum Safe Flaw Size. Maraging steel (350 grade) has a yield of approximately 2450 MPa (355 ksi) and a toughness of 55 MPa$\sqrt{\text{m}}$ (50 ksi$\sqrt{\text{in.}}$). A landing gear is to be fabricated from this material and the design stresses are 70 per cent of yield. Assuming that flaws must be 0.25 cm (0.1 in.) to be detectable, is this a reasonable stress at which to operate? Assume that small edge cracks are present. The stress-intensity parameter for this geometry is:

$$K = 1.12\,\sigma\sqrt{\pi a} \tag{Eq 12}$$

Solution. The flaw size at which fracture occurs is calculated by noting that at fracture $K = K_{\text{Ic}}$.

$$a_f = 0.797/\pi\,(K_{\text{Ic}}/\sigma_a)^2 = \frac{0.797}{\pi}\left[\frac{55}{1715}\right]^2$$

$$= 2.6 \times 10^{-4}\ \text{m} \tag{Eq 13}$$

Thus, critical flaws may escape detection even though the design stresses for the part are below the yield stress. Consequently, the stress is too high to ensure safe operation of the landing gear. It must be reduced to the point at which the critical flaw size is greater than the minimum detectable crack size (0.25 cm).

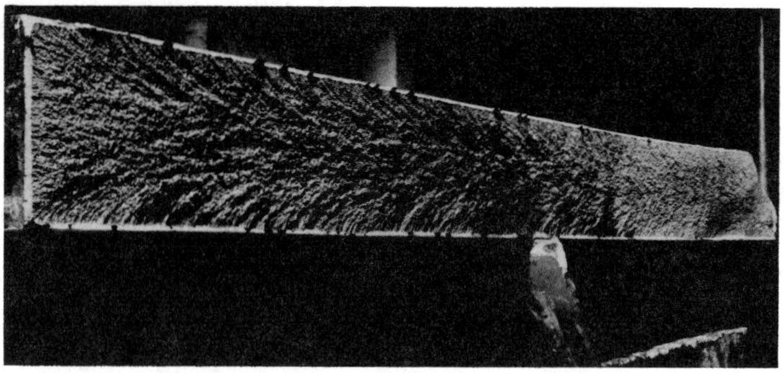

Relation of Fracture Toughness to Tensile Properties. As a practical matter, most microstructural modifications that increase the strength of materials cause corresponding decreases in the fracture toughness. It has been shown that the fracture toughness is related to other mechanical properties through the following expression (Ref 5):

$$K_{\text{Ic}} \propto n\,(E\,\sigma_{\text{ys}}\,\varepsilon_f)^{1/2} \tag{Eq 14}$$

where:

n = strain-hardening exponent
E = Young's modulus
σ_{ys} = yield strength
ε_f = the fracture strain

Considerable care must be exercised when using this equation. Strictly speaking, the critical stresses and strains should be determined under multi-axial conditions to replicate the crack-tip conditions. Generally, the fracture toughness decreases with increasing strength since both n and ε_f decrease with increasing strength. However, this trend is not always followed and noteworthy exceptions have been studied in high-strength steel systems (Ref 6-8).

Equation 14 should be viewed as being indicative of a trend more than a precise formula for computing the plane-strain fracture toughness. The direct calculation of fracture toughness from other mechanical properties is complicated by the fact that toughness is a property that manifests itself under the triaxial stress conditions that prevail at the crack tip. This means that the various quantities contained in Eq 14 would have to be measured under conditions of plane strain; such information is not usually available.

Another factor to consider is microstructural anisotropy that may develop as a result of processing. Table 1 shows that fracture toughness depends on orientation. When the crack plane is parallel to the rolling direction, segregated impurities and intermetallics that lie in these planes represent easy fracture paths and the toughness is low. When the crack plane is perpendicular to these weak planes, decohesion occurs and crack tip blunting or stress reduction takes place, thus effectively toughening the material. On the other hand, when the crack plane is parallel to the plane of these defects, the crack can propagate very easily and the toughness is reduced.

Application of Fracture Mechanics to Various Classes of Materials. Fracture mechanics was initially developed so that high-strength materials of limited ductility could be used safely in engineering situations. This formulation is referred to as linear elastic fracture mechanics (LEFM). In order to apply the LEFM methodology correctly, fracture must occur under essentially elastic conditions. Practically, this means that there can only be limited plastic deformation at the crack tip at the time of fracture. Additionally, plane-strain conditions should prevail and there should only be a single well-defined crack. The materials for which the LEFM approach works well are:

- High-strength metallic alloys such as heat-treated martensitic steels, precipitation-hardened aluminum alloys used in the aerospace industry, titanium alloys, and Ni-base alloys
- Ceramic materials which are essentially brittle
- Polymers of limited inelastic deformation

One important class of materials for which fracture mechanics is frequently not applicable is composites. In composites, cracks tend to be spatially distributed throughout the material. In this situation the usual fracture mechanics requirement of a single, well-defined crack is not met. Also, the requirements of homogeneous and isotropic materials are not met for most composite systems. Of course, fracture mechanics is very applicable to ceramics even at high temperatures because the ductility of these materials is limited.

Experimental Determination of Fracture Toughness. The plane-strain fracture toughness is a materials parameter of considerable engineering significance. The American Society for Testing and Materials (ASTM) has developed detailed procedures for determination of K_{Ic} (Ref 9).

Specimens are designed so as to minimize material and loading requirements. Frequently a compact type (CT) specimen, shown in Fig. 5(d), is used to experimentally determine the fracture toughness as well as other fracture properties which are discussed later in the chapter. The first step in carrying out a fracture toughness test is to

Fig. 9 Macroscopic chevron markings on the fracture surface pointing back to the fracture origin. ASTM A517H plate. Source: *Metals Handbook*, 9th ed., Vol 12, *Fractography*, 1987, p 347

Table 1 Selected fracture toughness values for some engineering alloys

Alloy	Yield strength		Orientation(a)	Temperature		Plane-strain fracture toughness(b)	
	MPa	ksi		°C	°F	MPa√m	ksi√in.
2042-T351	385	56	L-T	29	84	31	28
2024-T351	292	42	S-L	32	90	21	19
7075-T651	530	77	L-T	28	82	32	29
7075-T651	446	64.5	S-L	29	84	21	19
4140	1379	200	L-T	24	75	65	59
4140	1586	230	L-T	24	75	55	50
4340	1455	211	L-T	21	70	83	75.5
D6AC	1496	217	L-T	21	70	102	93
HP9-4-.20	1282	186	L-T	26	79	151	137
HP9-4-.20	1310	190	T-L	26	79	138	125.5
250 Maraging	1607	233	L-T	24	75	86	78
250 Maraging	1600	232	T-L	24	75	86	78
Ti-6Al-4V	889	129	L-T	24	75	64	58
Ti-6Al-4V	910	132	T-L	24	75	68	62
Ti-6Al-4V	883	128	S-L	24	75	75	68
Inconel 718	1041	151	T-L	24	75	87	79
Inconel 718	986	143	S-L	24	75	73	66

(a) The first letter gives the direction normal to the crack plane while the second letter gives the direction of crack propagation. In the L-T orientation, the crack is normal to the rolling direction (L) and propagates in the transverse direction (T). (b) All numbers are rounded off so as not to imply a greater precision than can be justified by the experimental procedure. Source: Data taken from Ref 10

Table 2 Calibration function for compact tension specimens

a/W	f(a/W)	a/W	f(a/W)
0.450	8.34	0.500	9.66
0.455	8.46	0.505	9.81
0.460	8.58	0.510	9.96
0.465	8.70	0.515	10.12
0.470	8.83	0.520	10.29
0.475	8.96	0.525	10.45
0.480	9.09	0.530	10.63
0.485	9.23	0.535	10.80
0.490	9.37	0.540	10.98
0.495	9.51	0.545	11.17
		0.550	11.36

Source: Ref 9

grow a fatigue starter crack at a low rate at the base of the machined notch. Such cracks are used to ensure reproducibility and to simulate service conditions. The specimen is then loaded to failure at a specified rate. The load vs. displacement curve is recorded. For the test to be valid, the load/displacement curve must be virtually linear and the crack front must be essentially straight. If these requirements are met, the toughness is calculated using expressions provided by the ASTM standard E-399. For the CT specimen, the expression for calculating the fracture toughness from experimentally measured quantities is:

$$K_{Ic} = \frac{P}{B\sqrt{W}} \cdot f(a/W) \qquad \text{(Eq 15)}$$

where:

K_{Ic} = plane-strain fracture toughness
P = appropriately chosen fracture load
B = specimen thickness
W = specimen width
$f(a/W)$ = calibration function

Values of the calibration function are provided in Table 2. It should be noted that K_{Ic} calculated in this manner must be subjected to various validation checks before it may be considered to be the plane-strain fracture toughness.

A considerable amount of data on fracture toughness has been analyzed for reliability and compiled by the Metals and Ceramics Information Center. These data are very useful for engineering calculations and are available in a handbook and additional supplements (Ref 10). Note that for metallic materials, as discussed previously, the fracture toughness often depends on the orientation of the specimen relative to processing directions. It is thus important to be sure that values of fracture toughness that are used are appropriate for the structure of the component being considered.

Fatigue Fracture

In a very schematic way, we may view fatigue as involving the following steps:

1. Cyclic deformation which causes internal damage in the form of dislocation debris.
2. The formation of a crack after sufficient damage has been caused by the cyclic loading.
3. Propagation of the crack to a size where it becomes unstable for the loads being applied.

While the "smooth bar" (S/N) approach to fatigue alluded to previously is undoubtedly useful, certain drawbacks arise in certain engineering situations:

- Most parts do not have smooth, highly polished surfaces.
- In general, the fatigue life is made up of an initiation phase and a propagation phase. However, in the smooth bar or S/N methodology, results are usually given in terms of the total life to failure with no indication of the fraction of life spent in the initiation and propagation phases.
- Most structures usually contain small pre-existing cracks. These cracks are frequently unavoidable, and arise from fabrication procedures and material defects. They will propagate under repeated loads until they become critical at which point fracture occurs. In such cases, the entire life is spent in the propagation phase.

For these reasons, a knowledge of the crack growth rate behavior is imperative if the overall fatigue life is to be computed.

In the fracture mechanics approach, the crack growth rate (i.e., the amount of crack extension per loading cycle) is correlated with the cyclic variation of the stress-intensity parameter K. This approach allows estimating useful safe life and inspection intervals, and it has been experimentally demonstrated that for many conditions, the rate of fatigue crack growth can be represented by the Paris equation, which has the form (Ref 11):

$$\frac{da}{dN} = C(\Delta K)^m \qquad \text{(Eq 16)}$$

where:

a = crack length
N = cycle
C, m = material parameters
$\Delta K = K_{max} - K_{min}$

The stress-intensity range is ΔK. This simple formula does not take into account the effect of the load (or stress) ratio R (i.e., P_{max}/P_{min}). As R increases the crack growth rate generally increases and this increase depends on the material being considered. For example, steels are relatively insensitive to load ratio effects while some aluminum alloys show a marked R ratio dependence.

More generally, the crack growth rate is represented by:

$$\frac{da}{dN} = f(\Delta K, R) \qquad \text{(Eq 17)}$$

Various formulations of the above equation have been proposed and one of these is discussed later.

The stress-intensity parameter range characterizes the cyclic stresses and strains ahead of the crack tip and uniquely characterizes the crack growth rate through a relationship such as Eq 16. Since the zone ahead of the crack which experiences cyclic plasticity (i.e., the fatigue plastic zone) is very small, plane-strain conditions can develop even for small thicknesses. This is an important conclusion since it means that data can be obtained from thin specimens and applied quite generally.

Example Problem: Calculation of Fatigue Lives. The crack growth rate of ferritic/pearlitic steels is given by the equation:

$$\frac{da}{dN} = 3.6 \times 10^{-10} (\Delta K)^{3.0} \qquad \text{(Eq 18)}$$

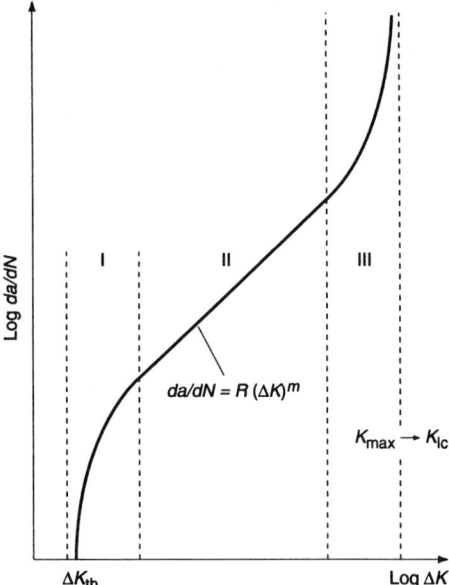

Fig. 10 Schematic representation of fatigue crack propagation behavior. In regime I the crack growth rate is low since the threshold for crack propagation is approached. In regime II the so-called Paris law is obeyed while in regime III the crack growth rate increases above that predicted by the Paris equation since the fracture toughness of the material is approached and there is local tensile overload fracture.

where ΔK is given in units of ksi$\sqrt{\text{in.}}$ and da/dN is in units of in./cycle. Assume that a part contains an edge crack that is 0.05 in. long. The stresses vary from 0 to 50 ksi and the fracture toughness is 100 ksi$\sqrt{\text{in.}}$. Compute the life of the part.

Solution. For this geometry, the stress-intensity parameter is given by:

$$\Delta K = 1.12 \Delta \sigma \sqrt{\pi a} \text{ (see footnote*)}$$

Using the information given in the problem statement and the above expression the crack growth rate is given by:

$$\frac{da}{dN} = 3.52 \times 10^{-4} \cdot a^{3/2}$$

This equation can be integrated from the initial condition of $N = 0$ and $a = a_0 = 0.05$ in. to the final condition of $N = N_f$ and $a = a_f$. The final crack length a_f occurs when $K_{max} = K_{Ic}$. Integrating yields:

$$N_f = 2.84 \times 10^3 \int_{a_0}^{a_f} \frac{da}{a^{3/2}}$$

Integration of the right-hand side of the equation gives:

$$N_f = 5.68 \times 10^3 \left[\frac{1}{\sqrt{a_0}} - \frac{1}{\sqrt{a_f}} \right]$$

*Rigorously, the expression should be $\Delta K = \Delta \sigma \cdot \sqrt{\pi a} + \sigma \sqrt{\pi} \cdot (1/2) \cdot a^{-1/2} \cdot \Delta a$. However, since $\Delta \sigma$ in a given cycle is much larger than Δa, to a very good approximation we may write $\Delta K = \Delta \sigma \sqrt{\pi a}$.

As mentioned above a_f depends on the fracture toughness and maximum stress. For the geometry being considered this is given by:

$$a_f = \frac{1}{\pi} \left(\frac{K_{Ic}}{\sigma} \right)^2 = \left[\frac{100}{1.12 \times 50} \right]^2 = 1.02 \text{ in.}$$

Substituting into the expression for life gives:

$$N_f = 5.68 \times 10^3 [1/\sqrt{0.05} - 1/\sqrt{1.02}] = 1.98 \times 10^4$$

The part is thus expected to last almost 20,000 cycles.

Other da/dN and ΔK Relationships. The representation of the crack growth rate (in Eq 16) is essentially correct for a reasonably wide domain of stress-intensity ranges. However, additional factors must be considered at both low and high stress-intensity ranges. When the stress-intensity range is low, a point will be reached where the average crack growth rate approaches the interatomic spacing. Physically, a crack cannot propagate a fraction of an interatomic spacing; however, we must remember that the measured crack growth rate in a typical experiment represents an *average* across an entire crack front. In some regions the crack may be stopped while in others it may be moving. For practical purposes, the threshold stress-intensity range (ΔK_{th}) occurs when the average crack growth rate is less than 10^{-12} m/cycle. For values of ΔK less than ΔK_{th} the crack growth rate is effectively zero. Equation 16 cannot be used to describe the crack growth behavior in the threshold regime. Also, when the stress-intensity parameter becomes so large that it approaches the fracture toughness, K_{Ic}, the rate of crack growth becomes much more rapid than that predicted by Eq 16 since, in addition to the fatigue process, a considerable amount of local tensile failure occurs during each cycle. Again, Eq 16 no longer provides a correct representation of the crack growth behavior. Over the broad range of stress-intensity parameters, fatigue crack growth rates may be represented by an equation of the form (Ref 12):

$$\log \frac{da}{dN} = C_1 \sinh \left\{ C_2 \log (\Delta K) + C_3 \right\} + C_4 \quad \text{(Eq 19)}$$

where:

C_1 = material constant, and

C_2, C_3, C_4 = functions of load ratio, frequency, and temperature

The above equation is represented schematically in Fig. 10. Numerous other empirical expressions have also been proposed to describe crack growth behavior from ΔK_{th} to values of ΔK when K_{max} approaches K_{Ic}. Regardless of the representation, the life may always be computed by integrating the crack growth curve as was done in the preceding example problem. It is worth noting that use of Eq 16 is usually conservative for the threshold to intermediate crack growth rate regime. While Eq 16 underestimates the growth rate for large ΔK values near the upper limit, the underestimate is small. This is because as the stress intensity approaches the fracture

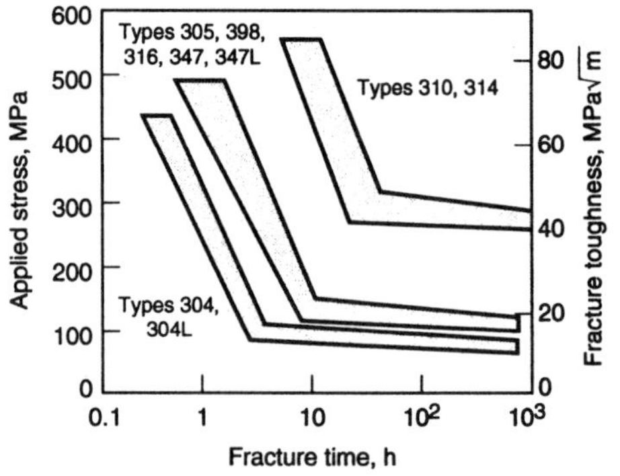

Fig. 11 Applied stress versus life of tensile type specimens in simulated sea water for stainless steels

toughness of the material, the crack is growing so rapidly that very few cycles are accumulated in this regime. Use of threshold data must be done very judiciously due to the fact that a small error in threshold can lead to significant overestimates of the fatigue life.

Summary of Fatigue Crack Propagation. The life of components containing pre-existing flaws can be computed in principle using fracture mechanics concepts. For complex load/crack geometries and stress patterns, the equations for ΔK are complex. In addition, the da/dN vs. ΔK relationships may be quite complicated (or perhaps not even available in functional form). Such conditions do not lead to easily evaluated integrals for the fatigue life as was the case in the example problem.

However, the life of a component can always be computed through step-wise numerical integration techniques similar in principle to the technique that was used in the example problem. In this scheme an increment of crack growth Δa is selected. The number of cycles associated with a growth increment Δa is obtained from $\Delta N = \Delta a(dN/da)$. All of the ΔN's are added up from the initial crack length to the final crack length. In such cases, it is most efficient to use computers to do the actual computations.

Environmental Effects

The presence of aggressive environment reduces the ability of a given material to bear load. In general, there are usually a number of environments for any given material which will lead to environmentally induced fracture. A particular environment does not attack all materials. Environmentally induced failures always occur as over a period of time as the deleterious environment reacts with the material to degrade material properties. In other words, if a specimen of a material is subjected to stress (frequently below the yield strength) in an environment in which it is susceptible to attack, for the most part failure does not occur instantaneously upon application of the load. Depending on the applied load level, the time to failure may vary considerably. This is shown in Fig. 11 for a high-strength steel in a simulated sea water environment consisting of

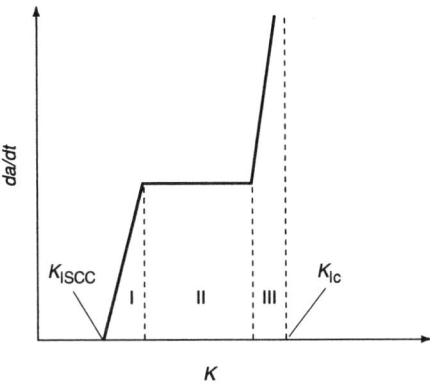

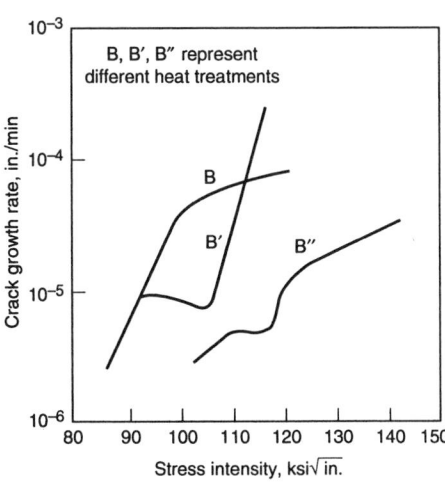

Fig. 12 Crack growth rate curve for stress-corrosion cracking. (a) Idealized crack growth rate curve where regime I is SCC enhancement of cracking above the threshold. In regime II the mechanical enhancement has attained a maximum and cracking is transport controlled. In regime III the process is mechanically dominated. Final fracture occurs very rapidly near K_{Ic}. (b) Experimental crack growth rate as a function of stress intensity for a high strength steel in humid air from Ref 14

3.5% salt (NaCl). This phenomenon is also called delayed failure. This example is particularly appropriate for selecting steels for marine applications such as ship hulls and offshore platforms. For both applications, it is attractive to consider the use of high-strength steels to save weight. However, the allowable stress levels in structural members of this high-strength steel in sea water environment are not dictated by the yield strength but rather by the level of stress above which environment-induced fracture occurs. Therefore the strength and environmental resistance of these steels have to be carefully balanced in order to optimize weight. In most large structures, it is necessary to consider the case of pre-existing cracks and to characterize the materials with such cracks present. These cracks usually result from processing and fabrication, although in some instances they may also occur during use. In this section we briefly review the fracture mechanics approach to environmentally-induced fracture.

Fracture Mechanics Approach. As for fatigue, stress-corrosion cracking can also be described within a fracture mechanics framework. In this approach, it is assumed that a crack already exists and it is the extension of the crack that governs the life of the component. Under such conditions it would be expected that the rate of crack growth would correlate with the magnitude of the applied stress-intensity parameter. In fact, it has been demonstrated that SCC rates of cracked specimens correlate *uniquely* with the stress-intensity parameter and that any attempt to correlate the crack growth rate with stress alone leads to logical inconsistencies (Ref 13).

In general there is a threshold for stress-corrosion cracking, denoted K_{Iscc}, below which crack growth is not observed. Above this level, increases in stress intensity produce increases in the real time crack growth rate da/dt. Depending on the material and the mechanism of crack extension, further increases in the stress intensity do not produce significant increases in the crack growth rate. This can be understood in terms of a transport-controlled step in which the *maximum* potential for acceleration of the crack growth rate by the crack tip stress field has been reached. At this point, the rate is governed by how rapidly the corrosive media can be transported to the crack tip or the limiting value of diffusion of a damaging species into the bulk material. The crack growth rates are more sensitive to parameters such as temperature, pressure, pH levels, etc. As K approaches K_{Ic}, tensile fracture mechanisms begin to appear, the process becomes mechanically dominated, and the crack growth rates again increase with K (Region III in Fig. 12a). Final failure occurs at K_{Ic} at which point the crack is moving at a substantial fraction of the velocity of sound and is essentially free of the environment. An idealized SCC curve is shown in Fig. 12(a) while actual experimental data are shown in Fig. 12(b).

The life of a cracked component in an aggressive environment may be calculated by an integration technique, similar in principle to that used to calculate fatigue lives. In order to compute the life, the crack growth rate as a function of K must be known as well as other factors such as operating stresses, initial crack length, fracture toughness of the material and the geometry of the part in which the crack is embedded. These concepts are illustrated in the following problem.

Example: Estimation of Life in Corrosive Environment. The crack growth rate of a high-strength steel in H_2O at 80 °C (175 °F) is given by:

$$\frac{da}{dt} = 9.32 \times 10^{-5} K - 2.08 \times 10^{-3}$$

Here K is in MPa$\sqrt{m}$, a is in meters and t is in days. Calculate the life of the part assuming that edge cracks of length 2.02×10^{-3} m are initially present. Assume that K_{Ic} is 55 MPa$\sqrt{m}$ and that the part is subjected to a constant stress of 345 MPa.

Solution. The initial stress intensity is given by:

$$K_i = 1.12 \times 345 \times (\pi \times 2.02 \times 10^{-3})^{1/2} = 30.8 \text{ MPa}\sqrt{m}$$
$$\text{(from Fig. 5)}$$

The threshold stress intensity may be calculated assuming:

$$\frac{da}{dt} = 0 \text{ at } K = K_{Iscc}$$

This gives a value of about 22.3 MPa$\sqrt{m}$ for K_{Iscc}. Since $K_i > K_{Iscc}$ the crack will grow by a SCC process.

The rate of growth is given by:

$$\frac{da}{dt} = 9.32 \times 10^{-5} \times 1.12 \times 345\sqrt{\pi a} - 2.08 \times 10^{-3}$$
$$= 0.0640 a^{1/2} = 2.08 \times 10^{-3}$$

The above equation can be integrated:

$$t_f = \int_{t=0}^{t=t_f} dt = \int_{a=a_0}^{a_f} \frac{da}{0.064 a^{1/2} - 2.08 \times 10^{-3}}$$

The crack length at fracture may be computed by noting that at this point $K = K_{Ic}$:

$$a_f = \frac{1}{\pi}\left[\frac{K_{Ic}}{1.12\sigma}\right]^2 = \frac{1}{\pi}\left[\frac{55}{386.40}\right]^2 \approx = 6.45 \times 10^{-3}$$

Using tables of integrals, $t_f \sim 2.5$ days.

Summary

This article has introduced a powerful methodology for computing the onset of fracture (i.e., the fracture stress given a crack of a certain size, or a critical crack size given an applied stress). In practical terms a material containing a crack will fracture when the value of the stress-intensity parameter reaches a critical value. This value is denoted by the symbol K_{Ic} and is called the plane-strain fracture toughness. The value of K_{Ic} is independent of the load/crack geometry. It can be measured experimentally and formulas for the stress-intensity parameter as a function of crack size, crack shape, applied stress (or load), and geometry of load application are readily available.

The rate of crack growth can be characterized by the stress-intensity parameter range for fatigue loading. The fatigue crack growth rate curves may be integrated to yield the fatigue life of an propagating crack.

Cracking in a corrosive environment may also be considered in a fracture mechanics framework. In this case the stress intensity is related to the temporal crack growth rate. The crack growth rate equation can be integrated to obtain the life of a cracked component in an aggressive medium.

In the following articles, fracture mechanics is used to characterize behavior in several important alloy systems and mechanisms of toughening, fatigue crack growth, and crack growth in corrosive media.

REFERENCES

1. A.A. Griffith, *Trans. Roy. Soc.*, Vol A221, 1924, p 163, *Trans. ASM,* Vol 61, 1968, p 871
2. E. Orowan, *Fatigue and Fracture of Metals,* MIT Press, Cambridge, 1950, p 139
3. P.C. Paris and G.C. Sih, ASTM STP 381, 1965, p 30
4. H. Tada, P.C. Paris, and G. Irwin, *The Stress Analysis of Cracks Handbook*, Del Research Corporation, 1973
5. G.T. Hahn and A.R. Rosenfield, ASTM STP 432, 1968, p 5
6. S.D. Antolovich and B. Singh, *Met. Trans.*, Vol 2, 1971, p 2135
7. S.D. Antolovich, A. Saxena, and G.R. Chanani, *Met. Trans.*, Vol 5, 1974, p 623
8. E.R. Parker and V.F. Zackay, *Engr. Fract. Mech.*, Vol 5, 1973, p 147
9. Standard Test Method for Plane-Strain Fracture Toughness of Metallic Materials, Annual ASTM Standards, Section 3, Vol 3.01
10. *Damage Tolerant Design Handbook*, MCIC-HB-01, Metals and Ceramics Information Center, Batelle Columbus Laboratories Including 1st and 2nd supplements, Jan 1975
11. P.C. Paris, *Proc. of 19th Sagamore Conference*, Syracuse University Press, Syracuse, N.Y., 1964, p 107
12. R.M. Wallace, C.G. Annis, Jr., and D. Sims, Report AFML-TR-76-176, Part II, 1976
13. H.R. Smith, D.E. Piper, and F.K. Downey, *Eng. Fract. Mech.*, Vol 1, 1968, p 123
14. S.D. Antolovich and G.R. Chanani, *Eng. Fract. Mech.*, Vol 4, 1972, p 765

Fracture Resistance of Structural Alloys

K.S. Ravichandran, The University of Utah, and A.K. Vasudevan, Office of Naval Research

FRACTURE MECHANICS is a multidisciplinary engineering topic that has foundations in both mechanics and materials science. From the perspective of a metallurgist, fracture mechanics often emphasizes mathematical mechanics, where the primary focus is on analytical methods. However, the microstructural aspects of fracture mechanics (quantified in terms of various measures of fracture toughness such as K_{Ic}, K_c, or K_{Id}) is important for several reasons. First, in many applications, fracture toughness is useful for design and/or as a quality control parameter. Secondly, fracture mechanics provides a more meaningful measure of fracture resistance in the presence of cracks or defects than other material properties such as ductility. Therefore, fracture mechanics plays a major role in both the application and the development and production of structural materials for petroleum, chemical, mining, aerospace, and naval applications.

The objective of this article is to summarize the microstructural aspect of fracture resistance in structural materials. The intent is to selectively compile and compare information on microstructure and fracture resistance of structural materials from literature and some of the author's work. The article begins with brief coverage on basic fracture principles, followed by material examples. Included in this text are examples of steels, aluminum alloys, titanium alloys, cermets, and composites. More detailed coverage is provided in other sections in this Volume.

Basic Fracture Principles

The mechanics of the fracture process (in either elastic or elastic-plastic conditions) is understood by considering a body with a crack length a subjected to an applied tensile stress σ. For purely brittle fracture originating from this crack, Griffith postulated that the critical rate of strain energy released during unstable crack extension, G_c, is related to the surface energy of the material, γ, as:

$$G_c = 2\gamma \tag{Eq 1}$$

$$G_c = \frac{\pi \sigma_c^2 a}{E} \tag{Eq 2}$$

where σ_c is the critical stress at the onset of fracture and E is the elastic modulus. From the linear elastic stress field ahead of a sharp crack, Irwin found that at regions very close to the crack tip, the stress normal to the fracture plane, σ_{yy}, is related to the stress-intensity factor, K, as:

$$\sigma_{yy} \approx \frac{K}{\sqrt{2\pi r}} \tag{Eq 3}$$

where $K = \sigma\sqrt{\pi a}$, σ is applied stress, and r is the distance from the crack tip on the crack plane. K is a measure of buildup or concentration of stress at the tip of a sharp crack. If fracture occurs from the crack, then the local critical value of σ_{yy} at which fracture occurs (e.g., by cleavage) is reflected as the critical value of K. Irwin further showed that K and the strain energy release rate, G, are related as:

$$G = \frac{K^2}{E} \tag{Eq 4}$$

It then follows that fracture will occur from the crack when the critical stress intensity at the crack tip, $K = K_c$, is reached, corresponding to $G = G_c$. Due to this simplicity, K_c has been accepted as a useful parameter representing fracture resistance of materials in engineering applications.

If the crack size is much smaller than the body dimensions, the relationship between fracture toughness, K_c, and fracture stress, σ_c, can be simply written as:

$$K_c = \sigma_c\sqrt{\pi a} \tag{Eq 5}$$

However, when the size of the crack from which fracture occurs is significant in relation to body dimensions:

$$K_c = \sigma_c\sqrt{\pi a}\, F\left(\frac{a}{W}\right) \tag{Eq 6}$$

where $F(a/w)$ is a function that accounts for the effects due to the finiteness of the body and is dependent on the shape of the body. The values of these factors can be found in handbooks on fracture mechanics as well as in the ASTM standard for fracture toughness testing, E 399. The appendix of this Volume also contains updated information on geometry factors, $F(a/W)$.

The above fracture relationships are strictly applicable to brittle materials in which the energy dissipation due to plastic deformation is almost negligible. Many structural materials, particularly those in the high-strength category, show evidence of plastic deformation and have fracture toughness levels higher than those that can be estimated from surface energy (Eq 1) alone. Hence, the modified form of Eq 1 to account for this additional contribution to fracture resistance due to plastic deformation can be written as:

$$G_c = \underset{(\text{elastic})}{2\gamma} + \underset{(\text{plastic})}{\Gamma\,\sigma_y\delta_c} \tag{Eq 7a}$$

or in terms of fracture toughness:

$$K_c = \sqrt{2E\gamma + \Gamma\, E\, \sigma_y\delta_c} \tag{Eq 7b}$$

where Γ is constant, σ_y is the material yield stress, and δ_c is the critical opening displacement at the crack tip at the onset of fracture. The first term is the energy consumed in the creation of two fracture surfaces during fracture and is considered to be independent of microstructure. The second term, $\sigma_y\delta_c$, represents approximately the energy consumed in plastic deformation accompanying fracture, a strong function of microstructure. Because the latter is several times higher than the former, the surface energy term is often ignored in the case of metallic structural materials.

From Table 1 it is evident that crack-tip plasticity accounts for much of the fracture resistance of structural metals. This fact makes it possible to alter microstructure for optimizing fracture resistance, through variables that influence strength and ductility, such as strain hardening, dislocation-particle interactions, and slip behavior. The mechanism of crack propagation depends on microstructural features as classified by Schwalbe (Fig. 1). This forms the basis of the discussion on individual alloys later. Further, due to an increase

Table 1 Fracture toughness values estimated from interfacial energies and the measured fracture toughness values of alloys

Material	Interfacial energy (2·γ), dynes/cm^2	G_{Ic}, MPa · m	E, MPa	K_{Ic}, MPa√m (a)	K_{Ic}, MPa√m (b)
α-Fe	4000	0.0004	200,000	8.9	20-150
Ti	2300	0.00023	100,000	4.8	30-80
Al	1350	0.00014	70,000	3.0	20-60

(a) Estimated. (b) Measured for alloys

in elastic modulus, the ranges of fracture toughness values obtainable by microstructure manipulation increases in the order of aluminum alloys, titanium alloys, and iron alloys (Fig. 2).

Fracture toughness values for materials are obtained by testing precracked compact tension or three-point bend specimens, following the ASTM E 399 test procedure. A measure of fracture resistance can also be obtained from Charpy specimens used in impact toughness testing. However, for comparison with linear elastic fracture mechanics specimens, precracked specimens are often used. Figure 3 illustrates the correlation between the critical strain energy release rates for fracture (K_{Ic}^2/E) and total energy absorbed per unit fracture area of precracked Charpy specimens, tested in slow bending, for different materials. This confirms that the critical strain energy release rate represents the energy required to cause fracture, even in the presence of plastic deformation. An approximate measure of this can also be obtained from simple Charpy specimens, as an alternative to the ASTM standard for fracture toughness test procedure.

Fracture Resistance of High-Strength Steels

Although engineering applications use many types of steels (such as mild steels, high-carbon steels, and alloy steels), only high-strength alloy steels are described in this article because of the general inverse relationship between strength and fracture toughness (Fig. 4). Examples of such steels include high-strength low-alloy steels, tool steels, dual-phase steels, maraging steels, and precipitation-hardenable stainless steels. The development of strength and toughness in these steels is linked to such factors as the size and distribution of carbides and nitrides, relative proportions of martensite and austenite phases, and grain size. In general, as shown in Fig. 4, high-fracture-toughness steels have ductile (low-carbon) martensite and retained metastable austenite as dominant phases in the microstructure. Steels that contain predominantly ferritic and pearlitic structures have relatively low fracture toughness. Table 2 summarizes the effects of microstructure on toughness.

Maraging (martensite-aging) steels, based on the Fe-Ni-Ti system, have very high strength and high fracture toughness and are designed with very low levels of carbon (<0.05%) and high levels of alloying elements (typically 18% Ni, 3% Ti, and 1 to 2% Co). The benefit of low carbon is twofold: crack-initiating carbides are absent and the martensite matrix is more ductile. In addition, maraging steels contain retained austenite (due to the high nickel content), which causes extensive plastic deformation and strain-induced martensitic transformation at the crack tip during fracture, thus increasing the fracture resistance. The strengthening lost by the elimination carbides is

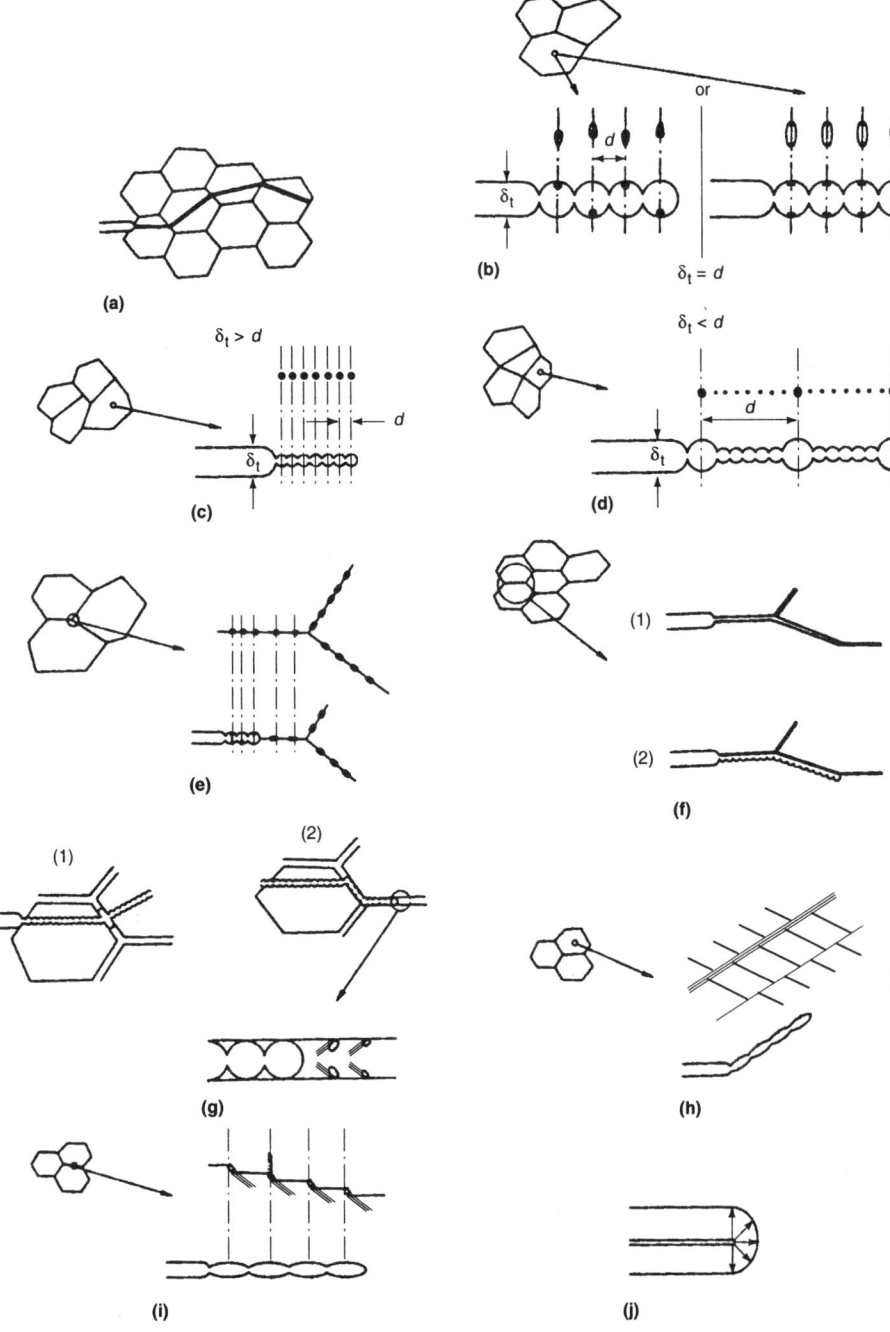

Fig. 1 Crack propagation mechanisms: (a) Cleavage crack propagation. (b) Dimple fracture due to coarse particles. (c) Dimple fracture due to fine particles. (d) Dimple fracture due to coarse and fine particles. (e) Intergranular crack propagation due to grain boundary precipitates. (f) Intergranular crack propagation due to a hard phase grain boundary film. (g) Crack propagation mechanisms when a soft phase grain boundary film is present. (h) Crack propagation by slip plane/slip plane intersection. (i) Crack propagation by slip plane/grain boundary intersection. (j) Crack propagation solely by plastic blunting

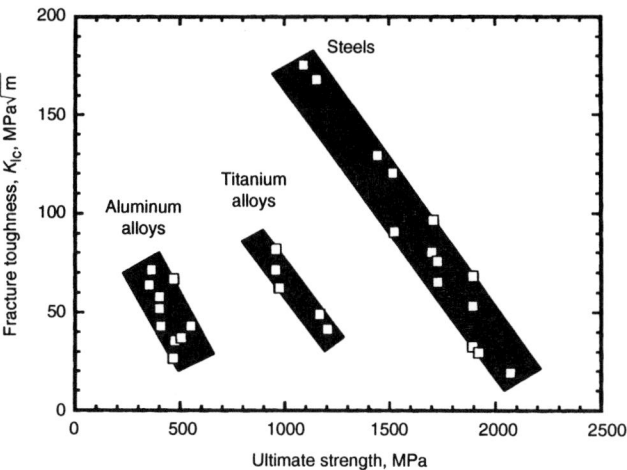

Fig. 2 Fracture toughness as a function of strength for high-strength structural alloys

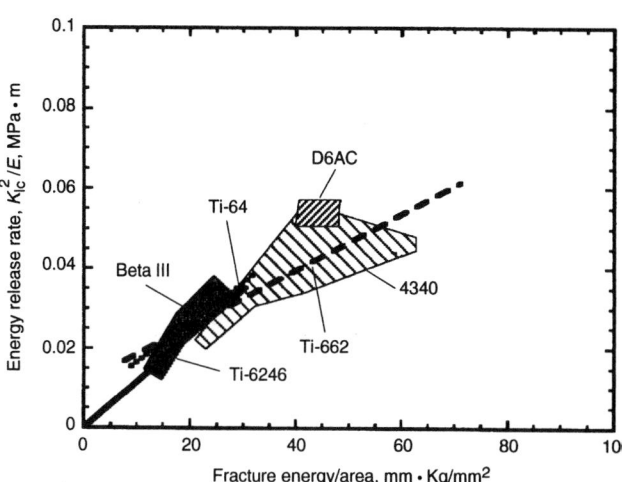

Fig. 3 The correlation between energy release rate and fracture energy of Charpy specimens for various materials

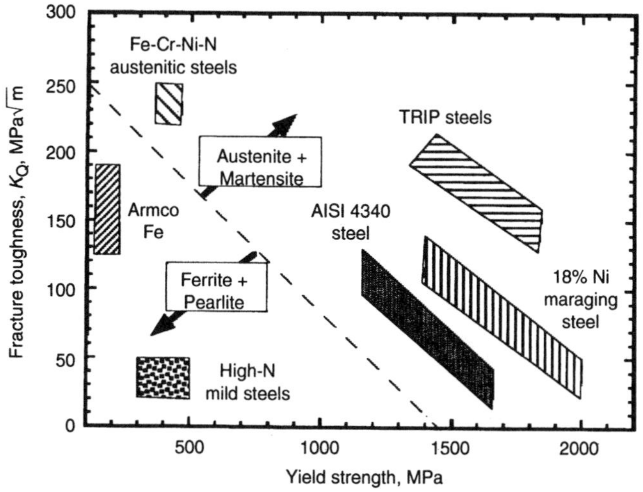

Fig. 4 Fracture toughness as a function of yield strength for structural steels. TRIP, transformation-induced plasticity

Table 2 Effects of microstructural variables on fracture toughness of steels

Microstructural parameter	Effect on toughness
Grain size	Increase in grain size increases K_{Ic} in austenitic and ferritic steels
Unalloyed retained austenite	Marginal increase in K_{Ic} by crack blunting
Alloyed retained austenite	Significant increase in K_{Ic} by transformation-induced toughening
Interlath and intralath carbides	Decrease K_{Ic} by increasing the tendency to cleave
Impurities (P, S, As, Sn)	Decrease K_{Ic} by temper embrittlement
Sulfide inclusions and coarse carbides	Decrease K_{Ic} by promoting crack or void nucleation
High carbon content (>0.25%)	Decrease K_{Ic} by easily nucleating cleavage
Twinned martensite	Decrease K_{Ic} due to brittleness
Martensite content in quenched steels	Increase K_{Ic}
Ferrite and pearlite in quenched steels	Decrease K_{Ic} of martensitic steels

replaced by precipitation of fine $(Fe,Ni)_3Ti$ phases from martensite, obtained by aging. However, under certain conditions, precipitation of phases at grain boundaries leads to deterioration in toughness. Hence, chemistry, processing, and heat treatment conditions should be designed to avoid the grain boundary precipitation. Typical heat treatment of a maraging steel involves austenitization and oil quenching or air cooling,

followed by aging at high temperature in the α + γ field.

Metastable Austenite-Based Steels. Transformation-induced plasticity (TRIP) steels with high contents of nickel and manganese retain high-temperature face-centered cubic austenite (γ) at room temperature, upon quenching. By a judicious choice of composition, this austenite is designed to be metastable after quenching and

transformable during deformation. The deformation and the volume change accompanying the austenite-to-martensite transformation increase the energy required for extension of a moving crack, resulting in high fracture toughness. The fracture toughness of such steels depends on the stability of austenite, measured in terms of the transformation coefficient, m, in the following relationship between the volume fraction of martensite formed (V_α) and tensile strain (ε):

$$V_\alpha = m \sqrt{\varepsilon} \tag{Eq 8}$$

The higher the tensile strain or crack opening displacement at the tip, the larger the volume fraction of transformed martensite and the higher the TRIP effect on toughness, as illustrated in Fig. 5. These steels possess the highest fracture toughness levels attainable in steels and hence are used in mining, drilling, and other applications requiring wear and erosion resistance.

Quenched-and-Tempered Steels. Fracture resistance in quenched-and-tempered steels is achieved by eliminating coarse alloy carbides, increasing hardenability to minimize ferrite formation, and alloying to retain austenite at room temperature.

An increase in the austenitization temperature, besides coarsening the grain size, dissolves carbides and nitrides present in steels. This eliminates the crack nucleation from carbides, thereby increasing the fracture toughness, as shown in Fig. 6 for two steels. The required austenitization time is also critical, and both time and temperature are chosen according to the amount of carbon and alloying elements in the steel. The effect of tempering temperature on fracture toughness depends on the type of prior austenitization treatment (Fig. 7). For example, the heat treatment involving austenitization at 1200 °C (2190 °F), followed by an ice brine quench and refrigeration in liquid nitrogen, results in high fracture toughness levels at all tempering temperatures, compared to austenitizing at 870 °C (1600 °F) and quenching in oil. The higher fracture toughness

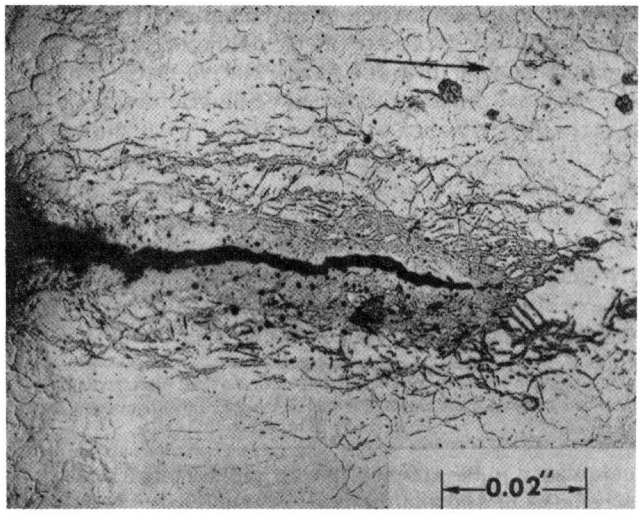

(a)

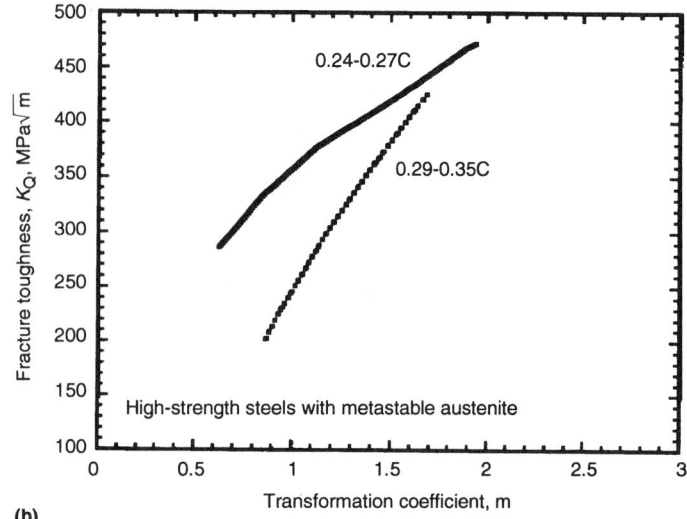

(b)

Fig. 5 TRIP steel crack and toughness. (a) Formation of martensite around a crack in a TRIP steel. (b) The effect of austenite transformation on the fracture toughness of metastable austenitic steels

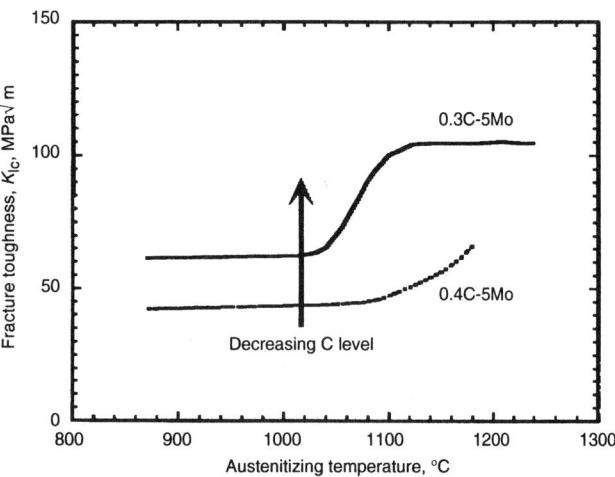

Fig. 6 The effect of austenitization temperature on the fracture toughness of two quenched-and-tempered steels

Fig. 7 The effect of tempering temperature on fracture toughness

levels of the former is attributed to the formation of 100% refined martensite upon quenching. Heat treatment at 870 °C (1600 °F) results in a mixture of blocky ferrite and upper bainite having continuous films of carbides at lath boundaries. This leads to low fracture toughness due to easy crack propagation along the lath boundaries at low tempering temperatures. However, at higher tempering temperatures, elimination of continuous carbide film by spheroidization increases the fracture toughness. In addition, in some alloy steels, retained austenite contributes to further increases in fracture toughness, by either crack-tip blunting or strain-induced transformation.

Fracture toughness in steels also depends on the nature of martensite (low-carbon ductile martensite or high-carbon twinned martensite). Supersaturated carbon in martensite increases the twinning to accommodate the strains in iron lattice due to carbon. The low fracture toughness

levels of some martensitic steels at low tempering temperatures are due to the presence of brittle twinned martensite. An increase in the twin density of martensite results in low fracture toughness, as shown in Fig. 8.

Inclusions decrease fracture toughness (Fig. 9) by promoting crack nucleation by inclusion fracture, void nucleation at the particle-matrix interface, and early coalescence. This reduces the extent of void growth before fracture, limiting the plastic energy absorption in the process zone at the crack tip. A decrease in sulfur level in steel increases the spacing between the inclusions, thereby increasing the size of the plastically deformed process zone. This contribution to increased fracture toughness can be rationalized in terms of Kraft's model:

$$K_{Ic} \approx \sqrt{2\,\pi\,E\,n\,d_T}$$ (Eq 9)

where n is the monotonic strain hardening exponent and d_T is the size of the process zone at the crack tip, proportional to the spacing between crack/void nucleating inclusions.

Fracture Resistance of Aluminum Alloys

Aluminum alloys based on Al-Cu ($2xxx$ series), Al-Mg-Si ($6xxx$ series), Al-Zn-Mg ($7xxx$ series), and, recently, Al-Li ($8xxx$ and $209x$ series) are the predominant age-hardenable alloys, used extensively in aerospace and other medium- to high-strength structural applications. Typical examples are alloys 2024, 2124, 6061, 7075, 7150, 7475, 8090, and 2091. Due to complex chemistry, precipitation, and intermetallic compound formation in aluminum alloys, control of the size and distribution of age-hardening coherent pre-

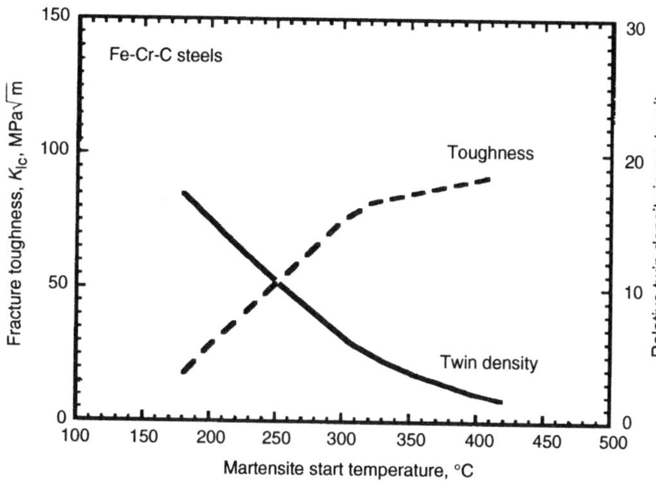

Fig. 8 Fracture toughness and martensite twin density as a function of martensite start temperature for an Fe-Cr-C steel

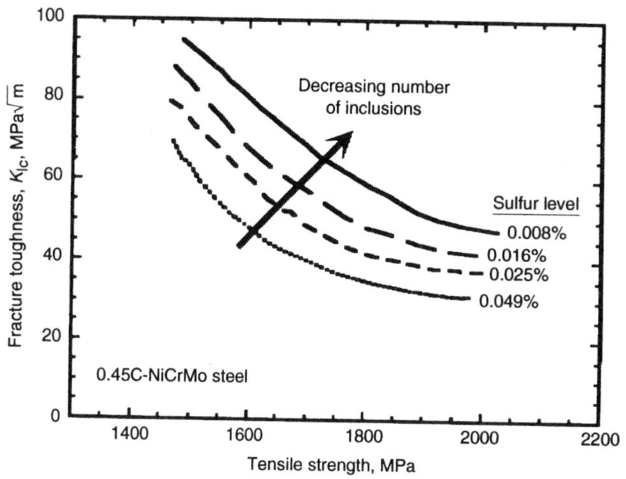

Fig. 9 The effect of sulfur on the fracture toughness vs. strength relationship

Table 3 A list of precipitates and intermetallic compounds in aluminum alloys

Precipitate phases (beneficial)	Intermetallic compounds (detrimental)
Al_2Cu	Al_7Cu_2Fe
Al_2CuMg	Mg_2Si
$MgZn_2, Mg_2Si$	$Al_{12}Mn_2Cr$
$Al_2Zn_3Mg_3$	$Al_{20}Cu_2Mn$
Al_3Zr, Al_3Li	$(Fe, Mn)Al_6$

Table 4 Strength and fracture toughness levels for selected aluminum alloys

Alloy	0.2% yield strength, MPa	K_{Ic}, MPa√m
2014-T6	436	20
2024-T851	443	21
2124-T851	435	26
7075-T7351	391	31
7079-T651	502	27

Table 5 Effects of processing/microstructural variables on the fracture toughness of aluminum alloys

Variable	Effect on fracture toughness
Quench rate	Decrease in K_{Ic} at low quench rates
Impurities (Fe, Si, Mn, Cr)	Decrease in K_{Ic} with high levels of these elements
Grain size	Decrease in K_{Ic} at large grain sizes due to coarse grain boundary precipitation
Grain boundary precipitates	Increase in size and area fraction decrease K_{Ic}
Underaging	Increases toughness
Peak aging	Increases fracture toughness
Overaging	Decreases fracture toughness
Grain boundary segregates (Na, K, S, H)	Lower fracture toughness in Al-Li alloys

cipitates and incoherent intermetallic phases is critical in achieving a balance of strength and resistance to fracture and stress-corrosion cracking. Table 3 lists some important precipitate phases and intermetallic compounds that form in aluminum alloys. While the precipitates control strength, intermetallic compounds that form during solidification primarily control ductility and fracture toughness.

Typical heat treatment of an aluminum alloy involves solution treatment, quenching in water, and aging at a suitable temperature for a specified period of time. Additionally, warm or cold working, such as stretching after solution treatment, is performed to control the size and distribution of precipitates. Such thermomechanical processing routes for high-strength aluminum alloys have been well developed to impart desirable combinations of strength, ductility, fracture toughness, and stress-corrosion cracking resistance.

Fracture resistance in aluminum alloys is strongly sensitive to purity, aging, the presence of

intermetallic compounds, thermomechanical treatment, grain size, and orientation or texture. Typical fracture toughness values of selected aluminum alloys are given in Table 4. Table 5 gives a list of variables and the nature of their effect on the fracture toughness of aluminum alloys.

Figure 10 illustrates the effects of intermetallic-forming elements (Cr, Zr, Fe, Si, and Mn) on fracture in terms of the relationship between fracture resistance (measured in terms of the unit propagation energy in fracturing a notched bar) and tensile yield strength for two different orientations of crack propagation. Zirconium addition is beneficial due to the grain-refining effect of Al_3Zr phase. Chromium and manganese primarily lead to intermetallic compound formation and hence must be reduced to achieve a combination of high fracture toughness and high strength. Most of the intermetallic compounds form at grain boundaries in wrought alloys and at interdendritic regions in cast alloys. This is the primary reason for lower levels of fracture toughness, especially when a crack propagates in the short transverse plane (plane of rolling), in which the grain boundary area intersected by the crack plane is high compared to other orientations.

The detrimental effect of intermetallic particles on fracture toughness can be understood from a simple relationship. The critical fracture strain, ε_c, of a ligament between the crack tip and a crack/void nucleating particle is related to the particle volume fraction, V_f, as:

$$\varepsilon_c = f\left(\frac{1}{V_f}\right) \tag{Eq 10}$$

The fracture toughness of aluminum alloys is related to the critical fracture strain as:

$$K_{Ic} \approx \sqrt{\frac{2 C E \varepsilon_c \sigma_y n}{(1 - v^2)}} \tag{Eq 11}$$

where C is a constant and v is the Poisson's ratio. As the volume fraction of brittle intermetallic particles is reduced, fracture toughness increases. For a given volume fraction of particles, fracture toughness also increases with increase in yield strength and the strain-hardening exponent of the matrix. From Eq 11, fracture toughness levels of several aluminum alloys can be predicted (Fig. 11) with good accuracy.

Figure 12 shows the effect of grain size on the fracture toughness of a 7xxx alloy tested with crack propagation in the long transverse direction. The decrease in fracture toughness is attributed to the increase in grain boundary fracture at large grain sizes. The increased intergranular fracture also coincides with the thickening of precipitates (e.g., the size of $MgZn_2$ precipitates in Fig. 13) at grain boundaries under prolonged aging. This behavior is also reflected in the change in fracture mode from transgranular to

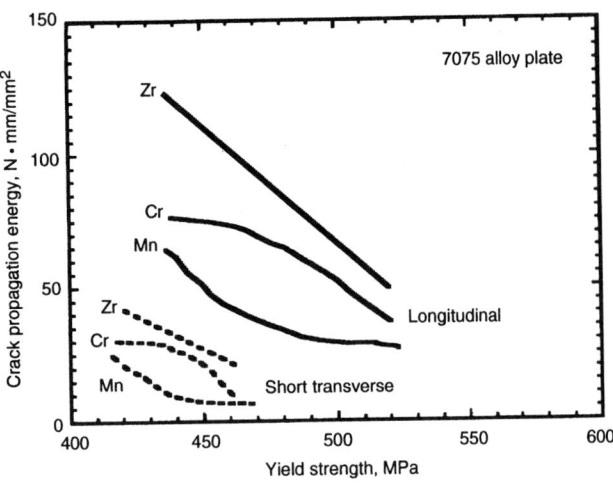

Fig. 10 The effect on toughness of elements that form intermetallic compounds

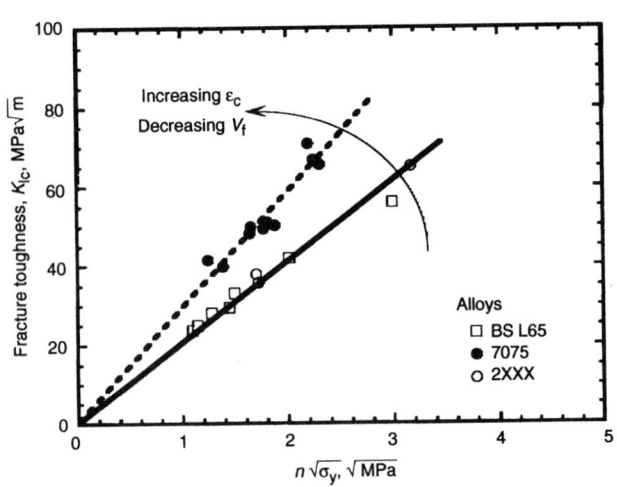

Fig. 11 The effect of parameter, $n\sqrt{\sigma_y}$ on the fracture toughness of aluminum alloys

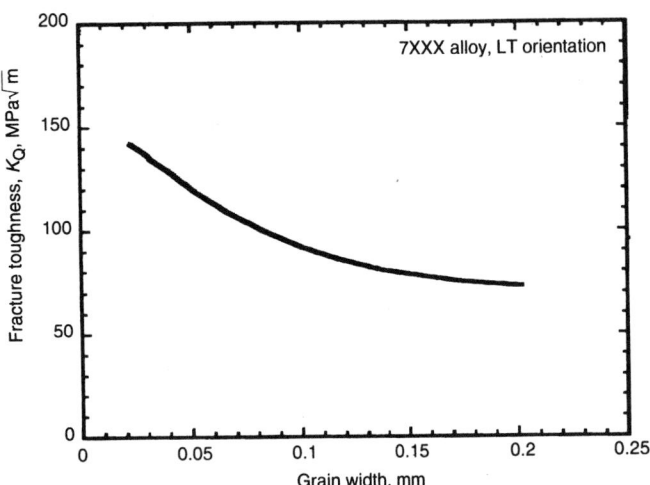

Fig. 12 The effect of grain size on the fracture toughness of a 7xxx alloy

Table 6 Fracture toughness levels of Ti-6Al-4V alloy in different microstructural conditions

Microstructure	0.2% yield strength, MPa	Elongation, %	K_{Ic}, MPa√m
β-processed (aligned lamellar α)	903	12	78
α+β processed (equiaxed α in aged β matrix)	917	16	53
Recrystallized (fully equiaxed α)	925	19	47

intergranular. This suggests that reduced grain boundary area accompanied by coarse grain boundary precipitation is detrimental to the fracture toughness of aluminum alloys.

Figure 14 shows the broad range of data of fracture toughness and yield strength for both 2xxx and 7xxx alloys as affected by the degree of aging. Overaging generally results in low fracture toughness levels for a given yield strength and alloy type, compared with underaging. This is attributed to the increased occurrence of intergranular failure, consistent with the observations illustrated in Fig. 12 and 13.

The variables influencing fracture toughness of Al-Li alloys are similar to those that affect other age-hardenable aluminum alloys. These include degree of aging, area fraction of grain boundary precipitates, impurities, and orientation. However, Al-Li alloys are more anisotropic due to strong texture formation, relative to aluminum alloys, and hence they show a much stronger sensitivity of fracture toughness to orientation.

Fracture Resistance of Titanium Alloys

Titanium alloys are primarily used in aerospace applications owing to their good combination of specific strength, ductility, and fracture toughness. As in steels and aluminum alloys, this combination is achieved by careful control of two-phase microstructures. Among the two phases (α and β), β is more ductile and is preferable in increasing the fracture toughness of titanium alloys. The three broad classes of titanium alloys are near-α, α + β, and β alloys, grouped according to the levels of α or β stabilizing elements. Typically, β content by volume is: near-α, <10%; α + β, 10-25%; and β, >25%. Figure 15 shows the fracture toughness/strength relationship maps for different titanium alloys. Metastable β alloys possess the highest combination of strength and toughness. This arises from a large volume fraction of β phase and fine aged-α precipitates.

Unlike steels and aluminum alloys, titanium alloys are generally free from inclusions and intermetallics that form during solidification. Neither is there precipitation and coarsening of brittle phases, so the control of microstructure for fracture toughness is less difficult. Microstructure plays a major role in controlling the fracture toughness of titanium alloys. Microstructure control is primarily achieved by mechanical processes, such as hot/cold working and heat treatment involving solution treatment followed by quenching and aging or slow cooling. Table 6 lists the fracture toughness values of a typical titanium alloy under different microstructural conditions. In general, for a given β phase content, fracture toughness increases with an increase in the amount of lamellar α as well as an increase in the aspect ratio of α phase.

The dominant variables that influence fracture toughness in titanium alloys are the interstitial elements, grain size, microstructural morphology, and relative proportions of α and β phases. Table 7 lists these variables and the nature of their effect on fracture toughness.

Figures 16 and 17 illustrate that increases in oxygen and hydrogen levels in Ti-6Al-4V alloy decrease fracture toughness. This is caused by an increase in the planarity slip, promoted by the ordering of Ti₃Al phase, which causes easy crack nucleation at grain and phase boundaries. This tendency to ordering is also increased at high

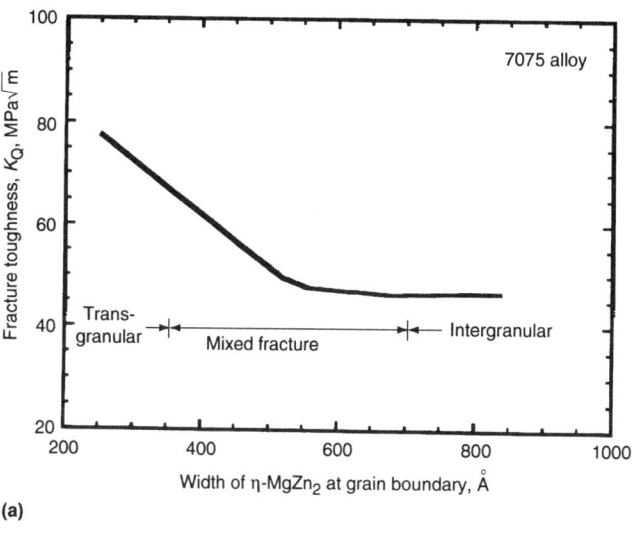

(a)

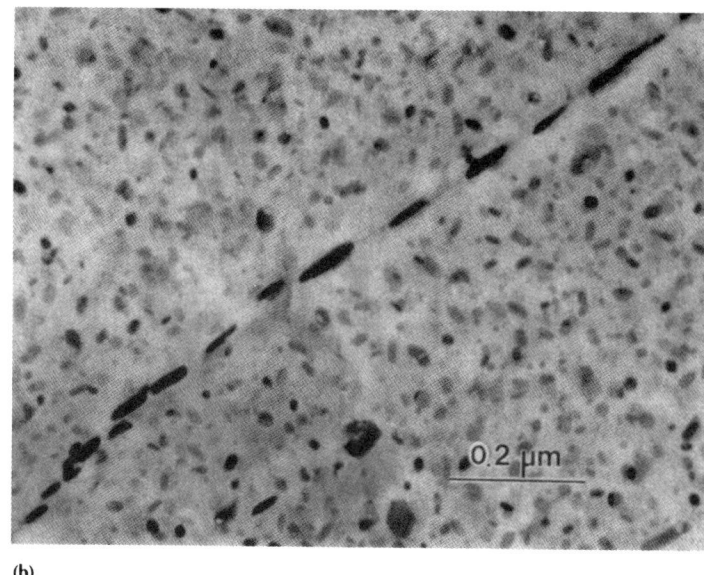

(b)

Fig. 13 7075 Al alloy. (a) The effect of grain boundary precipitate size on fracture toughness and fracture morphology. (b) Equilibrium grain boundary η-MgZn₂ precipitates at grain boundaries

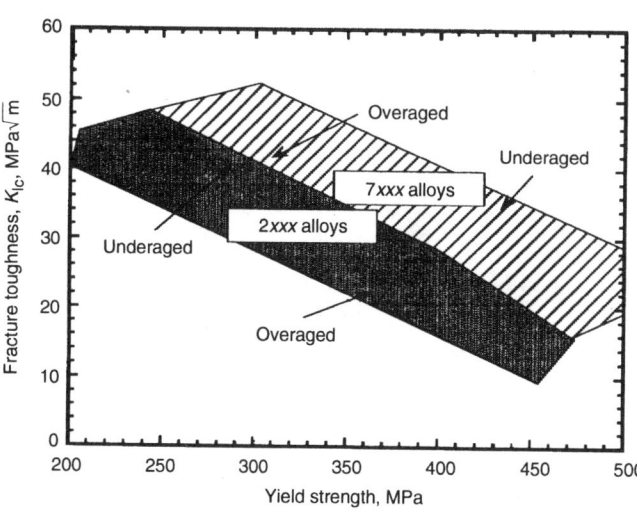

Fig. 14 The effects of alloy type and aged condition on the strength/fracture toughness relationship for aluminum alloys

Fig. 15 The relationship between fracture toughness and strength for different classes of titanium alloys and microstructures

Table 7 Effect of microstructural variables on the fracture toughness of titanium alloys

Variable	Effect on fracture toughness
Interstitials (O, H, C, N)	Decrease in K_{Ic}
Grain size	Increase in grain size decreases K_{Ic}
Lamellar colony size	Increase in colony size increases K_{Ic}
β phase	Increases in β volume fraction, continuity increase K_{Ic}
Grain boundary α phase	Increases in thickness and continuity increase K_{Ic}
Shape of α phase	Increase in aspect ratio of α phase increases K_{Ic}
Orientation	Crack oriented for easy cleavage along basal planes gives low K_{Ic}

aluminum contents, and hence compositions of commercial alloys rarely exceed 6% Al. Hydrogen causes cleavage and interface cracking due to the formation of hydrides (TiH₂). Alloys with high levels of β phase can dissolve more hydrogen, thereby preventing the decrease in fracture toughness due to hydrogen. The other interstitial elements, carbon and nitrogen, have low solid solubility in titanium and form fine TiC and TiN dispersions when the solubility level is exceeded. These particles drastically decrease the ductility as well as the fracture toughness of titanium alloys and hence must be eliminated.

The effect of β grain size on fracture toughness is illustrated in Fig. 18 for Ti-5.2Al-5.5V-1Fe-0.5Cu alloy. There is an inverse relationship of fracture toughness to β grain size. As the grain size increases, the tendency to intergranular fracture increases, due to the weakening effect of fine 0.2 μm thick particles at the β-β grain boundary. This is primarily due to the increased density of grain boundary precipitates as a result of the reduction in available grain boundary area.

The same alloy was heat treated differently to produce thick continuous α phase at the grain boundary, which increased fracture toughness (Fig. 19). However, for this to occur, the grain interior (aged β matrix) should be stronger than α.

The orientation of crack plane in a fracture toughness test with respect to the rolling direction of the titanium alloy plate has a significant effect, due to the preferred orientation of hexagonal close-packed crystal grains having limited slip systems, relative to body-centered cubic and face-centered cubic crystals. The effect of orientation on the fracture toughness/strength relationship is illustrated in Fig. 20. While a strong inverse relationship between fracture toughness and yield strength is seen for the longitudinal orientation, it is less strong in the transverse orientation. The effect of orientation on fracture

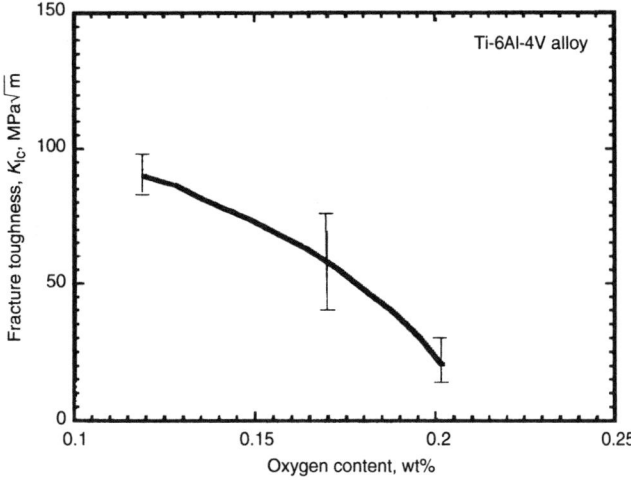

Fig. 16 The effect of oxygen level on the fracture toughness of Ti-6Al-4V alloy

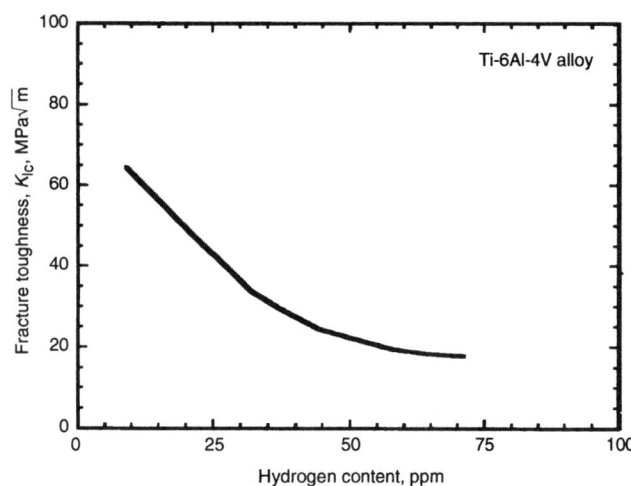

Fig. 17 The effect of hydrogen level on the fracture toughness of Ti-6Al-4V alloy

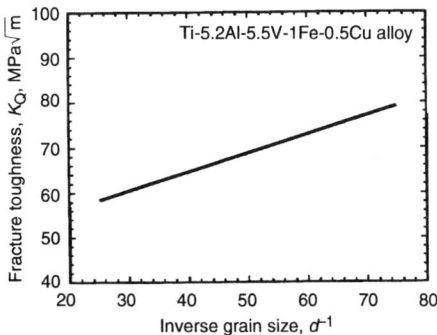

Fig. 18 The effect of grain size on the fracture toughness of a titanium alloy

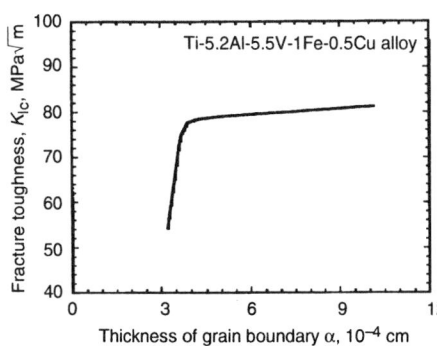

Fig. 19 The effect of the thickness of grain boundary α phase on the fracture toughness of a titanium alloy

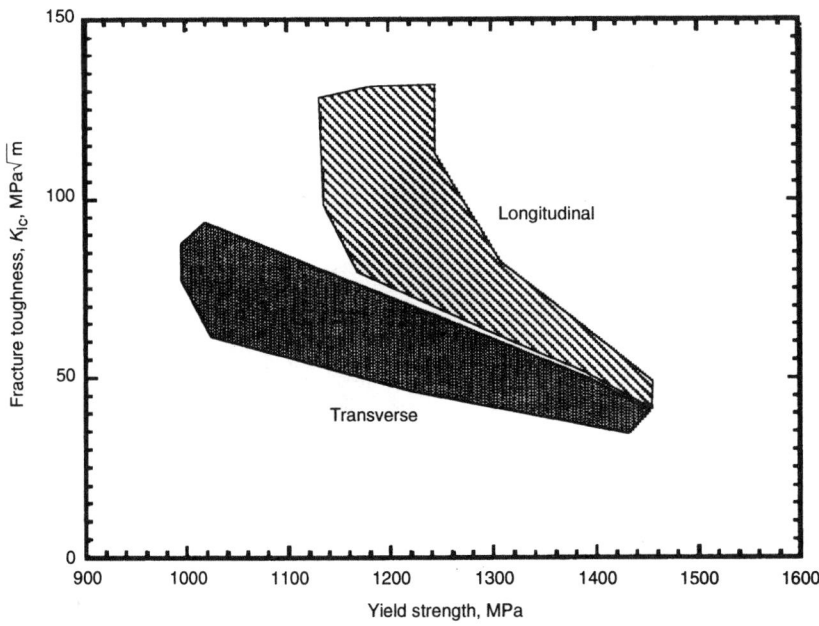

Fig. 20 The effect of crack plane orientation on the fracture toughness of Ti-6Al-4V alloy

toughness arises due to the relative orientation of slip systems such as {0001} <1120>, {1010} <1120> with respect to the crack plane. This is also evident from the variation of elastic modulus with orientation, presented in Table 8.

It is clear that the key to improving the combination of strength and toughness in titanium alloys is to increase β phase content, increase the lamellar α volume fraction and the aspect ratio of α phase, and reduce planarity of slip and interface embrittlement by reducing the levels of oxygen, hydrogen, carbon, and nitrogen.

Fracture Resistance of Composites

Brittle Matrix-Ductile Phase Composites. Ductile phases have been used to improve the fracture resistance of many structural materials, including ceramics, intermetallics, glasses, and other low-toughness materials, such as steels having a hard martensitic structure. Table 9 lists some brittle materials and the possible ductile phase reinforcements that can be used to improve the fracture toughness levels. Such an alloy design concept generally increases fracture toughness with little sacrifice in strength. The ductile phases absorb energy by plastic deformation during crack propagation. Bridging of the crack and the constrained deformation of ductile phase contribute to the increase in toughness. In general, the fracture toughness increases with an increase

in the volume fraction of ductile phase (Fig. 21). For such composites, the critical strain energy release rate, G_c, for unstable fracture can be expressed as the sum of fractional energy absorbed in fracturing the brittle and ductile phases:

$$G_c = \underbrace{(1 - V_f)\, G_m}_{\text{(matrix)}} + \underbrace{V_f\, \sigma_0 a_0\, \chi}_{\text{(ductile phase)}} \qquad \text{(Eq 12)}$$

Table 8 Effect of texture on the fracture toughness of Ti-6Al-2Sn-4Zr-6Mo alloy

Orientation	0.2% yield strength, MPa	E, MPa	K_{Ic}, MPa√m
Longitudinal	953	107,000	75
Transverse	1198	134,000	91
Short transverse	926	104,000	49

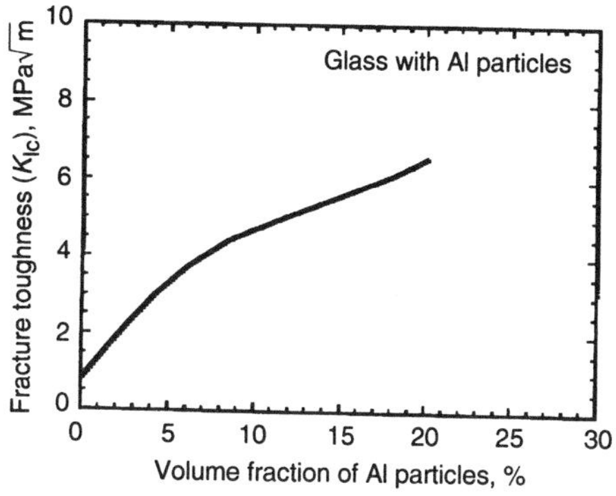

Fig. 21 Fracture toughness trend for glass having embedded ductile aluminum particles

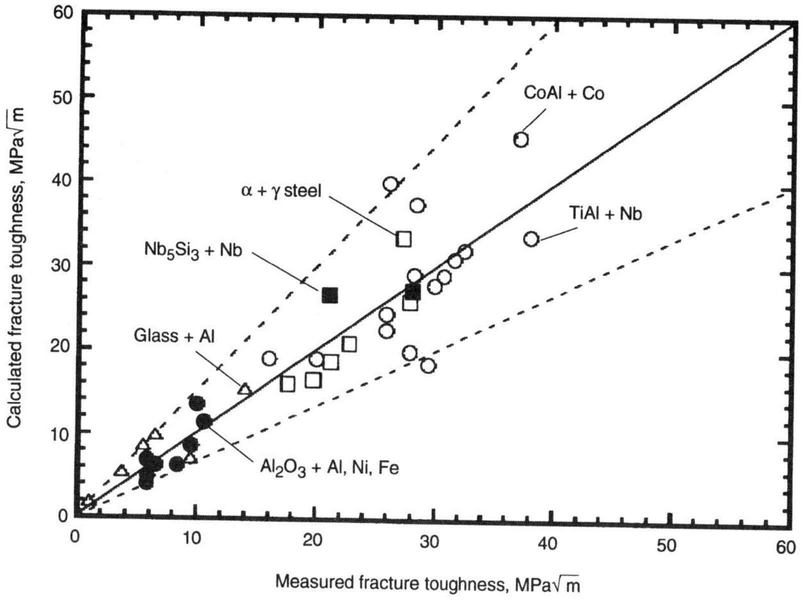

Table 9 Brittle materials and the possible ductile phases

Material	Ductile phase
Ceramics: Al_2O_3, glass, ZrO_2, SiC, Si_3N_4	Ti, Ni, Pb, Al, Fe
Intermetallics: TiAl, NiAl, CoAl, Nb_5Si_3, Cr_3Si	Ti, Nb, Co, Mo, Cr
Martensite	Austenite, ferrite

Fig. 22 The correlation between measured and calculated fracture toughness levels of several brittle materials having ductile phases as reinforcements

where σ_0 is the flow stress, a_0 is the radius of ductile phase, G_m is the energy release rate of the brittle matrix, and χ is a measure of microstructural constraint. A modified form for the above equation for the plane strain fracture toughness, K_{Ic}, can be written as:

$$K_{Ic} = \sqrt{\frac{E_c (1 - v_m^2) (1 - V_f) K_m^2}{(1 - v_c^2) E_m} + \frac{V_f E_c \sigma_0 a_0 \chi}{(1 - v_c^2)}}$$

(Eq 13)

where E_c and v_c are respectively the elastic modulus and the Poisson's ratio of the composite, and K_m, E_m, and v_m are respectively the fracture toughness, elastic modulus, and Poisson's ratio of the matrix material. For the case of plane stress, $E_c/(1 - v_c^2)$ and $E_m/(1 - v_m^2)$ are to be replaced by E_c and E_m, respectively. The parameter χ (usually in the range of 2 to 6) is a measure of the constraint experienced by the ductile phase in the elastic matrix during deformation. If χ is known, in addition to matrix and particle properties, fracture toughness of the composite can be estimated with reasonable accuracy.

Figure 22 shows the correlation between the measured fracture toughness and the calculated toughness, following Eq 13. The good agreement suggests that Eq 13 adequately represents functional dependence of fracture toughness on important microstructural parameters of the composite. Increases in the size, volume fraction, and

yield strength of the ductile phase, together with an increase in composite modulus, should significantly increase the fracture toughness. The effect of Young's modulus of matrix on the composite fracture toughness is not significant. A major contribution to fracture toughness is the constraint factor, which is a measure of increase in resistance to the in situ plastic deformation of ductile phase, as imposed by the surrounding elastic matrix.

WC-Co Cermets. The case of WC-Co cermets is similar to that of the above-described ductile phase composites, except that the ductile cobalt phase surrounds the brittle WC almost completely. Figure 23 illustrates the fracture path through the cermet microstructure. The volume fraction of cobalt in these cermets is usually be-

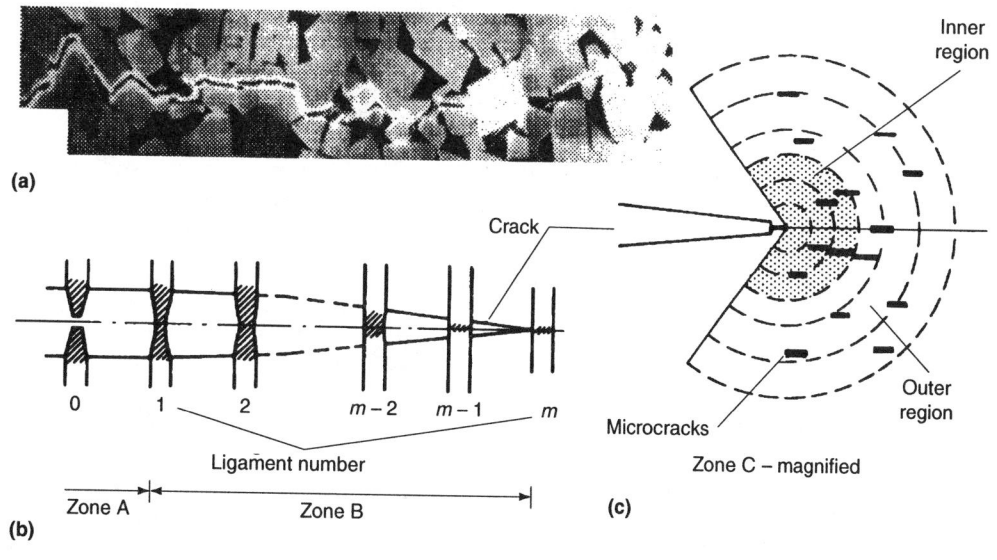

Fig. 23 Mechanisms of crack growth and fracture in WC-Co cermets

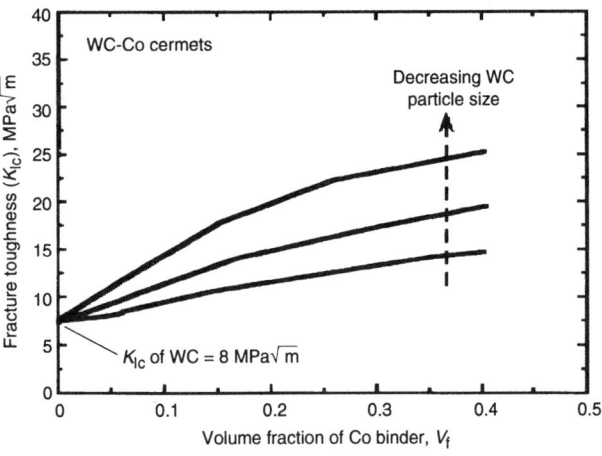

Fig. 24 The effect of ductile cobalt volume fraction and WC particle size on the fracture toughness of cermets

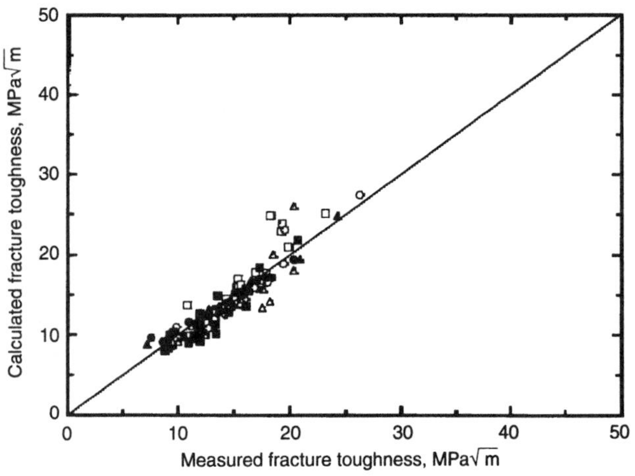

Fig. 25 The correlation between the measured and calculated fracture toughness levels of WC-Co cermets having varying sizes and volume fractions of WC and cobalt

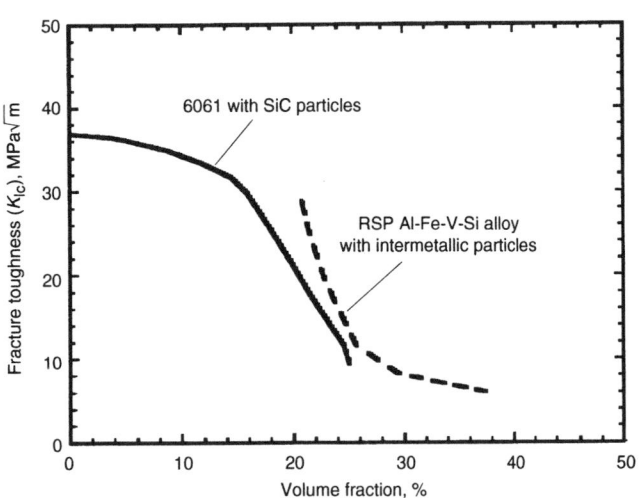

Fig. 26 The effect of dispersoids on the fracture toughness of aluminum-base metal-matrix composites

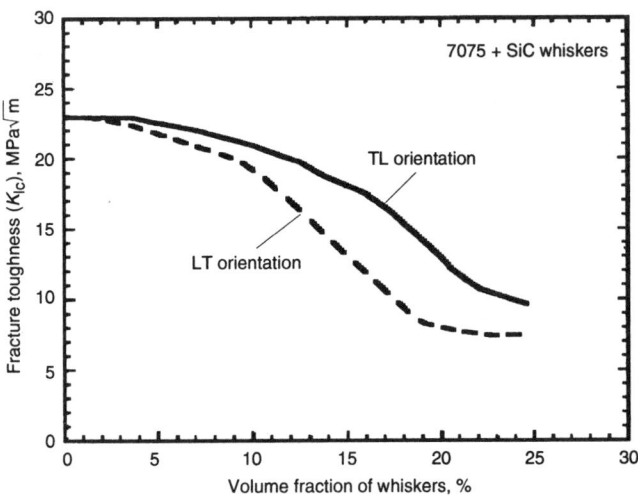

Fig. 27 The effect of orientation and whisker volume fraction on the fracture toughness of composites based on 7075 aluminum alloy

tween 0.1 and 0.3. The cermets are made by presintering WC to obtain a skeleton with continuous porosity and then infiltrating the skeleton with molten cobalt. In general, a decrease in WC particle size and an increase in cobalt volume fraction increases the fracture toughness of cermets (Fig. 24). Because of the thin layer of cobalt present between WC grains, its in situ deformation behavior during fracture is similar to the deformation of a thin ductile copper strip sandwiched between hard tool steel platens. In order to estimate the fracture toughness of cermets, this deformation analogy can be incorporated in Eq 13 for the constraint factor χ, through the relationship

$$\chi = \frac{\sigma_{eff}}{\sigma_0} = \left[1 + \frac{2k}{3} \left(\frac{d}{2h} \right) \right] \qquad \text{(Eq 14)}$$

where σ_0 is the bulk flow stress of the binder in the absence of any constraint and σ_{eff} is the flow stress of the binder in situ in the microstructure. The

constant k is defined as the maximum shear factor, which is taken as 0.577, and d and h are respectively the width of the rigid platen and the thickness of the ductile layer. In the case of cermets, as a first approximation, d and h can be considered to represent the mean WC particle diameter and the thickness of the cobalt binder, respectively. An increase in cobalt binder thickness and a decrease in WC particle size would therefore increase the constraint for deformation and hence the fracture toughness. Figure 25 compares experimental data with the theoretically calculated fracture toughness levels using Eq 13 and 14. The good correlation suggests that Eq 13 and 14 capture the effects of important microstructural parameters on the fracture toughness of cermets and can be used in the design of cermet composition and microstructure.

Metal-Matrix Composites. Light metals such as aluminum and magnesium are reinforced with particulates and whiskers based on SiC, Al₂O₃, TiC, and so on to increase the stiffness and high-temperature strength. These composites are made by dispersing reinforcements in liquid metal and

casting or by mixing with metal powder and hot pressing. In general, the size and spacing of particles, the strength of the interface between the particles, and the aspect ratio of the whiskers influence the strength and fracture toughness of composites. Figure 26 illustrates the fracture toughness levels of aluminum alloys reinforced with second-phase particles, showing a decrease in fracture toughness at large particle volume fractions.

The fracture toughness of metal-matrix composites can be estimated approximately from:

$$K_{Ic} \propto \sqrt{E \, \sigma_y \varepsilon_f \, l^*} \qquad \text{(Eq 15)}$$

where ε_f is the fracture strain of the ligament between the crack tip and the closest particle and l^* is the size of the process zone at the crack tip, usually taken as interparticle spacing. From this equation, it is clear that decreasing the interparticle spacing by increasing the volume fraction of dispersions reduces the fracture toughness of composites.

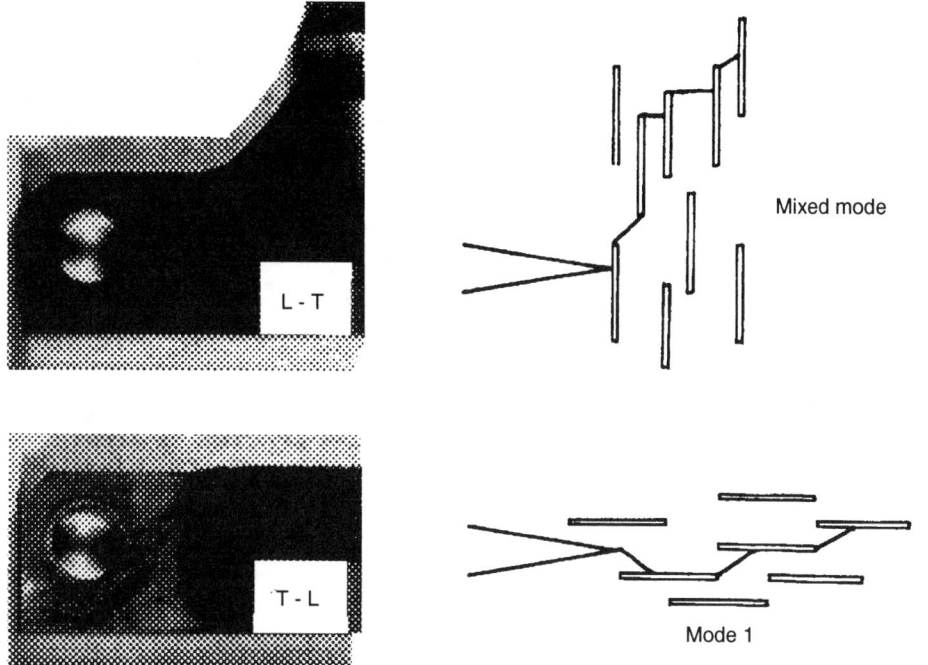

Fig. 28 Effect of whisker orientation on crack path and fracture in fracture toughness tests of 7075 + SiC-whisker composites

Figure 26 shows two composites, 6061 Al-Mg-Si alloy reinforced with various amounts of SiC particles and a rapidly solidified Al-Fe-V-Si alloy containing intermetallic particles. The trend is similar for both materials. There are two mechanisms by which reinforcements can affect fracture toughness. First, plastic flow localization at the interface and interface decohesion can significantly reduce the extent of void growth before ultimate failure, thus reducing fracture toughness. This is the case for the particulate composites. Alternatively, crack-tip blunting and crack path deviation around whiskers can increase fracture toughness by increasing the energy required for crack extension. However, for this to occur, the matrix-whisker interface must be strong. In reality, the interface is weaker due to reaction between the matrix and the whiskers during processing and the presence of oxides on the surfaces of whiskers. This effect is illustrated in Fig. 27, which shows that both orientations (whiskers oriented normal or parallel to crack plane) lead to a decrease in fracture toughness due to interface fracture. Figure 28 shows the crack path and whiskers are oriented differently with respect to the crack propagation direction in fracture toughness tests. The decrease in fracture toughness in the latter orientation is higher, due to increased weak interface area.

SELECTED REFERENCES

Basic Fracture Principles

- D. Broek, *Elementary Engineering Fracture Mechanics*, Kulewer Academic Publishers, 1986

- J.F. Knott, *Fundamentals of Fracture Mechanics*, Butterworths, London, 1973
- J.M. Kraft, Elastic-Plastic Fracture, *App. Mater. Res.*, Vol 3, 1964, p 88
- R.A. Wullaert, D.R. Ireland, and A.S. Telelman, Use of the Precracked Charpy Specimen in Fracture Toughness Testing, *Fracture Prevention and Control*, American Society for Metals, 1974, p 255

Steel

- A.J. Birkle, R.P. Wei, and G.E. Pellissier, *Trans. ASM*, Vol 59, 1966, p 981
- C.L.M. Cottrell, in *Fracture Toughness of High Strength Materials: Theory and Practice*, Publication 120, Iron and Steel Institute, London, 1970, p 112
- R.F. Decker, Alloy Design, Using Second Phases, *Metall. Trans.*, Vol 4, 1973, p 2495
- R.O. Ritchie and A.W. Thompson, On Macroscopic and Microscopic Analyses for Crack Initiation and Crack Growth Toughness in Ductile Alloys, *Metall. Trans.*, Vol 16A, 1985, p 233
- A.R. Rosenfield and A.J. McEvily, Some Recent Developments in Fatigue and Fracture, *Metallurgical Aspects of Fatigue and Fracture*, AGARD Report 610, NATO, 1973
- V.F. Zackay, Fundamental Considerations in the Design of Ferrous Alloys, *Alloy Design for Fatigue and Fracture Resistance*, AGARD Report 185, NATO, 1976, p 5.1
- V.F. Zackay and E.R. Parker, Fracture Toughness, *Alloy and Microstructure Design*, J.K. Tien and G.S. Ansell, Ed., Academic Press, 1976, p 213

- Z. Fan, The Grain Size Dependence of Ductile Fracture Toughness of Polycrystalline Metals and Alloys, *Mater. Sci. Eng.*, Vol A191, 1995, p 73
- S.D. Antolovich, A. Saxena, and G.R. Chanani, Increased Fracture Toughness in a 300 Grade Maraging Steel as a Result of Thermal Cycling, *Met. Trans.*, Vol 5, 1974, p 623

Aluminum Alloys

- J.D. Embury, Basic Microstructural Aspects of Aluminum Alloys and Their Influence on Fracture Behavior, *Alloy Design for Fatigue and Fracture Resistance*, AGARD Report 185, NATO, 1976, p 1.1
- G.C. Garrett and J.F. Knott, The Influence of Compositional and Microstructural Variations on the Mechanism of Static Fracture in Aluminum Alloys, *Metall. Trans.*, Vol 4A, 1978, p 1187
- G.T. Hahn and A.R. Rosenfield, Metallurgical Factors Affecting Fracture Toughness of Aluminum Alloys, *Metall. Trans.*, Vol 6A, 1975, p 653
- J.G. Kaufman, Design of Aluminum Alloys for High Toughness and High Fatigue Strength, *Alloy Design for Fatigue and Fracture Resistance*, AGARD Report 185, NATO, 1976, p 2.1
- A.K. Vasudevan, R.D. Doherty, and S. Suresh, Fracture and Fatigue Characteristics in Aluminum Alloys, *Aluminum Alloys-Contemporary Research and Applications*, A.K. Vasudevan and R.D. Doherty, Ed., Vol 31, *Treatise in Materials Science and Technology*, 1989, p 446

Titanium Alloys

- B.L. Averback, Microstructure and Fracture Toughness, *Fracture Prevention and Control*, American Society for Metals, 1974, p 97
- J.P. Hirth and F.H. Froes, Interrelations between Fracture Toughness and Other Mechanical Properties in Titanium Alloys, *Metall. Trans.*, Vol 8A, 1977, p 1165
- N.E. Paton, J.C. Williams, J.C. Chesnutt, and A.W. Thompson, The Effects of Microstructure on the Fatigue and Fracture of Commercial Titanium Alloys, *Alloy Design for Fatigue and Fracture Resistance*, AGARD Report 185, NATO, 1976, p 4.1
- K.H. Schwalbe, On the Influence of Microstructure on Crack Propagation Mechanisms and Fracture Toughness of Metallic Materials, *Eng. Fract. Mech.*, Vol 9, 1977, p 795
- C.A. Stubbington, Metallurgical Aspects of Fatigue and Fracture in Titanium Alloys, *Alloy Design for Fatigue and Fracture Resistance*, AGARD Report 185, NATO, 1976, p 3.1
- J.C. Williams, J.C. Chesnutt, and A.W. Thompson, The Effects of Microstructure on Ductility and Fracture Toughness of α+β Titanium Alloys, *Microstructure, Fracture Toughness and Fatigue Crack Growth Rate in Tita-*

nium Alloys, A.K. Chakrabarti and J.C. Chesnutt, Ed., TMS-AIME Publications, 1987, p 255

Composites and Cermets

- M.F. Ashby, F.J. Blunt, and M. Bannister, Flow Characteristics of Highly Constrained Metal Wires, *Acta Metall.*, Vol 37, 1989, p 1847
- A.G. Evans and R.M. McMeeking, *Acta Metall.*, Vol 34, 1988, p 2435
- K. Hirano, R & D Trends on Advanced Metal Matrix Composites and Fracture Mechanics Characterization, *ISIJ International*, Vol 32, 1992, p 1357
- F. Osterstock and J.L. Chermant, Some Aspects of the Fracture of WC-Co Composites, *Science of Hard Materials*, R.K. Viswanatham, D.J. Rowcliffe, and J. Gurland, Ed., Plenum Press, 1981, p 615
- K.S. Ravichandran, Fracture Toughness of Two Phase Composites based on WC-Co Cermets, *Acta Metall. Mater.*, Vol 42, 1994, p 143
- K.S. Ravichandran, A Survey of Toughness in Ductile Phase Composites, *Scripta Metall. Mater.*, Vol 26, 1992, p 1389
- K.S. Ravichandran and E.S. Dwarakadasa, An Overview of Structure Property Relationships in Advanced Aerospace Al Alloys, *J. Metals*, Vol 39, 1987, p 28
- V.V. Kristic, P.S. Nicholson, and R.G. Hoagland, Toughening of Glasses by Metallic Particles, *J. Am. Ceram. Soc.*, Vol 64, 1981, p 499

Fracture Toughness Testing

John D. Landes, University of Tennessee, Knoxville

FRACTURE TOUGHNESS is defined as a "generic term for measures of resistance to extension of a crack" (Ref 1). The term fracture toughness is usually associated with the fracture mechanics methods that deal with the effect of defects on the load-bearing capacity of structural components. Fracture toughness is an empirical material property that is determined by one or more of a number of standard fracture toughness test methods. In the United States, the standard test methods for fracture toughness testing are developed by the American Society for Testing and Materials (ASTM). These standards are developed by volunteer committees and are subjected to consensus balloting. This means that all objecting points of view to any part of the standard must be accounted for. Other industrial countries have equivalent standards writing organizations that develop fracture toughness test standards. In addition, international bodies such as the International Organisation for Standardisation (ISO) develop fracture toughness test standards that have an influence on products intended for the intentional market. In this review of fracture toughness testing, the ASTM approach will be emphasized to give the article a consistent point of view.

The standard fracture toughness test methods have been written mostly with metals in mind. Toughness testing of nonmetals is also an important consideration. For many nonmetals, the equivalent standard for metals is adapted with some possible modification. Fracture toughness test methods that are written specifically for a particular nonmetal are mostly in preparation. Therefore, this review emphasizes those standards written for metals without intending to make them apply exclusively to metals.

General Fracture Toughness Behavior. As a general background before discussing the details of fracture toughness testing and analysis, fracture toughness behavior and the parameters used to describe it are discussed. Fracture toughness is defined as resistance to the propagation of a crack. This propagation is often thought to be unstable, resulting in a complete separation of the component into two or more pieces. Actually, the fracture event can be stable or unstable. With unstable crack extension, often associated with a brittle fracture event, the fracture occurs at a well-

defined point and the fracture characterization can be given by a single value of the fracture parameter. With stable fracture, often associated with a ductile fracture process, the fracture is an ongoing process that cannot be readily described by a point (Ref 2). This fracture process is characterized by a crack growth resistance curve or R curve. This is a plot of a fracture parameter versus the ductile crack extension, Δa. An example K-based R curve is shown in Fig. 1. Sometimes a single point is chosen on the R curve to describe the entire process; this is mostly done for convenience and does not give a complete quantitative description of the fracture process.

Whether the fracture is ductile or brittle does not directly influence the deformation process that a component or specimen might undergo during the measurement of toughness (Ref 2). The deformation process is generally described as being linear-elastic or nonlinear. This determines which parameter is used in the fracture toughness test characterization. All loading begins as linear-elastic. For this, the primary fracture parameter is the well-known crack tip intensity factor, K, originally defined by Irwin (Ref 3) and described elsewhere in this Volume. If the toughness is relatively high, the loading may progress from linear-elastic to nonlinear during the toughness measurement and a nonlinear parameter is needed. The nonlinear parameters that are most often used in toughness testing are the J-integral (Ref 4), labeled J, and the crack tip opening displacement (CTOD), labeled δ (Ref 5). Because all loading starts as linear-elastic, the nonlinear parameters are all written as a sum of a linear component incorporating K and a nonlinear component. This is illustrated with the individual descriptions of the various methods in this article.

Test Methods Covered. The test methods covered include linear-elastic and nonlinear loading, slow and rapid loading, crack initiation, and crack arrest. The development of the test methods followed a chronological pattern; that is, a standard was written for a particular technology soon after that technology was developed. Standards written in this matter tend to become exclusive to a particular procedure or parameter. Because most fracture toughness tests use the same specimens and procedures, this exclusive nature of

each new standard did not allow much flexibility in the determination of a toughness value. The newer approach is to write standards to encompass all parameters and measures of toughness into a single test procedure. This approach is labeled the common method approach and is being developed by ASTM as well as by many standards organizations in other countries. The standards for fracture toughness testing are periodically revised, and several new ones are often in development. New methods under development are briefly discussed at the end of this article, and volume 3.01 of the *ASTM Annual Book of Standards* should be consulted for the current version of the tests mentioned in this article.

The fracture toughness test is generally conducted on a test specimen containing a preexisting defect; usually the defect is a sharp crack introduced by fatigue loading and called the precrack. The test is conducted on a machine that loads the specimen at a prescribed rate. Measurements of load and a displacement value are taken during the test. The data resulting from this are subjected to an analysis procedure to evaluate the desired toughness parameters. These toughness results are then subjected to qualification proce-

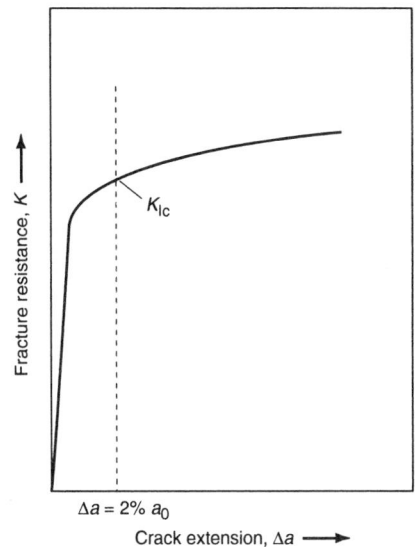

Fig. 1 Schematic of K-Based crack resistance (R) curve with definition of K_{Ic}

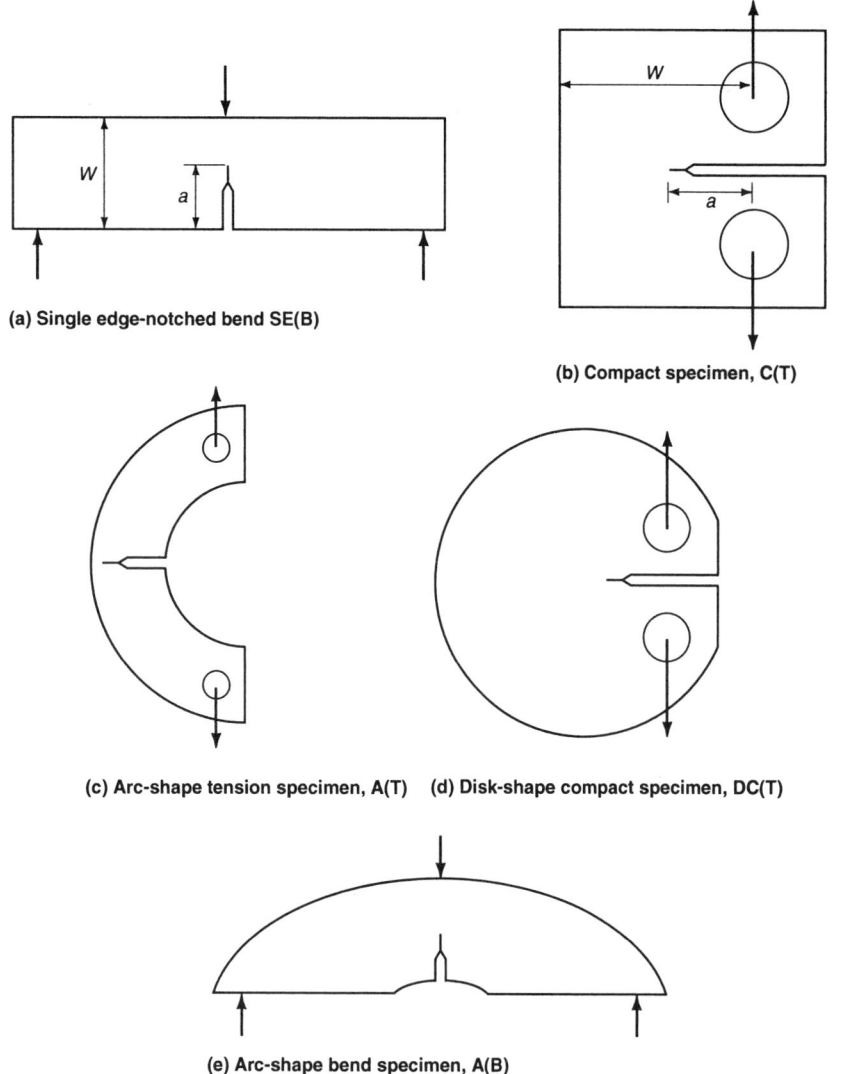

Fig. 2 Specimens types used in the K_{Ic} test (ASTM E 399). (a) Single edge-notched bend, SE(B). (B) Compact specimen, C(T). (c) Arc-shape tension specimen, A(T). (d) Disk-shape compact specimen, DC(T). (e) Arc-shape bend specimen, A(B)

dures (or validity criteria) to see if they meet the conditions for which the toughness parameters are accepted. Values meeting these qualification conditions are labeled as acceptable standard measures of fracture toughness. The standard fracture toughness test then has these ingredients: test specimens, types, and preparation; loading and instrumentation requirements; measurements taken; data analysis; and qualification of results. The following sections then discuss the various standard fracture toughness test methods following this format. The fracture toughness test methods written as ASTM standards follow a prescribed format. It is not always easy to determine the step-by-step procedure required to conduct the test from the standard. The sections below, which describe the various methods, follow a format of a step-by-step procedure rather than the format of the actual standards. The application of the fracture toughness result to the evaluation of structural components containing defects is not explicitly covered in the ASTM standard test

methods. Nor is it covered in this article. The description of the fracture toughness test methods follow a somewhat chronological outline, beginning with the methods that use the linear-elastic parameter K. After this, the methods that use the nonlinear parameters J and δ are discussed. Finally, the methods in development that should be completed and in the standards books before the end of the century are also discussed.

Linear-Elastic Fracture Toughness Testing

Fracture mechanics and fracture toughness testing began with a strictly linear-elastic methodology using the crack-tip stress-intensity factor, K. Later, nonlinear parameters were developed. However, the first test methods developed used the linear-elastic parameters and were based on K. These methods are described first in this article.

Plane-Strain Fracture Toughness (K_{Ic}) Test (ASTM E 399). The first fracture toughness test that was written as a standard was the K_{Ic} test method, ASTM E 399. This test measures fracture toughness that develops under predominantly linear-elastic loading with the crack-tip region subjected to near-plane-strain constraint conditions through the thickness. The test was developed for essentially ductile fracture conditions, but can also be used for brittle fracture. As a ductile fracture test, a single point to define the fracture toughness is desired. To accomplish this, a point where the ductile crack extension equals 2% of the original crack length is identified. This criterion is illustrated schematically with a K-R curve in Fig. 1. This criterion gives a somewhat size-dependent measurement, and so validity criteria are chosen to minimize the size effects as well as restrict the loading to essentially linear-elastic regime. The various elements of the K_{Ic} test are discussed in a little more detail than are some of the other tests for fracture toughness measurement. In this way, it can serve as a model for the other discussions. The details of this test can be found in Ref 6.

Test Specimen Selection. The first element of the test is the selection of a test specimen. Five different specimen geometries are allowed. These are the single edge-notched bend specimen, SE(B), compact specimen, C(T), arc-shape tension specimen, A(T), disk-shape compact specimen, DC(T), and the arc-shape bend specimen, A(B). Many of these specimen geometries are used in the other standards as well. They are shown schematically here in Fig. 2(a) through (e). The acronyms are standard ASTM nomenclature given in Ref 1. The bend and compact specimens (Fig. 2a and b, respectively) are traditional fracture toughness specimens used in nearly every fracture toughness test method. The other three are special geometries that represent special component structural forms. Therefore, most fracture toughness tests are conducted with either the edge-notched bend or compact specimens. The choice between the bend and compact specimen is based on:

- The amount of material available (the bend takes more)
- Machining capabilities (the compact has more detail and costs more to machine)
- The loading equipment available for testing (discussed next)

All of the specimens for the K_{Ic} test must be precracked in fatigue before testing. Refer to the standard, E 399 (Ref 6), for details on precracking.

The choice of the specimen also requires a choice of the size. Because the validity criteria depend on the size of the specimen, it is important to select a sufficient specimen size before conducting the test. However, the validity criteria cannot be evaluated until the test is completed; therefore, choosing the correct size is a guess that may turn out to be wrong. There are guidelines (Ref 6) for choosing a correct size, but no guarantee that the chosen size will pass the validity requirement. The test specimens must also be

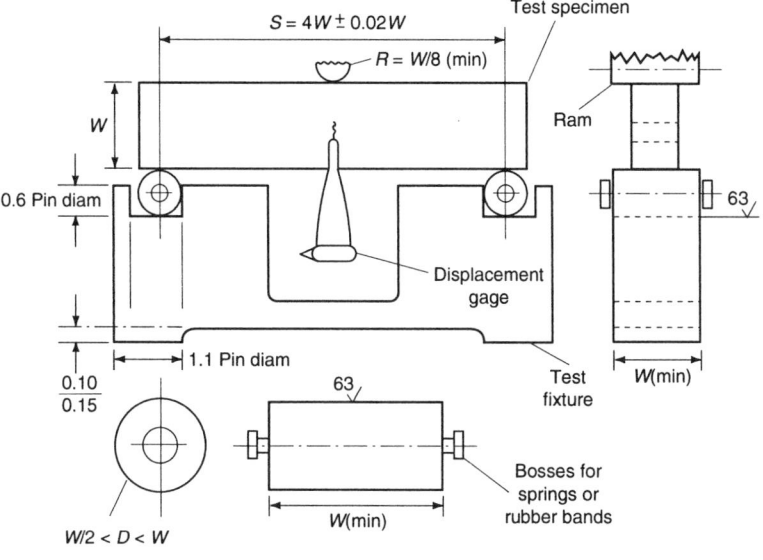

(a) Fixtures for the bend test

(b) Clevises for the compact specimen

Fig. 3 Test fixtures for the K_{Ic} test specimens. (a) Fixtures for the bend test. (b) Clevises for the compact specimen

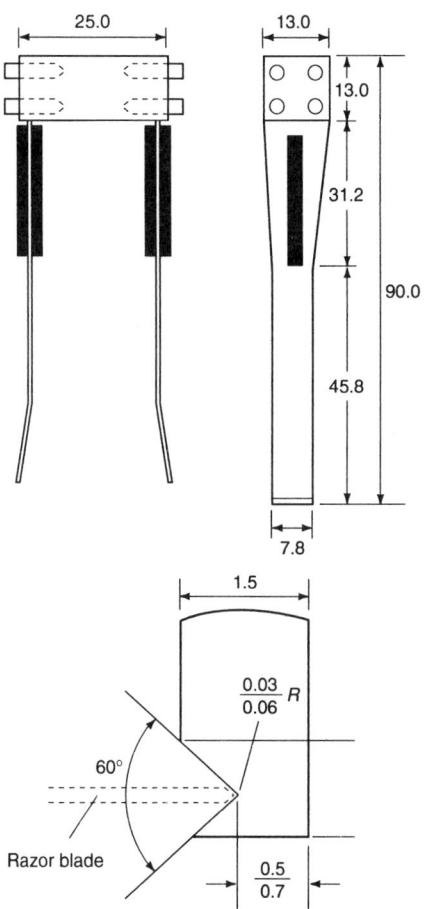

Fig. 4 An example clip gage for displacement measurement (all dimensions in mm)

which is usually done at a constant load range so load control is desired. The second type of loading machine is less expensive and may give more stability, but allows only crosshead control. Because this is required in most of the fracture toughness tests, this type of machine is quite satisfactory for the actual fracture toughness testing but is not so good for precracking.

Loading fixtures must be designed for the test. Two types can be used; choice of loading fixture depends on the test specimen chosen. The bend specimens SE(B) and A(B) use a bend fixture. The tension specimens C(T), DC(T), and A(T) require a pin-and-clevis loading. These two types of fixtures are illustrated in Fig. 3(a) and (b). Note in Fig. 3(a) that the bend loading is three point; this is the case for all bend-loaded specimens. Also note in Fig. 3(b) that the clevis has a loading flat at the bottom of the pin hole. This allows free rotation of the specimen arms during the test and is essential for getting good results.

For the K_{Ic} test, a continuous measurement of load and displacement is required during the test. The load is measured by a load cell, which should be on all loading machines. The measurement of displacement is usually done with a strain-gaged clip gage that is positioned over the mouth of the crack in the specimen. An example of a clip gage is shown in Fig. 4. Figure 3(a) shows the bend

chosen so that the proper material is sampled. This means that the location in the material source and the orientation of the sample must be correct and accounted for. The ASTM standards have a letter system to specify orientation (Ref 1). As the specimens are being prepared, requirements for tolerances on such things as locations of surfaces, size and location of the notch and pin holes, and surface finishes must be followed.

Loading Machines and Instrumentation. The next step in the test procedure is the choice of a loading machine and the preparation of loading fixtures and instrumentation for recording the test data. Most tests are conducted on either closed-loop servo-hydraulic machines or a constant-rate crosshead drive machine. The first allows load, displacement, or other transducer control, but is more expensive. It is preferred for precracking,

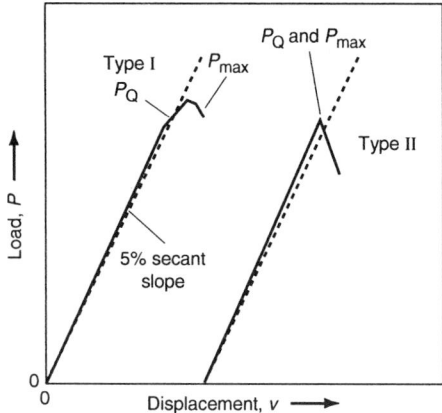

Fig. 5 Typical load-versus-displacement record for the K_{Ic} test, two types

specimen with a clip gage in place. The standards give guidelines for the working requirements of the load and displacement gages used in the tests.

The loading of the specimen is done at a prescribed rate. It must be done fast enough so that any environmental or temperature interactions are not a problem. On the other hand, it must be done slowly enough so that it is not considered a dynamically loaded test. For the K_{Ic} test, the load must be applied at a rate so that the increase in K is given by the range 0.55 to 2.75 MPa√m/s. The loading is done in displacement control, which usually means test machine crosshead control. During the loading, the load and displacement are measured continuously. This can be done autographically or digitally.

Test Data and Analysis. The load-and-displacement record provides the basic data of the test. The data are then analyzed to determine a provisional K_{Ic} value labeled K_Q. This provisional value is determined from a provisional load, P_Q, and the crack length. The P_Q value is determined with a secant slope on the load-and-displacement record (Fig. 5). The P_Q value is determined by drawing the original loading slope of the load-versus-displacement record. A slope of 5% less than the original slope is then drawn from the origin. For a monotonically increasing load, the P_Q is taken where the 5% secant intersects the load-versus-displacement curve; this is illustrated as type I in Fig. 5. For other records in which an instability or other maximum load is reached before the 5% secant, the maximum load reached up to and including the possible intersection of the 5% secant is the P_Q. Type II illustrated in Fig. 5 is an example of one of the other types of load-versus-displacement records. The 5% secant corresponds to about 2% ductile crack extension; this may be physical crack extension or effective crack extension related to plastic zone development. Unstable failure before reaching the 5% offset also marks a measurement point for P_Q at the maximum load reached at the point of instability.

The P_Q value is used to determine the corresponding K_Q value. This is calculated from the equation:

$$K = P \; f(a/W)/B\sqrt{W} \qquad \text{(Eq 1)}$$

where P is load, B and W are specimen thickness and width, and $f(a/W)$ is a calibration function that depends on the ratio of crack length to specimen width, a/W, and is given in the standard. For the calculation of K, a crack length value, a, is required. This comes from a physical measurement on the fracture surface of a broken specimen half. The specimen must be fractured into halves if it is not already that way from the test. The crack length is measured to the tip of the precrack using an averaging formula in the test standard. This value of crack length normalized with width, W, is used in the calibration function $f(a/W)$ to determine the K_Q value.

The K_Q is provisional K value that may be the K_{Ic} if it passes the validity requirements. The two major validity requirements are to ensure that crack resistance does not increase significantly with crack growth and that linear-elastic loading and plane-strain thickness are achieved. The first of these two requirements is quantified as:

$$\frac{P_{max}}{P_Q} \leq 1.10 \qquad \text{(Eq 2)}$$

which limits the R-curve behavior to an essentially flat trend and ensures some physical crack extension. The second requirement is:

$$a, B \geq 2.5 \left(\frac{K_Q}{\sigma_{ys}}\right) \qquad \text{(Eq 3)}$$

which guarantees linear-elastic loading and plane-strain thickness. P_{max} is the maximum value of load reached during the test. An example of P_{max} is shown in Fig. 5; σ_{ys} is the 0.2% offset yield strength. Values of K_Q that pass these validity requirements are labeled as valid K_{Ic} and are reported as such in a standard prescribed reporting of test results.

Fatigue precracking should be done in accordance with ASTM E 399. The K level used for precracking each specimen should not exceed about two-thirds of the intended starting K-value for a given environmental exposure. This prevents fatigue damage or residual compressive stress at the crack tip, which may alter the fracture toughness behavior, particularly when testing at a K-level near the K_Q value for the specimen.

Chevron notches are sometimes used to facilitate starting such mechanical precracks (see the Appendix at the end of this article). These modifications also may be necessary to control fatigue precracking of some materials.

Rapid-Load $K_{Ic}(t)$. A value of fracture toughness labeled $K_{Ic}(t)$ can be determined for a rapid-load test. Details of this method are given in a special annex to the method E 399 (Ref 6). For the static loading rate K_{Ic} value, the maximum loading rate is defined as 2.75 MPa√m/s. Anything faster than that is labeled as a rapid-load fracture toughness. The specimen apparatus and procedure are much the same as for the regular K_{Ic} test. Special instructions are given to ensure that the instrumentation can handle the rapidly changing signals. The interpretation of results

must be based on a dynamic value of the yield stress, σ_{YD}. An equation for σ_{YD} is given in the Annex to E 399. Results are reported as $K_{Ic}(t)$, where the loading time of the test is written in parentheses after the measured toughness value.

K-R Curve (ASTM E 561). Ductile fracture toughness behavior is measured by a crack growth resistance curve or "R-curve," which is defined as "a plot of crack-extension resistance as a function of slow-stable crack extension" (Ref 1). Although many ductile fracture processes can be measured as a single point, such as with K_{Ic}, the R-curve is a more complete description of the fracture toughness. When the R-curve increases significantly, a single-point measurement is even less descriptive of the actual fracture toughness. Steeply rising R-curves occur in many metallic materials but especially in thin plate or sheet materials. The steeply rising R-curve makes the single-point definition more size- and geometry-dependent and does not lend to structural evaluation.

The $K-R$ curve is a good method for fracture toughness characterization in cases where the R curve is steeply rising but the fracture behavior occurs under predominantly linear-elastic loading conditions. The $K-R$ curve procedure is given by ASTM E 561 (Ref 7). The objective of the method is to develop a plot of K, the resistance parameter, versus effective crack extension, Δa_e. The method allows three different test specimens, the compact, C(T), the center-cracked tension panel, M(T), and the crack-line-wedge-loaded specimen, C(W). The compact specimen is shown in Fig. 2(b). The center-cracked tension panel and the crack-line-wedge-loaded specimens are shown in Fig. 6(a) and (b), respectively. The first two specimens use a conventional loading machine with fixtures that are specified in the test method. The C(W) specimen is wedge loaded to provide a stiff, displacement-controlled loading system (Fig. 6b). This can prevent rapid, unstable failure of the specimen under conditions where the R curve toughness is low so that the R curve can be measured to larger values of Δa_e. All specimens must be precracked in fatigue.

The instrumentation required on the specimens is similar to that for the K_{Ic} test, except for the case of the C(W) specimen. The basic test result is a plot of load versus a displacement measured across the specimen mouth. From this, an effective crack length is determined from secant offset slopes to the load-versus-displacement record (Fig. 7). An effective crack extension is the difference between the original and effective crack lengths. Effective crack length is determined from the slope of the secant offset using the appropriate compliance function, which relates this slope to crack length. The K is determined as a function of the load and corresponding effective crack length. This is given by:

$$K = P \; f(a_e/W)/B\sqrt{W} \qquad \text{(Eq 4)}$$

The resulting plot of K versus effective crack length is the desired $K-R$ curve fracture toughness. The result is subjected to a validity requirement that

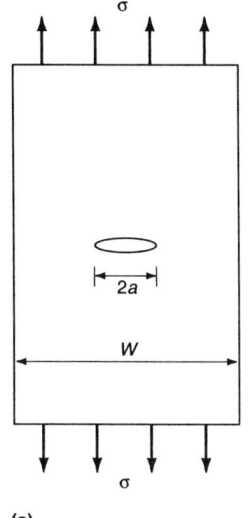

(a)

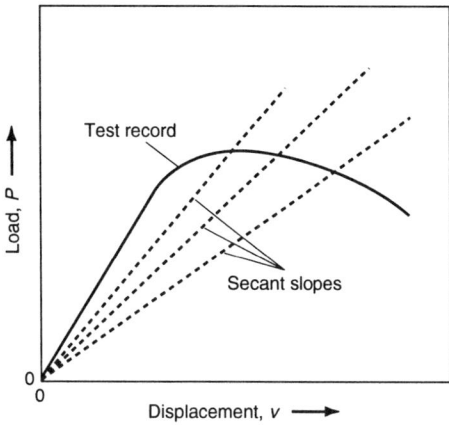

Fig. 7 Secant offset measurement of effective crack length

(b)

Fig. 6 (a) Center-cracked tension specimen, M(T). (b) Crack-line-wedge-loaded compact specimen, C(W), in loading fixture

limits the amount of plasticity. For the C(T) and C(W) specimens

$$b = (W - a) \geq (4/\pi)(K_{max}/\sigma_{ys})^2 \qquad \text{(Eq 5)}$$

where b is the uncracked ligament length, σ_{ys} is the 0.2% offset yield strength, and K_{max} is the maximum level of K reached in the test. For the M(T) specimen the net section stress based on the physical crack size must be less than the yield strength.

For the C(W) specimen, a load is not measured. The data collected are a series of displacement values taken at two different points along the crack line, one near the crack mouth and one nearer the crack tip. From the two different displacement values, an effective crack length can be determined from the ratio of the two displacement values and from calibration values given in a table in E 561. From the crack length and displacement a K value can be determined and the K-R curve constructed. The toughness result is then a curve of K versus Δa_e, somewhat similar to the one in Fig. 1.

Crack Arrest, K_{Ia} (ASTM E 1221). This procedure allows a toughness value to be determined based on the arrest of a rapidly growing crack, which may be lower than the initiation value. The specimen and procedure are somewhat different from the previously discussed toughness test methods that determine initiation toughness values only. The specimen for crack arrest testing is called the compact-crack-arrest compact specimen (Fig. 8). It is similar to the crack-line-wedge-loaded specimen, C(W), of the K-R curve method and requires wedge loading in order to provide a very stiff loading system to arrest the crack. The notch preparation is different from the other standards in that the specimen has a notch with no precrack. Generally, a brittle weld bead is placed at the notch tip to start the running crack, although other methods are allowed. The running crack advances rapidly into the test material and must be arrested by the test material to produce a K_{Ia} result. The only instrumentation on the specimen is a displacement gage. A load cell is placed on the loading wedge, but it does measure the

load on the specimen. The displacements at the beginning of the unstable crack extension and at the crack arrest position are measured and converted to K values. To eliminate effects of nonlinear deformation, which cannot be directly measured with only a displacement gage, a series of loads and unloads are conducted on the specimen until the unstable cracking occurs. When the specimen is unloaded, the crack tip can be marked by a procedure called "heat tinting." Heat tinting consists of marking the physical crack extension by heating the specimen until oxidation occurs on the crack. The specimen is then broken open, and the crack extension measured on the fracture surface.

The value of K_{Ia} is determined from a displacement value and the crack length at the arrest point. Validity is determined from the size criterion:

$$W - a \geq 1.25 (K/\sigma_{YD})^2 \qquad \text{(Eq 6)}$$

where σ_{YD} is a dynamic yield strength. To complete a successful K_{Ia} test, careful attention must be paid to the instructions in E 1221 (Ref 8).

Nonlinear Fracture Toughness Testing

Linear-elastic parameters are used to measure fracture toughness for relatively low toughness materials that fracture under or near the linear loading portion of the test. For many materials used in structures it is desirable to have high toughness, a value at least high enough so that the structure would not reach fracture toughness before significant yielding occurs. For these materials, it is necessary to use the nonlinear fracture parameters to measure fracture toughness properties. The two leading nonlinear fracture parameters are J and δ. For high toughness materials,

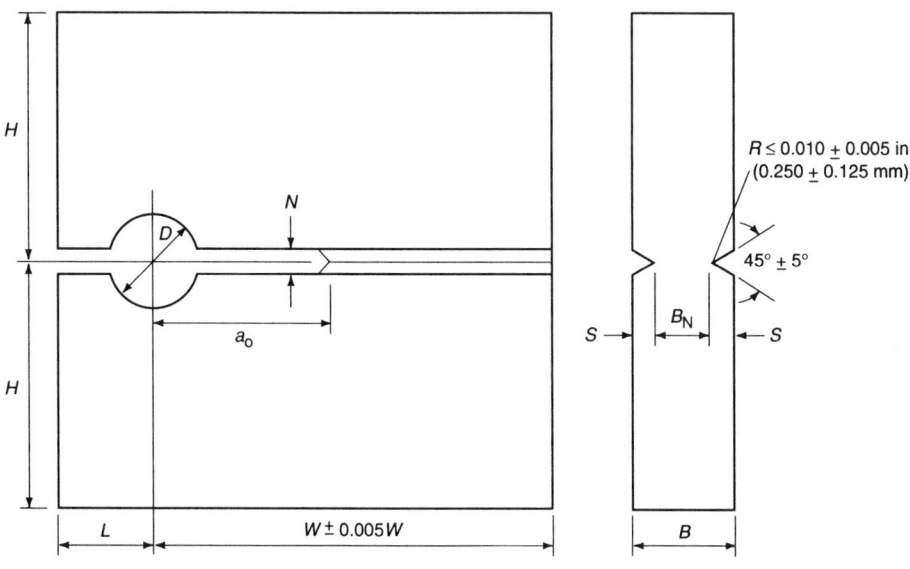

Fig. 8 Crack-line-wedge-loaded compact-crack-arrest specimen

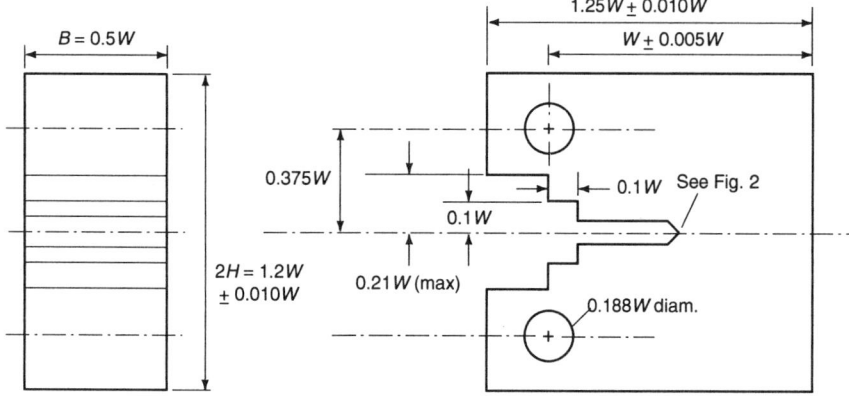

Fig. 9 J_{Ic} compact specimen with loadline cutout

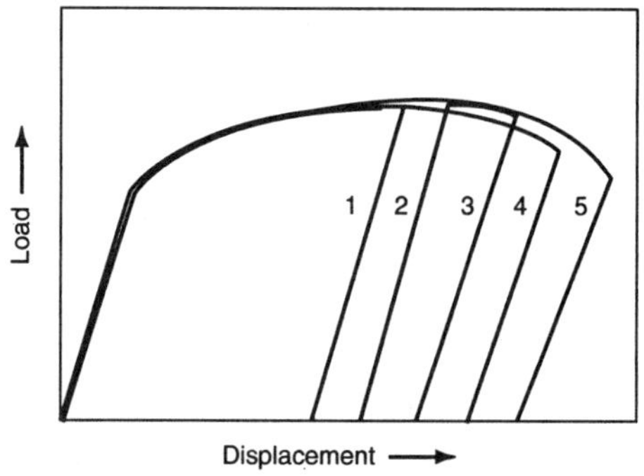

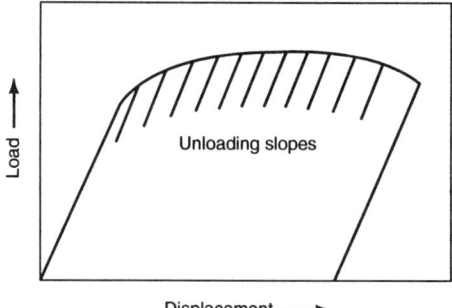

Fig. 10 Load versus displacement for multiple-specimen tests

fracture is often by a ductile mechanism, but this is not necessarily the case for all materials.

J_{Ic} Testing (ASTM E 813). One of the first tests developed using the J parameter is the J_{Ic} test per ASTM E 813 (Ref 9). In this test, an R curve is developed using J versus Δa pairs and a point near the beginning of the R curve is defined as J_{Ic} "a value of J near the onset of stable crack extension" (Ref 9). The specimens for the J_{Ic} test are the bend SE(B) and compact C(T). These specimens are similar to the ones used for K_{Ic} testing in Fig. 2(a) and (b); however, the compact specimen for J testing has a cutout on the front face so that a displacement gage can be mounted directly on the load line; that is, in the line of the applied loads (Fig. 9). The loading fixtures required are the bend fixture for the bend (Fig. 3a) and the pin-and-clevis for the compact (Fig. 3b). Again, a clevis with a loading flat at the bottom of the pin hole in the clevis is essential to get free rotation of the specimen. The instrumentation required is the load cell and a displacement measuring clip gage. If the electrical potential system is used additional instrumentation is required. The clip gage for the J_{Ic} test requires more resolution than that for the K_{Ic} test if a single specimen test method is used. For the bend specimen, a loadline clip gage is needed to measure J. Additionally, a second clip gage can be used over the crack mouth if a single specimen method is used.

J_{Ic} Test Procedures. The basic output of the test is a plot of J versus physical crack extension (Δa). (Unlike the K-R curve method, which uses effective crack extension, the J_{Ic} test uses physical crack extension.) To obtain the required J versus Δa data measurements of load, displacement and physical crack length are required during the test. There are two techniques used to develop these data. The first is the multiple-specimen test method, in which each specimen develops a single value of J and Δa but no special crack monitoring equipment is needed during the test. Crack extension is measured on the fracture surface at the conclusion of the test. However, for this technique a number of specimens are required to develop the plot of J versus Δa values needed for the result. The other method is the single specimen test from which all the J versus Δa values are

developed from one test. To accomplish this, a method of crack length monitoring is needed during the test. The primary method for crack length monitoring during the test is called the elastic unloading compliance method, in which crack length is measured from an elastic slope. The measurement of the elastic slope requires only a clip gage that measures displacement. The compact specimen uses the gage mounted on the load line. The bend specimen could use two gages: one on the loadline to measure J and one over the crack mouth to measure slope. The second single-specimen method is the electrical potential crack monitoring system, in which the electrical resistivity of the specimen is measured and correlated with crack length during the test. This is a secondary method to measure crack length and is not described in detail in the E 813. It requires some additional expertise on the part of the tester to use.

The test procedure depends on the method of crack length monitoring. For the multiple-specimen test, five or more specimens are loaded to prescribed displacement values that are thought to give some physical crack extension but not complete separation of the specimen. This results in a number of individual load-versus-displacement records, as shown in Fig. 10. When the prescribed displacement is reached, the specimen is unloaded and the crack tip is marked by heat tinting.

The single-specimen method using elastic compliance is initially loaded in the same way; however, during the test partial unloadings are taken to develop elastic slopes from which crack length can be evaluated using compliance relationships (Fig. 11). The compliance relationships are given in E 813 (Ref 9). For the electrical potential method, the load, displacement, and potential change are measured simultaneously. Potential changes are related to crack length through either an analytical or empirical correlation.

Data Evaluation. From these test results, J is evaluated from the load-versus-loadline displacement record. The J is calculated from a linear combination of an elastic term and a plastic term given as:

Fig. 11 Load versus displacement with unloading slopes

$$J = J_{el} + J_{pl} = \frac{K^2(1-\nu^2)}{E} + \frac{\eta_{pl}}{Bb}\int_0^{v_{pl}} P\,dv_{pl} \qquad \text{(Eq 7)}$$

where K is the stress-intensity factor, E is elastic modulus, ν is Poisson's ratio, P is load, v_{pl} is plastic displacement, B is specimen thickness, b is specimen uncracked ligament ($W - a$, where W is specimen width) and η_{pl} is a coefficient that has values of $\eta_{pl} = 2$ for the SE(B) specimen of Fig. 2(a) and $\eta_{pl} = 2 + 0.522b/W$ for the compact specimen of Fig. 9. Equation 7 is the basic J formula for the case of a nongrowing crack. It is based on a K equivalence for the elastic component of J and an area term for the plastic component of J. Alternate J formulas are given in E 813 (Ref 9) for the growing crack.

The crack length is used to determine $\Delta a = a - a_o$, where a_o is the original crack length at the beginning of the test. For the multiple specimen method, all Δa values are determined from measurements taken from the broken surface of the test specimen. For some metals, heat tinting does not oxidize the crack surfaces and another method of marking the crack extension, for example post-test fatiguing, can be used. The specimen is broken in two after the heat-tint procedure, usually at a low temperature to induce brittle fracture for easy reading of the ductile crack extension and to otherwise minimize plastic deformation during this procedure. Crack lengths a_o and a_f, original and final, are measured on the fracture surface. A nine-point measurement and averaging method is used because the crack front is usually not straight and regular. This procedure

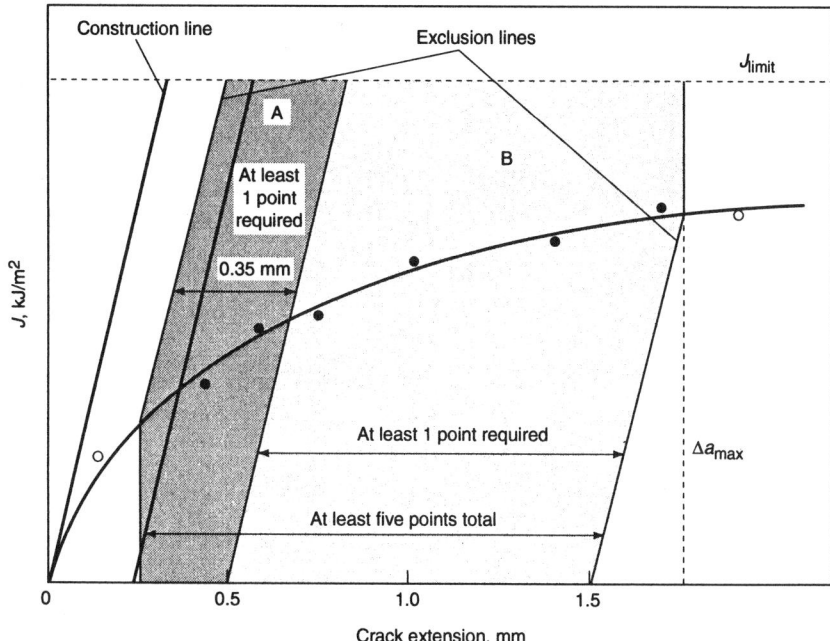

Fig. 12 J_{Ic} evaluation scheme

is described in E 813 (Ref 9). For the single-specimen methods for which crack length monitoring systems are used, the crack length is evaluated at prescribed points during the test. For example, in the elastic unloading compliance method, a crack length can be determined at each unload. Typically, about 15 of 30 data pairs, P, v, and a are evaluated for each test. For single-specimen tests, a physical measurement of the final crack length is made at the end of the test using the same procedure that is followed for the multiple-specimen test, so that this value can be used to compare with the final crack length evaluated by the crack monitoring system.

The J versus Δa results form a part of the J-R curve and are the basic data of the J_{Ic} method. The objective is to get J versus Δa values in a certain restricted range. These data are then subjected to a prescribed evaluation scheme to choose a point on the J-R curve that is near the initiation of stable cracking. The method for developing the J_{Ic} is somewhat complicated, and the details are given in E 813 (Ref 9). Basically, the J versus Δa pairs are evaluated to see which fall in a prescribed range. The pairs falling in the correct range are fitted with a power-law equation:

$$J = C_1 (\Delta a)^{C_2} \qquad \text{(Eq 8)}$$

where C_1 and C_2 are constants. A construction line is drawn, and the intersection of this with the fitted line, Eq 8, is the evaluation point for a candidate J_{Ic} value. This candidate value is labeled J_Q. A schematic of the process of J_{Ic} evaluation is shown in Fig. 12.

The candidate J_Q value is subjected to qualification criteria to see if it comprises an acceptable value. The basic one is to guarantee a sufficient specimen size:

$$b, B \geq 25(J_Q/\sigma_Y) \qquad \text{(Eq 9)}$$

where σ_Y is an effective yield strength and

$$\sigma_Y = (\sigma_{ys} + \sigma_{uts})/2 \qquad \text{(Eq 10)}$$

where σ_{ys} and σ_{uts} are the yield strength and ultimate tensile strength, respectively.

If the qualification requirements are met, the J_Q is J_{Ic} and the results are reported following the prescribed format in E 813 (Ref 9).

J-R Curve Evaluation (ASTM E 1152). A more complete evaluation of fracture toughness for ductile fracture based on J is the J-R curve procedure in ASTM E 1152 (Ref 10). This standard uses the same specimens, instrumentation, and test procedures as the J_{Ic} test. The J-R curve test cannot be conducted with the multiple-specimen test procedure; it must use a single-specimen procedure. The purpose of the J-R curve is to develop points of J versus Δa; these comprise the fracture toughness value, a single value of J is not required. The single-specimen methods are again the elastic unloading compliance method and the electrical potential method. Qualification criteria are given in the E 1152 (Ref 10). The J_{Ic} and J-R curve methods are very similar, hence a combined J standard is being prepared.

Crack Tip Opening Displacement (CTOD) (ASTM E 1290). The crack tip opening displacement method of fracture toughness measurement was the first one that used a nonlinear fracture parameter to evaluate toughness (Ref 5). The first CTOD standard was written by the British Standards Institution (Ref 11). Subsequently, ASTM E 1290 was written as the U.S. version of this test method (Ref 12). The basic idea of the test method is to evaluate a fracture toughness point for brittle fracture or to evaluate a safe point for

the case of ductile fracture. The primary measurements of toughness are at unstable fracture before significant ductile crack extension, labeled δ_c, unstable fracture after significant crack extension, δ_u, or the point of maximum load in the test, δ_m. The method originally had a point near the beginning of stable crack extension, δ_i, that was measured as a point on an R curve in a similar manner to J_{Ic}. This point was subsequently removed from the test method.

The CTOD standard uses the same bend and compact specimens that are used in the J_{Ic} test; thus the same loading fixtures are used. The method requires measurement of load and displacement during the test. The formulas for δ calculation use a combination of an elastic and a plastic component for δ:

$$\delta = K^2(1 - v^2)/2\,\sigma_{ys}E$$
$$+ r_p(W - a_0)v_p/[r_p(W - a) + a_0 + z] \qquad \text{(Eq 11)}$$

In this equation, the elastic component of δ is based on an equivalent K and the plastic component is based on a rigid plastic rotation of the specimen about a neutral stress point at $r_p(W - a_0)$ from the crack tip. In Eq 11, v is Poisson's ratio, σ_{ys} is the yield strength, r_p is a rotation factor, v_p is a plastic component of displacement, $W - a_0$ is the uncracked ligament length, and z is the position of the clip gage from the crack measurement position.

For many years the CTOD test was the only one that measured toughness for a brittle, unstable fracture event using a nonlinear fracture parameter. In addition, the method allows the measurement of toughness after a "popin," which is described as a discontinuity in the load-versus-displacement record usually caused by a sudden, unstable advance of the crack that is subsequently arrested.

Test Methods in Preparation

The development of standard fracture toughness test methods is not completed. By the end of 1995, four new test methods were being prepared for fracture toughness testing. They are:

- The combined J standard
- The common test method
- The transition fracture toughness standard
- The standard for testing of weldments

Because these methods will likely become test standards in the next few years, it is important to describe them briefly here as a likely reference in the future.

Combined J Standard (E1737-96). Because the J_{Ic} and J-R curve test standards are very similar in many respects, they are in the process of being combined into a single test standard for the ASTM standard books as ASTM E1737-96. The combined J standard will allow the measurement of both the J-R curve and the J_{Ic} point. Whereas the traditional J-based standards only allowed measurements of fracture toughness for ductile fracture (where stable cracking occurs with no unstable cracking and an *R* curve is used to measure toughness), the combined standard will have a single-point evaluation of unstable, brittle fracture based on *J*. The specimens and procedures for brittle fracture are the same as for the J_{Ic} and *J-R* curve tests; however, the test would end with sudden unstable crack advance. The parameter *J* is evaluated at the point of unstable cracking and labeled J_c if the unstable fracture is before significant stable cracking occurred and labeled J_u if it is after significant stable cracking.

The combined *J* test will use the same techniques as E 813 and E 1152. The single-specimen elastic unloading compliance method is the standard procedure. Electrical potential measurement of crack advance is allowed, and a detailed description of the technique is given in the standard. In addition, the multiple-specimen method is allowed, but only for measurement of a J_{Ic} point.

An additional feature of the new combined *J* standard is an initialization procedure to ensure that the initial portion of the *J-R* curve is aligned properly with the initial measured crack length. The E 813 method of J_{Ic} measurement did not specifically align the initial portion of the curve and could give artificially raised or lowered values of J_{Ic} reflecting the offset in the initial *J-R* curve alignment.

Common Fracture Toughness Test Method (E1820-96). The most universal of the test methods for fracture toughness will be the common test method approach. This allows a measurement of fracture toughness using the linear-elastic parameter, *K*, or the nonlinear parameters *J* or δ, whichever is appropriate to the result of the test. The idea of the method is that most of the fracture toughness tests use the same specimens, instrumentation, and test procedures. The exclusive nature of each test method described above was derived from the historical development of the fracture mechanics methodology. However, the way individual methods are written gives a good chance that a test can give an invalid or unqualified result with no way to use the analysis procedure of another test method to try to obtain an acceptable result. The common method approach is designed so that nearly every test can deliver some acceptable measurement of fracture toughness.

The common method will use the same specimens and loading fixtures as K_{Ic}, but will allow measurement of toughness for both brittle and ductile fracture mechanisms. The method allows the option of a single-specimen or a multiple-specimen procedure when crack extension must be measured for the development of an *R* curve. The common test method concept of combining all measurements of fracture toughness into a single standard rather than having many specialized standards has become popular around the world, and most of the standards writing organizations have either recently developed a version of the common test standard or are in the process of developing one. The ASTM common method for fracture toughness testing is in preparation as ASTM E1820-96.

Transition Fracture Toughness Testing (1997-1998). The transition fracture toughness has long been a problem area for fracture toughness testing. The transition fracture behavior is usually brittle, sometimes after an initial period of ductile crack extension. The toughness values show extensive scatter and size dependency that cause difficulty in the characterization of toughness for the evaluation of structures. The scatter and size dependency has been attributed to statistical influences and constraint differences (Ref 13). The characterization of toughness relies mainly on the statistical handling of the data. The test method in preparation concentrates mainly on these aspects. The specimens, fixtures, instrumentation, test procedures, and calculation of toughness parameters will follow existing standards. The evaluation of the statistical aspects are handled with a Weibull statistical distribution (Ref 13). From this, a median value of toughness is identified from the scatter; that median value of toughness is aligned on a master curve. From the master curve, the toughness distribution at other temperatures can be identified. Also from the statistical distribution, a percentage lower bound confidence level of toughness can be identified. For example, a 95% lower bound confidence level can be determined from the statistical distribution as a function of temperature in the transition. The transition fracture toughness standard is in development and should be available in 1997 or 1998. There are presently no documents to reference for this method, but it would be in the same volume as all of the other fracture toughness test standards.

Fracture Testing of Weldments (year of issue uncertain). The final test method in preparation is the fracture of weldments test method. Weldments do not require a different set of parameters, specimens, or equipment for toughness testing; however, there are special problems for the testing of weldments that are not contained in the other standards. Weldments have a composite of materials starting in the base metal, going through the heat-affected zone and into the weld metal. Such things as the placement of the notch for the sampling of the correct material, the precracking procedure to get the crack to grow in an acceptable manner, and handling of such things as distortion and residual stresses will be covered in the standard. The parts that are common with the other standards will not be covered in this method, but the tester will be referred to the other standards to complete the testing and analysis after the special problems inherent to the testing of weldments have been taken care of. This standard presently needs additional verification before it is proposed for standards balloting.

Appendix 1: Chevron Notched Specimens*

Chevron-notched specimens can be used to determine the fracture toughness of materials that are difficult to precrack, materials unavailable in large section sizes, and materials that are economically prohibitive to test using other specimen configurations. A natural crack initiates at the notch tip and extends in a stable manner as the load increases during testing. This eliminates fatigue precracking and simplifies interpretation of results.

As previously mentioned in this article, ASTM E 399 centers attention on the start of crack extension from a fatigue precrack. In contrast the chevron-notched specimen method makes use of a steady-state propagating crack or a crack at the initiation of a crack jump. Although both methods are based on the principles of linear elastic fracture mechanics, this distinction, coupled with differences in test procedures, causes fracture toughness values measured by the chevron-notched specimen method to be somewhat higher than those obtained using ASTM E 399.

Chevron-notched specimens were originally used for determining the fracture toughness of brittle materials that were difficult to fatigue precrack. These materials exhibited nearly ideal linear-elastic behavior and required only maximum load to failure for calculating fracture toughness.

Use of chevron-notched specimens was extended to ductile materials that required impractically large specimens for determination of fracture toughness by other methods. The thin chevron slots provide good plane-strain constraint at the crack ends, resulting in smaller specimen sizes for valid fracture toughness values from a given material. Thus, chevron-notched specimens have gained acceptance due to some economic advantages over other specimen configurations for determining fracture toughness of metallic engineering alloys and other engineered materials, such as ceramics and glass.

Specimen Description

Chevron-notched specimen geometries (Fig. 13) can be divided into two broad categories: specimens with rectangular cross sections (short bar) and those with round cross sections (short rod). Both straight and curved chevron slots are used with either specimen cross section, depending on the side-slot machining equipment available. A curved slot is made by plunge cutting with a circular blade, whereas a straight slot is cut by using an electrical discharge wire or by feeding a circular cutter through the specimen.

The very thin chevron slot significantly reduces the size of the nonplane-strain zone at the flank ends of the crack. This increased triaxial constraint allows valid fracture toughness deter-

---

* Adapted from article in 9th ed. *Metals Handbook*, Vol 8, *Mechanical Testing*.

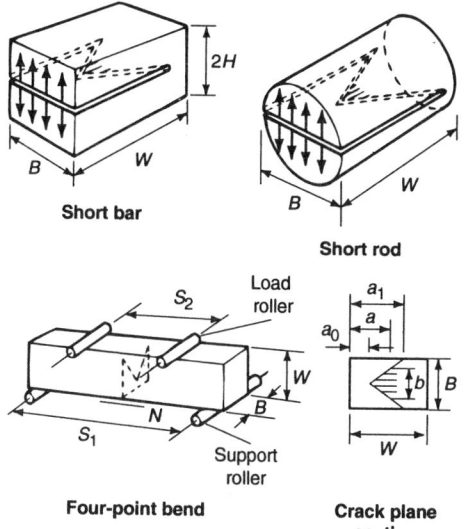

Fig. 13 Chevron notch test configuration for both bar and rod

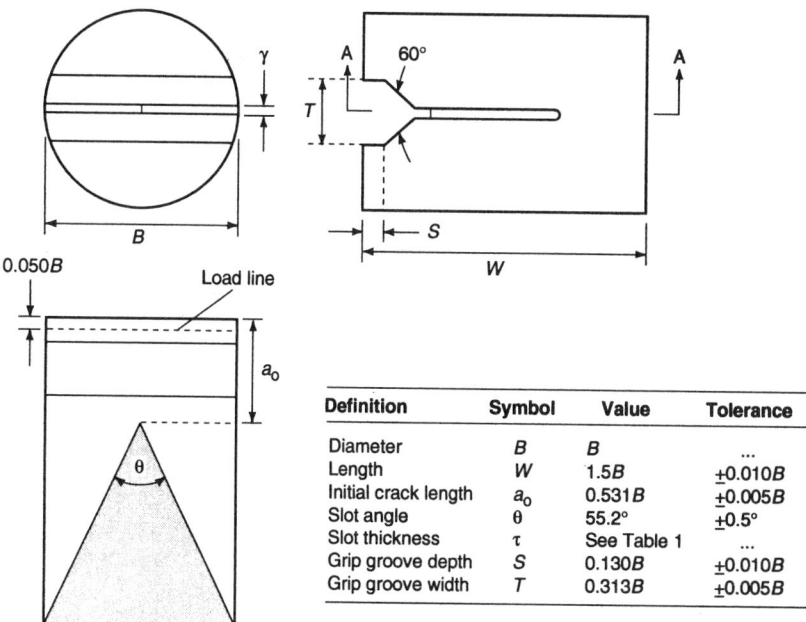

Definition	Symbol	Value	Tolerance
Diameter	B	B	...
Length	W	$1.5B$	$\pm 0.010B$
Initial crack length	a_0	$0.531B$	$\pm 0.005B$
Slot angle	θ	$55.2°$	$\pm 0.5°$
Slot thickness	τ	See Table 1	...
Grip groove depth	S	$0.130B$	$\pm 0.010B$
Grip groove width	T	$0.313B$	$\pm 0.005B$

Short rod specimen

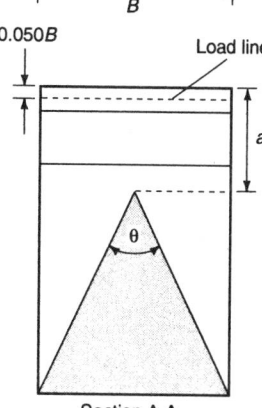

Definition	Symbol	Value	Tolerance
Breadth	B	B	...
Length	W	$1.5B$	$\pm 0.010B$
Height	H	$0.870B$	$\pm 0.005B$
Initial crack length	a_0	$0.531B$	$\pm 0.005B$
Slot angle	θ	$55.2°$	$\pm 0.5°$
Slot thickness	τ	See Table 1	...
Grip groove depth	S	$0.130B$	$\pm 0.010B$
Grip groove width	T	$0.313B$	$\pm 0.005B$

Short bar specimen

Fig. 14 Typical specimen geometries with straight chevron slots. The load line is the line along which the opening load is applied in the mouth of the specimen. Common sizes have B equal to 12.7 mm (0.5 in.), 19.08 mm (0.75 in.), and 25.4 mm (1.0 in.).

minations for a given metallic alloy to be made using much smaller specimens than required by ASTM E 399. For example, a 12-mm by 12-mm by 19-mm (0.5-in. by 0.5-in. by 0.75-in.) chevron-notched specimen of SAE 15B35 steel, quenched and tempered to 50 HRC with a yield strength of 1120 MPa (162 ksi) and $K \approx 80$ MPa$\sqrt{\text{m}}$ (72.8 ksi $\sqrt{\text{in.}}$), would weigh about 0.02 kg (0.05 lb). An ASTM E 399 compact tension specimen of the same material would be 25 mm by 64 mm by 61 mm (1.0 in. by 2.5 in. by 2.4 in.) and weigh 0.73 kg (1.62 lb).

Besides smaller specimen sizes, the chevron slot serves three additional critical functions. First, the slot defines the crack plane and propagation direction, providing a simple method for orienting the fracture with respect to the bulk material. Second, the slot forces crack initiation at the point of the "V" when the load is applied. Finally, the chevron slot configuration produces stable crack initiation and growth, because the crack front widens as it advances. This is advantageous when testing brittle alloys that are difficult to fatigue precrack.

Common specimen geometries are shown in Fig. 14 and 15. Slot thickness and slot configurations (Table 1) have been extensively tested and analyzed for various specimen geometries and accurate determination of fracture toughness (Ref 14-23). Figure 15 shows a curved chevron notch pattern for fracture toughness testing under development for aluminum alloys. Tests on aluminum alloys have shown that these parameters correlate closely with K_{Ic} (Ref 24) and that useful plane-strain fracture toughnesses can be measured with a specimen that is smaller than that used in the ASTM E 399 test. More recent references on the use of chevron-notched specimens are listed in the "Selected References" at the end of this article.

The Society of Automotive Engineers has adopted aerospace recommended procedure ARP

1704 for determination of fracture toughness using chevron-notched specimens, and the ASTM chevron-notched test method is specified in ASTM E 1304. Although the SAE recommended test practice exists for these specimen geometries, the SAE test has not been included in any aircraft company specifications to date. Further information in alternate geometries for chevron-notched specimens is described in Vol 8 of the 9th edition *Metals Handbook, Mechanical Testing.*

Brittle Materials

Chevron-notched specimens made from very hard, brittle materials, such as glass, ceramics, and carbides, exhibit nearly ideal linear elastic behavior. Crack initiation occurs at I, where the loading first deviates from linearity (Fig. 16a). The crack then extends in a stable manner throughout the test. The load required to advance the crack increases to a smooth maximum at the

Fig. 15 Short bar toughness aluminum-alloy specimens with curved chevron notches

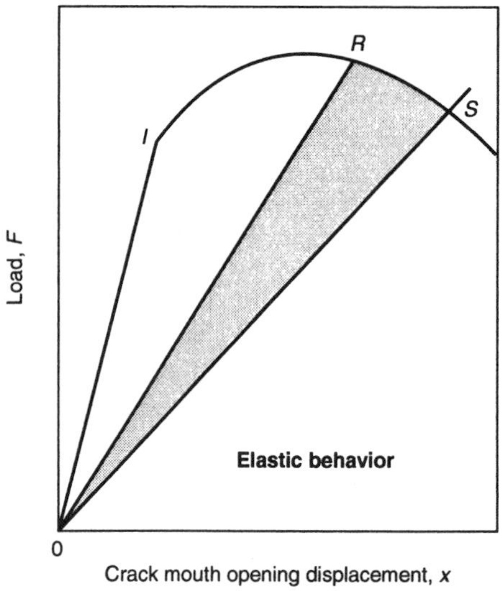

Table 1 Results of slot geometry study of chevron notches

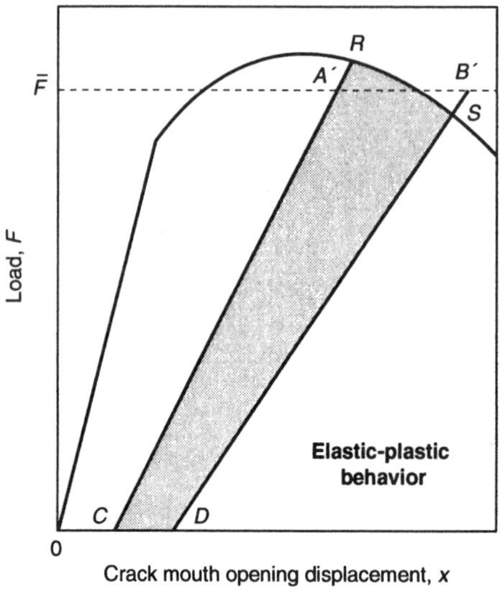

Slot thickness		Effect on specimen calibration, %	Plane-strain constraint(a)	Slot configuration
mm	in.			
0.38	0.015	0	Excellent	Sharp tip
0.8	0.031	−1	Excellent	Sharp tip
1.6	0.063	−3	Excellent	Sharp tip
0.38	0.015	0	Excellent	Round tip
0.8	0.031	−1	Good	Round tip
1.6	0.063	−3	Poor	Round tip
0.38	0.015	0	Good	Square tip
0.8	0.031	−1	Poor	Square tip
1.6	0.063	−3	Poor	Square tip

(a) Excellent, less than +2% effect on the measurement; good, less than +5% effect on the measurement; poor, more than +5% effect on the measurement

critical crack length. This maximum load is used to calculate fracture toughness. The critical crack length is dependent only on specimen geometry and is independent of material. Therefore, there is no need to measure crack length when determining fracture toughness.

The rationale for using linear elastic fracture mechanics principles for calculating fracture toughness of an ideal elastic material is illustrated in Fig. 16(a). Unloading from either R or S is linear to 0. Therefore, the shaded area represents the irrecoverable work to form the fracture surface (i.e., the fracture toughness of the material). The following equation is used to calculate the fracture toughness of the specimens shown in Fig. 14 (Ref 18-23):

$$K = \frac{(AF)}{(B)^{3/2}}$$

where K is the fracture toughness obtained from chevron-notched specimens, F is maximum load, B is specimen diameter for short rods or thickness for

short bars, and A is a dimensionless calibration constant dependent only on specimen geometry.

The currently recommended calibration constant for the specimens shown in Fig. 14 and 3 is 22.0. The equations and calibration constants for the specimens shown in Fig. 4 are given in Ref 14-17 and Ref 19.

Metal Alloys

Most chevron-notched specimens made from practical engineering alloys exhibit elastic-plastic behavior when tested. A plot of load versus mouth opening displacement for ideal elastic-plastic behavior is shown in Fig. 16(b). Unloading from R and S is offset from 0 to C and D, respectively. This offset is the result of yielding in two plastic zones in the specimen: the plane-stress plastic zone and the plane-strain plastic zone.

The plane-stress plastic zone occurs at the crack tip where the crack intersects the bottom of

the chevron notch. The size of this zone is dependent on chevron-notch acuity. Sufficient constraint usually is generated to adequately minimize the plane-stress plastic zone size when one of the following criteria is met:

- The chevron slot thickness is less than 1.5% of the short rod diameter or short bar thickness.
- The chevron slot bottoms are sharp pointed, as with a 60° included angle at the bottom of the slots.
- The plastic zone size is less than 0.2% of the specimen diameter or thickness.

However, the plane-strain plastic zone is located at the crack tip within the bulk of the specimen. Yielding in the plane-strain plastic zone partially relieves the stress at the crack tip. There-

Fig. 16 Schematic of (a) elastic specimen behavior, and (b) elastic-plastic specimen behavior. For the elastic-plastic behavior shown in (b) $\bar{F}$ is the average load between points R and S on the curve and locates both A' and B' on the unloading lines.

fore, the crack is apparently longer than is the actual case and thus the measured fracture toughness values are too low.

A plasticity factor, P, is used to adjust the fracture toughness to the proper value. The plasticity factor is measured experimentally from the load versus mouth opening displacement plot in Fig. 13 as follows:

$$P = \frac{(B' - A')}{(D - C)}$$

Note that the plasticity determination requires a plot of load versus displacement and two specimen unloadings and reloadings. Useful fracture toughness values are obtained when $-0.05 < P < 0.1$. When too large, P is reduced for a given material with sharper chevron slot acuity and/or larger specimen size.

There is no assurance that the peak load will occur at the critical crack length in elastic-plastic tests. However, for both linear elastic and elastic-plastic tests, the critical crack length occurs at a critical unloading slope ratio, r_c, which is a constant for any particular specimen configuration. The unloading slope ratio is defined as the un-loading slope divided by the slope of the initial elastic loading path. Therefore, the correct load, F_c, to use for calculating fracture toughness is located where an unloading line having the critical unloading slope intersects the load displacement curve. This line is found from vertical inter-polation between two actual unloading lines. This interpolation is done either graphically on the load versus displacement plot or automatically with a computer.

Fracture toughness for elastic-plastic behavior for the specimens shown in Fig. 14 is then calculated using the following equation:

$$K = \left[\frac{(AF_c)}{B^{3/2}}\right]\left[\frac{(1 + P)}{(1 - P)}\right]$$

where A is a dimensionless calibration constant dependent only on specimen geometry (the currently recommended calibration constant for the specimens shown in Fig. 14 is 22.0), F_c is the load at the critical unloading slope ratio, B is specimen diameter for a short rod or thickness for a short bar, and P is the plasticity factor.

more desirable because analysis is done following well known procedures. Therefore, assumptions, limitations, and conservatisms involved in the fracture prediction are all well known to the analysts.

To serve the needs of all fracture control analyses both the direct fracture mechanics approach to measuring fracture toughness and applying it with fracture parameters and the correlative approach of using qualitative fracture data and empirical correlation graphs plays an important role in designing for safety and reliability.

Appendix 2: Dynamic Fracture Toughness Testing

A rapid loading rate can influence the value of the measured fracture toughness. Usually the effect is to make toughness lower and the transition occur at a higher temperature when the fracture mode is brittle and to do just the opposite, that is, raise the toughness when the fracture mode is ductile. Many structural components can experience a high loading rate in service; therefore, it is often important to measure the toughness under high loading rates, a condition generally labeled dynamic. In the past, many requirements for fracture toughness data specified dynamic loading conditions. Before the quantitative approach to fracture toughness testing with the nonlinear parameters J and δ were developed, and fracture toughness measurement was limited to that characterized with the linear-elastic K, qualitative dynamic fracture testing had to be used whenever the linear-elastic K, qualitative dynamic fracture testing had to be used whenever the linear-elastic requirements could not be met. The Charpy V-notch impact test was one of the first of these types of tests and proved to give extremely valuable, though qualitative information about the toughness of a material. With the further development of the fracture mechanics methods, these qualitative tests could be replaced by more quantitative types of tests. However, new quantitative fracture toughness tests are slow to develop. Appendix A7 of method E399 allows a measurement of $K_{Ic}(t)$ which is a K_{Ic} test at a higher loading rate than the traditional quasi-static K_{Ic} test. Dynamic testing using the nonlinear J and δ parameters are under consideration but are not near to completion at this point. Therefore, the more qualitative dynamic tests are still being used for dynamic loading conditions. Two of these test methods are presented here in Appendix 2 to supply additional information about dynamic loading fracture testing. They are the dynamic tear test with its related application to fracture control analysis called the ratio-analysis diagram and the instrumented impact test. These tests do not provide direct fracture toughness results in terms of a fracture parameter that can be used to analyze structural safety and reliability. Rather they give information that can be used in a qualitative way. For example, to make material comparisons or to establish safe ranges of temperature for operation of a component. In most cases the information gathered with these tests can be used with empirical correlations to estimate actual fracture toughness values. For the dynamic tear test the ratio-analysis diagram can be sued to make the application to fracture control directly from these correlative graphs without a need to use the fracture mechanics calculations. The correlative approach has an appeal to the nonexpert in that fracture control information can be directly inferred without need for a special fracture mechanics education. For the more expert person the direct fracture mechanics approach may be

Dynamic Tear Tests

Many metals and alloys, especially at lower strength levels are too tough and too ductile to fracture under plane-strain conditions in the sizes normally used in structures. In an effort to obtain reliable values of fracture toughness of ductile metals and alloys, the Naval Research Laboratory introduced the dynamic tear (DT) test. This test is intended to evaluate metals and alloys over a wider range of fracture toughness than can other fracture-toughness tests. Correlation of DT toughness and K_{Ic} toughness has been published (Ref 25).

The standard DT-test specimen is similar to the Charpy specimen, but has greater depth and has a proportionately deeper notch, which is sharpened by a pressed knife edge. The DT-test specimen is broken by impact opposite the notch in a manner similar to the Charpy specimen, and the energy not absorbed is measured by the swing of the pendulum, if this type of machine is used, or by deformation of lead or aluminum plates if a drop-weight machine is used. Details on DT testing of metallic materials can be found in ASTM E 604 (Ref 26) and in Ref 27.

The DT test is a 1964 modification of the NRL drop-weight tear test (DWTT), which originally had a deep, sharp crack introduced by an electron-beam weld, rather than a notch. A different modification of the DWTT was introduced in 1963 by Battelle Memorial Institute. The Battelle DWTT, described in ASTM E 436 (Ref 28), uses a shallow notch pressed by a sharp chisel edge, rather than a deep crack or notch. In the Battelle DWTT, only the fracture appearance is recorded, while in the DT test, the energy to fracture the specimen is recorded.

Ratio-Analysis Diagram (RAD). From consideration of fracture mechanics and plastic-flow properties, and from numerous measurements, a RAD can be constructed to represent many of the conditions relating to material selection for fracture control. The conception and development of this tool has been discussed (Ref 29). Simplified versions of RADs for steels, aluminum alloys, and titanium alloys originally presented in Ref 29 are given here in Fig. 17-19. On the ordinate scales are values of DT energy and values of K_{Ic} that correspond to the DT values. The K_{Ic} scale is extended to the point corresponding to a 1-to-1 ratio of K_{Ic} to σ_{YS} (yield strength) for the best alloy of those to be considered. Sloping lines representing various ratios of K_{Ic} to σ_{YS} are

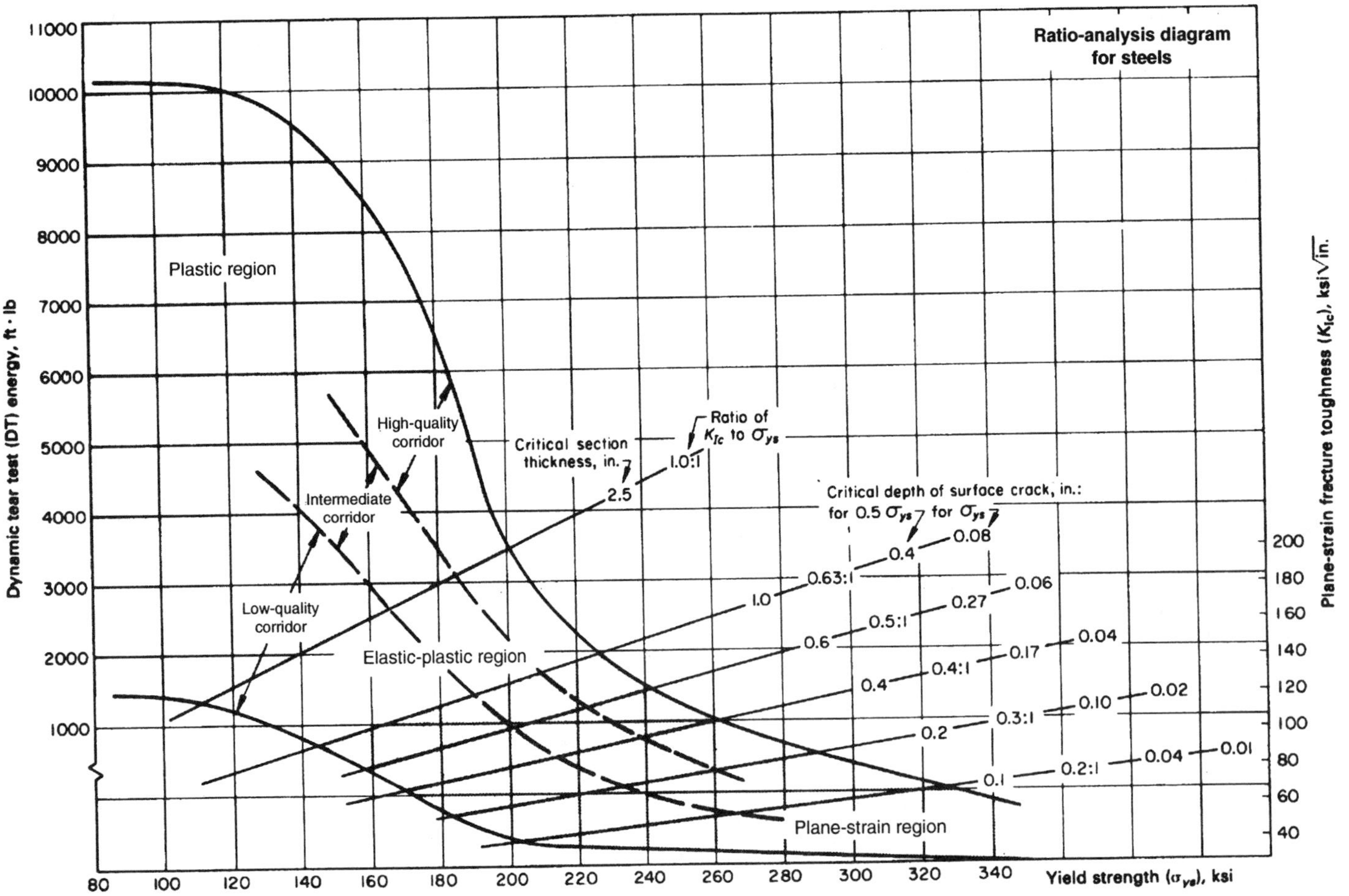

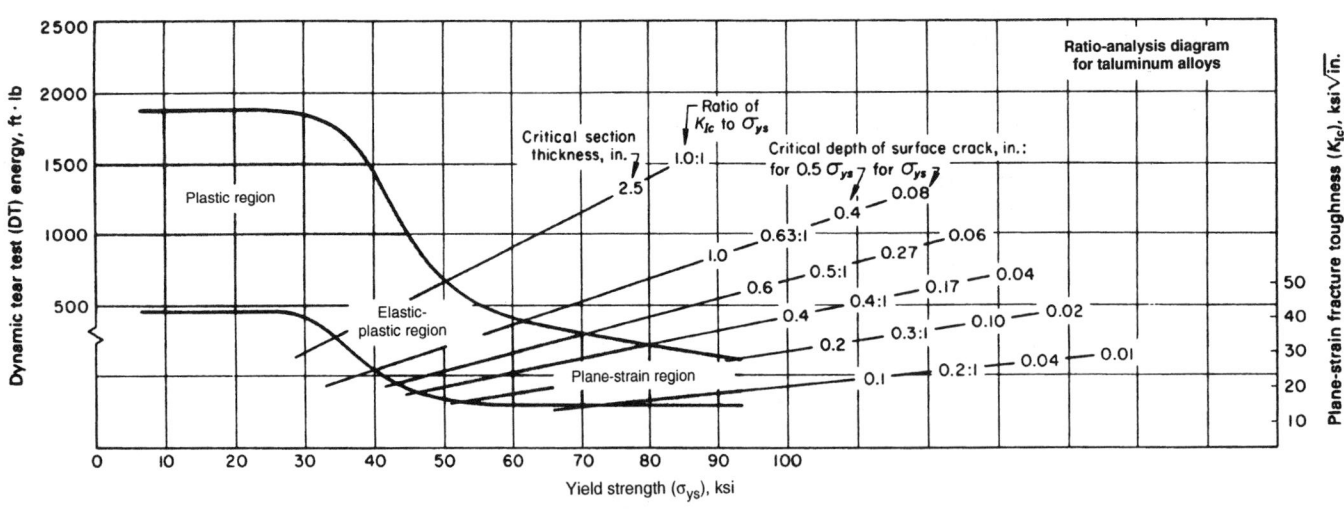

Fig. 17 RAD for steels, as prepared for trade-off analyses for components 25 mm (1.0 in.) thick. Data were determined using specimens 12.5 to 75 mm (0.5 to 3.0 in.) thick. Source: Ref 29

Fig. 18 RAD for aluminum alloys, as prepared for trade-off analyses for components 25 mm (1.0 in.) thick. Data were determined using specimens 25 to 100 mm (1.0 to 4.0 in.) thick. Source: Ref 29

drawn. By the equations of fracture mechanics, these ratios can be related to the critical flaw (crack) depth expected to initiate fracture in a plate of a particular thickness. Each sloping line is applicable to a certain critical thickness. Such a line can be constructed by plotting values of toughness versus yield strength for each alloy being considered for a design. If the thickness in the design is less than the critical thickness denoted by the sloping line, crack tips in the part will not be stressed under plane-strain conditions and either elastic-plastic or gross-yielding behavior can be expected. If design thickness is greater than critical thickness, the part can sustain even greater crack-tip stress intensity before crack extension will begin.

Because the modulus of elasticity is implicit in the relations between K_{Ic} and crack size, different diagrams are required for steel, aluminum alloys, and titanium alloys. Furthermore, at a given yield strength, different alloys have different fracture-

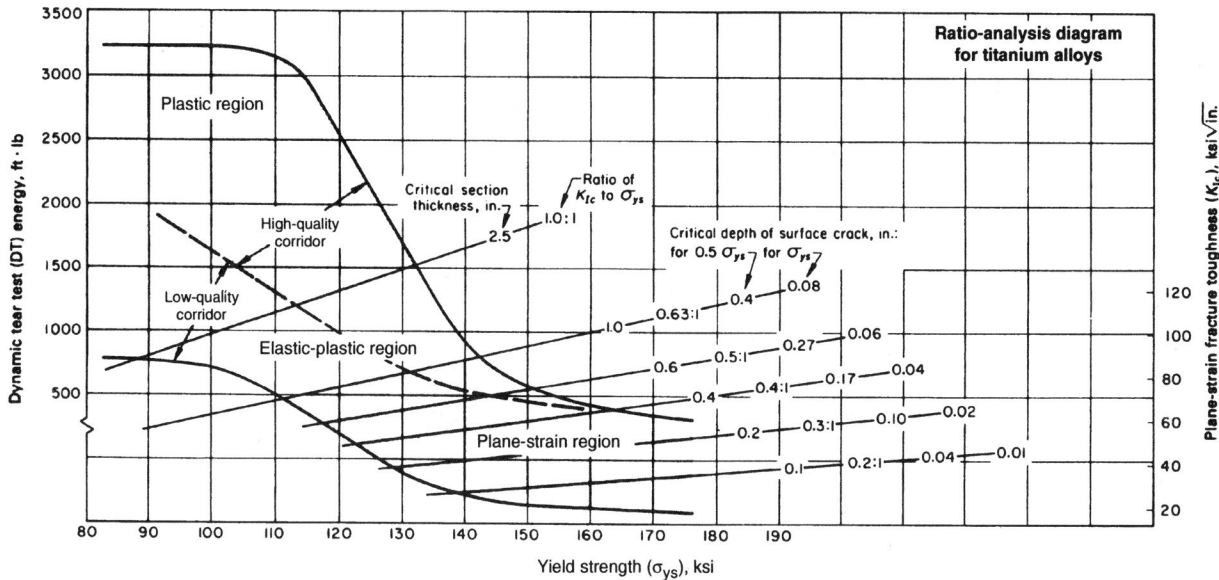

Fig. 19 RAD for titanium alloys, as prepared for trade-off analyses for components 25 mm (1.0 in.) thick. Data were determined using specimens 25 to 100 mm (1.0 to 4.0 in.) thick. Source: Ref 29

Table 2 Typical Charpy/K_{Ic} correlations for steels

Correlation	Transition temperature regime	Correlation	Transition temperature regime
Barsom (Ref 31)		**Marandet-Sanz—three steps (Ref 35)**	
$\dfrac{K_{Id}^2}{E} = 5 \text{ (CVN)}$		$T_{100} = 9 + 1.37\, T_{28}J$	$T_{100} = {}^\circ\text{C, for which } K_{Ic} = 100 \text{ MPa}\sqrt{\text{m}}$ $T_{28} = {}^\circ\text{C, for which CVN} = 28\,J$
Barsom-Rolfe (Ref 32)	$K_{Ic}, K_{Id} = \text{psi}\sqrt{\text{in.}}$	$K_{Ic} = 19 \text{ (CVN)}^{1/2}$	$K_{Ic} = \text{MPa}\sqrt{\text{m}}$ $\text{CVN} = J$
$\dfrac{K_{Ic}^2}{E} = 2 \text{ (CVN)}^{3/2}$	$E = \text{psi}$ $\text{CVN} = \text{ft} \cdot \text{lb}$		
Sailors-Corten (Ref 33)		Shift K_{Ic} curve through T_{100} point	
$\dfrac{K_{Ic}^2}{E} = 8 \text{ (CVN)}$		**Wullaert-Server (Ref 36)**	
		$K_{Ic,d} = 2.1\,(\sigma_y \text{ CVN})^{1/2}$	$K_{Ic,d} = \text{ksi}\sqrt{\text{in.}}$
$K_{Id}^2 = 15.873 \text{ (CVN)}^{3/8}$	$K_{Id} = \text{ksi}\sqrt{\text{in.}}$ $\text{CVN} = \text{ft} \cdot \text{lb}$		$\text{CVN} = \text{ft} \cdot \text{lb}$ $\sigma_y = \text{ksi corresponding to approximate loading rate}$
Begley-Logsdon—three points (Ref 34)		**Upper shelf region**	
$(K_{Ic})_1 = 0.45\,\sigma_y$ at 0% shear fracture temperature $(K_{Ic})_2$ From Rolfe-Novak Correlation at 100% shear fracture temperature $(K_{Ic})_3 = \frac{1}{2}[(K_{Ic})_1 + (K_{Ic})_2]$ at 50% shear fracture temperature	$K_{Ic} = \text{ksi}\sqrt{\text{in.}}$ $\sigma_y = \text{ksi}$	**Rolfe-Novak—$\sigma_y > 100$ ksi (Ref 37)** $\left(\dfrac{K_{Ic}}{\sigma_y}\right)^2 = 5 \text{ (CVN}/\sigma_y - 0.05)$	$K_{Ic} = \text{ksi}\sqrt{\text{in.}}$ $\text{CVN} = \text{ft} \cdot \text{lb}$ $\sigma_y = \text{ksi}$
		Ault-Wald-Bertolo—ultrahigh-strength steels (Ref 38) $\left(\dfrac{K_{Ic}}{\sigma_y}\right)^2 = 1.37 \text{ (CVN}/\sigma_y) - 0.045$	$K_{Ic} = \text{ksi}\sqrt{\text{in.}}$ $\text{CVN} = \text{ft} \cdot \text{lb}$ $\sigma_y = \text{ksi}$

1.0 ksi = 6.8948 MPa; 1.0 ksi$\sqrt{\text{in.}}$ = 1.099 MPa$\sqrt{\text{m}}$; 1.0 ft · lbf = 1.356 J. Note: CVN is the designation for Charpy impact energy; σ_y is the yield stress; and E is the Young's modulus.

toughness levels. Consequently, the diagrams may be divided into quality corridors based on test results for alloys with high, intermediate, and low toughness. With a RAD, it is then possible to guide the selection of alloys and working yield strengths so as to place the metal in either the ductile regime or the fracture mechanics regime, as may be appropriate for the cost or risk involved. Results obtained in DT tests extend fracture analysis beyond the linear-elastic fracture mechanics range.

Charpy and Dynamic Toughness

Correlations of Charpy Toughness with Fracture Toughness. Several empirical attempts have tried to correlate the Charpy impact energy with K_{Ic} to allow a quantitative assessment of critical flaw size and permissible stress levels. Most of these correlations are dimensionally incompatible, ignore differences between the two measures of toughness (in particular, loading rate

and notch acuity), and are valid only for limited types of materials and ranges of data. Additionally, these correlations can be widely scattered. However, some correlations can provide a useful guide to estimating fracture toughness; for example, some design criteria for nuclear pressure vessel and bridge steels are partially based on such correlative procedures (Ref 30, 31).

Some of the more common correlations are listed in Table 2 with appropriate units. Note that some of the correlations attempt to eliminate the

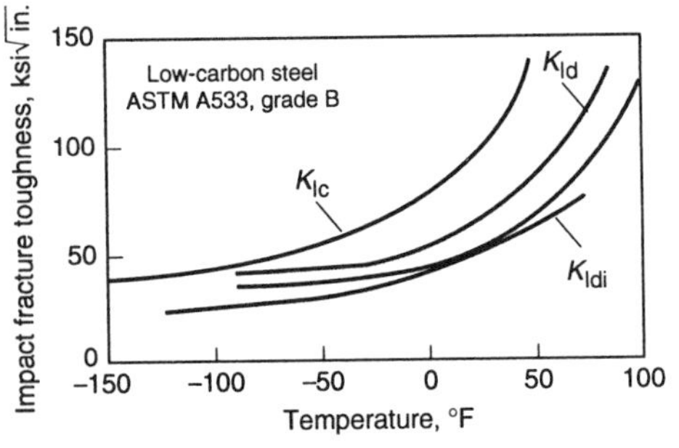

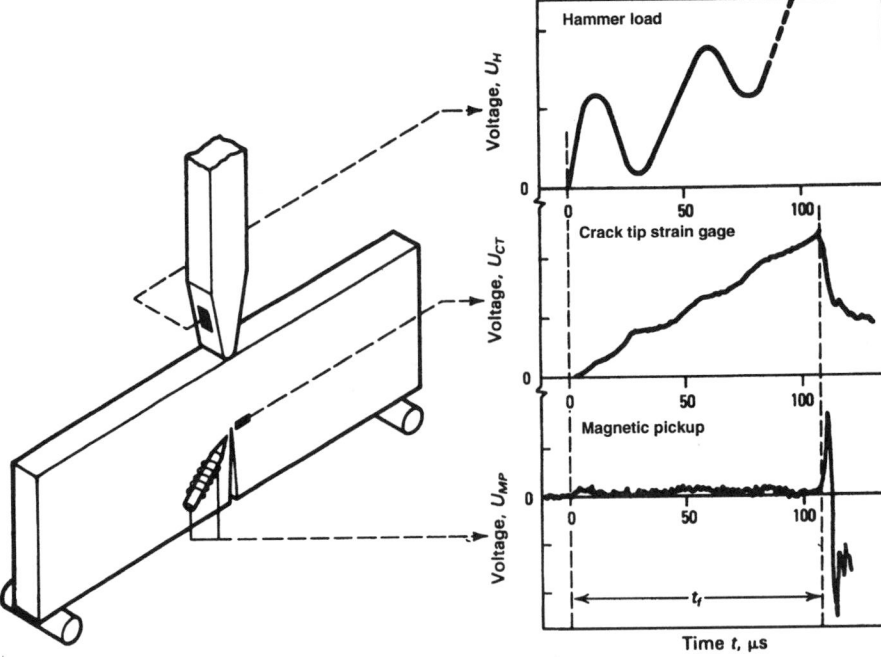

Fig. 20 Comparison of static (K_{Ic}), dynamic (K_{Id}), and dynamic-instrumented (K_{Idi}) impact fracture toughness of precracked specimens of ASTM A533, grade B, steel, as a function of test temperature. The stress-intensity rate was about 1.098×10^4 MPa$\sqrt{m} \cdot s^{-1}$ (10^4 ksi$\sqrt{in.} \cdot s^{-1}$) for the dynamic tests and about 1.098×10^6 MPa$\sqrt{m} \cdot s^{-1}$ (10^6 ksi$\sqrt{in.} \cdot s^{-1}$) for the dynamic-instrumented tests. Source: *Fracture Control and Prevention*, Use of Precracked Charpy Specimens, American Society for Metals, 1974, p 255-282

Fig. 21 Time-to-fracture measurement of a precracked Charpy specimen

effects of loading rate; the dynamic fracture toughness, K_{Id}, is correlated with Charpy energy.

Other attempts have been made to improve and explain some of the correlations (see, for example, Ref 39). A study has also been conducted using a portion of the Charpy energy to separate initiation and propagation components in the Charpy test (Ref 40). The results from this study for an upper shelf J_{Ic} correlation for pressure vessel steels were not significantly better than the Rolfe-Novak correlation listed in Table 2. A statistically based correlation for lower bound toughness has been developed for pressure vessel steels (Ref 41, 42), which may eventually replace the ASME K_{IR} curve described previously. Thus, simple and empirical correlations can be used as general guidelines for estimating K_{Ic} or K_{Id} within the limits of the specific correlation. Correlations using sharp-notch Charpy specimens have also been made and are discussed in *ASM Handbook*, Volume 8, *Mechanical Testing* (1985).

Instrumented Impact Testing

An instrumented-impact test is a modification of the Charpy test (and other impact tests) that allows recording of load and energy versus time, which correlates with deflection. From the record of load versus deflection, it is possible to determine the elastic portion of the stress-strain curve, the onset of crack extension, the energy for crack initiation, and the energy for crack propagation. With use of the instrumented impact test on precracked specimens, values of K_{Id} may be obtained under the same restrictions and in a similar manner as valid K_{Ic} values.

Instrumented Charpy Impact Test. The use of additional instrumentation with a standard Charpy impact machine allows augmented monitoring of the analog load/time response of Charpy V-notch specimen deformation and fracturing. The primary advantage of instrumenting the Charpy test is the additional information obtained

while maintaining low cost, small specimens, and simple operation. The most commonly used approach is application of strain gages to the striker to sense the load-time behavior of the test specimen.

One of the primary reasons for the development of the instrumented Charpy test was to apply existing notch bend theories (slow bend) to the dynamic three-point bend Charpy impact test. Obtaining load information during the standard Charpy V-notch impact test establishes a relationship between metallurgical fracture parameters and the transition temperature approach for assessing fracture behavior (Ref 43).

Interest in instrumented impact testing has expanded to include testing of different types of specimens (precracked, large bend), variations in test techniques (low blow, full-size components), and testing of many different materials (plastics, composites, aerospace materials, ceramics, etc.). The rapid growth in the application of instrumented impact testing has produced a correspondingly large demand for standardized test methods. However, no standard currently exists for instrumented Charpy testing. Typical dynamic fracture toughness of a structural steel is shown in Fig. 20.

General Test Requirements. The International Institute of Welding first attempted to standardize the instrumented Charpy test, but concluded that the test was not sufficiently documented, and the effort was discontinued (Ref 44). A few years later, two significant events prompted serious consideration of standardization. The development of the K_{IR} curve by the Pressure Vessel Research Committee and its inclusion in the ASME Code, Section III, created the need for dynamic initiation toughness, K_{Id}, data. The formation of an ASTM Task Group

(E24.03.03) to standardize precracked Charpy testing also encouraged standardization.

Simultaneously, two other related groups began formulating procedures and conducting interlaboratory round robins. A Pressure Vessel Research Committee/Metals Property Council Task Group on Fracture Toughness Properties for Nuclear Components developed procedures for measuring K_{Id} values from precracked Charpy specimens (Ref 45). Also, the Electric Power Research Institute (EPRI) funded work to develop the "EPRI Procedures" (Ref 46). Currently, only subtle differences exist between the two sets of procedures (Ref 47).

Precracked Charpy Test. By inducing a fatigue precrack in the Charpy specimen, the notch acuity and depth restrictions are eliminated. Early work concentrated on correlations with fracture toughness using only the total absorbed energy (i.e., uninstrumented testing). These energy values usually are normalized per unit area below the fatigue crack; the normalized energy values are designated as W/A.

Most of the correlations of W/A with fracture toughness have been conducted using slow-bend specimens. The basic problem in reaching an impact correlation is the difference in loading rates between the Charpy impact and the static K_{Ic} tests, particularly for loading rate sensitive materials (Ref 48). A general trend exists for a correlation between K_{Ic}^2/E and W/A, but the limited data and scatter make this difficult to utilize (Ref 38). A better correlation with K_{Id} may be possible. The reason for using K_{Ic}^2/E as the basis is the approximate proportionality between K_{Ic}^2/E and W/A, based on a presumed fracture mechanics relationship (Ref 48).

The precracked Charpy W/A values can also be used to estimate the nil-ductility transition tem-

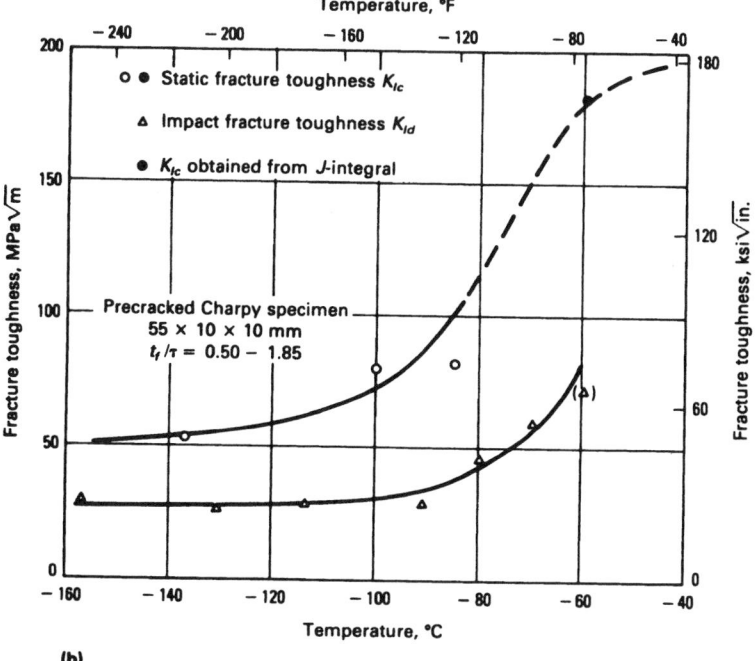

Fig. 22 Measured impact fracture toughness data for low-alloy steels tested at an impact velocity of 5 m/s (16.4 ft/s). (a) Alloy 30CrNiMo8. (b) Alloy StE 460. Data points in parentheses are invalid according to ASTM Standards. Source: *Metals Handbook*, 9th ed., Vol 8, 1985, p 272

cability of the test. To achieve sufficiently small times-to-fracture, the impact velocity must be limited. The maximum allowable velocity depends on the toughness of the material to be tested. These difficulties and restrictions are overcome by applying the concept of impact response curves for measurement of the impact fracture toughness K_{Id} (Ref 52-54).

The technique of measuring impact fracture toughness, K_{Id} with impact response curves and time-to-fracture measurements has several advantages over the conventional quasi-static technique described previously in this article in the section on precracked Charpy testing. The impact response curve technique represents a fully dynamic evaluation. Kinetic effects are correctly accounted for during the entire impact event. The method can thus be applied to all experimental test conditions, particularly in the short time-to-fracture range, i.e., when high impact velocities are used or brittle materials are tested.

This method does not require a calibrated instrumentation of the hammer, which is usually a prerequisite in impact experiments designed to determine the load at crack initiation. The data-measuring procedure consists of two separate tasks: determination of the impact response curve and the measurement of the time-to-fracture. The more complicated determination of the impact response curve need only be carried out once.

The time-to-fracture of a precracked Charpy specimen subjected to impact loading can be obtained from signals of two uncalibrated strain gages, one of which is located on the tup of the striking hammer and the other on the specimen to the side of the crack tip. The leading edge of the signal from the hammer strain gage marks the beginning of the impact event. The onset of crack propagation, on the other hand, is indicated by the rapid drop in load registered by the crack tip strain gage. The time-to-fracture, t_f, is the interval between the two signals. Typical oscillograms of time-to-fracture measurements are shown in Fig. 21. Typical data results are shown in Fig. 22.

perature. The typical technique defines an inflection point between lower shelf and transition region behavior as the estimated nil-ductility transition temperature (Ref 49). Some exceptions have been noted to this approach (Ref 50).

The types of data and test techniques used for instrumented precracked Charpy testing are the same as those for instrumented Charpy impact testing. The greatest advantage of precracking is the transformation of the Charpy V-notch specimen into a dynamic fracture mechanics test piece. The direct calculation of fracture toughness (within certain limitations) is now possible using the instrumented load-time information. See, for example, the calculation of fracture toughness of ferritic steels from precracked Charpy specimens (Ref 51).

Time-to-Fracture Tests

Interpretation of load-time records obtained in instrumented impact tests with precracked Charpy specimens can be difficult, particularly when the early time range is considered. This difficulty and the use of a quasi-static evaluation procedure for determining the dynamic impact fracture toughness K_{Id} restricts the range of appli-

REFERENCES

1. "Standard Terminology Relating to Fracture Testing," E 616, *Annual Book of ASTM Standards*, Vol 03.01, ASTM
2. J.D. Landes and R. Herrera, Micromechanisms of Elastic/Plastic Fracture Toughness, J_{Ic}, *Proc. 1987 ASM Materials Science Seminar*, ASM International, 1989, p 111-130
3. G.R. Irwin, Analysis of Stresses and Strains near the End of a Crack Traversing a Plate, *J. Appl. Mech.*, Vol 6, 1957, p 361-364
4. J.R. Rice, A Path Independent Integral and the Approximate Analysis of Strain Concentrations by Notches and Cracks, *J. Appl. Mech.*, Vol 35, 1968, p 379-386
5. A.A. Wells, Unstable Crack Propagation in Metals: Cleavage and Fast Fracture, *Proc. Cranfield Crack Propagation Symposium*, Vol 1, Paper 84, 1961
6. "Standard Method for Plane-Strain Fracture Toughness of Metallic Materials," E 399, *An-*

nual Book of ASTM Standards, Vol 03.01, ASTM

7. "Standard Practice for R-Curve Determinations," E 561, Annual Book of ASTM Standards, Vol 03.01, ASTM

8. "Standard Method for Determining Plane-Strain Crack Arrest Toughness, K_{Ia}, of Ferritic Steels," E 1221, Annual Book of ASTM Standards, Vol 03.01, ASTM

9. "Standard Method for J_{Ic}, a Measure of Fracture Toughness," E 813, Annual Book of ASTM Standards, Vol 03.01, ASTM

10. "Standard Method for Determining J-R Curves," E 1152, Annual Book of ASTM Standards, Vol 03.01, ASTM

11. "Methods for Crack Opening Displacement (COD) Testing," BS5762: 1979, The British Standards Institution, 1979

12. "Standard Method for Crack-Tip Opening Displacement (CTOD) Fracture Toughness Measurement," E 1290, Annual Book of ASTM Standards, Vol 03.01, ASTM

13. J.D. Landes and D.H. Shaffer, Statistical Characterization of Fracture in the Transition Region, Fracture Mechanics: Twelfth Conference, STP 700, ASTM, 1980, p 368-382

APPENDIX 1 REFERENCES: Chevron Notched

14. D. Munz, R.T. Bubsey, and J.L. Shannon, Jr., Performance of Chevron-Notch Short Bar Specimen in Determining the Fracture Toughness of Silicon Nitride and Aluminum Oxide, J. Test. Eval., Vol 8 (No. 3), 1980, p 103-107

15. D. Munz, R.T. Bubsey, and J.E. Srawley, Compliance and Stress Intensity Coefficients for Short Bar Specimens with Chevron Notches, Int. J. Frac., Vol 16 (No. 4), 1980, p 359-394

16. R.T. Bubsey, D. Munz, W.S. Pierce, and J.L. Shannon, Jr., Compliance Calibration of the Short Rod Chevron-Notch Specimen for Fracture Toughness Testing of Brittle Materials, Int. J. Frac., Vol 18 (No. 2), 1982, p 125-133

17. J.L. Shannon, Jr., R.T. Bubsey, W.S. Pierce, and D. Munz, Extended Range Stress Intensity Factor Expressions for Chevron-Notched Short Bar and Short Rod Fracture Toughness Specimens, Int. J. Frac., Vol 19, 1982

18. J.F. Beech and A.R. Ingraffea, Three-Dimensional Finite Element Calibration of the Short Rod Specimens, Int. J. Frac., Vol 18, 1982, p 217-229

19. A. Mendelson and L.J. Ghosen, Three-Dimensional Analysis of Short-Bar Chevron-Notched Specimens by Boundary Integral Method, in Chevron-Notched Specimens: Testing and Stress Analysis, STP 855, ASTM, Philadelphia, 1984

20. A.R. Ingraffea, R. Perucchio, T-Y Han, W.H. Gersthe, and Y.P. Huang, Three-Dimensional Finite and Boundary Element Calibration of the Short Rod Specimen, in Chevron-Notched Specimens: Testing and Stress Analysis, STP 855, ASTM, Philadelphia, 1984

21. L.M. Barker and R.V. Guest, "Compliance Calibration of the Short Rod Fracture Toughness Specimen," Report TR-78-20, Terra Tek, Inc., Salt Lake City, 1978

22. L.M. Barker, "Compliance Calibration of a Family of Short Rod and Short Bar Fracture Toughness Specimens," Report TR-81-07, Terra Tek, Inc., Salt Lake City, 1981

23. L.M. Barker, "Short Rod and Short Bar Fracture Toughness Specimen Geometries and Test Methods for Metallic Materials," Report TR-80-11, Terra Tek, Inc., Salt Lake City, 1980

24. K.R. Brown, "An Evaluation of the Chevron-Notched Short-Bar Specimen for Fracture Toughness Testing in Aluminum Alloys," Kaiser Aluminum & Chemical Co., Oct 1, 1981

APPENDIX 2 REFERENCES

25. W.S. Pellini, Evolution of Principles for Fracture-Safe Design of Steel Structures, NRL 6957, U.S. Naval Research Laboratory, Washington, Sept 1969

26. "Standard Test Method for Dynamic Tear Testing of Metallic Materials," E 604, Annual Book of ASTM Standards, Vol 03.01, ASTM, Philadelphia, 1984, p 641-652

27. E.A. Lange, P.P. Puzak, and L.A. Cooley, Standard Method for the ⅝" Dynamic Tear Test, NRL 7159, U.S. Naval Research Laboratory, Washington, July 1970

28. "Standard Method for Drop-Weight Tear Tests of Ferritic Steels," E 436, Annual Book of ASTM Standards, Vol 03.01, ASTM, Philadelphia, 1984, p 555-560

29. W.S. Pellini, Criteria for Fracture Control Plans, NRL 7406, U.S. Naval Research Laboratory, Washington, May 1970

30. "Pressure Vessel Research Committee Recommendation on Toughness Requirements for Ferritic Materials," Welding Research Council Bulletin 175, New York, Aug 1972

31. J.M. Barsom, The Development of AASHTO Fracture Toughness Requirements for Bridge Steels, Eng. Frac. Mech., Vol 7 (No. 3), Sept 1975, p 605-618

32. J.M. Barsom and S.T. Rolfe, Correlations Between K_{Ic} and Charpy V-Notch Test Results in the Transition Temperature Range, in Impact Testing of Materials, STP 466, ASTM, Philadelphia, 1979, p 281-302

33. R.H. Sailors and H.T. Corten, Relationship Between Material Fracture Toughness Using Fracture Mechanics and Transition Temperature Tests, in Fracture Toughness, Proceedings of the 1971 National Symposium on Fracture Mechanics, STP 514, Part II, ASTM, Philadelphia, 1972, p 164-191

34. J.A. Begley and W.A. Logsdon, "Correlation of Fracture Toughness and Charpy Properties for Rotor Steels," WRL Scientific Paper 71-1E7-MSLRF-P1, Westinghouse Research Laboratory, Pittsburgh, July 1971

35. B. Marandet and G. Sanz, Evaluation of the Toughness of Thick Medium-Strength Steels

by Using Linear Elastic Fracture Mechanics and Correlations Between K_{Ic} and Charpy V-Notch, in Flaw Growth and Fracture, STP 631, ASTM, Philadelphia, 1977, p 72-95

36. R.A. Wullaert, Fracture Toughness Predictions from Charpy V-Notch Data, in What Does the Charpy Test Really Tell Us?, Proceedings of the American Institute of Mining, Metallurgical and Petroleum Engineers, Denver, American Society for Metals, 1978

37. S.T. Rolfe and S.R. Novak, Slow-Bend K_{Ic} Testing of Medium-Strength High-Toughness Steels, in Review of Developments in Plane-Strain Fracture Toughness Testing, STP 463, ASTM, Philadelphia, 1970, p 124-159

38. "Rapid Inexpensive Tests for Determining Fracture Toughness," National Materials Advisory Board, National Academy of Sciences, Washington, DC, 1976

39. What Does the Charpy Test Really Tell Us?, Proceedings of the American Institute of Mining, Metallurgical and Petroleum Engineers, Denver, American Society for Metals, 1978

40. D.M. Norris, J.E. Reaugh, and W.L. Server, A Fracture-Toughness Correlation Based on Charpy Initiation Energy, in Fracture Mechanics: Thirteenth Conference, STP 743, ASTM, Philadelphia, 1981, p 207-217

41. W.L. Server et al., "Analysis of Radiation Embrittlement Reference Toughness Curves," EPRI NP-1661, Electric Power Research Institute, Palo Alto, CA, Jan 1981

42. Reference Fracture Toughness Procedures Applied to Pressure Vessel Materials, Metal Properties Council MPC-24, Proceedings of the Winter Annual Meeting of the American Society for Mechanical Engineers, New Orleans, ASME, New York, 1984

43. R.A. Wullaert, Application of the Instrumented Charpy Impact Test, in Impact Testing of Metals, STP 466, ASTM, Philadelphia, 1970, p 148-164

44. E.C.J. Buys and A. Cowan, Interpretation of the Instrumented Impact Test, Welding in the World, Vol 8 (No. 1), 1970, p 70-76

45. "Instrumented Precracked Charpy Testing, Report I—Recommended Testing Procedure," and "Report II—Associated Test Program," Pressure Vessel Research Committee/Metal Properties Council Working Group in Instrumented Precracked Charpy Testing, Westinghouse Research Laboratory, Pittsburgh, 1974

46. D.R. Ireland, W.L. Server, and R.A. Wullaert, "Procedures for Testing and Data Analysis," ETI Report TR-75-43, Effects Technology, Inc., Santa Barbara, CA, Oct 1975

47. W.L. Server, Impact Three-Point Bend Testing for Notches and Precracked Specimens, J. Test. Eval., Vol 6 (No. 1), 1978, p 29-34

48. T.M.F. Ronald, J.A. Hall, and C.M. Pierce, Usefulness of Precracked Charpy Specimens for Fracture Toughness Screening Tests of Titanium Alloys, Met. Trans., Vol 3, April 1972, p 813-818

49. G.M. Orner and C.E. Hartbower, Transition-Temperature Correlations in Construction Al-

loy Steels, *Weld. J.*, Vol 40 (No. 9), Oct 1961, p 459s

50. J.H. Gross, The Effect of Strength and Thickness on Notch Ductility, *Weld. J.*, Vol 48 (No. 10), Oct 1969, p 441s

51. *Metals Handbook*, 9th ed., Volume 8, *Mechanical Testing*, 1985, p 268

52. J.F. Kalthoff et al., "Measurements of Dynamic Stress Intensity Factors in Impacted Bend Specimens," Committee on Safety of Nuclear Installations Specialist Meeting on Instrumented Precracked Charpy Testing, EPRI NP-2102-LD, Electric Power Research Institute, Palo Alto, CA, Nov 1981

53. J.F. Kalthoff, S. Winkler, W. Böhme, and W. Klemm, Determination of the Dynamic Fracture Toughness K_{Id} in Impact Tests by Means of Impact Response Curves, in *Advances in Fracture Research*, D. Francois et al., Ed., Pergamon Press, New York, 1980, p 363-373

54. J.F. Kalthoff, "Time Effects and Their Influences on Test Procedures for Measuring Dynamic Material Strength Values," in *Proceedings of the International Conference on the Application of Fracture Mechanics to Materials and Structures*, G.C. Sih, E. Sommer, and W. Dahl, Ed., Freiburg, West Germany, June 20-24, Martinus Nijhoff Publishers, The Hague, 1983, p 107-136

SELECTED REFERENCES

• T.L. Anderson, *Fracture Mechanics, Fundamentals and Applications*, 2nd ed., CRC Press, 1995

• J.M. Barsom and S.T. Rolfe, *Fracture and Fatigue Control in Structures*, 2nd ed., Prentice-Hall, 1987

• J.A. Begley and J.D. Landes, The J Integral as a Fracture Criterion, *Fracture Toughness, Proc. 1971 National Symposium on Fracture Mechanics, Part II*, STP 514, ASTM, 1972, p 1-20

• D. Broek, *Elementary Fracture Mechanics*, 4th rev. ed., Martinus Nijhoff, 1986

• W.F. Brown, Jr. and J.E. Srawley, Plane-Strain Crack Toughness Testing of High Strength Metallic Materials, STP 410, ASTM, 1966

• G.A. Clarke, W.R. Andrews, P.C. Paris, and D.W. Schmidt, Single Specimen Tests for J_{Ic} Determination, *Mechanics of Crack Growth*, STP 590, ASTM, 1976, p 24-42

• G.A. Clarke, W.R. Andrews, J.A. Begley, J.K. Donald, G.T. Embley, J.D. Landes, D.E. McCabe, and J.H. Underwood, A Procedure for the Determination of Ductile Fracture Toughness Values Using J Integral Techniques, *J. Test. Eval.*, Vol 7 (No. 1), Jan 1979, p 49-56

• M.G. Dawes, Elastic-Plastic Fracture Toughness Based on CTOD and J-Contour Integral Concepts, *Elastic-Plastic Fracture*, STP 668, J.D. Landes, J.A. Begley, and G.A. Clarke, Ed., ASTM, 1979, p 307-333

• A. Joyce and J.P. Gudas, Computer Interactive J_{Ic} Testing of Navy Alloys, *Elastic-Plastic Fracture*, STP 668, J.D. Landes, J.A. Begley, and G.A. Clarke, Ed., ASTM, 1979, p 451-468

• J.D. Landes and J.A. Begley, The Effect of Specimen Geometry on J_{Ic}, *Fracture Toughness, Proc. 1971 National Symposium on Fracture Mechanics, Part II*, STP 514, ASTM, 1972, p 24-39

• J.D. Landes and J.A. Begley, Test Results from J_{Ic} Studies—An Attempt to Establish a J_{Ic} Testing Procedure, *Fracture Analysis*, STP 560, ASTM, 1974, p 170-186

• J.D. Landes and J.A. Begley, Recent Developments in J_{Ic} Testing, *Developments in Fracture Mechanics Test Methods Standardization*, STP 632, ASTM, 1977, p 57-81

• K.H. Schwalbe and D. Hellmann, Application of the Electrical Potential Method of Crack Length Measurement Using Johnson's Formula, *J. Test. Eval.*, Vol 9 (No. 3), 1981, p 218-221

• K. Wallin, Statistical Modeling of Fracture in the Ductile to Brittle Transition Region, *Defect Assessment in Components—Fundamentals and Applications*, J.G. Blauel and K.H. Schwalbe, Ed., ESIS/EGF, Mechanical Engineering Publications, 1991, p 1-31

• K. Wallin, Fracture Toughness Transition Curve Shape for Ferritic Structural Steels, *Proc. Joint FEFG/ICF International Conference on Fracture of Engineering Materials* (Singapore), 6-8 Aug, 1991, p 83-88

SELECTED REFERENCES: Chevron-Notched Specimens

• L.M. Barker, ASTM E 1304, The New Standard Test for Plane-Strain (Chevron-Notched) Fracture Toughness: Usage of Test Results, *Fracture Mechanics: Twenty-Second Symposium, Vol II*, ASTM, 1992, p 58-68

• J. Bray, Use of the Chevron-Notched Short Bar Test to Guarantee Fracture Toughness for Lot Release in Aluminum Alloys, *Chevron-Notch Fracture Test Experience: Metals and Non-Metals*, ASTM, 1992, p 131-143

• M.O. Lai and K.H. Lee, K_{Ic} Determination Using a Chevron Notched Compact Tension Specimen, *Engineering Fracture Mechanics*, Vol 41 (No. 3), Feb 1992, p 453-456

• H. Luo, et al., A Study of Fracture Toughness of Hardmetals by Chevron-Notching Method, *Powder Metallurgy Technology* (China), Vol 7 (No. 3), Aug 1989, p 165-171

• C.W. Marschall, et al., Using Chevron-Notch, Short-Bar Specimens for Measuring the Fracture Toughness of a Martensitic Stainless Steel at High Hardness, *Chevron-Notch Fracture Test Experience: Metals and Non-Metals*, ASTM, 1992, p 144-156

• J.M. Paddon and R. Morrel, Evaluation of the Chevron Notch Fracture Toughness Test for Brittle Materials, National Physical Laboratory Reports, Report DMM(A)72, 1992, p 102

• K.R. Raju, Application of Short Rod Chevron Notched Specimen for K_{Ic} Testing of Cast Iron, Transactions of the Indian Institute of Metals, Vol 43 (No. 5), Oct 1990, p 304-308

• K. Ray and S. Ray, Accelerated Fracture Toughness Testing for Quality Control of Micro-Alloyed Steels, *Fatigue and Fracture in Steel and Concrete Structures, ISFF '91*, Oxford & IBH Publishing Co., 1991, p 317-332

• A. Rosenfield and B. Majumdar, Fracture Toughness Evaluation of Ceramic Bonds Using a Chevron-Notch Disk Specimen, *Chevron-Notch Fracture Test Experience: Metals and Non-Metals*, ASTM, 1992, p 63-73

• P.A. Withey and P. Bowen, Fracture Toughness Determination by the Use of Chevron Notches, *Mechanical Behaviour of Materials—VI, Vol 4*, Pergamon Press, 1992, p 153-158

Concepts of Fracture Control and Damage Tolerance Analysis

David Broek, FractuREsearch

FRACTURE CONTROL is the concerted effort to ensure safe operations without catastrophic failure by fracture. Very seldom does a fracture occur due to an unforeseen overload on the undamaged structure. Fractures are usually the end result of crack growth from a small defect or flaw. Due to repeated or sustained "normal" service loads, a crack may develop (starting from a flaw or stress concentration) and slowly grow in size. Cracks and defects impair the strength of the component. Thus, during the continuing development of the crack, the structural strength decreases until it becomes so low that the service loads cannot be carried any more, and fracture ensues.

If fracture is to be prevented, the strength should not drop below a certain safe value. This means that cracks must be prevented from growing to a size at which the strength would drop below an acceptable limit. In order to determine which size of crack is admissible, one must be able to calculate how the structural strength is affected by cracks (as a function of their size), and in order to determine the safe operational life, one must be able to calculate the time in which a crack grows to the permissible size. Damage tolerance analysis is used to obtain this information.

Damage tolerance is the property of a structure to sustain defects or cracks safely, until such time that action is (or can be) taken to eliminate the cracks by repair or by replacing the cracked structure or component. In the design stage, one still has the options to select a more crack-resistant material or improve the structural design to ensure that cracks will not become dangerous during the projected economic service life. Alternatively, periodic inspections may be scheduled, so that cracks can be repaired or components replaced when cracks are detected. Either the time to retirement (replacement) or the inspection interval and type of inspection must follow from the crack growth time calculated in the damage tolerance analysis.

Inspections can be performed by means of any nondestructive inspection technique, but destructive techniques such as proof testing are essentially inspections. If a burst occurs during hydrostatic testing of a pipe line, for example, then there was apparently a crack of sufficient size to cause the burst. The proof test detects defects above a certain size under controlled circumstances (e.g., with water pressure) to prevent catastrophic failure during operation when the line is filled with oil or gas.

Fracture control is a systematic process to prevent fracture during operation. The extent of the fracture control measures depends on the criticality of the component, the economic consequences of the structure's being out of service, and the damage that would be caused by a fracture failure. Fracture control of a hammer may be as simple as selecting a material with sufficient fracture resistance. Fracture control of an airplane includes damage tolerance analysis, tests, and subsequent inspection and repair/replacement plans. Inspections, repairs, and replacements must be scheduled rationally using the information from the damage tolerance analysis.

The mathematical tool employed in damage tolerance analysis is fracture mechanics; it provides the concepts and equations used to determine how cracks grow and affect the strength of a structure. Acclaimed inaccuracies are due to inaccurate inputs much more often than to inadequacy of the concepts. Although further improvements of fracture mechanics concepts may well be desirable from a fundamental point of view, it is unlikely that damage tolerance analysis can be much improved, as its accuracy is determined mostly by assumptions and (predicted) loads and stresses. Therefore, the accuracy of all assumptions and input data must be clearly understood and identified, and the results of damage tolerance analysis must be used judiciously by engineers. In this context, damage tolerance analysis can give useful answers to questions that hitherto could not be answered at all. The answers may only be preliminary, but a reasonable answer is better than none. This and the following article describe the key principles and practical guidelines to obtain useful and reasonable answers from damage tolerance analysis.

Principles of Fracture Control

Establishment of a fracture control plan requires knowledge of the structural strength as it is affected by cracks, and knowledge of the time involved for cracks to grow to the permissible size. Thus, damage tolerance analysis has two objectives, namely, to determine:

- The effect of cracks on strength (margin against fracture)
- The crack growth as a function of time

Figure 1 shows the effect of crack size on strength, where the strength is expressed in terms of the load, P, the structure can carry before fracture occurs (fracture load). Supposing that a new structure has no significant defects ($a = 0$), then the strength of the structure is the design strength (P_u).

In every design a safety factor is used. This factor may be applied in different ways. Usually the safety factor is applied to load. For example, if the maximum anticipated service load is P_s, the structure is actually designed to sustain $j \cdot P_s = P_u$, where j is the safety factor. The designer sizes the structure in such a manner that the stress is equal

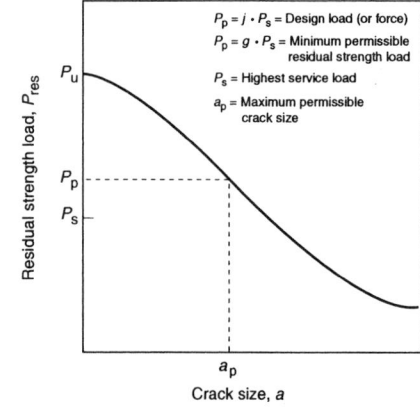

Fig. 1 Residual strength diagram in terms of load (or force). j, design safety factor; g, safety factor based on residual strength

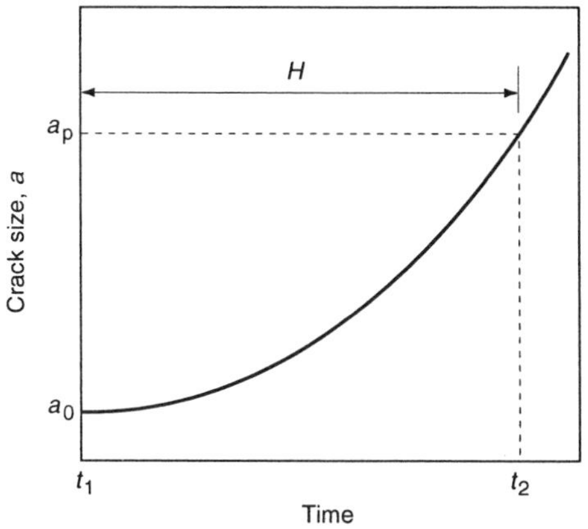

Fig. 2 General crack growth curve showing time interval, H, to reach maximum permissible crack size, a_p

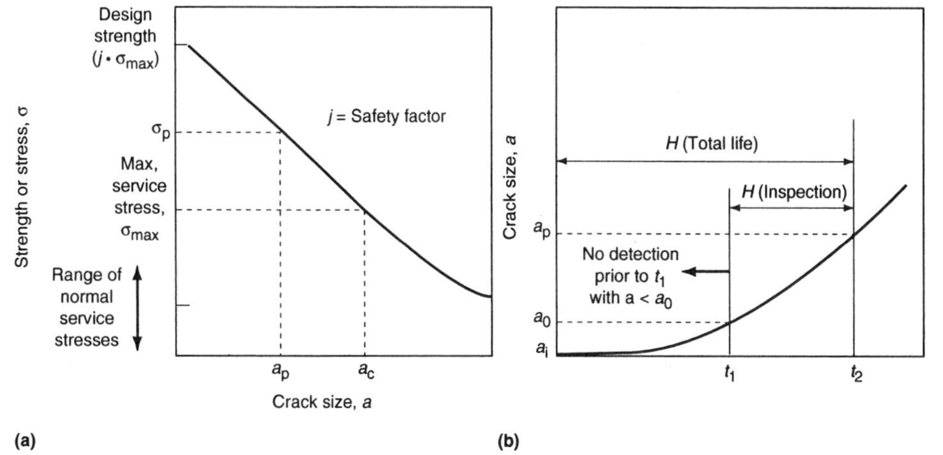

(a)

(b)

Fig. 3 Key parameters for fracture control. (a) Residual strength in terms of stress. (b) Crack growth and time period when inspection can be performed

to, or slightly less than, the tensile strength when the load is P_u. Alternatively, the safety factor can be applied to the allowable stress: if the actual material strength (tensile strength) is F_{tu}, the structure is sized in such a way that the stress at the highest service load, P_s, is less than or equal to F_{tu}/j. Because load and stress are usually proportional, the structural part is actually capable of carrying $j \cdot P_s = P_u$. The value of j is between 3 (many civil engineering structures) and 1.5 (aircraft).

It is emphasized that P_s is the highest service load. If the service load varies, the load may well be much less than P_s most of the time. For example, the loads on cranes, bridges, off-shore structures, ships, and airplanes are usually much less than P_s. Only in exceptional circumstances does the load reach P_s. At other times the load is only a fraction of P_s, so that the margin against failure is much larger than j, except in extreme situations.

The new structure has a strength P_u. If the load should reach P_u the structure fails. The probability of this occurring is not zero, but experience has shown that it is acceptably low if the prescribed safety factor is adhered to. If cracks are present, the strength is less than P_u. This remaining strength under the presence of cracks is generally referred to as the *residual strength* (P_{res}). Therefore, the diagram in Fig. 1 is called the residual strength diagram. With a residual strength $P_{res} < P_u$ the safety factor has decreased to $j = P_{res}/P_s$, which is less than $j = P_u/P_s$. In concert, the probability of fracture failure has become higher.

With a crack of size a, the residual strength is P_{res}. Should a load $P = P_{res}$ occur, then fracture takes place. If a load P_{res} does not occur and service loading continues at lower loads, the crack may continue to grow by fatigue. If continued crack growth occurs, then the residual strength decreases further. If nothing is done, fracture will eventually occur under normal service loads. This is what must be prevented.

The above discussion implies that the limit should be set somewhat above P_s. For example, one may require that the residual strength never

be less than $P_p = g \cdot P_s$, where g is the remaining safety factor and P_p is the minimum permissible residual strength. The design engineer (or user) does not usually decide what should be the initial safety factor j. The factor is prescribed by rules and regulations issued by engineering societies (e.g., ASME) or government authorities (e.g., FAA, DoD). These rules or requirements should also prescribe g. This has not been done for all types of structures yet, and the ASME code, for example, approaches the problem somewhat differently (see the section "Damage Tolerance Requirements" in this article).

Provided that the shape of the residual strength diagram is known and P_p has been prescribed, the maximum permissible crack size follows from the diagram. In order for damage tolerance analysis to determine the largest permissible crack, the first objective must be the calculation of the residual strength diagram of Fig. 1. If a_p can be calculated directly from P_p it may not be necessary to calculate the entire residual strength diagram, but only the point (a_p, P_p). However, this is seldom possible and rarely time saving. In general, calculation of the entire diagram is preferable. The residual strength diagram will be different for different components of a structure and for different crack locations; permissible crack sizes will be different as well.

Knowing that the crack may not exceed a_p is of little help, unless it is known when the crack may reach a_p. The second objective of the damage tolerance analysis is then the calculation of the crack growth curve, shown diagrammatically in Fig. 2. Under the action of normal service loading the cracks grow by fatigue at an ever faster rate, leading to the convex curve shown in Fig. 2.

Starting at some crack size a_0 the crack grows to a_p. Provided that one can calculate the curve in Fig. 2, one obtains the time H of safe operation (until a_p is reached). If a_0 is for example an (assumed or real) initial defect, then the component or structure must be repaired or replaced after a time H. Alternatively, a_0 may be the limit of crack detection by inspection. This crack a_0 will grow to a_p within a time of H. Because crack

growth is not allowed beyond a_p, the crack must be detected and repaired or otherwise eliminated before the time H has expired. Therefore, the time between inspections must be less than H; it is often taken as $H/2$. In any case, the time of safe operation by whatever means of fracture control follows from H.

Concepts of Damage Tolerance Analysis

Before any fracture control can be exercised, the residual strength diagram of Fig. 1 and the crack growth curve of Fig. 2 must be calculated. Although crack growth may occur by other mechanisms than fatigue (e.g., stress corrosion, creep, etc.), the discussion in this section will be limited to fatigue crack growth.

The first step in damage tolerance analysis is the calculation of a_p, or rather, of the residual strength diagram. Usually, the residual strength diagram is expressed in terms of stress rather than load, as in Fig. 3(a). Different fracture criteria may apply (K, J, or others), depending on the material used for the structure, so the calculations of the residual strength diagram are dealt with in detail in a separate article, "Residual Strength of Metal Structures." For the purpose of this discussion, an analysis based on K is sufficient.

Using K, the residual strength, σ_c, follows from:

$$\sigma_c = K_c/\beta \sqrt{\pi a} \qquad \text{(Eq 1)}$$

where K_c is the toughness of the material, β is the geometry factor defined by the details of the structure, and a is the crack size. Given the toughness, K_c, a residual strength diagram can be calculated easily by taking different values of a and calculating σ_c. This calculation is adequate if σ_c is sufficiently below yield stresses. For high stresses (as a approaches zero), the criterion of plastic collapse must be introduced (see "Residual Strength of Metal Structures"). Once the diagram is established, the permissible crack size a_p can be read from the resulting diagram (plot) for the required σ_p (Fig. 3a).

The second step is to calculate the crack growth curve (discussed in detail in the article "The Practice of Damage Tolerance Analysis" in this Volume). The rate of crack growth is a function (f) of ΔK and R (the stress ratio) such that:

$$da/dN = f(\Delta K, R) \qquad \text{(Eq 2)}$$

The precise nature of f (such as the Paris law or the Forman equation) is discussed elsewhere in this Volume. The problem is to obtain the crack growth curve (a vs. N) by integration of Eq 2, *fatigue crack growth rates* as follows:

$$N = \int_{a_0}^{a_p} da/f(\Delta K, R) \qquad \text{(Eq 3)}$$

where $\Delta K = \beta\Delta\sigma\sqrt{\pi a}$, and where β reflects the geometry of the particular structural detail. In the general case, solution of Eq 3 can be accomplished only by numerical integration performed by a computer. (Details are provided in the article "The Practice of Damage Tolerance Analysis" in this Volume.) For the purpose of the present discussion we shall assume that the resulting crack growth curve (Fig. 3b) can be so obtained.

Once the two curves are known, decisions on how to exercise fracture control can be made in accordance with the foregoing. The residual strength analysis provides the permissible crack size, a_p, and the crack growth analysis provides the value of H, the time available to exercise fracture control.

Fracture Control Measures

When the residual strength diagram has been calculated (Fig. 3a), the maximum permissible crack size a_p follows from the minimum permissible residual strength. The other information from analysis is the crack propagation curve. It shows how a crack develops by fatigue as a function of time. The maximum permissible crack size a_p, following from the residual strength analysis of Fig. 3(a), can be plotted on the calculated crack growth curve as in Fig. 3(b).

There are several ways in which this information can be used to exercise fracture control. In all cases, the time period H is the essential information needed. The following options are available for the implementation of fracture control:

- Periodic inspection
- Fail-safe features
- Durability design or mandated retirement (safe-life approach)
- Periodic (destructive) inspection by proof testing; repair after failure in proof test (if feasible)

Damage tolerance requirements sometimes prescribe the fracture control procedure. For example, commercial aircraft requirements prescribe nondestructive periodic inspection.

Periodic Inspection. Safety is ensured when cracks are eliminated before reaching a_p. Therefore, cracks must be discovered before that point by means of periodic inspection. Whatever the inspection method, there is a certain size of crack, a_0, that is not likely to be detected. This implies that discovery and repair must occur in between a_0 and a_p as shown in Fig. 3(b). For any inspection done before or at time t_1, the crack will be missed; should the next inspection be at t_2, the crack would already be too long (having reached a_p). Hence, the inspection interval, I, must be shorter than H (inspection) = $t_2 - t_1$. It is often taken as $I = H/2$, but a more rational procedure to determine inspection intervals can be employed. Naturally, a crack once discovered must be repaired at the operator's earliest convenience. Because a_p is a permissible and not a critical crack, and because detection commonly occurs at sizes less than a_p, immediate repair may not be necessary, but any complacency will defy all analysis and inspection efforts.

Regardless of how long or short the inspection interval, safety is maintained with some reservation. Whether inspections must be performed every day (i.e., $I = 1$ and $H = 2$ days) or every year (i.e., $I = 1$ and $H = 2$ years), there would always be two inspections between a_0 and a_p. Although a daily inspection may be cumbersome, the achieved safety is not really different in the cases of daily or yearly inspections. If a crack is missed in daily inspections, a potential fracture will occur sooner, but if a crack is missed in yearly inspections, fracture will occur nonetheless before the year is over. If short inspection intervals are undesirable, one has the option of selecting a more refined inspection procedure with a smaller a_0. Then H, and hence the inspection intervals, will be longer.

It does not matter either at which time the crack initiates (Fig. 4). Inspections scheduled at an $H/2$ interval will always give two opportunities for detection, regardless of when crack growth begins, provided that inspections are scheduled at $H/2$ intervals starting from hour zero (even if initially the chance of a crack is small). Similarly, if the interval is chosen as $H/3$, there will always be three inspections between a_0 and a_p.

Fail safety (in the context of fracture control and damage tolerance) is essentially a variation of fracture control by periodic inspection: cracks or failed members must be detected and repaired. The only difference is that the structure is designed to tolerate more damage than that which is observed. Fail safety can be achieved by means of crack arresters or multiple load paths.

For example, crack arresters are designed specifically to prevent an advancing crack from reaching critical crack size. Alternatively, a pipeline or pressure vessel can be designed so that cracks will cause leaks rather than breaks (see "Leak-Before Break" in the article "Operating Stress Maps for Failure Control"). Because a leak is presumably obvious, no special inspections are necessary other than frequent checks for leaks. The same can be accomplished by providing multiple load paths (e.g., parallel redundant members). When properly designed, the structure can still sustain σ_p when one member fails. Inspection for cracks would not be required, but regular checks for failed members would be. Of course, member failure must be obvious, because a second member will soon develop cracks when it must carry additional load.

Durability (or Safe-Life) Fracture Control. A safe-life approach to fracture control is applied to structures with hard-to-detect cracks (e.g., inaccessible locations) or when "initial" crack size limits are below inspectability thresholds. If no inspections can or will be done, a small initial crack (a_i) could be assumed to exist initially in the new structure. The time, H, for the crack to grow from a_i to a_p is then the available safe life. In that case the structure or component must be retired or replaced after, for example, $H/2$ hours. This is called the durability approach.

A durability analysis (based on S-N data) and/or tests are used to establish a "design life." To account for scatter or uncertainties in the operating conditions, retirement or replacement are then mandated at fraction (e.g., 1/5 to 1/3) of that design life. Crack growth and inspection intervals are not involved. The drawbacks of this approach are:

- The critical dependence on the durability analysis and safety factor (which may be inadequate)
- The high costs of retiring components without an obvious "cause"

Nonetheless, some components (such as landing gear) are designed to safe-life criteria, because cracks are noninspectable (the materials have low toughness) and proof tests are not practical.

Another problem with the durability approach is how the size of a_i is assumed. The assumption of an initial crack size may be difficult or unclear for high-quality components (that may be essentially defect free). In welded structures the assumption of an initial flaw is more realistic. Welds often contain defects such as porosity or lack of fusion. In particular, the latter is a sharp defect equivalent to a crack of about equal size.

Destructive Inspection by Proof Testing. If the toughness is very low, the maximum permissible flaw, a_p, may be smaller than the detectable

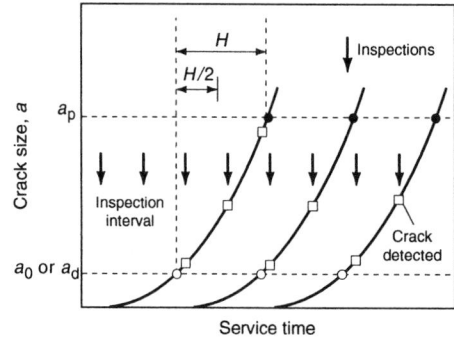

Fig. 4 Two possibilities for crack detection with $H/2$, regardless of when crack starts

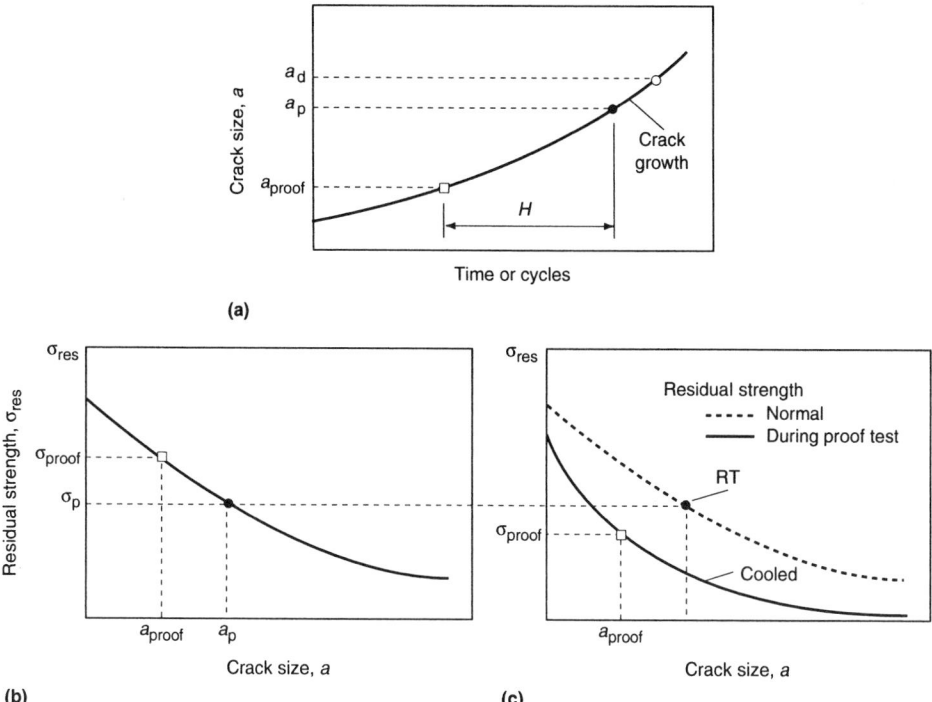

(a)

(b)   **(c)**

Fig. 5 Destructive inspection with proof test. (a) Crack growth. (b) Detection of cracks equal to or greater than proof size. (c) Lower proof load with cooling

crack, a_d (Fig. 5). This can also be the case when the structure is so large that inspections for cracks are impractical (e.g., a 1000 mile pipeline), so that the "detectable" crack size is effectively infinite. In such situations, proof testing may be another fracture control option.

As an example, consider a component that is subjected to a proof stress, σ_{proof}. Fracture will occur if a crack a_{proof} is present, as shown in Fig. 5(b). Conversely, if no fracture occurs, the maximum possible crack is a_{proof}, so that a safe operational period H (for growth from a_{proof} to a_p) is ensured. If the proof test is repeated every H hours, a period of at least H hours of safe operation is available after each successful proof test. Should failure occur during the proof test (i.e., a crack of size a_{proof} is detected), then a repair or replacement is made. The life can be extended

forever, provided that proof tests are always conducted at the proper interval, H.

Pipelines and pressure vessels are eminently suitable for the proof test approach. A line or vessel normally filled with gas or dangerous chemicals can be proof tested (hydrotested) with water. A failure during the proof test would happen under controlled circumstances, causing a water leak only. In many cases, hydrotests are already performed anyway. Selecting the proof stress level and interval on the basis of fracture mechanics analysis, to calculate H, would give these a rational foundation.

Proof tests on structures other than pressurized containers are often hard to perform. However, if the component can be easily loaded (or easily removed), the option is available and has been exercised (on wing hinges of F-111 aircraft).

Cooling the structure or component during the proof test causes a drop in toughness, which permits the use of lower proof stress to "detect" the same a_{proof} (as shown in Fig. 5c). After the test and warmup, the original toughness and residual strength are restored.

The Probability of Missing the Crack. All information needed to determine H and the inspection interval can be calculated, except for the detectable crack size. The latter must be obtained from inspection experience, which may be difficult to quantify. The length of the inspection interval (or the time period available for inspection, H) also is very sensitive to the (choice of) detectable crack size, because the slope of the crack growth curve is small for small cracks (Fig. 3b). Consequently, an *assumption with regard to the detectable crack size* may have more weight in determining the inspection interval than the painstaking and costly damage tolerance analysis. This is unsatisfactory. A more rational procedure for establishing inspection intervals is desirable.

Detection of cracks larger than the "detectable size" is not a certainty. It is affected by many factors: the skill of the inspector; the specificity of the assignment (e.g., one specific location, as opposed to a whole wing or bridge); the accessibility and viewing angles; exposure (e.g., part of a crack may be hidden behind other structural elements), possible corrosion products inside the crack; and so on. In general, discrimination of flaws depends on detection response levels (or crack size) and the level of the application noise (Fig. 6). Analysis of signal and signal plus noise are common in electronic devices, optics, and other discrimination processes. However, it is important to recognize that the dominant noise source in a nondestructive evaluation (NDE) process is not electronic noise that may be reduced by filtering, multiple sampling, and averaging techniques, but is instead the noise due to nonrelevant signals generated in applying the NDE procedure to a specific hardware element.

The outcome from an NDE measurement/decision process needs to be understood as a problem in conditional probability. When an NDE assessment is performed for the purpose of crack detection, the outcome is not a simple accept/reject

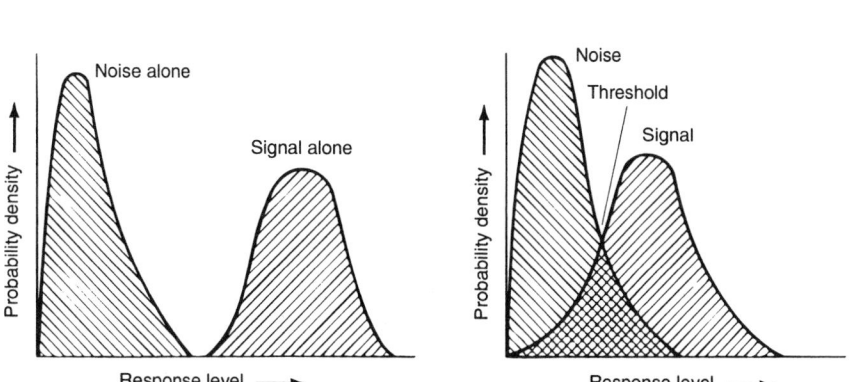

(a)   **(b)**   **(c)**

Fig. 6 Signal/noise density distribution for (a) large, (b) medium, and (c) small flaw size

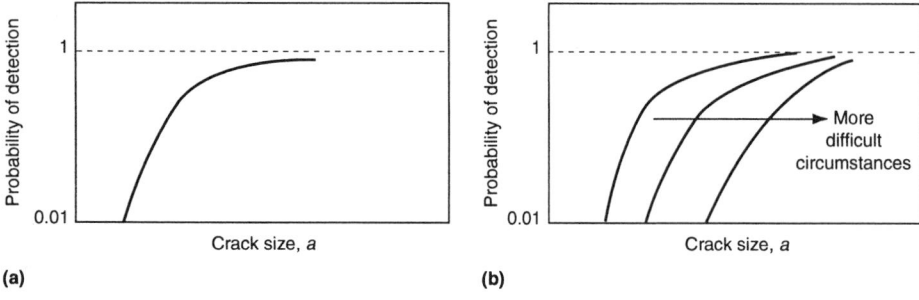

Fig. 7 Probability of crack detection in one inspection. (a) Basic curve. (b) Effect of accessibility, and specificity, or other difficulty factor

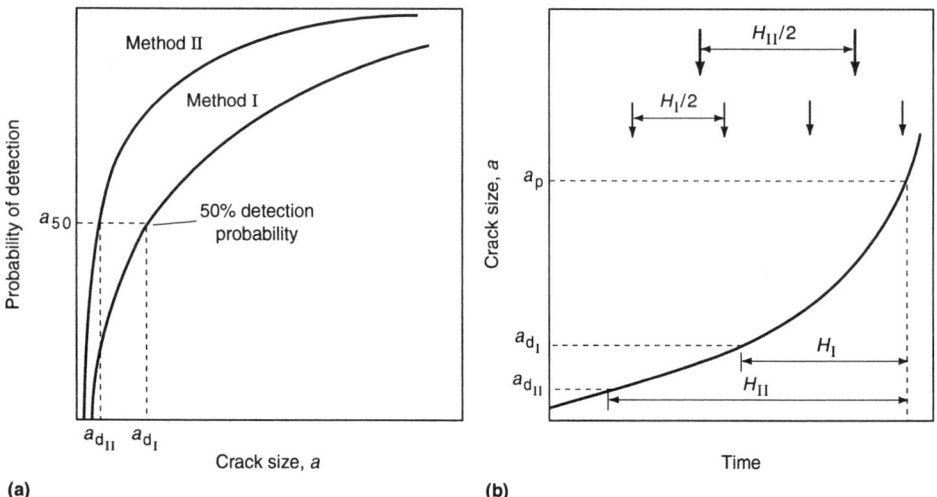

Fig. 8 Inspection intervals based on $H/2$. (a) Probability of detection for two inspection methods. (b) Inspection times on crack growth curve

(binary) process, as is frequently envisioned. It is actually the product of conditional acceptance due to the interdependence of the measurement and decision responses. In general, the four possible outcomes from an inspection process are:

- *True positive:* A crack exists and is detected.
- *False positive:* No crack exists but one is identified.
- *False negative:* A crack exists but is not detected.
- *True negative:* No crack exists and none is detected.

Typically, the probability of crack detection depends on crack size in the manner shown in Fig. 7. There is a certain crack size, a_0, below which detection is physically impossible. For example, for visual inspection this would be determined by the resolution of the eye, or for ultrasonic inspection by the wavelength. In reality, a_0 is larger than these physical limits. The probability of detection is never equal to 1, even for large cracks; any crack may be missed. It follows that the probability curve must have the general form shown in Fig. 7.

A crack is subject to inspection several times before it reaches the permissible size. At successive inspections, the crack will be longer, and the

probability of detection higher, but there is still a chance that it will go undetected. Consider 100 cracks growing at equal rate (same population), all in the same stage of growth (same size). Let the probability of detection at a certain inspection be $p = 0.2$. The probability that a crack will be missed is then $q = 1 - p = 0.8$. That means that 80 cracks will go undetected. At the next inspection the cracks are longer; let the probability of detection then be $p = 0.6$, so that $q = 0.4$. Thus of the remaining 80 cracks, $0.4 \times 80 = 32$ cracks will go undetected. Apparently the cumulative probability that a crack will be missed in successive inspection is $Q = q_1 * q_2 * \ldots * q_n$. In the above example, $Q = 0.8 \times 0.4 = 0.32$. Of the 100 cracks, 32 cracks remain undetected after two inspections. The cumulative probability of detection is $P = 1 - Q$. In the above example, $P = 0.68$. Of the 100 cracks, 68 were detected after two inspections, but 32 were missed.

Figure 8 shows what happens if inspection intervals are determined as $I = H/2$, where H is the time required for crack growth from a_d to a_p. The detectable crack size, a_d, might be selected as a crack with a certain probability of detection. For example, the detectable crack size could be defined as a_{50}, a crack with 50% probability of detection. Such a criterion certainly has appeal,

because it seems consistent, yet it still leads to inconsistencies.

For the case of Fig. 8 either method I or method II could be prescribed. The detectable crack sizes, a_{50}, lead to different inspection intervals, $H_I/2$ and $H_{II}/2$. Inspections would take place as indicated by arrows in Fig. 8(b). The two methods would be equivalent but certainly do not provide equal probability of detection, which is unsatisfactory. The probability-of-detection curves are different for the two procedures in Fig. 8. The cracks would be inspected for the first time when they still have a size smaller than a_{50}. At this first inspection, the probability of detection is not zero (unless $a < a_0$). There is a distinct probability, p_1, that the crack is already detected during that first inspection; the probability that it is missed is $q_1 = 1 - p_1$. At the next inspection the crack is larger, and the probability of detection is q_2, and so forth. By the time the crack reaches a_p, it has been inspected n times. The cumulative probability of the crack having been detected at any one of these inspections can be calculated as in the example above.

The length of the inspection interval can be established so as to provide a consistent safety level (cumulative probability of detection), independent of the shape of the crack growth curve, the accessibility, and the specificity of the inspection. The target cumulative probability of detection could be set, for example, at 95 or 98% and be specified in damage tolerance requirements. Given the calculated crack growth curve, the permissible crack size, and the probability of detection for the relevant specificity and accessibility, the cumulative probability of detection can be calculated for different lengths of the inspection interval. When the results are plotted, the interval for the desirable probability of detection can be obtained from the curve (Fig. 9). The interval will be different for different inspection methods, crack growth curves, accessibility, and specificity, but the cumulative probability of detection is always the same (equal safety). The problems with $H/2$ are then eliminated automatically (Ref 1-4).

A computer can perform the calculation for different interval lengths (Ref 5), provided that the crack growth curve calculated in the damage tolerance analysis and the applicable inspection parameters to the probability of detection curve are provided as input. Typical computer results are shown in Fig. 9(a) and 9(b) for the two crack propagation curves in Fig. 10. The criterion $I = H/2$ would assign the same inspection interval to both cracks. This would lead to different cumulative probabilities of detection for case 1 and case 2. In order to ensure the same probability in both cases, the intervals must be shorter in case 2. Although refinements can be made, the above provides a rational procedure to establish inspection intervals for which the probability of detection is independent of the inspection technique, the crack growth curve, and the assignment. The procedure is finding acceptance in the aircraft industry.

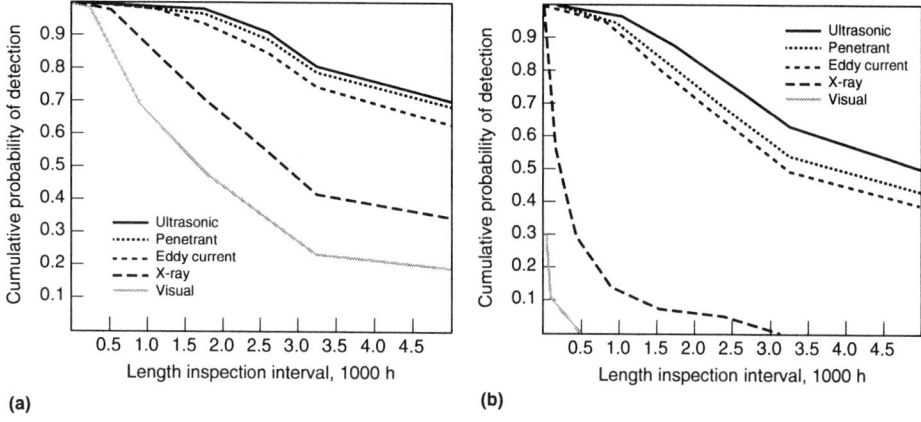

Fig. 9 Cumulative probability of crack detection as a function of the length of the inspection interval. (a) Case 1 of Fig. 10. (b) Case 2 of Fig. 10

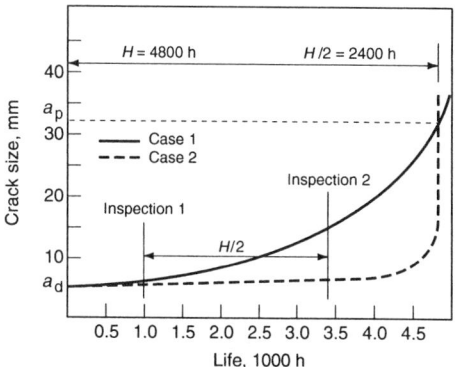

Fig. 10 Two hypothetical crack growth curves used in Fig. 9

Fracture Control Plans

The optimum fracture control plan depends on the consequences of a fracture. If the number of fractures experienced is considered to be an acceptable level with a certain fracture control plan at acceptable costs, the plan is close to optimum. Before implementation of a fracture control program, the objectives must be identified. If a structure can sustain assumed damage under an assumed loading condition, it is not necessarily safe, despite all analysis. Before defining the permissible residual strength or permissible crack size, the desired level of safety should be established, even if only qualitatively. It will appear that every component and structure calls for different fracture control requirements.

Fracture control measures must be in accordance with the acceptable risk. On the basis of rules and regulations issued by governments or engineering societies, the designer or manufacturer prescribes the details of the fracture control plan and the operator implements this plan through maintenance, inspection, repair, or replacement. The plan must be suitable for a particular structure and for the potential operators. Professional operators of pressure vessels, airplanes, and the like can implement more complex fracture control measures than the general public

operating automobiles. Here the concern is with those cases where materials selection alone does not provide adequate safeguards against fractures. Table 1 provides conceptual control plans for a variety of circumstances.

Detectable Cracks. Table 1 shows the ingredients of fracture control plans for structures in which cracks are detectable by inspection. If initial defects will not grow during service, plan I is applicable. If defects, whether initial or developing later, may grow under service loading, a crack eventually will become critical, unless it is readily discovered and repaired. Inspections should be scheduled in accordance with plan II.

Cracks Not Detectable by Inspection. Cracks may not be detectable, either because their permissible size is so small that it defies inspection, the location is not accessible, or the structure is so large that inspections are not feasible. Plan III is applicable in such cases. Plan IV involves destructive inspection by proof testing to show that no cracks larger than a_{proof} are present. If larger cracks are present, a failure will occur during the proof test, but this failure must not be catastrophic in its consequences.

After Crack Detection. Although fracture control calls for immediate repair or replacement when a crack is discovered, this is not always convenient. Large savings may be realized if remedial action can be scheduled for the next major overhaul or shutdown, or if operations can continue until a new part or component has been manufactured and received. Whether this is possible depends upon the fracture control plan in force. A well-conceived plan II already contains information on crack growth and residual strength. Using this information as an initial safeguard, operation can be continued, but the analysis should be updated and plan VI should be put into action. A crack may be discovered accidentally in a structure not subject to a fracture control plan. When no analysis is to be done, plan V is the only possible course, but it is recommended that analysis be done and that plan VI be followed. Recurrence of the incident can be prevented using plan VII.

Damage Tolerance Requirements

Requirements for Commercial Airplanes. The U.S. Federal Aviation Requirements [FAR.25b], enforced in a similar way in other countries, stipulate that the residual strength with damage present must not fall below the so-called limit load P_L, so that $P_p = P_L$. The limit load is, generally speaking, the load anticipated to occur once in the aircraft life. Given P_p, the residual strength diagram provides the maximum permissible crack size a_p.

In essence, the above is the complete requirement; little more is necessary. To satisfy the requirement, the manufacturer is obliged to design in such a manner that cracks can be detected before they reach a_p and to prescribe to the operator where and how often to inspect. Similarly, the operator is obliged to follow the manufacturer inspection instructions. Fracture control by these rules must be exercised by inspection. The excuse that some cracks are noninspectable is not justifiable (except, of course, if components are designed by safe-life criteria as in the case of landing gear). Every crack will become detectable if large enough. Thus, the requirement forces structural designs that are tolerant of damage large enough for detection, which promotes fail-safe design with multiple load path and crack arrest features.

Although there is a problem in the definition of detectable cracks, an (arbitrary) specification of detectable size would not improve the requirement, because detectability depends on the type of structure, its location, and its accessibility. The best way to determine inspection intervals is based on the cumulative probability of detection, as discussed. A useful improvement of the requirements would specify the desirable cumulative probability of inspection.

The U.S. Air Force requirements (adopted by some other forces as well but usually in a modified form) make a distinction between non-failsafe and fail-safe structures (crack arrest or multiple load path). More stringent requirements apply to non-fail-safe structures. The military requirements (MIL-A-87221, MIL-A-83444, and MIL-STD-1530) go into great detail with regard to the assumptions to be made in the analysis for initial crack size and crack development. Otherwise they are very similar to those for commercial aircraft.

ASME and Other Requirements. Other damage tolerance requirements exist for ships and offshore structures and for nuclear pressure vessels. Requirements for ships and offshore structures are issued by shipping bureaus, such as Lloyds of London, Veritas (Norway), and American Bureau of Shipping. Similar requirements exist for military ships. Those for ships are essentially preventive requirements; no analysis is necessary. Ships of a certain size and over must be equipped with so-called arrest strakes, which are located at the gunwale, at the bilge, and sometimes at mid-decks. They are longitudinal strakes of a higher-toughness material than the normal hull plating.

Table 1 Fracture control plans

Plan	Condition	Fracture control methods
Plans for anticipated cracks detectable by inspection		
Plan I	For initial defect not expected to grow by fatigue	• Calculate permissible size of defect. • If stress corrosion can occur, calculate which size of defect can be sustained indefinitely given the K_{Iscc} of the material. • Inspect once using a technique that can reliably detect defects of the calculated sizes. • Eliminate all detected defects larger than those of the calculated sizes.
Plan II	For all defects (initial or initiating later) that will grow during service—these will reach critical size	• If possible, show by analysis (or tests) that the structure can sustain without failure such large defects that the damage will be obvious (e.g., readily apparent leak or failed component; fail safety). Repair when damage is discovered. • If the above cannot be shown, calculate permissible crack size. • Establish crack size that can be detected reliably. • Calculate time for crack growth. • Implement periodic inspection based on crack growth calculation. • Start inspection immediately because time of crack initiation is not known. • Repair or replace when crack is detected.
Plan for anticipated cracks not detectable by inspection		
Plan III	For parts where a is so small that it defies inspection	• Make best estimate of possible initial defects. • Calculate permissible crack size a_p. • Calculate crack growth life from initial defect size to a_p. • Replace/retire after calculated life expires (using adequate factor).
Plan for structures for which inspection is not feasible, but proof testing is possible		
Plan IV	For components or structures that can be proof tested and where failure during proof testing is not a catastrophe (e.g., leak before break)	• Determine feasible proof test pressure or load. • Calculate maximum crack size a_{proof} that could be present after proof test. • Calculate maximum permissible crack a_p. • Calculate crack growth time, H, from a_{proof} to a_p. • Repeat proof test before H has expired (using adequate factor).
Plans for cracks discovered in service		
Plan V	For detected cracks for which no analysis is done	• Repair or replace unconditionally.
Plan VI	For detected cracks for which analysis is done (if immediate replacement is impractical)	• Show by analysis that a larger defect can be sustained. • Check growth daily; drill stop hole if permissible (as an interim measure). • Prepare for repair or replacement at earliest convenience. • Determine exact size and shape of crack. • Find materials data; if possible, cut test specimens from structure. • Obtain reliable load and stress information. • Calculate a_p. • Calculate time, H, for growth to a_p. • Prepare for repair or replacement before H (with adequate factor) expires. • If crack grows faster than calculated, update prognosis and speed up replacement or repair actions. If possible, reduce operational loads. Repair or replace as soon as possible.
Plan VII	For structures identical to those in which a crack was previously detected	• Use parts of cracked or failed structure to obtain material properties. • Implement Plan II, III, or VI.

The damage tolerance requirements for nuclear pressure vessels are contained in the ASME boiler and pressure vessel code, section XI and its appendix A. The requirements essentially provide acceptance limits for cracks detected in service. A variety of possible crack configurations and locations are identified. The requirements then provide the crack sizes for each case that may be left unattended. Should a detected crack exceed the prescribed limits, one has two options: repair weld or perform analysis.

If the option to perform analysis is selected, the following damage tolerance requirements apply: $K <$ Arrest toughness/$\sqrt{10}$ for upset conditions, and $K <$ Toughness/$\sqrt{2}$ for emergency and faulty conditions, where K is the stress intensity at these conditions. Strangely enough, these requirements are expressed in terms of the stress intensity and toughness. However, with the fracture condition of $K =$ Toughness (in the upset condition, at the stress σ_{cu}) fracture would occur if $\beta_p \sigma_p \sqrt{\pi a_p} =$ Toughness. If the actual stress intensity must be

smaller by a factor $\sqrt{10}$, it follows that $\sigma_{cu}/\sigma_p \approx \sqrt{10} = 3.16$, assuming $\beta_p \approx \beta_{cu}$. This means that a safety factor of 3.16 must remain with regard to upset conditions. Thus, the requirement can be stated in terms of the minimum permissible residual strength σ_p. If follows that $\sigma_p = \sigma_{cu}/3.16$, in accordance with the previous discussions. This case is displayed in Fig. 11(a) using the same nomenclature as before.

At the same time, the stress intensity must be less than the toughness divided by $\sqrt{2}$ for the emergency conditions (stress σ_{ce}). Using the same arguments as above, the minimum permissible residual strength, σ_p, must provide a safety factor of $\sqrt{2} = 1.41$ with regard to upset conditions. This is shown in Fig. 11(c).

Different toughnesses are used in the two cases, the arrest toughness and the regular toughness. Because the former is less than the latter, there are effectively two residual strength curves in play, as shown in Fig. 11. This does not change the principle of the analysis. In either case the permissible crack size a_p follows from the residual strength diagram as shown. Obviously, it is impossible to satisfy both requirements exactly. One is always more severe than the other. If it can be foreseen which of the two generally is the severest, the requirement can be simplified.

The requirement presents an alternative. Instead of the above, one may satisfy the following requirements:

- $a_p \leq a_{cu}/10$ upset and operating conditions
- $a_p \leq a_{ce}/2$ emergency conditions

where a_{cu} is the critical crack (causing fracture at upset conditions and a_{ce} is the critical crack in emergency conditions. (In the ASME code, a_p is denoted a_f.) Thus:

$$a_{cu} = 10a_p \tag{Eq 4}$$

$$K = \beta_{cu}\sigma_{cu}\sqrt{\pi \times 10a_p} = \beta_p\sigma_p\sqrt{\pi a_p} \tag{Eq 5}$$

so that

$$\beta_p\sigma_p/\beta_{cu}\sigma_{cu} = \sqrt{10} = 3.16 \tag{Eq 6}$$

This requirement will be identical to the one stated previously if $\beta_p = \beta_{cu}$. Because a_{cu} is a longer crack than a_p, in general $\beta_{cu} > \beta_p$, so that the requirement leads to a safety factor somewhat smaller than 3.16. The same arguments hold for a_p and a_{ce}. Both sets of requirements apparently attempt to set the same conditions, and only one is necessary.

Once a crack is detected and analysis is preferred instead of immediate repair, crack growth must be analyzed as well. Fatigue crack growth must be calculated starting at a_d, which is the crack actually present and discovered, and continuing over the period until the next inspection (shutdown). Over this period the crack may not grow beyond a_p as determined by the criteria discussed above. This condition is shown in Fig. 11(b) and 11(d). The requirements fully prescribe the analysis procedure as well as the toughness and rate data. (Use of other analysis and data is

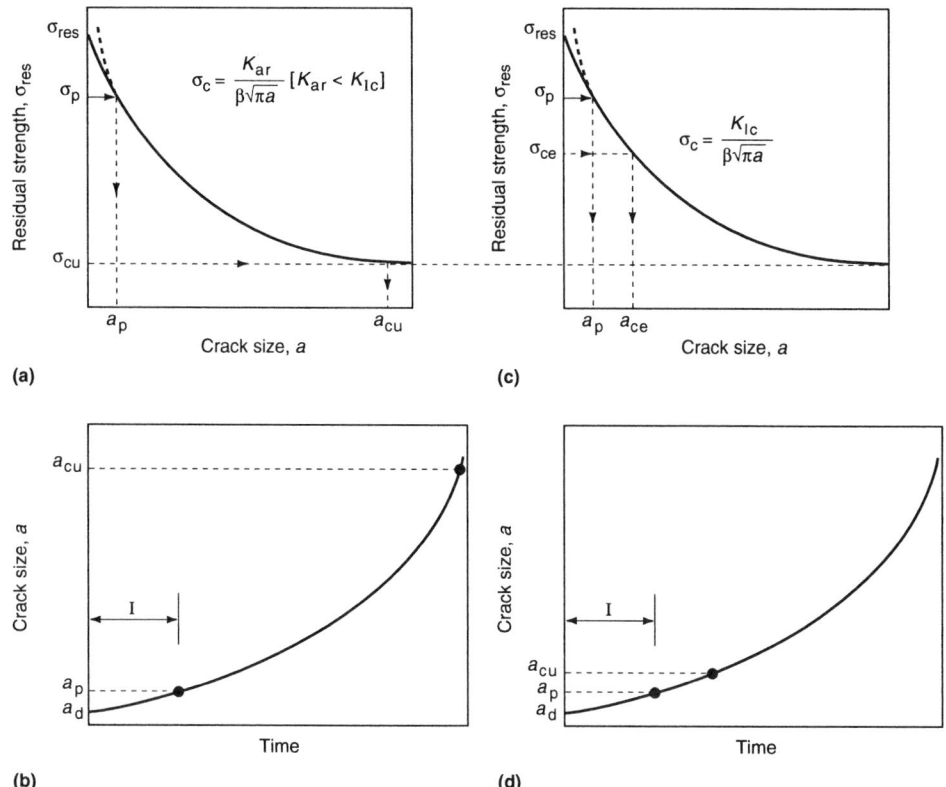

Fig. 11 ASME requirements for: (a) Permissible crack size (a_p) for upset condition (where K_{ar} is arrest toughness). (b) Crack growth for upset condition. (c) Permissible crack size for emergency condition (I = inspection interval). (d) Crack growth for emergency condition. (Note: Permissible crack size, a_p, is actually denoted as a_i in the ASME code.)

subject to approval by authorities.) The analysis procedure is mostly in agreement with general practice (Ref 6).

The requirement is specifically for a case where a crack is detected in service. One may then prove by analysis that this crack is not dangerous during further operation until the next shutdown. In the ASME requirements, the damage tolerance analysis is used to decide whether a structure with a *known crack* can be left in service without repair. Besides, analysis is not used to determine the inspection interval. The approach for aircraft has an important difference, in that cracks must be repaired and the analysis is used to ensure detection and repair a *potential crack*, not to determine whether it is safe to fly with a

known crack. This statement is not meant as criticism, but rather as a clarification of the difference in approach.

Fracture Mechanics and Fatigue Design

Fracture mechanics does not provide revolutionary new insights for fatigue design. It merely confirms what has been known for many years. Probably the most important rule in designing for high-fatigue performance is that stress concentrations should be kept at a minimum. Fracture mechanics merely emphasizes the significance of this. Without a stress concentration, the stress

intensity for a small crack ($\beta \approx 1$) is $K = \sigma\sqrt{\pi a}$. For a small crack emanating from a notch or fillet, the geometry factor approximately equals the stress concentration factor: $\beta \cong k_t$, putting the stress intensity at $K = k_t\sigma\sqrt{\pi a}$.

Because fatigue crack growth depends roughly on the third power of K, small cracks at stress concentrations grow faster by about k_t^3. Thus, a crack emanating from a simple hole grows about nine times faster than one of equal size starting at a smooth surface. Most service cracks start at stress concentrations, and the implications are self-evident.

The sole objective of damage tolerance analysis is to establish fracture control measures so that cracks can be eliminated before they become dangerous, by either repair or replacement of the component. There is no excuse for a fracture that results from known cracks, regardless of what analysis predicts.

If crack discovery demands repair, a new problem arises. Not only must the repair be adequate to restore strength, it must also be analyzed for damage tolerance again.

A repair cannot be treated too casually. A simple cover plate usually does not suffice; such a repair may rather aggravate the situation and cause new cracks in due time. Repairs must be designed to cause gradual transfer of loads to reduce the stress concentration. Fracture control measures must be reinstated for the repair.

The time available for fracture control is the time of safe operation, (H), which is governed by the residual strength and the crack growth curve. If H is short, frequent inspections must be scheduled, or the component must be replaced soon. As long as all fracture control decisions are indeed based on H, safety will be maintained. But long inspection intervals or replacement times are desirable from an economic point of view. The question then is which measures can be taken to improve the situation when H is too small to be economically acceptable. Possible options (Fig. 12) are summarized below.

Option (a): Use of a Material with Better Properties (Fig. 12b). A higher toughness will provide a somewhat larger a_p, but generally speaking it is not of great influence on H; most of the life is in the early phase of crack growth. Increasing toughness (a_p) affects only the steeper part of the curve, which in general has only a small effect.

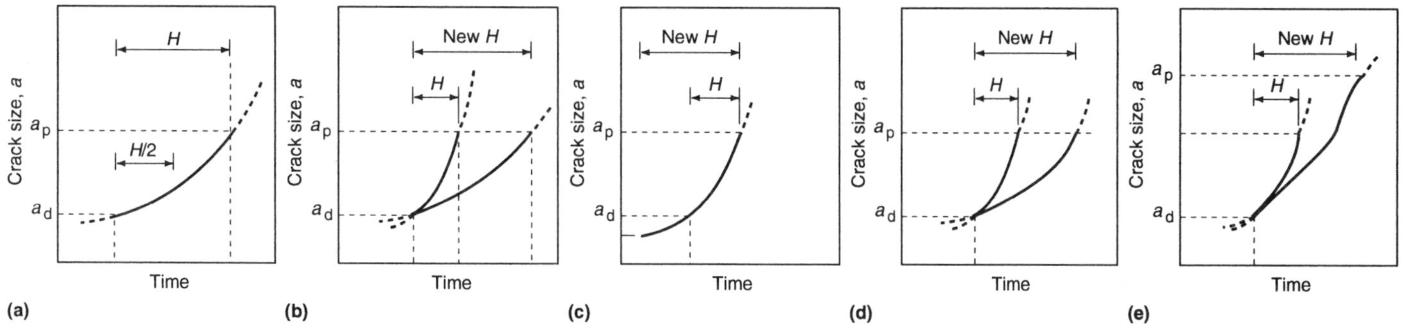

Fig. 12 Ways to increase inspection interval. (a) Base case. (b) Use of better material. (c) Use of more sensitive inspection method. (d) (Detail) design with lower stress. (e) Redundancy (fail-safe) or crack arresters

Option (b): Selection of a Better Inspection Procedure (Fig. 12c). Improving the inspection technique, (e.g., by selecting a more sophisticated inspection procedure) reduces detectable crack sizes. This usually has a significant effect on *H* because of the small slope of the initial part of the crack growth curve. The penalty will be a more difficult inspection, but the inspection interval will be longer, so fewer inspections will be needed.

Option (c): Redesign or Lower Stress (Fig. 12d). The crack growth curve is governed by the stress intensity. Reducing the stress by 15%, for example, will reduce *K* by 15%. Because crack growth rates are roughly proportional to K^3, a 15% reduction in stress will increase *H* by a factor $1.15^3 = 1.5$. Such a stress reduction seldom requires a general "beef up" of the structure; cracks occur where the local stresses are high, and the stress reductions are needed only locally. Reduction of stress concentrations, larger fillet radii, and less eccentricity will add hardly any material, cost, or weight.

Redesign may also affect *K*; a reduction in β is just as effective as a reduction in σ. In the above example of the reduced stress concentration, the effect is actually in $\beta(k_t)$ instead of in σ (the nominal stress does not change). The redesign may be in the production procedure, so that cracks occur in the cross-grain direction instead of along exposed grain boundaries.

Option (d): Providing Redundancy and Arresters (Fig. 12e). Building the structure out of more than one element provides multiple load paths. In a well-designed multiple load path structure, only inspections for a failed member may be necessary, provided that the fasteners or welds can transfer the load of the failed member by shear.

All of these options can be exercised during design. It is crucial, therefore, that damage tolerance analysis commence in the early design phase when modifications are still possible. Once the design is finalized, the options for improvement are drastically reduced. For finalized designs and existing structures, Option (b) (Fig. 12c) is often the only recourse, although doublers or arresters sometimes can be added later.

REFERENCES

1. D. Broek, Fracture Control by Periodic Inspection with Fixed Cumulative Probability of Crack Detection, *Structural Failure, Product Liability and Technical Insurance*, Interscience Enterprises Ltd., 1987, p 238-258
2. W.H. Lewis et al., *Reliability of Non-destructive Inspections*, SA-ALC/MME 76-6-38-1, 1978
3. E. Knorr, *Reliability of Detection of Flaws and of the Determination of Flaw Size*, AGARDograph 176, 1974, p 396-412
4. U. Goranson, paper presented at ICAF meeting, Toulouse, 1983
5. D. Broek, Damage Tolerance Analysis Software, Fracture REsearch Inc., 1995
6. D. Broek, *The Practical Use of Fracture Mechanics*, Kluwer Academic Publishers, 1986

SELECTED REFERENCES

1990-1994

- H. Ansell et al., "A Manufacturer's Approach to Ensure Long Term Structural Integrity," CP-3160, National Aeronautics and Space Administration, 1992, p 379-405
- J.M. Gaillardon et al., "Ageing Airplane Repair Assessment Program for Airbus A300," CP-3160, National Aeronautics and Space Administration, 1992, p 283-289
- R. Johnson, Aging Aircraft and Airworthiness, *Aerospace Engineering*, Vol 12 (No. 7), July 1992, p 23-30
- M.K.S. Madugula, Design against Fatigue and Fracture in Steel Structures: An Overview, *Fatigue and Fracture in Steel and Concrete Structures*, Vol 2, Conf. Proc. ISFF '91, Oxford & IBH Publishing Co., New Delhi, India, 1991, p 1261-1276
- H.I. McHenry and R.M. Denys, Measurement of HAZ Toughness in Steel Weldments, *Application of Fracture Mechanics to Life Estimation of Power Plant Components*, Conf. Proc., EMAS Ltd., 1990, p 211-222
- C. Miki, Applications of NDI in Fracture Control of Honshu-Shikoku Bridge Project, *Review of Progress in Quantitative Nondestructive Evaluation*, Conf. Proc., Plenum, 1994, p 2071-2082
- A.C. Pickard et al., Crack Detection and Crack Length Measurement in the Gas Turbine Industry, *Fatigue Crack Measurement: Techniques and Applications*, Engineering Materials Advisory Services Ltd., 1991, p 457-485
- T. Swift, Damage Tolerance Capability, *International Journal of Fatigue*, Vol 16 (No. 1), 1994, p 75-94

1985-1989

- J. Bernardi et al., Sizing Radial Cracks in Bore Holes by Eddy Current, *16th Symposium on Nondestructive Evaluation*, Southwest Research Institute, NTIAC, 1987, p 194-202
- P.M. Besuner et al., "Fracture Control for Fixed Offshore Structures," SSC-328, Ship Structure Committee, U.S. Coast Guard, 1985
- C.M. Branco, Critical Analysis of Flaw Acceptance Methods, *Fracture Mechanics Methodology: Evaluation of Structural Component Integrity*, 1984, p 151-168
- D. Broek, Fracture Control by Periodic Inspections with Fixed Cumulative Probability of Crack Detection: Rational Use of Fracture Mechanics Analysis, *Structural Failure, Product Liability and Technical Insurance*, Conf. Proc., Interscience Enterprises, Geneva, Switzerland, 1987, p 338-358
- T.R. Brussat et al., Damage Tolerance Assessment of Aircraft Attachment Lugs, *Eng. Fract. Mech.*, Vol 23 (No. 6), 1986, p 1067-1084
- A.G. Denyer, Aircraft Structural Maintenance Recommendations Based on Fracture Mechanics Analysis, *Case Histories Involving Fatigue and Fracture Mechanics*, Conf. Proc. ASTM, 1986, p 291-310
- L. Faria, Reliability in Probabilistic Design, Martinus Nijhoff Publishers, The Hague, The Netherlands, 1984, p 169-174
- D.W. Hoeppner, "Parameters That Input to Application of Damage Tolerance Concepts to Critical Engine Components," CP-393, AGARD (NATO), Oct 1985, p 4.1-416
- T.E. Kirchner et al., Crack Measurement Techniques for High-Temperature Applications, *Exp. Tech.*, Vol 12 (No. 12), Dec 1988, p 26-31
- R.D. Streit, Requirements of Quantitative NDE in Developing Fracture Control Plans, *Review of Progress in Quantitative Nondestructive Evaluation*, Conf. Proc., Plenum, 1985, p 1305-1313
- J.-M. Thomas and B. Heciak, Taking Account of Damage in the Sizing of Aircraft Parts: Prediction of Lifetime, *Mec. Mater. Electr.*, Jan-Feb 1987, p 59-62
- M. Torres and B. Plissonneau, "Repair of Helicopter Composite Structure: Techniques and Substantiations," CP-402, AGARD (NATO), Oct 1986, p 6.1 to 6.21

SUPPLEMENTARY REFERENCES*

Listed below are additional monographs, textbooks, and papers on fracture mechanics and damage tolerance. The main contents of each reference are summarized in brackets next to it. Though not mentioned here, the proceedings from annual or biannual conferences such as ICAF and ASIP are also excellent information sources.

- P.R. Albekis and C.M. Hudson, Ed., *Design of Fatigue and Fracture Resistant Structures*, ASTM STP 761, 1982 [Several papers on durability and damage tolerance, design and analysis; aerospace, nuclear, ground vehicle applications]
- C. Amzallag et al., Ed., *Low-Cycle Fatigue and Life Prediction*, ASTM STP 770, 1982 [Several papers on engine materials, temperature/environmental effects, short cracks, low-cycle fatigue and durability; aerospace and other applications]
- H.E. Boyer, Ed., *Atlas of Fatigue Curves*, American Society for Metals, 1986 [Fatigue and crack growth data collected from the literature]
- D.F. Bryan and J.M. Potter, Ed., *Effect of Load Spectrum Variables on Fatigue Crack Initiation and Propagation*, ASTM STP 714, 1980 [Papers dealing with damage tolerance, spectrum loading; mostly aerospace and aircraft engine applications]
- J.B. Chang and C.M. Hudson, Ed., *Methods and Models for Predicting Fatigue Crack Growth under Random Loading*, ASTM STP 748, 1981 [Papers dealing with spectrum loading, crack closure model, and crack retardation phenomena]

*Supplementary references provided by Antonio Rufin, Boeing, with assistance from Fatigue Technology Inc., Seattle.

- J.B. Chang and J.L. Rudd, Ed., *Damage Tolerance of Metallic Structures*, ASTM STP 842, 1984 [Papers on damage tolerance methodology, fatigue and fracture; aerospace applications]
- *Damage Tolerance Design Handbook*, Parts I and II, Metals and Ceramics Information Analysis Center, 1972 (updated 1973 and 1975) [Fracture and crack growth data for common aerospace structural materials]
- R.P. Gangloff et al., Fatigue Crack Propagation in Aerospace Aluminum Alloys, *AIAA Journal of Aircraft*, Vol 31 (No. 3), 1994 [Review of damage tolerance methods and data on aerospace aluminum alloys, reference materials]
- R.W. Hertzberg, *Deformation and Fracture Mechanics of Engineering Materials*, 3rd ed., Wiley, 1989
- L.F. Impellizzeri, Ed., *Cyclic Stress-Strain and Plastic Deformation Aspects of Fatigue Crack Growth*, ASTM STP 637, 1977 [Crack tip plasticity, small flaws, low-cycle fatigue, and overloads]
- J.F. Knott, *Fundamentals of Fracture Mechanics*, Butterworth, 1973
- A.S. Kobayashi, Ed., *Experimental Techniques in Fracture Mechanics*, Vol I and II, Society for Experimental Stress Analysis, 1973 [Several articles covering fundamentals of fracture mechanics, analysis methods, and experimental techniques]
- H. Liebowitz, Ed., *Fracture Mechanics of Aircraft Structures*, AGARD Monograph 176, 1976 [Introduction to fracture mechanics, damage tolerance, fail-safe concepts, non-destructive evaluation]
- *Metallic Materials and Elements for Aerospace Vehicle Structures*, MIL-HDBK-5F, 1990 [Fatigue and fracture data for aerospace structural materials]
- G.R. Neegard, "Guide to ASIAC Computer Programs," WL-TR-94-3127, Air Force Wright Labs, 1994 [Listing and descriptions of damage tolerance analysis computer programs available from the U.S. Air Force]
- J.M. Potter, Ed., *Fatigue in Mechanically Fastened Composite and Metallic Joints*, ASTM STP 927, 1986 [Several papers on joint durability]
- *Proceedings of the FAA/NASA International Symposium on Advanced Structural Integrity Methods for Airframe Durability and Damage Tolerance*, NASA CP 3274, 1994 [Several papers on durability and damage tolerance design and analysis; aircraft structures]
- T.P. Rich and D.J. Cartwright, Ed., *Case Studies in Fracture Mechanics*, MS 77-5, Army Mechanics and Materials Research Center, 1977 [Papers on fracture mechanics, joints, and materials; aerospace, pressure vessels, rotating machinery, and ground vehicles]
- M.S. Rosenfeld, Ed., *Damage Tolerance in Aircraft Structures*, ASTM STP 486, 1971 [Papers on fatigue crack propagation, damage tolerance, and fail-safe design]
- L. Schwarmann, *Material Data of High-Strength Aluminium Alloys for Durability Evaluation of Structures*, Aluminium-Verlag, Düsseldorf, 1985 [Fatigue and fracture data for aluminum alloys]
- Crack growth analysis computer programs include CRKGRO (USAF), NASA FLAGRO (NASA), NASCRAC (Failure Analysis Associates), FASTRAN (COSMIC/University of Georgia), and a crack growth program by J. Gallagher, University of Dayton Research Institute

The Practice of Damage Tolerance Analysis

David Broek, FractuREsearch

DAMAGE TOLERANCE ANALYSIS consists of three parts:

- Calculation of the residual strength diagram to obtain the permissible crack size
- Calculation of the crack growth curve
- Calculation of the inspection interval

The last item is discussed in the article "Concepts of Fracture Control and Damage Tolerance Analysis" in this Volume.

In most cases the residual strength analysis is performed with linear elastic fracture mechanics and is based on the stress-intensity factor, K, and the toughness (the plane strain fracture toughness, K_{Ic}, or the plane stress fracture toughness, K_c). Linear elastic conditions are assumed for this article. Residual strength analysis on the basis of elastic-plastic fracture mechanics and plastic fracture mechanics is discussed in the article "Residual Strength of Metal Structures" in this Volume.

In this article it is assumed that crack growth is predominantly by fatigue rather than by corrosion, creep, or other mechanisms. This does not exclude the combination of (mechanical) fatigue and stress corrosion (i.e., corrosion-assisted fatigue). After all, fatigue crack growth is almost always corrosion assisted, because at least one corrodent, humid air, is usually present. It is the occurrence of cyclic loading, rather than the presence of a corrodent, that puts crack growth in the category of fatigue. The response of the material, in terms of the crack growth rate, da/dN, includes the effect of corrosion.

Input to the Analysis

Damage tolerance analysis, whether of crack growth or residual strength (with the limitations expressed) is based on the stress-intensity factor, K:

$$K = \beta\sigma\sqrt{\pi a} \qquad \text{(Eq 1)}$$

where σ is the applied stress, a is the crack size, and β is the nondimensional geometry factor that incorporates all the geometrical complications of the structure at hand.

For residual strength analysis, the fracture criterion is usually:

$$K = K_{Ic} \text{ (or } K_c) \qquad \text{(Eq 2)}$$

in which K_{Ic} is the plane strain fracture toughness and K_c is the plane stress fracture toughness. To obtain the residual strength diagram, the following equation must be solved for a number of a values:

$$\sigma_c = K_{Ic}(\text{or } K_c)/\beta\sqrt{\pi a} \qquad \text{(Eq 3)}$$

However, in general the establishment of the residual strength diagram is more complicated than this, as explained in the article "Residual Strength of Metal Structures" in this Volume.

In fatigue, the crack growth data are expressed as a function of the stress-intensity range, ΔK, and the stress ratio, R, as:

$$da/dN = f(\Delta K, R) \qquad \text{(Eq 4)}$$

Calculation of the crack growth curve requires integration of Eq 4 with K expressed per Eq 1. The expression for ΔK includes $\Delta\sigma$ as per Eq 1, where $\Delta\sigma$ is the stress range in the cycle (or any particular cycle). Few structural parts are subjected to constant-amplitude loading. In general, the loading is random or semirandom, so that $\Delta\sigma$ is different from cycle to cycle. Thus, the following input is required for the analysis:

- Material data in terms of K_{Ic} (or K_c), rates in terms of rate diagram, $da/dN = f(\Delta K, R)$, and the yield strength, F_{ty}
- The geometry factor, β, as a function of crack size for the structure at hand
- The stress history (i.e., the sequence of stress values, $\Delta\sigma$, to which the detail in question will be subjected)

The following discussions briefly consider the crucial aspects of obtaining material data and geometry factors, then turn to the problem of crack growth analysis through the integration of Eq 4. This integration can seldom, if ever, be performed in closed form; therefore, numerical integration by computer is indicated. Most crack growth analysis computer programs contain extensive libraries of material data and geometry factors, which tend to give the user a false sense of security. *The selections made by the user* from these libraries are what make or break the analysis. The damage tolerance analyst must be thoroughly familiar with what follows in this article and would do well to become versant with other aspects of the problem (Ref 1).

Material Data

Damage tolerance analysis is based on material property data such as fracture toughness, K_{Ic} or K_c, and yield strength, F_{ty}. Fracture toughness is relevant for the calculation of residual strength, but otherwise it has no effect on the calculated crack growth curve (unless crack growth is described from curve-fitting models such as the Forman equation or the Collipriest equation).

Innocuous as it may seem, the yield strength has an overriding effect if the crack growth analysis must account for crack retardation after overloads or in low-cycle fatigue. This issue is so complicated that the reader is well advised to consult the literature (e.g., Ref 1, 2) before attempting to deal with it. Retardation may well change the result of the crack growth analysis by a factor of 2 or more, depending on the yield strength value used. However, most analysts prefer to ignore crack retardation because doing so is more conservative. That being the case, the only input of relevance is the rate diagram for the material, of which a typical example is shown in Fig. 1.

Curve Fitting da/dN Data. Data sets such as that shown in Fig. 1(b) must be interpreted before they can be used in analysis. Clearly, the "scat-

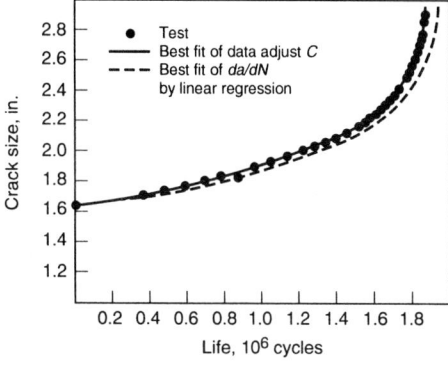

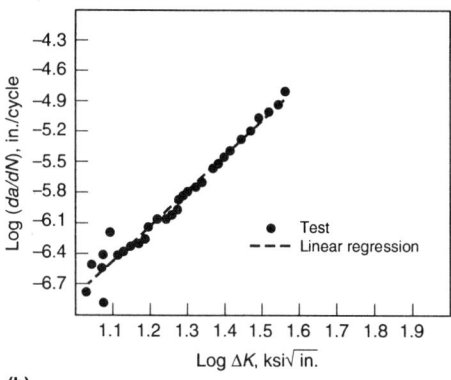

Fig. 1 Finding the best representation of rate data (ASTM A 533B steel at 26 °C, or 500 °F). (a) Re-prediction of crack growth. (b) Rate data

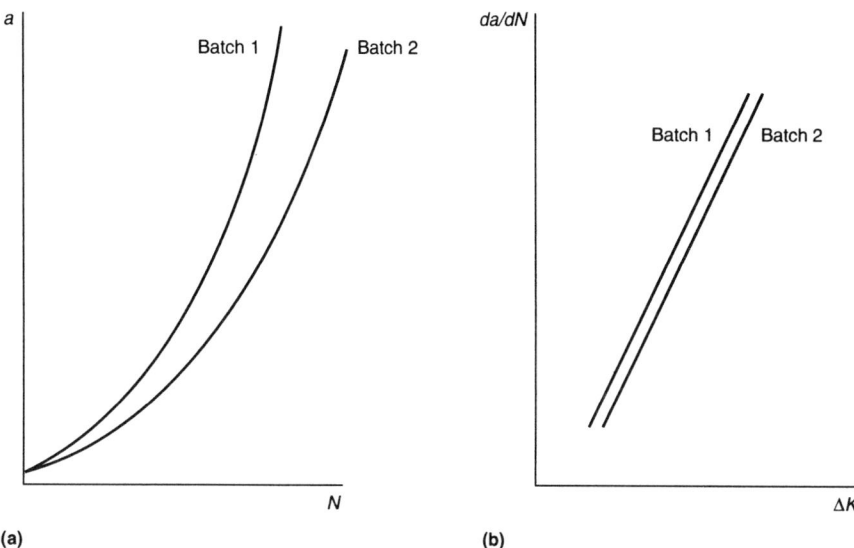

Fig. 2 Typical true scatter in fatigue crack growth. (a) Crack growth curves. (b) Rates

tered" data points cannot be used directly. In some cases a growth rate equation may be desirable; in particular, if the data appear to fall on straight lines, the Paris or Walker equation may be convenient (Ref 1, 2). Most computer programs have options for the use of a number of equations but also permit the use of tabular data. The latter eliminates the need for force-fitted (sometimes poorly fitted) equations.

All equations derive from "curve" fitting and have no physical basis. None of them is fundamentally better than any other, and none is more universally useful than any other. The most appropriate equation is the one that gives the best fit for the case at hand. Fits are often obtained by using a least squares fit of the $da/dN - \Delta K$ data. This is all that can be done if the original raw test data (a versus N) are not available (the original crack growth data usually are not reported in handbooks or other literature). The best fit through the $da/dN - \Delta K$ data must then be used, although the "best" fit may not give the best predictions if the analysis is required to cover a wide range of ΔK values.

A measured crack growth curve is shown in Fig. 1(a) and the da/dN data are shown in Fig. 1(b). In this case a straight line is appropriate (Paris equation, $da/dN = C_p(\Delta K)^{m_p}$. A least squares fit of these data provides $C_p = 6.496 \times 10^{-11}$ and $m_p = 3.43$. When these parameters are used in a crack growth analysis to re-predict the original crack growth curve, the result is as shown in Fig. 1(a), which is certainly not the best fit to the actual crack growth curve. In this case the curvature of the predicted curve seems appropriate (the value of m_p is correct), but there is a more or less proportional error, which can be corrected by adjusting C_p. The predicted life is too long as compared to the test life by a factor of 1,937,483/1,871,080 = 1.035. Multiplication of C_p by this factor ($C_p = 1.035 \times 6.496 \times 10^{-11} = 6.72 \times 10^{-11}$) will result in a better prediction, as shown in Fig. 1(a).

The regression fit of the da/dN data is not necessarily the best fit for analysis. In curve-fitting procedures, every data point gets equal weight, yet the points for high da/dN affect only a small portion of the life. The points for low da/dN are the relevant data during most of the life and should weigh more in the curve fitting. In the above example, the curvature of the predicted curves was appropriate, and only C_p needed adjustment. If the curvature is found to be incorrect, m_p must be adjusted first by trial and error. This usually causes wild gyrations in the predicted curve, but these can be ignored. The objective is to arrive at a line with the proper curvature, which will have the proper m_p. Once this m_p value is found, the curve can be adjusted by adjusting C_p in the manner discussed.

Clearly, fitting an equation is not trivial; it requires judgment and re-prediction of the original data using a predictive computer code. But if the original crack growth curve is not available, there is no other option than to fit the da/dN data as well as possible. This can be done by regression, but regression analysis may be a refinement and not necessarily an improvement over a hand-drawn best fit line.

Dealing with Scatter in Fatigue Crack Growth Data. Statistical procedures used to account for scatter are seldom in accord with physical reality. Most statistical procedures address data analysis without direct consideration of the physics and mechanics of the problem. For exam-

ple, the 90% confidence curves are commonly determined by using the individual da/dN data points as the statistical population sample, yet applying the correct mathematics does not lend credence to the physical result.

The three main sources of scatter are:

- Consistent differences between heats A and B of the same alloy
- Local differences due to inhomogeneities and "weak spots"
- Measurement errors

Essentially, only the first source of scatter is relevant; the other two have little bearing on the problem.

Consistent differences between various heats or batches of materials, and to a certain degree differences from location to location in one batch or plate, will be reflected in a more or less consistent difference in crack growth curves. As a consequence, the rate also will be consistently different, as shown in Fig. 2. This is true material scatter and must be accounted for in an analysis, because the exact properties of the batch used in the structure are not known.

Scatter due to inhomogeneities and "weak spots" may occur throughout the material, but only locally. At some locations the crack will accelerate, but soon afterward it will resume normal behavior. Another crack (in a different specimen) will encounter such "weak" spots at other locations and will speed up locally (at different stages in life than the first one), then resume normal growth. Similarly, it will sometimes slow down locally. On the whole, the two crack growth curves will be identical, as shown in Fig. 3(a). Also, the rate data will be identical, except that each test will provide a few outlying data points due to local higher or lower rates, as indicated in Fig. 3(b). The outlying data points occur at different locations in the two tests. If, instead of two tests, many tests are performed on specimens from the same plate, more and more outlying data

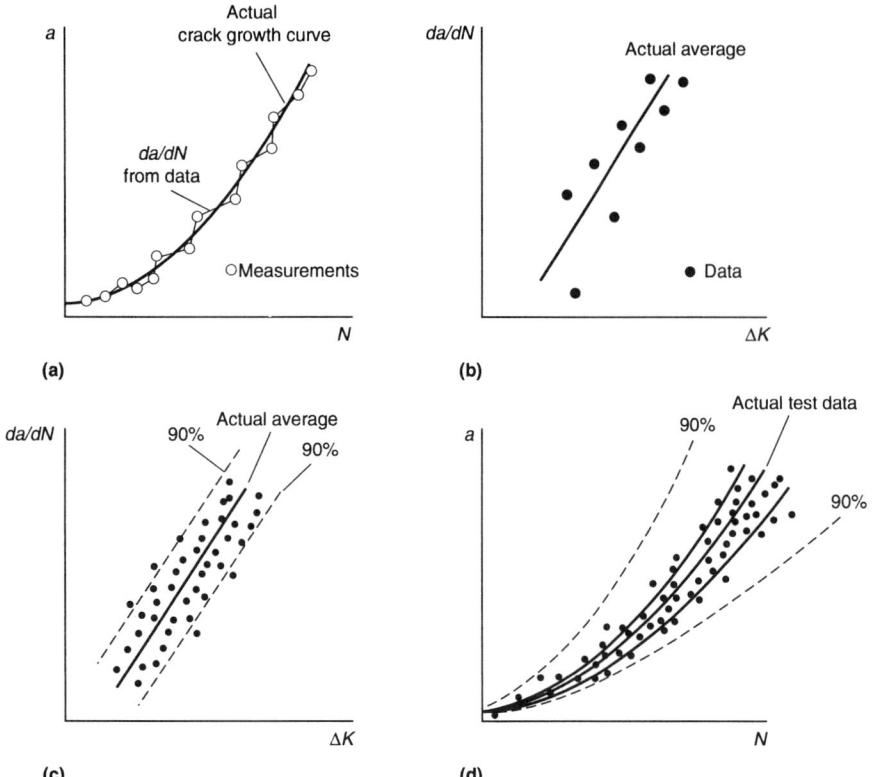

Fig. 3 Scatter due to inhomogeneities and measurement of fatigue crack growth. (a) Data. (b) Rates of crack growth. (c) Scatterband. (d) Re-prediction of crack growth curves versus cycles *(N)*

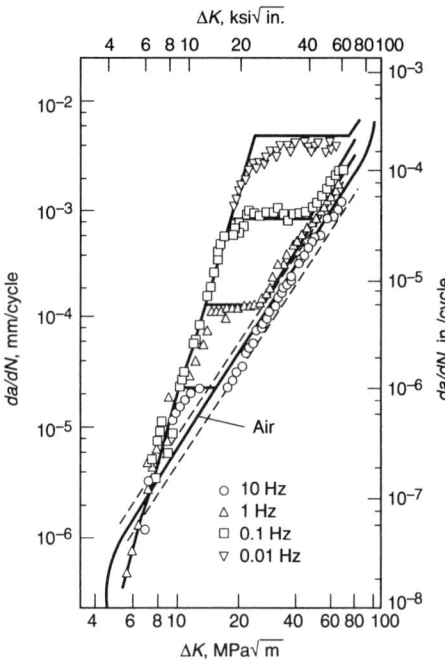

Fig. 4 Crack growth in X-65 line pipe steel in 3.5% salt water at –1.04 V (saturated calomel electrode) and at four frequencies. Source: Ref 3

points will appear and the scatter will seem to become well established (Fig. 3c).

Figure 3(d) results from taking the upper and lower (or 90% confidence) bounds of this scatter (Fig. 3c), then reconstructing crack growth curves on this basis. Clearly, the upper and lower bound data lead to unrealistic results, because all measured crack growth curves are essentially identical. This is caused by the implicit assumption that in some cases, all crack growth could be through a continuous string of weak spots (note that this scatter is caused by local weak spots). This is an untenable assumption. Weak spots are local; in each case only a few will be encountered. The entire bulk will not be one great weak spot. Hence, the average curve is the relevant one; the "scatter'" is only apparent, not real.

Measurement errors do occur, but in general, the crack size will not be consistently overmeasured or undermeasured. Even if the measurement were always over or under, it would result only in a shift of the crack growth curve. The error in the rates would be small:

$$[a_2 + \delta - (a_1 + \delta)] / \Delta N = (a_2 - a_1) / \Delta N$$

$$= \Delta a / \Delta N \qquad \text{(Eq 5)}$$

The main reason for the problem is that the differentiation (the procedure for obtaining Δa) tends to exaggerate measurement inaccuracies. For example, consider a measurement that is accurate within 0.005 in., about as high an accuracy as can be attained. If $a = 0.5$ in. nominally, its

value is between 0.495 and 0.505 in.; the possible error is only 1%. Let the next measurement be 0.52 in., indicating a crack size between 0.515 and 0.525, again with an accuracy of 1%. The value of Δa is then between $0.525 - 0.495 = 0.030$ and $0.515 - 0.505 = 0.010$, with an expected value of $0.52 - 0.50 = 0.020$. Hence, the error in rate will be as large as 50% (0.020 ± 0.010). This problem is not unique for crack growth; it always occurs where (numerical) differentiation is performed.

To counteract this problem, ASTM E647 recommends taking a moving average for the rate determination. Yet the differentiation will still exaggerate measurement inaccuracies, which will appear as exaggerated "scatter" in the rate diagram. If a sufficient number of such values are accumulated, the scatterband may become impressive, even if the other sources of scatter are completely absent. The resulting problem is the same as was discussed on the basis of Fig. 3. Upper and lower bounds would give crack growth curves bearing no relation to the tests, because all tests essentially showed the same crack growth curves. Determining a 90% confidence band with a statistical treatment that does not account for this reality is inappropriate.

Implications. Most of the scatter observed in data plots can be ignored, because it is apparent scatter only. The line representing the average of the data is the only realistic one (Fig. 3). However, it is prudent to account for the batch-to-batch, heat-to-heat, and manufacturer-to-manufacturer variations, because it is not known a

priori which batch of material will be used in the structure. As a rule of thumb, these effects can cause a difference of about a factor of 2 between worst and best crack growth rates.

Corrosion-Assisted Fatigue Crack Growth. Should the environmental effect result in data of the form given in Fig. 4, none of the common equations is applicable and a tabular representation of the data is indicated. In the case of variable-amplitude loading, ΔK varies considerably from one cycle to the next. In one cycle, the ΔK may call for rates at the "plateau" level of Fig. 4, whereas in the next cycle ΔK may be low and call for rates close to the threshold. The baseline data were obtained for gradually increasing ΔK. The environmental effect is time dependent and is measured at "chemical equilibrium" at the crack tip, but it is questionable whether the effect is the same in variable-amplitude loading. The equilibrium condition for a totally different ΔK cannot be reached immediately in one cycle. Computer codes do not consider this problem.

Should the environment change during crack growth, another problem arises. In winter the environment is cold, dry air; in summer it is warm, wet air. There may be 100% relative humidity in winter, but cold air contains much less moisture than warm air, even if saturated. For example, air with 100% relative humidity at room temperature contains approximately 30,000 ppm of water, but air with 100% relative humidity at –55 °C (–70 °F) contains only about 200 ppm of water. (Relative humidity is the fraction of the possible moisture content.)

This problem occurs for transport vehicles, bridges, airplanes, and many other structures exposed to weather. While the following example is for an airplane, it applies to other structures. A wing full of fuel may warm up considerably when the airplane is serviced, standing in the sun. During ascent, when many of the gust loadings are encountered, the material is warm, the air is warm, and the moisture content is high. During flying in the stratosphere, the structure and fuel cool to –55 °C (–70 °F). Cyclic loading still occurs, but crack growth rates in this environment may be less than those in wet air by a factor of 3. Upon descent, the structure is cold but the air is warm with much more water, and the air may contain pollutants, which cause much higher growth rates.

Clearly, such a changing environment defies any theory and any modeling. Only pragmatic engineering is of use. Pragmatism can be exercised in many ways, depending on goals, outlook, and the level of conservatism desirable. Assumptions must be made, and they can have far-reaching consequences. No categoric recipes can be given, only an example. Consider a case where the data set can be represented by a Paris or Walker equation, all with the same exponent(s), so that environment changes will affect only the coefficient C. By estimating the relative times spent in each environment, one can calculate a weighted average of C. Naturally, this assumption is disputable, but so are all other assumptions. Clearly, the time spent in the different environments can only be estimated, so more "refined" assumptions are as good as the above estimates.

Geometry Factors

In order to analyze any fracture or crack growth problem, the analyst must know the geometry factors for K or J (the crack growth energy release rate), or both, for the structural crack of interest. Geometry factors for many generic configurations have been compiled in handbooks (Ref 4-6). These can be used for generic loading and geometries, but actual structural details are unique, and ready-made handbook solutions cannot be expected to be available.

In such a case a solution can be obtained in principle by means of finite element analysis. However, this type of analysis may be too costly because of the complexity of the structural details, or for many other reasons. Fortunately, simple procedures can provide geometry factors quickly and with good accuracy (Ref 1):

- Use handbook solutions combined with superposition and compounding.
- Use Green's functions or weight functions, in combination with finite element stress analysis of the uncracked structure if necessary.

These procedures are included in some general-purpose fracture mechanics computer programs, but they can be done easily by hand in a short time using a handbook (Ref 4-6). The stress-intensity factor is defined as $K = \beta\sigma\sqrt{\pi a}$, in which σ is the nominal stress away from the crack. The geometry factor β accounts for the fact

that average stresses in the cracked section are higher, as well as for all free boundaries affecting the crack-tip stress as expressed by K. Thus $\beta = \beta(a/W, a/D, a/S, ...)$, where W, D, and S are relevant structural dimensions or in general $\beta = \beta(a/L)$, where L has a generalized length. Determining geometry factors requires deriving the function $\beta(a/L)$ for the specific loading and geometry details relevant to the crack to be analyzed. The way in which the function is to be used in crack growth and fracture analysis depends on the definition of stress as well. Additional information on geometry factors and new information not contained in handbooks is in the appendix "Summary of Stress Intensity Factors" in this Volume.

The Reference Stress. In the case of uniform applied stress, there is no problem in the definition of σ in the stress-intensity factor as described above. But if the stress distribution is nonuniform it may not be immediately obvious which stress should be used.

Before turning to the case of nonuniform stress distributions, it may be worthwhile to consider an example. For a central crack of size $2a$ in a plate of width W under uniform tension, it has been shown that $\beta = \sqrt{\sec(\pi a/W)}$ and that $K = \beta\sigma\sqrt{\pi a} = [\sec(\pi a/W)]^{1/2}\sigma\sqrt{\pi a}$. Should one insist on using a reference stress different from σ in this expression, one can legitimately do so. For example, one could use the average stress in the cracked section, σ_{net}, which is given by $\sigma_{net} = \sigma W/(W - 2a)$. Then K would become:

$$K = (1 - 2a/W)\sqrt{\sec\pi a/W}\ \sigma_{net}\sqrt{\pi a} = \beta\sigma_{net}\sqrt{\pi a} \quad \text{(Eq 6)}$$

where $\beta = (1 - 2a/W)\sqrt{\sec\pi a/W}$.

Obviously, the values of K so obtained are identical to those based on σ. Consistent use of Eq 6 in a residual strength analysis would provide as output the fracture stress in terms of σ_{net} (from which σ could be obtained). Similarly, cyclic stress input in a crack growth analysis should be in terms of σ_{net}, and the input for β should be in accordance with Eq 6. In this case a reference stress other than σ offers no advantage, but the example illustrates the principle. Any reference stress can be used, provided that β is adjusted such that the product $\beta\sigma$ remains unaffected. The β-input must be for the proper reference stress, and the input of stress ranges, exceedance diagrams, stress histories, and/or stress occurrences must be in terms of the reference stress. The output will be in terms of the reference stress as well and may require interpretation.

Compounding. In the general expression for the stress-intensity factor $K = \beta\sigma\sqrt{\pi a}$, the geometry factor β accounts for the effect of all boundaries. That is, $\beta = f(a/W, a/D, ...)$, where W, D, and so on are relevant dimensions of the structure. In many cases the individual effects of these boundaries can be found in handbooks. Their composite effect is obtained by compounding, which is multiplication of all individual effects. Possibly the most prominent example of compounding is demonstrated by the classical solution for the elliptical surface flaw (Fig. 5). (Other

solutions have since been obtained; see Ref 7 and 8.) The various boundary effects are due to the back free surface (BFS), front free surface (FFS), width (W), and crack front curvature (CFC):

$$K = \beta_{BFS}\beta_{FFS}\beta_W\beta_{CFC}\sigma\sqrt{\pi a} \quad \text{(Eq 7)}$$

or simply $K = \beta\sigma\sqrt{\pi a}$. If W is large, $\beta_W = 1$ and $\beta_{BFS} = 1.12$. Then the classical solution provides:

$$\beta_A = 1.12\ \beta_{FFS}/Q \quad \text{(Eq 8)}$$

where β_{FFS} and Q can be found in handbooks (Fig. 5).

In obtaining β for a structural crack in a complex geometry, the effect of the individual boundaries can be found in handbooks. By compounding these effects the "total" β is obtained. Rigorous compounding adheres to slightly different rules (Ref 9), but the procedure shown here is generally used and accepted.

Superposition is addition of stress-intensity factors due to various mode I loadings. For example, in a combination of bending (ben) and tension (ten), the total crack-tip stress is:

$$\sigma_{ij} = \frac{K_{ben}}{\sqrt{2\pi r}}f_{ij}(\theta) + \frac{K_{ten}}{\sqrt{2\pi r}}f_{ij}(\theta) = \frac{K_{total}}{\sqrt{2\pi r}}f_{ij}(\theta) \quad \text{(Eq 9)}$$

Because the solution of the crack-tip stress field is universal, the functions $f_{ij}(\theta)$ in both terms are identical, and the total stress intensity is:

$$K_{total} = K_{ten} + K_{ben} \quad \text{(Eq 10)}$$

First the separate β values must be obtained by compounding. Then, after the superposition is complete, the final β value must be obtained for a suitable reference stress, because damage tolerance analysis is based on $K = \beta\sigma\sqrt{\pi a}$.

$$K_{total} = \beta_{ben}\sigma_{ben}\sqrt{\pi a} + \beta_{ten}\sigma_{ten}\sqrt{\pi a} \quad \text{(Eq 11)}$$

In order to obtain β for the combination, a reference stress must be selected. This can be σ_{ten}, σ_{ben}, $\sigma_{total} = \sigma_{ben} + \sigma_{ten}$, or any other suitable stress. Selection of σ_{total} provides:

$$K = \left(\beta_{ben}\sigma_{ben}/\sigma_{total} + \beta_{ten}\sigma_{ten}/\sigma_{total}\right)\sigma_{total}\sqrt{\pi a}$$
$$= \beta\sigma_{total}\sqrt{\pi a} \quad \text{(Eq 12)}$$

Thus, β becomes the weighted average (with regard to the stresses) of those for the two loading conditions. In this manner geometry factors can be obtained easily, even for complicated geometries and loading conditions (Ref 1).

Integration of Crack Growth Rates

Constant-Amplitude Loading. The crack growth curve for a structural crack is obtained from integration of the rates:

$$da/dN = f(\Delta K, R) \text{ or } dN = da/f(\Delta K, R) \quad \text{(Eq 13)}$$

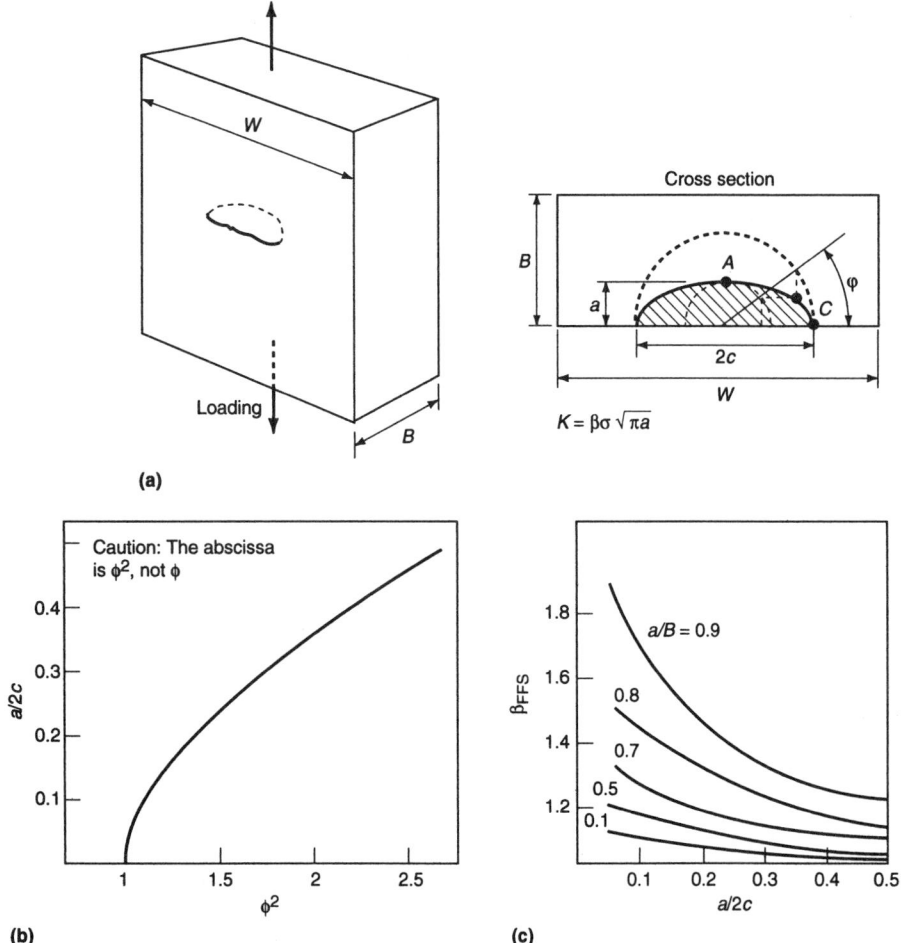

Fig. 5 Geometry factors for surface flaw. (a) Configuration. (b) Geometry factors for shape. (c) Geometry factor for front free surface

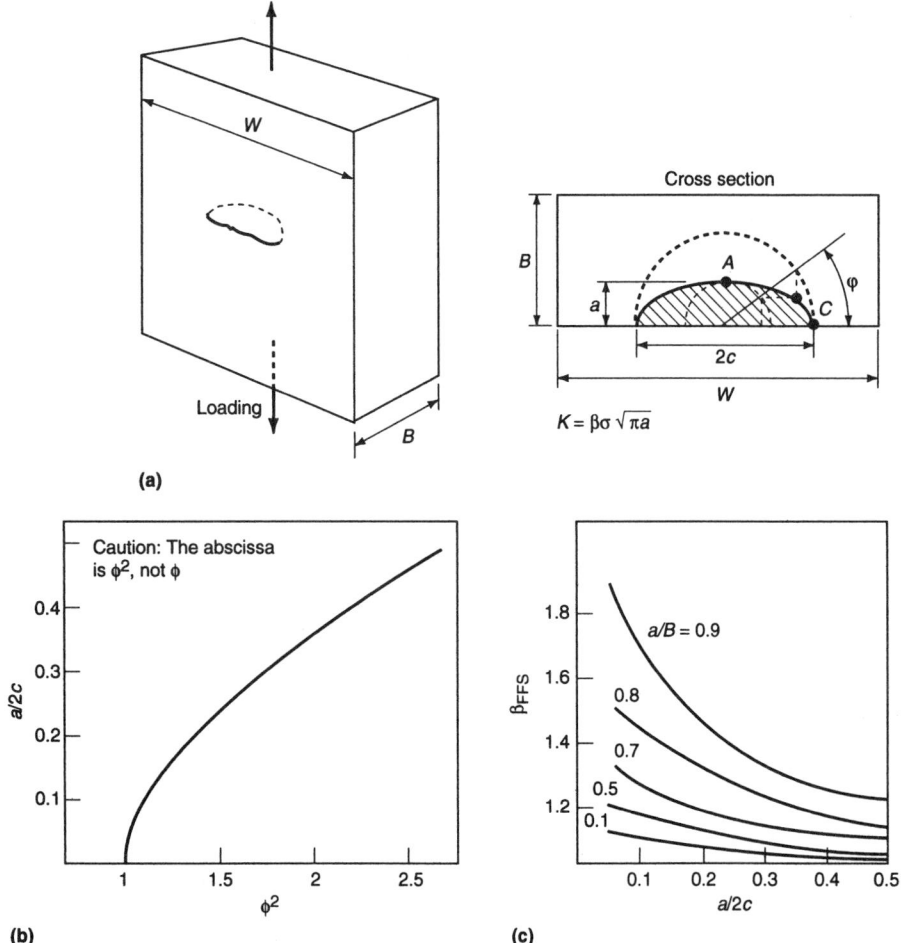

Fig. 6 Integration of fatigue crack growth

Integration provides:

$$N = \int_{a_0}^{a_p} da / f\,(\Delta K, R) \qquad \text{(Eq 14)}$$

Generally, the integration is done numerically. It can seldom be performed in closed form, because of the complexities of the function f, β in ΔK, and the stress history.

The β for a structural crack is usually a lengthy polynomial in a/W or is known only in tabular form. Numerical integration is indicated even if the function f is simple and $\Delta\sigma$ is constant (independent of a). If $\Delta\sigma$ is constant (constant amplitude), it can be taken out of the integration. The remaining parameters in ΔK all depend on the crack size a, so what must be integrated is $N = A \int F'(a)\,da = F(a)$. As in any integration, this is the area under the curve $F'(a)$, shown in Fig. 6. The very definition of integration is that this area is obtained as the sum of the area of many rectangles of height $F'(a)$ and width da, the area of each of which is height times width. In closed-form integration, the size of da vanishes to zero, which gives the true result.

In a numerical integration, the step size da is not infinitesimally small but finite. However, it is quite obvious from Fig. 6 that the result is accurate anyway, because the overshoot triangles at the top left of the rectangles very quickly compensate for the undershoot triangles at the top right of the rectangles. Indeed, numerical integration is a very "forgiving" procedure. Before the computer age, when numerical integrations had to be done by hand, one wanted to make large steps to get done quickly. The (small) inaccuracies thus introduced were compensated for by integration rules such as the Simpson rule. With the computer, the step size can be made so small that the error becomes negligible.

Based on Fig. 6, the integration procedure is then:

$$a_j = a_0; N = 0$$

$$a_i = a_j$$

$$\beta_i = f(a_i/w)$$

$$\Delta K_i = \beta_i \Delta\sigma \sqrt{\pi a_i}$$

$$da/dN = f(\Delta K_i, R)$$

$$\Delta a_i = \delta a_i \ (\text{e.g.} \delta = 0.01)$$

$$\Delta N_i = \Delta a_i/(da/dN)_i$$

$$N_j = N_i + \Delta N_i$$

$$a_i = a_j + \Delta a_i$$

if $a_j < a_p$ then return to beginning $\qquad$ (Eq 15)

The process also involves checks to stop it when the crack reaches the edge of the component, when $\Delta K = K_c$, or when β changes (e.g., a surface crack becomes a through-thickness crack).

Cycle-by-Cycle Counting with Variable-Amplitude Loading. When the loading is of variable amplitude, the stress in every cycle is different from the previous stress. The integration as shown above is then not possible, because ΔK may be very different from that in the previous cycle. We must then resort to cycle-by-cycle integration (i.e., the step size in the integration is one cycle only). For this cycle, ΔK is obtained for cycle i using the proper β value and the $\Delta\sigma$ value appropriate for that cycle. The crack growth, da, follows from Eq 13. It is added to the existing a value and the process is repeated for the next cycle.

If there is retardation, *the crack growth rate in any one cycle depends on that in previous cycles.* In that case the history must be remembered in order for the growth in the next cycle to be calculable at all. However, justified or not, most users prefer to ignore retardation because doing so is conservative. In that case, crack growth in any one cycle is independent of that in the previous cycle.

Hence, without retardation the integration often can be done as a quasiconstant-amplitude computation. This reduces the computation time from hours or minutes to a few seconds, depending on the complexity of the stress history. The computer applies all cycles in the history at each quasiconstant-amplitude step. It adds the crack growth of all these cycles and then determines an effective equivalent cycle that would provide the same crack growth, and it evaluates how many of those cycles are needed to extend the crack over an increment of, say, 1% of the current crack size. In essence, therefore, a weighted average of crack growth is determined for all cycles in the stress

Table 1 Error sources in damage tolerance analysis

Category	Cause of error	Comment	Possible factor on calculated life
Intrinsic shortcomings of fracture mechanics	Linear elastic fracture mechanics	Small error in a_p for small cracks or small parts	1.1-2
Data input	K_{Ic}, or K_c, J_R	Error in a_p	1.1-1.2
	da/dN data	Normal scatter	1.1-1.5
	Assumption 90% band	Apparent scatter	1-2
	Assumption variable environment	E.g., weighted averages	1-2
	Equations for da/dN	Unnecessary force fits	1.5
Assumptions	Direction (e.g., LT vs. SL)	Wrong data applied	1-2
	Initial flaw size	Very influential	1-3
	Shape	Surface flaws	1-2
	Development	E.g., multiple cracks, load transfer, etc.; continuing damage	1.1-2
Interpretation of stress history	Sequence	Semirandom vs. random	1.5
	Truncation	Improper truncation	1.1-1.3
	Clipping	Assumptions	1.5-2
Stress intensity	Loads/stresses	Measurement	1-1.5
		Analysis 15%	1-1.5
Retardation	Model	Small if calibrated	1.1-1.3
	State of stress	If not accounted for	1.1-1.5
	Yield strength	If, say, 10% too low	1.5-2.0

Note: If all errors were to work in the same direction, the total life could be miscalculated by a factor of 1-24,000.

history, recognizing that any one cycle in the history may occur at any one time.

Sources of Error

Sources of error in damage tolerance analysis can be classified as:

- Uncertainty and assumptions in data input
- Uncertainty due to assumptions about flaws
- Interpretations of, and assumptions in, stress history
- Inaccuracies in stress intensity
- Computer modeling errors

In each of these categories, a number of factors contribute to inaccuracies in the analysis. Table 1 provides rough estimates of particular errors, given as a factor on life (not as a percentage error). The numbers are from analysis experience.

Data Input. Retardation models are not ideal, but calibrated models provide results in which the error in life is generally only around 10% (given as a factor of 1.1 in Table 1), and in only a few exceptions does it run as high as 30% (given as a factor of 1.3). This is under conditions where β, da/dN, stress history, and calibration factors are known. Misinterpretation of scatter and force fitting of unsuitable equations may well introduce a factor of 2, and even careful assumptions may cause a factor of 1.5. The situation is worse for mixed environments where the data used are a weighted average for separate environments.

Flaws. By assuming a "conservative" circular flaw instead of an elliptical flaw, one may unwittingly introduce another factor of 2 (Ref 1). Moreover, many flaws are not elliptical, so the assumption of ellipticity can also cause errors. Assumptions about initial flaw size may introduce errors that are equally large. Assumptions about flaw development, continued cracking

when cracks run into holes, and so on are also influential.

Load history is always an approximation. Loads and number of occurrences must be approximated, and decisions have to be made about clipping and truncation. The simple clipping of a few loads can have dramatic effects (Ref 1), so the decision on clipping should be made by damage tolerance experts, not load experts. Improper sequencing is another error driver. Randomizing a load history that in reality was semirandom (mild weather and storms) can cause great differences (Ref 1).

Stress-intensity errors are drivers of intermediate importance. Crack growth is roughly proportional to the third power of K. Because $K = \beta\sigma\sqrt{\pi a}$, all errors in life are proportional to the errors in β and σ to approximately the third power. A 10% error in stress causes a factor on life of 1.1^3, or 1.33. Errors in stress stem from errors in loads and stress analysis. The stresses may be obtained within 10% for the given load, but the load also contains an error. Hence, the final stress may have an error larger than 10%, possibly 15%. This causes a factor on life between 0.85^3 and 1.15^3, or 0.61 and 1.5, the expected life being 1. The error in β is also included in the stress intensity. If this error can be reduced (e.g., from 5 to 3%), the gain is only from a factor of 1.05^3 to a factor of 1.03^3, or from 1.15 to 1.09, a small improvement indeed in comparison with other factors.

Computer modeling errors may be due to the integration scheme and/or rounding. A crack growth calculation is a simple numerical integration that does not give rise to large errors, but rounding can sometimes cause significant errors, especially in cycle-to-cycle integration where da (one cycle) is very small. This is an intrinsic problem of numerical computers. Personal computers provide eight significant figures in single precision and 16 in double precision. Double precision may be necessary in adding small num-

bers to large numbers, as when the small crack growth in one cycle is added to a large crack ($a + da$).

For example, assume that in a particular cycle da is evaluated as:

$$7.45 \times 10^{-8} = 0.000,000,074,500,000 \quad \text{(Eq 16)}$$

This occurs properly in eight significant figures; leading zeros do not count. If the crack size is 12, the results will be:

$$a + da = 12.000,000 + 0.000,000,0745$$
$$= 12.000,000 \quad \text{(Eq 17)}$$

Because 12.000,000 has eight significant figures, da will be rounded off and not be counted. It will appear as if there is no growth. This might occur in a similar way in 10 million successive cycles. The total growth would then have been:

$$10,000,000 \times (7.45 \times 10^{-8}) = 0.745 \quad \text{(Eq 18)}$$

so that $a + da = 12.745$. However, in each cycle the growth was rounded off, and after the 10 million cycles a is still 12. Double precision will mend this problem, but only to a degree:

$$a + da = 12.000,000,000,000,00$$
$$+ 0.000,000,0745 = 12.000,000,074,500,00 \quad \text{(Eq 19)}$$

Indeed, after 10 million cycles the size will be 12.745. However, if da appears to be 7.45×10^{-16}, this crack growth will still be ignored. Fortunately, the above problem seldom arises, but the use of double precision is recommendable at one place in the software, namely where $a + da$ is evaluated.

In the case that all errors discussed are active (which depends on the complexity of the problem), the total life (Table 1) could be miscalculated by a factor of 24,000. Naturally, errors generally will not all operate in the same direction, and some will compensate for others. However, the reliability of the result is affected much more by assumptions than by the shortcomings of fracture mechanics or computer software. It is not worthwhile to improve the strong links in a chain; the weak link must be improved. The weak links are the assumptions about rate data, clipping, flaw size, flaw shape, and so on, not the geometry factors, fracture mechanics concepts, and calibrated retardation models.

Implications. The magnitude of the inaccuracies due to assumptions should be assessed by repeating the analysis using different assumptions. It should be second nature to a damage tolerance analyst to perform calculations a number of times to evaluate the effects of assumptions with regard to stresses, loads, stress history, clipping levels, and so on. The common practice of making "conservative" assumptions everywhere assumes that all errors work in the same direction. Realism and sound judgment are necessary. Even with the best estimates the answer will be in error, but it will be closer to the truth. In the end, a safety factor should be applied as in a conventional design.

REFERENCES

1. D. Broek, *The Practical Use of Fracture Mechanics*, Kluwer Academic Publishers, 1988
2. D. Broek, *Elementary Engineering Fracture Mechanics*, Kluwer Academic Publishers, 4th ed., 1986
3. O. Vosikovsky, Fatigue Crack Growth in X-65 Line Pipe Steel at Low Frequency in Aqueous Environments, *ASME Trans.*, H97, 1975, p 298-305
4. D.P. Rooke and D.J. Cartwright, *Compendium of Stress Intensity Factors*, Her Majesty's Stationery Office, London, 1976
5. G.C. Sih, *Handbook of Stress Intensity Factors*, Lehigh University, 1973
6. H. Tada et al., *The Stress Analysis of Cracks Handbook*, Del Research, 1973-1986
7. J.C. Newman and I.S. Rajun, "Stress Intensity Factors for Cracks in Three-dimensional Finite Bodies," ASTM STP 791, 1983, p I-238
8. J.C. Newman and I.S. Rajun, "Analysis of Surface Cracks in Finite Plates under Tension and Bending Loads," NASA TP-1578, National Aeronautics and Space Administration, 1979
9. D.P. Rooke et al., "Simple Methods of Determining Stress Intensity Factors," AGARDograph 257, Chap. 10, 1980, Advisory Group for Aerospace Research and Development, NATO

Residual Strength of Metal Structures

David Broek, FractuREsearch

COMPLETE DAMAGE TOLERANCE ANALYSIS requires calculation of both the fatigue crack growth curve and the residual strength diagram. Calculation of the complete residual strength diagram is needed for determining the maximum permissible crack size (a_p), which may be smaller than the critical crack size (a_c). Calculation of the fatigue crack growth curve is needed for determining when the crack may reach a_p.

Problems can occur if either step is ignored. For example, often only the crack growth curve is calculated, because many computer programs have an automatic cutoff when the stress intensity, K, reaches the plane strain fracture toughness, K_{Ic}, or the plane stress fracture toughness, K_c (Ref 1, 2). This criterion allows crack growth to a_c instead of to the more conservative a_p. In addition, this criterion more often than not gives false information about a_c, even if linear elastic fracture mechanics (LEFM) applies to the material in question. This is discussed in more detail below.

It is also of little use (although it is often done) to calculate only the residual strength diagram and a_p. Knowing that a crack is smaller than permissible is of little comfort if further growth causes a fracture tomorrow. Hence, the crack growth curve must also be calculated. (See the article "The Practice of Damage Tolerance Analysis" in this Volume.)

The procedure to be followed for LEFM cases is reviewed in this article, along with elastic-plastic fracture mechanics (EPFM) and plastic fracture mechanics (PFM) procedures. The problem will be approached generally to show that LEFM and PFM are special cases of EPFM. Because the so-called "collapse condition" is crucial to the limitations of both LEFM and EPFM, collapse is briefly discussed first.

Collapse and Net Section Yield

Consider two parallel bars of the same size and of the same material, as in Fig. 1(a). Each carries half of the total load; the strain in the bars is equal, causing an elongation, ΔL. If the left bar is cut in two, the right bar will have to carry all the load. Assuming no bending, the stress, the strain, and the elongation will be twice as high as before. The left bar carries no stress at all, and the gap between the two halves of the left bar is $2\Delta L$.

Next, consider the case where the left and right bar are attached (e.g., welded), as in Fig. 1(b). If the bars are intact, the situation is identical to the previous case. However, if the left bar is cut, a different situation develops. When the top half of the right bar is strained, the top half of the left bar must necessarily undergo approximately the same strain. Both bars being of the same material (same modulus of elasticity, E), equal strain in the bars dictates equal stress. Thus, each bar still carries the same stress and each bar carries half the load. However, because the left bar is cut, the right bar alone must carry all the load across the cut. Below the cut, the two bars are again attached and must strain equally. Consequently, the bottom halves of the bar again share the load equally.

The attachment sets the condition for approximately equal strain and equal stress in both bars, almost all the way to the cut. Close to the cut, the load of the left bar must be transferred to the right bar, which will then carry the total load over a short distance. The load from the left bar must be transferred to the right and back over such a small distance that the additional load cannot be distributed evenly. Instead, most of the extra load will be carried by a small portion of the right half, so that higher stresses occur close to the cut (Fig. 1b): there is a stress concentration at the cut. It does not matter whether the bars are welded or are one piece. Transfer of the load from the cut half to the right half takes place by shear.

It is helpful to consider load path (load flow) lines, imaginary lines that indicate, for example,

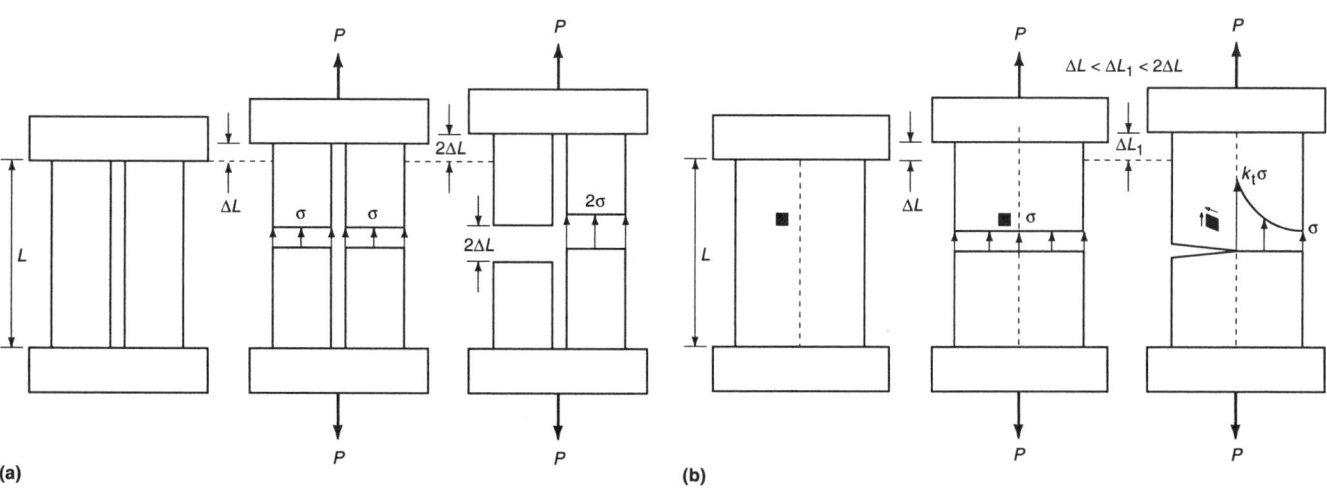

Fig. 1 Effect of cuts (cracks) on stress and strain distribution. (a) Two parallel bars. (b) Two welded bars or one piece

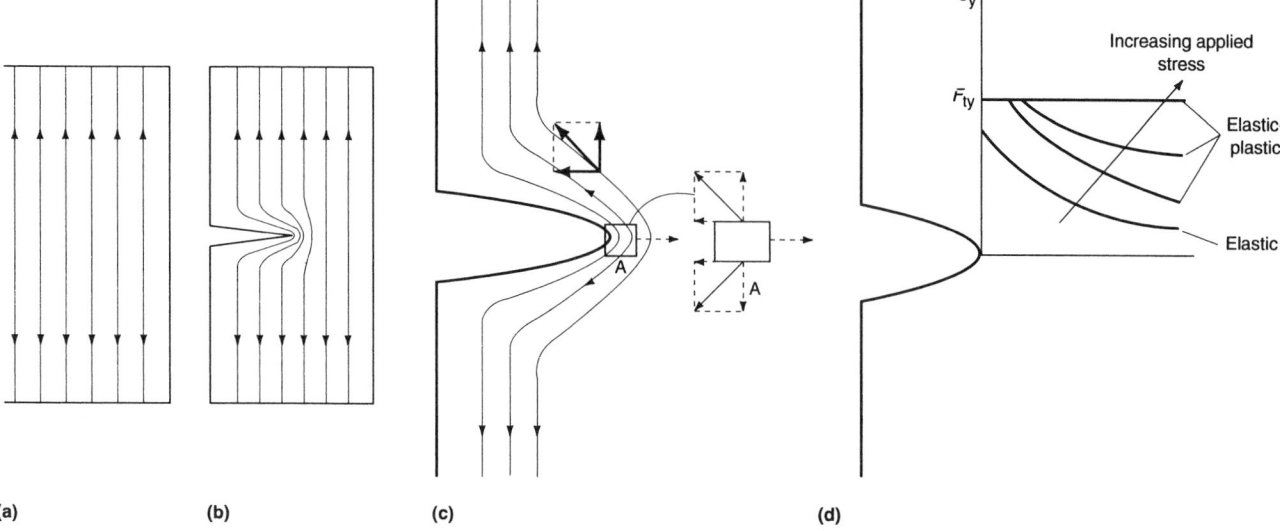

Fig. 2 "Load-flow" lines. (a) No crack. (b) With crack. (c) At crack. (d) Stress distribution for increasing load and plasticity

how one unit of load is transferred from one loading point to the other (Fig. 2). For uniform load, the flow lines are straight and evenly spaced, indicating that the load is evenly distributed (uniform stress). If the load path is interrupted by a cut, the flow lines must go around this cut within a short distance, as shown. At the tip of the cut the flow lines are closely spaced, indicating that more load is flowing through a smaller area, which means higher stress.

Assuming that the material does not exhibit strain hardening (horizontal stress-strain curve beyond yield), the stress cannot increase further after yielding occurs. Hence, by the time the entire section is yielding (fracture may occur before this can happen), the stress distribution in the section is as shown in Fig. 2(d). Note that even stress distributions may not be reached if fracture occurs before the entire ligament yields. Obviously, with the given stress-strain curve without work hardening, the cross section with the notch cannot carry any more load once the entire cross section is yielding, because the yielding will continue uninhibited until fracture results. This is called plastic collapse. Thus, the maximum load-carrying capability is:

$$P_{max} = B(W - a)F_{ty} \qquad \text{(Eq 1)}$$

where a is the crack depth, W is the total width, and B is the thickness. This failure load is called the collapse load or limit load. The nominal stress in the full-width part is $\sigma = P/BW$. Hence, the part fails when the nominal stress is:

$$\sigma_{fc} = P/W = F_{ty}(W - a)/W \qquad \text{(Eq 2)}$$

This is the equation of a straight line (as a function of the notch depth a). If fracture occurs as a consequence of collapse, the strength, σ_{fc}, is the residual strength.

If the material work hardens, the notched cross section can carry a higher load. In general, how-

ever, the entire cross section cannot carry a stress equal to the tensile strength. Apparently, collapse will occur at an average ligament stress somewhat higher than the material yield strength, F_{ty}, but less than the ultimate tensile strength, F_{tu}. The average ligament stress at which collapse occurs is called the collapse strength, F_{col}.

For a non-work-hardening material, $F_{col} = F_{ty}$, and at best $F_{col} = F_{tu}$. It is not easy to determine F_{ty} and F_{tu} other than by a tensile test, and similarly, it is not easy to determine F_{col} other than by a test on a cracked sample.

For a work-hardening material, Eq 2 changes into:

$$\sigma_{fc} = F_{col}(W - a)/W \qquad \text{(Eq 3)}$$

which is also the equation of a straight line as a function of notch depth.

This discussion applies only to the case of uniform applied loading. If there is bending (or other stress gradients in the applied stresses), the conditions for collapse are slightly more complicated (Ref 1).

The Energy Criterion for Fracture

The principle of energy conservation applies during every physical process. Whether a certain event can or will occur can often be determined by equating the energy available for the event and the energy required to make it happen.

Fracture requires energy. Experiments show that the energy is not constant but rather increases while fracture is in progress. It is possible to account for this rise; the problem becomes somewhat more complicated, but it can still be solved rationally (Ref 1, 2). However, in many practical cases, the rise can be ignored because the process cannot be stopped.

We will denote the fracture energy as W. Thus, we are considering the amount of energy ΔW to

propagate a crack of size a over a small distance Δa. Mathematically, this is dW/da.

The available energy, also called the driving energy or released energy, is the energy available to make fracture happen. Two sources can deliver the energy required: the work done by the applied loads and the strain energy stored in the material. If a crack of size a advances to size $a + da$, the applied loads may do some work, and the strain energy contained in the body will change. The net sum of these two types of energy is the available energy. It can be shown (Ref 1, 2) that the net result always equals the absolute change in strain energy, U, namely dU/da per unit thickness:

$$dU/da = (dU_{total}/da)/t \qquad \text{(Eq 4)}$$

where the subscript "total" stands for the full thickness and t is the thickness.

The strain energy per unit volume equals the area under the stress-strain curve. If the stress-strain curve is a straight line (linear elastic), this area equals $\frac{1}{2}\sigma\varepsilon$, where σ is the stress and ε is the strain. For a nonlinear stress-strain curve, the area is $C\sigma\varepsilon$, where the constant C depends on the curvature. The total strain energy in a body of length L, width w, and thickness t is $U_{total} = \alpha\sigma\varepsilon Lwt$, in which α may be a complicated function of the geometry and curvature of the stress-strain curve. Per unit thickness, $U = \alpha\sigma\varepsilon Lw$. Because α and ε are dimensionless, while σ has the dimension of force/length-squared and L and w have the dimension of length, the dimension of U is that of force. Differentiation to crack size effectively means division by a length, so the units of dU/da are force/length.

This means that dU/da and dW/da have the dimension of N/m or lb/in., which is equivalent to $N \cdot m/m^2$ or $in. \cdot lb/in.^2$ Because the unit of energy (work) is length times force, the unit of fracture energy is work per unit area. Thus, dU/da is the energy available for fracture per unit thickness and per unit crack extension. Similarly, dW/da is

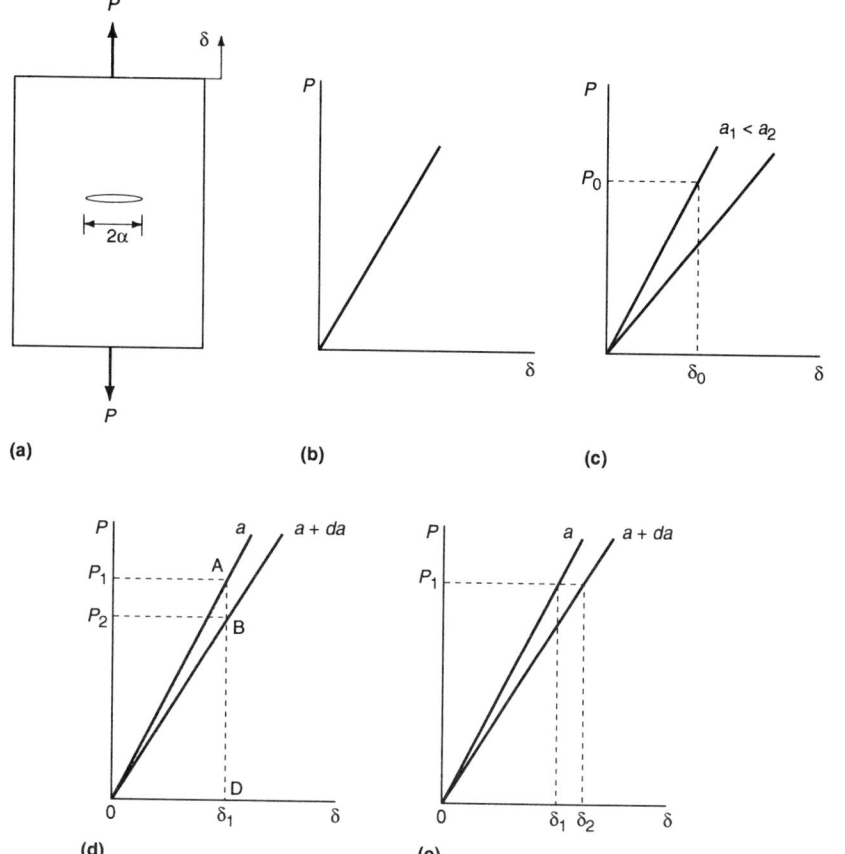

Fig. 3 Load displacement and energy release. (a) Plate with crack. (b) Load displacement without crack. (c) Load displacement for two crack sizes. (d) Energy release at constant displacement. (e) Work done by the load at constant load

the energy needed for fracture, again per unit thickness and per unit crack extension.

The Stress-Strain Curve and the Available Energy. It is necessary to find a quantitative expression for the available energy, dU/da. For uniaxial tension, this is possible with an equation (preferably a simple equation) for the stress-strain curve. The Ramberg-Osgood equation is useful:

$$\varepsilon = \sigma/E + \sigma^N/F \qquad (Eq\ 5)$$

where N is the so-called strain-hardening exponent. For lack of a better name, F is called the nonlinear modulus. The total strain consists of a linear and a nonlinear part:

$$\varepsilon = \varepsilon_{elastic} + \varepsilon_{plastic} \qquad (Eq\ 6)$$

Other common forms of the Ramberg-Osgood equations may be used instead. Equation 6 is used here to improve insight into the equations that follow.

The total strain energy in a body (per unit thickness) equals the area under the load-displacement curve (Fig. 3). The sample with the larger crack ($a + da$) has a lower overall stiffness, so its load displacement curve is lower (Fig. 3c, d). Suppose a sample with a crack of size a is loaded to P_1. The strain energy in that sample is

equal to the area OAD in Fig. 3(b). Further suppose that fracture would occur at load P_1, at constant displacement δ_1. After fracture to $a + da$, the total strain energy would equal the area OBD. The released strain energy equals the hatched area between the two curves (Fig. 3d), this being the available energy if fracture indeed occurs. Also, the fracture energy equals that area. While Fig. 3(d) considers the case of fracture at constant displacement, it can be shown (Ref 1, 2) that fracture under constant load would deliver energy equal to the absolute change in strain energy (dU/da per unit thickness).

Were the stress-strain curves to consist of a linear part only, the available energy would equal that shown in Fig. 3(d). In the case of nonlinear stress-strain curves, the difference with Fig. 3(d) is often quite small, so that LEFM applies.

Quantification of Fracture Energy. Fracture occurs when the available energy equals the energy required for fracture. Therefore, the criterion for fracture is:

$$dU/da = dW/da \qquad (Eq\ 7)$$

By implication, fracture will not occur when Eq 7 cannot be satisfied. There is no fracture when the area in Fig. 3(d) is smaller than the (required) fracture energy, and if there is no fracture, no energy will be released either. If we increase the load, more energy can be delivered,

and if this higher deliverable energy equals the required fracture energy, fracture will occur.

The energy criterion can be quantified easily for a linear stress-strain curve. The earliest attempt to incorporate the nonlinearity of the stress-strain curve was by Rice in 1968 (Ref 3). He expressed dU/da in terms of a complicated integral. (For this reason, $J = dU/da$ is often called the J-integral.) As shown below, an expression for J can be obtained without that integral. With hindsight and dimensional analysis, the problem can be solved in a simple way.

Equation 7 expresses the failure criterion. U contains the product $\alpha\sigma\varepsilon$, the strain energy per unit volume, where α is a dimensionless constant that depends on the shape of the stress-strain curve and the shape of the body. One is at liberty to define σ as the stress at any location. It can be the local stress at any place or the applied stress, as long as one recognizes that the dimensionless α depends on the choice. It is advantageous, but not necessary, to choose σ as the applied stress, away from the crack.

Because σ has units of force/area, and α and ε are dimensionless, while dU/da must have units of energy/unit area (or force · length/area), the proper dimension for dU/da can be obtained only if the quantity $\alpha\sigma\varepsilon$ is multiplied by one that has the unit of length. Clearly and unavoidably, dU/da must depend on the size of the crack, a, which has dimensions of length. Therefore, the expression for dU/da has to be:

$$dU/da = \alpha\sigma\varepsilon a \qquad (Eq\ 8)$$

which has unit dimensions of energy/unit area (e.g., N · m/m² or in. · lb/in.²). While this expression for dU/da is obvious from dimensional analysis, it follows from other considerations as well (Ref 1). Consequently, the criterion for fracture of Eq 7 becomes:

$$\alpha\sigma\varepsilon a = dW/da \qquad (Eq\ 9)$$

where α depends on the shape of the stress-strain curve and the geometry of the body. In its simplest form, $\alpha = f(a/w, N)$. With the Ramberg-Osgood equation, we can separate the linear and nonlinear parts of dU/da and arrive at:

$$\alpha_{linear}\sigma\varepsilon a + H_{nonlinear}\sigma\varepsilon a = dW/da \qquad (Eq\ 10)$$

Further, when we substitute $\varepsilon = \sigma/E$ and $\varepsilon = \sigma^N/F$ for the elastic and plastic parts (ignoring anisotropy), the final expression becomes:

$$\pi\beta^2(a/W)\sigma^2 a/E + H(a/W, N)\sigma^{N+1}a/F = dW/da \qquad (Eq\ 11)$$

where the fracture energy dW/da is a measure of toughness. The coefficients have been replaced by $\alpha = \pi\beta^2$ for the linear part and by H for the nonlinear part. The first term for the linear part is identical to the second term with $N = 1$ and $F = E$.

We have now arrived at a quantifiable expression for the fracture energy. The equation contains two unknown geometry factors, β and H. (E, F, and N can be obtained from the measured

Table 1 Values for H for center-cracked panels in plane stress

a/w	$N = 1 (\pi\beta^2)$ (c)	N = 2	N = 3	N = 5	N = 7	N = 10
			H (a)(b)			
0.0625	3.20 (3.20)	4.66	5.99	8.71	11.84	17.56
0.1250	3.38 (3.40)	5.28	7.44	13.48	23.30	50.79
0.1875	3.74 (3.77)	6.48	10.32	24.64	58.25	214.40
0.2500	4.42 (4.44)	8.80	16.48	57.92	208.64	1464.32
0.3125	5.65 (5.63)	13.58	32.05	190.13	1169.88	18365.75
0.3750	8.28 (8.21)	27.36	93.44	1239.04	17694.72	909115.39
0.4375	16.64 (16.10)	100.48	670.72	35389.44	...	...

(a) This table provides H for cases analyzed in Ref 4. For other values of a/w and N, use interpolation. (b) The H values were derived from so-called h_1 values (Ref 4) after sanitizing the expression of Eq 23 for J to that of Eq 17. (c) For the value of N = 1, H must equal $\pi\beta^2 = \pi$ sec $(\pi a/w)$. Source: Ref 4

Table 2 Examples of linear elastic fracture mechanics (LEFM), elastic-plastic fracture mechanics (EPFM), and collapse in center-cracked panels

Parameter	Case 1	Case 2	Case 3
Yield strength, N/mm² = MPa	380	210	380
E, GPa	69	207	69
$\sigma_f (= F^{1/N})$, MPa	639	660	639
N	12	5	12
Crack size, a, mm	50	50	50
Panel size, W, mm	600	600	200
Failure stress (test), MPa	200	160	190
β	1,017	1,017	1.19
H (rough approximation)	32	10	2000
Toughness, K_c, Eq 15, MPa√m	2550	2039	...
$J_{elastic}$, Eq 14, N/mm	94.2	20.1	...
$J_{plastic}$, Eq 17, N/mm	0.3(a)	67.1	...
$J_{total} = dW/da$, Eq 19, N/mm	94.5	87.2	...
Failure condition	LEFM	EPFM	Collapse

(a) Negligible compared to 94.2

stress-strain curve.) Both β and H can be calculated (see the appendix to this article, "Computation of Geometry Factors.") β can be easily established (Ref 1) for almost any geometry. For example, for a center crack in a rectangular body of width W, the expression for β is:

$$\beta = \sqrt{\sec(\pi a/W)} \qquad \text{(Eq 12)}$$

Determining H is more difficult, but it has been calculated for some cases (Ref 4). Table 1 shows values of H for center cracks.

Toughness and Geometry Factors

Measurement of Toughness. Provided that β and H are known, the toughness of any material can be measured by means of two tests. A normal tensile test yields elongation as a function of load, from which the stress-strain curve can be obtained. The stress-strain curve is curve fitted to Eq 5 to obtain E, F, and N. The second test is performed on a coupon with a crack of any geometry for which β and H are known, although ASTM has standardized certain specimen types. The length of this crack, a, is known, and the load to failure of the specimen is measured. Given width and thickness, the applied stress at fracture can be obtained from the load. Together the two tests provide all the information needed to calculate

the toughness: E, F, and N from the tensile test; β, H, and a from the size of the crack, and σ, the fracture stress measured in the second test. According to convention, dU/da and dW/da are evaluated per crack tip. The convention implies that a central crack is defined as 2a. If one wishes to use established values for β and H, the convention must be followed.

The definition of toughness is provided in Eq 11. Over time, the three terms in Eq 11 have been given a variety of shorthand names. For the two terms on the left side of the equation, the most common shorthand names have been G and J, respectively. Sometimes J is used for the sum of the two. Many researchers use $J_{elastic}$ and $J_{plastic}$. The shorthand expressions for the term on the right side have been R, J_R, J_c, and others. Hence, the following equalities all express the same thing as Eq 11:

$$G + J = J_c$$
$$G + J = J_R$$
$$J_{elastic} + J_{plastic} = J_c \text{ (or } J_{Ic}) \qquad \text{(Eq 13)}$$

The most important assumption underlying Eq 11 is that the material is nonlinear elastic (Ref 2). It is assumed that the same nonlinear curve is followed during unloading as during loading. Loading presents no problems, because the nonlinear curve is certainly followed during loading.

However, during unloading much of the deformation of the nonlinear part is not retrievable. This means that the associated nonlinear part of the strain energy is not retrievable either and is not available to deliver the required fracture energy. The validity of J breaks down after a small amount of unloading, even if this is due to crack extension. The point at which this happens is not fixed and can be determined only by finite element analysis (Ref 5). For this reason, variations of J have been proposed.

LEFM and EPFM. Table 2 shows typical results for center-cracked panels where H is known. In many cases the second term in Eq 11 is very small compared to the first. Under this condition (LEFM), it is justifiable to neglect the second (plastic) term:

$$\beta^2\sigma^2\pi\, a/E = dW/da \qquad \text{(Eq 14)}$$

This can be reduced to:

$$\beta\sigma\sqrt{\pi a} = K_{Ic}$$
$$K = K_{Ic} \qquad \text{(Eq 15)}$$

where K is the stress-intensity factor and K_{Ic} is the LEFM fracture toughness, defined as $K_{Ic} = \sqrt{EdW/da}$. Hence, in LEFM the residual strength can be calculated from:

$$\sigma_{res} = K_{Ic} / \beta\sqrt{\pi a} \qquad \text{(Eq 16)}$$

In Eq 16, K_c can be substituted for K_{Ic}.

In another case in Table 2, the situation is reversed: the first term is negligible with regard to the second. Under this condition (PFM), only the plastic term needs to be carried:

$$H\sigma^{H+1} a/F = dW/da \qquad \text{(Eq 17)}$$

Then the residual strength (σ_{res}) can be obtained as:

$$\sigma_{res} = (FJ_c/Ha)^{1/N+1} \qquad \text{(Eq 18)}$$

where $J_c = dW/da$.

In the remaining case (EPFM), neither term is negligible and both must be carried. The residual strength is obtained by solving the following equation for σ:

$$\pi\beta^2\sigma^2 a/E + H\sigma^{N+1} a/F = J_c \qquad \text{(Eq 19)}$$

For the LEFM and PFM cases (Eq 16 and 18), the equations can be solved directly for σ to obtain the residual strength. However, no solutions exist for algebraic equations with powers higher than four, so for the EPFM case (Eq 19), the residual strength can be obtained only by iteration. Fortunately, fast converging iteration schemes can be employed that obtain the solution quickly, especially if a computer is used.

It is now clear that LEFM and PFM are but limiting cases of the general equation, Eq 19, for EPFM. The use of these limiting cases is perfectly legitimate, provided that one term is negligibly small (Table 2). It is also clear that, for the purpose of tests and residual strength analysis, there is no need to express J as an integral. Undeniably, the integral has been historically useful in the development of EPFM, but it has little significance for practical toughness testing and damage tolerance analysis. Its remaining use is in the computation of the geometry factor H, as a convenience and not as a fundamental necessity. (See the appendix to this article, "Computation of Geometry Factors.")

The Geometry Factors β and H. Only one geometry factor, β, is needed for LEFM, and only one geometry factor, H, is needed for PFM, but both are required for EPFM. Values of β have been obtained in abundance (mostly by numerical analysis) and have been compiled in handbooks (Ref 6-8). Using the principles of superposition and compounding (Ref 1), these solutions can be used to obtain β for almost any conceivable case (see also the article "The Practice of Damage Tolerance Analysis" in this Volume).

So far, however, values for H have been obtained for only a few geometries, again by numerical analysis, and handbooks for H are small. The reason is that computation of H is expensive; not only must the analysis be nonlinear, but the dependence of H on N requires that H be computed for a range of N values. However, in time, handbooks for H may be expected to become as comprehensive as those for β.

Before this happens it is to be hoped that analysts start using Eq 14, because an unnecessary complication was introduced when presently available H values were computed (Ref 9). The developers of the first "H-handbook" (Ref 11, 12) chose to use a different expression for the Ramberg-Osgood equations than the one used in Eq 2, namely:

$$\varepsilon_{total}/\varepsilon_0 = \sigma/\sigma_0 + \alpha(\sigma/\sigma_0)^N \qquad (Eq\ 20)$$

There is no fundamental objection to this expression, but there is an important practical objection. Equation 5 contains three parameters (E, F, N), while Eq 20 contains four (ε_0, σ_0, α, N). If only three are needed, the use of four implies that one parameter is merely a function of the other three and is expendable. It also means that one of the parameters may be chosen freely (e.g., ε_0). The developers insist that ε_0 must be chosen because $\varepsilon_0 = \sigma_0/E$, where σ_0 is the yield strength of the material, but that is merely their choice (see pages 102-108 of Ref 1). Any value of σ_0 is permissible as long as $\varepsilon_0 = \sigma_0/E$. It would have been preferable if the superfluous parameter α had been eliminated for Eq 16 to read:

$$\varepsilon_{total}/\varepsilon_f = \sigma/\sigma_f + (\sigma/\sigma_f)^N \qquad (Eq\ 21)$$

where $\sigma_f = F^{1/N}$ as in Eq 2 and $\varepsilon_f = \sigma_f/E$ (where σ_f is the stress for which the plastic strain equals unity). This would only mean replacing F by σ_f^N in all the above equations, merely for mathematical convenience. Be that as it may, we are left with a superfluous parameter, α, which obviously depends on the other three (because of the assumptions made), by:

$$\varepsilon_0 = \sigma_0/E$$

$$\alpha = \alpha_0^N \varepsilon_0 F \qquad (Eq\ 22)$$

With the use of Eq 16 the equations for $J_{plastic}$ (Eq 17, 19) can be written as:

$$J_{plastic} = \alpha\sigma_0\varepsilon_0 ch_1 (P/P_0)^{N+1} \qquad (Eq\ 23)$$

In this equation P is the load and P_0 is the load at collapse, supposing that σ_0 were the collapse strength. Instead of the crack size, a, the unbroken ligament, c, is used. Finally, h_1 is the geometry factor provided in the handbook (Ref 4). Clearly, the load P is related to the stress, P_0 is related to σ_0, and the ligament c is related to the crack size, a:

$$P = g\sigma$$

$$P_0 = k\sigma_0$$

$$c = (W/2a - 1)$$

$$a = la \qquad (Eq\ 24)$$

where g, k, and l are just functions of the geometry. With this knowledge, the complicated Eq 23 turns into:

$$J_{plastic} = (\sigma_0^N/\varepsilon_0 F) \sigma_0\varepsilon_0 lah_1 (g/k)^{N+1} (\sigma/\sigma_0)^{N+1} \qquad (Eq\ 25)$$

which reduces to:

$$J_{plastic} = H\sigma^{N+1} a/F \qquad (Eq\ 26)$$

where $H = lh(g/k)^{N+1}$. Equation 26 is the basic form of the equation already presented as Eq 11. Equation 23 is just a complicated version of the same. Obviously, α, σ_0, and ε_0 can be divided out; they are superfluous. This should be expected; a collapse load cannot enter into J because the stress-strain equation used has no limit. The elastic energy release, G, could be expressed in the same manner by using $P = g\sigma$, $P_0 = k\sigma_0$, and $\sigma_0 = \varepsilon_0 E$:

$$G = \pi\beta^2\sigma_0\varepsilon_0 (k/g)^2 (P/P_0)^2 \qquad (Eq\ 27)$$

Introducing the collapse load does not make G dependent on same; it merely amounts to multiplying the numerator and denominator by the same number, which does not change the basic equation. Collapse does not enter LEFM or EPFM equations, and an artificial introduction does not change this fact. Collapse is a competing condition that must be assessed separately.

Two other objections can be raised against Eq 23. Instead of just one geometry parameter, H, one must use the four geometry parameters given in Eq 23 and 24: h, g, k, and l. Every time a calculation is performed, double work is necessary: parameters must be derived and are subsequently divided out. Naturally, one could, once and for all, calculate H from $H = lh(g/k)^{N+1}$ and from then on use Eq 17 and 19. This was done to obtain the H values in Table 1. The second objection is that J is expressed in the load P, while in engineering one works with stress. Therefore Eq

19 is more useful; all other fracture mechanics equations are expressed in stress for this very reason. For complicated structures the conversion from load to stress is done in the design stage, not at the time of fracture analysis. Once more, it is necessary to urge that future computations and compilations of H be based on Eq 19. Practical fracture mechanics has no need for unnecessary and irrelevant complications.

Calculation of the Residual Strength Diagram

Depending on the case (LEFM, EPFM, or PFM), the complete residual strength diagram is obtained by solving one of the equations (Eq 16, 18, or 19) for a range of a values. (Twelve to 16 values for a are usually sufficient to establish the curve for plotting.) If β and H are simple functions or available in tabular form, the computations for LEFM and PFM can be accomplished quickly with a hand calculator or a spreadsheet computer program. The iterative solution of Eq 19 must be programmed, so the use of a dedicated software program for residual strength analysis is advisable. Some damage tolerance software is equipped with these capabilities.

First, consider the use of LEFM and Eq 16. Obviously, solutions can be obtained for any geometry (any β), but for the purpose of this discussion assume that β is given by Eq 12 for a center-cracked plate. Then if $a = W/2$ (the whole plate is already cracked), β becomes infinite, so that Eq 19 provides $\sigma_{res} = 0$, as it should (Fig. 4a). Imag-

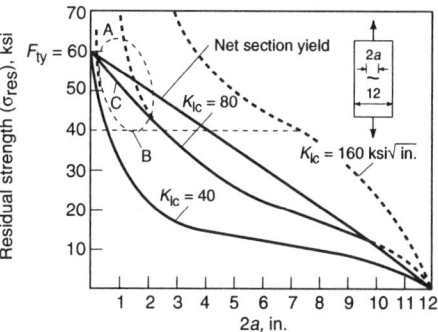

(a)

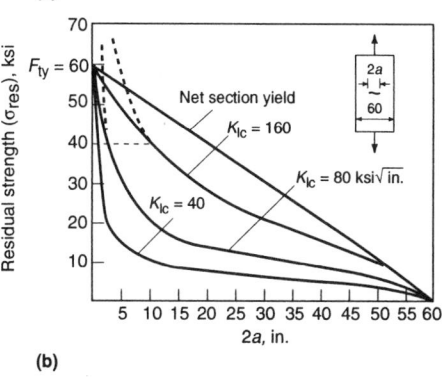

(b)

Fig. 4 Effect of panel size on residual strength with material yield strength (F_{ty}) of 415 MPa (60 ksi). (a) Panels 300 mm (12 in.) wide. (b) Panels 1525 mm (60 in.) wide

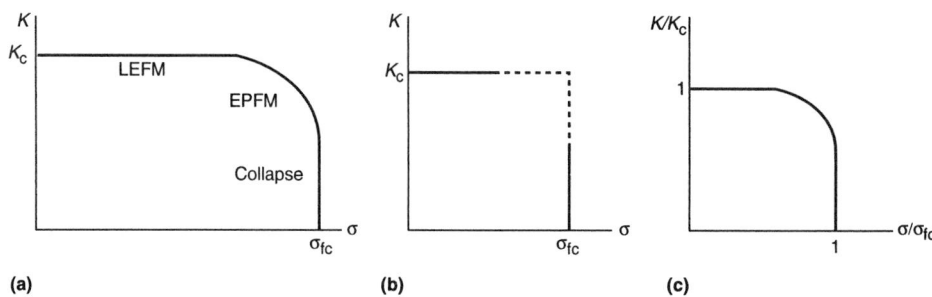

Fig. 5 Effect of plate size on failure and collapse. Residual strength (solid lines) is the lowest value of the superimposed curves for σ_{fr}, σ_t, and σ_{fc}.

Fig. 6 Failure analysis diagram. (a) Regimes of fracture mechanics. (b) Elastic-plastic region. (c) Normalized diagram

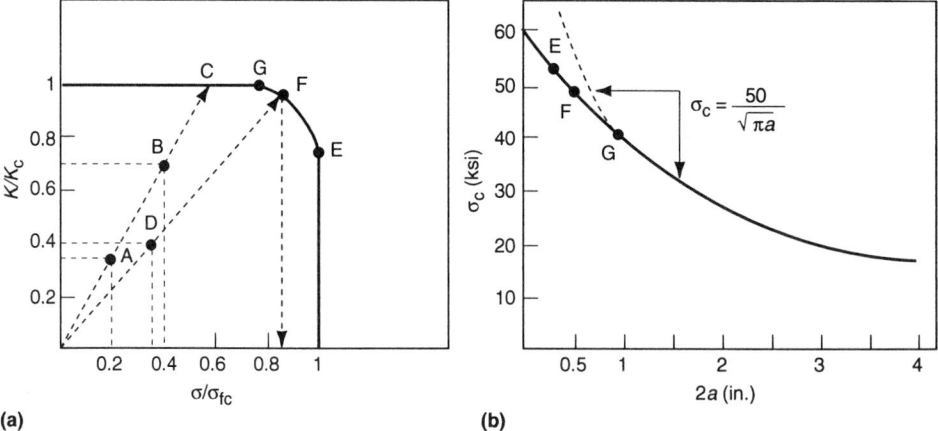

Fig. 7 Failure analysis diagram. (a) Construction (see text for details). (b) Comparable points in residual strength diagram (compare Fig. 4)

ine for a moment that β is universally equal to 1. Then Eq 16 represents the equation for an orthogonal hyperbola with the two axes as asymptotes. For $\beta = 1$ the resulting curve is therefore the modified hyperbola shown in Fig. 4(a).

For very small a the value of β is indeed equal to unity, so that the left-hand part of the curve tends to infinity (vertical dashed lines in Fig. 4b). Substitution of $a = 0$ in Eq 14 gives $\sigma = \infty$. Naturally, an infinite strength is impossible, because the (residual) strength of an uncracked plate or structure is equal to F_{tu}.

As discussed above, if the predicted fracture stress is higher than the stress causing failure by collapse, then collapse will prevail. This is the case when the result of Eq 3 is less than the result of Eq 16. The lower of the two is the actual residual strength discussed. It appears that there are three situations in which a collapse failure could prevail:

- The toughness is very high.
- The crack is very small.
- The width W is very small.

These situations are illustrated in Fig. 4 and 5. For the material with $K_{Ic} = 175$ MPa$\sqrt{m}$ (160 ksi$\sqrt{in}$.), a panel that is 300 mm (12 in.) wide always fails by collapse, but a panel of the same material that is 1500 mm (60 in.) wide fails by Eq 16. Apparently, for a particular structural configuration, failure may occur by collapse (unconditionally leading to fracture) in certain instances and by fracture in other instances. Collapse always prevails if the cracks are small. At any toughness, Eq 16 puts the fracture stress at infinity when the crack size approaches zero. Thus, the residual strength curve will always rise asymptotically to infinity for small a. This means that there will always be a point at a certain small a where collapse failures will prevail, regardless of how low the toughness is.

Thus, the left part of the curve will always be in error, whether the toughness is high or low. For $a = 0$, the strength is F_{tu} (or less), while for large a, Eq 16 applies (the curve in Fig. 4). Obviously, the behavior between $a = 0$, $\sigma \cong F_{ty}$, and the curve for large a cannot be as A or B in Fig. 4(a). Hence, if one assumes curve C (tangent to the curve), the assumption cannot be far from the truth; it certainly will be adequate for engineering analysis.

It is important to understand that two plates of different sizes, but of the same material, can fail differently: one by collapse, the other by fracture. This is depicted in Fig. 4 and 5. If, as in the case of Fig. 5, a plate of the smaller size W_3 is used in a test, the failure occurs by collapse. Hence, the value of the stress intensity at the time of failure would still be lower than the toughness, because K has not yet reached the toughness.

In the case of EPFM and PFM, the residual strength diagram is calculated in the same manner, through the use of Eq 18 and 19 instead of Eq 16. Some general damage tolerance software provides all options, includes libraries for β and H, and performs the necessary compounding and superposition. (See the article "The Practice of Damage Tolerance Analysis" in this Volume.)

Contrary to common belief, the same problem is encountered in PFM and EPFM, because Eq 18 and 19 also predict an infinite strength for $a \to 0$. Of course, artificialities, such as those leading to Eq 23 and 28, do not cure this ill. Thus, whether one uses LEFM, EPFM, or PFM, the competing condition for collapse must always be evaluated. The actual residual strength is the lower of the two. This is one of the reasons that the permissible crack size does not follow simply from a cutoff in crack growth analysis.

The Failure Analysis Diagram. The conditions discussed above are illustrated in a so-called failure analysis diagram (Ref 10) that was developed in Great Britain. The failure analysis diagram can be used to represent the whole gamut of fractures, from brittle to fully plastic, as shown in Fig. 6. Stress intensity is plotted along the ordinate; stress, along the abscissa. The stress is limited by collapse and the stress intensity is limited by the toughness.

The limits of K_c at one end and collapse at the other end require that there be a limiting line from K_c to σ_{fc}. The end portions of this contour must be straight, so that there is only a relatively small curved part. The top horizontal part is governed by LEFM and the vertical portion is governed by collapse. The curved part is the regime of EPFM. For many applications it may be permissible to approximate the diagram by two straight lines (Fig. 6b). The failure analysis diagram is usually presented in normalized form by plotting K/K_c and σ/σ_{fc} so that the intercepts with the axes are at 1 (Fig. 6c). This normalization makes the diagram universal.

The use of the failure analysis diagram can best be demonstrated by an example. Consider a material with $K_{Ic} = 55$ MPa$\sqrt{m}$ (50 ksi$\sqrt{in}$.) and $F_{col} = 415$ MPa (60 ksi). Assume that the "structure" is a center-cracked panel, 12 in. wide, with a crack $2a = 2$ in. subjected to a stress of 10 ksi. The nominal stress at collapse according to Eq 2 would be $\sigma_{fc} = 60 (12 - 2) = 50$ ksi. Thus, $\sigma/\sigma_{fc} = 10/50 = 0.20$. The stress intensity is $K = \beta\sigma\sqrt{\pi a}$, so that with $\beta \cong 1$ its value is $K = 10\sqrt{\pi \times 1} = 17.7$ ksi$\sqrt{in}$. Thus, $K/K_c = 0.35$. Now the point $\sigma/\sigma_{fc} = 0.20$ with $K/K_c = 0.35$ can be plotted in the diagram as point A in Fig. 7. If the stress is raised to 10 ksi, the stress intensity becomes $K = 20\sqrt{\pi \times 1} = 35.5$ ksi$\sqrt{\text{in.}}$, and $K/K_c = 35.5/50 = 0.71$. Further, $\sigma/\sigma_{fc} = 20/50 = 0.40$. This produces point B in the diagram.

Clearly, K/K_c and σ/σ_{fc} increase proportionally. Therefore, a straight line through the origin provides all combinations of K and σ. Fracture occurs where this line intersects the fracture locus. Judgment of the proximity of fracture can be made from the distance between a point and the contour. Extension of the line will also show whether fracture occurs by LEFM, EPFM, or collapse. In the present example, LEFM applies (point C).

In its normalized form, the failure analysis diagram is the same for all materials and structures, regardless of K_c and F_{col}. Given the scatter in material behavior, an approximation of the curved part will suffice for many purposes. A more precise diagram (curved part) can be drawn based on EPFM.

The use of the diagram for a particular application requires calculation of the stress at collapse and the stress intensity at a given stress and crack length. This permits calculation of K/K_c at that stress, which can be plotted. The diagram presents a means for assessing the proximity of fracture, and it shows what kind of fracture to expect, putting the three areas of fracture analysis (LEFM, EPFM, and collapse) in perspective. It is useful in conjunction with, not instead of, the residual strength diagram, as it can be derived from the latter. Its significance is that it illustrates, from a technical point of view, that the EPFM fracture criterion is not very sensitive, and also that in EPFM, collapse must be treated separately, as a competing condition.

REFERENCES

1. D. Broek, *The Practical Use of Fracture Mechanics*, Kluwer Academic Publishers, 1988
2. D. Broek, *Elementary Engineering Fracture Mechanics,* 4th ed., Kluwer Academic Publishers, 1986, p 5-16
3. J.R. Rice, A Path Independent Integral and the Approximate Analysis of Strain Concentrations by Notches and Cracks, *J. Appl. Mech,* 1968, p 379
4. V. Kumar et al., "An Engineering Approach for Elastic-Plastic Fracture Analysis," Report NP-1931, EPRI, 1981
5. R.M. McMeeking and D.M. Parks, "On Criteria for *J*-dominance of Crack-tip Fields in Large-scale Yielding," STP 668, ASTM, 1979, p 175-194
6. D.P. Rooke and D.T. Cartwright, *Compendium of Stress Intensity Factors*, Her Majesty's Stationery Office, London, 1976
7. G.C. Sih, *Handbook of Stress Intensity Factors*, Lehigh University, 1973
8. H. Tada et al., *The Stress Intensity of Cracks Handbook*, Del Research, 1973-1986
9. D. Broek, "*J* Astray and Back to Normalcy," *ECF Fracture Control of Structures,* Vol II, Engineering Materials Advisory Services, 1986, p 745-760
10. G.C. Chell, "A Procedure for Incorporating Thermal and Residual Stresses into the Concept of a Failure Analysis Diagram," STP 668, ASTM, 1979
11. V. Kumar et al., "An Engineering Approach for Elastic Plastic Fracture Analysis," Report NP-1931, EPRI, 1981
12. V. Kumar et al., "Advances in Elastic-Plastic Fracture Analysis," Report NP-3607, EPRI, 1984

Appendix: Computation of Geometry Factors

Because the geometry factor, β, that appears in both J and K is for the elastic case, it is independent of the stress-strain properties. For a given configuration and crack size, it can be obtained by several kinds of numerical analysis, finite element analysis presently being the most prominent. Given a geometry and crack size, a, two analyses are performed with an arbitrary load of, say, 1000 N. In one case, the crack size is a and in the second case it is $a + \Delta a$. In both cases, the strain energy, U, per unit thickness is calculated. From this, dU/da is obtained as $(U_1 - U_2)/\Delta a$. Then β can be obtained from $\beta\sigma\sqrt{\pi a} = \sqrt{E \times dU/da}$, where a, σ, and E are known. Alternatively, because $J = dU/da$, the J-integral can be calculated from the finite element analysis result, from which β can be obtained in the same manner as above.

The above computation needs to be done only once for a range of crack sizes. Many of the resulting geometry factors are presented in handbooks (Ref 6). Because the procedure is relatively easy, geometry factors have been obtained for many generic configurations. By using these handbook solutions, the geometry factors for many more configurations and loading conditions can be obtained by means of compounding and superposition (Ref 1).

The geometry factor, H, for J can be obtained in the same manner. Indeed, by obtaining two solutions for a and $a + \Delta a$, one arrives at $J = dU/da$ without the use of the J-integral. But it is more convenient to evaluate the J-integral from the results of the finite element analysis, in the same manner as can be done for K and β. For H, however, the computation is more expensive because the finite element analysis must be nonlinear, which increases computation time greatly. Besides, the computations must be repeated for a range of N values, because H depends on N. For each case, H is obtained from $H\sigma^N a/F = J$ ($= dU/da$), where σ, a, N, and F are known and J is calculated in the analysis.

The computation of H for one geometry (and the necessary range of N values) requires on the order of 100 times the effort needed to calculate β for the same case, so the "H-handbook" (Ref 4) is small. Besides, H values for other configurations cannot be obtained so easily by superposition and compounding, unless one uses estimation schemes (Ref 1).

Unfortunately, although the above procedure was used in compiling the H-handbook (Ref 4), it is based on a needlessly complicated definition of J (containing three unnecessary parameters, as given in Eq 23). The H values shown in Table 1 were derived from h_1 (the geometry factors presented in the handbook) by sanitizing the expression for J to that of Eq 17. The J-integral is useful for the computation of geometry factors, but for fracture mechanics applications, J can be derived without the integral.

Fatigue and Fracture Control of Weldments

Tarsem Jutla, Caterpillar Inc.

FAILURES in engineering structures are still common today, despite the fact that modern tools for designing structures are very sophisticated and readily available. Computer-aided design, finite-element stress analysis, computerized material property databases, and an array of process simulation tools are among the many design aids accessible to the engineer. In addition, over the past 20 to 30 years national and international codes and standards have been developed for many industries, and these provide explicit guidelines on such issues as material selection, design methods, standardized load histories, and safety factors (Ref 1-16).

In general, most of today's fabricated components and structures are welded, and invariably the weld joint is the most critical area from the performance perspective. An examination of structural and component failures documented in open literature over the past 50 years or so clearly indicates that failures predominantly start at connections, and in particular at weld joints. Placement of a weld onto a metallic material imparts many detrimental features, which include changes in microstructure and mechanical properties, introduction of welding residual stresses, local elevation of the applied stress, and introduction of weld imperfections. Depending on the operating environment and loading conditions, one or more of these features could lead to failure by a number of different mechanisms. Even though today's engineers are armed with good tools, and they comply with specific industry/regulatory codes or standards for design and manufacturing, failures are still inevitable. There are many reasons why in-service failures can occur, including:

- Lack of knowledge of service loads and cycles
- Lack of knowledge of the operating environment
- Improper specification of the design life
- Improper use of the design method and lack of consideration of key failure modes
- Use of incorrect material properties for design
- Improper selection of materials and welding procedures

- Lack of inspection during fabrication and variability in fabrication practice
- Operation of equipment/component beyond design specification

In welded structures, fatigue cracking is by far the most common failure mechanism, and unstable fracture (although rare) is perhaps the most dramatic, occurring without warning and often leading to serious consequences. This article discusses the various options for controlling fatigue and fracture in welded steel structures, the factors that influence them the most, and some of the leading codes and standards for designing against these failure mechanisms. It excludes high-temperature fracture, environmental cracking, and nonferrous materials.

Fracture Control Plans for Welded Steel Structures

Weld Imperfections. All welded structures contain imperfections to some level of examination, and the joint itself is a discontinuity in the structure. Welding imperfections (Fig. 1) fall into three broad categories: planar imperfections, volumetric imperfections, and geometrical imperfections.

Planar imperfections are sharp crack-like features that can substantially reduce the fatigue strength of a welded joint or cause initiation of brittle fractures. Examples include hydrogen cracks, lamellar tears, lack of fusion, reheat cracks, solidification cracks, and weld toe intrusions. The last group are nonmetallic intrusions at the weld toe region that act as crack starters for fatigue. They are present on a microscopic level, and on average they are typically 0.1 mm (0.004 in.) in depth and can be as much as 0.4 mm (0.016 in.) deep (Ref 17). These features are the primary reason why fatigue in welds is dominated by crack propagation.

Volumetric imperfections include porosity and slag inclusions. Because these types of imperfections tend to be nearly spherical in form, their notch effect is minor and they usually have little or no influence on the fatigue behavior. However, they do reduce the load-bearing area of the weld and hence reduce the static strength of the joint.

Geometrical imperfections include misalignment, overfill, stop/starts, undercut, and weld ripples. Geometric imperfections have the effect of locally elevating the stress over and above the general stress concentration due to the joint geometry. Methods for calculating the stress concentration factor due to misalignment are readily available (Ref 18). However, unintentional mis-

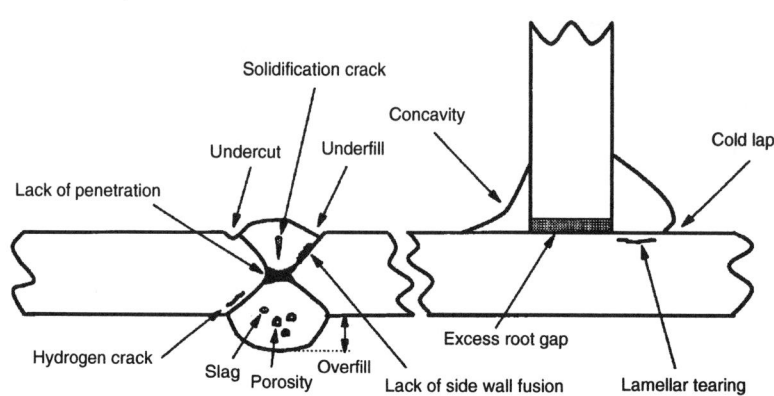

Fig. 1 Defects and discontinuities in welded joints

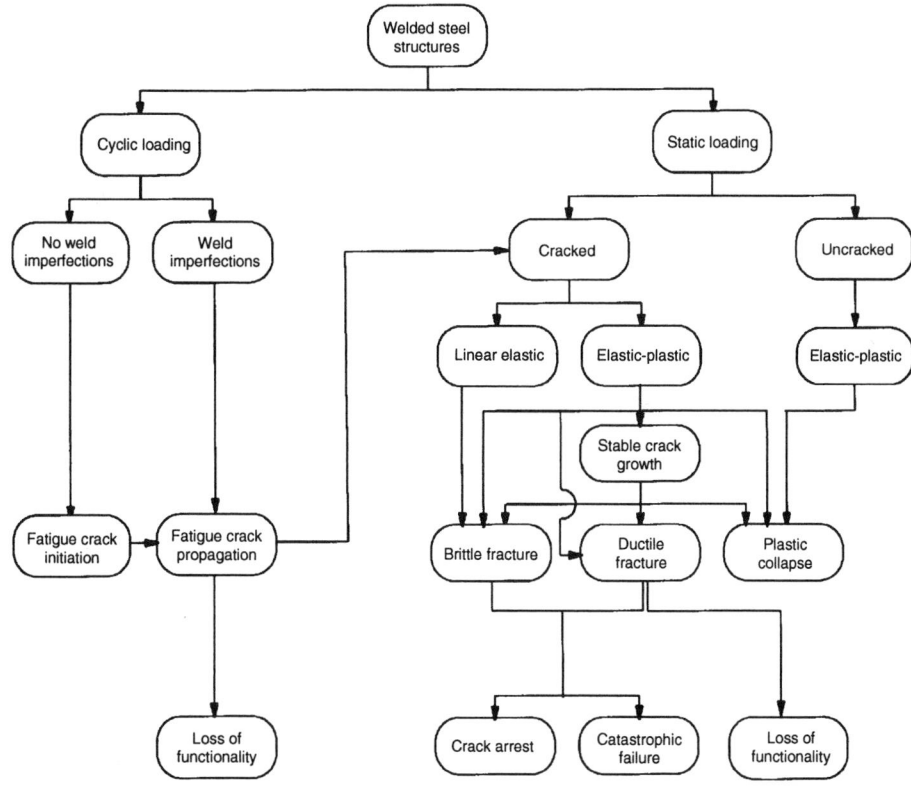

Fig. 2 Fracture paths for welded steel structures

Fig. 3 Schematic representation of a crack growing in service under fatigue loading. Failure occurs if $a_f \geq a_c$. a_0, start-of-life defect size; a_f, defect size due to fatigue crack growth; a_c, critical defect size due to unstable fracture or plastic collapse

alignment that occurs during fabrication cannot be allowed for during design, and this can only lead to early failures. Guidance is also available on acceptable levels for undercut, overfill, and location of stop/starts (Ref 19). Weld ripples become stress concentrators for fatigue only when the weld is loaded axially.

Failure Modes. Depending on the operating environment and the nature of the applied load, a steel structure or component operating under ambient conditions can fail in many different modes. The potential failure modes for welded fabrications are:

- Unstable fracture (brittle or ductile)
- Ductile fracture
- Plastic collapse
- Buckling
- Fatigue
- Corrosion fatigue
- Corrosion
- Stress-corrosion cracking
- Hydrogen-induced cracking

The first four modes of failure occur under static load. While brittle fracture can occur at nominal stresses substantially below the yield point and with negligible overall deformation, ductile fracture, plastic collapse, and buckling are often preceded by significant plasticity. The remaining five failure modes can be broken down into fatigue and environmental cracking.

Figure 2 summarizes the different loading paths that can result in failure of a statically or

cyclically loaded structure (assuming environmental effects are not significant). The loading paths range from initiation of fatigue cracks to crack propagation under cyclic loading, which could then potentially lead to brittle fracture under nominally elastic conditions (applied stresses well below yield) or to plastic collapse (overload of the remaining structure).

Fracture Control Methods. Figure 3 shows a schematic of a weld joint in a structure subjected to cyclic loading. As stated earlier, there are preexisting metallurgical discontinuities in welded joints to some degree. In Fig. 3 these are referred to as start-of-life defect size, a_0. If the applied loading conditions are favorable, fatigue cracks start from the initial starter cracks represented by a_0 and propagate into the material. The instantaneous fatigue crack size at any stage during its life is referred to as a_f, and the rate of crack growth is controlled by the material properties and by the applied loading conditions. During this stage the crack will continue to grow in a stable manner until critical conditions are reached for failure. These critical conditions could be loss in the functionality of the structure/component due to increase in compliance of the load-bearing member, or failure of the remaining ligament material ahead of the growing fatigue crack by unstable fracture or plastic collapse. The critical flaw dimension, a_c, at the onset of this condition is controlled by the applied loading conditions and by material properties such as fracture toughness and flow strength.

Extension of preexisting cracks or crack-like features in a stable manner (through fatigue cracking or ductile tearing) or in an unstable manner (through brittle fracture) depends on three critical parameters: the applied loading conditions, the resistance of the material to crack extension, and the size of the crack. This combination represents the classical basis for fracture mechanics. For a given crack, the applied loading conditions (primary stresses from application loads) and secondary stresses (e.g., welding residual stresses and stress concentration effects) provide the crack driving force, which is resisted by the material ahead of the crack to extension, whether this is under stable or unstable conditions. In fracture mechanics terms, ΔK_{th} is frequently used as the resistance of the material to crack extension for fatigue. For unstable fracture, toughness parameters such as K, J, and crack-tip opening displacement (CTOD) are used. The general principles described here are schematically shown in Fig. 4. The concept of arresting a propagating crack is also introduced. In certain industries, designing-in the ability to arrest a running crack can prevent serious consequences or reduce the amount of damage.

The above discussion provides a basis for defining the various fatigue and fracture control plans that are used in industry. These are schematically shown in Fig. 2, and they can be broken down as: approaches to fatigue control, approaches to fracture control, or approaches to crack arrest.

Approaches to fatigue control include: 1) avoid fatigue crack initiation, 2) avoid fatigue crack growth, and 3) assume that fatigue crack growth will occur and design to prevent unstable fracture. In welded joints, initiating a fatigue crack is rare. Most of the time is spent in propagating preexisting features in the weld area. Only in certain circumstances, such as machined joints or specially treated welds, can designing for crack initiation become an option.

Approaches to fracture control include: 1) avoid unstable fracture initiation, and 2) assume that unstable fracture may occur and design for crack arrest. For most welded fabrications, the second option is not acceptable. Catastrophic failures are avoided by specifying adequate "initiation" toughness for both the weld area and the base material at the minimum design temperature. However, for applications such as storage vessels and ship structures (particularly oil tank-

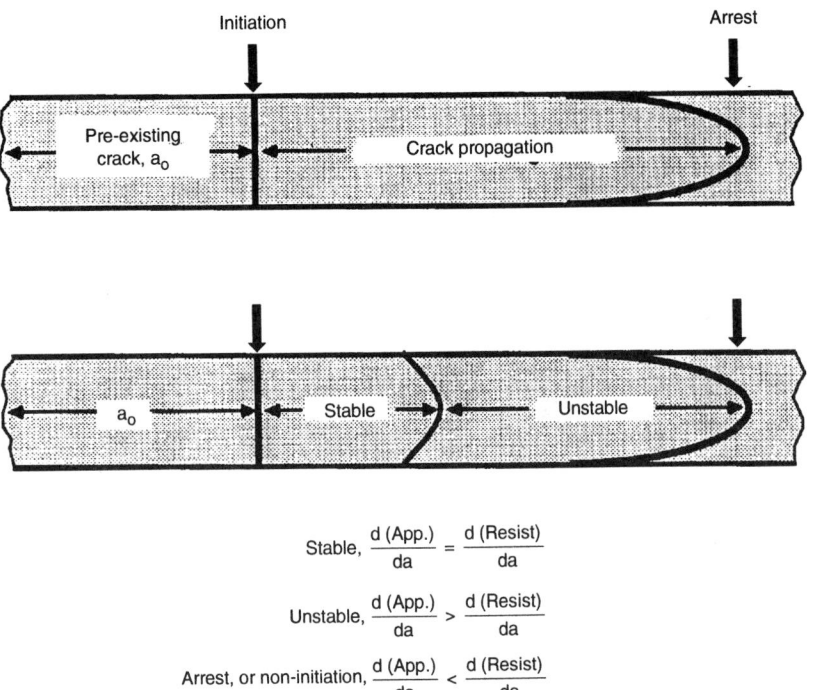

Stable, $\dfrac{d(\text{App.})}{da} = \dfrac{d(\text{Resist})}{da}$

Unstable, $\dfrac{d(\text{App.})}{da} > \dfrac{d(\text{Resist})}{da}$

Arrest, or non-initiation, $\dfrac{d(\text{App.})}{da} < \dfrac{d(\text{Resist})}{da}$

Fig. 4 Schematic representation of the conditions for stable and unstable fracture and crack arrest

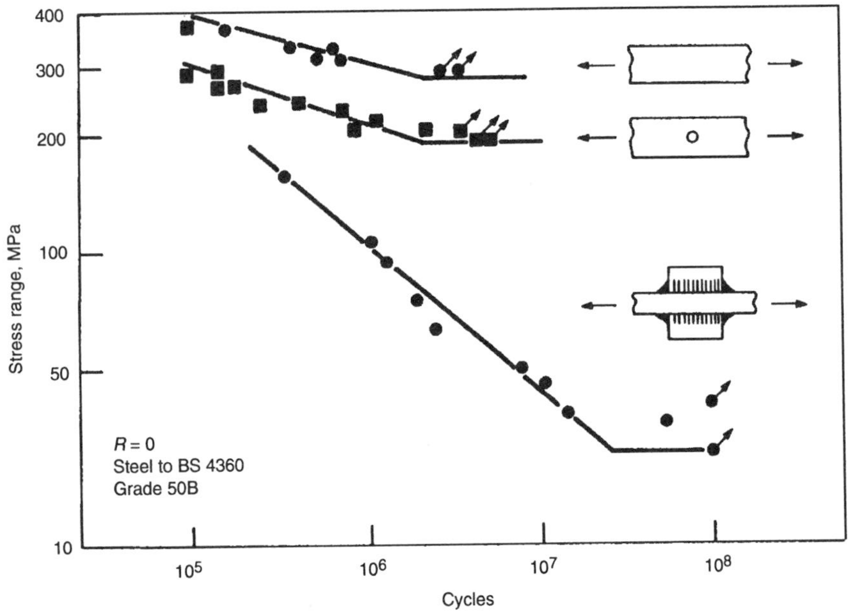

Fig. 5 Comparison of fatigue behavior of a welded joint and parent metal

significantly. This is often illustrated by comparing the fatigue lives of welded and plain components. As shown in Fig. 5, with all things being equal, attaching a weld to a load-carrying member can not only reduce the fatigue strength substantially, but also lower the fatigue limit. In this example, the fatigue limit of the welded component is one-tenth of that of the plain component. As a consequence of this phenomenon, it is frequently found that in cyclically loaded welded components the design stresses are limited by the fatigue strength of the welded joints.

There are several reasons why a weld reduces the fatigue strength of a component. These reasons fall into the following categories:

- Stress concentration due to weld shape and joint geometry
- Stress concentration due to weld imperfections
- Welding residual stresses

In order to be able to design against fatigue in welded structures, it is important to understand the influence of these factors on the performance of welded joints.

Stress Concentration due to Weld Shape and Joint Geometry. Making a welded joint either increases or decreases the local cross-sectional dimensions where the parent metal meets the weld. Generally, any change in cross-sectional dimensions in a loaded member will lead to concentration of stress. Thus, it is inevitable that the introduction of a weld will produce increase in the local stress. The precise location and the magnitude of stress concentration in welded joints depends on the design of the joint and on the direction of the load. Some examples of stress concentrations in butt welds and fillet welds are given in Fig. 6. In Fig. 6(b) the weld does not carry any load, but it nevertheless causes concentration of stress at the toes of the welds. Indeed, the weld toe is often the primary location for fatigue cracking in joints that have good root penetration. In situations where the root penetration is poor or the root gap is excessive, or in load-carrying fillet welds where the weld throat is insufficient, the root area can become the region of highest stress concentration. Fatigue cracks in these situations start from the root of the weld and generally propagate through the weld (Fig. 7).

The geometry/shape parameters that influence fatigue of welded joints by affecting the local stress concentration include plate thickness (T), attachment toe-to-toe length (L), attachment thickness (t), weld toe radius (r), weld angle (θ), and the profile of the weld surface (convex vs. concave). See Fig. 8. It is generally found that the fatigue strength of a welded joint decreases with increasing attachment length (Fig. 9a), increasing plate thickness (Fig. 9b), increasing weld angle (Fig. 9c, d), decreasing toe radius (Fig. 9e), and misalignment (Fig. 9f). The results given in Fig. 9 are experimental but have been confirmed by finite-element stress analysis of the local weld area. Thurlbeck and Burdekin (Ref 20) show that decreasing the ratio of attachment length to thickness (L/T) from 2.0 to 0.375 will increase the relative fatigue strength by a factor of about 1.2.

ers), it should be comforting to know that when a brittle fracture is initiated, the crack will not extend far before coming to rest. In this case, the second option is a secondary means to minimize the damage by arresting the brittle fracture close to the initiation site.

Approaches to crack arrest include: 1) crack arrestors built into the structures (e.g., "rat holes,") 2) change in section, or addition of stiffeners and 3) materials selection (use of materials with high crack arrest toughness).

In the next two sections, the above approaches are discussed in the context of fatigue and fracture separately. The important factors associated with each of these failure modes are identified.

Fatigue in Welded Joints

It is now commonly accepted that attaching a weld to a cyclically loaded structural member can change the service performance of that member

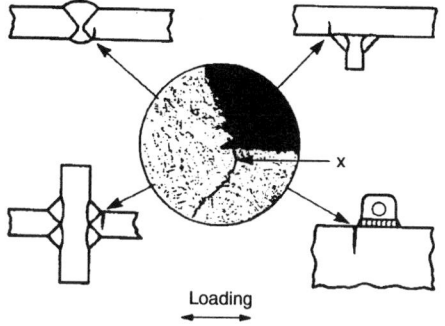

Fig. 6 Examples of stress concentrations in welded joints

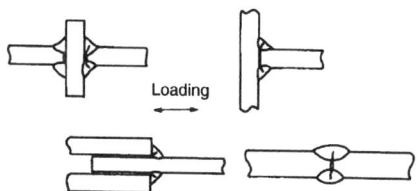

Fig. 7 Fatigue cracking from the weld root

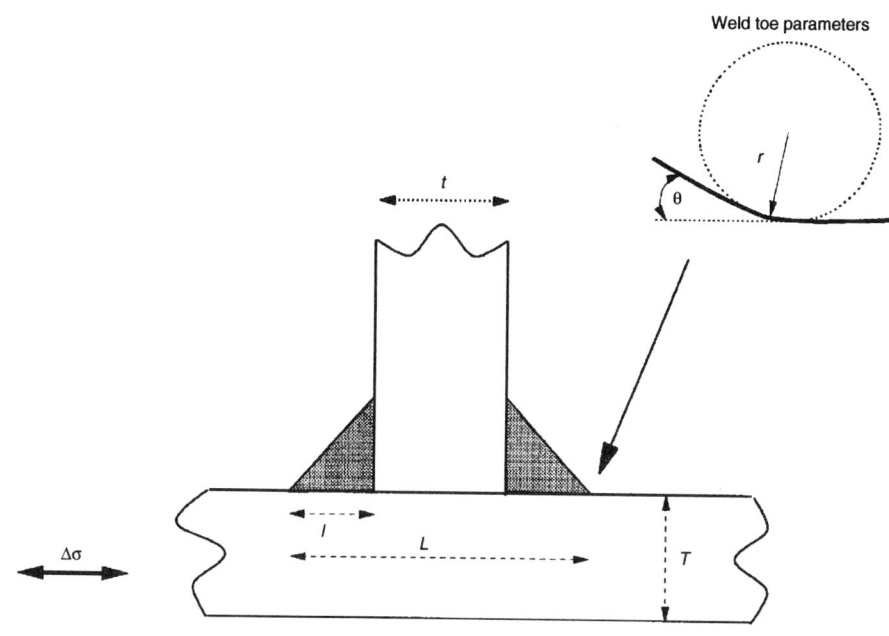

Fig. 8 Geometry parameters that affect weld toe stress concentration. *r*, weld toe root radius; θ, weld angle; *L*, toe-to-toe weld length

Similar results are obtained for changing the weld angle from 90° to 22.5°. By far the largest influencing parameter is the weld toe radius, which, for example, can increase the fatigue strength over the range $r = 1.0$ to 8.0 mm (0.04 to 0.3 in.) by a factor of 1.3, effectively doubling the life. The weld toe stress concentration factor, K_t, is a function of all these geometry variables, and a number of formulas are available for butt welds and cruciform joints subjected to bending or tensile load (Ref 21, 22). The following formula is for a T-joint subjected to bending load:

$$K_t = 1 + \left[\frac{1 - \exp\left[-0.90\,\theta\,\sqrt{\frac{T}{2l}+1}\right]}{1 - \exp\left[-0.45\pi\,\sqrt{\frac{T}{2l}+1}\right]} \right]$$

$$\left[\frac{0.13 + 0.65\left(1 - \frac{r}{T}\right)^4}{\left(\frac{r}{T}\right)^{1/3}} \right] \tanh\left[\frac{\left(\frac{2l}{T}\right)}{1 - \frac{r}{T}} \right]^{1/4} \quad \text{(Eq 1)}$$

with the validity boundaries $r/T = 0.02$ to 0.2, $l/T = 0.5$ to 1.2, and $\theta = 30$ to 80°.

Misalignment is another geometry-related parameter that can influence the fatigue performance of a welded joint. Misalignment can manifest itself in several ways, angular and axial misalignments being the most common. Generally fatigue strength decreases with increasing misalignment, as shown in Fig. 9(f). Formulas to calculate the local weld stress concentration factor due to misalignment are now well established (Ref 18).

The degree of influence of the geometry parameters cited above can vary significantly with the loading condition (e.g., tension vs. bending) and the joint type.

Stress Concentration due to Weld Discontinuities. It is found from stress analysis of an idealized model of a fillet weld loaded in the transverse direction that the stress concentration factor (K_t) at the weld toe is about 3. This is comparable to K_t for a hole in a plate. In view of this, it would be reasonable to expect that the fatigue behavior of a fillet weld is similar to that of a plate with a hole. However, as shown in Fig. 5, the fatigue performance of the welded joint is substantially lower, implying that other factors come into play for welded joints.

As stated above, weld imperfections are to some extent controllable and can be avoided during fabrication, or their effects can be included in the design. The difference observed in Fig. 5 has been attributed by some researchers to the presence of microscopic features at the weld toe. These features are small, sharp nonmetallic intrusions and are present in most, if not all, welds. The extent and distribution of these features vary with the welding process, and also possibly with the quality of the steel plate and its surface condition. The exact source of these intrusions is not precisely known, but it is believed that slag, surface scale, and nonmetallic stringers from a dirty steel are the primary causes. An example of a weld toe intrusion is given in Fig. 10. The combined effect of these sharp crack-like features and concentration of stress due to the weld geometry is that fatigue cracks initiate very early on, and most of the life is spent in crack propagation.

Planar weld imperfections (e.g., hydrogen cracks, lack-of-side wall fusion) are clearly to be avoided because they will substantially reduce the fatigue life. Volumetric imperfections such as slag inclusions and porosity can be tolerated to some extent (Ref 19), because the notch effect of these imperfections is generally lower than that of the weld toe (Fig. 11).

Welding Residual Stresses. In welded structures, it is normally found that residual stresses are present in the weldment area, and these can be high and can approach the yield strength of the material. These stresses occur as a result of the thermal expansion and contraction during welding, as a result of the constraint provided by the fabrication or by the fixtures, and as a result of distortion in the structure during fabrication (often referred to as reaction stresses). These stresses are localized to the weld zone and are self-balancing (i.e., both tensile and compressive stresses are present). Transverse to the weld toe the residual stress is typically tensile and can approach yield point. When a load cycle is applied to the structure, it is superimposed onto the residual stress field, and the effective stresses acting at the weld joint can fluctuate down from yield level (Fig. 12). The range of each cycle remains unchanged, but the effective mean stress can be significantly different from the applied mean stress. Because of the dominance of crack propagation in welded joints and the presence of high tensile residual stresses, the mean stress effect is negligible, and fatigue life is controlled by the stress range (Fig. 13).

Reducing residual stresses using postweld heat treatment can improve fatigue life, but only if the applied load cycles are partially or fully compressive. For fully tensile applied loads, postweld heat treatment does not improve the fatigue life (Fig. 13b). Thus, it is important to know the exact nature of applied loads before a decision is made on the need to heat treat a welded structure. Indeed, it should be noted that stress relief of welded joints is never fully effective. Residual

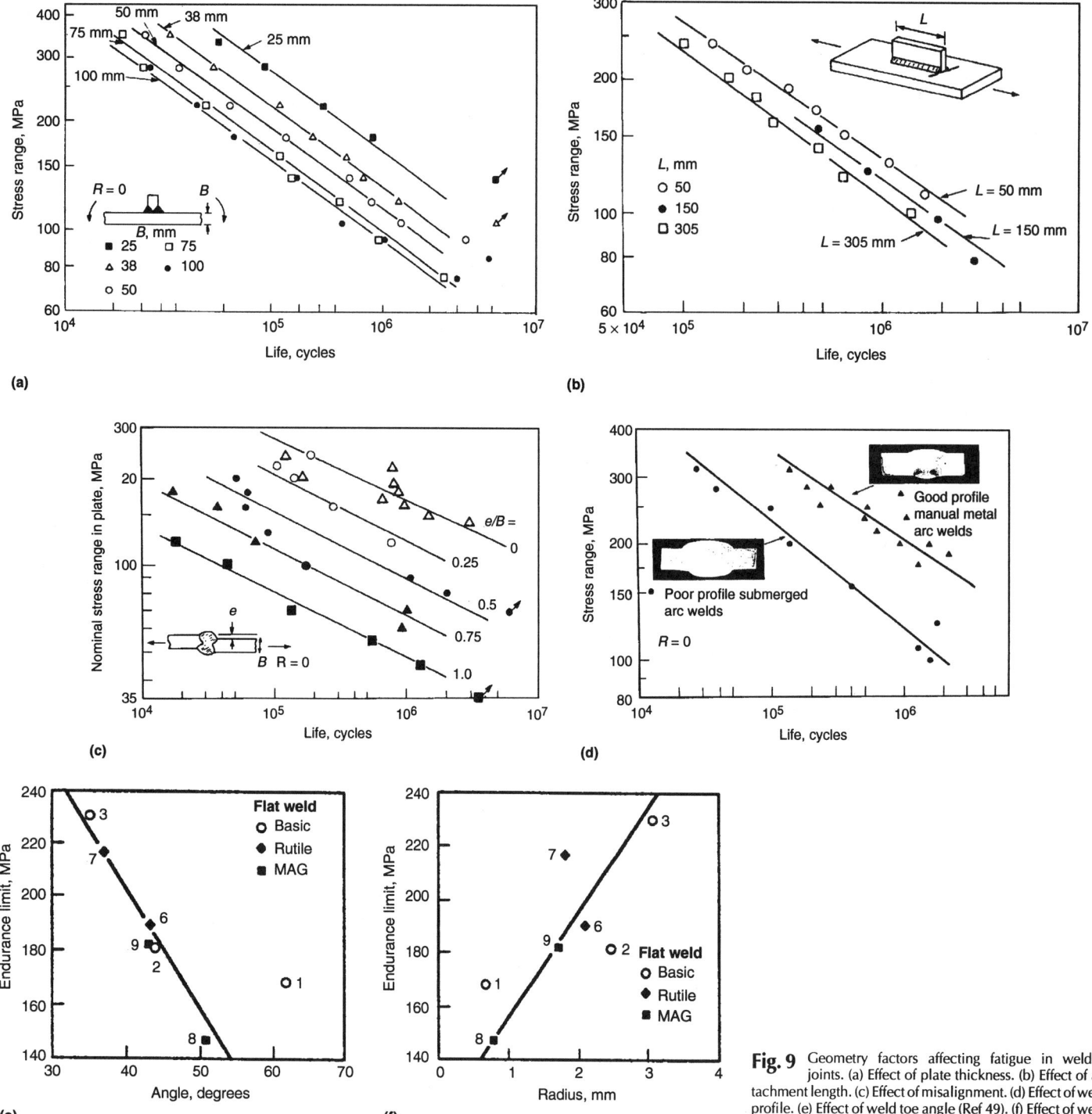

Fig. 9 Geometry factors affecting fatigue in welded joints. (a) Effect of plate thickness. (b) Effect of attachment length. (c) Effect of misalignment. (d) Effect of weld profile. (e) Effect of weld toe angle (Ref 49). (f) Effect of weld toe radius (Ref 49)

stresses up to yield point have been measured in stress-relieved pressure vessels and in up to 75% of yield in stress-relieved nodes in offshore structures. Due consideration has to be given to the complexity of the overall structure when stress relieving and to the time and temperature of the process.

Effect of Material Properties. Because crack propagation dominates the fatigue life of welded joints, material properties have no effect on fatigue strength. This is illustrated in Fig. 14, where

it can be seen that data points for steels with different strengths fall within the same scatter band. Thus, using a high-strength steel to improve fatigue life will not be beneficial for welded structures.

Microstructure. A fatigue crack starting at the weld toe will immediately grow into the heat-affected zone (HAZ) and then into the base metal. During this period it will propagate through a variety of microstructures, and as can be seen from Fig. 15, its growth rate will not be influ-

enced in any way. Thus, the variety of HAZ microstructures (and hardness levels) in the weldment area have little or no effect on the rate at which the crack grows.

Fracture Toughness. As with tensile strength, the fracture toughness of the weld metal, the HAZ, or the base metal does not influence the crack growth rate. This can be deduced from Fig. 15, which represents a variety of materials with different strength levels and toughness values. The only influence of fracture toughness is the

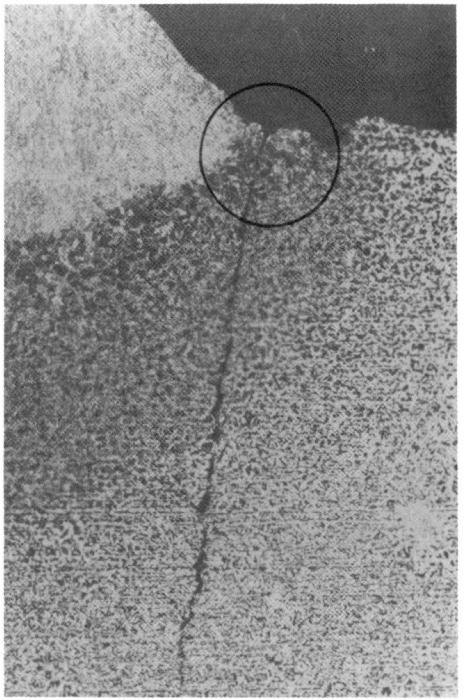

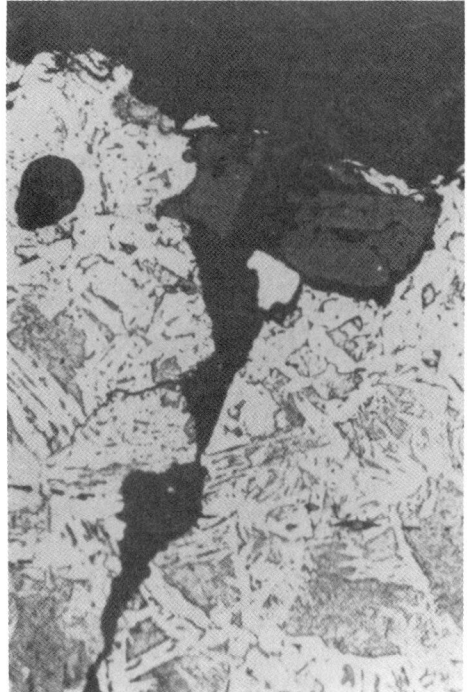

(a)

(b)

Fig. 10 Example of weld toe intrusions

limiting size the fatigue crack could reach before the material fails in an unstable manner. Tougher materials are able to tolerate bigger fatigue cracks.

Methods for Improving the Fatigue Life of Welded Joints. Postweld fatigue life improvement techniques are becoming popular for increasing the fatigue life of weld joints. From the preceding sections it can be seen that three factors influence fatigue of welded joints: stress concentration due to joint and weld geometry, stress concentration due to localized defects, and welding residual stresses. Improvement in fatigue life can be obtained by reducing the effects of one or more of these parameters. This is particularly true for cracks starting from the weld toe, which is by far the most common failure site. In load-carrying

fillet welds, however, cracks can start at the weld root and propagate through the weld throat until failure occurs. In this case, additional weld metal to increase the weld throat dimension will result in reduction in shear stress and, hence, a corresponding increase in fatigue strength.

Because the most common failure site is the weld toe, many postweld treatments for this region have been developed to improve the fatigue life. These techniques largely rely on removing the detrimental intrusions at the weld toe, reducing the joint stress concentration, or modifying the residual stress distribution. There are primarily two broad categories of postweld techniques: modification of the weld geometry and modification of the residual stress distribution (Ref 23, 24). A complete list is given in Fig. 16. Methods

that reduce the severity of the stress concentration or removal of weld toe intrusions include grinding, machining, or remelting. These techniques essentially focus on altering the local weld geometry by removing the intrusions and at the same time on achieving a smooth transition between the weld and the plate. Typical profiles of burr ground and tungsten inert gas (TIG) dressed welds are given in Fig. 17. Because weld toe intrusions can be up to 0.4 mm (0.016 in.) in depth, the general guideline is that the grinding or machining operation must penetrate at least 0.5 mm (0.02 in.) into the parent plate. In these procedures the depth, the diameter of the groove, and the direction of the grinding marks become important issues. In TIG and plasma dressing, the weld toe is remelted in order to improve the local profile and also to "burn" away or move the intrusions. With these techniques, the position of the arc with respect to the weld toe and the depth of the remelt zone are two critical variables. Controlling the shape of the overall weld profile has also been recognized as a method of improving the fatigue life. The American Welding Society (AWS) weld profile procedure using the "dime" test reduces the weld geometry K_t and hence increases the fatigue life (Ref 25, 26).

Methods that modify the residual stress field include heat treatment, hammer and shot peening, and overloading. Postweld heat treatment is known to reduce tensile residual stresses but does not eliminate them completely. The benefits of postweld heat treatment can be realized only if the applied loads are either partially or fully compressive. Overloading techniques rely on reducing the tensile residual stress field and/or introducing compressive stresses at the weld toe. Exact loading conditions for a complex structure are difficult to establish, so this technique is rarely used. To obtain significant improvement in fatigue strength, it is necessary to introduce compressive stresses in the local area in a consistent and repeatable manner. The three peening techniques (shot, needle, and hammer peening) aim to achieve this by cold working the surface of the weld toe. As with the grinding methods, it is necessary to penetrate the parent plate and deform to a depth of at least 0.5 mm (0.02 in.). Hammer peening is perhaps the best technique to

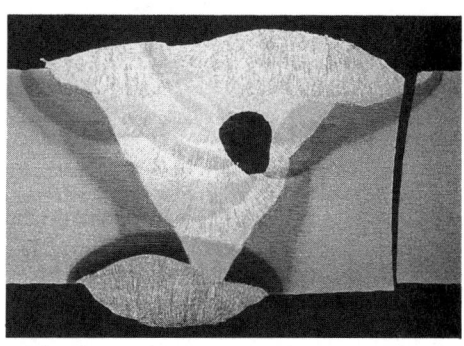

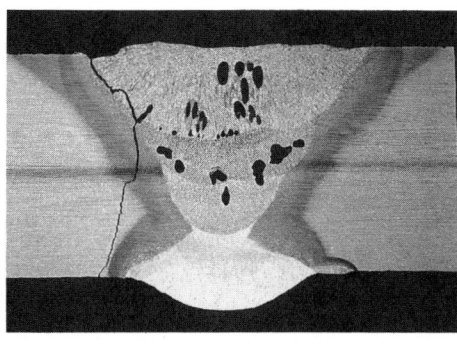

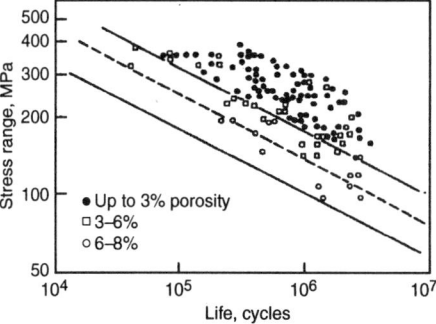

(a)

(b)

(c)

Fig. 11 Effect of volumetric defects on fatigue. (a) Slag inclusion in butt weld. Cracking from weld toe. (b) Porosity in butt weld. Cracking from weld toe. (c) Transverse groove welds containing porosity

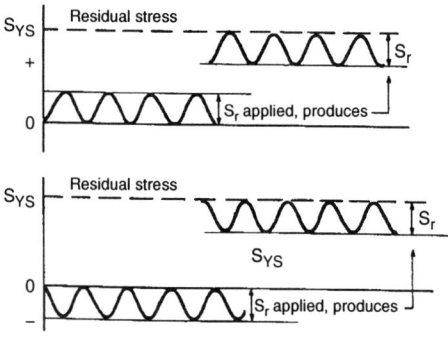

Fig. 12 Superposition effect of applied and local welding residual stresses

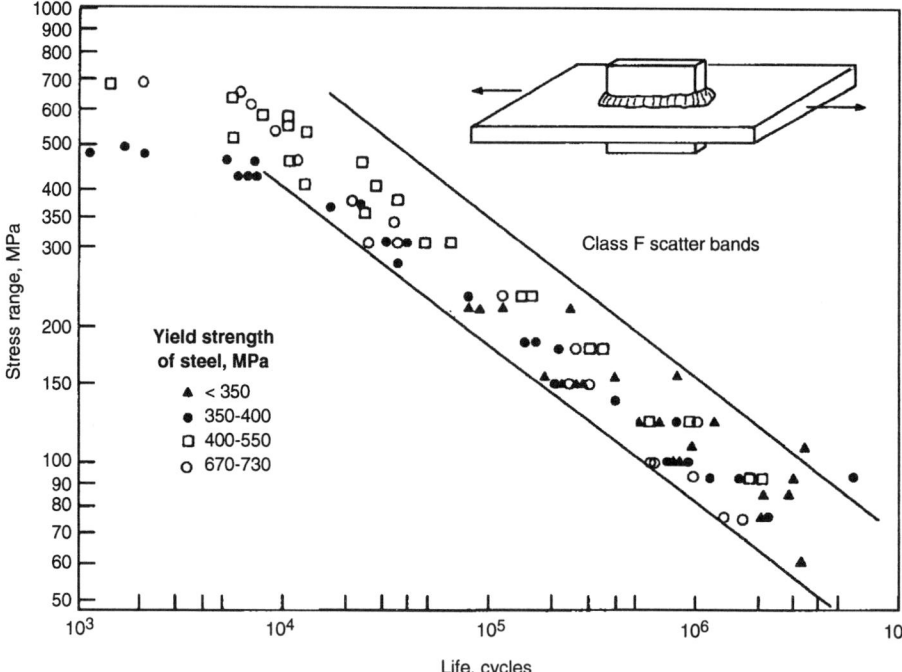

Fig. 14 Fatigue test results from fillet welds in various strengths of steel

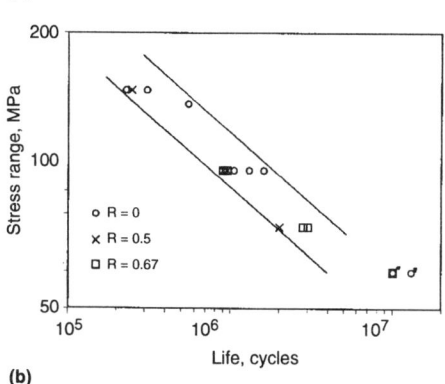

(a)

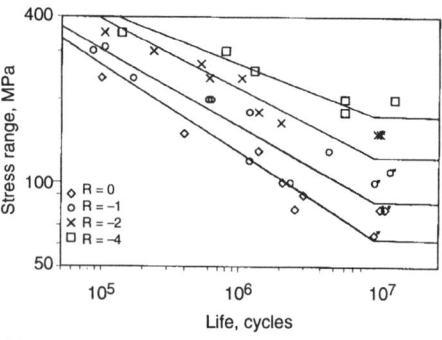

(c)

Fig. 13 Influence of welding residual stress on fatigue. (a) Effect of applied stress ratio on as-welded joints. (b) Effect of stress relief as a function of tensile load cycle. All specimens stress relieved. (c) Effect of stress relief as a function of tension-compression load cycle. All specimens stress relieved

do this, and it has the advantage of removing the weld toe intrusions by cold working them out. Peening techniques require special equipment and can present safety challenges for noise control. They are very difficult to automate and control in production.

A comparison of the improvement in fatigue strength obtained by some of these techniques is given in Fig. 18. It can be seen that hammer peening is perhaps the best technique, perhaps because it reduces or eliminates intrusions as well as introducing compressive stresses. Burr grinding, which is easier to implement, also produces significant improvement in fatigue life. Disc grinding and burr grinding are probably the most frequently used methods for improving the fatigue life of welded joints.

Fracture Control in Welded Structures

The brittle fracture of ships during World War II stimulated intensive research into avoiding catastrophic failures in fabricated structures. The nature of this type of failure provides no warning, and the aftermath is often spectacular (Fig. 19). These early failures led to many studies to identify the parameters that need to be controlled in order to avoid brittle fractures. Toughness of the material was one parameter that emerged from these studies. Much of the early work focused on measuring the toughness of steels used in ship structures, using notched-bar impact specimens. The most widely used impact specimen is the Charpy V-notch impact test, and today the test has become one of the necessary means of ensuring that the weld metal, the base metal, and the

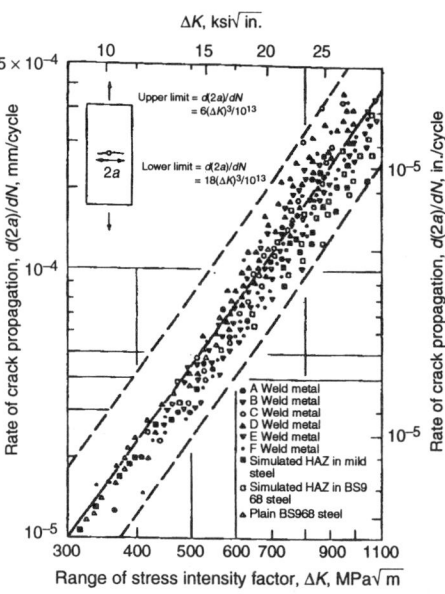

Fig. 15 Crack growth rate data showing no influence of weld metal, HAZ, or base metal

HAZ have sufficient toughness to avoid unstable fractures.

The Charpy test evaluates both the *initiation* of brittle fracture as well as its subsequent *propagation*. The significance attached to controlling fracture initiation or propagation has influenced much of the fracture research work since the early period, and it has resulted in two philosophies for fracture control.

U.S. researchers concluded that because of the highly heterogeneous nature of welds, it would

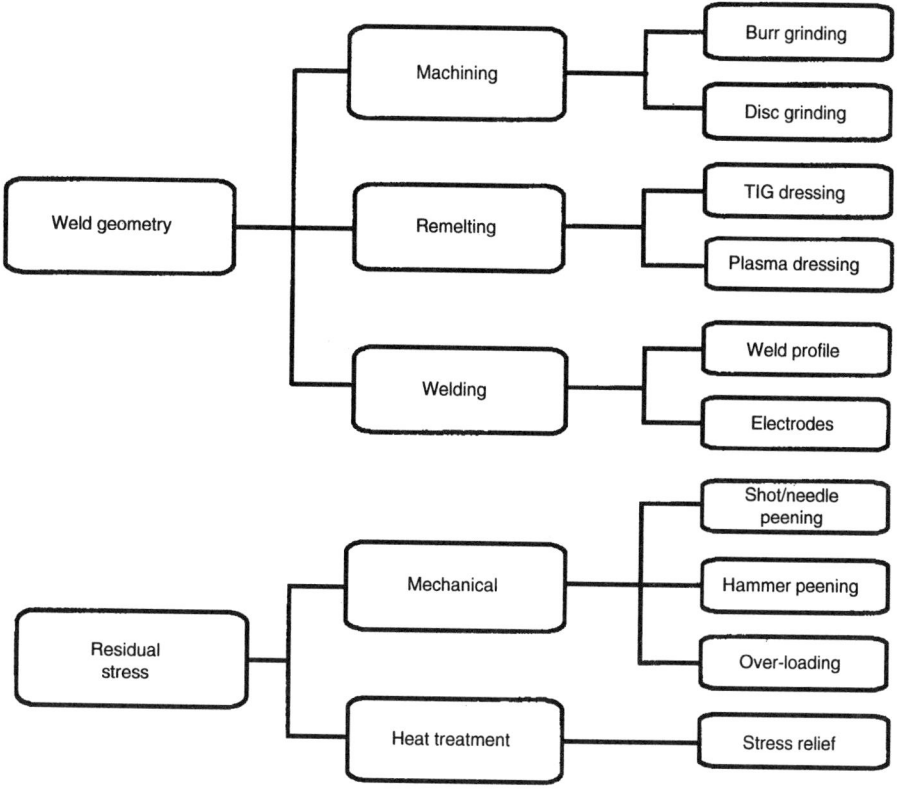

Fig. 16 Range of postweld fatigue life improvement techniques

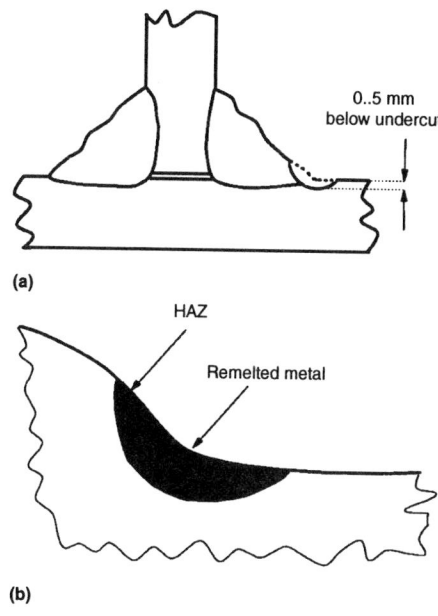

Fig. 17 Typical profiles for (a) burr grinding and (b) tungsten inert gas dressing of weld toe

be impractical and risky to try to control brittle fracture initiation in welded structures. It was argued that "pop-in" of fractures from brittle regions was likely and that the steel toughness should be sufficient to prevent a pop-in from propagating into a full-scale fracture. Pellini et al. (Ref 27) used this concept to propose the first practical design tool linking the required fracture toughness to temperature and design stress. The approach was presented in the form of a fracture analysis diagram (Fig. 20) that has a constant shape, where the position of the curve along the temperature axis is indexed to the "nil-ductility temperature" (NDT), which is obtained from the drop weight test. This approach has been widely used for fracture control of many types of structures, and as far as is known, there have been no brittle fractures in steels that meet the NDT criteria.

The competing approach is to try to control initiation of the fracture. This has been advocated mostly by the Europeans, and now more recently by the North Americans. The initiation approach is intimately linked to fracture mechanics concepts, and in many codes and standards it has been used to specify toughness requirements for weld and base metals and has been valuable in establishing flaw acceptance criteria for inspection. For welded joints it was recognized that initiation of fracture took place under the influence of high tensile residual stresses and stress concentration effects, which led to the use and advancement of postyield fracture criteria, pri-

marily through the development of fracture mechanics approaches for CTOD. The onset of brittle fracture requires the presence of a flaw of critical dimensions, low toughness, and stress (primary and secondary) of critical magnitude. Here the discussion focuses only on toughness and on the specification of toughness levels for

welded regions and parent materials to avoid fracture initiation. The factors that influence the initiation of unstable fracture are identified.

Factors Affecting Fracture Toughness. The toughness of a material can be measured using a standard Charpy V-notch specimen, as stated above, or by using a fracture mechanics test specimen. The commonly referenced results from Charpy impact tests include the impact energy required to fracture the specimen, the lateral expansion, and the percent shear fracture appearance at the specified temperature. The latter two

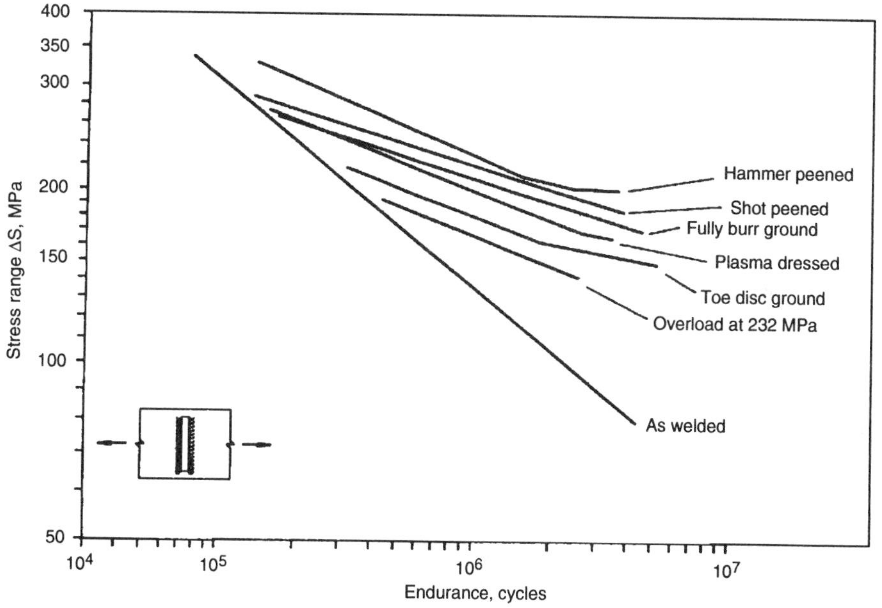

Fig. 18 Comparison of different improvement techniques

are intended as measures of ductility. In contrast to the Charpy test, fracture mechanics toughness tests fall into two groups: linear elastic fracture mechanics tests (K_{Ic}) and elastic-plastic fracture mechanics tests (CTOD, J-curves, and R-curves). The major advantage of using fracture mechanics specimens is that the data from these tests can be used directly in fracture mechanics fitness-for-service procedures to assess the tolerance of the structure to the presence of defects and cracks. The parameters that affect the toughness of a material include:

- Material grade and chemistry
- Strength and hardness
- Microstructure and level of impurities
- Grain orientation
- Size of structure or specimen (constraint)
- Notch/crack acuity
- Loading rate
- Temperature

Some of these are discussed below in the context of welded regions.

In conventional structural steels that have a ferritic microstructure, both Charpy and fracture mechanics toughness tests show a transition from ductile to brittle behavior with decreasing temperature. Figure 21 shows Charpy energy transition curves for a variety of steels and for a number of other materials. At higher temperatures, in the upper-shelf regime, a ferritic material behaves in a fully ductile manner and is capable of absorbing considerable amounts of energy when deformed. At these temperatures, ferritic materials are generally considered tough. As the temperature is reduced, a transition to lower-shelf behavior occurs. In this regime ferritic materials are brittle, and fracture occurs in a cleavage mode. The decrease in toughness (energy absorbed) from upper-shelf behavior can be substantial, and for some materials the transition temperature range is very small.

From Fig. 21 it is clearly evident that toughness is a function of the material type and the yield strength. Furthermore, the transition behavior is affected by the size and type of the specimen, the loading rate, and the notch acuity. Generally, the toughness is reduced with increasing loading rate, increasing specimen size, increasing notch acuity, increasing thickness, and increasing crack depth (Fig. 22). Thus, neither the fracture strength nor the toughness is a sole predictor or criterion for fracture control. This can present challenges to a design engineer who would like to be able to predict conditions for the onset of brittle fracture from some simple specifiable material properties. Also, this indicates that it is important to match the test specimen size and test conditions closely with the local region in the actual structure.

For welded joints, the challenge of generating toughness information is increased further by the nonhomogeneous nature of the region. Variations in notch location and orientation can result in significantly different toughness values. The weld HAZ presents particular difficulties in toughness measurements owing to its narrow

Fig. 19 Example of brittle fracture in ship structures

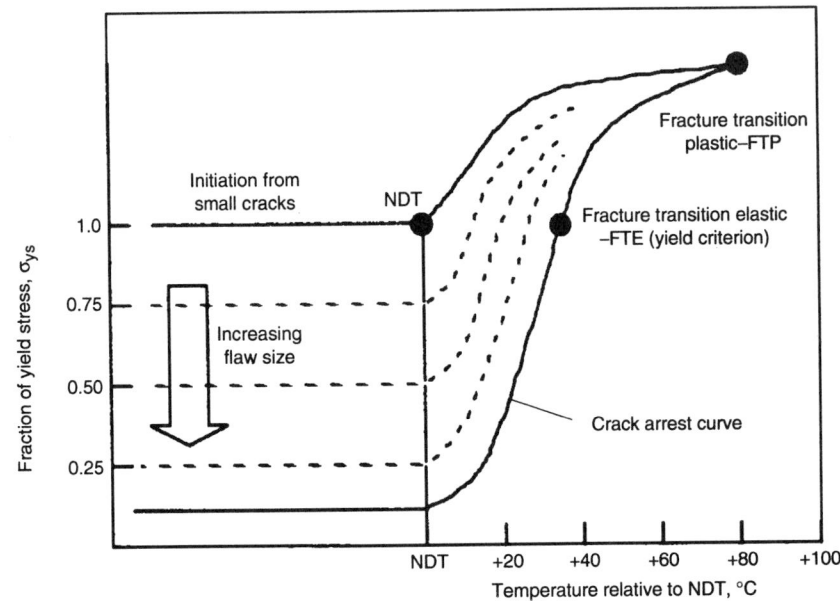

Fig. 20 Pellini's fracture analysis diagram

width. For example, in structural steels welded using typical arc energies, the transformed HAZ is usually no more than 3 to 4 mm (0.12 to 0.16 in.) wide. Within this distance there are wide ranges of microstructures and properties (Fig. 23), and the critical region of interest may be 1 mm (0.04 in.) or less. Despite their small size, brittle regions within the HAZ can have a significant influence on the integrity of a structure with respect to failure by brittle fracture. A number of catastrophic brittle fractures of engineering structures, including pressure vessels, storage tanks and bridges, in which the fracture started in the HAZ testify to this (Ref 28).

Figure 24 shows some examples of notch locations and orientations in the weld metal and the HAZ for measurement of fracture toughness. While placement of the notch in the weld metal is simple and straightforward, HAZ notch location can be difficult. Some industries have developed specific procedures that simplify Charpy testing of HAZ and ask for through-the-thickness notches located at the fusion boundary (FB), FB + 2 mm (0.08 in.), and FB + 5 mm (0.2 in.).

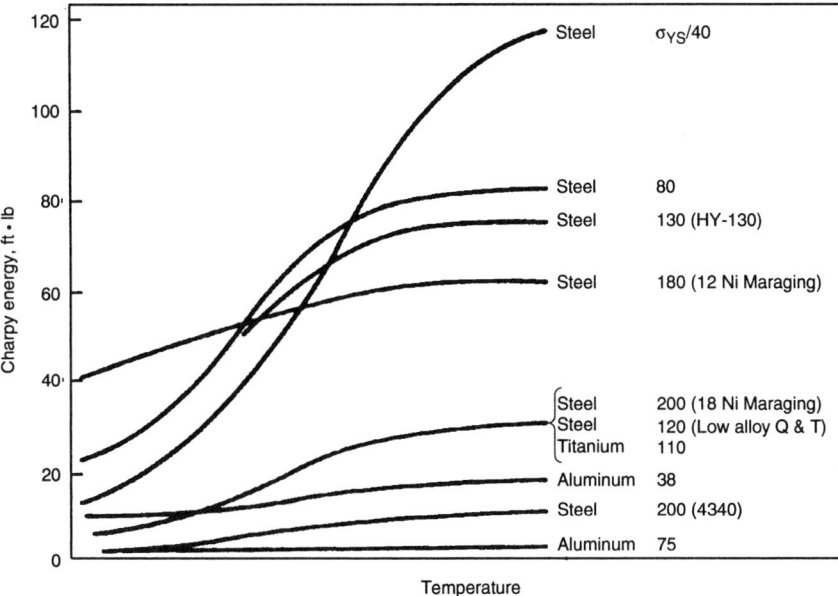

Fig. 21 Ductile-to-brittle transition curves for a variety of materials

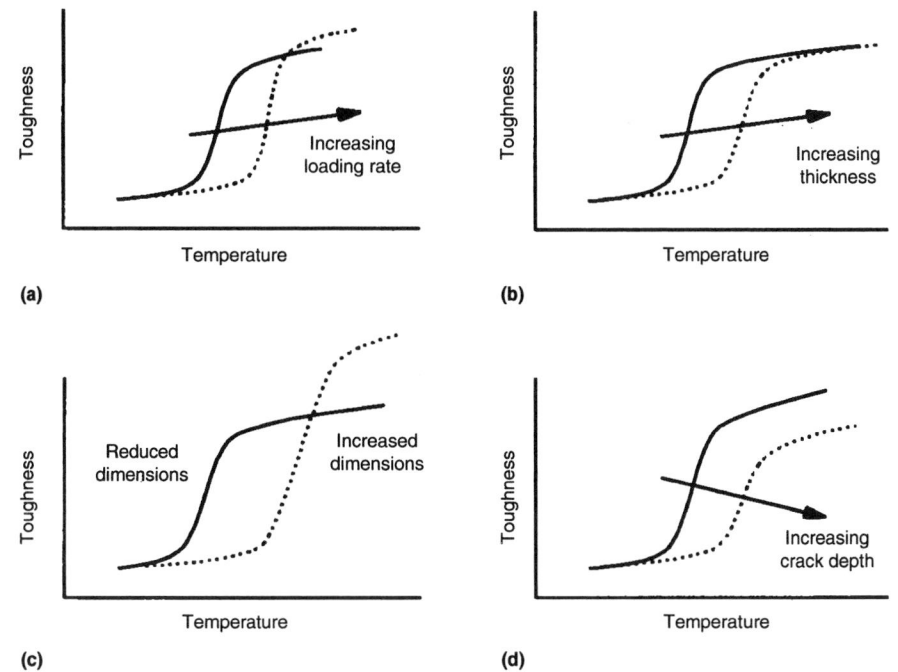

Fig. 22 Effects of loading rate, thickness, specimen dimensions, and crack depth on the ductile-to-brittle transition behavior

Although a certain amount of information on the variation in HAZ toughness can be obtained from these tests, their primary aim is to ensure that a certain level of quality in workmanship is maintained. Results obtained from Charpy testing of the HAZ have to be treated with caution, because it has been shown that the blunt notch of this type of test specimen can miss sampling local regions of low toughness (Ref 29). This caution also applies to Charpy testing of multipass welds, because it is well known that in the transition regime, depending on the orientation of the

Charpy notch, bimodal fracture behavior can be obtained at the same test temperature. This is because the notch of one test specimen may sample microstructure exhibiting upper-shelf behavior, and a specimen from the adjacent location may sample low-toughness microstructure (Ref 30). Thus, weldment toughness data can exhibit significant scatter, particularly in the transition regime, and care has to be taken in interpreting the results.

Although both Charpy and fracture mechanics toughness test standards are now well established

for parent material (Ref 31-33), the complex nature of the weld joint has hindered the development of standardized procedures for measuring the toughness. At the time of this writing, both ASTM and BSI were developing draft standards for weldment toughness testing, but restricted to weld metal. An essential requirement for tests on welded joints is that the test welds should be fully representative of the service structure of interest. This requirement is based on the knowledge that the fracture toughness of weld metals and weld HAZs may be critically dependent on such factors as:

- Welding process and consumables
- Base metal composition
- Joint thickness
- Preheat and interpass temperature
- Heat input
- Welding position
- Joint configuration
- Restraint
- Postweld heat treatment
- Time between welding and testing

Information on the effects of these welding factors can be obtained from Ref 34.

Embrittlement Mechanisms. Localized embrittlement in welded regions can occur by a number of mechanisms, and these can increase the risk of brittle fracture initiation. The three most common mechanisms are hydrogen embrittlement, strain aging, and temper embrittlement. Each of these mechanisms can reduce the toughness of the material, either soon after welding or during service. There are many catastrophic failures associated with these embrittlement mechanisms (Ref 35), and an effective fracture control plan has to include their potential effect.

Hydrogen Embrittlement. Hydrogen in steels and welded joints can be introduced in several ways. In welding, hydrogen from the arc atmosphere can dissolve into the liquid weld pool, and if the weld cools rapidly, a significant quantity may be retained in the weldment at low temperatures. Alternatively, in the petrochemical industry, hydrogen can be absorbed by the containment vessel during service when the medium in the vessel contains diffuseable hydrogen. Sufficient hydrogen levels will cause cracking in the weld metal and the HAZ, but lower levels of hydrogen can cause other problems, including the formation of "fish eyes" and embrittlement of the material. Hydrogen embrittlement in weld metal often manifests as a loss of ductility and fracture toughness. The susceptibility of a material to hydrogen embrittlement is dependent on the chemistry and microstructure, and therefore the different regions of the weldment will embrittle differently due to the presence of hydrogen.

The data given in Fig. 25(a) for a weld in a C-Mn steel shows that the fracture toughness transition curve shifts to the right in the presence of hydrogen, and the upper shelf is lowered. Hydrogen embrittlement therefore affects the complete transition curve, regardless of the failure mode. However, it should be noted that the effects of hydrogen embrittlement are sensitive to

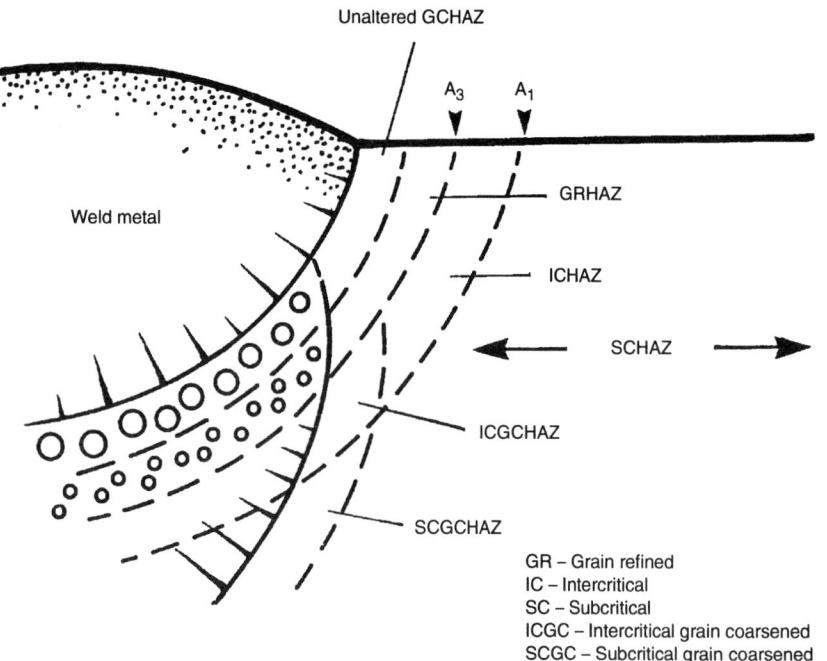

Fig. 23 Schematic diagram of the variety of microstructures that can be obtained in the HAZs of multipass welds. GR, grain refined; IC, intercritical; SC, subcritical; ICGC, intercritical grain coarsened; SCGC, subcritical grain coarsened

the strain rate. The trends observed in Fig. 25(a) may not be apparent with a Charpy test that is conducted at high strain rates, compared to a CTOD fracture toughness test. Thus, caution should be observed when attempting to measure the hydrogen embrittlement effects using a Charpy V-notch toughness test.

Earlier it was stated that the toughness of the weld metal or the HAZ is dependent on the time between welding and testing. This is largely due to the presence of hydrogen and its diffusion over time. For structures that will not be subjected to subsequent hydrogen environment during service, it would be misleading is test the weldment soon after fabrication. The presence of the diffuseable hydrogen could lower toughness, as shown in Fig. 25(b). This figure shows that a true value of toughness is obtained more than 100 h after welding, but this time is dependent on the size of the weld and the amount of hydrogen present. An alternate way to ensure that the hydrogen effects are dispelled is to soak the welded sample for a few hours at 150 °C (300 °F) before testing. This is referred to as hydrogen release heat treatment.

Strain Aging. The thermal and strain cycles that accompany welding can give rise to "dynamic" strain aging embrittlement. This is caused by the diffusion of carbides and free nitrogen into dislocations in the steel matrix. These dislocations, generated through the straining mechanism, become pinned (their movement is inhibited), thus increasing the yield strength (effectively the stress required to move the dislocations). Strain aging may also result in an increase in tensile strength, and it is usually accompanied by reduction in toughness. Strain aging in welds is often thought to occur in the root regions, which are highly restrained and receive complex thermal cycling. Strain age embrittlement has caused many failures in service, one example being the Ashland storage tank failure in January 1988 (Ref 36). An oil tank owned by Ashland Petrochemical Company ruptured as it was being filled for the first time after reconstruction, and as a consequence, approximately 4 million gallons (15×10^6 liters) of diesel fuel were released. The failure was attributed to a small defect left behind from an oxyfuel cutting process close to the circumferential seam. The material ahead of the defect had been strain aged by the welding process, thus lowering the local toughness. The material toughness away from the defect was found to be high. As a consequence of the presence of the defect, the low toughness at the crack tip (on a cold day), and high local weld residual stresses, the oxyfuel cutting defect triggered a brittle fracture that caused the complete collapse of the storage tank.

Temper embrittlement is the loss in toughness of alloy steels on exposure to temperatures in the range 325 to 575 °C (620 to 1065 °F). It occurs in coarse austenite grain microstructure (e.g., HAZs and weld metal) in which no ferrite is formed on the prior-austenite grain boundaries. It is caused by migration of elements such as Mn, Si, Sb, As, Sn, and P to high-angle austenite grain boundaries, which become embrittled. The risk of temper embrittlement can be reduced by promoting fine-grained austenite microstructure in weld and HAZ regions. The degree of embrittlement is

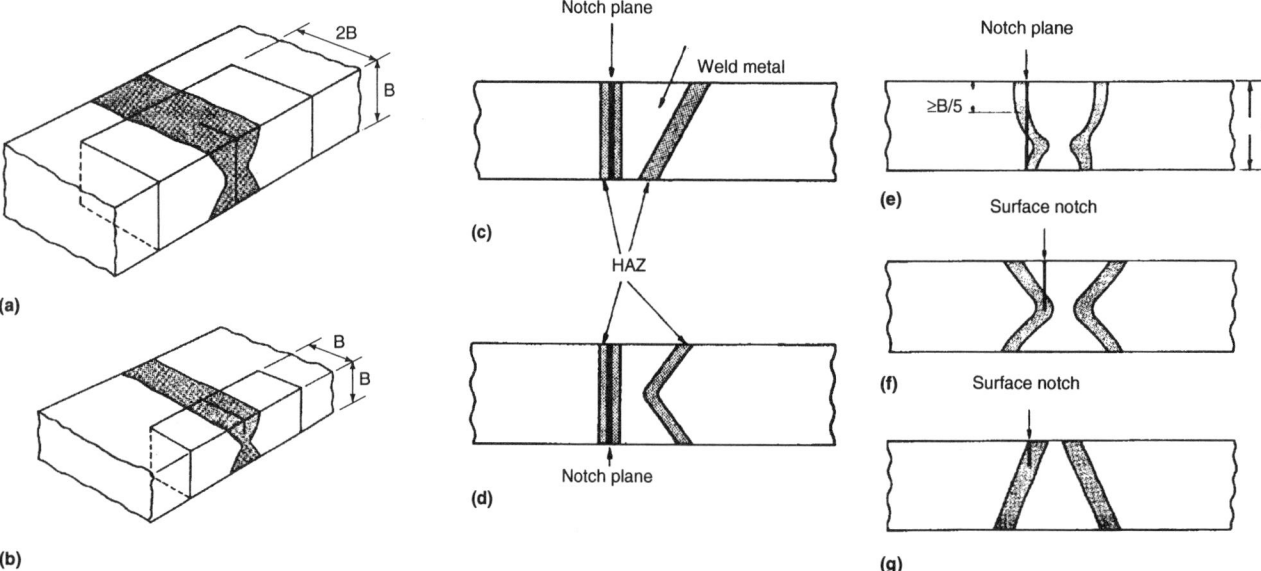

Fig. 24 Examples of notch locations for toughness testing of weld metals (a and b) and HAZ (c through g)

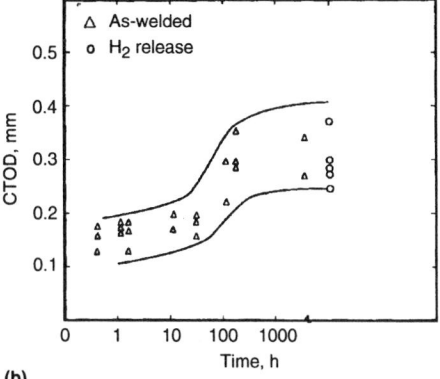

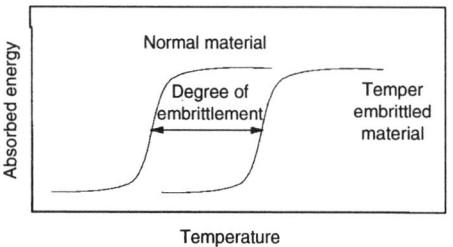

Fig. 25 Importance of time and hydrogen release treatment on fracture toughness. (a) Effect of hydrogen on fracture toughness immediately after welding. (b) Effect of time on as-welded specimens

Fig. 26 Schematic diagram showing the effect of temper embrittlement on toughness

often specified as the temperature shift in the Charpy transition curve (Fig. 26).

Toughness Requirements for Avoidance of Brittle Fracture. This section does not review the toughness requirements in different industries for avoidance of brittle fracture. Reference 37 provides excellent coverage of this topic.

Analysis of early Charpy V-notch data of ship steels indicated that initiation of brittle fracture occurred only in those steels showing less than 13.5 J (10 ft · lbf) Charpy energy, and cracks propagated only in those steels showing less than 27 J (20 ft · lbf). This gave rise to the 20 J (15 ft · lbf) criterion, which, although not universally applicable, has nevertheless been widely used. Today, most codes specify fracture toughness requirements that consider the influences of parameters such as tensile properties, heat treat-

ment condition, constraint, and material grade. All these factors influence the material ductility and the ductile-to-brittle transition temperature. Explicit specification of toughness requirements for weld metal and HAZ is rare, although some industry codes and standards assume that the toughness specification for base metal also applies to weld metal and HAZ. Material toughness requirements are generally part of comprehensive fracture control plans that include quality control, material selection and fabrication, and inspection. Account is taken of the in-service environment, degradation of properties during service (e.g., aging), applied loading condition, temperature cycles, and the consequences of catastrophic failure.

Toughness requirements differ widely in the broad spectrum of industries, but they generally reflect the degree of conservatism needed for the specific structure and the consequences of failure. An example of a code that embodies all the factors and provides a comprehensive fracture control plan, including crack arrest, is the Canadian Offshore Structures Standard (Ref 7). In this standard, the problem of fracture control is approached by considering both the susceptibility of a structural element to fracture initiation and the consequences of its failure. This leads to the definition of a 3 × 3 matrix that corresponds to three levels of failure consequences, based on risk to life and environment. There is a requirement of increasing crack arrest capability in the base metal as failure becomes less tolerable, this capability being measured by the Pellini drop weight test. Fracture initiation is primarily seen as a weld metal and HAZ problem, and CTOD fracture toughness requirements for cases of high initiation susceptibility reflect this. In the Canadian standard the following Charpy requirements for fracture initiation are based on correlations with CTOD:

Specified minimum yield strength (SMYS) of base metal, MPa	Charpy energy, J
≤270	27
270 < SMYS ≤ 410	SMYS/10
>410	41

Table 1 shows the risk matrix reproduced from the standard. For each of the boxes in Table 1, the corresponding toughness requirements and test temperatures are given, as shown in Table 2, where T is the toughness design temperature of the structural element. For crack arrest, no break performance in the drop weight test is required at the specified temperature. Note that the standard includes the effects of stress relief.

For more rigorous fracture toughness specifications, some codes allow the requirements to be specified on the basis of fracture mechanics analysis, where the effects of a known or a hypothesized flaw, the applied loading conditions, and the heat treatment condition are considered for the specific structure. The toughness specifications are usually based on parameters such as K, J, or CTOD, and they are generally required for the full plate or joint thickness (Ref 34).

Codes and Standards

Fitness-for-Service Codes and Standards for Fatigue and Fracture Control. Since the mid-1970s, a number of fracture mechanics assessment procedures have been developed that enable the significance of weld discontinuities to be assessed on a fitness-for-service basis. Using this concept, a structure is considered fit for the intended service if it can operate safely throughout its design life. The adoption of fitness-for-service concepts in several industry standards has resulted in the development of rigorous fatigue and fracture control plans. In addition to allowing the integrity of a structure to be assessed, a fitness-for-service code enables the remaining life to be computed.

The use of fitness-for-service technology in design and inspection can result in a number of significant advantages. The major benefit undoubtedly is that this technology allows individual flaws to be assessed and allows the operator to decide whether or not to repair. Although adoption of a fitness-for-service approach may allow weld discontinuities to remain in a structure, it should also lead to a safer structure, because parameters that could potentially cause the

Table 1 Risk matrix of the Canadian Offshore Structures Standard

Susceptibility to fracture initiation	Safety class of structural element		
		Safety class 1	
	Safety class 2	Highly redundant structure	Limited redundancy of structure
Low (e.g., element in compression or low tension)	Box 1 No toughness requirements	Box 2 Nominal toughness	Box 3 Some initiation toughness; moderate crack arrest toughness
Moderate (e.g., element in tension; no high stress concentrations; no hot spots; no plastic straining)	Box 4 Nominal toughness	Box 5 Moderate control of initiation; moderate crack arrest toughness	Box 6 Moderate control of initiation; good crack arrest toughness
High (e.g., element in tension; high stress concentrations; hot spots; plastic straining)	Box 7 Some initiation toughness; some crack arrest toughness	Box 8 Good control of initiation; moderate crack arrest toughness	Box 9 Good control of initiation; good crack arrest toughness

Note: Box numbers correspond to box numbers in Table 2. Source: Ref 7

Table 2 Toughness requirements of the Canadian Offshore Structures Standard

Upper band (Boxes 1, 2, 3):

Thickness, t, mm	Box 1 — Base metal: Test / Test temperature	Box 1 — HAZ and weld metal: Test / HAZ / AW / SR	Box 2 — Base metal: Test / Test temperature	Box 2 — HAZ and weld metal: Test / HAZ / AW / SR	Box 3 — Base metal: Test / Test temperature	Box 3 — HAZ and weld metal: Test / HAZ / AW / SR
t ≤ 26	Steel grades in accordance with clause 5.1.2	Welding consumables in accordance with clause s7.1.1 of the supplement to WS9	Base metal toughness to CAN/CSA-G40.21 category 1 and clause 6.2.1.4	Welding consumables in accordance with clause s7.1.1 of the supplement to WS9	NDT, CVN / T	CVN / T / T–5 / T
26 < t ≤ 40					CVN / T	CVN / T / T–10 / T
40 < t ≤ 52					NDT, CVN / T–10	CVN / T / T–20 / T
t > 52					CVN, NDT, CVN / T–20	CVN / T / T–20 / T

Lower band (Boxes 4, 5, 6):

Thickness, t, mm	Box 4 — Base metal: Test / Test temperature	Box 4 — HAZ and weld metal: Test / HAZ / AW / SR	Box 5 — Base metal: Test / Test temperature	Box 5 — HAZ and weld metal: Test / HAZ / AW / SR	Box 6 — Base metal: Test / Test temperature	Box 6 — HAZ and weld metal: Test / HAZ / AW / SR
t ≤ 26	Base metal toughness to CAN/CSA-G40.21 category 1 and clause 6.2.1.4	Welding consumables in accordance with clause s7.1.1 of the supplement to WS9	NDT, CVN / T	CVN / T / T–10 / T–5	NDT, CVN / T–10	CVN / T / T–10 / T–5
26 < t ≤ 40			CVN / T	CVN / T / T–20 / T–5	CVN / T–20	CVN / T / T–20 / T–5
40 < t ≤ 52			NDT, CVN / T–10	CVN / T–10 / T–30 / T–10	NDT, CVN / T–30	CVN / T–10 / T–30 / T–10
t > 52			CVN, NDT, CVN / T–20	CVN / T–10 / T–30 / T–10	NDT, CVN / T–35	CVN / T–10 / T–30 / T–10

Bottom table (Boxes 7, 8, 9):

Thickness, t, mm	Box 7: Test / Temp / Test / HAZ / AW / SR	Box 8: Test / Temp / Test / HAZ and weld metal (AW / SR)	Box 9: Test / Temp / Test / HAZ and weld metal (AW / SR)
t ≤ 26	NDT, CVN / T / CVN / T / T–5	NDT, CVN / T / CTOD (AW 0.1, SR 0.07); CVN (AW T, SR T)	NDT, CVN / T–10 / CTOD (AW 0.1, SR 0.07); CVN (AW T, SR T)
26 < t ≤ 40	NDT, CVN / T / CVN / T / T–10	NDT, CVN / T / CTOD (AW 0.15, SR 0.11); CVN (AW T, SR T)	NDT, CVN / T–20 / CTOD (AW 0.15, SR 0.11); CVN (AW T, SR T)
40 < t ≤ 52	NDT, CVN / T / CVN / T / T–20	NDT, CVN / T / CTOD (AW 0.2, SR 0.14); CVN (AW T–10, SR T–10)	NDT, CVN / T–30 / CTOD (AW 0.2, SR 0.14); CVN (AW T–10, SR T–10)
t > 52	NDT, CVN / T–20 / CVN / T / T–20	NDT, CVN / T–20 / CTOD (AW 0.2, SR 0.41); CVN (AW T–10, SR T–10)	NDT, CVN / T–35 / CTOD (AW 0.2, SR 0.14); CVN (AW T–10, SR T–10)

Note: Box numbers correspond to box numbers in Table 1. For T < –20 °C, steels and welding consumables may not be commercially available to satisfy the requirements. HAZ, heat-affected zone; AW, as-welded; SR, stress relieved; NDT, nil-ductility transition; CVN, Charpy V-notch; T, toughness design temperature of the structural element in °C; CTOD, crack-tip opening displacement in mm at T. Source: Ref 7

crack to become unsafe are taken into consideration.

Given that a flaw or a discontinuity is found in a weld, fracture mechanics analysis can be performed by computing the crack driving force (which is a function of the applied stresses, the flaw size, and the geometry) and comparing this with the resistance of the material to crack extension (which is the fracture toughness of the material in which the crack resides). In the sections that follow, procedures to calculate the crack driving force for fatigue and fracture are briefly discussed.

Fracture Mechanics

Fitness-for-Service Codes. A number of different fitness-for-service assessment methodologies for calculating allowable or critical flaw sizes are currently in use throughout the world. A comprehensive review can be found in Ref 38.

Common methodologies in use today include:

- BSI 7608:1993 (Ref 6)
- BSI PD 6493:1991 (Ref 16)
- CEGB R6 (Ref 39)
- EPRI/GE J and CTOD estimation scheme (Ref 10)
- Deformation plasticity failure diagram (Ref 40)
- The local approach (Ref 41)
- WES 2805 (Ref 42)
- IIW approach (Ref 8)

- API 579 approach (Ref 43)
- ASMI IX approach (Ref 44)

A survey conducted by Burdekin (Ref 45) under the auspices of Commission X of the International Institute of Welding (IIW) found that PD 6493:1980 (Ref 15) was the most widely used assessment method. This document has been revised, and the latest version (entitled "Guidance on Methods for Assessing the Acceptability of Flaws in Fusion Welded Structures') was released in August 1991. At the time of this writing, the third revision to PD 6493 was in progress. While some of the approaches listed above differ, their fundamental basis is the same, and in this article the intention is to point the reader to established and validated fracture mechanics procedures. In this context, some of the procedures outlined in PD 6493:1991 are briefly described.

Fracture Mechanics Methods for Fracture Control. BSI PD 6493:1991 includes three different fracture assessment routes, to enable structures to be assessed at a level of complexity appropriate to the problem under consideration. The Level 1 approach is based on the CTOD design curve method, which forms the basis of the elastic-plastic fracture assessment procedure included in BSI PD 6493:1980. This approach has a deliberate safety factor built into the calculations, and it results in computation of "tolerable" crack sizes.

In PD 6493:1991 the Level 1 approach is presented as a method for performing preliminary assessments. The Level 2 approach is based on a plane-stress plastic collapse modified strip yield model, and there is no inherent safety factor. This method is the preferred assessment level for the majority of applications. The Level 3 method is the most sophisticated assessment level in the proposed revisions and would normally be used only for the assessment of high-work-hardening materials when the Level 2 approach has proved too restrictive or when a tearing instability analysis is required. The fracture assessment model adopted in the Level 3 approach is based on the reference stress model proposed by Ainsworth (Ref 46). This assessment level requires extensive material characterization data, including details of the stress-strain behavior of the material in which the defect is located. This presents a challenge for the HAZ. All three levels can be undertaken in terms of the stress-intensity factor, K (or K derived from fracture resistance J) or the CTOD fracture mechanics parameters. Level 2 approach is based on procedures similar to CEGB R6 Rev. 2, and the Level 3 approach is based on CEGB R6 Rev. 3. These codes have been extensively used in the electric utilities industry in the U.K.

In all three levels of fracture assessment, the resistance of a structure to failure is determined using a failure assessment diagram (FAD). The y-axis of the FAD indicates the resistance of the structure to brittle fracture, while the x-axis assesses its resistance to plastic collapse. The failure assessment curve interpolates between these two limiting failure modes.

The Level 2 assessment procedure, which is recommended as the primary method, has a FAD that is defined by:

$$K_r, \sqrt{\delta_r} = S_r \left[\frac{8}{\pi^2} \ln \sec \left(\frac{\pi}{2} \right) S_r \right]^{-\frac{1}{2}} \qquad \text{(Eq 2)}$$

where K_r is the ratio of the applied stress-intensity factor to the fracture toughness K, $\sqrt{\delta_r}$ is the ratio of the applied CTOD to the material CTOD, and S_r is the stress ratio that is taken as the ratio between the net-section stress and the flow strength of the material.

For a given flaw, both K_r or $\sqrt{\delta_r}$ and S_r are determined using appropriate solutions, and these are plotted as a coordinate of a point onto the FAD (Fig. 27). If the point falls within the envelope of the failure locus, the flaw is considered safe. If the point falls outside the locus, the flaw is unacceptable.

The FAD to assess the integrity of a structure assumes that the size of the flaw is known. In order to compute limiting flaw size, say for inspection purposes, a point that lies on the failure assessment line and corresponds to the applied loading conditions has to be determined. This is achieved using iterative techniques by plotting a series of assessment points as a function of flaw size. The point that intersects the failure line gives the limiting size of a flaw. By performing these calculations for a series of crack aspect ratios, tolerable flaw size plots of the type given in Fig. 28 can be generated. This type of plot can be used either to set inspection levels or to allow known flaws to be assessed. In the same way, an iterative process has to be used to generate the critical fracture toughness for some hypothetical (or assumed) flaw size and assumed loading conditions. This computed value can then be directly used for material selection.

Another approach that is sometimes used to ensure that catastrophic failure does not occur is leak-before-break. This design philosophy has been developed primarily for applications where the crack extends in a stable manner by such mechanisms as stable tearing or fatigue crack growth. This approach can be invoked only if the crack shape at breakthrough can be precisely predicted. In most cases it is generally assumed that the defect length at breakthrough is 2 to 3 times the plate thickness. However, there have been instances where defects up to 10 times the plate thickness have been found during inspection of petrochemical plants. Leak-before-break becomes less likely as the defect length increases, so it should be invoked with great caution.

Method for Fatigue Control

The two most widely used approaches for fatigue control in welded joints are the S-N curve approach and fracture mechanics assessment methods. Both approaches are used extensively for design and in-service assessment of welded fabrications, and they are included in many industry and regulatory codes as standardized procedures.

The S-N curve approach for fatigue control in "nominally sound" welded joints is the simplest and most widely used in a variety of industries. The approach is based on the results of extensive welded joint fatigue tests during the early 1970s. It was found that welded joints could be classified into groups, with each group having a specific S-N curve. There are a number of design classification approaches in existence, for example AWS D1.1 (Ref 26), AASHTO-Bridge Code (Ref 6), and BS 7608 (Ref 11), but the fundamental basis for them is the same. In Fig. 29, a typical set of fatigue design curves, where BS 7608 is taken as a reference, each joint detail is placed into one of nine classes: A, B, C, D, E, F, F2, G, or W. Class A is the parent material S-N curve obtained from uniform, machined, and polished specimens; class W is the welded joint S-N curve based on nominal shear stress on the minimum weld throat area; and classes B through G are the welded joint S-N curves for joints of decreasing fatigue performance (or increasing joint stress concentration factor and notch severity). This approach is based on actual weld joint test data, so the joint classification takes into consideration the influence of the local stress concentration created by the weld geometry, effects of local discontinuities (e.g., weld toe intrusions), and residual stresses. The design stress, therefore, is the nominal stress adjacent to the weld detail under consideration. However, where joints are situated in regions of stress concentration resulting from the gross shape of the structure (e.g., adjacent to a hole), then it is necessary to consider the additional concentration of stress

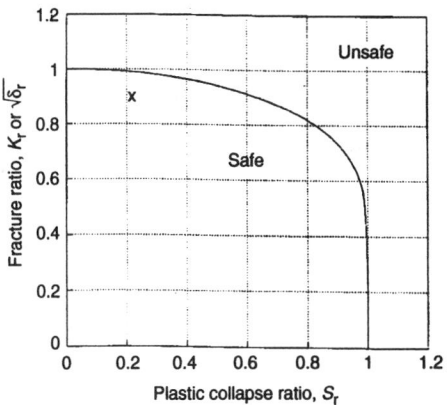

Fig. 27 Failure assessment diagram concept for assessing cracked components for brittle fracture and plastic collapse

when the nominal stress range is calculated. S-N curves are used in design either to determine the maximum allowable cyclic stress for a given fatigue design life or to estimate the design life for a given set of operating conditions.

Fracture Mechanics Approach. The second most widely used design approach for welded structures uses fracture mechanics as its fundamental basis. As stated earlier, the majority of the fatigue life of a welded joint is spent in propagating existing crack-like features. Consequently, the analysis of fatigue cracking in welded joints is well suited to fracture mechanics, which can be used to predict the fatigue strength of nominally sound welds by considering the propagation of a fatigue crack from the inherent discontinuities that exist at the weld toe and the weld root. Fracture mechanics also can be used to predict the life of a joint with a weld discontinuity.

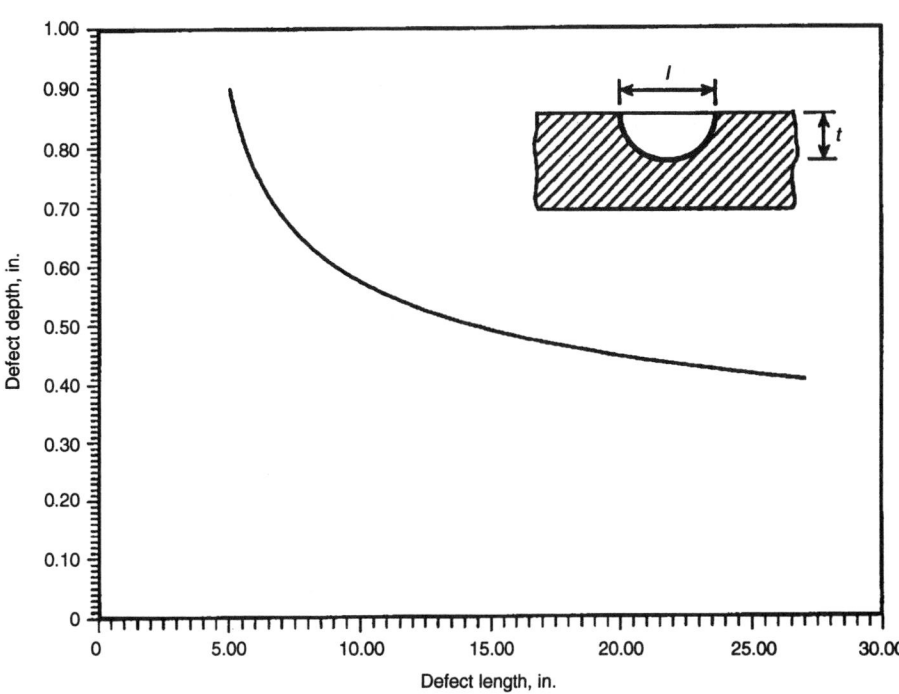

Fig. 28 Example of a "tolerable" flaw size plot derived using fracture mechanics principles

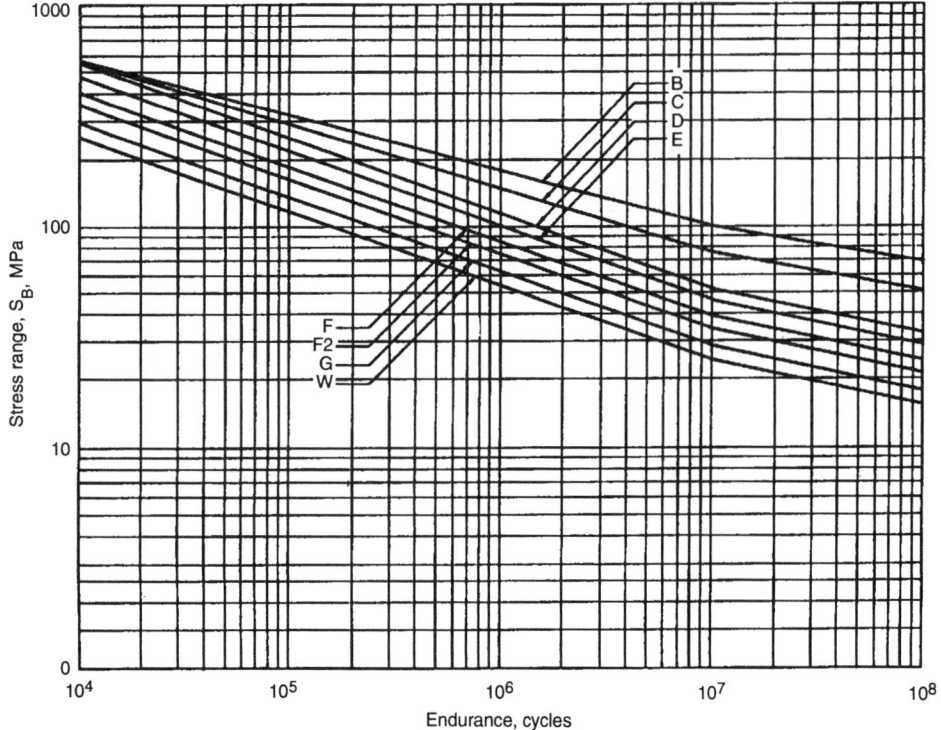

Fig. 29 Mean minus two standard deviation *S-N* curves for welded joints. See text for definitions of letter symbols.

Higher crack growth rates may occur in weld metals and HAZs when the normal crack growth (by striation mechanism) is accompanied by cleavage or microvoid coalescence. Under such circumstances the value for constant C can be increased to 6×10^{-13}. For structural ferritic steels operating in marine environments up to 20 °C (68 °F), the following crack growth constants are recommended:

$$da/dN = 2.3 \times 10^{-12} (\Delta K)^3 \text{ (N, mm units)} \qquad \text{(Eq 8)}$$

For design purposes, it is sometimes necessary to estimate the limiting flaw size that will not extend by fatigue during service. The limiting crack size below which fatigue crack growth will not occur can be calculated using threshold concepts based on ΔK_{th}. Near threshold da/dN becomes very sensitive to a number of variables including the applied stress ratio R (minimum stress/maximum stress). The following threshold stress-intensity factor ranges, representing the mean minus two standard deviations, are recommended for ferritic steels operating in air or seawater environments and also for austenitic stainless steels:

$$\Delta K_{th} = 170 - 214R \text{ (N, mm units) for } 0 \le R \le 0.5 \text{ (Eq 9)}$$

$$\Delta K_{th} = 63 \text{ (N, mm units) for } R > 0.5 \qquad \text{(Eq 10)}$$

Welding residual stresses must be taken into account when calculating the stress ratio. Using the above expressions, it is possible to calculate the limiting flaw size that will not extend by fatigue during service.

Using the fracture mechanics fatigue crack growth procedures described above, the user can determine the number of cycles (or time) needed for a crack to extend from an initial size to a final size. This information can be used in a powerful way to define inspection criteria, and in particular inspection intervals. Alternatively, it can be used to compute the remaining life of a welded component, before failure by unstable fracture or plastic collapse becomes a reality. Another important use of this technology is the establishment of flaw acceptance criteria for nondestructive examination (NDE) of welded joints. Here the process is reversed by first defining the total design life of the structure and the final flaw size for fracture, then computing the initial crack size for that life cycle. This computed value of the flaw size can be used directly for specifying NDE flaw acceptance levels. Adoption of this approach should enhance the safety of the structure, because all failure modes are considered and the flaw acceptance levels are specifically calculated rather than being based on workmanship criteria.

In documents such as PD 6493:1991, it is recommended that where possible, the fatigue crack growth relationship based on the Paris law should be used:

$$da/dN = C(\Delta K)^m \qquad \text{(Eq 3)}$$

where ΔK is the applied stress-intensity factor range, a is the crack size, N is the number of cycles, and C and m are constants that depend on the material and environment. This crack growth law is bounded by the threshold value for ΔK, ΔK_{th}, and the critical value of K_{max} for fracture toughness, K_c, as shown in Fig. 30. Experience with this crack growth law shows that it can be used to make reasonable estimates of the fatigue lives of welded joints, particularly when the initial crack size is relatively large. However, the actual sigmoidal shape may need to be represented in case of random loading assessments, and assessments for which K_{max} is close to K_c in the idealized relationship of Eq 3 can be nonconservative.

In general, the stress-intensity factor range, ΔK, can be expressed as:

$$\Delta K = YM_k (\Delta \sigma) \sqrt{\pi a} \qquad \text{(Eq 4)}$$

where $\Delta \sigma$ is the stress range, Y is a function of geometry and loading, and M_k is a factor incorporating the stress concentration effect of the weld. The relationship between $\Delta \sigma$ and the fatigue life N for an initial crack size a_0 and maximum crack size a_c is obtained by substituting Eq 4 into Eq 3, then integrating and rearranging to give:

$$\int_{a_0}^{a_c} \frac{da}{(YM_k \sqrt{\pi a})^m} = C(\Delta \sigma)^m N \qquad \text{(Eq 5)}$$

M_k factors are normally obtained by finite-element analysis of the type of joint and loading under consideration, and they are expressed in a nondimensional form. M_k factors are a function of the weld geometry parameters shown in Fig. 8:

$$M_k = f(r/T, L/T, \theta) \qquad \text{(Eq 6)}$$

General expressions for M_k factors have been produced by Thurlbeck and Burdekin (Ref 20), and these can be used with standard stress-intensity factor solutions to predict fatigue life.

Maddox (Ref 47) has carried out a series of fatigue crack propagation tests on various parent steels, weld metals, and HAZ microstructures (Fig. 15). He found no significant variation in crack propagation rates among any of the materials tested when the growth occurred by the striation mechanism. Based on such information, in BSI PD 6493:1991 firm recommendations are given for crack growth rate data. The option is available to use specific crack growth rate data for the material of interest, but otherwise the following upper-bound crack growth rate constants for ferritic steels (up to 600 MPa, or 87 ksi) operating in nonaggressive environments are recommended:

$$da/dN = 3 \times 10^{-13} (\Delta K)^3 \text{ (N, mm units)} \qquad \text{(Eq 7)}$$

The use of this crack growth relationship represents approximately 97.7% probability of survival.

REFERENCES

1. *Offshore Installations: Guidance on Design, Construction and Certification*, 4th ed., U.K. Department of Energy, HMSO, London, 1990 (last revisions Dec 1993)

2. "Design of Steel Structures," Eurocode 3, Draft for Development DD ENV 1993-1-1, British Standards Institution, London, 1992

3. "Specification for Unfired Fusion Welded Pressure Vessels," BS5500, British Standards Institution, London, Jan 1994

4. "Steel, Concrete and Composite Bridges," BS5400, British Standards Institution, London, 1982

5. *ABS Rules for Building and Classing Steel Vessels*, American Bureau of Shipping, 1992

6. *Guide, Specifications for Fracture Critical Non-redundant Steel Bridge Members*, American Association of State Highway and Transportation Officials, 1978 (last revisions 1991)

7. "Steel Structures (Offshore Structures)," CAN/CSA-S473-92, Canadian Standards Association, Ontario, 1992

8. "IIW Guidance on Assessment of the Fitness for Purpose of Welded Structures," Draft for Development, IIW/IIS-SST-1157-90, International Institute of Welding, 1990

9. "Unstable Fracture," Recommended Practice D404, *Veritas Offshore Standards*, Det Norske Veritas, Hovik, Norway, Aug 1988

10. V. Kumar, M.D. German, and C.F. Shih, "An Engineering Approach for Elastic-Plastic Fracture Analysis," Technical Report NP 1931, Electric Power Research Institute, July 1981

11. "Code of Practice for Fatigue Design and Assessment of Steel Structures," BS7608, British Standards Institution, 1993

12. *Offshore Installations: Guidance on Design, Construction and Certification*, 4th ed., U.K. Department of Energy, HMSO, London, 1990 (last revisions Dec 1993)

13. "Fatigue Strength Analysis for Mobile Offshore Units," Classification Note 302, Det Norske Veritas, Hovik, Norway, 1984

14. "Planning, Designing and Constructing Fixed Offshore Platforms," API RP2A, American Petroleum Institute, 1993

15. "Guidance on Some Methods for the Derivation of Acceptance Levels for Defects in Fusion Welded Joints," BSI PD 6493:1980, British Standards Institution, March 1980

16. "Guidance on Methods for Assessing the Acceptability of Flaws in Fusion Welded Structures," BSI PD 6493:1991, British Standards Institution, Aug 1991

17. E.J. Signes, R.G. Baker, J.D. Harrison, and F.M. Burdekin, Factors Affecting the Fatigue Strength of Welded High Strength Steels, *Br. Weld. J.*, Vol 14 (No. 3), 1967, p 108-116

18. S.J. Maddox, "Fitness-for-Purpose Assessment of Misalignment in Transverse Butt Welds Subject to Fatigue Loading," Document XIII-1180-85, International Institute of Welding, 1985

19. S.J. Maddox, Recent Advances in the Fatigue Assessment of Weld Imperfections, *Weld. J.*, July 1993, p 42-51

20. S.D. Thurlbeck and F.M. Burdekin, Effects of Geometry and Loading Variables on the Fatigue Design Curve for Tubular Joints, *Welding in the World*, Vol 30 (No. 7-8), July/Aug 1992, p 189-200

21. V.I. Makhnenko and R.Y. Mosenkis, Calculating the Coefficient of Concentration of Stresses in Welded Joints with Butt and Fillet Welds, *Automatic Welding*, Aug 1985

22. K. Iida, Y. Kho, J. Fukakura, M. Nihei, T. Iwadate, and M. Nagai, "Bending Fatigue Strength of Butt Welded Plate with Uranami Bead," IIW/IIS-XIII-1202-86

23. G.S. Booth, Ed., *Improving the Fatigue Performance of Welded Joints*, Welding Institute, 1983

24. *Fatigue Handbook—Offshore Steel Structures*, TAPIR, 1985

25. P.W. Marshall, Fatigue Design Rules in the USA, *Proc. Ninth Int. Conf. on Offshore Mechanics and Arctic Engineering*, Vol III, Part A, 1990, p 175-184

26. "Structural Welding Code—Steels," AWS D1.1, American Welding Society, 1994

27. W.S. Pellini and P.P. Puzak, "Fracture Analysis Diagram Procedures for the Fracture-Safe Engineering Design of Steel Structures," WRC Bulletin 88, May 1963, p 1-28

28. J.D. Harrison, "Why Does Low Toughness in the HAZ Matter?" paper presented at The Welding Institute Seminar, Coventry, England, 1983

29. S.E. Webster and E.F. Walker, The Significance of Local Brittle Zones to the Integrity of Large Welded Structures, *Proc. Offshore Mechanics and Arctic Engineering (OMAE) Seventh International Conference*, Vol 3, 1988, p 395-403

30. N. Bailey, "Bimodality Revised—Split Behavior of Weld Metal," TWI Bulletin 5, Sept/Oct 1991

31. "Standard Method for Notched Impact Testing of Metallic Materials," ASTM E 23

32. "Plain-Strain Fracture Toughness of Metallic Materials," ASTM E 399

33. "Draft British Standard Fracture Mechanics Toughness Tests," BS 7448 Part 1, British Standards Institution, London

34. S.J. Squirrell, H.G. Pisarski, and M.G. Dawes, "Recommended Procedure for the Crack-tip Opening Displacement (CTOD) Testing of Weldments," TWI Research Report 311, 1986

35. G.M. Boyd, "Brittle Fracture in Steel Structures," Butterworth & Co., 1970

36. C.W. Marshall, R.E. Mesloh, R.D. Buchheit, and J.F. Kiefner, Oil Tank Fracture Traced to

Embrittled Flaw Tip of New Weld, *Weld. J.*, April 1989, p 34-38

37. R. Phaal and C.S. Wiesner, "A Compendium of Toughness Requirements for Ferritic Materials in Various International Application Codes," TWI Report 458, 1992

38. J.R. Gordon, "A Review of Currently Available Fracture Analysis Methods and Their Applicability to Welded Structures," paper presented at Sixth Annual North American Welding Research Conference, Edison Welding Institute and American Welding Society, Oct 1990

39. I. Milne, R.A. Ainsworth, A.R. Dowling, and A.T. Stewart, "Assessment of the Integrity of Structures Containing Defects," R/H/R6-Rev. 3, Central Electricity Generating Board, 1987

40. J.M. Bloom, *Validation of a Deformation Plasticity Failure Assessment Diagram Approach to Flaw Evaluation*, STP 803, Vol II, ASTM, 1983, p 206

41. F. Mundry, paper presented at Seminaire International sur l'approche Locale de la Rupture, Moret-sur-Loing, June 1986, p 243

42. "The Method of Assessment for Defects in Fusion Welded Joints with Respect to Brittle Fracture," WES 2805, Japan Welding Engineering Society

43. "Recommended Practice for Fitness-for-Service of Refinery Equipment," API 579, American Petroleum Institute, to be published

44. *ASME Boiler & Pressure Vessel Code*, Section IX, American Society of Mechanical Engineers

45. F.M. Burdekin, "Final Report on Questionnaire on the Use of Fracture Mechanics Methods for the Assessment of the Significance of Weld Defects," Document X-1076-84, International Institute of Welding, 1984

46. R.A. Ainsworth, The Assessment of Defects in Structures of Strain Hardening Material, *Engineering Fracture Mechanics*, Vol 19, 1984, p 633

47. S.J. Maddox, *Weld. Res. Int.*, Vol 4 (No. 1), 1974, p 36-60

48. S.J. Maddox, "Fatigue Strength of Welded Structures," Abington Publishing, 1991

49. I. Huther, L. Promot, H.P. Lieurade, J.J. Janosch, D. Colchen, and S. Debiez, "Weld Quality and the Cyclic Fatigue Strength of Steel Welded Joints," IIW-DOC-XIII-1563-94

50. S.T. Rolfe and J.M. Barsom, "Fracture and Fatigue Control in Structures—Applications of Fracture Mechanics," Prentice-Hall, Inc.

51. S.J. Garwood, "Fatigue Crack Growth Threshold Determination," Welding Research Bulletin, Sept 1979, Vol 20, No. 9

52. "The Integrity of Offshore Pipeline Girth Welds," HMSO publication, UK, (ISBN 0114128707)

Fracture Mechanics in Failure Analysis

Alan R. Rosenfield

FRACTURE MECHANICS has developed into a useful tool in the design of crack-tolerant structures and in fracture control; it also has a place in failure analysis. Fracture mechanics can provide helpful quantitative information on the circumstances that led to the failure, and it can be used to prescribe preventive measures to avoid the recurrence of failures in similar components.

This article illustrates the role that fracture mechanics can play in failure analysis. It will be seen that the investigator has to draw on several disciplines (materials engineering plus analytical and experimental mechanics) and requires more expertise than has been traditionally needed for a failure analysis. This article describes the important failure criteria as relations between design (e.g., section geometry) and materials factors (e.g., fracture toughness), which are used to correlate fracture mechanics analysis to the observations of a failure analysis. Descriptions include an indication of how the factors are typically evaluated. Finally, a group of failure analysis examples illustrate how fracture mechanics parameters can be determined and how they may be fitted into an overall failure investigation.

It should be emphasized at the outset that a full fracture mechanics analysis is usually not required to understand failures that have occurred and to prevent similar failures, particularly in cases where the costs are relatively low, people have not been injured, and even when litigation is involved (Ref 1). Many failures can be corrected by appropriate specification of standard properties, good metallurgical practice, and proper design (see, e.g., three cases of brittle fracture described in Ref 2).

However, there are occasional failures involving large costs, extensive damage, and injuries, where it is not clear whether the best fracture control technology was used. Some questions that can be addressed by fracture mechanics are:

- Was the structure or operating component properly inspected periodically to detect potentially critical cracks?
- If cracks were found, was a proper analysis performed to determine whether continued safe operation was possible, and, if so, for how long?

- What modifications need to be made to similar structures to promote safe operation?

Approaches to such questions can be aided by the study of the fracture mechanics evaluations of real failures that have been investigated in the past.

Benefits and Limitations. While fracture mechanics is a well-developed approach to the study of failures, there is a gap between theory and practice, because it is often difficult to obtain exact values for many of the inputs into the analysis when structural failure occurs. If the failure conditions are well understood, the material is uniform, and the geometry is relatively simple, fracture mechanics can work well. But, in many situations, conditions are far less than ideal and fracture mechanics can only provide a reality check.

When fracture mechanics analysis indicates that failure should not have occurred based on the measurement of flaw size, material properties, and operating conditions, other factors that may have altered any of these conditions at the failure region must be investigated more thoroughly.

The benefits of uncovering the sources of significant differences between calculated and measured stresses are improvements in fracture control, such as use of tougher materials, reduced stress concentrations, and selection of better inspection procedures.

Using Fracture Mechanics in Failure Analysis

The investigator should first carry out a traditional failure analysis and determine the failure mechanism(s), the nature and location of the fracture origin, and the operating conditions.

The more detailed fracture mechanics evaluation requires numerical evaluation of a relation between mechanical properties, loads, and crack geometry. The first objective of this evaluation is to use the input data to determine whether the failure could have been predicted. If the occurrence of failure is not consistent with a fracture mechanics analysis, certain factors have not been considered and the investigator must search them out.

Such an analysis should include fractography to reveal whether the failed part contained serious defects and whether final failure was preceded by fatigue crack growth or stress-corrosion cracking. The information thus gained may be sufficient to explain the failure and to prevent future occurrences.

Analysis of Final Fracture Conditions

To perform the analysis of conditions leading to final fracture, all of the quantitative information must be collected that is necessary for the investigator to evaluate the failure condition:

$$K_I \geq K_{Ic} \qquad \text{(Eq 1)}$$

where K_I is the applied stress intensity (the demand placed on the cracked part by loading) and K_{Ic} is the fracture toughness (the resistance of the metal to fracture). If Eq 1 is satisfied, it can be deduced that a sound analysis has been performed. On the other hand, if poor agreement is obtained, some factors have not been properly accounted for (e.g., residual stress). There is no well-accepted criterion for reasonable agreement; probably a discrepancy of less than 25% between both sides of Eq 1 is acceptable, while a 50% discrepancy would certainly indicate further evaluation of the failure. The in-between region is uncertain.

Because the inputs into Eq 1 are never certain, the best that can be hoped for is approximate agreement between calculated and measured stresses. *Furthermore, because of these uncertainties, it is not possible to use a structural failure to back-calculate an unknown property or stress with any degree of reliability.* An additional potential complication is the occurrence of extensive local plasticity, which requires the use of elastic-plastic fracture mechanics, characterized by the J-integral, but this refinement is rarely necessary for analyzing real failures.

Evaluation of stress intensity can be done either experimentally (Ref 3, 4), such as by use of photoelasticity, or analytically using finite element methods (Ref 5). These techniques are particularly useful for complicated geometries and provide the quantity:

$$K_I / \sigma = g \tag{Eq 2}$$

where σ is the stress on the component, and g is defined as the ratio of stress intensity to stress. Equation 1 then becomes:

$$\sigma \geq K_{Ic} / g \tag{Eq 3}$$

and the analysis now becomes a comparison between the stress calculated from Eq 3 and the stress believed to be acting on the body.

Stress intensity factors are discussed in more detail elsewhere in this Volume for simple geometries. As an example, the following equation for stress intensity can be used for pressurized pipes:

$$K_I = f \sigma \sqrt{\pi a} \tag{Eq 4}$$

so that Eq 1 becomes:

$$\sigma \geq K_{Ic} / f \sqrt{\pi a} \tag{Eq 5}$$

where a is the crack length and the f factor depends on crack size, crack shape, component size, and loading pattern (tension, bending, pressure, etc.). For through-wall cracks, a is conventionally taken to be half the crack length. This is equivalent to calling a the total crack length and setting $f = 0.707$.

Evaluation of stress can be done using standard strength-of-materials equations if the configuration is simple (Ref 6) or by finite element analysis if the configuration is complex. However, errors can arise because the stress to be used in Eq 3 or 5 is not necessarily limited to the operating stress. Residual stresses and extraneous loads (such as impact) may also play an important role in fracture and must be taken into account.

Fatigue loadings complicate the picture. The stress used in Eq 3 or 5 must be the peak stress, σ_{max}, where the fluctuating stress $\Delta\sigma = \sigma_{max} - \sigma_{min}$.

Evaluation of fracture toughness requires measurements (by appropriate test methods as described in the article "Fracture Toughness Testing" in this Volume) using material from the failed part. Unfortunately, this recommended practice is not always followed; handbook values of fracture toughness are often used instead of toughness measurements on material in the vicinity of the failure.

The investigator must also compare fracture mechanisms of the part and any test samples. Note that the failure mechanism (such as brittle cleavage or ductile rupture) does not enter explicitly into evaluation of Eq 3 and 5. The investigator must verify that the fracture mechanics test specimen fails by the same mechanism as that of the actual failure. If the specimen fails by dimpled rupture and the component by cleavage, a serious overestimate of the K_{Ic} value will result.

Because fracture toughness varies with material orientation, the K_{Ic} specimen must be machined to ensure that the crack growth plane and direction are the same as those of the failure. The investigator must also recognize that the ASTM

procedure requires precise control of fatigue precracking and environment. The subcritical crack growth of the failed component is unlikely to lie within the ASTM envelope, so the effective K_{Ic} value of the failure can be different from the ASTM value. Finally, large errors can arise if the failure was in an atypical region (such as a weld, heat-affected zone, or a hot spot) and the toughness is measured on base metal.

An additional toughness error can arise in certain situations where the actual toughness is higher than K_{Ic}. This elevation is found in thin sections of low-strength alloys and is due to the ability of the metal to deform plastically and partially relieve the stresses at the crack tip. Reference 7 gives some examples of such situations.

Statistical Evaluation. Uncertainty in fracture toughness (Ref 8) can be very significant, particularly for cleavage fracture where the standard deviation of K_{Ic} values is typically 25% of the mean.

Ideally, a complete statistical evaluation of the properties of the material is desirable, but it is not generally available. The nearest approach to general agreement is for structural steels and weld metals in the ductile-brittle transition region, where a three-parameter Weibull distribution is used (Ref 9):

$$F = 1 - \exp\{-[(K_{Ic} - K_I)/K_0]^4\} \tag{Eq 6}$$

where F is the cumulative failure probability and K_I and K_0 are fitting parameters. Representative coefficients of variation (standard deviation/mean, expressed as a percentage) for a variety of plates and weldments are summarized in Table 1. Individual data sets for the MIL-HDBK-5E entries in Table 1 are unique combinations of alloy designation, heat treatment, product form, and orientation, while the other high-strength alloy entries are less restrictive. It would appear that the typical coefficient of variation for handbook values of K_{Ic} of these materials is on the order of 10 to 20%, depending on how well characterized the materials are.

An analogous situation holds for upper-shelf toughness values for structural steels. Note that this section of the table includes J-integral variability (J_{Ic}), which appears to be about twice that for K_{Ic}, as would be anticipated from the relation $K_{Ic} \propto \sqrt{J_{Ic}}$. For multiple base-plate data sets, the coefficient of variation of K_{Ic} is somewhat larger than 13%, compared to about 8% for single plates. Based on all of the upper-shelf data, a working hypothesis is that the coefficient from handbook values is also 10 to 20% for both plates and weld metals.

The first entry for the lower-shelf and transition regions of steel was calculated using Eq 6, which represents a very large database (Ref 10, 11). If the coefficient of variation of crack-tip opening displacement, in terms of which much of the weld data are reported, is also twice that of K_{Ic}, a typical coefficient for K_{Ic} data below the upper shelf would be on the order of 20% for both base plate and weld metal. The heat-affected zone data exhibit an even larger coefficient of variation, and this is a serious concern.

Table 1 Variability in fracture toughness

Material	Coefficient of variation, %		
	K_{Ic}	J_{Ic}	CTOD
Upper-shelf toughness			
High-strength alloys(a)			
Aluminum-base	3-33	...	...
Aluminum-base(b)	8.9-10.4	...	...
Aluminum-base (7075-T6)	32	...	...
Steels	7-23	...	...
Steel (4340)	22	...	...
Structural steels			
Base plate	7	13	...
Base plate		16	...
Base plate	8.8	15	...
Base plates(a)	>13(c)	...	...
Weldments(a)		>17	...
Weldment	20.5	36	...
Lower-shelf and transition region			
Base plates(a)	...	...	40 (est.)
Base plates & welds(a)	18-28(d)	...	...
Weldments(a)	25	47	...
Weld HAZ(a)	...	...	50-75(e)

Note: CTOD, crack-tip opening displacement; HAZ, heat-affected zone. (a) Multiple lots of material. Otherwise single plates or weldments are reported. (b) Data sets with more than 100 entries. (c) Values at 100-150 °C. (d) Based on three-parameter Weibull distribution, with exponent = 4; variability arises from variation in the K_I:K_0 ratio. (e) Estimate based on the tenth percentile of 485 tests. Source: Ref 8

Estimating Toughness. Several estimation methods for fracture toughness values have been developed. These include measurement of the crack-tip profile in the scanning electron microscope (Ref 12), correlations based on Charpy and tensile tests and on microstructure (Ref 13), and handbook values of toughness of an alloy nominally the same as that of the failed part. Estimates deduced using any of these methods should be treated with great caution, because the magnitude of the error is usually unknown and can be large.

Evaluation of crack size and shape may be done visually, because actual failures often involve catastrophic growth of cracks that are several centimeters in size. However, the fracture surface at the origin is often lost or badly damaged and crack geometry has to be reliably estimated. In the event that the crack is very small, microscopic measurements may be required to reveal crack geometry. Conventionally, the size of a through-thickness crack is taken to be its average length, while the size of a surface crack is taken as its maximum depth.

Evaluation of crack geometry and loading pattern (the f factor) can be made using handbook values (Ref 14-16), provided that both the loading pattern and the crack size and shape are known. Finite element calculations are required for complicated geometries but are expensive and are not often essential, because f can usually be estimated roughly to sufficient accuracy.

Stress Estimates from Fatigue Striation Spacings. If subcritical crack growth occurred by fatigue, striation spacings can be used to estimate the alternating component of stress intensity (Ref 17), designated $\Delta K = \Delta K_{max} - \Delta K_{min}$. Because ΔK is proportional to (striation spacing)$^{1/4}$, a large error in spacing measurement leads to a small error in ΔK. However, large errors in striation spacings and hence in ΔK are possible in materials such as high-strength steels, where striations may not be as distinct as they are for aluminum alloys, for example. Also, surface damage, such as by rubbing, can obliterate striations.

If striation measurements are used, the alternating stress can be calculated as follows:

$$\Delta\sigma = \Delta K/g \qquad (\text{Eq 7a})$$

$$\Delta\sigma = \Delta K/f\sqrt{\pi a} \qquad (\text{Eq 7b})$$

In Eq 7b, the subcritical crack length a is evaluated at the striation location, and is not the final crack size.

Subcritical Fracture Mechanics (SCFM)

The growth of the crack that eventually leads to fracture occurs by mechanisms entirely different from the fracture itself. During most of the cracking process, the crack is much smaller than the one that would cause fracture at the prevailing stress. Therefore, the crack-tip stress field is less severe, and the size of the plastic zone smaller than at the time of fracture. Due to this small plastic zone, SCFM can often be approached using elastic concepts. If the crack size is close to critical (fracture), this may no longer be true, but because by far the longest time is spent in the growth of much smaller cracks, it is justifiable to use LEFM to obtain the time for crack growth with good accuracy. A notable exception is creep crack growth.

Fatigue. In the case of cyclic loading, crack growth occurs by fatigue. The amount of growth in one cycle depends on the crack tip stress field, which is described by K. During a load cycle, K varies from a minimum $K_{min} = \sigma_{min} \beta\sqrt{\pi a}$ to a maximum $K_{max} = \sigma_{max} \beta\sqrt{\pi a}$, over a range $\Delta K = \Delta\sigma\beta\sqrt{\pi a}$, where σ_{min} is the minimum stress in the cycle, σ_{max} is the maximum stress in the cycle, and $\Delta\sigma$ is the stress range $\Delta\sigma = \sigma_{max} - \sigma_{min}$.

It must be expected that the amount of growth per cycle depends on ΔK and K_{max}. If one defines a stress ratio R as $R = K_{min}/K_{max}$, the above statement is equivalent to the assertion that the amount of growth per cycle depends on ΔK and R. Use of ΔK and R is often more convenient because $R = K_{min}/K_{max} = \sigma_{min} \beta\sqrt{\pi a}/\sigma_{max} \beta\sqrt{\pi a} = \sigma_{min}/\sigma_{max}$. Hence, in the case of constant-amplitude loading where σ_{min} and σ_{max} do not change from cycle to cycle, K remains constant, while K_{max} would depend on crack size.

The amount of growth per cycle is the rate of growth. This means that the above statement is equivalent to:

$$\frac{da}{dN} = \text{dependent on } \Delta K \text{ and } R \qquad (\text{Eq 8})$$

where N is the number of cycles, and da/dN is the rate of growth.

It cannot be known *a priori* how da/dN depends on ΔK and R; this information must be provided by a test of the material. By measuring the rate of growth in a test and plotting the growth rates as a function of ΔK (calculated from $\Delta K = \beta\Delta\sigma\sqrt{\pi a}$), the dependence on ΔK is obtained, and in a similar fashion, tests at different R stress ratios provide the dependence on R. Crack size as a function of time (number of cycles) is obtained by integration:

$$N = \int_{a_1}^{a_2} \frac{da}{da/dN} \qquad (\text{Eq 9})$$

The integration is typically performed numerically using a computer. If data for a fixed value of R fall nearly on a straight line on a log-log scale, the rate data can be represented by the Paris equation. Unfortunately, the Paris equation covers only one R value. If the lines for different R values are parallel, one could use the modified equation:

$$\frac{da}{dN} = \frac{C_0}{(1-R)^n}\Delta K^m \qquad (\text{Eq 10})$$

where n and C_0 are also empirical constants. By noting that $R = K_{min}/K_{max} = (K_{max} - \Delta K)/K_{max}$ so that $K_{max} = \Delta K/(1-R)$, Eq 10 becomes:

$$\frac{da}{dN} = C_0\frac{\Delta K^n}{(1-R)^n}\Delta K^{m-n} \qquad (\text{Eq 11})$$

substituting for R and letting $p = (m-n)$:

$$\frac{da}{dN} = C_0 K_{max}^n \Delta K^p \qquad (\text{Eq 12})$$

Equation 12 is known as the Walker equation. Its advantage over the Paris equation is that it covers all values of R.

Many other equations are used. It should be pointed out, however, that no equation has any physical meaning: all are curve-fitting equations. There is no objection to their use if they fit the data. However, if the numerical integration must be done by a computer, use of the original data in tabular form is as convenient as use of a sometimes poor fitting equation.

Stress-Corrosion Cracking. In the case of stress-corrosion cracking, the rate of growth will depend on K, by the same arguments used for fatigue. In this case, the rate of growth depends directly on time so that the growth rate da/dt is:

$$\frac{da}{dt} = f(k)$$

Again, the rates must be obtained from a test. Integration of the rate data for the loading and β of a structure will provide the crack size as a funtion of time.

For combined stress-corrosion and fatigue cracking, one obtains da/dN from a cyclic-loading test in the appropriate environment. Again, a numerical integration will provide the crack

growth curve for the structure regardless of the shape of the da/dN-ΔK data. Complications due to load interaction and load-environment interaction must be considered.

In the case of stress corrosion at sustained load, the stress intensity must exceed a certain minimum value for crack growth to occur. This minimum value is known as the stress-corrosion threshold, denoted by K_{ISCC}. The threshold value is best determined by performing tests on cracked specimens and by measuring the time to fracture. A plot of the stress intensity applied as a function of time to failure will show an asymptote, which is K_{ISCC}. If growth does occur, it will continue to occur because K increases, and a failure will result. Conversely, if no failure occurs, no crack growth occurs; therefore, the stress intensity at the asymptote is indeed the threshold.

Fatigue-Crack Growth. Fatigue-striation counts (if possible) generally provide a reasonable account of the crack-growth rates and crack-growth curve. If the crack-growth-rate behavior of the material and the stresses are known, the stress intensity can be calculated, and a comparison can be made between actual and anticipated properties for a conclusion about the adequacy of the material. Conversely, if the stresses are not known, the measured rates and the rate properties can be used to estimate what the acting stresses were. This procedure will at least provide the magnitude of the stresses.

From the amount of crack growth (crack size at fracture), known stresses, and growth-rate properties, a reasonable insight can always be obtained regarding the question of misuse, for example, continuous overloading. The time to failure and final crack size are determined using fracture mechanics, as discussed above. When the results are not in accordance with the observations, the analysis can be repeated to determine how much higher (or lower) the stresses would have had to have been to produce the cracking time and extent of cracking as observed.

Any change of loading or change in environment during the cracking process is likely to leave its mark. Such changes produce a change in crack-growth rate, which is usually associated with a change in microfracture topography (surface roughness). Because the roughness affects the reflection of light, a change in roughness will appear as a line (beach mark). Any beach mark is an indication of a change in circumstances during cracking. The beach marks on fatigue-crack surfaces are well known. Similar marks may occur on stress-corrosion-crack surfaces due to changes in loading or environment.

A beach mark is clearly the delineation of the crack front at some point during the cracking process. Thus, the crack size at the time of the change is known. If any information on the nature of possible changes is known, the crack size at which they occurred can be used to obtain information on rate properties or stresses in the manner discussed above. If there is no information about the nature of the changes, such information (when it is a change in loading in particular) can be obtained from known growth-rate properties:

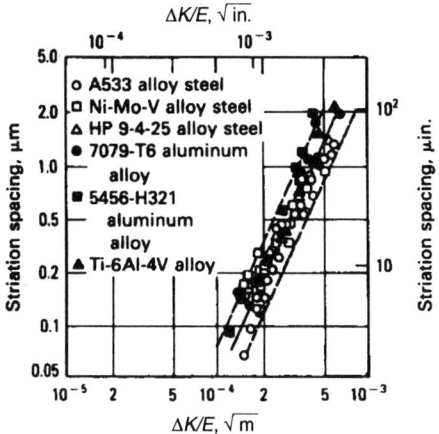

Fig. 1 Correlation of fatigue striation spacing with ΔK normalized by Young's modulus (E)

the time for growth between beach marks can be calculated, and the stress required to produce the observed crack sizes estimated.

Ductile striations are the most commonly observed fatigue features that occur in the microstructurally insensitive intermediate ΔK regime. The undulating or ripple-like fatigue striations appear to be correlated to a normalized stress-intensity range, ΔK. Although there are exceptions, such regular striations have been found to correlate to $\Delta K/E$ (stress-intensity range over Young's modulus), as shown in Fig. 1. A good correlation is:

$$\text{Striation spacing} \approx 6 \left(\frac{\Delta K}{E} \right)^2$$

Figure 1 illustrates the application of this equation to many material types. More quantitative relationships have interpreted this growth mechanism to be proportional to the crack tip displacement. Such displacements may be changed by microstructural variations of either the flow stress or crack path. Thus, within the scatter band of Fig. 1, and in some cases outside it, microstructure may play a role.

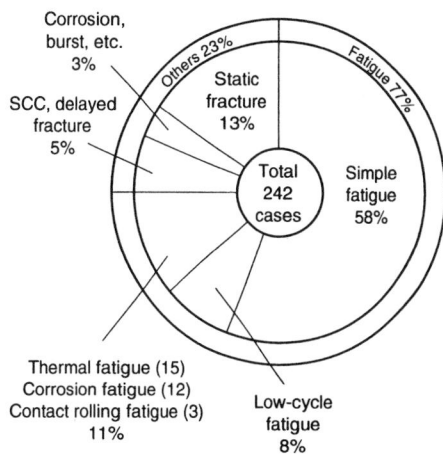

Fig. 2 Distribution of failures according to mechanism. Source: Ref 19

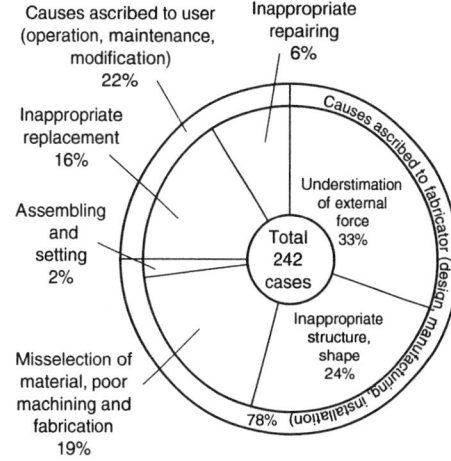

Fig. 3 Distribution of failures according to cause. Source: Ref 19

Case Studies of Failure Analyses Using Fracture Mechanics

Reports of a number of fracture-mechanics-based failure analyses have been published. Four examples (from Ref 18) are summarized in this section in order to illustrate the state of the art of the approach. Some simplifications and slight alterations have been made to the published reports in order to increase clarity. They were chosen to illustrate both typical procedures and the departures from ideal textbook procedures that can be encountered in practice. One striking point about these case studies is the very frequent need to estimate factors that are unknown or uncertain.

The "Selected References" at the end of this article also lists collections of case studies, which include examples involving fracture mechanics concepts. These collections provide insight into how to perform fracture mechanics analysis and also should give the reader insight as to how to decide whether such an analysis would be helpful.

Some of the collections of case studies discuss common characteristics of failures. For example, Fig. 2 shows that fatigue was the dominant mechanism for numerous industrial failures that the author investigated, while Fig. 3 points out that he found that errors in production (such as in design, material selection, and fabrication) were much more common than errors in operation (Ref 19). Section changes leading to stress concentrations are known to be a particular problem. There are three caveats regarding Fig. 2 and 3:

- Because service failures are very rare, it should not be concluded that use of defective parts is widespread.

- The possibility of service failures due to production deficiencies is limited by the use of safety factors and the practice of proof testing by overstressing.
- The distributions in Fig. 2 and 3 are not valid for every application. For example, gas-transmission pipeline failures are most commonly caused by accidental damage due to third-party impact.

Some insight has been obtained into the relative importance of fracture mechanics parameters when failure occurs from pre-existing cracks. The two criteria deduced from Ref 8 are that failure is most likely to occur when crack size is greater than 25 mm (1 in.) and the ratio of fracture toughness (K_{Ic}) to yield strength is less than $0.16\sqrt{m}$ ($1\sqrt{in}$.). Because of the relatively small number of failures discussed in Ref 8, it cannot be argued that these failures provide general rules as to which cracked structures are safe and which are likely to fail. However, they do point out the danger of allowing components made from low-toughness materials and containing large cracks to remain in service.

Failure of a Cryogenic Pressure Vessel

Background (Ref 18, p 43). This example is a rather straightforward application of fracture mechanics analysis with a satisfactory result, despite some uncertainties in the data. A cryogenic pressure vessel used in petrochemical manufac-

ture exploded. The vessel operated at –130 °C (–202 °F) and was made from a special fine-grained low-carbon steel. The failure initiated and first propagated in the neck or dome region of the vessel.

A traditional failure investigation using fractographic examination revealed evidence of fatigue crack growth, possibly corrosion assisted, at the origin. Several years after the explosion a more detailed investigation was undertaken to evaluate where safe operation could be ensured for similar vessels still in service. A "leak-before-break" philosophy was adopted; it assumed that operating vessels are safe if the critical crack length is greater than the wall thickness, under which conditions the subcritical crack would penetrate the wall and cause leakage before rupture would occur. Unfortunately, only a small piece of the failed vessel was available for examination.

Fracture Mechanics Analysis. Because of the complex geometry, a finite element calculation of *stress* was performed. The calculation revealed that the highest hoop stress was in the neck re-

Fig. 4 Ruptured liquid propane gas cylinder

gion, where the service failure did originate, and amounted to 80 MPa (11.6 ksi) at the outer wall and 60 MPa (8.7 ksi) at the inner wall. *Fracture toughness* is highly variable at the failure temperature; measurements on the same grade of steel from a similar vessel indicated a range from 39 to 67 MPa$\sqrt{m}$ (36 to 61 ksi$\sqrt{in}$). In the earlier investigation, critical *crack length* had been estimated to be in the range of 40 to 130 mm (1.6 to 5.1 in.). The crack was assumed to be a through-wall crack so that the *f* factor was 0.707 (see "Evaluation of Stress Intensity" in this article).

A range of hoop stress was calculated from Eq 5 by combining the largest estimated crack length with the smallest measured fracture toughness and vice versa. Because the result ranged from 87 to 267 MPa (12.6 to 38.8 ksi), the actual hoop stress of 80 MPa (11.6 ksi) could lead to failure if the crack was as large as the largest estimate and the fracture toughness had its lowest probable value. Moreover, the investigation confirmed that the critical crack size was greater than the wall thickness so that leak-before-break could be anticipated.

Evaluation. The investigators concluded that the failed vessel must have leaked prior to the rupture but that the leakage had not been detected. They urged more careful and extensive inspection of similar vessels still in service.

Failure of a Liquid Propane Gas Cylinder

Background (Ref 18, p 75). This is another example of straightforward analysis, but one in which the fracture mechanics evaluation did not markedly contribute to resolving the problem.

Five liquid propane gas (LPG) cylinders ruptured after about one year of service. The breaks were all longitudinal and about 400 to 500 mm (16 to 20 in.) long (Fig. 4). Visual observation of one of the failed cylinders revealed that the inner surface was scored with deep-drawing flaws of approximately the same length as the through crack. Examination of the fracture surface showed that a crack had initiated at one of these flaws and had propagated by cleavage almost to the outer surface. Final fracture involved formation of a small shear lip at the outer surface of the cylinder.

The cylinders were of welded construction and were made from a deep-drawn 0.13% C plain carbon steel with a yield strength of 326 MPa (47 ksi). Following postweld heat treatment, the cylinders were proof tested by pressurization at 3.1 MPa (450 psi). Operating pressure was 1.38 MPa (200 psi).

Fracture Mechanics Analysis. *Stress* was calculated using the simple equation for hoop stress of a pressurized cylinder; the result was 78 MPa (11.3 ksi). *Fracture toughness* was measured using a center-cracked panel machined from steel close to the location of the actual fracture. Of three specimens tested, only one failed by cleavage and its K_{Ic} value was 50 MPa$\sqrt{m}$ (45 ksi$\sqrt{in}$). The other specimens failed by dimpled rupture and could not be used to evaluate the toughness of the fractured cylinder. Because the crack propagated in the through-thickness direction, the critical *crack length* was taken to be the depth of the deep-drawing flaw, about 1.3 mm (0.05 in.). The *f factor* was calculated using a handbook of stress intensity values by treating the flaw as a straight-edge crack. This estimation is acceptable for a crack that is much wider than it is deep and led to a value of $f = 2.1$.

Using Eq 5, the predicted failure stress was found to be 372 MPa (54 ksi), which is about five times as high as the value of 78 MPa (11 ksi), calculated from the operating pressure. A test of the undamaged pressure relief valve demonstrated that it would be activated at 2.2 MPa (320 psi), corresponding to a hoop stress of 124 MPa (18 ksi), which is still well below the calculated failure stress, indicating that accidental overpressure could be eliminated as a primary failure cause.

Evaluation. Because the crack geometry was well established, the large error in predicted stress could not be attributed to errors in either crack length or the *f* factor. A hardness traverse was made around the circumference and it was found that there was a strong gradient, with a value of 170 HV in the region of the rupture and 90 HV at a location 180° from the rupture.

The investigators concluded that the cylinder had not been properly heat treated after welding. This error led to two contributions to the failure:

- The residual stresses were not relieved. Residual stresses added to the pressure stress could account for the difference in calculated and operating stress.
- Local embrittlement had occurred due to heating during welding. This suggests that the reported fracture toughness value was not representative, because the tip of the crack in the fracture toughness specimen was not close enough to the embrittled region.

While the published report of this investigation does not contain explicit recommendations, it is clear that the ruptures of the cylinders were caused by manufacturing deficiencies that could be corrected.

Fatigue Failure of a Large Fan

Background (Ref 18, p 37). This example illustrates analysis of a fatigue failure. A large fan, which provided draft for a coal-fired power plant, shattered. The fan had been in service for only ten days following periodic scheduled maintenance. The failure originated in a shroud plate, which showed evidence of extensive fatigue crack growth. The steel was described as low carbon, medium strength, with no details on manufacture being given.

The failure origin was close to a weld, but some details were not available because the mating surface to the origin was never found. Figure 5 shows the fracture surface with the initiating defect on the bottom side, a surface flaw 1.8 mm (0.07 in.) deep and 15 mm (0.6 in.) long. Most of the fracture surface exhibited fatigue markings indicating that the critical crack size at failure was much larger than the surface flaw.

Fracture Mechanics Analysis. Both static and alternating components of *stress* were evaluated using a finite element analysis. The static stress was 110 to 120 MPa (16 to 17.5 ksi), and the alternating stress was 10 to 15 MPa (1.5 to 2 ksi). *Fracture toughness* was measured using standard ASTM techniques and was found to be 65 ± 9 MPa$\sqrt{m}$ (59 ± 8 ksi$\sqrt{in}$). Critical *crack size* and *crack shape* were measured, but not reported, by the investigators. Figure 4 suggests that a good approximation for the onset of final failure would be a through crack about 110 mm (4.3 in.) long. The *f factor* for a through crack is 0.707 (see "Evaluating Stress Intensity" in this article).

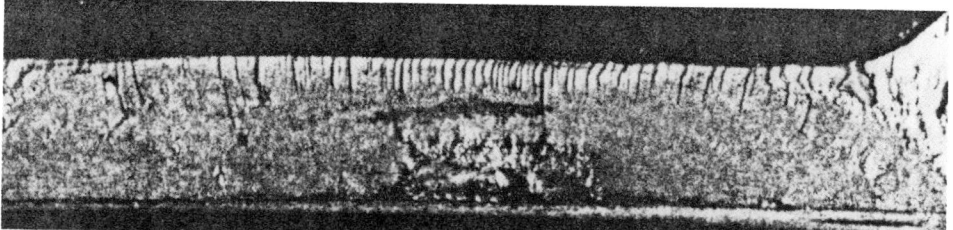

Fig. 5 Fracture surface of shroud plate. Approximately full scale

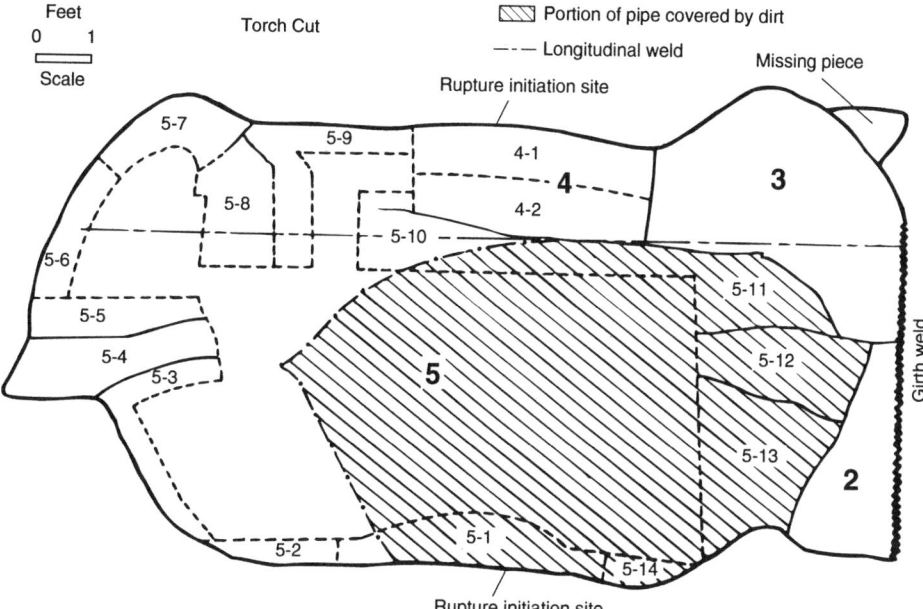

Fig. 6 Failure pattern in pipe. Dotted lines indicate subsequent torch cuts. Shaded region was partially protected from heat of fire by a dirt cover.

Using the above inputs, the failure stress was calculated to be in the range of 135 to 178 MPa (19.6 to 25.8 ksi), averaging somewhat greater than the combined steady-state and alternating stresses (120 to 135 MPa, or 17.5 to 19.6 ksi).

Evaluation. Although there was sufficient information at this point to provide remedies to prevent similar failures, the investigators decided to verify their stress evaluation by measuring fatigue striations on the small piece that was left from the failure. Using this technique they found that the cyclic stresses were greater than ±50 MPa (±7.25 ksi), which would add sufficiently to the steady-state stress to put the total stress firmly within the calculated failure range. However, strain gaging of a similar fan never revealed the presence of such unexpectedly high fluctuating stresses.

In stating their recommendations, the investigators pointed out that the steel was operating on its lower shelf and was probably too brittle for the application. In addition, they were troubled by the uncertainty in the magnitude of the fluctuating stress. They therefore suggested that similar fans were likely to fail. A number of improvements adopted as a result of this failure included use of a tougher steel, design changes, improved welding procedures, and more stringent inspections.

Failure of a Gas Transmission Pipeline

Background (Ref 18, p 173). This example illustrates the problem of performing a fracture mechanics analysis in failures where there is massive damage to the structure.

A gas-transmission pipeline ruptured, the gas was ignited, and the resulting fire burned for over one hour before it could be extinguished. The pipeline was made from a 0.3 wt% C plain carbon steel and was 760 mm (30 in.) in diameter with 9.5 mm (0.375 in.) wall thickness. It was buried to a depth of 1 m (40 in.) and had been cathodically protected, with the protection adjusted some months before the failure.

The failure was about 5.5 m (18 ft) long and was confined to a single section of pipe, arresting at a girth weld at one end and by tearing off (spiraling) at the other (see Fig. 6). When the fragments had been collected it was found that one face of the fracture at the origin in piece 5-1 was covered with dirt. This region had a microstructure of either quenched-and-tempered martensite or bainite, which would not be anticipated in this steel in the as-rolled condition. The uncovered opposite face (piece 4-1) showed metallographic evidence of reaustenization, indicating that the fire had been severe enough to generate very high temperatures. More important, the heat treatment caused by the fire had altered the steel in the region of the fracture origin so that its properties at the time of the explosion could not be determined experimentally. The maximum measured hardness near the origin of piece 5-1 was 300 HK, but the investigator concluded that this piece had been tempered and that the hardness was originally higher at the origin.

Fractographic examination revealed a semielliptical surface flaw at the origin that was 4.4 mm (0.175 in.) deep by 4.2 mm (0.165 in.) wide. While some intergranular fracture was observed, most of the fracture was by cleavage.

Fracture Mechanics Analysis. Using the standard equation for hoop stress in a cylinder pressurized to 3.9 MPa (560 psi), the operating *stress* was found to be 152 MPa (22 ksi). No direct estimate of *fracture toughness* at the origin could be made because of the heat treatment caused by the fire. The proper choice of critical *crack length* for this geometry is its depth of 4.4 mm (0.175 in.), while the *f factor* is 1.8.

Because fracture toughness was unknown, it had to be calculated from Eq 5, with the result being 32 MPa√m (29 ksi√in.). The investigator concluded that the only way that such a low toughness could be achieved in this pipeline steel was if the origin was at a local hard spot, whose hardness might have been as high as 600 HK (55 HRC).

Evaluation. The data available to the investigator suggested the possibility that either stress-corrosion cracking or hydrogen stress cracking was the cause of the failure. Both of these mechanisms require high stress, high strength, and hydrogen absorption.

Existence of a local hard spot could satisfy the strength requirement and contribute to the stress requirement. The presence of the hard spot would lead to a local residual stress and a flat region because of its greater resistance to deformation during fabrication. The local stress could then be higher than the calculated value because of residual stresses associated with nonuniform deformation (out-of-roundness of the pipe) and also because of the geometric irregularity.

The hardness and the metallographic observation of tempered martensite or bainite indicated that the strength of the hard spot was higher than the nominal strength of the steel. It is plausible that the hard spot region was susceptible to the embrittlement processes at the time of fracture.

Hydrogen access was found to be likely, in that deterioration of the protective coating was observed in the undamaged section of the pipeline away from the fracture.

In summary, the rupture was probably caused by access of hydrogen to a susceptible site on the pipeline, even though this scenario could not be definitely proved.

REFERENCES

1. L.E. Murr, *What Every Engineer Should Know about Material and Component Failure, Failure Analysis and Litigation*, M. Dekker, 1987
2. A.R. Rosenfield and C.W. Marschall, Ductile-to-Brittle Fracture Transition, *Metals Handbook*, 9th ed., Vol 11, American Society for Metals, 1986, p 66-71
3. J.W. Dally and W.F. Riley, *Experimental Stress Analysis*, McGraw-Hill, 1991
4. A.S. Kobayashi, *Handbook on Experimental Mechanics*, 2nd ed., VCH, 1993
5. O.C. Zinekiewick, *The Finite Element Method*, McGraw-Hill, 1977
6. W.C. Young, *Roark's Formulas for Stress and Strain*, McGraw-Hill, 1989
7. D.R.H. Jones, *Engineering Materials 3: Materials Failure Analysis—Case Studies and Design Implications*, Pergamon Press, 1993
8. A.R. Rosenfield and C.W. Marschall, *Engineering Fracture Mechanics*, Vol 45, Pergamon Press, 1993, p 333-338
9. D.E. McCabe, R.K. Nanstad, A.R. Rosenfield, C.W. Marschall, and G.R. Irwin, Investigation of the Bases for Use of the K_{Ic} Curve, *ASME PVP*, Vol 213, 1991, p 141-148
10. K. Wallin, Statistical Aspects of Constraint with Emphasis to Testing and Analysis of Laboratory Specimens in the Transition Region, ASTM STP 1171, 1993, p 264-288
11. D. Sienstra, T.L. Anderson, and L.J. Ringer, Statistical Inferences on Cleavage Fracture Toughness Data, *J. Eng. Mater. Technol.*, Vol 112, 1990, p 31-37
12. D. Broek, *The Practical Use of Fracture Mechanics*, Kluwer Academic Publishers, 1989, p 433-434
13. P.F. Timmins, *Fracture Mechanics and Failure Control for Inspectors and Engineers*, ASM International, 1994
14. Y. Murakami, *Stress Intensity Factors Handbook*, Pergamon Press, 1987
15. D.P. Rooke and D.J. Cartwright, *Compendium of Stress Intensity Factors*, The Hillingdon Press, 1976
16. H. Tada, R.C. Paris, and G.R. Irwin, *The Stress Analysis of Cracks Handbook*, Del Research Corp., 1985
17. R.M. Pelloux and A.S. Warren, Fatigue Striations and Failure Analysis, *Failure Analysis: Techniques and Applications*, J.I. Dickson et al., Ed., ASM International, 1992, p 45-49
18. V.S. Goel, Ed., *Analyzing Failures*, American Society for Metals, 1986
19. S. Nishida, *Failure Analysis in Engineering Applications*, Butterworth-Heinemann, 1992

SELECTED REFERENCES

- D. Broek, *The Practical Use of Fracture Mechanics*, Kluwer Academic Publishers, 1989
- J.I. Dickson, E. Abramovici, and N.S. Marchand, Ed., *Failure Analysis: Techniques and Applications*, ASM International, 1992
- K.A. Esakul et al., Ed., *Handbook of Case Histories in Failure Analysis*, ASM International, Vol 1, 1992; Vol 2, 1993
- V.S. Goel, Ed., *Analyzing Failures*, American Society for Metals, 1986
- C.M. Hudson and T.P. Rich, Ed., *Case Histories Involving Fatigue and Fracture Mechanics: A Symposium*, ASTM, 1986
- D.R.H. Jones, *Engineering Materials 3: Materials Failure Analysis—Case Studies and Design Implications*, Pergamon Press, 1993
- S. Nishida, *Failure Analysis in Engineering Applications*, Butterworth-Heinemann, 1992
- R.B. Tait and G.G. Garrett, Ed., *Fracture and Fracture Mechanics: Case Studies*, Pergamon Press, 1985

Operating Stress Maps for Failure Control

P.F. Timmins, President, Risk Based Inspection Inc.

FRACTURE MECHANICS CONCEPTS give engineers and inspectors the means to assess service situations for potential failure. An operating stress map (Fig. 1) is one way to assess potential fracture for an alloy of given toughness and operating stress. Operating stress maps are based on the same principle as a residual strength diagram, in that both diagrams are based on a plot of the following equation (Eq 16 in the article "Residual Strength of Metal Structures" in this Volume) for elastic fracture mechanics:

$$\sigma_c = K_{Ic}/\beta\sqrt{\pi a_c} \qquad \text{(Eq 1)}$$

Plotting of this equation illustrates the regions where different combinations of stress and crack size are of concern, or where fracture mechanics or net section yield applies for a given ratio of operating stress and yield stress.

This article describes the basis of "operating stress maps" based on Eq 1, as an alternative to the British CEGB (Central Electricity Generating Board) R6 method (Ref 1), which is based on failure assessment diagrams (FADs). The CEGB R6 method has received much attention (Ref 2), and it is useful in conjunction with (but not instead of) diagrams based on the critical stress

relation (Eq 1). The use of operating stress maps was developed by the author (Ref 3) as a tool for assessing potential fracture in new or existing structures. If operating equipment contains a crack of initial length a_1, and, during inspection intervals, it grows by some mechanism to length a_2 (less than the critical length), the growth rate may be low enough to allow operation until the next shutdown before the critical crack length is achieved, thus allowing time for repairs to be made during the shutdown. Alternatively, the operating stress may be reduced, giving a larger critical crack length according to Eq 1. This relation in Eq 1 between operating stress and crack length (or depth, depending on geometry) is the basis for deriving an operating stress map.

However, it must be emphasized that an analysis of damage tolerance cannot be based on Eq 1 alone. As discussed in the article "Residual Strength of Metal Structures" in this Volume, any analysis of damage tolerance requires information on both subcritical crack growth rates and the residual strength for a maximum permissible crack size (a_p). A crack smaller than a_p may in fact be unacceptable if subcritical crack growth rates are large enough to reach critical crack (a_c) and fracture before the next outage or scheduled maintenance. Therefore, calculation of a_p from a residual strength diagram must be supplemented by information on when a crack may reach a_p or a_c. Damage tolerance analysis must also account for net section yield (where fracture is governed by plastic deformation regardless of flaw size).

Constructing an Operating Stress Map

The process of constructing an operating stress map is based on the same principles used in constructing so-called "residual strength diagrams." Operating stress maps can be added to, depending on the particular service conditions or geometry. For example, a subcritical growth mechanism rate law can be incorporated to determine the time available between crack detection and growth to critical (i.e., unstable) size.

The method presented here provides maps for through-thickness and surface crack geometries. A number of maps are presented; however, the

intention is that engineers will be able to construct their own maps to meet specific needs. In addition, operating stress maps are compared to the use of FADs in the British CEGB R6 method.

Construction of an operating stress map is based on calculations of net section yield and fracture mechanics. For purposes of simplicity, this section describes the construction of a through-thickness crack (Fig. 2) in the linear elastic regime. For a through-thickness crack of length $2a$ in a plate (or a panel, a vessel wall, or line-pipe), the geometry factor β is one. For an edge crack in the same plate, $\beta = 1.12$. It is also important to note the convention used for crack length. If a crack has two tips (as in a through-thickness crack), then the total crack length is twice the crack length on one side ($2a$). The depth of an edge or surface crack, a, corresponds to the crack size. Other more complicated geometries are discussed in the section "Using Operating Stress Maps" in this article.

Plane strain fracture toughness (K_{Ic}) is the inherent minimum toughness limit and thus pro-

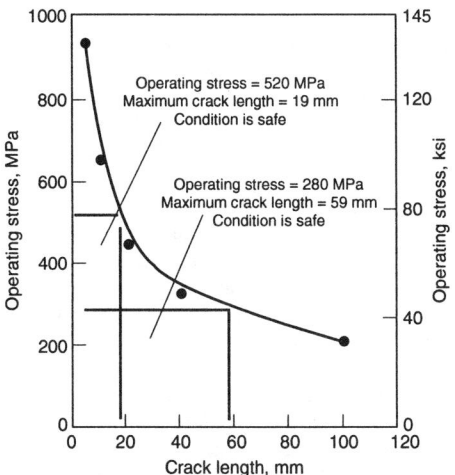

Fig. 1 Operating stress map of a through-thickness crack in forged plate of Ti-6Al-4V with yield strength (σ_y) of 790 MPa (115 ksi) and K_{Ic} of 83 MPa√m (75 ksi√in.)

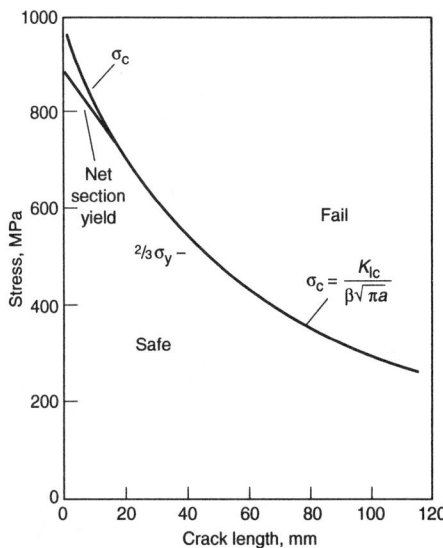

Fig. 2 Definition of quantities in constructing an operating stress map for a through-thickness crack ($\beta = 1$) in a plate of width W. When applied stress is at the critical level (σ_c), the crack of length $2a$ propagates across the plate of width W.

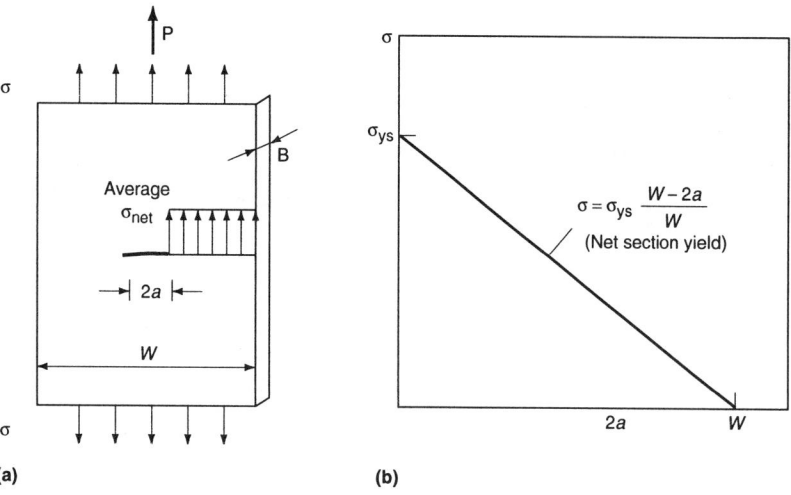

Fig. 3 Net section yield for small cracks. (a) Average net section stress. (b) Net section yield line

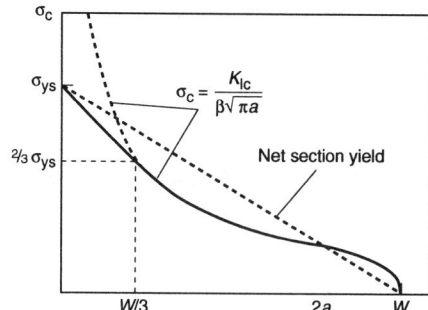

Fig. 4 Tangent approximation for short cracks with $\beta \approx 1$

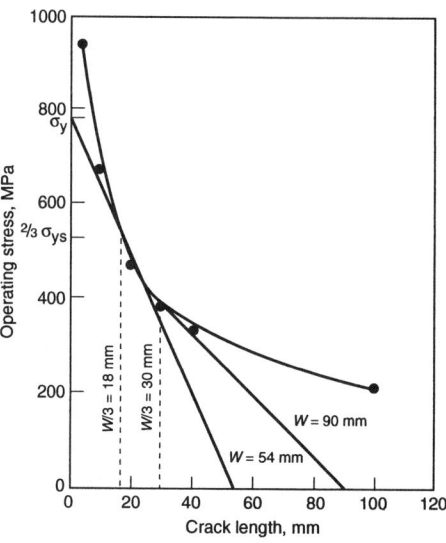

Fig. 5 Operating stress map of Ti-6Al-4V plate with a through-thickness crack and net section yield lines for two plate widths (W). Same material as in Fig. 1, $\sigma_y = 790$ MPa (115 ksi) and $K_{Ic} = 83$ MPa$\sqrt{m}$ (75 ksi$\sqrt{in.}$)

vides a conservative toughness value for plotting an operating stress map with Eq 1. Plane strain conditions represent maximum constraint and occur when section thickness (B) is:

$$B \geq \left(\frac{K_{Ic}}{\sigma_y}\right)^2 \cdot 2.5 \qquad \text{(Eq 2)}$$

$$B \leq \left(\frac{K_c}{\sigma_y}\right)^2 \cdot 0.3 \qquad \text{(Eq 3)}$$

The plastic zone at the crack tip is larger in plane stress than in plane strain, and less constraint and higher toughness occur in plane stress. Thus K_{Ic} is a conservative measure for constructing an operating stress map. Although Eq 2 and 3 define basic regimes of plane strain and plane stress toughness, there is no clearcut transition point between the two conditions. Cracks behave under conditions of either predominantly plane strain or predominantly plane stress.

Net Section Yield. When crack size becomes very small, Eq 1 is not adequate in calculating a safe limit for operating stress (or residual strength). As a_c approaches zero, σ_c approaches infinity, which is clearly erroneous. If the stress is much higher than the yield strength, then the entire plate is yielding and the plastic zone is as large as the plate. In this situation, the plastic zone would have become too large for K to be applicable much before the yield stress is reached.

In this situation of small crack sizes, the residual strength or safe operating stress is determined by calculation of net section yield (Fig. 3). If P is the applied load, and W and B are the width and thickness of the plate, the nominal remote stress is $\sigma = P/BW$, while the average stress over the net section is $\sigma_{net} = P/B(W - 2a)$, so that:

$$\sigma = \sigma_{net} \frac{W - 2a}{W} \qquad \text{(Eq 4)}$$

The whole net section yields when $\sigma_{net} = \sigma_y$, which occurs when:

$$\sigma = \sigma_y \frac{W - 2a}{W} \qquad \text{(Eq 5)}$$

This equation represents a straight line between the points ($\sigma = \sigma_y$; $2a = 0$) on the ordinate and ($\sigma = 0$; $2a = W$) on the abscissa (Fig. 3b). If the residual strength curve, $\sigma_c = K_{Ic}/\beta\sqrt{\pi a}$, is shown in the same diagram, it turns out that (usually) there are two regions in which the predicted residual strength is larger than the stress for net section yield at very large crack sizes and at small crack sizes (Fig. 4).

For small crack sizes, the general procedure recommended by Feddersen (Ref 4) is to draw a tangent from the yield strength value to the critical stress curve, $\sigma_c = K_{Ic}/\beta\sqrt{\pi a}$ (Fig. 4). For a through-thickness crack (where $\beta = 1$), the point of tangency is always at 2/3 σ_{ys} and W/3 (Fig. 4). As plate width becomes smaller, the net section yield line (Eq 5) moves left until it is tangent to the σ_c curve at 2/3 σ_y (Fig. 5). All operating stresses on the net section yield line represent a condition for failure, regardless of crack size.

Because the point of tangency for $\beta = 1$ is at 2/3 σ_y and $2a = W/3$, then the operating stress at this point can be calculated from Eq 1. From the fact that the operating stress is 2/3 the tensile yield stress and $a = W/6$, it follows that:

$$\frac{2}{3}\sigma_y = \frac{K_{Ic}}{\sqrt{\pi W/6}} \qquad \text{(Eq 6)}$$

or

$$W = \frac{27}{2\pi}\left(\frac{K_{Ic}}{\sigma_y}\right)^2 \qquad \text{(Eq 7)}$$

These relationships (Ref 5) give the minimum plate width for which K_{Ic} applies. This value of plate width is also the maximum that will fail at net section yield, as will all smaller plates, irrespective of the crack length.

A complete diagram is important when analyzing a structure for fracture criticality. Once the residual strength curve (σ_c in Eq 1) is drawn, the line for net section yield can be constructed, and it can be seen immediately whether or not the residual strength should be found from the curve (K_{Ic}) or from the tangent to the curve. The whole σ_c curve has to be determined before the tangent can be constructed. In any analysis, understanding of crack growth behavior is also required (see the section "Subcritical Crack Growth" in this article).

Empiricism and K_{Ic}. The main reason for using the stress intensity approach in failure control is that the equations relate crack size to the operating stress, so the units are of stress and dimension. The less meaningful but cheap tests that have been used for years to describe toughness behavior are notched-bar impact tests.

There are empirical relationships in the literature that relate impact data to the more meaningful, but more expensive, fracture toughness data. Unfortunately, there are about 36 different relationships available, and they must be used for specific conditions. However, two relationships

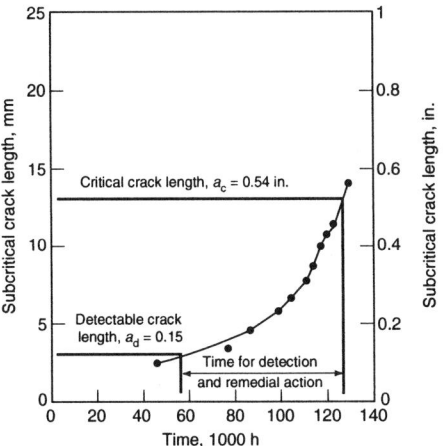

Fig. 6 Example of subcritical crack growth plot for fatigue crack growth of a surface crack in quenched-and-tempered steel

are used quite commonly (Ref 6). For conditions approaching upper shelf toughness values:

$$\left(\frac{K_{Ic}}{\sigma_y}\right)^2 = 4\left(\frac{CVN}{\sigma_y} - 0.1\right)\left(in., \frac{ft\ lb}{ksi}\right) \qquad (Eq\ 8a)$$

or

$$\left(\frac{K_{Ic}}{\sigma_y}\right)^2 = 0.52\left(\frac{CVN}{\sigma_y} - 0.02\right)\left(m, \frac{J}{MPa}\right) \qquad (Eq\ 8b)$$

and for conditions toward the lower shelf (Ref 7): $K_{Ic} = 12\sqrt{CVN}$ (all units), where CVN = Charpy toughness.

Subcritical Crack Growth

As previously noted, an analysis of damage tolerance in terms of the residual strength also requires information on crack growth prior to reaching the critical stress, $\sigma_c = K_{Ic}/\beta\sqrt{\pi a_c}$. Understanding of the crack growth behavior is necessary to estimate the time before critical crack length, a_c, is reached. For example, if subcritical crack growth behavior is known, crack length can be plotted versus time (Fig. 6). From a detected crack length, a_d, then the time to reach a_c can be estimated. This section briefly describes fatigue crack growth behavior. More detailed discussions of crack growth behavior of particular materials and service conditions (e.g., corrosion fatigue and stress corrosion cracking) are covered in other articles in this Volume.

$$da/dN = \frac{A(\Delta K)^n}{(1 - R)\ K_{Ic} - \Delta K} \qquad (Eq\ 9)$$

where A and n are empirical constants. This Forman equation (Ref 8) is general because R is accounted for. It has a vertical asymptote as $\Delta K = (1 - R)K_{Ic}$ and an asymptote at the fatigue crack growth thresh-

old, ΔK_{th}. However, neither this nor the Paris equation, nor any other of the 32 rate equations, are derived from first principles. The relations are, to a large extent, empirical curves from data.

More sophisticated relationships that consider crack growth retardation and a variety of geometries for a wide range of materials are currently available. Such a system is the NASA/FLAGRO computer program developed by Royce Forman at NASA, which provides a comprehensive package (Ref 9).

Fatigue Crack Growth Integration Example: SAE 1020 Steel. A wide plate of cold-rolled SAE 1020 steel has the following properties:

- Yield strength, $\sigma_y = 630$ MPa
- Young's modulus, $E = 207$ GPa
- Ultimate tensile strength, UTS = 670 MPa
- Plane strain fracture toughness, $K_{Ic} = 104$ MPa$\sqrt{m}$

The plate is subjected to constant-amplitude uniaxial cyclic loads from 200 MPa (σ_{max}) to -50 MPa (σ_{min}). To find the fatigue life that would be attained if an initial through-thickness edge crack were no greater than 0.5 mm in length:

$$K_{max} = (1.12)\ (200)\ \sqrt{\pi(0.0005)} = 9\ MPa\sqrt{m}$$

The final crack length, a_c, can be determined by setting K_{max} at fracture equal to K_{Ic}:

$$\Delta K = \Delta\sigma\beta\sqrt{\pi a}$$

$$\frac{da}{dN} = c(\Delta K)^m = c(\Delta\sigma\beta\sqrt{\pi a})^m = c(\Delta\sigma)^m(\pi a)^{m/2}\beta^m$$

which can define a critical crack size:

$$a_c = \frac{1}{\pi}\left(\frac{K_{Ic}}{\sigma_{max}\beta}\right)^2$$

Integrating the rate equation:

$$N_f = \int_0^{N_f} dN = \int_{a_i}^{a_c} \frac{da}{c(\Delta\sigma)^m\ (\pi a)^{m/2}\beta^m}$$

$$= \frac{1}{c(\Delta\sigma)^m(\pi)^{m/2}\beta^m}\int_{a_i}^{a_c} \frac{da}{a^{m/2}}$$

If $m \neq 2$, then:

$$\int_{a_i}^{a_c} \frac{da}{a^{m/2}} = \frac{a^{-(m/2)+1}}{-m/2 + 1}\bigg|_{a_i}^{a_c}$$

$$= \frac{a_c^{(-m/2)+1} - a_i^{(-m/2)+1}}{-m/2 + 1} \therefore N_f$$

$$= \frac{a_c^{(-m/2)+1} - a_i^{(-m/2)+1}}{(-m/2 + 1)c(\Delta\sigma)^m(\pi)^{m/2}\beta^m}$$

which is the general integration of the Paris equation when β is independent of crack length, a, and when m is not equal to 2. This equation is not correct if β is a function if a, which is the usual case.

Because specific crack growth data were not given for the SAE 1020 steel, a reasonable first approximation could use the conservative empirical equation for ferritic-pearlitic steels (see the

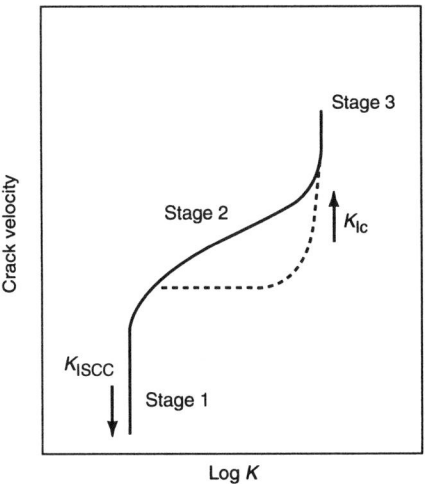

Fig. 7 Schematic plot of stress corrosion cracking velocity against stress intensity factor

article "Fracture Mechanic Properties of Carbon and Alloy Steels" in this Volume):

$$\frac{da}{dN} = c(\Delta K)^m = 6.9 \times 10^{-12}\ (\Delta K)^{3.0}$$

Although this equation was developed for $R = 0$, the small compressive stress, 50 MPa, will not have much effect on crack growth and can be neglected. Thus,

$$\Delta\sigma = 200 - 0 = 200\ MPa$$

Also,

$$a_c = \frac{1}{\pi}\left(\frac{K_{Ic}}{\sigma_{max}\beta}\right)^2 = \frac{1}{\pi}\left(\frac{105}{200 \times 1.12}\right)^2$$

Substituting,

$$N_f = \frac{(0.068)^{-3/2 + 1} - (0.0005)^{-3/2 + 1}}{(-3/2 + 1)(6.9 \times 10^{-12})(200)^3(\pi)^{3/2}(1.12)^3}$$

$$= \frac{(0.068)^{-0.5} - (0.0005)^{-0.5}}{-2.16 \times 10^{-4}} = 189,000\ cycles$$

Stress-corrosion cracking rates can depend on many factors, including temperature, chemistry of the environment, microstructure, and the stress intensity at the crack tip. Stress corrosion growth laws are similar to those described by fatigue crack extension (Fig. 7), but there is no single unified theory to explain SCC.

Stress-corrosion cracking degrades K_{Ic} to a level K_{ISCC}, below which crack growth does not occur. In between K_{ISCC} and K_{Ic}, a growth law will operate, but it will be peculiar to a given alloy and environment.

If circumstances arise for detection of SCC, it becomes prudent to generate operating stress maps for K_{ISCC} and K_{Ic}. Such maps are presented in Fig. 8 and 9 and for 4340 steel, from the data of Yokobori et al. (Ref 10). These maps are for surface crack geometries of various aspect ratios.

The operating stress map in Fig. 8 indicates that 4340 with a K_{ISCC} of 40 MPa$\sqrt{m}$ can be operated

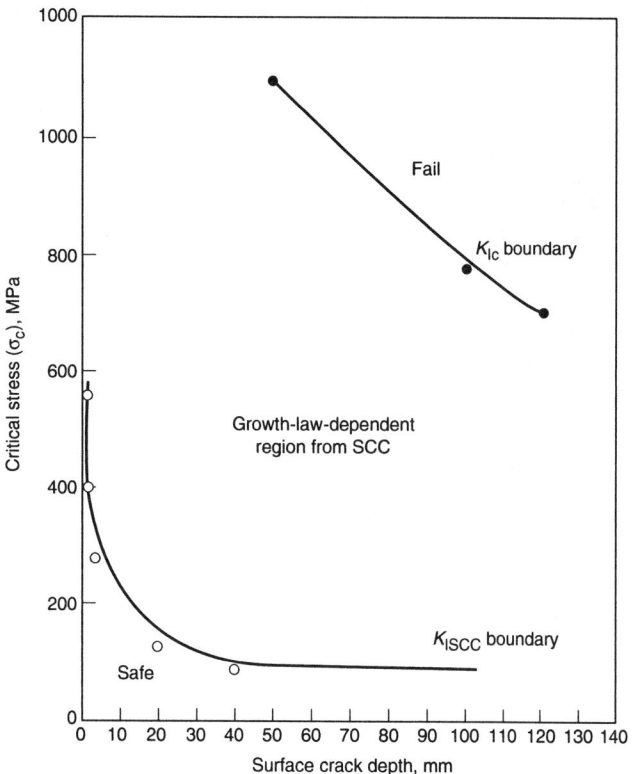

Fig. 8 Stress-corrosion cracking (SCC) operating stress map for 4340 steel with surface crack ($a/c = 0.6$). Material condition: quenched and tempered; $K_{Ic} = 127$ MPa$\sqrt{m}$; $\sigma_y = 1168$ MPa; $K_{ISCC} = 40$ MPa$\sqrt{m}$

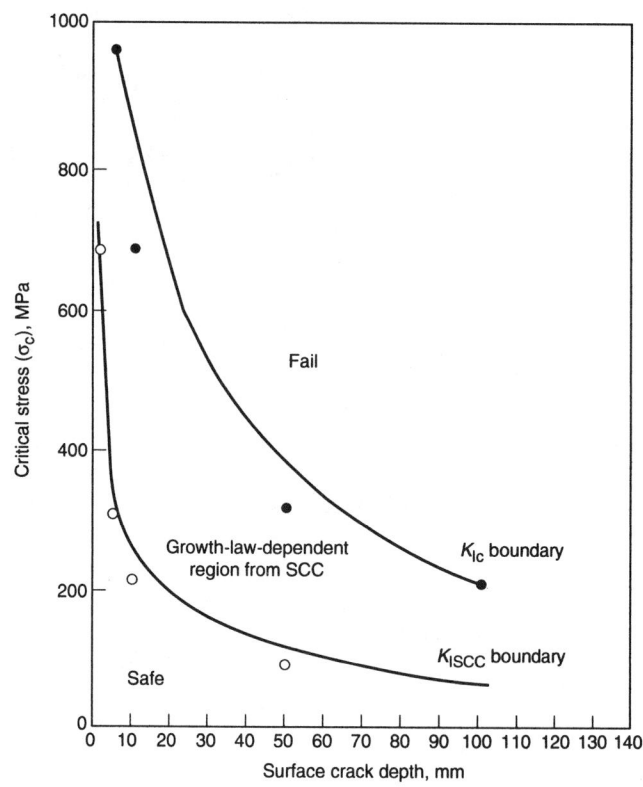

Fig. 9 Stress-corrosion cracking (SCC) operating stress map for 4340 steel with surface crack ($a/c = 0$). Same steel condition as in Fig. 8: $K_{Ic} = 127$ MPa$\sqrt{m}$; $\sigma_y = 1168$ MPa; $K_{ISCC} = 40$ MPa$\sqrt{m}$

safely at 25% of the yield strength, with surface cracks of aspect ratio $a/c = 0.6$ present up to maximum depth of 10 mm. The environment used must be one that gives rise to the K_{ISCC} value used.

Deeper cracks can be tolerated, but because the crack growth rate law is not known, the time-to-failure of such cracks cannot be predicted from the data available. A complete set of operating stress maps may be generated for surface cracks of various aspect ratios, through-thickness cracks, and edge cracks.

Corrosion Fatigue. There is not yet a generally accepted growth law for corrosion fatigue. The da/dN versus ΔK curves have a shape similar to those of fatigue and SCC curves. Testing in a similar environment under loading conditions similar to those expected in the application is needed to characterize corrosion fatigue crack growth rates.

Comparison with Failure Assessment Diagrams

As discussed in the article "Residual Strength of Metal Structures" in this Volume, the failure assessment diagram (FAD) is used to assess potential fracture in the whole range of conditions from brittle to fully plastic behavior. The FAD used in the CEGB R6 method is useful in conjunction with (but not instead of) the residual strength or operating stress diagram, because a

FAD can be derived from the latter. The example below also illustrates the simplicity of using an operating stress map in comparison to the FAD method. An operating stress map (or a residual strength diagram) relates operating stress directly to crack size, while the FAD method gives a proximity of fracture number as a "safety margin against failure" and requires several equations. Nonetheless, the FAD has significance because it illustrates from a technical view the competing conditions of plastic collapse and linear (elastic) fracture mechanics.

The CEGB R6 method (Ref 1) uses a failure assessment diagram (Fig. 10), which plots the ratio $K_r = K/K_c$ versus the ratio $S_r = \sigma/\sigma_{fc}$ (where σ_{fc} is the stress for plastic collapse). This normalization makes a universal diagram, where K_r represents the limit for elastic fracture and S_r represents the limit for plastic collapse. The diagram provides a graphical illustration of competing conditions and a basis for estimating the regime of elastic-plastic fracture between the two limits. Failure would occur in the region outside the envelope bounded by the axes and the assessment curve. This graphical approach is instructive, because it illustrates that elastic-plastic fracture mechanics (EPFM) does not require sensitive fracture criterion. This method of failure assessment is used in the example below.

Example 1: CEGB R6 Method in Failure Assessment. The FAD method can be illustrated with an example similar to that originally set by Spiekhout (Ref 2) in the R6 format. Consider a

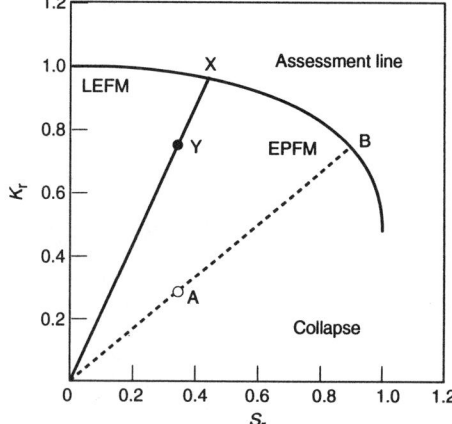

Fig. 10 Failure assessment diagram based on the stress intensity ratio ($K_r = K/K_c$) and stress ratio ($S_r = \sigma/\sigma_{fc}$) where σ_{fc} is the plastic collapse stress. General regions are shown for linear elastic fracture mechanics (LEFM) and elastic plastic fracture mechanics (EPFM).

cast steel plate with a surface flaw and the following conditions:

- Depth of surface flaw, $a = 7$ mm
- Length of surface flaw, $l = 35$ mm (1.38 in.)
- Yield strength, $\sigma_y = 430$ MPa (62 ksi)
- Ultimate tensile strength, UTS = 589 MPa (85 ksi)
- Plane strain fracture toughness, $K_{Ic} = 32$ MPa$\sqrt{m}$ (29 ksi$\sqrt{in.}$)

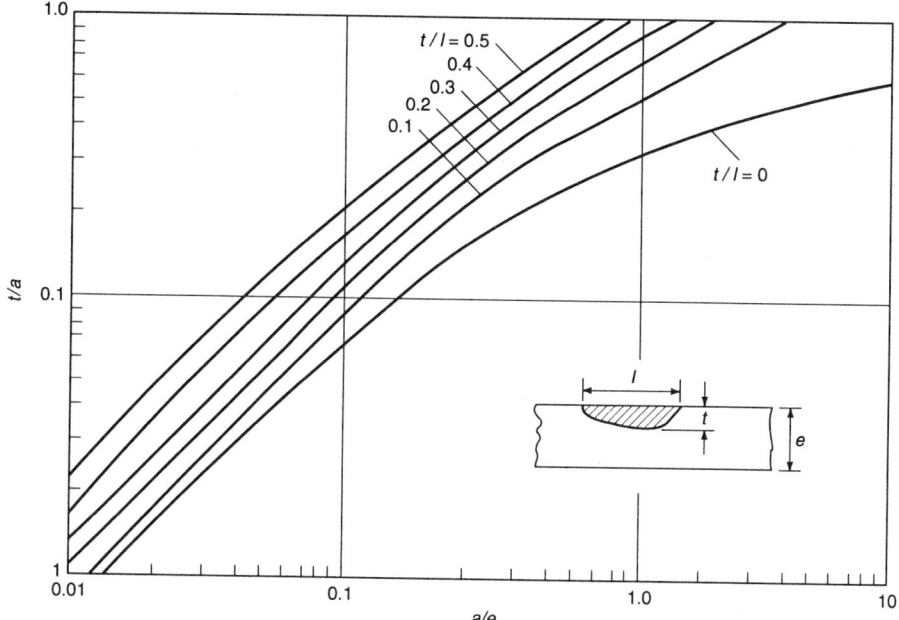

Fig. 11 Relationship between actual flaw dimensions and the parameter *a* for surface flaws

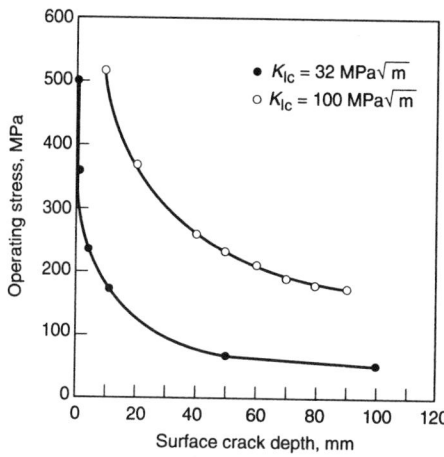

Fig. 12 Operating stress map with a plot of critical stress (σ_c, Eq 1) for cast steel plate with a surface crack ($a/c = 0.2$). Material condition: quenched-and-tempered steel, $\sigma_y = 430$ MPa (62 ksi), ultimate tensile strength = 590 MPa (85 ksi), $K_{Ic} = 32$ MPa$\sqrt{m}$ (29 ksi$\sqrt{in}$.), and plate thickness = 108 mm (4.25 in.). Higher toughnesses are inserted for comparison.

The operating stress (σ) is 170 MPa (25 ksi), and the stress for plastic collapse (σ_{fc}, or σ_{MC}) is estimated as follows:

$$\sigma_{fc} \approx (\sigma_y + UTS)/2 = 510 \text{ MPa}$$

This value for plastic collapse and the operating stress are used to define the value S_r as follows:

$$S_r = \sigma/\sigma_{fc} = 170/510 = 0.33$$

The ratio $K_r = K_I/K_{Ic}$ is then derived from British standard PD6493 (Fig. 11) (Ref 11) as follows:

$$\frac{\text{Crack depth}}{\text{Plate thickness}} = \frac{t}{e} = \frac{7}{108} = 0.065$$

and

$$\frac{\text{Crack depth}}{\text{Crack length}} = \frac{t}{l} = \frac{7}{35} = 0.2 \text{ aspect ratio}$$

Therefore:

$$\frac{a}{e} = 0.06$$

and the surface crack geometry is thus converted to a through-thickness geometry such that $a = 6.5$ mm. With $K_I = \beta\sigma_c\sqrt{\pi a}$ and with $\beta = 1$ for a through-thickness crack:

$$K_I = 170\sqrt{\pi \cdot 6.5} = 24.3 \text{ MPa}\sqrt{m}$$

Therefore the ratio K_r is defined as follows:

$$K_r = \frac{24.3}{32} = 0.76$$

From the FAD (Fig. 10), the point falls within the curve, so failure will not occur. The safety factor against failure is:

$$\frac{OX}{OY} = 1.3$$

In revision 3 of R6, S_r is changed to L_r and the limits imposed on L_r are reduced. In revision 2, S_r never exceeded 1.

Example 2: Operating Stress Map in Failure Assessment. Consider the same example of a cast steel plate with a surface flaw using an operating stress map. The operating stress map is shown in Fig. 12 for the data based on the following equation for an elliptical crack (see "Complex Geometry Factors" in the next section):

$$K_{Ic} = \frac{1.12 \, \sigma_c \sqrt{\pi a}}{\Phi}$$

With an aspect ratio of $7/35 = 0.2$, $\Phi = 1.016$ (since $a/2c = 0.2$). Solving for a_c, then:

$$a_c = \frac{32^2 \cdot 1.016^2}{1.12^2 \cdot 170^2 \cdot \pi} = \frac{1057.03}{113,904.29} = 0.00928 \text{ m}$$

$a_c = 9.3$ mm, or

$$a \cdot \sigma_c^2 = \frac{1057.03}{3.941} = 268$$

a and σ_c are varied to produce the operating stress map.

Changes in K_{Ic}. Increasing the value of K_{Ic} to 101.25 MPa$\sqrt{m}$ would change K_r for the FAD as follows:

$$K_r = \frac{24.3}{101.25} = 0.24$$

Thus, with $S_r = 0.33$ and $K_r = 0.24$, the point on the FAD (Fig. 10) is such that failure will not occur and the safety factor against failure is:

$$\frac{OB}{OA} = 2.64$$

With an operating stress map (Fig. 12):

$$a_c = \frac{101.25^2 \cdot 1.016^2}{1.12^2 \cdot 170^2 \cdot \pi} = 92.9 \text{ mm, or}$$

$$a \cdot \sigma_c^2 = \frac{10,582.24}{3.941} = 2685.15$$

Summary of Linear Elastic Fracture Mechanics Concepts

A brief summary of linear elastic fracture mechanics (LEFM) concepts is provided below as background for explanation of the application of LEFM in damage tolerance analysis. Additional information is also contained in the articles "XX" and "XXX" in this Volume.

The stress-intensity factor, K, is a scaling factor that relates applied mechanical stresses to the intensification of stresses at the tip of a crack. For a general body under tensile (mode I) loading (Fig. 13), the stress-intensity factor near the crack tip (K_I) (at $\theta = 0$ in Fig. 13) is:

$$\sigma_{yy} = K_I\sqrt{2\pi r} \quad (\theta = 0) \tag{Eq 10}$$

where r is the distance from the crack tip. All the effects of loading, geometry, and crack size are incorporated into one parameter, K_I. For different geometries, crack sizes, and loading, K_I will have a different value, but apart from the value of K, the stress field will be the same. Equation 10 is the general solution for all crack problems. The parameter K governs the field, and K is the only significant parameter for all crack problems. Once K is known, the near-crack-tip stress field is known in its entirety.

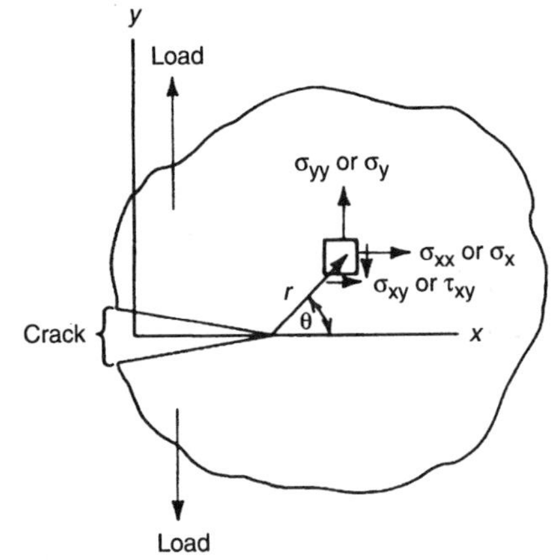

Fig. 13 General coordinate system for expression of crack-tip stresses where $\sigma_{ij} = (K_r/\sqrt{\pi 2r})f_{ij}(\theta)$ $+ C_1 r^0 + C_2 r^{1/2} +, \ldots$

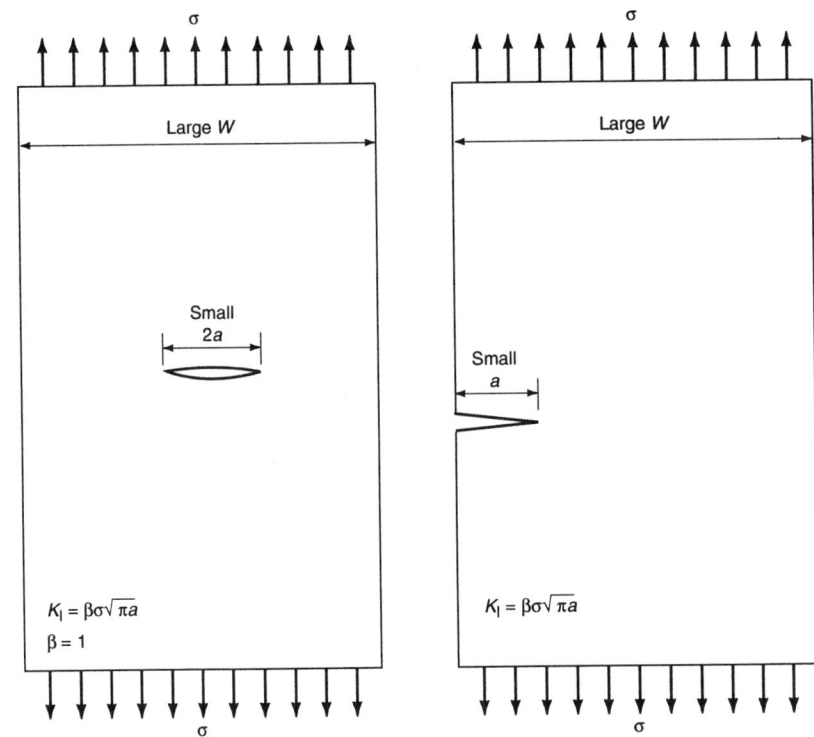

Fig. 14 Crack-tip stress intensity (K_I) in terms of applied stress (σ) for (a) a center-cracked plate and (b) an edge-cracked plate under uniform tension

As a general solution, Eq 10 can be expressed in terms of a geometric factor and crack size according to Eq 1, $K = \beta \sigma \sqrt{\pi a}$. All effects of geometry are reflected in one geometric parameter β. This geometric parameter has been calculated for many generic geometries, and the results have been compiled in various handbooks. (See also the appendix "Stress Intensity Factors" in this Volume.)

The stress intensity K_I represents the mechanical side of the equation; the toughness, K_c, is the material side. Fracture occurs when $K_I = K_c$. This is analogous to the statement that yielding will occur when the stress, σ, equals the yield stress, σ_y where σ is the mechanical side of the equation. Thus, if the material exhibits a fracture for a certain value of K, it will always exhibit fracture at that value of K (similitude for equal K).

These arguments remain valid even if some plastic deformation occurs. However, one condition must be satisfied: the plastic zone must be so small that its size is determined fully by K and by K only. This will be the case if the plastic zone does not extend beyond a value of r, at which the first term in the expansion series (Fig. 13) is still much larger than all other terms. Otherwise, the constants C_1, C_2, and so on (see caption equation in Fig. 13) will become significant. In general, this will not occur until the value of K_I/σ_y exceeds about 2 (this also depends on geometry). If it does occur, LEFM is no longer valid, and elastic plastic fracture mechanics (EPFM) must be used, although the use of LEFM can be stretched further with simple approximations (see the article "Residual Strength of Metal Structures" in this Volume). Some typical factors affecting K in the application of LEFM are briefly summarized below.

Geometry Factor β for Crack-Tip Stress Intensities (K_I). As long as stresses are elastic, then the stress in the region near the crack tip (σ_{yy} in Eq 10) is proportional to the applied stress (σ). Consider the case of a plate (Fig. 14a) with a large width (W) and a through-thickness center crack. It is expected that the crack-tip stress depends on the crack size. Because σ_{yy} depends on $1/\sqrt{r}$

according to Eq 10, it is inevitable that it depends on $\sqrt{a}$; otherwise, the dimensions would be wrong. Hence, a general relation for σ_{yy} is:

$$\sigma_{yy} = \frac{\beta \sigma \sqrt{\pi a}}{\sqrt{2\pi r}} \qquad \text{(Eq 11)}$$

where β is a dimensionless constant. The crack-tip stresses will be higher when W is smaller. Thus, β must depend on W. It is known that β must be dimensionless, yet β cannot be dimensionless and depend on W at the same time, unless β depends on W/a or a/W, that is, $\beta = f(a/W)$. Comparison of Eq 10 and 11, then, shows that:

$$K_I = \beta \left(\frac{a}{W}\right) \sigma \sqrt{\pi a} \qquad \text{(Eq 12)}$$

If the crack-tip stress is affected by other geometric parameters—for example, if a crack emanates from a hole, the crack-tip stress will depend on the size of the hole—the only effect on the stress-intensity factor (and the crack-tip stresses) will be in β. Consequently, β will be a function of all geometric factors affecting the crack-tip stress: $\beta = \beta(a/W, a/L, a/D)$. Crack-tip stresses are always given by Eq 10; the value of K in Eq 10 is always given by Eq 12.

For a through-thickness crack in a plate (Fig. 14a), $\beta = 1$ and the length is $2a$ when the crack has two tips. For a small edge crack, the length is a and $\beta = 1.12$ (Ref 12).

Complex geometry factors are summarized below and in Fig. 15. These approximations agree well with more complex expressions.

For the crack at a hole shown in Fig. 15(g), no rigorous solution currently exists and controversy exists over the variation of K_{Ic} along the crack front (Fig. 16), such that consensus dictates the K_{Ic} value chosen (Ref 13).

β values may be "compounded" (Ref 14) to give a total geometry factor. For example, a surface crack close to a weldment may have β compounded or a stress concentration (k_t) value built into it. (Compounding β values means multiplying them together. Building in a k_t value means adding to an extra factor to increase the β value.)

The k_t for fatigue notch is given by (Ref 15):

$$k_t = 1 + \frac{K - 1}{1 + (a/r)} \qquad \text{(Eq 13)}$$

where K = elastic stress concentration factor; r = root radius; and a = material parameter, which is a function of ultimate tensile strength. Other k_t values for different geometries are available in the literature (Ref 15, 16).

K_{Ic} values may be added or subtracted, depending on geometry. For example, K_{Ic} for a part of a geometry may be known, and K_{Ic} for the remaining part of the desired geometry may be known. The total value of K_{Ic} needed is:

$$K_{Ic} \text{ (total)} = K_{Ic} \text{ (part 1)} + K_{Ic} \text{ (part 2)}$$

The easy way to remember this is to add the K_{Ic} and multiply the β factors.

Embedded Elliptical Crack. For an embedded elliptical crack in the plane shown in Fig. 15(a), where $2a$ is the minor diameter and $2c$ is the

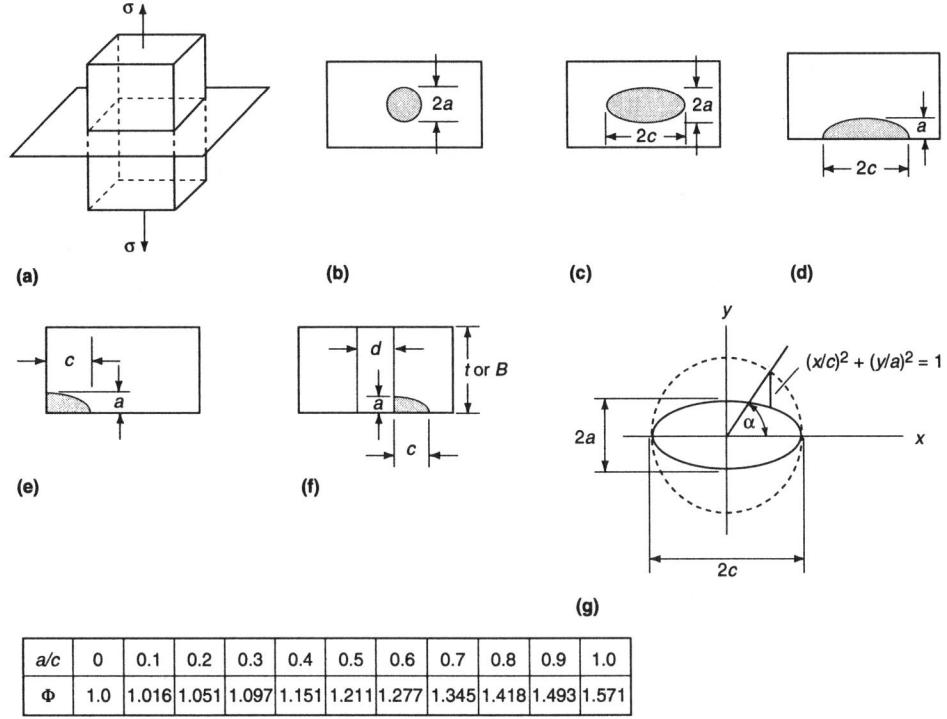

Fig. 15 Elliptical and circular cracks. (a) Tensile loading and crack plane. (b) Embedded circular crack. (c) Embedded elliptical crack. (d) Surface half-elliptical crack. (e) Quarter-elliptical corner crack. (f) Quarter-elliptical corner crack emanating from a hole. (g) Elliptical crack parameters. (h) Values of Φ

a/c	0	0.1	0.2	0.3	0.4	0.5	0.6	0.7	0.8	0.9	1.0
Φ	1.0	1.016	1.051	1.097	1.151	1.211	1.277	1.345	1.418	1.493	1.571

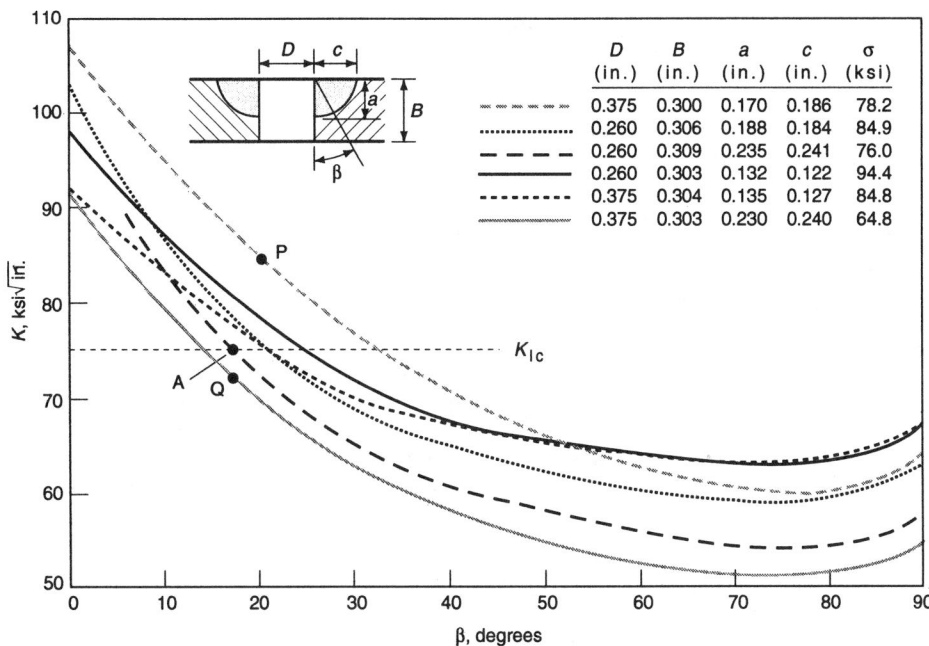

Fig. 16 Test data for corner cracks at holes

major diameter (Fig. 15c), the critical stress intensity factor (K_{Ic}) is given by (Ref 17):

$$K_{Ic} = \frac{\sigma_c\sqrt{\pi a}}{\Phi}\left[\sin^2\alpha + \left(\frac{a}{c}\right)^2\cos^2\alpha\right]^{1/4} \quad \text{(Eq 14)}$$

where α is the angle shown in Fig. 15(g) and Φ is the double elliptical integral, which depends on the crack aspect ratio, a/c. Values of Φ are shown in Fig. 15(h), where Φ varies from 1.0 to 1.51 for a/c ranging from zero (slender ellipse) to unity (circle). The stress intensity varies along the elliptical crack

tip: the maximum value at the minor axis and the minimum value at the major axis. In fatigue, this crack would grow into a circle with a uniform stress intensity at all points.

Circular Embedded Crack. For a circular embedded crack (Fig. 15b), K_{Ic} is:

$$K_{Ic} = \sigma_c\sqrt{\pi a}\left(\frac{2}{\pi}\right) \quad \text{(Eq 15)}$$

Surface elliptical cracks tend to grow to other elliptical shapes due to the free surface effect. A general expression for the semielliptical surface crack (Fig. 15d) is:

$$K_{Ic} = \frac{\sigma_c\sqrt{\pi a}}{\Phi}M_f\cdot M_b\,[\sin^2\alpha + \left(\frac{a}{c}\right)^2\cos^2\alpha]^{1/4} \quad \text{(Eq 16)}$$

where M_f is a front face correction factor and M_b is a back face correction factor: M_f and M_b are functions of α.

For a semielliptical crack in a thick plate, K_{Ic} at the deepest point is approximated by:

$$K_{Ic} = \frac{1.12\sigma_c\sqrt{\pi a}}{\Phi} \quad \text{(Eq 17)}$$

where an M_f factor of 1.12 is analogous to the free edge correction of the single edge crack.

Quarter-Circular Corner Crack. For the quarter-circular corner crack ($a/c = 1$, Fig. 15e) with two free edges, K_{Ic} is approximated by:

$$K_{Ic} = \frac{(1.12)^2\sigma_c\sqrt{\pi a}}{\Phi} \quad \text{(Eq 18)}$$

Crack-Tip Plasticity. Even when the analysis is based in LEFM, there is always a region of plasticity in the form of a "roll" at the leading edge of the crack tip (Fig. 17). The radius of this roll of plastic deformation or plastic zone depends on the condition of constraint at the crack front. The nature of this "roll" can be understood as a consequence from the theory of elasticity, where the crack-tip stress according to Eq 10 would be infinite for $r = 0$ regardless of the value of K. This is, of course, a nonphysical result. In reality, a material will exhibit plastic deformation that limits the stress (Fig. 17 and 18).

The radius of this roll of plastic deformation or plastic zone depends on the condition of constraint at the crack front. The most common situation is that of plane strain, and the plastic zone "size" or radius of the plastic roll is given by (Ref 18):

$$r_p = \frac{1}{6\pi}\left(\frac{K_{Ic}}{\sigma_y}\right)^2 \quad \text{(Eq 19)}$$

where $1/6\pi$ is determined experimentally. In plane stress:

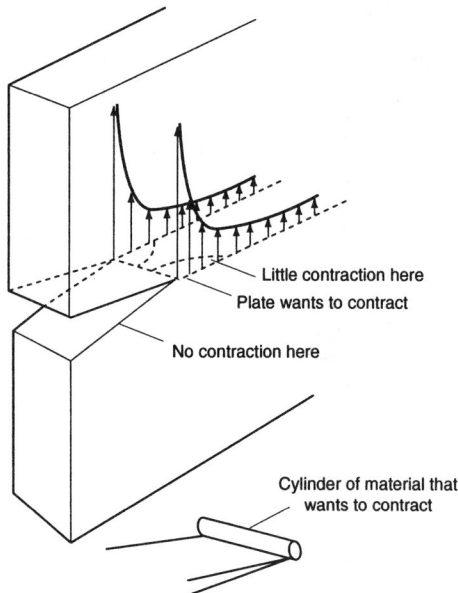

Fig. 17 Contraction and constraint from crack-tip plasticity. High stress and strain at the crack tip in the x- and y-directions cause a contraction in the z-direction. However, because the crack itself is stress free, no contraction occurs behind the crack front. Further away from the crack, the stresses are low; therefore, the contraction is small. Thus, there is a thin cylindrical-like zone of material at the crack tip wanting to contract a great deal, while the surrounding material does not need to contract. If this zone is long and thin, its contraction will be prevented because it is attached to the surrounding material. The surrounding material will constrain the contraction by exerting a tension stress upon the cylindrical zone in the z-direction so as to keep $\varepsilon = 0$.

$$r_p = \frac{1}{2\pi}\left(\frac{K_c}{\sigma_y}\right)^2 \qquad \text{(Eq 20)}$$

where $1/2\pi$ is determined experimentally.

Plane Strain and Plane Stress Constraint Conditions. The amount of constraint depends on the length-to-diameter ratio of the cylindrical zone of material at the crack tip that needs to undergo large contractions. If the zone is long, there is plane strain; if the zone is short, there is plane stress. For the through-thickness crack shown in Fig. 17, the length of the zone equals the thickness. Therefore, for this situation, the thickness determines whether there is plane strain or plane stress: large thicknesses will give plane strain, and small thicknesses will give plane stress. Therefore, the toughness will depend on thickness. Once there is complete constraint, the situation cannot worsen, so the toughness does not decrease further once it reaches K_{Ic}.

Because constraint is determined by the length-to-diameter ratio of the cylindrical zone defined above, section thickness determines constraint conditions for plane strain and plane stress toughness (Eq 2 and 3) of a through-thickness crack. However, for part-through cracks (surface flaws and corner cracks), the length of the zone does not depend on the thickness. In the case of part-through cracks the state of stress is always plane strain (unless the flaw is almost completely

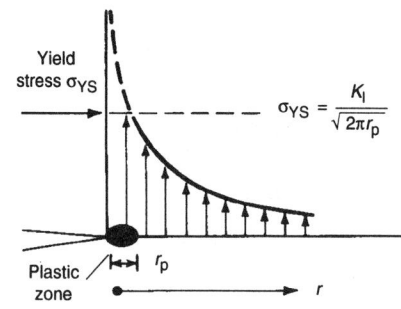

Fig. 18 Crack-tip stress distribution. (a) Elastic. (b) Elastic-plastic

through). Hence, one should use K_{Ic}, not K_c, regardless of thickness.

Evaluating Manufacturing or Fabrication Discontinuities. Numerous questions arise in equating fabrication discontinuities to solutions for elliptical cracks. For example, the ASME Section XI Boiler and Pressure Vessel Code formulation is an innately conservative one: that is, to define the effective fracture mechanics discontinuity as an ellipse that surrounds or fully encloses the fabrication discontinuities (Ref 19). Situations arise when this approach becomes impossible. In some cases, a center-cracked plate geometry may be assumed in the failure control of large-diameter, welded vessels (see the next section of this article).

Fracture Control Applications

Operating stress maps provide useful information for fracture control in a variety of applications. Real industrial applications are described in this section for both through-thickness cracks and surface cracks. Examples include determination of a safe level of pressure for a large-diameter vessel, criteria for "leak-before-break" operation, and the use of operating stress maps to evaluate surface cracks in weldments and castings.

Example 3: Safe Operating Pressure of a Large Pressure Vessel with a Surface Crack. As mentioned in the previous section of this article, it may not always be possible or necessary to analyze defects or discontinuities in terms of solutions for elliptical cracks. In some cases, frac-

ture control can be based on predicted behavior for a surface crack that grows into a through-thickness crack. When a crack of surface length $2a$ grows through a vessel wall, two outcomes are possible:

- The crack grows no farther and is stable.
- The crack grows catastrophically and the vessel breaks.

In some cases, fracture control may be verification of the first possibility.

For example, consider a large-diameter vessel with a torispherical head of 304 stainless steel. The outside surface of the vessel has evidence of some intergranular corrosion, and the plant manager wants to know if the vessel can be placed in operation at decreased pressure. The operating pressure, P, for a torispherical head (from ASME Section VIII, Div. 1) is:

$$P = \frac{SEt}{0.885L + 0.1t}$$

or:

$$SE = \frac{P(0.885L + 0.1t)I_v}{t}$$

where P = pressure in psig, S = stress in psig, E = joint efficiency = 1, L = inside crown radius in inches, t = minimum thickness in inches, and I_v = estimated stress concentration factor.

For a 3500 mm (138 in.) radius of the sphere with a minimum thickness of 9.5 mm (0.375 in.), the stress concentration is $I_v = 2$ for a 760 mm (30 in.) opening and the stress would be:

$$S = \frac{P(0.885 \times 138 + 0.1 \times 0.375)2}{0.375}$$

such that $S = 10,266$ psi (70 MPa) if the operating pressure is 15 psig (1 MPa).

The vessel dome is 304 stainless steel with a tensile yield strength of 240 MPa (35 ksi) and an operating stress map for through-thickness cracks as shown in Fig. 19. Inspection of the operating stress map reveals that for a wall stress of $S = 10$ ksi (70 MPa) for $P = 15$ psig (1 MPa), the critical crack length is 140 in. (3555 mm). With a head radius of 138 in. (3500 mm), operation below 15 psig (1 MPa) could be considered safe. If the surface flaw grew to a through-thickness crack, it would not be expected to grow catastrophically with an operating pressure below 15 psig.

Example 4: Leak-Before-Break Analysis. Operating stress maps of through-thickness cracks can also provide information on leak-before-break conditions. As in the preceding example, when a surface crack grows through a vessel wall, the crack may grow catastrophically, or the crack may be stable (in which case pressurized vessels would leak). Thus, an analysis of through-crack conditions can identify operating stresses for leak-before-break, as illustrated in the two following examples.

Case 1: Leak-Before-Break for a Longitudinal Seam Weld. The situation is a 304 stainless steel

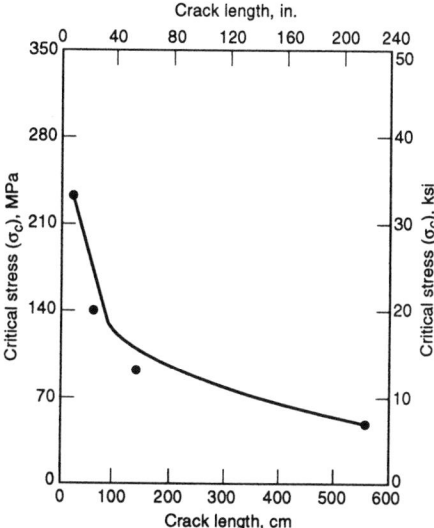

Fig. 19 Map of σ_c for cold-worked and annealed 304 stainless steel with a through-thickness crack ($\beta = 1$)

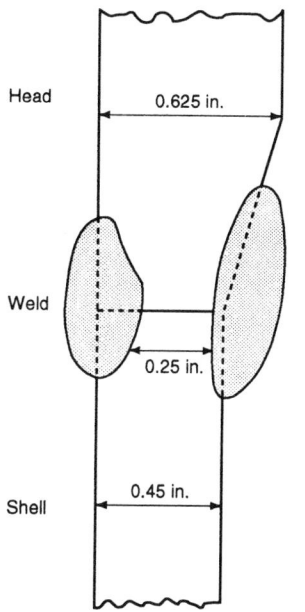

Fig. 20 Weld discontinuity in leak-before-break analysis (see text, Case 2 of Example 4)

vessel cylindrical section, which has a longitudinal seam weld. The vessel contains an innocuous liquid at room temperature. The vessel outer diameter ($2 R_o$) is 96 in. (2440 mm) with a 0.25 in. (6.35 mm) wall thickness. With an operating pressure (P) of 50 psig (345 kPa) the wall stress (S) is:

$$SE = \frac{PR_o - 0.4Pt}{t} \qquad E = 1$$

$$= \frac{50 \times 48 - 0.4 \times 50 \times 0.25}{0.25}$$

$$= 9580 \text{ psi}$$

From the operating stress map (Fig. 19), the crack length at 9.580 ksi is about 140 in., giving rise to a wall thickness of 70 in. The actual value of wall thickness is 0.25 in. The situation for unstable, catastrophic crack growth is two orders of magnitude greater than the true vessel thickness. This vessel will always leak before break under the conditions specified.

Case 2: Leak-Before-Break Conditions of a Weld Discontinuity. Consider a vessel in which there is a continuous discontinuity in the head in the head-to-shell weld (Fig. 20). The thickness in the weld region is 0.875 in. (22 mm). The discontinuity is 0.25 in. (6.35 mm). The minimum thickness is $0.625 - 0.25 = 0.4$ in., from the diagram. In this case, the material is A285C with a K_{Ic} of 77 ksi√in., and a yield stress of 30 ksi (205 MPa). The owner would like to operate the vessel safely with $P = 50$ psig, outer radius $R_o = 48$ in., $E = 1.7$, and $t = 0.4$ in. (minimum from Fig. 20). Thus:

$$t = \frac{PR_o}{2SE + 0.4P}$$

$$0.4 = \frac{50 \times 48}{2SE + 0.4 \times 50}$$

$$S = \frac{2392}{0.56}$$

$$= 4271.4 \text{ psi}$$

From the operating stress map for the ASTM A285C steel (Fig. 21), the net section yield line with $W = 0.875$ in. (plate width for through-thickness crack of 0.25 in.) indicates that elastic fracture mechanics does not apply in this case because, at the plate width and the operating stress of 4.271 ksi, failure will always be by net section yielding. Thus, with a crack length of $2 \times 0.25 = 0.5$ in., at the operating stress of 4.271 ksi, the vessel is safe to operate.

Consider now the case where one side of the weld is ground off. The wall thickness is now reduced to $0.45 - 0.25 = 0.2$ in. (5 mm). With:

$$t = \frac{PR_o}{2SE + 0.4P}$$

$$S = \frac{2396}{0.28} = 85,571 \text{ psi}$$

For the edge crack of 0.2 in., the operating stress of 8.557 ksi would be safe also.

This example indicates the need to always consider the net section yielding, where elastic fracture mechanics does not apply.

Example 5: Fracture Control with a Surface Crack. A casting made in 1.5Ni-Cr-Mo steel is to be operated at half of its yield stress (740 MPa) and one application of load. It has a surface crack, the major axis of which is ten times the length of the minor axis. The fracture toughness (K_{Ic}) of the steel is measured at 86 MPa√m. There are two ways to analyze the surface crack for fracture control: by calculations and by operating stress mapping. Fatigue problems lend themselves to solutions by numerical integration.

In operating stress mapping, the operating stress is normalized by the yield stress to produce maps for various aspect ratios. For example, Fig. 22 is an operating stress map for cast 1.5%Ni-Cr-Mo steels from the data in Table 1. Calculations based on complex formulas are examined and compared to the result predicted by the operating stress map for the Ni-Cr-Mo cast steel.

Method 1: Formula Calculations. The critical defect size may be calculated as:

$$a_{cr} = K_{Ic}^2 \left[\frac{\Phi - 0.212 \left(\dfrac{\sigma_c}{\sigma_y} \right)^2}{1.21 \, \pi \sigma_c^2} \right] \qquad \text{(Eq 21)}$$

where K_{Ic} is plane strain fracture toughness, σ_c is gross working stress normal to the major axis of the crack, a_{cr} is critical depth of a surface flaw (i.e., half the width of an embedded crack), σ_y is 0.2% proof stress, and Φ is double elliptical integral. As a simplification of Eq 21, let

$$Q = \left[\Phi^2 - 0.212 \left(\frac{\sigma_c}{\sigma_y} \right)^2 \right]$$

so then $\left[\dfrac{a}{Q} \right]_{cr} = \dfrac{K_{Ic}^2}{1.21 \, \pi \, \sigma_c^2}$ (Eq 22)

For embedded cracks, the coefficient on the denominator is taken as unity.

To define the shape of the crack, $a/2c$ can be considered to represent the crack size aspect ratio, where a is the minor and $2c$ is the major axis of an ellipse (when $a/c = 1$, the ellipse becomes a circle). The relationship between Φ and $a/2c$ is given in Fig. 23 for easy reference.

In this example, $a/2c = 0.1$ and $\sigma_c = 740/2$. Using these values in Eq 22:

$$\left(\frac{a}{Q} \right)_{cr} = \frac{K_{Ic}^2}{\sigma_c^2} \times \frac{1}{1.21 \, \pi}$$

$$= \frac{86^2}{370^2} \times \frac{1}{3.80}$$

$$= 0.0540 \times 2.263 = 0.0142$$

that is, $a_{cr} = Q \times 14.2$ mm. (Eq 23)

Now, for $a/2c = 0.1$ (from Fig. 23), $\Phi = 1.05$ and

$$Q = \Phi^2 - 0.212 \left(\frac{\sigma_c}{\sigma_y} \right)^2$$

$$= 1.05^2 - 0.212(0.5)^2$$

$$= 0.971$$

Inserting this value for Q in Eq 23, then $a_{cr} = 0.971 \times 14.2$ mm $= 13.78$ mm.

Method 2: Calculation with the Residual Strength Equation. Consider the same calculation using the residual strength equation (Eq 1) as follows:

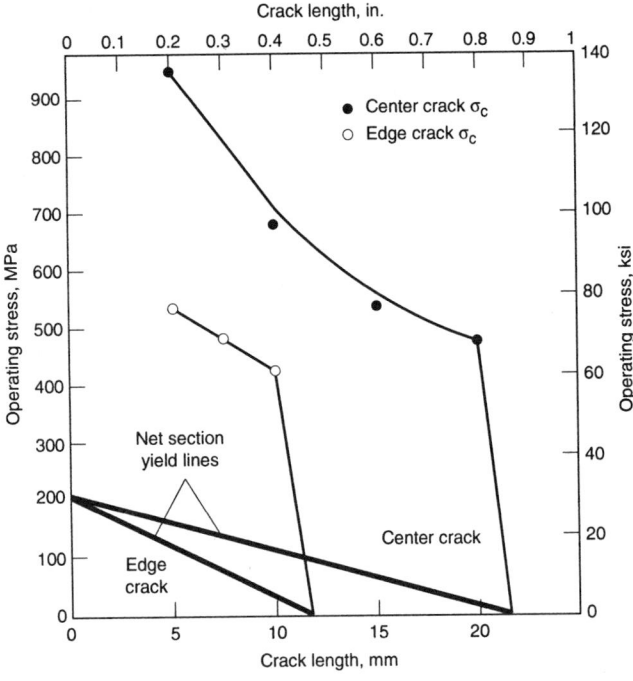

Fig. 21 Operating stress map with net section yield lines and critical stresses for ASTM A285C weldment

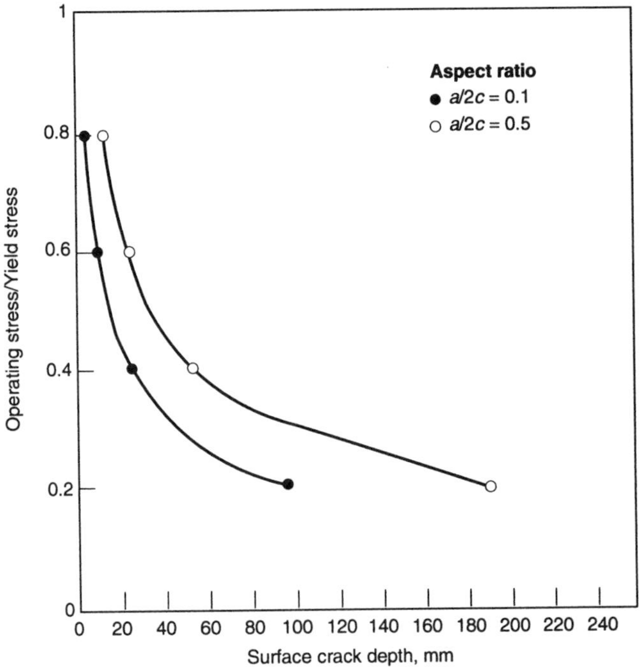

Fig. 22 Ratio of critical stress and yield strength for as-cast 1.5%Ni-Cr-Mo steel. Material condition: K_{Ic} = 86 MPa$\sqrt{m}$ (78 ksi$\sqrt{in.}$); σ_y = 740 MPa (107 ksi)

$$K_{Ic} = \frac{1.12\sigma_c\sqrt{\pi a}}{\Phi}$$

where K_{Ic} = 86 MPa$\sqrt{m}$, σ_c = 370 MPa, and Φ = 1.05 (from Fig. 23, $a/2c = 0.1$). Therefore:

$$a = \frac{86 \times 1.05}{1.12 \times 370 \times 3.142}$$

$$= 15 \text{ mm}$$

Compared to the value of 13.78 mm from the preceding calculation, there is a difference of 1.22 mm or 8.1%.

In Fig. 22, the operating stress map yields a value for a of about 14 mm for σ_c of 370 MPa. Also, if section thickness, B, is considered and the back surface correction M_b is used, then for crack depths of up to a/B of 0.25 with $a/2c$ still at 0.1 (Fig. 24), the preceding equation holds true.

Example 6: Simplified Calculations for Fracture Control with Surface Cracks. As shown in Example 5, the residual strength equation (Eq 1) is a basis for fracture control analysis. Other calculations are described below.

Estimating Effect of Transformation Stress on Fracture. During the water quenching of Christmas tree steel valve cast component, 30 mm in section, the transformation stress generated was 120 MPa. The measured K_{Ic} was 25 MPa$\sqrt{m}$, and the 0.2% proof stress was 600 MPa.

The maximum size of surface defect specified in production was 0.50 mm. It is necessary to determine the tolerable crack size (with the aspect ratio of the crack, $a/2c = 1/10$) and how transformation stresses affect crack growth.

Table 1 Critical crack depths for two cast steels

$a/2c$	for σ_c/σ_Y =			
	0.2	0.4	0.6	0.8
0.5C-1Cr steel				
K_{Ic} = 46 MPa $\sqrt{m}$; σ_y = 480 MPa; a_{crit}, mm				
0.1	67	16.3	6.9	3.7
0.2	79	19.4	8.3	4.5
0.25	88	21.5	9.3	5.0
0.3	101	25.0	10.5	5.8
0.4	121	30.0	13.0	7.0
0.5	129	37.0	16.3	9.0
1.5Ni-Cr-Mo steel				
K_{Ic} = 86 MPa$\sqrt{m}$; σ_y = 740 MPa; a_{crit}, mm				
0.1	98	23.9	10.2	5.4
0.2	116	28.6	12.3	6.5
0.25	129	31.8	13.7	7.3
0.3	149	36.6	15.4	8.5

If the transformation stress is 120 MPa, then from the residual strength equation:

$$K_{Ic} = \frac{1.12\sigma_c\sqrt{\pi a}}{\Phi}$$

and with $a/2c = 0.1$, $\Phi = 1.05$, then:

$$a = \frac{1.05^2 \cdot 25^2}{1.12^2 \cdot 120 \cdot \pi}$$

$$= \frac{689}{56,548.6} = 12.2 \text{ mm}$$

The size specified in production is therefore safe, because cracks would have to grow to a depth of

12.2 mm to reach critical levels for catastrophic fracture. However, if the transformation stress approaches the yield stress, then:

$$a = \frac{1.05^2 \cdot 25^2}{1.12^2 \cdot 600^2 \cdot \pi} = 0.48 \text{ mm}$$

This critical depth is too close to the specified size (0.5 mm) to be safe.

Fracture of a Pump Casing. The fracture of a pump casing during pressure testing gave rise to the fact that an internal crack had extended 5 mm by 2 mm. The fracture stress was 1000 MPa.

The steel had been subjected to heat treatment that led to a 0.2% proof stress of 1200 MPa and

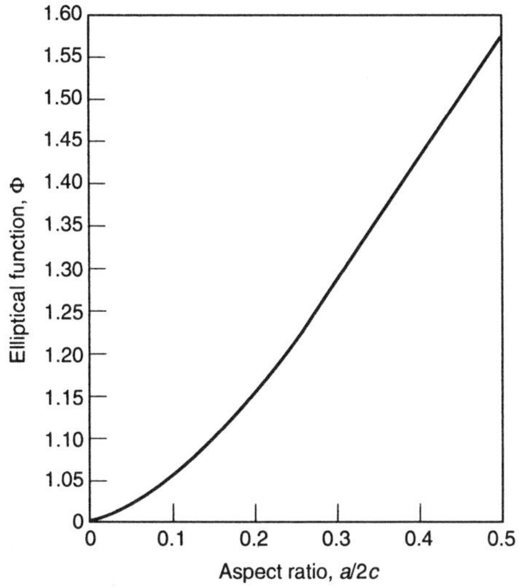

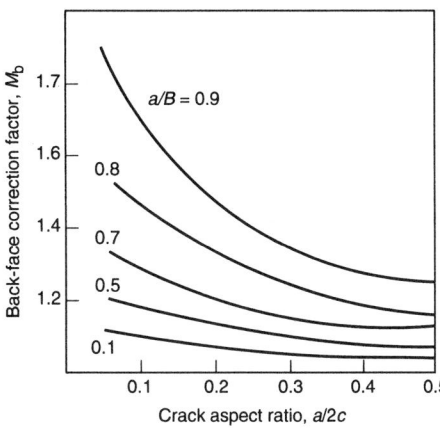

Fig. 23 Elliptical function vs. aspect ratio (a/2c)

Fig. 24 Effect of back-face correction on aspect ratio

$K_{Ic} = 50$ MPa$\sqrt{m}$. Determine the operating stress necessary to cause fracture using the relationship for an embedded elliptical crack:

$$K_{Ic} = \frac{1.12\sigma_c\sqrt{\pi a}}{\Phi}$$

$$\sigma_c^2 = \frac{50^2 \cdot 1.05^2}{1.12^2 \cdot 0.001 \cdot \pi} = 838 \text{ MPa}$$

This critical stress estimate of 838 MPa is within 16% of the observed 1000 MPa fracture stress.

REFERENCES

1. R.P. Harrison, K. Loosemore, I. Milne, and A.R. Dowling, "Assessment of the Integrity of Structures Containing Defects," Central Electricity Generating Board, R/H/R6-Rev. 2
2. J. Spiekhout, "Fitness-for-Purpose Assessment of Weld Flaws—Application of Various Fracture Mechanics Codes," *Welding Journal*, Sept 1988, p 55
3. P.F. Timmins, *Fracture and Failure Control for Inspectors and Engineers*, ASM International, 1994
4. C.E. Feddersen, "Evaluation and Prediction of the Residual Strength of Center Cracked Tension Panels," ASTM STP 486, 1971, p 50-78
5. C.E. Feddersen, "Evaluation and Prediction of Residual Strength of Center Cracked Tension Panels," ASTM STP 486, 1971, p 50-78
6. R. Roberts and C. Newton, "Interpretive Report on Small-Scale Test Correlations with K_{Ic} Data," WRC Bulletin No. 265
7. S.T. Rolfe and J.M. Barsom, *Fracture and Fatigue Control in Structures*, Prentice Hall, 1977
8. R.G. Forman, V.E. Kearney, and R.M. Engle, "Numerical Analysis of Crack Propagation in Cyclic-Loaded Structure," *ASME Transactions, Journal of Basic Engineering*, Vol 89 (No. 3), 1967, p 459
9. R.G. Forman et al., "Development of the NASA/FLAGRO Computer Program," ASTM STP 945, 1988, p 781
10. T. Yokobori et al., "Evaluation of KISCC Testing Procedure by Round Robin Tests on Steels," ASTM STP 945, 1988, p 843
11. PD6493.1980, "Guidance on Some Materials for the Derivation of Acceptance Levels for Defects in Fusion Welded Joints"
12. P.C. Paris and G.C. Sih, ASTM STP 381, 1965, p 30
13. L.R. Hall and W.L. Engstrom, "Fracture and Fatigue-Crack-Growth Behavior of Surface Flaws and Flaws Originating at Fastener Holes," AFFDL-TR-7447, 1974
14. D. Broek, *Elementary Engineering Fracture Mechanics*, Noordhoff, 1978
15. R.E. Peterson, *Metal Fatigue*, Sines and Waisman, Ed., McGraw-Hill, 1959
16. R.E. Peterson, "Analytical Approach to Stress Concentration Effect in Fatigue of Aircraft Structures," WADS Symposium, Wright Air Development Center, Dayton, OH, Aug 1959
17. J.M. Svoboda, Steel Founders' Society of America Research Report 94A, Oct 1982
18. J.F. Knott, *Fundamentals of Fracture Mechanics*, Butterworths, 1973
19. "Rules for Inservice Inspection of Nuclear Reactor Coolant Systems," ASME Boiler and Pressure Vessel Code, Section XI, American Society of Mechanical Engineers, 1980

Failure Control in Process Operations

P.F. Timmins, Risk Based Inspection Inc.

FAILURE CONTROL METHODS for piping and process systems encompass the inspection, monitoring, life assessment, repair, maintenance, and life extension of various engineering components used in power plants, chemical processing plants, and refineries. Major costs are associated with each of these steps, and cost-effective implementation of failure control can be viewed, in general terms, from the standpoint of economic loss and the risk of asset loss. For the various types of equipment in process systems, the significance of failure can be ranked as a function of failure probability and the cost of failure (Fig. 1). The types of process equipment most important in controlling risk of asset loss, listed in decreasing order of importance, appear to be piping, reactors, tanks, and process towers (Fig. 1). This ranking suggests the priority areas for inspection, monitoring, life assessment, repair, and maintenance.

Another key factor is the definition of failure. While complete breakage or rupture may be the ultimate and self-evident criterion of failure, more conservative definitions are invariably employed to retire a component prior to such unforeseen and catastrophic failure. Failure of a component may generally be defined as the inability of

the component to perform its intended function reliably, economically, and safely. For example, various definitions of failure for high-temperature process equipment are listed in Table 1.

Component-retirement decisions are often based on economic justification rather than on technical need. A logical and technically based decision may, for instance, involve a sequence of steps such as remaining-life calculations based on operating history, inspections, material testing, assessment of remaining life, and final disposition of the component in terms of continued service, repair, or replacement. Unfortunately, there are major cost factors associated with each of these steps. The cost of the component itself is usually a small fraction of the cost of disassembling the unit as necessary and performing all of the above operations. If, after all of this, a wrong decision is made and the component fails in service, the economic penalties are severe. The owner of the plant has to weigh all of these economic factors carefully and not make decisions purely on a technical basis. A conservative but not uncommon approach has therefore been simply to replace critical components in a plant after 30 to 40 years, regardless of the technical merits of such action.

Nonetheless, maintenance can have a dramatic impact on profitability (Fig. 2). Corrective maintenance or unplanned maintenance is the most costly. In Fig. 2, this form of maintenance is reflected as unity on the unit cost basis shown. Preventive maintenance (planned maintenance on a fixed time scale) is about 60% of the cost of corrective maintenance. Predictive maintenance (maintenance on a sliding time scale) is about 40% of the cost of corrective maintenance. Clearly, predictive maintenance or proactive maintenance is desirable. If inspection, maintenance and operations departments are organized to run in the predictive or proactive mode, then their costs will be reduced and the profitability of the process operation will increase.

This article focuses on the subject of proactive or predictive maintenance, with particular emphasis on the control and prediction of corrosion damage for life extension and failure prevention. Corrosion is a prime source of failure in process systems, and corrosion-related problems are among the most expensive and hazardous in various process industries. Predictive or proactive

Table 1 Failure criteria and definitions of high-temperature component creep life

History-based criteria

30 to 40 years have elapsed
Statistics of prior failures indicate impending failure
Frequency of repair renders continued operation uneconomical
Calculations indicate life exhaustion

Performance-based criteria

Severe loss of efficiency indicating component degradation
Large crack manifested by leakage, severe vibration, or other malfunction
Catastrophic burst

Inspection-based criteria

Dimensional changes have occurred, leading to distortions and changes in clearances
Inspection shows microscopic damage
Inspection shows crack initiation
Inspection shows large crack approaching critical size

Criteria based on destructive evaluation

Metallographic or mechanical testing indicates life exhaustion

Source: R. Viswanathan and R.B. Dooley, *Creep Life Assessment Techniques for Fossil Plant Boiler Pressure Parts*, Proceedings of International Conference on Creep, JSME-IME-ASTM-ASME, Tokyo, Apr 1986, p 349-359

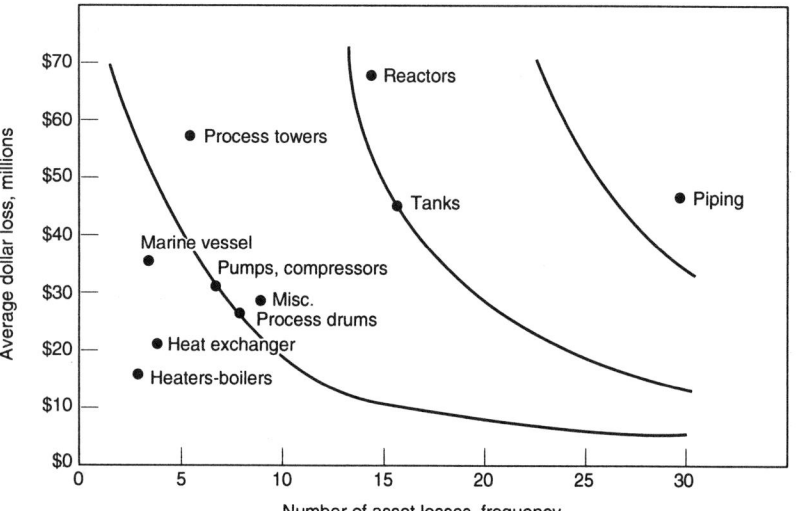

Fig. 1 Asset loss risk as a function of equipment type. Source: W.G. Garrison, Loss Prevention, *Hydrocarbon Processing*, Sept 1988

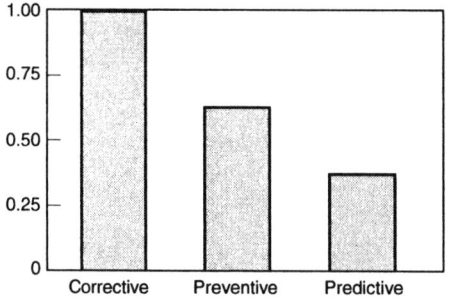

Fig. 2 Relative cost of maintenance approaches. Source: Power Plant Diagnostics Go On-Line, *Mechanical Engineering,* Dec 1989

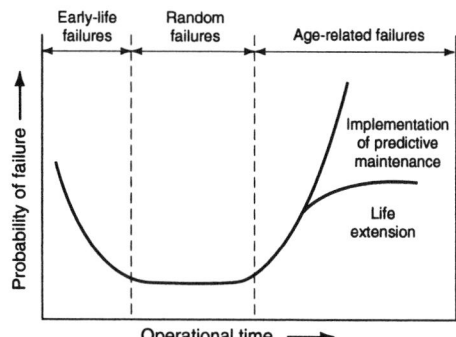

Fig. 3 Typical reliability curve illustrating the conceptual objective of life extension

maintenance programs are used for life extension. However, if a predictive maintenance program is not in place, then condition assessment is required, because it is fundamental to the continued safe operability and reliability of refinery or process equipment. Condition or life assessment is also discussed in this article.

Predictive Maintenance

Predictive maintenance differs from traditional preventive maintenance in that key equipment parameters are monitored as preliminary indications of failure. Vibration monitoring and signature analysis of rotating or reciprocating equipment is one industrial example of a predictive maintenance system. Another example is leak detection systems with acoustic emission sensors. The objectives of this maintenance approach are to monitor the performance and reliability of critical components or subcomponents and to reduce age-related failures in the traditional "bath tub" reliability curve (Fig. 3).

Nondestructive testing and inspection techniques are being improved and used more often to improve the reliability and safety of chemical and refinery plants. Early meaningful detection of deterioration can provide the basis for repairs or changes in operating conditions that may extend equipment life. However, some conditions can be difficult to assess by nondestructive inspection. For example, embrittlement from hydrogen, carburization, and strain aging can be difficult to determine. Nonetheless, useful nondestructive methods (Table 2) include not only conventional methods such as ultrasound, radiography, magnetic particles, and dye penetrants, but also eddy current testing, capacitive strain gage testing, replication metallography, and borescope examination of internal piping. More extensive discussion is contained in the article "Nondestructive Testing and Life Assessment" in this Volume.

Implementation of a predictive maintenance system requires the identification of critical components or weak links that could justify continuous monitoring or detailed time-based examination. For example, an asset risk diagram (Fig. 1) is one way of identifying equipment priorities for failure control. Each major piece of equipment is then critically examined in terms of weak links and their failure mechanisms. Corrosion is a prime source of failure in process systems, and corrosion-related problems are probably the most expensive and hazardous in the process industries today. Nonetheless, many failures involve more than one factor; that is, one process may initiate cracking and other processes may eventually lead to failure. For example, fatigue failures that cause tubes or pipes to leak may start at corrosion pits.

Failure mechanisms for a given component or process must be understood for effective implementation of predictive maintenance. Failure mechanisms typically are classified according to temperature. At high temperature, damage mechanisms include embrittlement, creep, thermal fatigue, hot corrosion, oxidation, and erosion. At lower temperatures, corrosion, erosion, pitting, corrosion fatigue, stress corrosion, and hydrogen embrittlement can play major roles.

Following the work of Stoneburg et al. (Ref 1), evaluation of defects and the type of analysis applied often depend on the failure mechanism or mode of operation that led to the defect. For instance, a manufacturing defect that was found some years later would require a remaining-life analysis based on fracture mechanics. A defect produced by misuse of equipment may require a different kind of analysis. It is very important to determine the root cause of the defect in order to apply the appropriate analysis, so an understanding of the applicable failure categories is necessary.

All plants operating with corrosive elements in the process, or in corrosive environments such as salt air or acids from nearby processes, have an obvious concern with process equipment degradation, originating either internally or externally. However, a noncorrosive process with high-velocity flow or unseen cavitation is also a major concern, because wall-thickness degradation is usually unexpected and visually undetectable from the exterior of the process equipment. Other service-related defects leading to failure include cracking, embrittlement, creep, and mechanical and thermal fatigue. Insulation that becomes wet may cause and hide corrosion. Many of these defects are visually undetectable from the exterior of process equipment until failure has occurred (i.e., through-wall propagation or excessive distortion).

Table 2 Inspection and measurement techniques for corrosion control

Technique	Description
Acoustic emission	Measures the location, initiation, and propagation of cracks and defects under stress in metals
Dye penetrant	Simple procedure for locating surface cracks. Requires shutdown for internals
Magnetic particle	Surface and subsurface crack and defect location, seams, and inclusions
Radiography	X-ray and gamma-ray (Co^{60}, Ir^{192}) wall-thickness measurement and crack and defect location
Thermography	Identifies lack of bond, hot spots, local thinning, and temperature changes due to poor/wet lagging
Ultrasonics	Indicates internal defects, porosity, lack of fusion, crack location, and wall thickness
Visual exam	Identifies localized corrosion, erosion pitting, deposit scaling and fouling problems, and staining and corrosion leakage due to cracking
Sentinel holes	Small telltale holes drilled from outside of carbon steel pipework to depth of corrosion allowance. Weepage indicates replacement is required
Weight loss coupons	Traditional method of limited sensitivity but used in all environments. Type of corrosion observed is an important indicator
Electrical resistance	Measures change in probe resistance. Widely used for carbon steel fabrications in gas and liquid phases. Automated readings
Corrosion potential	Identifies corrosion conditions (i.e., active, passive, pitting, stress-corrosion cracking regimes) in ionically conductive media
Linear polarization resistance	An electrochemical dc method used for measurement of uniform corrosion. Standard electrochemical technique. Typically requires a conductive electrolyte, but some new probes have a conductive separator between the metal probe elements
Zero resistance ammetry	Established method for assessing galvanic corrosion between dissimilar metals, but can be used with nominally "identical" electrodes in some applications
Hydrogen probes	Measures rate of diffusion of hydrogen through steels, either by means of a pressure gage or by electrochemical techniques
Thin layer activation	Measures change of radioactivity as a local irradiated area corrodes
Electrochemical impedance	An ac method used for general corrosion measurements similar to linear polarization resistance. More versatile and accurate than dc measurements
Electrochemical noise	A more recent technique used for assessing general corrosion and potential fluctuations associated with localized corrosion

Over the years, experience and advancements in nondestructive examination (NDE) techniques and equipment have led to an improvement in the ability to detect defects in process equipment. This improvement has led to the detection of original manufacturing defects in existing process equipment that may not have been detected at the time of manufacture. This does not necessarily indicate a lack of quality in older process equipment, but it may indicate that many manufacturing defects have no effect on the integrity or service life of the equipment and that a high level of confidence can be placed in the current NDE technology.

Once a piece of equipment is in service, it cannot be assumed that it will not require further evaluation. It is imperative to obtain and maintain all the design documents for process equipment, and to never accept shipment of new equipment or changes without those documents. It is far easier to verify design drawings and engineering changes than to destructively examine equipment in order to resolve an issue. A typical example of this is weld backing rings that are left in place on the inside of a vessel. These backing rings assist the manufacturer in making the closing weld on process equipment. If the backing ring is discovered during an in-service radiographic examination, the analyst must determine whether the material is compatible with the process fluid and whether the ring is integral or protected from coming into contact with the process fluid. All of this may be resolved when thorough documentation is available; otherwise, it may be impossible to determine without destructive examination. Other improper designs may make use of the incorrect material for the service or process that the equipment will support. Poor construction details can lead to in-service failure. For instance, rough edges or joints may collect corrosive material that leads to pitting and accelerated degradation.

During examinations, indications should be characterized and recorded as to their type and size. Each indication should then be evaluated. The first step in classifying indications is to compare them to the acceptance criteria of the applicable consensus code (e.g., ASME Section VIII). All indications that meet the code acceptance criteria (e.g., Ref 2 and 3) are considered adequate, and no further evaluation is required. Indications that exceed the code acceptance criteria are normally considered defects and require additional evaluation. Code acceptance criteria are established based on quality control levels that are arbitrary yet conservative. As such, indications that do not exceed the code acceptance criteria are accepted without further consideration. Indications that exceed the code acceptance criteria are rejectable per that code, but they are not necessarily unacceptable. Further analysis as to their fitness for continued service is required.

Corrosion Monitoring

Corrosion monitoring has become an important aspect of the design and operation of modern industrial plants because it alerts plant engineering and management personnel to damage caused by corrosion and the rate of the deterioration. A large variety of techniques are available for corrosion monitoring in plant corrosion tests. The most widely used and simplest method involves the exposure and evaluation of the corrosion in actual test coupons (specimens). The ASTM standard G 4 was designed to provide guidance for this type of testing.

In the selection of a corrosion monitoring method, a variety of factors should be considered. First, the purpose of the test should be understood by everyone concerned with the corrosion monitoring program. The cost and applicability of the methods under consideration should be known, and it is important to consider the reliability of the method selected. In many cases, it will be desirable to include more than one method in order to provide more confidence in the information generated.

Monitoring Locations. A principal part of any corrosion monitoring program is deciding where to locate the monitoring devices. Because corrosion will probably not occur uniformly throughout the plant, it is desirable to find sites at which the highest corrosion rates will be experienced.

The problems involved in developing corrosion monitoring programs for a plant are illustrated in the example of a distillation column. The most logical points for corrosion monitoring in a distillation column are the feed point, the overhead product receiver, and the reboiler or bottoms product line. These points are the locations at which the highest and lowest temperatures are encountered, as well as the points at which the most and least volatile products are concentrated. However, these points are usually not the locations of the most severe corrosion.

The species causing the corrosion often concentrate at an intermediate point in the column because of chemical changes within the column. Therefore, if there is a possibility of concentrating a corrosive species within the distillation column, several monitoring points would be required throughout the column (Fig. 4). A monitoring program can be restricted to the most corrosive location within the column once this area has been identified.

Another problem with distillation columns is that the liquid on trays tends to be frothy, which creates difficulties for electrochemical methods. One solution is to install bypass loops that remove liquid from the column, pass it over the corrosion monitoring probes, and reinject it at a lower point in the column. This practice avoids the problem of foam and froth and provides a more controlled flow rate over the corrosion monitoring equipment. Use of a bypass loop also allows removal of liquid samples at times of high corrosion rates.

Redundancy is also important in designing corrosion monitoring programs. The use of at least two different types of monitoring devices at any location is often desirable. For example, the use of an electrical resistance probe with a polarization resistance probe allows measurement of

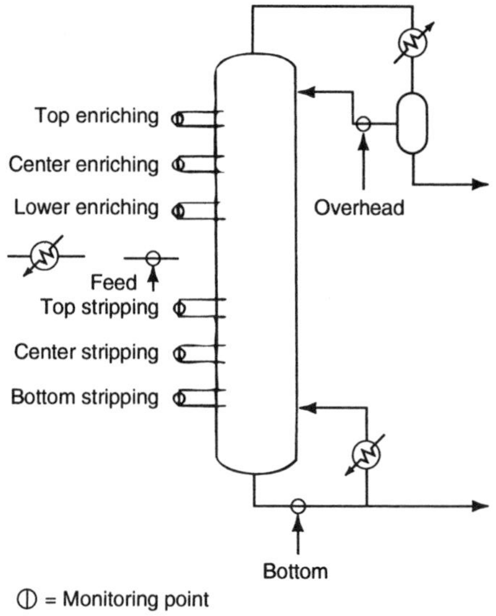

Fig. 4 Distillation column showing preferred locations of monitoring probes or other devices

both instantaneous corrosion rates and an average corrosion rate. The data thus obtained can be correlated, and this is very helpful in identifying spurious or inaccurate readings.

In another approach, the polarization resistance probe is weighed before and after the test in order to correlate mass loss with the average corrosion rate that the probe suffered. There is reason to expect that the electrochemical value is in error if the average corrosion rate and the mass loss of the probe do not agree. Also, to obtain time-averaged corrosion rate values, independent coupons can be installed together with polarization resistance probes.

A variety of corrosion monitoring approaches must be used when designing pilot or demonstration plants. Coupon tests can be very helpful in selecting optimum materials for processes based on pilot plant experience. Polarization resistance monitoring is very useful for determining whether certain processing conditions cause corrosive situations to develop. Because the corrosion mechanisms are often not well understood, and because the result of erroneous information can be serious overdesign or exposure to unexpected hazards, redundancy in the design of such monitoring systems is important in pilot plants.

The process stream can be sampled in locations other than distillation columns. A sample tap can be helpful in conjunction with polarization resistance devices. The alarm on polarization resistance monitoring equipment can be used to signal the need to remove samples. This is particularly desirable in the case of pilot plant operation, in which wide variations in processing conditions are encountered. It is also helpful in plants that produce different products in the same equipment.

High-velocity gas streams in pipes may cause problems with conventional monitoring systems. In this case, the presence of an aqueous phase is usually restricted to a thin layer on the surface of the pipe. A probe that protrudes into the pipe may miss the liquid layer that is present only close to the pipe wall. A flush-mounted surface probe can be used in such cases. This probe permits the measurement of polarization resistance in order to estimate the corrosion rate of the pipe wall.

Probe location is also critical in storage tanks containing nonaqueous liquids. The most corrosive location in these tanks may be at the liquid level if the liquid in the tank has a density exceeding that of water. In this case, the corrosion monitoring probe should be mounted on a floating platform so it can detect the presence of a corrosive aqueous phase. However, when the liquid stored in the tank is less dense than the water, the probe should probably be positioned at the bottom of the tank.

Interpretation and Reporting. In-service monitoring, more than any other type of corrosion testing, requires the utmost skill in interpretation and reporting. Important economic decisions are often based on the test results. A number of standards provide guidelines for certain procedures, but none is comprehensive. To plan an appropriate test program, the investigator must know or have good advice on both the chemistry and the mechanics of the processes involved; that is, the investigator must understand the entire corrosion system. There must be a strong emphasis on the strategy of the program and a searching analysis of the test results. It is important to prepare detailed records of what was done as part of the experiment, and it is also important to document any unplanned changes that occur in the process stream or the equipment during the investigation. Without a valid interpretation and effective (timely) reporting, the price of the work can be significantly greater than the cost of time and materials. However, the consequences of corrosion failures go beyond additional costs. Also involved are personal safety risks (and liability), hazard potentials, and product quality and pollution problems.

Corrosion Monitoring Methods

The primary purpose of corrosion monitoring is to estimate the condition of the equipment so that planned turnarounds can be scheduled and replacement of equipment can be anticipated. Frequent measuring also helps minimize the duration of corrosive upset conditions.

Corrosion Coupons. If properly located, the coupon is a reasonably accurate tool. In addition to producing corrosion data in measurable terms (mpy or mm/yr), the coupon is helpful in identifying the type of corrosion activity present at the point where the coupon is located in the system. Pitting on the coupon surface indicates activity from a strong acid or acid salt, such as hydrochloric acid or ammonium chloride. General thinning indicates the activity of hydrogen sulfide on the coupon.

Coupons are widely used to monitor inhibitor programs in, for example, water treatment or refinery overhead streams. With retractable coupon holders, the coupons can be extracted from the process without having to shut down in order to determine the corrosion rate. Coupons can be designed to detect such phenomena as crevice corrosion, pitting, and dealloying corrosion. For example, some pulp mill bleach plant washer drums are electrochemically protected to mitigate crevice corrosion. Specifically designed crevice corrosion test coupons are used to monitor the effectiveness of the electrochemical protection program. These coupons are periodically removed from the equipment and examined for evidence of crevice corrosion.

Coupon testing does have important limitations. First, coupon testing cannot be used to detect rapid changes in the corrosivity of a process. Second, localized corrosion cannot be guaranteed to initiate before the coupons are removed, even with extended test durations. Coupons only show corrosion that has already taken place, and a single coupon will not show whether corrosion was uniform or occurred all at once. Generally, a retractable coupon must be left in the system for 20 to 30 days before reliable information can be obtained. The coupon will not record upset conditions rapidly or measure the true corrosion activity in the system if it is improperly placed.

Third, the calculated or measured corrosion rate of the coupon may not translate directly to that of the equipment. Despite every effort to achieve equivalence, differences in mass and coupon area/solution volume ratio are usually sufficient to render direct comparison meaningless. For example, the metal surface temperature of the coupon is governed by the process stream that surrounds it. Where a cooling medium is present on the other side of the metal surface, the skin temperature of the tube is lower than the temperature of the surrounding hydrocarbon. The cool tube skin could cause surface water condensation at temperatures above the anticipated water dewpoint in the process fluid. Coupons would not indicate corrosion activity of this nature. Useful correlations can be established by monitoring the corrosion rate of the equipment with ultrasonic thickness monitoring and by comparing this corrosion rate with the calculated rate for equivalent coupons.

Lastly, certain forms of corrosion cannot be detected with coupons. The principal limitation is the simulation of erosion-corrosion and heat transfer effects. Careful placement of the coupons in the process equipment can slightly offset these weaknesses.

Erosion-corrosion is related to process turbulence, and process turbulence is often a function of equipment design. Because coupons tend to shield one another from the effects of process turbulence, field coupon testing is not reliable as a method of simulating erosion-corrosion.

For heat transfer effects, specially designed coupons are required that simulate effects such as those found in heating elements or condenser tubes. Coupons range in design from thermowell-shaped devices to sample tubes in a test heat exchanger. Thermowell-shaped devices are heated or cooled on the inside and project into the process stream. Heat transfer tests can also be conducted in the laboratory. In this environment, the coupon forms part of the wall of the test vessel and can therefore be heated or cooled from one side. Because of the cost involved, heat transfer coupon tests are usually carried out on only one (or perhaps two) alloys that have been selected from a larger group.

Preparing, Installing, and Interpreting Coupons. Corrosion coupons may be flat or cylindrical and may be installed in any accessible location. It must be remembered that coupons measure corrosion only where they are placed. Coupons show corrosion that has already taken place, and a single coupon will not show whether the corrosion was uniform or occurred all at once.

Different types of coupon holders or chucks are used, depending on the system, the pressure, the location, or other factors. Most coupons are run in a 25 or 50 mm (1 or 2 in.) threaded plug. Flat coupon holders hold two coupons, while cylindrical coupon chucks may contain eight or more. The multichuck coupons allow a coupon to be pulled at intervals to see if the corrosion rate is uniform or not.

High-pressure systems require a special coupon check and insertion device. The insertion tool fits into a special attachment on the pipe or vessel that has a high-pressure chamber with a valve on each end. The inner valve is closed, the retrieval tool inserted, and the inner valve opened. The tool is then run in and left. The procedure is reversed to remove the coupon.

The industry guide for preparing, installing, and interpreting coupons is NACE standard RP-07-75. The primary consideration is that all coupons be treated exactly alike. A method of preparation that does not alter the metallurgy of the coupon is required. Grinding and sanding of coupons should be controlled to avoid metallurgical changes and to provide a consistent and reproducible surface finish.

Coupons should be handled carefully and stored in noncorrosive envelopes until they are installed. Rust spots caused by improper handling, fingerprints, and so on may initiate a pit that is not representative of the system being evaluated. Prior to installation, the weight, serial number, date installed, name of system, location of coupon, and orientation of the coupon and holder should be recorded. The coupons are left in the system for a predetermined number of days and then removed.

When the coupons are removed, the serial number, date removed, observations of any erosion or mechanical damage, and appearance should be recorded. A photograph of the coupon may be valuable in some cases. The coupons should then be placed in a moisture-proof envelope impregnated with a vapor phase inhibitor and taken immediately to the laboratory for cleaning and weighing. The coupons can be blotted (not wiped) dry prior to being placed in the envelope.

The laboratory receives the coupon and inspects, cleans, and weighs it. A report is issued showing the thickness loss, any pitting observations, and any other observations of interest.

Electrical resistance probes are specially designed corrosion coupons. Their corrosion rate is calculated from measurement of electrical resistance rather than mass loss. These measurements are made by installing a wire or other device fabricated from the material in question in such a way that its electrical resistance can be conveniently measured. Corrosion reduces the cross section of the exposed element; therefore, its electrical resistance will increase with exposure time if corrosion is taking place. A temperature-compensating element should be incorporated in such a probe, because the resistance of the probe is also influenced by the temperature.

Electrical resistance probes measure the remaining average metal thickness. To obtain the corrosion rate, measurements are made over a period of time, and the results are plotted as a function of exposure time. The corrosion rate can be determined from the slope of the resulting plot. There are several advantages to this approach. Because probes are relatively small, they can be installed easily. For determination of the metal remaining, the probe can be wired directly to a control room location or to a portable resistance bridge at the probe location.

There are also some disadvantages to this approach. Vessel and piping walls must be penetrated in order to install the probes; such penetration results in the potential for leaks. It is expensive to direct-wire the probe to a control room location, and such work must be carried out with care to avoid spurious signals and errors. On the other hand, it is time-consuming and sometimes impossible to take measurements at the probe site with a portable bridge. The temperature compensation device reacts slowly, and it can be a source of error if the temperature varies when the measurement is taken. Corrosion rate measurements obtained in short periods of time can be inaccurate because the method measures only the remaining metal, not the rate of attack; this increases the signal-to-noise ratio in short exposures. This method provides no information on localized attack.

Ultrasonic thickness measurements can be used to monitor corrosion rates in situ. Ultrasonic thickness measurements involve placement of a transducer against the exterior of the vessel in question. The transducer produces an ultrasonic signal. This signal passes through the vessel wall, bounces off the interior surface, and returns to the transducer. The thickness is calculated by using the time that elapses between emission of the signal and its subsequent reception, along with the velocity of sound in the material. To obtain a corrosion rate, a series of measurements must be made over a time interval, and the metal loss per unit time must be determined.

Measurement errors can occur when vessel walls are at high or low temperatures. Serious problems may exist in equipment that has a metallurgically bonded internal lining, because it is not obvious from which surface the returning signal will originate. Despite these drawbacks, the ultrasonic thickness approach is widely practiced where it is necessary to evaluate vessel life and suitability for further service. It must be kept in mind, however, that depending on the type of transducer used, the ultrasonic thickness method can overestimate metal thicknesses when the remaining thickness is under approximately 1.3 mm (0.05 in.).

Polarization Resistance Measurement. Unlike the previously discussed methods, which provide information on remaining thickness, the technique of polarization resistance provides an estimate of the corrosion rate. The theory behind the technique is that the corrosion rate of a probe is inversely proportional to its polarization resistance, that is, the slope of the potential-current response curve near the corrosion potential. It is necessary in a plant situation to use a probe that enters the vessel in the area where the corrosion rate is desired. The electrodes of the probe are fabricated from the material in question. An electronic power supply polarizes the specimen about 10 mV from the corrosion potential. The resulting current is recorded as a measure of the corrosion rate. Polarization resistance yields an instantaneous estimate of corrosion rate.

There are several limitations to this approach. The corroding environment must be an electrolyte with reasonably low resistivity. High-resistivity electrolytes produce erroneously low corrosion rates. The vessel wall must be penetrated, and this involves concerns regarding leaks, personnel safety, and other problems. The ability to use direct wiring from the probe location to a remote control room is desirable, but the installation of these wiring systems is costly. In addition, these systems do not provide information on localized corrosion, such as pitting and stress-corrosion cracking. Also, the corrosion rate values are approximate at best, and the method is best suited for use during periods when substantial corrosion rate changes occur.

Measurement of Corrosion Potentials. The use of corrosion potential measurements for in-service corrosion monitoring is not as widespread as the use of polarization resistance. However, this approach can be valuable in some cases, particularly where an alloy could show both active and passive corrosion behavior in a given process stream. For example, stainless steels can provide excellent service as long as they remain passive. However, if an upset occurs that would introduce either chlorides or reducing agents into the process stream, stainless alloys may become active and may exhibit excessive corrosion rates. Corrosion potential measurements would indicate the development of active corrosion, and they may be coupled with polarization resistance measurements as additional confirmation of high corrosion rates.

The success of corrosion potential measurements depends on the long-term stable performance of a standard reference electrode. Such electrodes have been developed for continuous pH monitoring of process streams, and their application for measuring corrosion potentials is straightforward. However, the conditions of temperature, pressure, electrolyte composition, pH, and other possible variables can limit the applications of these electrodes for corrosion monitoring service.

Hydrogen Probes. The concept of the hydrogen probe is based on the fact that one of the cathodic reaction products in nonoxidizing acidic systems is hydrogen. The hydrogen atoms thus generated diffuse through the thickness of the vessel and are liberated at the exterior surface.

Hydrogen probe analysis measures the corrosion rate, unlike ultrasonic thickness measurements and other techniques that measure remaining wall thickness. However, hydrogen probe analysis is limited to systems in which the temperature is close to ambient and the diffusion rate of hydrogen is high. Gas pipeline service is the most common application. In this case, corrosion can occur when hydrogen sulfide, water, and sometimes carbon dioxide are present. This approach has another variation that consists of simply attaching a chamber to the exterior of the pipe and monitoring hydrogen liberation through increasing pressure.

Commercial devices are available for corrosion monitoring by this technique, although it is questionable whether such devices could be positioned and allowed to operate unattended for extended periods of time. In addition, these units have all of the problems associated with any electrochemical measuring device, namely the need for complex electronic equipment and wiring and the need for operators and installers with a sensitivity to the requirements of such equipment. Also, this method is in practice limited to steel, which has a high hydrogen diffusivity and low solubility of hydrogen. Exterior hydrogen monitoring does not supply a quantitative measurement of hydrogen damage.

Analysis of Process Streams. Another useful in-service corrosion monitoring technique is analysis of the process streams for the presence of corrosion products. This straightforward approach usually does not require the installation of specialized equipment. For example, process streams from the bottom of the distillation column can be routinely sampled, and atomic absorption analysis techniques can be used to determine such heavy metals as iron, nickel, and chromium at very low levels. The concentration of such impurities is then directly proportional to the corrosion rate multiplied by the area of metal corroding, if the only source of metal ions is corrosion and if the corrosion products are not precipitating. One problem is that the corroding area may not be known with certainty; if not, the results are relative. However, they do help to determine whether conditions have improved.

Sentry Holes. Small sentry holes can be drilled into the outside of a vessel or pipe at areas that are considered particularly susceptible to corrosion. The holes are drilled to the pressure design thickness. Thus, when corrosion has almost consumed the corrosion allowance, the appearance of a small leak indicates that action must be taken to prevent a major failure. Sentry holes may be threaded, or they may have nipples attached to facilitate plugging. Nondestructive testing is frequently performed in the area near the

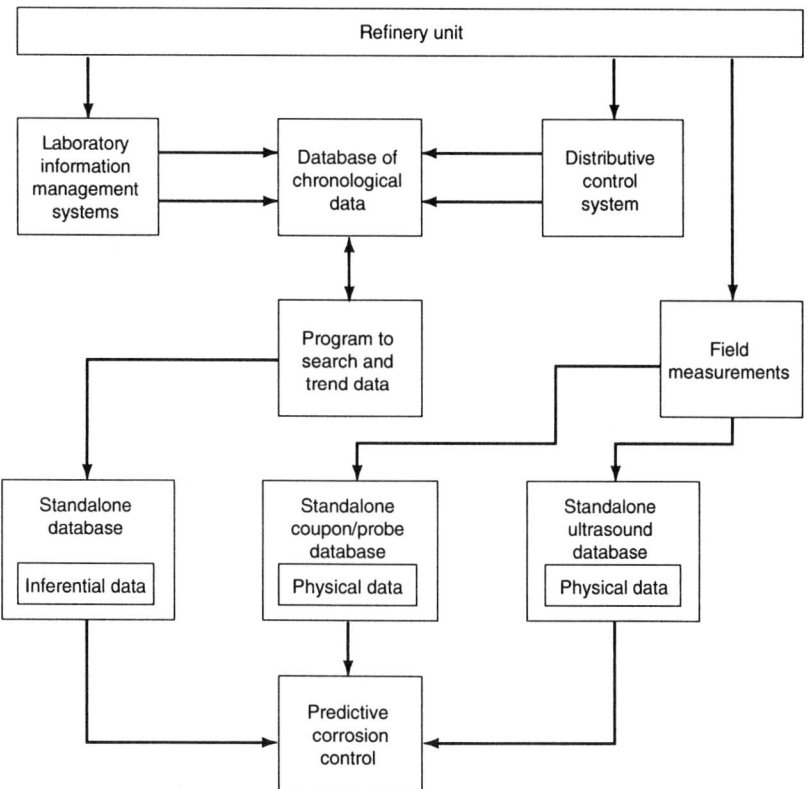

Fig. 5 Flowchart of a typical predictive corrosion control system

Predictive Corrosion Control

Corrosion is a key source of failure in process systems, and predictive corrosion control follows the predictive maintenance approach of extending plant reliability by reducing age-related failures (Fig. 3). The general approach involves the collection and systematic analysis of operational corrosion behavior and data on key plant equipment that is critical for safety or reliability. A general example of a predictive control flow chart is shown in Fig. 5.

A key step in developing a predictive corrosion control program is to identify critical components in a given process unit, based on a risk assessment such as that in Fig. 1 or some other prioritization criteria. The first step is to first subdivide each process unit into functional sections (e.g., preheat train, heater, reactor, overheads, pumparounds, etc.). For a given unit, it is helpful to sketch the system under consideration and draw in areas of concern. Figure 6 is an example. Key components can be identified with an asset risk analysis, which in the case of Fig. 1 ranks piping high, followed by reactors. Which piping has the greatest risk in a given unit or section remains to be determined.

The next step is to analyze the corrosion mechanisms and failures and determine critical areas for monitoring. Sometimes sections are di-

leak to determine the extent of the damage before repair or shutdown.

vided according to the mechanisms of aqueous corrosion or high temperature. For example, aqueous corrosion is predominant in overheads and pumparounds, while high-temperature corrosion is the main concern in a preheat train, reactor, or bottoms. High-temperature corrosion is generally controlled by metallurgy, because the mechanisms are typically hot sulfur attack and hot hydrogen attack. Aqueous phase corrosion is generally controlled by process control and chemical inhibition, because the mechanisms are typically wet hydrogen sulfide, ammonium bisulfide, carbonates, hydrogen chloride, and ammonium chloride.

Complete understanding of the corrosion mechanisms and the system is essential for further development of a predictive corrosion control program with a given unit or section. To illustrate the complexity of such a program, corrosion factors for a crude unit overhead are summarized in the following section.

Corrosion Mechanisms in a Crude Unit Overhead

A typical crude unit overhead system is representative of the aqueous corrosion experienced in many refinery systems. In a crude unit, the most aggressive corrosion is typically from concentrated hydrochloric acid (HCl). A substantial portion of the acid comes from ammonium chloride near the dewpoint of water in a system. Few metallurgies will withstand the combination of corrosives that are found near the dewpoint of water in a sour crude unit overhead system. Con-

sequently, mitigation and corrosion control of carbon steel handling ammonium chloride (NH_4Cl) is generally by filming inhibitors and sustained wash water rates.

Oxygen and Oxidizing Agents. Oxygen and other materials present in the feedstock (e.g., ferric chloride and cupric chloride) act as oxidizing agents and can play a role in the corrosion process that occurs on any unit. In an environment such as the crude unit, free oxygen is generally not a source of corrosion attack, because oxygen readily reacts with hydrogen sulfide (H_2S) to form elemental sulfur:

$$O_2 + 2H_2S \rightarrow 2H_2O + 2S\downarrow$$

The presence of free sulfur in an overhead system confirms that oxygen is getting into the system. Elemental sulfur cannot distill up a crude tower. Sulfur has a boiling point of 444 °C (831 °F) and would be in the bottoms from the fractionator if it were present in the crude.

It is widely recognized that the presence of oxygen or oxidizing agents rapidly accelerates corrosion in a process unit. In one laboratory study, the following relationship between oxygen and H_2S was observed:

Corrodent, 6.0 pH	Corrosion rate of metal loss, mm/yr (mils/yr)
H_2S only	0.76 (30)
O_2, 1 ppm	0.35 (14)
O_2, 1 ppm, plus H_2S	3.0 (120)

As can be seen from these data, there is a synergistic influence on corrosion when the corrosives are present together. In this case, the rate of metal loss was increased by four times over the highest corrosion shown by either corrosive. In a second study, corrosion activity increased from 0.7 to 2.4 mm/yr (27 to 95 when oxygen was added to H_2S). A similar increase in corrosion activity would be present with other oxidizing agents, such as ferric and cupric chloride.

Oxygen contamination sources include water washes. In one instance, corrosion rates more than tripled when accumulator wash water rates were recycled ten times rather than three. Oxygen and other oxidizing agents can be introduced with crude feedstock, slop oils, and desalter wash water. Oxygen levels in water must be controlled to be below 50 parts per billion (wt).

Hydrochloric acid is a major cause of corrosion in most crude units. The acid is formed in the crude preheat exchanger and in the crude atmospheric furnace by hydrolyzing magnesium chloride and calcium chloride present in the crude to form hydrogen chloride. The reactions are of the type:

$$MgCl_2 + 2H_2O \rightarrow Mg(OH)_2 + 2HCl$$

In the overhead, tower top, or top pumparound near the water dewpoint, a low pH condition is produced by the corrosion mechanism:

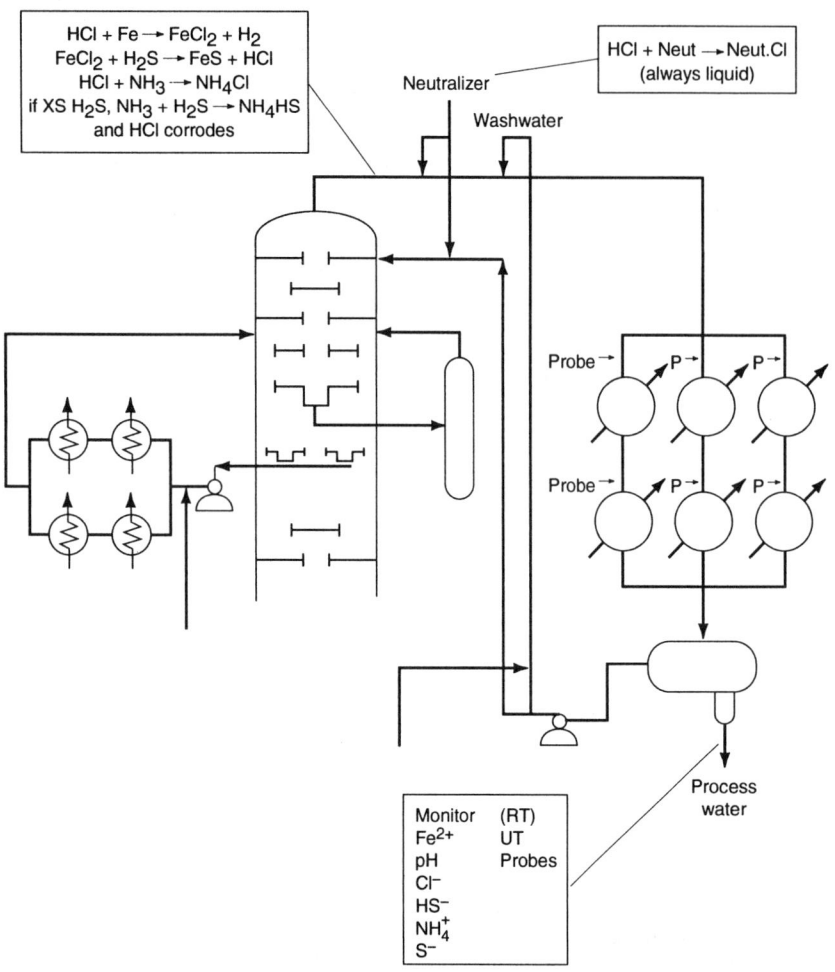

$$HCl + Fe \rightarrow FeCl_2 + H_2$$
$$FeCl_2 + H_2S \rightarrow FeS + HCl$$
$$HCl + NH_3 \rightarrow NH_4Cl$$
if XS H_2S, $NH_3 + H_2S \rightarrow NH_4HS$
and HCl corrodes

Neutralizer

$$HCl + Neut \rightarrow Neut.Cl$$
(always liquid)

Washwater

Probe →

Probe →

Process water

Monitor   (RT)
Fe^{2+} UT
pH   Probes
Cl^-
HS^-
NH_4^+
S^-

Fig. 6   Example of system schematic used to define critical points for corrosion control

$$2HCl \overset{H_2O}{\rightarrow} 2H^+ + 2Cl^-$$

$$2HCl + Fe \overset{H_2O}{\rightarrow} FeCl_2 + H_2\uparrow$$

At lower temperatures, or at an elevated pH in the presence of H_2S, the corrosion mechanism is:

$$FeCl_2 + H_2S \overset{H_2O}{\leftrightarrow} Fe\downarrow + 2HCl$$

After hydrogen chloride dissolves in the water near the point of initial condensation, the strong acid attacks the metal surface, producing a pitted appearance. The corrosion products are metal chloride salts. Further, when the pH of the water in this region is below 5.0, HCl also dissolves any metal sulfide corrosion products that are present. These corrosion products tend to be carried away from the site of the corrosion attack. Hydrogen sulfide, a weak acid that is present in most systems, dissolves in the water where it reacts with the metal chloride, that is, ferrous chloride, producing iron sulfide. Iron sulfide (FeS) is insoluble in water and precipitates from the aqueous phase, causing fouling on the surface of the heat transfer equipment. Because FeS tends to precipitate from the water phase, HCl is

then free to react a second or third time with the metal surface.

It is important to understand that the system is extremely dynamic, such that there is a tendency for small pockets to form where each corrosion reaction is occurring. All corrosion reactions tend to occur simultaneously in these systems. Mitigation and corrosion control of carbon steels handling HCl is by neutralization and/or inhibition by chemical treatments, in addition to sustained wash water rates.

Ammonium Chloride. Reactions involving ammonia and hydrogen chloride result in a vapor phase reaction between ammonia and hydrogen chloride gas:

$$NH_3 + HCl \leftrightarrow NH_4Cl \text{ solid}$$

In the presence of heat and a small amount of water:

$$NH_4Cl \overset{H_2O}{\rightarrow} NH_4^+ + Cl^-$$

$$NH^+ + Cl^- \underset{Heat}{\overset{H_2O}{\rightarrow}} NH_3\uparrow + H^+Cl^-$$

This HCl is free to attack the metal surface:

$$HCl + Fe \overset{H_2O}{\rightarrow} FeCl_2 + H_2\uparrow$$

and in the presence of H_2S:

$$FeCl_2 + H_2S \overset{H_2O}{\rightarrow} FeS\downarrow + 2HCl$$

Ammonium chloride (NH_4Cl) is formed between a strong acid and a moderately weak base. For this reason, the salt is an acid salt. Ammonium chloride (NH_4Cl) has the ability to absorb water from steam at temperatures above the dewpoint of water. This can produce a saturated solution of NH_4Cl when at the boiling point of water, and the salt can easily have a pH as low as 3.3. Also, slightly moistened NH_4Cl has good adhesive qualities. Because of this, the salt has a tendency to stick to wetted metal surfaces.

When the hygroscopic NH_4Cl contacts water, it ionizes into ammonium ions and chloride ions. Once the chloride ion enters the water it is very stable and stays in the aqueous phase. The ammonium is not nearly as stable and decomposes to a more stable hydrogen ion and ammonia gas (NH_3). Ammonia has a very high vapor pressure and tends to escape from the aqueous phase back into the vapor.

As ammonia flashes into the vapor phase, it leaves a concentrated hydrochloric solution behind. Hydrochloric acid attacks the metal surface, producing a metal chloride corrosion product. The strong acid will dissolve any metal sulfide or metal oxides that may be present on the metal surfaces of the equipment. This type of secondary attack on sulfide-oxide metal films is particularly aggressive at a pH below 5.0.

Other Corrosives. Some types of crude release small quantities of sulfur trioxide when they are heated. Sulfur trioxide is very hygroscopic and forms sulfuric acid readily. Sulfuric acid is a strong acid and can rapidly accelerate corrosion. In a crude unit, this acid is not usually found in high concentrations in the atmospheric tower, but high concentrations have been found on occasion in the vacuum tower overhead water. This is probably because the vacuum furnace operates at 70 to 100 °C (130 to 185 °F) higher than the atmospheric furnace. Control of corrosion from this acid is handled in a manner which is very similar to control over corrosion from HCl and ammonium chloride.

In some oilfield operations, carbon dioxide is injected into the wells to increase crude oil recovery and production. When carbon dioxide dissolves in water it becomes carbonic acid. This acid is weak and can drive the pH of a system down to only about 4.5 at the temperatures present in the system. However, the addition of H_2S increases the aggressive nature of the weak acid. Neutralization and inhibition with a filming inhibitor provide acceptable control over the acid.

Ammonia can also be classified as a corrosive in systems that have been alloyed with copper-base metallurgies. This type of corrosion attack accelerates rapidly with increasing pH and is especially aggressive at pH ranges above 8.5.

Corrosion Control in a Crude Unit Overhead

Crude Oil Desalting. The first step to failure control on a crude unit is optimizing the desalter unit. The major corrosive coming from the desalting unit is magnesium chloride, which is hydrolyzed to form HCl in the desalter, preheat exchangers, and furnace. The desalter can also be a source of ammonia and other corrosives that can influence overhead corrosion control.

Chloride-containing salts in crude are found in three forms: sodium chloride (NaCl), calcium chloride ($CaCl_2$), and magnesium chloride ($MgCl_2$). The chloride distribution may be different for each crude type, but in general the distribution is NaCl, 75%; $CaCl_2$, 15%; and $MgCl_2$, 10%.

Magnesium chloride starts to hydrolyze at temperatures above 120 °C (250 °F). Calcium chloride starts to hydrolyze to form hydrogen chloride by the mechanism given below:

Salt	Salt hydrolysis to acid
$MgCl_2$	$MgCl_2 + 2H_2O \rightarrow Mg(OH)_2 + 2HCl\uparrow$
$CaCl_2$	$CaCl_2 + 2H_2O \rightarrow Ca(OH)_2 + 2HCl\uparrow$
NaCl	$NaCl + H_2O \rightarrow NaOH + HCl\uparrow$

Suppose the desalted crude from a unit contains sufficient salt to generate a concentration of 115 ppm of chlorides in the condensed water in the overhead. If the system is sour and ammonia is used for pH control, the level of HCl found in the water that initially condenses could be as high as 1150 ppm. This level of hydrochloric acid could easily drive the pH of the water at dewpoint down to 1.5. No metallurgy could withstand this level of acidity at the temperatures present at dewpoint. Thus, HCl concentrations in the overhead water require the optimization of desalting.

Another, less apparent reason to optimize the desalting operation is the cost associated with poor crude dehydration or water separation from the crude. Poor water separation increases the quantity of water that must be condensed in the overhead system, and this costs the refinery extra money to vaporize and then cool the water. In addition, steam can occupy 7 to 10 times more volume than crude in the preheat. An increase in steam could increase pressures in the atmospheric tower, which could reduce the quantity of light ends that are separated from the flashed crude.

From the standpoint of corrosion control, increased quantities of water in the desalted crude increase the quantity of steam in the overhead. This raises the dewpoint temperature of water in the overhead. Suppose that a dewpoint of 107 °C (224 °F) is in the overhead. Water carryover in the desalter could easily raise the dewpoint of the water to 125 °C (260 °F) in the overhead. If the tower top temperature is 116 °C (242 °F), a relatively dry, noncorrosive tower top could become a very wet tower top. This could result in an aggressive level of corrosion attack on the top tray in the tower. Corrosion in the tower top is aggressive dewpoint corrosion.

Ammonia contamination in the desalter stabilizes crude emulsions and can upset overhead corrosion control. Some refiners have injected sulfuric acid into the desalter with the wash water. This procedure consumes ammonia, because ammonium sulfate is formed. Ammonium sulfate is extremely stable and will not sublime or dissociate at the temperatures present in the preheat. The problem with sulfuric acid is that the material can be carried over, causing sulfuric acid corrosion activity in the overhead. This type of attack can be extremely aggressive.

The best pH for good desalting is between 5.0 and 6.5. If bases are introduced, emulsion stability in the desalter is increased.

The primary source of ammonia in the desalter is from the sour water strippers that process sour water from delayed cokers, thermal crackers, and fluid cracking units. These units produce large quantities of ammonia and H_2S. Phenolic compounds produced in the cracking operation are partially soluble in high-pH waters. Phenols are pollutants that are closely monitored in discharge water from the refinery. Because of the overload that phenols place on water treatment facilities, many plants use the desalter as a waste disposal unit for phenols. If the sour water stripper is operated to minimize both the H_2S content and the ammonia content in the stripped water, there is no problem with using this water in the desalter. Unfortunately, few refineries operate their strippers to optimize ammonia removal. When this situation is allowed to exist, corrosion control becomes increasingly difficult.

Caustic Injection. Good corrosion control in a crude unit overhead system is relatively easy to obtain if overhead water chloride concentrations are maintained in the range of 40 to 50 ppm, measured as NaCl. When chlorides exceed this level, control becomes increasingly difficult. At a chloride level above 100 ppm in the overhead water, corrosion control is very difficult. If the desalting unit has been optimized, and overhead chlorides are above the target of 40 to 50 ppm, the use of caustic soda should be considered to reduce overhead chloride concentrations.

Sodium chloride is a stable salt that does not hydrolyze in appreciable quantities until a temperature of about 525 °C (980 °F) is reached. Sodium hydroxide converts the salts that hydrolyze easily into more stable NaCl:

$$MgCl_2 + 2NaOH \rightarrow Mg(OH)_2 + 2NaCl$$

$$CaCl_2 + 2NaOH \rightarrow Ca(OH)_2 + 2NaCl$$

$$HCl + 2NaOH \rightarrow H_2O + NaCl$$

$$FeCl_3 + 3NaOH \rightarrow Fe(OH)_3 + 3NaCl$$

Note that caustic neutralizes any acid in the crude and, being a strong base, also displaces weaker bases in the salts found in crude. It is almost impossible to stoichiometrically calculate the amount of caustic needed to reduce overhead concentrations to acceptable levels.

Disadvantages of sodium hydroxide injection into desalted crude include the following:

- The reaction products, hydroxides, are insoluble in crude and contribute to exchanger fouling and furnace coking in atmospheric and vacuum furnaces.
- Unreacted caustic can increase fouling and coking tendencies in the preheat system.
- On occasion, unreacted caustic has caused furnace tube stress cracking and embrittlement.
- Caustic can influence the foaming and emulsification characteristics of the crude.
- The strong base and the reaction products can cause foaming in the atmospheric and vacuum tower flash zones.
- Sodium contents in residual fuels may be increased.
- Sodium content in coke produced at the delayed coker increases.
- Sodium is a poison on equilibrium catalysts.
- Caustic can reduce the activity of many antifoulants.

An extremely conservative approach should be taken to injecting caustic into any preheat system.

Filming Inhibitors. Adequate control over the corrosive process is not possible without adequate neutralization of the acids in the system and proper use of a filming inhibitor. Inhibition and neutralization are particularly important, because without the proper application of both, the corrosion process will continue.

Corrosion is inhibited if the metal is separated from its environment by an inert and impervious barrier. The barrier, or film, must satisfy a number of requirements to function as a suitable inhibitor. For example, it must firmly adhere to metal surfaces; be continuous, dense, and nonporous; and be relatively nonreactive.

Two types of inhibitor are used in process units: oil soluble and water soluble. The filming mechanism is slightly different for each type, although both inhibitors are polar compounds.

The oil-soluble inhibitor uses the hydrocarbon in the process stream to provide the protective barrier. The inhibitor molecule is polar. One end of the molecule is called an oil-soluble tail. Its function is to dissolve in the hydrocarbon that is present in the stream. The chemical composition of this portion of the inhibitor determines in what hydrocarbon streams the inhibitor will be soluble. Many inhibitors are soluble in butane and heavier hydrocarbon. Extremely hydrocarbon-soluble inhibitors generally provide better filming inhibition.

At the opposite end of the filming inhibitor is the polar head, the portion of the inhibitor that is attracted to the metal surface. The exact mechanism of this attractive force is not completely understood; however, what probably occurs is that the polar head forms a coordinate bond with the metal surface. The attraction is called absorption. The bond is pH-dependent, and high-pH conditions break the bond between the metal surface and the polar head. When such a condition exists, the polar head portion of the inhibitor may be preferentially attracted toward water, as opposed to the metal surface.

Measurement

Corrosion Coupons. If properly located, the coupon is a reasonably accurate tool. The primary purpose of measuring the corrosion activity on a unit is to provide an estimate of the condition of the equipment so that planned turnarounds can be scheduled and replacement of equipment anticipated. Frequent measuring also helps minimize the duration of corrosive upset conditions. It is well documented that short upset conditions can cause very serious corrosion activity in an overhead system.

In addition to producing corrosion data in measurable terms (mpy or mm/yr), the coupon is helpful in identifying the type of corrosion activity present at the point where the coupon is located in the system. Pitting on the coupon surface would indicate activity from a strong acid or acid salt, such as hydrochloric acid or ammonium chloride. General thinning indicates the activity of hydrogen sulfide on the coupon.

However, where a cooling medium is present on the other side of the metal surface, the skin temperature of the tube is lower than the temperature of the surrounding hydrocarbon. The cool tube skin could cause surface water condensation at temperatures above the anticipated water dewpoint in the hydrocarbon. Coupons would not indicate corrosion activity of this nature. This is especially important in units which use cold crude as a heat exchange medium for hot overhead vapors.

Corrosometer Probe. If properly placed, the corrosometer probe is a fast, accurate method for determining relative corrosion activity in a system. The probe produces data in meaningful terms (mpy or mm/yr) for periods as short as one week. Upset conditions can be observed in even shorter periods. The major limitation of the probe lies in the placement. Also, the probe is not sensitive to cool tube skin condensation conditions, and it is not a reliable corrosion measuring tool in systems that are extremely sour because the probe element measures corrosion product laydown as metal thickness. If sulfiding corrosion activity accelerates, the corrosometer probe tends to show increasing metal thickness rather than metal loss.

Sample Analysis. One of the more useful measuring methods available is sample analysis. Unlike the probe or coupon, the sample indicates corrosion activity through the entire overhead system at one point in time, or continuously, depending on the type of sampling (i.e., remote or online). Sample analysis is used to identify the level of corrosives in the system and the products produced as a result of their presence.

The level of metal corrosion product found in overhead water from a process unit is a function of corrosion activity, filmer inhibitor emulsification tendencies, the pH of the water, and the H_2S concentration. Inhibitors that emulsify hydrocarbon badly tend to increase the level of metal corrosion products found in hydrocarbon streams. In most units, the principal corrosion product is a metal sulfide. Inhibitors that cause excessive emulsification tend to hold these corrosion products up in the hydrocarbon stream. Also, as the pH rises above 6.0, FeS may preferentially stay with the hydrocarbon phase, or it may concentrate at the interface between the water and hydrocarbon in the accumulator water leg. When this condition exists, sample analyses of overhead water alone are not very significant.

Samples of overhead hydrocarbon product should be run to supplement data generated from analyses of water samples. Typical sample analyses from an overhead system should include:

Sample	Analyses
Overhead water	pH
	Chlorides
	Total iron
	Ammonia
	Hydrogen sulfide
Overhead hydrocarbon	Total iron
	Inhibitor residual
	Water content

Sample analysis data are difficult to interpret and can be misleading, so sample data should be subjected to trend analysis. The major value of the sample is that it provides the ability to identify and correct upset conditions rapidly on the unit. This capability is extremely important. It has ensured the success of many corrosion control programs.

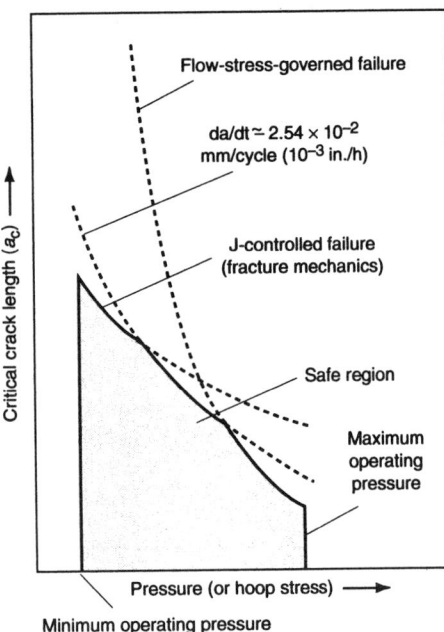

Fig. 7 Combined failure criteria for a pipe under internal pressure. Source: R. Viswanathan, *Damage Mechanisms and Life Assessment of High-Temperature Components*, ASM International, 1989, p 247

Remaining-life determination is generally confined to equipment that is degraded by creep, fatigue, hot hydrogen attack, or combinations of these mechanisms. In the absence of a predictive corrosion control program (or, more generally, predictive maintenance for life extension), condition assessment is required, because it is fundamental to the continued safe and reliable operation of refinery or process equipment. For the most part, however, life assessment is based on educated guesses from various approaches, generally nondestructive, deterministic, or phenomenological approaches. A recent review by Scasso (Ref 4) of the Italian Welding Institute focuses on the regulatory approach adopted by Europe and the U.S. The findings are summarized in Tables 3 to 6.

The Finnish standard SFS 3280:E provides a credible deterministic approach (Ref 5). The scope of SFS 3280:E is to provide guidance in the monitoring of creep and the assessment of the degree of creep in tubes, pipe systems, and headers, that have been designed on the basis of creep strength, and in comparable steel constructions. The objectives of assessing the degree of creep are to establish the intervals between inspections and to estimate the residual service life of the constructions.

Critical Crack Sizes

The critical crack size (a_c) can be defined in a number of ways, based on fracture toughness, ligament size, crack-growth-rate transitions, or other considerations as appropriate. A common definition for many heavy-section components is based on the fracture toughness of the material.

Condition and Life Assessment

Condition and life assessment involve defining a critical crack size for a given component and measuring existing crack sizes by NDE or other inspection techniques. Life assessment requires an understanding of the factors affecting critical crack size and the methods for crack measurement. Under certain circumstances, any defect or flaw observable in a component by visual or other NDE methods constitutes grounds for retirement. Under other circumstances, NDE observations are combined with crack growth analysis to determine remaining life. The conservative approach is to replace components based on crack initia-

tion. However, with increasing awareness of fracture mechanics considerations, it is possible to consider the crack tolerance of components.

Techniques that use crack initiation as a failure criterion include calculations based on history, extrapolations of failure statistics, strain measurements, accelerated mechanical testing, microstructural evaluations, oxide scale growth, hardness measurements, and advanced NDE techniques. For crack-growth-based analysis, the NDE information, results from stress analysis, and crack growth data are integrated and evaluated with reference to a failure criterion.

Table 3 Regulations and standards on life assessment

Country	Regulation or standard
Denmark	Instructions for manufacturing and operation of high-pressure piping issued by the Danish Power Companies
Finland	SFS 3280:E (1984-10-22). Inspection of pressure vessels. Assessment of degree of creep
Germany	TRD 508 (1978-10). Additional tests on components calculated with time-dependent design strength values—inspection and testing. TRD 508, Annex 1. Method for the calculation of components having time-dependent design strength values. VdTuV MB 451 (1986-3). Investigation of the surface structure of building components subjected to creep rupture according to TRD 508
Italy	ISPESL (1992). Components of steam generators and components of vessels under steam or gas pressure operating in creep condition of materials—calculations and testing
The Netherlands	TO1o2V (1985-2), Appendix 1. Rules for pressure vessels—periodic inspection: parts in the creep range
Portugal	ISQ-AVR-002. General procedure for the remanent life assessment of static components that have undergone high temperature
United Kingdom	BSI-PD 6510: 1982. A review of the present state of the art of assessing remanent life of pressure vessels and pressurized systems designed for high-temperature service
United States	API Recommended Practice 530, Appendix E (1988-89). Calculation of heater tube thickness in petroleum refineries—estimation of remaining tube life. ASME Code Case N-47-26, Appendix T (1986-2-23)(a). Class 1 components in elevated-temperature service (Section III, Division 1)—rules for strain, deformation, and fatigue limits at elevated temperatures

(a) The ASME Code Case N-47-26 is a design document and therefore is not aimed at assessing the remanent life; however, the creep-fatigue damage can be calculated according to its rules

Several alternative criteria can be used to establish a value for the critical crack size: a flow-stress-governed failure criterion, a J_{Ic}-controlled failure criterion, and a limiting creep-crack growth rate criterion.

The first two criteria are employed in a scenario in which rupture occurs during or immediately following a startup transient, in the absence of creep. The third criterion is used in a scenario where failure occurs by creep-crack growth under operating conditions. The lowest value of critical crack size determined by use of these criteria is then used for remaining-life analysis. The use of combined failure criteria to define the safe operating pressure for a pipe is illustrated in Fig. 7.

Stress level is not the only factor that affects the definition of critical crack size. Often critical crack sizes decrease with time due to embrittlement or aging. For example, severely embrittled bolts, vessels, and rotors may have critical crack lengths below the detection limits of conventional testing techniques.

In other instances, a_c may be large but the rate of crack growth may be so high that once a crack initiates, it reaches critical size rapidly. Many environmentally induced failures in highly stressed components exhibit this behavior. For instance, in generator retaining rings and in steam turbine blades where crack growth under corro-sive conditions is encountered, the presence of a pit or pitlike defect is cause for retirement.

With respect to material behavior, the major problem is the unavailability of data pertaining to crack growth and toughness in the service-degraded condition specific to the component. While considerable data may be available on materials in the virgin condition, the data bank on service-exposed materials is very small. Nondestructive methods are needed to determine those properties with specific reference to a given component.

In welded components, the problem is further compounded by the fact that a weldment contains a complex microstructure of many zones with varying material properties. Failure can occur through any of these zones or at the interfaces between them. In cases where crack growth rates might be rapid, conventional NDE techniques are often inadequate to detect the initial crack.

The uncertainties in interpretation of NDE results can sometimes be overwhelming. Difficulties in distinguishing between innocuous versus harmful flaws and identifying their orientations can lead to uncertainties in life assessment. If there are numerous indications closely spaced, the manner of treating them in terms of a linkup analysis could be very crucial. Geometric discontinuities such as fillets, section transitions, and weld backing rings interfere with NDE signals and mask flaws. Guidance on the assessment of cracks and flaws may be obtained from the article "Operating Stress Maps for Failure Control" in this Volume.

Creep Life Assessment

Failure due to creep can be classified as resulting either from widespread bulk damage or from localized damage. The structural components that are vulnerable to bulk damage (e.g., boiler tubes) are subjected to uniform loading and uniform temperature distribution during service. If a sample of material from such a component is examined, it will truly represent the state of damage in the material surrounding it. The life of such a component is related to the creep-rupture properties.

On the other hand, components that are subjected to stress (strain) and temperature gradients (typical of thick-section components) may not fail by bulk creep rupture. It is likely that at the end of the predicted creep-rupture life, a crack will develop at the critical location and propagate to cause failure. A similar situation exists where failure originates at a stress concentration or at pre-existing defects in the component. In this case, most of the life of the component is spent in crack propagation, and creep-rupture-based criteria are of little value. Therefore, this section briefly reviews creep life assessment from the perspective of creep-rupture properties and creep crack growth. Practical methods based on replication and parametric approaches are described at the end of this article.

Creep Rupture Life

Although it is relatively easy to quantify damage in laboratory creep tests conducted at constant temperature and stress (load), components in service hardly ever operate under constant conditions. Start-stop cycles, reduced power operation, thermal gradients, and other factors result in variations in stresses and temperatures. Procedures are needed that will permit estimation of the cumulative damage under changing exposure conditions.

Damage Rules. The most common approach to calculation of cumulative creep damage is to compute the amount of life expended by using

Table 4 Nondestructive evaluation requirements of regulations and standards given in Table 3

Country	Visual exam(a)	Dimensional check(b)	Magnetoscopic test	Penetrants	Replica exams	Ultrasound	Radiography	Magnetic permeability(c)
Denmark		X	X	X	(d)	X	X	
Finland	X	X	X	X	X	X		
Germany	X	X	X	X	X	X		
Italy	X	X	X	X	X	X	X	
The Netherlands(e)	X	X	X	X	X			
Portugal	X	X	X	X	X	X		X
United Kingdom	X	X	X	X	X	X		

Note: Not all the mentioned inspections are always demanded; on the contrary, other examinations, different from those indicated, can be required according to the results of the previous one. (a) Supported by appropriate devices such as endoscopes. (b) Thickness measurements included. (c) In chemical plants, for tubes subject to carburization. (d) Metallurgical examination on parts cut out of the piping. (e) The minimum extent of nondestructive examinations is indicated in the document.

Table 5 Inspection frequencies of regulations and standards given in Table 3

Country	Timing of inspection	
	First check(a)	Subsequent checks
Denmark	Approx 1 yr after commissioning of the plant	Every third year, counting from first measurement until registration of 1% creep. Then every year until registration of 2% creep. After that the pipe system must be scrapped or the Inspection Authority must be approached.
Finland	TLC < 60% TTL ST < 80% DTB PS expected >1%	According to inspection results, but at least every 4 yr
Germany	TLC < 60% TTL TLC by fatigue >50% TTL (at least at the first periodic inspection after the mentioned limits have been attained)	According to inspection results; after the 100% TTL or 1% permanent set has been attained, the test intervals can be reduced.
Italy	TLC < 60% TTL(b) ST < 100% DTB(c) TLC ≥ 60% TTL(d)	According to inspection results, but: If ST <100% DTB, at least every 50,000 h or 20% TTL, whichever is smaller and if ST >100% DTB, at least every 25,000 h
The Netherlands	TLC ≥ 60% TTL(e)	Planned schedule (if there is no reason not to use it)
Portugal	ST < 80% DTB	According to inspection checks
United Kingdom	Not specified	Not specified

TLC, total life consumption; TTL, total theoretical life; ST, service time; DTB, design time base; PS, permanent strain. (a) The results should be compared with the results of the fabrication checks, if available. (b) Only for dimensional checks. (c) Only for visual inspections, dimensional checks, penetrant tests, magnetic tests, ultrasound, or radiographic tests. (d) Only for replica examinations. (e) At least at the first periodic inspection after 60% TTL has been attained

Table 6 Rejection criteria of regulations and standards given in Table 3

Country	Rejection criteria
Denmark	Attainment of a permanent set of 2%
Finland	Not specified (according to the calculation and testing results)
Germany	Presence of nonrepairable cracks Consumption of total theoretical creep life and/or total theoretical fatigue life, unless safe, continued operation can be demonstrated Attainment of a permanent set of 2%, if reference measurement results were available from the beginning of service Attainment of a permanent set of 1%, if reference measurements were not available after 60% of the total theoretical life had been reached
Italy	According to the calculation and testing results Consumption of total theoretical life Attainment of a permanent set of 2%, if reference measurement results were available from the beginning of service Attainment of a permanent set of 1%, if reference measurement results were not available after 60% of the total theoretical life had been reached
The Netherlands	According to the calculation and testing results, particularly in the presence of nonrepairable cracks
Portugal	Not specified (according to the calculation and testing results)
United States	API Recommended Practice 530, Appendix E: not specified (according to the calculation and testing results) ASME Code Case N-47-26, Appendix T: not applicable

time or strain fractions as measures of damage. When the fractional damages add up to unity, then failure is postulated to occur. The most prominent rules are as follows:

- *Life-fraction rule (LFR)* (Ref 6):

$$\Sigma \frac{t_i}{t_{ri}} = 1$$

- *Strain-fraction rule* (Ref 7):

$$\Sigma \frac{\varepsilon_i}{\varepsilon_{ri}} = 1$$

- *Mixed rule* (Ref 8):

$$\Sigma \left(\frac{t_i}{t_{ri}}\right)^{1/2} \left(\frac{\varepsilon_i}{\varepsilon_{ri}}\right)^{1/2} = 1$$

- *Mixed rule* (Ref 9):

$$k\Sigma\left(\frac{t_i}{t_{ri}}\right) + (1-k)\Sigma\left(\frac{\varepsilon_i}{\varepsilon_{ri}}\right) = 1$$

where k is a constant; t_i and ε_i are the time spent and strain accrued at condition i; and t_{ri} and ε_{ri} are the rupture life and rupture strain under the same conditions.

Example: Life-Fraction Rule Calculation. The purpose of this example is to illustrate the use of the LFR. A piping system, made of 1¼Cr-½Mo steel designed for a hoop stress of 7 ksi, was operated at 540 °C (1000 °F) for 42,500 h and at 550 °C (1025 °F) for the next 42,500 h. From the minimum curve of Larson-Miller parameter for the steel, it is found that, at σ = 48 MPa (7 ksi):

t_r at 1000 °F = 220,000 h

t_r at 1025 °F = 82,380 h

Life fraction expended, t/t_r, at 1000 °F

$$= \frac{42,500}{220,000} = 0.19$$

Life fraction expended, t/t_r, at 1025 °F

$$= \frac{42,500}{82,380} = 0.516$$

The total life fraction expended is 0.71.

Validity of Damage Rules. Goldhoff and Woodford (Ref 10) studied the Robinson life-fraction rule and determined that for a Cr-Mo-V rotor steel it worked well for small changes in stress and temperature. Goldhoff (Ref 11) assessed strain-hardening, life-fraction, and strain-fraction rules under unsteady conditions for this steel. While all gave similar results, the strain-fraction rule was found to be the most accurate.

From careful and critical examination of the available results, the following overall observations can be stated (Ref 12).

- Although several damage rules have been proposed, none has been demonstrated to have a clearcut superiority over any of the others. The LFR is therefore the most commonly used.
- The LFR is clearly not valid for stress-change experiments. Under service conditions where stress may be steadily increasing due to corrosion-related wastage (e.g., in boiler tubes), application of the LFR will yield nonconservative life estimates; that is, the actual life will be less than the predicted life. On the other hand, residual-life predictions using postexposure tests at high stresses will yield unduly pessimistic and conservative results.

- The LFR is generally valid for variable-temperature conditions as long as changing creep mechanisms and environmental interactions do not interfere with test results. Hence, service life under fluctuating temperatures and residual life based on accelerated-temperature tests can be predicted reasonably accurately by use of the LFR.
- The possible effects of material ductility (if any) on the applicability of the LFR need to be investigated. A major limitation in applying the LFR is that the properties of the virgin material must be known or assumed. Postexposure tests using multiple specimens often can obviate the need for assuming any damage rule.

Creep Crack Growth

Gross and uniform creep deformation of components is usually the exception rather than the rule. Localized defects and stress concentrations often play decisive roles in failure. Under these circumstances, the growth of cracks and defects is governed by the creep ductility of the material.

Extensive creep crack growth data pertaining to Cr-Mo piping steels have been collected, analyzed, and consolidated (Ref 13,14). It has been observed that a crack tip driving force parameter, C_t, that takes time-dependent creep deformation into account correlates much better with crack growth rates, da/dt, than the traditionally used elastic stress intensity factor, K. The relation between da/dt and C_t can be expressed as:

$$\frac{da}{dt} = bC_t^m$$

To perform a remaining-life assessment of a component under creep-crack growth conditions, two principal ingredients are needed: an appropriate expression for relating the driving force C_t to the nominal stress, crack size, material constants, and geometry of the component being analyzed; and a correlation between this driving force and the crack growth rate in the material, which has been established on the basis of prior data or by laboratory testing of samples from the component. Once these two ingredients are available, they can be combined to derive the crack size as a function of time. The general methodology for setting inspection intervals using this approach has been described elsewhere (Ref 12, 15).

A number of variables have been identified as affecting the crack growth rate by modifying b, C_t, or m (Ref 16):

- In-service degradation increases da/dt by increasing C_t in the case of ductile materials and by increasing m and/or b for brittle material.
- Crack growth rates in welds, fusion lines, and heat-affected zone (HAZ) materials are at least a factor of 5 higher compared to those in base metal.
- The presence of localized chains of inclusion, assisted by segregation of impurities to interfaces such as grain boundaries and fusion lines, causes significant increases in creep-crack growth rates.
- The presence of large amounts of impurities in the steel accelerates crack growth by increasing m.
- All material and experimental variables that reduce creep ductility result in higher crack growth rates.
- Temperature can have mixed effects on crack growth. In cases where the effect of temperature is merely to increase creep rate, the da/dt increases with increase in temperature due to increase in C_t. On the other hand, if a transition from a brittle to ductile condition is involved, increase in temperature may actually decrease the crack growth rates.
- Crack tip constraint has a pronounced effect on crack growth. Assumptions regarding plane stress or plane strain conditions can have a pronounced effect on da/dt.
- Inclusion of primary creep, in addition to the secondary creep, in calculating C_t results in larger value of da/dt and reduced remaining life.

Additional information on creep-crack growth is contained in the article "Creep-Crack Growth in Metals" in this Volume.

Hydrogen Cracking in Wet H₂S

Cracking resistance of steels is a major concern in refining and petrochemical industries where aqueous H_2S is present. The generally accepted theory of the mechanism for hydrogen damage in wet H_2S environments is that monatomic hydrogen is charged into steel as a result of sulfide corrosion reactions that take place on the material surface. The primary source of atomic hydrogen available at internal surfaces of pipeline and vessel steels is generally the oxygen-accelerated dissociation of the H_2S gas molecule in the presence of water. The basic reaction is:

$$Fe + H_2S \xrightarrow{H_2O} FeS + 2H$$

The FeS formed on the surface of the steel is readily permeated by atomic hydrogen, which diffuses further into the steel.

This diffusion of atomic hydrogen into steel is associated with three distinct forms of cracking:

- Hydrogen-induced cracking
- Stress-oriented hydrogen-induced cracking
- Hydrogen stress cracking (also known as sulfide stress cracking and sulfide stress-corrosion cracking)

Hydrogen-induced cracking (HIC) and stress oriented hydrogen-induced cracking (SOHIC) are both caused by the formation of hydrogen gas (H_2) blisters in steel. Hydrogen-induced cracking, also called stepwise cracking or blister cracking, is primarily found in lower-strength steels, typically with tensile strengths less than about 550 MPa (80 ksi). It is primarily found in line pipe steels.

In contrast, hydrogen stress cracking does not involve blister formation, but it does involve cracking from the simultaneous presence of high stress and hydrogen embrittlement of the steel. Hydrogen stress cracking occurs in higher-strength steels or at localized hard spots associated with welds or steel treatment. As a general rule of thumb, hydrogen stress cracking can be expected to occur in process streams containing in excess of 50 ppm H_2S (although cracking has been found to occur at lower concentrations).

The basic factors of these cracking modes include temperature, pH, pressure, chemical species and their concentration, steel composition and condition, and welding or the condition of the weld HAZ. This section briefly reviews HIC, SOHIC, and hydrogen stress cracking of line pipe and pressure vessel steels in aqueous H_2S environments with respect to differences between these types of cracking and important variables for failure control.

Hydrogen-Induced Cracking

Hydrogen-induced cracking, also known as stepwise or blister cracking, manifests itself in low-strength steels in the form of small cracks and/or blisters. This type of cracking is typically oriented parallel to the rolling plane of the steel and is associated with inclusions and segregation bands in the material. These cracks can appear in the absence of an applied stress and propagate by linking up in a stepwise manner (Fig. 8), leading to component failure by reducing the effective thickness of the material. The driving force for

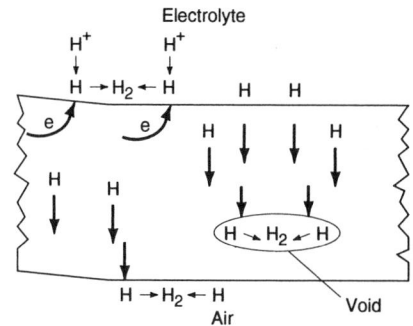

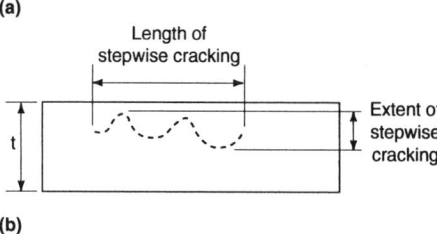

Fig. 8 Hydrogen blistering and stepwise cracking in steel. (a) Schematic of blister formation process. (b) Schematic of stepwise cracking. Source: *International Metals Review*, Vol 30 (No. 6), 1985, p 291-301

crack propagation is the buildup of hydrogen pressure in the cracks.

The internal hydrogen blisters form quickest at internal discontinuities in the steel, which can be hard spots of low-temperature transformation products or laminations. However, manganese sulfide (MnS) inclusions are the primary sites for this to occur. The formation of H_2 at these sites is facilitated by the gap that exists around MnS due to the favorable variation in expansion coefficients between the steel and MnS. Elongated sulfides are particularly attractive sites for formation of H_2.

Increasing H_2 pressure at these sites leads to hydrogen damage, such as hydrogen blistering, longitudinal cracking and, through interaction of plastic zones at the ends of these sites, a delayed shear reaction frequently referred to as stepwise cracking. As cracks initiate and propagate, they begin to link up with others, and a series of stepwise cracks can propagate through the material. An applied stress is not required.

Controlling HIC of Pipe Steel. There are three methods of controlling HIC in new and replacement plant. One method is to prevent the H_2S molecule from undergoing the dissociation process by controlling the gathered gas composition. Secondly, the creation of a tenacious, impervious film on the steel surface would inhibit corrosion.

The elimination of favorable sites for H_2 formation in the steel is another preventive measure. Shape control of sulfide inclusions is perhaps the best way to minimize the tendency toward HIC in line pipe steels. Elongated MnS inclusions promote crack initiation and propagation due to the high stresses at the tips of the inclusions. However, the addition of calcium or rare earths to the steel makes the sulfides spherical, and because of their hardness, they remain spherical after processing. Reduction of the sulfur content is also beneficial in reducing the susceptibility of steels to HIC.

Copper and Cobalt Alloying. Other alloying additions that reduce hydrogen permeation, such as copper up to about 0.25% (Ref 17) and cobalt (Ref 18), are also beneficial. A Pacific Rim supplier produced an ASTM A 516 GR70 steel with a 1 wt% Co addition, which resulted in greatly reduced absorbed hydrogen at pH levels around 3 (Ref 18). Further, a similar steel from the same supplier, with a 1 wt% Co + 0.3 wt% Cu addition, had a significant benefit in that the observed hydrogen permeation rate in the NACE solution + H_2S showed a downward trend after 200 hours. Steel with only the 0.3 wt% Cu addition appeared to follow an increasing rate after this time (Ref 18).

HIC Control for Existing Plant. Selecting appropriate steel grades or suppliers is satisfactory for both new and replacement units, but for existing plant components, replacement cost, particularly in terms of production downtime, is extremely high. The problem, then, is inspecting an existing plant that is prone to hydrogen blistering and defining when unsafe conditions prevail such that maintenance is cost-effective. Important in defining these conditions are the inclusion species present and the types of blistering present.

Inclusions Species. Existing plant units are generally constructed from C-Mn steels, which meet the required compositional codes but have a high distribution of MnS and alumina (Al_2O_3) inclusions. These steels, which are generally fully killed, tend to present a twofold problem in terms of hydrogen behavior as it diffuses through.

MnS inclusions are such that the difference in expansion coefficients between these inclusions and the C-Mn steel matrix creates a gap around these inclusions. Atomic hydrogen readily forms molecular hydrogen in these interfaces.

Al_2O_3 inclusions are such that the inclusion exerts a stress on the surrounding matrix to the extent that hydrogen in the region of these inclusions produces hydrogen embrittlement of the matrix in the inclusion vicinity. This is due to the stress filed, coupled with the formation of H_2 at the interfaces. The embrittled matrix allows easy propagation of wedges created around these inclusions by H_2 such that longitudinal cracking in these Al_2O_3 stringer regions is an easy process.

Similarly, Al_2O_3 stringers in close proximity exhibit earlier delayed shear cracks, or stepwise cracks, than MnS inclusions. Determining total residual aluminum and sulfur is a way to identify the predominant inclusion species within a given grade of C-Mn steel. This can serve as a guide to the early nature of hydrogen blistering.

Types of Blistering. In aluminum-killed steels of high sulfur content, the predominant inclusion species are Al_2O_3 stringers and probably duplex MnS-Al_2O_3 inclusions. Blisters in the form of longitudinal cracks occur in the Al_2O_3 stringers initially. Subsequent formation of an internal cavity is at these sites and, as is frequently observed, at MnS-Al_2O_3 inclusions.

Further developments ensue. The linking of cavities occurs by the delayed shear or stepwise cracking process. This type of cracking tends to form between internal blisters and the internal surfaces of pipe or vessels.

Blisters themselves are classified into two categories (Ref 19). In type I, the steel on either side of the blister is intact. In type II, the steel between the blister and the internal surface is degraded by corrosion to the extent that the effective wall thickness of the pipe or vessel is the ligament that exists between the external wall surface and the blister surface.

It may be supposed that modeling of these four hydrogen-diffusion-controlled processes is easy using an elastic-plastic fracture mechanics approach to determine plastic zone sizes and assumed inclusion distributions, based on the Jernkontorets chart, for example. However, such modeling fails because no acceptable value for the diffusion coefficient of hydrogen in C-Mn steel is available.

Nondestructive Evaluation. For large problem plant units, which are invariably covered with insulating material and then an aluminum alloy jacket, the practical forms of NDE that can be used to detect blisters and similar change are ultrasonic testing, radiography, and acoustic emission.

For towers and large vessels, a common practice is to cut holes in the insulation material at the areas most likely to be damaged and examine these regions with an ultrasound device, usually a thickness meter or a shear-wave scope. The area is then mapped, and repeat inspections may be carried out to provide a growth rate and pattern for any blisters detected. Various objections based on statistical validity can be raised against this approach. However, there is no other cost-effective method.

Four types of HIC defects result from H_2 formation at internal surfaces in C-Mn steel: longitudinal cracking, stepwise cracking or delayed shear, and type I and type II blistering. Pointers as to the early form that blistering may take can be derived from residual aluminum and sulfur levels in the steels. The more complex type I and type II blistering may be differentiated using, for example, shear wave ultrasonics. In any event, it is probably safer to always assume type II blistering and monitor the remaining wall thickness.

Stress-Oriented Hydrogen-Induced Cracking

Stress-oriented hydrogen-induced cracking is a form of classical HIC in which the cracking has a specific orientation with respect to an applied and/or residual stress. Similar to the HIC mechanism, SOHIC tends to stack up in the wall thickness direction, typically in the HAZs of welds where residual stresses are high and at areas of high applied stress or areas of stress concentration. SOHIC is characterized by the stacking of HIC in a direction perpendicular to the axis of principal applied stress and by the microscopic interlinking of the "stacked" hydrogen-induced cracks (Ref 20-23). The interlinking cracks are

both perpendicular to the stress and parallel to the axis defined by the nonmetallic inclusions (Fig. 9).

This type of cracking has often been observed in the base metal adjacent to weld HAZ, which typically are areas of highest residual stress. SOHIC is often important in wet H_2S, but it also has been observed in anhydrous hydrogen fluoride environments (Ref 24). SOHIC can occur in HIC-susceptible materials stressed to as little as 30% of the specified minimum yield strength (SMYS) in the absence of a weld (Ref 21). The implication of SOHIC is that it causes through-wall hydrogen stress cracking of a material that otherwise would be resistant.

Failure control of SOHIC is best achieved by postweld heat treatment to reduce residual stresses, but this method alone cannot guarantee the elimination of SOHIC, because the operating stress may preclude stress concentration reduction.

Hydrogen Stress Cracking

Hydrogen stress cracking was first identified in the production of sour crude oils when high-strength steels used for well-head and down-hole equipment cracked readily after contacting produced water that contained H_2S. Hydrogen stress cracking was not experienced by refineries and petrochemical plants until the introduction of high-pressure processes that required high-strength bolting and other components in gas compressors. With the increased use of submerged arc welding for pressure vessel construction, it was found that weld deposits significantly harder and stronger than the base metal could be produced. This led to transverse cracking in the weld deposit.

Hydrogen stress cracking is typically transgranular and contains sulfide corrosion products. It should not be confused with hydrogen-induced stepwise cracking. The mechanism of hydrogen stress cracking has been the subject of many investigations, most of which attempted to address the cracking seen in high-strength steels instead of the lower-strength steels used in refinery and petrochemical plant equipment. It occurs primarily at ambient temperature.

Mechanisms. In general terms, hydrogen stress cracking occurs in the same corrosive environments that lead to hydrogen embrittlement. As in the case of hydrogen embrittlement and hydrogen blistering, hydrogen stress cracking of steels in refinery and petroleum plants often requires

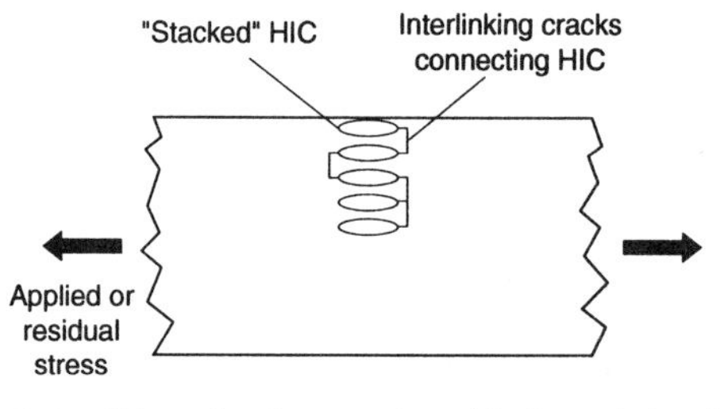

Fig. 9 Schematic of stress-oriented hydrogen-induced cracking. Source: Ref 23

Table 7 Neubauer classification of creep damage

Damage level (Neubauer)	Description	Recommended action
1	Undamaged	No creep damage detected
2	Isolated	Observe
3	Oriented	Observe, fix inspection intervals
4	Microcracked	Limited service until repair
5	Macrocracked	Immediate repair

Source: Ref 30

Table 8 Correlation of damage level and life fraction consumed

Damage level	Consumed life fraction Range X	Remaining life factor $(1/X-1)$ Minimum	Maximum
1	0.00-0.12	7.33	Unknown
2	0.04-0.46	1.17	24.00
3	0.3-0.5	1.0	2.33
4	0.3-0.84	0.19	2.33
5	0.72-1.00	0 = failed	0.39

Source: Ref 31

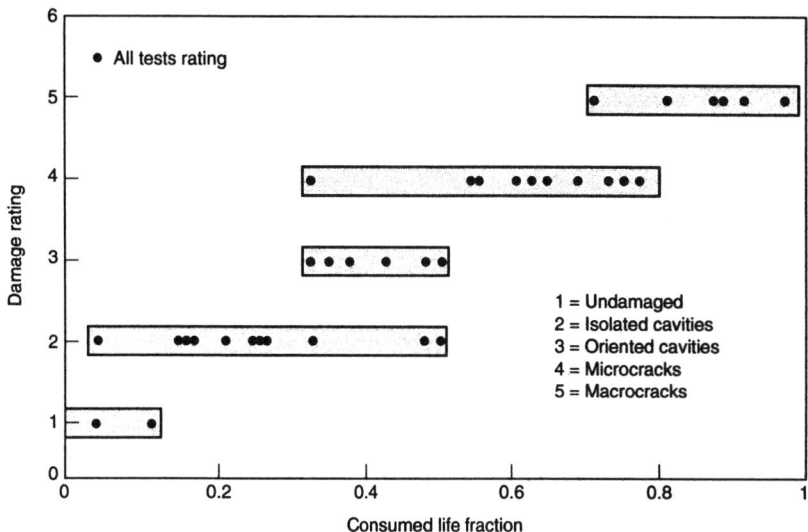

Fig. 10 Relation between Neubauer damage rating (Table 7) and consumed life fraction. Source: Ref 30, 31

the presence of cyanides. Hydrogen sulfide affects corrosion rates and hydrogen absorption and directly affects the maximum allowable hardness of the HAZ or the cracking threshold stress. For example, the allowable maximum hardness value decreases 30 HB, and the allowable threshold stress decreases by 50%, for a tenfold increase in H_2S concentration (Ref 25).

Prevention. The most effective way of preventing hydrogen stress cracking is to ensure proper metallurgical condition of the steel. For example, line pipe failures of this type may be prevented in several ways if the hard spots can be located. Success has been achieved by using internal NDE devices based on the magnetic flux-leakage principle to locate hard spots. Once the hard spots are located, they can be removed, shielded to prevent the cathodic current from reaching them, or tempered to reduce the hardness. Failures tend to occur only in areas with hardnesses exceeding 30 HRC. Hard spots also are now prohibited by API 5L, "Specification for Line Pipe" and must be inspected for during the production of pipe. One of the best inspection techniques is to examine the pipe surface visually for flat spots.

For welds, hardness is limited to 200 HB (Ref 26). Because hard zones can also form in the HAZs of welds and shell plates from hot forming, the same hardness limitation should be applied in these areas. Guidelines for dealing with the hydrogen stress cracking that occurs in refineries and petrochemical plants are given in API 942 (Ref 27) and NACE RP-04-72 (Ref 28).

Postweld heat treatment of fabricated equipment will greatly reduce the occurrence of hydrogen stress cracking. The effect is twofold: First, there is the tempering effect of heating to 620 °C (1150 °F) on any hard microstructure, and second, the residual stresses from welding or form-

ing are reduced. The residual stresses represent a much larger strain on the equipment than internal pressure stresses.

A large number of the ferrous alloys, including the stainless steels, as well as certain nonferrous alloys are susceptible to hydrogen stress cracking. Cracking may be expected to occur with carbon and low-alloy steels when the tensile strength exceeds 620 MPa (90 ksi). Because there is a relationship between hardness and strength in steels, the above strength level approximates the 200 HB hardness limit. For other ferrous and nonferrous alloys used primarily in oil field equipment, limits on hardness and/or heat treatment have been established in NACE MR-01-75 (Ref 29). Although oil field environments can be more severe than those encountered during refining, the recommendations can be used as a general guide for material selection.

With the grades of steels typically used in the oil and gas industry, hydrogen stress cracking is typically observed in the HAZs of welds. Susceptible areas can be identified by high hardness values. However, small localized hard zones can be present in the HAZ and can initiate cracking even though the macrohardness (Rockwell C or Brinell hardness) is low. Therefore, at times microhardness testing may be prudent.

Practical Life Assessment in Creep Regime

Replication Methods. Surface replication is a well-known sample preparation technique that can be used to assess the condition of high-temperature power plant and petrochemical components from creep damage. The usual method of metallographic investigation involves cutting large pieces from components, which thus renders the component unfit for service. In contrast, surface replication allows examination of microstructural damage without cutting sections from

the component (see the article "Replication Microscopy Techniques for NDE" in the *ASM Handbook*, Volume 17, *Nondestructive Evaluation and Quality Control*, 1989).

Replication techniques are sufficiently sophisticated to allow classifications of microstructural damage (such as in Table 7, for example) that can be directly correlated to life fractions (Fig. 10). A distinct correlation exists for these data, such that a minimum and maximum remaining life fraction can be specified such as in Table 8, for example (Ref 30, 31). For assessed consumed life fraction, X, after exposure time, T_{exp}, the remaining life (T_{rem}) is

$$T_{rem} = T_{exp}(1/X - 1)$$

The qualitative-quantitative relation is advantageous because data from surface replication can be predictive in terms of generating a conservative minimum- and a maximum-life estimate. The maximum life is useful in predictive maintenance environment, as it would dictate the planning of future repairs or replacement.

Parametric Methods. Extrapolation to service stress and temperature are performed using the appropriate extrapolation rule of the general form

$$P(\sigma) = C(T, T_{rup})$$

where the stress (σ), temperature (T), and rupture life (T_{rup}) are related by various functional form P and C, such as the frequently used Larson-Miller parametric relation (Ref 32) where

$$P = (T_d + 460)(C + \log L_d) \times 10^{-3}$$

where C is a constant and P is the Larson-Miller parameter, T_d is the design steel temperature, and L_d is the design life. By adopting the preceding approaches, either singularly or combined (depending on the availability of data), useful practical assessments of the remaining life of refinery equipment in the creep regime can be achieved.

REFERENCES

1. D.H. Stoneburg et al., Ed., *Practical Considerations for Engineering Evaluation and Analysis Applied to Process Equipment*; conference proceedings, Improving Reliability in Petroleum Refineries and Chemical and Natural Gas Plants, 1993
2. National Board Inspection Code (ANSI NB-23), The National Board of Boiler and Pressure Vessel Inspectors (U.S.)
3. Pressure Vessel Inspection Code (API 510: Maintenance Inspection, Rating, Repair and Alteration), American Petroleum Institute
4. M. Scasso, "Comparison Among the Present Regulations Concerning the Remanent Life Assessment of Components Designed for High Temperature Service," Italian Institute of Welding, 1993
5. "Inspection of Pressure Vessels: Assessment of Degree of Creep," Finnish Pressure Vessel Commission, SFS 3280:E, 1986
6. E.L. Robinson, Effect of Temperature Variation on the Creep Strength of Steels, *Trans. ASME*, Vol 160, 1938, p 253-259
7. Y. Lieberman, Relaxation, Tensile Strength and Failure of E1 512 and Kh1 F-L Steels, *Metalloved Term Obrabodke Metal*, Vol 4, 1962, p 6-13
8. H.R. Voorhees and F.W. Freeman, "Notch Sensitivity of Aircraft Structural and Engine Alloys," Wright Air Development Center Technical Report, Part II, Jan 1959, p 23
9. M.M. Abo El Ata and I. Finnie, "A Study of Creep Damage Rules," ASME Paper 71-WA/Met-1, American Society of Mechanical Engineers, Dec 1971
10. R.M. Goldhoff and D.A. Woodford, *The Evaluation of Creep Damage in a CrMoV Steel*, STP 515, American Society for Testing and Materials, 1982, p 89
11. R.M. Goldhoff, Stress Concentration and Size Effects in a CrMoV Steel at Elevated Temperatures, *Joint International Conference on Creep*, Institute of Mechanical Engineers, London, 1963
12. R. Viswanathan, *Damage Mechanisms and Life Assessment of High Temperature Components*, ASM International, 1989, p 73, 87-103
13. A. Saxena, J. Han, and K. Banergi, *Creep Crack Growth in Boiler and Steam Pipe Steels*, Report CS-5583, Electric Power Research Institute, Jan 1988
14. A. Saxena, *Creep Crack Growth in CrMoV Rotor Steels*, Report RP2481-5, Electric Power Research Institute
15. P.K. Liaw and A. Saxena, *Remaining Life Estimation of Boiler Pressure Parts—Crack Growth Studies*, Report CS-4688, Electric Power Research Institute, July 1986
16. R. Viswanathan and S. Gehl, Creep Life Assessment Techniques for Piping, *Mechanical Behavior of Materials, VI*, Vol 2, p 117-122
17. G.J. Biefer, The Stepwise Cracking of Line-Pipe Steels in Sour Environments, *Mater. Perform.*, Vol 21, 1982, p 19
18. A. Ikeda et al., "Influence of Environmental Conditions and Metallurgical Factors on Hydrogen Induced Cracking of Line Pipe Steel," Paper 80, Corrosion '80, Chicago, 1980
19. P.F. Timmins, Assessing Hydrogen Damage in Sour-Service Lines and Vessels Is Key to Plant Inspection, *Oil and Gas Journal*, 5 Nov 1984
20. R.T. Hill and M. Iino, Correlation between Hydrogen-Induced Blister Cracking of Stressed and Unstressed Specimens, *Current Solutions to Hydrogen Problems in Steels*, American Society for Metals, 1982, p 196-199
21. R.D. Kane et al., "Review of Hydrogen Induced Cracking of Steels in Wet H_2S Refinery Service," paper presented at the International Conference of Interaction of Steels with Hydrogen in Petroleum Industry Pressure Vessel Service (Paris, France, 28-30 March, 1989), Materials Properties Council, Inc.
22. R.D. Merrick and M.L. Bullen, "Prevention of Cracking in Wet H_2S Environments," Paper 269, Corrosion '89, New Orleans, 1989
23. J.P. Ribble et al., The Effect of Metallurgical and Environmental Variables on Hydrogen-Induced Cracking of Steels, *1990 Mechanical Working and Steel Processing Proceedings*, Vol XXVIII, p 499-505
24. C.C. Seastrom, *1990 Mechanical Working and Steel Processing Proceedings*, Vol XXVIII, p 507-515
25. T.G. Gooch, Hardness and Stress Corrosion Cracking of Ferritic Steel, *Weld. Inst. Res. Bull.*, Vol 23 (No. 8), 1982, p 241-246
26. D.J. Kotecki and D.G. Howden, Wet Sulfide Cracking of Submerged Arc Weldments, *Proc. API*, Vol 52 (III), 1972, p 631-653
27. "Controlling Weld Hardness of Carbon Steel Refinery Equipment to Prevent Environmental Cracking," Recommended Practice 942, 2nd ed., American Petroleum Institute, 1983
28. "Methods and Controls to Prevent In-Service Cracking of Carbon Steel (P-1) Welds in Corrosive Petroleum Refinery Environments," NACE RP-04-72 (1976 Revision), National Association of Corrosion Engineers, 1976
29. "Sulfide Stress Cracking Resistant Metallic Materials for Oil Field Equipment," NACE MR-01-75 (1980 Revision), National Association of Corrosion Engineers, 1980
30. B. Neubauer, "Creep and Fracture of Engineering Materials and Structures," Pineridge, U.K., 1981, p 617
31. J.M. Brear et al, "Possibilistic and Probabilistic Assessment of Creep Cavitation," ICM 6 Pergamon, 1991
32. F.R. Larson and J. Miller, *Trans. ASME*, Vol 74, 1952, p 765

Stress-Corrosion Cracking and Hydrogen Embrittlement

Gerhardus H. Koch, CC Technologies, Inc.

STRESS-CORROSION CRACKING (SCC) is a cracking phenomenon that occurs in susceptible alloys, and is caused by the conjoint action of a tensile stress and the presence of a specific corrosive environment. For SCC to occur on an engineering structure, three conditions must be met simultaneously, namely, a specific crack-promoting environment must be present, the metallurgy of the material must be susceptible to SCC, and the tensile stresses must be above some threshold value. This cracking phenomenon is of particular interest to users of potentially susceptible structural alloys, because SCC occurs under service conditions, which can result, often without any prior warning, in catastrophic failure. Many different mechanisms for SCC have been proposed, but in general these mechanisms can be divided into two general groups, namely the anodic dissolution mechanisms and cathodic mechanisms. The parameters that control SCC can be divided into materials, environmental, and mechanical parameters. In this article, an overview of the SCC behavior of different engineering materials is presented with emphasis on carbon and low-alloy steels, high-strength steels, stainless steels, nickel-base alloys, aluminum alloys, and titanium alloys. Although these materials do not encompass all materials susceptible to SCC, they comprise the most commonly used materials in a wide range of industries.

Key Factors of SCC

Materials Factors. The alloy composition and microstructure have a great effect on the susceptibility of a material to SCC in a particular environment. The bulk alloy composition may affect the formation and stability of a protective film on the surface. The alloy composition includes the nominal composition, the presence of constituents, and the presence and composition of impurities or trace elements. The metallurgical condition, which affects the susceptibility to SCC, includes the strength level, the presence of phases in the matrix and at the grain boundaries, the composition of the phases, the grain size and orientation, grain-boundary segregation, and residual stresses.

An example of strong influence of alloy composition and microstructure on the susceptibility to SCC is given by austenitic stainless steels, where chromium and molybdenum promote the formation of passive films on the surface. Trace elements such as carbon at concentrations greater than 0.03 wt%, may cause sensitization by forming chromium carbides at the grain boundaries and depleting zones around the carbides of chromium, thereby rendering the steel susceptible to intergranular SCC (IGSCC). Austenitic stainless steels will fail transgranularly in high-temperature chloride solutions.

Similarly, the susceptibility of aluminum alloys to SCC strongly depends on the microstructure, which can be modified by heat treatment. The 7000 series aluminum alloys are precipitation-hardening alloys, and the peak-aged microstructure (T6) is the most susceptible to SCC. Overaging to the T76 or T73 condition usually reduces or eliminates the susceptibility to cracking. Peak aging of this alloy results in a fine distribution of coherent precipitates, which give strength to the alloy. However, the heat treatment also results in the formation of large incoherent precipitates at the grain boundaries and the depletion of solute in the region adjacent to the grain boundaries.

Environmental Factors. Stress-corrosion cracking of susceptible alloys is environment specific. The environmental effects can simply be summarized by listing the alloy/environment combinations in which SCC has been observed. Table 1 (Ref 1) shows a partial list of alloy/environment combinations, which has, in recent years, increased in number. For example, transgranular SCC of copper and SCC of stainless steels and nickel-base alloys in high-purity water can be added to the list. Although a list such as shown in Table 1 can be used as a general guideline for materials selection, it should be realized that SCC depends on a great many factors other than the bulk environment. Environments that cause SCC are usually but not necessarily aqueous, and specific environmental parameters must be in specific ranges for cracking to occur. These include, but are not limited to:

- Temperature
- pH
- Electrochemical potential
- Solute species
- Solute concentration
- Oxygen concentration

Changing any of these environmental parameters may significantly affect the crack nucleation process or the rate of crack propagation. Although the parameters listed above are important in controlling the rate of SCC, conditions inside a propagating crack and at the crack tip, which actually control the crack propagation process, are often quite different from the so-called bulk environmental parameters. The pH inside cracks often differs from that in the bulk environment. In low-alloy steels containing about 1 wt% Cr, dissolution and hydrolysis of chromium can result in a lowering of the pH to values near 4 (Ref 2).

In the case of stainless steels, the pH value in cracks can range from 0 to 3, with the lowest pH values associated with concentrated salt solutions containing chromium and ferrous ions. The pH inside cracks of aluminum and aluminum alloys

Table 1 Alloy/environment systems exhibiting SCC

Alloy	Environment
Carbon steel	Hot nitrate, hydroxide, and carbonate/bicarbonate solutions
High-strength steels	Aqueous electrolytes, particularly when containing H_2S
Austenitic stainless steels	Hot, concentrated chloride solutions; chloride-contaminated steam
High-nickel alloys	High-purity steam
α-brass	Ammoniacal solutions
Aluminum alloys	Aqueous Cl^-, Br^-, and I^- solutions
Titanium alloys	Aqueous Cl^-, Br^-, and I^- solutions; organic liquids; N_2O_4
Magnesium alloys	Aqueous Cl^- solutions
Zirconium alloys	Aqueous Cl^- solutions; organic liquids; I_2 at 350 °C (660 °F)

Source: Ref 1

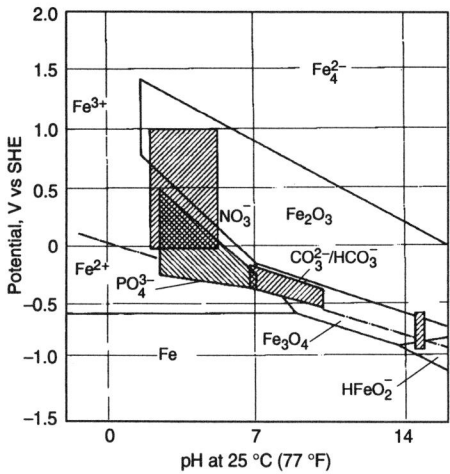

Fig. 1 Relationship between pH-potential conditions for SCC susceptibility of carbon steel in various environments and the stability regions for solid and dissolved species on the electrochemical equilibrium diagram. Ref 1

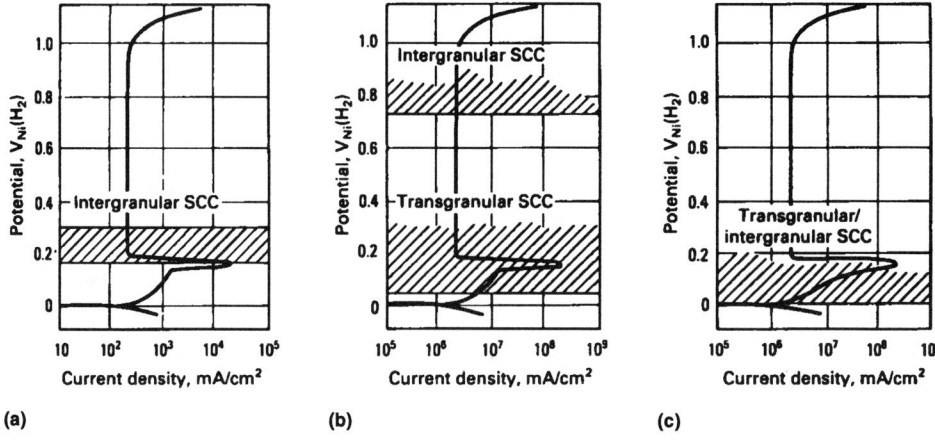

Fig. 2 Potentiodynamic polarization curves and potential values at which intergranular and transgranular SCC occurs in a 10% NaOH solution at 288 °C (550 °F). (a) Alloy 600, (b) Alloy 800, (c) Type 304 stainless steel. Source: Ref 1

is generally in the range of 3 to 4, while the pH inside propagating cracks in titanium alloys can be as low as 1 (Ref 2).

Stress-corrosion cracking consists of a crack nucleation and propagation phase. Very little is known about the conditions that control the nucleation of a crack, other than that the thermodynamic and kinetic conditions must be right for the crack to nucleate. For example, for anodically assisted SCC, metal dissolution and subsequent formation of a protective oxide film must be thermodynamically possible. The thermodynamic requirement of simultaneous dissolution and film formation has led to the identification of critical potentials at which SCC can occur. The thermodynamic conditions at which dissolution and film forming occurs is described by potential-pH (Pourbaix) diagrams. For example, the Pourbaix diagram in Fig. 1 describes the conditions at which metal dissolution and film formation on carbon steel can occur in different environments such as phosphate, nitrate, and carbonate/bicarbonate solutions. The effects of many of the external parameters listed above, such as pH, temperature, potential, and solute and oxygen concentration can have a great effect on thermodynamic stability and thus on the susceptibility to SCC. The diagram indicates that severe susceptibility to SCC is encountered when a protective film such as carbonate, phosphate, or magnetite is thermodynamically stable.

In addition to the thermodynamic stability requirements for crack nucleation and propagation, kinetic requirements also need to be met. As in the thermodynamic requirements for SCC, environmental parameters such as potential, pH, solute and oxygen concentration, temperature, and crack-tip chemistry have a strong effect on the crack nucleation and crack growth kinetics. Figure 2 shows examples of potentiodynamic polarization curves for alloy 600, alloy 800, and type 304 stainless steel in a 10% NaOH solution at 288 °C (550 °F), indicating the various potential re-

gions for susceptibility to SCC. The figure shows that for all three alloys the active-passive transition region presents the critical potential range for SCC to occur. However, the critical potentials, as well as the mode of cracking, are different for each alloy.

Mechanical Factors. Threshold stresses and stress-intensity factors, the presence of a stress-independent crack-growth regime, and the dependence of cracking to strain rate are important features in determining the susceptibility of alloys to SCC. The threshold stress is typically the stress value obtained from constant-load testing below which SCC does not occur and can serve as a simple measure for susceptibility of a material to SCC in a certain environment.

The stress-intensity factor (K) is a parameter that describes the relationship between the applied stress and crack length for specific specimen geometries. Figure 3 shows the stress-intensity factor K as a function of the crack propagation rate da/dt. The threshold is defined in this figure by the minimum detectable crack growth rate. The threshold stress intensity is generally associated with the development of a plastic zone at the crack tip. Stage I crack growth shows a rapid increase in crack growth rate, while in Stage II the crack growth rate is independent of the stress intensity.

The stress-corrosion crack propagation is usually studied with linear elastic fracture mechanics (LEFM), which assumes little plasticity at the tip of the propagating crack, such that the stress state is triaxial or plane strain. When the plastic zone size exceeds a certain value, either by increased stress, by propagation in a ductile material, or by crack propagation in a thin member, the stress state becomes biaxial or plane stress, and LEFM is not applicable. Then the more fundamental parameters, the energy release rate or J-integral, can be applied to describe the propagating stress-corrosion crack (Ref 3).

The slow-strain-rate technique provides an excellent way to determine the susceptibility of an alloy to SCC (Ref 4, 5). However, the strain-rate behavior strongly depends on the alloy/environ-

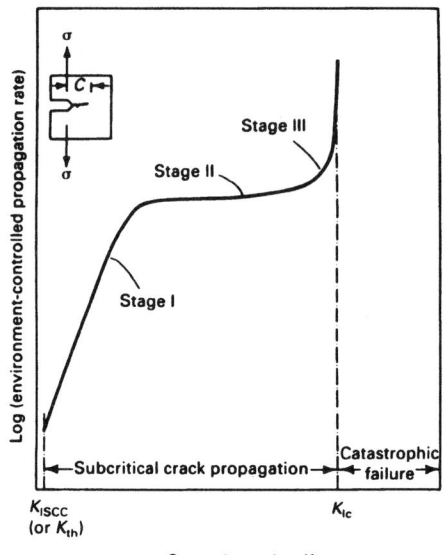

Fig. 3 Schematic diagram of stress-corrosion crack velocity as a function of stress-intensity factor K

ment combination. For example, for most materials the critical strain rate, at which the maximum susceptibility is obtained, is 10^{-6}/s. This critical strain rate points to a cracking mechanism whereby the rate of anodic dissolution is equal to the rate of protective film formation. If a higher strain rate is applied, the mechanical fracture will be more rapid than the rate of anodic dissolution. On the other hand, when a lower strain rate is applied, anodic dissolution will continue to blunt the crack, and SCC cannot occur. When other SCC mechanisms are predominant, the critical strain rate may be at a higher value, as is often the case with internal hydrogen embrittlement, or there may be no critical value, which occurs when the susceptibility decreases with decreasing strain rate. This has been observed in cases where the mechanism of SCC is thought to be hydrogen embrittlement.

Stress-Corrosion Cracking Mechanisms

It is unlikely that a single mechanism for SCC exists. The specific mechanism that is operative depends on the type of material, environment, and loading conditions. Although many models have been proposed, they can be divided into two main groups, namely those based on anodic dissolution and those that involve mechanical fracture.

Anodic Dissolution Models. The SCC mechanisms, which fall under the anodic dissolution model include the active path intergranular SCC and the film-rupture model. The active-path intergranular SCC results from a difference in alloy composition at the grain boundary, and the crack velocity can be described by Faraday's equation:

$$da/dt = i_a M/zF\rho$$

where i_a is the anodic current density, M is the atomic weight, z is the valence, F is Faraday's constant, and ρ is the material density. This equation assumes that the crack tip remains bare, and that the crack walls remain relatively inactive. It has been shown (Ref 6) that the Faradaic relationship is applicable to various material/environment combinations (see Fig. 4).

The film-rupture SCC model assumes that the stress at the crack tip acts to open the crack tip and rupture the protective surface film. The bare metal then dissolves rapidly, resulting in crack growth. There are essentially two schools of thought, one assuming that the crack tip remains bare because the rate of dissolution is higher than the rate of repassivation (Ref 7, 8). The second assumes that the crack tip repassivates completely and is periodically fractured by the emergence of slip steps (see Fig. 5) (Ref 9-12). Although considerable evidence has been found for both mechanisms, crack arrest markings or striations have been found on both intergranular and transgranular fracture surfaces, supporting the notion that crack propagation in both cases is discontinuous. However, because transgranular SCC fracture surfaces have very sharp cleavage markings, which match precisely on opposite fracture surfaces, a film rupture and dissolution model is not a likely mechanism for transgranular SCC (Ref 13).

Mechanical Fracture Models. There are several proposed SCC models that fall under this category. These include the corrosion tunnel model, the adsorption-enhanced plasticity model, the tarnish rupture model, the film-induced cleavage model, the adsorption-induced brittle fracture model, and the hydrogen embrittlement model.

The corrosion tunnel model assumes that tunnels of corrosion form at the crack tip until the remaining ligaments fracture in a ductile manner. The crack would thus propagate by alternating corrosion and ductile rupture, which would result in a grooved fracture surface with evidence of microvoid coalescence on the peaks, as is illustrated in Fig. 6(a). Silcock and Swann (Ref 14)

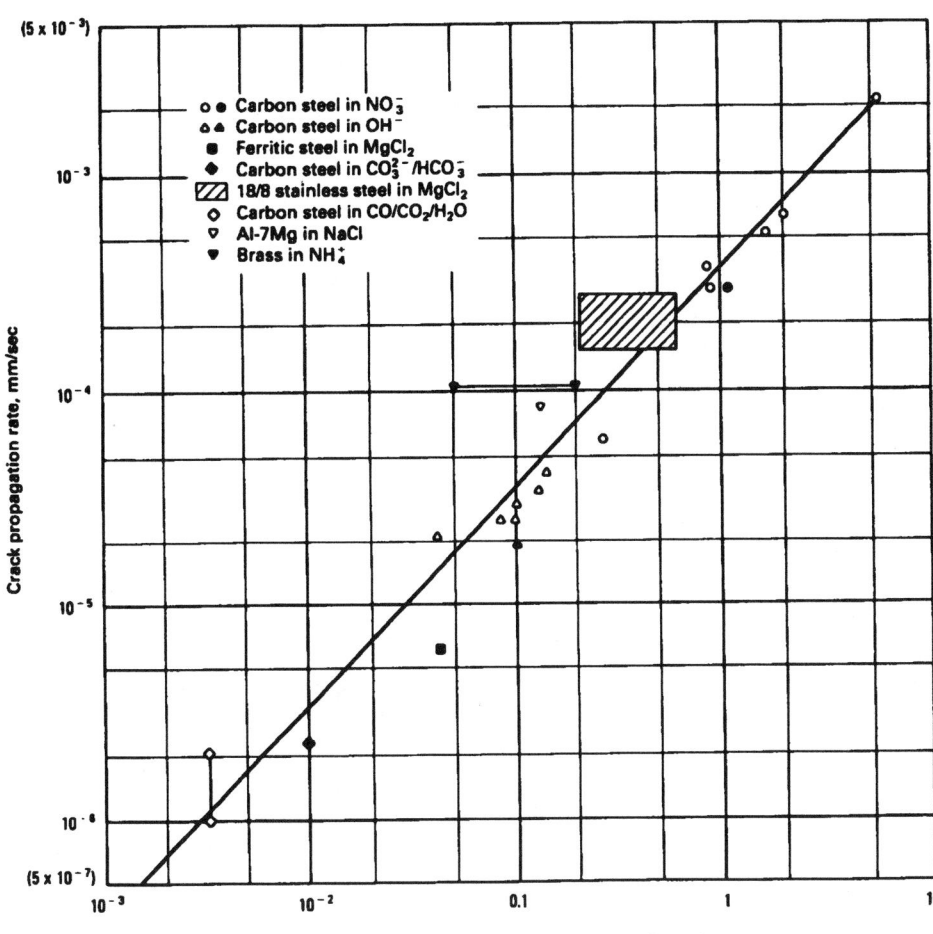

Fig. 4 Relationship between the average crack-propagation rate and the oxidation (i.e., dissolution and oxide growth) kinetics on a straining surface for several ductile alloy/aqueous environment systems (Ref 6)

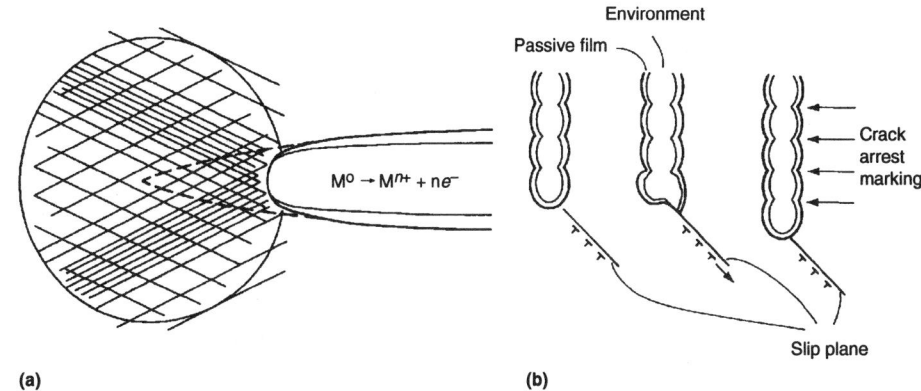

Fig. 5 Schematic representation of crack propagation by the film-rupture model. (a) Crack tip stays bare as a result of continuous deformation (Ref 7, 8). (b) Crack tip passivates and is ruptured repeatedly (Ref 9-12).

pointed out that this mechanism is not consistent with fractographic features on stainless steels and suggested that the stress at the crack tip changes the morphology of the corrosion damage, such that the tunnels become flat slots (Fig. 6b). It was concluded that transgranular SCC of austenitic stainless steels could be explained in terms of this model and the formation and mechanical separation of corrosion slots (Ref 14).

The adsorption-enhanced plasticity model for SCC assumes that certain species from the environment adsorb to the crack tip by chemisorption (Ref 15-17). Fractographic studies were used to demonstrate that transgranular cleavage occurs by slip at the crack tip in conjunction with the formation of microvoids ahead of the crack. Further, the chemisorption supposedly facilitates the nucleation of dislocations at the crack tip, pro-

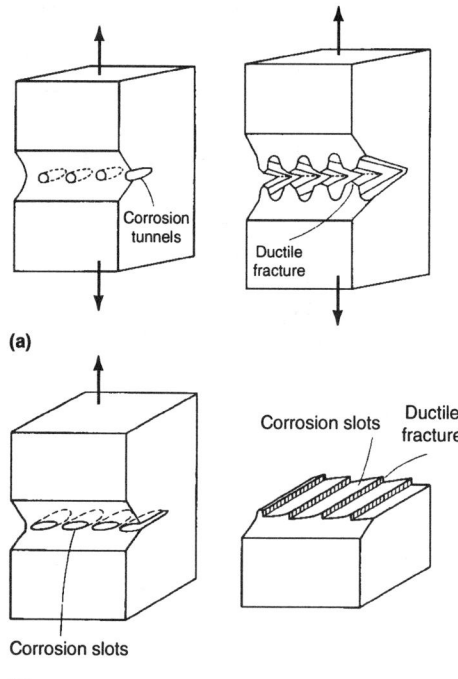

Fig. 6 Corrosion tunnel models. (a) Schematic of tunnel model showing the nucleation of a crack by the formation of corrosion tunnels at slip steps and ductile deformation and fracture of the remaining ligaments. (b) Schematic diagram of the tunnel mechanism of SCC and flat-slot formation as proposed in Ref 14

Table 2 Effect of alloying within normal limits on the SCC resistance of martensitic low-alloy steels to chloride

Element	AISI 4120 (Yield strength = 1034 MPa, or 150 ksi)	AISI 4340 (Yield strength = 1172-1448 MPa, or 170-210 ksi)
Carbon	Decrease	Decrease
Manganese	No effect	Decrease
Nickel	Increase	No effect
Chromium	Increase	No effect
Molybdenum	Increase	No effect
Vanadium	Increase	...
Niobium	Increase	...
Titanium	Increase	...
Zirconium	Increase	...
Boron	No effect	...
Copper	No effect	...
Silicon	No effect	...
Sulfur	Beneficial	No effect
Phosphorus	Decrease	No effect
Oxygen	Decrease	...
Nitrogen	Decrease	No effect

Source: Ref 24

moting the shear process responsible for the cleavagelike fracture. Although fractography has not offered any direct evidence of the validity of this model, similarities between the different types of failure, SCC, liquid metal embrittlement (LME), and hydrogen embrittlement (HE), may be explained by the adsorption-enhanced plasticity model.

The tarnish rupture model explains both transgranular and intergranular SCC by the formation of a brittle surface film that fractures under applied stress (Ref 18, 19). When the film fractures, bare metal is exposed to and reacts rapidly with the environment to form a new surface film. Thus, by this mechanism, the crack propagates by alternating film growth and fracture. This model predicts discontinuous crack propagation, which would result in crack-arrest markings on the fracture surface and also in penetration of the film ahead of the crack tip.

The film-induced cleavage model proposes that dealloying and/or vacancy injection could induce brittle fracture (Ref 20). Sieradzki and Newman (Ref 21) developed this concept into a model where a surface film could induce cleavage fracture by assuming that a thin film forms on the surface. According to this model, a brittle crack nucleates in this film and propagates through the film across the film-substrate interface into the ductile metal substrate. Once the crack enters the ductile metal, it will continue to propagate in a brittle manner for some time and will eventually

blunt and stop, after which the process of film formation and brittle fracture repeats itself.

The adsorption-induced brittle fracture or stress-sorption model is based on the assumption that adsorption of species from the environment can lower the interatomic bond strength and the stress required for cleavage fracture (Ref 21, 22). Similar mechanisms have been proposed for liquid metal embrittlement and hydrogen embrittlement (Ref 23). According to this model the crack propagates continuously, with the crack propagation rate controlled by the rate at which the embrittling species arrive at the crack tip. The model does not explain how a sharp crack tip can be maintained in an otherwise ductile material.

The hydrogen embrittlement model has been proposed as an operating mechanism for several material/environment systems. Hydrogen can enter an alloy lattice from both gaseous and aqueous phases. In an aqueous phase where corrosion reactions take place, both anodic and corresponding cathodic reactions occur. In many cases, hydrogen ion reduction is the cathodic reaction where hydrogen atoms are formed on the surface. While some of the hydrogen atoms combine to form hydrogen gas, other hydrogen atoms remain adsorbed to the surface. Because of the high partial hydrogen pressure or fugacity (thousands of psi), there exists a driving force for the hydrogen atoms to be absorbed into the lattice.

Hydrogen-induced cracking has been proposed as the SCC mechanism for carbon and high-strength ferritic steels, nickel-base alloys, titanium alloys, and aluminum alloys.

SCC of Carbon and Low-Alloy Steels

Stress-corrosion cracking of carbon and low-alloy steels is a significant problem in several industries, including power generation, oil and gas production, gas transmission, oil refining and processing, and pulp and paper processing.

Materials Factors. Alloying has a significant effect on the susceptibility of carbon steel to SCC

in various environments. Table 2 gives an overview of the effect of alloying elements on the susceptibility to SCC of two low-alloy steels (Ref 24). It should be noted that the elements listed in the table not only affect SCC behavior, but also affect other properties such as strength and welding characteristics. The carbon content generally has a significant effect on the susceptibility to SCC, with the SCC resistance decreasing as the carbon content increases (Ref 25-28). However, the SCC resistance of alloyed steel appears to be less dependent on the carbon content (Ref 29). Manganese has an adverse effect on the SCC resistance of medium-carbon steels in chloride environments, as well as in hydrogen sulfide and nitrate environments (Ref 25, 30, 31). High sulfur and phosphorus alloy concentrations have reduced the SCC resistance of carbon steel in hydrogen sulfide, as well as in nitrate solutions (Ref 32, 33). Nitrogen levels below 0.01% increase the susceptibility to SCC in nitrate solutions, but higher levels appear to be beneficial (Ref 34). The other alloying elements listed in Table 2 have generally a beneficial effect on SCC resistance.

Low-carbon steels can be divided into ferritic-pearlitic and quenched-and-tempered types. Quenched-and-tempered heat treatments are applied to increase the yield strength above 621 MPa (90 ksi). Fine grain size generally increases the resistance of quenched-and-tempered steels to SCC in a number of environments (Ref 35, 36). Further, the presence of twinned martensite in these steels has an adverse effect on SCC resistance, as do the presence of ε carbides and high dislocation densities (Ref 37-40). Martensitic microstructures generally provide the highest SCC resistance, particularly if fine spheroidized carbides are uniformly dispersed in the ferrite. A bainite structure also has a beneficial effect on SCC resistance (Ref 41), while untempered martensite is considered detrimental (Ref 42).

Grain-boundary segregations, which contain elements such as phosphorus and sulfur, have an adverse effect on the intergranular SCC of carbon steel in several environments (Ref 43). Other elements that promote intergranular SCC include arsenic, tin, and antimony (Ref 44). Finally, intermetallic inclusions, such as sulfides, have been shown to act as nucleation sites for cracking and also to accelerate crack propagation (Ref 45, 46).

Mechanical Factors. The strength level has a significant effect on the resistance of carbon steels to SCC, with the resistance decreasing as the strength is increased (Ref 47). Steels with yield strengths less than 1241 MPa (180 ksi) are often considered resistant to aqueous chloride SCC (Ref 48). However, because SCC has occasionally been observed in lower-strength steels, the assumption has been made that steels should have yield strengths below 689 MPa (100 ksi) to be resistant to SCC (Ref 49).

Environmental Factors. There are several environments that can induce SCC in carbon and low-alloy steels. Some of these environments, which are specific to certain industrial applications, will be discussed in the following paragraphs.

Aqueous Chlorides. Stress-corrosion cracking is common to various industries in which aqueous chlorides are encountered, including marine, aerospace, power generation, oil and gas production and refinement, and construction. Although SCC is generally associated with the presence of an aqueous phase, it has been reported in vapor, with crack growth rates increasing with an increase in relative humidity (Ref 50).

Although chloride concentration has little effect on the K_{Iscc} (Ref 51), it does increase crack propagation rates, with the effect more pronounced at the lower chloride concentrations (Ref 52, 53). Lowering the pH and increasing the temperature were found to have significant detrimental effects on the SCC resistance (Ref 54-56).

Hydrogen Sulfide. Stress-corrosion cracking in environments that contain hydrogen sulfide commonly occurs in the production, transmission, and refining of oil and gas, and failures have been reported on gas and oil well tubulars, wellhead equipment, pipelines, process piping, and pressure vessels. It is generally assumed that water must also be present, and that the susceptibility to cracking increases with increasing hydrogen sulfide concentration.

The pH level in solutions that contain hydrogen sulfide has been found to have a significant effect on the susceptibility of steels to SCC. Although SCC can occur at pH levels up to about 9, and up to about 12 for steels with high strengths, reduced pH values will result in a decrease in SCC threshold levels (Ref 57-60). Increased temperatures have been found to decrease the susceptibility to SCC in hydrogen sulfide, with the maximum susceptibility to SCC near room temperature (Ref 61, 62).

Stress-corrosion cracking in hydrogen sulfide, also termed hydrogen sulfide cracking, has been attributed to a hydrogen embrittlement mechanism.

Sulfuric Acid. Stress-corrosion cracking of steel in sulfuric acid is generally associated with high strength levels and is suggested to be analogous to SCC in aqueous chloride solutions (Ref 63).

Hydrogen Gas. Cracking of carbon steels has been reported in pressure vessels containing high-pressure hydrogen gas and has been associated with weld and nozzle forgings of medium-strength steels (Ref 64, 65). It was demonstrated that the susceptibility to cracking in hydrogen gas increases with increasing strength of the steel (Ref 66, 67). Although cracking has been observed over a wide range of temperatures, the maximum susceptibility occurs at or near room temperature (Ref 68). The presence of small amounts of oxygen or sulfur dioxide was shown to inhibit cracking in gaseous hydrogen (Ref 69, 70).

Caustic Solutions. Stress-corrosion cracking of carbon steel in sodium hydroxide or caustic solution has been well documented and has been most commonly associated with steam boilers. Caustic cracking has also been identified as the cause of cracking of continuous digesters used in the pulp and paper industry (Ref 71). Cracking generally occurs in highly stressed areas where the caustic can concentrate. Plastic deformation is considered to be a prerequisite for caustic cracking (Ref 72, 73). Stress-corrosion cracking occurs over a wide range of hydroxide concentrations, between 5 and 70%, and in a temperature range of 100 to 349 °C (212 to 660 °F) (Ref 72, 74, 75).

Caustic cracking has been shown to occur in a very narrow range of electrochemical potentials near the active-passive transition on polarization diagrams (Ref 76, 77). This potential range is though to be associated with the presence of a protective magnetite film. Small additions of oxygen and/or chlorides promote SCC, while addition of larger amounts promote passivation by having a strong oxidizing effect and hence inhibit cracking (Ref 78, 79).

Ammonia. Carbon steels have been used widely for transportation and storage of ammonia. Recent surveys have shown that a large number of leaks could be attributed to SCC (Ref 80). Stress-corrosion cracking in this environment has been found in cold-formed and welded steel (Ref 80), and the presence of oxygen and carbon dioxide is required for cracking to occur (Ref 81).

Carbonate/Bicarbonate Solutions. Stress-corrosion cracking of natural gas transmission pipelines has been attributed to aqueous solutions of carbonate/bicarbonate (CO_3-HCO_3) and occurs over a very limited potential range from about –670 to –770 mV versus Cu/CuSO$_4$ at 75 °C (Ref 82, 83). This potential range is associated with the active-passive transition in a potentiodynamic polarization curve. Stress-corrosion cracking also occurs over a limited pH range around 9. The mechanism of this specific form of SCC is anodic dissolution-film rupture. Plastic deformation at the crack tip ruptures the protective passive film and exposes fresh metal which undergoes anodic dissolution. Depending on the dissolution rate and the crack extension rate, the crack tip may repassivate or continue to propagate. Passivation of the crack walls maintains a sharp aspect ration

of the crack, which is necessary to concentrate plasticity at the crack tip.

Recently, transgranular SCC on pipelines was discovered, which was correlated with low-pH (pH <8) dilute CO_2-containing environments (Ref 84, 85). This form of SCC has been termed low-pH or nonclassical SCC. In contrast with the high-pH SCC, low-pH SCC does not appear to be sensitive to potential, and in fact it was demonstrated that the severity of low-pH SCC does not increase with increasing negative potential. Comparing low-pH with high-pH SCC also has demonstrated that while the high-pH cracking occurs under constant load or constant displacement testing, the low-pH cracking does not occur under these loading conditions. The latter exhibits crack growth only under cyclic loading conditions (Ref 86). The most plausible mechanism for crack propagation in the low-pH environment is a hydrogen-related mechanism, with the most likely source of hydrogen being carbonic acid, which is formed by the dissolution of CO_2 in the ground water.

SCC of High-Strength Steels

Steels with yield strengths greater than 1240 MPa (180 ksi) fall under the general group of high-strength steels (Ref 54). The steels in this group include quenched-and-tempered low-alloy steels and maraging steels. These steels are all highly susceptible to SCC, even in benign environments, such as water vapor, which would only cause minimal corrosion. The quenched-and-tempered steels have a tempered martensitic microstructure, where carbon is an important alloying element. These high-strength steels come in a low-alloy and a high-alloy variety. Maraging steels obtain their strength from martensite formation, followed by the growth of precipitates, such as Ni$_3$Mo and Ni$_3$Ti, during aging, and unlike in the case of the quenched-and-tempered steels, carbon is not an alloying element but rather an impurity, which reduces the strength and fracture toughness by forming TiC (Ref 87).

Material Factors. Although alloying elements have a significant effect on the resistance of a steel to SCC, these effects are difficult to assess, because they also may influence the yield strength, which in turn affects the susceptibility to SCC. For example, titanium is known to increase the resistance of maraging steel to SCC, but also increases the yield strength contributing to a decrease in resistance to SCC (Ref 88). The effect of carbon is detrimental to the SCC resistance (K_{Iscc}) of AISI 4340 steel up to a concentration of 0.4% (Ref 89, 90). Likewise, manganese addition in amounts up to 3% lowers the resistance to SCC. Other elements investigated include phosphorus (≤0.05%), sulfur (≤0.03%), chromium (≤2.5%), molybdenum (≤1.2%), cobalt (≤3.5%), and silicon, none of which were found to affect the resistance to SCC. Finally, nickel has been shown to have little effect on SCC when present in concentrations of up to 9%.

The effect of silicon addition on the SCC susceptibility has been researched in detail (Ref 91,

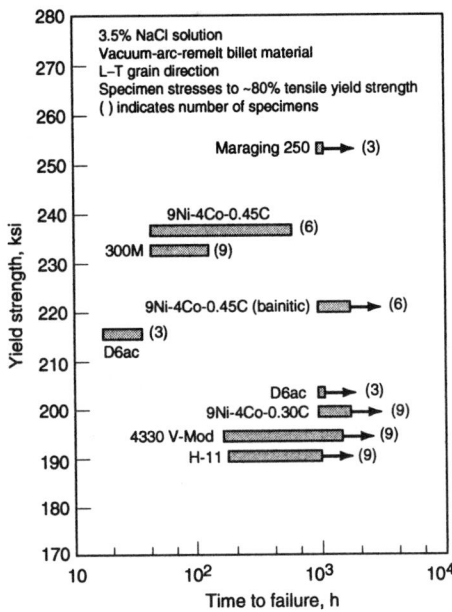

Fig. 7 Alternate-immersion SCC data from smooth bend specimens (Ref 94)

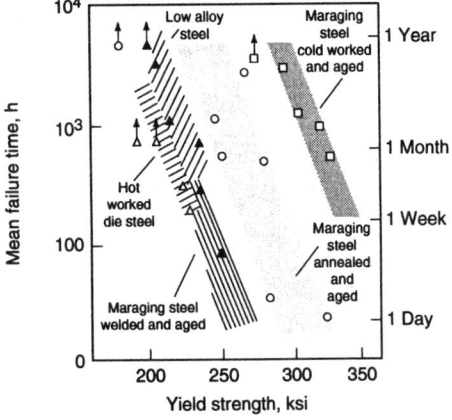

Fig. 8 Results of bent-beam tests in aerated distilled water. Specimens were exposed at 75% of the yield stress (Ref 94).

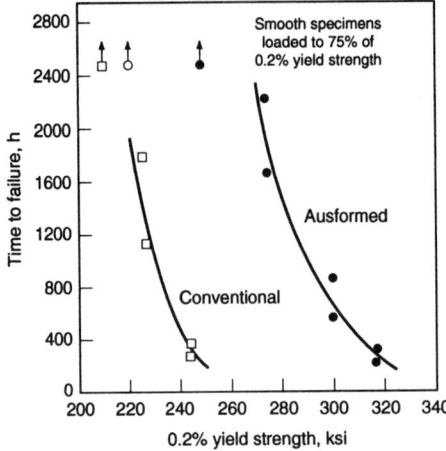

Fig. 9 Effect of ausforming on SCC of D6AC steel in distilled water (Ref 97)

92). Although small amounts of silicon do not affect SCC, concentrations above 1.5% have been found to decrease crack growth and increase the time to failure. On the other hand, the combined effect of silicon and manganese was found to lower the K_{Iscc}, by promoting phosphorus segregation during austenitization (Ref 92).

Maraging and quenched-and-tempered steels exhibit different susceptibilities to SCC, with the former having high resistance and the latter low resistance to SCC (Ref 93). Figure 7 shows the time-to-failure data for bend specimens subjected to alternate immersion in a 3.5% NaCl solution (Ref 94). While the time-to-failure is only hours or days for the D6-Air-Cooled (D6AC) steel at a yield strength of 100 MPa (220 ksi), the time to failure for the 18Ni-250 maraging steel is of the order of months for a yield strength of 1720 MPa (250 ksi). An explanation for the susceptibility to SCC of the quenched-and-tempered steels can be given in terms of a combination of factors, namely, the presence of ε carbide, high dislocation density, and coherency strains between twinned martensitic platelets, which lower the yield strength (Ref 93).

Because SCC in high-strength steels is generally intergranular, it is obvious that the compositional variations at the grain boundaries are important. In all types of steel, residual elements tend to precipitate at the prior-austenite grain boundaries and exert their deleterious effect on SCC at these boundaries. Thus, if the prior-austenite grain size is made as fine as possible, a maximum in SCC resistance can be achieved. Austenitizing treatments of 18Ni-300 grade maraging steel were found to have little effect on K_{Iscc}, but the rate of stress-corrosion crack growth was found to decrease with decreasing prior-austenite grain size (Ref 95).

Different annealing and aging treatments in 18Ni-300 maraging steel, reaching the same yield strength, do not affect the value of K_{Iscc} (Ref 96). However, the crack growth rate does increase with decreasing aging temperatures. Although SCC in high-strength steels is generally intergranular, the grain size was found to have no influence on the K_{Iscc} of AISI 4340 steel, but a significant increase in crack velocity with increasing grain size was found.

Figure 8 demonstrates that cold working of 18Ni maraging steels prior to aging decreases the susceptibility to SCC (Ref 94). The figure further shows a ranking of the different types and treatments of steel. Ausforming and light tempering result in a significant increase in K_{Iscc} of H11 steel (Ref 97). Similar improvement in resistance to SCC has been shown in quenched-and-tempered D6AC steel (see Fig. 9).

Environmental Factors. High-strength steels are generally susceptible to most aqueous environments, including environments containing water vapor. The most common damaging species is chloride, while those environments that contain cathodic poisons are extremely harmful as well (Ref 93, 98). These include species such as arsenic, antimony, selenium, tellurium, and most importantly, sulfur. The poisons prevent the recombination of atomic hydrogen into molecular form, facilitating hydrogen to absorb into the steel matrix. Sulfur occurs in the form of H_2S, which is naturally found in sour oil wells.

Other environments that may promote SCC include aqueous solutions containing sulfate, phosphate and nitrate ions, and various organic compounds (Ref 54).

The electrochemical potential and pH of a solution have a significant effect on the SCC of high-strength steels (Ref 47). Although an impressed cathodic potential has little effect on the K_{Iscc} of AISI 4340 steel in seawater or 3.5% NaCl aqueous solutions, it significantly affects the crack growth rate (see Fig. 10) (Ref 54, 88, 99, 100). Generally, the effect of pH on K_{Iscc} in SCC of quenched-and-tempered steel is small, except when it is below 1 or above 9. Highly acidic conditions promote cracking, and highly basic conditions reduce or even prevent cracking, presumably by preventing pitting corrosion. It is generally believed that the pH and potential at a propagating crack tip promote the local hydrogen ion reduction reaction and hence hydrogen absorption into the matrix and hydrogen embrittlement.

Mechanisms of SCC in High-Strength Steels. Extensive research has confirmed that hydrogen embrittlement is the most likely mechanism of SCC of high-strength low-alloy steel. Local environments, such as potential and pH at the tips of propagating cracks, were found conducive to conditions for hydrogen embrittlement. Similarities between fracture appearance of stress-corrosion cracks and cracks formed in gaseous hydrogen further confirmed a hydrogen-related fracture mechanism for SCC.

The four principal fracture mechanisms that have been proposed are:

- The pressure theory (Ref 101)
- The surface energy theory (Ref 102)
- The decohesion theory (Ref 103, 104)
- The enhanced plasticity theory (Ref 106–108)

The pressure theory (Ref 101) is based on the concept that molecular hydrogen precipitates in preexisting voids in the metal. When the pressure is built up in the voids, the resulting stress reduces the applied stress force fracture. This theory has not been supported by experimental evidence.

The surface energy theory (Ref 102) postulates that hydrogen, adsorbed at the metal surface, acts to lower the surface free energy, lowering the stress required for cracking. Again no experimental support for this model has been offered.

The decohesion model has been developed from Troiano's original theory that electrons donated from dissolved hydrogen enter the incompletely filled d-bond (Ref 103). The cohesive strength of the metal is then reduced by the increased electron density, which causes an increase in interatomic spacing. Experimental work by Fu and Painter (Ref 106) on iron aluminide has suggested that hydrogen embrittlement results from depletion of d-bonding charge from iron sites in the ordered aluminide lattice.

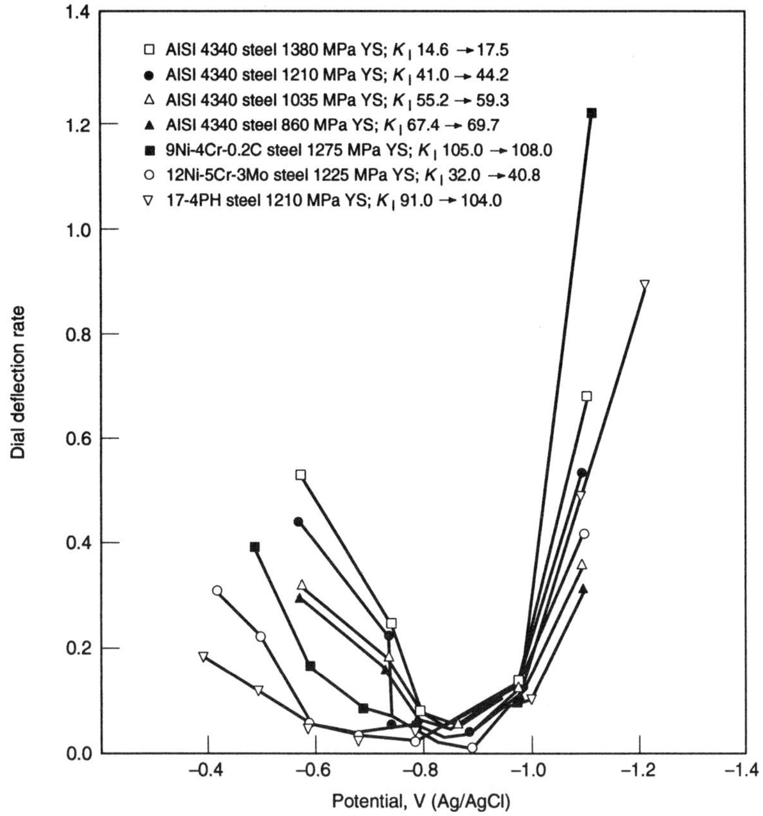

Fig. 10 legend:
- □ AISI 4340 steel 1380 MPa YS; K_I 14.6 → 17.5
- ● AISI 4340 steel 1210 MPa YS; K_I 41.0 → 44.2
- △ AISI 4340 steel 1035 MPa YS; K_I 55.2 → 59.3
- ▲ AISI 4340 steel 860 MPa YS; K_I 67.4 → 69.7
- ■ 9Ni-4Cr-0.2C steel 1275 MPa YS; K_I 105.0 → 108.0
- ○ 12Ni-5Cr-3Mo steel 1225 MPa YS; K_I 32.0 → 40.8
- ▽ 17-4PH steel 1210 MPa YS; K_I 91.0 → 104.0

Fig. 10 Effect of applied potential on crack growth rates of various stainless steels in an aqueous chloride environment (Ref 54)

The hydrogen-enhanced localized plasticity (HELP) model suggests that hydrogen in the lattice reduces the shear stress required for dislocation motion at the crack tip, easing the local plastic flow associated with crack propagation. This theory of hydrogen-induced SCC has been supported experimentally by Birnbaum's work on thin foils of several metals, including steel, using an environmental cell in a high-voltage transmission electron microscope (Ref 107, 108).

SCC of Stainless Steels

Stainless steels are a group of iron-base alloys that contain a minimum of 11% Cr to prevent general corrosion in unpolluted atmospheric environments. The susceptibility of stainless steels to SCC strongly depends on the composition, microstructure, and heat treatment of the steels. Figure 11 shows a composite diagram of the alloys in the stainless steel family, starting from the basic Fe-19Cr-10Ni (304) stainless steel (Ref 109). The figure indicates four separate groups, namely, austenitic, ferritic, duplex, and martensitic and precipitation-hardening stainless steels. These different stainless steels can be susceptible to SCC in various environments, including aqueous chlorides, caustic, high-temperature pure water, polythionic acid thionates, sulfides, and radiation fields. The stainless steels of the four basic groups—austenitic, ferritic, duplex, and martensitic and precipitation-hardening stainless steels—will be discussed in some detail in the following sections.

Austenitic stainless steels, based on the Fe-18Cr-10Ni alloy, are particularly susceptible to intergranular SCC in several environments. Sensitization of the microstructure of austenitic stainless steels is a feature that plays an important role in SCC of these steels, and considerable effort has gone into the development of alloys that resist sensitization. Carbon is present in austenitic stainless steels as austenite stabilizer. At concentrations between 0.03 and 0.7% C, the equilibrium structure at room temperature should contain austenite, α-ferrite, and carbide ($M_{23}C_6$). Carbon concentrations up to 0.03% C should be soluble in austenite up to a temperature of 800 °C (1470 °F). Austenite that contains more than 0.03% C precipitates $M_{23}C_6$ upon cooling below the solubility line (Ref 110). However, at relatively rapid cooling rates, the precipitation reaction is suppressed, and the austenite is supersaturated with carbon. If this supersaturated austenite is heated into the austenite + $M_{23}C_6$ field, precipitation of chromium-rich $M_{23}C_6$ will take place at the austenite grain boundaries. When the chromium does not diffuse back into the austenite adjacent to the grain-boundary carbides, a zone depleted of chromium will be formed around the carbides, which has lower resistance to corrosion. This type of structure, which is known as sensitized structure, can be obtained by slow cooling, heat treatment, elevated-temperature service, or welding (Ref 111).

In order to minimize sensitization, various steps can be taken: (1) reduce the carbon concen-

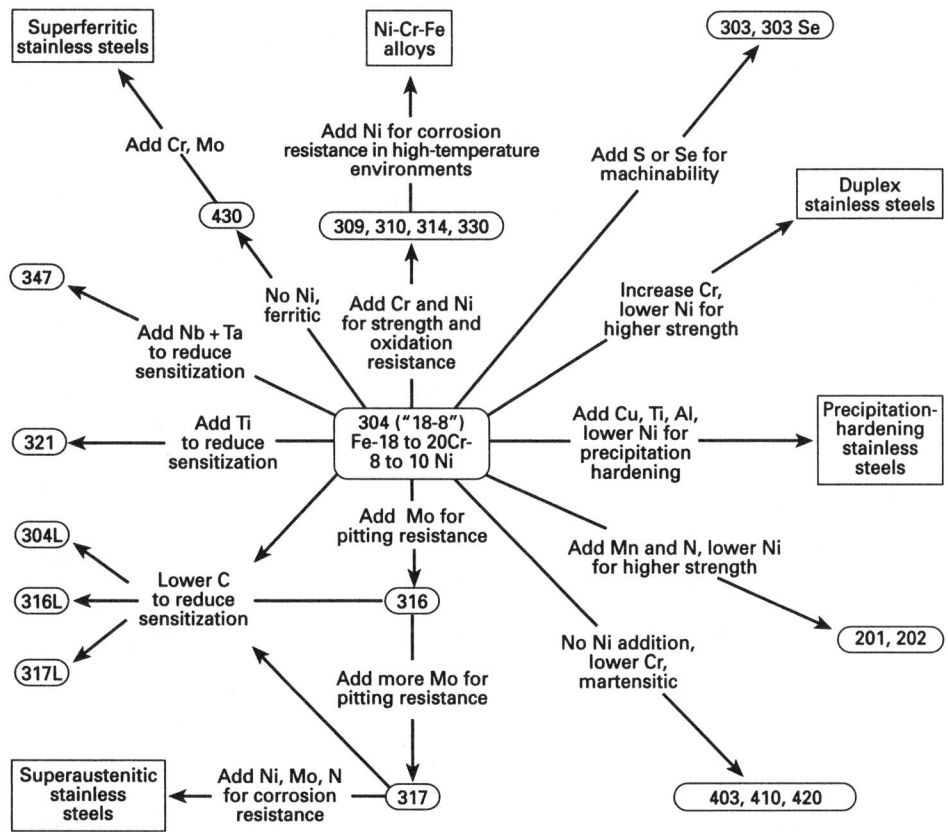

Fig. 11 Schematic diagram showing compositional and property linkages in the stainless steel family of alloys (Ref 109)

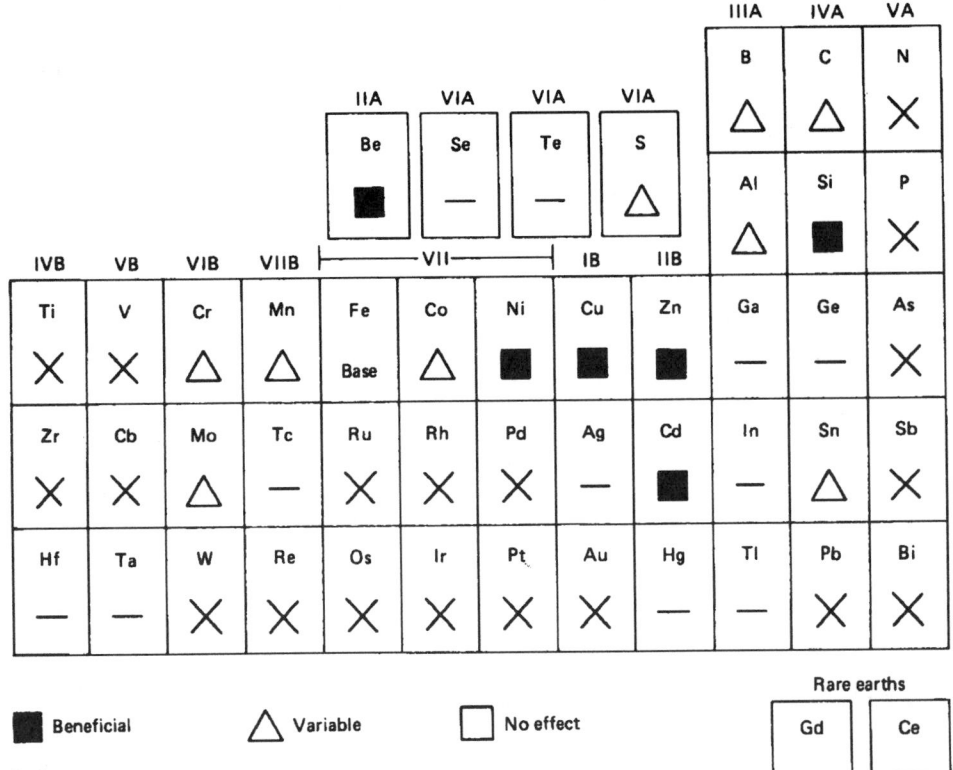

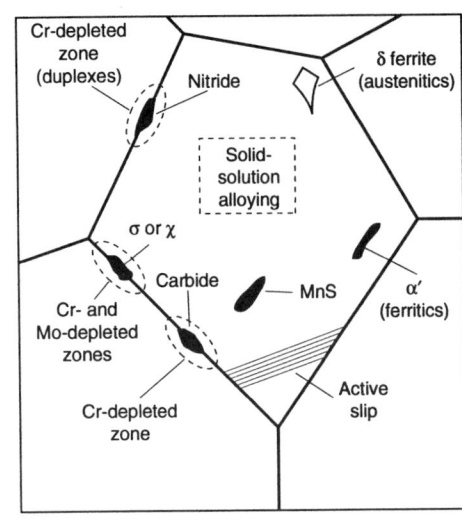

Fig. 13 Schematic drawing of the microstructural features found in stainless steels (Ref 115)

Fig. 12 Effect of various alloying elements on the resistance of austenitic stainless steels to SCC in chloride solutions (Ref 111)

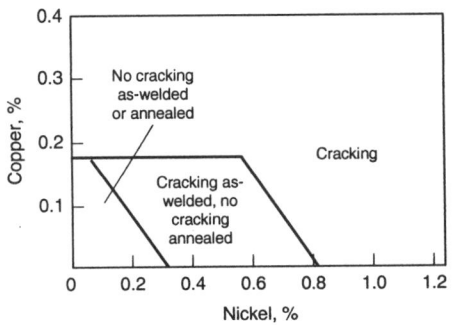

Fig. 14 Effect of copper and nickel contents on the SCC resistance of U-bend specimens of ferritic Fe-18Cr-2Mo-0.35Ti-0.015C-0.015N stainless steels exposed to a magnesium chloride solution boiling at 140 °C (284 °F) (Ref 122)

tration below 0.03% (for example, types 304L, 316L, and 317L stainless steels), (2) heat treat to rediffuse chromium back into the chromium-depleted zone, and (3) add titanium (type 321 stainless steel) or niobium (type 347 stainless steel) to precipitate carbides at higher temperature, so that no carbon is left to precipitate as chromium-rich carbide. The latter treatments result in so-called stabilized grades, which may suffer from another form of corrosion attack, known as knifeline attack (Ref 111). Carbon strengthens the matrix, and when its concentration is lowered, the strength may be reduced. To compensate for this loss in strength, nitrogen has been added as an alloying element (Ref 112).

In addition to carbon, titanium, and niobium, which affect the degree of sensitization of austenitic stainless steels and hence their susceptibility to intergranular SCC, there are other alloying elements that affect the susceptibility to transgranular SCC (Fig. 12) (Ref 111, 113, 114). The elements designated "beneficial" are nickel, cadmium, zinc, silicon, beryllium, and copper. It should be noted that the beneficial effect of nickel only applies to austenitic stainless steels. Figure 12 further shows many elements that have a detrimental effect on SCC.

The microstructure of austenitic stainless steels has a significant effect on the susceptibility to chloride SCC. The various phases and microstructural features found in stainless steels are shown schematically in Fig. 13 (Ref 115). The presence of δ ferrite improves the resistance to

SCC by interference with cracks propagating across austenite grains. In order to be effective, a significant amount of δ ferrite is required, such that the alloy composition of duplex stainless steels is reached. The σ or χ phases have an effect on sensitization of grain boundaries, similar to that of carbides. However, few attempts have been made to study the effect of these phases on chloride SCC.

Ferritic stainless steels have been reported to have high resistance to SCC (Ref 116-118). However, SCC of ferritic stainless steels has been reported for specific alloy/environment combinations (Ref 119-121). Materials factors that have been identified as detrimental to the chloride SCC cracking of ferritic stainless steels include the presence of certain alloying elements, sensitization, cold work, high-temperature embrittlement, and the precipitation of α' (at 475 °C, or 887 °F) (Ref 109). Although it is not clear whether these effects are directly related to chloride SCC or represent manifestations of other phenomena such as hydrogen embrittlement or stress-aided intergranular corrosion, the effects can be detected in the standard boiling magnesium chloride test.

The alloying elements that are detrimental to the resistance of ferritic stainless steel to chloride SCC, include copper, nickel, molybdenum (in the presence of nickel), cobalt (in the presence of molybdenum), ruthenium, sulfur, and carbon (Ref 109). Figure 14 shows that the concentrations of nickel and copper needed for cracking are

relatively low (Ref 122). Similarly low concentrations for molybdenum and cobalt are needed for chloride SCC of ferritic stainless steels.

Any structural features that reduce the ductility of ferritic stainless steels such as the formation of carbonitrides or α', and cold work can reduce the resistance of these steels to SCC (Ref 122, 123).

Hydrogen embrittlement has been identified as the cause of SCC of ferritic stainless steels when they are cathodically polarized in aqueous environments (Ref 124, 125). For example, the super ferritic stainless steels 29-4C and Sea-Cure exhibit hydrogen embrittlement when they are cathodically polarized to potentials in the range of –0.9 to –1.4 V (SCE) (Ref 124). Hydrogen embrittlement of these alloys can also occur when they are exposed to acidic environments, or in the presence of hydrogen-ion recombination poisons. The fracture mode of hydrogen embrittlement in these superferritic stainless steels is transgranular cleavage.

Duplex stainless steels with 10 to 28% Cr and 4 to 8% Ni contain both austenite and ferrite. An attractive feature of these alloys is that they have much higher strength than either the austenitic or the ferritic stainless steels. Although duplex

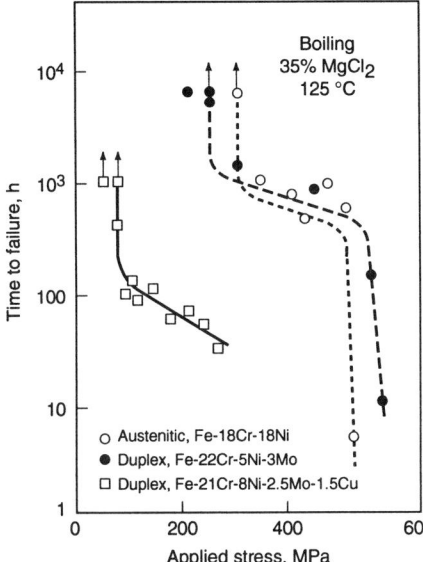

Fig. 15 Effect of applied stress on the times to failure of various stainless steels in a boiling magnesium chloride solution (Ref 126)

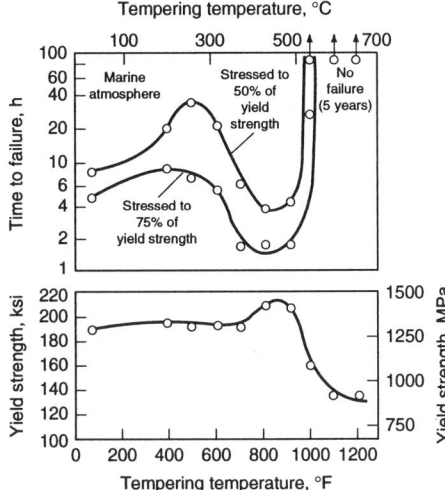

Fig. 16 Effect of tempering temperature on the cracking resistance and yield strength of an Fe-12Cr-1Mo-0.33V-0.25C martensitic stainless steel (Ref 130)

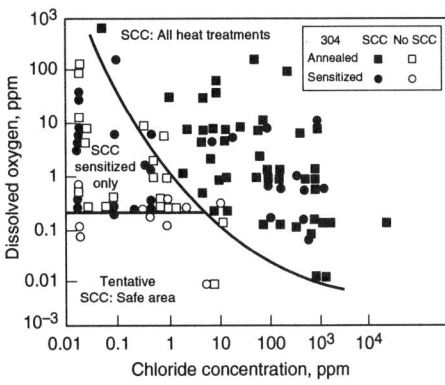

Fig. 17 Concentration ranges of dissolved oxygen and chloride that may lead to SCC of type 304 in high-purity water at temperatures ranging from 260 to 300 °C (500 to 570 °F). The applied stresses are greater than the yield strength and test times are greater than 1000 h, or strain rates are greater than 10^{-5}/s (Ref 132).

stainless steels are susceptible to chloride SCC, the threshold stress for SCC is generally much higher for duplex stainless steels (Fig. 15) (Ref 126). The microstructural parameters that have a significant effect on the susceptibility to chloride SCC are the presence of σ phase and the ferrite content. The presence of σ phase has been shown to be detrimental to SCC resistance in boiling 35% $MgCl_2$ at 125 °C (257 °F) (Ref 126). Ferrite has been shown to affect the resistance to chloride SCC in boiling 42% $MgCl_2$, with the resistance reaching a maximum at 40% and decreasing on either side of this value (Ref 127, 128).

Welding of duplex stainless steels may have a significant effect on the resistance to chloride SCC (Ref 111, 126). Welding can destroy the duplex structure of an annealed material by form-ing continuous regions of ferrite along the weld HAZ. In addition, welding can result in the formation of carbonitrides, α′, and precipitate-free zones, which contribute to the susceptibility to SCC.

Martensitic and precipitation-hardening stainless steels are used in heat-treated conditions when high strength is required. These steels are susceptible to SCC in aqueous chloride environments with or without sulfides present. The mechanism of SCC in aqueous chloride environments such as marine environments is generally accepted to be hydrogen embrittlement (Ref 121). Martensitic and precipitation-hardening alloys are susceptible to hydrogen embrittlement in marine environments at yield strengths above 1035 MPa (150 ksi) (Ref 129). In precracked specimens, hydrogen embrittlement may occur at lower yield strengths. Tempering, in the case of martensitic stainless steels, and overaging, in the case of precipitation-hardening steels, can significantly increase the resistance to hydrogen embrittlement, but also lower the yield strength (Ref 130); see Fig. 16 for an Fe-12Cr-1Mo-0.33V-0.25C martensitic stainless steel.

Environmental Factors in SCC of Stainless Steels. Austenitic stainless steels are susceptible to intergranular and transgranular SCC in aqueous chloride environments. Intergranular cracking occurs in sensitized stainless steels in chloride environments such as NaCl or seawater environments, while transgranular SCC occurs in nonsensitized steels (chloride SCC). Most of the early evaluations of chloride SCC was conducted in boiling magnesium chloride (Ref 131). This test has been standardized as ASTM G 36.

Other environments that induce SCC in austenitic stainless steels are high-temperature water and various sulfur compounds. High-temperature water SCC denotes SCC in the cooling water of nuclear reactors and has been used to describe intergranular SCC of weld heat-affected zones (HAZs) in type 304 stainless steel piping exposed to oxygenated boiling water reactor (BWR) coolants. Figure 17 shows a compilation of data on the stress-corrosion susceptibility of type 304 stainless steels in high-temperature water at various dissolved oxygen and chloride concentrations (Ref 132). The figure indicates that sensitized type 304 stainless steel can undergo SCC at dissolved oxygen concentrations between 0.15 and 0.3 ppm and chloride concentrations between 0.02 and 0.5 ppm. These oxygen and chloride ranges span the BWR operating ranges. A quantitative and predictive model developed by Ford and coworkers (Ref 133-135) is based on the slip-dissolution/film-rupture model. In this model, a thermodynamically stable oxide film is ruptured by an increase in the strain in the steel matrix. Faraday's law is then used to relate the amount of subsequent crack propagation to the oxidation charge density associated with dissolution and oxide growth. This model predicts stress-corrosion crack propagation for austenitic stainless steels in 288 °C (550 °F) water within a factor of two and is therefore used in the nuclear power industry as a crack-propagation prediction tool.

Intergranular SCC of austenitic stainless steels occurs in various sulfur-containing environments. For example, SCC occurs in catalytic reformers used in the oil-refinery industry, when the steels are in contact with the condensates of reforming furnaces, containing polythionic acids ($H_2S_xO_6$, where x = 3, 4, or 5) (Ref 136, 137). More recent studies (Ref 138, 139) have shown that of all the polythionic acids, only tetrathionic acid ($H_2S_4O_6$) could induce SCC. A laboratory test for resistance to polythionic cracking has been standardized as ASTM G 35.

Other sulfur-containing compounds that induce intergranular SCC include sulfides, thiosulfates, and sulfate solutions. Stress-corrosion cracking in hydrogen sulfides (H_2S) is of interest in downhole environments, while SCC in thiosulfate solutions is of interest to the nuclear industry. Borate solutions are stored in stainless steel tanks for use as spray in case of nuclear accidents. In addition to borate, the solution also contains thiosulfate anions, which react with the iodine fissure products. Stress-corrosion cracking of stainless steels in these borate solutions has been attributed to the presence of the thiosulfate (Ref 140).

Austenitic stainless steels are also susceptible to SCC in caustic solutions (Ref 132). The resistance of stainless steels to caustic cracking is of interest to the power, chemical process, and pulp and paper industries. Data have shown that austenitic stainless steels are particularly susceptible to caustic cracking when the temperature approaches 100 °C (212 °F). The presence of oxygen in caustic solutions has been reported to have a detrimental effect on the susceptibility to SCC. Specifically, it has been shown that caustic cracking of austenitic stainless steel is very severe in aerated solutions of sodium hydroxide and that high levels of both nickel and chromium in the alloy appear to be necessary to resist caustic cracking (Ref 141). Caustic SCC can be mitigated by the addition of phosphates, chlorides, and chromates to the environment (Ref 142-145).

Ferritic stainless steels generally have a high resistance to chloride SCC. However, SCC of

Table 3 Nominal chemical compositions of various commercial nickel-base alloys used for aqueous corrosion resistance

Common alloy designation	UNS No.	Chemical composition, wt %								
		C(a)	Cr	Cs	Fe	Mo	Nb	Ni	Ti	Others
Nickel										
200	N02200	0.1	...	0.3(a)	0.4(a)	...	...	99.0(b)	0.1(a)	...
201	N02201	0.02	...	0.3(a)	0.4(a)	...	...	99.0(b)	0.1(a)	...
Ni-Cu										
400	N04400	0.15	...	31.5	1.25	...	...	66.6	...	...
Ni-Mo										
B-2	N10665	0.001	1.0(a)	...	2.0(a)	28	...	71	...	...
B	N10001	0.05	1.0(a)	...	5.0	28	...	66	...	0.3 V
Ni-Cr-Fe										
600	N06600	0.08	16.0	0.5(a)	8.0	...	...	75	0.3(a)	...
800	N08800	0.1	21.0	0.8(a)	44.0	...	...	32.5	0.4(a)	...
Nimonic 75	...	0.15	19.5	0.5	5.0	...	...	73	0.4	...
690	N06690	0.02	29.0	...	10.0	...	...	61	0.3	...
800H	N09910	0.06-0.1	21.0	0.8(a)	44.0	...	...	32.5	0.4(a)	Al + 0.8 Ti(b)
Ni-Cr-Fe-Mo										
825	N00825	0.05	21.5	2.0	29.0	3.0	...	42	1.0	...
Sanicro	N08028	0.02	27	1.2	36	3	...	31.4	...	...
G-3	N06985	0.015	22.0	2.0	19.5	7.0	0.8(a)	44	...	...
2550	N06975	0.03	24.5	1.0	20.0	6.0	...	48	...	1.5 W(a)
G-30	N06030	0.03	29.5	2.0	15.0	5.5	0.8	43	...	2.5 W
G-50	N06950	0.02	20.0	0.5(a)	17.0	9.0	...	50(b)	...	...
Ni-Cr-Mo-W										
N	N10003	0.06	7.0	...	5.0(a)	16.5	...	71	...	...
W	N10004	0.12	5.0	...	6.0	24.0	...	63	...	...
625	N06625	0.1	21.5	...	5.0(a)	9.0	4.0	62	...	...
C-276	N10276	0.01	15.5	...	5.5	16.0	...	57	...	4.0 W
C-4	N06455	0.01	16.0	...	3.0(a)	15.5	...	65	0.7(a)	...
C-22	N06022	0.015	22.0	...	3.0(a)	13.0	...	56	...	3.0 W
ALLCORR	N06110	0.15	30.0	...	...	10.0	...	53	1.5(a)	4 W(a)
Age-hardenable alloys										
K-500	N05500	0.25	...	29	2.0(a)	...	...	86	0.6	2.7Al
X-750	N07750	0.08	15.5	...	7.0	...	0.9	70(b)	2.5	0.7 Al
718	N07718	0.05	18.0	...	19.0	3.0	5.0	52.5	0.4(a)	0.006 B
Nimonic 80	N07080	0.10	19.5	...	3.0	...	...	73	2.25	1.4 Al
Nimonic 105	...	0.20	14.5	...	2.0	5.0	...	56	1.2	4.7Al; 20 Co
R-41	N07041	0.09	19.0	...	5.0(a)	10.0	...	52	3.1	1.5 Al; 11 Co
925	N09925	0.03	21.0	2.0	22(b)	3.0	...	43	2.1	...
625 Plus	N07716	0.03	21	...	5.0	8.5	3.5	61	1.3	0.35 Al(a)
725	N07725	0.03	21	...	7.5	8.0	3.5	57	1.4	0.35 Al(a)

(a) Maximum. (b) Minimum. Source: Ref 147

ferritic stainless steels have been reported for type 434, type 430, and Fe-18Cr-2Mo in lithium chloride solutions (Ref 119), for sensitized type 446 in boiling magnesium chloride and sodium chloride solutions (Ref 120), for type 430F in marine atmospheres (Ref 121), and for the nickel-containing superferritics, AL 29-4-2, Monit, and Sea-Cure, in boiling magnesium chloride (Ref 146). As in the case of austenitic stainless steels, cathodic polarization can prevent chloride SCC of ferritic stainless steels.

Ferritic stainless steels have been reported to be resistant to caustic SCC. Rather, these alloys exhibit uniform corrosion or intergranular corrosion in caustic environments. However, SCC of type 446 and E-Brite 26-1 heat treated at 871 °C (1600 °F) was reported after exposure to 50% NaOH at 316 °C (600 °F) (Ref 141).

SCC of Nickel-Base Alloys

Nickel-base alloys are often used because of their high resistance to corrosion and SCC in various environments. However, SCC can occur under specific combinations of environment, microstructure, and applied stress. The number of environments that have been recognized to cause SCC in nickel-base alloys has increased over the years, as shown in Table 3 (Ref 147). The table shows that depending on their composition, these alloys can be susceptible to SCC in environments ranging from aqueous halide solutions to various acids, high-temperature pure water, steam, to sulfur compounds.

Materials Factors. The composition of nickel-base alloys can range from binary systems to multicomponent systems with several alloying elements serving different applications. Major alloying systems are Ni-Cu, Ni-Mo, Ni-Cr-Fe, Ni-Cr-Fe-Mo, and Ni-Cr-Mo-W (Ref 148). In addition, impurities such as carbon, nitrogen, sulfur, phosphorus, tin, and silicon may be present in various concentrations. Copper and molybdenum are added to increase the resistance to reducing environments. Chromium is added to provide resistance to oxidizing environments. Molybdenum and tungsten are added to chromium-containing alloys to increase the resistance to pitting and crevice corrosion in halide environments. Niobium and titanium are added to prevent the formation of chromium carbides. In precipitation-hardenable alloys niobium, aluminum, and titanium also combine with nickel to form coherent intermetallic phases.

Commercially pure nickel (200) is primarily used in the handling of caustic solutions. The Ni-Cu alloys (alloy 400), and the age-hardenable Ni-Cu alloy (alloy K-500) are primarily used for seawater applications. The Ni-Mo alloys (alloys B and B-2) are used in the chemical processing industry for handling reducing acids. The Ni-Cr-Fe alloys (alloys 600, 800, Nimonic 75, and alloy 690) find wide application in the nuclear power industry as steam generator tubing and turbine wheel components. The Ni-Cr-Fe-Mo alloys (alloys 825, G3) are generally used in industries, where corrosion resistance is required. The precipitation-hardenable Ni-Cr-Fe-Mo alloy 718, first developed for aerospace applications, is now used in the nuclear and oil and gas industries. Finally, the Ni-Cr-Mo-W alloys, such as alloys C276 and C22, are used mainly in the chemical process, oil and gas and pollution control industries because of their high resistance to corrosion in contaminated acids.

As in the case of austenitic stainless steels, the precipitation of carbides plays an important role in the susceptibility of nickel-base alloys to SCC. In nickel-base alloys, carbides can be classified as primary and secondary carbides (Ref 147, 148). The primary carbides form during solidification and include the MC types, where M is niobium, tantalum, or titanium, and the M_6C types, where M is usually molybdenum or tungsten. These precipitates are difficult to dissolve and have little effect on the susceptibility to SCC. The secondary carbides precipitate as a result of heat treatment and welding, typically along the grain boundaries, and are detrimental to intergranular corrosion and SCC resistance because of their associated chromium-depletion zones. Carbides in nickel-base alloys can be classified as those rich in chromium (Cr_7C_3 or $Cr_{23}C_6$), and those that are rich in refractory elements such as $Cr_{21}(Mo\text{-}W)_2C_6$.

In addition to carbides, intermetallic phases form that may influence the susceptibility of nickel-base alloys to SCC. The γ, γ', and γ'' and δ are coherent precipitates that are desirable for increased strength (Ref 149-151). The γ phase is a face-centered cubic (fcc) precipitate, while the γ'' is a DO_{22}-ordered coherent precipitate (metastable Ni_3Nb). The γ' is a coherent precipitate ($L1_2$), which is a metastable version of $Ni_3(Al,Ti,Si)$ and normally present in age-hardenable alloys. The δ phase has, like the γ'' phase, an ideal composition of Ni_3Nb, and has an orthorhombic crystal structure. The detrimental intermetallic phases are generally the σ, μ, and Laves phases. The σ phase has a complex tetragonal layered structure and tends to nucleate from $M_{23}C_6$ carbides, to which it is structurally related (Ref 152). The molybdenum-rich μ phase has a complex rhombohedral structure and occurs in chromium-containing nickel-base alloys with large amounts of molybdenum and tungsten. The Laves phase is an ordered (A_3B) structure that forms if the local concentration of niobium is high.

An extensive study was conducted by Streicher on the effects of microstructure on Ni-Cr-Mo alloys (C and C-276) on localized corrosion and SCC (Ref 153). It was demonstrated that these alloys have experienced varying degrees of susceptibility to SCC as a result of carbide and μ-phase precipitation. The study by Streicher (Ref 153) indicated that solution-annealed alloys were not susceptible to SCC in boiling 45% $MgCl_2$, and that the precipitation of M_6C was much more detrimental to the SCC resistance than the μ phase.

Environmental Factors in SCC of Nickel-Base Alloys. Generally, nickel-base alloys are not susceptible to SCC in aqueous halide solutions. Stress-corrosion cracking has been observed in these solutions, however, when the temperature is high enough. At temperatures slightly above the boiling point of chloride solutions, no cracking is observed in alloys with nickel content above 35% (Ref 154). Figure 18 shows the effect of nickel content on the crack growth rate and K_{ISCC} for Fe-Ni-Cr alloys in hot chloride solutions. The figure clearly shows that the alloy is the most susceptible at a nickel concentration of about 20 wt%. At lower and higher concentrations, the resistance to SCC increases dramatically.

Cracking has been observed in several precipitation-hardening nickel-base alloys (e.g., alloy

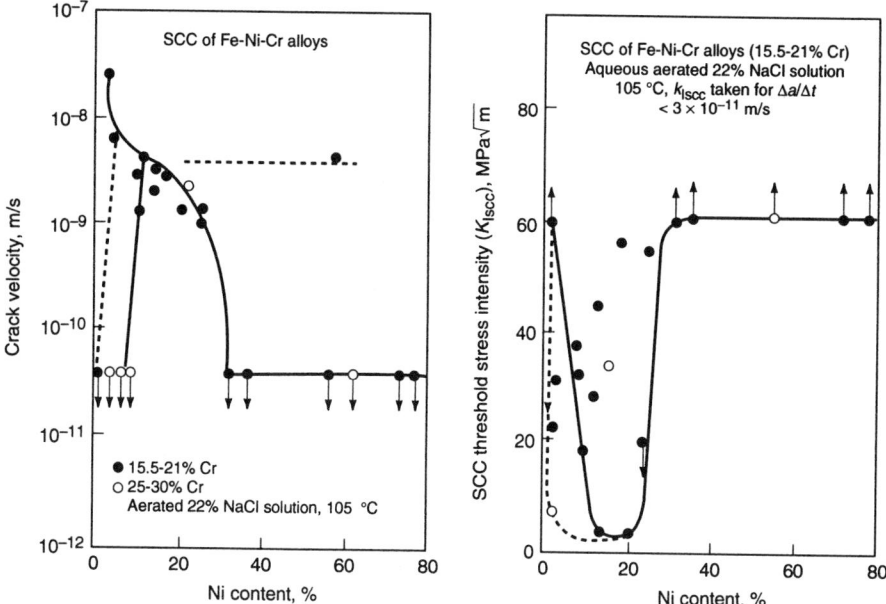

Fig. 18 Effect of nickel content on the maximum crack growth rate and threshold stress-intensity factor for Fe-Ni-Cr alloys in hot chloride solutions (Ref 154)

718) exposed to hot HCl, whereas no cracking was found in solid-solution-strengthened alloys (Ref 155). It was speculated that the combination of changes in distribution of alloying elements between the γ and γ' phases, chromium depletion at the grain boundaries, and changes in the deformation mode may have contributed to the increased susceptibility of alloy 718 to SCC.

Generally, nickel-base alloys are resistant to SCC in fluoride, bromide, and iodide environments. However, in a few special cases SCC has been observed. Graf and Wittich (Ref 156) demonstrated that aqueous solutions containing cupric fluoride would cause SCC of Nickel 400. An Ni-Cu alloy with 33% Cu was found to be the most susceptible to cracking. Nickel alloys with less than 30% Cu or more than 40% Cu are resistant. The reason that SCC is limited to the vapor phase rather than liquid was said to result from the enrichment of a thin layer of aqueous hydrogen fluoride with copper fluoride corrosion products. The presence of oxide was found to cause the formation of cupric fluoride, accelerating SCC. Nickel-chromium alloys, such as alloy 600, have been used in components exposed to hydrogen fluoride vapors to minimize SCC (Ref 147).

Stress-corrosion cracking of nickel-base alloys has been observed in concentrated bromide solutions used as packer fluids (Ref 157). The susceptibility to SCC in these environments has been attributed to the presence of H_2S. Transgranular SCC of nickel-base alloys in iodide environments has been encountered in the manufacturing of acetic acid (CH_3COOH), where potassium iodide is used as promoter.

Nickel-base alloys have found wide use in sulfur-containing environments, which are encountered in oil and gas production and in refinery operations (Ref 158-160). These environments include the sulfide (H_2S) environments that may

also include chloride, carbon dioxide, and elemental sulfur, as well as sulfur-oxyanions such as thiosulfate, tetrathionate, and polythionic acid. Early investigations of Ni-Cr-Mo alloys found that under cathodic polarization by galvanic coupling with carbon steel, intergranular cracking could occur in an $H_2S + CO_2 + Cl^-$ environment (Ref 161). When increasing the temperature, the susceptibility to SCC increases, but in the absence of galvanic coupling with carbon steel, the cracking is transgranular. Elemental sulfur (S_8) is often present, and recent studies of nickel-base alloys in sour environments have indicated that S_8 is extremely detrimental to localized corrosion and SCC of nickel-base alloys in brine solutions (Ref 162). No explanation has been offered in the literature with regard to the mechanism by which S_8 affects cracking in these environments. Acidification by HCl or CH_3COOH has been shown to increase the susceptibility of nickel-base alloys to both H_2S and $H_2S + S_8$ environments (Ref 157, 163-165), while increasing the pH by adding NaOH has been shown to prevent SCC in these environments (Ref 157, 164).

Metastable oxyanions, such as thiosulfates and polythionates, can also induce intergranular SCC in sensitized Ni-Cr-Fe alloys (Ref 111, 137, 166). Much work has been done on intergranular cracking of alloy 600 because of its extensive use as steam generator tubing in pressurized water reactors (PWRs) (Ref 167). From this research it has become apparent that intergranular SCC of Ni-Cr-Fe alloys in the presence of metastable sulfur oxyanions can be avoided by using prolonged heat treatments (e.g., 15 h at 700 °C, or 1290 °F, for alloy 600), which replenish the chromium-depleted zone.

Stress-corrosion cracking failures have been experienced in other nuclear plant environments, including high-temperature pure-water environments and caustic environments. It is well known that solid-solution strengthened alloys, such as alloy 600, and precipitation-hardened alloys, such as alloys 750 and 718, are susceptible to intergranular SCC in deaerated high-purity water at temperatures of about 300 to 350 °C (570 to 660 °F) (Ref 168-171). The effect of dissolved oxygen on the intergranular SCC susceptibility of alloy 600 in high-temperature water has been reviewed in detail (Ref 172, 173). The accelerating effect of oxygen is well documented and is even more significant in creviced structures (Ref 174).

Stress-corrosion cracking of the solid-solution-strengthened alloy 600 in hot caustic environments at temperatures of 350 °C (660 °F) has been studied extensively, because of the use of this alloy for steam generator tubing in PWRs (Ref 141, 175-179). All the austenitic Ni-Cr-Fe alloys are susceptible to SCC, and alloy 690 is the most resistant of the commercial alloys. Only pure nickel seems to be immune to cracking.

SCC of Aluminum Alloys

Stress-corrosion cracking of aluminum alloys can occur in several different environments, ranging from water condensate to salt water solutions. Only aluminum alloys that are precipitation-hardening alloys and contain soluble alloying elements, such as copper, magnesium, silicon, and zinc, are susceptible to SCC. The susceptibility of these alloys to SCC strongly depends on the heat treatment, which results in specific microstructures. Thus, because the susceptibility depends so strongly on the microstructure resulting from heat treatment, heat treatments have been developed to reduce and even eliminate the susceptibility to SCC. However, as will be discussed in later sections, such heat treatments also can result in a loss of strength.

The susceptibility of aluminum alloys is strongly dependent on alloy composition. There are basically five commercial alloy types, namely Al-Cu-Mg (2000 series) alloys, Al-Cu-Li (2000 and 8000 series) alloys, Al-Mg (5000 series) alloys, Al-Mg-Si (6000 series) alloys, and Al-Zn-Mg (7000 series) alloys. The Al-Cu and Al-Zn-Mg alloys are precipitation-hardening alloys, whereas the Al-Mg alloys are not considered heat treatable and do not develop their strength through heat treatment.

Water or water vapor is the key environmental factor in SCC of aluminum alloys. Halide ions (Cl^-, I^-, and Br^-) have a significant effect in accelerating SCC, with chloride having the greatest effect. In general, the susceptibility to SCC is greater in neutral solutions than in alkaline solutions and greater still in acidic solutions.

Aluminum-Copper-Magnesium (2000 Series) Alloys. The Al-Cu-Mg alloys were the first heat-treatable, high-strength aluminum alloys and are still widely used. Al-Cu and Al-Cu-Mg alloys derive their strength from precipitation hardening, which is achieved by solution heat treatment, followed by rapid cooling, and either natural aging at room temperature (T4 temper) or artificial aging at elevated temperatures (T6 tem-

per). Cold working after the quench further increases the strength, resulting in the T3 temper; artificially aging results in the T8 temper.

Thick-section products of the 2000 series alloys, such as 2024, 2014, and 2219 in the T3 and T4 tempers have a low resistance to SCC in the short-transverse direction (Ref 180-182). If these alloys in the T3 and T4 tempers are heated for short periods in the temperature range used for artificial aging, selective precipitation along the grain boundaries may further decrease the resistance to SCC. Coarse Al-Cu precipitates form along the grain boundaries, depleting the regions adjacent to the grain boundaries of solute (Ref 183). At longer heating times, as specified for the T6 and T8 tempers, the precipitation becomes more homogeneous, and the resistance to SCC increases.

The precipitates are formed within the grains at a greater number of nucleation sites during treatment to the T8 temper. This temper requires cold working after quenching from the solution annealing treatment, and it provides the highest strength and resistance to SCC before artificial aging in the 2000 series alloys.

There is a consensus among the majority of researchers (Ref 184) that the mechanism of both intergranular corrosion and SCC is crack-tip anodic dissolution along the grain boundaries. The anodic or dissolving phase could be the solute-depleted zone along the grain boundary, the grain boundary itself, or the coarse grain boundary precipitates.

Aluminum-Copper-Lithium (2000 and 8000) Series Alloys. The Al-Cu-Li alloys, such as 2090, and Al-Li-Cu-Mg alloys, such as 8090, have higher strength than the typical Al-Cu alloys, but are more susceptible to intergranular SCC (Ref 185-187). The alloys have their highest resistance to SCC at or near the peak-age tempers. While underaging increases the susceptibility significantly, overaging has only a small effect. The susceptibility of the underaged alloys has been attributed to the precipitation of intermetallic constituent particles (Al_2CuLi) at the grain boundaries. These constituent particles are believed to be anodic to the copper-rich matrix, resulting in crack-tip anodic dissolution along the grain boundaries. With increasing aging time, copper-containing precipitates form inside the grains. It has been speculated that the formation of these precipitates may increase the anode-cathode area ratio in the microstructure such that preferential grain boundary attack is avoided. Similar behavior has been observed with SCC of Al-Li-Cu-Mg alloys such as 8090 (Ref 188).

Aluminum-Magnesium (5000) Series Alloys. Aluminum-magnesium alloys are not considered heat treatable and do not develop their strength by precipitation hardening. However, these alloys can be processed to H3 tempers, which require a final thermal stabilizing treatment to eliminate age softening, or to H2 tempers, which require a final partial annealing (Ref 182). The H116 or H117 tempers are also used for high-magnesium alloys and involve special temperature control during fabrication to achieve

a fine distribution of Mg_5Al_8 precipitates that increases the resistance to intergranular corrosion and SCC (Ref 189).

Although 5000 series aluminum alloys are not heat treatable, they can develop good strength through solution hardening by magnesium in solid solution, dispersion hardening by precipitates (Mg_2Al_3), and strain-hardening effects.

Cold-rolled and stabilized tempers of Al-Mg alloys with magnesium contents above 5% are usually very susceptible to SCC. Because the solid solutions in these alloys are more highly supersaturated, the excess magnesium tends to precipitate out as Mg_2Al_3, which is anodic to the alloy matrix (Ref 182). The precipitation of a continuous string of this phase along the grain boundaries, accompanied by little or no precipitation inside the grains, results in a highly susceptible alloy. Alloys with relatively low magnesium content, such as 5052 and 5454 (2.5 and 2.7% Mg, respectively), do not form continuous grain boundary precipitates and are therefore resistant to intergranular SCC. On the other hand, alloys exceeding magnesium concentrations of approximately 3%, such as 5083, when in strain-hardened tempers, may develop susceptible microstructures as a result of heating or even after long times at room temperature.

Although anodic dissolution appears to be the mechanism for SCC in the 5000 series aluminum alloys, some authors (Ref 190, 191) have postulated that SCC is caused by hydrogen embrittlement of strings or seams of β-phase (Al_8Mg_5), which form along the grain boundaries. If the β-phase is continuous, the susceptibility to SCC is much greater.

Aluminum-Magnesium-Silicon (6000) Series Alloys. Alloys of the 6000 series alloys are used in applications requiring intermediate strength and high resistance to SCC (Ref 192). The most common commercially used alloy is 6061. The 6000 series aluminum alloys are strengthened by precipitation hardening. Although there have been no reported cases of SCC of this group of alloys, certain abnormal thermal treatments, such as a high solution annealing temperature, followed by a slow quench, can make these alloys susceptible to SCC in the naturally aged T4 condition.

Aluminum-Zinc-Magnesium-Copper (7000) Series Alloys. The 7000 series aluminum alloys are precipitation-hardening alloys and are used in applications requiring the highest strength. One of the oldest and most commonly used alloys is alloy 7075, which was introduced as a high-strength aircraft alloy in 1943.

The resistance of most 7000 series alloys in the peak-age temper to SCC is very low when loaded in the short-transverse direction. Particularly, the low-copper-content alloys 7079-T69 and 7039-T61 and T64 are highly susceptible to SCC. Because the heat treatment has a significant effect on the susceptibility of the 7000 alloys, the various stages of heat treating have been investigated to obtain more resistant microstructures (Ref 182, 193-198). The rate of cooling from the solution treatment temperature through a critical temperature range of 398 to 288 °C (750 to 550 °F) has a pronounced effect on the susceptibility of the copper-bearing 7000 series alloys, such as 7075, to intergranular corrosion (Ref 93). Although high quenching rates increase the resistance to intergranular corrosion of copper-bearing alloys, increasing the quenching rate has no effect on the susceptibility of these alloys to SCC in the short-transverse direction. However, in the case of copper-free 7000 series alloys, the resistance to SCC in sodium chloride solutions increases with decreasing cooling rate (Ref 194).

In order to increase the resistance to SCC, a number of overaging treatments have been developed (Ref 195-198). The T73 overaging treatment, which is generally achieved by a two-stage heat treatment, provides a high degree of SCC resistance. However, this heat treatment also results in a loss of strength of about 14% compared with the T6 temper. An alternative temper is the T76 temper, which was originally developed to improve the resistance to intergranular and exfoliation corrosion. Moreover, some increase in resistance to SCC with no reduction in strength can be achieved with this temper.

Since the early development of the 7075 alloy, several 7000 series alloys have been developed to improved mechanical, fracture, corrosion, and SCC properties. These alloys include 7050, 7175, 7475, 7017, 7018, 7178, and 7079. The latter two alloys were early modifications of 7075 with the intent to improve mechanical and fracture behavior. These alloys demonstrated extreme and untreatable susceptibility to corrosion and SCC.

The SCC mechanism of 7000 alloys is extensively studied, and the majority of researchers considers SCC to be caused by hydrogen-induced cracking (Ref 184). However, there are several explanations as to the mechanism of hydrogen involvement. These include:

- Increased localized plasticity (Ref 199, 200)
- Brittle hydride formation (Ref 201-203)
- Fissures introduced in the passive film
- Reduced cohesive strength (Ref 190, 203)
- Nucleation of microvoids (Ref 204)
- Void pressurization with hydrogen bubbles (Ref 205)
- Crack blunting with hydrogen bubbles (Ref 206)

Other mechanisms that have been considered for SCC of 7000 alloys include the brittle rupture of the passive film, anodic dissolution of either the grain boundary, or the precipitates at the grain boundary.

Mechanical Factors. Many wrought aluminum alloy products have highly directional grain structures. These directional structures result in anisotropic behavior with respect to susceptibility to intergranular SCC (Ref 182). The resistance to intergranular SCC is the highest when the stress is applied in the longitudinal direction, the lowest in the short-transverse direction, and intermediate in the other directions. The differences in directional susceptibility are the most noticeable in the more susceptible tempers, such as T3 for 2000 series alloys and T6 for 7000 series alloys,

but are usually much lower in temper produced by extended precipitation (overaging) treatments, such as T6 and T8 for the 2000 series alloys and T73, T736, and T76 for the 7000 series alloys.

The component configuration and the anticipated direction and magnitude of the stress often control the selection of alloy and temper. For example, for thin section products, peak hardness tempers may be selected because there are no stresses in the short-transverse direction. On the other hand, more resistant tempers may be required for thick sections or forgings, where significant stresses may be present in the short-transverse direction.

SCC of Titanium Alloys

Titanium alloys are well known for their excellent resistance to corrosion in many aggressive environments. This resistance results from the ability of titanium to form and maintain a coherent tenacious and self-healing surface oxide film, typically rutile (TiO_2) (Ref 207, 208). In the early application of titanium alloys catastrophic failures were experienced, which were attributed to SCC. For example, in the mid-1950s titanium alloys were found to be susceptible to SCC in red-fuming nitric acid (Ref 209, 210), and in the early 1960s titanium alloy missile tanks containing anhydrous methanol or N_2O_4 experienced catastrophic stress-corrosion failures (Ref 209-211). In more recent years, titanium and its alloys have found broader application, ranging from supersonic aircraft and deep-sea submersibles to downhole application for geothermal brine and deep sour gas well service (Ref 207, 208). These particular environments could promote SCC of certain titanium alloys.

Material Factors

Titanium exists in two forms. At low temperature the metal has a close-packed hexagonal (HCP) structure—the α phase—while at temperatures above 880 °C (1650 °F) the metal has a body-centered cubic (bcc) structure—the β phase (Ref 210). The commercial titanium alloys are typically categorized into four basic types, namely: α, near-α, $\alpha + \beta$, and β. A list of selected commercial alloys is given in Table 4 (Ref 208). The α alloys include commercially pure or unalloyed titanium, and those alloys that contain α-stabilizing alloying elements, such as aluminum, tin, and the interstitials, oxygen, nitrogen, and carbon. The strength of the α alloys is derived from solid-solution hardening. The β phase is stabilized by either molybdenum, vanadium, niobium, tantalum, manganese, or chromium.

The α, near-α, and ($\alpha + \beta$) alloys are particularly susceptible to SCC in a variety of environments. In these alloys the α phase is the susceptible phase. The SCC behavior of this phase is particularly sensitive to the aluminum and oxygen content. The susceptibility of α-phase containing alloys increases at aluminum concentrations of ≥ 5 wt%, which results from the increased tendency to form Ti_3Al (α_2), which has an or-

Table 4 Designations and nominal compositions of selected commercial titanium alloys

Common alloy designation	Nominal composition, %	Alloy type
Grade 1	Unalloyed titanium	α
Grade 2	Unalloyed titanium	α
Grade 3	Unalloyed titanium	α
Grade 4	Unalloyed titanium	α
Ti-Pd	Ti-0.15Pd	α
Grade 12	Ti-0.3Mo-0.8Ni	Near α
Ti-3-2.5	Ti-3Al-2.5V	Near α
Ti-6-4	Ti-6Al-4V	$\alpha + \beta$
Ti-6-2-1-8	Ti-6Al-2Nb-1Ta-0.8Mo	Near α
Ti-5-2.5	Ti-5Al-2.5Sn	α
Ti-8-1-1	Ti-8Al-1V-1Mo	Near α
Ti-6-2-4-2	Ti-6Al-2Sn-4Zr-2Mo	Near α
Ti-4-3-1	Ti-4Al-3Mo-1V	$\alpha + \beta$
Ti-550	Ti-4Al-2Sn-4Mo-0.5Si	$\alpha + \beta$
Ti-6-6-2	Ti-6Al-6V-2Sn-0.6Fe-0.6Cu	$\alpha + \beta$
Ti-7-4	Ti-7Al-4Mo	$\alpha + \beta$
Ti-6-2-4-6	Ti-6Al-2Sn-4Zr-6Mo	$\alpha + \beta$
Ti-8Mn	Ti-8Mn	β
Beta III	Ti-11.5Mo-6Zr-4.5Sn	β
Ti-3-8-6-4-4	Ti-3Al-8V-6Cr-4Zr-4Mo	β
Ti-13-11-3	Ti-3Al-13V-11Cr	β
Ti-8-8-2-3	Ti-8V-8Mo-3Al-2Fe	β
Ti-15-5-3	Ti-15Mo-5Zr-3Al	β

Source: Ref 208

dered hexagonal structure (DO_{19}) (Ref 210, 212). This ordered phase forms in the temperature range of 400 to 700 °C (750 to 1290 °F). The microstructure has little influence on the susceptibility of α alloys, whether it is equiaxed or martensitic.

In near-α and ($\alpha + \beta$) alloys, the α phase is also the phase susceptible to SCC. Unlike in the case of α alloys, the microstructure of the near α and ($\alpha + \beta$) alloys, which results from the various heat and processing treatments, has a significant effect on the susceptibility to SCC (Ref 210). The degree of susceptibility is directly related to the grain size, volume fraction, and mean free path of the susceptible α phase. Because the β phase is ductile and resistant to SCC, it can act as crack arrester (Ref 213, 214). Thus, structures that have an increased volume fraction of β phase, or where the β phase forms a continuous network breaking up the α phase, are more resistant to SCC. Figure 19 shows the K_{Ic} and K_{Iscc} values for the ($\alpha + \beta$) alloy Ti-6Al-4V in various metallurgical conditions in a 3.5% NaCl aqueous solution (Ref 210). A fine equiaxed structure in which the β phase is dispersed throughout the α matrix, and which results from processing in the $\alpha + \beta$ field, is much more susceptible to SCC than the structure in which the β phase is lamellar, resulting from processing in the β field. A particularly SCC-resistant microstructure is the transformed β microstructure, which results in either a martensitic or so-called Widmanstätten or basketweave structure. Fine α platelets surrounded by a continuous β phase and smaller prior-β grains are more resistant to SCC than long α platelets in transformed β structures. The martensitic phases α' and α'', which are formed upon quenching from the β region, are also resistant to SCC (Ref 213, 214).

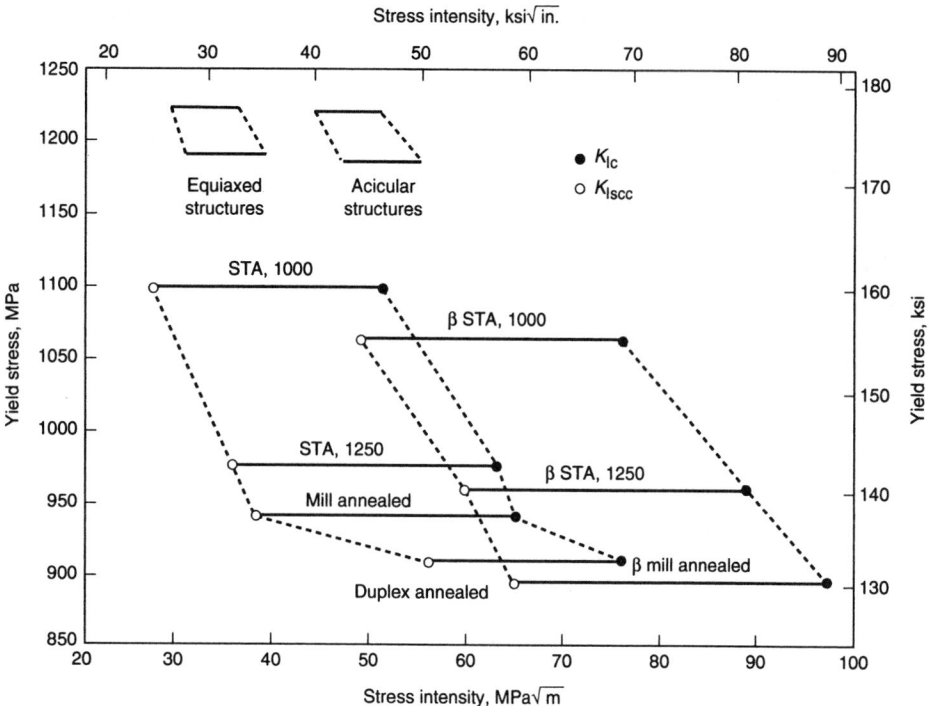

Fig. 19 Range of K_{Ic} and K_{Iscc} values for Ti-6Al-4V in various metallurgical conditions in a 3.5% aqueous NaCl solution at 24 °C (75 °F) (Ref 210)

Table 5   Environments known to promote stress-corrosion cracking of commercial titanium alloys

Medium	Temperature		Titanium alloys with known susceptibility
	°C	°F	
Oxidizers			
Nitric acid (red-fuming)	RT	RT	Ti, Ti-8Mn, Ti-6Al-4V, Ti-5Al-2.5Sn
Nitrogen tetroxide (no excess NO)	30-75	86-167	Ti-6Al-4V
Organic compounds			
Methyl alcohol (anhydrous)	RT	RT	Ti-6Al-4V, grade 2, grade 4, Ti-4Al-3Mo-1V, Ti-3Al-8V-6Cr-4Zr-4Mo, Ti-8Al-1Mo-1V, Ti-13V-11Cr-3Al, Ti-5Al-2.5Sn
Methyl chloroform	370	700	Ti-8Al-1Mo-1V, Ti-6Al-4V, Ti-5Al-2.5Sn, Ti-13V-11Cr-3Al
Ethyl alcohol (anhydrous)	RT	RT	Ti-8Al-1Mo-1V, Ti-5Al-2.5Sn
Ethylene glycol	RT	RT	Ti-8Al-1Mo-1V
Trichloroethylene	370, 620, 815	700, 1150, 1500	Ti-8Al-1Mo-1V, Ti-5Al-2.5Sn
Trichlorofluoroethane	788	1450	Ti-8Al-1Mo-1V, Ti-5Al-2.5Sn, Ti-6Al-4V, Ti-13V-11Cr-3Al
Chlorinated diphenyl	315-370	600-700	Ti-5Al-2.5Sn
Hot salt			
Chloride and other halide salts/residues	230-430	450-805	Most commercial alloys except grades 1, 2, 7, 11, 12, and 9
Metal embrittlement			
Cadmium (solid + liquid)	25-600	77-1110	Ti-8Mn, Ti-13V-11Cr-3Al, grade 2, Ti-6Al-4V
Mercury (liquid)	RT	RT	Grade 4, Ti-6Al-4V
	370	700	Ti-13V-11Cr-3Al, Ti-8Al-1Mo-1V
Silver (solid) and AgC	232-480	450-900	Ti-7Al-4Mo, Ti-5Al-2.5Sn
Ag-5Al-2.5Mn (braze alloy)	340	645	Ti-6Al-4V, Ti-8Al-1Mo-1V
Miscellaneous			
Seawater/NaCl solution	RT	RT	Unalloyed Ti (with >0.3% O), Ti-2.5Al-1Mo-11Sn-5Zr-0.2Si (IMI-679), Ti-3Al-11Cr-13V, Ti-5Al-2.5Sn, Ti-8Mn, Ti-6Al-4V, Ti-6Al-6V-2Sn, Ti-7Al-2Nb-1Ta, Ti-4Al-3Mo-1V, Ti-8Al-1Mo-1V, Ti-6Al-2Sn-4Zr-6Mo
Distilled water	RT	RT	Ti-8Al-1Mo-1V, Ti-5Al-2.5Sn, Ti-11.5Mo-6Zr-4.5Sn
Chlorine gas	288	550	Ti-8Al-1Mo-1V
10% HCl	35, 340	95, 645	Ti-5Al-2.5Sn, Ti-8Al-1Mo-1V
LiCl, KBr, and Na$_2$SO$_4$ solution (0.6M)	RT	RT	Ti-6Al-4V, Ti-6Al-6V-2Sn
Molten chloride/bromide salts	300-500	570-930	Ti-8Al-1Mo-1V

RT, room temperature. Source: Ref 208

Beta Alloys. Although the β titanium alloys are more resistant to SCC, they are somewhat susceptible to either transgranular or intergranular SCC, depending on the alloy composition and microstructure (Ref 215, 216). Intergranular cracking has been observed in a few β alloys where fine α precipitates have formed at lower aging temperatures. Beta-phase alloys where the β phase is stabilized with molybdenum, vanadium, niobium, or tantalum are resistant to SCC, whereas alloys stabilized by the eutectic element manganese and chromium are susceptible to transgranular SCC. Transgranular SCC of the β phase is known to occur in Ti-8Mn with a β + α structure and in solution-treated Ti-13V-11Cr-3Al (Ref 213).

Environmental Factors

Although titanium and titanium alloys are resistant to SCC in many common operating environments, there are several environments that promote SCC. The environments that are known to promote SCC of specific titanium alloys are listed in Table 5 (Ref 208). Some of the more common cracking environments include aqueous solutions and organic compounds such as methanol.

Aqueous Environments. The α and (α + β) titanium alloys are susceptible to SCC in aqueous environments, and the degree of susceptibility depends on the species present in the solution, the pH, temperature, and viscosity of the solution. In certain metallurgical conditions, some highly susceptible titanium alloys have exhibited susceptibility to SCC in distilled water (Ref 210). These alloys include Ti-8Al-1V-1Mo, Ti-5Al-2.5Sn, and Ti-11.5Mo-6Zr-4.5Sn. Additions of halides, chloride, bromide, and iodide ions increase the susceptibility and make alloys that are otherwise not susceptible to SCC in distilled water susceptible to cracking. Chloride ions have been found to have the greatest effect on the susceptibility to SCC, and hence much of the data available have been generated in 3.5% NaCl and seawater at ambient temperature (Ref 217-219). An example of the effect of chloride ion concentration on the cracking velocity of a susceptible alloy is given in Fig. 20 (Ref 219).

The addition of other anions does not increase the susceptibility to SCC and may in some cases inhibit SCC. Examples of such neutral or inhibiting ions are NO_3^-, SO_4^{2-}, OH^-, CrO_4^{2-}, and PO_4^{3-} (Ref 217-218).

Solution pH and temperature and electrochemical potential have a significant effect on the susceptibility of titanium alloys to SCC. Increasing acidity increases the susceptibility to cracking, while increasing alkalinity appears to have no obvious or significant effect. At high hydroxide concentrations, greater than $1M$, inhibition may be expected. Little data are available on the effect of temperature. While the critical stress intensity for crack initiation (K_{Iscc}) in Ti-8Al-1V-1Mo in a neutral 3.5% NaCl solution does not vary with temperature, the crack velocity is strongly temperature dependent (Fig. 21) (Ref 219).

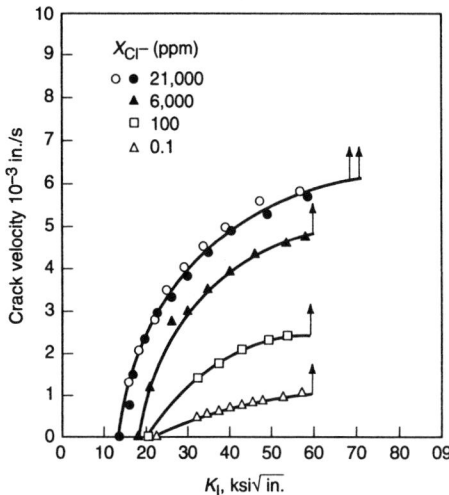

Fig. 20 Influence of chloride concentration on the SCC behavior of Ti-8Al-1Mo-1V in aqueous chloride solutions at 25 °C (77 °F) (Ref 219)

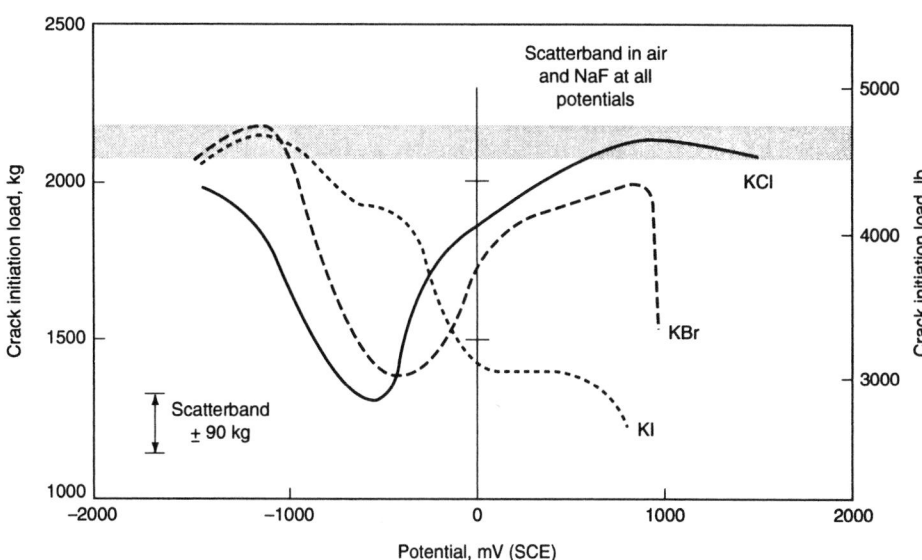

Fig. 22 Effect of potential on crack initiation stress for $\alpha + \beta$ titanium alloys in various halide solutions at 25 °C (77 °F) (Ref 220)

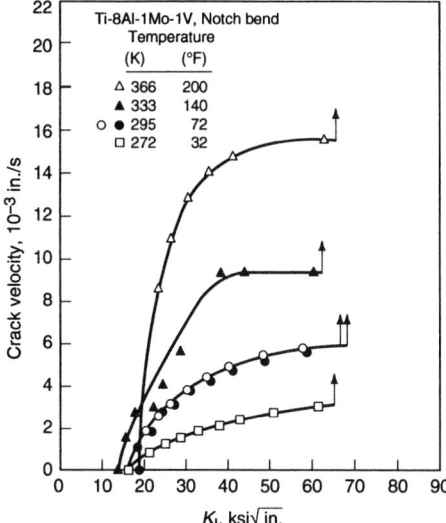

Fig. 21 Effect of temperature on SCC velocity of Ti-8Al-1Mo-1V (notch-bend specimens) in a 3.5% aqueous NaCl solution (Ref 219)

The effects of potential must be considered because titanium alloy components are often coupled to other metals when incorporated into a structure. Anodic or cathodic polarization tends to inhibit stress-corrosion crack initiation and increase the K_{Iscc} of several susceptible alloys, as is illustrated in Fig. 22 (Ref 220). This effect has not been observed for the highly susceptible alloys, such as Ti-8Al-1V-1Mo.

Three basic mechanisms have been proposed for SCC in aqueous environments (Ref 213-214). The first mechanism is a film rupture model, which claims that crack nucleation results from planar slip and the formation of wide slip steps that rupture the protective surface oxide film. The crack nucleation is highly dependent on (1) the degree of slip planarity and slip-step width, which concentrates the slip, (2) the oxide film repassivation kinetics, and (3) the strain rate.

A second mechanism is based on anodic dissolution at highly localized sites of stress concentration. This model assumes a balance between the rate of dissolution at the crack tip, the crack-tip stress intensity, and the crack-tip environment. Fractographic evidence has demonstrated that SCC in the α phase is mechanical in nature, and therefore a third mechanism may be more appropriate.

The third mechanism is based on a hydrogen-assisted cracking phenomenon (Ref 221). Hydrogen that results from the corrosion reaction is absorbed into the matrix and can migrate and concentrate near the crack tip. Localized hydrogen embrittlement ahead of the crack tip may then promote crack propagation. Only the α phase in α and $(\alpha + \beta)$ alloys are subject to hydrogen embrittlement, because local precipitation of brittle titanium hydride forms ahead of the crack tip, promoting crack propagating by cleavage through the hydride. The β phase in $(\alpha + \beta)$ alloys is not susceptible to hydrogen embrittlement, but has high hydrogen solubility and can therefore act as means of transport for hydrogen.

Methanol. Titanium alloys are highly susceptible to SCC on methanol liquid and vapor (Ref 210). The fracture mode can be intergranular or transgranular, depending on alloy composition and stress intensity (Ref 222). Intergranular failure in methanol is observed in alloys that are not susceptible to SCC in aqueous solutions, such as the grade 1 and 2 alloys, and β alloys such as Ti-13V-11Cr-3Al. Intergranular SCC in methanol generally involves anodic dissolution and requires little or no stress to propagate. The level of impurities in methanolic solutions has a significant effect on this failure mode. In fact, intergranular SCC in methanolic solution requires traces of halides or halogen (Ref 223). Furthermore, water represents a cracking inhibitor when added above a certain level. For example, a safe minimum water content to prevent intergranular SCC of Ti-6Al-4V in methanol is 1.5 wt%.

Transgranular SCC in methanol is generally observed in those alloys that are susceptible to SCC in aqueous solutions. As in the case of intergranular SCC, a trace level of halides or halogens is required for transgranular SCC to occur, and water can act as an inhibitor when added in sufficient quantity. However, SCC of alloys that are susceptible to cracking in distilled water, such as Ti-8Al-1V-1Mo cannot be inhibited by adding water. Numerous other species have been identified that inhibit transgranular SCC in methanolic solutions (Ref 223). These include nitrate and sulfate ions, and metallic ions such as Al^{3+}, Zr^{4+}, Cd^+, and Sn^{2+}.

The SCC mechanism is a mixture of oxide film rupture, dissolution, and hydrogen embrittlement (Ref 208). Crack nucleation is associated with rupture of the passive oxide film. Because TiO_2 formation is not possible in a water-free or water-lean environment, titanium continues to dissolve, forming a nonprotective titanium methylate, or methoxide, surface film:

$$\text{no } H_2O: Ti + 4CH_3OH \rightarrow Ti(OCH_3)_4 + 4H$$

$$\text{with } H_2O: Ti + 3CH_3OH + H_2O \rightarrow TiOH(OCH_3)_3 + 4H$$

In both reactions atomic hydrogen is a by-product that can be absorbed into the matrix and promote crack propagation by hydrogen embrittlement, such as hydride formation in the α matrix.

Other Environments. In addition to the environments discussed above, titanium alloys have demonstrated various degrees of susceptibility to SCC to several environments. These environments include:

- Organic solvents, such as carbon tetrachloride (CCl_4), methylene chloride (CH_2Cl_2), methylene iodide (CH_2I_2), trichlorethane (CH_3CCl_3), and trichlorethylene (C_2HCl_3) (Ref 210)
- Hot chloride salt (Ref 210, 224)

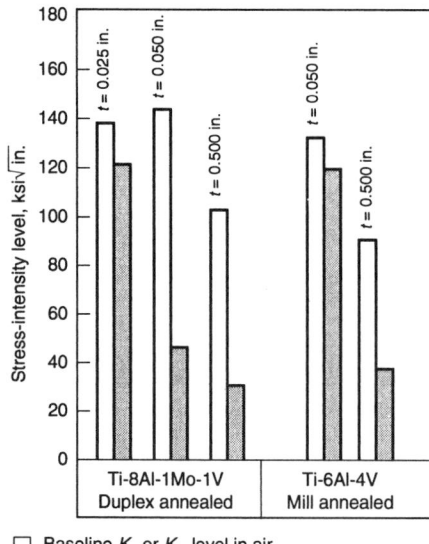

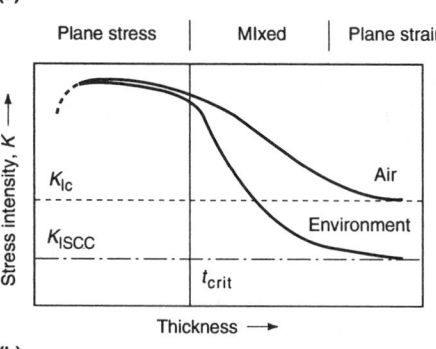

Fig. 23 Effect of specimen thickness on the SCC susceptibility of titanium alloys. (a) Fracture toughness
of duplex-annealed Ti-8Al-1Mo-1V and mill-annealed Ti-
6Al-4V, tested in air and in 3.5% NaCl. (b) Variation of fracture toughness with specimen thickness. t_{crit}, specimen
thickness below which SCC does not occur

- Oxidizing nitrogen-oxide compounds, such as
  nitrogen tetroxide (N_2O_4), and red-fuming nitric acid (Ref 225-227)
- Molten salts (Ref 228)
- Liquid and solid metals, such as cadmium (Ref
  210, 229-231)
- Gaseous environments, including hydrogen
  gas (Ref 221) and moist chlorine gas (Ref 232)

Mechanical Factors in Ti-Alloys

Most commercial titanium alloys demonstrate
resistance to SCC in loaded smooth or notched
configurations but become susceptible at very
high stress intensities, such as those associated
with highly stressed precracked components. In
addition to the stress-concentration effects, material thickness, orientation and loading mode, and
rate can have a significant effect on the SCC
behavior (Ref 233, 234). Figure 23 shows a diagram that indicates that the susceptibility to SCC
decreases with decreasing thickness and that
there is a critical thickness below which SCC
does not occur (Ref 234). This critical thickness

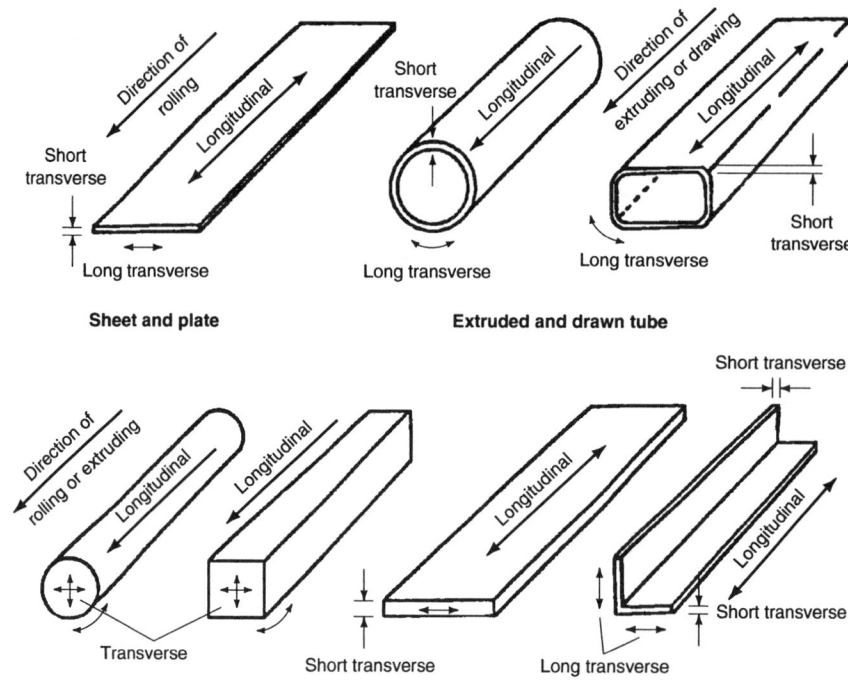

Fig. 24 Grain orientations in standard wrought forms of alloys (Ref 236)

is dependent on the alloy composition, heat treatment, orientation, and loading rate. Apparently,
the critical thickness relates to the transition from
plane strain to plane stress.

The mode of SCC in α, near-α, and (α + β)
alloys is transgranular cleavage, with the fracture
plane along {1017}, which is a 15° angle from
the basal plane (0001) (Ref 221, 235). The basal
plane is usually aligned parallel to the rolling
direction, and therefore, the susceptibility of the
α-titanium alloys is strongly orientation dependent, particularly in processed material. Figure 23
shows that the resistance of these alloys to transgranular SCC is the least in the transverse direction, resistance to SCC: TL < SL < LT (Ref 230).

Because the cleavage plane for the bcc β phase
is typically {100}, transgranular SCC of susceptible alloy is less orientation dependent.

Evaluation of Stress-Corrosion Cracking

In order to determine the susceptibility of alloys to SCC, several types of testing are available.
If the objective of testing is to predict the service
behavior or to screen alloys for service in a specific environment, it is often necessary to obtain
SCC information in a relatively short period of
time, which requires acceleration of testing by
increasing the severity of the environment or the
critical test parameters. The former can be accomplished by increasing the test temperature or
the concentration of corrosive species in the test
solution and by electrochemical stimulation. Test
parameters that can be changed to reduce the
testing time include the application of higher

stresses, continuous straining, and precracking,
which allows by-passing of the crack nucleation
phase of the SCC process.

Stress-corrosion specimens can be divided into
two main categories, namely smooth, and precracked or notched specimens. Further distinction can be made in the loading mode, such as
constant deflection, constant load, and constant
extension or strain rate. These different loading
modes will be discussed in more detail in the
following sections.

During alloy processing operations used in the
production of wrought alloys, the metal is forced
in a predominant direction, so that the grains are
elongated in the direction of flow. Because it is
important to relate the application of stress and
the grain flow direction, two conventions are
used to relate the two parameters. In one system,
which is primarily used for smooth specimens,
the three stressing directions are designated by
indicating the direction of the stress, namely longitudinal (L), long-transverse (LT), transverse
(T), and short-transverse (Fig. 24) (Ref 236).

A second system, which is particularly useful
for precracked specimens, indicates both the
cracking plane and the direction of crack propagation. The system uses three letters (L, T, and
W) to indicate three perpendicular directions,
namely L for the longitudinal direction, T for the
thickness direction, and W for the width direction. The crack plane is indicated by the direction
normal to the crack, and the crack propagation is
indicated by one of the directions L, T, or W.
Figure 25 demonstrates the various orientations
for a double-cantilever-beam (DCB) specimen
(Ref 237).

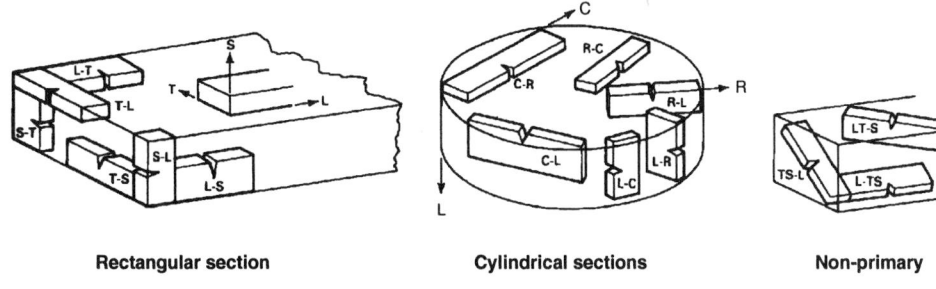

Fig. 25 Fracture plane identification. L, direction of grain flow; T, transverse grain direction; S, short transverse grain direction; C, chord of cylindrical cross section; R, radius of cylindrical cross section; first letter, normal to the fracture plane; second letter, direction of crack; propagation in fracture plane (Ref 237)

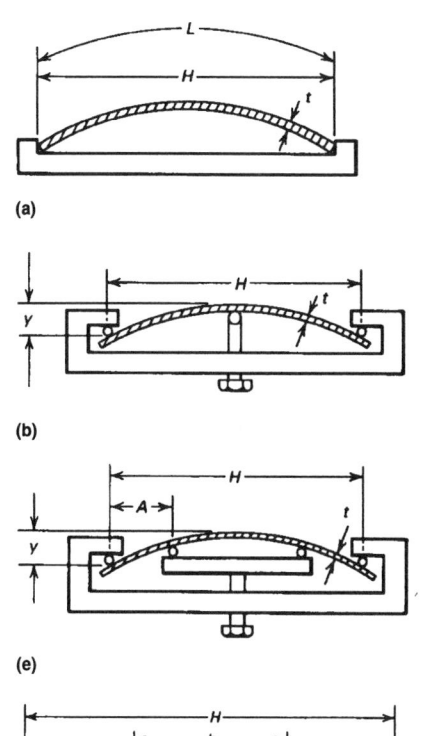

(a)

(b)

(e)

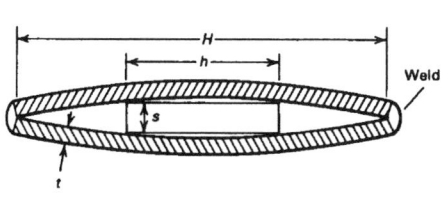

(d)

Fig. 26 Schematic specimen and holder configurations for bent-beam specimens. (a) Two-point loaded specimen. (b) Three-point loaded specimen. (c) Four-point loaded specimen. (d) Welded double-beam specimen (Ref 239)

Other parameters that play an important role in SCC testing are surface condition and residual stress. The nucleation of stress-corrosion cracks strongly depend on initial surface reactions, and thus the surface condition of the test specimens, particularly smooth specimens, has a significant effect on the test results. Smooth test specimens are often tested with a mechanically (machined or abraded) or (electro)chemically treated surface. It is very important to avoid or to remove machining marks or scratches perpendicular to the loading direction (Ref 238).

Smooth Specimen Testing

Smooth SCC specimens allow for the evaluation of the total SCC life, which includes crack nucleation and propagation. Testing can be conducted under constant extension or strain, constant load, and constant extension or strain rate. The selection of a specific test method for SCC strongly depends on the particular service application, and the time allowed for testing.

Constant Extension Testing. Constant extension or constant strain tests on smooth specimens are widely used and do not require elaborate testing fixtures. Depending on the specific configuration of the test articles, different types of constant extension tests are being used, the most common being bent-beam, U-bend, C-ring, and tensile type specimens.

Bent-Beam Specimens. The different types of bent-beam specimens are illustrated in Fig. 26 (Ref 239). These specimens may be used to test sheet plate and flat extruded material, or wires and extrusions with a circular cross section. The figure shows that the bending can be accomplished in several ways depending on the dimensions of the specimen. Stressing of the specimen is accomplished by bending the specimen in a stressing device, while restraining the ends. During stress-corrosion testing both specimen and stressing device are exposed to the test environment. The most simple loading arrangement is the two-point loaded bent-beam, which can only be used on relatively thin sheet or wire material. The elastic stress at the mid-point of the specimen can be estimated from the following equation:

$$L = (ktE/\sigma)\sin^{-1}(H/ktE)$$

where L is the specimen length, σ is the maximum stress, E is the elastic modulus, H is the length of

holder, t is the specimen thickness, and k is the empirical constant (1.280).

Three-Point Bend Specimens. Three-point bend tests are commonly used because of the ease of load application and the ability to use the same loading rigs for different stresses. The load is applied by turning a bolt in the rig, deflecting the specimen. The elastic stress at the mid-point of the specimen is calculated from the following equation:

$$\sigma = 6Ety/H^2$$

where σ is the maximum tensile stress, E is the elastic modulus, t is the specimen thickness, y is the maximum deflection, and H is the length of holder.

This test has a number of disadvantages. First, dissimilar metal corrosion and/or crevice corrosion can occur under the bolt. Secondly, once the crack has formed, the stress condition changes such that the outer layer of the specimen is not subject to a tensile stress only, but to a complex combination at tensile and bending stresses. The propagating crack will then deviate from the centerline. Thus, the three-point bend test can only be used as a qualitative test to assess the susceptibility to stress-corrosion cracking. With the four-point bend test, described in the next paragraph, tensile stresses can be maintained during the growth of the crack.

Four-Point Bend Specimens. Four-point bend testing provides a uniform tensile stress over a relatively large area of the specimen. The elastic stress in the outer layer of the specimen between the two inner supports can be calculated from the following equation:

$$\sigma = 12Ety/(3H^2 - 4A^2)$$

where σ is the maximum tensile stress, E is the elastic modulus, t is the specimen thickness, y is the maximum deflection, H is the distance between outer supports, and A is the distance between outer and inner supports.

U-bend specimens are prepared by bending a strip 180° around a mandrel with a predetermined radius (Fig. 27). The figure shows that bends less than 108 degrees are also used. Standardized test methods are described in ASTM G 30 (Ref 240). Because of the ease of fabrication, a large amount of specimens can be fabricated, and this test is therefore widely used to qualitatively evaluate the

susceptibility of alloy and heat treatment to stress-corrosion cracking.

A good approximation of the strain at the apex of the U-bend is:

$$\varepsilon = t/2R, \text{ when } t < R$$

where t is the specimen thickness and R is the radius of the bend.

Then, an appropriate value for the maximum stress can be obtained from the stress-strain curve of the test material.

C-ring specimens are commonly used to determine the susceptibility to stress-corrosion cracking of alloys in different product forms (Ref 241). This test is particularly useful for testing of tubing, rod, and bar in the short-transverse direction, as is illustrated in Fig. 28. The specimens are typically bolt loaded to a constant strain or constant load per ASTM G 338 (Ref 241) and if the stresses in the outer layers of the apex of the C-rings are in the elastic region, the stresses can be accurately calculated using the following equations:

$$D_f = D - \Delta$$

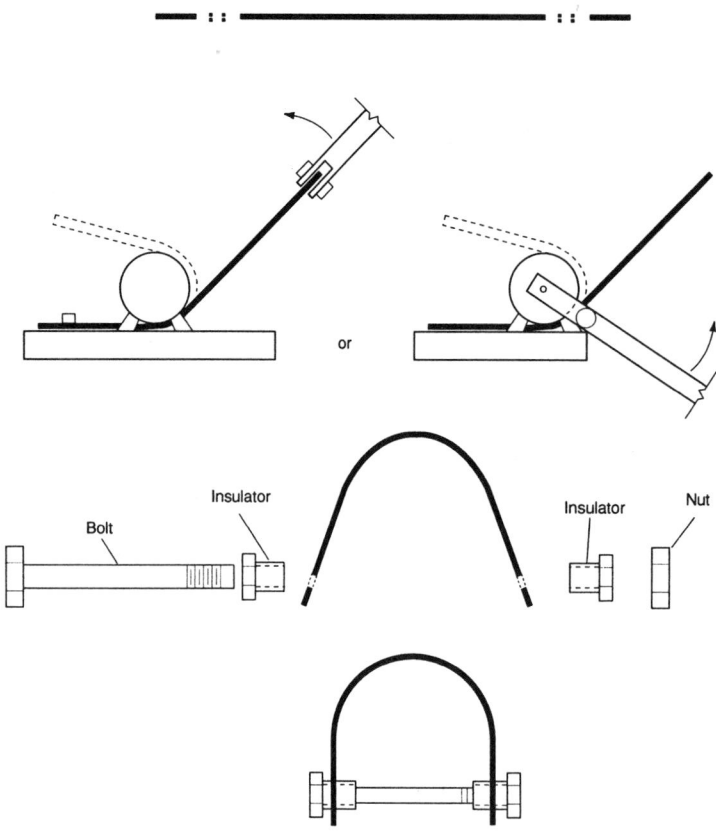

Fig. 27 Schematic two-stage stressing of a U-bend specimen (Ref 240)

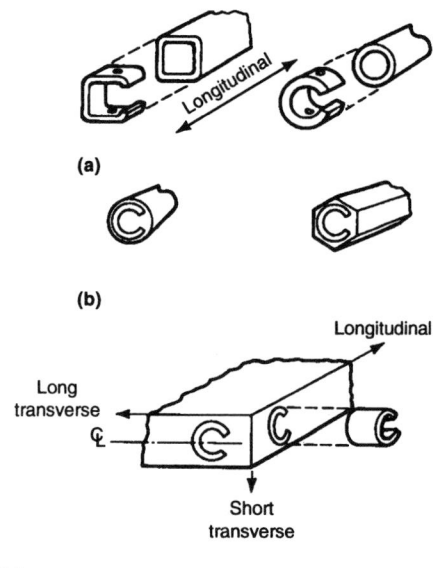

Fig. 28 Sampling procedure for testing various products with C-rings. (a) Tube. (b) Rod and bar. (c) Plate (Ref 241)

$$\Delta = \sigma d^2/4EtZ$$

where D is the outer diameter of the C-ring before stressing, D_f is the outer diameter of stressed C-ring, σ is the elastic stress, Δ is the change of D at the desired stress, d is the mean diameter $(D - t)$, t is the wall thickness, E is the elastic modulus, and Z is the correction factor for curved beam.

The stress on C-ring specimens can be more accurately determined by attaching circumferential and transverse strain gages to the stressed surface. The circumferential (σ_C), and transverse (σ_T) elastic stresses can be calculated with (Ref 242, 243):

$$\sigma_C = E/(1-\mu^2) \times (\varepsilon_C + \mu\varepsilon_T)$$

$$\sigma_T = E/(1-\mu^2) \times (\varepsilon_T + \mu\varepsilon_C)$$

where E is the elastic modulus, μ is the Poisson's ratio, ε_C is the circumferential strain, and ε_T is the transverse strain.

Tensile Specimens. For specific purposes, such as alloy development, a large number of stress-corrosion specimens need to be evaluated. Tensile specimens have been used for this purpose where specimens used to determine tensile properties in air are adapted to SCC, as discussed in ASTM G 49. When uniaxially loaded in tension, the stress pattern is simple and uniform, and the magnitude of the applied stress can be accurately determined. Specimens can be quantitatively stressed by using equipment for application of either a constant load, a constant strain, or an increasing load or strain.

This type of test is one of the most versatile methods of SCC testing because of the flexibility permitted in the type and size of the test specimen, the stressing procedures, and the range of stress level. It allows the simultaneous exposure of unstressed specimens (no applied load) with stressed specimens and subsequent tension testing to distinguish between the effects of true SCC and mechanical overload.

A wide range of test specimen sizes can be used, depending primarily on the dimensions of the product to be tested. Stress-corrosion test results can be significantly influenced by the cross section of the test specimen. Although large specimens may be more representative of most structures, they often cannot be prepared from the available product forms being evaluated. They also present more difficulties in stressing and handling in laboratory testing.

Smaller cross-sectional specimens are widely used. They have a greater sensitivity to SCC initiation, usually yield test results rapidly, and permit greater convenience in testing. However, the smaller specimens are more difficult to machine, and test results are more likely to be influenced by extraneous stress concentrations resulting from nonaxial loading, corrosion pits, and so on. Therefore, use of specimens less than about 10 mm (0.4 in.) in gage length and 3 mm (0.12 in.) in diameter is not recommended, except when testing wire specimens.

Tension specimens containing machined notches can be used to study SCC and hydrogen embrittlement. The presence of a notch induces a triaxial stress state at the root of the notch, in which the actual stress will be greater by a concentration factor that is dependent on the notch geometry. The advantages of such specimens include the localization of cracking to the notch region and acceleration of failure. However, unless directly related to practical service conditions, the results may not be relevant.

Tension specimens can be subjected to a wide range of stress levels associated with either elastic or plastic strain. Because the stress system is intended to be essentially uniaxial (except in the case of notched specimens), great care must be exercised in the construction of stressing frames to prevent or minimize bending or torsional stresses.

The simplest method of providing a constant load consists of a dead weight hung on one end of the specimen. This method is particularly useful for wire specimens. For specimens of larger cross section, however, lever systems such as those used in creep-testing machines are more practical. The primary advantage of any dead-weight loading device is the constancy of the applied load.

Constant-strain SCC tests are performed in low-compliance tension-testing machines. The specimen is loaded to the required stress level, and the moving beam is then locked in position. Other laboratory stressing frames have been used, generally for testing specimens of smaller cross section.

Constant Load Testing. Although the constant extension tests are widely used for evaluating the susceptibility of alloys to stress-corrosion cracking because of the ease of specimen prepa-

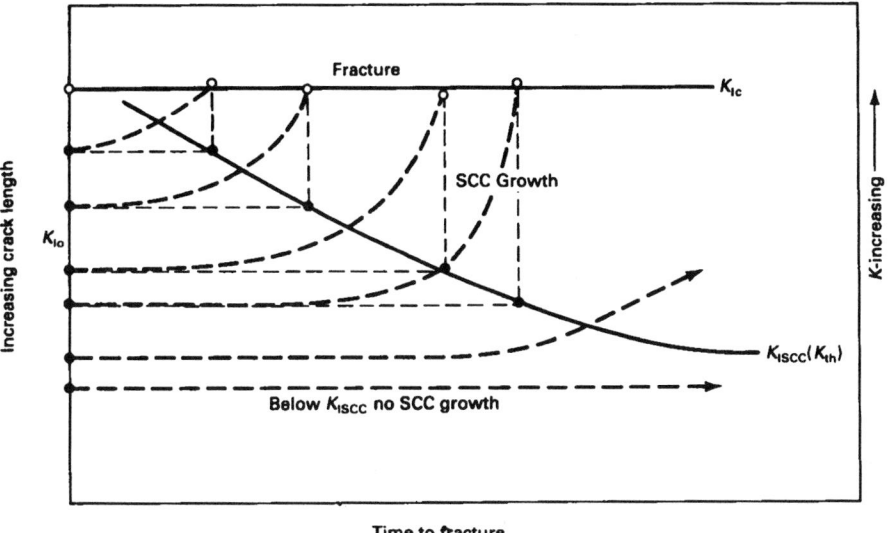

(a)

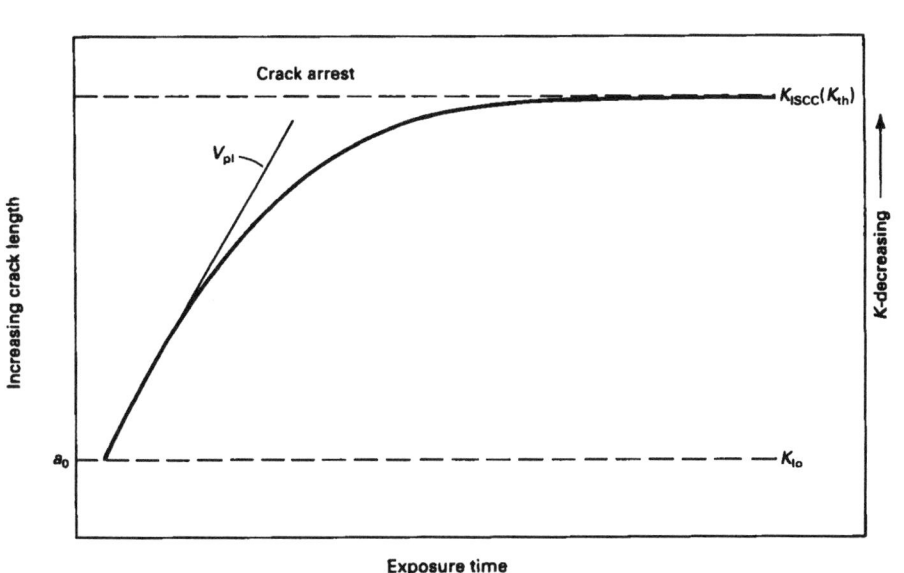

(b)

Fig. 29 Schematic comparison of determination of threshold stress integrity factor (K_{Iscc} or K_{th}). (a) Constant-load (K-increasing) test. (b) Constant crack opening displacement (K-decreasing) test (Ref 244)

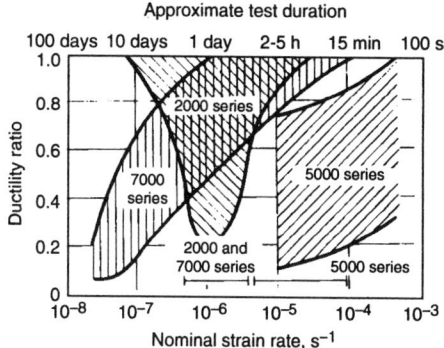

Fig. 30 Strain rate regimes for SCC of 2000, 5000, and 7000 series aluminum alloys in a 3% aqueous NaCl solution plus 0.3% H_2O_2 (Ref 245)

indicates that the principal mechanism for cracking of the 2000 series alloys is film rupture—anodic dissolution model, while the predominant mechanism for cracking of the 7000 series alloys is hydrogen embrittlement.

The parameters that are typically measured in slow strain rate testing to determine the susceptibility to SCC are:

- Time to failure
- Percent elongation
- Percent reduction in cross-sectional area at the fracture surface
- Reduction in ultimate (UTS) and yield (YTS) tensile stress
- Presence of secondary cracking on the specimen gage section
- Appearance of the fracture surface

In order to assess the susceptibility of a material to SCC, the results of the slow strain rate test in a particular environment must be compared with those in an inert environment, such as dry nitrogen gas.

Precracked Specimen Testing

The use of precracked specimens in the evaluation of SCC is based on the engineering concept that all structures contain cracklike flaws (Ref 246, 247). Moreover, precracking can contribute to the susceptibility to SCC of alloys such as titanium alloys, and this susceptibility may not always be evident from smooth specimens.

Precracking eliminates the uncertainties that are associated with crack nucleation and can provide a flaw geometry for which a stress analysis is available through fracture mechanics. Expressing stress-corrosion characteristics in terms of fracture mechanics provides a relationship between applied stress, crack length, and crack growth in a corrosive environment. When the plasticity can be ignored, or in other words, when the plastic zone ahead of the propagating crack is below a certain value and a triaxial or plane strain stress state exists at the crack tip, linear elastic fracture mechanics (LEFM) can be applied to describe the relationship between crack length (a) and the applied stress (σ) by the stress intensity factor K:

ration and the ability to test a large number of specimens at one time, there is one major drawback. Once stress-corrosion cracks have formed, the gross cross-section stress decreases, which will eventually cause the crack to stop. Application of a constant or a static load provides an alternative test method that represents some actual field conditions that can provide threshold values. It should be cautioned, however, that such threshold values are strongly dependent on the method of loading (i.e., dead weight or spring) and the specimen size and cannot be considered a materials property. Moreover, Fig. 29 (Ref 244) shows that as a crack develops, the stress at the crack tip increases, possibly decreasing the time-to-failure.

Constant Strain Rate Testing. Constant or slow strain rate testing is a very useful technique to evaluate the susceptibility of materials to SCC in a relatively short period of time (Ref 244, 245). Typical strain rates range between 10^{-5}/s and 10^{-7}/s, but for most materials the typical strain rate is at 10^{-6}/s. The strain sensitivity to SCC can change for different alloys, even of the same metal. Figure 30 (Ref 245) shows that for the 2000 series aluminum alloys, the critical strain rate for the highest susceptibility to cracking is 10^{-6}/s, whereas no such critical strain rate exists for the 7000 series aluminum alloys. This difference in slow strain rate behavior of the two alloys may indicate different mechanisms for stress-corrosion cracking. The slow strain rate behavior

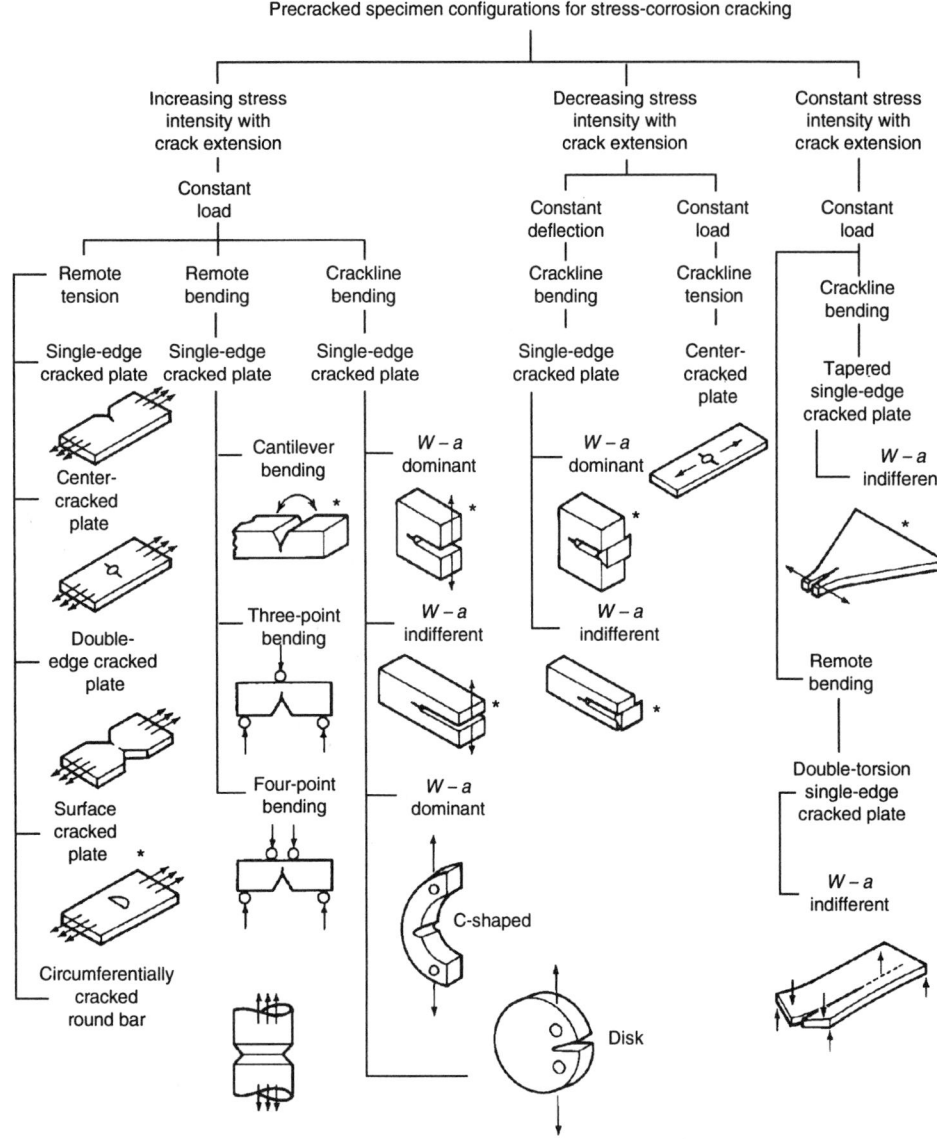

Fig. 31 Classification of precracked specimens for SCC testing (Ref 248)

$$K = \sigma\sqrt{a} \cdot F$$

where F is a polynomial factor that accounts for the specimen geometry. Linear elastic fracture mechanics, and thus the K factor, cannot be used to describe the relationship between applied stress and the crack length when there is significant plasticity or when the stress state at the crack tip is biaxial or plane stress. Then, a more fundamental parameter, the J integral, is used.

Almost all standard plane strain fracture mechanics test specimens can be adapted to SCC testing. Several examples are illustrated schematically in Fig. 31 (Ref 248). ASTM Standard E 399 describes the allowable specimen dimensions and test procedures for precracked specimens.

Specimen Preparation. When using precracked fracture mechanics specimens, specific dimensional requirements need to be considered, as well as crack configuration and orientation.

The basic dimensional requirement for application of linear elastic fracture mechanics is that dimensions are such that plane strain condition can be maintained. In general, for a valid K measurement, neither that crack length nor the specimen thickness should be less than $2.5\,(K_{Ic}/\sigma_Y)^2$.

Several designs of initial crack configuration are available. ASTM E 399 recommends that the notch root radius is not greater than 0.127 mm (0.005 in.), unless a chevron notch is used, in which case it may be 0.25 mm (0.01 in.). In order to start out with a crack as sharp as possible, ASTM E 399 describes procedures for precracking. The K level used for precracking should not exceed about two-thirds of the intended initial K value. This procedure prevents the forming of compressive stresses at the crack tip, which may alter the SCC behavior of the alloys.

Aluminum alloys can also be precracked by the pop-in method, where the wedge-opening method is used to the point of tensile overload.

This method cannot be used for steels and titanium alloys, because of the strength of these alloys.

Loading Procedures. Stress-corrosion crack growth in precracked specimens can be studied in K-increasing and K-decreasing tests (Ref 244). In constant load or K-increasing tests, crack growth results in increased crack opening, which keeps the environment at the crack tip and corrosion products from interfering with crack growth. One of the problems with this mode of loading is that with increasing K, the plastic zone ahead of the crack tip may increase and at some point interfere with crack propagation. Moreover, for this type of testing bulky and relatively expensive equipment is required.

Constant displacement (K-decreasing) tests do not have the problems of the K-increasing tests indicated above. The plastic zone ahead of the crack tip does not increase with increasing crack size, so that the stress condition always remains in the plane strain mode. Also, the constant displacement tests can be self-loaded, and thus external testing equipment is not needed. Because in these tests the stress-intensity factor decreases with increasing crack growth, the stress-corrosion threshold stress intensity factor (K_{Iscc}) can be easily determined by exposing a number of specimens loaded to different initial K_1 values. This can even be accomplished by crack arrest in one specimen.

A major problem with this test method occurs when corrosion products form in the crack, blocking the crack mouth and interfering with the environment at this crack tip. Moreover, the oxide can wedge open the crack and change the originally applied displacement and load.

Measurement of Crack Growth. In order to quantify the crack growth behavior in precracked stress-corrosion specimens, the crack length needs to be monitored, so that the crack velocity (da/dt) can be calculated, and the relationship between the increasing K and the crack velocity can be determined. There are basically three methods to monitor the growth of stress corrosion cracks: (1) visual/optical measurements, (2) measurement of the crack-opening displacement using clip gages, and (3) the potential drop measurement, which monitors the increase in resistance across two on either side of the propagating crack (Ref 249).

REFERENCES

1. R.H. Jones and R.E. Ricker, Mechanism of Stress-Corrosion Cracking, *Stress-Corrosion Cracking: Materials Performance and Evaluation*, R.H. Jonas, Ed., ASM International, 1992

2. A. Turnbull, *Advances in Localized Corrosion*, NACE-9, H. Isaacs, et al., Ed., National Association of Corrosion Engineers, 1990

3. J.A. Begley and J.D. Landes, *Proc. 1971 National Symposium on Fracture Mechanisms, Part III*, STP 514, ASTM, 1972, p 1

4. G.M. Ugianski and J.H. Payer, Ed., *Stress-Corrosion Cracking—The Slow Strain-Rate Technique*, STP 665, ASTM, 1979

5. J.A. Beavers and G.H. Koch, "Limitations of the Slow Strain Rate Test for Stress-Corrosion Cracking," Publication No. 39, Materials Technology Institute of the Chemical Process Industries (MTI), 1995

6. R.N. Parkins, *Br. Corrosion J.*, Vol 14, 1979, p 5

7. H.J. Engle, in *Theory of Stress-Corrosion Cracking in Alloys*, NATO, 1971, p 86

8. J.C. Scully, *Corros. Sci.*, Vol 15, 1975, p 207

9. D.A. Vermilyea, *J. Electrochem. Soc.*, Vol 119, 1972, p 405

10. D.A. Vermilyea, in *Stress-Corrosion Cracking and Hydrogen Embrittlement of Iron Base Alloys*, NACE, 1977, p 208

11. R.W. Staehle, Theory of Stress-Corrosion Cracking in Alloys, NATO, 1971, p 223

12. R.W. Staehle, in *Stress-Corrosion Cracking and Hydrogen Embrittlement of Iron Base Alloys*, National Association of Corrosion Engineers, 1977, p 37

13. E.N. Pugh, *Corrosion*, Vol 41 (No. 9), 1985, p 517

14. J.M. Silcock and P.R. Swann, *Environment-Sensitive Fracture of Engineering Materials*, Z.A. Foroulis, Ed., The Metallurgical Society, 1979, p 133

15. S.P. Lynch, *Hydrogen Effects in Metals*, A.W. Thompson and I.M. Bernstein, Ed., The Metallurgical Society, 1981, p 80

16. S.P. Lynch, *Mater. Sci.*, Vol 15 (No. 10), 1981, p 403

17. S.P. Lynch, *J. Mater. Sci.*, Vol 20, 1985, p 3329

18. A.J. Forty and P. Humble, *Philos. Mag.*, Vol 8, 1963, p 247

19. E.N. Pugh, *Stress Corrosion Cracking and Hydrogen Embrittlement of Iron Based Alloys*, National Association of Corrosion Engineers, 1977, p 37

20. A.J. Forty, *Physical Metallurgy of Stress Corrosion Fracture*, T.N. Rhodin, Ed., Interscience, 1959, p 99

21. K. Sieradzki and R.C. Newman, *Philos. Mag. A*, Vol 5 (No. 1), 1985, p 95

22. H.H. Uhlig, *Physical Metallurgy of Stress Corrosion Fracture*, T.N. Rhodin, Ed., Interscience, 1959, p 1

23. N.S. Stoloff, *Environment-Sensitive Fracture of Engineering Materials*, Z.A. Foroulis, Ed., The Metallurgical Society, 1979, p 486

24. S.W. Ciaraldi, Stress Corrosion Cracking of Carbon and Low-Alloy Steels Strength Less Than 1241 MPa, *Stress-Corrosion Cracking: Materials Performance and Evaluation*, R.H. Jones, Ed., ASM International, 1992

25. G. Sandoz, *Metall. Trans.*, Vol 2, 1971, p 1055

26. M.A. Eaglesham et al., *Theoret. Appl. Frac. Mech.*, Vol 10, 1988, p 97

27. R.N. Parkins, *Conf. Theory of Stress Corrosion Cracking in Alloys*, J.C. Scully, Ed., NATO Science Committee Research Evaluation Conference (Portugal), 1971, p 167

28. L.M. Long and H.H. Uhlig, *J. Electrochem. Soc.*, Vol 112, 1965, p 964

29. J. Flis and J.C. Scully, *Corrosion Sci.*, Vol 8, 1968, p 235

30. J.P. Frazer and G.G. Eldredge, *Corrosion*, Vol 14, 1958, p 524t

31. L.M. Long and N.A. Lockington, *Corros. Sci.*, Vol 7, 1967, p 447

32. S.W. Ciaraldi, *Proc. 1981 Mechanical Working and Steel Processing Conf.*, AIME, 1981, p 605

33. E. Snape, *Corros. Anticorros.*, Vol 24, 1968, p 261

34. A.E. Schuetz and W.D. Robertson, *Corrosion*, Vol 13, 1957, p 437t

35. N. Bailey, *Met. Constr. Brit. Weld. J.*, Vol 2, 1970, p 339

36. M. Henthorne and R.N. Parkins, *Brit. Corros. J.*, Vol 5, 1967, p 186

37. S. Das and G. Thomas, *Trans. ASM*, Vol 62, 1969, p 55

38. F. Watkinson et al., *Brit Weld. J.*, Vol 10, 1963, p 54

39. D.A. Vaughan and D.I. Phalen, Reactions Contributing to the Formation of Susceptible Paths for Stress Corrosion Cracking, STP 429, ASTM, 1967, p 209

40. M.T. Wang and R.W. Staehle, *Proc. Hydrogen in Metals Int. Conf.* (Paris), 1972, p 342

41. W.M. Cain and A.R. Troiano, *Petrol. Eng.*, Vol 37, May 1965, p 78

42. R.L. McGlasson et al., *Corrosion*, Vol 16, 1960, p 113

43. R.H. Jones et al., *Corrosion*, Vol 45 (No. 6), 1989, p 494

44. C. Lea and E.D. Hondros, *Proc. R. Soc. (London)*, Vol 377A, 1981, p 477

45. R.A. Davis, *Corrosion*, Vol 19, 1963, p 45t

46. N.A. Tiner and G.A. Gilpin, *Corrosion*, Vol 22, 1966, p 271

47. E.H. Phelps, *Proc. Conf. Fundamental Aspects of Stress Corrosion Cracking*, Ohio State University, 1967

48. E.H. Phelps, *Proc. 7th World Petroleum Congr.*, Elsevier, 1967, p 27

49. C.S. Carter and M.V. Hyatt, *Stress Corrosion Cracking and Hydrogen Embrittlement in Iron Base Alloys*, Vol 6, National Association of Corrosion Engineers, 1977, p 524

50. H.H. Johnson and A.M. Wilmer, *Appl. Mater. Res.*, Vol 4, 1965, p 33

51. A.H. Freedman, "Development of an Accelerated Stress Corrosion Test For Ferrous and Nickel Base Alloys," NOR 68-58, NASA Contract NAS-20333, Northrup Corp., 1968

52. S. Yamamoto and T. Fujita, *Fracture, 1969, Proc. Int. Conf. Fracture*, P.L. Pratt, Ed., Chapman and Hall, 1969, p 425

53. S. Fukui and A. Asada, *Trans. Iron Steel Inst. Jpn.*, Vol 9, 1969, p 448

54. G. Sandoz, High Strength Steels, *Stress Corrosion Cracking in High Strength Steels and in Titanium and Aluminum Alloys*, B.F. Brown, Ed., Naval Research Laboratory, 1969

55. M.J. May and A.H. Priest, "The Influence of Solution pH on Stress Corrosion Cracking Resistance of a Low Alloy High Strength Steel," MGA/A/45/68, BISRA, British Steel Corp., 1968

56. J.B. Greer et al., *Corros. Sci.*, Vol 28, 1972, p 328

57. E. Snape, *Corrosion*, Vol 23, 1967, p 154

58. C.M. Hudgins, et al., *Corrosion*, Vol 22, 1966, p 238

59. J.D. Gilchrist and R. Narayan, *Corros. Sci.*, Vol 11, 1971, p 281

60. D.R. Johnston and T.G. McCord, *Proc. NACE 26th Conf.*, National Association of Corrosion Engineers, 1970

61. L.M. Dvorecek, *Corrosion*, Vol 26, 1972, p 177

62. H.E. Townsend, *Corrosion*, Vol 28, 1972, p 99

63. H.T. Effinger, et al., *Oil Gas J.*, No. 2, 1951, p 99

64. W.T. Chandler and R.T. Walter, "Effects of High Pressure Hydrogen on Steels," paper presented to ASM and AWS, American Society for Metals, 1967

65. J.S. Laws, et al., "Hydrogen Gas Pressure Vessel Problems in the M-1 Facilities," NASA CR-1305, National Aeronautics and Space Administration, 1969

66. W. Haufman and W. Rouls, *Weld. J.*, Vol 44, 1965, p 225s

67. R.J. Walter and W.T. Chandler, *Mater. Sci. Eng.*, Vol 8, 1971, p 90

68. D.P. Williams and H.G. Nelson, *Metall. Trans.*, Vol 1, 1970, p 63

69. G.G. Hancock and H.H. Johnson, *Trans. Metall. Soc. AIME*, Vol 326, 1966, p 513

70. H.W. Liu and F.J. Ficalora, *Int. J. Fract. Mech.*, Vol 8, 1972, p 223

71. R.A. Yeske, Corrosion by Kraft Pulping Liquors, *Metals Handbook*, Vol 13, 9th ed., *Corrosion*, ASM International, 1987, p 1210

72. F.P.A. Robinson and L.G. Neil, *2nd Int. Congr. Metallic Corrosion*, National Association of Corrosion Engineers, 1963

73. K. Bohenkamp, *Proc. Conf. Fundamental Aspects of Stress Corrosion Cracking*, R.W. Staehle, Ed., National Association of Corrosion Engineers, 1967, p 374

74. A.A. Berk and W.F. Waldeck, *Chem. Eng.*, Vol 57 (No. 6), 1950, p 235

75. H.W. Schmidt, et al., *Corrosion*, Vol 7, 1951, p 295

76. H. Grafen, *Corros. Sci.*, Vol 7, 1967, p 177

77. G. Herbsleb and W. Schwenk, *Stahl Eisen.*, Vol 90, 1970, p 903

78. W. Radeker and H. Grafen, *Stahl Eisen.*, Vol 76 (No. 11), 1956, p 1616

79. J.E. Reinoehl and W.E. Berry, *Corrosion*, Vol 28, 1972, p 151

80. T.J. Dawson, *Weld J.*, Vol 35, 1956, p 50

81. J.M. Suttcliffe, et al., *Corrosion*, Vol 28, 1972, p 313

82. R.N. Parkins, *Conf. Theory of Stress Corrosion Cracking in Alloys*, J.C. Scully, Ed., NATO Science Committee Research Evaluation Conference (Portugal), 1971, p 167

83. J.A. Beavers and R.N. Parkins, *7th Symposium On-Line Pipe Research*, American Gas Association (AGA), 1986

84. J.T. Justice and J.D. Mackenzie, *Proc. NG-18/E PRG 7th Biennial Joint Tech. Mtg. Line Pipe Research*, Paper No. 28, 1988

85. J.A. Beavers and N.G. Thompson, "Electrochemical Studies On-Line Pipe Steel in NS3 Groundwater," Final Report to TransCanada Pipelines Ltd., 1989
86. B.A. Harle and J.A. Beavers, *Corrosion*, Vol 49 (No. 10), 1993, p 861
87. K. Rohrbach and M. Schmidt, Maraging Steels, *Metals Handbook*, Vol 1, 10th ed., *Properties and Selection: Iron, Steels, and High-Performance Alloys*, ASM International, 1990
88. J.W. Kennedy and J.A. Whittaker, *Corros. Sci.*, Vol 8 (No. 6), 1968, p 359
89. G. Sandoz, *Metall. Trans.*, Vol 2, 1971, p 1055
90. G. Sandoz, *Metall. Trans.*, Vol 3, 1972, p 1169
91. C.S. Carter, *Corrosion*, Vol 25 (No. 10), 1969, p 423
92. S.K. Banerji, et al., *Metall. Trans. A*, Vol 9, 1978, p 237
93. G.E. Kerns, M.T. Wong, and R.W. Staehle, *Stress-Corrosion Cracking and Hydrogen Embrittlement of Iron Base Alloys*, NACE-5, R.W. Staehle, et al., National Association of Corrosion Engineers, 1971, p 70
94. D.P. Dautovich and S. Floreen, *Stress Corrosion Cracking and Hydrogen Embrittlement of Iron Base Alloys*, NACE-5, R.W. Staehle, et al., Ed., National Association of Corrosion Engineers, 1971, p 798
95. R.P.M. Proctor and H.W. Paxton, *Trans. ASM*, Vol 62, 1969, p 989
96. C.S. Carter, *Metall. Trans.*, Vol 2, 1971, p 1621
97. R.T. Ault, et al., *Trans. ASM*, Vol 60, 1967, p 79
98. J.J. DeLuccia, *Hydrogen Embrittlement: Prevention and Control*, STP 962, L. Raymond, Ed., ASTM, 1988, p 17
99. H.P. Leckie, *Proc. of Conf. on Fundamental Aspects of Stress Corrosion Cracking*, National Association of Corrosion Engineers, 1969, p 411
100. W.A. Van Der Sluys, *Eng. Fract. Mech.*, Vol 1, 1969, p 447
101. R.A. Oriani, *Hydrogen Effects in Environment-Induced Cracking of Metals*, NACE-10, R.P. Gangloff and M.B. Ives, Ed., National Association of Corrosion Engineers, 1990, p 439
102. N.J. Petch and P. Stables, *Nature*, Vol 169, 1952, p 842
103. A.R. Troiano, *Trans. ASM*, Vol 52, 1960, p 54
104. R.A. Oriani, *Ber Bunsenges. Phys. Chem.*, Vol 76, 1972, p 848
105. C.D. Beacham, *Metall. Trans.*, Vol 3, 1972, p 437
106. C.L. Fu and G.S. Painter, *J. Mater. Res.*, Vol 6 (No. 4), 1991, p 719
107. H.K. Birnbaum, *Atomistics of Fracture*, R.M. Latanision and J.R. Pickens, Ed., Plenum Press, 1983, p 733
108. P. Sofronis and H.K. Birnbaum, Hydrogen Localized Plasticity: A Mechanism for Hydrogen Related Fracture in AD, Vol 36, *Fatigue and Fracture of Aerospace Structural Materials*, American Society of Mechanical Engineers, 1993, p 15
109. A.J. Sedricks, Stress Corrosion Cracking of Stainless Steels, Chapter 4, *Stress Corrosion Cracking: Materials Performance and Evaluation*, R.H. Jones, Ed., ASM International, 1992
110. V.N. Krivonbok, *The Book of Stainless Steels*, E.E. Thum, Ed., American Society for Steel Treating, 1933
111. A.J. Sedriks, *Corrosion of Stainless Steels*, John Wiley and Sons, 1979
112. J.E. Alexander, "Alternative Alloys for BWR Piping Applications," Final Report, NP-2671-LD, General Electric Co., Oct 1982
113. R.M. Latanison and R.W. Staehle, *Proc. Conf. Fundamental Aspects of Stress Corrosion Cracking*, National Association of Corrosion Engineers, 1969, p 214
114. G.J. Theus and R.W. Staehle, *Stress Corrosion Cracking and Hydrogen Embrittlement of Iron Base Alloys*, National Association of Corrosion Engineers, 1977, p 845
115. A.J. Sedriks, *Corrosion*, Vol 42 (No. 7), 1986, p 376
116. M.A. Scheil, *Symposium on Stress Corrosion Cracking of Metals*, STP 64, ASTM, 1945, p 395
117. A.S. Couper, *Mater. Protect.*, Vol 8, 1969, p 17
118. C. Edeleanu, *J. Iron Steel Inst.*, Vol 173, 1953, p 140
119. L. Bedman, *Corrosion*, Vol 35, 1979, p 96
120. M.A. Streicher, "Stress Corrosion of Ferritic Stainless Steels," Paper No. 68, presented at Corrosion/75, National Association of Corrosion Engineers, 1975
121. B.F. Brown, *Stress Corrosion Cracking Control Measures*, Monograph 156, National Bureau of Standards, 1977, p 55
122. R.F. Steigerwald, A.P. Bond, H.J. Dundas, and E.A. Lizlovs, *Corrosion*, Vol 33, 1977, p 279
123. I.L.W. Wilson, F.W. Pement, and R.G. Aspden, "Stress Corrosion Studies on Some Stainless Steels in Elevated Temperature Aqueous Environments," Paper No. 136 presented at Corrosion/77, National Association of Corrosion Engineers, 1977
124. J.F. Grubb and J.R. Maurer, "Use of Cathodic Protection With Superferritic Stainless Steels," Paper No. 28 presented at Corrosion/84, National Association of Corrosion Engineers, 1984
125. L.S. Redmerski, J.J. Eckenrod, K.E. Pinnow, and C.W. Kovach, "Cathodic Protection of Seawater Cooled Power Plant Condensors Operating With High Performance Ferritic Stainless Steel Tubing," Paper No. 208, presented at Corrosion/85, National Association of Corrosion Engineers, 1985
126. H. Spaehn, *Environment-Induced Cracking of Metals*, R.P. Gangloff and M.B. Ives, Ed., National Association of Corrosion Engineers, 1990, p 449
127. J.E. Truman, *Int. Met. Rev.*, Vol 26 (No. 6), 1981, p 301
128. S. Shimodaira, et al., *Stress Corrosion Cracking and Hydrogen Embrittlement of Iron Base Alloys*, R.W. Staehle, et al., Ed., National Association of Corrosion Engineers, 1977, p 1003
129. R.J. Schmitt and E.H. Phelps, *J. Met.*, March 1970, p 47
130. J.F. Bates and A.W. Loginow, *Corrosion*, Vol 20, 1964, p 189t
131. M.A. Scheil, *Symposium on Stress Corrosion Cracking of Metals*, STP 64, ASTM, 1945, p 395
132. M.O. Speidel, "Stress Corrosion Cracking of Austenitic Stainless Steels," Report to the Advanced Research Project Agency, ARPA Order No. 2616, Contract No. N00014-75-C-0703, Ohio State University, Aug 1977
133. F.P. Ford, D. Taylor, P.L. Andersen, and R.G. Ballinger, "Environmentally Controlled Cracking of Stainless and Low Alloy Steels in Light Water Reactors," Final Report, EPRI Contract No. RP2000-6, Electric Power Research Institute, 1986
134. F.P. Ford and M. Silverman, *Corrosion*, Vol 36 (No. 10), 1980, p 558
135. F.P. Ford and P.L. Andersen, "The Theoretical Prediction of the Effect of System Variables on the Cracking of Stainless Steels and Its Use in Design," Paper No. 83, presented at Corrosion/87, National Association of Corrosion Engineers, 1987
136. A. Dravnieks and C. Samans, *Proc. Am. Pet. Inst.*, Vol 37 (No. 111), 1957, p 100
137. C. Samans, *Corrosion*, Vol 20 (No. 8), 1964, p 256t
138. S. Ahmad, M.L. Mehta, S.K. Saraf, and I.P. Saraswat, *Corrosion*, Vol 38 (No. 6), 1964, p 347
139. S. Ahmad, M.L. Mehta, S.K. Saraf, and I.P. Saraswat, *Corrosion*, Vol 37 (No. 7), 1981, p 412
140. I.S. Isaacs, B. Vyas, and M.W. Kendig, *Corrosion*, Vol 38 (No. 3), 1982, p 130
141. A.R. McIlree and H.T. Michels, *Corrosion*, Vol 33, 1977, p 60
142. G.C. Wheeler and E. Howells, *Power*, Sept 1960, p 86
143. E. Howells, *Corros. Technol.*, 1960, p 368
144. A.V. Ryabchenkov, V.I. Gerasinov, and V.P. Sidoroc, *Prof. Met. (USSR)*, Vol 2, May-June 1966, p 217
145. R.W. Staehle and A.K. Agrawal, "Corrosion, Stress Corrosion Cracking and Electrochemistry of the Fe and Ni Base Alloys in Caustic Environments," Report to ERDA, Contract E(11-1)-2421, Ohio State University, 1976
146. R.M. Davidson, T. Debold, and M.J. Johnson, *Metals Handbook*, Vol 13, 9th ed., *Corrosion*, ASM International, 1987, p 547
147. N. Sridhar and G. Cragnolino, Stress Corrosion Cracking of Nickel-Base Alloys, Chapter 5, *Stress-Corrosion Cracking: Materials Performance and Evaluation*, R.H. Jones, Ed., ASM International, 1992
148. A.I. Asphahani, Corrosion of Nickel-Base Alloys, *Metals Handbook*, Vol 13, 9th ed., *Corrosion*, ASM International, 1987
149. Y.C. Fayman, *Mater. Sci. Eng.*, Vol 82, 1986, p 203

150. J.R. Crum, M.E. Adkins, and W.G. Lipscomb, *Mater. Perform.*, Vol 25, 1986, p 27

151. M.T. Miglin, J.V. Monter, C.S. Wade, and J.L. Nelson, *Proc. of 5th Int. Symposium on Environmental Degradation of Materials in Nuclear Power Systems—Water Reactors*, American Nuclear Society, Aug 1991, p 279

152. H.J. Wernik, *Topologically Close Packed Structures in Intermetallic Compounds*, John Wiley and Sons, 1967

153. M.A. Streicher, *Corrosion*, Vol 32, 1976, p 79

154. M.O. Speidel, *Metall. Trans. A*, Vol 12, 1981, p 779

155. J. Kolts, "Heat Treatment of Environmental Embrittlement of High Performance Alloys," Paper No. 407, presented at Corrosion/86, National Association of Corrosion Engineers, 1986

156. L. Graf and W. Wittich, *Werkst. Korros.*, Vol 5, 1966, p 385

157. J. Kolts, "Laboratory Evaluation of Corrosion Resistant Alloys for the Oil and Gas Industry," Paper No. 323, presented at Corrosion/86, National Association of Corrosion Engineers, 1986

158. R.N. Tuttle and R.D. Kane, Ed., "H2S Corrosion in Oil and Gas Production—A Completion of Classic Papers," National Association of Corrosion Engineers, 1981

159. R.D. Kane, *Int. Met. Rev.*, Vol 30, 1985, p 291

160. P.R. Rhodes, "Stress Cracking Risks in Corrosive Oil and Gas Wells," Paper No. 322, presented at Corrosion/86, National Association of Corrosion Engineers, 1986

161. A.I. Asphahani, "High Performance Alloys for Deep Sour Gas Wells," Paper No. 42, presented at Corrosion/78, National Association of Corrosion Engineers, 1988

162. M. Watkins, H.E. Chaung, and G.A. Vaughn, "Laboratory Testing of the SCC Resistance of Stainless Alloys," Paper No. 283, presented at Corrosion/87, National Association of Corrosion Engineers, 1987

163. S.W. Ciaraldi, "Stress Corrosion Cracking of Corrosion-Resistant Alloy Tubulars in Highly Sour Environments," Paper No. 284, presented at Corrosion/87, National Association of Corrosion Engineers, 1987

164. S.M. Wilhelm, "Effect of Elemental Sulfur on Stress Corrosion Cracking of Nickel Base Alloys in Deep Sour Gas Well Production," Paper No. 77, presented at Corrosion/87, National Association of Corrosion Engineers, 1987

165. N. Shridhar and S.M. Corey, "The Effect of Elemental Sulfur on Stress Cracking of Ni-Base Alloys," Paper No. 12, presented at Corrosion/89, National Association of Corrosion Engineers, 1989

166. G.A. Gragnolino and D.D. Macdonald, *Corrosion*, Vol 38, 1982, p 406

167. R.C. Scarberry, S.C. Peatman, and J.R. Crum, *Corrosion*, Vol 32, 1976, p 401

168. P. Berge, *3rd Int. Symp. Environmental Degradation of Materials in Nuclear Power Systems—Water Reactors*, G.J. Theus and J.R. Weeks, Ed., The Metallurgical Society

169. C.S. Welty, Jr. and J.C. Blomgren, Steam Generator Issues, *Proc. 4th Int. Symp. Environmental Degradation of Materials in Nuclear Power Systems—Water Reactors*, D. Cubicciotti, Ed., National Association of Corrosion Engineers, 1990, p 127

170. A.R. McIlree, *Proc. Int. Symp. Environmental Degradation of Materials in Nuclear Power Systems—Water Reactors*, National Association of Corrosion Engineers, 1985, p 838

171. H. Hanninen and I. Aho-Mantila, *Proc. 3rd Int. Symp. Environmental Degradation of Materials in Nuclear Power Systems—Water Reactors*, G.J. Theus and J.R. Weeks, Ed., The Metallurgical Society, 1988, p 77

172. R.L. Cowan and G.M. Gordon, Intergranular Stress Corrosion Cracking and Grain Boundary Composition in Fe-Ni-Cr Alloys, *Stress Corrosion Cracking and Hydrogen Embrittlement of Iron Base Alloys*, R.W. Staehle, et al., Ed., National Association of Corrosion Engineers, 1977, p 1023

173. D. Van Rooyen, *Corrosion*, Vol 31, 1975, p 327

174. R. Copson and G. Economy, *Corrosion*, Vol 24, 1968, p 55

175. I.L.W. Wilson and R.G. Aspden, Caustic Stress Corrosion Cracking of Iron-Nickel Chromium Alloys, *Stress Corrosion Cracking and Hydrogen Embrittlement of Iron Base Alloys*, R.W. Staehle, et al., Ed., National Association of Corrosion Engineers, 1977, p 1109

176. G. Theus, *Nucl. Technol.*, Vol 28, 1976, p 388

177. J.R. Crum, *Corrosion*, Vol 38, 1982, p 40

178. K.H. Lee, G. Cragnolino, and D.D. Macdonald, *Corrosion*, Vol 41, 1985, p 540

179. J.R. Crum, *Corrosion*, Vol 42, 1986, p 368

180. D.O. Sprowls and R.H. Brown, *Resistance of Wrought High-Strength Aluminum to Stress Corrosion*, Technical Paper 17, Aluminum Company of America, 1962

181. F.H. Haynie and W.K. Boyd, "Stress Corrosion Cracking of Aluminum Alloys," DMIC Report 228, Battelle Mem. Inst., July 1966

182. M.V. Hyatt and M.O. Speidel, *High Strength Aluminum Alloys in Stress Corrosion Cracking in High Strength Steels and in Titanium and Aluminum Alloys*, B.F. Brown, Ed., Naval Research Laboratories, 1972

183. E.H. Dix, *Trans. AIME*, Vol 137, 1940, p 11

184. T.D. Burleigh, *Corrosion*, Vol 47 (No. 2), 1991, p 89

185. J.G. Rinker, M. Marek, and T.H. Sanders, Jr., *Microstructure, Toughness, and SCC Behavior of 2020 in Aluminum-Lithium Alloys*, T.H. Sanders, Jr. and E.A. Starke, Jr., Ed., American Institute of Mining, Metallurgical, and Petroleum Engineers, 1983, p 597

186. A.K. Vasudevan, P.R. Ziman, S.C. Jha, and T.H. Sanders, Jr., *Stress Corrosion Resistance of Al-Cu-Li-Zr Alloys in Aluminum-Lithium Alloys*, Vol III, G. Baker, P.J. Gregson, S.J. Harris, and P.J. Peel, Ed., The Institute of Metals, 1986, p 303

187. E.L. Calvin, S.J. Murtha, and R.K. Wyss, The Effect of Aging Time on the Stress-Corrosion Cracking Resistance of 2090-T8E41, *Proc. of*

Int. Conf. Aluminum Alloys (Charlottesville, VA), 15-20 June 1986

188. N.J.H. Holroyd, A. Gray, G.M. Scamans, and R. Herman, *Environ-Sensitive Fracture of Al-Li-Cu-Mg Alloys in Aluminum Lithium Alloys*, Vol III, C. Balia, P.J. Gregson, S.J. Harris, and C.J. Peel, Ed., The Institute of Metals, 1986, p 310

189. C.L. Brooks, *Nav. Eng. J.*, Vol 82 (No. 40), 1970, p 29

190. W. Gruhl, *Z. Metallkd.*, Vol 75 (No. 11), 1984, p 819

191. Z. Cui, *Acta. Metall. Sin.*, Vol 20 (No. 6), 1984, p B323

192. D.O. Sprowls and R.H. Brown, *Proc. Conf. on Fundamental Aspects of Stress Corrosion Cracking*, National Association of Corrosion Engineers, 1969, p 466

193. H.Y. Hunsicker, The Metallurgy of Heat Treatment, *Aluminum V.I., Properties, Physical Metallurgical, and Phase Diagrams*, D.R. Van Horn, Ed., American Society for Metals, 1967, p 109

194. W.D. Vernan and W.A. Anderson, "Thermal Treatment of Aluminum Base Alloy Products," U.S. Patent 3,171,760, 1965

195. G.H. Koch and D.T. Kolijn, *J. Heat Treat.*, Vol 1 (No. 2), 1979, p 3

196. M.O. Speidel, *Proc. Conf. Fundamental Aspects of Stress Corrosion Cracking*, 1969, p 561

197. D.O. Sprowls and J.A. Nock, Jr., "Thermal Treatment of Aluminum Base Alloy Articles," U.S. Patent 3,198,676, 1965

198. J.A. Vaccari, *Mater. Eng.*, Vol 71, 1970, p 22

199. S.P. Lynch, *Acta. Metall.*, Vol 36 (No. 10), 1988, p 2639

200. G.M. Bond, I.M. Robertson, and W.K. Birnbaum, *Acta. Metall.*, Vol 36 (No. 8), 1988, p 2193

201. C.D.S. Tuck, *Hydrogen Effects in Metals*, I.M. Bernstein and A.W. Thompson, Ed., American Institute of Mining, Metallurgical, and Petroleum Engineers, 1981, p 503

202. S.W. Ciaraldi, J.L. Nelson, R.A. Yeske, and E.N. Pugh, *Hydrogen Effects in Metals*, I.M. Bernstein and A.W. Thompson, American Institute of Mining, Metallurgical, and Petroleum Engineers, 1981, p 437

203. Z.-X. Tong, S. Lin, and C.M. Hsiao, *Metall. Trans. A*, Vol 20A (No. 5), 1989, p 925

204. T. Ohnishi and K. Higashi, *J. Jpn. Inst. Light Met.*, Vol 34 (No. 11), 1984, p 850

205. H.P. Van Leeuwen, *Corrosion*, Vol 29, 1973, p 197

206. K. Rajan, W. Wallace, and J.C. Beddoes, *J. Mater. Sci.*, Vol 17, 1982, p 2817

207. R.W. Schutz, *Titanium, Process Industries Corrosion—The Theory of Practice*, National Association of Corrosion Engineers, 1986, p 503

208. R.W. Schutz, Stress Corrosion of Titanium Alloys, *Stress Corrosion Cracking—Materials Performance and Evaluation*, R.H. Jones, Ed., ASM International, 1995

209. *Proc. Int. Symposium on Stress Corrosion Mechanisms in Titanium Alloys*, Jan 1971,

Georgia Institute of Technology and National Association of Corrosion Engineers

210. M.J. Blackburn, W.H. Smyrl, and J.A. Feeney, Titanium Alloys in Stress-Corrosion Cracking, *High Strength Steels and in Titanium and Aluminum Alloys*, B.F. Brown, Ed., Naval Research Laboratories, 1972

211. "Accelerated Crack Propagation of Titanium in Methanol Halogenated Hydrocarbons, and Other Solutions," DMIC Memorandum 228, DMIC, Battelle Memorial Institute, March 1967

212. M.J. Blackburn, *Trans. Amer. Soc. Met.*, Vol 62, 1969, p 147

213. R.J.H. Wanhill, *Br. Corros. J.*, Vol 10 (No. 2), 1975, p 69

214. J. Brettle, *Met. Mater.*, 1972, p 442

215. D.T. Powell and J.C. Scully, *Corrosion*, Vol 24 (No. 6), 1968, p 151

216. T.R. Beck, *J. Electrochem. Soc.*, Vol 115, 1968, p 890

217. T.R. Beck, M.J. Blackburn, W.H. Smyrl, and M.O. Speidel, "Stress Corrosion Cracking of Titanium Alloys: Electrochemical Kinetics, SCC Studies with Ti-8-1-1, SCC and Polarization Curves in Molten Salts, Liquid Metal Embrittlement, and SCC Studies with Other Titanium Alloys," Contract NAS 7-409, Quarterly Progress Report 14, Boeing Scientific Research Laboratory, 1969

218. N.G. Feige and T. Murphy, *Met. Eng. Q.*, Vol 7 (No. 1), 1967, p 53

219. J.D. Boyd, P.J. Moreland, W.K. Boyd, R.A. Wood, D.N. Williams, and R.I. Jaffe, "The Effect of Composition on the Mechanism of Stress Corrosion Cracking of Titanium Alloys in N2O4 and Aqueous and Hot-Salt Environments," Contract NAS-100 (09), Battelle Memorial Institute, 1969

220. T.R. Beck and M.J. Blackburn, *AIAA J.*, Vol 6 (No. 2), 1968, p 326

221. G.H. Koch, A.J. Bursle, R. Lui, and E.N. Pugh, *Metall. Trans. A.*, Vol 12A, 1981, p 1833

222. A.J. Sechiles, J.A.S. Green, and P.W. Slattery, *Corrosion*, Vol 24 (No. 6), 1968, p 172

223. E.G. Haney and P. Fugassi, *Corrosion*, Vol 27, 1971, p 99

224. S.P. Rideout, R.S. Ondrejan, and M.R. Louthan, Hot Stress-Corrosion Cracking of Titanium Alloys, *The Science Technology and Application of Titanium*, Pergamon Press, 1970

225. J.D. Jackson and W.K. Boyd, "Corrosion of Titanium," DMIC Memorandum 218, DTIC Battelle Memorial Institute, Sept 1966

226. W.K. Boyd, Stress Corrosion of Titanium and Its Alloys, *Proc. Int. Symposium on Stress Corrosion Mechanisms in Titanium Alloys*, Georgia Institute of Technology and National Association of Corrosion Engineers, Jan 1991

227. H.B. Bomberger, *Corrosion*, Vol 13 (No. 5), 1957, p 17

228. H.L. Logan, *Proc. Conf. Fundamental Aspects of Stress-Corrosion Cracking*, National Association of Corrosion Engineers, 1969, p 662

229. R.E. Stolz and R.H. Stulen, *Corrosion*, Vol 35 (No. 4), 1979, p 165

230. D.N. Fager and W.F. Spurr, *Corrosion*, Vol 26 (No. 10), 1970, p 409

231. D.A. Meyn, *Corrosion*, Vol 29 (No. 5), 1973, p 192

232. R.E. Adams and E. Von Tieschenhausen, *Proc. Conf. Fundamental Aspects of Stress Corrosion Cracking*, National Association of Corrosion Engineers, 1969, p 691

233. R.E. Curtis, R.R. Boyer, and J.C. Williams, *Trans. ASM*, Vol 62, 1969, p 457

234. D.E. Piper, S.H. Smith, and R.V. Carter, *Met. Eng. Q.*, Vol 8 (No. 30), 1968, p 50

235. D.A. Meyn, "A Study of the Crystallographic Orientation of Cleavage Facets Produced by Stress-Corrosion Cracking of Ti-7Al-2Nb-1Ta in Water," Report of NRL Progress, National Research Laboratory, 1965, p 21

236. D.B. Franklin, "Design Criteria for Controlling Stress Corrosion Cracking," George C. Marshall, Space Flight Center Report MSFC-SPEC-522, National Aeronautics and Space Administration, Jan 1977

237. "Standard Method Test for Plane-Strain Fracture Toughness Testing of Metallic Materials," E 399, ASTM

238. "Standard Practice for Preparing, Cleaning, and Evaluating Corrosion Test Specimens," G 1, ASTM, 1979

239. "Standard Practice for Preparation and Use of Bent-Beam Stress Corrosion Test Specimens," G 39, ASTM, 1979

240. "Standard Recommended Practice for Mailing and Using U-Bend Stress Corrosion Test Specimens," G 30, ASTM, 1979

241. "Standard Recommended Practice for Making and Using C-Ring Stress Corrosion Test Specimens," G 338, ASTM, 1979

242. *Stress Corrosion Testing*, STP 425, ASTM, 1967, p 3

243. H.L. Craig, D.O. Sprowls, and D.E. Piper, *Handbook on Corrosion Testing and Evaluation*, W.H. Ails, Ed., John Wiley and Sons, 1976, p 213

244. G. Vogt, *Werkstoff. Korros.*, Vol 29, 1978, p 721

245. N.J.H. Holroyd and G.M. Scamans, Slow-Strain Rate Stress Corrosion Testing of Aluminum Alloys, *Environment-Sensitive Fracture*, S.V. Dean, E.N. Pugh, and G.M. Ugianski, Ed., STP 821, ASTM, 1984, p 202

246. B.F. Brown, *Metall. Rev.*, Vol 13, 1968, p 171

247. R.P. Wei, *Proc. Int. Conf. Fundamental Aspects of Stress Corrosion Cracking*, R.W. Staehle et al., Ed., National Association of Corrosion Engineers, 1969, p 104

248. D.O. Sprowls, "Evaluation of Stress-Corrosion Cracking," *Stress-Corrosion Cracking: Materials Performance and Evaluation*, R.H. Jones, Ed., ASM International, 1992

249. "Standard Test Method for Measurement of Fatigue Crack Growth Rates," E 647, ASTM, 1991

Elevated-Temperature Crack Growth

Richard H. Norris, Parmeet S. Grover, B. Carter Hamilton, and Ashok Saxena, Mechanical Properties Research Laboratory, School of Materials Science and Engineering, Georgia Institute of Technology

HIGH-TEMPERATURE operating requirements for parts and equipment have drastically increased over the past 20 to 30 years, and businesses such as the utility, aircraft, and chemical industries are greatly dependent on the safe and efficient operation of such equipment. In the power-generation industry, for example, components operate for extended periods of time at temperatures between 0.3 to 0.5 of their absolute melting temperature and have design lives that are limited by creep. Defense and aerospace applications rely heavily on materials that maintain their integrity in the presence of combinations of high temperature and stress (Ref 1). Also, in other cases, as the original design lives are expiring, assessment of the remaining life of components currently in operation with the objective of extending the component service life has become an economic and safety consideration (Ref 2). Many facilities are now relying on retirement-for-cause philosophy to determine the end of service life for parts (Ref 3-5).

Despite the sophisticated methods of flaw detection that are available, defects and impurities are commonly present in all large components and can potentially escape detection. In the high-temperature regime, components fail by the accumulation of time-dependent creep strains at these defects, which with time evolve into cracks causing eventual failure. This results in financial losses in repair and downtime costs as well as possible loss of human life (Ref 6). This article focuses on the concepts for characterizing and predicting elevated-temperature crack growth in structural materials. Both creep and creep-fatigue crack growth will be considered, focusing mainly on test methods. For a discussion on the application of the data in life prediction, refer to Ref 1, 5, and 7.

Creep and Creep-Fatigue

Failures that are attributed to creep result from either widespread or localized creep damage (Ref 8). If the part is subjected to uniform stress and temperatures, the damage is likely to be widespread and failure by creep rupture is apt to result. This is most commonly observed in thin section components such as steam pipes. Components with localized damage, which is a result of nonuniform stress and temperature distribution found most commonly in large structures, are more prone to fail as a result of creep crack propagation rather than stress rupture.

Service conditions experienced by components can also involve cyclic loading and unloading at elevated temperatures. Hence, in these situations, crack growth occurs not only under static loading (creep conditions), but creep-fatigue interactions play a major role in the initiation and growth of cracks. Components operating at high temperatures experience changes in conditions from beginning to end of each operating cycle, resulting in transient temperature gradients. Considering the case of steam turbines as an example, cracks in castings are typically located at the steam inlets of high-pressure and intermediate-pressure turbine sections because the local thermal stresses are higher. The primary cause of crack initiation and propagation in turbine casings is fatigue and creep-fatigue and occasionally brittle fracture due to high transient thermal stresses (Ref 4). Thermal stresses are responsible for fatigue and creep-fatigue crack growth (CFCG) in the lower-temperature regions, while creep contributes to crack growth in regions where temperature exceeds 427 °C (800 °F) (Ref 7).

Creep-Ductile Materials. Creep and creep-fatigue crack growth is a common occurrence in most engineering materials operating at high temperatures. Materials classified as creep-ductile have the ability to sustain significant amounts of crack growth prior to failure. Also, crack growth in these materials is accompanied by significant amounts of creep deformation at the crack tip. Therefore, a complete understanding of crack growth mechanics and damage mechanisms is required for accurate predictions of life of high-temperature components made from such materials. A typical flow diagram of this methodology is shown in Fig. 1. Examples of such materials include Cr-Mo steels, stainless steels, and Cr-Mo-V steels.

This damage in creep-ductile materials at high temperatures is usually in the form of grain-boundary cavitation. It has been most commonly observed that this cavitation initiates at second-phase particles or defects on the grain boundaries (Ref 9). Nucleation and growth of these cavities lead to coalescence of these voids, eventual crack formation, and growth (Ref 10), which is the primary mechanism of creep crack growth. Failure due to creep-fatigue interaction can be described from two points of view (Ref 11): influence of cyclic loading on cavitation damage and influence of cavitation on cyclic initiation and propagation. These mechanisms are illustrated in Fig. 2, adapted from Ref 11. Three prominent mechanisms for fatigue crack growth at elevated temperatures, in the presence of hold times, are (1) alternating slip mechanism (crack-tip blunting mechanism), (2) fatigue crack growth caused by grain-boundary cavitation, and (3) influence of corrosive environment.

Creep-Brittle Materials. A second class of high-temperature structural materials is known as creep-brittle materials, which include high-temperature aluminum alloys, titanium alloys, nickel-base superalloys, intermetallics, and ceramic materials. The primary difference between this class of materials and the creep-ductile materials discussed previously is that creep crack growth in these materials is usually accompanied by small-scale creep deformation and by crack growth rates that are comparable to the rate at which creep deformation spreads in the cracked body. As discussed later in this article, this has a significant influence on the crack tip parameters that characterize crack growth rates. Because of these differences, the time-dependent fracture mechanics (TDFM) concepts described in the subsequent discussion will address these two types of material behavior separately.

Time-Dependent Fracture Mechanics

Stationary-Crack-Tip Parameters. As previously mentioned, crack growth in creep-ductile materials lags considerably behind the spreading of the creep zone. Therefore, a practical assumption of a stationary crack tip is made when searching for crack-tip parameters (although the limitations and validity of this assumption in defining crack-tip parameters are discussed later in the

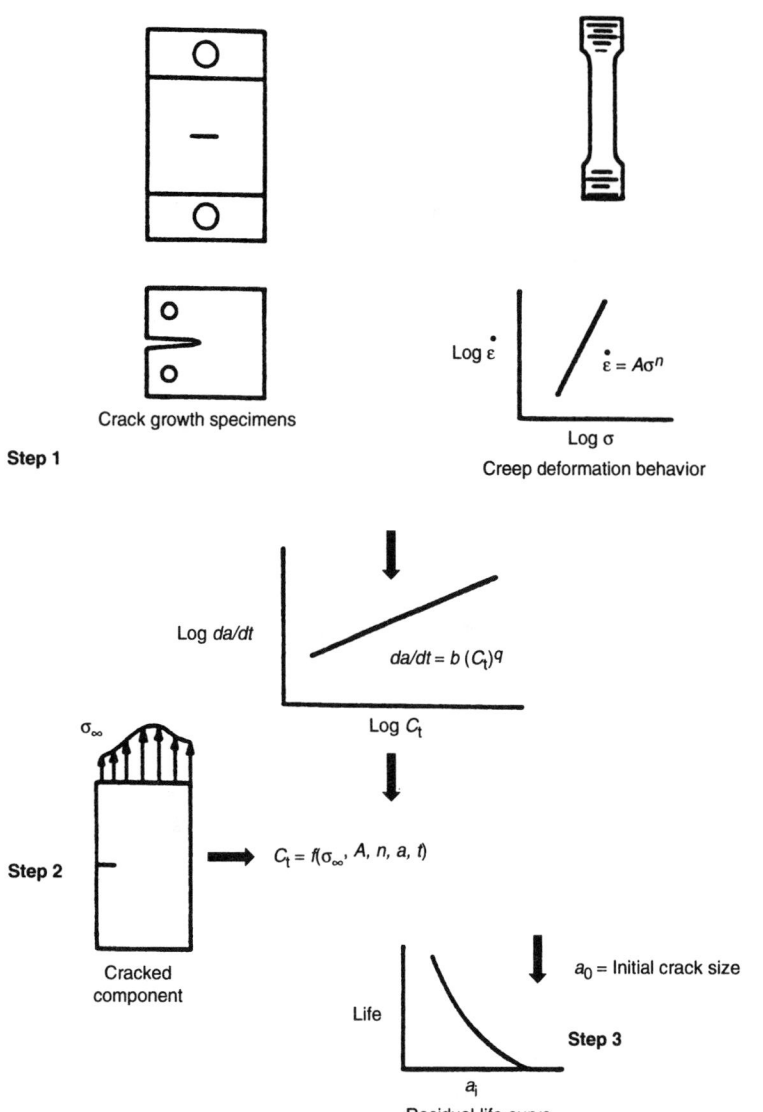

Fig. 1 Methodology for predicting crack propagation life using time-dependent fracture mechanics

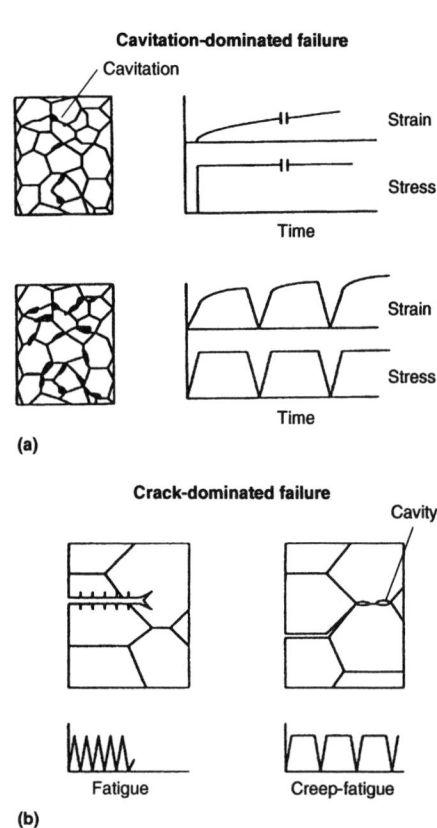

Fig. 2 Schematic representation of mechanistic aspects of creep-fatigue. (a) Effect of cycling on cavitation damage. (b) Effect of cavitation on cyclic crack growth. Source: Ref 11

section "Conditions with Growing Cracks.") In order to describe the mechanics of creep and creep-fatigue crack growth, an understanding of the stress-strain-time response at the crack tip of a body subjected to a load in creep-temperature regime must be developed first. The stages of the evolving deformation zone ahead of a crack tip when a member is subjected to a load in the creep regime is shown in Fig. 3. The initial response of the body is elastic-plastic, and the crack-tip stress field is proportional to the stress-intensity factor, K, if the scale of plasticity is small compared with the crack size (Ref 12, 13). If the plastic zone is not small, the J-integral characterizes the instantaneous crack tip stresses and strains (Ref 14). With increasing time, creep deformation causes the relaxation of the stresses in the immediate vicinity of the crack tip, resulting in the formation of the creep zone, which continually increases in size with time. Because the parameters K and J are independent of time, they are not able to uniquely characterize the crack-tip stresses and

strains within the creep zone. The parameters C^*, $C(t)$, and C_t have been developed to describe the evolution of time-dependent creep strains in the crack-tip region (Ref 15-18) and will be discussed later in this section. Within these creep regions, the crack-tip stress and strain fields resemble the Hutchinson-Rice-Rosengren (HRR) fields noted in elastic-plastic fracture mechanics (Ref 19, 20).

For a body undergoing creep, the uniaxial stress-strain-time response for a material that exhibits elastic, primary, secondary, and tertiary creep is given by:

$$\dot{\varepsilon} = \frac{\dot{\sigma}}{E} + A_1 \, \varepsilon^{-p} \sigma^{n_1(1+p)} + A\sigma^n$$

$$+ A_3 \sigma^{n_3} (\varepsilon - A\sigma^n t)^{p_3} \qquad \text{(Eq 1)}$$

where ε and σ are the strain and stress, respectively, and $\dot{\varepsilon}$ and $\dot{\sigma}$ denote their time derivatives. The val-

ues of A, A_1, A_3, p, p_3, n, n_1, and n_3 are the creep regression constants derived from creep deformation data. The terms on the right-hand side of the equation represent the elastic, primary, secondary, and tertiary creep strain contributions, respectively. This equation is convenient for analyzing the creep deformation behavior of cracked bodies under creep loading conditions.

The crack-tip stress and strain behavior for a creeping body change with time, as a result of continuous evolution of the creep zone. The changes in the creep deformation zone follow the progression shown in Fig. 4. During the initial stage of small-scale creep, the creep zone is small compared with the crack size and the remaining ligament. Primary creep strains accumulate at a faster rate than the secondary creep strains; therefore, the primary creep strains initially dominate this region. Next, the primary creep zone continues to expand, and the secondary creep zone begins to evolve within the primary creep zone. Then the primary creep zone envelopes the entire remaining ligament, while the secondary creep zone continues to grow in size within the primary creep zone. Eventually, the secondary creep zone engulfs the entire remaining ligament. In heavily cavitating materials, the tertiary creep zone begins as a small zone near the crack tip, but can eventually cover the entire remaining ligament. In chromium-molybdenum steels, cavitation is

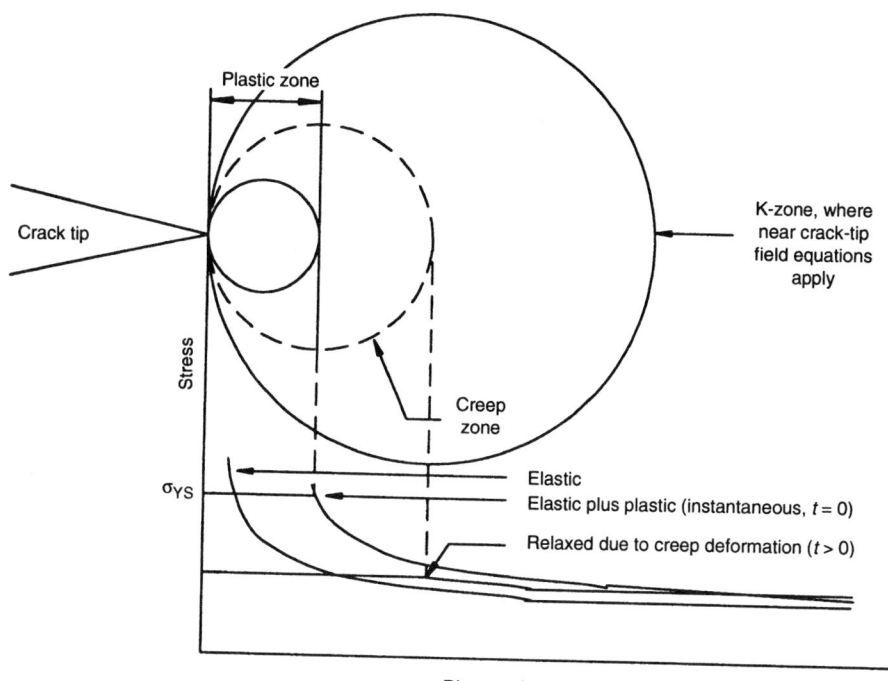

Fig. 3 Formation of deformation zones ahead of crack tip upon initial loading in the creep regime

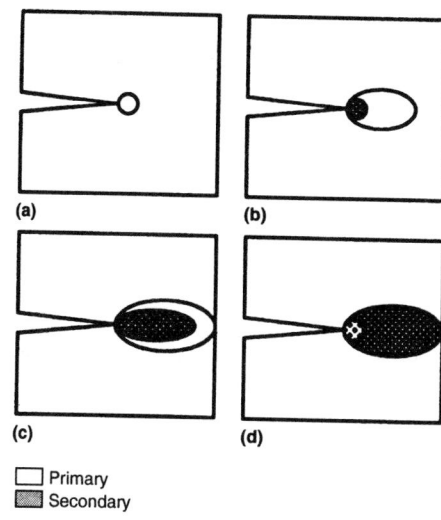

Fig. 4 Creep zone evolution. (a) Small scale primary creep conditions. (b) Secondary zone evolving within the primary zone. (c) Secondary zone becoming comparable in size with the extensive primary zone. (d) Extensive secondary zone enveloping entire ligament (steady-

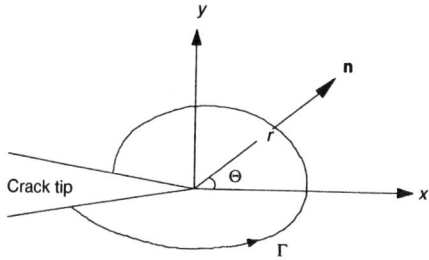

Fig. 5 Schematic of the contour integral in terms of crack-tip coordinate system used to define C^*. **n** is the unit normal vector.

usually limited to a small region near the crack tip, and the consideration of tertiary creep strains is not relevant in estimating the crack tip parameters.

Cracked Body Deforming under Steady-State Creep Conditions. When steady-state creep conditions dominate, as shown in Fig. 4(d), the relationship between stress and strain rate, Eq 1 simplifies to the so-called Norton relation:

$$\dot{\varepsilon} = A\,\sigma^n \qquad \text{(Eq 2)}$$

For these conditions, the crack-tip parameter, C^*, was defined by Landes and Begley (Ref 15) and Nikbin, Webster, and Turner (Ref 21) by analogy to the J-integral. The C^*-integral is defined as follows:

$$C^* = \int_{\Gamma} W^* \, dy - T_i\left(\frac{\partial \dot{u}_t}{\partial x}\right) ds \qquad \text{(Eq 3)}$$

where

$$W^* = \int_0^{\dot{\varepsilon}_{mn}} \sigma_{ij} \, d\dot{\varepsilon}_{ij} \qquad \text{(Eq 4)}$$

is the strain energy rate density associated with the point stress and strain rate. In Eq 3, Γ is an arbitrary counterclockwise line contour starting at the lower crack surface and ending on the upper crack surface enclosing the crack tip and no other defect, T_i is the component of the traction vector in the direction of the outward normal, and ds is the increment in the contour path. Figure 5 illustrates this integral on a reference crack-tip coordinate system. A more detailed account of these notations can be found in Ref 22 to 24.

C^* is a path-independent integral whose value can be obtained by calculation of the integral along an arbitrary path, as mentioned before. The C^*-integral is also related to the energy rate or power difference between two identically loaded specimens having incrementally different crack lengths; therefore, C^* can be measured at the loading pins of the specimen and defined in that way by:

$$C^* = -\frac{1}{B}\frac{dU^*}{da} \qquad \text{(Eq 5)}$$

where B is the specimen thickness and U^* is the steady-state energy rate (or stress-power) difference between two specimens in which the crack lengths differ by an incremental amount da, but are otherwise identical. The C^*-integral also describes the strength of the crack-tip stress and strain-rate singularities (Ref 25):

$$\sigma_{ij} = \left(\frac{C^*}{I_n A r}\right)^{1/(n+1)} \tilde{\sigma}_{ij}(\Theta) \qquad \text{(Eq 6)}$$

$$\dot{\varepsilon}_{ij} = A\left(\frac{C^*}{I_n A r}\right)^{1/(n+1)} \tilde{\dot{\varepsilon}}_{ij}(\Theta) \qquad \text{(Eq 7)}$$

where r is the distance from the crack tip, Θ is the angle from the plane of the crack, and $\tilde{\sigma}_{ij}(\Theta)$ and $\tilde{\dot{\varepsilon}}_{ij}(\Theta)$ are angular functions specified in Ref 26. A is the Norton law coefficient in the relation between stress and steady-state creep rate. I_n is a constant dependent on the steady-state creep exponent n, whose values may be found in tables (Ref 26). For most values of n of practical interest, I_n can be expressed approximately as 3 for plane-stress

conditions and 4 for plane-strain conditions. Thus, C^* characterizes the strength of the crack-tip-stress singularity commonly known as the Hutchinson-Rosengren-Rice (HRR) singularity.

Using the load-line deflection rate, which can be measured directly from the specimen, the applied load and crack length, C^* can be determined by (Ref 27):

$$C^* = \frac{P\dot{V}_{SS}}{BW}\,\eta\left(\frac{a}{W}, n\right) \qquad \text{(Eq 8)}$$

where W is the specimen width, B is the specimen thickness, V_{SS} is the steady-state load-line deflection rate, and η is a geometric function that is also dependent on the crack length to width ratio, a/W, and the secondary creep exponent, n. This method of determining C^* has been successfully used under laboratory conditions with test specimens, most commonly center crack panels and compact tension specimens. For the compact tension geometry, the value of η can be approximated as follows (Ref 2, 28-30):

$$\eta = \frac{n}{n+1} \left[\frac{2 + 0.522 \left(1 - \frac{a}{W}\right)}{1 - \frac{a}{W}} \right] \qquad \text{(Eq 9)}$$

where the term $n/n+1$ on the right-hand side of the equation is strictly valid for secondary creep only and is replaced by $n_1/n_1 + 1$ when most of the test time is spent under primary creep conditions (Ref 31). The C^*-integral can also be determined from expressions analogous to the analytical expressions for estimating the fully plastic portion of the J-integral. This method is useful when the experimental values of the load-line deflection rates are not available and either plane-strain or plane-stress conditions prevail. When planar conditions are prevalent (assumption for most thick or very thin in-service members), C^* can be calculated as follows for compact tension specimens (Ref 12, 24, 32):

$$C^* = A \, (W - a) h_1 \left[\frac{P}{\beta \zeta B (W - a)} \right]^{n+1} \qquad \text{(Eq 10)}$$

where h_1 depends on a/W, n, and the state of stress (Ref 33), β equals either 1.455 or 1.071 for plane-strain or plane-stress conditions, respectively, and:

$$\zeta = \left[\left(\frac{2a}{W-a} \right)^2 + 2 \left(\frac{2a}{W-a} \right) + 2 \right]^{1/2}$$
$$- \left[\left(\frac{2a}{W-a} \right) + 1 \right] \qquad \text{(Eq 11)}$$

Elastic plus Secondary Creep Conditions. As previously mentioned, the creep deformation zone changes with time and these zones evolve from small-scale to extensive creep conditions. For a cracked body to reach extensive secondary creep conditions, a characteristic transition time, t_T, has been proposed for when C^* becomes valid from the time of initial loading (Ref 32, 34):

$$t_T = \frac{K^2 (1 - v^2)}{(n+1) \, EC^*} \qquad \text{(Eq 12)}$$

For times less than the calculated value of t_T, stress redistribution in the crack-tip region cannot be ignored. Thus Eq 2 must be modified to include the elastic term in addition to the power-law creep term. Under these circumstances, C^* is path-dependent and it no longer uniquely determines the crack-tip stress fields given by Eqs 6 and 7. The size of the secondary creep zone (r_c) can be determined by the relationship (Ref 32):

$$r_c (\theta, t, n) = \alpha K^2 (EAt)^{2/(n-1)} \tilde{r}(\theta, n) \qquad \text{(Eq 13)}$$

where n is the creep exponent and α is a function of the state of stress and n, and is given by:

$$\alpha = \frac{1}{2\pi} \left[\frac{(n+1) I_n}{2\pi (1 - v^2)} \right]^{-2/(n-1)}$$

and I_n is a dimensionless parameter related to the HRR stress field (Ref 26).

For several applications, the transition times may be large in comparison with the average operating time between start-up and shutdown for components. If operational shutdown is accepted as a part of normal operating mode, it is reasonable to infer that some components may never actually reach steady-state conditions, thereby spending their service life in the small-scale and transition creep regimes. Thus C^* is not applicable for characterizing creep crack growth in these components. Furthermore, primary creep behavior must be incorporated in the above analysis for it to be generally useful.

Even under small-scale creep, in the immediate vicinity of the crack tip, the creep strain rates exceed the elastic strain rates; therefore, selection of any integration path in the creep-dominated region will yield path-independence for the C^*-integral. The C^*-integral taken along a path near the crack tip has been called the $C(t)$ parameter and has been shown to characterize the amplitude of the crack-tip stress and strain fields (Ref 13, 35). For small-scale secondary creep (SSC) conditions, $C(t)$ can be approximated by the following equation (Ref 32, 34):

$$[C(t)]_{SSC} \cong \frac{(1 - v^2) \, K^2}{(n+1) \, Et} \qquad \text{(Eq 14)}$$

As the extensive creep conditions become prevalent, $C(t)$ becomes equal to C^*, and it also becomes completely path-independent. An interpolation formula for analytically estimating $C(t)$ during small-scale to extensive creep conditions for elastic-secondary creep conditions is given by (Ref 35):

$$C(t) \cong \frac{(1 - v^2) \, K^2}{(n+1) \, Et} + C^* \qquad \text{(Eq 15)}$$

Consideration of Primary Creep in Crack-Tip Parameters. Primary creep can be present in the small-scale as well as extensive creep conditions and can be of considerable importance in many elevated-temperature components such as chromium-molybdenum steels (Ref 24). Under extensive primary creep conditions, the second term in Eq 1 becomes dominant. Integrating this term and solving for ε results in $\varepsilon = \sigma^{n_1} [A_1 t (1 + p)]^{1/1+p}$. Because for a given material, A_1, p, and n_1 are constants, the accumulated strain is a function of stress and time. Furthermore, the strain and strain-rate dependence on stress and time is separable. Because of this property, the C^*-integral is path-independent for extensive primary creep; however, its value changes with time. Primary creep can also be included in the estimation of $C(t)$ under small-scale creep conditions. The transition time for the progression of small-scale primary creep conditions to evoke extensive primary creep conditions, t_1, is defined by (Ref 36):

$$t_1 = \frac{1}{n_1 + 1} \left(\frac{J}{C^*} \right)^{p+1} \qquad \text{(Eq 16)}$$

where J is the J-integral (Ref 14). For $t > t_1$, extensive primary creep conditions prevail and as mentioned earlier, C^* is path-independent at a fixed time and thus defined as $C^*(t)$, whose value changes with time. It uniquely characterizes the instantaneous crack-tip stresses. $C^*(t)$-integral can be related to another path-independent integral, C_h^*, which is independent of time, by the following equation (Ref 36, 37):

$$C^*(t) = \frac{C_h^*}{(1 + p) t^{p/(1+p)}} \qquad \text{(Eq 17)}$$

Thus, the time dependence of $C^*(t)$ can be separated from the crack-size and load-dependent parameter, C_h^*, which can be determined analytically much like J and C^*. For compact specimens (Ref 16):

$$C_h^* = [A_1 (1 + p)]^{1/(1+p)} (W - a) h_1 \left[\frac{P}{\beta \zeta B (W - a)} \right]^{n_1 + 1} \qquad \text{(Eq 18)}$$

where A_1, p, and n_1 are the primary creep constants from Eq 1. With continuing evolution of the secondary creep zone within the primary creep zone, the elapsed time for the secondary zone to overtake the primary-zone boundary and engulf the remaining ligament is derived by (Ref 37):

$$t_2 = \left[\frac{(n+p+1) \, C_h^*}{(n+1) \, (1+p) \, C^*} \right]^{-(1+p)/p} \qquad \text{(Eq 19)}$$

In a manner similar to the determination of the secondary creep zone size, the extent of the primary creep zone during small-scale creep can also be determined:

$$r_c (\theta, t, n) = \alpha' \, K^2 \, [E (A_1 t)^{1/(1+p)}]^{2/(n_1 - 1)} \tilde{r}(\theta, n_1) \qquad \text{(Eq 20)}$$

where α' is a function of the state of stress and the primary creep exponent n_1.

A condition commonly observed is one in which both primary and secondary creep strains occur simultaneously in the ligament. The $C^*(t)$-integral in this regime can be approximated by the following relationship (Ref 22):

$$C^*(t) \approx \frac{C_h^*}{(1 + p) t^{p/(1+p)}} + C_s^* \qquad \text{(Eq 21)}$$

where C_s^* is the steady-state value of the C^*-integral. The $C(t)$-integral also characterizes the amplitude of the HRR fields under these conditions and a wide range expression for $C(t)$ is approximated by (Ref 16):

$$C(t) \approx [C(t)]_{\text{SSC}} + C^*(t) = [C(t)]_{\text{SSC}} + C^* \left[\left(\frac{t_2}{t} \right)^{p/(p+1)} + 1 \right]$$

$$\text{(Eq 22)}$$

The parameter $C(t)$ is useful for characterizing the creep crack growth for the small-scale and steady-state regimes. However, one significant disadvantage of $C(t)$ is that it cannot be measured in the small-scale (transient) region and can only be calculated analytically. In the extensive creep regime, $C(t) = C^*$ so it can be measured from the load-line displacement readings directly from a test specimen as given earlier in Eq 8.

The C_t Parameter. Because $C(t)$ cannot be measured at the load-line under small-scale conditions, another parameter, C_t, has been proposed and shown to characterize creep-crack growth rates under a wide range of creep conditions (Ref 18, 38). The C_t parameter is defined as the instantaneous rate at which stress-power is dissipated and can be measured at the loading pins in the entire regime from small-scale to extensive creep. Thus by definition, C_t is equal to $C^*(t)$ and $C(t)$ in the extensive regime (Ref 24) and is given by (Ref 18):

$$C_t = \lim_{\Delta a \to 0} \frac{\Delta U_t^*}{B \Delta a} = -\frac{1}{B} \frac{\partial U_t^*}{\partial a} \Big|_{V_c} \quad \text{(Eq 23)}$$

where B is the specimen thickness, a is crack length, and ΔU_t^* is the instantaneous difference in the stress power between two cracked bodies that have incrementally differing crack lengths of Δa but are otherwise identical.

For small-scale creep conditions, the Irwin concept of effective crack length has been modified to define a stationary crack to accommodate the expression for C_t (Ref 18, 38):

$$a_{\text{eff}} = a + \chi \dot{r}_c \quad \text{(Eq 24)}$$

In this equation, χ is the scaling factor, which is approximately equal to $\frac{1}{3}$ as determined by finite element analysis (Ref 39, 40), $\dot{r}_c$ is the creep zone size, a_{eff} is the effective crack length, and a is the physical crack length. This leads to an expression for C_t in the small-scale creep regime (Ref 18, 24) in which the load-line deflection can be directly measured during the test:

$$(C_t)_{\text{SSC}} = \frac{P \left(\dot{V}_c \right)_{\text{SSC}}}{BW} \frac{F'}{F} \quad \text{(Eq 25)}$$

where

$$F \left(\frac{a}{W} \right) = \frac{KB\sqrt{W}}{P}; \quad F' \left(\frac{a}{W} \right) = \frac{dF}{d \left(\frac{a}{W} \right)}$$

Analytically, the small-scale creep deflection rate can be determined by (Ref 38):

$$(\dot{V}_c)_{\text{SSC}} = \frac{2B(1 - \nu^2)}{EP} K^2 \chi \dot{r}_c \quad \text{(Eq 26)}$$

Substituting Eq 26 into Eq 25, an equation which directly relates $(C_t)_{\text{SSC}}$, K and $\dot{r}_c$ is determined:

$$(C_t)_{\text{SSC}} = 2(1 - \nu^2) \chi \left(\frac{F'}{F} \right) \frac{K^2 \dot{r}_c}{EW} \quad \text{(Eq 27)}$$

Using Eq 27, an analytical expression for a stationary crack can be derived for $(C_t)_{\text{SSC}}$ in which knowledge of load-line deflection is not needed, assuming that constants in the appropriate creep constitutive laws are available (Ref 38). For example, for elastic, secondary creep $(C_t)_{\text{SSC}}$ can be given by:

$$(C_t)_{\text{SSC}}$$

$$= \frac{4\alpha\chi\tilde{r}_c (1 - \nu^2)}{E(n - 1)} (EA)^{2/(n-1)} \frac{K^4}{W} \frac{F'}{F} \left(\frac{1}{t} \right)^{(n-3)/(n-1)} \quad \text{(Eq 28)}$$

where α has been previously defined.

An expression for estimating C_t for a wide range of creep conditions from small-scale creep to extensive creep and also including primary creep has been derived that is very similar to the way in which $C(t)$ is derived (Ref 2, 18, 41):

$$C_t = (C_t)_{\text{SSC}} + C^*(t) \quad \text{(Eq 29)}$$

where the value of $(C_t)_{\text{SSC}}$ can be either experimentally determined from Eq 25 or analytically determined from Eq 27. If Eq 25 is used, the expression to measure C_t experimentally over the entire secondary creep range is given as (Ref 38):

$$C_t = \frac{P \left(\dot{V}_c - \dot{V}_{SS} \right)}{BW} \left[\frac{F'}{F} \right] + C^* \quad \text{(Eq 30)}$$

where the load-line deflection rate in extensive creep conditions is subtracted from the total rate of deflection caused by creep.

Furthermore, a wide range expression for determining C_t in the presence of primary and secondary creep conditions has been determined in a similar manner as Eq 22 (Ref 41, 42):

$$C_t = \frac{P \left[\dot{V}_c - \dot{V}^*(t) \right]}{BW} \frac{F'}{F} - C^* \left[\left(\frac{t_2}{t} \right)^{p/(1+p)} + 1 \right] \quad \text{(Eq 31)}$$

where $\dot{V}_c^*(t)$ is the load-line deflection rate under extensive primary-secondary creep conditions.

Conditions with Growing Cracks. As previously stated, all the crack-tip parameters discussed thus far are based on the assumption of a stationary crack tip. Crack growth can significantly alter the crack-tip stress fields if the rate of crack growth is comparable to the rate at which creep deformation spreads at the crack tip. For a crack progressing with a velocity $\dot{a}$, the stress

fields are dependent on the crack velocity by (Ref 43):

$$\sigma_{ij} = \alpha \left[\frac{\dot{a}}{AEr} \right]^{1/(n-1)} \tilde{\sigma}_{ij}(\Theta) \quad \text{(Eq 32)}$$

This stress field is alluded to as the Hui-Riedel (HR) field. By solving for r in Eq 6 and 32 and then setting them equal, it can be shown that for steady-state creep, stress distribution within the HR field is (Ref 44):

$$\sigma_{ij} = \left(\frac{EC^*}{\dot{a}} \right)^{1/2} \beta_1(\Theta) \quad \text{(Eq 33)}$$

It is intuitive from this equation that the extent of the HR field must be small in comparison to the extent of the HRR field for C^* (or C_t) to uniquely characterize the creep crack growth rate. By equating Eq 32 and 33, an estimation of the HR field size can be obtained. For materials in which the crack growth behavior is characterized by C^* (or C_t), the size of the HR zone has been estimated to be on the order of 0.01 Å which is negligible (Ref 43). This implies that C^* is a viable parameter even in the presence of growing cracks, provided the crack growth is slow in comparison to the rate of spread of creep deformation. This condition can be ensured by applying deflection-rate partitioning as shown below (Ref 27, 29):

$$\dot{V} = \dot{V}_e + \dot{V}_p + \dot{V}_c$$

$$= \dot{a} \left[\left(\frac{\partial V_e}{\partial a} \right)_P + \left(\frac{\partial V_p}{\partial a} \right)_P \right] + \dot{V}_c \quad \text{(Eq 34)}$$

The term on the left side of Eq 34 is the total load-line deflection rate at constant load, while the first two terms on the right side are the deflection rates due to crack growth as a result of elastic and plastic compliance change, and the third term is due to creep deformation. The contribution of deflection due to creep is found by subtracting the deflection rates attributed to the change in elastic and plastic compliances from the total deflection rate:

$$\dot{V}_c = \dot{V} - \dot{a} \frac{B}{P} \left[\frac{2K^2}{E} + (m + 1)J_p \right] \quad \text{(Eq 35)}$$

where J_p is the plastic portion of J and m is the plasticity exponent. Stationary crack tip parameters can only be used as long as the second term in the right-hand side contributes negligibly to $\dot{V}_c$. This condition is ensured by allowing only those data for which $\dot{V}_c/\dot{V} \geq 0.8$ (Ref 45).

Creep-Brittle Materials

In many situations, the crack growth rate is comparable to the rate of expansion of the creep zone, and the crack can no longer be assumed to be effectively stationary within an expanding creep field. These conditions are typical of creep-brittle materials, where the rate of creep strain accumulation at the crack tip is comparable to crack extension rates and where crack growth significantly perturbs the crack-tip stress field. Stated in another way, the HR field is no longer small in comparison to the extent of the HRR fields. Thus, the stationary-crack-tip parameters no longer characterize the crack-tip conditions and can no longer be expected to correlate uniquely with creep crack growth rate.

In creep-brittle materials, the rate of deflection caused by creep deformation represents only a small percentage of the total deflection rate; therefore, in the absence of significant plasticity, the rate of deflection caused by change in elastic compliance is comparable to the total deflection rate. Because the elastic contribution is analytically determined and the total deflection rate is experimentally measured, Eq 35 can sometimes erroneously yield negative creep deflection rates due to experimental error in the total deflection rate measurement. The negative creep deflection rates result in negative C_t values, which have no clear physical interpretation. Negative C_t values can also sometimes result if the creep zone size decreases (Ref 46), which is often the case in creep-brittle materials following incubation or toward the end of the test even in creep-ductile materials when stable crack growth sets in and the crack grows very rapidly. This aspect will be discussed in more detail in the next section.

Because Eq 35 is accurate only when the creep deflection rate dominates the total deflection rate, the equation lacks precision for creep contributions less than 80% of the total deflection (Ref 45). Therefore, the use of this equation and also the crack-tip parameters C_t and C^* is not suitable in creep-brittle materials.

Under special circumstances, time-independent fracture parameters, such as the J-integral (Ref 19, 20) or K, may correlate with the creep crack growth rate in creep-brittle materials. At elevated temperatures, some aluminum alloys have exhibited such creep-brittle behavior. For example, correlations between the creep crack growth rate and K have been established for aluminum alloys 2219-T851 (Ref 47, 48), 2519-T87 (Ref 49), and to a limited extent for 8009 (Ref 50). Similar correlations have also been demonstrated for other creep-brittle materials such as Ti-6242 (Ref 51) and for carbon-manganese steels at temperatures of 360 °C (680 °F) as discussed later in this article.

The precise conditions under which K or J characterize the crack growth behavior of creep-brittle materials are not yet well understood. However, the creep deformation resistance of creep-brittle materials is believed to be a significant factor. The accumulation of creep strain ahead of the crack tip is impeded in creep-brittle materials by microstructural features such as precipitates or dispersoids, and simultaneously decreasing rupture ductility increases the crack growth rate. As a result, the creep crack growth rate and the rate of creep strain accumulation in the crack-tip region are comparable. The movement of the crack perturbs the crack-tip stress

fields, and it is no longer possible to represent the crack-tip fields using the Riedel-Rice formulation (Ref 32). In an idealized situation, one can imagine that the creep zone boundary and the crack tip in the plane of the crack move at equal speeds. Thus, the creep zone size at the crack tip remains constant, and using a coordinate system that moves with the crack tip as reference, the crack-tip stress field is also constant and completely determined by K. If it is assumed that the creep zone size remains small with respect to the pertinent length dimensions of the specimen and the shape also remains constant, creep crack growth is expected to correlate uniquely with K. In practice, all these conditions are most likely seldom met. Nevertheless, correlations between creep crack growth rates and K have been observed. However, the limitations of such correlations are not well understood. This remains an area of active research.

Incubation Period. When a cracked specimen is first loaded at elevated temperature, the material ahead of the crack tip is free from prior creep damage and, therefore, time-dependent crack growth does not begin instantly. Creep crack growth studies have shown that crack extension occurs following a specific time period, which has been termed incubation time. Incubation models based on ductility exhaustion and creep cavitation concepts have been developed (Ref 10, 52, 53); however, incubation time currently lacks a precise definition among researchers because it is unknown if the crack actually remains stationary during this period or if the crack grows at an indiscernible rate. Some researchers have defined the incubation period as the time required for the crack to grow through the initial creep zone size (Ref 54), while others have defined it based on a certain increase in the direct current potential drop output when utilizing the potential drop technique to monitor crack extension (Ref 55).

The incubation period is more pronounced in creep-brittle materials than in creep-ductile materials, and it can comprise nearly 90% of the total test time during creep crack growth testing of creep-brittle specimens. Thus, incubation period is vitally important when considering the creep crack growth characteristics of creep-brittle materials. The correlation between the creep crack growth rate and the stress-intensity factor displays a steep slope in creep-brittle materials, indicating that the crack growth rate rapidly increases for very small increases in K. The K-range, therefore, over which creep crack extension occurs prior to specimen fracture is small. Thus, once the crack begins to grow, the test material cannot sustain stable crack extension for a significant duration before failure occurs, a characteristic that limits the engineering application of the material. Some studies, however, indicate that the incubation period required prior to crack extension quickly increases for small decreases in the initial K-level (Ref 49). For some critical K-levels and below, the incubation time may be so large that the crack effectively remains stationary within the testing time frame. Total time to failure, therefore, must be considered to

be a combination of the incubation period prior to crack initiation and the crack growth period following crack initiation. Evaluation of the test material for engineering applications cannot solely rely on the da/dt-K correlation, but must take into account the time required to initiate creep crack growth.

Crack-Tip Parameters for Creep-Fatigue Crack Growth

From previous sections, C_t appears to be the most appropriate crack-tip parameter for correlating creep crack growth over the regime from small-scale creep to extensive creep conditions for creep-ductile materials. The average value of C_t, $(C_t)_{avg}$, is used for characterizing the average crack growth rate, $(da/dt)_{avg}$, in creep-fatigue experiments (Ref 56-60).

The value of $(C_t)_{avg}$ is determined by two methods. The first method is suitable for specimens when load-line deflections during hold time are available, while the second is well suited for components when the load-line deflection must be calculated using the deformation laws for the material. Expressions for calculating $(C_t)_{avg}$ from laboratory experiments and in components have been developed (Ref 56, 61-63). The discussion here focuses on calculating $(C_t)_{avg}$ from experimental measurements because the primary emphasis in this article is on test methods for characterizing creep and creep-fatigue crack growth.

For materials whose deformation behavior is characterized by elastic, secondary creep, $(C_t)_{avg}$ can be determined as follows (Ref 63):

$$(C_t)_{avg} = \frac{\Delta P \, \Delta V_c}{B \, W \, t_h} \frac{F'}{F} - C^* \left(\frac{F'}{F} \frac{1}{\eta} - 1 \right) \qquad (Eq\ 36)$$

where ΔP is the applied load range, ΔV_c is the load-line deflection change due to creep during hold time t_h. Because the amount of crack extension during hold times is small, the change in total deflection during hold times is approximately equal to ΔV_c. The remaining variables have been previously defined. As previously noted for compact specimens, the ratio $(F'/F)/\eta \approx 1$, thus Eq 36 can be written as (Ref 57):

$$(C_t)_{avg} = \frac{\Delta P \, \Delta V_c}{B \, W \, t_h} \frac{F'}{F} \qquad (Eq\ 37)$$

It is also noted that the ΔV_c value in Eq 37 includes both primary and secondary creep contributions to the load-line deflection change. Thus Eq 37 also accounts for primary creep deformation in the estimation of $(C_t)_{avg}$. The $(da/dt)_{avg}$ is calculated as follows:

$$\left(\frac{da}{dt} \right)_{avg} = \frac{1}{t_h} \left[\left(\frac{da}{dN} \right) - \left(\frac{da}{dN} \right)_{cycle} \right] \qquad (Eq\ 38)$$

where $(da/dN)_{cycle}$ is the cyclic crack growth rate and has to be obtained from a continuous cycling

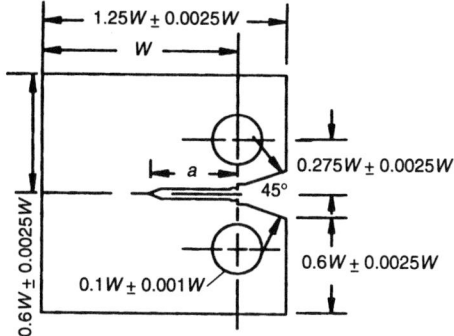

Fig. 6 Typical compact tension specimen dimensions (in inches) for a 1.00 in. thick (1T) specimen per ASTM E 1457. Specimen thickness (B) is related to width (W) as $B = 0.5W \pm 0.005W$.

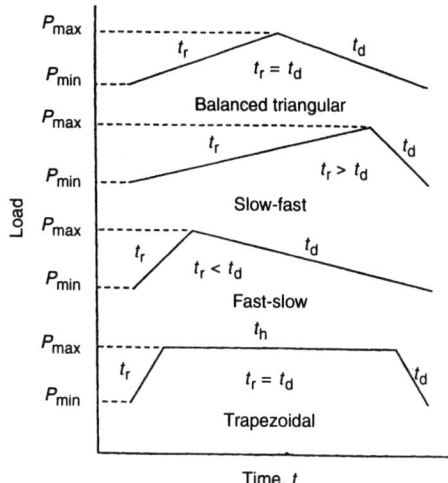

Fig. 7 Typical loading waveforms used during creep-fatigue crack growth testing. Source: Ref 66

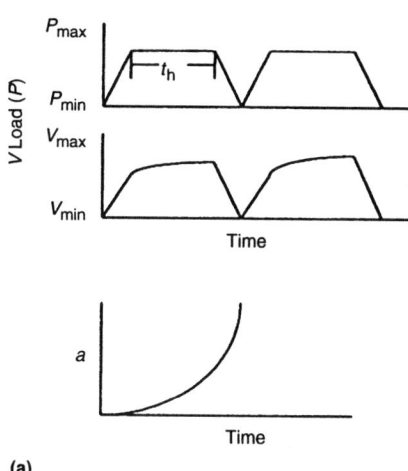

(a)

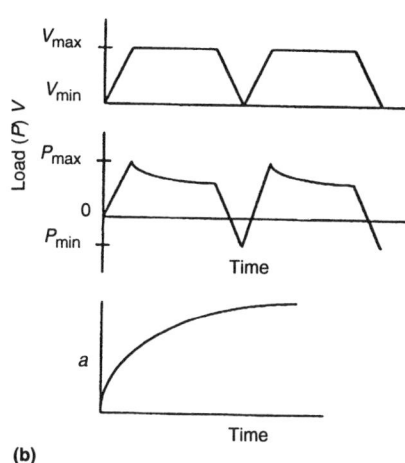

(b)

Fig. 8 Schematic comparison of load-controlled (a) and displacement-controlled (b) testing under trapezoidal loading. Source: Ref 66

fatigue crack growth rate test for which the rise and decay times are usually the same as the trapezoidal waveform used in the creep-fatigue crack growth tests. The overall crack growth rate, da/dN, is calculated from the crack length measurements during the creep-fatigue crack growth test.

Creep Crack Growth Testing

Creep crack growth tests are performed in accordance with ASTM E 1457, "Measurement of Creep Crack Growth Rates in Metals" (Ref 45) using the compact specimen geometry. The test procedure involves heating precracked specimens to the prescribed test temperature and applying a constant load until significant crack extension or specimen failure occurs. During the test, the crack length and load-line deflection are constantly monitored with time. The reduced test data compare the time rate of crack growth, da/dt, in terms of an elevated-temperature crack growth parameter described in the earlier section of this article.

Equipment

Specimen Configuration. The most widely used specimen configuration for creep and creep-fatigue crack growth testing is the compact-type (CT) specimen as shown in Fig. 6. Certain special dimensional requirements such as notch configuration and specimen width-to-thickness ratios can be found in ASTM specifications (Ref 45). Other configurations such as the center-cracked tensile (CCT) panel and single-edge notch (SEN) specimens have also been used; however, due to several reasons of convenience, the CT specimen geometry remains the most suitable geometry for creep and creep-fatigue crack growth testing. First, the transition time for extensive creep conditions to develop in CT specimens is longer than in CCT specimens for the same K and a/W for specimens of same width (W) (Ref 64). Because of the longer transition times in the CT specimens, the condition that $t_c/t_1 \ll 1$ during creep-fatigue testing, where t_c is the cycle time, is more easily satisfied. Another advantage of the CT specimen is that a clip-gage can be conveniently

placed at the load-line to measure the deflection change. This is important for reliable and direct measurement of load-line deflection change, which is the measured quantity required to calculate the crack-tip parameters. Perhaps the most important advantage of CT specimens is that the magnitude of the applied load for the same applied value of K is significantly lower than for CCT specimens. Thus, smaller load capacity machines and smaller fixturing can be used for testing.

Test Machine. Creep crack growth testing should be performed in either deadweight or servomechanical test machines that are capable of maintaining a constant load over an extended period of time. During the duration of the testing, variations in loading must not exceed ±1.0% of the nominal value at any time. When using cantilever-type deadweight creep machines, it is important that the lever remain as close to horizontal as possible. The cantilever loading conditions are performed such that the loading ratios are 20 to 1 or 10 to 1, depending on the level length. Significant deviation of the cantilever from the level position results in significant variations in the load on the specimen. ASTM E 4 "Practices for Load Verification of Testing Machines" (Ref 65) details the required accuracy for the test equipment.

Creep-fatigue tests can be conducted under either load-controlled or displacement-controlled conditions. Figure 7 (Ref 66) shows various waveforms used for load-controlled testing in which the specimen is cycled between a fixed maximum and minimum load value. The displacement versus time and crack size versus time responses in these tests are shown schematically in Fig. 8(a). Displacement-controlled tests involve cycling between fixed displacements as shown schematically in Fig. 8(b) (Ref 66) along with the corresponding changes in the load-line displacement and the load versus time and crack size versus time responses.

The majority of the creep-fatigue testing performed to date has been under load-control con-

ditions, largely because it is more convenient. Displacement-controlled testing requires specially designed grips that ensure that the specimen does not buckle or experience any torsional stresses during the compressive portion of the loading. However, this type of testing has some advantages over load-controlled testing, as discussed below.

In load-controlled tests, the net section stress ahead of the crack rises continuously as the crack grows. Thus, the stress-intensity factor continuously increases with crack growth while the size of the remaining ligament decreases. The scale of plasticity or creep in the specimen increases as the test progresses causing ratcheting in the specimen with accumulation of inelastic deflection after each cycle. In displacement-controlled tests, the applied load decreases with crack extension, as shown in Fig. 8(b), and the specimen deflection is forced to the minimum value at the end of each cycle. Therefore, ratcheting of displacement cannot occur (Ref 67), and data can be collected for greater crack extensions than in load-controlled tests. Load-controlled tests are suitable for low crack growth rates, and displacement amplitude tests are more suitable for higher crack growth rates ($>4 \times 10^{-6}$ mm/cycle) and longer hold time tests. Gladwin et al. (Ref 68) have noted that added complication due to static modes of fracture (e.g., tearing) may influence the creep-fatigue damage process in positive R load-controlled tests. Therefore, the apparent accelerations in crack growth rate associated with the hold time (creep damage) may be a result of, for example, stable tearing effects due to large deformations caused by ratcheting effects rather than the result of true creep-fatigue interaction (i.e., accelerated fatigue crack growth because of creep damaged material) (Ref 68). This problem is avoided when testing is performed under displacement range control.

Specimen heating is usually performed in electric resistance furnaces or convection laboratory ovens. Temperature control should be within ± 2 °C (± 3 °F) for tests performed at temperatures up to 1000 °C (1800 °F) and ± 3 °C (± 5 °F) for tests above 1000 °C (1800 °F). Specimens are usually tested in air at atmospheric pressure; however, inert atmosphere environments and vacuum conditions have been used and aid in the reduction of the effects of oxidation and other forms of corrosion. Thermocouples are used to monitor the specimen temperature. The thermocouples should be located in the uncracked ligament within a 2 to 5 mm (0.08 to 0.2 in.) region around the crack plane. Thermocouples should be in intimate contact with the specimen. Ceramic insulation is recommended for covering the individual lines to prevent shorting of the temperature circuit.

Fixturing. Pin-and-clevis assemblies are used to support test specimens in the load frames. This type of fixturing is used on both top and bottom specimen faces and allows in-plane rotation during specimen loading and during subsequent crack extension. Materials used for these fixtures must be creep resistant and able to withstand the loading and temperatures employed. The fixtures can be fabricated from grades 304 and 316 stainless steel, grade A286 steel, Inconel 718 and Inconel X-750 in the annealed or solution-treated condition. After fabrication, hardenable parts should be heat treated so that they develop resistance to creep deformation.

Fatigue precracking is performed on creep test specimens to eliminate any effects of the machined notch and to provide a sharp crack for initial crack growth. The methodology for fatigue precracking is described extensively in ASTM E 399, "Test Method for Plane-Strain Fracture Toughness of Metallic Materials" (Ref 69) and also briefly in the article "Fracture Toughness Testing" in this Volume. The precracking is to be carried out on the material in the same condition in which it is to be tested for creep crack growth behavior. The precracking can be performed between room temperature and the anticipated test temperature. The equipment used to precrack should allow symmetric load distribution in reference to the machined notch, and the maximum stress-intensity factor during the operation, K_{max}, should be controlled within $\pm 5\%$. The procedure may be carried out at any frequency that allows accurate load application, and the specimen

should be precracked to at least a length of 2.54 mm (0.100 in.).

The initial precracking is conducted at stress intensities high enough to allow crack initiation and growth out of the machined starter notch and with growth of the precrack, the load is decreased to avoid transient effects and to allow lower stress intensities for the creep test. The load values for precracking, P_f, are determined so as not to exceed the following value (Ref 45):

$$P_f = \frac{0.4 B_N (W - a_0)^2 \sigma_{YS}}{(2W + a_0)} \qquad \text{(Eq 41)}$$

where B_N is the corrected specimen thickness, W is the specimen width, and a_0 is the initial crack length measured from the load-line and σ_{YS} is the yield strength. During the final 0.64 mm (0.025 in.) of fatigue precrack extension, the maximum load should not be larger than P_f as determined from above or a load such that the ratio of the stress-intensity factor range to the Young's modulus ($\Delta K/E$) is equal to or less than 0.0025 mm$^{1/2}$ (0.0005 in.$^{1/2}$) (Ref 45), whichever is smaller. In doing so it is ensured that the final precrack loading does not exceed the initial load used during creep crack growth testing.

The crack length during precracking can be measured optically with a traveling microscope if the test is being performed at room temperature, and if the precracking is performed at elevated temperature, the electric potential drop technique can be used. Measurement of the fatigue precrack should be accurate within 0.1 mm (0.004 in.). Measurements should be made on both surfaces, and the value should not vary more than 1.25 mm (0.05 in.). If the surface cracks exceed this range, further extension is necessary until these criteria are met.

Required Measurements

Crack Length Measurement (Electric Potential Method). In monitoring the crack propagation during the elevated-temperature creep test, crack extensions must be resolved to at least 0.1 mm (0.004 in.). Because optical measurement techniques within an enclosed furnace are not feasible and the through-thickness crack fronts sometime differ significantly from observed surface lengths, crack length measurements are most commonly made during an elevated-temperature creep test using the electric potential drop technique. By applying a fixed electric current, any increase in crack length (a corresponding decrease in uncracked ligament) results in an increase in electric resistance, which is noted as an increase in output voltage across the output locations.

The current input and voltage output lead locations for typical CT and CCT specimens are shown in Fig. 9. These leads can either be welded onto the specimen or connected with screws. The choice of application method is dependent on material and test temperature. For softer materials tested at lower temperatures, threaded connections would be acceptable, but for harder materi-

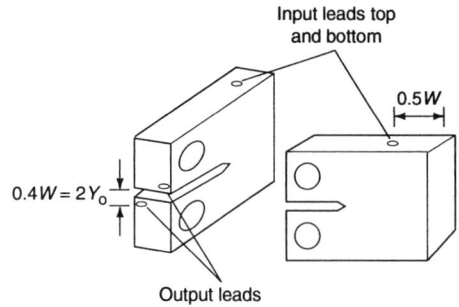

(a)

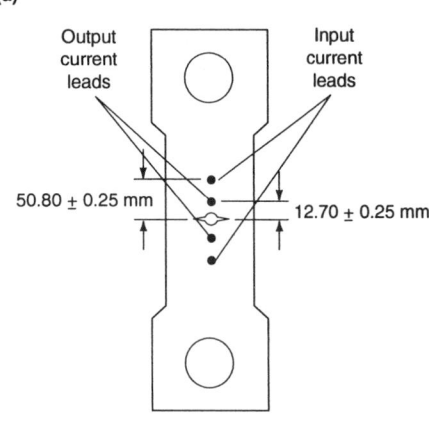

(b)

Fig. 9 Input current and voltage output lead locations for typical compact (a) and center-cracked tension specimens (b)

als and especially at higher temperatures, welding of leads is recommended. The leads should be sufficiently long to allow current input devices and output voltage measuring instruments to be well away from the furnace to avoid excessive heating. The leads should also be approximately the same length and contain similar junctions to avoid excessive lead resistance, which contributes to the thermal voltage, V_{th} as described below. Use of 2 mm (0.08 in.) diameter stainless steel wires have been shown to work well because of superior oxidation resistance at elevated temperature; however, any oxidation-resistant material capable of carrying a current that is stable at the test temperature may be utilized as lead connectors. Nickel and copper wires have been successfully used as lead material for lower temperature tests.

When using the direct current electric potential method, the instantaneous voltage V and the initial voltage V_0 usually deviate from the indicated readings. This is due to the thermal voltage, V_{th}, which is caused by several factors, such as differences in the junction properties of the connectors used, differences in the resistance of the output leads, differing output lead lengths, and temperature differences in output leads themselves. Measurements should be taken of the V_{th} prior to the load application and at various times during the test. These measurements are made by turning off the current source and recording the output voltage. Before analyzing the crack length data, the values of V_{th} should be subtracted from the respective V and V_0 in order to determine the actual crack extension.

Knowing the corrected original voltage, V_0, and the corrected instantaneous change in voltage during crack extension, V, the crack length in a CT specimen can be computed by using the following closed form equation (Ref 55, 70):

$$a_i = \frac{2W}{\pi} \cos^{-1} \left\{ \frac{\cosh\left(\dfrac{\pi Y_0}{2W}\right)}{\cosh\left[\dfrac{V}{V_0}\cosh^{-1}\left(\dfrac{\cosh\dfrac{\pi Y_0}{2W}}{\cos\dfrac{\pi a_0}{2W}}\right)\right]} \right\} \quad \text{(Eq 42)}$$

where a_i is the instantaneous crack length, a_0 is the original crack length after precracking, Y_0 is the half separation distance between the voltage output leads, V_0 is the initial output voltage before load application, V is the instantaneous output voltage, and W is the specimen width.

Materials with high electrical conductivities can experience fluctuations in V_{th}. These fluctuations can be of the same magnitude as the voltage changes that accompany crack extension and could mask this information. Because of the potential variation in thermal voltage, the direct current electric potential method should not be the only nonvisual technique for crack length measurement. The use of more sophisticated electric potential setups, such as the reversing potential method, is recommended.

Crack lengths, both initial and final, are required to differ by no more than 5% across the specimen thickness. Maintaining a straight crack front is sometimes dependent on material and material thickness. Upon post test examination, thicker specimens have been noted to experience crack tunneling, or nonstraight crack extension. Crack tunneling (thumbnail-shaped crack fronts) is common in non-side-grooved (or parallel-sided) specimen configurations (Ref 8). This occurrence is a direct result of the conditions being closer to plane stress near the surfaces of the specimen and plane strain in the center, which results in higher crack growth rates near the specimen center. Side grooving of test specimens on the crack plane has been shown to greatly reduce this problem. Side grooves up to 25% in reduction are acceptable, but reductions of approximately 20% have been found to work well for many materials. The included angle of the grooves is typically less than 90° with a root radius less than or equal to 0.4 ± 0.2 mm (0.016 ± 0.008 in.). It is prudent to perform the side grooving after fatigue precracking because precracks are hard to see when located in the grooves.

Load-Line Deflection Measurements. Continuous displacement measurements are needed in the determination of the crack-tip parameters for the duration of the testing. These displacement readings should be taken directly from the load-line as much as possible. For CT specimens, the measuring device should be attached on the machined knife edges of the specimens. For CCT specimens, the deflection is to be measured on the load line at points that are ±35 mm (±1.40 in.) from the crack centerline. The measurement of the displacement can be directly measured by placing an elevated-temperature clip gage (either strain gages for temperatures up to approximately 150 °C (300 °F) or capacitance gages for higher temperatures) on the specimen and placing the entire assembly inside the furnace. If this type of device is not available, the displacements may be transferred outside the furnace with a rod-and-tube assembly that is connected to a displacement transducer—either a direct current displacement transducer (DCDT), linear variable displacement transducer (LVDT), or capacitance gage—outside the furnace. In these transfer-type displacement devices the transfer rod and tubes should be fabricated of material that experiences low thermal expansion and is thermally stable (Ref 45). The resolution of these deflection measurement devices should be a minimum of 0.01 mm (0.0004 in.) or less.

If measurements cannot be taken directly on the specimen load-line, deflection can be measured from the test machine crosshead movement with the use of dial gages and/or the displacement transducers noted above. Under the constant loading conditions, the crosshead/load-train deflection will be primarily due to the test specimen with the exception of the initial deflection, which will contain some elastic contributions.

With smaller-sized specimens, the reduced notch dimensions will not accommodate a clip-on load-line gage. Past research has used a modified rod-and-tube extensometer connected to an external DCDT, the arms of which were attached on the outer surface (top and bottom faces) of the specimens on the load-line just above the pinholes (Ref 71). Clevises with deep throats can be used to accommodate this extensometry.

Data Acquisition. The measurements taken during the testing, electric potential voltage, load-line displacement, temperature, and cycles for creep-fatigue tests can be recorded continuously with the use of strip chart recorders, voltmeters, or digital data acquisition systems. The resolution of these acquisition systems should be at least one order of magnitude better than the measuring instrument. However, no matter which technique is used, it is important to remember the thermal voltage in the electric potential readout should be measured at least once every 24 h as described previously.

Typical Test Procedures

Test Setup. Because of the inherent scatter observed in most test situations, more than one test per condition is suggested so that data confidence intervals can be obtained. In creep and creep-fatigue crack growth testing, variables such as microstructural differences, load precision, environmental control, and, to a lesser degree, data processing contribute to scatter (Ref 45).

Prior to installation in the testing machine, the test specimen should be fitted with electric potential leads. The exposed surface of the potential leads that are on the interior of the furnace can be

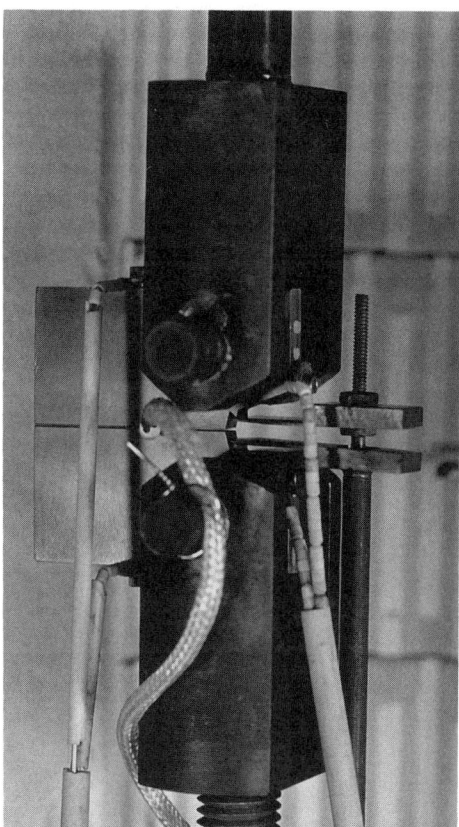

Fig. 10 Typical installed specimen ready for creep crack growth testing

covered with ceramic insulators or other shielding to avoid direct contact with the furnace elements and other components (thermocouples, extensometry, etc.) inside the furnace. After securing of the test specimen in the clevises with clevis pins, a slight load not exceeding 10% of the intended test load may be applied to improve the axial stability of the load train. The extensometer is then placed on the load line of the specimen, and care is taken to ensure that the knife edges are securely in contact. The thermocouples are then placed in contact with the specimen on the crack plane in the uncracked ligament region. Figure 10 displays a close-up of a fully fitted 0.25T-CT test specimen prior to furnace heating. The furnace is then placed into position, sealed, and started. Specimen heating should be performed gradually to avoid overshooting the test temperature because the aforementioned temperature control limits also apply to specimen heating. The current in the electric potential system should also be on during the furnace heating because resistance heating of the specimen occurs by the applied current. Once the test temperature is achieved and stable, the specimen is allowed to "soak" for at least 1 h per 25 mm (1 in.) of specimen thickness at the test temperature prior to applying load. Once sufficient time at temperature has been achieved, a set of measurements are recorded in the no-load condition for the reference conditions. Then the load is carefully applied in order to avoid inertial loading. The choice of load or K

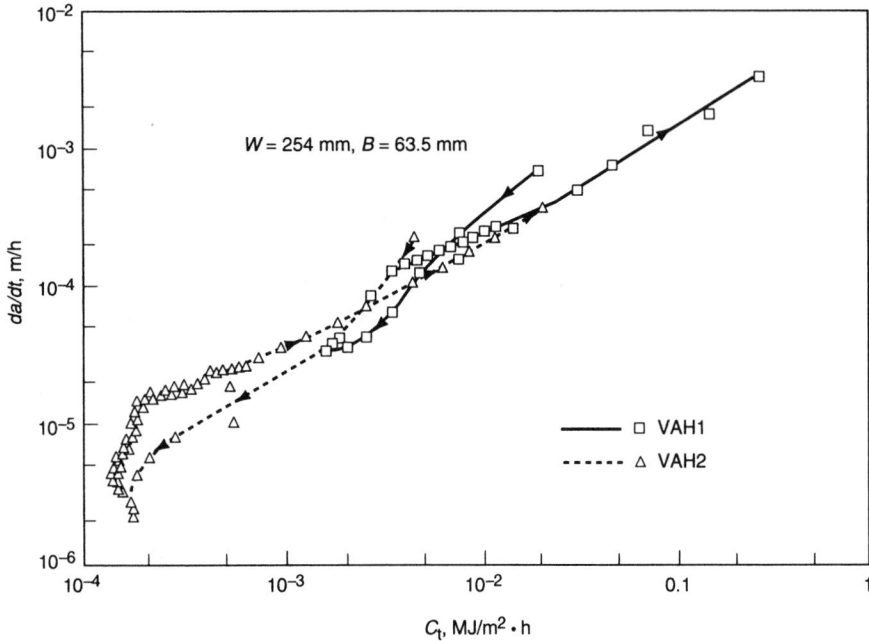

Fig. 11 Creep crack growth rate from specimens of 63.5 mm (2.5 in.) thickness in which crack growth occurred in small-scale, transition, and extensive creep regimes. Source: Ref 51

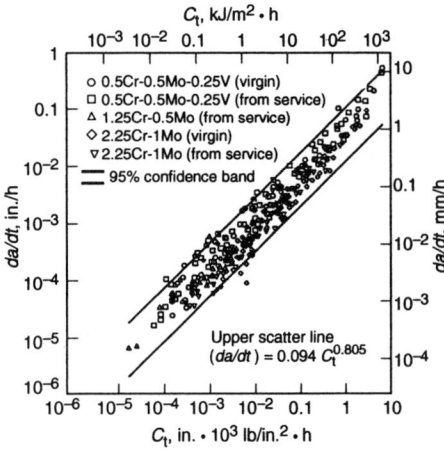

Fig. 12 Creep crack growth rate for chromium-molybdenum steels (tested at 1000 to 1022 °F) compiled from various laboratories. Source: Ref 7

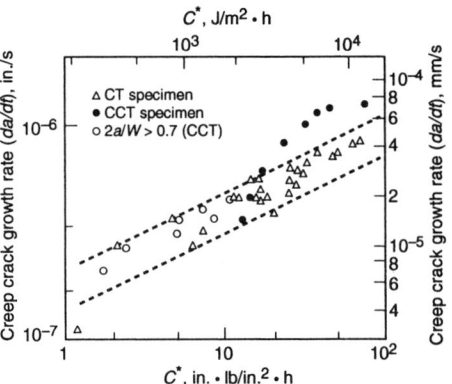

Fig. 13 Creep crack growth rate for 304 stainless steels at 594 °C (1100 °F) with differing specimen geometries. Source: Ref 27

level is dependent on the crack growth rates required during the test. Ideally, crack growth rates should be the same as those encountered by the material during service. The time of specimen loading should be as short as possible, and another set of measurements of electric potential and displacement are recorded immediately upon completion of loading as the initial loading condition (time = 0).

For creep-fatigue crack growth testing, additional consideration is given to specimen loading because cyclic tests are required. The hold time should be selected in conjunction with the K-level such that crack extensions of approximately 5 mm (0.2 in.) are obtained during the planned duration of the test. The hold time should also be selected such that the deflection that accumulates during the hold time is approximately three to five times the sensitivity of the displacement gage/amplifier system used. The loading waveshape should simulate the service loading conditions. As mentioned above, load-controlled testing can be performed under a variety of waveforms. For power-plant components and gas turbines used in airplanes, a trapezoidal waveform is a good approximation. The rise-decay and hold times should be representative of the relative times of fatigue and creep loading conditions in service.

Post Test Measurements. Once the test is completed, either due to specimen failure or by attainment of sufficient crack growth, the load is removed, the furnace is turned off, and the specimen is allowed to naturally cool and is then removed from the loading clevises. The original crack length (after precracking) and the final crack length (resulting from creep crack growth) are measured at nine equally spaced locations along the crack front. All the data are processed using computer programs that utilize either the

secant method or seven-point polynomial method to calculate the deflection rates, dV/dt, crack growth rates, da/dt and the crack-tip parameters discussed previously. The details of these methods can be found in the ASTM E 1457 (Ref 45).

Crack Growth Correlations

Creep Crack Growth. The experimental creep crack growth rate data for creep-ductile materials have been shown to correlate with the C^*-integral and the C_t parameter. As demonstrated in previous discussion, C_t and C^* are identical in the extensive creep regime. However, C_t is also valid in the small-scale and transition regimes where C^* is no longer path-independent and therefore does not uniquely characterize the crack-tip stress fields. Because C_t is more general than C^*, it has been chosen as the primary parameter for correlating creep crack growth in the discussion that follows. Figure 11 shows the creep crack growth rate of 1Cr-1Mo-¼V steel obtained from specimens that were 254 mm (10 in.) wide and had nominal thicknesses of 63.5 mm (2.5 in.) (Ref 72). These experiments were performed at the National Research Institute for Metals (NRIM) in Japan (Ref 72). In these tests, the initial crack growth rate occurred in the small-scale and the transition creep regions, while in the latter part of the test extensive creep conditions prevailed. The arrows mark the direction in which data were collected. Note that a high C_t value was obtained in the initial portion of the tests that progressively decreased to a minimum value and then subsequently increased. This was the pattern in both the tests identified by the specimen numbers VAH1 and VAH2. The authors note that the crack growth rates uniquely correlate with C_t during the increasing and decreasing portions of

the tests, lending support to the theory that C_t can correlate crack growth rates over the wide range of conditions from small-scale to extensive creep.

Figure 12 shows the creep crack growth rates for several chromium-molybdenum steels correlated with C_t (Ref 73). These data were consolidated from several experimental studies, essentially proving that consistency in the data can be obtained from one laboratory to another during creep crack growth testing. It is also observed that when da/dt is correlated with C_t (or C^* for that matter, in the case of extensive creep), a first-order normalization of temperature effects are obtained. Because C_t and C^* are obtained from the product of the load, P, and the creep displacement rate, V_c, where one is applied and the other is a response dependent on temperature, it is expected that to some extent the correlation between da/dt and C_t is independent of temperature (Ref 73). However, when changing the temperature will result in fundamental changes in creep deformation and damage mechanisms, such normalization of temperature effects is not expected.

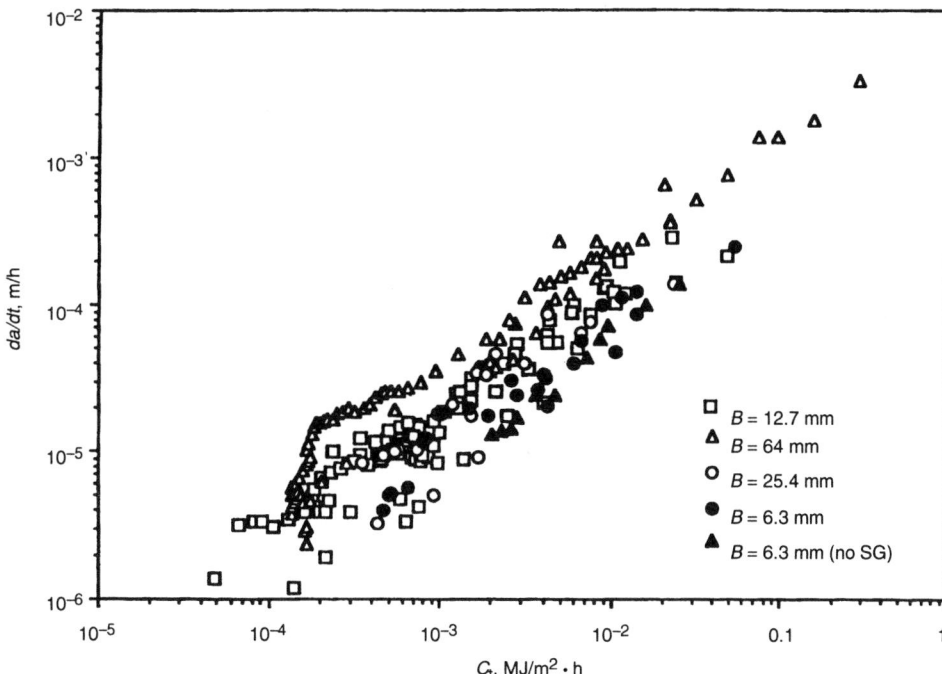

Fig. 14 Creep crack growth rate for 1Cr-1Mo-¼V steels with differing specimen sizes. SG, side grooved. Source: Ref 51

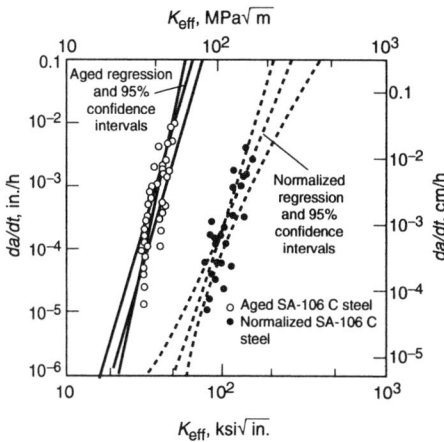

Fig. 15 Creep crack growth rate for a typical creep-brittle material. Source: Ref 74

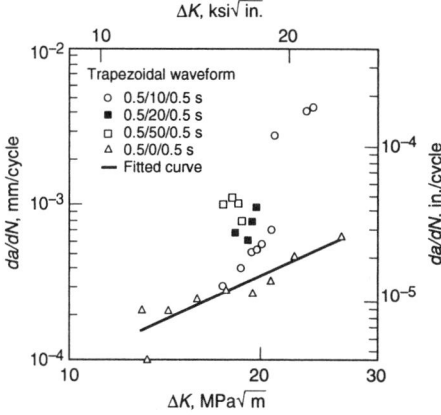

Fig. 16 da/dN versus ΔK for a 2¼Cr-1Mo steel at 594 °C (1100 °F) tested with and without hold times. Source: Ref 61

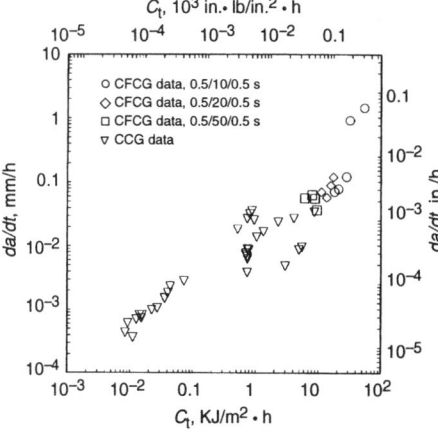

Fig. 17 Correlation of measured crack growth rates with the C_t calculated from experimental measurements (Ref 61) for 2.25Cr-1.0Mo steel at 594 °C (1100 °F). (Note da/dt versus C_t plotted for the creep crack growth data and $(da/dt)_{avg}$ with $(C_t)_{avg}$ for the creep-fatigue data)

Figure 13 shows the correlation between da/dt and C^* for 304 stainless steel where compact as well as center crack tension geometries were used for obtaining the data (Ref 12). This demonstrates the geometry independence of such data. All tests in this study were in the extensive creep conditions, thus no distinction is made between C_t and C^*.

Figure 14 shows creep crack growth rate data from specimens of different sizes from the 1Cr-1Mo-¼V steel (Ref 37). It is noted that the data from the specimens 6.25 mm (0.25 in.) thick seem to lie at the lower end of the scatterband while the data from the specimens 63.5 mm (2.5 in.) thick seem to lie on the upper end of the scatterband. There seems to be a systematic effect of thickness indicating a state-of-stress effect. Therefore, in generating creep crack growth rate data, it is essential to give proper consideration to the thickness of the specimen, depending on the end use of the data.

Figure 15 shows the creep crack growth rate as a function of K for a highly cold-worked carbon-manganese steel tested at 360 °C (680 °F) (Ref 74). At this temperature, this material exhibits creep-brittle behavior. These correlations are very sensitive to the effects of temperature. Similar correlations have been shown for nickel-base alloys (Ref 75, 76), cold-worked 316 stainless steel (Ref 29), Ti-6242 alloy (Ref 51), and for 2519 Al alloy (Ref 49). Conditions for the unique correlation between da/dt and K are not well understood for creep-brittle materials, and this continues to be an area of considerable research.

Creep-Fatigue Crack Growth Correlations. Figure 16 (Ref 61) shows plots of da/dN versus ΔK for a 2.25Cr-1.0Mo steel at 595 °C (1100 °F). The regression line through the elevated-tem-perature fatigue test ($t_h = 0$) is used to get the cycle-dependent part in modeling the creep-fatigue data. The lack of correlation between da/dN and ΔK for the creep-fatigue tests is evident from the data scatter in this figure. Such lack of correlation has also been shown in Cr-Mo-V steel (Ref 77). An increase in the da/dN with increasing hold time for fixed ΔK has also been reported by Saxena and Bassani (Ref 78). This is due to the increasing contribution of time-dependent crack growth (Ref 79). Creep damage at the crack tip, influence of the environment, or microstructural changes such as formation of cavities that occur during loading at elevated temperatures could be responsible for this behavior (Ref 58).

The average time-dependent crack growth rates, $(da/dt)_{avg}$, are correlated with $(C_t)_{avg}$. Figure 17 (Ref 17) is a plot of $(da/dt)_{avg}$ versus $(C_t)_{avg}$ for 2.25Cr-1.0Mo steel, for various hold times at elevated temperatures. All data show a clear trend and fall into a narrow scatterband despite the range of hold times used. Similar trends have also been shown for 1.25Cr-0.5Mo steel as shown in Fig. 18 (Ref 60). This strongly indicates the usefulness of $(C_t)_{avg}$ in characterizing creep-fatigue rates. Furthermore, the creep crack growth (CCG) data for each of these materials has also been plotted on these graphs. However, da/dt has been correlated with C_t for the creep crack growth data. All the creep and creep-fatigue crack growth rate data show the same trend. This has the important implication that life prediction procedures for these materials would be considerably simplified because CCG data could be used to predict the life of components under creep-fatigue conditions and vice versa. In comparing a $(da/dt)_{avg}$ versus $(C_t)_{avg}$ relation of creep-fatigue with a da/dt versus C_t

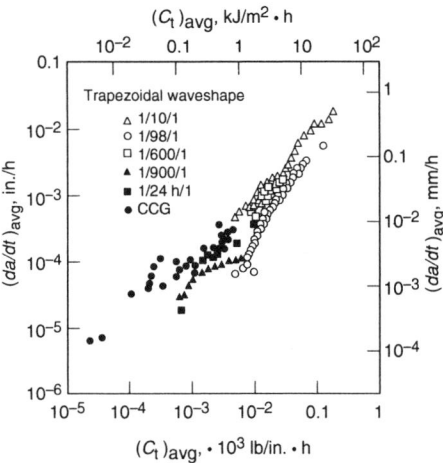

Fig. 18 Comparison between creep and creep-fatigue crack growth data in terms of the estimated $(C_t)_{avg}$ for 1.25Cr-0.5Mo steel at 538 °C (1000 °F). Source: Ref 59, 60

relation of CCG, it must be kept in mind that although C_t and $(C_t)_{avg}$ are equivalent parameters with the same physical interpretation, their exact values may differ slightly in the small-scale creep regime by a constant factor for a given material (Ref 60). The value of this constant ranges from 1 to 1.3 for different materials (Ref 60).

The time dependence of the life-prediction model is obtained by generating a regression line through all the data. The total fatigue crack growth rate per cycle is a linear summation of the cycle and time-dependent crack growth rates. Such an expression obtained for 2.25Cr-1.0Mo steel at 595 °C (1100 °F) under trapezoidal loading waveshapes is given in the following equation (Ref 57) for SI units (mm/h, MPa$\sqrt{m}$, and C_t in kJ/m$^2 \cdot$ h):

$$\frac{da}{dN} = 1.08 \times 10^{-6} \, \Delta K^{1.94}$$
$$+ 1.46 \times 10^{-2} \left[(C_t)_{avg} \right]^{0.722} t_h \quad \text{(SI)}$$

In English units (in./h, ksi$\sqrt{in.}$, and C_t in units 10^3 lb/in. $\cdot$ h) the relation is:

$$\frac{da}{dN} = 5.11 \times 10^{-8} \, \Delta K^{1.94}$$
$$+ 2.40 \times 10^{-2} \left[(C_t)_{avg} \right]^{0.722} t_h \quad \text{(English)}$$

The first term in the equations above represents the cycle-dependent crack growth rate and the other term represents the time-dependent crack growth rate. These equations can be effectively used to predict the service life of high-temperature components made of 2.25Cr-1.0Mo steel under both creep and creep-fatigue crack growth conditions at 595 °C (1100 °F). An upper and lower scatterband can also be generated for the data in Fig. 17 and 18 for design purposes. This model has been established under the assumption

that the crack growth during hold time is only due to creep deformation. Any other time-dependent effects like oxidation at the crack tip have not been considered (Ref 57). Neither have any synergistic effects due to any complicated interactions of the creep and fatigue mechanisms of crack growth during unloading/reloading been incorporated. However, with the assumption that the unloading/reloading times are much smaller compared with the hold times, their exclusion seems justified (Ref 57). If such effects were to be considered, depending on the material, an equation of the type presented above would be too simplistic in its description of the creep-fatigue behavior of a material. This remains a subject of future research.

REFERENCES

1. J.A. Harris, Jr., D.L. Sims, and C.G. Annis, Jr., "Concept Definition: Retirement for Cause of F100 Rotor Components," AFWAL-Tr-80-4118, Air Force Wright Aeronautical Laboratories, Sept 1980
2. A. Saxena and P.K. Liaw, "Remaining Life Estimations of Boiler Pressure Parts—Crack Growth Studies," Final Report, CS 4688 per EPRI Contract RP 2253-7, 1986
3. Proc. EPRI Conf. on Life Extension and Assessment of Fossil Plants, June 1986, Electric Power Research Institute (Washington, D.C.)
4. W.A. Logsdon, P.K. Liaw, A. Saxena, and V. Hulina, *Eng. Fract. Mech.*, Vol 25, 1986, p 259
5. A. Saxena, P.K. Liaw, W.A. Logsdon, and V. Hulina, *Eng. Fract. Mech.*, Vol 25, 1986, p 290
6. The National Board of Boiler and Pressure Vessel Inspectors, *Natl. Board Bull.*, Vol 43 (No. 2), 1985
7. A. Saxena, "Life Assessment Methods and Codes," EPRI TR-103592, Electric Power Research Institute, Jan 1996
8. A. Saxena, "Recent Advances in Elevated Temperature Crack Growth and Models for Life Prediction," *Advances in Fracture Research: Proc. Seventh Int. Conf. on Fracture*, ICF-7, March 1989 (Houston, TX)
9. J.T. Staley, Jr., "Mechanisms of Creep Crack Growth in a Cu-1 wt.% Sb Alloy," M.S. thesis, Georgia Institute of Technology, March 1988
10. J.L. Bassani and V. Vitek, *Proc. Ninth National Congress of Applied Mechanics—Symposium on Non-Linear Fracture Mechanics*, L.B. Freund and C.F. Shih, Ed., ASME, 1982, p 127-133
11. R. Raj, *Flow and Fracture at Elevated Temperatures*, R. Raj, Ed., American Society for Metals, 1983, p 215-249
12. A. Saxena, *Fracture Mechanics—12*, STP 700, ASTM, 1980, p 131
13. J.L. Bassani and F.A. McClintock, Creep Relaxation of Stress around a Crack Tip, *Int. J. Solids Struct.*, Vol 7, 1981, p 479-492
14. J. Rice, A Path Independent Integral and the Approximate Analysis of Strain Concentration

by Notches and Cracks, *J. Appl. Mech.*, Vol 35, 1986, p 379-386
15. J.D. Landes and J.A. Begley, A Fracture Mechanics Approach to Creep Crack Growth, *Mechanics of Crack Growth*, STP 590, ASTM, 1976, p 128-148
16. C.P. Leung, D.L. McDowell, and A. Saxena, "Influence of Primary Creep in the Estimation of C_t Parameter," Topical Report on Contract 2253-10, Electric Power Research Institute, 1988
17. H. Riedel and V. Detampel, Creep Crack Growth in Ductile, Creep Resistant Steels, *Int. J. Fracture*, Vol 24, 1987, p 239-262
18. A. Saxena, Creep Crack Growth under Non-Steady-State Conditions, *Fracture Mechanics—17*, STP 905, ASTM, 1986, p 185-201
19. J.W. Hutchinson, Singular Behavior at the End of a Tensile Crack in a Hardening Material, *J. Mech. Phys. Solids*, Vol 16, 1968, p 13-31
20. J.W. Rice and G.F. Rosengren, Plane Strain Deformation near Crack Tip in a Power Law Hardening Material, *J. Mech. Phys. Solids*, Vol 16, 1968, p 1-12
21. K.M. Nikbin, G.A. Webster, and C.E. Turner, Relevance of Nonlinear Fracture Mechanics to Creep Cracking, *Cracks and Fracture*, STP 601, ASTM, 1976, p 47-62
22. H. Riedel, *Fracture at High Temperatures*, Springer-Verlag, Berlin, 1987
23. M.F. Kanninen and C.H. Popelar, *Advanced Fracture Mechanics*, Oxford University Press, 1985, p 437
24. A. Saxena, Mechanics and Mechanisms of Creep Crack Growth, *Fracture Mechanics: Microstructures and Micromechanisms*, ASM International, 1989, p 283-334
25. N.L. Goldman and J.W. Hutchinson, Fully Plastic Crack Problems: The Center-Cracked Strip under Plane Strain, *Int. J. Solids Struct.*, Vol 11, 1975, p 575-591
26. C.F. Shih, "Table of Hutchinson-Rice-Rosengren Singular Field Quantities," Technical Report MRL E-147, Brown University, June 1983
27. A. Saxena and J.D. Landes, *Advances in Fracture Research*, ICF-6, Pergamon Press, 1984, p 3977-3988
28. D.J. Smith and G.A. Webster, *Elastic-Plastic Fracture*, Vol I, *Inelastic Crack Analysis*, STP 803, ASTM, 1983, p 654
29. A. Saxena, H.A. Ernst, and J.D. Landes, *Int. J. Fracture*, Vol 23, 1983, p 245-257
30. A. Saxena, T.T. Shih, and H.A. Ernst, *Fracture Mechanics—15*, STP 833, ASTM, 1984, p 516
31. A. Saxena, *Eng. Fract. Mech.*, Vol 40 (No. 4/5), 1991, p 721-736
32. H. Riedel and J.R. Rice, *Fracture Mechanics—12*, STP 700, ASTM, 1980, p 112
33. H.A. Ernst, *Fracture Mechanics—14*, Vol I, *Theory and Analysis*, STP 791, ASTM, 1983, p I-499
34. K. Ohji, K. Ogura, and S. Kubo, *Jpn. Soc. Mech. Eng.*, No. 790-13, 1979, p 18
35. R. Ehlers and H. Riedel, *Advances in Fracture Research*, ICF-5, Vol 2, Pergamon Press, 1981, p 691

36. H. Riedel, *J. Mech. Phys. Solids*, Vol 29, 1981, p 35

37. H. Riedel and V. Detampel, *Int. J. Fracture*, Vol 33, 1987, p 239

38. J. Bassani, D.E. Hawk, and A. Saxena, *Nonlinear Fracture Mechanics*, Vol I, *Time Dependent Fracture*, STP 995, ASTM, 1989, p 7

39. A. Saxena, *Mater. Sci. Eng. A*, Vol 108, 1988, p 125

40. C.P. Leung, D.L. McDowell, and A. Saxena, *Nonlinear Fracture Mechanics*, Vol I, *Time Dependent Fracture*, STP 995, ASTM, 1989, p 55

41. C.P. Leung and D.L. McDowell, *Int. J. Fracture*, Vol 46, 1990, p 81-104

42. C.P. Leung, Ph.D. dissertation, School of Mechanical Engineering, Georgia Institute of Technology, 1988

43. C.Y. Hui and H. Riedel, *Int. J. Fracture*, Vol 17, 1981, p 409-425

44. H. Riedel and W. Wagner, *Advances in Fracture Research*, ICF-5, Vol 2, Pergamon Press, 1985, p 683-688

45. "Standard Test Method for Measurement of Creep Crack Growth Rates in Metals," E 1457, ASTM, 1992

46. D.E. Hall, Ph.D. dissertation, School of Mechanical Engineering, Georgia Institute of Technology, 1995

47. P.L. Bensussan and R.M. Pelloux, Creep Crack Growth in 2219-T851 Aluminum Alloy: Applicability of Fracture Mechanics Concepts, *Advances in Fracture Research*, ICF-6, Vol 3, Pergamon Press, 1986, p 2167-2179

48. P.L. Bensussan, D.A. Jablonski, and R.M. Pelloux, *Metall. Trans. A*, Vol 15, 1984, p 107-120

49. B.C. Hamilton, M.S. thesis, School of Materials Science and Engineering, Georgia Institute of Technology, 1994

50. K.A. Jones, M.S. thesis, School of Materials Science and Engineering, Georgia Institute of Technology, 1993

51. B. Dogan, A. Saxena, and K.H. Schwalbe, *Mater. High Temp.*, Vol 10, 1992, p 138-143

52. P. Bensussan, *High Temperature Fracture Mechanisms and Mechanics: Proc. MECAMAT Int. Seminar on High Temperature Fracture Mechanisms and Mechanics,* P. Benussan and J. Mascavell, Ed., Mechanical Engineering Publications, Vol 3, 1990, p 1-17

53. P. Bensussan, G. Cailletaud, R. Pelloux, and A. Pineau, *The Mechanisms of Fracture*, V.S. Goel, Ed., American Society for Metals, 1986, p 587-595

54. T.S.P. Austin and G.A. Webster, *Fat. Fract. Eng. Mat. Struct.*, Vol 15 (No. 11), 1992, p 1081-1090

55. H.H. Johnson, *Mater. Res. Stand.*, Vol 5 (No. 9), 1965, p 442-445

56. P.S. Grover and A. Saxena, *Sādhnā, Integrity Engineering Components*, Vol 20, Part I, 1995, p 53-85

57. P.S. Grover and A. Saxena, Characterization of Creep-Fatigue Crack Growth Behavior in $2\frac{1}{4}$Cr-1Mo Steel using $(C_t)_{avg}$, *Int. J. Fracture*, Vol 73, No. 4, 1995, p 273-286

58. A. Saxena, *JSME Int. J. Series A*, Vol 36 (No. 1), 1993, p 1-20

59. K.B. Yoon, Ph.D. dissertation, School of Mechanical Engineering, Georgia Institute of Technology, June 1990

60. K.B. Yoon, A. Saxena, and P.K. Liaw, *Int. J. Fracture*, Vol 59, 1993, p 95-114

61. P.S. Grover, M.S. thesis, School of Materials Science and Engineering, Georgia Institute of Technology, 1993

62. N. Adefris, Ph.D. dissertation, School of Materials Science and Engineering, Georgia Institute of Technology, 1993

63. K.B. Yoon, A. Saxena, and D.L. McDowell, *Fracture Mechanics—22*, STP 1131, ASTM, 1992, p 367-392

64. A. Saxena, Limits of Linear Elastic Fracture Mechanics in the Characterization of High-Temperature Fatigue Crack Growth, *Basic Questions in Fatigue: Vol II,* R. Wei and R. Gangloff, Ed., STP 924, ASTM, 1989, p 27-40

65. Practices of Load Verification of Testing Machines, E4-94, *Annual Book of Standards*, Vol 3.01, ASTM, 1994

66. P.S. Grover, Ph.D. dissertation, unpublished research, School of Materials Science and Engineering, Georgia Institute of Technology, 1995

67. A. Saxena, R.S. Williams, and T.T. Shih, *Fracture Mechanics—13*, STP 743, ASTM, 1981, p 86

68. D.N. Gladwin, D.J. Miller, and R.H. Priest, *Mater. Sci. Technol.*, Vol 5, Jan 1989, p 40-51

69. "Test Method for Plane-Strain Fracture Toughness of Metallic Materials," E 399, *Annual Book of ASTM Standards*, Vol 03.01, ASTM, 1994, p 680-714

70. K.H. Schwalbe and D.J. Hellman, *Test. Eval.*, Vol 9 (No. 3), 1981, p 218-221

71. R.H. Norris, Ph.D. dissertation, School of Materials Science and Engineering, Georgia Institute of Technology, 1994

72. A. Saxena, K. Yagi, and M. Tabuchi, *Fracture Mechanics: Vol 24*, STP 1207, ASTM, 1992, p 481-497

73. A. Saxena, J. Han, and K. Banerji, *J. Pressure Vessel Technol.*, Vol 110, 1988, p 137-146

74. Y. Gill, Ph.D. dissertation, School of Materials Science and Engineering, Georgia Institute of Technology, 1994

75. K. Sadananda and P. Shahinian, *Fracture Mechanics*, N. Perrone, et al., Ed., 1978, p 685-703

76. R.M. Pelloux and J.S. Huang, *Creep-Fatigue-Environment Interactions*, R.M. Pelloux and N.S. Stoloff, Ed., TMS-AIME, 1980, p 151-164

77. C.B. Harrison and G.N. Sandor, *Eng. Fract. Mech.*, Vol 3, 1971, p 403-420

78. A. Saxena and J.L. Bassani, *Fracture: Interactions of Microstructure, Mechanisms and Mechanics*, TMS-AIME, 1984, p 357-383

79. P.S. Grover and A. Saxena, *Structural Integrity: Experiments, Models and Applications,* ECF-10, K. Schwalbe and C. Bergin, Ed., Engineering Materials Advisory Services, 1994, p 1-21

High-Temperature Life Assessment*

A.F. Liu, Rockwell International (retired)

CURRENT FRACTURE MECHANICS theory treats cyclic crack growth as a linear elastic phenomenon. The residual strength of a test coupon, or a structural component, is frequently computed based on linear elastic fracture indexes. Elastic-plastic, or fully plastic analysis such as the J-integral approach is used when large scale yielding occurs. All the existing crack growth analysis methods for spectrum life prediction basically deal with using material constant amplitude crack growth rate data to compute crack growth history of a structural element. For crack growth at high temperature, the conventional crack growth methodology that was based on material room temperature behavior will no longer be applicable.

The need for an updated fracture mechanics technology that can handle the combined effects of temperature, stress amplitude, cyclic frequency, and dwell time was first recognized by researchers in the nuclear and aircraft engine industries and government agencies. Substantial research efforts have been made since the mid-1970s. Summaries of the accomplishments have been documented in a number of review papers, listed here as Ref 1 through 9. Application of these new technologies to damage tolerance analysis of aircraft structures is discussed in Ref 10 through 13.

The key products developed during this period (mid-1970 to 1991) include:

- New fracture mechanics indexes for characterizing material residual strength and sustained load crack growth at high temperature, namely, the steady-state creep parameter C^*, and the transient creep parameter, C_t (see previous article)
- A large quantity of test data revealing the various aspects of high-temperature crack growth behavior
- Computer models for constant amplitude crack growth at high temperature (Ref 14, 15)
- An updated computer code for spectrum crack growth life prediction (Ref 16)

*Portions of this article were reprinted from the book *Damage Mechanisms and Life Assessment of High-Temperature Components*, by R. Viswanathan. ASM International, 1989.

This article discusses the variables affecting the material crack growth rate behavior and those essential elements in making spectrum crack growth life prediction. In addition, life assessment for bulk creep damage is briefly reviewed. More extensive discussions of these methods are presented in Ref 17 on a component-specific basis for boilers, turbines, pressure vessels, and advanced steam plants.

Assessment of Bulk Creep Damage

The current approaches to creep damage assessment of components can be classified into two broad categories: (1) history-based methods, in which plant operating history in conjunction with standard material property data is employed to calculate the fractional creep life that has been expended, using the life-fraction rule or other damage rules and (2) methods based on postservice evaluation of the actual component.

In history-based methods, plant records and the time-temperature history of the component are reviewed. The creep-life fraction consumed for each time-temperature segment of the history can then be calculated and summed up using the lower-bound ISO data and the life-fraction rule, or other damage rules.

The most common approach to calculation of cumulative creep damage is to compute the amount of life expended by using time or strain fractions as measures of damage. When the fractional damages add up to unity, then failure is postulated to occur. The most prominent rules are as follows:

- Life-fraction rule, $\Sigma\, t_i/t_{ri} = 1$
- Strain-fraction rule, $\Sigma\, \varepsilon_i/\varepsilon_{ri} = 1$
- Mixed rule, $\Sigma\, (t_i/t_{ri})^{1/2}\, (\varepsilon_i/\varepsilon_{ri})^{1/2} = 1$
- Mixed rule, $k\, \Sigma\, (t_i/t_{ri}) + (1-k)\, \Sigma\, (\varepsilon_i/\varepsilon_{ri}) = 1$

where k is a constant, t_i and ε_i are the time spent and strain accrued at condition i, and t_{ri} and ε_{ri} are the rupture life and rupture strain under the same conditions.

The life-fraction procedure usually is inaccurate because of errors in assumed history, in material properties, and in the life-fraction rule itself. The temperature-history information may be somewhat refined by supplemental information concerning the current oxide-scale thickness and microstructural details. In spite of such refinements, only gross estimates of creep damage are obtained using the calculation technique.

Direct postservice evaluations represent an improvement over history-based methods, because no assumptions regarding material properties and past history are made. Unfortunately, direct examinations are expensive and time-consuming. The best strategy is to combine the two approaches. A history-based method is used to determine if more detailed evaluations are justified and to identify the critical locations, and this is followed by judicious postservice evaluation. Table 1 summarizes the techniques that are in use for life assessment and some of the issues pertaining to each technique (Ref 17).

Current postservice evaluation procedures include conventional nondestructive evaluation (NDE) methods (e.g., ultrasonics, dye-penetrant inspection, etc.), dimensional (strain) measurements, and creep-life evaluations by means of accelerated creep testing. All of these methods have limitations. Normal NDE methods often fail to detect incipient creep damage and microstructural damage, which can be precursors of rapid, unanticipated failures. Due to unknown variations in the original dimensions, changes in dimensions cannot be determined with confidence. Dimensional measurements fail to provide indications of local creep damage caused by localized strains such as those in heat-affected zones of welds and regions of stress concentrations in the base metal. Cracking can frequently occur without manifest overall strain. Furthermore, the critical strain accumulation preceding fracture can vary widely with a variety of operational material parameters and with stress state.

A common method of estimating the remaining creep life is to conduct accelerated rupture tests at temperatures well above the service temperature. The stress is kept as close as possible to the service stress value, because only isostress-varied temperature tests are believed to be in compliance with the life-fraction rule. The time-to-rupture results are then plotted versus test temperature. By extrapolating the test results to the service temperature, the remaining life under service conditions is estimated.

Table 1 Life-assessment techniques and their limitations for creep-damage evaluation for crack initiation and crack propagation

Crack initiation		
Technique	Issues	Crack propagation
Calculation	Inaccurate	
Extrapolation of past experience	Inaccurate	
Conventional NDE	Inadequate resolution	
High-resolution NDE:	Not sufficiently developed at this time	
Acoustic emission		
Positron annihilation		
Barkhausen noise analysis		
Strain (dimension) measurement	Uncertainty regarding original dimensions	
	Lack of clear-cut failure criteria	
	Difficulty in detecting localized damage	
Rupture testing	Difficulty in sample removal	
	Difficulty in using as a monitoring technique	
	Validity of life-fraction rule	
	Effects of oxidation and specimen size	
	Uniaxial-to-multiaxial correlations	**Issues:**
Microstructural evaluation:	Quantitative relationships with remaining life are lacking	Uncertainties in interpretation of NDE results
Cavitation measurement		Lack of adequate crack growth data in creep and creep-fatigue
Carbide-coarsening measurements		Lack of methods for characterizing crack growth rates specific to the degraded components
Lattice parameter		Lack of a clear-cut end-of-life criterion under creep conditions
Ferrite chemistry analysis		Difficulty in assessing toughness of in-service components
Hardness monitoring		
Oxide scale measurements for tubes	Need data on oxide scale growth in steam	
	Kinetics of hot-corrosion and constant-damage curves	

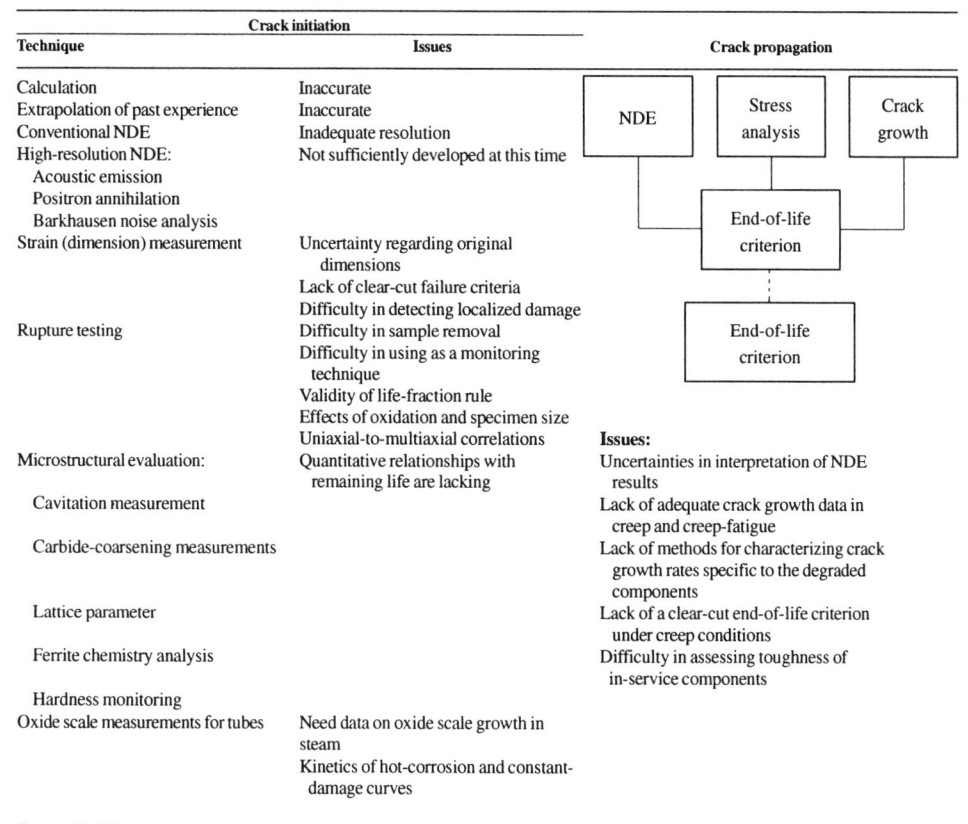

Source: Ref 17

Creep Crack Growth

As described in previous sections of this article, creep crack growth can be characterized in terms of fracture mechanic crack growth by the

Implementation of the above procedure requires a reasonably accurate knowledge of the stresses involved. For cyclic stressing conditions, and in situations involving large stress gradients, selection of the appropriate stress for the isostress tests is uncertain. Furthermore, the procedure involves destructive tests requiring removal of large samples from operating components. There are limitations on the number of available samples and the locations from which they can be taken. Periodic assessment of the remaining life is not possible. The costs of cutting out material, machining specimens, and conducting creep tests can add up to a significant expenditure. These costs are further compounded by the plant outage during this extended period of evaluation and decision making. Development of nondestructive techniques, particularly those based on metallographic and miniature-specimen approaches, has therefore been a major focus of the programs aimed at predicting crack initiation (see the article "Failure Control in Process Operations" in this Volume).

steady-state (large-scale crack growth regime) parameter C^* and the transient parameter C_t. The parameter C_t is theoretically equivalent to K (the crack-tip stress-intensity factor) under small-scale creep. It becomes C^* when the amount of creep deformation approaches steady-state creep condition. Therefore, C_t can be used for the entire range of creep deformations. Because K is a more convenient parameter and all the crack growth analysis methodologies have been developed based on K, the use of K (in lieu of C_t) is preferable when situation permits. If a material is creep-resistant, its creep zone size will be smaller than those creep zones in the creep ductile materials. Upon reviewing and analyzing creep crack growth data (Table 2), most high-temperature superalloys for aircraft belong to this creep-resistant category of materials. In such cases, the linear elastic index K may, in some cases, adequately characterize the high-temperature crack growth behavior in these alloys.

For heavy-walled process equipment, initiation criteria can be combined with crack-growth data to perform fracture-mechanic analysis of localized damage by creep crack growth. To estimate critical crack sizes for end-of-life under brittle-fracture conditions, nondestructive methods are needed for characterizing the toughness of the in-service components. Several methods, including Auger analysis, chemical analysis, miniature-specimen testing, chemical etching, electron mi-

croscopy, and others, have been explored for ferritic steels.

To perform a remaining-life assessment of a component under creep crack growth conditions, two principal ingredients are needed: (1) an appropriate expression for relating the driving force K, C^*, or C_t to the nominal stress, crack size, material constants, and geometry of the component being analyzed; and (2) a correlation between this driving force and the crack growth rate in the material, which has been established on the basis of prior data or by laboratory testing of samples from the component. Once these two ingredients are available, they can be combined to derive the crack size as a function of time. The general methodology for doing this is illustrated below, assuming C_t to be the driving crack-tip parameter.

Estimating C_t. The generalized expression for calculating C_t from measurements of load versus deflection rates on laboratory samples is defined (in Eq 36 of the previous article). Under extensive creep conditions, C_t can be simply calculated from the C^* expression (Eq 8 in previous article).

Another analytical expression for calculating C_t has been given as (Ref 2):

$$C_t = \frac{4\,\alpha\beta\,(1-\nu^2)}{E(n-1)} \frac{K^4}{W}$$

$$\times (EA)^{2/(n-1)}\, t^{-(n-3)/(n-1)} \frac{F'}{F} + C^* \qquad \text{(Eq 1)}$$

where β has a value of approximately $1/7.5$. In Eq 1, the first term denotes the contribution from small-scale creep and the second term denotes the contribution from steady-state, large-scale creep. The first term is time-variant whereas the second term is time-invariant. In the limit of $T \to 0$, approaching small-scale creep conditions, the first term dominates, implying that K is the controlling parameter in crack growth, with time also explicitly entering the relationship. In the limit $t \to \infty$, the first term becomes zero and C_t becomes identical with C^*. In Eq 36, F is a function of (a/W) and F' is given by $dF/d(a/W)$. In Eq 43, α is a constant whose value is a function of n, and A and n are the Norton law coefficients.

Equation 1 can be used to estimate C_t from an applied load (stress) and from a knowledge of the elastic and creep behavior of the material, the K calibration expression, and the C^* expression for the geometry of interest. The K and C^* expressions can be found in handbooks—at least for selected geometries. The material properties A and n can be obtained from creep tests. The C^* expressions (for example, in Ref 39) are not as abundantly available for different geometries as the K expressions. At the present time, this is viewed as a limitation of the technology. More detailed descriptions of the derivations of the C^* and C_t expressions, and the manner of obtaining some of the constants and calculating their values, are presented in the previous article. Procedures for estimating C^* based on the reference-stress approach also have been described by Ainsworth et al. (Ref 40).

Life Assessment under Creep Crack Growth with C_t. The general expression for C_t given in Eq 1 essentially reduces to the form

$$C_t = \sigma\dot{\varepsilon}(A, n)aH \text{ (geometry, } n) \qquad \text{(Eq 2)}$$

where σ is the stress far from the crack tip, obtained by stress analysis, $\dot{\varepsilon}$ is the strain rate far from the crack tip, which is a function of the constants A and n in the Norton relation (see Eq 2 in the previous article "Elevated-Temperature Crack Growth"), a is the crack depth from NDE measurements, and H is a tabulated function of geometry and the creep exponent n. The values of A and n are either assumed from prior data or generated by creep testing of samples. By assembling all the constants needed, the value of C_t can be calculated.

Once C_t is known, it can be correlated to the crack-growth rate through the constants b and m in the following relation:

$$\dot{a} = bC_t^m \qquad \text{(Eq 3)}$$

Values of the constants b and m for all the materials analyzed by Saxena et al. are listed in Table 3. It can be shown (from Eq 6 and 7 of the previous article) that m should have the approximate value $n/(n + 1)$, where n is the creep-rate exponent.

Combining Eq 2 and 3 provides a first-order differential equation for crack depth, a, as a function of time, t. Theoretically, this equation can be solved by separating variables and integrating. However, the procedure is complicated by the time dependency of C_t and the a (crack size) dependency of the term H in Eq 3. To circumvent this, crack growth calculations are performed with the current values of a and the corresponding values of da/dt to determine the time increment required for incrementing the crack size by a small amount Δa; that is, $\Delta t = \Delta a/\dot{a}$. This provides new values of a, t, and C_t, and the process is then repeated. When the value of a reaches the critical size a_c as defined by K_{Ic}, J_{Ic} wall thickness, remaining ligament thickness, or any other appropriate failure parameter, failure is deemed to have occurred.

Although this procedure appears complex at first sight, the calculations are relatively easy once the principles are understood. Computer programs have been developed that perform the entire analysis on personal computers. The only judgment involved is in selecting proper values for the constants A, n, b, and m, because large scatter in creep and crack growth data necessitates subjective choices. If actual creep and/or crack growth tests could be performed, more accurate results could be obtained. Several case histories are available in the literature to acquaint the reader with the procedures involved (Ref 17).

A sample output may be in the form of a table of crack depth versus time or a plot of crack size versus remaining life (Fig. 1). This plot was generated for a thick-wall cylinder under internal pressure containing a longitudinal crack. The outside radius and wall thickness of the cylinder were assumed to be 45.7 and 7.62 cm (18 and 3 in.), respectively, and the hoop stresses were calculated for internal pressures of 8.96 and 13.79 MPa (1.3 and 2 ksi). Material properties in the degraded condition (hot region) as well as in the undergraded condition were considered. The results show that the remaining life is a function of the stresses as well as of prior degradation. Plots of this type could be used to determine remaining life or to set inspection criteria and inspection

Table 2 Controlling parameters for creep crack growth analysis

Alloy	Temperature, °C	Controlling parameter	Other parameters attempted	Remarks	Ref
Aluminums					
RR58	150-200	C^*	K		31
2219-T851	150	K			32
2219-T851	175	K	σ_{nom}, σ_{net} σ_{ref}, C^*		48
Low-alloy steels					
A470 Class 8	482, 538	C^*, C_t	K		33, 49
1.0Cr-0.5Mo	535	C^*	K		35
1.0Cr-0.5Mo	525	C^*	K, J, σ_{ref}		37
1.0Cr-1.0Mo-0.25V	427-538	C_t	K, $C(t)$, C^*		23
0.5Cr-0.5Mo-0.25V	540	C^*, C_t	$C(t)$	(b)	50
2.25Cr-1.0Mo	540	C^*, C_t	$C(t)$	(b)	50
1.25Cr-0.5Mo	482, 538	C_t		(b)	36
0.16C	400, 500	C^*	σ_{net}	Tested in vacuum and in air	34
Stainless steels					
800H	535	C^*	K, σ_{net}		27
316	740	σ_{net}	K		20
316	593	C^*	K, σ_{ref}		21, 22
316	593	J	C^*	(a)	33
316	600, 650	C^*			34
304	593	C^*			18, 19
304	650	C^*	K, σ_{net}, σ_{ref}		38
304	593	C^*	C_t		33
304	650	C^*	σ_{net}	Also correlated with C_t (Ref 33)	34
Superalloys					
Udimet 700	850	K	J, C^*		26
Discaloy	649	C^*	K, σ_{nom}		51
Astroloy	704	K			24
Waspaloy	704	K			24
Nimonic 80A	650	K	C^*		35
Nimonic 115	704	K			24
René 95	704	K			24
Inco 718	538	K	C^*		25, 30
Inco 718	538	K		Tested in vacuum	29
Inco 718	593	K	C^*		25, 30
Inco 718	649	K	C^*		25, 30
Inco 718	649	K		Tested in vacuum	29
Inco 718	704	K			24
Inco 718	704	K		Tested in vacuum	24
IN 100	732	K, C^*, σ_{net}		C^* and σ_{net} correlated with CT specimens only	28

(a) Net section failure per Ref 2. (b) Both the as-processed and the used materials were tested. Sources: As listed.

Table 3 Creep crack growth constants b and m for various ferritic steels

Material	b				m	
	Upper scatter line		Mean		Upper scatter	Mean
	BU(a)	SI(b)	BU(a)	SI(b)		
All base metal	0.094	0.0373	0.022	0.00874	0.805	0.805
2¼Cr-1Mo weld metal	0.131	0.102	0.017	0.0133	0.674	0.674
1¼Cr-½Mo weld metal	(c)	(c)	(c)	(c)	(c)	(c)
2¼Cr-1Mo and 1¼Cr–½Mo heat-affected-zone/fusion-line material	0.163	0.0692	0.073	0.031	0.792	0.792

(a) BU = British units: da/dt in in./h; C_t in in. · lb/in. · h × 10^3. (b) SI = Système International units: da/dt in mm/h; C_t in kJ/m² · h. (c) Insufficient data; creep crack growth rate behavior comparable to that of base metal. Source: Saxena, Han, and Banerji, "Creep Crack Growth Behavior in Power Plant Boiler and Steam Pipe Steels," EPRI Project 2253-10, published in Ref 17

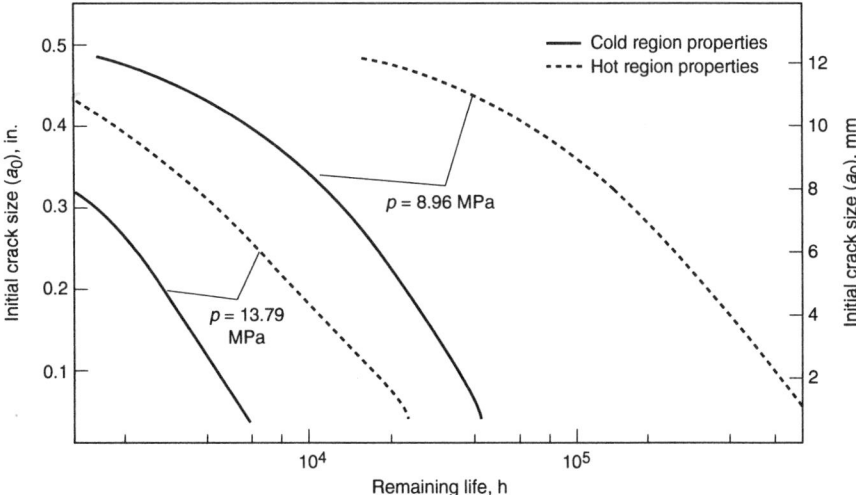

Fig. 1 Typical output from crack growth analysis showing remaining life versus initial crack size for an internally pressurized cylinder of 1.25Cr-0.5Mo steel at 538 °C (1000 °F). Source: Ref 17

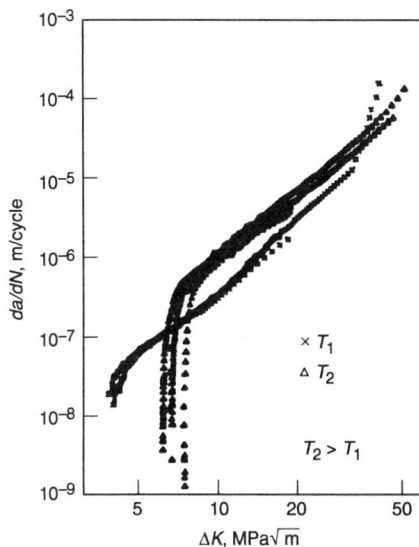

Fig. 2 Schematic of temperature effect on fatigue crack threshold and growth rates

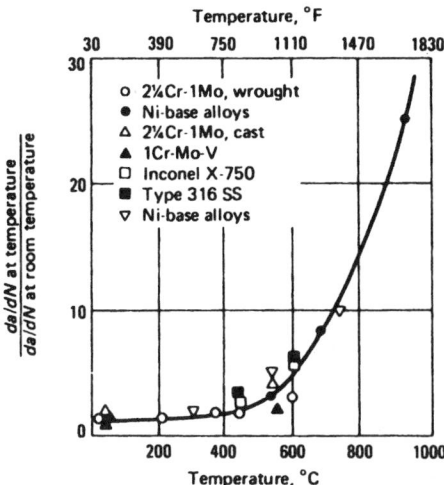

Fig. 3 Variation of fatigue crack growth rates as a function of temperature at $\Delta K = 30$ MPa√m (27 ksi√in.). Source: Ref 17

intervals. Examples of remaining-life analyses are presented in Ref 17 on boilers and rotors.

Ainsworth et al. have recently described a unified approach for structures containing defects (Ref 40). This approach incorporates structural failure by rupture, incubation behavior preceding crack growth, and creep crack growth in a single framework. Service life is governed by a combination of time to rupture, time of incubation, and time of crack growth. All of these quantities are calculated using a reference stress that is specifically applicable to the geometry of the component and is derived analytically or based on scale-model tests. If the desired service life exceeds the calculated rupture time, retirement may be necessary. In the opposite situation, further analysis is carried out to calculate the incubation time during which no crack growth is expected to occur. If the calculation indicates that the incubation time t_i is less than the desired service life, then a crack growth analysis is performed to calculate the crack growth life t_g. If the total life, $t_i + t_g$, is less than the desired service life, safe operation beyond that point would be considered undesirable. This approach seems very promising and is the subject of further investigation.

High-Temperature Fatigue Crack Growth

As previously mentioned in the section "Creep Crack Growth," the use of the linear elastic stress intensity factor (K) may be adequate for analyzing high-temperature fracture resistance of creep-resistant superalloys. This section briefly summarizes the factors affecting high-temperature fatigue crack growth in the context of traditional K-factor analysis.

If cyclic crack growth testing at high temperature is done in a traditional way (i.e., with a sinusoidal or symmetrically triangular waveform

at a moderately high frequency), the crack growth rates are functions of ΔK and R similar to those at room temperature with the following exceptions: (1) for a given R, the value of the crack threshold ΔK_{th} is higher at higher temperature, and (2) for a given R, the terminal ΔK value is higher at higher temperature because K_c is usually higher at a higher temperature (due to the fact that the material tensile yield strength is lower at a higher temperature). A schematic representation of these temperature influences, on da/dN is shown in Fig. 2. It should be noted that crack growth rates are not always higher at temperatures higher than room temperature as implied in Fig. 2. Depending on frequency and ΔK range, some material (particularly those sensitive to environment) may exhibit slower crack growth rates at intermediate temperatures. Moisture, which might have acted similarly to a corrosive medium, was vaporized by heat; therefore, the magnitude of the environmental fatigue component (which inherently associates with crack propagation) would be reduced (Ref 41).

In the power-law (Paris equation) crack growth regime, the effects of temperature, stress ratio (R), and hold times have been investigated for many high-temperature alloys. Typical behavior and crack growth results for specific alloys are covered elsewhere in this Volume. However, a general comparison of temperature effects on fatigue crack growth of several different high-temperature alloys is shown in Fig. 3. Because the reported data are obtained at various ΔK ranges and temperature ranges, the general comparison is based on a constant ΔK (arbitrarily chosen as 30 MPa√m, or 27 ksi√in.). A clear trend of crack growth rate increase with increasing temperature can be seen as shown in Fig. 3. At temperatures up to about 50% of the melting point (550 to 600 °C, or 1020 to 1110 °F), the growth rates are relatively insensitive to temperature, but the sensitivity increases rapidly at higher temperatures. The crack growth rates for all the materials at temperatures up to 600 °C relative to the room-

temperature rates can be estimated by a maximum correlation factor of 5 (2 for ferritic steels).

Besides temperature, cyclic frequency, or duration of a stress cycle (e.g., with hold time), is a key variable in high-temperature crack growth. At high frequency—that is, fast loading rate with short hold time (or no hold time)—the crack growth rate is cycle dependent and can be expressed in terms of crack growth per cycle (da/dN). At low frequency (or with long hold time), however, the crack growth rate is time dependent; that is, da/dN is in proportion to the total time span of a given cycle. For tests of different cycle times, all crack growth rate data points are collapsed into a single curve of which da/dt is the dependent variable. A mixed region exists in between the two extremes. The transition from one type of behavior to another depends on material, temperature, frequency, and R (Ref

42). For a given material and temperature combination, the transition frequency is a function of R. The frequency range at which the crack growth rates remain time dependent increases as R increases (Ref 42). The limiting case is R approaching unity. It is equivalent to crack growth under sustained load, for which the crack growth rates at any frequency will be totally time dependent.

To further understand the complex interaction mechanisms of stress, temperature, time, and environmental exposure, a vast amount of experimental and analytical data was compiled (from a bibliography of 42 references) and reviewed. Crack growth behavior for 36 types of loading profiles, which were in excess of 60 combinations in material, temperature, frequency, and time variations, were examined. A compilation of the results is presented in Ref 12. The evaluation method was to classify the data into groups representing a variety of isolated loading events. In this way, the phenomenological factors that influence load/environment interaction mechanisms can be determined. It also enables micromechanical modeling to be made to account for the contribution of each variable to the total behavior of crack growth. A composite of these load segments also provides a basis for spectrum loading simulation.

At room temperature, cyclic frequency and the shape of a stress cycle have an insignificant effect on either constant amplitude or spectrum crack growth behavior. The magnitude and sequential occurrences of the stress cycles are the only key variables affecting room-temperature crack growth behavior. Therefore, an accurate representation of the material crack growth rate data as a function of the stress amplitude ratio (i.e., the so-called crack growth law or crack growth rate equation), and a load interaction model for monitoring the load sequence effects on crack growth (the commonly called crack growth retardation/acceleration model), are the only two essential elements in cycle-dependent crack growth life predictive methodology. However, the time factor in a given stress cycle (whether it is associated with hold time or low frequency), which promotes time-dependent crack growth behavior, might play a significant role in high-temperature crack growth. Therefore, the section below briefly reviews cycle-dependent versus time-dependent crack growth.

Cycle-Dependent Versus Time-Dependent Crack Growth. Research conducted on conventional high-temperature superalloys, Inco 718 in particular, has shown that sustained load creep crack growth rate data can be used to predict cyclic crack growth in the time-dependent regime (Ref 43, 44). In those regions in which the cycle-dependent and the time-dependent phenomena are both present, implementation of a semiempirical technique may be required. A summary on formulating a procedure to predict crack growth in the time-dependent regime is given in this section.

Applying the Wei-Landes superposition principle for subcritical crack growth in an aggressive environment (Ref 45), crack growth rate for a given stress cycle can be treated as the sum of three parts:

1. The uploading part (i.e., the load rising portion of a cycle)
2. The hold time
3. The downloading (unloading) portion of a cycle

Therefore

$$\left(\frac{da}{dN}\right) = \left(\frac{da}{dN}\right)_r + \left(\frac{da}{dN}\right)_H + \left(\frac{da}{dN}\right)_d \qquad \text{(Eq 4)}$$

It has been shown frequently, by experimental tests, that the amount of da for the unloading part is negligible unless the stress profile is unsymmetric, and the uploading time to the unloading time ratio is significantly small (that is, the unloading time, compared to the uploading time, is sufficiently long). For simplicity, this discussion is limited to those cases based on two components only, that is, by setting $(da/dN)_d$ equal to zero. However, the $(da/dN)_r$ term may be cycle-dependent, or time-dependent, or mixed. This term consists of two parts; one part accounts for the cyclic wave contribution, and another part accounts for the time contribution. Consequently, Eq 4 can be rewritten as:

$$\frac{da}{dN} = \left(\frac{da}{dN}\right)_c + \left(\frac{da}{dN}\right)_t + \left(\frac{da}{dN}\right)_H \qquad \text{(Eq 5)}$$

The first term on the right-hand side of Eq 5 represents the cycle-dependent part of the cycle. It comes from the conventional crack growth rate data at high frequency; that is, it follows those crack growth laws cited in the literature (such as the Paris and Walker equations). In reality, when a stress cycle is totally cycle dependent, the magnitude of the second term on the right-hand side of Eq 5 will be negligibly small. On the other hand, when a stress cycle is totally time dependent, the contribution of $(da/dN)_c$ to the total da/dN is negligible; thereby the validity of Eq 5 in respect to full frequency range is maintained.

When a crack growth rate component exhibits time-dependent behavior, it is equivalent to crack growth under a sustained load of which the crack growth rate description is defined by da/dt (instead of da/dN) as:

$$\frac{da}{dt} = C(K_{max})^m \qquad \text{(Eq 6)}$$

This quantity is obtained from a sustained-load test. To express the second term on the right-hand side of Eq 5 in terms of da/dt, consider a generalized function that can describe K at any given time in a valley to peak cycle. That is:

$$K(t) = R \cdot K_{max} + 2K_{max} \cdot (1-R) \cdot t_r \cdot f \qquad \text{(Eq 7)}$$

where t_r is the time required for ascending the load from valley to peak, and f is the frequency of the cyclic portion of a given load cycle. For symmetric loading, Eq 7 gives $K(t) = K_{min}$ at $t_r = 0$, and $K(t) =$

K_{max} at $t_r = 1/2f$. The amount of crack extension over a period t_r can be obtained by replacing the K_{max} term of Eq 6 by $K(t)$, and integrating, that is,

$$\left(\frac{da}{dN}\right)_t = C \int_0^{t_r} [K(t)]^m \, dt \qquad \text{(Eq 8)}$$

For any positive value of m, Eq 8 yields

$$\left(\frac{da}{dN}\right)_t = C \, (K_{max})^m \, (t_r \, R_m) \qquad \text{(Eq 9)}$$

where

$$R_m = (1 - R^{m+1})/[(m+1)(1-R)] \qquad \text{(Eq 10)}$$

Therefore, for a given K_{max}, $(da/dN)_r$ increases as R increases in the time-dependent regime. This trend is opposite to that commonly observed in the high-frequency (cycle-dependent) regime.

The third term on the right-hand side of Eq 5 simply equals da/dt times the time at load. Recognizing that the first term on the right-hand side of Eq 9 is actually equal to da/dt, finally, Eq 5 can be expressed as:

$$\frac{da}{dN} = \left(\frac{da}{dN}\right)_c + \frac{da}{dt}(t_r \, R_m) + \frac{da}{dt} t_H \qquad \text{(Eq 11)}$$

where t_H is the hold time.

The applicability of Eq 11 is demonstrated in Fig. 4 and 5. In these figures, the test data were generated from the Inco 718 alloy, at 649 °C, having various combinations of ΔK, R, t_H, and f. The test data, which were extracted from the open literature (Ref 43, 44), are presented in the figures along with the predictions.

One of the two data sets in Fig.4 was generated using trapezoidal stress cycles with a frequency of 1 Hz (i.e., 0.5 s for uploading, and 0.5 for unloading) and varying $t_H = 1$ to 500 s. The data points for the other data set were obtained by conducting tests at various frequencies without hold time. A constant ΔK level (either 25 MPa$\sqrt{m}$ or 36 MPa$\sqrt{m}$ with $R = 0.1$) was applied to all the tests. Crack growth rate per cycle was plotted as a function of total time per cycle. For example, for a total cycle time of 100 s, it would mean that the test was conducted at a frequency of 1 Hz with $t_H = 99$ s, or $f = 0.01$ Hz without hold time. The predictions were made by using Eq 6 and 11 with $C = 2.9678 \times 10^{-11}$ m/s and $m = 2.65$. The value for the $(da/dN)_c$ term was set to those experimental data points for $f = 10$ Hz. It is seen that the correlation between Eq 11 and the trapezoidal load test data is quite good.

It is also shown in Fig.4 that Eq 11 correlates with those triangular load test data in the time-dependent region ($f \leq 0.02$ Hz, or total time ≥ 50 s) but fails to predict the crack growth rates in the mixed region ($0.02 < f < 10$ Hz). For this group of data, a better correlation was obtained by using the latest version of the Saxena equation (Ref 46):

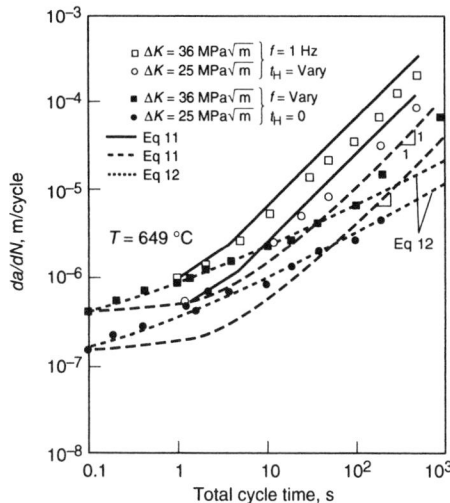

Fig. 4 High-temperature fatigue crack growth rates of Inco 718 (actual and predicted rates, $R = 0.1$)

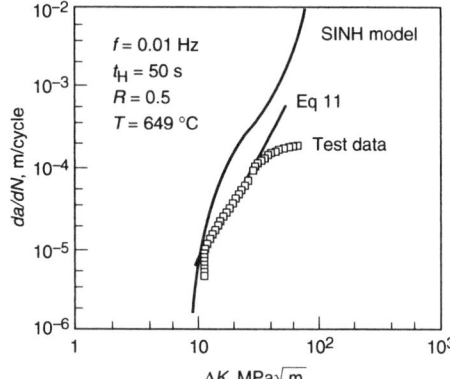

Fig. 5 High-temperature fatigue crack growth rates of Inco 718 (actual and predicted results, $R = 0.5$)

$$\frac{da}{dN} = \left(\frac{da}{dN}\right)_c + C_4 \, \Delta K^\alpha \, [1/\sqrt{f} - 1/\sqrt{f_0}] \qquad \text{(Eq 12)}$$

where f_0 is the characteristic frequency, which separates the cycle-dependent and the mixed regions. For those data sets in Fig. 4, the value for f_0 was assumed to be 10 Hz. Using a procedure given by Saxena (Ref 47), it was determined that $C_4 = 1.075 \times 10^{-10}$, $\alpha = 2.35$.

The example case shown in Fig. 5 involves all three loading variables, t_H, t_r, and R (as compared to those data sets shown in Fig. 4, of which the crack growth rates were functions of t_H and R or f and R). The test condition for this data set was: $R = 0.5$, $f = 0.01$ Hz (i.e., $t_r = 50$ s), $t_H = 50$ s, $K_{max} = 20$ to 140 MPa√m ($\Delta K = 10$ to 70 MPa√m). A very good match was obtained (up to $\Delta K = 35$ MPa√m). It thus appears that Eq 11 is superior to the other crack growth models. A comparison with the SINH model (Ref 14), a model that is widely used by the engine industry, is shown in Fig. 5.

Table 4 Comparison of fracture mechanics elements for room-temperature and high-temperature crack growth

	Room temperature	High temperature
Dependent variables	da/dN	da/dN, da/dt, mixed
Functions of	R-ratio	R-ratio, temperature, frequency, waveform
Controlling parameter	K, J	K, C_t, C^*
Crack-tip deformation mode	Plasticity (F_{ty}, n)	Plasticity (F_{ty}, n)
		Stress relaxation due to creep (E, A, n, t)
		Environmental diffusion coefficients (Q, R, t)
Spectrum life	Single mode for load interaction	Multiple modes for load/temperature/time interactions

In conclusion, crack growth behavior of a stress cycle having a trapezoidal wave form can be predicted by using the combination of conventional high-frequency da/dN data, sustained load data (da/dt), and Eq 11. For these stress cycles having a triangular wave form, test data for a specific frequency in question may be required. Otherwise, a set of test data containing several frequencies is needed for developing those empirical constants in the Saxena equation. It should be noted that Eq 12 is basically an empirical function for curve fitting and data interpolation; it is not a scientific rule that dictates the frequency effect on crack growth behavior. Therefore, although not essential, it is desirable to have an all-around method that can describe the da/dN behavior in the mixed region.

Summary. As long as load/environment interactions are absent, the total crack growth rate for a loading block containing both triangular and trapezoidal stress cycles will be

$$\left(\frac{da}{dN}\right)_{total} = \sum_i \left(\frac{da}{dN}\right)_i$$

where i denotes the ith loading step in the entire group of loads under consideration. The amount of da for each loading step is determined by using Eq 11 or 12.

An attempt to extend the existing load interaction models (for room temperature) to handle high-temperature crack growth under variable amplitude and variable waveform loadings involving all three types of crack-tip deformation modes (i.e., plasticity, creep, and environment-induced damage) is more speculative and beyond the scope of this article. Significant modifications on characterization of material properties and the load interaction and damage accumulation models are required due, in part, to the numbers of variables involved in defining the high-temperature crack growth behavior. The complexity of high-temperature crack growth is summarized in a comparison of all the major elements involved in room-temperature and high-temperature crack growth (Table 4).

REFERENCES

1. A. Saxena, Mechanics and Mechanism of Creep Crack Growth, *Fracture Mechanics: Microstructure and Micromechanisms*, ASM International, 1989
2. H. Riedel, Creep Crack Growth, *Flow and Fracture at Elevated Temperatures*, American Society for Metals, 1983, p 149-177
3. H. Ghonem, T. Nicholas, and A. Pineau, "Analysis of Elevated Temperature Fatigue Crack Growth Mechanisms in Alloy 718," Annual Winter Meeting, American Society of Mechanical Engineers, 1991
4. J.M. Larsen and T. Nicholas, Cumulative-Damage Modeling of Fatigue Crack Growth in Turbine Engine Materials, *Eng. Fract. Mech.*, Vol 22, 1985, p 713-730
5. T. Nicholas, J.H. Laflen, and R.H. Van Stone, *Proc. Conf. on Life Prediction for High Temperature Gas Turbine Materials*, Syracuse University Press, 1986, p 4.1-4.61
6. L.S. Fu, *Eng. Fract. Mech.*, Vol 13, 1980, p 307-330
7. K. Sadananda and P. Shahinian, Review of the Fracture Mechanics Approach to Creep Crack Growth in Structural Alloys, *Eng. Fract. Mech.*, Vol 15, 1981, p 327-342
8. K. Sadananda and P. Shahinian, Creep Crack Growth Behavior and Theoretical Modeling, *Met. Sci.*, Vol 15, 1981, p 425-432
9. H.P. Van Leeuwen, "The Application of Fracture Mechanics to the Growth of Creep Cracks," AGARD Report No. 705, presented at the 56th Meeting of the Structures and Materials Panel, North Atlantic Treaty Organization, April 1983 (London, UK)
10. A. Nagar, "A Review of High Temperature Fracture Mechanics for Hypervelocity Vehicle Application," AIAA Paper No. 88-2386, AIAA/ASME/ASCE/AHS/ASC 29th Structures, Structural Dynamics and Materials Conference, 1988 (Williamsburg, VA)
11. D.M. Harmon, C.R. Saff, and J.G. Burns, "Development of an Elevated Temperature Crack Growth Routine," AIAA Paper No. 88-2387, AIAA/ASME/ASCE/AHS/ASC 29th Structures, Structural Dynamics and Materials Conference, 1988 (Williamsburg, VA)
12. A.F. Liu, "Element of Fracture Mechanics in Elevated Temperature Crack Growth," AIAA Paper No. 90-0928, Collection of Technical Papers, part 2, AIAA/ASME/ASCE/AHS/ASC 31st Structures, Structural Dynamics and Materials Conference, 2-4 April 1990 (Long Beach, CA), p 981-994
13. A.F. Liu, "Assessment of a Time Dependent Damage Accumulation Model for Crack Growth at High Temperature," Paper No. ICAS-94-9.7.1, *ICAS Proc. 1994*, Vol 3, p

2625-2635 (19th Congress of the International Council of the Aeronautical Sciences, 18-23 Sept 1994 (Anaheim CA)

14. J.M. Larsen, B.J. Schwartz, and C.G. Annis, Jr., "Cumulative Damage Fracture Mechanics under Engine Spectra," Report AFML-TR-79-4159, Air Force Materials Laboratory, 1980

15. A. Utah, "Crack Growth Modeling in Advanced Powder Metallurgy Alloy," Report AFWAL-TR-80-4098, Air Force Wright Aeronautical Laboratories, 1980

16. D.M. Harmon and C.R. Saff, "Damage Tolerance Analysis for Manned Hypervelocity Vehicles," Vol I, Final Technical Report, WRDC-TR-89-3067, Flight Dynamics Laboratory, Wright Research and Development Center, Sept 1989

17. R. Viswanathan, *Damage Mechanisms and Life Assessment of High-Temperature Components*, ASM International, 1989

18. A. Saxena, Evaluation of C^* for the Characterization of Creep-Crack-Growth Behavior in 304 Stainless Steel, *Fracture Mechanics—12*, STP 700, ASTM, 1980, p 131-151

19. T.T. Shih, A Simplified Test Method for Determining the Low Rate Creep Crack Growth Data, *Fracture Mechanics—14*, Vol II, *Testing and Applications*, STP 791, ASTM, 1983, p II-232 to II-247

20. R.D. Nicholson and C.L. Formby, *Int. J. Fracture*, Vol 11, 1975, p 595-604

21. K. Sadananda and P. Shahinian, Evaluation of J* Parameter for Creep Crack Growth in Type 316 Stainless Steel, *Fracture Mechanics—14*, Vol II, *Testing and Applications*, STP 791, ASTM, 1983, p II-182 to II-196

22. K. Sadananda and P. Shahinian, Parametric Analysis of Creep Crack Growth in Austenitic Stainless Steel, *Elastic-Plastic Fracture—2*, Vol I, *Inelastic Crack Analysis*, STP 803, ASTM, 1983, p I-690 to I-707

23. A. Saxena and B. Gieseke, "Transients in Elevated Temperature Crack Growth," *Proc. ME-CAMAT*, Int. Seminar on High Temperature Fracture Mechanisms and Mechanics, 13-15 Oct 1987 (Dourdan, France), p III/19 to III/36

24. S. Floreen, The Creep Fracture of Wrought Nickel-Base Alloys by a Fracture Mechanics Approach, *Metall. Trans. A*, Vol 6, 1975, p 1741-1749

25. K. Sadananda and P. Shahinian, Creep Crack Growth in Alloy 718, *Metall. Trans. A*, Vol 8, 1977, p 439-449

26. K. Sadananda and P. Shahinian, *Metall. Trans. A*, Vol 9, 1978, p 79-84

27. M. Welker, A. Rahmel, and M. Schutze, Investigations on the Influence of Internal Nitridation on Creep Crack Growth in Alloy 800 H, *Metall. Trans. A*, Vol 20, 1989, p 1553-1560

28. R.C. Donath, T. Nicholas, and S.L. Fu, *Fracture Mechanics—13*, STP 743, ASTM, 1981, p 186-206

29. M. Stucke, M. Khobaib, B. Majumdar, and T. Nicholas, Environmental Aspects in Creep Crack Growth in Nickel Base Superalloy, *Advances in Fracture Research*, Vol 6, Pergamon Press, 1986, p 3967-3975

30. G.K. Haritos, D.L. Miller, and T. Nicholas, Sustained-Load Crack-Growth in Inconel 718 under Nonisothermal Conditions, *J. Eng. Mater. Technol. (Trans. ASME)*, Series H, Vol 107, 1985, p 172-179

31. K.M. Nikbin and G.A. Webster, Temperature Dependence of Creep Crack Growth in Aluminum Alloy RR58, *Micro and Macro Mechanics of Crack Growth*, TMS-AIME, 1982, p 137-147

32. J.G. Kaufman, K.O. Bogardus, D.A. Mauney, and R.C. Malcolm, Creep Cracking in 2219-T851 Plate at Elevated Temperatures, *Mechanics of Crack Growth*, STP 590, ASTM, 1976, p 149-168

33. A. Saxena and J.D. Landes, Characterization of Creep Crack Growth in Metals, *Advances in Fracture Research*, Vol 6, Pergamon Press, 1986, p 3977-3988

34. S. Taira, R. Ohtani, and T. Kitamura, Application of J-Integral to High-Temperature Crack Propagation, Part I—Creep Crack Propagation, *J. Eng. Mater. Technol. (Trans. ASME)*, Series H, Vol 101, 1979, p 154-161

35. H. Riedel and W. Wagner, Creep Crack Growth in NIMONIC 80A and in a 1Cr-½Mo Steel, *Advances in Fracture Research*, Vol 3, Pergamon Press, 1986, p 2199-2206

36. S. Jani and A. Saxena, Influence of Thermal Aging on the Creep Crack Growth of a Cr-Mo Steel, *Effects of Load and Thermal Histories on Mechanical Behavior of Materials*, TMS-AIME, 1987, p 201-220

37. H.P. Van Leeuwen and L. Schra, Fracture Mechanics and Creep Crack Growth of 1%Cr-1/2%Mo Steel with and without Prior Exposure to Creep Conditions, *Eng. Fract. Mech.*, Vol 127, 1987, p 483-499

38. R. Koterazawa and T. Mori, Applicability of Fracture Mechanics Parameters to Crack Propagation under Creep Condition, *J. Eng. Mater. Technol. (Trans. ASME)*, Series H, Vol 99, 1977, p 298-305

39. V. Kumar, M.D. German, and C.F. Shih, "An Engineering Approach for Elastic-Plastic Fracture Analysis," EPRI Report NP 1931, Electric Power Research Institute, Palo Alto, CA, 1981

40. R.A. Ainsworth et al., CEGB Assessment Procedure for Defects in Plant Operating in the Creep Range, *Fatigue Fract. Eng. Mater. Struct.*, Vol 10 (No. 2), 1987

41. T.T. Shih and G.A. Clarke, Effect of Temperature and Frequency on the Fatigue Crack Growth Rate Properties of a 1950 Vintage CrMoV Rotor Material, *Fracture Mechanics*, STP 677, ASTM, 1979, p 125-143

42. T. Nicholas and N.E. Ashbaugh, Fatigue Crack Growth at High Load Ratios in the Time-Dependent Regime, *Fracture Mechanics—19*, STP 969, ASTM, 1988, p 800-817

43. G.K. Haritos, T. Nicholas, and G.O. Painter, Evaluation of Crack Growth Models for Elevated-Temperature Fatigue, *Fracture Mechanics—18*, STP 945, ASTM, 1988, p 206-220

44. T. Nicholas and T. Weerasooriya, Hold-Time Effects in Elevated Temperature Fatigue Crack Propagation, *Fracture Mechanics—17*, STP 905, ASTM, 1986, p 155-168

45. R.P. Wei and J.L. Landes, Correlation Between Sustained-Load and Fatigue Crack Growth in High-Strength Steels, *Materials Research and Standards*, TMRSA, Vol 9, 1969, p 25-28

46. A. Saxena and J. Bassani, Fracture: Interactions of Microstructures, Mechanisms, and Mechanics, TMS, 1984, p 357-383

47. A. Saxena, A Model for Predicting the Effect of Frequency on Fatigue Crack Growth Behavior at Elevated Temperature, *Fat. Eng. Mater. Struct.*, Vol 3, 1981, p 247-255

48. P.L. Bensussan and R.M. Pelloux, Creep Crack Growth in 2219-T851 Aluminum Alloy: Applicability of Fracture Mechanics Concepts, *Advances in Fracture Research*, Vol. 3, Pergamon Press, NY, 1986, p 2167-2179

49. A. Saxena, Creep Crack Growth under Non-Steady State Conditions, *Fracture Mechanics—17*, STP 905, ASTM, 1986, p 185-201

50. H. Riedel and V. Detampel, Creep Crack Growth in Ductile, Creep-Resistant Steels, *International Journal of Fracture*, Vol 33, 1987, p 239-262

51. J.D. Landes and J.A. Begley, "A Fracture Mechanics Approach to Creep Crack Growth," *Mechanics of Crack Growth*, ASTM STP 590, American Society for Testing and Materials, 1976, p 128-148

Thermal and Thermomechanical Fatigue of Structural Alloys

Huseyin Sehitoglu, Department of Mechanical and Industrial Engineering, University of Illinois

STRUCTURAL ALLOYS are commonly subjected to a variety of thermal and thermomechanical loads. If the stresses in a component develop under thermal cycling without external loading, the term *thermal fatigue* (TF) or *thermal stress fatigue* is used. This process can be caused by steep temperature gradients in a component or across a section and can occur in a perfectly homogeneous isotropic material. For example, when the surface is heated it is constrained by the cooler material beneath the surface, and thus the surface undergoes compressive stresses. Upon cooling, the deformation is in the reverse direction, and tensile stresses could develop. Under heat/cool cycles, the surface will undergo TF damage. Examples of TF are encountered in railroad wheels subjected to brake-shoe action, which generates temperature gradients and, consequently, internal stresses (Ref 1, 2).

On the other hand, TF can develop even under conditions of uniform specimen temperature, instead caused by internal constraints such as different grain orientations at the microlevel or anisotropy of the thermal expansion coefficient of certain crystals (noncubic). Internal strains and stresses can be of sufficiently high magnitude to cause growth, distortion, and surface irregularities in the material (Ref 3). Consequently, thermal cycling results in damage and deterioration of the microstructure. This behavior has been observed in pure metals such as uranium, tin, and cadmium-base alloys and in duplex steels with ferritic/martensitic microstructures.

The term *thermomechanical fatigue* (TMF) describes fatigue under simultaneous changes in temperature and mechanical strain (Ref 4, 5). Mechanical strain is defined by subtracting the thermal strain from net strain, which should be uniform in a specimen. The mechanical strain arises from external constraints or externally applied loading. For example, if a specimen is held between two rigid walls and subjected to thermal cycling (and is not permitted to expand), it undergoes "external" compressive mechanical strain. Examples of TMF can be found in pressure vessels and piping; in the electric power industry, where structures experience pressure loadings

and thermal transients with temperature gradients in the thickness direction; and in the aeronautical industry, where turbine blades and turbine disks undergo temperature gradients superimposed on stresses due to rotation. In the railroad application discussed earlier, when external loading due to rail/wheel contact is considered, then the material undergoes the more general case of TMF. The temperature rise on the surfaces of cylinders and pistons in automotive engines combined with applied cylinder pressures also represents TMF. Based on the mechanical strain range, the results of TF and TMF tests should correlate well.

A distinction must be drawn between isothermal high-temperature fatigue as cyclic straining under constant nominal temperature conditions versus TMF. As such, isothermal fatigue (IF) can be considered a special case of TMF. In most cases, the deformation and fatigue damage under TMF cannot be predicted based on IF information. Therefore, TMF experiments have been considered in studies of both stress-strain representation and damage evolution.

Sometimes the term *low-cycle thermal fatigue* or *low-cycle thermomechanical fatigue* is used. Low-cycle fatigue (LCF) can be identified two ways: (1) high-strain cycling where the inelastic strain range in the cycle exceeds the elastic strain range and (2) where the inelastic strains are of sufficient magnitude that they are spread uniformly over the microstructure. Fatigue damage at high temperatures develops as a result of this inelastic deformation where the strains are nonrecoverable. In low-cycle cases, the material suffers from damage in a finite (short) number of cycles. Thermomechanical fatigue is often a low-cycle fatigue issue. For example, in railroad wheels only severe braking applications—occurring infrequently over thousands of miles—contribute to damage, fewer than 10,000 cycles take place during a wheel's lifetime. Similarly, the largest thermal gradients and transients in jet engines develop during startup and shutdown. The total number of takeoffs and landings for an aircraft is fewer than 30,000 cycles over the lifetime of an aircraft. In the laboratory, investigations often are conducted under low-cycle conditions to complete the experiments in a reasonable period of time.

The inability to predict TMF damage from the IF database continues to challenge engineers and researchers. Thermomechanical fatigue encompasses several mechanisms in addition to "pure" fatigue damage, including high-temperature creep and oxidation, which directly contribute to damage. These mechanisms differ, depending on the strain-temperature history. They are different from those predicted by creep tests (with no reversals) and by stress-free (or constant-stress) oxidation tests. Microstructural degradation can occur under TMF in the form of (1) overaging, such as coarsening of precipitates or lamellae; (2) strain aging in the case of solute-hardened systems; (3) precipitation of second-phase particles; and (4) phase transformation within the temperature limits of the cycle. Also, variations in mechanical properties or thermal expansion coefficients between the matrix and strengthening particles (present in many alloys) result in local stresses and cracking. These mechanisms influence the deformation characteristics of the material, which inevitably couple with damage processes.

Because of the importance of TMF in real-world applications, considerable attention has been devoted to the problem via workshops and symposiums. Ever since the early 1950s and 1960s, TMF experiments have been reported by research groups in the United States, Europe, and Japan. A number of books, review articles, and symposia proceedings on the subject have been published (Ref 6-20). The advent of computer control and servohydraulic testing equipment has allowed simultaneous, accurate control of temperature and strain. Consequently, research in the field has flourished.

The database of TMF research, however is small compared to the IF database. Experiments involving TMF remain difficult and expensive. The use of IF data to predict the performance of a material under TMF has been demonstrated to have drawbacks. The use of isothermal LCF and mechanical strain-range results at the maximum temperature end (or at a temperature with low IF resistance) may still be nonconservative. Attempts have been made to relate TMF crack growth to IF crack growth using linear elastic fracture mechanics (LEFM) concepts, but further refinements incorporating elastic-plastic fracture mechanics (EPFM) are needed. Many of the existing models do not account for the interaction of the mechanical strain with temperature. This interaction is rather complex and not well understood.

A distinction must be drawn between TMF and thermal shock (Ref 6, 21). Thermal shock involves a very rapid and sudden application of temperature (due to surface heating or internal heat generation), and the resulting stresses are often different from those produced under slow heating and cooling (i.e., quasi-static) conditions. Physical properties, such as specific heat and conductivity (which do not appear in low-strain-rate cases), appear explicitly in the thermal shock case. The rate of strain influences the material response and should be considered in damage due to thermal shock or in selection of materials for better thermal shock resistance.

Finally, if the body is subjected to thermal cycling conditions with superimposed net section loads, the component will undergo thermal ratcheting, which is the gradual accumulation of inelastic strains with cycles (Ref 22, 23). Failure due to thermal ratcheting involves both fatigue and ductile rupture mechanisms. The "two-bar structure" will be used to illustrate thermal ratcheting in a later section. Thermal ratcheting sometimes occurs unintentionally in thermomechanical tests when a region of the specimen is hotter than the surrounding regions, resulting in a bulge in the hot region.

This article provides an overview of the experimental methods in TF and TMF and presents experimental results on structural materials that have been considered in TF and TMF research. Life prediction models and constitutive equations suited for TF and TMF are also covered.

Mechanical Strain and Thermal Strain

Free (unrestrained) thermal expansion and contraction produce no stresses. When the thermal expansion of a body is restrained upon uniform heating, thermal stresses develop. Consider the case where a bar is held between two rigid walls and subjected to thermal cycling. The length of the bar cannot change during heating and cooling. Let T_0 be the reference temperature at which the bar was placed under *total constraint*. The compatibility equation for this bar is given as:

$$\varepsilon_{net} = \varepsilon_{th} + \varepsilon_{mech} = \alpha (T - T_0) + \varepsilon_{mech}$$

In this case the net strain is zero, and all of the thermal strain is converted to mechanical strain. The thermal strain is defined as the product of coefficient of thermal expansion and the temperature range $T - T_0$, where T is the current temperature. Then,

$$\varepsilon_{mech} = -\alpha (T - T_0)$$

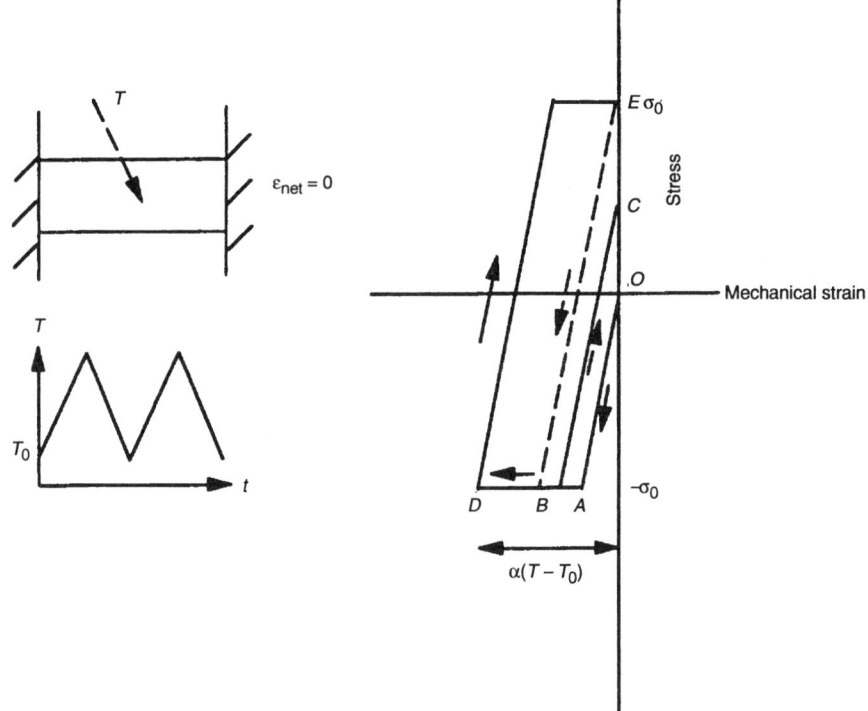

Fig. 1 Idealized stress-strain behavior under total constraint

Sometimes the total constraint case is identified as $\dot{\varepsilon}_{th}/\dot{\varepsilon}_{mech} = -1$. When this ratio is larger than -1, some free expansion and contraction occur, and the term *partial constraint* is used. If the $\dot{\varepsilon}_{th}/\dot{\varepsilon}_{mech}$ ratio is lower than -1, the condition is known as *overconstraint* (Ref 24). Therefore, the constraint influences the mechanical strain for a given thermal strain. Mechanical strain comprises elastic strain and inelastic strain (once yield stress is reached) and is the key parameter in TMF studies. The stress/mechanical-strain behavior shown in Fig. 1 is highly idealized; the material exhibits no hardening after yielding, the tension and compression strength are the same, and elastic modulus is independent of temperature. Upon heating, the bar is elastic and follows the stress-strain curve along OA. At A, the bar yields in compression, and upon further increase in temperature the mechanical strain on the bar increases along AB. The bar accumulates inelastic strain along AB. If the bar is cooled from B, it will deform in the reverse (i.e., tensile) direction. When the initial temperature is reached, the bar will return to zero mechanical strain, but a residual tensile stress will exist in the bar at point C. If the bar is again heated to the maximum temperature, the material will cycle between the stress point B and C. The bar is operating within the "shakedown" regime. It is unlikely that the bar will fail under these conditions because there is no plastic flow after the first reversal.

Next, consider the case when the thermal strain in the first heating portion of the cycle exceeded twice the elastic strain and a mechanical strain corresponding to point D is reached. Upon cooling back to the initial temperature, T_0, the bar will yield in tension and inelastic flow will occur until

point E is reached. Upon reheating, the bar will deform in the reverse direction (dashed line) until it reaches point D in compression. A hysteresis loop develops as a result of this thermal cycle. Under alternate heat/cool cycles, forward and reverse yielding will occur every cycle, resulting in failure in a finite number of cycles.

The constrained bar model is conceptually easy to visualize, but in real structures the condition can be different from total constraint. This will be analyzed later in this article.

Experimental Techniques in TF and TMF

Table 1 summarizes different heating methods for TF and TMF. The advantages and disadvantages of each technique are listed, as are the materials examined. In early work, experimenters subjected specimens alternately to high and low temperatures with no external loading. One way to accomplish this procedure was by immersing the specimens in cold and hot fluidized beds, which can be operated up to 1150 °C (2100 °F) (Ref 49). Over the years, various wedge-shape specimen geometries have been used.

The crack growth can be observed and the data presented as a function of maximum temperature. If the results are to be compared to IF or TMF tests, the stresses and strains should be calculated (Ref 51, 55) with the finite-element method (FEM) or other numerical methods. The geometry and strain-temperature variation for the wedge specimen are shown in Fig. 2. Note that as the temperature increases, the strain-temperature variation is out of phase (OP). The minimum

Table 1 Summary of TMF and TF test methods

Type of test	Heating method	Advantages	Disadvantages	Materials studied	Reference
TF	Immersion in hot and cold oil bath	Simplicity of the experiment	Transient stress strain could be present and should be calculated	Noncubic crystals, including tin, zinc, cadmium	3
TMF	Direct resistance	Rapid heating; allows space to mount the extensometer and pyrometer for crack growth measurements	Electric isolation of grips; local heating of crack tips	Conductive materials, stainless steel	25-31
TMF and TF	Induction (10-450 kHz, 5-40 kW capacity)	Rapid heating; complex specimen geometries permitted; inert environment testing using bellows	Experience with coil design required; electric noise in the strain signal due to high-frequency magnetic fields; high cost of unit	Aluminum, copper, steels, nickel-base superalloys	24, 32-45
TMF and TF	Quartz lamp (radiation)	Inexpensive; uniform temperature over different zones of the specimen	Shadow effects; slow cooling rates; enforced cooling needed	Nickel- and cobalt-base superalloys, metallic composites	46-48
TF	Fluidized bed	Good for screening TF resistance of materials	Stress-strain temperature transients must be calculated and surface oxidation removed	Nickel-base superalloys	49-52
TF	Burner heating; flame heating	Good for screening TF resistance of materials; surface hot corrosion damage representative of service	Stress-strain temperature transients must be calculated	Nickel-base superalloys, steels	53
TF	Thermal fatigue under bending	Under reversed bending one surface undergoes OP, the other undergoes IP	Stress-strain gradients must be calculated	Nimonic alloys	54
TF	Dynamometer (friction heating)	Very high temperatures on surface reached; representative of service	Oxides are wedged into cracks; friction characteristics change with time	0.5 to 0.7% C steels	1, 2

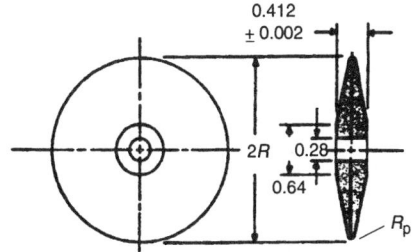

Part	$R_p \pm 0.001$	$2R \pm 0.01$
1	0.010	2.38
2	0.020	2.40
3	0.030	2.42
4	0.040	2.44

(a)

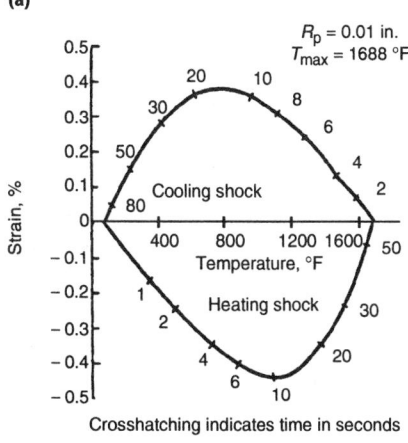

(b)

Fig. 2 (a) Wedge geometry for TF studies. Dimensions given in inches. (b) Strain-temperature variation in the fluidized-bed experiments. Source: Ref 13

Fig. 3 Disk specimen, showing radial cracks larger than 6.35 mm, used by Simovich (Ref 34) in thermal fatigue studies on steels. The dark spot at the right is used for temperature sensing.

strain is reached within 10 s. At times beyond 10 s, the strain-temperature variation is in-phase (IP). Upon cooling, the reverse behavior was observed. Relative thermal fatigue resistance of many alloys can be classified with fluidized-bed experiments. This technique has proved to be of considerable value in examining the role of directional solidification, grain size, and γ′ size and morphology in superalloys (Ref 51, 52, 56). The ε–T variation in Fig. 2(b) resembles the TMF diamond counterclockwise (DCCW) history that will be discussed later.

Instead of the fluidized-bed technique, burner heating and quartz lamp heating can be applied to the specimens. More recently, Remy and colleagues (Ref 47) used the quartz heating method to study the thermal fatigue behavior of nickel- and cobalt-base superalloys. Their specimen geometry was slightly different from the wedge used in early studies, but the principles of the method were the same. The use of quartz lamps is considerably more economical than other thermal fatigue heating methods. Simovich (Ref 34) developed a different specimen design: 5 cm diam disk (Fig. 3). This specimen was heated

axisymmetrically using an induction heater with no external load, and a temperature gradient was developed in the radial direction. Cooling water was pumped through the large hole; the dark spot at the right marks a typical location for thermocouples. The maximum temperature considered was 650 °C (1200 °F) and the cycle time was approximately 60 s (controlled by induction heating). Using an axisymmetric model, Simovich calculated the circumferential stresses upon cooling and compared these results to experimental measurements. Under these conditions, cracks near 7 mm (0.3 in.) appeared in less than 2000 cycles in 0.7% C class steels.

All the experiments discussed thus far involved no external loading. Thermomechanical fatigue experiments with externally imposed strain were pioneered by Coffin (Ref 4, 25), who plotted the results versus plastic strain range. Both hollow and solid specimen designs were used. The hollow design allows more rapid heating and cooling. On the other hand, the solid specimen design lowers the possibility of buckling. Most IF experiments have been conducted on solid specimens; to obtain meaningful comparisons, such specimens should also be used in TMF studies.

Currently used techniques include resistance heating (Ref 4, 25-29), quartz lamp heating (Ref 46-48), and induction heating (Ref 24, 32-38). Induction heating is preferred, and the actual temperature gradient in the specimen should be known. Temperature measurements have been accomplished with spot-welded thermocouples, strapped-on thermocouples, or pyrometers. The temperature must be continuously monitored throughout the test. Infrared pyrometers are preferred in order to avoid potential failure originating from thermocouple beads or oxides formed at the thermocouple/specimen intersection. If thermocouples are chosen, a backup thermocouple is advised in case one should break off. A different temperature profile at different specimen locations could result in specimen barreling or instability effects. Depending on the thermal mass

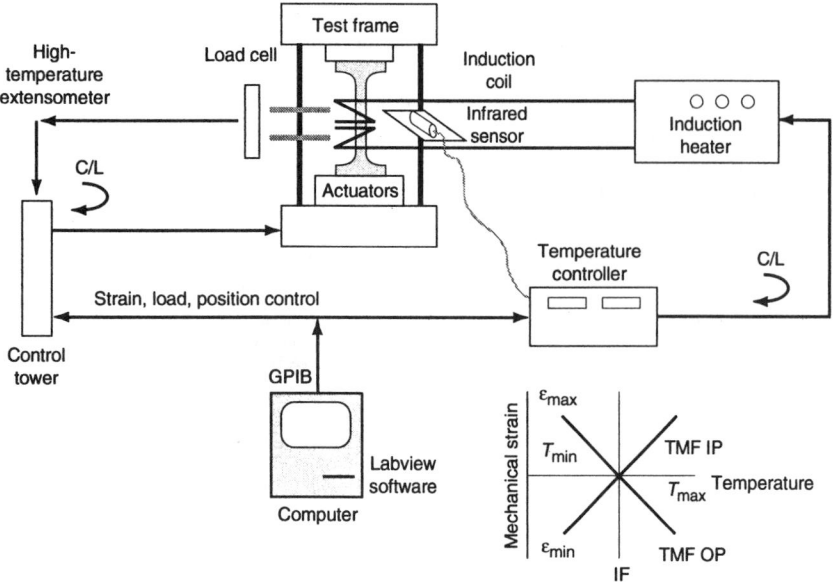

Fig. 4 Schematic of a modern TMF system

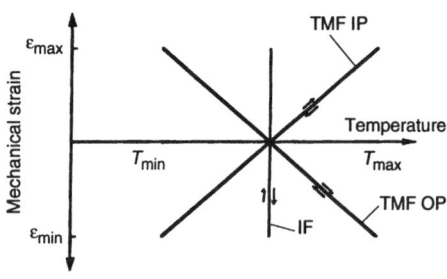

(a)

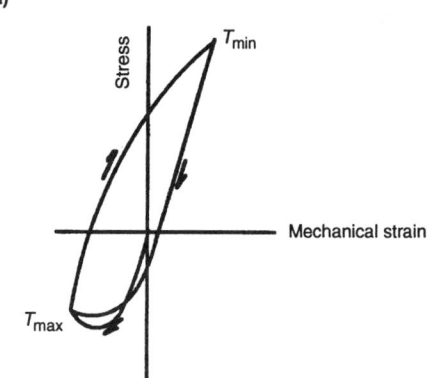

(b)

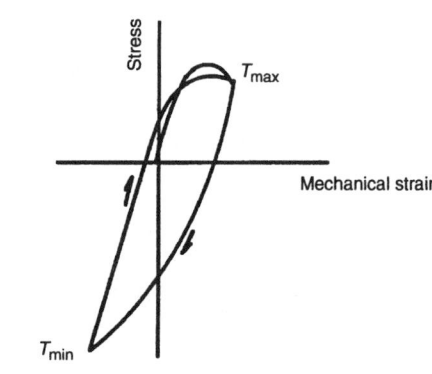

(c)

Fig. 5 (a) Mechanical strain/temperature variation in TMF OP, TMF IP, and IF. (b) TMF OP stress-strain response. (c) TMF IP stress-strain response

(i.e., grips) at the ends of the specimen and the "chimney" effect with induction or quartz heating, the coils or the lamp power in different zones of the specimen should be adjusted to avoid temperature gradients more than 5 °C (9 °F). Coffin (Ref 57) has observed progressive thickening of the sample cross section at one region and progressive thinning at another region. Manson (Ref 6) has shown that if a local region of the specimen undergoes higher temperatures relative to the major length of the specimen, localized plastic strains and creep will occur in this region due to reduced yield stress. In some cases, when localized deformation as described above occurred, experimenters accounted for it in their analysis; interpretation of the results, however, is rather complex. Optimizing the dynamic rather than the static temperature profiles circumvents this problem and should be completed before a serious TMF research program is undertaken.

Quartz or alumina rod extensometers are used to control and measure the net strain during TMF experiments. Net strain is defined as the deflection divided by the initial gage length. Special attention should be paid in mounting the extensometer in the presence of an induction coil. The ends of the rods can be conical or chisel edged. At high temperatures, the spring load on the rods should be reduced to avoid penetration and notching of the specimen. In early studies, diametral strain measurements were made on hourglass specimens and converted to axial strain (Ref 4, 25, 28). The conversion requires Poisson's ratio and modulus of elasticity as a function of temperature and could have caused some errors in strain determination.

Thermal strain compensation is achieved by cycling the temperature at zero load before the test and determining the thermal strain as a function of temperature and time. Thermal strain can be defined using the coefficient of thermal expansion (CTE). Mechanical strain that produces

stresses is defined by subtracting the thermal strain from the net strain. A good calibration and a good extensometry are required, because in TMF the mechanical strain range could be much lower than the thermal strain range.

Figure 4 shows a schematic of a modern TMF test system (currently used at the University of Illinois at Urbana-Champaign). The test machine is a digital-control servohydraulic test frame. There are two close loops (C/L) in the control system. The control tower receives axial strain, position, and load signals from the test frame and sends them to a Macintosh computer fitted with a general-purpose instrumentation bus (GPIB) board. The computer, using Labview software, generates strain and temperature histories, which are transmitted to the temperature controller and to the control tower. Data collection is performed with the Labview software, and the results are displayed on the monitor during the experiment. A noncontact infrared pyrometer device has been used for temperature measurements. Specimens were heated using a high-frequency induction heater with a 15 kW capacity. The test system can perform TMF IP and OP tests, IF tests, and other complex strain-temperature variations.

TMF IP versus TMF OP

Mechanical strain/temperature waveform is classified according to the phase relation between mechanical strain and temperature. In-phase TMF means that peak strain coincides with maximum temperature; out-of-phase TMF means that peak strain coincides with minimum temperature. These two cases are shown in Fig. 5(a), along with the IF case. Generic hysteresis loops corresponding to the TMF OP and TMF IP cases are shown in Fig. 5(b) and (c), respectively. For a TMF cycle, the hysteresis loops are "unbalanced" in tension versus compression. In the TMF OP case, considerably more inelastic strains develop in compression relative to tension. The opposite

behavior occurs in the TMF IP case. Some TMF experiments have been conducted under $R_\varepsilon = -1$ (i.e., completely reversed) conditions. Other TMF experiments have been conducted under $R_\varepsilon = -\infty$ (maximum mechanical strain is zero; see Fig. 1 [Ref 24] and $R_\varepsilon = 0$ conditions (minimum mechanical strain is zero [Ref 4]).

The inelastic strain is defined by subtracting the elastic strain from the mechanical strain:

$$\varepsilon_{in} = \varepsilon_{mech} - \frac{\sigma}{E\{T(t)\}} \qquad \text{(Eq 1)}$$

For computational purposes, pairs of stress and temperature data points are needed. The variation in elastic modulus, $E\{T(t)\}$, as a function of temperature should be determined from isothermal experiments. A stress/inelastic strain hysteresis loop can be constructed using Eq 1. If there are hold periods

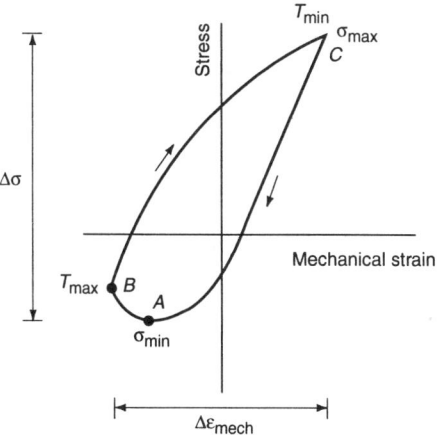

Fig. 6 Definitions of stress range and mechanical strain range in TMF

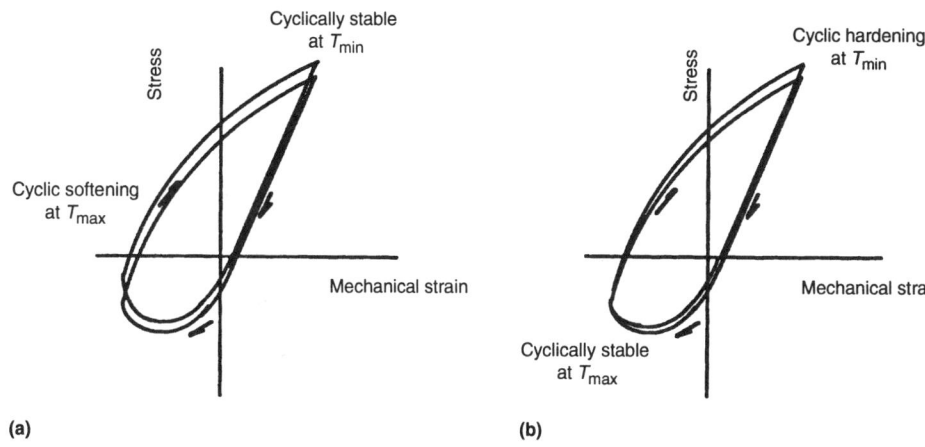

Fig. 7 Stress-strain response under cyclic softening (a) or cyclic hardening (b) conditions

during the TMF cycle, the equation will still be valid. The mechanical strain range, $\Delta\varepsilon_{mech}$, is shown in Fig. 6. The stress range in a TMF cycle is also shown for the OP case. The loop for the IP case is similar, but reversed. Note that at the minimum strain (point B) the stress is not necessarily a minimum. Inelastic deformation with softening due to decrease in strength with increasing temperature is observed during AB. At B the maximum temperature is reached. Upon cooling, the behavior is elastic, followed by plastic deformation at the low-temperature end.

For engineering purposes, the inelastic strain range of a thermomechanical cycle can be determined to a first approximation by subtracting the elastic strains computed at the maximum and minimum strain levels. This gives:

$$\Delta\varepsilon_{in} \approx \Delta\varepsilon_{mech} - \left(\frac{|\sigma_B|}{E_B} + \frac{|\sigma_C|}{E_C} \right) \qquad \text{(Eq 2)}$$

where E_C is the elastic modulus at the maximum strain and E_B is the elastic modulus corresponding to the minimum strain. Equation 2 slightly underestimates the inelastic strain range compared to the more exact equation. Note that the inelastic strain range includes the plastic strain, creep strain, and other strain components (e.g., transformation strain). Separation of plastic and creep strains in a TMF cycle is not straightforward. If needed, it can be done experimentally (Ref 58) by stress hold at selected points of the hysteresis loop or via constitutive models including plasticity and creep. Several constitutive models have been proposed for thermomechanical cyclic loadings and will be discussed later.

Just as in IF conditions, the TMF response of engineering materials involves cyclic hardening, cyclic softening, or cyclically stable behavior, depending on the microstructure, the maximum temperature level, and the phasing of strain and temperature. However, the behavior can be somewhat complex because of strain-temperature interaction. A material can harden, soften, or be cyclically stable at T_{max} of the cycle; likewise, at T_{min} the material can cyclically soften, harden, or

be stable. Two possibilities are shown in Fig. 7. In Fig. 7(a), the material softens at T_{max} and remains cyclically stable at T_{min}. The material can cyclically soften at high temperature due to thermal recovery, causing coarsening of the microstructure, and in this case the hysteresis loops appear to "climb" in the tensile direction. Therefore, the tensile mean stress increases with increasing number of cycles. The microstructural coarsening could subsequently affect the strength at T_{min}, with the maximum stress in the cycle dropping with increasing number of cycles. Thereafter, the climbing of the hysteresis loops stops and the range of stress in the cycle decreases. This behavior has been documented in Ref 59. In the second example (Fig. 7b), stable behavior is observed at T_{max}, but the strength at T_{min} increases because of dynamic or static strain-aging effects. In this case, the hysteresis loops also climb in the tensile direction and, at the same time, overall stress range increases. Examples of this are discussed in Ref 60.

Other Strain-Temperature Variations in TF and TMF

Diamond (or baseball) TMF strain variation is obtained by changing the mechanical strain and temperature 90° or 270° out of phase. The diamond path should be specified as clockwise (DCW) or counterclockwise (DCCW), which could influence TMF life. The strain-temperature variation and the generic hysteresis loops for the DCCW case are shown in Fig. 8. In many structural alloys studied, the DCW and DCCW were not as damaging as TMF IP or TMF OP, because at the maximum temperature neither the strains nor the stresses were at a maximum. It is important to study the diamond TMF histories; they are encountered in many practical situations, such as turbine blades. Examples of more complex strain-temperature histories observed in service will be discussed in a later section.

A variation of the diamond history was proposed in the early 1970s (Ref 28). The term *bithermal fatigue* was coined in the mid-'80s by NASA researchers (Ref 30). In this case, the tensile portion of the loop is applied at one tem-

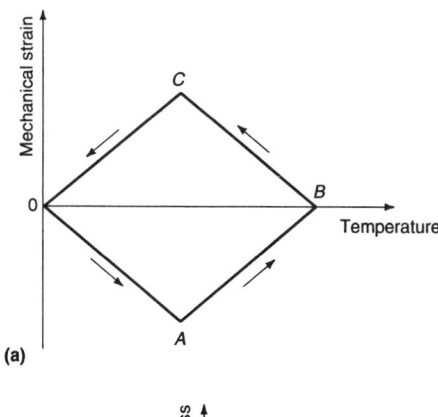

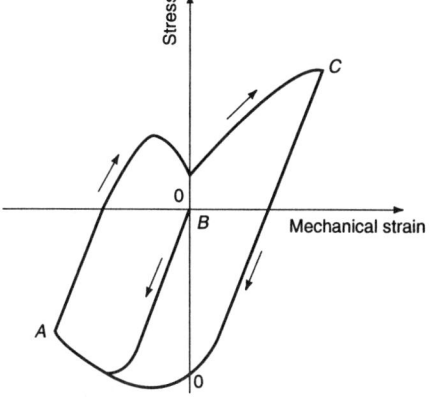

Fig. 8 Strain-temperature variation (a) and schematic of stress-strain response (b) for the DCCW case

perature, T_1, and the compressive portion of the loop is conducted at a different temperature, T_2. The temperature is changed, $T_1 \rightleftarrows T_2$, at zero stress. Advantages of this technique are that the tests can be conducted without the need for TMF computer software and the results more readily related to IF tests. If the thermal strains are large, however, the extensometer must have the range and the resolution to handle strain control at both temperature extremes. Also, some creep recovery due to internal stresses could occur during the zero stress temperature excursions.

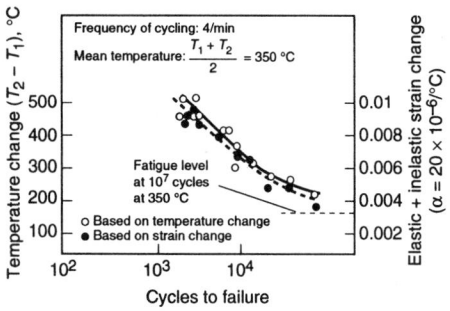

Fig. 9 Results of TMF experiments by Coffin (Ref 4) under total constraint for annealed type 347 stainless steel

TF and TMF of Structural Alloys

Carbon Steels, Low-Alloy Steels, and Stainless Steels

One of the early laboratory investigations of thermal fatigue in steels was conducted at the University of Illinois to further understand TMF in railroad wheels (Ref 1, 2). During the brake-shoe action on a railroad wheel, the rim is constrained by the surrounding cooler hub and the plate. Upon heating, circumferential compressive stresses develop; upon cooling, yielding in the tensile direction can occur. Under repeated brake applications, TF cracks can develop. This is simulated in the laboratory with a wheel dynamometer and a brake-shoe heating. Thermal cracks can grow to a size sufficient to exceed the fracture toughness of the material, resulting in catastrophic fracture.

Later experiments were conducted by Simovich (Ref 4), who subjected disk specimens to TF with induction heating. In this case, disks approximately 50 mm (2 in.) in diameter were heated and cooled with induction, generating a temperature gradient in the radial direction. The steel developed radial cracks that grew to sizes near 10 mm (0.4 in.). Simovich conducted a thermal analysis and stress analysis of the disk and correlated the fatigue results as a function of mechanical strain and stress range.

Well-controlled TMF experiments using direct resistance heating were conducted by Coffin at General Electric. The effect of prestrain on the fatigue life of type 347 stainless steel under thermal cycling has been established (Ref 4), as has the maximum temperature effect (up to 600 °C, or 1110 °F). The role of strain hold period in reducing TMF lives was established for periods of from 6 to 180 s. The influence of the hold period was explained as an increase in the inelastic strain range of the cycle. Coffin also examined the effect of thermal cycling on the subsequent stress-strain response and noted strain hardening of the material. Papers by Coffin (Ref 4) and Manson (Ref 5) were the first to propose a relationship between plastic strain range and life was proposed; this was later coined as the low-cycle fatigue, or Coffin-Manson, equation.

Coffin's results on type 347 stainless steel are shown in Fig. 9. The specimens were subjected to total constraint ($\varepsilon_{net} = 0$). The mean temperature was maintained constant at 350 °C (660 °F), and the maximum temperature considered was as high as 650 °C (1200 °F). The horizontal axis (log scale) is the fatigue life, defined as fracture of the specimen. The vertical axis (linear coordinates) is given in terms of temperature range and mechanical strain. In the experiments of Coffin, these two quantities are not exactly equal because of deformation of supports and temperature distribution along the length of the tube. Therefore, the mechanical strain range is slightly lower than the thermal strain range in these experiments. The specimen is hot in tension and cold in compression and is undergoing TMF OP loading. Whether the specimen is clamped at the minimum temperature or the maximum temperature influences only the mean strain and has very little, if any, influence on fatigue life.

Coffin used resistance heating and developed a cam-and-lever mechanism to apply independent strains on the sample. This design has been duplicated in a number of subsequent TMF investigations in Japan, the United States, and the former Soviet Union. For example, research on constrained specimens was conducted on railroad wheel material, and hysteresis loops were established for carbon steels (0.4 to 0.7% C) at temperatures reaching 500 °C (930 °F) (Ref 39).

In Great Britain, the thermal fatigue resistance of carbon steels, alloy steels, and cast irons was investigated by Baron and Bloomfield (Ref 38) in the early 1960s. They used induction heating of an edge of a cold specimen. The strains were not measured or calculated for this case, but results have been displayed using T_{max} versus cycles to form a crack of finite size. The maximum temperature considered was 900 °C (1650 °F), where most steels transform to austenite, and martensite formed upon rapid cooling, resulting in rapid formation of cracks. At high temperatures austenitic stainless steels were found to be superior to other steels, while some of the nodular irons approached the TMF resistance of plain carbon steels.

Thermomechanical fatigue research on steels has attracted considerable attention in Japan. Kawamoto et al. (Ref 29) conducted TMF experiments on 0.7% C steels and 18-8 stainless steels. They confirmed that the hold periods reduced fatigue life and suggested that hold periods allow formation of metal carbides and oxides at grain boundaries. They found no considerable difference between IP and OP cycling when the results were compared based on mechanical strain range. They made the noteworthy observation that under TMF the lives were shorter than under IF, even when the IF test was conducted at the maximum temperature of the thermal cycle.

Taira and colleagues (Ref 35-37) have authored a number of key TMF papers covering a range of steels, including 1016 steel, chromium-molybdenum steels, and type 304 stainless steel. They used the mechanical strain range and plastic strain range to compare their data obtained under thermal cycling of 1016 steel in the temperature range of 100 to 600 °C (210 to 1110 °F).

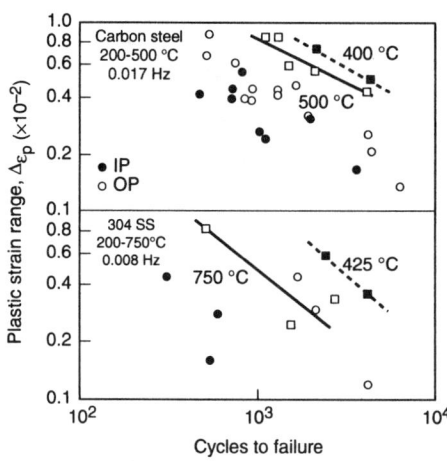

Fig. 10 Results of Fujino and Taira (Ref 53) on carbon steel and type 304 stainless steel, showing more damage in TMF IP relative to TMF OP

It is expected that considerable creep and oxidation effects are present in these steels at temperatures exceeding 500 °C (930 °F). Taira et al. (Ref 61, 62) also conducted thermal ratcheting tests under TMF OP conditions with stress control. Thermomechanical fatigue damage was predicted from creep-rupture data for these experiments.

Fujino and Taira (Ref 63) demonstrated that for type 304 stainless steel (at 200 to 750 °C, or 390 to 1380 °F) the TMF IP lives were shorter than the TMF OP case by nearly a factor of four. Their results are shown in Fig. 10 for carbon steel and for type 304 stainless steel. Isothermal fatigue data at 425 and 750 °C (800 and 1380 °F) for type 304 and at 400 and 500 °C (750 and 930 °F) are given. The TMF OP lives were lower than IF lives in carbon steel, whereas for the stainless steel they were similar to the IF lives at maximum temperature of the cycle. The researchers made measurements of grain-boundary sliding and found evidence of it in TMF IP cases, but not in TMF OP or IF loadings. As the maximum temperature was lowered from 750 to 600 °C (1380 to 1110 °F) in TMF experiments (Ref 36), the TMF OP, TMF IP, and IF results at T_{max} of the cycle converged. It is clear that grain-boundary sliding due to unbalanced displacements at the microlevel becomes more pronounced as the maximum temperature in the cycle is increased.

Other studies in Japan were conducted by Udoguchi and Wada (Ref 31), who considered H46 martensitic stainless steel and type 347 stainless steel under TMF OP conditions with a maximum temperature of 700 °C (1290 °F) and 1040 steel with a maximum temperature of 400 °C (750 °F). Their results also confirmed that the TMF resistance is inferior to IF even when the results are compared at T_{max}. One of the most systematic investigations of TMF of steels was undertaken by Kuwabara and Nitta in the mid to late 1970s (Ref 40, 41). They conducted TMF OP and TMF IP experiments on type 304 stainless steel under continuous cycling and also in the presence of tensile or compressive hold periods (300 to 600 °C, or 570 to 1110 °F). The TMF IP

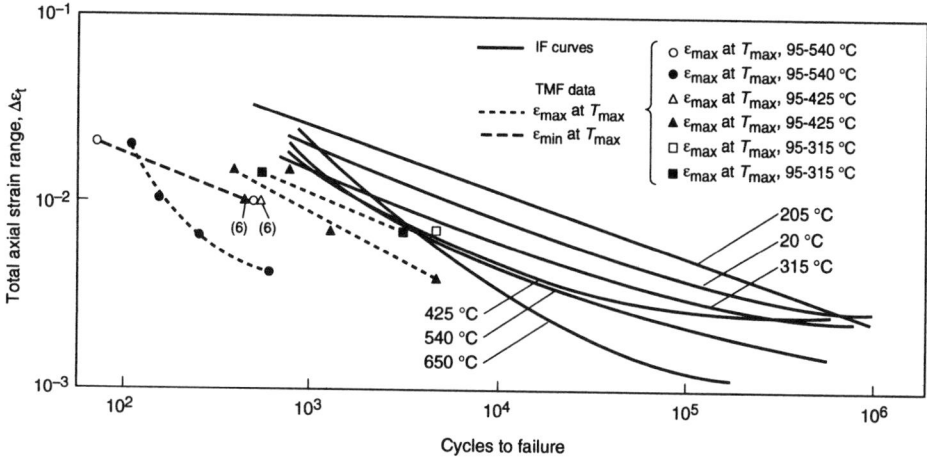

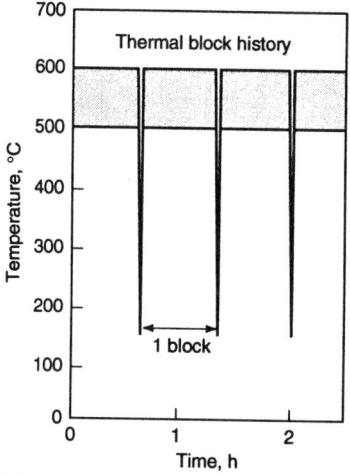

Fig. 11 Results of Jaske (Ref 66) on TMF of 1010 carbon steel. Note: (6) indicates a 6 min hold time at maximum temperature.

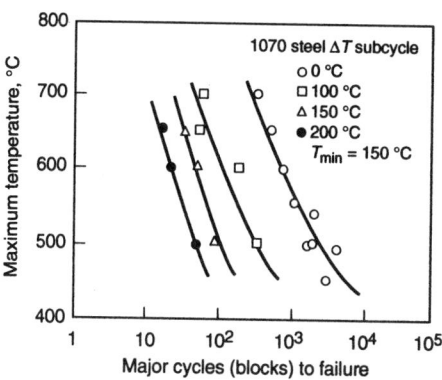

Fig. 12 (a) Temperature-time history under block loading. (b) Thermomechanical fatigue lives under TMF OP block loading. Source: Ref 59

lives were shorter than TMF OP lives, but comparison with IF lives at 600 °C (1110 °F) showed that IF tests were more damaging. In another set of experiments on type 304, they showed that TMF IP and TMF OP were comparable when the maximum temperature was only 550 °C (1020 °F) (Ref 42); still, more intergranular cracking was observed for the IP case relative to the IF and OP cases. A hold period on these steels drastically reduced the TMF IP lives, but had little effect on the TMF OP behavior. The trends were somewhat reversed when chromium-molybdenum-vanadium steels were investigated; 1Cr-Mo-0.25V (Ref 43) (examined between 300 and 550 °C, or 570 and 1020 °F) and nickel-molybdenum-vanadium forged steel (Ref 44) exhibited shorter lives in TMF OP compared to TMF IP. This behavior is consistent with the propensity of these alloys to suffer from considerably higher oxidation damage relative to stainless steels. For both alloys, higher surface crack density was measured in the TMF OP case. At high strains TMF OP and TMF IP results converged. In early studies, Manson et al. (Ref 64) demonstrated the significant surface cracking due to oxidation effects in low-alloy steels; the TMF results were consistent with the shorter lives observed in PC (plasticity in tension reversed by creep in compression) type cycling relative to CP (creep in tension reversed by plasticity in compression) type cycling for this class of alloys.

Research at NASA (Ref 58) considered type 316 stainless steel subjected to temperature cycling in the range of 230 to 815 °C (445 to 1500 °F). Considerable creep strains were measured, and thermal recovery was present in these TMF experiments. Later, Halford and colleagues (Ref 30) conducted bithermal IP and bithermal OP tests on type 316, which showed good agreement with the earlier TMF OP and TMF IP tests. Their results showed that TMF IP was more damaging than TMF OP, a finding confirmed by Miller and Priest (Ref 19) on the same class of stainless steel. Sheffler (Ref 28) conducted bithermal TMF OP and TMF IP tests on type 304 stainless steel and demonstrated that both OP and IP lives were shorter than the IF data at maximum temperature

of the cycle. Similarly, for the A-286 alloy the bithermal IP lives were shortest compared to OP and IF results. Sheffler made the TMF and IF comparisons for tests conducted at the same frequency and in ultrahigh vacuum. An interesting observation was that cavities formed due to unreversed grain-boundary displacements cannot be fully accommodated by intergranular sliding.

In Canada, Westwood (Ref 27, 65) conducted TMF tests on type 304 in the temperature range of 350 to 700 °C (660 to 1290 °F). The results showed good agreement between TMF IP and IF tests conducted at T_{max} of the thermal cycle, while the lives under TMF OP were longer. Although a larger difference between TMF IP and IF is expected at these temperatures, the lifetimes are generally consistent with previous data reported for similar materials.

Hysteresis response and life was studied by Jaske (Ref 66) on 1010 steel subjected to thermal cycling in the range of 95 to 540 °C (200 to 1000 °F). The cyclic hardening phenomenon was noted when the maximum temperature was below 425 °C (800 °F), possibly due to strain-aging effects. The TMF lives were significantly shorter than IF lives, even when the IF results from the maximum temperature were considered. The results of this study are shown in Fig. 11. The mechanical strain range versus life is plotted for TMF OP, TMF IP, and IF cases. There appears to be crossover in lives between TMF OP and TMF IP slightly below a strain range of 0.02. All the TMF data shown fall below the IF curves. This is consistent with the findings of Fujino and Taira (Ref 63). Similarly, Laub (Ref 67) studied 1010 steel used in heat exchangers and subjected specimens to total constraint TMF OP cycling where the mean temperature of the cycle was maintained constant. The most severe case examined was a mean temperature of 315 °C (600 °F) and a maximum temperature of 760 °C (1400 °F). Considerable oxidation of crack tips has been noted at temperatures exceeding 480 °C (900 °F).

Finally, in the work of Sehitoglu on 1070 steel, the stress-strain response was determined under total/partial and overconstraint TMF OP conditions with a minimum temperature of 150 °C (300

°F) and a maximum temperature of 700 °C (1290 °F). In later studies, Sehitoglu and his students investigated variable amplitude effects in TMF (Ref 59), environment effects in TMF (Ref 68, 69), phasing effects (TMF OP versus TMF IP) (Ref 68), strain-temperature changes conducive to strain aging (Ref 60), and notch effects and crack growth behavior under TMF (Ref 24, 60). A temperature-time history for a two-step TMF OP loading (total constraint) is shown in Fig. 12(a). One block includes one major cycle plus 100 minor (sub) cycles (Ref 33). In this case the major cycle underwent 150 ⇄ 600 °C (300 ⇄ 1110 °F) cycling under total constraint and the minor cycle experienced 500 ⇄ 600 °C (930 ⇄ 1110 °F). Because of considerable coarsening of the microstructure due to high-temperature exposure, the strength of the material at 150 °C (300 °F) is considerably lowered and the inelastic strain range of the cycle increases. The fatigue lives for these types of histories, where $\Delta T_{sub} = 0$, 100, 150, and 200 °C (30, 210, 300, and 390 °F), are shown in Fig. 12(b). This diagram indicates the dramatic deterioration of fatigue life in the presence of subcycles. On the same class of steels, Neu and Sehitoglu (Ref 68, 69) observed a typical crossover of the fatigue lives: At high strains IP tests were more damaging than OP, whereas the trend reversed at small strains. In tests conducted in a helium environment, the TMF IP experiments were more damaging than the TMF OP. The results of TMF OP and TMF IP experiments in air and in helium environment are shown in Fig. 13. The use of maximum-temperature IF data obtained for strain rates comparable to the TMF test predicted the trends, but a more sophisticated TMF life model has been proposed. Some of these studies will be discussed.

Mughrabi and his group in Germany (Ref 70) recently conducted both TMF IP and TMF OP tests on type 304L stainless steels. They observed a higher stress amplitude in TMF relative to IF when the TMF cycle coincides with temperatures near 450 °C (840 °F), where maximum dynamic strain aging occurs. Similar to the work of Sehitoglu et al. (Ref 33), they found that as the maximum temperature is increased, the maximum stress in the cycle occurs before the maximum temperature and maximum strain are reached. Thermomechanical fatigue IP tests revealed shorter lives when creep damage became more pronounced, whereas cavitation damage was not observed in the TMF OP case.

It is difficult to compare the results of one investigator to another, especially in TMF loading cases. This is because the TMF strain rates or frequency is dictated by the heating and cooling system, which is unique to the investigator. Even if the same heating method is used, there are no standards for TMF specimen geometry or for test control software, and the tests are often slowed down to ensure proper agreement between temperature and strain. Improved hardware and software would lead to greater reliability and consistency among different laboratories.

Environmental Effects. Coffin (Ref 71, 72) was the first to emphasize the significance of oxide damage in steels. At temperatures exceeding 500 °C (930 °F), an oxide layer forms on the surface of iron-base alloys. The iron oxides that form are brittle and facilitate crack advance into the substrate. This layer experiences a mechanical strain, which can result from one or a combination of the following: (1) strain from the applied mechanical loading in the material, (2) mismatch in the thermal expansion coefficients among the different stoichiometries of the oxides and substrate (Ref 33 and 73), (3) load due to the volume difference between the substrate and the various oxides (e.g., Fe_2O_3, Fe_3O_4, and FeO) (Ref 74, 75), and (4) other factors discussed in Ref 68. These mechanisms could affect the morphology of the surface oxide as well as the growing oxide-induced crack. Tensile oxide fracture facilitates crack initiation and crack growth, because the repeated oxide fracture can channel crack growth into the substrate. In TMF OP, the oxide forms near maximum temperature and upon cooling undergoes tension and fractures locally. Skelton (Ref 76) has shown that on 0.5Cr-Mo-V steels the crack growth rate in air is nearly an order of magnitude faster than in vacuum, with crack growth rates in steam environment falling between these two extremes.

One way to separate environmental damage from fatigue and creep damage is by performing tests in an inert or nonoxidizing atmosphere. Although a number of studies have been conducted on LCF under nonoxidizing environments (Ref 77-79), only two studies have been made on TMF of steels under an inert environment (Ref 28, 68). Sehitoglu and Neu devised a unique method of testing the specimen surrounded by bellows in which helium is trapped. The experiments were conducted under both TMF IP and TMF OP conditions. The increase in life relative to air results in the TMF OP case was nearly a factor of five, whereas in the TMF IP case the lives were not significantly influenced. The results are shown in Fig. 13.

Under conditions where creep mechanisms are dominant compared to environmental interaction effects, the fatigue life in air is about the same as in an inert atmosphere (Ref 78). However, when an environmental contribution exists, the fatigue life of smooth specimens is increased by a factor of 2 to 20 in a nonoxidizing atmosphere compared to tests performed in air (Ref 68, 76, 77, 79).

Strain Rate and Temperature Effects. Sehitoglu and his students conducted numerous investigations of 0.7% C steels (used in railroad wheels) and established the stress-strain behavior over the temperature range of 150 to 700 °C (300 to 1290 °F) (Ref 80-82). The effects of maximum temperature, strain aging, and thermal recovery due to spheroidization effects on stress-strain response have been identified. The effects of alloying were also examined. Early experiments have been reported on TMF behavior of carbon steels; unfortunately, the hysteresis loops have not been provided in these cases.

The influence of strain rate and temperature on life has been examined by Majumdar (Ref 83), who conducted experiments with a minimum temperature of 425 °C (800 °F) and a maximum temperature of 595 °C (1100 °F). They showed that at strain rates equal to or higher than 10^{-4}/s, the fractures are predominantly transgranular. For strain rates lower than 10^{-4}/s, however, the fractures become intergranular and the lives are shorter than the isothermal lives at maximum temperature. Majumdar also investigated the effect of hold time and demonstrated that hold periods reduce the cycles to failure.

The strain rates used in TMF studies vary in a narrow range. In his original study, Coffin (Ref 4) considered the influence of hold time and demonstrated that the cycles to failure decreased by a factor of three when the hold time increased from 6 to 180 s. Several high-strain-rate experiments were conducted by Taira and colleagues (Ref 61, 62) and Udoguchi and Wada (Ref 31) in their work on steels. Strain rates on the order of 10^{-4}/s were considered, which correspond to cycle times of 60 s. On the other extreme cycle, times near 30 min were considered (Ref 58). In most TMF research, the cycle time is on the order of 2 to 4 min, which corresponds to 5×10^{-5}/s. Direct resistance and induction heating methods can readily be used to produce strain rates on the

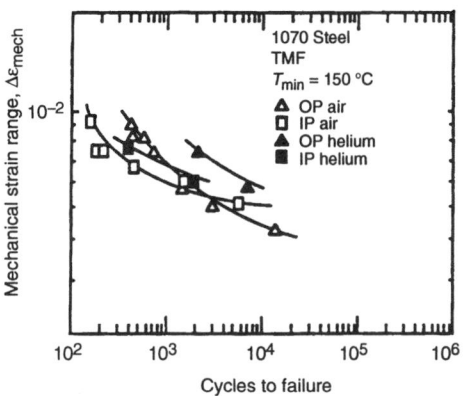

Fig. 13 Comparison of TMF OP and TMF IP experiments in air and in helium. Source: Ref 68, 69

order of 5×10^{-5}/s. Higher strain rates and the accompanying rapid temperature changes could produce temperature gradients in the specimen and make interpretation of the results difficult. Kuwabara and Nitta (Ref 84) examined the relationship between cycle time and TMF life in the range of 2 to 20 min. As the cycle time was increased, the fraction of intergranular cracks increased in the TMF IP case; however, the TMF OP results were not sensitive to strain rate. Commensurate with this finding is that TMF IP lives decreased with increasing cycle time while TMF OP lives remained constant.

Carbon steels undergo metallurgical changes in the form of coarsening of the pearlite lamellae and, ultimately, spheroidization at temperatures exceeding 400 °C (750 °F). Strain plays an important role in the spheroidization of pearlite. Deformation sets up subboundaries within the cementite, which are then rounded by diffusion driven by chemical potential gradients at the interface. This rounding of the interface edges leads to a complete band of ferrite separating the cementite. Many of these divisions occurring throughout the cementite break the lamellae up into segments, which then spheroidize. Strain-accelerated spheroidization can greatly reduce the time necessary to spheroidize a specimen at a given temperature.

In Fig. 14 the maximum stress in the TMF OP cycle is plotted versus the mechanical strain range in the cycle. Because the T_{min} was maintained constant in these experiments, higher strain ranges were achieved with higher maximum temperature. As the temperatures exceed the 500 °C (930 °F) value (or when the mechanical strain amplitude exceeds 0.003), the maximum stress decreases gradually with increasing mechanical strain. There are two main implications of this result: (1) the maximum stress cannot be used as a predictor of fatigue damage because for the same maximum stress there are two corresponding strain levels, one in the low-temperature and the other in the high-temperature regime (see discussion in Ref 33), and (2) the softening of the material at 150 °C (300 °F) means that the resistance of the material to deformation has decreased and the inelastic strain in the cycle has increased, producing enhanced damage.

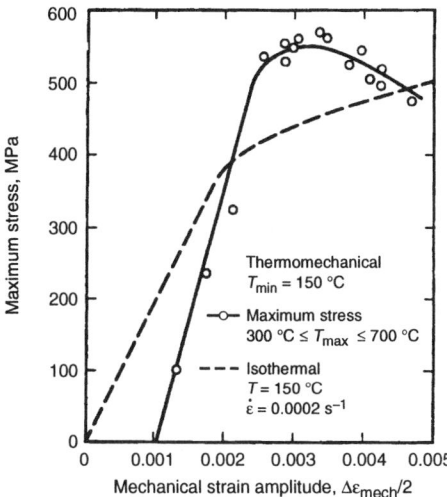

Fig. 14 Decrease in maximum stress in TMF OP case due to thermal recovery. Source: Ref 45

Microstructural Changes in Steels. The mechanical properties below the transformation temperature (body-centered cubic, or bcc, phase) and above the transformation temperature (austenite face-centered cubic, or fcc, phase) are considerably different (Ref 85). Two series of creep tests were performed under constant stress and temperature. Creep tests were conducted at temperatures of 400, 450, 500, and 550 °C (750, 840, 930, and 1020 °F) on 1070 steel well below the transformation temperature of 660 °C (1220 °F); the second series was conducted to investigate creep at temperatures above the austenitic transformation at temperatures of 700 and 800 °C (1290 and 1470 °F). It was found that both the transient and steady-state creep strain rates in the fcc phase were higher than the creep rates predicted with bcc phase properties by two orders of magnitude.

The second series of relaxation experiments (Ref 85) differed from the previous set in that, before each experiment, the specimen was heated to above its austenitic transformation temperature, to 925 °C (1700 °F), and was held at this temperature for 1 h. Then the specimen was rapidly cooled in air to the desired temperature of 500 or 550 °C (930 or 1020 °F), and the experiment proceeded as outlined above. Results indicate that the final stresses are similar for experiments with preheating and experiments without preheating to 925 °C (1700 °F), considering that preheat experiments indicate an initial stress about 50% lower than experiments without preheating.

The effect of phase transformations during thermal fatigue has been explored by Nortcott and Baron (Ref 86), who noted that repeated formation of austenite and martensite during the TF cycle generally leads to cracking of the material. Similarly, Sehitoglu (Ref 24) considered TMF experiments beyond 650 °C (1200 °F) on 1070 steels where the hysteresis loops recorded displayed the transformation effect. Thermal expansion characteristics are influenced by the na-

ture of austenitic transformation. The mean value of α for 1070 steel below the transformation temperature is 8.34×10^{-6} 1/°F, whereas the mean value for α above the transformation temperature is 1.52×10^{-5} 1/°F. The coefficient of thermal expansion can be defined two ways: (1) tangent to the thermal strain-temperature curve, or (2) as a secant modulus, the slope of the line connecting the thermal strain point to the origin. It is important to specify whether the CTE is a tangent or a secant value. When phase transformations occur, it is advisable to use a secant modulus; this avoids the problem of rapid changes in the tangent modulus upon phase transformation.

Many steels undergo strain aging, which results in considerable hardening in a TMF test or a test that involves exposure of the material to temperatures below 400 °C (750 °F). Certain temperature-strain histories in solute-hardened materials produce strain aging, and thus strengthening, of the material. Thermomechanical fatigue studies under strain-aging conditions for steels and nickel-base superalloys have been discussed in Ref 20. Strengthening is caused by interstitial solute atoms, which anchor the dislocation motion. If the pinning of the dislocations occurs during deformation, the term *dynamic strain aging* is used (Ref 87). If the aging occurs under a constant load (after some plastic deformation), it is called *static strain aging* (Ref 87).

Static strain-aging experiments have been conducted on both 1020 and 1070 steels (Ref 80). The material was cycled at 20 °C (70 °F), but was exposed to the aging temperature time at zero stress every reversal. The experimental results are shown in Fig. 15. In these experiments, the deformation is at 20 °C (70 °F) but with intermittent exposure to 300 °C (570 °F) (up to 30 min) at zero load of the cycle. Increase in room-temperature

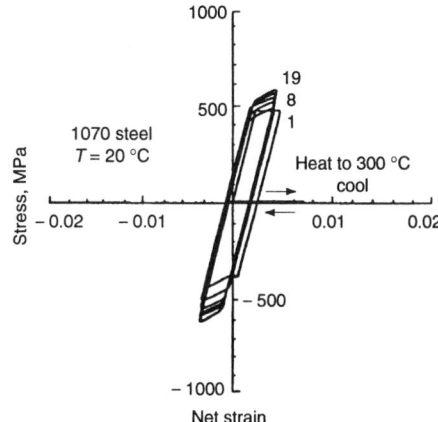

Fig. 15 Experimental σ–ε response under TMF strain-aging conditions. Source: Ref 80

strength as high at 30% has been measured after 40 reversals. The strain range of the hysteresis loops is $0.005 + \Delta\sigma/2E$, where $\Delta\sigma$ is the stress range and E is the elastic modulus at 20 °C (70 °F). Since this material is cyclically stable at room temperature, the observed strengthening is attributed to strain aging. The specimen is cycled at a strain rate of 2×10^{-3}/s.

Thermal recovery effects have been observed in steels when temperatures exceed 500 °C (930 °F). In the case of pearlitic steels, the lamellae structure coarsens and, ultimately, a spheroidized material results. Examples of the microstructural change are illustrated in Fig. 16. Consequently, these changes alter the stress-strain response of the material. Figure 16(a) shows the mean lamellae thickness to be nearly 145 Å. As the microstructure becomes spheroidized (Fig. 16b), the mean spheroidite diameter is much larger than

Table 2 Summary of microstructural damage mechanisms in steels

Material	TMF IP	TMF OP	Reference
1070 steel	Crack growth at pearlite colony boundaries; ferrite-pearlite interfaces Internal oxygen attack of MnS particles; coarsening of pearlite lamellae; spheroidization Phase transformation, bcc-fcc; recrystallization	Strain-aging effect due to exposure at elevated temperature, followed by low temperature Formation and repeated fracture of oxides; internal oxygen attack of MnS particles Coarsening of pearlite lamellae; spheroidization	20, 24, 38, 45
Type 304 stainless steel	Strengthening due to strain aging; creep damage (grain-boundary triple points) in tensile stress part of the cycle Higher dislocation density compared to IF Mixture of dislocation arrangements compared to IF High density of intergranular cracks Higher grain-boundary sliding in tension relative to compression, resulting in ratcheting at the microlevel Grain-boundary residual stresses at low temperature relax at high temperatures, resulting in cavity nucleation	Strengthening due to strain aging; higher dislocation density compared to IF; mixture of dislocation arrangements compared to IF Lowest density of intergranular cracks compared to IF and IP	70, 63, 28, 19
1Cr-1Mo-0.25V steel	...	Higher density of crack formation in OP relative to IP and IF, possibly due to fracture of oxide scale	44, 76
Type 347 stainless steel	Creep damage at grain boundaries produced shortest lives for IP	...	31
1016 steel	Integral breadth of x-ray diffraction profiles as a measure of subgrain evolution during TMF	...	88

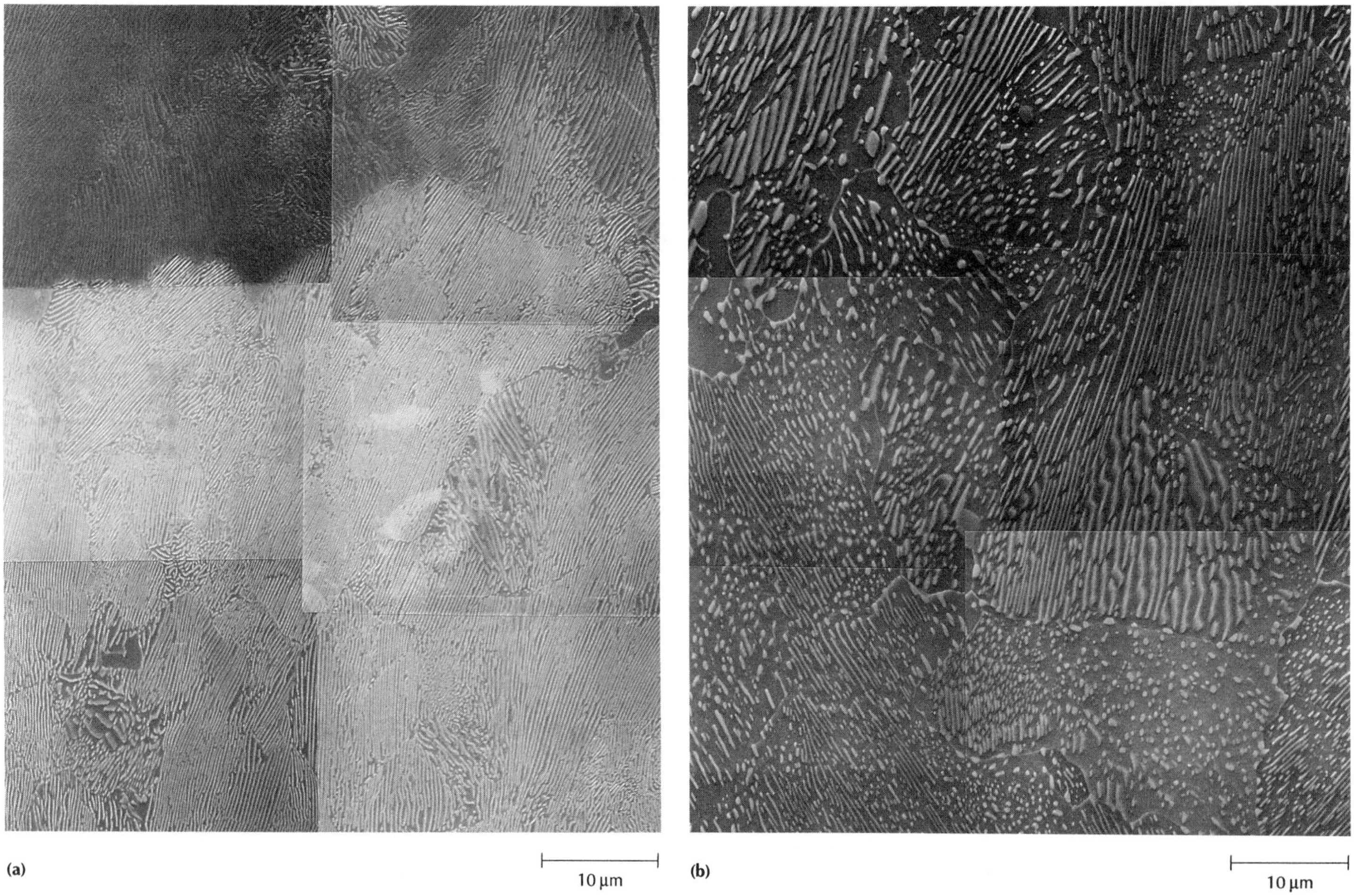

(a)   10 µm   (b)   10 µm

Fig. 16 Change in lamellae morphology upon exposure of steel at high temperature. Source: Ref 82

any of the cementite thicknesses. The coarsening process takes the form of an early breaking-up of the lamellae, followed by spheroidite growth. This difference in size is apparent in Fig. 16. This phenomenon was documented in early studies by Sehitoglu (Ref 25, 34, 82). Table 2 summarizes the microstructural damage mechanisms identified by various experiments on steels.

TMF of Aluminum Alloys

Only a handful of experiments has been reported on the elevated-temperature behavior of aluminum alloys. At temperatures exceeding 150 °C (300 °F), aluminum alloys undergo creep damage in the form of grain-boundary cavitation (Ref 89) and intergranular crack growth (Ref 90). Under TMF conditions with $T_{mean} = 200$ °C (390 °F), creep damage is expected to occur under both OP and IP conditions. Extensive studies on the fatigue of aluminum (Ref 91-93) at room temperature revealed accelerated fatigue damage in air relative to a vacuum environment. At elevated temperatures the environment (oxidation) effect is expected to be more pronounced (Ref 94-96). Figure 17 compares TMF OP and TMF IP lives for aluminum alloys 2xxx-T4, a powder metallurgy material with minimal porosity level. In the experiments, $R_\varepsilon = -1$, the minimum tem-

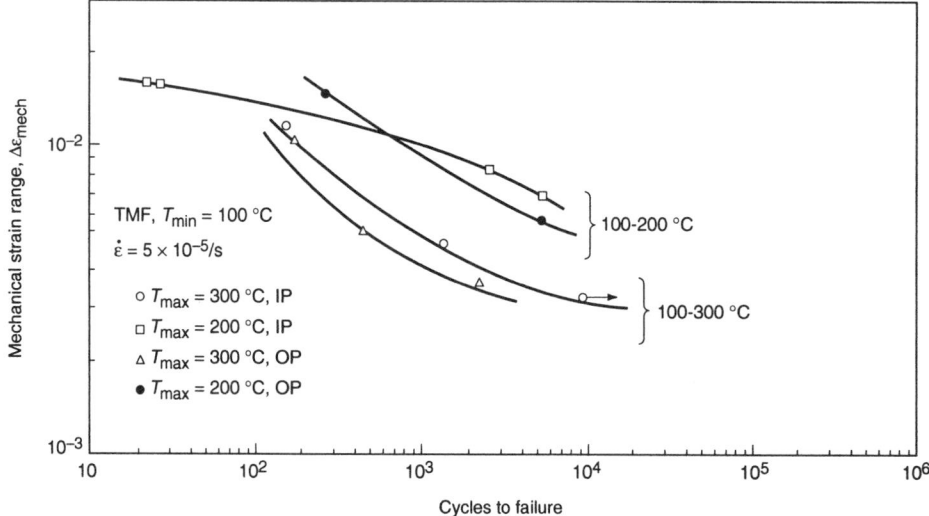

Fig. 17 Comparison of TMF OP and TMF IP lives for Al 2xxx-T4. Source: Ref 95

perature was 100 °C (210 °F), and the maximum temperatures were 200 and 300 °C (390 and 570 °F). A crossover in lives occurred for the 100 ⇄ 200 °C case, but there was no such crossover for the 100 ⇄ 300 °C case (Ref 90). Two other

studies on TMF of aluminum alloys have been reported (Ref 97, 98). In Ref 97, a cast Al-Si-10Mg alloy was studied under total constraint TMF OP conditions; the mechanical strain increased proportionally with increasing maximum temperature. The minimum temperature was maintained constant at 50 °C (120 °F). The most severe case studied was under 50 ⇄ 350 °C (120 ⇄ 660 °F) conditions, and considerable cyclic softening was observed both at the low- and high-temperature ends of the cycle. In Ref 98, the cast alloys A1319 and Al356 were considered. This work studied the role of dendrite arm spacing, porosity level, composition, and heat treatment.

Relatively very few studies have been conducted on the TMF aluminum alloys. The major issues are the following:

- Oxidation has an influence on fatigue damage both at room temperature and at elevated temperatures. A number of fundamental studies have been made of IF of aluminum at room temperature and a few vacuum tests at elevated temperatures (Ref 99), but there are no reported experiments under TMF loading. Based on the work of Bhat and Laird (Ref 99), considerable oxidation damage is present in polycrystalline aluminum alloys.

- If the maximum temperature exceeds the aging temperature, then considerable softening can be observed in TMF due to changes in the shape and size of the precipitates. The aging temperatures can vary from 150 to 200 °C (300 to 390 °F).

- Creep damage has been observed at temperatures exceeding 200 °C (390 °F) in TMF experiments. The creep damage is in the form of distributed cracks. When creep damage with diffuse cracks occurs, continuum damage mechanics concepts would be appropriate. In this case, the σ-ε behavior of a damaged material can be described by using effective stress and hydrostatic stress integrated over the cycle (Ref 100).

TMF of Nickel-Base High-Temperature Alloys

Much has been published on the high-temperature behavior of nickel-base superalloys—the development of which is ongoing, including the use of coating treatments. The major advantage of nickel-base superalloys over other metals is their useful TMF operating range, which extends to $T_{max}/T_m = 0.8$, where T_{max} is the maximum temperature in the cycle and T_m is the melting temperature in degrees Kelvin. On the other hand, when T_{max}/T_m values exceed 0.5, the TMF strength of steels is considerably lowered.

Depending on the test temperature, the superalloys may show either cyclic softening or cyclic hardening behavior. The hardening behavior is attributed to dislocation buildup at the precipitate/matrix interface. On the other hand, softening behavior is considered to be due to precipitate shearing, increased dislocation climb facilitated by increased diffusion rates, and reduced dislocation densities caused by recovery processes (for a review, see Ref 101 and 102).

Similar to IF tests, TMF experiments exhibit a temperature and strain range dependence of cyclic stress response. Castelli et al. (Ref 103) observed cyclic hardening of Hastelloy X under OP

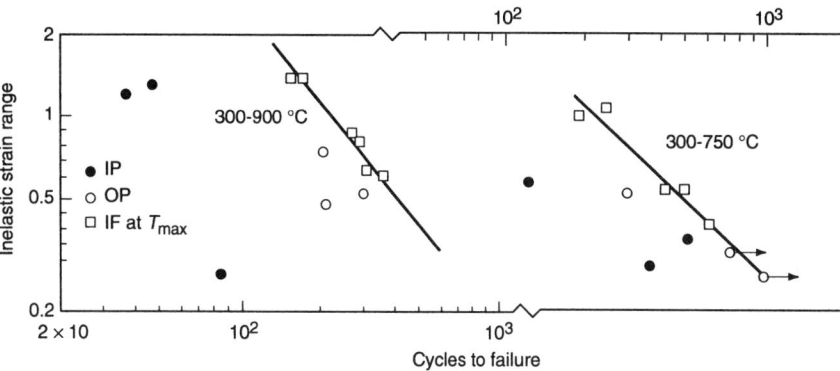

Fig. 18 Comparison of TMF OP and TMF IP lives for Hastelloy X. Source: Ref 36

cycling at $\Delta\varepsilon_m = 0.006$ over a temperature range of 600 to 800 °C (1110 to 1470 °F). When the temperature range was increased to 800 to 1000 °C (1470 to 1830 °F), cyclic softening was observed. Marchand et al. (Ref 104) tested B-1900+Hf under TMF OP and TMF IP at a temperature range of 400 ⇄ 925 °C (750 ⇄ 1700 °F). Cyclic stress-strain curves revealed cyclic hardening at low strain ranges and cyclic softening at high strain ranges when compared to cyclic stress-strain response at T_{max}. Sehitoglu and Boismier (Ref 105), working with polycrystalline Mar-M247 (500 ⇄ 870 °C), found gradual cyclic softening for most of the life at small strains, whereas cyclic hardening was observed at high strains. Stress-strain behavior has been found for René 80 (Ref 106), Mar-M247 (Ref 105), and Mar-M246 (Ref 107), CMSX-6 single crystals (Ref 108), and on Inconel 617 by Macherauch and colleagues (Ref 109). They studied Inconel 617 and reported significant hardening at T_{max} of 750 to 850 °C (1380 to 1560 °F). When the maximum temperature was higher than 950 °C (1740 °F), the response was stable.

A number of Japanese investigators have published results on TMF of superalloys. An extensive study of superalloys has been undertaken by Kuwabara et al. (Ref 42), who considered Inconel 718, Inconel 738LC, Inconel 939, Mar-M247, and René 80. For Inconel 718, the temperatures were 300 ⇄ 650 °C (570 ⇄ 1200 °F); for the other alloys, 300 ⇄ 900 °C (570 ⇄ 1650 °F). For Inconel 718, Inconel 939, and Mar-M247, shorter lives were demonstrated for the TMF IP case in the high strain range and a crossover in life at small strain levels. The Inconel 738LC and René 80 exhibited shorter lives for the TMF OP case relative to TMF IP. Taira et al. (Ref 36) considered Hastelloy X in the temperature ranges 300 ⇄ 900 °C and 300 ⇄ 750 °C (570 ⇄ 1650 °F and 570 ⇄ 1380 °F) and found that TMF IP lives are considerably shorter than the TMF OP case. These results are shown in Fig. 18.

In Great Britain, extensive thermal fatigue studies have been reported by Glenny and Taylor (Ref 110) on Nimonic and directionally solidifed nickel alloys. The duration of the thermal cycle (immersion times) and the maximum temperature effect (up to 920 °C, or 1690 °F) have been examined, and intergranular cracking has been

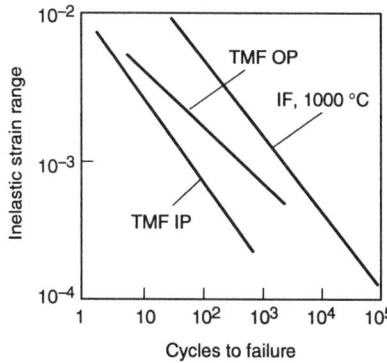

Fig. 19 Comparison of TMF OP and TMF IP lives for polycrystalline Mar-M200. Source: Ref 112

noted. Tilly (Ref 111) used the tapered-disk geometry with the fluidized-bed technique under 20 ⇄ 920 °C (70 ⇄ 1690 °F) conditions. This author also conducted reverse bend tests with temperature cycling of 350 ⇄ 1000 °C (660 ⇄ 1830 °F) in air and in vacuum and showed an increase in fatigue life in vacuum relative to air of nearly a factor of two. The lives were lower than those predicted based on IF data at T_{max}.

Other TF experiments have been conducted by Woodford and Mowbray (Ref 50) on the nickel-base superalloys Inconel 738 and René 77 using the tapered-disk specimen. The hysteresis behavior was calculated at the disk periphery, and the temperature-strain phasing was similar to the DCW type. Crack length was monitored as a function of cycles. The temperature range was 22 ⇄ 920 °C (72 ⇄ 1690 °F) for both materials. These investigators made the important observation of γ′-depleted zones in the vicinity of crack tips. In these regions, aluminum is depleted and the γ′ structures break down. Bizon and Spera (Ref 51) considered 22 nickel-base superalloys using the fluidized-bed technique. They noted the positive role of directional solidification and coatings on TF life.

Most nickel-base superalloys exhibit a crossover of the TMF IP and TMF OP mechanical strain-life curves. In this case, TMF IP fatigue lives are shorter than TMF OP lives at high mechanical strain ranges, but are greater than TMF OP lives at low mechanical strain ranges. The crossover occurs at approximately $\Delta\varepsilon_m = 0.0045$.

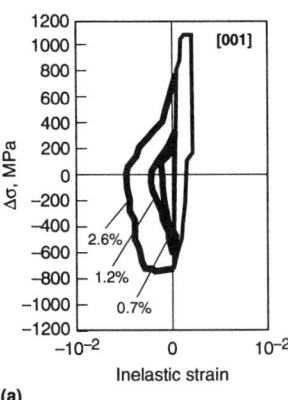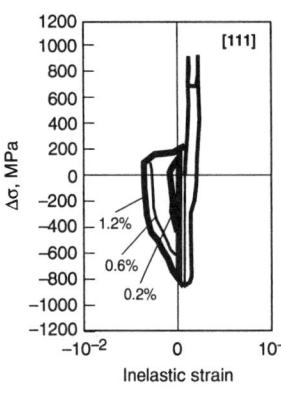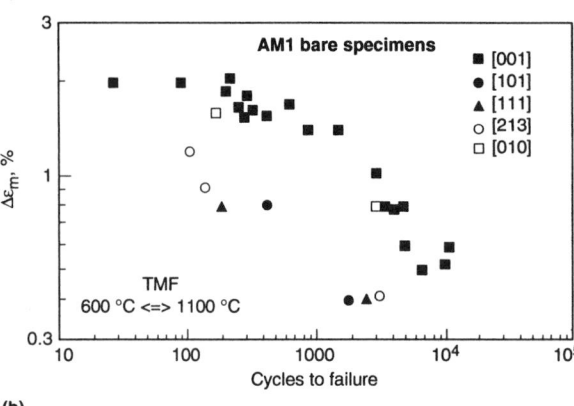

Fig. 20 Effect of crystallographic orientation on TMF behavior of AM1. Source: Ref 120

Kuwabara et al. (Ref 42) observed the crossover in life curves for Inconel 718 and Mar-M247 under temperature cycling of $300 \rightleftarrows 650$ °C and $300 \rightleftarrows 900$ °C ($570 \rightleftarrows 1200$ °F and $570 \rightleftarrows 1650$ °F), respectively. Bill et al. (Ref 112) investigated Mar-M200 at mechanical strain ranges greater than 0.01 and over a temperature range of $495 \rightleftarrows 1000$ °C ($925 \rightleftarrows 1830$ °F). These results are shown in Fig. 19.

Nelson et al. (Ref 113) studied B-1900+Hf at a temperature range of $540 \rightleftarrows 870$ °C ($1000 \rightleftarrows 1600$ °F) and also observed a crossover corresponding to a mechanical strain range of 0.0045. Ramaswamy and Cook (Ref 114) conducted tests on Inconel 718 at $\Delta\varepsilon_m = 0.015$ and over temperature ranges of $345 \rightleftarrows 565$ °C and $345 \rightleftarrows 650$ °C ($650 \rightleftarrows 1050$ °F and $650 \rightleftarrows 1200$ °F) and found TMF IP to be more damaging than TMF OP. However, they noted that René 80 ($760 \rightleftarrows 870$ °C, or $1400 \rightleftarrows 1600$ °F) also showed a crossover in the TMF life curves. The TMF OP lives were shorter than the TMF IP cases (Ref 115); this has been attributed to high mean stresses in the TMF OP case. Gayda et al. (Ref 116) found that for coated PWA 1480 at inelastic strain ranges of less than 0.2%, the TMF OP lives are significantly longer than TMF IP, whereas in the high strain regime the TMF OP and TMF IP lives are comparable.

In the past, TMF life has been approximated by IF life at the maximum temperature of the TMF cycle using the same mechanical strain range. This appears to be applicable for TMF OP conditions. Experiments conducted by Bill et al. (Ref 112) on Mar-M200 revealed that IF life at T_{max} was slightly longer than TMF OP life ($500 \rightleftarrows 1000$ °C, or $930 \rightleftarrows 1830$ °F). The IF lives may have been greater because the IF tests were conducted at a frequency 100 times greater than that of the TMF tests. Nelson et al. (Ref 113) also found a correlation between IF life at T_{max} and OP life (40 to 870 °C, or 1000 to 1600 °F) for B-1900+Hf. Malpertu and Rémy (Ref 117) conducted TMF experiments on Inconel 100 utilizing a cycle similar to a counterclockwise 135° cycle over a temperature range of 600 to 1050 °C (1110 to 1920 °F). Initiation life was the same as the IF initiation life at T_{max}. There is a correlation between TMF OP life and IF life at T_{max}, but

there are discrete differences in the damage mechanisms.

A few TMF studies have included nonproportional phasing of the mechanical strain and temperature. Nonproportional loading cycles are very important because they more closely approximate a service-induced strain-temperature history of an actual component. An example is the diamond-shape history, where the maximum and minimum mechanical strain occur at the median of the temperature range. Embley and Russell (Ref 115) conducted DCW history tests on Inconel 738 and found lives approaching two orders of magnitude longer than TMF OP and TMF IP lives over the same temperature range (425 to 870 °C, or 800 to 1600 °F). Nelson et al. (Ref 113) conducted TMF experiments on B-1900+Hf utilizing a counterclockwise elliptical strain-temperature cycle at 540 to 870 °C (1000 to 1600 °F). They discovered a fivefold increase in life over TMF OP lives.

Guedou and Honnorat (Ref 48) considered three alloys—Inconel 100, AM1, and DS 200—subjected to DCW and DCCW histories with a $650 \rightleftarrows 1100$ °C ($1200 \rightleftarrows 2010$ °F) temperature range. The AM1 is a single-crystal alloy, DS-200 is directionally solidified, and Inconel 100 is polycrystalline. Based on mechanical strain range, AM1 exhibited the best properties. Isothermal fatigue data at T_{mean} was closest to the TMF mechanical strain life curve. Bernstein et al. (Ref 118) considered Inconel 738LC both in the coated and uncoated state under $425 \rightleftarrows 870, 915,$ and 980 °C ($800 \rightleftarrows 1600, 1680,$ and 1800 °F) conditions. They found shorter lives for the coated material relative to the uncoated material. They discussed the turbine blade strain-temperatures extensively (in particular, the role of the startup and shutdown) and showed that the TMF OP cycle best describes the engine conditions.

Recently Halford et al. (Ref 30) have proposed the use of bithermal fatigue cycles as a simple alternative to TMF testing. Bithermal results have been interpreted with strain range partitioning (SRP) as a predominantly PC or CP type of loading. Bithermal experiments conducted on B-1900+Hf were directly related to TMF results by use of an appropriate damage rule. Other research on TMF OP of Hastelloy X has been reported by

Kaufman and Halford (Ref 119) in the ranges $505 \rightleftarrows 905$ °C and $425 \rightleftarrows 925$ °C.

Recent research on TMF in Europe centers around Remy et al. at École de Mines (Ref 117, 120), Guedou at Snecma (Ref 48), Bressers and various colleagues at Petten (Ref 121), and Mughrabi and his students in Germany (Ref 70, 108). Remy and coworkers (Ref 120) studies the crystallographic orientation effect on the cyclic σ-ε behavior of AM1 superalloy in the temperature range of $600 \rightleftarrows 1100$ °C. These results are summarized in Fig. 20(a) for the [001] and [111] orientations. The inelastic strain range in the cycle was found to be strongly orientation dependent, with [001] producing smaller inelastic strains than [111] (Fig. 20a). This is evident when the stress/inelastic-strain loops are compared for the case of mechanical strain range of 1.2%. The longest fatigue lives among five crystal orientations were found for the [001]-oriented specimens (Fig. 20b).

Bressers et al. (Ref 121) used a 135° OP cycle (i.e., diamond counter clockwise, DCCW) and studied the TMF behavior of SRR99 in the coated and uncoated condition. They considered both $R = 0$ and $R = -\infty$ cases and monitored the crack length as a function of cycles. The role of oxidation is emphasized in their model. Mughrabi and coworkers (Ref 108) studied CMSX-6 single-crystal superalloys of [001] orientation under $600 \rightleftarrows 1100$ °C conditions and documented the coarsening of precipitates during TMF. They confirmed that the mean stress in nickel-base superalloys play a considerable role. The life under TMF OP was considerably shorter than under TMF IP (five times), with DCW (diamond clockwise) and DCCW cases between these two extremes. This study also confirms that inelastic strain range is not a good correlator of life when failure occurs in a finite number of cycles with very small $\Delta\varepsilon_{in}$ components. They noted that when the γ' structure rafts, soft γ-matrix channels permit unconstrained dislocation motion. Also, during the high-temperature phase of the cycle dislocation climb and during the low-temperature end of the cycle, cutting of the particles has been observed.

Other work from Europe includes Marchionni et al. (Ref 122), who have been studying an oxide

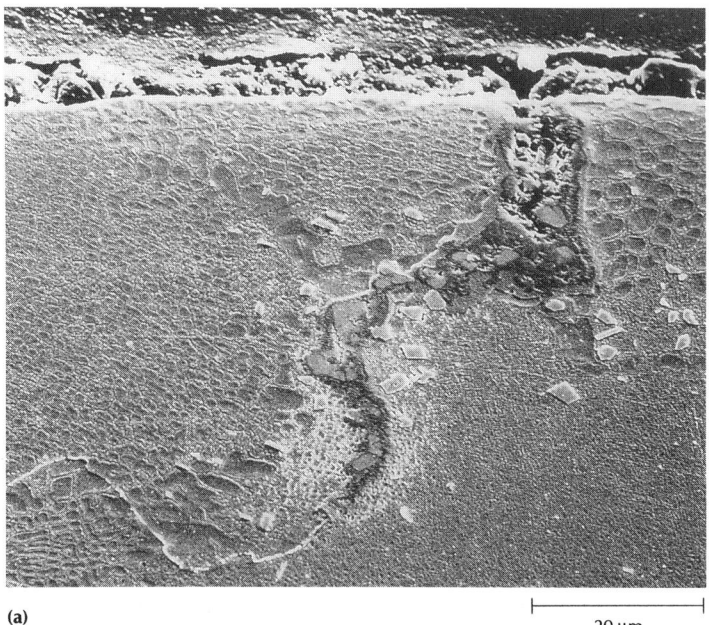

(a)　　　　　　　　20 µm　　　(b)　　　　　　　　50 µm

Fig. 21 Grain-boundary oxidation and cracking in Mar-M247. Source: Ref 105

dispersion (Y_2O_3) Inconel alloy. The TMF OP and TMF IP results are similar and lie within the scatter of data. Macherauch and his group at Karlsruhe (Ref 109) have reported TMF OP and TMF IP experiments on Inconel 617, showing TMF IP damage to be more significant than TMF OP. The maximum temperature was in the range of 850 to 1050 °C, while the minimum temperature was 600 °C.

Strain Rate and Frequency Effects. Strain rate can affect cyclic stress-strain response as well as fatigue life. In studies conducted on René 80 (Ref 123), Hastelloy X (Ref 103), Mar-M246 (Ref 49), and Mar-M200 (Ref 124), it has been reported that decreasing the frequency resulted in a decrease in the stress range; no change in the relative hardening and softening behavior has been observed. There is abundance of information on the strain-rate effects under IF conditions, including decreasing frequency, lowering strain rates, or introducing hold times (Ref 118, 124-126). These effects are attributed to increased environmental and creep damage (Ref 126-128). Only in rare cases do the strain rate or hold times have no effect (Ref 112, 129) or does decreasing the strain rate or introducing hold periods increase fatigue life (Ref 124, 130, 131). This latter behavior can be explained based on a reduced creep component caused by reduced cyclic stresses when γ' precipitate coarsening occurs.

There have been no systematic attempts to alter the strain rate (analogous to IF experiments), but some TMF experiments have introduced a hold period at maximum temperature. The effect of compressive hold periods on TMF of Inconel 738 has been established by Bernstein et al. (Ref 118), who reported shortened cycles to failure. In nickel-base superalloys, if the hold period in TMF OP results in stress relaxation in compression, high tensile mean stresses develop upon

reversed loading—which is detrimental to fatigue life.

Environmental Effects. The effects of the environment on nickel-base superalloys at elevated temperatures are very complex. Environmental damage can affect both crack initiation and crack propagation and has a detrimental effect on fatigue life. Crack nucleation often originates from preferentially oxidized grain boundaries (Ref 126, 130, 132, 133). Grain boundaries are preferentially oxidized because they are paths of rapid diffusion and their composition may differ from that of the matrix (Ref 105, 107, 134, 135). An example of oxidation at grain boundaries and intergranular initiation in Mar-M247 subjected to TMF IP conditions is shown in Fig. 21. The experiment was conducted under TMF IP 500 ⇄ 870 °C conditions.

Rémy et al. (Ref 136) oxidized precracked specimens and compared the crack growth with that of virgin specimens. These experiments revealed crack growth rates as much as three orders of magnitude higher than the virgin samples. They proposed a modified fracture mechanics approach to handle the crack growth under repeated oxide fracture; the oxidation constants were determined via integration over the cycle. The different TMF strain/temperature waveforms (600 ⇄ 1050 °C) were predicted with their model.

Under elevated-temperature conditions, a protective oxide scale forms on the surface of the specimen, separating the substrate from the environment. However, spalling and cracking of the protective oxide scale occur due to stresses developed in the scale. The principal sources of stress in the oxide scale are the thermal stresses due to the difference between the thermal expansion coefficients of the oxide and the matrix. Although there is zero thermal stress at the oxide formation

temperature, upon cooling by ΔT, a stress is generated in the oxide layer. Oxide spikes penetrate from the surface toward the inside of the substrate. The oxide spike morphology could form at the surface or at the coating/substrate interface upon failure of the coating. The problem of stress fields associated with oxide spikes has been studied by Kadioglu and Sehitoglu (Ref 107, 135). In their work, an oxide spike was modeled as a semispherical surface inhomogeneity. The stress field in the vicinity of the oxide spike was calculated using a technique based on Eshelby's method. Then the calculated strain at the tip of the oxide spike was used in the life prediction model. Strains at the oxide tips increase considerably as the ratio of oxide elastic modulus to metal elastic modulus decreases. The stresses under different levels of thermal mismatch also were shown. Sample results are presented in Fig. 22. The geometry of the oxide intrusion (spike) is shown in Fig. 22(a). The variation of $\sigma_{ij}/(E_m\varepsilon^{th})$, which is the stress tensor normalized by the product of matrix modulus and thermal mismatch strain, as a function of distance measured from the surface is shown in Fig. 22(b). The term $\Gamma = E_{ox}/E_m$ is the oxide to matrix modulus ratio, and X_3/c represents the normalized distance normal to the free surface. The $X_3/c > 1$ represents the matrix region ahead of the oxide intrusion, while $X_3/c < 1$ represents the oxide. The critical parameter extracted from these studies is $\Delta\varepsilon_m^{ox}$, the mechanical strain range at the tip of the oxide. This result was used in the life prediction model described in Ref 135.

Oxidation characteristics of nickel-base superalloys can vary widely. The oxidation products formed vary with alloy composition, temperature, and time at temperature. A general oxidation characterization for nickel-base superalloys at 870 °C (1600 °F) can be drawn from Ref 137 and 138. Initially, a continuous film of Al_2O_3 forms.

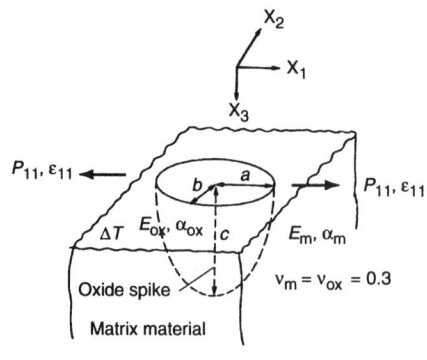

(a)

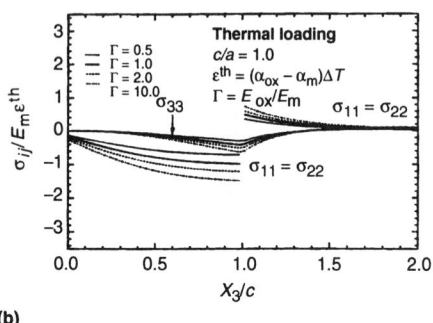

(b)

Fig. 22 (a) Geometry of an oxide spike (intrusion). (b) Oxide stresses as a function of α and E mismatch. Source: Ref 107

Diffusional mass transport of chromium through the Al_2O_3 layer alloys the formation of an outer layer of Cr_2O_3. Eventually, spinels of $Ni(Cr,Al)_2O_4$ are formed. Some TiO_2 may also be formed. This sequence of events is a specific case. The oxides formed will vary from alloy to alloy and with variations in temperature and time of exposure.

Oxidized surfaces usually are associated with an adjacent zone of alloy depletion. This is characterized by a zone depleted of γ' precipitates. Several studies have reported the existence of γ'-depleted zones (Ref 105, 107, 113, 126, 131-133). The γ'-depleted zone is caused by the loss of aluminum to the formation of oxides. This zone may also be depleted of solid-solution-strengthening elements such as chromium. The fatigue characteristics of such a layer may be markedly different in the initial cracking stages. Deformation bands develop in precipitate-free areas and lead to premature microcracking and fatigue failure (Ref 139). Due to oxidation of the crack tip, a region depleted of oxide-forming elements will be formed ahead of a fatigue crack. As a result, the crack will propagate into a region having changed mechanical properties. Crack growth in each cycle may be controlled by the size of the environmentally affected zone at the crack tip (Ref 126).

Steady-state formation of the oxide and alloy-depleted layers is governed by parabolic rate kinetics (Ref 128, 130, 138, 140). The rate of oxidation and alloy depletion is considered to be affected by the application of stress. It has been shown that oxidation and alloy depletion increases when stress is applied (Ref 108, 138). The effect of stress on environmental attack may vary with alloy composition and exposure conditions.

Coating Effects in Superalloys. Environmental degradation due to oxidation and corrosion may be prevented by using protective coatings. Various types of coatings have been used to reduce the deleterious effect of the environment (Ref 33, 45). Many of the coatings developed fulfill their protection role against oxidation or corrosion of the base material. Three main types of coatings have been used to protect superalloys: (1) diffusion coatings, (2) overlay coatings (MCrAlY, where M is nickel or cobalt), and (3) thermal barrier coatings (TBCs).

The predominant oxide formed on the coating is Al_2O_3. The overlay coatings consist of Ni(Co,Fe),Cr, Al, and Y, and are called MCrAlY type. Finally, TBCs have been used to limit the heat flow into the base alloy. The materials most commonly considered as TBCs are general oxides such as ZrO_2 and Al_2O_3. The low thermal expansion coefficient of the ceramic coatings and the relatively high thermal expansion coefficient of the base alloy result in a large mismatch strain, which encourages the propagation of cracks in the coating. The presence of porosity, discontinuities, random microcracking, and a columnar structure with grain boundaries to the surface, known as coating segmentation, has also been observed.

Under TMF conditions, coatings undergo complex stress-strain changes and at the low-temperature end of the cycle could fracture. Various researchers have found coatings to provide benefit, depending on the temperature (Ref 107, 128, 135, 141-144). In some cases, however, a reduction in the fatigue lives of some directionally solidified alloys (Ref 141) and other nickel-base superalloys (Ref 145, 146) has been noted. Goward (Ref 147) has investigated the TMF behavior of aluminide and CoCrAlY coatings. The tests were conducted under fully reverse condition by cycling the temperature between 425 and 925 °C. In these tests, the low-aluminum CoCrAlY coating exhibited a much higher resistance to crack initiation than the higher-aluminum CoCrAlY.

The effects of protective coatings on TF of superalloys have been studied by alternately immersing a variety of tapered disk and wedge-type specimens into hot and cold fluidized beds (Ref 148, 149). However, the strain/temperature cycle was not precisely known in these experiments. Thermomechanical fatigue tests in which the temperature and strain can be controlled separately have been used to investigate the effects of coatings on superalloys (Ref 150-152). Among the various forms of strain/temperature phase relations, the most damaging is TMF OP. In fact, it was reported that coated superalloys exhibited shorter lives under TMF OP than under TMF IP (Ref 150-151).

Under service conditions, additional strains on the coatings may arise due to thermal expansion mismatch, elastic moduli mismatch, diffusion between coating and substrate, phase transformation, or chemical reaction with the environment. These additional strains and stresses alter the crack initiation and propagation resistance of the materials, resulting in spallation of protective oxide scales and/or coating or early crack formation, which allows oxidation attack into the base alloy. To improve the fatigue performance of the coated components, these strains should be minimized by adjusting the mechanical and metallurgical properties of the coatings (provided that oxidation/corrosion resistance capability is maintained). Therefore, a life prediction methodology that will relate coating performance to the mechanical/physical properties of coating/substrate systems and environmental conditions is needed.

Coating cracking lives have been successfully correlated with total strain, which is the summation of the thermal expansion mismatch strain and the mechanical strain (Ref 152, 153). A fatigue crack growth model has also been proposed by Strangman (Ref 154). In this work, the penetration of a coating crack into the base metal has been analyzed using the fracture mechanics approach. In another approach (Ref 155), the life of a coated system was considered as the summation of the number of cycles to initiate a crack through the coating, the number of cycles for the coating crack to penetrate a small distance into the substrate, and the number of cycles to propagate the substrate to failure. In a recent study (Ref 151), the mechanical damages for the coating and the substrate have been calculated separately and then combined to produce an optimum prediction damage parameter. The predicted lives for coated superalloys were within a factor of two for TMF OP tests. The two-bar model representing the coating and the substrate has been used to determine the constitutive stress-strain loop for the coating under TMF conditions. Then, the number cycles to initiate a crack in the coating has been estimated by the hysteretic energy method (Ref 156). With this approach, the fatigue life of overlay coatings was estimated within a factor of 2.5 in the case of TMF conditions (Ref 155). Although the two-bar model is one dimensional, it captures the first-order effects of the coatings on the behavior of base alloys. To study the effect of biaxiality requires nonlinear FEM due to the highly nonlinear behavior of the coating/substrate system (Ref 157).

Swanson et al. (Ref 156) have conducted isothermal fatigue tests at 760, 925, and 1040 °C and TMF tests by cycling the temperature between 425 and 1040 °C on PWA 286 (Ni-CoCrAlY+Si+Hf) overlay and PWA 273 (NiAl, outward diffusion) aluminide-coated single-crystal PWA 1480 alloy. Their tests used hollow tubes as test specimens. They found that, in many cases, coating cracks had progressed into the PWA 1480 alloy and directly caused failure. In some specimens, however, the coating cracks did not extend into the substrate, and failure was caused by a crack initiated from the uncoated inner surface. The coating cracks penetrated into the substrate in both OP and IF tests for specimens coated with PWA 273, but only in the OP tests for specimens with overlay coating. In this

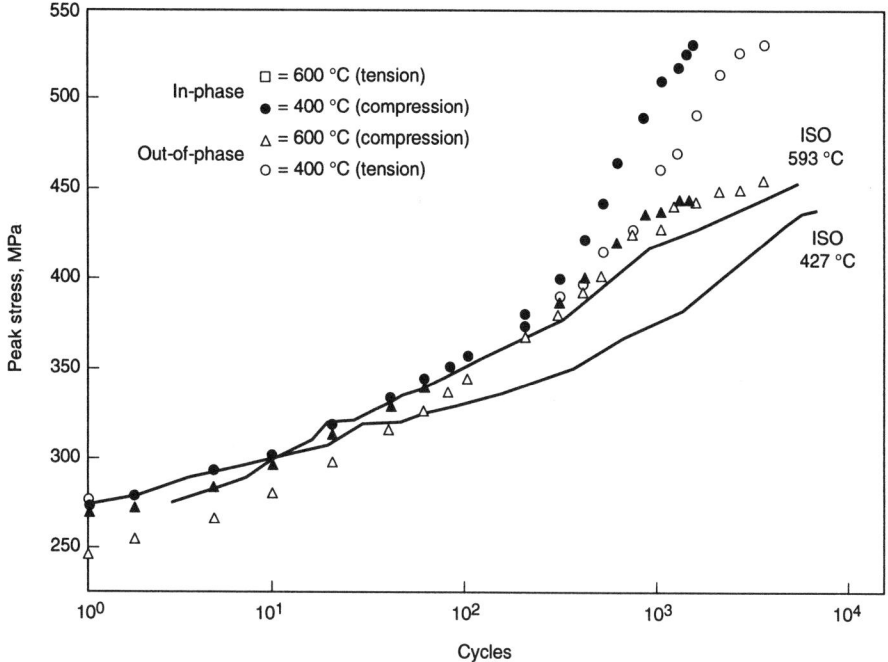

Fig. 23 Cyclic hardening for Hastelloy X under TMF conditions. Source: Ref 103

Table 3 Reported damage mechanisms for TMF in-phase and TMF out-of-phase loadings

Material	TMF in-phase	TMF out-of-phase	Reference
Coating Ni-base superalloys		Fracture of coating upon cooling below its ductile brittle transition	158
Mar-M 247 (uncoated polycrystalline)	Intergranular crack initiation and growth	Repeated oxide damage rafting of the γ' structure different than IF or TMF IP	105
Mar-M 247 (coated and uncoated)	Intergranular crack initiation and growth, internal crack initiation	Fracture of coating upon cooling below its ductile brittle transition	107, 135
	Rafting of the γ' structure different than IF or TMF OP	Rafting of the γ' structure different than IF or TMF IP	
Hastelloy X (solution strengthened Ni-base superalloy)	Strain rate dependent dynamic strain aging	Strain rate dependent dynamic strain aging	103
	Precipitation hardening due to Cr-rich $M_{23}C_6$	Precipitation hardening due to $M_{23}C_6$	
AM1 single crystals		Environment initiated damage in TMF differs from casting defect initiated damage in IF	48
IN 738 LC (coated and uncoated)		Lower lives for coated OP case because of coating fracture	118
CMSX-6 (single crystals)	Soft γ' matrix formation, cutting of particles at low temperatures, dislocation climb during high-temperature portion of the cycle		108

stresses) developed is lower than in other directions; consequently, for a given mechanical strain range the plastic strain range is lowest among all possible directions. Directionally solidified alloys remove the grain boundaries transverse to the principal stress and also lower the elastic modulus in that direction relative to polycrystalline materials. For polycrystalline nickel-base superalloys, grain size and coatings influence TMF lives.

- Because of the unbalanced nature of inelastic deformation in TMF, mean stresses are sustained and do not relax. The mean stresses play a considerable role at finite lives, because the plastic strain range is smaller than the elastic strain range.
- Complex chemistries of oxides form with properties different from those of the substrate, resulting in internal stresses and oxide fracture that channels the crack into the material. Considerable depletion in the vicinity of oxides has been measured.
- At small strains and long lives, oxidation damage persists. Depending on stress and temperature, creep damage appears to be more significant at short lives.
- For the majority of nickel-base alloys at temperatures above 700 °C, TMF results display strain-rate sensitivity. Generally, as the strain rate is reduced or hold periods are introduced, the cycles to failure are lowered.
- For most nickel-base superalloys, TMF IP damage is larger than TMF OP damage at high strain amplitudes, whereas the trend is reversed at long lives. The diamond cycle often produces lives that fall between the TMF IP and TMF OP extremes.

Microstructural Changes. Under TMF conditions considerable changes in nickel-base superalloy microstructure have been known to occur, including changes in the size and morphology of γ' precipitates and the formation of dislocation networks around precipitates. For polycrystalline nickel-base superalloys exposed to temperatures above 800 °C, TMF OP loading results in transgranular propagation and TMF IP results in intergranuiar propagation.

Castelli et al. (Ref 103) and later Castelli and Ellis (Ref 32) observed cyclic hardening of Hastelloy X under TMF OP at $\Delta\varepsilon_m = 0.006$ over a temperature range of 600 to 800 °C. In this alloy dynamic strain aging occurs in the region from 200 to 700 °C, and precipitation of chromium-rich precipitates also produces hardening. When the temperature range was increased to 800 to 1000 °C, cyclic softening was observed. Since this is a solute-hardened superalloy, it undergoes considerable dynamic strain aging when the temperature is at 600 °C. Considerable precipitation of $M_{23}C_6$ carbides also occurs in the vicinity of dislocations, which coarsens at high temperatures and loses its effectiveness as the additional hardening mechanism. Examples of extensive hardening for Hastelloy X in TMF are shown in Fig. 23. Both TMF IP and TMF OP results are shown for 400 $\rightleftarrows$ 600 °C conditions. The results are

case, coating-initiated cracking was the dominant failure mode. It was difficult to draw a general conclusion about the effects of coatings on fatigue life from this work due to variation of specimen design and orientations, frequencies used, cycle type, and strain ranges applied.

Wright (Ref 128) has examined the oxidation-fatigue interactions in René N4. Isothermal tests were performed on uncoated, aluminide-coated, and preoxidized alloy at 1095 °C. The test results showed that although there were no differences in the fatigue life of coated and uncoated specimens in low-frequency tests ($f = 1/2$ cycle/min), fatigue life increased significantly during high-frequency tests ($f = 20$ cycle/min). Glenny and Tay-

lor (Ref 56), Bizon and Spera (Ref 51), and Woodford and Mowbray (Ref 50) have investigated the thermal fatigue characteristics of uncoated and coated superalloys using a variety of tapered-disk and wedge-type specimens. However, the available data from these studies are difficult to interpret due to the large variety of specimen shapes, thermal cycle shapes, and differences in the definition of failure criteria.

In summary, the following general rules apply for the TMF of nickel-base superalloys:

- For single crystals, the best TMF resistance has been obtained in the [001] direction. In this direction, the elastic modulus (and thus the

obtained at strain rates of 5×10^{-5}/s conditions. Strengthening increases by more than a factor of two over several thousand cycles. The IF results at 595 and 425 °C are also shown for comparison.

Thermal recovery processes in TMF have been documented by Sehitoglu and Boismier (Ref 105) and by Kadioglu and Sehitoglu (Ref 135) in the form of γ' coarsening and eventual rafting of the γ' microstructure.

Research on TF and TMF has described several microstructural changes that influence deformation (stress-strain) behavior as well as fatigue lifetime. Some, but not all, of the findings are listed in Table 3. The most important mechanisms are:

- Aging of the microstructure when exposed to high temperatures
- Repeated oxide rupture due to mismatch in mechanical and physical properties of matrix and oxide
- Strain aging in the case of solute-hardened materials at the elevated-temperature end, resulting in considerable hardening
- Enhanced grain-boundary damage due to unequal deformation during the cycle
- Carbide precipitation at grain boundaries at high temperature

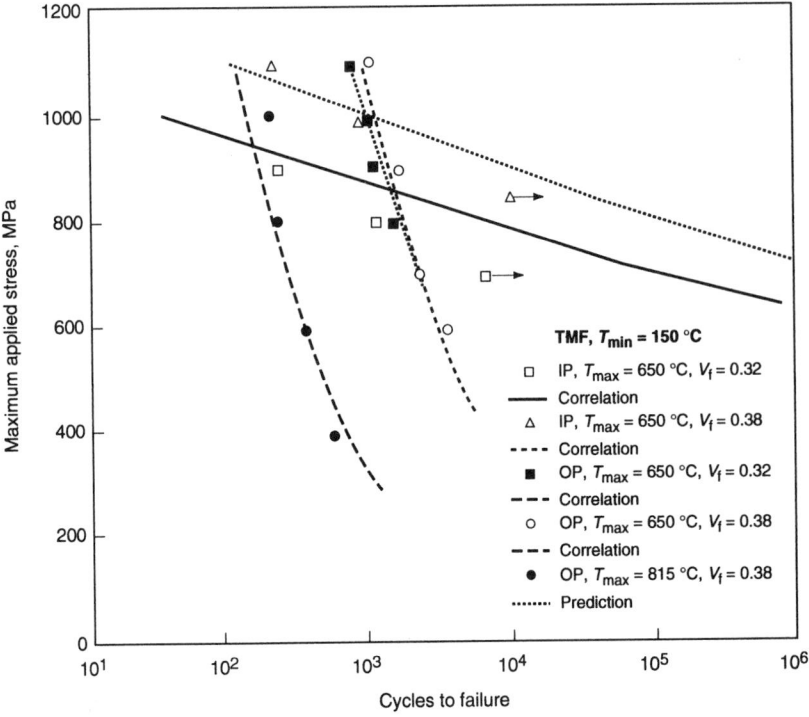

Fig. 24 TMF OP and TMF IP for SiC/titanium aluminide composite. Source: Ref 169

Other Structural Alloys

Several classes of advanced materials have been investigated under TMF loading conditions. Thermomechanical fatigue of titanium aluminide has been investigated by Wei et al. (Ref 159) under total constraint 25 °C $\rightleftarrows$ 750 °C, 25 °C $\rightleftarrows$ 900 °C conditions. The role of hydrogen and helium environment was investigated and the lives were ×2 higher in helium environment relative to air. On a similar material (Ti$_3$Al) Mall et al. (Ref 160) studied TMF crack growth under OP and IP cases.

TMF data on cobalt-base superalloys have been published by Reuchet and Rémy (Ref 140) and Kalluri and Halford (Ref 161). Early research by Sheffler and Doble (Ref 162) on TMF of tantalum alloys showed that the TMF IP lives are considerably shorter than the TMF OP case. These materials have been investigated as an alternative to nickel-base superalloys, but their cost and/or performance characteristics have not been superior to nickel-base superalloys. Undoubtedly, they will find some specialty applications.

Thermomechanical fatigue is of considerable interest in the electronics industry. Failure mechanisms are currently being investigated in aluminum thin films in integrated circuits and in lead-tin solders undergoing temperature, current density, and mismatch in thermal expansion conditions. Recent TMF data on 63Sn-37Pb solder alloys (Ref 163) have been published. In these experiments the material is subjected to simultaneous shear loading and temperature cycling. It is important to note that if the specimen is subjected to shear and temperature, a set of material planes will experience TMF OP loading and the other orthogonal planes will undergo TMF IP loadings.

Starting in mid-1980, interest grew in metal-matrix composites (metal reinforced with ceramic particulates whiskers or fibers). These materials, although expensive, have been touted for lower thermal expansion coefficient, higher elastic modulus relative to matrix, and better high-temperature properties. Although some properties have improved with these new classes of 'advanced' materials, the TMF resistance of these materials is not superior to that of the monolithic alloy. This is partly due to difficulties in the processing uniformity and detrimental residual stresses. Karayaka and Sehitoglu (Ref 164, 165), VanArsdell et al. (Ref 166, 167), and Sehitoglu (Ref 168) have authored a number of papers on TMF OP and IP of Al2024 reinforced with SiC particulates with volume fractions of SiC in the range 15% to 30%. Recent work (Ref 169, 170) focused on TMF of Timetal 21s (titanium alloy) reinforced with SiC (SCS-6) composites studied under TMF OP and TMF IP conditions (stress-control). Under TMF IP fiber damage was dominant and under TMF OP environment damage in the matrix or at interfaces was found to be most important. Sample results are shown in Fig. 24 for TMF OP and TMF IP loading conditions under stress control. The experimental techniques developed in Ref 46 were utilized. The results are plotted in an S_{max}-N_f format. Under TMF IP, conditions. the lives are controlled by the fiber failure while for TMF OP case the damage mechanism is a combination of environmental and fatigue processes. We note that at long lives the TMF OP is far more damaging. At high stresses there is crossover and TMF IP damage exceeds the TMF OP damage.

Multiaxial Effects in TF and TMF

The multiaxial effect is one of the least explored aspects of TF and TMF loadings. When a surface is heated the stresses are generally biaxial and this phenomenon has been discussed by Manson (Ref 6). Manson discusses thermal shock as well as slow heating cases where a two-dimensional stress state develops. Taira and Inoue (Ref 171) considered the biaxial stress fields in their TF analysis. In their experimental work they cooled a solid cylinder at one end, and plotted their TF results using the von Mises equivalent strain range and showed good agreement with uniaxial tests on .16% steel. Recent research considers TMF under multiaxial loadings (Ref 172) where axial-torsional loading is applied simultaneously with temperature.

Crack Initiation and Crack Propagation

Crack Initiation in TF and TMF. Crack initiation within nickel-base superalloys can occur intergranularly at oxidized surface exposed grain boundaries or transgranularly. Transgranular initiation can be caused by heterogeneous planar slip which produces initiation along persistent slip bands at free surfaces (Ref 125, 139, 173). Transgranular initiation can also occur at pores, inclusions, and carbides (Ref 117, 139). Transgranular crack initiation is more prominent at low temperatures and high frequencies. This is because the contributions from the creep and environmental components of damage are minimal. Gell and Leverant (Ref 126) indicated that one of the first observed effects of an increased creep component is a transition from transgranular to intergranular initiation. Runkle and Pelloux (Ref 174) found under IF conditions, as temperature is increased, initiation changes from transgranular to intergranular for Astroloy. A similar transition was observed for a decrease in strain rate by Nazmy (Ref 127) for IN 738 under IF loading at 900 °C (1650 °F).

At high temperatures crack initiation is predominantly intergranular (Ref 124, 130, 131, 174). This is attributed to increased damage contributions from creep and environmental attack. Environmental attack appears to be the more dominant of the two damage mechanisms. Intergranular crack initiation typically occurs at oxidized surface connected grain boundaries which are often accompanied by an adjacent zone of γ' depletion (Ref 50, 102, 118, 130, 131). Preferential grain boundary environmental attack occurs because of the easier path of oxygen diffusion. The decrease in fatigue life with increased temperature and decreased frequency can partially be attributed to the transition from transgranular to intergranular crack initiation. In general, intergranular crack initiation occurs at a faster rate than transgranular initiation (Ref 126). In the work of Kadioglu and Sehitoglu (Ref 107) the Mar-M246 exhibited intergranular crack initiation followed by a switch to transgranular crack growth.

Crack Propagation in TF and TMF. It is important to know the mode of crack propagation to develop physically meaningful fracture mechanics or micro-mechanical models to characterize the crack driving force. The interaction of creep damage and environmental attack and their effects on crack propagation under TMF conditions can be very complicated. In general, TMF OP produces transgranular crack propagation while intergranular fracture is observed for TMF IP case, within representative service temperature ranges (300-1000 °C) at moderate strain rates (10^{-3} to 10^{-6}/s). TMF experiments conducted on B-1900+Hf at temperature ranges of 427-925 °C (Ref 153) and 538-871 °C (Ref 113) revealed predominantly transgranular OP crack propagation and intergranular IP propagation. Kuwabara and Nitta (Ref 42) observed intergranular crack propagation for IN 718 (300-650 °C) and Mar-M247 (300-900 °C) under TMF IP loading. Milligan and Bill (Ref 124) performed TMF experiments on Mar-M200 (500-1000 °C) and found intergranular crack propagation and internal grain boundary cracking for TMF IP case, while TMF OP produced mixed transgranular and intergranular cracking. Ramaswamy and Cook (Ref 175) also observed transgranular cracking under TMF OP and intergranular cracking under TMF IP tests conducted on IN 718 (343-565 °C and 343-649 °C) and René 80 (760-871 °C). Intergranular crack propagation in TMF IP case is attributed to a tensile creep component that results in ratcheting at the microlevel resulting in weakening of the grain boundaries. In a study conducted on Mar-M247 similar results have been observed (Ref 105).

Rau et al. (Ref 176) performed TMF crack growth tests on Mar-M200 DS (427-927 °C) and B-1900+Hf (316-927 °C) under strain control. The Mar-M200 DS out-of-phase crack growth rates were greater than in-phase. The T_{max} IF crack growth rates for B-1900+Hf were the same as out-of-phase TMF crack growth rates. Leverant et al. (Ref 153) also tested B-1900+Hf (427-927 °C) under strain control. It was observed that out-of-phase and in-phase TMF crack propagation rates were the same. Heil et al. (Ref 177) conducted TMF crack growth tests on IN 718 under load control at a temperature range of 427-649 °C (800-1200 °F). They discovered that in-phase loading produced greater crack growth rates than out-of-phase loading. The subject of TMF crack growth rates needs to be investigated further. Figure 25 shows their result compared on the basis of the stress intensity range. These experiments were conducted under load control R = 0.1 conditions. The 90° refers to DCW and the 270° refers to DCCW load-temperature history. The 0° refers to TMF IP and 180° refers to TMF OP. They found that TMF IP is more damaging than TMF OP in 718 based on ΔK. Heil et al. forwarded a simple model incorporating crack growth from cycle and time dependent damage. Their results are intuitively correct in view of the fact that the crack closure phenomenon will be minimum for the TMF IP case.

Okazaki and Koizumi (Ref 178, 179) in Japan used the J-Integral to correlate crack growth rates under TMF IP conditions for low-alloy ferritic steel (Co-Mo steel). Sehitoglu (Ref 24) correlated crack growth rates in TMF OP (150 °C $\rightleftarrows$ 400 °C to 600 °C) and IF results from room temperature using the range in crack opening displacement. Gemma et al. (Ref 180) considered crack growth in both DS Mar-M200 and B-1900 alloys under TMF OP conditions with 427 °C $\rightleftarrows$ 1038 °C and showed that lower crack growth rates were obtained when the loading axis and the grain growth direction coincided confirming the benefit of DS alloys. In the UK, Skelton (Ref 181) studied crack growth in steels (Alloy 316, Alloy 800, 1/2Cr-Mo-V) and used both the linear elastic and elasto-plastic fracture mechanics parameters. The elasto-plastic crack growth parameter was $C(\varepsilon^{in})^n a Q$ where C, n, and Q are constants, a is crack length, and ε^{in} is inelastic strain. Skelton used this parameter for small cracks both for TMF and thermal shock conditions.

Recent research from Wright Patterson Air Force Base (Ref 160) has investigated TMF crack growth in titanium aluminide. Similar to IN718 the TMF IP crack growth rates were higher than TMF OP. The LEFM parameter ΔK was used. The authors noted a distinct difference between superalloys and aluminides in that the crack tip blunts in aluminide and sustained load cracking do not occur.

Jordan and Meyers (Ref 182) conducted TMF OP, TMF IP experiments as well as a DCCW type cycle. Higher crack growth rates were noted for the TMF OP case relative to IF test at T_{max} for the case 427 °C $\rightleftarrows$ 871 °C. Several elasto-plastic fracture mechanics parameters have been used

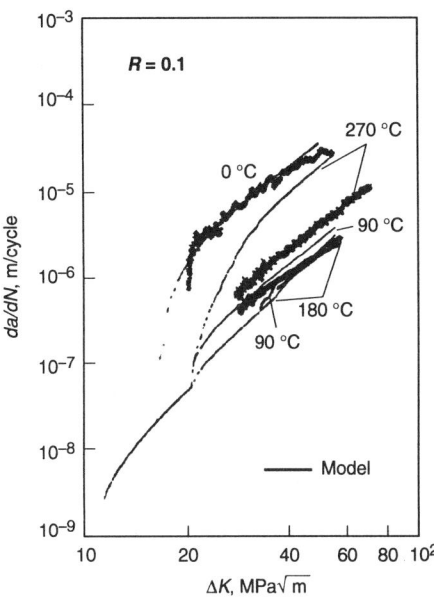

Fig. 25 Crack growth rates in TMF OP and TMF IP cases for Inconel 718. Source: Ref 177

including the strain-intensity range, stress-intensity range and the range in J-integral.

Temperature-Strain Variations in Service

McKnight et al. (Ref 183) conducted finite element analysis of turbine blades with a simplified mission cycle. The purpose was to analyze tip cracking due to thermomechanical fatigue. In their analysis of a hollow, air cooled turbine blade they did focus at a tip cap designated as a 'squealer tip' just below the leading edge. They showed that at this critical location, the blade experiences a TMF OP type of cycle with a small amount of ratcheting in the compressive direction. The analysis indicated that the mechanical strain range (the principal strain component) in the critical location was near 0.31% after several cycles and the maximum temperature was as high as 1100 °C. Both the maximum and the minimum mechanical strains were negative. The strain temperature was predominantly a TMF OP cycle with an initial temperature of 300 °C. Under these conditions the lifetime of René 80 simulated on a laboratory sample was 3000 cycles as established by Kaufman and Halford (Ref 119).

Embley and Russell (Ref 115) studied strain temperature phasings that are rather complex and representative of those experienced by turbine blades. Bernstein and colleagues (Ref 118) also studied 'faithful' histories typical of leading edge of turbine blades. A schematic of this history is given in Fig. 26. The schematics in Embley and Russell (Ref 115) work and Bernstein's paper are similar. We note that during the startup the behavior is of TMF OP type. Higher strains in the compressive direction are achieved for 'fast' acceleration relative to 'normal' acceleration. The temperature does not reach a maximum at the acceleration peak; the temperature reaches a

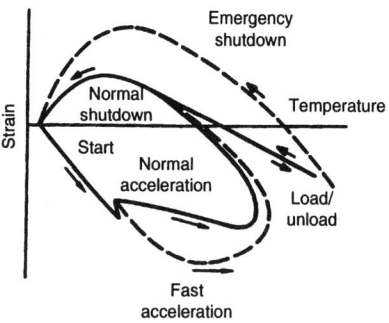

Fig. 26 Strain/temperature history on a turbine blade. Source: Ref 118

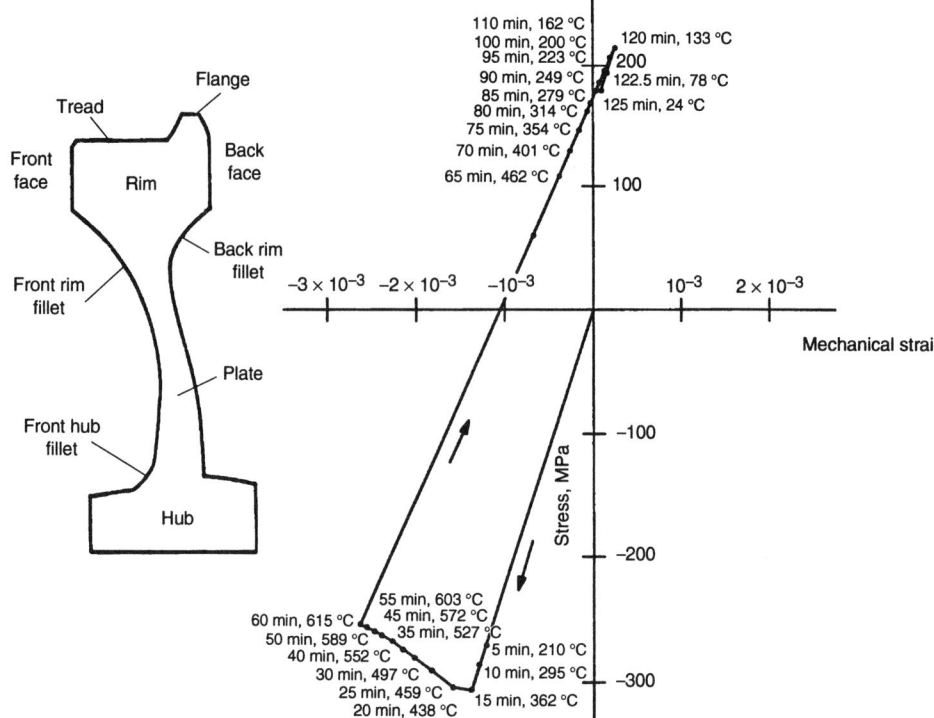

Fig. 27 Railroad wheel design and stress/mechanical strain

maximum during the steady state operation (the load/unload portion). Upon normal shutdown the strain increases in the tensile direction as temperature is reduced. In the case of an emergency shutdown the tensile mechanical strains reached can be considerably higher.

Heating and cooling of the surface of a thick structure under the impingement of steam has been considered by Skelton (Ref 76). When the surface is exposed to the steam at 550 °C (1020 °F) it undergoes compression (OP behavior) because it is surrounded by the cooler material. With time, the temperature front moves into the structure, the temperature gradient reduces, and the surface unloads elastically at temperature. During steady state conditions the strain remains constant and some relaxation of the stress can occur. When the steam is shut off, the surface wants to contract but is again constrained by the warmer surrounding material. The surface then undergoes tension and the maximum stress in the cycle is reached corresponding to the minimum temperature.

The current railroad wheel design has evolved as a result of TF experiments on actual wheels and also from thermomechanical elasto-plastic analysis of stresses under braking conditions. The original wheel design had a straight plate region which produced higher stresses (due to the constraint) than the curved plate design. The curved plate design is shown in Fig. 27(a). Localized heating due to friction occurs at the tread area. The results of a FEM analysis are given in Fig. 27(b). The circumferential stress-mechanical strain behavior under the brake-shoe (at the tread region of the wheel) is depicted in Fig. 27(b). The times and temperatures are also shown during the different stages of the heating process (Ref 184). The analysis was conducted for a 50HP application to the surface for a period of 1 h followed by 1 h cooling, then 5 min cool down to room temperature. The minimum stress develops 15 minutes into the 50HP application and upon subsequent rise in temperature the material softens. At the end of the heating period (60 min) the peak temperature at the surface has reached 615 °C (1139 °F). We note that the tensile stresses upon cooling are near 200 MPa (29 ksi).

Often in engineering applications an approximate measure of inelastic strains and stresses are needed without the execution of an elasto-plastic

FEM analysis. If the results of an elastic FEM analysis are available, the results of such an analysis can be used to estimate the inelastic strains. As pointed out earlier by Manson (Ref 6), the elastic strains calculated from FEM are assumed as total strains. Then, given the total strains, corresponding stresses and inelastic strains can be determined. This approach is referred to as 'strain invariance principle' and has been successfully demonstrated for TF analysis of the wedge specimen (Ref 13).

Thermal Ratcheting

To understand thermal ratcheting, a two bar structure shown schematically in Fig. 28 is considered. The two bars are in series and are subjected to a net section load, P_n, and one of the bars undergoes thermal cycling. In the present model, Bar 2 remains at a steady temperature of T_0 and Bar 1 undergoes thermal cycling such that the maximum temperature is T_0 and the minimum temperature is $T_0 - \Delta T$. Several regimes of material behaviors can be identified. These are (i) elastic (high-cycle fatigue), (ii) elastic shakedown (high-cycle fatigue), (iii) reversed plasticity (thermomechanical fatigue, low-cycle fatigue, and (iv) ratcheting regimes (shown as progressive and excessive distortion regions). In the first mode, all strains are elastic. In the second mode, one of the two bars yields during the first half of the cycle followed by elastic response in both bars.

In the third mode, called plasticity, one of the bars yields plastically while the second bar remains elastic. TMF occurs in this regime. In the

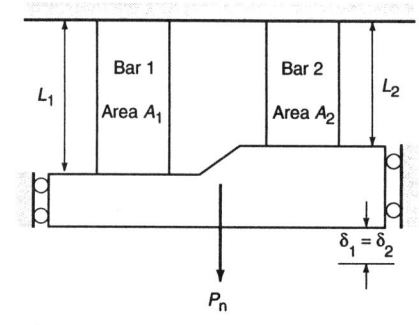

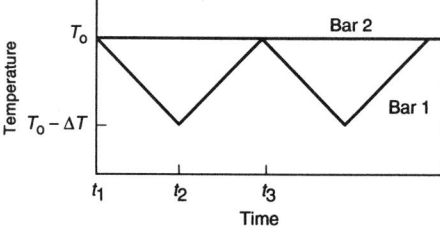

Fig. 28 Schematic of a two-bar structure and the temperature history on bars 1 and 2. Source: Ref 185

ratcheting case, one bar yields during the cooling half of the cycle while the second bar yields in the same direction during the heating portion of the cycle. This results in accumulation of strains in the tensile direction and ultimately failure due to a combined fatigue and ductility exhaustion mechanism. The equilibrium equation is

$$P_n = \sigma_1 A_1 + \sigma_2 A_2$$

and the compatibility equation ($\delta_1 = \delta_2$) is

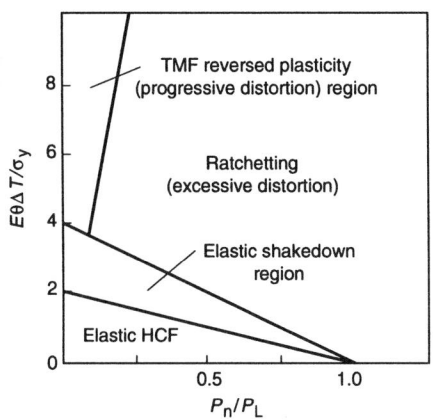

Fig. 29 Diagram showing the operating regimes of component behavior. Source: Ref 185

$$\varepsilon_1 L_1 = \varepsilon_2 L_2$$

where ε_1 is the net strain on Bar 1 given as

$$\varepsilon_1 = \varepsilon_1^e + \varepsilon_1^{in} + \varepsilon_1^{th}$$

$$\varepsilon_2 = \varepsilon_2^e + \varepsilon_2^{in}$$

where ε_1^e, ε_1^{in}, ε_1^{th} represent elastic, inelastic, and thermal strain components.

A common way of representing the different deformation regimes is to plot the dimensionless parameter $E\theta\Delta T/\sigma_y$ versus the P_n/P_L where P_L is the limit load determined from the yield strength σ_y at T_0 and θ is the thermal expansion coefficient. We note that similar results would be obtained by considering a pressure vessel subjected to a temperature gradient across the wall thickness and subjected to an internal pressure. The regime of 'thermal ratcheting' and TMF are shown in Fig. 29. The experimental results obtained from various tests are also shown in the diagram. The letter "*P*" denotes plasticity, "*R*" denotes ratcheting, "*S*" stands for shakedown, and HCF stands for high-cycle fatigue. The stress-net strain response of a two-bar structure with $P_n/P_L = 0.5$ and $\Delta T = 150\ °C$ (270 °F) are shown in Fig. 30 for a 304 stainless steel. We note that Bar 1 undergoes tension upon cooling. Therefore, the response depicted in Fig. 30 is termed "OP" (temperature is a minimum when stress is a maximum).

These experiments were conducted by using two servohydraulic test systems with command signals from a microcomputer which enforces equilibrium and compatibility of the two-bar structure. Similar experiments have been reported by Swindeman and Robinson (Ref 186) using 2.25Cr-1Mo steel. In addition to thermal cycling they also considered compression hold period effects on ratcheting.

Thermal ratcheting experiments have been conducted under stress control cycling or constant stress with superimposed thermal cycling (Ref 61, 62). Taira (Ref 88) has proposed constitutive models for predicting the transient and steady-state stage of deformation under these conditions and forwarded creep rupture as the mechanism of failure under this type of cycling.

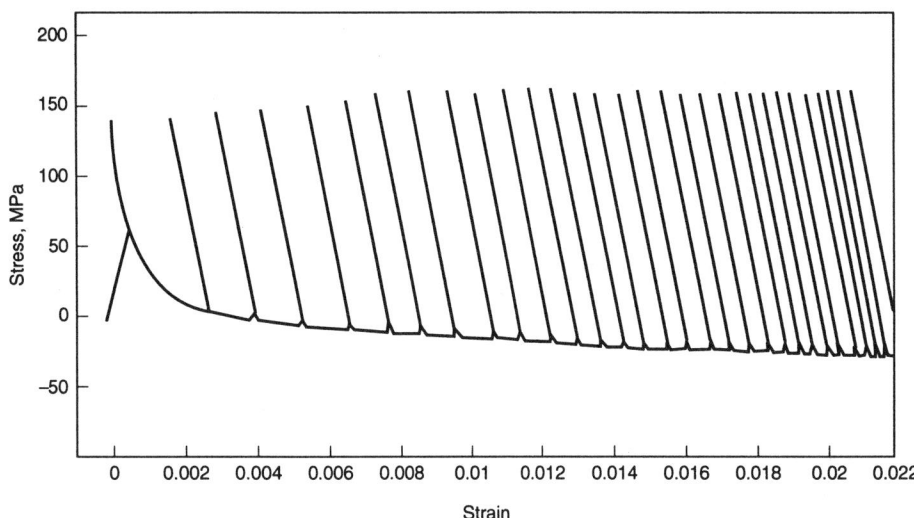

Fig. 30 Ratcheting of the two-bar structure for type 304 stainless steel. Source: Ref 185

Also, Russ et al. at WPAFB (Ref 187) conducted stress control TMF experiments which resulted in thermal ratcheting in titanium aluminides.

Thermal Shock

Thermal shock occurs under rapid temperature change conditions and results in thermal stresses that cause fracture of the material. There are three main differences between thermal shock and thermal fatigue: (i) thermal shock is a single application of a temperature change while thermal fatigue means repeated thermal cycling, (ii) physical properties such as specific heat and thermal conductivity influence the stresses developed which is not the case in the case of quasistatic thermal stress because the strain rates are high in the thermal shock case, the sensitivity of the material to strain rate is important, and (iii) the stresses generated in thermal shock are high enough that fracture toughness level is important while it does not enter in thermal fatigue considerations. In this section, we will be only concerned with thermal shock behavior of structural metallic materials and not the more well studied thermal shock in ceramics. Thermal shock in metals can be encountered in hot working manufacturing applications, such as the rapid temperature rise in the bore of a gun barrel, friction heating conditions such as in disk brakes, and rapid startups such as that in space shuttle main engines. The thrust chamber of the space shuttle has an inner liner with axial flow coolant channels constructed from copper or a copper-zirconium-silver alloy. During the firing cycle liquid hydrogen enters the channels to cool the chamber (or the hot gas wall) and a severe temperature rise occurs within a second.

The flame heating has been used to simulate engine environment and involves very rapid heating of the surface (Ref 21). The Association of American Railroads used flame heating on the rim of a rotating wheel of 0.7% C steel which was then subjected to a water quench when the same

region rotated through a water pool. In this case, because the surface temperatures were above the transformation temperature, thermal cracks developed upon quenching to room temperature. Baron (Ref 188) employed a tapered disk specimen similar to Glenny and Taylor (Ref 56) and using induction heating, reached heating of the order of 1 s. Rapid heating of a surface while the bulk remains cool can produce thermal shock conditions. For more details of thermal shock, the reader is referred to the text by Manson (Ref 6). More recently, Skelton (Ref 181) has studied crack growth under 'thermal shock' conditions.

Another variation of thermal shock is called 'thermal striping.' In thermal striping fatigue the surface temperature fluctuations develop due to mixing of fluid streams which impinge on the surface of structures. A temperature gradient develops through the component wall and the frequencies are generally much higher than in classical TF tests.

Life Prediction under TF and TMF

The question has been raised when correlating IF and TMF results as to what equivalent temperature should be used for meaningful comparisons. The maximum stress in a TMF experiment may not be at the maximum temperature and also the maximum strain may not coincide with the maximum temperature. Even though conventional IF data at the maximum temperature of a TMF cycle have been used to predict TMF lives, many researchers have shown that nonconservative predictions can still result (Ref 18, 40-44, 59, 60, 66-83) with this approach. It is essential to compare TMF and isothermal fatigue data at similar strain rates so that time-dependent damage mechanisms in both cases correspond.

Both environmental damage and creep damage on the material contribute to fatigue damage at elevated temperatures. A number of oxidation damage mechanisms have been proposed. These include (a) enhanced crack nucleation and crack growth by brittle surface oxide scale cracking (Ref 130, 189, 190), (b) grain-boundary oxidation which results in intergranular cracking (Ref 130, 134), and (c) preferential oxidation of second-phase particles (Ref 140). Oxidation and fatigue can interact resulting in a life much less than either of them acting alone. TMF OP loading is more damaging whenever the rate of oxide fracture damage exceeds that of creep damage. Oxidation damage accelerates with increasing $\Delta\varepsilon$, T_{max}, or increasing thermal coefficient mismatch of metal and oxide. The low ductility of oxides promotes cracking. If repeated oxidation is very severe for a given alloy, this could result in shorter TMF OP lives compared with TMF IP lives over a wide range of life regime. Low-carbon steels and low-alloy Cr-Mo steels exhibit shorter lives under TMF OP compared with TMF IP. These materials are known to be susceptible to oxidation damage. The TMF OP curve will be also lowered if the mean stress effects become significant. In the TMF OP case, tensile mean stresses develop which are conducive to minimal contact between crack surfaces and rapid crack growth rates.

Many mechanisms have been proposed to explain creep-induced damage and creep-fatigue interactions. These include (a) coalescence of intergranular voids ahead of an advancing crack (Ref 191, 192), (b) a greater crack tip plastic zone resulting from the summation of the plastic zones of voids ahead of a crack (Ref 193), (c) grain-boundary sliding initiating wedge-type cracks at grain boundaries (Ref 89), and at hard second-phase particles on the grain boundaries (Ref 194), (d) grain boundaries acting as weak paths for flow localization and crack growth (Ref 195), and (e) the modification of the crack tip strain fields in the absence of cavities (Ref 196). It would be expected that a number of these creep mechanisms would operate under both isothermal and TMF conditions depending on the alloy.

The models developed for TMF studies can broadly be divided into two categories: the continuum-based models (using parameters such as strain range, stress range, product of maximum stress and strain range, and plastic work) and physical damage equations (incorporating oxidation damage kinetics, microstructural observations, and creep damage mechanism). The major issue is how these equations perform at log lives where TMF data are scarce. Extrapolation of short-life data to long lives has limitations because the severity of the mechanisms will change. Therefore, models based on physical mechanisms and not empirical fits to data are preferred when predictions in the long-life regime are considered. We finally note that microstructural coarsening and other metallurgical instabilities are encountered in alloys at high temperatures and this could result in a decrease in lives for both the TMF OP and TMF IP cases.

Oxidation Damage and Oxidation Fatigue Laws. Specific modeling of oxide failure processes has been attempted by a number of investigators and are given in Ref 197 to 203. The range

of mechanisms include separation of an oxide film at a slip step (Ref 197), failure due to mismatch of the coefficient of thermal expansion (Ref 198), failure due to growth stresses (Ref 199, 200, 201, 203), interfacial shear failure (Ref 202), failure due to oxide buckling (Ref 203), and fatigue failure of oxide (Ref 68, 69, 135). These models are listed in Table 4.

The majority of the models proposed characterize failure due to scale growth stresses or temperature change conditions. The effect of mechanical strains on oxidation is not well understood. Many of the models consider uniform oxide thickness. Experimental observations, however, indicate nonuniform thickness in the form of oxide intrusions (Ref 135).

Table 4 Summary of oxide failure mechanisms and models

Ref	Mechanisms of failure	Equation
197	Separation of an oxide film at a slip step	$\dfrac{\sigma_f}{E_{ox}} = [\varepsilon_f/0.38(1-K)]\,(h_c/a)^{1/4}\mathrm{Sin}(w)$, Tensile failure $\dfrac{\tau_f}{E_{ox}} = [\varepsilon_f/(2(1+n))^{1/2}](h_c/a)^{1/2}\mathrm{Cos}(w)$, Shear failure h = oxide thickness, E_{ox} = oxide modulus σ_f = fracture strength, τ_f = shear strength a = interfacial layer thickness, K = dimensionless parameter, w = geometry constant
198	Tensile failure due to α mismatch	$\sigma_f = \dfrac{E_{ox}\,\Delta T(\alpha_{ox}-\alpha_m)}{1+\dfrac{E_{ox}}{E_m}\left(\dfrac{h_c}{h_m}\right)}$ E = elastic modulus h_c = critical rupture thickness of oxide h_m = thickness of metal ΔT = uniform temperature change
199	Failure due to compressive hoop strain during growth	$h_c = \dfrac{d\,\varepsilon\,\rho_m(56+16x)}{112\,\rho_{ox}}$ FeO_x ρ_m = density of metal ρ_{ox} = density of oxide h_c = critical rupture thickness
200	Scale displacement (M) causing compressive stress on convex surfaces	metal loss $= H = \dfrac{12\gamma}{E_{ox}}\left[\dfrac{R^2}{M^2\,\phi}\right]^{1/3}$ R = radius of metal surface $m = \phi(1-\alpha)-(1-V)$ ϕ = oxide/metal ratio
200	Loss of scale integrity when oldest part of oxide sustains strains average stored energy	$S = \dfrac{1}{2}\,h\,\dfrac{E\,\varepsilon^2}{2\gamma}$ h = metal thickness lost ε = strain on oxide failure
201	Growth stresses	$\sigma_f = \dfrac{E_m\,(h_m)^2}{6\,r\,h_c}$ r = radius of curvature h_c = critical rupture thickness E_m = elastic modulus of metal h_m = thickness of base metal
202	Interfacial shear failure	$\tau_f = \dfrac{h_c E_{ox}\,\Delta T(\alpha_m-\alpha_{ox})}{L+\left(1+\dfrac{E_m}{E_{ox}}\dfrac{h_c}{h_m}\right)}\,(1-v)$
203	Buckling of oxide	$\alpha\Delta T = \left[\dfrac{(h_c)^4}{a^4}+\dfrac{w}{E_{ox}h_c}\right]^{1/2}$ $\alpha\,\Delta T$ = thermal strain in oxide h_c = critical oxide rupture thickness a = crack length
68, 69	Fatigue failure of oxide	$h = \dfrac{B\,K_{peff}\,t^\beta}{\bar{h}_f}$ h = oxide intrusion depth $\bar{h}_f = \dfrac{\delta_0}{(\Delta\varepsilon_m)^2\phi\,\dot{\varepsilon}^a}$ B,β,δ_0,a = constants; K_{peff} = parabolic oxidation constant h_f = critical rupture thickness; $\Delta\varepsilon_m$ = applied mechanical strain range ϕ = phasing factor for thermomechanical loading
135		$\bar{h}_f = \dfrac{\delta_0}{(\Delta\varepsilon_m^{ox})^2\phi\,\dot{\varepsilon}^a}$ $\Delta\varepsilon_m^{ox}$ = Local (oxide tip) mechanical strain range

Table 5 Summary of oxidation-fatigue laws

Ref	Experiments/mechanism	Material	Equation
72, 126	Isothermal in air and vacuum, surface and crack tip oxidation	A286 steel	$\Delta\varepsilon = A(N_f)^b \nu^m$ A,b,m = constants N_f = cycles to failure $\Delta\varepsilon$ = strain range ν = cycle frequency
130	Isothermal in air, preoxidized samples, hold periods/surface and grain boundary oxidation	René 80	$\Delta\varepsilon_p = A \left[\dfrac{n}{1 + nt_h^{1/n}} \right] \exp\left(-\dfrac{Q}{RT}\right)(N_f)^b$ A,n,b,Q,R = constants T = temperature, t_h = hold period $\Delta\varepsilon_p$ = plastic strain range n = cycle frequency
134	Isothermal in air at different frequencies and hold periods, crack tip oxidation	In 718, In 750, Astroloy, 304 SS, 1/2 CrMoV steel	$\dfrac{da}{dN} = \dfrac{D_{gb}}{n} f(K_{max})$ D_{gb} = oxygen diffusivity at grain boundary n = cycle frequency da/dN = fatigue crack growth rate K_{max} = maximum stress intensity
140	Isothermal in air and in vacuum, crack tip oxidation of matrix and carbides	Mar-M509	$\dfrac{da}{dN} = 0.51\,\Delta\varepsilon_p[\mathrm{Sec}\,\dfrac{\pi}{2}\dfrac{s}{s_0} - 1]\,a$ $+ (1 - f_c)\,a_M \left(1 + \dfrac{\Delta\varepsilon_p}{2}\right) t_c$ $+ f_c a_c \exp(bs) t_c^{1/4}$ $\Delta\varepsilon_p$ = plastic strain range s,s_0 = applied stress, flow stress b,a_M,a_c = constants f_c = carbide fraction t_c = cycle period a = crack length
68, 69, 105, 107, 135	TMF in air and in helium, strain rates, oxide intrusions, repeated oxide rupture	0.7% C steel, Mar-M246, Mar-M247, Tib21s, Al2xxx-T4	$\dfrac{1}{N_f^{ox}} = \left[\dfrac{h_{cr}\delta_0}{B\Phi^{ox}K_p^{eff}} \right]^{-1/\beta} \dfrac{2(\Delta\varepsilon_{mech})^{(2/\beta)+1}}{\dot{\varepsilon}^{1-(a/\beta)}}$ $h_{cr}, \delta_0, B, a, \beta$ = constants $\dot{\varepsilon}$ = strain rate ϕ^{ox} = phasing factor $\Delta\varepsilon_m$ = mechanical strain range, $1/N_{ox}$ = oxidation damage K_{peff} = parabolic oxidation constant
188	Isothermal in air with hold period, oxide rupture in tension	2¼Cr-1Mo steel	$A\Delta\varepsilon^m \left(1 - \dfrac{t_h}{t_c}\right) = N_f$ t_c = cycle period, t_h = hold period A,m = constants, $\Delta\varepsilon$ = strain range N_f = cycles to failure

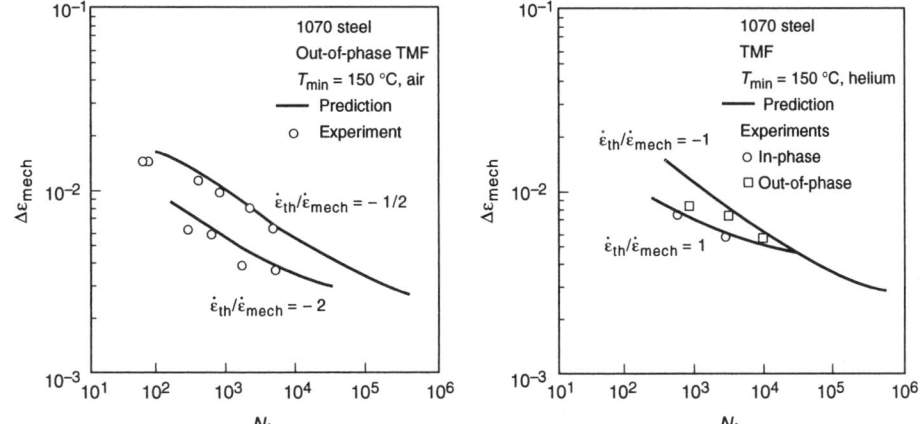

Fig. 31 Thermomechanical fatigue OP life prediction for steels under $\dot{\varepsilon}_{th}/\dot{\varepsilon}_{mech} = -1/2$ and $\dot{\varepsilon}_{th}/\dot{\varepsilon}_{mech} = -2$ conditions. Source: Ref 68, 69

A summary of oxidation-fatigue laws is given in Table 5 (Ref 1-4, 25-29, 34, 46-48). The models depicted in Table 5 specifically address fatigue plus oxidation damage. Despite the lack of quantitative understanding of oxidation-fatigue, the models provide a qualitative description of the oxidation-fatigue process. The earliest oxidation-fatigue model is that proposed by Coffin (Ref 72, 126) and is referred to as the frequency modified life approach. Antolovich and coworkers (Ref 130) recognized the formation of oxide spikes at preferential grain boundaries in Ni alloys and proposed a strain-life relation incorporating oxidation kinetics. Liu and Oshida (Ref 134) modified fracture mechanics parameters and accounted for accelerated crack growth upon oxide film rupture at crack tips. Rémy and Reuchet's (Ref 140) model incorporates crack growth according to a Dugdale type of model modified to account for oxidation of metal and carbides. The model proposed by Sehitoglu and coworkers (Ref 68, 69, 105, 107, 135) incorporates an oxidation phasing factor and accelerated oxidation under mechanical straining conditions. This is based on measurement of oxide thicknesses and observations of oxide failure at the surface and at crack tips. The details of this model will be illustrated later.

Several observations should be made on these models. Since the experimental data that form their foundations have been obtained primarily under IF conditions, the use of these relations for thermomechanical loading is not recommended. The Neu-Sehitoglu model incorporates a varying oxidation severity depending on the strain-temperature history and can handle TMF cases.

Depending on the material, OP or IP loading condition, and whether or not tensile or compressive hold periods are encountered, oxidation damage could be significantly greater than creep damage. By performing experiments in controlled atmospheres, the mechanisms of oxide failure can be isolated and more easily interpreted.

Creep Damage and Creep Fatigue Models. A summary of creep laws is given in Table 6. The strain range partitioning method (Ref 50, 62) separates the inelastic strain range into four generic components (pp = plastic-plastic, pc = plastic-creep, cp = creep-plastic, cc = creep-creep). Plastic-plastic stands for plasticity in tension reversed by plasticity in compression; plastic-creep stands for plasticity in tension reversed by creep in compression; etc. Recently, the method has been modified to handle time effects on the strain components, and a total strain range version has also been proposed. The application of the model to TMF has been outlined where stress hold experiments conducted at various points around the stress-strain hysteresis loop are needed (Ref 50).

The time-cycle fraction rule (adopted as an ASME Code, Ref 204) involves linear summation of fatigue and creep damage, where the fatigue damage is expressed as cycle ratio and the creep damage is written as a time ratio. Creep damage is determined from stress-rupture diagrams, and fatigue damage is obtained from the strain-life equation. Since experimental results indicate that predictions based on the time-cycle fraction rule can be nonconservative, a modified time-cycle fraction rule has been proposed by

Table 6 Summary of creep-fatigue laws

Ref	Experiments/mechanism	Equation
58, 64	Isothermal: creep damage in tension versus compression	$\dfrac{1}{N_{pred}} = \dfrac{F_{pp}}{N_{pp}} + \dfrac{F_{pc}}{N_{pc}} + \dfrac{F_{cp}}{N_{pp}} + \dfrac{F_{cc}}{N_{cc}}$ N_{pred} = predicted life
204	Isothermal: time-cycle fraction rule, Miner-Robinson Rule	$D = \Sigma\dfrac{N}{N_f} + \Sigma\dfrac{t}{t_f}$ N_f = cycles to failure, t_f = time to failure
205	Isothermal: nonlinear damage	$D_f = 1 - (1 - N/N_f)^{P+1}$ $D_c = 1 - (1 - N/N_c)^{q+1}$
206	Isothermal	$D = \Sigma\dfrac{N}{N_f} + \Sigma\dfrac{\dot\varepsilon}{\varepsilon_f}$ $\dot\varepsilon$ = creep strain rate, ε_f = creep ductility p,q = constants, N_f = cycles to failure
207	Isothermal: void growth at crack tips	$\dfrac{da}{dt} = \alpha\left(\dfrac{K^2}{t}\right)^{\beta/n+1}$ β = void growth exponent, n = stress exponent K = stress intensity, t = time, da/dt = crack growth rate
83, 191	Isothermal: void growth in tension, void healing in compression	$\dfrac{dD_c}{dt} = \left(\dfrac{\varepsilon^m c}{\hat\varepsilon_c}\right)\left\|\dot\varepsilon^{in}\right\|$ $\dfrac{dD_f}{dt} = (1 + \alpha D_c)(\dot\varepsilon_{f0})^{-(m+1)}\varepsilon^m\left\|\dot\varepsilon^{in}\right\|$ $\dot\varepsilon^{in}$ = creep strain rate, $\dot\varepsilon_{f0}$ = creep ductility $\alpha,m,m_c,\dot\varepsilon_{f0},\hat\varepsilon_c$ = constants, D_c = creep damage, D_f = fatigue damage
208	Isothermal: decrease in crack closure load	$da/dN = C(U\Delta K)^m$ $U = U_0 + (1 - 0.5/F(S_{max}/\sigma_0)\{1 - \exp(-6\times 10^3(S_{max}/\sigma_0)^m\frac{t_h\cdot\varepsilon_0}{(\sigma_c/\sigma_0)^m})\})^2$ U_0 = time independent effective stress ratio, σ_0 = yield strength, σ_c = creep strength, m = creep exponent S_{max} = maximum stress, t_h = hold period in tension
68, 69	Thermomechanical: intergranular, cracking	$D^{creep} = \phi^{creep}\displaystyle\int_0^{t_c} A e^{(-\Delta H/RT)}\left(\dfrac{\alpha_1\bar\sigma + \alpha_2\sigma_H}{K}\right)^m dt$ (Terms defined in the text)
118	TMF life model	$N_f = C_0(\Delta\varepsilon)^{C_1}(t_h)^{C_2}\exp\left(\dfrac{C_3}{A}\right)$
209	Microcrack propagation in TMF	$\dfrac{da}{dN} = C_f(\Delta J)^{m_f} + C_c\hat c^{m_c} + C_0(\Delta J)^{m_0}(\Delta t)^\Psi$ (Summation of crack growth from fatigue, creep, and oxidation contributions)

Table 7 Nonunified plasticity model

Elastic strain rate	$\dot\varepsilon^e_{ij} = ((1-\nu)\dot\sigma_{ij} - \nu\dot\sigma_{kk}\delta_{ij})/E - ((1-\nu)\sigma_{ij} - \nu\sigma_{kk}\delta_{ij})\dfrac{\partial E}{\partial T}\cdot\dfrac{\dot T}{E^2}$
Thermal strain rate	$\dot\varepsilon^{th}_{ij} = \theta\dot T\delta_{ij}$
Yield criteria	$f = \dfrac{1}{2}\left(S_{ij} - S^c_{ij}\right)\left(S_{ij} - S^c_{ij}\right) - k^2 = 0$
Plastic strain rate	$\dot\varepsilon^p_{ij} = H\dfrac{\partial f}{\partial S_{ij}}\left[\dfrac{\partial f}{\partial S_{kl}}\dot S_{kl} + \dfrac{\partial f}{\partial T}\dot T\right]$
Creep strain rate	$\dot\varepsilon^c_{ij} = B(\bar\sigma)^{S-1}S_{ij}\exp(-\Delta H/RT)$
Plastic modulus	$H = A' J_2^{N/2}/(1 \pm \beta(\exp - p/p_0))$
Accumulated plastic strain	$p = \displaystyle\int_t\left(\dfrac{2}{3}\dot\varepsilon^p_{ij}\dot\varepsilon^p_{ij}\right)^{1/2}dt$

Source: Ref 81

Lemaitre and Plumtree (Ref 205) with applications involving cumulative damage. Recognizing that creep ductility is a function of strain rate, a ductility exhaustion approach has also been proposed which defines creep damage as strain rate to ductility ratio (Ref 206).

Several researchers considered creep-fatigue to be a crack propagation-controlled problem where the creep mechanism is assumed to influence the fatigue crack growth or vice versa. Majumdar and Maiya (Ref 83) considered the influence of creep cavity growth ahead of a crack growing by fatigue in their damage-rate equations. In their model, sintering of cavities occurs in compression, effectively reversing the creep damage occurring in tension. The model is suitable for materials that exhibit copious cavitation. These damage-rate equations have recently been applied to thermomechanical fatigue loadings (Ref 191).

The model of Saxena and Bassani (Ref 207) accounts for cavity formation ahead of a crack tip and modified the crack tip stress fields; therefore, modified fracture mechanics parameters have been used to handle fatigue-creep crack growth. In the absence of cavities, the crack tip stress-strain fields are still modified; this is the basis for a modified crack closure model proposed by Sehitoglu and Sun (Ref 208). The model by Neu and Sehitoglu (Ref 68, 69) incorporates effective stress and hydrostatic stress components and a creep phasing factor which depends on the thermomechanical history.

By performing tests in a helium environment, oxidation damage is eliminated leaving only the fatigue-creep damage. This allows the constants in the fatigue-creep damage term to be formulated directly.

In the Neu-Sehitoglu model (Ref 68, 69), to characterize the oxidation rate at different partial pressures of oxygen, the $(PO_2)^{1/q}$ term should be placed on the right-hand side of the growth law where q is a constant and PO_2 is the partial pressure of oxygen in the testing environment. In general, K_P will not be constant for a cycle which undergoes a varying temperature history. Therefore, an effective oxidation constant, K_P^{eff} is defined. The growth of voids and intergranular cracks occur predominantly under tensile loading. Consequently, to take into account the load asymmetry, the creep damage term is a function of effective and hydrostatic stress components (see Ref 68, 69, and 105 for details). The constitutive model proposed by Slavik and Sehitoglu (Ref 80), which utilizes two state variables, was used in the simulations, but other constitutive models can also be used. In Table 6 we did not show some of the commonly used life prediction models including the $\sigma_{max}\Delta\varepsilon^{in}/2$, a quantity related to tensile hysteresis energy, and the ΔW_p, plastic work or the total hysteresis energy parameters. As pointed out by Halford (Ref 18) the use of these parameters in TMF places rather unrealistic restrictions on the flow (σ–ε) and failure behavior of existing structural materials.

Prediction of TMF OP lives for 1070 steel are shown in Fig. 31. The minimum temperature is maintained at 150 °C (300 °F) in these experiments and the mechanical strain increases proportionally with maximum temperature with $\dot\varepsilon_{th}/\dot\varepsilon_{mech} = -1/2$ in the first case and $\dot\varepsilon_{th}/\dot\varepsilon_{mech} = -2$ in the second case. The predictions with the model are shown as solid lines. The $\dot\varepsilon_{th}/\dot\varepsilon_{mech} = -2$ is termed *partial constraint* and the case $\dot\varepsilon_{th}/\dot\varepsilon_{mech} = -1/2$ is termed *overconstraint* (Ref 24, 68, 69).

Constitutive Equations Suitable for TMF

Recent research on high-temperature deformation has produced a considerable number of constitutive models. Two classes of constitutive models suitable for thermomechanical loading have been reviewed by Slavik and Sehitoglu (Ref 81).

Nonunified Creep Plasticity Models and Their Use in Thermal Loading. In the first class

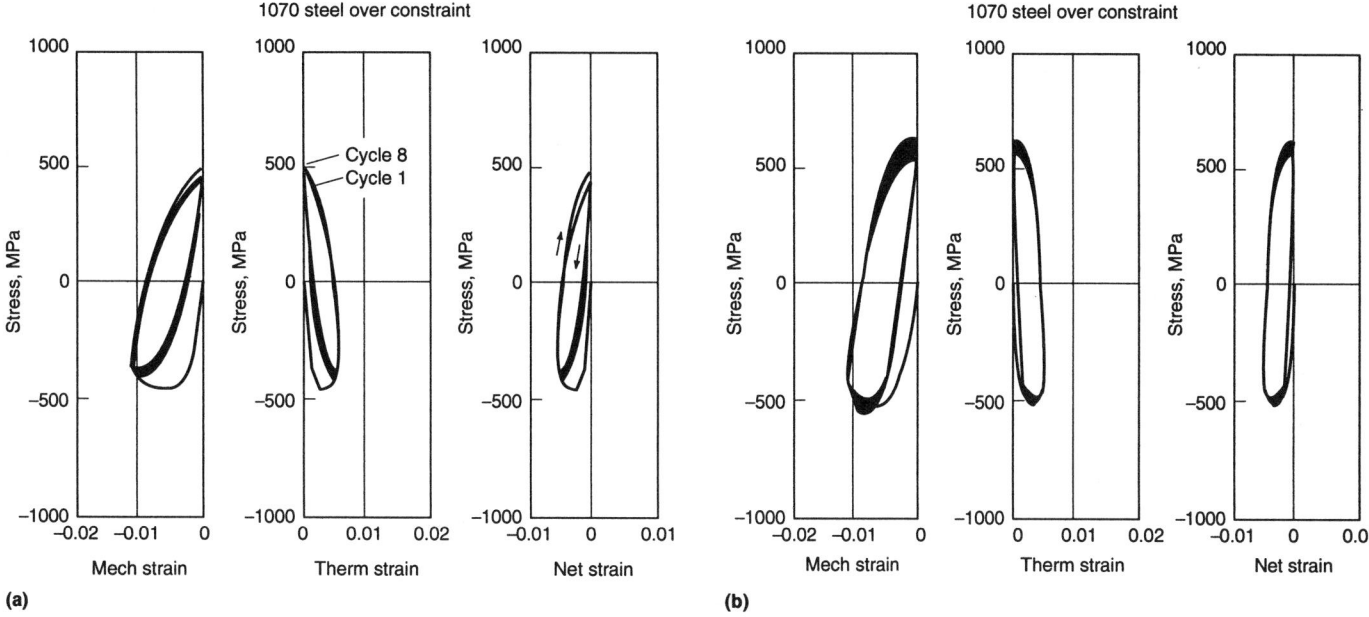

Fig. 32 Thermomechanical fatigue (out-of-phase) stress strain of 1070 steel. (a) Experimental. (b) Prediction using nonunified equations. Source: Ref 81

of models a time-dependent creep strain is added to the plastic strain resulting in the so-called nonunified models. The plastic strain can be described by the classical von Mises yield criteria and Prager or Ziegler rules. In the paper by Slavik and Sehitoglu (Ref 81), the Drucker-Palgen model (Ref 210) was used. The different strain rate components and other important parameters are given in Table 7.

In Table 7 the first row gives the elastic strain rate of an isotopic material where the elastic modulus is a function of temperature. The second row is the thermal strain rate for an isotropic material. The yield criteria is the von Mises type where k is the yield stress in shear and the center of the yield surface is permitted to move in stress space. The equations for this translation are given in Ref 81. The plastic strain rate is normal to the yield surface and also changes with temperature as shown in Table 7. The creep strain rate is expressed as a function of equivalent stress and the creep rate is in the same direction as the deviatoric stress rate. The plastic modulus, the slope of the stress-plastic strain rate, varies with plastic strain history such that it exponentially approaches a steady state value. The plastic strain history is described by the accumulated plastic strain which is a scalar positive quantity.

The advantages of this model are the following: (1) it can use the existing database of plastic and creep properties, (2) it conforms to existing FEM codes, and (3) it does not require special integration schemes. Experimental results and predictions of TMF OP behavior of 1070 steel with the above equations are given in Fig. 32(a) and (b) (Ref 81), respectively. In this case, 150 °C $\rightleftarrows$ 450 °C cycling under 'overconstraint' ($\dot{\varepsilon}_{mech} = -2, \dot{\varepsilon}_{th}$) is shown. "Overconstraint" means that the mechanical strain is larger than the thermal strain because during heating the material is also

Table 8 Two unified models used for TMF $\sigma - \varepsilon$ prediction

Ref	Flow rule	Back stress	Drag stress
80	$\dot{\varepsilon}_{ij}^{in} = \dfrac{3}{2} f(\bar{\sigma}/K) \dfrac{S_{ij} - S_{ij}^c}{\bar{\sigma}}$ $f\left(\dfrac{\bar{\sigma}}{K}\right) = \begin{cases} A\left(\dfrac{\bar{\sigma}}{K}\right)^n & \dfrac{\bar{\sigma}}{K} \leq 1 \\ A\exp\left[\left(\dfrac{\bar{\sigma}}{K}\right)^m - 1\right] & \dfrac{\bar{\sigma}}{K} > 1 \end{cases}$	$\dot{S}_{ij}^c = \dfrac{2}{3} h_\alpha \dot{\varepsilon}_{ij}^{in} - r_\alpha S_{ij}^c$ $h_k = B(k_{sat} - K)\,\dot{\bar{\varepsilon}}^{in.}$ $r_k = C(K - K_{rec})$	$\dot{K} = h_k - r_k + \Theta\dot{T}$
211, 212	$\dot{\varepsilon}_{ij}^{in} = f\left(\dfrac{K^2}{J_2}\right)S_{ij}$ $f\left(\dfrac{K^2}{J_2}\right) = \dfrac{D_0^2}{J_2}\exp\left[-\left(\dfrac{n+1}{n}\right)\left[\dfrac{K^2}{3J_2}\right]^n\right]^{0.5}$		$\dot{K} = m(K_1 - K) \cdot \dot{W}^{in.} - A\left(\dfrac{K - K_i}{K_1^*}\right)^r \exp\left(\dfrac{-\Delta H}{RT}\right)$

Table 9 Constants for the unified model for selected materials

Material	E (MPa)	θ (1/°C)	A (sec^{-1})	n	m
1070 steel (Ref 80)	$202{,}250 - 31.0\,T$ $T \leq 440$ °C $309{,}990 - 275.7\,T$ $T > 440$ °C	1.7×10^{-5}	$4.0 \times 10^9 \exp[-25{,}300/(T + 273)]$	5.4	8.3
René 80 (Ref 215)	$192{,}170 - 60.7\,T$ $T < 871$ °C $310{,}990 - 197.1\,T$ $T \geq 871$ °C	1.6×10^{-5}(a)	$2.33 \times 10^{-10}, T \leq 650$ °C $8.14 \times 10^{15} \exp[-54{,}288/(T + 273)], T > 650$ °C	8.06	5.69
Mar-M247 (Ref 105)	$253{,}900 - 107.8\,T$	1.6×10^{-5}	$1.33 \times 10^{23} \exp[-64{,}515/(T + 273)]$	11.6	17.5
Al2xxx-T4 (Ref 96, 164)	$72{,}750 - 50\,T$ $T < 150$ °C $82{,}000 - 90\,T$ $T \geq 150$ °C	3.0×10^{-5}	$9.8 \times 10^{11} \exp[-18{,}722/(T + 273)]$	4.6	10.1
Ti-β21S (Ref 214)	$114{,}000 - 42.3\,T$ $T \leq 483$ °C $151{,}500 - 120\,T$ $T > 483$ °C	$8.24 \times 10^{-6} + 3.64$ $\times 10^{-9}\,T$	$7.7 \times 10^5 \exp[-18{,}300(T + 273)]$	2.7	8.2

(a) Not determined from experiments

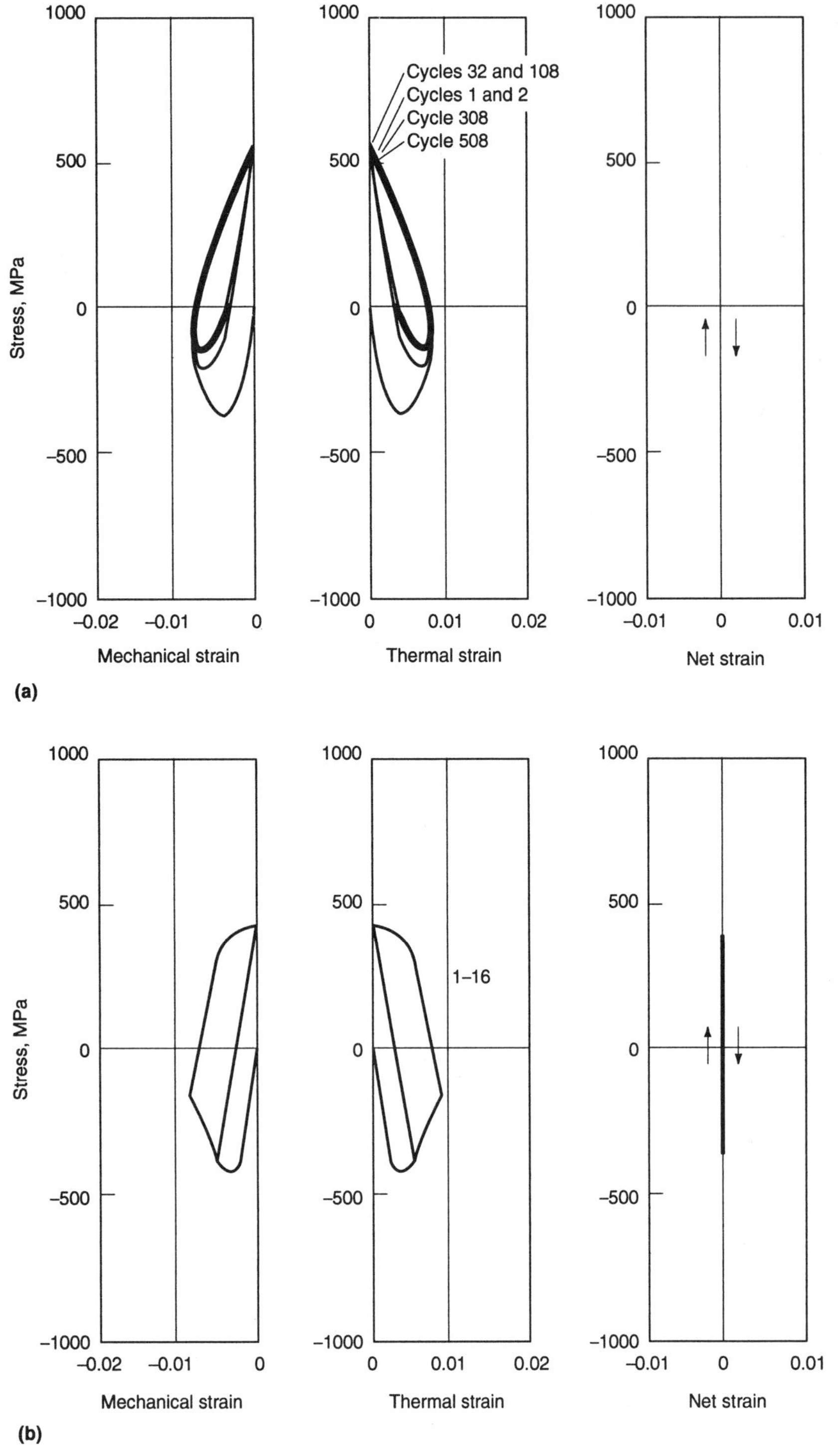

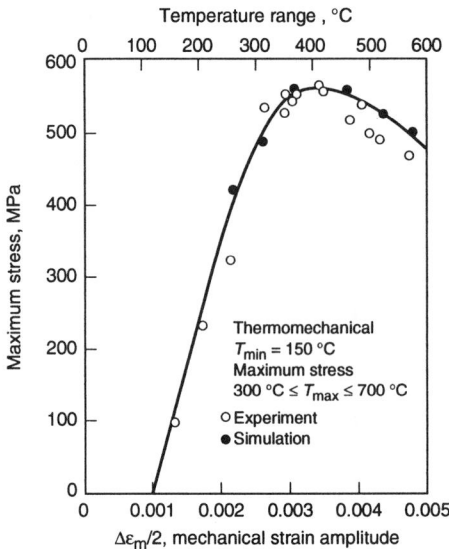

Fig. 34 Prediction of maximum stress in the TMF OP case in 1070 steel. Source: Ref 80

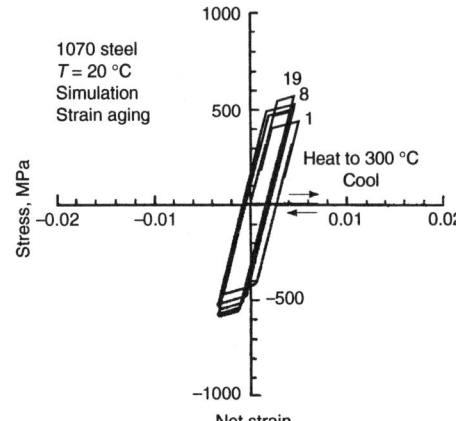

Fig. 35 Prediction of 1070 steel response under strain aging conditions. Source: Ref 80

(a)

(b)

Fig. 33 Thermomechanical fatigue (out-of-phase) stress strain of 1070 steel. (a) Experimental. (b) Prediction using Bodner's model. Source: Ref 81

subjected to a negative net strain as shown in Fig. 32(a). The prediction of the stress-strain response with the nonunified model is satisfactory (Fig. 32b).

Unified Creep Plasticity Models and Their Use in Thermal Loading. These models, termed *unified models*, have the potential to predict creep-plasticity interactions and strain rate effects more accurately than the nonunified models. A unified model that has been widely used was proposed by Bodner and Partom (Ref 211) and later modified for cyclic loading (Ref 212). The simulations of TMF OP with this model are given in Fig. 33 (Ref 81). Comparison of experimental results shown in Fig. 33(a) and predictions shown in Fig. 33(b) prove that the model is satisfactory. In Fig. 33(a) the stress-net strain response is shown under 150 °C $\rightleftarrows$ 600 °C and total con-

straint ($\dot{\varepsilon}_{mech} = -\dot{\varepsilon}_{th}$) conditions. In this case, the net strain is zero and the mechanical strain is equal but opposite of thermal strain.

In the unified models, there is no yield surface assumption. Inelastic strain rates are permitted at small levels of effective stress which is usually the deviatoric stress-deviatoric backstress. The unified models are often composed of two state variables, the deviatoric back stress, S_{ij}^c and the drag stress, K. The back stress can be used to predict the Bauschinger effect in room-temperature loading and also the transient and steady state creep response at high temperatures. The drag stress, K, accounts for cyclic hardening or softening, and the influence of plasticity on creep or vice versa. The strain rate sensitivity is determined by the flow rule. The general form of these unified relations is given in Table 8. We note that Bodner's original form did not include a back stress.

The first equation in Table 8 is the flow rule where $\dot{\varepsilon}_{ij}^{in}$ is the inelastic strain rate (which is the

Table 10 Constants for the unified model for the drag stress term

Material	K_0 (MPa)	K_{sat} (MPa)	B	C	K_{rec} (MPa)
1070 steel	$262.7 - 0.04T$, $T \leq 440\,°C$ $403.0 - 0.36T$, $T > 440\,°C$	$256.0 + 1.4 \times 10^{-3}T^2$, $T \leq 304\,°C$ $568.0 - 0.6T$, $T > 304\,°C$	5.0	0, $T < 300\,°C$ $10^8 \exp[-20,000/(T+273)]$, $T \geq 300\,°C$	$548 - 0.62\,T$
René 80	384.0, $T \leq 60\,°C$ $2.66 \times 10^{-3}\,{*}E$, $T > 760\,°C$	NA	0	0	NA
Mar-M247	N/A	$886.1 - 0.376\,T$	NA(a)	0	NA
Al2xxx – T4	$226 - 0.15T$ $T < 150\,°C$ $256 - 0.28T$ $T \leq 150\,°C$	$620 - 1.66T$, $T < 150\,°C$ $420 - 0.30T$, $T \geq 150\,°C$	5	$4.9 \times 10 - 10 + 4.0 \times 10^{-6}T$ $- 3.2 \times 10^{-8}\,T^2 + 1.2 \times 10^{-10}$ $T^3 - 2.3 \times 10^{-13}\,T^4 + 1.7 \times$ $10^{-16}\,T^5$	$260 - 0.8T$, $T < 150\,°C$ 20 $T \geq 150\,°C$
Ti-β21S	$730 - 0.271T$ $T \leq 483\,°C$ $970 - 0.768T$ $T > 483\,°C$	$506.5 - 0.291T$	5(b)	0	NA

(a) $K = K_{sat}$ in this study. Constants were not determined for transient behavior. (b) Not determined from experiments

Table 11 Constants for the unified model for the back stress evolution term

Material	h_α	b	r_α
1070 steel	$b(17,200 - 20\bar{\alpha})$ $\alpha_{ij}\,\dot{\varepsilon}_{ij}^{in} \geq 0$ $b(17,200)$ $\alpha_{ij}\,\dot{\varepsilon}_{ij}^{in} < 0$	$0.89 + 2.58 \times 10^{-3}T + 1.53 \times 10^{-5}T^2 - 6.4$ $\times 10^{-8}T^3 + 5.15 \times 10^{-11}T^4$	$128.6 \exp[-19,460/(T+273)]\,(\bar{\alpha})^{3.3}$
René 80	$(2.096 \times 10^{-5} + 1.50 \times 10^{-15}\,|\bar{\alpha}|^b)^{-1}$	$5.02 + 2.25 \times 10^{-4}T$, $T \leq 760\,°C - 109.58$ $+ 0.414T - 4.99 \times 10^{-4}T^2 + 2.005 \times 10^{-7}T^3$, $T > 760\,°C$	0
Mar-M247	Not determined	Not determined	Not determined
Al2xxx-T4	$b(20,000 - 1000\bar{\alpha})$ $\alpha_{ij}\,\dot{\varepsilon}_{ij}^{in} \geq 0$ $b(20,000)$ $\alpha_{ij}\,\dot{\varepsilon}_{ij}^{in} < 0$	$1.013 - 8.8 \times 10^{-4}T - 2.7 \times 10^{-6}T^2$	0
Ti-β21S	$b(14,500 - 422\,\bar{\alpha} + 3.0\,\bar{\alpha}^2)$ $\alpha_{ij}\,\dot{\varepsilon}_{ij}^{in} \geq 0$ $b(14,500)$ $\alpha_{ij}\,\dot{\varepsilon}_{ij}^{in} < 0$	$1 + 0.0125\,(T-23) - 3.38 \times 10^{-5}\,(T-23)^2 +$ $2.09 \times 10^{-8}\,(T-23)^3$	0

combination of plastic and creep strains), $\bar{\sigma}$ is the effective stress, S_{ij} is the deviatoric stress, S_{ij}^c is the deviatoric back stress, and K is the drag stress. The second equation describes the evolution of back (internal) stress where h_α is the hardening function for the back stress and r_α is the recovery function for the back stress. The third equation depicts the evolution of drag stress where h_k is the hardening function for the drag stress, r_k is the recovery function for the drag stress, and Θ is the thermally activated drag stress change term (defined as $\partial K_0/\partial T$). The term that distinguishes different models is the choice of the flow rule, $f(\sigma/K)$ and the manner in which the hardening and recovery functions are determined. Different deformation mechanisms (plasticity, power law creep, diffusional creep) have been identified in deformation mechanism maps (Ref 47) but have not been explicitly considered. Sehitoglu and Slavik confirmed that at high strain rates the strain rate effective stress relation has the form of the exponential function (i.e., rate insensitive behavior) while at lower strain rates the relation between inelastic strain rate and effective stress is consistent with the power law creep relation. An experimentally based unified constitutive model has been proposed earlier by Sehitoglu and Slavik (Ref 80, 81). A list of functions and corresponding experiments to determine these constants follow:

Function	Experimental determination
Flow rule, f	High and low strain rates, yield strength measurements in tensile or compressive monotonic tests
Hardening of back stress	High strain rate (room and high temperature)
Recovery of back stress	Low strain rate or creep (high temperature)
Hardening of drag stress	High strain rate cycling (room and high temperature)
Recovery of drag stress Θ	Rest periods (high temperature) Change of K_O with temperature

The material constants for the Slavik-Sehitoglu model for different class of materials are listed in Tables 9 to 11. These are 1070 steel (Ref 80), Mar-M247 (Ref 105), René 80 (Ref 213), Al2xxx-T4 (Ref 96, 164) and Ti-β21S (Ref 214). These tables are also summarized in Ref 215 and explained in Ref 80.

Predictions of material response utilizing this relation are given in Fig. 34 (Ref 80). The experimental results were shown earlier in Fig. 14. In these experiments (total constraint TMF OP), the minimum temperature was constant at 150 °C (300 °F). Because of thermal recovery effects, the drag stress decreases upon exposure to high temperatures and this results in loss of strength at the minimum temperature. Note that the model (solid rectangular points) predicts the decrease in maximum stress with increasing mechanical strain

very accurately. Because of the thermal recovery effects, the maximum stress in the cycle decreases with increasing mechanical strain range. The agreement between experiment and prediction is excellent. It is possible to incorporate additional terms to predict such phenomena such as strain aging and dynamic recovery into the constitutive models. Other TMF and IF simulations with this unified model can be found in Ref 80, 96, 105, 164, and 213. Simulations for the case of strain aging for the 1070 steel have been demonstrated in Ref 80 and are further discussed below.

Certain temperature-strain histories in solute-hardened materials produce strain aging, hence strengthening of the material. Thermomechanical fatigue studies (under strain aging conditions for steels and nickel-base superalloys) have been discussed in Ref 60 and 215. Strengthening is caused by interstitial solute atoms which anchor the dislocation motion. If the pinning of the dislocations occurs during deformation, the term 'dynamic strain aging' is used (Ref 87). If the aging occurs under a constant load (after some plastic deformation), it is called 'static strain aging' (Ref 87).

Static strain aging experiments were conducted on 1070 steel. The material is cycled at 20 °C but is exposed to the aging temperature time at zero stress every reversal. A schematic of the strain-temperature history and the results for 1070 steel was shown earlier in Fig. 15. The predictions using the unified formulation are given in Fig. 35. The prediction of the strengthening under these conditions is very favorable. The details can be found in Ref 80 and 215.

Finally, we note that other unified models have been tested under TMF loadings but not as extensively as the Slavik and Sehitoglu model. For example, Miller (Ref 216) proposed a unified model which he checked against experimental TMF IP data on PWA 663. His predictions were satisfactory. Walker (Ref 217) proposed a model for high-temperature/time-dependent loadings that was also applied to predict TMF behavior.

REFERENCES

1. H.J. Schrader, The Friction of Railway Brake Shoes at High Speed and High Pressure, *Eng. Exp. Station Bull.*, Univ. Ill. Bull., Vol 35 (No. 72), May 1938

2. H.R. Wetenkamp, O.M. Sidebottom, and H.J. Schrader, The Effect of Brake Shoe Action on Thermal Cracking and on Failure of Wrought Steel Railway Car Wheels, *Eng. Exp. Station Bull., Univ. Ill. Bull.*, Vol 47 (No. 77), June 1950

3. W. Boas and R.W.K. Honeycombe, The Deformation of Tin-Base Bearing Alloys by Heating and Cooling, *Inst. Met. J.*, Vol 73, 1946-1947, p 33-444

4. L.F. Coffin, Jr., A Study of the Effects of Cyclic Thermal Stresses on a Ductile Metal, *Trans. ASME*, Vol 76 (No. 6), 1954, p 931-950

5. S. Manson, Behavior of Materials under Conditions of Thermal Stress, *Heat Transfer*

Symp., Univ. Mich. Eng. Res. Inst., Vol 27-38, 1953: see also NACA TN-2933, 1953

6. S.S. Manson, *Thermal Stress and Low-Cycle Fatigue*, McGraw-Hill, 1966

7. *Thermal and High-Strain Fatigue*, Monograph and Report Series No. 32, Institute of Metals, London, 1967

8. D.J. Littler, Ed., *Thermal Stresses and Thermal Fatigue*, Butterworths, London, 1971

9. R.P. Skelton, *Fatigue at High Temperature*, Applied Science Publishers, London, 1983

10. R.P. Skelton, *High Temperature Fatigue: Properties and Prediction*, Applied Science Publishers, London, 1983

11. A. Weronski and T. Hejwoski, *Thermal Fatigue of Metals*, Marcel Dekker, 1991

12. A.E. Carden, A.J. McEvily, and C.H. Wells, Ed., *Fatigue at Elevated Temperatures*, STP 520, ASTM, 1973

13. D.A. Spera and D.F. Mowbray, Ed., *Thermal Fatigue of Materials and Components*, STP 612, ASTM, 1976

14. H. Solomon, G. Halford, L. Kaisand, and B. Leis, Ed., *Low Cycle Fatigue*, STP 942, ASTM, 1988

15. H. Sehitoglu, Ed., *Thermo-Mechanical Fatigue Behavior of Materials*, STP 1186, ASTM, 1991

16. H. Sehitoglu and S.Y. Zamrik, Ed., *Thermal Stress, Material Deformation and Thermo-Mechanical Fatigue*, American Society of Mechanical Engineers, 1987

17. J. Bressers and L. Rémy, Ed., *Symp. Fatigue under Thermal and Mechanical Loading*, Kluwer Academic Publishers, May 1995

18. G. Halford, Low-Cycle Thermal Fatigue, TM 87225, National Aeronautics and Space Administration, 1986; see also Low-Cycle Thermal Fatigue, *Thermal Stress II*, R.B. Hetnarski, Ed., Elsevier, 1987

19. D.A. Miller and R.H. Priest, Material Response to Thermal-Mechanical Strain Cycling, *High Temperature Fatigue: Properties and Prediction*, R.P. Skelton, Ed., Applied Science Publishers, 1983, p 113-176

20. H. Sehitoglu, *Thermo-Mechanical Fatigue Life Prediction Methods*, STP 1122, ASTM, 1992, p 47-76

21. H.G. Baron, Thermal Shock and Thermal Fatigue, *Thermal Stress*, P.P. Benham and R.D. Hoyle, Ed., Pitman, London, 1964, p 182-206

22. D.R. Miller, Thermal-Stress Ratchet Mechanism in Pressure Vessels, *J. Basic Eng. (Trans. ASME)*, Vol 81, No. 2, 1959, p 190-196

23. J. Bree, Elastic-Plastic Behavior of Thin Tubes Subjected to Internal Pressure and Intermittent Heat Fluxes with Application to Fast-Nuclear-Reactor Fuel Elements, *J. Strain Anal.*, Vol 2 (No. 3), 1967, p 226-238

24. H. Sehitoglu, Constraint Effect in Thermo-Mechanical Fatigue, *J. Eng. Mater. Technol. (Trans. ASME)*, Vol 107, 1985, p 221-226

25. L.F. Coffin, Jr. and R.P. Wesley, Apparatus for Study of Effects of Cyclic Thermal Stresses on Ductile Metals, *Trans. ASME*, Vol 76 (No. 6), Aug 1954, p 923-930

26. A. Carden, Ed., *Thermal Fatigue Evaluation*, STP 465, ASTM, 1970, p 163-188

27. H.J. Westwood, High Temperature Fatigue of 304 Stainless Steel under Isothermal and Thermal Cycling Conditions, *Fracture 77: Advances in Research on the Strength and Fracture of Materials*, D.M.R. Taplin, Ed., Pergamon Press, 1978, p 755-765

28. K.D. Sheffler, Vacuum Thermal-Mechanical Fatigue Testing of Two Iron Base High Temperature Alloys, *Thermal Fatigue of Materials and Components*, STP 612, D.A. Spera and D.F. Mowbray, Ed., 1976, p 214-226

29. M. Kawamoto, T. Tanaka, and H. Nakajima, Effect of Several Factors on Thermal Fatigue, *J. Mater.*, Vol 1 (No. 4), 1966, p 719-758

30. G. Halford, M.A. McGaw, R.C. Bill, and P. Fanti, Bithermal Fatigue: A Link between Isothermal and Thermomechanical Fatigue, *Low Cycle Fatigue*, STP 942, H. Solomon, G. Halford, L. Kaisand, and B. Leis, Ed., ASTM, 1988

31. T. Udoguchi and T. Wada, Thermal Effect on Low-Cycle Fatigue Strength of Steels, *Thermal Stresses and Thermal Fatigue*, D.J. Littler, Ed., Butterworths, London, 1971, p 109-123

32. M.G. Castelli and J.R. Ellis, Improved Techniques for Thermo-Mechanical Testing in Support of Deformation Modeling, *Thermo-Mechanical Fatigue Behavior of Materials*, STP 1186, H. Sehitoglu, Ed., ASTM, 1991, p 195-211

33. M. Karasek, H. Sehitoglu, and D. Slavik, Deformation and Damage under Thermal Loading, *Low Cycle Fatigue*, STP 942, H. Solomon, G. Halford, L. Kaisand, and B. Leis, Ed., ASTM, 1988, p 184-205

34. T.R. Simovich, "A Study of the Thermal Fatigue Characteristics of Several Plain Carbon Steels," M.S. thesis, University of Illinois, 1967

35. S. Taira, Relationship between Thermal Fatigue and Low-Cycle Fatigue at Elevated Temperature, *Fatigue at Elevated Temperatures*, STP 520, A.E. Carden, A.J. McEvily, and C.H. Wells, Ed., ASTM, 1973, p 80-101

36. S. Taira, M. Fujino, and R. Ohtani, Collaborative Study on Thermal Fatigue of Properties of High Temperature Alloys in Japan, *Fatigue Eng. Mater. Struct.*, Vol 1, 1979, p 495-508

37. S. Taira, M. Fujino, and S. Maruyama, Effects of Temperature and the Phase between Temperature and Strain on Crack Propagation in a Low Carbon Steel during Thermal Fatigue, *Mechanical Behavior of Materials*, Society of Materials Science, Kyoto, 1974, p 515-524

38. H.G. Baron and B.S. Bloomfield, Resistance to Thermal Stress Fatigue of Some Steels, Heat Resisting Alloys and Cast Irons, *J. Iron Steel Inst.*, Vol 197, 1961, p 223-232

39. D.R. Adolphson, "Stresses Developed in Uniaxially Restrained Railway Car Wheel Material When Subjected to Thermal Cycles," M.S. thesis, University of Illinois, 1957

40. K. Kuwabara and A. Nitta, Thermal-Mechanical Low Cycle Fatigue under Creep-Fatigue

Interaction on Type 304 Stainless Steel, *Proc. ICM 3*, Vol 2, 1979, p 69-78

41. K. Kuwabara and A. Nitta, Effect of Strain Hold Time of High Temperature on Thermal Fatigue of Type 304 Stainless Steel, *ASME-MPC Symp. Creep-Fatigue Interaction*, American Society of Mechanical Engineers, 1976, p 161-177

42. K. Kuwabara, A. Nitta, and T. Kitamura, Thermal Mechanical Fatigue Life Prediction in High Temperature Component Materials for Power Plant, *ASME Int. Conf. Advances in Life Prediction*, D.A. Woodford and J.R. Whitehead, Ed., American Society of Mechanical Engineers, 1983, p 131-141

43. K. Kuwabara and A. Nitta, "Isothermal and Thermal Fatigue Strength of Cr-Mo-V Steel for Turbine Rotors," Report E277005, Central Research Institute of Electric Power Industry, Tokyo, 1977

44. A. Nitta, K. Kuwabara, and T. Kitamura, "Creep-Fatigue Damage in Power Plant Materials, Part 2: The Behavior of Elevated Temperature Low Cycle Fatigue Crack Initiation and Propagation in Steam Turbine Rotor Steels," Report E282002, Central Research Institute of Electric Power Industry, Tokyo, 1982

45. H. Sehitoglu and M. Karasek, Observations of Material Behavior under Isothermal and Thermo-Mechanical Loading, *J. Eng. Mater. Technol. (Trans. ASME)*, Vol 108 (No. 2), 1986, p 192-198

46. G. Hartman, III, A Thermal Control System for Thermal Cycling, *J. Test. Eval.*, Vol 13 (No. 5), 1985, p 363-366

47. A. Koster, E. Chataigner, G. Cailletaud, and L. Rémy, Development of a Thermal Fatigue Facility to Simulate the Behavior of Superalloy Components, *Symp. Fatigue under Thermal and Mechanical Loading*, J. Bressers and L. Rémy, Ed., European Commission, Petten, May 1995

48. J. Guedou and Y. Honnorat, Thermo-Mechanical Fatigue of Turbo-Engine Blade Superalloys, *Thermo-Mechanical Fatigue Behavior of Materials*, STP 1186, H. Sehitoglu, Ed., ASTM, 1991, p 157-175

49. E. Glenny, J.E. Nortwood, S.W.K. Shaw, and T.A. Taylor, A Technique for Thermal-Shock and Thermal-Fatigue Testing, based on the Use of Fluidized Solids, *J. Inst. Met.*, Vol 87, 1958-59, p 294-302

50. D.A. Woodford and D.F. Mowbray, Effect of Material Characteristics and Test Variables on Thermal Fatigue of Cast Superalloys, *Mater. Sci. Eng.*, Vol 16, 1974, p 5-43

51. P.T. Bizon and D.A. Spera, Thermal-Stress Fatigue Behavior of Twenty-Six Superalloys, *Thermal Fatigue of Materials and Components*, STP 612, D.A. Spera and D.F. Mowbray, Ed., ASTM, 1976, p 106-122

52. M.A.H. Howes, Evaluation of Thermal Fatigue Resistance of Metals Using the Fluidized Bed Technique, *Fatigue at Elevated Temperatures*, STP 520, A.E. Carden, A.J. McEvily, and C.H. Wells, Ed., ASTM, 1973, p 242-254

53. F.K. Lampson, I.C. Isareff, Jr., and A.W.F. Green, Thermal Shock Testing under Stress of Certain High Temperature Alloys, *Proc. ASTM*, Vol 57, 1957, p 965-976

54. P.G. Forrest and K.B. Armstrong, The Thermal Fatigue Resistance of Nickel-Chromium Alloys, *Proc. Inst. Mech. Eng.*, 1963, p 3.1-3.7

55. D.F. Mowbray and J.E. McConnelee, Nonlinear Analysis of a Tapered Disk Specimen, *Thermal Fatigue of Materials and Components*, STP 612, D.A. Spera and D.F. Mowbray, Ed., ASTM, 1976, p 10-29

56. E. Glenny and T.A. Taylor, A Study of the Thermal Fatigue Behavior of Materials, *J. Inst. Met.*, Vol 88, 1959-60, p 449-461

57. L.F. Coffin, Instability Effects in Thermal Fatigue, *Thermal Fatigue of Materials and Components*, STP 612, D.A. Spera and D.F. Mowbray, Ed., ASTM, 1976, p 227-238

58. G.R. Halford and S.S. Manson, Life Prediction of Thermal-Mechanical Fatigue Using Strain Range Partitioning, *Thermal Fatigue of Materials and Components*, STP 612, D.A. Spera and D.F. Mowbray, Ed., ASTM, 1976, p 239-254

59. M. Karasek, H. Sehitoglu, and D. Slavik, Deformation and Damage under Thermal Loading, *Low Cycle Fatigue*, STP 942, H. Solomon, G. Halford, L. Kaisand, and B. Leis, Ed., ASTM, 1988, p 184-205

60. H. Sehitoglu, Crack Growth Studies under Selected Temperature-Strain Histories, *Eng. Fract. Mech.*, Vol 26 (No. 4), 1987, p 475-489

61. S. Taira and M. Ohnami, Fracture and Deformation of Metals Subjected to Thermal Cycling Combined with Mechanical Stress, *Joint Int. Conf. Creep*, Institution of Mechanical Engineers, 1963, p 57-62

62. S. Taira, M. Ohnami, and T. Kyogoku, Thermal Fatigue under Pulsating Thermal Stress Cycling, *Bull. Jpn. Soc. Mech. Eng.*, Vol 6, 1963, p 178-185

63. M. Fujino and S. Taira, Effect of Thermal Cycling on Low Cycle Fatigue Life of Steels and Grain Boundary Sliding Characteristics, *Proc. ICM 3*, Vol 2, 1979, p 49-58

64. S.S. Manson, G.R. Halford, and M.H. Hirschberg, Creep-Fatigue Analysis by Strain-Range Partitioning, *Design for Elevated Temperature Environment*, S.Y. Zamrik, Ed., American Society of Mechanical Engineers, 1971, p 12-28

65. J.J. Westwood and W.K. Lee, Creep-Fatigue Crack Initiation in 1/2 Cr-Mo-V Steel, *Proc. Int. Conf. Creep. Fract. Eng. Mater. Struct.*, Pineridge Press, 1981, p 517-530

66. C.E. Jaske, Thermal-Mechanical, Low-Cycle Fatigue of AISI 1010 Steel, *Thermal Fatigue of Materials and Components*, STP 612, D.A. Spera and D.F. Mowbray, Ed., ASTM, 1976, p 170-198

67. J.S. Laub, Some Thermal Fatigue Characteristics of Mild Steel for Heat Exchangers, *Thermal Fatigue of Materials and Components*, STP 612, D.A. Spera and D.F. Mowbray, Ed., ASTM, 1976, p 141-156

68. R. Neu and H. Sehitoglu, Thermo-Mechanical Fatigue Oxidation, Creep, Part I: Experiments, *Metall. Trans. A*, Vol 20A, 1989, p 1755-1767

69. R. Neu and H. Sehitoglu, Thermo-Mechanical Fatigue Oxidation, Creep, Part II: Life Prediction, *Metall. Trans. A*, Vol 20A, 1989, p 1769-1783

70. R. Zauter, F. Petry, H.-J. Christ, and H. Mughrabi, Thermo-Mechanical Fatigue of the Austenitic Stainless Steel AISI 304L, *Thermo-Mechanical Fatigue Behavior of Materials*, STP 1186, H. Sehitoglu, Ed., ASTM, 1993, p 70-90

71. L.F. Coffin, Jr., Cyclic Strain-Induced Oxidation of High Temperature Alloys, *Trans. ASM*, Vol 56, 1963, p 339-344

72. L.F. Coffin, Fr., The Effect of High Vacuum on the Low Cycle Fatigue Law, *Metall. Trans.*, Vol 3, July 1972, p 1777-1788

73. J.K. Tien and J.M. Davidson, Oxide Spallation Mechanisms, *Stress Effects and the Oxidation of Metals*, J.V. Cathcart, Ed., TMS-AIME, 1974, p 200-219

74. N.B. Pilling and R.E. Bedworth, The Oxidation of Metals at High Temperatures, *J. Inst. Met.*, Vol 29, 1923, p 529-582

75. M. Schutze, Deformation and Cracking Behavior of Protective Oxide Scales on Heat-Resistant Steels under Tensile Strain, *Oxid. Met.*, Vol 24 (No. 3/4), 1985, p 199-232

76. R.P. Skelton, Environmental Crack Growth in 0.5 Cr-Mo-V Steel During Isothermal High Strain Fatigue and Temperature Cycling, *Mater. Sci. Eng.*, Vol 35. 1978, p 287-298

77. E. Renner, H. Vehoff, H. Riedel, and P. Neumann, Creep Fatigue of Steels in Various Environments, *Low Cycle Fatigue and Elasto-Plastic Behavior of Materials*, 2nd Int. Conf. Low Cycle Fatigue and Elasto-Plastic Behavior of Materials (Munich), Elsevier, 1987, p 277-283

78. P.S. Maiya, Effects of Wave Shape and Ultrahigh Vacuum on Elevated Temperature Low Cycle Fatigue in Type 304 Stainless Steel, *Mater. Sci. Eng.*, Vol 47, 1981, p 13-21

79. C.R. Brinkman, High-Temperature Time-Dependent Fatigue Behavior of Several Engineering Structural Alloys, *Int. Met. Rev.*, Vol 30 (No. 5), 1985, p 235-258

80. D. Slavik and H. Sehitoglu, A Constitutive Model for High Temperature Loading, Part I: Experimentally Based Forms of the Equations; Part II: Comparison of Simulations with Experiments, *Thermal Stress, Material Deformation, and Thermo-Mechanical Fatigue*, H. Sehitoglu and S.Y. Zamrik, Ed., American Society of Mechanical Engineers, 1987, p 65-82

81. D. Slavik and H. Sehitoglu, Constitutive Models Suitable for Thermal Loading, *J. Eng. Mater. Technol. (Trans. ASME)*, Vol 108 (No. 4), 1986, p 303-312

82. H. Sehitoglu, Changes in State Variables at Elevated Temperatures, *J. Eng. Mater. Technol. (Trans. ASME)*, Vol 111, 1989, p 192-203

83. S. Majumdar, Thermo-Mechanical Fatigue of Type 304 Stainless Steel, *Thermal Stress, Material Deformation, and Thermo-Mechanical*

Fatigue, H. Sehitoglu and S.Y. Zamrik, Ed., American Society of Mechanical Engineers, 1987, p 31-36

84. K. Kuwabara and A. Nitta, Thermal-Mechanical Low-Cycle Fatigue under Creep-Fatigue Interaction on Type 304 Stainless Steels, *Fatigue Fract. Eng. Mater. Struct.*, Vol 2, 1979, p 293-304

85. D. Mikrut, Elevated Temperature Time Dependent Behavior of 0.7% Carbon Steels, M.S. thesis, University of Illinois, 1989

86. L. Nortcott and H.G. Baron, The Craze Cracking of Metals, *J. Iron Steel Inst.*, 1956, p 385-408

87. J.D. Baird, Strain Aging of Steel—A Critical Review, *Iron Steel*, Vol 36, 1963, p 186-192, 368-374, 400-405

88. S. Taira, Relationship between Thermal Fatigue and Low Cycle Fatigue at Elevated Temperature, *Fatigue at Elevated Temperatures*, STP 520, A.E. Carden, A.J. McEvily, and C.H. Wells, Ed., ASTM, 1973, p 80-101

89. S. Baik and R. Raj, Wedge Type Creep Damage in Low Cycle Fatigue, *Metall. Trans. A*, Vol 13A, 1982, p 1207-1214

90. M. Karayaka and H. Sehitoglu, Thermomechanical Fatigue of Particulate Reinforced Aluminum 2xxx-T4, *Metall. Trans. A*, Vol 22A, 1991, p 697-707

91. D.A. Meyn, The Nature of Fatigue Crack Propagation in Air and in Vacuum for 2024 Aluminum, *Trans. ASM*, Vol 61, 1968, p 52-61

92. C.Q. Bowles and J. Schijve, Crack Tip Geometry for Fatigue Cracks Grown in Air and in Vacuum, STP 811, *Fatigue Mechanisms: Advances in Quantitative Measurement of Physical Damage*, ASTM, 1983, p 400-426

93. M.J. Hordon, Fatigue Behavior of Aluminum in Vacuum, *Acta Metall.*, Vol 14, 1966, p 1173-1178

94. M. Karayaka and H. Sehitoglu, Thermomechanical Fatigue of Al-SiCp Composites, *Proc. 4th Conf. Fatigue and Fracture Thresholds*, Vol 3, Materials and Components Engineering Publications, 1990, p 1693-1698

95. M. Karayaka and H. Sehitoglu, Thermomechanical Fatigue of Particulate Reinforced Aluminum 2xxx-T4, *Metall. Trans. A*, Vol 22A, 1991, p 697-707

96. M. Karayaka and H. Sehitoglu, Thermo-mechanical Deformation Modeling of Al2xxx-T4 Composites, *Acta Metall.*, Vol 41 (No. 1), 1993, p 175-189

97. B. Flaig, K.H. Lang, D. Lohe, and E. Macherauch, Thermal-Mechanical Fatigue of the Cast Aluminum Alloy GK-AlSi10Mgwa, *Symp. Fatigue under Thermal and Mechanical Loading*, J. Bressers and L. Rémy, Ed., European Commission, Petten, May 1995

98. R.B. Gundlach, B. Ross, A. Hetke, S. Valterra, and J.F. Mojica, Thermal Fatigue Resistance of Hypoeutectoid Aluminum Silica Casting Alloys, *AFS Trans.*, Vol 141, 1994, p 205-223

99. S. Bhat and C. Laird, Cyclic Stress-Strain Response and Damage Mechanisms at High Temperature, *Fatigue Mechanisms*, STP 675, Jeffrey Fong, Ed., ASTM, 1979, p 592-623

100. H. Sehitoglu and M. Karayaka, Prediction of Thermomechanical Fatigue Lives in Metal Matrix Composites, *Metall. Trans. A*, Vol 23A, 1992, p 2029-2038

101. R.V. Miner, Fatigue, *Superalloys II*, C.T. Sims, N.S. Stoloff, and W.C. Hagel, Ed., John Wiley & Sons, 1987, p 263-289

102. W.W. Milligan, E.S. Huron, and S.D. Antolovich, Deformation, Fatigue and Fracture Behavior of Two Cast Anisotropic Superalloys, *Fatigue '87*, Vol 3, 3rd Int. Conf. Fatigue and Fatigue Thresholds, Engineering Materials Advisory Services, 1987, p 1561-1591

103. M.G. Castelli, R.V. Miner, and D.N. Robinson, Thermo-Mechanical Deformation of a Dynamic Strain Aging Alloy, *Thermo-Mechanical Fatigue Behavior of Materials*, STP 1186, H. Sehitoglu, Ed., ASTM, 1991, p 106-125

104. N. Marchand, G.L. Espérance, and R.M. Pelloux, Thermal-Mechanical Cyclic Stress-Strain Responses of Cast B-1900+Hf, *Low Cycle Fatigue*, STP 942, H. Solomon, G. Halford, L. Kaisand, and B. Leis, Ed., ASTM, 1988, p 638-656

105. H. Sehitoglu and D.A. Boismier, Thermo-Mechanical Fatigue of Mar-M247, Part 1: Experiments, *J. Eng. Mater. Technol. (Trans. ASME)*, Vol 112, 1990, p 68-80; see also Thermo-Mechanical Fatigue of Mar-M247, Part 2: Life Prediction, *J. Eng. Mater. Technol. (Trans. ASME)*, Vol 112, 1990, p 80-90

106. T.S. Cook, K.S. Kim, and R.L. McKnight, Thermal Mechanical Fatigue of Cast René 80, *Low Cycle Fatigue*, STP 942, H. Solomon, G. Halford, L. Kaisand, and B. Leis, Ed., ASTM, 1988, p 692-708

107. Y. Kadioglu and H. Sehitoglu, Thermomechanical and Isothermal Fatigue Behavior of Bar and Coated Superalloys, *J. Eng. Mater. Technol. (Trans. ASME)*, Vol 118, 1996, p 94-102

108. S. Kraft, R. Zauter, and H. Mugrabi, Investigations on the High Temperature Low Cycle Thermomechanical Fatigue Behavior of the Monocrystalline Nickel Base Superalloys CMSX-6, ASTM STP 1263, *Symposium on Thermo-Mechanical Fatigue Behavior of Materials*, ASTM, 1996

109. Y. Pan, K. Lang, D. Lohe, and E. Macherauch, Cyclic Deformation and Precipitation Behavior of NiCr22Co12Mo9, *Phys. Status Solidi (a)*, Vol 138, 1993, p 133-145

110. E. Glenny and T.A. Taylor, A Study of the Thermal-Fatigue Behavior of Metals: The Effect of Test Conditions on Nickel-Base High-Temperature Alloys, *J. Inst. Met.*, Vol 88, 1959-60, p 449-461

111. G.P. Tilly, Laboratory Simulation of Thermal Fatigue Experienced by Gas Turbine Blading, *Thermal Stresses and Thermal Fatigue*, D.J. Littler, Ed., Butterworths, London, 1971, p 47-65

112. R.C. Bill, M.J. Verrilli, M.A. McGaw, and G.R. Halford, "Preliminary Study of Thermomechanical Fatigue of Polycrystalline MAR-M 200," TP-2280, National Aeronautics and Space Administration, Feb 1984

113. R.S. Nelson, J.F. Schoendorf, and L.S. Lin, "Creep Fatigue Life Prediction for Engine Hot Section Materials (Isotropic)—Interim Report," CR-179550, National Aeronautics and Space Administration, Dec 1986

114. V.G. Ramaswamy and T.S. Cook, "Cyclic Deformation and Thermomechanical Fatigue Model of Nickel Based Superalloys, Abstracts," presented at ASTM Workshop on Thermo-Mechanical Fatigue and Cyclic Deformation (Charleston, SC), ASTM, 1986

115. G.T. Embley and E.S. Russell, "Thermal-Mechanical Fatigue of Gas Turbine Bucket Alloys," presented at 1st Parsons Int. Turbine Conf. (Dublin), June 1984

116. J. Gayda, T.P. Gabb, and R.V. Miner, paper presented at NASA 4th TMF Workshop, National Aeronautics and Space Administration, 1987

117. J.L. Malpertu and L. Rémy, Thermomechanical Fatigue Behavior of a Superalloy, *Low Cycle Fatigue*, STP 942, H. Solomon, G. Halford, L. Kaisand, and B. Leis, Ed., ASTM, 1988, p 657-671

118. H. Bernstein, T.S. Grant, R.C. McClung, and J. Allen, Prediction of Thermal-Mechanical Fatigue Life for Gas Turbine Blades in Electric Power Generation, *Thermo-Mechanical Fatigue Behavior of Materials*, H. Sehitoglu, Ed., STP 1186, ASTM, 1991, p 212-238

119. A. Kaufman and G. Halford, "Engine Cyclic Durability by Analysis and Material Testing," TM-83557, National Aeronautics and Space Administration, 1984

120. E. Chataigner, E. Fleury, and L. Rémy, Thermo-Mechanical Fatigue Behavior of Coated and Bare Nickel-Base Superalloy Single Crystals, *Symp. Fatigue under Thermal and Mechanical Loading*, J. Bressers and L. Rémy, Ed., European Commission, Petten, May 1995

121. A. Arana, J.M. Martinez-Esnaola, and J. Bressers, Crack Propagation and Life Prediction in a Nickel Based Superalloy under TMF Conditions, *Symp. Fatigue under Thermal and Mechanical Loading*, J. Bressers and L. Rémy, Ed., European Commission, Petten, May 1995

122. M. Marchionni, D. Ranucci, and E. Picco, Influence of Cycle Shape and Specimen Geometry on TMF of an ODS Ni-Base Superalloy, *Symp. Fatigue under Thermal and Mechanical Loading*, J. Bressers and L. Rémy, Ed., European Commission, Petten, May 1995

123. L.F. Coffin, Jr., The Effect of Frequency on the Cyclic Strain and Fatigue Behavior of Cast René at 1600 °F, *Metall. Trans.*, Vol 5, May 1974, p 1053-1060

124. W.W. Milligan and R.C. Bill, "The Low Cycle Fatigue Behavior of Conventionally Cast MAR-M 200 at 1000 °C," TM-83769, National Aeronautics and Space Administration, Sept 1984

125. J. Gayda and R.V. Miner, Fatigue Crack Initiation and Propagation in Several Nickel-Based Superalloys at 650 °C, *Int. J. Fatigue*, Vol 5, July 1983, p 135-143

126. M. Gell and G.R. Leverant, Mechanisms of High Temperature Fatigue, *Fatigue at Elevated Temperatures*, ASTM 520, A.E. Carden, A.J. McEvily, and C.H. Wells, Ed., ASTM, 1973, p 37-67

127. M.Y. Nazmy, High Temperature Low Cycle Fatigue of IN 738 and Application of Strain Range Partitioning, *Metall. Trans. A*, Vol 14A, March 1983, p 449-461

128. P.K. Wright, Oxidation-Fatigue Interactions in a Single-Crystal Superalloy, *Low Cycle Fatigue*, STP 942, H. Solomon, G. Halford, L. Kaisand, and B. Leis, Ed., ASTM, 1988, p 558-575

129. L. Rémy, F. Rezai-Aria, R. Danzer, and W. Hoffelner, Evaluation of Life Prediction Methods in High Temperature Fatigue, *Low Cycle Fatigue*, STP 942, H. Solomon, G. Halford, L. Kaisand, and B. Leis, Ed., ASTM, 1988, p 1115-1132

130. S.D. Antolovich, R. Baur, and S. Liu, A Mechanistically Based Model for High Temperature LCF of Ni Base Superalloys, *Superalloys 1980*, American Society for Metals, 1980, p 605-613

131. S.D. Antolovich, S. Liu, and R. Baur, Low Cycle Fatigue Behavior of René 80 at Elevated Temperature, *Metall. Trans. A*, Vol 12A, March 1981, p 473-481

132. W.W. Milligan, E.S. Huron, and S.D. Antolovich, Deformation, Fatigue and Fracture Behavior of Two Cast Anisotropic Superalloys, *Fatigue '87*, Vol 3, 3rd Int. Conf. Fatigue and Fatigue Thresholds (Charlottesville, VA), 1987, p 1561-1591

133. C.J. McMahon and L.F. Coffin, Jr., Mechanisms of Damage and Fracture in High-Temperature, Low-Cycle Fatigue of a Cast Nickel-Based Superalloy, *Metall. Trans.*, Vol 1, Dec 1970, p 3443-3450

134. Y. Oshida and H.W. Liu, Grain Boundary Oxidation and an Analysis of the Effects of Oxidation on Fatigue Crack Nucleation Life, *Low Cycle Fatigue*, STP 942, H. Solomon, G. Halford, L. Kaisand, and B. Leis, Ed., ASTM, 1988, p 1199-1217

135. Y. Kadioglu and H. Sehitoglu, Modeling of Thermo-Mechanical Fatigue Damage in Coated Alloys, *Thermo-Mechanical Fatigue Behavior of Materials*, STP 1186, H. Sehitoglu, Ed., ASTM, 1991, p 17-34

136. L. Rémy, H. Bernard, J.L. Malpertu, and F. Rezai-Aria, Fatigue Life Prediction under Thermal-Mechanical Loading in a Nickel Base Superalloy, *Thermo-Mechanical Fatigue Behavior of Materials*, STP 1186, H. Sehitoglu, Ed., ASTM, 1991, p 3-16

137. S.T. Wlodek, The Oxidation of René 41 and Udimet 700, *Trans. Met. Soc. AIME*, Vol 230, Aug 1964, p 1078-1090

138. G.E. Wasielewski and R.A. Rapp, High-Temperature Oxidation, *The Superalloys*, C.T. Sims and W.C. Hagel, Ed., John Wiley & Sons, 1972, p 287-316

139. J.D. Varin, Microstructure and Properties of Superalloys, *The Superalloys*, C.T. Sims and

W.C. Hagel, Ed., John Wiley & Sons, 1972, p 231-257

140. J. Reuchet and L. Rémy, Fatigue Oxidation Interaction in a Superalloy—Application to Life Prediction in High Temperature Low Cycle Fatigue, *Metall. Trans. A*, Vol 14A, Jan 1983, p 141-149

141. K. Schneider and H.W. Gruling, Mechanical Aspects of High Temperature Coatings, *Thin Film Solids*, Vol 107, 1983, p 395-416

142. R. Lane and N.M. Geyer, Superalloy Coatings for Gas Turbine Components, *J. Met.*, Vol 18, Feb 1966, p 186-191

143. G.F. Paskeit, D.H. Boone, and C.P. Sullivan, Effect of Aluminide Coating on the High-Cycle Fatigue Behavior of a Nickel Base High Temperature Alloy, *J. Inst. Met.*, Vol 100, Feb 1972, p 58-62

144. C.H. Wells and C.P. Sullivan, Low Cycle Fatigue of Udimet 700 at 1700 °F, *Trans. ASM*, Vol 61, March 1968, p 149-155

145. R.S. Bartocci, Behavior of High Temperature Coatings for Gas Turbines, in *Hot Corrosion Problems Associated with Gas Turbines*, STP 421, ASTM, 1967, p 169-187

146. G. Liewelyn, Protection of Nickel Base Alloys against Sulfur Corrosion by Pack Aluminizing, in STP 421 (ibid), ASTM, 1967, p 3-20

147. G.W. Goward, Current Research on Surface Protection of Superalloys for Gas Turbines, *J. of Metals*, Oct 1970, p 31-39

148. D.H. Boone and C.P. Sullivan, in *Fatigue at Elevated Temperatures*, STP 520, A.E. Carden, A.J. McEvily, and C.H. Wells, Ed., ASTM, 1972, p 401-415

149. A.J. Santhanam and C.G. Beck, *Thin Solid Films*, Vol 73, 1980, p 387-395

150. G.A. Swanson, I. Linask, D.M. Nissley, P.P. Norris, T.G. Meyer, and K.P. Walker, Report 179594, National Aeronautics and Space Administration, 1987

151. J.E. Heine, J.R. Warren, and B.A. Cowles, "Thermomechanical Fatigue of Coated Blade Materials," Final Report, Wright Research Development Center, 27 June 1989

152. K.R. Bain, The Effects of Coatings on the Thermomechanical Fatigue Life of a Single Crystal Turbine Blade Material, *AIAA/SAE/ASME/ASEE 21st Joint Propulsion Conf.*, 1985, p 1-6

153. G.R. Leverant, T.E. Strangman, and B.S. Langer, in *3rd Int. Conf. Superalloys 1976*, Claitors, Vol 75, 1976, p 285

154. T.E. Strangman, "Thermal Fatigue of Oxidation Resistant Overlay Coatings for Superalloys," Ph.D. thesis, University of Connecticut, 1978: see also T.E. Strangman and S.W. Hopkins, Thermal Fatigue of Coated Superalloys, *Ceram. Bull.*, Vol 55 (No. 3), 1976, p 304-307

155. G.R. Halford, T.G. Meyer, R.S. Nelson, and D.M. Nissley, paper presented at 33rd Int. Gas Turbine and Aeroengine Congress (Amsterdam), ASME, June 1988

156. G.A. Swanson, I. Linask, D.M. Nissley, P.P. Norris, T.G. Meyer, and K.P. Walker, Report 179594, National Aeronautics and Space Administration, 1987

157. D.M. Nissley, Fatigue Damage Modeling for Coated Single Crystal Superalloys, *NASA Conf. Publ. 3003*, Vol 3, 1988, p 259-270

158. G.R. Leverant, T.E. Strangman, and B.S. Langer, Parameters Controlling the Thermal Fatigue Properties of Conventionally-Cast and Directionally-Solidified Turbine Alloys, *Superalloys: Metallurgy and Manufacture*, B.H. Kear, et al., Ed., Claitors, 1976, p 285-295

159. W. Wei, W. Dunfee, M. Gao, and R.P. Wei, The Effect of Environment on the Thermal Fatigue Behavior of Gamma Titanium Aluminide, *Symp. Fatigue under Thermal and Mechanical Loading*, J. Bressers and L. Rémy, Ed., European Commission, Petten, May 1995

160. S. Mall, T. Nicholas, J.J. Pernot, and D.G. Burgess, Crack Growth in a Titanium Aluminide Alloy under Thermal Mechanical Cycling, *Fatigue Fract. Eng. Mater. Struct.*, Vol 14 (No. 1), 1991, p 79-87

161. S. Kalluri and G. Halford, Damage Mechanisms in Bithermal and Thermo-Mechanical Fatigue of Haynes 188, *Thermo-Mechanical Fatigue Behavior of Materials*, STP 1186, H. Sehitoglu, Ed., ASTM, 1991, p 126-143

162. K.D. Sheffler and G.S. Doble, Thermal Fatigue Behavior of T-111 and Astar 811C in Ultrahigh Vacuum, *Fatigue at Elevated Temperatures*, STP 520, A.E. Carden, A.J. McEvily, and C.H. Wells, Ed., ASTM, 1973, p 482-489

163. P. Hacke, A Sprecher, and H. Conrad, Modeling of the Thermomechanical Fatigue of 63Sn-37Pb Alloy, *Thermo-Mechanical Fatigue Behavior of Materials*, STP 1186, H. Sehitoglu, Ed., ASTM, 1991, p 91-105

164. M. Karayaka and H. Sehitoglu, Thermomechanical Cyclic Deformation of Metal Matrix Composites: Internal Stress-Strain Fields, *ASTM Cyclic Deformation, Fracture and Nondestructive Evaluation of Advanced Materials*, ASTM, 1990

165. M. Karayaka and H. Sehitoglu, Thermomechanical Fatigue of Metal Matrix Composites, *Low Cycle Fatigue and Elasto-Plastic Behavior of Materials*, Vol 3, T.T. Rie, Ed., Elsevier, 1992, p 13-18

166. W. VanArsdell, "The Effect of Particle Size on the Thermo-mechanical Fatigue of Metal Matrix Composites," M.S. thesis, University of Illinois at Urbana, 1993

167. W. VanArsdell, H. Sehitoglu, and M. Mushiake, "The Effect of Particle Size on the Thermo-mechanical Fatigue of Metal Matrix Composites," *Fatigue '93*, EMAS, 1993

168. H. Sehitoglu, The Effect of Particle Size on the Thermo-Mechanical Fatigue Behavior of Metal Matrix Composites, *Symp. Fatigue under Thermal and Mechanical Loading*, J. Bressers and L. Rémy, Ed., Kluwer Academic Publishers, 1996

169. R. Neu, Thermo-Mechanical Fatigue Damage Mechanism Maps for Metal Matrix Composites, STP 1263, *2nd Symposium on Thermo-Mechanical Fatigue Behavior of Materials*, M. Verilli and M. Castelli, Ed., ASTM, 1996

170. R. Neu, A Mechanistic-Based Thermomechanical Fatigue Life Prediction Model for Metal Matrix Composites, *Fatigue Fract. Eng. Mater. Struct.*, Vol 16 (No. 8), 1993, p 811-828

171. S. Taira and T. Inoue, Thermal Fatigue under Multiaxial Thermal Stresses, *Thermal Stresses and Thermal Fatigue*, D.J. Littler, Ed., Butterworths, London, 1971, p 66-80

172. J. Meersman, J. Ziebs, H.-J. Kuhn, R. Sievert, J. Olscewski, and H. Frenz, The Stress-Strain Behavior of In 738LC under Thermo-mechanical Uni- and Multiaxial Fatigue Loading, *Symp. Fatigue under Thermal and Mechanical Loading*, J. Bressers and L. Rémy, Ed., European Commission, Petten, May 1995

173. M. Gell and D.J. Duquette, The Effects of Oxygen on Fatigue Fracture of Engineering Alloys, *Corrosion Fatigue: Chemistry, Mechanics and Microstructure*, NACE, 1971, p 366-378

174. J.C. Runkle and R.M. Pelloux, Micromechanisms of Low-Cycle Fatigue in Nickel-Based Superalloys at Elevated Temperatures, *Fatigue Mechanisms*, STP 675, ASTM, 1979, p 501-527

175. V.G. Ramaswamy and T.S. Cook, "Cyclic Deformation and Thermomechanical Fatigue Model of Nickel Based Superalloys, Abstracts," presented at ASTM Workshop on Thermo-mechanical Fatigue and Cyclic Deformation (Charleston, SC), ASTM, 1986

176. C.A. Rau, Jr., A.E. Gemma, and G.R. Leverant, Thermal-Mechanical Fatigue Crack Propagation in Nickel- and Cobalt-Base Superalloys under Various Strain-Temperature Cycles, *Fatigue at Elevated Temperatures*, STP 520, A.E. Carden, A.J. McEvily, and C.H. Wells, Ed., ASTM, 1973, p 166-178

177. M.L. Heil, T. Nicholas, and G.K. Haritos, Crack Growth in Alloy 718 under Thermal-Mechanical Cycling, *Thermal Stress, Material Deformation, and Thermo-Mechanical Fatigue*, H. Sehitoglu and S.Y. Zamrik, Ed., American Society of Mechanical Engineers, 1987, p 23-29

178. M. Okazaki and T. Koizumi, Crack Propagation During Low Cycle Thermal-Mechanical and Isothermal Fatigue at Elevated Temperatures, *Metall. Trans. A*, Vol 14A (No. 8), Aug 1983, p 1641-1648

179. M. Okazaki and T. Koizumi, Effect of Strain Waveshape on Thermal-Mechanical Fatigue Crack Propagation in a Cast Low Alloy Steel, *J. Eng. Mater. Technol (Trans. ASME)*, A.E. Carden, A.J. McEvily, and C.H. Wells, Ed., ASTM, Vol 81-87, 1983, p 1641-1648

180. A.E. Gemma, B.S. Langer, and G.R. Leverant, Thermo-Mechanical Fatigue Crack Propagation in an Anisotropic (Directionally Solidified) Nickel Base Superalloy, *Thermal Fatigue of Materials and Components*, STP 612, D.A. Spera and D.F. Mowbray, Ed., ASTM, p 199-213

181. R.P. Skelton, Crack Initiation and Growth in Simple Metal Components during Thermal Cycling, *Fatigue at High Temperature*, Applied Science Publishers, London, 1983, p 1-63

182. E.H. Jordan and G.J. Meyers, Fracture Mechanics Applied to Nonisothermal Fatigue Crack Growth, *Eng. Fract. Mech.*, Vol 23 (No. 2), 1986, p 345-358

183. R.L. McKnight, J.H. Laflen, and G.T. Spamer, "Turbine Blade Tip Durability Analysis," CR 165268, National Aeronautics and Space Administration, 1982

184. M.R. Johnson, private communication

185. D. Morrow, "Stress-Strain Response of a Two-Bar Structure Subject to Cyclic Thermal and Steady Net Section Loads," M.S. thesis, University of Illinois, 1982; see also E. Abrahamson, "Modeling the Behavior of Type 304 Stainless Steel with a Unified Creep-Plasticity Theory," Ph.D. thesis, University of Illinois, 1983

186. R.W. Swindeman and D.N. Robinson, Two-Bar Thermal Ratcheting of Annealed 2.25 Cr-1Mo Steel, *Thermal Stress, Material Deformation, and Thermo-Mechanical Fatigue*, H. Sehitoglu and S.Y. Zamrik, Ed., American Society of Mechanical Engineers, 1987, p 91-98

187. S.M. Russ, C.J. Boehlert, and D. Eylon, Out-of-Phase Thermomechanical Fatigue of Titanium Composite Matrices, *Mater. Sci. Eng.*, Vol A192/193, 1995, p 483-489

188. H.G. Baron, Thermal Shock and Thermal Fatigue, *Thermal Stress*, P.P. Benham and R.D. Hoyle, Ed., Pitman, London, 1964, p 182-206

189. K.D. Challenger, A.K. Miller, and C.R. Brinkman, An Explanation for the Effects of Hold Periods on the Elevated Temperature Fatigue Behavior of 2.25Cr-1Mo Steel, *J. Eng. Mater. Technol. (Trans. ASME)*, Vol 103, Jan 1981, p 7-14

190. J. Bressers, U. Schusser, and B. Ilschner, Environmental Effects on the Fatigue Behavior of Alloy 800H, *Low Cycle Fatigue and Elasto-Plastic Behavior of Materials*, 2nd Int. Conf. Low Cycle Fatigue and Elasto-Plastic Behaviour of Materials (Munich), Elsevier Applied Sciences, 1987, p 365-370

191. S. Majumdar and P.S. Maiya, A Mechanistic Model for Time-Dependent Fatigue, *J. Eng. Mater. Technol. (Trans. ASME)*, Vol 102, Jan 1980, p 159-167

192. B. Kirkwood and J.R. Weertman, Cavity Nucleation During Fatigue Crack Growth Caused by Linkage of Grain Boundary Cavities, *Micro and Macro Mechanics of Crack Growth*, K.

Sadananda, B.B. Rath, and D.J. Michel, Ed., TMS-AIME, 1982, p 199-212

193. J. Wareing, Creep-Fatigue Interaction in Austenitic Stainless Steels, *Metall. Trans. A*, Vol 8A, May 1977, p 711-721

194. B.K. Min and R. Raj, Hold-Time Effects in High Temperature Fatigue, *Acta Metall.*, Vol 26, 1978, p 1007-1022

195. B. Tomkins, Fatigue: Mechanisms, *Creep and Fatigue in High Temperature Alloys*, J. Bressers, Ed., 1981, p 111-143

196. J.K. Tien, S.V. Nair, and V.C. Nardone, Creep-Fatigue Interaction in Structural Alloys, *Flow and Fracture at Elevated Temperatures*, R. Raj, Ed., American Society for Metals, 1985, p 179-213

197. J.C. Grosskreutz and M.B. McNeil, *J. Appl. Phys.*, Vol 40, 1969, p 355

198. D. Bruce and P. Hancock, Mechanical Properties and Adhesion of Surface Oxide Films on Iron and Nickel Measured during Growth, *J. Inst. Met.*, Vol 97, 1969, p 148-155

199. D. Bruce and P. Hancock, Influence of Specimen Geometry on the Growth and Mechanical Stability of Surface Oxides Formed on Iron and Steel in the Temperature Range 570 °C-800 °C, *J. Iron Steel Inst.*, Nov 1970, p 1021-1024

200. M.I. Manning, Geometrical Effects on Oxide Scale Integrity, *Corros. Sci.*, Vol 21 (No. 4), 1981, p 301-316

201. J. Stringer, Stress Generation and Relief in Growing Oxide Films, *Corros. Sci.*, Vol 10, 1970, p 513

202. J.K. Tien and J.M. Davidson, Oxide Spallation Mechanisms, *Stress Effects and the Oxidation of Metals*, J.V. Cathcart, Ed., TMS-AIME, 1974, p 200-219

203. C.H. Wells, P.S. Follansbee, and R.R. Dils, Mechanisms of Dynamic Degradation of Surface Oxides, *Stress Effects and the Oxidation of Metals*, J.V. Cathcart, Ed., TMS-AIME, 1974, p 220-244

204. ASME Boiler and Pressure Vessel Code, Case N-47-23, "Class 1 Components in Elevated Temperature Service," Section III, Division 1, American Society of Mechanical Engineers, 1986

205. J. Lemaitre and A. Plumtree, Application of Damage Concepts to Predict Creep-Fatigue Failures, *J. Eng. Mater. Technol. (Trans. ASME)*, Vol 101, 1979, p 284-292

206. R. Hales, A Method of Creep Damage Summation Based on Accumulated Strain for Assessment of Creep-Fatigue Endurance, *Fatigue Fract. Eng. Mater. Struct.*, Vol 6, 1983, p 121

207. A. Saxena and J.L. Bassani, Time-Dependent Fatigue Crack Growth Behavior at Elevated Temperature, *Fracture: Microstructure, Mechanisms and Mechanics*, J.M. Wells and J.D. Landes, Ed., TMS-AIME, 1984, p 357-383

208. H. Sehitoglu and W. Sun, The Significance of Crack Closure under High Temperature Fatigue Crack Growth with Hold Periods, *Eng. Fract. Mech.*, Vol 33, 1989, p 371-388

209. M. Miller, D.L. McDowell, R. Oehmke, and S. Antolovich, A Life Prediction Model for Thermomechanical Fatigue Based on Microcrack Propagation, *Thermo-Mechanical Fatigue Behavior of Materials*, STP 1186, H. Sehitoglu, Ed., ASTM, 1991, p 35-49

210. D. Drucker and L. Palgen, On the Stress-Strain Relations Suitable for Cyclic and Other Loading, *J. Appl. Mech. (Trans. ASME)*, Vol 48, 1981, p 479-485

211. S. Bodner and Y. Partom, Constitutive Equations for Elastic Viscoplastic Strain Hardening Materials, *J. Appl. Mech. (Trans. ASME)*, Vol 42, 1975, p 385-389

212. S. Bodner, I. Partom, and Y. Partom, Uniaxial Cyclic Loading of Elastic Viscoplastic Material, *J. Appl. Mech. (Trans. ASME)*, Vol 46, 1979, p 805-810

213. D. Slavik and T.S. Cook, A Unified Constitutive Model for Superalloys, *Int. J. Plast.*, Vol 6, 1990, p 651-664

214. R. Neu, private communication, 1995

215. H. Sehitoglu, "Thermomechanical Deformation of Engineering Alloys and Components—Experiments and Modeling," *Mechanical Behavior of Materials at High Temperature*, Kluwer, 1996, p 349-380

216. A.K. Miller, A Realistic Model for the Deformation Behavior of High Temperature Materials, *Fatigue at Elevated Temperatures*, STP 520, A.E. Carden, A.J. McEvily, and C.H. Wells, Ed., ASTM, 1973, p 613-624

217. K.P. Walker, "Research and Development Program for Nonlinear Structural Modeling with Advanced Time-Temperature Dependent Constitutive Relationships," CR-165533, National Aeronautics and Space Administration, 1981

Life Extension and Damage Tolerance of Aircraft

Mitchell P. Kaplan and Timothy A. Wolff, Willis & Kaplan, Inc.

THE LIFE of a structural component is limited by its ability to resist the effects of its usage history, which may consist of cyclic loads, fluctuations in temperature, or a corrosive environment. Even with anticipated usage, the material properties of a component degrade over time, and that degradation can eventually result in failure. One of the mechanisms by which this process occurs, fatigue, is the initiation and growth of cracks. Fatigue is a primary cause of failure in aircraft structures. For this reason, a reliable prediction of component fatigue life is critical to aircraft safety.

Life extension using the damage tolerance design method has become more prominent in the aviation industry. As the average age of the world's transport fleet continues to rise, this analysis technique becomes more widespread. With the increasing cost of new aircraft, operators find it more economical to refurbish and maintain their existing aircraft than to purchase new ones. The resulting increase in the number of aircraft that are being operated beyond their expected service life raises important safety questions regarding the structural integrity of these aircraft. To answer these questions, several approaches have been developed that attempt to provide some measure of the useful life of aircraft structural components. These approaches are either based upon fatigue methods, which can employ statistical data and testing to arrive at an estimate of component life, or damage tolerance methods which arrive at a reasonable inspection interval to ensure component safety. This latter method is based upon the growth rate of cracks assumed to exist in the component as a result of fatigue, accidental damage, or material and manufacturing defects.

The effect of corrosion, and the degradation it imposes upon materials is complex and not readily deterministic as a function of time. Aircraft manufacturers select materials and processes to impede the effect of corrosion, and users of the aircraft attempt to preclude the destructive nature of corrosion as a regular part of their inspection program. This article provides a general overview of these approaches to corrosion identification and prevention, and discusses their application to the process of extending the life of aircraft structural components.

Analysis Methods

The analysis methods that are used to determine the life of aircraft are those using fatigue methods and those using the methods of fracture mechanics. The similarities in these methods are the input information, i.e., discrete localized stresses and the stress time history (stress spectrum) of the component being analyzed. The differences in these methods are the initial condition and the inspection interval. The initial condition in fatigue is a pristine (unflawed) condition in the component, while fracture mechanics assumes an initial flaw of a given size and geometry. The fatigue test is complete at specimen failure. Two pieces of information are known in a fatigue test; the number of cycles to failure, and the loading that the specimen (component) experienced. In a fracture mechanics test, these two are known as well as the number of cycles (flight) for the crack to grow a given distance. This additional information allows the analyst to determine recurring inspection intervals as the growth behavior of the flaw is identified.

Fatigue life methods are empirical in nature, using statistical techniques to establish safe estimates of fatigue life from $S\text{-}N$ or $\varepsilon\text{-}N$ test data. A safe estimate is one in which the probability of component failure within a specified time frame is very remote. Component life estimates using these methods are typically based upon material and load data gathered from axial loading tests, material component tests, and full-scale testing. In axial loading tests, test samples of the material are subjected to tension-compression or tension-tension cycles until failure. The data are then presented in the form of an S/N curve, which may depict maximum stress, minimum stress, or stress amplitude versus the number of cycles to failure. The fatigue life obtained from an S/N curve represents the time for a flaw to initiate and propagate to failure. Because the loading is cyclical, the

stress ratios are defined to be a function of the minimum and maximum stresses for a given test. The two most commonly used ratios are the R factor (the ratio of minimum stress to maximum stress) and the A factor (the ratio of alternating stress to mean stress) such that:

$$R \text{ ratio} = \frac{\sigma_{min}}{\sigma_{max}}$$

$$A \text{ ratio} = \frac{\sigma_a}{\sigma_m}$$

where the stress amplitude $(\sigma_a) = (\sigma_{max} - \sigma_{min})/2$ and the mean stress $(\sigma_m) = (\sigma_{max} + \sigma_{min})/2$. The stress factors are related directly:

$$R = (1 - A)/(1 + A)$$

$$A = (1 - R)/(1 + R)$$

and thus $R = -1$ and $A = \infty$ under fully reversed loading (zero mean stress). The R factor is normally used in airframe design, while the A factor is used in engine design.

S/N curves are created by averaging the data points from tests conducted on an appropriate sample size of the material. In most cases, a test is run to failure at three or four different stress levels to determine the shape of the S/N curve. Remaining samples are then tested to failure at the same selected stress levels and the results are averaged to produce a curve that reflects a 50% probability of failure. If a large enough number of data points are collected for various stress levels, a probability-stress-life plot can be constructed as shown in Fig. 1. For a given stress level, a fatigue life can be chosen corresponding to an acceptable probability of failure. Statistical models commonly used in these plots are the Weibull and log normal distributions (Ref 1).

Constant-Amplitude Fatigue Limit. Testing a material under constant amplitude (often at $R = -1$, $A = \infty$) can be used to define a fatigue limit. For low-strength steels, the data points on an S/N curve tend to approach an asymptote as the number of cycles becomes large. The stress value

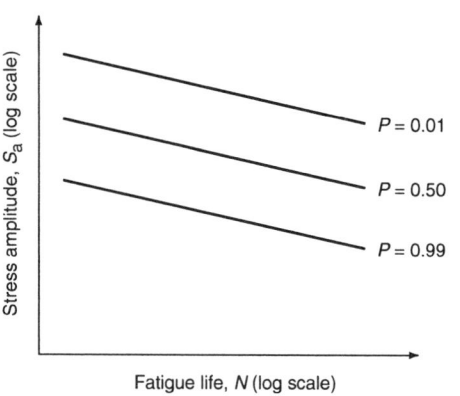

Fig. 1 Typical probability-stress-life plot

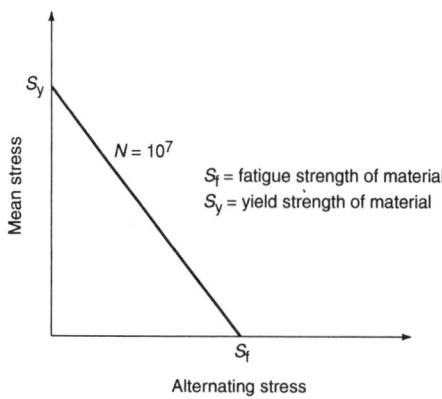

Fig. 2 Plot of the Soderberg relationship

corresponding to this region is referred to as the fatigue limit of the material. For those materials which do not exhibit this region, such as high-strength steels and aluminum alloys, the fatigue limit often is defined as the stress value corresponding to 10^7 cycles (Ref 2). A fatigue limit is useful in designing components for 'infinite' life, but in practice the mean stress is not always zero. Constant-life diagrams, such as the Soderberg line shown in Fig. 2, serve as a guide in fatigue design by relating non-zero mean stress cases to the fatigue limit and other properties of a material (Ref 3). Combinations of mean and alternating stress which fall on or beneath the line correspond to an indefinite fatigue life for a component. The Soderberg line provides conservative estimates for ductile materials. Other relationships, such as the Goodman line (Ref 4) or Gerber parabola (Ref 5), provide comparable results.

Variable Amplitude Loading. In reality, aircraft components are likely to be subjected to variable amplitude loading rather than the simplified constant amplitude loading assumed so far. Determining the fatigue life of a component under these conditions requires a method that can relate the variations in mean and alternating stress to fatigue data obtained from constant amplitude loading tests. One such method is the Palmgren-Miner Rule (Ref 6, 7), which has the mathematical form:

$$\sum_{i=1}^{m} \frac{n_i}{N_i} = 1$$

where n_i represents the number of cycles of damage and N_i represents the life of a component at a number of levels of mean and alternating stress. For each combination of stresses in the spectrum, the fraction of damage done by each is added together to obtain a total damage fraction. The reciprocal of this fraction provides a measure of the number of loading spectra that a component can sustain before failure. This estimate is based upon the assumptions that the relationship between n and N is linear and that the fatigue life is independent of the order in which the stresses are applied. In most cases, these assumptions are only marginally valid, and fatigue life estimates obtained using the Palmgren-Miner Rule

are adjusted through testing or by dividing by a suitable safety factor (Ref 8).

Fracture mechanics methods are concerned with the initiation and propagation of cracks. Fracture mechanics is based upon the fundamental assumption that all structures, whether new or in service for many years, contain minute flaws which, under the proper conditions, can cause fracture. These flaws can be in the form of material irregularities, such as an inclusion or void, or they can be created during the manufacturing process. A scratch on the inside surface of a fastener hole caused by a dull drill bit is an example of a manufacturing defect. In-service structures can sustain flaws from their environment and maintenance practices. Manufacturing and maintenance procedures are closely regulated in the aviation industry to prevent the creation of flaws, especially in the vicinity of stress concentrations. However, it is an accepted fact that flaws will always exist.

Residual Strength and Griffith Criterion. Once a crack initiates, it will continue to grow longer under the action of repeated loads or due to the service environment, until failure occurs. As the crack lengthens, the rate at which it grows will increase. Failure occurs when the magnitude of the stress concentration produced by the crack and at the crack tip exceeds the residual strength of the structure. The residual strength is a measure of the highest stress a structure can carry for a given crack size before failing.

In order to determine the residual strength of a structure, the relationship between the crack tip stress field and crack length must be established. In 1920, Griffith proposed that a crack can propagate in a solid only if the energy available for crack growth equals or exceeds the energy required (Ref 9). The energy required for crack growth, referred to as the crack resistance, represents the energy necessary to break atomic bonds in the formation of new crack surfaces. Griffith arrived at his theory by experimenting with glass rods, where the crack resistance consisted solely of surface energy as a result of the brittle nature of glass. For ductile materials, such as metals, the crack resistance consists primarily of plastic en-

ergy, as the material at the crack tip must yield before crack growth can occur.

Griffith's theory is essentially a statement of conservation of energy. The potential energy contained within a solid before an increment of crack growth must equal the potential energy remaining in the solid after crack growth plus any energy increases or decreases resulting from external work done on the solid or energy lost to the surroundings. By analyzing a uniformly stressed plate with a preexisting crack in its center and applying the principle of minimum potential energy, Griffith arrived at an equation relating applied stress to crack length. The general form of Griffith's equation is:

$$G = \frac{\pi\sigma^2 a}{E}$$

where G is the elastic energy release rate, σ is the stress applied to the plate, a is half the crack length, and E is Young's modulus. The elastic energy release rate represents the energy available for crack growth. When G equals or exceeds a certain critical value, crack growth will occur.

Stress Intensity Factor. Irwin expanded upon Griffith's theory by determining the crack tip stress field for an edge crack in a thin plate (Ref 10). When the crack is small compared to the dimensions of the plate, Irwin found the following relationship:

$$K_I = \sigma\sqrt{\pi a}$$

where K_I is defined as the stress intensity factor, σ is the stress applied at the edges of the plate, and a is the crack length. The subscript, I, denotes mode I crack loading. Three modes of loading are possible for a crack in a solid: mode I, referred to as a tensile or 'opening mode,' mode II, referred to as shear or a 'sliding mode,' and mode III, referred to as torsion or a 'tearing mode.' For a crack in a plate, mode I loading is created by normal stresses on the edges of the plate which tend to separate the opposing surfaces of the crack. Mode II loading is created by shear stresses in the plane of the plate and mode III by shear stresses perpendicular to the plane of the plate. Stress intensity factors can be defined for all three loading conditions, but mode I tends to be the most critical condition for component life and failure.

The stress intensity factor is a useful parameter that integrates material properties, loading conditions, and geometry into the relationship between structural strength and crack size. It is not a stress concentration factor. For finite geometries, the equation for the stress intensity factor takes the form:

$$K_I = \sigma\sqrt{\pi a}\,\beta$$

The dimensionless factor, β, is a function of geometry and crack size. For a plate with a central crack subject to uniform tension, β is a function of the ratio of crack size to plate width. For a plate subject to uniform tension with a hole in its center and a crack emanating from one side of the hole, β is a function of the ratio of crack size to hole radius. Stress intensity factor solutions exist for many combinations of loading conditions and geometry. Since these solutions are based upon linear elastic stress analysis, the principle of superposition can be used to solve more complex cases.

The stress intensity factor is related to the elastic energy release rate by the equation:

$$G = \frac{K_I^2}{E} \text{ for plane stress (i.e., thin plates)}$$

$$G = \frac{(1 - v^2) K_I^2}{E} \text{ for plane strain (i.e., thick plates)}$$

where v is Poisson's ratio. From these relationships, it becomes clear that if a critical value of G exists, above which fracture will occur, a critical stress intensity factor, K_{Ic}, must also exist. This critical value is referred to as the fracture toughness.

Fracture toughness is a material property that is determined through load-displacement tests. Typically, the crack opening displacement is plotted as a function of applied load, and the fracture toughness is calculated from the load corresponding to a certain percentage of crack extension. The results of these tests are dependent upon many factors, such as temperature, material temper, crack geometry, and specimen geometry, and must be applied carefully. The American Society for Testing and Materials (ASTM) has written a specification defining this test and its parameters (Ref 11). Plane strain fracture toughness tests have well-defined criteria for specimen and crack geometry. Compact tension and bend specimens are generally used with their dimensions carefully selected to ensure plane strain conditions. Plane stress fracture toughness tests are not as clearly defined and usually require much larger specimens (Ref 12). Knowledge of fracture toughness for a given set of material conditions, coupled with the equation for stress intensity factor, allows the critical crack length, the largest crack size a structure can withstand before failure, to be calculated for any applied load. This value is important in the design of damage tolerant structures.

Crack Growth Relationships. Once the critical crack length is known, the next question that arises is how long will it take an existing crack to propagate from a certain initial crack size to the critical crack length? This question can be answered by developing an equation that describes the behavior of a crack under conditions of stable growth. The term stable growth refers to crack growth that occurs when the elastic energy release rate is just equal to the crack resistance ($G = R$). When the elastic energy release rate exceeds the crack resistance, relatively rapid, unstable crack growth, referred to as fracture instability, will result. Fracture instability occurs at the critical crack length. One of many relationships that are assumed to describe stable crack growth is the Paris equation:

$$\frac{da}{dN} = C(\Delta K)^n$$

where C and n are material constants, and ΔK is the alternating stress intensity factor (Ref 13). This equation states that the change in crack length per cycle is a function of the difference between the maximum and minimum stress intensity factors. Since stress intensity is a function of stress as well as crack length, ΔK relates the stress spectrum directly to the crack growth rate. Tests have shown that below a certain value of ΔK, referred to as the threshold stress intensity factor (ΔK_{th}), the crack growth rate is very small. Therefore, crack growth is assumed to occur only when ΔK exceeds the threshold.

Figure 3 shows a double logarithmic plot of da/dN data for 2024-T3 aluminum at $R = 0$. The Paris equation predicts that the data points should lie on a straight line of slope n. As can be seen, the plot clearly consists of several slopes, pointing out the limitations of the Paris equation.

Other equations have been developed that attempt to account for the variations in slope and R factor (Ref 14-16). While these equations are applicable over certain ranges and R factors, none of them is applicable over the full range of ΔK values for all cases. Nonetheless, these equations are useful for estimating the time required for an existing crack in a structure to grow from its initial size to the critical crack length, providing some measure of the life of the structure. In most instances, a crack growth equation can be found that conforms to the test data over a limited range of ΔK values. For this reason, many computer programs contain several crack growth models, allowing the user to select the applicable crack growth equation. Additionally, these crack growth programs have as an option, an algorithm that allows the user to input raw test data. The crack growth program then interpolates the crack growth rate as a function of ΔK from the input data.

Life Assessment Procedures

Safe-Life Approach. The relevance of fatigue in aircraft design was not always readily understood. During the 1950s, the loss of several Comet aircraft within a short period of time prompted an investigation into the structural integrity of the Comet's fuselage. A pressurization test on a sample aircraft revealed that the combination of fatigue and stress concentration at the corner of a window was the probable cause of the failures (Ref 17). These accidents as well as others prompted the development of fatigue methods for aircraft design. Using these methods, the safe-life approach predicts a replacement time for aircraft components, usually specified as a number of allowable landings or flight hours. The replacement time is based upon the time required for failure, which is obtained from component fatigue tests. In most cases, a component is designed so that the replacement time for that component exceeds the expected service life of the aircraft. However, critically loaded components, such as engine rotors or landing gear, can have much shorter replacement times. Once a component reaches its replacement time, its safe-life is

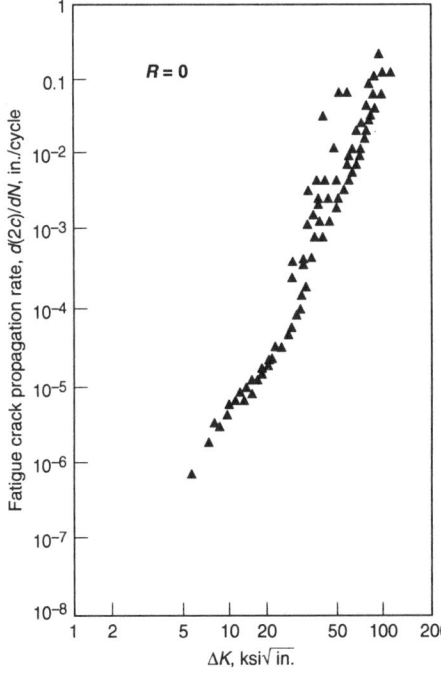

Fig. 3 Crack growth data for 2024-T3 aluminum. 2024-T3 Al, 0.090 in. sheet, L-T direction, 70 °F, lab air; frequency, 0.3 to 20 Hz; CC specimens, 0.090 in. thick, 12.0 in. wide. Source: *Damage Tolerant Handbook*, MCIC Handbook, HB-01

considered to be used up and it is retired, whether or not any fatigue cracks are present. Ideally, a component designed according to safe-life principles will be replaced before it develops a fatigue crack. There were, however, two significant problems inherent in this method:

- The safety of an aircraft was not protected if it contained a manufacturing or maintenance induced defect.
- Retirement times were not based upon statistically based safety factors. To protect safety, the selected safety factor was conservative. As a result, many components were prematurely retired.

Fail-Safe Approach. The fail-safe approach to aircraft fatigue design was developed during the 1960s and implemented in a number of commercial aircraft (Ref 18). The goal of the fail-safe philosophy is to design multiple load path structures, such that if an individual element should fail, the remaining elements would have sufficient structural integrity to carry the additional loads from the failed element until the damage is detected through scheduled maintenance inspections. In addition to multiple load path structures, crack stoppers are also commonly used in fail-safe designs. Crack stoppers, which typically consist of materials with a high fracture toughness, are used to supplement the residual strength of the surrounding structure and prevent cracks from propagating to failure. An example of a crack stopper is a stringer in a pressurized fuselage. The stringer reduces the amount of energy available for crack growth, slowing or stopping the advance of a crack that crosses it. Ideally, an aircraft designed according to fail-safe principles can sustain damage and remain airworthy until the damage is detected and repaired. This necessitates periodic inspection to examine the structure to determine if the primary load carrying member contains cracks. The frequency of these inspections is typically assigned by the manufacturer based upon service experience. This method does not consider the initiation and growth (linking up) of small cracks at fastener holes. As a result, the loss of several 'fail-safe' aircraft in the mid 1970s emphasized the need to locate cracks and repair damage before failure occurred (Ref 19).

Damage Tolerance Approach. In the early 1970s, the U.S. Air Force initiated a number of standards and specifications that led to a new method of fatigue design-damage tolerance (Ref 20-21). Based upon fracture mechanics techniques, the damage tolerance approach redefined the basis for analyzing fatigue cracks in aircraft structures. With economic and safety advantages over the previous methods, the damage tolerance philosophy was eventually adopted by the commercial aircraft industry as well (Ref 22-23). The objective of the damage tolerance approach is to detect cracks in principal structural elements before they propagate to failure. A principal structural element (PSE) is defined as any aircraft structure carrying flight, ground, or pressurization loads, whose failure could result in the loss of the aircraft. By establishing inspection intervals for these elements based upon the time it takes a crack to grow from an initial detectable size to the critical crack length, the objective of the damage tolerance approach can be achieved.

Before inspection intervals can be established for an aircraft, its usage profile must be defined. This profile describes the various flight conditions, such as taxi, climb, cruise, descent, and landing impact, and the amount of time spent at each gross weight, speed, and altitude. The usage profile is then used to create a load factor spectrum at the center of gravity of the aircraft, based on the frequency of occurrence of each increment of load factor, and to identify the principal structural elements of the aircraft. An inspection interval must be established for each principal structural element. This task is accomplished by converting the load factor spectrum into a stress spectrum for each location and incorporating the effects of the service environment. Using crack growth equations, such as the Forman equation, the stress spectrum is combined with material properties data and stress intensity factor solutions applicable to each principal structural element to determine the number of cycles or flight hours to failure. This number is usually divided by a factor of two to arrive at the inspection interval. Dividing by a factor of two ensures that, should a principal structural element develop a crack, it will be inspected at least once before the crack propagates to failure.

Unlike the safe-life approach where components are retired whether or not they are damaged, damage tolerant components are only replaced if a crack is found during an inspection. It is important to note that, although the detectable crack size dictates the time between inspections to a certain degree, any size crack found during an inspection mandates replacement of the damaged component. In addition to its economic advantages, the damage tolerance approach also allows for a reduction of safety factors in design. Since fracture mechanics provides a more precise characterization of crack behavior, the large scatter factors typical of fatigue methods are not required.

Damage tolerant components are designed on the basis of residual strength as shown in Fig. 4. The residual strength of a cracked component decreases as the crack grows longer. After a certain amount of time, the residual strength of the component will fall below the maximum expected service stress, at which time failure of the component becomes probable. The inspection interval must be made short enough to locate the damage before a crack reaches a length likely to result in failure. The task of the designer is to produce components that have sufficient residual strength to withstand the maximum stresses expected, while at the same time, maximizing the critical crack length. Some of the ways in which this task can be achieved are by selecting a material with good fracture toughness, eliminating stress concentrations, or integrating crack stoppers into the design.

The above discussion has focused on the deterministic aspects of damage tolerance. Stochastic methods, such as risk analysis, also play a role. Combining the data generated from deterministic analysis with stochastic factors such as manufacturing quality, risk analysis attempts to assess the probability of fracture of principal structural elements. An initial flaw in a principal structural element is represented by an Equivalent Initial Flaw Size Distribution (EIFSD), which is obtained from tests that determine the distribution of times it takes a crack of some initial size to reach a specified reference size (Ref 24). This distribution of times is referred to as the Time to Crack Initiation (TTCI) distribution. This approach differs from a deterministic analysis which assumes a single value for the initial flaw size. Once the TTCI distribution is known, the EIFSD may be obtained for any service time. Inspection intervals are then established based upon the amount of time it takes the crack sizes at the lower end of the distribution to reach the critical crack length. These inspection intervals may later be modified to account for the quality of maintenance and inspection capability (the probability of detecting cracks).

Life Limiting Factors

To extend the life of an aircraft, one first must understand those factors which limit its life. It is the purpose of the following discussion to briefly discuss some of those aspects that control the durability of the aircraft. The life of an aircraft is

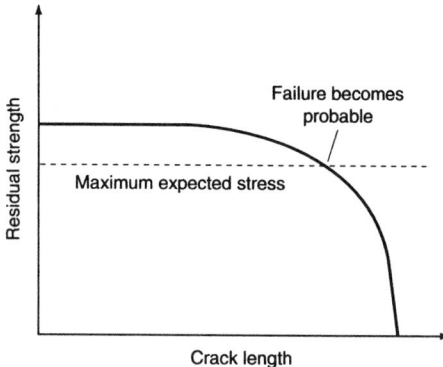

Fig. 4 Schematic of a residual strength diagram

dependent upon a number of variables, such as flight profiles, usage rate, external environment, material selection, and design philosophy. The first two of these (flight profile and usage rate) affect the design life of the aircraft, i.e., the number of expected safe flight hours. The external environment affects corrosion, and therefore component life, while material selection (composite versus metal structure) and design philosophy affect individual components. In a study conducted by the NTSB, they found that in the fifteen years between 1970 and 1985, corrosion or fatigue was involved in 77 out of 579 aircraft incidents (Ref 25). These many incidents may be greatly reduced through improved design methods and inspection procedures.

It is interesting to note that while the aircraft is operational, the dynamic factors (fatigue, flutter, vibration) are the responsible parameters for determining the life aspects of selected components, but while the aircraft is sitting on the ground, corrosion behavior is normally the dominant driver. It should be noted that this is not the case in the galley, lavatories, and bilge areas. There are factors, however, that exacerbate both these conditions. These are manufacturing and maintenance actions which allow for both the premature initiation and propagation of cracks and also enhance the onset and continuing damage of corrosion. These may be dings, scratches, grinding marks, improper cleaning, non-removal of drilled or reamed material, etc.

Manufacturing defects may be subdivided into two broad categories. These are material defects and mechanical defects. Both these type of defects generally exist in the aircraft at the time the product leaves the manufacturer. The material defect may manifest itself as a result of a metallurgical anomaly. This anomaly or metallurgical defect may occur as a result of faulty furnace control during heat treating, improper plating which may result in unwanted hydrogen in high-strength steel, processing problems (forging laps, shuts, undesirable grain structure, etc.), welding defects, etc. A second type of material defect may occur in composite materials. Defects that occur in composite materials include debonding, poor ply adhesion, improper cleaning, and non-removal of ply separation material.

For example, a material defect caused failure of an F-111 aircraft which used D6AC steel in the wing box. Material allowables were obtained through a testing program. Unfortunately, the test specimens to determine the toughness of the steel had a section thickness different from the material placed upon the aircraft. The material located on the aircraft had a toughness substantially lower, approximately half, than the toughness of the material used in the test program due to the difference in thermal behavior during heat treatment. Additionally, a forging lap was present in the material used for the joint. The crack was not observed during inspection, and the material was placed onto the aircraft with the forging lap present. The aircraft was flown for 107 flights safely, at which time catastrophic failure occurred causing the destruction of the aircraft and the death of the pilot (Ref 26).

Subsequent inspection noted the presence of the flaw, and a rigorous test program indicated a significant difference in material toughness between the coupons used in the original test program, and the forging used on the aircraft. As a result of this failure, means to periodically inspect the components were required. After much deliberation, the method utilized was a proof test. Chambers were built that housed the F-111, and a system was designed and fabricated that would lower the temperature of the chamber below nil ductility transition (NDT) for the steel. The aircraft was placed into the chamber, the temperature stabilized, and fixtures would load the wing box as required. If a crack equal to or greater than critical existed in the structure, failure would occur. If no failure occurred, the aircraft would be able to safely fly another flight interval. This inspection technique was repeated for each aircraft as appropriate.

There are a number of examples where failures of structural components initiated from manufacturing defects. For military aircraft, these include T-38 and F-5 (improper use of a router caused these aircraft to lose wings as a result of fatigue cracking) (Ref 27), civilian aircraft include United Flight 232 which failed near Sioux City, Iowa in July, 1989 (Ref 28), and failures of titanium discs on an RB211 engine (Ref 29). The Sioux City mishap was due to a manufacturing problem in a titanium fan disc, while the RB211 failures were due to tensile residual stresses due to processing of the titanium. These residual stresses when added to the normal operating stresses caused an effective operating stress that had sufficient magnitude to lower the fatigue life of the structure. It can be seen that these manufacturing defects occur as a result of both manufacturing assembly and material processing.

Corrosion is a source of time-dependent metal damage that requires protection of structure according to FAA regulations (Ref 30). In addition, the FAA has issued an Advisory Circular (AC) on corrosion (Ref 31). This document discusses the types of corrosion, those areas on the aircraft susceptible to corrosion, inspection techniques, the agents which cause corrosion, and methods of corrosion removal and component rework. The

AC is not meant to be all-inclusive, but is a useful primer for those interested in corrosion. In the *Metals Handbook*, Volume 13, an article "Corrosion in the Aircraft Industry" discusses the many types of corrosion found in aircraft along with many illustrative examples (Ref 32).

In most cases, corrosion itself does not lead to structural failure, but it is the most expensive aspect of maintaining structural integrity. One operator of a fleet of aircraft indicated that 95% of the parts that were removed on his aircraft were removed as a result of corrosion (Ref 25). The damage that is caused to the aircraft can manifest itself as corrosion and paint degradation. Both of these are exacerbated by the presence of salt water, acid rain, and sunshine. As a result, the down time of the aircraft, whether it is kept in a hanger or on the flight line, and its primary geographical location all become important variables in the determination of the inspection intervals and types of inspections that are required.

The effect of corrosion needs to be addressed by denoting the types of corrosion and where they can occur, inspection techniques to find the corrosion in situ, and repair or replacement methods to remove the offending material. Each of these areas is an extensive technical topic, which is beyond the scope of this article. In abbreviated form, corrosion may be broken into two broad categories: enhanced fatigue crack growth due to corrosion and general corrosion. General corrosion includes the degradation of coatings due to atmospheric conditions. Several types of corrosion (Table 1) occur in aircraft, and these different types of corrosion must be defined and located. As an example, Boeing developed a Corrosion Task Force to accumulate and review all corrosion problems existing on their aircraft (Ref 33). Boeing developed a program that identified, to a specific usage baseline, the location, type, and inspection intervals for their aircraft. If a particular aircraft were based in an area that has a more severe environment, these intervals would be decreased. As a result of this task force, Boeing now has corrosion prevention and control programs on their aircraft, and has developed a Corrosion Design Handbook. The feedback between the users of the aircraft and its manufacturers has led to designs which decrease corrosion.

Table 1 Types of corrosion

Bonded Structure Corrosion
Concentration Cell Corrosion
Crevice Corrosion
Exfoliation Corrosion
Fretting Corrosion
High Temperature Corrosion
Intergranular Corrosion
Pitting Corrosion
Cavitation Corrosion
Corrosion Fatigue
Erosion Corrosion
Filiform Corrosion
Galvanic Corrosion
Hydrogen Embrittlement
Microbiological Corrosion
Stress Corrosion Cracking
Uniform Corrosion

Environmentally Enhanced Fatigue Crack Growth. Crack growth analysis is normally performed using material properties obtained in laboratory air. It is well known that the introduction of active environments (sump water, salt water, and fuel) increase crack growth rate (Ref 34-37). Along with the environment, other important factors include the cyclic frequency and shape. It has been shown that the greater the time the material is exposed to a high stress in an aggressive environment, the greater the crack growth rate (Ref 38, 39). Analysts have developed many techniques to account for the enhanced crack growth rate in hostile environments, but no universal method has yet been determined. Generally, the analyst uses the increased crack growth rate of the particular environment (cyclic, chemical, and temperature), and it has been found that the inspection intervals decrease. However, the alternating stresses that cause the crack to grow occur in flight, where the effect of salt water (a ground environment condition) is absent. Additionally, at high altitudes, the temperatures are decreased thereby slowing the chemical reactions. In wet wings, the effect of the chemical environment is always present, however the temperature decreases as altitude increases. The complexity of accurately determining the advance of the crack is a substantive problem. An approach that is used by the Air Force to aid in solving this problem is through the use of a *Corrosivity Factor* (Ref 40). This method documents the air bases at which the particular aircraft was assigned, and then each base was assigned a particular Corrosivity Factor. This empirical approach allows determination of crack growth rate, and thus inspection intervals as a function of the particular location of the aircraft and its corrosive behavior as a function of that location.

Improper Maintenance. The manufacturer of the aircraft system presents many documents with that system as part of its delivery. These include pilots handbooks, structural repair manuals, corrosion manuals, etc. It is imperative that these manuals be adhered to in a rigid fashion. It is easy to devise methods that may be shorter and easier than those discussed in the manual, but the manufacturer selected that particular procedure for sound reasons. It is a danger to deviate from these methods for the deviation itself may place undue stress or cause latent damage to components apart from those being examined. This high stress or latent damage can lead to premature failure of these other components, and that may result in undesirable consequences. Additionally, the personnel performing the particular functions must take special care to ensure that the maintenance action itself does not result in damage to the aircraft. American Airlines Flight 191 crashed in Chicago in May, 1979 as a result of improper maintenance actions (Ref 41).

The previously mentioned crash of a U.S. Air Force T-38 trainer occurred as a result of a router mark on the lower skin of the wing. The mark was quite small, but as the stresses were significant, a crack initiated and grew until failure occurred. The fatigue crack, at failure, was only 0.097 in.

deep. A similar mark also caused the crash of an F-5 aircraft. From these examples, it may be seen that extreme care during maintenance and the inspection following maintenance must be addressed to ensure that proper procedures and techniques are rigorously followed.

Multiple Site Damage. The effect of fatigue crack growth on the life of aircraft has been much studied in recent years. Since the accident involving Aloha Airlines Flight 243 in April, 1988, the interest regarding multiple site damage has increased significantly (Ref 42). The Air Force, through their specifications, account for multiple site damage by assuming that each fastener has a crack of a particular size, and the most critical location has a crack with a significantly larger size. As the aircraft is utilized, both these types of cracks grow, albeit at different rates. There are methods that are commonly used to retard multiple site damage. These include fastener selection, metal working, and component geometry. Additionally, much analytical and test work has been accomplished to study the effect of many small cracks and their linking together. Methods and procedures that are used are discussed in this section.

The onset of multiple site damage (MSD) especially in fuselage structure in the first row of rivets in joints is a serious concern. It is well known that in a riveted, load carrying structure, the load is transferred from one panel to another through the rivets. If there are a number of rows of these rivets (assuming a lap joint shown in Fig. 5), the highest load is in the first row of rivets, the second highest load in the second row of rivets, etc. The loads in the holes adjacent to these rivets are twofold. One needs to consider both the far field stress (the pressure loads), and the local loading of the rivets on the skin (the bearing load). It is the superposition of these separate loads that must be considered in this type of structure. Much work has been accomplished in determining these loads, and a paper by Swift discusses a solution that is based upon an experimental study (Ref 43).

The localized or bearing loads are quite high, but the length over which they are dominant is quite small. However, the initiation of a crack by these loads has much significance. Because a crack may be initiated in a short time period, the crack growth life of the structure must be carefully analyzed. If an existing crack is located on an airframe, or if the user wishes to make a change or addition which necessitates additional cutouts in the fuselage structure, a doubler must be added. The design of the doubler, and the method of attachment of the doubler to the existing structure must be ascertained. Because the loads from the skin are placed into the doubler through the fastener, and Newton's Laws dictate displacement compatibility of the fastener/doubler/skin system, the thickness of the doubler greatly affects the load on the fastener. As the doubler thickness increases, the load induced on the fastener increases, with a resulting decrease in crack initiation time. This decrease in crack initiation time greatly affects the life of the overall

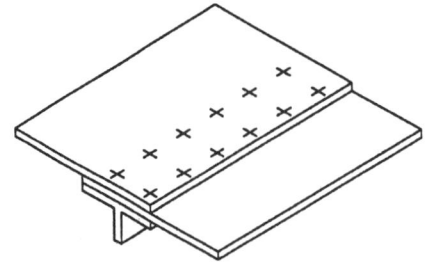

Fig. 5 Lap splice joint

structure. Other aspects of the fastener design also play a large role in structural life. This includes fastener spacing and type, and hole preparation.

Life Extension Techniques

Methods of extending aircraft life are dependent upon a multiplicity of factors. The damage tolerance portion of life extension depends primarily upon three independent variables. These are the material properties, usage history, and the damage intrinsic to the material component itself. Knowledge of these variables and their interaction, if any, allows determination of component life. In other instances, fatigue initiation or crack growth may determine the life of a particular part or component. In these latter instances, knowledge of the geometry and location of the flaw, and the loading and usage of the aircraft become increasingly important. To extend the life of the structure it is necessary to quantify (either deterministically or statistically) the information obtained from field experience on either these or like aircraft, and then have analysis techniques that allow the user to determine inspection procedures and techniques, and their appropriate timing to extend the life of the selected components. In those instances where corrosion and its growth may be the driving factor behind the life of the particular structural member, the determination of component life is much less straightforward. There is no known method for determining either corrosion initiation or propagation times. As a result, periodic inspections to seek out the existence of corrosion is the only viable method. This is accomplished by preventative means. What becomes important is the geographical location of the airport where the aircraft is parked, material selection, surface preparation, surface protection, material removal and cleaning methods. It is these factors that dictate component replacement, and thus may be the relevant factor in aircraft life.

The FAA has issued Advisory Circular 91-56, "Supplemental Inspection Program for Large Transport Category Airplanes," dated May 6, 1981, with its purpose being the issuance of a supplemental inspection document (SID) (Ref 44), and Advisory Circular 91-60, "The Continued Airworthiness of Older Airplanes," dated June 13, 1983 with its purpose to expand the scope of the previous advisory circular to other airplanes (Ref 45). The SID informs the user of

the changes required in the maintenance actions as the aircraft accumulates an increasingly larger number of service hours. The inspection program defined by the SID requires the user to continually re-evaluate the inspection procedures and intervals based upon usage, service experience of the particular fleet of aircraft, and advances in technology (analyses, testing, inspection procedures, etc.). The SID emphasized obtaining this information on principal structural elements. The interested reader should examine these documents, and additionally Ref 46 and the article "Damage Tolerance Certification of Commercial Aircraft" in this Volume.

Corrosion Protection. The inspection of components to ensure that corrosion has not initiated and grown to a condition where part cleanup or replacement is necessary must be accomplished. But prior to this, it is important to design the particular components in a fashion that impedes this insidious growth. There are several different approaches to counteract the onset and growth of corrosion. Many of these methods are applicable to only one type of component, but there are methods that lend themselves to several portions of the aircraft. The methods that the manufacturer may use to attain a longer corrosion initiation time can fall into several categories. These include appropriate selection of materials, judicious choices for materials that are used in contact with one another, the design of the metallurgy of the material for the intended loading, and the coating or surface treatment of the material itself.

Aluminum, as an alloy, has been used on aircraft for many years. However, the alloy itself and its heat treatment have changed substantially. Manufacturers have used duralumin (2024) in the 1940s, 7178, 7079, and 7075 in the 1960s. The first two of these alloys, 7178 and 7079, were both of higher strength than the 2000 series alloys, but the stress-corrosion cracking susceptibility was substantially greater. As a result of the increased degradation in active environments, these alloys were no longer used in aircraft design. The alloy 7075 was first used in its high-strength temper, T6. It was found that the alloy in this temper was susceptible to stress-corrosion cracking. Experiments determined that if this alloy was overaged, and used in the T76 or T73 condition, its resistance to stress-corrosion cracking was enhanced. There was, of course, a trade-off. By going to the overaged condition, the strength of the alloy was lessened.

Aluminum alloys were also susceptible to surface corrosion, pitting corrosion, and exfoliation. It was found that by cladding the alloy with pure aluminum, the resistance to these types of corrosion was enhanced. For many aircraft usages, these alloys are now being used in the clad condition. It should be noted that the clad is removed at fastener holes due to the machining process. The fasteners are usually countersunk, and as a result, the clad material is removed in their vicinity. Rivets for aircraft are generally manufactured from 2117 aluminum, therefore, no galvanic reaction occurs at the skin/rivet interface. When steel bolts are used, they are usually cadmium plated to minimize galvanic corrosion. Other surface treatments of aluminum, in addition to priming and painting the surface, include Alodine and anodizing.

Another material commonly used in the manufacture of aircraft is titanium. This alloy is stiffer and stronger than aluminum, yet not as dense as steel. Its corrosion resistance is quite good, although it is susceptible to chloride. As a result, both sea water and methyl chloride can cause corrosion in titanium.

The remaining metal that is used widely on aircraft is steel. As low-alloy steels are commonly used on aircraft, the effect of corrosion must be continually addressed. A primary location of steel is in the landing gear. In this usage, steel is the preferred material because of its high strength and stiffness. To minimize the amount of material used, the selected alloy is generally used in its high-strength condition. As the strength levels are increased, the material becomes chemically more active. For this reason, many of the steel alloys (4340, 4140, 300M, etc.) are coated or plated, usually with cadmium, to prevent exposure with the atmosphere.

Mechanical Techniques of Controlling Multiple Site Damage. A typical lap splice joint, shown in Fig. 5, can attach a stringer to the outer skin of the aircraft, and have a crack arrest strip integral to the design. Since these rivets are highly loaded due to both pressurization and fuselage bending loads, and additionally having superimposed localized bearing loads, small cracks can initiate and propagate. And, as the Aloha accident has shown, a sudden failure in a panel can cause several additional panels to fail almost instantly. The MSD is the existence and linking of these small cracks into a large panel crack that can lead to catastrophic consequences.

The design of the doubler helps determine both the stress concentration at the doubler/skin interface, and the inspectability of the structure itself. These two considerations are both important in evaluating MSD. The location of the skin/stringer or the skin/longeron interface attachment has a known load. The known load at the first rivet away from the attachment is substantially similar, but the thickness at that location is substantially smaller. The smaller cross sectional thickness results in higher stresses. Because the length over which the stress increase occurs is limited, a stress concentration results. To avoid these stress concentrations, designs that alleviate the stress increase may be considered. These designs may include a tapered skin, a tapered doubler, or the tying in of the first rivet at different rows. To aid in inspecting the structure, the designer may include a doubler designed with tabs that allow inspection of the skin material in a straightforward manner. Figure 6 illustrates these designs.

The bearing stress on a rivet is dependent upon its pitch (the distance between adjacent rivets). For an aircraft skin that is 0.040 in. thick and subject to a stress of 15 ksi, the shear flow would be 600 lb/in. For a basic five row doubler joint (0.040 in. skin and 0.050 in. doubler) with a rivet

spacing of 1.00 in. and a rivet diameter of 0.125 in., a displacement compatibility analysis would indicate a maximum shear load of 176 lb/rivet on the outer row of rivets. The bearing stress from these rivets would be 35 ksi. As the allowable bearing stress (for yielding conditions) on 2024-T3 alloy is 88 ksi, these stresses are acceptable. If the rivet spacing were increased to 2 in. (50 mm), then the bearing stress on the same rivets would become 55 ksi (380 MPa). Even though the static margin is high, the change in stress due to the fastener spacing is significant.

Another important factor in improving component life is hole preparation. If a hole is drilled through the skin and doubler, the resulting stress concentration factor is three. However, if the hole is cold worked by expanding the material through the use of a mandrel and split sleeve, then a residual compressive stress is placed into the material. This residual compressive stress greatly reduces the effect of the bearing stress at the edge of the hole, and crack initiation time is greatly lengthened (Ref 47). Furthermore, small cracks or discontinuities near the edge of the hole are closed, and component life is greatly enhanced.

A second method which can be used to improve component behavior through hole preparation and component attachment is fastener selection. There are several types of fasteners that are used to attach structural members (rivets, bolts, hi-loks), while hi-tigue fasteners actually deform the material adjacent to the hole and impart a residual compressive stress. The effect of the residual compressive stress is to reduce the localized stress in the vicinity of the hole thus improving component life.

Inspection. The inspection programs for each particular aircraft are described in its SID. The rationale for these inspections, on the many

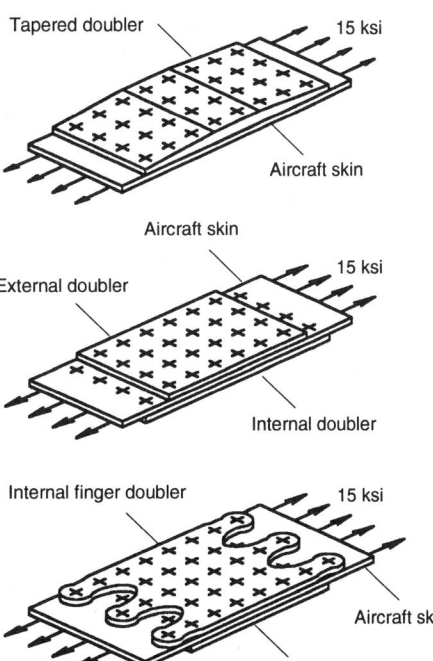

Fig. 6 Doubler designs that reduce first fastener load

PSEs, may include both a deterministic and a stochastic approach. As the aircraft ages, the number of inspection locations based upon initial flaw assumptions using EIFSD usually grows. As a result, the aging of the aircraft causes an increase in the inspection time. The determination of when and where to inspect becomes increasingly complicated. Using a lead-the-fleet aircraft, or a full scale test article gives good indication of the growing flaw population. The continuous analysis of the data from the lead the fleet aircraft, and all other aircraft inspected can be used to tune the inspection intervals. A goal of this iterative testing and analysis procedure is to determine the economic life of the aircraft itself. It is at this juncture that the aircraft is retired, because further usage can no longer be economically justified.

The documentation that is necessary to accomplish this complex undertaking is lengthy. At McDonnell Douglas, as stated in Ref 46, three volumes of information are prepared. These are:

- Volume I, General Information, PSE Description, Inspection Requirements
- Volume II, Inspection Methods
- Volume III, Inspection Planning and Status Report

McDonnell Douglas plans to annually update and republish Volume III by a computer program specifically written for this purpose (Ref 46). It will contain information on fleet status, PSE population, projected supplemental inspection start dates and intervals for each PSE, inspection planning data, and a record of past inspection, according to Ref 46.

Inspection Procedures for Corrosion. Inspection procedures used on aircraft cover the gamut of inspection methods used in other industries. They include visual, radiography, ultrasonic, acoustic emission, eddy current, magnetic particle, and dye penetrant (Ref 48). The type of inspection procedure is dependent upon the location of the component tested, its material, and the type of corrosion. In many instances, it is necessary to gain access to the area of interest.

The preparation of a Corrosion Prevention and Control Plan, as described above, is paramount. The user must know where to inspect, how to inspect, and how often to inspect. These inspection documents are normally quite detailed. A typical mechanism is the definition and use of PSE's. These are structural components that can be grouped together in terms of corrosion susceptibility or inspectability. For each PSE, and that may consist of many components, the user is informed of all aspects of corrosion susceptibility and inspection procedures. This plan generally will call out many other specifications (ASTM, SAE, and Military) which discuss particular methods and procedures of accomplishing the necessary inspection. In many instances, it is necessary to remove plating or painting, and after the inspection to repaint or replate the affected areas.

Repair Limits for Corrosion Damaged Structure. The user of the aircraft receives from the manufacturer details on maintainability or repair of the aircraft. The user is given informa-

methods of first identifying, and secondly removing the corrosion from the aircraft. In some instances, e.g., pitting corrosion, the removal method may be as straightforward as sanding or by using chemicals to remove the offending material. If one gets to a more severe corrosion such as exfoliation, the removal of segments of material must be accomplished. If the removal of the material is within the prescribed limits of the manufacturer, the manufacturer's repair manual is sufficient. However, if the repair falls outside of the limits set by the manufacturer, approval to exceed these limits must be obtained from the FAA.

Additionally, corrosion is generally insidious. The user does not know of the existence of corrosion until it is seen, and then the type and destruction it causes cannot be deterministically evaluated. The repair of corrosion is generally an iterative process of clean-up and evaluation. Again, depending upon the type of corrosion, the amount required to clean-up cannot be determined, and portions of the structure must be removed. Stress-corrosion cracking follows grain boundaries in aluminum alloys, and its depth and severity cannot be accurately determined in a nondestructive fashion.

After the corroded area has been identified and the corrosion product removed, the repaired area should have the same surface treatment as the unaffected area adjacent to it. This may necessitate painting, plating, or using some other surface treatment on the material. As stated above, the manufacturer's repair manual usually discusses these requirements in detail.

Ongoing Programs

A document entitled "A Review of Research in Aeronautical Fatigue in the United States, 1993-1995" was presented at the ICAF meeting in May, 1995 (Ref 49). A summary of the programs that investigated corrosion and multiple site damage is given below.

Analytical and Test Studies on Corrosion. There were ten programs discussed that were in the area of corrosion and corrosion fatigue, and thirteen papers were referenced. The studies consisted of examining the formation of fatigue cracks in 2024 and 7075 aluminum. Initiating mechanisms included pits that were formed as a result of pitting corrosion, wear that may cause fretting, especially in joints and around rivets, and areas that had experienced prior corrosion such as exfoliation and crevice corrosion. Additional studies determined novel means (shot peening) to determine the extent of exfoliation corrosion. Finally, there were a number of studies that examined the crack growth behavior of aluminum alloys exposed to hostile chemical environments.

Analytical and Test Studies of Multiple Site Damage. The document listed nine research programs that were conducted in the U.S. in this time period on MSD, and they reference fourteen papers. The research consisted of analytical studies encompassing the loss of residual strength as a result of MSD, the study of crack opening displacement (COD) and the critical crack tip opening angle (CTOA) to determine the onset of MSD. Another study was centered upon the development of a finite element program that contained techniques that allow the investigator to quickly and easily conduct analysis at the crack tip with many different structural geometries which included stringers, lap joints, tear straps, frames, etc. Additions to this program include the use of elastic-plastic FEA to better define the effect of growing cracks on residual strength.

Experimental studies that were accomplished included the teardown inspection of fuselage structure that has experienced 60,000 pressurization cycles. The locations and the size of the cracks, and whether they could be identified using nondestructive techniques are goals of the study. Other studies included the effect of crack arrest straps in fuselage structure, and large scale testing of actual fuselage structure. Finally, a study on multiple site damage in fuselage lap splices was conducted.

The conclusions from many of these efforts could be formulated on a statistical basis, i.e., a risk analysis. The results of the risk analysis would give the user information regarding the ultimate safety of the aging aircraft fleet. It may be seen that the study of MSD is a fertile area, and that Aloha Flight 243 initiated several well-needed studies.

REFERENCES

1. Richard C. Rice, Fatigue Data Analysis, *Metals Handbook*, Vol 8, ASM, 1985, p 695-720
2. S. Suresh, *Fatigue of Materials*, Cambridge University Press, Cambridge, 1991
3. C.R. Soderberg, Factor of Safety and Working Stress, *Transactions of the American Society of Mechanical Engineers*, Vol 52, 1939, p 13-28
4. J. Goodman, *Mechanics Applied to Engineering*, Longmans Green, London, 1899
5. H. Gerber, Bestimmung der zulassigen Spannungen in Eisen-konstructionen, *Zeitschrift des Bayerischen Architeckten und Ingenieur-Vereins*, Vol 6, 1874, p 101-110
6. A. Palmgren, Die Lebensdauer von Kugellagern, *Zeitschrift des Vereins Deutscher Ingenieure*, Vol 68, 1924, p 339-341
7. M.A. Miner, Cumulative Damage in Fatigue, *Journal of Applied Mechanics*, Vol 12, 1945, p 159-164
8. W.J. Crichlow, On Fatigue Analysis and Testing for the Design of the Airframe, *Fatigue Life Prediction for Aircraft Structures and Materials*, LS 62, AGARD, 1973, p 6-1-6-36
9. A.A. Griffith, The Phenomenon of Rupture and Flow in Solids, *Philosophical Transactions of the Royal Society*, Vol A221, 1920, p 163-198
10. G.R. Irwin, Analysis of Stresses and Strains Near the End of a Crack Traversing a Plate, *Journal of Applied Mechanics*, Vol 24, 1957, p 361-364
11. Standard Test Method for Plane-Strain Fracture Toughness of Metallic Materials, E399, *Annual Book of ASTM Standards*, Vol 03.01, ASTM, Philadelphia, 1985, p 547-582
12. David Broek, *Elementary Engineering Fracture Mechanics*, Kluwer Academic Publishers, Dordrecht, 1991
13. P.C. Paris, M.P. Gomez, and W.P. Anderson, A Rational Analytic Theory of Fatigue, *The Trend in Engineering*, Vol 13, 1961, p 9-14
14. J.E. Collipriest and R.M. Ehret, A Generalized Relationship Representing the Sigmoidal Distribution of Fatigue Crack Growth Rates, *Rockwell International Report SD74-CE-0001*, 1974
15. R.G. Forman, V.E. Kearney, and R.M. Engle, Numerical Analysis of Crack Propagation in Cyclic-Loaded Structures, *Journal of Basic Engineering*, Vol 89, 1967, p 459-464
16. E.K. Walker, The Effect of Stress Ratio During Crack Propagation and Fatigue for 2024-T3 and 7075-T6 Aluminum, *Effects of Environment and Complex Load History for Fatigue Life*, STP 462, ASTM, Philadelphia, 1970, p 1-14
17. T. Swift, Damage Tolerance in Pressurized Fuselages, *New Materials and Fatigue Resistant Aircraft Design*, 14th Symposium of the International Committee on Aeronautical Fatigue, Ottawa, 1987, Douglas Paper 7768, p 4-11
18. M. Stone, Fatigue and Fail-Safe Design of a New Jet Transport Airplane, *Fatigue Design Procedures*, Proceedings of the 4th Symposium of the International Committee on Aeronautical Fatigue, E. Gassner and W. Schutz, Ed., Pergamon Press, London, 1969, p 1-65
19. U.S. Department of Transportation, *Damage Tolerance Assessment Handbook*, Report No. DOT-VNTSC-FAA-93.13.I, 1993, p 1-11
20. MIL-STD-1530, *Aircraft Structural Integrity Requirements*, U.S. Air Force
21. MIL-A-83444, *Airplane Damage Tolerance Requirements*, U.S. Air Force
22. 25.571 Subpart C, Part 25-Airworthiness Standards: Transport Category Aircraft, *Federal Aviation Regulations*, Federal Aviation Administration, Washington, D.C.
23. Advisory Circular 25.571-1A, *Damage Tolerance and Fatigue Evaluation of Structure*, Federal Aviation Administration, Washington, D.C.
24. A.P. Berens, J.C. Burns, and J.L. Rudd, Risk Analysis for Aging Aircraft Fleets, *Structural Integrity of Aging Airplanes*, S.N. Alturri et al., Ed., Berlin, 1991, p 37-51
25. Proceedings of the International Conference on Aging Airplanes, June 1988, DOT-TSC-FAA890-88-26
26. U.A. Hinders, "F-111 Design Experience— Use of High Strength Steel," AIAA 2nd Aircraft Design and Operations Meeting
27. M.P. Kaplan, TAC T-38 Wing Life Assessment, Program Management review, Wright Patterson Air Force Base, OH, 1973
28. Aircraft Accident Report, United Airlines, Inc., N1819U, NTSB-AAR-90-06, Nov 1990
29. U.S. Air Force Personnel, private discussion
30. CFR Title 14, Part 25.609, Protection of Structure

31. "Corrosion Control for Aircraft, Federal Aviation Administration Advisory Circular AC 43-4A, July 1991
32. M.L. Bauccia, "Corrosion Control in the Aircraft Industry," *Metals Handbook*, 9th ed., Vol 13, *Corrosion*, ASM, 1987
33. U.G. Goranson, "Damage Tolerance: Fact and Fiction," 14th Platema Memorial Lecture, presented at ICAF, Stockholm, Sweden, June 1993
34. A. Hartman, "On the Effect of Water Vapor and Oxygen on the Propagation of Fatigue Cracks in an Aluminum Alloy," *Int. J. Fract. Mech.*, Vol 1, 1965, p 167-183
35. M.R. Achter, "Effect of Environment on Fatigue Cracks," *Fatigue Crack Propagation*. STP 415, ASTM, 1967, p 181-204
36. P. Hartman and J. Schijve, "The Effects of Environment and Frequency on the Crack Propagation Laws for Macrofatigue Cracks," *Eng. Fract. Mech.*, Vol 1, 1970, p 615-631
37. T.T. Shih and R.P. Wei, "Load and Environment Interactions in Fatigue Crack Growth," Prospects of Fracture Mechanics," Noorthoff, 1974, p 237-250
38. J. Schijve and D. Broek, "The Effect of Frequency on the Propagation of Fatigue Cracks," NLR TR-M-2094, National Aerospace Institute, Amsterdam, 1961
39. J. Schijve and P. DeRijk, "The Effect of Temperature and Frequency on Fatigue Crack Propagation in 2024-T3," NLR TM-M-2138, National Aerospace Institute, Amsterdam, 1963
40. R.N. Miller and F.H. Meyer, "Computerized Corrosion Forecasting Model for C-5 Aircraft," *Proceedings of the 1987 ASIP/ENSIP Conference*, AFWAL-TR-88-4128
41. Aircraft Accident Report, American Airlines, Inc., DC-10-10, N110AA, NTSB-AAR-79-17, May 1979
42. W.R. Hendricks, "The Aloha Accident—A New Era for Aging Aircraft, *Structural Integrity of Aging Aircraft*, S.N. Alturri, et al., Ed., Springer-Verlag, 1991, p 153
43. T. Swift, "Report to Damage Tolerant Aircraft," op.cit., p 433
44. "Supplemental Structural Inspection Program for Large Transport Category Airplanes," Federal Aviation Administration Advisory Circular AC 91-56, May 1981
45. "The Continued Older Airplanes," Federal Aviation Administration Advisory Circular AC-91-60, June 1983
46. P.R. Abelkis, et al., "Use of Durability and Damage Tolerance Concepts in the Development of Transport Aircraft Continuing Structural Integrity Programs," presented at ICAF, Pisa, Italy, May 1985
47. B.K. Young, Jr., "Cold Expansion Techniques for Fatigue Enhancement," Durability of Metal Aircraft Structures, S.N. Alturri, et al., Ed., Atlanta Technology Publications, 1992, p 405
48. *Metals Handbook*, 9th ed., Vol 17, *Nondestructive Evaluation and Quality Control*, ASM, 1989
49. J.L. Rudd, "A Review of Research on Aeronautical Fatigue in the United States, 1993-1995," presented at ICAF, Melbourne, Australia, May 1995

SELECTED REFERENCES

- David Broek, *Elementary Engineering Fracture Mechanics*, Kluwer Academic Publishers, Dordrecht, 1991
- Department of Defense, *Metallic Materials and Elements for Aerospace Vehicle Structures*, MIL-HDBK-5F
- J.P. Gallagher, F.J. Giessler, A.P. Berens, and R.M. Engle, *USAF Damage Tolerant Design Handbook: Guidelines for the Analysis and Design of Damage Tolerant Aircraft Structures*, Report No. AFWAL-TR-82-3073, Wright Patterson Air Force Base, Ohio, 1984
- S. Suresh, *Fatigue of Materials*, Cambridge University Press, Cambridge, 1991

Damage Tolerance Certification of Commercial Aircraft

T. Swift, Federal Aviation Administration

DAMAGE TOLERANCE is the current philosophy used for maintaining the structural safety of commercial transport aircraft. The requirements for damage tolerance certification of these structures are contained in the Federal Airworthiness Requirements (FAR) 25.571 (Ref 1). Guidance for complying with FAR 25.571 is given in Advisory Circular 25.571-1A from the Federal Aviation Administration (FAA) (Ref 2). This article describes the structural evaluations necessary to comply with the regulation.

For a complete explanation of damage tolerance certification of commercial transport aircraft, it is appropriate to review the historical evolution of the damage tolerance philosophy. This is followed by discussion of design philosophies and a summary of evaluation tasks for damage tolerance certification.

Historical Background

After many major fatigue failures in the 1950s on both military and commercial aircraft, the most notable of which were the DeHavilland Comet failures in early 1954, the U.S. Air Force (USAF) initiated the Aircraft Structural Integrity Program (ASIP) in 1958. The fatigue methodology adopted in the ASIP was the reliability approach, which became known as the "Safe-Life" method. (See the section "Structural Design Philosophies" in this article.) This approach, used in the development of USAF aircraft in the 1960s, involved analysis and testing to four times the anticipated service life. On the commercial scene, another philosophy, "Fail Safety," was introduced in the early 1960s, and a choice between Safe-Life and Fail-Safe methods was allowed by commercial airworthiness requirements. However, it was found that the Safe-Life method did not prevent fatigue cracking within the service life, even though the aircraft were tested to four lifetimes to support one service life (i.e., scatter factor of 4). Some notable examples were:

- *KC-135 Safe-Life 13,000 h:* 14 cases of unstable cracking in lower wing skins between 1800 and 5000 h. Lower wing surfaces were reskinned at 8500 h.
- *F5 Safe-Life 4000 h:* Fatigue failure of the lower wing skin root region at 1900 h caused loss of aircraft.
- *F-111 Safe-Life 4000 h:* Fatigue failure of the center wing box at 105 h caused loss of aircraft.

Almost without exception, initial manufacturing or in-service accidental damage is the cause of fatigue cracking that occurs within one service lifetime on aircraft that has been substantiated both analytically and by test to four lifetimes. In fact, studies performed by the Air Force Flight Dynamics Laboratory in the late 1960s (reported in AFFDL TR-70-149) indicated that more than 56% of all in-service fatigue problems were caused by pre-existing material and fabrication quality deficiencies (i.e., initial manufacturing damage not simulated in the fatigue test or accounted for analytically).

The fatigue failure of the F-111 aircraft cited above was caused by a defect in the D6AC steel center wing box fitting, which propagated over 105 h of service to a critical size at tension stresses induced by a 4.0 g maneuver. The aircraft was designed for a load factor of 11.0 g. Of course the fatigue test was free of this defect. The failure resulted in the largest single investigation of a structural alloy ever undertaken and precipitated investigations into the history of USAF accidents related to fatigue. These investigations culminated in a complete change in design criteria for USAF aircraft. Many of the design specifications were changed and others were introduced. The most important document to come out of the investigations was MIL-A-83444, "Airplane Damage Tolerance Requirements," issued in July 1974. This document specified the fracture mechanics principles to be used in the design of all future military aircraft. Since the release of MIL-A-83444, all USAF aircraft previously designed to the Safe-Life philosophy have been assessed using damage tolerance principles.

While the USAF was developing its change in design philosophy, the commercial large transport industry was also considering changes in design philosophy. After the first generation of fail-safe aircraft had been in service for some time, the Civil Aviation Authority (CAA) in the United Kingdom became concerned about the loss of fail safety with time due to widespread fatigue damage. The CAA considered it risky that on U.S. designs, the fatigue testing of large components and full-scale structures was not continued for as long as on equivalent structures in Europe. Most of the full-scale testing was done to two lifetimes or less, compared to five lifetimes in Europe. In addition, the CAA believed that U.S. inspection programs were not based on a sound engineering evaluation. In the early 1970s, with these concerns in mind, the CAA limited the service life of the Boeing 707 aircraft on their registry to 60,000 h.

At about the same time, the FAA became concerned about the fact that old first-generation fail-safe aircraft were being sold to new operators. It believed that new operators of old aircraft might not be familiar with the scope of required maintenance and the means that had been used to keep the aircraft in a safe and airworthy condition. The FAA held a series of meetings with the industry about aging aircraft issues, and as a result it issued a number of Advisory Circulars containing Maintenance Inspection Notes for several of these aircraft.

In late 1976, as the life limit of 60,000 h approached, U.K. operators of Boeing 707 aircraft became concerned about the economic impact of their aircraft's being grounded. However, the CAA was convinced that its policy was well founded, and it decided to seek confirmation of its policy from the Airworthiness Requirements Board (ARB). The ARB asked its technical committee to consider the matter, and a number of meetings took place starting in March 1977. These meetings included manufacturers from the

United States and the United Kingdom as well as representatives from the FAA.

The life-limiting policy of the CAA had been substantiated by the loss of a Hawker Siddley 748 in Argentina, one of the first aircraft designed by Hawker Siddley Manchester Division on fail-safe principles. Wing bending material, instead of being concentrated in front and rear spar caps (as had been the case with the Manchester and Lancaster bombers, Shakletons, Tudors, and the Argossy), was distributed to form a multiplicity of stiffening elements on a multiple-load-path fail-safe principle. Even so, on 14 April 1976, separation of the right wing occurred due to undetected fatigue. The crack had originated at rivet holes of a reinforcing plate near the external rib of an engine, and it had propagated undetected because of the lack of adequate inspections based on engineering evaluation (i.e., damage tolerance). The aircraft had accumulated 25,760 h.

Another important incident that took place during the discussion period about aging aircraft was the loss of a Boeing 707 at Lusaka, Zambia in May 1977. This aircraft suffered the loss of pitch control following the in-flight separation of the right-hand horizontal stabilizer and elevator, which resulted from fatigue failure of the rear spar upper cap. This incident was important because the structure had been designed on fail-safe multiple-load-path principles similar to those used for the 748. The spar had an upper and lower cap plus a fail-safe center member. These members extended to three sets of lugs for fail safety. The propagation life of the upper cap from initiation to failure was approximately 7,200 flights. A further 100 flights took place with the upper spar cap failed, but the failure was undetected prior to separation of the stabilizer. The fail-safe multiple-load-path philosophy did not prevent catastrophic failure, again because of the lack of proper inspections based on engineering evaluation. At the time of failure the aircraft had accumulated 47,621 h and 16,723 flights. This accident reinforced the CAA position, and it was probably the most important incident during the discussions because it happened only two months after the start of the meetings.

After considerable discussion among authorities and manufacturers, a policy was adopted in January 1978. This policy is set out in CAA Airworthiness Notice 89, issued on 23 August 1978 with Information Leaflet AD/IL/0067/I-5. As an alternative to "life-ing" the aircraft, these documents require a structural review that results in a supplemental inspection program. As an example, the 707 limits enforced by the CAA could be eliminated pending the completion of a structural audit (damage tolerance assessment) that resulted in inspections in addition to the existing inspection program. Such supplemental inspections were made mandatory by the CAA.

The activity in Europe in the late 1970s was paralleled in the United States. In November 1976, the FAA gave notice of a Fatigue Regulatory Review Program for commercial transport aircraft and invited interested persons to submit proposals to amend the requirements of FAR

25.571 and 25.573. These proposed amendments were precipitated by significant developments in the state of the art and in industry practice regarding fatigue and damage tolerance evaluation of commercial transport aircraft. The FAA subsequently held a Fatigue Regulatory Review Conference in March 1977 to obtain the views of all concerned. (Approximately 13 meetings had been held with industry and other organizations prior to this time.) This joint effort by industry and government led to the deletion of FAR 25.573 and a revision and expansion of FAR 25.571. Amendment 45 was issued in December 1978, slightly preceded by FAA Advisory Circular 25.571-1A, which outlines an acceptable means of compliance. Thus, all commercial transport aircraft designed after December 1978 are required to meet the damage tolerance requirements of FAR 25.571.

The issue of Airworthiness Notice 89 for the aging commercial fleet in Europe was equaled in the United States. A number of meetings with the Aircraft Industries Association had taken place by the late 1970s, and a rough draft Advisory Circular was formulated as a result of committee action. Eventually, in May 1981, FAA Advisory Circular 91-56 was released. It contains advisory material related to the aging aircraft problem, and it requires evaluation of the structure using modern fracture technology and a Supplemental Inspection Document. The philosophy incorporated into the Advisory Circular is identical to that of Airworthiness Notice 89: continued safe operation of aircraft beyond the intended life can be ensured by a continuing inspection program that is supplemental to the basic maintenance program.

Structural Design Philosophies

Definitions

Damage tolerance is the attribute of the structure that permits it to retain its required residual strength for a period of use after the structure has sustained specific levels of fatigue, corrosion, or accidental or discrete source damage. The term "specific levels of damage" refers to damage between the levels of in-service detectability and critical at-limit load. This damage will vary, depending on the inspection method and the limit load capability for each principal structural element (PSE). A PSE is an element that contributes significantly to carrying flight, ground, and pressurization loads and whose failure, if remained undetected, could eventually lead to loss of the aircraft. FAA Advisory Circular 25.571-1A lists examples of PSEs.

Fail-safe is the attribute of the structure that permits it to retain its required residual strength for a period of use after the failure or partial failure of a PSE, prior to repair.

The safe life of a structure is the number of events (e.g., flights, landings, or flight hours) during which there is a low probability that the strength will degrade below its designed ultimate value due to fatigue cracking.

General Requirements

In general, an evaluation of the structure under typical load and environmental spectra must show that catastrophic failure due to fatigue, corrosion, or accidental damage will be avoided throughout the operational life of the aircraft. This can be ensured only with an adequate inspection program, and therefore the evaluation must result in inspection and maintenance procedures for each PSE. The evaluation must show that the structure is damage tolerant for fatigue, corrosion, or accidental or discrete source damage unless the manufacturer can show, to the satisfaction of the FAA, that damage tolerance is impractical. In the latter case a safe-life evaluation must be made using appropriate scatter factors. Impracticality should be related to geometrical constraints such as in-service non-inspectability rather than to manufacturing cost differences between safe-life and damage-tolerant elements.

Philosophy of the Damage Tolerance Evaluation. Because of the large number of areas to be considered in the damage tolerance evaluation, a high degree of reliance must be placed on analysis. It is recognized that to test every PSE for damage tolerance characteristics would generally be economically unfeasible. Thus, the evaluation is considered to be an analytical evaluation. However, sufficient testing must be performed to ensure that analysis methods are not unconservative. This is particularly true in the field of fracture mechanics, the primary tool used in damage tolerance evaluation. Each manufacturer must provide a separate report describing the analytical methodology used in the evaluation. The report must contain sufficient analysis/test correlation to assure the FAA, with a high degree of confidence, that methods are not unconservative.

The FAA does not dictate how many locations should be evaluated. A plan is normally submitted to the FAA Certification Office for approval. Some manufacturers have evaluated as many as 500 elements, while others have considered less than 100. This variability is primarily due to differences in how each manufacturer interprets the extent of each PSE.

Flyable Crack Philosophy. Damage tolerance evaluation has been interpreted in the past as a means to allow continued safe operation in the presence of known cracking. This interpretation is incorrect. No regulations allow the strength of the structure to be *knowingly* degraded below ultimate strength (1.5 times limit). The damage tolerance evaluation is merely a means of providing an inspection program for an aircraft that is not expected to crack under normal circumstances but may crack in service due to inadvertent circumstances. If cracks are found in primary structure, they must be repaired before further flight. The only allowable exception is through an engineering evaluation, which must show that the strength of the structure will never be degraded below ultimate strength during flight with known cracking. This evaluation must take into account slow, stable tearing during an ultimate load application.

Damage Tolerance Evaluation Tasks

The objective of damage tolerance evaluation is to provide a crack growth and residual strength analysis, based on the expected use of the aircraft, so that an inspection program can be developed to maintain safety during service. The primary tasks to be accomplished by the damage tolerance evaluation are:

- Define aircraft utilization
- Develop load factor spectra at the aircraft center of gravity
- Select critical locations for evaluation
- Develop stress spectra for each location
- Establish the environment for each location
- Develop crack growth rate (da/dN) data for each location
- Validate basic crack growth analysis methods
- Obtain fracture toughness data for each material and geometry
- Determine the extent of damage for each location at limit load
- Validate residual strength analysis methods
- Determine the structural category for each location
- Produce a crack growth curve for each location
- Convene meetings between airline operators, manufacturers, and authorities
- Collectively decide on inspection methods, thresholds, and frequencies consistent with operational economics

A flowchart of the first 12 tasks, which involve an engineering evaluation, appears in Fig. 1. Details about the tasks are provided below.

Aircraft utilization specifications should account for the different flight types expected in service, the flight lengths in terms of hours, and the percentage of time spent flying each flight type. Flight types may include long versus short flights, cargo versus passenger flights, low-altitude versus high-altitude flights, and the percentage of time spent in training. For each flight type it is important to consider the distribution of time spent at each gross weight, speed, and altitude.

Consideration also needs to be given to the structural elements involved in the damage tolerance evaluation. For example, low-level flying is extremely severe on lower wing surface crack growth life. However, for the majority, but not all, of fuselage critical elements, this effect may be secondary to the cabin pressure. In the case of longitudinal cracking in the fuselage, the internal cabin pressure is the most important loading condition. For circumferential cracking on the crown of the fuselage skin, the primary loading would most likely be fuselage downbending due to inertial loads.

Flight profiles for each flight type should be compiled that reflect each segment of the flight, such as taxi, climb, cruise, descent, landing impact, and landing roll. The number of different flight types considered may vary from about 25

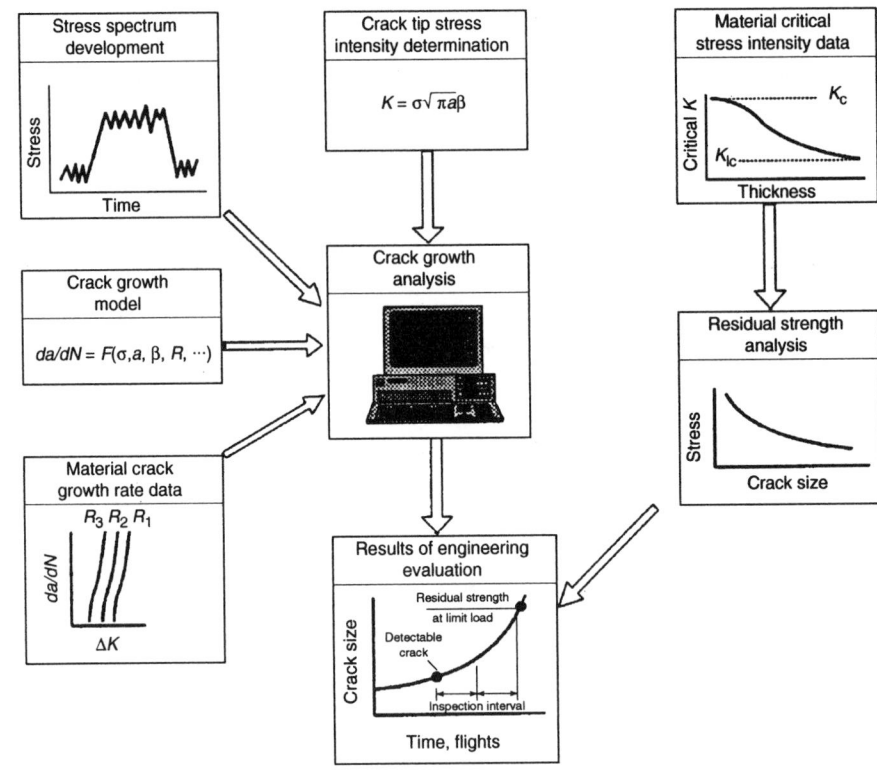

Fig. 1 Crack growth and residual strength evaluation for each principal structural element

for a complex military aircraft to as low as five or six for a commercial transport aircraft.

Load factor spectra at the aircraft center of gravity are usually developed in the form of data about cumulative frequency of occurrence of incremental load factor. These data, obtained from recorded flight histories for similar aircraft, are developed for taxi, maneuver, gust, and landing impact. All loads in the spectra must be initially considered. The spectra may be simplified in some cases by eliminating loads that are not damaging. This must be done with extreme care when considering crack growth. In the past, nondamaging loads have been eliminated by ensuring that they produce stresses below the endurance limit for crack initiation. For spectra to be used in crack growth analysis and testing, considerable care must be exercised when eliminating cycles; otherwise, the crack growth life will be unconservatively affected.

The load exceedance data referred to above is available from various sources, such as Engineering Sciences Data Unit Ltd. (ESDU) (Ref 3). Both gust and maneuver exceedances are included in ESDU data sheets. Manufacturers obtain in-flight recorded history from many other references, such as Ref 4 to 6. One of the more recent programs to obtain in-flight data was the Digital VGH Program of the National Aeronautics and Space Administration (NASA), which used data from existing flight data recorders on L-1011, B-727, B-747, and DC-10 aircraft. This program, originally terminated by NASA before completion, was later completed under a NASA-

FAA interagency agreement. This work is summarized in Ref 7.

Critical Locations For Evaluation. Following are some considerations for choosing the location where damage is to be simulated in the PSE:

The static stress analysis should be reviewed to determine areas where static margins are at a minimum (primarily areas subjected to tension and shear loading).

Locations of high stress concentration should be reviewed, particularly those at which a number of surfaces intersect each other, creating a feather edge that is prone to accidental damage. These areas usually occur in forgings and are often not easy to detect on drawings.

Locations of high spectrum severity should be reviewed not only for areas of high stress (1 g), but also for areas where a large number of cycles may occur during every flight. One such area, not usually given enough attention, is flap support structure. On most aircraft this structure is subjected to aerodynamic buffeting loads that may occur thousands of times per flight. The stresses induced by these loads, although often low in range, are usually at high stress ratios and can contribute significantly to crack growth damage.

Locations where stresses would be high in secondary members after failure of a primary member should be carefully reviewed. This is particularly necessary for each area that is not inspectable for less than primary member failure. In such a case the inspection interval will depend on the life of the secondary member under redistributed loads after primary member failure.

Locations where crack propagation rates are high and fracture toughness values are low should be reviewed.

The results of full-scale fatigue testing can be valuable for locating critical fatigue areas of the structure. This testing is normally conducted too late in the design development of the aircraft structure to be useful in the initial selection of critical locations. However, maintenance programs can easily be modified to include these areas. In fact, it is expected that the results of the damage tolerance evaluation submitted prior to certification will be updated as service and full-scale test experience is gathered.

Locations of likely fatigue damage and crack propagation paths, particularly when the crack path might be affected by multiple-site damage, may be most accurately determined by component fatigue testing to complete failure.

The number of areas considered should ensure that adequate coverage exists to maintain safety. This is a general statement, of course, that means little when a manufacturer is trying to judge the scope of the damage tolerance evaluation task. Generally speaking, the initial number of areas considered is approximately 150 for the complete airframe. During the course of the evaluation, this may be reduced to about 90 PSEs for which crack growth and residual strength results are documented. The number varies with aircraft type, of course, but is not necessarily a function of aircraft size because all aircraft have the same major features.

PSE Stress Spectra Development. Stress spectra must be developed for each PSE. This is done by converting the load factor spectra at the aircraft center of gravity (discussed above) using stress transfer functions. These transfer functions are required for each segment of the flight profile. In certain parts of the structure, such as the outer wing, dynamic magnification factors are often necessary. Load factor exceedance data are converted to stress exceedance data for each PSE, then into flight-by-flight stress spectra for use in a crack growth computer program. Most of the major manufacturers use proprietary computer programs to perform this conversion, but one commercially available program is LICAFF (Ref 8).

Because the damage tolerance evaluation of the structure is essentially analytical, it is important to develop spectra that minimize crack growth life consistent with aircraft use. From an economic standpoint, not all cycles that may be applied to a structure in service need to be considered, but all cycles that contribute to crack propagation damage must be considered.

In the past, stress spectra were developed in the form of blocks of cycles at various stress levels, to be used in either crack initiation analysis or testing. The analysis was based on linear cumulative damage laws. The spectra were simplified by eliminating nondamaging cycles with stresses below the endurance limit. This was possible because crack initiation damage is not as sensitive to low-level cycles as crack propagation damage. However, for crack propagation analysis that sat-

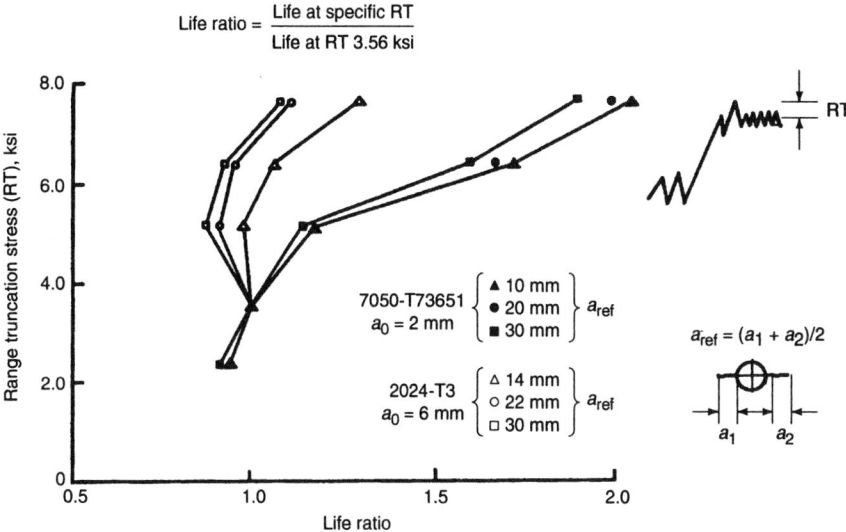

Fig. 2 Effect of stress range truncation on crack growth life. Source: Ref 9. These data were included in Ref 9 by permission of the DeHavilland Aircraft Co. of Canada.

isfies the requirements of FAR 25.571, it is better to arrange the spectra in flight-by-flight order. Spectrum simplification by elimination of cycles should be performed with extreme caution, because low-level cycles that may be nondamaging during the crack initiation phase may be quite damaging during the crack propagation phase. In order to ensure inclusion of all cycles that contribute to crack growth damage, a stress range truncation study must be performed for each aircraft type. An element most sensitive to the gust environment, such as the lower wing surface, is normally chosen. The results of such studies have been quite unexpected to some manufacturers.

Figure 2 shows minimum stress range truncation data for damage tolerance evaluation of the Dash 8 aircraft. The data presented are for 2024-T3 (lower wing surface skin material) and 7050-T73651 (lower wing surface stiffener material). Stress spectra were generated at four stress range truncation levels for the 2024-T3 material: 3.56, 5.18, 6.4, and 7.66 ksi (24.5, 35.7, 44.1, and 52.8 MPa). For the 7050-T73651 material, an additional stress range of 2.35 ksi (16.2 MPa) was considered. Figure 2 also defines the stress range truncation level. Taking the first stress range of 3.56 ksi (24.5 MPa) as an example, all cycles with a stress range less than 3.56 ksi were eliminated from the spectrum. In the second case, all cycles with a stress range less than 5.18 ksi (35.7 MPa) were eliminated, and so on. Each data point in Fig. 2 is an average of three specimen tests.

In the case of the 2024-T3 material, the initial crack length (a_0), an average of two through cracks at a fastener hole, was 6 mm (0.24 in.). Three final crack lengths were chosen: 14, 22, and 30 mm (0.55, 0.87, and 1.18 in.). For the 7050-T73651 material, the average a_0 was 2 mm (0.08 in.), and the three final crack lengths considered were 10, 20, and 30 mm (0.4, 0.8, and 1.2 in.). Figure 2 represents the life to grow the cracks from initial to final size, but it is plotted against a life ratio equal to the life at the particular stress

range truncation level divided by the life at a stress range truncation level of 3.56 ksi (24.5 MPa). The objective is to fix the stress range truncation level that results in the lowest crack growth life. For the Dash 8 wing spectra, this value was 5.18 ksi (35.7 MPa) for the 2024-T3 skin material and 2.35 ksi (16.2 MPa) for the 7050-T73651 stiffener material. A weighted mean stress range truncation level of 3.56 ksi (24.5 MPa) was chosen for the Dash 8 spectrum.

It can be seen that there is considerable variation in crack growth life, depending on the lower stress range truncation level used. The question of which level to use is often difficult, because one needs to balance safety with economics. Figure 2 illustrates the importance of this decision from a safety standpoint. The case for economics is illustrated by Fig. 5 of Ref 9 (which is identical to Fig. 5-4 of Ref 10), which shows that for a short-takeoff-and-landing transport aircraft spectrum, there are six times more cycles in a spectrum at a stress range truncation level of 2.0 ksi (14 MPa) than at 4.0 ksi (28 MPa). The economic issue is most important when developing spectra to be used for full-scale fatigue testing.

The highest loads in the spectrum should be those that occur once in one-tenth the projected life of the aircraft. Including higher loads can increase the projected life due to retardation effects, which may not be realistic for all aircraft in a fleet. Reference 10 provides details of the effect of including high loads that occur less frequently than once in one-tenth the projected life of the aircraft.

PSE Environment. FAR 25.571(a)(1)(i) states that "The typical loading spectra, temperatures, and humidities expected in service" apply. This should be interpreted to mean that any environment that can affect the crack growth life or residual strength should be accounted for.

Low temperatures can considerably reduce the residual strength of cracked structures. This is true particularly for 7000-series aluminum alloys and some 2000-series alloys. Crack growth life for these alloys generally improves with decreasing temperature, but care should be exercised in taking full advantage of this, because the majority of crack growth damage for some elements is experienced at low altitudes and during landing and takeoff.

Increased humidity and moisture usually increase crack growth. Salt content in the atmosphere can also increase crack growth rates, particularly for aircraft taking off and landing over the ocean through mist or fog.

Sump tank water containing chlorides can subject lower wing surface elements to increased crack growth rates.

Increased temperatures: Components subjected to increased temperatures, such as those surrounding engine installations, should receive careful attention. Crack growth rates are usually much higher in materials subjected to temperatures higher than room temperature.

Corrosive environments: Care should be exercised when developing crack growth rate data in a corrosive environment. This is true particularly in the very low ΔK region at low stress ratios such as zero. Crack tip blunting due to corrosion can reduce crack growth rates and create misleading results. The effect of a corrosive atmosphere is more pronounced at high stress ratios, and care should be taken to account for this effect.

Every effort should be made to specify the environment that each PSE will be subjected to in service, so that the crack growth and residual strength data will not be unconservative.

Crack Growth Rate Data. The primary material property required for crack growth analysis is the crack growth rate (*da/dN*) versus the crack-tip stress-intensity factor range (ΔK). Paris et al. (Ref 11) were the first to observe that *da/dN* plotted against ΔK is a straight line on log-log paper. The stress-intensity factor range on a cycle is defined as $\Delta K = K_{max} - K_{min}$, where K_{max} and K_{min} are the maximum and minimum applied stress-intensity factors, respectively. Crack-tip stress intensity is defined as:

$$K = \sigma(\pi a)^{1/2}\beta \qquad (Eq\ 1)$$

where σ, a, and β are the applied gross stress, crack size, and effect of geometry, respectively.

It is essential to develop data for *da/dN* versus ΔK by testing for each material, product form, and heat treatment considered in the damage tolerance evaluation. This data should be generated for the specific environment that the PSE will be subjected to in service. The loading direction and crack orientation, relative to the material grain direction, should also be considered.

Developing *da/dN* data is probably the most important task in the damage tolerance evaluation. The data collected in the past have produced a wide variety of results, largely because there has been no standard procedures for collecting and reporting them. The American Society for Testing and Materials (ASTM) has developed a standard test method (Ref 12) that may reduce some of the scatter that apparently exists in the

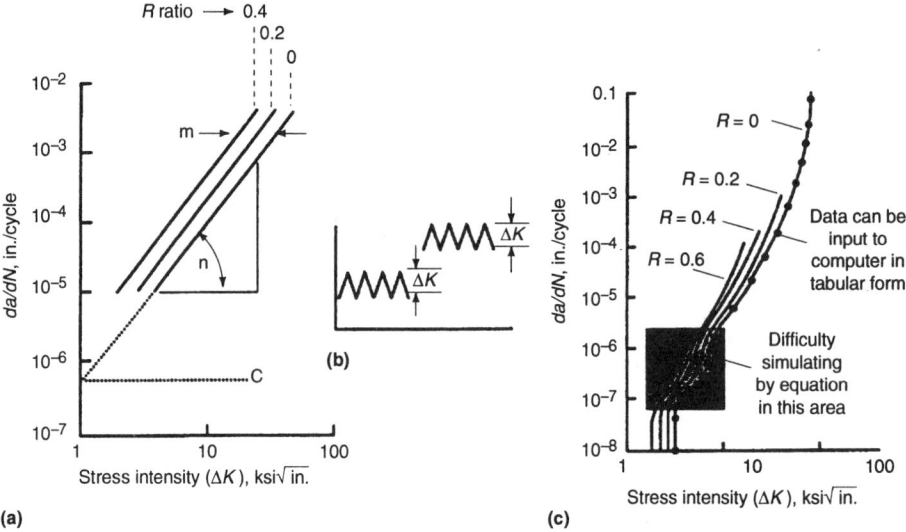

Fig. 3 Crack growth rate data simulation

literature. It is still acceptable to use data from the literature, provided that the origin of the data and the means of collection are known, and provided that assurance can be given that the data are reliable and applicable to the element in question. The data generation standards should be compared to current standards to ensure reliability. Reference 12 contains code for a FORTRAN computer program that generates a polynomial fit to the crack growth data. The polynomial is mathematically differentiated to provide *da/dN* data. Use of this program will also help reduce apparent scatter in the data.

There are a number of ways to represent *da/dN* versus ΔK, and each manufacturer has its own method. The simplest form is through the Paris equation (Ref 11):

$$da/dN = C(\Delta K)^n \qquad (Eq\ 2)$$

where *C* is the intercept of the vertical axis at $\Delta K = 1.0$ and *n* is the slope of the line. This equation gives a constant slope on log-log paper, as shown in Fig. 3(a).

The Paris equation was modified by Walker (Ref 13) to account for the stress ratio *(R)*, as indicated in Fig. 3(b)

$$da/dN = C[\Delta K/(1-R)^{1-m}]^n \qquad (Eq\ 3)$$

The Walker equation enables lines to be plotted at each *R*-ratio, as shown in Fig. 3(a). The value *m* adjusts the width of the line spread, as shown in this figure.

The Walker equation is widely used throughout the world, but many other equations are also in constant use. These include the Forman equation (Ref 14), the Collipriest equation (Ref 15), and the modified Collipriest equation (Ref 16). The latter is probably the most sophisticated, but none of these equations can be considered exact, and they were developed to fit test data. This can be done for most materials over the range of *da/dN*

data shown in Fig. 3(a), but difficulty is experienced with some materials at lower *da/dN* levels, as shown in Fig. 3(c).

Many manufacturers now represent *da/dN* data in tabular form and input the data for each *R*-ratio into a computer, which performs logarithmic interpolation between each data point and each *R*-ratio. This procedure has been proven to give the most accurate results. Irrespective of the method used to represent *da/dN* data, it should be considered a requirement to validate the approach using a small element test program.

Crack Growth Analysis Model Validation. As mentioned above, each manufacturer has developed its own methodology for crack propagation analysis. These methods vary from extremely simplistic to extremely sophisticated. The simplistic methods are usually inexpensive for the manufacturers but tend to produce conservative results, which may be burdensome to the airline operator. The more sophisticated methods are more expensive to the manufacturer in terms of computer time, but the results are usually more accurate and the inspection frequencies specified may not be as burdensome to the operator.

There are many elements to the basic crack propagation computer program:

- Method of representing *da/dN*
- Basic *da/dN* integrating routine ($N = \int da/F(\Delta K$, *M*, *R*, etc.)
- Method of cycle representation
- Retardation model
- Acceleration model
- Method of calculating *K*

Each of these elements interacts with the others, and the combination of the elements must be validated by testing.

The method of representing da/dN varies, as discussed above. In the opinion of this author, a tabular input of the data yields the most accurate results.

Basic da/dN Integration Scheme. Some programs incorporate complex integration routines that tend to be costly in terms of computer time. Attempts are often made to reduce the cost of cycle-by-cycle integration by assuming constant growth rates over groups of similar cycles within a flight. This reduces the effective number of cycles required in cycle-by-cycle integration.

The method of cycle representation in the crack growth program has presented problems in the past. Some manufacturers consider each cycle and assume that crack growth occurs on increasing load, so that the inputs are the maximum stress and the *R*-ratio for each cycle. Other manufacturers "rain flow count" the cycles. The two representations give different crack growth lives using the same *da/dN* data. This problem is briefly discussed in Ref 9.

Retardation Model. There are a number of crack growth retardation models in use today. Some of the more common ones are the Wheeler model, the Willenborg model, the generalized Willenborg model, and the closure model. The Wheeler model depends on test data to establish an exponent term in the model, whereas the other three models are analytical. This author has had considerable success using the generalized Willenborg model. Some manufacturers have not incorporated retardation models into their crack growth programs and are therefore providing conservative crack growth curves.

Acceleration Model. Very few programs currently make use of acceleration models. Unless aircraft use will include extremely high sink rates on landing (which cause high compressive stresses in the lower wing surface), an acceleration model may not be necessary.

The method of calculating K is of course extremely important for accurate analysis. A number of classical solutions are usually incorporated into the computer program in analytical form. However, to account for the complex effects of geometry, such as in stiffened structure, it is essential to include a tabular routine to input β in the stress-intensity factor equation, Eq 1.

Validation. There is significant variability between different manufacturers' programs for obtaining data about crack length versus time. Therefore it is essential to validate the basic elements of the program by performing crack growth tests on simple unstiffened panels under flight-by-flight spectrum loading. For the spectrum cycle elimination discussed above, the specimens used to determine the maximum stress range truncation level are ideal for such a study.

Fracture toughness data should be obtained for the full range of material thickness used. This is necessary because the critical stress-intensity factor varies with material thickness, from the higher plane-stress values (K_c) in thin materials through a mixed-mode region to the lower plane-strain values (K_{Ic}) in thicker parts. Critical stress-intensity factors in the mixed-mode region are influenced by the percentage of plane-strain fracture, which is a function of material thickness. As the material thickness is reduced, a greater por-

tion of the fracture surface will be plane-stress (45°) fracture.

Standard testing methods are available for plane-strain fracture toughness to obtain K_{Ic} values (Ref 17). However, the determination of K_c is difficult because many phenomena influence plane-stress fracture, including panel width effects, crack buckling, and slow stable tearing.

Panel Width Effects. If the plane-stress critical stress-intensity factor is plotted against panel width for ductile materials, it becomes apparent that critical *K* is affected by panel width, even though fracture toughness is supposed to be a material parameter. For example, K_c values cannot be fully developed for 2024-T3 sheet unless the test panel is at least about 60 in. wide, whereas full K_c values can be developed for 7075-T6 sheet using a panel width of 16 in. This is due to gross net section yielding of the uncracked ligament. This effect can mislead designers who conduct parametric studies during candidate material selection.

Crack buckling, which is caused by induced compressive stress parallel to and along the edge of the crack, reduces the effective critical stress-intensity factor at failure. This is due to either local bending at the crack tip or induced Mode III fracture. Antibuckling guides are commonly used to stabilize the crack edge during center crack panel testing.

Slow stable tearing is another effect that needs careful consideration in thin sheet ductile materials. Crack extension can occur during application of a single load cycle. This phenomenon is usually represented by an *R*-curve of applied *K* plotted against slow stable crack extension, Δa.

The fracture toughness of many aluminum alloys is affected by temperature. For example, at –65 °F (–55 °C), considered to be the soaked-out temperature at altitude, the fracture toughness of most 7000-series alloys is approximately 28% lower than at room temperature. This is also true of some 2000-series alloys, particularly 2014-T6. The effects of temperature should be given careful consideration in residual strength calculations, particularly in candidate material selection for tension critical structure.

Fracture toughness in either plane-stress or plane-strain regions is affected by grain direction for a number of alloys, so data appropriate for the correct grain orientation should be used.

PSE Limiting Residual Strength. It is important to determine the maximum extent of damage that each PSE can tolerate at the limit load conditions specified in FAR 25.571(b). This damage extent is considered the limit for crack growth calculations, and an inspection program must be established to detect damage before these proportions are reached. Two-bay skin damage with central broken stiffening elements is usually inspectable by walk-around if it is easily observable from the ground.

Discrete Source Damage. FAR 25.571(e) requires a discrete source damage evaluation. The aircraft must be capable of successfully completing a flight during which structural damage is likely to occur as a result of:

- Impact with a 4 lb bird at cruise velocity (V_c) at sea level or at 0.85 V_c at 8000 ft, whichever is greater
- Uncontained fan blade impact
- Uncontained engine failure
- Uncontained high-energy rotating machinery failure

Immediately obvious damage from discrete sources should be determined, and the remaining structure should be shown to have static strength for the maximum loads (considered ultimate loads) expected during completion of the flight. According to FAA Advisory Circular 25.571-1A, the load conditions during and subsequent to the incident should be as follows:

- *At the time of the incident:* The maximum normal operating differential pressure multiplied by a factor of 1.1 (valve tolerance), plus the expected external aerodynamic pressures during 1 g level flight combined with 1 g flight loads. The aircraft, assumed to be in 1 g level flight, should be shown to be able to survive any maneuver or any flight path deviation caused by certain incidents listed in FAA Advisory Circular 25.571-1A, taking into account any likely damage to the flight controls and normal corrective action by the pilot.
- *Following the incident:* 70% limit maneuver loads and, separately, 40% of the limit gust velocity (vertical or lateral) at the specified speeds, each combined with the maximum appropriate cabin differential pressure (including the expected external aerodynamic pressure). The aircraft must be shown by analysis to be free from flutter up to the maximum dive speed (defined by the dive speed/Mach number envelope), with any change in structural stiffness resulting from the incident.

The federal regulations do not specify damage sizes or acceptable risk levels in the event of uncontained engine failure. In the recent past, manufacturers have conducted a quantitative risk assessment using the interpretive material in European Joint Airworthiness Regulation (ACJ) 25.903(d)(1) to show compliance with FAR 25.571(e). Briefly, for the more critical case this assessment involves showing that there is not more than a 1 in 20 chance of catastrophe resulting from the release of a single one-third piece of disc plus 1/3 height of a blade. Trajectories for this engine debris are outlined in FAA Advisory Circular 20-128 (Ref 18).

In the case of the fuselage, considered to be the most critical element for discrete source damage, the risk assessment described in ACJ 25.903(d)(1) involves an evaluation of fuselage bending capability in the presence of the discrete source damage. However, what is most important is to prevent explosive decompression failure due to longitudinal damage. Many of the larger manufacturers have a history of wedge penetration testing: a pressurized fuselage section, subjected to downbending loads, is pierced with a sharp wedge blade to simulate engine disc damage. Such testing is described in Ref 19.

It is the opinion of this author that, although not required by regulations, wise manufacturers will design the pressure cabin to sustain a full two-bay skin crack with a broken frame if there is a tendency for skin fast fracture within two bays at the applied principal stress resulting from skin hoop tension and fuselage downbending shear stresses. This is usually easy to achieve in a fuselage with crack stopper straps, as shown in Fig. 4. Assuming that the applied principal stress is given by line AD and that the damage size is as shown, skin fast fracture will occur at B, and the damage will be arrested at C. If the fuselage design does not include crack stopper straps and the peak of the residual strength curve is at point E, fast fracture at B will not be arrested, and there will be fuselage explosive decompression failure. However, if the combination of skin fracture toughness and applied stress level is such that fast fracture will not occur within two bays, as indicated by line FG, then it is not normally necessary to require capability for full two-bay damage.

Residual Strength Analysis Validation. Although nonlinear analysis is necessary for some configurations, linear elastic fracture mechanics has been used with a high degree of success to calculate residual strength in the presence of fatigue damage for complex structures. A wide variety of methods are available:

• Lumped parameter finite element analysis
• Energy release rate finite element analysis
• Cracked element finite element analysis
• Displacement compatibility analysis

It is generally understood that the maximum extent of damage at limit load cannot be determined for every PSE by test. Thus, this task is considered primarily analytical. However, a background of enough testing must be performed on complex structures to provide assurance that the methods used are reliable. As a minimum, the types of structures to be considered are:

• Lower wing surface stiffened components
• Horizontal stabilizer upper surface stiffened components
• Fuselage stiffened components subjected to pressure for longitudinal cracks
• Fuselage stiffened components (may be flat panels) for circumferential cracks
• Any other principal structure for which analysis is considered unreliable

The two-bay skin crack criterion was introduced with wide-body aircraft. Although not specifically required by regulation, it is wise to design the vast expanse of basic structure in the fuselage, lower wing surface, horizontal stabilizer upper surface, and so on to sustain a two-bay skin crack with a broken central stiffener at limit load. This damage scenario is consistent with expected cracking scenarios, and in many cases it relieves airline operators of the need for considerable sophisticated nondestructive inspection (NDI).

Figure 5(a) illustrates a fatigue-critical location in a fuselage skin at the first fastener in a circum-

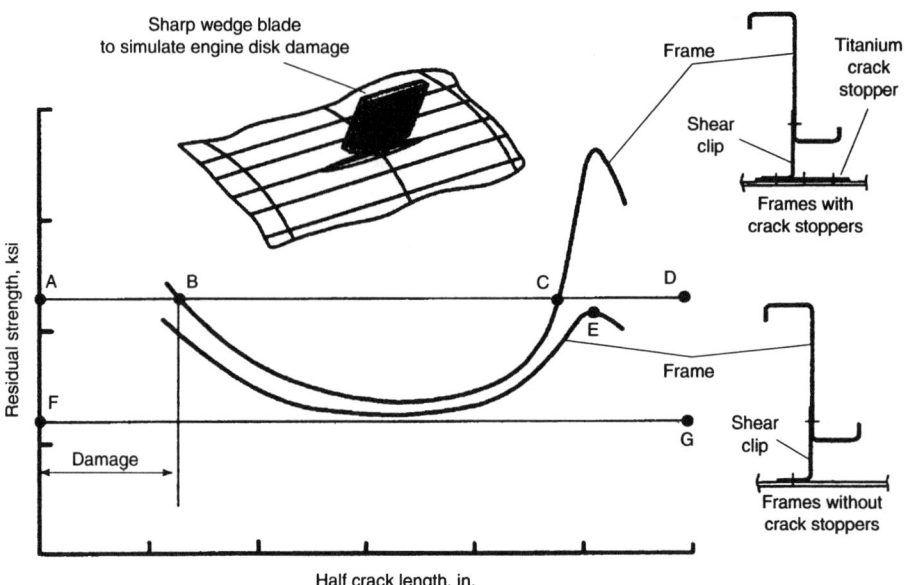

Fig. 4 Discrete source damage residual strength for longitudinal damage

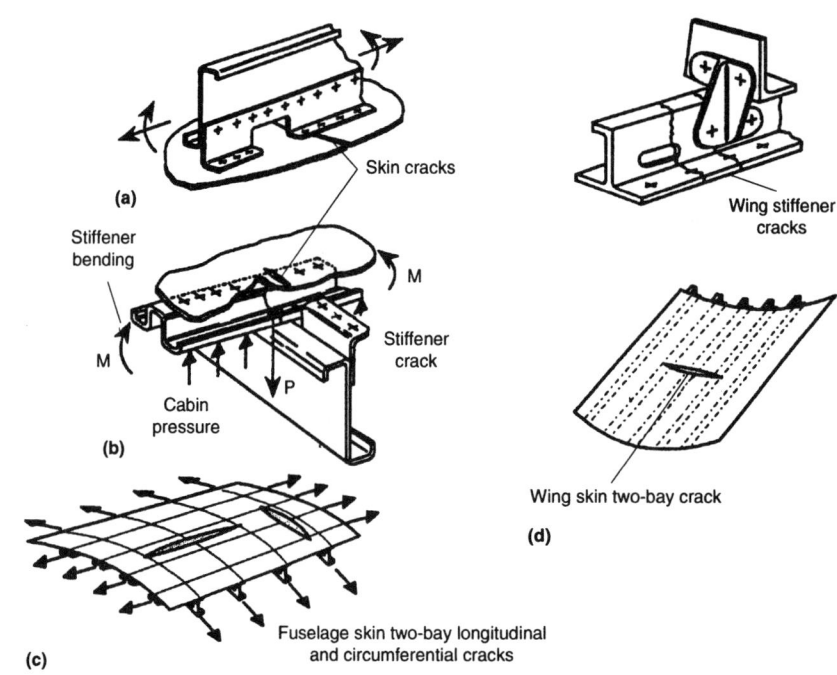

Fig. 5 Cracking scenarios leading to the two-bay skin crack criteria

ferential frame shear clip cutout. The cutout, which allows continuous axial stiffeners to pass through the frame, creates a stress concentration in the skin, which is aggravated by increased skin stress due to frame bending at some locations. Fatigue cracks at this location will propagate into both adjacent skin bays, creating the two-bay skin crack scenario. In addition, discrete source damage, discussed below, can create the potential for stiffening element failure in addition to two-bay skin damage.

Figure 5(b) illustrates a fatigue-critical location in the fuselage in the axial stiffeners at a circumferential frame connection. Radial loading due to internal cabin pressure causes stiffener bending *(M)* and concentrated load transfer *(P)* from the stiffener into the frame. A combination of these loads creates high local stresses in the stiffener, creating a potential for stiffener fatigue cracking. After failure of the stiffener, the skin becomes overloaded at the skin-to-stiffener rivets on each side of the stiffener crack. This creates the poten-

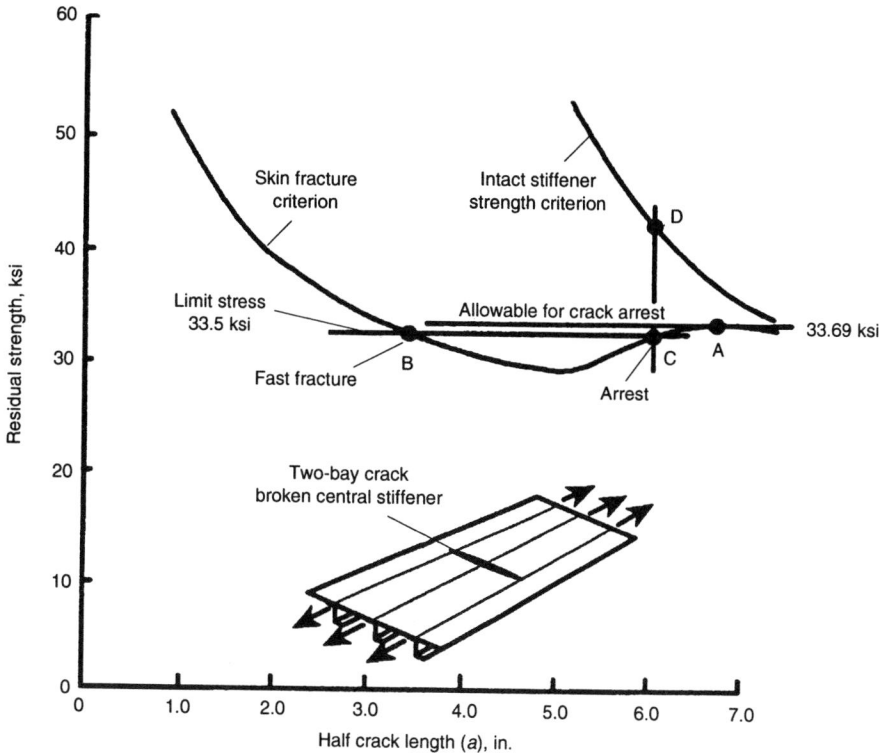

Fig. 6 Typical lower wing surface residual strength diagram

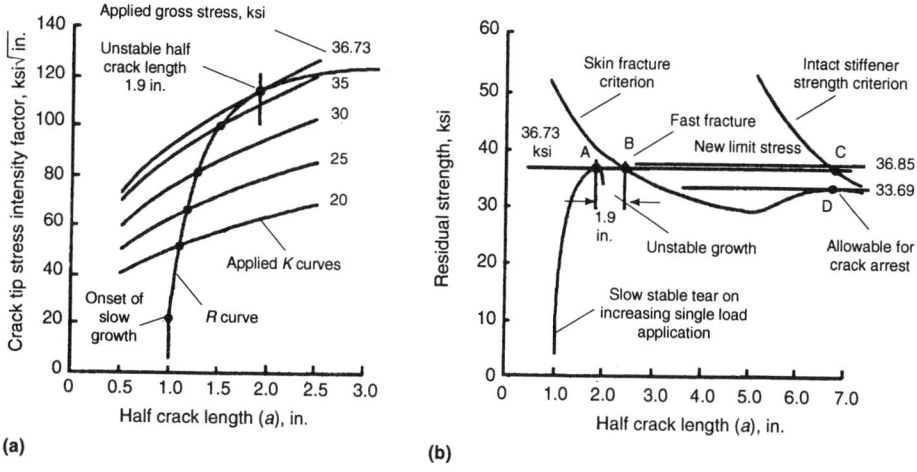

Fig. 7 Typical *R*-curve with applied *K*-curves for lower wing surface residual strength crack resistance

tial for skin cracks that would propagate into two adjacent bays, creating the potential for a two-bay skin crack with a broken central stiffener. Figure 5(c) summarizes these damage scenarios.

Figure 5(d), another example of basic structure, shows potential cracking of a lower wing surface stiffener at fuel transfer holes or at rib attachment fastener holes. After stiffener failure, the first fastener holes on each side of the stiffener crack become highly loaded, causing skin stress concentration and potential skin cracking. This skin cracking can propagate into both adjacent bays, creating the potential for a two-bay skin crack with a broken central stiffener.

In many areas of the airframe basic structure (e.g., in the lower wing surface and for the fuselage skin circumferential cracking case), it is essential to design for the two-bay skin crack with a broken central stiffener at limit load to relieve the inspection burden created for the airline operator. Figure 6, a typical example, shows a residual strength curve for a two-bay skin crack with a central broken stiffener in a lower wing surface. Assume that the wing has been designed such that the limit gross stress is 33.5 ksi (241 MPa). The allowable stress for the two-bay condition is given by point A on the skin fracture curve. Any skin fast fracture below this point will be arrested.

Fast fracture at stress levels above this point will cause failure. If limit load is applied during a particular flight (regulatory level), a skin fast fracture at point B will be arrested at point C. This damage would be considered detectable by walk-around visual inspection, and it would most likely be found.

Suppose then that a decision was made to make the limit stress level 10% higher, or 36.85 ksi (254 MPa), to save weight. This requires abandonment of the two-bay skin crack criterion if the same structural geometry is to be used. Figure 7(a) illustrates a resistance curve analysis for the typical lower wing surface element considered. This figure shows that if a skin half crack length of 1.0 in. exists at a broken stiffener, then slow stable growth will occur on a limit load application and the crack will start to become unstable at a stress of 36.73 ksi (253 MPa), which is below the limit stress level. Figure 7(b) indicates that fast fracture will occur and that the structure will fail before the limit stress is reached.

The implication is that the critical half crack size in the skin is less than 1.0 in. (25 mm). In order to establish a reasonable inspection interval, the detectable crack size has to be very small. In addition, when the skin crack is very small, the crack-tip stress intensity is high, because load transfer out of the broken stiffener causes rapid crack growth rates. Sophisticated NDI is then required over a wide expanse of structure, which creates a considerable burden on the airline operator. The alternative is to use an inspection method capable of finding the broken stiffener on the inside (either internal inspection of the wing tank or external inspection with low-frequency eddy current), which again is a burden on the operator. A more detailed discussion of this topic can be found in Ref 20.

Structural Category Description. Aircraft structures can be divided into five major categories.

Safe-life: In this type of structure, flaws can become critical at limit load before they are readily detectable in-service. Under FAR 25.571, a structure cannot be qualified in the safe-life category unless it is established that damage tolerance is impractical. Currently, only landing gear are considered safe-life structures, and these must generally be qualified by testing that incorporates appropriate scatter factors.

Single-load-path, damage-tolerant: This type of structure is allowed by regulations, provided that a reasonable inspection threshold and frequency can be established that are economically feasible for the airline operator.

Multiple-load-path, externally inspectable: This category is recommended for the vast expanse of basic structure. If the structure is designed such that a reasonable safe period for inspection is feasible (considering skin crack growth after failure of an internal member), the component will be externally inspectable.

Multiple-load-path, not inspectable for less than load path failure: In this type of structure, a crack in the primary member may not be readily inspectable without disassembly. However, the

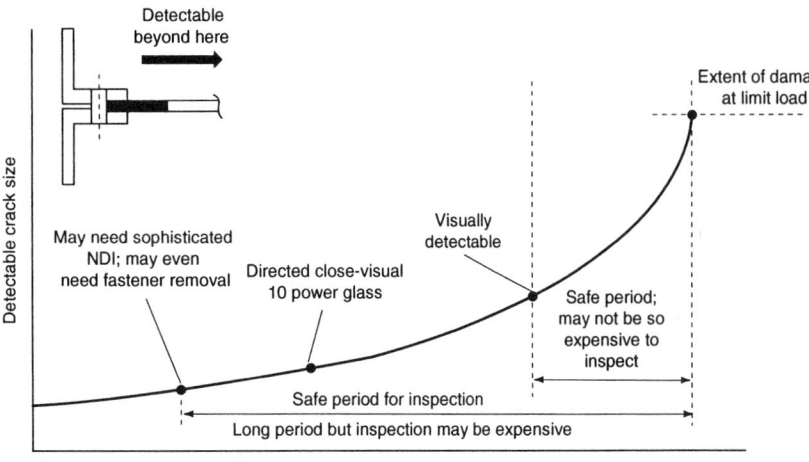

Fig. 8 Tradeoff between detectable crack size and safe inspection

structure may be easily inspectable for complete failure of the primary member. In this case, inspection frequency is based on the remaining life of the secondary structure. Further explanation can be found in Ref 9.

Multiple-load-path, inspectable for less than load path failure: In this type of structure, the primary member is inspectable, so crack growth in the primary member followed by growth in the secondary member can be considered in establishing inspection frequency.

These categories are described in more detail in Ref 9.

A crack growth curve must be provided for each PSE. The curve should be plotted for *detectable* crack growth, as shown in Fig. 8. The most critical crack position and orientation should be assumed.

Inspection Threshold Philosophy. None of the regulatory material for the design of new aircraft discusses a means of establishing the threshold for detailed inspection of fatigue-critical areas. When the damage tolerance philosophy was conceived, it was intended that the threshold be equal to the interval. This was objected to by airline operators because it meant introducing detailed inspections much earlier than before. Consequently, each manufacturer established its own methods, some of them based on the growth of 0.05 in. corner cracks at fastener holes (as indicated by Fig. 44 of Ref 9). Other manufacturers based the threshold on fatigue analysis methods using an appropriate scatter factor. However, the latter approach does not account for possible initial manufacturing damage, and for some structural configurations, fatigue cracks could become critical prior to the threshold. This problem is discussed in detail, along with a typical example, in Ref 20 in a section entitled "Threshold for Detailed Inspection."

Industry and regulatory authorities have agreed to include threshold criteria in a revised version of FAA Advisory Circular 25.571-1B. As of October 1995 the document has not been officially released, but the draft form reads as follows: "(1)

Where it can be shown by observation, analysis, and/or test that a load path failure in multiple load path 'fail-safe' structure, or partial failure in crack arrest 'fail-safe' structure, will be detected and repaired prior to failure of the remaining structure, the thresholds can be established using either: (i) fatigue analysis and tests with appropriate scatter factor, or (ii) slow crack growth analysis and tests, based on appropriate initial manufacturing damage and an appropriate scatter factor. (2) For single load path structure and for multiple load path and crack arrest 'fail-safe' structure, where it cannot be demonstrated that load path failure, partial failure, or crack arrest will be detected and repaired prior to failure of the remaining structure, the thresholds shall be established based on crack growth analyses and tests, assuming the structure contains an initial flaw of maximum probable size that could exist as a result of manufacturing or service induced damage."

As mentioned above, a number of manufacturers are already using this philosophy in establishing inspection thresholds.

Inspection Frequency Philosophy. The frequency of inspection is based on crack growth life, starting with a crack size that has a high probability of being detected using the specified inspection methodology and ending with the extent of damage at limit load. The regulation does not specify probability of detection, but 90% probability with 95% confidence appears to be a generally accepted value in a deterministic approach. The crack growth life is generally divided by a factor to provide an inspection interval. The minimum value of this factor has been 2.0 for multiple-load-path structure and 3.0 for single-load-path structure. Because the probability of finding a crack increases with crack length, some manufacturers have used a joint probability calculation that determines the number of inspections required to provide a constant probability of survival. One such calculation is based on the computer program IPOCRE (Ref 21).

Inspection cost is influenced by the inspection interval as well as by the choice of method. Looking for small cracks using NDI may be expensive, but the interval is long, as illustrated in Fig. 8. However, it is the opinion of this author that frequent visual inspection is safer than NDI at extended intervals. Care should be exercised not to make the interval too long under the assumption that extremely small cracks will be found, because the probability of detecting them might be low.

If NDI will be used, it is usual for the fracture mechanics analyst to mark up a drawing, during development of the aircraft, with the most critical crack locations for each PSE. The NDI specialist then determines the inspection method and specifies any NDI instrumentation calibration specimens needed. These data are presented in an NDI manual for the aircraft.

Full-Scale Fatigue Testing. It is well known in the industry that FAA requirements do not include a mandate to perform full-scale fatigue testing for aircraft previously designated "fail-safe" and currently designated "damage tolerant." However, most manufacturers in recent years have completed a minimum of two lifetimes of simulated fatigue loading on full-scale aircraft. The two primary reasons for full-scale fatigue testing are to determine the locations of "hot spots" where early fatigue cracking might occur and to determine the point in the life at which widespread, multiple-site fatigue cracking may occur.

With industry agreement, FAR 25.571 is being modified to include the following terminology: "The evaluation must include repeated loads and static analyses supported by full-scale fatigue test evidence. It must be demonstrated with sufficient full-scale fatigue test evidence that widespread multiple-site damage will not normally occur within the design service goal of the airplane."

It has become customary with some manufacturers to perform residual strength testing at the termination of the full-scale fatigue test by introducing saw cuts in critical locations. In this case it is advisable to conduct a small element test program to validate that the same residual strength will result from saw cuts relative to fatigue cracks. This subject is discussed in more detail in Ref 9.

Teardown Inspection Philosophy. At the conclusion of full-scale fatigue testing, it is advisable to conduct teardown inspection of selected areas of the structure that may be susceptible to multiple-site damage (MSD). This inspection is not considered merely a visual check for cracks. It involves cutting out sections of the structure, breaking open the sections through fastener holes, and fractographically examining the fracture surfaces to determine if any sign of MSD has occurred. Extremely small cracks, even of less than detectable size, can substantially reduce the residual strength capability for which the aircraft was designed and certified. Information about MSD is needed to help determine the onset of widespread fatigue damage.

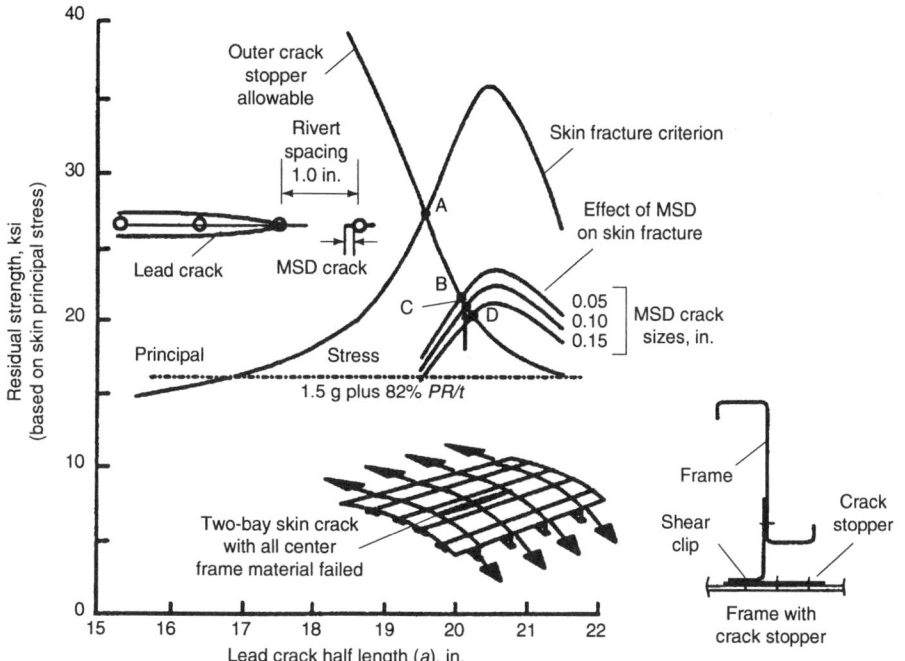

Fig. 9 Effect of multiple-site damage on lead crack residual strength frames with crack stoppers. Rivet spacing 1.0 in. (25 mm)

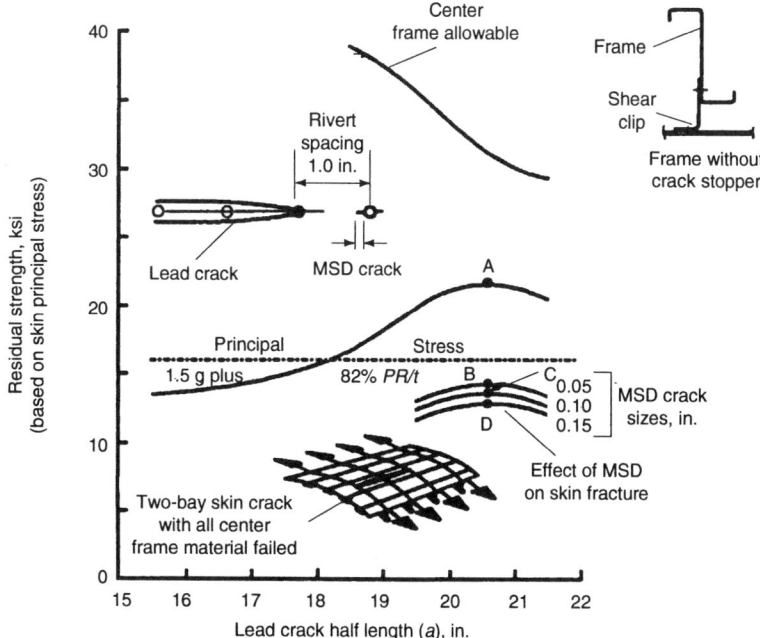

Fig. 10 Effect of multiple-site damage on lead crack residual strength frames without crack stoppers. Rivet spacing 1.0 in. (25 mm)

Widespread fatigue damage is the point in the aircraft life when MSD and multiple-element damage (MED) have degraded the originally certified residual strength capability below the regulatory level of limit load. The critical damage size can be influenced by the condition of the surrounding structure. If the structure is young, it is unlikely that MSD is present to the extent it would affect the lead crack residual strength. However, if the structure is operating beyond the life substantiated by fatigue testing, there is a strong possibility that MSD may affect lead crack residual strength. This condition is particularly important in the event of discrete source damage, as in the case of engine burst.

MSD and MED have become the most important issues in discussions of aging aircraft, and many references deal with this subject (Ref 9, 19, 20, 22, 23). References 9 and 19 were published in 1983 and 1987, respectively, and Ref 20, 22, and 23 were.published after the Aloha 737 accident in 1988. Detailed discussion of MSD and MED is beyond the scope of this article, but some guidance can be offered:

- The best way to handle MSD is to make sure that it never occurs within the operational life of the aircraft. Sufficient full-scale fatigue data should be gathered, followed by a detailed teardown inspection as described above. As mentioned, for new aircraft this will eventually be a requirement.
- For existing aircraft that may not have had the benefit of full-scale fatigue testing and teardown inspection, it is essential to predict the onset of widespread fatigue damage so that aircraft can be modified in areas susceptible to MSD and MED.

Loss in lead crack residual strength appears to be a function of structural geometry. Figure 9 shows the result of finite element analysis for the case of a two-bay longitudinal skin crack with a broken central frame. The frame cross section includes a titanium crack stopper strap. Figure 10 shows residual strength for the same cracking scenario, but in this case the frame does not include a crack stopper strap. In both figures, the residual strength is given by point A. The effects of MSD have been determined based on an intuitive failure criterion that is described in Ref 23.

In the case where titanium crack stopper straps are present, considerable MSD is tolerable and the residual strength capability is still above the required stress. However, in the case where no crack stoppers are present, even undetectable MSD can degrade the residual strength below the required value. Thus, aircraft needs to be evaluated in all areas susceptible to MSD in order to determine the extent of MSD that will degrade the residual strength below limit.

Steering Committee Activity. When the crack growth and residual strength have been determined for each PSE, manufacturers should convene a series of meetings, known as steering committee meetings, with airline operators and government authorities. Because the operators are responsible for performing the inspections, they should be involved in considering the trade-off between inspection method and frequency, based on the engineering data provided by the manufacturers. Some operators prefer frequent inspections with less sophistication in the method, while others prefer a more sophisticated and usually more expensive NDI method but with reduced frequency. Not involving operators at an early stage has created problems with extended service life programs, resulting in the need for Supplemental Inspection Documents (SIDs) for mature fleets of commercial aircraft.

Airworthiness Limitations. The inspection program developed from the damage tolerance evaluation to prevent catastrophic failure must be included in the Airworthiness Limitation Section of the Instruction for Continuous Airworthiness required by FAR 25.1529.

Repairs and Modifications. Any repair or modification to an airframe structure has the potential to degrade the fatigue life and damage tolerance capability. Inspection thresholds and frequencies can be drastically affected by changes in structural geometry due to repairs or modifications. In view of this, each repair or modification should be evaluated for damage tolerance, and new inspection thresholds and frequencies should be established. This applies to aircraft initially certified as damage tolerant as well as to older aircraft operating under SIDs.

REFERENCES

1. "Fatigue Evaluation, Damage Tolerance and Fatigue Evaluation of Structure," Sec. 25.571, *Airworthiness Standards: Transport Category Airplanes*, DOT Part 25, Federal Aviation Administration
2. "Damage Tolerance and Fatigue Evaluation of Structure," Advisory Circular 25.571-1A, Federal Aviation Administration
3. "Average Gust Frequencies: Subsonic Transport Aircraft," Data Item 69023, Engineering Sciences Data Unit Ltd., London
4. P.A. Hunter, "An Analysis of VGH Data from One Type of a Four Engine Turbojet Transport Airplane during Commercial Operations," Report TN D-4330, National Aeronautics and Space Administration, Feb 1968
5. "Operational Experience of Turbine-Powered Commercial Transport Airplanes," Report TN D-1392, National Aeronautics and Space Administration, Oct 1962
6. P.A. Hunter, "Summary of CG Acceleration Experienced by Commercial Transport Airplanes in Landing Impact and Ground Operations," Report TN D-6124, National Aeronautics and Space Administration, April 1971
7. "The NASA Digital VGH Program—Exploration of Methods and Final Results," Contractor Report 181909, National Aeronautics and Space Administration, 1989
8. D. Broek, *LICAFF: An Interactive Fatigue Crack Growth Integration Program for Applications without Retardation, Using Effective Cycles*, FractuREsearch Inc., Galena, Ohio
9. T. Swift, "Verification of Methods for Damage Tolerance Evaluation of Aircraft Structures to FAA Requirements," paper presented to 12th Symposium of the International Committee on Aeronautical Fatigue (Toulouse, France), May 1983, Document 1336, Centre d'Essais Aeronautique de Toulouse
10. "Effect of Transport/Bomber Loads Spectrum on Crack Growth," AFFDL-TR-78-134, Air Force Flight Dynamics Laboratory, Nov 1978
11. P.C. Paris, M.P. Gomez, and W.E. Anderson, "A Rational Analytical Theory of Fatigue," *The Trend in Engineering*, Vol 13, p 9-14
12. "Standard Test Method for Constant-Load-Amplitude Fatigue Crack Growth Rates Above 10^{-8} M/cycle," ASTM E 647-81
13. E.K. Walker, "The Effect of Stress Ratio during Crack Propagation and Fatigue for 2024-T3 and 7075-T6," *The Effects of Environment and Complex Load History on Fatigue Life*, ASTM STP 462, 1970
14. R.G. Forman, V.E. Kearney, and R.M. Engle, "Numerical Analysis of Crack Propagation in Cyclic Loaded Structures," *J. Basic Engineering*, Vol 89, *Trans. ASME*, Series D, 1967, p 459
15. J.E. Collipriest and R.M. Ehret, "A Generalized Relationship Representing the Sigmoidal Distribution of Fatigue Crack Growth Rates," Report SD74-CE-0001, Rockwell International, March 1974
16. C.E. Jaske, C.E. Fedderson, K.B. Davies, and R.C. Rice, "Analysis of Fatigue, Fatigue Crack Propagation and Fracture Data," Report CR-132332, National Aeronautics and Space Administration, Nov 1973
17. "Test Method for Plane-Strain Fracture Toughness of Metallic Materials," ASTM E 399-81
18. "Design Considerations for Minimizing Hazards Caused by Uncontained Turbine Engine and Auxiliary Power Unit Rotor and Fan Blade Failures," Advisory Circular 20-128, Federal Aviation Administration
19. T. Swift, "Damage Tolerance in Pressurized Fuselages," 11th Plantema Memorial Lecture, presented at 14th Symposium of the International Committee on Aeronautical Fatigue (Ottawa, Ontario, Canada), June 1987
20. T. Swift, "Damage Tolerance Capability," *International Journal of Fatigue*, Vol 16 (No. 1), 1994
21. D. Broek, *IPOCRE Software Module for the Determination of Inspection Intervals for Certain Cumulative Probability of Crack Detection*, FractuREsearch, Inc., Galena, Ohio
22. J.R. Maclin, "Performance of Fuselage Pressure Structure," paper presented at the 1991 International Conference on Aging Aircraft and Structural Airworthiness (Washington, D.C.), Nov 1991
23. T. Swift, "Widespread Fatigue Damage Monitoring—Issues and Concerns," paper presented at the Fifth International Conference on Structural Airworthiness of New and Aging Aircraft (Hamburg, Germany), June 1993

The U.S. Air Force Approach to Aircraft Damage Tolerant Design

Mitchell P. Kaplan, Willis & Kaplan, Inc.
John W. Lincoln, Aeronautical Systems Center, Wright Patterson Air Force Base

DAMAGE TOLERANT DESIGN in U.S. Air Force aircraft has evolved over the last twenty-five years since fracture mechanics was first used in the 1970s as a tool for analyzing structural problems of aircraft. Fracture mechanics originated in the 1920s, and as a tool became viable in the 1950s. It was then a reasonably short time before fracture mechanics became the methodology used by the U.S. Air Force to ensure safe design of their aircraft. In the 1970s, the B-1A airframe was designed using damage tolerance techniques. Prior to this time, fracture mechanics was used to determine the rationale causing the failure of the TFX (F-111) aircraft and to determine a method for ensuring that the problem underlying the crash did not occur again.

It was also during the 1970s that it became apparent that significant structural problems existed in the airframe of the C-5A aircraft, which was designed using stress analysis and fatigue as the primary design tools. Fracture mechanics was used as a tool to determine the extent of these problems and to define "a fix" for them. During the 1970s and 1980s, the U.S. Air Force established and evaluated a set of standards that would change the method of design for their aircraft and structurally reassessed their existing fleet of aircraft. Finally, the U.S. Air Force expanded the use of fracture mechanics technology to encompass not only airframes, but engines and mechanical components as well.

The chronology and rationale underlying this change is discussed in detail in this article. It is important to understand the historical bases for the change as well as the technological advances that underlie them. During this same time period, the 1970s and 1980s, the U.S. Air Force expended significant effort in a number of areas to include fracture mechanics in design. These included, as stated above, a series of standards that required contractors to use damage tolerance procedures as the primary design tool to ensure safety for new aircraft, a structural reassessment

of in-service aircraft, and laboratory support from the Wright Laboratory. Funding by several areas of the U.S. Air Force in fundamental research was also provided. Additional funding included expenditures by several system programs to design new materials, obtain material data, and further examine analysis procedures that would be useful in determining component life. Finally, this data was published in a fashion that would hasten its application.

The use of fracture mechanics thus has evolved into the U.S. Air Force design program for aircraft that are damage tolerant, that is, aircraft designed to operate with manufacturing and in-service induced defects. This change in ensuring aircraft structural integrity has manifested itself in a number of ways, including:

- Fewer structural problems (surprises) occur.
- Life-cycle cost of a system can be determined.
- Economic life of a system can be determined.
- Fleet utilization is enhanced.
- Inspections procedures are known.
- Inspection confidence is improved.
- Inspection intervals can be scheduled.
- Change in usage can be determined in terms of component and aircraft life.
- Advances in instrumentation allow (1) measurement of additional relevant parameters, (2) enhanced (real time) measurement of parameters, and (3) embedding of crack gages into structure (smart structure).
- Advances in software allow (1) calculations of severity change on a per flight basis, (2) improved inspection schedule, and (3) an improved maintenance schedule.

In short, the inclusion of damage tolerance design and a systematic review of design procedures allows the U.S. Air Force to design, manufacture, and maintain systems that are structurally safe and economically prudent. After a brief introduction of fracture mechanics, this article describes those particular aspects that relate to aircraft design.

Fracture Mechanics

Fracture mechanics provides the framework for the interrelationship among the toughness of the material, the applied stress, and the presence of a defect or crack within the material itself. It should be noted that the primary difference between fracture mechanics analysis and the type of analyses conducted previously is the a priori assumption that a crack is present. Stress analysis, fatigue analysis, corrosion, creep, and so forth do not assume the immediate presence of a crack. Many of these analyses do not consider the presence of a crack until failure occurs. Fracture mechanics analysis procedures assume that a crack exists immediately subsequent to manufacture of the component.

The study of fracture mechanics may be subdivided in a number of ways. For the purpose of this section, the subdivision consists of critical crack growth and subcritical crack growth. Critical crack growth entails those situations where the combination of a particular flaw size and applied stress lead instantly to catastrophic failure. The stress may be tensile, shear, or torsional and is not applied in a cyclical manner. As this discussion considers fracture mechanics as the primary method of material failure, the flaw is assumed to be an infinitely sharp crack. Subcritical crack growth may occur as a result of cyclic loading, corrosion, or creep. The crack grows until the combination of applied stress and crack size causes catastrophic failure. Generally, in the case of airframes, the temperature is insufficiently high for creep to become a meaningful problem and is usually ignored. This is not true in engines. Cyclic loading will be discussed, and corrosion will be addressed in conjunction with both cyclic or sustained loading.

Critical Crack Growth. Fracture mechanics as understood today began in 1920 with A.A. Griffith (Ref 1). At that time, Griffith was working with brittle materials such as glass and the conditions that caused them to fracture. By using basic thermodynamic principles, he was able to derive an equation that led to the energy required to create new surfaces. Through quantification of thermodynamics and the introduction of solid mechanics, he was able to derive an equation that was based on measurable invariant properties of the materials involved. These measurable properties included the surface energy of the material and Young's modulus, and additionally, the geometry of the initiation site was required. Griffith was able to show that additional new surface would be created when the elastic energy available to form a new surface was equal to or greater than the reduction of elastic energy due to the cohesive strength of the material.

The method used by Griffith to express the relationship between the cohesive strength of the material and the elastic energy was based on work done previously by C.E. Inglis (Ref 2). Inglis showed, using mathematical techniques for ellipses with many different aspect ratios, the variation of the stress in a flat, infinite plate. It is apparent that an ellipse was chosen because of its ability to mirror several different flaws that occur in nature. If we consider an elliptical hole that has equal major and minor axes (aspect ratio equal to one), then the ellipse degenerates to a circle. If the major and minor axis differ by a thousand (a very high aspect ratio), the shape of the ellipse approaches a crack.

By using Inglis's methods, Griffith was able to show

$$\sigma \sqrt{\pi a} = \sqrt{2 E \gamma_e} \qquad \text{(Eq 1)}$$

where σ is the applied stress, a is the crack length, E is Young's modulus, and γ_e is the elastic surface energy of the material.

It was noted that the left side of Eq 1 contains measured parameters (applied stress and crack length), while the right side of the equation contains material constants (Young's modulus and the elastic surface energy). In 1948, Irwin (Ref 3) and Orowan in 1949 (Ref 4), suggested that the elastic surface energy can be rewritten to include both the elastic and plastic surface energy. In a material that is completely brittle, the plastic portion of the surface energy would be quite small. The material constants (surface energy and Young's modulus) are grouped on the right side of the equation, and they can be replaced with a new material constant, the stress-intensity factor. Rewriting the equation in this form yields

$$\sigma \sqrt{\pi a} = \sqrt{2 E \left(\gamma_e + \gamma_p \right)} = K_I \qquad \text{(Eq 2)}$$

where γ_p is the plastic strain energy, and K_I is the stress-intensity parameter (mode I, tension).

As a result of Eq 2, a method came into existence that allowed the determination of the material behavior as a function of the applied stress and the crack length. In contrast to the method

proposed by Griffith in Eq 1, this method does not require the independent knowledge of the surface energy. The energy required to create new surface (elastic and plastic) is combined within the equation with Young's modulus to give a new parameter. This is the stress-intensity factor. What is required is a careful determination of the stress at the tip of the crack and of the relationship between the crack-tip stress and the applied stress of the material being tested. This was accomplished in the paper written by Irwin in 1957 (Ref 5).

The results obtained for mode I loading (tension) by Irwin are:

$$\sigma_x = \frac{K_I}{(2\pi r)^{1/2}} \cos \frac{\theta}{2} \left(1 - \sin \frac{\theta}{2} \sin \frac{3\theta}{2} \right)$$

$$\sigma_y = \frac{K_I}{(2\pi r)^{1/2}} \cos \frac{\theta}{2} \left(1 + \sin \frac{\theta}{2} \sin \frac{3\theta}{2} \right)$$

$$\tau_{xy} = \frac{K_I}{(2\pi r)^{1/2}} \sin \frac{\theta}{2} \cos \frac{\theta}{2} \cos \frac{3\theta}{2}$$

$$\sigma_z = \nu(\sigma_x + \sigma_y) \quad \tau_{xy} = \tau_{yz} = 0 \qquad \text{(Eq 3)}$$

In this instance, σ is the normal stress, with the subscript showing the direction, τ is the shear stress, and ν is Poisson's ratio.

These equations, and the similarly derived equations for mode II (shear) and mode III (torsion), show that the stress level and the geometry surrounding the crack (both the shape of the crack and the component in which the crack is situated) have a role in the quantitative value of the stress-intensity factor, but do not affect the nature of the distribution of the stress field. As Irwin stated (Ref 5):

...in situations such that a generalized plane stress or a plane stress analysis is appropriate, the influence of the test configuration, loads, and crack length upon the stresses near an end of the crack may be expressed in terms of two parameters. One of these is an adjustable uniform stress parallel to the direction of a crack extension. It is shown that the other parameter, called the stress intensity factor, is proportional to the square root of the force tending to cause crack extension.

There now exists the necessary mathematical relationships supported by Newtonian mechanics that allow the investigator to relate the applied stress and the crack length in a fashion that demonstrates the onset of fracture. The material parameter, K_{Ic}, may be experimentally obtained. The stress at the crack tip is related to the gross stress by the equations derived by Irwin, and the crack length is measured.

Subcritical Crack Growth. The primary area of subcritical crack growth discussed in this section is the extension of cracks due to cyclic loading. Cyclic loading (fatigue) as a failure mechanism has been studied for more than 130 years (Ref 6). The life of components experiencing fatigue loading has been analyzed using Miner's Rule (Ref 7), which does not account for stress sequence, and the Neuber method (Ref 8), which

considers the effect of the stress sequence. In both cases, it is assumed that no crack exists until final failure occurs.

In 1961, a paper by Paris et al. (Ref 9) investigated the effect of crack growth using fracture mechanics criteria. These criteria included the initiation of a crack into the component prior to the start of the test, and the study of the crack as it was exposed to cyclic loading.

They found that

$$\frac{da}{dN} = f(K_{max}, R) \qquad \text{(Eq 4)}$$

where da/dN is the crack growth per cycle, K_{max} is the maximum stress-intensity factor, and R is the ratio of minimum to maximum stress-intensity factor.

As the crack-growth behavior was affected by a number of different test and environmental conditions, the analysts had to go to more complicated equations that would best represent the data. These investigators developed a number of equations that could represent the crack-growth rate as a function of stress-intensity factor (Ref 10-17).

One of the more important parameters that affected the crack-growth rate was the stress ratio, or R ratio. This was the ratio obtained, during cyclic loading, when the minimum stress was divided by the maximum stress. Other material behavior characteristics that had to be included in the crack-growth equation to best represent the data were the existence of a stress-intensity threshold below which cracking would not occur, followed by a straight-line portion of the crack-growth curve, and finally a portion of the curve that would be increasing at an increasing rate. The curve representing this data is shown in Fig.

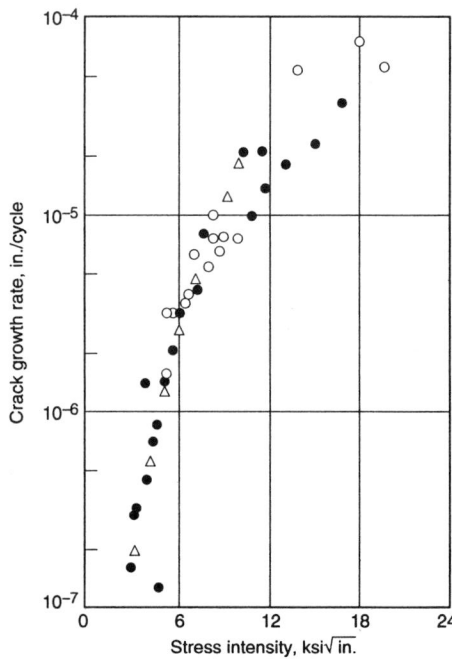

Fig. 1 Typical crack growth curve

1. The crack-growth-rate data itself were found to be affected by environment, microstructure, heat treatment, texture, material form (sheet, plate, casting, forging, etc.) cyclic rate, waveform, R ratio, stress history, and so forth (Ref 13-24).

Another set of equations was used in tandem with the crack-growth-rate equations. These dealt with retardation of crack growth. As stated above, Neuber had found that fatigue life is affected by the previous stress history that the particular component has experienced. Several investigators developed equations that account for crack-growth retardation behavior (Ref 25-29). These equations calculate the residual compressive stress that exists after plastic deformation occurs at the crack tip due to a high tensile load. This residual compressive stress decreases as the distance from the applied high stress increases. This varying compressive stress field is calculated, and the residual compressive stress is subtracted from the next set of minimum and maximum stresses. As the minimum and maximum stresses are now changed as a result of the compressive loading, change also occurs with both the stress ratio and the difference between the maximum and minimum stress-intensity factor (ΔK).

These parameters are the primary driving forces that cause the crack to grow, thereby affecting the crack-growth rate of the particular component. There are, as previously stated, a number of mathematical approximations to determine the crack-growth-rate behavior of the material. All of these formulas use crack-growth data derived from constant amplitude cyclic applied stresses as the basis for their approximations. However, in many instances the stress-time history of a particular component does not behave as if the stresses were applied in this fashion. An exception to that statement may be the skin of a fuselage that is subjected primarily to pressurization stresses. When the aircraft is on the ground, the stresses due to pressurization of the fuselage skin is zero. At maximum altitude, the pressure differential may be 8 to 10 psi (55 to 70 kPa). Upon landing, the pressurization differential is again zero. For other areas on the aircraft, this is not true. The lower surface of the wing on a fighter aircraft sees an almost random stress-time history for each segment of the flight. Much work has been accomplished relating the life of a component to whether the stresses are applied from the lowest stress to the highest stress, the highest stress to the lowest stress, or low stress to high stress to low stress. Results indicate that randomly ordering the stresses gives the lowest component life (Ref 30, 31).

There is yet another dilemma that exists between the analysis and the testing of a particular component. The component can be tested by placing the stresses in a random fashion, but the component cannot be analyzed in the same manner. Because the analysis is based on constant amplitude data, the stress cycle goes from minimum to maximum to minimum, and this is not true for the test spectrum. To account for this effect, stress-counting methods such as rain flow or range pair (Ref 32) must be used. These methods examine the stresses and combine the maximum half cycle with the minimum half cycle, the second maximum half cycle with the second minimum half cycle, and so forth. There is now a collection of full cycles that can be analytically examined. The order of the stresses is not identical; however, the results obtained more accurately represent the test results (Ref 30, 31).

The user can determine component life once the following information is obtained:

- Material properties, static and dynamic
- Component geometry
- Loadings
- Load-time history
- Stress-to-load ratios
- Stress-intensity factors
- Stress counting computer routines
- Crack growth equations

By selecting an initial flaw size, the component life can be determined based on that initial flaw size. An inspection interval of one-half that life is usually specified as the inspection interval. It is at this time in the service history that the particular component is inspected.

Application of Fracture Mechanics

In the 1960s, a transition occurred in fracture mechanics. It was taken from the laboratory and the research journals and placed in the daily venue of design engineering. This transition was hastened as a result of the space program (in particular, for its widespread use of pressure vessels) and the F-111 aircraft accident.

Fracture Mechanics as a Design Tool in the Space Program. In 1965, Tiffany and Masters wrote an article (Ref 33) that was instrumental in promoting the use of fracture mechanics as a technique that would be useful in aircraft design. This article discusses the use of fracture mechanics as a method of predicting failure, understanding failure mechanisms, and suggesting inspection methods to protect against failure in pressure vessels. The materials from which the pressure vessels were manufactured included steel, titanium, and aluminum.

At the time, Tiffany and Masters worked at a major aircraft company, and the article was an outgrowth of a portion of their ongoing tasks. The importance of the article, however, supersedes this. It was the first time that representatives of a major manufacturer formally addressed the notion that the onset of cracking may occur as a result of manufacturing defects and worked toward a design philosophy that accounted for this occurrence in a quantitative manner.

Tiffany and Masters examined pressure vessels and the cracks that may occur as a result of their manufacturing processes, especially welding, and stated:

Clearly, it must be recognized that fabricated structures and, indeed, even the raw materials contain defects and flaws of various kinds. The lives of these structures are controlled by the flaw sizes required to cause fracture at the operating stress levels, *the initial flaw size,* and the subcritical flaw growth characteristics of the materials.

This was a vast departure in design philosophy. Never before had industry discussed the existence of flaws (cracks) in new structure. Because fracture mechanics is based on the existence of cracks, the need for fracture mechanics in design not only became viable, but necessary. The previous methods used to validate structural integrity was through either strength (static failure), or fail-safe and safe-life design (fatigue). Both of these methods assumed that no cracks exist, and more importantly, had no method to deal with cracks. Only fracture mechanics had that capability.

The article written by Tiffany and Masters dealt with the application of fracture mechanics. Because a rocket motor case is basically a pressure vessel, it was necessary to get information (materials data) regarding the critical stress-intensity factor (K_{Ic}), and the crack-growth behavior (da/dN versus ΔK). In addition, the mathematical relationships for a part-through flaw (a semielliptical embedded crack) and inspection criteria were needed. The inspection criteria that were needed included the area(s) to be inspected, how to inspect the particular location, and how often the inspection should be accomplished. Here is yet another significant difference between fracture mechanics design and strength design. In the previous design systems, the structure would be retired (taken from service) after the appropriate amount of time. In damage tolerant design, the component may be used to the point of onset of widespread fatigue damage if proper inspection procedures are utilized.

An inspection method suggested in the paper is that of proof testing. If the maximum stress the component experiences is assumed to be X lb/in.^{2}, then the component is tested at some higher level, for example, 1.5X lb/in.^{2}. If failure does not occur, then the analyst knows that the maximum crack size existing in the component is 0.44 times the crack size that causes failure. By performing a crack-growth analysis, the investigator determines the number of cycles (flights) for the crack to grow to a size where failure occurs during normal operations. After that number of cycles is reached, the component can be proof tested yet again. In this manner, the life can be extended indefinitely. It should be noted that crack growth can occur as a result of the proof test, and that must be taken into account.

Fracture Mechanics as an Engineering Tool in Failure Investigation. During the late 1960s, the U.S. Air Force and Navy were jointly funding the development of the F-111 aircraft. This aircraft had a swing wing and was designated as a fighter bomber. The wing box of the aircraft was manufactured from D6ac steel. During a test flight that occurred in December 1969, the wing failed in the steel wing pivot fitting and the aircraft crashed, killing the pilots. An investigation was undertaken to determine the cause of failure. Recovered from the wreckage was a portion of the wing box containing the flaw that initiated the crash sequence. The semicircular flaw was caused by a forging lap. The flaw was 1 in. long and 0.25 in. deep. The area where the flaw was located had been inspected several times, and the inspectors had not noticed it (Ref 34).

During the investigation, it was found that D6ac steel is susceptible to cooling rate. Tests indicated the toughness of the D6ac steel was significantly high, but tests of the failed material indicated a substantially lower toughness. This quench sensitivity of the D6ac steel resulted in a difference in fracture toughness by a factor of two. It was found that the testing conducted during the development program was on thin specimens, and the cooling rate was dissimilar to the cooling rate of the material in "as-forged" condition. This difference in the cooling rate led to the difference in material toughness. The combination of these factors lowered toughness, and the existence of the manufacturing defect led to catastrophic results.

The U.S. Air Force was faced with a problem of significant proportions. They had an aircraft that was an integral part of their fleet, yet had the potential to contain a flaw that would render the aircraft unable to perform its mission. It was decided that periodic proof testing of the aircraft would allow the aircraft to become a viable member of the U.S. Air Force contingency. It is well known that steel, when subjected to lower temperatures, becomes brittle. If a steel alloy is tested below its nil-ductility transition (NDT) temperature, the toughness of the steel drops considerably. It was decided that the aircraft should be proof tested below the NDT temperature of the D6ac steel in the wing box. To allow this test to occur, it was necessary to build rooms (hangars) of sufficient size and with the appropriate thermal characteristics to allow the temperature of the room to stay below the NDT temperature. The wing box structure would be tested. If failure did not occur during testing, the aircraft could be flown safely until the next inspection. After the next inspection, the proof test would be repeated. This inspection method increases the life of the aircraft until failure occurs during test. If no crack is present, the aircraft may be flown indefinitely.

An important aspect of this approach to inspection and aircraft utilization was that retirement of the aircraft would now be done for cause. Previously, when aircraft components arrived at a particular life, the component would be retired. No inspections looking for cracks were required, and none were conducted. When the life of an aircraft (component) was reached, the aircraft (component) was removed from service. With this new methodology, for a particular part of the aircraft to be retired, a tangible reason such as a crack would be required. Safety would be based on the results of inspections and not on the results of analyses and component tests. As a result of the success using fracture mechanics to uncover the reason for the failure of the F-111 and to allow the implementation of a periodic inspection technique that would ensure flight safety, the U.S. Air Force decided to incorporate fracture mechanics as a design procedure.

Damage Tolerance in Aircraft Design

With the success of fracture mechanics both in design and in failure analysis, the U.S. Air Force presented the idea to industry to use this discipline as the design methodology that would protect the safety of aircraft. A damage tolerance design approach that would incorporate fracture mechanics, periodic inspections, and a maintenance policy based on these inspections was initiated in the B-1A program, and a structural assessment of the C-5A was initiated using this procedure.

The B-1A Aircraft. In the early 1970s, the U.S. Air Force began procurement of the B-1A aircraft. That aircraft was the first design that used damage tolerance as a primary method for material selection, detail design, and inspection methodology. Using damage tolerance as the design technology for the B-1A required a significant increase in analysis methods and materials data. Fracture mechanics had previously been a tool to aid the researcher in understanding the failure mechanism. Its use as a design methodology required the unleashing of significant resources to obtain the methodologies that could be used to apply it to normal design procedures on future aircraft systems.

Development was required to both obtain and develop crack-growth programs, retardation models, standard crack solutions, nonstandard crack solutions unique to specific B-1A geometries, materials data, inspection methods, and so forth. The gathering and classification of the methods and materials was in itself significant. The need to gather the necessary information changed in a nontrivial fashion. The cascade of questions seemed unending. How was the aircraft to be used? What would be the stress-time history of the aircraft? How would all the parts of the aircraft react to these loadings? What inspection intervals would be used? What initial flaw size should be used? What should be the initial flaw geometry? How would inspections occur during fabrication to ensure that no flaw greater than that assumed existed? Maintenance actions became an increasingly important topic. How should the inspections be accomplished? How does the user know the flight profiles, the severity of the flights, and the damage done during the flights? How do the inspection intervals change over time? What is the dependency of the inspection intervals on the inspection technique? The B-1A program became both a development program for an advanced bomber and a testing and proving ground for an exercise in solid mechanics.

The C-5A Aircraft. Also during the early 1970s, the U.S. Air Force was experiencing structural problems with the C-5A aircraft. As a result of these structural problems, a team of engineers and scientists was assigned to work on the C-5A. These individuals were selected from the U.S. Air Force, Lockheed (the designer and manufacturer of the aircraft), and selected experts from other aircraft manufacturers. The C-5A aircraft had been designed using fatigue criteria. The function

of the assembled team was to structurally reassess the aircraft using damage tolerance criteria. One year was allocated for this project. During this time, material tests were initiated and completed, analyses were done, and the project was completed. Critical areas were identified, and the anticipated usage was determined. The analyses consisted of determining the external loads, the stresses, and the crack-growth life of the critical components. The results of this program illustrated to the U.S. Air Force that the structural integrity of the C-5A wing was questionable, and actions were required if a life of 30,000 h was needed for the airframe.

The B-1A and the C-5A programs, combined with the success of the F-111 accident investigation, led the U.S. Air Force to initiate a new method of structural design. The issuance of two new specifications, MIL-STD 1530 (Ref 35) and MIL-A-83444 (Ref 36) was the beginning for this new approach. The documents led to the utilization of the damage tolerance procedure as the desired method for designing U.S. Air Force aircraft. A new standard, MIL-A-87221, has been issued (Ref 37). This standard incorporates many of the requirements of the previous standards.

Damage Tolerance Design

At the time of the B-1A development and the C-5A investigation, specifications were also being prepared to apply damage tolerance criteria more in airframe design. MIL-A-83444 was written after a series of meetings between the U.S. Air Force and representatives from all the major airframe manufacturers. Prior to the publication of the document, drafts were written and comments solicited from all present. The purpose of the specification is to present to airframe manufacturers a document that would contain the requirements for protecting the flight safety of aircraft. The specification discusses:

- Initial flaw size requirements
- In-service flaw size requirements
- Residual strength
- Design definitions

It is not the purpose of the document to present a method of analysis, but to levy requirements that would result in structural aircraft safety. Because of this, the U.S. Air Force is careful not to describe a desired design methodology.

Initial Flaw Size. The initial flaw requirements were based on both structural design philosophies and component geometries. It is stated in Section 3.12.1 of the Appendix to MIL-A-87221 (Ref 37) that "Initial flaws shall be assumed to exist as a result of material and structural manufacturing and processing operations." This sentence in the requirement precludes the use of other design methods, for example, static strength and fatigue analysis. As stated earlier, neither of these analysis techniques has the capability to deal with the existence of flaws. The same paragraph further states, "Small imperfections equivalent to a 0.005 in. radius corner flaw

Table 1 Flaw sizes prescribed in MIL-A-87221 for slow crack and fail-safe design

Structural design	Flaw location	Material thickness	Flaw geometry	Initial flaw size
Slow crack and fail safe	Holes	>0.05 in.	Semicircular	0.05 in. (radius)
	Holes	≤0.05 in.	Through crack	0.05 in. long
	Not holes	≤0.125 in.	Through crack	0.25 in. long
	No holes	>0.125 in.	Semicircular	0.25 in. (length)
				0.125 in. deep

resulting from these operations shall be assumed to exist in each hole of each element of the structure . . ." The contractor is now on notice that this cracking is widespread. The use of fracture mechanics to ensure safe design is now mandatory in U.S. Air Force airframes.

The U.S. Air Force, however, does not categorically select the initial flaw size. The manufacturer has the opportunity to select a flaw size compatible with his inspection methodology. But in order to choose a flaw size smaller than that discussed in the specification, it is essential that the airframe manufacturer have an nondestructive inspection (NDI) program that can reliably *not miss* flaws of the selected size. The inspection methodology that is then selected, tested, and approved is used during the manufacturing process. The significant difference between *finding* flaws and *not missing* flaws should be noted.

The third aspect of the initial flaw size is the location of the initial flaw. The specification requires that the initial flaw be present and located in the most critical location on each "separate element of the structure." However, all remaining holes (and other critical locations) must be assumed to contain a smaller flaw. This is to ensure the continuing damage portion of the requirement.

If the assumed initial flaw is smaller than that stated in the specification, then 100% inspection of these flaw locations subsequent to manufacturing is required. If the initial flaw sizes chosen are that suggested in the specification, 100% inspection of the assumed flaws is unnecessary. The sizes of the flaws prescribed in MIL-A-87221 are shown in Table 1. The definition of "slow crack" and "fail safe" are:

- "Slow crack growth structure consists of those design concepts where flaws or defects are not allowed to attain the critical size required for unstable crack propagation. Safety is assured through slow crack growth for specified periods of usage depending upon the degree of inspectability."
- "Fail-safe structure is a structure designed and fabricated such that unstable crack propagation will be stopped within a confined area of structure prior to complete failure (crack-arrest fail-safe structure), or structure designed and fabricated in segments where each segment contains localized damage and thus prevents complete loss of the structure (multiple-load-path fail-safe structure)."

Regardless of the type of fail-safe structure, "the strength and safety will not degrade below a speci-

fied level for a specified period of unrepaired service usage."

There is a benefit for using either interference fit fasteners or cold expanded holes that would reduce the maximum tensile stress at the hole/fastener interface. The benefits of the desired fastener system need to be demonstrated through laboratory testing using joint design similar to that found in the aircraft. If the desired fastener system shows beneficial results, the assumed initial flaw size may be as low as 0.005 in. (0.125 mm). This reduction in initial flaw size significantly increases the safety limit (twice the inspection interval) of the component.

The advent of widespread cracking, and a means of preventing it, is an important part of this specification. It should be noted that each hole (located in the noncritical location) also has an initial flaw requirement of 0.005 in. (0.125 mm), and that flaw will grow during normal aircraft usage. As that flaw increases in length, a time will occur when inspection of those flaws becomes necessary. If, for example, the flaw for a particular critical location (initial size of 0.05 in., or 1.25 mm) were growing from its location, and grew into another hole, then the new flaw size would be the distance between the holes, the two hole diameters, and the length the crack grew from the second hole assuming the initial flaw size for that hole was 0.005 in. (0.125 mm). One begins to see both the complexity of the analysis procedure and the implicit safety in the continuing damage philosophy.

In-Service Flaw Size. It is appreciated that the personnel maintaining the aircraft may not have either the same tools available to them or the same means of inspection capability (or both) as the manufacturer of the aircraft. As a result, the flaw size subsequent to initial inspection depends on field methods. In the event that the methods and access to the components are identical to those used during manufacture, then MIL-A-87221 requires that the in-service flaw size be the same as the initial flaw size. If the component (or fastener) cannot be removed, and dye penetrant, magnetic particle (ferrous alloys only), eddy current, or ultrasonic techniques are used, then the initial flaw size at fasteners would be 0.25 in. (6.35 mm) of uncovered length. For critical location other than fasteners, the flaw size would be 0.50 in. (12.5 mm). If visual inspection techniques are required, the assumed flaw size (uncovered length) would be 2 in. (50 mm).

Residual Strength Requirements. The residual strength of a member is "the minimum load an aircraft must sustain with damage present and without endangering safety of flight or degrading

performance of the aircraft for the specified minimum period of unrepaired service usage." This definition contains a number of terms that must be explained. *Damage present* is the length of the crack in the critical structural component. A *critical structural component* is an aircraft structure whose failure will either cause the plane to crash or not successfully complete its mission. If the aircraft is a fighter aircraft on an air-to-ground mission, it may need to perform flight sequences that are extremely severe. If the aircraft cannot fly these sequences without component failure, then aircraft performance has been degraded. *Unrepaired service usage* is a time period dependent on the categorization of the inspectability of the particular structure. If an aircraft has a wet wing (fuel is stored in the wing) and a spar fails, the wing will leak fuel. This condition is observable to either the flight crew or the maintenance crew prior to the next flight. The unrepaired service usage would be one flight. If a failure occurred in a frame member, and that frame was inspected on a periodic basis, then the uninspected service usage would be that time period.

When calculating the residual strength, the particular time period regarding the inspection of the component must be defined (see the article "Residual Strength of Metal Structures" in this Volume). It is this time period that determines the maximum assumed flaw size. This is accomplished, in part, by finding the maximum expected load during this time frame. With this maximum assumed flaw size and the material properties, the minimum failure load can be determined. If that load is higher than the expected maximum load for that level of inspectability (time period), then the residual strength of the structure has been defined. If that load is less than the expected maximum load, then either the level of inspectability or inspection procedure must be redefined for the structure to meet this particular requirement.

As stated above, one part of the information that must be obtained is the maximum expected load. The loads the aircraft experiences are generally obtained through an exceedance function (see Fig. 2). The data that define the exceedances may be obtained from specifications (Ref 38) or directly from instrumented aircraft. It is well known that the load will increase as a function of time as exceedance curves are generally plotted in number of load exceedances per 1000 h. The higher loads have a lower number of occurrences. For example, there may be a load occurrence of 0.1 per thousand hours, or once every 10,000 h. To increase conservatism, the load required to determine the residual strength of a particular component is analyzed to a higher load. If the cracks for a particular component can be seen (or become evident, i.e., fuel leakage) when the aircraft is parked on the ground or during flight, one would expect that the load to determine the residual strength for that component would need to occur every flight. However, a multiplication factor of 100 is present. This means that the maximum load that occurs once every 100 flights is the actual load used, and that load is higher than

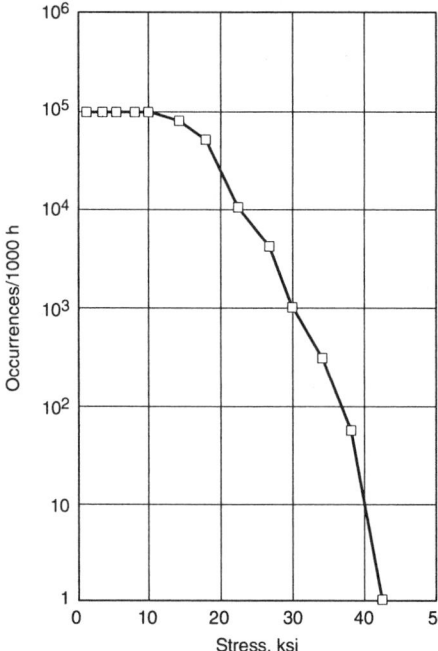

Fig. 2 Typical stress exceedance curve

the maximum load that occurs every flight. If walk-around visual inspection is necessary to see the damage, then the residual strength load would be 1000 h. If disassembly is required to visually observe the crack, then 50 years of flight (about 12,000 h of flight time) is assumed to determine the appropriate load. The remaining two-component design categories, depot level inspection and uninspectable structure, have significantly longer time periods and therefore higher loads. The base-level inspection period is five lifetimes of the aircraft, and the load required for uninspectable structure would be equal to 20 design lifetimes. To perform the crack-growth analysis, the data obtained must be placed into a stress-time history. References 39 to 44 discuss various methods of accomplishing this transformation.

Design Definitions. Included in the specification are several definitions, which are discussed

in detail in Ref 37. Other than those already discussed, they will not be presented here.

Airframe Structural Integrity Programs (ASIP)

The Airframe Structural Integrity Program (ASIP) began in 1958 as a result of fatigue failures in the B-47 aircraft. These failures were important in the issuance of a directive from the Chief of Staff of the U.S. Air Force (Ref 45). The objectives of the structural integrity program were to control structural failure of operational aircraft, to determine methods of accurately predicting aircraft service life, and to provide a design and test approach that would avoid structural fatigue problems in future weapon systems.

The original ASIP used a reliability-based method to establish the operational life. This was commonly called the "safe-life approach." The safe-life approach relied on the results of the laboratory fatigue test of a full-scale aircraft. The time-load history of the test aircraft was carefully selected to emulate the actual loadings that aircraft would experience during its anticipated lifetime. The "safe life" of the aircraft was established by dividing the number of successful test-simulated flight hours by a factor (four was commonly used by the U.S. Air Force). The purpose of the factor was to account for the many uncertainties that exist in materials and manufacturing quality. This concept was the basis for all new U.S. Air Force aircraft designed during the 1960s.

There was success in using this approach. The C-141 had an excellent safety record and good overall service history. There were, however, problems that could not be solved using this method, for example, those with the F-111 and the C-5A, which have already been discussed. Other problems began to occur as well. Losses of aircraft such as the F-5, B-52, and T-38, and the widespread cracking found on the KC-135 pointed to a deficiency in the criterion. Additionally, the ability to inspect these aircraft was impacted by the design of the aircraft itself. Because

inspectability was not required as a part of the safe life approach, many details were found to be outside the inspection capability of the 1960s. Relatively small cracks and defects could escape detection with dramatic consequences.

The shortcoming of the safe life process, illustrated by several service incidents, demanded that a fundamental change be made in the approach to design, qualification, and inspection of aircraft. An improved reliability method based on a damage tolerance approach emerged as the candidate for this change. This approach assumed the aircraft was subject to a wide range of initial quality from both the manufacturing process and service-induced damage, and the aircraft had to be inspectable. To ensure the design could be safely operated in the presence of these initial flaws, it was necessary to design the structure to be tolerant of these defects for a selected period of service usage prior to inspection. The damage tolerance approach is used to determine the period of safe operation, the safety limit (the time for the crack to grow from its assumed initial size to failure), and the requisite inspection intervals. These inspection intervals are usually taken as half the safety limit.

In 1975, the damage tolerance approach was formally made a part of ASIP. During the 1970s and 1980s, the U.S. Air Force performed an assessment on every major aircraft weapon system using a damage tolerance approach to develop the appropriate inspection and modifications that were deemed necessary to ensure flight safety.

ASIP now includes five separate tasks that cover all aspects of development and support of an aircraft structure. These tasks, described in detail in Table 2, are:

- *Task I:* Design information
- *Task II:* Design analyses and development tests
- *Task III:* Full-scale testing
- *Task IV:* Force management data package
- *Task V:* Force management

Design Information. The first ASIP task consists of obtaining the information that allows the

Table 2   USAF Aircraft Structural Integrity Program tasks

Task I Design information	Task II Design analyses and development tests	Task III Full-scale testing	Task IV Force management data package	Task V Force management
ASIP master plan	Materials and joint allowables	Static tests	Final analyses	Loads/environment spectra survey
Structural design criteria	Load analysis	Durability tests	Strength summary	Individual airplane tracking data
Damage tolerance durability control plans	Design service load spectra	Damage tolerance tests	Force structural maintenance plan	Individual airplane maintenance times
Selection of materials, processes, and joining methods	Design chemical/thermal environment spectra	Flight and ground operations tests	Load/environment spectra survey	Structural maintenance records
Design service life and design usage	Stress analysis	Sonic tests	Individual airplane tracking program	
	Damage tolerance analysis	Flight vibration tests		
	Durability analysis	Flutter tests		
	Sonic analysis	Interpretation and evaluation of test results		
	Vibration analysis			
	Flutter analysis			
	Nuclear weapons effects analysis			
	Non-nuclear weapons effects analysis			
	Design development tests			

contractor to perform a preliminary design. For completion of this task to become a reality, it is necessary for the contractor and the contracting agency to work closely to define the critical parameters. The performance criteria of the aircraft must be rigorously defined by the contracting agency. The desired life of the structure, its operating envelope, the missions, desired aircraft configuration(s), and so forth must be so stated.

ASIP Master Plan. Given this information, the contractor can develop the structural integrity plan. This master plan will be, in essence, a PERT or critical path method (CPM) chart for the entire life cycle of the aircraft. This includes the design and analysis phase, the test phase, the manufacturing phase, and the force-management phase. These charts include all requirements, specifications, design criteria, analyses, coupon and component tests, full scale tests—both ground testing (stress and fatigue/fracture) and flight testing, definition of inspection criteria, and individual aircraft tracking techniques. Long-lead-time items are identified, critical paths defined, and possible problem areas identified. Cost and schedule impacts for any of the items contained in the chart should be noted, and alternative paths are identified to minimize these problems.

Structural Design Criteria. The second portion of this task is the identification of the structural design criteria. The contracting documents define these criteria. The contractor ensures that his interpretation and the contracting agency's definition agree in all matters. Subservient specifications are identified, and all problems concerning definitions and criteria should be resolved by issuance of this document.

Damage Tolerance and Durability Control Plan. In this subtask, a damage tolerance and durability control plan is issued and the various fracture critical components are defined. In addition, the design criteria for each of these fracture critical components need to be identified. These include, but are not limited to, hole preparation, manufacturing inspection criteria and procedures, fail-safe design, inspectable and uninspectable structure. It is from these definitions that inspection intervals and methodology are determined. Also, this subtask involves the determination of the structure that is required to meet the durability requirements of the applicable specifications. This requirement addresses structure that is susceptible to fatigue loading and whose failure may be costly in terms of excessive maintenance or functional problems. The economic life of these components are defined, and must be no less than design service of the aircraft.

The fracture control plan must include the basic experimental data that are used to analytically define the life of the parts, the issuance of a fracture control parts list, traceability of materials, identification on the drawings describing the part as fracture critical, special processing controls, quality control procedures, vendor certification, and so forth. The durability control plan is similar; however, the controls one must maintain on the various components are less stringent.

Selection of Materials, Processes, and Joining Methods. The fourth portion of this task is the selection of the materials, the processing of the materials, and the methods used to join these materials. Trade-off studies are used in preliminary design before final selection of materials are made. The final choice of materials and their processing history is justified. The materials meet the design criteria as required by the contract. Joining the materials becomes increasingly important. Many failures occur where structures are joined together. This includes the use of fasteners, welding, brazing, diffusion bonding, and adhesive bonding. The inspectability, inherent residual stresses caused by the processing itself, environmental degradation, and other factors for these material joints must be discussed to ensure that the optimal choice is made, and that the requirements as set forth in the contract can be met.

Design Service Life and Usage. The final part of defining design information in task I of ASIP is the design service life and design usage portion. These data, in the case of ASIP, provided by the U.S. Air Force, are the anticipated usages that the contractor needs to know to determine the life of the designed structure. It is also known that initial usage data may not be sufficiently severe, and that measured usage later in the program can establish whether provided data are not severe enough.

Design Analyses and Development Tests (Task II). The purpose of this ASIP task is to determine the environment in which the aircraft will operate and then design the various structural components to withstand the loads and stresses for the design life for each of the components. As stated in the various specifications, it is unnecessary for every component to be designed with a life equal to that of the system. It is important that the life of every component be known and that each component be inspectable as its design life requires. The types of analyses that are required are described below.

Design Allowables for Materials and Joints. For the material and joint allowables, ASIP requires the contractor to use the data available in the various handbooks. These handbooks include *MIL-HDBK-5* (Ref 46), *MIL-HDBK-017* (Ref 47), *MIL-HDBK-23* (Ref 48), and *MCIC-HDBK-01* (Ref 49). Other data can be procured from other sources or generated by the contractor; however, they must be approved by the contracting agency.

Load Analysis. An analysis of loading must be conducted by the contractor and must be accomplished in accordance with the appropriate specification. Both the static and dynamic loads that the aircraft and aircraft component will encounter must be described. In this instance, a load analysis implies utilizing the external aircraft parameters such as roll, roll rate, roll acceleration, yaw, yaw rate, yaw acceleration, airspeed, weight and weight distribution, vertical and horizontal acceleration (n_z and n_y), and so forth, and obtaining from these parameters the shear and bending moment on different locations of the airframe. Loads

for all flight and landing conditions must be obtained, and these include air loads, ground loads, weapon effect loads, and so forth. In addition, the analysis should include the effects of temperature, frequency, chemical environment, dynamic response of the structure, and so forth.

Design Load Spectra. An anticipated service-load spectrum must be defined for design purposes in conjunction with the specification that best defines the anticipated use of the system. This is an anticipated load-time distribution that can be expected over the service life of the aircraft. A difficulty inherent in new systems is that the actual use has not been determined. All that exist are service load spectra for aircraft that have flown in roles similar to that which is presently being designed and built. The load spectrum must take into account maneuver, gust, landing, taxi and take-off loads, and must consist of flight-by-flight type loading. A spectrum with a small and discrete number of loads that is ordered from low to high (or high to low) with a number of cycles in each of these "blocks" is generally not allowed. A random flight by flight spectrum is generally desirable.

The chemical/thermal environment spectrum that is assumed for design must also be taken into account. Again the appropriate specification that presents these data for similarly used aircraft must be utilized. Parameters of interest include environment, waveform, and frequency.

Structural Analysis. The next eight subtasks in this task discuss the various types of analyses that are required to structurally design a viable aircraft. These analyses are used in conjunction with the load analysis discussed above. These analyses include stress analysis, damage tolerance analysis, durability analysis, sonic fatigue analysis, vibration analysis, flutter analysis, nuclear weapons effect analysis, and non-nuclear weapons effect analysis.

The stress analysis that is to be conducted must be in agreement with the details of the contract requirements, and the non-nuclear weapons effects analysis must follow the guidelines of the design handbook. All the other analyses must be conducted in concert with the U.S. Air Force specification that describes that particular analysis. Some of these specifications are very detailed in nature, while others are more generic. For example, one may discuss the methodology and procedure for a particular analysis, and another may discuss only the initial conditions that must be addressed and an array of various design concepts that may be used for particular components.

An important part of these subtasks is the interrelationships they have with the testing portions of this effort. For example, there is a requirement for a full-scale fatigue test, durability test, and damage tolerance test. Which ultimate load conditions to be tested, the load distribution desired, and how to obtain this distribution must be determined as the result of the stress analysis. A similar type of analysis needs to be conducted for the durability and damage tolerance test article. In this instance, there is a description of the loading spectrum, where to inspect for flaws during the durability test phase, and where to place the flaws during the damage tolerant test phase.

Design Development Tests. The last step in task II of ASIP is the testing of decisions or results from the design and development process. This step may be:

- Tests required to obtain the material data
- Coupon and component tests to ensure that the stresses and the fatigue and fracture characteristics are known
- Tests to allow an independent verification for both the stresses and the fatigue and fracture behavior of the material
- Tests to get material allowables for joints, splices, fittings, and so forth

The size of the tests ranges from coupons to large full-scale components. A complete set of plans for these tests must be prepared showing the purpose, scope, scheduling, correlation with analyses, impact with design, possible trade-off of designs, and so forth. In addition, these plans must be approved by the U.S. Air Force.

Full-Scale Testing (Task III). The purpose of this task is to ensure the "structural adequacy" of the completed airframe. This is accomplished by ground testing and flight testing. It is important to determine the structural adequacy as early as possible in the contract. For this reason, these tests are conducted on the first production article that is released. The tests that are conducted in task III are identical to those that are analyzed in the previous task. These are the ground test articles that include the static test article and a damage tolerant and durability test article. The flight test articles include those from which external load analyses may be obtained, and sonic, vibration, and flutter tests.

Ground Tests. The static test again will comply with the specification that has been written to describe this test task. The contractor must submit a plan that will fully describe this test, and this test plan must be approved by the U.S. Air Force. This includes plans, procedures, and schedules. The significant design limit and ultimate design conditions will be tested to allow a thorough evaluation of the airframe. Special environmental factors such as heat and chemical environment must be included as appropriate. The scheduling of this test must be accomplished so as not to impact the schedule for the later flight tests.

The durability and damage tolerance test requires a second airframe (the first airframe being used for the static test). These tests are run consecutively with the durability test being conducted first. Both of these tests will be conducted in accordance with the appropriate specifications. The contractor must prepare a test procedure, plans, and test schedule that must be approved by the U.S. Air Force prior to the start of any testing. It is incumbent that the test spectrum be a random flight-by-flight spectrum and that all durability critical locations be tested. One of the purposes of this test is to determine any additional areas that have not previously been identified as critical. Also, as a result of this test, inspection intervals and modification times for critical structure may

be obtained. Therefore, realistic loads and loading geometry are absolutely essential.

As mentioned previously, the article selected for the durability test should be an early airframe, but as representative of actual production design and manufacturing techniques as possible. One lifetime of testing including an inspection of selected components, if necessary, should be completed on this article prior to release of the production contract by the U.S. Air Force. A second lifetime of testing and inspection should be completed prior to delivery of the first production aircraft. If the airframe being tested does not complete two lifetimes without unacceptable cracking, a complete evaluation of the durability design may be required. If this be the case, the original production schedule would be altered. This decision would be made at the appropriate time. The contractor may, at the discretion of the U.S. Air Force, continue the test for an additional two lifetimes. This would allow for possible life extension of the airframe. The contractor must prepare a cost and schedule that would allow for this possibility.

Subsequent to the durability test, damage tolerance testing would be initiated on the same article. In addition, full-scale test components may be used for additional selected tests. Again, the test plan, procedures, and schedules are required to be approved by the U.S. Air Force prior to testing. Flaws of known characteristics would be physically introduced into the structure prior to testing. The flaw growth behavior would be monitored to ensure that the predictions from earlier analyses and testing are correct.

Flight Tests. The primary purpose of the flight and ground operations tests is to ensure that the external load analysis results are in agreement with the measured results. This aircraft will be heavily instrumented and flown in certain mission and flight conditions to compare the results of the analyses with the tests. The ground operations test is primarily a vibration test whose purpose is to verify mass, stiffness, and damping characteristics that would be used in the flutter analysis. The flutter test determines whether the aircraft is free from aeroelastic instabilities (flutter) and has acceptable damping in the area encompassed by the flight envelope.

If any problem is uncovered in these tests, the contractor has to determine the cause, corrective actions, fleet impact, and cost to fix the problem. All problems and fixes along with their schedule and cost impact and fleet consequence have to be submitted to the U.S. Air Force for approval.

Force Management Data Package (Task IV). After the aircraft is developed and manufactured, it must be maintained. Again, the maintenance of these aircraft is the responsibility of the U.S. Air Force Matériel Command. To ensure appropriate maintenance actions are conducted, it is important that additional information be developed. It is in task IV that such information is determined. That information consists of five primary items: the final analyses (load analyses, stress analyses, damage tolerance, and durability analyses, etc.), a strength summary of the air-

frame, a force structural maintenance plan, a load/environment stress survey, and an individual aircraft tracking program. Each of these items is discussed below.

After the various tests are completed, for example, the full scale static test article or the load environment/spectra survey, it may be seen that a discrepancy exists between the stresses determined from the test at certain locations on the aircraft and those obtained from analysis. An investigation is required to determine wherein the inconsistency lies. If the analysis requires a reevaluation, then the equations must be appropriately modified. After the modification occurs, the changes must be approved by the U.S. Air Force. It is this final analysis that would be used by The Air Material Command to determine repairs, modifications, and inspection intervals.

In addition to the final analyses, a strength summary is submitted to the U.S. Air Force. This summary is to be written in a manner that would consolidate the many volumes of data that exist on the particular airframe and allow the user to quickly ascertain the areas that require remedial action if an unforeseen event occurs. It is this type of information that would be made readily available to the user through this document.

The third portion of this task, the force structural maintenance plan, is written to inform the user and the maintenance organization of the inspection and modification requirements for the airframe and its estimated economic life. The information in this plan should be as complete as possible so that the U.S. Air Force can rely on this document for determining long-term costs for the airframe. The initial release of this plan will be based on the design and analysis portions of the program. As more data are obtained from the testing, especially the load and durability portions of the program, the plan would be updated as appropriate. The U.S. Air Force assumes that there would be both an initial and a final plan written into the statement of work for the program, and any additional updates would require a separate negotiation. The contents of this plan, its organization, and structure must be approved by the U.S. Air Force.

The next portion, the load/environment spectra survey, is the largest and most extensive portion of this task. In this subtask, the subcontractor is required to instrument 10 to 20% of the fleet for three years or one design lifetime to determine how the aircraft is actually being flown. This is for later comparison with the assumed spectrum used during design. The parameters that must be recorded include weight, airspeed, altitude, various rates, and accelerations such as pitch, roll, and yaw, "g" levels, and so forth. These data must be collected, synthesized, and analyzed in a manner that allows the maintenance organization to determine the inspection intervals for the various fracture critical parts and to ascertain the economic life vis-à-vis the actual aircraft usage. The contractor has to define the number of aircraft to be instrumented, the instrumentation used, the parameters collected, the algorithms, and so forth. Finally, the actual usage spectra for the

Table 3 Engine Structural Integrity Program tasks

Task I Design information	Task II Design analysis material characterization and development tests	Task III Component and core engine tests	Task IV Ground and flight engine tests	Task V Engine life management
Development plans		Component tests	Ground engine tests	
ENSIP master plan	Design duty cycle	Strength	Thermal survey	Updated analyses
Durability and damage tolerance control plans	Material characterization	Vibration	Ground vibration strain and flutter boundary	Engine structural maintenance plan
Material process characterization plan	Design development tests	Durability	Unbalanced rotor vibration	Operational usage survey
Corrosion prevention and control	Analyses	Damage tolerance	Strength	Individual engine tracking
Inspection and diagnostics plan	Sensitivity	Containment	Impedance	Durability and damage tolerance control actions (production)
	Critical parts list		Clearance	
	Thermal		Containment	
	Strength		Ingestion	
	Containment		Accelerated mission tests (AMT)	
	Vibration/flutter		Damage tolerance	
Operational requirements	Stress/environment spectra	Core engine tests	Flight engine test	
Design service life and design usage	Durability	Thermal survey	Fan strain survey	
Design criteria	Damage tolerance	Vibration strain and flutter boundary survey	Nacelle temperature survey	
	Creep		Installed vibration	
	Installed engine inspectability		Deterioration	
	Manufacturing, Process, and Quality Controls			
	VSR			
	NDI demonstration			

aircraft need to be defined using the collected data as a baseline. From these newly developed spectra, algorithms would be developed that allow the user to utilize these data to determine safety limits and inspection intervals for the aircraft. This entire package of requirements needs to be proposed, using the appropriate U.S. Air Force and government standards, as required, and submitted to the U.S. Air Force for approval.

The last portion of this task is the individual aircraft tracking. This subtask is actually the report aspect of the previous subtask. It is known that various components will need inspections or replacement at different times. In addition, some aircraft will be flown more severely than others. As a result, scheduling of maintenance actions needs to be addressed. It is in this subtask that this will be accomplished.

Force management (task V) addresses the required operations that the U.S. Air Force must perform to ensure safety and operational readiness of the individual aircraft and the fleet of aircraft. The U.S. Air Force will rely heavily on the documents and methods previously produced by the contractor to accomplish these operations. There are four parts to this task. Two of these, the load/environment spectra survey and individual airplane tracking data, are identical to those in task IV. The primary difference is that it will be the responsibility of the U.S. Air Force to accomplish them, that is, to obtain the equipment and train personnel both to use the equipment and to obtain valid data from the equipment. The usage of that data and the training of Air Force personnel are paramount.

The final two sections of task V, individual aircraft maintenance times and structural maintenance records, are self-explanatory. They simply say that the U.S. Air Force will determine the inspection intervals and adhere to them, and documentation regarding them is required.

Engine Structural Integrity Program (ENSIP)

The Engine Structural Integrity Program (ENSIP) (Ref 50) began in the 1970s as a result of a number of structural problems that were occurring on U.S. Air Force turbine engines. These problems not only affected the safety of the aircraft, but also the durability of the engine itself. The cost of maintaining and modifying the engines was escalating at an alarming rate. Both of these sets of problems manifested themselves in fleet readiness. The purpose of ENSIP was to design a procedure that would systematically evaluate the structural integrity of a turbine engine with respect to safety, economic life, and total life-cycle cost. Many papers were written by U.S. Air Force personnel discussing ENSIP. A paper written by Tiffany and Cowie (Ref 51) discusses ENSIP in more detail.

Engine structural integrity is accomplished by ensuring structural safety through the use of damage tolerance design. By using the principles of fracture mechanics, the crack propagation rate can be determined, and periodic inspection ensures safe operation. The components of the engine itself, dependent on the aircraft role and the selected component, have a particular life. The durability requirement of ENSIP dictates that the economic life of each cold component be greater than the design service life. To ensure that the engine can be maintained throughout its service life, ENSIP insists that new and old parts be interchangeable. In addition, the inspectability and repairability of the various parts must be defined. The environment to which the components are exposed must be defined. This would include thermal (steady-state and transient), pressure, chemical, steady-state and dynamic stresses, and so forth). Because of the unusually harsh environment to which the engine compo-

nents are exposed, the contractor must prepare a material and process characterization plan that is valid throughout engine development. To ensure that the engine meets its durability criteria, ground tests on a completed engine must be performed. Maintenance actions, when required, must be accomplished. Finally, a policy instituting usage and tracking criteria must be formulated.

As discussed above in ASIP, there are five major tasks associated with ENSIP. Each of these tasks is divided into several subtasks. The major tasks, described in detail in Table 3, are as follows:

- *Task I:* Design information
- *Task II:* Design analysis, component and material characterization
- *Task III:* Component and core engine tests
- *Task IV:* Ground and flight engine tests
- *Task V:* Production quality control and engine life management

Rather than discuss the intricacies of the specification, it would be beneficial to demonstrate the specific specification requirements for selected areas. The particular functional areas to be discussed are damage tolerance, durability, maintainability, material characterization, environmental definition, ground test methods, and tracking philosophy.

Damage Tolerance. According to the specification, damage tolerance criteria must be used on fracture critical parts. A fracture critical part is defined as one "whose failure would result in probable loss of the aircraft as a result of noncontainment or, for single engine aircraft, power loss preventing sustained flight either due to direct part failure or by causing other progressive part failures" (Ref 52).

The specification emphasizes component inspectability. The more inspectable the component, the more often it may be inspected and the shorter the inspection interval will be. The worst case would be a particular component that may be inspected only during engine teardown in the depot. The initial flaw size would be large (the maximum nondetectable flaw size), and the inspection interval would be long (two depot inspection intervals). In those cases where the inspection interval is long, the crack may grow substantially. As a result, adherence to design detail to minimize stress concentrations factors and appropriate material selection is paramount.

Durability Requirements. The durability of a particular component is closely tied to its economic life. For cold parts, the primary failure mode is structural cracking. To protect against incipient damage, fracture mechanics methods are used. But to prevent generalized cracking prior to the component attaining its design life, low-cycle fatigue (LCF) methods are generally used. The use of LCF in design and analysis, in conjunction with appropriate material selection and substantiation tests, is the method required to ensure that generalized cracking does not occur during the component design life. The LCF design life is usually equal to the design life of the component based on minimum material properties. For hot parts (those parts in the hot gas stream), the primary failure mode becomes more complicated. The durability of the component is adversely affected by stress rupture and erosion in addition to LCF. The minimum design life for hot parts is one-half the design service life for cold parts.

The durability of the engine and its components is usually obtained by test. Similar to that discussed previously for ASIP, a full-scale test of the production article is desired. A usage history of the engine is determined. The test can be accelerated by omitting many of the times at cruise and stress cycles that would cause no damage. The resultant accelerated mission test (AMT) is equivalent to a lifetime of usage but can be accomplished in a reasonable time. The test article is heavily instrumented and monitored closely, and changes that would either improve mission performance or enhance life are instituted into the design. The economic life of the engine is attained when the cost of repairing the article is unacceptable and continued usage of the unrepaired article impairs its operational capability.

Design, Analysis, and Test Procedures. As engine development is an evolutionary process, the manufacturer will continue to use those methods and materials that provide structurally durable and reliable engines. Because the engine will have unique requirements, additional studies will be required to ensure that the fully designed engine has been optimized for its particular role. The basis for this optimization study is definition of the combination of the power setting and times this engine will experience in its particular role. This anticipated usage, the engine duty cycle, is formulated by the contractor and approved by the U.S. Air Force. This duty cycle is ultimately used to provide confidence that the engine design and structure meet the mission requirements and de-

sired economic life requirements of the U.S. Air Force.

The material allowables that are used by the contractor must be verified experimentally. Sufficient data are obtained to allow a statistical substantiation of the material properties. The allowables themselves that are used in design will be -3σ data. The processing of the material must be accomplished from full-size heats that are subsequently machined into test components and specimens. Allowable stress levels, crack growth behavior, inspection capability, and so forth all need to be verified by tests.

A set of tests must be accomplished on the core engine. It is through these tests that thermal and vibration data can be obtained. Other tests will be accomplished in spin pits and on test rigs to gather information that allows the manufacturer to ensure that the desired structural integrity is obtained. Low-cycle fatigue tests and damage tolerance testing may be accomplished on the same components. First these components are tested without flaws, and then flaws are induced and the components tested to another lifetime. One result of all these tests is a list of critical parts, failure modes, and failure times. Inspection techniques will be enhanced as a result of the knowledge obtained from accomplishment of this test program.

Ground and Test Flight Verification. A significant inclusion in the ENSIP requirements is the need to perform ground testing on the production design. The AMT testing gives important information regarding the LCF characteristics of the engine, including the location of previously unknown problems, as well as the damage tolerance behavior of a production engine. Furthermore, as the AMT engine is essentially a "lead-the-force" engine in terms of time, structural and performance behavior can be monitored and information regarding actual engine behavior can be accurately determined.

It is important to schedule the ground testing early enough in the program to allow both the contractor and the U.S. Air Force to determine if the engine meets its performance and structural requirements. As a result, it is necessary for one lifetime of AMT testing to be completed before production approval. AMT testing is also required on a second engine, identical to production, and a third lifetime of testing is necessary using an updated AMT usage history. It is assumed that flight data will then be available for inclusion in the derivation of the test spectrum. Other information that may be obtained from these ground test engines is determination of methods and procedures for inspection that can be used in the field.

Prior to flight test, it is necessary to obtain data that ensure anomalous damage does not occur due to excessive temperature or overspeeding of the engine. Knowing the results of this test plus the results of resonant searches over the entire flight regime then allows the contractor to place the engine in an aircraft for flight test. Instrumentation is placed in the flight engine, the aircraft is flown to several "points in the sky," and data are

collected. The data are evaluated and compared with ground test data and analysis. Correlations are made, and techniques are optimized to ensure proper prediction of engine behavior.

With these data and the improved analyses, all locations on the engine that indicated a problem in test or performance will be critically examined. An understanding of all structural problems regarding crack initiation and failure mode will be determined. The result of these studies will culminate in a plan that will reduce the occurrences of these undesirable events. One aspect of this plan can be feedback to production and design to reduce or eliminate the cause of this cracking.

Engine Life Management. The purpose of engine life management is to determine where each engine is in its particular life history and to perform the required maintenance work in a timely fashion. To perform the first function, an engine tracking program is required. Engine tracking requires the use of instrumentation on the aircraft to provide knowledge of critical parameters that determine the life of an engine to the maintenance organization. This could include flight hours, rotor speeds, temperatures, take-offs, and so forth. Software programs must be available to take this information from individual engines and process it quickly and succinctly. Additionally, maintenance organizations must be trained in the necessary skills to disassemble, inspect, replace, and repair the various components. After the engine is repaired, testing must be accomplished to ensure proper operation. This portion of the effort is the primary responsibility of the U.S. Air Force. However, the information obtained as a result of ENSIP allows this task to occur.

Mechanical Subsystems Structural Integrity Program (MECSIP)

The Mechanical Subsystems Structural Integrity Program (MECSIP) (Ref 52) is a program that was initiated by the U.S. Air Force in the late 1980s to expand the integrity process that had been started with ASIP and ENSIP to mechanical subsystems and equipment. The methodology and the approach to MECSIP is similar to the earlier programs. MECSIP was developed to fill a need to have an organized approach with the discipline of the earlier programs to ensure safety, reliability, and durability in mechanical subsystems. The typical subsystems that MECSIP would be used to evaluate include landing gear, crew escape modules, flight control equipment, hydraulic equipment, engine controls and accessories, and so forth. The equipment itself could consist of pumps, valves, actuators, fittings, tubings, and pressure vessels. Because of the breadth of components and systems to which MECSIP is applicable, the specification is broadly written, yet the particular program can be tailored to fit an individual system. The tailoring of the specification to fit the particular system is one of the first tasks in MECSIP.

Table 4 Mechanical equipment and subsystems program tasks

Task I Preliminary planning and evaluation	Task II Design information	Task III Design analyses and development tests	Task IV Component development and system functional tests	Task V Integrity management data package	Task VI Integrity management
Program strategy	MECSIP master plan	Load analyses	Functional tests	Final analyses	Operational usage survey
Trade studies	Design criteria	Design stress	Strength testing	Maintenance planning and task	Maintenance records service
Development and refinement of	Design service life/design usage	environment/spectra	Durability testing	development	reporting
requirements	Critical parts analyses and class	development	Vibration dynamics/acoustic	Individual systems tracking	Individual subsystem
Preliminary integrity analysis	M & P selection/characterization	Performance and functional	testing		maintenance times
	Product integrity control plan	sizing analysis	Damage tolerance testing		
	Corrosion prevention and	Thermal/environment analysis	Thermal and environment		
	control	Stress/strength analysis	survey		
		Durability analysis	Maintainability/repairability		
		Damage tolerance analysis	demonstration		
		Vibration/dynamics/acoustics	Evaluation and interpretation		
		analysis	Integrated test plan		

A significant difference between MECSIP and ASIP/ENSIP is the customer. The contractors who develop the system using ASIP/ENSIP have the U.S. Air Force as their customer. The contractor who utilizes MECSIP generally has as the customer the airframe or engine manufacturer. If the manufacturer is designing and producing a gear pump, the contract is with the engine manufacturer. If the manufacturer is designing and producing landing gear, the customer is the airframe manufacturer. It is these contractors who supply the data needed for design to the mechanical subsystems and equipment manufacturer.

Similar to ASIP and ENSIP, there are six major tasks in MECSIP, each containing several subtasks. These are described in detail in Table 4 and are listed below:

- *Task I:* Preliminary planning and evaluation
- *Task II:* Design information
- *Task III:* Design analyses and development tests
- *Task IV:* Component development and system functional tests
- *Task V:* Integrity management data package
- *Task VI:* Integrity management

Analyses Required. One of the first needs of the contractor is to understand his subsystem and equipment and the interrelationship between his mechanical equipment and/or subsystem and the aircraft in which it will be installed. The subsystem manufacturer must be given information regarding the requirements levied on his equipment or subsystem by the system(s) they service. An important aspect of this is to understand the failure modes of his system and the effects the failure of his system would have on the operation and performance of the aircraft itself.

After this information is known, and the requirements for the aircraft become solidified, the equipment and subsystems manufacturer can refine his requirements. Trade studies regarding materials and processes as well as approaches to solving particular needs will have been accomplished. The design of the product may then begin with the contractor performing life and performance goals.

The next step is to gather the necessary design information. This task relates to the manufacturer determining the needs of the system and then tailoring his subsystem to fill those needs. To accomplish this, there are several subtasks that should be completed. One of the most important is the MECSIP Master Plan. This plan formalizes the different steps the manufacturer must accomplish during this program. This is a living document throughout the program and is updated as frequently as necessary.

The design criteria for the system must be defined. The specific requirements for the subsystem, as determined by the airframe or engine manufacturer, are translated into design requirements. Materials, processes, geometries, operating environment, life requirements, planned maintenance actions, weight, performance, and so forth are determined. The next step would be the classification of parts. Some of the components of the particular subsystem, if they fail, could adversely impact the mission. Other parts may adversely impact the longevity of the system, that is, cause a durability problem. Special considerations must be determined to ensure complete knowledge of the realistic life behavior, and the ramification of failure must be understood. Fracture of gears on a fuel pump can cause immediate system failure, while transistor life as a function of time at temperature is predictable. A failure effects and modes analysis (FEMA) improves the completeness of these plans and increases the utility of the particular analyses.

After the design has been finalized, the various analyses that are required for ASIP and ENSIP are required for MECSIP as well (see Table 4). It should be noted that all aspects of the plans are not applicable to each particular component. For solder connections, material selection may be well defined, whereas for joints vibration analyses may be more significant.

Testing Required. The testing that is required demonstrates the accuracy of the analyses and the life and inspection capabilities of the components. Additionally, nondestructive evaluation methodology is demonstrated in this phase of the program. Again the types of tests required are predicated on the components and/or subsystem, their utility, and their failure mode. Because the testing required is extensive, the plans regarding the test design, the test duration, fixtures, and life requirements should be accomplished early in the program to ensure one lifetime of testing prior to production go-ahead. As in the case of ASIP and ENSIP, production articles and/or methods should be used as much as possible. Because these components depend on their location in the engine or airframe, special environmental factors such as temperature, acoustical fatigue, vibration, and chemical environment, can play important roles. These conditions must be properly accommodated during testing.

Life Management. The function of life management again falls to the U.S. Air Force. However, the contractor needs to prepare plans regarding inspection locations, inspection intervals, inspection techniques, and so forth. The U.S. Air Force personnel has to know the particular usage spectrum for which these inspections were designed. The software the manufacturer used to develop these inspection intervals should be part of the data package the U.S. Air Force receives. If changes need to be made as a result of in-flight tracking, the U.S. Air Force must have the capability of performing these analyses. In short, life management is a U.S. Air Force action item based on the completeness and accuracy of the tasks performed by the contractor.

REFERENCES

1. A.A. Griffith, The Phenomena of Rupture and Flow in Solids, *Philos. Trans. R.Soc.*, Vol 221, 1920, p 163-198
2. C.E. Inglis, Stresses in Plates Due to the Presence of Cracks and Sharp Corners, *Trans. Inst. Navel Arch.*, Vol 60, London, 1913, p 219-230
3. G.R. Irwin, Fracture Dynamics, *Fracturing of Metals*, American Society for Metals, 1948
4. E. Orowan, Fracture and Strength of Solids, *Rep. Progr. Physics*, Vol 12, 1949
5. G.R. Irwin, Analysis of Stresses and Strains Near the End of a Crack Traversing a Plate, *J. Appl. Mech.*, Sept 1957, p 361-364
6. A. Wöhler, Versuche über die Festigkeit der Eisenbahnwagenachsen, *Z. Beuwessen*, Vol 10, 1860, English Summary 1867 Engineering, Vol 4, p 160-161
7. M.A. Miner, Cumulative Damage in Fatigue, *J. Appl. Mech.*, Vol 12, 1945, p 159-164
8. H. Neuber, Theory of Stress Concentration for Shear-Strained Prismatical Bodies with Arbi-

trary Non-Linear Stress-Strain Law, *J. Appl. Mech.,* Vol 28, 1961, p 544-550

9. P.C. Paris, et al., A Rational Analytic Theory of Fatigue, *Trend Eng.,* Vol 13 (No. 1), Jan 1961, p 9-14

10. P.C. Paris, The Fracture Mechanics Approach to Fatigue, *Fatigue—An Interdisciplinary Approach,* Syracuse University Press, 1964, p 107-132

11. E.K. Walker, The Effect of Stress Ratio during Crack Propagation and Fatigue for 2024-T3 and 7075-T6 Aluminum, *Effects of Environment and Complex Load History on Fatigue Life,* STP 462, M.S. Rosenfeld, Ed., ASTM, 1970, p 1-14

12. R.G. Forman, V.E. Kearney, and R.M. Engle, Numerical Analysis of Crack Propagation in a Cyclic-Loaded Structure, *J. Basic Eng.,* Vol 89, Series D, 1967, p 459-464

13. J. Schijve and P. DeRijk, "Fatigue-Crack Propagation in 2024-T3 Alclad Sheet Materials of Seven Different Manufacturers," NLR TR-M-2162, National Aerospace Laboratory,1966

14. D. Broek, "The Effect of Sheet Thickness on the Fatigue Crack Propagation in 2024-T3 Sheet," NLR-TR-M-2129, National Aerospace Institute Amsterdam, 1963

15. D.R. Donaldson and W.E. Anderson, Crack-Propagation Behavior of Some Airframe Materials, Royal Aircraft Technical Report 73183, Dec 1973 and *Cranfield Symposium,* Vol II, 1960, p 375-441

16. S.H. Smith, T.R. Porter, and W.D. Sump, "Fatigue-Crack Propagation and Fracture Toughness Characteristics of 7079 Aluminum Alloy Sheets and Plates in Three Aged Conditions," NASA CR-996, National Aeronautics and Space Administration, 1968

17. A. Hartman, On the Effect of Water Vapour and Oxygen on the Propagation of Fatigue Cracks in an Aluminum Alloy, *Int. J. Fract. Mech.,* Vol 1, 1965, p 167-188

18. M.R. Achter, Effect of Environment on Fatigue Cracks, *Fatigue Crack Propagation,* STP 415, ASTM, 1967, p 181-204

19. P. Hartman and J. Schijve, The Effects of Environment and Frequency on the Crack-Propagation Laws for Macrofatigue Cracks, *Eng. Fract. Mech.,* Vol 1, 1970, p 615-631

20. T.T. Shih and R.P. Wei, Load and Environment Interactions in Fatigue-Crack Growth, *Prospects of Fracture Mechanics,* Noordhoff, 1974, p 237-250

21. J. Schijve and D. Broek, "The Effect of the Frequency on the Propagation of Fatigue Cracks," NLR TR-M-2094, National Aerospace Institute Amsterdam, 1961

22. J. Schijve and P. DeRijk, "The Effect of Temperature and Frequency on Fatigue-Crack Propagation in 2024-T3," National Aerospace Laboratory (NRL, Amsterdam), NLR TR-M-2138, 1963

23. J.P. Gallagher and R.P. Wei, Corrosion Fatigue Crack Propagation Behavior in Steels, *Corrosion Fatigue: Chemistry, Mechanisms, and Microstructure,* A.J. McEvily and R.W. Staehle, Ed., National Association of Corrosion Engineers, 1972, p 409-423

24. H.D. Dill and C.R. Saff, "Environment-Load Interaction Effects on Crack Growth," AFFDL-TR-78-137, Air Force Flight Dynamics Laboratory, Nov 1978

25. O.E. Wheeler, Spectrum Loading and Crack Growth, *J. Basic Eng.,* Vol 94D, 1972, p 181

26. J.D. Willenborg, R.M. Engle, and H.A. Wood, "A Crack-Growth-Retardation Model Using an Effective Stress Concept," AFFDL-TM-71-1 FBR, Air Force Flight Dynamics Laboratory,1971

27. J.P. Gallagher, "A Generalized Development of Yield Zone Models," AFFDL-TM-FBR 74-28, Air Force Flight Dynamics Laboratory,1974

28. J.P. Gallagher and T.F. Hughes, "Influence of the Yield Strength on Overload Affected Fatigue-Crack-Growth Behavior of 4340 Steel," AFFDL-TR-74-27, Air Force Flight Dynamics Laboratory,1974

29. H.D. Dill and C.R. Saff, Analysis of Crack Growth Following Compressive Loads Based on Crack Surface Displacements and Contact Analysis, *Cyclic Stress-Strain and Plastic Deformation Aspects of Fatigue Crack Growth,* STP 637, ASTM, 1977, p 141-152

30. J.A. Reiman, M.A. Landy, and M.P. Kaplan, Effect of Spectrum Type on Fatigue Crack Growth Life, *Fatigue Crack Growth under Spectrum Loads,* STP 595, ASTM, 1976, p 187-202

31. M.A. Landy, M.P. Kaplan, and J.A. Reiman, *Derivation and Analysis of Loading Spectra for USAF Aircraft,* ASD-TR-76-1, Aeronautical Systems Division, Wright Patterson, 1976

32. N.E. Dowling, Fatigue Failure Prediction for Complicated Stress-Strain Histories, *J Mater.,* Vol 7, 1972, p 71-87

33. C.F. Tiffany and J.N. Masters, Applied Fracture Mechanics, *Fracture Toughness Testing and Its Application,* STP 381, ASTM, 1965, p 249-278

34. U.A. Hinders, "F-111 Design Experience—Use of High Strength Steel," AIAA 2nd Aircraft Design and Operations Meeting, 1970

35. *MIL-STD-1530, Aircraft Structural Integrity Program Requirements,* U.S. Air Force

36. *MIL-A-83444, Airplane Damage Tolerance Requirements,* U.S. Air Force

37. *MIL-A-87221, Aircraft Structures,General Specification for,* U.S. Air Force

38. *MIL-A-8866, Airplane Strength and Rigidity, Reliability Requirements, Repeated Loads and Fatigue,* U.S. Air Force

39. J.P. Gallagher, Estimating Fatigue-Crack Lives for Aircraft: Technique, *Exp. Mech.,* Vol 16 (No. 11), Nov 1976, p 425-433

40. T.R. Brussat, An Approach to Predicting the Growth to Failure of Fatigue Cracks Subjected to Arbitrary Uniaxial Cyclic Loading, *Damage Tolerance in Aircraft Structures,* STP 486, ASTM, 1971, p 122-143

41. J.P. Gallagher and H.D. Stalnaker, Predicting Flight-by-Flight Fatigue Crack Growth Rates, *J. Aircraft,* Vol 12 (No. 9), Sept 1975, p 699-705

42. L.R. Hall and W.I. Engstrom, "Fracture and Fatigue-Crack-Growth Behavior of Surface Flaws and Flaws Originating at Fastener Holes," AFFDL-TR-74-47, Air Force Flight Dynamics Laboratory, 1974

43. S.H. Smith, F.A. Simonen, and W.S. Hyler, "C-141 Wing Fatigue Crack Propagation Study," Final Report to Warner Robins ALC; BCL Report G-2954-1, Battelle, 1975

44. S.H. Smith, "Fatigue-Crack-Growth Behavior of C-5A Wing Control Points," ASD-TR-74-18, Aeronautical Systems Division, Wright Patterson, 1974

45. "Detail Requirements of Structural Fatigue Certification Program," Technical Memorandum WCLS-TM-58-4, Aircraft Laboratory Directorate of Laboratories, Wright Air Development Center, Air Research and Develop Command, U.S. Air Force, June 1958

46. *MIL-HDBK-5, Aerospace Vehicle Structures, Metallic Materials and Elements,* U.S. Air Force

47. *MIL-HDBK-17, Plastic for Aerospace Vehicles,* U.S. Air Force

48. *MIL-HDBK-23, Structural Sandwich Composites,* U.S. Air Force

49. *MCIC-HDBK-01, Damage Tolerant Design Handbook,* U.S. Air Force

50. *MIL-STD-1783, Engine Structural Integrity Program (ENSIP),* U.S. Air Force

51. C.F. Tiffany and W.D. Cowie, "Progress on the ENSIP Approach to Improved Structural Integrity in Gas Turbine Engines/An Overview," Paper 78-WA/GT-13, American Society of Mechanical Engineers, 1978

52. *MIL-STD-1798, Mechanical Equipment and Subsystems Integrity Program,* U.S. Air Force

Section 5: Fatigue and Fracture Resistance of Ferrous Alloys

Fracture and Fatigue Properties of Structural Steels

Alexander D. Wilson, Lukens Steel Company

THE USE OF FRACTURE MECHANICS in the design of steel structures has been a growing discipline over the past 30 years. Steel structures can cover a diverse number of markets, such as construction of bridges and buildings, pressure vessels, construction and mining equipment, and offshore platforms and ships. Both carbon, high-strength low-alloy (HSLA) and alloy steels are used in these applications. Although some of the end users of these structures have their own steel specifications, the vast majority rely on ASTM specifications, particularly for the basic strength properties and chemical composition. An end user who does have a steel specification often bases it on an ASTM grade with some modification.

Within the ASTM structural and pressure vessel specifications, there are close to 1000 that might be specified. In reality, only a few dozen are actually applied in the industry. A specification is chosen for an application based on the design stresses, which dictate the strength and thickness of the product; the temperature and environment of the application; weldability concerns during fabrication; and of course cost. However, beyond specifying the ASTM grade for an application, often other variables must be considered, particularly when "generic" fracture mechanics properties are applied in the design of the structure. Furthermore, more basic properties such as Charpy V-notch (CVN) impact can often provide some general indication of how these variables affect a key fracture mechanics parameter.

Fracture Toughness

Materials and Testing. Steels used in structural applications can cover a wide variety of chemistries, processing characteristics, and thicknesses. These all can have a significant effect on the fracture mechanics properties. Table 1 lists several popular steel grades for which fracture mechanics data are presented in this article. These grades include low-strength carbon steels and higher-strength alloy steels.

Test methods for characterizing the fracture mechanics properties of steels have been in development for over 30 years. Prior to the widespread use of fracture mechanics, traditional toughness tests such as the CVN were widely used to characterize steels and establish guidelines for their fitness for service. Many of these approaches are still widely used today. In fact, today the CVN test is the most widely used quality control test for evaluating steel toughness. Therefore, it is important to consider the new fracture test methods in comparison to the traditional ones, such as the CVN. Table 2 lists the popular traditional and newly developed fracture tests in use today. Comparisons between the traditional methods, such as dynamic tear (DT) testing and drop-weight testing (to determine the nil-ductility transition temperature) are useful. This is shown schematically in Fig. 1. Because the DT test specimen is larger and has a sharper notch than the CVN specimen, it tends to shift the transition curve to higher temperatures. In fact, it has been shown that the lower knee of the DT transition curve comes at the nil-ductility transition temperature determined by the drop-weight test.

Steel toughness depends on several metallurgical and fabrication factors, such as whether the steel is rolled, control rolled, normalized, or hardened and tempered. These basic factors are exten-sively discussed in the literature, along with the important issue of steel cleanliness, particularly as measured by sulfur level (Ref 1). Today structural steel grades can be obtained with sulfur levels limited to as low as 0.001% or as high as 0.040%. Historically, it has been established that "low"-sulfur steels have 0.010% maximum sulfur or better and have inclusion shape control, most popularly obtained by calcium treatment. Low-sulfur steels have fewer nonmetallic inclusions,

Table 1 Popular plate steels used in structural applications

ASTM steel specification	Type(a)	Yield strength, min(b)		Heat treatment(c)	Microstructure(d)
		MPa	ksi		
A516 grade 70	Carbon, C-Mn	262	38	N+	F-P
A588 grade A	HSLA, Cu-Cr-V	345	50	N+	F-P
A633 grade C	HSLA, C-Mn-Cb	345	50	N	F-P
A808	HSLA, C-Mn-Cb	345	50	CR	F-P
A514 grade F	Alloy, Ni-Cr-Mo-B	690	100	Q&T	M
A533 grade B class 1	Alloy, Mn-Mo-Ni	345	50	Q&T	B
A710 grade A class 3	Alloy, Cu-Ni	517	75	Q&T	F-P, B, M

(a) HSLA: high strength, low alloy. (b) Specification minimum. Lower minimums may be applicable for thicker plates. (c) N, normalized; N+, normalized, but may be produced as-rolled; CR, control rolled; Q&T, quenched and tempered. (d) Most typical microstructure: F-P, ferrite-pearlite; M, martensitic; B, bainitic

Table 2 ASTM testing specifications used in fracture testing

ASTM	Standard test methods for:	Common name
E 23	Notched Bar Impact Testing of Metallic Materials	CVN
E 604	Dynamic Tear Testing of Metallic Materials	DT
E 208	Conducting Drop-Weight Test to Determine Nil-Ductility Transition Temperature of Ferritic Steels	NDT
E 399	Plane-Strain Fracture Toughness of Metallic Materials	K_{Ic}
E 813	J_{Ic}, A Measure of Fracture Toughness	J_{Ic}
E 1152	Determining J-R Curves	...
E 1290	Crack-Tip Opening Displacement (CTOD) Fracture Toughness Measurement	CTOD
E 647	Measurement of Fatigue Crack Growth Rates	$da/dN - \Delta K$

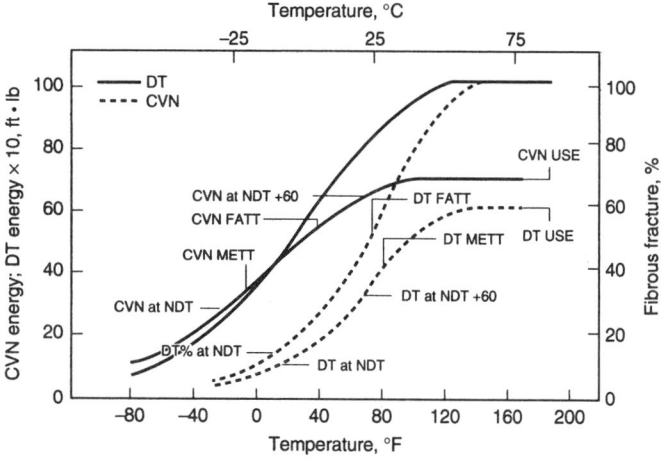

Fig. 1 Comparison of absorbed energy and fracture appearance for Charpy V-notch (CVN) and dynamic tear (DT) testing of an A533B steel. METT, mid-energy transition temperature; FATT, 50% fracture appearance transition temperature; NDT, nil-ductility transition

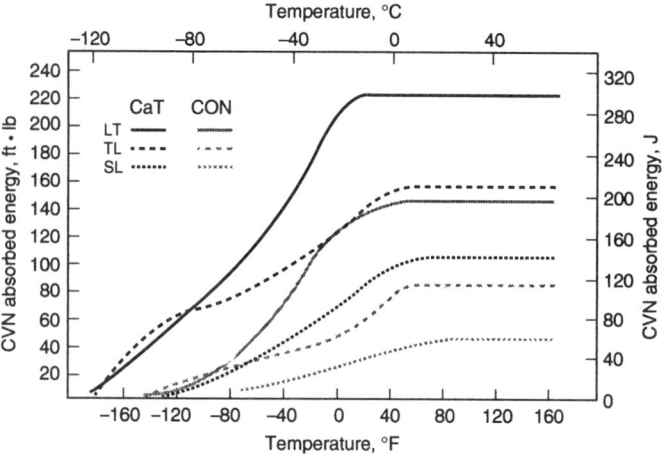

Fig. 2 The effect of processing of two 102 mm (4 in.) A633C plate steels on the full Charpy V-notch (CVN) transition curve in the three major testing orientations. CON, conventional; CaT, calcium treatment

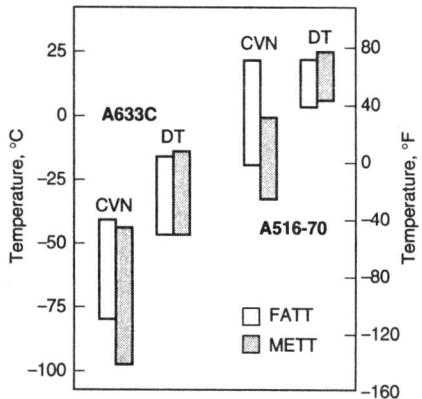

Fig. 3 Comparison of the Charpy V-notch (CVN) and dynamic tear (DT) transition temperatures for a number of A516-70 plates and A633C plates. The improved transition temperatures for A633C are demonstrated. FATT, 50% fracture; METT, mid-energy transition temperature

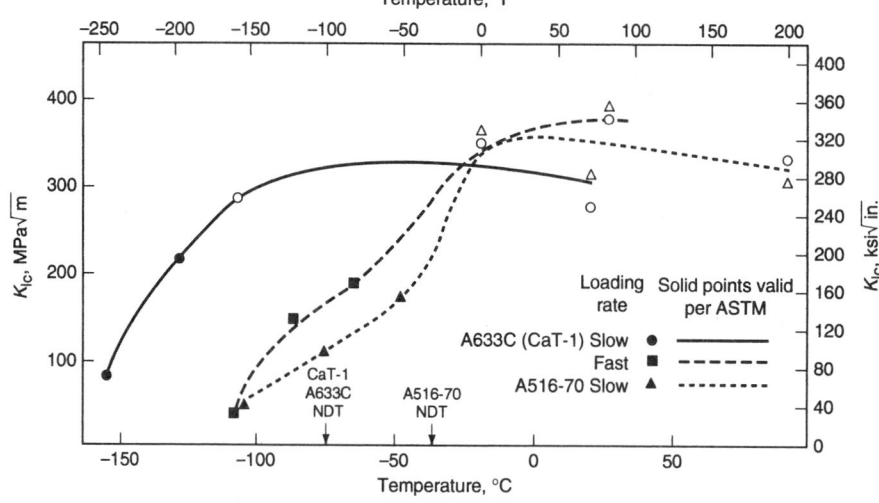

Fig. 4 Fracture toughness (K_{Ic} established from J_{Ic}) comparison for A633C and A516-70 plate. Both slow and fast loading rates are shown for A633C plate. CaT, calcium treatment; NDT, nil-ductility transition

and calcium treatment further gives inclusion shape control. The influence of sulfur control is very dramatic in its effect on toughness, particularly on the upper shelf, as shown in Fig. 2. Sulfur levels also affect the change in toughness level with different testing orientations, which are commonly designated per ASTM E 389 as LT (longitudinal), TL (transverse), and SL (short transverse, through-thickness).

Fracture Mechanics Toughness. Most structural steels have a combination of strength and toughness, which makes the use of elastic-plastic toughness testing (J_{Ic}; crack-tip opening displacement, or CTOD) more popular than plane-strain fracture toughness (K_{Ic}). For example, when comparing the two steel grades A633C and A516-70 in Fig. 3, traditional tests methods (CVN, DT) show the improved behavior of A633C because of its lower carbon levels and the niobium addition. The same benefit is also demonstrated in J_{Ic} testing in Fig. 4.

It is also important to quantify the transition to brittle fracture from toughness testing. However, J_{Ic} and CTOD testing in the transition region can show significant variability in results. Scatter of fracture mechanic toughness in the transition regime is discussed elsewhere in this Volume.

The influence of inclusion control discussed previously can also be characterized by J_{Ic} and J-R curve evaluations (Fig. 5 and 6). Improved-cleanliness calcium-treated steels also show both better initiation toughness and propagation toughness. Most structural steels would show a similar benefit, with the benefit increasing as the strength level increases.

Often it is useful to develop empirical correlations between the various toughness tests. These correlations are easiest to develop for upper-shelf testing where only ductile fracture by microvoid coalescence is the fracture mode. Such empirical

comparisons are commonly made between J and CVN, J and CTOD, and CVN and DT. Figure 7 is an example comparing upper-shelf energy (USE) for CVN and DT test values (Ref 3, 4).

Fatigue Crack Growth

The fatigue crack propagation of structural steels is often considered relatively insensitive to changes among grades. However, steel cleanliness has been shown to have a significant influence on fatigue crack growth rate behavior within a steel. For example, significant anisotropy or directionality of behavior between conventional and calcium-treated A588A steels is detailed in Fig. 8. It has been demonstrated that nonmetallic inclusions can accelerate fatigue crack growth, particularly in the through-thickness (SL, ST)

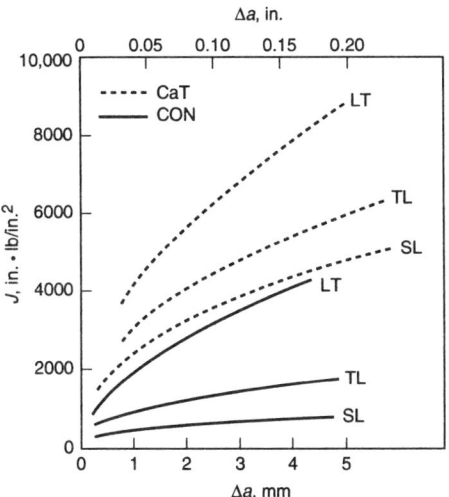

Fig. 5 *J-Δa* resistance curves for two A588A steels in noted quality levels. CON, conventional; CaT, calcium treatment. The effects of quality and testing orientation are noted. Source: Ref 2

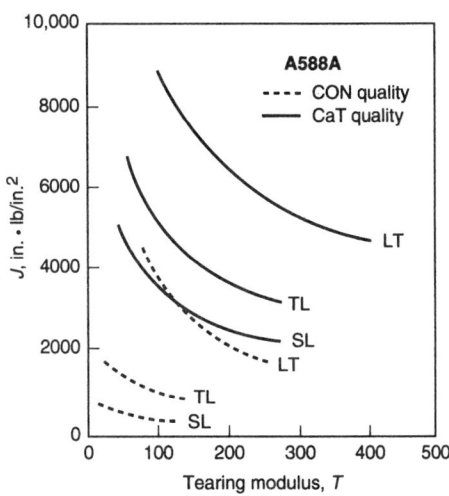

Fig. 6 *J-T* curves for the two A588A steels from Fig. 5. CON, conventional; CaT, calcium treatment. The effects of quality and testing orientation are noted. Source: Ref 2

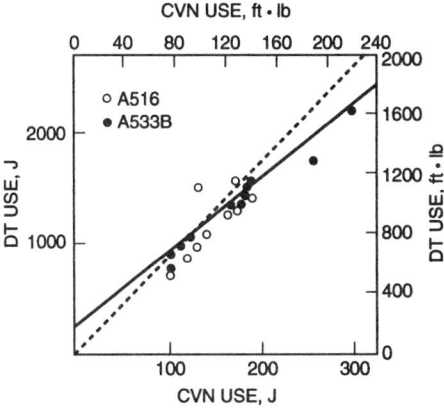

Fig. 7 Comparison plot of DT and CVN upper-shelf energy (USE) data. Solid line represents statistical correlation of DT USE = 250 + 6.59 CVN USE with R^2 = 0.821. Dashed line represents a rough approximation of DT USE = 8.5 CVN USE (correlations in joules). Source: Ref 3, 4

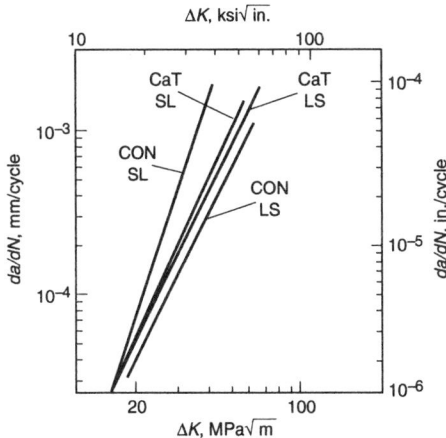

Fig. 8 Fatigue crack growth rate results for two A588A steels showing comparison of LS and SL testing orientations. CON, conventional; CaT, calcium treatment. Improved isotropy of the calcium-treated steel is noted.

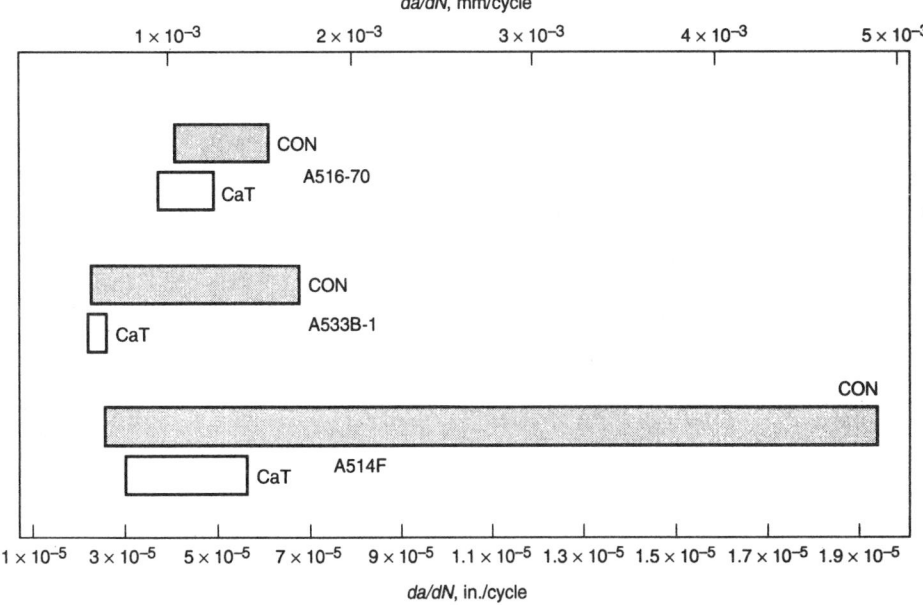

Fig. 9 Range of fatigue crack growth notes at ΔK = 55 MPa√m (50 ksi√in.) in six testing orientations for conventional (CON) and calcium-treated (CaT) quality plates of A516-70, A533B-1, and A514F. Improved isotropy of quality levels is demonstrated for calcium-treated plate. Source: Ref 5

and transverse (TL) testing orientations, and particularly at higher ΔK levels. Figure 9 (Ref 5) demonstrates this for three other steel grades. These results indicate that cleanliness and loading orientation must be considered when using fatigue crack growth rate data.

All structural steel data in air can be summarized in a band of data using the Paris equation ($da/dN = C\Delta K^n$), as shown in Fig. 10 and 11. If the environment of the application is other than benign air, there can be a further acceleration of fatigue crack growth (Ref 6). Salt water, for example, can have a strong effect, particularly at slower testing frequencies (Fig. 12). The modeling of this data with the Paris equation still is appropriate, although the data now are all outside the correlation shown earlier (Fig. 13). More detailed coverage of fatigue crack growth behavior and thresholds is provided in the article "Fracture

Properties of Carbon and Alloy Steels" in this Volume.

Fatigue Life Behavior

In general, fatigue is classified as either low-cycle or high-cycle. Low-cycle fatigue is more closely associated with the modeling and analysis of localized plastic deformation, which is essentially the cause of fatigue crack initiation. As such, low-cycle fatigue behavior is typically determined from strain-based fatigue testing. This method is useful because it provides a basis to

relate component fatigue behavior to localized regions of plastic deformation at notches or other design details. This is briefly discussed in the section "Axial Strain-Life Fatigue" in this article. More detailed information on strain-based fatigue is covered in the article "Fundamentals of Modern Fatigue Analysis for Design" in this Volume.

High-cycle fatigue of steels is often expressed by a fatigue limit, which is the value of stress below which the steel can presumably endure an infinite number of cycles. The fatigue limit should not be confused with fatigue strength, which is the endurance stress under fatigue loading for a specified number of cycles. Fatigue

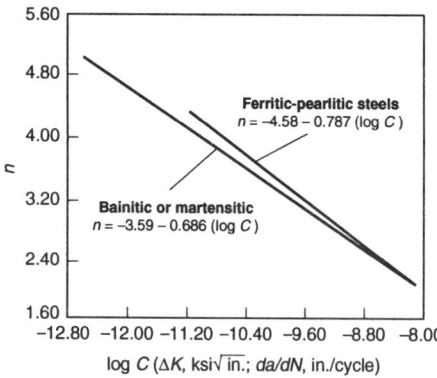

Fig. 10 Plots of n and C from the Paris fatigue crack propagation equation, comparing noted microstructures in a number of steels with various quality levels and testing orientations

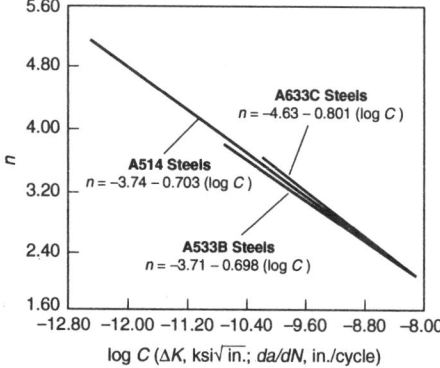

Fig. 11 Comparison of n and C values for the Paris fatigue crack propagation equation comparing three grades of steel with various quality levels and testing orientations. No effect of strength level is shown.

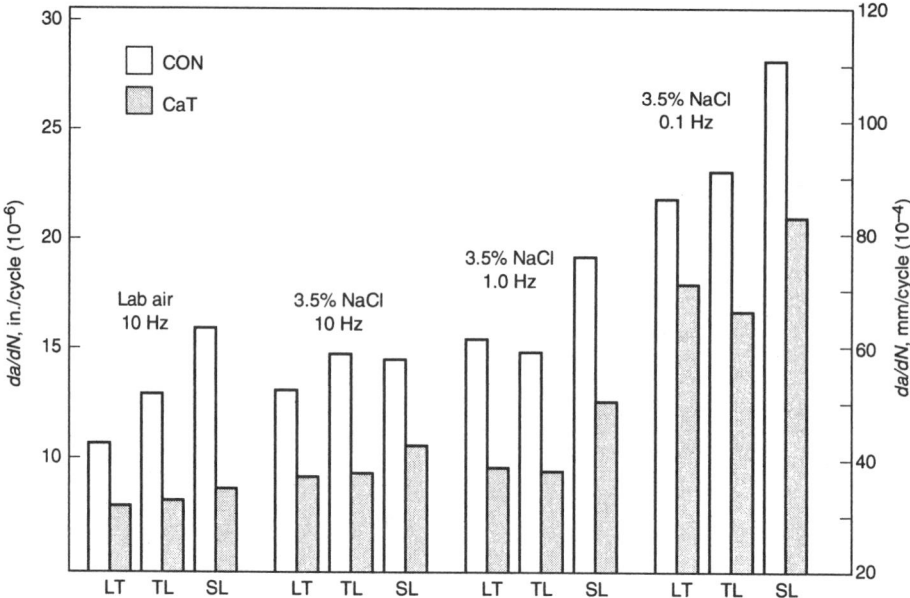

Fig. 12 Bar graphs comparing fatigue crack growth rates at $\Delta K = 55$ MPa$\sqrt{m}$ (50 ksi$\sqrt{in.}$) for conventional and calcium-treated A633C steels. CON, conventional; CaT, calcium treatment. The effect of testing frequency in salt water and steel quality is demonstrated.

strength is used when materials do not have an apparent fatigue limit.

In general, the stress-life (S-N) curve of steels with tensile strengths below about 1100 MPa (160 ksi) has a definite fatigue limit. To a large extent, the fatigue limit depends on the tensile strength (irrespective of whether strengthening is achieved through cold work, transformation hardening, solid-solution alloying, or precipitation hardening). However, the fatigue limit as a ratio of ultimate tensile strength depends on several factors, such as material, loading, and environmental conditions. Moreover, under some conditions (e.g., corrosion fatigue, certain material conditions, or surface stress concentrations), a fatigue limit may not be observed in a steel. If a fatigue limit is observed, it typically occurs above 10^6 cycles between the range of 10^7 and 10^8 cycles.

Unnotched fatigue limits of various steels are shown in Fig. 14 as a function of tensile strength from a systematic of fatigue for ferritic, martensitic, and austenitic steels. A broad band encompasses fatigue limits for 11 martensitic (quenched-and-tempered) steels, while the

austenitic grade (type 304 stainless) and the ferritic grade (type 430) have a distinct band when correlating fatigue limits to tensile strength. This implies a general basis for estimating fatigue limits. However, typical tensile strengths (such as those reported on mill sheet reports) do not necessarily correlate with actual fatigue limits, because typical mechanical properties may be based on heat treatment and test coupon properties that may not apply to the actual situation.

In addition fatigue limits at a given tensile strength are subject to scatter that becomes more pronounced with the presence of inclusions. The relation illustrated in Fig. 14 is based on special tests with steels that contain fewer inclusions than generally expected and thus less scatter. If inclusions or defects are more pronounced, or if testing encompasses steels with widely varying quality, then the scatter of fatigue limits for a given tensile strength should tend to become more asymmetric, because more data would occur in the lower region of the fatigue limit/tensile strength ratio. For the data comparisons in Fig. 14, the relative effects of inclusion size are shown in Fig. 15, where the decrease in relative fatigue strength seems to begin at a defect size of 45 μm and greater (Ref 7).

Notch Effects. A systematic analysis of notch effects/sensitivity on the fatigue strength of various steels has been performed by The National Research Institute for Metals in Japan (Ref 8) in an ongoing fatigue data sheet project, in which fatigue properties have been examined systematically for engineering materials manufactured in Japan. The effects of temperature and stress concentration factor on high-cycle fatigue properties for nine kinds of engineering steels (Fig. 16) were obtained, and the following conclusions were drawn:

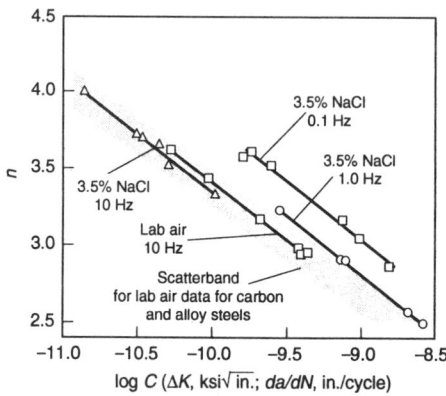

Fig. 13 Plots of n and C from the Paris fatigue crack propagation equation for A633C steels in salt water in conventional and calcium-treated quality levels and three testing orientations (LT, TL, SL). Comparison is made to overall scatterband of all carbon and alloy steels in air testing (37 plates for 8 steel grades).

- The ratio of the fatigue strength of the smooth specimens at 10^8 cycles to the tensile strength at the same temperature varied from 0.4 to 0.6, depending on temperature and material conditions. The mean value was about 0.5 (Fig. 17).
- The fatigue strength of notched specimen testing had less scatter than test results for unnotched specimens (Fig. 17), and the fatigue strength of notched specimens (expressed in terms of "crack initiation" at 10^8 cycles) also corresponded well to the tensile strength, independent of temperature and material conditions.

A notch sensitivity index [$\eta = (K_f - 1)/(K_t - 1)$] for the results had a tendency to become small with increasing stress concentration factor or

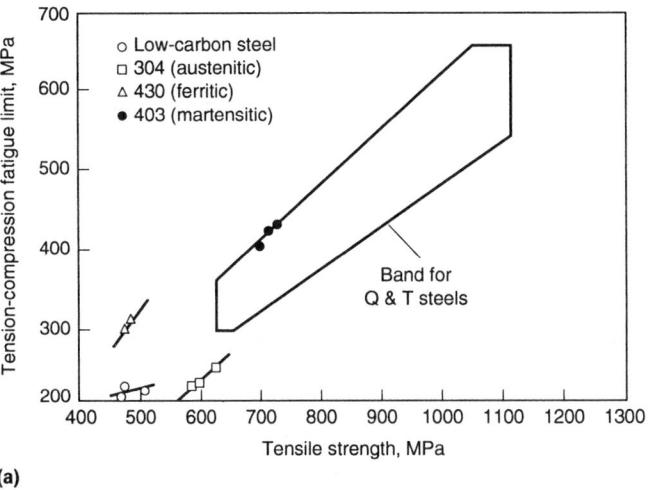

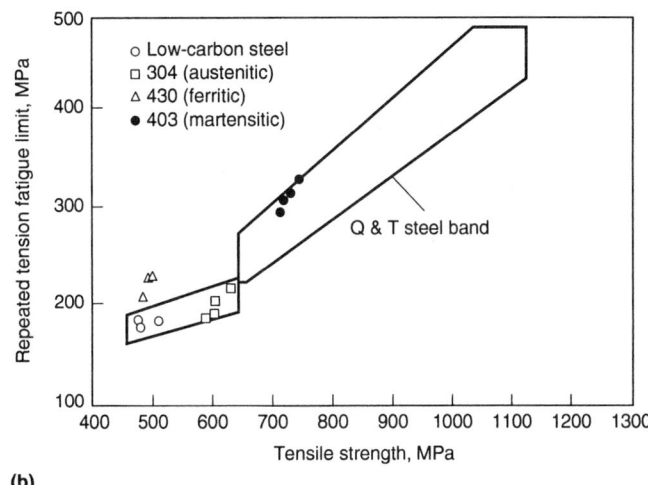

(a)

(b)

Fig. 14 Relation of tensile strength and fatigue limit for stainless steels and quenched-and-tempered (Q&T) structural steels. (a) Tensile strengths vs. reversed tension-compression ($R = -1$) fatigue. (b) Repeated tension fatigue ($R = 0$) vs. tensile strength. Source: Ref 7

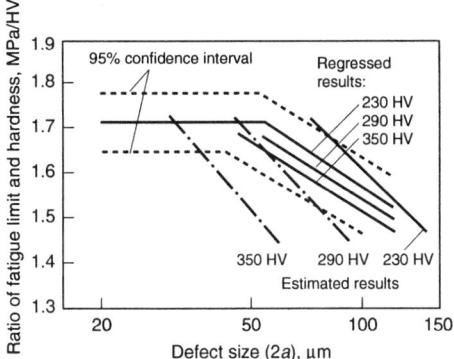

Fig. 15 Relation of relative fatigue strength ratio and size of nonmetallic inclusions at fatigue crack initiation site. The relative fatigue strength ratio for reversed bending fatigue limit and the Vickers hardness (HV) are plotted against defect size. Defect size was evaluated by averaging the largest three diameters of nonmetallic inclusions found for a given heat/lot. The data are plotted tentatively at the 20 μm position for steels revealing no inclusion at crack initiation sites. Horizontal lines indicate regions with no inclusion at the initiation site, while inclined lines indicate regions where defects greater than about 45 μm decrease relative fatigue limits. Source: Ref 7

with decreasing tensile strength (Fig. 18), where K_f is the fatigue notch factor, defined as the ratio of unnotched fatigue strength to notched fatigue strength. A slight decrease in η is noted for the lower-strength steel and austenitic grades (Fig. 18).

Axial Strain-Life Fatigue. The effect of notches on fatigue life can also be considered in terms of strain-life fatigue. Although most engineering components are designed for application in the elastic regime, stress concentrations or notches often cause plastic strains to develop in the vicinity of notches. Fatigue damage is known to be caused by plastic deformation, and analysis of fatigue crack initiation at the root of the notch can be modeled in terms of plastic strain. This use of strain-life fatigue is known as "equivalent fatigue damage," where the plastic deformation of material at the root of the notch is considered equivalent to the plastic strain during cyclic, strain-controlled axial loading of a smooth fatigue specimen. Therefore, strain life data provides useful information in evaluating and understanding the effect of notches on fatigue (see the article "Estimating Fatigue Life" in this Volume).

Strain-life data and the typical fatigue life parameters (expressed in terms of fatigue life exponents and coefficients) for both plastic and elastic strain fatigue are also detailed in the appendix "Parameters for Estimating Fatigue Life" in this Volume. In the high-cycle regime, where plastic strain is minor, fatigue life can be characterized by either the stress- or strain-controlled methodology.

In cases where only total strain life ($\Delta\varepsilon = \Delta\varepsilon_p + \Delta\varepsilon_e$) is available, the plastic strain range can often be correlated with the total strain range through an expression of the form:

$$\Delta\varepsilon_t = B\Delta\varepsilon_p^\gamma \qquad \text{(Eq 1)}$$

where B and γ are experimentally determined constants (Ref 9, 10). Equation 1 is only an approximation of the behavior at intermediate strain ranges and breaks down for small values of $\Delta\varepsilon_t$ and large values of $\Delta\varepsilon_p$ where $\Delta\varepsilon_t \approx \Delta\varepsilon_p$. In addition, some austenitic stainless steels (especially type 301 with the lower nickel content as an austenite stabilizer) may develop martensite in their microstructures during plastic deformation. This metastable aspect of the

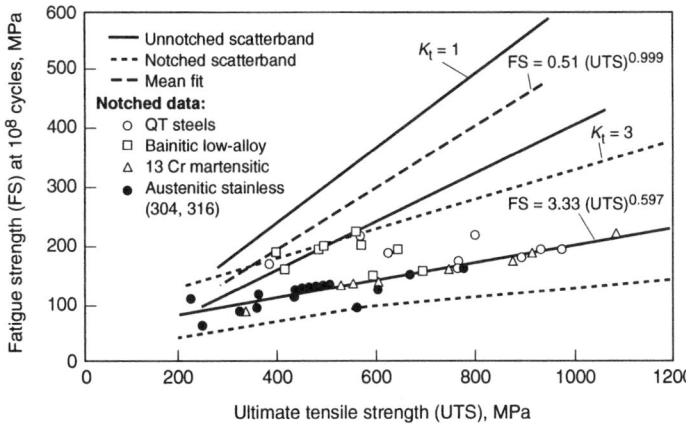

Fig. 16 Relation between tensile strength and fatigue strength of smooth specimens and notched specimens

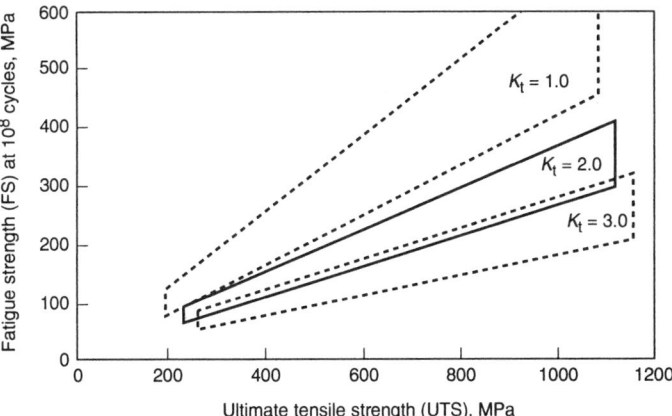

Fig. 17 Relation between tensile strength and fatigue strength for various notch factors. Source: Ref 8

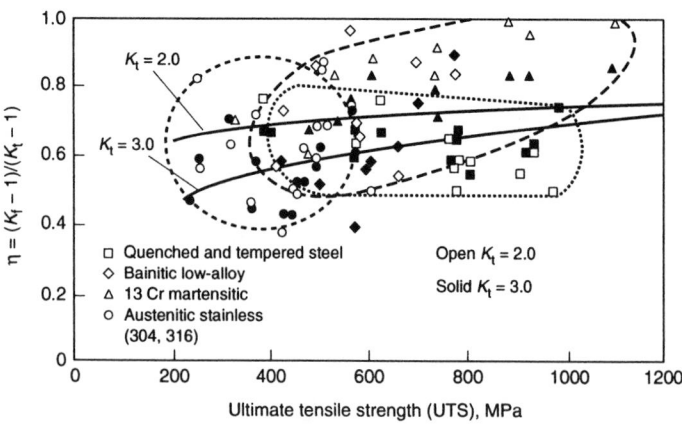

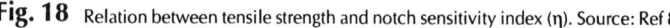

Fig. 18 Relation between tensile strength and notch sensitivity index (η). Source: Ref 8

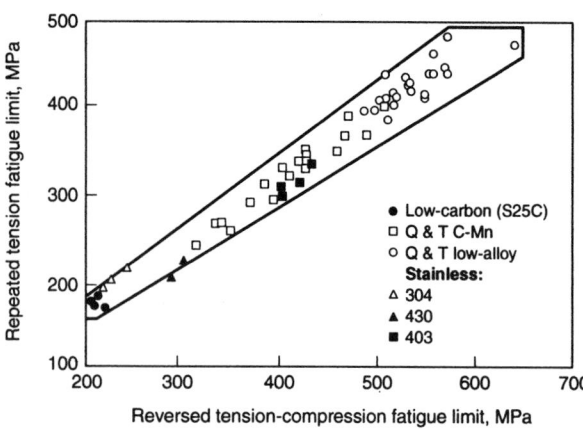

Fig. 19 Relation of fatigue limit under repeated tension to fatigue limit under reversed tension-compression. Q&T, quenched and tempered. Source: Ref 7

austenitic grades can have a nonlinear effect on plastic strain fatigue life and is a subject of research.

Mean stress effects on fatigue limits of various machinery structural steels are shown in Fig. 19 and 20. Figure 19 gives the relation between fatigue strengths under repeated tension and under reversed tension-compression. A general ratio of 0.78 is found for quenched-and-tempered steels of higher strengths, while it is variable for steels of lower strengths and higher ductilities.

The same data are expressed as Haigh's diagram (Fig. 20), which gives the relation of amplitude to mean stress of fatigue limit for different strength levels of steels. In this diagram, fatigue limit lines are combined to yield limit lines, indicating that the material can be used without failure in zones under each curve. The lowest curve, labeled 490 MPa (71 ksi) in tensile strength, represents the trend for low-carbon (S25C) steels in the test program.

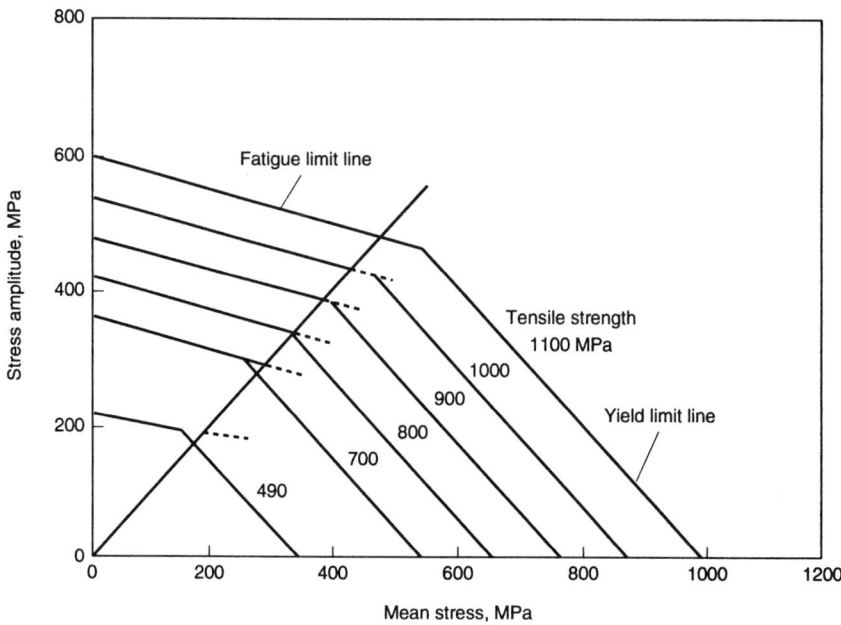

Fig. 20 Fatigue limit diagram relating stress amplitude and mean stress of materials at different tensile strength levels. Source: Ref 7

Corrosion Fatigue Strength

Corrosion or electrochemical surface effects are key variables in both fatigue crack initiation and fatigue crack growth. The current emphasis in corrosion fatigue testing is to measure rates of fatigue crack propagation in different environments rather than to specify the total life to failure. Time-to-failure data are typical of *S-N* tests, while crack growth rate testing has stemmed from the change to a "fail safe" criterion from the use of a "safe life" criterion. The fail safe criterion generally assumes that either cracks are present in the material before service or that cracks are initiated very early in the service history. These assumptions suggest that crack growth controls fatigue behavior in most practical applications.

Nonetheless, evaluation of the effect of corrosion on fatigue life is important and useful. In particular, the simultaneous combination of corrosion and fatigue loading can cause large reductions of fatigue strength at long lives and even the elimination of fatigue limits. For example, the fatigue limit of mild steel is eliminated in aqueous environments, depending on the electrochemical potential (Fig. 21). In this case, corrosion fatigue

represents the elimination of fatigue crack thresholds (ΔK_{th}) from the presence of an electrolyte.

Corrosion fatigue strength can also be interpreted as a "notch effect" from the stress concentrations caused by corrosion pits. This effect can be simulated by prior corrosion, where precorroded samples are subsequently fatigue tested in air or another passive environment. In this case, fatigue limits are observed, but at a reduced level relative to that for uncorroded (smooth) specimens (Fig. 22). The notch effect from corrosion pitting can also be illustrated when fatigue limits are observed for stainless steels during fatigue testing in aqueous solutions (Fig. 23). Fatigue endurance results for smooth and notched specimens of a 13% Cr ferritic steel in distilled water and salt water are shown in Fig. 23. These environments are seen to cause a large reduction in the endurance limit of the smooth specimens but only a relatively modest effect on the notched specimens. The importance of surface finish and

the adverse effect of corrosion, even when it is allowed to occur only before a fatigue test, is of course well known and included in standard texts on fatigue design.

Corrosion fatigue depends on the material, the environment, and how the corrosive environment is applied. For example, specimens submerged in fresh or salt water have better corrosion fatigue resistance than those subjected to a water spray, drip, or wick or those submerged in continuously aerated water. This is due to the great importance of oxygen and the formation of oxide films in the corrosion fatigue process.

The typical fatigue strengths of various structural alloys are summarized in Table 3 for aqueous environments. In general, corrosion fatigue tests on smooth specimens of high-strength steel indicate that very large reductions in fatigue strength or fatigue life can occur in salt water or spray (Fig. 24). For instance, the fatigue strength at 10^6 cycles could be reduced to as little as 10%

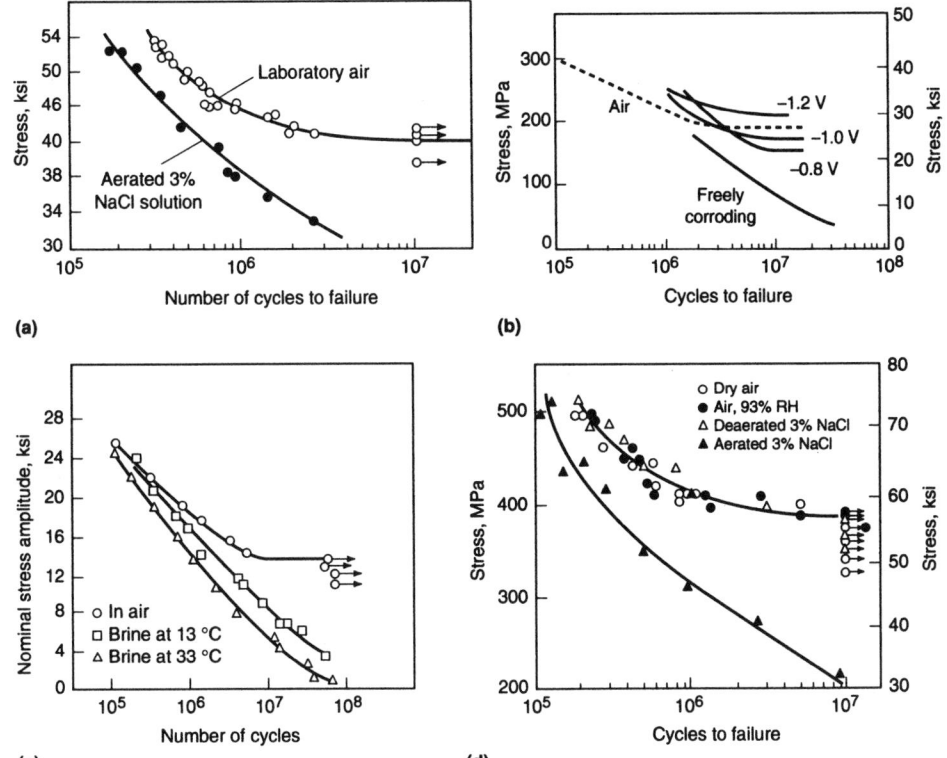

(a)

(b)

(c)

(d)

Fig. 21 Effect of environment on the fatigue limits of various steels. (a) Elimination of fatigue limit for 1018 steel in salt solution. (b) Fatigue behavior of carbon steel in seawater as a function of specimen potential. (c) Effect of temperature on corrosion fatigue behavior of notched specimens of mild steel. (d) The deleterious effect of aerated aqueous chloride on the high-cycle fatigue life of tempered martensitic AISI

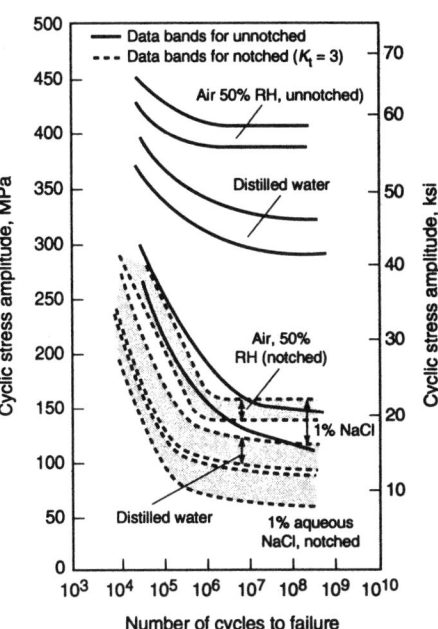

Fig. 23 Corrosion fatigue endurance data for specimens of 13% Cr steel in rotating-bending testing (mean load zero) at a frequency of 50 Hz and temperature of 23 °C (73 °F). Notched specimens $k_t \approx 3$. Source: *ASM Handbook,* Vol 13, p 296

of that in dry air. In these tests, the main role of the environment was corrosive attack of the polished surface, creating local stress raisers that initiated fatigue cracks. Stainless steels also have a lower fatigue strength in seawater than in fresh water, probably because of the presence of chloride ions in seawater. Chlorides are known to attack the protective oxide surface films on stainless steels and therefore expose the underlying material to the environment and affect fatigue strength.

Corrosion fatigue crack initiation in notched specimens is most effectively characterized by the notch-root plastic strain range calculated by Neuber's method, elastic-plastic finite elements, or fracture mechanics approximation (Ref 13, 14). Figure 25 shows the fracture mechanics method from the results of over 100 experiments with C-Mn and alloy steels in aqueous chloride compared to moist air. The load cycles to produce 1 mm of fatigue crack extension increase with decreasing $\Delta K/\sqrt{R}$ (where R is the notch tip radius) and are reduced by chloride (free corrosion) relative to moist air. An endurance limit is observed for moist air, but not for corrosion fatigue. Information on other methods (e.g., Neuber's method) is given in the article "Estimating Fatigue Life" in this Volume.

Mechanisms of Corrosion Fatigue Crack Initiation. Although the mechanisms of corrosion fatigue are not well understood, it is known to be an electrochemical process dependent on the environment/material/stressing interaction. Surface

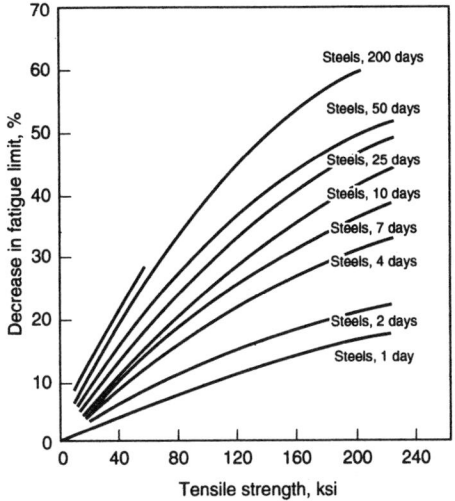

Fig. 22 Relation between tensile strength and the percentage decrease in fatigue limit of steels from prior corrosion. Source: Ref 19.

features at origins of corrosion fatigue cracks vary with the alloy and with specific environmental conditions. In carbon steels, cracks often originate at hemispherical corrosion pits and often contain significant amounts of corrosion products (Fig. 26). The cracks are often transgranular and may exhibit a slight amount of branching. Surface pitting is not a prerequisite for corrosion fatigue cracking of carbon steels, nor is the transgranular fracture path; corrosion fatigue

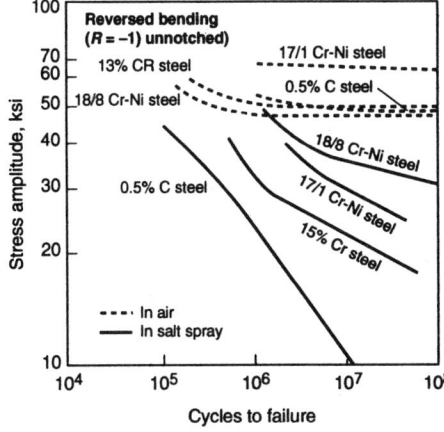

Fig. 24 Effect of salt spray on corrosion fatigue of various structural steels (unnotched). Source: Ref 19

cracks sometimes occur in the absence of pits and follow grain boundaries or prior-austenite grain boundaries.

Environmental effects on the nucleation process of fatigue cracks are more extensive for aqueous environments than for gases. The mechanisms of environmentally assisted crack nucleation in aqueous conditions include (Ref 15):

- Pitting, where nucleation of cracks takes place at pits formed through corrosive attack
- Preferential dissolution, where highly deformed material acts as a local anode with the surrounding undeformed material acting as a local cathode. The anodic regions then serve as crack nucleation sites.

Table 3 Corrosion fatigue behavior above 10^7 cycles for selected structural alloys

Material	Process description	S_u, MPa	ksi	Frequency, Hz	Corrosive medium	Fatigue limit in air MPa	ksi	Corrosion fatigue strength(a) MPa	ksi	Ratio of corrosion fatigue strength and air fatigue limit
Mild steel	Normalized			...	River water	260	38	32	4.6	0.12
0.21% C steel	Annealed	490	71	22	Sea water	220	32	30	4.3	0.13
1035		600	87	29	6.8 percent salt water;	280	41	170	25	0.61
1050		650	94	...	complete immersion	220	32	140	20	0.63
1050	Q&T	900	130	...		415	60	170	25	0.42
4130	Q&T	880	128	...		480	70	190	27	0.38
9260	Normalized	985	143	...		500	72	170	25	0.35
5Cr steel	Q&T	990	130	...		510	74	365	53	0.71
Wrought iron		325	47	...		210	30	130	19	0.64
12.5Cr steel	Annealed	1000	146	6	Fresh water in torsion	225	37	125	18	0.49
18/8 stainless	Annealed	1300	188	...		195	28	85	12	0.44
18.5Cr steel	Annealed	770	112	...		240	35	195	28	0.79
Aluminum	Annealed	90	13	24	River water with	19	2.70	7.6	1.1	0.41
Copper	Annealed	215	31	...	saline solution	69	10	70	10.5	1.04
Monel	Annealed	565	82	...	about one-half	250	36	200	29	0.81
60/40 brass	Annealed	365	53	...	water of sea	150	22	130	19	0.85
Nickel	Annealed	530	77	...		235	34	160	23	0.68
Phosphor bronze	Normalized	430	62	37	3% salt spray	150	22	180	26	1.20
Aluminum-3Mg	Heat treated	...	...	83	3% salt spray	125	18	48	7	0.38
Aluminum-7Mg	Heat treated	...	...	83	3% salt spray	110	16	48	7	0.45
Aluminum-Cu Mg	Heat treated	...	...	83	3% salt spray	180	26	85	12	0.47
Mg-Al-Zn AZG		...	...	...	Tap water	75	11	40	6	0.49
Mg-Al-Mn AZ537		...	...	...	Tap water	75	11	55	8	0.68
Mg-Al-Mn AM503		...	...	...	3% salt water	55	8	20	3	0.36
Mg-Al-Mn AZM		...	...	...	3% salt water	150	22	14	2	0.08

Note: Rotating bending specimens unless specified. These results have been collected from various sources and illustrate corrosion fatigue effects only; they are not to be used as design values. Source: Ref 19

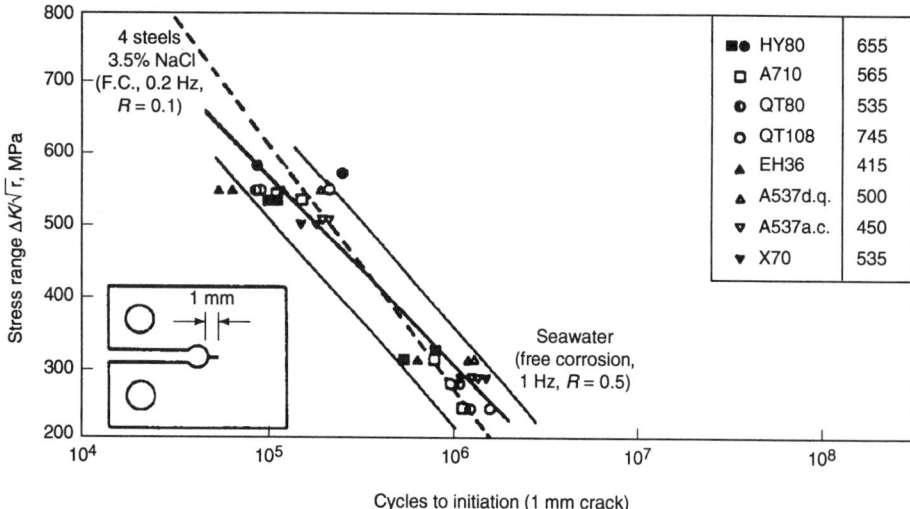

Fig. 25 The effect of chloride on the corrosion fatigue crack initiation resistance of notched steel specimens. *r*, notch tip radius. Source: Ref 20

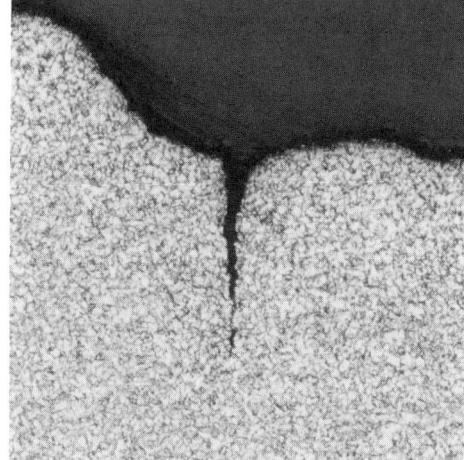

Fig. 26 Initiation of fatigue cracks at corrosion pit from a failed steel tubing in a drilling application (Example 1). 50×

• Rupture of protective oxide films, where localized corrosion occurs in the metal immediately below deformation-induced rupture of the protective oxide. Fatigue then starts at the regions of localized attack.
• Surface energy reduction, where adsorption of environmental species promotes Griffith-type crack growth by reducing the surface energy.

Localized preferential dissolution is unlikely in gaseous environments because development of local electrochemical cells requires the presence of an electrolyte. However, mechanisms have been devel-

oped for pitting and for adsorption-induced surface energy reductions in gaseous environments.

Gas Environments. Pitting in gases is generally restricted to oxidizing environments. However, even non-oxidizing gases that do not induce pitting may still significantly enhance corrosion fatigue initiation. The pitting model is therefore applicable to specific metal-environment systems. Gas environments can alter the mechanisms of fatigue crack nucleation by (Ref 15):

• Obstruction of reversible slip
• Inhibition of crack rewelding in slip bands

• Reduction of surface energy at the crack tip
• Formation of bulk oxide

Aqueous Corrosion Fatigue. Important variables in aqueous corrosion fatigue include alloy composition and solution chemistry. In general, corrosion fatigue behavior parallels environment corrosivity, with increasing corrosion rates resulting in decreased fatigue resistance. However, a number of exceptions to this observation have been noted and observed. When improved corrosion fatigue resistance has been observed to be inversely related to corrosion rate, the improve-

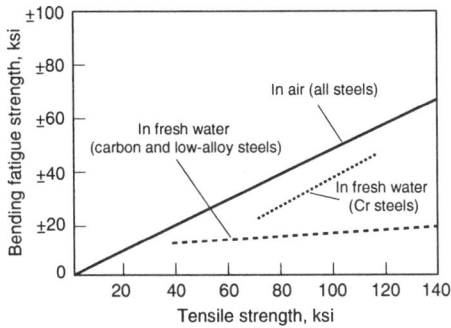

Fig. 27 The influence of strength and chemical composition on the corrosion fatigue strength of steels. Unnotched specimens, rotating bending fatigue strength, 20 × 10⁶ cycles, 1450 cycles per min. Source: Ref 19

ment has generally been ascribed to delays in crack initiation or early propagation caused by dissolution of microcracks or stress concentrators.

Early experiments with mild steel specimens in deaerated 3% sodium chloride solutions showed that a critical corrosion rate was necessary to initiate corrosion fatigue failures (Ref 16). When the same material was exposed to artificial sea water at elevated temperatures, shallow and uniformly distributed pits were found over the surface. These pits served as initiation sites for corrosion fatigue failures. Both of these results thus suggested that corrosion-induced defects could enhance fatigue crack initiation.

However, as previously mentioned, pitting does not always occur in corrosion fatigue. For example, low-carbon steels have been susceptible to corrosion fatigue in acid solutions in spite of a total lack of formation of pits (Ref 17). Similar susceptibility to corrosion fatigue has been observed in steel samples tested in deaerated solutions while under small impressed anodic currents (Ref 18). Thus, the pitting model cannot be a generalized mechanism for corrosion fatigue of metals, and any pitting model is only applicable to a specific metal-environment combination.

Prevention of Corrosion Fatigue. Both temporary and permanent solutions for corrosion fatigue involve reducing or eliminating cyclic stress, increasing the corrosion-fatigue strength of the material, and/or reducing or eliminating corrosion. These objectives are accomplished by changes in material, design, or environment.

Changes in Material or Heat Treatment. Although heat treatment can be a factor, alloying with chromium is the material change that most effectively reduces the effects of corrosion fatigue on steel. Figure 27 summarizes a comprehensive air and corrosion fatigue program with carbon, low-alloy, and chromium steels, using rotating-bending specimens subjected to fresh water spray (Ref 19). The results in air show the usual behavior, the fatigue limit being approximately proportional to the tensile strength. For the results in water, however, the fatigue strength of the carbon steels and low-alloy steels is almost independent of the tensile strength. All the results on carbon steels, covering a range of carbon con-

tent from 0.03 to 1.09%, both annealed and hardened and tempered, showed corrosion fatigue strengths between ±80 and ±150 MPa (±12 and ±22 ksi). For both carbon and low-alloy steels, the corrosion fatigue strength was usually higher in the annealed condition than in the hardened-and-tempered condition. The corrosion-resistant steels, containing ≥5% Cr, showed much greater resistance to corrosion fatigue, approximately proportional to the tensile strength.

Although annealing has been beneficial in a few cases, alloying and heat treatment of steels usually have little effect on corrosion fatigue characteristics unless the alloying is undertaken specifically to improve corrosion resistance. In some cases, deleterious heat treatments such as those leading to sensitization in stainless steels have resulted in lower corrosion fatigue resistance.

Component Design Changes. Sometimes it is possible to reduce operating stress to increase corrosion fatigue strength. Operating stress may be lowered by reducing either the mean stress or the amplitude of the cyclic stress. This almost always involves a change in component design. Sometimes only a minor change is required, such as increasing a fillet radius to reduce the amount of stress concentration at a critical location. In other instances, more extensive changes are required, such as significantly increasing the cross-sectional area or adding a strengthening rib.

Corrosion effects can be lessened by alloying, by providing galvanic protection, or by altering or removing a corrosive environment. Galvanic protection by sacrificial anodes or applied cathodic currents has been successful in reducing the influence of corrosion on fatigue of metals and alloys exposed to aqueous environments, except in such alloys as high-strength steels that are subject to hydrogen-induced delayed cracking. In similar instances, anodic polarization has been used for protection of stainless steels. With passive alloys or alloys that can be polarized to produce passive behavior, crevices must be avoided, because corrosion within crevices may actually proceed more rapidly due to the anode-cathode relationship. For example, in a part made of low-carbon steel, an area under an O-ring can fail by corrosion fatigue in a caustic solution with a pH of 12 at a location where only low cyclic stresses exist. Where pitting corrosion has occurred, corrosion pits may act as stress raisers and thus accelerate fatigue failure.

Inhibitors are sometimes added to the environment or included in organic coatings to eliminate corrosion fatigue. The effect of inhibitors is believed to depend solely on their ability to reduce corrosion rates to acceptable values. However, corrosion fatigue is a complex electrochemical process. For example, experiments performed on boiler steels at 275 °C showed that additions of 0.7 g/L NaOH to distilled water improved the fatigue limit by approximately 20%, but that increasing the NaOH concentration to 200 g/L lowered the fatigue limit by approximately 10% (Ref 16). At the higher alkaline concentrations, intergranular as well as transgranular cracking was

observed, indicating that caustic cracking (stress-corrosion cracking) was also occurring during normal fatigue cracking.

There is considerable evidence that dissolved oxygen is important to the mechanism of corrosion fatigue. For example, an improvement in the fatigue life of steel specimens was noted when the specimens were completely immersed at 96 °C, suggesting that this improvement was due to the limited solubility of oxygen at this temperature. Sodium chloride solution dripped through air proved to be extremely damaging to fatigue specimens, whereas a solution dripped through a commercial hydrogen atmosphere resulted in higher fatigue life, with still further improvement as the purity of hydrogen was increased (Ref 16).

Electrochemical polarization may have a marked effect on corrosion fatigue resistance. For example, anodic polarization of carbon steels and of copper alloys has been shown to result in decreases in fatigue lives. For some cases, however, severe anodic polarization actually results in increases in fatigue resistance. Examination of specimens subjected to large amounts of anodic dissolution reveals that surface-initiated cracks are blunted by the corrosive environment, leading to a decrease in stress concentration.

Fatigue Failure Examples

Simple rules for fatigue-resistance design cannot be stated because of the diversity in function, loads, stresses, materials, and environments. The adverse effect of stress risers and discontinuities within a metal, either at the surface or subsurface, may also arise from primary or secondary working of the material. Some of these factors are illustrated with the following case histories of fatigue failure of various steel components.

Effect of Thread Design and Low-Frequency Loading on Corrosion Fatigue Failure of a Threaded Low-Carbon Steel Rod: Submitted by *J.M. Gallardo, Universidad de Sevilla, Spain.* A threaded rod component failed after 10,000 hours of discontinuous service as an anchoring and tensioning element for the traction ropes of a ladle-carrying car on a sea-side harbor crane. One-half of the element was square in section (75 × 75 mm); the other half was a square-threaded 70 mm diam rod. The fracture occurred in the second thread nearest to the section change. The fastener was made of normalized UNE F1120 steel, which is similar in composition and properties to AISI-SAE 1029.

Investigation. The fracture surface was typical of a fatigue fracture. Nearly 90% of the section showed beach marks, the remaining section displayed a final, catastrophic fracture surface. Fracture origin was identified at the second thread root. Cracks were also detected at the first thread root. Chemical analysis (0.27 C, 0.75 Mn, 0.18 Si, 0.011 S, 0.023 P), heat treatment, thread shape, surface roughness, general dimensions, etc. fulfilled initial design.

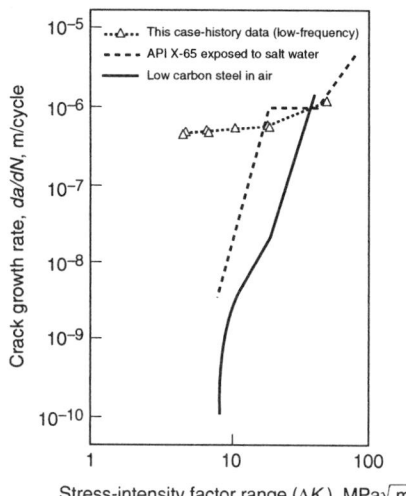

Fig. 28 Measured rate of crack growth in low-carbon steel (F1140) in seaside environment compared with crack growth of low-carbon steel in air and API X-65 steel in salt water

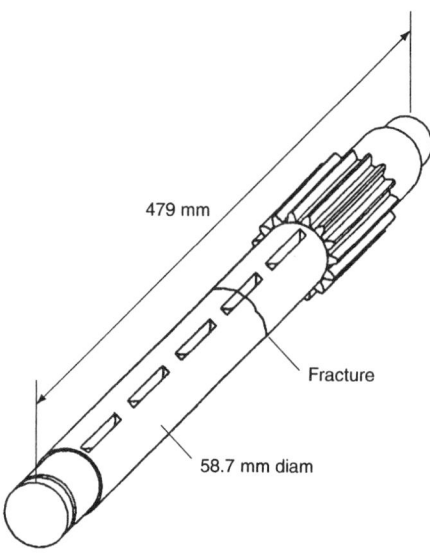

Fig. 29 Fracture location at the keyway that was loaded during shaft straightening

cycle frequency is low enough to permit corrosion at the crack tip between two successive jumps (as seen in Fig. 28 for API X-65 in salt water). Nevertheless, corrosion (3% aqueous solution of sodium chloride) is said to produce only slight effects in fatigue behavior of lower-strength steels such as A36 and A588 Grade A, at frequencies higher than 0.2 Hz (Ref 26). However, corrosion fatigue can affect crack growth rates at a lower load-cycle frequency, which in this case was only 0.04 Hz. Furthermore, corrosion products, of high chloride content, were identified inside non-propagated cracks at the first thread.

The very high crack propagation rate measured may explain why only one out of four similar fasteners failed. Integration of Paris equation, between an initial crack size of 100 μm (calculated value of crack depth for the estimated ΔK_{th}) and the measured crack size before final fracture (52.7 mm) gives a steady crack-growth period of $N = 82300$ cycles. This figure corresponds to 16 months service time, i.e., 13% of the total 10 years that crane was in operation. Minute differences in fatigue initiation process may be responsible for differences in beginning the sustained crack propagation cycle, which, on the other hand, is very fast.

Conclusions. The threaded rod failed by corrosion-fatigue. Main factors associated to fracture initiation were thread profile-square, at constant flank height in the first threads and minimum root radius-, very high machining roughness and coastal saline environment. Rapid fatigue crack propagation was attributed to low-frequency cycling in a sea-side corrosive environment. Corrective measures include a recommended 12-month inspection interval and a new fastener design. New fasteners should be designed with higher fillet root radius (0.5 mm), lower surface roughness (ISO N6/N7), gradual transition between square and round sections and diminishing flank height for the first threads.

Fatigue Failure of Transmission Shaft from Straightening after Heat Treatment: Submitted by *F. Carrion, Instituto Mexicano del Transporte, Mexico.* The secondary transmission shaft of a heavy weight truck failed after 10,000 Km (or 5% guaranteed life). The piece was AISI 8620 steel with five keyways and a gear (Fig. 29) and a carburized case (0.5 mm deep hard surface of 60-62 HRC) and an interior hardness of 20-25 HRC after heat treatment. The failure always happened at the same point on the second keyway to an estimated 5% of the total manufactured shafts.

Investigation. From the analysis of the manufacturing process, a wide dispersion of the deformation of the shafts after heat treatment was obtained with an eccentricity ranging from 0.0076 mm to 0.294 mm with an average of 0.122 mm, due to the incorrect positioning of the shafts during the heat treatment. Despite of the degree of deformation, all shafts went through a straightening process in which force was applied at the center of the shaft on the second keyway.

Macroscopic fractography revealed typical beach marks from a tension-compression fatigue failure at low nominal stress. The starting zone was located at the corner of the keyway shaft and the final fracture zone covered about 30% of the fracture surface. Electron microscopic analysis of the crack initiation site revealed two zones: the first corresponding to an impact fracture with the size of the hardened case (0.5 mm), and the second corresponding to the initiation of the typical fatigue fracture. The hardened case was martensite and bainite, and the interior was ferrite and bainite. Grain size in both cases was 7-8 ASTM throughout, and the tensile yield strength of the material was 549 MPa with an ultimate strength of 942 MPa.

A fatigue crack propagation test was done with a modified three-point bending specimen from the ASTM E-648 standard (63 mm long, 6 mm wide, 15 mm high and with a 3 mm groove for the initial crack). Although the specimen size criteria was not met with this width, the tests found that the difference between the length of the crack at the center and the length of the crack at the edge was always less than 10%. This size of specimen was selected due to the interest of investigating the spatial distribution of the fracture mechanics properties of the material across the transversal area of the shaft.

The average of the fracture toughness K_Q (as an approximate value of K_{Ic}) was 102 MPa√m. From a load ratio (R) of 0.1 and a constant DP of 1000 kN, the relation between the fatigue crack growth rate (da/dN) and the applied stress intensity was obtained using the ASTM E-399 standard formula for the three-point bending specimen. The coefficients for Paris' equation (stage II) were found to be $n = 2.3$ and $C = 1.664 \times 10^{-11}$. With an initial crack of 0.5 mm, a remaining life of 2000 Km was calculated. However, this calculation did not consider the increase of

Unfortunately, the original design was lacking in fatigue-prevention measures. The very first threads, near the section change, were not machined with diminishing flank height to reduce stress concentrations at thread root (Ref 21). A similar reduction can be obtained by using a large thread root radius (Ref 22) such as 0.5 mm (Ref 23) instead of the mean measured value of 70 μm. The afore-mentioned advisable radius value is referred to a trapezoidal thread of the same dimensions, which has similar load-carrying capabilities and an improved fatigue behavior (Ref 23), compared to the square thread used. Finally, surface roughness, in compliance with design drawings specifications, was 21.4 μm (N11, ISO 1032), but a preferable value is 0.8/1.6 μm (N6/N7) (Ref 24). Three other contiguous rods were tested by magnetic particle; no flaws were found at the threads.

Fastener design and fabrication did not favor fatigue resistance, but three other fasteners working in parallel had shown no flaws. To explain this, a fracture mechanics approach was used to characterize the fatigue process. A scanning electron microscopy survey was performed by measuring the striations spacings (da/dN) at several distances (a) from fracture origin to determine the crack-growth rate. Stress intensity factor K_I values were calculated at the crack front center. Load cycle data $\Delta\sigma$ were obtained from design calculations and the geometric factor for threaded rods (Ref 25).

The resulting Paris-model relationship is presented in Fig. 28. Along with this case-history experimental data, the behavior of a ferritic low-carbon steel (0.15-0.20 C, 0.60-0.90 Mn, 0.04 max P, 0.04 max Si) in air (Ref 26) and API X-65 (0.24 C, 1.35 Mn, >0.02% V and/or 0.005 Nb) in salt water (Ref 26) are compared. Measured crack growth rate is very high in comparison with crack propagation in air. Corrosive media may increase growth rate by a factor of 10^2-10^3, if the loading

life caused by the variable loads, the change of direction of the crack (initial crack is parallel to the longitudinal axis of the shaft), and the contribution of the initial propagation of the stage I.

Conclusions. Crack initiation was due to the impact force during straightening after heat treatment. Once the crack initiated, it grew due to fatigue during operation. The remaining life after cracking of the hardened case is not acceptable. The heat treatment was improved to reduce the dispersion of the deformation of the shafts. The geometry of the keyway also was changed to reduce the stress concentration at that point. An analysis based in fracture mechanics and plasticity theory was done to determine an acceptable limit on the eccentricity of 0.211 mm for the straightening process to avoid cracking.

Intergranular Fracture with Continuous Carbide Network in a Steel Transmission Gear: Submitted by *M. Pepi, U.S. Army Proving Ground, Maryland.* A spiral bevel transmission gear from an Army cargo helicopter failed during a training flight leading to an immediate landing. Inspection of the number 2 engine transmission revealed that an 8-tooth segment (out of a 35-tooth gear) had fractured and penetrated the transmission housing. The failed component was fabricated from X-2 steel (a modified H11 tool steel), and carburized to a required case hardness of 59-64 HRC. The broken parts were subject to visual examination, magnetic particle inspection, light optical microscopy, chemical analysis, hardness testing, case depth measurement, X-ray diffraction, scanning electron microscopy (SEM), and energy dispersive spectroscopy (EDS).

Investigation. Visual examination and light optical microscopy of the fracture surfaces revealed characteristics consistent with a fatigue failure, including smoothness of fracture and beach marks. The beach marks and radial lines on the fracture surface revealed that the origin was located in the damping ring groove portion of the gear at a darkened half-moon shaped defect. The pre-existing crack was oriented perpendicular to the direction of grinding, suggesting the possibility of a grinding crack/burn. The darkened region had a featureless topography resembling that of a steel surface which has been exposed to a heat treatment atmosphere. Energy dispersive spectroscopy (EDS), in conjunction with a scanning electron microscope (SEM), within the darkened area revealed the presence of sodium. This finding was considered evidence that a pre-existing crack was present during the manufacturing process, since sodium nitrate and sodium dichromate are widely used to black oxide finish steel components. EDS spectra outside this darkened region failed to detect the presence of sodium. Further visual examination and magnetic particle inspection of the remaining component revealed small grinding cracks located in the same area as the fracture origin. A sample was sectioned through the origin to confirm that the pre-existing defect was a grinding crack. The sample was metallographically prepared, and showed evidence of rehardening (white region) and retempering (dark region). In addition, metallography revealed what appeared to be continuous carbide networks (CCN's) in the region of the gear which contained the origin. These networks were not continuous in high stress areas, such as the gear root and flank. Continuous carbide networks are generally caused by an excessive carbon potential during the carburization process, and can act to embrittle the case and reduce the fatigue limit of the component under bending fatigue conditions. Also, these networks render a surface sensitive to grinding, and if cracking occurs, it usually follows the path of the networks.

Metallography combined with microhardness testing of the damping ring groove area (fracture origin location) revealed a greater case depth than specified. The case depth was required to be 0.030-0.050 in., however, the average effective case depth was 0.075 in. It was later learned that the damping ring groove was subject to both the first and second carburization treatments, rather than the required first treatment only (the region was left inadvertently unmasked after the first treatment). The root, fillet, and flank effective case depths all conformed to the governing requirement of 0.030-0.050 in. Chemical analysis was performed to determine the elemental composition of the alloy. The carbon content was determined through combustion-infrared detection, and the sulfur content was found utilizing combustion-automatic titration. The remaining elements (silicon, manganese, phosphorus, tungsten, chromium, vanadium, and molybdenum) were determined using direct current plasma emission spectroscopy. The composition conformed to the governing specification. A 0.013 in. section was sliced from the top of the damping ring groove region, and subject to analysis for carbon content. The weight percent of carbon was 1.31%, suggesting the part was subject to a higher than nominal carbon potential during carburization. Hardness testing was performed on both the case and core of the component. The case hardness met the governing requirements of 59-64 HRC with an average of 61 HRC. The core had an average hardness of 40 HRC which fell within the required range of 36-44 HRC. X-ray diffraction was performed on sections of the fractured gear to determine the stress profile and percent retained austenite. The stress profile was consistent with a properly shot-peened X-2 steel. The percent retained austenite was 8.9%, which met the governing requirement (20% maximum). Fractographic examination of a secondary through fracture revealed intergranular fracture within the case of the damping ring groove area (Fig. 30). The remaining morphology consisted of ductile dimples, indicative of overload fracture. No evidence of fatigue was noted. This finding was evidence that the continuous carbide networks can contribute to an intergranular fracture.

Conclusions. The primary fracture origin was located at a pre-existing processing defect, which was a half-moon shaped darkened region. This defect was determined to be a grinding crack/burn, and was considered a contributing

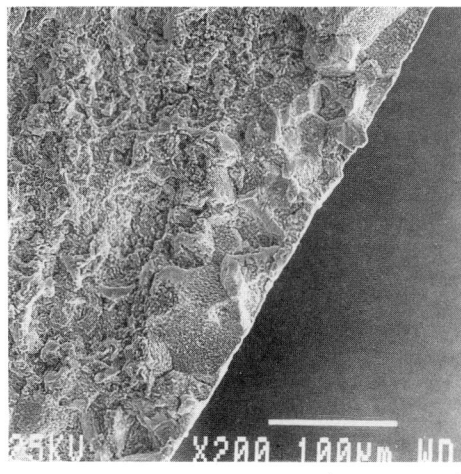

Fig. 30 Scanning electron micrograph of a secondary fracture through the damping ring groove region showing evidence of intergranular cracking at the surface. This region coincided with the origin area on the primary fracture. Intergranular morphology was attributed to the carbide network noted within the carburized case.

factor to the failure of this component. The grinding crack/burn was characterized by an orientation perpendicular to the direction of grinding, a half-moon shape, a darkened, featureless topography, the presence of sodium in the darkened area, and a rehardened and retempered structure adjacent to the fracture origin. Although the morphology of the primary fracture origin (darkened area) was featureless, it was possible that the continuous carbide networks within the carburized case facilitated grinding crack propagation.

Corrective Measures. This failure was the second spiral bevel transmission gear to fail in three years. The first failure occurred when a 9-tooth segment of the gear fractured. Although major fire damage made that investigation difficult, it was determined that a fatigue crack had grown before final fracture. The origin of that failure was also determined to be the damping groove region. Steps were taken by the contractor to ensure this type of failure would not recur, including a more thorough magnetic particle inspection in the groove region. In addition, grinding in the damping ring groove has been eliminated and replaced with a turning procedure which should keep heat generation to a minimum and ensure that defects related to grinding will not occur. In addition, a nital etch with 10× magnification step has been added to the inspection process, which should also help to detect future defects. The contractor also agreed to consider material specifications with micrographs of unacceptable case microstructures. This would aid the subcontractor in determining rejectable parts before they are placed into service.

Failure of an Overhead Crane Drive Shaft Due to Rotating-Bending Fatigue: Submitted by *J.R. Emmons, Rocketdyne Division, Rockwell International, Canoga Park, CA.* A drive shaft

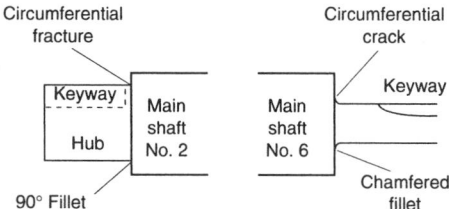

Fig. 31 Locations of observed fractures in a shaft assembly

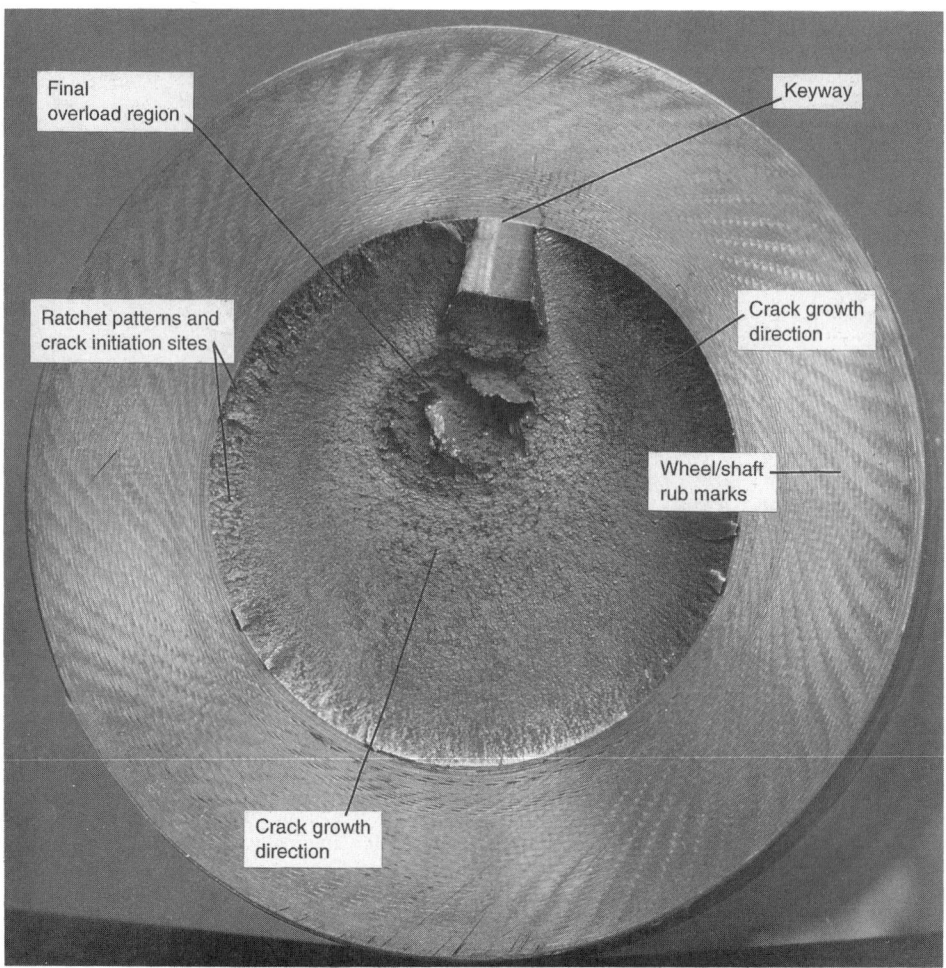

Fig. 32 Fracture section of steel shaft

used as part of the drive train for an overhead facilities crane fractured due to low-cycle rotating-bending fatigue, initiating and propagating by combined torsional and reverse bending stresses. The part submitted for examination was a principal drive shaft which fractured in the location of a 90 degree fillet where a wheel hub is machined down to 1.375 in. diameter from a 2.375 in. diameter main shaft. A 0.375 in. wide by 0.125 in. deep keyway was machined into the entire length of the hub, ending approximately 0.062 in. away from the 90 degree fillet (Fig. 31). The hub had fractured, separated from the main shaft, and lodged in the shaft hole of the wheel when the wheel fell away from the shaft. The wheel, together with the fractured hub and the shaft, were submitted for analysis.

Investigation. The pre-service and post-failure radiographic and penetrant inspection records were reviewed for the shaft and were found to be satisfactory. No additional cracks or indications were evidenced. The crack which led to the failure was not present when the shaft was installed. It was estimated that installation had been 1-1/2 to 2 years prior to the failure. The shaft material was verified as tempered 4150 steel by using wet chemical analysis and metallographic techniques. The hardness and the microstructure were surveyed by preparing metallographic specimens from the fractured hub section. The microstructure was confirmed to be a tempered martensite structure. The surface of the crack was examined at low and high magnifications utilizing optical and scanning electron microscopy.

Two additional shafts had not fractured to failure, but were radiographically and penetrant inspected as part of a verification analysis to determine if the remaining drive shafts were serviceable. One of these shafts was found to have a radial penetrant indication, this time in the chamfered fillet in the one-inch diameter machined motor drive end of the shaft, which is the end opposite to that which fractured in the principal shaft (Fig. 31). This shaft was sectioned diametrically to expose a polished cross section of the indication. One quarter of the radial indication was broken open to expose the crack fracture features for observation in order to determine the mode of fracture. Because of the resultant similarity between this crack and the fracture in the principal shaft, no further forensic work other than basic fractography, was performed on this second shaft.

Figure 32 shows a magnified view of the fractured hub which caused the wheel to break off the principal drive shaft of the overhead crane. The concentric circles revolving around the final overload, along with the ratchet marks around the outer circumference pointing towards the overload, are typical of a crack growth induced by rotating-bending stresses. The ratchet marks are exposing the fatigue initiation sites around the outer perimeter, as verified by high magnification scanning electron microscopy performed later in the investigation. The individual cracks propagated toward a single crack front and then toward the final overload area. Figure 32 shows the darker ductile dimples of the final overload region slightly off-center of the hub. The final overload was probably off-center due to the rotating-bending stresses which initiated the fracture on one side of the hub (most likely the side opposite the final overload) while at the same time driving the crack around the outer circumference producing multiple inward growing crack fronts.

Heavy corrosion was present traversing the fracture surface, extending for an inward distance of 0.125 in. from the outer edge, as identified in Fig. 32. The corrosion was confirmed to be rust. This corrosion may signify an older portion of the entire crack which may have been growing for quite some time. It was originally suspected that

the overall fracture had initiated at the keyway, which is usually the case in shafts where keyways are present, but the forensic analysis provided the contributing evidence that the fatigue cracks actually propagated towards the keyway rather than initiating from it.

Conclusions. A drive shaft used to assist in driving an overhead crane for the movement of medium to heavy weight components inside a manufacturing building, fractured at the mating interface with the drive wheel, in the location of a reduced diameter wheel hub machined at a sharp 90 degree fillet from the main bulk shaft. The sharp 90 degree fillet co-existed with a keyway in the vicinity of the fracture, although the keyway did not appear responsible for the failure. A second shaft was cracked in a sharply chamfered region where the main shaft was machined to a reduced diameter to form the motor drive hub opposite the wheel hub end of the shaft. The fracture mode for both shafts was low-cycle rotating-bending fatigue initiating and propagating by combined torsional and reverse bending stresses. All drive shafts were replaced with new designs which eliminated the sharp 90 degree chamfers in favor of a more liberal chamfer which reduced the stress concentration in these areas.

Fatigue of Coiled Tubing

Terry McCoy and Jerry Foster, Halliburton Energy Services, Duncan, OK

Coiled tubing (CT) is used in the oil and energy industries to provide a number of production enhancement and maintenance services. The operational concept of a CT system is to run a continuous string of small-diameter tubing into a well to perform specific well servicing operations without disturbing existing completion tubulars and equipment. When servicing is complete, the small-diameter tubing is retrieved from the well and spooled onto a large reel for transport to and from work locations. Coiled tubing offers many advantages over conventional jointed tubing including time savings, pumping flexibility, fluid placement, reduced formation damage, and safety.

However, the combination of low yield strength, thin wall, small diameter, and plastic strain cycling events limit CT applications. The continuous length of tubing experiences plastic deformation before entering the wellbore during the process of unwinding from the reel and passing through the surface machinery. The plastic strains superposed with high tangential stresses introduces the low-cycle fatigue failure mechanism. Tubing subjected to these operating conditions has a finite life. Many researchers have proposed methods of quantifying the effects of the low-cycle fatigue (Ref 27, 28). The loading conditions inside the wellbore are also complex (but not in the plastic strain regime) and of a dynamic nature. Tubing mechanics algorithms are used to calculate tubing forces, indicate the buckling state, and compute operating stresses.

Tubing integrity (Ref 29) is the limiting factor in establishing the operating envelope (Ref 30). Factors such as yield strength, working depth, working pressure, physical dimensions, and condition of the tubing combine to limit its application. The accepted industry operating boundary of 5000 psi (34.5 MPa) and 15,000 ft (4,572 m) depth evolved from practical experience based on the material properties available in this time period. The operating envelope has been expanded since the introduction of CT as a result of improvements in metallurgy and manufacturing process control. The continuous milling technique improved homogeneity by eliminating the butt welds previously used to join segments of tubing. Continuous milling removed the potential flaws associated with welds. The perfection of the longitudinal seam weld process also contributed to the acceptance of coiled tubing. The manufacturing process includes 100% nondestructive inspection of the product for flaw detection plus full hydrostatic test to 80% of the yield pressure.

The condition of tubing after it has entered service is monitored using visual and analytical tools. The fatigue state is calculated from the records of pressures, forces, plastic-bend events, and the physical parameters of the equipment.

Operating techniques are used to mitigate the effects of corrosion by using inhibitors and to minimize stress by proper selection of equipment. Imperfections introduced from the environment such as mechanical damage, corrosion pitting, and cracking may reduce the service life below expectations.

Three basic grades of steel tubing have proven to be acceptable for CT workover operations:

- 70-grade with minimum yield strength of 70 ksi (483 MPa)
- 80-grade with minimum yield strength of 80 ksi (552 MPa)
- 100-grade with minimum yield strength of 100 ksi (690 MPa)

The 70- and 80-grade materials are usually selected for hostile environments because of their inherent resistance to cracking in the presence of wet hydrogen sulfide and other aggressive chemicals.

Coiled tubing is available in outside diameters from 0.75 through 3.50 in. with wall thickness from 0.087 through 0.203 in., and lengths exceeding 25,000 ft (7620 m). The most common OD sizes are 1.25 in. and 1.50 in. The trend is toward greater use of larger tubing sizes, higher strength, and greater wall thickness (W).

Example 1: Failure of Coiled Tubing in a Drilling Application. Coiled tubing is used in drilling operations because it eliminates the need for a drilling rig and a reduction in equipment requirements and time to completion. Coiled tubing for drilling applications is subjected to multiple plastic-strain cycles while under load in the presence of potentially corrosive fluids, either encountered downhole or pumped from the surface. In one application, coiled tubing (2.000 in. OD × 0.156 in. W) with a yield strength of 483 MPa (70 ksi) was used to drill 68 steam injection wells in California for a major oil company. Depth of the wells ranged from 850 to 1000 ft (259 to 305 m). The tubing had used about 50% of its estimated fatigue life when failure occurred.

Investigation. The tubing failure was transverse to the tubing axis. Visual examination showed a flat fracture surface (arrow in Fig. 33) that was about 1/2 in. (13 mm) in length with the rest of the fracture showing shear lips indicative of tensile overload. The fracture was not associated with the longitudinal weld seam. The field report indicated that a pinhole leak in the tubing was noticed when the tubing was being spooled over the reel and then the tubing "snapped."

The flat portion of the fracture surface is typical for fatigue cracking. Fatigue striations were visible when the sample was examined with a scanning electron microscope (SEM). Closer examination showed corrosion pitting on the tubing ID from which the fatigue crack had propagated. The

corrosion pits were essentially flat-bottomed and were about 0.010 to 0.012 in. (0.25 to 0.30 mm) deep. Other areas adjacent to the primary fatigue crack also showed small fatigue cracks initiating from corrosion pits (Fig. 26).

The corrosion pitting on the tubing ID was primarily on one side of the tubing. X-ray diffraction analysis of the scale taken from the area in question showed only amorphous iron compounds with a moderate amount of quartz and small amounts of various clays. The exact cause of the corrosion pitting was not determined but may have occurred when the tubing was idle and fluids accumulated at the bottom of the tubing wraps.

Chemical and mechanical property testing performed verified that the CT met specifications. Hardness of the tubing was 92 HRB, which met the specified requirement of 22 HRC maximum. The microstructure of the tubing was typical for this grade of tubing having a fine-grain, ferritic structure. Tubing dimensions showed ovality of 0.082 in. (2.1 mm), indicating that the tubing had been significantly plastically deformed.

Conclusion. The CT failed prematurely due to low-cycle fatigue, initiating at ID corrosion pitting. Multiple fatigue cracks initiated at the bottom of ID corrosion pits, one of which predominated and propagated to the tubing OD, causing a pinhole leak. When the tubing was spooled onto the reel, the crack length was sufficient to allow the tubing to "snap" when bent over onto the reel. Although the exact cause of the corrosion was not determined, it is speculated that corrosion occurred during tubing lay-up since the pitting corrosion was isolated to one side of the tubing.

Corrective Measures. Corrosive attack on the coiled tubing can be reduced by completely removing fluids or modifying the fluids in the tubing. If water cannot be totally removed from the tubing, then an alkaline solution (to bring pH to 8 or 9) and an oxygen scavenger can be added. The tubing can also be purged by flowing dry nitrogen through the tubing to dry it out, then capping the ends to maintain a noncorrosive atmosphere in the tubing.

Example 2: Failure of Coiled Tubing due to Hydrogen Sulfide (H_2S) Exposure. Coiled tubing with 80 ksi (552 MPa) yield strength is manufactured to a maximum hardness of 22 HRC,

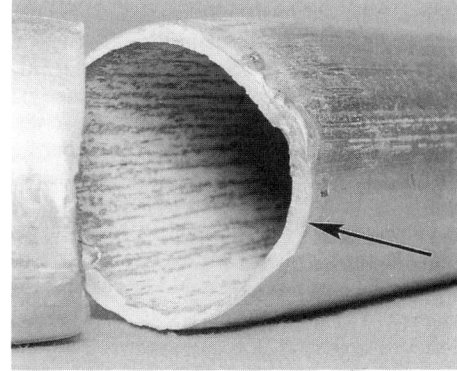

Fig. 33 Fracture surface of coiled steel tubing showing a flat fracture surface (arrow) indicative of fatigue

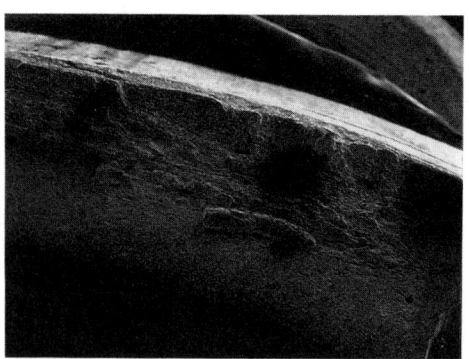

Fig. 34 Fractograph of tubing with a brittle zone near the OD indicative of sulfide SCC

thereby meeting NACE Standard MR0175 requirement for sour gas service. However, extra precautions must be taken when using this tubing in sour (H2S containing) wells because of the plastic cycling the tubing experiences in service. Coiled tubing (1.500 in. OD × 0.095 in. W) with length of 10,000 ft (3,048 m) was being used in Red Deer, Alberta, Canada area for various types of jobs, including sand cleanout, acid wash, and nitrogen lifts. The tubing had been on 38 jobs when the failure occurred. Seventeen of the 38 jobs involved wells containing H2S gas. H2S concentration varied from 0.1% to 7%. The tubing had seen relatively high usage with 262,630 total running feet (80,050 m). Detailed analysis of the jobs revealed that at the area of failure, the tubing had experienced 70% of its estimated fatigue life.

Investigation. The primary failure point was a 1/2 in. (13 mm) transverse crack where a leak occurred. Visual examination of the tubing also showed two other transverse cracks that were approximately 0.33 in. (8.4 mm) and 0.18 in. (4.6 mm) in length. Examination under a low-power microscope revealed numerous OD surface fissures, some of which were adjacent to the primary failure area. Most of these fissures or small cracks were associated with longitudinal gouges found on the tubing OD but some were located in areas showing no surface damage.

The fracture surface of the 1/2-in. (13 mm) crack was examined on the scanning electron microscope (SEM). This showed a brittle zone near the OD, identified as a sulfide stress crack (SSC), followed by fatigue cracking (Fig. 34). Sulfide stress cracking is defined as brittle failure by cracking under the combined action of tensile stress and corrosion in the presence of water and hydrogen sulfide (H2S). The cracking mode was ascertained through comparison of this failure to known laboratory tested samples. Numerous other small cracks were also noted.

Conclusion. Mechanical property testing of the tube sample showed that it met design requirements. Hardness was measured to be 96 to 97

HRB, which is below 22 HRC. Failure occurred initially due to sulfide stress cracking on the tubing OD followed by fatigue crack propagation through to the tubing ID. Although the CT was below the 22 HRC maximum required by NACE Standard MR0175 for use in sour service, the amount of plastic working of the tubing contributed to occurrence of failure before expectations, although the CT did achieve 70% of its expected life.

Corrective Measures. In the presence of known corrosive environments, it was recommended that tubing should not be used above 50% of its theoretical fatigue life unless the tubing is closely evaluated. In addition, a lower strength, 70 ksi (483 MPa) yield strength grade may perform better since it should have inherently better resistance to sulfide stress cracking than the higher-strength grades.

REFERENCES

1. A.D. Wilson, Ongoing Challenges for Clean Steel Technology, *Clean Steel Technology*, ASM/TMS, 1992, p 135-143
2. A.D. Wilson, Influence of Inclusions on the Fracture Properties of A588A Steel, *Fracture Mechanics: Fifteenth Symposium*, STP 833, ASTM, 1984, p 412-435
3. A.D. Wilson, Comparison of Dynamic Tear and Charpy V-Notch Impact Properties of Plate Steels, *Trans. ASME*, Vol 100, April 1978, p 204-211
4. A.D. Wilson and J.K. Donald, Evaluating Steel Toughness Using Various Elastic-Plastic Fracture Toughness Parameters, *Non Linear Fracture Mechanics: Volume II—Elastic-Plastic Fracture*, STP 995, ASTM, 1989, p 144-168
5. A.D. Wilson, Fatigue Crack Propagation in Steels: The Role of Inclusions, *Fracture: Interaction of Microstructure, Mechanisms and Mechanics*, TMS AIME, 1984, p 235-254
6. A.D. Wilson, Corrosion Fatigue Crack Propagation Behavior of a C-Mn-Cb Steel, *J. Eng. Mater. Technol.*, Vol 106, July 1984, p 233-241
7. S. Nishijima, "Basic Fatigue Properties of JIS Steels for Machine Structural Use," Report SR-93-02, National Research Institute for Metals, Tokyo, 1993
8. K. Kanazawa, K. Yamaguchi, M. Sato, and S. Nishijima, Effects of Temperature and Stress Concentration Factor on High Cycle Fatigue Properties of Engineering Steels, *J. Soc. Mater. Sci. Jpn.*, Vol 37, 1988, p 254-260
9. R.P. Skelton, The Prediction of Crack Growth Rates from Total Endurances in High Strain Fatigue, *Fatigue Eng. Mater. Struct.*, Vol 2, 1979, p 305-319
10. R.P. Skelton, Crack Initiation and Growth during Thermal Cycling in Fatigue at High Temperatures, *Fatigue at High Temperatures*, R.P.
11. O.H. Burnside et al., *Long-Term Corrosion Fatigue of Welded Marine Steels*, SSC-326, U.S. Coast Guard Ship Structure Committee, 1984
12. H. Lee and H. Uhlig, *Met. Trans.*, Vol 3, 1972, p 2949-2957
13. N.E. Dowling, *Mechanical Behavior of Materials*, Prentice-Hall, 1993
14. J.M. Barsom and S.T. Rolfe, *Fatigue and Fracture Control in Structures*, 2nd ed., Prentice-Hall, 1987
15. T. Sudarshan and M. Louthan, *Inter. Metals Review*, Vol 32, 1987, p 121-151
16. D. Duquette, in *Fatigue and Microstructure*, American Society for Metals, 1979, p 335-364
17. D.J. Duquette, in *Mechanisms of Environment Sensitive Cracking of Materials*, P.R. Swann et al., Ed., The Metals Society, London, 1977, p 305
18. H. Masuda and D.J. Duquette, *Metall. Trans.*, Vol 6A, 1975, p 87
19. F. Forrest, *Fatigue of Metals*, Pergamon, 1962
20. S.S. Rajpathak and W.H. Hartt, in *Environmentally Assisted Cracking: Science and Engineering*, W.B. Lisagor, T.W. Crooker, and B.N. Leis, Ed., ASTM, 1990, p 425-446
21. J.H. Bickford, An Introduction to the Design and Behaviour of Bolted Joints, Marcel Dekker, Inc., New York, 1990, p 485-492
22. Failure Analysis: The British Engine Technical Reports, compiled by F.R. Hutchings and P.M. Unterweiser, American Society for Metals, Metals Park, 1981, p 110
23. A. Ferraresi, Disegno di Costruzioni Meccaniche e Studi di Fabbricazione, Vol I, Hoepli, Milano, 1991, p 108-109
24. N. Larburu, Máquinas, Prontuario, Paraninfo, Madrid, 1989, p 278
25. Toribio et al., Factor de Intensidad de tensiones en un tornillo fisurado sometido a tracción y flexión, IV Encuentro del Grupo Español de Fractura, Domínguez, García-Lomas y Navarro, Ed., Sevilla, Marzo, 1989, p 50-56
26. Application of Fracture Mechanics for Selection of Metallic Structural Materials, J.E. Campbell, W.W. Gerberich, and J.H. Underwood, Ed., ASM, Metals Park, 1982, p 41-103
27. V.A. Avakov, J.C. Foster, and E.J. Smith, Coiled Tubing Life Prediction, Offshore Technology Conference Paper No. 7325, Houston, May 1993
28. K.R. Newman and D.A. Newburn, Coiled Tubing Life Modeling, Society of Petroleum Engineers Paper No. 22820, SPE Annual Technical Conference, Proceedings, Dallas, Oct 1991
29. R.D. Kane and M.S. Cayard, Factors Affecting Coiled Tubing Serviceability, Petroleum Engineer International, Jan 1993
30. K.R. Newman, Coiled Tubing Pressure and Tension Limits, Society of Petroleum Engineers Paper No. 23131, Offshore Europe Conference, Aberdeen, Scotland, Sept 1991

Skelton, Ed., Applied Science Publishers, London, 1983, p 1-62

Fatigue Resistance and Microstructure of Ferrous Alloys*

Ronald W. Landgraf, Virginia Polytechnic Institute and State University

THE BROAD RANGE OF PROPERTY COMBINATIONS available in ferrous alloy systems provides a unique opportunity to control fatigue resistance through microstructural manipulation. Considerable progress has been made in improving the understanding of microstructural effects on fatigue behavior (Ref 1-3). Such knowledge has provided the basis for a more fundamental approach to designing and processing alloys for increased fatigue performance.

The traditional approach to fatigue design with ferrous alloys, based on endurance limits and infinite life criterion, has been supplanted by approaches based on finite-life behavior that emphasize the cyclic deformation aspects of the fatigue process (Ref 4, 5). Central to these approaches for predicting structural performance in service situations is a set of properties that characterize the cyclic deformation and fracture behavior of an alloy. These properties, which are related to alloy strength, ductility, and strain-hardening characteristics, provide a useful quantitative basis for assessing and interpreting the influence of various microstructural features on material fatigue resistance.

This article will first review general trends in cyclic response for representative commercial alloys to establish the spectrum of cyclic properties attainable through microstructural alteration. Individual alloy classes are then examined in greater detail to assess the current understanding of relationships between microstructure and fatigue resistance. Finally, the role of internal defects and selective surface processing in influencing fatigue performance is discussed.

The following ferrous alloy systems are covered:

- Ferritic-pearlitic steels
- Martensitic steels
- Maraging steels

- Austenitic stainless steels
- Cast iron and steel

Strengthening mechanisms considered include:

- Grain refinement
- Cold working
- Solid-solution strengthening
- Precipitation strengthening
- Transformation strengthening
- Thermomechanical processing
- Selective service processing

Fatigue resistance is evaluated in terms of:

- Cyclic stability
- Resistance to crack initiation and propagation

- Notch sensitivity

General Trends in Cyclic Behavior

In this section, the spectrum of properties achievable in ferrous systems is surveyed using data for representative commercial alloys. Important characterizing parameters are introduced and key behavioral responses noted.

Cyclic Deformation Behavior. Because material flow properties may be altered, often significantly, by reversed deformation, it is necessary to use a *cyclic* stress-strain relation in fatigue studies (Ref 4). Such curves are conveniently characterized in the form of a power-law hardening relation between strain amplitude, ε_a, and

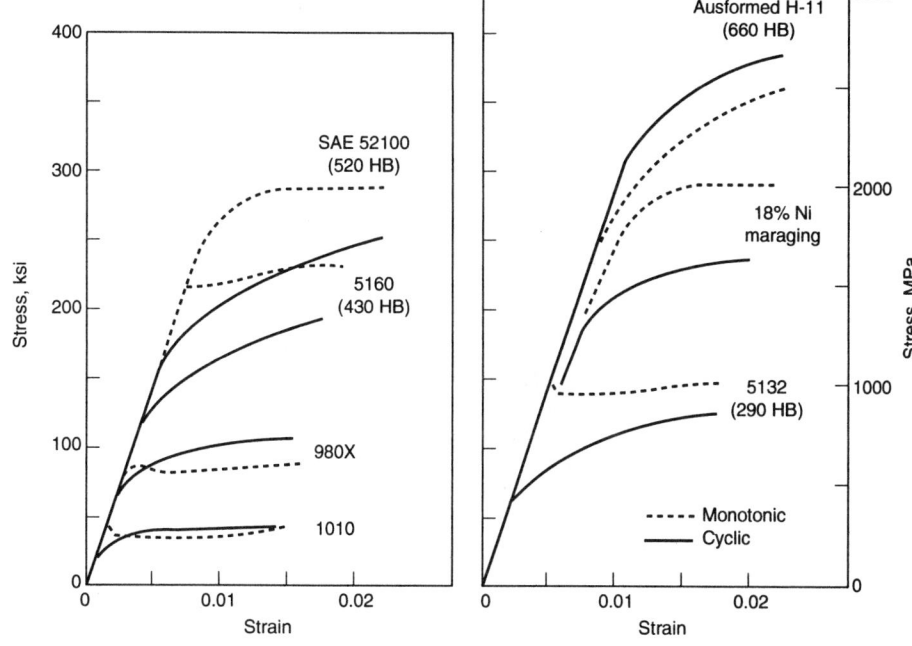

Fig. 1 Monotonic and cyclic stress-strain behavior of representative ferrous alloys

* Adapted from: R. W. Landgraf, Control of Fatigue Resistance through Microstructure—Ferrous Alloys, *Fatigue and Microstructure*, American Society for Metals, 1979, p 439-466

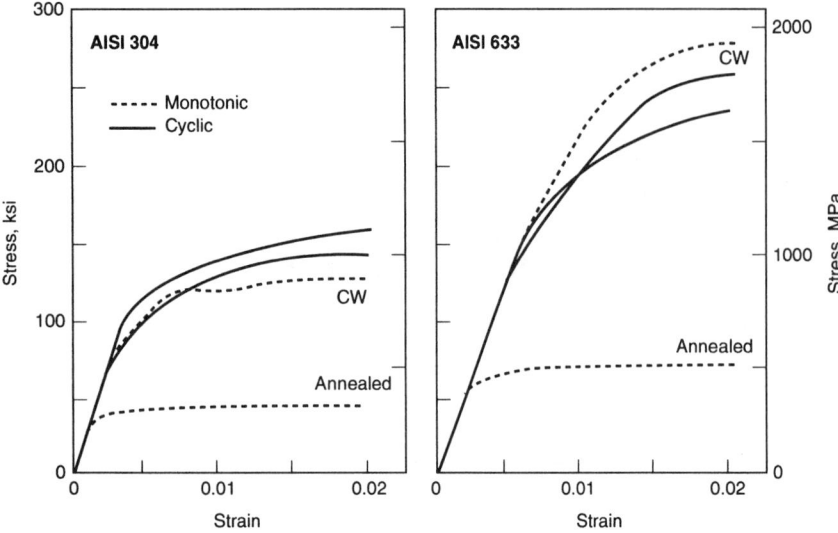

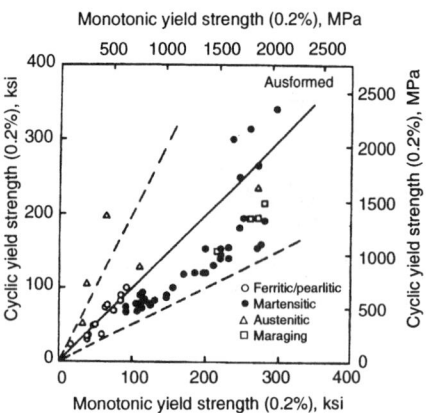

Fig. 3 Cyclic versus monotonic yield strengths for various classes of ferrous alloys

Fig. 2 Stress-strain responses for metastable austenitic stainless steels. Ann., annealed; CW, cold worked

stress amplitude, σ_a, as is often used for monotonic curves:

$$\varepsilon_a = \frac{\sigma_a}{E} + \left(\frac{\sigma_a}{K'}\right)^{1/n'} \quad \text{(Eq 1)}$$

where E is the elastic modulus and K' and n' are cyclic deformation properties. This formulation allows for direct comparison with monotonic tension curves so that cyclic-hardening or cyclic-softening responses can be quickly assessed. Monotonic and cyclic stress-strain curves for a series of representative steels are compared in Fig. 1. Ferritic-pearlitic steels, such as hot-rolled, low-carbon SAE 1010 and microalloyed SAE 980X, exhibit some cyclic softening at low strains followed by modest cyclic hardening at higher strains. Quenched-and-tempered martensitic steels (SAE 5132, 5160, and 52100) all show significant cyclic softening. Such behavior is typical of a broad range of low- and intermediate-hardness martensitic alloys (Ref 6). Similar softening trends are observed for the precipitation-strengthened 18% Ni maraging steel. Ausformed H-11 tool steel, on the other hand, is seen to cyclic harden above its already high monotonic value. Such high strength levels, obtained by severe hot working of austenite prior to quenching, result in the highest long-life fatigue strengths yet achieved (Ref 7, 8).

In contrast to these trends is the response of two austenitic stainless steels shown in Fig. 2. In the annealed condition, these steels exhibit pronounced cyclic hardening as a result of a deformation-induced martensitic transformation (Ref 9). Prior cold working also causes transformation hardening as evidenced by substantial increases in monotonic yield strengths. In this condition, cyclic response is reasonably stable.

A general view of the cyclic responses for a range of ferrous systems is shown in Fig. 3 as a plot of cyclic versus monotonic yield strengths; points above the 45° line indicate cyclic-harden-

ing alloys; those below, cyclic-softening alloys. Ferritic-pearlitic microstructures are seen to be reasonably stable. The pronounced softening of low- and intermediate hardness martensites and maraging alloys is clearly apparent. As-transformed or slightly-tempered martensites exhibit stable to hardening tendencies, as do the metastable austenitic stainless steels.

The practical significance of these responses depends on the particular application. For example, in notched members, local softening at a notch root could result in redistribution of stress, thus alleviating the stress-concentration effect. In surface-treated members, on the other hand, the

effects of favorable residual stresses will be maximized by a high-strength, cyclically stable alloy.

Fatigue-Life Behavior. Contemporary fatigue design methods, which are based on local stress-strain concepts, use the following relationship between strain amplitude, ε_a, and reversals to failure, $2N_f$, to quantify the fatigue resistance of a material:

$$\varepsilon_a = \frac{\sigma'_f}{E}(2N_f)^b + \varepsilon'_f (2N_f)^c \quad \text{(Eq 2)}$$

where E is the elastic modulus, and σ'_f, b, ε'_f, and c are cyclic properties related to material strength, ductility, and strain-hardening behavior. The total strain resistance is viewed as the summation of an elastic plus a plastic-strain term. The correspondence between the strain-life curve (Eq 2) and the

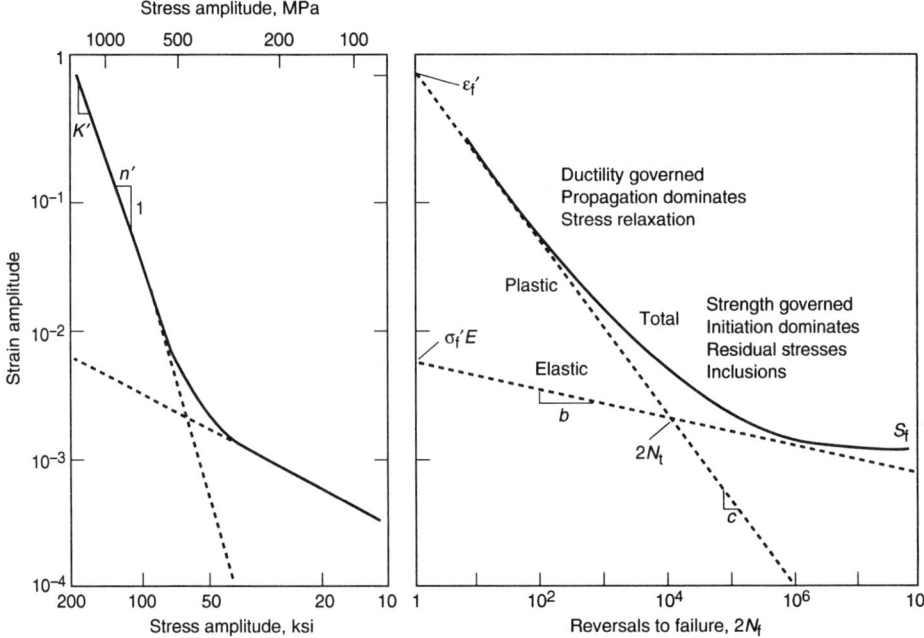

Fig. 4 Representation of stress-strain and strain-life behavior

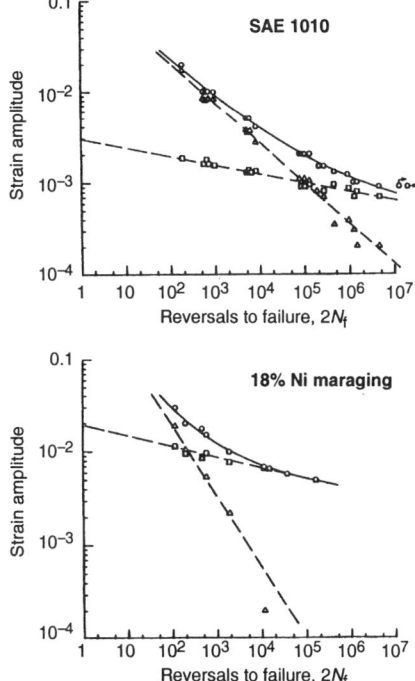

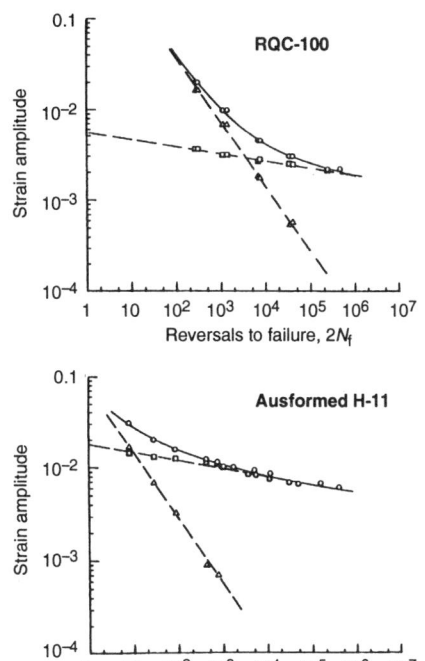

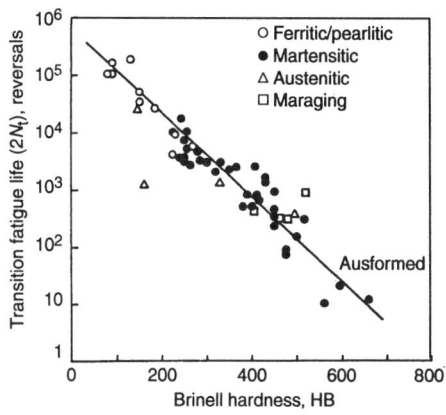

Fig. 6 Transition fatigue life as a function of hardness for ferrous alloys

Fig. 5 Strain-life curves for representative steels

cyclic stress-strain curve (Eq 1), obtained by plotting each curve on a common strain axis, is shown in Fig. 4. This construction allows direct determination of the steady-state stress amplitude associated with a given strain amplitude. Note that an endurance limit, S_f, can be estimated from a point on the cyclic curve where the plastic strain becomes vanishingly small—a small plastic-strain offset.

While conveying useful quantitative information, this representation also offers a conceptual scheme for identifying regimes wherein particular mechanisms and behavioral patterns are operative. These regimes are determined primarily by the relative levels of plastic and elastic strain rather than by the absolute fatigue life. At high plastic strains, for example, material ductility is the prime consideration governing fatigue resistance. Cracks are found to initiate relatively early in the life, and crack growth is a dominant failure mode. Conversely, at low plastic strains, strength governs behavior and crack initiation becomes the increasingly dominant event. It is in this regime that material defects, such as inclusions, play a major role. Also, residual-stress effects will be most important here. The transition fatigue life, $2N_t$ (the life at which the plastic and elastic strain amplitudes are equal), provides a useful zone of demarcation between these two regimes and can be determined from material properties as follows:

$$2N_t = \left(\frac{\varepsilon_f' E}{\sigma_f'}\right)^{1/(b-c)} \quad \text{(Eq 3)}$$

Strain-life plots for four different steels are compared in Fig. 5; elastic, plastic, and total

strain components are shown. The low-carbon SAE 1010 steel is seen to be ductility-governed over most of its life, with a transition fatigue life of about 10^5 reversals. The higher-strength martensitic RQC-100 steel shows improved long-life resistance and a shorter transition life. Increasing strength, first by maraging and then by ausforming, further increases long-life resistance and decreases short-life resistance, with an attendant decrease in transition life.

Transition fatigue life as a function of hardness for a variety of ferrous alloys is shown in Fig. 6. A difference of four orders of magnitude in transition life is noted between low-strength alloys and ausformed steel. Decreasing transition life can be associated with the increase of crack-initiation resistance, a factor that accounts for the

increased sensitivity of hardened steels to notches and internal defects.

As further evidence of the influence of strain regime on failure mode, fatigue crack development in an SAE 1045 steel is shown in Fig. 7. Figure 7(a) illustrates behavior near the transition life. Here, slip bands formed early in the life result in multiple crack origins. Inclusions may or may not serve as initiation sites. Crack growth occurs by a linking process through previously slipped material. At a lower strain amplitude (Fig. 7b), slip markings are not observed and isolated cracks form late in life, invariably at inclusions. By way of contrast, the behavior of a low-carbon steel, at an even lower strain level and longer life, is shown in Fig. 8. This photomicrograph, obtained at about 10% of the fatigue life, shows well-developed slip bands and the beginning of microcrack linking. Thus, crack development in low-carbon steel at long lives is similar to crack development observed for hardened steel at short lives. Again, it is the plastic-strain level, rather than life regime, that determines behavior.

These patterns of crack development have important implications for irregular loading histories. Occasional overstrains may initiate cracks in a material that can then propagate at lower strain

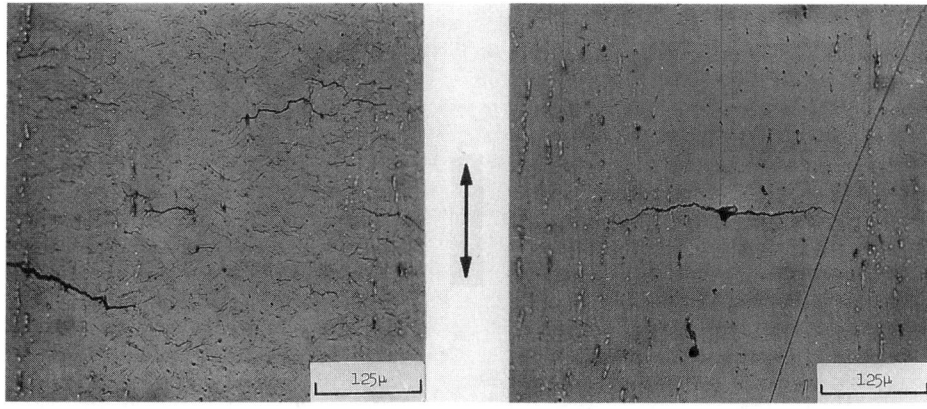

Fig. 7 Fatigue crack development in a heat-treated steel (SAE 1045, hardness: 450 HB) at high (a) and low (b) cyclic strains. (a) $\Delta\varepsilon/2 = 0.011$, $2N_f = 2000$. (b) $\Delta\varepsilon/2 = 0.004$, $2N_f = 53,300$

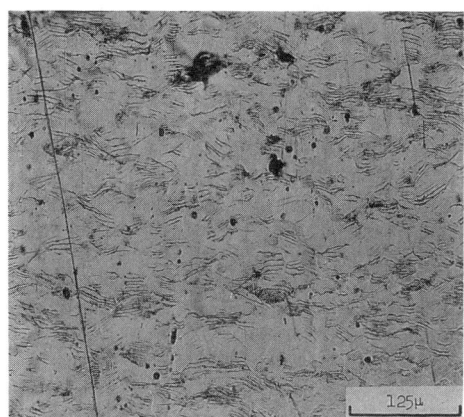

Fig. 8 Fatigue crack development in a low-carbon steel (SAE 1010). $\Delta\varepsilon/2 = 0.002$, $2N_f = 100,000$

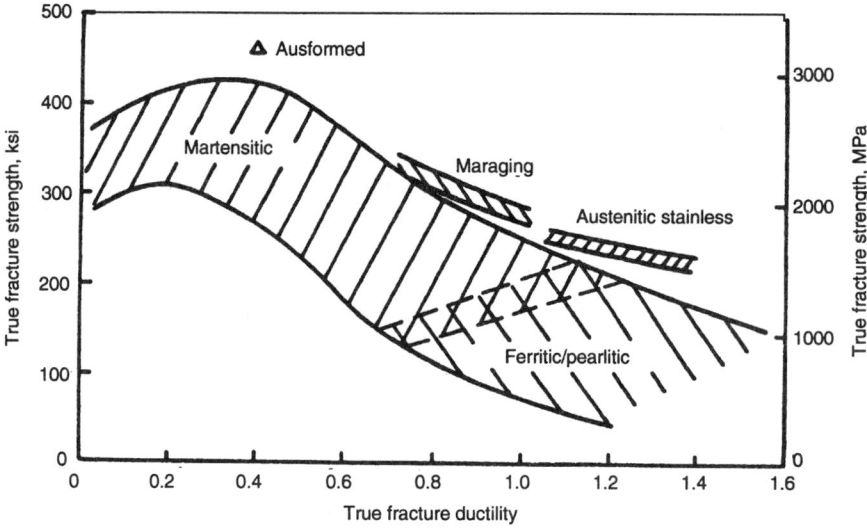

Fig. 10 Fracture strength-ductility combinations attainable in commercial steels

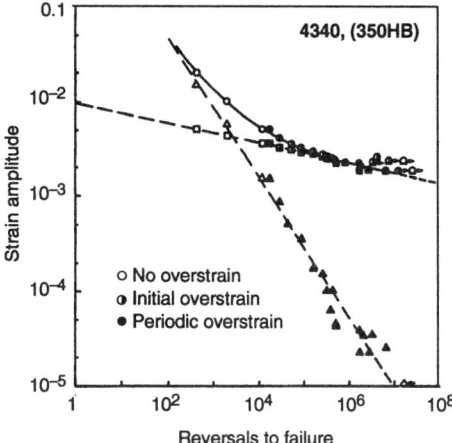

Fig. 9 Influence of overstrains on the fatigue life of a heat treated steel (SAE 4340, hardness: 350 HB) (Ref 10)

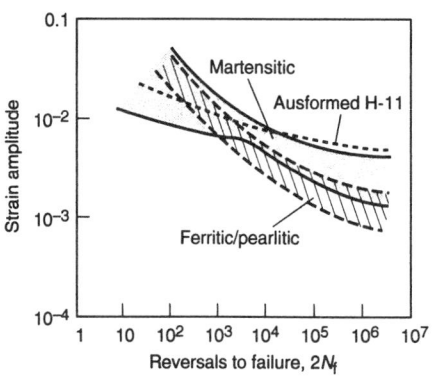

Fig. 11 Representative strain-life behavior for ferritic-pearlitic and martensitic steels

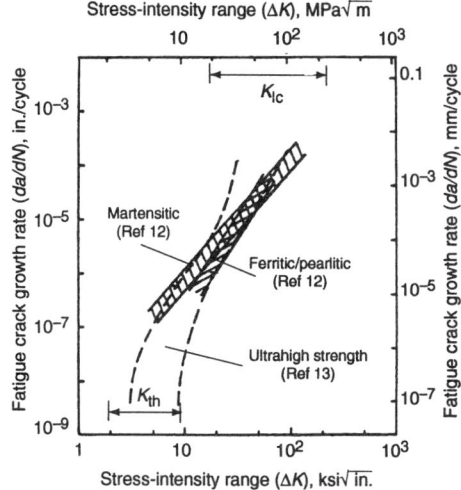

Fig. 12 Fatigue crack growth behavior for representative classes of steel (Ref 12, 13)

levels. Dowling (Ref 10) has demonstrated this effect in an SAE 4340 steel, as shown in Fig. 9. Data are compared for tests performed with no overstrains, with initial overstrains, and with periodic overstrains. The first two test conditions result in runouts (no failure) at comparable strain levels. Specimens subjected to periodic overstrains, however, show no runouts at these strain levels; in fact, the data points fall on extrapolations of the finite-life strain curves. This behavior suggests that endurance limits will be lowered, if not eliminated, under spectrum loading conditions.

Strength-Ductility Considerations. The foregoing discussion emphasizes the important interplay between strength and ductility in determining overall fatigue resistance. The true fracture strength and ductility, as determined in a monotonic tension test, are found to be sensitive to many of the metallurgical factors that affect fatigue behavior. Consequently, they provide a useful basis for comparing material characteristics (Ref 11); trends for various alloy classes are shown in Fig. 10. The inverse nature of strength-ductility relations is apparent. The broad

scatterbands observed for the conventional ferritic-pearlitic and martensitic alloys suggest that there is considerable latitude available for adjusting these properties. Also, it is seen that improvements are attainable by judicious alloying and processing (for example, maraging and ausforming).

These trends in strength and ductility are clearly reflected in the strain-life curves shown in Fig. 11. The higher-strength martensitic alloys are found to give superior long-life resistance, whereas the more ductile alloys are superior at short lives. At intermediate lives, because the curves cross one another, less variation is seen in fatigue resistance. The curve for ausformed H-11 steel is included to indicate the upper limit on long-life resistance currently achievable by bulk processing; the fatigue strength of ausformed steel is higher by a factor of seven than the bottom of the ferritic-pearlitic band.

Representative fatigue crack growth trends are shown in Fig. 12 for ferritic-pearlitic and martensitic alloys (Ref 12) and for an ultrahigh-strength steel (Ref 13). Similarities with Fig. 11 are appar-

ent. In the middle region, there is little variation among materials. The major differences are at the two extremes—that is, in the threshold levels at low growth rates and in fracture toughness levels (indicating unstable growth). The bands shown in Fig. 12 represent the extremes observed in commercial alloys. A point of particular significance here is that microstructures resulting in good long-life fatigue resistance will generally give lower thresholds for crack growth. Thus, it is important to distinguish between resistance to crack initiation and resistance to crack propagation.

Proceeding with the comparison of materials, Fig. 13 shows that maraging steels, with an uncommonly high combination of strength and ductility, offer high-cycle resistance approaching that of ausformed steel while maintaining excellent low-cycle resistance. The austenitic stainless

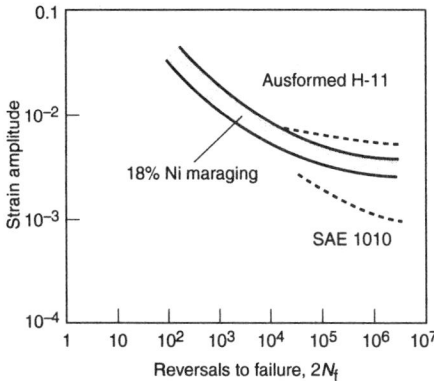

Fig. 13 Strain-life behavior of 18% Ni maraging steel

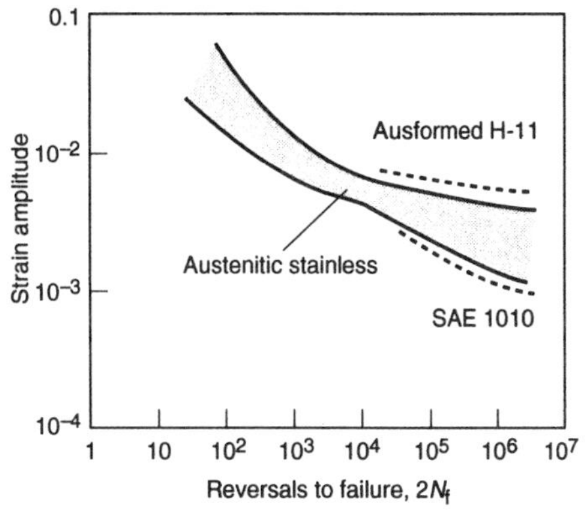

Fig. 14 Strain-life behavior of austenitic stainless steels

steels (Fig. 14) offer a wide range of properties, depending on their relative stability to reversed plasticity.

General Microstructural Effects. As a final part of this section, we identify microstructural features that can alter, often significantly, the foregoing trends. These can be classified as effects caused by directionality and inhomogeneity.

Directionality effects are particularly apparent in ausformed steel (see Fig. 15). Because of the high level of deformation during processing, an extremely textured structure is obtained. This results in a tendency for cracks to branch and follow weaker longitudinal planes, serving as an effective crack-arrest mechanism in certain applications (Ref 14). If, however, such an alloy were subjected to transverse straining, greatly reduced fatigue performance could be anticipated. In this regard, it is noted that the yield and ultimate strengths of ausformed steel are nearly the same in longitudinal and transverse directions (Ref 15). Fracture strength and ductility, however, are significantly reduced in the transverse direction—further evidence that fracture properties are more

reliable indicators of fatigue performance. In general, alloys strengthened by cold working or thermomechanical processing will exhibit direction-dependent properties that may be important in many applications.

Inhomogeneity effects in a maraging steel are illustrated in Fig. 16. Here, low-hardness, alloy-depleted zones serve as sites for early crack initiation. These zones can appreciably reduce long-life fatigue performance. Such observations are equally relevant to decarburized steels and multiphase alloys, such as dual-phase steels in which the low-carbon ferritic zones serve as preferential crack initiation sites.

Microstructural Considerations

Ferritic-Pearlitic Alloys. Details of the fatigue process in iron and iron-carbon alloys are well documented in terms of the structural alterations accompanying cyclic deformation (Ref 16-18), strain-aging effects (Ref 19), and the influence of microstructure and composition on

fatigue properties (Ref 20). Considerable attention has been directed toward increasing the useful strength of low-carbon steels through microalloying and thermomechanical processing (Ref 21). One can see notable progress by comparing, in Fig. 1, the stress-strain curve for SAE 980X with the conventional SAE 1010 steel. Strength has been increased by a factor of three by a combination of grain refinement and precipitation strengthening (Ref 22, 23).

These strengthening effects are examined in more detail in Fig. 17, where monotonic and cyclic yield strengths for a series of model alloys are presented in the form of a Hall-Petch plot. For carbon and carbon-manganese alloys, monotonic yield strengths are found to be much more sensitive to grain refinement than cyclic yield strengths. Large-grain alloys exhibit cyclic hardening, whereas small-grain alloys cyclically soften. This is consistent with observations in iron (Ref 18) showing that a characteristic dislocation-cell structure (that is essentially independent of prior condition) is formed during fatigue. The addition of niobium (columbium) and the resulting precipitation strengthening increase monotonic strength by about 100 MPa (15 ksi); more importantly, cyclic stability is achieved. (The softening observed for the intermediate grain size in Fig. 17 is a result of an underaged condition.)

Long-life fatigue data on a carbon-manganese steel (Ref 5) are presented in Fig. 18. Not surprisingly, the effect of grain refinement on the endurance limit, S_f, closely parallels trends in cyclic yield strength. This suggests that a small cyclic offset yield strength would reliably predict long-life fatigue resistance in these alloys. Also shown in Fig. 18 is the influence of grain size on the threshold-intensity range for crack growth. Grain refinement is seen to degrade crack-growth resistance somewhat.

Cyclic stress-strain curves for a series of low-alloy steels are shown in Fig. 19. Using iron as a baseline, this representation clearly shows the effects of alloying and grain refinement on cyclic

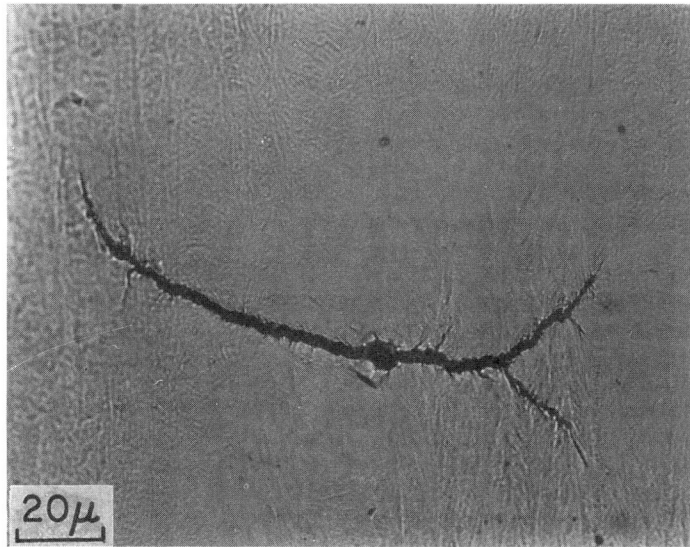

Fig. 15 Fatigue crack development in a strongly textured ausformed H-11 tool steel

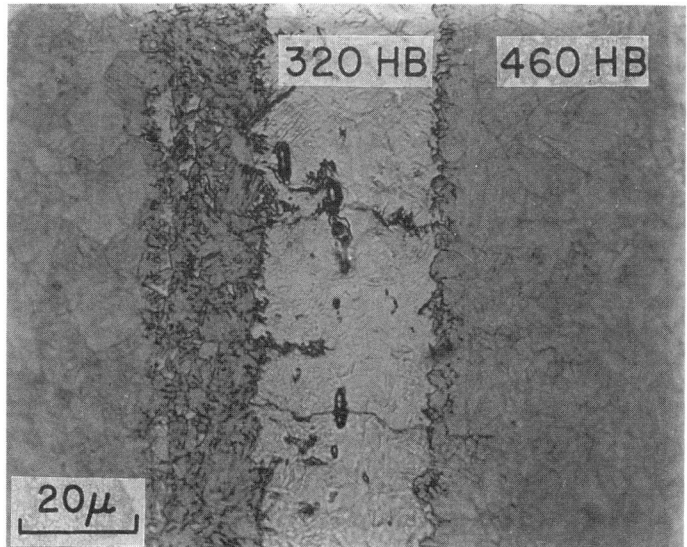

Fig. 16 Fatigue crack development in alloy-depleted zones in 18% Ni maraging steel

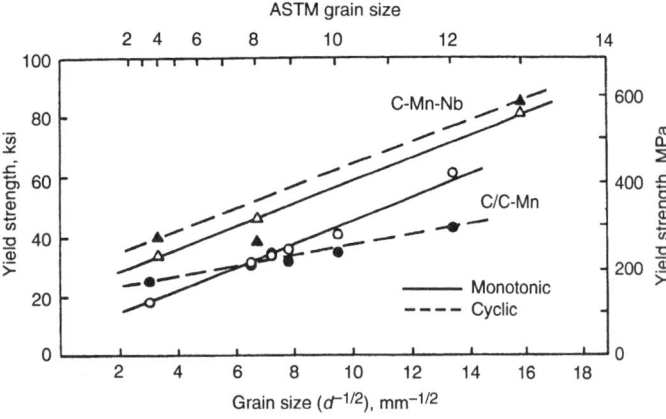

Fig. 17 Effects of grain size on yield strengths of low-carbon steels (Ref 24)

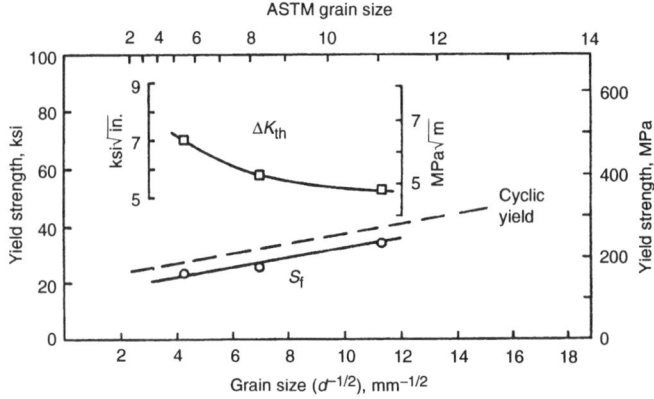

Fig. 18 Effect of grain size on endurance limit and crack growth threshold behavior of low-carbon steel (Ref 25)

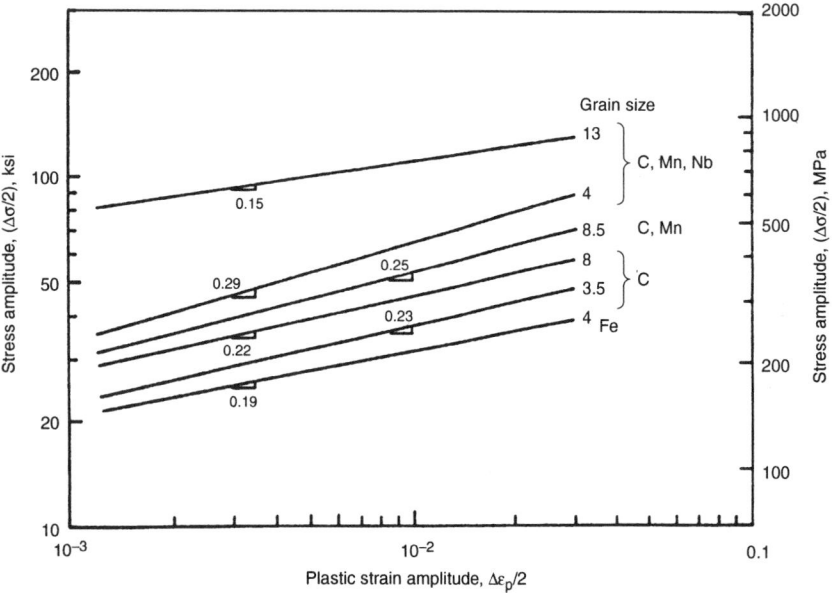

Fig. 19 Effect of grain size and alloying on cyclic stress-strain response

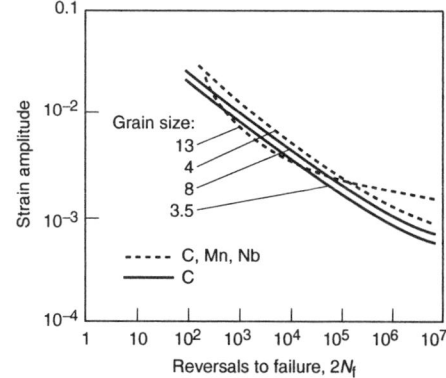

Fig. 20 Strain-life behavior as influenced by grain size and alloying (Ref 24)

deformation resistance. Strain-life curves for selected conditions, shown in Fig. 20, confirm the correlation between cyclic stress-strain and strain-life response. Grain refinement and precipitation strengthening are seen to contribute about equally to improvements in long-life fatigue resistance.

Martensitic Alloys. Quenched-and-tempered martensitic steels, as a class, offer the greatest latitude in tailoring properties for fatigue resistance. Monotonic strength, ductility, and fracture toughness properties for SAE 1045 steel are shown in Fig. 21 as a function of hardness. Of note are the observed optimum in strength properties and the precipitous drop in ductility at about 600 HB. This trend has been attributed to the influence of inclusions and a transition in failure mode (Ref 26). Fracture toughness exhibits a similar drop at a lower hardness. Interestingly, ausformed steel provides an important exception to these trends.

The propensity for tempered martensites to cyclically soften is a point of practical concern. As shown by data for SAE 1045 steel in Fig. 22, maximum softening occurs at intermediate hardnesses, while cyclic stability is approached at both hardness extremes. Stability can be associated with stabilization of the heavily dislocated martensite during quenching . A minimum carbon content of 0.05 to 0.1 wt% is required to achieve stability in as-transformed iron-nickel-carbon martensites (Ref 27). From these observations, it is possible to correlate cyclic responses with carbon content and degree of tempering. The region of cyclic stability or hardening correlates well with the first stage of tempering (that is, up to tempering temperatures of about 200 °C, or 390 °F). In studies of a 4142 steel, Theilen et al. (Ref 28) have attributed such hardening to strain aging. Cyclic softening, observed at higher tempering temperatures, is associated with rearrangement of the dislocation substructure and reduc-

tion of the dislocation density. In this respect, behavior is similar to a cold-worked material. Finally, the mixed response at low hardnesses is attributed to removal of yield point at low strains followed by bulk strain hardening at higher strains.

Strain-life curves as a function of hardness are presented in Fig. 23 to demonstrate the range of properties available by tempering. This family of intersecting curves provides clear evidence of the crucial interplay of strength and ductility in determining overall fatigue resistance. This information, used in conjunction with life-prediction models, provides a quantitative basis for optimizing material processing for specific applications.

Thermomechanical Processing. The effectiveness of ausforming in achieving notable improvements in fatigue performance is well established. The success of this process is apparently due to the formation of a well-stabilized substructure (Ref 8, 19). Other techniques involving deformation of martensite have been less effective (Ref 29). The cyclic behavior of a quenched-and-deformed SAE 4142 steel is compared with a conventionally heat-treated alloy in Fig. 24. Although the monotonic tensile curve is elevated, the compressive curve is lowered, and identical cyclic curves are obtained for the two conditions.

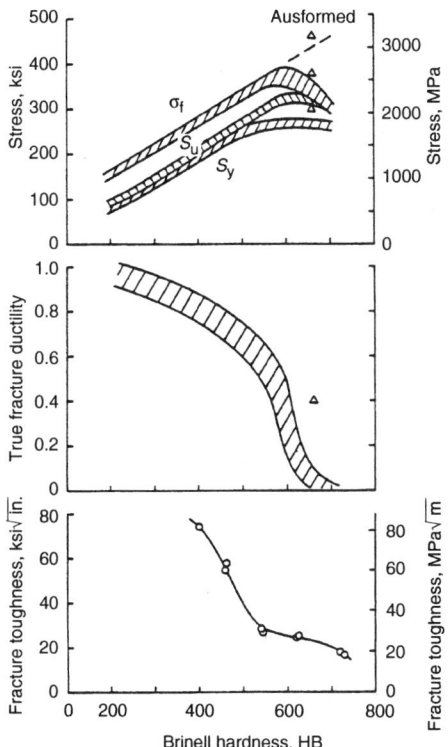

Fig. 21 Monotonic strength, ductility, and fracture toughness as a function of hardness for a medium-carbon steel (SAE 1045). σ_f is true fracture strength, S_y is yield strength.

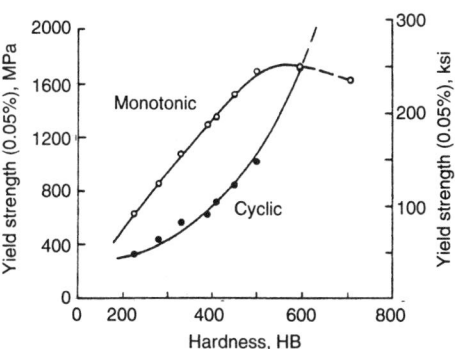

Fig. 22 Monotonic and cyclic yield strength as a function of hardness for a medium-carbon steel (SAE 1045)

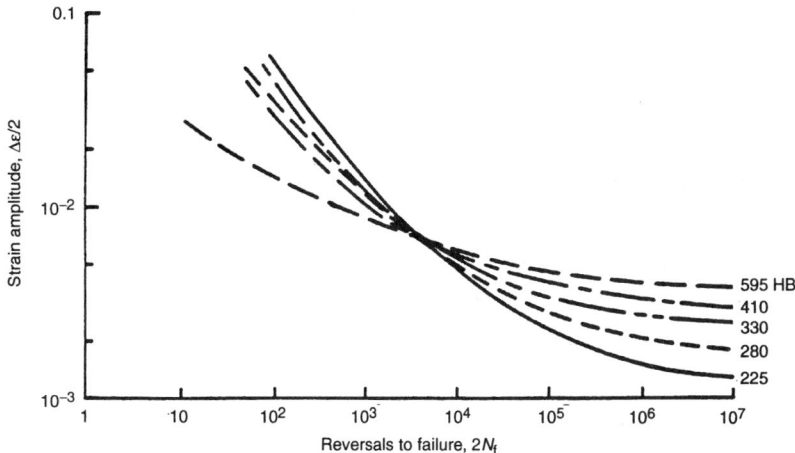

Fig. 23 Strain-life behavior as a function of hardness (HB) for a medium-carbon steel (SAE 1045)

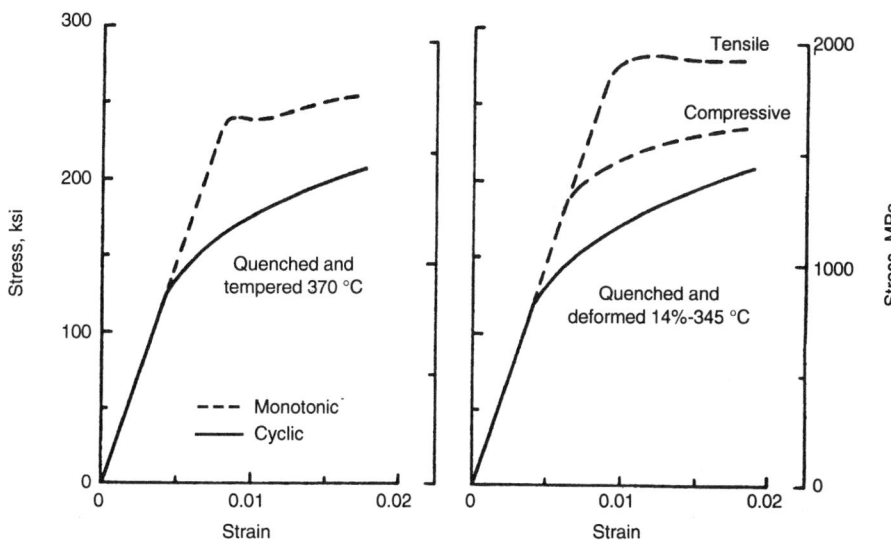

Fig. 24 Effect of deformation processing on deformation response of a medium-carbon steel (SAE 4142, hardness: 450 HB)

It is the directional nature of such strengthening that limits its effectiveness in reversed deformation situations. Deformation processing could improve resistance under unidirectional loading, however.

Maraging. Although the resistance of maraging steel to high-cycle fatigue is somewhat lower than might be expected (because of cyclic softening), its overall fatigue characteristics suggest that it would perform well in spectrum-loading situations. Studies of the cyclic responses in maraging steel in both the annealed and aged conditions indicate that cyclic softening of the annealed structure is associated with the formation of a dislocation-cell structure, whereas softening of the aged structure is attributed to the rearrangement of dislocation networks around precipitates (Ref 30).

Metastable Austenitic Alloys. Austenitic stainless steels (see Fig. 2) are of interest due to their intriguing capability of transforming to martensite when cyclically strained (Ref 31-34). Results indicate that such transformations degrade low-cycle resistance; however, improvements in high-cycle and crack growth resistance have been noted. It is possible that this mechanism can be used to improve the performance of notch members through preferential transformation at stress concentrations. In this regard, it is significant that the degree of instability can be controlled through appropriate alloying.

Notch Sensitivity and Internal Defects

Microstructure plays a particularly prominent role in determining the sensitivity of ferrous alloys to small geometric stress concentrations and internal defects. Mitchell has studied the behavior of cast irons and steels (Ref 35) and has developed an approach for predicting fatigue performance in internally defective materials based on consideration of matrix properties and the size, shape, and distribution of defects. A schematic of microdiscontinuities treated is shown in Fig. 25; the microdiscontinuities include various graphite morphologies in cast irons, porosity in cast steels, and nonmetallic inclusions in cast and wrought steels. This approach provides guidelines for adjusting matrix properties to best accommodate particular types of defects as well as quantitatively assessing the benefits of improving steel cleanliness.

The sensitivity of a particular ferrous alloy to internal defects is strongly dependent on their size. When the notch size approaches the size of the microstructural unit (for example, grain size or dislocation-cell size), the material can no longer sustain the stress gradient predicted by continuum theory. Thus, small notches have less influence on behavior than do large notches. Figure 26 shows the effect of notch or defect size on

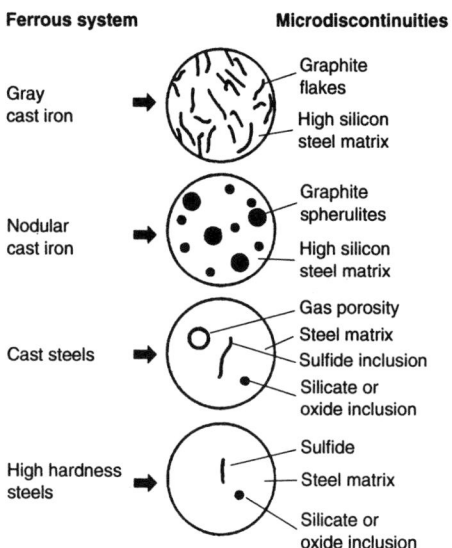

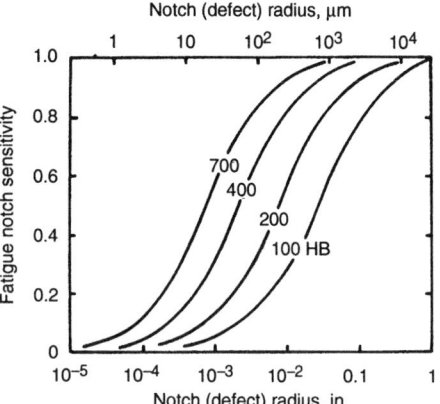

Fig. 25 Examples of microdiscontinuities that affect fatigue behavior of ferrous systems (Ref 35)

Fig. 26 Effect of notch size on fatigue notch sensitivity of steels as a function of hardness (HB)

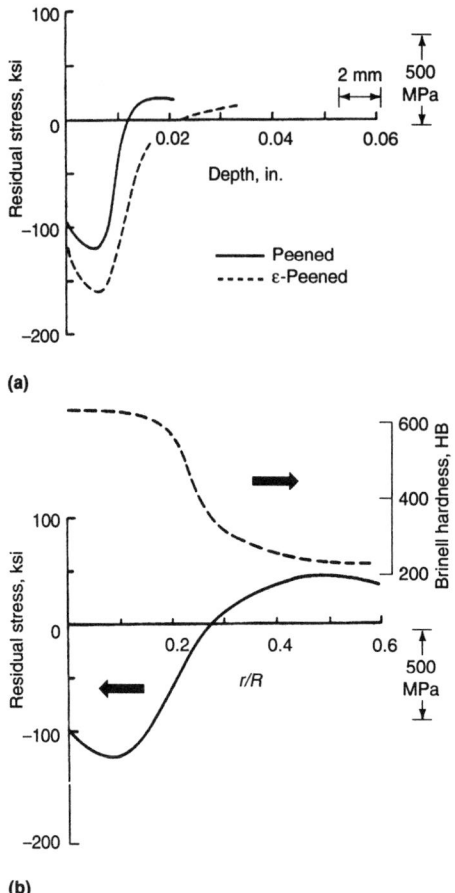

Fig. 27 Typical residual stress patterns obtained by shot peening (a) and induction hardening (b)

fatigue notch sensitivity for steels at different hardness levels. Notch sensitivity ranges from 0 (no notch effect) to 1 (full theoretical effect). Hardened steel is sensitive to defects two orders of magnitude smaller than is a normalized steel. This explains why inclusions can severely limit the performance of high-strength steel and also why threshold stress intensities are observed to decrease at increasing strength levels. Considerable effort is expended to remove inclusions from high-carbon bearing steels (SAE 52100) for this reason. In addition, the lower-strength, ductile steels are able, through local yielding, to redistribute the high local stresses, thus ameliorating the notch effect.

These effects should be considered when selecting materials because the perceived benefits of a high-strength steel may be diminished due to a lower tolerance for notches and internal defects. Also, as mentioned previously, the presence of occasional high cyclic strains will often dictate the choice of a lower-strength, higher-ductility

condition. High strength does not necessarily lead to high fatigue resistance. In addition, while inclusions are known to lower fatigue strength, not all inclusions are the same as to their effects. For example, oxides are more deleterious than sulfides. In addition, from a practicing engineers' perspectives, it is possible to specify the acceptable inclusion ratings as well as steel making routes including lowering the sulfur level or treating with rare earth elements to globularize the sulfides.

Selective Surface Processing

Surface-processing treatments, such as shot peening, induction hardening, carburizing, and nitriding, are commonly used to improve the fatigue resistance of steel components subjected to bending or torsional loading. The influence of these processes on fatigue performance depends, in a complex way, on local material properties, the service loading and, importantly, on the magnitude, distribution, and stability of residual stresses (Ref 36, 37).

Residual Stress. The development of compressive residual stress at the surface of components can often result in dramatic improvements in fatigue performance (Ref 38). Residual stress

profiles obtained by shot peening a spring steel are shown in Fig. 27. Strain peening—in which a component, such as a spring, is deformed during processing—is seen to impart higher compressive stresses and a deeper pattern of penetration than does conventional peening. For induction hardening, the residual stress pattern, associated with volume changes during transformation, depends on the depth of hardening and, in general, is related to the hardness profile as indicated in Fig. 27(b). Here the favorable surface stresses are coupled with a higher-strength surface layer which, at long lives, offers further performance improvements. Fatigue failures in peened or induction-hardened components are often found to initiate below the surface.

Cycle-Dependent Stress Relaxation. It is common to assess residual stress effects on fatigue by treating them as mean stresses (Ref 36). Their effectiveness will depend on their stability under service loading; cyclic-dependent stress relaxation may negate any potential benefits. The resistance of a steel to cyclic stress relaxation can be determined by subjecting axial specimens to biased strain cycling and observing the cyclic change in mean stress. Residual (mean) stress can be expected to relax whenever the applied loading results in reversed plastic straining. Because of the tendency for many steels to exhibit cycle-dependent softening, this may occur at lower stresses than would be anticipated from monotonic yield strengths.

The cyclic stress relaxation behavior of an SAE 1045 steel at three hardnesses is shown in Fig. 28. The ratio of instantaneous mean stress to the first cycle mean stress is plotted against cycles on logarithmic coordinate. The relaxation rate is characterized by the slope of this line, r, resulting in the following relation:

$$\sigma_{mN} = \sigma_{m1}(N)^r \qquad \text{(Eq 4)}$$

where σ_{mN} is the mean stress on the Nth cycle, and σ_{m1} is the mean stress on the first cycle. The relaxation exponent, r, a function of steel hardness and applied strain amplitude, can be estimated using procedures in Ref 36.

Fatigue Life Prediction. In order to achieve maximum benefits from surface-processing treatments, it is necessary to consider both surface and subsurface failure modes. Using the previously established analogy between mean and residual stresses, this is conveniently accomplished using the following life relation developed by Morrow (Ref 39):

$$2N_f = \left(\frac{\sigma_a}{\sigma'_f - \sigma_m - \sigma_r} \right)^{1/b} \qquad \text{(Eq 5)}$$

where fatigue life in reversals, $2N_f$, is a function of material properties, σ'_f and b, stress amplitude, σ_a, mean stress, σ_m, and residual stress, σ_r.

In Fig. 29, the individual stress gradients resulting from processing and applied loading of an induction-hardened shaft are plotted from surface to centerline. These are then combined and plot-

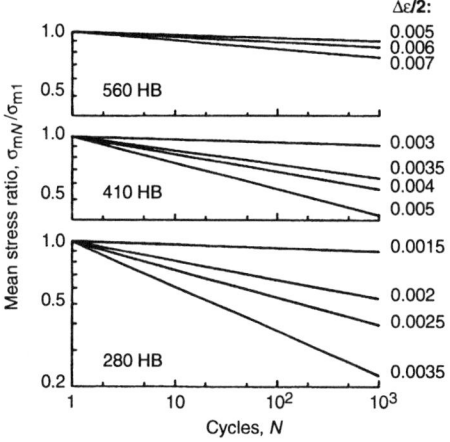

Fig. 28 Cyclic stress relaxation rates as a function of hardness and strain level for SAE 1045 (Ref 36)

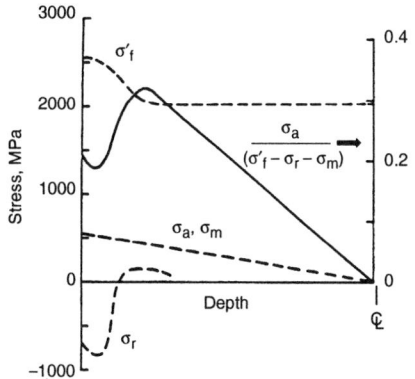

Fig. 29 Stress and damage parameter profiles for a surface-processed bending member (Ref 36)

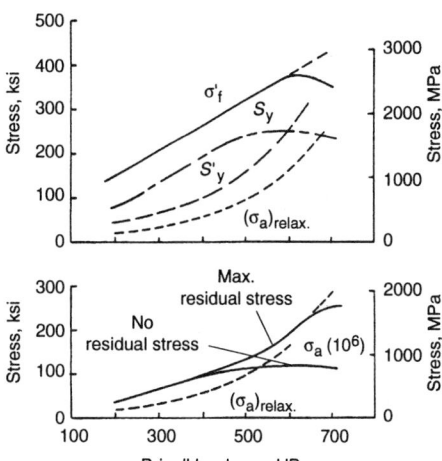

Fig. 30 Residual stress effects on long-life fatigue as a function of hardness (Ref 36)

ted as a damage parameter profile as indicated. From Eq 5, the higher the magnitude of this parameter, the shorter the fatigue life. In Fig. 29, the parameter attains a maximum below the surface, thus predicting subsurface crack initiation. This is often observed in surface processes members and, in this instance, is the result of a shallow case depth and the tensile residual stress field subsurface. Life improvements are possible by deepening the case. The optimal processing would result in equal failure likelihood at the surface and subsurface.

Processing for Fatigue Performance. To gain additional insight into material characteristics conducive to surface-processing response, it is informative to consider the variation of key properties with hardness. In Fig. 30, trends representative of a range of medium- to high-carbon steels are shown. Fatigue strength, as indicated by σ'_f, increases linearly with hardness, attains a maximum around 600 HB, and decreases slightly thereafter. The decrease is due, in part, to inclusion effects: cleaner steels show less effect. With regard to residual stresses, note that the cyclic yield strength, S_y', falls below the monotonic value, S_y, at hardnesses below 550 HB. Such cycle-dependent softening promotes rapid residual stress relaxation.

The stress amplitude threshold for relaxation, $(\sigma_a)_{relax}$, follows the trend of the cyclic yield strength. Furthermore, at hardnesses below 550 HB, stress relaxation can be expected at lives of 106 reversals or longer. The influence of compressive residual stresses on fatigue resistance at 10^6 is largest at the highest hardnesses and diminishes rather rapidly with decreasing hardness. Material characteristics promoting strong residual stress effects include: a monotonic yield strength to achieve a high initial residual and cyclic stability in order to avoid relaxation. These trends suggest that an improvement of approximately a factor of three in long-life fatigue strength is attainable through hardening, with another factor of two possible through residual stress effects.

Nelson et al. (Ref 40) have dramatically demonstrated the powerful effect of residual stress on fatigue. By means of a severe quenching procedure, they were able to obtain a compressive residual stress of 250 ksi (1720 MPa) at the surface of an as-transformed SAE 1045 steel specimen. This resulted in a million-cycle fatigue strength of 230 ksi (1590 MPa).

REFERENCES

1. C.E. Feltner and P. Beardmore, *STP 467*, ASTM 1970, p 77
2. J.C. Grosskreutz, *STP 495*, ASTM 1971, p 5
3. C. Laird, *Alloy and Microstructural Design*, Academic Press, 1976, p 175
4. *Fatigue Design Handbook*, Society of Automotive Engineers, 1988
5. J.A. Bannantine, J.J. Comer, and J.L. Handrock, *Fundamentals of Metal Fatigue Analysis*, Prentice-Hall, Englewood Cliffs, NJ, 1990
6. R.W. Landgraf, *Work Hardening in Tension and Fatigue*, AIME, 1977, p 240
7. F. Borik, W.M. Justusson, and V.F. Zackey, *Trans. ASM*, Vol 5, 1966, p 327
8. R. Phillips and W.E. Duckworth, *Appl. Mater. Res.*, Vol 5, 1966, p 13
9. R.W. Smith, M.H. Hirschberg, and S.S. Manson, TN D-1574, National Aeronautics and Space Administration, 1963
10. N.E. Dowling, *J. Test. Eval.*, Vol 1, 1973, p 271
11. R.W. Landgraf, *STP 467*, ASTM 1970, p 3
12. J.M. Barsom, *J. Eng. Ind., (Trans. ASME)*, Vol 6, 1971, p 1190
13. R.O. Ritchie, *J. Eng. Mater. Technol (Trans. ASME)*, Vol 99, 1977, p 195
14. W.M. Justusson and R. Clark, *Mechanical Working and Steel Processing*, Vol IV Gordon & Breach, 1969, p 481
15. R.H. Bush, A.J. McKevily, Jr., and W.M. Justusson, *Trans. ASM*, Vol 57, 1964, p 57
16. M. Klesnil, M. Holzmann, P. Lukas, and P. Rys, *J. Iron Steel Inst.*, Vol 203, 1966, p 47
17. M. Klesnil and P. Lukas, *J. Iron Steel Inst.*, Vol 205, 1967, p 746
18. H. Abdel-Raouf, P.P. Benham, and A. Plumtree, *Can. Metall. Q.*, Vol 10, p 87
19. D.V. Wilson, *Met. Sci.*, Vol 11, 1977, p 321
20. J.D. Grozier and J.H. Bucher, *J. Mater.*, Vol 2, 1967, p 393
21. *Microalloying 75*, Union Carbide, 1977
22. A.M. Sherman, *Metall. Trans. A*, Vol 6A, 1975, p 1035
23. D.J. Quesnel and M. Meshii, *Mater. Sci. Eng.*, Vol 30, 1977, p 223
24. R.W. Landgraf, A.M. Sherman, and J. Sprys, *2nd Int. Conf. Mechanical Behavior of Materials*, American Society for Metals, 1976, p 513
25. S. Taira, K. Tanaka, and M. Hoshina, *STP 675*, ASTM, 1979, p 135
26. J. Morrow, G.R. Halford, and J.F. Millan, *Proc. 1st Int. Conf. Fracture*, Vol 3, 1965, p 1611
27. P. Beardmore, *2nd Int. Conf. Mechanical Behavior of Materials*, American Society for Metals, 1976, p 1898
28. P.N. Theilen, M.E. Fine, and R.A. Fournelle, *Acta Metall.*, Vol 24, 1976, p 1
29. B. Mintz and D.V. Wilson, *J. Iron Steel Inst.*, Vol 204, 1966, p 91
30. L.F. Van Swam, R.M. Pelloux, and N.J. Grant, *Metall. Trans. A*, Vol 6A, 1975, p 45
31. G.R. Chanai, S.D. Antolovich, and W.W. Gerberich, *Metall. Trans.*, Vol 3, 1972, p 2661
32. G.R. Chanai and S.D. Antolovich, *Metall. Trans.*, Vol 5, 1974, p 217
33. D. Hennessy, G. Steckel, and C. Altstetter, *Metall. Trans. A*, Vol 7A, 1976, p 415
34. G. Baudry and A. Pineau, *Mater. Sci. Eng.*, Vol 28, 1977, p 229
35. M.R. Mitchell, *J. Eng. Mater. Technol. (Trans. ASME)*, Vol 99, 1977, p 329
36. R.W. Landgraf and R.A. Chernenkoff, *STP 1004*, ASTM, 1988, p 1
37. R.W. Landgraf and R.H. Richman, *STP 569*, ASTM, 1975, p 130
38. R.W. Landgraf and R.C. Francis, *SAE Trans.*, Vol 88, Sect. 2, 1979, p 1485
39. J. Morrow, *Fatigue Design Handbook*, Society of Automotive Engineers, 1968, p 27
40. D.V. Nelson, R.E. Ricklefs, and W.P. Evans, *STP 467*, ASTM, 1970, p 228

Fracture Mechanics Properties of Carbon and Alloy Steels*

THE STEELS discussed in this article range from the less expensive mild steels used in bridge construction to the very expensive 18Ni maraging steels used in high-strength pressure vessels and aircraft components. Such steels may differ by more than an order of magnitude in both yield strength and alloy additions as well as in cost per unit weight. It has been clear for some time that selection of ultrahigh-strength steels for critical structures requires application of fracture mechanics principles if high degrees of reliability are to be achieved. In fact, such principles have been applied to selection of steels since the early 1960s in developing rocket motor chambers with improved performance. These concepts have been adopted by the aerospace industry and have since been incorporated into standards for determining fracture toughness of metals and for design of airframes and engine components.

For example, the U.S. Air Force has established damage tolerance requirements (MIL-A-83444) that are applied in the design, fabrication, and inspection of military aircraft. This resulted because of recurring structural failures in military aircraft, such as those caused by problems in material selection, fabrication, and heat treatment of the D6ac steel wing carrythrough structure for the F-111 aircraft. Other government agencies have included damage tolerance requirement specifications, and it is expected that this trend will continue. These concepts have already been extended to medium-strength rotor forging steels in the electric power industry and are gradually being applied to such problem areas as dynamic fracture toughness of roll bars for off-road vehicles and corrosion fatigue of line pipe steel in aggressive hydrogen sulfide environments.

Because these concepts are being applied to a broad spectrum of steels with extreme variations in strength, composition, and microstructure, this article attempts to present typical data for several classes of steels (e.g., ferrite-pearlite steels, Ni-Cr-Mo generator rotor steels, and ultrahigh-strength steels). Within the scope of this article, it is not possible to present sufficient data for the

user to apply this as a design manual. Rather, an attempt is made to identify those effects of alloy composition, microstructure, and heat treatment that are most critical to the various types of fracture behavior encountered due to variations in loading, temperature, and environment. More detailed data are covered in other articles in this Volume.

This article is organized into four major sections that address fracture toughness, fatigue crack propagation, sustained-load crack propagation, and fundamental aspects of fracture in steel. Where possible, these major sections are further subdivided to illustrate the effects of variations in alloy chemistry, microstructure, temperature, strain rate (or frequency), and environment on the various fracture toughness or crack growth rate parameters. In the following sections, the metallurgy of steel is given only a summary treatment, being followed by the major mechanical property sections dealing with static fracture toughness, K_{Ic}; dynamic fracture toughness, K_{ID}; fatigue or sustained-load crack growth rates, da/dN or da/dt; and fatigue or sustained-load thresholds, ΔK_{th} or K_{th}.

Common classes of steels are austenitic (face-centered-cubic) steels, ferrite-pearlite or bainitic (body-centered-cubic) steels, and martensitic (body-centered-tetragonal) steels. The austenitic stainless steels are discussed in a separate article in this Volume. In this article, nearly all of the data deal with ferrite-pearlite steels and steels with tempered-martensite microstructures, although some of the larger components of heat-treated or slowly cooled alloys invariably contain some bainite. Particular emphasis is given to maraging-type steels and high-strength low-alloy (HSLA) steels. The HSLA steels contain niobium (columbium), vanadium, or titanium as microalloying constituents in the form of finely dispersed carbides. The high-nickel maraging steels contain a low-carbon martensite matrix with precipitates of various compositions, after age hardening.

Typical Compositions and Alloying. Nominal compositions of some of the steels considered in this article are given in Table 1. Table 1 does not include the compositions for the HSLA and Cr-Mo families of steel, which are large and are

covered in separate articles in this Volume. However, this article does provide some comparative data on these major steel types.

With regard to strength, ferrite-pearlite steels generally achieve their strength from carbon in the form of Fe_3C (pearlite content) or silicon and manganese as solid-solution strengtheners. The exceptions are the HSLA grades, which may be strengthened by niobium, vanadium, or titanium carbides. In both instances, the standard contribution of strength from the inverse square-root grain size relationship is most important. For tempered martensitic steels, carbon is the major strengthener for low tempering temperatures, whereas chromium, tungsten, molybdenum, or vanadium alloy carbides contribute to secondary hardening of alloy steels tempered in the range from 500 to 600 °C (930 to 1110 °F). General comparison of strength is shown in Table 2.

Alloying additions to low-strength ferrite-pearlite steels are minimal. However, for heavy steel sections of higher strength and for ultrahigh-strength steels, alloying additions are of primary consideration for hardenability, strength, and toughness. To expand the austenite stability range, additions of elements such as boron, carbon, chromium, nickel, manganese, molybdenum, and vanadium are most common, whereas copper, tungsten, niobium, and titanium may be added to specialty steels, and silicon and aluminum are added as deoxidants. The latter element may also be added as a grain refiner, because AlN pins grain boundaries and retards ferrite grain growth. Such alloying additions, except boron and carbon, are readily dissolved in ferrite, but some tend to form strong carbides, as originally outlined by Bain and Paxton (Ref 1).

With regard to impact energy and fracture toughness of ferritic steels, manganese or nickel is most commonly added to reduce the ductile-to-brittle transition temperature. In addition, grain refiners such as aluminum may cause net improvements in low-temperature toughness. Rare-earth additions such as cerium often are made for inclusion shape control to improve toughness at higher temperatures in the upper-shelf region. These factors are described in more detail later in this article.

* Developed from ASM source articles acknowledged at the end of this article.

Table 1 Nominal compositions of carbon and alloy steels referred to in this article

Type or designation	UNS designation	Nominal composition, %									Nominal yield strength or range, MPa
		C	Mn	Si	Cr	Ni	Mo	V	Co	Other	
Ferrite-pearlite steels											
A285, grade C	K02801	0.17	0.52	0.06	...	...	...	...	...	...	240
A36	...	0.25	1.0	...	...	...	...	...	...	...	≥250
St 37-3	...	0.08	0.45	0.17	...	...	...	...	...	...	320
A588, grade A	K11430	0.15	1.1	0.20	0.5	...	...	0.05	...	0.3Cu	>290
A588, grade F	K11541	0.15	0.8	0.20	0.2	0.8	0.15	0.05	...	0.6Cu	≥345
A515, grade 70 (ASTM A212)	K03101	0.27	0.71	0.19	...	...	...	...	...	...	≥450
A302, grade B	K12022	0.20	1.29	0.18	...	0.08	0.55	...	...	...	470
A572 HSLA	...	0.26 max	1.35 max	...	...	...	...	(a)	...	(a)	350-450
API X-65	...	0.24	1.35	...	...	...	...	(b)	...	(b)	≥450
Q & T low-alloy (medium strength) steels											
A514, grade F	K11576	0.15	0.80	0.25	0.5	0.9	0.5	0.05	...	0.3Cu, 0.004B	690
A533, grade B	K12539	0.25	1.25	0.20	...	0.50	0.5	...	...	...	730
A517, grade F	K11576	0.17	0.85	0.20	0.56	0.81	0.5	0.04	...	0.31Cu	805
HY 80	...	0.12	0.30	0.22	1.46	2.49	0.43	...	...	...	590
HY 100	...	0.18	0.45	0.25	1.40	2.60	0.32	0.01	...	...	730
HY 130	...	0.12	0.75	0.30	0.55	5.0	0.50	0.08	...	...	950
QT 35	...	0.18	1.2	0.20	0.90	1.1	0.45	0.07	...	...	600
API M N80	...	0.41	1.15	0.28	...	...	0.09	...	...	...	605
API N 110	...	0.43	1.19	0.26	...	...	0.09	...	...	...	920
Ni-Cr-Mo-V rotor steel	...	0.30	0.40	0.20	1.70	3.5	0.60	0.15	...	...	700
3340	...	0.40	0.45	...	1.52	3.33	...	...	...	...	890
Ultrahigh-strength steels (large strength ranges due to tempering)											
4130	G41300	0.30	0.50	0.25	1.0	...	0.20	...	...	...	900-1700
4140	G41400	0.40	0.83	0.25	0.93	...	0.20	...	...	...	900-1800
4340	G43400	0.40	0.70	0.25	0.80	1.9	0.25	...	...	...	900-1800
300M	K44315	0.42	0.76	1.60	0.76	1.76	0.41	0.10	...	...	1070-1740
D6ac	K24728	0.45	0.80	0.25	1.15	0.60	1.0	0.06	...	...	900-1800
H-11	T20811	0.40	0.30	0.90	5.00	...	1.3	0.5	...	...	900-1800
HP9-4-20	K91401	0.20	0.30	0.10	0.80	9.0	1.0	0.1	4.5	...	1100-1300
HP9-4-25	...	0.27	0.21	0.02	0.39	9.48	0.51	0.08	3.9	...	1200-1400
HP9-4-45	...	0.45	0.10	0.10	0.30	8.5	0.20	0.10	3.75	...	1300-1800
10Ni-Cr-Mo-Co	...	0.11	0.2	0.10	2.20	10.0	1.0	...	8.0	...	1310
AF1410	...	0.15	0.06	0.10	2.0	10.0	1.0	...	14.0	...	1480
12Ni-5Cr-3Mo	K91890	0.02	...	...	5.0	12.1	3.2	...	...	0.25Ti, 0.30Al	...
18Ni, grade 200	K92810	0.01	0.02	0.10	...	18.0	3.0	...	8.5	0.20Ti, 0.1Al	1380
18Ni, grade 250	K92890	0.006	0.02	0.02	...	18.7	4.8	...	8.5	0.40Ti, 0.15Al	1720
18Ni, grade 300	K93120	0.013	0.02	...	...	18.6	4.8	...	8.9	0.59Ti, 0.1Al	2070
18Ni, grade 350	...	0.004	0.02	0.02	...	18.0	4.2	...	12.5	1.70Ti, 0.1Al	2400
12Cr-Mo-V	...	0.20	1.0	0.5	12.0	0.6	0.5	0.3	...	...	900-1450

(a) 0.01-0.1 V and/or 0.005-0.05 Nb. (b) 0.02 V min and/or 0.005 Nb min

Table 2 General comparison of mild (low-carbon) steel with various high-strength steels

Steel	Chemical composition, % (a)				Minimum yield strength		Minimum tensile strength		Minimum ductility (elongation in 50 mm, or 2 in.), %
	C (max)	Mn	Si	Other	MPa	ksi	MPa	ksi	
Low-carbon steel	0.29	0.60-1.35	0.15-0.40	(b)	170-250	25-36	310-415	45-60	23-30
As-hot rolled carbon-manganese steel	0.40	1.00-1.65	0.15-0.40	...	250-400	36-58	415-690	60-100	15-20
HSLA steel	0.08	1.30 max	0.15-0.40	0.02 Nb or 0.05 V	275-450	40-65	415-550	60-80	18-24
Heat-treated carbon steel									
Normalized(b)	0.36	0.90 max	0.15-0.40	...	200	29	415	60	24
Quenched and tempered	0.20	1.50 max	0.15-0.30	0.0005 B min	550-690	80-100	660-760	95-110	18
Quenched-and-tempered low-alloy steel	0.21	0.45-0.70	0.20-0.35	0.45-0.65 Mo, 0.001-0.005 B	620-690	90-100	720-800	105-115	17-18

(a) Typical compositions include 0.04% P (max) and 0.05% S (max). (b) If copper is specified, the minimum is 0.20%.

Structural Steels

Before discussing the fracture mechanics properties of ferritic and quenched-and-tempered structural steels, this section briefly reviews some basic types and metallurgical aspects of strengthening mechanisms and fracture toughness for structural steels. Like most materials, toughness generally decreases with increasing yield strength, as shown in Fig. 1 and 2 for HSLA (ferritic-pearlitic) steels and quenched-and-tempered steels, respectively. Improvements in both strength and toughness can be achieved by grain refinement (Fig. 1) with heat treatment or thermomechanical processing methods such as controlled rolling. In general, quenched-and-tempered steels provide a better combination of strength and toughness when compared to as-

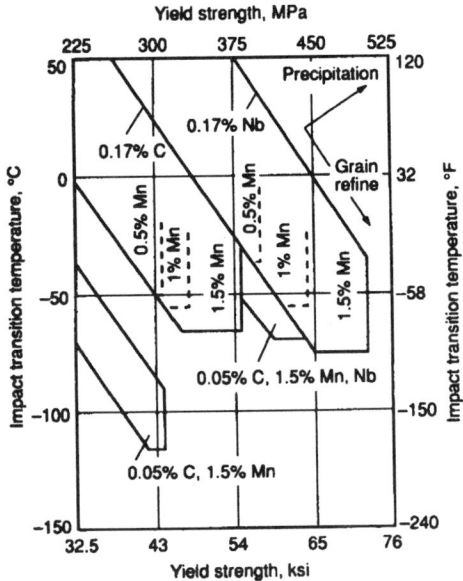

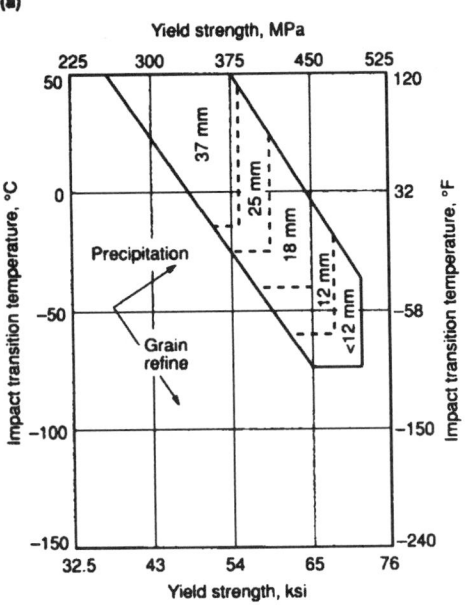

(a)

(b)

Fig. 1 General combinations of yield stress and impact transition temperatures available in controlled-rolled steels of various (a) compositions and (b) section sizes

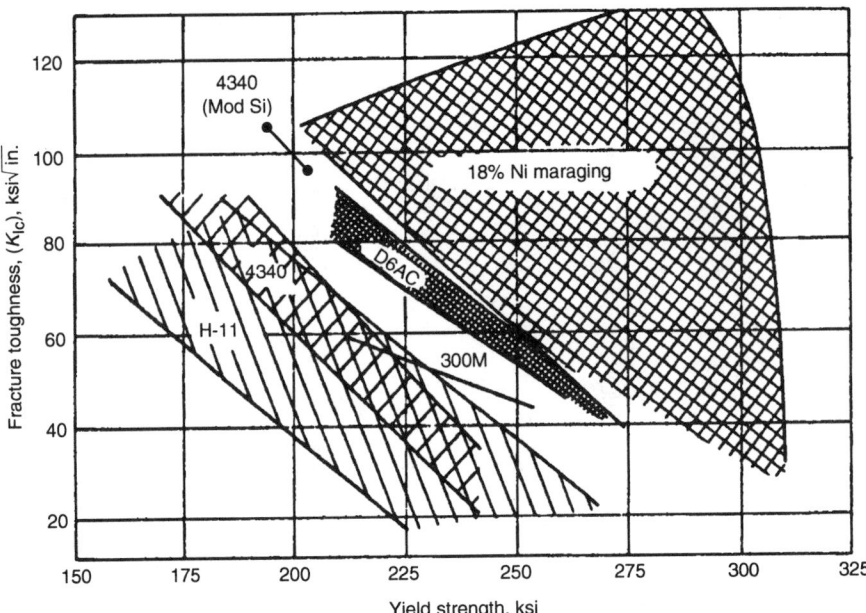

Fig. 2 Correlation between fracture toughness and yield strength for various high-strength steels

with about 0.4 to 0.7% Mn, 0.1 to 0.5% Si, and some residuals of sulfur, phosphorus, and other elements. These steels are not deliberately strengthened by alloying elements other than carbon; they contain some manganese for sulfur stabilization and silicon for deoxidation. Mild steels are mostly used in the as-rolled, forged, or annealed condition and are seldom quenched and tempered.

The largest category of mild steels is the low-carbon (<0.08% C, with ≤0.4% Mn) mild steels used for forming and packaging. Mild steels with higher carbon and manganese contents have also been used for structural products such as plate, sheet, bar, and structural sections. Typical examples include:

Steel	Minimum yield strength	
	MPa	ksi
Hot-rolled SAE 1010 steel sheet	207	30
ASTM A 283, grade D	228	33
ASTM A 36	250	36

Before the advent of HSLA steels, these mild steels were commonly used for the structural parts of automobiles, bridges, and buildings. In automotive applications, for example, hot-rolled SAE 1010 sheet has long been used as a structural steel. However, as lighter weight automobiles became more desirable during the energy crisis, there was a trend to reduce weight by using higher-strength steels with suitable ductility for forming operations.

The trend for structural steels used in the construction of bridges and buildings has also been away from mild steels and toward HSLA steels. For many years, ASTM A 7 (now ASTM A 283, grade D) was widely used as structural steel. In about 1960, improved steelmaking methods resulted in the introduction of ASTM A 36, with

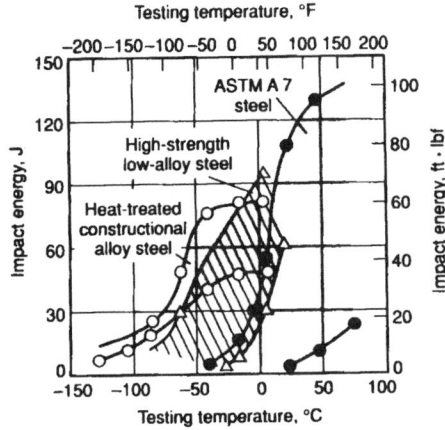

Fig. 3 General comparison of Charpy V-notch toughness for a mild-carbon steel (ASTM A 7, now ASTM A 283, grade D), an HSLA steel, and a heat-treated constructional alloy steel

improved weldability and slightly higher yield strength. Now, however, HSLA steels often provide a superior substitute for ASTM A 36, because HSLA steels provide higher yield strengths without adverse effects on weldability. Weathering HSLA steels also provide better atmospheric corrosion resistance than carbon steel.

Hot-Rolled Carbon-Manganese Structural Steels. For rolled structural plate and sections, one of the earliest approaches in achieving higher strengths involved the use of higher manganese contents. Manganese is a mild solid-solution strengthener in ferrite and is the principal strengthening element when it is present in amounts over 1% in rolled low-carbon (<0.20% C) steels. Manganese can also improve toughness properties (Fig. 4b).

rolled carbon or HSLA ferritic-pearlitic steels (Table 2, Fig. 3). However, quenching and tempering is a more involved process than the production of as-rolled steels, which is one reason the as-rolled HSLA steels are an attractive alternative. Compared to mild and heat-treated steels (Table 2), the as-rolled HSLA steels also have lower carbon contents, which enhance not only toughness but also weldability for structural fabrication.

Structural Steel Types

Mild (low-carbon) steels are normally considered to have carbon contents up to 0.25% C

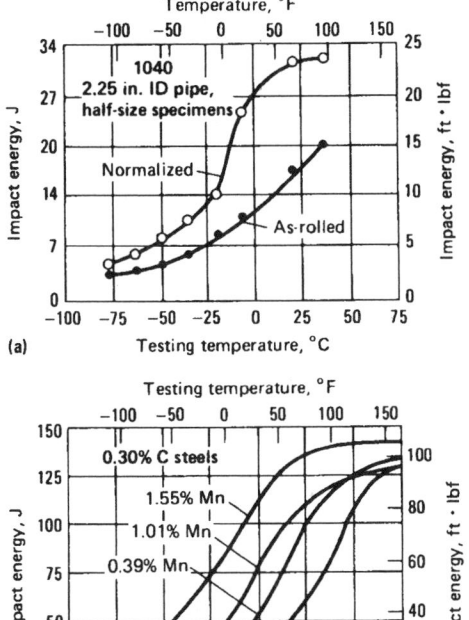

Fig. 4 Effect of (a) normalizing and (b) manganese content on the Charpy V-notch impact energy of normalized carbon steels. (a) Impact energy and transition temperature of 1040 steel pipe, deoxidized with aluminum and silicon. (b) Charpy V-notch impact energy for normalized 0.30% C steels containing various amounts of manganese

Before World War II, strength in hot-rolled structural steels was achieved by the addition of carbon up to 0.4% and manganese up to 1.5%, giving yield strengths of the order of 350 to 400 MPa (50 to 58 ksi). The strengthening of these steels relies primarily on the increase in carbon content, which results in greater amounts of pearlite in the microstructure and thus higher tensile strengths. However, the high carbon contents of these steels greatly reduce notch toughness and weldability. Moreover, the increase of pearlite contents in hot-rolled carbon and alloy steels has little effect on yield strength, which, rather than tensile strength, has increasingly become the main strength criterion in structural design.

Nevertheless, carbon-manganese steels with suitable carbon contents are used in a variety of applications. If structural plate or shapes with improved toughness are required, small amounts of aluminum are added for grain refinement. Carbon-manganese steels are also used for stampings, forgings, seamless tubes, and boiler plates.

High-strength structural carbon steels have yield strengths greater than 275 MPa (40 ksi) and are available in various product forms:

- Cold-rolled structural sheet
- Hot-rolled carbon-manganese steels in the form of sheet, plate, bar, and structural shapes

- Heat-treated (normalized or quenched-and-tempered) carbon steels in the form of plate, bar, and, occasionally, sheet and structural shapes

Heat-treated carbon structural steels typically attain yield strengths of 290 to 690 MPa (42 to 100 ksi). The heat treatment of carbon steels consists of either normalizing or quenching and tempering. These heat treatments can be used to improve the mechanical properties of structural plate, bar, and, occasionally, structural shapes. Structural shapes (such as I-beams, channels, wide-flange beams, and special sections) are primarily used in the as-hot-rolled condition because warpage is difficult to prevent during heat treatment. Nevertheless, some normalized or quenched-and-tempered structural sections can be produced in a limited number of section sizes by some manufacturers.

Normalizing involves air cooling from austenitizing temperatures and produces essentially the same ferrite-pearlite microstructure as that of hot-rolled carbon steel, except that the heat treatment produces a finer grain size. This grain refinement makes the steel stronger, tougher, and more uniform throughout. Charpy V-notch impact energies at various temperatures are given in Fig. 4(a) for a normalized carbon steel.

Quenching and tempering, that is, heating to about 900 °C (1650 °F), water quenching, and tempering at temperatures of 480 to 600 °C (900 to 1100 °F) or higher, can provide a tempered martensitic or bainitic microstructure that results in better combinations of strength and toughness. An increase in the carbon content to about 0.5%, usually accompanied by an increase in manganese, allows the steels to be used in the quenched-and-tempered condition. For quenched-and-tempered carbon-manganese steels with carbon contents up to about 0.25%, low hardenability restricts the section sizes to about 150 mm (6 in.).

Low-alloy steels contain alloy elements, including carbon, up to a total alloy content of about 8.0%. Low-alloy steels with suitable alloy compositions have greater hardenability than structural carbon steel and thus can provide high strength and good toughness in thicker sections by heat treatment. Their alloy contents may also provide improved heat and corrosion resistance. However, as the alloy contents increase, alloy steels become more expensive and more difficult to weld. Quenched-and-tempered structural steels are primarily available in the form of plate or bar products.

Ferritic nickel steels (5% and 9% Ni steels) are too tough at room temperature for valid K_{Ic} data to be obtained on specimens of reasonable size, but limited K_{Ic} data have been obtained on these steels at subzero temperatures by the J-integral method. The 5% Ni steel retains relatively high fracture toughness at −162 °C (−260 °F), and the 9% Ni steel retains relatively high fracture toughness at −196 °C (−320 °F). These temperatures approximate the minimum temperatures at which these steels may be used.

High-strength low-alloy steels are a group of low-carbon steels that have small amounts of alloying elements and attain yield strengths greater than 275 MPa (40 ksi) in the as-rolled or normalized condition. These steels have better mechanical properties and sometimes better corrosion resistance than as-rolled carbon steels. Moreover, because the higher strength of HSLA steels can be obtained at lower carbon contents, the weldability of many HSLA steels is comparable to or better than that of mild steel.

High-strength low-alloy steels are primarily hot rolled into the usual wrought product forms (sheet, strip, bar, plate, and structural sections) and are commonly furnished in the as-hot-rolled condition. However, the production of hot-rolled HSLA products may also involve special hot-mill processing that further improves the mechanical properties of some HSLA steels and product forms. These processing methods include:

- *The controlled rolling* of precipitation-strengthened HSLA steels to obtain fine austenite grains and/or highly deformed (pancaked) austenite grains, which during cooling transform into fine ferrite grains that greatly enhance toughness while improving yield strength.
- *The accelerated cooling* of, preferably, controlled-rolled HSLA steels to produce fine ferrite grains during the transformation of austenite. These cooling rates cannot be rapid enough to form acicular ferrite, nor can they be slow enough so that high coiling temperatures result and thereby cause the overaging of precipitates.
- *The quenching or accelerated air or water cooling* of low-carbon steels (≤0.08% C) that possess adequate hardenability to transform into low-carbon bainite (acicular ferrite). This microstructure offers an excellent combination of high yield strengths (275 to 690 MPa, or 60 to 100 ksi), excellent weldability and formability, and high toughness (controlled rolling is necessary for low ductile-to-brittle transition temperatures).
- *The normalizing* of vanadium-containing HSLA steels to refine grain size, thereby improving toughness and yield strength.
- *The intercritical annealing* of HSLA steels (and also carbon-manganese steels with low carbon contents) to obtain a dual-phase microstructure (martensite islands dispersed in a ferrite matrix). The usefulness and cost effectiveness of these processing methods are highly dependent on product form and alloy content.

High-strength low-alloy steels include many standard and proprietary grades designed to provide specific desirable combinations of properties such as strength, toughness, formability, weldability, and atmospheric-corrosion resistance. These steels are not considered alloy steels, even though their desired properties are achieved by the use of small alloy additions. Instead, HSLA steels are classified as a separate steel category, which is similar to as-rolled mild-carbon steel with enhanced mechanical properties obtained by the judicious (small) addition of alloys and, perhaps, special processing techniques such as controlled rolling. This separate product recognition of HSLA steels is reflected by the fact that HSLA steels are generally priced from the base price for carbon steels, not from the base price for alloy steels. Moreover, HSLA steels are often sold on the basis of minimum mechanical properties, with the specific alloy content left to the discretion of the steel producer.

Although HSLA steels are available in numerous standard and proprietary grades, they can be divided into seven categories:

- *Weathering steels*, which contain small amounts of alloying elements such as copper and phosphorus for improved atmospheric corrosion resistance and solid-solution strengthening
- *Microalloyed ferrite-pearlite steels*, which contain very small (generally, less than 0.10%) additions of strong carbide or carbonitride-forming elements such as niobium, vanadium, and/or titanium for precipitation strengthening, grain refinement, and possibly transformation temperature control
- *As-rolled pearlitic steels*, which may include carbon-manganese steels but which may also have small additions of other alloying elements to enhance strength, toughness, formability, and weldability
- *Acicular ferrite (low-carbon bainite) steels*, which are low-carbon (<0.08% C) steels with an excellent combination of high yield strengths, weldability, formability, and good toughness
- *Dual-phase steels*, which have a microstructure of martensite dispersed in a ferritic matrix and provide a good combination of ductility and high tensile strength for sheet forming
- *Inclusion-shape-controlled steels*, which provide improved ductility and through-thickness toughness by small additions of calcium, zirconium, titanium, or perhaps rare-earth elements so that the shape of the sulfide inclusions are changed from elongated stringers to small, dispersed, almost spherical globules
- *Hydrogen-induced cracking-resistant steels* with low carbon, low sulfur, inclusion shape control, and limited manganese segregation, plus copper contents greater than 0.26%

These seven categories are not necessarily distinct groupings, in that an HSLA steel may have characteristics from more than one grouping. For example, all the above types of steels can be inclusion shape controlled. Microalloyed ferrite-pearlite steel may also have additional alloys for corrosion resistance and solid-solution strengthening. A separate category might also be considered for the HSLA 80 (Navy) Ni-Cu-Nb steel (0.04% C, 1.5% Mn, 0.03% Nb, 1.0% Ni, 1.0% Cu, and 0.7% Cr).

Maraging steels comprise a special class of high-strength steels that differ from conventional steels in that they are hardened by a metallurgical reaction that does not involve carbon. Instead, these steels are strengthened by the precipitation of intermetallic compounds at temperatures of about 480 °C (900 °F). The term *maraging* is derived from martensite age hardening and denotes the age hardening of a low-carbon, iron-nickel lath martensite matrix.

Commercial maraging steels are designed to provide specific levels of yield strength from 1030 to 2420 MPa (150 to 350 ksi). Some experimental maraging steels have yield strengths as high as 3450 MPa (500 ksi). These steels typically have very high nickel, cobalt, and molybdenum contents and very low carbon contents. Carbon, in fact, is an impurity in these steels and is kept as low as commercially feasible in order to minimize the formation of titanium carbide (TiC), which can adversely affect strength, ductility, and toughness. Other varieties of maraging steel have been developed for special applications.

The absence of carbon and the use of intermetallic precipitation to achieve hardening produce several unique characteristics that set maraging steels apart from conventional steels. Hardenability is of no concern. The low-carbon martensite formed after annealing is relatively soft—about 30 to 35 HRC. During age hardening, there are only very slight dimensional changes. Therefore, fairly intricate shapes can be machined in the soft condition and then hardened with a minimum of distortion. Weldability is excellent. Fracture toughness (Fig. 2) is considerably better than that of conventional high-strength steels. This characteristic in particular has led to the use of maraging steels in many demanding applications.

Fracture Modes

In steels, fractures occur by any of three basic mechanisms—cleavage, intergranular fracture, and dimpled rupture—or by a mixture of these three mechanisms. The first two generally occur below the ductile-to-brittle transition temperature, whereas dimpled rupture occurs above the transition temperature. These mechanisms are briefly described below. Additional information on fracture mechanisms is described in the article "Micromechanisms of Monotonic and Cyclic Crack Growth" in this Volume.

Cleavage is a transgranular fracture mode by which fracture occurs along crystallographic planes and after which the fracture surface appears as a series of flat planes. The critical event in cleavage fracture is the nucleation of a cleavage facet as it crosses crystallographic boundaries, causing formation of a new facet. One way to improve the toughness of a steel which fails by cleavage is by reduction of the microstructural unit that controls the facet size. In ferritic steels, the facet size is the ferrite grain size (Ref 2), whereas in the multiphase structures of pearlite and bainite the fracture facet size is controlled by the prior austenite grain size (Ref 3).

Low-temperature intergranular fracture can occur by two basic mechanisms: trace-element grain-boundary segregation, or precipitation of films or finely dispersed phases along grain boundaries. Impurity-segregation-induced intergranular fracture is most often called "tem- per embrittlement"; fracture generally occurs along prior-austenite grain boundaries. Fracture by this mode occurs most frequently in alloy steels containing nickel and chromium that have been tempered in the temperature range from 400 to 560 °C (750 to 1050 °F) or cooled slowly through this range.

Intergranular fracture due to grain-boundary precipitation generally results from improper processing. An example is precipitation of grain-boundary films of titanium carbonitrides in maraging steels on slow cooling through the range from 1000 to 750 °C (1830 to 1380 °F) (Ref 4). Also, in low-alloy steels, high-temperature austenitization above 1300 °C (2370 °F) results in manganese sulfide precipitation along the austenite grain boundaries. These particles then act as void nucleation sites for intergranular dimpled rupture (Ref 5). This phenomenon can also occur in overheated welds.

Dimpled rupture is a mode of fracture that occurs above the ductile-to-brittle transition temperature in low-strength steels and over a wide temperature range in high-strength steels. In this type of fracture, voids are nucleated at inclusions, carbides, or other dispersed phases. With increasing plastic strain these voids grow and coalesce, resulting in a fracture surface covered with cuplets of "dimples"—hence the term "dimpled rupture." This fracture mode is also known as "microvoid coalescence" or "plastic fracture."

Dimpled rupture commences by void nucleation at inclusions such as manganese sulfides. Void nucleation events have also been observed at oxides in welds and at titanium carbonitrides in maraging steels. In some materials, such as maraging steel, final fracture occurs by growth and impingement of these voids, resulting in a relatively equiaxed dimple fracture surface. In other materials, the void growth stage is interrupted by the formation of a sheet of voids at small dispersed-phase particles. This mechanism is discussed further in "Effects of Microstructure and Heat Treatment" in this section.

Embrittlement

Several forms of embrittlement in steel can occur during heat treatment or elevated-temperature service. Types of embrittlement include:

- Temper embrittlement
- Tempered martensite embrittlement
- Blue brittleness
- Quench-age embrittlement
- Graphitization
- Strain-age embrittlement
- Aluminum nitride embrittlement

Temper Embrittlement and Tempered Martensite Embrittlement. In many classes of steels, two of the most important metallurgical embrittlement phenomena are temper embrittlement and tempered martensite embrittlement. Temper embrittlement is quite common in slowly cooled heavy sections of steels tempered in the range from 400 to 560 °C (750 to 1050 °F). Here,

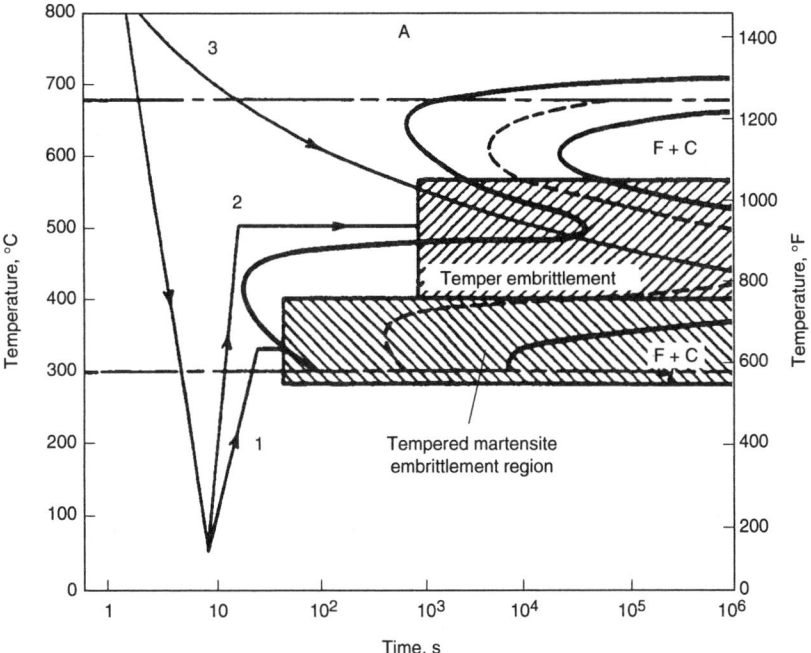

Fig. 5 Heat treatment cycles that could produce (1) tempered martensite embrittlement or (2) and (3) temper embrittlement in a 3340 steel. A, austenite; F, ferrite; C, cementite

segregation of very small amounts of tramp elements such as arsenic, phosphorus, sulfur, antimony, tin, and tellurium may cause weakening of prior-austenite grain boundaries and reductions in notch toughness. For ultrahigh-strength steels, tempered martensite embrittlement arises when tempering is done in the range from 300 to 400 °C (570 to 750 °F), usually because platelike carbides are precipitated along prior-austenite grain boundaries that have already been weakened by impurity segregation during the prior austenitizing treatment (Ref 6). These two types of embrittlement could be observed in the same steel, depending on the thermal cycle. Such a case is indicated for AISI 3340 steel in Fig. 5. Here, thermal heat treatment cycle 1 could produce tempered martensite embrittlement, whereas cycle 2 or cycle 3 could produce temper embrittlement.

Blue Brittleness. Carbon steels generally exhibit an increase in strength and a reduction of ductility and toughness at temperatures around 300 °C (570 °F). Because such temperatures produce a bluish temper color on the surface of the specimen, this problem has been called blue brittleness. It is generally believed that blue brittleness is an accelerated form of strain-age embrittlement. Deformation in the blue-heat range followed by testing at room temperature produces an increase in strength that is greater than when the deformation is performed at ambient temperature. Blue brittleness can be eliminated if elements that tie up nitrogen are added to the steel (e.g., aluminum or titanium).

Quench-Age Embrittlement. If a carbon steel is heated to a temperature slightly below its lower critical temperature and then quenched, it will become harder and stronger but less ductile. This problem has been called quench aging or quench-

age embrittlement. The degree of embrittlement is a function of time at the aging temperature. Aging at room temperature requires several weeks to reach maximum embrittlement. Lowering the quenching temperature reduces the degree of embrittlement. Quenching from temperatures below about 560 °C (1040 °F) does not produce quench-age embrittlement. Carbon steels with a carbon content of 0.04 to 0.12% appear to be most susceptible to this problem. Quench aging is caused by the precipitation of carbide and/or nitride from solid solution.

Temper Embrittlement of Steels

Temper embrittlement is a major cause of degradation of toughness of ferritic steels. Numerous components otherwise in sound condition become candidates for retirement if they are severely embrittled. The problem is encountered as a result of exposure of a steel in the temperature range 345 to 540 °C (650 to 1000 °F). Tempering, postweld heat treatments, or service exposure in this range must be avoided. The problem may be avoided by heat treating above this range, followed by rapid cooling. Unfortunately, in the case of massive components such as rotors, no rate of cooling is fast enough and some residual embrittlement may be inevitable. Subsequent to heat treatment, exposure of the component during service in the critical range can also lead to embrittlement.

Many steel components in a plant invariably are exposed to the critical temperature range during service and hence embrittlement cannot be avoided. Some examples are boiler headers, steam pipes, turbine casings, pressure vessels,

blades, rotors, and combustion turbine disks. In the case of components that are subject to embrittlement, such as HP/IP rotors and pressure vessels, restrictions are imposed on the startup and shutdown procedures. To avoid the risk of brittle failure during these transients, loading is avoided until a certain temperature has been reached. For instance, rotors are prewarmed up to a certain temperature before loading. Similarly, pressure vessels may sometimes need to be depressurized during shutdown prior to reaching a certain temperature. These requirements result in additional operational and maintenance costs and loss of production. Temper embrittlement phenomena thus adversely affect the longevity, reliability, cycling ability, efficiency, and operating costs of high-temperature equipment. An excellent review of this phenomenon and its characteristics has been published by McMahon (Ref 7).

It is well established at this time that segregation of tramp elements (antimony, phosphorus, tin, and arsenic) to prior-austenite grain boundaries in steel is the principal cause of temper embrittlement. Until the advent of the Auger electron spectroscope in the mid-1960s, no conclusive evidence of such segregation could be obtained. In decreasing order of effect, temper embrittlement is caused by impurity contents of antimony, phosphorus, tin, and arsenic. Antimony is generally not present in large quantities in commercial steels and is therefore neglected from consideration. Arsenic is not a potent embrittler and hence is not very important. Phosphorus and tin are, therefore, the major residual elements of concern.

Among the alloying elements, manganese, silicon, nickel, and chromium are known to exacerbate the effects of impurities. When these elements are present in combination, the effect is further increased. It is well known that nickel and chromium in combination increase embrittlement significantly more than either element alone. For this reason, Ni-Cr-Mo-V steel rotors are considered to be much more susceptible to embrittlement than Cr-Mo-V steel rotors, as shown in Fig. 6 (Ref 8).

For a given class of steels, manganese and silicon have the major influence (Fig. 7). The data show considerable synergism among manganese, silicon, phosphorus, and tin. The maximum embrittlement is observed when all these elements are present together. Because sulfur levels formerly could not be minimized, the presence of manganese was always necessary for sulfur control. Silicon was generally added for deoxidation.

In modern practice, silicon can be eliminated by replacing it as a deoxidant by alternate deoxidation processes such as vacuum carbon deoxidation. Manganese levels can be reduced commensurate with lower sulfur levels. Control of phosphorus and tin to much lower levels can be achieved by careful selection of scrap iron and better steelmaking practices.

A combination of all these improvements has brought the temper-embrittlement problem under greater control in recent years. Various compositional factors for prediction of temper-embrittle-

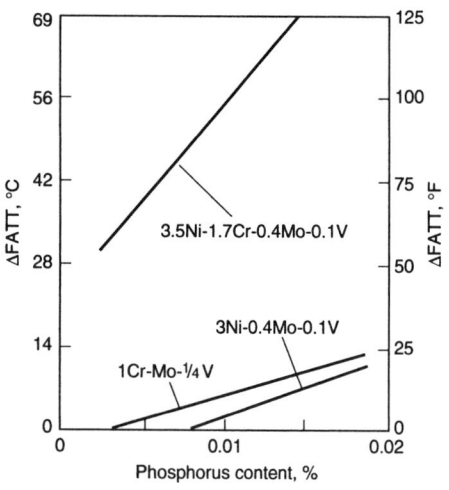

Fig. 6 Effect of phosphorus content on the temper embrittlement (ΔFATT) of three step-cooled forging steels. Source: Ref 8

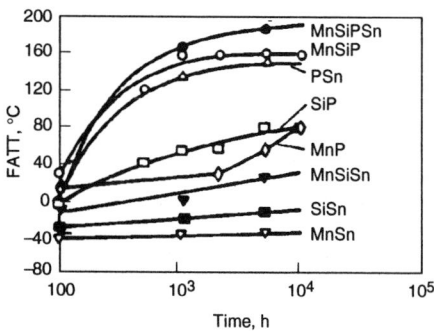

Fig. 7 Effects of manganese, silicon, phosphorus, and tin on the kinetics of temper embrittlement at 480 °C. (895 °F) for a 2¼Cr-1Mo steel. Source: Ref 9, 10

ment susceptibilities have evolved over the years (Ref 11-13). The effects of various alloying elements and their potential roles in temper embrittlement are summarized in Table 3.

Effects of Microstructural Factors. In heavy-section components such as turbine rotors, casings, and steam chests, inhomogeneities in chemical composition as well as thermal gradients during heat treatment often result in nonuniformities in the microstructure. Hence, the risk of temper embrittlement can vary with location in the component. There have been few systematic studies of these effects. Another important variable affecting susceptibility to temper embrittlement is grain size. Unfortunately, it is difficult to isolate grain size effects, because variations in grain size often result in other microstructural changes due to the effect of grain size on hardenability.

Failure Analysis of Temper-Embrittled Components. Identification of temper embrittlement as a failure mechanism in a failed component is relatively easy. The presence of large amounts of phosphorus, tin, manganese, and silicon in the steel is an indication that temper embrittlement

Table 3 Role of composition in temper embrittlement of steels

Important impurities

Tin, phosphorus	Ni-Cr-base steels (e.g., 3.5Ni-Cr-Mo-V)
Phosphorus	Cr-Mo-base steels (e.g., 2¼Cr-1Mo; Cr-Mo-V)

Major effects of alloying elements

Nickel	Raises inherent resistance of steel to brittle fracture; promotes segregation of tin and silicon (and antimony, if present)
Chromium	Imparts hardenability; imparts some resistance to softening at elevated temperatures; promotes segregation of phosphorus
Manganese	Imparts hardenability; scavenges sulfur; promotes segregation of phosphorus
Silicon	Deoxidizes; promotes segregation of phosphorus
Molybdenum	Imparts (bainitic) hardenability; imparts resistance to softening; scavenges phosphorus and tin
Vanadium	Imparts resistance to softening; aids in grain refinement
Niobium	Imparts resistance to softening; scavenges phosphorus; aids in grain refinement

Source: Ref 14

may be involved. Fractography generally indicates an intergranular fracture with little evidence of ductile dimples on the fracture surface. The extent of the intergranular fracture present is a function of the temperature at which the fracture is produced and the microstructure. It has been observed that a plot of intergranular fracture (%) versus Charpy test temperature for embrittled Cr-Mo-V steels resembles a bell-shape curve with

the maximum intergranular fracture being observed at 50% of the fracture-appearance transition temperature (Ref 15). For a given degree of embrittlement, more intergranular fracture is observed in martensitic structures than in bainitic structures (Ref 15).

Time-Temperature Relationships for Temper Embrittlement. Temper embrittlement obeys a C-curve behavior in time-temperature plots. At high temperatures, the kinetics of impurity diffusion to grain boundaries are rapid, but the tendency to segregate is low because the matrix solubility for the element increases with temperature. Hence, embrittlement occurs rapidly but to a small degree. At low temperatures, the tendency to segregate is high, but the diffusion kinetics are not rapid enough to reach maximum embrittlement. The optimum combination of thermodynamic and kinetic factors favoring embrittlement occurs at some intermediate temperature, called the "knee" of the C-curve. For commercial steels of interest, the knee occurs in the temperature range from 455 to 510 °C (850 to 950 °F) but can be shifted up or down depending on the composition, grain size, and microstructure of the steel.

Instability of the microstructure complicates the picture further. With increasing exposure, the carbide structures and compositions evolve into more stable configurations with concomitant changes in the ferrite matrix. It is therefore difficult to represent temper embrittlement kinetics in terms of rate processes with unique activation energies. Attempts to predict long-time behavior based on short-term evaluations have met with little success. For practical purposes, however, it is not uncommon to assume parabolic kinetics, under isothermal conditions.

Fracture Toughness

The concepts of fracture mechanics are concerned with the basic methods for predicting the load-carrying capabilities of structures and components that contain cracks. The fracture mechanics approach is based on a mathematical description of the characteristic stress field that surrounds any crack in a loaded body. When the region of plastic deformation around a crack is small compared to the size of the crack (as is often true for large structures and high-strength materials), the magnitude of the stress field around the crack is commonly expressed as the stress-intensity factor, K, such that:

$$K = \sigma \sqrt{(a)} \, Y \qquad \text{(Eq 1a)}$$

where σ is remotely applied stress, a is characteristic flaw size dimension, and Y is a geometry factor determined from linear elastic stress analysis. The stress-intensity factor, K, thus represents a single parameter that includes both the effect of the stress applied to a sample and the effect of a crack of a given size in a sample. For example, in mode I

loading (i.e., tensile loading perpendicular to the plane of the crack), the stress-intensity factor, K_I, for a crack tip in any body is given by the relationship:

$$K_I = \sigma \sqrt{\pi a} \, f(g) \qquad \text{(Eq 1b)}$$

where a is crack length and $f(g)$ is a function that accounts for crack geometry and structural configuration. This general relationship makes it possible to translate laboratory results into practical design information without the need for extensive service experience or correlations. Other relations between the stress-intensity factor and various body configurations; crack sizes, shapes and orientations; and loading conditions are available in the published literature. In this way, linear elastic analysis of small-scale yielding can be used to define a unique factor, K, that is proportional to the local crack-tip stress field outside the small crack-tip plastic zone.

An underlying principle of fracture mechanics is that unstable fracture occurs when the stress-intensity factor at the crack tip reaches a critical value, K_c. The fracture toughness, K_c, varies with the degree of localized constraint to plastic flow along the tip of the fatigue crack. Thus, cracks in very thick members are subjected to higher constraints than cracks in thinner members. The maximum constraint, as defined in ASTM E399, occurs under plane-strain conditions and results in the lowest value of fracture toughness, K_{Ic} (Fig. 8). Under identical test conditions, the K_c values for thinner plates are usually higher than those observed under plane-strain conditions (i.e., $K_c > K_{Ic}$).

By knowing the critical value of K_I at failure (K_c, K_{Ic}, or K_{Id}) for a given material of a particular thickness and at a specific temperature and loading rate, the designer can determine flaw sizes that can be tolerated in structural members for a given design stress level. Conversely, the designer can determine the design stress level that can be safely used for an existing crack that may be present in a structure.

In general, fracture-toughness property varies with constraint such that:

- K_c = critical stress-intensity factor for static loading and plane-stress conditions of variable constraint (i.e., this value depends on specimen thickness and geometry, as well as on crack size)
- K_{Ic} = critical stress-intensity factor for static loading and plane-strain conditions of maximum constraint
- K_{Id} = critical stress-intensity factor for dynamic (impact) loading and plane-strain conditions of maximum constraint
- K_c, K_{Ic}, or $K_{Id} = C\sigma\sqrt{a}$

where C is a constant that is a function of specimen and crack geometry, σ is nominal stress in ksi, and a is flaw size. Each of these values (K_c, K_{Ic}, and K_{Id})

is also a function of temperature, particularly for those structural materials exhibiting a transition from brittle to ductile behavior.

The parameter K_{Ic} is a true material property in the same sense as is the yield strength of a material. The value of K_{Ic} determined for a given material is unaffected by specimen dimensions or type of loading, provided that the specimen dimensions are large enough relative to the plastic zone to ensure plane-strain conditions around the crack tip (strain is zero in the through-thickness or z-direction). Tests on precracked specimens of a wide variety of materials have shown that the critical K value at the onset of crack extension

approaches a constant value as specimen thickness increases, as shown in Fig. 8 for 4340 steel. In general, when the specimen thickness and the inplane dimensions near the crack are large enough relative to the size of the plastic zone, then the K-value for the onset of crack growth approaches a constant, minimum value, K_{Ic}, of the material. The thickness, B, required to ensure valid plane-strain behavior is given by:

$$B \geq 2.5\left(\frac{K_{Ic}}{\sigma_{ys}}\right)^2 \qquad \text{(Eq 2)}$$

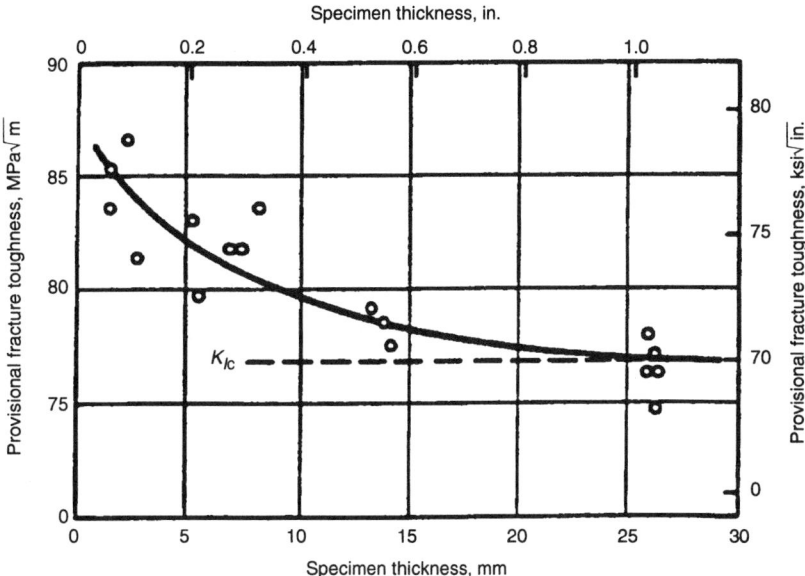

Fig. 8 Critical fracture stress vs. section size of 4340 steel tempered at 400 °C (750 °F) to a yield strength of 1470 MPa (213 ksi). Source: Ref 16

Table 4 Room-temperature fracture toughness of various structural steels

Alloy name (nominal, wt %)	Processing and product form	Condition or heat treatment	Ultimate tensile strength, MPa	Tensile yield strength, MPa	K_c, MPa√m	K_{Ic}, MPa√m	Specimen and load type	Specimen thickness	Source (a)
Mn-Cr-Mo-V steel (1.5% Mn)	...	...	...	372 (54 ksi)	...	63 (b)	Tension on 75 × 19 mm edged crack specimen	9 mm (0.354 in.)	GPC
Mn-Cr-Mo-V steel (1.5% Mn)	...	...	...	372 (54 ksi)	...	79 (c)	Tension on 75 × 19 mm edged crack specimen	9 mm (0.354 in.)	GPC
1340 (0.4 C, 1.75 Mn, 0.28 Si)	...	...	1517 (220 ksi)	...	...	65	...	...	SAH
1340	...	...	1920 1862 1715 1568 1500	...	50-63 39 50 57-82 65	...	...	...	GPC
2340	...	...	1960 1852 1715 1578 1509	...	77 65-76 62 76 87	...	...	...	GPC
3140	...	...	1862 1646	...	56 54-62	...	...	...	GPC
4140	Forged bar 1.6 mm (0.62 in.) oil quenched from 870 °C (1600 °F) and 1 h temper	205 °C temper 395 °C temper 280 °C temper	...	1448 1517 1585	...	44 55 55	CT (L-T)	15 mm (0.6 in.)	DTH

(continued)

(a) Sources: GPC, C.P. Cherepanov, Mechanics of Brittle Fracture, McGraw-Hill, 1979; SAH, *Structural Alloys Handbook*, Volume 1, Metals and Ceramics Information Center, 1987; DTH, *Damage Tolerance Handbook*, Volume 2, MCIC-HB-018, Metals and Ceramics Information Center, Dec 1983. (b) Crack parellel to fibers. (c) Crack perpendicular to fibers. CT, compact tension specimen with standard nomenclature for specimen orientation (L-T, S-L, T-L)

Table 4 Continued

Alloy name (nominal, wt %)	Processing and product form	Condition or heat treatment	Ultimate tensile strength, MPa	Tensile yield strength, MPa	K_c, MPa√m	K_{Ic}, MPa√m	Specimen and load type	Specimen thickness	Source (a)
4140	Plate 25 mm (1 in.) 870 °C (1600 °F) 1 h + 843 °C (1550 °F) 1 h, oil quench, 1 h temper	482 °C temper	...	1213	...	75	CT(T-L)	25 mm (1 in.)	DTH
		425 °C temper		1365		44			
		482 °C temper		1096		93			
		425 °C temper		1207		64.5			
4330 (modified)	...	232 °C temper	...	...	...	57	...	...	GPC
4340	...	260 °C temper	...	1640 (238 ksi)	...	48.5 ± 2	35 bend specimens	...	Ref 18
4340	...	425 °C temper	...	1420 (206 ksi)	...	87 ± 4	37 bend specimens	...	Ref 18
4340	...	260 °C temper	...	1640 (238 ksi)	...	50 ± 2.75	46 CT specimens	...	Ref 18
4340	...	425 °C temper	...	1420 (206 ksi)	...	87 ± 3	48 CT specimens	...	Ref 18
4340	Plate 16 mm (0.62 in.)	Heat treated to 51 HRC	...	1517 (220 ksi)	...	57	CT (L-T)	25 mm (1 in.)	DTH
4340	Round bar 115 mm (4.5 in.) diam	...	1240 (180 ksi)	1330 (193 ksi)	...	117.5	CT (L-T)	25 mm (1 in.)	DTH
4340	Plate 25 mm (1 in.) thick, oil quenched from 843 °C (1550 °F)	425 °C temper	...	1420	...	84	CT (L-T)	20 mm (0.8 in.)	DTH
		260 °C temper		1640		49			
4340	255 mm (10 in.) billet oil quenched and tempered	425 °C temper	...	1360-1455	...	83-89	CT (L-T)	25 mm (1 in.)	DTH
4340	Forged bar 16 mm (0.62 in.) oil quenched from 870 °C (1600 °F) and 1 h temper	205 °C temper	...	1345	...	65	CT (T-L)	15 mm (0.6 in.)	
		395 °C temper		1448		100			
		345 °C temper		1495		87.7			
		280 °C temper		1503		67			
4340	...	426 °C temper	...	...	...	60	...	...	GPC
4340	Vacuum melting with consumable electrode	149 °C temper	2332	1450	51	...	...	2.54 mm (0.1 in.)	GPC
		204 °C temper	2038	1470	100				
		260 °C temper	1813	1540	161				
4340	Vacuum melting with consumable electrode	316 °C temper	1764	1530	201	...	...	2.54 mm (0.1 in.)	GPC
		371 °C temper	1636	1558	>254				
		427 °C temper	1430	1333	>225				
4340	...	...	1920	...	94	...	...	...	GPC
			1783		218				
			1646		196				
4340	Oil quenched bar, 25 mm (1 in.)	540 °C temper	1260	1172	...	110	...	...	SAH
		425 °C temper	1530	1380		75			
		205 °C temper	1950	1640		53			
5140	...	...	1920	...	55	...	...	...	GPC
			1852		58				
			1607		71				
			1509		72				
5150	...	...	1517 (220 ksi)	...	...	71	...	...	SAH
HP9-4-0.2	Forging 75 mm (3 in.) thick	Annealed	...	1303 (189 ksi)	...	118-145 132 (avg)	CT (L-T)	25-50 mm (1-2 in.)	DTH
HP9-4-0.2	Forging 85-180 mm (3.4-7 in.) thick	Heat treated	...	1275-1365 (185-198 ksi)	...	148-165 (154 avg)	CT (L-T)	38 mm (1.5 in.)	DTH
HY TUF	Forging 165 mm (6.5 in.) thick	Heat treated	...	1365 (198 ksi)	...	124-132 (122 avg)	CT (L-T)	50 mm (2 in.)	DTH
Super Hy Tuf	...	...	1764	...	57	...	...	...	GPC
			1617		32-43				
HY140	25 mm (1 in.) plate, quenched and tempered	540 °C temper	1027 (149 ksi)	980 (142 ksi)	...	~275	...	...	SAH
H-11	Vacuum melted	Quenched-and-tempered bar	1793	1448	...	38	...	...	SAH
			1930	1585		30			
H-11	Vacuum melted	Quenched-and-tempered sheet	1793	1448	...	77	...	...	SAH
			1930	1517		60			
300M	Quenched-and-tempered plate	425 °C temper	1793	1585	...	46	...	...	SAH
		540 °C temper	1585	1448		66			
300M	Quenched-and-tempered sheet	425 °C temper	1793	1585	...	104	...	...	SAH
		540 °C temper	1585	1413		...			
300M	...	...	1920	...	145	...	...	...	GPC
			1783		212				
D6ac	Plate	Heat treated to 46 HRC	...	1420 (206 ksi)	...	86	Three-point bending (T-L)	17.78 mm (0.7 in.)	DTH
D6ac plate	900 °C(1650 °F) ausbay quench, 525 °C (975 °F) slack quench, 163 °C (325 °F)	Aging at 540 °C (1000 °F)	...	1495 (217 ksi)	...	49-91 (L-T) (73.5 avg)	CT(L-T)	38 mm (1.5 in.)	DTH
D6ac forging	900 °C(1650 °F) ausbay quench, 525 °C (975 °F) slack quench, 205 °C (400 °F)	Aging at 540 °C (1000 °F)	...	1475 (214 ksi)	...	55-105 (72.5 avg)	CT(L-T)	16.5 mm (0.65 in.)	DTH

(continued)

(a) Sources: GPC, C.P. Cherepanov, *Mechanics of Brittle Fracture*, McGraw-Hill, 1979; SAH, *Structural Alloys Handbook*, Volume 1, Metals and Ceramics Information Center, 1987; DTH, *Damage Tolerance Handbook*, Volume 2, MCIC-HB-018, Metals and Ceramics Information Center, Dec 1983. (b) Crack parellel to fibers. (c) Crack perpendicular to fibers. CT, compact tension specimen with standard nomenclature for specimen orientation (L-T, S-L, T-L)

Table 4 Continued

Alloy name (nominal, wt %)	Processing and product form	Condition or heat treatment	Ultimate tensile strength, MPa	Tensile yield strength, MPa	K_c, MPa√m	K_{Ic}, MPa√m	Specimen and load type	Specimen thickness	Source (a)
D6ac plate	925 °C (1700 °F) ausbay quench, 525 °C (975 °F) oil quench, 60 °C (140 °F)	Aging at 540 °C (1000 °F)	...	1495 (217 ksi)	...	77-111 101 (avg)	CT(L-T)	38 mm (1.5 in.)	DTH
D6ac	...	...	...	1645 (238.5 ksi)	108-229	...	...	2.0 mm (0.787 in.)	GPC
D6ac	...	...	2058 1920 1783 1646	...	65-84 64-76 107 142	...	...	...	GPC
D6ac	Quenched-and-tempered plate (L)	260 °C temper 540 °C temper	2000 1585	1724 1482	...	33 66	...	...	SAH
AF1410	Water quenched 50 mm (2 in.) plate	510 °C age for 5 h, air cooled	...	1572 (228 ksi)	...	140(L-T) 137(T-L)	CT	44.45 mm (1.750)	DTH
AM 355	...	...	...	...	61-63	...	...	...	GPC
AMS 6434	...	...	2058 1920 1783 1646	...	98 131 131 218	...	...	...	GPC
PH15-7Mo	...	...	1646 (239 ksi)	...	...	84	...	...	GPC
AGCX-7 (0.38 C, 0.65 Mn, 1.65 Si, 1.38 Ni, 1.78 Cr, 0.64 Mo, 0.075 V)	...	...	1764-1842	1568	245-260	...	...	2.0 mm (0.0787 in.)	GPC
VL-1D steel (0.30 C, 0.90 Si, 0.92 Mn, 0.56 Cr, 1.0 Ni, 0.49 Mo, 1.07 W)	Hardened from 930 °C with air cooling, tempered 3 h	210 °C temper 450 °C temper 600 °C temper No temper	1675 1490 1078 1695	1372 1254 960 1430	184 173 170 163	...	Tension on specimen with central surface crack	2.0 mm (0.0787 in.)	GPC
VKS-1 steel (0.39 C, 1.08 Si, 0.83 Mn, 0.65 Ni, 0.53 Mo, 0.07 V)	Hardened from 940 dc with air cooling, tempered 3 h	270 °C temper No temper	1813 2097	1538 1656	77.5 32	...	Tension on specimen with central surface crack	2.0 mm (0.0787 in.)	GPC
40KhN steel	...	...	...	1391 (200 ksi)	...	70	...	...	GPC
35GS steel	...	...	...	882 (128 ksi)	...	56	...	...	GPC
U8 high-carbon steel (0.8 C, 0.25 Mn, 0.25 max Ni, 0.2 max Cr)	Hardened from 810 °C with oil	190 °C temper	...	...	...	18	Tension on 360 × 180mm specimen with central crack	2.0 mm (0.0787 in.)	GPC
9Ni-4Co-25	Quenched-and-tempered plate	315 °C temper 540 °C temper	1482 1380	1345 1310	...	82 132	...	...	SAH
9Ni-4Co-45	Quenched-and-tempered plate	315 °C temper 540 °C temper	1930 1482	1758 1413	...	49 104	...	...	SAH
N18 maraging steel	...	Aged 4 h at 490 °C	1773 1862	1695 1803	...	84(b) 117(c)	Three-point bending above edge crack	19 mm (0.75 in.)	GPC
Ni-Co-Mo maraging steel (18% Ni)	Two melts	Aged 3 h at 482 °C	...	2048 (297 ksi)	...	136 (melt 1) 104 (melt 2)	Tension on 25 × 305 mm specimen with edge crack	3.6 mm (0.14 in.)	GPC
Ni-Co-Mo maraging steel (18% Ni)	Two melts	Aged 3 h at 482 °C	...	2048 (297 ksi)	...	123 (melt 1) 80 (melt 2)	Tension on 25 × 305 mm specimen with central crack	3.6 mm (0.14 in.)	GPC
Ni maraging steel (18% Ni)	...	...	2058 1920 1783 1715 1646	...	190 237 174 204 221			...	GPC
N18 maraging steel	...	...	...	1920	...	85	Tension on 76 mm wide specimen with central crack	4.0 mm (0.157 in.)	GPC
N18 maraging steel	...	...	...	1920	...	86	Extension, 76 mm wide with edge crack	4.0 (0.157)	GPC
N18 maraging steel	...	...	...	1920 (278 ksi)	...	92	Three-point bending above edge crack	4.0 mm (0.157 in.)	GPC
N18 maraging steel	...	...	...	1920 (278 ksi)	...	90	Pure (four-point) bending with edge crack	4.0 mm (0.157 in.)	GPC
N18 maraging steel	Vacuum melted	...	2058 (298 ksi)	...	217	93	...	...	GPC
18Ni-Marage 200 (0.1Al, 8.5 Co, 3.25 Mo, 18.5 Ni, 0.2 Ti, 0.10 Mn, 0.1 Si)	...	Plate aged at 480 °C	1550	1482	...	120-154	...	...	SAH

(continued)

(a) Sources: GPC, C.P. Cherepanov, *Mechanics of Brittle Fracture*, McGraw-Hill, 1979; SAH, *Structural Alloys Handbook*, Volume 1, Metals and Ceramics Information Center, 1987; DTH, *Damage Tolerance Handbook*, Volume 2, MCIC-HB-018, Metals and Ceramics Information Center, Dec 1983. (b) Crack parellel to fibers. (c) Crack perpendicular to fibers. CT, compact tension specimen with standard nomenclature for specimen orientation (L-T, S-L, T-L)

Table 4 Continued

Alloy name (nominal, wt %)	Processing and product form	Condition or heat treatment	Ultimate tensile strength, MPa	Tensile yield strength, MPa	K_c, MPa√m	K_{Ic}, MPa√m	Specimen and load type	Specimen thickness	Source (a)
18Ni-Marage 250 (18 Ni, 7.75 Co, 4.9 Mo, 0.40 Ti, 0.10 Al)	Vacuum melted sheet and plate aged at 480 °C	Plate (T)	1634	1565	...	75	...	...	SAH
		Plate (L)	1765	1634		91			
		Sheet (T)	1855	1813		77-134			
18Ni-Marage 300	Vacuum melted sheet and plate aged at 480 °C	Plate (T)	1980	1910	...	75	...	...	SAH
		Plate (L)	1985	1925		79			
		Sheet (T)	1958	1875		92			
18Ni-Marage 350	Vacuum melted	25 mm (0.5 in.) plate aged at 480 °C	2427	2380	...	49	...	...	SAH

(a) Sources: GPC, C.P. Cherepanov, Mechanics of Brittle Fracture, McGraw-Hill, 1979; SAH, *Structural Alloys Handbook*, Volume 1, Metals and Ceramics Information Center, 1987; DTH, *Damage Tolerance Handbook*, Volume 2, MCIC-HB-018, Metals and Ceramics Information Center, Dec 1983. (b) Crack parellel to fibers. (c) Crack perpendicular to fibers. CT, compact tension specimen with standard nomenclature for specimen orientation (L-T, S-L, T-L)

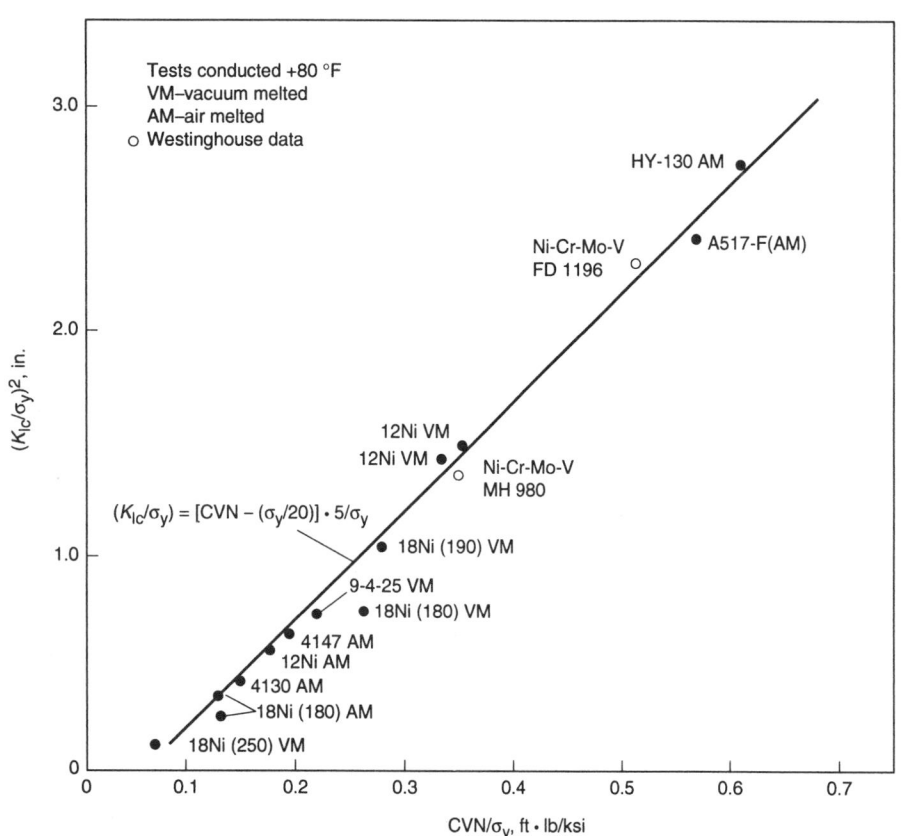

Fig. 9 Relation between K_{Ic} and Charpy V-notch values in the upper-shelf region. Source: Ref 17

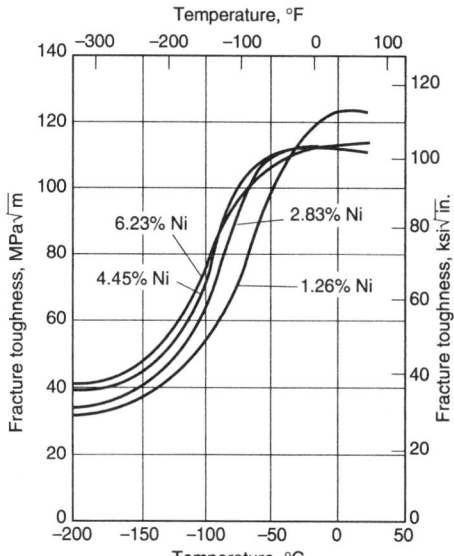

Fig. 10 Effect of nickel content on fracture toughness of high-strength steels containing (in wt%) 0.35 C, 0.65 Mn, 0.35 Si, 0.80 Cr, 0.30 Mo, 0.10 V, and various amounts of nickel; all steels hardened to a yield strength of 1175 MPa (170 ksi). Source: Ref 19, 20

Typical room-temperature toughness of various structural steels is summarized in Table 4.

Traditionally, however, the notch toughness of low- and intermediate-strength steels is described in terms of Charpy V-notch (CVN) test results. The CVN impact specimen is the most widely used specimen for material development, specifications, and quality control. The relationship between K_{Ic} and CVN impact toughness has been developed empirically by Barsom and Rolfe (Ref 17) for steel with room-temperature yield strengths higher than about 760 MPa (110 ksi). The correlation is shown in Fig. 9 and is given by the equation:

$$\left(\frac{K_{Ic}}{\sigma_{ys}}\right)^2 = \frac{5}{\sigma_{ys}}\left(CVN - \frac{\sigma_{ys}}{20}\right) \qquad \text{(Eq 3)}$$

where K_{Ic} is in ksi · in.$^{1/2}$; σ_{ys} is in ksi; and CVN is energy absorption, in ft · lb, for a Charpy V-notch impact specimen tested in the upper-shelf (100% shear fracture) region.

Effect of Steel Composition and Condition

Among the metallurgical variables that affect toughness, the significant ones are yield strength, microstructure, grain size, content of inclusions, and impurities. Generally, an increase in yield strength results in a decrease in K_{Ic}. Among the various transformation products, tempered martensite exhibits the highest toughness, followed by bainite, followed by ferrite-pearlite structures.

It is important to note that for a given alloy chemistry or microstructural condition, the fracture toughness measured is highly reproducible for tests at different laboratories and for tests with different specimens, as shown for 4340 in Table 4 (Ref 18). The general effects of alloy chemistry and microstructural condition are described below. The next section describes important test variables such as temperature and strain rate.

Effect of Nickel on Toughness. As indicated for maraging steels, nickel improves toughness, even though it can be very difficult to separate the effects of chemistry on fracture toughness from the effects of other variables (e.g., hardenability).

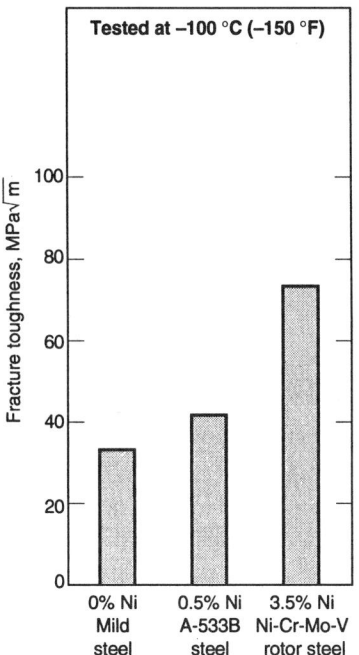

Fig. 11 Effect of nickel content on fracture toughness of lower-strength steels with yield strengths ranging from 500-700 MPa (73-102 ksi) at –100 °C (–150 °F). Source: Ref 21

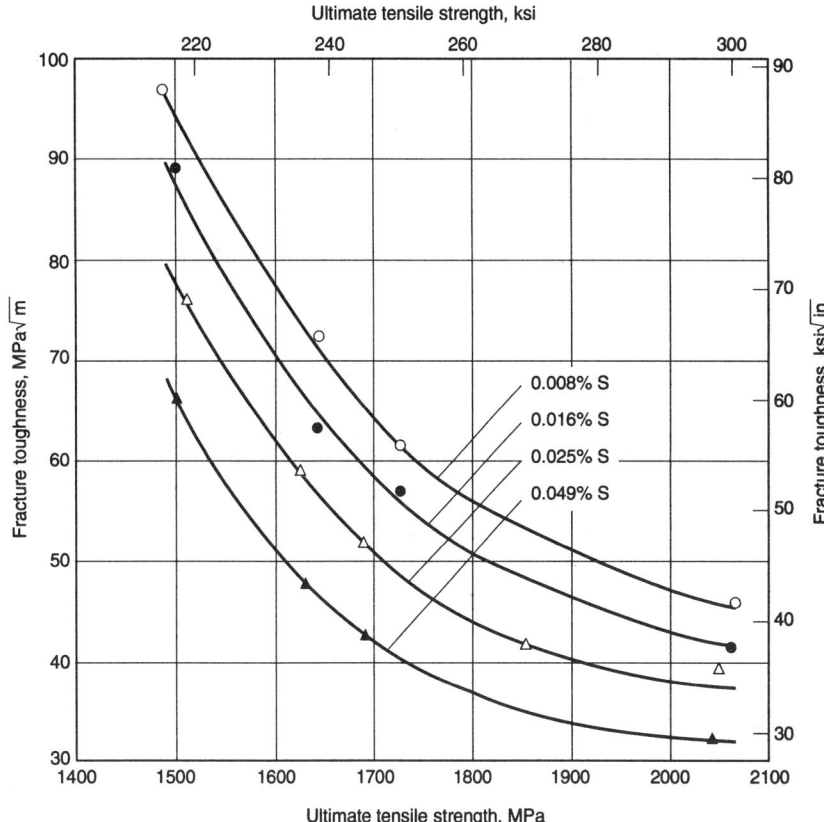

Fig. 13 Effect of sulfur content on fracture toughness of 4345 steel hardened and tempered to various strength levels. Source: Ref 22

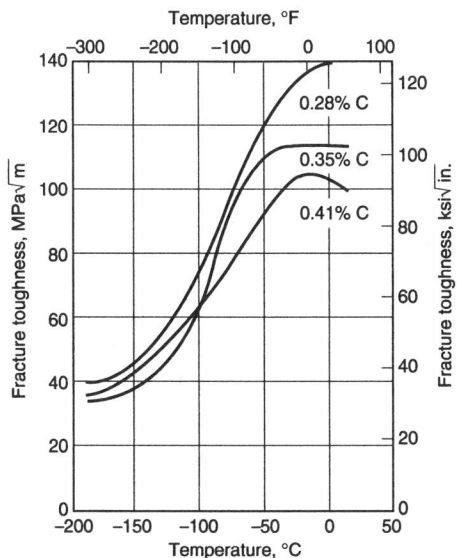

Fig. 12 Effect of carbon content and testing temperature on fracture toughness of alloy steels. Data are for steels containing (in wt%) 0.65 Mn, 0.35 Si, 0.80 Cr, 3.00 Ni, 0.30 Mo, 0.10 V, and various amounts of carbon; all steels were hardened and tempered to a yield strength of approximately 1175 MPa (170 ksi). Source: Ref 19, 20

In one series of experiments in which strength was held constant (Ref 19, 20), an increase in nickel content of about 5% increased fracture toughness by about 50% in the transition temperature range, for a steel with a yield strength of 1175 MPa (170 ksi) at room temperature (Fig. 10). On the other hand, for lower-strength steels with yield strengths ranging from 500 to 700 MPa

(73 to 102 ksi) at –100 °C (–150 °F), addition of 3.5% Ni increased fracture toughness by more than 100% (Ref 21), as illustrated in Fig. 11. It should be noted that moderate additions of nickel mainly improve low-temperature fracture toughness through resistance to cleavage, and that little effect may be seen at higher temperatures at which microvoid coalescence is the fracture mode. Such is the case in Fig. 10 at temperatures in the upper-shelf region.

Effect of Carbon on Ductile-to-Brittle Transition Temperature. It is well known that decreasing the carbon content decreases the transition temperature in ferrite-pearlite steels (*ASM Handbook*, Vol 1). This trend is less prevalent in ultrahigh-strength steels if strength is maintained at a constant level by tempering. The effect of carbon content on the fracture toughness transition curves for alloy steels containing 0.28, 0.35, and 0.41% C, with strength maintained at 1175 MPa (170 ksi), is illustrated in Fig. 12. There is little effect in the transition temperature range; however, there is a substantial effect in the upper-shelf region, where microvoid coalescence is the fracture mode.

Nonmetallic inclusions (single inclusions or clusters of inclusions) reduce fracture toughness, particularly at high strength levels. The effect of sulfide inclusion content (which is a function of sulfur content) on fracture toughness of 4345 steel is shown in Fig. 13 (Ref 22). Because of this effect, special melting and processing require-

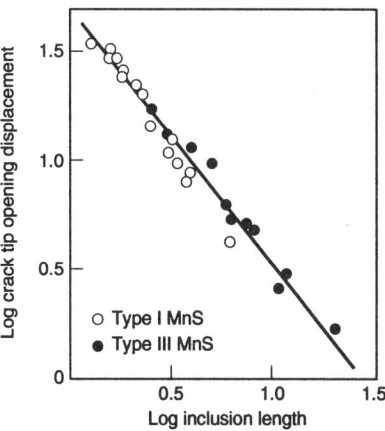

Fig. 14 Relationship between projected inclusion length per unit area and crack-tip opening displacement to fracture in sulfur-bearing steels

ments are often specified when alloy steels are selected for certain critical applications, such as aircraft landing gear. Vacuum arc remelting, vacuum induction remelting, and electroslag remelting are three of the special processes that normally produce cleaner steels than more common steelmaking processes.

Not only does sulfur affect ultrahigh-strength steels, but it may also affect grades of low to medium strengths (Ref 23). This is largely a mi-

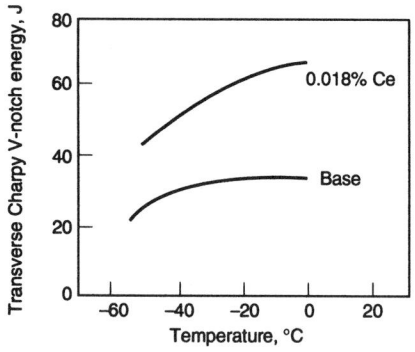

Fig. 15 Effect of rare-earth additions on impact properties of Al-Si killed X65 pipeline steel.

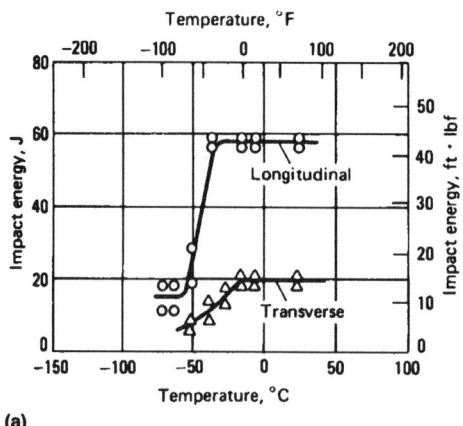

(a)

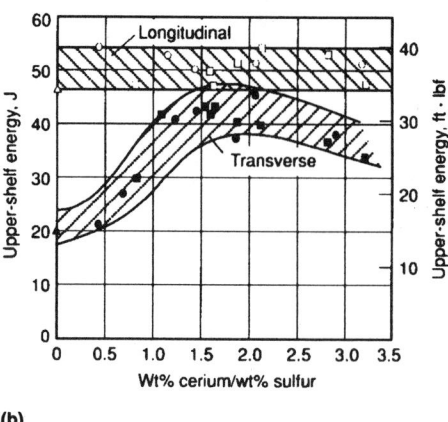

(b)

Fig. 17 Effect of sulfide shape control on transverse toughness of structural steels. (a) Typical transition behavior for HSLA steel without inclusion shape control. Data determined on half-size Charpy V-notch test specimens. (b) Effect of cerium-to-sulfur ratio on upper-shelf impact energy for HSLA steel. Circles, steel treated with mischmetal; squares, steel treated with rare-earth silicides

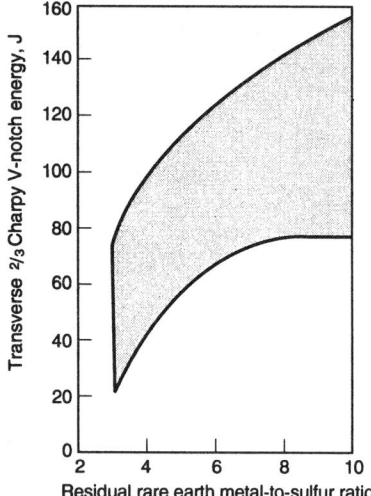

Fig. 16 Relationship between transverse Charpy V-notch energy and rare-earth metal: sulfur ratio (⅔-size Charpy specimens) at –18 °C (–1 °F). Source: Ref 25

crovoid nucleation mechanism, and if the sulfide inclusion length is plotted versus crack-tip opening displacement, an inverse relationship is revealed (Fig. 14). The critical crack-tip opening displacement, δ_c, is a measure of fracture toughness. Under plain-strain conditions, an approximate relation between δ_c and K_{Ic} is:

$$\delta_c \cong \frac{K_{Ic}^2}{2\sigma_{ys}E} \qquad (Eq\ 4)$$

where σ_{ys} is yield strength and E is Young's modulus. Thus, one means of improving toughness is to reduce the projected inclusion length by controlling the shape of the inclusions (Ref 23). This can be achieved by adding rare earths, which combine with MnS to form hard-to-deform particles comprised of sulfides, oxides, and/or oxysulfides.

Sulfide inclusion shape control performs several important roles in HSLA steels. It improves transverse impact energy, and it can minimize lamellar tearing in welded structures by improving through-thickness properties that are critical in constrained weldments. The main ob-

jective of inclusion shape control is to produce sulfide inclusions with negligible plasticity at even the highest rolling temperatures. Sulfide inclusions, which are plastic at rolling temperatures and thus elongate and flatten during rolling, adversely affect ductility in the short transverse (through-thickness) direction.

The preferred method for sulfide shape control involves calcium-silicon ladle additions. However, sulfide shape control is also performed with small additions of rare-earth elements, zirconium, or titanium that change the shape of the sulfide inclusions from elongated stringers to small, dispersed, almost spherical globules. This change in the shape of sulfide inclusions substantially increases transverse impact energy and improves formability. Inclusion shape control was introduced with the advent of hot-rolled sheet and light plate having a yield strength of 550 MPa (80 ksi) in the as-rolled condition. This technology has also been extended to include grades with lower yield strengths ranging from 310 to 550 MPa (45 to 80 ksi). The improved formability of these grades is recognized in ASTM A 715.

Inclusion shape control with rare-earth elements is seldom used because rare-earth elements produce relatively dirty steel. Nonetheless, representative improvements in transverse toughness with rare-earth additions are shown in Fig. 15 and 16 (Ref 24, 25). With regard to shape control, Luyckx et al. (Ref 26) demonstrated that when there was a complete change from elongated to globular sulfides, at a cerium-to-sulfur ratio of two, the upper-shelf impact energy became a maximum. This is seen in Fig. 17. As might be expected, there was little effect of rare-earth metal additions on longitudinal properties, because the shape of sulfides is not particularly detrimental to longitudinal fracture toughness in steels of low to medium strength.

Chemistry Effects on Temper Embrittlement. As previously mentioned, chemistry also has important influence on temper embrittlement. The use of Auger electron spectroscopy has shown that steels that fail by this mode have

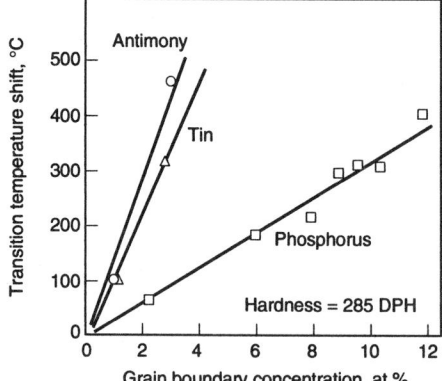

Fig. 18 Change in ductile-to-brittle transition temperature as a function of grain boundary impurity concentration. Data obtained on 3340 steel doped individually with 0.06% P, 0.06% Sn, or 0.06% Sb. a/o, atomic percent. 285 DPH = 890 MPa (129 ksi) ultimate strength. Source: Ref 27

severe segregation of trace elements such as antimony, arsenic, tin, and phosphorus along prior-austenite grain boundaries, even though the bulk concentration is in the parts per million range. This segregation exists for only a few atom layers. In Fig. 18, the results of an investigation (Ref 27) on 3340 steel indicate that the transition temperature shift was proportional to the grain boundary concentration of the embrittling trace impurity.

Temper embrittlement is a complex phenomenon, because it is dependent not only on the trace elements present but also on the major alloying additions, such as nickel and chromium. For example, in steels containing the same amounts of trace impurities, Ni-Cr alloys are more susceptible than those having either nickel or chromium separately (Fig. 6 and Ref 28). In either case, the degree of embrittlement can be reduced by lowering the trace element content of the steel. To avoid temper embrittlement, steels with low trace

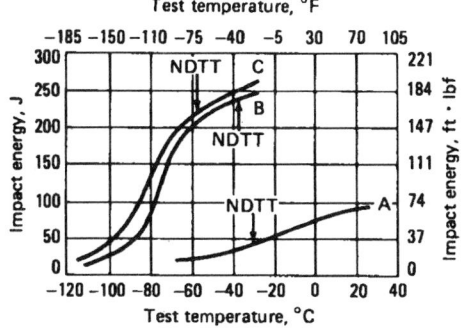

Fig. 19 Charpy V-notch impact energy with nil-ductility transition temperature for three steels: A, 60 mm (2⅜ in.) thick old carbon-manganese steel (0.21% C) with a yield strength of 355 MPa (51 ksi); B, 70 mm (2¾ in.) thick modern carbon-manganese steel (0.114C-0.29Ni-0.025Nb-0.022Cu) with a yield strength of 369 MPa (54 ksi); and C, 50 mm (2 in.) thick thermomechanically controlled processing steel (0.11C-0.23Ni-0.03Nb-0.24Cu) with a yield strength of 506 MPa (73 ksi)

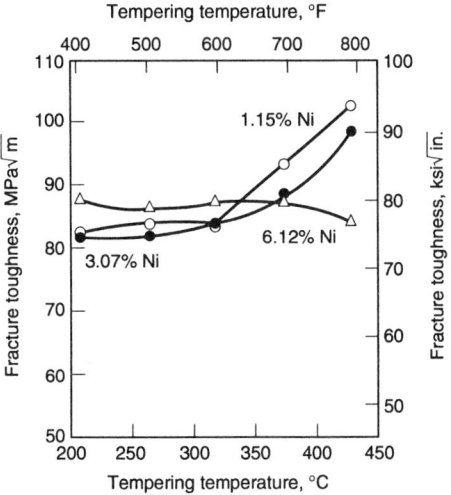

Fig. 20 Variation of room-temperature notch-bend fracture toughness with tempering temperature for steel containing (in wt%) 0.35 C, 0.65 Mn, 0.35 Si, 0.80 Cr, 0.30 Mo, 0.10 V, and various amounts of nickel. Source: Ref 20

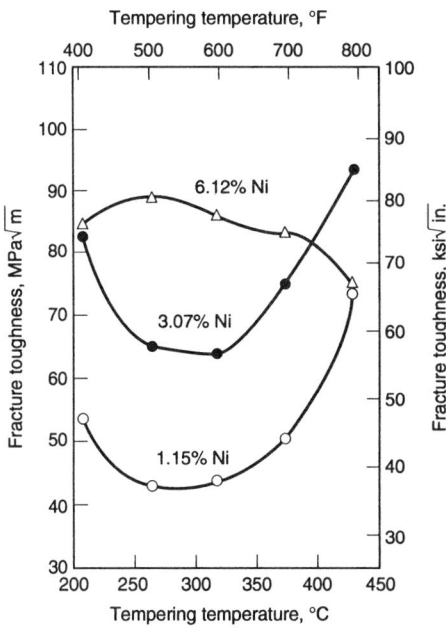

Fig. 21 Variation of notch-bend fracture toughness at –73 °C (–100 °F) with tempering temperature for steels containing (in wt%) 0.35 C, 0.65 Mn, 0.35 Si, 0.80 Cr, 0.30 Mo, 0.10 V, and various amounts of nickel. Source: Ref 19

element concentrations and proper ratios of major alloying additions should be used. It is also advisable to avoid service in the embrittling temperature range. A small austenite grain size also inhibits temper embrittlement. It should be noted that this behavior is reversible, that is, it can be eliminated by heating above the temper embrittlement range provided that the material is not subsequently maintained at temperatures in the embrittlement range for extended periods of time.

Comparison of Modern HSLA Steels with Older Structural Steels. As previously described, several advances in steelmaking technology have allowed the development of tough, low-carbon HSLA steels for high-strength structural applications. As a general comparison of HSLA steel improvements with older structural steels, Fig. 19 compares impact toughness curves of old and new steels using transverse subsurface specimens. Steel A is a 60 mm (2⅜ in.) carbon-manganese steel that was used in 1975. It has a carbon level of 0.21%. Steel B is a modern 70 mm (2¾ in.) thick normalized carbon-manganese steel with a carbon level of 0.114% and some microalloying (0.29% Ni, 0.025% Nb, and 0.022% Cu). Steel C is a modern 50 mm (2 in.) thick controlled-rolled and accelerated-cooled thermomechanically controlled processing steel with a carbon level of 0.11% and some microalloying (0.23% Ni, 0.03% Nb, and 0.24% Cu). The yield strengths of steels A, B, and C are 355, 369, and 506 MPa (51, 54, and 73 ksi), respectively. Figure 19 shows the improved fracture toughness of modern steels as indicated by a decrease in the transition temperature and an increase in the upper-shelf energy.

Quenched-and-Tempered Steels. Typical microstructural changes on heat treatment of high-strength steels are accomplished by changing the austenitizing temperature and time at temperature, the quenching rate, and the tempering temperature and time at temperature. Increasing the austenitizing temperature and time has a twofold effect in that it increases the grain size and/or increases the solutionizing of alloy carbide form-

ers. The former effect may produce mixed results in that austenitizing at 1200 °C (2200 °F) may increase the sharp-crack fracture toughness of an AISI 4340 steel over that obtained using conventional heat treatments; however, the large prior-austenite grains may produce a lower impact toughness (Ref 29). As discussed in the next section, increased prior-austenite grain size in high-strength steel may also have a detrimental effect on resistance to fatigue cracking. On the other hand, it is well known that reasonable austenitizing temperatures and times are necessary to dissolve the carbide-forming elements prior to quenching.

Variations in quenching rates also influence fracture toughness of alloy steels by causing variations in the as-quenched microstructures. Quenching alloy steels in oil to obtain nearly 100% martensite on quenching will result in higher fracture toughness in the tempered condition than slack quenching and tempering at the same temperature. This condition is discussed by Peterman and Jones (Ref 30) for heat-treated aircraft structures of D6ac steel. Briefly, austenitizing at 900 °C (1650 °F), quenching in salt at 200 °C (400 °F), and tempering at 200 °C resulted in an average fracture toughness of 57 MPa√m (52 ksi√in.), whereas austenitizing at 925 °C (1700 °F), quenching in oil at 60 °C (140 °F), and tempering at 200 °C (400 °F) resulted in an average fracture toughness of 101 MPa√m (92 ksi√in.).

The effects of tempering ultrahigh-strength steels at temperatures in the upper portion of the tempering range are to increase fracture toughness and to reduce yield strength (Fig. 2). In the intermediate tempering temperature regime between 200 and 400 °C (400 and 750 °F), these effects are not always apparent. This is shown in Fig. 20, where fracture toughness at 25 °C (75 °F) is relatively independent of tempering temperature for three different nickel additions (Ref 20). For tests at –73 °C (–100 °F), however, severe

tempered martensite embrittlement is partly reduced by addition of 3% Ni and is completely suppressed by addition of 6% Ni (Fig. 21). The effect of carbon level on the degree of tempered martensite embrittlement in a 3% Ni ultrahigh-strength steel is illustrated in Fig. 22. For low-temperature service, it is clear that tempering of such steels in the range from 250 to 350 °C (480 to 660 °F) is to be avoided.

Grain Refinement of Low-Carbon and HSLA Steels. For lower-strength steels, useful microstructural modifications to improve low-temperature fracture toughness involve decreasing the ferrite or prior austenite grain size. Results from two investigations presented in Fig. 23 show that, for temperatures above –150 °C (–240 °F), a substantial improvement in fracture toughness can be achieved through grain refinement (Ref 20, 31, 32). Here, it is indicated that there may be a relationship between K_{Ic} and grain size, d. This is shown for both ferrite grain size and prior austenite grain size in Fig. 24 and 25, respectively (Ref 33, 34). Such data have been analyzed by Stonesifer and Armstrong (Ref 34) in terms of a critical plastic zone at the crack tip, with the grain size dependence of K_{Ic} arising from effects of grain size on yield strength. More simply, it can be said that the plastic constraint in the triaxially stressed zone at the crack tip raises the yield stress to the cleavage fracture stress, σ_f. This has been used by others (Ref 35) and is essentially the Orowan criterion given by:

$$pcf \cdot \sigma_{ys} = \sigma_f \qquad (Eq\ 5)$$

where σ_{ys} is the uniaxial yield stress at the temperature and strain rate of interest and pcf is the plastic constraint factor. The simplest empirical determina-

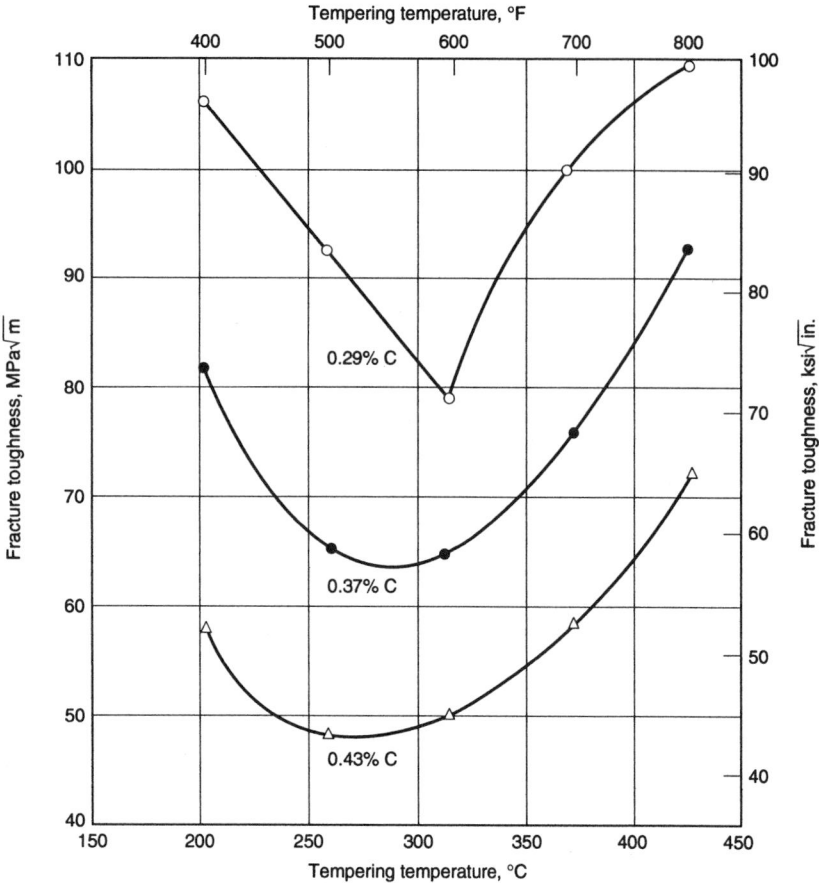

Fig. 22 Effect of carbon content and tempering temperature on fracture toughness of alloy steels containing (in wt%) 0.65 Mn, 0.35 Si, 0.80 Cr, 3.00 Ni, 0.30 Mo, 0.10 V, and various amounts of carbon. Source: Ref 19

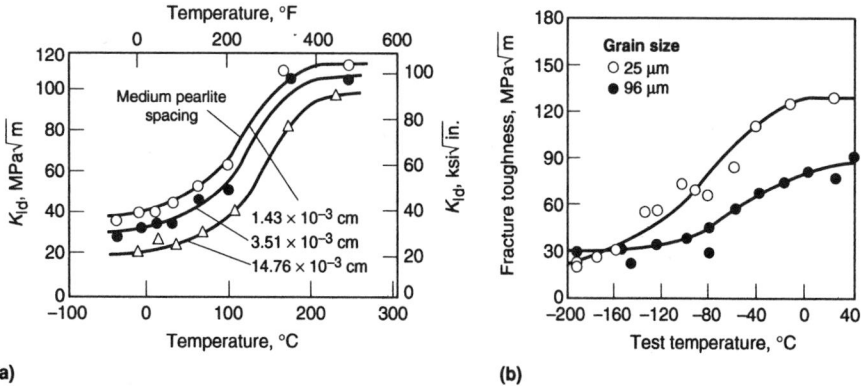

(a)　　　　　　　　　**(b)**

Fig. 23 Effect of grain size on fracture toughness. (a) Dynamic fracture toughness (K_{ID}) curves for fully pearlitic steels as a function of temperature for three prior-austenite grain sizes. (b) Fracture toughness as a function of temperature for St 37-3 steel in two grain sizes. Source: Ref 20, 31, 32

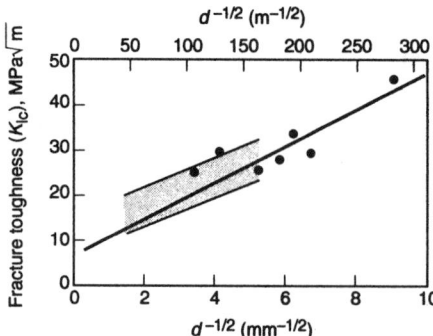

Fig. 24 Relationship of fracture toughness to inverse square root of grain size. Fracture toughness of several plain carbon steels vs. reciprocal square root of ferrite grain size at −120 °C (−184 °F). Source: Ref 33

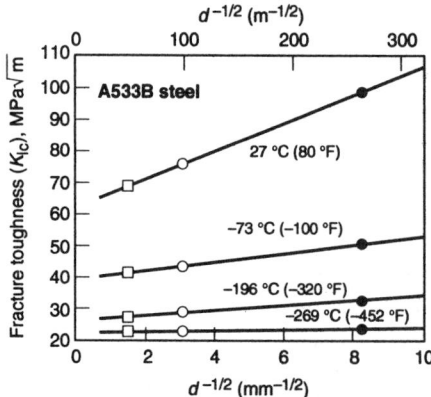

Fig. 25 Relationship of fracture toughness to inverse square root of grain size. Dependence of fracture toughness on prior-austenite grain size at four different temperatures. Source: Ref 34

tion of pcf imposed by a plane-strain crack was originally proposed by Hahn and Rosenfield (Ref 36) to be:

$$pcf = 1 + \alpha \{K_{Ic}/\sigma_{ys}\} \qquad \text{(Eq 6)}$$

For strain hardening in mild steels, it may be assumed that the value of α is approximately $20\sqrt{m}$, because this would give pcf on the order of 2.5 to 3.0, values that are predicted from elas-

tic-plastic analysis (Ref 37). Relationships for yield and fracture stress dependence on grain size have been independently determined by Curry and Knott (Ref 33) and by Rosenfield et al. (Ref 38) to be, for mild steel:

$$\sigma_{ys}(-120\,^\circ C) = 210\,\text{MPa} + (0.73\,\text{MPa}\sqrt{m})d^{-1/2} \qquad \text{(Eq 7a)}$$

and

$$\sigma_f = 343\,\text{MPa} + (3.3\,\text{MPa}\sqrt{m})d^{-1/2} \qquad \text{(Eq 7b)}$$

By combining Eq 5, 6, and 7, the following equation describing the effect of grain size on K_{Ic} is obtained:

$$K_{Ic}(\text{at} -120\,^\circ C)$$

$$\cong 6.6\,\text{MPa}\sqrt{m} + 0.13\,\text{MPa}\cdot\text{m}(d^{-1/2}) \qquad \text{(Eq 8)}$$

It can be seen that this equation fits the data in Fig. 24, and gives a good rationalization for the $1/\sqrt{d}$ dependence of fracture toughness. Such a relationship can also be approximated for A533B steel at room temperature, because Knott (Ref 39) has estimated that $\sigma_f \cong 2600$ MPa for an A533B steel with a 15 μm grain size. This, along with Stonesifer and Armstrong's determination of the relationship between yield strength and grain size (Ref 34), gives for A533B steel:

$$\sigma_{ys}(\text{at } 25\,^\circ C) = 572\,\text{MPa} + (0.11\,\text{MPa}\sqrt{m})d^{-1/2} \qquad \text{(Eq 9a)}$$

and

$$\sigma_f = 1750\,\text{MPa} + (3.3\,\text{MPa}\sqrt{m})d^{-1/2} \qquad \text{(Eq 9b)}$$

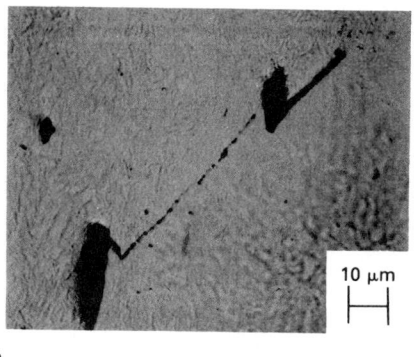

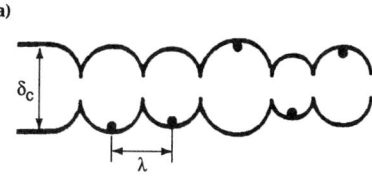

(a)

(b)

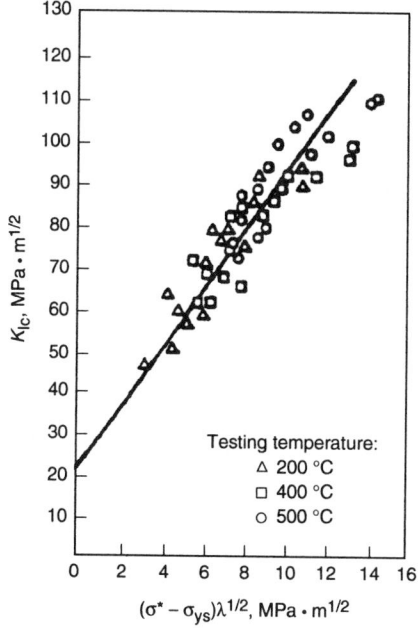

(c)

Fig. 26 Effect of particles on fracture toughness. (a) Void sheets linking three inclusions in a sectioned tensile sample of 4340 steel. (b) Schematic relationship between crack tip opening displacement and inclusion spacing. Source: Ref 42. (c) K_{Ic} for a martensitic 0.45C-Ni-Cr-Mo-V steel as a function of inclusion spacing and yield strength. Source: Ref 43

Following the above procedure, this leads to:

K_{Ic}(at 25 °C)

$$\approx 58.8\,\mathrm{MPa}\sqrt{\mathrm{m}} + 0.16\,\mathrm{MPa}\cdot\mathrm{m}(d^{-\frac{1}{2}}) \qquad \text{(Eq 10)}$$

and gives a curve nearly identical to the line drawn through the room-temperature data in Fig. 25 for A533B steel.

It is clear, then, that one goal for improved low-temperature fracture toughness in steels of low to medium strength is a refined grain size such as has been achieved in HSLA steels. The development of fine grain size in these steels through the use of minor alloying additions and lower hot rolling finishing temperatures has resulted in large improvements in fracture toughness (Ref 40).

Inclusions and Void Nucleation. Reduced ductility and toughness due to particle nucleation of microvoids is most often observed in high-strength steels. Figure 26(a) shows a metallographic section of a specimen of AISI 4340 steel strained just short of fracture, exhibiting a "void sheet" between large voids that nucleated at lower strains at manganese sulfides. In this case, the smaller voids nucleated at the cementite particles that formed during tempering. Void sheet formation results in duplex dimple size distributions. This process aborts further void growth and tends to reduce fracture toughness. Cox and Low (Ref 41) concluded that void sheet formation is easier in materials with larger-size strengthening precipitates. This is supported by the absence of void sheets in maraging steel having a yield strength of 1380 MPa (200 ksi) and containing precipitates a few hundred angstroms in length and by the presence of void sheets in 4340 steel containing cementite particles 0.17 μm in length. Cox and Low also suggest that this is why maraging steels are tougher than quenched-and-tempered steels of similar strength levels (Fig. 2).

The key event in dimpled rupture is the void nucleation event, which can be delayed to larger plastic strains by reductions in inclusion size (Ref 41). Toughness can also be improved by increasing the spacing between inclusions so that voids must grow to larger sizes during the fracture process. Reductions in impurities that occur as inclusions reduce the inclusion volume fraction and increase the inclusion spacing, thereby improving K_{Ic}.

The effect of particles has been treated at great length in the literature (Ref 42), but one of the simplest quantitative relationships that has been derived is that of fracture toughness as a function of inclusion spacing. Schwalbe (Ref 42) suggests that the critical crack-tip opening displacement, δ_c, should be proportional to the distance between inclusions, λ, as shown schematically in Fig. 25(b). According to Eq 4, this would result in K_{Ic} being proportional to the square root of λ. Such a semiempirical relationship has been found by Priest (Ref 43) for a 0.45C-Ni-Cr-Mo-V steel similar to 4340 steel but with somewhat higher chromium, molybdenum, and vanadium contents. This relationship is given by:

$$K_{Ic} = 23\,\mathrm{MPa}\sqrt{\mathrm{m}} + 7(\sigma^* - \sigma_{ys})(\lambda)^{\frac{1}{2}} \qquad \text{(Eq 11)}$$

with σ^* equal to 2000 MPa (290 ksi). In Fig. 26(c), this relationship, shown by the dashed curve, is seen to fit the data for a large variation in λ and for three different test temperatures where dimpled rupture is the microstructural fracture mode. Such observations demonstrate the advantage of providing inclusion-controlled steels for high fracture toughness performance.

Temperature and Strain-Rate Effects

A proper use of fracture-mechanics methodology for fracture control of structures necessitates the determination of fracture toughness for the material at the temperature and loading rate representative of the intended application. The fracture toughness of structural steels can vary significantly with temperature and loading rates, particularly if loading and/or temperature causes a transition from ductile to brittle fracture modes. These general effects on fracture modes and fracture toughness are summarized in Fig. 27 for static and high-strain-rate (impact) loading. This

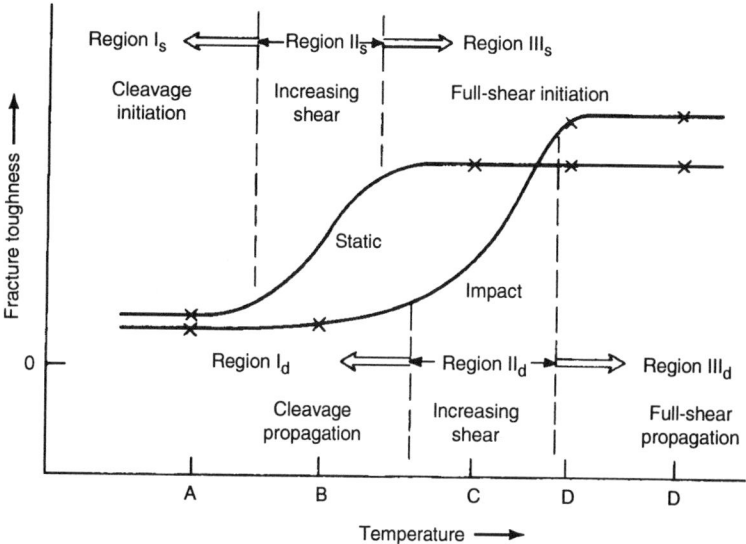

Fig. 27 Fracture-toughness transition behavior of steel under static and impact loading. The static fracture-toughness transition curve depicts the mode of crack initiation at the crack tip. The dynamic fracture-toughness transition curve depicts the mode of crack propagation.

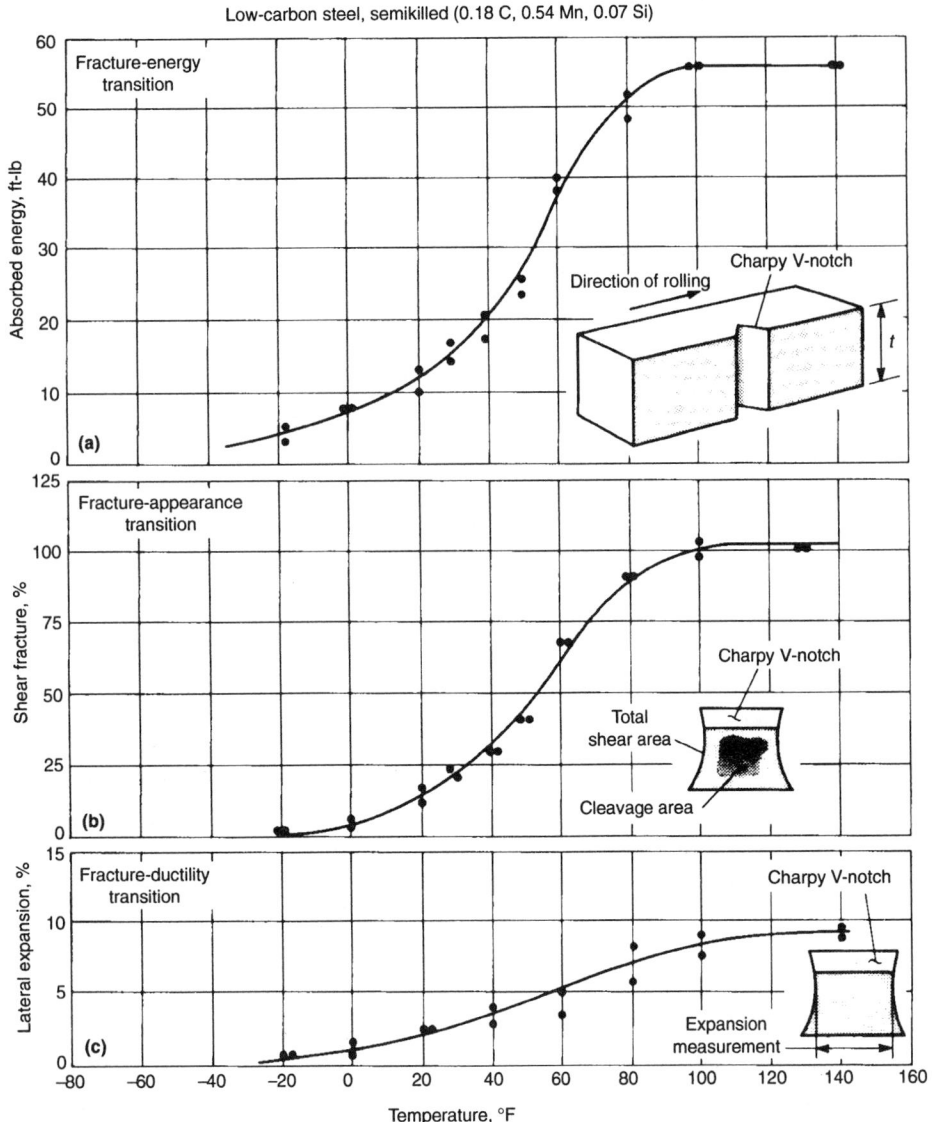

Low-carbon steel, semikilled (0.18 C, 0.54 Mn, 0.07 Si)

Fig. 28 Characteristics of the transition-temperature range for Charpy V-notch testing of low-carbon steel plate, as determined by (a) fracture energy, (b) fracture appearance, and (c) fracture ductility. The drawings at lower right in the graphs indicate: (a) orientation of the specimen notch with plate thickness, t, and direction of rolling; (b) location of the total shear area on the fracture surface; and (c) location of the expansion measurement in this series of tests—all illustrated for a Charpy V-notch specimen. Percentage of shear fracture and lateral expansion were based on the original dimensions of the specimen.

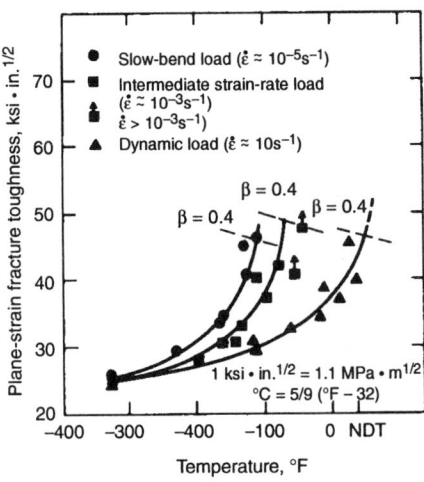

Fig. 29 Effect of temperature and strain rate on plane-strain fracture-toughness behavior of ASTM type A36 steel. Source: Ref 44

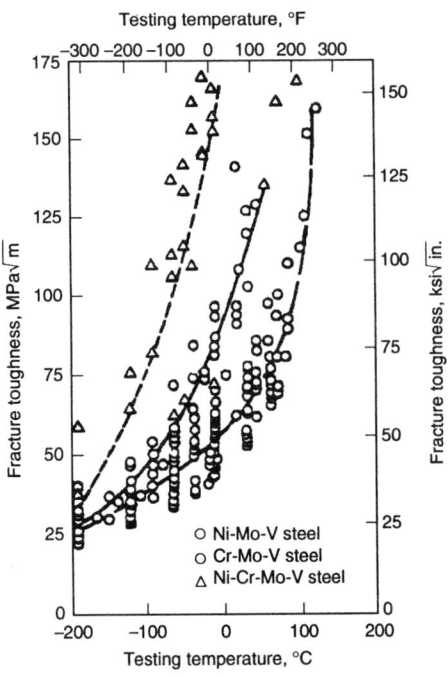

Fig. 30 Effect of temperature on fracture toughness of three alloy steels. Source: Ref 45

well-known transition from ductile to brittle fracture depends on several parameters. Most structural steels can fail in either a ductile or brittle manner depending on conditions such as temperature, loading rate, and constraint (e.g., section size). Strength and the composition of the material also affect toughness and the transition rate from ductile to brittle fracture. Thus, the effects of testing temperature and loading rate on fracture toughness vary from one grade of steel to another.

The effects of temperature and load rates can be measured with a variety of specimens, such as CVN impact specimens, the dynamic tear specimen, a plane-strain fracture toughness specimen under static loading (K_{Ic}), or a plane-strain specimen under dynamic loading (K_{Id}). Traditionally,

the fracture toughness and ductile-to-brittle transition behavior of structural steels is based on CVN specimen testing (ASTM test standards A23 and A370). CVN specimens are tested at different temperatures, and the impact notch toughness at each test temperature is determined from the energy absorbed during fracture, the percent shear (fibrous) fracture on the fracture surface, or the change in the width of the specimen (lateral expansion). An example of the ductile-to-brittle transition with temperature for each of these parameters is presented in Fig. 28. The actual values for each parameter and the locations of the curves along the temperature axis are usually different for different steels and even for a given steel composition. Because the change from ductile to brittle behavior occurs over a

range of temperatures, it has been customary to define a single temperature within the transition range that reflects the behavior of the steel under consideration. Several equally useful definitions are in use, including the 15 ft · lb temperature, the 15 mil temperature, and the 50% shear temperature.

Similar transitions can be observed in plane-strain fracture toughness for structural steel plate (Fig. 29) and several rotor steels (Fig. 30). The rate of increase of K_{Ic} with temperature does not remain constant, but rather increases markedly above a given test temperature. This transition in plane-strain fracture toughness is related to a

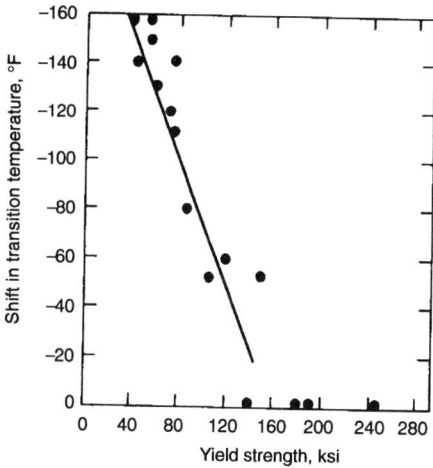

Fig. 31 Effect of yield strength on the shift in transition temperature between impact and static plane-strain fracture-toughness curves. Source: Ref 44

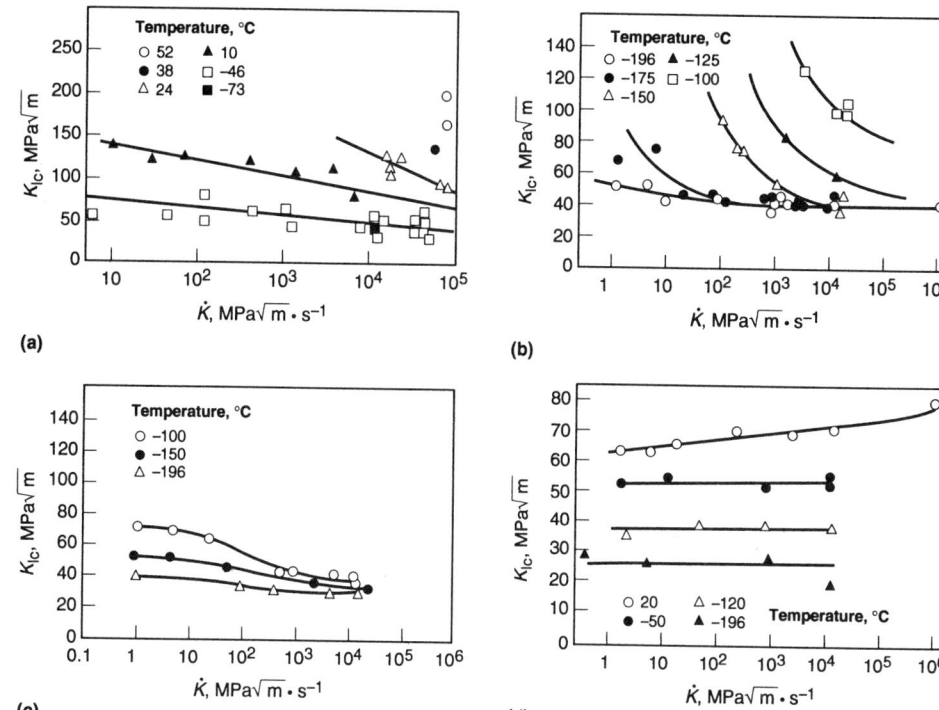

Fig. 32 Influence of temperature and loading rate on fracture toughness. (a) A533B steel; $\sigma_{ys} \approx 450$ MPa (65 ksi). (b) QT35 steel; $\sigma_{ys} \approx 600$ MPa (87 ksi). (c) Ni-Cr-Mo-V rotor forging steel; $\sigma_{ys} \approx 670$ MPa (97 ksi). (d) 0.4C-9Ni-4Co steel; $\sigma_{ys} \approx 970$ MPa (140 ksi). (Yield strengths reported are at room temperature.) Source: Ref 21

change in the microscopic mode of crack initiation at the crack tip from cleavage to increasing amounts of ductile tearing.

Fractures modes and toughness under various conditions of temperature and load rates can be understood by considering toughness-transition behavior under static and impact loading (Fig. 27). The static fracture toughness transition curve depicts the mode of crack initiation at the crack tip. The dynamic fracture toughness transition curve depicts the mode of crack propagation.

The fracture toughness curve for either static or dynamic loading can be divided into three regions as shown in Fig. 27. In region I_s for the static curve, the crack initiates in a cleavage mode from the tip of the fatigue crack. In region II_s, the fracture toughness to initiate unstable crack propagation increases with increasing temperature. This increase in crack initiation toughness corresponds to an increase in the size of the plastic zone and in the zone of ductile tearing (shear) at the tip of the crack prior to unstable crack extension. In this region, the ductile-tearing zone is usually very small and is difficult to delineate by visual examination. In region III_s, the static fracture toughness is quite large and somewhat difficult to define, but the fracture initiates by ductile tearing (shear).

Once a crack has initiated under a static load, the morphology (cleavage or shear) of the fracture surface for the propagating crack is determined by the dynamic behavior and degree of plane strain at the temperature. Regions I_d, II_d, and III_d in Fig. 27 correspond to cleavage, increasing ductile tearing (shear), and full-shear crack propagation, respectively. Thus, at temperature A, the crack initiates and propagates in cleavage. At temperatures B and C, the crack exhibits ductile initiation but propagates in cleavage. The only difference between the behaviors at temperatures B and C is that the ductile-tearing zone for crack initiation is larger at temperature C than at temperature B. At temperature D, cracks initiate and propagate in full shear. Consequently,

full-shear fracture initiation and propagation occur only at temperatures for which the static and dynamic (impact) fracture behaviors are on the upper shelf.

Strain Rate Effects. The effect of loading rate is obviously important because of its effect on not only toughness but also the transition temperature (Fig. 27, 29). Strain rate effects are even more important in low- and medium-strength steels such as A36 (Fig. 29), which are strain-rate sensitive. An analysis of plane-strain fracture-toughness data for constructional steels (Ref 44) shows that the fracture toughness transition curve is translated (shifted) to higher temperature values as the rate of loading is increased. Thus, at a given temperature, fracture toughness values measured at high loading rates are generally lower than those measured at lower loading rates. Also, the fracture toughness values for constructional steels decrease with decreasing test temperatures to a minimum K_{Ic} value of about 27 MPa$\sqrt{m}$ (25 ksi$\sqrt{in}$.). This minimum fracture toughness value is independent of the rate of loading used to obtain the fracture toughness transition curve.

The magnitude of shifts in transition temperatures from strain-rate effects depends on yield strength and has been the subject of several studies (Ref 44, 46-48). From the work of Barsom (Ref 44), data for steels having yield strengths between 250 and 1725 MPa (36 and 250 ksi), such as those presented in Fig. 31, show that the shift between static and impact plane-strain fracture-toughness curves is given by the relationship:

$$T_{shift} \text{ (in °F)} = 215 - 1.5\sigma_{ys} \text{ (in ksi)} \qquad \text{(Eq 12a)}$$

or

$$T_{shift} \text{ (in °C)} = 119 - 0.12\sigma_{ys} \text{ (in MPa)} \qquad \text{(Eq 12b)}$$

for steels with yield strengths between 195 and 895 MPa (28 and 130 ksi) where σ_{ys} is the room-temperature yield strength. There is no transition shift from loading rates for steels with yield strengths greater than about 960 MPa (139 ksi).

Similar results have also been verified by Priest (Ref 121) in a large number of tests of low- to medium-strength steels, some of which are illustrated in Fig. 32. Here, up to six test temperatures and six orders of magnitude of loading rate (in K, MPa$\sqrt{m} \cdot s^{-1}$) were used to determine fracture toughness. It can be seen that there are substantial shifts for low-strength steels. However, for the steel with a yield strength of 970 MPa (140 ksi) in Fig. 32(d), although there is a change in fracture toughness with test temperature, there is no shift with loading rate.

The combined effect of yield strength and strain rate for steels with yield strengths less than 965 MPa (139 ksi) is given by the following relations for strain rates ($\dot{\varepsilon}$, in s^{-1}) between 10^{-3} s^{-1} and 10^{-1} s^{-1}:

$$T_{shift} \text{ (in °F)} = (150 - \sigma_{ys})\dot{\varepsilon}^{0.17} \qquad \text{(Eq 13a)}$$

where σ_{ys} is room-temperature yield strength in ksi, or where

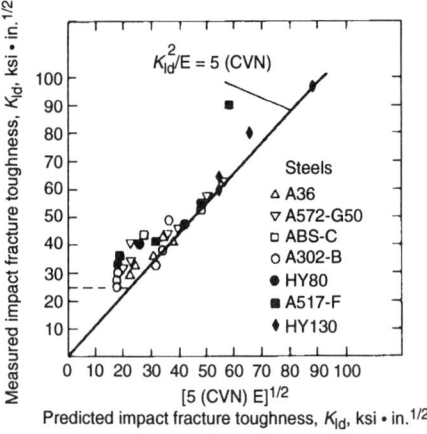

Fig. 33 Correlation of plane-strain impact fracture toughness and impact Charpy V-notch energy absorption for various grades of steel

$$T_{shift} \text{ (in °C)} = (83 - 0.08\sigma_{ys})\dot{\varepsilon}^{0.17} \qquad \text{(Eq 13b)}$$

with σ_{ys} in MPa. These relations provide the difference between static and any intermediate or impact plane-strain toughness curves. The strain rate is calculated for a point on the elastic-plastic boundary according to the equation:

$$\dot{\varepsilon} = \frac{2\sigma_{ys}}{tE} \qquad \text{(Eq 14)}$$

where t is the loading time for the test and E is the elastic modulus.

Dynamic plane-strain fracture toughness (K_{Id}) can be estimated from CVN toughness or the upper-shelf static toughness (K_{Ic}). In the upper shelf, the effects of loading rate and notch acuity are not as critical as in the transition region. The effect of loading rate is to elevate the yield strength by about 170 MPa (25 ksi). Thus, Eq 3 may be used to calculate K_{Id} values by replacing σ_{ys} with the dynamic yield strength, σ_{yd}, where $\sigma_{yd} \approx \sigma_{ys} + 25$ ksi. This use of Eq 3 to calculate K_{Id} is consistent with the observation that, in the upper-shelf region, the dynamic fracture toughness of steels is higher than the static fracture toughness.

In the transition region, a correlation between CVN toughness and impact fracture toughness (K_{Id}) is given by the equation (Ref 17, 44):

$$\frac{(K_{Id})^2}{E} = 5(CVN) \qquad \text{(Eq 15)}$$

where K_{Id} is in ksi$\sqrt{\text{in.}}$, E is in ksi, and CVN toughness is in ft · lb. The validity of this correlation is apparent from the data presented in Fig. 33 for various grades of steel ranging in yield strength from about 36 to about 140 ksi and in Fig. 34 for eight heats of SA 533B, class 1 steel. Consequently, a given value of CVN impact energy absorption corresponds to a given K_{Id} value (Eq 15), which in turn corresponds to a given toughness behavior at lower rates of loading.

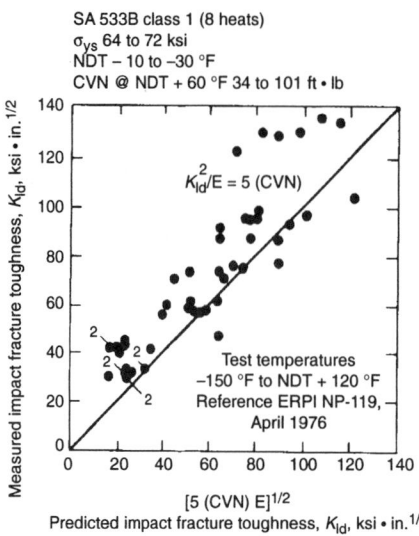

Fig. 34 Correlation of plane-strain impact fracture toughness and impact Charpy V-notch energy absorption for SA 533B, class 1 steel

The behavior for rates of loading less than impact are established by shifting the K_{Id} value to lower temperatures by using Eq 12 or 13. Conversely, for a desired behavior at the minimum operating temperature and maximum in-service rate of loading, the corresponding behavior under impact loading can be established by using Eq 12 or 13, and the equivalent CVN impact value can be established by using Eq 15.

Scatter of Toughness Data in Ductile-to-Brittle Transition Regime. Fracture toughness testing of ferrous materials in the ductile-to-brittle transition region is complicated by an extreme amount of scatter in toughness data. Data for

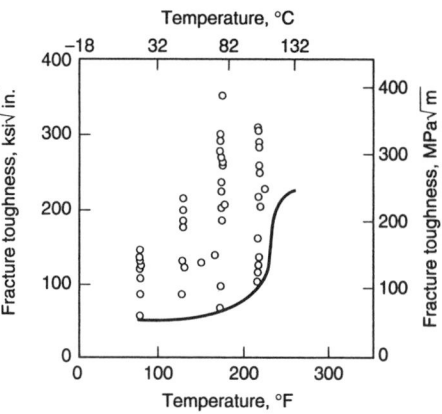

Fig. 35 Typical fracture toughness scatter in the ductile-brittle transition region for a carbon steel forging. Pressure vessel steel (ASTM A508 C12) with test material was taken from a forged hollow cylinder normalized at 893 °C (1640 °F) for 13 h and air cooled, then austenitized at 857 °C (1575 °F) for 8.75 h and water quenched. Tempering was done at 523 °C (973 °F) for 9.5 h, followed by a postweld heat treatment at 561 °C (1041 °F) for 12 h. The yield strength at room temperature is 620 MPa (90 ksi) and the ultimate strength is 779 MPa (113 ksi). Source: Ref 49

pressure vessel steels, for example, typically covers half an order of magnitude (Fig. 35). Some of the scatter is, of course, attributable to experimental errors and specimen-to-specimen differences. Scatter is also caused by the occurrence of different fracture mechanisms (cleavage, tearing, or ductile tearing followed by cleavage) in the transition region. Some experimental work has investigated the use of dynamic fracture to promote cleavage initiation and suppress prior ductile tearing (Ref 49).

Fracture Mechanics of Steel Fatigue

Although a considerable amount of life data are available for many steels and other structural materials, the existence of surface irregularities and cracklike imperfections can have a profound effect on the prevention of fatigue failure. As such, fracture mechanics offers a useful method for understanding not only fatigue crack propagation but also the factors that may eliminate or reduce the crack initiation portion of fatigue. Therefore, this section briefly reviews notch effects on crack initiation prior to the more traditional focus of fracture mechanics on fatigue crack propagation.

Notch Effects on Crack Initiation

Initiation of fatigue cracks in structural and equipment components occurs in regions of stress concentrations, such as notches, as a result of stress fluctuation. The material element at the tip of a notch in a cyclically loaded component is subjected to the maximum stress range, $\Delta\sigma_{max}$.

Consequently, this material element is most susceptible to fatigue damage and is, in general, the origin of fatigue crack initiation. It can be shown that for sharp notches, the maximum-stress range on this element can be related to the stress-intensity-factor range, ΔK_I (Eq 16), as follows (Ref 17):

$$\Delta\sigma_{max} = \frac{2}{\sqrt{\pi}}\frac{\Delta K_t}{\sqrt{\rho}} = \Delta\sigma(k_t) \qquad \text{(Eq 16)}$$

where ρ is the notch-tip radius, $\Delta\sigma$ is the range of applied nominal stress, and k_t is the stress-concentration factor.

Fatigue crack initiation behavior of various steels is presented in Fig. 36 for specimens subjected to zero-to-tension bending stress and containing a smooth notch that resulted in a stress-concentration factor of about 2.5. The data show that $\Delta K_I/\sqrt{\rho}$, and therefore $\Delta\sigma_{max}$, is the primary parameter that governs fatigue crack initiation

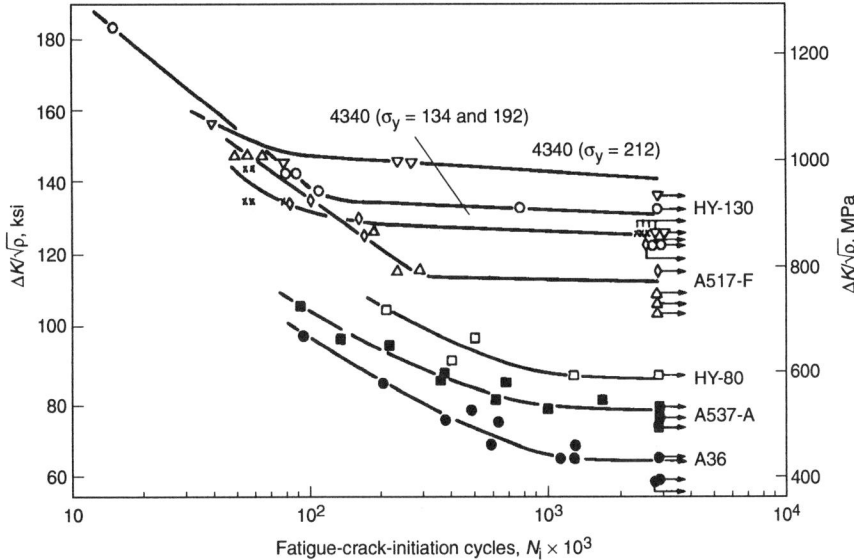

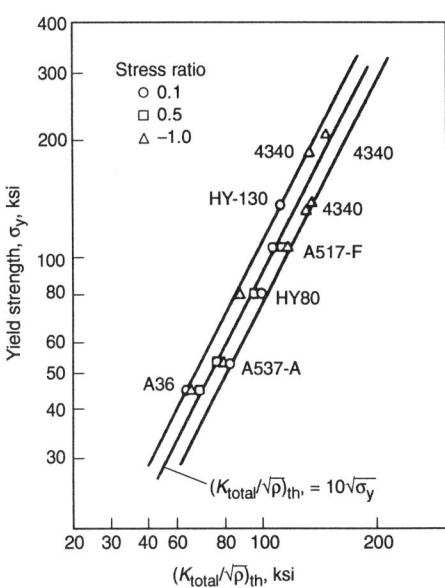

Fig. 36 Fatigue crack initiation behavior of various steels at a stress ratio of +0.1. Source: Ref 17

Fig. 37 Dependence of fatigue crack initiation threshold on yield strength. Source: Ref 51

behavior in regions of stress concentration for a given steel tested in a benign environment. The data also indicate the existence of a fatigue crack initiation threshold, $\Delta K_{\mathrm{I}}/\sqrt{\rho}\,)_{\mathrm{th}}$, below which fatigue cracks would not initiate at the roots of the tested notches. The value of this threshold is characteristic of the steel and increases with increasing yield or tensile strength of the steel. The data show that the fatigue crack initiation life of a component subjected to a given nominal-stress range increases with increasing strength. However, this difference in fatigue crack initiation life among various steels decreases with increasing stress-concentration factor (Ref 17).

Finally, fatigue crack initiation data for various steels subjected to stress ratios (ratio of nominal minimum applied stress to nominal maximum applied stress) ranging from –1.0 to +0.5 indicate that fatigue crack initiation life is governed by the total maximum stress (tension plus compression) range at the tip of the notch (Ref 50). The data presented in Fig. 37 indicate that the fatigue crack initiation threshold, $(\Delta K_{\mathrm{I}}/\sqrt{\rho})_{\mathrm{th}}$, for various steels subjected to stress ratios ranging from –1.0 to +0.5 can be estimated from:

$$\frac{\Delta K_{\mathrm{total}}}{\sqrt{\rho}} = 10\sqrt{\sigma_{ys}} \qquad (\mathrm{Eq}\ 17)$$

where $\Delta K_{\mathrm{total}}$ is the stress-intensity factor range calculated by using the tension-plus-compression stress range, and σ_{ys} is the yield strength of the material.

Fatigue Crack Propagation

Cyclic loading can cause crack propagation in certain alloy steels at stress intensities as low as one-twentieth of the K_{Ic} value. Threshold stress intensities may range from 3 to 20 MPa$\sqrt{\mathrm{m}}$ (2.7 to 18 ksi$\sqrt{\mathrm{in.}}$). This demonstrates that factors other than continuum plasticity considerations are important.

The general nature of fatigue crack propagation using fracture mechanics techniques is summarized in Fig. 38. A logarithmic plot of the crack growth per cycle, da/dN, defines the rate of subcritical crack growth from fatigue loading, da/dN, in terms of the stress-intensity range, $\Delta K = K_{\max} - K_{\min}$. The three regions extend from ΔK levels associated with almost zero crack-growth rate to conditions approaching fast fracture.

The fracture mechanics approach for characterizing fatigue crack growth can be used in design applications to estimate maximum flaw sizes that allow a part to reach its design life. This approach is also very useful for conducting failure analyses.

Because predicting service fatigue life often involves integrating crack growth rates over a range of ΔK values, the da/dN versus ΔK relationship can be represented by:

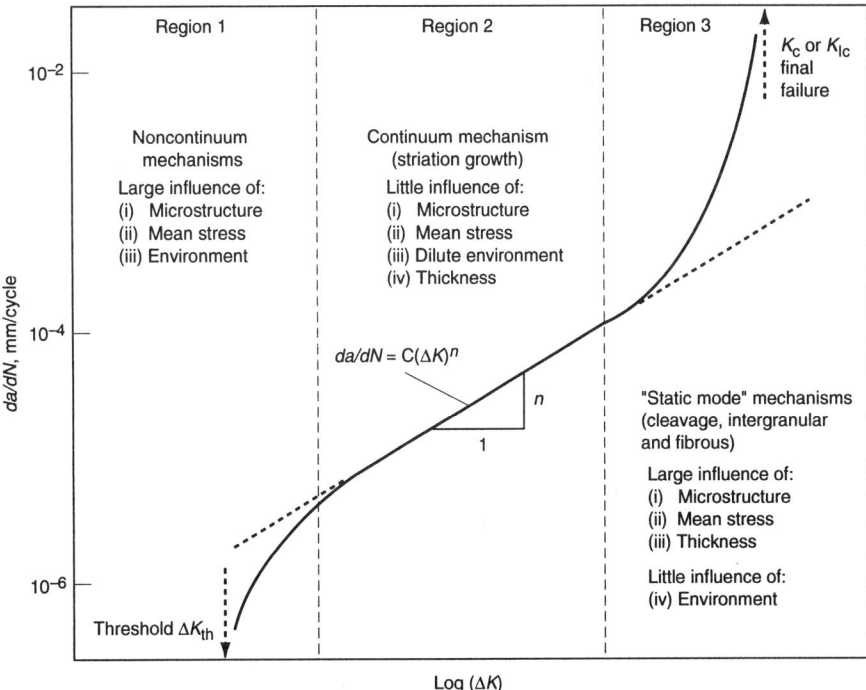

Fig. 38 Schematic illustration of variation of fatigue crack growth rate, da/dN, with alternating stress intensity, ΔK, in steels, showing regions of primary crack-growth mechanisms

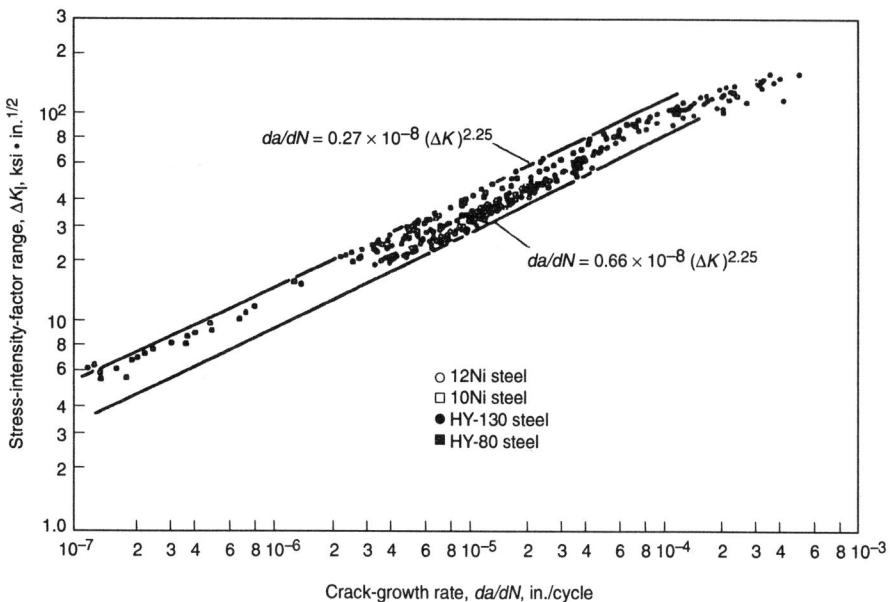

Fig. 39 Summary of fatigue crack growth data for martensitic steels. Source: Ref 17

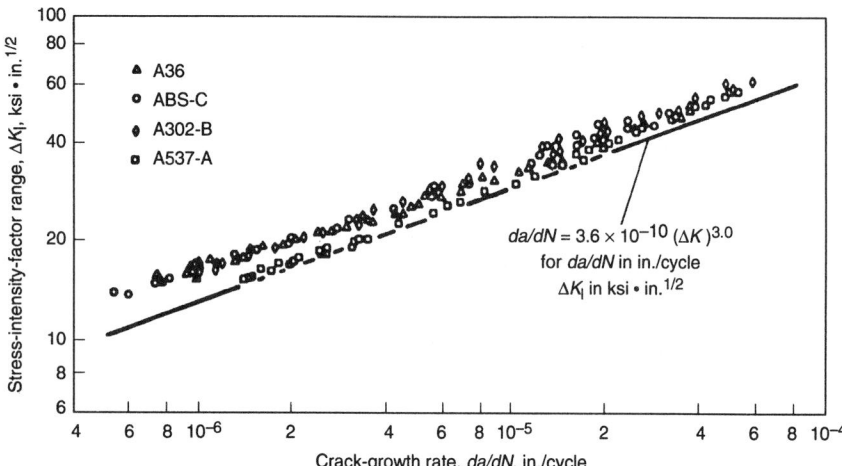

Fig. 40 Summary of fatigue crack growth data for ferrite-pearlite steels. Source: Ref 17

$$\left(\frac{da}{dN}\right)^{-1} = \frac{A_1}{(\Delta K)^{n1}}$$

$$+ A_2 \left[\frac{1}{(\Delta K)^{n2}} - \frac{1}{[(1-R)K_c]^{n2}} \right] \quad \text{(Eq 18)}$$

where $n1$, $n2$, A_1, A_2, and K_c are best-fit material constants determined by regression analysis. Equation 18 is general and applies to all three regions of crack growth.

Variables that influence fatigue crack growth behavior are stress ratio, loading frequency (not a factor in benign environments), material chemistry, heat treatment, test temperature, and environment. Nonetheless, results of fatigue crack growth rate tests for nearly all metallic structural

materials have shown that the da/dN versus ΔK curves have three distinct regions (Fig. 38). These three regions are briefly described below, followed by sections covering the testing, environmental, and material factors that affect fatigue crack growth rates.

Crack Threshold Region. The behavior in region 1 of Fig. 38 exhibits a fatigue crack growth threshold, ΔK_{th}, which corresponds to the stress-intensity factor range below which cracks do not propagate, or in which the crack growth rate becomes diminishingly small. The fatigue crack growth threshold is an important design parameter for such applications as rotating shafts involving low-stress, high-frequency fatigue loading where no crack extension during service can be permitted. In region 1, crack growth is negligible for an almost unlimited number of cycles.

An analysis of experimental results published on nonpropagating fatigue cracks shows that conservative estimates of ΔK_{th} for various steels subjected to different stress ratios, R, can be predicted (Ref 17) from

$$\Delta K_{th} = 6.4(1-0.85R) \text{ for } R \geq +0.1 \quad \text{(Eq 19a)}$$

and

$$\Delta K_{th} = 5.5 \text{ for } R < +0.1 \quad \text{(Eq 19b)}$$

where ΔK_{th} is in ksi$\sqrt{m}$.

Equation 19 indicates that the fatigue crack propagation threshold for steels is primarily a function of the stress ratio and is essentially independent of chemical or mechanical properties. Other examples of stress ratio effects on crack thresholds are provided in the section "Mean Stress Effects" in this article.

Linear Fatigue Crack Growth Region. At intermediate values of ΔK (region 2 in Fig. 38), a straight line usually is obtained on a log-log plot of ΔK versus da/dN. This is described by the power-law relationship

$$\frac{da}{dN} = C(\Delta K)^n \quad \text{(Eq 20)}$$

where C and n are constants for a given material and stress ratio. The Paris crack growth equation is generally valid within the ΔK range of 300 to 1800 MPa$\sqrt{mm}$ (9 to 52 ksi$\sqrt{in.}$). Values of K below about 300 MPa$\sqrt{mm}$ (9 ksi$\sqrt{in.}$) fall in the threshold range where crack propagation does not occur, and values above about 1800 MPa$\sqrt{mm}$ (52 ksi$\sqrt{in.}$) fall in the range where the static mode of fracture occurs as the fracture toughness limit of the material is approached.

Extensive fatigue crack growth rate data for various steels show that the primary parameter affecting growth rate in region 2 is the stress-intensity factor range, and that the mechanical and metallurgical properties of these steels have negligible effects on the fatigue crack growth rate in a room-temperature air environment. The data for martensitic steels fall within a single band, as shown in Fig. 39, and the upper bound of scatter can be obtained (Ref 17) from

$$\frac{da}{dN} \text{ (m/cycle)} = 1.35 \times 10^{-10} (\Delta K \text{ MPa}\sqrt{m})^{2.25} \text{(Eq 21a)}$$

$$\frac{da}{dN} \text{ (in./cycle)} = 6.6 \times 10^{-9} (\Delta K \text{ ksi}\sqrt{in.})^{2.25} \quad \text{(Eq 21b)}$$

Likewise, for ferritic-pearlitic steels (Fig. 40), data for region 2 crack growth rates fall within a single band (different from the band for martensitic steels), and the upper bound of scatter can be calculated from

$$\frac{da}{dN} \text{ (m/cycle)} = 6.9 \times 10^{-12} (\Delta K \text{ MPa}\sqrt{m})^{3.0} \quad \text{(Eq 22a)}$$

$$\frac{da}{dN} \text{ (in./cycle)} = 3.6 \times 10^{-10} \, (\Delta K \text{ ksi}\sqrt{\text{in.}})^{3.0} \quad \text{(Eq 22b)}$$

Similar results are obtained for a variety of weld metals and heat-affected zone microstructures (Ref 52). The crack growth parameters C and n are estimated in Ref 52 at a C value of 3×10^{-13} and a value of 3, in units of newtons (load) and millimeters (length), for ferritic steels with yield strengths up to 600 MPa (87 ksi). These values are based on the upper limit of air fatigue data shown in Fig. 41.

Region 2 behavior is relevant to design situations involving a finite number of fatigue cycles. The stress ratio and mean stress have negligible effects on the rate of crack growth in region 2. Also, the frequency of cyclic loading and the wave form (sinusoidal, triangular, square, trapezoidal) do not affect the rate of crack propagation per cycle of load for steels in benign environments (Ref 17).

Austenitic stainless steels, described elsewhere in this Volume, have a similar crack growth rate relation in region 2, as follows:

$$\frac{da}{dN} \text{ (m/cycle)} = 5.6 \times 10^{-12} \, (\Delta K \text{ MPa}\sqrt{\text{m}})^{3.25} \quad \text{(Eq 23a)}$$

$$\frac{da}{dN} \text{ (in./cycle)} = 3.0 \times 10^{-10} \, (\Delta K \text{ ksi}\sqrt{\text{in.}})^{3.25} \quad \text{(Eq 23b)}$$

At high ΔK values (region 3 in Fig. 38), unstable behavior occurs, resulting in a rapid increase in the crack growth rate just prior to complete failure of the specimens. The acceleration of fatigue crack growth rates that determines the transition from region 2 to region 3 appears to be caused by the superposition of a brittle or ductile tearing mechanism onto the mechanism of cyclic subcritical crack extension, which leaves fatigue striations on the fracture surface. These mechanisms occur when the strain at the tip of the crack reaches a critical value (Ref 17). Thus, the fatigue-rate transition from region 2 to region 3 depends on the maximum stress-intensity factor, the stress ratio, and the fracture properties of the material (Ref 17).

Region 3 behavior for rapid crack growth is significant only for applications that may experience very few (of the order of ten or fewer) load and unload cycles (e.g., pressure vessels in which pressure may be discharged only a few times during the service life).

Crack growth behavior and ultimate failure in region 3 can occur in one of two ways (Ref 54-56). The first possibility is operative for high-strength, low-toughness metals, in which specimen sizes normally used for fatigue crack growth rate testing behave in a linear elastic manner at K levels equal to K_{Ic}. In this case, increasing crack length during constant load testing causes the peak stress intensity to reach the fracture toughness, K_{Ic}, of the material, and the unstable behavior is related to the early stages of brittle fracture.

The second possibility, plastic limit load behavior, is common for ductile metals, particularly if K_{Ic} is high. In this case, the growing crack reduces the uncracked area of the specimen sufficiently for the peak load to cause fully plastic limit load behavior. When plastic limit load behavior causes unstable crack growth, ΔK values have no meaning, because the limitations of linear elastic fracture mechanics have been exceeded. Here, the use of the J-integral concept, crack opening displacement, or some other elastic-plastic fracture mechanics approach is more appropriate than ΔK for correlating the data.

Fatigue Crack Propagation Modes. Various microstructural modes in fatigue crack growth are briefly reviewed here prior to detailed discussions on environmental and testing effects. Although it is usual to consider ductile fatigue striations as the prime fatigue fracture mode, it is

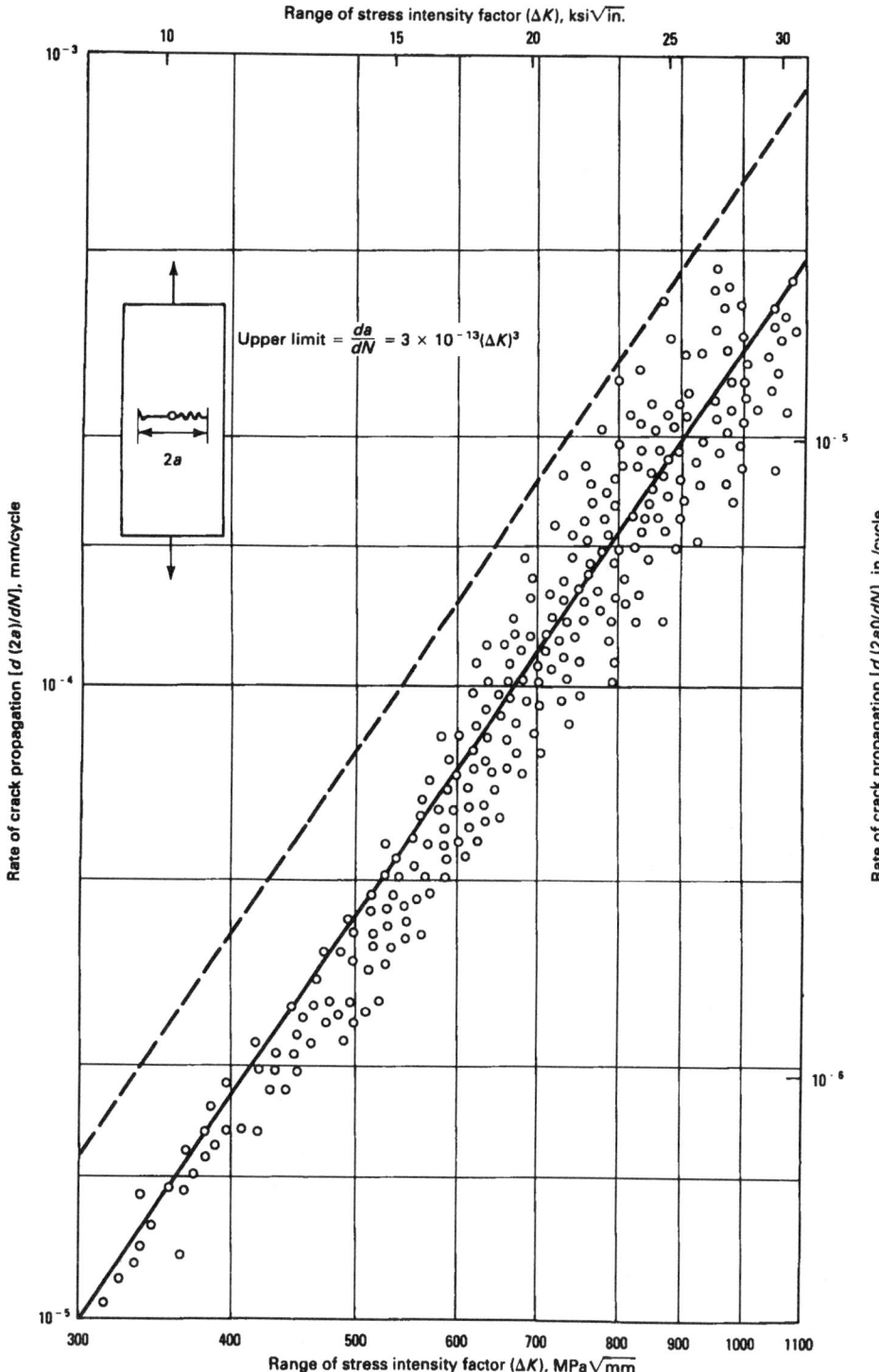

Fig. 41 Fatigue crack growth in the weld metal and heat-affected zones of carbon-manganese steel base plates in an air environment. Source: Ref 53

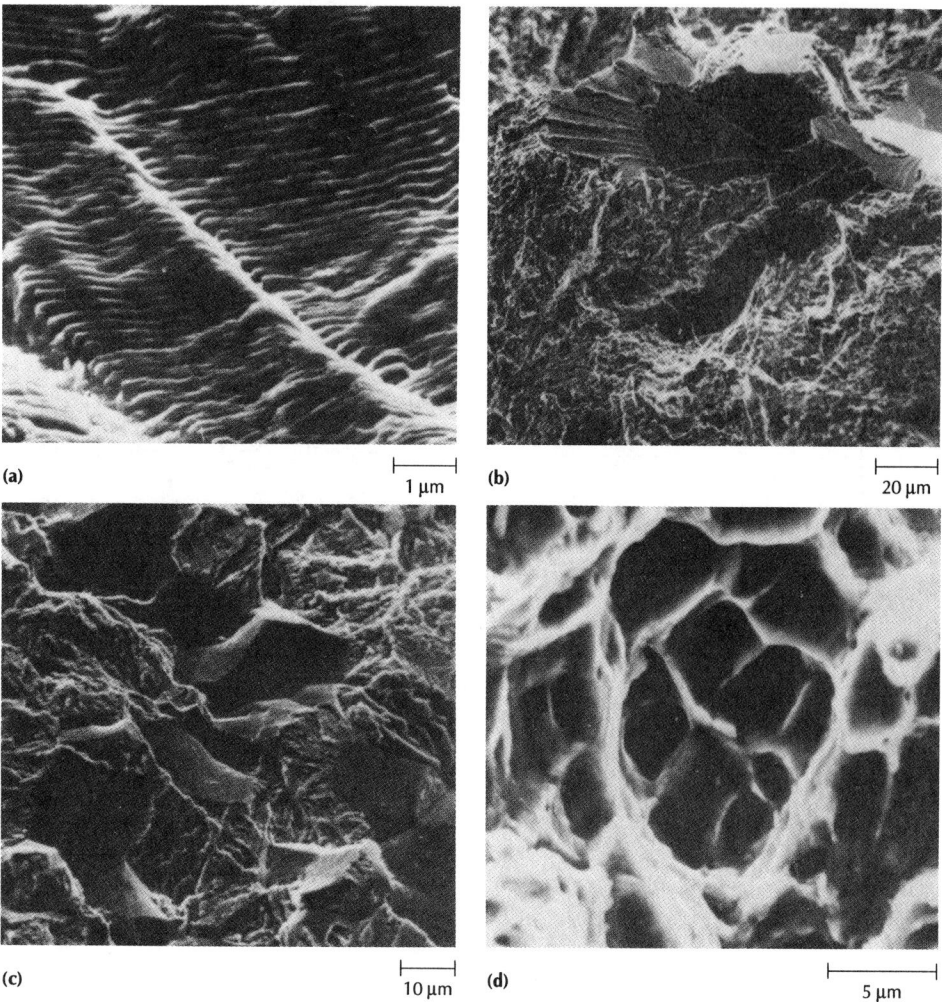

Fig. 42 Fractography of fatigue crack propagation at intermediate (region 2) and high (region 3) growth rates in steels tested in moist air at $R = 0.1$. See text for details about regions. (a) Ductile striations in 9Ni-4Co steel at $\Delta K = 30$ MPa$\sqrt{m}$. (b) Additional cleavage fracture in mild steel at $\Delta K = 40$ MPa$\sqrt{m}$. (c) Additional intergranular fracture in 4Ni-1.5Cr steel at $\Delta K = 40$ MPa$\sqrt{m}$. (d) Microvoid coalescence in 9Ni-4Co steel at $\Delta K = 70$ MPa$\sqrt{m}$. Source: Ref 57

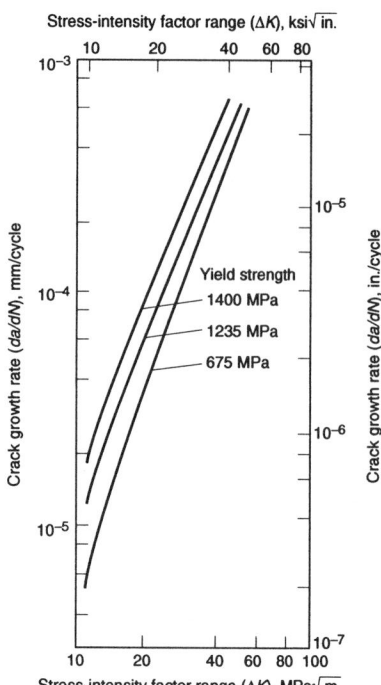

Fig. 43 Effect of strength level on fatigue crack growth rates. Fatigue crack growth rate behavior of 9Ni-4Co-0.30C steel hardened and tempered to indicated strength levels. Source: Ref 60

apparent that other fracture modes often occur, based on reviews (Ref 57, 58) of results from a large number of investigators. As previously mentioned, data indicate that, in general, crack growth in region 2 is relatively insensitive to variations in either microstructure or crack shape. In the threshold region (region 1), the rate of crack growth per cycle drops off rapidly as ΔK is decreased. Behavior in this region may exhibit considerable sensitivity to the microstructure of the material. In region 3, the growth rate may become large because the maximum value of the stress-intensity factor at maximum cyclic load approaches the fracture toughness. Here the local stresses are sufficiently large to activate fracture by other microstructural modes, such as intergranular fracture, cleavage fracture, or dimple rupture. In Fig. 42, a series of microstructural fatigue failure modes in steels are shown. Although striations are shown at a ΔK of 30 MPa$\sqrt{m}$ (27 ksi$\sqrt{in}$.), at higher stress intensities it is seen that cleavage fracture, intergranular fracture or microvoid coalescence may occur in

addition to striation formation (see Fig. 42b, c, and d). Such modes (Ref 58, 59), particularly intergranular and cleavage fracture, may also appear at lower stress intensities.

Because heat treatment, alloy additions, and grain size may affect the tendency toward any microstructural fracture mode, these factors must be considered in addition to yield strength and external loading variables.

Effects of Microstructure and Heat Treatment

In ultrahigh-strength steels, the effect of tempering is relatively small for region 2 growth rates, but the effect may be large near the threshold. An investigation of HP-9-4-30 steel, quenched and tempered to yield strengths of 675, 1235, and 1400 MPa (95, 180, and 200 ksi), showed increases in crack growth rate with in-

creasing yield strength, (Fig. 43). On the other hand, Imhof and Barsom (Ref 60) evaluated AISI 4340 steel at yield strengths ranging from 895 to 1515 MPa (130 to 220 ksi) and found no significant variation in region 2. Whereas the effect on da/dN may only constitute a factor of two or three (or less) in region 2, there may be order-of-magnitude differences in growth rate near the threshold. This is shown for a 300M steel tempered at 100 to 650 °C (212 to 1200 °F) in Fig. 44. This is a superimposed environmental effect that shows up in laboratory air with as little as 40% relative humidity (Ref 57).

Besides tempering effects, microstructure may also have relatively little effect on region 2 growth (Fig. 45). However, it has been found recently that modified microstructures may produce large variations in region 1 growth and may even extend to region 2 (Ref 62). This is shown in Fig. 46 for a low-carbon steel, where a mixture of about 38% martensite in ferrite produced about the same resistance to crack growth as that indicated in Fig. 45 for ferrite-pearlite microstructures. This is for heat treatment A, as indicated in Fig. 46, where the ferrite phase is continuous. If the thermal cycle is now varied to make the martensitic phase continuous, order-of-magnitude improvements are obtained at low ΔK with heat treatment B. Thus, it would appear that substantial improvements in fatigue crack growth resistance near the threshold are possible with microstructural control.

Another means of microstructural control is the grain size effect on fatigue thresholds, ΔK_{th} (Ref 54). One investigation (Ref 63) shows the grain effect on mild steel thresholds (Fig. 47). Al-

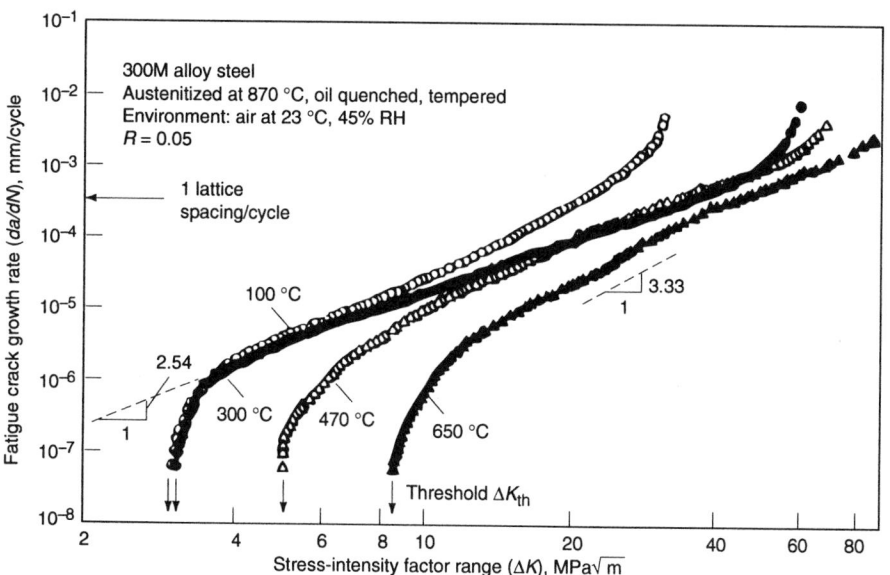

Fig. 44 Effect of strength level on fatigue crack growth rates. Variation of fatigue crack propagation in moist air at R = 0.05 with ΔK for ultrahigh-strength 300-M martensitic steel, quenched and tempered at temperatures from 100-650 °C (212-1200 °F) to produce tensile strengths from 2300-1190 MPa (330-275 ksi), respectively. Source: Ref 57

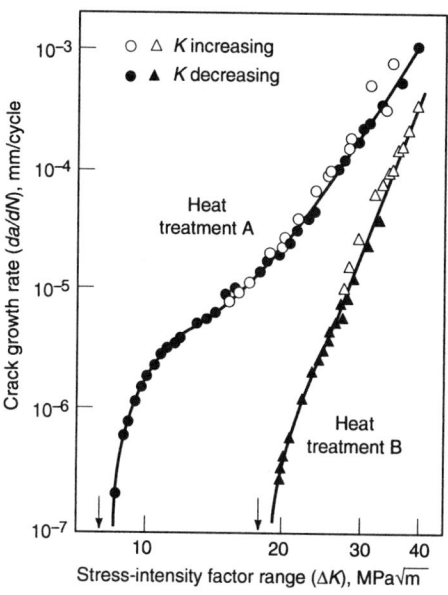

Property	Heat treatment A (ferritic)	Heat treatment B (martensitic)
0.2% proof stress, MPa (ksi)	293 (43)	452 (66)
Ultimate tensile strength, MPa (ksi)	543 (79)	750 (109)
Elongation, % in 12.7 mm	…	9.9
Reduction in area, %	32.6	…

Fig. 46 Rate of crack growth vs. ΔK for two microstructures in low-carbon steel (0.15-0.20% C, 0.60-0.90% Mn, 0.04% max P, 0.04% max S). Source: Ref 62

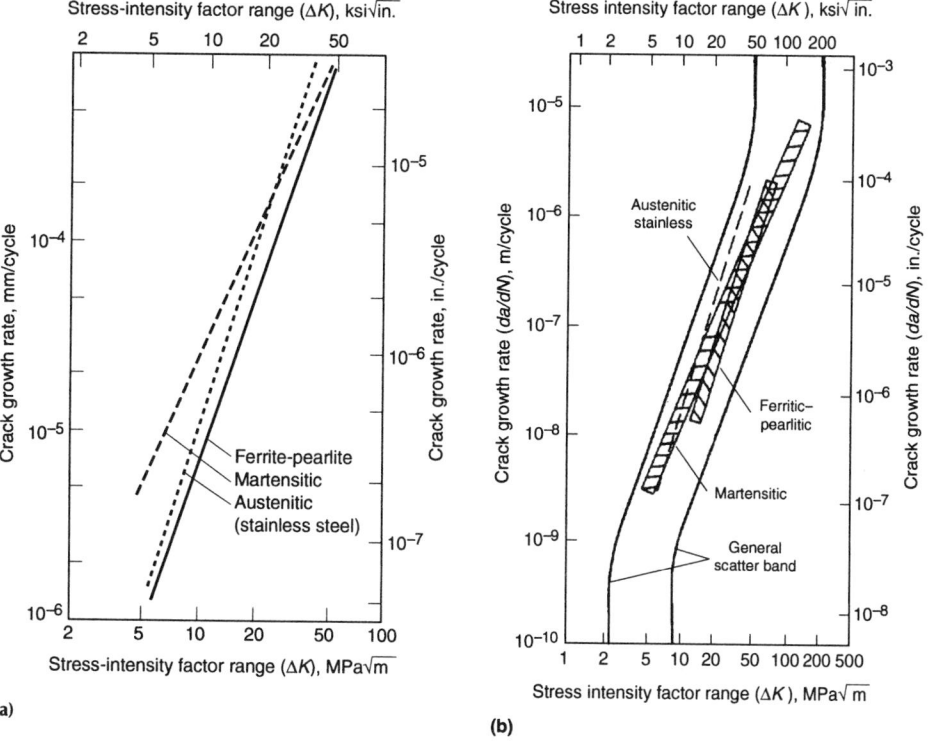

(a)

(b)

Type of microstructure	Yield strength MPa	ksi	Tensile strength MPa	ksi	Strain-hardening exponent
Austenitic (stainless steel)	205-345	30-50	515-655	75-95	>0.30
Ferrite-pearlite	205-550	30-80	345-755	50-110	0.15-0.30
Martensitic	>480	>70	>620	>90	<0.15

Fig. 45 Fatigue crack growth rates for ferritic, martensitic, and austenitic steel microstructures. (a) Upper limits of fatigue crack growth rates for three types of steel microstructures. Source: Ref 17. (b) Superposition of scatterbands on general scatterbands for steels

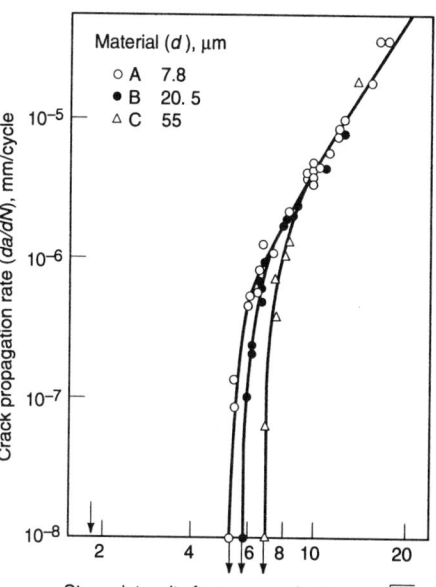

Fig. 47 Relationship between crack propagation rate and stress-intensity factor range for R = −1. Steels are ferrite-pearlite mixtures containing (in wt%) 0.2 C, 0.92 Mn, 0.26 Si, 0.11 P, 0.15 S, and 0.009 N, with ferrite grain sizes of 7.8, 20.5, and 55 μm. Source: Ref 63

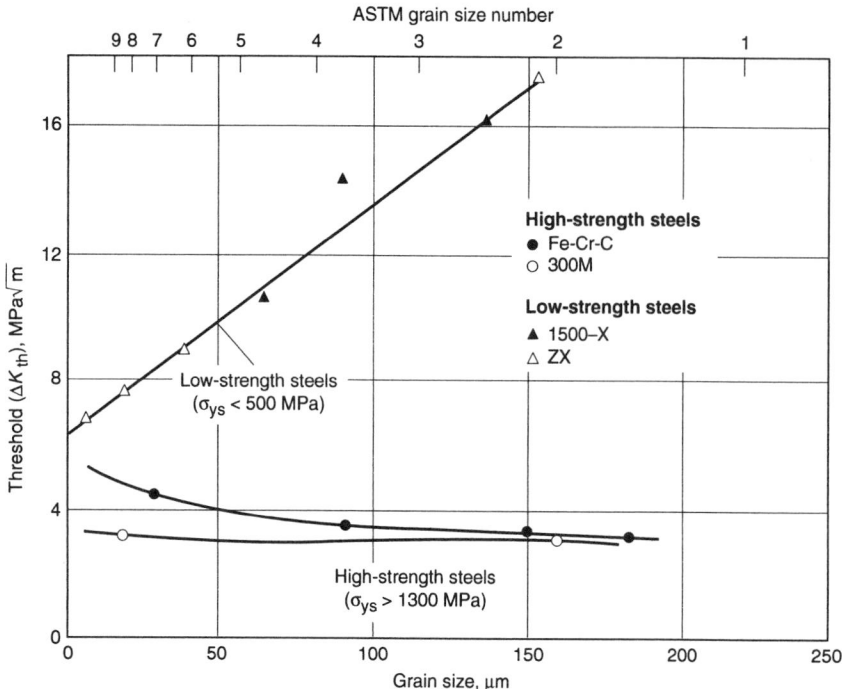

Fig. 48 Variation of threshold ΔK_{th} with grain size for steels at $R = 0.05$. For ferritic-pearlitic low-strength steels, grain size refers to ferritic grain size; for martensitic high-strength steels, grain size refers to prior-austenite grain size. Source: Ref 57

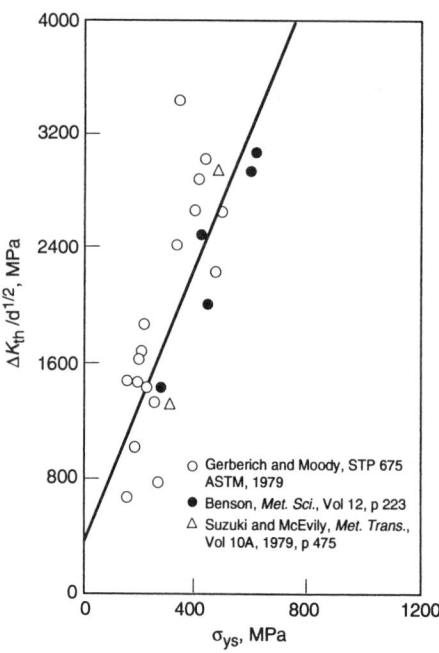

Fig. 49 Effect of yield strength on normalized threshold stress-intensity range

though there was no difference in region 2 growth, the threshold values varied from 5.3 to 7.0 MPa√m (4.8 to 6.4 ksi√in.) as grain size increased from 7.8 to 55 μm. A collection of data from two reviews (Ref 57, 58) shows this effect very well. Figure 48 shows the beneficial effect of large grain size on thresholds for low-strength steels and the detrimental effect of large grain size for high-strength steels. The former is considered to be related to a cyclic slip or microstructurally sensitive crack path that is controlled by the grain diameter, whereas for the high-strength steel, the effect is considered to be environmentally related (Ref 57). For the low- to medium-strength steels, it is useful to compare the grain

size with the reversed plastic zone size, $R_{p\pm}$, because many studies (Ref 64-66) have claimed that this is the controlling microstructural unit. Using twice the plane-strain plastic zone radius and twice the yield stress due to the stress reversal gives:

$$R_{p\pm} \cong \frac{2\Delta K^2}{6\pi(2\sigma_{ys})^2} = \frac{\Delta K^2}{12\pi(\sigma_{ys})^2} \qquad \text{(Eq 24)}$$

One way of understanding this is to consider that general cyclic slip will not proceed if the grain size is greater than the reversed plastic zone size. Thus,

a description of the fatigue threshold could be obtained by substitution of $d = R_{p\pm}$

$$K_{th} = \sigma_{ys}(12\pi d)^{1/2} = 6.14\sigma_{ys}(d)^{1/2} \qquad \text{(Eq 25)}$$

If threshold values were normalized by grain size and plotted against yield strength, the slope of this curve could then be compared with Eq 25. A collection of data in Fig. 49 plotted in this manner predicts a least-squares slope of 5.0 instead of 6.14. However, if the curve is forced through the origin, the best fit gives a slope of 6.06, which verifies Eq 25 and demonstrates that both ferrite grain size and yield strength have strong effects on low-strength threshold stress intensities.

This *does not* necessarily mean that a designer should select large-grained steels for perform-

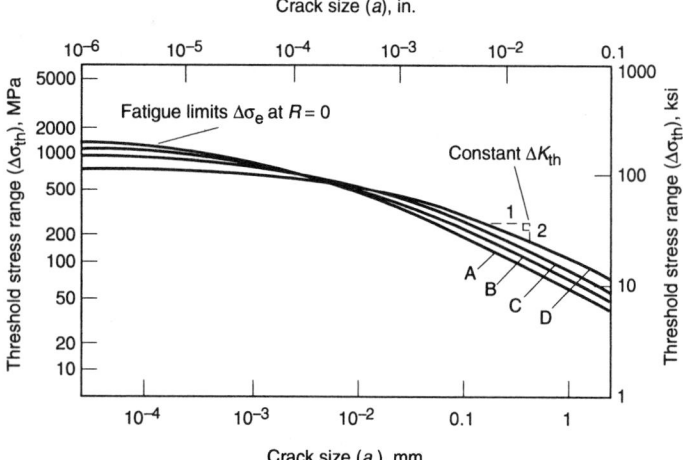

Fig. 50 Predicted variation of threshold stress $\Delta\sigma_{th}$ at $R = 0$ with crack size a. Based on data for 300M ultrahigh-strength steel tempered at temperatures from 100 to 650 °C (212 to 1200 °F) to produce a variety of tensile strengths. Source: Ref 57

Curve	Tempering temperature		Ultimate tensile strength	
	°C	°F	MPa	ksi
A	100	212	2238	324
B	300	570	1737	252
C	470	880	1683	244
D	650	1200	1186	172

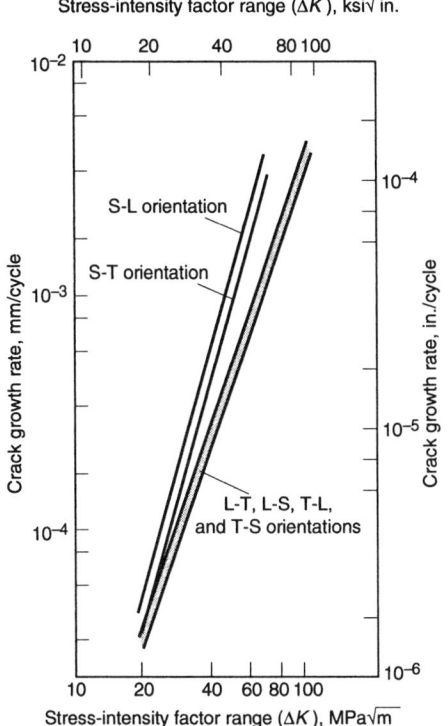

Fig. 51 Fatigue crack growth rates for ASTM A533B steel tested for each of six orientations. Source: A.D. Wilson, Fatigue Crack Propagation in A533B Steel, *Trans. ASME, J. Pressure Vessel Technol.*, Vol 99, 1977, p 459-469

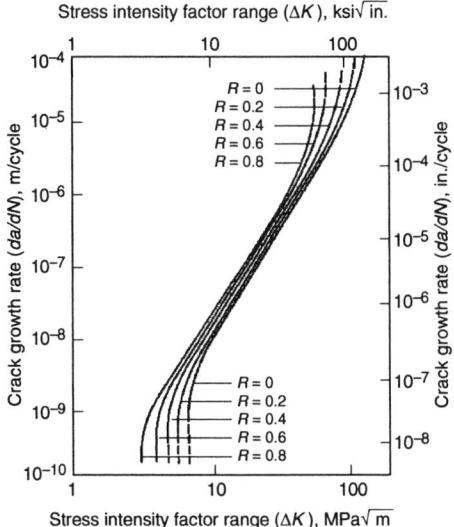

Fig. 52 General effect of mean stress influence on fatigue crack growth rates

ance. The number of cycles to produce a crack that leads to failure also depends heavily on grain size, with the finest grains giving the best performance. Thus, fine grains appear to be best for resistance to fatigue initiation, and coarse grains appear to be best for resistance to crack propagation for long cracks. In fact, it may be shown that there is a crossover in endurance limit for very

short cracks versus long cracks. The short-crack endurance limit is highest for fine-grained steel, and the long-crack endurance limit is highest for coarse-grained steel. A similar crossover has been postulated for high-strength steels, except that the critical parameter is yield strength. This is shown in Fig. 50 for a 300M steel quenched and tempered to wide variations in strength level. Whereas the low-strength material would offer the best performance for long cracks, the high-strength material would be superior for short cracks. This becomes important because the optimization of microstructure depends on whether the structural engineer is designing for crack initiation or crack propagation from a pre-existing flaw.

Crack growth resistance also may be different for cracks that propagate along different directions measured relative to the rolling direction. Figure 51 presents data on crack growth rate for specimens of conventionally melted ASTM A533 type B steel tested for each of six orientations. In this study, the effects of three processing techniques—conventional melting, calcium treatment, and electroslag remelting—were investigated. The electroslag remelted material produced the lowest crack growth rates and the least sensitivity to orientation.

Mean Stress Effects

In both fatigue life and fatigue crack propagation testing, the mean stress intensity is an important variable. Because K is proportional to stress (S), the standard ratio ($R = K_{min}/K_{max} = S_{min}/S_{max}$) is the variable most often used in fatigue crack growth testing. Unlike stress or strain life (S-N or ε-N) testing (which are usually done in the fully reversed, $R = -1$, stress condition), fatigue crack growth data is usually done under cyclic tension, $R = 0$. Fatigue crack growth testing at $R = 0$ or approximately zero is based on the concept that the crack would close during compression loading. In this case, the stress-intensity factor, K, would vanish. Thus, the compression loads should have little influence on constant-amplitude fatigue crack growth behavior. In general, this is fairly realistic, but under variable-amplitude loading, compression cycles can be important to fatigue crack growth (see the section "Variable-Amplitude Loading" in this article).

The general influence of mean stress effects is shown in Fig. 52. Mean stress effects in region II are small, while larger effects occur in regions I and III. The most commonly used equation to model stress effects in regions II and III is the Forman equation (Ref 67):

$$\frac{da}{dN} = \frac{A(\Delta K)^n}{(1 - R)K_c - \Delta K} \qquad \text{(Eq 26)}$$

where A and n are empirical fatigue material constants and K_c is the applicable fracture toughness for the material and thickness. The Forman equation is a modification of the Paris equation. Typical mean

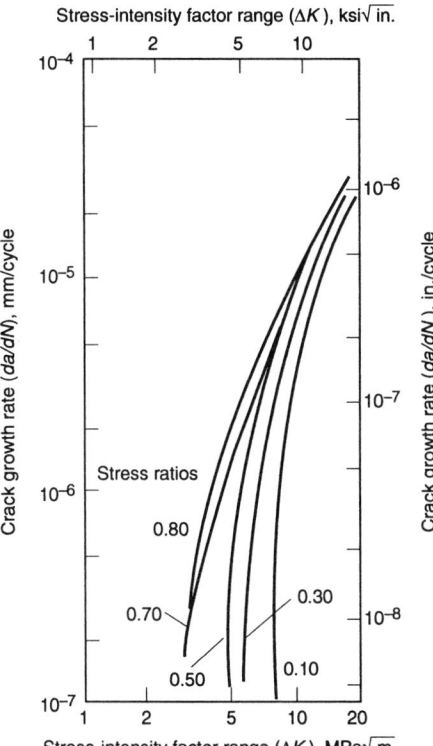

Fig. 53 Effect of stress ratio on fatigue threshold stress-intensity factor range, ΔK_{th}, for A533B-1 steel. Data are for A533B-1 steel in region 1, tested at various stress ratios, at 60 Hz and 25 °C (75 °F) in air. Source: Ref 17

stress effects in the threshold region are shown in Fig. 53 for manganese-molybdenum-nickel pressure vessel steel (ASTM A533B, class 1, 80 to 100 ksi tensile strength). Figure 54 at $R = 0.1$ is for comparison.

The effect of negative R-ratios, which includes compression in the cycle, has not been sufficiently investigated, particularly at the threshold levels. The results of many negative R-ratio tests on wrought and cast steels, cast irons, and aluminum alloys subjected to constant-amplitude conditions in regions II and III indicate that crack growth rates based on ΔK values (which neglect compressive nominal stresses) are similar to $R = 0$ results, or are increased by not more than a factor of 2 (Ref 68).

Mean Stress Effects on Fatigue Crack Growth Thresholds. As shown in Fig. 63(b) and Fig. 64, mean stress effects have a significant influence on fatigue crack growth thresholds. Empirical relations (such as Eq 19) have been considered for the effect of stress ratios on thresholds. Two examples are described below.

Example 1. In a class of low- to medium-strength steels, the effect of stress ratio on the threshold stress-intensity range, ΔK_{th}, apparently is similar for seven different materials, as shown in Fig. 55. For values of R greater than 0.1, ΔK_{th} can be given by:

$$\Delta K_{th} = C_1(1 - 0.85R) \qquad \text{(Eq 27)}$$

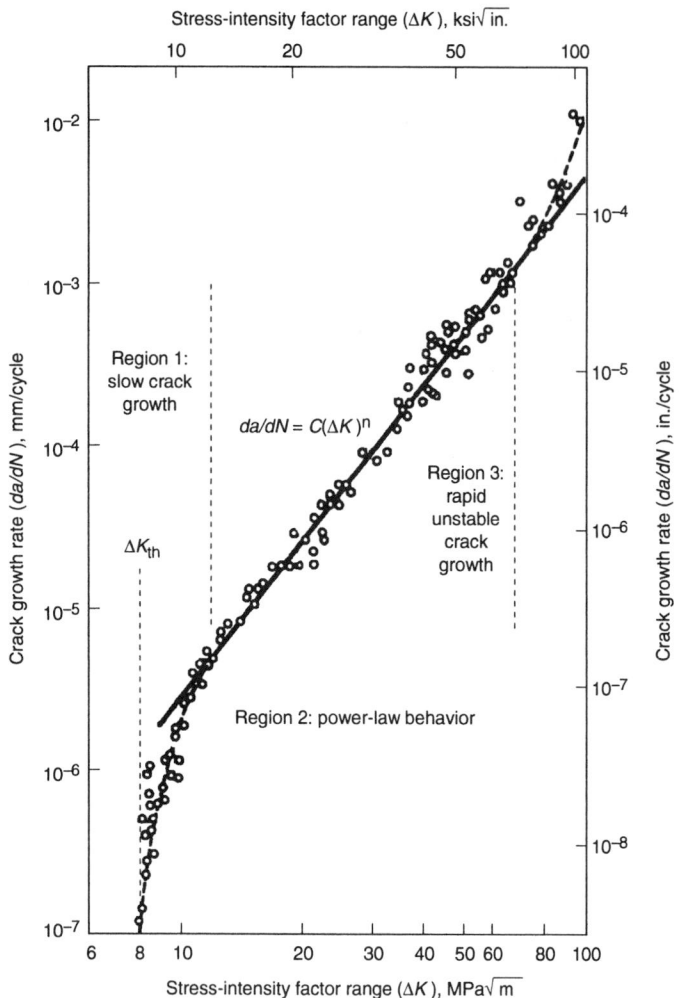

Fig. 54 Fatigue crack growth behavior of ASTM A533 B1 steel. Yield strength of 470 MPa (70 ksi). Test conditions: $R = 0.10$; ambient room air, 24 °C (75 °F)

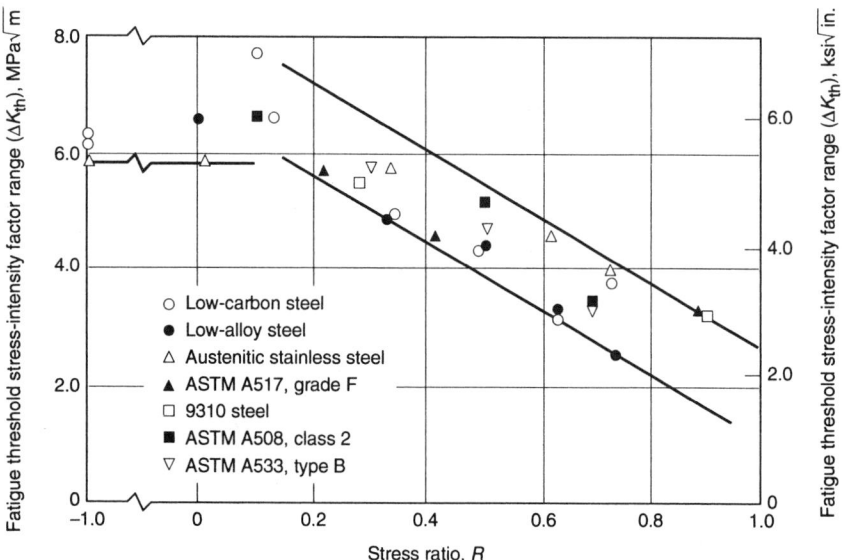

Fig. 55 Effect of stress ratio on fatigue threshold stress-intensity factor range, ΔK_{th}, for several steels. Source: Ref 17

where C_1 is a constant whose value is 7.0 MPa$\sqrt{m}$ (6.4 ksi$\sqrt{in}$). Because of the narrow grouping of data in Fig. 55, it must be assumed that these were relatively fine-grained steels without the microstructural influences previously discussed. Where greater influences of strength, grain size, microstructure, or environment are encountered, the constant C_1 would necessarily change and the R-effect might even be qualitatively different. For example, in vacuum test conditions, the threshold may be independent of R (Ref 57).

Example 2. Another relationship between ΔK_{th} and the applied stress ratio, R, is provided in BS PD6493:

$$\Delta K_{th} = 190 - 144\,R\ \text{MPa}\sqrt{mm} \qquad \text{(Eq 28)}$$

This relationship provides the lower bound to all published threshold data for British grade 50D steel in air and seawater (Ref 69). It has been suggested that other data for similar steels and for austenitic steels lie below the PD6493 line (Ref 70). Including these data, the following relationship, based on a 97.7% probability of survival for the data in Fig. 56, has been recommended (Ref 70):

$$\Delta K_{th} = 170 - 214\,R\ \text{MPa}\sqrt{mm}\ \text{for}\ 0 \le R < 0.5$$

$$= 63\ \text{MPa}\sqrt{mm}\ \text{for}\ R \ge 0.5 \qquad \text{(Eq 29)}$$

Variable-Amplitude Loading

Fatigue testing under constant-amplitude loading is substantially different from variable-amplitude service loads, and many methods have been developed to predict or model variable-amplitude performance from data generated in constant-amplitude testing. Different methods are used for fatigue life (*S-N*) data and crack propagation. More information on various methods is described in the article "Estimating Fatigue Life" in Section 3 and the article "Fatigue Crack Growth under Variable Amplitude Loading" in Section 2 of this Volume.

Prediction of fatigue crack growth rates under variable-amplitude loading can be complicated when retardation or acceleration of crack growth is affected by overloads. It is well known that tensile overstressing during fatigue crack propagation usually causes retardation of crack propagation, its effect being beneficial rather than detrimental (Ref 71). Experimental evidence and theoretical analyses have been accumulated to show that the retardation is due to compressive residual stresses developed within the plastic zone ahead of the fatigue crack by the overstressing (Ref 71).

However, it has also been shown that compressive applied stress relieves the residual stress, resulting in substantial reduction of the retardation effect (Ref 71, 72). In fully reversed tension-compression tests, the retardation was completely suppressed and even some acceleration was observed in the transient period immediately after the overloading (Ref 73, 74). This effect leads to the significant acceleration.

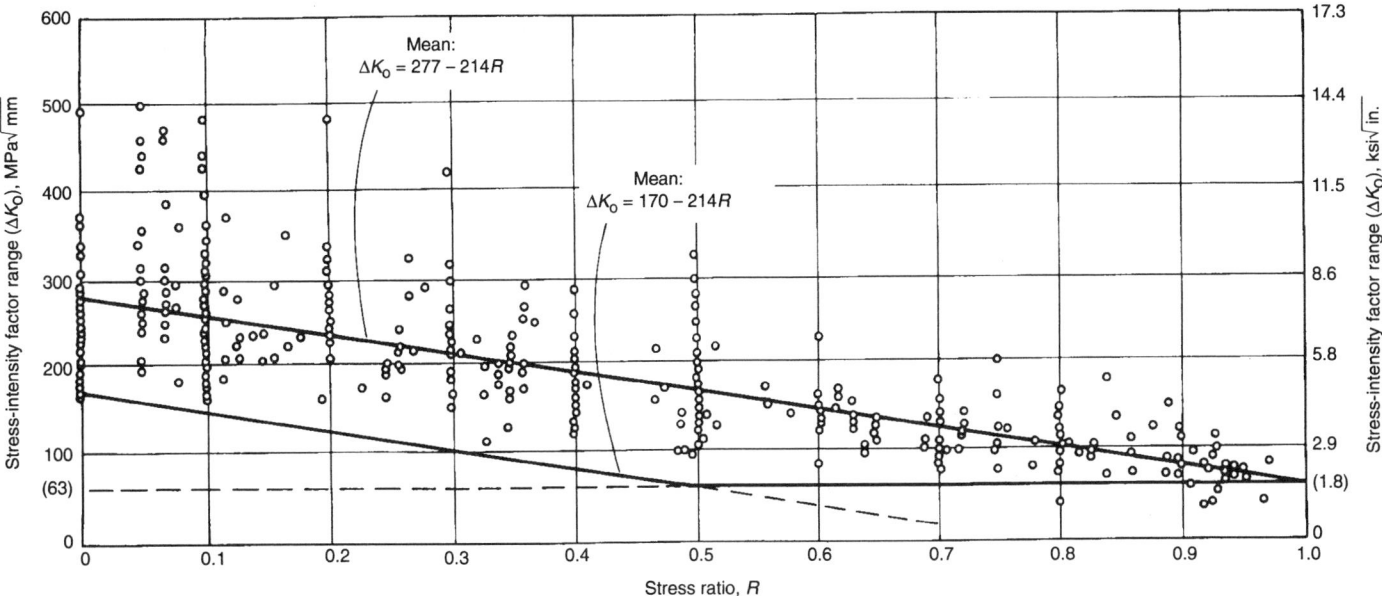

Fig. 56 Fatigue crack growth threshold data for ferritic steels with yield strengths up to 600 MPa (87 ksi). ΔK_o is the threshold intensity (ΔK_{th}).

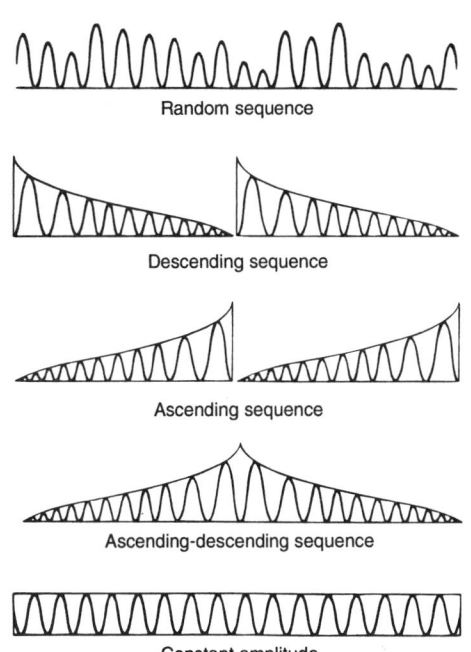

Fig. 57 Schematic representation of load sequences for fatigue crack growth rate testing. Source: Ref 17

In one case, for example, significant acceleration (more than one hundred times) occurred where a very small number of cycles (e.g., 2 cycles) of overstress were applied intermittently between very large numbers of cycles (of the order of 10^6 cycles) of understress below the threshold stress intensity, ΔK_{th} (Ref 75). These results indicate that ΔK_{th}, which has been thought to be an important design criterion in fracture mechanics, may have less significance as a threshold for fatigue crack propagation under such variable stress conditions. This acceleration

can be a vital factor that has to be considered in the design of such components.

When retardation or acceleration of crack growth rates is significant, the prediction of fatigue crack growth rate under variable amplitude loading is complex. Many complex cycle-by-cycle models have been developed and proposed in the literature (Ref 76-82) to take these factors into account. Some textbooks (Ref 83, 84) also provide introductory descriptions of sequence effects. These models require knowledge of the order of appearance of the peaks and valleys of the loading history, information which is not always available. However, even in simulation loading, for which cycle interaction effects are expected to be important, a simple linear accumulation model can produce good predictions as compared to much more complex models (see, e.g., Ref 76, p 103-114).

Summation of Crack Increments. If stress overloads or load history (sequence effects) do not significantly affect fatigue crack growth, then crack growth (Δa) in each individual cycle of variable-amplitude loading can be estimated from the da/dN versus ΔK curve. Summing the values of Δa, while keeping track of the number of cycles, is a straightforward method of estimating fatigue life under variable-amplitude loading. However, this method does not provide accurate estimates if sequence effects or overstresses cause retardation or acceleration of fatigue crack growth rates.

Block or Spectrum Loading. Another approach is to approximate the actual service load with repeated application of different loading sequences of finite length. This procedure is numerically equivalent to the summation of crack increments if the crack length, a, does not change by a large amount during one repetition of the sample load history.

Block or spectrum loading is often used to predict the life of structural components. To correlate variable-amplitude data with constant-amplitude data, some method of normalizing the varying stress-intensity factor ranges in the spectrum is necessary. One such normalization scheme is to use the root mean square (rms) value of the stress-intensity factor. Here ΔK_{max} would replace ΔK in the Paris power-law relationship (Eq 20) where ΔK_{rms} is given by:

$$\Delta K_{rms} = \sqrt{\frac{\sum_{i=1}^{N} (\Delta K_I)^2}{N}} \qquad \text{(Eq 30)}$$

Barsom (Ref 17) used the loading sequences shown in Fig. 57 to establish a database for evaluation of the ΔK_{rms} normalization procedure. The results shown in Fig. 58 show an excellent correlation between the constant-amplitude data (ΔK) and the variable-amplitude data (ΔK_{rms}). It should be emphasized that such an independence of loading sequence may not apply to region 1 crack growth or to situations involving aggressive environments.

Temperature Effects

High-Temperature Fatigue Crack Growth Rates. Increasing temperature increases fatigue crack growth rates. A clear trend of crack growth rate increase with increasing temperature can be seen in Fig. 59. In this figure it can be seen that at temperatures up to about 50% of the melting point (550 to 600 °C, or 1020 to 1110 °F), the growth rates are relatively insensitive to temperature, but that the sensitivity increases rapidly at

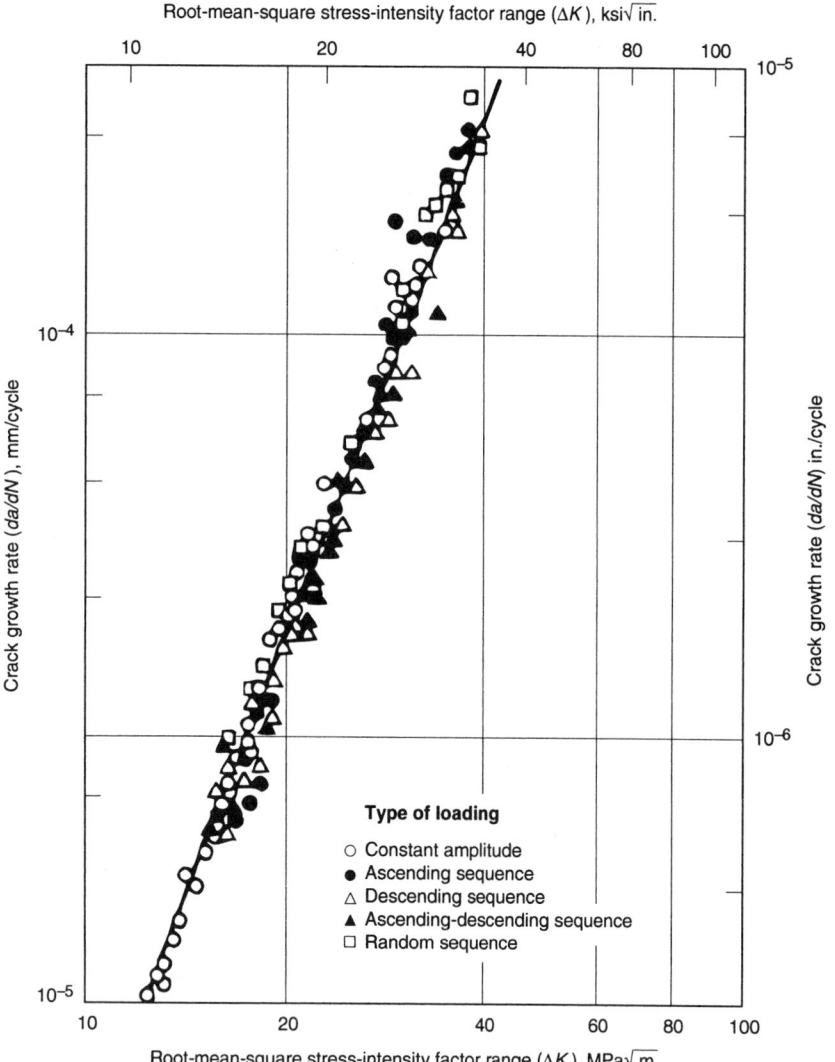

Root-mean-square stress-intensity factor range (ΔK), ksi$\sqrt{}$ in.

Type of loading

○ Constant amplitude
● Ascending sequence
△ Descending sequence
▲ Ascending-descending sequence
□ Random sequence

Root-mean-square stress-intensity factor range (ΔK), MPa$\sqrt{}$ m

Fig. 58 Effect of loading sequence on fatigue crack growth rate. The loading patterns shown in Fig. 57 were imposed on specimens of A514B steel; rationalizing the stress-intensity factor range by a root-mean-square method caused the data to fall within a single band, independent of loading pattern. Source: Ref 17

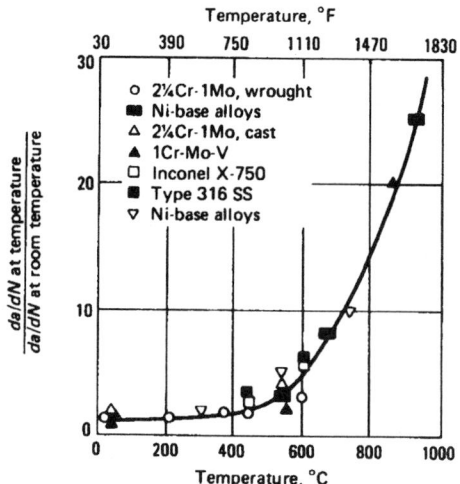

○ 2¼Cr-1Mo, wrought
■ Ni-base alloys
△ 2¼Cr-1Mo, cast
▲ 1Cr-Mo-V
□ Inconel X-750
■ Type 316 SS
▽ Ni-base alloys

Temperature, °C

Fig. 59 Variation of fatigue crack growth rates as a function of temperature at ΔK = 30 MPa$\sqrt{}$m (27 ksi$\sqrt{}$in.)

higher temperatures. The crack growth rates for all the materials at temperatures up to 600 °C (1110 °F) relative to the room-temperature rates can be estimated by a maximum correlation factor of 5 (2 for ferritic steels). Additional information on high-temperature steels (such as Cr-Mo steels and austenitic stainless steels) is provided in other articles in this Volume.

Temperature can affect crack growth rates in several ways. At high temperatures, ease of dislocation motion, dynamic strain aging, or oxidation can be sufficiently enhanced so that creep-fatigue or environmental-fatigue interactions result. McHenry and Pense (Ref 85) found that the crack growth behavior of HP 9-4-35 steel was not appreciably affected by temperatures as high as 350 °C (650 °F) or changes in frequency from 0.02 to 10 Hz. However, both ASTM A212, grade B (now ASTM A515, grade 70), and A517, grade F steels showed that fatigue crack growth rates increased with increasing testing temperature, as illustrated in Fig. 60. Decreasing the frequency of

loading for specimens of A517, grade F steel also increased the rate of crack growth.

Low-Temperature Fatigue Crack Growth. The fatigue crack growth behaviors of ferritic steels show dramatic changes at low temperatures, which correlate with their fracture toughness behavior. In general, reduced temperatures have insignificant effects or are slightly beneficial as long as ductile cracking modes are operative. The onset of cleavage, however, ultimately causes a dramatic acceleration of fatigue crack growth at the lowest temperatures. The behavior of 9% Ni steel (Fig. 61) is typical: as the temperature is reduced, there is a slight improvement between 295 and 111 K, followed by a rapid loss of fatigue crack growth resistance between 111 and 4 K (Ref 86). The temperature dependence of fatigue crack growth resistance in this case closely parallels the temperature dependence of fracture toughness.

At low temperatures, the yield strength may be sufficiently increased so that cleavage is pro-

moted at the fatigue crack tip. In Fig. 62, Stonesifer (Ref 87) shows that very low temperatures produce very high crack growth rates at low ΔK levels in A533B steel. It is clear that the slope of the –196 °C (–320 °F) curve is such that n in the power-law relationship $da/dN = C(\Delta K)^n$ is an order of magnitude larger for the –196 °C data than it is for the 20 °C (68 °F) data. This effect of testing temperature on the fatigue crack propagation exponent has been reviewed, for a large number of iron-base alloys and steels (Ref 58). The results in Fig. 63 demonstrate the increase in the value of n with decreasing testing temperature.

However, from these results, one should not assume that crack growth rates are going to be higher at any temperature below room temperature. In fact, at temperatures only slightly below room temperature, improved resistance to fatigue crack propagation may be observed if severe microcleavage can be avoided. This implies that there is some low temperature where there is a crossover in resistance; this has been discussed elsewhere for iron-base systems (Ref 88).

The important point is that one should avoid extrapolating room-temperature behavior to temperatures below room temperature. To avoid microcleavage phenomena, one solution is to use Fe-Ni steels. The nickel additions decrease the ductile-to-brittle transition and result in low fatigue crack growth exponents, even at –196 °C (–320 °F) (Fig. 61, 63).

Aggressive Environments

Aggressive environments affect fatigue crack growth rates in all regions of growth but most often in regions 1 and 2, where exposure times are much greater. Three patterns of corrosion fatigue behavior that might be observed in steels are indicated in Fig. 64. Figure 65(a) illustrates the

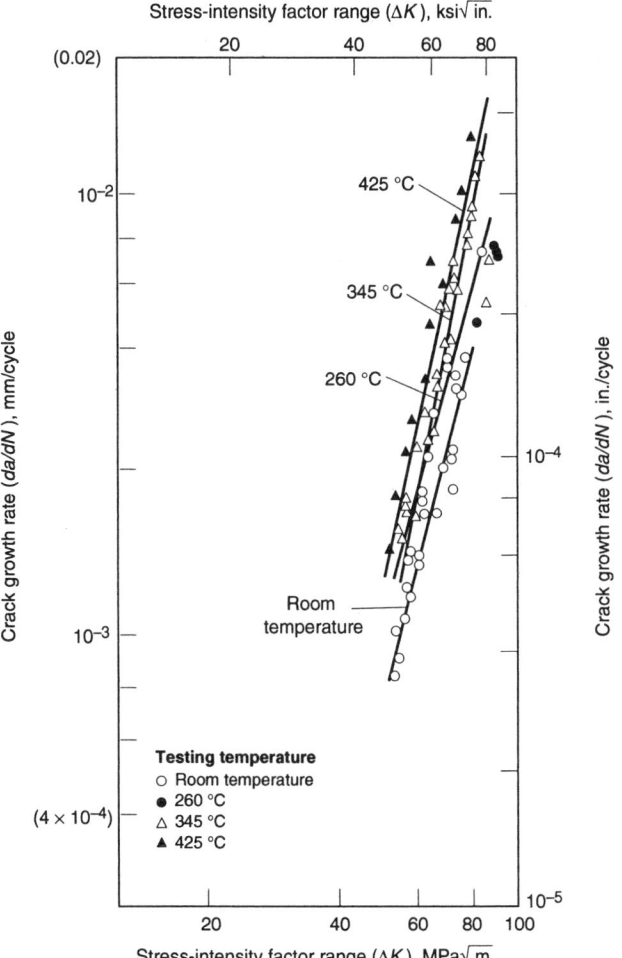

Fig. 60 Effect of elevated testing temperature on fatigue crack growth rates in ASTM A212B and A517F steels. Source: Ref 85

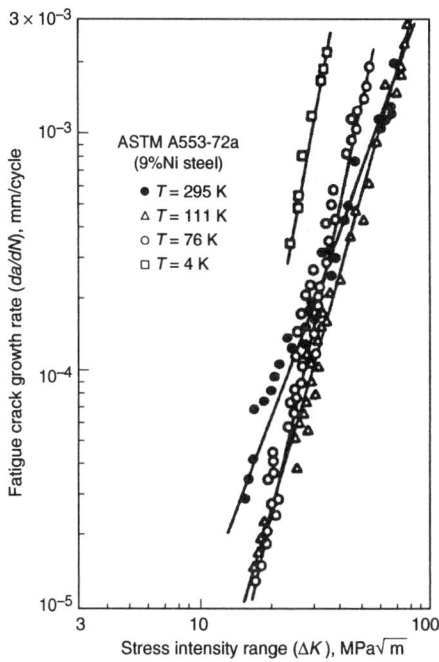

Fig. 61 Effect of temperature on the fatigue crack growth rates of a 9% Ni ferritic steel. Source: Ref 86

Table 5 Selected corrosion fatigue crack growth data for steels

Steel	Solution	Frequency dependent?	$\dfrac{(da/dN)_{sol}}{(da/dN)_{air}}$	Stress-intensity range, MPa $\sqrt{m}$	Reference
A212B	Liquid Na	Yes	3	50<ΔK<80	85
0.5Cr-0.5Mo-0.25V	Distilled H_2O	?	10	12<ΔK<24	93
A302B	200 °C H_2O	Yes	4	50<ΔK<100	94
X-65 line pipe	1 ppm H_2S crude oil	No	1	10<ΔK<25	90
X-65 line pipe	1 ppm H_2S crude oil	Yes	5	25<ΔK<70	90
X-65 line pipe	4700 ppm H_2S crude	No	20	10<ΔK<25	90
X-65 line pipe	4700 ppm H_2S crude	Yes	20	25<ΔK<70	90
9Ni-4Co-0.25C	3½%NaCl	No	1	50<ΔK<100	95
12Ni (1240 MPa) maraging	3½% NaCl	?	4	30<ΔK<100	95
18Ni (1720 MPa) maraging	Dehumidified hydrogen	Yes	10	20<ΔK<45	96
12Ni-5Cr-3Mo	3% NaCl	Yes	4	20<ΔK<80	60
12Cr-8Ni-2Mo	Distilled H_2O	Yes	10	30<ΔK<90	97

corrosion fatigue behavior of AISI 4340 steel quenched and tempered to a yield strength of 1700 MPa (245 ksi) (Ref 89). In distilled water, a 1000-fold increase in the crack growth rate is produced by a 1000-fold decrease in the cyclic frequency. This behavior falls somewhere between types B and C behavior in Fig. 64. Similar behavior is observed for X-65 line pipe steel tested in salt water with a superimposed cathodic potential (Ref 90). However, as shown in Fig. 65(b), the same change in frequency does not produce quite as large a change in fatigue crack growth rate. In addition, the cathodic potential increases the growth rate substantially over the open-circuit case for X-65 steel. Both the effects in Fig. 65 have been associated with hydrogen entry, and in fact most corrosion fatigue of high- and ultrahigh-strength steels is considered to oc-cur by a combined mechanism of hydrogen embrittlement and fatigue.

Effects of alloying additions are shown in Fig. 66 and 67. First, it can be seen in Fig. 66 that for the identical test frequency, the susceptibility of a 12Ni-5Cr-3Mo maraging steel to corrosion fatigue in a 3% aqueous sodium chloride solution is nearly identical to that of 4340M steel in distilled water. It should be noted that the K_{Iscc} value for this alloy is 60 MPa $\sqrt{m}$ (55 ksi $\sqrt{in}$.), so that nearly all of the environmental effects show up well below the static stress-corrosion threshold, K_{Iscc}. This corresponds most closely to type A behavior in Fig. 64. The same data for 12Ni-5Cr-3Mo at 0.1 Hz are compared with those for two other alloys in Fig. 67. The lowest corrosion-fatigue crack growth rates were achieved in specimens of the 10Ni-Cr-Mo-Co steel.

For lower-strength steels tested in a 3% solution of aqueous sodium chloride, only slight effects were observed in A36, A588 grade A, and A514 grade F steels at frequencies of 0.2 and 1 Hz (Ref 91). However, this was for relatively high growth rates in region 2. It is not clear what effects would occur in region 1 or at lower cyclic frequencies. Results of fatigue crack growth rate tests on specimens of A508-2, A533B, A516, and SA333 in reactor-grade water indicate that the growth rate curves tend to lie in three basic regimes, depending on test conditions (Ref 92).

Corrosion fatigue data for several alloy steels in a variety of environments are given in Table 5. In general, corrosion fatigue depends on cyclic frequency and testing temperature and may occur over a large stress-intensity range and in solutions of any pH value.

Inhibiting Corrosion Fatigue. There is some evidence that inhibitors provide effective passive films or modify the pH at the crack tip (Ref 98). The result in high-strength AISI 4340 steel, shown in Fig. 68, is that growth rates in air at 90% relative humidity are decreased by a factor of about three at a cyclic frequency of 0.167 Hz. These inhibitors were added to the humidified test chamber by means of an organic complex wherein the inorganic inhibitors were incorporated into a quaternary ammonium salt. The result is that the amount of intergranular fracture decreases and the environmental susceptibility to hydrogen embrittlement is reduced.

Fatigue Crack Growth in Hydrogen Sulfide. Crude oil with hydrogen sulfide (H_2S) is a common aggressive environment for application of structural steels. The crack growth rates in sour crude oil with low (~1 ppm) and high (4,700 ppm) H_2S content are presented in Fig. 69. The results show a predominant effect of H_2S. In crude oil with low H_2S content, the growth rates in the low-ΔK range are lower than in air. This indicates that crude oil alone inhibits the crack growth at low ΔK and shifts the threshold to a higher ΔK value. Frequency affects growth rate in a similar way to salt water at free corrosion, but to much less a degree. Compared to air, maximum acceleration of growth rate in low-H_2S crude oil is only half of that in salt water under free corrosion at the same frequency (0.1 Hz).

In crude oil saturated with H_2S, contrary to the results in other environments, high acceleration of growth rate persists up to high ΔK and is affected by frequency only at high stress intensities. The growth rate curve exhibits a frequency-independent decrease in slope at a growth rate of 10^{-3} mm/cycle. From this value and above, the curve is parallel to the air data, and the crack grows 20 times faster than in air.

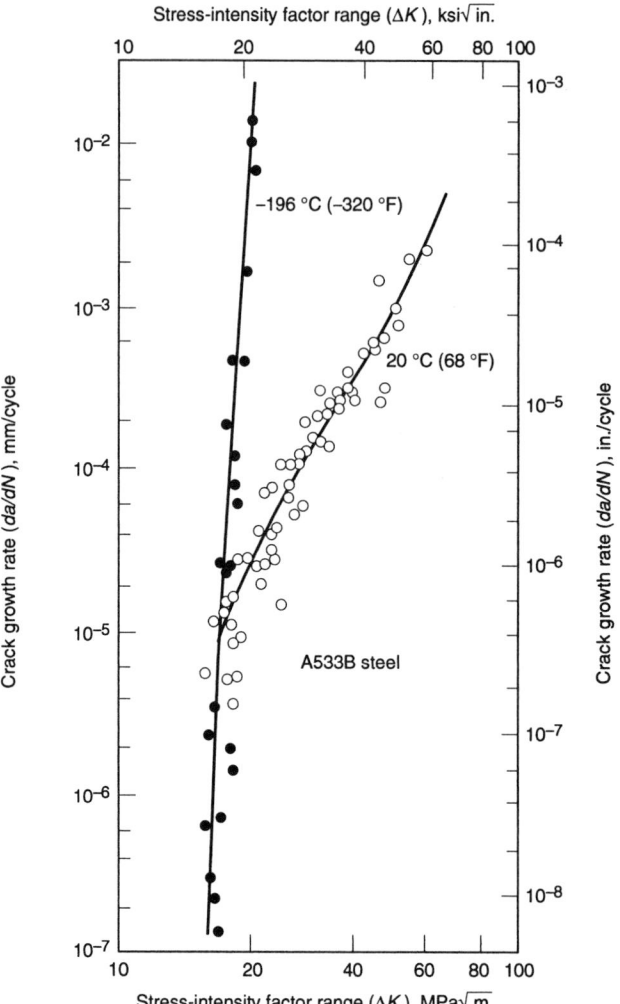

Fig. 62 Effect of low testing temperature on fatigue crack growth rates in A533B steel. Source: Ref 87

Sustained-Load Crack Propagation

Environmentally assisted cracking may occur in any class of steel, given the right combination of environment, tensile stress, temperature, and presence or absence of a superimposed electrical potential. For a given combination, delayed failure may occur within minutes, as in the case of ultrahigh-strength steel fasteners coupled to aluminum in the presence of water, or it may occur years later, as in the case of steam turbine disks exposed to high-purity steam.

Typical results for static loading of precracked specimens of four high-strength steels tested in distilled water are shown in Fig. 70. Each point represents a test of one specimen loaded to a particular level of stress intensity. Failure time periods depend on composition and strength of the steel and on stress intensity, provided that K_I is above a certain threshold level. The threshold stress intensity here is for tests of relatively thin sheet (plane stress). For considerably thicker

plate test conditions (plane strain), the threshold value would be the critical value of K_{Iscc}, which occurs at a significantly lower K-level than for thin sheet. This thickness transition has been discussed in detail elsewhere (Ref 100), but one should be very careful to evaluate the material in the thickness to be used or have data for a thicker test condition that would provide a conservative estimate. Unfortunately, no simple thickness relationship in terms of (K_{Iscc}/σ_{ys}) is available that would provide a basic index for K_{Iscc} in the same manner as for K_{Ic}. Plane-strain versus plane-stress effects also have been described (Ref 105).

Microstructural Fracture Modes. Examinations of environmentally assisted fractures in specimens of both low- and high-strength steels indicate that cracking may proceed from intergranular fracture to transgranular cleavage to transgranular dimple rupture with increasing applied stress intensity. One or more of these modes

may be absent, depending on the microstructure or the strength level of the material as well as on the aggressiveness of the environment.

Steel Selection

The most reliable method of preventing stress-corrosion cracking (SCC) of carbon and low-alloy steels is proper materials selection. Strength is perhaps the greatest influence on the SCC resistance of steels under a variety of conditions, with resistance decreasing as strength is increased. Steels with good fracture toughness also resist SCC. For example, a good correspondence between fracture toughness and SCC resistance in aqueous chlorides has been reported for 1034 MPa (150 ksi) steels (Ref 101). A similar correlation between Charpy fracture toughness and SCC in synthetic seawater was observed for steels with yield strengths of 690 to 1380 MPa (100 to 200 ksi) (Ref 102). By contrast, no significant correlation between toughness and SCC resistance in aqueous hydrogen sulfide has been reported for AISI 4xxx series steels (Ref 101).

Because of the predominant influence of strength level on SCC, limitations on steel strengths (or hardnesses) are the main selection criterion. For example, limiting the yield strength of steels used in aqueous chlorides to 690 MPa (100 ksi) is a typical requirement (Ref 103). In contrast, high-yield-strength steels are susceptible to SCC in most aqueous environments (Ref 104, 105). Figure 71 demonstrates that moisture (represented by moist argon) is alone capable of producing cracking in H-11 steel at the 1580 MPa (230 ksi) yield-strength level (Ref 106). Moisture is believed to cause condensation at the crack tip.

The most harmful environments are those that contain cathodic poisons (Ref 104, 105). These prevent the association of atomic hydrogen to molecular hydrogen and thus increase the supply of atomic hydrogen. Elements in groups V and VI of the periodic table, such as arsenic, selenium, tellurium, and sulfur, all act as poisons. These elements are listed in order of decreasing potency.

The most common cathodic poison is probably sulfur in the form of H_2S, as found naturally in "sour" oil wells. For exposure to these environments, hardness in steels is normally limited to 22 HRC (Ref 107). Because high-yield-strength steels usually have hardnesses in excess of 40 HRC, their use is inappropriate.

Of the various forms of SCC of carbon and low-alloy steels, SCC in aqueous chlorides is perhaps the most common. This is due largely to the variety of industries and environments in which aqueous chlorides are encountered. Interestingly, K_{Iscc} for AISI 4340 steel with yield strengths of 1380 and 1720 MPa (200 and 250 ksi) are very similar, regardless of chloride ion content (Ref 105). There is little variation in K_{Iscc} between distilled water and aqueous solutions containing varying amounts of chlorides. Similar behavior is found for the high-alloyed HP 9-4-45 steel (Ref 105). For 18Ni-300 maraging steel, K_{Iscc} is the same in distilled water, 3% NaCl solution, and H_2SO_4 (Ref 105).

Other aqueous solutions that may promote SCC include those containing sulfate, phosphate, or nitrate ions (Ref 105). High-yield-strength steels are also susceptible to organic compounds. This behavior has been attributed to the effect of impurities, similar to the response of titanium alloys to methanol.

The general pattern of behavior appears to be that K_{Iscc} is consistent in aqueous solutions, provided that a poison is not introduced. Relative humidity should be treated as if it were an aqueous solution. The K_{Iscc} for a maraging steel will be consistently higher than that of quenched-and-tempered steels at the same strength level. In general, maraging steels have better SCC resistance than other types of steels with comparable yield strengths (Ref 105, 108). The superiority of maraging steels in terms of K_{Iscc} and crack velocity for chloride-containing and nonchloride aqueous environments is evident from Fig. 72. Steel selection and SCC prevention for some common environments are described below.

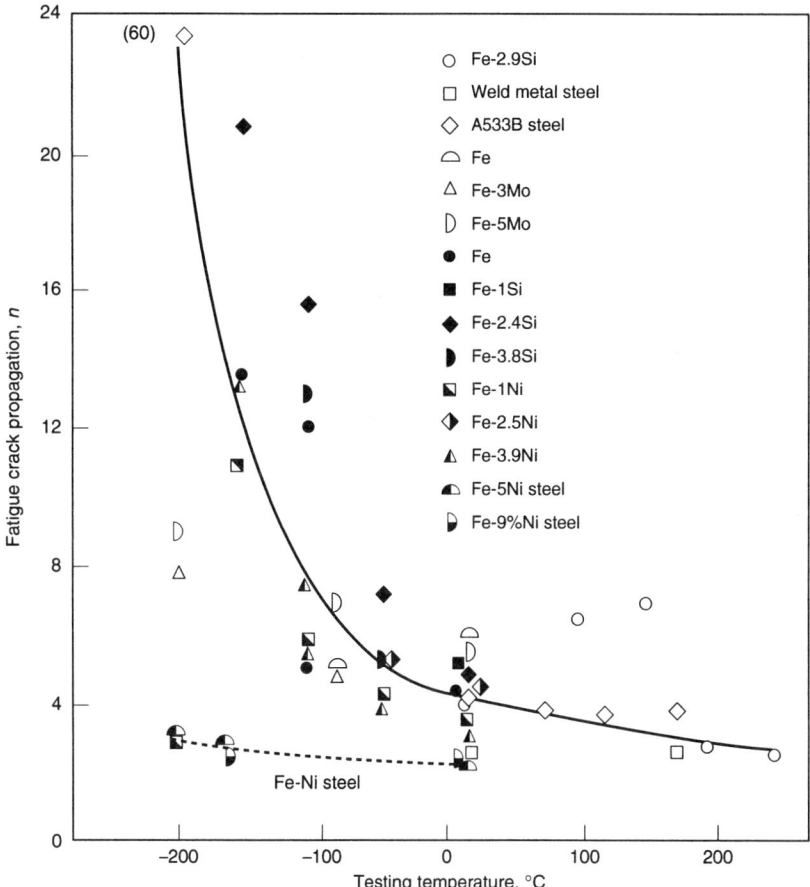

Fig. 63 Influence of testing temperature on fatigue crack propagation exponent for iron-base alloys. Source: Ref 68

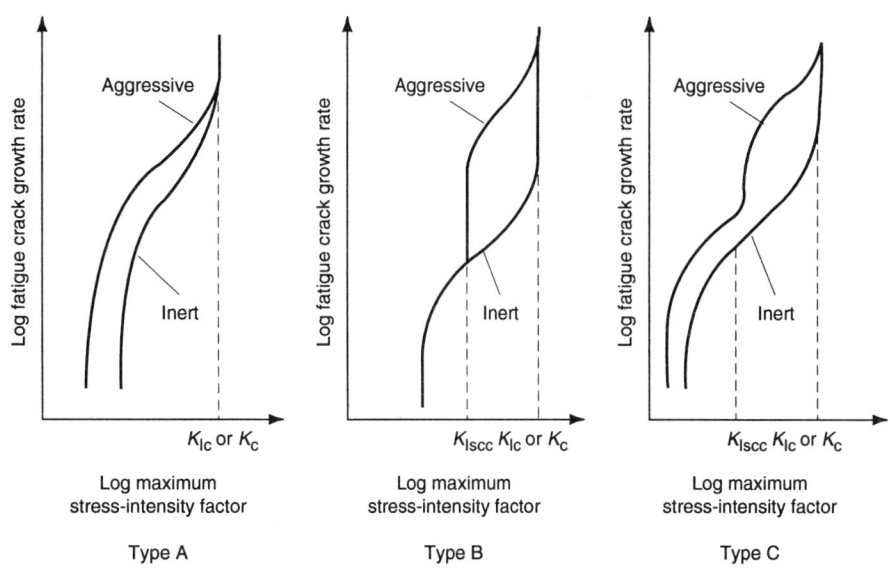

Fig. 64 Schematic diagrams showing three types of corrosion fatigue behavior. Source: A. McEvily and R. Wei, Fracture Mechanics and Corrosion Fatigue, *Proceedings—International Conference on Corrosion Fatigue*, NACE, 1971, p 381-395

Aqueous Chlorides. Although a clearly defined yield strength threshold for SCC in aqueous chlorides is not apparent, it can be generalized that steels with yield strengths of less than 690

MPa (100 ksi) are resistant (Ref 103). Above this value, SCC in aqueous chlorides can occur, with SCC resistance decreasing with increased yield strength. This strength level corresponds to a

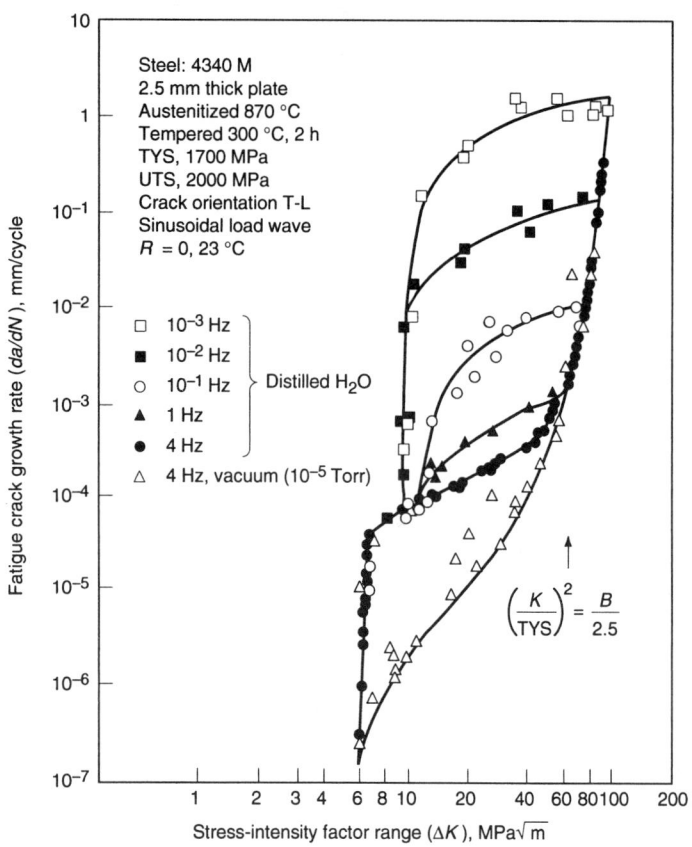

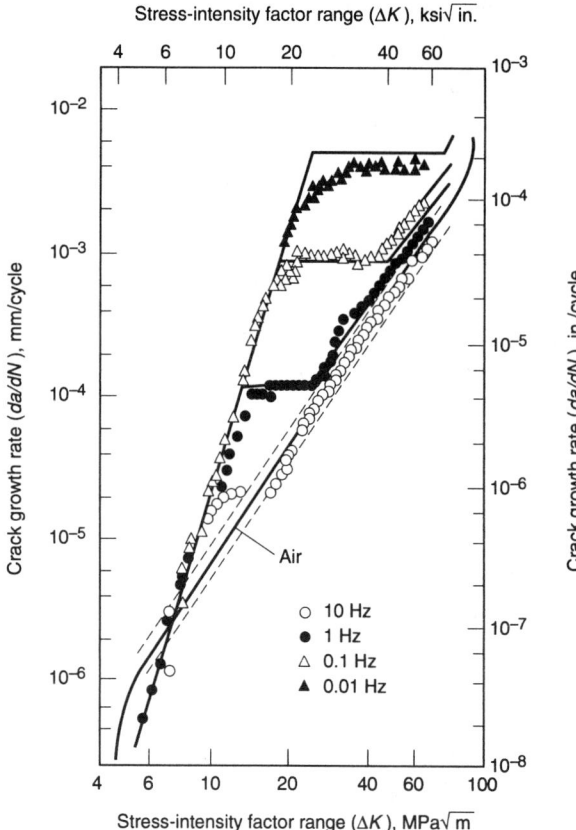

Fig. 65 Effect of cyclic frequency on corrosion fatigue. (a) High-strength steel 4340 M exposed to water and vacuum. Source: Ref 89. (b) X-65 line pipe steel exposed to air and salt water with a superimposed cathodic potential. Source: Ref 90

Table 6 Effect of cold work on stress-corrosion cracking (SCC)

Prestress, (% elongation)	Applied stress (% yield strength)(a)	
	130	80
0	No SCC	No SCC
1	No SCC	No SCC
2	No SCC	No SCC
3	SCC	No SCC
5	SCC	No SCC

(a) Pipeline steel tested in 5% NaCl saturated with hydrogen sulfide. (a) Source: *Corrosion*, Vol 22, 1966, p 238

threshold stress intensity for SCC of about 110 MPa√m (100 ksi√in.).

For steels with yield strengths of less than 1241 MPa (180 ksi), increasing chloride content of SCC-conducive environments apparently increases crack growth rates (Ref 109-111). This effect is more pronounced at lower chloride levels. However, chloride concentration has been reported to have little influence on the threshold stress level of precracked specimens (Ref 112).

In aqueous chlorides, plastic prestrain to about 5% has little effect on the SCC of low- and medium-strength steels, but as little as 1% prestrain can adversely affect higher-strength steels (Ref 102, 113). Plastic prestrain can, however, result in compressive residual stresses. In such cases, SCC resistance may increase due to

inhibiting effects on SCC initiation (Ref 114). Thus, at similar strength levels, cold-drawn wires have been reported to have better SCC resistance than heat-treated wires (Ref 103).

Hydrogen Sulfide. As previously noted, H_2S is one of the more aggressive and common SCC agents. The effect of H_2S on the toughness of 4140 steel is shown in Fig. 73. SCC problems in H_2S are more prevalent for steels with higher strength levels and are often associated with hard zones caused by cold working or welding. In H_2S environments, cold working as little as 1% has a strongly detrimental effect on SCC resistance. As shown in Table 6, cold working can impart SCC susceptibility even to low-strength steels that would not otherwise exhibit SCC. Such effects are believed to be due to increases in both available sites for SCC initiation and hydrogen solubility.

Additions of acetic acid to H_2S-containing solutions promote SCC, probably by both reducing pH and removing protective sulfide films. This latter effect has also been suggested for cyanides. Carbon dioxide has generally been observed to promote SCC both in the laboratory and in service environments. Increasing amounts of chlorides have also been reported to enhance SCC; however, at fixed H_2S and pH levels, the effects have been minimal (Ref 115).

Perhaps the most comprehensive SCC prevention and control guidelines have been developed

for H_2S service. These guidelines have been summarized in the document commonly referred to as MR-01-75, published by the National Association of Corrosion Engineers (NACE). In general, the NACE guidelines have been supported by service experience, even though SCC can be obtained for some of the recommended materials in laboratory tests.

With respect to carbon and low-alloy steels, NACE MR-01-75 recommendations include limits on hardness levels to avoid hydrogen sulfide SCC. For most steels, hardness is restricted to a maximum of 22 HRC. Certain AISI 4xxx-series steels are allowed at higher hardnesses in the quenched-and-tempered condition, but the user is cautioned that SCC testing of such materials is advisable. Additionally, MR-01-75 sets limits with respect to nickel content (1% maximum) and degree of cold work, and it specifies required stress-relief heat treatments.

Although not strictly applicable, MR-01-75 is often used for other environments, generally those that, like H_2S, are believed to cause SCC by hydrogen embrittlement. Examples include gaseous hydrogen service and applications in which steel contact with an acidic aqueous phase could cause hydrogen adsorption.

Sulfuric Acid. Stress-corrosion cracking of steel in sulfuric acid solutions is generally associated with high strength levels, although some cases of hydrogen blistering have been reported

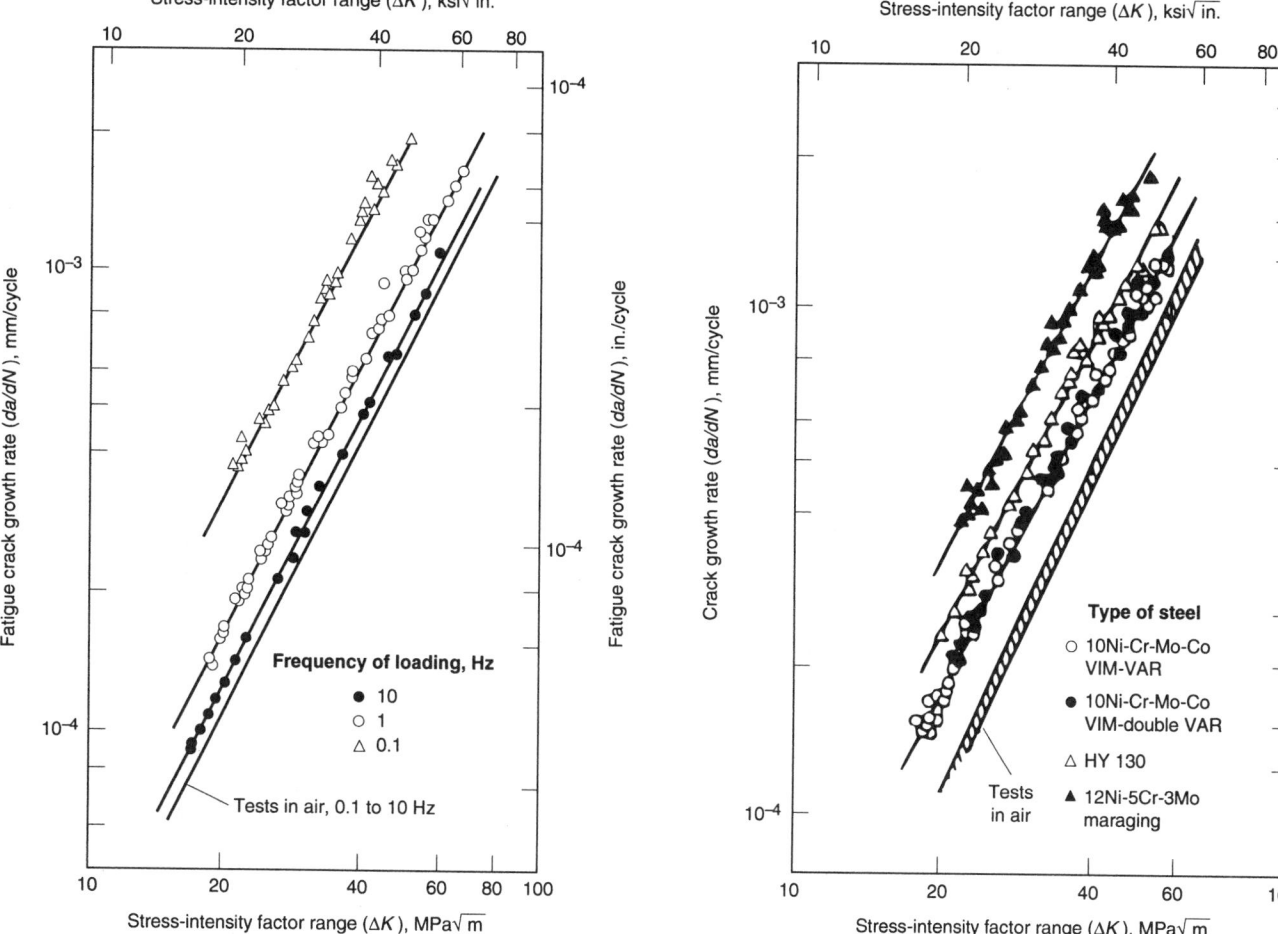

Fig. 66 Effect of cyclic frequency on corrosion fatigue for 12Ni-5Cr-3Mo maraging steel. The steel was tested in air and in a 3% aqueous solution of sodium chloride with sinusoidal loading. Source: Ref 17

Fig. 67 Effect of composition on corrosion fatigue crack growth rates for three different high-strength steels. The steels were tested in air and in a 3% aqueous solution of sodium chloride at 0.1 Hz. Source: Ref 17

in mild steels. Stress-corrosion cracking in sulfuric acid solutions has been suggested to be analogous to SCC in aqueous chlorides, producing many similar effects (Ref 116).

Nitrates. Reported SCC service failures by nitrates are generally associated with welded steel equipment and/or high-strength steels. To study such failures, laboratory investigations have sometimes used concentrated nitrate solutions. However, SCC can occur at quite low nitrate concentrations in boiling solutions. In these solutions, SCC may be insensitive to concentration increases above a certain level (Ref 117).

For low-carbon steels, cold working reportedly improves SCC resistance in nitrates (Ref 118, 119). At intermediate carbon levels (0.09 to 0.19%), SCC resistance initially decreases and then increases with increasing amounts of cold work (Ref 119). Cold working has also been reported to be more beneficial than heat treatment to a similar strength level for the SCC of steel wires in nitrate solutions (Ref 120). Introduction of residual compressive stresses by shot peening improves the nitrate SCC resistance of steels (Ref 121). Plastic straining appears to be a prerequisite for SCC in hydroxide environments (Ref 122).

Alloying Effects

Because strength is the dominant factor in steel selection for SCC resistance, the effects of chemistry and alloying can be masked by their effect on strength or hardness. It is also difficult to generalize on the effects of alloying additions on sustained-load cracking because of the vast number of other variables. Nevertheless, a few cases for specific environments have been isolated (Ref 123, 124).

For HSLA steels (Ref 125) in aqueous environments, manganese may lower resistance by producing either untempered martensite or twinned martensite. Nickel may also produce undesirable effects if it causes an increase in retained austenite that leads to formation of untempered martensite after tempering. If no untempered martensite is present, nickel is considered to produce no significant effects on sustained-load cracking. Molybdenum additions on the order of 0.5% enhance resistance. For other elemental additions, either studies have not been made or results have been controversial. Effects of impurity elements are discussed at the end of this section.

For 18-Ni maraging steels, an extensive study (Ref 124) was conducted on 14 alloy heats with varying additions of cobalt, molybdenum, titanium, and aluminum. An alloy parameter, shown as the horizontal axis in Fig. 74, was devised that predicted the K_{Iscc} resistance to an aqueous 3.5% NaCl solution with reasonable accuracy. A linear regression analysis indicated a correlation coefficient of −0.88, which means that this alloy parameter explains about 77% (r^2) of the variation in K_{Iscc}. Because the yield strengths range from 1420 to 1960 MPa (206 to 284 ksi) and there were additional possibilities of microstructural influence, it is difficult to account for all variables.

With regard to medium-strength steels, a very extensive study (Ref 126) was conducted on a molybdenum-modified 4130 steel to determine the effect of molybdenum additions on resistance to H_2S cracking. Here, 12 heats of steel with varying molybdenum contents were tempered to a total of 56 conditions. The resulting K_{Iscc} values in an aqueous 0.5% acetic acid solution saturated with H_2S are shown in Fig. 75. In this study the yield strength level was held constant at 760 MPa (110 ksi). It was observed that molybdenum additions of about 0.75% maximized resistance to

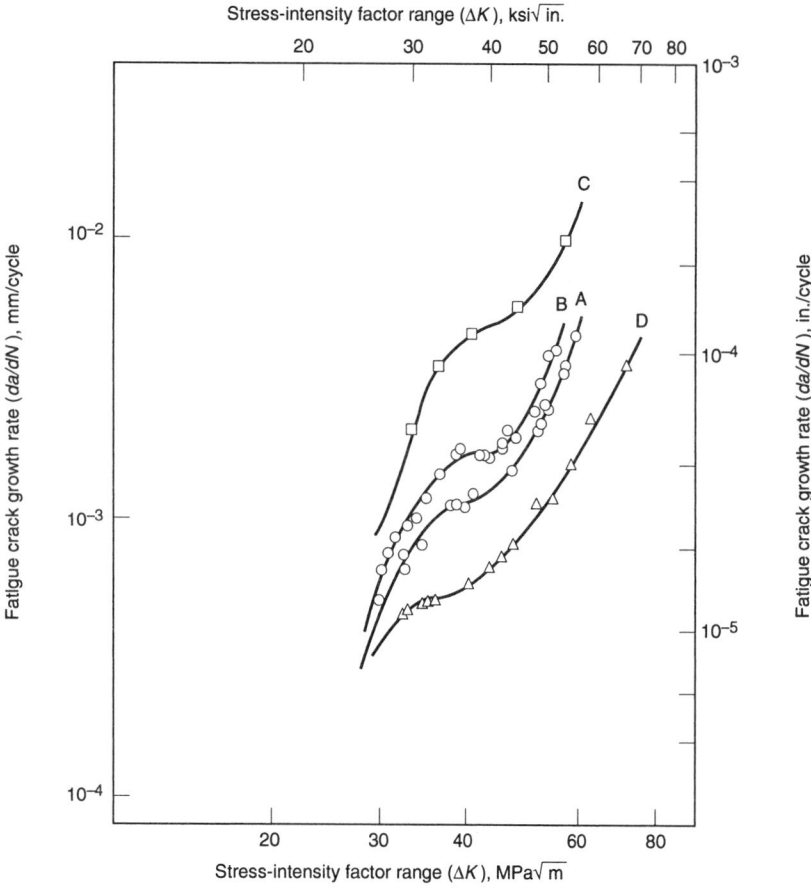

Fig. 68 Fatigue crack growth rate vs. stress-intensity factor range, ΔK, for type 4340 steel. Curve a: 90% relative humidity + $Na_2Cr_2O_7$ + $NaNO_2$ + $Na_2B_4O_7$. Curve b: 90% relative humidity + $Na_2Cr_2O_7$. Curve c: 90% relative humidity only. Curve d: dry air; relative humidity 15%. Source: Ref 98

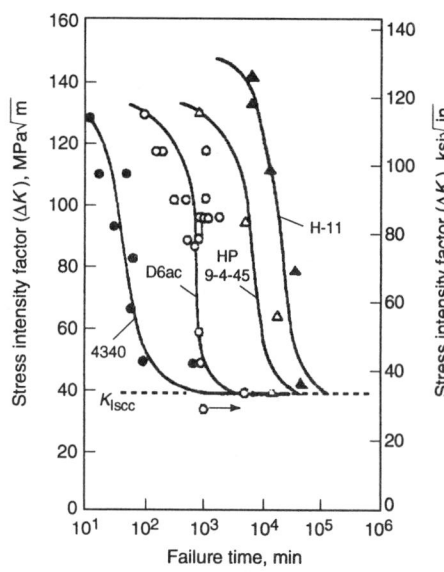

Fig. 70 Stress-corrosion cracking in four high-strength steels. Precracked specimens of four high-strength steels were subjected to sustained loading in an environment of distilled water. Each steel had been hardened and tempered to a tensile strength of about 1650 MPa (240 ksi). Source: Ref 99

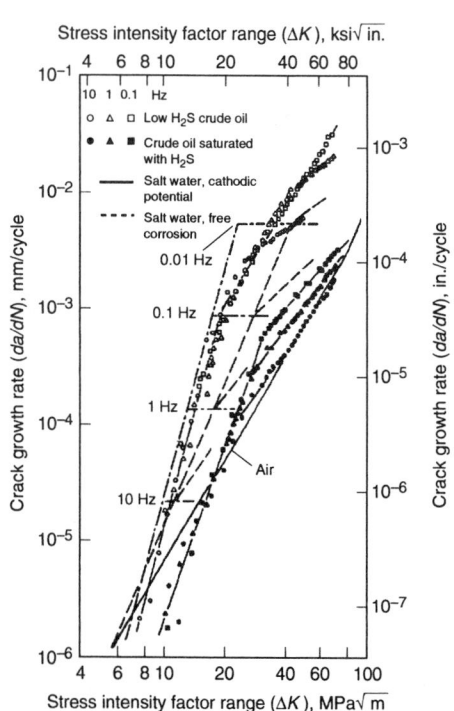

Fig. 69 Fatigue crack growth rates in crude oil with two levels of hydrogen sulfide content at three frequencies

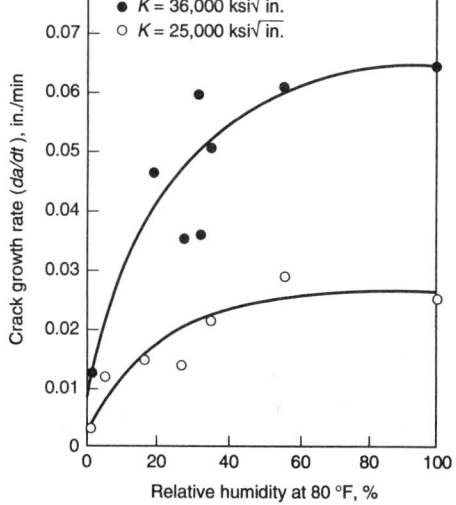

Fig. 71 Effect of relative humidity on crack velocity for H-11 steel with a yield strength of 1580 MPa (230 ksi) in moist argon environments. Source: *Applied Materials Research*, Vol 4, 1965, p 34

cracking. Smooth-bar bend tests on the same heats and in the same solution indicated that 0.9% Mo maximized threshold stresses. Many possible suggestions for this improvement were offered, including the control of tramp elements by segregation of phosphorus, arsenic, antimony, and tin to carbide interfaces instead of to prior-austenite grain boundaries.

The conclusion of a previous review (Ref 123) was that alloy chemistry probably has little effect on SCC resistance except for its influence on microstructure and strength. This is possibly true, with the one major exception that tramp elements segregated to prior-austenite grain boundaries seriously decrease threshold stress-intensity factors. Just as it has been shown that phosphorus, sulfur, tin, antimony, arsenic, and tellurium lower fracture toughness, combined effects of temper embrittlement and hydrogen embrittlement during SCC can be very detrimental.

Effects of Microstructure and Heat Treatment

Fine grain size has been shown to increase the SCC resistance of quenched-and-tempered steels in H_2S solutions (Ref 127, 128). Similar effects have been observed for carbon steels in nitrates, where increased austenitizing temperature, re-sulting in grain growth, has adversely affected SCC resistance (Ref 129). Increased cooling rate after austenitizing has also been shown to decrease SCC resistance to nitrate environments (Ref 130). Prolonged aging at temperatures below about 593 °C (1100 °F) reportedly improves SCC resistance in nitrates, but aging at higher temperatures promotes SCC susceptibility (Ref 131-133).

For quenched-and-tempered steels, the presence of twinned martensite is thought to reduce SCC resistance as it does fracture toughness and

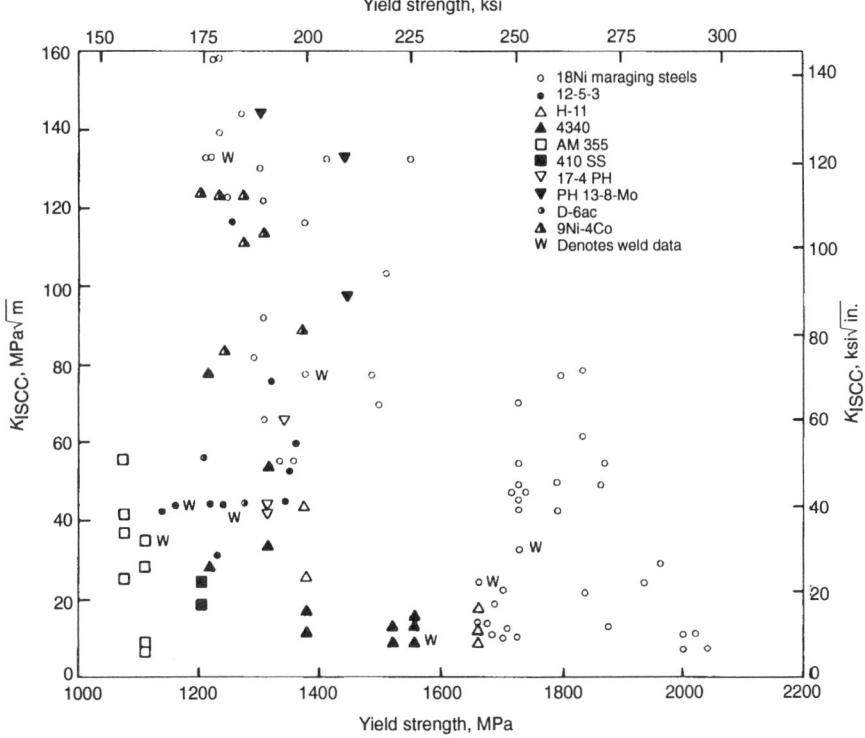

Fig. 72 Threshold stress intensity (K_{Iscc}) values for maraging steels and other high-strength steels as a function of yield strength

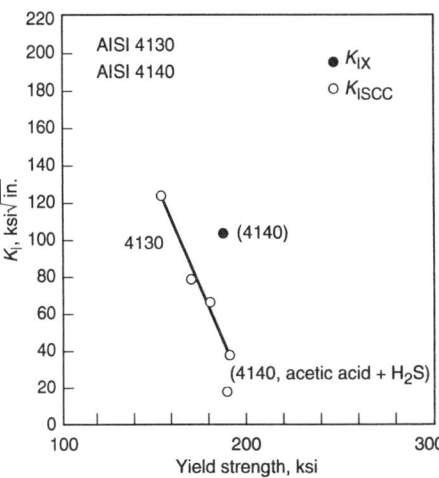

Fig. 73 Stress-corrosion resistance and fracture toughness of AISI 4130 and 4140 steels. Source: Ref

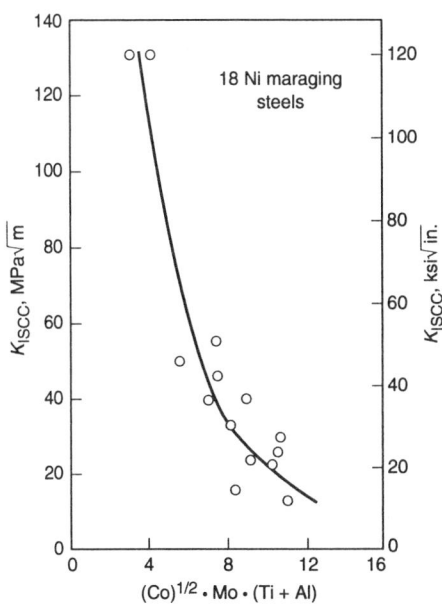

Fig. 74 Effect of alloying element parameter on K_{Iscc} for 18Ni maraging steels in an aqueous solution of sodium chloride. Source: Ref 124

Table 7 Observation of stress-corrosion cracking in low-strength steels

Boiling solution	Steel	Fracture mode	Reference
20% NH$_4$NO$_3$	0.19% C	Intergranular	150
Ca(NO$_3$)$_2$NH$_4$NO$_3$	0.02 to 0.22% C	Intergranular	150
KNO$_3$ or NaNO$_3$	0.02 to 0.22% C	Intergranular	150
4N NH$_4$NO$_3$	Mild steel	Intergranular	151
4N NH$_4$NO$_3$	0.32C-0.5Mo-3Cr	Intergranular	152
9N NaOH	0.32C-0.5Mo-3Cr	Intergranular	152
35% NaOH	Mild steel	Intergranular	151
KOH, NaOH, or LiOH (300 °C)	Mild steel	Intergranular	153
Anhydrous NH$_3$	ASTM A-212, A-285	Intergranular	150
Anhydrous NH$_3$	0.14C-1.73Mn	Intergranular + cleavage	151
Na$_2$CO$_3$-NaHCO$_3$	Mild steel (variable Ti, C)	Intergranular	154
NH$_4$CO$_3$ or Na$_2$CO$_3$-NaHCO$_3$	Mild steel (variable C, Si, Ni, Cr)	Intergranular	155
45% MgCl$_2$	1 to 6% Ni steel	Intergranular + cleavage	151

hydrogen-embrittlement resistance. The presence of ε carbides and high dislocation densities also reduces SCC resistance (Ref 134, 135).

In aqueous H$_2$S, quenched-and-tempered steels have been shown to have superior SCC resistance compared with normalized-and-tempered steels of similar strength levels (Ref 136). Martensitic microstructures generally give the best SCC resistance at a given strength level, especially if fine spheroidized carbides uniformly dispersed in ferrite (such as is obtained by high-temperature tempering) are present (Ref 137). Bainitic structures have approximately equivalent SCC resistance (Ref 138). The presence of untempered martensite is considered detrimental. Coarse or globular carbides that result from slow cooling and tempering produce an intermediate level of SCC resistance. In nitrate solutions,

steels with pearlitic microstructures are superior in SCC resistance compared with steels having fine spheroidized carbides (Ref 139).

Manganese segregation in tempered steels can produce localized regions of increased hardness. This results in a preferential path for SCC in aqueous H$_2$S and a corresponding drop in SCC resistance (Ref 140, 141).

Grain-boundary segregants can significantly influence the SCC of steels in environments in which intergranular cracking predominates. These effects have been systematically studied in elevated-temperature nitrate solutions (Ref 142-144). Following aging treatments to promote segregation, enhanced SCC susceptibility of steels to calcium nitrate was attributed to locally high phosphorus and sulfur levels at grain boundaries (Ref 142). Somewhat mixed results have been

reported for silicon and tin in similar environments (Ref 143). In ammonium nitrate solutions, phosphorus, sulfur, arsenic, tin, and antimony (when present with silicon, germanium, selenium, tellurium, and bismuth) strongly promoted intergranular SCC (Ref 144).

Inclusions in steels act as SCC initiation sites and accelerate crack propagation in the major-axis direction of elongated inclusions such as sulfides (Ref 145, 146). Resultant anisotropy of H$_2$S SCC has been reported in both plate and tubular products (Ref 147, 148).

The general effect of increased tempering temperatures in many alloy steels is to lower the strength and to increase the K_{Iscc} value from about 10 MPa$\sqrt{m}$ (9 ksi$\sqrt{in.}$) to 70 MPa$\sqrt{m}$ (64

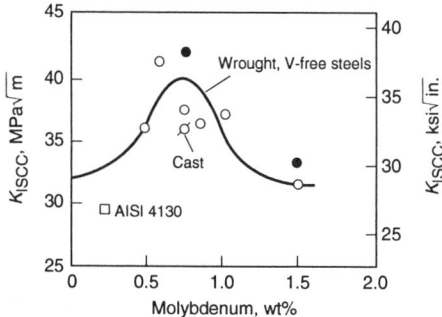

Fig. 75 Effect of molybdenum content on K_{Iscc} for alloy steels. These data are for steel with a yield strength of 760 MPa (110 ksi) in an aqueous 0.5% acetic acid solution saturated with hydrogen sulfide. Solid circles denote heats with vanadium additions. Source: Ref 126

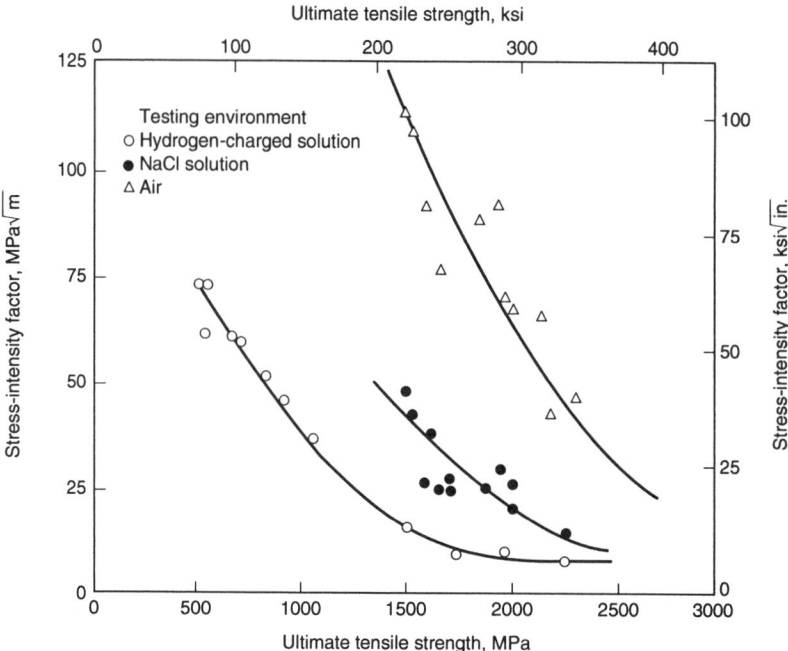

Fig. 76 Correlation between tensile strength and stress-intensity factor for crack propagation in alloy steels in several environments. Reported values of stress-intensity factor are: K_{ic} values for testing in air; K_{Iscc} values for testing in a 3.5% aqueous sodium chloride solution; and K values for threshold cracking of specimens electrolytically charged with hydrogen. Source: *Fracture 1977*, Vol 2, University of Waterloo Press, 1977, p 255

ksi√in.) (Fig. 76). In this instance, the behavior parallels fracture toughness. However, it should be emphasized that resistance to SCC does not always parallel fracture toughness. For example, 18-Ni maraging steel typically undergoes a continuous improvement in resistance to SCC with increased aging temperature, while fracture toughness goes through a minimum (Ref 124).

For lower-strength conditions, quenched-and-tempered steels, normalized steels, and isothermally transformed steels respond quite differently to sulfide cracking. A series of heat treatments that produced six different types of mixtures of martensite, ferrite, and untempered martensite demonstrated the superiority of quenched-and-tempered microstructures (Fig. 77). Even though these results were for notched bars loaded in bending, the general ranking with regard to microstructural resistance to SCC should be the same. It was found that microstructures that had fine spheroidizing carbides uniformly distributed throughout the ferrite, which are typically achieved in quenched-and-tempered steels, had superior resistance to sulfide cracking (Ref 149).

Effects of Temperature and Environment

Temperature and environment may produce unexpected results with regard to predictions of how a material will behave. Even though corrosion attack usually increases as temperature increases, it does not follow that SCC will necessarily be worse. Thus, data obtained for one combination of material, temperature, environment, and loading should not be used for making predictions concerning any other combination. This is aptly illustrated in Fig. 78(a), where it is shown that for AISI 4130 steel tested at one hydrogen pressure, the lower the temperature, the less resistant the material near the threshold. However, there is a crossover effect at high crack velocities, and the material is more resistant at temperatures above and below 53 °C (127 °F). Thus, data obtained as a function of either temperature or stress intensity should not be extrapo-

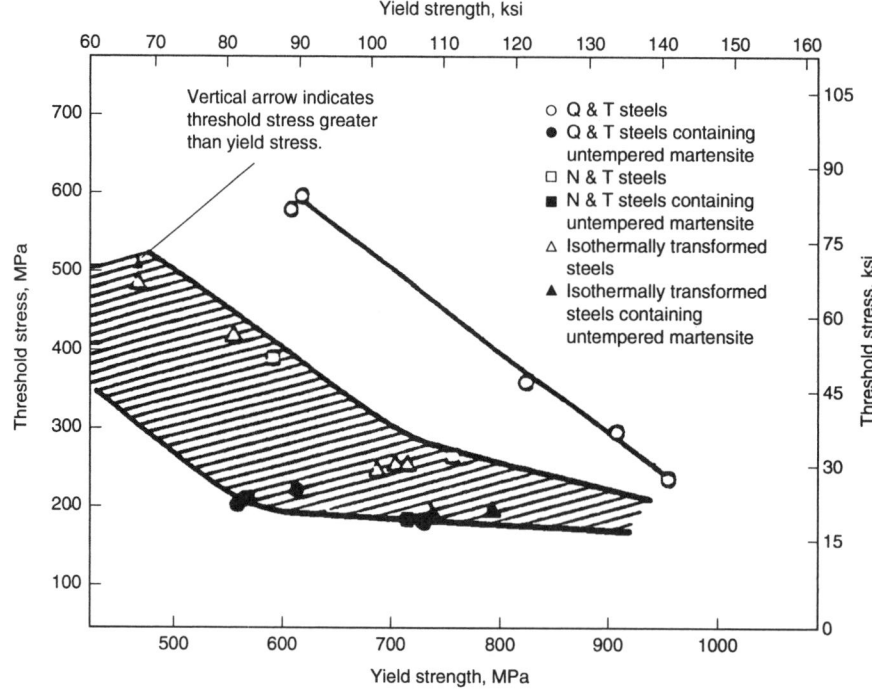

Fig. 77 Influence of microstructure on crack resistance for API N80 and AISI 4140 and 4340 steels. The data are for notched specimens loaded in bending in 5% NaCl-0.5% acetic acid solution saturated with hydrogen sulfide. Q&T, quenched and tempered; N&T, normalized and tempered. Source: *Corrosion*, Vol 24, 1968, p 261-282

lated. The effect of hydrogen concentration is more orderly, as shown in Fig. 78(b), where an increase in hydrogen pressure causes an increase in the crack growth rate.

In general, any environmental species or electrochemical situation that promotes hydrogen entry will accelerate SCC in medium- and high-strength steels. For low-strength steels, there is a major categorical exception to this in that SCC occurs by anodic dissolution in boiling solutions of nitrates, carbonates, chlorides, and hydroxides as well as anhydrous ammonia. Various combinations of materials and boiling solutions that resulted in SCC, along with the fracture mode observed for each combination, are listed in Table 7 (Ref 150-155).

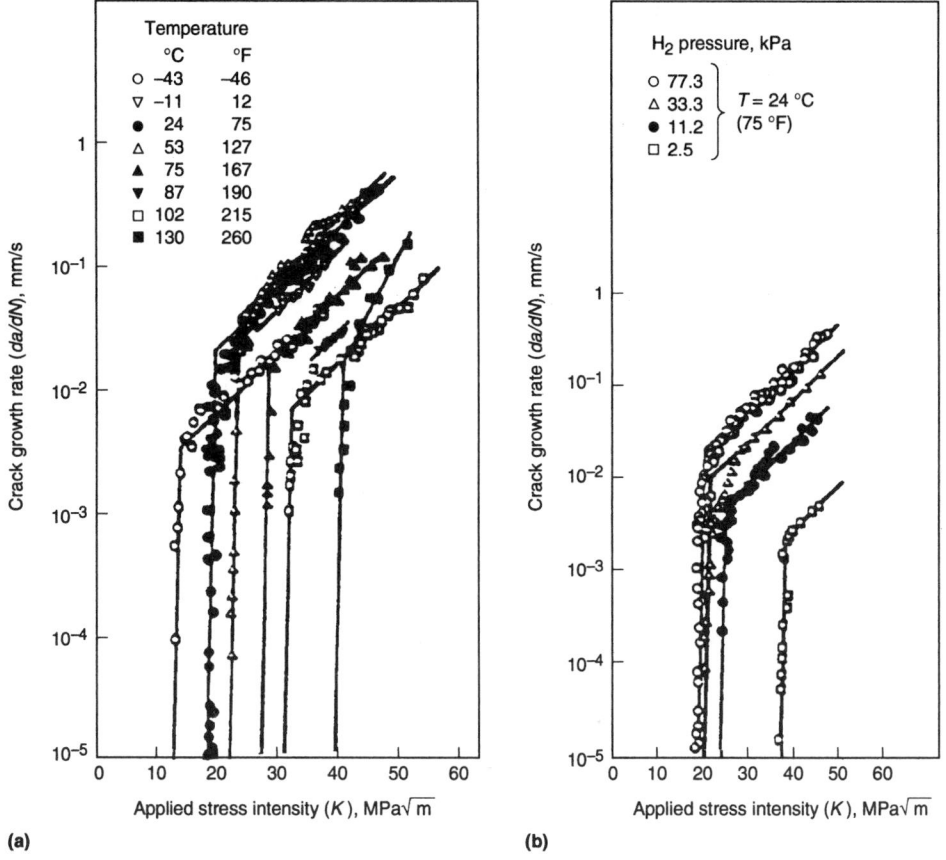

Fig. 78 Dependence of crack growth rate, *da/dt*, on applied stress intensity, *K*, for 4130 steel with a yield strength of 1330 MPa (193 ksi). (a) At various temperatures in hydrogen at a pressure of 77.3 kPa. (b) At various hydrogen pressures at 24 °C (75 °F). Source: Ref 149

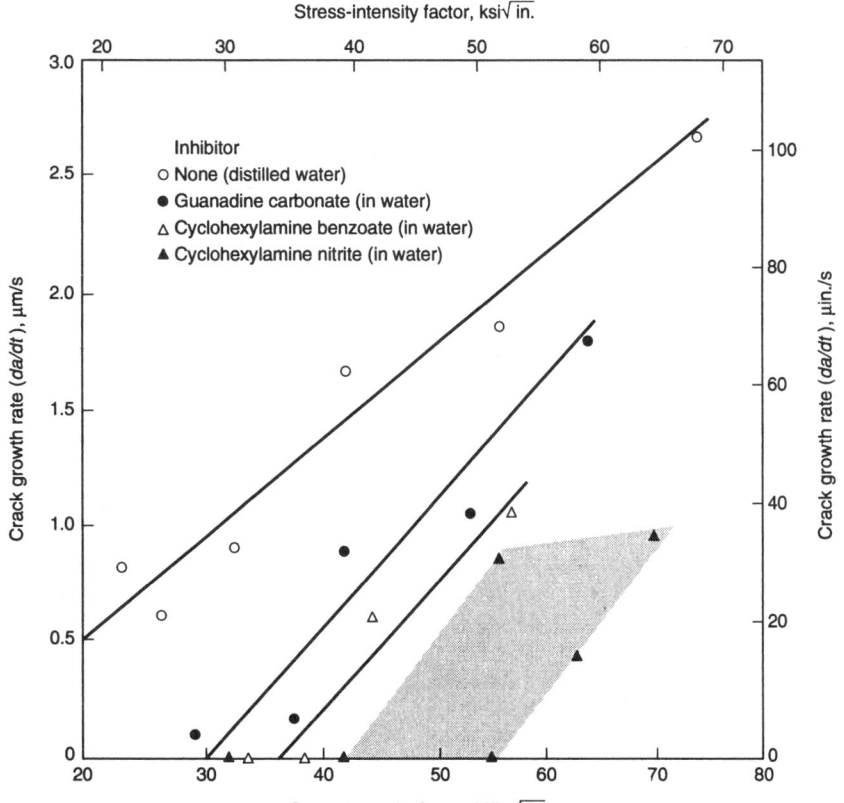

Fig. 79 Effect of inhibitors on crack growth rate for 300M steel. Source: Ref 156

With regard to failure prevention, the effects of SCC can be minimized by choice of steel composition, by cathodic protection against caustic and nitrate embrittlement, or by use of protective coatings or inhibitors. Some successful applications have been: inhibited epoxy coatings on 18-Ni maraging steel exposed to aqueous solutions containing 3% NaCl; 0.2% water additions to vessels containing agricultural ammonia; and film-forming amines added to hydrocrackers containing H_2S, ammonia, or hydrocyanic acid. Most of these studies were not performed on systems containing cracks, so it is not known how successful such coatings and inhibitors are for dynamic or precracked specimens. As shown in Fig. 79, crack velocities in high-strength 300M steel tested in distilled water can be markedly reduced by adding organic inhibitors, possibly by forming a chemisorbed double layer to prevent hydrogen entry.

ACKNOWLEDGMENTS

This article was developed from source material contained in the following ASM publications:

- W.W. Gerberich, R.H. VanStone, and A.W. Gunderson, Fracture Properties of Carbon and Alloy Steels, *Application of Fracture Mechanics in the Selection of Metallic Structural Materials*, American Society for Metals, 1982
- J. Barsom, Fracture Toughness, Fracture Mechanics—Fatigue and Fracture, *Metals Handbook Desk Edition*, American Society for Metals, 1985, p 32-2 to 32-8
- R. Viswanathan, *Damage Mechanisms and Life Assessment of High-Temperature Components*, ASM International, 1989
- S.W. Ciaraldi, Stress Corrosion Cracking of Carbon and Low-Alloy Steels, *Stress Corrosion Cracking Materials Performance and Evaluation*, ASM International, 1992, p 41-62
- P.G. Marsh and W.W. Gerberich, Stress Corrosion Cracking of High Strength Steels, *Stress Corrosion Cracking Materials Performance and Evaluation*, ASM International, 1992, p 63-90

REFERENCES

1. E.C. Bain and H.W. Paxton, *Alloying Elements in Steel*, American Society for Metals, 1961
2. J.R. Low, Jr., The Fracture of Metals, *Progress in Materials Science*, Vol 12, 1963, p 1-96
3. A.M. Turkalo, The Morphology of Brittle Fracture in Pearlite, Bainite and Martensite, *Trans. AIME*, Vol 218, 1960, p 24-30
4. W.C. Johnson and D.F. Stein, A Study of Grain Boundary Segregants in Thermally Embrittled Maraging Steel, *Met. Trans.*, Vol 5, 1974, p 549-554
5. T.S. Brammer, A New Examination of the Phenomena of Overheating and Burning of Steels, *J. Iron Steel Inst.*, Vol 201, 1963, p 752-761
6. C.L. Briant and S.K. Banerji, Tempered Martensite Embrittlement in a High Purity

Steel, *Met. Trans. A*, Vol 10, 1979, p 1151-1155

7. C.J. McMahon, Temper Embrittlement—An Interpretive Review, *Temper Embrittlement in Steel*, STP 407, ASTM, 1968, p 127-167

8. C.J. McMahon, Jr., Problems of Alloy Design in Pressure Vessel Steels, *Fundamental Aspects of Structural Alloy Design*, R.I. Jaffee and B.A. Wilcox, Ed., McGraw-Hill, 1977, p 295-322

9. C.J. McMahon et al., The Effect of Composition and Microstructure on Temper Embrittlement in 2¼Cr-1Mo Steels, *ASME J. Eng. Mater. Tech.*, Vol 102, 1980, p 369

10. R. Viswanathan and R.I. Jaffee, 2¼Cr-1Mo Steels for Coal Conversion Pressure Vessels, *ASME J. Eng. Mater. Tech.*, Vol 104, July 1982, p 220-226

11. J. Watanabe et al., Temper Embrittlement of 2¼Cr-1Mo Pressure Vessel Steel, paper presented at ASME 29th Petroleum Mechanical Engineering Congress, Dallas, 15-18 Sept 1974

12. R. Bruscato, *Weld. Res. Suppl.*, Vol 49, 1973, p 1485

13. B.J. Shaw, "The Effect of Composition on Temper Embrittlement of Low Carbon Rotor Steels," Scientific Paper 77-1D9-GRABO-P1, Westinghouse Research Laboratories, Pittsburgh, 1977

14. C.J. McMahon et al., "The Elimination of Impurity Induced Embrittlement in Steels, Part I," Report NP 1501, Electric Power Research Institute, Sept 1980

15. R. Viswanathan and A. Joshi, The Effect of Microstructure on the Temper Embrittlement of CrMoV Steels, *Met. Trans. A*, Vol 6, 1975, p 2289-2297

16. J. Campbell, W. Gerberich, and J. Underwood, *Application of Fracture Mechanics for Selection of Metallic Structural Materials*, ASM, 1982

17. S.T. Rolfe and J.M. Barsom, *Fracture and Fatigue Control in Structures—Applications of Fracture Mechanics*, Prentice-Hall, 1977

18. D.E. McCabe, Evaluation of the Compact Tension Specimen for Determining Plane-Strain Fracture Toughness of High Strength Materials, *J. Mater.*, Vol 7 (No. 4), Dec 1972, p 449-454

19. C. Vishnevsky, Effect of Alloying Elements on Tempered Martensite Embrittlement and Fracture Toughness of Low Alloy High Strength Steels, Report CR 69-18(F), Army Materials and Mechanics Research Center, Watertown, MA, Jan 1971

20. C. Vishnevsky and E.A. Steigerwald, Influence of Alloying Elements on the Toughness of Low Alloy Martensitic High Strength Steels, Report CR 68-09(F), Army Materials and Mechanics Research Center, Watertown, MA, Nov 1968

21. A.H. Priest, Influence of Strain Rate and Temperature on the Fracture Toughness and Tensile Properties of Several Metallic Materials, *Dynamic Fracture Toughness*, M.G. Dawes,

Ed., Welding Institute, Cambridge, U.K., 1977, p 95-111

22. R.P. Wei, Fracture Toughness Testing in Alloy Development, *Fracture Toughness and Its Applications*, STP 381, ASTM, 1965, p 279-289

23. P.E. Waudby, Rare Earth Additions to Steel, *International Metals Reviews*, Review 229 (No. 2), 1978, p 74-98

24. W.G. Wilson and G.J. Klems, Impact Toughness of Fuel Pipelines Demands Steel Cleanliness, *Industrial Heating*, Oct 1974, p 12-16

25. H.J. Kirsching, H.J. Hornbeck, H. Schenk, and C. Carius, Effect of Cerium on the Properties of a Stainless Chromium Nickel Steel, *Arch. Eisenhuttenwes*, Vol 34 (No. 4), 1963, p 269-277

26. L. Luyckx, R.J. Bell, A. McLean, and M. Korchynsky, Sulfide Shape Control in High-Strength Low-Alloy Steels, *Met. Trans.*, Vol 1 (No. 12), 1970, p 3341-3350

27. C.J. McMahon, Jr., Intergranular Fracture in Steels, *Mater. Sci. Eng.*, Vol 25, 1976, p 233-239

28. J.R. Low, Jr., D.F. Stein, A.M. Turkalo, and R.P. Laforce, Alloy and Impurity Effects on Temper Brittleness of Steel, *Trans. AIME*, Vol 242, 1968, p 14-24

29. R.O. Ritchie and R.M. Horn, Further Considerations on the Inconsistency in Toughness Evaluation of AISI 4340 Steel Austenitized at Increasing Temperatures, *Met. Trans. A*, Vol 9 (No. 3), March 1978, p 331-341

30. G.L. Peterman and R.L. Jones, Effects of Quenching Variables on Fracture Toughness of D6ac Steel Aerospace Structures, *Met. Eng. Q.*, Vol 15 (No. 2), May 1975, p 59-64

31. J.M. Hyzak and I.M. Bernstein, The Role of Microstructure on the Strength and Toughness of Fully Pearlitic Steels, *Met. Trans. A*, Vol 7, 1976, p 1217-1224

32. W. Dahl and W.B. Kretzschmann, Influence of Precracking and Grain Size on Fracture Toughness of Structural Steels, *Fracture 1977*, Vol 2A, D.M.R. Taplin, Ed., Pergamon Press, 1977, p 17-21

33. D.A. Curry and J.F. Knott, The Relationship between Fracture Toughness and Microstructure in the Cleavage Fracture of Mild Steel, *Met. Sci.*, Vol 10, 1976, p 1-6

34. F.R. Stonsifer and R.W. Armstrong, Effect of Prior Austenite Grain Size on the Fracture Toughness Properties of A 533 B Steel, *Fracture 1977*, Vol 2A, D.M.R. Taplin, Ed., Pergamon Press, 1977, p 1-6

35. R.O. Ritchie, J.F. Knott, and J.F. Rice, On the Relationship between Critical Tensile Stress and Fracture Toughness in Mild Steel, *J. Mech. Phys. Solids*, Vol 21, 1973, p 395-410

36. G.T. Hahn and A.R. Rosenfield, Experimental Determination of Plastic Constraint Ahead of a Sharp Notch under Plane-Strain Conditions, *Trans. ASM*, Vol 59, 1966, p 909-919

37. J.R. Rice and M.B. Johnson, The Role of Large Crack Tip Geometry Changes in Plane Strain Fracture, *Inelastic Behavior of Solids*, M.F. Kanninen et al., Ed., McGraw-Hill, 1970, p 641-672

38. A.R. Rosenfield, G.T. Hahn, and J.D. Embury, Fracture of Steels Containing Pearlite, *Met. Trans.*, Vol 3, 1972, p 2797-2804

39. J.F. Knott, *Fundamentals of Fracture Mechanics*, Butterworths, 1979

40. *Low Alloy High Strength Steels*, conf. proc., Metallurg Companies, Nuremberg, BRD, 21-23 May 1970

41. T.B. Cox and J.R. Low, Jr. *Met. Trans.*, Vol 5, 1974, p 1457-1470

42. K.H. Schwalbe, On the Influence of Microstructure on Crack Propagation Mechanisms and Fracture Toughness of Metallic Materials, *Eng. Fract. Mech.*, Vol 9, 1977, p 795-832

43. A.H. Priest, *Effect of Second-Phase Particles on the Mechanical Properties of Steel*, Iron and Steel Institute, London, 1971

44. J.M. Barsom, Effect of Temperature and Rate of Loading on the Fracture Behavior of Various Steels, *Dynamic Fracture Toughness*, M.G. Davies, Ed., Welding Institute, Cambridge, U.K., 1979, p 113-125

45. H.D. Greenberg, E.T. Wessel, and W.H. Pryle, Fracture Toughness of Turbine Generator Rotor Forgings, *Eng. Fract. Mech.*, Vol 1, 1970, p 653-674

46. J.M. Barsom and S.T. Rolfe, The Correlations between K_{Ic} and Charpy V Notch Test Results in the Transition Temperature Range, *Impact Testing of Metals*, STP 466, ASTM, 1970, p 281-302

47. J.M. Barsom, Development of the AASHTO Fracture-Toughness Requirements for Bridge Steels, *Eng. Fract. Mech.*, Vol 7 (No. 3), 1975, p 605-618

48. R. Roberts, Fracture Toughness of Bridge Steels, Phase II Report, Report FHWA-RD-74-59, Federal Highway Administration, Sept 1974

49. J. Joyce, *Elastic Plastic Fracture Test Methods*, STP1114, ASTM, 1991, p 275

50. M.E. Taylor and J.M. Barsom, Effect of Cyclic Frequency on the Corrosion-Fatigue Crack-Initiation Behavior of ASTM A 517 Grade F Steel, *Fracture Mechanics: Thirteenth Conference*, STP 743, ASTM 1981

51. R. Roberts, J.M. Barsom, J.W. Fisher, and S.T. Rolfe, Fracture Mechanics for Bridge Design, Report FHWA-RD-78-69, Federal Highway Administration, July 1977

52. "Guidance on Some Methods for the Derivation of Acceptance Levels for Defects in Fusion Welded Joints," PD6493, British Standards Institution, 1980

53. S.J. Maddox, *Weld Res. Inst.*, Vol 4 (No. 1), 1974

54. N.E. Dowling and J.A. Begley, Fatigue Crack Growth During Gross Plasticity and the *J*-integral, *Mechanics of Crack Growth*, STP 590, ASTM, 1976, p 82-103

55. C.M. Carmen and J.M. Katlin, *Trans. ASME, J. Basic Eng.*, 1966, p 792-800

56. J.M. Barsom and E.J. Imhof, in *Progress in Flaw Growth and Fracture Toughness Testing*, STP 536, ASTM, 1973, p 182-205

57. R.O. Ritchie, Near-Threshold Fatigue-Crack Propagation in Steels, *International Metals Reviews*, Vol 24 (No. 5-6), 1979, p 205-230

58. W.W. Gerberich and N.R. Moody, A Review of Fatigue Fracture Topology Effects on Threshold and Growth Mechanisms, *Fatigue Mechanisms*, J.F. Fong, Ed., STP 675, ASTM, 1979, p 292-341

59. C.E. Richards and T.C. Lindley, The Influence of Stress Intensity and Microstructure on Fatigue Crack Propagation in Ferritic Materials, *Eng. Fract. Mech.*, Vol 4, 1972, p 951-978

60. E.J. Imhof and J.M. Barsom, Fatigue and Corrosion-Fatigue Crack Growth of 4340 Steel at Various Yield Strengths, *Progress in Flaw Growth and Fracture Toughness*, STP 536, ASTM, 1973

61. S.T. Rolfe and J.M. Barsom, *Fracture and Fatigue Control in Structures*, Prentice-Hall, 1977

62. H. Suzuki and A.J. McEvily, Microstructural Effects on Fatigue Crack Growth in a Low Carbon Steel, *Met. Trans. A*, Vol 10, 1979, p 475-481

63. S. Taira, K. Tanaka, and M. Hoshina, Grain-Size Effect on Crack Nucleation and Growth in Long-Life Fatigue of Low-Carbon Steel, *Fatigue Mechanisms*, J.F. Fong, Ed., STP 675, ASTM, 1979, p 135-162

64. C.J. Beevers, Fatigue Crack Growth Characteristics at Low Stress Intensities of Metals and Alloys, *Met. Sci.*, Aug/Sept 1977, p 362-367

65. A. Yuen, S.W. Hopkins, G.R. Leverant, and C.A. Rau, Correlations between Fracture Surface Appearance and Fracture Mechanics Parameters for Stage II Fatigue Crack Propagation in Ti-6Al-4V, *Met. Trans.*, Vol 5, 1974, p 1833-1842

66. R.J. Cooke, P.E. Irving, G.S. Booth, and C.J. Beevers, The Slow Fatigue-Crack Growth and Threshold Behavior of a Medium-Carbon Alloy Steel in Air and Vacuum, *Eng. Fract. Mech.*, Vol 7, 1975, p 69-77

67. R.G. Forman, *Trans. ASME, J. Basic Eng.*, Vol 89 (No. 3), 1967, p 459

68. H.O. Fuchs and R.I. Stephens, *Metal Fatigue in Engineering*, John Wiley and Sons, 1980, p 89

69. "Offshore Installations: Guidance on Design and Construction," U.K. Department of Energy, 1985

70. "Rules for the Design, Construction and Inspection of Offshore Structures," Det Norske Veritas, 1977

71. For example, *Fatigue Crack Growth under Spectrum Loads*, STP 595, ASTM, 1976

72. J. Schijve, *Advances in Aeronautical Sciences*, Vol 3-4, Pergamon Press, 1962, p 387-408

73. R.I. Stephens, D.K. Chen, and B.W. Hom, STP 595, ASTM, 1976, p 27-40

74. K. Koterazawa, M. Mori, T. Matsui, and D. Shimo, *Trans. ASME*, Series H. 95, 1973, p 202-212

75. R. Koterazawa, Acceleration of Fatigue and Creep Crack Propagation under Variable Stresses, *Fatigue Life: Analysis and Prediction*, American Society for Metals, 1986, p 187-196

76. J.B. Chang and C.M. Hudson, Ed., *Methods and Models for Predicting Fatigue Crack Growth under Random Loading*, STP 748, ASTM, 1981

77. D.F. Socie, *Eng. Fract. Mech.*, Vol 9, 1977, p 849-865

78. A.U. DeKoning, *Fracture Mechanics: Thirteenth Conference*, STP 743, ASTM, 1981, p 63-85

79. R. Sunder, *Eng. Fract. Mech.*, Vol 12, 1979, p 155-165

80. S. Willenborg, R.M. Engle, and H.A. Wood, Report AFFDL-TM-71-1-FBR, Air Force Flight Dynamics Laboratory, 1971

81. O.E. Wheeler, *Trans. ASME, J. Basic Eng.*, Vol 94, 1972, p 181-186

82. F.G. Hamel and J. Masounave, paper presented at Ninth Polytechnique, 1-3 June 1983

83. N. Dowling, *Mechanical Behavior of Materials*, Prentice Hall, 1993

84. J. Bannantine, J. Comer, and J. Handrock, *Fundamentals of Metal Fatigue Analysis*, Prentice Hall, 1990

85. H.I. McHenry and A.W. Pense, *Fatigue Crack Propagation in Steel Alloys at Elevated Temperatures*, STP 520, ASTM, 1973, p 345-354

86. *Materials at Low Temperature*, American Society for Metals, 1983, p 287

87. F.R. Stonsifer, "Effects of Grain Size and Temperature on Sub-Critical Crack Growth in A533 Steel," Memorandum Report 3400, Naval Research Laboratory, Nov 1976

88. N.R. Moody and W.W. Gerberich, Fatigue Crack Propagation in Iron and Two Iron Binary Alloys at Low Temperatures, *Materials Science and Engineering*, Vol 41, 1979, p 271-280

89. M.O. Speidel, Corrosion Fatigue in Fe-Ni-Cr Alloys, *Stress Corrosion Cracking and Hydrogen Embrittlement of Iron Base Alloys*, NACE-5, National Association of Corrosion Engineers, 1977, p 1071-1094

90. O. Vosikovsky, Fatigue-Crack Growth in an X-65 Line-Pipe Steel at Low Cyclic Frequencies in Aqueous Environments, *Trans. ASME*, Series H, Vol 97, 1975, p 298-305

91. J.M. Barsom, Effect of Cyclic-Stress Form on Corrosion-Fatigue Crack Propagation below K_{Iscc} in a High-Yield-Strength Steel, *Corrosion Fatigue*, NACE-2, National Association of Corrosion Engineers, 1972, p 424-436

92. W.H. Cullen and K. Torronen, "A Review of Fatigue Crack Growth of Pressure Vessel and Piping Steels in High-Temperature, Pressurized Reactor-Grade Water," Memorandum Report 4298, Naval Research Laboratory, 19 Sept 1980

93. T. Misawa, N. Ringshall, and J.F. Knott, Fatigue Crack Propagation in Low Alloy Steel in a De-aerated Distilled Water Environment, *Corrosion Science*, Vol 16 (No. 11), Nov 1976, p 805-818

94. T. Kondo, T. Kikuyama, H. Nakajima, M. Shindo, and R. Nagasaki, Corrosion Fatigue of ASTM A-302 B Steel in High Temperature Water, the Simulated Nuclear Reactor Environment, *Corrosion Fatigue*, National Association of Corrosion Engineers, 1972, p 539

95. T.W. Crooker and E. Lange, The Influence of Yield Strength and Fracture Toughness on Fatigue Design Procedures for Structural Steels, *Conference on Fatigue of Welded Structures*, Vol 2, Welding Institute, Cambridge, U.K., 1971, p 243-256

96. R.P. Wei and J.D. Landes, Correlation between Sustained Load and Fatigue Crack Growth in High Strength Steels, *Mater. Res. Stand.*, Vol 9 (No. 7), July 1969, p 25-27

97. W.W. Gerberich, J.P. Birat, and V.F. Zackay, On the Superposition Model for Environmentally Assisted Fatigue Crack Propagation, *Corrosion Fatigue*, NACE-2, National Association of Corrosion Engineers, 1972, p 396-408

98. V.S. Agarwala and J.J. DeLuccia, New Inhibitors for Crack Arrestment in Corrosion Fatigue of High Strength Steels, *Corrosion*, Vol 36 (No. 4), 1980, p 208-212

99. W.D. Benjamin and E.A. Steigerwald, Effect of Composition on the Environmentally Induced Delayed Failure of Precracked High-Strength Steel, *Met. Trans.*, Vol 2, 1971, p 606-608

100. W.W. Gerberich and Y.T. Chen, Hydrogen-Controlled Cracking—An Approach to Threshold Stress Intensity, *Met. Trans. A*, Vol 6, 1975, p 271-278

101. W.S. Pellini, "Advances in Fracture Toughness Characterization Procedures and in Quantitative Interpretations to Fracture Safe Design for Structural Steels," Memorandum Report 6713, Naval Research Lab, April 1968

102. S.R. Novak, "Comprehensive Investigation of K_{Iscc} Behavior of Candidate HY180/210 Steel Weldments," Report 89.021-024(1) B63105, U.S. Steel Corp., 31 Dec 1970

103. C.S. Carter and M.V. Hyatt, Review of Stress Corrosion Cracking in Low Alloy Steels with Yield Strengths below 150 ksi, *Stress Corrosion Cracking and Hydrogen Embrittlement in Iron Base Alloys*, Vol 6, National Association of Corrosion Engineers, 1977, p 524-600

104. G.E. Kerns, M.T. Wang, and R.W. Staehle, Stress Corrosion Cracking and Hydrogen Embrittlement, *Stress Corrosion Cracking and Hydrogen Embrittlement of Iron Base Alloys*, NACE-5, R.W. Staehle, J. Hochmann, R. McCright, and J. Slater, Ed., National Association of Corrosion Engineers, 1971, p 700-735

105. B.F. Brown, Ed., *Stress Corrosion Cracking in High Strength Steels and in Titanium and Aluminum Alloys*, Naval Research Laboratory, 1972, p 80-145

106. H.H. Johnson and M. Willner, *Appl. Mater. Res.*, Vol 4, 1965, p 34

107. "Sulfide Stress Cracking Resistant Metallic Material for Oil Field Equipment," MR-01-75, National Association of Corrosion Engineers, 1980

108. D.P. Dautovich and S. Floreen, The Stress Corrosion and Hydrogen Embrittlement Behavior of Maraging Steels, *Stress Corrosion Cracking*

and Hydrogen Embrittlement of Iron Base Alloys, NACE-5, R.W. Staehle, J. Hochmann, R. McCright, and J. Slater, Ed., National Association of Corrosion Engineers, 1971, p 798-815

109. S. Yamamoto and T. Fujita, Delayed Failure Properties of High Strength Steels in Water, Fracture 1969: Proc. Int. Conf. Fracture, P.L. Pratt, Ed., Chapman and Hall, 1969, p 425

110. V.J. Colangelo and M.S. Fergusen, "Susceptibility of Gun Steels to Stress Corrosion Cracking," WVT-7012, Watervliet Arsenal, U.S. Government Report AD717553, 1970

111. S. Fukui and A. Asada, Trans. Iron Steel Inst. Jpn., Vol 9, 1969, p 448

112. A.H. Freedman, "Development of an Accelerated Stress Corrosion Test for Ferrous and Nickel Base Alloys," NOR 68-58, NASA Contract NAS-20333, Northrup Corp., 1968

113. S.R. Novak, "Effect of Plastic Strain on the K_{Iscc} of High Strength Steels in Seawater," paper presented at Symp. Fracture and Fatigue, George Washington University, May 1972

114. C.S. Carter, Met. Trans., Vol 3, 1972, p 584

115. L.M. Dvorecek, Corrosion, Vol 26, 1970, p 177

116. M.J. May and A.H. Priest, "The Influence of Solution pH on Stress Corrosion Cracking Resistance of a Low Alloy High Strength Steel," MG/A/45/68, BISRA, British Steel Corp., 1968

117. J.D. Gilchrist and R. Narayan, Corros. Sci., Vol 11, 1971, p 281

118. H.H. Uhlig and J. Sava, Trans. ASM, Vol 56, 1963, p 361

119. M.J. Humphries and R.N. Parkins, Effects of Cold Work on Stress Corrosion Cracking of Mild Steel, Proc. Conf. Int. Congr. Corrosion, National Association of Corrosion Engineers, 1969, p 151

120. G.G. Hancock and H.H. Johnson, Trans. Met. Soc. AIME, Vol 326, 1966, p 513

121. F.H. Cocks and J. Bradspies, Corrosion, Vol 28, 1972, p 192

122. K. Bohenkamp, Caustic Cracking of Mild Steel, Proc. Conf. Fundamental Aspects of Stress Corrosion Cracking, R.W. Staehle, Ed., National Association of Corrosion Engineers, 1967, p 374

123. C.S. Carter and M.V. Hyatt, Review of Stress Corrosion Cracking in Low Alloy Steels with Yield Strengths below 150 ksi, Stress Corrosion Cracking and Hydrogen Embrittlement of Iron Base Alloys, NACE-5, National Associa-

tion of Corrosion Engineers, 1977, p 524-600

124. D.P. Dautovich and S. Floreen, The Stress Corrosion and Hydrogen Embrittlement Behavior of Maraging Steels, Stress Corrosion Cracking and Hydrogen Embrittlement of Iron Base Alloys, NACE-5, National Association of Corrosion Engineers, 1977, p 798-815

125. G. Sandoz, The Effect of Alloying Elements on the Susceptibility to Stress Corrosion Cracking of Martensitic Steels in Salt Water, Met. Trans., Vol 2, 1971, p 1055-1063

126. P.J. Grobner, D.L. Sponseller, and D.E. Diesburg, Effect of Molybdenum Content on the Sulfide Stress Cracking Resistance of AISI 4130 Steel with 0.357% Cb, Corrosion, Vol 35 (No. 4), 1979, p 175-185

127. N. Bailey, Met. Constr. Brit. Weld. J., Vol 2, 1970, p 339

128. E.E. Hofmann et al., Arch. Eisenhüttenwes., Vol 39, 1968, p 677

129. M. Henthorne and R.N. Parkins, Brit. Corros. J., Vol 5, 1967, p 186

130. R.N. Parkins, Stress Corrosion Cracking of Low Carbon Steels, Proc. Conf. Fundamental Aspects of Stress Corrosion Fracture, National Association of Corrosion Engineers, 1967, p 361

131. E. Houdremont et al., Stahl Eisen, Vol 60, 1940, p 757

132. M. Smialowski et al., Bull. Acad. Polon. Sci. Lett., A1 (No. 2 suppl), 1950, p 163

133. G. Athavale and W. Eilender, Korros-Metalschutz, Vol 16, 1940, p 127

134. M.T. Wang and R.W. Staehle, Effect of Heat Treatment and Stress Intensity Parameters on Crack Velocity and Fractography of AISI 4340 Steel, Proc. Hydrogen in Metals Int. Conf., 1972, p 342

135. D.A. Vaughan and D.I. Phalen, Reactions Contributing to the Formation of Susceptible Paths for Stress Corrosion Cracking, STP 429, ASTM, 1967, p 209

136. J.T. Waber and H.J. McDonald, Stress Corrosion Cracking of Mild Steels, Corrosion Publishing, 1947

137. E. Snape, Corros. Anticorros., Vol 24, 1968, p 261

138. W.M. Cain and A.R. Troiano, Petrol. Eng., Vol 37, May 1965, p 78

139. R.L. McGlasson et al., Corrosion, Vol 16, 1960, p 113

140. R.N. Tuttle, Mater. Protect., Vol 9 (No. 4), 1970, p 11

141. S.W. Ciaraldi, Corrosion, Vol 40, 1984, p 77

142. R.H. Jones et al., Corrosion, Vol 45 (No. 6), 1989, p 494

143. K.L. Moloznik et al., Corrosion, Vol 35 (No. 7), 1979, p 331

144. C. Lea and E.D. Hondros, Proc. R. Soc. (London), Vol 377A, 1981, p 477

145. R.A. Davis, Corrosion, Vol 19, 1963, p 45t

146. N.A. Tiner and G.A. Gilpin, Corrosion, Vol 22, 1966, p 271

147. P.C. Hughes et al., J. Iron Steel Inst., Vol 203, 1965, p 154

148. S.W. Ciaraldi, Anisotropy of Sulfide Stress Cracking in C-90 Grade Oil Country Tubular Goods, Proc. 1981 Mechanical Working and Steel Processing Conf., AIME, 1981, p 605

149. H.G. Nelson and D.P. Williams, Quantitative Observations of Hydrogen-Induced, Slow Crack Growth in a Low Alloy Steel, Stress Corrosion Cracking and Hydrogen Embrittlement of Iron Base Alloys, NACE-5, National Association of Corrosion Engineers, 1977, p 390-404

150. H.L. Logan, The Stress Corrosion of Metals, J. Wiley and Sons, 1966

151. B. Poulson, The Fractography of SCC in Carbon Steel, Corros. Sci., Vol 15 (No. 8), Aug 1976, p 469-477

152. de G. Jones, J.F. Newman, and R.P. Harrison, Stress Corrosion Cracking of Low Alloy Steels, Proceedings of the Fifth International Congress on Metal Corrosion, National Association of Corrosion Engineers, 1974, p 434-438

153. de G. Jones and M.J. Humphries, Preliminary Experiments on the Stress Corrosion of Mild Steel in Lithium Hydroxide Solutions, Proceedings of the Fifth International Congress on Metal Corrosion, National Association of Corrosion Engineers, 1974, p 410-413

154. R.D. Armstrong and A.C. Coates, A Correlation between Electrochemical Parameters and Stress Corrosion Cracking, Corros. Sci., Vol 16 (No. 7), July 1976, p 423-433

155. D. Hixson and H.H. Uhlig, Stress Corrosion Cracking of Mild Steel in Ammonium Carbonate Solution, Corrosion, Vol 32 (No. 2), 1976, p 56-59

156. W.W. Gerberich, N.R. Moody, and B. Miksic, University of Minnesota and Cortec Corp., unpublished data

Fatigue and Fracture Properties of Cast Steels

S.W. Becker and G.F. Carpenter, NACO Technologies

CASTINGS offer unique cost advantages over other manufacturing methods for many components, especially those having complex three-dimensional geometry. Unlike built-up assemblies, castings can be designed for function rather than for ease of assembly. However, castings are not specified by designers as often as their advantages would imply, because many designers mistakenly believe that the casting process inherently produces components which contain flaws that deleteriously affect properties and part performance. This situation is changing. The casting process is well understood, and high-integrity castings are being made with the same reliability as forgings. Indeed, castings prove their quality every day in applications as demanding as the rotating hardware in gas turbine engines and primary aircraft structures.

With more demanding structural applications, the fatigue and fracture resistance of castings is of considerable interest to designers of castings. In the past, material selection and product design were based on the results of tension tests, stress-life fatigue tests, and impact tests. This method of design against fatigue and fracture was not always satisfactory. In recent years, design philosophies based on low-cycle fatigue strain versus life behavior, crack propagation, and fracture toughness have emerged as being sound and successful.

This article summarizes the general fatigue and fracture properties of cast steels, with emphasis on general comparisons with wrought steel and usefulness design data based on strain-life and fatigue crack growth testing. Topic coverage includes:

- Cyclic stress-strain behavior and low- and high-cycle fatigue life behavior
- Plane-stress fracture toughness
- Plane-strain fracture toughness
- Constant-amplitude fatigue crack initiation and growth
- Variable-amplitude fatigue crack initiation and growth

The subjects of sustained-load cracking and stress-corrosion cracking, creep crack growth, creep-fatigue behavior, and thermal fatigue are also important to designers; however, these topics are beyond the scope of this article. Reference 1 contains more information on these topics.

Comparisons with wrought steels are emphasized when possible because most available data are for wrought steels at room temperature. However, a significant portion of this article is based on a very comprehensive study of the fatigue and fracture properties of cast steels at room and low climatic temperatures done by R.I. Stephens and his students (Ref 2, 3). The study included fatigue behavior under both constant-amplitude and variable-amplitude loading, for the following grades of cast steel:

- *SAE 0030:* normalized and tempered (NT) at 137 HB (average)
- *SAE 0050A:* normalized and tempered (NT) at 192 HB
- *Carbon-Manganese Steel:* normalized, quenched and tempered (NQT) at 174 HB
- *Manganese-Molybdenum Steel:* normalized, quenched and tempered (NQT) at 206 HB
- *AISI 8630:* normalized, quenched, and tempered at 305 HB

The room-temperature ultimate tensile strength and yield strength ranges are 500 to 1150 MPa (72 to 166 ksi) and 300 to 1000 MPa (44 to 143 ksi), respectively. Brinell hardnesses range from 137 to 305 HB and both ferritic-pearlitic and tempered martensitic microstructures are involved. These cast steels are representative of the carbon and low-alloy cast steels used for structural applications.

Structure and Property Correlations

Carbon and low-alloy steel castings are produced to a great variety of properties because composition and heat treatment can be selected to achieve specific combinations of properties, including hardness, strength, ductility, fatigue, and toughness. Although selections can be made from a wide range of properties, it is important to recognize the interrelationships among these properties. These general relations are summarized below for toughness, fatigue, and component design factors such as section size and discontinuities.

Strength and Toughness. Several test methods are available for evaluating the toughness of steels or the resistance to sudden or brittle fracture. These include the Charpy V-notch impact test, the drop-weight test, the dynamic tear test, and specialized procedures to determine plane-strain fracture toughness. Higher toughness is obtained when a steel is quenched and tempered, rather than normalized and tempered; quenching, followed by tempering, produces superior toughness.

Charpy impact toughness is the most common measure of toughness with several types of specimen configurations such as the V-notch or keyhole specimen. The loss in toughness at lower temperatures is more distinct in Charpy V-notch data as compared to keyhole-notch specimen testing, which is why V-notch data are more commonly used to evaluate ductile-brittle transition temperatures.

Typical Charpy V-notch data are shown in Fig. 1 for five common structural cast steels and one cast austenitic stainless steel (CF8). Austenitic steels retain considerable toughness at cryogenic temperatures. Comparative data for other cast stainless steels are summarized in Fig. 2. The lower carbon duplex cast stainless steel grades CF3M, CF8, and CF8M and the austenitic CN7M and CF20 exhibit significantly higher toughness than the stabilized grade of CF8C, and the free-machining grade CF8F (a lower carbon version of C-16). The higher carbon austenitic grades CH-20 and CK-20 show lower impact resistance than the other grades listed.

Nil ductility transition temperatures (NDTT) for cast carbon and cast low-alloy steels range from 38 °C (100 °F) to as low as –90 °C (–130 °F) for normalized-and-tempered cast carbon and

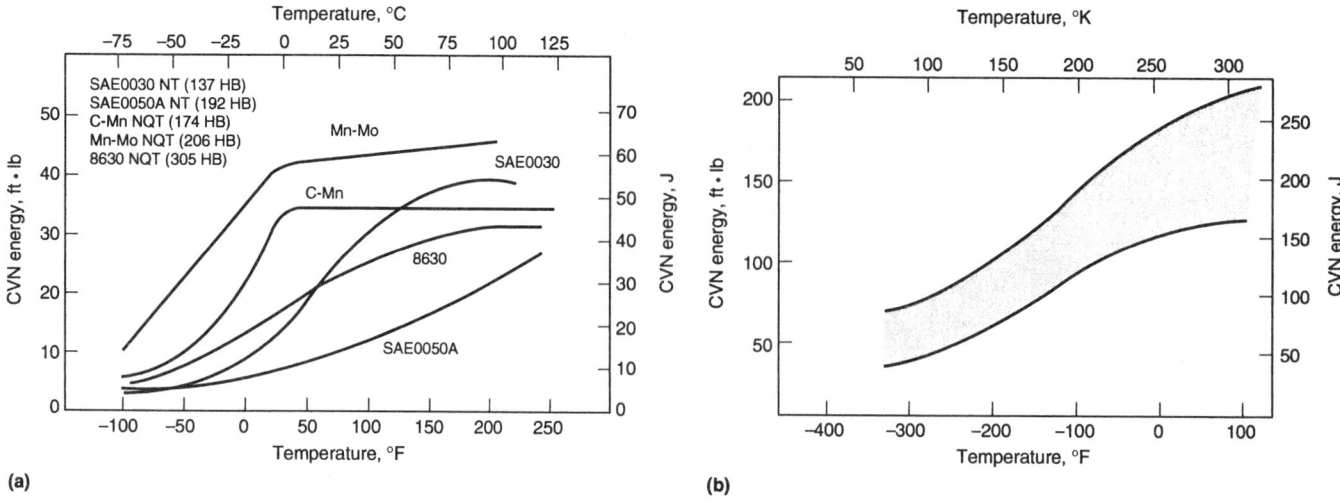

Fig. 1 Charpy V-notch (CVN) impact toughness of various cast steels (a) carbon and low-alloy steels with ferritic-pearlitic (NT treatments) or tempered martensite (NQT treatments) microstructures. (b) Low-temperature Charpy energy band of an austenitic cast stainless steel (CF-8, solution treated and quenched)

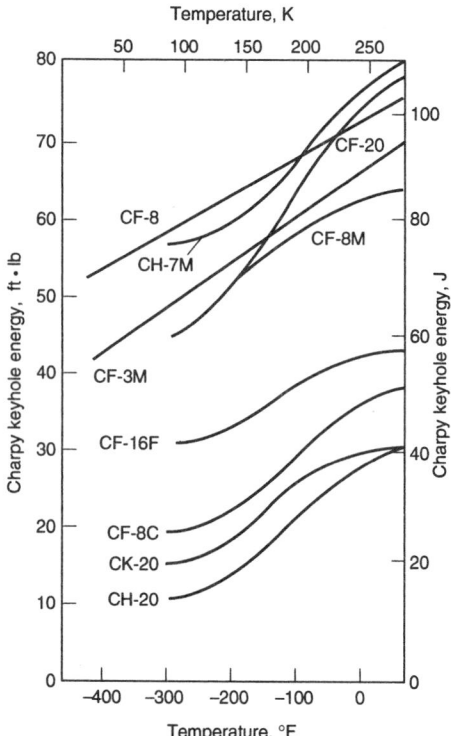

Fig. 2 Charpy keyhole impact toughness of various cast stainless steels (solution annealed and quenched). Source: Ref 1

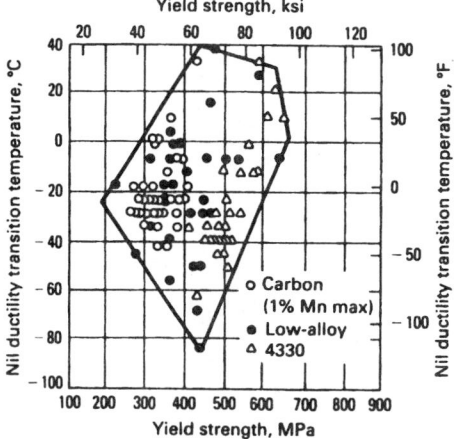

Fig. 3 Nil ductility transition temperatures and yield strengths of normalized-and-tempered commercial cast steels

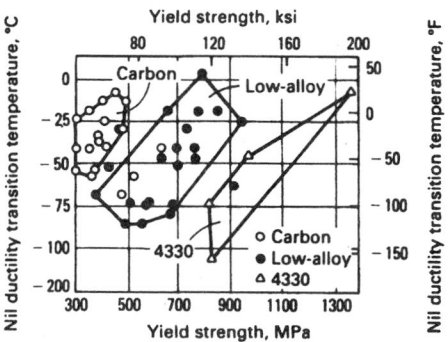

Fig. 4 Nil ductility transition temperatures and yield strengths of quenched-and-tempered commercial cast steels

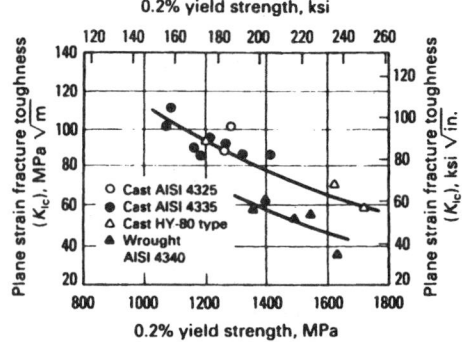

Fig. 5 Plane-strain fracture toughness K_{Ic} and strength relationships at room temperature for quenched-and-tempered nickel-chromium-molybdenum steels

low-alloy steels in the yield strength range of 207 to 655 MPa (30 to 95 ksi) (Fig. 3). Comparison of the data in Fig. 3 with those of Fig. 4 shows the superior toughness values at equal strength levels that low-alloy steels offer compared to carbon steels. When cast steels are quenched and tempered, the range of strength and of toughness is broadened. Depending on alloy selection, NDTT values of as high as 10 °C (50 °F) to as low as –107 °C (–160 °F) can be obtained in the yield strength range of 345 to 1345 MPa (50 to 195 ksi) (Fig. 4).

An approximate relationship exists between the Charpy V-notch impact energy-temperature behavior and the NDTT value. The NDTT value frequently coincides with the energy transition temperature determined in Charpy V-notch tests.

Plane-strain fracture toughness (K_{Ic}) data for a variety of steels reflect the important strength-toughness relationship. Fracture mechanics tests have the advantage over conventional toughness tests of being able to yield material property values that can be used in design equations.

Plane-strain fracture toughness (K_{Ic}) data for the various cast steels are in Fig. 5. For quenched-and-tempered nickel-chromium-molybdenum steels, Fig. 5 indicates high K_{Ic} values of about 110 MPa√m (100 ksi√in.). at a 0.2% offset yield strength level of 1034 MPa (150 ksi). At a yield strength level of 1655 MPa (240 ksi), K_{Ic} values decrease to about 66 MPa√m (60 ksi√in.). Data are also plotted in Fig. 5 for wrought plates made of comparable steel of somewhat higher carbon content.

Tensile Strength and Fatigue Strength Limits. Cast and wrought steels have similar fatigue (or endurance) limits for notched specimens with comparable tensile strength (e.g., Fig. 6). The endurance ratio (endurance limit divided by the tensile strength) of cast carbon and low-alloy steels as determined by rotating-beam bending fatigue tests (mean stress = 0) is generally taken

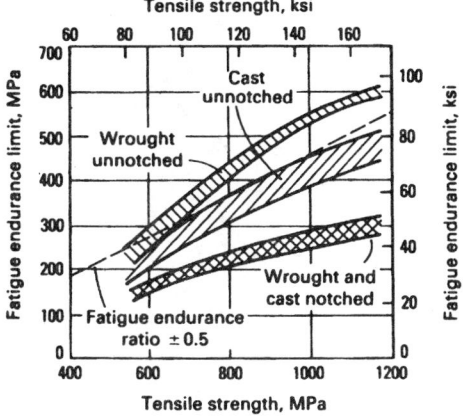

Fig. 6 Fatigue endurance limit versus tensile strength for notched and unnotched cast and wrought steels with various heat treatments. Data obtained in R.R. Moore rotating beam fatigue tests of nine steels (K_t = 2.2). Source: *Metals Handbook*, Volume 1, 8th ed., 1961

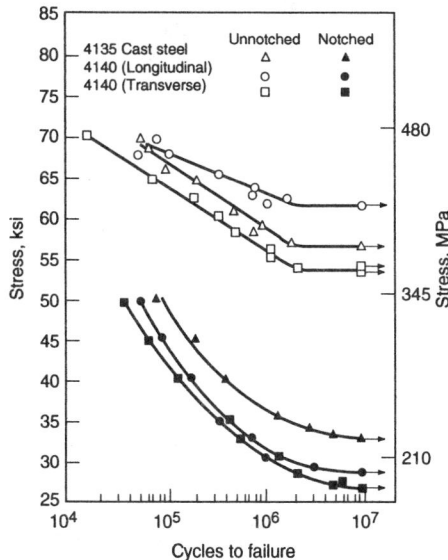

Fig. 7 *S-N* curves (R = –1) of a normalized and tempered AISI 4140 wrought steel in the longitudinal and transverse direction and cast 4135 steel normalized and tempered. Tensile strength for wrought steel: longitudinal, 110.0 ksi (758 MPa); transverse, 110.7 ksi (763 MPa); cast steel: 112.7 ksi (770 MPa)

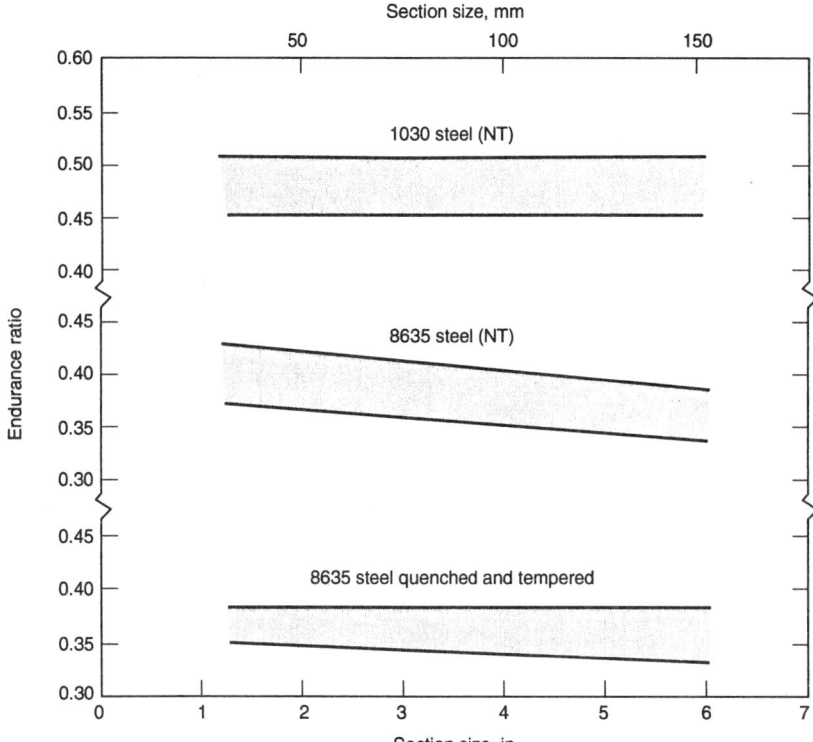

Fig. 8 Fatigue endurance ratios of three cast steels versus section size. Bands show the range for specimens taken at different depths from 32 mm, 76 mm, and 152 mm squares.

Table 1 Fatigue notch sensitivity of several cast and wrought steels (R = –1)

			Endurance limit						Fatigue notch
	Tensile strength		Unnotched		Notched		Fatigue endurance ratio		sensitivity
Steel	MPa	ksi	MPa	ksi	MPa	ksi	Unnotched	Notched	factor, q
Normalized and tempered									
1040 cast	648	94.2	260	37.7	193	28	0.40	0.30	0.29
1040 wrought	620	90	...	...	...	...	...	...	0.50
1330 cast	685	99.3	334	48.4	219	31.7	0.49	0.32	0.44
1330 cast	669	97	288	41.7	215	31.2	0.43	0.32	0.28
4135 cast	777	112.7	353	51.2	230	33.3	0.45	0.30	0.45
4335 cast	872	126.5	434	63	241	34.9	0.50	0.28	0.68
8630 cast	762	110.5	372	54	228	33.1	0.49	0.30	0.53
8640 wrought	748	108.5	...	...	...	...	...	...	0.85
Quenched and tempered									
1330 cast	843	122.2	403	58.5	257	37.3	0.48	0.31	0.48
1340 wrought	836	121	...	...	...	...	...	...	0.73
4135 cast	1009	146.4	423	61.3	280	40.6	0.42	0.28	0.43
4335 cast	1160	168.2	535	77.6	332	48.2	0.46	0.29	0.51
8630 cast	948	137.5	447	64.9	266	38.6	0.47	0.27	0.57
8640 wrought	953	138.2	...	...	...	...	...	...	0.90
Annealed									
1040 cast	576	83.5	229	33.2	179	26	0.40	0.31	0.23
1040 wrought	561	81.4	...	...	...	...	...	...	0.43

(a) Notched tests run with theoretical stress concentration factor of 2.2. (b) $q = (K_f - 1)/(K_t - 1)$, where K_f is the endurance limit notched/endurance limit unnotched and K_t is the theoretical stress concentration factor rotating beam bending tests (mean stress = 0)

to be approximately 0.40 to 0.50 for smooth bars. The data given in Table 1 indicate that this endurance ratio is largely independent of strength, although the endurance ratio tends to decrease at higher tensile strengths.

The fatigue notch sensitivity factor, q, determined in rotating-beam bending fatigue tests is related to the microstructure of the steel (composition and heat treatment) and the strength. Table 1 shows that q generally increases with strength—from 0.23 for annealed carbon steel at a tensile strength of 576 MPa (83.5 ksi) to 0.68 for the higher-strength normalized-and-tempered low-alloy steels. The quenched-and-tempered steels with a martensitic structure are less notch sensitive than the normalized-and-tempered

steels with a ferrite-pearlite microstructure. Similar results and trends in notch sensitivity have been reported for tests with sharper notches. Because of the inherent microdiscontinuities in castings, cast steels suffer less degradation of fatigue properties due to notches than equivalent wrought steels.

Endurance limits of cast steels also are less sensitive to the testing direction than wrought steels. For wrought steels, the endurance limit is lower in the transverse direction than in the longitudinal direction of prior working (Fig. 7). This is attributed to the elongation of inclusions in the direction of prior working and the stress concen-

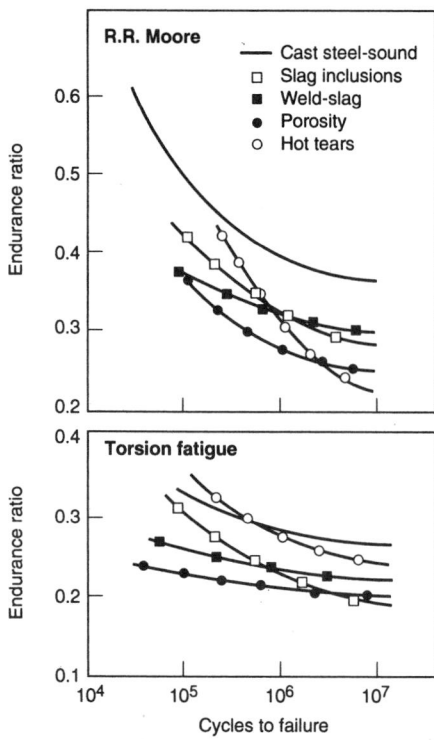

Fig. 9 Effect of various defects on endurance ratio (fatigue strength/monotonic strength) of quenched-and-tempered 8630 cast steel. (a) R.R. Moore rotating beam. (b) Torsion fatigue

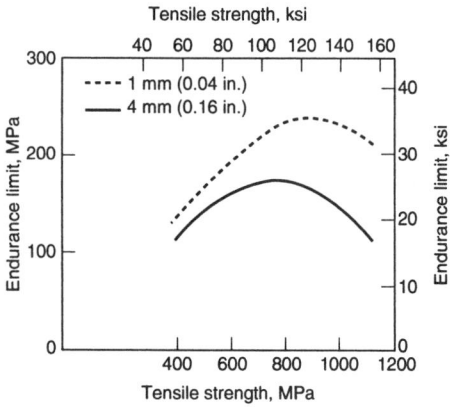

Fig. 10 Effect of defect size on fatigue endurance limits of cast steel. Source: Ref 5

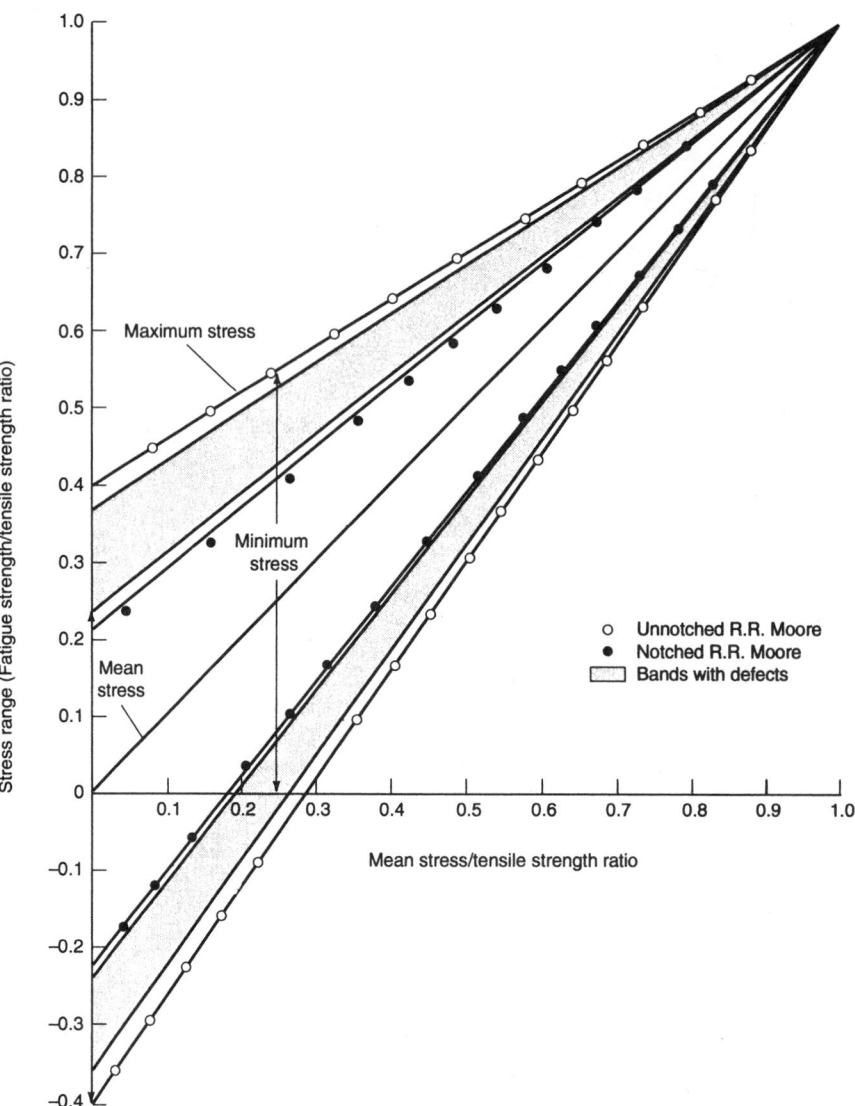

Fig. 11 Goodman diagram for the bending fatigue (R.R. Moore) of 8630 cast steel (normalized and tempered) for determining the fatigue limits in terms of cyclic stress range. The stress range (which is the difference between maximum and minimum stress) for unnotched specimens at zero mean stress is about ±0.4 UTS (line A), while at zero minimum stress ($R = 0$) for unnotched specimens, the stress range is reduced by the amount shown by line B. The ranges for castings with defects and discontinuities are shown as bands for the maximum and minimum stresses, and the bands include discontinuities such as incomplete penetration of welds, weld undercut, weld-slag, sound welds, slag inclusion, hot tears, and cast surface porosity. Notched specimens were machined. Notch of R.R. Moore specimen: 60° included angle, 0.0015 in. (0.0381 mm) root radius

trating effect of loading normal to the major axis of an essentially elliptical void. Cast steels, however, do not exhibit this directionality since the microdiscontinuities generally have no preferential orientation. It is also common practice to liken the fatigue resistance of cast steels to those of a comparable wrought steel tested in a transverse direction.

Fatigue of Cast Steel

Besides the effect of strength and external notches (Fig. 6 and 7), the fatigue of cast steels is influenced principally by the following factors:

- Section size
- Defect size
- Stress modes and waveform type

These factors are briefly described below with additional information on the strain-life behavior of cast steels. Because fatigue cracks initiate from plastic deformation in regions of stress concentration or inherent discontinuities, strain-life data provide useful information for design analysis. A brief review of inherent notch effects from internal microdiscontinuities is also summarized.

Section Size Effects. Increasing section size reduces tensile strength and thus fatigue strength. However, if the fatigue endurance is just a function of tensile strength, then the *endurance ratio* (fatigue limit/tensile strength) would be constant

versus section size. This is not strictly true for all castings. In some castings, a slight decrease in the endurance ratio may be observed. In Fig. 8, for example, a slight decrease in endurance ratio is observed for 8635, when compared to 1030 castings.

Effect of Defects. Figure 9 illustrates the typical effect of various defects on fatigue strength of a cast steel. The reduction in fatigue strength depends on defect size (Fig. 10), with the largest defects being the weakest links. The notch effect from defects is described by various relations in terms of the fatigue notch factor (K_f). Analysis by de Kazinczy (reported in Ref 4) specifies:

$$K_f = 1 + 0.16 \sqrt{d} \qquad (\text{Eq 1})$$

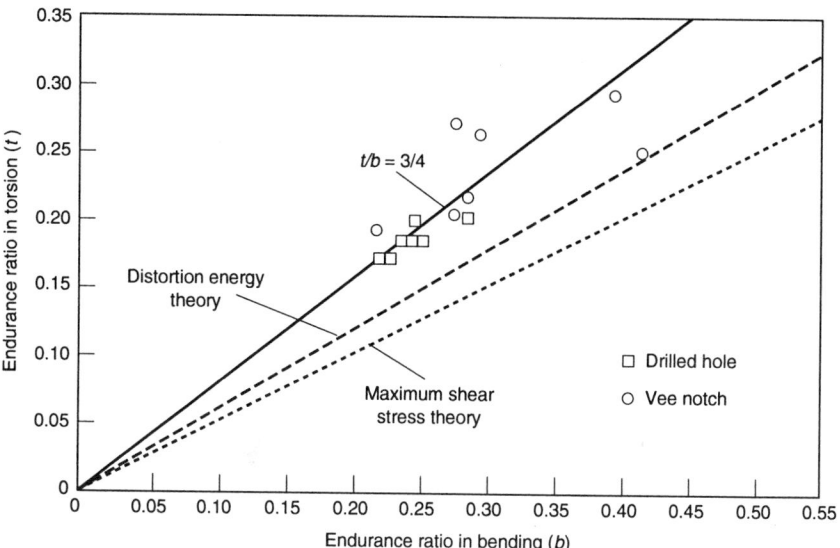

Fig. 12 Torsion and bending endurance ratios (fatigue endurance/monotonic strength) with drilled holes and notched values compared with theory

Table 2 "Endurance ratios" in bending and torsion

Type of specimen	Endurance ratio in bending	Endurance ratio in torsion	$\left(\dfrac{t}{b}\right)$(a)
Q & T			
Cast steel-sound(a)	0.310	0.298	0.96
Weld-machine-sound	0.251	0.230	0.92
Slag inclusions	0.246	0.246	1.00
As welded-sound	0.241	0.221	0.92
Weld-slag	0.243	0.184	0.75
Weld-undercut	0.233	0.195	0.84
Cavities	0.117	0.100	0.86
Hot tears	0.274	0.146	0.53
N & T			
Cast steel-sound(a)	0.361	0.270	0.75
Weld-machine-sound	0.352	0.261	0.74
As welded-sound	0.345	0.250	0.73
Weld-slag	0.314	0.234	0.75
Weld-undercut	0.280	0.230	0.82
Cavities	0.235	0.195	0.83
Slag inclusions	0.292	0.208	0.71
Hot tears	0.245	0.241	0.98

Note: "Endurance ratio" using R.R. Moore specimen. Q&T unnotched 0.390, (0.015 in. R) 0.255; Q&T notched (0.001 in. R) 0.174; N&T unnotched 0.395, (0.015 in. R) 0.252; N&T notched (0.001 in. R) 0.225. (a) $\left(\dfrac{t}{b}\right) = \dfrac{\text{"Endurance ratio" in torsion}}{\text{"Endurance ratio" in bending}}$; Average for Q&T = 0.85.

where d is the defect diameter in mm. However, this relation does not account for the shape of the defect or the matrix strength. Another method (Ref 4) is based on the method of Peterson, which relates K_f to defect as follows:

$$K_f = 1 + \frac{K_t - 1}{1 + a/r} \qquad \text{(Eq 2)}$$

where K_t is the theoretical stress concentration factor, r is the tip radius of the surface, and a is a material constant, which for most ferrous alloys is estimated in mm as

$$a = 0.0254 \, [2070/\text{UTS (in MPa)}]^{1.8} \, \text{mm} \qquad \text{(Eq 3)}$$

For example, with a hemispherical surface discontinuity ($K_t \cong 2.5$) and an average diameter of 0.40 mm ($r = 0.20$ mm) for the *largest* microdiscontinuities in a cast steel with a Brinell hardness of 160 HB (and UTS = 3.5 HB = 560 MPa), then:

$$a \cong 0.0254 \, [3.7]^{1.8} \cong 0.27 \, \text{mm}$$

$$K_f \cong 1 + \frac{(2.5 - 1)}{(1 + 0.27/0.20)} = 1.6 \qquad \text{(Eq 4)}$$

Applied stress effects include variations in mean stress, variable amplitude loading, and the differences between torsion and bending. Mean stress effects are commonly addressed with modified Goodman diagrams such as the one shown in Fig. 11 for 8630 steel. Mean stress effects should not differ substantially from notched wrought steels. Variable amplitude fatigue is discussed later with respect to fatigue crack growth.

Differences in fatigue behavior for torsional and bending loads are typically described with the assumption of homogeneous and isotropic materials. However, because castings have internal discontinuities, comparison between bending

and torsional fatigue is based on a "drilled hole" assumption, where the hole accounts for inherent notch effects from microdiscontinuities (Ref 4). Figure 12 shows a plot of "endurance" ratio in torsion (t) to that in bending (b) for notched, wrought steels. Dashed lines represent the distortional energy theory and maximum shear stress theory lines for $t/b = 0.577$ and 0.5, respectively. The data points for the "drilled hole" show best agreement to $t/b = 3/4$ developed in the previous argument. V-notch specimens (55° circumferential) deviate slightly, due to a difference in stress concentration, but are still in better agreement with $t/b = 3/4$.

Cast metals have t/b ratios which vary between 0.7 and 1.05 (Ref 5) with an average of 0.85. As such, the concept of a "drilled hole" method seems reasonable. However, an analysis by Vishnevsky (Table 2) concludes that "all the points except one lie above the Maxwell-von Mises criterion indicating that discontinuities present in torsion are not nearly so damaging as they are in bending." Table 2 shows that endurance ratios in torsion are lower than those in bending and that the t/b ratio average of 0.85 and 0.79 for the QT and HT condition are also in agreement with the drilled-hole argument.

Cyclic Stress-Strain Behavior. Cyclic stress-strain behavior provides a measure of the steady-state cyclic deformation resistance and thus a basis for understanding resistance to crack nucleation. Cyclic stress-strain curves (using the incremental step method and the companion specimen method) are shown in Fig. 13 and 14 for five cast steels at room temperature and low temperature. The cyclic stress-strain curves are initially identical to the monotonic curves and then become nonlinear at stresses or strains of about 20 to 50% below that for monotonic curves.

The monotonic upper yield point found in all but the 8630 steel (Fig. 14) was eliminated under the cyclic conditions. At higher strains, the cyclic curves intersect or converge with the monotonic

curves for these four steels. These steels cyclic strain soften at the lower strain levels and then cyclic strain harden at the higher strain levels. The cyclic curves for 8630 steel were always equal to or below the monotonic curve and exhibited cyclic strain softening.

A review of available data on wrought steels (Ref 6) indicates that cyclic strain hardening exponent (n' values) for the cast steels are similar to those for wrought steels. Moreover, the strain hardening and softening behaviors of the cast steels are similar to those for wrought steels. Cyclic strain hardening was found for cast steels when the tensile/yield strength ratio was greater than 1.5. When this ratio was less than 1.3, cyclic strain softening occurred. At intermediate ratios, mixed cyclic strain softening and hardening were observed. These ratios are about 0.1 higher than those for wrought materials (Ref 7). The monotonic yield and tensile strengths, cyclic stress-strain curves, and cyclic yield strengths at the low temperature were about 10% higher than room-temperature results except for 8630 steel.

Low-Cycle Axial Fatigue Behavior. Low-cycle axial fatigue tests were conducted on the five steels (Ref 2 and 3) at room and low temperature with $R = -1$ at constant strain amplitudes from 0.0013 to 0.015, which gave fatigue lives of 10^2 to 10^6 cycles. Half-life stable hysteresis loops were used to obtain elastic and plastic strain amplitudes. With Young's modulus, E, obtained from the first-quarter cycle, the elastic strain amplitude was calculated from:

$$\Delta\varepsilon_e/2 = \Delta\sigma/2E \qquad \text{(Eq 5)}$$

and the plastic strain amplitude was calculated as

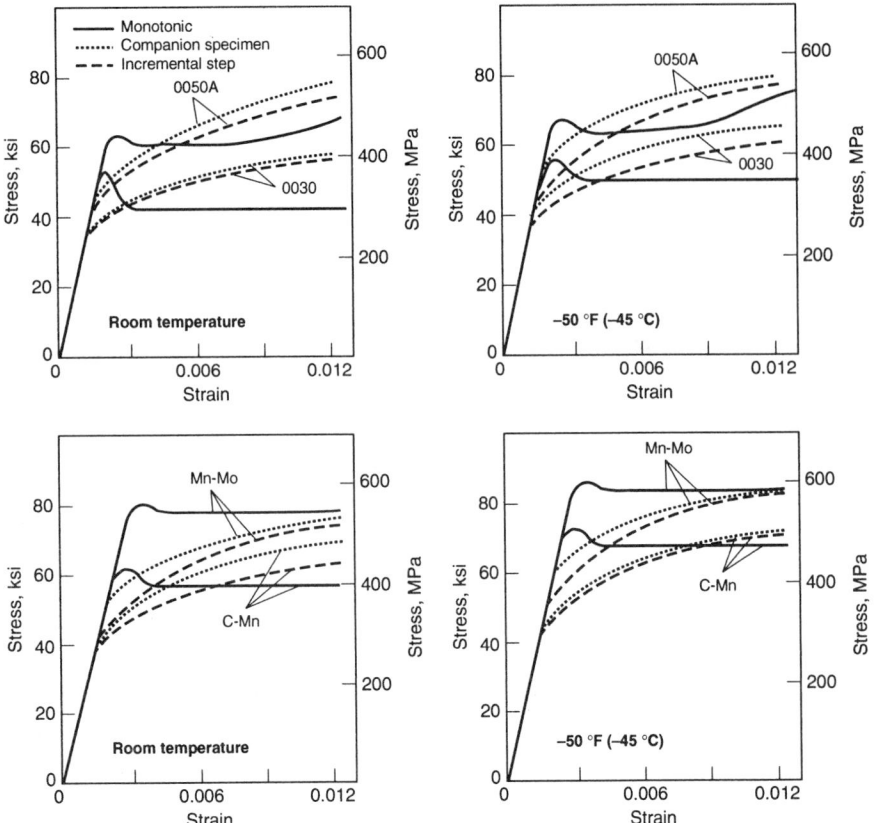

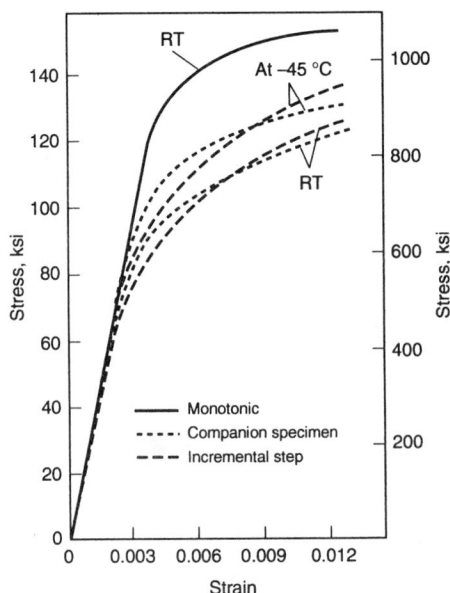

Fig. 14 Cyclic stress-strain of cast 8630 steel (NQT, 305 HB average) at room temperature and –45 °C (–50 °F). Source: Ref 2

Fig. 13 Room-temperature and low-temperature cyclic stress-strain curves for two ferritic-pearlitic cast steels (SAE 0050A and 0030 normalized-and-tempered to average hardness of 192 and 137 HB, respectively) and two quenched-and-tempered low-alloy cast steels (C-Mn and Mn-Mo steels at 174 and 206 average HB, respectively). Stable hysteres of cyclic stress-strain were obtained from both companion test specimens and from a single specimens subjected to incremental strain steps. For the incremental step tests, each strain block contained 79 reversals with the magnitude ranging from 0 to 0.012. The strain rate was constant and equal to 0.01/s. Source: Ref 2

Table 3 Monotonic and low-cycle fatigue properties

Steel	S_u MPa	ksi	S_Y (0.2%) MPa	ksi	ε_f	ε_f'	σ_f MPa	ksi	σ_f' MPa	ksi	b	c	S_u/S_y
At room temperature													
0030	496	72	303	44	0.62	0.28	752	109	655	95	–0.083	–0.552	1.6
0050A	787	114	415	60	0.21	0.30	869	126	1338	194	–0.127	–0.569	1.9
C-Mn	583	85	402	58	0.34	0.15	703	102	869	126	–0.101	–0.514	1.5
Mn-Mo	702	102	542	79	0.38	0.78	752	109	1117	162	–0.101	–0.729	1.3
8630	1144	166	995	143	0.35	0.42	1268	184	1936	281	–0.121	–0.693	1.2
At –45 °C (–50 °F)													
0030	542	79	320	46	0.36	0.18	621	90	834	121	–0.089	–0.506	1.7
0050A	834	121	436	63	0.17	0.32	924	134	1282	186	–0.111	–0.582	1.9
C-Mn	612	89	464	67	0.26	0.07	634	92	717	104	–0.067	–0.439	1.3
Mn-Mo	758	110	562	81	0.36	0.47	862	125	1096	159	–0.090	–0.671	1.3
8630	1178	171	999	145	0.33	0.35	1254	182	1785	259	–0.099	–0.659	1.2

Source: Ref 2

$$\Delta\varepsilon_p/2 = \Delta\varepsilon/2 - \Delta\varepsilon_e/2 \qquad \text{(Eq 6)}$$

$$= \Delta\varepsilon/2 - \Delta\sigma/2E \qquad \text{(Eq 7)}$$

where $\Delta\varepsilon/2$ is the controlled total strain amplitude. These elastic, plastic, and total strain amplitudes are plotted versus number of applied reversals to fracture, $2N_f$, on log-log coordinates in Fig. 15 for four

of the five cast steels at room temperature (with total life at –45 °C (–50 °F) included for comparison). The elastic and plastic components are represented by the following linear relationships:

$$\Delta\varepsilon/2 = \Delta\varepsilon_e/2 + \Delta\varepsilon_p/2 \qquad \text{(Eq 8)}$$

$$= \sigma_f'/E \times (2N_f)^b + \varepsilon_f' \ (2N_f)^c \qquad \text{(Eq 9)}$$

where σ_f' is the fatigue strength coefficient, b is the fatigue strength exponent, ε_f' is the fatigue ductility coefficient, c is the fatigue ductility exponent, E is Young's modulus, and $2N_f$ is the number of reversals to failure.

The low-cycle fatigue material properties are listed in Table 3 along with yield and tensile strengths obtained from the monotonic tension tests. Low-cycle fatigue behavior at the low temperature was equal to or better than at room temperature for lives greater than 5×10^5 reversals. Mixed behavior existed at shorter lives. The fatigue strengths at 10^6 reversals were from 0 to 30% better at the low temperature. Mixed low-cycle fatigue behavior at low temperatures has also been reported in the literature for wrought steels (Ref 8). In addition, values of low-cycle fatigue material properties for the five cast steels were within the ranges published for wrought steels (Ref 6). The room-temperature fatigue strengths of the five cast steels at 10^6 reversals ranged from 208 MPa (30 ksi) for 0030 steel to 370 MPa (53.7 ksi) for 8630 steel. Room-temperature fatigue strengths were within 30 to 40% of the ultimate tensile strength (32 to 46% at the low temperature).

High-Cycle Axial Fatigue Behavior. The high-cycle axial fatigue tests were conducted in the same manner as for the low-cycle tests, except that load control was used instead of strain control and the test frequency was 23 to 37 Hz. Failures ranged from 10^5 reversals to runouts at 2×10^7 reversals. High-cycle (10^7 cycle) fatigue data for 0030 and 8630 cast steels are superimposed with the low-cycle fatigue data in Fig. 15(a) and 15(b), respectively, for the 0030 carbon steel and 8630 steel. The solid circles represent total strain amplitudes for the strain-controlled low-cycle fatigue tests, and the solid squares represent total strain amplitudes from the load-controlled

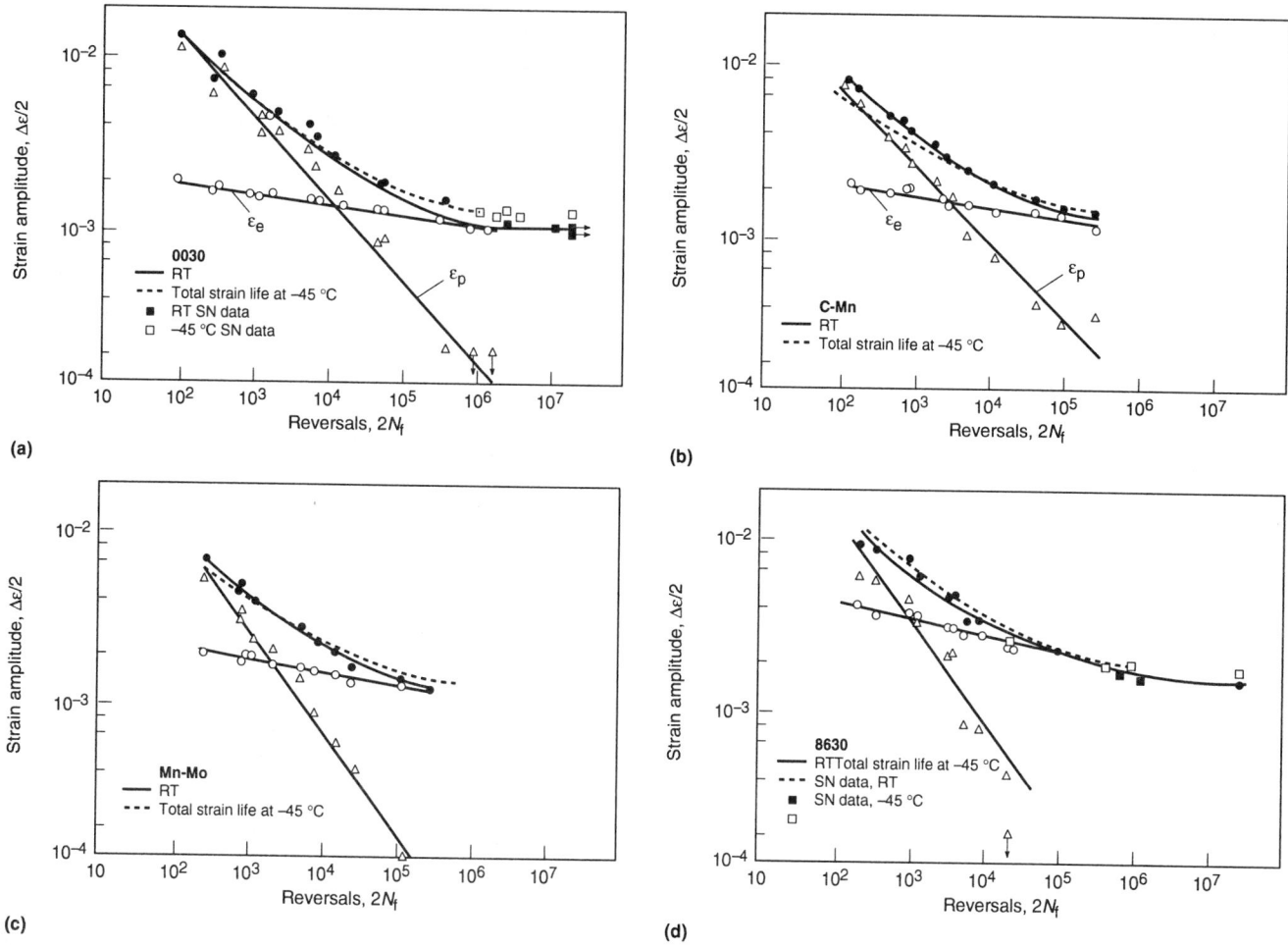

Fig. 15 Axial strain life and the elastic (ε_e) and plastic (ε_p) components of four cast steels. (a) SAE 0030 (NT, 137 HB). (b) C-Mn steel (NQT, 174 HB). (c) Mn-Mo steel (206 HB), and (d) 8630 cast steel at 305 HB. Total strain life at –45 °C (–50 °F) is included. Smooth hourglass specimens with a minimum diameter of 6.4 mm (0.25 in.) over 17.5 mm ($^{11}/_{16}$ in.) length. Source: Ref 2

Table 4 Axial fatigue limits and fatigue ratios

Steel	At room temperature Fatigue limit (S_f) MPa	ksi	Fatigue ratio, S_f/S_u	At –45 °C (–50 °F) Fatigue limit (S_f) MPa	ksi	Fatigue ratio, S_f/S_u
0030	196	28.5	0.40	241	35	0.44
0050A	237	34.4	0.30	243	35.2	0.29
C-Mn	248	36	0.42	255	37	0.42
Mn-Mo	232	33.7	0.33	269	39	0.35
8630	293	42.4	0.26	365	53	0.31

Source: Ref 3

high-cycle fatigue tests. The latter strains were calculated by dividing the applied stress amplitude by Young's modulus because these tests were run in the elastic region. All five cast steels exhibited typical fatigue limits around 5×10^6 to 10^7 reversals for both test temperatures. Values of the fully reversed fatigue limits, S_f, obtained from such plots for all five cast steels at both test temperatures, are listed in Table 4 along with the fatigue ratios, S_f/S_u. The fatigue limits vary from 196 to 293 MPa (28.5 to 42.4 ksi) at room temperature and from 241 to 365 MPa (35 to 53 ksi) at –45 °C (–50 °F). The fatigue ratios vary from

0.26 to 0.42 at room temperature and 0.29 to 0.44 at the low temperature. The fatigue ratios for the 0050A and 8630 steels are lower than those for the other cast steels. Axial fatigue ratios for wrought steels range generally from about 0.37 to 0.43 (75 to 85% of the typical rotating bending fatigue ratio of 0.5).

Fracture Mechanics

The treatment of discontinuities in design has undergone major changes as a result of the wider

use of fracture mechanics in the industry. If the plane-strain fracture toughness K_{Ic} of a material is known at the temperature of interest, designers can determine the critical combination of flaw size and stress required to cause failure in one load application. In addition, designers can calculate the remaining life of a component having a discontinuity, or they can compute the largest acceptable flaw from a knowledge of the crack growth properties (da/dN) and fracture mechanics parameters (K, n, and C) of a material.

Fracture toughness depends on the volume of plastic deformation near the crack. For thin specimens, the plastic region at the crack tip is constrained and is under plane stress conditions, which produces a relatively high toughness, K_c. As section thickness increases, fracture behavior approaches plane-strain conditions, and the toughness approaches the inherent lower limit of toughness (K_{Ic}) for the material. In the absence of suitable plane-strain fracture mechanics data, approximations can be made on the basis of test results obtained from a variety of tensile, impact, and fatigue tests.

Plane-stress fracture toughness (K_c) is usually difficult to obtain because of the complexity of the elastic-plastic behavior at the crack tip.

Table 5 Plane-strain fracture toughness of cast low-alloy steels at room temperature

Alloy type	Heat treatment(a)	Yield strength, 0.2% offset MPa	ksi	Plane-strain fracture toughness, K_{Ic} MPa√m	ksi√in.
Fe-1.25Cr-0.5Mo	SRANTSR	275	40	88	80
Cast 1030	NT	303	44	127	116
Fe-0.5Cr-0.5Mo-0.25V	NT	367	53	55	50
Fe-0.5C-1.5Mn	NT	412	59	107	98
Fe-0.5C-1Cr	NT	413	60	58	53
Fe-0.5C	NT	425	61	65	59
Cast 9535	NT	614	89	67	61
Fe-0.35C-0.6Ni-0.7Cr-0.4Mo	NT	683	99	64	58
Cast 4335	SLQT	747	108	69	63
Cast 9536	NT	752	109	59	54
Fe-0.3C-1Ni-1Cr-0.3Mo	NT	787	114	66	60
Cast 4335	SLQT	814	118	97	87
Cast 4335	SLQT	903	131	105	96
Cast 4335	QT	1090	158	115	105
Cast 4335	QT	1166	169	92	84
Ni-Cr-Mo	QT	1207	175	98	89
Cast 4340	QT	1207	175	115	105
Cast 4325	QT	1263	183	75	82
Cast 4325	QT	1280	186	104	95
Cr-Mo	QT	1379	200	84	76
Cast 4340	QT	1450	210	67	61

(a) SR, stress relieved; A, annealed; N, normalized; T, tempered; Q, quenched; SLQ, slack quenched

Table 6 J_{Ic}, J_c, K_{Ic}, and K_c toughness of cast steels

Steel	J_{Ic} kJ/m²	in. · lb/in.²	J_c kJ/m²	in. · lb/in.²	K_{Ic} MPa√m	ksi√in.	K_c MPa √m	ksi√in.
At room temperature								
0030	73	415	...	...	130	118	...	...
0050A	...	...	37(a)	209(a)	...	...	92(a)	84(a)
			25(b)	145(b)			77(b)	70(b)
C-Mn	84	479	...	...	139	126	...	...
Mn-Mo	139	794	...	...	179	163	...	...
8630	80	456	...	...	135	123	...	...
At –45 °C (–50 °F)								
0030	...	...	49(a)	282(a)	...	...	108(a)	98(a)
			44(b)	215(b)			93(b)	85(b)
0050A	...	...	17(a)	95(a)	...	...	61(a)	56(a)
			14(b)	78(b)			56(b)	51(b)
C-Mn	75	428	...	...	132	120	...	...
Mn-Mo	118	674	...	...	166	151	...	...
8630	...	...	38(a)	218(a)	...	...	95(a)	86(a)
			30(b)	174(b)			85(b)	77(b)

(a) Average value. (b) Lowest value. Source: Ref 2

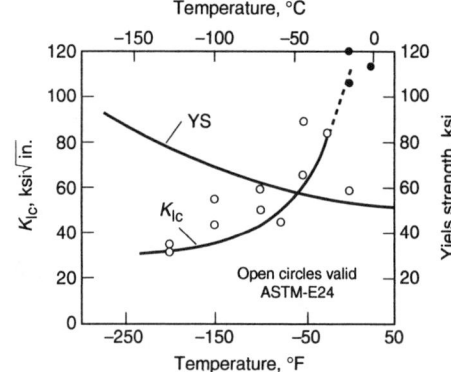

Fig. 16 Plane-strain fracture toughness of ASTM A 216 steel casting (Grade WCC). Source: Ref 1

Moreover, the slow stable crack extension prior to fracture makes it difficult to determine the maximum stress-intensity level at crack instability because the crack length is uncertain.

The R-curve analysis (Ref 9) is applicable for determining K_c for intermediate-strength materials where the plastic zone size, r_y, is small relative to the specimen dimensions, where:

$$r_y = [1/(2\pi)] \times (K/S_y)^2 \qquad (Eq~10)$$

An R-curve is a continuous record of toughness development in terms of crack growth resistance, K_R, plotted against crack extension as the crack is driven under a continuously increasing stress-intensity factor, K. During slow-stable fracturing, the developing crack growth resistance is equal to the applied stress-intensity factor.

To have a valid plane-stress fracture toughness test, the uncracked ligament should satisfy the following:

$$(w - a) \geq 8\,r_y \qquad (Eq~11)$$

where w is the specimen width and a is the crack length (or plastic zone corrected length). Values of r_y near unstable crack extension for five cast steels (Ref 2) ranged from 8 to 18 mm (0.3 to 0.7 in.) and at low temperature ranged from 2.5 to 18 mm (0.1 to 0.7 in.). The validity equation was satisfied only with the 8630 steels at low temperature. The localized plasticity was too large for the other test conditions. However, R-curves were established with K_R equal to K without a plastic zone correction.

Plane-Strain Fracture Toughness. The critical value of toughness (K_{Ic}) is reached when specimen thickness is large enough to produce plane-strain conditions at the tip of a crack. As the thickness increases, the small plastic zone relative to the thickness produces a stress parallel to the crack tip as the plastic zone attempts to contract in this direction but is prevented from doing so by the large volume of surrounding elastic material. The resulting high degree of triaxial stresses leads to lower toughness and a predominantly flat fracture. This critical K value, termed plane-strain fracture toughness, or K_{Ic}, is the minimum toughness and an inherent property of a material.

Typical K_{Ic} values for various cast steels are summarized in Table 5. Fracture toughness of steels is rarely below 27 MPa√m (25 ksi√in.) even at low temperature (Fig. 16). However, this depends on loading rates. If the K_{Ic} values are known or reliable estimates are available, the flaw tolerance of a given structure can be computed. Except for special cases the K_{Ic} level should be considered as the onset of rapid fracture.

J-Integral Method. When alloys have appreciable ductility and low yield strengths, large specimens are required to obtain valid K_{Ic} results. In these instances, the path-independent J-integral is used to estimate fracture toughness by extending fracture mechanics concepts from linear-elastic behavior to elastic-plastic behavior (Ref 10). For linear-elastic behavior, the J-integral is identical to G, the strain energy release rate per unit crack extension, which is directly related to the stress intensity factor, K. Thus, linear-elastic fracture criteria based on K_{Ic}, G_{Ic}, or J_{Ic} are identical. For mode I linear-elastic plane-strain conditions:

$$J_{Ic} = G_{Ic} = K_{Ic}^2/E \times (1 - v^2) \qquad (Eq~12)$$

where E is Young's modulus and v is Poisson's ratio. Because J_{Ic} characterizes the toughness of materials at or near the onset of crack extension from a preexisting sharp crack, it can be used as a conservative estimate of K_{Ic} on specimens that contain appreciable ductility but lack sufficient thickness to be tested for K_{Ic}. Procedures for J_{Ic} tests are described in Ref 11 and elsewhere in this Volume.

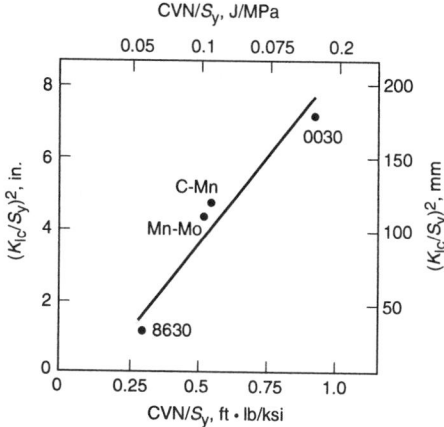

Fig. 17 Correlation between room-temperature K_{Ic}, yield strength (S_y), and upper-shelf Charpy V-notch (CVN) toughness for four cast steels. In SI units, $(K_{Ic}/S_y)^2 = 1170 (CVN/S_y - 0.022)$. Source: Ref 2

Table 7 Constant-amplitude fatigue crack growth constants for $da/dN = C(\Delta K)^n$

Steel	$R \approx 0$ C m/cycle	$R \approx 0$ C in./cycle	n	$R = \frac{1}{2}$ C m/cycle	$R = \frac{1}{2}$ C in./cycle	n
At room temperature						
0030	3.34×10^{-14}	1.98×10^{-13}	4.33	...	...	...
0050A	2.24×10^{-13}	1.19×10^{-11}	3.88	1.99×10^{-12}	1.08×10^{-10}	3.40
C-Mn	7.74×10^{-13}	4.18×10^{-11}	3.35	2.34×10^{-12}	1.25×10^{-10}	3.21
Mn-Mo	1.12×10^{-12}	5.99×10^{-11}	3.28	7.15×10^{-12}	3.7×10^{-10}	2.89
8630	2.63×10^{-12}	1.42×10^{-10}	3.03	1.39×10^{-11}	7.55×10^{-10}	2.67
At −45 °C (−50 °F)(a)						
0030	3.48×10^{-15}	2.14×10^{-13}	4.72	...	...	...
0050A	1.79×10^{-14}	1.08×10^{-12}	4.53	8.36×10^{-14}	4.94×10^{-12}	4.30
C-Mn	5.4×10^{-14}	3.11×10^{-12}	4.01	6.36×10^{-13}	3.49×10^{-11}	3.52
Mn-Mo	9.76×10^{-14}	5.52×10^{-12}	3.84	3.13×10^{-12}	1.64×10^{-10}	3.04
8630	6.38×10^{-13}	3.45×10^{-11}	3.38	1.97×10^{-12}	1.05×10^{-10}	3.22

(a) At −34 °C (−50 °F) for 0030. Source: Ref 2

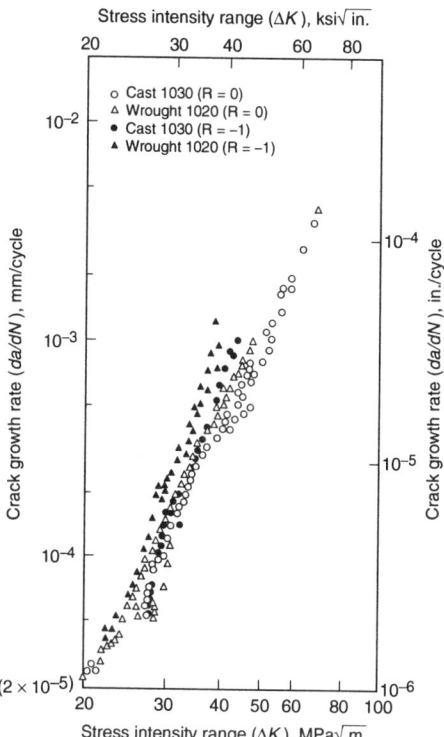

Fig. 18 Constant amplitude fatigue crack growth of cast and wrought steels at $R = 0$ and $R = -1$

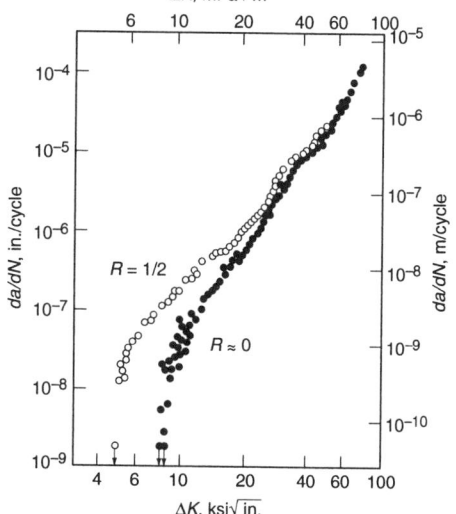

Fig. 19 Fatigue crack threshold behavior of cast steel (SAE 0030 carbon steel) at room temperature. Source: Ref 3

Valid values for J_{Ic} for six test conditions are listed in Table 6. Also listed are values of J_c, both lowest and average, for the invalid J_{Ic} tests. The values of K_{Ic} and K_c given in Table 6 were obtained from J_{Ic} and J_c values (with an assumed Poisson ratio of $v = \frac{1}{3}$). The manganese-molybdenum and 1050A cast steels exhibited the highest and lowest fracture toughness, respectively, at both test temperatures. The three martensitic cast steels (carbon-manganese, manganese-molybdenum, and 8630) had better fracture toughness at room temperature compared with the two ferritic-

pearlitic cast steels (0030 and 0050A). However, at the low temperature, the 8630 steel had J_c less than that of 0030. The 8630 steel had the largest decrease in fracture toughness as the temperature was lowered. The fracture toughness of the carbon-manganese and manganese-molybdenum steels were lowered only 10 to 15% as the temperature was lowered.

Correlation of K_{Ic} and Charpy V-notch upper-shelf impact energy was found for the four cast steels with valid J_{Ic} results at room temperature, as shown in Fig. 17. A similar correlation was not found for the steels tested at the low temperature.

Fatigue Crack Growth Rates (Constant Amplitude). Given the similar notched fatigue strength of cast and wrought steels, similar fatigue crack growth rates should also be expected (Fig. 18). Typical Paris-law parameters for intermediate crack growth rates (Region II) are summarized for five cast steels in Table 7 (Ref 2). At low temperature (−45 °C, or −50 °F) a decrease in

fatigue crack growth rates was observed by amounts of from 1.2 to 3 for $R \approx 0$ and from 1 to 2 for $R = \frac{1}{2}$ at the lower ΔK values (Ref 2, 3). At the higher ΔK values, the room- and low-temperature data tended to converge or crossover. As R was increased from 0 to $\frac{1}{2}$, the fatigue crack growth rate increased by factors from 1.5 to 3. This increase was slightly higher than that suggested in the literature for wrought steels (Ref 12).

For $R \approx 0$ at room temperature, the fatigue crack growth rates for the five cast steels are 2 to 5 times lower than conservative equations proposed by Barsom (Ref 13) for wrought steels. Similar results for similar cast steels were reported by Greenberg et al. (Ref 14) and Kapadia and Imhoff (Ref 15). Thus the fatigue crack growth behavior of cast steels appears to be equal to or better than that of similar wrought steels.

Threshold Crack Growth Behavior. Threshold and near-threshold (Region I) fatigue crack growth behavior of the five cast steels at room and low temperature was determined at $R \approx 0$ and $R = \frac{1}{2}$ for growth rates between 10^{-8} and 10^{-10} m/cycle (4×10^{-7} and 4×10^{-9} in./cycle). Compact (Chevron notched) specimens were tested per ASTM E647. The incremental load-shedding method was used to obtain the fatigue crack growth rate data in the near-threshold region. The fatigue crack growth rate of 10^{-10} m/cycle (4×10^{-9} in./cycle) was used to define the threshold ΔK.

A typical plot of the region I and region II data is shown in Fig. 19. It is evident that the threshold ΔK value for $R = \frac{1}{2}$ is about half that for $R \approx 0$. All of the steels exhibited similar behavior. The values of ΔK_{th} are listed in Table 8. The room-temperature $R \approx 0$ values were higher than the majority of values reported in the literature for wrought steels (Ref 16). The ΔK_{th} values were higher for the low temperature, except for the 8630 steel at $R \approx 0$, for which the values were the same. The ΔK_{th} values showed no correlation with other properties such as yield strength, ultimate strength, or fatigue limit.

Variable Amplitude Fatigue Crack Initiation and Growth. Constant-amplitude fatigue

Table 8 Fatigue crack thresholds (ΔK_{th}) for five cast steels

Steel	ΔK_{th}, $R = 0.05$		ΔK_{th}, $R = \frac{1}{2}$	
	MPa√m	ksi√in.	MPa√m	ksi√in.
At room temperature				
0030	9.1	8.3	5.3	4.8
0050A	10.2	9.3	5.2	4.7
C-Mn	8.3	7.6	3.9	3.5
Mn-Mo	8.1	7.4	4.1	3.7
8630	9.4	8.6	4.1	3.7
At –45 °C (–50 °F)				
0030	14.2	12.9	9.3	8.5
0050A	12.3	11.2	6.8	6.2
C-Mn	14.4	13.1	7.1	6.5
Mn-Mo	10.7	9.7	6.5	5.9
8630	9.4	8.6	5.7	5.2

Source: Ref 3

behavior does not provide information on sequential and interactive effects with variable loading. A more realistic comparison of material or component fatigue behavior can be determined from load or strain histories that closely duplicate or simulate actual service history and environment. SAE International (formerly Society of Automotive Engineers) has developed several service spectra indicative of real-life loading (Ref 2). Variable amplitude tests of five cast steels at room and low temperatures with an as-drilled keyhole notch compact specimen are presented in Ref 2. Eliminating the compression at the four temperatures increased fatigue crack initiation life by factors of 3 to 9, increased short crack growth life by factors of 1 to 4, and increased long crack growth life by factors of 1 to 3. Except for 0050A cast steel, all the average low-temperature fatigue lives for the three different life criteria discussed in Ref 2 were equal to or better than those at room temperature. The increase in crack initiation life was within a factor of 2.5 as was the total fatigue life to fracture. Crack growth life, however, tended to increase as the temperature was lowered and then this trend reversed. The 0050A steel had a continuous decrease in crack growth life for lower temperatures. In general, the three martensitic cast steels (carbon-manganese,

manganese-molybdenum, and 8630) had better fatigue resistance with variable amplitude loading at the four temperatures than the ferritic-pearlitic cast steels (0030 and 0050A). The 0030, carbon-manganese, manganese-molybdenum, and 8630 steels are satisfactory for low-temperature conditions, but 0050A is not.

Lee (Ref 3, 17) used constant-amplitude material properties developed for the five steels (Table 6) at room temperature and at –45 °C (–50 °F) and several mathematical models to calculate variable amplitude fatigue crack initiation and growth. Comparison of the calculated and experimentally determined fatigue lives showed that total fatigue life calculations were generally within a factor of ±2 for all tests. However, greater scatter (+7 to –4) was noted for crack initiation life calculations. These results appear to confirm that the models developed for wrought steels can be used for cast steels.

ACKNOWLEDGMENTS

The authors are grateful for the work done by Professor Stephens and his students at the University of Iowa, which formed the basis for much of this article. In addition, the authors wish to thank the Steel Founders' Society of America for sponsoring the original work.

REFERENCES

1. *Steel Castings Handbook*, 6th ed., Steel Founders' Society of America and ASM International, 1995
2. R.I. Stephens, "Fatigue and Fracture Toughness of Five Carbon or Low Alloy Cast Steels at Room or Low Climatic Temperature," Research Report No. 94A, Steel Founders' Society of America, Oct 1982
3. R.I. Stephens, "Fatigue and Fracture Toughness of Five Carbon or Low Alloy Cast Steels at Room or Low Climatic Temperatures (Part II)," Research Report No. 94B, Steel Founders' Society of America, May 1983
4. M.R. Mitchell, *J. Engr. Materials and Tech.*, Vol 99, Oct 1977, p 329-343
5. C. Vishnevsky et al., ASME paper No. 67-WA/Met-17, 1967
6. Technical Report on Fatigue Properties, SAE J1099, Society of Automotive Engineers, Feb 1975
7. R.W. Smith, M.W. Hirsberg, and S.S. Manson, "Fatigue Behavior of Materials under Strain Cycling in the Low and Intermediate Life Range," TND-1574, National Aeronautics and Space Administration, April 1963
8. R.I. Stephens, J.H. Chung, and G. Glinka, Low Temperature Fatigue Behavior of Steels—A Review, *SAE Trans.*, Vol 88, 1980
9. "Standard Practice for R-Curve Determination," E 561, *Annual Book of ASTM Standards*, ASTM, E.T. Wessel, State of the Art of the WOL Specimen for K_{Ic} Fracture Toughness Testing, *Eng. Fract. Mech.*, Vol 1 (No. 1), 1968
10. J.R. Rice, A Path Independent Integral and the Approximate Analysis of Strain Concentration by Notches and Cracks, *J. Appl. Mech.*, Vol 35, June 1968, p 379
11. "Standard Test Method for J_{Ic}, A Measure of Fracture Toughness," E 813, *Annual Book of ASTM Standards*, ASTM
12. S.T. Rolfe and J.M. Barsom, *Fracture and Fatigue Control in Structures—Applications of Fracture Mechanics*, Prentice-Hall, 1977
13. J.M. Barsom, Fatigue-Crack Propagation in Steels of Various Yield Strengths, *J. Eng. Ind. (Trans. ASME)*, Series B, No. 4, Nov 1971
14. H.D. Greenberg and W.G. Clark, Jr., A Fracture Mechanics Approach to the Development of Realistic Acceptance Standards for Heavy Walled Steel Castings, *Met. Eng. Quart.*, Aug 1969
15. B.M. Kapadia and E.J. Imhoff, Jr., Fatigue-Crack Growth in Cast Irons and Cast Steels, *Cast Metals for Structural and Pressure Containing Applications*, MPC-11, ASME, 1979
16. K. Tanaka, A Correlation of ΔK_{th} Values with the Exponent, m, in the Equation of Fatigue Crack Growth for Various Steels, *Int. J. Fract.*, Vol 15 (No. 1), Feb 1979
17. S.G. Lee, "Estimating Fatigue Crack Initiation and Propagation Life of Cast Steels Under Variable Loading History," Ph.D. thesis, University of Iowa, Dec 1982

Fatigue and Fracture Properties of Cast Irons

S. Lampman, ASM International

CAST IRONS have several well-known manufacturing and engineering advantages over cast steel, including 30 to 40% lower manufacturing costs relative to cast steel and more desirable performance characteristics, such as better wear resistance and vibration damping. These advantages over cast steel are based on several metallurgical characteristics of cast irons. First, cast iron has melting temperatures (and therefore pouring temperatures) that are 300 to 350 °C (540 to 630 °F) lower than those of cast steel. Second, because of a larger concentration of free carbon and higher silicon, graphitic (gray and ductile) cast iron has the greatest fluidity and the least shrinkage of any ferrous metal. For example, cast steel will generally experience a volume shrinkage of more than 4% during solidification, whereas gray and ductile cast iron shrinkage can be less than 1%, depending on the composition and the processing conditions. This difference in shrinkage allows products to be made to exact dimensions much more easily using gray and ductile cast iron, with very little problem in obtaining pressure tightness as a result of reduced interdendritic shrinkage. In addition, cast irons are more machinable than cast steels, and they are relatively more wear resistant because the graphite acts as a self-lubricating system. Graphite also attenuates sound and mechanical vibration, which makes cast iron ideal for many mechanical applications, such as brakes.

However, the higher carbon concentration responsible for several desirable physical properties and ease of manufacturing with cast iron is, unfortunately, also responsible for the degradation of the ductility and fracture toughness. The carbon, usually present principally as graphite, serves to nucleate fatigue and fracture processes at relatively low strain levels, significantly impairing the fatigue and fracture resistance of cast iron compared with that of cast or wrought steel. Considerable work has been done to characterize better the fatigue and fracture of cast iron as well as to understand better the micromechanisms that control the fracture process in these systems. The purpose of this article is to summarize the fatigue and fracture behavior of the various types of cast iron as a function of chemical composition, matrix microstructure, and graphite morphology.

The five types of commercial cast iron considered in this article are gray, ductile, malleable, compacted graphite, and white iron. With the exception of white cast iron, all cast irons have in common a microstructure that consists of a graphite phase (usually 8 to 14 vol%) in a matrix that may be ferritic, pearlitic, bainitic, tempered martensitic, or combinations thereof. The four types of graphitic cast irons are roughly classified according to the morphology of the graphite phase. Gray iron has flake-shaped graphite, ductile iron has nodular or spherically shaped graphite, compacted graphite iron (or "vermicular" graphite iron) is intermediate between these two (with graphite in a vermicular or "wormlike" coral shape), and malleable iron has irregularly shaped globular or "popcorn"-shaped graphite that is formed during tempering of white cast iron. These categories of graphite shape have a strong influence on fatigue and fracture properties, although microstructure (tensile strength), component/specimen size, surface condition,

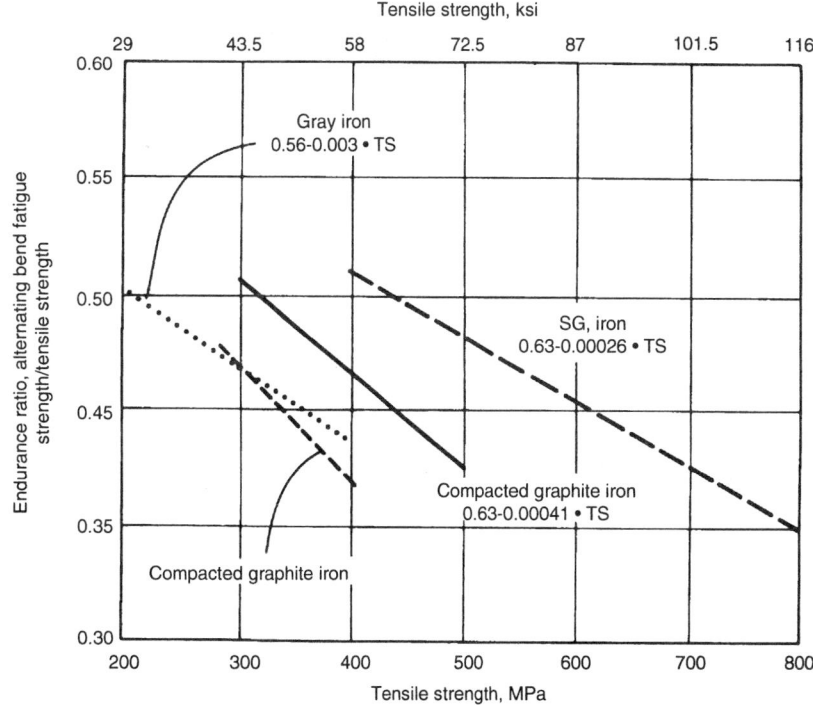

Fig. 1 Ratio of alternating bend fatigue strength/tensile strength of gray cast iron, compacted graphite iron, and spheroidal graphite (SG) iron (i.e., ductile iron). Statistical analysis of a number of experimental data allowed the calculation of a relationship between fatigue strength (FS) at or above 10^6 cycles and tensile strength (TS) of compacted graphite irons, which is: FS (in MPa) = (0.63 – 0.00041 · TS), where TS is also in MPa. Values calculated with this equation fit well between those of gray and ductile irons, as shown. Source: Ref 6

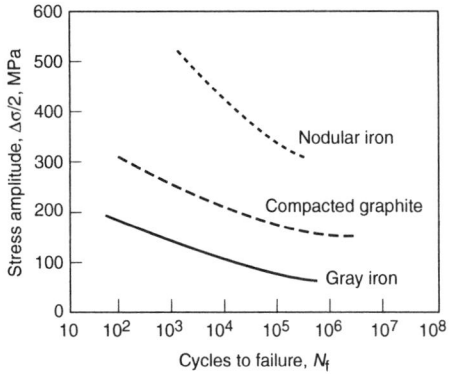

Fig. 2 Comparison of typical stress-life fatigue curves for nodular (ductile), gray, and compacted graphite cast irons. Source: Ref 2

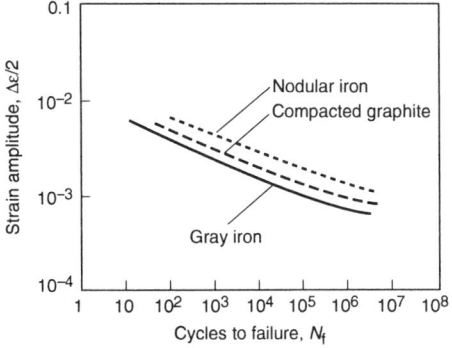

Fig. 3 Comparison of typical strain-life fatigue curves for nodular (ductile), gray, and compacted graphite cast irons. Source: Ref 2

Table 1 Typical fatigue endurance of cast irons

Class or type	Ultimate tensile strength(a), ksi	Hardness, HB	Fatigue endurance strength(b) Unnotched, ksi	Notched, ksi
Gray cast irons				
Medium section (type 20)	20-25	140-187	9.5-10	9.5
Medium section (type 25)	25-30	170-229	12-15	
Medium section (type 30)	30-35	170-241	13.7-15.5	13.5
Medium section (type 35)	35-40	187-269	16-17.5	
Medium section (type 40)	40-48	207-269	17.5-19.5	17.5
Medium section (type 45)	45-52	217-269	21.5-25.5	
Medium section (type 50)	50-57	215-260	24.5-27.5	21.5
Medium section (type 60)	60-66	230-290	24.5-29.5	
Medium section (type 70)	70		29.5-31.5	25.5
Nickel alloyed gray iron (NCI 30)	30 min	200	15	12
Nickel alloyed gray iron (NCI 40)	40	230	19	14
Nickel alloyed gray iron (NCI 50)	50	260	23	16
Nickel alloyed gray iron (NCI 60)	60	290	27	18
Nickel alloyed gray iron (NCI 70)	70	320	31	20
Nickel alloyed gray iron (NCI 80)	80	350	35	22
Malleable cast irons				
Ferritic malleable (32510)	50-58	110-156	28	
Ferritic malleable (35018)	53-60	110-156	31	
Pearlitic malleable (43010)				
Pearlitic malleable (45010)	65	163-207		
Pearlitic malleable (45007)	65	163-217		
Pearlitic malleable (48004)	70	163-228		
Pearlitic malleable (48005)				
Pearlitic malleable (50007)	75	179-228	37	30
Pearlitic malleable (53004)		197-241		
Pearlitic malleable (60003)	80	197-255	39	
Pearlitic malleable (70002)				
Martensitic malleable (80002)	100	241-269	40	
White and alloy cast irons				
Unalloyed white	20-50	300-575		
High Si (Duriron, Durichlor)	13-18	480-520		
High Cr (corrosion resistant)	30-90	290-400		
Medium Si (Silal)	25-45	170-250		
Ni-Cr-Si	20-45	80-150		
High Al	34-90	180-350		
Ni-resist (type 1)	25-30	130-170	12	
Ni-resist (type 2)	25-30	125-170	12	
Ni-resist (type 2b)	30-35	170-250	18	
Ni-resist (type 3)	25-35	120-160	13.5	
Ni-resist (type 4)	25-30	150-210	9	
Ni-resist (type 5)	20-25	100-125	9.9	
Ni-hard (type 1 sand cast)	40-50	550 (s)		
Ni-hard (type 1 chill cast)	50-60	600 (s)		
Ni-hard (type 2 sand cast)	45-55	525 (s)		
Ni-hard (type 2 chill cast)	60-70	575 (s)		
Almanite (Meehanite type WS)	60-80	400-525		
Ductile irons				
Standard ductile 60-40-18	60-80	149-187	30.5	18-23
Standard ductile 80-55-06	80-100	179-248	39-40	21-24
Standard ductile 100-70-03	100-120	217-269		
Standard ductile 120-90-02	120-175	240-300	49	30
Ni-Resist ductile (austenitic) D-2	55-69	140-200	30	20
Ni-Resist ductile (austenitic) D-2B	58-70	150-210		
Ni-Resist ductile (austenitic) D-2C	55-65	130-170		
Ni-Resist ductile (austenitic) D-3	55-67	140-200		
Ni-Resist ductile (austenitic) D-3A	55-65	130-190		
Ni-Resist ductile (austenitic) D-4	60-72	170-240		
Ni-Resist ductile (austenitic) D-5	55-60	130-180		
Ni-Resist ductile (austenitic) D-5B	55-65	140-190		

(a) 0.750 in. diameter specimens machined from 1.2 in. diameter test bar. (b) Rotating cantilever beam tests; strength specified is between 10^6 and 10^7 cycles. Source: Ref 4

chemistry, and temperature also affect the general fatigue and fracture performance of cast irons (Ref 1-5).

Fatigue of Cast Irons

The fatigue performance of cast irons, in general, is influenced by graphite morphology, matrix microstructure and tensile strength, specimen size, surface condition, surface degradation such as corrosion, and the type of loading stress (e.g., axial, bending, reversed bending, torsion, multiaxial, variable amplitude). The effect of graphite shape on cast iron fatigue has received considerable attention and is a key variable. The free graphite in cast iron acts as an inherent notch that increases stress concentrations for fatigue crack initiation. Therefore, the fatigue performance of cast irons is influenced greatly by the quantity, size, and shape of the graphite phase as well as its interaction with the matrix.

The effect of different graphite morphologies on fatigue is illustrated in Fig. 1 to 3 for three types of cast iron. The flake form of graphite in gray iron increases stress concentrations and thus lowers fatigue strength relative to that of irons that have more spheroidal or compacted graphite

forms. In contrast, ductile irons have a nodular graphite, which reduces the concentration of internal stresses and improves the relatively fatigue performance at a given tensile strength. However, gray irons are also less notch sensitive than ductile irons (Table 1).

Most of the available data on cast iron fatigue are based on rotating bending testing where the specimen diameter is generally less than 13 mm (0.5 in.). Results are often reported in terms of fatigue endurance limits (Fig. 1), although testing in the finite life range (below 10^6 cycles) is also

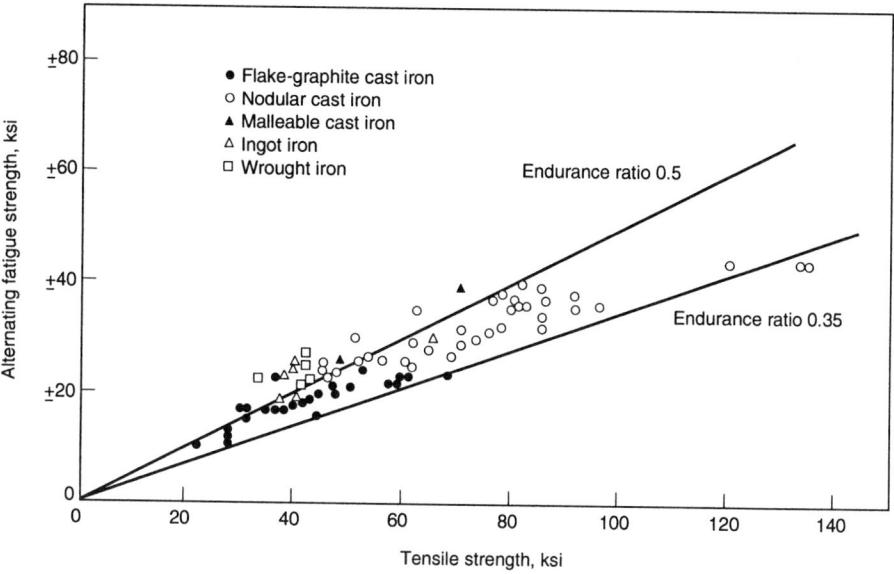

Fig. 4 Relation between rotating-bending fatigue strength and tensile strength of iron, based on fatigue limit or failure in 10^7 cycles. Source: See Angus (1960) and Grover (1960) in "Selected References: Prior to 1980"

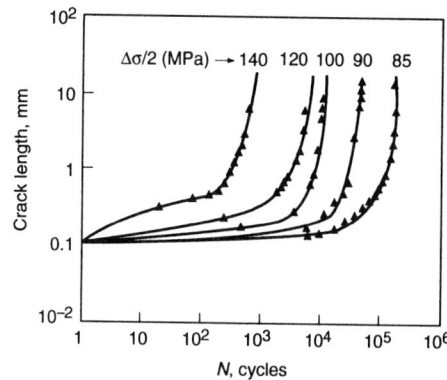

Fig. 6 Crack development in gray iron. Source: Ref 2

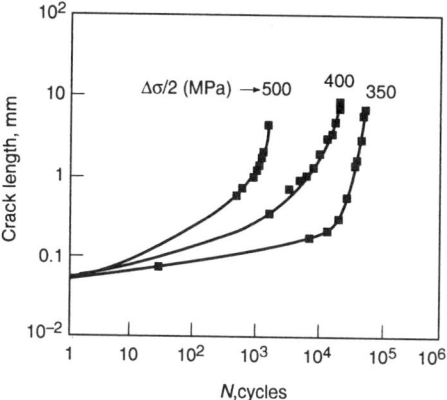

Fig. 5 Crack development in ductile iron. Source: Ref 2

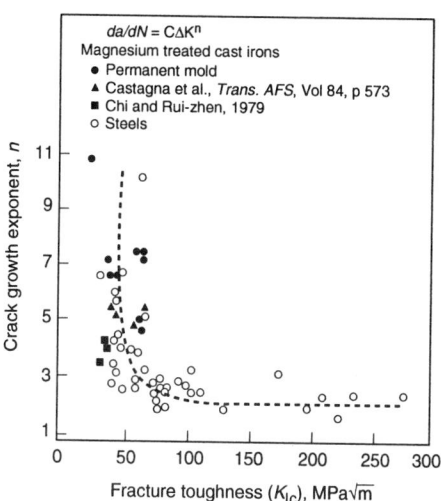

Fig. 7 Variation of crack growth exponent with fracture toughness in magnesium-treated cast irons and steels. Steels data is from J. Knott, *Fundamentals of Fracture Mechanics*, Butterworth, 1973. Source: Ref 7

Fatigue Crack Growth. Fatigue crack initiation is the limiting fatigue mechanism when fatigue life approaches about 10^6 cycles or the fatigue-limit regime. Below about 10^6 cycles, the majority of fatigue life tends to be consumed by fatigue crack growth for gray, ductile, and compacted graphite irons (Ref 2). For practical purposes, crack initiation in the low-cycle regime (10^2 to about 10^6 cycles) can be assumed to occur on the first cycle of loading. Thus, life prediction models for low-cycle fatigue of cast irons should be based on crack growth behavior (Fig. 5 and 6).

The Paris relation, $da/dN - C(\Delta K)^n$, is valid for cast iron in the regime of intermediate stress intensities. Fatigue crack growth studies on permanent mold ductile iron castings of a hypereutectic composition (3.6C-3.0Si) indicate that it may be possible to improve the fatigue crack resistance by lowering the pearlite content and increasing the nodule count. Crack growth parameters from this study are summarized in Table 4. Results of various investigations on fatigue crack growth rate are shown as a function of the fracture toughness (K_{Ic}) in Fig. 7. The crack growth exponent values in permanent mold ductile iron castings are significantly higher than in others. Further, slight reductions in fracture toughness lead to large increases in exponent values.

Fracture Appearance. In flake graphite cast irons, it is not generally possible to determine from fracture appearance whether failure has been due to fatigue or to a single overload in tension, because no distinctive fatigue fracture is obtained. This is particularly so with relatively weak flake graphite cast irons. In nodular irons, however, fatigue failures frequently have a distinctive appearance different from that associated with failure by other means.

Load Variables. Axial loading or torsional loading cycles are frequently encountered in designing parts of cast iron, and in many instances these are not completely reversed loads. Types of

being emphasized more (Fig. 2, 3). The fatigue endurance limit (or the "knee" in the fatigue curve) usually occurs between 10^6 and 10^7 cycles. For many tests, if a specimen endures 10^7 cycles of a given stress, then it is often assumed that the material would withstand such a stress indefinitely. However, fatigue failures do occur above 10^7 cycles in rotating bending fatigue test. Thus, some fatigue limits are based on 2×10^7 cycles (Ref 1).

Typical cast iron fatigue data generated in the 1950s and 1960s are summarized in Fig. 4 and Tables 2 and 3. The general relation between tensile strength and fatigue limit (Fig. 4) is determined to a large extent by the strength/microstructure of the matrix, as discussed in more detail later in this article for each particular type of cast iron. Endurance ratios may also be affected by alloying and graphite form, also discussed for particular types of cast iron. More general fatigue characteristics are summarized below.

regularly repeated stress variation can usually be expressed as a function of a mean stress and a stress range. Wherever possible, the designer should use actual data from the limited information available. Without precisely applicable test data, an estimate of the reversed bending fatigue limit of machined parts may be made by using about 35% of the minimum specified tensile strength of the particular grade of gray iron being considered. This value is probably more conservative than an average of the few data available on the fatigue limit for gray iron.

An approximation of the effect of range of stress on fatigue limit may be obtained from diagrams such as Fig. 8. Tensile strength is plotted on the horizontal axis to represent fracture strength under static load (which corresponds to a 0 stress range). Reversed bending fatigue limit is plotted on the ordinate for 0 mean stress, and the two points are joined by a straight line. The resulting diagram yields a fatigue limit (maximum value of alternating stress) for any value of mean stress.

Table 2 Reversed bending ($R = -1$) fatigue strength of cast irons (circa 1960)

Material	Comment	UTS, ksi	Type	Size, in.	K_t	10^4	4×10^4	10^5	4×10^5	10^6	4×10^6	10^7	10^8
Gray cast iron 3.32 total C, 0 combined C, 1.16Si, 0.102S, 0.38P, 0.58Mn	Large casting	21	Unnotched	0.35d	…	…	16	13	10	8	7.0	7.0	…
Cast iron 3.68 total C, 0.55 combined C, 2.17Si, 0.03S, 0.646P, 0.54Mn	Large bar	23	Unnotched	0.33d	…	…	16	14	12	12	12.0	12.0	…
Gray cast iron 3.35 total C, 0.55 combined C, 1.10Si, 0.095S, 0.51P, 0.63Mn	Large casting	28	Unnotched	0.35d	…	…	17	14	12	11	10.0	10.0	10.0
	Large casting	28	Transverse hole	0.055d, 0.35D	3.0	…	…	12	8	6	6.5	6.5	6.5
Cast iron 3.51 total C, 0.69 combined C, 1.58Si, 0.088S, 0.72P, 0.54Mn, 0Ni, 0.01Cr, 0.05V, 0.06Ti	Large bar	30	Unnotched	0.33d	…	23	19	17	15	15	15.0	15.0	…
Gray cast iron 3.44 total C, 0.68 combined C, 1.10Si, 0.093S, 0.51P, 0.62Mn	Large casting	32	Unnotched	0.35d	…	…	17	15	13	11	11.0	11.0	…
	Large casting	32	Unnotched	0.30d	…	20	19	18	17	16	16.0	16.0	…
	Large casting	32	Unnotched	0.35d	…	…	18	15	13	11	10.0	10.0	10.0
	Large casting	32	V-notch	60° ω, 5/64r, 0.74D, 0.35d	1.5	…	13	12	11	10	10.0	10.0	…
	Large casting	32	V-notch	60° ω, 5/64r, 0.501D, 0.342d	3.0	…	14	13	11	10	9.0	9.0	…
	Large casting	25	Unnotched	0.40d	…	…	…	14	13	12	11.0	10.0	…
Gray cast iron 3.44 total C, 0.68 combined C, 1.10Si, 0.093S, 0.51P, 0.62Mn	Large casting	12	Unnotched	0.40d	…	…	14	12	10	9	8.0	7.0	…
Cast iron 3.23 total C, 0.57 combined C, 1.96Si, 0.072S, 0.44P, 0.54Mn	Large casting	32	Unnotched	0.40d	…	22	18	16	12	12	12.0	12.0	…
Cast iron 3.24 total C, 0.40 combined C, 2.85Si, 0.101S, 0.168P, 0.57Mn	Large casting	32	Unnotched	0.33d	…	23	20	18	15	13	13.0	13.0	…
Cast iron 3.57 total C, 0.58Mn, 1.58Si, 0.66P, 0.06S	Cast 7/8-in. bar	34	Unnotched	0.33d	…	31	27	25	23	22	22.0	22.0	…
	Cast 7/8 in. bar	34	Kommers notch		5+	…	22	19	17	16	16.0	16.0	…
Austenitic cast iron 2.63 total C, 0.67 combined C, 2.14Si, 1.23Mn, 6.94Cu, 2.09Cr, 14.9Ni, 0.022Ti, 0.005V, 0.012As, 0.16P, 0.065S	Plate (RT data)	35	Unnotched	0.30d	…	…	…	…	18	16.0	14.0	…
	Plate (RT data)	35	Transverse hole	0.032d, 0.32D	2.2	…	…	…	15	12	11.0	11.0	…
	Plate (400 °F data)	31	Unnotched	0.30d	…	21	18	17	17	17	17.0	17.0	…
	Plate (400 °F data)	31	Transverse hole	0.032d, 0.32D	2.2	17	15	13	12	11	11.0	11.0	…
	Plate (700 °F data)	30	Unnotched	0.30d	…	…	21	18	17	15	15.0	15.0	…
	Plate (700 °F data)	30	Transverse hole	0.032d, 0.32D	2.2	…	…	…	15	14	13.0	13.0	…
	Plate (900 °F data)	26	Unnotched	0.30d	…	…	15	13	12	12.0	10.0	…	…
	Plate (900 °F data)	26	Transverse hole	0.032d, 0.32D	2.2	…	…	14	12	12	12.0	12.0	…
Austenitic cast iron 2.63 total C, 0.67 combined C, 2.14Si, 1.23Mn, 6.94Cu, 2.09Cr, 14.9Ni, 0.022Ti, 0.005V, 0.012As, 0.16P, 0.065S	Plate (1160 °F data)	20	Unnotched	0.30d	…	…	14	13	12	12.0	12.0	…	…
	Plate (1160 °F data)	20	Transverse hole	0.032d, 0.32D	2.2	…	…	14	13	12	10.0	10.0	…
Cast iron 3.30 total C, 0.64 combined C, 1.12Si, 0.09S, 0.21P, 0.62Mn	Large bar	36	Unnotched	0.33d	…	28	24	22	19	19	19.0	19.0	…
Cast iron 3.35 total C, 0.50 combined C, 2.25Si, 0.084S, 0.26P, 0.60Mn	Large bar	36	Unnotched	0.33d	…	…	24	21	19	19.0	19.0	…	…
Cast iron 3.46 total C, 0.44 combined C, 2.35Si, 0.078S, 0.27P, 0.62Mn, 0.63Ni, 0.22Cr	Large bar	37	Unnotched	0.33d	…	…	28	26	23	22	20.0	20.0	…
Cast iron 3.31 total C, 0.61 combined C, 1.45Si, 0.106S, 0.37P, 0.86Mn	Large bar	40	Unnotched	0.33d	…	…	25	22	19	19	19.0	19.0	…
Cast iron 2.86 total C, 0.65 combined C, 2.51Si, 0.079S, 0.057P, 0.62Mn	Large bar	43	Unnotched	0.33d	…	…	…	28	24	24	24.0	24.0	…
Cast iron 3.03 total C, 0.58 combined C, 1.93Si, 0.64Mn, 0.023S, 0.076P	Large bar	45	Unnotched	0.33d	…	…	30	27	24	22	20.0	20.0	…
	Large bar	45	Kommers notch		5+	26	23	21	18	17	17.0	17.0	…
High-strength cast iron 2.84 total C, 0.74 combined C, 1.52Si, 1.05Mn, 0.37Cu, 0.31Cr, 0.20Ni, 0.07P, 0.124S (gravity), 0.106S (H$_2$S)	Slab (RT data)	48	Unnotched	0.30d	…	…	…	…	25	23	20.0	20.0	…
	Slab (RT data)	48	Transverse hole	0.032d, 0.32D	2.2	…	24	22	19	18	17.0	17.0	17.0
High-strength cast iron 2.84 total C, 0.74 combined C, 1.52Si, 1.05Mn, 0.37Cu, 0.31Cr, 0.20Ni, 0.07P, 0.124S (gravity), 0.106S (H$_2$S)	Slab (300 °F data)	45	Transverse hole	0.032d, 0.32D	2.2	23	20	19	18	17	17.0	17.0	…
	Slab (500 °F data)	46	Unnotched	0.30d	…	30	24	21	19	18	18.0	18.0	…
	Slab (500 °F data)	46	Transverse hole	0.032d, 0.32D	2.2	23	22	20	19	16	16.0	16.0	…
	Slab (700 °F data)	50	Unnotched	0.30 d	…	…	30	25	23	22	22.0	22.0	…
	Slab (700 °F data)	50	Transverse hole	0.032d, 0.32D	2.2	…	25	22	21	20	20.0	20.0	…
	Slab (850 °F data)	45	Unnotched	0.30d	…	…	27	25	23	23	23.0	23.0	…
	Slab (850 °F data)	45	Transverse hole	0.032d, 0.32D	2.2	25	20	18	17	17	17.0	17.0	…
	Slab (1000 °F data)	33	Unnotched	0.30d	…	25	18	17	16	15	15.0	15.0	…
	Slab (1000 °F data)	33	Transverse hole	0.032d, 0.32D	2.2	17	15	14	14	14	14.0	14.0	…
	Slab (1200 °F data)	16	Unnotched	0.30d	…	15	10	8	7	7	7.0	7.0	…
	Slab (1200 °F data)	16	Transverse hole	0.032d, 0.32D	2.2	…	13	10	9	8	7.0	7.0	…

UTS, ultimate tensile strength; RT, room temperature; d, diameter; D, outer diameter; r, radius. Source: H.J. Grover, S.A. Gordon, and L.R. Jackson, *Fatigue of Metals and Structures*, Department of the Navy, 1960, p 278-279

Few data available are applicable to design problems involving dynamic loading where the stress cycle is predominantly compressive rather than tensile. Other work done on aluminum and steel indicates that for compressive (negative) mean stress, the behavior of these materials could be represented by a horizontal line beginning at the fatigue limit in reversed bending, as indicated in Fig. 8. Gray iron is probably at least as strong as this for loading cycles resulting in negative mean stress, because it is much stronger in static compression than in static tension. It is therefore a natural assumption that the parallel behavior shown in Fig. 8 is conservative.

If, prior to design, the real stress cycle can be predicted with confidence and enough data are available for a reliable S-N diagram, the casting might be dimensioned to obtain a minimum safety factor of 2 based on fatigue strength. (Some uses may require more conservative or more liberal loading.) The approximate safety factor is best illustrated by point P in Fig. 8. The safety factor is determined by the distance from the origin to the fatigue limit line along a ray through the cyclic-stress point, divided by the distance from the origin to that point. In Fig. 8 this is *OF/OP*.

On this diagram, point P' represents a stress cycle having a negative mean stress. In other words, the maximum compressive stress is greater than the tensile stress reached during the loading cycle. In this instance, the safety factor is the distance *OF'/OP'*. However, this analysis assumes that overloads will increase the mean stress and alternating stress in the same proportion. This may not always be true, particularly in

Table 3 Effect of a circumferential V-notch on the bending fatigue strength of cast steel and cast iron

Material	Tensile strength, ksi	Bending fatigue strength, ksi Plain	Bending fatigue strength, ksi Notched	K_f	K_t (Neuber)	Angle, deg	Root radius, in.	Minimum diameter, in.	Depth, in.
Cast 0.4% C steel	75	29.6	23	1.28	2.2	60	0.015	0.22	0.035
	84	33.6	25	1.34	2.2	60	0.015	0.22	0.035
Cast alloy steel	100	45.8	30	1.54	2.2	60	0.015	0.22	0.035
	113	56.2	31	1.80	2.2	60	0.015	0.22	0.035
	130	54.6	36	1.51	2.2	60	0.015	0.22	0.035
	150	69.2	43	1.61	2.2	60	0.015	0.22	0.035
Cast alloy steel	130	39	26	1.50	3.1	55	0.0086	0.300	0.0375
	150	44	29	1.52	3.1	55	0.0086	0.300	0.0375
	164	46	30	1.53	3.1	55	0.0086	0.300	0.0375
	170	47	32	1.47	3.1	55	0.0086	0.300	0.0375
Graphitic cast steel	100	32	18	1.78	3.1	55	0.0086	0.300	0.0375
Flake-graphite gray cast iron	46	22	14	1.57	3.1	55	0.0086	0.300	0.0375
	44	16	16	1.00	3.1	55	0.0086	0.300	0.0375
	52	24	22	1.09	3.1	55	0.0086	0.300	0.0375
	48	20	20	1.00	3.1	55	0.0086	0.300	0.0375
	50	21	21	1.00	3.1	55	0.0086	0.300	0.0375
	41	18	15	1.20	3.1	45	0.010	0.331	0.071
	60	23	21	1.09	3.1	45	0.010	0.331	0.071
Nodular cast iron (as cast)	80	38	22	1.73	3.1	45	0.010	0.301	0.086
	86	37	21	1.76	3.1	45	0.010	0.301	0.086
	91	38	26	1.46	3.1	45	0.010	0.301	0.086
	78	38	20	1.90	3.5	45	0.010	0.417	0.142
	79	35	19	1.84	3.5	45	0.010	0.417	0.142
	81	37	21	1.76	3.5	45	0.010	0.417	0.142
	85	34	16	2.13	3.5	45	0.010	0.417	0.142
Nodular cast iron (heat treated)	46	24	21	1.14	3.1	45	0.010	0.301	0.086
	61	29	17	1.70	3.1	45	0.010	0.301	0.086
	73	30	17	1.77	3.1	45	0.010	0.301	0.086
	77	32	17	1.88	3.1	45	0.010	0.301	0.086
	42	22	15	1.47	3.5	45	0.010	0.417	0.142
	46	23	18	1.28	3.5	45	0.010	0.417	0.142
	51	26	16	1.63	3.5	45	0.010	0.417	0.142

Source: *Fatigue of Metals*, National Physics Lab, 1962, p 154-155

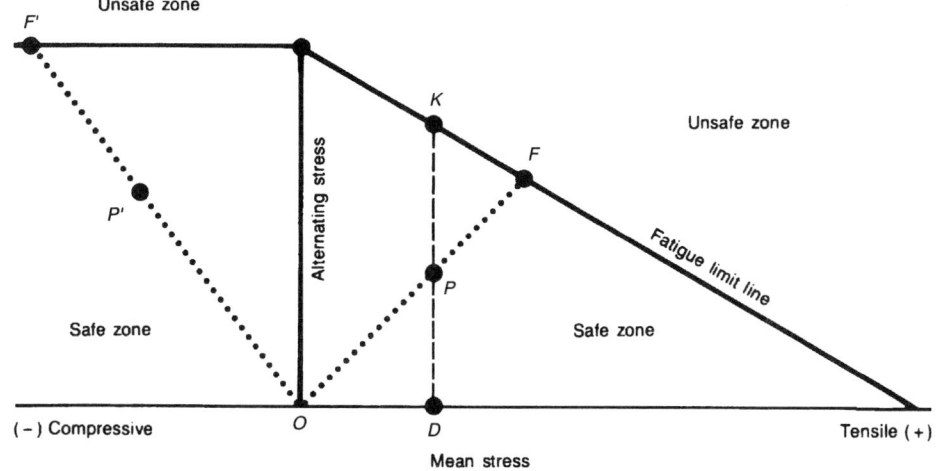

Fig. 8 Diagram showing safe and unsafe fatigue zones for cast iron subjected to ranges of alternating stress superimposed on a mean stress. Example point P shows conditions of tensile (positive) mean stress; P' shows compressive (negative) mean stress. The safety factor is represented by the ratio of OF to OP or OF' to OP'. For conditions of constant mean tensile stress, DK/DP is the safety factor.

systems with mechanical vibration in which the mean stress may remain constant. For this condition, the vertical line through P would be used; that is, DK/DP would be the factor of safety.

Most engineers use diagrams such as Fig. 8 mainly to determine whether a given condition of mean stress and cyclic stress results in a design safe for infinite life. The designer can also determine whether variations in the anticipated mean stress and alternating stress will place the design in the unsafe zone. Usually the data required to analyze a particular set of conditions are obtained experimentally (e.g., Table 5). It is emphasized that the number of cycles of alternating stress implied in Fig. 8 is the number normally used to determine fatigue limits, that is, approximately

10 million. Fewer cycles, as encountered in infrequent overloads, will be safer than indicated by a particular point plotted on a diagram for infinite life. Too few data are available to draw a diagram for less than infinite life.

Fatigue Notch Sensitivity. Gray irons are less notch sensitive than ductile irons. Because the graphite flakes in gray iron can be thought of as internal notches, gray irons have little or no sensitivity to the presence of additional notches resulting from design features. The strength-reducing effect of the internal notches is included in the fatigue limit values determined by conventional laboratory tests with smooth bars. High-strength irons usually exhibit greater notch sensitivity, but probably not the full theoretical value represented by the stress concentration factor. Normal stress concentration factors (see Ref 9) are probably suitable for high-strength gray irons. Relative notch factors (the ratio of unnotched to notched fatigue strength) for cast irons are in the range of about (Ref 6):

- 1.7 to 1.8 for compacted graphite irons
- Less than 1.5 for gray iron
- Greater than 1.85 for ductile iron

Thermal Fatigue. When castings are used in an environment where frequent changes in temperature occur, or where temperature differences are imposed on a part, thermal stresses occur in castings and may result in elastic and plastic

Table 4 Toughness and fatigue crack growth data for various ductile irons

Property	As cast(a)	As cast(a)	As cast(a)	Austenitized at(b): 800 °C	900 °C	1000 °C	900 °C
Graphite shape(c)	S	S	I	S	S	S	S
Nodule count, mm^{-2}	201	221	...	107	95	101	269
Pearlite, %	73.8	45.2	37.5	[<2%, annealed ferritic]			
Ferrite grain size, mm	...	...	...	0.021	0.032	0.039	0.034
K_{Ic}, MPa√m	25.6	38.5	37.5	61.5	66.3	67.3	64.5
σ_y, MPa	612.2	562.1	437.6	409.4	405.3	402.7	401.3
C (Paris law coefficient)	1.58	1.83	1.10	9.86	4.54	8.51	2.09
n (Paris law exponent)	10.75	6.66	7.14	7.44	7.41	7.27	5.12

(a) As cast (chill free); pearlite variation was achieved through changes in mold preheating temperature and mold wall thickness. (b) After soaking at the indicated temperature for 12 h, all the castings were furnace cooled to 690 °C, held for 24 h, then air cooled to room temperature. (c) S, spheroidal graphite; I, irregular (vermicular) graphite. Source: Ref 7

Table 5 Gray iron fatigue strength in bending and torsion

Parameter	Gray iron No. 1	2	3	4	5
Tensile strength, ksi	44.0	48.0	55.0	56.5	76.5
Endurance limit in complete reversed bending, ksi	19.0	21.0	22.0	22.0	25.0
Endurance limit in half-cycle bending, 0-max, ksi	23.0	32.0	27.0	33.0	38.0
Endurance limit in complete reversed torsion, ksi	16.0	16.5	21.0	20.0	22.0
Endurance limit in half-cycle torsion, 0-max, ksi	23.0	25.0	26.0	33.0	29.0

Source: Ref 8

strains and finally in crack formation. Changes in microstructure, associated with stress-inducing volume changes, as well as surface and internal oxidation, may also be associated with stresses induced by temperature differences.

The interpretation of thermal fatigue tests is complicated by the many different test methods employed by various investigators. The two widely accepted methods are constrained thermal fatigue and finned-disk thermal shock tests (Ref 10, 11).

In general, for good resistance to thermal fatigue, cast irons should have high thermal conductivity, low modulus of elasticity, high strength at room and elevated temperatures, and, for use above 500 to 550 °C (930 to 1020 °F), resistance to oxidation and structural change. The relative ranking of irons varies with test conditions. When high cooling rates are encountered, experimental data and commercial experience show that thermal conductivity and a low modulus of elasticity are most important. Consequently, gray irons of high carbon content (3.6 to 4%) are superior (Ref 10, 11). When intermediate cooling rates exist, ferritic ductile and compacted graphite irons have the highest resistance to cracking but are subject to distortion. When low cooling rates exist, high-strength pearlitic ductile irons or ductile irons alloyed with silicon and molybdenum are best with regard to cracking and distortion (Fig. 9).

A rather detailed analysis of the behavior of various irons at elevated temperatures is given in Ref 6. Extensive experimental work on cylinder heads is reviewed. A critical analysis of most of the accepted criteria for assessing the quality of irons for castings used at elevated temperatures is also included.

Constrained Thermal Test. In the constrained thermal fatigue test (Ref 11), a specimen is mounted between two stationary plates that are held rigid by two columns, heated by high-frequency (450 kHz) induction current, and cooled by conduction of heat to water-cooled grips. The thermal stress that develops in the test specimen is monitored by a load cell installed in one of the grips holding the specimen. As thermal cycling continues, the specimen accumulates fatigue damage in a fashion similar to that in mechanical fatigue testing; ultimately, the specimen fails by fatigue.

Experimental results (Fig. 10) point to higher thermal fatigue for compacted graphite iron than for gray iron and also indicate the beneficial effect of molybdenum. In fact, regression analysis of experimental results indicates that the main factors influencing thermal fatigue are tensile strength (TS) and molybdenum content:

$$\log N = 0.934 + 0.026 \cdot \text{TS} + 0.861 \cdot \text{Mo}$$

where N is the number of thermal cycles to failure, tensile strength is in kips (1000 lbs) per square inch (ksi), and molybdenum is in percent.

Finned-Disk Test. In the finned-disk thermal shock test (Ref 11), the specimen is cycled between a moderate-temperature environment and a high-temperature environment, which causes thermal expansion and contraction. Because thermal conductivity plays a significant role in this type of test, gray iron shows much greater resistance to cracking than compacted graphite iron. Major cracking occurred in less than 200 cycles in all compacted graphite iron specimens, while the unalloyed gray iron developed minor cracking after 500 cycles and major cracking after 775 cycles (Ref 11). Because of its higher elevated-temperature strength, alloyed gray iron was even more resistant to thermal fatigue.

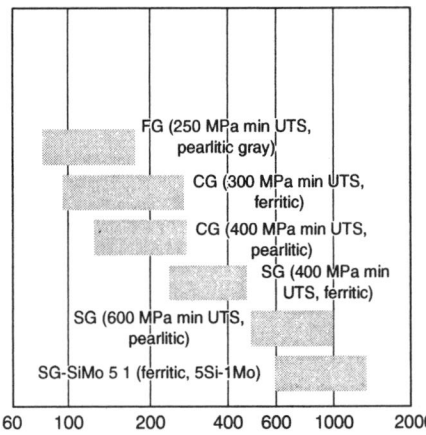

Composition					
C	Si	Mn	P	Mg	Other
2.96	2.90	0.78	0.66	...	0.12 Cr
3.52	2.61	0.25	0.051	0.015	...
3.52	2.25	0.40	0.054	0.015	1.47 Cu
3.67	2.55	0.13	0.060	0.030	...
3.60	2.34	0.50	0.053	0.030	0.54 Cu
3.48	4.84	0.31	0.067	0.030	1.02 Mo

Fig. 9 Results of thermal fatigue tests on various cast irons; specimens cycled between 650 and 20 °C (1200 and 70 °F). FG, flake graphite; CG, compacted graphite; SG, spheroidal graphite; UTS, ultimate tensile strength. Source: Ref 10

Corrosion Fatigue of Cast Irons. A limited amount of information is available in the literature on the corrosion fatigue of cast irons. From a literature review of a few tests in tap and salt water, the fatigue reduction factors due to corrosion were from 1.28 to 1.59 in tap water and 1.55 to 2.47 in sea water for gray irons (Ref 1). Fatigue strengths were reduced by about 7 MPa (1 ksi) when tested in tap water.

Additional corrosion fatigue tests carried out by the British Cast Iron Research Association (Ref 1) were conducted in various environments, as summarized in Tables 6 and 7. The fatigue strength in air, in demineralized water, and in inhibitor-containing water was based on a life of 50×10^6 cycles of stress, and the tests in sodium chloride solution were based on 100×10^6 cycles of stress. The stress frequency was approximately

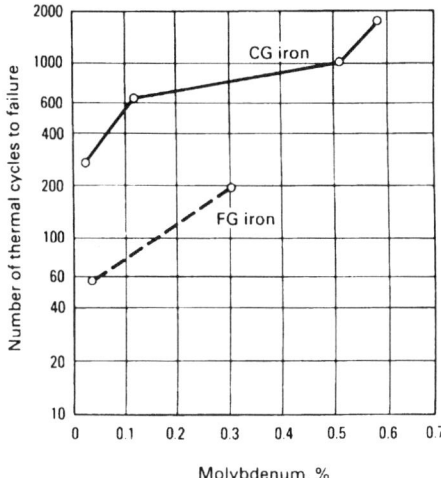

Fig. 10 Results of constrained thermal fatigue tests conducted between 100 and 540 °C (212 and 1000 °F). CG, compacted graphite; FG, flake graphite. Source: Ref 11

Table 6 Corrosion fatigue of cast irons

	Pearlitic flake		Pearlitic nodular		Ferritic nodular	
Environment	Fatigue strength, ksi	Fatigue strength reduction factor	Fatigue strength, ksi	Fatigue strength reduction factor	Fatigue strength, ksi	Fatigue strength reduction factor
Air	16	…	35	…	27	…
Water	13	1.23	29	1.21	23	1.18
3% sodium chloride	5	3.20	6	5.83	6	4.50
1% borax	14	1.14	25	1.40	20	1.35
3% "Sobenite"(a)	16	1.00	29	1.21	24	1.13
3% sodium carbonate	16	1.00	32	1.09	25	1.08
3% soluble oil	16	1.00	27	1.30	24	1.13
0.25% potassium chromate	16	1.00	34	1.03	28	0.97

(a) "Sobenite" is a mixture of 10 parts sodium benzoate to 1 part sodium nitrite. Source: Ref 2

Table 7 Corrosion fatigue strengths of cast irons and steels

		Fatigue strength, ksi		Tensile strength, ksi
Material	Corrosive medium	In air	In corrosive medium	
Flake graphite iron	Water	16	13	40
Mild steel	Water	33	13	52
Nodular graphite iron	Water	35	29	94
1.3Ni-0.6Cr steel	Water	57	30	114
18Cr-8Ni stainless steel	Water	29	26	84
Flake graphite iron	3% salt water	16	5	40
0.2% C steel	Sea water	29	4	64
0.5% C steel (as-drawn)	3% salt water	49	7	130
Nodular graphite iron	3% salt water	35	6	94
3Ni-0.7Cr steel	Sea water	46	10	96
18Cr-8Ni stainless steel	Sea water	29	10	84

Source: Ref 1

3,000 cycles/min. The following conclusions from Ref 1 are noted:

- "Corrosion fatigue of flake graphite (gray) irons in demineralized water can be prevented by the addition to the water of known alkaline inhibitors except borax which is not completely effective for pearlitic irons. Soluble oils are also effective for flake graphite irons."
- "Tests in corrosion-inhibited demineralized water on both pearlitic and ferritic nodular (ductile) irons revealed that the only satisfactory inhibitor giving more or less complete protection against corrosion fatigue was potassium chromate. Some protection was obtained with sodium carbonate for pearlitic irons and this, together with tri-sodium orthophosphate and sodium nitrite, was effective for ferritic irons. Soluble oils are not effective in preventing corrosion fatigue of nodular graphite irons."
- "All inhibitors prevented visible corrosion of the surface, but this is no indication of the effectiveness of an inhibitor in preventing corrosion fatigue."

See Ref 1 for more details.

Fracture Toughness of Cast Irons

Like the fracture resistance of structural steels and other alloys, the measurement of the fracture resistance of cast irons can be broadly divided into two groups:

- Measurement of fracture energy curves based on dynamic tear tests (ASTM E604-83) or Charpy impact tests
- Measurement of resistance to crack propagation using a fracture mechanics method, such as linear elastic fracture mechanics (ASTM E 399), elastic-plastic fracture mechanics with the J integral (ASTM E 813), crack-tip opening displacement (BS 5762) measurements, or

dynamic fracture toughness parameters such as K_{Id} or J_{Id}

Fracture energy curves and dynamic fracture toughness (K_{Id} or J_{Id}) both can be used to evaluate transition temperature behavior, where fracture behavior changes from ductile to brittle. However, comparison of transition temperatures measured in the different dynamic tests (e.g., Charpy impact, dynamic tear energy tests, or K_{Id} or J_{Id} tests) is inappropriate. Cast iron yield strengths are rate dependent, and thus transition temperatures are affected when loading rates change from slow bending to impact. Therefore, valid comparison can only be made when loading rate, specimen size, and notch acuity are similar. These different types of fracture toughness tests are discussed elsewhere in this Volume (see the article "Fracture Toughness Testing"). A comparison of the different test methods is also reviewed for cast iron toughness in Ref 5. In general, test bars for the Charpy or Izod impact testing of cast irons are larger than those used for testing steels. Increases in the size of as-cast specimens cause a relative reduction in impact strength and tensile strength.

In metallurgical terms, the general effects of graphite morphology and matrix microstructure are shown in Fig. 11 and 12 for cast irons. Like fatigue fracture, gray iron has the lowest fracture resistance, due to the inherent notch effect from the flake morphology of its graphite. Gray iron is

also less notch sensitive, with little or no change in notch toughness at lower temperatures. Ductile-to-brittle transition temperatures are more pronounced for ductile and compacted graphite irons (Fig. 11, 12). The effect of graphite morphology is more pronounced for upper shelf toughness, but below the transition temperature, graphite morphology has little effect on toughness.

Most alloying elements or impurities affect toughness indirectly through their effect on graphite morphology and matrix microstructure. However, silicon and phosphorous are two exceptions. Both of these impurities have a strong influence on toughness (Fig. 13). The effect of microstructure and alloying on toughness is summarized for each specific type of cast iron in the next sections.

Table 8 is a general summary of room-temperature Charpy toughness for various cast irons. Charpy testing is the most widely used method to evaluate the fracture toughness of cast irons, and additional information can be found in the *Metals Handbook* (8th and 9th editions), the more recent *ASM Handbook* (Volume 1), and other reference books on cast irons (Ref 3, 4).

Gray Cast Iron

Fatigue Strength. Notches and surface imperfections have little effect on the fatigue strength

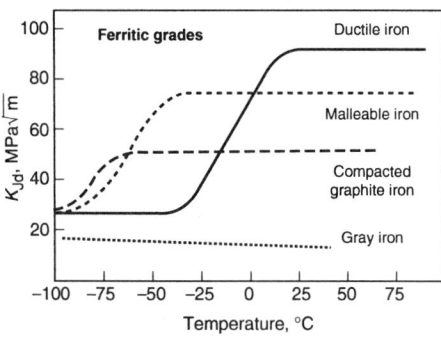

Fig. 11 Effect of graphite shape on dynamic fracture toughness, (K_{Jd}) of ferritic grades of various cast irons. These results are generally based on data from the literature, with some approximation where data in literature is incomplete. Source: Ref 5

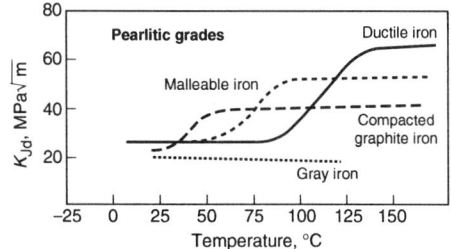

Fig. 12 Effect of graphite shape on fracture toughness of pearlitic grades of various cast irons. These results are generally based on data from the literature with some approximation in regions where data is incomplete. Source: Ref 5

of gray cast irons, which is one reason for its use in automobile crankshafts and similar applications. Minimum tensile and fatigue properties of gray cast iron per ISO 185 are summarized in Table 9. The ratio of fatigue endurance limits to ultimate tensile strength has been reported to vary from 0.33 to 0.62 for gray cast irons (Ref 1). Graphite flake size, matrix microstructure (strength), and phosphorous content all influence the endurance ratio. Some general reported values for the endurance ratio have been:

- 0.4 to 0.7 for repeated loads in compression only (Ref 4)
- 0.45 for irons with tensile strengths of 55 to 70 MPa (8 to 10 ksi) (Ref 1)
- 0.38 for tensile strengths from 80 to 100 MPa (12 to 15 ksi) with acicular-matrix microstructures (Ref 1)

Typical values for different loading conditions are summarized in Table 5. The ratio of endurance limits between torsion loading versus bending loads is reported to vary from 0.75 to 1.25 (Ref 4). In general, the fatigue endurance limit ratio with ultimate tensile strength is smaller at higher tensile strengths.

Impact Strength and Toughness. Gray cast irons are not notch sensitive and do not have a pronounced ductile-to-brittle transition temperature (Fig. 11, 12). However, the impact toughness of unnotched gray iron is reported (Ref 4) to

decrease about 10 to 20% at –75 °C (–100 °F) and 50% at –185 °C (–300 °F). Gray irons with flake graphite have Charpy V-notch upper-shelf toughness at around 1.4 to 6.8 J (1 to 5 ft · lbf), with higher values obtained with higher-strength-ma-

trix microstructures (Ref 5). This unusual behavior (toughness generally decreases with higher strength) is attributed to the fact that fracture occurs through the graphite flakes rather than through matrix deformation (Ref 5).

Table 8 Typical impact strength of cast irons

Class or Type	Ultimate tensile strength(a), ksi	Hardness, HB	V-notch Charpy, ft · lbf	Unnotched Charpy, ft · lbf	Unnotched Izod 1.2 in. diam, ft · lbf	Unnotched Charpy(b) 1.2 in. diam, ft · lbf
Gray cast irons:						
Medium section (type 20)	20-25	140-187	...	...	21	55
Medium section (type 25)	25-30	170-229	...	...	22	55
Medium section (type 30)	30-35	170-241	...	...	23	60
Medium section (type 35)	35-40	187-269	...	...	25	60
Medium section (type 40)	40-48	207-269	...	...	31	70
Medium section (type 45)	45-52	217-269	...	...	36	...
Medium section (type 50)	50-57	215-260	...	...	65	80
Medium section (type 60)	60-66	230-290	...	...	75	115(c)
Medium section (type 70)	70	...	...	...	120	...
Nickel alloyed gray iron (NCI 30)	30 min	200	...	...	26	...
Nickel alloyed gray iron (NCI 40)	40	230	...	...	35	...
Nickel alloyed gray iron (NCI 50)	50	260	...	...	43	...
Nickel alloyed gray iron (NCI 60)	60	290	...	...	52	...
Nickel alloyed gray iron (NCI 70)	70	320	...	...	61	...
Nickel alloyed gray iron (NCI 80)	80	350	...	...	70	...
Malleable cast irons:						
Ferritic malleable (32510)	50-58	110-156	14-17	50-60	...	70-90
Ferritic malleable (35018)	53-60	110-156	16.5	...	...	70-90
Pearlitic malleable (43010)	...	...	...	...	...	...
Pearlitic malleable (45010)	65	163-207	9	...	...	...
Pearlitic malleable (45007)	65	163-207	...	...	...	...
Pearlitic malleable (48004)	70	163-228	...	...	...	...
Pearlitic malleable (48005)	...	...	6	...	...	...
Pearlitic malleable (50007)	75	179-228	...	...	...	22-35
Pearlitic malleable (53004)	...	197-241	4	...	...	...
Pearlitic malleable (60003)	80	197-255	...	...	...	22-35
Pearlitic malleable (70002)	...	...	...	...	...	22-35
Martensitic malleable (80002)	100	241-269	...	...	...	22-35
White and alloy cast irons						
Unalloyed white	20-50	300-575	...	...	...	3.5-10.0
High Si (Duriron, Durichlor)	13-18	480-520	...	...	...	2-4(d)
High Cr (corros. resistant)	30-90	290-400	...	...	...	20-35(d)
Medium Si (Silal)	25-45	170-250	...	...	...	15-23(d)
Ni-Cr-Si	20-45	80-150	...	...	...	110-210(d)
High Al	34-90	180-350	...	...	...	180-350(d)
Ni-Resist (type 1)	25-30	130-170	...	...	100	...
Ni-Resist (type 2)	25-30	125-170	...	...	100	...
Ni-Resist (type 2b)	30-35	170-250	...	...	60	...
Ni-Resist (type 3)	25-35	120-160	...	...	150	...
Ni-Resist (type 4)	25-30	150-210	...	...	80	...
Ni-Resist (type 5)	20-25	100-125	...	...	150	...
Ni-Hard (type 1 sand cast)	40-50	550 (s)	...	...	20-30	...
Ni-Hard (type 1 chill cast)	50-60	600 (s)	...	...	25-40	...
Ni-Hard (type 2 sand cast)	45-55	525 (s)	...	...	25-35	...
Ni-Hard (type 2 chill cast)	60-70	575 (s)	...	...	35-55	...
Almanite (Meehanite type WS)	60-80	400-525	...	...	to 180	...
Ductile irons:						
Standard ductile 60-40-18	60-80	149-187	10-15	75-115	120+	...
Standard ductile 80-55-06	80-100	179-248	2-5	15-65	120+	...
Standard ductile 100-70-03	100-120	217-269	2-6	35-50	...	...
Standard ductile 120-90-02	120-175	240-300	2-6	25-40	...	...
Ni-Resist ductile (Austenitic) D-2	55-69	140-200	12	26	...	...
Ni-Resist ductile (Austenitic) D-2B	58-70	150-210	10	...	...	...
Ni-Resist ductile (Austenitic) D-2C	55-65	130-170	28	...	...	...
Ni-Resist ductile (Austenitic) D-3	55-67	140-200	7	...	...	...
Ni-Resist ductile (Austenitic) D-3A	55-65	130-190	14	...	...	...
Ni-Resist ductile (Austenitic) D-4	60-72	170-240	...	...	...	...
Ni-Resist ductile (Austenitic) D-5	55-60	130-180	17	...	...	...
Ni-Resist ductile (Austenitic) D-5B	55-65	140-190	6	...	...	...

(a) 0.750 in. diameter specimens machined from 1.2 in. diameter test bar. (b) 18 in. between supports. (c) 6 in. between supports, 1.125 in. diameter. (d) 1.2 in. diameter, 6 in. between supports. Source: Ref 4

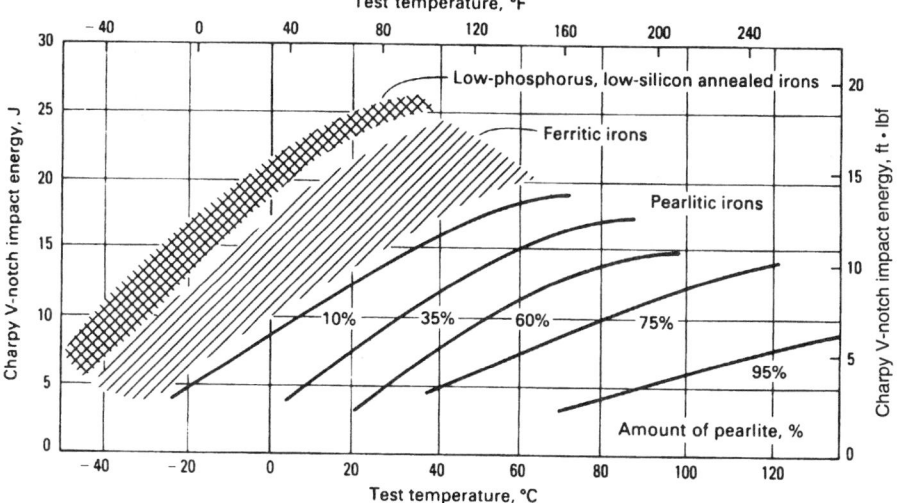

Fig. 13 Charpy V-notch impact energies of ductile irons. Source: Pellini, Sandoz, and Bishop, Noten Ductility of Nodular Irons, *Trans. ASM*, Vol 46, 1954, p 418-445

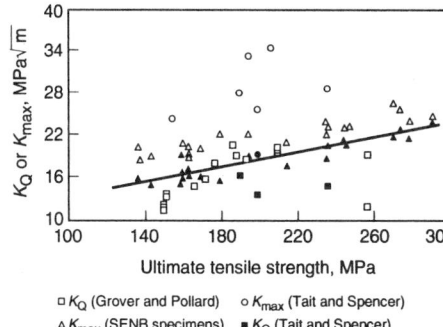

Fig. 14 Relationship between tensile strength and fracture toughness parameters K_Q and K_{max} for gray cast iron. Source: Ref 5

Several attempts have been made to measure the fracture toughness of gray iron using fracture mechanics methods (Ref 5). Some results are summarized in Fig. 14. As mentioned above for Charpy toughness, there is a trend of higher toughness at higher strengths. The values of fracture toughness reported in the literature for gray cast iron almost always give either K_Q or K_{max}, rather than K_{Ic}, because of the ambiguity in determining accurately the moment of crack extension. A question then arises as to which of the two is the better measure of the fracture toughness. Bradley observed more than 1.6 mm of crack advance before reaching maximum load in class 35 gray iron (Ref 12). In view of this result and the discussion presented in Ref 5, K_Q values are considered reasonable and conservative estimates of K_{Ic}. K_{max} values would tend to overestimate the fracture toughness of gray cast iron.

Ductile Irons

Fatigue Strength. Minimum tensile strength and fatigue limits for ductile irons per ISO 1083 are summarized in Table 10. In fatigue applications, ductile irons are generally used in the pearlitic condition (as-cast or normalized). Austempered ductile iron and ferritic ductile iron are also used. Ductile iron is frequency sensitive, and the cycle frequency in testing should not exceed the expected frequency for service.

Fatigue endurance ratios for ductile iron are affected primarily by the matrix microstructure/strength (Fig. 15) and by notch factors (Fig. 16) or surface condition. Table 11 summarizes typical reversed bending fatigue results for three standard grades of ductile iron. Results of rotating bending tests carried out by the British Cast Iron Research Association are summarized in Table 12 and are discussed in more detail in Ref 1. Ductile irons follow the expected trend of decreasing fatigue endurance ratios for higher strength of the matrix microstructure. However, heat treatments that result in an increase of tensile strength do not show a proportional increase in the fatigue limit (Table 11). References 1 to 8 and the "Selected References" listed at the end of this article are additional sources of property data on the fatigue of ductile iron.

Charpy Impact Toughness. Upper-shelf Charpy V-notch energy for ferritic ductile iron is about 14 to 24 J (10 to 18 ft · lbf) as compared to 12 to 19 J (9 to 14 ft · lbf) for malleable iron and 1 to 6 J (0.7 to 4.5 ft · lbf) for gray iron. Data from a comprehensive study show that increasing pearlite content decreases impact toughness (Fig. 13) and that the transition temperature is significantly affected by silicon and phosphorous content.

The effects of various heat treatments on Charpy V-notch impact energy are shown in Fig. 17 for a ductile iron alloyed with about 0.75% Ni. Curve F shows that austempering heat treatment

Table 9   Minimum tensile and fatigue properties of gray cast irons to ISO 185 (1988)

	Minimum properties					
Property	Grade 150	Grade 200	Grade 250	Grade 300	Grade 350	Grade 400(a)
Tensile strength, MPa (ksi)	150 (21.75)	200 (29.0)	260 (37.7)	300 (43.5)	350 (50.75)	400 (58.0)
Fatigue limit (Wohler)						
Unnotched(b), MPa (ksi)	68 (9.85)	90 (13.05)	117 (16.95)	135 (19.55)	149 (21.6)	152 (22.0)
V-notched(c), MPa (ksi)	68 (9.85)	87 (12.6)	108 (15.65)	122 (17.7)	127 (18.7)	129 (18.4)

(a) Not in ISO 185 but included in some national specifications. (b) 8.4 mm diameter. (c) Circumferential 45° V-notch with 0.25 mm root radius

Table 10   Minimum tensile and fatigue properties of ductile irons produced to ISO 1083 (1987)

	Ferritic irons		Intermediate grades			Pearlitic as-cast and normalized irons			Hardened-and-tempered irons		
Property	Grades 350-22 350-22L	Grade 400-18 400-18L	Grade 450-10	Grade 500-7	Grade 600-3	Grade 700-2	Grade 800-2	Grade 900-2	Grade 700-2	Grade 800-2	Grade 900-2
Tensile strength, MPa (ksi)	350 (50.75)	400 (58.0)	450 (65.25)	500 (72.5)	600 (87.0)	700 (101.5)	800 (116)	900 (130)	700 (101.5)	800 (116)	900 (130)
0.19 proof strength, MPa (ksi)	203 (29.45)	247 (35.8)	293 (42.5)	323 (46.85)	346 (50.2)	385 (55.8)	440 (63.8)	495 (71.8)	525 (76.1)	600 (87)	675 (97.9)
Elongation, %	22	18	10	7	3	2	2	2	2	2	2
Fatigue limit (Wohler)											
Unnotched (10.6 mm), MPa (ksi)	180 (26.1)	195 (28.3)	210 (30.5)	224 (32.5)	248 (36)	280 (40.6)	304 (44.1)	317 (46)	280 (40.6)	304 (44.1)	317 (46)
Notched(a), MPa (ksi)	114 (16.5)	122 (17.7)	128 (18.6)	134 (19.4)	149 (21.6)	168 (20)	182 (26.4)	190 (27.5)	168 (24.35)	182 (26.4)	190 (27.55)

(a) Diameter at notch root, 10.6 mm; circumferential 45° V-notch with 0.25 mm root radius and a notch depth of 3.6 mm

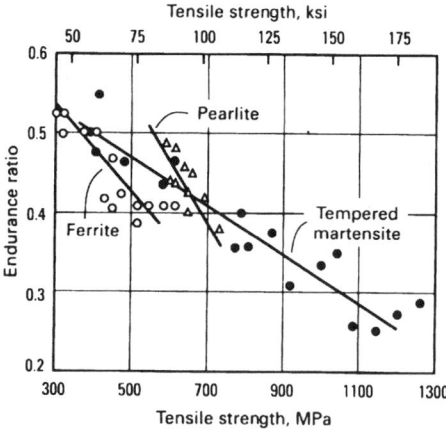

Fig. 15 Effect of tensile strength and matrix structure on endurance ratio for ductile iron. Source: *Metals Handbook*, 9th ed., Vol 1, p 45

Table 11 Reversed-bending fatigue properties of three standard grades of ductile iron

	Unnotched(a)					45° V-notched			
	Tensile strength, S_t		Endurance limit, S_e		Endurance ratio(b)	Endurance limit, S_n		Endurance ratio(c)	Notch sensitivity factor(d)
Type	MPa	ksi	MPa	ksi		MPa	ksi		
64-45-12	490	71	210	30.5	0.43	145	21	0.30	1.4
80-55-06	620	90	275	40	0.44	165	24	0.27	1.7
120-90-02(e)	930	135	338	49	0.36	207	30	0.22	1.6

(a) Polished specimens with 10.6 mm diameter. (b) S_e/S_t. (c) S_n/S_t. (d) S_e/S_n. (e) Oil quenched from 900 °C (1650 °F) and tempered at 595 °C (1100 °F). Source: *ASM Handbook*, Vol 1, p 48

Table 12 Effect of heat treatment on fatigue properties of ductile iron

Treatment	Tensile strength, ksi	Unnotched fatigue limit, ksi	Endurance ratio	Notched fatigue limit, ksi
As-cast	85	39	0.46	27
Normalized	134	44	0.33	27
As-cast	86	36	0.42	23
Oil-quenched, tempered 2 h 600 °C	120	44	0.37	27
Oil-quenched, tempered 2 h 550 °C	133	44	0.33	25

Source: Ref 1

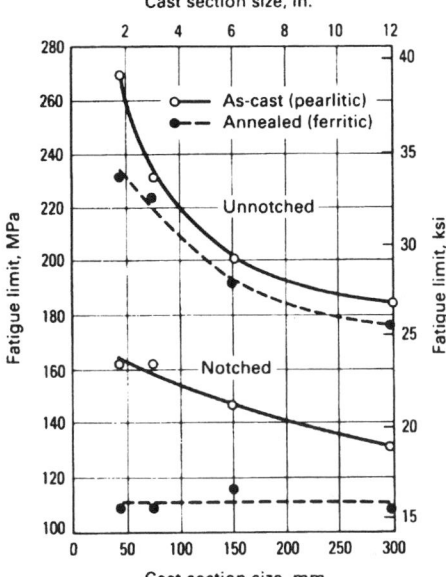

Fig. 16 Effect of cast section size on the fatigue properties of ductile irons. Source: *Metals Handbook*, 9th ed., Vol 15, p 662

not only improves elongation at high strength but also produces the highest room-temperature impact energy of any of the conventional heat treatments, with the exception of surface-hardened as-cast ductile iron, curve *A*, with a hardness of 207 HB.

The use of Charpy V-notch data is generally inappropriate to determine dynamic fracture toughness of ductile irons (Ref 5). Because graphite debonding dominates the fracture process in ductile iron, Charpy V-notch specimens may be acceptable only when the internodular spacing is less than the standard notch root radius of 0.25 mm. This would require a nodular count of less than 20 mm^{-2}, which is below the typical nodular counts in commercial ductile irons.

Dynamic Tear Energy. Upper-shelf dynamic tear energy for ferritic ductile iron is about 175 J (129 ft · lbf) for a standard 15.9 mm specimen, as compared to 100 J (73 ft · lbf) for malleable iron (Ref 5). Greater amounts of pearlite in the matrix microstructure reduce the upper-shelf toughness (Fig. 18b). With higher levels of pearlite, the upper-shelf toughness is comparable to that of compacted graphite cast iron (Fig. 18a). Similar comparisons are expected for upper-shelf toughness derived from Charpy testing.

Fracture Toughness. Early measurements of room-temperature fracture toughness of ferritic ductile irons were almost all incorrect. Systematic increases occurred over time as better test and evaluation methods were developed (Ref 5). Linear elastic fracture mechanics is not suitable for measuring the upper-shelf toughness of ferritic ductile irons unless very large specimens are used. Certain lower-strength grades of ductile iron do not fracture in a brittle manner when tested under nominal plane-strain conditions in a standard fracture toughness test. In the low-strength ductile irons, plane-strain conditions are established only at temperatures low enough to embrittle the ferrite. Otherwise, an increase in the size of the fracture toughness test specimens is necessary to provide the degree of mechanical constraint necessary to obtain a valid measurement of K_{Ic}.

Plane-strain fracture toughness (K_{Ic}) at ambient temperature and upper-shelf regime is estimated at 88 to 90 MPa√m (80 to 81 ksi√in.) for ferritic ductile irons based on the review of Bradley and Srinivasan (Fig. 19). Upper-shelf K_{Ic} values of 100 MPa√m (90 ksi√in.) or more are also considered possible for ferritic ductile irons, as compared with 75 MPa√m (68 ksi√in.) for malleable irons (Ref 5). Typical fracture toughness results are shown in Fig. 19 and 20. These

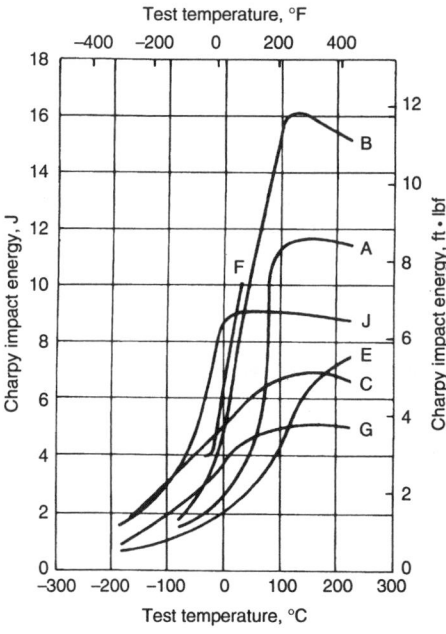

Fig. 17 Effect of heat treatment on Charpy V-notch impact properties of ductile iron. Source: *Metals Handbook*, 9th ed., Vol 1, p 42

fracture toughness estimates are higher than earlier results (e.g., Table 13); as discussed in Ref 5, the data in Table 13 are not accurate because the specimen was too small for proper toughness results on a ductile material.

Lower-shelf fracture toughness is more consistent. Typical results for fully ferritic ductile iron and ferritic ductile iron with about 15% pearlite are presented in Fig. 20. The conclusion from review of data available (Ref 5) is that both fer-

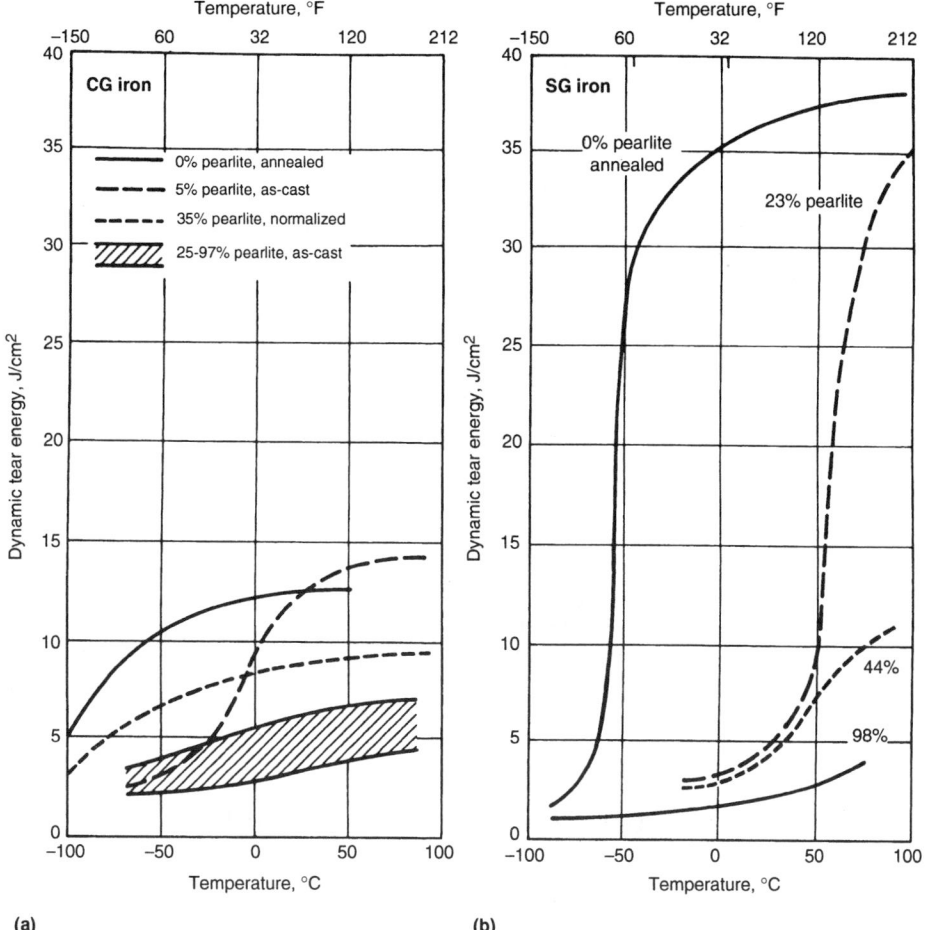

(a)　　　　　　　　　　　　　　　　　　　**(b)**

Fig. 18　Effect of pearlite content on the toughness of (a) compacted graphite (CG) cast iron and (b) spheroidal graphite (SG) cast iron (i.e., ductile iron). This comparison for several different matrix microstructures indicates that the deleterious effect of vermicular graphite on fracture toughness in CG iron is much more pronounced for a ferritic matrix than for a pearlitic matrix. Furthermore, the compacted graphite cast iron seems to have a somewhat lower ductile-to-brittle transition temperature than ductile iron, possibly because the ductile fracture has been made easier relative to cleavage by the formation of compacted graphite. Source: Ref 13

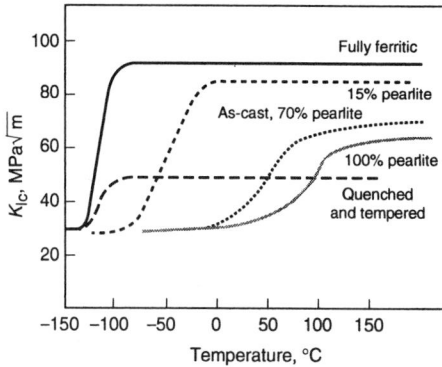

Fig. 19　Effect of microstructure on fracture toughness of ductile iron. These results are generally based on data from the literature, with some approximation in regions where data were unavailable. Source: Ref 5

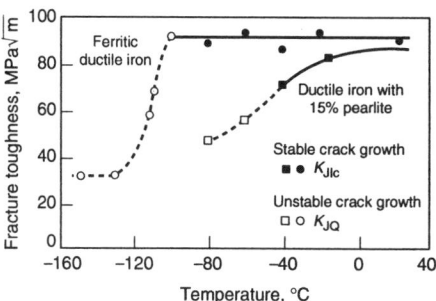

Fig. 20　Comparison of fracture toughness of fully ferritic ductile iron (GGG-40) with ferritic ductile iron with 15% pearlite in matrix (S-45). Source: Ref 5

Table 13　Fracture toughness of ductile iron from early work

Type of iron	Condition	Ultimate tensile strength		Yield strength		Elongation, %	K_{Ic}, MPa · m$^{0.5}$ (ksi · in.$^{1/2}$) at:		
		MPa	ksi	MPa	ksi		20 °C (70 °F)	–40 °C (–40 °F)	–105 °C (–160 °F)
Ferritic									
3.0% Si	As-cast	521	75.6	427	62.0	11.0	...	35.1 (32.0)	30.2 (27.5)
3.5% Si	As-cast	547	79.4	471	68.3	9.0	...	27.0 (24.6)	...
Pearlitic									
2.5% Si	As-cast	703	102.0	374	54.2	7.5	...	37.1 (33.8)	...
	Normalized	918	133.2	552	80.0	3.6	45.3 (41.3)	...	...
	Austempered	...	...	620	90.0(a)	...	36.5 (33.3)	...	...

Note: These fracture toughness data are from early test results, which are probably not representative of inherent K_{Ic} values based on more current evaluations. See text. (a) Estimated. (b) Estimated yield strength. Source: Ref 14, reported in *ASM Handbook*, Vol 1, p 47

ritic and pearlitic ductile irons have similar lower-shelf fracture toughness values of about 23 to 28 MPa√m (21 to 25 ksi√in.). At temperatures above –110 °C, the fracture toughness of ferritic ductile iron increases sharply with increasing temperature. The presence of as little as 15%

pearlite in ductile iron will cause a significant increase in the ductile-to-brittle transition temperature (Fig. 20). For temperatures around 0 °C, the fracture toughness K_{Jc} is a very sensitive function of the percentage of pearlite in the matrix, the latter affecting the fracture toughness by

lowering the upper-shelf value and, more importantly, by shifting the ductile-to-brittle transition temperature.

Dynamic Fracture Toughness. Figures 11 and 12 summarize a review (Ref 5) of dynamic fracture toughness evaluations for ferritic and pearlitic grades of various cast irons. As expected, pearlite reduces upper-shelf toughness and increases transition temperatures.

There are only a few reliable results for K_{Id} and K_{Jd} for ferritic ductile iron published in the literature. The upper-shelf dynamic fracture toughness of ferritic ductile iron (with 2.3% silicon and having less than 100 mm^{-2} nodule count) as measured by K_{Jd} is about 90 to 95 MPa√m (82 to 86 ksi√in.). This value decreases with increasing nodule count, decreasing nodularity, and increasing pearlite content in the matrix (Ref 5).

Malleable Cast Irons

There are two basic types of malleable iron: blackheart and whiteheart. Blackheart malleable iron is the only type produced in North America and is the most widely used throughout the world. Whiteheart malleable iron is the older type and is essentially decarburized throughout in an extended heat treatment of white iron. Unless other-

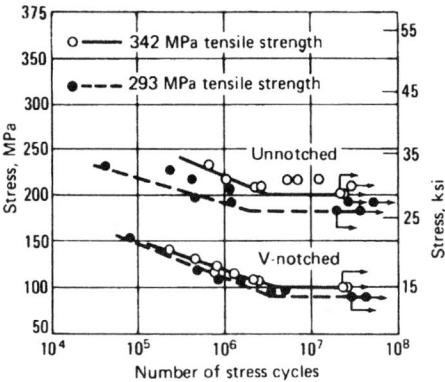

(a)

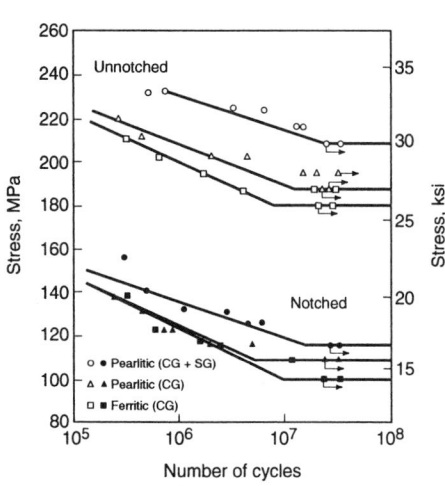

(b)

Fig. 21 Rotating-bending fatigue strength of malleable cast iron and compacted graphite (CG) cast iron. (a) Unnotched and notched fatigue curves of two ferritic malleable irons (25 mm diameter specimens). (b) Unnotched and notched fatigue curves for ferritic, pearlitic, and higher-nodularity CG irons. SG, spheroidal graphite. Sources: Ref 26, 27

Table 14   Minimum tensile and fatigue properties of blackheart malleable irons produced to BS 310 (1972)

Property	Minimum properties		
	Grade B290/6	Grade B310/10	Grade B340/12
Tensile strength, MPa (ksi)	290 (42)	310 (45)	340 (49.3)
0.2% proof stress, MPa (ksi)	191 (27.7)	205 (29.7)	224 (32.5)
Elongation, %	6	10	12
Fatigue limit (Wohler)			
Unnotched (10 mm diam), MPa (ksi)	174 (25.2)	186 (27.0)	204 (29.6)
Notched(a) (10 mm diam at root), MPa (ksi)	87 (12.6)	93 (13.5)	102 (14.8)

(a) Circumferential 45° V-notch with 0.25 mm root radius and a notch depth of 2 mm

wise specified, this article considers only the blackheart type.

Malleable iron, like medium-carbon steel, can be heat treated to produce a wide variety of mechanical properties. The different grades and mechanical properties are essentially the result of the matrix microstructure, which may be a matrix of ferrite, pearlite, tempered pearlite, bainite, tempered martensite, or a combination of these (all containing nodules of temper carbon). This matrix microstructure is the dominant factor influencing the mechanical properties.

Minimum tensile properties and fatigue limits for whiteheart and blackheart malleable irons are summarized in Tables 14 and 15. Fatigue curves for a ferritic malleable iron are shown in Fig. 21 with fatigue curves for various compacted graphite irons. Fatigue curves for the ferritic grades of these two iron types are comparable. Notch radius generally has less effect on fatigue strength than notch depth for ferritic malleable irons (Fig. 22).

Charpy Impact Toughness. Like compacted graphite cast irons, malleable cast irons have a lower ductile-to-brittle transition than ductile irons with a comparable matrix microstructure. Malleable cast irons are generally regarded as having the best low-temperature toughness of the cast irons due to the lower silicon content (about

1% less silicon than ductile irons), which thus lowers the yield strength and the ductile-to-brittle transition temperature. Ferritic malleable cast iron has an upper-shelf Charpy energy of 13 J (9.5 ft · lbf), compared with 9 J (6.5 ft · lbf) for pearlitic malleable cast iron (Fig. 23). Other investigators have reported a lower transition temperature and a somewhat higher upper shelf Charpy energy of 19 to 20 J for ferritic malleable cast iron and 12 J for pearlitic malleable cast iron (Ref 15, 16). Dramatic reduction in upper-shelf toughness and increase in the ductile-to-brittle transition temperature result from phosphorus concentrations of 0.15% or greater in both ferritic and pearlitic malleable cast iron (Ref 17, 18). The phosphorus not only forms a very low-melting-point phosphide but also degenerates the graphite particle shape.

Dynamic Tear Energy. Figure 24 compares the tear energy of malleable iron with that of various ductile irons. As shown previously for compacted graphite irons (Fig. 18), ferritic malleable iron has a lower ductile-to-brittle transition than ductile iron. These data also show that annealing above the critical temperature reduces upper-shelf toughness compared to subcritical annealing.

Fracture Toughness. Some published data (such as those reported in Table 16 and Ref 19-

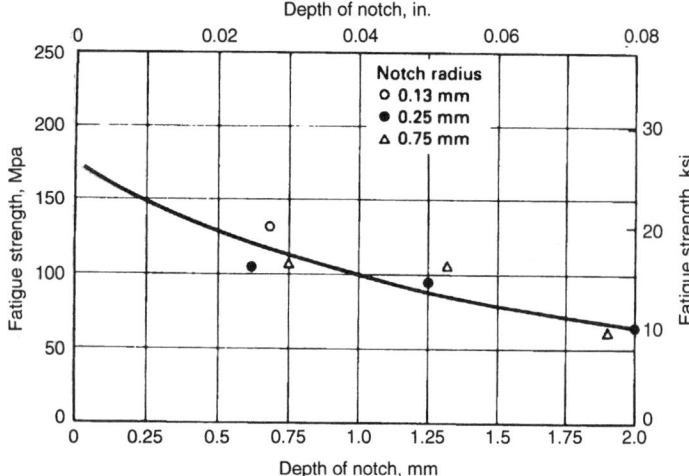
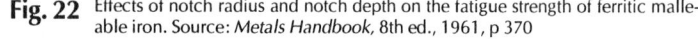

Fig. 22 Effects of notch radius and notch depth on the fatigue strength of ferritic malleable iron. Source: *Metals Handbook*, 8th ed., 1961, p 370

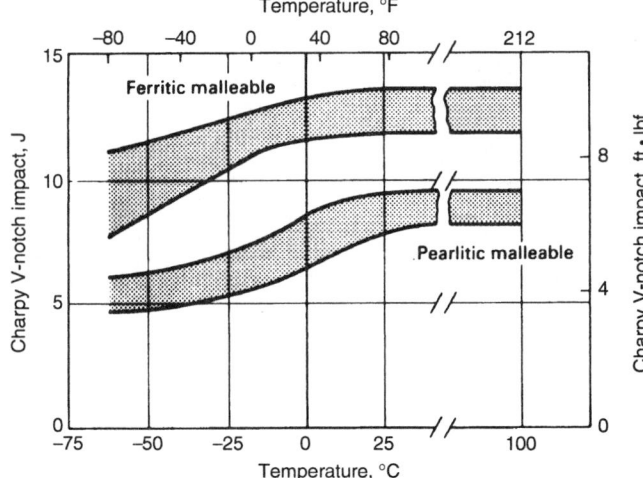

Fig. 23 Charpy V-notch transition curves for ferritic and pearlitic malleable irons. Source: Ref 3

Table 15 Minimum tensile and fatigue properties of whiteheart malleable irons produced to BS 309 (1972)

Property	Grade W340/4 section thickness 9 mm	Grade W340/4 section thickness 12 mm	Grade W340/4 section thickness 15 mm	Grade W410/4 section thickness 9 mm	Grade W410/4 section thickness 12 mm	Grade W410/4 section thickness 15 mm
Tensile strength, MPa (ksi)	270 (39.15)	310 (45)	340 (49.3)	350 (50.75)	390 (56.55)	410 (59.5)
0.2% proof stress, MPa (ksi)	149 (21.6)	171 (24.8)	187 (27.1)	193 (28)	215 (31.15)	226 (32.75)
Elongation, %	7	4	3	10	6	4
Fatigue limit (Wohler)						
Unnotched (10 mm diam), MPa (ksi)	122 (17.7)	140 (20.3)	153 (22.2)	158 (22.9)	176 (25.5)	185 (26.8)
Notched(a) (10 mm diam at root), MPa (ksi)	73 (10)	84 (12)	92 (13)	95 (13.8)	106 (15.4)	111 (16.1)

(a) Circumferential 45° V-notch with 0.25 mm root radius and notch depth of 2 mm

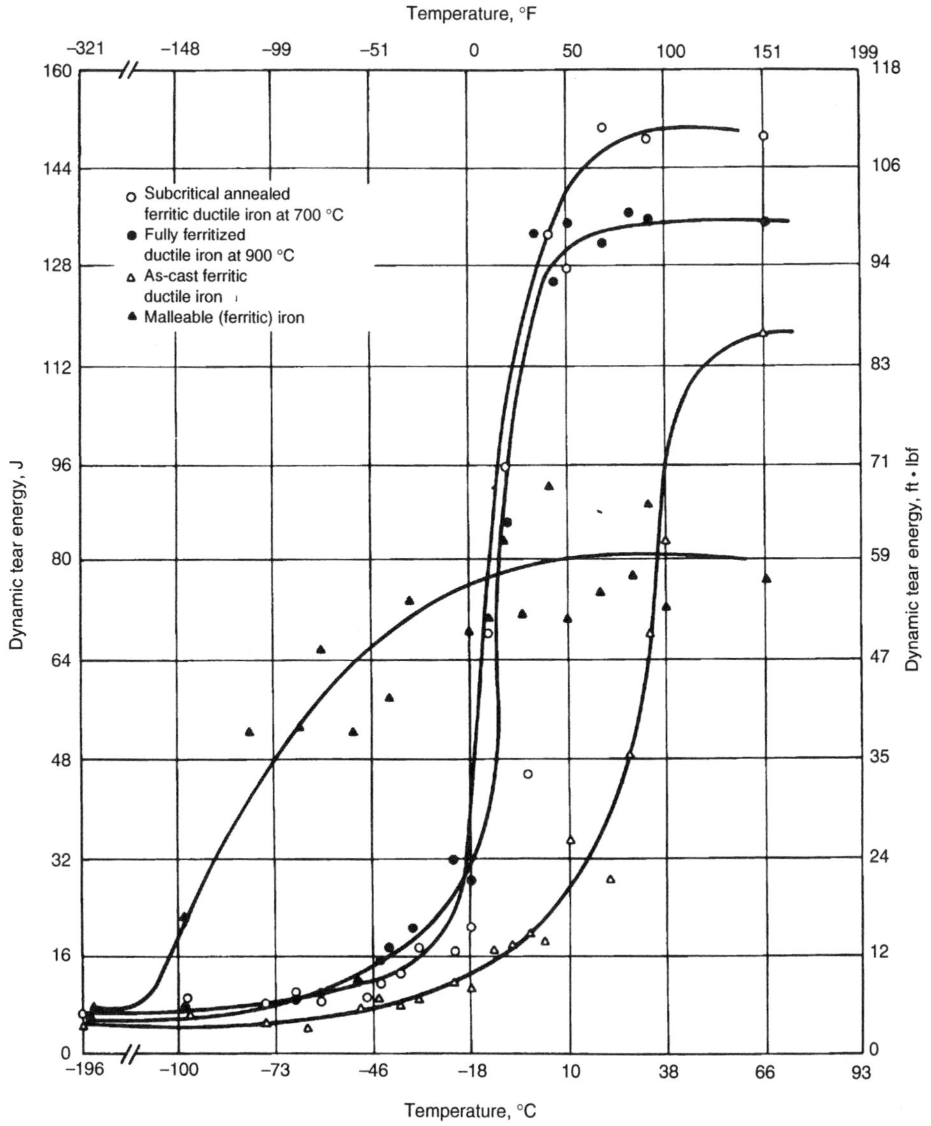

Fig. 24 Dynamic tear energy versus temperature for four ferritic cast irons. Specimen size: 190 mm (7½ in.) long, 130 mm (5⅛ in.) wide, and 41 mm (1⅝ in.) thick; 13 mm (½ in.) notch depth. Source: Ref 28

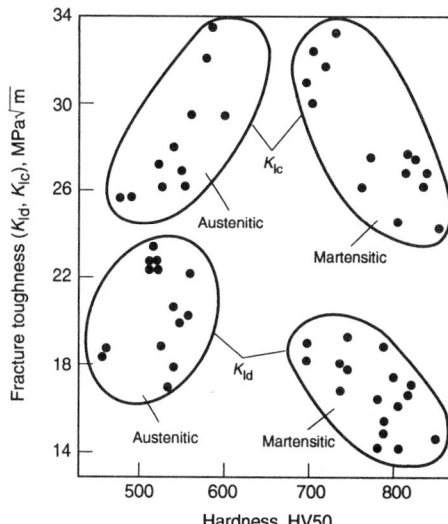

Fig. 25 Relationship between fracture toughness and hardness for white cast irons of different microstructures. Source: Ref 5

22) have attempted to establish linear elastic (K_{Ic}) fracture toughness limits for malleable irons. However, as discussed in Ref 5, the fracture toughness of malleable irons needs to be evaluated by an elastic-plastic criterion such as K_{Jc}. Upper-shelf K_{Jc} values range from 53 to 73 MPa$\sqrt{m}$ (48 to 66 ksi $\sqrt{in.}$, as summarized in Table 17.

Linear elastic fracture toughness values reported in the literature (K_{Ic} in Table 16 and K_{max} or K_Q in Ref 19-23) are invalid in their indication of increases in upper-shelf toughness with increasing strength and/or lowering of temperature (Ref 5). While this may be a reasonable correlation for gray cast iron, it is not so for malleable cast iron. The reason is that a *J*-integral approach is needed to calculate the critical *J* value at or just before maximum load where crack extension begins (Ref 5). It is inadvisable to use K_Q or K_{max} for toughness ranking of malleable irons (Ref 5).

White Cast Iron

A number of studies have been conducted to determine the fracture toughness of white cast iron (Fig. 25). White cast iron is generally used in applications where abrasion resistance is desirable. Gahr and Scholz (Ref 24) have found that the dynamic fracture toughness (K_{Id}) values, measured on 10 × 10 × 55 mm specimens with a 0.2 mm wide slot, were in the range of 14 to 24 MPa$\sqrt{m}$ (13 to 22 ksi$\sqrt{in.}$), while the quasistatic fracture toughness values, measured on fatigue precracked compact tension specimens, were in the range of 25 to 33 MPa$\sqrt{m}$ (23 to 30 ksi$\sqrt{in.}$).

Table 16 Fracture toughness of malleable irons from early results

Malleable iron grade	Test temperature °C	Test temperature °F	Yield strength MPa	Yield strength ksi	K_{Ic} MPa$\sqrt{m}$	K_{Ic} ksi$\sqrt{in.}$
Ferritic						
M3210	24	75	230	33	44	40
	−19	−3	240	35	42	38
	−59	−74	250	36	44	40
Pearlitic						
M4504 (normalized)	24	75	360	52	55	50
	−19	−2	380	55	48	44
	−57	−70	390	57	30	27
M5503 (quenched and tempered)	24	75	410	60	45	41
	−19	−3	440	64	52	47
	−58	−73	455	66	30	27
M7002 (quenched and tempered)	24	75	520	75	54	49
	−19	−3	550	80	38	35
	−58	−72	570	83	40	36

Note: These fracture toughness data are from early test results, which are not considered representative of K_{Ic} values from more recent evaluations (especially in the ferritic condition). See text. Source: Reported in *ASM Handbook*, Vol 1, p 80

Table 17 Fracture toughness of various malleable irons

Material	Yield strength, MPa	Temperature, °C	K_{Jmax} (new)(a), MPa$\sqrt{m}$	K_{Jc} (previously reported), MPa$\sqrt{m}$
M3210 (ferritic malleable iron)	230	24	76	44
		−19	76	41
		−57	71	43
M4504 (ferritic/pearlitic malleable iron)	359	24	70	55
		−19	72	48
		−59	30	30
M5503 (tempered martensitic malleable iron)	411	24	57	45
		−18	53	52
		−58	34	29
M7002 (tempered martensitic malleable iron)	540	24	50	50
		−19	39	39
		−58	39	39
80-60-03 (pearlitic/ferritic ductile iron)	433	24	33	27
		−19	25	25
		−48	26	26
DQ & T (SAE grade) (tempered martensitic ductile iron)	717	24	48	52
		−19	51	48
		−59	50	51
D5B (SAE grade) (austenitic ductile iron)	304	24	89	64
		−47	85	…
		−59	91	67

(a) K_{Jmax} values were recalculated from earlier work and are assumed to equal K_{Jc}. Source: Ref 5

Under dynamic loading, austenitic grades of white cast iron had better impact strength than did martensitic grades, though no such clear pattern was noted for K_{Id} values.

Eriksson (Ref 25) measured a fracture toughness value of 22 MPa$\sqrt{m}$ (20 ksi$\sqrt{in.}$) on a white cast iron sample with martensitic matrix corresponding to a Vicker's hardness value of 633. A comparison of his results (Fig. 25) with those of Gahr and Scholz indicates both lower hardness and lower toughness. This may be due to the presence of some flake graphite in samples tested by Eriksson.

REFERENCES

1. K.B. Palmer, *Fatigue Properties of Cast Iron*, British Cast Iron Research Association

2. D. Socie and J. Fash, Fatigue Behavior and Crack Development in Cast Irons, *Trans. AFS*, Vol 90, 1982, p 385-392
3. C.F. Walton and T.J. Opar, *Iron Castings Handbook*, Iron Casting Society, 1981
4. D. Maykuth, *Structural Alloys Handbook*, Mechanical Properties Data Center, published originally by Battelle, now offered by CINDAS
5. W.L. Bradley and M.N. Srinivasan, Fracture and Fracture Toughness of Cast Irons, *International Materials Reviews*, Vol 35 (No. 3), 1990, p 129-159
6. E. Nechtelberger, *The Properties of Cast Iron up to 500 °C*, Technicopy Ltd., 1980
7. S. Seetharamu and M.N. Srinivasan, Fatigue Crack Propagation in Permanent Mold Ductile Iron Castings, *AFS Transactions*, Vol 94, 1986, p 265-270

8. C. Walton, *Gray and Ductile Iron Castings Handbook*, Iron Casting Society, 1965
9. R.E. Peterson, *Stress Concentration Factors*, John Wiley, 1974
10. K. Roehrig, *Trans. AFS*, Vol 86, 1978, p 75
11. Y.J. Park, R.B. Gundlach, R.G. Thomas, and J.F. Janowak, *Trans. AFS*, Vol 93, 1985, p 415
12. W. Bradley, *Trans. AFS*, Vol 89, 1981, p 837-848
13. K. Cooper and C. Loper, *Trans. AFS*, Vol 86, 1978, p 241
14. R.K. Nanstad, F.J. Worzala, and C.R. Loper, Jr., Static and Dynamic Toughness of Ductile Cast Iron, *Trans. AFS*, Vol 83, 1975
15. G. Yu. Schulte, *Metalloved. Term. Obrab. Met.*, Nov 1973, p 35-37
16. F.W. Jacobs and F.L. Preston, *Trans. AFS*, Vol 83, 1975, p 263-270
17. P.B. Burgess, *Trans. AFS*, Vol 75, 1967, p 665-675
18. T. Okumato and K. Kondo, *Imono (J. Jpn Foundrymen's Soc.)*, Vol 34, 1962, p 34
19. G.N.J. Gilbert, *Engineering Data on Malleable Cast Irons*, British Cast Iron Research Association, 1968
20. F.J. Worzala, R.W. Heine, and Y.W. Cheng, *Trans. AFS*, Vol 84, 1976, p 675-682
21. A. Little and H.J. Heine, in *The Metallurgy of Cast Iron*, B. Lux et al., Ed., Georgi Publishing, Switzerland, 1975, p 767-788
22. V.I. Ovchinnikov, V.A. Kurdyukov, and S.G. Girenkov, *Metalloved. Term. Obrab. Met.*, Feb 1982, p 13-14
23. *ASM Handbook*, Vol 1, p 80
24. K.Z. Gahr and W.G. Scholz, *J. Met.*, Vol 32, Oct 1980, p 38-44
25. K. Eriksson, *Scand. J. Metall.*, Vol 2, 1973, p 197-203
26. L.W.L. Smith et al., *Properties of Modern Malleable Irons*, BCIRA International Center for Cast Metals Technology, 1987
27. K.B. Palmer, *BCIRA J.*, Report 1213, Jan 1976, p 31
28. *First International Conference on Austempered Ductile Iron*, American Society for Metals, 1984

SELECTED REFERENCES

1980 to 1995

- A. Alagarsamy and J. Janowak, Fatigue Strength of Commercial Ductile Irons, *Trans. AFS*, Vol 98, 1990, p 511-518
- P. Bhandhubanyong, I. Umeda, and Y. Kimura, Fatigue Strength and Crack Propagation in Spheroidal, Compacted Vermicular and Flake Graphite Cast Irons, *Trans. Jpn. Foundrymen's Soc.*, Vol 3, 1984, p 40-44
- S.B. Biner, The Role of Eutectic Carbide Morphology on the Fracture Behavior of High-Chromium Cast Irons, Part I: Austenitic Alloys, *Can. Metall. Q.*, Vol 24 (No. 2), 1985, p 155-162
- J.H. Bulloch, Fractographic Analysis of Fatigue Cracking in Spheroidal Graphite Cast Irons, *Theoretical and Applied Fracture Mechanics*, Vol 17 (No. 1), 1992, p 19-45

- G. Faubert, D. Moore, and K. Rundman, Heavy Section ADI: Fatigue Properties in As-Cast and Austempered Specimens, *Trans. AFS*, Vol 97, 1990, p 759-764
- R. Higuchi and J. Wallace, Fatigue Properties of Gray Iron, *Trans. AFS*, Vol 89, 1981, p 483-494
- K.H. Kloos and J. Adelmann, Effect of Deep Rolling on Fatigue Properties of Cast Irons, *J. Mech. Behav. Mater.*, Vol 2, 1989, p 75-86
- H. Sachar and J. Wallace, Effect of Microstructure and Testing Modes on the Fatigue Properties of Gray Cast Iron, *Trans. AFS*, Vol 90, 1982, p 777-793
- M. Srinivas and G. Malakondaiah, Effect of Grain Size on Threshold Stress Intensity for Fatigue Crack Growth in Armco Iron, *Scr. Metall.*, Vol 20 (No. 5), 1986, p 689-692
- Z. Tao and J. Luo, Influence of Surface Hardening on Contact Fatigue Life of Nodular Cast Iron, *Acta Metallurg. Sinica*, Vol 3A (No. 3), 1990, p 188-193
- A. Tiziani, E. Ramous, and A. Molinari, Fracture Morphology in Compacted Graphite Cast Iron, *Advances in Fracture Research*, Vol 2, Pergamon Press, 1984, p 1489-1496
- A.V. Vesnitskii, Criteria of the Contact Strength, Ductility, and Fracture of Materials in Shear with Buckling, *Strength of Materials*, Vol 24 (No. 6), 1993, p 402-408
- S. Wu and G. Shu, Thermal Fatigue Property of Vermicular Graphite Cast Iron, *Fatigue 90*, Materials and Component Engineering Publications, Birmingham, U.K., 1990, p 1705-1710

Prior to 1980
- H.T. Angus, Physical and Engineering Properties of Cast Iron, British Cast Iron Research Association, 1960
- D.R. Askeland and R.F. Fleischman, Effect of Nodule Count on the Mechanical Properties of Ferritic Malleable Iron, *Trans. AFS*, Vol 86, 1978, p 373-378
- A.G. Fuller, Effect of Graphite Form on Fatigue Properties of Pearlitic Ductile Irons, *Trans. AFS*, Vol 85, 1977, p 527-536
- G.N.J. Gilbert, *Engineering Data on Malleable Cast Irons*, British Cast Iron Research Association, 1968
- Forrest, *Fatigue of Metals*, National Physics Lab, 1962
- H. Morrogh, Fatigue of Cast Irons, *Fatigue of Metals*, Chapman-Hall, 1959
- H. Grover, S. Gordon, and C. Jackson, *Fatigue of Metals and Structures*, Department of the Navy, 1960
- M. Sofue, S. Okada, and T. Sasaki, High-Quality Ductile Iron with Improved Fatigue Strength, *Trans. AFS*, Vol 86, 1978, p 173-183

Bending Fatigue of Carburized Steels

George Krauss, Colorado School of Mines

BENDING FATIGUE of carburized steel components is a result of cyclic mechanical loading. The bending produces stresses, which are tensile at the surface, decrease with increasing distance into the component, and at some point become compressive. Such loading is a characteristic of rotating shafts and the roots of gear teeth. Carburizing produces a high-carbon, high-strength surface layer, or high-strength case, on a low-carbon, low-strength interior or core and is therefore an ideal approach to offset the high surface tensile stresses associated with bending. Thus when the design of a component maintains operating stress gradients below the fatigue strength of the case and core microstructures, excellent bending fatigue resistance is established.

There are, however, many alloying and processing factors that produce various microstructures, and therefore variable strength and fracture resistance, of the case regions of carburized steels. When applied surface stresses exceed the surface strength, surface fatigue crack initiation and eventual failure will develop (Ref 1). When the surface strength is adequate, depending on the steepness of the applied stress gradients in relationship to the case/core strength gradient, subsurface fatigue cracking may develop. The purpose of this article is to review the alloying and processing factors that influence the microstructures and bending fatigue performance of carburized steels.

The Carburizing Process and Microstructure

Bending fatigue performance of carburized steels can vary significantly. One study reported values of experimentally measured endurance limits ranging from 200 to 1930 MPa (29 to 280 ksi), with most values between 700 and 1050 MPa (100 and 152 ksi) (Ref 2). This wide variation in fatigue performance is a result of variations in specimen design and testing, alloying, and processing interactions that produce large variations in carburized microstructures and the response of the microstructures to cyclic loading. A number of books and reviews have addressed the general relationships of processing to the microstructure and properties of carburized steels (Ref 3-9).

Gas carburizing is the most widely used heat-treating process for parts that must be produced in high volumes, but carburizing is also performed in liquid salt baths (Ref 3, 10), in solid carbonaceous material (Ref 3), by vacuum (Ref 11), and by plasma carburizing (Ref 12). Gas-carburized steels are exposed to atmospheres that contain partial pressures of oxygen-containing components, such as CO, CO_2, and H_2O, and therefore generally are subject to some degree of surface oxidation. The resulting surface oxides constitute another microstructural feature that may influence fatigue performance. Vacuum- or plasma-carburized parts that are carburized with partial pressures of only carbon-containing components, such as propane or methane, are not subject to surface oxidation.

The object of all carburizing processes is to expose the surface of a low-carbon steel to a carbon-containing atmosphere where austenite, with its high solubility for carbon, is stable, typically at temperatures between 850 and 950 °C (1560 and 1740 °F). The carbon diffuses into the steel, to a depth determined by steel composition, temperature, and time (Ref 13, 14). The carburized part is then quenched and the austenite transforms to martensite. The quenched parts are most often subjected to a low-temperature tempering treatment at temperatures between 150 and 200 °C (300 and 400 °F), and very fine transition carbides precipitate in the martensite (Ref 9, 15, 16). The high-carbon, low-temperature-tempered (LTT) martensite that forms at the surface of a carburized part has the very high strength required for excellent bending and rolling contact fatigue (Ref 16). In addition, the sequence of phase transformations during quenching of a carburized part produces surface residual compressive stresses that greatly enhance resistance to bending fatigue (Ref 17, 18). However, microstructural features other than the LTT martensite, as discussed below, frequently affect fatigue performance.

Two major types of microstructures may form in the near-surface case regions of carburized steels (Ref 15, 19). Within both types, there are additional microstructural features (e.g., retained austenite, prior austenite grain size, carbide parti-

cles, nonmetallic inclusions, and surface oxides) that, depending on their densities and magnitudes, can influence fatigue performance. The first major microstructure type, illustrated in Fig. 1, consists of relatively coarse plates of LTT high-carbon martensite within a matrix of retained austenite. Typically it is formed by direct quenching of the austenite after lowering the temperature of the part from the carburizing temperature to about 850 °C in order to minimize distortion.

The second major type of microstructure, illustrated in Fig. 2, consists of lower-carbon martensite, retained austenite, and dispersed carbide particles. This type of microstructure is produced by reheating carburized parts to a temperature between the A_1 and A_{cm} temperatures of the high-carbon case. At this temperature, carbides form, lowering the carbon content of the austenite. Upon cooling, the martensite that forms in the lower-carbon austenite is fine, with the morphology shifting from plate to lath martensite. The amount of retained austenite between the marten-

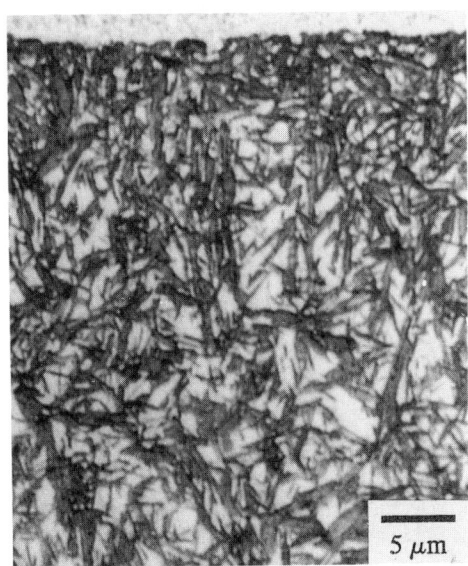

Fig. 1 Typical near-surface case microstructure of direct-quenched carburized steel. Martensite plates etch dark and retained austenite appears white. Gas-carburized AISI 8719 steel (1.06% Mn, 0.52% Cr, 0.5% Ni, 0.17% Mo). Light micrograph, nital etch

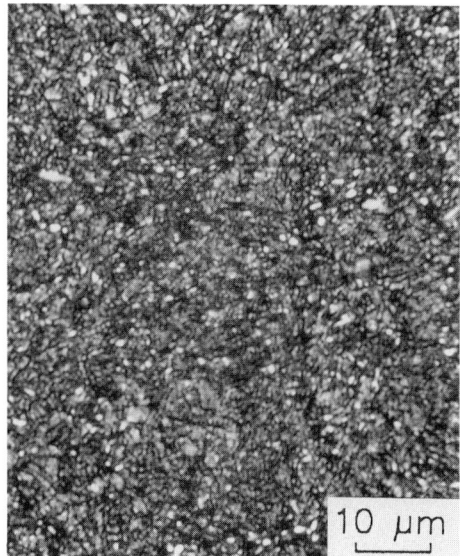

Fig. 2 Typical near-surface case microstructure of carburized steel (SAE 8620: 0.81% Mn, 0.19% Mo, 0.48% Ni) reheated between A_1 and A_{cm}. Retained carbides are small, white spherical particles, and matrix consists of a dark etching of mixture of martensite and austenite too fine for resolution in the light microscope. Light micrograph, nital etch

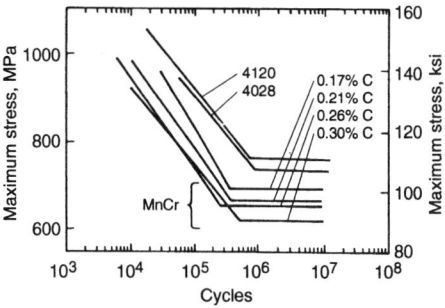

Fig. 3 Typical maximum stress (S) vs. number of cycles (N) bending fatigue plots for 6 carburized steels. R = –1. Source: Ref 6

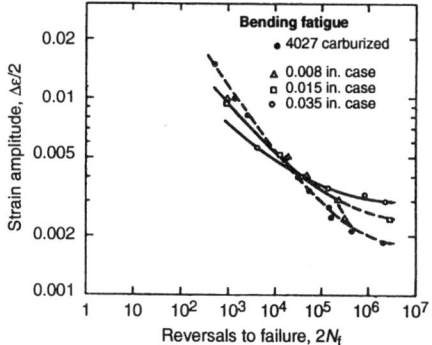

Fig. 4 Strain amplitude vs. reversals to failure for uncarburized (solid symbols) and carburized (open symbols) 4027 steel (0.80% Mn, 0.28% Si, 0.27% Mo). Source: Ref 22

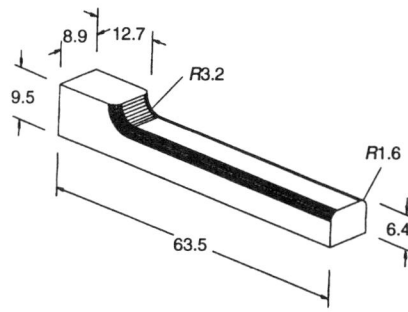

Fig. 5 Example of a cantilever specimen used to evaluate bending fatigue of carburized steels. Specimen edges are rounded and maximum stress is applied at the location shown in Fig. 6. Dimensions in millimeters. Courtesy of R.S. Hyde

site crystals is reduced by virtue of the higher M_s temperatures of the lower-carbon parent austenite. The case LTT martensite-retained austenite microstructure of an intercritically reheated carburized part, in fact, is so fine that it cannot be clearly resolved in the light microscope.

Bending Fatigue Testing

Most bending fatigue data for carburized steels are presented as plots of maximum stress, S, versus the number of cycles, N, to fracture for a specified stress ratio (R), which is the ratio of minimum (or compressive) stress to maximum tensile stress (R = min-S/max-S). Figure 3 shows an example of typical S-N plots for a series of carburized alloy steels (Ref 6). The S-N curve consists of two parts: a straight section with negative slope at low cycles and a horizontal section at high cycles (Ref 20). The horizontal line defines the fatigue limit or endurance limit, which is taken to be the maximum applied stress below which a material is assumed to be able to withstand an infinite number of stress cycles without failure. Pragmatically, the endurance limit is taken as the stress at which no failure occurs after a set number of cycles, typically on the order of 10 million cycles. The low-cycle portion of the S-N plot defines various fatigue strengths or the stresses to which the material can be subjected for a given number of cycles. The more cycles at a given strength, the better the low-cycle fatigue resistance of a material.

Analysis of bending fatigue behavior of carburized steels based on S-N curves represents a stress-based approach to fatigue and assumes that the carburized specimens deform nominally only

in an elastic manner (Ref 20). This assumption is most valid at stresses up to the endurance limit and is useful when machine components are designed for high-cycle fatigue. However, as maximum applied stresses increase above the endurance limit, plastic strain becomes increasingly important during cyclic loading, and fatigue is more appropriately analyzed by a strain-based approach. In this approach, described in detail in the article "Fundamentals of Modern Fatigue Analysis for Design" in this Volume (and in many other books such as Ref 20 and 21), the total strain range is the sum of the applied elastic and plastic strains, and the strain amplitude is plotted as a function of the strain reversals required for failure at the various levels of strain.

According to the strain-based approach, low-cycle fatigue behavior is determined by plastic strains, while high-cycle fatigue behavior is determined by elastic strains. In particular, ductile materials with microstructures capable of sustaining large amounts of plastic deformation have better low-cycle fatigue resistance, while high-strength materials with high elastic limits and high yield strengths have better high-cycle fatigue resistance. Figure 4 shows the results of strain-based bending fatigue testing of uncarburized and carburized 4027 steel (Ref 22). The more ductile, low-strength uncarburized specimens show better fatigue resistance at low cycles than do the carburized specimens. The performance is reversed at high cycles, where the carburized specimens with their high-strength surfaces show better fatigue resistance, especially those specimens with the deeper cases.

Specimen Design

Many types of specimens have been used to evaluate bending fatigue in carburized steels. Rotating beam (Ref 23), unnotched four-point bend (Ref 24), notched four-point bend (Ref 25, 26), and cantilever beam (Ref 27-29) specimens have all been used, and they have in common a maximum applied surface tensile stress and decreasing tensile stress with increasing distance into the specimen. Axial fatigue testing of carburized specimens, which applies the maximum tensile stresses uniformly over the cross section of a specimen, invariably results in subsurface initia-

tion and propagation of fatigue cracks in the core of carburized specimens (Ref 22, 30), and therefore it does not permit evaluation of the resistance of case microstructures to fatigue.

Brugger was the first to use cantilever bend specimens to evaluate the fracture and fatigue of carburized steels (Ref 27). Figure 5 shows a cantilever bend specimen that has evolved from the Brugger specimen. The radius between the change in section simulates the geometry at the root of gear teeth and results in maximum applied surface stresses just where the cross section begins to increase, as shown in Fig. 6 (Ref 29). An important feature of this specimen is the rounding of the corners of the beam section. If the corners are square, the carbon introduced into the corner surfaces cannot readily diffuse into the interior of the specimen. As a result, the corner microstructures may have significantly elevated levels of retained austenite and coarse carbide structures, both microstructural features that influence bending fatigue resistance (Ref 29, 31).

The maximum applied surface tensile stress is the testing parameter plotted in S-N curves that characterize bending fatigue. However, the applied stress ranges between maximum and minimum values during a fatigue cycle, and two other parameters, mean stress and stress ratio, are important for the characterization of fatigue. The

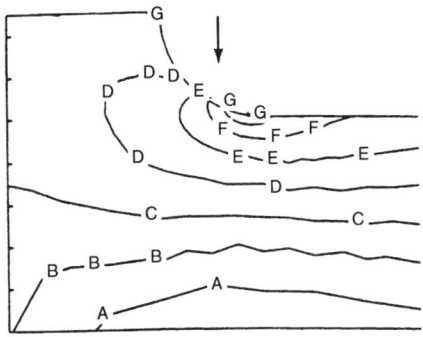

Fig. 6 Location of maximum stress on the cantilever bend specimen shown in Fig. 5 as determined by finite element modeling. Courtesy of K.A. Erven

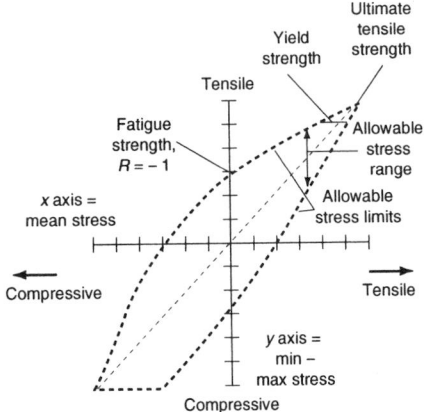

Fig. 7 Typical allowable-stress-range diagram. Source: Ref 32

mean stress is the algebraic average of the maximum and minimum stresses in a cycle, and as discussed above, the stress ratio, R, is the ratio of the minimum stress to the maximum stress in a cycle. Thus R values may range from −1, for fully reversed loading that ranges between equal maximum tensile and compressive stresses, to positive values where the stress is cycled between two tensile values (Ref 20). Much of the bend testing of cantilever specimens described below is performed with R values of 0.1 in order to preserve details of the fracture surface.

Figure 7 shows a typical allowable-stress diagram that plots fatigue strength versus mean stress for a given material (Ref 32). The diagram shows that the most severe condition for fatigue is for fully reversed testing with $R = -1.0$. As the mean stress increases, the fatigue strength in terms of maximum applied stress increases, but the allowable stress range decreases. Zurn and Razim (Ref 33) have examined the effect of notch severity and retained austenite on allowable-stress/mean-stress diagrams of carburized steels, and they conclude that carburizing, relative to the use of through-hardened steels, is especially effective for parts with sharp notches. In the absence of notches, carburizing is most suitable for parts subjected to fatigue loading at low values of mean stress.

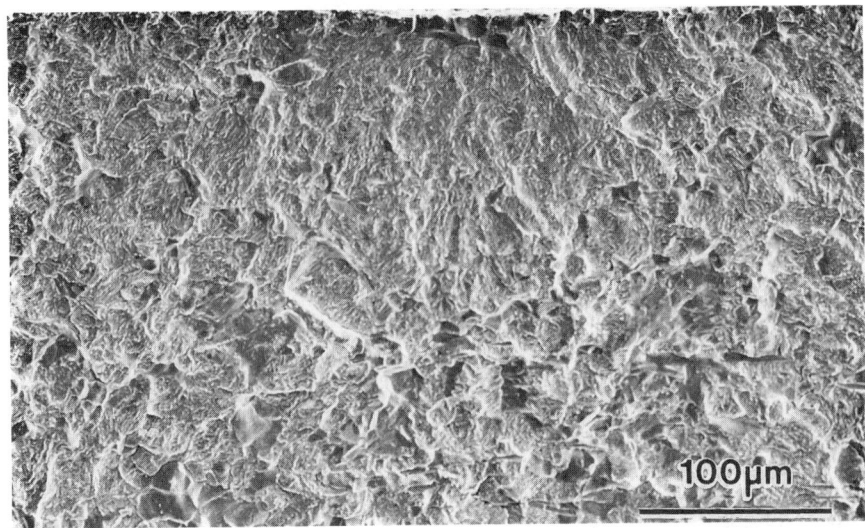

(a)

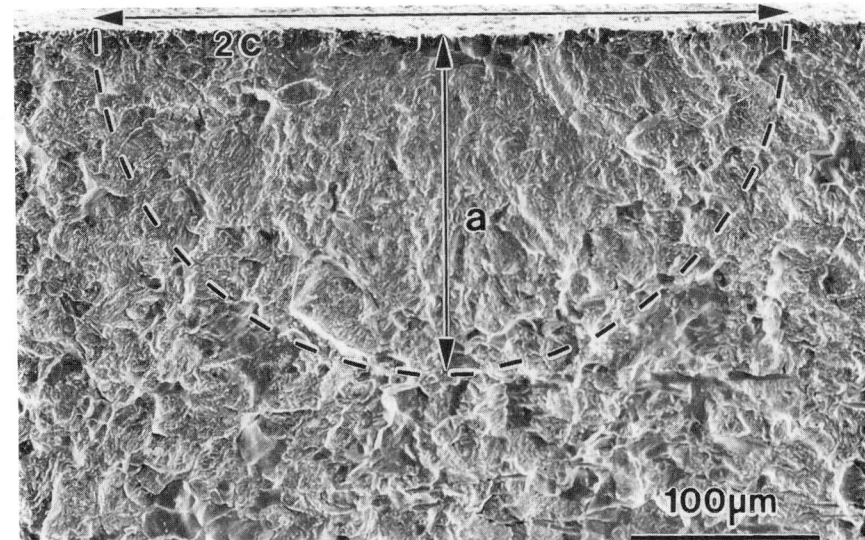

(b)

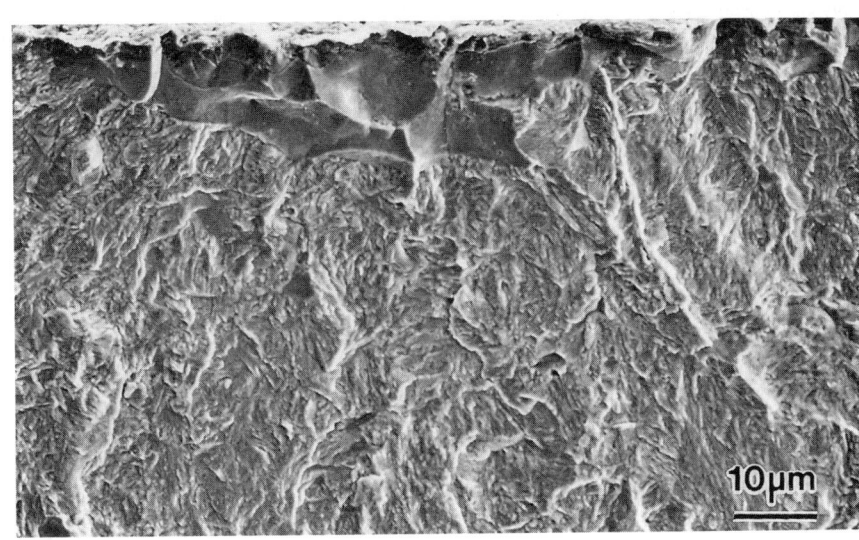

(c)

Fig. 8 Fatigue fracture in gas-carburized and modified 4320 steel. (a) Overview of initiation, stable crack propagation, and unstable crack propagation. (b) Same area as shown in (a), but with extent of stable crack indicated by dashed line. (c) Higher magnification of intergranular initiation and transgranular stable crack propagation areas. SEM micrographs. Source: Ref 36

Table 1 Fracture toughness results for carburized SAE 4320 bending fatigue specimens

Max stress, MPa	Cycles to failure	Depth, a, µm	Width, c, µm	a/c	M	K_{Ic}, MPa√m, Eq 1	K_{Ic}, MPa√m, Eq 2	K_{Ic}, MPa√m, Eq 3
1370	6400	175	388	0.45	0.87	30	29	22
	16,900	170	355	0.48	0.86	29	28	21
	15,300	200	210	0.95	0.78	18	18	12
1285	17,400	230	300	0.77	0.81	22	22	15
	18,700	230	295	0.78	0.81	22	22	15
1235	34,100	210	300	0.70	0.81	22	22	15
1160	21,700	230	295	0.78	0.81	19	19	13
	32,300	220	295	0.75	0.81	19	19	14

Equations used to determine the above are defined in text. Source: Ref 36

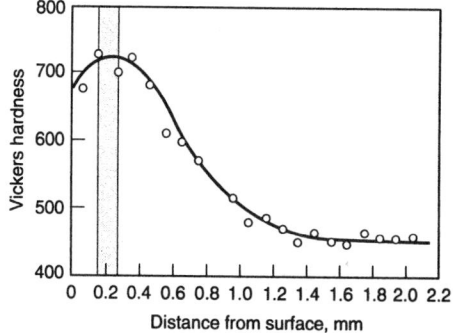

Fig. 9 Hardness vs. distance from the surface of direct-cooled gas-carburized SAE 4320 steel. Superimposed on the hardness profile is the range of critical depths (vertical dashed band) at which stable fatigue cracks became unstable in bending fatigue of similarly processed steels. Source: Ref 36

The testing of actual machine components is another important approach to the fatigue evaluation of carburized steels. An example of component testing is the bending fatigue testing of single teeth in gears (Ref 34). Gears are fabricated, carburized, and mounted in a fixture so that one tooth at a time is subjected to cyclic loading. Recently, identically carburized specimens of the same steel were subjected to cantilever bend and single tooth bending fatigue testing (Ref 35). The mechanisms of fatigue failure, based on fracture surface examination, were found to be the same, but the single tooth testing showed higher levels of fatigue resistance than did the cantilever testing, a result attributed to the higher surface compressive stresses that were measured in the gear tooth specimens.

Stages of Fatigue and Fracture

Bending fatigue fractures of carburized steels consist of well-defined stages of crack initiation, stable crack propagation, and unstable crack propagation. The fracture sequence is strongly influenced by the gradients in strength, microstructure, and residual stress that develop in carburized steels. Figure 8 shows a series of scanning electron microscope (SEM) fractographs that characterize the typical fracture sequence of a direct-quenched carburized steel with a near-surface case microstructure similar to that shown in Fig. 1. The cantilever bend specimen from which the fractographs of Fig. 8 were taken was a 4320 steel carburized to a 1 mm case depth at 927 °C (1700 °F), quenched from 850 °C (1560 °F) into oil at 65 °C (150 °F), and tempered at 150 °C (300 °F) for 1 h. The specimen was tested in bending fatigue with an R value of 0.1 (Ref 36). Figures 8(a) and (b) show a low-magnification overview of the initiation, stable propagation, and unstable fracture surfaces, and Fig. 8(c) shows the intergranular initiation and transgranular stable crack propagation zones of the fracture at a higher magnification.

Intergranular cracking at prior-austenite grain boundaries is an almost universal fracture mode in the high-carbon case of direct-quenched carburized steels (Ref 19, 23, 24, 36, 37). Not only do the fatigue cracks initiate by intergranular cracking, but also the unstable crack propagates largely by intergranular fracture until it reaches the lower-carbon portion of the case, where ductile fracture becomes the dominant fracture mode. In fact, sensitivity of the case microstructures to intergranular fracture makes possible the quantitative characterization of the size and shape of the stable fatigue crack, as shown in Fig. 8(b). The transition from the transgranular fracture of the stable crack to the largely intergranular fracture of the unstable fracture is identified by the dashed line.

A replica study of carburized specimens subjected to incrementally increasing stresses showed that surface intergranular cracks initiated when the applied stresses exceeded the endurance limits (Ref 36). Thus, it appears that in direct-quenched carburized specimens, intergranular cracks are initiated as soon as the applied surface bending stress reaches a level sufficient to exceed the surface compressive residual stress and the cohesive strength of the prior-austenite grain boundary structures. The surface intergranular cracks are shallow, typically on the order of two to four austenite grains, and are arrested, perhaps because of a plastic zone smaller than the grain size at the tip of the sharp intergranular cracks, and the fact that strain-induced transformation of retained austenite in the plastic zone ahead of the crack introduces compressive stresses (Ref 38, 39). The fatigue crack then propagates in a transgranular mode, and when the stable crack reaches critical size, as defined by the fracture toughness, unstable fracture occurs.

The initiation and stable crack zones of carburized steels are quite small and are often difficult to identify. Figure 9, based on the measurement of critical crack sizes in a number of direct-quenched carburized 4320 steels, shows that the size of the unstable cracks ranges from 0.170 to 0.230 mm, and that the cracks therefore become unstable well within the high-carbon portion of the carburized specimens. The small critical crack sizes are consistent with the low fracture toughness of high-carbon steel LTT microstructures susceptible to intergranular fracture (Ref 40). When the critical crack sizes and the stresses at which the cracks become unstable are used to calculate the fracture toughness of the case microstructures of carburized steels (Ref 36), the results show good agreement with the range of fracture toughness, 15 to 25 MPa√m, that has been measured from through-hardened specimens with high-carbon LTT martensitic micro-

structures (Ref 40). Table 1 shows the data used to calculate the various case fracture toughness values in gas-carburized 4320 steel and the fracture toughness values calculated according to three different fracture toughness equations (Ref 36) as follows:

$$K_{IC} = \frac{1.2\sigma_a\sqrt{a\pi}}{Q} \tag{Eq 1}$$

$$K_{IC} = \frac{\left[1.2\sigma_a + 0.683\left(\dfrac{\partial\sigma}{\partial x}\right)a\right]\sqrt{a\pi}}{Q} \tag{Eq 2}$$

and

$$K_{IC} = \frac{M\sigma_a\sqrt{a\pi}}{Q} \tag{Eq 3}$$

where

$$Q = \phi^2 - 0.212\left(\frac{\sigma_a}{\sigma_{ys}}\right)^2$$

with ϕ the aspect ratio of crack depth (a) and crack length (c) such that:

$$\phi^2 = 1 + 1.464\left(\frac{a}{c}\right)^{1.65}$$

The unstable crack that proceeds through the high- and medium-carbon martensitic portions of the case may be arrested when the sensitivity to intergranular fracture decreases at a case carbon content between 0.5 and 0.6% (Ref 41). The continued application of cyclic loading at this point then may cause a secondary stage of stable fatigue crack propagation, characterized by transgranular fracture, resolvable fatigue striations, and secondary cracking (Ref 23, 35, 37). This stage of low-cycle, high-strain fatigue is short and gives way to ductile overload fracture of the core.

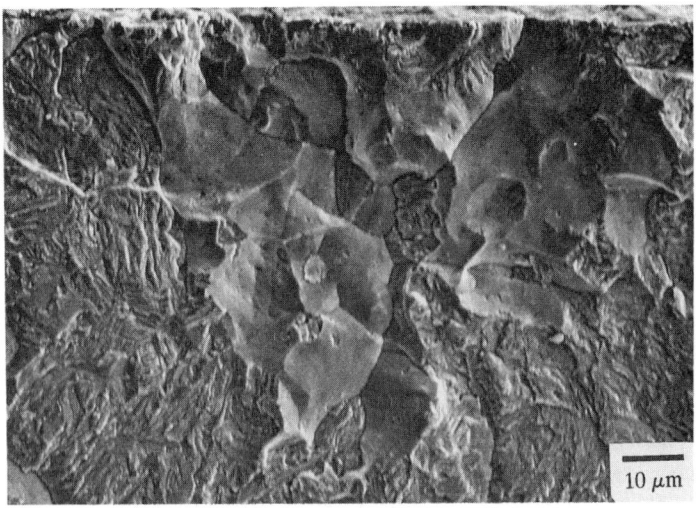

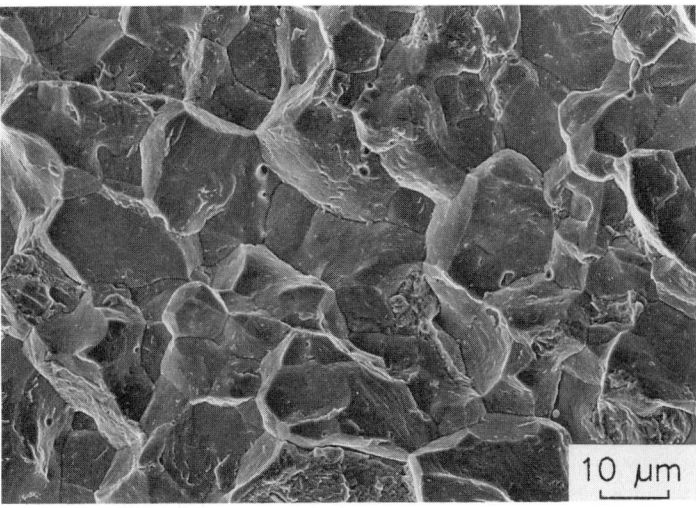

Fig. 10 Intergranular bending fatigue crack initiation at the surface of a gas-carburized and direct-cooled SAE 8219 steel specimen. Source: Ref 42

Fig. 11 Intergranular fracture in case unstable crack propagation zone in gas-carburized and direct-cooled SAE 4320 steel. Courtesy of A. Reguly

Intergranular Fracture of Carburized Steels

As noted above, intergranular fracture at the prior-austenite grain boundaries of high-carbon case microstructures dominates bending fatigue crack initiation and unstable crack propagation of direct-quenched carburized steels. The intergranular cracking may be associated with other microstructural features, such as the surface oxides generated by gas carburizing, but it generally extends much deeper into a carburized case than the oxide layer. Several studies have documented bending fatigue crack initiation by intergranular fracture even in the absence of surface oxidation, where for example the oxidized surface has been removed by chemical or electropolishing (Ref 24, 25) or no oxidation is present because the specimens were vacuum or plasma carburized (Ref 28).

Figure 10 shows an example of intergranular fatigue crack initiation in a direct-quenched specimen of gas-carburized type 8719 steel (Ref 42). There is a shallow zone of surface oxidation, about 10 μm in depth, but the intergranular cracking extends much deeper into the specimen. Figure 11 shows extensive intergranular cracking in the unstable crack propagation zone in the case of a direct-quenched, gas-carburized 4320 steel. Auger electron spectroscopy shows that such intergranular fracture surfaces have higher concentrations of phosphorus and carbon, in the form of cementite, than do transgranular fracture surfaces removed from prior-austenite grain boundaries (Ref 19, 43, 44). Thus, the brittle intergranular fracture that occurs in stressed high-carbon case microstructures of carburized steels is associated with the combined presence of segregated phosphorus and cementite at prior-austenite grain boundaries. These grain boundary structures are present in as-quenched specimens and do not require tempering for cementite formation, as is typical in the intergranular mode of tempered martensite embrittlement in medium-carbon

steels (Ref 15). This embrittlement, termed "quench embrittlement," is found in quenched steels with carbon contents as low as 0.6% (Ref 16). There is evidence that phosphorus segregation stimulates the formation of the grain boundary cementite (Ref 43, 44).

The higher the phosphorus content of a carburized steel, the lower its bending fatigue resistance and case fracture toughness. Figure 12 shows S-N curves for a series of gas-carburized and direct-quenched modified 4320 steels with systematic variations in phosphorus content from 0.031 to 0.005% (Ref 43). Endurance limits and low-cycle fatigue resistance increase with decreasing phosphorus content, but little difference is noted between the performance of the 0.005 and 0.017% phosphorus specimens. All of the specimens, even those with the lowest phosphorus content, failed by intergranular initiation of fatigue cracks.

The bending endurance limits of gas-carburized specimens in which fatigue is initiated by intergranular fracture typically range between 1050 and 1260 MPa (Ref 45, 46). This range is based on studies of cantilever bend specimens with good surface finish, rounded specimen corners, nominal amounts of surface oxidation, and loading at R = 0.1. Variations within this range may be due to variations in austenitic grain size, inclusion contents, retained austenite content, or residual stresses, as discussed below. Nevertheless, the common mechanism of bending fatigue crack initiation of direct-quenched specimens is intergranular fracture at embrittled grain boundaries in a microstructure of LTT martensite and retained austenite, as described in Fig. 1.

Carburized steels with high nickel content do not appear to be as susceptible to intergranular cracking as steels with low nickel content (Ref 14, 47, 48). Also, major changes in the case microstructures of carburized steel, such as those produced by reheating and described relative to Fig. 2, result in bending fatigue crack initiation sites other than embrittled prior-austenite grain boundaries. Microstructural conditions that pro-

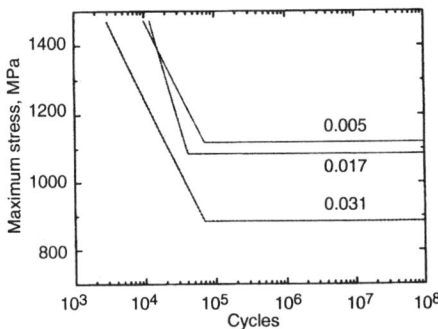

Fig. 12 Effect of phosphorus content on the bending fatigue of direct-quenched, gas-carburized modified 4320 steel with 0.005, 0.017, and 0.031 wt% phosphorus, as marked. Source: Ref 43

duce fracture initiation other than by intergranular cracking are also described below.

Austenitic Grain Size and Fatigue

Prior-austenite grain size of carburized steels correlates strongly with bending fatigue resistance. Generally, the finer the prior-austenite grain size, the better the fatigue performance. For example, Fig. 13 shows a direct relationship of bending fatigue endurance limit on prior-austenitic grain size, plotted as the inverse square root of the grain size, for several sets of carburized 4320 steels (Ref 49).

The refinement of austenite grain size has several effects on the case microstructure of carburized steels. A finer prior-austenite grain size produces a finer martensitic microstructure on quenching and therefore raises the strength of the carburized case. Increases in strength are beneficial to high-cycle fatigue resistance, as discussed above in the section on bending fatigue testing.

Another very important consequence of fine austenitic grain size is the dilution of the grain boundary segregation of phosphorus. In fact,

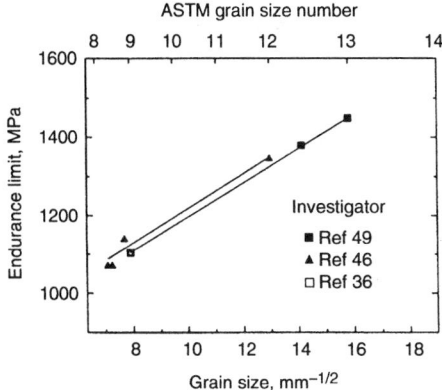

Fig. 13 Endurance limits as a function of prior-austenite grain size from various studies of bending fatigue of gas-carburized 4320 steels. Source: Ref 49

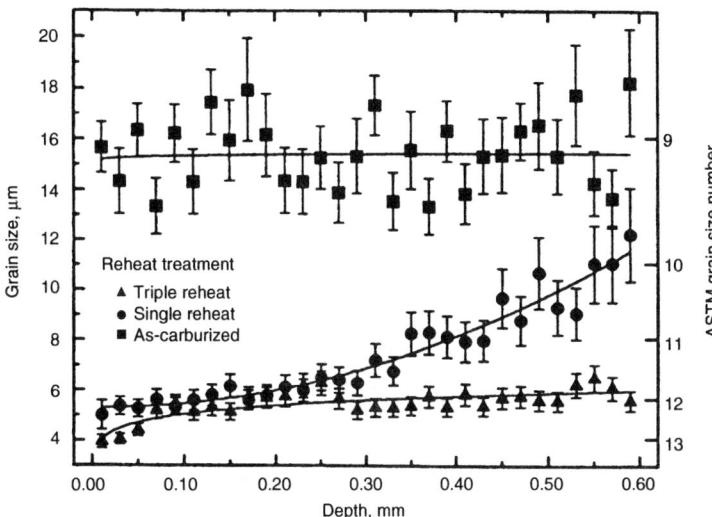

Fig. 14 Prior-austenite grain size as a function of depth from the surface of gas-carburized 4320 specimens in the as-carburized, direct-quenched condition and reheated conditions. Source: Ref 49

very fine austenitic grain sizes can eliminate the sensitivity of high-carbon case microstructures to intergranular fracture. As a result, other mechanisms of fatigue crack initiation replace intergranular cracking, generally to the benefit of fatigue performance. The high values of endurance limits shown for the very fine grain specimens in Fig. 13 were associated with fatigue crack nucleation at surface oxidation, not at embrittled prior-austenite grain boundaries. Although, as discussed below, surface oxides form on austenite grain boundaries and fatigue cracks may nucleate on the oxide-covered austenite boundaries, fine-grain specimens show no intergranular fracture below the oxidized surface layers.

The most effective way to produce very fine grains in carburized steels is by slow cooling and reheating of carburized parts at temperatures below the A_{cm} where austenite and cementite are stable. The cementite particles effectively retard austenite grain growth and reduce the carbon content of the austenite, causing the type of microstructure shown in Fig. 2 to form on quenching. Specimens reheated to above the A_{cm} may show grain refinement, depending on the temperature of heating, but because all carbides are dissolved, grain size refinement is not as effective as in specimens heated below the A_{cm}, and the type of microstructure shown in Fig. 1 develops upon quenching (Ref 24). Figure 14 shows austenite grain size as a function of distance from the carburized surface of gas-carburized 4320 steel specimens in the direct-quenched condition and after one and three reheating treatments (Ref 49). The reheat treatments very effectively reduce the near-surface case grain size where carbon content is the highest and therefore the greatest density of carbide particles is retained during intercritical reheating.

Intercritical-temperature reheat treatments of carburized steels produce very fine austenitic grain sizes and high endurance limits. Typically the endurance limits are above 1400 MPa (Ref 24, 28, 49). However, the beneficial effects of the reheating on bending fatigue resistance are not due to grain size refinement alone. The reduced

carbon content of the austenite when carbides are retained raises M_s temperatures and reduces the amount of retained austenite in the as-quenched case microstructures. Reduced levels of retained austenite raise the strength of case microstructures and therefore also may contribute significantly to the improved high-cycle fatigue performance of fine-grain, intercritically reheated carburized steels.

Surface Oxidation and Fatigue

Elements such as chromium, silicon, and manganese, commonly found in low-alloy carburizing steels, readily oxidize during gas carburizing as a result of H_2O/H_2 and CO/CO_2 equilibria in the carburizing gas atmospheres (Ref 4, 50). Other elements, such as molybdenum, nickel, and iron, do not oxidize. Figures 15 and 16 show the characteristics of the surface oxidation in gas-carburized specimens of 20MnCr5 steel containing 1.29% Mn, 0.44% Si, 1.25% Cr, and 0.25% Ni (Ref 51). The oxides grow into the steel and form two zones. In the deeper zone, silicon- and manganese-containing oxides grow on austenite grain boundaries. Figure 16, an SEM fractograph, shows that the grain boundary oxides grow with a lamellar morphology. In the shallower zone, chromium-rich oxides nucleate and grow as dispersed particles within austenite grains. The depths of the oxidized zones depend on the composition of the carburizing gas, the time and temperature of carburizing, and the chemistry of the steel (Ref 52, 53). Because the oxides penetrate into the steel, especially along austenite grain boundaries, surface oxidation is also referred to as internal oxidation or intergranular oxidation (IGO).

The surface oxidation produced during gas carburizing may or may not significantly reduce bending fatigue resistance. The most severe effects of such oxidation are associated with a reduction in near-surface case hardenability, which

results from the removal of chromium, manganese, and silicon from solution in the austenite by the oxide formation (Ref 54-56). The reduced case hardenability can cause nonmartensitic microstructures, such as ferrite, bainite, and pearlite, to form at the surface of the carburized steel. Not only is the surface hardness reduced, but the residual surface stresses may become less compressive or even tensile.

Figures 17 and 18 show the effects of surface oxidation with reduced hardenability on the bending fatigue and residual stresses of 8620 and 4615 gas-carburized specimens (Ref 57). The 4615 steel has higher hardenability by virtue of higher nickel and molybdenum contents and a lower sensitivity to surface oxidation by virtue of reduced manganese and chromium contents (Ref 57). As a result of the different chemistries, the 8620 steel formed pearlite in the near-surface regions of the case while the microstructure of the 4615 steel, despite some oxidation, consisted only of plate martensite and retained austenite at the surface (Ref 57). These differences in microstructures due to surface oxidation and reduced case hardenability are consistent with the differences in bending fatigue performance and residual stresses shown between the two steels in Fig. 17 and 18. This study illustrates the importance of steel chemistry on controlling surface oxidation and the associated formation of nonmartensitic microstructures in gas-carburized steels. Another approach used to reduce surface oxide formation in steels with low hardenability is to use more severe quenching with higher cooling rates.

If the hardenability of a steel is sufficient to prevent the formation of nonmartensitic microstructures for a given gas carburizing and quenching schedule, surface oxidation has a much reduced effect on bending fatigue performance. In direct-quenched specimens, as discussed above and demonstrated in Fig. 10, intergranular fracture to depths much deeper than the oxidized layers dominates fatigue crack initiation. How-

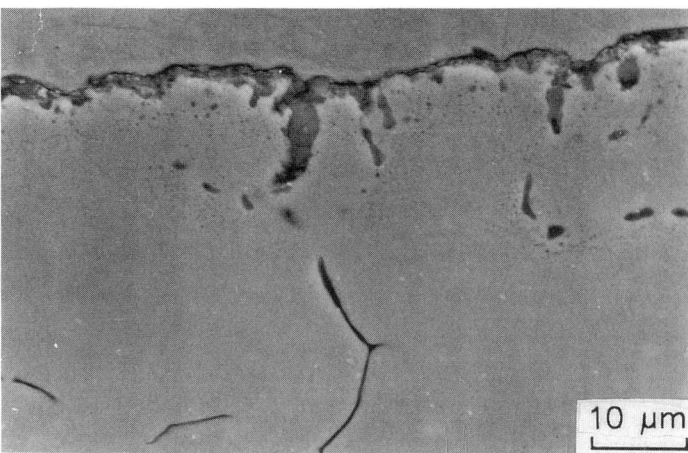

Fig. 15 Oxides (dark features) at surface of a gas-carburized 20 MnCr 5 steel containing 1.29% Mn, 0.44% Si, 1.25% Cr, 0.25% Ni, and 0.0015% B. SEM micrograph,

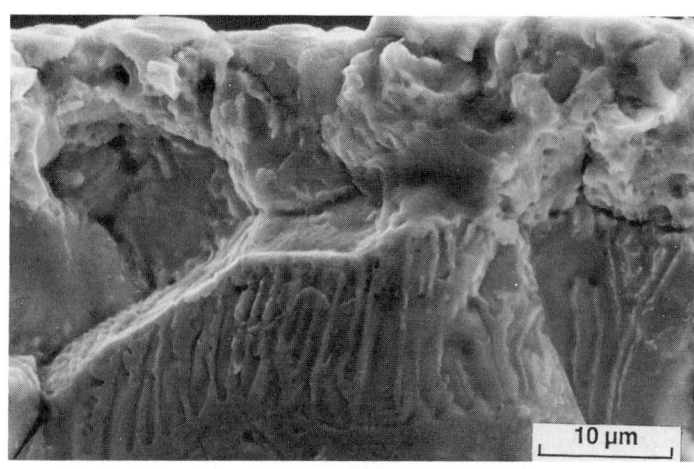

Fig. 16 Surface oxides on a fracture surface from a specimen of the steel identified in Fig. 15. Source: Ref 51

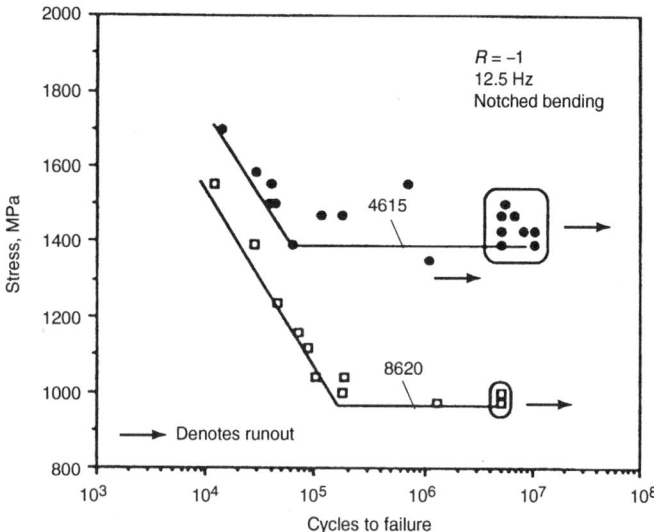

Fig. 17 S-N curves for direct-quenched gas-carburized 4615 and 8620 steels, notched 4-point bend specimens. Compositions of the steels are given in Table 2. Nonmartensitic transformation products were present on the surfaces of the 8620 steel specimens and absent on the 4615 steel specimens. Source: Ref 57

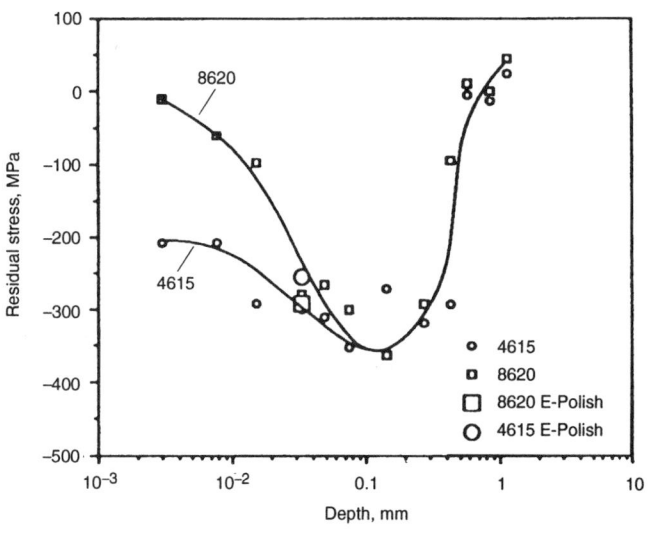

Fig. 18 Residual stress as a function of depth below the surface of the direct-quenched gas-carburized 4615 and 8620 steel specimens described in Fig. 17. Source: Ref 57

ever, when the conditions for intergranular crack initiation are minimized, as for example by reheating (Ref 24, 49) or shot peening (Ref 58), the surface oxide layers become a major location for bending fatigue crack initiation. Figure 19 shows crack initiation in the oxidized zone of a gas-carburized and reheated specimen of 4320 steel. The initiation is confined to the oxidized zone, and stable transgranular fatigue propagation proceeds directly below the oxidized zone with no evidence of intergranular fracture.

Retained Austenite and Fatigue

Next to LTT martensite, retained austenite is the most important microstructural component in the case of carburized steels. The amounts of retained austenite vary widely, depending on carbon and alloy content, heat-treating condi-

tions, and special processing steps such as shot peening and subzero cooling (Ref 9, 15). Generally, the higher the carbon and alloy content, the lower the M_s temperature and the higher the retained austenite content in the microstructure. The low-temperature tempering applied to carburized steels, generally performed at temperatures below 200 °C (400 °F), is not high.enough to cause the retained austenite to transform, and therefore retained austenite remains an important component of the microstructure. At higher tempering temperatures, retained austenite transforms to cementite and ferrite with attendant decreases in hardness and strength as the martensitic microstructure coarsens (Ref 15).

The role that retained austenite plays in the bending fatigue performance of carburized steels has been difficult to identify because of the variable loading conditions that may be applied to carburized machine components and the compli-

cating effects of other factors, such as residual stresses, grain boundary embrittling structures, and surface oxidation. With respect to loading conditions, it appears that higher amounts of retained austenite are detrimental to high-cycle fatigue and reduce endurance limits (Ref 28, 33, 38), while higher amounts of retained austenite are beneficial for low-cycle, high-strain fatigue (Ref 38, 59, 60).

Reduced retained austenite contents of LTT martensite/austenite composite microstructures increase elastic limits and yield strengths (Ref 61) and therefore benefit stress-controlled, high-cycle fatigue. One of the approaches to reducing the retained austenite content in the case microstructures of carburized steels, as noted above, is to reheat carburized specimens to temperatures below the A_{cm} and quench to produce the type of microstructure shown in Fig. 2. Invariably such reheating and quenching significantly increases

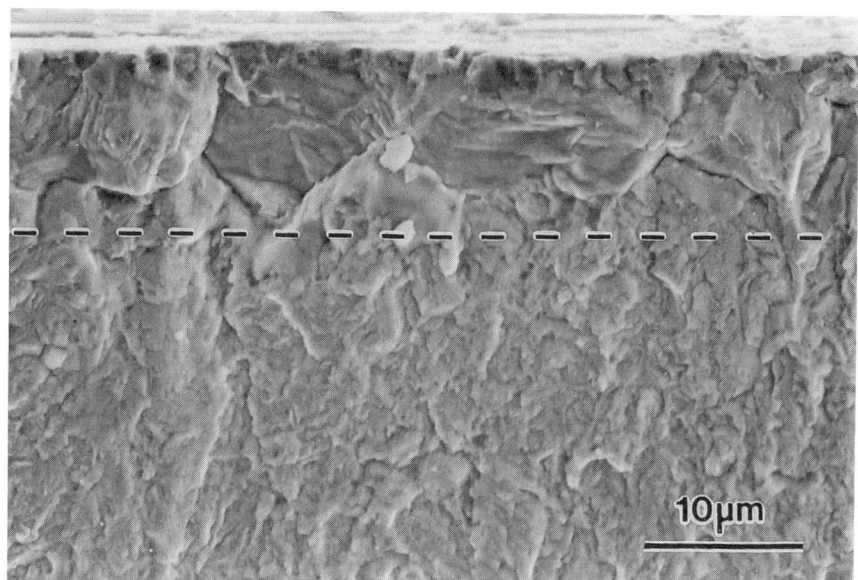

Fig. 19 Bending fatigue crack initiation in gas-carburized and reheated 4320 steel. The dashed line corresponds to maximum depth of surface oxidation, and all fracture below dashed line is transgranular. Source: Ref 49

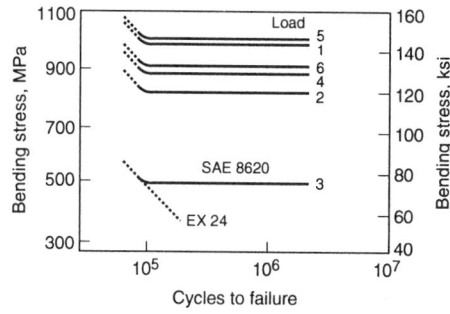

Fig. 20 S-N curves of vacuum-carburized 8620 and EX 24 (0.89% Mn, 0.24% Mo, 0.55% Cr) steels. The lower curves were obtained from specimens subzero cooled to –196 °C and the upper curves were obtained from specimens not subjected to subzero cooling. Source: Ref 31

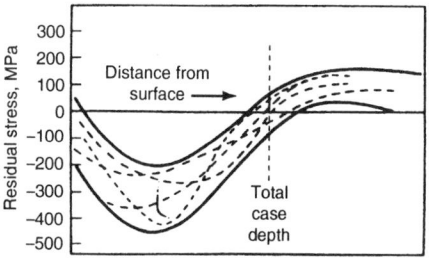

Fig. 21 Ranges and patterns of residual stresses as a function of depth for 70 carburized steels. Source: Ref 5

bending fatigue endurance limits compared to direct-quenched specimens of identically carburized specimens (Ref 24, 49). The reheating not only reduces retained austenite but also refines the austenitic grain size, refines the martensitic structure, and reduces susceptibility to intergranular fracture, all features that are known to improve fatigue resistance. Therefore, improved high-cycle fatigue resistance of reheated and quenched specimens is related to a combination of microstructural changes, including low retained austenite contents.

The benefit of retained austenite to strain-controlled, low-cycle bending fatigue is related to the improved ductility and reduced strength and hardness that retained austenite contributes to a composite LTT martensite/austenite case microstructure. In addition, retained austenite, at sufficiently high applied strains and stresses, undergoes deformation-induced transformation to martensite (Ref 62). The volume expansion associated with the strain-induced formation of martensite creates compressive stresses (Ref 39) that lead to reduced rates of fatigue crack growth, accounting for the enhanced low-cycle fatigue performance that is observed in carburized steels with high amounts of retained austenite in the case (Ref 60).

Subzero Cooling and Fatigue

Cooling carburized steel below room temperature is a processing approach sometimes used to reduce the retained austenite content in the case regions of carburized steels. The transformation of austenite to martensite is driven by temperature changes, and the low M_s temperatures of high-carbon case regions of carburized alloy steels limit the temperature range between M_s and room temperature over which martensite forms (Ref 15). Therefore, the temperature range for

martensite formation and the reduction of retained austenite is extended by cooling below room temperature. The cooling treatments are variously referred to as subzero cooling, refrigeration treatments, or deep cooling.

In addition to the effects of retained austenite on bending fatigue, as discussed above, any deformation-induced transformation of retained austenite during cyclic loading in service, because of the volume expansion that accompanies the transformation of austenite to martensite, may change the dimensions of a carburized component. Therefore, subzero cooling is one approach to reduce retained austenite in parts that require high precision and stable dimensions throughout their service life. However, several studies show that subzero cooling lowers the bending fatigue resistance of carburized steels. Nevertheless, high-quality, high-performance aircraft and helicopter gears are routinely subjected to subzero cooling without apparent detrimental effects (Ref 63). For example, a commonly used carburizing steel for aircraft gears is 9310, which contains about 3 wt% nickel. The high nickel content lowers the M_s temperature and increases the amount of austenite at room temperature. The austenite content can be reduced by subzero cooling, probably with adverse effects on localized residual stress, as discussed below. However, the latter adverse effect of subzero cooling may be offset by fine austenite grain size, and high nickel content may improve the fracture toughness and fatigue resistance of carburized steels (Ref 14, 47, 48, 64) to a level where fatigue resistance is not adversely affected by subzero cooling.

In the low-alloy steels commonly used for carburizing, subzero cooling used to reduce retained austenite may reduce the bending fatigue resistance of carburized components. According to Parrish and Harper (Ref 5), refrigeration treatments should be "considered as a last resort," and if they are applied, parts should be tempered both

before and after. Figure 20 shows an example of the detrimental effects of subzero cooling on the bending fatigue resistance of carburized specimens. The data were produced in an experimental study of vacuum-carburized specimens of 8620 and EX 24 steel that were deep cooled to –196 °C in liquid nitrogen (Ref 31). The overall fatigue performance in this study was complicated by high retained austenite contents and coarse carbide particles at square specimen corners, but the detrimental effect of subzero cooling on bending fatigue performance is clearly demonstrated in Fig. 20.

The detrimental effect of subzero cooling on the bending fatigue of carburized specimens has been related to changes in residual stress by several investigations (Ref 23, 65, 66). The overall surface residual stresses become increasingly compressive, as measured from the martensite in the case and as expected from the constraint of the expansion that accompanies the transformation of austenite to martensite as temperature is decreased. However, the residual stresses in the austenite phase are measured to be tensile, especially at the surface of the carburized specimens. These tensile stresses would then be expected to lower the surface tensile stresses applied in bending to initiate fatigue cracks. Microcrack formation within martensite plates and at plate/austenite interfaces (Ref 67) may be enhanced by the localized residual stresses induced by subzero cooling (Ref 66), but they could be minimized by maintaining a fine prior-austenite

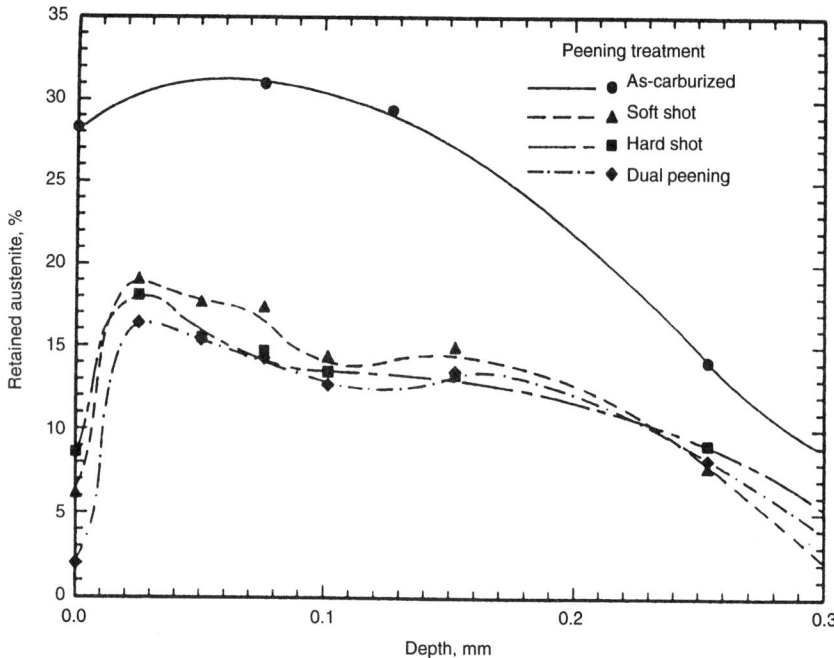

Fig. 22 Retained austenite as a function of depth below the carburized surface for gas-carburized 4320 specimens in the as-carburized, direct-quenched and various shot-peened conditions after direct quenching. Source: Ref 58

and they are positively modified (made locally more compressive) by the strain-induced transformation of austenite to martensite. Tempering lowers residual compressive stresses because of dimensional changes that accompany the recovery and coarsening of the martensitic microstructure during tempering (Ref 75).

Shot peening is an effective way to increase the case compressive residual stresses in carburized steels (Ref 52, 76-78) and as a result improve the bending fatigue performance. Shot peening causes deformation-induced transformation of case retained austenite, and the constraint of the associated volume expansion causes the development of additional compressive stresses. Figures 22 to 24 show, relative to unpeened specimens, the decrease in retained austenite, the increase in the case compressive stresses, and the increased bending fatigue performance, respectively, that are associated with shot peening of direct-quenched carburized 4320 specimens (Ref 58).

grain size and applying reheating treatments (Ref 68, 69).

Residual Stresses and Shot Peening

Compressive residual stresses are formed in the case microstructures of carburized steels as a result of transformation and temperature gradients induced by quenching (Ref 17, 18). The magnitude and distribution of the residual stresses therefore are complex functions of the temperature gradients induced by quenching (Ref 70), which in turn are dependent on specimen size and geometry, the hardenability of the steel, the carbon gradient, and the case depth. The residual stresses as a function of case depth are routinely

measured by x-ray diffraction (Ref 71), and considerable effort has been applied to modeling residual stress profiles in carburized steels as a function of cooling and hardenability (Ref 72-74).

Figure 21 shows the range and pattern of compressive residual stresses typically formed in the case regions of direct-cooled carburized steels (Ref 5). The compressive residual stresses offset the adverse effects of factors such as quench embrittlement and intergranular fracture to which high-carbon microstructures are susceptible (Ref 16), and they increase the fracture and fatigue resistance of direct-quenched parts to levels that provide good engineering performance. As discussed above, case residual stresses in carburized steels are adversely modified by subzero cooling,

REFERENCES

1. A.K. Sharma, G.H. Walter, and D.H. Breen, An Analytical Approach for Establishing Case Depth Requirements in Carburized Gears, *J. Heat Treating*, Vol 1 (No. 1), 1979, p 48-57
2. R.E. Cohen, P.J. Haagensen, D.K. Matlock, and G. Krauss, "Assessment of Bending Fatigue Limits for Carburized Steel," Technical Paper 910140, SAE International, 1991
3. ASM Committee on Gas Carburizing, *Carburizing and Carbonitriding*, American Society for Metals, 1977
4. G. Parrish, *The Influence of Microstructure on the Properties of Case-Carburized Components*, American Society for Metals, 1980
5. G. Parrish and G.S. Harper, *Production Gas Carburizing*, Pergamon Press, Oxford, 1985
6. D.V. Doane, Carburized Steel—Update on a Mature Composite, *Carburizing: Processing and Performance*, G. Krauss, Ed., ASM International, 1989, p 169-190

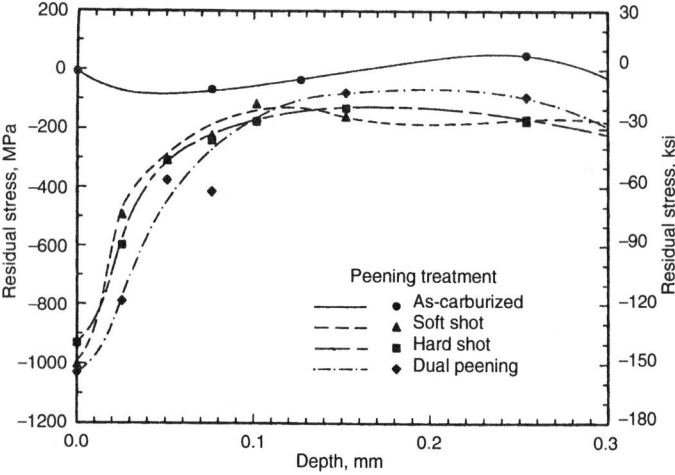

Fig. 23 Residual stress profiles for the specimens described in Fig. 22. Source: Ref 58

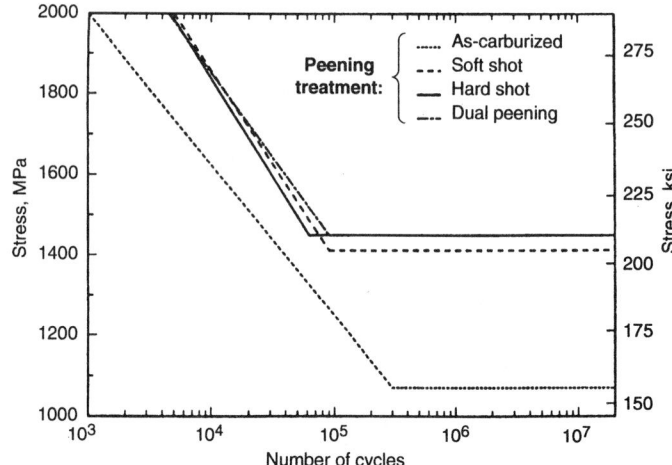

Fig. 24 *S-N* curves for the specimens described in Fig. 22. Source: Ref 58

7. G. Krauss, Microstructure and Properties of Carburized Steels, *Heat Treating*, Vol 4, *ASM Handbook*, ASM International, 1991, p 363-375

8. J. Grosch, H. Bomas, D. Liedtke, and H. Streng, *Einsatzharten*, Expert Verlag, Renningen-Malmsheim, Germany, 1994

9. G. Krauss, "Martensite," "Austenite," "Austenite and Fatigue," and "Oxidation and Inclusions," a 4-part series of articles published in *Advanced Materials and Processes*, May, July, September, and December, 1995

10. G. Wahl, Development and Application of Salt Baths in the Heat Treatment of Case Hardening Steels, *Carburizing: Processing and Performance*, G. Krauss, Ed., ASM International, 1989, p 41-56

11. J. St. Pierre, Recent Developments in Vacuum Carburizing, *Carburizing: Processing and Performance*, G. Krauss, Ed., ASM International, 1989, p 31-39

12. W.L. Grube and J.G. Gay, High-Rate Carburizing in a Glow-Discharge Methane Plasma, *Met. Trans. A*, Vol 9, 1978, p 1421-1429

13. C.A. Stickels and C.M. Mack, Overview of Carburizing Processes and Modeling, *Carburizing: Processing and Performance*, G. Krauss, Ed., ASM International, 1989, p 1-9

14. J. Grosch, Fundamentals of Carburizing and Toughness of Carburized Components, *Quenching and Carburising*, The Institute of Metals, London, 1993, p 227-249

15. G. Krauss, *Steels: Heat Treatment and Processing Principles*, ASM International, 1990

16. G. Krauss, Heat Treated Martensitic Steels: Microstructural Systems for Advanced Manufacture, *ISIJ International*, Vol 35 (No. 4), 1995, p 349-359

17. D.P. Koistinen, The Distribution of Residual Stresses in Carburized Cases and Their Origin, *Trans. ASM*, Vol 50, 1958, p 227-241

18. L.J. Ebert, The Role of Residual Stresses in the Mechanical Performance of Case Carburized Steel, *Met. Trans. A*, Vol 9, 1978, p 1537-1551

19. G. Krauss, The Microstructure and Fatigue of a Carburized Steel, *Met. Trans. A*, Vol 9, 1978, p 1527-1535

20. H.O. Fuchs and R.I. Stephens, *Metal Fatigue in Engineering*, John Wiley & Sons, 1980

21. J. Bannantine, J. Comer, and J. Handrock, *Fundamentals of Metal Fatigue Analysis*, Prentice-Hall, 1990

22. R.W. Landgraf and R.H. Richman, Fatigue Behavior of Carburized Steel, *Fatigue of Composite Materials*, STP 569, American Society for Testing and Materials, 1975, p 130-144

23. M.A. Panhans and R.A. Fournelle, High Cycle Fatigue Resistance of AISI E9310 Carburized Steel with Two Different Levels of Surface Retained Austenite and Surface Residual Stress, *J. Heat Treating*, Vol 2 (No. 1), 1981, p 54-61

24. C.A. Apple and G. Krauss, Microcracking and Fatigue in a Carburized Steel, *Met. Trans.*, Vol 4, 1973, p 1195-1200

25. L. Magnusson and T. Ericsson, Initiation and Propagation of Fatigue Cracks in Carburized Steel, *Heat Treatment '79*, The Metals Society, London, 1979, p 202-206

26. R. Sieber, "Bending Fatigue Performance of Carburized Gear Steels," Technical Paper 920533, SAE International, 1992

27. H. Brugger and G. Kraus, Influence of Ductility on the Behavior of Carburizing Steel during Static and Dynamic Bend Testing, *Archiv Eisenhuttenwesen*, Vol 32, 1961, p 529-539

28. J.L. Pacheco and G. Krauss, Microstructure and High Bending Fatigue Strength in Carburized Steel, *J. Heat Treating*, Vol 7 (No. 2), 1989, p 77-86

29. R.E. Cohen, D.K. Matlock, and G. Krauss, Specimen Edge Effects on Bending Fatigue of Carburized Steel, *J. Materials Engineering and Performance*, Vol 1 (No. 5), 1992, p 695-703

30. L. Magnusson and T. Johannesson, Fatigue of Case Hardened Low Alloyed Steel, *Scandinavian J. Metallurgy*, Vol 6, 1977, p 40-41

31. K.D. Jones and G. Krauss, Effects of High-Carbon Specimen Corners on Microstructure and Fatigue of Partial Pressure Carburized Steels, *Heat Treatment '79*, The Metals Society, London, 1979, p 188-193

32. D.H. Breen and E.M. Wene, Fatigue in Machines and Structures—Ground Vehicles, *Fatigue and Microstructure*, American Society for Metals, 1979, p 57-99

33. Z. Zurn and C. Razim, On the Fatigue Strength of Case Hardened Parts, *Carburizing: Processing and Performance*, G. Krauss, Ed., ASM International, 1989, p 239-248

34. M.B. Slane, R. Buenneke, C. Dunham, M. Semenek, M. Shea, and J. Tripp, "Gear Single Tooth Bending Fatigue," Technical Paper 821042, SAE International, 1982

35. D. Medlin, G. Krauss, D.K. Matlock, K. Burris, and M. Slane, "Comparison of Single Gear Tooth and Cantilever Beam Bend Fatigue Testing of Carburized Steel," Technical Paper 950212, SAE International, 1995

36. R.S. Hyde, R.E. Cohen, D.K. Matlock, and G. Krauss, "Bending Fatigue Crack Characterization and Fracture Toughness of Gas Carburized SAE 4320 Steel," Technical Paper 920534, SAE International, 1992

37. R.D. Zipp and G.H. Walter, A Fractographic Study of High Cycle Fatigue Fractures in Carburized Steel, *Metallography*, Vol 7, 1975, p 77-81

38. M.A. Zaccone, J.B. Kelley, and G. Krauss, Strain Hardening and Fatigue of Simulated Case Microstructures in Carburized Steel, *Carburizing: Processing and Performance*, G. Krauss, Ed., ASM International, 1989, p 249-265

39. M.M. Shea, "Impact Properties of Selected Gear Steels," Technical Paper 780772, SAE International, 1978

40. G. Krauss, The Relationship of Microstructure to Fracture Morphology and Toughness of Hardened Hypereutectoid Steels, *Case-Hardened Steels: Microstructural and Residual Stress Effects*, D.E. Diesburg, Ed., TMS-AIME, 1984, p 33-58

41. R.S. Hyde, "Quench Embrittlement and Intergranular Oxide Embrittlement: Effects on Bending Fatigue Initiation of Gas-Carburized Steel," Ph.D. dissertation, Colorado School of Mines, 1994

42. K.A. Erven, D.K. Matlock, and G. Krauss, Effect of Sulfur on Bending Fatigue of Carburized Steel, *J. Heat Treating*, Vol 9, 1991, p 27-35

43. R.S. Hyde, G. Krauss, and D.K. Matlock, Phosphorus and Carbon Segregation: Effects on Fatigue and Fracture of Gas-Carburized Modified 4320 Steel, *Met. Trans. A*, Vol 25, 1994, p 1229-1240

44. T. Ando and G. Krauss, The Effect of Phosphorus Content on Grain Boundary Cementite Formation in AISI 52100 Steel, *Met. Trans. A*, Vol 12, 1981, p 1283-1290

45. K.A. Erven, D.K. Matlock, and G. Krauss, Bending Fatigue and Microstructure of Gas-Carburized Alloy Steels, *Materials Science Forum*, Vol 102-104, 1992, p 183-198

46. R.E. Cohen, G. Krauss, and D.K. Matlock, "Bending Fatigue Performance of Carburized 4320 Steel," Technical Paper 930963, SAE International, 1993

47. D. Wicke and J. Grosch, Das Festigkeitsverhalten von Legierten Einsatzstahlen bei Schlagbeanspruchung, *Harterei-Tech. Mitt.*, Vol 32, 1977, p 223-233

48. B. Thoden and J. Grosch, Crack Resistance of Carburized Steel under Bend Stress, *Carburizing: Processing and Performance*, G. Krauss, Ed., ASM International, 1989, p 303-310

49. R.S. Hyde, D.K. Matlock, and G. Krauss, "The Effect of Reheat Treatments on Fatigue and Fracture of Carburized Steel," Technical Paper 940788, SAE International, 1994

50. R. Chatterjee-Fischer, Internal Oxidation during Carburizing and Heat Treatment, *Met. Trans. A*, Vol 9, 1978, p 1553-1560

51. C. Van Thyne and G. Krauss, A Comparison of Single Tooth Bending Fatigue in Boron and Alloy Carburizing Steels, *Carburizing: Processing and Performance*, G. Krauss, Ed., ASM International, 1989, p 333-340

52. K. Namiki and A. Hatano, High Tough and High Fatigue Strength Carburizing Steel, *Heat and Surface '92*, Japan Technical Information Service, Tokyo, 1992, p 361-364

53. B. Edenhofer, Progress in the Technology and Applications of In-situ Atmosphere Production in Hardening and Case-Hardening Furnaces, *1995 Carburizing and Nitriding with Atmospheres*, J. Grosch, J. Morral, and M. Schneider, Ed., ASM International, 1995, p 37-42

54. B, Hildenwall and T. Ericsson, Residual Stresses in the Soft Pearlite Layer of Carburized Steel, *J. Heat Treat.*, Vol 1 (No. 3), 1980, p 3-13

55. S. Gunnarson, Structure Anomalies in the Surface Zone of Gas-Carburized Case-Hardened Steel, *Met. Treat. Drop Forg.*, Vol 30, 1963, p 219-229

56. T. Naito, H. Ueda, and M. Kikuchi, Fatigue Behavior of Carburized Steel with Internal Oxides and Nonmartensitic Microstructures near the Surface, *Met. Trans. A*, Vol 15, 1984, p 1431-1436

57. W.E. Dowling, Jr., W.T. Donlon, W.B. Copple, and C.V. Darragh, Fatigue Behavior of Two Carburized Low Alloy Steels, *1995 Carburizing and Nitriding with Atmospheres*, J. Grosch, J. Morral, and M. Schneider, Ed., ASM International, 1995, p 55-60

58. J.A. Sanders, "The Effects of Shot Peening on the Bending Fatigue Behavior of a Carburized SAE 4320 Steel," M.S. thesis, Colorado School of Mines, 1993

59. C. Razim, Uber den Einfluss von Restaustenit auf des Festigkeitsverhalten Einsatzgeharteter Probenkorper bei Schwingender Beanspruchung, *Harterei-Tech. Mitt.*, Vol 23, 1968, p 1-8

60. R.H. Richman and R.W. Landgraf, Some Effects of Retained Austenite on the Fatigue Resistance of Carburized Steel, *Met. Trans. A*, Vol 6, 1975, p 955-964

61. M. Zaccone and G. Krauss, Elastic Limit and Microplastic Response of Hardened Steels, *Met. Trans. A*, Vol 24, 1993, p 2263-2277

62. G.B. Olson, Transformation Plasticity and the Stability of Plastic Flow, *Deformation, Processing, and Structure*, G. Krauss, Ed., ASM International, 1984, p 391-424

63. Final Report, Advanced Rotorcraft Transmission Program, National Aeronautics and Space Administration, Cleveland, Ohio

64. R.J. Johnson, The Role of Nickel in Carburizing Steels, *Metals Engineering Quarterly*, Vol 15

(No. 3), 1975, p 1-8

65. C. Kim, D.E. Diesburg, and R.M. Buck, Influence of Sub-zero and Shot-Peening Treatment on Impact and Fatigue Fracture Properties of Case Hardened Steels, *J. Heat. Treat.*, Vol 2 (No. 1), 1981, p 43-53

66. J. Grosch and O. Schwarz, Retained Austenite and Residual Stress Distribution in Deep Cooled Carburized Microstructures, *1995 Carburizing and Nitriding with Atmospheres*, J. Grosch, J. Morral, and M. Schneider, Ed., ASM International, 1995, p 71-76

67. A.R. Marder, A.O. Benscoter, and G. Krauss, Microcracking Sensitivity in Fe-C Plate Martensite, *Met. Trans.*, Vol 1, 1970, p 1545-1549

68. R.P. Brobst and G. Krauss, The Effect of Austenite Grain Size on Microcracking in Martensite of an Fe-1.22C Alloy, *Met. Trans.*, Vol 5, 1975, p 457-462

69. M.G. Mendiratta, J. Sasser, and G. Krauss, Effect of Dissolved Carbon on Microcracking in Martensite of an Fe-1.39 pct C Alloy, *Met. Trans.*, Vol 3, 1972, p 351-352

70. B. Liscic, State of the Art in Quenching, *Quenching and Carburizing*, The Institute of Materials, London, 1993, p 1-32

71. "Residual Stress Measurement by X-ray Diffraction," SAE Handbook Supplement J784a, SAE, 1971, p 19

72. T. Ericsson, S. Sjostrom, M. Knuuttila, and B.

Hildenwall, Predicting Residual Stresses in Cases, *Case-Hardened Steels: Microstructural and Residual Stress Effects*, D.E. Diesburg, Ed., TMS-AIME, 1984, p 113-139

73. A.K. Hellier, M.B. McGirr, S.H. Alger, and M. Stefulji, Computer Simulation of Residual Stresses during Quenching, *Quenching and Carburizing*, The Institute of Materials, London, 1993, p 127-138

74. *Quenching and Distortion Control*, G. Totten, Ed., ASM International, 1992

75. G. Krauss, Microstructure, Residual Stresses and Fatigue of Carburized Steels, *Quenching and Carburizing*, The Institute of Materials, London, 1993, p 205-225

76. B. Scholtes and E. Macherauch, Residual Stress Determination, *Case-Hardened Steels: Microstructural and Residual Stress Effects*, D.E. Diesburg, Ed., TMS-AIME, 1984, p 141-159

77. K. Naito, T. Ochi, T. Takahashi, and N. Suzuki, Effect of Shot Peening on the Fatigue Strength of Carburized Steels, *Proc. Fourth International Conference on Shop Peening*, The Japan Society of Precision Engineering, Tokyo, 1990, p 519-526

78. A. Inada, H. Yaguchi, and T. Inoue, The Effects of Retained Austenite on the Fatigue Properties of Carburized Steels, *Heat and Surface '92*, Japan Technical Information Service, Tokyo, 1992, p 409-412

Contact Fatigue of Hardened Steel

R. Scott Hyde, The Timken Company

CONTACT FATIGUE is the cracking of a surface subjected to alternating Hertzian stresses (Ref 1) produced under combined rolling and sliding loading conditions. In addition to cracking, contact fatigue can result in microstructural alterations, including changes in retained austenite, residual stress, and martensite morphology. When traction forces (frictional) are negligible, this phenomenon is known as rolling contact fatigue. However, frictional forces are always present and thus contact fatigue will be the term used in this article to describe this phenomenon. Contact fatigue is encountered most often in rolling-element bearings and gears, where the surface stresses are high due to concentrated loads that are repeated many times during normal operation. The mechanism of contact fatigue can be understood in terms of several sources of stress concentration, or stress raisers, within the macroscopic Hertzian stress field.

The location of contact fatigue initiation is dependent on the applied normal load, the magnitude of surface tractive forces, and type and location of stress raisers, resulting in surface or subsurface fatigue initiation. Surface initiation is usually controlled by asperities, oxides, defects, and geometry, whereas subsurface initiation is due to inclusions, carbides, defects, and grain boundaries. Other factors that influence fatigue initiation and propagation are microstructure/hardness and residual stress profile.

It is useful to classify contact fatigue damage according to a system that is related to origin, propagation mode, and macroscopic appearance of advanced stages of damage. Material and environmental factors coupled with load, contact geometry, surface microtopography, and conditions of lubrication are the primary variables that combine to determine the predominant mode(s) of contact fatigue damage.

Component damage due to contact fatigue can be reduced through enhancements in lubrication, steel cleanness, alloy selection, surface finish, geometry and design, and manufacturing techniques. These enhancements focus on providing greater resistance to contact fatigue, thus translating into a more power-dense component or system by improving life or increasing load-carrying capabilities (Ref 2). Improved contact fatigue resistance of bearings and gears depends on the

mechanisms of contact fatigue, the controlling metallurgical and tribological factors, and the use of improved, cleaner materials that have been properly heat treated to the optimal hardness and correct microstructure.

Cyclic Contact Stresses

Contact stresses are caused by the pressure of one solid in contact with another solid over limited areas of contact (Ref 3). Most machine components are designed on the basis of stress in the main body of the member, that is, in portions of the body not affected by the localized stresses. In other words, damage to most mechanical components is associated with stresses and strains in portions of the component far removed from the points of application of the loads. However, in certain situations the contact stresses between the surfaces of two externally loaded bodies (e.g., meshing gear teeth or contact between the roller or ball and its race in a bearing) can be the significant stresses; that is, the stresses on or somewhat below the surface of the contact are the major causes of damage to one or both of the bodies. Therefore, analysis of cyclic (Hertzian) stresses provides important information about surface and subsurface stresses for static loading of concentrated contacts, particularly the location of the maximum shear stress below the surface. The following review on dynamic loads and stress variation under rolling/sliding conditions is thus important to the understanding of cyclic stresses that cause contact fatigue.

Hertzian Shear Stresses at and below the Contact Surfaces. Damage caused by contact fatigue is the result of cyclic shear stresses developed at or near contact surfaces during operation. The effects on the shear stress field below the surface of a bearing raceway as a cylindrical roller travels across the surface is shown schematically in Fig. 1 (Ref 4). The initial location of the roller is shown at position 1 (Fig. 1a). The material directly below the roller experiences no shear stresses parallel to the coordinate axes, τ_{zy} and τ_{yz}, but the maximum shear stress, τ_{max}, is depicted on planes at 45° to the coordinate axes. The τ_{zy} and τ_{yz} stresses are referred to as orthogonal shear stress ($\tau_0 = \tau_{zy} = \tau_{yz}$) and the τ_{max} is

referred to as the octahedral shear stress. However, the adjacent material is subjected to shear stresses parallel to the coordinate axes, τ_{zy} and τ_{yz}, and the material more distant from the roller experiences no applied shear stresses. As the roller moves to position 2 in Fig. 1(b), the shear stresses below the roller become a maximum in the planes oriented 45° to the surface and τ_{zy} and τ_{yz} tend toward zero. Similar to Fig. 1(a), the adjacent material exhibits shear stresses τ_{zy} and

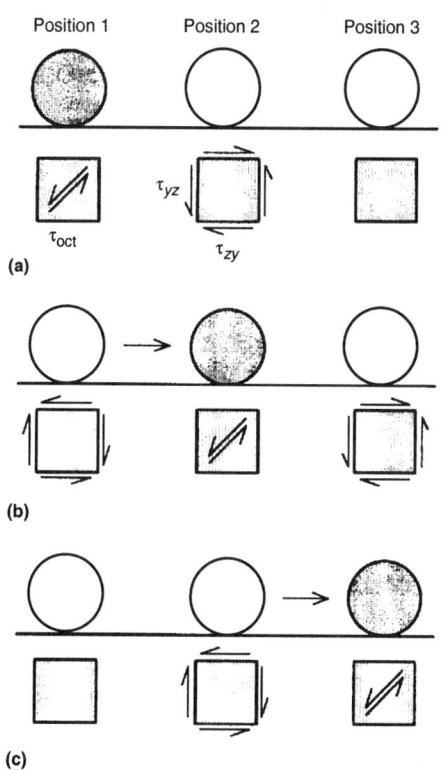

Position 1   Position 2   Position 3

(a)

(b)

(c)

Fig. 1 Schematic of the variation of subsurface shear stresses at one depth for three locations immediately below, ahead, and behind the rolling element; this is a two-dimensional view for "line contact" and also applies to a plane containing the rolling direction, through the center of a point contact. At different depths below the contact surface, the variation of shear stresses will be similar, but of different magnitude for pure rolling. When sliding is superimposed with the rolling, the tangential friction forces at the surface modify the range of shear stresses from the pure rolling situation at any given location at and below the surface.

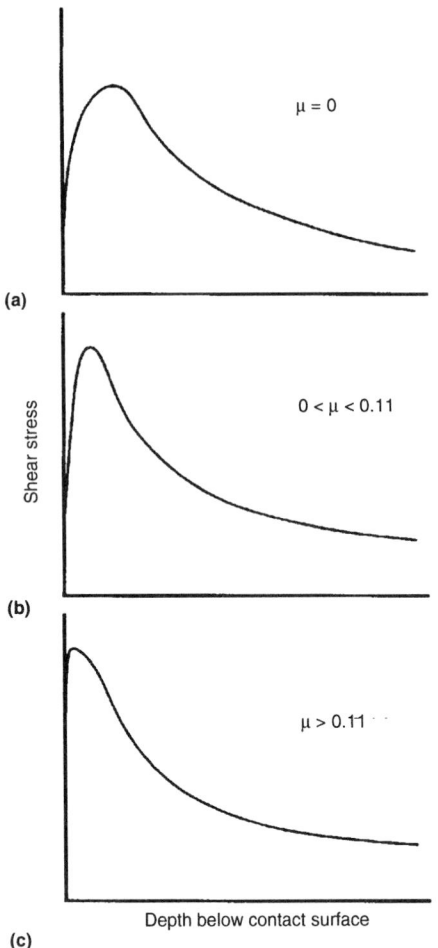

Fig. 2 Variations in maximum shear stress as a function of depth for three different friction coefficients: (a) $\mu \approx 0$, (b) $0 \le \mu \le 0.11$, and (c) $\mu \ge 0.11$. This figure is for static loading, but the range of shear stress versus depth is also at or very near the surface for tangential forces from rolling/sliding contact.

τ_{yz} equal to a finite value. With the roller in position 3 (Fig. 1c), the shear stresses below the roller again reach a maximum τ_{max} on 45° planes and τ_{zy} and τ_{yz} approach zero. Meanwhile, the adjacent material reverts back to only τ_{zy} and τ_{yz} having finite values, and the material away from the roller experiences no applied stresses. It should be noted that the shear stress field in position 2 changes direction from Fig. 1(a) to (c). Although τ_0 is always smaller than τ_{max}, for a given point in a contacting body, τ_0 changes sign as the rolling element approaches and leaves the region above the point. Therefore, the range of the maximum orthogonal shearing stress is $2\tau_0$, and for most applications this range is greater than the range of the octahedral shearing stress. These alternating shear stresses cause fatigue and are believed to be the primary contributor to bearing damage. Rolling bearing fatigue is based on the Lundberg-Palmgren theory (Ref 5). The Lundberg-Palmgren theory postulates that the stress responsible for fatigue damage is the orthogonal shear stress, which can be derived from the Hertzian pressure distribution in the contact surface and the subsurface. The orthogonal shear

stress amplitude reaches a maximum simultaneously in two planes, one parallel and one perpendicular to the contact surface. Damage resulting from contact stresses starts as a localized, inelastic cyclic deformation (localized yielding or distortion) followed by the initiation and propagation of a crack. Inelastic deformation theories are based on material deformation occurring when either the maximum shearing stress (Tresca's criterion) or the maximum octahedral shearing stress (von Mises criterion) at any point in a component reaches a critical value that causes slip of the crystal along specific crystallographic planes (Ref 3). The maximum shearing stress is given by:

$$\tau_{max} = \frac{1}{2}(\sigma_1 - \sigma_3)$$

where σ_1 and σ_3 are the maximum and minimum values of the principal stresses at a point. The maximum octahedral shearing stress is given by the equation:

$$\tau_{oct\,(max)} = \frac{1}{3}\sqrt{(\sigma_1 - \sigma_2)^2 + (\sigma_2 - \sigma_3)^2 + (\sigma_3 - \sigma_1)^2}$$

where σ_1 and σ_3 are defined as before and σ_2 is the third principal stress.

If normal forces act alone, that is the friction coefficient is zero, the stresses would be (Ref 6):

$$\sigma_{1max} = -p_o, \ \sigma_{2max} = p_o, \ \text{and} \ \sigma_{3max} = -0.5p_o$$

$$\tau_{max} = 0.30\,p_o$$

$$\tau_{oct(max)} = 0.27\,p_o$$

and these values of τ_{max} and $\tau_{oct(max)}$ occur on the z-axis at a certain distance below the contact surface. This distance is a function of only contact geometry. Figure 2(a) shows a schematic of stress variations with respect to depth in a solid subjected to static contact stresses only.

However, if tangential forces (frictional) act in addition to the normal forces, both principal and shear stresses increase. It has been found that tangential forces caused by a coefficient of friction equal to 0.33 increases the maximum principal stress by 39%, the maximum shearing stress by 43%, and the maximum octahedral shearing stress by 37% (Ref 6). Furthermore, the location of these maximum shearing stresses changes with variations in friction coefficient, such that, as the friction coefficient increases from zero, depth of the shear stress maxima decreases. Investigation shows that when the coefficient of friction is greater than 0.11 shearing stress maxima are located on the contact surface, but below 0.11 these stress maxima are found a distance beneath the surface (Ref 6). This distance is now a function of both contact geometry and the degree of tangential forces. Figures 2(b) and (c) show schematics of stress variations with respect to depth in a solid subjected to contact stresses and tangential forces.

In order to better understand the mechanism of contact fatigue, an understanding of the design characteristics of the application (rolling-element

bearings and gears), the basic applied stresses, and the generation of operating stresses is required. Therefore, bearings and gear characteristics are briefly summarized below. In addition, variations of normal and shear stresses also need to be considered in three dimensions. Three-dimensional variations of normal and shear stresses for point contact and for the ends of line contact in bearings and gears are important under conditions of inadequate crowning or end geometry modifications.

Rolling-element bearings, ball and roller, use rolling elements interposed between two raceways and relative motion is permitted by the rotation of these elements. Bearing raceways that conform closely to the shape of the rolling elements are normally used to house the rolling elements. The rolling elements are usually positioned within the bearing by a retainer, cage, or separator. However, in some ball, cylindrical roller, and needle bearings, the elements occupy the available space and locate themselves by contact with each other. Fatigue damage of rolling-element bearings is usually termed rolling contact fatigue due to the negligible tractive forces present in most bearings of this type. Roller bearings develop line-contact loading due to their cylindrical shape, and ball bearings tend to develop point contact loading. This difference, in conjunction with the double-axis rotation inherent in ball bearings, results in variations in contact fatigue fracture appearance (Ref 7).

Ball bearings can be divided into three categories depending upon the direction of applied load: radial contact, angular contact, and thrust. Radial contact ball bearings are designed for applications in which loading is primarily radial with only low thrust (axial) loads. Angular contact bearings are used in applications that involve combinations of radial and high axial loads and require precise positioning of shafts. Thrust bearings are used primarily in applications involving axial loads. Roller bearings have higher load-carrying capacities than ball bearings for a given envelope size (outer dimensions) and are usually used in moderate speed, heavy-duty applications. However, improvements in materials, design, and manufacturing have now made the use of these bearings in high-speed applications practical (Ref 4). Principal types of roller bearings are cylindrical, needle, tapered, and spherical.

Gears transmit motion and force from one machine component to another. This transmission of motion and force may be in either the same plane and direction or in a different plane and/or direction. Similar to rolling-element bearings, there are many types of gears: spur, worm, helical, and bevel (straight, spiral, and hypoid) (Ref 8). Spur and helical gears transmit motion in the same plane, whereas bevel gears change the plane of motion. The design and function of gears are associated, since the design is based on a specific function. The basic design of gears results in line or point contact of properly aligned teeth, which results in the development of contact stresses. However, because of the occurrence of elastic deformation (deflection) on the surface of loaded teeth and misalignment in service, contact occurs along narrow bands or in small areas instead of along the expected line contact. The radius of curvature of the tooth profile has an effect on the amount of deformation, the width of the resulting contact bands, and the overall contact stress field.

As a contacting gear tooth moves up the profile of the loaded tooth, a sliding-rolling action takes place at the profile interface. At the pitch line, tractive forces are negligible and loading conditions resemble those developed in roller bearings (Ref 9). Above the pitch line, a rolling-sliding action takes over, but the sliding is in the opposite direction. The action on the profile of the contacting tooth is exactly the same as on the loaded tooth, except in reverse order.

In addition to the sliding and contact stresses, gear teeth experience tensile and compressive stresses that are greatest within the root area (Ref 10). However these stresses develop along the tooth profile up to the point of contact and therefore play a role in contact fatigue damage. The main differences between gear types are the degree and direction of sliding action that takes place as a tooth cycles in and out of mesh. The degree of sliding, and thus of tractive forces, are more significant in gears than in rolling-element bearings, resulting in the maximum shear stresses that are much closer to or on the contact surfaces.

Gear and Bearing Materials

Although this article focuses primarily on the rolling contact fatigue of hardened bearing steels, this section provides a brief overview of the key types of gear and bearing steels. In addition, a brief discussion of nonferrous gear and bearing materials is also included as general background.

For both gear and bearing steels, it is important to recognize the differences between carburized steels and through-hardened steels in terms of material selection, fatigue performance, and failure analysis or fracture appearance. In terms of fatigue behavior, for example, it is understood that through-hardened and case-hardened parts are very different in initiation and propagation periods. Spalls (or more generally macropitting) may propagate slower and sometimes seem to stop for periods in carburized material, possibly because of the usually higher residual stresses near the surface than in through-hardened steel. Through-hardened steels probably have better dimensional stability and accept higher contact stresses because of lower retained austenite. However, carburized steels may be less sensitive to surface-initiated failure and run-in or changes of the surface roughness may be different, occurring in different, perhaps shorter, time periods.

Differences between these two different types of steels may also be a factor in selection of a specific steel dictated by the primary performance characteristics required for specific applications. In other words, specific factors listed in the section "Factors Influencing Contact Fatigue Life" in this article may be more important than others, depending on the specific failure mode(s) expected, as described in the section "Effect of Contact Stresses on Fracture Appearance" in this article and based on the operating and environmental conditions of a Hertzian contact.

Gear Materials. A variety of cast irons, powder-metallurgy materials, nonferrous alloys, and nonmetallic materials are used in gears, but steels, because of their high strength-to-weight ratio and relatively low cost, are the most widely used gear materials. Most gears are made of carbon and low-alloy steels, including carburizing steels and the limited number of low-alloy steels that respond favorably to nitriding.

Among the through-hardening steels in wide use are 1040, 1552, 1060, 4140, and 4340. These steels can also be effectively case hardened by induction heating. Among the carburizing steels used in gears are 1018, 1524, 4026, 4118, 4320, 4620, 4820, 8620, and 9310 (AMS 6260). Many high-performance gears are carburized. Some special-purpose steel gears are case hardened by either carbonitriding or nitriding. Other special-purpose gears, such as those used in chemical or food-processing equipment, are made of stainless steels or nickel-base alloys because of their corrosion resistance, their ability to satisfy sanitary standards, or both. Gears intended for operation at elevated temperatures may be made of tool steels or high modifications of either through-hardened or carburized carbon steel grades.

Because resistance to fatigue failure is partly dependent on the cleanliness of the steel and on the nature of allowable inclusions, melting practice may also be a factor in steel selection and may warrant selection of a steel produced by vacuum melting or electroslag refining. The mill form from which a steel gear is machined is another factor that may affect its performance. Many heavy-duty steel gears are machined from forged blanks that have been processed to provide favorable grain flow consistent with load pattern rather than being machined from blanks cut from mill-rolled bar.

Rolling-element bearing materials include through-hardened steel, such as 52100, for ball bearings and carburized materials, such as 8620, for roller bearings. A commonly accepted minimum surface hardness for most bearing components is 58 HRC. The carburizing grades have a core hardness range of 25 to 45 HRC.

At surface-hardness values below the minimum 58 HRC, resistance to pitting fatigue is reduced, and the possibility of brinelling (denting) of bearing raceways is increased. Because hardness decreases with increasing operating temperature, the conventional materials for ball and roller bearings can be used only to temperatures of approximately 150 °C (300 °F). Although ball bearings made of high-temperature materials, such as M50 (Fe-0.80C-4Cr-1V-4.25Mo), or roller bearings made of CBS1000M (Fe-0.13C-0.5Mn-0.5Si-1.05Cr-3Ni-4.5Mo-0.4V) are usable to approximately 315 °C (600 °F), the practical limit is actually determined by the breakdown temperature of the lubricant, which is 205 to 230 °C (400 to 450 °F) for the synthetic lubricants that are widely used at elevated temperatures.

Molybdenum high-speed tool steels, such as M1, M2, and M10, are suitable for use to about 425 °C (800 °F) in oxidizing environments. Grades M1 and M2 maintain satisfactory hardnesses to about 480 °C (900 °F), but the oxidation resistance of these steels becomes marginal after a long exposure at this temperature. An important weakness of these highly alloyed materials is their fracture toughness. Also, regardless of operating temperature, bearings require adequate lubrication for satisfactory operation.

For bearings that operate in moderately corrosive environments, AISI type 440C stainless steel should be considered. Its maximum obtainable hardness is about 62 HRC, and it is recommended for use at temperatures below 175 °C (350 °F). However, the dynamic-load capacity of bearings made from type 440C stainless steel is not expected to be comparable to that of bearings made from 52100 steel. The carbide structure of 440C is coarser, the hardness generally lower, and the fracture toughness about one-half that of 52100 steel.

Materials for Sliding Bearings. Almost all materials, such as carbon, ceramics, white cast iron, and metal alloys of all kinds, can be used as bearings. These materials may be chosen primarily for corrosion resistance, for prevention of sparking, for resistance to high volumes of abrasives, or simply because they are available. High-performance sliding bearing materials include tin-base or lead-base babbitts, copper-lead alloys, and aluminum alloys containing tin and lead.

Polymers, gray cast irons (with large amounts of graphite), and bronzes are selected as bearing materials because of their ability to survive even though they may be made to poor tolerances or may be operated with occasional lubricant starvation or without maintenance. Poor treatment of these bearings will often appear as adhesive grooving and galling. Where these materials are run against a harder countersurface, there may be metal transfer to that surface.

The lowest cost bearings are bushings of molded polymers (plastics) in which steel shafts rotate. The lowest cost polymers are the nylons and acetals. At low sliding speeds, they usually wear by abrasion due to dirt or the roughness of the surface of the shaft.

For low friction, the more expensive polytetrafluoroethylene (PTFE) is available. Polytetrafluoroethylene is sometimes added to acetals to reduce friction. Graphite and molybdenum disulfide (MoS_2) are also added; although they decrease friction, these solid lubricants usually do not prevent wear.

Microstructural Changes from Contact Fatigue

Cyclic contact stresses cause microstructural changes, which are important because they are related to stress levels and the number of cycles. The microstructural changes usually observed in martensitic microstructures from contact fatigue are known as dark etching areas, butterflies, and

(a)

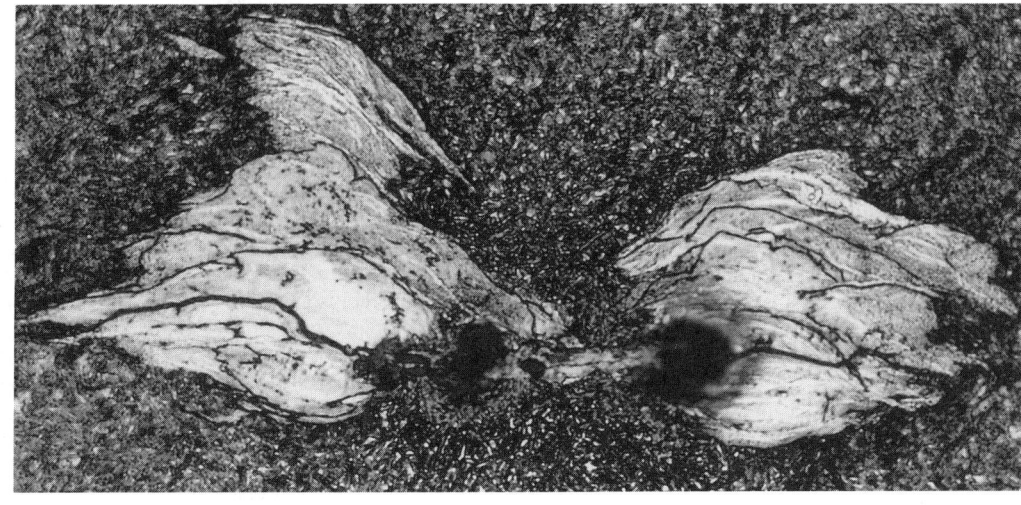

(b)

Fig. 3 Micrograph showing a butterfly on a plane perpendicular to the rolling direction within a martensite/austenite matrix. (a) A subsurface fatigue crack is visible to the right of the butterfly. 100×. (b) 750×. Etched in nital

white bands. The development of these structural changes depends on the local magnitude of the applied shear stresses and the number of cycles such that, below a certain stress limit or number of cycles, these structural changes are not observed. For example, of these three microstructural changes, the so-called white bands occur from many cycles (approximately 100 million total) after dark etching and butterfly formation. These microstructural changes from cyclic contact stress therefore can be useful because the influence of stress level and number of cycles allows one to infer what loads and stresses have been present in the examination of a damaged or failed bearing or component.

Microstructural alterations from contact fatigue have been reported in the literature since the work published by Jones (*Steel*, Vol 119, p 68-100) in 1946. Different investigators have used different terms for identifying and describing the various structural changes, but most are likely describing only a few different microstructural changes. In addition, most researchers have restricted their methods to light optical microscopy, and the principal problems are reproducibility and standardization of etching and microscopy techniques. The earliest work on the transmission electron microscopic evidence of dislocation cell formation in fatigue-damaged microstructures was provided by O'Brien et al. (Ref 11).

The distinction between through-hardened and carburized steels is also important here. Microstructural alterations are different for 52100 type steels than for carburized steels. The undissolved carbides in 52100 and the higher level of retained austenite in the carburized steels contribute to these differences. Additional differences such as hardness/strength and residual stress gradients found in carburized steels may influence the type and location of the microstructural alterations with respect to load levels and stress cycles. The above-mentioned alterations have been observed in both types of materials and thus, material selection should be based upon manufacturing considerations and operating environments.

Dark etching areas (DEA) appear in the region of maximum Hertzian shear stress after a high number of cycles, depending on load (Ref 12). In through-hardened 52100 steel DEA has been initially observed in as few as 10^6 cycles at contact pressures exceeding 3500 MPa and at cycles as long as 10^9 with contact pressures of 2700 MPa (Ref 19). Many researchers have studied this dark etching phase and have suggested many terms to describe the microstructure, such as "troostite," tempered martensite, low-temperature bainite, and ferrite (Ref 13).

Dark etching areas are produced when a material volume is subjected to contact stresses in which the subsurface flow stress is locally exceeded and the material plastically deforms (Ref 14, 15). Plastic deformation caused by contact stresses occurs by dislocation slip along certain crystallographic planes (Ref 16). For this mechanism, it is typical that slip lines are located orthogonal to each other and form at an angle of about 45° with respect to the contact surface corresponding to the maximum shear stress. Areas appear dark after polishing and etching where slip has occurred in the tempered martensite crystals. The darker these areas appear, the more plastic deformation has occurred in the tempered martensite. From the position and darkness of the DEA, conclusions can be drawn about the heaviest stresses of the material under the contact surface (Ref 17). These areas are present in very localized regions beneath the contact surface, approximately corresponding in depth to the maximum shear stress, which suggests a stress dependence of these microstructural alterations.

The DEAs have somewhat the same microstructure as that of martensite tempered to relatively high temperatures, and some investigators have even theorized such a structural change during contact fatigue is caused by heat generated by cyclic loading (Ref 13). However, distribution of the dark areas is not uniform, but discontinuous, and if heat were the main cause of the structural alteration, distribution of the constituents would be more uniform. Sugino et al. (Ref 18) surmised the possibility that very fine carbides present in the original tempered martensite would dissolve into the matrix by cyclic stressing. This phenomenon is associated with the diffusion of carbon atoms toward heavily dislocated regions, which are caused by inhomogeneous microplastic deformation during the fatigue process (Ref 12). Also, moving dislocations would break the coherence of the carbides with the martensite, or even cut the carbides. This supersaturation of the martensitic matrix due to carbide dissolution is not stable and the reprecipitation of incoherent carbides would occur at such favorable sites as grain boundaries. Furthermore, Sugino et al. postulated that the dissolution of carbides is a function of the degree of slip on discrete crystallographic planes, thus causing the dependence of structural alteration on the orientation of martensite crystals. This dependency is consistent with features of the dark etching areas (Ref 18).

Butterflies, also known as white etching areas, is the term given to frequently observed structural changes that occur due to contact stresses. Figure 3 shows an example of a butterfly in which white areas extend outward from a dark center region. One characteristic feature of butterflies is that they take a definite inclination to the contact surface due to the static shear stresses that develop (Ref 18). The center region of the butterfly, which etches dark, is usually a nonmetallic inclusion. According to Sugino et al. (Ref 18), butterflies originate at almost all kinds of nonmetallic inclusions. However, the statistical result of many observations shows a trend that these structural changes originate most preferentially at alumina type inclusions (Ref 18). Alumina inclusions form a hard, brittle phase relative to the steel that exhibits an incoherent interface with the matrix

Table 1 List of inclusion type and occurrence associated with butterfly formation

Type of nonmetallic inclusion	Appearance of nonmetallic inclusion	Frequency of butterflies
MnS	Elongated (<3 μm)	Rare
	Elongated (>3 μm)	Few
	Very thin and long	None
Al_2O_3	Finely dispersed stringers	Many
TiN		None
MnS + Al_2O_3		Many

microstructure, whereas sulfide inclusions are soft, ductile particles that have a semicoherent interface with the surrounding steel matrix. Table 1 shows inclusion type and appearance correlated with associated frequency to butterfly formation (Ref 18). Sulfides combined with alumina are occasionally found, but sulfides alone are usually not associated with the butterfly structure.

From the orientation of the white etching areas, it can be assumed that butterflies are generated due to the maximum shear stress, and the depth at which most butterflies are found coincides with the zone of maximum contact stress. Many theories for butterfly formation have been postulated. One theory by Schlicht et al. (Ref 19) explains the formation of butterflies as caused by tensile and shear stresses that develop due to contact stresses in the vicinity of nonmetallic inclusions. Local stresses develop in the vicinity of these inclusions as a result of the elastic modulus differences between the inclusion and the martensite-retained-austenite matrix and/or the weak interfacial energy between the inclusion and the matrix. If the material is overstressed, the elastic energy concentrated in the area of an inclusion is transformed into deformation energy and energy needed to initiate cracks that propagate in the direction of maximum shear stress. The cracked surfaces are pressed together during propagation due to hydrostatic pressure and are plastically deformed. The plastic deformation is associated with adiabatic heating, and consequently the microstructure is reaustenitized. During the reaustenitization, carbides dissolve, and upon subsequent quenching a highly deformed martensite is formed. Other investigations postulate that carbide dissolution is caused by severe deformation of the carbides, resulting in diffusion of carbon from the carbide to a microcrack or interface (Ref 19). This hypothesis does not require the formation of high temperatures and localized reaustenitization of the steel microstructure. However, both theories require that a microcrack be a prerequisite for the formation of the butterflies. The mechanism of butterfly formation is similar to the generation of adiabatic shear bands observed with high impact loading (Ref 19). Also, the martensitic microstructure formed by this method of rehardening resembles a microstructure that is formed during a marquenching process. In both of the above examples, the microstructure that forms exhibits highly deformed martensite, similar to what is observed with butterflies.

The distribution of the butterfly constituent is dependent on various factors, such as the number of cycles and other structural alterations (Ref 20). While the number of butterflies and the size of the zone in which they generate increase with repeated cycles, their density becomes reduced at certain depths. In addition, as the size of butterflies increases with cycles, their growth is suppressed in these same areas. It has been shown that the formation of the DEAs influences the generation and growth of the butterflies such that the region in which butterfly generation is suppressed corresponds to the location of the DEA and the number of cycles at which this suppression occurs also corresponds to the number of cycles needed to generate the DEAs. These dependencies can be explained by considering that plastic deformation, the cause of DEAs, reduces the stress-raising effect around a nonmetallic inclusion and consequently the formation of a microcrack, which is required for the formation of the butterfly. In fact, butterflies have been found at the depth of maximum shear stress prior to the formation of the DEAs. Therefore, the interaction of the two structural changes is not contradictory to the idea that shear stress is essential for their formation.

White bands occur within DEAs after a high number of cycles following DEA and butterfly formation. In through-hardened 52100 steel white bands have been initially observed in as few as 10^8 cycles at contact pressures approaching 4000 MPa and at cycles as long as 10^{11} with contact pressures of 2500 MPa (Ref 19). White bands have had different descriptives, such as gray lines, nonetching bands, 30° bands, and deformation bands (Ref 13).

White etching bands are observed at angles between 20 and 30° (shallow bands) with respect to the contact surface (Ref 17). Their inclination is not constant; slight changes are observed with variations in contact conditions, lubrication, and tractional forces (Ref 21). With increased cycles, additional white bands are observed. This second set of white bands form at angles of 70 to 80° (steep bands) with respect to the contact surface and develop within the zone of previous white band formation (Ref 17). These steep bands form after a greater number of cycles and are located closer to the contact surface. With a reverse of motion with respect to the contact surfaces, these bands will develop in a corresponding new orientation, which suggests a stress-related mechanism for their formation (Ref 12).

The white band microstructure was shown by Österlund and Vingsbo (Ref 13) to consist of carbide disks sandwiched between an extremely fine-grain, ferritelike phase, free of resolvable carbides. The mechanism of white band formation can be thought of as a form of mechanical tempering in which dislocation motion adds to the energy needed for carbon diffusion, as opposed to a traditional tempering process in which elevated temperatures and times are required. Due to extensive plastic deformation resulting from contact stresses, a dislocation structure of fine cells develops within the ferritelike phase,

thus resembling heavily cold-rolled ferrite (Ref 13). The dislocation density is similar to that of martensite, whereas the dislocation structure is similar to that of heavily strained ferrite. Because the decay of the original tempered martensite is oriented (i.e., specific white band orientation), it can be concluded that white band formation caused by shearing stresses developed during contact loading.

Fracture Appearance

The fatigue of hardened steel under cyclic contact stress has been the subject of extensive research using analytical and experimental methods. A rapidly growing number of publications have revealed a large variation in the contact fatigue strength of similar materials when tested under different conditions of contact geometry, speed, lubrication, temperature, and sliding in combination with rolling. These variations result from a poor understanding of the basic mechanism of contact fatigue.

The existence of several distinctly different modes of contact fatigue damage has been recognized, but the factors that control the nucleation and propagation of each type are only partially understood. In addition to a lack of understanding of the basic mechanism, the bearing and gear industries have, in the past, used completely different nomenclature to describe identical fatigue damage modes. The basic mechanism of contact fatigue initiation and propagation is the same, regardless of the application such that bearings, gears, and other contact-stressed components fail in the same manner.

Although bearing and gear industries have used completely different nomenclature to describe contact fatigue in the past, current nomenclature used by both industries is very similar. American National Standard ANSI/AGMA 1010-E96 published in 1996 represents several years' work that emphasized standardized nomenclature for failure modes that is compatible with both gear and bearing industries. Reference 9 (C.A. Moyer, Comparing Surface Failure Modes in Bearings and Gears) uses nomenclature for bearings that is similar to that used in ANSI/AGMA 1010-E96 for gears. In order to improve communication, it is imperative to maintain compatible nomenclature per ANSI/AGMA 1010-E96 such that:

- The term macropitting should be used instead of "spalling." Spalling has been used in the past to describe several different failure modes and therefore is confusing. It should be avoided where possible.
- The term micropitting should be used instead of "peeling." Peeling is not a good name for micropitting because it does not describe either the appearance or the mechanism of the failure mode. Micropitting is a much better name because it describes both appearance and mechanism and links the failure mode to macropitting, which is similar except for scale.

Table 2 List of contact fatigue terminology used to describe the different fatigue mechanisms

Macroscale	Microscale
Macropitting	**Micropitting**
Pitting	Microspalling
Initial pitting	Frosting
Destructive pitting	Glazing
Flaking	Gray staining
Spalling	Surface distress
Scabbing	Peeling
Shelling	
Fatigue wear	
Subcase Fatigue	
Case crushing	

Fracture origin and appearance of a component damaged by contact fatigue is a function of the applied stresses and material selection, with variations in those applied stresses resulting from environmental factors. Table 2 lists terminology used by the bearing and gearing industries to describe the fracture appearance of contact fatigue-failed components.

Macropitting

As previously noted, macropitting is a preferred general term that includes spalling (Ref 22) and other forms of macroscale damage (Table 2) caused by Hertzian contact fatigue. Spalling (or macropitting) of bearing raceways or gear teeth is generally due to contact fatigue, which occurs from localized plastic deformation, crack initiation, and finally macropitting from crack propagation in and near the contact surface.

Macropitting or spalling results from the *subsurface* growth of fatigue cracks, which may have a surface or subsurface origin. Figure 4 shows macropit formation by subsurface growth of surface-origin pitting on a gear tooth. The bottom of inclusion-origin macropits form first, and so forth. When circumstances cause surface-origin pitting, crack growth often occurs within and beyond the range of maximum shear stresses in the contact stress field (so-called hydraulic-pressure propagation crack paths that may be due to chemical reactions at the crack tip as well as to lubricant viscosity effects).

When fully developed, the craters exist at a depth comparable to that of the maximum alternating Hertzian shear stress. A macropit does not result from the gradual enlargement of a small cavity, but rather results from the subsurface growth of a fatigue crack, which eventually separates from the main body of material (Ref 23). In the example of a gear damaged by crater formation due to surface-origin macropitting (Fig. 4), the craters are generally steep walled and essentially flat bottomed, as the walls and bottom are formed by fracture surfaces. The bottoms of macropits usually form first, and then the crack extends through the contact surface, creating the walls of the crater. However, there are circumstances in which macropits are initiated at or near the contact surface and the crack propagates through the Hertzian shear stress field, forming

the walls and then the bottom of the crater (Ref 22).

Macropitting fatigue life is inherently statistical because defect severity and location are randomly distributed among macroscopically identical contact components. Two major classes of macropitting damage are distinguished according to the location of the initiating defect: subsurface- and surface-origin macropitting.

Subsurface-origin macropitting results from defects in the bulk material subjected to the Hertzian cyclic stress field. Subsurface-origin macropits (or spalls) are characterized by a smooth area parallel to the contact surface with a steep wall exit (inclined by more than 45° with respect to the contact surface). Sometimes the subsurface defect that caused the damage remains visible after the macropit forms. The defects that cause subsurface-origin macropits are inclusions and/or material microstructure alterations. The defects that most likely produce this type of macropit are located above the depth of maximum alternating Hertzian shear stress. Variables governing the life of a component with respect to this type of macropit are Hertzian shear stress level, material matrix fatigue resistance, defect severity, and defect location.

Inclusion origin fatigue is normally initiated at nonmetallic inclusions below the contact surface and is the primary mode of contact fatigue damage in antifriction bearings (Ref 24-26). The initiation and propagation of fatigue cracks are the result of cyclic stresses that are locally intensified by the shape, size, and distribution of nonmetallic inclusions in steels. Oxide inclusions from deoxidation, reoxidation, or refractory sources are the most frequently observed origins of inclusion-origin fatigue damage (Ref 26). Sulfide inclusions alone are rarely associated with contact fatigue cracks (Ref 18). The ductility of sulfide inclusions at hot-working temperatures and a relatively low interfacial energy make them less effective stress concentrators than the harder, incoherent oxide inclusions. In addition, oxide inclusions are often present as "stringers" or elongated aggregates of particles, which provide a much greater statistical probability for a point of

stress concentration to be in an area of high contact stress.

When a nonmetallic inclusion as described above results in significant stress concentration, two situations are possible: (1) localized plastic flow occurs, and/or (2) a fatigue crack begins to propagate with no plastic flow as observed microscopically. If localized plastic deformation occurs, microstructural alterations begin to develop and the effective stress concentration of the inclusion is reduced. The probability of fatigue crack initiation/propagation is less than before the plastic flow occurred so a macropit may not form. However, if a fatigue crack forms due to the inclusion, prior to any plastic deformation, the stress concentration is increased and crack propagation results in macropitting or spalling fatigue. This hypothesis includes the possibility of a fatigue crack originating from an inclusion that is associated with some sort of microstructural alteration, but also proposes that microstructural alterations are not a necessary step in the initiation/propagation of inclusion origin fatigue.

Early macroscopic propagation of inclusion-origin fatigue takes on a characteristic pattern. Many branching subsurface cracks are formed around the inclusion with propagation often being most rapid perpendicular to the rolling direction, with a resulting elliptical shape for the incipient macropit (or spall). In some instances, the cracked material around the inclusion spalls out early and further propagation is relatively slow, leaving a small macropit with very few subsurface cracks to cause further damage. However, when material has spalled from the contact surface, the result may be debris bruising and denting of other contact areas with subsequent surface origin macropitting.

Surface-origin macropitting (or spalling) is caused by defects in the immediate subsurface material subject to asperity scale cyclic stress fields and aggravated by surface tractive forces. These defects are either preexisting defects, such as nicks, dents, grinding furrows, surface discontinuities, and so forth, or microscale pits, which are described later. The distinguishing features of a surface-origin macropit are in the entrance zone of the macropit. The entrance zone may exhibit a shallow-angle entry wall (inclined less than 30° to the contact surface), an arrowhead configuration, the presence of a visible surface defect, or an association with a stress concentration due to design geometry. Surface-origin macropits have been classified by the nature of the defect: geometric stress concentration (GSC) and point surface origin (PSO).

Geometric stress concentration macropits are distributed on the contact surface at the ends of line contact. When the contact geometry, deflection under applied loads, and alignment cause the contact stress to be higher at the end of line contact, fatigue occurs within a narrow band in which the contact stresses are more severe than those associated with inclusions. Figure 5 shows an example of several tapered roller bearing cones that were damaged by GSC macropitting. The GSC macropits can propagate more rapidly

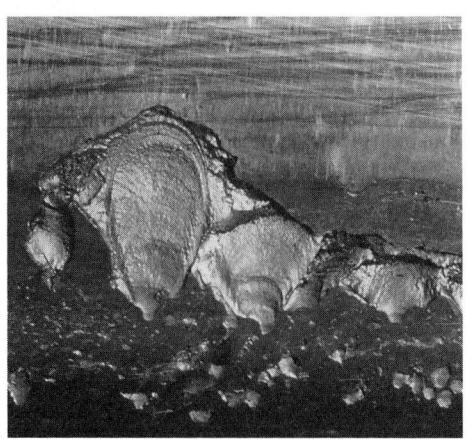

Fig. 4 Fractograph showing the advanced stages of macropitting fatigue for a helical pinion tooth. Courtesy of GEARTECH

Fig. 5 Fractograph showing various advanced stages of geometric stress concentration with macropitting (spalling) fatigue of bearing raceways. Relative life numbers are indicated to show that propagation is not related to life.

Fig. 6 Fractograph showing the advanced stages of point-surface-origin macropitting fatigue of a bearing raceway in which the origin is still visible

in either direction—parallel or transverse to the rolling direction. This mode of contact fatigue is more frequently observed in life tests accelerated due to overload conditions, because contact geometry is designed to produce uniform loading under expected service loads.

Crowning of bearing rollers and gear teeth in line contact situations have been shown to be an effective method of prolonging life or preventing the GSC mode of macropitting. The fact that GSC macropits can occur simultaneously at both ends of a line contact indicates that this mode of contact fatigue is truly the result of end stress concentrations and not simply misalignment. However, misalignment can aggravate the situation by increasing end contact stresses.

Early stages of GSC macropitting exhibit cracks extending to the surface, but it is not clear whether the first cracks always originate at the surface or slightly subsurface. However, the origins are rarely associated with a nonmetallic inclusion, but occasionally an inclusion at the end of line contact will serve as the nucleus for a GSC fatigue macropit.

Point surface origin (PSO) macropits occur randomly distributed on the contact surface, similar to inclusion-type macropits. However, PSO macropits are different in two important aspects: First, there is no consistent association with non-

metallic inclusions, and second, the origin is located at or near the contact surface, whereas with inclusion-origin macropits, the initiation site is located subsurface. These characteristics result in a distinct arrowhead appearance, in which the tip of the arrowhead is the origin and the advanced stages of the macropit develop in a fanned-out appearance with respect to the tip in the rolling direction. Figure 6 shows an example of a tapered roller bearing cone that was damaged due to PSO macropitting.

In the advanced stages of propagation, PSO macropits exhibit a brittle appearance and propagation is rapid in both area and depth. However, the depth of cracking is not related to the subsurface contact stress profile. Although PSO macropits do not show a tendency to be associated with inclusions, when conditions are favorable for PSO fatigue, inclusion-type damage sometimes propagates rapidly, and after some macropitting are similar in appearance to PSO fatigue. Point-surface macropitting is often seen propagating from nicks, dents, and debris bruises. A PSO classification indicates the discrete origin of fatigue cracking at or very near the contact surface. A GSC classification indicates a geometric stress concentration due to contact surface profile normal to the rolling/sliding direction, or end-of-contact geometry, in combination with elastic deflections or misalignment in the bearing system.

However, the aforementioned surface anomalies for a PSO classification are common in many contact-loaded applications where PSO macropitting is not encountered and therefore that are not always associated with PSO. Thus, PSO macropitting is better described as a combination of a point-surface origin or initiation followed by hydraulic-pressure propagation (HPP). The occurrence of HPP can be linked to lubricant viscosity and chemical effects—for example, dis-

solved water and/or aggressive additives (some extreme pressure or EP additives)—that influence fatigue fracture behavior at the crack tip. This type of propagation is due to lubrication trapped within a fatigue crack (see the section "Lubrication and Surface Finish" in this article) and has been shown to nearly always promote PSO macropitting or spalling.

Point-surface-origin macropitting below the pitch line on gear teeth has been observed and duplicated in a test rig by combined rolling and sliding forces imposed by phasing gears between the test specimen and a crowned test roller. Without the phasing gears, such that loading conditions were nearly pure rolling, the fatigue was reported to be subsurface inclusion origin. Thus, tangential forces caused by friction in combination with rolling are believed to be the most likely cause of PSO macropitting fatigue. Higher frictional forces are encountered under low lubricant viscosity and/or thin-film conditions and thus results in the presence of PSO macropits.

Micropitting

As previously noted, micropitting is the preferred term for peeling fatigue which is defined as microscale spalling fatigue. It is damage of rolling/sliding contact metal surfaces by the formation of a glazed (burnished) surface, asperity scale microcracks, and asperity scale micropit craters. Micropitting damage is a gradual type of surface fatigue damage that is a complex function of surface topography and its interaction with the lubricant (even under high contact loads, a thin film of lubricant between the contacting surfaces from elastohydrodynamic (EHD) lubrication is present).

Cumulative plastic deformation occurs early in cycling and leads to flattening of asperity tips as the surface-finishing topography is gradually leveled and only smooth, low frequency waviness and scratches at the bottom of deeper furrows remain. The resultant surface has a mirrorlike reflection with finishing lines nearly obliterated and is known as a glazed surface. The surface and immediate subsurface material of a glazed surface are heavily deformed and with continued cyclic stressing, the ductility of the material decreases and microcracks form (Ref 27). The cracks tend to propagate parallel to the contact surface at depths comparable to that of asperity scale shear stresses. A microcracked surface is not usually distinguishable from a glazed surface by the unaided eye, but with microscopy, microcracks opening to the surface may be visible in glazed areas (Ref 22). As the microcracks multiply and propagate, the surface becomes undermined (in asperity dimensions) and multiple microscopic pits form. A micropitted surface appears frosted to the unaided eye with black spots representing the micropits.

Two controlling variables for the occurrence of micropitting are EHD film thickness to roughness ratio and surface microgeometry (Ref 28). In the presence of a sufficiently high EHD film thickness to surface roughness height ratio, micropitting does not occur because the film prevents high-contact microstress in asperity interactions. However, film thickness may be sufficient to prevent generalized micropitting, but not local micropitting in the vicinity of surface defects such as nicks and dents that locally depressurize and thereby thin the EHD film (Ref 22, 26). Tribology features that affect micropitting are both the height and sharpness (RMS height and slope) of the asperities. There is also some evidence that the normal wear in boundary-lubricated gear contacts removes incipient surface-origin pitting, which is one reason why gears may be less prone to spalling (macropitting) from surface-origin spalling under some circumstances.

Micro-Hertzian contact stress fields exist in the vicinity of each asperity ridge in unlubricated or insufficiently lubricated systems (Ref 27). The absolute maximum shear stress along the axis of symmetry of the micro-Hertzian asperity contact is shown to occur at a depth approximately equal to twice the peak-to-valley height of the asperity. The value of this shear stress is many times greater than the macro-Hertzian octahedral shear stress and the range of orthogonal shear stress.

Micropitting (or peeling) is characterized by the limited depth of cracking and propagation over areas rather than propagation in depth. Micropitting fatigue cracks usually propagate parallel to the surface at depths from 2.5 to 13 μm (0.1 to 0.5 mil) and rarely as deep as 25 μm (1 mil) (Ref 26). The high stress gradient that exists in the vicinity of asperities explains why such cracks do not propagate to greater surface depths of the macro-Hertzian contact stress field. In essence, micropitting removes the asperity ridges and relieves the high stress gradient. An additional visual characteristic is that the micropitting occurs more or less in the rolling direction, following finishing lines.

Shear stresses generated due to asperity contact do not affect the location at which the maximum octahedral and orthogonal shear stresses occur (Ref 27). The micro-Hertzian asperity contact field merely increases these stresses above the classical Hertzian theory. Thus, the depth at which subsurface fatigue originates remains unaffected by surface finish (Ref 27). However, traction forces are not considered in classical Hertzian theory and will decrease the depth at which the maximum contact shear stresses occur.

If micro-Hertzian contact stresses exist in the vicinity of asperity ridges, then similar, but larger contact stresses exist around surface defects such as grinding furrows, nicks, dents, and scratches. Because these surface defects raise the absolute micro-Hertzian shear stress, one would expect damage to occur at greater depths than those associated with asperity ridges. Results show that surface origin fatigue commonly occurs at a depth of approximately 25 to 50 μm (1 to 2 mils) below the contact surface (Ref 29). These types of fatigue mechanisms were discussed earlier and referred to as PSO and GSC type fatigue.

Macropitting (or spalling) fatigue is the most hazardous consequence of micropitting (or peeling) (Ref 29). In the absence of preexisting surface defects, surface-origin pitting initiates from micropits through the process of asperity-scale fatigue. More often, the presence of micropits can lead to premature macropitting (or spalling) (Ref 7). The principal operating factor influencing the rate of macropitting from micropits is tractive interfacial stress. When friction is high, macropits form rapidly from the micropits on the surface. If early macropitting does not terminate the operation of a component that has suffered micropitting, then surface material delamination may occur (Ref 27). The original surface is thereby lost over wide areas, and the component becomes inoperative through loss of dimensional accuracy, noise, or secondary damage.

In many cases, micropitting (or peeling) is not destructive to the bearing raceway or gear tooth surface. It sometimes occurs only in patches and may stop after the tribological conditions have been improved by run-in. Run-in is defined as an initial transition process occurring in newly established wearing contacts, often accompanied by transients in coefficient of friction, wear rate, or both, that are uncharacteristic of the long-term behavior of the given tribological system (Ref 30). Gears appear less sensitive to the consequences of surface distress than rolling-element bearings (Ref 7). This may be due to the greater ductility of lower-hardness materials. However, micropitting has been observed in gears as well as bearings.

Subcase Fatigue

Subcase fatigue can also be defined as a macro-Hertzian contact fatigue mechanism. Subcase fatigue is fracture of case-hardened components by the formation of cracks below the contact surface within the Hertzian stress field (Ref 7). However, the depth at which the cracks form is much greater than those that cause macropitting fatigue, and it is a function of material strength in conjunction with the alternating Hertzian shear stresses. Subcase fatigue is also sometimes referred to as case crushing, but since it results from a fatigue crack that initiates below the effective case depth or in the lower-carbon portion of the case, the former nomenclature will be used throughout this article.

In case-hardened components—produced by carburizing, nitriding, or induction hardening—a gradient of decreasing hardness (shear strength) exists from the case to the core. If the material shear strength gradient is steeper than the gradient in Hertzian shear stress, then the applied-stress-to-strength ratio is least favorable at some depth below the normal maximum Hertzian shear stress region. This depth is usually located at (or near) the case/core interface due to shear strength and stress gradients, in addition to unfavorable tensile residual stresses. The reversal in the sign of the residual stress from compression in the case to tension in the core occurs at or very near the case/core interface, and therefore the Hertzian shear stresses are increased due to the tensile residual stresses.

Once formed, such cracks usually propagate parallel to the case-hardened region until branch-

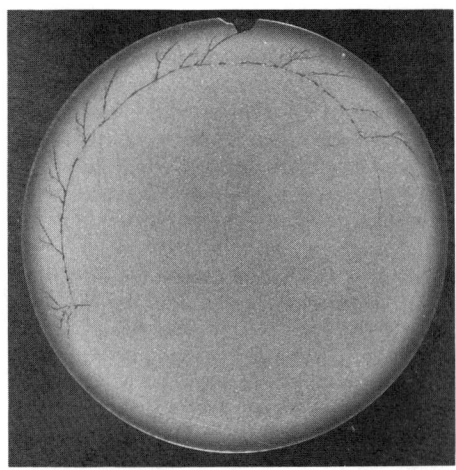

(a)

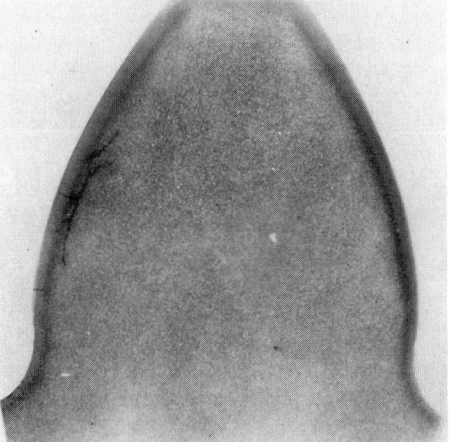

(b)

(c)

Fig. 7 Micrographs showing etched cross-sections of (a) a carburized cylindrical test specimen (1.9×) and (b) a carburized gear tooth in which subcase fatigue cracks initiated and propagated during testing. Subcase fatigue, called case crushing by Pederson and Rice, shows the early stages of fatigue cracking in the lower carbon portion of the carburized case. (c) The diagram indicates how the stress and strength gradients combine to cause the weakest condition at the subcase location where fatigue cracking begins. Source: Ref 26, 31, and 33

ing-cracks form and propagate toward the surface, resulting in macropit formation (Ref 31). Figure 7 shows micrographs of a gear tooth and a roller specimen in which subcase fatigue cracks

Table 3   List of contact fatigue modes and their controlling factors

Mode of failure	Factors that control occurrence
Subsurface	
Inclusion origin	Size and density of oxide or other hard inclusions
	Absence of other modes of failure
Subcase fatigue	Low core hardness
	Thin case depth relative to radius of curvature and load
Surface	
Point-surface origin	Low lubricant viscosity
	Thin EHD film compared with asperities in contact
	Tangential forces and/or gross sliding
Geometric stress concentration	End contact geometry
	Misalignment and deflections
	Possible lubricant film thickness effects
Micropitting	Low lubricant viscosity
	Thin EHD film compared with asperities in contact
	Loss of EHD pressure
	Slow operating speeds

had formed and propagated during testing. Note that the main crack formed in a direction parallel to the case-hardened layer with cracks branching towards the surface. Subcase fatigue is usually caused by too thin a case, such as on a nitrided gear tooth, by insufficient core hardness on carburized or induction-hardened parts, or in heavily loaded case-hardened rolling-element bearings (Ref 32).

Factors that tend to promote the various types of contact fatigue are listed in Table 3. It should be noted that three of the five modes of fatigue damage are influenced by lubrication, including all surface-origin damage modes. Figures 8(a) to (d) show Hertzian shear stresses and material shear strengths as a function of depth below the contact surface. These are an adaptation of the subcase fatigue analysis by Pederson and Rice (Ref 31). As noted earlier, fatigue occurs when the applied shear stress is greater than the shear strength of the material. Thus, the different modes of contact fatigue can be described by determining where these stress profiles cross each other and are shown schematically in Fig. 8.

Factors Influencing Contact Fatigue Life

Factors influencing contact fatigue life of hardened steel bearings and gears can be roughly classified into four categories: material, lubrication and surface finish, dimensional precision, and environmental conditions. The latter two factors are concerned more with machine element design and will not be discussed in any great detail in this article.

Most factors discussed in this section are dependent upon and different for the type of material in question, through-hardened or carburized. Therefore, material usage should be a function of

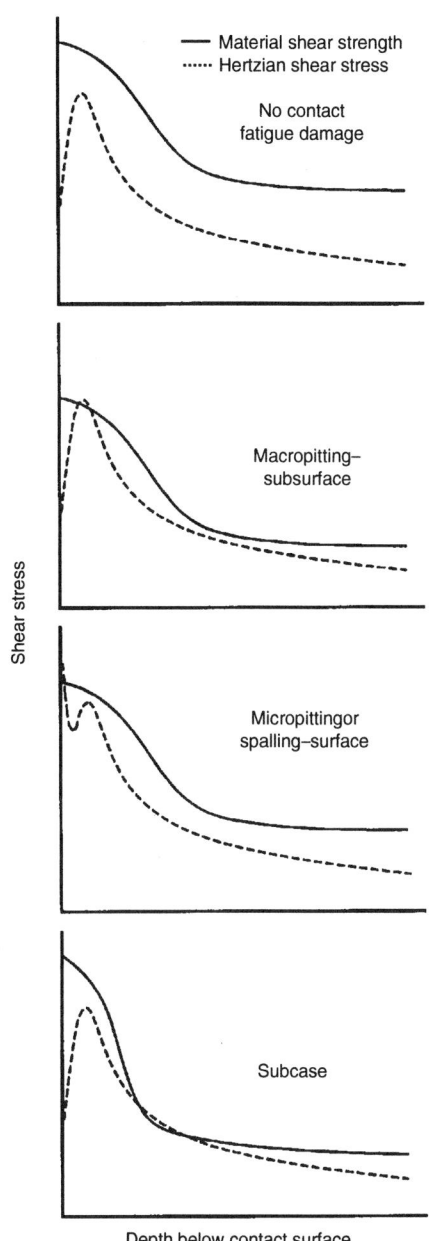

Fig. 8 Applied shear stress and material shear strength as a function of depth representing types of fatigue damage. (a) No damage. (b) Subsurface-origin, macropitting fatigue. (c) Micropitting or surface-origin macropitting fatigue. (d) Subcase fatigue

manufacturing capabilities, operating environments, and more often that not, cost. The influence of several factors, such as retained austenite and hardness, are difficult to resolve. Typically, as retained austenite increases, hardness decreases. This is important in debris denting, in which high surface hardness is desirable to prevent debris denting, but higher levels of retained austenite are desirable to prevent fatigue cracking from debris dents. Therefore, material selection is primarily based upon the desired performance issues, such as wear, fatigue, dimensional stability, debris contamination, and past experience with a given steel.

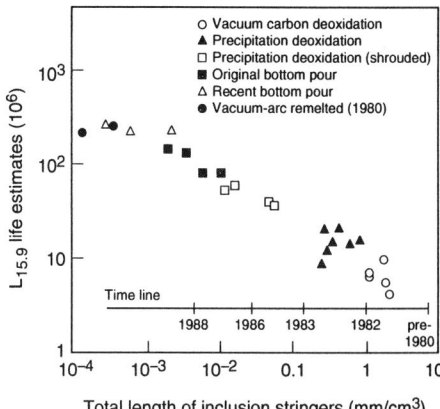

Fig. 9 Plot of relative contact fatigue life for tapered roller bearings versus steel cleanness for several steel heats produced by various electric-arc furnace steelmaking practices

Hardness. As demonstrated earlier in this article, several distinctly different modes of contact fatigue can cause cracks to initiate and propagate independently at various rates; thus the phenomenon appears to be influenced more by highly localized conditions than by the general properties of the bulk material. Early research on contact fatigue utilized mechanical engineering and elastic stress analysis with little emphasis on materials, until an understanding of how material hardness influences contact fatigue life (Ref 33). Contact fatigue measurements on hardened steels indicate an increasing life with increasing component hardness, although there is an optimal hardness range that is application-dependent to ensure maximum contact fatigue life (Ref 34, 35). In addition to hardness, some important metallurgical factors other than a predominantly martensitic microstructure that influence contact fatigue life are inclusions, carbides or other hard second-phase precipitates, retained austenite content, nonmartensitic transformation products (NMTP), grain size, residual stress, heat-treat methods, and steel type (Ref 36-49).

Inclusions. Numerous research investigations have been reported concerning the metallurgical variables that affect the contact fatigue life of bearings and gears. However, nonmetallic inclusions have been shown to be the single most influential factor on the contact life of hardened steel (Ref 29, 33). Of the nonmetallic inclusions present in steel, the most detrimental of these is alumina, followed by silicates (Ref 18). Types and sizes of inclusions present in steel are governed mainly by the steelmaking practice, and their frequency or severity is referred to as cleanliness. The effect of steel cleanliness (based on electric-arc furnace manufacturing practice) on the contact fatigue life of tapered-roller bearings is shown in Fig. 9. The introduction of ladle refining, inert gas stirring, and inert gas shrouding has improved the life of roller bearings two orders of magnitude and improved steel cleanness to levels approaching that of vacuum-arc remelted steel (Ref 41).

Another type of oxide that has proved to be detrimental to contact fatigue resistance is known as intergranular oxidation (Ref 50). These oxides form along the surface of a steel during gas carburizing due to the oxygen potential present in the carburizing atmosphere in combination with the presence of elements that have a high affinity for oxygen such as silicon, chromium, and manganese. These oxides are important in applications such as gears for which the final contact surface remains in the as-carburized condition.

Although total oxygen content is generally accepted as an indication of steel cleanness, it is not necessarily exclusive with regard to contact fatigue life. In some instances, a steel with low oxygen content and only a few large inclusions will not give the expected increase in fatigue life. Based on observations of inclusions in 52100 and M1 steels, it was concluded that although inclusion location and type are of primary importance to contact fatigue life, size and orientation are also important and oxides and larger carbides/nitrides are more harmful than softer sulfide inclusions (Ref 51).

Carbides. Although there have been many studies performed to investigate the effect of inclusions, there has been little research showing the effects of carbides on contact fatigue. Carbides of interest are primarily those found in tool steels. The transition from using 52100 to tool steels in contact-loaded machine elements has taken place because of dimensional stability, retention of hot hardness, wear resistance, and oxidation resistance at elevated temperatures. The larger, primary carbides found in high-alloy steels act as stress raisers to initiate the micropitting mode of contact fatigue under thin lubrication films and contribute to subsurface microstructural alterations (Ref 52). Research has shown that contact fatigue life increases as the size of carbides decrease (Ref 53). A double-quench procedure following a carburizing cycle is used to keep both residual carbides and retained austenite at reasonable levels (Ref 54). A double-quench process involves carburizing followed by an initial quench and subsequent reaustenitization and quenching. This procedure is done to control matrix carbon concentration to avoid excessive retained austenite and massive carbides. In addition, double quenching produces a refined microstructure that also increases material toughness and fatigue resistance.

Retained Austenite. The influence of retained austenite has most commonly been referred to as beneficial in contact fatigue, most notably under debris-contaminated lubrication conditions and/or low lubrication film thickness (Ref 55, 56). However, the issue of retained austenite is application dependent because increased retained austenite content results in a lower material hardness for a given heat treatment and carbon content. Nevertheless, a certain amount of properly distributed retained austenite is usually desirable for contact fatigue resistance (Ref 57).

The optimal level of retained austenite depends on expected fatigue modes. For example, the plastic deformation of retained austenite can re-lax stress concentrations around oxide inclusions. However, with recent improvements in steelmaking practices, oxide concentrations have decreased such that variations in retained austenite offer fewer beneficial effects from inclusion-origin damage. Fatigue damage modes associated with contaminated lubrication are micropitting caused by debris and/or surface-initiated macropitting caused by debris dents, such that fatigue life can be increased by optimizing surface hardness and retained austenite (Ref 58). Increased retained austenite content appears to decrease crack sensitivity (increased fracture toughness) and therefore reduces the damaging effects of debris denting (Ref 59). Furthermore, it is postulated that the plastic deformation or transformation of the retained austenite results in favorable compressive residual stresses that diminish the actual applied cyclic stress.

Although retained austenite has been shown to benefit contact fatigue life, there are situations in which austenite can be a detriment, as related to dimensional stability and surface hardness. During operation, metastable retained austenite will transform under stress and strain to untempered martensite, which results in an associated volume expansion. This volume expansion can create distortion and may result in decreased life, through misalignment and noise. Also, excessive amounts of retained austenite will lower material hardness and result in a decreased resistance to fatigue initiation. To attain a given stability with a higher hardness, a subzero treatment between quenching and tempering is sometimes applied in order to control retained austenite content. However, with a subzero treatment, the mechanical properties of the steel, especially toughness and fatigue resistance, may be significantly reduced (Ref 60). Therefore, the optimal level of retained austenite is application dependent and should be varied based on expected damage mechanism.

Retained austenite content can be controlled by heat treatment, multiple quenching, tempering, cryogenics, and/or alloying. Alloying elements such as carbon, manganese, nickel, molybdenum, and chromium lower the martensite start (M_s) temperature of iron and thus produce greater levels of retained austenite. Variations on the concentrations of these elements is reflective of steel grade, and variation of carbon content produces steel types suitable for carburizing (e.g., 8620, 4320, 9310), induction-hardening (e.g., 4140, 1552, 5160), and through-hardening (e.g., 52100, M1, A2) heat treatments. Basically, a steel must have sufficient quantities of the correct alloying elements to produce a component with proper surface and core hardness, in addition to avoiding unwanted nonmartensitic transformation products (NMTP), such as quenching pearlite. Carbon content controls surface hardness, and the other elements aid in controlling core hardness and the amount of NMTP. The NMTP microconstituents—ferrite, pearlite, and bainite—are softer than the martensite matrix, reducing the contact fatigue resistance of steel and thus should be avoided.

Compared with through-hardened steels, carburized and induction-hardened steels produce residual compressive stresses and relatively large amounts of retained austenite in the subsurface region of a component. In general, resistance to contact fatigue is greatest in carburized steels, followed by through-hardened steels, and the induction-hardened steels. In the case of small components, the life of a carburizing steel is slightly longer than that of a through-hardening steel and considerably longer than that of an induction-hardened steel (Ref 56, 57). The fatigue lives of both induction-hardened and through-hardened steels for large components are nearly the same, but carburizing steels last approximately three times longer (Ref 57). The main differences among these steel types exist in hardness, retained austenite, and residual stress distributions resulting from variations in carbon concentration and heat treatment.

Residual Stresses. Favorable compressive residual stress profiles, like those found in carburized components, increase the contact fatigue life of bearings and gears (Ref 61). Residual stresses can be formed during heat treatment, subsequent peening processes, and/or cyclic loading. Residual stress can either increase or decrease the maximum Hertzian shear stress according to the following equation (Ref 34):

$$\tau_{oct}^{r} = -\tau_{oct} - \frac{1}{2}(\pm S_r)$$

where τ_{oct} is the maximum shearing stress, τ_{oct}^{r} is the maximum shearing stress modified by the residual stress, and S_r is the residual stress. The positive/negative sign ($\pm$) indicates that the residual stress may be tensile or compressive, respectively. For the case of compressive residual stresses, fatigue crack initiation and propagation rates are decreased due to a decrease in the effective applied stress and results in longer life.

Lubrication and tribology also play important roles in contact fatigue life. As shown in Table 3, it is apparent that lubrication and surface finish can be influential in three out of the five modes of contact fatigue, including all surface-initiated modes. Therefore, it is important to have a basic understanding of lubrication and surface texture variables (roughness, waviness, directionality, and microtopography) to better comprehend the phenomenon of contact fatigue.

These variables include lubricant properties such as absolute viscosity, viscosity index, pressure viscosity coefficient, additives, film thickness, surface roughness, surface tension, and operating temperatures (Ref 62). The absolute viscosity, μ_o, is resistance to fluid flow, and the viscosity index and the pressure viscosity coefficient, φ, are measures of fluid viscosity change with temperature and pressure, respectively. Additives are materials incorporated in lubricants to impart new or enhance existing properties, such as anticorrosive, antifoam, detergent, antiscuff, and so forth (Ref 63). The film thickness refers to the distance between two contacting surfaces under lubricated conditions and can be approximated by the empirical Grubin equation (Ref 64):

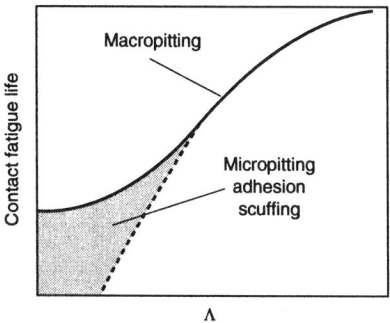

Fig. 10 Relations between contact fatigue life, macropitting, and the film parameter (Λ) for EHD lubrication

$$h_o = 0.5 \, (\mu_o \, \varphi \, v_r)^{8/11} \, (P/\vartheta)^{-1/11} \, (\Sigma 1/R)^{-4/11}$$

where h_o is the nominal film thickness, v_r is the relative velocity of the contacting surfaces, P is the normal load, ϑ is the length of contact, and R is the radius of curvature for the contacting surfaces. The shape of the contact zone and the resultant stresses depend on the magnitude of the load and the characteristics of the EHD film (Ref 65). From the above equation, one can conclude that the most important factors influencing lubricant film thickness are absolute viscosity, pressure viscosity coefficient, and contact surface velocity. Although it is advantageous to know the nominal film thickness, variations will exist from the oil inlet side to the oil exit side and in the vicinity of surface defects that result in local variations in stress distribution and adds to the inherent statistical nature of fatigue damage.

In a surface free of flaws, the surface texture becomes critical as the EHD film thickness approaches the dimensions of the surface roughness. Thus, the calculated film thickness might be less than the combined surface roughness of the contacting surfaces, resulting in asperity contact and the initiation of micropitting fatigue. In addition to surface roughness, it has been suggested that the radius of curvature or slopes of the asperities play an important role in film thickness and effects on micropitting. A surface roughness criterion for determining the extent of asperity contact is based on the ratio of the EHD film thickness to a composite surface roughness. The film parameter is given as Λ (Ref 62):

$$\Lambda = \frac{h_{min}}{\left(\sigma_1^2 = \sigma_2^2\right)^{1/2}}$$

where h_{min} is the minimum film thickness and σ_1 and σ_2 are the root-mean-square (RMS) roughness of the two surfaces in contact. At Λ less than 1, operation is in the boundary lubrication regime where macropitting life is relatively short. In fact, macropitting may not occur if it is preempted by micropitting, adhesion, or scuffing. For Λ between 1 and 3, macropitting life is increased, and micropitting, adhesion, and scuffing usually do not occur. For Λ greater than 3, macropitting may still occur, but the contact fatigue life is relatively long (Fig. 10).

Contact fatigue life has been shown to increase with increases in viscosity, pressure viscosity co-

efficient, and/or velocity. With mineral oils, investigations have shown that macropitting life increases with increasing viscosity (Ref 35). These factors again imply that contact fatigue life increases with increasing EHD film thickness, as shown in Fig. 11. However, many environmental factors influence lubricant film thickness by altering base lubricant properties or modifying the chemical activity of the additive packages. One such factor is temperature, both bulk and local. Changes in bulk temperature may significantly alter lubricant viscosity and thus affect EHD film thickness and overall fatigue performance of surfaces in contact. In addition to bulk temperature, variations in temperature within contact areas may locally impose stress gradients due to thermal expansion (Ref 66). These stress gradients can also alter lubricant viscosity and EHD film thickness. Such stresses are insignificant except under conditions of appreciable sliding, where they become more damaging as Λ decreases (Ref 67). Temperature also affects the rate of all relevant chemical reactions between lubricant and contact surface, affecting both crack initiation and propagation. Temperatures that aid in preventing crack initiation may at the same time promote mechanisms that increase crack propagation rates. High temperatures may degrade lubricants by oxidation in air, thus forming acids that have been shown to decrease contact fatigue strength. In addition to oxidation of lubricants, the physical adsorption of oxygen and water by the lubricant can result in chemical reaction (changes in surface tension) with the contact surfaces and may aid in crack propagation. In addition, the adverse effects of dissolved or entrained contamination on surface tension may also occur from physical adsorption and chemical reaction with the fresh fracture surface at the crack tip, affecting fatigue fracture toughness, including possible hydrogen embrittlement effects.

The propagation of cracks extending to the contact surface (surface or subsurface initiated) can be accelerated by hydrodynamic effects created by lubrication. This phenomenon is referred to as hydraulic pressure propagation (HPP) and was first suggested as a damage mode in 1935 (Ref 23). It has been shown that, during each contact cycle, oil under pressure can open and enter surface cracks of great length and accelerate crack propagation (Ref 68). The HPP mode of fatigue occurs when lubricant viscosity and operating speeds are low and when the contact surface has a relatively high fracture toughness (Ref 33). However, two distinctly different mechanisms for HPP have been proposed: crack propagation due to the development of stresses resulting from hydraulic oil pressure and crack propagation due to chemical interactions between the oil and crack tip.

The characteristics of surface-origin macropits (or spalls) formed due to combined rolling/sliding action, macropits formed due to nominally pure rolling action with low Λ ratios, and macropits formed by propagation of subsurface inclusion origin damage due to HPP are remarkably similar (Ref 33). However, upon metallographic

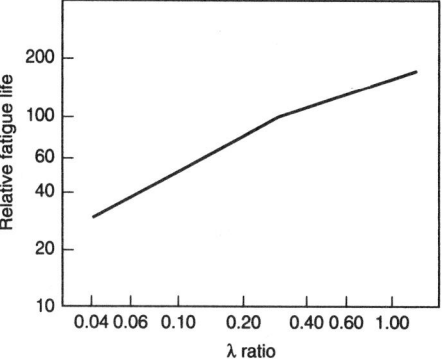

Fig. 11 Plot of relative contact fatigue life for tapered roller bearings versus Λ (the ratio between the minimum elastohydrodynamic film thickness and the root-mean-square roughness of the contacting surfaces)

sectioning, HPP macropits show a crack path that is unrelated to the known distribution of shear stresses or any other calculable stresses except for those generated by the lubricant trapped in the crack. Cracks that propagate by HPP are those inclined downward in the rolling direction at a small angle with respect to the contact surface. These observed crack propagations support the mechanism of HPP due to the development of stresses resulting from hydraulic pressure. Alternative mechanisms are possible and may operate in conjunction with the previous mechanism. These mechanisms involve a chemical reaction between the lubricant, additives, and/or impurities in the microstructure of the bearing or gear. One such mechanism is called the Rehbinder Effect after the author of publications on the subject. The Rehbinder Effect is the effect of surface-active materials on the mechanical behavior of crystalline materials. This mechanism reduces crack-extension forces (stresses) by the interaction of polar molecules with the motion of dislocations or reductions in interfacial energy, although it is not considered a corrosion-type mechanism. Increased concentrations of EP-additive packages have been shown to result in reduction of contact fatigue life and have revealed some surface attack. Penetration of more active additives into fine surface cracks led to rapid crack propagation and premature component damage (Ref 68). However, in the case of contaminated lubricants, corrosion of the crack tip may occur and result in crack growth or hydrogen generation. Hydrogen embrittlement is widely prevalent in steels with carbon contents and hardness levels typical of contact-loaded machine elements and may also explain the mechanism of HPP (Ref 69).

REFERENCES

1. H. Hertz, *Miscellaneous Papers*, Macmillan, 1896
2. R.L. Leibensperger, "Power Density: Product Design for the 21st Century," IJTM Special Publication on the Role of Technology in Corporate Policy, Int. J. Technology Management, 1991

3. A.P. Boresi and O.M. Sidebottom, *Advanced Mechanics of Materials*, John Wiley & Sons, 1985

4. R.L. Widner, Failures of Rolling Element Bearings, *Failure Analysis and Prevention*, Vol 9, 9th ed., *Metals Handbook*, American Society for Metals, 1986

5. G. Lundberg and A. Palmgren, Dynamic Capacity of Rolling Bearings, *Acta Polytech.*, Vol 1 (No. 3), 1947

6. J.O. Smith and C.K. Liu, Stresses Due to Tangential and Normal Loads on an Elastic Solid with Application to Some Contact Stress Problems, *J. Applied Mech.*, 1953

7. T.E. Tallian, *Failure Atlas for Hertz Contact Machine Elements*, American Society of Mechanical Engineers, 1992

8. "Appearance of Gear Teeth—Terminology of Wear and Failure," ANSI/AGMA 1010-E96, 1996

9. C.A. Moyer, "Comparing Surface Failure Modes in Bearings and Gears: Appearances versus Mechanisms," AGMA Tech. Paper, 91 FTM 6, 1991

10. L.E. Alban, *Systematic Analysis of Gear Failures*, American Society for Metals, 1985

11. J.L. O'Brien and A.H. King, *J. Iron Steel Inst.*, Vol 204, 1966, p 55

12. A.P. Voskamp, Material Response to Rolling Contact Loading, *J. Tribology*, Vol 107, 1985, p 359-364

13. R. Österlund and O. Vingsbo, Phase Changes in Fatigued Ball Bearings, *Metall. Trans. A*, Vol 11 (No. 5), 1980, p 701-707

14. A.P. Voskamp, R. Österlund, P.C. Becker, and O. Vingsbo, Gradual Changes in Residual Stress and Microstructure during Contact Fatigue in Ball Bearings, *Metals Tech.*, Jan 1980, p 14-21

15. E.V. Zaretsky, R.J. Parker, and W.J. Anderson, A Study of Residual Stress Induced during Rolling Contact, American Society of Mechanical Engineers, 1968

16. O. Zwirlein and H. Schlicht, Rolling Contact Fatigue Mechanisms—Accelerated Testing versus Field Performance, *Rolling Contact Fatigue Testing of Bearing Steels*, STP 771, J.J.C. Hoo, Ed., ASTM, 1982, p 358-379

17. H. Swahn, P.C. Becker, and O. Vingsbo, Martensite Decay during Rolling Contact Fatigue in Ball Bearings, *Metall. Trans. A*, Vol 7, 1976, p 1099-1110

18. K. Sugino, K. Miyamoto, M. Nagumo, and K. Aoki, Structural Alterations of Bearing Steels under Rolling Contact Fatigue, *Trans. ISIJ*, Vol 10, 1970, p 98-111

19. H. Schlicht, E. Schreiber, and O. Zwirlein, Effects of Material Properties on Bearing Steel Fatigue Strength, *Effect of Steel Manufacturing Processes on the Quality of Bearing Steels*, STP 987, J.J.C. Hoo, Ed., ASTM, 1988, p 81-101

20. M. Nagumo, K. Sugino, K. Aoki, and K. Okamoto, Initiation and Propagation of Cracks under Rolling Contact Fatigue in Bearing Steels, *Fracture*, 1969, p 587-597

21. V. Bhargava, G.T. Hahn, and C.A. Rubin, Rolling Contact Deformation, Etching Effects, and

22. T.E. Tallian, *Failure Atlas for Hertz Contact Machine Elements*, American Society of Mechanical Engineers, 1992

23. S. Way, Pitting Due to Rolling Contact, *J. Applied Mech.*, 1934, p A49-A58

24. R.L. Widner and J.O. Wolfe, Analysis of Tapered Roller Bearing Damage, No. C 7-11.1, American Society for Metals, 1967

25. C.A. Moyer, H.P. Nixon, and R.R. Bhatia, "Tapered Roller Bearing Performance for the 1990's," No. 881232, Society of Automotive Engineers, 1988

26. W.E. Littmann and R.L. Widner, Propagation of Contact Fatigue from Surface and Subsurface Origins, No. 65-WA/CF-2, *Trans. ASME*, 1966, p 624-636

27. R.L. Leibensperger and T.M. Brittain, Shear Stresses below Asperities in Hertzian Contact as Measured by Photoelasticity, *J. Lubr. Technol. (Trans. ASME)*, July 1973, p 277-286

28. M. Tokuda, M. Nagafuchi, N. Tsushima, and H. Muro, Observations of the Peeling Mode of Failure and Surface Originated Flaking from a Ring to Ring Rolling Contact Fatigue Test Rig, *Rolling Contact Fatigue Testing of Bearing Steels*, STP 771, J.J.C. Hoo, Ed., ASTM, 1982, p 150-165

29. T.E. Tallian, On Competing Failure Modes in Rolling Contact, *ASLE Trans.*, Vol 10, 1967, p 418-439

30. *Friction, Lubrication, and Wear Technology*, Vol 18, *ASM Handbook*, ASM International, 1992

31. R. Pederson and S.L. Rice, Case Crushing of Carburized and Hardened Gears, *SAE Trans.*, Vol 68, 1960, p 187

32. K. Maeda, H. Kashimura, and N. Tsushima, Investigation on the Fatigue Fracture of Core in Carburized Rollers of Bearings, *ASLE Trans.*, Vol 29, 1984, p 85-90

33. W.E. Littmann, The Mechanism of Contact Fatigue, Conf. Proc., *An Interdisciplinary Approach to the Lubrication of Concentrated Contacts*, National Aeronautics and Space Administration, 1969, p 309-377

34. E.V. Zaretsky and W.J. Anderson, Material Properties and Processing Variables and Their Effect on Rolling-Element Fatigue, Tech. Memo. TM X 52227, National Aeronautics and Space Administration, 1966

35. D. Scott, The Effect of Material Properties, Lubricant, and Environment on Rolling Contact Fatigue, *Inst. Mech. Eng.*, Vol 181, 1966, p 103-115

36. J. Brown, Big Chill to Extend Gear Life, *Power Trans. Design*, March 1995, p 59-61

37. "Rolling Contact Fatigue Life of Various Types of Carburizing Steels," Internal Publication, NTN Toyo Bearing Co., 1983

38. N. Tsushima, K. Maeda, and H. Nakashima, Influence of Improved Steel Quality on Rolling Contact Fatigue, *ASME Trans.*, 1991, p 51-57

39. K. Maeda, H. Nakashima, H. Kashimura, N. Tsushima, and S. Ito, "Rolling Contact Fatigue

Failure of High-Strength Bearing Steel, *Metall. Trans. A*, Vol 21 (No. 7), 1990, p 1921-1931

Test with Large-Sized Specimen," No. 850763, Society of Automotive Engineers, 1985

40. T.W. Morrison, T. Tallian, H.O. Wolp, and G.H. Baile, The Effect of Material Variables on the Fatigue Life of AISI 52100 Steel Ball Bearings, *ASLE Trans.*, Vol 5, 1962, p 347-364

41. J.D. Stover and U. Muralidharan, "Effect of Residual Titanium on Rolling Contact Fatigue Resistance of Bearing Steels," Conf. Proc., Steel in Motor Vehicle Manufacture International Conference, Sept 1990

42. R.A. Baughman, "Effect of Hardness, Surface Finish, and Grain Size on Rolling Contact Fatigue Life of M50 Bearing Steel," No. 59-LUB-11, American Society of Mechanical Engineers, 1959

43. S. Enekes, Effects of Some Metallurgical Characteristics on the Fatigue Life of Bearing Steels, *J. Iron Steel Inst.*, 1972, p 83-88

44. H. Yamada and N. Tsushima, Evaluation of Non-Metallic Inclusions of Steels Used for Rolling Bearings from Fracture Surface by Rotating Ring Fatigue Fracture Test, *Wear*, Vol 118, 1987, p 305-317

45. R.F. Johnson and J.R. Blank, Fatigue in Rolling Contact: Some Metallurgical Aspects, *Inst. Mech. Eng.*, 1962, p 95-104; and *Proceedings of The Symposium on Fatigue of Rolling Contact*, Institution of Mechanical Engineers, 1963

46. F. Hengerer, U. Brockmüller, and P.O. Sörström, Through Hardening or Case Hardening for Tapered Roller Bearings, *Creative Use of Bearing Steels*, STP 1195, J.J.C. Hoo, Ed., ASTM, 1993, p 21-33

47. R. Baum, K. Böhnke, J. Otto, and H.D. Pflipsen, Improved Properties of Bearing Steels by Advanced Metallurgical Processing, *Creative Use of Bearing Steels*, STP 1195, J.J.C. Hoo, Ed., ASTM, 1993, p 252-270

48. J. Monnot, B. Heritier, and J.Y. Cogne, Relationship of Melting Practice, Inclusion Type, and Size with Fatigue Resistance of Bearing Steels, *Effect of Steel Manufacturing Processes on the Quality of Bearing Steels*, STP 987, J.J.C. Hoo, Ed., ASTM, 1988, p 149-165

49. E.V. Zaretsky, Selection of Rolling Element Bearing Steels for Long Life Applications, *Effect of Steel Manufacturing Processes on the Quality of Bearing Steels*, STP 987, J.J.C. Hoo, Ed., ASTM, 1988, p 5-43

50. S.L. Rice, "Pitting Resistance of Some High Temperature Carburized Cases," Tech. Paper Series, 780773, Society of Automotive Engineers, 1978

51. R.F. Johnson and J.F. Sewell, The Bearing Properties of 1%C-Cr Steel as Influenced by Steelmaking Practice, *J. Iron Steel Inst.*, Vol 196, 1960, p 414-444

52. P.K. Pearson and T.W. Dickinson, The Role of Carbides in Performance of High-Alloy Bearing Steels, *Effect of Steel Manufacturing Processes on the Quality of Bearing Steels*, STP 987, J.J.C. Hoo, Ed., ASTM, 1988, p 113-131

53. J.L. Chevalier and E.V. Zaretsky, "Effect of Carbide Size, Area, and Density on Rolling-Element Fatigue," Tech. Note TN D-6835, Na-

tional Aeronautics and Space Administration, 1972

54. H. Muro, Y. Sadaoka, S. Ito, and N. Tsushima, The Effect of Retained Austenite on the Rolling Fatigue of Carburized Steels, *Proc. J. Congress Mater. Res.*, 1967, p 74-77

55. R.S. Sayles and P.B. Macpherson, Influence of Wear Debris on Rolling Contact Fatigue, STP 771, J.J.C. Hoo, Ed., ASTM, 1982, p 255-274

56. N. Tsushima, H. Nakashima, and K. Maeda, Improvement of Rolling Contact Fatigue Life of Carburized Tapered Roller Bearings, No. 860725, Society of Automotive Engineers, 1986

57. N. Tsushima, K. Maeda, and H. Nakashima, Rolling Contact Fatigue Life of Various Kinds of High-Hardness Steels and Influence of Material Factors on Rolling Contact Fatigue Life, STP 987, J.J.C. Hoo, Ed., ASTM, 1988, p 132-148

58. K. Toda, T. Mikami, and T.M. Johns, "Development of Long Life Bearing in Contaminated Lubrication," Tech. Series 921721, Society of Automotive Engineers, 1992

59. I. Sugiura, O. Kato, N. Tsushima, and H. Muro, "Improvement of Rolling Bearing Fatigue Life under Debris-Contaminated Lubricant by Decreasing the Crack Sensitivity of the Material," No. 81-AM-1E-2, American Society of Lubrication Engineers, 1981

60. A. Kroon and H. Nützel, Bearing Steel Development, *Ball Bearing J.*, Special '89, 1989, p 40-47

61. X. Hogbin, C. Qing, S. Eryu, W. Dengzhen, C. Zhaohong, and W. Zhengle, The Effect of Shot Peening on Rolling Contact Fatigue Behavior and Its Crack Initiation and Propagation in Carburized Steel, *Wear*, Vol 151, 1991, p 77-86

62. E.V. Zaretsky and W.J. Anderson, How To Use What We Know About EHD Lubrication, *Mach. Des.*, 1968, p 167-173

63. H. Winter and P. Oster, Influence of Lubrication on Pitting and Micropitting Resistance of Gears, *Gear Tech.*, March/April 1990, p 16-23

64. B. Fitzsimmons and A.J. Jenkins, "Bearing Damage Analysis and the Influence of Environmental Factors Relevant to Accelerated Testing Programs," ASAE, No. 71-634, 1971

65. L.B. Sibley and F.K. Orcutt, Elasto-Hydrodynamic Lubrication of Rolling-Contact Surfaces, *ASLE Trans.*, Vol 4, 1961, p 234-249

66. B.W. Kelley, The Importance of Surface Temperature to Surface Damage, *Handbook of Mechanical Wear*, University of Mich. Press, 1961, p 155

67. J.A. Jefferis and K.L. Johnson, Traction in Elastohydrodynamic Contacts, *Proc. Inst. Mech. Eng., 1967-68*, Vol 182 (No. 14), 1968, p 281-291

68. B. Michau, D. Berthe, and M. Godet, Observations of Oil Pressure Effects in Surface Crack Development, *Tribol. Int.*, June 1974, p 119-122

69. D. Scott, B. Loy, and G.H. Mills, Metallurgical Aspects of Rolling Contact Fatigue, *Proc. Inst. Mech. Eng.*, Vol 181, 1966, p 94-102

Fatigue and Fracture Resistance of Heat-Resistant (Cr-Mo) Ferritic Steels

R.W. Swindeman and W. Ren, Oak Ridge National Laboratory

The Cr-Mo steels are preferred in the construction of high-temperature components because they possess excellent strength, toughness, and corrosion resistance relative to carbon steels and most low-alloy steels. Components fabricated from Cr-Mo steels serve in the petroleum, petrochemical, fossil power, and pulp and paper industries. The steels are most often used at temperatures of 316 °C (600 °F) and higher, when creep effects, graphitization, or hydrogen attack may be significant in carbon steels.

Like most steels, the Cr-Mo steels respond to heat treatment, and a single composition may be produced in several strength classes. The low-chromium Cr-Mo steels have moderate hardenability. Slow cooling and isothermal annealing during cooling produce ferrite-pearlite microstructures that have long-time stability but relatively low tensile strength and toughness. Rapid air cooling and enhanced cooling produce bainitic steels, which tend to be stronger and tougher. Medium-chromium Cr-Mo steels have moderate hardenability, and air cooling is generally sufficient to produce bainite through-thicknesses of 150 mm (6 in.) and more. The bainitic steels are usually tempered. The high-chromium Cr-Mo steels have excellent hardening ability, and air cooling is sufficient to produce martensite through most thicknesses. Accelerated cooling of the steels is permitted, but care must be taken to avoid cracking. The martensitic steels are always tempered.

Compositions and Metallurgy. Nominal compositions for the heat-resistant, Cr-Mo steels are provided in Table 1. The steels range from ½ to 12% Cr and ½ to 1% Mo. Product forms include tubes, pipes, forgings, fittings, plates, and castings. Welded tubing and pipes are included. Depending on the product, the maximum use temperature for pressure-boundary applications ranges from 480 to 650 °C (900 to 1200 °F) (Ref 1).

Table 1 Summary of Cr-Mo steels used in high-temperature service

Nominal composition	Specification	Product form	Minimum or range of ultimate strength MPa	ksi	Minimum yield strength MPa	ksi	Maximum use temperature °C	°F	Typical product forms	Typical UNS numbers
½Cr–½Mo	A 335 P2	Tube	380	55	205	30	540	1000	Pipe, plate	K11547, K12143
	A 230 T2	Tube	414	60	205	30	540	1000	Tube	K11547
	A 182 F2	Forgings	485	70	275	40	540	1000	Forgings	K12122
1Cr-½Mo	A 387 Gr12/Cl1	Plate	380-550	55	220	33	650	1200	Plate	K11757
	A 213 T12	Tube	415	60	220	32	650	1200	Tube, fittings, pipe	K11562, K12062
	A 387 Gr12/Cl2	Plate	450-585	65	275	40	650	1200	Plate	K11757
	A 182 F12/Cl2	Forgings	485	70	275	40	650	1200	Forgings	K11564
1¼Cr–½Mo-Si	A 182 F11/Cl1	Forgings	415	60	205	30	650	1200	Forgings, tubes, fittings	K11597
	A 387 Gr11/Cl1	Plate	415-585	60	245	35	650	1200	Plate	K11789
	A 187 F11/Cl2	Forgings	485	70	275	40	650	1200	Forgings	K11572
	A 387 Gr11/Cl2	Plate	515-690	75	310	45	650	1200	Plate	K11787
2¼Cr-1Mo	A 182 F22	Forgings	415	60	205	30	650	1200	Forgings, tubes, plate	K21590
	A 217 WC9	Castings	485-655	70	275	40	650	1200	Castings	J21890
	A 182 F22/Cl3	Forgings	515	75	310	45	650	1200	Forgings, plate, bar	K21590, K21390
2¼Cr–1Mo–¼V	A 182 F22V	Forgings	585-760	85	415	60	480	900	Forgings, plate	K31835
2¼Cr-1.6W	A 213 T23	Tube		74	400	58	650	1200	Tube	
3Cr-1Mo	A 199 T21	Tube	415	60	170	25	650	1200	Tube	K31545
	A 213 T21	Tube	415	60	205	30	650	1200	Tube, forgings, plate	K31545
	A 336 F21/Cl3	Forgings	515	75	310	45	650	1200	Forgings, plate	K31545
3Cr-1Mo-¼V-Ti-B	A 182 F3V	Forgings	585-760	85	415	60	480	900	Forgings, plate	K31830
5Cr-½Mo	A 213 T5	Tube	415	60	205	30	650	1200	Tube, pipe	K41545
	A 336 F5A	Forgings	550-725	80	345	50	650	1200	Forgings	K42544
	A 182 F5a	Forgings	620	90	450	65	650	1200	Forgings	K42544
5Cr-½Mo-Si	A 213 T5b	Tube	415	60	205	30	650	1200	Tube, pipe	K51545
5Cr-½Mo-Ti	A 213 T5c	Tube	415	60	205	30	650	1200	Tube, pipe	K41245
9Cr-1Mo	A 213 T9	Tube	415	60	205	30	650	1200	Tube, fittings, pipe	K81590, K90941
	A 182 F9	Forgings	585	85	380	55	650	1200	Forgings	K81590
	A 212	Castings	620	90	415	60	650	1200	Castings	J82090
9Cr-1Mo-V	A 182 F91	Forgings	585	85	415	60	650	1200	Forgings, plate, tube	S50460
9Cr-2W	A 213	Tube	620	90	440	64	650	1200	Tube, pipe	...
12Cr-2W	A 213	Tube	620	90	400	58	650	1200	Tube, pipe	...

The ½Cr–½Mo steel is produced to three strength levels and is approved in the ASME Boiler and Pressure Vessel Code (BPVC) for service to 540 °C (1000 °F). The 1Cr-½Mo and 1¼Cr–½Mo steels are produced in three strength levels and are approved for service to 650 °C (1200 °F). The 1¼Cr–½Mo steel is the most widely used of the Cr-Mo steels, but the greatest body of information is available for 2¼Cr-1Mo steel. Wrought 2¼Cr-1Mo steel is produced to several strength levels. The annealed steel (class 1) has a largely ferrite-pearlite microstructure, although bainite is often present. The class 1 steel is used in the power industry as superheater/reheater tubing, headers, and main steam line piping for service temperatures to 650 °C (1200 °F). The normalized-and-tempered steel (class 2) is used for heavy-wall pressure vessels in the petroleum and petrochemical industries for temperatures to 480 °C (900 °F). In Europe and Asia, the class 2 material is often used for tubing and piping in the power industry for service to 650 °C (1200 °F). Quenched-and-tempered versions of 2¼Cr-1Mo steels exist. These steels are intended for heavy-wall pressure vessels but are limited to service at temperatures below the creep range because of their susceptibility to hydrogen attack (Ref 2). The 3Cr-1Mo steel is available in two strength levels and has strength properties comparable to those of 2¼Cr-1Mo steel, but improved resistance to hydrogen attack.

Recently, 2¼Cr-1Mo and 3Cr-1Mo steels with vanadium additions have been commercially developed for high-temperature pressure vessels and piping (Ref 3-8). The vanadium-bearing steels have better strength and are more resistant to hydrogen attack than steels without vanadium additions. The use for vanadium-modified 2¼Cr-1Mo steel and 3Cr-1Mo steel has been for heavy-wall pressure vessels (Ref 4, 5). Further modifications of the 3Cr-1Mo-¼V steels have been produced by adding titanium, niobium, calcium, and the like (Ref 5, 6). The strength and toughness of the steels are comparable to that of the steels listed in Table 1.

The straight 9Cr-1Mo steel is approved for service in the BPVC to 650 °C (1200 °F), but usage has not been widespread. However, the 9Cr-1Mo-V steel developed in the late 1980s (Ref 8) has been installed as tubing, piping, forgings, and castings in a variety of applications for the power, petroleum, and petrochemical industries and has accumulated up to 15 years of exposure at temperatures in the creep regime. Even more recently, tungsten has been used as a replacement for some of the molybdenum to improve the weldability and toughness of Cr-Mo steels. Three steels included in the grouping include a 2¼Cr-1.6W steel (Ref 9), a 9Cr-2W steel (Ref 10), and a 12Cr-2W steel (Ref 11). The tungsten-bearing steels contain ½% Mo and ¼% V, and they are intended for superheater/reheater tubing and piping for steam lines. Steels that contain high carbon or manganese and nickel for improved hardenability are not considered here. These steels are usually restricted to lower-tem-

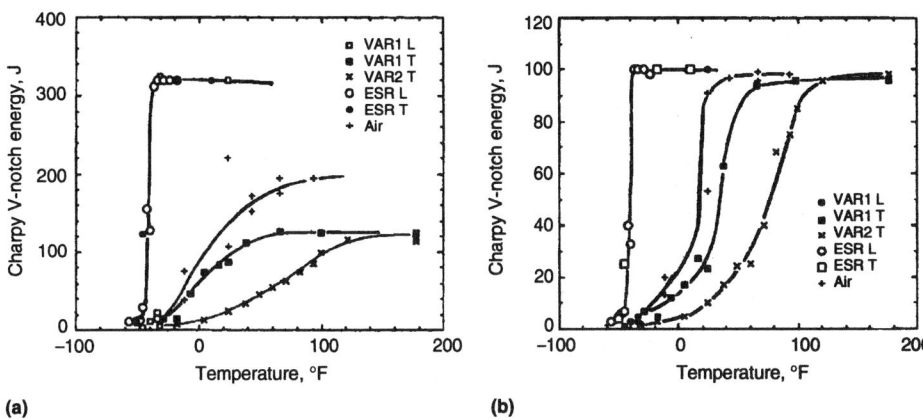

Fig. 1 Comparison of the Charpy V-notch energy and fracture appearance of 2¼Cr-1Mo steel produced by air, vacuum-arc remelting (VAR), and electroslag remelting (ESR) practices. Source: Ref 12

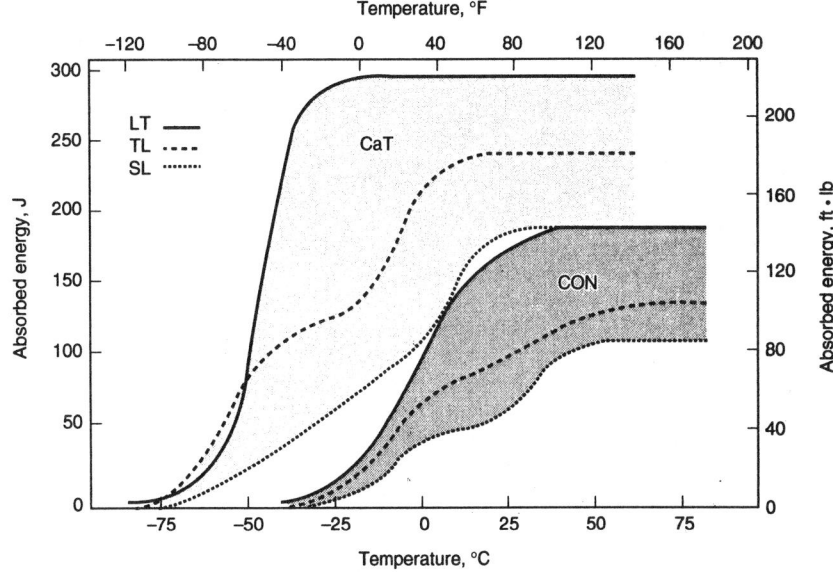

Fig. 2 Comparison of the range of Charpy V-notch energy for 2¼Cr-1Mo steel (class 2) produced by conventional (CON) melting and calcium treatment (CaT) melting. Source: Ref 13

perature service. Neither are the high-strength Cr-Mo-V steels developed for bolting, turbine casings, rotors, and disks covered here. However, these steels are very susceptible to creep, fatigue, and creep-fatigue damage, so many of the fracture and high-temperature fatigue issues discussed below apply to them.

Charpy V-Notch (CVN) Toughness

Clean steelmaking practices have significantly improved the toughness of Cr-Mo steels over the last 15 years by reducing the content of residual elements responsible for some types of embrittlement. Refinement of grain size and shape control of inclusions have produced additional improvements in toughness. Typically, the procurement specifications for the Cr-Mo steels require that the room-temperature elongations meet or exceed values in the range of 18 to 22%, depending on the strength and other factors related to the spe-

cific grade. A maximum ultimate strength or hardness is sometimes specified to ensure adequate toughness in the delivered product. For thick-section products, such as plates, a supplementary requirement for toughness is available when needed. Toughness criteria derived from the CVN or other types of toughness tests are often invoked for nuclear applications. For most Cr-Mo steels, toughness data are available that include CVN energy, fracture appearance, and lateral expansion over a range of test temperatures that include the brittle-to-ductile transition. The as-normalized or as-tempered toughness of Cr-Mo steels varies with composition, strength, and microstructure, so it is difficult to produce curves that can be considered typical of a grade or class of steel. Curves for various grades and classes overlap substantially. However, there is a tendency for toughness concerns to increase with increased alloy content and strength level. As the alloy content is increased for better high-tem-

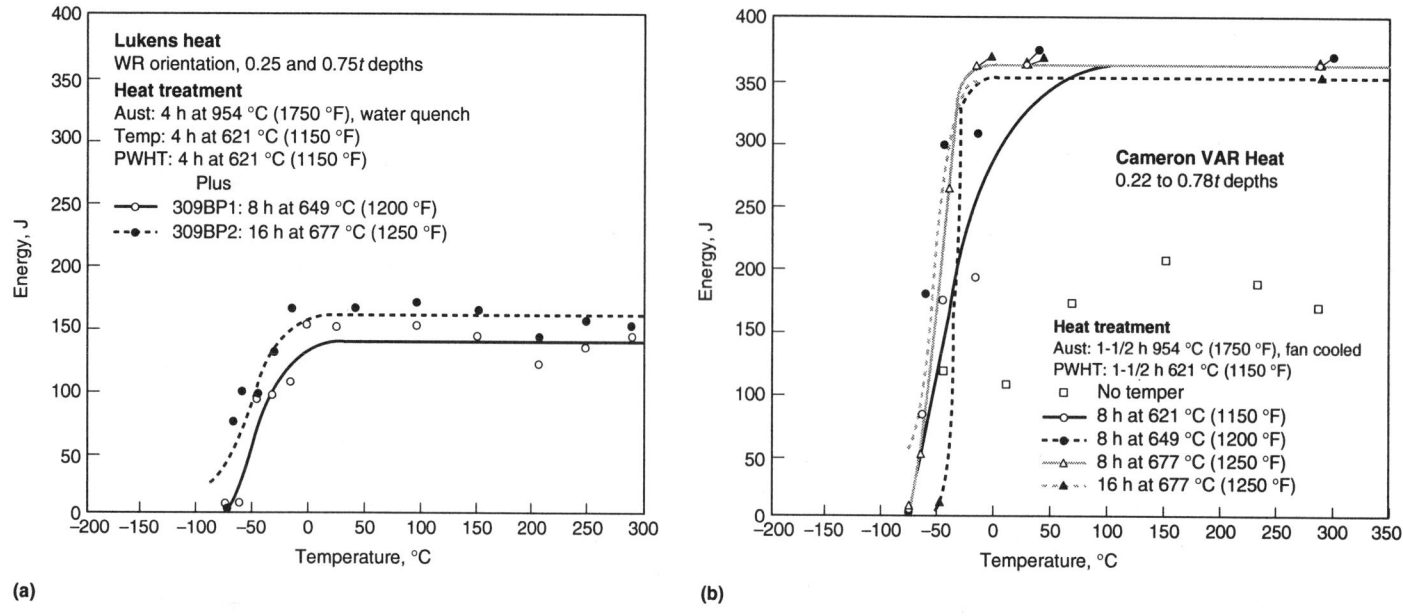

Fig. 3 Comparison of the effects of tempering on the Charpy V-notch (CVN) energy of 2¼Cr-1Mo steel (class 2), conventional and vacuum-arc remelted (VAR). (a) 152 mm thick plate. (b) 51 mm thick VAR, forged, and hot-rolled plate after a postweld heat treatment of 1.5 h at 621 °C. Source: Ref 14

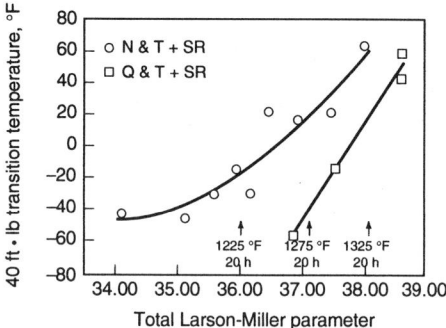

Fig. 4 Effect of postweld heat treatment on the Charpy V-notch 54 J transition temperature for 2¼Cr-1Mo steel (A387, grade 11). N&T, normalized-and-tempered; Q&T, quenched-and-tempered; SR, stress relieved. Source: Ref 13

perature strength, the susceptibility to reheat cracking and temper embrittlement increases. Such concerns promote more testing and evaluation.

The preponderance of toughness data reported in the literature for pressure-bearing, high-temperature Cr-Mo steels addresses 2¼Cr-1Mo steel. Typical CVN toughness data for four heats of annealed 2¼Cr-1Mo steel plate (SA-387, grade 22, class 1) are plotted in Fig. 1. One heat was air melted, two were vacuum-arc remelted (VAR), and one was electroslag remelted (ESR) (Ref 12). All heats were annealed and furnace cooled, then subjected to a simulated postweld heat treatment (PWHT) of 20 h at 730 °C (1350 °F). In Fig. 1(a), the CVN energies are plotted against temperature. Relative to the air-melted material, the ESR heat exhibited very low transi-

tion temperatures and high upper-shelf energies. The VAR heats exhibited good toughness in the longitudinal orientation but possessed higher transition temperatures and upper-shelf energies of only 120 J. The fracture appearance data plotted in Fig. 1(b) followed the same trends. The superior toughness in the ESR steel was attributed to the reduced inclusion content.

Charpy V-notch energies for the normalized-and-tempered plates (SA-387, grade 22, class 2) are shown in Fig. 2. Ranges are plotted for steels produced by conventional melting and calcium treatment melting practices (Ref 13). Trends for three sample orientations are included. Substantial improvements were made in both the transition temperature and the upper-shelf energy values by means of the calcium treatment, which reduces sulfur content and refines the grain size.

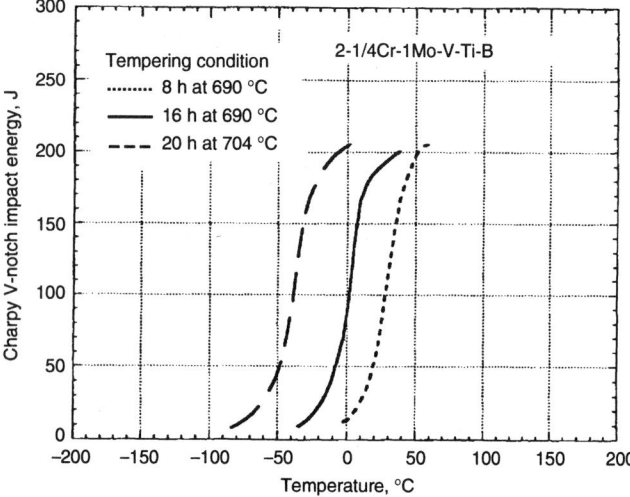

Fig. 5 Effect of tempering conditions on the Charpy V-notch energy curve for 2¼Cr-1Mo-V-Ti-B steel. Source: Ref 4

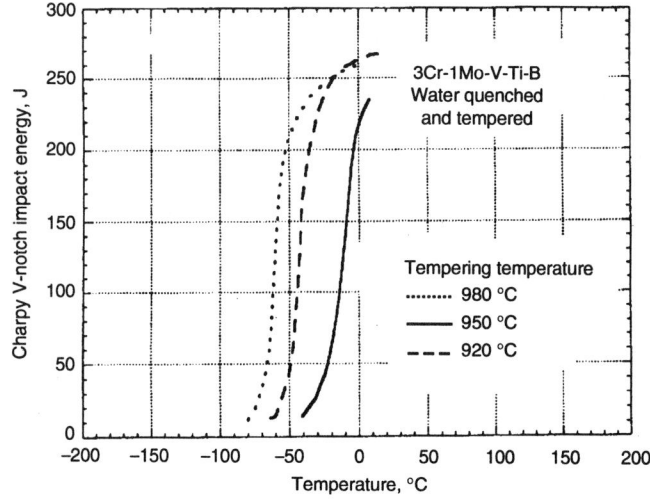

Fig. 6 Effect of normalizing temperature on the Charpy V-notch energy curve for 3Cr-1Mo-V-Ti-B steel. Source: Ref 6

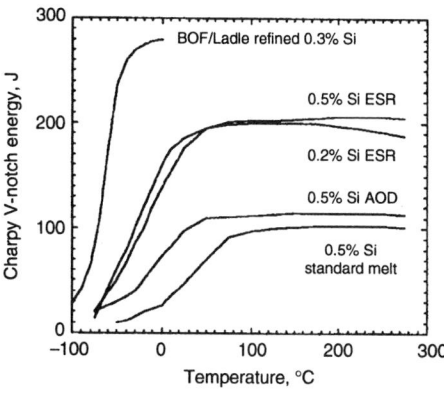

Fig. 7 Effect of silicon and melting practice on the Charpy V-notch energy curve for 3Cr-1Mo-V steel. Source: Ref 15

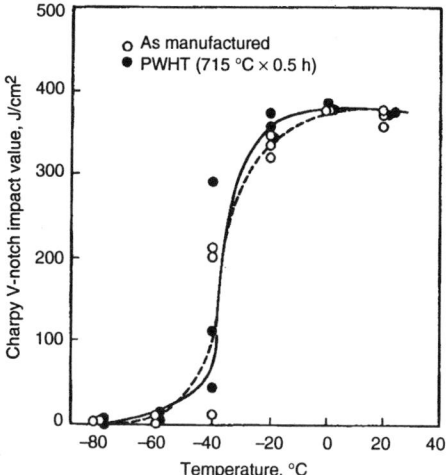

Fig. 8 Charpy V-notch energy curves for 2¼Cr-1.6W-V-Nb steel. PWHT, postweld heat treatment. Transition temperature, –39 °C. Source: Ref 9

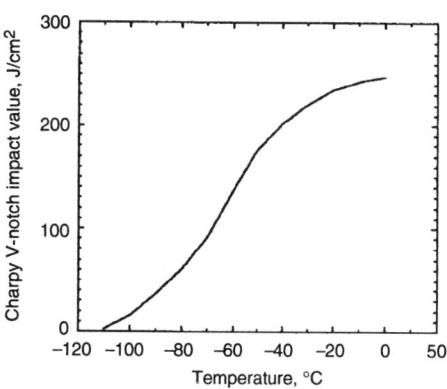

Fig. 9 Charpy V-notch energy curves for 9Cr-W-V steel as tempered

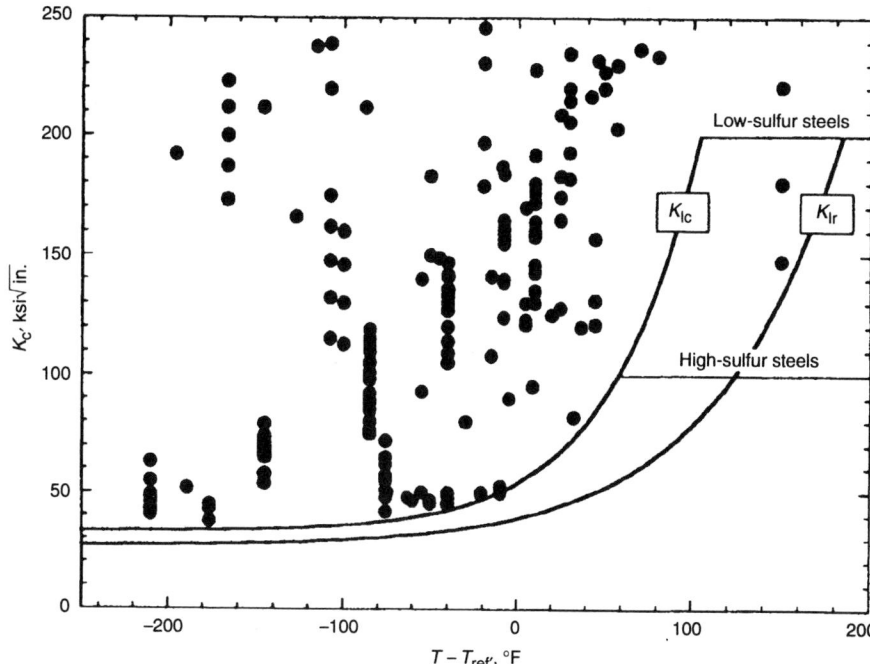

Fig. 10 Fracture toughness data for carbon and Cr-Mo steels compared with lower-bound curves developed for ASME Section III, Appendix G

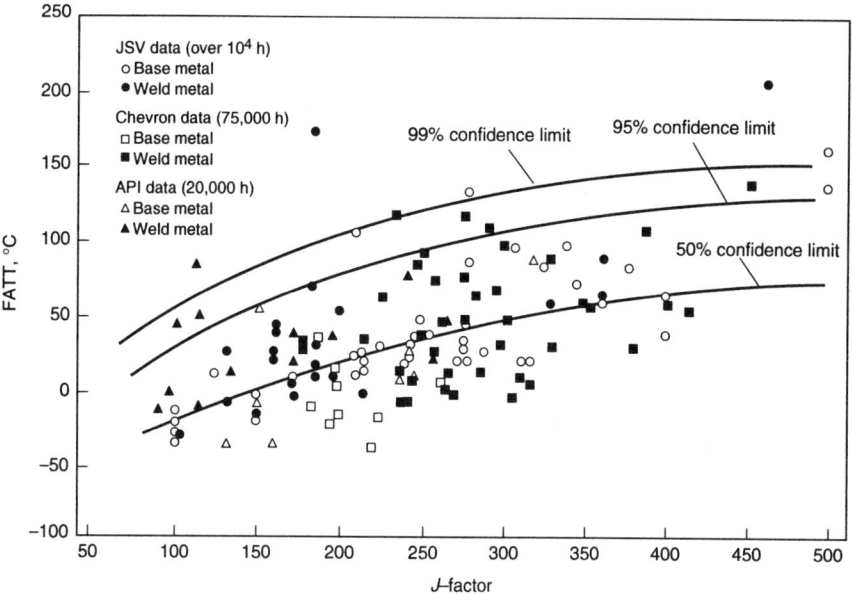

Fig. 11 Correlation of the Charpy V-notch fracture appearance transition temperature (FATT) with the J-factor for temper embrittlement of Cr-Mo steels

The toughness of 2¼Cr-1Mo steel is also affected by the normalizing temperature (through its influence on the grain size and hardenability), the cooling rate from the normalization temperature (through its effect on microstructure), and the tempering parameter (which determines the strength level and microstructure). Tempering and PWHT are important to convert the Fe₃C to more thermodynamically stable carbides and to soften the hardened heat-affected zone (HAZ) of welds. In the bainitic steels, toughness improves rapidly with tempering. Some examples are shown in Fig. 3, which plots CVN energies against testing temperature for an air-melted heat (Fig. 3a) and a VAR heat (Fig. 3b).

Wilson et al. (Ref 13) have observed that excessive PWHT may reduce toughness substantially, and they have shown examples for both 2¼Cr-1Mo and 1¼Cr-1Mo steels. Figure 4 provides a correlation between the 54 J (40 ft · lb) transition temperature and the tempering parameter for normalized-and-tempered and quenched-and-tempered grade 11 steels. Here, the tempering parameter, TP, is equivalent to the Larson-Miller parameter and is given by:

$$TP = T(20 + \log t)/1000$$

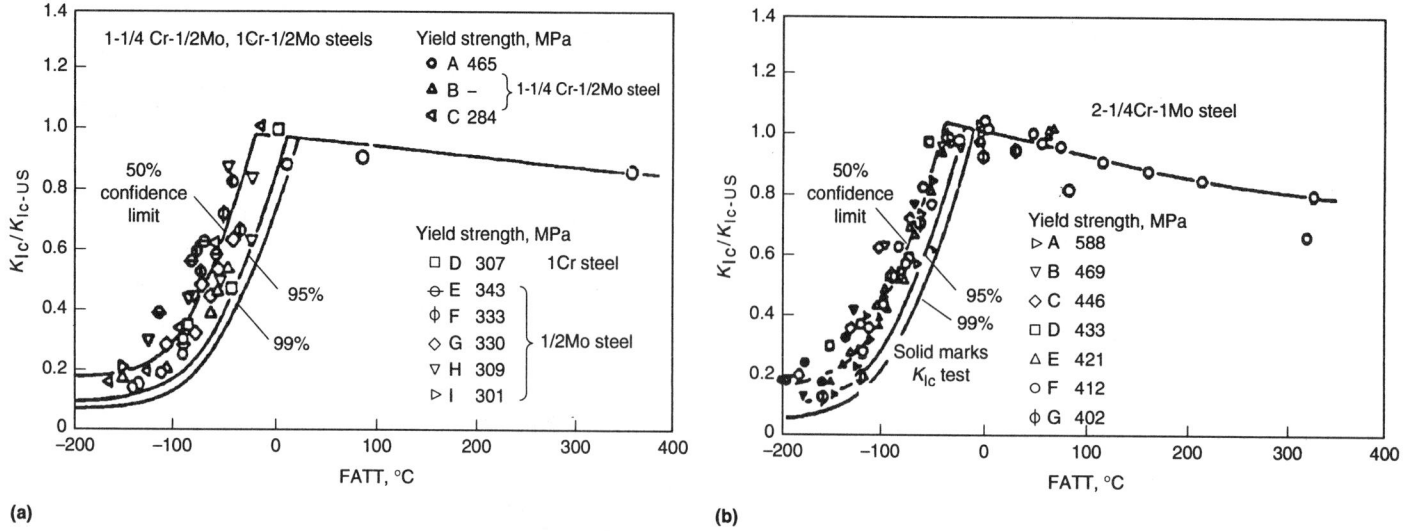

Fig. 12 Master curves for the prediction of the K_{Ic} transition curve for Cr-Mo steels exposed to embrittlement conditions. (a) $1\frac{1}{4}$Cr–$\frac{1}{2}$Mo and 1Cr-$\frac{1}{2}$Mo steel. (b) $2\frac{1}{4}$Cr-1Mo steel

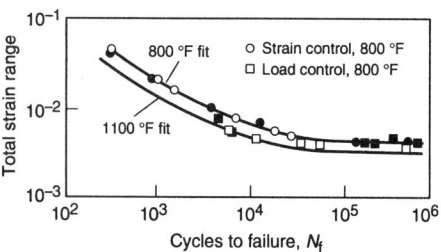

Fig. 13 Strain-fatigue curves for $2\frac{1}{4}$Cr-1Mo steel (class 1) at 425 °C (800 °F) with comparison of strain life at 595 °C (1100 °F). Open points, annealed; solid points, isothermally annealed. $R = -1$. Source: Ref 18

Fig. 14 Strain-fatigue for $2\frac{1}{4}$Cr-1Mo steel (class 1) at three temperatures. Strain rate, 4×10^{-3} per second. Source: $R = -1$. Ref 19

where T is temperature in Kelvins and t is time in hours.

The compositions and heat treatments for the new vanadium-modified Cr-Mo steels have been optimized to ensure good toughness and strength. Some typical CVN energy curves are provided in Fig. 5 for the $2\frac{1}{4}$Cr-1Mo-V-Ti-B steel (Ref 4) and in Fig. 6 for 3Cr-1Mo-V-Ti-B steel (Ref 6). The two vanadium-modified steels have similar tempering behavior, but Fig. 5 was chosen to show how the ductile-to-brittle transition temperature is decreased by increased tempering time or temperature, and Fig. 6 was chosen to show how it is increased by increased normalizing temperature, which increases grain size.

Data for the CVN toughness of martensitic 9Cr-1Mo-V steel are included in Fig. 7. Air-melted, high-silicon heats exhibit low upper-shelf energies and high transition temperatures (Ref 15). With reductions in silicon and utilization of clean steelmaking practice, very low transition temperatures and high upper-shelf energies can be achieved for 9Cr-1Mo-V steel.

The Cr-W-Mo steels are intended for tube and pipes, so extensive characterization of toughness properties has not been undertaken. Nevertheless, they appear to have good properties. The CVN energy for the $2\frac{1}{4}$Cr-1.6W steel (SA-213, grade 23) is shown as a function of temperature in Fig. 8. Toughness is excellent in both the as-normalized and PWHT conditions (Ref 9). Similarly, in the normalized-and-tempered condition the toughness of the 9Cr-2W steel (SA-213, grade 92) is excellent, as indicated in Fig. 9 for a tube heat. Reported energies at 0 °C range from 206 to 279 J/cm^2 for tubes and 116 to 162 J/cm^2 for pipes. At 0 °C, the 12Cr-2W steel has CVN energy values in the range of 65 to 330 J/cm^2 for tubes and 66 to 98 J/cm^2 for pipes.

Fracture Mechanics

Several degradation mechanisms embrittle Cr-Mo and Cr-W-Mo steels and their weldments: temper embrittlement during long-time exposure, aging embrittlement due to precipitation of brittle phases, creep damage, hydrogen attack, and hydrogen embrittlement. Traditionally, CVN data for exposed material have been used to assess the severity of some degradation mechanisms in regard to structural integrity. A more recent approach has been to perform J-R testing on "de-graded" material and derive fracture mechanics properties. As part of the development of a fitness-for-service assessment methodology, for example, the Materials Properties Council has collected fracture toughness data for carbon and Cr-Mo steels. These data have been compared to lower-bound curves more or less comparable to the ASME Section III, Appendix G curve for nuclear pressure vessel steels. Data for several grades of Cr-Mo steels are shown in Fig. 10 (Ref 16).

An approach taken by Iwadate et al. (Ref 17) to address temper embrittlement is to correlate the CVN fracture-appearance transition temperature (FATT) against an embrittlement factor. For $2\frac{1}{4}$Cr-1Mo and 3Cr-1Mo steels they chose the J factor, defined by:

$$J = (\%Si + \%Mn)(\%P + \%Sn) \times 10^4$$

For $1\frac{1}{4}$Cr–$\frac{1}{2}$Mo steel, they used the X factor, defined by:

$$X = (10P + 5Sb + 4Sn + As)/100$$

where the element content is in parts per million. With such a correlation and knowledge of the upper-shelf energy and yield strength, master curves were constructed that could be used to predict K_{Ic}. Curves for the correspondence of the FATT with the J factor are shown in Fig. 11 for the Cr-Mo steels exposed up to 75,000 h. Curves for K_{Jc} versus temperature are shown in Fig. 12 for Cr-Mo steels. Here the term K_{Ic-US} is derived from the Rolfe-Barsom correlation:

$$(K_{Ic-US}/\sigma_{0.2})^2 = 0.6478\,[(CVN_{US}/\sigma_{0.2}) - 0.0098]$$

where CVN_{US} and $\sigma_{0.2}$ are the impact energy (in joules) and the 0.2% offset yield strength (in megapascals) at the upper-shelf temperature. The correlation holds for base metal and weld metal.

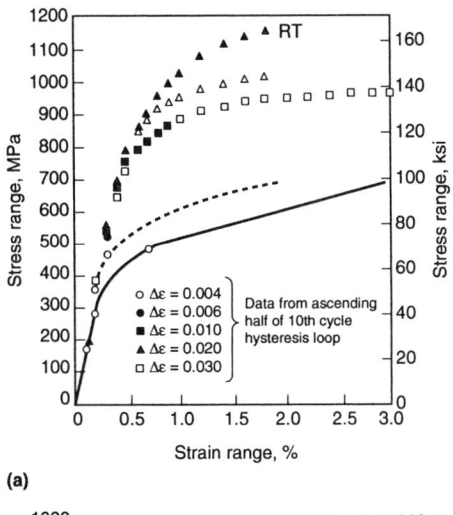

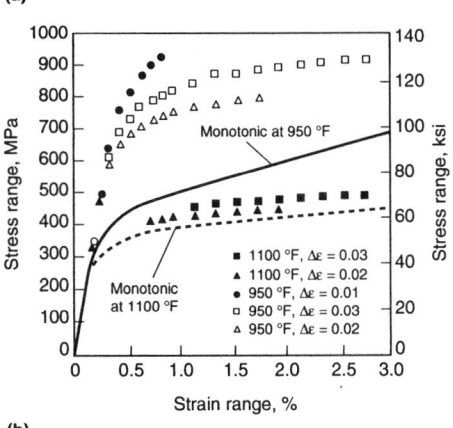

Fig. 15 Cyclic hardening curves for 2¹⁄₄Cr-1Mo steel (class 1) derived from the hysteresis loop after ten cycles. (a) Data at room temperature (RT) and 425 °C (800 °F). (b) Data at 510 and 595 °C (950 and 1100 °F). (c) Data at RT and 595 °C (1100 °F). *R* = –1. Source: Ref 20

Low-Cycle Fatigue

The Cr-Mo steels have a long history of satisfactory performance in high-temperature compo-

nents. Because the steels exhibit low thermal expansion and high thermal conductivity, relative to austenitic stainless steels, they are less susceptible to fatigue failures induced by thermal-mechanical loading. With continued operation of equipment, increased cyclic operation, and higher service temperatures, however, the risk of fatigue-related failures in Cr-Mo steels increases. At temperatures up to 370 °C (700 °F) and in the low-cycle fatigue (LCF) region, the lower-strength Cr-Mo alloys with better tensile ductility exhibit better LCF resistance. Generally, this is true when the cyclic strain range exceeds 1%. At low strain ranges corresponding to the high-cycle fatigue (HCF) regime, the steels with higher ultimate strengths exhibit better fatigue resistance. Above 370 °C (700 °F), the LCF resistance of Cr-Mo steels is affected by temperature, strain rate, load histogram, and environment. The temperature effect is relatively weak. For some combinations of strain range and strain rate, an improvement in life with increasing temperature is observed, and the improvement appears to result from dynamic strain aging. Above 370 °C (700 °F), the LCF damage mechanism in the Cr-Mo steels involves the formation of cracks in the surface oxide that lead to the penetration of transgranular cracks. These cracks appear as oxide-filled wedges.

The strain-controlled (LCF) of 2¹⁄₄Cr-1Mo steel and its weldments has been studied extensively. Annealed (class 1), normalized-and-tempered (class 2), and quenched-and-tempered (A-542) steels have been examined. One collection of LCF data for the annealed steel is provided in Fig. 13 (Ref 18). These data represent three heats for annealed steel (pipe and plate) and isothermally annealed steel (tubing). Testing temperatures ranged from 427 °C (800 °F) to 593 °C (1100 °F). Both load-controlled and strain-controlled data are included, and lives extend over four decades. Additional data for three heats of plate products are plotted in Fig. 14 for temperatures of 371, 427, and 538 °C (700, 800, and 1000 °F) (Ref 19). One low-carbon heat is included.

These trends are applicable to continuous cycling at high strain rates (0.001/s and greater) and have formed the basis for the design fatigue curves found in ASME Section III, Division 1, Subsection NH. Although the LCF curves exhibit minimal temperature effects, the cyclic strength varies with temperature, strain rate, and processing history, so the elastic and plastic components of the strain range for a given life may vary substantially. A comparison of the monotonic and cyclic hardening curves for annealed material is shown in Fig. 15 (Ref 20). The cyclic data were obtained from the rising half of the tenth-cycle hysteresis loop and correspond to ranges rather than amplitudes. Strain rates are low (0.005/min.) compared to the rate used for fatigue testing, but the trends suggest that the annealed material is a strain-hardening material rather than a strain-softening material.

The LCF for class 2 steels has been well characterized to 600 °C (1110 °F) for tubing by the National Research Institute for Metals (Ref 21). The LCF data for several temperatures and strain rates are shown in Fig. 16. These data indicate a slight decrease in fatigue life with increasing temperature and decreasing strain rate. These steels exhibit cyclic softening. Comparisons of the monotonic and cyclic stress versus strain curves are shown in Fig. 17 (Ref 22).

Unpublished data for the 2¹⁄₄Cr-1Mo-V and 3Cr-1Mo-V steels indicate LCF behavior similar to that of 2¹⁄₄Cr-1Mo (class 2) steel at temperatures in the range of 370 to 540 °C (700 to 1000 °F).

The fatigue behavior of 9Cr-1Mo-V steel has been well characterized from research in the U.S., Asia, and Europe. The LCF curves for three heats and four temperatures are shown in Fig. 18 (Ref 23). Data are tightly grouped at all temperatures.

For conditions that involve creep-fatigue interaction, the load histogram is important. Hold time at the compressive or tensile strain limit usually produces shorter fatigue lives. Often, researchers have reported that compressive hold times are more damaging than tensile hold times and pro-

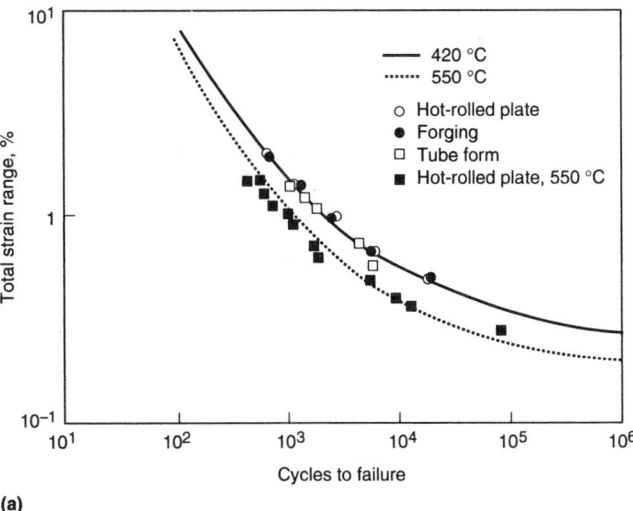

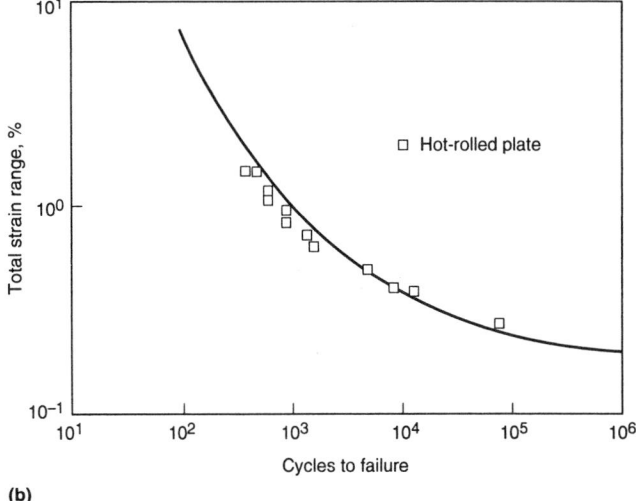

Fig. 16 Strain-fatigue curves for 2¹⁄₄Cr-1Mo (class 2). Strain rate, 0.1% per second. (a) Data at 420 °C. (b) Data at 550 °C. *R* = –1. Source: Ref 21

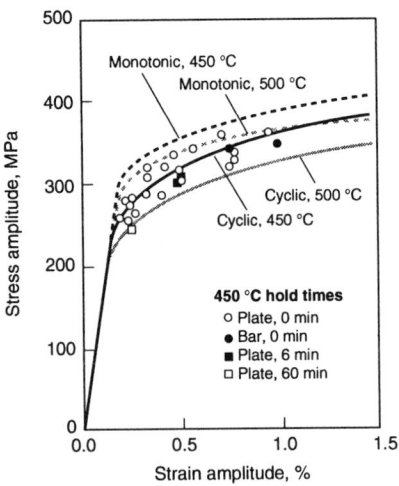

Fig. 17 Cyclic stress vs. strain curves for 2¼Cr-1Mo steel (class 2), derived from the stress amplitude at half life, at 450 and 550 °C. Strain rate 0.1% per second. $R = -1$. Source: Ref 22

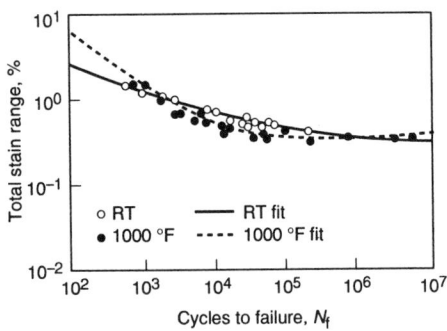

Fig. 18 Strain-fatigue curves for 9Cr-1Mo-V steel at room temperature and 540 °C (1000 °F). Strain rate, 4×10^{-3} per second. $R = -1$. Source: Ref 23

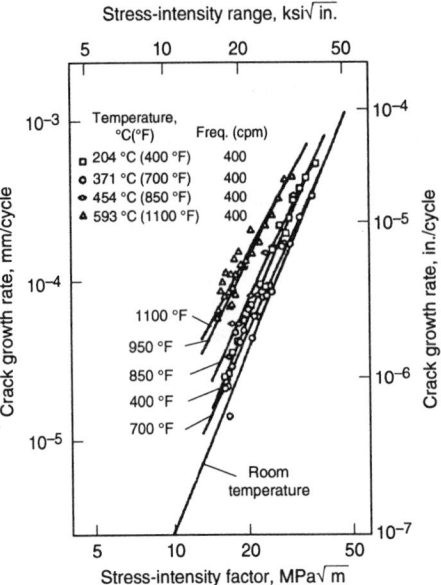

Fig. 19 Effect of temperature on the fatigue crack growth rate of 2¼Cr-1Mo steel (class 1). $R = 0.05$ in air. Source: Ref 24

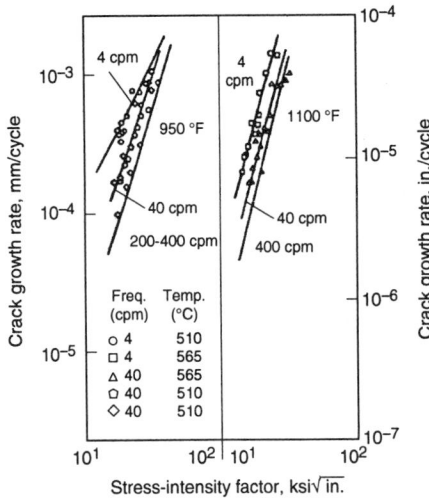

Fig. 20 Effect of frequency on the fatigue crack growth rate of 2¼Cr-1Mo steel (class 1) at 510 and 593 °C (950 and 1100 °F). $R = 0.05$ in air. Source: Ref 24

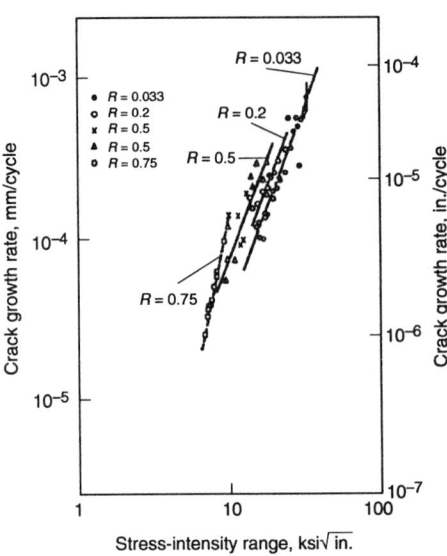

Fig. 21 Effect of mean stress on the fatigue crack growth rate of 2¼Cr-1Mo steel (class 1) at 510 °C (950 °F). Tested in air at 40 to 50 cpm. Source: Ref 24

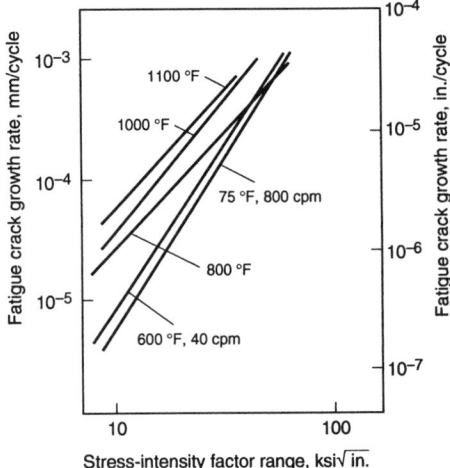

Fig. 22 Fatigue crack propagation behavior of modified 9Cr-1Mo steel (ASTM A387, grade 91), tested in air at room temperature (800 cpm) and at elevated temperatures (40 cpm). $R = 0.05$. Source: Ref 25

duce shorter lives. Explanations for the hold time effects have been offered that relate to mean stress effects and to the oxide cracking mechanism, but no consensus exists in regard to the best method to evaluate creep-fatigue damage in Cr-Mo steels. Linear damage calculations based on cycle fractions and time fractions require continuous cycling data and stress rupture data, and fatigue damage calculations require the type of fatigue data presented above. Most other approaches, such as strain-range partitioning, ductility exhaustion, and frequency separation, require supplementary data that are not provided in this article.

Fatigue Crack Growth

Often, LCF curves are used to estimate the cycles to fatigue crack initiation in relatively thick sections. Thereafter, subcritical fatigue crack growth procedures are used to estimate remaining life. Typical fatigue crack growth data for annealed 2¼Cr-1Mo steel are plotted in Fig. 19 to 21 (Ref 24). The effect of temperature is

illustrated in Fig. 19, the effect of frequency is illustrated in Fig. 20, and the effect of stress ratio is shown in Fig. 21. Fatigue crack growth curves for 9Cr-1Mo-V steel are provided in Fig. 22 (Ref 25).

Weldments

Although the fatigue behavior of weldments is covered elsewhere in this Volume, it should be mentioned that the strength, toughness, fatigue, and creep-fatigue behavior of Cr-Mo steel weld metal and weldments is of major concern in the

selection of materials and the manufacture of pressure-bearing components. For new construction, the testing requirements for weld metal and weldments are specified in the BVPC Section IX. Filler metals are available that exhibit strength properties comparable or superior to those of base metal, but the hardenability of most Cr-Mo steels is sufficient to produce hard, low-toughness regions in the HAZ of thick-section weldments. A proper PWHT restores toughness properties. Often, the HAZ contains microstructures developed from heating the base metal to the intercritical zone. Very fine-grain material in this zone tends to have poor high-temperature strength. Under fatigue and creep-fatigue loading, failures often occur in the HAZ of weldments. Because composition, metallurgical structure, loading history, and section geometry strongly influence the life under cyclic loading, high-temperature test data on the fatigue, creep-fatigue, and fatigue-crack-growth of Cr-Mo weldments are not well enough defined to be included here.

REFERENCES

1. Boiler and Pressure Vessel Code, Section II, Part D, American Society of Mechanical Engineers, 1995
2. W.E. Erwin and J.G. Kerr, The Use of Quench and Tempered 2¼Cr-1Mo Steel for Thick Wall Reactor Vessels in Petroleum Refinery Processes: An Interpretive Review of 25 Years of Research and Application, Bulletin 275, Welding Research Council, Feb 1982
3. R.W. Swindeman and M. Gold, Developments in Ferrous Alloy Technology for High-Temperature Service, *J. Pressure Vessel Technol.*, Vol 113, May 1991, p 113-140
4. T. Ishiguro, Y. Murakami, K. Ohnishi, and J. Watanabe, A 2¼Cr-1Mo Pressure Vessel Steel with Improved Creep Rupture Strength, *Application of 2¼Cr-1Mo Steel for Thick-Wall Pressure Vessels*, STP 755, American Society for Testing and Materials, 1982, p 129-147
5. A.G. Imgram, S. Ibarra, and M. Prager, A Vanadium Modified 2¼Cr-1Mo Steel with Superior Performance in Creep and Hydrogen, *New Alloys for Pressure Vessel and Piping*, PVP Vol 210, American Society of Mechanical Engineers, 1990, p 1-28
6. T. Ishiguro, Y. Murakami, K. Ohnishi, S. Mima, and J. Watanabe, Development of a 3Cr-1Mo-¼V-Ti-B Pressure Vessel Steel for Enhanced Design Stress, *Research in Chrome-Moly Steels*, MPC-21, American Society of Mechanical Engineers, 1984, p 43-51
7. M. Yamada, T. Sakai, and S. Nose, Development of Cr-Mo-V-Cb-Ca Steel for High Pressure and High Temperature Hydrogenation Reactors, *Fitness-for-Service and Decisions for Petroleum and Chemical Equipment*, PVP Vol 315, American Society of Mechanical Engineers, 1995, p 385-396
8. V.K. Sikka, C.T. Ward, and K.C. Thomas, Modified 9Cr-1Mo Steel—An Improved Alloy for Steam Generator Application, *Ferritic Steels for High-Temperature Applications*, American Society for Metals, 1983
9. F. Masuyama, T. Yokoyama, Y. Sawaragi, and A. Asada, Development of Tungsten Strengthened Low Alloy Steel with Improved Weldability, *Service Experience and Reliability Improvement: Nuclear, Fossil, and Petrochemical Plants*, PVP Vol 288, American Society of Mechanical Engineers, 1994, p 141-146
10. H. Matsumoto et al., Development of a 9Cr-0.5Mo-1.8W Steel for Boiler Tubes, *First International Conference on Improved Coal-Fired Power Plants*, EPRI, 1986, p 5.203-5.218
11. A. Iseda, Y. Sawaragi, S. Kato, and F. Masuyama, Development of a New 0.1C-11Cr-2W-0.4Mo-1Cu Steel for Large Diameter and Thick Wall Pipe for Boilers, *Creep: Characterization, Damage, and Life Assessments*, ASM International, 1992, p 389-397
12. H.P. Offer and P.J. Ring, "An Evaluation of Vacuum Arc Remelted Tubesheet Forging Properties for CRBRP Steam Generators," GEFR-00245, General Electric Co., Sunnyvale, CA, Sept 1977
13. A.D. Wilson, C.R. Roper, K.E. Orie, and F.B. Flecher, Properties and Behavior of Modern A387-22 Cr-Mo Steels, *Serviceability of Petroleum, Process, and Power Equipment*, PVP Vol 239, American Society of Mechanical Engineers, 1992, p 69-79
14. R.W. Swindeman, R.K. Nanstad, J.F. King, and W.J. Stelzman, "Effect of Tempering on the Strength and Toughness of 2¼Cr-1Mo Steel Weldments," ORNL/TM-9307, Oak Ridge National Laboratory, Oct 1984
15. C.R. Brinkman, D.J. Alexander, and P.J. Mazaisz, "Influence of Long-Term Thermal Aging (50,000 h) on the Charpy V-Notch Properties of Modified 9Cr-1Mo Steel," ORNL-6506, Oak Ridge National Laboratory, Oct 1988
16. T.L. Anderson, Probabilistic Establishment of Fracture Toughness Distributions for Fitness-for-Service Evaluations, *Fitness-for Service and Decisions for Petroleum and Chemical Equipment*, PVP Vol 315, American Society of Mechanical Engineers, 1995, p 485-490
17. T. Iwadate, Y. Tanaka, and H. Takemata, Prediction of the Fracture Toughness KII Transition Curves of Pressure Vessel Steels from Charpy V-Notch Impact Test Results, *Serviceability of Petroleum, Process, and Power Equipment*, PVP Vol 239, American Society of Mechanical Engineers, 1992, p 95-101
18. J.R. Ellis, M.T. Jakub, C.E. Jaske, and D.A. Utah, Elevated Temperature Fatigue and Creep-Fatigue Properties of Annealed 2¼Cr-1Mo Steel, *Structural Materials for Service at Elevated Temperatures in the Nuclear Power Generation*, MPC-1, American Society of Mechanical Engineers, 1975, p 213-246
19. M.P. Booker, J.P. Strizak, and C.R. Brinkman, "Analysis of the Continuous Cycling Fatigue Behavior of 2¼Cr-1Mo Steel," ORNL-5593, Oak Ridge National Laboratory, Dec 1979
20. C.E. Jaske, B.N. Leis, and C.E. Pugh, Monotonic and Cyclic Stress-Strain Response of Annealed 2¼Cr-1Mo Steel, *Structural Materials for Service at Elevated Temperatures in the Nuclear Power Generation*, MPC-1, American Society of Mechanical Engineers, 1975, p 191-212
21. "Data Sheets on Elevated-Temperature, Low-Cycle Fatigue Properties of SCMV4 (2.25Cr-1Mo) Steel Plate for Pressure Vessels," Fatigue Data Sheet 7, National Research Institute for Metals, Tokyo, 1978
22. K. Aoto, H. Kawasaki, Y. Wada, and I. Nihel, Analytical Study on Correlationship between Stress Relaxation and Creep Deformation Behavior of Ferritic Steels, *Collection and Uses of Relaxation Data in Design*, PVP Vol 172, American Society for Mechanical Engineers, 1989, p 9-14
23. C.R. Brinkman, J.P. Strizak, and M.K. Booker, "Smooth- and Notched-Bar Fatigue Characteristics of Modified 9Cr-1Mo Steel," ORNL-6330, Oak Ridge National Laboratory, Feb 1987
24. C.R. Brinkman, W.R. Corwin, M.K. Booker, T.L. Hebble, and R.L. Klueh, Time Dependent Mechanical Properties of 2¼Cr-1Mo Steel for Use in Steam Generator Design, ORNL-5125, Oak Ridge National Laboratory, March 1976
25. L.A. James and K.W. Carlsen, The Fatigue Crack Growth and Ductile Fracture Toughness Behavior of ASTM A387 Gr91 Steel, *Residual-Life Assessment, Nondestructive Examination, and Nuclear Heat Exchanger Materials*, PVP Vol 98-1, American Society of Mechanical Engineers, 1985, p 97-107

Fatigue and Fracture Properties of Stainless Steels

S. Lampman, ASM International

STAINLESS STEELS are used in a wide variety of structural applications. Most of the structural applications occur in the chemical and power engineering industries, which account for more than a third of the market for long and flat stainless steel products (circa 1993, Ref 1). These applications include an extremely diversified range of use, including nuclear reactor vessels, heat exchangers, oil industry tubulars, components for the chemical processing and pulp and paper industries, furnace parts, and boilers used in fossil fuel electric power plants.

Stainless steels of all types contain chromium in amounts ranging from 10 to 30%. The stainlessness and oxidation resistance of stainless steels is primarily dependent on chromium content. Other alloying elements, including silicon, nickel, molybdenum, copper and aluminum, contribute to the general corrosion resistance, but their influence is limited compared to that of chromium. Standard compositions of stainless steels (Table 1) are generally divided by microstructure into the following groups: austenitic types, ferritic types, duplex types, martensitic types, and precipitation hardening types. The following sections briefly summarize the key mechanical characteristics of these stainless steel types with particular emphasis on fracture properties and corrosion fatigue. Typical room-temperature mechanical properties and fatigue endurance limits are summarized in Table 2. Additional information on corrosion and environmentally assisted fracture of stainless steels is contained in Ref 2 and elsewhere in this Volume.

Stainless Steel Types

Ferritic stainless steels are so named because their body-centered-cubic (bcc) crystal structure is the same as that of iron at room temperature. These alloys are magnetic and cannot be hardened by heat treatment. Ferritic stainless steels contain between 11 and 30% Cr, with only small amounts of austenite-forming elements, such as carbon, nitrogen, and nickel. Their general use depends on their chromium content for corrosion resistance.

In general, ferritic stainless steels do not have particularly high strength (Fig. 1). Their annealed yield strengths range from 275 to 350 MPa (40 to 50 ksi), and their poor toughness and susceptibility to sensitization limit their fabricability and the usable section size. Their chief advantages are their resistance to chloride stress-corrosion cracking, atmospheric corrosion, and oxidation at a relatively low cost.

Ferritic stainless steels can be a viable alternative to austenitic steels for some high-temperature structural applications. In view of their lower alloy content and lower cost, ferritic steels have an economic advantage over austenitic steels, stress conditions permitting. Ferritic stainless steels can also be desirable in applications with thermal cycling because ferritic steels have higher thermal conductivities and lower thermal expansion coefficients than austenitic steels. Ferritic steels may therefore allow reductions in thermal stresses and improved thermal-fatigue resistance.

In addition various grades of high-chromium heat-resistant ferritic steels are important alloys for high-temperature structural applications. High-chromium heat-resistant ferritic steels were first developed for use in gas and steam turbine applications. In the early 1940s, a need for improved high-strength, corrosion-resistant materials for gas turbine disks and steam turbine blades, operating at temperatures near 540 °C (1000 °F), led metallurgists in England to increase development work on 12% Cr heat-resistant steels. This research produced two steels (H-46 and FV448) and other alloy steels in the 1950s with improved creep-rupture strengths. During the same time period, other 12% Cr alloys such as AISI 422 (UNS S42200) and Lapelloy were introduced in the United States; they have been used for steam turbine blades and bolting materials for turbine cylinders. Although many of the high-chromium heat-resistant ferritics are not strictly considered stainless steels, the chromium contents are high enough to exhibit corrosion resistance comparable to that of typical stainless steels. These alloys are considered in the article "Fatigue and Fracture Resistance of Heat-Resistant (Cr-Mo) Ferritic Steels" in this Volume.

Austenitic stainless steels constitute the largest stainless family in terms of number of alloys and usage. Like the ferritic alloys, they cannot be hardened by heat treatment. Austenitic alloys are nonmagnetic, and their structure is face-centered-cubic (fcc). They possess excellent ductility, formability, and toughness, even at cryogenic temperatures. In addition, they can be substantially hardened by cold work.

Austenitic stainless steels can be subdivided into two categories: chromium-nickel alloys, such as S30400 and S31600, and chromium-manganese-nitrogen alloys, such as S20100 and

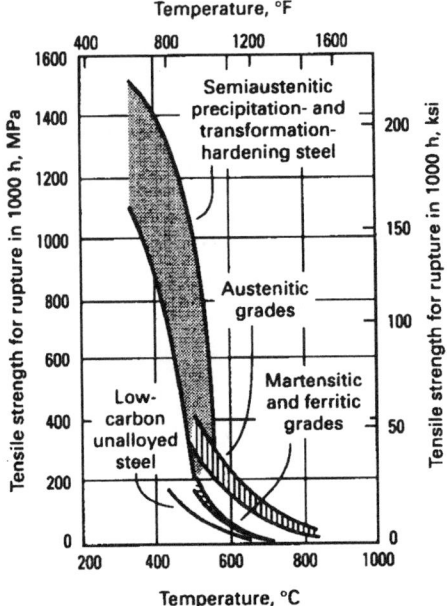

Fig. 1 General comparison of the hot-strength characteristics of austenitic, martensitic, and ferritic stainless steels with those of low-carbon unalloyed steel and semiaustenitic precipitation and transformation-hardening steels

Table 1 Compositions of common structural stainless steels

Type	UNS number	C	Mn	Si	Cr	Ni	P	S	Others	Yield strength(b), MPa (ksi)
Austenitic types										
301	S30100	0.15	2.00	1.00	16.0-18.0	6.0-8.0	0.045	0.03	...	205 (30) ann
304	S30400	0.08	2.00	1.00	18.0-20.0	8.0-10.5	0.045	0.03	...	205 (30) ann
304L	S30403	0.03	2.00	1.00	18.0-20.0	8.0-12.0	0.045	0.03	...	170 (25) ann
304LN	...	0.03	2.00	1.00	18.0-20.0	8.0-10.5	0.045	0.03	0.10-0.15N	205 (30) ann
308	S30800	0.08	2.00	1.00	19.0-21.0	10.0-12.0	0.045	0.03	(Welding electrode composition)	205 (30) ann
309S	S30908	0.08	2.00	1.00	22.0-24.0	12.0-15.0	0.045	0.03	...	205 (30) ann
310S	S31008	0.08	2.00	1.50	24.0-26.0	19.0-22.0	0.045	0.03	...	205 (30) ann
316	S31600	0.08	2.00	1.00	16.0-18.0	10.0-14.0	0.045	0.03	2.0-3.0Mo	205 (30) ann
316N	S31651	0.08	2.00	1.00	16.0-18.0	10.0-14.0	0.045	0.03	2.0-3.0Mo, 0.10-0.16N	240 (35) ann
321	S32100	0.08	2.00	1.00	17.0-19.0	9.0-12.0	0.045	0.03	(5 × %C) min Ti	
348	S34800	0.08	2.00	1.00	17.0-19.0	9.0-13.0	0.045	0.03	0.2Cu, (10 × %C) min (Nb + Ta)(c)	
21-6-9(d)	S21900	0.08	8.0-10.0	1.00	19.0-21.5	5.0-7.0	0.06	0.03	0.15-0.40N	
22-13-5(e)	S20910	0.06	4.0-6.0	1.00	20.5-23.5	11.5-13.5	0.04	0.03	1.5-3.0Mo, 0.1-0.3Nb, 0.1-0.3V, 0.2-0.4N	345 (50) ann 380 (55) ann
Kromarc 58	...	0.03	9.3	0.05	15.5	23.0	0.005	0.005	2.2Mo, 0.02Al, 0.16V, 0.008Zr, 0.016B, 0.17N	371 (54) ann
Ferritic types										
405	S40500	0.08	1.0	1.0	11.5-14.5	...	0.04	0.03	0.1-0.3Al	170 (25) ann
409	S40900	0.08	1.0	1.0	10.5-11.75	0.5	0.045	0.045	...	205 (30) ann
430	S43000	0.12	1.0	1.0	16-18	...	0.04	0.03	0.6Mo	205 (30) ann
434	S43400	0.12	1.0	1.0	16-18	...	0.04	0.03	0.75-1.25Mo	365 (53) ann
442	S44200	0.20	1.0	1.0	18-23	...	0.04	0.03	...	275 (40) ann
446	S44600	0.20	1.5	1.0	23-27	...	0.04	0.03	0.25N	275 (40) ann
Martensitic types										
403	S40300	0.15	1.00	0.50	11.5-13.0	...	0.04	0.03	...	550 (80) ht
410	S41000	0.15	1.00	1.00	11.5-13.0	...	0.04	0.03	...	550 (80) ht
420	S42000	0.15 min	1.00	1.00	12.0-14.0	...	0.04	0.03	...	1480 (215) ht
422	S42200	0.20-0.25	1.00	0.75	11.0-13.0	0.5-1.0	0.025	0.025	0.75-1.25Mo, 0.75-1.25W, 0.15-0.30V	760 (110) ht
431	S43100	0.20	1.00	1.00	15.0-17.0	1.25-2.50	0.04	0.03	...	1030 (149) ht
Precipitation hardening types(f)										
15-5PH	S15500	0.07	1.00	1.00	14.0-15.5	3.5-5.5	0.04	0.03	2.5-4.5Cu, 0.15-0.45 (Nb + Ta)	1000 (145) ht
17-4PH	S17400	0.07	1.00	1.00	15.5-17.5	3.0-5.0	0.04	0.03	3.0-5.0Cu, 0.15-0.45 (Nb + Ta)	1000 (145) ht
PH13-8Mo	S13800	0.05	0.10	0.10	12.25-13.25	7.5-8.5	0.01	0.008	2.0-2.5Mo, 0.90-1.35Al, 0.01N	1310 (190) ht
Custom 455 (XM-16)	S45500	0.05	0.50	0.50	11.0-12.5	7.5-9.5	0.04	0.03	0.5Mo, 1.5-2.5Cu, 0.8-1.4Ti, 0.1-0.5Nb	1410 (205) ht
17-7PH	S17700	0.09	1.00	1.00	16.0-18.0	6.5-7.75	0.04	0.03	0.75-1.5Al	1030 (150) ht
PH15-7Mo	S15700	0.09	1.00	1.00	14.0-16.0	6.5-7.75	0.04	0.03	2.0-3.0Mo, 0.75-1.5Al	1100 (160) ht
AM355	S35500	0.10-0.15	0.5-1.25	0.50	15.0-16.0	4.0-5.0	0.04	0.03	2.5-3.25Mo	1030 (150) ht
A-286	K66286	0.04	1.30	0.40	15.0	26.0	...	...	1.3Mo, 2.0Ti, 0.2Al, 0.3V, 0.005B	755 (110) ht

(a) Compositions are in weight percent. Single values are maximum values except for Kromarc 58 and A-286, for which nominal compositions are shown. (b) Ann-annealed, ht-heat treated; yield strengths are minimum values for annealed material and typical values for heat treated material. (c) Optional Nb + Ta. (d) Nitronic 40 (XM-10). (e) Nitronic 50 (XM-19). (f) 15-5PH, 17-4PH, PH13-8Mo and Custom 455 are martensitic precipitation hardening types; 17-7PH, PH15-7Mo and AM355 are semiaustenitic precipitation hardening types; and A286 is an austenitic precipitation hardening type.

S24100. The latter group generally contains less nickel and maintains the austenitic structure with high levels of nitrogen. Manganese (5 to 20%) is necessary in these low-nickel alloys to increase nitrogen solubility in austenite and to prevent martensite transformation. The addition of nitrogen also increases the strength of austenitic alloys. Typical chromium-nickel alloys have tensile yield strengths from 200 to 275 MPa (30 to 40 ksi) in the annealed condition, whereas the high-nitrogen alloys have yield strengths up to 500 MPa (70 ksi).

The most common austenitic stainless steel is Type 304, which has been used for critical elevated-temperature applications such as high pressure piping in power generating stations and for very low temperature applications such as liquid helium transfer piping and storage vessels. The alloy is generally used in the annealed condition (heated at 1060 °C, or 1950 °F, to dissolve carbides, and quenched in water or cooled in air at a rate high enough to prevent carbide reprecipitation). Type 304 has high ductility and toughness in this condition over a wide range of service temperatures. Because of its high toughness and relatively low yield strength, plane-strain conditions cannot be obtained on monotonic loading up to the point of unstable crack propagation in precracked fracture toughness specimens of reasonable size. The same limitation applies to all austenitic stainless steel in the annealed condition. Therefore, available fracture mechanics data for these alloys generally are limited to fatigue crack growth rate data (da/dN). However, valid monotonic fracture mechanic properties of austenitic stainless steels are an area of active engineering interest (see the article "Fracture Toughness of Austenitic Stainless Steels and Their Welds" in this Volume).

All of the austenitic stainless steels in Table 1 are modifications of Type 304. Increasing the nickel content increases the austenite stability and reduces the work hardening effect. Increasing the chromium content increases the resistance to deformation at elevated temperatures, and also increases the oxidation and corrosion resistance. Addition of molybdenum increases corrosion resistance. Titanium, niobium and tantalum are added to stabilize carbide formation and minimize grain boundary sensitization following welding. Up to 0.30% nitrogen and small amounts of other elements may be added to heats of stainless steel by melting under a nitrogen atmosphere and by additions of master alloys.

Table 2 Typical tensile properties and fatigue limits of stainless steels

Alloy	Form or thickness(a)	Condition	Ultimate tensile strength, ksi	Tensile yield strength 0.2%, ksi	Elongation in 2 inches, %	Reduction of area (for bars), %	Hardness	Izod, ft-lb	Charpy V, ft-lb	Smooth specimen fatigue strength(b), ksi
Austenitic types										
201	Sheet	Annealed	110	55	55	65	95 HRB	110	100	
		10% CR	130	90	36		25 HRC			
		40% CR	190	165	10		41 HRC			
202	Sheet	Annealed	105	50	55		87 HRB		85	
		10% CR	128	98	31					
		40% CR	184	155	8					
203 EZ	1 inch bar	Annealed	87	45	53	63	86 HRB	184	85	
		CD	115	80	15	35				
		Full hard	170	140	17	47	35 HRC			
301	Sheet	Annealed	120	40	75	65	80 HRB	110	114	35
		10% CR	150	88	55	40	22 HRC	100		
		40% CR	195	170	23	10	45 HRC			80
		60% CR	235	230	10					80 (R)
302	Sheet	Annealed	93	40	68	65	80 HRB	110	118	34 (bar)
		10% CR	108	92	43	41	10 HRC	65		
		40% CR	151	132	13	8	35 HRC			75
302B	Plate	Annealed	90	40	50	65	82 HRB	90	90	
303	1 inch bar	Annealed	90	37	65	60	80 HRB	85		34
		10% CW	105	45	45	57				48
		40% CW	173	80	13	45				
303 Se	1 inch bar	Annealed	90	35	50	65	160 HB	80		
		Half hard	137	105	25					
		Full hard	175	137	14		32 HRC			
304	Bar	Annealed	87	34	57	67	80 HRB	110	110	35
		10% CW	98	70	35		10 HRC			60
		40% CW	146	135	12		35 HRC			92
304L	Bar	Annealed	85	32	57	70	77 HRB	110	110	40
		10% CW	106	73	35					
		40% CW	154	137	9		25 HRC			
305	Sheet	Annealed	86	33	62	73	80 HRB	110		
		10% CR	94	70	45	48				
		40% CR	145	130	10	25	30 HRC			
308	Sheet	Annealed	80	33	58	57	82 HRB	107	100 min	35 (bar)
		10% CR	93	64	38		10 HRC			
		40% CR	153	128	11		35 HRC			
309	Sheet	Annealed	80	37	55	60	77 HRB	110		
		10% CR	87	57	45					
		40% CR	135	124	8					
310	Sheet	Annealed	85	45	47	70	77 HRB	107	89	31.5
		10% CR	108	67	28					
		40% CR	155	138	5		30 HRC			
314	Sheet	Annealed	100	50	40	60	85 HRB			
316	Sheet	Annealed	85	38	61	67	77 HRB	107	78	39
		10% CR	95	70	40		10 HRC			
		40% CR	143	128	8		30 HRC			
316L	Sheet	Annealed	77	37	55	70	77 HRB	107		
		Half hard	125	90	25					
		Full hard	165	127	12		31 HRC			
317	1 inch bar	Annealed	87	38	55	70	77 HRB	107		38
		10% CW	104	57	37					
		40% CW	148	134	7					
321	Sheet	Annealed	90	32	55	60	80 HRB	107	110	38
		10% CR	100	70	45		10 HRC		108	
		40% CR	150	133	7		35 HRC		80	
347	Sheet	Annealed	95	40	50	60	80 HRB	107	120	39
		10% CR	105	77	33		10 HRC		70	
		40% CR	158	138	5		35 HRC		33	88
348	Bar	Annealed	90	37	50	65	150 HB		110	39
Ferritic types										
400	Sheet	Annealed	62	32	32		68 HRB			
405	Sheet	Annealed	67	47	27		77 HRB	27		
		CR	90	65	20	60				
		CR	130	125	10					
430	Bar	Annealed	75	52	25	65	82 HRB	30		40
		CR	90	65	20	60				
		CR	130	125	2	45				
442	Sheet	Annealed	80	52	25	52	82 HRB	10		
		CR	90	70	20	45				
		CR	110	100	15	30				

(continued)

Unless otherwise stated, properties shown apply to room-temperature conditions for longitudinal specimens. (a) Properties from transverse specimens are identified by (T). (b) Fatigue endurance limits shown apply at from 10^6 to 10^8 cycles. Fatigue results are from rotating beam tests except as noted as follows: (A), Axial; (B), Flexural bending

Table 2 Continued

Alloy	Material Form or thickness(a)	Condition	Ultimate tensile strength, ksi	Tensile yield strength 0.2%, ksi	Elongation in 2 inches, %	Reduction of area (for bars), %	Hardness	Impact strength Izod, ft-lb	Impact strength Charpy V, ft-lb	Smooth specimen fatigue strength(b), ksi
Ferritic types (continued)										
446	Sheet	Annealed	80	52	25	52	82 HRB	2		47
		CR	90	70	20	45				
		CR	110	110	15	30				
18 SR	Sheet	Annealed	85	65	27		90 HRB			
E-Brite	Strip	Annealed	65	46	44	88	83 HRB		240	47
261		20% CR	96	88	8		97 HRB			
		60% CR	124	116	3		104 HRB			
Martensitic types										
403	Bar	Annealed	75	45	30	70	155 HB	90		40
		Temper 1200F	110	85	23	65	97 HRB	75		55
		Temper 600F	180	140	15	55	39 HRC	35		
410	1 inch bar	Annealed	65	35	27	80	80 HRB	85		40
		Temper 1100F	118	104	22	66	241 HB	75	38	62
		Temper 600F	181	143	17	62	361 HB		38	
		Temper 300F	188	148	17	60	388 HB		42	
414	Bar	Annealed	115	90	15		220 HB	50		
		Temper 1200F	120	105	20	65	247 HB	50		
		Temper 800F	200	150	17	60	395 HB			
416	Bar	Annealed	65	35	30	75	135 HB	70		40
		Temper 1200F	110	90	19	54	21 HRC	25		55
		Temper 600F	180	140	13	40	39 HRC		23	
420	1 inch bar	Annealed	95	50	25	65	190 HB	70		
		Temper 1200F	110	70	25	60	250 HB	60		
		Temper 400F	250	215	8	25	50 HRC	10		
422	Bar	Annealed	97	74	26					
		Temper 1400F	121	90	19	53	26 HRC		38	
		Temper 1200F	149	125	18	52	34 HRC		15	90
		Temper 800F	237	168	15	52	49 HRC		7	
431	1 inch bar	Annealed	120	95	25	60	240 HB	50		
		Temper 1100F	140	115	19	57	302 HB	48	50	
		Temper 500F	198	149	16	55	415 HB	40		
440A	Bar	Annealed	105	60	20	45	95 HRB	2		
		Temper 600F	260	240	5	20	51 HRC	10		
440B	Bar	Annealed	107	62	18	35	96 HRB	2		
		Temper 600F	280	270	3	15	55 HRC	3		
440C	Bar	Annealed	110	65	14	25	97 HRB	2		52
		Temper 600F	285	275	2	10	57 HRC	4		
		As quenched					60 HRC			117
440F	Bar	Temper 600F	285	275			56 HRC			
12 MoV	Sheet	Annealed	100	65	22		22 HRB			
		Temper 900F	247	195	10.5					135 (A)
		Temper 700F	236	188	9.5					
Greek Ascoloy	Bar	Temper 1000F	165	137	17	59	38 HRC	37	19	75
		Temper 500F	209	175	18	53			19	
Precipitation hardening types										
17-14 Cu-Mo	Bar	1350F, WQ	89	41	45	59			26	
22-4-9	Bar	ST + age	162	102	9	9	344 HB		11	68
17-7PH	Sheet (T)	Annealed	130	40	35		85 HRB			
		TH 1050	200	185	9	26	43 HRC		6	80
		RH 950	235	220	6		48 HRC		6	106, 85 (A)
		CH 900	265	260	2		49 HRC			82
PH15-7Mo	Sheet (T)	Annealed	130	55	35		88 HRB			
		TH 1050	210	200	7		44 HRC		4	120 (A)
		RH 950	240	225	6		48 HRC		4	160 (A)
		CH 900	265	260	2		50 HRC			
PH14-8Mo	Air melt sheet	Annealed	125	55	25		88 HRB			
		SRH 1050	215	205	5		46 HRC			130 (A)
		SRH 950	235	220	5		49 HRC			
		CH 950	298	289	1.5		52 HRC			
AM 350	Sheet	Annealed	161	61	38		95 HRB			
		DA (850F)	179	154	12	28	42 HRC			84
		SCT (850F)	206	175	12		46 HRC		14	86
		CRT	205	175	19		46 HRC			115 (A)
AM 355	Sheet	Annealed	160	57	26		100 HRB			
		DA (850F)	190	160	12		43 HRC		33	
		SCT (850F)	216	182	19	38.5	48 HRC	17	17	135 (A)
		CRT (25%)	230	210	12		52 HRC			
		XH	350	330	0.5		56 HRC			

(continued)

Unless otherwise stated, properties shown apply to room-temperature conditions for longitudinal specimens. (a) Properties from transverse specimens are identified by (T). (b) Fatigue endurance limits shown apply at from 10^6 to 10^8 cycles. Fatigue results are from rotating beam tests except as noted as follows: (A), Axial; (B), Flexural bending

Table 2 Continued

Alloy	Material Form or thickness(a)	Condition	Ultimate tensile strength, ksi	Tensile yield strength 0.2%, ksi	Elongation in 2 inches, %	Reduction of area (for bars), %	Hardness	Impact strength Izod, ft-lb	Charpy V, ft-lb	Smooth specimen fatigue strength(b), ksi
Precipitation hardening types (continued)										
17-4PH	Bar	Annealed	150	110	10	45				
		H 1150	145	125	19	60	33 HRC		50	87
		H 1075	165	150	16	58	36 HRC		40	
		H 1025	170	165	15	56	38 HRC		35	84
		H 900	200	185	14	54	44 HRC		17	90
15-5PH	Bar	Annealed								
		H 1150	145	125	19	60	33 HRC		50	100
		H 1075	165	150	16	58	36 HRC		40	100 (A)
		H 1025	170	165	15	56	38 HRC		35	133 (A)
		H 900	200	185	14	50	44 HRC		15	96
PH13-8Mo	Bar	Annealed					363 HB			
		H 1100	170	150	16	60	36 HRC		60	
		H 1050	190	180	15	55	43 HRC		50	
		H 1000	215	205	13	50	45 HRC		30	100
		H 950	225	205	12	40	47 HRC		20	135 (A)
W	Sheet	Annealed	135	95	11		25 HRC	108		
		Age 1050F	190	170	11		40 HRC			
		Age 950F	210	195	11		43 HRC		4	85
AM 363	Strip	Annealed	123	107	12.2			135 (bar)		90 (A)
		CD + 900F	206	206	11					
Custom 450	1 inch bar	Annealed	141	118	13	50	28 HRC		95	70
		Age 1150F	141	90	24	70	28 HRC		97	80
		Age 1000F	172	169	18	65	39 HRC		51	
		Age 900F	196	187	15	55	43 HRC		40	76
Custom 455	1 inch bar	Annealed	145	115	14	60	31 HRC		70	
		Age 1000F	205	195	14	55	45 HRC		20	86
		Age 950F	230	220	12	50	48 HRC		14	93
		Age 900F	245	235	10	45	49 HRC		9	109
14-4	Bar	Temper 1100					52 HRC			
		Temper 900					57 HRC			
		Temper 300					63 HRC			

Unless otherwise stated, properties shown apply to room-temperature conditions for longitudinal specimens. (a) Properties from transverse specimens are identified by (T). (b) Fatigue endurance limits shown apply at from 10^6 to 10^8 cycles. Fatigue results are from rotating beam tests except as noted as follows: (A), Axial; (B), Flexural bending

These additions, as in Kromarc 58, increase the strength of the alloy. Compositions of other austenitic stainless steels have not been included in Table 2 because little or no fracture mechanics data have been available for them.

Of the austenitic grades, most fatigue crack growth rate data have been obtained on Types 304 and 316 because these two stainless steels have been shown to have good fatigue crack growth rate resistance in nuclear reactor environments.

Martensitic stainless steels are quenched and tempered for increased strength and good ductility and toughness. These alloys are magnetic, and they have a tensile yield strength of about 275 MPa (40 ksi) in the annealed condition. The strength obtained by heat treatment depends on the carbon content of the alloy. Increasing carbon content increases strength but decreases ductility and toughness. The most commonly used alloy in this family is S41000, which contains about 12% Cr and 0.1% C. This alloy is tempered to a variety of hardness levels, from 20 to 40 HRC. Both chromium and carbon contents are increased in alloys S42000, S44002, S44003, and S44004. The first of these contains 14% Cr and 0.3% C and has a hardness capability of 50 HRC. The other three alloys contain 16% Cr and from 0.6 to 1.1% C. These alloys are capable of 60 HRC and a tensile yield strength of 1900 MPa (280 ksi).

Fatigue strength of martensitic steels (such as in Fig. 2 for Type 410) is similar to that of carbon and low-alloy steels at the same hardness/strength level. As expected, the surface condition is a key variable in crack initiation or high-cycle (SH) fatigue. The sensitivity to surface condition usually increases with higher hardness. Martensitic stainless steels are selected for certain applications in which good corrosion resistance is required along with strength and/or hardness greater than may be obtained in austenitic or ferritic grades.

Precipitation hardening (PH) stainless steels have been developed to meet a need for constructional components with good corrosion resistance, with high strength, and with better toughness than that of the martensitic stainless steels discussed above. Many different types of precipitation hardening stainless steels have been developed, and many of them have been used in certain special applications. Common compositions presented in Table 1 have fracture mechanics data available.

The precipitation hardening grades may be divided into three classes. Of those shown in Table 1, 15-5PH, 17-4PH, PH13-8Mo and Custom 455 are martensitic; 17-7PH, PH15-7Mo and AM355 are semiaustenitic; and A-286 is austenitic. The class depends on the microstructure of the steel when it is cooled from the annealing temperature to room temperature, and this in turn depends on the relative amounts of ferrite-promoting and austenite-promoting elements in the chemical composition. Applications of the martensitic and semi-austenitic classes are usually limited to room-temperature and elevated-temperature service. Applications for the austenitic class (face centered cubic) include both cryogenic and elevated-temperature service.

Many variations of multiple-cycle heat treatments may be used in developing a range of strength and toughness properties in the precipitation hardening steels. The steels in the martensitic class (15-5PH, 17-4PH, PH13-8Mo and Custom 455) are usually supplied by the mill in condition A—i.e., in the solution treated condition. Solution treating temperatures range from 815 to 1040 °C (1500 to 1900 °F) depending on composition, and aging temperatures range from 480 to 620 °C (900 to 1150 °F). The microstructures of these steels are martensitic after solution treating (similar to annealed 18Ni maraging steels). Aging produces precipitates of intermetallic compounds, resulting in precipitation hardening. Within the age hardening temperature range, the highest strengths are obtained after aging at 480 °C (900 °F). Aging at 540 or 565 °C (1000 or 1050 °F) results in lower strength but better ductility and fracture toughness.

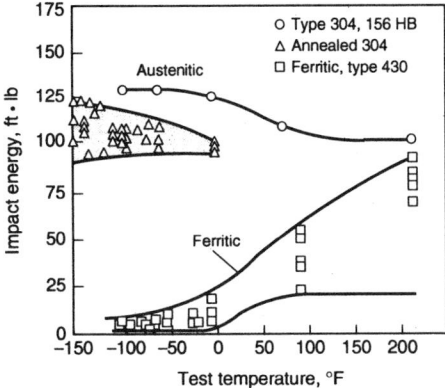

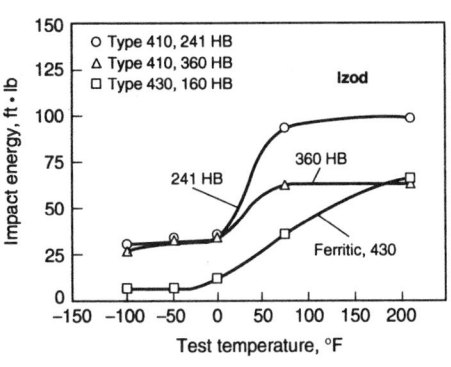

Fig. 2 Impact toughness comparison of stainless steels. (a) Ferritic Type 430 and austenitic Type 304. (b) Ferritic type 430 and martensitic Type 410. Each point represents an average of five tests, Izod specimens. Source: Adapted from *Metals Handbook*, 8th ed., Vol 1, 1961

Heat treatments for the semiaustenitic grades (17-7PH, PH15-7Mo and AM355) require solution annealing at 1040 to 1065 °C (1900 to 1950 °F) and an additional thermal conditioning step prior to aging to develop the desired properties. Heat treatment of A-286 stainless steel is not as effective in increasing its strength as for the martensitic and semiaustenitic grades because of the relatively low strength of the austenitic matrix. Mill forms of A-286 are usually obtained in the annealed condition, annealed at 980 °C (1800 °F). For some applications, annealed A-286 may be re-solution treated at 900 °C (1650 °F) and quenched in oil prior to further processing. After fabrication, components of A-286 stainless steel are aged in the range from 700 to 760 °C (1300 to 1400 °F). On aging, precipitates of intermetallic compounds are formed in the matrix austenite. Aging increases the yield strength from 248 MPa (36 ksi) in the annealed condition to 690 MPa (100 ksi) in the aged condition.

Duplex stainless steels are chromium-nickel-molybdenum alloys that are balanced to contain a mixture of austenite and ferrite and are magnetic, as well. Their duplex structure results in improved stress-corrosion cracking resistance, compared with the austenitic stainless steels, and improved toughness and ductility, compared with the ferritic stainless steels. They are capable of tensile yield strengths ranging from 550 to 690 MPa (80 to 100 ksi) in the annealed condition, which is approximately twice the strength level of either phase alone. These alloys are discussed in the article "Fatigue and Fracture Properties of Duplex Stainless Steels" in this Volume.

Embrittlement

Stainless steels are susceptible to embrittlement during thermal treatment or elevated-temperature service. These thermally-induced forms of embrittlement of stainless steels include sensitization, 475 °C (885 °F) embrittlement, and σ-phase embrittlement as briefly described below. Stainless steels are also susceptible to various forms of environmentally-assisted crack growth by hydrogen embrittlement. This is discussed separately by alloy type.

Sensitization. Stainless steels become susceptible to localized intergranular corrosion when chromium carbides form at the grain boundaries during high-temperature exposure. This depletion of chromium at the grain boundaries is termed "sensitization" because the alloys become more sensitive to localized attack in corrosive environments. Sensitization and intergranular corrosion can occur in austenitic, duplex, and ferritic stainless steels, depending on the alloy content and time-temperature exposures required for carbide precipitation.

Austenitic stainless steels become susceptible to intergranular corrosion when subjected to temperatures in the range of 480 to 815 °C (900 to 1500 °F). Several approaches have been taken to minimize or prevent the sensitization of austenitic stainless steels. If sensitization results from welding heat and the component is small enough, solution annealing will dissolve the precipitates and restore immunity. However, in many cases this cannot be done because of distortion problems or the size of the component. In these cases, a low-carbon version of the grade or a stabilized composition should be used. Complete immunity requires a carbon content below about 0.015 to 0.02%. Additions of niobium or titanium to tie up the carbon are also effective in preventing sensitization as long as the ratio of these elements to the carbon content is high enough. Stabilizing heat treatments are not very effective.

Ferritic stainless steels can be susceptible to sensitization, but ferritic grades with less than 15% Cr can be immune from sensitization. Reducing the carbon and nitrogen interstitial levels improves the intergranular corrosion resistance of ferritic stainless steels.

Sensitization can occur in titanium-stabilized ferritic stainless steels. The thermal treatment that causes sensitization, however, is altered by the addition of titanium.

Duplex stainless steels are resistant to intergranular corrosion when aged in the region of 480 to 700 °C (895 to 1290 °F). It has been recognized for some time that duplex grades with 20 to 40 vol% ferrite exhibit excellent resistance to intergranular corrosion.

475 °C Embrittlement. Iron-chromium alloys containing 13 to 90% Cr are susceptible to embrittlement when held within or cooled slowly through the temperature range of 550 to 400 °C (1020 to 750 °F). This phenomenon, called 475 °C (885 °F) embrittlement, increases tensile strength and hardness and decreases tensile ductility, impact strength, electrical resistivity, and corrosion resistance.

475 °C embrittlement occurs with iron-chromium ferritic and duplex ferritic-austenitic stainless steels, but not with austenitic grades. Aging at 475 °C (885 °F) can cause a rapid rate of hardening with aging between about 20 and 120 h because of homogeneous precipitation. The rate of hardening is much slower with continued aging from 120 to 1000 h. During this aging period, precipitation increases. Aging beyond 1000 h produces little increase in hardness because of the stability of the precipitates.

Even for a severely embrittled alloy, 475 °C embrittlement is reversible. Properties can be restored within minutes by reheating the alloy to 675 °C (1250 °F) or above. The degree of embrittlement increases with chromium content; however, embrittlement is negligible below 13% Cr. Carbide-forming alloying additions, such as molybdenum, vanadium, titanium, and niobium, appear to increase embrittlement, particularly with higher chromium levels. Increased levels of carbon and nitrogen also enhance embrittlement and, of course, are detrimental to nonembrittled properties as well. Cold work prior to 475 °C (885 °F) exposure accelerates embrittlement, particularly for higher-chromium alloys.

Sigma-Phase Embrittlement. The existence of σ-phase in iron-chromium alloys has been identified in over fifty binary systems and in other commercial alloys. Sigma phase has a tetragonal crystal structure and a hardness equivalent to approximately 68 HRC (940 HV). Because of its brittleness, σ often fractures during indentation.

In general, σ forms with long-time exposure in the range of 565 to 980 °C (1050 to 1800 °F), although this range varies somewhat with composition and processing. Sigma formation exhibits C-curve behavior with the shortest time for formation (nose of the curve) generally occurring between about 700 and 810 °C (1290 and 1490 °F); the temperature that produces the greatest amount of σ with time is usually somewhat lower.

In commercial austenitic and ferritic stainless steels, even small amounts of silicon markedly accelerate the formation of σ. In general, all of the elements that stabilize ferrite promote σ formation. Molybdenum has an effect similar to that of silicon; aluminum has a lesser influence. Increasing the chromium content also favors σ formation. Small amounts of nickel and manganese increase the rate of σ formation, but large amounts, which stabilize austenite, retard σ formation. Carbon additions decrease σ formation by forming chromium carbides, thereby reducing the amount of chromium in solid solution. Additions of tungsten, vanadium, titanium, and niobium also promote σ formation.

As might be expected, σ forms more readily in ferritic than in austenitic stainless steels. Coarse grain sizes from high solution-annealing tem-

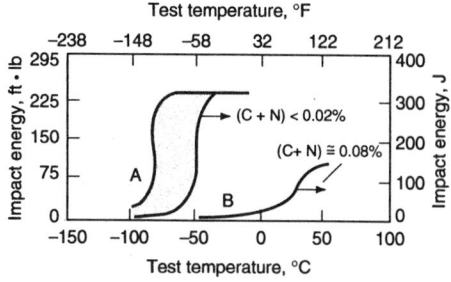

Fig. 3 Charpy V-notch transition temperature range for A-commercially produced, electron-beam-melted ferritic 26Cr-1Mo. B-quarter-size V-notch impact specimens of an air-melted 26Cr-1Mo steel

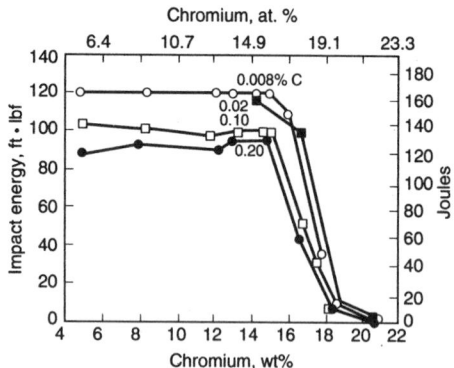

Fig. 4 Effect of variation in chromium and carbon content on the notch impact toughness of commercial ferritic stainless steels. Source: Ref 3

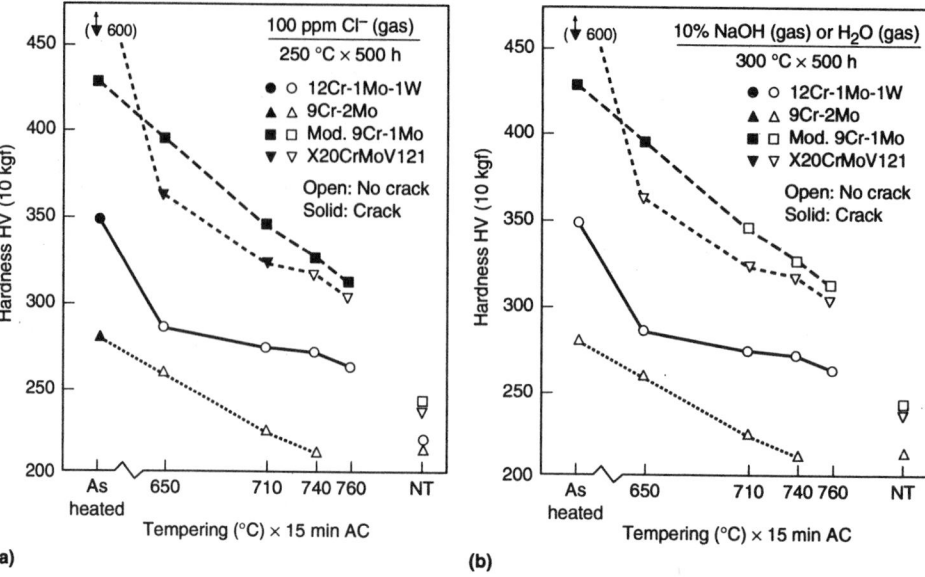

(a)

(b)

Fig. 5 Relationship between SCC properties and hardness of simulated HAZ before and after tempering of four high-chromium ferritic steels. Maximum hardness of 600 HV is observed in X20CrMoV121 steel heated at 1250 °C because of a higher carbon content of 0.2 wt%. Hardness distributions for fully martensitic X20CrMoV121 and Modified 9Cr-1Mo steels are higher than those of two-phase steels. 9Cr-2Mo steel is softer than 12Cr-1Mo-1W-V-Nb steel due to lower carbon content with 0.06 wt% and absence of strengthening elements such as V and Nb. Source: Ref 4

peratures retard σ formation, and prior cold working enhances it. The influence of cold work on σ formation depends on the amount of cold work and its effect on recrystallization. If the amount of cold work is sufficient to produce recrystallization at the service temperature, σ formation is enhanced. If recrystallization does not occur, the rate of σ formation may not be affected. Small amounts of cold work that do not promote recrystallization may actually retard σ formation.

The most sensitive room-temperature property for assessing the influence of σ is the impact strength. High-temperature exposure can produce a variety of phases, and embrittlement is not always due solely to σ formation. Therefore, each situation must be carefully evaluated to determine the true cause of the degradation of properties.

Ferritic Stainless Steels

Fracture Toughness. Ferritic stainless steels have comparatively low ductility (Table 2) and toughness (Fig. 2). Available toughness data for ferritic stainless steels are limited primarily to impact toughness energy. Very little fracture mechanics data are available for ferritic stainless steels.

Ferritic stainless steels are notch sensitive similar to low-carbon and low-alloy steels. Impact toughness of ferritic grade 430 is compared with two other common stainless steels (Fig. 2). The toughness and transition temperatures are influenced by carbon contents, chromium levels, and hardness. When carbon contents exceed about 0.018 wt%, a large increase in transition temperature occurs. Typical effects of composition and test temperature are shown in Fig. 3 and 4.

Stress-Corrosion Cracking (SCC). Ferritic stainless steels such as Type 405 and 430 have high SCC resistance in chloride solutions (Ref 1-3), and ferritic grades UNS S44400 (Remanit 4522) and S44800 (Remanit 4575) can be fully resistant to chloride SCC at stress-intensity levels of 60 MPa√m (55 ksi√in.) in 22% NaCl solutions at 105 °C (220 °F).

Test data from U-bend also confirm good SCC resistance in chloride environments (Ref 1, 2), although SCC of the ferritics has been reported in applications such as:

- Type 434, type 430, and Fe-18Cr-2Mo in lithium chloride solutions
- Sensitized type 446 in boiling magnesium chloride and sodium chloride solutions
- Type 430 F in the marine atmosphere
- Nickel-containing superferritics AL 29-4-2 (UNS S44800), Monit (UNS S44635), and Sea-Cure (UNS S44660) in boiling magnesium chloride

Chloride SCC resistance of ferritic stainless steels can be reduced by the presence of certain alloying elements, sensitization (induced by heat treatment or welding), cold work, high-temperature embrittlement. For example, the producers of AL 29-4C (UNS S44735), which is shown to be SCC-resistant in U-bend tests, caution that the addition of 1% Ni to the alloy composition will introduce susceptibility to chloride SCC in boiling magnesium chloride environments.

Hardness and microstructure also influence SCC resistance as shown in Fig. 5 from an evaluation of weldment SCC. In the case of 10 vol% NaOH solution and pure water, SCC occurs at hardnesses above 380 HV. In the case of the more severe condition in 100 ppm Cl⁻ solution, SCC occurs in all steels after 1250 °C as-heated condition. However, no cracking is observed for two-phase steels (ferrite and δ-ferrite) after post-weld heat treatment (PWHT) (at least above 650 °C for 15 min). The fully martensitic steels had SCC depending on hardness after PWHT. Critical hardness to prevent SCC in 100 ppm Cl⁻ solution is about 310 HV. In all test conditions, no SCC is observed for the normalized-and-tempered base metals. Thus, depending on hardness, microstructure, and environment, SCC occurs even in the ferritic steels. The SCC resistance of 12Cr-1Mo-1W-V-Nb steel is superior to conventional 9-12Cr steels, because δ-ferrite decreases the hardness of HAZ and prevents crack propagation.

Corrosion Fatigue. Ferritic stainless steels have similar fatigue behavior as ferritic low-carbon and low-alloy steels in nonaggressive environments. In corrosive environments such as seawater, however, chromium steels have better high-cycle fatigue strength than low-carbon steels (see the article "Fatigue and Fracture Properties of Structural Steels" in this Volume). For example, tests on higher-chromium, low-interstitial 26-1 ferritic stainless steel containing molybdenum (Fe-26.21Cr-1.03Mo-0.001C-0.007N) show a relatively high fatigue strength of 350 MPa (50 ksi) at 3×10^7 cycles in an aerated 3% NaCl solution (Ref 2). Fatigue crack growth rates for this alloy is compared in Fig. 6 (Ref 5).

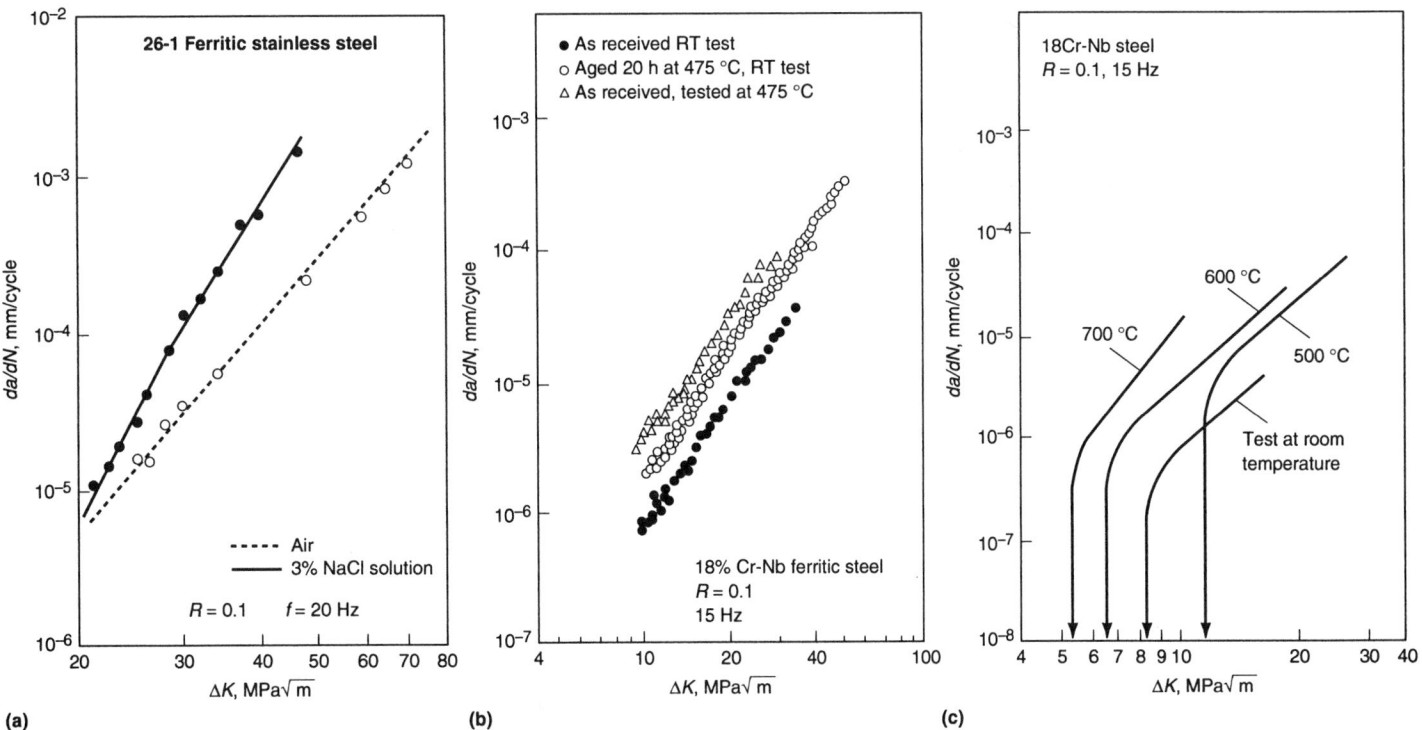

Fig. 6 Fatigue crack growth rates of ferritic stainless steels under various conditions. Source: Ref 5 and K. Makhlouf and J.W. Jones, *Int. Journal of Fatigue*, Vol 15, 1993, p 163-171

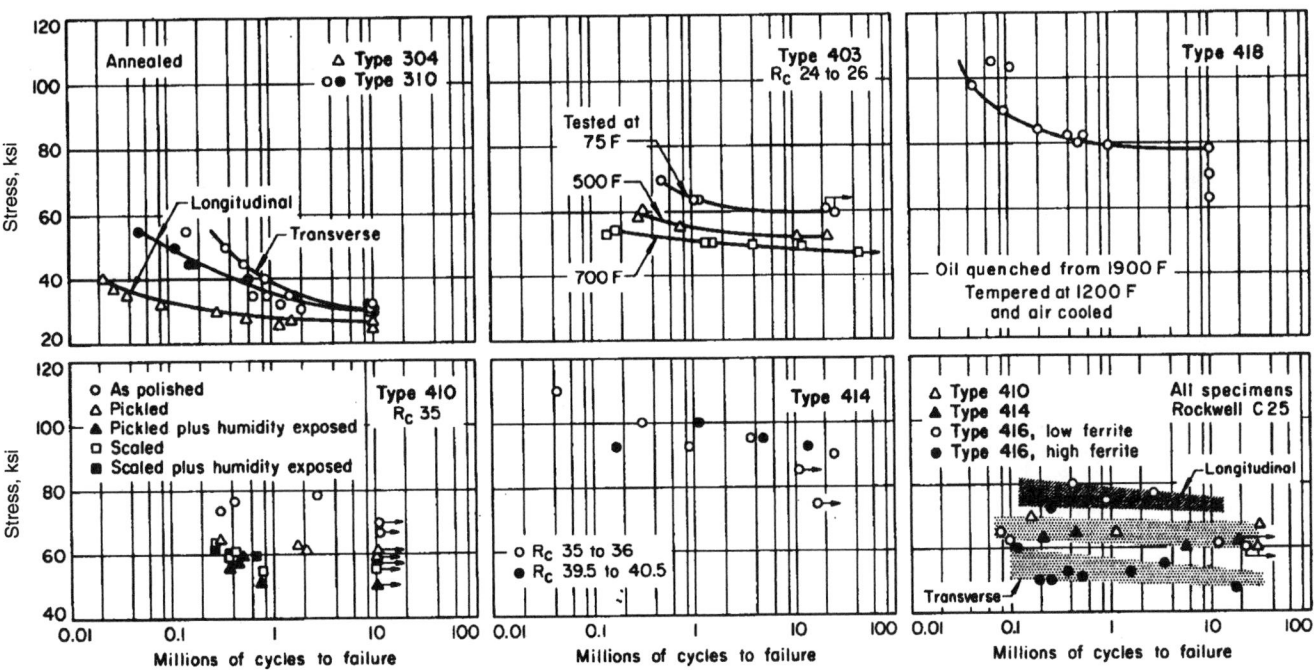

Fig. 7 Factors affecting fatigue properties of stainless steels. Source: *Metals Handbook*, 8th ed., Vol 1

Martensitic Stainless Steels

Types 403 and 410 stainless steels represent the basic compositions of the martensitic stainless steels. These alloys are used in various fracture-critical structural applications. For example, Type 403 martensitic stainless steel is used extensively for steam turbine rotor blades and rotors that operate at temperatures up to 480 °C (900 °F). For this type of application, the components are tempered at 590 °C (1100 °F) or higher, after which embrittlement at service temperatures is negligible.

The fatigue and fracture resistance of martensitic stainless steels is similar to low-alloy martensitic steels with comparable strength and hardness. Typical fatigue strength of various martensitic stainless steels are shown in Fig. 7.

Mean stress effects for Type 422 (AISI 616) are given in Fig. 8 (Ref 6).

Fracture Toughness and SCC Thresholds. Fracture toughness data on martensitic grades are available from various tests on specimens of Types 403, 403 modified, 420, 422, and 431. Table 3 summarizes some valid K_{Ic} test results for Type 420 in air and 3.5% NaCl solution. The specimens tempered at 450 °C (840 °F) had the

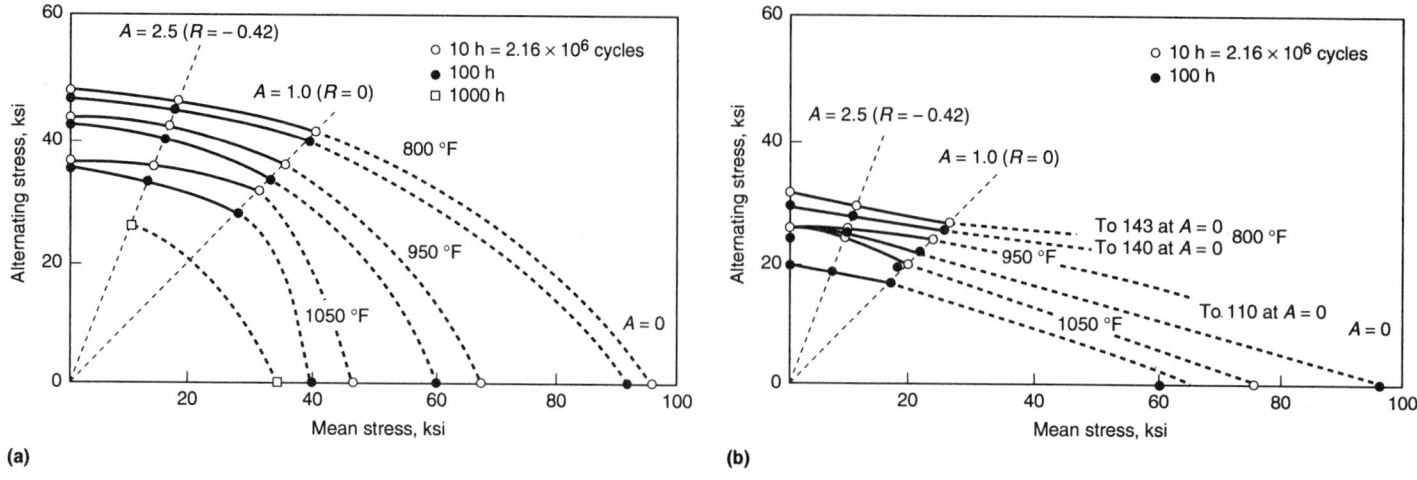

Fig. 8 Stress range diagrams for AISI 616 (type 422) martensitic stainless steel. (a) Unnotched; (b) notched. A = stress amplitude/mean stress, or $R = (1 - A)/(1 + A)$. Source: Ref 6

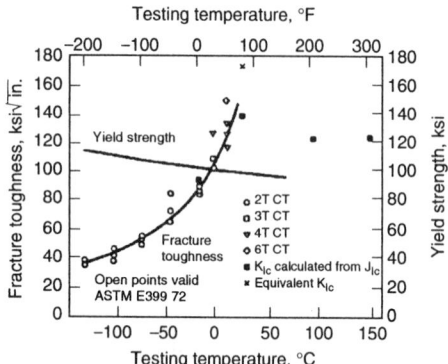

Fig. 9 Fracture toughness data obtained over ranges of temperature and specimen thickness for Type 403 modified stainless steel. $K_{Ic}(J)$ specimens were 1 in. thick. Source: Ref 9

Fig. 10 Fatigue behavior of Type 410 martensitic stainless steel in air and in a 0.03% NaCl solution. Source: Ref 2

lowest fracture toughness and the lowest K_{Iscc} values because of the 450 °C embrittlement. Such embrittlement is caused by formation of fine carbide precipitates and Fe_3C plates, whereas the low corrosion resistance is caused by chromium depletion through precipitation of Cr_7C_3. For the specimens tempered at 250 °C (480 °F) and 650 °C (1200 °F), those that were quenched in oil had the highest fracture toughness and the highest K_{Iscc} values. The highest K_{Iscc} value was obtained after quenching in oil and tempering at 650 °C (1200 °F). During tempering at 650 °C, $M_{23}C_6$ precipitates are formed which reduce the hardness but do not cause chromium depletion.

Valid K_{Ic} data at room temperature for relatively tough materials like martensitic steels require unusually thick specimens. For example, fracture toughness test data have been obtained by Logsdon (Ref 7) on three heats of Type 403 modified stainless steel using compact specimens from 50 to 200 mm (2 to 8 in.) thick for tests in the temperature range from –196 to +80 °C (–320 to +175 °F). The thicker specimens were used to obtain valid K_{Ic} data at higher temperatures. In a later study, (Ref 8), 25-mm (1-in.)-thick compact specimens were tested at temperatures from –18

to +150 °C (0 to 300 °F) using the J-integral method. From –18 to +24 °C (0 to 75 °F), the $K_{Ic}(J)$ data points were close to the fracture toughness curve defined by the K_{Ic} data points for the larger specimens from the same heat. At higher temperatures, the $K_{Ic}(J)$ data points apparently define a plateau in toughness values (Fig. 9). Yield strength and fracture toughness are also compared in Fig. 9.

Type 431 Fracture Toughness. Different combinations of yield strength and fracture toughness were obtained by heat treating Type 431 speci-

mens cut from rolled bar (Ref 9). One set of specimens was tempered at 635 °C (1175 °F) for two hours (HT125) and the other was tempered at 290 °C (550 °F) for two hours, cooled in air, and retempered at 290 °C for two hours (HT200). Yield strength and fracture toughness (K_{Ic} for 2 in. specimens from T-L orientation) were as follows:

- 717 MPa (104 ksi) yield strength and 93 MPa$\sqrt{m}$ (85 ksi$\sqrt{in.}$) fracture toughness for HT125 specimens

Table 3 Fracture toughness and SCC thresholds of martensitic stainless steel, Type 420

Quenching treatment	Tempering temperature		Average hardness, HV	K_{Ic}		K_{Iscc}(a)	
	°C	°F		MPa$\sqrt{m}$	ksi$\sqrt{in.}$	MPa$\sqrt{m}$	ksi$\sqrt{in.}$
In oil	250	482	465	62.4	56.8	25.8	23.5
Martempered in oil	250	482	425	35.5	32.3	25.3	23.0
Martempered in air	250	482	305	48.1(b)	43.8(b)	23.1	21.0
In oil	450	842	365	49.7	45.2	18.2	16.6
Martempered in oil	450	842	435	45.1	41.0	17.0	15.5
Martempered in air	450	842	402	40.4	36.8	17.0	15.5
In oil	650	1202	267	83.1(b)	75.6(b)	73.4	66.8
Martempered in air	650	1202	251	78.5(b)	71.4(b)	66.5	60.5

(a) K_{Iscc} in 3.5% NaCl solution. (b) Did not meet all requirements of ASTM Method E399

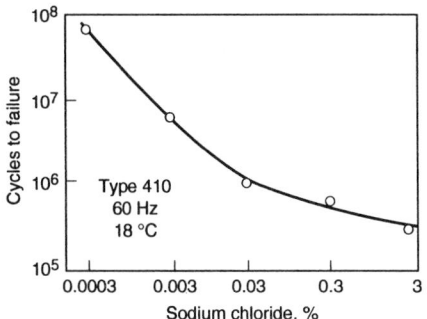

Fig. 11 Effect of NaCl concentration on the cycles to failure at an alternating stress of 340 MPa for a Type 410 martensitic stainless steel. Source: Ref 2

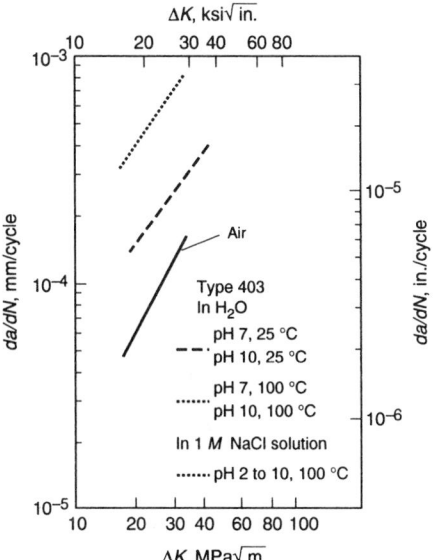

Fig. 12 Fatigue crack growth rates in Type 403 stainless steel in air, water, and a 1 M NaCl solution at 10 Hz and an R ratio of 0.5. Compact specimens (0.5 in. thick) obtained from L-T orientation of plate that had been austenitized at 950 °C (1750 °F), cooled in air, and tempered at 650 °C (1200 °F) for one hour to obtain a yield strength of 682 MPa (100 ksi). All fatigue crack growth rate tests were conducted at room temperature or at 100 °C (212 °F), at an R ratio of 0.5 and at frequencies of 0.1 to 40 Hz. The specimens were completely submerged in the solutions during testing. Source: Ref 11

- 1137 MPa (165 ksi) yield strength and 75 MPa√m (68 ksi√in.) fracture toughness for HT200 specimens

The Type 431 specimens were also tested for SCC thresholds. Results of K_{Iscc} tests in 20% sodium chloride, for compact specimens (T-L) bolt loaded to 95% of the K_{Ic} value, were 47 MPa√m (43 ksi√in.) for the HT125 condition and 13 MPa√m (12 ksi√in.) for the HT200 condition.

Corrosion fatigue resistance is dependent on the chromium content. Types 403, 410, and 422 have lower chromium content, which thus lowers pitting and fatigue crack initiation resistance in chloride solutions. Corrosion effects on high-cycle fatigue resistance of Type 410 are shown in Fig. 10 and 11.

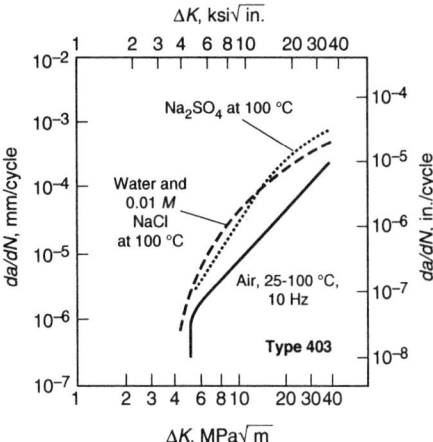

Fig. 13 Fatigue crack growth rates in Type 403 stainless steel in air, water, 0.01 M NaCl solution, and 0.01 and 1.0 M Na$_2$SO$_4$ solutions. Tests in the 0.01 M (molar) and 1.0 M sodium chloride solutions were made with the solutions at pH levels of 2, 7, and 10 and with an open circuit. Source: Ref 11

Fatigue crack growth rates of martensitic stainless steels are less sensitive to environmental effects at room temperature. For example, Clark (Ref 10) reported nearly identical crack growth rates of modified 403 in air, distilled water, seawater, and sulfurous acid (pH 73) at room temperature. In these tests, compact specimens from two heats of Type 403 modified were tested in air, distilled water, seawater and sulfurous acid at 24 and 93 °C (75 and 200 °F) and in high-oxygen (40 ppm) and low-oxygen (1 ppm) steam at 100 °C (212 °F). The specimens had been austenitized at 960 °C (1760 °F), quenched in oil, and tempered at 663 °C (1225 °F) for four hours. For these tests, R ratios were from 0 to 0.1 at a testing frequency of 30 Hz—except at low ΔK levels, when a testing frequency of 160 Hz was used. At room temperature, environmental effects on crack growth were insignificant.

The fatigue crack growth rate properties in air at room temperature and at 93 °C (200 °F) were also the same. At 93 °C, however, distilled water, seawater and sulfurous acid environments increased the rate of fatigue crack growth by factors of approximately 2.5, 3, and 5, respectively, over that for the air environment. The oxygen content of the steam did not influence the rate of fatigue crack growth. In distilled water at 270 °C (520 °F), a slight increase in crack growth rates was also observed for modified Type 403 (Ref 7).

Somewhat different results were obtained from crack growth testing of Type 403 in water at room temperature (Ref 11). Exposure to water at 25 °C resulted in intermediate crack growth rates between those in air and those in water at 100 °C (Fig. 12). However, fatigue crack growth rates in 0.01 M sodium chloride at pH 10 and 100 °C were the same as those in water at 100 °C. At lower cyclic frequencies, the fatigue crack growth rates were higher than at 40 Hz at ΔK values above 20 MPa√m (18 ksi√in.). For tests in the 1.0 M sodium chloride solution at 100 °C (212 °F), fatigue crack growth rates were the

same as for water at the same temperature (Fig. 13). At 100 °C (212 °F), fatigue crack growth rates in 1.0 M sodium phosphate solution at pH 10 and at 10 and 40 Hz and in 1.0 M sodium silicate at pH 10 and at 10 Hz were practically the same as those in air (Fig. 13).

Martensitic PH Stainless Steels

Martensitic (maraging) precipitation hardening stainless steels are usually purchased in the solution annealed condition with a microstructure of low-carbon martensite. After fabrication, the components are age hardened (maraged) in the temperature range from 480 to 590 °C (900 to 1100 °F) and cooled in air. Aging develops the precipitation hardened properties in these steels. Heat treatments and corresponding tensile properties are practically the same for Types 15-5PH and 17-4PH. On aging, precipitates of submicroscopic copper compounds form in the martensitic matrix, causing substantial increases in strength and hardness. The composition of 15-5PH is such that no delta ferrite occurs in this alloy, and thus it has good short transverse centerline ductility in wrought products. A small percentage of delta ferrite may be observed in microstructures of 17-4PH stainless steel. An aluminum-containing intermetallic compound precipitates in the martensitic matrix during aging of PH13-8Mo, while intermetallic compounds of copper and titanium are precipitated on aging of Custom 455.

Plane strain fracture toughness (K_{Ic}) data for 15-5PH, 17-4PH, PH13-8Mo and Custom 455 martensitic precipitation hardening stainless steels are presented in Tables 4 and 5. For each of these steels, the fracture toughness increases as the aging temperature is increased. Of these four steels, Custom 455 has the highest strength in the H900 and H950 conditions. For forged and rolled bars in the H1000 condition, however, the strength and toughness of Custom 455 are comparable to those of PH13-8Mo. For many applications, either the H1000 or the RH1000 condition represents the best combination of strength and toughness.

Results of threshold stress corrosion (K_{Iscc}) tests on martensitic precipitation hardening stainless steels are presented in Table 6. Sump tank residue water and aqueous sodium chloride solutions may influence the threshold stress-intensity factors of specimens obtained in the L-T orientation in these steels depending on the steel condition and the environment, but the effect is more noticeable for specimens of the T-L orientation and probably still more damaging for specimens of the short transverse orientation. However, no K_{Iscc} data were located for specimens of the short transverse orientation.

Fatigue crack growth rates for specimens of 15-5PH and 17-4PH (H1050 and H1100) are shown in Fig. 14 for various conditions and environments at room temperature. For specimens in the H1050 condition, increasing the R ratio from 0.05 to 0.67 and incorporating a one-minute holding period at maximum load in each cycle substantially increased the crack growth rates at ΔK

Table 4   Fracture toughness (K_{Ic}) data for martensitic PH stainless steels

Type(a)	Condition(b)	Testing temperature °C	Testing temperature °F	Yield strength MPa	Yield strength ksi	Tensile strength MPa	Tensile strength ksi	Orientation	Fracture toughness K_{Ic} MPa√m	Fracture toughness K_{Ic} ksi√in.
15-5PH (VAR)	H900	RT	RT	1280	185	1380	200	L-T	96	87
	H900	RT	RT	1210	175	1320	192	...	81	74
	H900	RT	RT	1180	171	1330	193	T-L	81	74
	H1080	22	72	1030	149	1040	151	Random	115-122	104-111
	H1080	0	32	1044	151	1052	152	Random	96-114	87-104
	H1080	−20	−4	1041	151	1054	153	Random	89-101	81-92
17-4PH	H900	RT	RT	1210	176	1380	200	T-L	48	44
	H975	RT	RT	1160	168	1230	178	L-T	93	85
	H1100	RT	RT	883	128	972	141	T-L	153(c)	139(c)
17-4PH (AM)	H900	RT	RT	1170	170	1310	190	L-T	53	48
	H900	RT	RT	1210	176	1340	195	...	57	52
PH13-8Mo	H950	RT	RT	1360	197	1550	225	T-L	70	64
	H1050	RT	RT	1230	178	1320	192	T-L	112	102
Custom 455 (VAR)	H900	RT	RT	1760	255	...	...	L-T	51	46
	H950	RT	RT	1700	246	...	...	L-T	79	72
	H1000	RT	RT	1365(d)	198(d)	...	...	L-T	110	100

(a) Heat treatments: 15-5PH and 17-4PH were austenitized at 1040 °C (1900 °F), AC; PH13-8Mo was austenitized at 1000 °C (1825 °F), AC; Custom 455 was annealed at 980 °C (1800 °F), WQ, reheated to 815 °C (1500 °F), OQ. (b) Aging treatments: H900 at 480 °C (900 °F), AC; H950 at 510 °C (950 °F), AC; H975 at 525 °C (975 °F), AC; H1000 at 540 °C (1000 °F), AC; H1050 at 565 °C (1050 °F), AC; H1080 at 580 °C (1080 °F), AC; H1100 at 595 °C (1100 °F), AC. (c) $K_{Ic}(J)$ data. (d) Typical. Source: Ref 9

Table 5   Fracture toughness for compact specimens of PH13-8Mo stainless steel

Product form	Condition	Yield strength (L) MPa	Yield strength (L) ksi	Tensile strength (L) MPa	Tensile strength (L) ksi	Average fracture toughness, K_{Ic} L-T orientation MPa√m	L-T orientation ksi√in.	T-L orientation MPa√m	T-L orientation ksi√in.
Forged bar	H950	1410	204	1490	216	66	60	63	57
Rolled bar	RH950	1500	217	1630	236	68	62	...	...
		1510	219	1630	237	64	58	...	...
Rolled bar	RH975	1490	216	1610	233	79	72	...	...
		1510	219	1590	231	72	66	...	...
Forged bar	H1000	1390	201	1460	212	104	95	99	90
		1320	191	1430	208	87	79	89	81
		1460	212	1510	219	113	103	99	90
Rolled bar	H1000	1430	208	1490	216	96	87	82	75
	Welded joint(b)	...	...	...	...	91	83	...	...
	Welded joint(c)	...	...	...	...	97	88	...	...
Rolled bar	RH1000	1480	215	1530	222	122	111	...	...
		1500	218	1560	226	104	95	...	...
Extruded bar	H1000	1480	214	1520	221	74	67	72	66

(a) Heat treatments: H950 and H1000—austenitized at 925 °C (1700 °F), AC; RH950, RH975 and RH1000—austenitized at 925 °C (1700 °F), AC, cooled to −73 °C (−100 °F) for 5 h; H950 and RH950—aged at 510 °C (950 °F) for 4 h; RH975—aged at 525 °C (975 °F) for 4 h; H1000 and RH1000—aged at 540 °C (1000 °F) for 4 h. (b) Weld metal. (c) Heat-affected zone. Source: Ref 9

Table 6   Stress-corrosion threshold for martensitic PH stainless steels at room temperature

Type	Condition	Environment	K_{Ic} MPa√m	K_{Ic} ksi√in.	K_{Iscc}(a) MPa√m	K_{Iscc}(a) ksi√in.
15-5PH	H900	3.5% NaCl	81	74	62	56
	H900	20% NaCl	79	72	36 (T-L)	33
17-4PH	H900	3.5% NaCl	57	52	57	52
PH13-8Mo	H950	3.5% NaCl	81	74	81	74
Custom 455	H950	3.5% NaCl	79	72	79	72

(a) L-T orientation except as noted. Source: Ref 9

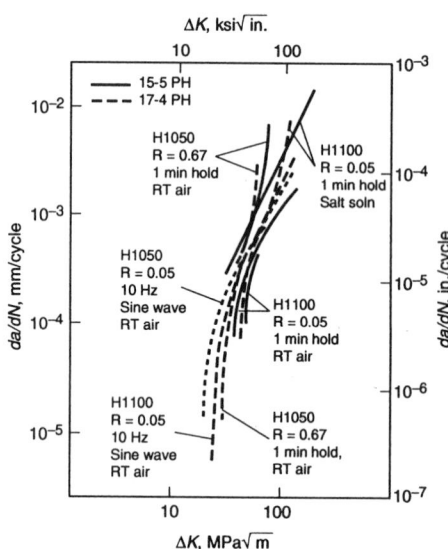

Fig. 14 Fatigue crack growth rates in WOL specimens of 15-5 PH and 17-4 PH stainless steel in the H1050 and H1100 conditions in room-temperature air and in a 3.5% NaCl solution. Adapted from Ref 9

values over 40 MPa√m (36 ksi√in.). For specimens in the H1100 condition, exposure to a salt solution environment during tests with a one-minute holding period at maximum load increased the fatigue crack growth rates over those of specimens tested in air with one-minute holding time or with continuous cycling (Ref 12).

Fatigue crack growth rate data collected by Crooker, Hasson, and Yoder on specimens of 17-4PH (H1050 and H1150) also show the marked increases in fatigue crack growth rates that accompany increases in load ratio over the range from 0.04 to 0.80 (Ref 13). In comparing fatigue crack growth rates for specimens of 17-4PH stainless steel in a dry argon environment and in a 100% humid argon environment, Rack and Kalish (Ref 14) have shown that high humidity increases fatigue crack growth rates in the absence of oxygen. The specimens were in the H900, H1000, and H1100 conditions, the frequency was 10 Hz, and the R ratio was 0.1.

Effects of increasing the load ratio, R, on fatigue crack growth rates of PH13-Mo (H1000) in low humidity air (LHA) and in sump tank residue water (STW) are shown in Fig. 15 (Ref 15). The highest fatigue crack growth rates in this series were obtained on specimens tested at an R ratio of 0.3 in STW. Increasing the load ratio from 0.08 to 0.3 had a marked effect on the growth rates at all ΔK values in the range covered by these tests. Specimen orientation also influenced the results. Fatigue crack growth rates for PH13-8Mo (H1100) specimens tested in air at 0.17 Hz and a stress ratio of zero (Ref 16) are lower at given ΔK values than the rates for corresponding specimens in the H1000 condition (Fig. 15).

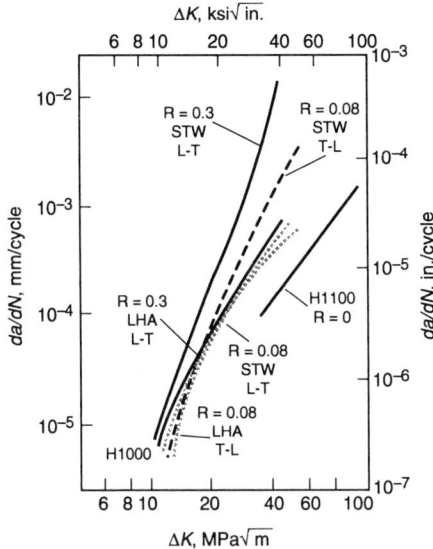

Fig. 15 Fatigue crack growth rates in compact specimens of PH 13-8 Mo stainless steel in the H1000 condition for room-temperature tests at 1 Hz, R ratios of 0.08 and 0.3, L-T and T-L orientations, in low-humidity air (LHA) or sump tank water (STW). Crack growth rates for H1100 condition at R = 0 included for comparison. Source: Ref 15 and 16

Table 7 Fracture toughness and stress-corrosion threshold for semiaustenitic PH stainless steels at room temperature

Type	Condition(a)	Yield strength MPa	Yield strength ksi	Orientation	Fracture toughness K_{Ic} MPa√m	Fracture toughness K_{Ic} ksi√in.	K_{Iscc} MPa√m	K_{Iscc} ksi√in.
17-7PH	RH950	1180	171	L-T	35	32	<21(b)	<19(b)
	TH1050	...	...	L-T	43	39	17.5(b)	16(b)
	RH1050	1310	190	T-L	52	47	<20(c)	<18(c)
PH15-7Mo	RH950	1405	204	T-L	34	31	<16(c)	<15(c)
	RH950	1350	196	L-T	35	32	15(b)	14(b)
	RH1050	1345	195	T-L	44	40	<22(c)	<20(c)
	TH1050	1160	168	L-T	37	34	20(b)	18(b)
	TH1080	...	...	L-T	55	50	...	...
AM355	SCT850	1240	180	L-T	65	59	35(b)	32(b)
	SCT850	1240	180	T-L	53	48	9(c)	8(c)
	SCT1000	1170	170	T-L	115	105	41(c)	37(c)

(a) **RH heat treatments for 17-7PH and PH15-7Mo:** solution annealed at 1065 °C (1950 °F) and air cooled; conditioned by heating at 955 °C (1750 °F) for 10 min, air cooling, subzero cooling to –75 °C (–100 °F) for 8 h, and warming in air; then aged at 510 °C (950 °F) for 1 h (RH950) or aged at 565 °C (1050 °F) for 1 h (RH1050). **TH heat treatments for 17-7PH and PH15-7Mo:** solution annealed at 1065 °C (1950 °F) and air cooled; conditioned by heating at 760 °C (1400 °F) for 1½ h, cooling to 16 °C (60 °F) within 1 h of removal from furnace, and holding for 30 min; then aged at 565 °C (1050 °F) for 1½ h and air cooled (TH1050) or aged at 580 °C (1080 °F) for 1½ h and air cooled (TH1080). **SCT heat treatments for AM355:** solution annealed at 1040 °C (1900 °F), water quenched, subzero cooled to –75 °C (–100 °F), held for 3 h, reheated to 955 °C (1750 °F), air cooled or water quenched, subzero cooled to –75 °C (–100 °F), and held for 3 h; then aged at 455 °C (850 °F) for 3 h (SCT850) or aged at 540 °C (1000 °F) for 3 h (SCT1000). (b) In 3.5% NaCl solution. (c) In 20% NaCl solution. Source: Ref 9

Fatigue life of two maraging steels with comparable tensile strength is given in Fig. 16. Reference 17 provides SN curves for several other PH steels at various tensile strengths and stress ratios.

Semi-Austenitic and Austenitic PH Steels

Semi-Austenitic PH Steels. Only limited fracture toughness data have been published on the semiaustenitic precipitation hardening stainless steels. Typical K_{Ic} data are presented in Table 7. The highest toughness was obtained for specimens of AM355 (SCT1000) at a yield strength of 1170 MPa (170 ksi). Increasing the aging temperature from 455 °C (850 °F) to 540 °C (1000 °F) reduced the yield strength only slightly but increased the fracture toughness substantially.

Threshold stress-corrosion cracking (K_{Iscc}) data also are presented in Table 7 for these steels. Apparently these steels have relatively low resistance to crack growth in aqueous sodium chloride solutions. The highest value for K_{Iscc} in this series was obtained for the AM355 (SCT1000) specimens.

Austenitic precipitation hardening stainless steel A-286 is the main representative in this category. It contains titanium and small amounts of vanadium and aluminum, which precipitate as intermetallic compounds such as Ni_3(Al, Ti) and Ni_4Mo(Fe, Cr) Ti on aging. Various mill forms of the alloy are usually supplied in the annealed condition—Condition A (980 °C, or 1800 °F, for one hour followed by quenching in oil or water). Precipitation hardening occurs on aging in the range from 700 to 760 °C (1300 to 1400 °F) for 16 hours. Other combinations of heat treatments may be used depending on the application. One variation is to re-solution treat at 900 °C (1650 °F) for two hours, quench in oil or water, and age at 700 °C (1300 °F) for 16 hours. This variation results in improved room-temperature properties but less desirable stress-rupture properties.

Because of the high toughness of A-286 stainless steel, even at –269 °C (–452 °F), available

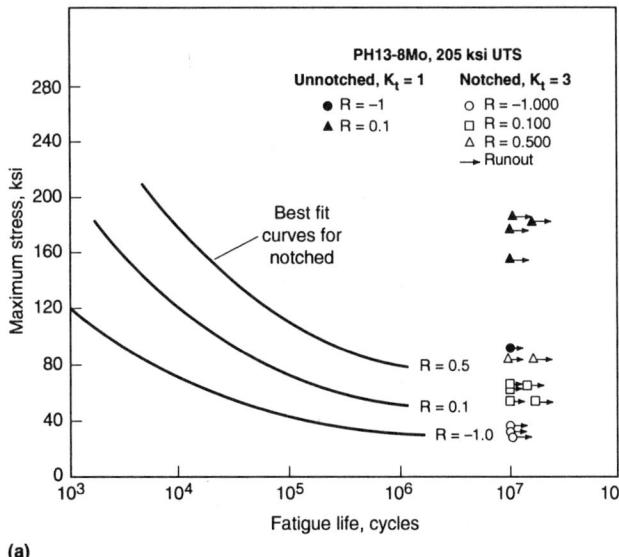

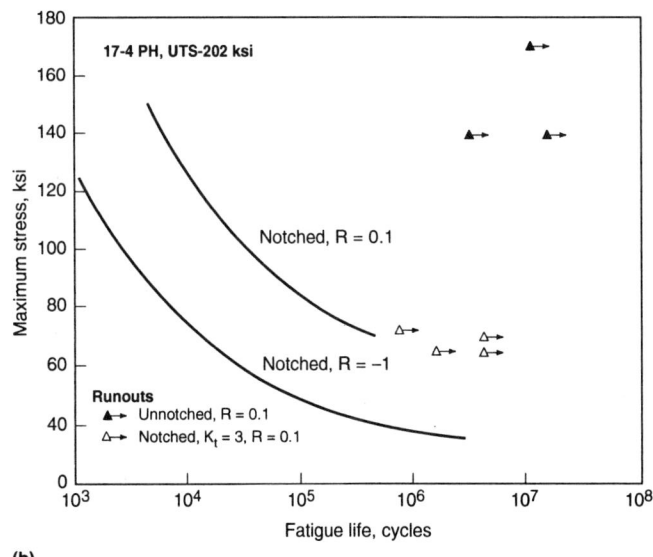

Fig. 16 Room-temperature axial fatigue curves of two maraging (martensitic) grades of precipitation hardening stainless steels with comparable tensile strength. Solid symbols indicate runout for unnotched (K_t = 1) specimens. Best-fit SN curves are shown for notched specimens (K_t = 3) with runouts. (a) PH13-8Mo (H1000) forged bar in longitudinal and long transverse directions. (b) 17-4PH (H900) bar in longitudinal direction. Source: MIL HDBK 5

Table 8 Fracture toughness of A-286 austenitic precipitation hardening stainless steel based on the J-integral method

Heat treatment	Room-temperature yield strength		Specimen orientation	Specimen thickness		Testing temperature		J_{Ic}		$K_{Ic(J)}$	
	MPa	ksi		mm	in.	°C	°F	kJ/m²	in.-lb/in.²	MPa√m	ksi√in.
980 °C (1800 °F) ½ h, WQ, 720 °C (1325 °F) 16 h	769	112	T-L	12.6	0.5	25	77	133	758	167	152
						430	800	92	524	139	126
						540	1000	81	463	130	119
980 °C (1800 °F) ½ h, WQ, 720 °C (1325 °F) 16 h	722	105	...	3.05	0.12	25	77	120	686	159	144
						540	1000	99	563	144	131
STA (solution treated and aged)	...	...	L-T	...	...	25	77	121	692	159	145
900 °C (1650 °F) 2 h, OQ, 730 °C (1350 °F) 16 h	607	88	T-S	38	1.5	25	77	75	426	125	114
						−196	−320	67	385	123	112
						−269	−452	61	350	118	107
900 °C (1650 °F) 5 h, OQ, 718 °C (1325 °F) 20 h	822	119	...	12.7	0.5	24	75	121	692	161	146
						−269	−452	143	815	180	163

Source: Ref 9

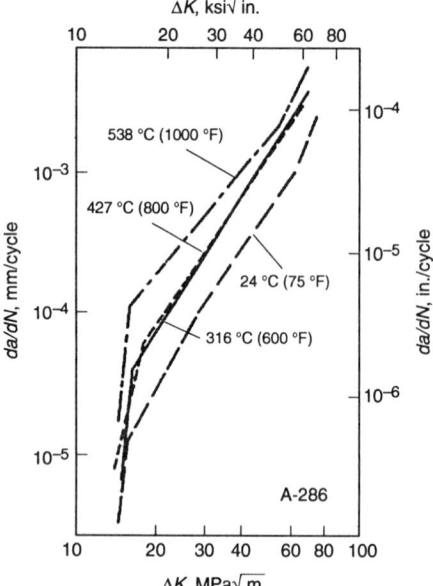

Fig. 17 Fatigue crack growth rates for specimens of A-286 stainless steel at room temperature and elevated temperatures for tests in air at 3 Hz (RT) and 0.67 Hz (elevated temperatures), an *R* ratio of 0.05, and at L-T, T-L, R-L, and R-C orientations. Source: Ref 9

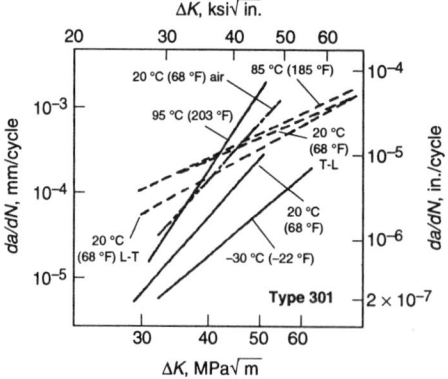

Fig. 18 Fatigue crack growth rates for Type 301 stainless steel in the annealed and warm worked conditions, in air and argon environments, and at temperatures from −30 to +95 °C (−22 to +203 °F). These results were obtained on compact specimens 7 mm (0.28 in.) thick at a cyclic frequency of 20 Hz with a sinusoidal wave form at a load ratio (*R*) of 0.01. All specimens were tested in dry argon except one series that was tested in laboratory air. Source: Ref 9

fracture toughness data have been obtained only by the J-integral method. Results from several sources are presented in Table 8. Results shown for the series of tests performed by Reed, Tobler, and Mikesell (Ref 18) are lower than the others at room temperature. For this series, the heat treatment and the specimen orientation were not the same as for the others. Results of the J_{Ic} tests done by Wells et al. (Ref 19) show that toughness increases as the testing temperature is decreased to −269 °C (−452 °F).

Fatigue crack growth rate data have been reported by James (Ref 20) for compact specimens from A-286 stainless steel flat bar 12.7 mm (0.5 in.) thick and round bar 38 mm (1.5 in.) in diameter which had been annealed at 980 °C (1800 °F), quenched in water, and aged at 720 °C (1325 °F) for 16 hours. These specimens were obtained in the L-T, T-L, R-L, and R-C orientations and were tested at room temperature and at 316, 427, and

538 °C (600, 800, and 1000 °F) at an *R* ratio of 0.05 in air. Testing frequencies were 3 Hz at room temperature and 0.67 Hz at the elevated temperatures. The summary curves in Fig. 17 show that fatigue crack growth rates tend to increase as the exposure temperature is increased. Variations in the orientation of the specimens from the two product forms had no effect on the fatigue crack growth rates. For these specimens of A-286 alloy, the trend of the fatigue crack growth rates at each temperature was similar to that for 20% cold worked Type 316 stainless steel tested under similar conditions.

Results of fatigue crack growth rate tests on compact and single-edge-notch specimens of A-286 stainless steel also have been reported by Gamble and Paris (Ref 21). Tests were made at room temperature and at 482 °C (900 °F) in air at *R* ratios of 0 and −1. At room temperature, variations in load ratio had no effect on fatigue crack growth rates. At 482 °C (900 °F), however, growth rates at *R* = −1 were approximately three times greater than for the corresponding *R* = 0 data. Specimen geometry did not influence the results. Results of the fatigue crack growth rate

tests were used in predicting the allowable number of service cycles for gas turbine disks subjected to cyclic thermal stresses.

Fatigue crack growth rates reported by Tobler and Reed (Ref 22) for solution treated and aged A-286 stainless steel specimens tested at −196 and −269 °C (−320 and −452 °F) were lower, at each Δ*K* value, than rates obtained at room temperature. In this respect, A-286 is similar to the other stable stainless steels. Fatigue crack growth rate data reported by Wells et al. (Ref 19) also showed that crack growth rates at −269 °C (−452 °F) were lower than those at room temperature for A-286 in the solution treated and aged condition.

Austenitic Stainless Steels

Fatigue crack growth (FCG) rates of austenitic stainless steels have been extensively studied, given their widespread use in high-temperature structural parts with cyclic stressing over a wide range of frequencies and load ratios. As previously noted, static fracture mechanics is discussed elsewhere in this Volume.

The general effect of cold work and annealing on the FCG rates of Types 301 and 304 are summarized in Fig. 18 and 19. For annealed Type 301 specimens tested in argon, fatigue crack growth rates at a given Δ*K* value increased as the temperature increased over the testing temperature range (Fig. 18). Fatigue crack growth rates in laboratory air at 20 °C (68 °F) were higher than for corresponding conditions in argon, indicating that the humidity and/or oxygen in the air influenced the growth rates. The warm worked specimens were reduced 65% at 450 to 500 °C (840 to 930 °F), resulting in a substantial increase in strength. Fatigue crack growth rates for the warm worked specimens (see Fig. 18) indicate that the fatigue crack propagation properties of the warm worked alloy are different than those of the annealed alloy. This effect of warm working has been observed for other austenitic stainless steels. These differences are attributed to the extent of the strain-induced transformation at the crack tip. This transformation effect would be most noticeable in Type 301, because it is less stable than the other alloys in the UNS S3*xxxx* series.

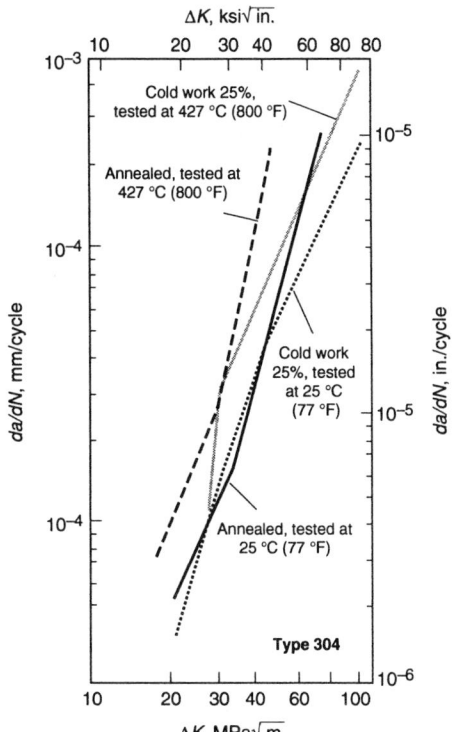

Fig. 19 Fatigue crack growth rates for annealed and cold worked Type 304 stainless steel at 25 and 427 °C (77 and 800 °F), 0.17 Hz, and an *R* ratio of 0. Source: Ref 16

Type 304 stainless steel is also used in the cold worked condition for improved strength. A comparison of fatigue crack growth rate data (Fig. 19) shows that the high-ΔK crack growth rates were lower for the cold worked specimens than for the annealed specimens. Cold working of Type 316 also reduces crack growth rates at comparable ΔK values (Fig. 20).

Because the expected service lives of most components of austenitic stainless steels are many years, an evaluation of the effect of long-time aging at service temperatures is important. Results of fatigue crack growth rate tests on specimens that were tested in the unaged and aged conditions (5000 hours at 593 °C, or 1100 °F) are shown in Fig. 21 (Ref 23). After aging for 5000 hours at this temperature, precipitation of $M_{23}C_6$ carbides is essentially complete. These results indicate that at 593 °C (1100 °F) there are no deleterious effects of aging on the crack growth rates of specimens that are continuously cycled. When a holding time of 0.1 or 1.0 minute is included in each loading cycle, there tends to be a slight increase in the fatigue crack growth rate at a given ΔK level.

Aging of cold worked specimens of Type 304 at 593 °C (1100 °F) for 5000 hours tends to increase slightly the fatigue crack growth rates of specimens that are continuously cycled at 593 °C (1100 °F) (Ref 23). With holding times of 0.1 and 1.0 minute, the fatigue crack growth rates also increased (Ref 23). However, an opposite effect is observed in aged 316 with hold times (see "Aging and Hold Time Effects").

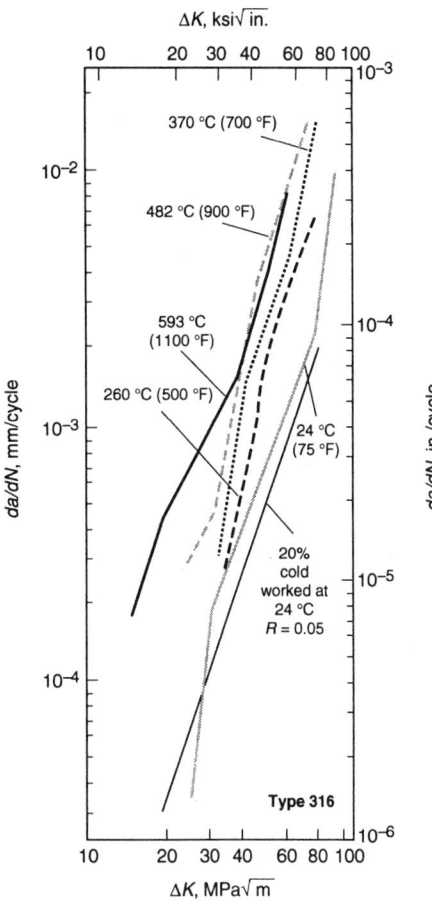

(a)

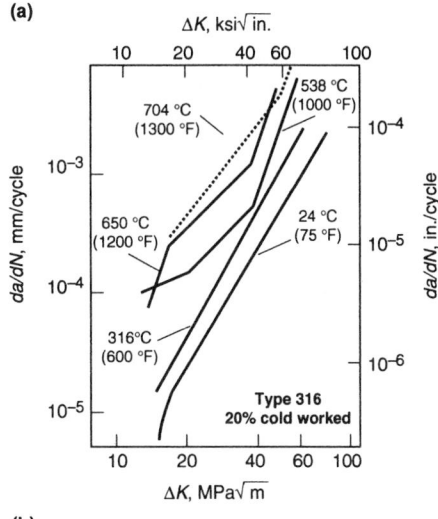

(b)

Fig. 20 Effect of testing temperature on fatigue crack growth of Type 316. (a) Annealed Type 316 stainless steel tested in air at 0.17 Hz and an *R* ratio of 0. (b) Fatigue crack growth rates of 20% cold worked Type 316 stainless steel for L-T and T-L specimens at each temperature in air; 3 Hz at 24 °C, 0.67 Hz at elevated temperatures: *R* = 0.05.

Effect of Stress Ratio on Fatigue Crack Growth Rates. Tests on austenitic stainless steels have shown that FCG rates tend to increase as the *R* ratio increases, when compared at a given ΔK range. For example, Fig. 22 shows fatigue

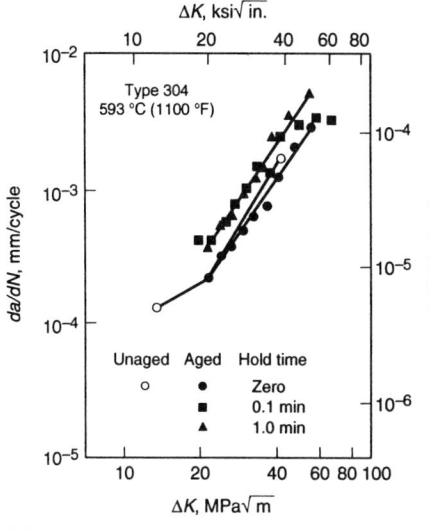

(a)

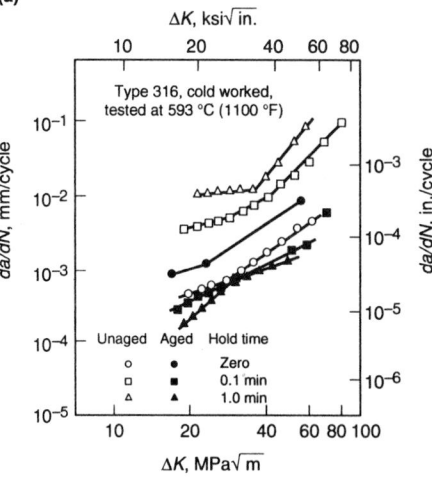

(b)

Fig. 21 Effect of aging and hold times on FCG rates. (a) Effect of aging at 593 °C (1100 °F) for 5000 h, and hold times of 0.1 and 1.0 min for each cycle, on fatigue crack growth rates of L-T oriented specimens of Type 304 stainless steel tested in air at 0.17 Hz and an *R* ratio of 0. (b) Effect of exposure at 593 °C (1100 °F) for 5000 h, and hold times during cycling, on fatigue crack growth rate of 20% cold worked Type 316 stainless steel at 593 °C in air. Source: Ref 23

crack growth rates of Type 304 at several stress ratios. In austenitic stainless steels such as 304 and especially Type 301, martensitic transformation can take place as the fatigue crack propagates. This complicates the effect of *R* on *da/dN*, because the crack opening stress intensity factor relative to the maximum stress intensity factor (K_{op}/K_{max}) increases and, consequently, *da/dN* decelerates with K_{max} particularly under the condition that σ_{max}/σ_y is small. These results are very different from those of other materials, and these fatigue crack growth behaviors correspond to the state of the formation of martensite at the vicinity of fatigue crack surfaces.

The effect of stress ratio variations can be based on the "effective stress intensity factor," K_{eff}, rather than on ΔK, to account for the effect

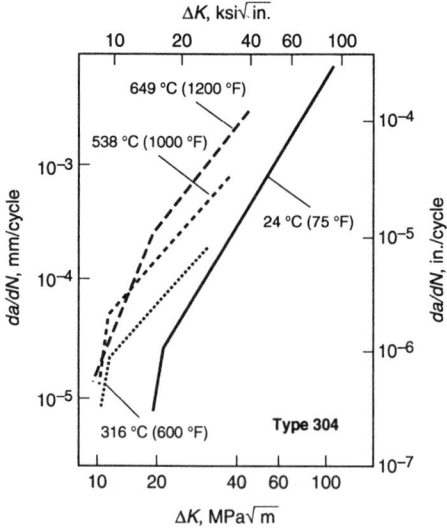

(a)

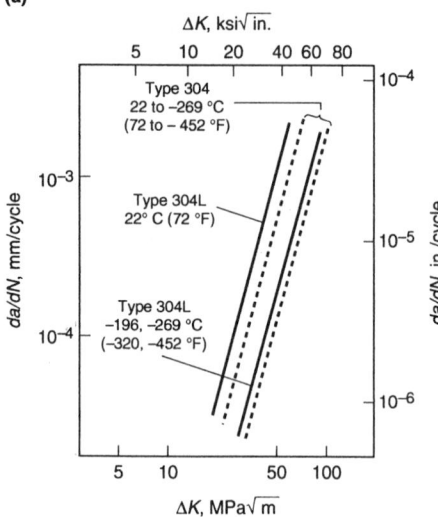

(b)

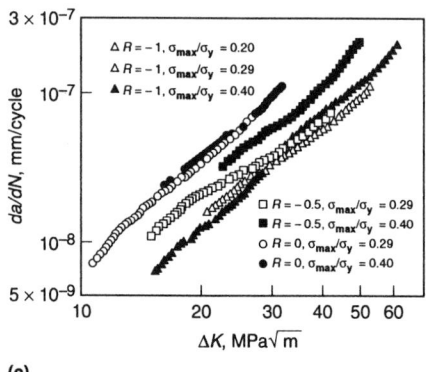

(c)

Fig. 22 Fatigue crack growth rates of 304 under various conditions. (a) Effect of testing temperature on fatigue crack growth rates for annealed Type 304 stainless steel tested in air at 0.066 Hz and an *R* ratio of 0 to 0.05. (b) Fatigue crack growth rates for annealed Type 304 and 304L stainless steels at room and cryogenic temperatures, 20 to 28 Hz, and an *R* ratio of 0.1. (c) Effect of stress ratio on Type 304. Source: Ref 9 and 24

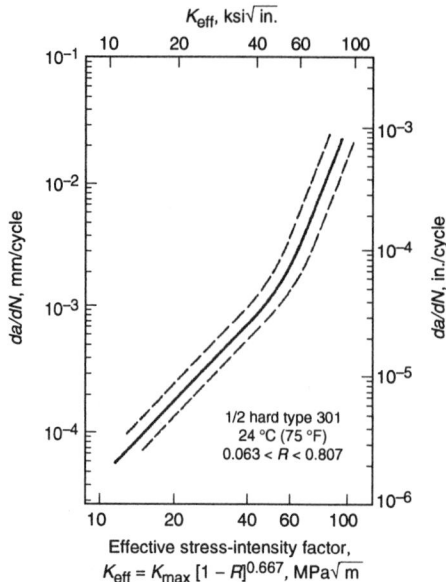

Fig. 23 Scatter band of fatigue crack growth rates for $^1/_2$-hard Type 301 stainless steel, tested at 24 °C (75 °F), 10 Hz, and *R* ratios of 0.063 to 0.807 based on effective stress-intensity factor, K_{eff}. Source: Ref 9

of the range of stress ratios. K_{eff} is defined as follows:

$$K_{eff} = K_{max} (1 - R)^m \qquad (Eq\ 1)$$

where *m* is determined empirically and *R* is the load ratio (minimum load/maximum load) on cyclic loading. The crack growth rate law then becomes:

$$da/dN = C[K_{max}(1 - R)^m]^n \qquad (Eq\ 2)$$

Results of fatigue crack growth rates tests on Type 301 ($^1/_2$ hard) at different *R* ratios are summarized in Fig. 23 based on K_{eff}. This empirical method only applies in the Paris regime (Stage II crack growth rate).

Another empirical method for stress ratio effects is based on the relation (Ref 25):

$$\Delta K_R = \Delta K_{R=0}/V(R) \qquad (Eq\ 3)$$

where *V(R)* is a correlation function that is derived from empirical data. Figure 24 shows derived values of *V(R)* for various materials including Type 304.

Results of fatigue crack growth rate tests also have been reported by James over a range of stress ratios from –0.150 to +0.750 for compact specimens of Type 304 stainless steel at 538 °C (1000 °F) and at a frequency of 6.67 Hz. For these tests, the parameter $K_{max}(1 - R)^m$, or K_{eff}, where *m* = 0.5 (Eq 1), again provided a much better correlation of results than the parameter ΔK (Ref 26).

Environmental Effects

Effect of Temperature and Moisture on FCG Rates. Higher temperatures increase crack growth rates as shown in Fig. 24 for various austenitic grades. The curves in Fig. 25 show that, at room temperature, the fatigue crack growth rates for Types 304, 316, 321, and 348 all fall in a narrow band. For tests at 593 °C (1100 °F), however, specimens of Type 316 had the least fatigue crack propagation resistance, whereas specimens of Type 348 had the highest fatigue crack propagation resistance, over the ΔK range studied. Results of tests on specimens of Type 304 and 321 were nearly the same at 593 °C (1100 °F) in air.

The presence of moisture and oxygen cause an increase in crack growth rates as shown in Fig. 25 for various grades. Fatigue crack growth rate data at 25 °C (77 °F) show that crack growth rates increased slightly with increased humidity when oxygen was present but that high humidity in an inert gas had no significant effect. Fatigue crack growth rates in room air at room temperature were the same for Types 316 and 321 stainless steels. Furthermore, in tests at 649 °C (1200 °F) in dry nitrogen, fatigue crack growth rates for Types 316 and 321 also were the same. In air, however, fatigue crack growth rates in Type 316 specimens increased by a factor of about 22 over rates in an inert environment at the same temperature. The corresponding increase in fatigue crack growth rates for specimens of Type 321 was about 5 times that for the inert environment at 649 °C (1200 °F). If components of these stainless steels are exposed to inert environments instead of to air or oxygen-containing environments, fatigue crack growth rates will be substantially lower than those expected on the basis of tests in air.

The effects of humid air environments on the room-temperature fatigue crack growth rates of specimens of annealed Type 304 stainless steel also are shown in Fig. 26 for specimens cycled at 0.17 Hz with an *R* ratio of zero. At the lower end of the ΔK range, fatigue crack growth rates in humid air are substantially greater than crack growth rates in dry air. However, fatigue crack growth rates of specimens of Type 304 stainless steel tested in a pressurized water reactor environment at 260 to 315 °C (500 to 600 °F) with *R* ratios of 0.2 and 0.7 were no greater than the fatigue crack growth rates in air at the same temperature with an *R* ratio less than 0.1 (Ref 27). However, variations in *R* ratios influenced the fatigue crack growth rates in the pressurized water reactor environment.

Heat-to-heat variations in high-temperature FCG rates do not appear significant for 304 and 316 stainless steels. From FCG tests at 538 °C in air (*R* = 0.05, 0.00138 Hz) on specimen from various heats (73-85 HRB and 0.024-0.064 wt% C variations), results indicate that, for the conditions employed, there is no effect of heat-to-heat variations on fatigue crack growth rates for commercially produced Types 304 and 304L stainless steels (Ref 28). Furthermore, there is no apparent effect of grain size and crack orientation (L-T vs. T-L) on fatigue crack growth rates at 538 °C (1000 °F) for the commercial product in the annealed condition. A similar conclusion was reached for Type 316 (Ref 28).

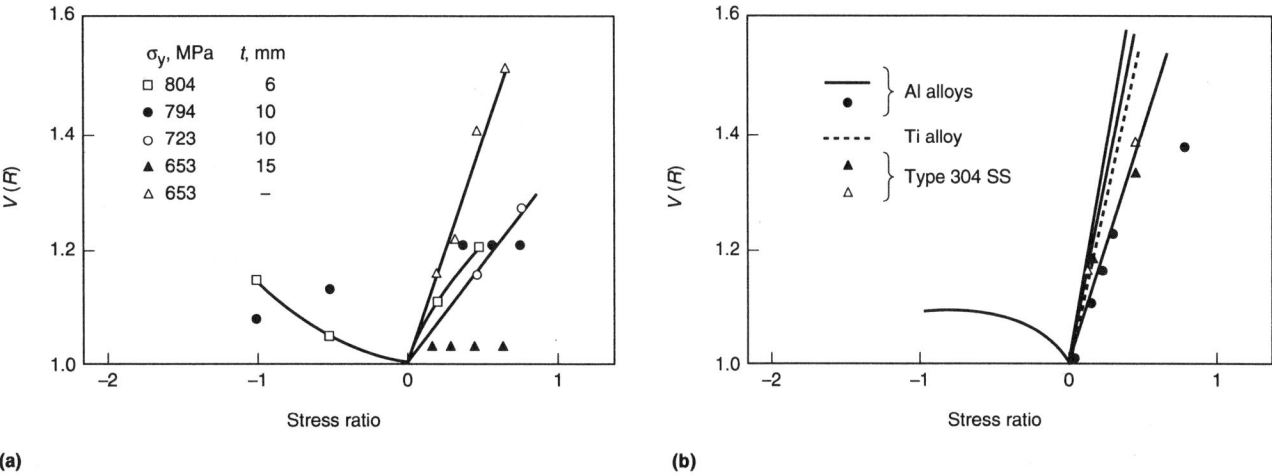

Fig. 24 Stress ratio effects from correlation function $V(R)$ from published fatigue crack growth rate curves for (a) Quenched and tempered steels at growth rates between 10^{-8} and 10^{-7} m/cycles and (b) for aluminum alloys, Type 304 stainless steel, and a titanium alloy. Source: Ref 25

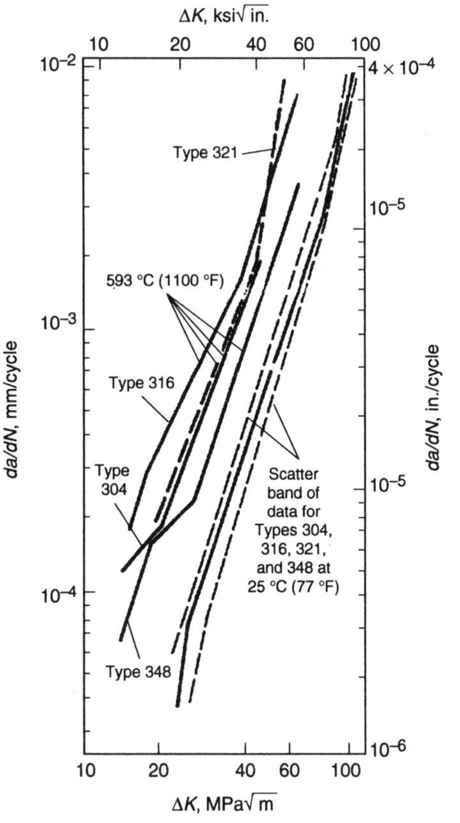

Fig. 25 Fatigue crack growth rates for annealed Types 304, 316, 321, and 348 stainless steels in air at room temperature and 593 °C (1100 °F), L-T orientation, 0.17 Hz, and an R ratio of 0. Tests were made on single-edge-notch cantilever specimens of Types 321 and 348 stainless steels from the L-T orientation at 0.17 Hz with an R ratio of zero at room temperature and at elevated temperatures to 593 °C (1100 °F).

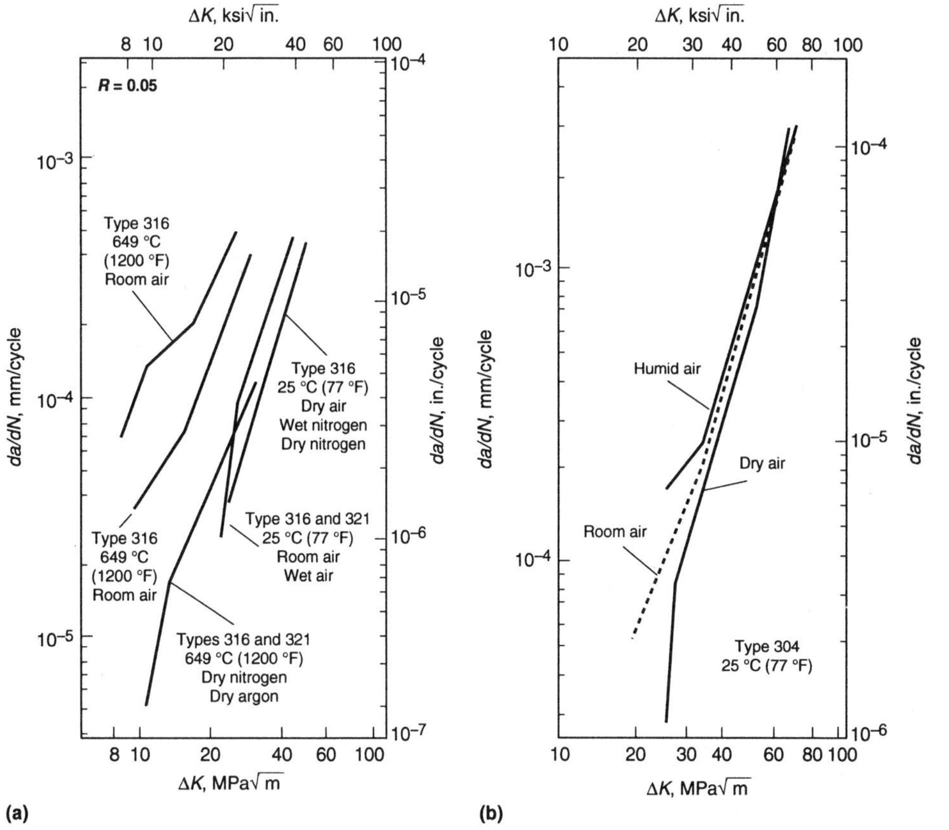

Fig. 26 Effect of environments on fatigue crack growth rates. (a) Types 316 and 321 stainless steels at 25 and 649 °C (77 and 1200 °F). Compact specimens were tested in fatigue loading according to a sine wave loading pattern at 5 Hz with an R ratio of 0.05 in room air, dry air, humid air, dry nitrogen, wet nitrogen, and dry argon. (b) Effect of humidity on fatigue crack growth rates for Type 304 stainless steel tested at room temperature, 0.17 Hz, and an R ratio of 0. Source: Ref 9

Aging and Hold Time Effects. The effect of long-time aging at service temperatures is an important factor for austenitic stainless. As previously noted, aging of Type 304 has little or no effect on FCG rates with continuous (zero hold time) loading (Fig. 21a) or causes a slight increase in FCG rates of cold worked Type 304.

Hold times tend to increase FCG rates of aged Type 304 (Fig. 21a), while hold times have a different effect on aged 316 stainless steel. Fatigue crack growth rates for specimens of Type 316 stainless steel aged at 649 °C (1200 °F) for 6000 hours and tested at 538 °C (1000 °F) were either lower than, or within the scatterband for, specimens tested at 538 °C without aging (Ref 29). Other results are summarized below.

Cold Worked Type 316. Effects of holding times on cyclic loading of unaged and aged specimens of 20% cold worked Type 316 are shown in Fig. 21(b) for tests at 593 °C (1100 °F). The frequency for specimens cycled with zero holding time was 0.17 Hz, and the R ratio was zero. Aging was done for 5000 hours at 593 °C (1100 °F), and testing was done in air. For the unaged specimens, increasing the holding time signifi-

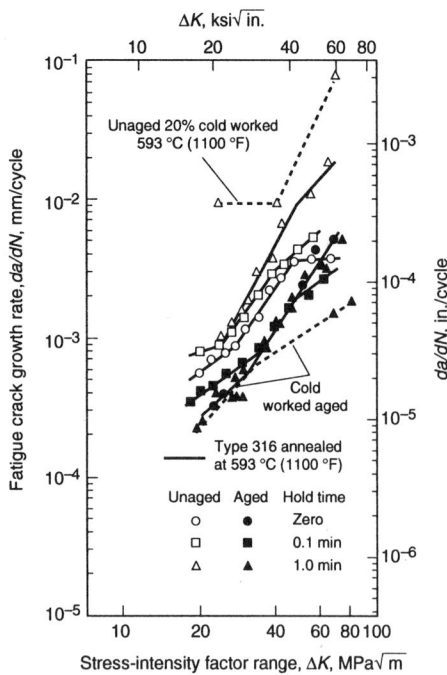

Fig. 27 Effect of exposure in air at 593 °C (1100 °F) for 5000 h, and hold times, on fatigue crack growth rates for annealed Type 316 stainless steel at 593 °C in air

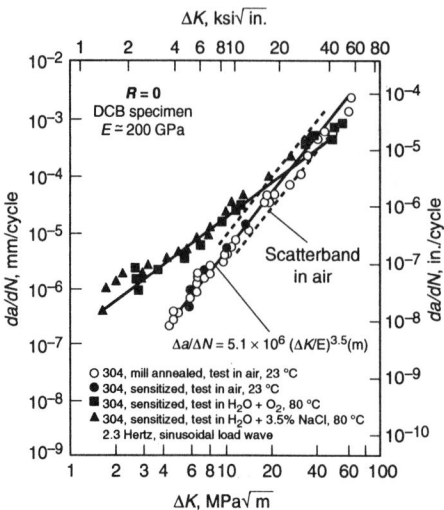

Fig. 28 Effect of environment and cyclic stress intensity range on the growth rate of fatigue cracks in Type 304 stainless steel. Source: Ref 39

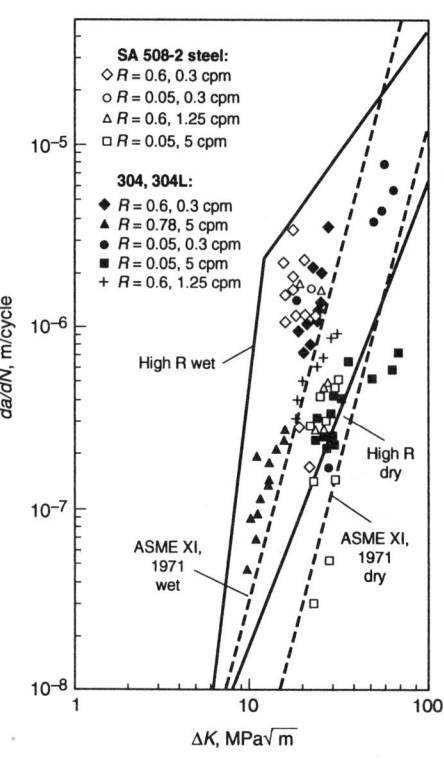

Fig. 29 Corrosion fatigue crack growth data for two austenitic stainless steels in normal BWR water compared with A508 steel. The solid and dashed lines are from Section 11 of the ASME Boiler and Pressure Vessel Code and are intended to represent the upper bounds for fatigue crack growth data for the conditions indicated. These upper bounds are modified periodically as more information becomes available. Source: Ref 40

cantly increased the fatigue crack growth rates as shown. For the aged specimens, holding at maximum load for 0.1 or 1.0 minute for each loading cycle reduced the fatigue crack growth rates over those obtained with no holding time. These data indicate that cold working and aging at 593 °C (1100 °F) before or during service exposure can lead to improved fatigue crack growth resistance and that short hold times at maximum load reduce FCG rates.

Annealed Type 316. Similar results are obtained for annealed Type 316 after aging (Fig. 27). Long-time exposure (5000 hours of aging) at 593 °C (1100 °F) in air on the fatigue crack growth rates of specimens of annealed Type 316 substantially reduced the fatigue crack growth rates at ΔK levels from 18 to 55 MPa$\sqrt{m}$ (16 to 50 ksi$\sqrt{in}$.) for the continuous cycling tests and over the whole testing range for specimens cycled with 0.1- and 1.0-minute holding times for each cycle. Fatigue crack growth rates for specimens tested without prior exposure and with holding times of 0.1 and 1.0 minute for each cycle were higher than those for specimens cycled continuously under the same conditions. However, the effect of hold time was less significant for specimens that had been aged at 593 °C (1100 °F) before testing at the same temperature. In other work (Ref 30), holding times of up to 8 minutes did not cause significant increases in fatigue crack growth rates at 593 °C (1100 °F) in aged specimens. However, holding for 16 minutes caused marked increases in fatigue crack growth rates.

Annealed Type 321 stainless steel aged at 593 °C (1100 °F) for 5000 hours and then tested at

593 °C has shown that long-time exposure at the service temperature does not reduce the fatigue crack propagation resistance in air (Ref 31). Aged specimens tested with zero holding time had lower crack growth rates than corresponding specimens that were not aged.

Fatigue Crack Growth in Aqueous Environments. Aggressive environments accelerate crack growth rates of most materials including austenitic stainless steels in aqueous solutions (Ref 32-38). Figures 28 and 29 are examples of environmentally sensitive crack growth in oxygenated water and chloride solutions at different stress ratios. Figure 28 (Ref 39) illustrates the effect of oxygenated and chloride solutions on crack growth in the near threshold regime, while Fig. 29 (Ref 40) compares FCG rates of 304 and 304L with A508 pressure vessel steels in normal boiling water reactor (BWR) water. Crack growth rates of 304 are accelerated in BWR environments, but they are not significantly different from A508 vessel steel. In contrast, PWR environments have little or no effect on 304 and similar austenitic steel FCG. For example, effects of an environment of pressurized reactor water on the fatigue crack growth rates of a Type 316N stainless steel forging, of cast 316 stainless steel, and of Type 316 welds at 288 °C (550 °F) were evaluated by Bamford (Ref 41). Effects of variations in load ratio, frequency, specimen orientation, and heat-to-heat properties for wrought, cast and weld metal in air and in the reactor water were determined. Results of the study showed that fatigue crack growth rates in the reactor water were not significantly different from FCG rates in air. This suggests the importance of dissolved oxygen in promoting crack growth, as BWR environments have more dissolved oxygen than PWR environments. From early studies on aqueous corrosion fatigue crack growth behavior of pressure vessel steels (SA533-B plate or SA508 forgings), the three most important variables are the cyclic frequency, the applied R ratio,

and the temperature. The influence of a BWR or PWR coolant on fatigue crack growth only becomes evident at cyclic frequencies below 1.0 Hz, is a maximum at a temperature of 200 °C, and increases with increasing positive R ratio.

Mechanisms of Corrosion Fatigue. Corrosion fatigue involves complex interactions between loading, environmental, and metallurgical variables. The influences of potential and solution pH, and the contributions of individual micromechanisms to crack growth of austenitic stainless steels are discussed in Ref 42 to 45. The mechanisms (i.e., anodic dissolution vs. hydrogen embrittlement) for corrosion fatigue of steels in aqueous environments is briefly summarized in Ref 43 with some modeling of chemical/electrochemical reaction control of corrosion fatigue crack growth from the perspective of hydrogen embrittlement for steels HT70, HY130, 304 stainless, and NiCrMoV. Variables that control crack growth are considered in relation to improved design and fracture control procedures.

To assist in the understanding of micromechanisms for corrosion, fatigue crack growth in metastable austenitic steels, the relationships between the crack paths and the underlying microstructure were investigated for annealed and cold-rolled (CR) 304 stainless steels that had been tested in a deaerated 3.5% NaCl solution, air, and vacuum (Ref 44). Corrosion fatigue in the deleterious environments (3.5% NaCl and air) was brittle. In

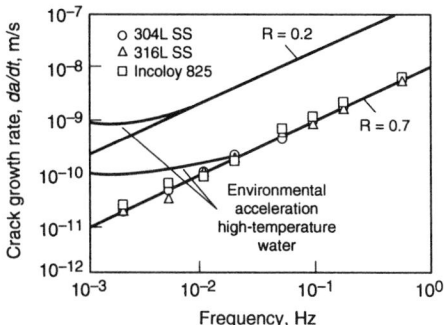

Fig. 30 Time-based crack growth rate vs. frequency of stainless steels in 93 °C simulated well water. Source: Ref 47

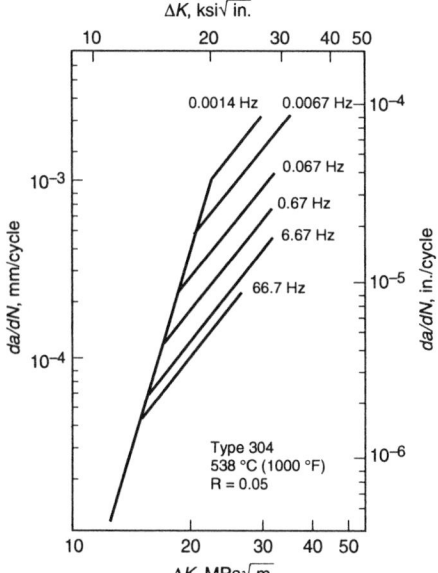

(a)

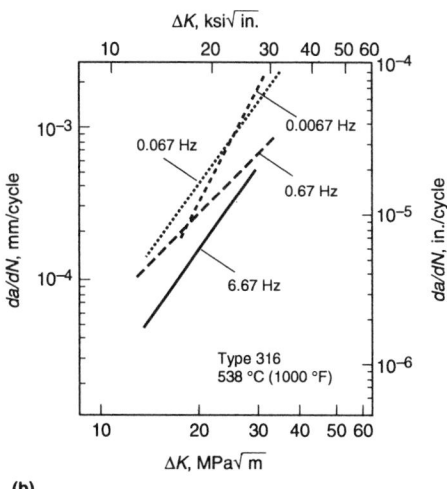

(b)

Fig. 31 Effect of variation in cyclic frequency on fatigue crack growth rates. (a) Annealed Type 304 stainless steel at 538 °C (1000 °F) for an R ratio of 0.05 in air with a sawtooth waveform. (b) Annealed Type 316 stainless steel in air at 538 °C (1000 °F) and an R ratio of 0.05. Source: Ref 9

contrast, fatigue cracking in vacuum was ductile, fully transgranular, and noncrystallographic. These results, taken in conjunction with the crack growth and electrochemical reaction data, support hydrogen embrittlement as the mechanism for corrosion fatigue crack growth in 304 stainless steels in 3.5% NaCl solution. Martensitic transformation appears not to be the only responsible factor for embrittlement. Other microstructural components, such as twin and grain boundaries, slip bands, and cold work-induced lattice defects, may play more important roles in enhancing crack growth rates.

Load History Effects. Environmentally assisted fatigue crack growth is a complex time-dependent process, which is influenced by not only electrochemical factors but also by load history factors such as rise-time, hold-time, unload-time, and cyclic frequency. Load history factors are important because they affect the allowable time for exposure of growing surface cracks. Therefore, lower frequency loading or increased hold times tend to increase FCG rates.

For example, fatigue crack growth behavior of weld heat-affected zone of Type 304 stainless steel in high-temperature water was examined in Ref 46 to clarify the effects of welding residual stress, cyclic frequency, and thermal aging on crack growth rate. A lower crack growth rate of the HAZ than of the base metal was observed in both the high-temperature water and the ambient air caused by the compressive residual stress. With the effect of the welding residual stress evaluated separately from the environmental effect of the high-temperature water, the crack growth rate increased at a cyclic frequency of 0.0167 Hz but did not increase at 3 and 5 Hz. The crack growth behavior of the thermally aged HAZ at 400 °C for 1800 h was almost the same as that of the unaged material tested at 0.0167 and 5 Hz in the high-temperature water.

In another analysis (Ref 47) fracture-mechanics crack growth tests were conducted on 25.4 mm thick compact tension specimen of Types

304L and 316L stainless steel and Incoloy 825 at 93 °C and 1 atm of pressure in a simulated well-water environment. Crack growth rates were measured under various load conditions: load ratios of $R = 0.2$ to 1.0 and frequencies from 2×10^{-4} to 1 Hz with rise times of 1-5000 s and peak stress intensities of 25-40 MPa$\sqrt{m}$. The measured crack growth rates are bounded by the predicted rates from the current ASME Section XI correlation for fatigue crack growth rates of austenitic stainless steel in air. Environmentally accelerated crack growth was not evident in any of the three materials when exposed to the room-temperature simulation of water. This is in contrast with the region of accelerated crack growth of these materials (Fig. 30).

Time-dependent crack growth encompasses several load history factors that affect crack growth rates in either corrosive or high-temperature conditions. In general, environmental factors become more significant as the cyclic loading frequency decreases. For example, Fig. 31 compares FCG rates at different frequencies for annealed 304 and 316 stainless steels at 538 °C (1000 °F). The pattern is different between 304 and 316, but FCG rates for both steels increase with decreasing frequency. The effect of hold time is more complicated, as previously discussed.

Creep-Fatigue Crack Growth. Time-dependent effects become more pronounced when high-temperature creep occurs. In recent years, a large number of models describing fatigue crack growth at high temperatures has been developed. In general, the process of crack growth is usually divided into two parts, that of fatigue and creep. Creep crack growth at hold times may be divided into global and local sections. In the global part, only the continuum damage process is considered; no reference is made to the crack tip stress field. The local part of creep growth includes the nucleation as well as subsequent crack propagation; this is analogous to the process of pure fatigue. These models are discussed in more de-

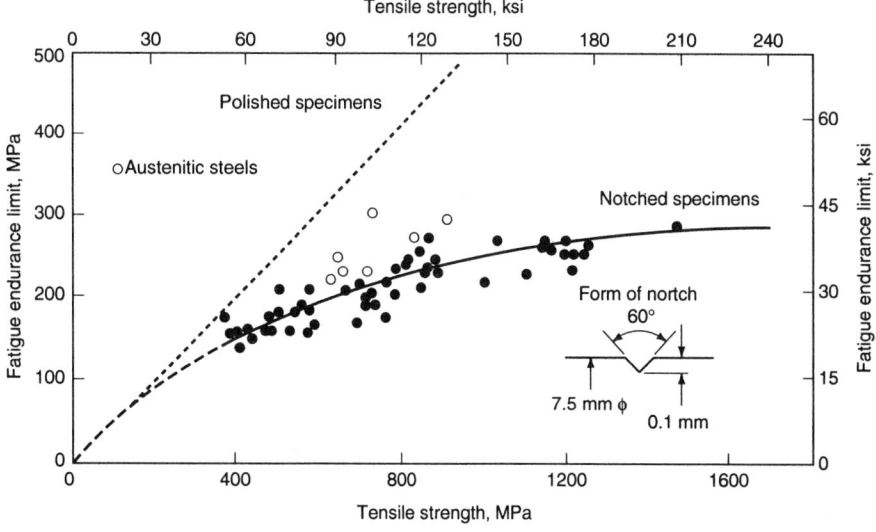

Fig. 32 Relation between endurance limit and tensile strength for polished and notched steel specimens. Source: F. Sisco, *Alloys of Iron and Carbon*, Vol II, *Properties*, McGraw-Hill, 1937

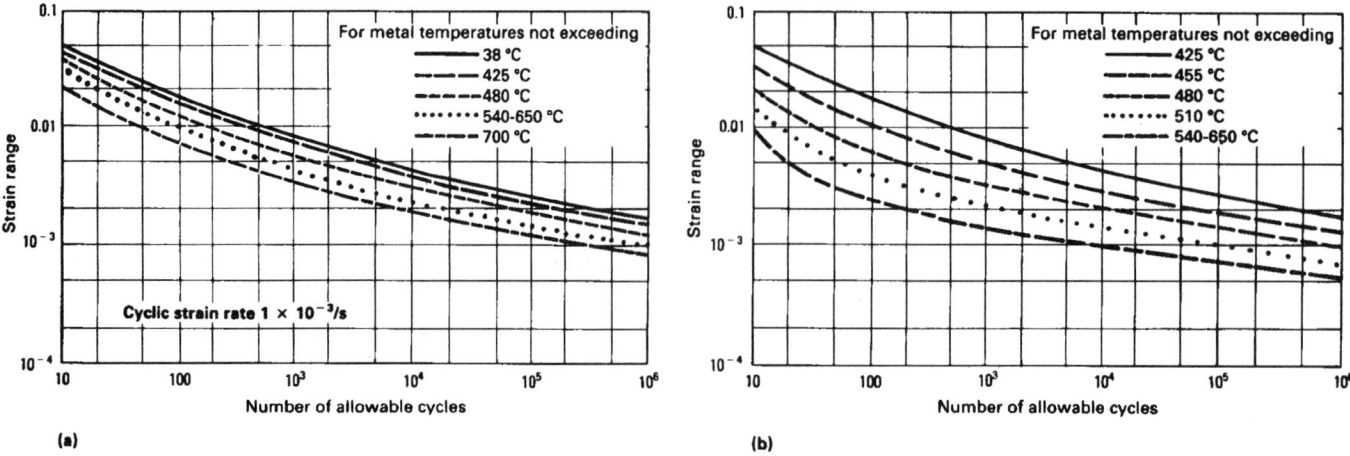

Fig. 33 Design fatigue-strain range curves for 304 and 316 stainless steel. (a) Design curves with continuous cycling (pure fatigue). (b) Design curves with hold times (creep-fatigue interaction)

tail in the article "Elevated-Temperature Crack Growth" in this Volume. References 48 and 49 describe creep-fatigue behavior and fractography of austenitic stainless steels in air and vacuum.

Corrosion Fatigue. Crack growth rate under cyclic loads in a corrosive environment (da/dt) may be expressed as a sum of contributions by:

- Stress-corrosion cracking, $(da/dt)_{SCC}$
- Corrosion fatigue, $(da/dt)_{CF}$, representing the additional crack growth under cyclic loading due to the environment; and
- Mechanical fatigue, $(da/dt)_{air}$, representing the fatigue growth in air

The net result is

$$(da/dt) = (da/dt)_{SCC} + (da/dt)_{CF} + (da/dt)_{air} \qquad \text{(Eq 4)}$$

The first two terms on the right side of the equation are environment sensitive and depend on loading history variables. In oxygenated-water environments, the environment-sensitive terms can contribute significantly to crack growth rates of austenitic stainless steels (Ref 36-47). Under low-R and high-frequency loading, mechanical fatigue dominates. Environmental contributions become more significant as the frequency decreases as previously noted in Fig. 29-31. Modeling of time-dependent effects on aqueous high-temperature FCG rates is still an active area of research.

Fatigue Crack Growth Rates at Cryogenic Temperatures. Fatigue crack growth rate data were obtained by Tobler and Reed (Ref 22) on specimens of Type 304, 316, and 304L stainless steels (annealed) at temperatures in the range from room temperature to liquid helium temperature (–269 °C, or –452 °F). The data for Type 304 were scattered over a range while for Type 304L the data at room temperature described one curve and the data at the cryogenic temperatures were more distinct. These results indicate that cryogenic fatigue crack growth rates for Type 304 do not deviate significantly from room-temperature fatigue crack growth rates over the ΔK range studied. Furthermore, if design calculations for

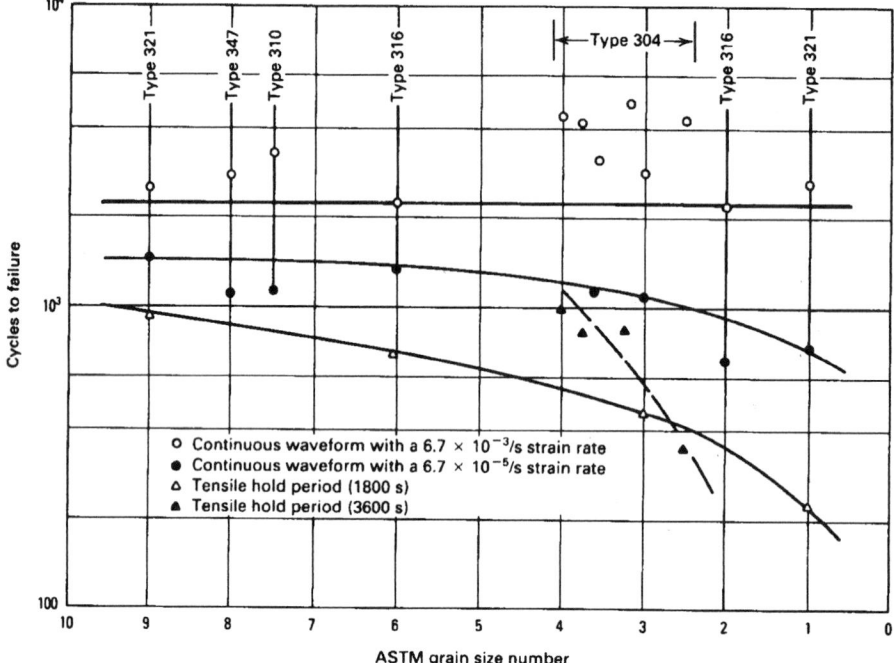

Fig. 34 Effect of strain rate and grain size on the fatigue life of various stainless steels at elevated temperatures. Grain size has the greatest influence on fatigue life when hold times are increased. Test conditions: total strain range = 1.0%; test temperature, 593 to 600 °C (1100 to 1110 °F). Source: Ref 50

Type 304L are based on room-temperature fatigue crack growth rates, the calculations will be conservative for cryogenic exposure.

Fatigue strength of austenitic stainless steels is well documented in the literature. Austenitic stainless steels have a distinct fatigue limit and similar endurance ratio as other unnotched steel specimens of similar tensile strength. However, austenitic steels (which are more ductile and resistant to crack growth) appear to be less notch sensitive than other steels (see the article "Fracture and Fatigue Properties of Structural Steels" in this Volume). A similar result of less notch sensitivity for lower-strength austenitic grades is

also observed in early work reported by Sisco (Fig. 32).

Fatigue properties at elevated temperatures are dependent on several variables including strain range, temperature, cyclic frequency, hold times, and the environment. The fatigue design curves in Fig. 33(a) show the simple case of pure fatigue (that is, continuous cycles without hold times) for 304 and 316 stainless steel. These design curves (from Code Case N-47 in the ASME Boiler Code) have a built-in factor of safety and are established by applying a safety factor of 2 with respect to strain range or a factor of 20 with respect to the number of cycles, whichever gives

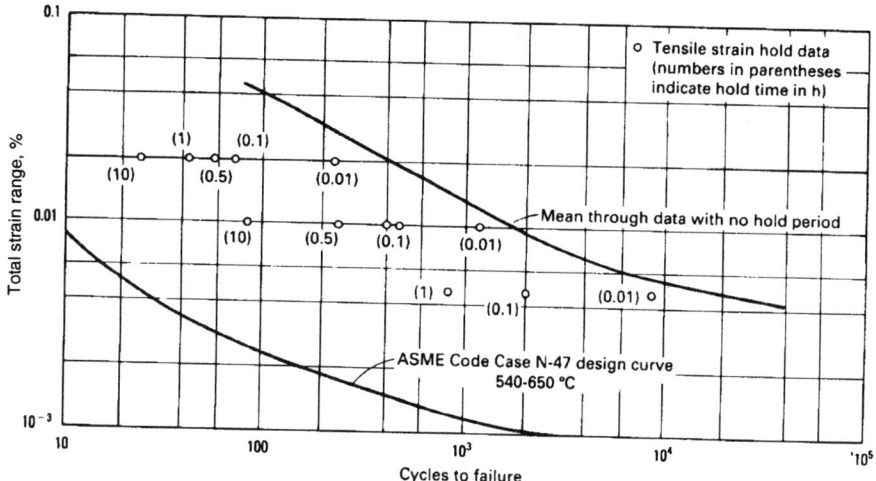

Fig. 35 Influence of tensile hold times at peak strain on failure life of a single heat of Type 316 stainless steel tested at 593 °C (1100 °F). Source: Ref 50

the lower value. The creep-life fraction is determined by the time-life fraction per cycle using assumed stresses 1.1 times the applied stress and the minimum stress-rupture curves incorporated in the code.

The design curves in Fig. 33(a) are based on a strain rate of 1×10^{-3}/s. If the strain rate decreases, fatigue life also decreases. In Fig. 34, for example, the fatigue lives of several stainless steels are shown for continuous cycling at two different strain rates. Fatigue life is reduced with a lower strain rate, while grain size has little effect on fatigue life when life is determined from pure fatigue (or continuous cycling).

When hold times are introduced, a different set of design curves is used (Fig. 33b) to determine the allowable fatigue-life fraction (creep-life fraction is determined the same way as for continuous cycling). These allowable fatigue-life curves are a more conservative set of curves than those of Fig. 33(a). They incorporate the effect of creep damage by applying a fatigue life reduction factor, which includes hold time effects in addition to the factor of safety (2 in strength and 20 in cycles, whichever gives the lower value). Figure 35 (Ref 50) compares the 540 to 650 °C (1000 to 1200 °F) design curve in Fig. 33(b) with actual fatigue life results from testing 316 stainless at 593 °C (1100 °F) and various hold times. When hold times are introduced, the influence of grain size may also be more pronounced (Fig. 34). More detailed information on creep-fatigue behavior is in *ASM Handbook*, Vol 8, 1985, p 346-359.

REFERENCES

1. J. Davis, *ASM Specialty Handbook-Stainless Steels*, ASM International, 1993
2. A.J. Sedriks, *Corrosion of Stainless Steels*, John Wiley & Sons, 1996
3. D. Peckner and I.M. Bernstein, *Handbook of Stainless Steels*, McGraw-Hill, 1977
4. *Heat-Resistant Materials*, ASM International, 1991, p 577-586
5. *Corrosion-Fatigue Technology*, ASTM STP 642, American Society for Testing and Materials, 1978, p 117
6. R.G. Matters and A.A. Blatherwick, "High-Temperature Rupture, Fatigue and Damping Properties of AISI 616 (Type 422 Stainless Steel)," *Trans. ASME, J. Basic Eng.*, June 1965
7. W.A. Logsdon, *Engr. Fracture Mechanics*, Vol 7, March 1975, p 23-40
8. W.A. Logsdon, STP 590, *American Society for Testing and Materials*, 1976, p 43-60
9. *Application of Fracture Mechanics*, American Society for Metals, 1982, p 105-169
10. W.G. Clark, Jr., The Fatigue Crack Growth Rate Properties of Type 403 Stainless Steel in Marine Turbine Environments, in *Corrosion Problems in Energy Conversion*, Electrochemical Society, 1974, p 368-383
11. L. Abrego and J.A. Begley, Fatigue Crack Propagation of 403 Stainless Steel in Aqueous Solutions at 100 °C, Paper No. 235 for the International Corrosion Forum, Palmer House, Chicago, 3-7 March 1980
12. K.R. Kondas, Influence of Microstructural and Load Wave Form Control on Fatigue Crack Growth Behavior of Precipitation Hardening Stainless Steels, Doctor's degree dissertation, University of Missouri, July 1976
13. T.W. Crooker, D.F. Hasson, and G.R. Yoder, A Fracture Mechanics and Fractographic Study of Fatigue-Crack Propagation Resistance in 17-4 PH Stainless Steel, NRL Report 7910, Naval Research Laboratory, Washington, July 1975
14. H. Rack and D. Kalish, The Strength, Fracture Toughness and Low Cycle Fatigue Behavior of 17-4 PH Stainless Steel, *Metallurgical Transactions*, Vol 5 (No. 7), 1974, p 1595-1604
15. R. Berryman, AFML-TR-76-137, 1976
16. P. Shahinian, H. Watson, and H. Smith, Fatigue Crack Growth in Selected Alloys for Reactor Applications, *Journal of Materials*, Vol 7 (No. 4), Dec 1972, p 527-535
17. *MIL Handbook 5F*, U.S. Government
18. R. Reed et al., The Fracture Toughness and Fatigue Crack Growth Rate of an Fe-Ni-Cr Superalloy at 298, 76, and 4K, *Advances in Cryogenic Engineering*, K.D. Timmerhaus, Ed., Vol 22, Plenum Press, 1977, p 68-79
19. J. Wells et al., Evaluation of Weldments in Austenitic Stainless Steels for Cryogenic Applications, *Advances in Cryogenic Engineering*, K.D. Timmerhaus, Ed., Vol 24, Plenum Press, 1978, p 150-160
20. L. James, The Effect of Temperature on the Fatigue-Crack Propagation Behavior of A-286 Steel, Report HEDL-TME-75-82, Westinghouse Hanford Co., Richland, WA, Jan 1976
21. R. Gamble and P. Paris, Cyclic Crack Growth Analysis for Notched Structures at Elevated Temperatures, STP 590, American Society for Testing and Materials, Philadelphia, 1976, p 345-367
22. R. Tobler and R. Reed, Fatigue Crack Growth Resistance of Structural Alloys at Cryogenic Temperatures, *Advances in Cryogenic Engineering*, K.D. Timmerhaus, Ed., Vol 24, Plenum Press, 1978, p 82-90
23. D.J. Michel and H.H. Smith, NRL-MR-3627, Naval Research Laboratory, Oct 1977
24. H. Ogiyama et al., Effect of Negative Stress Ratio on Fatigue Crack Growth Rate in 2017-T3 and SUS 304 Steel, *J. Society Materials Science*, Japan, Vol 40, 1991, p 575-580
25. I.M. Robertson, *Fatigue Fract. Engng. Mater. Struc.*, Vol 17, 1994, p 327-338
26. L.A. James, The Effect of Stress Ratio on the Elevated Temperature Fatigue-Crack Propagation of Type 304 Stainless Steel, *Nuclear Technology*, Vol 14 (No. 2), May 1972, p 163-167
27. W. Bamford, Trans ASME, *Journal of Pressure Vessel Tech.*, Vol 101, 1979, p 73-79
28. L. James, Effect of Heat-to-Heat and Melt Practice Variations Upon Fatigue Crack Growth in Two Austenitic Steels, STP 679, American Society for Testing and Materials, Philadelphia, 1979, p 3-16
29. L. James, *Met. Trans.*, Vol 5, 1974, p 831-838
30. D. Michel and H. Smith, *Acta Met.*, Vol 28, 1980, p 999
31. D. Michel and H. Smith, Effect of Hold Time and Thermal Aging on Elevated Temperature Fatigue Crack Propagation in Austenitic Stainless Steels, Report NRL-MR-3627, Naval Research Laboratory, Washington, Oct 1977
32. L.A. James and D.P. Jones, "Fatigue Crack Growth Correlation for Austenitic Stainless Steels in Air," *Proc. of the Conference on Predictive Capabilities in Environmentally-Assisted Cracking*, R. Rungta, Ed., PVP Vol 99, American Society of Mechanical Engineers, 1985, p 363-414
33. T. Kawakubo et al., "Crack Growth Behavior of Type 304 Stainless Steel in Oxygenated 290 °C Pure Water under Low Frequency Cyclic Loading," *Corrosion 36*, 1980, p 638-647

34. J.D. Gilman et al., "Corrosion-Fatigue Crack Growth Rates in Austenitic Stainless Steels in Light Water Reactor Environments," *Int. J. Pres. Ves. Piping 31*, 1988, p 55-68

35. W.J. Shack, Corrosion Fatigue Curves for Austenitic Stainless Steels in Light Water Reactor Environments, in *Environmentally Assisted Cracking in Light Water Reactors*, Semiannual Report, Oct 1990-March 1991, *NUREG/CR-4667*, Vol 12, ANL-91/24, 1991, p 30-36

36. T.S. Srivatsan, in *Environmental Degradation of Engineering Materials*, Penn State University Press, 1987

37. P.K. Liaw et al., *Scripta Metall.*, Vol 17, 1983, p 611

38. T. Magnin and L. Coudereuse, *Acta Metall.*, Vol 35, 1987, p 2105-2113

39. M.O. Speidel, Stress Corrosion Cracking of Austenitic Stainless Steels, Advanced Research Projects Agency (ARPA) N00014-75-C-0703, Ohio State University, 1977

40. P.M. Scott and B. Tomkins, in *Corrosion Fatigue*, R.N. Parkins and Ya.M. Kolotyrkin, Ed., Metals Society, 1980, p 156-164

41. W. Bamford, Fatigue Crack Growth of Stainless Steel Piping in a Pressurized Water Reactor Environment, *Transactions of ASME, J. Pres. Ves. Technology*, Vol 101 (No. 1), Feb 1979, p 73-79

42. R.P. Wei and A. Alavi, *J. Electrochem. Soc.*, Vol 138 (No. 10), Oct 1991, p 2907-2912

43. R.P. Wei and M. Gao, Micromechanism for Corrosion Fatigue Crack Growth in Metastable Austenitic Stainless Steels, Conference: Corrosion—Deformation Interactions, Les Editions de Physique, 1993, p 619-629

44. R.P. Wei and M. Gao, Crack Paths, Microstructure, and Fatigue Crack Growth in Annealed and Cold-Rolled AISI 304 Stainless Steels, *Met. Trans. A*, Vol 23A (No. 1), 1992, p 355-371

45. R.P. Wei, Mechanistic Considerations of Corrosion Fatigue of Steels, Retroactive Conference: International Conference on Evaluation of Materials Performance in Severe Environments, Vol 1, The Iron and Steel Institute of Japan, 1989, p 71-85

46. M. Itatani et al., *Nucl. Eng. Des.*, Vol 153 (No. 1), Dec 1994, p 27-34

47. J. Park, W.J. Shack, and D.R. Diercks, Crack Growth Behavior of Candidate Waste Container Materials in Simulated Underground Water, *Application of Accelerated Corrosion Tests to Service Life Prediction of Materials*, STP 1194, 1994, p 188-203

48. R. Koterazawa, *Fatigue Fract. Mater. Struc.*, Vol 16, 1993, p 619-630

49. R. Koterazawa and T. Nosho, *Fatigue Fracture Mater. Struc.*, Vol 14, 1991, p 1-9

50. C. Brinkman, *International Metals Review*, Vol 30, 1985, p 235-258

Fracture Toughness of Austenitic Stainless Steels and Their Welds

W.J. Mills, Member of ASM International, Upper St. Clair, PA

Austenitic stainless steel alloys are used extensively in heat-resistant structural components in power-generating and chemical industries as a result of their metallurgical stability, excellent corrosion resistance, and good creep and ductility properties at elevated temperatures. In addition, AISI type 300-series stainless steels are the most widely used structural alloys for cryogenic applications, because they exhibit high strength, ductility, and fracture toughness properties as well as low thermal expansion and low magnetic permeability.

Extensive fracture toughness testing of this class of alloys has been conducted to predict the failure conditions for flawed components. This article summarizes the fracture toughness behavior of wrought base metals and welds. Minimum expected toughness values, based on statistical analyses of literature data, are provided for use in fracture mechanics evaluations.

Elastic-plastic fracture mechanics methods are required to characterize fracture properties for austenitic stainless steels due to their ductile response even after high levels of cold work and neutron irradiation. The J-integral resistance (J-R) curve approach is generally used to characterize the fracture behavior of ductile materials; however, the standard ASTM test methods for determining J_{Ic} and J-R curves are not generally applicable to this class of high-ductility, high-toughness materials. Modified testing and analysis methods, which are detailed in the next section, have been developed and successfully applied to these materials.

Fracture Toughness Determination

The fracture toughness of stainless steels is determined by both single- and multiple-specimen J-R curve techniques, but the analysis procedures differ slightly from those described in ASTM E 813 for J_{Ic} determination and E 1152 for J-R curve determination. ASTM procedures and size requirements are generally not applicable to these alloys due to their exceptionally high toughness, ductility, and strain hardening capacity. The primary difference involves the crack-tip blunting relationship. The ASTM blunting line is given by:

$$J = 2\sigma_f(\Delta a) \qquad \text{(Eq 1)}$$

where σ_f is the flow strength, which is equal to the average of the 0.2% offset yield strength and ultimate tensile strength [$\sigma_f = \frac{1}{2}(\sigma_{YS} + \text{UTS})$], and Δa is the crack extension.

For austenitic stainless steels, Eq 1 significantly overpredicts the apparent crack extension associated with the blunting process, which results in nonconservative J_{Ic} values. The appropriate blunting line for this class of alloys is (Ref 1):

$$J = 4\sigma_f(\Delta a) \qquad \text{(Eq 2)}$$

The factor of 2 difference in Eq 1 and 2 is related to differences in the constraint factor, m, relating J to crack-tip opening displacement, CTOD (Ref 2):

$$J = m\sigma_{YS}(\text{CTOD}) \qquad \text{(Eq 3)}$$

The m value for intermediate- to high-strength, low strain hardening materials (i.e., Ramberg-Osgood strain hardening exponents, N, greater than 5) is close to unity (Ref 2), whereas low-strength, high-strain-hardening materials exhibit values of m on the order of 2 (Ref 1, 3, 4). Because the apparent crack extension associated with blunting is approximately one-third the CTOD (Ref 1, 4) and the yield strength is approximately two-thirds the flow strength (i.e., $\sigma_{YS} \approx \frac{2}{3}\,\sigma_f$) for this class of materials (Ref 1), the crack-tip blunting relationship is given by:

$$J = m\sigma_{YS}\,(\text{CTOD}) \approx 2\left(\frac{2\sigma_f}{3}\right)(3\Delta a) = 4\sigma_f(\Delta a) \qquad \text{(Eq 4)}$$

This equation has been shown to adequately represent the crack-tip blunting behavior for type 304 (Ref 1, 4-10), 316 (Ref 10-13), 321 (Ref 14), and 347 (Ref 13) stainless steels and their welds (Ref 15-17).

The J-R curve is constructed by plotting J-integral values against the corresponding crack extension values and fitting the J-Δa data with either a linear or power-law regression line (see Fig. 1). The J_c fracture toughness is then determined at the intersection of the linear regression line with the blunting line, per ASTM E 813-81, or the intersection of the power-law curve with the 0.2 mm offset blunting line, per ASTM E 813-87. (Initiation toughness values are termed J_c, rather than J_{Ic}, because they do not strictly meet the analysis methods and size requirements of ASTM E 813.) For modest amounts of crack extension ($\Delta a < 4$ mm), the two regression analyses provide reasonable fits of the data and yield very similar J_c values for both welds (Ref 18) and base metals (Ref 19).

Austenitic stainless steels typically exhibit tremendous tearing resistance after cracking initiates, as reflected by a steep J-R curve slope, dJ/da. Average dJ/da values reported herein are

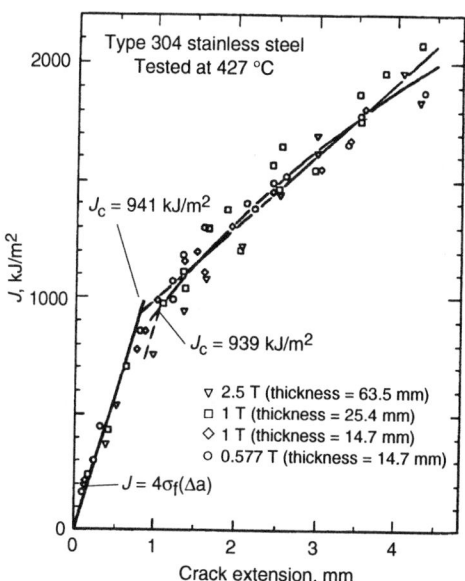

Fig. 1 Multiple-specimen J-R curves for 2.5T, 1T, and 0.577T compact tension specimens. The T factor denotes the planar dimension proportionality, relative to the standard specimen in ASTM E 399. Both the linear and power-law J-R curves adequately represent the data and produce comparable J_c values. Source: Ref 21

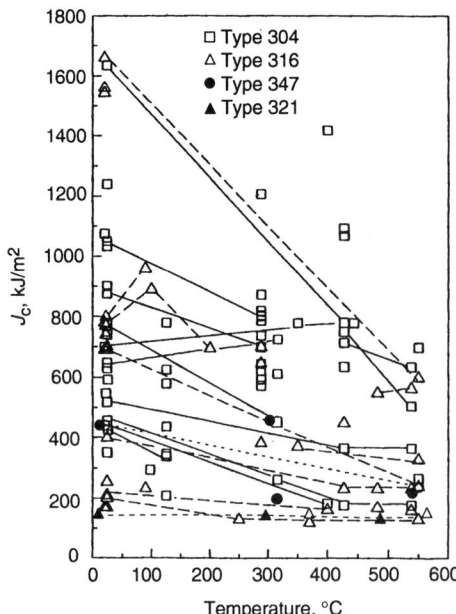

Fig. 2 Effect of test temperature on J_c fracture toughness. J_c values for the same heat are connected by a line. Source: Ref 5–10, 22, 24–46 (type 304); Ref 4, 10–13, 25, 29, 33, 34, 38, 43, 47–55 (type 316); Ref 13, 56 (type 347); Ref 14 (type 321)

determined based on the slope of either the linear J-R curve or the power-law J-R curve at approximately 2 to 3 mm of crack extension. Values of the tearing modulus (T), which represents a dimensionless form of the J-R curve slope, are computed from the following equation (Ref 20):

$$T = \frac{dJ}{da}\frac{E}{\sigma_f^2} \qquad \text{(Eq 5)}$$

where E is elastic modulus.

For modest amounts of crack extension ($\Delta a <$ 4 mm), conventional J-R curves are relatively independent of specimen size (Ref 21–23). For example, Fig. 1 (Ref 21) shows that the overall fracture behavior is not radically different for the different size compact tension, C(T), specimens of type 304, so data for all specimen sizes are combined in the linear and power-law regressions to establish J_c. The two regression analyses provide good fits to the data and result in comparable J_c and dJ/da values. Figure 1 also shows that Eq 2 accurately predicts the blunting behavior for type 304, regardless of specimen size.

Fracture Toughness of Base Metal

Fracture Properties at Ambient and Elevated Temperatures

Extensive fracture toughness testing of type 304 and 316 stainless steels shows that they are extremely resistant to fracture. Both types exhibit a ductile fracture response under a wide variety of conditions, but J_c values are highly variable, as shown in Fig. 2. It is seen that J_c values range from 169 to 1660 kJ/m^2 at room temperature and from 130 to 1420 kJ/m^2 at approximately 400 °C.

Some of this scatter is due to the lack of a standard J_c test method, but this effect is small in comparison to heat-to-heat variability and orientation effects. For example, common testing and analysis methods produced a factor of 4 difference in J_c values for five heats of type 304 (Ref 25) and three heats of type 316 (Ref 52). These findings demonstrate that significant heat-to-heat variability is inherent to austenitic stainless steels.

Figure 2 also shows that increasing the test temperature tends to decrease fracture toughness, with the highest toughness heats showing the greatest effect. Low toughness heats show only a very small toughness decrease between 24 and 550 °C. The apparent increase in toughness exhibited by two type 316 heats between room temperature and about 100 °C is believed to be associated with data scatter, rather than an inherent temperature effect. In addition, the type 304 heat showing a slight increase in J_c between 24 and 443 °C exhibits a factor of 2 reduction in dJ/da, such that the overall fracture resistance decreases with increasing temperature, consistent with the trends exhibited by the other heats. The decrease in fracture toughness reflects a decrease in both flow strength from 430 to 300 MPa and total elongation from 70 to 40% as temperature increases from 24 to 538 °C.

To account for heat-to-heat variability at ambient and elevated temperatures, a statistical analysis of post-1981 literature toughness data for types 304 and 316 was performed (Ref 19); results are summarized in Table 1. Both J_c and dJ/da were found to be distributed lognormally (i.e., logarithms of J_c and dJ/da were normally distributed) and were independent of material type. Although median J_c and dJ/da values are very high at room temperature (672 kJ/m^2 and 292 MPa, respectively), the large degree of scatter indicates the likelihood that a significant fraction of heats have values that are considerably lower. Lower bounds for 90%/95% tolerance limits bracketing 90% of the total population at a 95% confidence level are summarized in Table 1. The lower-bound J_c value of 215 kJ/m^2 and dJ/da value of 59 MPa, corresponding to a tearing modulus of 75, are relatively high, demonstrating that even the lower toughness heats are resistant to fracture.

While the median J_c value of 421 kJ/m^2 at 400 to 550 °C is about 35% lower than the room-temperature value, the median dJ/da value of 263 MPa is essentially the same as its room-temperature counterpart. This indicates that the tearing

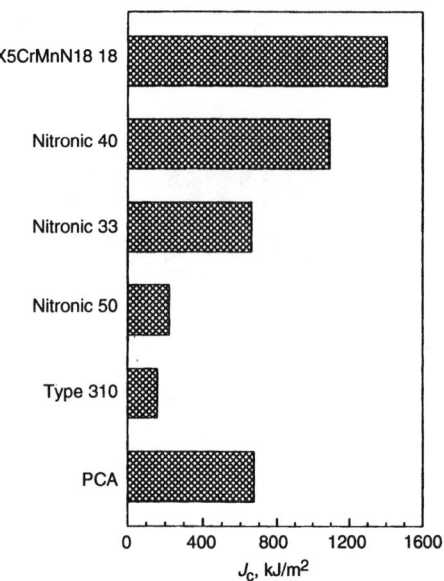

Fig. 3 J_c values for various stainless steels at room temperature. Source: Ref 51 (type 310); Ref 52 (PCA); Ref 57, 58 (nitrogen-strengthened high-manganese stainless steels)

resistance is not degraded at elevated temperatures. The 90%/95% lower-bound J_c and dJ/da values at elevated temperatures are 96 kJ/m^2 and 79 MPa, respectively. The lower-bound J-R curve slope, which corresponds to a tearing modulus of 120, indicates that types 304 and 316 retain excellent fracture resistance at temperatures up to 550 °C.

Figure 2 also provides the J_c response for types 347 and 321 at ambient and elevated temperatures. Toughness values range from 150 to 450 kJ/m^2 and fall in the lower portion of the toughness interval for types 304 and 316. Nevertheless, these toughness values indicate that type 347 has good fracture resistance between 24 and 550 °C.

Limited fracture toughness testing of other stainless steels has been performed at room temperature, and the results are summarized in Fig. 3. The materials represented in this figure are type 310, PCA, and various nitrogen-strengthened high-manganese alloys. Type 310 exhibits a relatively low J_c value of 158 kJ/m^2 (Ref 51). The high-manganese alloys, including Nitronic 33 (Fe-18Cr-3Ni-13Mn), Nitronic 40 (Fe-21Cr-6Ni-9Mn), Nitronic 50 (Fe-21Cr-12Ni-5Mn), and X5CrMnN18 18, exhibit intermediate to high toughness values, ranging from 230 to 1400

Table 1 Summary of J_c and dJ/da values for stainless steel base metals and welds

Alloy	Condition	Temperature, °C	Mean J_c, kJ/m^2	90%/95% L.B. J_c, kJ/m^2	Mean dJ/da, MPa	90%/95% L.B. dJ/da, MPa
304, 316	Wrought	20–125	672	215	292	59
304, 308, 316, 16-8-2	GTA weld	20–125	492	192	390	139
304, 308, 316, 16-8-2	SA weld	20–125	147	67	150	72
304, 316	Wrought	400–550	421	96	263	79
304, 308, 316, 16-8-2	GTA weld	400–550	293	180	307	107

90/95 L.B., lower bound for tolerance limits bracketing 90% of the total population at a 95% confidence level. GTA, gas-tungsten arc; SA, submerged arc. Source: Ref 19

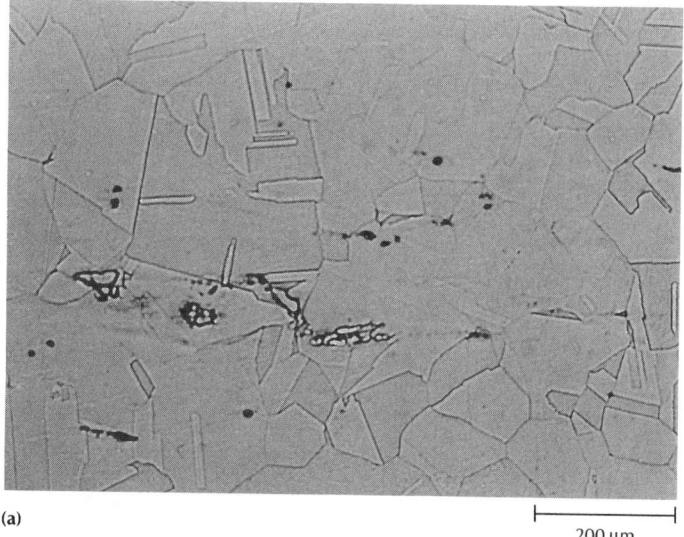

(a)

200 µm

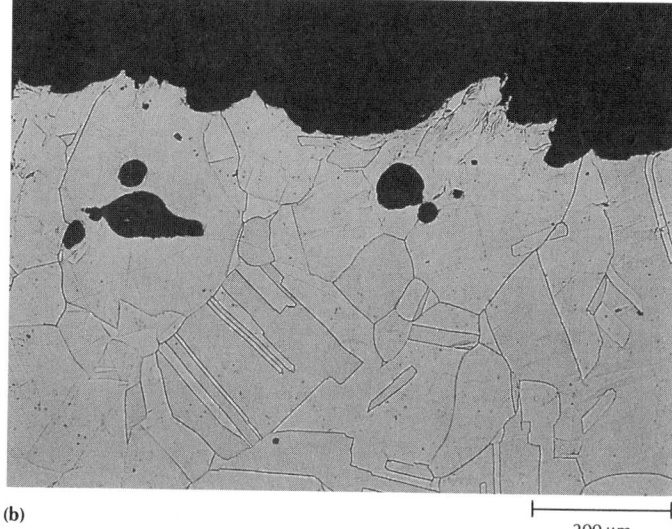

(b)

200 µm

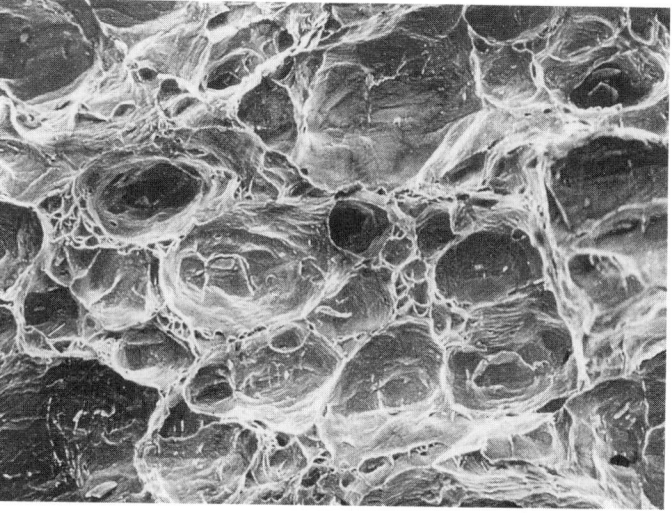

(c)

130 µm

Fig. 4 Microstructure and fracture surface morphology for a low-toughness type 304 heat (J_c = 178 kJ/m²). (a) Typical microstructure with MC inclusion clusters. (b) Fracture profile showing that MC-nucleated microvoids are localized along the fracture plane. (c) SEM fractograph showing shallow dimples nucleated by MC inclusion clusters. Source: Ref 25

kJ/m² (Ref 57, 58). PCA, a titanium-modified stainless steel (Fe-14Cr-16Ni-2Mo-0.25Ti) that is the most promising candidate austenitic material for the first wall component of fusion devices, has a J_c of 670 kJ/m² (Ref 52).

The reported J_c and dJ/da values demonstrate that stainless steels have exceptionally high fracture resistance at ambient and elevated temperatures; failure requires extensive plastic strains and long critical crack lengths. Therefore, fracture control via a fracture mechanics approach is generally not an important design requirement. Conventional stress and strain design limits, such as those provided in the American Society of Mechanical Engineers (ASME) Boiler and Pressure Vessel Code, are sufficient to guard against fracture because they preclude excessive plastic deformation. For applications where margins against ductile fracture must be quantified or where components are subjected to large plastic strains, elastic-plastic J-integral methods can be used to predict fracture conditions. These methods include the J-estimation schemes developed by General Electric/Electric Power Research Institute (Ref 59-61) and Novetech/Electric Power

Research Institute (Ref 62), or the Nuclear Regulatory Commission leak-before-break prediction method (Ref 63, 64). Calculation of applied J values for cracked components requires knowledge of the strain hardening capacity of the material in terms of the Ramberg-Osgood strain hardening relationship:

$$\frac{\varepsilon}{\varepsilon_{YS}} = \frac{\sigma}{\sigma_{YS}} + C\left(\frac{\sigma}{\sigma_{YS}}\right)^N \qquad \text{(Eq 6)}$$

where σ is the true stress, ε is the true strain, ε_{YS} is the yield strain at the 0.2% offset yield strength value, and C and N are the strain hardening coefficient and exponent, respectively. Values of C and N for type 304 and 316 stainless steels and their welds are provided in the literature (Ref 17, 25).

Fracture Mechanisms in Base Metals

The operative cracking mechanism in stainless steel is typically microvoid coalescence, regardless of test temperature. Heat-to-heat variability is associated with differences in both the matrix strength and the density and morphology of in-

clusions that serve as microvoid nucleation sites. Large inclusions fail during the early stages of plastic straining, so a high inclusion density drastically reduces the plastic energy required for microvoid coalescence. Typical inclusions include MC-type carbides, calcium aluminates, and manganese sulfides. The strength of the matrix generally depends on the effectiveness of solution annealing in removing plastic deformation introduced during thermomechanical processing. Retention of 1 to 2% cold work is sufficient to restrict plastic deformation during the fracture process.

Figures 4 and 5 compare the inclusion and dimple morphologies for low- and high-toughness heats of type 304 (Ref 25). In the low-toughness heat with a J_c of 178 kJ/m² (Fig. 4), primary dimples are shallow and confined to the immediate vicinity of the crack plane. Inclusion clusters aligned in the primary working direction fail early in the plastic straining process, creating a preferred crack path. In addition, the relatively high strength exhibited by this heat (σ_{YS} = 206 MPa) reduces the plastic energy required for microvoid coalescence. This combination of a rela-

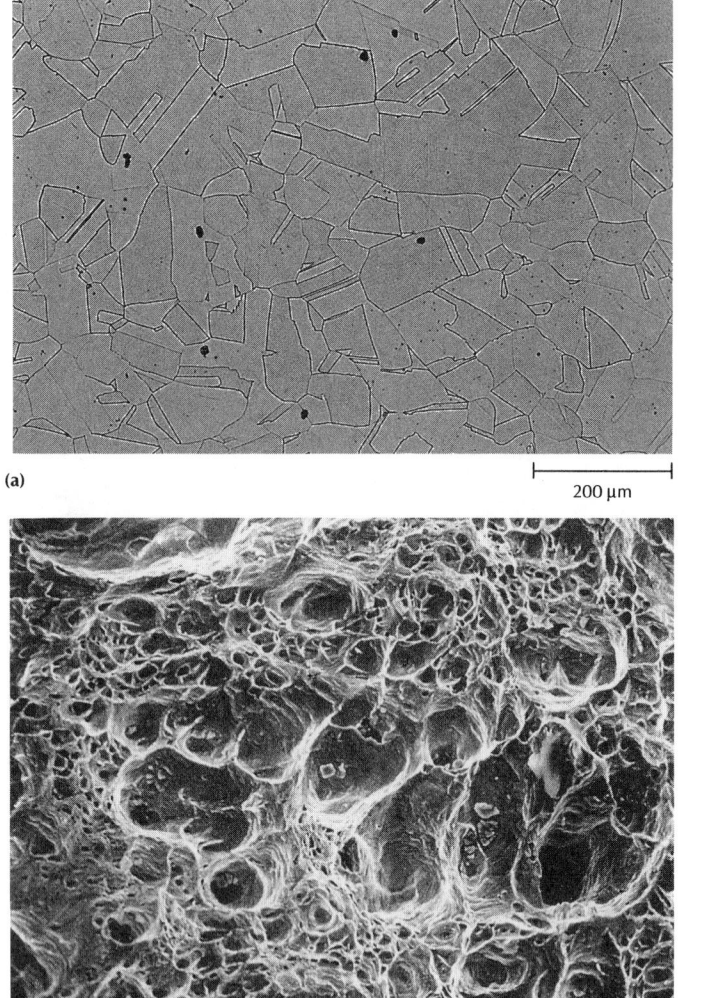

(a)

200 µm

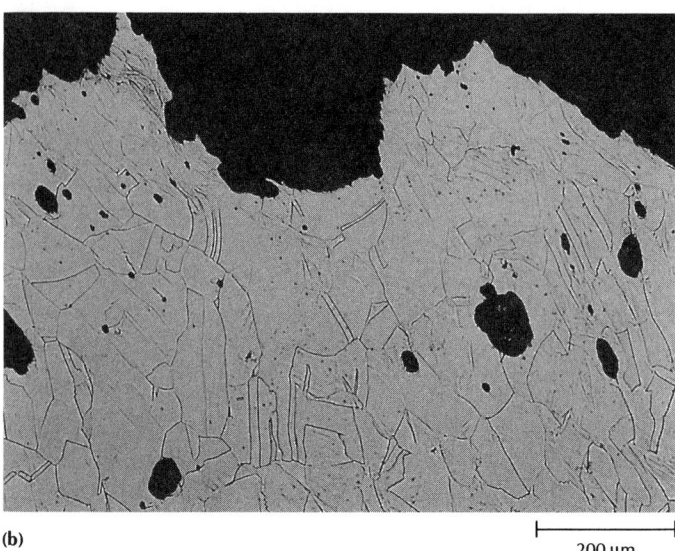

(b)

200 µm

(c)

100 µm

Fig. 5 Microstructure and fracture surface morphology for a high-toughness type 304 heat (J_c = 751 kJ/m^2). (a) Uniform distribution of relatively small MC inclusions and fine $M_{23}C_6$ carbides. (b) Fracture profile showing evidence of gross plasticity and MC-nucleated microvoids away from the fracture plane. (c) SEM fractograph showing duplex population of well-defined dimples nucleated by MC and $M_{23}C_6$ carbides. Source: Ref 25

tively high-strength matrix and aligned inclusion clusters accounts for both the shallow dimples confined to the primary fracture plane and the low fracture toughness.

In the high-toughness heat with a J_c of 751 kJ/m^2 (Fig. 5), the deep primary dimples and large microvoids away from the crack plane demonstrate that the fracture process involves gross plasticity. The smaller MC-type inclusions in this heat are more resistant to fracture, and the lower-strength matrix (σ_{YS} = 148 MPa) promotes increased plastic deformation. Because there is no preferred crack path, microvoids are nucleated away from the primary fracture plane. Gross plastic deformation is then required for microvoid coalescence, which accounts for the superior fracture toughness.

Fracture Toughness of Welds

Although stainless steel welds are predominantly austenitic, they typically contain a δ-ferrite phase that has a body-centered cubic (bcc) structure. This phase is needed to control the weld

solidification behavior and inhibit the formation of low-melting-point compounds, such as sulfides and phosphides, that promote microfissuring. Because the ferrite phase is brittle at low temperatures, stainless steel welds exhibit a ductile-brittle transition temperature phenomenon, as shown in Fig. 6 for type 308 welds (Ref 65-69). At ambient and elevated temperatures, the ferrite phase behaves in a ductile manner, so the welds are resistant to fracture.

Ambient Temperature Fracture Toughness

Figure 7 summarizes the J_c fracture toughness of types 304/308, 316/16-8-2, and 330 welds at 20 to 125 °C. Five welding processes are represented, including gas-tungsten arc (GTA), shielded-metal arc (SMA), submerged arc (SA), gas-metal arc (GMA), and flux-cored arc (FCA) processes. The fracture toughness is seen to be dependent on weld process, but not composition. This is consistent with findings that type 308 and 16-8-2 welds fabricated using the same welding process and welding parameters yield identical J_c

and dJ/da values, demonstrating that toughness is independent of filler material (Ref 17).

Figure 7 shows that GTA welds consistently exhibit the highest J_c values, while SA welds yield the lowest. The mean J_c value for GTA welds is about 25% lower than that for types 304 and 316, represented by the broken line on Fig. 7. The mean J_c and dJ/da values for SA welds are about three times lower than their GTA counterpart. Table 1 shows that the 90%/95% lowerbound J_c and dJ/da values for the GTA and SA welds are 192 versus 67 kJ/m^2 and 139 versus 72 MPa, respectively.

The SMA and FCA welds tested at room temperature have low to intermediate toughness values. Cryogenic (Ref 78) and high-temperature (Ref 17) data show that the SMA weld process results in tremendous variability, covering the full range of toughness values exhibited by SA and GTA welds. The excessive scatter is attributed to the variability associated with a manual welding process. For design purposes, it is reasonable to assume that the lower-bound toughness for SMA welds is comparable to that for SA welds. Cryogenic results (Ref 78-80) reveal that

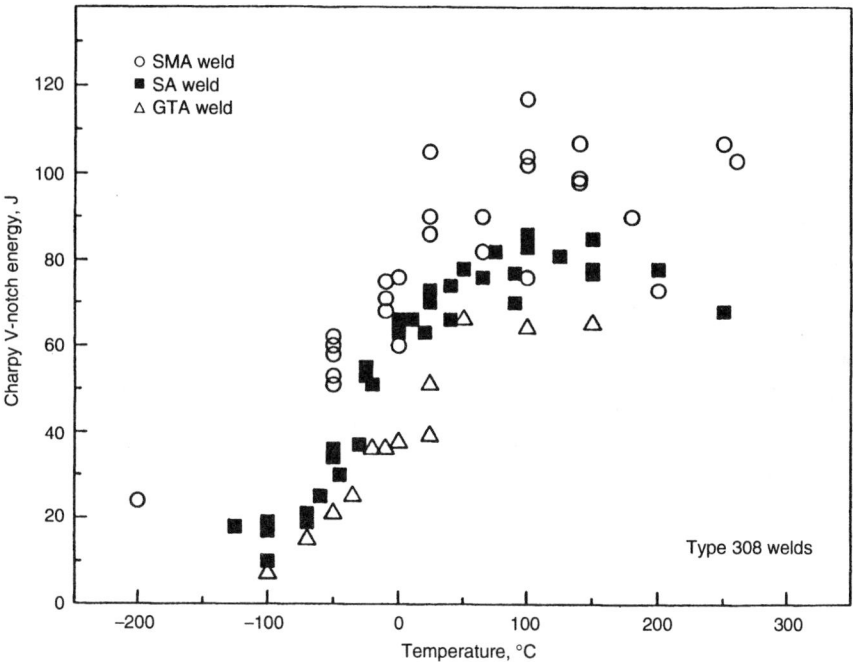

Fig. 6 Charpy impact energy vs. test temperature for type 308 welds showing the ductile-brittle transition temperature phenomena. SMA, shielded-metal arc; SA, submerged arc; GTA, gas-tungsten arc. Half-size Charpy specimens (5 × 5 × 25.4 mm with a 0.76 mm notch) were used to characterize the GTA weld. Source: Ref 65-67 (SMA), Ref 68 (SA), Ref 69 (GTA)

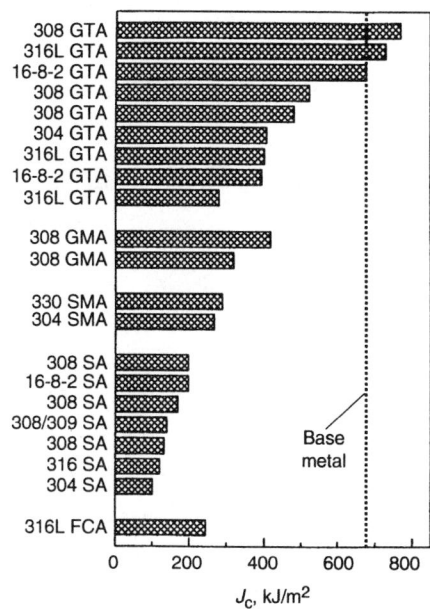

Fig. 7 Summary of J_c fracture toughness at 20-125 °C for welds. The mean J_c value for types 304 and 316 base metal is shown for comparison. GTA, gas-tungsten arc; GMA, gas-metal arc; SMA, shielded-metal arc; SA, submerged arc; FCA, flux-cored arc. Source: Ref 15, 22, 25, 34, 52, 69-72 (GTA); Ref 35, 36 (GMA); Ref 34, 73 (SMA); Ref 17, 34, 68, 74-76 (SA); Ref 77 (FCA)

FCA welds can exhibit lower toughness values than SA welds.

The two GMA welds tested at room temperature have high J_c values. This is consistent with cryogenic data (Ref 78, 79, 81-83) that show that J_c values for GMA welds are comparable to those for GTA welds.

Elevated-Temperature Fracture Toughness

The same ordering of fracture toughness levels for the various weld processes is observed at elevated temperatures, as shown in Fig. 8. At 427 to 538 °C, GTA welds consistently possess the highest J_c values (266 to 373 kJ/m²), SMA welds have intermediate toughness values (89 to 190 kJ/m²), and both SA welds tested in this regime have a J_c of 76 kJ/m². Table 1 provides mean and lower-bound values of J_c and dJ/da for GTA welds at elevated temperatures. The sparse elevated-temperature data base precludes meaningful statistical analysis for SA and SMA welds, so a simple lower-bound approach must be used to establish minimum expected toughness values. The lowest observed J_c and dJ/da values at high temperatures are 55 kJ/m² and 62 MPa, which are about 15% less than the lower-bound values for SA welds at room temperature.

The fracture resistance of electron-beam (EB) welds is high. While no specific fracture toughness values are available, Josefsson (Ref 71) reported that J_c values for EB welds are about twice those for GTA welds (550 kJ/m² at 250 °C and 330 kJ/m² at 450 °C).

The exceptionally high toughness for GTA and EB welds precludes rapid fracture concerns in most engineering structures. Although J_c values

for SMA and SA welds can be low, their relatively high tearing resistance indicates that unstable fracture is unlikely except after significant plastic deformation. Consequently, standard stress and strain design limits generally provide adequate protection against failure, and sophisticated elastic-plastic fracture mechanics assessments are not routinely needed.

Fracture-Mechanisms in Welds

Stainless steel welds fail exclusively by a dimple rupture mechanism, where the microvoids nucleate at inclusions and δ-ferrite particles. Thus, the overall fracture resistance is controlled by the density and morphology of second-phase inclusions. Type 304/308 and 316/16-8-2 welds have a duplex austenite-ferrite structure with about 4 to 12% δ-ferrite. Because δ-ferrite is ductile at ambient and elevated temperatures, its volume fracture and morphology do not control the fracture response. Fracture properties are controlled by the density of inclusions rich in manganese and silicon and believed to be manganese silicides and silicates. SA welds have a high density of coarse inclusions (Ref 17, 18, 89) as shown in Fig. 9. SMA welds contain a modest amount of coarse inclusions (Ref 17, 18, 55, 82), while GMA and GTA welds have a much lower inclusion density (Ref 17, 82, 90). O'Donnell et al (Ref 55) found that the volume fraction of inclusions for an SMA weld is an order of magnitude higher than that in a GTA weld. High silicon contents are generic to the SA weld processes due to silicon pickup from the flux. In SA and SMA welds, the deoxidation process produces oxides rich in manganese and silicon that are sometimes trapped in the molten pool during

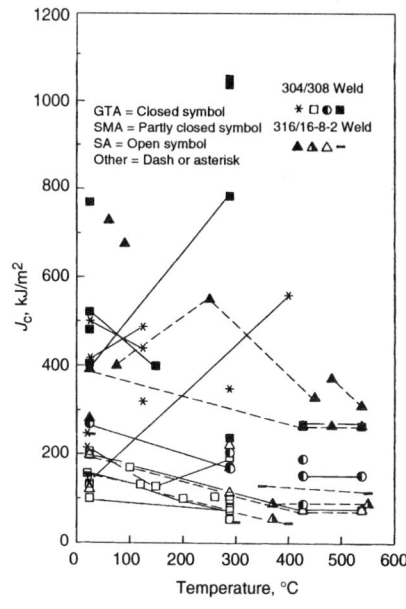

Fig. 8 Effect of test temperature on J_c for types 304/308 and 316/16-8-2 welds. J_c values for the same weld are connected by a line. GTA, gas-tungsten arc; SMA, shielded-metal arc; SA, submerged arc. A dash or asterisk represents either another welding process or a weld where the process was not identified. Source: Ref 10, 16-18, 22, 24, 29, 34-36, 38, 40, 43, 46, 54, 68-70, 74-76, 84-88 (types 304/308); Ref 15, 17, 18, 29, 34, 47, 50, 52, 55, 71, 72, 86 (types 316/16-8-2)

solidification, thereby accounting for the high density of inclusions. The low density of coarse inclusions in GTA and GMA welds arises from the inert shielding gas protecting the molten pool

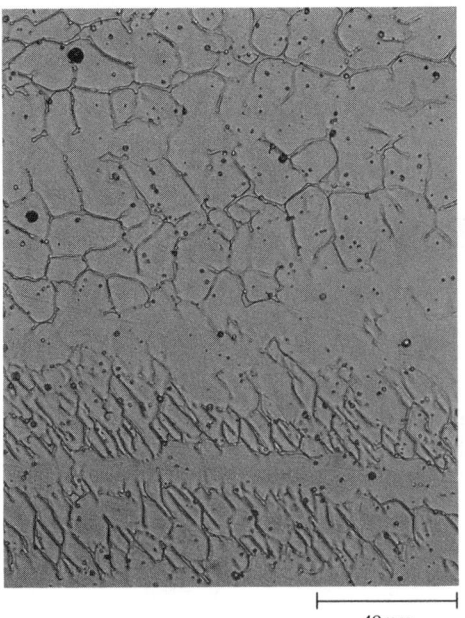

Fig. 9 Duplex austenite-ferrite microstructure for type 308 submerged arc weld showing very high density of inclusions rich in silicon and manganese. Source: Ref 17

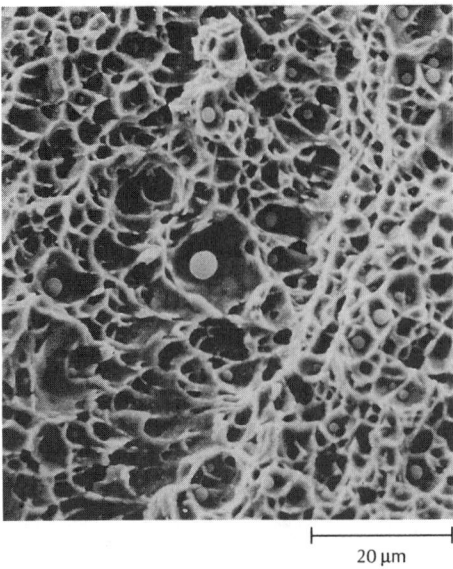

Fig. 10 SEM fractograph of type 16-8-2 submerged arc weld where the majority of dimples were nucleated by spherical inclusions rich in silicon and manganese. Source: Ref 17

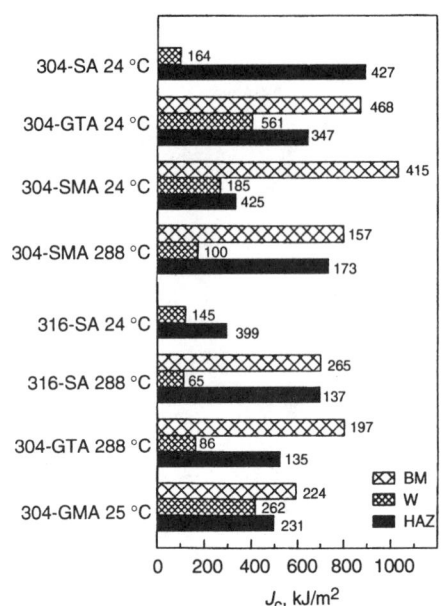

Fig. 11 Comparison of J_c values for heat-affected zone (HAZ), weld fusion zone (W), and base metal (BM). Values of dJ/da, in MPa, are provided beyond each bar. Cracks are oriented parallel to the welding direction. SA, submerged arc; GTA, gas-tungsten arc; SMA, shielded-metal arc; GMA, gas-metal arc. Source: Ref 22, 34, 36

from oxygen and the lack of a flux. GTA welds tend to have a few small particles, but they resist fracture and do not adversely affect properties (Ref 17).

The fracture surface morphology for GTA welds involves a microvoid coalescence mechanism. Microvoids are nucleated by localized failure or decohesion of the ferrite phase and small precipitates, but only after extensive deformation of both the ferrite and austenite matrix. The ability of the ferrite phase to accommodate plastic deformation and resist fracture accounts for the exceptional fracture toughness displayed by GTA welds. Although ferrite content and morphology have little effect on fracture properties at ambient and high temperatures, significant ductility loss can occur at cryogenic temperatures or after long-term thermal aging. Under these conditions, the ferrite can degrade the fracture resistance of the weld, as discussed later.

The coarse inclusions in SA and SMA welds cannot accommodate plastic deformation, so high secondary stresses develop at the particle-matrix interface. The high stresses cause inclusions to decohere from the matrix at low plastic strains, and the resulting microvoids serve as effective dimple nucleation sites. Figure 10 shows that most of the dimples in SA welds are nucleated by the spherical inclusions. Relatively little plastic deformation is required to initiate microvoids in SA welds, which accounts for their inferior fracture toughness. The modest inclusion density in SMA welds produces intermediate toughness values.

The superior toughness displayed by EB welds is associated with the absence of coarse inclusions and a refined microstructure produced by rapid cooling rates.

Fracture Toughness of Welding-Induced Heat-Affected Zones

Welding of types 304 and 316 produces heat-affected zones where the grain boundaries are often decorated with ferrite or chromium-rich $M_{23}C_6$ carbides. Immediately adjacent to the fusion zone, a thin ferrite layer sometimes forms along austenite grain boundaries and extends one to two grain diameters into the heat-affected zone. The ferrite is beneficial because it restricts grain growth and prevents liquation cracking by limiting impurity element diffusion and inhibiting wetting of liquid films (Ref 91, 92). Because the intergranular ferrite is ductile at ambient and elevated temperatures, it does not have an adverse effect on fracture resistance. Welding-induced precipitation of $M_{23}C_6$ carbides occurs up to 3 mm from the fusion boundary, depending on the carbon content of the base metal, weld travel speed, and heat input (Ref 93). While chromium depletion adjacent to the carbides creates a sensitized structure that is susceptible to corrosion and stress-corrosion cracking, the overall fracture resistance remains high because the intergranular precipitates have little effect on fracture properties (Ref 94). In Fig. 11, fracture toughness values for type 304 and 316 heat-affected zones (Ref 22, 34, 36) are compared with their weld and base metal counterparts. The heat-affected zone toughness is superior to the weld toughness and generally commensurate with that of the base metal. This behavior is observed for welds made with very different heat inputs, including GTA, SMA, and SA welds. In cases where J_c for the heat-affected zone is reduced relative to the base metal, values of dJ/da remain exceptionally high, ranging from 135 to 427 MPa. Thus, the limiting toughness for welded joints is typically control-

led by the weld fusion zone, not the heat-affected zone.

Effect of Crack Orientation on Base Metal and Weld Toughness

Thermomechanical processing of stainless steels can produce anisotropic microstructures with inclusion stringers aligned in the primary working direction. Because these inclusions fail and nucleate microvoids early in the plastic straining process, they degrade fracture toughness when aligned in the crack propagation direction. In some heats, δ-ferrite stringers are aligned in the primary working direction, but this phase is ductile, so it has no adverse effect on toughness. This is supported by the exceptional fracture resistance observed in heats with δ-ferrite stringers, even when the crack is parallel to the stringers (Ref 39, 42, 52).

The effect of crack orientation on J_c for types 304 and 316 and their welds is shown in Fig. 12. Crack orientation is defined in accordance with the ASTM E 616 two-letter code (L = longitudinal [i.e., primary working direction], T = transverse, S = short transverse, R = radial, and C = circumferential). The first letter designates the direction normal to the crack plane and the second is the expected crack propagation direction.

Figure 12 shows that the toughness of the type 304 plate in the T-L orientation is 40% lower than the L-T toughness, while the two type 316L plates show little or no difference in toughness in the L-T and T-L orientations. The small difference in J_c for the L-T and T-L orientations in some plates arises because plates are often cross-

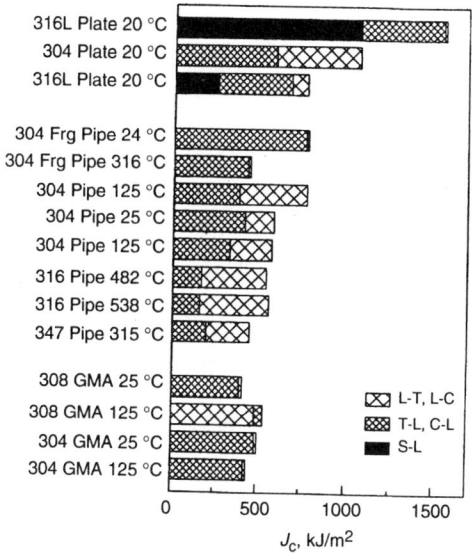

Fig. 12 Effect of crack orientation on the fracture toughness of base metal and welds. GMA, gas-metal arc welded; L, longitudinal (primary working direction); T, transverse; C, circumferential; S, short transverse. For the GMA welds, the welding direction is parallel to T-L oriented cracks and perpendicular to L-T oriented cracks. Source: Ref 8, 10, 11, 24, 27, 36, 56

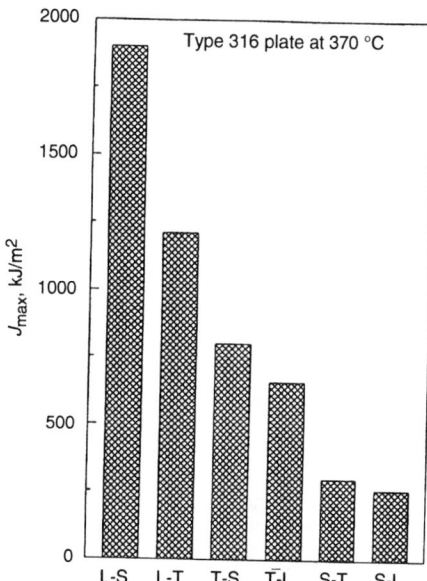

Fig. 13 Effect of crack orientation on J_{max}, which corresponds to the value of J at maximum load. L, longitudinal (primary working direction); T, transverse; S, short transverse. Source: Ref 95

rolled to minimize inclusion alignment in any one particular direction. Nevertheless, inclusions in cross-rolled plates are smeared out normal to the short-transverse direction, which accounts for the significant reduction in toughness for the S-L orientation. The low toughness in this orientation is generally not a concern, because most components are not subjected to significant stresses in the through-thickness direction.

Garwood (Ref 95) characterized the 370 °C fracture toughness of a type 316 plate in six different orientations; the results are summarized in Fig. 13. The toughness is characterized in terms of the J value at maximum load (J_{max}). For stainless steels, J_{max} is well above J_c, so it reflects the tearing resistance as well as the initiation toughness. Nevertheless, the trends shown in Fig. 13 provide an overview of crack orientation effects. Although the L-S and L-T orientations have a common crack plane, the different cracking directions yield a 40% difference in toughness. Cracks with an L-S orientation represent a crack-arrester geometry where failed inclusion stringers blunt the crack as it propagates in the through-thickness direction. Hence, the tremendous plastic energy required to extend the crack accounts for the exceptionally high J_{max} of 1900 kJ/m². The L-T orientation represents a crack-divider

geometry where the dimples nucleated by inclusions also blunt the crack, but not as effectively as the L-S orientation. The T-S and T-L orientations have a common crack plane that contains the primary axis of the stringers. Hence, the elongated dimples nucleated by the stringers provide a preferred crack path that accounts for the intermediate toughness for both of these orientations. Through-thickness loading for the S-T and S-L orientations produces by far the worst fracture properties. Because inclusion clusters are parallel to the crack plane, they provide low-energy crack paths, resulting in inferior fracture resistance.

Crack orientation effects for piping are summarized in Fig. 12. While the forged pipe has comparable toughness values in the C-L and L-C orientations (Ref 27), the other pipes show significant orientation effects with J_c in the C-L orientation, being 30 to 70% lower than its L-C counterpart. This orientation effect is attributed to the large stringers aligned in the longitudinal direction (Ref 10), as shown in Fig. 14. When these stringers are perpendicular to the cracking direction (L-C orientation), they nucleate deep, equiaxed microvoids that blunt the crack tip and enable the adjacent material to undergo gross plastic deformation prior to separation (Fig. 14b). In the C-L orientation, stringers aligned in the crack growth direction initiate elongated microvoids ahead of the crack (Fig. 14c). Coalescence of these microvoids causes premature crack advance without significant plastic deformation and results in a 70% lower J_c and 60% lower tearing modulus (Ref 10).

Figure 12 also shows that the fracture toughness of welds is independent of crack orientation: cracks parallel and normal to the welding direction yield equivalent J_c values. Second-phase particles in the weld fusion zone are not preferentially aligned, which accounts for the lack of an orientation effect.

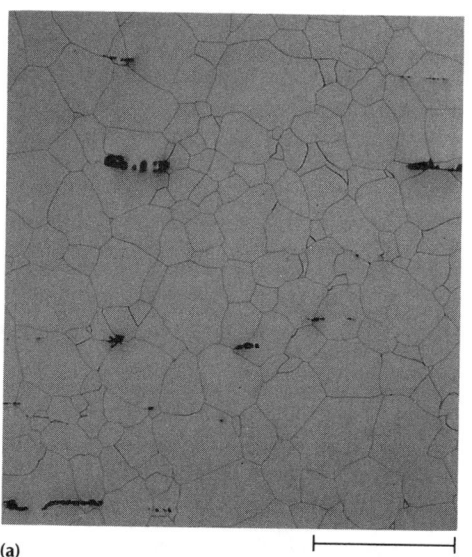
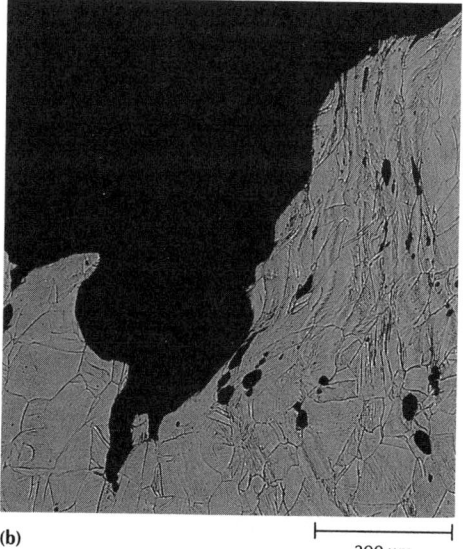
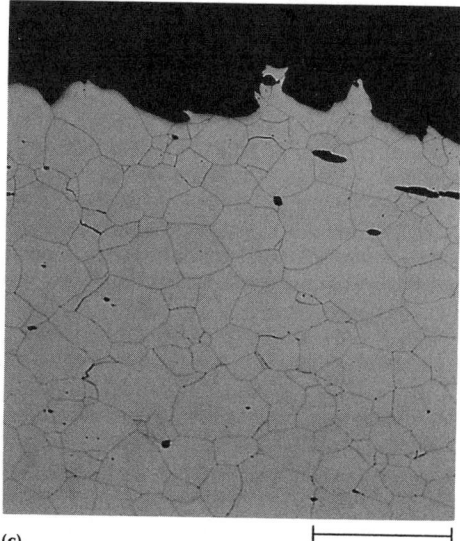

Fig. 14 Microstructure of compact tension specimens from type 316 piping. (a) Stringers aligned in axial direction. (b) Longitudinal-circumferential (L-C) orientation. Extensive plastic deformation and deep microvoids nucleated by stringers. (c) Circumferential-longitudinal (C-L) orientation. Shallow microvoids and limited plastic deformation. Source: Ref 10

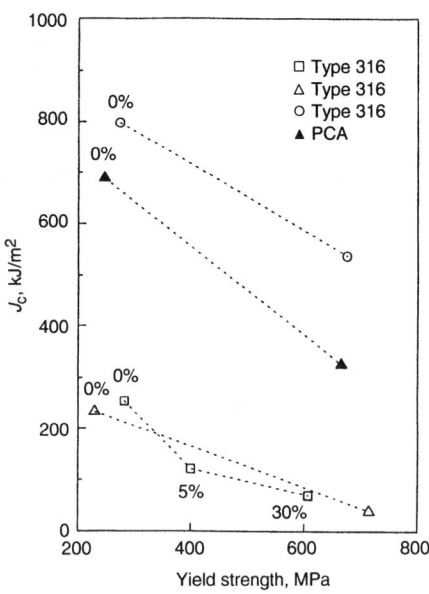

Fig. 15 Effect of cold-work-induced strengthening on fracture toughness. The percentage of cold work is provided next to data points. Source: Ref 49, 52

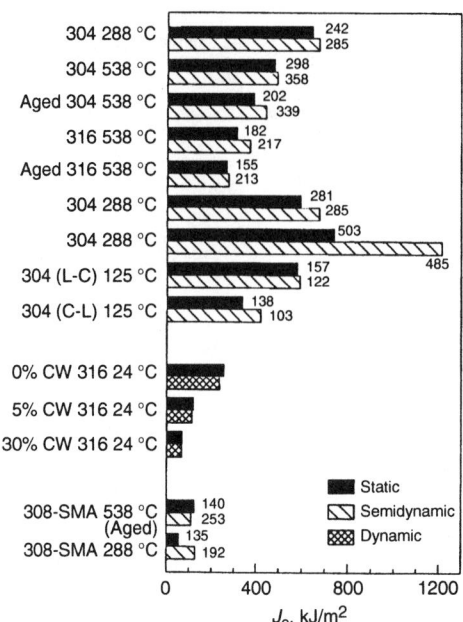

Fig. 16 Effect of strain rate on J_c and on dJ/da (provided next to each bar in MPa). L, longitudinal (primary working direction); C, circumferential; CW, cold work; SMA, shielded-metal arc welded. Source: Ref 46, 49, 94, 96, 97

Fracture Toughness of Cold-Worked Stainless Steels

While cold work is used to increase the strength of stainless steels, it significantly reduces ductility and fracture toughness. Annealed type 316 has yield strengths between 200 and 300 MPa, but after 5 to 30% cold work, yield strength levels are typically 400 to 600 MPa. The strength is increased at the expense of ductility; total elongation values are reduced from 60-75% in the annealed condition to 20-30% after 20% cold work. The reduced ductility is responsible for the decrease in toughness. Figure 15 summarizes the effect of cold-work-induced strengthening on fracture toughness. Chipperfield (Ref 49) found that the toughness of type 316 is reduced from 254 kJ/m² in the annealed condition to 121 kJ/m² after 5% cold work and to 70 kJ/m² after 30% cold work. Pawel (Ref 52) also found that cold work significantly reduces the toughness of two type 316 heats and a PCA heat. For the low-toughness type 316 heat (represented by open triangles), cold work severely degrades J_c from 233 to 39 kJ/m². By contrast, the high-toughness heat (represented by circles) shows a much smaller effect, and cold-worked PCA shows an intermediate loss in toughness. Strength levels for these cold-worked materials are similar (σ_{YS} = 675 to 715 MPa, σ_{UTS} = 725 to 770 MPa), so strength level does not correlate with toughness. The different J_c responses are associated with variability in inclusion density and morphology. Low-toughness heats generally contain a high density of inclusions, and cold work exaggerates their alignment in the primary working direction. When the inclusion stringers are aligned in the cracking direction, they nucleate elongated microvoids ahead of the crack front, thereby provid-

ing a low-energy crack path that significantly degrades fracture resistance.

Effect of Strain Rate on Base Metal and Weld Toughness

Figure 16 shows that semidynamic and dynamic strain rates do not adversely affect fracture resistance and that in some cases, higher loading rates have a beneficial effect. Semidynamic strain rates (~600 mm/min corresponding to crack initiation times of less than 0.5 s) have a small beneficial effect on both J_c and dJ/da (Ref 46, 94). In addition, load-displacement curves obtained under dynamic conditions (15 m/min) are identical to the semidynamic curves, indicating that the beneficial effects are maintained at very high strain rates (Ref 94). One of the two type 304 heats studied by Marschall (Ref 96) shows a similar response, with J_c increasing by 15% under semidynamic conditions. The second heat shows a much larger effect, with J_c increasing by 65%. Chipperfield (Ref 49) characterized the static and dynamic toughness of annealed and cold-worked type 316 and found that in both conditions the toughness response was essentially unaffected by loading rate: dynamic J_c values were within 7% of their static counterparts.

The toughness of an aged type 308 SMA weld at semidynamic strain rates is also superior to that under static strain rates (Ref 94). While J_c values for the two strain rates are similar, the 80% increase in dJ/da under semidynamic conditions results in a substantial improvement in overall fracture resistance. Moreover, dynamic testing of this weld produces load-displacement curves that are identical to the semidynamic curves, here

again demonstrating that both the semidynamic and dynamic toughness responses are superior to the static response. Marschall (Ref 96) found even more dramatic strain rate effects for a type 308 SA weld: semidynamic strain rates cause a 130% increase in J_c and a 40% increase in dJ/da.

Effect of Thermal Aging on Base Metal and Weld Toughness

Aging-Induced Microstructural Changes

Austenitic stainless steels are commonly used in high-temperature applications where aging-induced microstructural changes degrade fracture resistance. High-temperature aging of base materials results in the formation of carbides and intermetallics, including sigma [FeCr, FeMo, Fe(CrMo)], chi ($Fe_{18}Cr_6Mo_5$) and Laves (Fe_2Mo) phases (Ref 98, 99). Between 500 and 650 °C, chromium-rich $M_{23}C_6$ precipitation is dominant occurring successively at grain boundaries, twin boundaries, and finally intragranularly. At higher temperatures, $M_{23}C_6$ carbides and σ-phase form in both types 304 and 316, while molybdenum additions in the latter also result in Laves and χ precipitation.

The δ-ferrite phase in duplex austenitic-ferritic welds is unstable at elevated temperatures. Long-term aging above 500 °C causes the ferrite to transform into austenite, σ-phase, and $M_{23}C_6$ carbides that precipitate along the ferrite-austenite interface (Ref 100, 101). At lower temperatures, 885 °F (474 °C) embrittlement, which is common to ferritic steels with high chromium (Ref 102), is operative in the δ-ferrite because its composition is similar to that of a ferritic steel. This mechanism, also known as α' embrittlement, results from the spinodal decomposition of δ-ferrite into low-chromium (α) and high-chromium (α') regions (Ref 103, 104). In addition, low-temperature aging produces a nickel- and titanium-rich silicide (G phase: $Ti_6Ni_{16}Si_7$) in the ferrite (Ref 101), some carbide precipitation along the ferrite-austenite interface (Ref 101), and limited M_6C in the matrix.

Fracture Toughness of Aged Base Metal

The effect of long-term thermal exposure on the fracture toughness of stainless steels is summarized in Fig. 17. Aging at 450 °C for 50,000 h has no effect on fracture toughness (Ref 55). Aging at 550 and 566 °C causes only a modest degradation of fracture resistance, with J_c being reduced by about 20% after 10,000 h (Ref 10, 94) and by 35% after 50,000 h (Ref 55). The reduction in toughness results from aging-induced $M_{23}C_6$ carbides along grain boundaries that initiate a localized intergranular cracking mechanism. Intergranular cracking is confined to a few isolated regions, while the dominant fracture mechanism is transgranular dimple rupture nucleated by primary MC carbides (Fig. 18). The dominance of the dimple rupture mechanism in both the

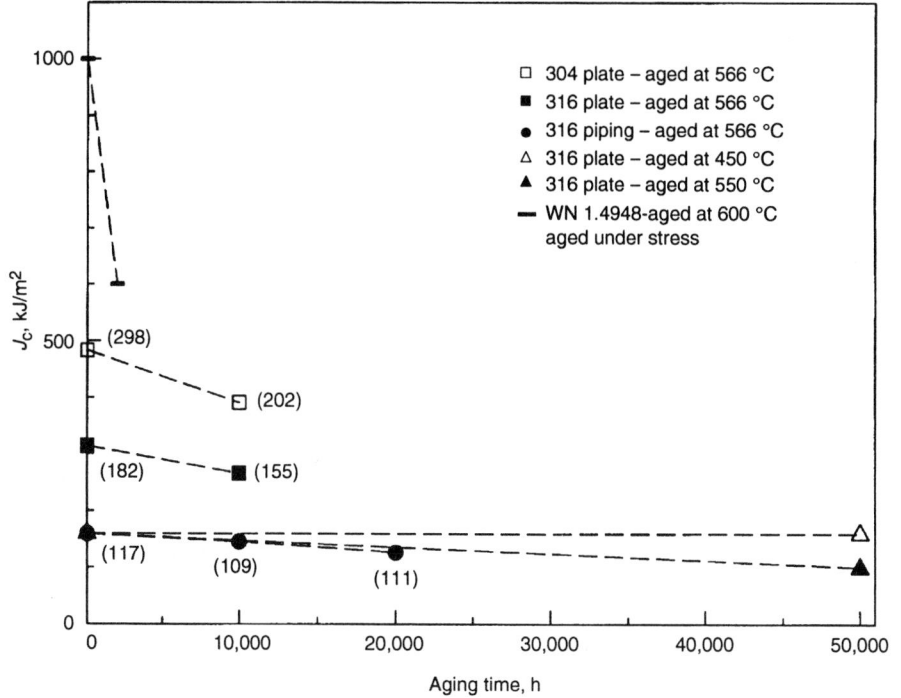

Fig. 17 Effect of aging on the fracture toughness of base metals. Values of dJ/da (in MPa) are provided in parentheses. Source: Ref 8, 10, 94

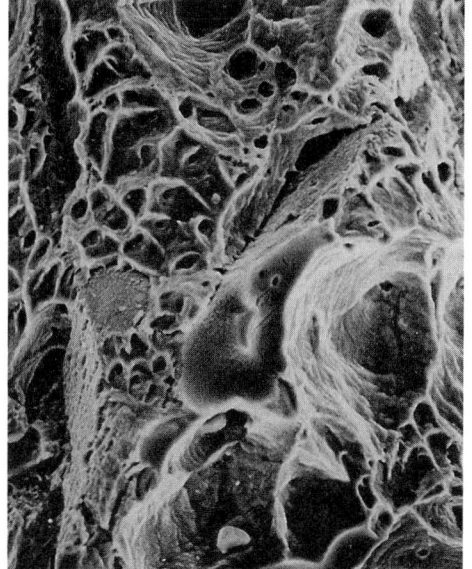

Fig. 18 SEM fractograph for type 304 aged at 566 °C. Transgranular dimple rupture is dominant, but limited intergranular cracking (denoted by arrows) is also observed. Source: Ref 94

Table 2 Fracture toughness of type 321 pipe before and after in-service aging at 450 to 550 °C for 51,000 h

Test temperature, °C	Condition	J_c, kJ/m^2	dJ/da, MPa
20	Unaged	162	49
20	Aged	107	48
300	Unaged	152	230
300	Aged	144	165
500	Unaged	152	260
500	Aged	129	194

Source: Ref 14

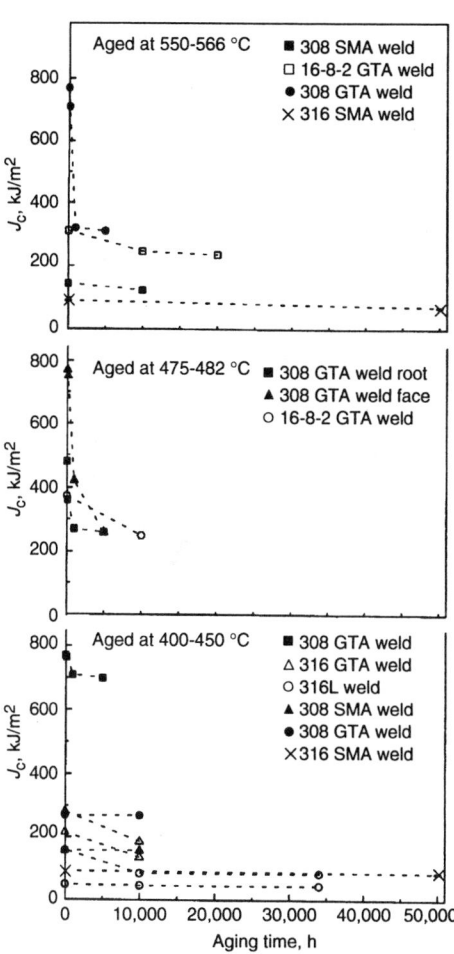

Fig. 19 Effect of aging on the fracture toughness of welds. SMA, shielded-metal arc; GTA, gas-tungsten arc. Range of aging temperatures: (a) 550 (Ref 55, 70) to 566 °C (Ref 10, 94). (b) 475 (Ref 69, 70) to 482 °C (Ref 10). (c) 400 (Ref 15, 70, 86) to 450 °C (Ref 10, 55)

unaged and aged conditions accounts for the relatively small effect of thermal aging on J_c.

Figure 17 also indicates that aging under stress accelerates toughness loss. Specifically, aging of WN 1.4948 (similar in composition to type 304) for 2000 h at 600 °C while stressed to 65% of the yield strength causes a 40% reduction in J_c (Ref 8). The loss of toughness is associated with intergranular and intragranular carbide precipitation

and development of a stress-induced dislocation substructure characteristic of secondary creep.

Table 2 summarizes the effect of in-service exposure to 450-550 °C for 51,000 h on the fracture toughness of type 321 pipe (Ref 14). The service-exposed pipe displayed a modest reduction in toughness, with the degree of degradation being dependent on test temperature. At room temperature, aging reduced J_c by 35% but had no effect on dJ/da. At 300 and 500 °C, aging produced a 5-15% reduction in J_c and 25-30% reduction in dJ/da.

Fracture Toughness of Aged Welds

Figure 19 shows considerable variability in the degree of degradation produced by long-term thermal aging of welds. The greatest toughness loss occurs in the higher-toughness welds. For example, exposing the low-toughness type 316 SMA weld to 550 °C for 50,000 h is seen to have only a modest effect on toughness, as J_c is re-

duced from 90 to 70 kJ/m^2 (Ref 55). Aging at 566 °C for 10,000 h causes a 14% reduction in J_c for the type 308 SMA weld with intermediate toughness (as-welded toughness J_c of 144 kJ/m^2) and 21% reduction in J_c for the type 16-8-2 GTA weld with high toughness (as-welded J_c of 311 kJ/m^2) (Ref 10). The fracture-resistant type 308 GTA weld (as-welded J_c of 770 kJ/m^2) shows a 60% decrease in toughness after aging at 550 °C for 5000 h (Ref 70). It is noteworthy that most of the toughness loss occurs within the first 1000 h of aging at 550 °C.

While fracture surfaces for both the unaged and aged welds are dominated by dimples produced by microvoid coalescence, the dimple size is slightly smaller for the welds aged at 550 and 566 °C. The smaller dimples arise from an increase in the number of microvoid nucleation sites due to aging-induced $M_{23}C_6$ carbides and σ-phase, as shown in the metallographic-fractographic profiles in Fig. 20 (Ref 94). These profiles are obtained by electropolishing selected areas of the fracture surface so that its topography and underlying microstructure can be characterized simultaneously. It is seen that microvoids in the unaged

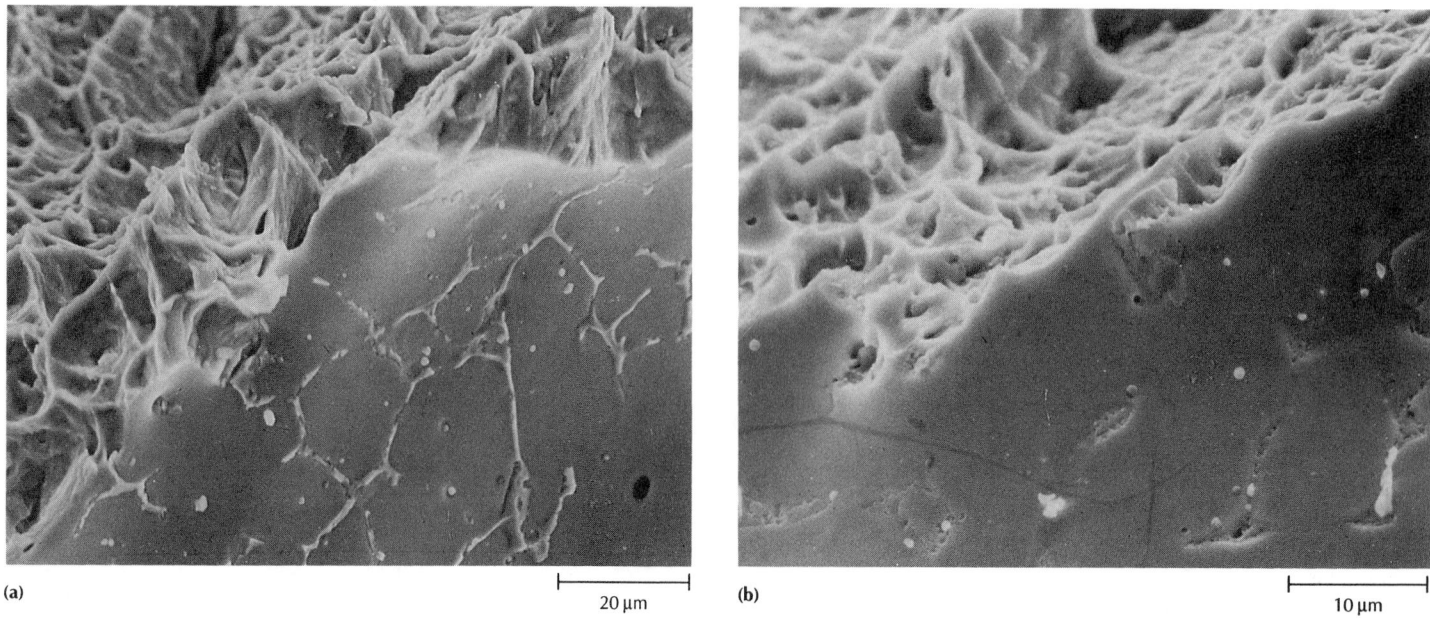

Fig. 20 SEM metallographic/fractographic interface for type 308 shielded-metal arc welds. (a) In the unaged condition, microvoids are nucleated by δ-ferrite and silicon-rich inclusions. (b) After aging δ-ferrite transformation products, σ-phase (white phase) and $M_{23}C_6$ carbides (on prior austenite-ferrite boundaries) provide additional microvoid nucleation sites. Source: Ref 94

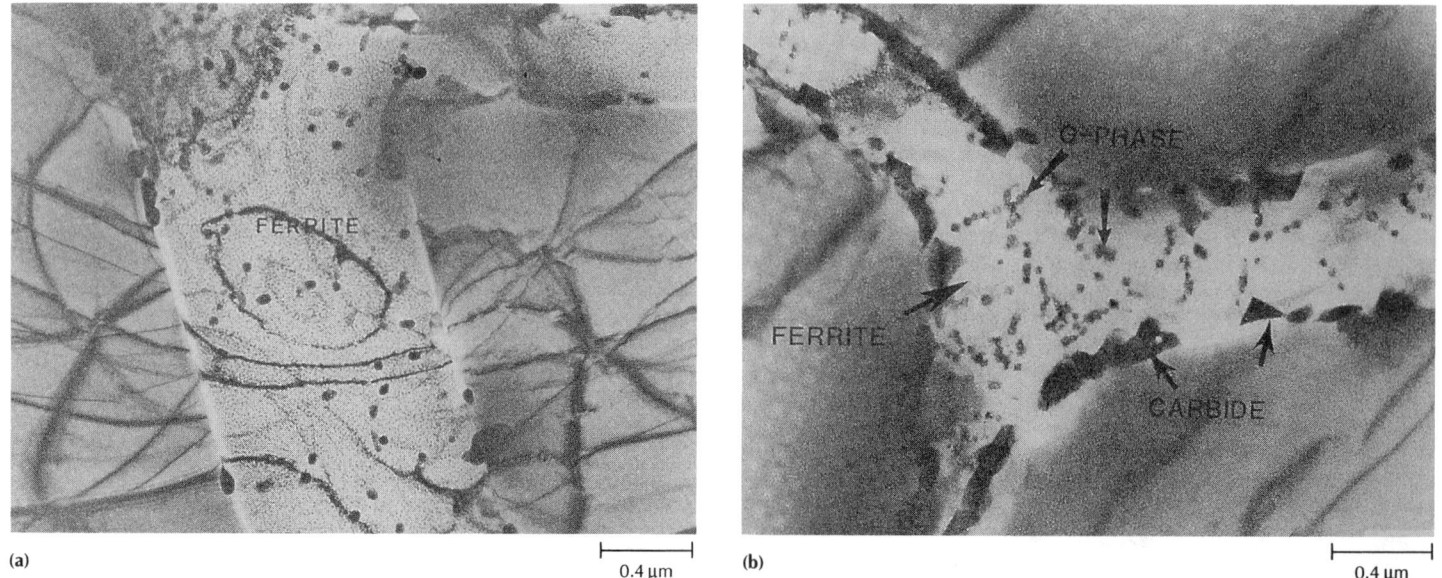

Fig. 21 Electron micrographs of aged type 308 weld. (a) Aged at 475 °C for 1000 h, showing mottled structure indicative of spinodal decomposition of the δ-ferrite and extensive G-phase precipitation. (b) Aged at 475 °C for 4950 h, showing $M_{23}C_6$ carbides at austenite-ferrite interface in addition to G-phase precipitates within the δ-ferrite. Source: Ref 101

weld are nucleated by inclusions and δ-ferrite particles (Fig. 20a), whereas microvoids in the aged weld are nucleated by inclusions, σ-phase, and $M_{23}C_6$ carbides (Fig. 20b).

The different aging responses exhibited by the various welds are attributed to the relative effectiveness of microvoid nucleation sites. While carbides and σ-phase provide additional nucleation sites in SMA welds, their effectiveness in nucleating microvoids is less than that associated with the inclusions already present in the as-welded

condition. This accounts for the modest aging effects for SMA welds. In high-toughness GTA welds, the small inclusions and unaged δ-ferrite resist failure, so they nucleate microvoids only after extensive plastic deformation. In this case, aging-induced carbides and σ-phase have a pronounced effect on toughness because they are more effective microvoid nucleation sites. It is noteworthy that even though aging of the GTA welds reduces fracture toughness by up to 60%, J_c values after aging (236 and 300 kJ/m²) are still

substantially higher than the J_c value of 144 kJ/m² for the unaged SMA weld. This comparison clearly shows that aging-induced carbides and σ-phase are less effective microvoid nucleation sites than the inclusions in SMA welds.

Aging of type 308 (Ref 69) and 16-8-2 (Ref 94) GTA welds at 475 to 482 °C (Fig. 19b) causes about the same degree of toughness loss as 550 to 566 °C aging, although the embrittlement mechanisms are different. During long-term exposure at temperatures below 500 °C, spinodal decomposi-

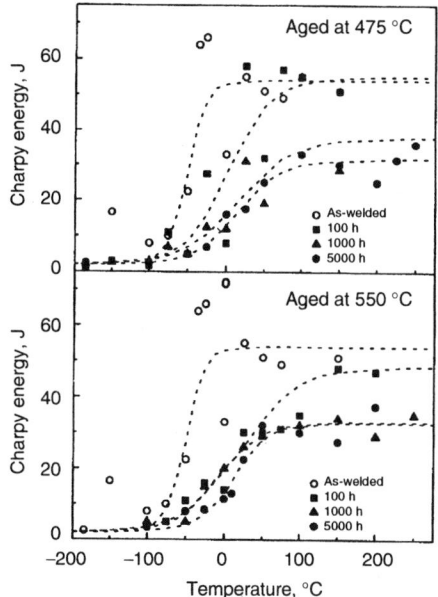

Fig. 22 Charpy V-notch energy as a function of test temperature for a type 308 gas-tungsten arc weld, aged at 475 and 550 °C. Source: Ref 70

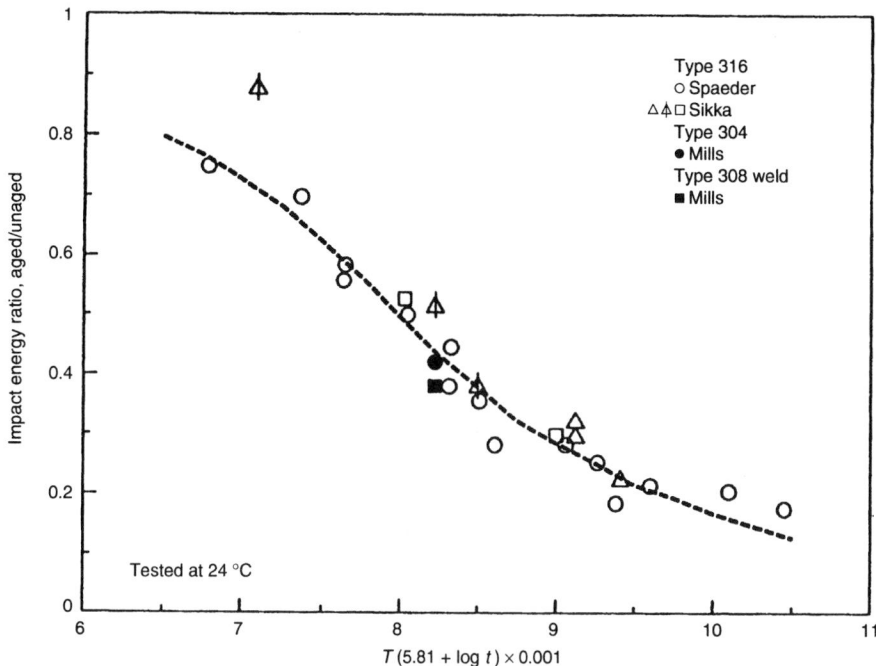

Fig. 23 Charpy V-notch impact energy ratio for aged vs. unaged materials as a function of the Larson-Miller type parameter developed by Sikka (see text and Ref 106). Source: Ref 106, 107 (type 316); Ref 94 (types 304 and 308)

tion of the δ-ferrite coupled with G-phase and carbide precipitation are responsible for the aging effects. Electron micrographs of the 475 °C aged weld (Fig. 21) reveal G-phase within the ferrite, $M_{23}C_6$ carbides along the ferrite-austenite interface, and a mottled ferrite appearance indicative of spinodal decomposition into α′ (Ref 69, 70). Atom probe field ion microscopy of the aged ferrite confirms the spinodal decomposition into iron-rich α regions and chromium-rich α′ regions with a periodicity of 6.5 nm (Ref 66). The α′ embrittlement hardens the ferrite, making it an effective microvoid nucleation site that reduces upper-shelf fracture energy. The dominant role of the α′ embrittlement mechanism was demonstrated using reversion heat treatment experiments where the aged weld was heat treated at 550 °C for 1 h (Ref 105). Because 550 °C is above the α + α′ miscibility gap and well below the G-phase and $M_{23}C_6$ solvus temperatures, the ferrite is homogenized while the precipitates are essentially unaffected. This treatment results in a substantial but not full recovery of the notch toughness, demonstrating that the spinodal reaction is responsible for most of the embrittlement while the second-phase precipitates play a minor role.

Embrittlement rates at 400 to 450 °C are relatively slow due to the sluggish kinetics of spinodal decomposition and G-phase precipitation. In this temperature regime (Fig. 19c), some welds show little or no aging effects after 5000 to 50,000 h, while others show a modest toughness loss. Type 308 GTA and SMA welds aged for 10,000 h at 427 °C and type 316 SMA weld aged for 50,000 h at 400 to 450 °C (Ref 55) show essentially no reduction in J_c or tearing modulus. Aging the type 308 GTA weld at 400 °C for times up to 5000 h causes only a 10% drop in J_c (from 770 to 700 kJ/m²), whereas aging at 475 °C for

the same time produces a 60% decrease in toughness (Ref 70). Ould et al. (Ref 86) found that 400 °C exposure for 10,000 to 34,000 h has essentially no effect on the low-toughness weld but causes a 50% decrease in J_c for an intermediate-toughness weld. Faure et al. (Ref 15) also reported a modest effect of 10,000 h aging at 400 °C for a type 316 GTA weld: J_c was reduced by 35%.

Although long-term thermal exposure degrades the toughness of welds, the high post-aging J_c values indicate that sufficient toughness is retained to preclude nonductile fracture. Ductile tearing of aged materials occurs only after substantial plastic deformation, so fracture control of aged stainless steels is typically not a critical engineering issue.

Fracture Toughness/Charpy Energy Correlations for Aged Stainless Steels

A wealth of data (Ref 65, 69, 70, 105-112) shows that thermal aging causes up to a 90% degradation in Charpy V-notch energy for both base metals and welds. For duplex welds, aging not only degrades the upper-shelf energy but also increases the ductile-brittle transition temperature. The Charpy energy response for a type 308 GTA weld aged at 475 and 550 °C is shown in Fig. 22 (Ref 70). At both temperatures, 1000 to 5000 h agings decrease the upper-shelf energy by about 50% and increase the transition temperature by almost 100 °C.

Sikka (Ref 106) developed the Larson-Miller type relationship in Fig. 23 that predicts the reduction in Charpy energy for type 316 in terms of exposure time (t in hours) and temperature (T in K) using:

$$P = T(5.81 + \log t) \times 10^{-3} \qquad \text{(Eq 7)}$$

where P is the Larson-Miller parameter. Although this correlation cannot account for the effects of composition, orientation, and δ-ferrite content in all heats and welds (Ref 94, 112), it provides satisfactory predictions of the degree of embrittlement for many materials. Sikka's observation that equivalent time-temperature combinations per Eq 7 produce similar intergranular carbide microstructures suggests that carbide precipitation is responsible for the aging-induced toughness degradation in type 316. This is consistent with the fractographic observations in Fig. 18(b), where aging-induced $M_{23}C_6$ precipitates along grain boundaries reduce toughness by nucleating localized intergranular cracks.

In contrast to the large degradation in Charpy energy, some fracture mechanics studies (Ref 10, 94, 113) reveal that thermal aging produces only a modest reduction in J_c and tearing resistance. This behavior is illustrated in Fig. 24 for type 304 and type 308 SMA weld aged at 566 °C for 10,000 h (Ref 94). The 60% reduction in blunt-notch Charpy impact energy for both materials, represented by the solid symbols in Fig. 23, agrees with the Larson-Miller type correlation developed by Sikka. In direct contrast to this response, however, precracked Charpy and fracture mechanics specimens show a much smaller aging effect. Aging reduces J_c values by about 20% and dJ/da values by 30%. This is consistent with the 25 to 35% reduction in precracked Charpy impact energy but is much less than the 60% reduction in Charpy V-notch impact energy.

Fractographic examinations reveal that the large degradation in Charpy V-notch energy after aging is due to differences in the energy required to initiate a crack from a blunt notch (Ref 94). The

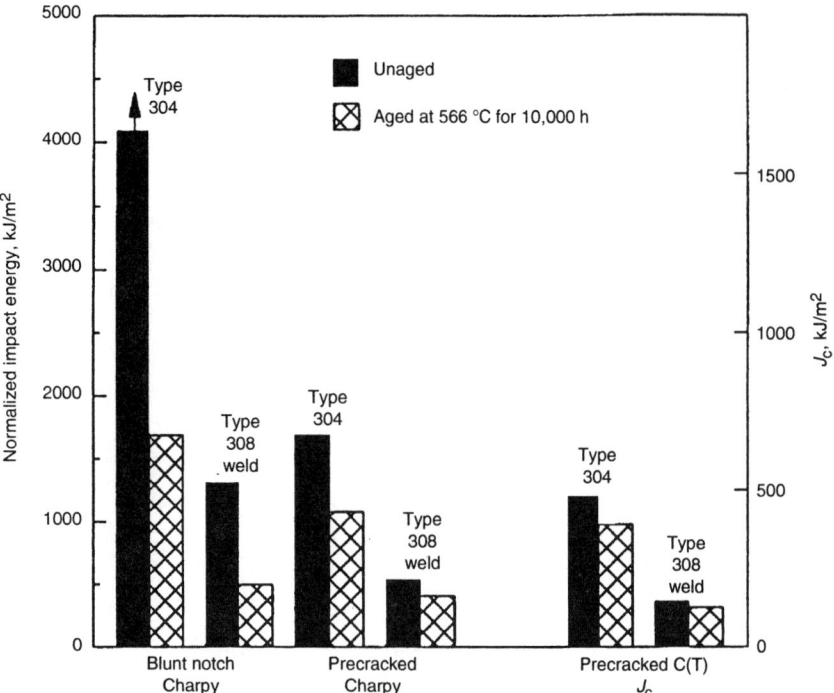

Fig. 24 Effect of thermal aging on the normalized Charpy V-notch impact energy, precracked Charpy impact energy, and J_c. Charpy tests were performed at 24 °C; J_c tests were performed at 538 °C. The unaged type 304 Charpy V-notch specimen did not fracture at a normalized impact energy of ~4000 kJ/m². Source: Ref 94

key fracture surface feature on unaged Charpy V-notch specimens is a series of tear-arrest markings, where cracks initiate but are quickly arrested by the fracture-resistant matrix. The tremendous amount of energy required for crack reinitiation accounts for the high blunt-notch Charpy impact energy in the unaged condition. After aging, there is little or no evidence of arrest markings. In the aged base metal, intergranular $M_{23}C_6$ carbides render the grain boundaries susceptible to cracking. During the early stages of plastic straining, formation of an intergranular crack ahead of the notch serves as a starter crack that drastically reduces fracture energy. In the aged weld specimens, the $M_{23}C_6$ carbides and σ-phase provide additional microvoid nucleation sites that enhance cracking and thereby preclude formation of tear-arrest markings. These findings demonstrate that the large reduction in blunt-notch Charpy energy is due to aging-induced microstructural changes that markedly decrease the energy required to initiate a sharp crack. Because conventional Charpy data reflect the energy associated with nucleating a crack, rather than extending a pre-existing crack, they do not necessarily correlate with fracture mechanics properties. In many cases, blunt-notch Charpy data are an overly pessimistic indicator of thermal aging effects for stainless steel components containing cracks or crack-like defects.

In contrast to this behavior, Alexander (Ref 70) found a good correlation between Charpy V-notch energy and fracture toughness for an aged type 308 GTA weld. For this weld, aging-induced microstructural changes appeared to have the same effect on the crack propagation energy and the energy required to initiate a crack from a blunt notch. Although Charpy and fracture mechanics data correlate well for some welds, caution must be exercised when using conventional Charpy tests to infer the fracture toughness for this class of materials, because results are dominated by the energy required to initiate a sharp crack from a blunt notch, rather than the energy required to propagate a pre-existing crack.

Effect of Neutron Irradiation on Base Metal and Weld Toughness

Irradiation-Induced Microstructure

Neutron irradiation displaces atoms from normal lattice positions, thereby creating point defects. Although most point defects recombine, survivors diffuse and cluster to form an irradiated microstructure consisting of cavities, precipitates, and dislocations. Microstructural evolution is dependent on irradiation temperature, neutron fluence, flux, and energy spectrum (Ref 114-119).

At temperatures below 350 °C, irradiation produces a very fine defect substructure consisting of small Frank interstitial loops and "black spot" damage, corresponding to nanometer-diameter vacancy or interstitial clusters. Between 350 and 600 °C, the defect structure coarsens. Black spots are replaced by Frank loops, and eventually network dislocations and cavities evolve at high neutron exposures. Cavities are three-dimensional clusters of vacancies (voids), gas atoms (bubbles), or a combination of the two. Above 400 °C, irradiation induces precipitation of vari-

ous phases, including G-phase, Ni_3Si, nickel- and silicon-rich M_6C, M_2P, M_3P, and Laves phase (Ref 114-116, 120, 121).

The lattice defect structure and precipitates serve as obstacles to dislocation motion that strengthen the matrix. In addition, irradiation damage causes a large ductility loss and greatly diminished strain hardening capacity as the yield strength approaches the ultimate strength. The irradiation-produced microstructure also promotes a planar slip mechanism, termed dislocation channeling. Specifically, dislocation glide on a narrow band of slip planes sweeps irradiation-produced dislocation loops from the band, creating a defect-free channel with highly deformed voids (Ref 122). Because the channels offer much less resistance to dislocation motion than the unswept regions, subsequent dislocation activity is confined to the dislocation channels. This produces a very heterogeneous deformation mode. The irradiation-induced strengthening and enhanced planar slip cause a pronounced degradation in fracture resistance.

Irradiation at temperatures above 600 °C does not cause significant displacement damage due to removal of lattice defects by annealing. Some degree of strengthening arises from second-phase precipitation, while significant reductions in ductility and toughness result from helium embrittlement. Helium, which is highly insoluble in metals, results from (n,α) transmutation of ^{10}B per the $^{10}B(n,\alpha)^7$ Li reaction, and the two-step reaction with natural nickel: $^{58}Ni(n,\gamma)^{59}Ni$ followed by $^{59}Ni(n,\alpha)^{56}Fe$. It causes high-temperature embrittlement, regardless of whether the material is irradiated at high or low temperatures. During high-temperature exposure, helium diffuses to grain boundaries, rendering them susceptible to intergranular cracking, because helium atoms reduce the grain boundary cohesive strength and helium bubbles serve as local crack nuclei (Ref 118, 123). At temperatures above 600 °C, the helium embrittlement mechanism dramatically degrades the ductility and fracture toughness, but whenever the temperature falls below 600 °C this mechanism is not operative.

The two parameters used to represent irradiation damage are neutron fluence in neutrons per square centimeter and the average number of displacements experienced by each atom, in displacements per atom (dpa). Displacements per atom, calculated as the product of total fluence and spectrum-averaged displacement cross section (Ref 124), is preferred because it accounts for the effectiveness of neutrons at various energies in damaging the lattice.

Intermediate Temperature Irradiation of Base Metal

The effect of intermediate temperature irradiation on the fracture toughness of types 304 and 316 is shown in Fig. 25. It is seen that irradiation damage can be separated into three regions: a threshold neutron exposure below which there is no loss in toughness, intermediate exposures where toughness decreases rapidly with neutron dose, and a saturation exposure above which in-

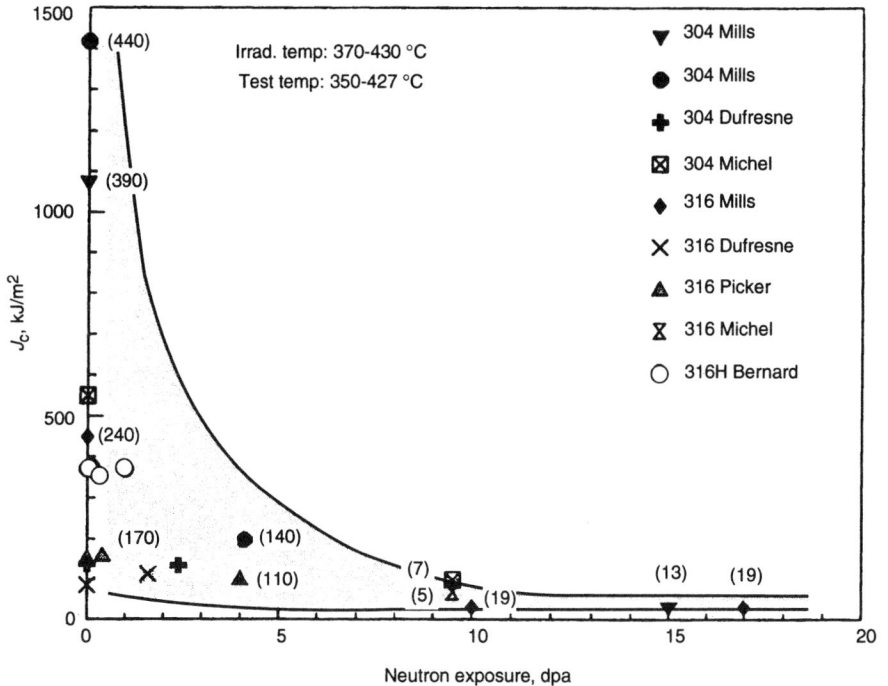

Fig. 25 J_c as a function of neutron exposure for base metals irradiated at intermediate temperatures. Values of dJ/da (in MPa) are provided in parentheses. Source: Ref 29, 42, 43, 47, 85, 125

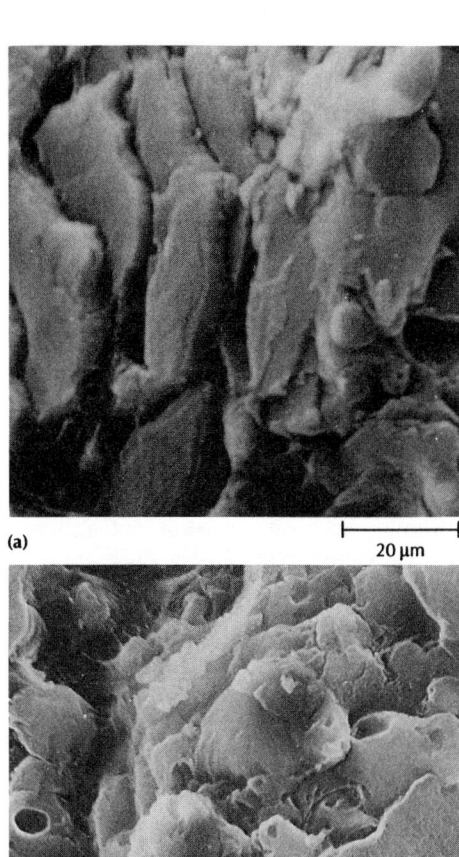

Fig. 26 SEM fractographs showing the stepped facets indicative of channel fracture. (a) Type 316 irradiated to 17 dpa. (b) A type 308 shielded-metal arc weld irradiated to 14 dpa. Note the presence of microvoids superimposed on crystallographic facets in the weld. Source: Ref 43

creasing irradiation damage produces little or no additional reduction in toughness. The threshold neutron exposure for toughness degradation appears to be about 1 dpa, which is consistent with the threshold exposure of 1 dpa for tensile ductility degradation (Ref 126). Therefore, the tensile and fracture toughness responses for stainless steel components subjected to neutron exposures less than 1 dpa are not significantly affected.

Irradiation exposures between 1 and 10 dpa cause a substantial decrease in fracture resistance, with the largest effects occurring in high-toughness heats. At approximately 10 dpa, the degradation in fracture properties saturates at a J_c value of about 30 kJ/m^2 and at dJ/da values less than 20 MPa, corresponding to tearing moduli less than 10 (Ref 43). The low tearing modulus indicates that highly irradiated stainless steels have little resistance to tearing at J-values above J_c. As a result of the dramatic decrease in toughness and increase in yield strength, postirradiation failures occur in or near the linear-elastic regime. The equivalent critical stress-intensity factor (K_{Jc}) computed from the saturation J_c value is approximately 70 MPa$\sqrt{m}$, based on the following equation (Ref 127):

$$K_{Jc} = (EJ_c)^{1/2} \qquad \text{(Eq 8)}$$

With this toughness level and postirradiation yield strengths of 600 to 700 MPa, 2 to 3 cm thick components provide sufficient constraint to satisfy plane-strain conditions. Under these conditions, acceptable load-crack length relationships can be computed using linear-elastic fracture mechanics concepts, based on critical K_{Jc}-levels calculated from Eq 8.

Limited postirradiation testing of other stainless steels reveals that they are also susceptible to irradiation embrittlement. Irradiation of type 348 to a fluence of 3×10^{22} neutrons/cm^2 ($E > 1$ MeV) results in J_c values ranging from 20 to 24 kJ/m^2 (Ref 128), which is slightly lower than the saturation toughness of 30 kJ/m^2 for types 304 and 316. Instrumented impact testing of precracked type 321 specimens irradiated at 230 and 400 °C to doses from 20 to 54 dpa reveals that J_{max} values are reduced to as low as 50 kJ/m^2 (Ref 129, 130). For highly irradiated materials, initiation of cracking occurs shortly before maximum load, so J_{max} values slightly overestimate J_c.

The degradation in toughness results from irradiation-induced strengthening and ductility loss, coupled with increased heterogeneous slip. At high neutron exposures, a fracture mechanism transition from dimple rupture to channel fracture is responsible for the dramatic reduction in toughness. The channel fracture mechanism (Ref 131) involves localized separation along intense dislocation channels, which produces the crystallographic faceted appearance shown in Fig. 26(a) (Ref 43). The stepped facets are steeply inclined to the overall crack plane, demonstrating that their formation involves a shear process rather than cleavage. When channel fracture becomes the dominant mechanism at about 10 dpa, the toughness degradation saturates at about 30 kJ/m^2 (Ref 43). The reduced data scatter in the saturation regime reflects a tendency of the irradiated alloys to evolve toward a saturation microstructure, where the density of irradiation-produced defects saturates with continued exposure. Because the saturation microstructure is independent of the starting condition, the variability

in postirradiation toughness is greatly diminished.

Intermediate Temperature Irradiation of Welds

The effect of intermediate temperature irradiation on the toughness of welds is demonstrated in Fig. 27. As with the base metal, irradiation effects can be categorized into the threshold, transition, and saturation regimes. Exposures up to 1 dpa have no significant effect on fracture resistance. In the transition region beyond 1 dpa, fracture resistance diminishes more rapidly than in the base metal because the embrittled δ-ferrite serves as an effective microvoid nucleation site. At exposures above 10 dpa, saturation J_c values range from 10 to 30 kJ/m^2 and dJ/da values are less than 20 MPa (Ref 43, 55, 85). Equivalent K_{Jc} values range from 40 to 70 MPa$\sqrt{m}$. At these toughness levels, 1 to 2 cm thick welded components possess sufficient constraint to induce plane-strain fracture conditions.

Fracture surfaces for highly irradiated welds, shown in Fig. 26(b), exhibit channel fracture with small microvoids superimposed on the crystallographic facets (Ref 43). The microvoids are nucleated by failure of embrittled δ-ferrite particles, but they cannot develop into dimples due to restricted plastic deformation capabilities after irradiation. These small holes act as stress concentrators and prematurely nucleate channel fracture, which causes saturation J_c levels for welds to fall below base metal values.

Low-Temperature Irradiations

The fracture toughness responses for base metals, heat-affected zones, and welds irradiated at low temperatures are summarized in Fig. 28 and 29. Irradiation of types 304, 316, and PCA to 2 or 3 dpa causes a 25 to 60% reduction in J_c and less than a 40% reduction in dJ/da. The diminished fracture resistance is due to formation of black spot damage and faulted Frank loops that increase the yield strength from 230 to 650 MPa (Ref 52). The overall fracture resistance remains high after irradiation, except for the type 316 heat with a postirradiation J_c of 155 kJ/m^2 and dJ/da of 30 MPa (Ref 52). This air-melted heat contains an unusually high volume of nonmetallic inclusions that promote microvoid coalescence. Hence, the inferior postirradiation toughness reflects the poor tearing resistance of this air-melted heat, coupled with a 90% increase in flow strength after irradiation. The other heats possess good tearing resistance, with postirradiation dJ/da values ranging from 90 to 290 MPa.

Figure 29 reveals that the heat-affected zone and welds are more sensitive to low-temperature irradiation than the base metal, which is consistent with the intermediate-temperature response. Irradiation to 2 or 3 dpa causes a 50 to 60% decrease in J_c and a 40 to 60% decrease in dJ/da for type 308 GMA and 316 GTA welds and the type 304 heat-affected zone. While the PCA gas-tungsten arc weld shows the greatest irradiation sensitivity, its toughness after irradiation to 3 dpa remains very high, with a postirradiation J_c of 317 kJ/m^2 and dJ/da of 130 MPa. The high postirradiation dJ/da values, ranging from 150 to 260 MPa, indicate that the GTA and GMA welds retain good fracture resistance at intermediate neutron exposures. Moreover, the postirradiation fracture toughness of the PCA electron beam weld is exceptionally high, with a J_c of 882 kJ/m^2 and dJ/da of 290 MPa.

Irradiation of the type 304 heat-affected zone to 2 dpa reduces J_c from 150 to 75 kJ/m^2 and dJ/da from 102 to 33 MPa. The inferior fracture properties for the irradiated heat-affected zone result from a combination of rather low toughness in the nonirradiated condition and a 50% increase in flow strength after irradiation.

Postirradiation Fracture Toughness of Cold-Worked Material

Figure 30 shows that intermediate-temperature irradiation of 20% cold-worked type 316 to flu-

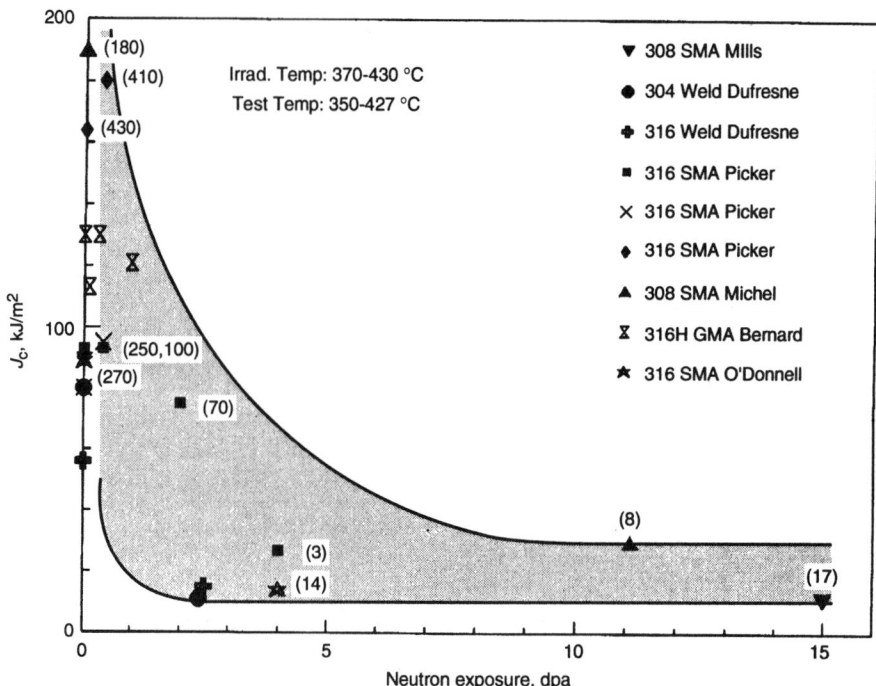

Fig. 27 J_c as a function of neutron exposure for welds irradiated at intermediate temperatures. Values of dJ/da (in MPa) are provided in parentheses. Source: Ref 29, 43, 47, 55, 85, 125

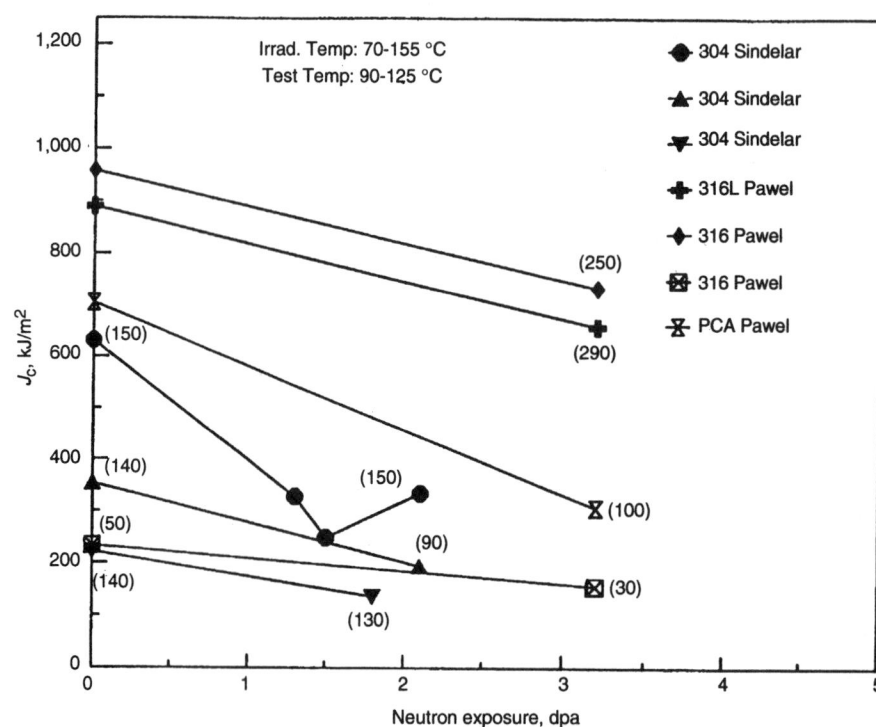

Fig. 28 J_c as a function of neutron exposure for base metals irradiated at low temperatures. Values of dJ/da (in MPa) are provided inside parentheses. Source: Ref 35, 52

ences greater than 10×10^{22} neutrons/cm^2 ($E >$ 0.1 MeV), corresponding to neutron exposures greater than 57 dpa (Ref 132), causes a 50% decrease in J_c and an order of magnitude reduction in dJ/da (Ref 133-135). Postirradiation J_c values are relatively independent of test temperature between 24 and 538 °C, ranging from 25 to

45 kJ/m^2, but drop to 6 kJ/m^2 at 649 °C. Tearing resistance is also dependent on test temperature: dJ/da approaches zero at 24 °C and ranges from 10 to 70 MPa at 200 to 538 °C.

While the nonirradiated material exhibits ductile dimple rupture at all test temperatures, operative fracture mechanisms in the irradiated condi-

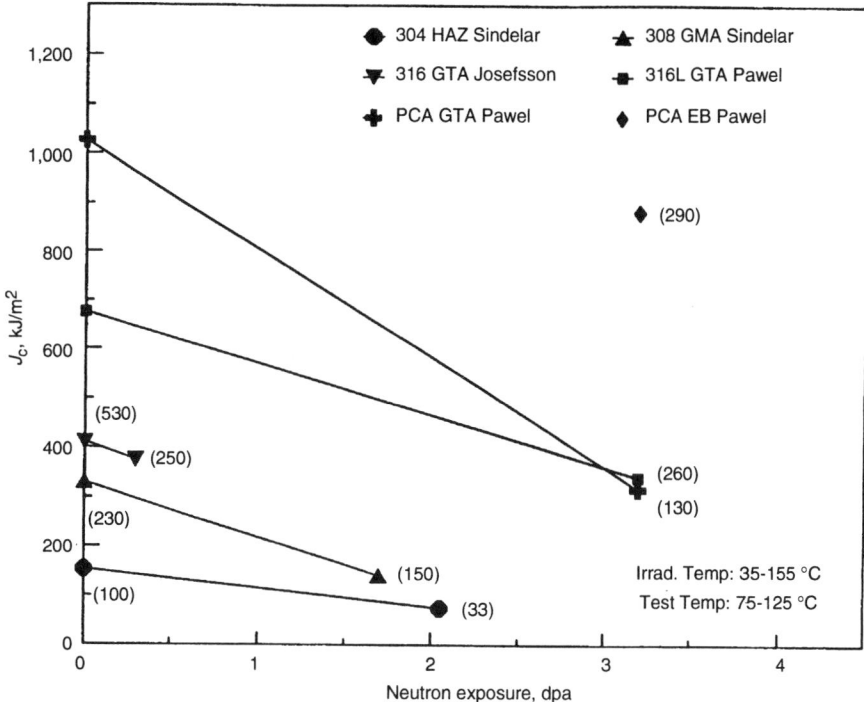

Fig. 29 J_c fracture toughness as a function of neutron exposure for welds and heat-affected zone irradiated at low temperatures. Values of dJ/da (in MPa) are provided in parentheses. Source: Ref 35, 52, 71

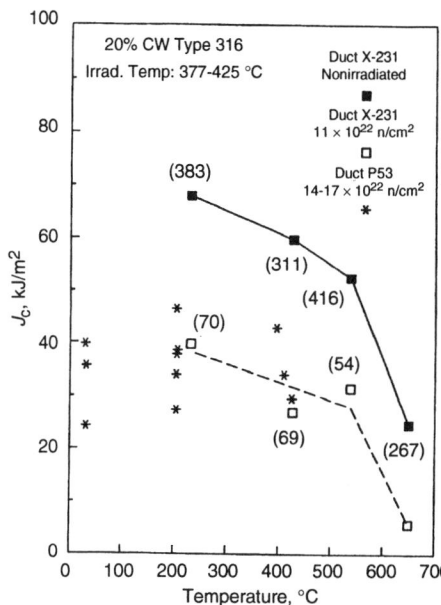

Fig. 30 Temperature dependence of J_c for nonirradiated and irradiated 20% cold-worked type 316. Values of dJ/da (in MPa) are provided in parentheses. Source: Ref 133-135

tion are dependent on test temperature (Ref 134, 135). Interphase fracture between the austenite (γ) matrix and deformation-induced ε (hexagonal close-packed) martensite platelets results in the exceptionally low tearing resistance at 24 °C (Ref 135). Channel fracture is the dominant cracking mechanism between 205 and 538 °C, and intergranular fracture due to a helium embrittlement mechanism is responsible for the severe toughness loss at 649 °C (Ref 134).

At intermediate neutron exposures, between 3 and 8 dpa, cold-worked type 316 shows less irradiation sensitivity than its solution-annealed counterpart. Michel and Gray (Ref 85, 136) report that intermediate temperature irradiation of 20% cold-worked type 316 to 8 dpa results in only a 10% reduction in J_c and has no effect on dJ/da, as shown in Fig. 31. Well-defined microvoid coalescence is observed in both the nonirradiated and irradiated materials, which is consistent with the irradiation-insensitive response. Although fracture properties are unaffected by irradiation, the flow strength is increased from 585 to 760 MPa.

The effect of 90 °C irradiation to 3 dpa on the fracture toughness of two heats of cold-worked type 316 and one heat of cold-worked PCA is also shown in Fig. 31 (Ref 52). The cold-worked PCA and high-toughness heat of cold-worked type 316 show a modest irradiation effect. A more pronounced effect is displayed by the low-toughness heat of cold-worked type 316: J_c is reduced from 39 to 21 kJ/m^2 and the tearing modulus is zero in both the nonirradiated and irradiated conditions. The inferior fracture resistance for this air-melted heat is due to an exceptionally high density of inclusions. The abundance of large

inclusions aligned in the cracking direction, coupled with the cold-worked matrix, promote microvoid nucleation and growth, which causes a low J_c and zero tearing resistance in the nonirradiated condition. Irradiation strengthening further restricts plastic deformation capabilities and degrades J_c to 21 kJ/m^2, even though the neutron exposure is only 3 dpa.

The lower-bound postirradiation J_c values of 20 to 25 kJ/m^2 for cold-worked type 316 tested at temperatures up to 538 °C correspond to K_{Jc} values of 55 to 65 MPa$\sqrt{m}$. These low toughness levels, coupled with yield strengths on the order of 800 MPa, indicate that fracture of cold-worked stainless steels in the highly irradiated condition is an important design issue and linear-elastic fracture mechanics can be used to predict failure conditions. At 649 °C, helium embrittlement degrades J_c to 6 kJ/m^2, corresponding to a K_{Jc} of only 30 MPa$\sqrt{m}$. This low toughness indicates that brittle fracture of highly irradiated stainless steels is an issue at temperatures above 600 °C, where the helium embrittlement mechanism is operative.

Cryogenic Fracture Toughness

Base Metals

Austenitic stainless steels, which are used extensively at cryogenic temperatures, can be separated into two categories: 1) metastable alloys (e.g., types 304 and 316 and their welds) that undergo a deformation-induced partial transformation of the austenite (face-centered cubic) to α' (bcc) martensite and ε-martensite, and 2) fully

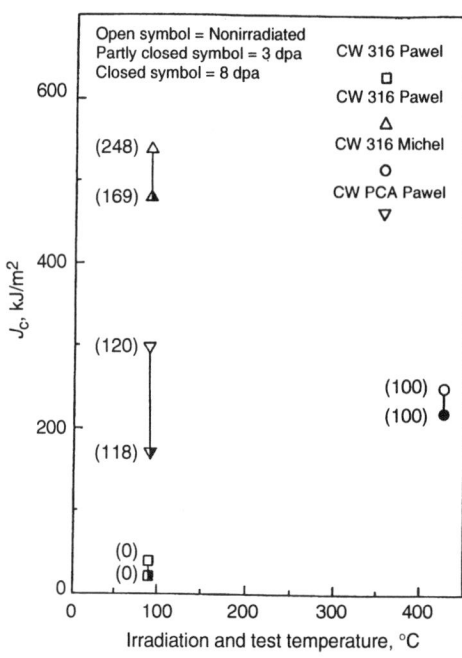

Fig. 31 Effect of irradiation to 3 and 8 dpa on the fracture toughness of cold-worked type 316. Specimens were irradiated and tested at the same temperature. Values of dJ/da (in MPa) are provided in parentheses. Source: Ref 52, 85, 136

stable alloys (e.g., types 310, 330, and the Nitronic series of alloys) that resist phase transformations at cryogenic temperatures as low as –269 °C. (The deformation-induced α' martensite transformation occurring at –269 °C is different than the α' embrittlement due to spinodal decomposition at approximately 475 °C. Unfortunately, the literature uses the term α' for both phases.)

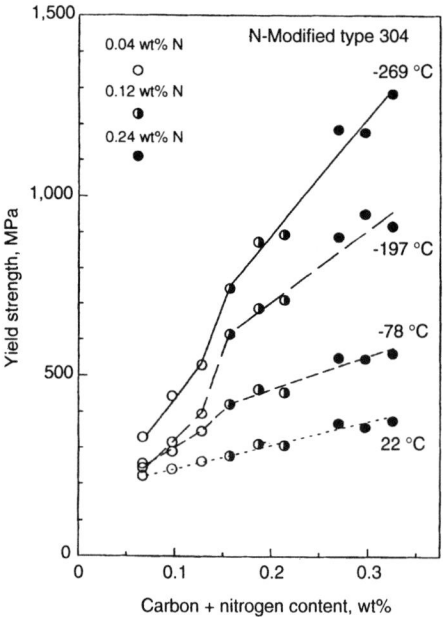

Fig. 32 Effect of carbon plus nitrogen content on the yield strength of type 304. Source: Ref 142

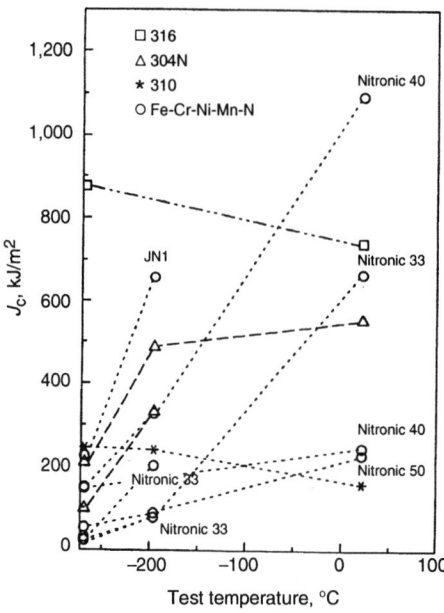

Fig. 33 Effect of cryogenic temperature on J_c for stainless steels. Source: Ref 51 (type 316); Ref 140, 141 (type 304N); Ref 51 (type 310); Ref 51, 57, 143 (Fe-Cr-Ni-Mn-N)

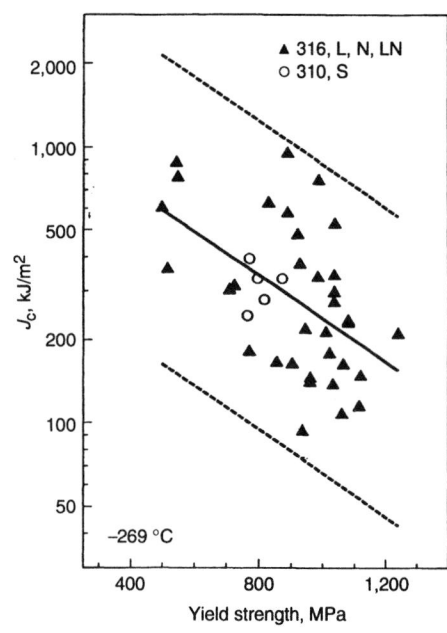

Fig. 34 Summary of –269 °C toughness as a function of yield strength. The solid line represents a least-squares exponential regression of the data, and the dashed lines correspond to linearized 90/95% tolerance limits. Source: Ref 51, 144 (type 316); Ref 145 (type 316L); Ref 146 (type 316N); Ref 139, 143, 144, 147-155 (type 316LN); Ref 51, 143 (type 310); Ref 138, 156, 157 (type 310S)

Evidence of martensitic phases has been detected in the crack-tip plastic zones of type 304 and 316 specimens tested at –269 °C (Ref 51, 137-141) but not at 24 °C except after high irradiation exposure.

Nitrogen-strengthened stainless steels, including AISI types 304N, 304LN, 316N, 316LN, and Nitronic alloys, are attractive because nitrogen stabilizes the austenite with respect to the martensite transformation and is an effective strengthener, particularly at –269 °C. The effect of interstitial carbon plus nitrogen on the yield strength of type 304 at temperatures between 22 and –269 °C is shown in Fig. 32 (Ref 142). Increasing the carbon plus nitrogen content from 0.1 to 0.3 wt% increases strength by 70% at room temperature and by 200% at –269 °C. At room temperature the effects of carbon and nitrogen are similar, whereas nitrogen is a more potent strengthener with decreasing temperature. Hence, the strength of these alloys can be tailored for demanding cryogenic applications by adjusting the nitrogen concentration.

The vast majority of austenitic stainless steel base metals do not exhibit a ductile-brittle transition temperature phenomenon, but their fracture toughness may tend to decrease at cryogenic temperatures due to increased strength and reduced ductility. This effect is most pronounced in nitrogen-strengthened alloys where the hardened matrix restricts plastic deformation. In extreme cases, such as in Nitronic alloys with low nickel and high nitrogen tested near absolute zero, a ductile-brittle transition associated with slip band cracking (see below) can occur. But most austenitic stainless steels are inherently ductile alloys and microvoid coalescence is the dominant cracking mechanism at all temperatures from –269 to 550 °C.

Extensive fracture toughness testing of 300-series stainless steels and their welds has revealed that these materials retain good fracture resistance at cryogenic temperatures. Figure 33 shows the effect of decreasing temperature on J_c for type 304N, 316, 310, and Fe-Cr-Ni-Mn-N stainless steels. The alloys with deliberate nitrogen additions show reduced toughness with decreasing temperature due to significant strengthening and ductility loss. Decreasing the temperature from 24 to –269 °C causes a two- to three-fold increase in yield strength and at least a factor of 2 decrease in elongation. The most dramatic temperature effect occurs between –196 and –269 °C, where yield strength increases by about 30 to 40% and elongation decreases by 10 percentage points. For types 316 and 310 without nitrogen additions, J_c increases slightly as temperature is decreased. Their yield strengths at –269 °C (545 and 765 MPa) are less than those for type 304N (794-986 MPa) and Fe-Cr-Ni-Mn-N alloys (986-1540 MPa). Thus, the smaller degree of strengthening experienced by types 316 and 310 accounts for their improved cryogenic fracture resistance.

The –269 °C fracture toughness values for types 316 and 310 are plotted in Fig. 34 as a function of yield strength. The higher yield strengths are achieved by increasing the nitrogen content. While J_c tends to decrease with increasing strength, there is considerable variability at any particular strength level. To account for data scatter, 90%/95% global tolerance limits bracketing 90% of the population at a 95% confidence level were developed (Ref 19) assuming that J_c values are log-normally distributed. The best-fit regression line and tolerance limits are represented by the solid and dashed lines. The order-of-magnitude scatter in J_c, which is consistent with the degree of scatter observed at room tem-

perature (Fig. 2), is attributed to differences in inclusion density and morphology, chemistry, and crack orientation. At strength levels below 700 MPa, the mean J_c values of 400 to 600 kJ/m^2 at –269 °C are somewhat lower than the mean J_c of 672 kJ/m^2 at room temperature. This indicates that cryogenic temperatures cause à modest toughness degradation in low-strength, low-nitrogen stainless steels. As nitrogen content and strength increase, mean J_c values are significantly lower because these materials show a greater temperature dependence.

Cryogenic fracture toughness values for type 310 and for 310S, a low-carbon version of type 310, are also represented in Fig. 34. J_c values for these high-nickel stainless steels are consistent with the mean toughness of nitrogen-strengthened type 316, with a yield strength of approximately 800 MPa.

The cryogenic fracture toughness for type 304, summarized in Fig. 35, shows the same trends as that for type 316. Both alloys exhibit an order-of-magnitude scatter in J_c and comparable reductions in J_c with increasing strength. It is seen that mean and lower-bound J_c values for type 304 are about 40% lower than those for type 316.

Minimum-expected J_c levels exceed 100 kJ/m^2 for type 304 at strength levels less than 600 MPa and for types 316 and 310 at strength levels less than 800 MPa; hence, fracture is not a primary design issue. As strength levels increase, however, the decrease in J_c causes increased fracture concerns. In the high-strength regime, conventional stress and strain design limits should be supplemented with fracture mechanics analyses to provide adequate protection against fracture

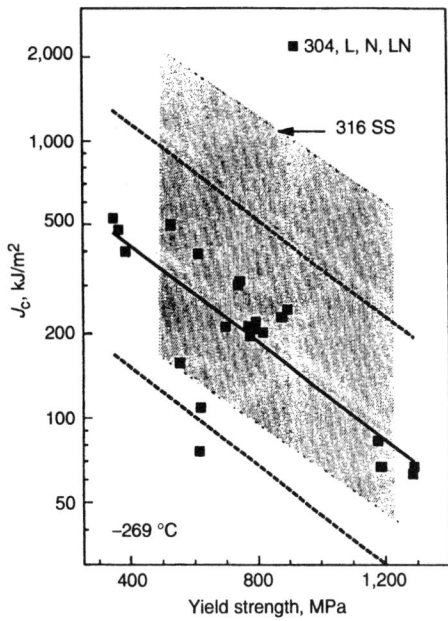

Fig. 35 Summary of –269 °C toughness as a function of yield strength. The solid line represents a least-squares exponential regression of the data, and the dashed lines correspond to linearized 90/95% tolerance limits. Source: Ref 137, 158, 159 (type 304); Ref 138, 145 (type 304L); Ref 140, 146, 159 (type 304N); Ref 57, 159 (type 304LN)

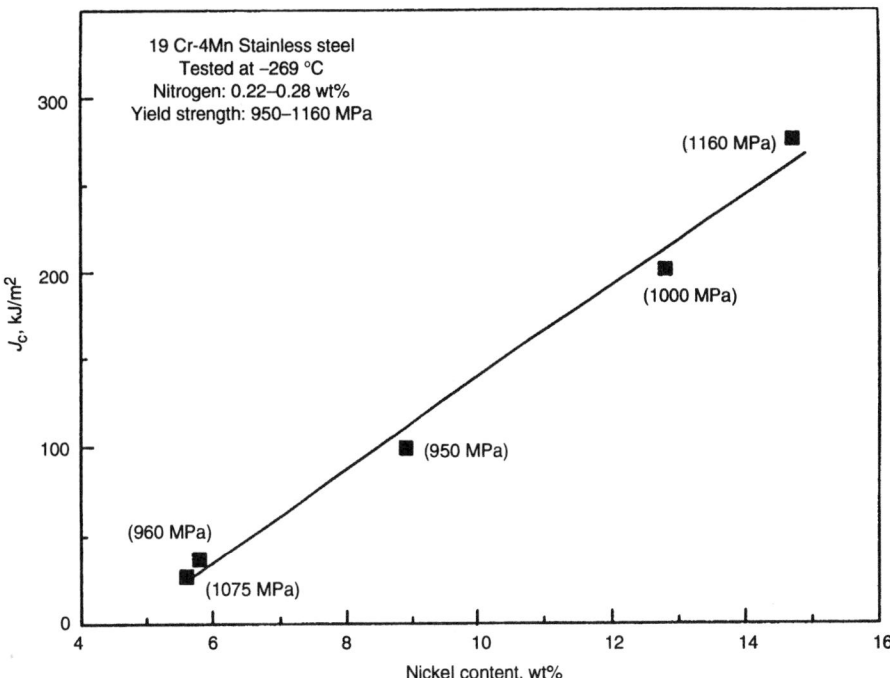

Fig. 36 Effect of nickel content on J_c for 19Cr-4Mn stainless steel at –269 °C. The yield strength for each heat is provided in parentheses. Source: Ref 160

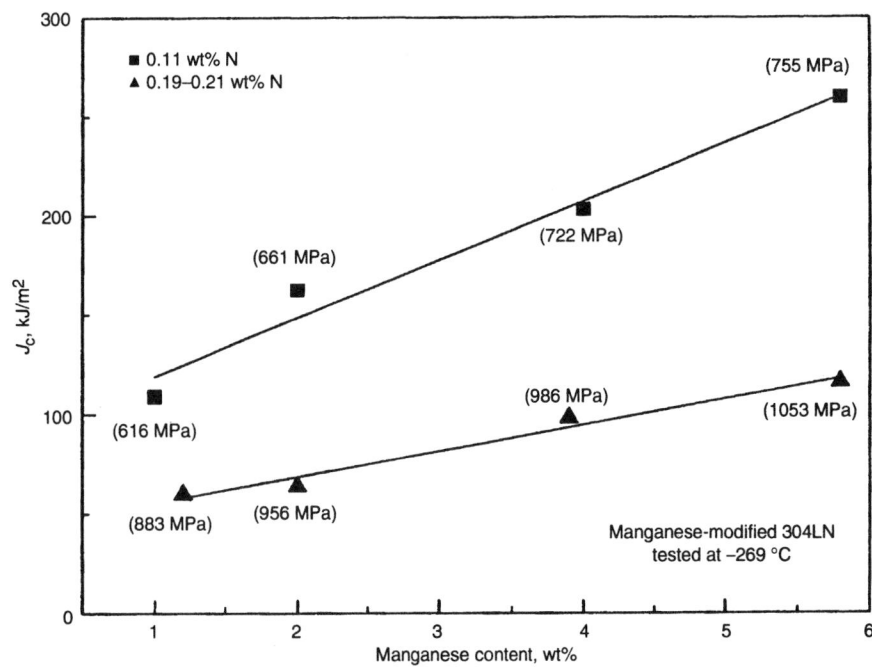

Fig. 37 Effect of manganese and nitrogen contents on J_c for modified type 304LN at –269 °C. The yield strength for each heat is provided in parentheses. Source: Ref 160

for critical engineering applications. Minimum expected toughness values at yield strengths of approximately 1000 MPa are 40 to 60 kJ/m², corresponding to K_{Jc} values of 90 to 110 MPa√m. At these toughness levels, 2 to 3 cm thick components provide sufficient constraint to induce plane-strain fracture conditions. As a result, a linear-elastic fracture mechanics methodology can be used to compute critical flaw sizes for large cryogenic components with yield strengths in excess of 1000 MPa.

The difference in fracture toughness behavior between types 304 and 316 is attributed to differences in nickel content. Nickel is beneficial to fracture toughness because it stabilizes the austenite and promotes dislocation cross-slip by increasing the stacking fault energy. Purtscher and Reed (Ref 160) systematically varied the nickel content in 19Cr-4Mn stainless steel and found that increasing it from 9 to 14 wt% more than doubled J_c without sacrificing strength, as shown in Fig. 36. Types 304 and 304N contain between 8 and 10.5 wt% Ni, while types 316 and 316N contain 10 to 14 wt% Ni. Hence, the lower nominal nickel content in type 304 is responsible for its 40% lower J_c values relative to type 316. In addition, nickel content partly accounts for the observed heat-to-heat variability. The type 304 heat falling below the lower tolerance limit in Fig. 36 contains 8.0 wt% Ni (Ref 137), the minimum allowable level, and the two points just above the tolerance limit represent only 8.3 and 8.7 wt% Ni (Ref 138, 159).

Manganese and molybdenum additions also improve the overall strength-to-toughness response (Ref 141, 153, 160). The beneficial effect of manganese has been clearly demonstrated by

systematically varying the manganese and nitrogen contents in manganese-modified type 304LN (Ref 141, 160). Figure 37 shows that increasing manganese from 1 to 6 wt% doubles J_c while increasing yield strength by 20%. The role of molybdenum is less clear. Purtscher et al. (Ref 153, 160) found that molybdenum additions up to 3 wt% significantly increased strength while causing a slight decrease in J_c. The overall

strength-to-toughness response, however, tends to improve. The beneficial influence of manganese and molybdenum, like that of nickel, is believed to be associated with enhanced stability of the austenite and increased stacking fault energy (Ref 160).

Figure 37 also shows that increasing the nitrogen from 0.1 to 0.2 wt% increases the yield strength of type 304LN at –269 °C by 250 to 300

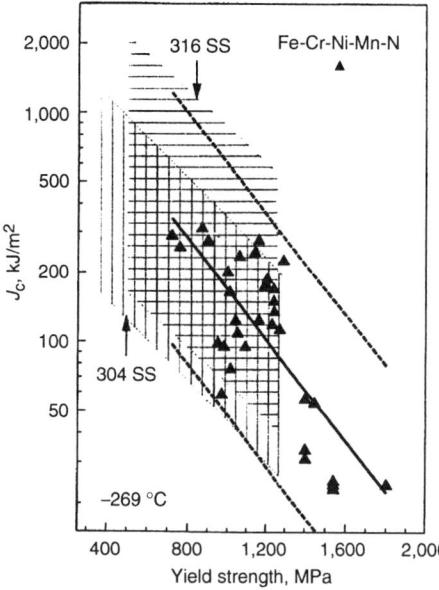

Fig. 38 J_c as a function of yield strength for Fe-Cr-Ni-Mn-N stainless steels tested at –269 °C. The scatter bands represent the range of toughness-strength values for types 316 and 304. Source: Ref 57, 143 (Nitronic 33); Ref 51, 57 (Nitronic 40); Ref 57 (Nitronic 50); Ref 143, 161 (JN1); Ref 162-169 (other Fe-Cr-Ni-Mn-N alloys)

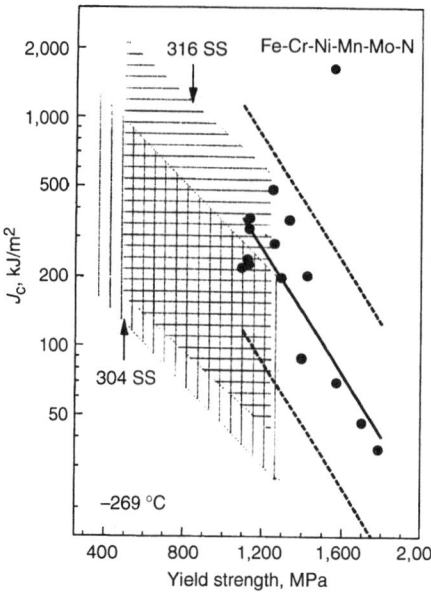

Fig. 39 J_c as a function of yield strength for Fe-Cr-Ni-Mo-N stainless steels tested at –269 °C. The scatter bands represent the range of toughness-strength values for types 316 and 304. Source: Ref 156 (Kromarc 58); Ref 145 (JK1); Ref 161 (JJ1); Ref 170, 171 (other Fe-Cr-Ni-Mn-Mo-N stainless steels)

Fig. 40 Crystallographic facets indicative of slip-band cracking in Fe-18Cr-3Ni-13Mn-0.37N tested at –269 °C. Source: Ref 168

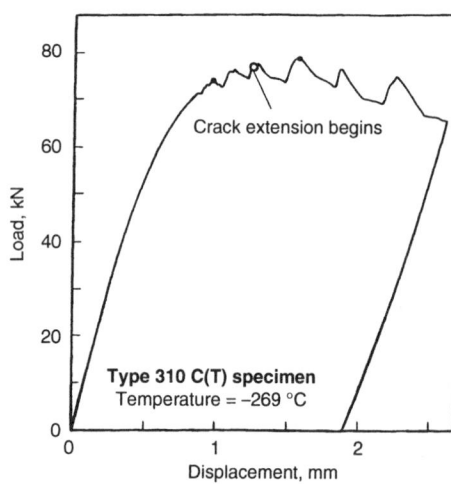

Fig. 41 Serrated load-displacement curve for type 310S at –269 °C. Source: Ref 157

MPa. But unlike nickel and manganese, nitrogen increases strength at the expense of ductility. Consequently, nitrogen additions significantly degrade fracture resistance.

Nitrogen-strengthened stainless steels with high nickel, manganese, and molybdenum have been developed to take advantage of the improved fracture resistance and yield strengths up to 1800 MPa. High-manganese and high-nickel stainless steels include Nitronic 33, Nitronic 40, Nitronic 50, JN1 (Fe-25Cr-15Ni-4Mn), and others with 13-25 wt% Mn. The strength-to-toughness behavior for these alloys is compared with that for types 304 and 316 in Fig. 38. For yield strengths up to 1200 MPa, the toughness range for the manganese-modified alloys and type 304N are similar. In addition, the rate at which toughness decreases with increasing strength is essentially the same for all materials represented in Fig. 38, but the manganese additions allow yield strength levels up to 1800 MPa to be achieved.

Figure 39 shows the strength-to-toughness relationship for stainless steels with high nickel, manganese, and molybdenum, including Kromarc® 58 (23Ni-16Cr-9Mn-2Mo), JK1 (13Ni-2Mo-1.3Mn), JJ1 (12Ni-10Mn-5Mo), and other Fe-Cr-Ni-Mn-Mo-N stainless steels with 13 wt% Ni, 3 to 7.5 wt% Mo, and 1 to 22 wt% Mn. Where strength levels overlap, toughness values for these high-alloy materials correspond to the upper portion of the range for type 316. The lower 90%/95% tolerance limit for Fe-Cr-Ni-Mn-Mo-N alloys is higher than those for the other alloys, indicating that Fe-Cr-Ni-Mo-N alloys possess superior fracture resistance.

Microvoid coalescence is the dominant cracking mechanism in most austenitic steels at cryo-genic temperatures, but nitrogen-strengthened alloys also exhibit slip-band cracking (Ref 137, 162, 163, 168, 169) where separation along planar slip bands results in the crystallographic faceted appearance shown in Fig. 40 (Ref 168). Accumulation of dislocation-induced defects along slip bands leads to localized cracking under the influence of resolved tensile stresses. Increasing amounts of slip-band cracking are observed in low-toughness, high-strength alloys, at extreme temperatures. In certain alloys of the Nitronic steels, slip band cracking is the predominant failure mechanism at –269 °C, so that a ductile-brittle transition occurs in the extreme cryogenic regime (Ref 168).

Stable and metastable stainless steels exhibit serrated load-displacement curves during tensile and fracture toughness testing at –269 °C, but not at temperatures above –230 °C. A typical example of a serrated load-displacement curve for type 310S is shown in Fig. 41 (Ref 157). The first serrations occur shortly after specimens experience appreciable plasticity, but prior to the onset of measurable crack extension. Their magnitude and periodicity are very reproducible. These features are not associated with phase transformations because they occur in both stable and metastable alloys. Neither are they associated with a fracture mechanism transition, because microvoid coalescence is the dominant fracture mechanism. Basinski (Ref 172) and Tobler (Ref 157) proposed that serrations result from an avalanche of dislocation motion associated with an adiabatic temperature increase. Adiabatic heating has a much greater impact at –269 °C because the specific heats are about 200 to 300 times lower than at room temperature. Typical specific heat values (Ref 173-176) for types 304 and 316 are

450 to 500 J kg^{-1} K^{-1} at room temperature, 140 to 280 J kg^{-1} K^{-1} at –196 °C, and 2 to 4 J kg^{-1} K^{-1} at –269 °C. Therefore, small amounts of thermal energy at –269 °C create a substantial adiabatic temperature rise that causes thermal softening due to the lower strength at higher temperatures. Because thermal conductivities are also low at this temperature, the region experiencing higher temperatures is very localized. During the load drop, plastic strain rates associated with an avalanche of dislocation motion in the thermally softened region exceed applied strain rates. Eventually the plastic strain instability arrests when the deformation process becomes stable due to local strain hardening and an increase in specific heat that precludes continued temperature increase. The flow process is then stable until the entire specimen cools to –269 °C and additional plastic straining reinitiates this process, thereby accounting for the reproducibility of the serrations.

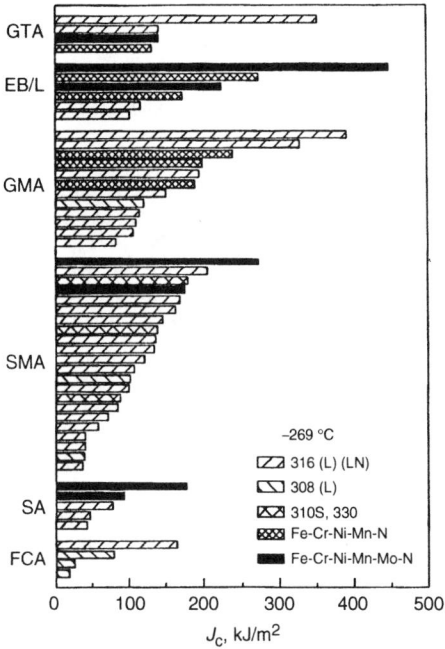

Fig. 42 Fracture toughness of welds fabricated using different processes. GTA, gas-tungsten arc; EB, electron beam; L, laser; GMA, gas-metal arc; SMA, shielded-metal arc; SA, submerged arc; FCA, flux-cored arc. Source: Ref 81, 177 (type 316); Ref 78-81, 83, 89, 177 (type 316L); Ref 82, 154 (type 316LN); Ref 81 (type 308); Ref 73, 79, 81 (type 308L); Ref 169 (type 310S); Ref 73 (type 330); Ref 90, 166, 170, 178, 179 (Fe-Cr-Ni-Mn-N); Ref 90, 170, 178, 180, 181 (Fe-Cr-Ni-Mn-Mo-N)

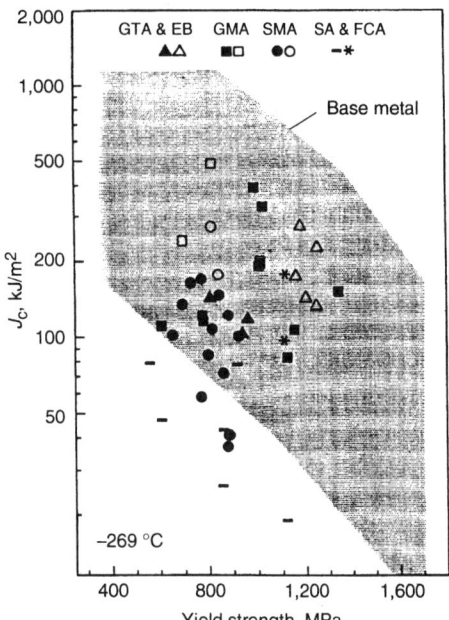

Fig. 43 J_c vs. yield strength at –269 °C for welds. The scatter band represents the range of toughness and strength values for base metals. Solid symbols represent types 308 and 316; open symbols and asterisks represent Fe-Cr-Ni-Mn-N and Fe-Cr-Ni-Mn-Mo-N. GTA, gas-tungsten arc; EB, electron beam; GMA, gas-metal arc; SMA, shielded-metal arc; SA, submerged arc; FCA, flux-cored arc. Source: Ref 73, 79, 81 (type 308); Ref 78-83, 89, 154, 177, 179 (type 316); Ref 90, 166, 170, 178, 179 (Fe-Cr-Ni-Mn-N); Ref 90, 170, 178, 180, 181 (Fe-Cr-Ni-Mn-Mo-N)

Welds

At cryogenic temperatures, welds typically exhibit higher strength and lower toughness than their base metal counterparts. The inferior toughness is associated with the presence of nonmetallic inclusions and δ-ferrite, which undergoes a ductile-brittle transition. Under cryogenic conditions, cleavage fracture of the δ-ferrite creates a network of microcracks ahead of a crack front. Subsequent cracking of the austenite matrix then causes an overall crack advance. Because cracks preferentially seek out the brittle δ-ferrite, the relative amount of cleavage fracture observed on the fracture surface tends to be much greater than the percentage of δ-ferrite present in the weld. In addition, as the volume fraction of δ-ferrite increases, its interconnected nature provides a low-energy crack path that reduces the overall cracking resistance. For optimum toughness, weld compositions are adjusted to produce the minimum amount of δ-ferrite that precludes microfissuring. Accordingly, cryogenic grades of filler metal are often specially formulated to produce low ferrite levels (i.e., ferrite numbers less than 2) and low-impurity concentrations to minimize the potential for microfissuring.

Figure 42 summarizes the toughness response of welds at –269 °C. The GTA, EB, laser, GMA, SMA, SA and FCA welding processes are represented. The wide variation in toughness is associated with differences in welding process, nonmetallic inclusion density, δ-ferrite content, and strength. Although toughness responses for the

various filler metals overlap, the limited data base indicates that J_c values for type 308 welds tend to be slightly lower than the nominal values for type 316 welds. This difference is probably related to variations in nickel content: type 308 welds contain 9 to 11 wt% Ni, whereas type 316 welds contain 11 to 14 wt% Ni. There is no systematic difference in J_c among type 316, 310S, and 330 welds, whereas experimental nitrogen-strengthened welds with high nickel, manganese, and molybdenum typically exhibit higher J_c values.

The trends associated with welding process are consistent with ambient- and elevated-temperature findings. GTA and GMA welds consistently show the highest fracture resistance, while SA and FCA welds possess the lowest. Two FCA welds were found to be very brittle, with valid K_{Ic} levels of 66 and 79 MPa√m, corresponding to J_c values of 20 and 25 kJ/m² (Ref 79). SMA welds display low to intermediate toughness values, with the lowest values (about 40 kJ/m²) being similar to the limiting toughness for SA welds.

Laser and EB welds generally exhibit high fracture toughness values, comparable to those for GTA and GMA welds. Although the type 316 EB and laser welds show only intermediate toughness values of 120 and 100 kJ/m², these values are matched reasonably well to the base metal toughness of 160 kJ/m² (Ref 154). Because the EB and laser welding processes remelt the base metal, use of high-toughness base metal heats is expected to increase the weld toughness.

The inferior toughness of SA welds and some SMA welds is associated with a high density of

nonmetallic inclusions. These processes tend to produce significant amounts of silicon and oxygen-rich inclusions that serve as effective microvoid nucleation sites, comparable to the behavior at ambient and elevated temperatures. GTA, GMA, EB and laser welding processes produce cleaner welds with superior fracture resistance. Although GMA welds are significantly cleaner than SMA and FCA welds, oxygen pickup from argon-oxygen shield gas increases the amount of oxide inclusions (Ref 82, 83). Thus, when high-oxygen shield gases are used to fabricate GMA welds, the fracture toughness is degraded due to an increase in inclusion density. The low toughness of the FCA welds is attributed to high nitrogen contents (Ref 79). Because this process is self-shielded without an inert shielding gas, it tends to pick up nitrogen from the atmosphere, which increases strength but reduces toughness. Based on these observations, optimum cryogenic fracture resistance can be achieved by selecting: 1) weld processes that minimize inclusion contents, 2) inert shielding gases that minimize oxygen and nitrogen uptake, and 3) filler metal compositions that minimize δ-ferrite while guarding against microfissuring.

In Fig. 43, comparison of the –269 °C toughness-to-strength behavior for type 308 and 316 welds (solid symbols) with the base metal response reveals that the weld toughness is often undermatched. At comparable strength levels, J_c values for SA welds are consistently below the lower-bound toughness for the base metal, while J_c values for SMA welds straddle the lower-bound toughness limit for the base metal. Toughness values for GMA and GTA welds are within the base metal scatter band.

The toughness of GTA, GMA, EB, and laser welds, with J_c values typically above 100 kJ/m² at –269 °C, is sufficiently high to avoid fracture concerns for most cryogenic applications. Significant plastic deformation is required to cause rapid crack extension, so conventional stress and strain design limits adequately guard against failure. The lower toughness of SA and SMA welds, with lower-bound J_c values on the order of 40 kJ/m², demonstrates that they are more susceptible to cracking. Hence, for critical cryogenic applications involving SA and SMA welds, fracture is a design concern and fracture mechanics evaluations should be used to demonstrate adequate performance. The 308L FCA welds exhibit a brittle fracture response with valid K_{Ic} values as low as 66 MPa√m. Because this type of weld is susceptible to brittle fracture, linear-elastic fracture mechanics analysis should be an integral part of the design to provide protection against unstable fracture.

Microfissure-free and δ-ferrite-free welds with high alloy contents are currently being developed to improve both strength and toughness. This includes Fe-Cr-Ni-Mn-N welds with 5 to 20 wt% Ni and 5 to 22 wt% Mn, Fe-Cr-Ni-Mo-N welds with 12 to 25 wt% Ni and 2 to 5 wt% Mo, and Fe-Cr-Ni-Mn-Mo welds with 5 to 20 wt% Ni, 4 to 14 wt% Mn, and 1 to 3 wt% Mo. Strength levels for these welds are typically controlled by nitrogen additions. The high nickel, manganese, and molybdenum contents are designed to improve strength and toughness by enhancing austenite stability, increasing stacking fault energy, and eliminating δ-ferrite from the weld fusion zone. As shown in Fig. 43, tensile and fracture toughness data for these welds (open symbols and asterisks) show increased strength and significantly improved fracture resistance for each welding process. The increase in SA weld toughness is particularly noteworthy. It is seen that fracture properties for these developmental welds are closely matched with the base metal toughness.

Charpy Impact Energy/Fracture Toughness Correlations at –269 °C

A significant amount of work has been performed to explore the possibility of establishing a correlation between Charpy impact energy and J_c at cryogenic temperatures, so that fracture toughness can be estimated from a simple and inexpensive Charpy test. While limited success has been made at –196 °C (Ref 73, 80, 143, 177), there is no general correlation between J_c and Charpy impact energy at –269 °C due to adiabatic heating of the specimen during impact loading (Ref 143, 182-184). As a result of the exceptionally low specific heats at –269 °C, adiabatic heating is unavoidable and can cause a large rise in the local temperature. In fact, the actual fracture temperature for Charpy specimens with an initial temperature of –269 °C can be above –200 °C. Thus, Charpy impact energy results for base metal and welds should not be used to estimate J_c or even infer fracture trends at –269 °C, because such data may they significantly overestimate fracture energy and incorrectly rank materials with different temperature sensitivities.

REFERENCES

1. W.J. Mills, On the Relationship between Stretch Zone Formation and the J Integral for High Strain-Hardening Materials, *J. Test. Eval.*, Vol 9, 1981, p 56-62
2. J.N. Robinson, An Experimental Investigation of the Effect of Specimen Type on the Crack Tip Opening Displacement and J-Integral Fracture Criteria, *Internat. J. Fract.*, Vol 12, 1976, p 723-737
3. S.A. Paranjpe and S. Banerjee, Interrelation of Crack Opening Displacement and J-Integral, *Eng. Fract. Mech.*, Vol 11, 1979, p 43-53
4. P.H. Davies, An Elastic-Plastic Fracture Mechanics Study of Crack Initiation in 316 Stainless Steel, *Fracture Resistance Curves and Engineering Applications*, Vol II, *Elastic-Plastic Fracture: Second Symposium*, STP 803, ASTM, 1983, p II-611 to II-631
5. H. Kobayashi, H. Nakamura, and H. Nakazawa, Evaluation of Blunting Line and Elastic-Plastic Fracture Toughness, *Fracture Resistance Curves and Engineering Applications*, Vol II, *Elastic-Plastic Fracture: Second Symposium*, STP 803, ASTM, 1983, p II-420 to II-438
6. M.I. de Vries and B. Schaap, Experimental Observations of Ductile Crack Growth in Type 304 Stainless Steel, *Elastic-Plastic Fracture Test Methods: The User's Experience*, STP 856, ASTM, 1985, p 183-195
7. N. Nakajima, S. Shima, H. Nakajima, and T. Kondo, The Fracture Toughness Measured on Sensitized 304 Stainless Steel in Simulated Reactor Water, *Nuclear Eng. Design*, Vol 93, 1986, p 95-106
8. G.T.M. Janssen, J.V.D. Eikhoff, and H.J.M. van Rongen, Prediction of Ductile Crack Growth in the Austenitic Steel WN 1.4948 at Ambient and Elevated Temperature, *Sixth International Conference on Structural Mechanics in Reactor Technology*, Commission of the European Communities, Vol L, Paper L5/2, 1981
9. G.S. Kramer and V. Papaspyroulos, A Study of the Initiation and Growth of Complex Cracks in Nuclear Piping under Pure Bending, *Elastic-Plastic Fracture*, Vol II, *Nonlinear Fracture Mechanics*, STP 995, ASTM, 1989, p 433-453
10. W.J. Mills, Fracture Toughness of Aged Stainless Steel Primary Piping and Reactor Vessel Materials, *J. Pressure Vessel Technol.*, Vol 109, 1987, p 440-448
11. P. Balladon, J. Heritier, and P. Rabbe, The Influence of Microstructure on the Ductile Rupture Mechanisms of a 316L Steel at Room and Elevated Temperatures, *Fracture Mechanics: 14th Symposium*, STP 791, ASTM, 1983, p II-496 to II-513
12. P. Balladon, J. Heritier, and C. Jarboui, Strain Hardening Effects on Fracture Toughness and Ductile Crack Growth in Austenitic Stainless Steels, *Fracture Mechanics: 16th Symposium*, STP 868, ASTM, 1985, p 293-307
13. P. Balladon and J. Heritier, Comparison of Ductile Crack Growth Resistance of Austenitic, Niobium Austenitic, and Austeno-Ferritic Stainless Steels, *Fracture Mechanics: 17th Symposium*, STP 905, ASTM, 1986, p 661-682
14. Y. Nakajima, Y. Iino, and M. Suzuki, Fracture Toughness of Service Exposed Type 321 Stainless Steel at Room and Elevated Temperature under Normal and Low Straining Rates, *Engr. Fract. Mech.*, Vol 33, 1989, p 295-307
15. F. Faure, P. Ould, and P. Balladon, Effect of Long Term Aging on the Mechanical Properties of Stainless Steel Welds in PWR, *International Trends in Welding Science and Technology*, ASM International, 1993, p 563-568
16. R.A. Hays, M.G. Vassilaros, and J.P. Gudas, Investigation of Ductile Fracture Properties of Welded Type 304 Stainless Steel Pipe and Large Plan Compact Specimens, *Fracture Mechanics: 18th Symposium*, STP 945, ASTM, 1988, p 134-150
17. W.J. Mills, Fracture Toughness of Stainless Steel Welds, *Fracture Mechanics: 19th Symposium*, STP 969, ASTM, 1988, p 330-355
18. R.M. Horn, H.S. Mehta, W.R. Andrews, and S. Ranganath, "Evaluation of the Toughness of Austenitic Stainless Steel Pipe Weldments," NP-4668, EPRI, June 1986
19. W.J. Mills, Fracture Toughness of Type 304 and 316 Stainless Steels and Their Welds, 1996, submitted for publication
20. P.C. Paris, H. Tada, A. Zahoor, and H.A. Ernst, The Theory of Instability of the Tearing Mode for Elastic-Plastic Crack Growth, *Elastic-Plastic Fracture*, STP 668, ASTM, 1979, p 5-36
21. W.J. Mills, "Effect of Specimen Size on the Fracture Toughness of Type 304 Stainless Steel—Interim Report," Report HEDL-TME 81-52, Westinghouse Hanford, 1982
22. M.G. Vassilaros, R.A. Hays, and J.P. Gudas, Investigation of the Ductile Fracture Properties of the Type 304 Stainless Steel Plate, Welds, and 4-Inch Pipe, *Proc. 12th Water Reactor Safety Research Information Meeting*, NUREG/CP-0058, U.S. Nuclear Regulatory Commission, Vol 4, 1984, p 176-189
23. D.E. McCabe and W.A. Logsdon, Plane Stress Fracture Toughness Evaluation of Removable Radial Shield Assembly Materials, *Trans. Seventh International Conf. on Structural Mechanics in Reactor Technology*, Commission of the European Communities, Vol E, Paper E8/4, 1983, p 333-340
24. W.H. Cullen, A.L. Hiser, J.R. Hawthorne, G.R. Caskey, and G.A. Abramczyk, Fractographic and Microstructural Aspects of Fracture Toughness Testing in Irradiated 304 Stainless Steel, *Properties of Stainless Steels in Elevated Temperature Service*, PVP Vol 132, ASME, 1987, p 129-156
25. W.J. Mills, Heat-to-Heat Variations in the Fracture Toughness of Austenitic Stainless Steels, *Eng. Fract. Mech.*, Vol 30, 1988, p 469-492
26. W.H. Bamford and J.A. Begley, Techniques for Evaluating the Flaw Tolerance of Reactor Coolant Piping, Paper 76-PVP-48, ASME, 1976
27. W.H. Bamford and A.J. Bush, Fracture Behavior of Stainless Steel, *Elastic-Plastic Fracture*, STP 668, ASTM, 1979, p 553-577
28. J.A. Begley, A.A. Sheinker, and W.K. Wilson, "Crack Propagation Testing for LMFBR Piping—Phase II Final Report," Research Report 76-8E7-Elbow-R2, Westinghouse, 1976
29. J. Dufresne, B. Henry, and H. Larsson, Fracture Toughness of Irradiated AISI 304 and 316L Stainless Steel, *Effects of Radiation on Structural Materials*, STP 683, ASTM, 1979, p 511-528
30. R. Herrera and J.D. Landes, Direct J-R Curve Analysis: A Guide to the Methodology, *Fracture Mechanics: 21st Symposium*, STP 1074, ASTM, 1990, p 24-43
31. M.F. Kanninen, A. Zahoor, G. Wilkowski, I. Abou-Sayed, C. Marshall, D. Broek, S. Sampath, H. Rhee, and J. Ahmad, *Instability Predictions for Circumferentially Cracked Type-304 Stainless Steel Pipes under Dynamic Loading*, Vol 1, EPRI NP-2347, Battelle Columbus Laboratories, 1982
32. M.A. Khan, T. Shoji, H. Takahashi, and H. Niitsuma, Combined Elastic-Plastic and

Acoustic Emission Methods for the Evaluation of Tearing and Cleavage Crack Extension, *Fracture Resistance Curves and Engineering Applications,* Vol II, *Elastic-Plastic Fracture: Second Symposium,* STP 803, ASTM, 1983, p II-506 to II-528

33. J.D. Landes, Size and Geometry Effects on Elastic-Plastic Fracture Characterizations, *Proc. U.S. Nuclear Regulatory Commission CSNI Specialists Meeting Tearing Instability,* NUREG/CP-0010, U.S. Nuclear Regulatory Commission, 1979, p 194-225

34. J.D. Landes, D.E. McCabe, and H.A. Ernst, "Elastic-Plastic Methodology to Establish R Curves and Instability Criteria—Eighth Semi-annual Report," EPRI Contract RP1238-2, Westinghouse R&D Center, Pittsburgh, PA, July 1984

35. R.L. Sindelar, G.R. Caskey, J.K. Thomas, J.R. Hawthorne, A.L. Hiser, R.A. Lott, J.A. Begley, and R.P. Shogan, Mechanical Properties of 1950's Vintage Type 304 Stainless Steel Weldment Components After Low Temperature Neutron Irradiation, *Effects of Radiation on Materials: 16th International Symposium,* STP 1175, ASTM, 1993, p 714-746

36. R.L. Sindelar and G.R. Caskey, Jr., Orientation Dependency of Mechanical Properties of 1950's Vintage Type 304 Stainless Steel Weldment Components Before and After Low Temperature Neutron Irradiation, *Microstructures and Mechanical Properties of Aging Materials,* TMS, 1993, p 361-369

37. G.M. Wilkowski, J.O. Wambaugh, and K. Prabhat, Single Specimen J-Resistance Curve Evaluations Using the D.C. Electric Potential Method and a Computerized Data Acquisition System, *Fracture Mechanics: 15th Symposium,* STP 833, ASTM, 1984, p 553-576

38. G.M. Wilkowski, J. Ahmad, C.R. Barnes, F. Brust, D. Guerrieri, J. Kiefner, G. Kramer, G. Kulhowvick, M. Landow, C.W. Marschall, W. Maxey, M. Nakagaki, P. Papaspyropoulos, V. Pasupathi, and P. Scott, *Degraded Piping Program—Phase II,* NUREG/CR-4082, Vol 4, U.S. Nuclear Regulatory Commission, March 1986

39. P. Balladon, J-R Curves on Austenitic Stainless Steels—Comparison of Unloading Compliance and Interrupted Test Methods—Effect of Specimen Size—Notch Effect, *Ductile Fracture Test Methods,* Organization for Economic Cooperation and Development (OCDE), Paris, 1983, p 110-114

40. J.W. Cardinal and M.F. Kanninen, Stable Crack Growth and Fracture Instability Predictions for Type 304 Stainless Steel Pipes with Girth Weld Cracks, *Elastic-Plastic Fracture,* Vol II, *Nonlinear Fracture Mechanics,* STP 995, ASTM, 1989, p 320-329

41. M.I. de Vries, J-Integral Fracture Toughness of Low Fluence Neutron-Irradiation Stainless Steel DIN 1.4948, *Influence of Radiation on Material Properties: 13th International Symposium (Part II),* STP 956, ASTM, 1986, p 162-173

42. W.J. Mills, L.A. James, and L.D. Blackburn, "Results of Fracture Mechanics Tests on PNC SUS 304 Plate," Report HEDL-7544, Westinghouse Hanford, Aug 1985

43. W.J. Mills, Fracture Toughness of Irradiated Stainless Steel Alloys, *Nuclear Technol.,* Vol 82, 1988, p 290-303

44. N. Miura, K. Kashima, K. Michiba, and T. Shimakawa, Ductile Fracture of Stainless Steel Welds at Elevated Temperature, *Fatigue, Fracture, and Risk—1991,* PVP Vol 215, ASME, 1991, p 81-86

45. P. Scott, R.J. Olson, and G.M. Wilkowski, The IPIRG-1 Pipe System Fracture Tests: Experimental Results, *Fatigue, Flaw Evaluation and Leak-Before-Break Assessments,* ASME, 1994, p 135-151

46. P. Scott, R.J. Olson, M. Wilson, and G.M. Wilkowski, The Effect of Inertial Loading on Circumferentially Cracked Pipe: Experimental Results, *Fatigue, Flaw Evaluation and Leak-Before-Break Assessments,* ASME, 1994, p 183-197

47. J. Bernard and G. Verzeletti, Elasto-Plastic Fracture Mechanics Characterization of Type 316H Irradiated Stainless Steel Up to 1 dpa, *Effects of Radiation on Materials: 12th International Symposium,* STP 870, ASTM, 1985, p 619-641

48. C.G. Chipperfield, Detection and Toughness Characterization of Ductile Crack Initiation in 316 Stainless Steel, *Internat. J. Fract.,* Vol 12, 1976, p 873-886

49. C.G. Chipperfield, "A Method for Determining Dynamic JQ and Ji Values and Its Application to Ductile Steels," paper presented at the International Conference on Dynamic Fracture Toughness, The Welding Institute /American Society for Metals, London, 1977

50. C.G. Chipperfield, A Toughness and Defect Size Assessment of Welded Stainless Steel Components, *Institution of Mechanical Engineers,* 1978, p 145-159

51. R L. Tobler, Fracture of Structural Alloys at Temperatures Approaching Absolute Zero, *Proc. Fourth International Conf. on Fracture,* 1977, p 279-285

52. J.E. Pawel, D.J. Alexander, M.L. Grossbeck, A.W. Longest, A.F. Rowcliffe, G.E. Lucas, S. Jitsukawa, A. Hishinuma, and K. Shiba, Fracture Toughness of Candidate Materials for ITER First Wall, Blanket, and Shield Structures, *J. Nuclear Mater.,* Vol 212-215, 1994, p 442-447

53. P.K. Liaw and J.D. Landes, Effects of Monotonic and Cyclic Prestrain on Fracture Toughness: A Summary, *Fracture Mechanics: 18th Symposium,* STP 945, ASTM, 1988, p 622-646

54. F.J. Loss and R.A. Gray, Jr., "Toughness of Irradiated Type 316 Forging and Weld Metal Using the J-Integral," Memorandum Report 2875, U.S. Naval Research Laboratory, July 1974, p 23-30

55. I.J. O'Donnell, H. Huthmann, and A.A. Tavassoli, *International J. Pressure Vessel Piping,* Vol 65, 1996, p 209-220

56. T.P. Magee and C.L. Hoffmann, Characterization of Elastic-Plastic Fracture Behavior of Type 347 Stainless Steel Piping Materials, *Fatigue and Fracture Mechanics in Pressure Vessels and Piping,* PVP Vol 304, ASME, 1995, p 267-275

57. D.T. Read and R.P. Reed, Toughness, Fatigue Crack Growth, and Tensile Properties of Three Nitrogen-Strengthened Stainless Steels at Cryogenic Temperatures, *The Metal Science of Stainless Steel,* AIME, 1979, p 92-121

58. O. Speidel and P.J. Uggowitzer, High Manganese, High Nitrogen Austenitic Stainless Steel: Their Strength and Toughness, *High Manganese High Nitrogen Austenitic Steels,* ASM International, 1993, p 135-142

59. V. Kumar, M.D. German, and C.F. Shih, "An Engineering Approach for Elastic-Plastic Fracture Analysis," Report NR-1931, EPRI, July 1981

60. V. Kumar, M.D. German, W.W. Wilkening, W.R. Andrews, H.G. deLorenzi, and D.F. Mowbray, "Advances in Elastic-Plastic Fracture Analysis," Report NR-3607, EPRI, Aug 1984

61. V. Kumar and M.D. German, "Elastic-Plastic Fracture Analysis of Through-Wall and Surface Flaws in Cylinders," Report NR-5596, EPRI, Jan 1988

62. A. Zahoor, "Ductile Fracture Handbook, Vol 1—Circumferential Throughwall Cracks," Report NP-6301-D, EPRI, June 1989

63. P.C. Paris and H. Tada, "The Application of Fracture Proof Design Methods Using Tearing Instability Theory to Nuclear Piping Postulating Circumferential Through Wall Cracks," Report NUREG/CR-3464, U.S. Nuclear Regulatory Commission, Sept 1983

64. F.W. Brust, "Approximate Methods for Fracture Analyses of Through-Wall Cracked Pipes," Report NUREG/CR-4853, U.S. Nuclear Regulatory Commission, Feb 1987

65. S. Yukawa, Effect of Long-Term Thermal Exposure on the Toughness of Austenitic Stainless Steels and Nickel Alloys, *Fracture Mechanics—Applications and New Materials,* PVP Vol 260, ASME, 1993, p 115-125

66. D.J. Alexander, K.B. Alexander, M.K. Miller, and R.K. Nanstad, The Effect of Aging at 343 °C on the Mechanical Properties and Microstructure of Type 308 Stainless Steel Weldments, *Microstructures and Mechanical Properties of Aging Materials,* TMS, 1993, p 263-269

67. D.J. Alexander, K.B. Alexander, M.K. Miller, and R.K. Nanstad, The Effect of Aging at 343 °C on the Type 308 Stainless Steel Weldments, *Fatigue, Degradation, and Fracture—1990,* PVP Vol 195, ASME, 1990, p 187-192

68. F.M. Haggag, W.R. Corwin, and R.K. Nanstad, Effects of Irradiation on the Fracture Properties of Stainless Steel Weld Overlay Cladding, *Nuclear Eng. Design,* Vol 124, 1990, p 129-141

69. J.M. Vitek, S.A. David, D.J. Alexander, J.R. Keiser, and R.K. Nanstad, Low Temperature Aging Behavior of Type 308 Stainless Steel

Weld Metal, *Acta Metall. Mater.*, Vol 39, 1991, p 503-516

70. D.J. Alexander, J.M. Vitek, and S.A. David, Long-Term Aging of Type 308 Stainless Steel Welds: Effects on Properties and Microstructure, *International Trends in Welding Science and Technology*, ASM International, 1993, p 557-561

71. B. Josefsson and U. Bergenlid, Tensile, Low Cycle Fatigue and Fracture Toughness Behaviour of Type 316L Steel Irradiated to 0.3 dpa, *J. Nuclear Mater.*, Vol 212-215, 1994, p 525-529

72. A.B. Poole, J.A. Clinard, R.L. Battiste, and W.R. Hendrich, Fracture Mechanics and Full Scale Pipe Break Testing for the Department of Energy's New Production Reactor-Heavy Water Reactor, *Nuclear Eng. Design*, Vol 152, 1994, p 57-65

73. H.I. McHenry, D.T. Read, and P.A. Steinmeyer, Evaluation of Stainless Steel Weld Metals at Cryogenic Temperatures, Report NBSIR 79-1609, National Bureau of Standards Boulder, 1979

74. S.M. Graham, W.R. Lloyd, and W.G. Reuter, Experience with J Testing of Type 304/308 Stainless Steel Weldment, *Elastic-Plastic Fracture Test Methods: The User's Experience*, Vol 2, STP 1114, ASTM, 1991, p 213-224

75. F.M. Haggag, Degradation of Mechanical Properties of Stainless Steel Cladding Due to Neutron Irradiation and Thermal Aging, *Effects of Radiation on Materials: 16th International Symposium*, STP 1175, ASTM, 1993, p 363-370

76. D.E. McCabe, "Fracture Evaluation of Surface Cracks Embedded in Reactor Vessel Cladding," Report NUREG/CR-5207, U.S. Nuclear Regulatory Commission, Sept 1988

77. D.J. Alexander and G.M. Goodwin, Thick-Section Weldments in 21-6-9 and 316LN Stainless Steel for Fusion Energy Applications, *Adv. Cryogenic Eng. (Mater.)*, Vol 38, 1992, p 101-107

78. G.M. Goodwin, Welding Process Selection for Fabrication of a Superconducting Magnet Structure, *Weld. J.*, Vol 64, 1985, p 19-26

79. H.I. McHenry and T.A. Whipple, Weldments for Liquid Helium Service, *Materials Studies for Magnetic Fusion Energy Applications at Low Temperatures—III*, NBSIR 80-1627, National Bureau of Standards, 1980, p 155-165

80. T.A. Whipple, H.I. McHenry, and D.T. Read, Fracture Behavior of Ferrite-Free Stainless Steel Welds in Liquid Helium, *Weld. J.*, Vol 60, 1981, p 72s-78s

81. T.A. Whipple and D.J. Kotecki, Weld Process Study for 316L Stainless Steel Weld Metal for Liquid Helium Service, *Materials Studies for Magnetic Fusion Energy Applications at Low Temperatures—IV*, NBSIR 81-1645, National Bureau of Standards, 1981, p 303-321

82. C.N. McCowan and T.A. Siewert, Inclusions and Fracture Toughness in Stainless Steel Welds at 4 K, *Adv. Cryogenic Eng. (Mater.)*, Vol 34, 1988, p 335-342

83. C.N. McCowan and T.A. Siewert, Fracture Toughness of 316L Stainless Steel Welds with Varying Inclusion Contents at 4 K, *Adv. Cryogenic Eng. (Mater.)*, Vol 36, 1990, p 1331-1337

84. J.P. Gudas and D.R. Anderson, J_I-R Curve Characteristic of Piping Material and Welds, *Proceedings of the U.S. Nuclear Regulatory Commission Ninth Water Reactor Safety Research Information Meeting*, NUREG/CP-0024, U.S. Nuclear Regulatory Commission, March 1982

85. D.J. Michel and R.A. Gray, Fracture Toughness of Irradiated FBR Structural Materials, *Proc. Conf. Environmental Degradation of Engineering Materials III*, Pennsylvania State University, 1987, p 619-626

86. P. Ould, P. Balladon, and Y. Meyzaud, Fracture Toughness Properties of Austenitic Stainless Steel Welds, *Bull. Cercle Etud. Metaux*, Vol 15, 1988, p 31.1 to 31.12

87. P.M. Scott and J. Ahmad, "Experimental and Analytical Assessment of Circumferentially Surface-Cracked Pipes under Bending," Report NUREG/CR-4872, U.S. Nuclear Regulatory Commission, April 1987

88. M. Nakagaki, C.W. Marschall, and F.W. Brust, Elastic-Plastic Fracture Mechanics Evaluations of Stainless Steel Tungsten/Inert-Gas Welds, *Elastic-Plastic Fracture*, Vol II, *Nonlinear Fracture Mechanics*, STP 995, ASTM, 1989, p 214-243

89. T.A. Whipple and H.I. McHenry, Evaluation of Weldments and Castings for Liquid Helium Service, *Materials Studies for Magnetic Fusion Energy Applications at Low Temperatures—IV*, NBSIR 81-1645, National Bureau of Standards, 1981, p 273-287

90. T.A. Siewert and C.N. McCowan, Joining of Austenitic Stainless Steels for Cryogenic Applications, *Adv. Cryogenic Eng. (Mater.)*, Vol 38, 1992, p 109-115

91. V.P. Kujanpaa, S.A. David, and C.L. White, Characterization of Heat-Affected Zone Cracking in Austenitic Stainless Steel Welds, *Weld. J.*, Vol 66, 1987, p 221s-228s

92. J.C. Lippold, W.A. Baeslack III, and I. Varol, Heat-Affected Zone Liquation Cracking in Austenitic and Duplex Stainless Steels, *Weld. J.*, Vol 71, 1992, p 1s-14s

93. N.S. Tsai and T.W. Eagar, The Size of the Sensitization Zone in 304 Stainless Steel Welds, *J. Mater. Energy Systems*, Vol 6, 1984, p 33-37

94. W.J. Mills, Effect of Loading Rate and Thermal Aging on the Fracture Toughness of Stainless Steel Alloys, *Fracture Mechanics: Perspectives and Directions (20th Symposium)*, STP 1020, ASTM, 1989, p 459-475

95. S.J. Garwood, Fracture Toughness of Stainless Steel Weldments at Elevated Temperatures, *Fracture Mechanics: 15th Symposium*, STP 833, ASTM, 1984, p 333-359

96. C.W. Marschall, M.P. Landow, G.M. Wilkowski, and A.R. Rosenfield, Comparison of Static and Dynamic Strength and J-R Curves of Various Piping Materials from the IPIRG-1

Program, *Internat. J. Pressure Vessels Piping*, Vol 62, 1995, p 49-58

97. K.J. Stoner, R.L. Sindelar, N.G. Awadalla, J.R. Hawthorne, A.L. Hiser, and W.H. Cullen, *Mechanical Properties of 1950's Vintage Type 304 Stainless Steel Weldment Components, Fatigue, Degradation, and Fracture—1990*, PVP Vol 195, ASME, 1990, p 149-156

98. B. Weiss and R. Stickler, Phase Instabilities during High Temperature Exposure of 316 Austenitic Stainless Steel, *Metall. Trans.*, Vol 3, 1972, p 851-866

99. Y. Minami, H. Kimura, and Y. Ihara, Microstructural Changes in Austenitic Stainless Steels during Long-Term Aging, *Mater. Sci. Technol.*, Vol 2, 1986, p 796-806

100. J.M. Vitek and S.A. David, The Sigma Phase Transformation in Austenitic Stainless Steels, *Weld. J.*, Vol 65, 1986, p 106s-111s

101. J.M. Vitek and S.A. David, The Aging Behavior of Types 308 and 308 CRE Stainless Steels and Its Effect on Mechanical Properties, *Properties of Stainless Steels in Elevated Temperature Service*, PVP Vol 132, ASME, 1987, p 157-171

102. P.J. Grobner, The 885 °F (475 °C) Embrittlement of Ferritic Stainless Steels, *Metall. Trans.*, Vol 4, 1973, p 251-260

103. A. Trautwein and W. Gysel, Influence of Long-Term Aging of CF8 and CF8M Cast Steel at Temperatures between 300 and 500 °C on Impact Toughness and Structural Properties, *Stainless Steel Casting*, STP 756, ASTM, 1982, p 165-189

104. M.K. Miller, J.M. Hyde, M.G. Hetherington, A. Cerezo, G.D.W. Smith, and C.M. Elliott, Spinodal Decomposition in Fe-Cr Alloys— Experimental Study at the Atomic Level and Comparison with Computer Models, Part I: Introduction and Methodology, *Acta Metall. Mater.*, Vol 43, 1995, p 3385-3401

105. D.J. Alexander and R.K. Nanstad, The Effects of Aging for 50,000 Hours at 343 °C on the Mechanical Properties of Type 308 Stainless Steel Weldments, *Proc. Seventh International Symp. on Environmental Degradation of Materials in Nuclear Power Systems—Water Reactors*, National Association of Corrosion Engineers, 1995, p 747-758

106. J.A. Horak, V.K. Sikka, and D.T. Raske, Review of Mechanical Properties and Microstructure of Type 304 and 316 Stainless Steel After Long-Term Thermal Aging, *Proc. IAEA Specialists Meeting on the Mechanical Properties of Structural Materials Including Environmental Effects*, IWGFR-49, Vol 1, International Atomic Energy Agency, 1983, p 179-213

107. C.E. Spaeder, Jr. and K.G. Brickner, Modified Type 316 Stainless Steel with Low Tendency to Form Sigma, *Advances in the Technology of Stainless Steels and Related Alloys*, STP 369, ASTM, 1964, p 43

108. N.L. Mochel, C.W. Ahlman, G.C. Wiedersum, and R.H. Zong, Performance of Type 316 Stainless Steel Piping at 5000 psi and 1200 °F,

Proc. American Power Conf., Vol 28, 1966, p 556-568

109. C.L. Clark, J.J.B. Rutherford, A.B. Wilder, and M.A. Cordovi, Metallurgical Evaluation of Superheater Tube Alloys After 12 and 18 Months' Exposure to Steam at 1200, 1350, and 1500 °F, *J. Eng. Power (Trans. ASME),* 1962, p 258-288

110. J.R. Hawthorne and H.E. Watson, Notch Toughness of Austenitic Stainless Steel Weldments with Nuclear Irradiation, *Weld. J.,* 1973, p 255s-260s

111. T.D. Parker, Strength of Stainless Steels at Elevated Temperatures, *Source Book on Stainless Steel,* American Society for Metals, 1976, p 80-99

112. J.J. Smith and R.A. Farrar, Influence of Microstructure and Composition on Mechanical Properties of Some AISI 300 Series Weld Metals, *Internat. Mater. Rev.,* Vol 38, 1993, p 25-51

113. W.H. Bamford, E.I. Landerman, and E. Diaz, Thermal Aging of Cast Stainless Steel, and Its Impact on Piping Integrity, *J. Eng. Mater. Technol.,* Vol 107, 1985, p 53-60

114. P.J. Maziasz, Overview of Microstructural Evolution in Neutron-Irradiated Austenitic Stainless Steels, *J. Nuclear Mater.,* Vol 205, 1993, p 118-145

115. P.J. Maziasz, Temperature Dependence of the Dislocation Microstructure of PCA Austenitic Stainless Steel Irradiated in ORR Spectrally-Tailored Experiments, *J. Nuclear Mater.,* Vol 191-194, 1992, p 701-705

116. P.J. Maziasz and C.J. McHargue, Microstructural Evolution in Annealed Austenitic Steels during Neutron Irradiation, *Internat. Mater. Rev.,* Vol 32, 1987, p 190-219

117. F.A. Garner, Irradiation Performance of Cladding and Structural Steels in Liquid Metal Reactors, *Materials Science and Technology,* Vol 10A, VCH Publishers Inc., 1994, p 419-543

118. E.E. Bloom, Irradiation Strengthening and Embrittlement, *Radiation Damage in Metals,* American Society for Metals, 1976, p 295-329

119. M.L. Grossbeck, P.J. Maziasz, and A.F. Rowcliffe, Modeling of Strengthening Mechanisms in Irradiated Fusion Reactor First Wall Alloys, *J. Nuclear Mater.,* Vol 191-194, 1992, p 808-812

120. P.J. Maziasz, Microstructural Stability and Control for Improved Irradiation Resistance and for High-Temperature Strength of Austenitic Stainless Steels, *MiCon 86: Optimization of Processing, Properties, and Service Performance through Microstructural Control,* STP 979, ASTM, 1988, p 116-161

121. T.M. Williams, Precipitation in Irradiated and Unirradiated Austenitic Steels, *Effects of Radiation on Materials: 11th Conference,* STP 782, ASTM, 1982, p 166-185

122. R.L. Fish, J.L. Staalsund, J.L. Hunter, and J.J. Holmes, Swelling and Tensile Property Evaluations of High-Fluence EBR-II Thimbles, *Effects of Radiation on Substructure and Mechanical Properties of Metals and Alloys,* STP 529, ASTM, 1973, p 149-164

123. M.L. Grossbeck, J.O. Stiegler, and J.J. Holmes, Effects of Irradiation on the Fracture Behavior of Austenitic Stainless Steels, *Radiation Effects in Breeder Reactor Structural Materials,* AIME, 1977, p 95-116

124. D.G. Doran, Neutron Displacement Cross Sections for Stainless Steel and Tantalum Based on a Lindhard Model, *Nuclear Sci. Eng.,* Vol 49, 1972, p 130-144

125. C. Picker, A.L. Stott, and H. Cocks, Effects of Low Dose Fast Neutron Irradiation on the Fracture Toughness of Type 316 Stainless Steel and Weld Metal, *Proc. IAEA Specialists Meeting on Mechanical Properties of Fast Reactor Structural Materials,* Paper IWGFR 49/440-4, International Atomic Energy Agency, 1983

126. L.D. Blackburn, A.L. Ward, and J.M. Steichen, Ductility of Irradiated Type 304 and 316 Stainless Steels, *Radiation Effects in Breeder Reactor Structural Materials,* AIME, 1977, p 317-326

127. J.A. Begley and J.D. Landes, The J Integral as a Fracture Criterion, *Fracture Toughness,* STP 514, ASTM, 1974, p 1-20

128. F.M. Haggag and A.K. Richardson, Precracking and Computerized Single-Specimen J_{IC} Determination for Irradiated Three-Point Bend Specimens, *Fracture Mechanics: 18th Symposium,* STP 945, ASTM, 1988, p 390-404

129. E.A. Little, The Fracture Toughness of Fast Reactor Irradiated Type 321 Stainless Steel and Nimonic PE16, *Dimensional Stability and Mechanical Behaviour of Irradiated Metals and Alloys,* British Nuclear Energy Society, London, 1983, p 139-142

130. E.A. Little, Dynamic J-Integral Toughness and Fractographic Studies of Fast Reactor Irradiated Type 321 Stainless Steel, *Effects of Radiation on Materials: 12th International Symposium,* STP 870, ASTM, 1985, p 563-579

131. C.W. Hunter, R.L. Fish, and J.J. Holmes, Channel Fracture in Irradiated EBR-II Type 304 Stainless Steel, *Trans. Am. Nuclear Soc.,* Vol 15, 1972, p 254-255

132. F.H. Huang, *Fracture Properties of Irradiated Alloys,* Avante Publishing, Richland, WA, 1995, p 219-246

133. F.H. Huang, Irradiation Effects on the Fracture Toughness of 20% Cold Worked Type 316 Stainless Steel, *Dimensional Stability and Mechanical Behaviour of Irradiated Metals and Alloys,* Vol 1, British Nuclear Energy Society, London, 1983, p 135-138

134. F.H. Huang, The Fracture Characterization of Highly Irradiated Type 316 Stainless Steel, *Internat. J. Fract.,* Vol 25, 1984, p 181-193

135. M.L. Hamilton, F.H. Huang, W.J.S. Yang, and F.A. Garner, Mechanical Properties and Fracture Behavior of 20% Cold-Worked 316 Stainless Steel Irradiated to Very High Neutron Exposures, *Influence of Radiation in Material Properties: 13th International Symposium (Part II),* STP 956, ASTM, 1987, p 245-270

136. D.J. Michel and R.A. Gray, Effects of Irradiation on the Fracture Toughness of FBR Structural Materials, *J. Nuclear Mater.,* Vol 148, 1987, p 194-203

137. J.W. Chan, J. Glazer, Z. Mei, and J.W. Morris, Jr., 4.2K Fracture Toughness of 304 Stainless Steel in a Magnetic Field, *Adv. Cryogenic Eng. (Mater.),* Vol 36, 1989, p 1299-1306

138. J.W. Chan, D. Chu, A.J. Sunwoo, and J.W. Morris, Jr., Metastable Austenites in Cryogenic High Magnetic Field Environments, *Adv. Cryogenic Eng. (Mater.),* Vol 38, 1992, p 55-58

139. H. Krauth and A. Nyilas, Fracture Toughness of Nitrogen Strengthened Austenitic Steels at 4K, *Fracture and Fatigue,* Pergamon Press, 1980, p 119-128

140. R.L. Tobler, D.T. Read, and R.P. Reed, Strength/Toughness Relationship for Interstitially Strengthened AISI 304 Stainless Steel at 4K Temperature, *Fracture Mechanics: 13th Conference,* STP 743, ASTM, 1980, p 250-268

141. R.L. Tobler and R.P. Reed, Tensile and Fracture Properties of Manganese-Modified AISI 304 Type Stainless Steel, *Adv. Cryogenic Eng. (Mater.),* Vol 28, 1982, p 83-92

142. R.L. Tobler and R.P. Reed, Interstitial Carbon and Nitrogen Effects on the Tensile and Fracture Parameters of AISI 304 Stainless Steel, *Materials Studies for Magnetic Fusion Energy Applications at Low Temperatures—III,* NBSIR 80-1627, National Bureau of Standards, 1980, p 17-48

143. I.S. Hwang, M.M. Morra, R.G. Ballinger, H. Nakajima, S. Shimamoto, and R.L. Tobler, Charpy Absorbed Energy and J_{IC} as Measures of Cryogenic Fracture Toughness, *J. Test. Eval.,* Vol 20, 1992, p 248-258

144. R.P. Reed, N.J. Simon, P.T. Purtscher, and R.L. Tobler, Alloy 316LN for Low Temperature Structures: A Summary of Tensile and Fracture Data, *Materials Studies for Magnetic Fusion Energy Applications at Low Temperatures—IX,* NBSIR 89-3050, National Bureau of Standards, 1986, p 15-26

145. M. Shimada, Fatigue Crack Growth at 4 K of Aged Austenitic Stainless Steels, *Adv. Cryogenic Eng. (Mater.),* Vol 36, 1990, p 1217-1224

146. D.T. Read and R.P. Reed, Fracture and Strength Properties of Selected Austenitic Stainless Steels at Cryogenic Temperatures, *Cryogenics,* 1981, p 415-417

147. T. Ogata, K. Ishikawa, T. Yuri, R.L. Tobler, P.T. Purtscher, R.P. Reed, T. Shoji, K. Nakano, and H. Takahashi, Effects of Specimen Size, Side-Grooving, and Precracking Temperatures on J-Integral Test Results for AISI 316LN at 4 K, *Adv. Cryogenic Eng. (Mater.),* Vol 34, 1988, p 259-266

148. T. Ogata, K. Nagai, K. Ishikawa, K. Shibata, and E. Fukushima, VAMAS Interlaboratory Fracture Toughness Test at Liquid Helium Temperature, *Adv. Cryogenic Eng. (Mater.),* Vol 36, 1990, p 1053-1060

149. T. Ogata, K. Nagai, and K. Ishikawa, VAMAS Tests of Structural Materials at Liquid Helium

Temperature, *Adv. Cryogenic Eng. (Mater.)*, Vol 40, 1994, p 1191-1198

150. E.S. Drexler, N.J. Simon, and R.P. Reed, Strength and Toughness at 4K of Forged, Heavy-Section 316LN, *Adv. Cryogenic Eng. (Mater.)*, Vol 40, 1994, p 1199-1206

151. W.J. Muster and J. Elster, Low Temperature Embrittlement After Ageing Stainless Steels, *Cryogenics*, Vol 30, 1990, p 799-802

152. A. Nyilas and H. Yanagi, 4K Fracture Toughness Investigations of 316 LN Stainless Steel Plate and Forging Materials, *Cryogenics*, Vol 29, 1989, p 191-195

153. P.T. Purtscher, R.P. Walsh, and R.P. Reed, Effect of Chemical Composition on the 4 K Mechanical Properties of 316LN-Type Alloys, *Adv. Cryogenic Eng. (Mater.)*, Vol 34, 1988, p 191-198

154. T.A. Siewert, D. Gorni, and G. Kohn, High-Energy-Beam Welding of Type 316LN Stainless Steel for Cryogenic Applications, *Adv. Cryogenic Eng. (Mater.)*, Vol 34, 1988, p 343-350

155. M. Shimada and S. Tone, Effects of Niobium on Cryogenic Mechanical Properties of Aged Stainless Steels, *Adv. Cryogenic Eng. (Mater.)*, Vol 34, 1988, p 131-139

156. W.A. Logsdon, J.M. Wells, and R. Kossowsky, Fracture Mechanics Properties of Austenitic Stainless Steels for Advanced Cryogenic Applications, Paper 75-9D4-CRYMT-P3, Westinghouse Research Laboratories, Pittsburgh, PA, Dec 1975

157. R.L. Tobler, Ductile Fracture with Serrations in AISI 310S Stainless Steel at Liquid Helium Temperatures, *Fracture Resistance Curves and Engineering Applications*, Vol II, *Elastic-Plastic Fracture: Second Symposium*, STP 803, ASTM, 1983, p II-763 to II-776

158. M. Shimada, R.L. Tobler, T. Shoji, and H. Takahashi, Size, Side-Grooving, and Fatigue Precracking Effects on J-Integral Test Results for SUS 304 Stainless Steel at 4 K, *Adv. Cryogenic Eng. (Mater.)*, Vol 34, 1988, p 251-257

159. N.J. Simon and R.P. Reed, Strength and Toughness of AISI 304 and 316 at 4 K, *Materials Studies for Magnetic Fusion Energy Applications at Low Temperatures—IX*, NBSIR 89-3050, National Bureau of Standards, 1986, p 27-42

160. P.T. Purtscher and R.P Reed, The Toughness of Austenitic Stainless Steels at 4 K, *Materials Studies for Magnetic Fusion Energy Applications at Low Temperatures—IX*, NBSIR 89-3050, National Bureau of Standards, 1986, p 53-76

161. K. Yoshida, H. Nakajima, M. Oshikiri, R.L. Tobler, S. Shimamoto, R. Miura, and J. Ishizaka, Mechanical Tests of Large Specimens at

4 K: Facilities and Results, *Adv. Cryogenic Eng. (Mater.)*, Vol 34, 1988, p 225-232

162. R.P. Reed, P.T. Purtscher, and L.A. Delgado, Low-Temperature Properties of High-Manganese Austenitic Steels, *High Manganese High Nitrogen Austenitic Steels*, ASM International, 1993, p 13-22

163. R.L. Tobler and M. Shimada, Warm Precracking at 295K and Its Effect on the 4-K Toughness of Austenitic Steels, *J. Test. Eval.*, Vol 19, 1991, p 312-320

164. H. Nakajima, K. Yoshida, S. Shimamoto, R.L. Tobler, P.T. Purtscher, and R.P. Reed, Round Robin Tensile and Fracture Test Results for an Fe-22Mn-13Cr-5Ni Austenitic Stainless Steel at 4K, *Adv. Cryogenic Eng. (Mater.)*, Vol 34, 1988, p 241-249

165. S. Tone, M. Shimada, T. Horiuchi, Y. Kasamatsu, H. Nakajima, and S. Shimamoto, The Development of a Nitrogen-Strengthened High-Manganese Austenitic Stainless Steel for a Large Superconducting Magnet, *Adv. Cryogenic Eng. (Mater.)*, Vol 30, 1984, p 145-152

166. S. Tone, M. Hiromatsu, J. Numata, T. Horiuchi, H. Nakajima, and S. Shimamoto, Cryogenic Properties of Electron-Beam Welded Joints in a 22Mn-13Cr-5Ni-0.22N Austenitic Stainless Steel, *Adv. Cryogenic Eng. (Mater.)*, Vol 32, 1986, p 89-96

167. R. Ogawa and J.W. Morris, Jr., The Influence of Processing on the Cryogenic Mechanical Properties of High Strength High Manganese Stainless Steel, *Adv. Cryogenic Eng. (Mater.)*, Vol 30, 1984, p 177-184

168. R.L. Tobler and D. Meyn, Cleavage-Like Fracture along Slip Planes in Fe-18Cr-3Ni-3Mn-0.37N Austenitic Stainless Steel at Liquid Helium Temperature, *Metall. Trans.*, Vol 19A, 1988, p 1626-1631

169. H. Takahashi, T. Shoji, and R.L. Tobler, Acoustic Emission and Its Application to Fracture Studies of Austenitic Stainless Steels at 4 K, *Adv. Cryogenic Eng. (Mater.)*, Vol 34, 1988, p 387-395

170. N. Yamagami, Y. Kohsaka, and C. Ouchi, Mechanical Properties of Welded Joints in Mn-Cr and Ni-Cr Stainless Steels at 4 K, *Adv. Cryogenic Eng. (Mater.)*, Vol 34, 1988, p 359-366

171. P.T. Purtscher, M.C. Mataya, L.M. Ma, and R.P. Reed, Effect of Processing of 4-K Mechanical Properties of a Microalloyed Austenitic Stainless Steel, *Adv. Cryogenic Eng. (Mater.)*, Vol 36, 1990, p 1291-1298

172. Z.S. Basinski, *Proc. Royal Soc.*, Vol 240, 1957, p 229-242

173. K. Nohara and Y. Habu, Strengthened High Manganese Non-Magnetic Steel: Application to Superconducting Magnet, *High Manganese*

High Nitrogen Austenitic Steels, ASM International, 1992, p 33-42

174. H.I. McHenry, Properties of Austenitic Stainless Steel at Cryogenic Temperatures, *Materials Studies for Magnetic Fusion Energy Applications at Low Temperatures—VI*, NBSIR 83-1690, National Bureau of Standards, 1983, p 127-155

175. S.D. Washko and G. Aggen, Wrought Stainless Steels, *ASM Handbook*, Vol 1, ASM International, 1990, p 841-907

176. M. Rouby and P. Blanchard, Physical and Mechanical Properties of Stainless Steels and Alloys, *Stainless Steels*, Les Editions de Physique, 1993, p 111-158

177. D.T. Read, H.I. McHenry, P.A. Steinmeyer, and R.D. Thomas, Jr., Metallurgical Factors Affecting the Toughness of 316L SMA Weldments at Cryogenic Temperatures, *Weld. J.*, Vol 59, 1980, p 104s-113s

178. J.W. Elmer, H.I. McHenry, and T.A. Whipple, Strength and Toughness of Fully Austenitic Stainless Steel Filler Metals at 4K, *Materials Studies for Magnetic Fusion Energy Applications at Low Temperatures—IV*, NBSIR 81-1645, National Bureau of Standards, 1981, p 289-302

179. C.N. McCowan, T.A. Siewert, and R.L. Tobler, Tensile and Fracture Properties of an Fe-18Cr-20Ni-5Mn-0.16N Fully Austenitic Weld Metal at 4 K, *J. Eng. Mater. Technol.*, Vol 108, 1986, p 340-343

180. R.L. Tobler, R.E. Trevisan, and R.P. Reed, Tensile and Fracture Properties of an Fe-14Mn-8Ni-1Mo-0.7C Fully Austenitic Weld Metal at 4 K, *Cryogenics*, Vol 25, 1985, p 447-451

181. R.L. Tobler, R.E. Trevisan, T.A. Siewert, H.I. McHenry, P.T. Purtscher, C.N. McCowan, and T. Matsumoto, Strength, Fatigue, and Toughness Properties of an Fe-18Cr-16Ni-6Mn-2Mo Fully Austenitic SMA Weld at 4K, *Adv. Cryogenic Eng. (Mater.)*, Vol 34, 1988, p 351-358

182. R.L. Tobler, R.P. Reed, I.S. Hwang, M.M. Morra, R.G. Ballinger, H. Nakajima, and S. Shimamoto, Charpy Impact Tests Near Absolute Zero, *J. Test. Eval.*, Vol 19, 1991, p 34-40

183. R.L. Tobler, A. Bussiba, J.F. Guzzo, and I.S. Hwang, Charpy Specimen Tests at 4 K, *Adv. Cryogenic Eng. (Mater.)*, Vol 38, 1992, p 217-223

184. H. Nakajima, K. Yoshida, H. Tsuji, R.L. Tobler, I.S. Hwang, M.M. Morra, and R.G. Ballinger, The Charpy Impact Test as an Evaluation of 4 K Fracture Toughness, *Adv. Cryogenic Eng. (Mater.)*, Vol 38, 1992, p 207-215

Fatigue and Fracture Properties of Duplex Stainless Steels

R. Johansson, Avesta Sheffield AB

WROUGHT DUPLEX STAINLESS STEELS has been used for more than 60 years (Ref 1). Over the years much effort has been devoted to production and welding metallurgy, as well as corrosion research of the duplex stainless steels. Therefore, duplex stainless steels are today established in a wide product range. In most cases duplex stainless steels are selected because they combine high strength and excellent corrosion resistance.

Typical chemical compositions of wrought duplex stainless steels are given in Table 1 (Ref 2), where a general description of the properties of duplex stainless steels is also given. The welding metallurgy of the duplex stainless steels is reviewed in Ref 3. During production and use, many different phases can be present in the duplex stainless steels due to high levels of chromium, molybdenum, and nitrogen. An overview of the metallurgy of the duplex stainless steels has been made in Ref 4.

Fatigue Crack Initiation

Fatigue is usually initiated at inclusions, at surface scratches, or at persistent slip bands emerging at the surface. In a corrosive environment the fatigue crack initiation is also influenced by slip bands breaking the thin passive layer at the surface and by corrosion pits acting as notches on the surface.

Crack Initiation from Local Strains. The influence of local strains on the initiation of fatigue cracks in duplex stainless steels has been demonstrated by low-cycle fatigue tests (Ref 5-7). At low strains, below 10^{-3}, the deformation in the austenitic phase controls the fatigue properties of the duplex alloy 22Cr-7Ni-2.5Mo-1.7Mn-0.07N, whereas at higher strains the deformation in the ferritic phase controls the fatigue properties. Cracks nucleate only in the austenitic phase at low strains. Because of the large reversibility of slip in the planar slip mode in the austenitic phase, good fatigue resistance of the duplex alloy is observed. Crack nucleation is delayed and takes place in the austenitic phase. At higher

strains, twinning and pencil glide promote early crack initiation in the ferritic phase and thus reduce the fatigue life of the duplex alloy.

The fatigue crack initiation in duplex stainless steels can thus occur first in either the ferritic phase or the austenitic phase. In Ref 8 it is shown on polished specimens of 2205 with 0.11% nitrogen that the crack initiation starts at persistent slip bands in the ferrite. The persistent slip bands were generally more numerous in the ferritic grains. Slip markings in the austenitic grains were less intensive, and only a few cracks were initiated in the austenite.

Higher nitrogen content in duplex stainless steels improves fatigue resistance, as highlighted in Ref 9 and 10 for duplex 2205 with 0.11 and

0.18% nitrogen. Nitrogen is dissolved mostly in the austenite, thus hardening the austenite and leading to an increase in the monotonic and cyclic stress-strain response of the steel. As a consequence the fatigue strength will be increased with increased nitrogen content. In Fig. 1 there is a comparison with the results from a duplex 2205 with a nitrogen content of 0.07%, denoted AF7 (Ref 9). The steel with the lowest nitrogen content has the shortest strain life at low strains.

Depending on strain levels and nitrogen content, crack initiation takes place in both the ferritic and austenitic phases, or only mainly in the ferritic phase. In both tested steels (nitrogen contents of 0.11 and 0.18%) the ferritic phase is the softest, while the hardness of the austenitic phase

Table 1   Typical chemical composition of some wrought duplex stainless steels

UNS number W. number AFNOR number SIS number	Trademarks	Typical chemical composition					PREN(a)
		Cr	Ni	Mo	N	Other	
Mo-free grades(b)							
UNS S32304 WNr1.4362 Z3CN 2304AZ	SAF 2304 UR 35 N	23	4	0.2	0.1	...	25
Standard grade(c)							
UNS S31803 WNr 1.4462 NFA 36209 Z3CND2205AZ	UR45N SAF 2205 AVESTA 2205 REMANIT 4462	22	5	2.7	0.14	...	33
	UR 45N+	23	6.2	3.2	0.19		36
Superduplex(d)							
UNS S32760 UNS S32550 Z3 CNDU 2506 UNS S32750	ZERON 100 UR 52 N+ FERRALIUM SD40 SAF 2507 URANUS 47N+	25 25 25	7 6.5 7	3.6 3.8 3.8	0.24 0.26 0.27	0.7 Cu, 0.7 W 1.5 Cu	41 41 41

(a) Pitting resistance equivalent number, $Cr + 3.3Mo + 16N$. The superduplex grades are characterized by a value of PREN > 40. (b) The 23Cr-4Ni-0.10N, molybdenum-free grade can be used to replace the austenitic grades AISI 304 and/or 316. (c) The composition 22Cr-5Ni-3Mo-0.17N can be considered the standard duplex stainless steel. Its nitrogen content has recently been increased to further improve its corrosion resistance in oxidizing chloride-rich acid media (essentially an increase in pitting resistance). Its corrosion resistance lies between those of the austenitic grade AISI 316 and the 4-5% Mo super austenitic alloys (904L). The chemical analysis of the grade can be optimized in order to obtain a PREN value between 33 and 36. (d) The superduplex grade with a pitting index PREN > 40 (with or without copper and/or tungsten additions) is especially designed for marine, chemical, and oil engineering applications requiring both high mechanical strength and resistance to corrosion in extremely aggressive environments (chloride-containing acids, etc.). The corrosion resistance is equivalent to that of the super austenitic steels containing 5-6% Mo. Source: Ref 2

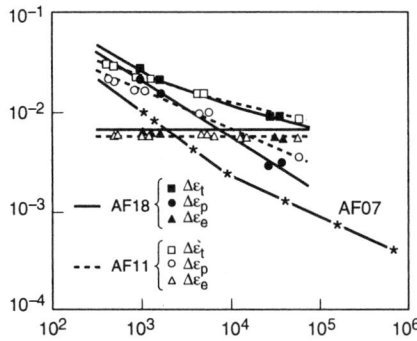

Fig. 1 Fatigue strength curves for duplex stainless steels AF11 and AF18, and Manson-Coffin curve for steel AF07. Source: Ref 9

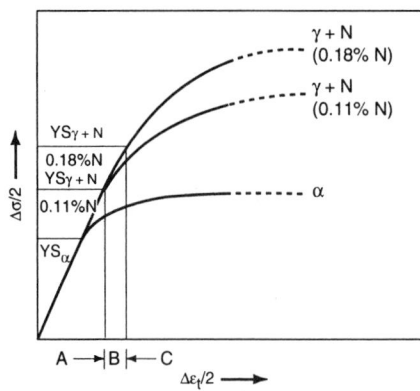

Fig. 2 Schematic interpretation of the effect of nitrogen on the low-cycle fatigue behavior of duplex stainless steel

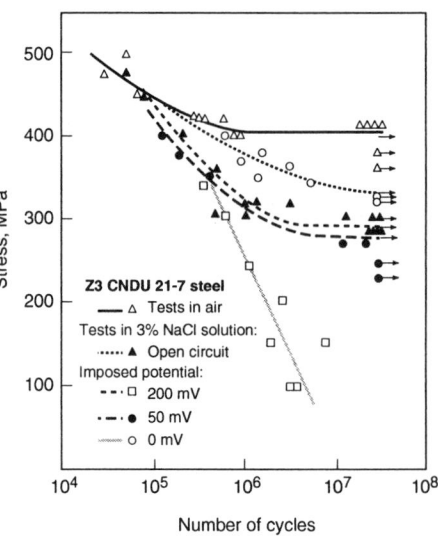

Fig. 3 Rotating bending fatigue strength of duplex stainless steel in salt solution

increases with the nitrogen content. The lower strength of the ferritic phase can result in a concentration of the deformation to the ferrite. This effect can reduce the positive effect of nitrogen at low plastic strains, and the fatigue life might be reduced.

In Ref 9 it is shown that at a plastic strain $\Delta\varepsilon_t = 0.8 \times 10^{-2}$, the slip lines and the crack initiation at the surface appear mostly in the ferritic phase. At $\Delta\varepsilon_t = 1.4 \times 10^{-2}$, slip lines appear in the 0.11% nitrogen steel mainly in the ferritic phase, but in the 0.18% nitrogen steel only in the ferritic phase. At $\Delta\varepsilon_t = 2 \times 10^{-2}$, the slip lines in the low-nitrogen steel are observed in both ferrite and austenite, while in the higher-nitrogen steel only a few slip lines are observed in the austenite. This behavior has been explained by the relative hardness of the two phases (Fig. 2). In both steels the ferritic phase is the softest. Depending on stress levels and nitrogen content, ferrite and austenite will to a certain extent take part in the deformation.

Crack Initiation with Corrosion Fatigue. Corrosion fatigue can on an electrochemical basis be categorized into four types (Ref 11):

- *Corrosion fatigue in the active state:* This type is characterized by corrosion throughout the entire fatigue life of the metal. Many cracks are formed, which frequently start from the bottom of corrosion pits. The fracture surface is rough and full of fissures, and the cracks are filled with corrosion products.
- *Corrosion fatigue in the passive state:* There are no pits and just a few cracks, which makes this type difficult to distinguish from conventional fatigue.
- *Corrosion fatigue under unstable passivity:* This type of corrosion fatigue occurs when the metal is passive in the beginning and becomes activated after a certain number of cycles by protruding glide steps or extrusions.
- *Corrosion fatigue under disturbed passivity:* Typical examples of this type include the superposition of corrosion fatigue and stress-corrosion cracking, pitting, or intergranular corrosion.

A stainless steel, resistant to corrosion fatigue cracking, should therefore have good resistance to different forms of localized corrosion such as stress-corrosion cracking, pitting, and intergranular corrosion. Many duplex stainless steels have a high resistance to local corrosion and therefore also a high resistance to corrosion fatigue.

An investigation of the corrosion fatigue mechanisms of a 22Cr-5Ni-3Mo-0.19N duplex stainless steel (Ref 12, 13) showed that at high stress levels the crack initiation occurred mainly at persistent slip bands in the austenite or near the phase boundaries. At lower stress levels, the cracks initiated mainly at nonmetallic inclusions in the surface region. The fatigue limit in air was 240 MPa (34 ksi). When tested in concentrated chloride solutions at 80 °C (175 °F), the fatigue limit was decreased only to 200 MPa (29 ksi). Under these stable, passive corrosion conditions the duplex steel had a pronounced fatigue limit. When the temperature was raised to 150 °C the corrosion fatigue strength was reduced significantly due to superposition of pitting corrosion.

The corrosion fatigue properties of the duplex alloy in a 3.5% NaCl solution also depends on whether the strain is localized to the austenitic phase or to the ferritic phase (Ref 7). At low stresses the austenitic phase is cathodically protected by the undeformed ferritic phase, and the risk of corrosion fatigue damage is reduced. At higher strains under a free corrosion potential, the anodic dissolution is localized at twins and at ferrite grain boundaries, causing a premature fatigue failure. If a cathodic potential is imposed on the specimen, the heterogeneous deformation in the ferritic phase will promote hydrogen entry, causing hydrogen embrittlement and consequently reducing the fatigue life even more. The presence of strain concentrations encountered as an effect of reversed strains is one reason why corrosion fatigue behavior sometimes differs from the stress-corrosion cracking behavior.

The corrosion fatigue crack initiation and propagation of a duplex stainless steel under different potentials was compared to that of an austenitic steel and a ferritic steel (Ref 14). The crack initiation was tested with a smooth specimen under rotating bending stresses, and crack propagation tests were performed on standard precracked compact tension specimens. Tests

in air were compared to tests in an aerated 3% solution of NaCl at pH = 6. In air, the fatigue limit for the duplex 21Cr-7Ni-2.5Mo steel was higher than that for the austenitic type 316 steel and the ferritic 26Cr-1Mo steel tested. When tested in sodium chloride, the fatigue limit for the ferritic steel was higher than that of the other two steels because of the higher chromium content of the ferritic steel, giving either a tough passive film or at least a passive film that would recover its passivity very rapidly if damaged.

During the corrosion fatigue test, the variation in either specimen potential was measured (Ref 14). The free potential for the duplex steel depended on the stress state. In one test it started at −100 mV, then rose to +100 mV before failure, a mixed potential as a result of the anodic and cathodic processes taking place during the test. The potential decreased more rapidly at higher stresses due to either the increase of the rate of anodic dissolution or the increase of anodic areas. The current potential curve for the duplex steel shows that the metal is passive below +100 mV and that pits are not repassivated above +100 mV. Corrosion fatigue tests under rotating bending stress and different imposed potentials in sodium chloride (Fig. 3) showed that the fatigue limit was drastically reduced when the imposed potential was +200 mV (Ref 14). Another result from the test is also shown in Fig. 3, where the fatigue limit at a potential of 0 mV is higher than at a free potential.

In a corrosion fatigue study for three different stainless steels (Ref 15), the time to initiation of the fatigue crack was longer for the duplex steel (22Cr-5Ni-3Mo-0.14N) than for a ferritic or an austenitic stainless steel. The test was performed in an aqueous sulfuric acid (0.05M or 2M) at 30 °C (86 °F) in the passive range under potentiostatically controlled potential. The main reasons for the longer time to crack initiation in the duplex steel were the thicker passive layer and the

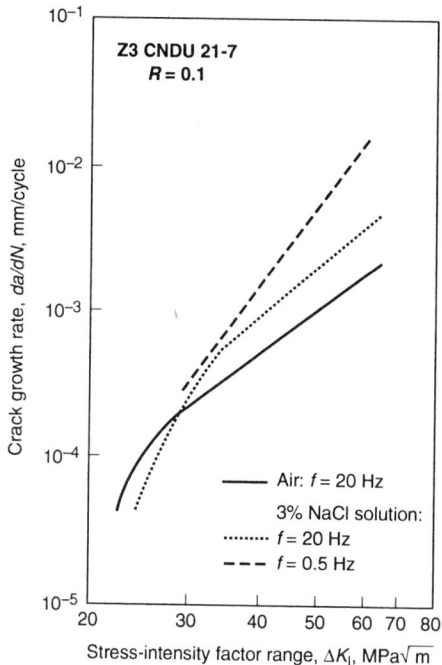

Fig. 4 Crack growth rates of AFNOR Z3 CNDU 21-7. Source: Ref 21

Fatigue Crack Growth

Fatigue crack growth is in most investigations measured on compact tension specimens or three-point bend specimens according to a standard such as the ASTM standard. One early investigation in Ref 21 showed that at high stress-intensity factors (and at $R = 0.1$), the crack propagation rate in a 3% NaCl solution is higher than in air. When the testing frequency was decreased from 20 to 0.5 Hz, the fatigue crack propagation rate increased further (Fig. 4). A similar result was found for a ferritic steel, while an austenitic steel was not influenced. The ferritic and the duplex steels have an extremely passive film. When this film is destroyed by a fatigue crack, an acceleration of the metal dissolution rate is observed, which leads to an increase in the rate of fatigue crack propagation. The austenitic steel, however, has a lower passivity, which facilitates the initiation of a corrosion fatigue crack, particularly at corrosion pits. Once a crack has initiated, the repassivation of the metal or a low dissolution rate leads to propagation rates close to the rates obtained in air.

The corrosion fatigue crack growth rate was measured in Ref 22 on 25Cr-6Ni-2.4Mo duplex stainless steel at a frequency of 0.17 Hz ($R = 0.1$), with the specimens protected by a cathodic potential in synthetic sea water. The propagation rates in the test with the cathodic polarization were much higher than those in the test with natural corrosion potential, because hydrogen embrittlement accelerates the propagation.

In a test program with several participating organizations (Ref 23), duplex stainless steels were tested in air and in ASTM synthetic sea water. The crack propagation rate for cast and forged materials from a 25Cr-6Ni-3Mo alloy in synthetic sea water ($R = 0.6$, frequency 10 Hz) were almost identical at 10^{-5} mm/cycle with $K = 10$ MPa$\sqrt{m}$. However, the fatigue limit under rotating bending stresses in 25 °C synthetic sea water at 30 Hz and 10^7 cycles was 350 MPa (50 ksi) for the forged specimens and roughly 200 MPa (29 ksi) for the cast specimens.

The fatigue crack growth of a duplex stainless steel is affected by the deformation status and the presence of brittle and hard phases. In Ref 24 it is shown that the fatigue crack growth threshold, ΔK_{th}, increases after 50% cold rolling and even more after 50% cold rolling followed by an annealing at 475 °C (885 °F) for 120 h. The fatigue crack growth, da/dN, as a function of the stress-intensity amplitude ΔK is shown in Fig. 5. The annealing produces a hardening through spinodal

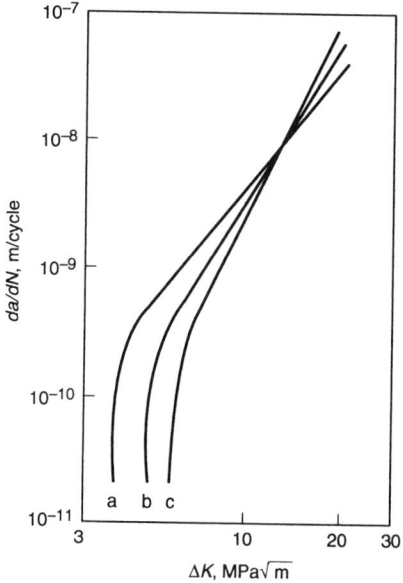

Fig. 5 da/dN vs. ΔK for the different conditions tested: (a) Hot-rolled, as-received. (b) Cold-rolled (50%). (c) Cold-rolled (50%) and annealed (120 h, 475 °C). Source: Ref 24

decomposition of the ferrite, thus increasing the threshold value.

One further way of demonstrating the crack propagation rate in duplex stainless steels is to compare the behavior of the duplex alloy with its tie-line single-phase ferritic and austenitic alloys, as has been done in Ref 25. The materials tested were laboratory melts of the duplex alloy 2205 and its tie-line single-phase alloys. The material was tested in its hot-rolled/quenched-and-annealed condition and after a tensile deformation of 8%. Fatigue crack growth data are shown in Fig. 6 and in Table 2. During fatigue crack growth testing the crack is closed during a part of the loading cycle, causing the effective ΔK to be less than the nominal ΔK. The crack closure, K_{cl}, can be induced by the crack surface roughness, oxides, or plasticity. In Table 2, $\Delta K_{th} = \Delta K_{th,eff} + K_{cl}$. H is a surface roughness parameter measured using quantitative fractography (Ref 26). From Fig. 6 and Table 2, it is evident that the threshold values are decreased after the tensile prestraining. The crack closure values are decreased for the single-phase alloys but not for the duplex alloy. The fracture surface roughness of the ferritic and duplex alloys is not affected by the prestraining. However, there is a smoother fatigue crack surface in the austenitic alloy after prestraining. This means that there is no simple relation between the

higher Cr-Fe relation in the passive layer as compared to the other steels tested.

Suggested Models of Corrosion Fatigue. A survey of suggested mechanisms for corrosion fatigue initiation and propagation for steels in general is made by Austen in Ref 16. These suggested mechanisms have in some cases led to the formulation of models for corrosion fatigue and computer programs for life prediction. However, the complete modeling of corrosion fatigue, all the way from initiation through notches and short cracks to the crack propagation of long cracks, still remains to be done.

One model for crack initiation (Ref 17, 18) is based on the film rupture mechanism (e.g., Ref 19, 20). The model assumes slip step dissolution and repassivation, and it incorporates mechanical and electrochemical factors. The model does not include general active dissolution, except at slip steps that break through the passive film. Pitting is also neglected. Included in the model are the current decay curve, the bare metal corrosion rate, and the critical slip step height. One example is shown where the calculated fatigue curves are in good agreement with experimental S-N data for an austenitic stainless steel. Duplex stainless steels have so far not been included in the experiments.

An attempt has been made, however (Ref 18) to express the critical penetration depth at the surface in terms of fracture mechanics, which could be used when looking at fatigue mechanisms for duplex steels. The fatigue crack growth threshold of small cracks is described by using the ΔK concept. A relation is obtained between the critical penetration depth, the critical crack depth, the applied stress, and the fatigue limit of the steel.

Table 2 Fatigue crack growth data ($R = 0$)

Grade and composition	ε	ΔK_{th}, MPa$\sqrt{m}$	$\Delta K_{th,eff}$, MPa$\sqrt{m}$	K_{cl} MPa$\sqrt{m}$	$\overline{H}$, μm
Duplex (0.018C, 2.21Cr, 5.8Ni, 3.0Mo, 0.17N)	0	4.6	3.3	1.3	2.6
	8%	3.1	1.9	1.2	3.2
Ferrite (0.005C, 23.3Cr, 4.4Ni, 3.8Mo, 0.005N)	0	4.8	2.4	2.4	5.9
	8%	3.3	2.2	1.1	6.7
Austenite (0.028C, 21.4Cr, 9.1Ni, 2.7Mo, 0.30N)	0	4.2	2.5	1.7	5.0
	8%	3.2	2.1	1.1	3.3

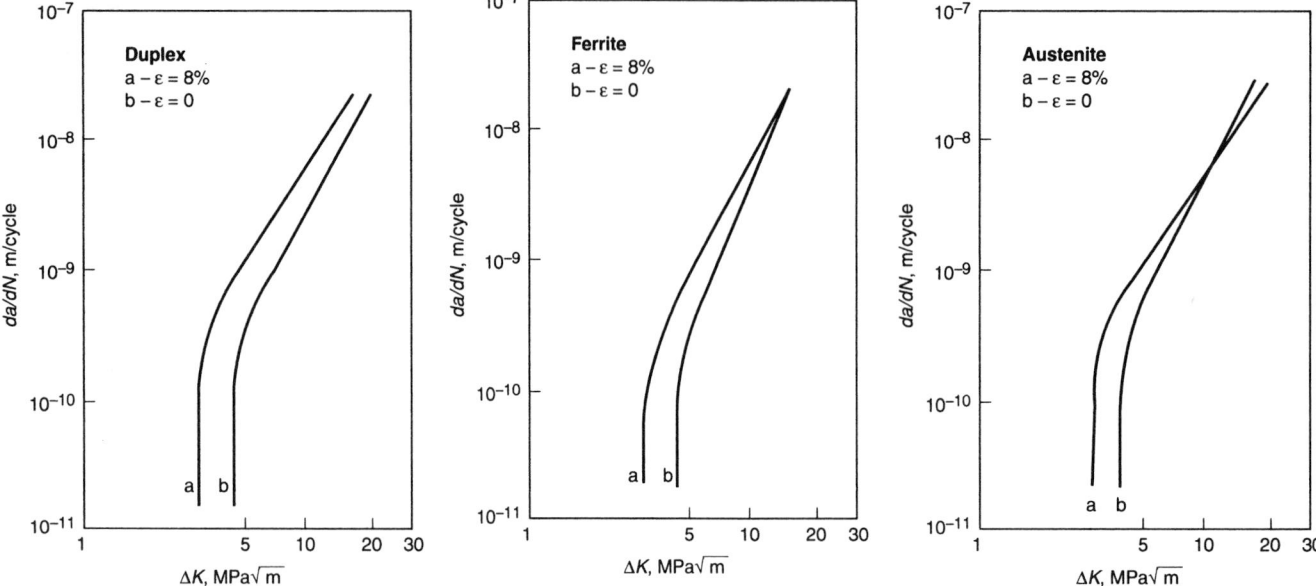

Fig. 6 da/dN vs. ΔK for duplex, ferritic, and austenitic steels in undeformed and prestrained conditions. Average curves for each state. Source: Ref 25

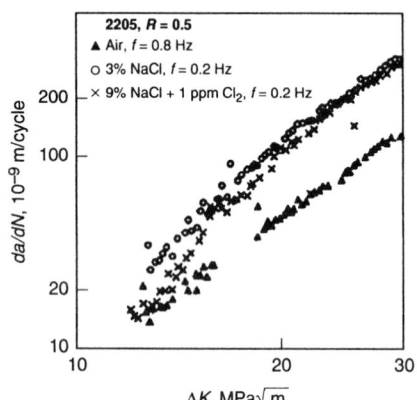

Fig. 7 Crack growth rate for 2205 in air, 3% NaCl, and 3% NaCl with 1 ppm chlorine. Frequency 0.8 Hz in air and 0.2 Hz in the aqueous environments, R = 0.5. Results with chlorine additions are from single tests only. Source: Ref 27

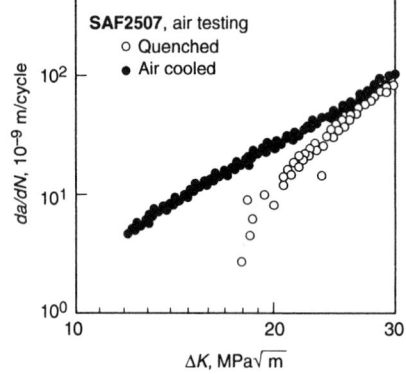

Fig. 8 Fatigue crack propagation results at 0.8 Hz in air with R = 0.5, for SAF 2507 in quenched and air-cooled conditions. Source: Ref 28

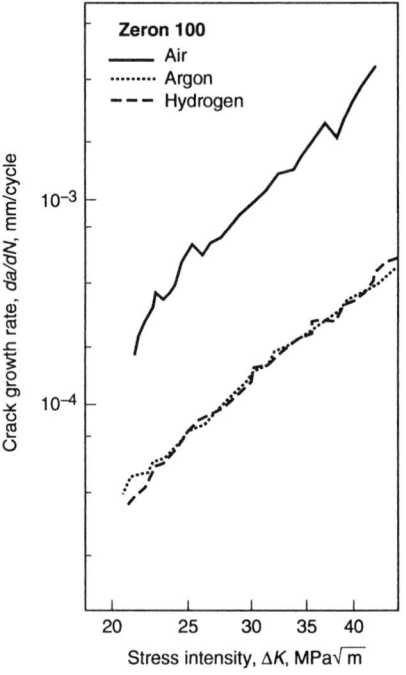

Fig. 9 Fatigue crack growth rates of a superduplex stainless steel (Zeron 100) in air, argon, and hydrogen

surface roughness and the crack closure level for any of the alloys in this investigation.

The crack growth rate of duplex alloy 2205 at room temperature in air and in a 3% NaCl solution is shown in Fig. 7 from Ref 27. In 3% NaCl the crack growth rate was about a factor of 2 greater than the crack growth in air. When the testing temperature for the NaCl test was increased to 80 °C (175 °F), no influence on the crack propagation rate was experienced (Ref 28). The same was true for the superduplex steel SAF 2507 (Ref 28). It was shown that the threshold value for SAF 2507 was higher than for the 2205 duplex steel and several austenitic and super austenitic steels. However, when air-cooled material of SAF 2507 was compared to quenched material, the threshold value for the air-cooled material was as low as for the other steels tested

(Fig. 8). This behavior was claimed to depend on differences in residual stresses.

Another superduplex stainless steel, Zeron 100, was embrittled in fatigue by gaseous hydrogen and water (Ref 29). The ferrite matrix failed by cyclic cleavage, while the austenite phase remained unaffected. The embrittled zone size was controlled by the rate of entry of hydrogen to the crack tip. The fatigue crack propagation rate was increased by a factor of 8 in hydrogen as compared to that in air (Fig. 9). All duplex steels were embrittled after annealing in the temperature range of 400 to 500 °C (750 to 930 °F), as was the superduplex steel investigated in Ref 30. After aging for 1000 h at 400 °C (750 °F), the superduplex steel underwent a fatigue crack propagation test at temperatures between 30 and 305 °C (86 and 580 °F). Figure 10 shows that there is an influence of temperature and specimen orientation on the fatigue crack growth: the higher the

testing temperature, the higher the crack propagation rate. Fatigue crack propagation in the aged superduplex steel was increased by initiation of cleavage cracks in the ferrite matrix. Further tests on thermally aged Zeron 100 (Ref 31) show that the embrittlement is caused by cleavage in the age-hardened ferrite. A model put forward suggests a strong influence of the ferrite volume fraction and predicts that heat-affected zones with large grain sizes are sensitive to enhanced fatigue crack propagation rates.

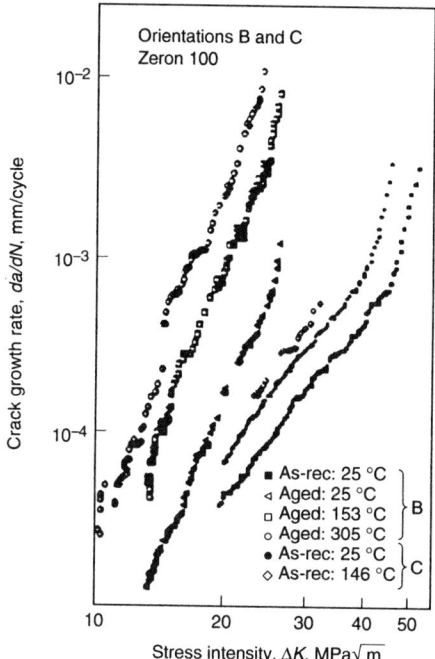

Fig. 10 Effect of temperature and aging on fatigue crack growth of a superduplex stainless steel (Zeron 100). Orientation B is crack growth in the transverse direction, and orientation C is cracking in the rolling direction. Source: Ref 30

Table 3 Fatigue strength under rotating bending stress

Grade	Environment	Yield strength, MPa	Ultimate tensile strength, MPa	Fatigue strength, MPa	Ref
2205	Air	495	730	±450	33, 34
2205	3% NaCl, pH=7	495	730	±425	33, 34
2205	3% NaCl, pH=3	495	730	±325	33, 34
2205	3% NaCl, pH=1	495	730	±180	33, 34
329	Air	538	760	±490	35
329	3% NaCl	538	760	±400	35
2205	Air	479	744	±500	35
2205	3% NaCl	479	744	±400	35
25,6,3	ASTM sea water	549	705	±350	36
2205	Air, longitudinal	494	741	±350	37
2205	Air, transversal	520	750	±325	37
2205	2M H$_2$SO$_4$	494	741	±100	37
3RE60	Distilled water	...	742	±320	38
3RE60	Sea water(a)	...	742	±305	38
3RE60	White water	...	735	±265	38
2205	White water	...	714	±265	38

(a) German Industrial Standard

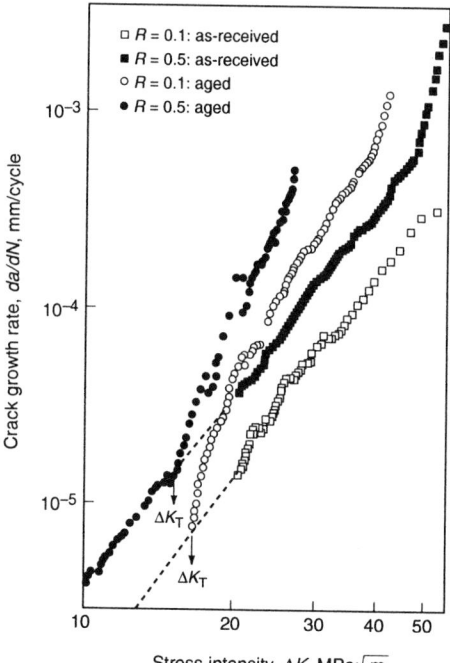

Fig. 11 Fatigue crack propagation rates as a function of stress-intensity factor range (ΔK) in Zeron 100 aged at 400 °C for up to 5000 h, tested at room temperature (22 °C) at R = 0.5 and R = 0.1. Source: Ref 31

Fatigue crack propagation rates for different aging times and testing temperatures at different R-values are shown in Fig. 11. The fatigue crack growth rate in Zeron 100 in air varied with crack

orientation in wrought structures, as expected (Ref 32). The fracture surface roughness showed little variation and was only somewhat affected by the specimen orientation. The only parameter that varied with crack orientation and crack growth was the number of austenite-ferrite grain boundaries crossed during crack propagation.

Fatigue Strength

Fatigue strength data useful for designing in duplex stainless steels are available in the literature to some extent. The data shown in the literature are mainly from tests performed in air at room temperature and in NaCl solutions, but some data are related to tests in other specific environments. The literature shows that the fatigue limit in air for the duplex stainless steels is very high and almost equal to the yield strength. In corrosive environments, the fatigue strength for the duplex steels is reduced but remains high. Fatigue tests also show that welded components of duplex stainless steels can withstand high levels of fatigue stresses. The data given in the literature are described as the fatigue limit or the fatigue strength at a specified number of cycles to failure. Nothing is said about confidence levels, but as a first approximation the data could be interpreted as the mean fatigue strength or the mean fatigue limit with a 50% probability of fracture at a low confidence level.

Fatigue Data for Rotating Bending Stresses. Duplex stainless steels usually have higher yield stress and tensile strength than the standard austenitic stainless steels. In Ref 33 and 34, duplex stainless steel of grade 2205 (22Cr-5Ni-3Mo-N) with tensile strength of 720 to 740 MPa and plate thickness 15 mm was tested in a rotating bending fatigue machine at 11, 100, and 200 Hz and was compared with austenitic stainless steels. The testing temperature was 40 °C (105 °F).

The testing frequency had no influence on the fatigue level, but the influence of the environment was significant, as shown in Table 3. The fatigue strength for 2205 was higher than the fatigue

strength for the austenitic stainless steels. At pH = 1, the austenitic steel without molybdenum (304LN) had the lowest fatigue strength, 40 MPa. This steel is in its active state in this environment, whereas the duplex steel is in its passive state. There were no signs of pitting corrosion on any steel after the different fatigue tests performed. The testing stresses were in all tests lower than the stresses giving stress-corrosion cracking in the same environment. Tests performed on welded specimens showed that the fatigue strength for the duplex steel was higher than in the austenitic steels investigated, one example being a fatigue strength of 150 to 210 MPa for the duplex steel in 3% NaCl with pH = 1.

An extensive study of over 40 types of stainless steels was performed in a rotating bend testing machine (Ref 35). One duplex type 329 steel and one duplex type 2205 steel with a yield stress of about 500 MPa reached a fatigue limit at 10⁸ cycles (70 Hz) of 490 to 500 MPa in air and 400 MPa in 3% NaCl at room temperature. The main conclusion of the investigation was that materials that have both high yield strength and a good corrosion resistance also have the highest fatigue strength.

Wrought and Cast Comparison (Ref 36). In a test program with several participating organizations, duplex stainless steels were tested in air and in ASTM synthetic sea water (Ref 36). The fatigue limit under rotating bending stresses in 25 °C synthetic sea water at 30 Hz and 10⁷ cycles was 350 MPa for the forged specimens and roughly 200 MPa for the cast specimens.

Transverse and Longitudinal Specimens in Sulfuric Acid (Ref 37). Rotating bending fatigue of 2205 was investigated under various conditions. At a testing frequency of 50 Hz, the fatigue limit at 10⁷ cycles was ±350 MPa for longitudinal specimens and ±325 MPa for transverse specimens when tested in air. Specimens with a stress concentration factor of α_k = 4.9 showed a reduced fatigue limit of 170 MPa (24.5 ksi) for longitudinal specimens and 145 MPa (21 ksi) for transverse specimens. Testing in 2M sulfuric acid lowered the fatigue strength to 100 MPa (14.5

Table 4 Fatigue strength under reversed bending stresses

Grade	Environment	Yield strength, MPa	Ultimate tensile strength, MPa	Fatigue strength, MPa	Ref
DP3	3% NaCl	610	815	±340	39
2304	Air	460	695	±445	40
2304	3% NaCl	460	695	±360	40
18-4-3-3	Air	548	821	±410	41

Table 5 Fatigue strength under pulsating tensile stresses

Grade	Environment	Yield strength, MPa	Ultimate tensile strength, MPa	Fatigue strength, MPa	Ref
2205	Air	470	700	250 ± 240	42, 43
2205	22% NaCl, 80 °C	...	...	250±200	42, 43
2205	22% NaCl, 150 °C	359 (150 °C)	645 (150 °C)	250±70	42, 43
26-4-6-2	Air	...	...	275±275	44
26-4-6-2	4N NaCl, 80 °C, pH=7	...	...	250±250	44
26-4-6-2	4N NaCl, 80 °C, pH=2	...	...	160±160	44
2205	Air	489	...	268±219	45
2205	3% NaCl, 45 °C	...	...	244±200	45
2205	3% NaCl, 80 °C	430 (80 °C)	...	232±190	45
2507	Air	558	...	295±240	45
2507	3% NaCl, 80 °C	515 (80 °C)	...	270±220	45
2507	3% NaCl, 90 °C	515 (80 °C)	...	222±181	45
3RE60	Air	448	720	284±234	38, 46
2205	Air	483	714	260±210	38, 46
2205	Air	519	750	275±225	38, 46
2507	Air	565	800	303±248	38, 46
2205	Air	578	722	245±245	47, 48
2205	3% NaCl	578	722	190±190	47, 48
DP 9	Air	436	752	180±180	49
DP 9	3N HNO₃, boiling	436	752	80±80	49

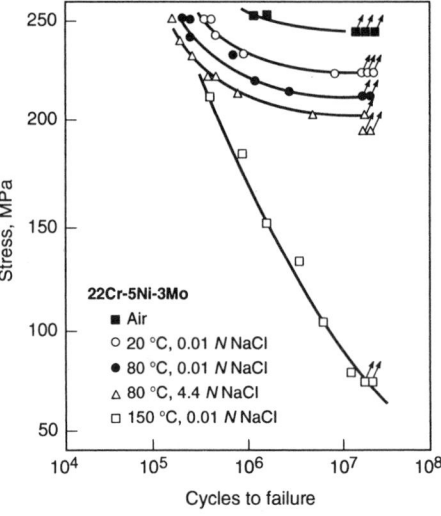

Fig. 12 Air and corrosion fatigue behavior of duplex stainless steel 22Cr-5Ni-3Mo. 250 MPa mean stress at 50 Hz. Source: Ref 43

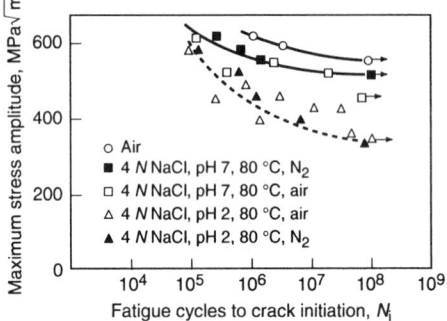

Fig. 13 Number of cycles to crack initiation for X4 CrMnNiMoN2664 in NaCl solution, pH 7 and pH 2. Source: Ref 44

ksi) or lower at 10^7 cycles. The lower values were found for transverse specimens and for the lower testing frequency of 6.5 Hz, as compared to 50 Hz in the main testing program. Stress concentrations can further reduce the fatigue strength to 50 MPa (7.25 ksi).

Fatigue Strength in Water (Ref 38). Fatigue testing was performed in different corrosive environments. The tested material (transverse specimens) were cut out of hot-rolled plates, 15 to 85 mm thick. A four-point rotating bending machine was used at a testing frequency of 25 Hz. The environments examined were distilled water, synthetic sea water (DIN 50905:4), and synthetic white water (pH = 3.5, Cl⁻ = 400 ppm, SO₄²⁻ = 250 ppm). The fatigue strength for 3RE60 at 10^7 cycles decreased with increased severity of the environment, from 320 to 265 MPa.

Reversed Bending Fatigue. Fatigue strength under reversed bending stresses in air or in 3% NaCl is summarized in Table 4 from three sources. In a low-frequency test (0.5 Hz) in a 3% NaCl solution the fatigue behavior of DP 3 (25Cr-6Ni-3Mo-0.3W) was investigated (Ref 39). The fatigue limit in reversed bending at 10^7 cycles was not decreased in the solution as compared to the result in air. This fatigue limit was ±340 MPa (±49 ksi) for specimens with a yield stress of 610 MPa (88.5 ksi). In the low-cycle range, however, the duplex steel showed a shorter fatigue life in salt water than in air. This behavior could be related to the finding that the crack propagation rate in salt water is larger than that in

air. In the tested environment the fatigue limit for stainless steels is increased with increased tensile strength, in contrast to the behavior of carbon and low-alloy steels.

The duplex stainless steel SAF 2304 (23Cr-4Ni-N) was fatigue tested under reversed bending stresses (Ref 40). In air the fatigue strength at 2×10^6 cycles and 20 Hz was ±445 MPa (64.5 ksi) for specimens with a yield strength of 460 MPa (66.7 ksi). The fatigue strength was decreased to ±360 MPa (52 ksi) when the test was performed in 3% NaCl.

A new high-strength duplex stainless steel intended for railway cars (Ref 41) is 18Cr-4Ni-3Si-3Mn-1Cu with a yield strength above 500 MPa (72.5 ksi) and a fatigue limit in bending fatigue of ±410 MPa (59 ksi) at 10^7 cycles in air. The fatigue limit for gas metal arc weldments is reduced to ±375 MPa (54 ksi).

Fatigue strength under pulsating tensile stresses in air or in 3% NaCl solution is summarized in Table 5 (Ref 42-49). An investigation of the corrosion fatigue mechanisms of a 22Cr-5Ni-3Mo duplex stainless steel showed that at high stress levels the crack initiation occurred mainly at persistent slip bands in the austenite or near the phase boundaries (Ref 42, 43). At lower stress levels, the cracks initiated mainly at nonmetallic inclusions in the surface region. The fatigue limit in air was 240 MPa (mean stress 250 MPa, 2×10^7 cycles, 50 Hz) (Table 5). When testing was done in concentrated chloride solutions at 80 °C (175 °F), the fatigue limit was decreased only to

200 MPa (29 ksi). Under these stable passive corrosion conditions the duplex steel had a pronounced fatigue limit (Fig. 12). When the temperature was raised to 150 °C (300 °F), the corrosion fatigue strength was reduced significantly (to 70 MPa, or 10 ksi) due to superposition of pitting corrosion.

Duplex stainless steel 26Cr-3.5Ni-5.6Mn-2.2Mo-0.3N was tested in air in a neutral chloride solution (pH = 7) and in an acid solution (pH = 2) (Ref 44). The fatigue strength at 80 °C (60 Hz, 10^8 cycles, R = 0) in the neutral chloride solution is 250 MPa (36 ksi), which is not much lower than in air. However, there is a significant decrease when the pH is lowered to 2 (see Table 5 and Fig. 13). Experiments on stress corrosion in the same media did not show any tendency to stress-corrosion cracking. In a pitting test, however, there was an increase in corrosion current of one order of magnitude when the pH decreased from values between 9 and 4 down to a value of 2. The decrease in fatigue strength with increased acidity was attributed to a decreased stability of the passive film and not to an increased tendency to pitting.

Alloys 2205, SAF 2507, and 3RE60. The fatigue strength of other duplex stainless steels (Ref 45) was tested in air and in a 3% NaCl solution at 45, 80, and 90 °C (110, 175, and 195 °F). The pulsating tensile fatigue strength at 2×10^6 cycles for grade 2205 was 487 MPa (70 ksi) (max at $R = 0.1$) in air at 20 °C, 444 MPa (64.4 ksi) in the NaCl solution at 45 °C, and 422 MPa (61 ksi) in the NaCl solution at 80 °C. The test frequency in air was 20 Hz and in the NaCl solution 5 Hz. The decrease in fatigue strength with increasing temperature could be explained by the same decrease in the yield strength. The initiation of fatigue cracks occurred mainly at surface scratches. Only in a few specimens fractured at 80 °C did the initiation of fatigue cracks start at corrosion pits, despite the fact that the fatigue test was performed at a temperature well above the critical pitting temperature.

In the same project some tests were also performed on SAF 2507 (25Cr-7Ni-4Mo-N) in air at 20 °C (68 °F) and in a 3% NaCl solution at 80 and 90 °C (175 and 195 °F). The pulsating fatigue strength in air was 535 MPa (77.5 ksi) and 490 MPa (71 ksi) in the NaCl solution at 80 °C. This decrease could also be referred to the decrease in yield strength with increasing temperature, and there was no evidence of corrosion pits in the specimens after the fatigue test.

Table 5 also lists data from fatigue testing in air at 20 Hz for 3RE60, 2205, and 2507 (Ref 38, 46). Fatigue strength at 2×10^6 cycles was 518 MPa (75 ksi) for 3RE60, 470-500 MPa (68-72.5 ksi) for 2205, and 551 MPa (80 ksi) for 2507.

The fatigue limit in air for alloy 2205 under cyclic tensile load at 25 Hz and 10^7 cycles was 490 MPa, and it was 380 MPa in 3% NaCl. Different stress concentration factors lowered these figures, for example from 490 to 285 MPa (71 to 41 ksi) in air for a stress concentration factor of 3.9, and from 380 to 240 MPa (55 to 35 ksi) for the same stress concentration factor in 3% NaCl. Values of the fatigue strength under tension-compression loads were also given, ±175 MPa (25 ksi) in air at 10^7 cycles for smooth specimens Ref 47, 48).

The duplex stainless steel DP 9 (23Cr-11Ni-3.3Si) was tested in air and in boiling $3N$ HNO$_3$ (Ref 49). The axial fatigue strength was 180 MPa (26 ksi) in air (max amplitude) and 80 MPa (11.6 ksi) in the boiling acid (104 °C) at 10^6 cycles. The crack propagation rate was somewhat higher in the boiling acid than in air.

Influence of Cold Work. The effect of cold working on the fatigue properties of the duplex steel A905 (26Cr-6Mn-4Ni-2.5Mo-0.15N) has been investigated (Ref 50). In a pulsating tensile test at 10 Hz in air, the fatigue limit at 5×10^7 cycles increased from 520 to 750 MPa (75 to 109 ksi) after 8% cold work (specimens from cold-drawn rods), whereas in $4N$ NaCl at 80 °C and pH = 5, the corrosion fatigue strength increased from 500 to 700 MPa (72.5 to 101.5 ksi) after 8% cold work. The tensile strength increased from 890 to 1000 MPa (129 to 145 ksi) due to the 8% cold work.

Environmental Effects and Influence of Testing Frequency. Tests performed in air and in NaCl in the laboratory are today more or less standard, while tests performed in real process environments are more rare. Rotating bending fatigue tests were performed at 50 Hz in geothermal steam from different geothermal fields (Ref 51). For notched specimens of alloy 2205 the fatigue limit in air was ±250 MPa (36 ksi) at 5×10^6 cycles. The fatigue strength was reduced to 200 MPa (29 ksi) in one geothermal environment and to 120 MPa (17 ksi) in another environment. The differences in fatigue strength were related to different contents of hydrogen and chlorides. Alloy 2205 was more influenced by the geothermal environments than the austenitic 316 steel.

In the literature cited above on fatigue data, some authors have claimed that there is no influence of the testing frequency on the fatigue limit. This can of course be true for tests performed in rather mild environments and tests that last for shorter time periods. Depending on the type of environment, and whether the steel is in its passive or active corrosion state, there will be an influence of the testing frequency. One example (Ref 38, 52) is duplex steels tested under rotating bending stresses in synthetic white water at test frequencies of 25 and 5 Hz. The reduction in testing frequency decreases the fatigue strength at 10^7 cycles from ±250 to ±210 MPa (±36 to ±30 ksi) due to the longer exposure times in the corrosion solution (Fig. 14).

Welds and Weldment Geometries. Usually, when welds are tested, the test specimens are cut out transversal to a weld, and the specimens are then machined and polished before the testing. If the welding has been performed according to the state of the art, there should not be any decrease in fatigue strength for the weld as compared to that of the base material. For example, in Ref 52 the base material of 2205 has roughly the same fatigue strength at 10^7 cycles as the submerged arc weld and the electroslag weld.

Fatigue tests carried out in air on joints of a duplex steel (23Cr-4Ni) in the as-welded condition examined joint geometries of butt welds and plates with longitudinal fillet-welded gussets on both sides (Ref 53). Fatigue test results were compared with International Institute of Welding design recommendations and results from the literature. The fatigue strengths of the joint geometries analyzed were as good as those of identical joints in structural steel. The very limited data for duplex stainless steel welded joints indicate fatigue performance comparable to that of C-Mn steel joints (Ref 54). Therefore, the current design S-N curves for C-Mn steels may be used for duplex joints.

Fracture Toughness

The fracture toughness of duplex stainless steels has been evaluated through impact testing and fracture mechanic testing methods such as crack-tip opening displacement (CTOD), K_{Ic},

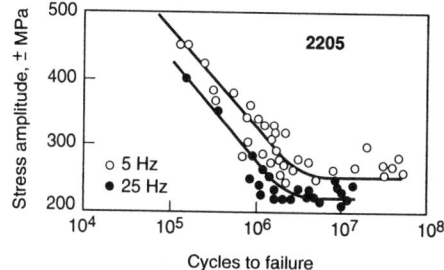

Fig. 14 Fatigue testing of 2205 SRG Plus in synthetic white water at 5 and 25 Hz, 50% probability of fracture

and J_{Ic}. The fracture toughness as measured by the standard testing methods for K_{Ic} and J_{Ic} often give very high values, but the requirement for plane strain is not fulfilled. Nevertheless, it is often claimed that the figures given could be interpreted as representative for the plate thickness that has been tested.

Fracture Modes. In the solution-annealed condition, fracture surfaces of duplex stainless steels have a ductile appearance with dimples (Ref 55). After aging alloy 2205 at 800 °C (1470 °F), the fracture surface contains numerous delaminations and microvoids, where the latter are much smaller than in the solution-annealed specimens (Ref 55). Aging alloy 2205 at 475 °C for 100 h will lead to unstable crack propagation and a fracture surface of quasicleavage type (Ref 56). Depending on the specimen orientation in a fracture toughness test, crack growth behavior can be evaluated along the rolling direction or perpendicular to it. Parallel to the rolling direction, voids nucleate preferentially in the ferrite phase or at ferrite-austenite phase boundaries (Ref 57). The crack growth perpendicular to the rolling direction takes place across ferrite and austenite grains with some cracking branching at the crack tip (Ref 57).

Base Material in the Solution-Annealed Condition. For solution-annealed materials, J_{Ic} values range from 200 to 800 kJ/m^2 (Ref 55-60). In Ref 55, for example, J_{Ic} values of 0.4 to 0.5 MJ/m^2 for single-edge notched bending (SENB) specimens and 0.7 MJ/m^2 for single-edged notched tension (SENT) specimens have been obtained. The duplex steel (solution-annealed 10 mm plate, 18Cr-5Ni-2.7Mo) was tested according to ASTM E 813. The initial value was low in comparison with the resistance to subsequent crack propagation. Other tests on solution-annealed 2205 (Ref 59) show $J_{Ic} = 300$ kJ/m^2.

In the solution-annealed condition, the standard duplex steels have a very high toughness as tested at room temperature. The duplex steels are rather insensitive to variations in solution annealing temperature. In one study, solution annealing at 1050, 1175, or 1300 °C for 1 h respectively and subsequent water quenching did not affect the crack resistance properties of the steel when it was tested at room temperature (Ref 55). The minor change in the relative amounts of ferrite and austenite phases did not in this case have any greater impact on the fracture toughness. How-

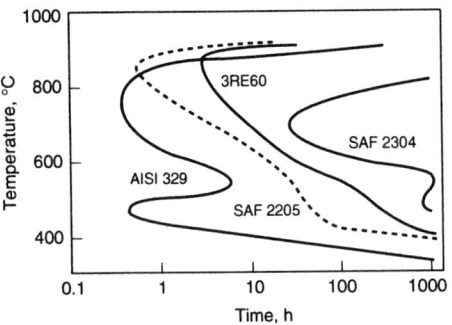

Fig. 15 Time-temperature embrittlement curves for some duplex stainless steels. The curves represent the 27 J transition at room temperature for standard Charpy V-notch specimens. Source: Ref 62

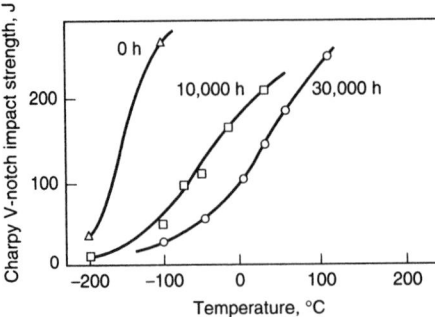

Fig. 16 Charpy V-notch impact curves for quench-annealed SAF 2205 aged at 325 °C. Source: Ref 62

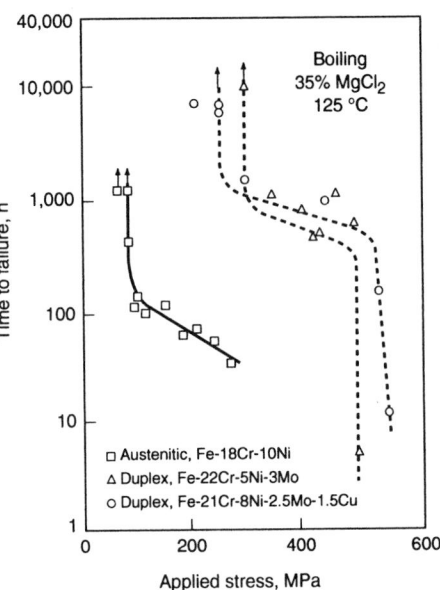

Fig. 17 Effect of applied stress on the times to failure of various stainless steels in a magnesium chloride solution. Source: Ref 75

ever, when the relation between the two phases is varied over a wider range (0 to 100% of ferrite), there is a marked influence on the elongation values and on the impact toughness. In Ref 61, the alloy with the highest ferrite content had the lowest values. The microstructure in the duplex stainless steels is usually elongated in the rolling direction. This of course has an influence on the fracture toughness, which is higher for test specimens with a crack extension transverse to the rolling direction (Ref 60).

Influence of Brittle Phases and Precipitations. The toughness behavior of duplex stainless steels can be summarized in time-temperature embrittlement diagrams of the type shown in Fig. 15 (Ref 62). There is a marked decrease in ductility for many of the duplex steels in the interval between 600 and 900 °C (1110 and 1650 °F) and also at 475 °C (885 °F) or in the range of 450 to 500 °C (840 to 930 °F). Cold-rolled duplex steels will get an even lower impact toughness after aging at 475 °C (Ref 63). When using the duplex steels in applications exposed to elevated temperatures during longer periods, the behavior at temperatures as low as 300 °C (570 °F) must be examined.

An example of aging treatment and resulting impact toughness is shown in Fig. 16 (Ref 62) for the standard duplex steel 2205. At longer times at 325 °C (615 °F), the impact toughness will decrease, thus limiting the application temperature to 300 °C. The maximum temperature for long-term use of superduplex grades is limited to 280 °C (535 °F) while molybdenum-free duplex grades may be considered for applications up to 325 °C (615 °F) (Ref 64). A summary of all the embrittling phases and precipitations found in duplex and superduplex steels are given in Ref 65 and 66.

Fracture toughness data for duplex steels decrease after different heat treatments in the temperature range between 475 and 1200 °C (885 and 2190 °F), the decrease being most pronounced after heating in the interval 475 to 900 °C (885 to 1650 °F) (Ref 67). The decrease at 675 to 900 °C follows 2 h of heat treatment, while at 475 °C the decrease in toughness is slower and reaches the lower values after 24 h of heat treatment. The J_{Ic} value is decreased from 500 to

below 100 kJ/m^2 after only 1 h of heat treatment at 800 °C (Ref 55). Aging of the duplex steel 2205 at 475 °C for 100 h decreased the J-value from 800 to 200 kJ/m^2 (Ref 56).

Samples of 2205 were sensitized for 2 and 24 h at 675 and 825 °C (Ref 68). Regeneration treatment for 0.5 to 6 h at 1050 °C almost recovered the toughness properties as measured by CTOD, while the stress-corrosion properties were fully recovered only for specimens which were sensitized for the shorter period (2 h).

The fracture toughness is also influenced to some extent by the volume fraction of inclusions and precipitations and the distances between them (Ref 56, 69).

Welds. The precipitation of brittle phases influences the ductility of welds. Short-time aging to simulate heat-affected zones has been carried out on grade 2205 (Ref 58, 59). Aging times of 15 min at 875 °C are long enough to have an embrittlement effect, given as a J_{Ic} value. However, the embrittlement effect after 1 h at 475 °C seems to be rather limited.

The influence of 60 to 85% ferrite on welds in duplex steel 2205 was measured (Ref 70). An increase in ferrite content decreased the fracture toughness values, measured as CTOD.

Weld metals for grade 2205 typically contain 9% Ni. Embrittlement occurs more rapidly in the weld metal than in the base metal. The impact toughness will decrease below the acceptance value of 27 J after 3 to 4 min at 850 to 950 °C (Ref 71).

The heat-affected zone in the superduplex steel UNS 32760 has been simulated (Ref 72, 73). Charpy V-notch values remained at a high level except in tests with very long cooling times. The superduplex steel UNS 32750 was submerged arc welded with different heat inputs and then impact tested (Ref 74). The impact toughness values were acceptable if the heat input was maximized to 2.5 kJ/mm and if the interpass temperature was maximized to 150 °C (300 °F).

Stress-Corrosion Cracking

Duplex stainless steels are used in environments that may produce anodic or cathodic stress-corrosion cracking (SCC). The behavior of duplex stainless steels under SCC conditions is

summarized in Ref 75. As in the case of austenitic stainless steels, the duplex stainless steels have a high protection potential in concentrated chloride solutions. σ-phase formation lowers the resistance against anodic SCC. Welding may also impair the SCC resistance when the cooling rate between 1200 and 800 °C (2190 and 1470 °F) is too high. Builtup weld seams or repair welds may be susceptible to SCC because of the high cooling rate involved that suppresses the reformation of austenite from high-temperature δ-ferrite.

The SCC mechanism is closely coupled to mechanical and electrochemical phenomena. In the ferrite, depassivation is essentially determined by mechanical twinning, whereas slip on the ferrite surfaces is insufficient to cause localized depassivation. At lower stress levels, the SCC of the duplex steels seems to be determined by local depassivation of the austenite, while at higher stresses the twinning in the ferrite is responsible for the cracking. At lower stresses the ferrite may in some duplex steels be less deformed than the austenite and may cathodically protect the plastically deformed austenite. At higher stresses, when twinning occurs in the ferrite, the ferrite will depassivate and localized corrosion and cracking will take place.

Stress-corrosion cracking of duplex stainless steels occurs in concentrated chloride solutions such as boiling magnesium chloride, as well as in sour gas environments that contain chloride and hydrogen sulfide (Ref 76). Threshold stresses for SCC of duplex stainless steels in chlorides are generally higher than those of austenitic stainless steels (Fig. 17), although this does depend in part on the test methods used in evaluation (see the next section, "Evaluating SCC Test Data"). There are also significant variations in chloride SCC resistance among the different commercial grades (Fig. 18) (Ref 77). The data of Fig. 18

Table 6 Compositions of the alloys investigated for chloride-induced stress-corrosion cracking

Steel	Austenite, %	Composition, %									
		C	Si	Mn	P	S	Cr	Ni	Mo	N
2304	44	0.013	0.39	1.41	0.022	0.005	22.54	4.64	0.24	0.083
2205	51	0.016	0.30	1.48	0.022	0.001	21.73	5.50	2.95	0.150
2507	52	0.017	0.37	0.64	0.018	0.002	24.90	7.30	4.00	0.260
F	0	0.015	0.35	1.30	0.011	0.003	24.50	4.15	3.77	0.029
FA	30	0.007	0.49	1.48	0.004	0.003	22.39	5.22	3.19	0.070
AF	64	0.017	0.44	1.63	0.004	0.003	21.43	6.27	2.74	0.203
A	98	0.023	0.40	1.86	0.004	0.003	20.47	9.09	2.29	0.196

F, ferritic; FA, ferritic-austenitic; AF, austenitic-ferritic; A, austenitic

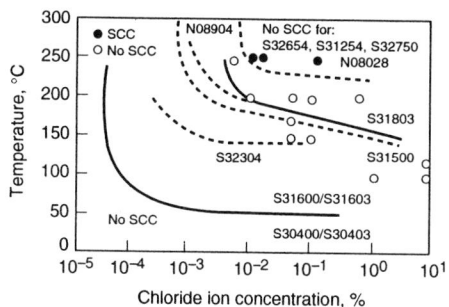

Fig. 18 Results of autoclave tests in aerated neutral chloride ion solutions for various stainless steels. Source: Ref 77, 83

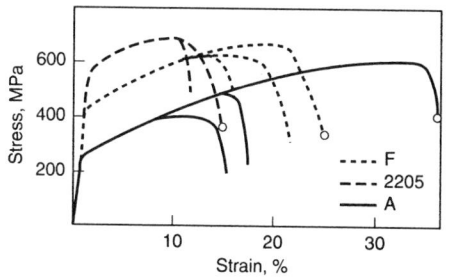

Fig. 19 Stress-strain curves for austenitic (A) and ferritic (F) alloys and alloy 2205 in oil (O) and in 50% LiCl at 100 °C, illustrating the necessity of parameter normalization for stress-corrosion cracking evaluation

Table 7 Stress-corrosion cracking test results on duplex steels in chloride environments

Alloy	Test temperature, °C	Hours to failure in U-bend test(a)	Slow strain rate testing(b) at:		
			$4 \times 10^{-6} \cdot s^{-1}$	$2 \times 10^{-6} \cdot s^{-1}$	$1 \times 10^{-6} \cdot s^{-1}$
43.5% MgCl₂					
2304	150	8.4	12 (7)	...	...
2205	150	1.8	12 (19)	...	...
50% LiCl					
2304	150	4.8	12 (13)(c)	...	8 (11)
2205	150	56	15 (60)(c)	...	7 (30)
63% CaCl₂					
2304	150	22	...	7 (12)	...
2205	150	2.2	...	12 (20)	...
40% MgCl₂					
2304	100	...	59 (44)	...	...
2205	100	...	33 (48)	...	...
2507	100	...	...	...	...
50% LiCl (deaerated)					
2304	100	...	74 (66)(c)	62	...
2205	100	...	68 (64)(c)	37 (59)	...
2507	100	...	59 (49)(c)	30 (53)	...
40% CaCl₂					
2304	100	...	51 (44)(c)	33 (34)(c)	...
2205	100	...	72 (88)(c)	45 (59)(c)	...
2507	100	...	62 (111)	84 (88)	...

(a) For U-bend testing the time to failure is taken as the time at which the first cracks were observed and is the average of four specimens.
(b) For slow strain rate testing two evaluation parameters are given: the normalized plastic elongation to failure and, in parentheses, the normalized area reduction. (c) Results are averages of at least two tests.

suggest that increasing the total chromium, nickel, molybdenum, and nitrogen contents results in increased chloride SCC resistance. These data do not differentiate between effects due to individual alloying elements. In fracture mechanics tests (Ref 78) using precracked fatigue specimens in a 22% NaCl solution at 105 °C (220 °F), two variants of type 329 duplex stainless steel were found to be resistant to chloride SCC at stress-intensity levels of 60 MPa√m (55 ksi√in.).

Evaluating SCC Test Data. Relative SCC rankings of steels depend on both the testing environment and the test method used. For example, data for chloride-induced SCC (Ref 79) were determined for three commercial alloys (type 2304, 2205, and 2507) and some tie-line alloys with 2 to 100% ferrite (Table 6). The results from the stress-corrosion testing are shown in Table 7. The relative ranking of the steel types depends on

both the testing environment and the test method used. In the U-bend test, 2205 has a higher susceptibility to cracking than the leaner alloy 2304 in both magnesium chloride and in calcium chloride, while in lithium chloride 2304 is more prone to cracking than 2205.

Stress-strain curves (Fig. 19) were recorded as a way to normalize the parameters for the stress-corrosion testing. In Table 7, the slow strain rate test has been evaluated in two ways: by judging the normalized plastic elongation to failure and by judging the normalized area reduction. In the lithium chloride test at 150 °C (300 °F), there was a clear strain rate dependence, but it was difficult to separate the different alloys when elongation to failure was used as a failure criterion. However, when the area reductions were measured, alloy 2304 had a much lower resistance to SCC in all three environments as compared to alloy 2205. In the slow strain rate test at 100 °C (212 °F), the more highly alloyed steel 2507 had better crack resistance in the calcium chloride environment

than the two leaner alloyed steels. However, in lithium chloride, 2507 had lower crack resistance.

The stress-corrosion resistance of the tie-line alloys is also dependent on the test type and the testing environment. In the U-bend test in calcium chloride, the ferritic alloy has a much higher resistance to cracking than the austenitic alloy and the duplex alloys have by far the lowest resistance (Ref 79). In slow strain rate testing in lithium chloride, there is a poor differentiation between the alloys, but there is a tendency for the two-phase alloys to have a somewhat higher crack resistance than the others.

The different rankings of the three duplex alloys cannot be explained by fracture path or electrochemical relations. Fracture surfaces from the testing of the tie-line alloys in lithium chloride have been compared at 100 °C. The cracking was transgranular in the ferritic alloy and intergranular in the austenitic alloy. In the two-phase alloys, the surfaces contained parts with transgranular

Table 8 Results of stress-corrosion tests and minimum yield stresses

UNS number	Material composition	Designation	CaCl₂ test(a), %	DET(b), %	Critical temperature at 0.05% Cl⁻(c), °C	Minimum 0.2% yield strength, MPa
Austenitic stainless steels						
S32654	24Cr-22Ni-7Mo-0.5N	654 SMO™	>90	>100	>300	430
S31254	20Cr-18Ni-6Mo-0.2N	254 SMO™	>90	≥90	>300	300
N08028	27Cr-31Nl-3.5Mo-1Cu	Sanicro 28	90	...	230	220
N08904	20Cr-25Ni-4.5Mo-1.5Cu	904L	90	≥70	183	220
N08825	22Cr-40Ni-3Mo-2Cu	Sanicro 41	90	...	...	240
S31050	25Cr-22Ni-2Mo	2RE69	38	...	...	270
S31603	17Cr-12Ni-2.5Mo	3R60	35	≤10	...	220
S31000	25Cr-20Ni	2RE10	20	...	52	210
S30403	18Cr-9Ni	3R12	12	≤10	52	210
Duplex stainless steels						
S32750	25Cr-7Ni-4Mo-0.3N	SAF 2057®	90	≥70	>300	550
S31803	22Cr-5Ni-3Mo-0.15N	SAF 2205®	90	≥40	183	450
S31500	19Cr-5Ni-2.5Mo-0.1N	3RE60	90	...	170	450
S32900	26Cr-5Ni-1.5Mo	10RE51	70(d)	...	...	485
S32304	23Cr-4Ni-0.1N	SAF 2304®	85	≥40	138	400

(a) Ratio between threshold stress and UTS. Time to failure is 500 h unless specified. (b) Ratio between threshold stress and 0.2% yield strength in drop evaporation tests. (c) Critical temperature at 0.05% chloride ion concentration in autoclave tests. (d) Time to failure is greater than 1000 h. Source: Ref 83

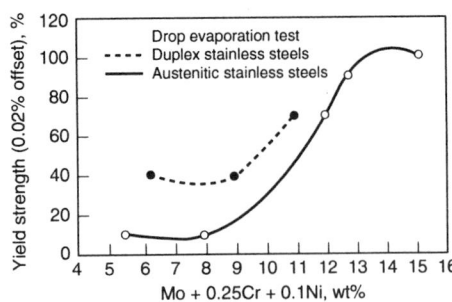

Fig. 20 Dependence of threshold stress in percentage of 0.2% yield strength on molybdenum equivalent for austenitic and duplex stainless steels

cracking in the ferrite. The crack in some areas followed the ferrite-austenite phase boundaries, revealing entire grains of austenite. In all three duplex alloys, tested in different environments, the fracture appearance was transgranular in the ferrite and interphase or intergranular in the austenite. In addition, the corrosion potential of the ferritic alloy was in all cases cathodic to that of the austenite, which resulted in cathodic protection of the austenite in almost all studied cases. Thus, neither the electrochemical relation between the ferrite and the austenite nor the behavior of the crack path could explain the different rankings of the three duplex alloys in the tested environments.

Another attempt to understand why different test methods give different rankings of duplex and austenitic alloys has been reported in Ref 80. The conventional monotonic slow strain rate test was compared to a test where a straining was followed by a relaxation period at constant deflection for 23 h. In the first cycle, the steels were strained to 90% of their yield stress, and corresponding deflection was held constant during relaxation. In the coming next cycles the stress was increased by 15 MPa per cycle. The testing was performed in lithium chloride at 100 °C. The stress-corrosion resistance of the 2304 and 2205 duplex steels was superior to that of the standard 316 austenitic stainless steel but inferior to that of the super austenitic stainless steel 254 SMO. Both testing methods yielded the same ranking. It was difficult, however, to see any clear differences between the two duplex steels. The main difference between the austenitic and duplex steels was found in the measurement of the corrosion potential. In the cyclic strain relaxation test, the measured potential for the austenitic alloys had a drop for every loading cycle, while the duplex steel had a pronounced drop in potential only just before final failure. This finding may to

some extent explain the different rankings of stainless steels in different stress-corrosion tests. The fracture surface of the duplex steels was transgranular in the ferrite and intergranular around the austenite phase, while the austenitic steels had both transgranular and intergranular cracking regions.

The influence of creep behavior on SCC was reported in Ref 81. The temperature creep dependence of the duplex stainless steel 2205 was much lower than that of the austenitic 316 steel in the temperature range tested. This behavior, together with the above-mentioned strain relaxation tests, may explain the different ranking of steels in constant-loading tests and in slow strain rate tests.

SCC in Caustic Solutions. A range of duplex and austenitic stainless steels were tested in caustic solutions at 200 °C (Ref 82). The solution used in the U-bend test and in the slow strain rate test contained 200 g/L sodium hydroxide and 10 g/L NaCl. The best behavior was shown by the molybdenum-free 23Cr-4Ni steel, which outperformed both the 22Cr-5Ni-3Mo and the 25Cr-7Ni-4Mo-N duplex steels and some austenitic stainless steels.

SCC Comparison of Austenitic and Duplex Stainless Steels. With the development of a database, experimental data have been compared for duplex and austenitic stainless steels tested with three different testing methods: the calcium chloride test, the autoclave test, and the drop evaporation test (Ref 83). In the calcium chloride test (Table 8), the duplex steels were ranked as high as highly alloyed austenitic stainless steels. In the autoclave test (Table 8 and Fig. 18), the lower-alloy duplex steels had a better stress-corrosion resistance than standard austenitic steels, but the superduplex steel and the highly alloyed austenitic stainless steels were more resistant. In the drop evaporation test (Fig. 20), the duplex

steels are performing better than standard austenitic stainless steels but were outperformed by the austenitic stainless steels S32654 and S31254. It appears, by this author, that tests in autoclaves agree well with actual service conditions and that the drop evaporation test is useful in simulating applications where evaporation may occur and cyclic stresses are involved. An additional summary of stress-corrosion data for duplex stainless steels can be found in Ref 84.

Fractography. The fracture behavior of the duplex stainless steels during SCC has been investigated for duplex 25Cr-6Ni-3.6Mo steel with slow strain rate testing in boiling 8 M lithium chloride plus 0.025 M thiourea (Ref 85). Fracture surface specimens revealed two different fracture modes (complex and cleavage-like) in adjacent grains. Like most polycrystalline materials, different fracture paths result from different grain orientations, making different cleavage planes active in the different grains. Engineering alloys with polycrystalline structure commonly change from cleavage to complex type of fracture, which is governed by a combination of grain orientation, local microstructure, and stressing conditions. By using the transmission electron microscope, a porous sponge-like film with a thickness of 1000 Å was found between the parent metal and the outer corrosion product. This film was enriched with chromium, nickel, molybdenum, and copper. The study suggests that the dealloying of the ferritic phase and the formation of a sponge-like region is associated with the SCC of the ferritic phase.

Elevated-Temperature Properties

Duplex stainless steels are generally used in applications up to 300 °C (570 °F). The steels will embrittle if used in the temperature range of 350 to 550 °C (660 to 1020 °F). If long exposure times can be avoided in this region, there are some applications at temperatures above 550 °C (1020 °F) where duplex alloys are a better choice than austenitic alloys.

The thermal cycling properties of a duplex stainless steel, 23Cr-5Ni-1.5Mo, was compared with those of a standard austenitic steel of type 316 (Ref 86). Different thermal cycling tests up to 1125 °C (2060 °F) showed remarkable acceleration of failures in the duplex steel as compared with the austenitic steel. This is related to the large differences in internal stresses between the two phases in the duplex steel and to an extensive grain growth during thermal cycling.

Thermal cycling of a 24Cr-4Ni-1.3Si duplex steel between room temperature and 900 °C (1650 °F) raised microstresses varying from grain to grain (Ref 87). The plastification caused an accumulating change of the shape of the specimen, which gave high residual stresses and internal cracks and damage.

The creep-fatigue behavior of the duplex steel 2205 was studied in a sulfur-containing environment of Ar + 3% SO_2 at 700 °C (Ref 88). Severe sulfidation attack occurred at the external surface of both 2205 duplex and 316 austenitic stainless steel under a combination of creep-fatigue loading and atmosphere. However, the attack on the duplex steel was less severe than the attack on the austenitic stainless steel.

REFERENCES

1. J. Olsson and M. Liljas, Sixty Years of Duplex Stainless Applications, Paper 209, *NACE 1994*

2. J. Charles, Composition and Properties of Duplex Stainless Steels, *Welding in the World*, Vol 36, 1995, p 43-54

3. M. Liljas, The Welding Metallurgy of Duplex Stainless Steels, *Proc. Duplex Stainless Steels '94*

4. J.-O. Nilsson, Super Duplex Stainless Steels, *Materials Science and Technology*, Vol 8, 1992, p 685-700

5. T. Magnin, L. Coudreuse, and J.M. Lardon, *Basic Questions in Fatigue*, Vol II, STP 924, ASTM, 1988, p 128-144

6. T. Magnin and J.M. Lardon, *Mater. Sci. Eng.*, Vol 104A, 1988, p 21

7. T. Magnin, J.M. Lardon, and L. Coudreuse, in *Low-Cycle Fatigue*, STP 942, ASTM, 1988, p 812-823

8. J. Polak, S. Degallaix, and G. Degallaix, The Role of Cyclic Slip Localization in Fatigue Damage of Materials, *J. Phys. (France)*, Vol 4 (No. 3), Nov 1993, p 679-684

9. S. Degallaix, A. Seddouki, J.-O. Nilsson, and J. Polak, Influence of Nitrogen on Monotonic and Cyclic Mechanical Properties of Duplex Stainless Steels, *High Nitrogen Steels 93*, Institute for Metal Physics, Ukraine, p 240-245

10. S. Degallaix, A. Seddouki, G. Degallaix, T. Kruml, and J. Polak, Fatigue Damage in Austenitic-Ferritic Duplex Stainless Steels, *Fatigue Fract. Engng. Mater. Struct.*, Vol 18, 1995, p 65-77

11. H. Spähn, *Metal Progress*, Feb 1979, p 32

12. Kh.G. Schmitt-Thomas, H. Meisel, and R. Mathis, *Arch. Eisenhüttenwes*, Vol 53, 1982, p 321

13. R. Mathis, *J. Materials Science*, Vol 22, 1987, p 907

14. C. Amzallag, P. Rabbe, and A. Desestret, *Corrosion-Fatigue Technology*, STP 642, ASTM, 1978, p 117

15. R. Spähn, *Werkstoffe und Korrosion*, Vol 42, 1991, p 109

16. I. Austen, paper presented at VTT Symposium 121, Tech. Res. Centre of Finland, 1991, p 31

17. M.A. Daeubler, A.W. Thompson, and I.M. Bernstein, *Met. Trans.*, Vol 22A, 1991, p 513

18. M.A. Daeubler, G.W. Warren, I.M. Bernstein, and A.W. Thompson, *Met. Trans.*, Vol 22A, 1991, p 521

19. M. Mueller, *Corrosion*, Vol 38, 1982, p 431

20. M. Mueller, *Met. Trans.*, Vol 13A, 1982, p 649

21. C. Amzallag, P. Rabbe, and A. Desestret, Corrosion-Fatigue Behavior of Some Special Stainless Steels, *Corrosion-Fatigue Technology*, STP 642, ASTM, 1978, p 117-132

22. T. Misawa, Effects of Cathodic Protection Potential on Corrosion Fatigue Crack Growth for a Duplex Stainless Steel in Synthetic Seawater, *Corrosion Engineering*, Vol 38, 1989, p 315-318

23. Data on Corrosion Fatigue and Stress Corrosion of Steels under Marine Environment, *Committee on Environmental Strength of Steels*, Iron and Steel Institute, Japan, Sept 1986

24. J. Wasén, B. Karlsson, and M. Nyström, Fatigue Crack Growth in a Duplex Stainless Steel, *Fatigue '90*, p 1167-1172

25. M. Nyström, B. Karlsson, and J. Wasén, Fatigue Crack Growth of Duplex Stainless Steels, *Duplex Stainless Steel '91*, Les Editions de Physique, 1991, p 795-802

26. B. Karlsson and J. Wasén, in *Proc. ECF 7*, EMAS, Warley, 1988, p 573

27. J. Linder, Corrosion Fatigue of Austenitic and Duplex Stainless Steels in 3% NaCl, *Proc. Fatigue under Spectrum Loading and in Corrosive Environments*, EMAS, 1993, p 371-385

28. J. Linder, Corrosion Fatigue of Duplex and Austenitic Stainless Steels in 3% NaCl at 80 °C and Room Temperature, *Swedish Institute for Metals Research*, IM-3212, 1995

29. T.J. Marrow, J.E. King, and J.A. Charles, Hydrogen Effects on Fatigue in a Duplex Stainless Steel, *Fatigue '90*, p 1807-1812

30. T.J. Marrow and J.E. King, Temperature Effects on Fatigue Crack Propagation in a Thermally Aged Super Duplex Stainless Steel, *Duplex Stainless Steels '91*, Les Editions de Physique, 1991, p 169-176

31. T.J. Marrow and J.E. King, Fatigue Crack Propagation Mechanisms in a Thermally Aged Duplex Steel, *Material Science and Engineering*, Vol A183, 1994, p 91-101

32. T.J. Marrow and J.E. King, Microstructural and Environmental Effects on Fatigue Crack Propagation in Duplex Stainless Steels, *Fatigue Fract. Engng. Mater. Struct.*, Vol 17, 1994, p 761-771

33. W. Wessling and H.E. Bock, *Stainless Steel*, Vol 77, 1977, p 217

34. E. Bock, *Technischer Bericht*, Fried. Krupp Hüttenwerke AG, 1978

35. J.E. Truman and G.A. Honeyman, in *Proc. 11th ICC*, Vol 5, 1990, p 553

36. Data on Corrosion Fatigue and Stress Corrosion of Steels under Marine Environment, *Committee on Environmental Strength of Steels*, Iron and Steel Institute, Japan, Sept 1986

37. G. Blümmel, doctoral thesis, Darmstadt, 1986

38. R.E. Johansson and H.L. Groth, Corrosion Fatigue and Fatigue Data for Duplex Stainless Steels, *Conf. Duplex Stainless Steels '91*, Vol 1, Les Editions de Physique, p 283-294

39. H. Hirakawa and I. Kitaura, The Sumitomo Search, No. 26, Nov 1981, p 136

40. J.-O. Nilsson, Internal Technical Report 5043, AB Sandvik Steel, 1985

41. S. Kakuchi, M. Aoki, H. Kondou, S. Fujiyama, Y. Takahashi, K. Hirahara, M. Takahashii, and K. Ogawa, *Proc. Stainless Steel '91*, ISIJ, 1991, p 700

42. Kh.G. Schmitt-Thomas, H. Meisel, and R. Mathis, *Arch. Eisenhüttenwes*, Vol 53, 1982, p 321

43. R. Mathis, *J. Materials Science*, Vol 22, 1987, p 907

44. A. Atrens, *Metals Technology*, Vol 9, 1982, p 117

45. Project JK-4314, Jernkontoret, Stockholm, 1991

46. R.E. Johansson and H. Groth, *Proc. Nordic Symp. on Mechanical Properties of Stainless Steels*, Avesta Research Foundation and Jernkonforet, Oct 1990, p 88-96

47. E. Roeder, R. Selzer, and J. Vollmar, *Steel Research*, Vol 62, 1991, p 459-464

48. J. Vollmar and E. Roeder, *Werkstoffe und Korrosion*, Vol 45, 1994, p 378-386

49. K. Toyama, N. Konda, and K. Suzuki, Paper 11, *Proc. of Symposium No. 900-50*, The Japan Soc. of Mech. Eng., Aug 1990

50. M.P. Muller, A. Atrens, and J.E. Allison, *Mat. und Technik*, No. 1, 1987, p 3

51. I. Torbjörnsson and T. Tomasson, Fatigue under Spectrum Loading and in Corrosive Environments, EMAS, 1993, p 459-472

52. J.-O. Andersson, R. Johansson, and E. Alfonsson, Avesta 2205 SRG Plus: A New Grade for Paper Mill Suction Rolls, *Proc. Stainless Steel '92*, Avesta Research Foundation and Jernkonforet, p 379-389

53. M. Kosimäki and E. Niemi, *Stainless Steel '92*, Avesta Research Foundation and Jernkonforet, p 889-898

54. R. Razmjoo, *Design Guidance on Fatigue of Welded Stainless Steel Joints*, The Welding Institute, 1994

55. R. Roberti, G.M. La Vecchia, and S. Rossi, Effect of Specimen Geometry on the Fracture Toughness of an Austenitic-Ferritic Stainless Steel, *Proc. ELVAMAT 89*, ISIJ, p 679-686

56. L. Iturgoyen, J. Alcalá, and M. Anglada, The Influence of Ageing at 475 °C on the Fracture Resistance of a Duplex Stainless Steel, *Proc. Conf. Duplex Stainless Steels '91*, Les Editions de Physique, 1991, p 757-764

57. R.E. Johansson and J.-O. Nilsson, Fracture Toughness of Austenitic and Duplex Stainless Steels, *Proc. Stainless Steels '84*, Institute of Metals, p 446-451

58. R. Roberti, W. Nicodemi, G.M. La Vecchia, and Sh. Basha, *Proc. Stainless Steel 1991*, ISIJ, p 700-707

59. R. Roberti, W. Nicodemi, G.M. La Vecchia, and Sh. Basha, *Proc. Duplex Stainless Steels '91*, Les Editions de Physique, p 993-1000

60. W. Nicodemi, R. Roberti, and G.M. La Vecchia, Duplex Stainless Steel Microstructure and Toughness, *Proc. Stainless Steel '92*, p 270-279

61. A. Hoshino, K. Nakano, and M. Kanao, Influence of Austenite on Toughness of Two Phase Stainless Steels, *Trans. National Research Institute for Metals*, Vol 2 (No. 4), 1980, p 185-194

62. P. Norberg, Applicability of Duplex Stainless Steels above 300 °C, *Conf. Duplex Stainless Steels '86*, Nederlands Instituut voor Lastechniek, p 298-302

63. M. Nyström, B. Karlsson, and J. Wasén, The Influence of Prestraining and 475 °C on the Mechanical Properties of a Duplex Stainless Steel (SAF2205), *Proc. Stainless Steel 1991*, ISIJ, p 738-745

64. J. Charles, Composition and Properties of Duplex Stainless Steels, *Welding in the World*, Vol 36, 1995, p 43-54

65. B. Josefsson, J.-O. Nilsson, and A. Wilson, Phase Transformations in Duplex Steels and the Relation between Continuous Cooling and Isothermal Heat Treatment, *Conf. Duplex Stainless Steels 1991*, Les Editions de Physique, 1991, p 67-78

66. J.-O. Nilsson, Super Duplex Stainless Steels, *Materials Science and Technology*, Vol 8, 1992, p 685-700

67. A.M. Irisarri, E. Erauzkin, F. Santamaria, and A. Gil-Negrete, Influence of the Heat Treatment on the Fracture Toughness of a Duplex Stainless Steel, *ECF 8, Fracture behaviour and design of materials and structures*, EMAS, 1990, p 373-376

68. A.M. Irisarri and E. Erauzkin, Effect of the Heat Treatment on the Fracture Toughness and Corrosion Embrittlement of a Duplex Stainless Steel 2205, *Proc. Duplex Stainless Steels '91*, Les Editions de Physique, 1991, p 779-785

69. R. Roberti, Fracture Behaviour of Austenitic and Duplex Stainless Steels, *Innovation Stainless Steel 1993*, Associazione Italiana di Metallurgia, p 2.129 to 2.137

70. F. Santamaria and A.M. Irisarri, Influence of Some Metallurgical Variables on Fracture Toughness of Duplex S.S. Weldments, *ECF 10, Structural Integrity: Experiments, Models and Applications*, EMAS, 1994, p 1139-1144

71. L. Karlsson, L. Ryen, and S. Pak, Precipitation of Intermetallic Phases in 22% Cr Duplex Stainless Weld Metals, *Welding Research Supplement*, Jan 1995, p 28-40

72. S. Atamert and J.E. King, The Effect of Heat-Affected-Zone Microstructures on Fracture Toughness of Duplex Stainless Steels, *ECF 8, Fracture Behaviour and Design of Materials and Structures*, 1990, p 394-399

73. J.-J. Dufrane, Heat Affected Zone Simulation of Super Duplex Stainless Steel UNS S 32760—ZERON 100, *Duplex Stainless Steels '91*, Les Editions de Physique, 1991, p 967-975

74. G. Björkman, L. Ödegård, and S.-Å. Fager, Submerged-Arc Welding of Sandvik SAF 2507 with Different Heat Inputs: Mechanical Properties and Corrosion Resistance, *Duplex Stainless Steels '91*, Les Editions de Physique, 1991, p 1035-1042

75. H. Spähn, Stress Corrosion Cracking and Corrosion Fatigue of Martensitic, Ferritic and Ferritic Austenitic (Duplex) Stainless Steels, *Int. Corros. Conf. Ser., Environ.-Induced Cracking Met.*, NACE, 1990, p 449-487

76. A.J. Sedriks, Stress Corrosion Cracking of Stainless Steels, *Stress Corrosion Cracking: Materials Performance and Evaluation*, ASM International, 1992, p 111-113

77. "Sandvik SAF 2507: A High-Performance Duplex Stainless Steel," Sandvik Steel, Sweden, March 1990

78. M.O. Speidel, *Met. Trans.*, Vol 12, 1981, p 779

79. R.F.A. Jargelius, R. Blom, S. Hertzman, and J. Linder, Chloride Induced Stress Corrosion Cracking of Duplex Stainless Steels in Concentrated Chloride Environments, *Proc. Third Int. Conf. Duplex Stainless Steels*, Vol 1, Les Editions de Physique, 1991, p 211-220

80. R.F.A. Jargelius and J. Linder, Use of Slow Strain Rate Technique to Assess the Stress Corrosion Resistance of Duplex and Austenitic Stainless Steels, *Proc. Applications of Stainless Steels '92*, Avesta Research Foundation and Jernkonforet, p 477-484

81. J. Linder, S. Hertzman, and R.F.A. Jargelius, Creep Behaviour of Stainless Steels at 100 °C to 325 °C, and Its Implications for Stress Corrosion Cracking, *Proc. Applications of Stainless Steels '92*, Avesta Research Foundation and Jernkonforet, p 1049-1058

82. G. Rondelli, B. Vincentini, M.F. Brunella, and A. Cigada, Effect of Alloy Element Contents on Caustic Stress Corrosion Cracking of Several Stainless Steels, *Werkstoffe und Korrosion*, Vol 44, 1993, p 57-61

83. L.-Z. Jin, The Chloride Stress-Corrosion Cracking Behavior of Stainless Steels under Different Test Methods, *J. Mater. Eng. Perform.*, Vol 4, 1995, p 734-739

84. P. Kangas and J.M. Nicholls, Chloride-Induced Stress Corrosion Cracking of Duplex Stainless Steels: Models, Test Methods and Experience, *Materials and Corrosion*, Vol 46, 1995, p 354-365

85. W.J. Nisbet, G.W. Lorimer, and R.C. Newman, A Transmission Electron Microscopy Study of Stress Corrosion Cracking in Stainless Steels, *Corrosion Science*, Vol 35, 1993, p 457-469

86. K. Kamachie et al., Thermal Fatigue by Impact Heating and Stresses of Two Phase Stainless Steel at Elevated Temperature, *Progress in Science and Engineering of Composites, ICCM-IV*, Tokyo, 1982, p 1383-1389

87. F.D. Fischer, F.G. Rammerstorfer, and F.J. Bauer, Fatigue and Fracture of High-Alloyed Steel Specimens Subjected to Purely Thermal Cycling, *Met. Trans.*, Vol 21A, April 1990, p 935-948

88. E. Aghion and C.A. Molaba, Creep-Fatigue Failure of SAF 2205 and 316 Stainless Steels in Ar + 3% SO₂ Environment at 700 °C, *J. Mater. Sci.*, Vol 29, 1994, p 1758-1764

Section 6: Fatigue and Fracture Resistance of Nonferrous Alloys

Selecting Aluminum Alloys to Resist Failure by Fracture Mechanisms*

R.J. Bucci, ALCOA Technical Center, G. Nordmark (retired), and E.A. Starke, Jr., University of Virginia, Department of Materials Science and Engineering

Though virtually all design and standard specifications require the definition of tensile properties for a material, these data are only partly indicative of mechanical resistance to failure in service. Except for those situations where gross yielding or highly ductile fracture represents limiting failure conditions, tensile strength and yield strength are usually insufficient requirements for design of fracture-resistant structures. Strength by itself may not be sufficient if toughness, resistance to corrosion, stress corrosion, or fatigue are reduced too much in achieving high strength.

The achievement of durable, long-lived structural components from high-strength materials requires consideration of severe stress raisers for which possible failure mechanisms are likely to be fatigue, brittle fracture, or fracture from some combination of cyclic and static loading in corrosive environments. Good design, attention to structural details, and reliable inspection are of primary importance in controlling corrosion-fatigue and fracture. Accordingly, designers have traditionally considered the minimization of stress raisers as more important than alloy choice.

However, proper alloy selection does represent an important means of minimizing premature fracture in engineering structures. Obviously, high tensile strength is potentially detrimental in parts containing severe stress raisers for which possible failure mechanisms are likely to be fatigue, brittle fracture, or fracture in combination with corrosion, static loads, and/or cyclic loading. Likewise, selecting ductile alloys of low enough strength to ensure freedom from unstable fracture is limited by economic or technical pressures to increase structural efficiencies. Therefore, optimum alloy selection for fracture control requires careful assessment and balance of trade-offs among the mechanical properties and corrosion behavior required for a given application.

*Adapted with permission from the article by R.J. Bucci in *Engineering Fracture Mechanics*, Vol 12, 1979, p 407-441 and from information contained in *Fatigue and Microstructure* (ASM, 1979, p 469-490) and from *Application of Fracture Mechanics for Selection of Metallic Structural Materials* (ASM, 1982, p 169-208)

In the aluminum industry, significant progress has been achieved in providing "improved" alloys with good combinations of strength, fracture toughness, and resistance to stress-corrosion cracking. Optimum selection and use of fatigue-resistant aluminum alloys also has become more of a factor for designers and materials engineers for extending fatigue life and/or structural efficiency. This emphasis on alloy development and selection is due, in part, to the greatly enhanced understanding of fatigue processes from the disciplines of strain control fatigue and fracture mechanics. The strain control approach is aimed primarily at fatigue crack initiation and early fatigue crack growth, while fracture mechanics concepts address the propagation of an existing crack to failure. This combination of knowledge from cyclic strain testing and fracture mechanics provides a basis for understanding of fatigue processes beyond the historical emphasis on crack nucleation studies from stress-controlled (stress to number of cycles, or *S-N*) fatigue testing. In this context, this article provides a brief overview on fatigue and fracture resistance of aluminum alloys.

Characteristics of Aluminum Alloy Classes

A wide variety of commercial aluminum alloys and tempers provide specific combinations of strength, toughness, corrosion resistance, weldability, and fabricability. The relatively high strength-to-weight ratios and availability in a variety of forms make aluminum alloys the best choice for many engineering applications. Like other face-centered cubic materials, aluminum alloys do not exhibit sudden ductile-to-brittle transition in fracture behavior with lowering of temperature (Ref 1, 2). Tensile test results indicate that almost all aluminum alloys are insensitive to strain rates between 10^{-5} mm/mm/sec and 1 mm/mm/sec (~10^5 MPa/sec) at room and low temperatures (Ref 2). Therefore, aluminum is an ideal material for structural applications in a wide range of operating temperatures and loading rates.

Aluminum alloys are classified in several ways, the most general according to their strengthening mechanisms. Some alloys are strengthened primarily by strain hardening (-H) while others are strengthened by solution heat treatment and precipitation aging (-T). A second commonly used system of classification is that of the Aluminum Association where the principal alloying element is indicated by the first digit of the alloy designation. Grouping of wrought aluminum alloys by strengthening method, major alloying element, and relative strength are given in Table 1. Another classification system established by the International Standards Organization (ISO) utilizes the alloying element abbreviations and the maximum indicated percent of element present (Table 2). This article utilizes the Aluminum Association system, which is described in more detail in *The Aluminum Association Standards and Data Handbook* (Ref 3) and in Volume 2 of the *ASM Handbook*.

Commercial aluminum products used in the majority of structural applications are selected from 2XXX, 5XXX, 6XXX, and 7XXX alloy groups, which offer medium-to-high strengths. Of these, 5XXX and 6XXX alloys offer medium-to-relatively high strength, good corrosion resistance, and are generally so tough that fracture toughness is rarely a design consideration. The 5XXX alloys provide good resistance to stress corrosion in marine atmospheres and good welding characteristics. Notably, this class of alloys has been widely used in low-temperature applications that satisfy the most severe requirements of liquefied fuel storage and transportation at cryogenic temperatures (Ref 2, 4, 5). Alloys of the 6XXX class, with good formability and weldability at medium strengths, see wide use in conventional structural applications. The 2XXX and 7XXX alloys generally are used in applications involving highly stressed parts. Certain alloys and tempers within these classes are promoted for their high toughness at high strength. Stress-corrosion cracking resistance of 2XXX and 7XXX

Table 1 Wrought aluminum and aluminum alloy designation system

Aluminum Association series	Type of alloy composition	Strengthening method	Range of tensile strength MPa	ksi
1XXX	Al	Cold work	70-175	10-25
2XXX	Al-Cu-Mg (1-2.5% Cu)	Heat treat	170-310	25-45
2XXX	Al-Cu-Mg-Si (3-6% Cu)	Heat treat	380-520	55-75
3XXX	Al-Mn-Mg	Cold work	140-280	20-40
4XXX	Al-Si	Cold work (some heat treat)	105-350	15-50
5XXX	Al-Mg (1-2.5% Mg)	Cold work	140-280	20-40
5XXX	Al-Mg-Mn (3-6% Mg)	Cold work	280-380	40-55
6XXX	Al-Mg-Si	Heat treat	150-380	22-55
7XXX	Al-Zn-Mg	Heat treat	380-520	55-75
7XXX	Al-Zn-Mg-Cu	Heat treat	520-620	75-90
8XXX	Al-Li	Heat treat	280-560	40-80

Table 2 ISO equivalents of wrought Aluminum Association designations

Aluminum Association international designation	ISO designation(a)
1050A	Al 99.5
1060	Al 99.6
1070A	Al 99.7
1080A	Al 99.8
1100	Al 99.0 Cu
1200	Al 99.0
1350	E-Al 99.5
...	Al 99.3
1370	E-Al 99.7
2011	Al Cu6BiPb
2014	Al Cu4SiMg
2014A	Al Cu4SiMg(A)
2017	Al Cu4MgSi
2017A	Al Cu4MgSi(A)
2024	Al Cu4Mg1
2030	Al Cu4PbMg
2117	Al Cu2.5Mg
2219	Al Cu6Mn
3003	Al Mn1Cu
3004	Al Mn1Mg1
3005	Al Mn1Mg0.5
3103	Al Mn1
3105	Al Mn0.5Mg0.5
4043	Al Si5
4043A	Al Si5(A)
4047	Al Si12
4047A	Al Si12(A)
5005	Al Mg1(B)
5050	Al Mg1.5(C)
5052	Al Mg2.5
5056	Al Mg5Cr
5056A	Al Mg5
5083	Al Mg4.5Mn0.7
5086	Al Mg4
5154	Al Mg3.5
5154A	Al Mg3.5(A)
5183	Al Mg4.5Mn0.7(A)
5251	Al Mg2
5356	Al Mg5Cr(A)
5454	Al Mg3Mn
5456	Al Mg5Mn
5554	Al Mg3Mn(A)
5754	Al Mg3
6005	Al SiMg
6005A	Al SiMg(A)
6060	Al MgSi
6061	Al Mg1SiCu
6063	Al Mg0.7Si
6063A	Al Mg0.7Si(A)
6082	Al Si1MgMn
6101	E-Al MgSi
6101A	E-Al MgSi(A)
6181	Al Si1Mg0.8
6262	Al Mg1SiPb
6351	Al Si1Mg0.5Mn
7005	Al Zn4.5Mg1.5Mn
7010	Al Zn6MgCu
7020	Al Zn4.5Mg1
7049A	Al Zn8MgCu
7050	Al Zn6CuMgZr
7075	Al Zn5.5MgCu
7178	Al Zn7MgCu
7475	Al Zn5.5MgCu(A)
...	Al Zn4Mg1.5Mn
...	Al Zn6MgCuMn

Note: The proposed ISO chemical composition standard for aluminum and its alloys references Aluminum Association equivalents as well as its own identification system. The ISO system is based on the systems that have been used by certain European countries. The main addition element is distinguished by specifying the required content (middle of range) rounded off to the nearest 0.5. If required, the secondary addition elements are distinguished by specifying the required content rounded off to the nearest 0.1, for two elements at most. (a) The chemical symbols for addition elements should be limited to four. If an alloy cannot otherwise be distinguished, a suffix in brackets is used: 6063 = Al Mg0.7Si; 6463 = Al Mg0.7Si(B); and international alloy registration, 6063A = Al Mg0.7Si(A). Note that suffixes (A), (B), and so on should not be confused with suffixes of the Aluminum Association.

alloys is generally not as great as in other aluminum alloy groups; however, service failures are avoided by good engineering practices and proper selection of alloy and temper or a suitable protective system. The 2XXX and 7XXX alloys see widespread use in aerospace applications. Certain 2XXX and 7XXX alloys provide good welding characteristics at high strength.

Alloys of the 1XXX class are used primarily in applications where electrical conductivity, formability, ductility, and resistance to stress corrosion are more important than strength. The 3XXX alloys, widely used in piping applications, are characterized by relatively low strengths and very good toughness, ductility, formability, brazing, and welding characteristics. The 4XXX alloys are used mainly for welding wire and brazing applications where it is desired to have a lower melting point than in the wire without producing brittleness in the weldment.

Alloy Selection Concepts. In any design plan, priority must be given to alloy properties. Optimum alloy choice involves evaluation and decision based on rating characteristics of a material that quantitatively measure resistance to failure by foreseeable failure mechanisms. In some instances, trade-off will be necessary among these material characteristics and among other factors, such as cost, fabricability, availability, expected service life, and maintainability. Relatively few generalizations can be made that will be valid for all material selection problems; individual problems must be treated separately or on the basis of closely related experience.

An important consideration to the relative ranking of importance of properties to prevent failure is the particular application and basic design strategy to which the selected alloy will be applied. It is pertinent to review basic design philosophies by which aluminum alloys are selected to resist failure by fracture mechanisms. Later discussion will treat alloy selection concepts related to the specific areas of fracture toughness, corrosion, stress-corrosion cracking (SCC), and fatigue.

Design Philosophies. In general, design philosophies for the prevention of fracture-type failures are of two basic types: safe life and damage-tolerant (or fail-safe). Neither approach is meant to be used as an extreme, nor is either approach meant to replace need for full-scale design verification tests. Many applications require a "fracture-control plan" to arrive at rational and cost-effective criteria for design, fabrication, and maintenance of reliable structures.

Safe Life Design Approach. Traditionally, component life has been expressed as the time (or number of fatigue cycles) required for a crack to be initiated and grow large enough to produce catastrophic failure. Prior to development of reliable crack detection techniques and fracture mechanics technology, little attempt was made to separate component failure into initiation and propagation stages. It was assumed that total life of a part consisted primarily of initiation of a crack, generally by fatigue or stress corrosion. Time for a minute crack to grow and produce failure was considered a minor portion of the service life. In the safe life approach, which is an outgrowth of this assumption, the designer seeks long, safe life by preventing cracks of significant size from occurring during the service life of the structure. In this approach it is the incubation period leading to development of a significant crack that is of major concern.

Small coupon-type specimens, though useful for rating materials and establishing sensitivity of various load and fabrication parameters, are not suitable for establishing the life of the part. A safe-life evaluation of a structure requires a reasonably accurate experimental simulation of the particular item of hardware. Under this procedure, accurately described loads are applied to the structure, life is determined, and a scatter factor is applied to establish the safe life of the structure. Structural "hot spots" are retrofitted as necessary. Generally, such elaborate tests prohibit evaluation of a large number of candidate materials and structural arrangements, since testing of each option may not be feasible because of economic and time constraints. Therefore, design and, consequently, material selection by this approach rely heavily on experience to eliminate need for excessive structural maintenance and retrofit.

Damage-Tolerant (Fail-Safe) Design Approach. Damage tolerance describes features of design that prohibit catastrophic loss of structural integrity. Damage tolerance evaluation of structure is intended to ensure that, should serious cracking or damage occur, the remaining structure can withstand reasonable loads without excessive structural deformation until the damage is detected. Consideration must be given to the probable existence of flaws (cracks) in the structure. These flaws could be initiated in service or be present as undetected initial material or fabrication defects. Given a crack-like flaw corresponding to the maximum size escaping reliable detection, life of the part is assumed to be spent propagating this flaw to the critical size that results in unstable fracture. The general design strategy is to select stress levels, configurations, and materials to provide a controlled slow rate of crack propagation with high residual strength. The designer thereby seeks to limit the rate of flaw growth so the largest flaw missed at one inspection will not cause catastrophic failure before one or more later inspections. Analysis procedures depend heavily on the use of crack growth rates and fracture toughness combined with fracture mechanics principles for prediction of crack growth life and fracture strength. Moreover, inspection is an integral part of the fracture control plan. Recognition of these principles and their implications for the safety, reliability, and durability of engineering structures has resulted in engineering standards and codes that impose requirements of fracture mechanics analyses and control of crack behavior. Perhaps the most notable of these is the Air Force structural integrity requirement (Ref 6). With this plan, use of fracture toughness and stress corrosion testing in material procurement is required to ensure that materials with properties lower than those used in design do not appear in the final structure.

Fracture Mechanics and Toughness

Fracture mechanics is concerned with catastrophic failure associated with crack-like flaws, regardless of how the flaw originated. Important parameters are crack size, local stress in the absence of the crack, yield strength, and materials fracture toughness. Use of these parameters allows prediction of the terminal flaw size in a part. The fracture mechanics approach is based on analysis of crack tip stress-strain fields. When stresses are below the yield stress, the critical stress concentration for fracture lies in the domain of linear elastic fracture mechanics and is an inherent material property K_{Ic} (plane-strain fracture toughness).

The concepts of fracture mechanics are concerned with the basic methods for predicting the load-carrying capabilities of structures and components containing cracks. The fracture mechanics approach is based on a mathematical description of the characteristic stress field that surrounds any crack in a loaded body. When the region of plastic deformation around a crack is small compared to the size of the crack (as is often true for large structures and high-strength materials), the magnitude of the stress field around the crack is commonly expressed as the stress intensity factor, K, where:

$$K = \sigma \sqrt{(a)}\, Y \qquad \text{(Eq 1)}$$

where σ = remotely applied stress, a = characteristic flaw size dimension, and Y = geometry factor, determined from linear elastic stress analysis. The stress-intensity factor, K, thus represents a single parameter that includes both the effect of the stress applied to a sample and the effect of a crack of a given size in a sample. The stress-intensity factor can have a simple relation to applied stress and crack length, or the relation can involve complex geometry factors for complex loading, various configurations of real structural components, or variations in crack shapes. In this way, linear elastic analysis of small-scale yielding can be used to define a unique factor, K, that is proportional to the local crack tip stress field outside the small crack tip plastic zone.

These concepts provide a basis for defining a critical stress-intensity factor (K_c) for the onset of crack growth, as a material property independent of specimen size and geometry for many conditions of loading and environment. For example, if a combination of σ and a were to exceed a critical value K_c, then the crack would be expected to propagate.

Tests on precracked specimens of a wide variety of materials have shown that the critical K value at the onset of crack extension approaches a constant value as specimen thickness increases. Figure 1 shows this effect in tests with 7075 aluminum alloy specimens over a range of thickness. In general, when the specimen thickness and the inplane dimensions near the crack are large enough relative to the size of the plastic zone, then the value of K at which growth begins is a constant and generally minimum value called the plane-strain fracture toughness factor, K_{Ic}, of the material. The parameter K_{Ic} is a true material property in the same sense as is the yield strength of a material. The value of K_{Ic} determined for a given material is unaffected by specimen dimensions or type of loading, provided that the specimen dimensions are large enough relative to the plastic zone to ensure plane-strain conditions around the crack tip (strain is zero in the through-thickness or z-direction).

Plane-strain fracture toughness, K_{Ic}, is also directly related to the energy required for the onset of crack propagation by the formula

$$K_{Ic} = \sqrt{\frac{EG_{Ic}}{1 - \nu^2}} \qquad \text{(Eq 2)}$$

where E is the elastic modulus (in MPa or psi), ν is Poisson's ratio (dimensionless), and G_{Ic} is the critical plane-strain energy release rate for crack extension (in kJ/m^2 or in.-lb/in.^{2}). In simplified concept, G_{Ic} is the critical amount of strain energy that is released from the elastic stress field of the specimen per unit area of new cracked surface for the first small increment of crack extension. The concepts of K_{Ic} and G_{Ic} are essentially interchangeable; K_{Ic} is generally preferred because it is more easily associated with the stress or load applied to a specimen. The value of K_{Ic} is measured directly using test methods described in ASTM E-399.

In the plane-strain state, a material is at its lowest point of resistance to unstable fracture. The onset of fracture is abrupt and is most clearly

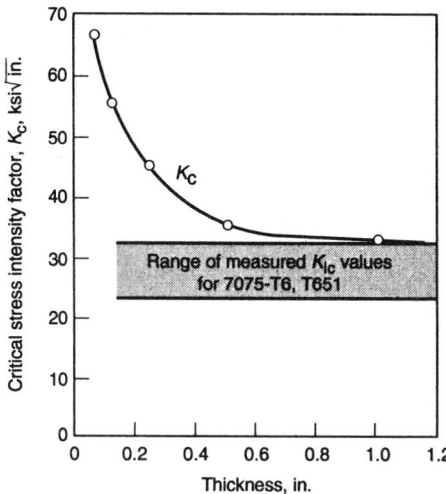

Fig. 1 Fracture toughness of 7075-T6, T651 sheet and plate from tests of fatigue-cracked center-notched specimens (transverse). Source: J.G. Kaufmann in *Review of Developments in Plane Strain Fracture Toughness Testing*, ASTM STP 463, 1970, p 7

observed in thick sections of low-ductility (high-strength) alloys, when the elastic stress state in a flawed component is highly constrained to that of plane strain. However, when stresses approach or exceed yield values, the elastic stress field surrounding the crack departs from that of plane strain (from the development of an enlarged crack tip plastic zone which generally enhances fracture toughness). With increasing load, slow stable crack extension (tearing) may accompany the increasing plastic zone size. Onset of rapid fracture occurs when increase in crack tip stress field, measured by K (increase in K due to increased nominal stress and crack length), equals or exceeds resistance to crack extension (due to an increase in plastic zone size, crack tip blunting, and change from flat to slant fracture). This behavior is most clearly seen in fracture of relatively tough thin plate and sheet alloys. Unstable fracture under these conditions cannot be described as a material property since events leading to rapid fracture are specimen configuration and size dependent. One standardized method for describing elastic-plastic fracture involves the resistance-curve or R-curve concept described in ASTM E561. Briefly, the resistance-curve concept involves measurement of the K values at which various amounts of crack growth occur in a thin-plate laboratory specimen. Then a plotted curve of K versus crack growth from the laboratory specimen can be used to predict crack-growth behavior in a structural component of the same material. Limitations of the method are that the component must have the same thickness as the laboratory specimen and that K relations must be known for both component and specimen. However, once a resistance curve is obtained for a given material and thickness, it can be used to predict the crack-growth and crack-instability behavior of other components of the same material.

J-Integral Method. Another concept for use in the analysis of elastic-plastic fracture is the

Table 3 Typical room-temperature yield strength and plane-strain fracture toughness values for several high-strength aluminum alloys

Product	Alloy	Temper	Yield strength(a) MPa	ksi	L-T MPa√m	ksi√in.	T-L MPa√m	ksi√in.	S-L MPa√m	ksi√in.
Plate	2014	T651	440	64	24	22	22	20	19	17
	2024	T351	325	47	36	33	33	30	26	24
	2024	T851	455	66	24	22	23	21	18	16
	2124	T851	440	64	32	29	25	23	24	22
	2219	T851	435	63	39	35	36	33	…	…
	7050	T73651	455	66	35	32	30	27	29	26
	7075	T651	505	73	29	26	25	23	20	18
	7075	T7651	470	68	30	27	24	22	20	18
	7075	T7351	435	63	32	29	29	26	21	19
	7475	T651	495	72	43	39	37	34	32	29
	7475	T7651	460	67	47	43	39	35	31	28
	7475	T7351	430	62	53	48	42	38	35	32
Die forgings	7050	T736	455	66	36(b)	33(b)	25(c)	23(c)	25(c)	23(c)
	7149	T73	460	67	34(b)	31(b)	24(c)	22(c)	24(c)	22(c)
	7175	T736	490	71	33(b)	30(b)	29(c)	26(c)	29(c)	26(c)
Hand forgings	2024	T852	430	62	29	26	21	19	18	16
	7050	T73652	455	66	36	33	23	21	22	20
	7075	T7352	365	53	37	34	29	26	23	21
	7079	T652	440	64	29	26	25	23	20	18
	7175	T736	470	68	37	34	30	27	26	24
Extrusions	7050	T7651x	495	72	31	28	26	24	21	19
	7050	T7351x	459	65	45	41	32	29	26	24
	7075	T651x	490	71	31	28	26	24	21	19
	7075	T7351x	435	63	35	32	29	26	22	20

(a) At 0.2% offset (longitudinal). (b) Parallel to grain flow. (c) Nonparallel to grain flow. Source: R. Bucci, *Engr. Fracture Mech.*, Vol 12, 1979, p 407-441

J-integral concept, where *J* is a nonlinear generalization of *G* (the elastic strain energy release rate). *J* can be thought of as the amount of elastic-plastic strain energy per unit area of crack growth which is applied toward extending the crack in a specimen under load. A critical value of *J*, called J_{Ic}, is the value required for the start of crack extension from a pre-existing crack. For material having a sufficiently high yield strength or for specimens of sufficient size, elastic stresses control the crack extension, and J_{Ic} is equal to G_{Ic}.

An important advantage of the J_{Ic} test method is that it can accommodate a significant amount of crack-tip blunting and general plastic deformation in the specimen. If the amount of plastic deformation is small enough, J_{Ic} will be identical to G_{Ic}, and thus J_{Ic} can be converted to an approximately equivalent measure of K_{Ic} (see Eq 2). For large amounts of plastic deformation, a size requirement limits the size of the specimen and, indirectly, the amount of plastic deformation which can be allowed. The specimen size requirement allows a significantly smaller specimen, often ten times smaller, to be tested with the J_{Ic} procedure than with the K_{Ic} procedure. So, although the J_{Ic} test is relatively time consuming due to multiple tests, it can be used over a wider range of material properties and specimen sizes than the K_{Ic} test. In addition, single-specimen J_{Ic} test procedures, such as incremental unloading methods, can reduce both testing time and the number of specimens required to obtain J_{Ic} test data. Another advantage of the J_{Ic} approach is that it makes possible the prediction of the failure load of cracked high-toughness, medium-strength alloys (with more tendency toward plastic deformation) for fracture-critical applications.

Alloy Selection for Fracture Toughness

Plane-strain fracture toughness, K_{Ic}, is particularly pertinent in materials selection because, unlike other measures of toughness, it is independent of specimen configuration. For comparison, the notch toughness of a material, which is most commonly measured by Charpy testing, does depend on the configuration of the specimen. Changes in the size of the specimen or in the root radius of the notch will affect the amount of energy absorbed in a Charpy test. The main reason for this is that the total energy required for initiation of the crack from the notch, for propagation of the crack across the specimen, and for complete fracture of the specimen is measured in a Charpy test. In contrast, a K_{Ic} test measures only the critical load required for a small extension of a pre-existing crack. Even though K_{Ic} is more difficult to measure than notch toughness, because of the requirements of a pre-existing crack and a specimen large enough for plane-strain conditions, it is a constant material property and can be more generally applied to materials selection.

In selection of structural materials, the single most important characteristic of K_{Ic} for nearly all materials is that it varies *inversely* with yield strength for a given alloy. Table 3 is a summary of plane-strain fracture toughness values for a wide variety of aluminum alloys—including 2024 and 7075, which have been used for components which do not require a high level of fracture toughness. Typical data for a lot of one high-toughness aluminum alloy (7475-T7351) are shown in Table 4.

Table 4 Typical fracture toughnesses measured on the same lot of a high-toughness 7475-T7351 aluminum alloy plate

Specified test method	Typical L-T toughness(a) MPa√m	ksi√in.
ASTM E 399, K_{Ic}	55	50
ASTM E 561, *R* curve		
5% secant	45-65	40-60
25% secant	75-110	70-100
Center-cracked panel 150 mm (6 in.) wide	130	120
Center-cracked panel 400 mm (16 in.) wide	200	180

Note: Test sample thicknesses are those typically specified. Data are approximate and are presented to contrast different test results, not for use for design purposes. *R*-curve testing also requires tensile data for test validity checks, and these should be provided to the testing laboratory. (a) Test sample orientation code is described in ASTM E 399 (Ref 43). The first letter represents the direction of applied tensile stress: the second letter is the direction of crack growth. L, longitudinal; T, transverse; S, thickness direction

Tough aluminum alloys such as those from the 1XXX, 3XXX, 4XXX, 5XXX, and most 6XXX series do not normally exhibit elastic unstable fracture, either in test panels or in real structures. These alloys are so tough that fracture toughness is rarely a design criterion. Because of this consideration and the relative difficulty of measuring toughness in a design-oriented manner by current methods, fracture toughness information for these alloys is rather limited. Much of the data available for these alloys (Fig. 2) has been developed by extrapolation of correlations from simpler tests such as the simple tear specimen of the design shown in Fig. 3, where the resistance of a

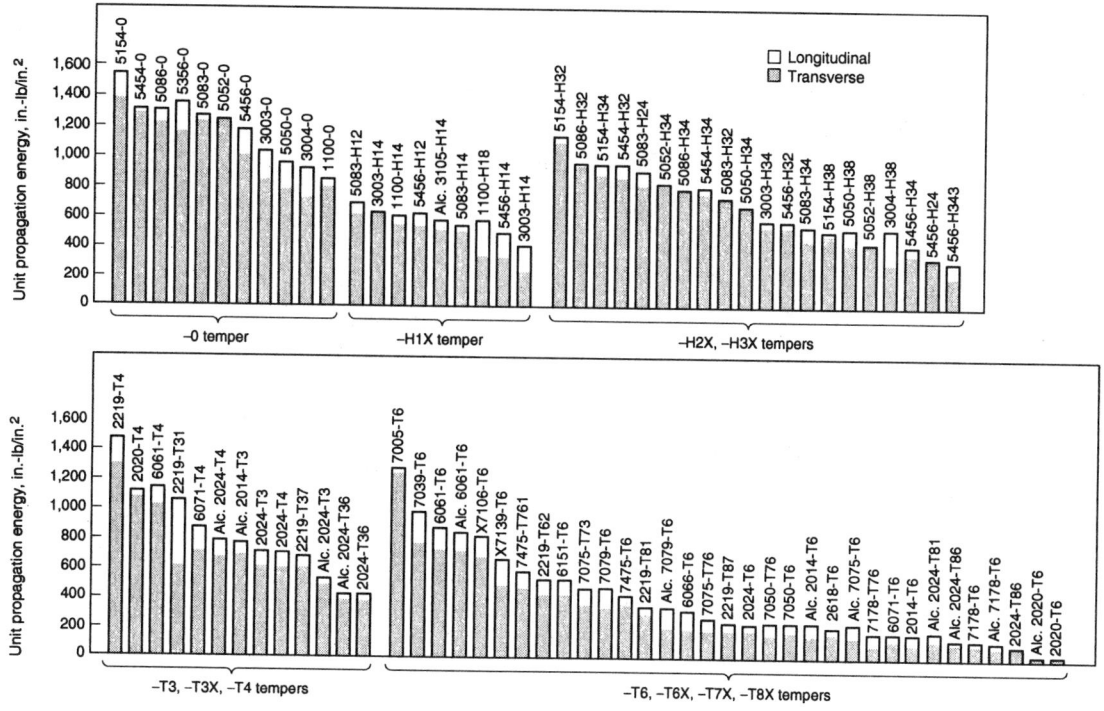

Fig. 2 Ratings of 1.6 mm (0.063 in.) aluminum sheet based on unit propagation energy (1 in. · lb/in.² = 0.175 kJ/m²)

material to crack growth in a nonuniform stress field is evaluated by measurement of appropriate areas under autographic load deformation records. A relative ranking of thin-section fracture toughness by tear tests is shown in Fig. 2 for a number of aluminum sheet alloys and tempers. Particularly, the unit propagation energy (UPE) has been found to be correlated with resistance to stable crack growth in thin sections, measured as K_{Ic} from wide-panel tests (Ref 1). These alloys are excluded from further consideration in this section.

Alloys for which fracture toughness is a meaningful design-related parameter fall into two categories:

* Controlled-toughness, high-strength alloys (i.e., those alloys developed primarily for their high fracture toughness at high strength)
* Conventional high-strength alloys, tempers, and products for which fracture toughness is a meaningful design parameter, but which are not promoted or used for fracture-critical components

The above categories are composed primarily of 2XXX and 7XXX alloys. Controlled-toughness, high-strength commercial products include 2124-T3 and T8-type sheet and plate; 2419-T6 and T8-type sheet, plate, extrusion, and forgings; 7050-T7-type plate, forgings, and extrusions; 7149-T7-type forgings; 7175-T6 and T7-type extrusions and forgings; and 7475-T6 and T7-type sheet and plate.

Recognized conventional high-strength alloys that are not produced to minimum toughness include 2014, 2024, 2219, 7075, 7079, and 7178. Typical strength and fracture toughness proper-

$$\text{Tear strength, psi} = \frac{P}{A} + \frac{MC}{I} = \frac{P}{bt} + \frac{3P}{bt} = \frac{4P}{bt}$$

$$\text{Unit propagation energy, in.-lb per square in.} = \frac{\text{energy to propagate a crack}}{bt}$$

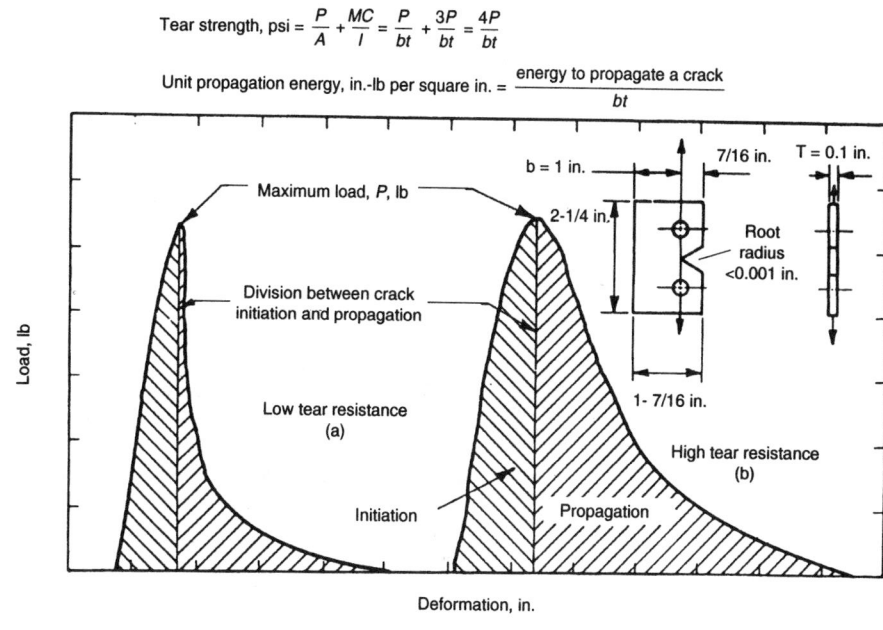

Fig. 3 Tear-test specimen and representative tear test curves

ties of several high-strength aluminum products are presented in Table 5. Evaluations have shown (Ref 7) that the fracture toughness of high-strength, precipitation-hardened 2XXX and 7XXX alloys is not adversely affected by high strain rate or moderate temperature reduction.

In general, fracture toughness decreases with increasing yield strength, as indicated by scatter bands of notched yield strength ratio (NYR) and

UPE established for a wide variety of commercial 2XXX and 7XXX products. To develop high toughness, the microstructure must accommodate significant plastic deformation, and yet a microstructure that resists plastic deformation is needed for high strength. As indicated by Fig. 4(a) and 4(b), 7XXX alloys have the highest combination of strength and toughness of any family of aluminum alloys. In 7XXX alloys,

Table 5 Typical room-temperature yield strength and plane-strain fracture toughness values of several high-strength aluminum alloys

| Product | Alloy | Temper | Yield strength at 0.2% offset (longitudinal) | | Plane-strain fracture toughness, K_{Ic} | | | | | |
| | | | | | L-T | | T-L | | S-L | |
			MPa	ksi	MPa√m	ksi√in.	MPa√m	ksi√in.	MPa√m	ksi√in.
Plate	2014	T651	440	64	24	22	22	20	19	17
	2024	T351	325	47	36	33	33	30	26	24
	2024	T851	455	66	24	22	23	21	18	16
	2124	T851	440	64	32	29	25	23	24	22
	2219	T851	435	63	39	35	36	33	…	…
	7050	T73651	455	66	35	32	30	27	29	26
	7075	T651	505	73	29	26	25	23	20	18
	7075	T7651	470	68	30	27	24	22	20	18
	7075	T7351	435	63	32	29	29	26	21	19
	7475	T651	495	72	43	39	37	34	32	29
	7475	T7651	460	67	47	43	39	35	31	28
	7475	T7351	430	62	53	48	42	38	35	32
Die forgings	7050	T736	455	66	36	33(a)	25	23(b)	25	23(b)
	7149	T73	460	67	34	31(a)	24	22(b)	24	22(b)
	7175	T736	490	71	33	30	29	26	29	26
Hand forgings	2024	T852	430	62	29	26	21	19	18	16
	7050	T73652	455	66	36	33	23	21	22	20
	7075	T7352	365	53	37	34	29	26	23	21
	7079	T652	440	64	29	26	25	23	20	18
	7175	T736	470	68	37	34	30	27	26	24
Extrusions	7050	T7651X	495	72	31	28	26	24	21	19
	7050	T7351X	459	65	45	41	32	29	26	24
	7075	T651X	490	71	31	28	26	24	21	19
	7075	T7351X	435	63	35	32	29	26	22	20

(a) Parallel to grain flow. (b) Nonparallel to grain flow

highest strength is associated with the T6 peak aged temper. Decreasing strength to acceptable levels by overaging provides a way to increase toughness (Fig. 5) as well as resistance to exfoliation, SCC, and fatigue crack growth in some 7XXX alloys. Alloys of the 2XXX class are used in both the naturally aged and artificially aged conditions. Commercial naturally aged 2XXX alloys (viz. T3- and T4-type tempers) provide good combinations of toughness and strength. Artificial aging to precipitation-hardened T8 tempers produces higher strength with some reduction in toughness, but in addition offers greater stability of mechanical properties at high temperatures and higher resistance to exfoliation and SCC.

The strength-fracture toughness interaction has been postulated to be the consequence of void link-up created by slip-induced breakdown of submicron strengthening particles, which occurs more readily at high strength levels (Ref 8). If the strengthening (matrix) precipitates are shearable they may promote strain localization which leads to premature crack nucleation and low fracture toughness. Whether or not the strengthening precipitates are sheared or looped and bypassed by dislocations depends on alloy composition and aging treatment. During aging, heterogeneous precipitation usually occurs at grain and subgrain boundaries resulting in soft, solute-denuded PFZ's in the matrix adjacent to the boundaries. The combination of these soft zones, that can localize strain, and grain boundary precipitates, that can aid in microvoid nucleation, also has an adverse effect on fracture toughness. Though this hypothesis remains unproven, it has been clearly demonstrated that the amounts, distribution, and

morphology of alloy phases and second-phase particles in alloy microstructure have a large influence on toughness (Ref 9-11). Developed understanding of the interrelationships of alloy microstructure and fracture mechanisms has led to design of new commercial aluminum alloys offering optimum high strength and high toughness. Primarily, the alloy improvements have evolved through microstructural control obtained by increased purity, modified compositions, and better homogenization, fabrication, and heat treatment practices (Ref 10-15).

The balance between strength and toughness is greatly affected by a variety of processing parameters, including solution heat treatment, quenching efficiency, deformation prior to aging (for 2XXX alloys) and aging treatment. The solution heat treatment determines the amount of solute in solid solution and the vacancy content, which affects subsequent aging kinetics. Quenching affects both the microstructure and properties by determining the amount of solute that precipitates during cooling and that which is available for subsequent age hardening. It also affects the level of residual stresses which can influence manufacturing costs, fatigue and corrosion behavior. After quenching, methods to obtain a balance of properties include cold working before aging, when practical (T8 temper), and selecting aging times and temperatures to minimize grain boundary precipitates and precipitate-free zones (PFZ). The deformation prior to aging aids in the nucleation and growth of the matrix precipitates which decreases the time to reach peak strength. This, along with low-temperature aging, minimizes the amount of grain boundary precipitates

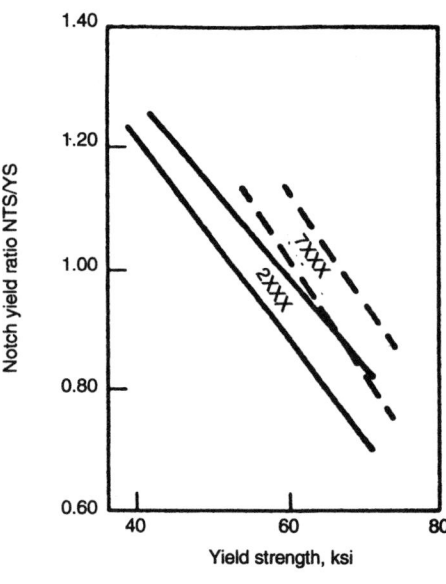

(a)

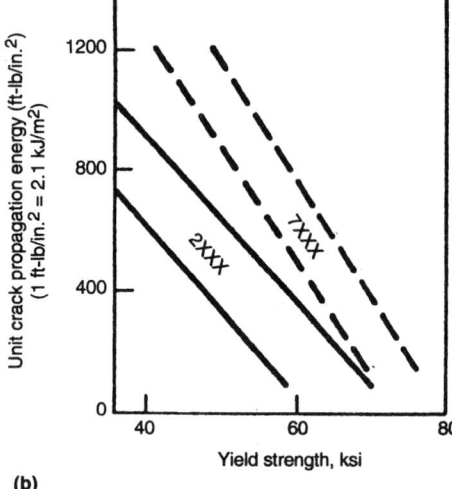

(b)

Fig. 4 Comparison of 2XXX and 7XXX commercial aluminum alloys (a) Notch toughness vs. yield strength. (b) Unit propagation energy vs. yield strength

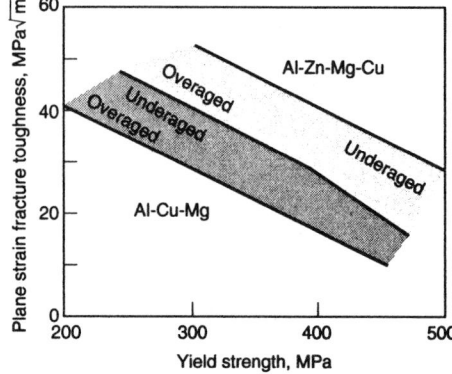

Fig. 5 Relationships of plane-strain fracture toughness to yield strength for the 2XXX and 7XXX series of aluminum alloys. Source: R. Develay, *Metals and Materials*, Vol 6, 1972, p 404

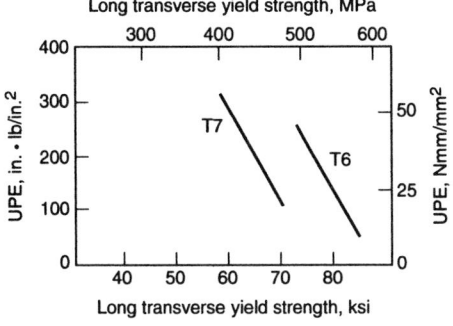

(a)

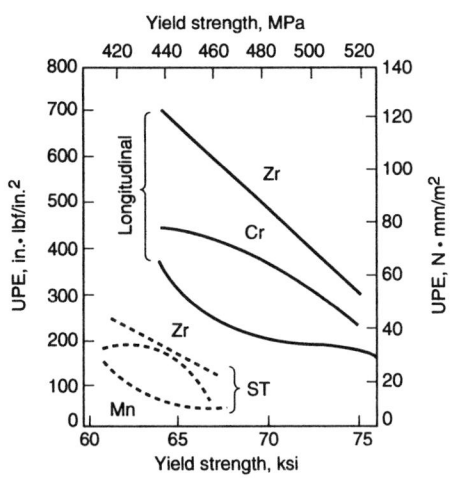

(b)

Fig. 6 Unit crack propagation energies (UPE), (a) commercial 7XXX aluminum alloy plate in peak strength and overaged tempers (Source: Ref 10) and (b) effects of dispersoid type on toughness of 75 mm (3 in.) 7075 plate (Source: J. Staley in ASTM STP 605)

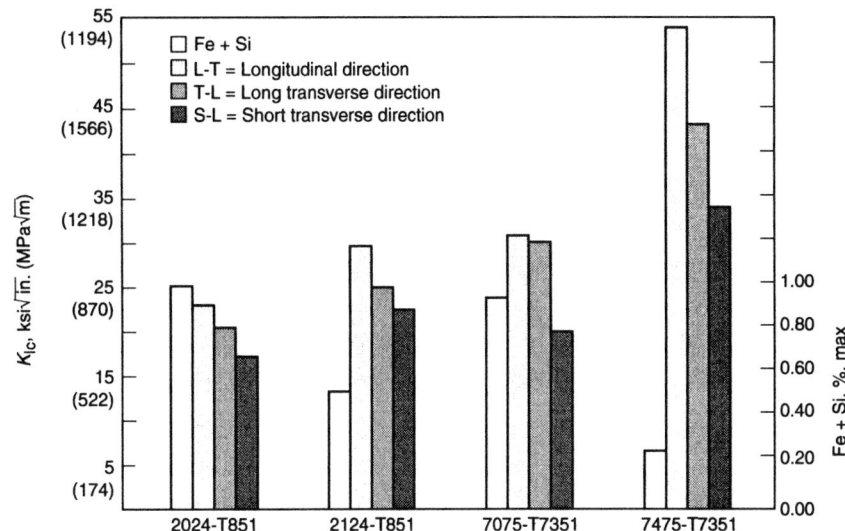

Fig. 7 Aluminum alloys 2124 and 7475 are tougher versions of alloys 2024 and 7075. High-purity metal (low iron and silicon) and special processing techniques are needed to optimize toughness in these materials. Source: Ref 14

Table 6 Effect of purity on the fracture toughness of some high-strength wrought aluminum alloys

Alloy and temper	% Fe, %, max	% Si, %, max	0.2 % proof stress, MPa	Tensile strength, MPa	Fracture toughness, MPa√m	
					Longitudinal	Short transverse
2024-T8	0.50	0.50	450	480	22-27	18-22
2124-T8	0.30	0.20	440	490	31	25
2048-T8	0.20	0.15	420	460	37	28
7075-T6	0.50	0.40	500	570	26-29	17-22
7075-T73	0.50	0.40	430	500	31-33	20-23
7175-T736	0.20	0.15	470	540	33-38	21-29
7050-T736	0.15	0.12	510	550	33-39	21-29

Source: M.O. Speidel, *Met. Trans.*, Vol 6A, 1975, p 631

Table 7 Nominal compositions of aluminum alloys used in low-temperature service

Alloy designation	Nominal composition, %								
	Si	Cu	Mn	Mg	Cr	Zn	Ti	Zr	Others
Wrought alloys									
1100	...	0.12	...	...	...	...	...	...	...
2014	0.8	4.4	0.8	0.5	...	...	...	...	...
2024	...	4.4	0.6	1.5	...	...	...	...	...
2219	...	6.3	0.3	...	...	...	0.06	0.18	0.1V
3003	...	0.12	1.2	...	...	...	...	...	...
5083	...	...	0.7	4.4	0.15	...	...	...	...
5456	...	...	0.8	5.1	0.12	...	...	...	...
6061	0.6	0.28	...	1.0	0.20	...	...	...	...
7005	...	...	0.45	1.4	0.13	4.5	0.04	0.14	...
7039	0.1	0.05	0.25	2.8	0.20	3.0	0.05	...	0.2Fe
7075	...	1.6	...	2.5	0.23	5.6	...	...	...
Cast alloys									
355	5.0	1.2	...	0.5	...	...	...	...	...
C355	5.0	1.3	...	0.5	...	...	...	...	...
356	7.0	...	...	0.3	...	...	...	...	...
A356	7.0	...	...	0.3	...	...	...	...	...

and PFZ's (which adversely affect fracture toughness) at the desired strength level.

Alloy 2124 was the first 2XXX alloy developed to have high fracture toughness. The principal contribution to high toughness was increased purity (low iron and silicon), which minimizes formation of relatively large insoluble constituents (>1 μm). The detrimental effect of large constituent phases on the fracture toughness of aluminum alloys has been documented by many investigators. Constituent particles participate in the fracture process through void formation at particle/matrix interfaces or by fracturing during primary processing. Their volume fraction can be minimized by reducing impurity elements, e.g., iron and silicon, and excess solute. The detrimental effect of dispersoids also depends on their size and the details of their interface with the matrix. For example, the strength-toughness relationships in Fig. 6(b) were determined for 7075 variants containing different dispersoid-forming elements. Because Zr particles are small and coherent with the matrix (strong interface), they are usually not involved in the fracture process.

Resultant improvement for production materials is shown in Fig. 7 (Ref 14). Minimization of insoluble constituents by process control was used to develop 2419 and 2214 as higher-tough-

ness versions of 2219 and 2014, respectively. Biggest gains in fracture toughness of 2XXX alloys by process control have been to the precipitation-hardened T8 tempers which are widely used in applications requiring good resistance to

exfoliation corrosion and SCC. The effect of impurity on toughness of other alloys is shown in Table 6.

Grain size and degree of recrystallization can have a significant effect on fracture toughness.

The desired degree of recrystallization depends on product thickness, i.e., whether the part is under plane stress or plane strain. In thin products under plane stress, fracture is controlled by plasticity and a small recrystallized grain size is preferable. If the grain size is small enough, plasticity will be enhanced without detrimental, low energy, intergranular fracture. However, for thick products under plane strain, fracture is usually controlled by coarse particles and an unrecrystallized grain structure is preferable.

Alloy 7475 represents one of the most successful applications of alloy design techniques. Its composition and properties are modified from those of alloy 7075 by

- Reducing iron and silicon contents
- Optimizing dispersoids
- Altering precipitates
- Controlling quenching rate
- Controlling grain size

These modifications result in the toughest aluminum alloy available commercially at high strength levels. For designers this influence is shown most clearly by information on crack lengths for unstable crack growth at specific design stresses, such as that shown in Fig. 8. The crack tolerance of the 7475-T761 alloy sheet is almost three times greater than that of conventional 7075-T6. Similar effects have been noted for plate.

Alloy 7475 represents the highest strength-toughness combinations available in a commercial aluminum alloy. However, patented process controls (in addition to controlling the purity of iron and silicon) are necessary to achieve highest toughness levels in 7475. In comparison to conventional high-strength alloys, the effectiveness of alloy 7475 in developing high toughness at high strength is shown by plane-strain fracture toughness (K_{Ic}) data (Fig. 9), plane-stress fracture toughness (K_c) data from wide center crack panel tests (Fig. 10), and crack resistance curves (Fig. 11). This advantage is demonstrated by the critical stress-flaw size relationships in Fig. 8. The effect of heat treatment on crack propagation energy is shown in Fig. 6. Controls on production processes for high-toughness alloys 2124 and 7475 should also improve fatigue crack growth resistance.

7150-T77 Plate. In response to a need for improved corrosion resistance, another temper was developed for 7150. Alloy 7150-T77 plate develops the same mechanical properties as does 7150-T6 with significantly improved resistances to both exfoliation corrosion and SCC. The first application was on the C17 cargo transport. This saved a considerable amount of weight because corrosion performance of 7150-T6 and T61 was deemed to be inadequate by the Air Force for this application, and strength of 7050-T76 is considerably lower. The combination of strength and corrosion characteristics of 7150-T77 is attributed to proprietary processing. This processing promotes the development of a precipitate structure which effectively resists the passage of dislocations equivalent to that provided by the T6 temper and simultaneously minimizes the elec-

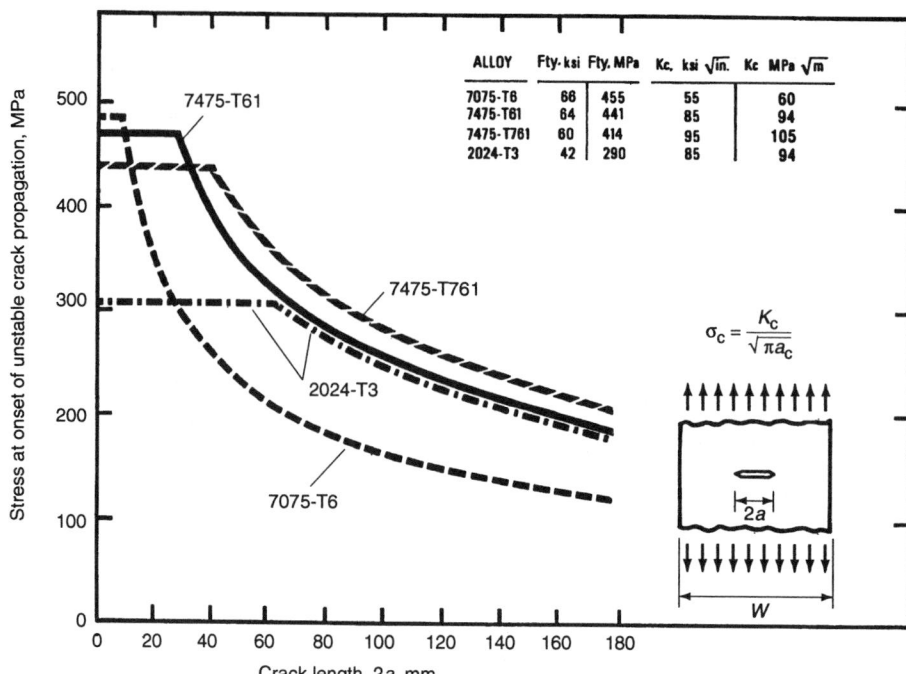

ALLOY	Fty, ksi	Fty, MPa	Kc, ksi √in.	Kc, MPa √m
7075-T6	66	455	55	60
7475-T61	64	441	85	94
7475-T761	60	414	95	105
2024-T3	42	290	85	94

$$\sigma_c = \frac{K_c}{\sqrt{\pi a_c}}$$

Fig. 8 Gross section stress at initiation of unstable crack propagation vs. crack length for wide sheet panels of four aluminum alloy/temper combinations. Source: Ref 13

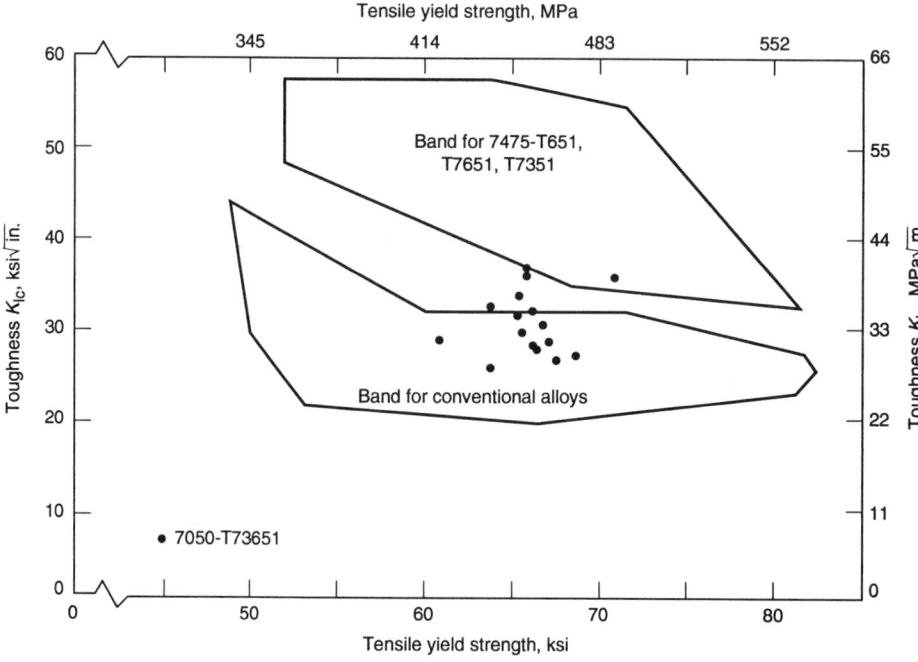

Fig. 9 Fracture toughness vs. yield strength for high-strength aluminum alloy plate (L-T orientation)

trochemical differences between the matrix and grain boundaries. Extruded products in T77 do not develop the 70 MPa strength advantage.

7055-T77 Plate. The implementation of the T77 temper for 7150 was followed by development of a proprietary material for compressively loaded structures. Alloy 7055-T77 plate offers a strength increase of about 10% relative to that of 7150-T6 (almost 30% higher than that of 7075-

T76). It also provides a high resistance to exfoliation corrosion similar to that of 7075-T76 with fracture toughness and resistance to the growth of fatigue cracks similar to that of 7150-T6. In contrast to the usual loss in toughness of 7XXX products at low temperatures, fracture toughness of 7055-T77 at −65 °F (220 K) is similar to that at room temperature. Resistance to SCC is intermediate to those of 7075-T6 and 7150-T77. The

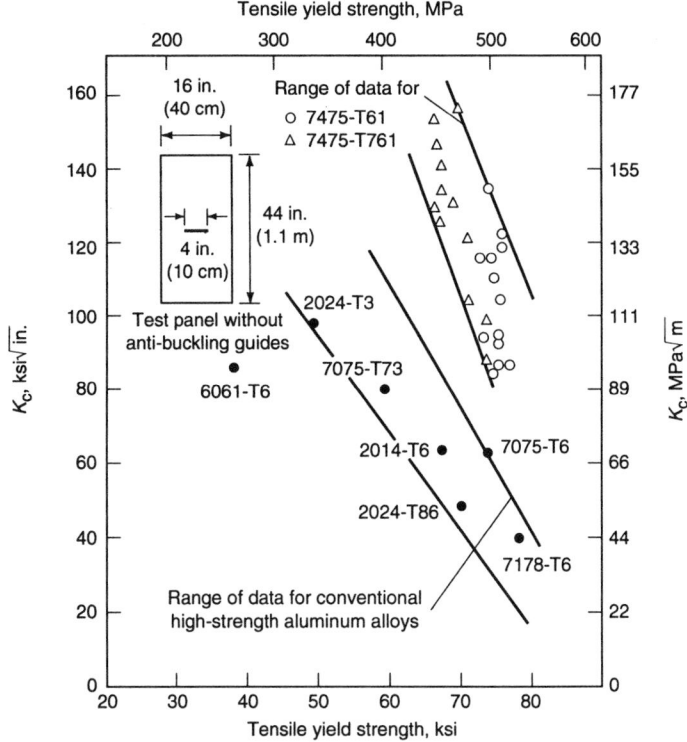

Fig. 10 Critical stress intensity factor, K_c, vs. tensile yield strength for 1.0 to 4.7 mm (0.040 to 0.188 in.) aluminum alloy sheet. Improved alloy 7475 is compared to other commercial alloys. Source: Ref 10

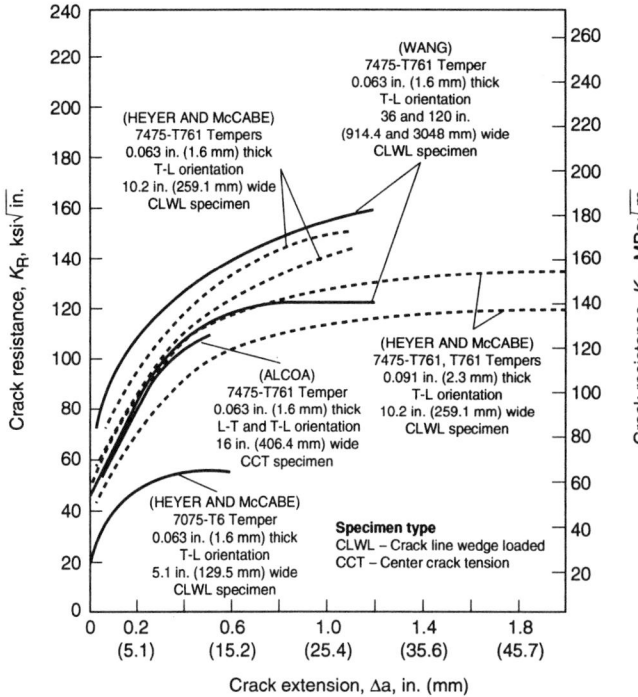

Fig. 11 Crack resistance curves for aluminum alloy 7475 sheet

Table 8 Fracture toughness of aluminum alloy plate

Alloy and condition	Room-temperature yield strength		Specimen design	Orientation	Fracture toughness, K_{Ic} or $K_{Ic}(J)$ at:							
					24 °C (75 °F)		−196 °C (−320 °F)		−253 °C (−423 °F)		−269 °C (−452 °F)	
	MPa	ksi			MPa√m	ksi√in.	MPa√m	ksi√in.	MPa√m	ksi√in.	MPa√m	ksi√in.
2014-T651	432	62.7	Bend	T-L	23.2	21.2	28.5	26.1	…	…	…	…
2024-T851	444	64.4	Bend	T-L	22.3	20.3	24.4	22.2	…	…	…	…
2124-T851(a)	455	66.0	CT	T-L	26.9	24.5	32.0	29.1	…	…	…	…
	435	63.1	CT	L-T	29.2	26.6	35.0	31.9	…	…	…	…
	420	60.9	CT	S-L	22.7	20.7	24.3	22.1	…	…	…	…
2219-T87	382	55.4	Bend	T-S	39.9	36.3	46.5	42.4	52.5	48.0	…	…
			CT	T-S	28.8	26.2	34.5	31.4	37.2	34.0	…	…
	412	59.6	CT	T-L	30.8	28.1	38.9	32.7	…	…	…	…
5083-O	142	20.6	CT	T-L	27.0(b)	24.6(b)	43.4(b)	39.5(b)	…	…	48.0(b)	43.7(b)
6061-T651	289	41.9	Bend	T-L	29.1	26.5	41.6	37.9	…	…	…	…
7039-T6	381	55.3	Bend	T-L	32.3	29.4	33.5	30.5	…	…	…	…
7075-T651	536	77.7	Bend	T-L	22.5	20.5	27.6	25.1	…	…	…	…
7075-T7351	403	58.5	Bend	T-L	35.9	32.7	32.1	29.2	…	…	…	…
7075-T7351	392	56.8	Bend	T-L	31.0	28.2	30.9	28.1	…	…	…	…

(a) 2124 is similar to 2024, but with higher-purity base and special processing to improve fracture toughness. (b) $K_{Ic}(J)$. Source: *Metals Handbook*, 9th ed., Vol 3, American Society for Metals, 1980, p 746, compiled from several references

attractive combination of properties of 7055-T77 is attributed to its high ratios of Zn/Mg and Cu/Mg. When aged by the proprietary T77 process this composition provides a microstructure at and near grain boundaries that is resistant to intergranular fracture and to intergranular corrosion. The matrix microstructure resists strain localization while maintaining a high resistance to the passage of dislocations. The extruded products in T77 do not develop the 70 MPa strength advantage.

Low-Temperature Toughness. Aluminum alloys represent a very important class of structural

metals for subzero-temperature applications. Aluminum and aluminum alloys have face-centered-cubic (fcc) crystal structures. Most fcc metals retain good ductility at subzero temperatures. Aluminum can be strengthened by alloying and heat treatment while still retaining good ductility along with adequate toughness at subzero temperatures. Nominal compositions of aluminum alloys that are most often considered for subzero service are presented in Table 7.

Data on fracture toughness of several aluminum alloys at room and subzero temperatures are summarized in Table 8. The room-temperature

yield strengths for the alloys in this table range from 142 to 536 MPa (20.6 to 77.7 ksi), and room-temperature plane-strain fracture toughness values for both bend and compact tension specimens range from 22.3 to 39.9 MPa√m (20.3 to 36.3 ksi√in.). This range in numerical values is not as impressive as actual service performances.

Of the alloys listed in Table 8, 5083-O has substantially greater toughness than the others. Because this alloy is too tough to obtain valid K_{Ic} data, the values shown for 5083-O were converted from J_{Ic} data. The fracture toughness of this alloy increases as exposure temperature decreases. Of the other alloys, which were evaluated in various heat-treated conditions, 2219-T87 has the best combination of strength and fracture toughness, both at room temperature and at –196 °C (–320 °F); this alloy can be readily welded.

Alloy 6061-T651 is another weldable alloy. It has good fracture toughness at room temperature and at –196 °C (–320 °F), but its yield strength is lower than that of alloy 2219-T87. Alloy 7039 also is weldable and has a good combination of strength and fracture toughness at room temperature and at –196 °C (–320 °F). Alloy 2124 is similar to 2024 but with a higher-purity base and special processing for improved fracture toughness. Tensile properties of 2124-T851 at subzero temperatures can be expected to be similar to those for 2024-T851.

Several other aluminum alloys, including 2214, 2419, 7050, and 7475, have been developed in order to obtain room-temperature fracture toughness superior to that of other 2000 and 7000 series alloys. Information on subzero properties of these alloys is limited, but it is expected that these alloys would have improved fracture toughness at subzero temperatures as well as at room temperature.

Stress-Corrosion Cracking of Aluminum Alloys

Stress-corrosion cracking (SCC) is a complex synergistic interaction of corrosive environment and sustained tensile stress at an exposed surface of metal resulting in cracking and premature failure at stresses below yield. In high-strength aluminum alloys, SCC is known to occur in ordinary atmospheres and aqueous environments. Both initiation and crack propagation may be accelerated by chlorides, temperature, and certain other chemical species. Susceptibility to SCC places a limitation on use of high-strength materials in certain applications. However, proper alloy/temper selection, good design and assembly practices, and environmental protection, combined with regular inspections, have proved to be highly successful techniques for the prevention of SCC failure in high-strength parts (Ref 15, 16).

The exact mechanisms responsible for SCC of a susceptible aluminum alloy in a particular environment remains controversial. However, most proposed mechanisms are variations of two basic theories: crack advance by anodic dissolution or hydrogen embrittlement. The controlling factors in these two SCC models are as follows:

Anodic dissolution is characterized by:

- Grain boundary precipitate size, spacing, and/or volume fraction
- Grain boundary PFZ width, solute profile or deformation mode
- Matrix precipitate size/distribution and deformation mode
- Oxide rupture and repassivation kinetics,

while *hydrogen embrittlement* is characterized by:

- Hydrogen absorption leading to grain boundary or transgranular decohesion
- Internal void formation via gas pressurization
- Enhanced plasticity (adsorption and absorption arguments exist)

An important fact to remember is that pure aluminum does not stress corrode, and for any given system, susceptibility usually increases with solute content. This fact, coupled with data and the controlling factors of the two models, suggests that microstructural alterations may influence SCC behavior for a given composition.

It is possible that hydrogen may contribute in the SCC of certain alloys and tempers of aluminum, although a detailed mechanistic understanding of SCC in aluminum alloys still requires more research (Ref 17). Recent literature surveys indicate considerable dispute as to how much, if at all, high-strength Al alloys are embrittled by hydrogen (Ref 18, 19, 20). There has not been enough evidence of hydrogen embrittlement to restrict commercialization of high-strength Al alloys (Ref 21).

SCC Resistance Ratings

An important step in controlling SCC by proper alloy selection is the SCC ranking of candidate materials. To establish performance that can be expected in service, it is necessary to compare candidate materials with other materials for which either long-term service experience or appropriate laboratory test data are available. Such comparisons, however, can be influenced significantly by test procedures (Ref 22-24). Laboratory stress-corrosion tests are generally of two types: constant deflection tests of smooth tensile bars or *C*-ring specimens loaded in aggressive environments, or crack propagation tests of precracked fracture mechanics specimens in aggressive environments. Commonly used criteria for SCC resistance from these tests include:

- Stress threshold (σ_{th}) below which laboratory specimens do not fail in aggressive environments
- Stress intensity threshold (K_{th}) below which crack propagation does not occur in precracked specimens
- Crack velocity measurements (*da/dt*) versus stress intensity in aggressive environments

There presently are no foolproof stress-corrosion test methods that are free of special limitations on test conditions and free of problems on interpretation of test results. However, a system of ratings of resistance to SCC for high-strength aluminum alloy products based on σ_{th} of smooth test specimens has been developed by a joint task group of ASTM and the Aluminum Association to assist alloy and temper selection, and it has been incorporated into ASTM G64 (Ref 25). Definitions of these ratings, which range from A (highest resistance) to D (lowest resistance), are as follows (adapted from G64-91):

- *A: Very high.* No record of service problems; SCC is not anticipated in general applications.
- *B: High.* No record of service problems; SCC is not anticipated at stresses of the magnitude caused by solution heat treatment. Precautions must be taken to avoid high sustained tensile stresses (exceeding 50% of the minimum specified yield strength) produced by any combination of sources including heat treatment, straightening, forming, fit-up, and sustained service loading.
- *C: Intermediate.* Stress-corrosion cracking is not anticipated if total sustained tensile stress is maintained below 25% of minimum specified yield strength. This rating is designated for the short-transverse direction in products used primarily for high resistance to exfoliation corrosion in relatively thin structures, where appreciable stresses in the short-transverse direction are unlikely.
- *D: Low.* Failure due to SCC is anticipated in any application involving sustained tensile stress in the designated test direction. This rating is currently designated only for the short-transverse direction in certain products.

Ratings are based on service experience, if available, or on standard SCC tests (ASTM G47, Ref 26) as required by many materials specifications. This exposure represents a severe control environment commonly used in alloy development and quality control. To rate a new material and test direction, according to G47, tests are performed on at least ten random lots and the test results must have 90% compliance at a 95% level of confidence for one of the following stress levels:

- A: Up to and including 75% of the specified minimum yield strength
- B: Up to and including 50% of the specified minimum yield strength
- C: Up to and including 25% of the specified minimum yield strength
- D: Fails to meet the criterion for rating C

It is cautioned, however, that these generalized SCC ratings may involve an oversimplification in regard to the performance in unusual chemical environments. In this rating system, a quantitative (numerical) ranking was avoided because current SCC test methods do not justify finite values. Table 9 contains a tabulation of alloys and tempers, product forms, and stressing directions,

Table 9 Relative stress-corrosion cracking ratings for high-strength wrought aluminum products

The associated stress levels for rankings A, B, C, D (see text) are not to be interpreted as threshold stresses and are not recommended for design. Documents such as MIL-HANDBOOK-5, MIL-STD-1568, NASC SD-24, and MSFC-SPEC-552A should be consulted for design recommendations. Resistance ratings are as follows: A, very high; B, high; C, intermediate; D, low (see text)

Alloy and temper(a)	Test direction(b)	Rolled plate	Rod and bar(c)	Extruded shapes	Forgings
2011-T3, -T4	L	(d)	B	(d)	(d)
	LT	(d)	D	(d)	(d)
	ST	(d)	D	(d)	(d)
2011-T8	L	(d)	A	(d)	(d)
	LT	(d)	A	(d)	(d)
	ST	(d)	A	(d)	(d)
2014-T6	L	A	A	A	B
	LT	B(e)	D	B(e)	B(e)
	ST	D	D	D	D
2024-T3, -T4	L	A	A	A	(d)
	LT	B(e)	D	B(e)	(d)
	ST	D	D	D	(d)
2024-T6	L	(d)	A	(d)	A
	LT	(d)	B	(d)	A(e)
	ST	(d)	B	(d)	D
2024-T8	L	A	A	A	A
	LT	A	A	A	A
	ST	B	A	B	C
2048-T851	L	A	(d)	(d)	(d)
	LT	A	(d)	(d)	(d)
	ST	B	(d)	(d)	(d)
2124-T851	L	A	(d)	(d)	(d)
	LT	A	(d)	(d)	(d)
	ST	B	(d)	(d)	(d)
2219-T3, -T37	L	A	(d)	A	(d)
	LT	B	(d)	B	(d)
	ST	D	(d)	D	(d)
2219-T6	L	(d)	(d)	(d)	A
	LT	(d)	(d)	(d)	A
	ST	(d)	(d)	(d)	A
2219-T87, -T8	L	A	A	A	A
	LT	A	A	A	A
	ST	A	A	A	A
6061-T6	L	A	A	A	A
	LT	A	A	A	A
	ST	A	A	A	A
7005-T53, -T63	L	(d)	(d)	A	A
	LT	(d)	(d)	A(e)	A(e)
	ST	(d)	(d)	D	D
7039-T63, -T64	L	A	(d)	A	(d)
	LT	A(e)	(e)	A(e)	(d)
	ST	D	(d)	D	(d)
7049-T73	L	A	(d)	A	A
	LT	A	(d)	A	A
	ST	A	(d)	B	A
7049-T76	L	(d)	(d)	A	(d)
	LT	(d)	(d)	A	(d)
	ST	(d)	(d)	C	(d)
7149-T73	L	(d)	(d)	A	A
	LT	(d)	(d)	A	A
	ST	(d)	(d)	B	A
7050-T74	L	A	(d)	A	A
	LT	A	(d)	A	A
	ST	B	(d)	B	B
7050-T76	L	A	A	A	(d)
	LT	A	B	A	(d)
	ST	C	B	C	(d)
7075-T6	L	A	A	A	A
	LT	B(e)	D	B(e)	B(e)
	ST	D	D	D	D
7075-T73	L	A	A	A	A
	LT	A	A	A	A
	ST	A	A	A	A

(continued)

(a) Ratings apply to standard mill products in the types of tempers indicated and also in TX5X and TX5XX (stress-relieved) tempers. They may be invalidated in some cases by use of nonstandard thermal treatments, or mechanical deformation at room temperature, by the user. (b) Test direction refers to orientation of direction in which stress is applied relative to the directional grain structure typical of wrought alloys, which for extrusions and forgings may not be predictable on the basis of the cross-sectional shape of the product: L, longitudinal; LT, long transverse; ST, short transverse. (c) Sections with width-to-thickness ratios equal to or less than two, for which there is no distinction between LT and ST properties. (d) Rating not established because product not offered commercially. (e) Rating is one class lower for thicker sections; extrusions, 25 mm (1 in.) and thicker; plate and forgings, 38 mm (1.5 in.) and thicker

with the classification of each into one of four categories from ASTM G64-91.

Precracked specimens and linear elastic fracture mechanics (LEFM) methods of analysis have also been widely used for SCC testing in recent years. It was anticipated that this new technique would provide a more quantitative measure of the resistance to the propagation of SCC of an alloy in the presence of a flaw. The test results are generally presented in a graph of the crack velocity versus the crack driving force in terms of a stress-intensity factor, K. Although the full diagram is required to describe the performance of an alloy, numbers derived from the diagram such as the "plateau velocity" and the "threshold stress intensity" (K_{th} or K_{Iscc}) can be used to compare materials. Effective use of the precracked specimen testing procedures, however, have proven very difficult to standardize, and there currently is no commonly accepted rating system for rating the resistance to SCC based on these descriptors. It is noteworthy that ranking of alloys by these criteria corresponds well with the ratings obtained with smooth specimens in ASTM G64.

Alloy Selection for SCC Resistance

In general, high-purity aluminum and low-strength aluminum alloys are not susceptible to SCC. Occurrence of SCC is chiefly confined to higher-strength alloy classes, such as 2XXX and 7XXX alloys and 5XXX Al-Mg alloys containing 3% or more Mg, particularly when loaded in the short-transverse orientation. Historically, in higher-strength alloys (e.g., aircraft structures) most service failures involving SCC of aluminum alloys have resulted from assembly or residual stresses acting in a short-transverse direction relative to the grain flow of the product (Ref 15, 18, 21). This is generally more troublesome for parts machined from relatively thick sections of rolled plate, extrusions, or forgings of complex shape where short-transverse grain orientation may be exposed. The specific alloy/temper combinations 7079-T6 (now obsolete), 7075-T6, and 2024-T3 have contributed to 90% of all service SCC failures of aluminum alloy products.

Within the high-strength alloy classes (2XXX, 7XXX, 5XXX), broad generalizations that relate susceptibility to SCC and strength or fracture toughness do not appear possible (Fig. 12). However, for certain alloys useful correlations of these properties with SCC resistance may be made over restricted ranges of the alloy's strength capability. For example, progressively overaging 7075 products from the T6 peak strength temper to T76 and T73 lowers strength but increases SCC resistance. However, "underaging" 7075 plate to T76 and T73 strength levels does not improve resistance to SCC.

Controls on alloy processing and heat treatment are key to assurance of high resistance to SCC without appreciable loss in mechanical properties and great accomplishments have been made. General developments are discussed below in several alloy classes.

Table 9 (continued)

Alloy and temper(a)	Test direction(b)	Rolled plate	Rod and bar(c)	Extruded shapes	Forgings
7075-T74	L	(d)	(d)	(d)	A
	LT	(d)	(d)	(d)	A
	ST	(d)	(d)	(d)	B
7075-T76	L	A	(d)	A	(d)
	LT	A	(d)	A	(d)
	ST	C	(d)	C	(d)
7175-T74	L	(d)	(d)	(d)	A
	LT	(d)	(d)	(d)	A
	ST	(d)	(d)	(d)	B
7475-T6	L	A	(d)	(d)	(d)
	LT	B(e)	(d)	(d)	(d)
	ST	D	(d)	(d)	(d)
7475-T73	L	A	(d)	(d)	(d)
	LT	A	(d)	(d)	(d)
	ST	A	(d)	(d)	(d)
7475-T76	L	A	(d)	(d)	(d)
	LT	A	(d)	(d)	(d)
	ST	C	(d)	(d)	(d)
7178-T6	L	A	(d)	A	(d)
	LT	B(e)	(d)	B(e)	(d)
	ST	D	(d)	D	(d)
7178-T76	L	A	(d)	A	(d)
	LT	A	(d)	A	(d)
	ST	C	(d)	C	(d)
7079-T6	L	A	(d)	A	A
	LT	B(e)	(d)	B(e)	B(e)
	ST	D	(d)	D	D

(a) Ratings apply to standard mill products in the types of tempers indicated and also in TX5X and TX5XX (stress-relieved) tempers. They may be invalidated in some cases by use of nonstandard thermal treatments, or mechanical deformation at room temperature, by the user. (b) Test direction refers to orientation of direction in which stress is applied relative to the directional grain structure typical of wrought alloys, which for extrusions and forgings may not be predictable on the basis of the cross-sectional shape of the product: L, longitudinal; LT, long transverse; ST, short transverse. (c) Sections with width-to-thickness ratios equal to or less than two, for which there is no distinction between LT and ST properties. (d) Rating not established because product not offered commercially. (e) Rating is one class lower for thicker sections; extrusions, 25 mm (1 in.) and thicker; plate and forgings, 38 mm (1.5 in.) and thicker

2XXX Alloys (Ref 28). Thick-section products of 2XXX alloys in the naturally aged T3 and T4 tempers have low ratings of resistance to SCC in the short-transverse direction. Ratings of such products in other directions are higher, as are ratings of thin-section products in all directions. These differences are related to the effects of quenching rate (largely determined by section thickness) on the amount of precipitation that occurs during quenching. If 2XXX alloys in T3 and T4 tempers are heated for short periods in the temperature range used for artificial aging, selective precipitation along grain or subgrain boundaries may further impair their resistance.

Artificial aging of 2XXX alloys to precipitation-hardened T8 tempers provides relatively high resistance to exfoliation, SCC, and superior elevated-temperature characteristics with modest strength increase over their naturally aged counterparts (Ref 27). Longer heating, as specified for T6 and T8 tempers, produces more general precipitation and significant improvements in resistance to SCC. Precipitates are formed within grains at a greater number of nucleation sites during treatment to T8 tempers. These tempers require stretching, or cold working by other means, after quenching from the solution heat treatment temperature and before artificial aging. These tempers provide the highest resistance for SCC and the highest strength in 2XXX alloys. This significant progress in improving fracture toughness of 2XXX alloys in T8 tempers is demonstrated by alloy 2124-T851 (also known as Alcoa 417 Process 2024-T851), which has had over 30 years of experience in military aircraft with no record of SCC problems. Typical data on 2XXX alloys are shown in Fig. 13.

Aluminum-Lithium Alloys. Some studies on aluminum-copper-lithium alloys indicate that these alloys have their highest resistance to SCC at or near peak-aged tempers. Underaging of these alloys (e.g., 2090) is detrimental; overaging decreases resistance only slightly. The susceptibility of the underaged microstructure has been attributed to the precipitation of an intermetallic constituent, Al_2CuLi, on grain boundaries during the early stages of artificial aging. This constituent is believed to be anodic to the copper-rich matrix of an underaged alloy, causing preferential dissolution and SCC. As aging time increases, copper-bearing precipitates form in the interior of the grains, thus increasing the anode-cathode area ratio in the microstructure to a more favorable value that avoids selective grain-boundary attack. Similar studies of stress-corrosion behavior are being conducted on aluminum-lithium-copper-magnesium alloys (e.g., 8090).

Newer Al-Li alloys have been developed that have lower lithium concentrations than 8090,

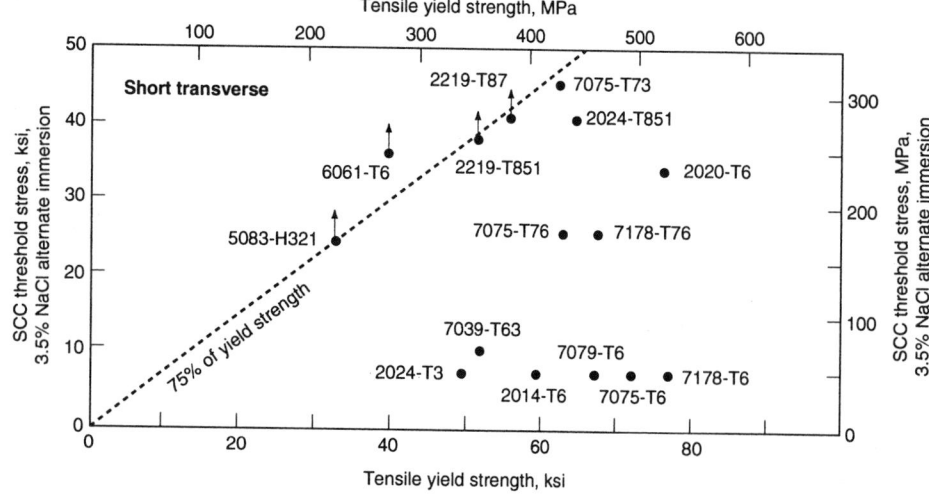

(a)

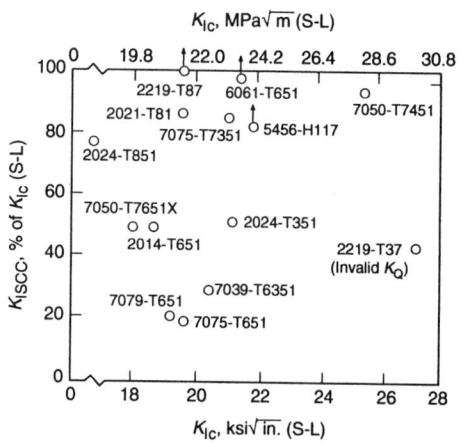

(b)

Fig. 12 Relationship between estimated stress-corrosion cracking "threshold stresses" and the tensile yield strength (a) and fracture toughness (b) of a wide variety of aluminum alloys and tempers. Data show that there is no general correlation. Source: Ref 16

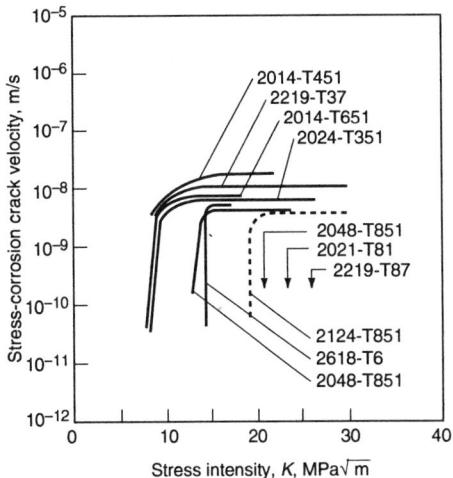

Fig. 13 Crack propagation rates in stress corrosion tests using precracked specimens of high-strength 2XXX series aluminum alloys, 25 mm thick, double cantilever beam, TL(S-L) orientation of plate, wet twice a day with an aqueous solution of 3.5% NaCl, 23 °C

2090, and 2091. These alloys do not appear to suffer from the same technical problems. The first of the newer generation was Weldalite 049 which can attain a yield strength as high as 700 MPa and an associated elongation of 10%. A refinement of the original alloy, 2195, is being considered for cryogenic tanks for the U.S. Space Shuttle. Alloy 2195 offers many advantages over 2219 for cryogenic tanks. Its higher strength coupled with higher modulus and lower density can lead to significant weight savings. Alloy 2195 also has good corrosion resistance, excellent fatigue properties, has a higher strength and fracture toughness at cryogenic temperatures than at room temperature, can be near-net shaped formed, and can be welded with proper precautions. However, further development work is required to identify optimum processing conditions that will ensure that the required combination of strength and fracture properties is obtained in the final product.

Other alloys containing less than 2% Li are being considered. Preliminary work indicates that new Al-Li alloy plate can be developed to provide a superior combination of properties for the bulkheads of high-performance aircraft, and analyses indicate that new Al-Li alloy flat-rolled products and extrusions would be competitive with polymer matrix composites for the horizontal stabilizer of commercial jetliners.

5XXX Alloys (Ref 28). These strain-hardening alloys do not develop their strength through solution heat treatment; rather, they are processed to H3 tempers, which require a final thermal stabilizing treatment to eliminate age softening, or to H2 tempers, which require a final partial annealing. The H116 or H117 tempers are also used for high-magnesium 5XXX alloys and involve special temperature control during fabrication to achieve a microstructural pattern of precipitate that increases the resistance of the alloy to intergranular corrosion and SCC. The alloys of the 5XXX series span a wide range of magnesium

contents, and the tempers that are standard for each alloy are primarily established by the magnesium content and the desirability of microstructures highly resistant to SCC and other forms of corrosion.

Although 5XXX alloys are not heat treatable, they develop good strength through solution hardening by the magnesium retained in solid solution, dispersion hardening by precipitates, and strain-hardening effects. Because the solid solutions in the higher-magnesium alloys are more highly supersaturated, the excess magnesium tends to precipitate out as Mg_2Al_3, which is anodic to the matrix. Precipitation of the phase with high selectivity along grain boundaries, accompanied by little or no precipitation within grains, may result in susceptibility to SCC.

The probability that a susceptible microstructure will develop in a 5XXX alloy depends on magnesium content, grain structure, amount of strain hardening, and subsequent time/temperature history. Alloys with relatively low magnesium contents, such as 5052 and 5454 (2.5 and 2.7% Mg, respectively), are only mildly supersaturated; consequently, their resistance to SCC is not affected by exposure to elevated temperatures. In contrast, alloys with magnesium contents exceeding about 3%, when in strain-hardened tempers, may develop susceptible structures as a result of heating or even after very long times at room temperature. For example, the microstructure of alloy 5083-O (4.5% Mg) plate stretched 1% is relatively free of precipitate (no continuous second-phase paths), and the material is not susceptible to SCC. Prolonged heating below the solvus, however, produces continuous precipitate, which results in susceptibility.

6XXX Alloys (Ref 28). The service record of 6XXX alloys shows no reported cases of SCC. In laboratory tests, however, at high stresses and in aggressive solutions, cracking has been demonstrated in 6XXX alloys of particularly high alloy content, containing silicon in excess of the Mg_2Si ratio and/or high percentages of copper.

7XXX Alloys Containing Copper (Ref 28). The 7XXX series alloy that has been used most extensively and for the longest period of time is 7075, an aluminum-zinc-magnesium-copper-chromium alloy. Introduced in 1943, this aircraft construction alloy was initially used for products with thin sections, principally sheet and extrusions. In these products, quenching rate is normally very high, and tensile stresses are not encountered in the short-transverse direction; thus, SCC is not a problem for material in the highest-strength (T6) tempers. When 7075 was used in products of greater size and thickness, however, it became apparent that such products heat treated to T6 tempers were often unsatisfactory. Parts that were extensively machined from large forgings, extrusions, or plate were frequently subjected to continuous stresses, arising from interference misfit during assembly or from service loading, that were tensile at exposed surfaces and aligned in unfavorable orientations. Under such conditions, SCC was encountered in service with significant frequency (Ref 29).

The problem resulted in the introduction (in about 1960) of the T73 tempers for thick-section 7075 products. The precipitation treatment used to develop these tempers requires two-stage artificial aging, the second stage of which is done at a higher temperature than that used to produce T6 tempers. During the preliminary stage, a fine, high-density precipitation dispersion is nucleated, producing high strength. The second stage is then used to develop resistance to SCC and exfoliation. The additional aging treatment required to produce 7075 in T73 tempers, reduces strength to levels below those of 7075 in T6 tempers. Excellent test results for 7075-T73 have been confirmed by extensive service experience in various applications. Environmental testing has demonstrated that 7075-T73 resists SCC even when stresses are oriented in the least favorable direction, at stress levels up to 300 MPa (44 ksi). Under similar conditions, the maximum stress at which 7075-T6 resists cracking is about 50 MPa (7 ksi).

Utilizing T7-type overaged tempers is a primary way to ensure improved resistance to exfoliation and SCC in 7XXX alloys. The T73 temper for alloy 7075 was the first aluminum alloy temper specifically developed to provide high resistance to stress-corrosion cracking with acceptable strength reduction from the T6 temper. Favorable evidence of this alloy's high resistance covers over 35 years of testing experience and extensive use in critical applications with no reported instances of failure in service by stress-corrosion cracking. This experience surpasses that of all other high-strength aluminum alloys and has become a standard of comparison for rating newer alloys and tempers (Ref 30).

Several commercial 7XXX alloys (7049-T73 and T76, 7175-T74 and 7050-T73, T74, and T76) offer combinations of strength, fracture toughness, and resistance to SCC superior to those combinations provided by conventional high-strength alloys, such as 7075-T6 and 7079-T6 (Ref 27). Alloys 7x49 and 7x50 were developed specifically for optimum combinations of the above properties in thick sections. Increased copper content provided good balance of strength and SCC resistance, while restriction of the impurity elements iron and silicon provided high toughness. Of particular note are 7149-T7451 and 7150-T7451 plate alloys, which offers optimum combinations of toughness, SCC resistance, and strength. Certain high-strength 7XXX alloys with lower copper content, such as 7079 and weldable 7005, exhibit excessive strength reduction when overaged to a T73-type temper, and a commercial stress-corrosion-resistant temper does not exist for these alloys. When using these alloys in existing commercial tempers, appreciable short-transverse tensile stresses, about 10 ksi (69 MPa) or above, should be avoided where exposure to an aggressive environment is of concern.

Alloy 7175, a variant of 7075, was developed for forgings. In the T74 temper, 7175 alloy forgings have strength nearly comparable to that of 7075-T6 and has better resistance to SCC (Fig.

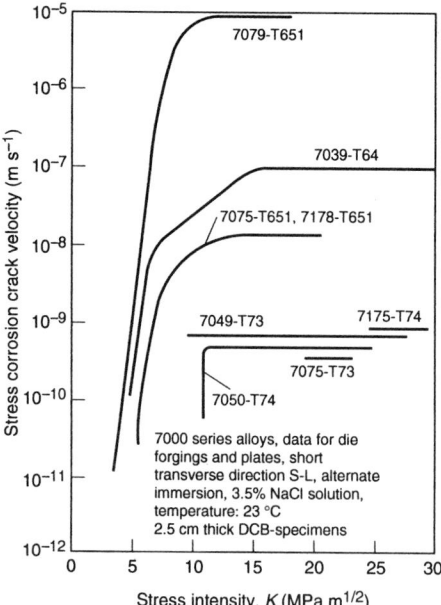

Fig. 14 Crack propagation rates in stress corrosion tests using 7XXX series aluminum alloys, 25 mm thick, double cantilever beam, short-transverse orientation of die transverse orientation of die forgings and plate, alternate immersion tests, 23 °C. Source: M.O. Speidel, *Met. Trans.*, Vol 6A, 1975, p 631

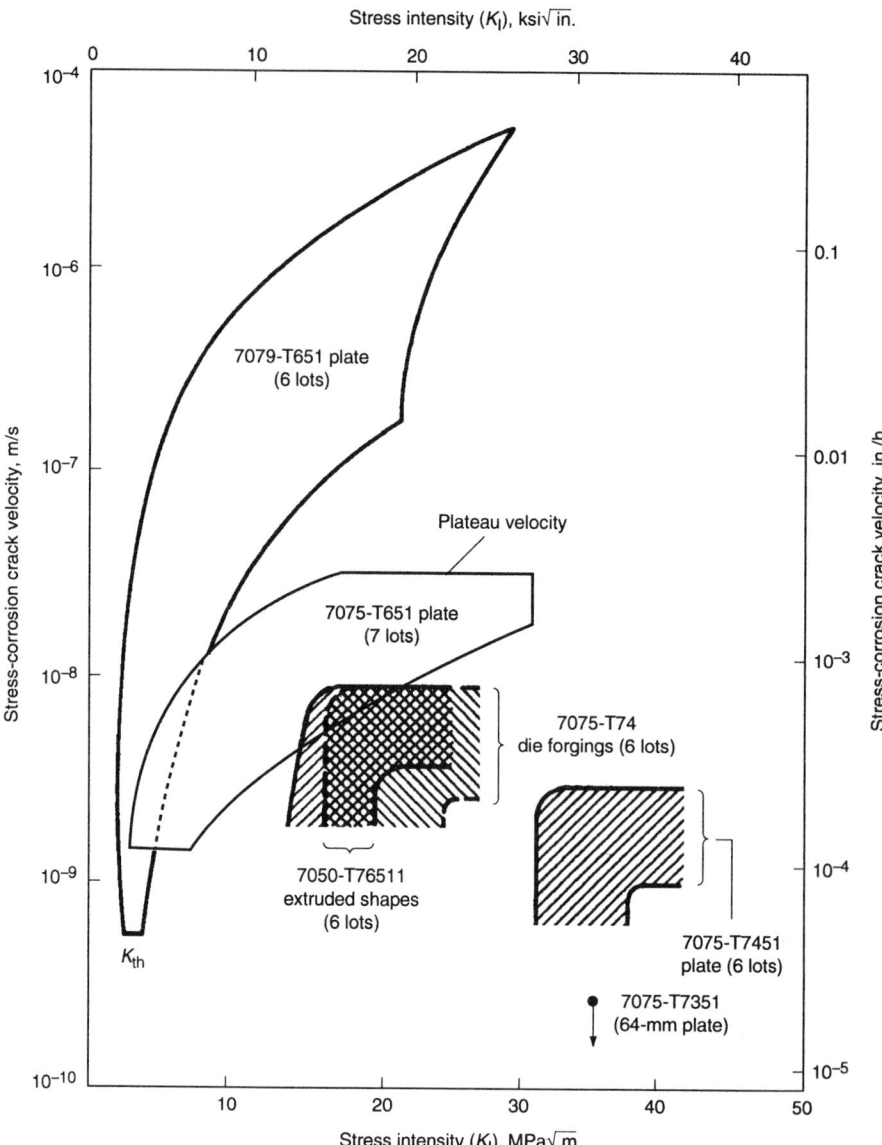

Fig. 15 SCC propagation rates for various aluminum alloy 7050 products. Double-beam specimens (S-L) bolt-loaded to pop-in and wetted three times daily with 3.5% NaCl. Plateau velocity averaged over 15 days. The right-hand end of the band for each product indicates the pop-in starting stress intensity (K_{Io}) for the tests of that material. Data for alloys 7075-T651 and 7079-T651 are from Ref 31. Source: Ref 32

14). Newer alloys—such as 7049 and 7475, which are used in the T73 temper, and 7050, which is used in the T74 temper—couple high strength with very high SCC resistance and improved fracture toughness. The superior performance is evident for alloys in the T7 tempers (Fig. 15) (Ref 31, 32).

The T76 tempers, which also require two-stage artificial aging and which are intermediate to the T6 and T73 tempers in both strength and resistance to SCC, are developed in copper-containing 7XXX alloys for certain products. Comparative ratings of resistance for various products of all these alloys, as well as for products of 7178, are given in Table 9.

The microstructural differences among the T6, T73, and T76 tempers of these alloys are differences in size and type of precipitate, which changes from predominantly Guinier-Preston (GP) zones in T6 tempers to η', the metastable transition form of $\eta(MgZn_2)$, in T73 and T76 tempers. None of these differences can be detected by optical metallography. In fact, even the resolutions possible in transmission electron microscopy are insufficient for determining whether the precipitation reaction has been adequate to ensure the expected level of resistance to SCC. For quality assurance, copper-containing 7XXX alloys in T73 and T76 tempers are required to have specified minimum values of electrical conductivity and, in some cases, tensile yield strengths that fall within specified ranges. The validity of these properties as measures of resistance to SCC is based on many correlation studies involving these measurements, laboratory and

field stress-corrosion tests, and service experience.

T77 Tempers. Until recently, overaging to T76, T74, and T73 tempers increased exfoliation resistance with a compromise in strength; strength was sacrificed from 5 to 20% to provide adequate resistance. The T77 temper, however, provides resistance to exfoliation with no sacrifice in strength, and resistance to SCC superior to that of 7075-T6 and 7150-T6. The highest strength aluminum alloy products, 7055 plate and extrusions, are supplied primarily in the T77 temper. Alloy 2024 products are also resistant to intergranular corrosion in the T8 temper, but fracture toughness and resistance to the growth of fatigue cracks suffer relative to 2024-T3.

New processing for 7150, resulting in 7150-T77, offers a higher strength with the durability and damage tolerance characteristics matching or

exceeding those of 7050-T76. Extrusions of 7150-T77 have been selected by Boeing as fuselage stringers for the upper and lower lobes of the new 777 jetliner because of the superior combination of strength, corrosion and SCC characteristics, and fracture toughness. Alloy 7150-T77 plate and extrusions are being used on the new C17 cargo transport. Use of this material saved considerable weight because corrosion performance of 7150-T6 was deemed to be inadequate.

The implementation of the T77 temper for 7150 was followed by development of new 7XXX products for compressively loaded structures. Alloy 7055-T77 plate and extrusions offer a strength increase of about 10% relative to that of 7150-T6 (almost 30% higher than that of 7075-T76). They also provide a high resistance to exfoliation corrosion similar to that of 7075-T76 with fracture toughness and resistance to the

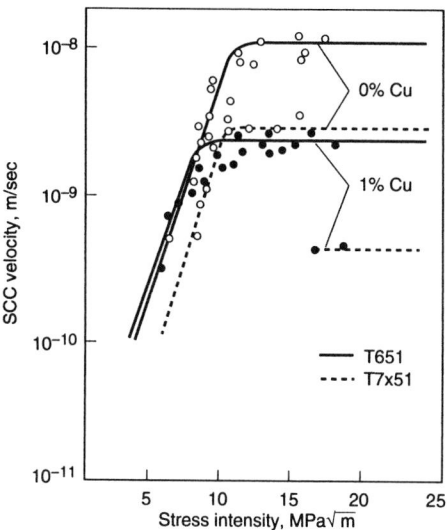

Fig. 16 Effect of overaging and copper content on SCC resistance of an Al-Zn-Mg alloy in 3.5% NaCl solution. Source: Ref 33

7079, a considerable amount of overaging is required with severe strength penalty to improve the stress-corrosion resistance. In general, increasing the copper content decreases the crack velocity (Fig. 16) (Ref 33). The effect can be mainly attributed to the change in the electrochemical activity of the precipitates as a function of their copper content. In the 7XXX series alloys the η phase is very active and anodic with respect to the film-covered matrix. If the alloy contains copper, copper both dissolves in the matrix and enters the η phase, making both more noble. As a result, the mixed potential at the crack tip shifts to a more noble value. The decrease in the crack velocity can then be attributed to the reduced rate of dissolution of the more noble precipitates, or reduced rate of hydrogen ion reduction and hydrogen adsorption at the crack tip at the more noble potential.

Casting Alloys (Ref 28). The resistance of most aluminum casting alloys to SCC is sufficiently high that cracking rarely occurs in service.

The microstructures of these alloys are usually nearly isotropic; consequently, resistance to SCC is unaffected by orientation of tensile stresses.

Accelerated laboratory tests, natural-environment testing, and service experience indicate that alloys of the aluminum-silicon 4XX.X series, 3XX.X alloys containing only silicon and magnesium as alloying additions, and 5XX.X alloys with magnesium contents of 8% or lower have virtually no susceptibility to SCC. Alloys of the 3XX.X group that contain copper are rated as less resistant, although the numbers of castings of these alloys that have failed by SCC have not been significant.

Significant SCC of aluminum alloy castings in service has occurred only in the highest-strength aluminum-zinc-magnesium 7XX.X alloys and in the aluminum-magnesium alloy 520.0 in the T4 temper. For such alloys, factors that require careful consideration include casting design, assembly and service stresses, and anticipated environmental exposure.

growth of fatigue cracks similar to that of 7150-T6. In contrast to the usual loss in toughness of 7XXX products at low temperatures, fracture toughness of 7055-T77 at –65 °F (220 K) is similar to that at room temperature. Resistance to SCC is intermediate to those of 7075-T6 and 7150-T77 products. The attractive combination of properties of 7055-T77 is attributed to its high ratios of Zn/Mg and Cu/Mg. When aged to T77 this composition provides a microstructure at and near grain boundaries that is resistant to intergranular fracture and to intergranular corrosion.

Copper-free 7XXX Alloys (Ref 28). Wrought alloys of the 7XXX series that do not contain copper are of considerable interest because of their good resistance to general corrosion, moderate-to-high strength, and good fracture toughness and formability. Alloys 7004 and 7005 have been used in extruded form and, to a lesser extent, in sheet form for structural applications. More recently introduced compositions, including 7016, 7021, 7029, and 7146, have been used in automobile bumpers formed from extrusions or sheet.

As a group, copper-free 7XXX alloys are less resistant to SCC than other types of aluminum alloys when tensile stresses are developed in the short-transverse direction at exposed surfaces. Resistance in other directions may be good, particularly if the product has an unrecrystallized microstructure and has been properly heat treated. Products with recrystallized grain structures are generally more susceptible to SCC as a result of residual stress induced by forming or mechanical damage after heat treatment. When cold forming is required, subsequent solution heat treatment or precipitation heat treatment is recommended. Applications of these alloys must be carefully engineered, and consultation among designers, application engineers and product producers, or suppliers is advised in all cases.

Overaging (T7x tempers) improves the SCC resistance of copper-containing alloys such as 7075, whereas for the low-copper alloys, like

Fatigue Life of Aluminum Alloys

Although high-strength aluminum alloys with high toughness have drastically lowered the probability of catastrophic failures in high-performance structures, corrosion fatigue requirements will continue to bear the burden for low-maintenance, durable, long-life structures. Good design, attention to structural details, and reliable inspection are of primary importance to controlling fatigue, and designers have traditionally considered these factors more important than alloy choice. However, a primary challenge facing designers and the materials engineer alike is extension of fatigue life and/or increased structural efficiency through optimum selection and use of fatigue-resistant alloys.

Differences in the fatigue performance of engineering materials can be translated into longer life, reduced weight, and reduced maintenance costs of present engineering structures. Fatigue improvements in aluminum, titanium, and steel alloys have been demonstrated through modifications to alloy composition, fabricating practice, and processing controls. Better understanding of fatigue mechanics has led to new hypotheses having the potential to lead to commercialization of improved alloys for fatigue.

Early work at Alcoa on large numbers of smooth and notched specimens demonstrated that wide variations in commercial aluminum alloys caused little or no detectable difference in fatigue strengths (Ref 34, 35). When early fatigue crack growth experiments categorized fatigue crack growth rates of aluminum alloys into one band, for example Fig. 17, it was generalized that fatigue resistance of all aluminum alloys were alike (Ref 36, 37). As a consequence of these early beliefs, further efforts to develop fatigue-resistant aluminum alloys were minimized, and

though several conceptual improvements have been advanced in laboratory experiments (Ref 13), none to date have reached commercial levels. For alloys developed to provide improved combinations of properties such as strength, corrosion resistance, and fracture toughness, fatigue resistance was determined as a last step before products were offered for sale, only to ensure that fatigue resistance was not degraded.

Despite early conclusions from laboratory data, users discovered that certain aluminum alloys performed decidedly better than others in

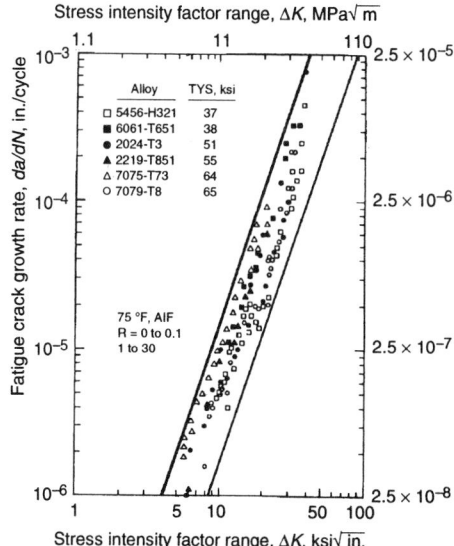

Fig. 17 Scatter band limits for fatigue crack growth rate behavior for a range of aluminum alloys. Source: Ref 26

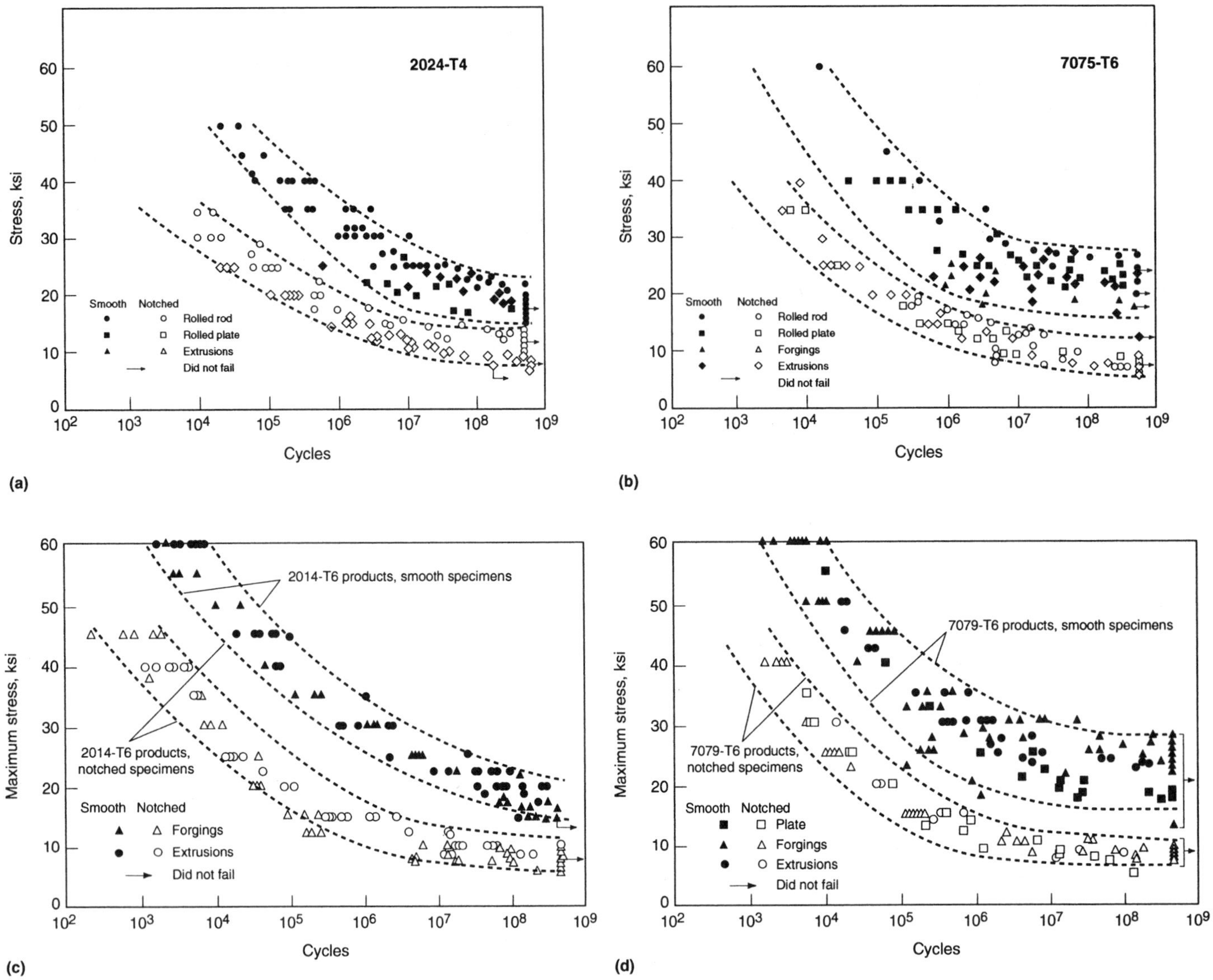

Fig. 18 Variation in rotating-beam fatigue for (a) 2024-T4, (b) 7075-T6, (c) 2014-T6, and (d) 7079-T6 alloys. Notches (60°) were very sharp ($K_t > 12$) with a radius of about 0.0002 in. Results are from over a thousand rotating-beam tests performed in the 1940s. Sources: R. Templin, F. Howell, and E. Hartmann, "Effect of Grain-Direction on Fatigue Properties of Aluminum Alloys," Alcoa, 1950 and *ASTM Proceedings*, Vol 64, p 581-593

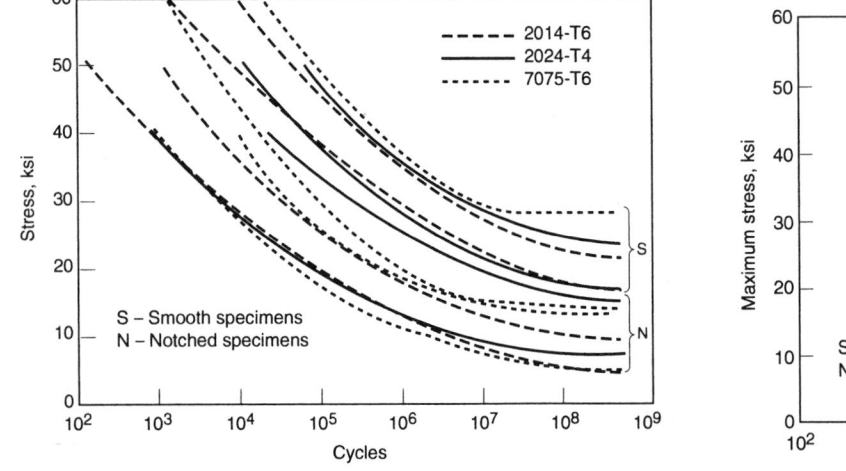

Fig. 19 Comparison of fatigue strength bands for 2014-T6, 2024-T4, and 7075-T6 aluminum alloys for rotating-beam tests. Source: R. Templin, F. Howell, and E. Hartmann, "Effect of Grain-Direction on Fatigue Properties of Aluminum Alloys," Alcoa, 1950

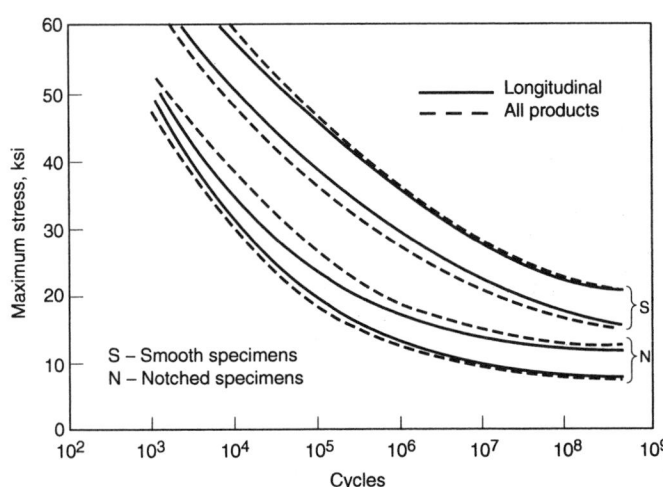

Fig. 20 Comparison of fatigue strength bands for 2014-T6 aluminum alloy products, showing effects of direction. Source: *ASTM Proceedings*, Vol 64, p 581-593

service when fluctuating loads were encountered, and therefore any generalization that all aluminum alloys are alike in fatigue is not wholly appropriate for design use. For example, alloy 2024-T3 has long been recognized as a better fatigue performer in service than alloy 7075-T6. In part, this may be explained by designers using higher design stresses on the basis of higher static strength of 7075-T6. However, results of Fig. 18(b) show alloy 7075-T6 to have broader scatter for smooth specimens and a lower bound of performance for severely notched specimens that is below that for alloy 2024-T4 (Fig. 18a). Broader scatter is also evident for 7079 compared to 2014 (Fig. 18c and d).

Most fatigue data were obtained from the basic stress-controlled cycling of notched and unnotched coupons in rotating-beam, axial, and flexure-type sheet tests. Test results from coupon specimens are useful for rating fatigue resistance of materials. However, material selection by the traditional *S-N* approach requires large numbers of material characterization tests for each material to simulate a myriad of possible service conditions. *S-N* data are also strongly influenced by many factors, such as specimen configuration, test environment, surface condition, load type, and stress ratio. Therefore, caution is required when translating coupon test results to a particular application. Evaluation of more than one material in component testing is needed to assist in final accurate material selection.

Nonetheless, extensive efforts have led to major improvements in the ability to characterize cyclic behavior and fatigue resistance of materials. Recognition of the importance of controlling basic elements of test procedure have led to development of recommended practices for establishing basic *S-N* fatigue data (Ref 38). The emerging disciplines of strain control fatigue and fracture mechanics have greatly enhanced understanding of fatigue processes. The strain control approach is aimed primarily at low-cycle fatigue crack initiation and early fatigue crack growth, while fracture mechanics concepts address the propagation of an existing crack to final failure. Each of these approaches is reviewed in the following sections.

S-N Fatigue

High-cycle fatigue characteristics commonly are examined on the basis of cyclic *S-N* plots of rotating-beam, axial, or flexure-type sheet tests. Many thousands of tests have been performed, and a collection of aluminum alloy *S-N* data is contained in the publication *Fatigue Data Book: Light Structural Alloys* (ASM, 1995). Early work on rotating-beam tests is summarized in Fig. 19. There seems to be greater spread in fatigue strengths for unnotched specimens than for notched specimens. This appears to be evidence that the presence of a notch minimizes differences, thus suggesting similar crack propagation after crack initiation with a sharp notch. In this context, the spread in smooth fatigue life is partly associated with variations in crack initiation

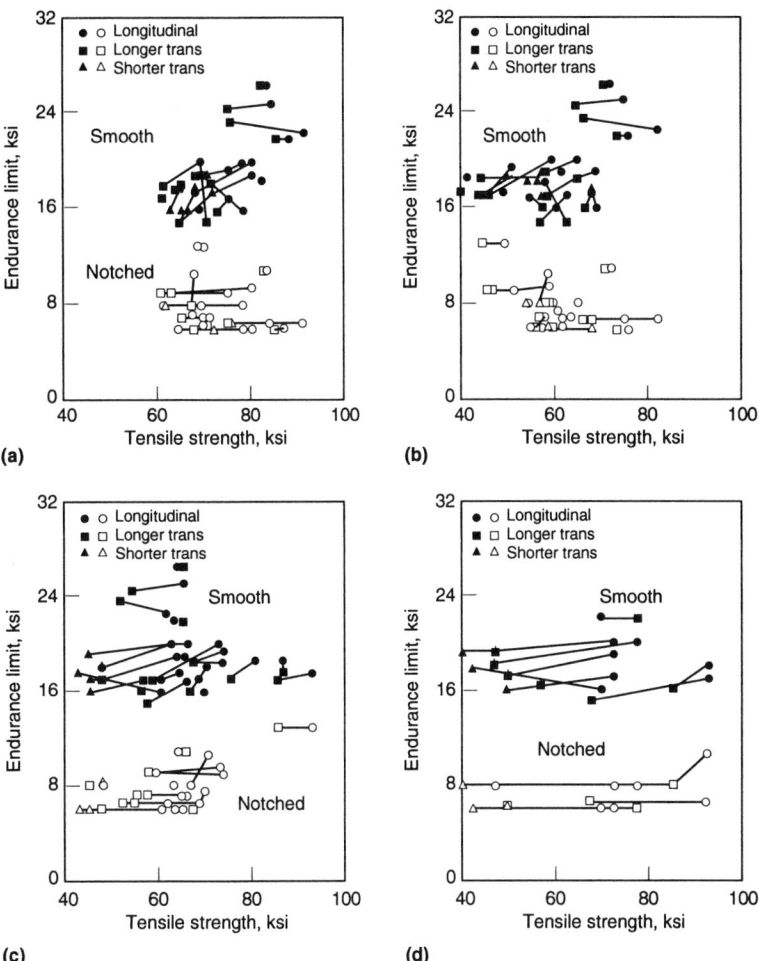

Fig. 21 Plots of fatigue with static mechanical properties for 2014, 2024, and 7075 aluminum alloys. (a) Endurance limit vs. tensile strength. (b) Endurance limit vs. yield strength. (c) Endurance limit vs. Elongation. (d) Endurance limit vs. reduction of area. Sharp notches ($K_t > 12$). Source: R. Templin, F. Howell, and E. Hartmann, "Effect of Grain-Direction on Fatigue Properties of Aluminum Alloys," Alcoa, 1950

sources (at surface imperfections or strain localizations). In general, however, the *S-N* approach does not provide clear distinctions in characterizing the crack initiation and crack propagation stages of fatigue.

The *S-N* response curves for rotating-beam fatigue strength of unnotched aluminum alloys tend to level out as the number of applied cycles approaches 500 million. This allows some rating of fatigue endurance, and estimated fatigue limits from rotating-beam tests have been tabulated for many commercial aluminum alloys (Table 10). Fatigue limits should not be expected in aggressive environments, as *S-N* response curves don't tend to level out when corrosion fatigue occurs. Rotating-beam strengths determined in the transverse direction are not significantly different from test results in the longitudinal direction. The scatter band limits in Fig. 20 show relatively small effects attributable to working direction, particularly for the notched fatigue data.

Rotating-beam data have also been analyzed to determine whether fatigue strength can be correlated with static strength. From a plot of average endurance limits (at 5×10^8 cycles) plotted against various tensile properties (Fig. 21), there

does not appear to be any well-defined quantitative relation between fatigue limit and static strength. This is consistent with results for most nonferrous alloys. It should be noted that proportionate increases in fatigue strength from tensile strengths do appear lower for age-hardened aluminum alloys than for strain-hardened alloys (Fig. 22). A similar trend appears evident for fatigue strength at 5×10^7 cycles (Fig. 23).

Effect of Environment. A key source of variability in *S-N* data is environment (Ref 39-41). Even atmospheric moisture is recognized to have a little corrosive effect on fatigue performance of aluminum alloys. Much high-cycle *S-N* testing has been carried out in uncontrolled ambient lab air environments, thereby contributing to scatter in existing data. This factor should be recognized when comparing results of different investigations.

Most aluminum alloys experience some reduction of fatigue strength in corrosive environments such as seawater, especially in low-stress, long-life tests (e.g., Fig. 24). Unlike sustained-load SCC, fatigue degradation by environment may occur even when the direction of principal loading with respect to grain flow is other than short-

Table 10 Typical tensile properties and fatigue limit of aluminum alloys

Alloy and temper	Ultimate tensile strength		Tensile yield strength		Elongation in 50 mm (2 in.), %		Fatigue endurance limit(a)	
	MPa	ksi	MPa	ksi	1.6 mm (1/16 in.) thick specimen	1.3 mm (1/2 in.) diam specimen	MPa	ksi
1060-O	70	10	30	4	43	…	20	3
1060-H12	85	12	75	11	16	…	30	4
1060-H14	95	14	90	13	12	…	35	5
1060-H16	110	16	105	15	8	…	45	6.5
1060-H18	130	19	125	18	6	…	45	6.5
1100-O	90	13	35	5	35	45	35	5
1100-H12	110	16	105	15	12	25	40	6
1100-H14	125	18	115	17	9	20	50	7
1100-H16	145	21	140	20	6	17	60	9
1100-H18	165	24	150	22	5	15	60	9
1350-O	85	12	30	4	…	(d)	…	…
1350-H12	95	14	85	12	…	…	…	…
1350-H14	110	16	95	14	…	…	…	…
1350-H16	125	18	110	16	…	…	…	…
1350-H19	185	27	165	24	…	(e)	50	7
2011-T3	380	55	295	43	…	15	125	18
2011-T8	405	59	310	45	…	12	125	18
2014-0	185	27	95	14	…	18	90	13
2014-T4, T451	425	62	290	42	…	20	140	20
2014-T6, T651	485	70	415	60	…	13	125	18
Alclad 2014-0	175	25	70	10	21	…	…	…
Alclad 2014-T3	435	63	275	40	20	…	…	…
Alclad 2014-T4, T451	420	61	255	37	22	…	…	…
Alclad 2014-T6, T651	470	68	415	60	10	…	…	…
2017-0	180	26	70	10	…	22	90	13
2017-T4, T451	425	62	275	40	…	22	125	18
2018-T61	420	61	315	46	…	12	115	17
2024-0	185	27	75	11	20	22	90	13
2024-T3	485	70	345	50	18	…	140	20
2024-T4, T351	470	68	325	47	20	19	140	20
2024-T361(b)	495	72	395	57	13	…	125	18
Alclad 2024-0	180	26	75	11	20	…	…	…
Alclad 2024-T3	450	65	310	45	18	…	…	…
Alclad 2024-T4, T351	440	64	290	42	19	…	…	…
Alclad 2024-T361(b)	460	67	365	53	11	…	…	…
Alclad 2024-T81, T851	450	65	415	60	6	…	…	…
Alclad 2024-T861(b)	485	70	455	66	6	…	…	…
2025-T6	400	58	255	37	…	19	125	18
2036-T4	340	49	195	28	24	…	125(c)	18(c)
2117-T4	295	43	165	24	…	27	95	14
2125	…	…	…	…	…	…	90	13(d)
2124-T851	485	70	440	64	…	8	…	…
2214	…	…	…	…	…	…	103	15(d)
2218-T72	330	48	255	37	…	11	…	…
2219-0	175	25	75	11	18	…	…	…
2219-T42	360	52	185	27	20	…	…	…
2219-T31, T351	360	52	250	36	17	…	…	…
2219-T37	395	57	315	46	11	…	…	…
2219-T62	415	60	290	42	10	…	105	15
2219-T81, T851	455	66	350	51	10	…	105	15
2219-T87	475	69	395	57	10	…	105	15
2618-T61	440	64	370	54	…	10	125	18
3003-0	110	16	40	6	30	40	50	7
3003-H12	130	19	125	18	10	20	55	8
3003-H14	150	22	145	21	8	16	60	9
3003-H16	180	26	170	25	5	14	70	10
3003-H18	200	29	185	27	4	10	70	10
Alclad 3003-0	110	16	40	6	30	40	…	…
Alclad 3003-H12	130	19	125	18	10	20	…	…
Alclad 3003-H14	150	22	145	21	8	16	…	…
Alclad 3003-H16	180	26	170	25	5	14	…	…
Alclad 3003-H18	200	29	185	27	4	10	…	…
3004-0	180	26	70	10	20	25	95	14
3004-H32	215	31	170	25	10	17	105	15
3004-H34	240	35	200	29	9	12	105	15
3004-H36	260	38	230	33	5	9	110	16
3004-H38	285	41	250	36	5	6	110	16
Alclad 3004-0	180	26	70	10	20	25	…	…
Alclad 3004-H32	215	31	170	25	10	17	…	…

(continued)

(a) Based on 500,000,000 cycles of completely reversed stress using the R.R. Moore type of machine and specimen. (b) Tempers T361 and T861 were formerly designated T36 and T86, respectively. (c) Based on 10 cycles using flexural type testing of sheet specimens. (d) Unpublished Alcoa data. (e) Data from CDNSWRC-TR619409, 1994, cited below. (f) T7451, although not previously registered, has appeared in literature and some specifications as T73651. (g) Sheet flexural. Sources: *Aluminum Standards and Data*, Aluminum Association, and E. Czyryca and M. Vassilaros, *A Compilation of Fatigue Information for Aluminum Alloys*, Naval Ship Research and Development Center, CDNSWC-TR619409, 1994

Table 10 (Continued)

Alloy and temper	Ultimate tensile strength MPa	Ultimate tensile strength ksi	Tensile yield strength MPa	Tensile yield strength ksi	Elongation in 50 mm (2 in.), % 1.6 mm (1/16 in.) thick specimen	Elongation in 50 mm (2 in.), % 1.3 mm (1/2 in.) diam specimen	Fatigue endurance limit(a) MPa	Fatigue endurance limit(a) ksi
Alclad 3004-H34	240	35	200	29	9	12	…	…
Alclad 3004-H36	260	38	230	33	5	9	…	…
Alclad 3004-H38	285	41	250	36	5	6	…	…
3105-0	115	17	55	8	24	…	…	…
3105-H12	150	22	130	19	7	…	…	…
3105-H14	170	25	150	22	5	…	…	…
3105-H16	195	28	170	25	4	…	…	…
3105-H18	215	31	195	28	3	…	…	…
3105-H25	180	26	160	23	8	…	…	…
4032-T6	380	55	315	46	…	9	110	16
4043-0	…	…	…	…	…	…	40	6(d)
4043-H38	…	…	…	…	…	…	55	8(d)
5005-0	125	18	40	6	25	…	…	…
5005-H12	140	20	130	19	10	…	…	…
5005-H14	160	23	150	22	6	…	…	…
5005-H16	180	26	170	25	5	…	…	…
5005-H18	200	29	195	28	4	…	…	…
5005-H32	140	20	115	17	11	…	…	…
5005-H34	160	23	140	20	8	…	…	…
5005-H36	180	26	165	24	6	…	…	…
5005-H38	200	29	185	27	5	…	…	…
5005-0	145	21	55	8	24	…	85	12
5050-H32	170	25	145	21	9	…	90	13
5050-H34	195	28	165	24	8	…	90	13
5050-H36	205	30	180	26	7	…	95	14
5050-H38	220	32	200	29	6	…	95	14
5052-0	195	28	90	13	25	30	110	16
5052-H32	230	33	195	28	12	18	115	17
5052-H34	260	38	215	31	10	14	125	18
5052-H36	275	40	240	35	8	10	130	19
5052-H38	290	42	255	37	7	8	140	20
5056-0	290	42	150	22	…	35	140	20
5056-H18	435	63	405	59	…	10	150	22
5056-H38	415	60	345	50	…	15	150	22
5083-0	290	42	145	21	…	22	160	23
5083-H11	303	44	193	28	…	16	150	22(e)
5083-H112	295	43	160	23	…	20	150	22(e)
5083-H113	317	46	227	33	…	16	160	23(e)
5083-H32	317	46	227	33	…	16	150	22(e)
5083-H34	358	52	283	41	…	8	…	…
5083-H321, H116	315	46	230	33	…	16	160	23
5086-0	260	38	115	17	22	…	145	21(e)
5086-H32, H116	290	42	205	30	12	…	50	22(e)
5086-H34	325	47	255	37	10	…	…	…
5086-H112	270	39	130	19	14	…	…	…
5086-H111	270	39	170	25	17	…	145	21(e)
5086-H343	325	47	255	37	10-14	…	160	23(e)
5154-0	240	35	115	17	27	…	115	17
5154-H32	270	39	205	30	15	…	125	18
5154-H34	290	42	230	33	13	…	130	19
5154-H36	310	45	250	36	12	…	140	20
5154-H38	330	48	270	39	10	…	145	21
5154-H112	240	35	115	17	25	…	115	17
5252-H25	235	34	170	25	11	…	…	…
5252-H38, H28	285	41	240	35	5	…	…	…
5254-0	240	35	115	17	27	…	115	17
5254-H32	270	39	205	30	15	…	125	18
5254-H34	290	42	230	33	13	…	130	19
5254-H36	310	45	250	36	12	…	140	20
5254-H38	330	48	270	39	10	…	145	21
5254-H112	240	35	115	17	25	…	115	17
5454-0	250	36	115	17	22	…	140	20(e)
5454-H32	275	40	205	30	10	…	140	20(e)
5454-H34	305	44	240	35	10	…	…	…
5454-H111	260	38	180	26	14	…	…	…
5454-H112	250	36	125	18	18	…	…	…
5456-0	310	45	160	23	…	24	150	22(e)
5456-H112	310	45	165	24	…	22	…	…

(continued)

(a) Based on 500,000,000 cycles of completely reversed stress using the R.R. Moore type of machine and specimen. (b) Tempers T361 and T861 were formerly designated T36 and T86, respectively. (c) Based on 10 cycles using flexural type testing of sheet specimens. (d) Unpublished Alcoa data. (e) Data from CDNSWRC-TR619409, 1994, cited below. (f) T7451, although not previously registered, has appeared in literature and some specifications as T73651. (g) Sheet flexural. Sources: *Aluminum Standards and Data*, Aluminum Association, and E. Czyryca and M. Vassilaros, *A Compilation of Fatigue Information for Aluminum Alloys*, Naval Ship Research and Development Center, CDNSWC-TR619409, 1994

Table 10 (Continued)

Alloy and temper	Ultimate tensile strength MPa	Ultimate tensile strength ksi	Tensile yield strength MPa	Tensile yield strength ksi	Elongation in 50 mm (2 in.), % 1.6 mm (1/16 in.) thick specimen	Elongation in 50 mm (2 in.), % 1.3 mm (1/2 in.) diam specimen	Fatigue endurance limit(a) MPa	Fatigue endurance limit(a) ksi
5456-H321, H116, H32	350	51	255	37	...	16	160	23(e)
5457-0	130	19	50	7	22	...	...	...
5457-H25	180	26	160	23	12	...	...	...
5457-H38, H28	205	30	185	27	6	...	...	...
5652-0	195	28	90	13	25	30	110	16
5652-H32	230	33	195	28	12	18	115	17
5652-H34	260	38	215	31	10	14	125	18
5652-H36	275	40	240	35	8	10	130	19
5652-H38	290	42	255	37	7	8	140	20
5657-H25	160	23	140	20	12	...	...	...
5657-H38, H28	195	28	165	24	7	...	...	...
6061-0	125	18	55	8	25	30	60	9
6061-T4, T451	240	35	145	21	22	25	95	14
6061-T6, T651	310	45	275	40	12	17	95	14
Alclad 6061-0	115	17	50	7	25	...	...	...
Alclad 6061-T4, T451	230	33	130	19	22	...	...	...
Alclad 6061-T6, T651	290	42	255	37	12	...	...	...
6063-0	90	13	50	7	...	...	55	8
6063-T1	150	22	90	13	20	...	60	9
6063-T4	170	25	90	13	22	...	...	...
6063-T5	185	27	145	21	12	...	70	10
6063-T6	240	35	215	31	12	...	70	10
6063-T83	255	37	240	35	9	...	...	...
6063-T831	205	30	185	27	10	...	...	...
6063-T832	290	42	270	39	12	...	...	...
6066-0	150	22	85	12	...	18	...	...
6066-T4, T451	360	52	205	30	...	18	...	...
6066-T6, T651	395	57	360	52	...	12	110	16
6070-T6	380	55	350	51	10	...	95	14
6101-H111	95	14	75	11	...	...	...	...
6101-T6	220	32	195	28	15	...	...	...
6151-T6	...	...	...	...	...	...	83	12
6201-T81	...	...	...	...	...	...	105	15
6262-T9	...	...	...	...	...	...	95	14
6351-T4	250	36	150	22	20	...	...	...
6351-T6	310	45	285	41	14	...	90	13
6463-T1	150	22	90	13	20	...	70	10
6463-T5	185	27	145	21	12	...	70	10
6463-T6	240	35	215	31	12	...	70	10
7002-T6	440	64	365	53	9-12	...	...	...
7039-T6	415	60	345	50	14	...	...	...
7049-T73	515	75	450	65	...	12	...	...
7049-T7352	515	75	435	63	...	11	...	...
7050-T73510, T73511	495	72	435	63	...	12	...	...
7050-T7451(f)	525	76	470	68	...	11	...	...
7050-T7651	550	80	490	71	...	11	...	...
7075-0	230	33	105	15	17	16	117	17(e)
7075-T6, T651	570	83	505	73	11	11	160	23
7072-H14	...	...	...	...	...	...	35	5(g)
7075-T73	503	73	435	63	13	...	150	22(e)
7076-T6	...	...	...	...	...	...	138	20(d)
Alclad 7075-0	220	32	95	14	17	...	...	...
Alclad 7075-T6, T651	525	76	460	67	11	...	...	...
7079-T6	490	71	428	62	10	...	160	23(e)

(a) Based on 500,000,000 cycles of completely reversed stress using the R.R. Moore type of machine and specimen. (b) Tempers T361 and T861 were formerly designated T36 and T86, respectively. (c) Based on 10 cycles using flexural type testing of sheet specimens. (d) Unpublished Alcoa data. (e) Data from CDNSWRC-TR619409, 1994, cited below. (f) T7451, although not previously registered, has appeared in literature and some specifications as T73651. (g) Sheet flexural. Sources: *Aluminum Standards and Data*, Aluminum Association, and E. Czyryca and M. Vassilaros, *A Compilation of Fatigue Information for Aluminum Alloys*, Naval Ship Research and Development Center, CDNSWC-TR619409, 1994

transverse. Fatigue response to environment varies with alloy, so final alloy selection for design should address this important interaction. When accumulating data for this purpose, it is recommended that any testing be conducted in a controlled environment, and preferably the environment of the intended application. However, an environment known to be more severe than that encountered in service is often used to conservatively establish baseline data and design guidelines. Because environmental interaction with fatigue is a rate-controlled process, interaction of time-dependent fatigue parameters such as frequency, waveform, and load history should be factored into the fatigue analysis (Ref 39-41).

Typically, the fatigue strengths of the more corrosion-resistant 5XXX and 6XXX aluminum alloys and tempers are less affected by corrosive environments than are higher-strength 2XXX and 7XXX alloys, as indicated by Fig. 25. Corrosion fatigue performance of 7XXX alloys may, in general, be upgraded by overaging to the more corrosion-resistant T7 tempers (Ref 42-47), as indicated by results shown in Fig. 26 and 27. With 2XXX alloys, the more corrosion-resistant, precipitation-hardened T8-type tempers provide a better combination of strength and fatigue resistance at high endurances than naturally aged T3 and T4 tempers. However, artificial aging of 2XXX alloys is accompanied by loss in toughness with resultant decrease in fatigue crack growth resistance at intermediate and high stress intensities (Ref 45, 46).

Table 11 Summary of the 7050 plate materials used in the study of the effect of microporosity on fatigue

Material	Product thickness, in.	Key microstructural features
Old-quality plate	5.7	Large porosity
New-quality plate	5.7	Porosity
Low-porosity plate	6.0	Small porosity, constituent particles
Low-particle plate	6.0 (T/4)	Small constituents, thick plate grain structure
Thin plate	1.0	Refined grain size and constituent particles

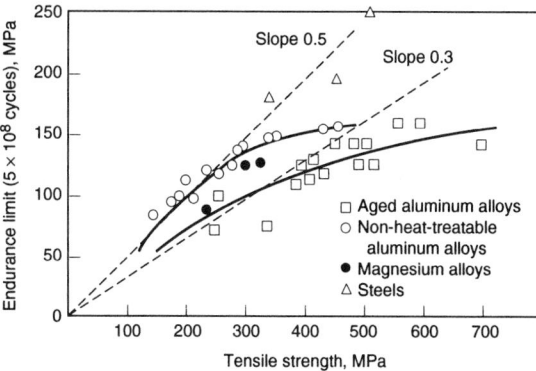

Fig. 22 Fatigue ratios (endurance limit/tensile strength) for aluminum alloys and other materials. Source: P.C. Varley, *The Technology of Aluminum and Its Alloys*, Newnes-Butterworths, London, 1970

Interaction of a clad protective system with fatigue strength of alloys 2024-T3 and 7075-T6 in air and seawater environments is shown in Fig. 24. In air, the cladding appreciably lowers fatigue resistance. In seawater, benefits of the cladding are readily apparent.

Reduced Porosity Materials*. The size of microporosity in commercial products is affected by the forming processes used in their production. A recent program was undertaken to determine whether the fatigue strength could be improved by the control of microporosity. Five variants of 7050 plate were produced to provide a range of microstructures to quantify the effects of intrinsic microstructural features on fatigue durability (Table 11). The first material, designated "old-quality" material, was produced using production practices typical of those used in 1984. The material is characterized by extensive amounts of centerline microporosity. Despite the centerline microporosity, this material still meets all existing mechanical property specifications for thick 7050 plate. Current quality production material, designated "new-quality" material, was also used, characterized by reduced levels of centerline microporosity compared to the old-quality material. The new-quality material represents the current benchmark for commercially available material. The processing methods used in the production of the new-quality material are a result of a statistical quality control effort to improve 7050 alloy thick plate (Ref 48). Material taken from two plant-scale production lots of each quality level provided the material for this program. Both materials are 5.7 in. thick 7050-T7451 plate. Static mechanical property characterization of the two 7050 plate pedigrees showed no significant differences in properties other than an increase in short transverse elongation for the new-quality material (Ref 49), and both materials meet the AMS material specification minimums. The fact that both materials meet the property requirements of the AMS specification underscores the limitation of existing specifications in that they do not differentiate intrinsic metal quality.

Effect of Microporosity on Fatigue. Smooth axial stress fatigue tests were performed for both

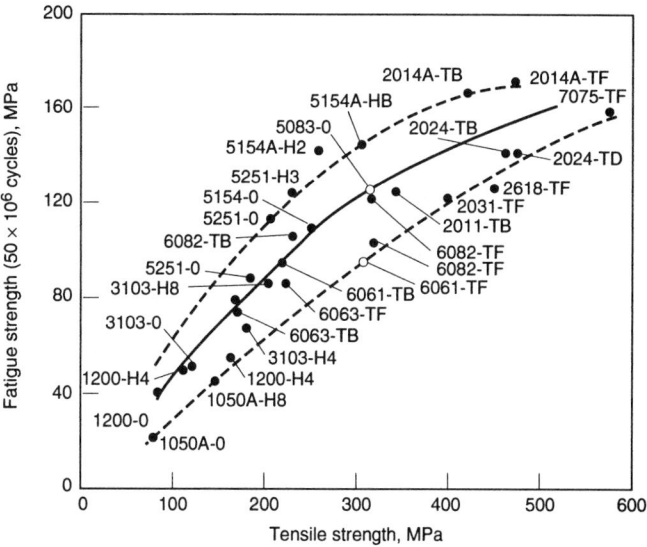

Fig. 23 Relationships between the fatigue strength and tensile strength of some wrought aluminum alloys

the old-quality and the new-quality plate materials. The tests were done on round bars with a gage diameter of 12.7 mm (0.5 in.). Gage sections were sanded longitudinally to remove circumferential machining marks. Testing was done at a maximum stress of 240 MPa (35 ksi), a stress ratio $R = 0.1$, and cyclic frequency of 10 Hz in laboratory air. The specimen orientation was long-transverse (L-T) relative to the parent plate. The specimens were removed from the midthickness (T/2) plane of the plate where microporosity concentration is the greatest (Ref 49). The lifetimes of the specimens are plotted in Fig. 28 on a cumulative failure plot, where the data are sorted in order of ascending lifetime and ordinate is the percentile ranking of the specimens relative to the total number of tests. Thus, the lifetime corresponding to the 50% point on the ordinate represents the median lifetime, where half of the specimens failed prior to that lifetime and half failed at longer lifetimes. The data show that the cumulative distribution of fatigue lifetimes for the new-quality material is substantially longer than for the old-quality material.

Fatigue tests were also performed for the old- and new-quality materials using flat specimens

containing open holes. Tests were performed at four stress levels for each material pedigree at a stress ratio of $R = 0.1$ and cyclic frequency of 25 Hz in laboratory air. As with the round specimens, L-T specimens were removed from the T/2 plane of the plate. The holes were deburred by polishing with diamond compound only on the corners and not in the bore of the hole; this resulted in slight rounding of the corners. The fatigue lifetime data are plotted in Fig. 29 as an *S-N* plot. Also plotted for both materials are the 95% confidence limits for the *S-N* curves. The confidence limits were obtained from a Box-Cox analysis of the data, which enables statistical determination of the mean *S-N* response and the 95% confidence limits (Ref 39). The data clearly show that, at equivalent stresses, the new-quality material exhibited longer lifetimes than the old-quality material.

Strain Control Fatigue

Considerable evidence suggests that failure data are more usefully presented in the form of strain-life curves, and that strain-based cumula-

*"Effect of Porosity" is adapted from J.R. Brockenbrough, R.J. Bucci, A.J. Hinkle, J. Liu, P.E. Magnusen, and S.M. Mixasato, "Role of Microstructure on Fatigue Durability of Aluminum Aircraft Alloys," Progress Report, ONR Contract N00014-91-C-0128, 15 April 1993

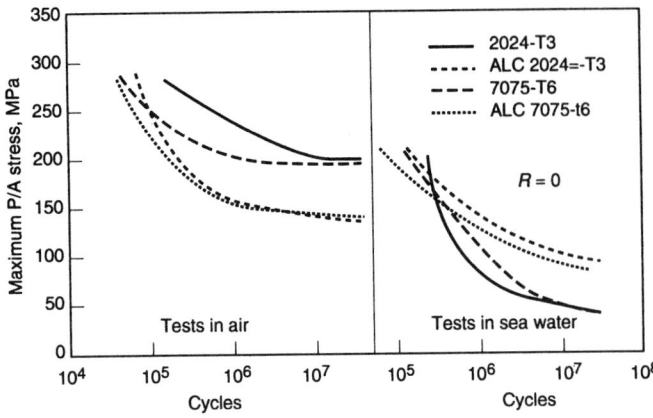

Fig. 24 Axial stress fatigue strength of 0.8 mm 2024, 7075, and clad sheet in air and seawater, $R = 0$. Source: Ref 33

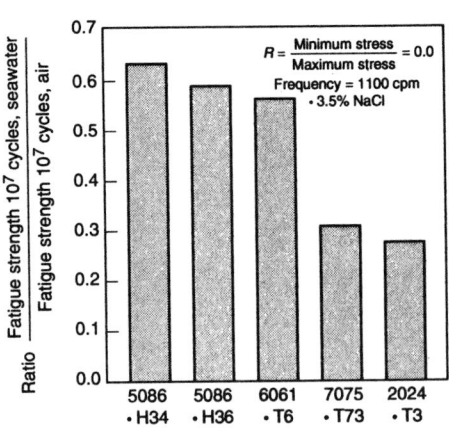

Fig. 25 Comparison of axial-stress fatigue strengths of 0.032 in. aluminum alloy sheet in seawater and air. Source: Ref 33

tive-damage life predictions are generally more reliable than conventional stress-based approaches (Ref 51-53). Strain-based prediction methods are capable of addressing interaction effects of variable load history and are better suited to handle "what if" situations than traditional stress approaches. In addition, they require a significantly reduced number of material characterization and component verification tests to make a material selection and/or design decision. Strain control fatigue is also essential in the understanding of crack initiation because, without localized plastic strain at areas of stress concentration in a structure, failure cannot occur. At high plastic strains, fatigue experiments on aluminum alloys (Ref 54) have shown that homogeneous slip (i.e., distribute plastic strain and avoid strain concentration sites) prolongs fatigue life to crack initiation. Recognized factors that promote homogeneous slip and/or increase low-cycle fatigue life are decreased coherency of strengthening

particles, increased magnesium content, and minimization and more uniform distribution of second-phase particles, which serve as initiation sites. Effects of alloy microstructure on fatigue initiation life depend on the level of strain.

In general, strain-life fatigue is based on the division of cyclic stress-strain response into plastic and elastic components (Fig. 30a), where the relation between stress and strain depends on the strength-ductility properties of the material (Fig. 30b) and also the cyclic hardening or softening of the material. For most metals, stress-strain hysteresis behavior (Fig. 30) is not constant, as cyclic softening or hardening can occur by reversed loading and cyclic straining. Generally (Ref 55-57), materials that are initially soft exhibit cyclic hardening, and materials that are initially hard undergo cyclic softening.

With strain-life fatigue, the elastic and plastic components may be separated and plotted on a strain life curve (Fig. 31). A plot on logarithmic

coordinates of the plastic portion of the strain amplitude (half the plastic strain range) versus the fatigue life often yields a straight line, described by the equation

$$\frac{\Delta \varepsilon_p}{2} = \varepsilon'_f (2N_f)^c \qquad \text{(Eq 3)}$$

where ε'_f is the fatigue ductility coefficient, c is the fatigue ductility exponent, and N_f is the number of cycles to failure ($2N_f$ is the number of load reversals). In contrast, elastic strains influence fatigue behavior under long-life conditions, where a stress-based analysis of fatigue is charted by plotting stress amplitude (half the stress range) versus fatigue life on logarithmic coordinates. The result is a straight line having the equation

$$\frac{\Delta \sigma}{2} = \sigma'_f (2N_f)^b \qquad \text{(Eq 4)}$$

where σ'_f is the fatigue strength coefficient and b is the fatigue strength exponent.

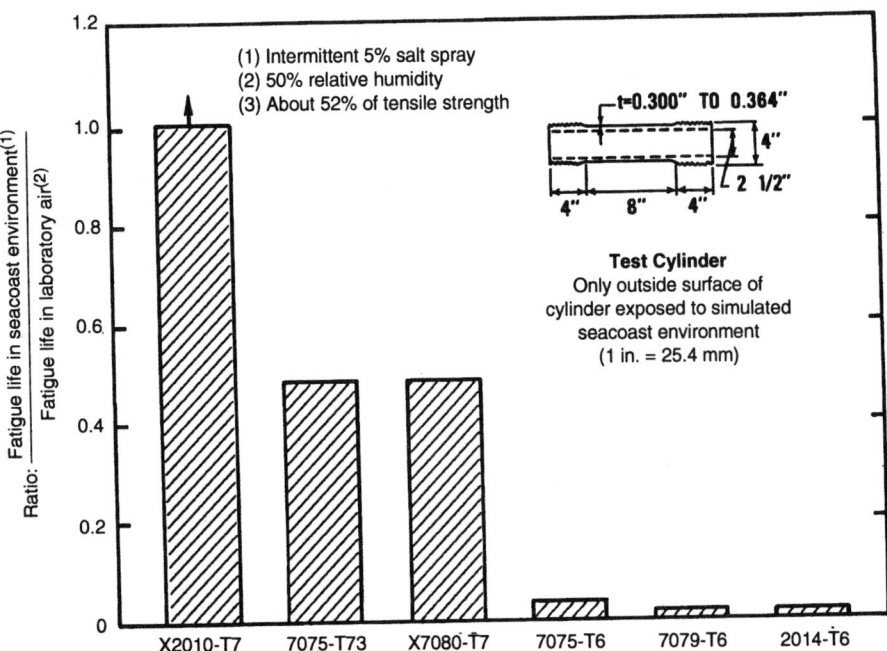

Fig. 26 Comparisons of fatigue lives of pressurized hydraulic cylinders in laboratory air and simulated seacoast environments at 80% design stress. Sources: Ref 3, 42

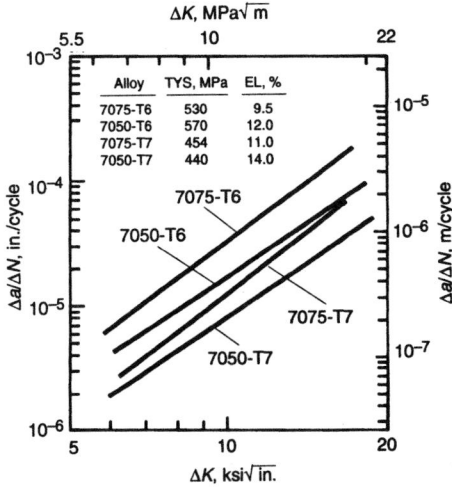

Fig. 27 Cyclic stress intensity range, ΔK, vs. cyclic fatigue crack growth rate, $\Delta a/\Delta N$, of laboratory-fabricated high-strength 7XXX aluminum alloys

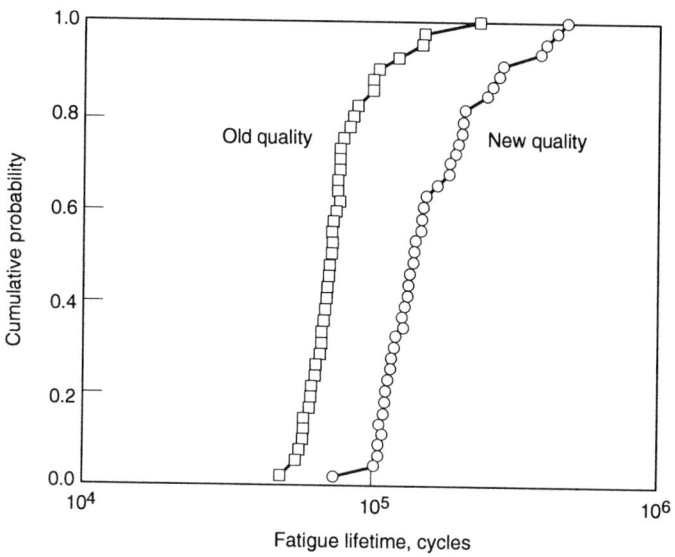

Fig. 28 Cumulative smooth fatigue lifetime distributors for old-quality and new-quality plate (see text for definitions). Tests conducted at 240 MPa (35 ksi) max stress, $R = 0.1$

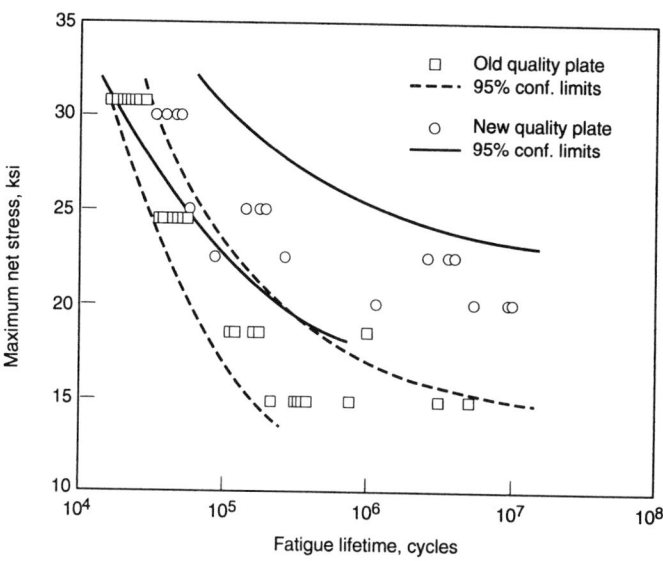

Fig. 29 Open-hole fatigue lifetimes for new-quality and old-quality plate (see text for definitions). Tests conducted at $R = 0.1$

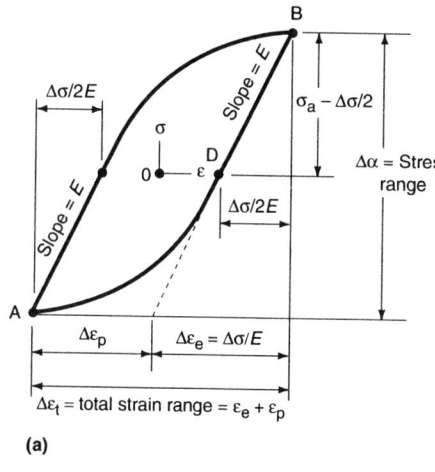

(a)

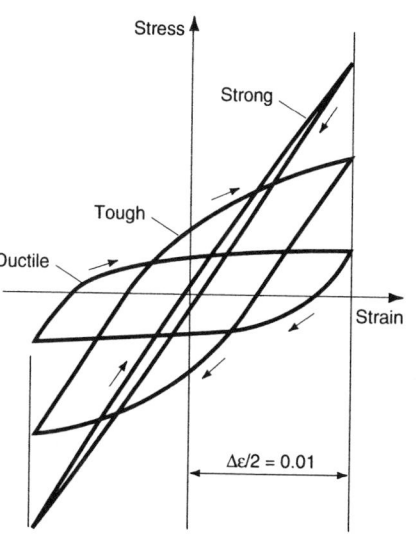

(b)

Fig. 30 Stress-strain hysteresis loop under cyclic loading. (a) Elastic and plastic strain range. (b) Hysteresis loops showing idealized stress-strain behavior for different types of materials

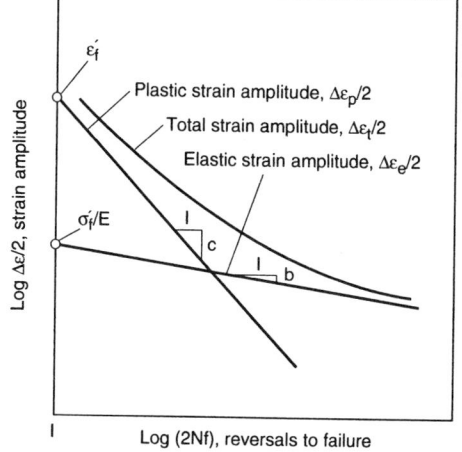

$$\frac{\Delta\varepsilon_t}{2} = \frac{\Delta\varepsilon_e}{2} + \frac{\Delta\varepsilon_p}{2}$$

$$= \frac{\sigma'_f}{E} (2Nf)^b + \varepsilon'_f (2Nf)^c$$

Where: σ'_f = Fatigue strength ductility coefficient
ε'_f = Fatigue ductility coefficient
b = Fatigue strength exponent
c = Fatigue ductility exponent

Fig. 31 Strain control fatigue life as a function of elastic-, plastic-, and total-strain amplitude

The elastic strain range is obtained by dividing Eq 4 by Young's modulus E:

$$\frac{\Delta\varepsilon_e}{2} = \frac{\sigma'_f}{E} (2N_f)^b \qquad \text{(Eq 5)}$$

The total strain range is the sum of the elastic and plastic components, obtained by adding Eq 3 and 5 (see Fig. 31):

$$\frac{\Delta\varepsilon}{2} = \varepsilon'_f (2N_f)^c - \frac{\sigma'_f}{E} (2N_f)^b \qquad \text{(Eq 6)}$$

For low-cycle fatigue conditions (frequently fewer than about 1000 cycles to failure), the first term of Eq 6 is much larger than the second; thus, analysis and design under such conditions must use the strain-based approach. For long-life fatigue conditions (frequently more than about 10,000 cycles to failure), the second term dominates, and the fatigue behavior is adequately described by Eq 4. Thus, it becomes possible to use Eq 4 in stress-based analysis and design.

This approach offers the advantage that both high-cycle and low-cycle fatigue can be characterized in one plot. From this relationship it is seen that long-life fatigue resistance is governed by the elastic line, while short-life fatigue resistance is governed by the plastic line (Ref 58). Within a bounded range of alloy types and microstructures, controlled strain fatigue lives greater than 10^4 cycles typically increase with increasing strength. On the other hand, low-cycle controlled strain fatigue lives for the same alloys generally increase with increasing ductility where ductility can be defined as $\ln(1/1-RA)$, RA being reduction of area determined from the standard tension test. The reciprocal strength-ductility relationship implies that materials selected on the basis of long-life resistance may not perform as well in low-cycle applications, and vice-versa. This is illustrated in Fig. 32 by the crossover in the strain-life relationships of X7046, a high-strength alloy, and 5083, a moderate-strength/high-ductility alloy. Results of strain control fatigue experiments on high-strength 7XXX laboratory-fabricated microstructures (Fig. 33) show similar crossover trends that can be correlated with strength and ductility. The observed crossovers imply that alloy selection, dependent on estimation of fatigue initiation life, requires identification of the most damaging cycles in the component fatigue spectrum for proper interpretation of mechanical property tradeoffs. This is accomplished using knowledge of the component strain spectrum, from strain gaged parts and/or stress analysis, and cumulative damage assessment of strain-life data. A compilation of fatigue strain-life parameters for various aluminum alloys is given in Table 12 and the appendix "Parameters for Estimating Fatigue Life" in this Volume. Corresponding monotonic properties are given in Table 13. Additional details on state-of-the-art fatigue analysis methods are given in Ref 59-62 and Section 3 "Fatigue Strength Prediction and Analysis" in this Volume.

Microstructure and Strain Life*

Plastic strain has been recognized as a controlling parameter in fatigue, and microstructures that homogeneously distribute the strain are desirable. Any microstructural feature that concentrates plastic strain or that results in an inhomogeneous distribution of plastic strain leads to undesirable local stress concentrations and large slip offsets at surfaces. Mechanisms representing these effects are illustrated schematically in Fig. 34, which shows two microstructural features that can result in strain localization: shearable precipitates, Fig. 34(a) and precipitate free zones (PFZs), Fig 34(b). These can lead to early crack nucleation and enhanced metal/environment interactions.

The following discussions briefly review the effect of shearable precipitates and PFZs on the strain life of aluminum alloys. These two microstructural features are considered because of their importance in commercial alloys. Inhomogeneous deformation similar to that in Fig. 34(a) can also occur in irradiated materials in which glide dislocations remove radiation defects, forming cleared channels of defect-free material. Low-stacking-fault-energy materials may also exhibit planar slip, but inhomogeneous deformation is not prevalent because softening in the slip plane does not occur. Inhomogeneous deformation similar to that in Fig. 34(b) may occur in two-phase materials having a soft and a hard phase. To some extent, localized deformation occurs in *all* materials at low stress and strain amplitudes.

The effect of shearable precipitates and PFZs on plastic-strain localization can be reduced by microstructural modification to improve fatigue life (see Table 14). The degree of plastic strain localization is primarily determined by the slip length and degree of age hardening. Because extensive age hardening and corresponding high yield strength are desirable, focus is placed on ways of improving the fatigue life by reducing the slip length. A reduction in grain size seems to be the most effective method for alloys that can contain both shearable precipitates and PFZs.

To definitively determine the influence of microstructure on fatigue life, it may be necessary to test in the low-cycle fatigue (LCF) regime under stress as well as strain control. Some microstructural features, through their effect on cyclic deformation behavior and resulting softening and/or

*Adapted from *Fatigue and Microstructure*, ASM, 1979, p 469-490

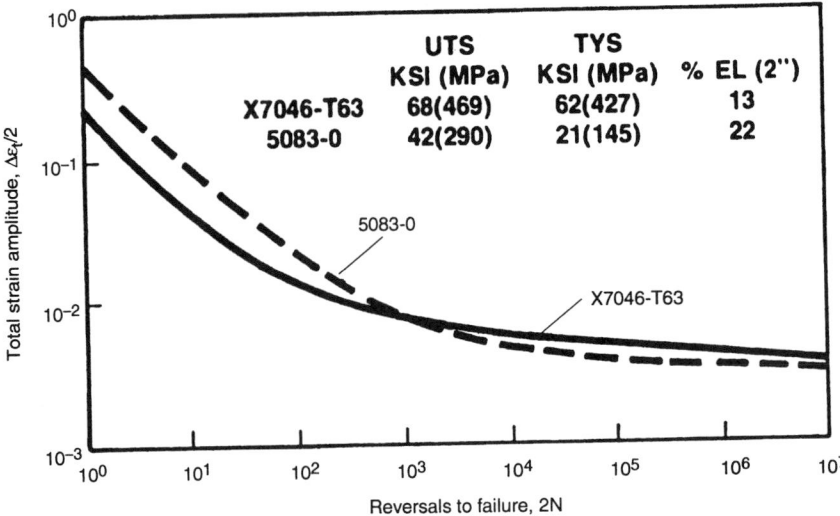

Fig. 32 Cyclic strain vs. life curve for X7046-T63 and 5083-O aluminum alloys

	UTS KSI (MPa)	TYS KSI (MPa)	% EL (2")
X7046-T63	68(469)	62(427)	13
5083-0	42(290)	21(145)	22

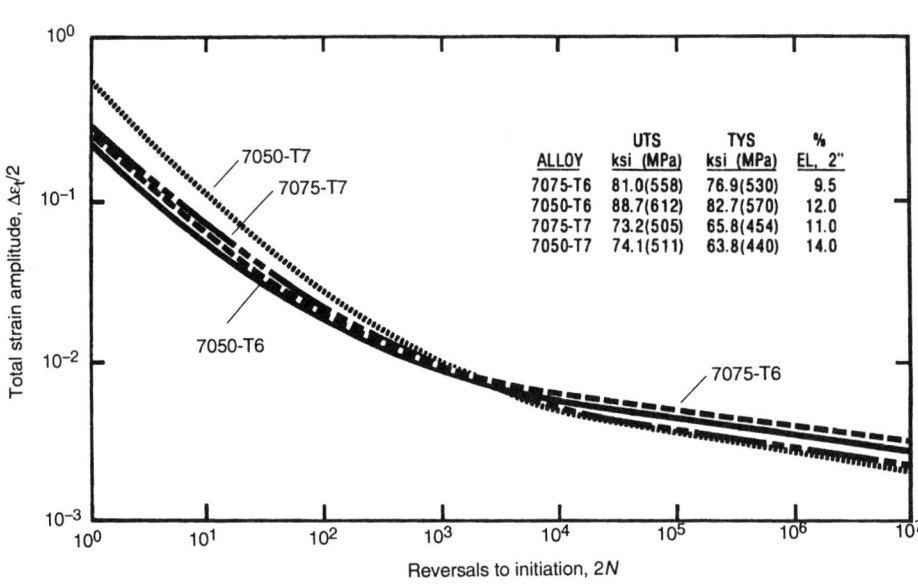

Fig. 33 Cyclic strain vs. initiation life for laboratory-fabricated high-strength 7XXX aluminum alloys. Fatigue resistance at low total strain amplitude is governed by the elastic-strain amplitude. Fatigue lives for total strain amplitudes less than about 5×10^{-3} generally increase with increasing strength. On the other hand, fatigue lives for total strain amplitude greater than about 10^{-2} generally increase with increasing ductility. Source: T.H. Sanders, Jr. and J.T. Staley, "Review of Fatigue and Fracture Research on High-Strength Aluminum Alloys," *Fatigue and Microstructure*, American Society for Metals, 1979, p 472

ALLOY	UTS ksi (MPa)	TYS ksi (MPa)	% EL, 2"
7075-T6	81.0(558)	76.9(530)	9.5
7050-T6	88.7(612)	82.7(570)	12.0
7075-T7	73.2(505)	65.8(454)	11.0
7050-T7	74.1(511)	63.8(440)	14.0

hardening, may improve or reduce the observed fatigue life, depending on the control mode. In addition, high-cycle fatigue (HCF) tests are important because in this region the influence of the yield stress usually dominates. It should also be emphasized that the microstructural parameters that accelerate or delay fatigue crack nucleation may have the opposite effect on fatigue crack propagation.

Precipitate Shearing

Overaging homogenizes slip and increases fatigue resistance in the low-cycle region where "ductility-controlled" fatigue dominates. This behavior, as it relates to the formation of nonshearable precipitates, alters fatigue properties, as shown in the Coffin-Manson life plots of Fig. 35. The two curves are for an underaged (with shearable precipitates) and an overaged (with nonshearable precipitates) 7050 alloy having identical yield strengths and strain to fracture (Ref 63). The fatigue life of the overaged alloy is consistently longer than that of the underaged alloy. The curves converge at low and high plastic strain amplitudes in these strain-controlled tests for the following reasons: For large strain amplitudes, all slip is homogeneous, regardless of the deformation mechanism, primarily due to multiple-slip activation. For small strain amplitudes, the sample with nonshearable precipitates hardens more extensively (due to the generation of geometrically necessary dislocations) than the sample with shearable precipitates (which normally softens). Consequently, for a strain-controlled test, failure occurs earlier than anticipated for the samples with nonshearable precipitates.

Table 12　Room-temperature cyclic parameters of various aluminum alloys (strain control, $R = -1$, unnotched)

Alloy/ temper	Form	Condition	Fatigue failure criterion	Ultimate tensile strength, MPa (ksi)	Tensile yield strength, MPa (ksi)	Fatigue strength coefficient, σ_f, MPa (ksi)	Fatigue strength exponent, b	Fatigue ductility coefficient, ε_f	Fatigue ductility exponent, c	Cyclic strain hardening coefficient, K', MPa (ksi)(a)	Cyclic strain hardening exponent, n' (a)
99.5% Al	Sheet	Cold rolled	Crack initiation(b)	73 (25)	19 (2.75)	95 (13.8)	-0.088	0.022	-0.328	255 (37)	0.265
99.5% Al	Sheet	Cold rolled	Crack initiation(c)	73 (25)	19 (2.75)	117 (17)	-0.109	0.017	-0.315	453 (65.7)	0.337
1100	Bar stock	As received	Rupture	110 (16)	97 (14)	159 (23)	-0.092	0.467	-0.613	184 (26.6)	0.159
2014-T6	Bar stock	As received	Rupture	511 (74)	463 (67)	776 (112.5)	-0.091	0.269	-0.742	704 (102)	0.072
2024-T3	Sheet	As received	5% load decrease	490 (71)	345 (50)	835 (121)	-0.096	0.174	-0.644	843 (122)	0.109
2024-T3	Sheet	5% cold formed	Crack initiation at 1 mm depth	490 (71)	476 (69)	891 (129)	-0.103	4.206	-1.056	669 (97)	0.074
2024-T3	Sheet	...	Crack initiation, 0.5 mm length	486 (70.5)	378 (55)	1044 (151)	-0.114	1.765	-0.927	590 (85.5)	0.040
2024-T4	Rod	Heat treated	...	476 (69)	304 (44)	764 (110.8)	-0.075	0.334	-0.649	808 (117)	0.098
2024-T351	Plate	Solution heat treated and cold worked(d)	...	455 (66)	380 (55)	927 (134)	-0.1126	0.4094	-0.7134	1067 (155)	0.1578
5454-H32	...	...	...	275 (40)	175 (26)	537 (77.8)	-0.0920	0.324	-0.6596	628 (91.1)	0.1394
5456-H311	Bar stock	As received	Rupture	400 (58)	235 (34)	702 (101.8)	-0.102	0.200	-0.655	635 (92)	0.084
6061-T6	...	ASTM grain size 3 to 5	...	328 (48)	300 (44)	654 (94.8)	-0.100	4.2957	-1.0072	566 (82)	0.0993
7075-T6	...	...	...	578 (84)	469 (68)	971 (140.8)	-0.072	0.7898	-0.9897	987 (143.2)	0.0728
7075-T6	Sheet	As received	5% load decrease	572 (83)	512 (74)	1048 (152)	-0.106	3.1357	-1.045	1500 (217.5)	0.186
7075-T6	Plate	As received	5% load decrease	572 (83)	512 (74)	776 (112.5)	-0.095	2.565	-0.987	521 (75.5)	0.045
7075-T6	Rod	Heat treated	...	580 (84)	470 (68)	886 (128.5)	-0.076	0.446	-0.759	913 (132)	0.088
7075-T7351	Plate	...	Crack initiation, 0.5 mm length	462 (67)	382 (55)	989 (143)	-0.140	6.812	-1.198	695 (100)	0.094
7475-T761	Sheet	As received	5% load decrease	475 (69)	414 (60)	983 (142.5)	-0.107	4.246	-1.066	675 (98)	0.059

(a) Stress-strain behavior at half-failure life. (b) Strain control, initiation criterion not specified. (c) Stress control, initiation criterion not specified. (d) Stress relieved by stretching 1.5% to 3% permanent set.
Sources: MarTest Inc., test data for Materials Properties Council; *J. of Materials*, Vol 4, 1969, p 159; and *Materials Data for Cyclic Loading Part D; Aluminum and Titanium Alloys*, Elsevier, 1987

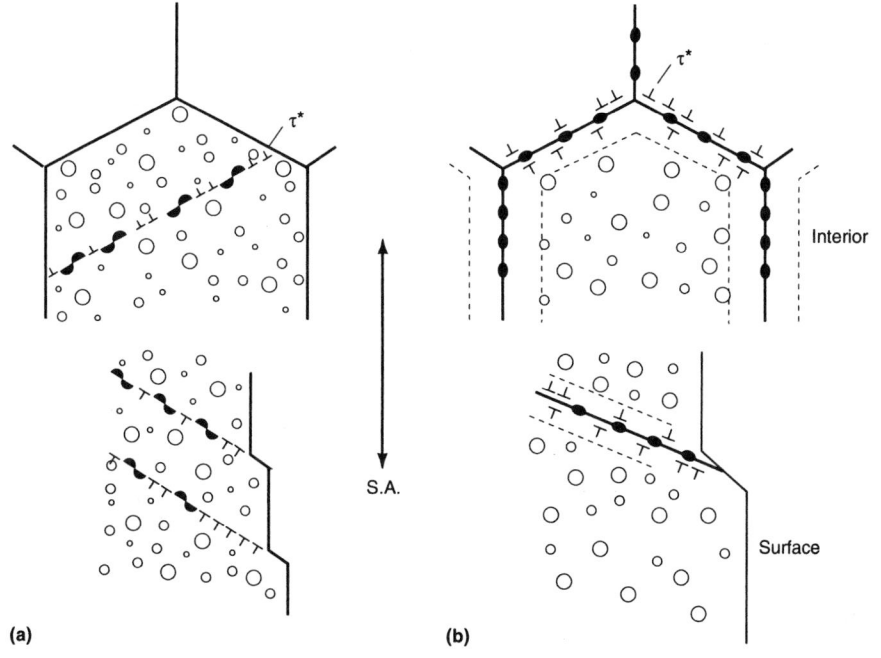

(a)　　**(b)**

Fig. 34　Schematic representation of two microstructural features that result in strain localization. τ^* represents stress concentration at indicated areas of grain boundaries. (a) Shearable precipitates. (b) Precipitate-free zones

Larger differences between the two heat treatments would occur under stress-controlled conditions, because the samples that harden would resist plastic deformation and those that soften would not.

Aggressive environments enhance the differences in fatigue life when one compares alloys having shearable precipitates (inhomogeneous deformation) with alloys having nonshearable precipitates (homogeneous deformation). This is illustrated by the Coffin-Manson life plots of LCF samples cycled in dry air and distilled water (Fig. 36). The aggressive H_2O environment decreases the fatigue life of the alloy with shearable precipitates by almost an order of magnitude when compared with the inert environment for the same plastic strain amplitude. The aggressive environment has little or no effect on the alloy with nonshearable precipitates. The degree of coherency in these Al-Zn-Mg-Cu alloys (Ref 64) was modified by changing the copper concentration. Increasing the copper content in the strengthening precipitates of 7XXX alloys results in earlier loss of coherency (Ref 63) and increases the probability of dislocation looping when compared with alloys containing lesser amounts of

Table 13 Room-temperature monotonic properties of various aluminum alloys

Alloy/ temper	Form	Condition	Ultimate tensile strength, MPa (ksi)	Tensile yield strength, MPa (ksi)	Elongation (EL) /reduction in area (RA), %	Static strain hardening coefficient, K, MPa (ksi)	Static strain hardening exponent, n	Cyclic strain hardening coefficient, K', MPa (ksi)(a)	Cyclic strain hardening exponent, n'(a)
99.5% Al	Sheet	Cold rolled	73 (25)	19 (2.75)	43% EL in 5D	42 (6)	0.117	255 (37)(b)	0.265(b)
99.5% Al	Sheet	Cold rolled	73 (25)	19 (2.75)	43% EL in 5D	42 (6)	0.117	453 (65.7)(c)	0.337(c)
1100	Bar stock	As received	110 (16)	97 (14)	87.6% RA	...	...	184 (26.6)	0.159
2014-T6	Bar stock	As received	511 (74)	463 (67)	25% RA	610 (88.5)	0.043	704 (102)	0.072
2024-T3	Sheet	As received	490 (71)	345 (50)	19% EL in 5D	...	...	843 (122)	0.109
2024-T3	Sheet	5% cold formed	490 (71)	476 (69)	16% EL in 5D/ 16% RA	476 (69)	0.0	669 (97)	0.074
2024-T3	Sheet	...	486 (70.5)	378 (55)	17.3% EL in 5D	627 (91)	0.074	590 (85.5)	0.040
2024-T4	Rod	Heat treated	476 (69)	304 (44)	35% RA	...	0.20	808 (117)	0.098
2024-T351	Plate	Solution heat treated and cold worked(d)	455 (66)	380 (55)	24.5% RA	455 (66)	0.032	1067 (155)	0.1578
5454-H32	...	...	275 (40)	175 (26)	28% RA	238 (34.5)	0.0406	628 (91.1)	0.1394
5456-H311	Bar stock	As received	400 (58)	235 (34)	34.6% RA	591 (85.7)	0.166	635 (92)	0.084
6061-T6	...	ASTM grain size 3 to 5	328 (48)	300 (44)	51.8%	...	...	566 (82)	0.0993
7075-T6	...	...	578 (84)	469 (68)	33% RA	827 (120)	0.1130	987 (143.2)	0.0728
7075-T6	Sheet	As received	572 (83)	512 (74)	10.8% EL in 5D	...	...	1500 (217.5)	0.186
7075-T6	Plate	As received	572 (83)	512 (74)	10.8% EL in 5D	...	...	521 (75.5)	0.045
7075-T6	Rod	Heat treated	580 (84)	470 (68)	33% RA	...	0.113	913 (132)	0.088
7075-T7351	Plate	...	462 (67)	382 (55)	8.4% EL in 5D	633 (91.8)	0.055	695 (100)	0.094
7475-T761	Sheet	As received	475 (69)	414 (60)	13.5% EL in 5D	...	...	675 (98)	0.059

(a) Stress-strain behavior at half-failure life, see accompanying table with fatigue characteristics. (b) Strain control, initiation criterion not specified. (c) Stress control, initiation criterion not specified. (d) Stress relieved by stretching 1.5% to 3% permanent set. Sources: MarTest Inc., test data for Materials Properties Council; *J. of Materials*, Vol 4, 1969, p 159; and *Materials Data for Cyclic Loading Part D; Aluminum and Titanium Alloys*, Elsevier, 1987

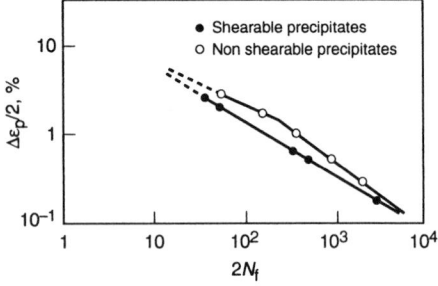

Fig. 35 Strain-life curves for samples of 7050 alloy with shearable precipitates (4 h at 120 °C, or 250 °F) and nonshearable precipitates (96 h at 150 °C, or 300 °F)

Table 14 Effect of microstructural modifications on the fatigue resistance of alloys containing shearable precipitates and PFZs

Modification to microstructure	Shearable precipitates	Precipitate-free zones
Overaging	Improves	No effect
Dispersoids	Improves	No effect
Unrecrystallized structures	Improves	Improves
Reduction of grain size	Improves	Improves
Steps in grain boundaries	No effect	Improves
Alignment of grain boundaries	No effect	Improves

copper with the same aging treatment. Cyclic deformation of the lower-copper-content alloys with shearable precipitates produced localized slip bands (Ref 64), which intensified metal/environment interactions. The nonshearable precipitates of the high-copper-content alloy prevented the occurrence of such inhomogeneous deformation.

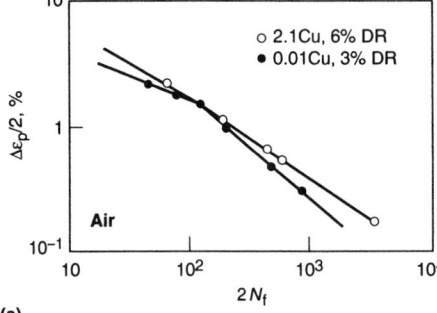

(a)

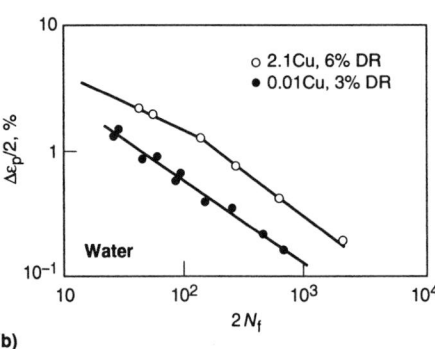

(b)

Fig. 36 Strain-life curves for samples of Al-Zn-Mg-*x* Cu alloys with shearable precipitates (0.01% Cu) and nonshearable precipitates (2.1% Cu). DR, degree of recrystallization. (a) Cycled in dry air. (b) Cycled in distilled water. Source: Ref 64

Addition of Nonshearable Precipitates. Although overaging homogenizes slip and increases the resistance of an alloy to fatigue crack nucleation, it normally results in a reduction in

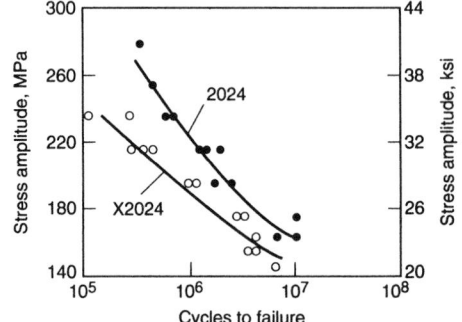

Fig. 37 S-N curves for commercial and experimental 2024 alloys with comparable tensile strengths. Both alloys contained a distribution of 5 μm diam iron- and silicon-rich inclusions; the commercial alloy also contained 0.1 to 0.2 m diam manganese-rich inclusions. Experimental alloy X2024 was free of the manganese inclusions and exhibited lower fatigue strength due to high crack density from sharp slip bands. Source: Pelloux and Stoltz Ref 65

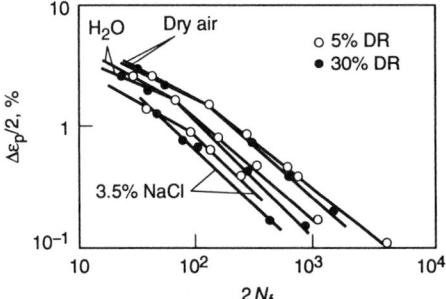

Fig. 38 Influence of degree of recrystallization (DR) and environment on the strain-life behavior of an Al-Zn-Mg-1.6 Cu alloy with shearable precipitates

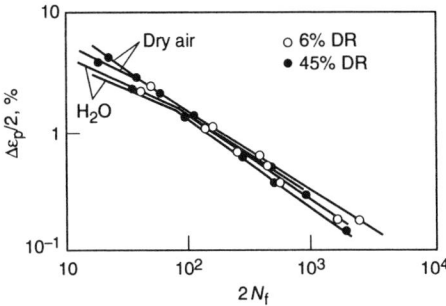

Fig. 39 Influence of degree of recrystallization (DR) and environment on the strain-life behavior of an Al-Zn-Mg-2.1 Cu alloy with nonshearable precipitates

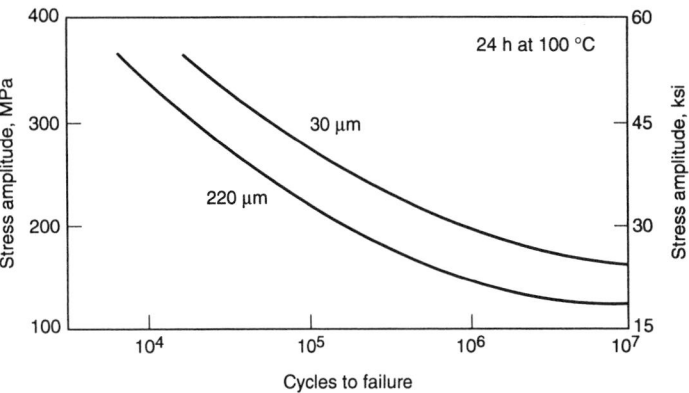

Fig. 41 Effect of grain size on the stress-life behavior of an X7075 alloy with shearble precipitates. Source: Ref 69

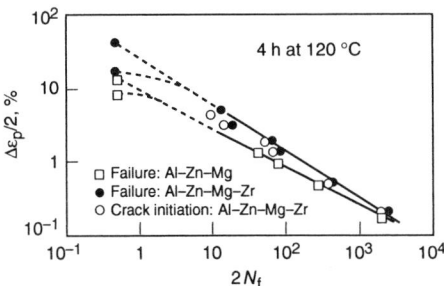

Fig. 40 Effect of grain size on the strain-life behavior of an alloy with shearable precipitates. The Al-Zn-Mg alloy had large grain size; the Al-Zn-Mg-Zr alloy, small grain size. Source: Ref 68

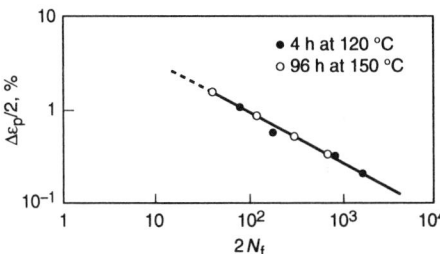

Fig. 42 Strain-life curves of large-grained Al-Zn-Mg alloy with shearable precipitates when underaged (4 h at 120 °C, or 250 °F) and nonshearable precipitates plus PFZs when overaged (96 h at 150 °C, or 300 °F). Source: Ref 64

static strength. Consequently, it is sometimes beneficial to have a dispersion of nonshearable precipitates intermixed with shearable precipitates. Some commercial alloys have alloying additions (for example, manganese and chromium in 2XXX and 7XXX aluminum alloys, respectively) that form small (0.1 to 0.2 μm), incoherent dispersoids during high-temperature homogenization treatments. The primary purpose of these small intermetallic compounds is to control grain size and shape. However, they also disperse slip and inhibit the formation of intense slip bands. Therefore, plastic deformation is more homogeneous, and early crack nucleation due to intense slip bands is avoided.

Figure 37 shows the results of a stress-controlled test of two 2XXX alloys—one (X2024) contains only shearable precipitates and the other (2024) both shearable and nonshearable precipitates. At all stress levels, alloy 2024 has a much longer fatigue life than X2024. It is important to note that the alloys have comparable tensile strengths—a necessity for a valid comparison in a stress-controlled test. The X2024 alloy having only shearable precipitates developed sharp, intense slip bands and a higher density of crack nuclei earlier in the fatigue life than did the alloy containing nonshearable dispersoids.

Unrecrystallized Structures. Figure 36 demonstrates that an alloy containing shearable precipitates has lower fatigue strength than a similar alloy containing nonshearable precipitates. Those results were obtained on material having a low (3 to 6%) degree of recrystallization. A larger difference in fatigue lives would have been ob-

served if the alloys were fully recrystallized. Unrecrystallized structures also promote homogeneous deformation and reduce the influence of the type (shearable or nonshearable) of precipitates.

Figure 38 shows Coffin-Manson life plots of an Al-Zn-Mg-Cu alloy with shearable precipitates (Ref 64). Two different degrees of recrystallization were tested in three different environments. The specimens having the largest volume fraction of unrecrystallized structure showed the greatest fatigue resistance in each environment. However, as expected, an aggressive environment enhanced the difference observed between specimens having a mostly unrecrystallized structure (homogeneous deformation) and those with a more recrystallized structure (inhomogeneous deformation). The dislocation substructure in the unrecrystallized regions and the nonshearable precipitates along subgrain boundaries reduce slip lengths and thus homogenize deformation. On the other hand, localized planar slip occurred in the recrystallized grains, resulting in an enhanced environmental effect and early crack nucleation.

For alloys with nonshearable precipitates, the degree of recrystallization has no effect on fatigue life, regardless of environment (Fig. 39). Further, the effect of the environment was small. Slip distances, which are controlled by the spacings of the nonshearable precipitates, are much smaller than the mean intercept length between subgrain boundaries.

Grain Size. A reduction in grain size results in beneficial effects that delay crack nucleation in alloys containing shearable precipitates. Reduced grain size reduces the slip length, and thus the stress concentration, by reducing the number of dislocations in a pileup. A reduction in slip length also reduces the number of dislocations that can egress at a free surface (and thus the slip-step height and extrusion/intrusion size). Another beneficial effect of grain-size reduction involves the volume of material needed to satisfy the von Mises criterion (Ref 66). In essence, this criterion requires multiple slip to occur in polycrystalline materials in order to preserve the external form of the specimen and maintain cohesion at the grain boundaries. However, as Calnan and Clews (Ref 67) have suggested, multiple-slip systems need only operate in the immediate vicinity of the grain boundary, whereas slip may occur on either

duplex or single systems in the body of the grains. Consequently, the smaller the grain size, the larger the volume fraction of material deformed by multiple slip and the more homogeneous the overall deformation.

Figure 40 illustrates the beneficial effect of reducing grain size for an alloy containing shearable precipitates (Ref 68). The ternary alloy had an equiaxed grain structure with a mean intercept length of 0.5 mm. Coarse planar slip and intense slip bands, which were later sites for crack nuclei, occurred early in the life of the large-grained material. The Al-Zn-Mg-Zr alloy had smaller elongated grains with mean grain dimensions of approximately 0.03 by 0.05 by 0.10 mm. Slip in the fine-grained material was less intense, and crack initiation was delayed. This is further illustrated in Fig. 40 by the fact that cycles to initiation for the fine-grained material exceeded cycles to failure for the coarse-grained material under the same plastic strain amplitude.

Again it is noted that the two curves converge at low plastic amplitudes (long life) for this strain-controlled test. As mentioned previously, this is due to differences in cyclic-hardening behavior. The strain-hardening exponent, n', of the Al-Zn-Mg-Zr alloy is approximately twice that of the ternary alloy, a fact attributed to a larger degree of multiple slip and more frequent dislocation-dislocation interactions in the Al-Zn-Mg-Zr alloy than in the Al-Zn-Mg alloy. The convergence would not have been observed in a stress-controlled test.

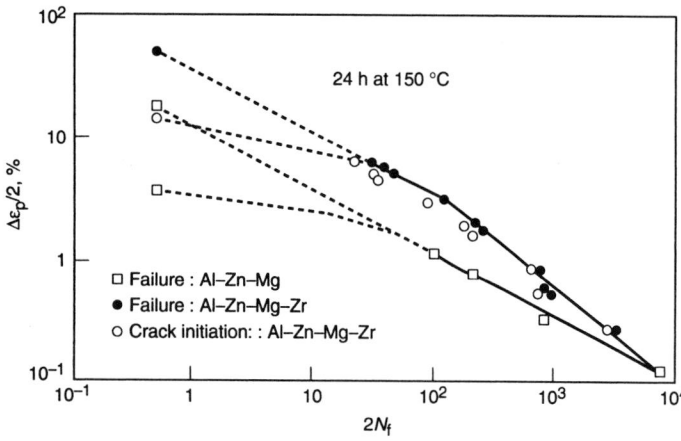

Fig. 43 Effect of grain size on the strain-life behavior of an alloy with nonshearable precipitates plus PFZs. The Al-Zn-Mg alloy had large grain size; the Al-Zn-Mg-Zr, small grain size. Source: Ref 68

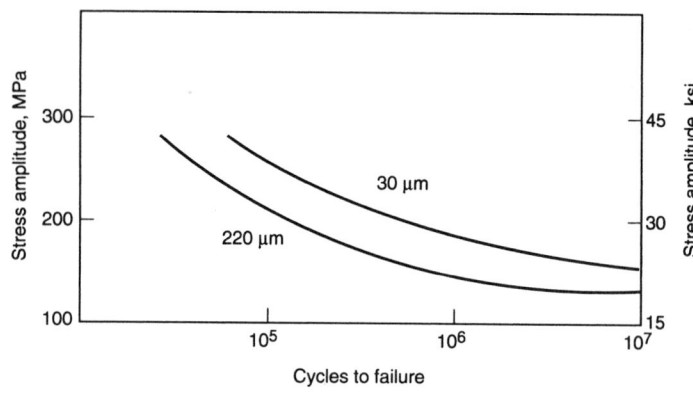

Fig. 44 Effect of grain size on the stress-life behavior of an X7075 alloy with nonshearable precipitates plus PFZs. Source: Ref 69

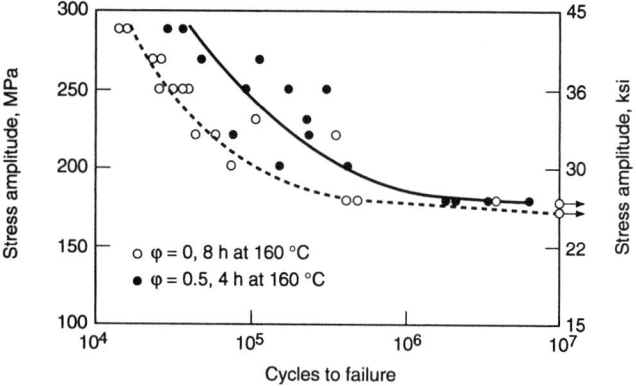

Fig. 45 Effect of grain-boundary ledges on the stress-life behavior of an X7075 alloy containing nonshearable precipitates and PFZs

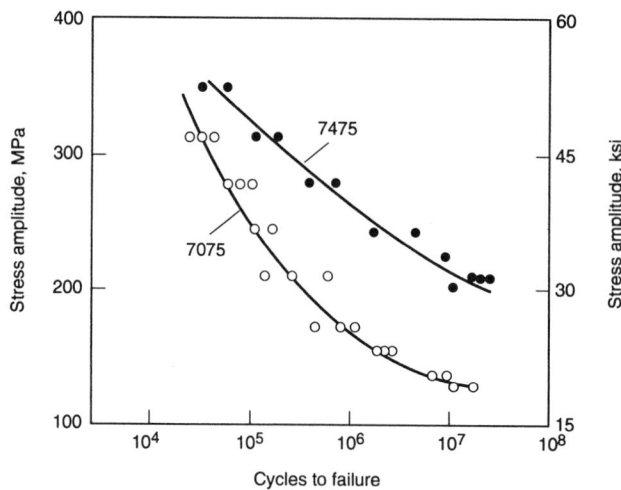

Fig. 46 Effect of inclusion density on the stress-life behavior of two 7XXX alloys: high inclusion density, alloy 7075; low inclusion density, alloy 7475

Figure 41 shows the grain-size effect in a stress-controlled test for a high-purity 7075 alloy (X7075) aged to contain shearable precipitates (Ref 69). Since the flow stress is determined by the interaction of dislocations with the coherent precipitates, the yield stress is approximately the same for both alloys. Optical examinations of the specimen surfaces show that cracks nucleate much earlier in specimens having the large grain size. Cracks nucleated at intense slip bands for both grain sizes. However, the slip bands were much more pronounced in specimens with a large grain size of 200 μm. For specimens with small grain size (30 μm), cracks at slip bands could be detected only in grains that were statistically larger than average.

Precipitate-Free Zones

A solute-depleted PFZ is weaker than the matrix and can be the site of preferential deformation. This preferential plastic deformation leads to high stress concentrations at grain-boundary triple points (Fig. 34) and to early crack nuclea-

tion. The magnitude of the stress concentrations will be a function of the grain-boundary length and the difference in shear strength of the age-hardened matrix and the soft PFZ.

Because the strain localization occurs in a region free of solute, overaging the matrix precipitates or adding dispersoids does not homogenize the deformation. This is clearly illustrated by comparing results for underaged and overaged specimens of large-grained Al-Zn-Mg alloy (Fig. 42). The tensile yield strength and strain to fracture are approximately the same for both specimens. As mentioned previously, the underaged alloy has shearable precipitates, which results in strain localization, the formation of intense slip bands, and early crack nucleation under cyclic loading. Overaging was one method described for homogenizing deformation; however, this method is not effective for large-grained material. Preferential deformation in the PFZ also leads to strain localization and results, for this particular case, in the same fatigue life. For the same reason dispersoids distributed throughout the matrix would not inhibit strain localization in the PFZ.

Reduction of grain size is a very effective method of reducing early crack nucleation due to preferential deformation in the PFZ. This reduces the slip distance and lowers the stress concentrations at grain-boundary triple points. The fracture mode can likewise change from a low-energy intergranular to a higher-energy transgranular mode.

The effectiveness of reducing the grain size is illustrated in Fig. 43, which shows Coffin-Manson life plots of two overaged Al-Zn-Mg alloys, described previously (Ref 68). The small-grained Al-Zn-Mg-Zr alloy has a much longer life than does the large-grained Al-Zn-Mg alloy. The improvement in life is attributed to increasing the cycles to crack initiation. For the lower plastic strain amplitudes, a convergence is noted for long lives (10^4 cycles) for this strain-controlled test. Since the fine-grained material hardens more at low strains, the stress to enforce the applied strain is greater at long lives, and this affects the life improvement due to the fine grains.

No such convergence is observed for a stress-controlled test (Fig. 44) for a similar alloy

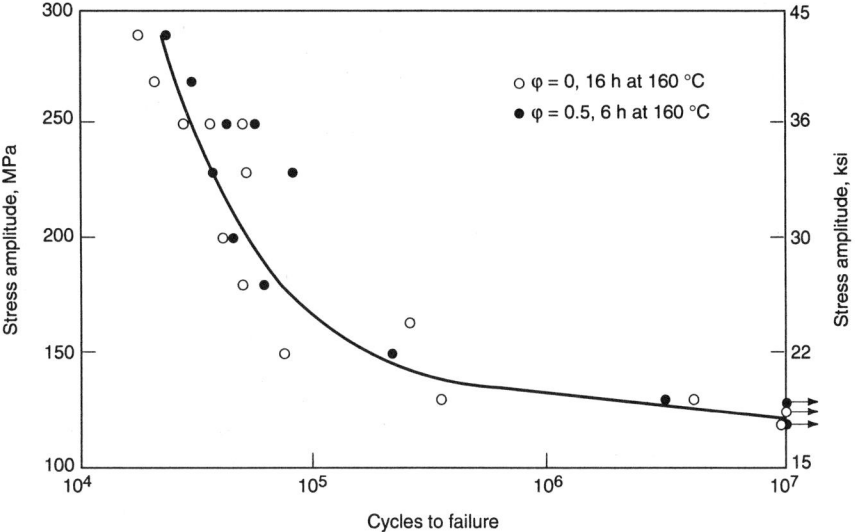

Fig. 47 Effect of alignment of grain boundaries—and alignment plus steps in grain boundaries—on the stress-life behavior of a 7475 alloy containing nonshearable precipitates and PFZs

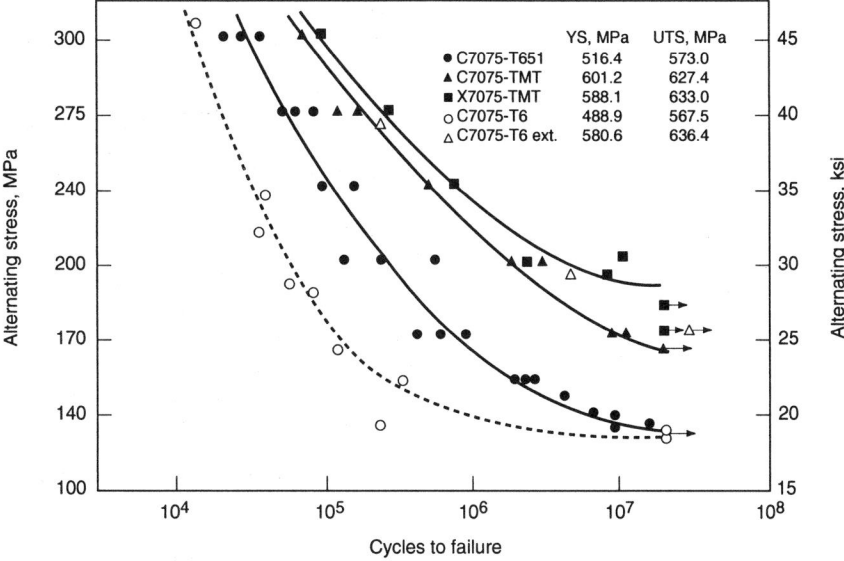

Fig. 48 S-N curves of 7075 aluminum alloys with and without TMTs. Axial loading with $R = 1$. Threaded 13 mm (0.5 in.) round, 75 mm (3 in.) long hourglass specimens (50 mm, or 2 in., radius) with 5 mm (0.2 in.) net section diameter were machined and longitudinally polished. A TMT was given to both commercial and high-purity bars by solution annealing at 460 °C (860 °F) for 1 h, water quenching, aging at 100 °C (212 °F) for 1 h, swaging at room temperature, and aging at 120 °C (250 °F) for 16 h. The commercial alloy, C7075-TMT, was reduced 30% in cross section, whereas the high-purity alloy, X7075-TMT, was swaged only 10% because of specimen size limitations. Source: Ref 72

(X7075) and heat treatment (Ref 69). Optical examination revealed that cracks were nucleated at grain boundaries parallel and perpendicular to the stress axis for the large-grained material, but only at grain boundaries perpendicular to the stress axis for the fine-grained material. This is a direct result of reducing the slip length and thus the local stress concentration. Cracks appearing parallel to the stress axis are a result of the tension-compression employed and the high stress concentrations at triple points in the large-grained material (Ref 69).

Steps in Grain Boundaries. The previous section described the use of grain-size reduction as a means of decreasing the slip length in the PFZ and thus the local stress concentration. This re-

sulted in improved resistance to fatigue crack nucleation and increased fatigue life. Thermomechanical processing is another method that can be used to reduce the slip length in the PFZ. If enough cold deformation is employed to introduce steps (or "ledges") into the grain boundaries, the effective slip length within the PFZ is drastically reduced (similar to a small grain size), with corresponding improvement in resistance to fatigue crack nucleation. Figure 45 shows the results of a stress-controlled test for two high-purity 7075 alloys, one cold worked 50% to produce grain-boundary steps. The cold work drastically reduced the incidence of grain-boundary cracking and improved the fatigue life at high stress amplitudes. At low stress amplitudes and long

fatigue lives, crack nucleation occurred at inclusions for both alloys. This effect is most likely due to lower stress concentration at inclusions.

This raises another important point about microstructure. Many alloys have large inclusions, which may concentrate strain during cyclic deformation and lead to early crack nucleation. This detrimental effect can be reduced substantially by lowering the impurity levels. This is illustrated in Fig. 46, which shows that a significant improvement in the HCF life of 7075 alloy is obtained by lowering the iron and silicon content (7475 alloy).

Alignment of Grain Boundaries. Like many other commercial alloys, high-strength aluminum alloys have dispersoids that inhibit grain growth during high-temperature processing and subsequent heat treatment. For these alloys, the resulting grain shape is characteristic of the processing treatment; for rolled plate it has a pancake shape. If these alloys are aged to contain nonshearable precipitates and have a solute-denuded PFZ, detrimental strain localization could occur only in the PFZ parallel to the long grain dimension and only if the PFZ is inclined to the stress axis. If the stress axis is parallel or perpendicular to the long grain dimension, there will be no shear stress parallel to the grain boundary, and preferential deformation within the PFZ will be restricted. Grain-boundary alignment is then as effective in restricting deformation in the PFZ as are steps produced by thermomechanical treatment (TMT), as shown by the stress-life curves in Fig. 47.

Thermomechanical Processing. Fatigue strength of age-hardened aluminum alloys can be improved in some cases by TMT involving cold work before or during aging. McEvily et al. (Ref 70, 71) found that the fatigue life of Al-Mg and Al-Zn-Mg alloys is increased marginally by cold working prior to aging, perhaps because of partial elimination of grain-boundary PFZs. Ostermann (Ref 72) showed that the long-life fatigue strength and fatigue ratio of smooth 7075 aluminum specimens were increased about 25% by cold working in the partially aged condition (Fig. 48). On the other hand, Reimann and Brisbane (Ref 62) found that the fatigue-life curves for notched 7075 ($K_t = 3$) were essentially unchanged by TMT, and suggested that TMT may affect crack initiation rather than crack growth.

The benefit of TMT is, however, not necessarily limited to crack initiation retardation. Crack growth retardation also has been observed as a result of cold working 2024 aluminum samples prior to aging (Ref 74). DiRusso and coworkers (Ref 75) compared the behavior of T6 and TMT 7075 aluminum and concluded that smooth TMT specimens have lower strength than T6, whereas notched samples may have higher strength. Other results (Ref 76) also demonstrate a significant improvement in fatigue strength in the long-life regime for both smooth and notched ($K_t = 8$) specimens of 7075 as a result of TMT (Fig. 49). The underlying cause of improvement is probably refinement and homogenization of microstructure as a result of TMT, and the consequent

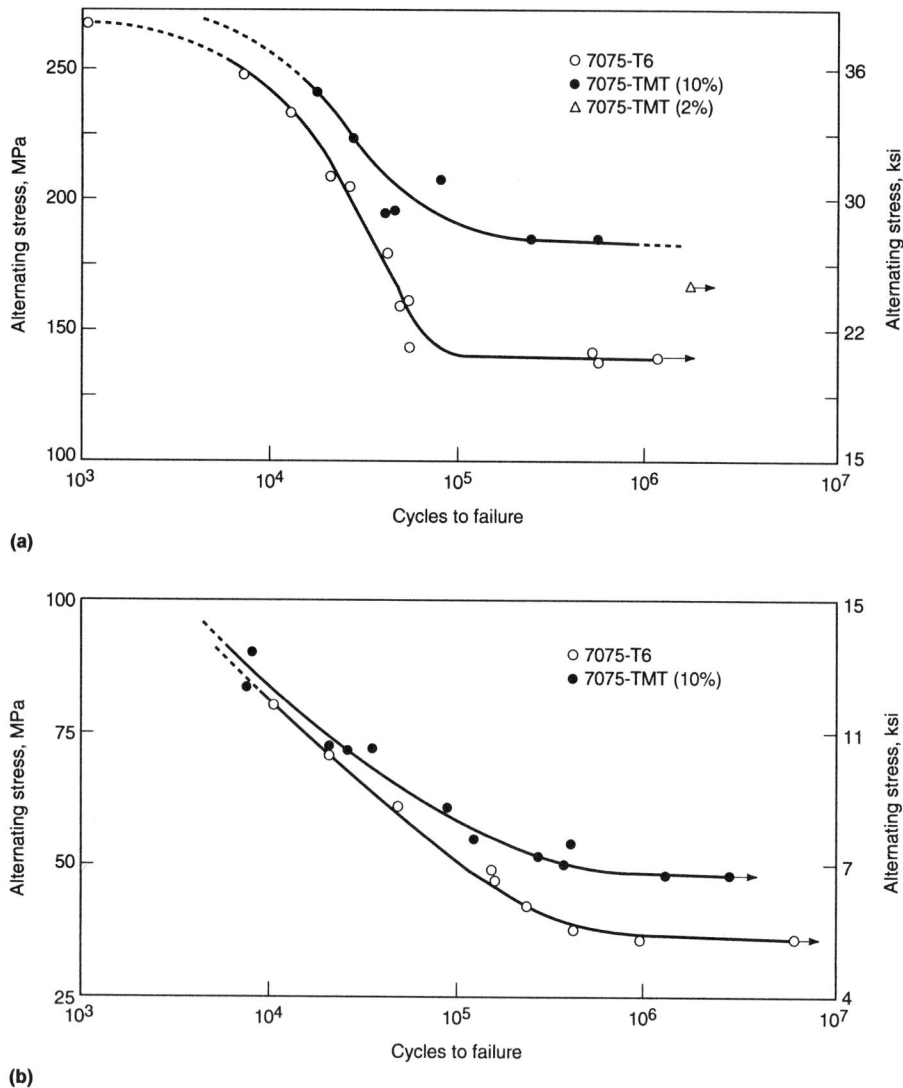

Fig. 49 Fatigue-life curves for 7075-T6 and 7075-TMT. (a) Unnotched. (b) Notched, $K_t = 8$

deformation by dispersed slip during cyclic loading (Ref 72). Other work (Ref 77) suggests that fatigue crack propagation rates in both 2024 and 2124 aluminum alloys depend on precipitate type and dislocation density. The substantial improve-

ment in fatigue strength of notched samples tested at $R = 0$ strongly suggests that microstructural changes due to TMT can promote increased resistance to crack propagation as well as crack initiation during fatigue cycling of 7075.

Fatigue Crack Growth of Aluminum Alloys

A material's resistance to stable crack extension under cyclic loading is generally expressed either in terms of crack length, a, versus number of cycles, N, or as fatigue crack growth rate, da/dN, versus crack tip cyclic stress intensity factor range, ΔK, using fracture mechanics concepts. The latter approach is particularly useful in damage-tolerant design for estimating the influence of fatigue crack growth on the life of structural components. Baseline data for flaw growth predictions is usually established from constant load-amplitude cyclic loading of precracked specimens. Crack length is measured as a function of elapsed cycles, and these data are subjected to numerical analysis to establish rates of crack growth. Crack growth rates are expressed as function of the applied cyclic range in stress intensity factor, ΔK, calculated from expressions based on linear elastic stress analysis. Fracture mechanics assumes that fatigue crack growth in an engineering structure occurs at the same da/dN of the precracked specimen when the range and

mean stress intensity factors for both configurations are the same. Component crack propagation life may therefore be estimated by numerical integration of crack growth rates established from the laboratory coupon specimen.

The typical relationship between fatigue crack growth rate and ΔK observed for most alloys when tested in a non-hostile environment is often classified by the three regions (Ref 78) as shown in Fig. 50. Within region A, crack growth rates become vanishingly small (approximately less than 10^{-5} mm/cycle) with decreasing ΔK, and there exists, within this region, a fatigue stress intensity threshold below which pre-existing cracks do not appear to grow. For many long and infinite life applications, growth of fatigue cracks at very slow rates comprise a major portion of component life, yet low fatigue crack growth rate data on aluminum alloys (and other structural alloys) are rather limited due to the relative high cost and time required to establish this information. Designers using fracture mechanics concepts are interested in low ΔK fatigue crack growth rate information since these rates correspond to early stages of crack formation and propagation where remedial measures can be instituted. In region B, behavior is often characterized by a linear relationship between log da/dN and log ΔK. Region B rates are of great practical interest, since they are generally associated with damage sizes for in-service inspection of high-performance parts. Final stages of fatigue crack propagation are characterized by region C as ΔK (or more specifically K_{max}) approaches the critical stress intensity, K_{Ic} or K_c. Region C growth rates are highly dependent on stress ratio, alloy toughness, and specimen thickness (if not plane strain). Tougher alloys exhibit better constant amplitude fatigue crack growth resistance in regions B and C (Ref 79-82) as indicated by 7075-T6 and high-toughness alloy 7475-T6 data of Fig. 51 (Ref 83).

In examining fatigue crack growth rate curves for many materials exhibiting very large differences in microstructure, the striking feature is the similarities between these curves, not the differences. This point is illustrated by Fig. 52, a compilation of data for 2XXX and 7XXX series aluminum alloys. The differences in crack growth rate between these alloys are important from the viewpoint of integrating along any one of them to obtain the lifetime of a structure, but from a mechanistic point of view, these differences are small. A larger range of metals can be represented by a single curve if the driving force (ΔK) is normalized by modulus. These data exclude the effect of environment (mainly water vapor) which is a major factor affecting fatigue crack growth rates.

The considerable use of the fracture mechanics approach in the evaluation of fatigue crack growth rates in aluminum alloys is evident from a four-part Compendium of Sources of Fracture Toughness and Fatigue-Crack Growth for Metallic Alloys published in the *International Journal of Fracture* (Ref 85-88). Another key reference is the *Damage Tolerant Design Handbook* (Ref 89).

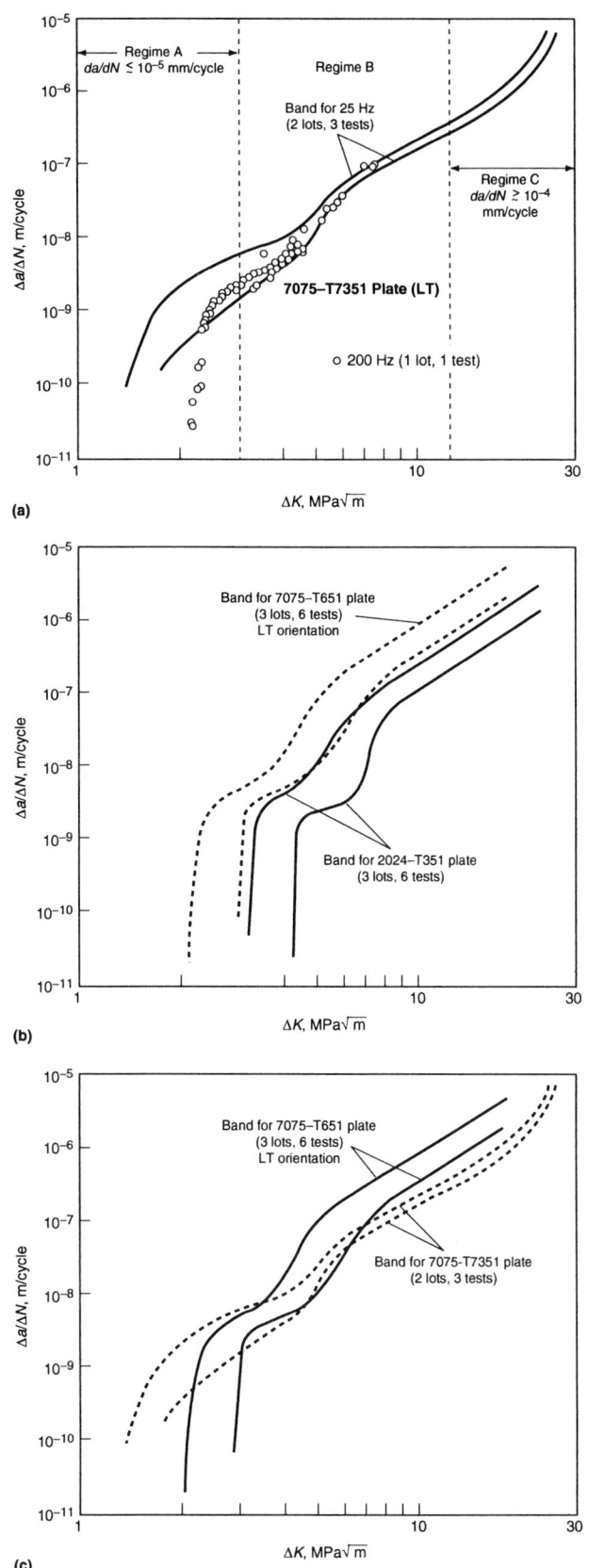

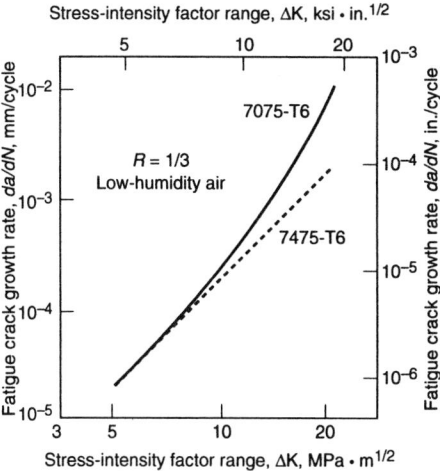

Fig. 51 Benefit of high-toughness alloy 7475 at intermediate and high stress intensity. Source: Ref 83

Fig. 50 Fatigue crack growth of 7075 and 2024 plate in moist air, $R = 0.33$ (a) 25 vs 200 Hz with crack growth regimes (b) and (c) typical scatterbands

Effect of Composition, Microstructure, and Thermal Treatments. In general, fatigue crack growth rates in non-hostile environments fall within a relatively narrow scatter band, with only small systematic effects of composition, fabricating practice or strength, as illustrated by Fig. 52 to 54. There are many sources of fatigue crack growth rate data which show the effects of various physical and microstructural variables on fatigue life of aluminum alloys (Ref 92-105), but there is little agreement on the key variables and there are few significant approaches for improving the fatigue crack growth resistance of these alloys. However, some generalizations can be made.

As discussed in the section "Microstructure and Strain Life" in this article, metallurgical microstructures that distribute plastic strain and avoid strain concentration help reduce crack initiation. Those metallurgical factors which contribute to increased fracture toughness also generally contribute to increased resistance to fatigue crack propagation at relatively high ΔK levels. For example, as illustrated in Fig. 51, at low stress intensities the fatigue crack growth rates for 7475 are about the same as those for 7075. However, the factors that contribute to the higher fracture toughness of 7475 also contribute to the retardation of fatigue crack growth, resulting in two or more times slower growth for 7475 than for 7075 at ΔK levels equal to or greater than about 16 MPa $\cdot \sqrt{m}$ (15 ksi $\cdot \sqrt{in.}$). A similar trend has been observed for 2124-T851, which exhibits slower growth than 2024-T851. Smooth specimens of alloys 2024 and 2124 exhibit quite similar fatigue behavior. Because fatigue in smooth specimens is dominated by initiation, this suggests that the large insoluble particles may not be significant contributors to fatigue crack initiation. However, once the crack is initiated, crack propagation is slower in material with relatively few large particles (2124) than in material with a greater number of large particles (2024).

Staley (Ref 90) summarized the role of particle size in influencing fatigue crack growth in aluminum alloys, as shown in Fig. 55 (Ref 106). The influence of alloy composition on dispersoid effect is shown in Fig. 56. The general trend in Fig.

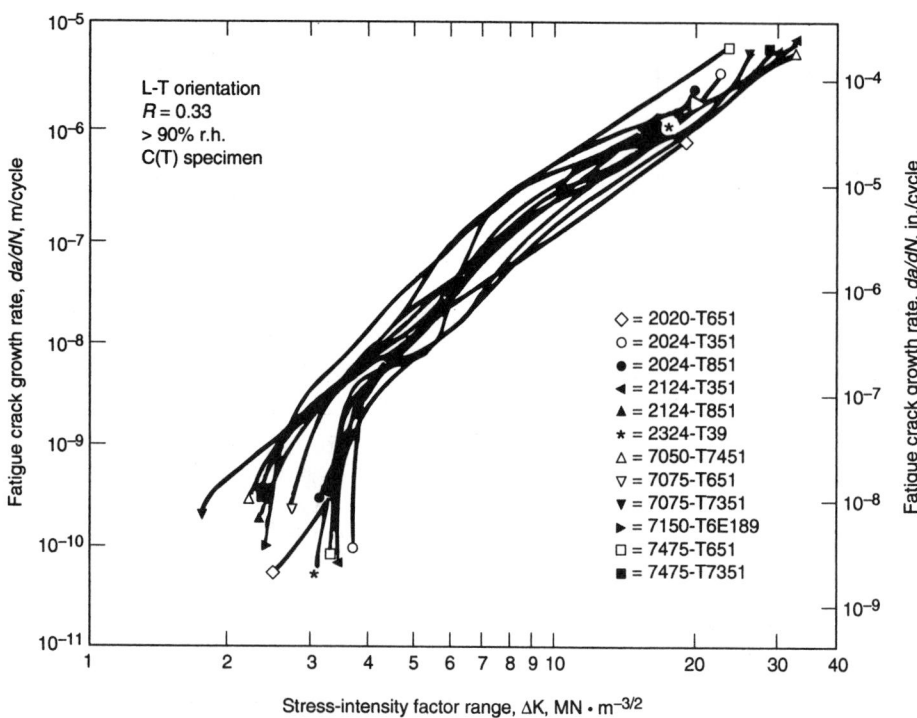

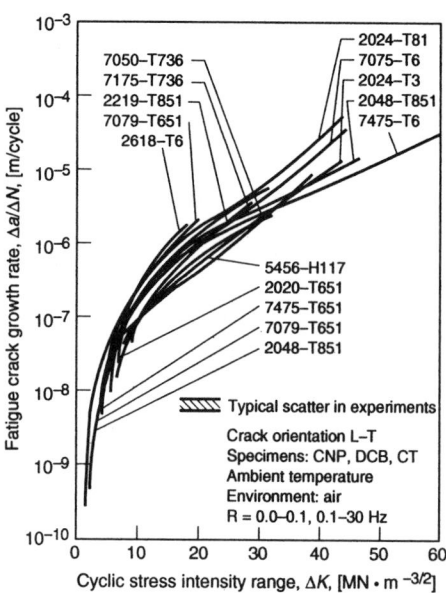

Fig. 52 Minor influences of differing microstructures on fatigue crack growth rate curves: data from twelve 2XXX and 7XXX aluminum alloys with different heat treatments. Source: Ref 73

Fig. 53 Crack growth comparison. Many commercial aluminum alloys show similar fatigue crack propagation rates in air, as indicated above. Source: Ref 90

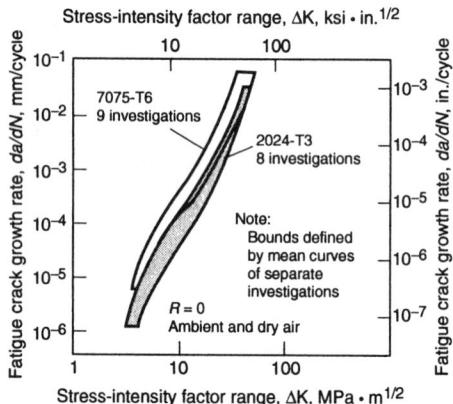

Fig. 54 Summary of fatigue crack growth rate data for aluminum alloys 7075-T6 and 2034-T3. Source: Ref 91

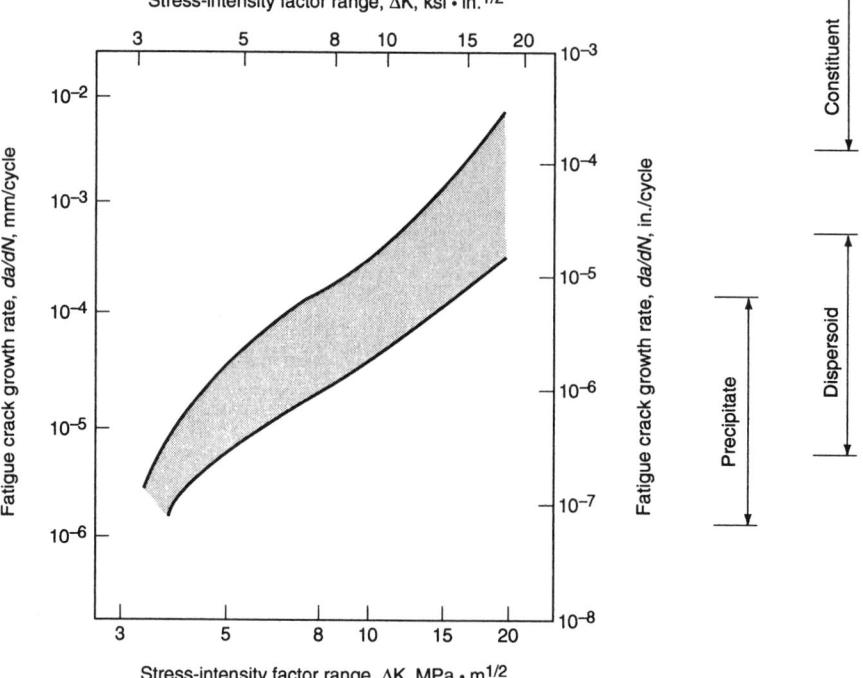

Fig. 55 Comparison of typical particle sizes in aluminum alloys with crack advance per cycle on fatigue loading. Source: Ref 106

56 is that for more finely dispersed particles, the fatigue crack propagation life is increased. Whereas dispersoid type appears to have a relatively small effect on mean calculated life, the smaller precipitates provided by aging produce a much larger effect.

Effect of Processing and Microstructure

The extensive use of age-hardenable aluminum alloys at high strength levels, i.e., greater than 520 MPa (75 ksi), has been hampered by poor secondary properties of toughness, stress-corrosion resistance, and fatigue resistance, particu-
larly in the short transverse direction. Some secondary property improvements have been obtained by employing slight changes in alloy chemistry (Ref 107, 108), different grain refining elements (Ref 92), or removal of the impurity elements Fe and Si (Ref 109-111). Such research has led to the development of alloys with improved fracture toughness and stress-corrosion resistance compared to the extensively used 7075. However, significant improvements in fatigue resistance have not been realized with these methods.

Microstructure control through modification of conventional primary processing methods has

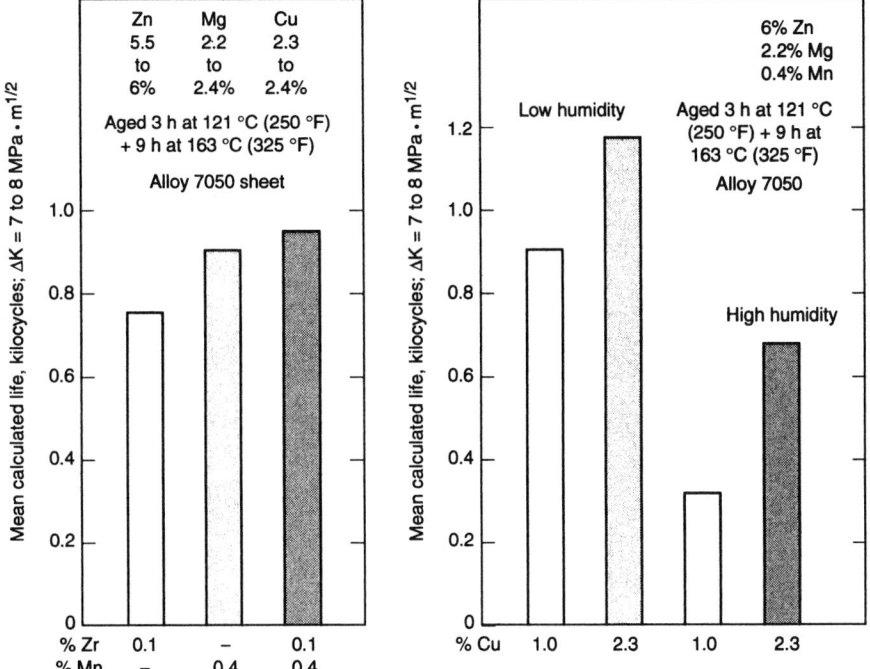

Fig. 56 Effect of dispersoid type (based on composition) on fatigue crack propagation life of 7050 alloy sheet. Source: Ref 106

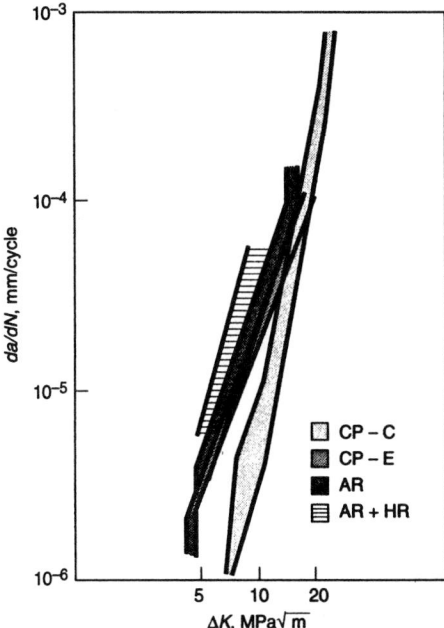

Fig. 57 Crack growth data for compact tension specimens from commercially processed (CP) plate and experimental intermediate thermomechanical treatment (ITMT) material in the as-recrystallized (AR) condition and the as-recrystallized plus hot-rolled (AR + HR) condition. The CP 7050 material was partially recrystallized (<50%) and specimens were from the center (CP-C) and the bottom or top edge (CP-E) of the plate. Scatter bands include data for specimens of both L-T and T-L orientations. Tests conducted at 10 cps and 20 cps in dry air with R = 0.1. Source: Ref 112

been examined as a way of upgrading the fatigue properties of these alloys. These methods, called thermomechanical treatments (TMT), include thermomechanical aging treatments (TMA) and intermediate thermomechanical treatments (ITMT), which are specialized ingot processing techniques applied before the final working operation. In general, for high-strength aluminum alloys, a fine grain structure produced by ITMT improves fatigue-crack-initiation resistance but reduces fatigue-crack-propagation resistance when compared with a typical pancake-shape, partially recrystallized, hot worked structure (Fig. 57) (Ref 112). This effect is more pronounced when the strengthening precipitates are shearable and the grain size determines the slip length. Figure 58(a) compares the LCF curves of ITMT and commercially pure (CP) 7XXX alloys. The ITMT material shows a significant increase in reversals to initiation for all strain amplitudes. The ITMT fine grain structure homogenizes the deformation, and the decrease in strain localization improves the resistance to fatigue-crack initiation. The convergence of the curves at high strain amplitudes results from homogenization of deformation by high strains.

Unfortunately, homogeneous deformation increases the rate of crack propagation because it allows single straight-running cracks during subcritical crack growth. The planar slip and inhomogeneous deformation of CP material enhance crack branching, increase the total crack path, and lower the effective stress intensity at the tip of the crack, all of which lower crack-growth rates (Ref 114). Figure 58(b) compares the fatigue crack propagation curves of the same material and shows the detrimental effect that a fine grain

structure has on the fatigue crack propagation rate.

Combined effects of grain size, deformation mode, and environment on propagation behavior are shown in Fig. 59 (Ref 115) for ITMT-7475. Both aging treatment and grain size significantly affect the fatigue crack growth rates (FCGRs) measured in vacuum (Fig. 59a). Decreasing the grain size by ITMT and overaging, both of which homogenize deformation and decrease the reversibility of slip, increase FCGRs. Although the same trends are observed in air (Fig. 59b), the

magnitude of the effect is considerably reduced due to environment-enhanced growth.

The results in Fig. 59 are consistent with other studies that show that slip character and grain size can have a pronounced effect on the fatigue crack growth behavior of age-hardenable aluminum alloys. When the strengthening precipitates are co-

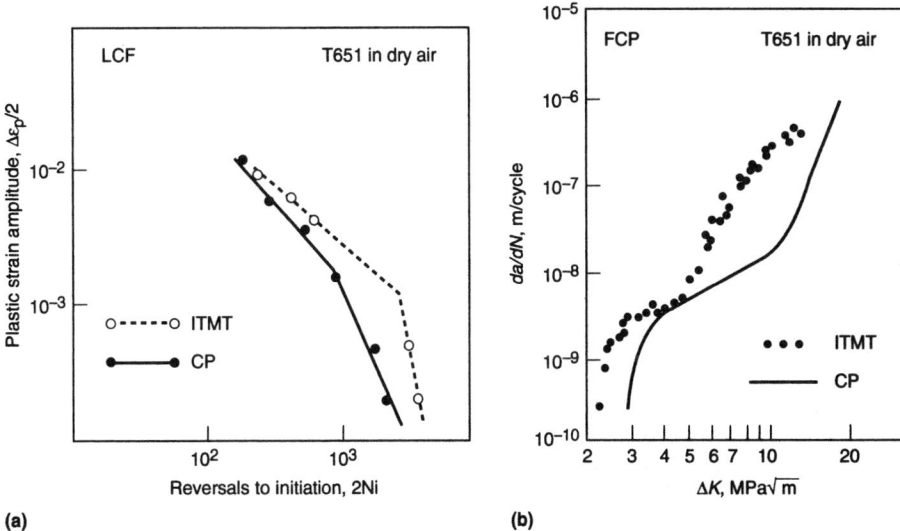

Fig. 58 Effects of intermediate thermomechanical treatments (ITMT) on (a) fatigue crack initiation and (b) fatigue crack propagation of 7XXX aluminum alloys. Source: Ref 113

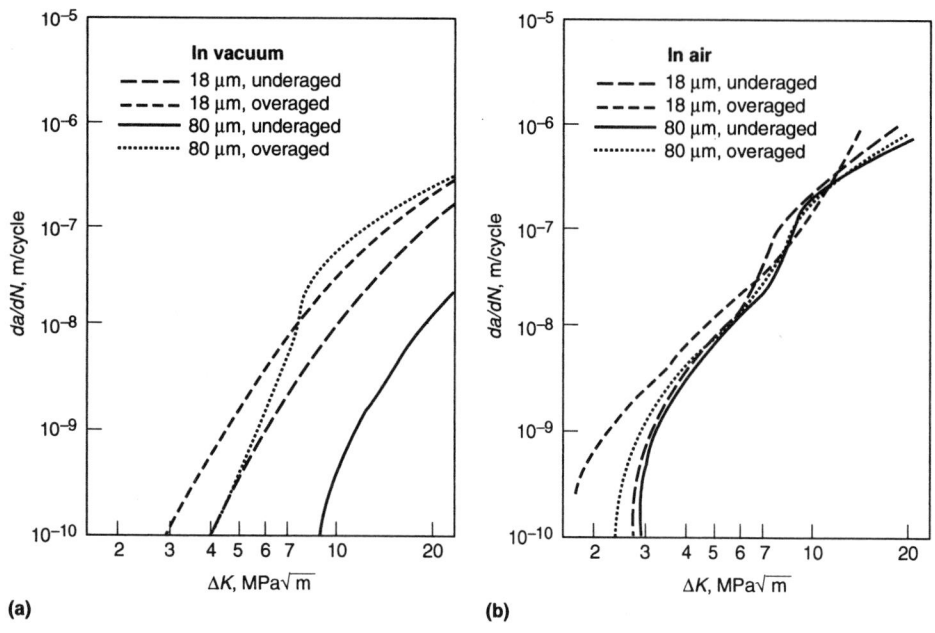

(a) **(b)**

Fig. 59 Effects of grain size and aging treatment on the FCGR of intermediate thermomechanical treatment (ITMT) alloy 7045: (a) tests in vacuum, and (b) tests in laboratory air. Differences in a vacuum could not be accounted for by closure effects. Source: Ref 115

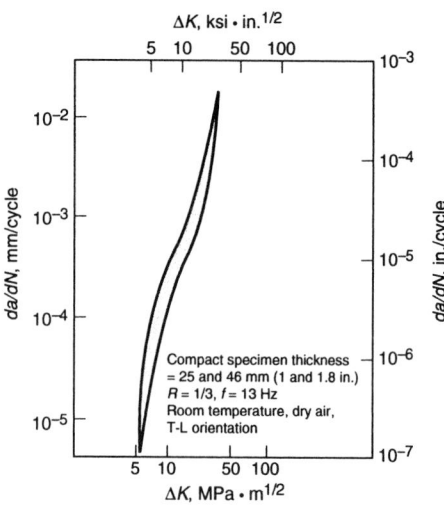

Fig. 61 Fatigue crack growth rates for 5083-O plate in thicknesses of 25, 70, and 178 mm (1, 1.8, and 7 in.). Source: Ref 116

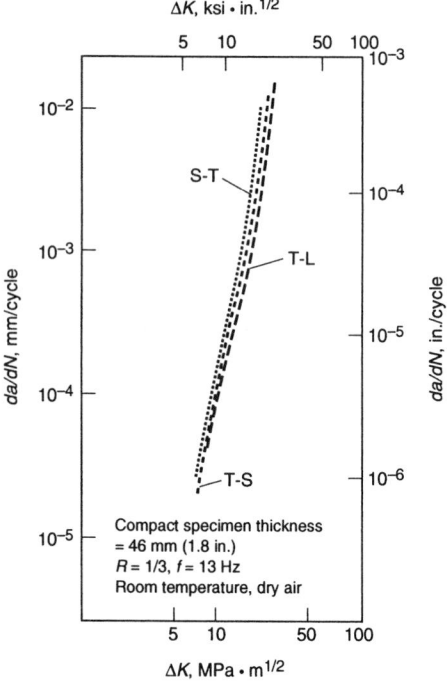

Fig. 60 Effect of orientation on fatigue crack growth rates in 180 and 196 mm (7.0 and 7.7 in.) 5083-O plate. Source: Ref 116

they are looped and bypassed by dislocations promoting more homogeneous deformation and reducing crack tortuosity. A reduction in grain size (by enhancing multiple slip at low ΔK values) and an aggressive environment (by decreasing the plasticity needed for fracture) can also reduce crack tortuosity although the oxides formed in air can have an opposite effect on crack growth rates by increasing crack closure. The slower crack growth rates associated with planar slip and large grains have been attributed to:

(a) Slip being more reversible
(b) The tortuosity of the crack path
(c) The ΔK of zigzag and branched cracks being smaller than the ΔK calculated assuming a single crack normal to the stress axis
(d) Enhanced closure associated with increased surface roughness

A reduction in grain size and overaging reduce the reversibility of slip and crack tortuosity. Consequently, it is not surprising that the 18 μm grain size in Fig. 59(b), overaged material had the fastest crack growth of all the conditions studied. The different fatigue crack growth rates for the various materials may be related to the difference in the extent of crack closure that they exhibited in the air environment, as discussed in Ref 115. However, in a vacuum, differences in growth rates for the various materials could not be accounted for by closure effects. The influence of environment and in particular the improvement in fatigue crack growth resistance in vacuum is well known. The extent of the improvement depends on aging condition and grain size, with the most significant improvements derived for coarse-grained material in an underaged condition. One factor which may account for this is the marked

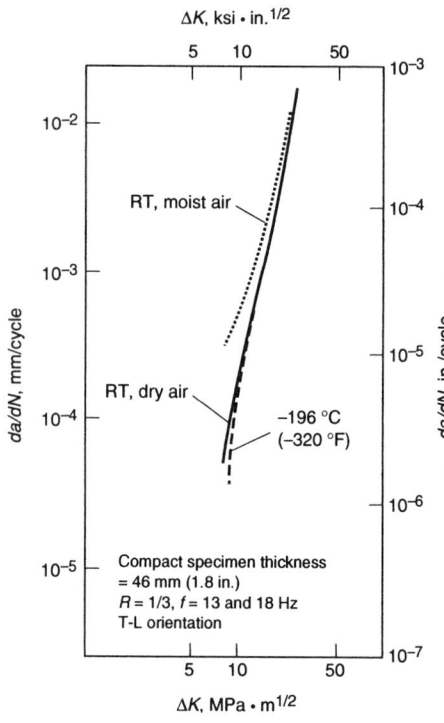

Fig. 62 Effect of temperature and humidity on fatigue crack growth in 180 mm (7.0 in.) 5083-O plate. Source: Ref 116

extent of slip reversibility in the underaged material compared with the multiple slip situation in the overaged material.

Effects of Product Form and Orientation. The rate of fatigue crack propagation in aluminum alloys is relatively insensitive to product form and orientation. This is illustrated in Fig. 60 for thick 5083-O plate that was evaluated for use in

herent with the matrix (underaged condition) they are sheared by dislocations promoting coarse planar slip and inhomogeneous deformation. This favors fracture along slip planes and the occurrence of zigzag crack growth and crack branching. When the strengthening precipitates are incoherent with the matrix (overaged condition),

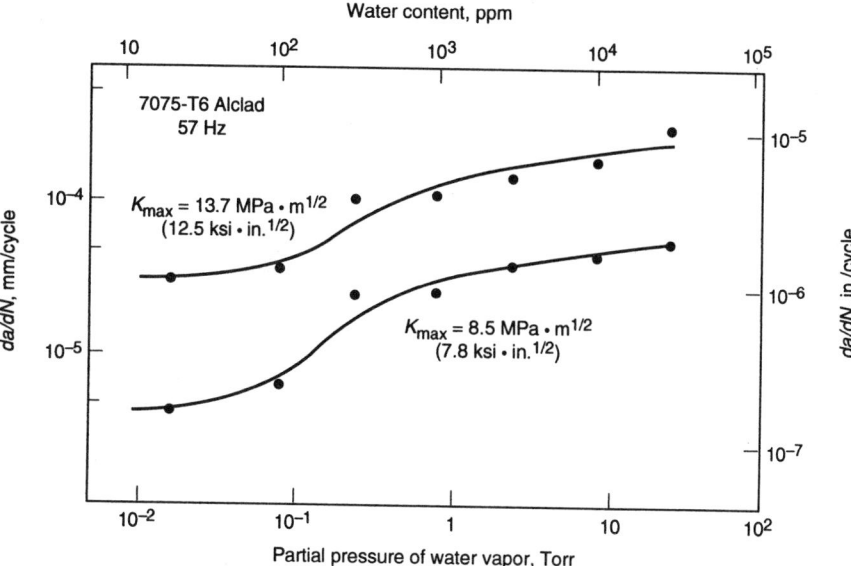

Fig. 63 Effects of moisture on fatigue crack growth rates in aluminum alloys. Source: Ref 101

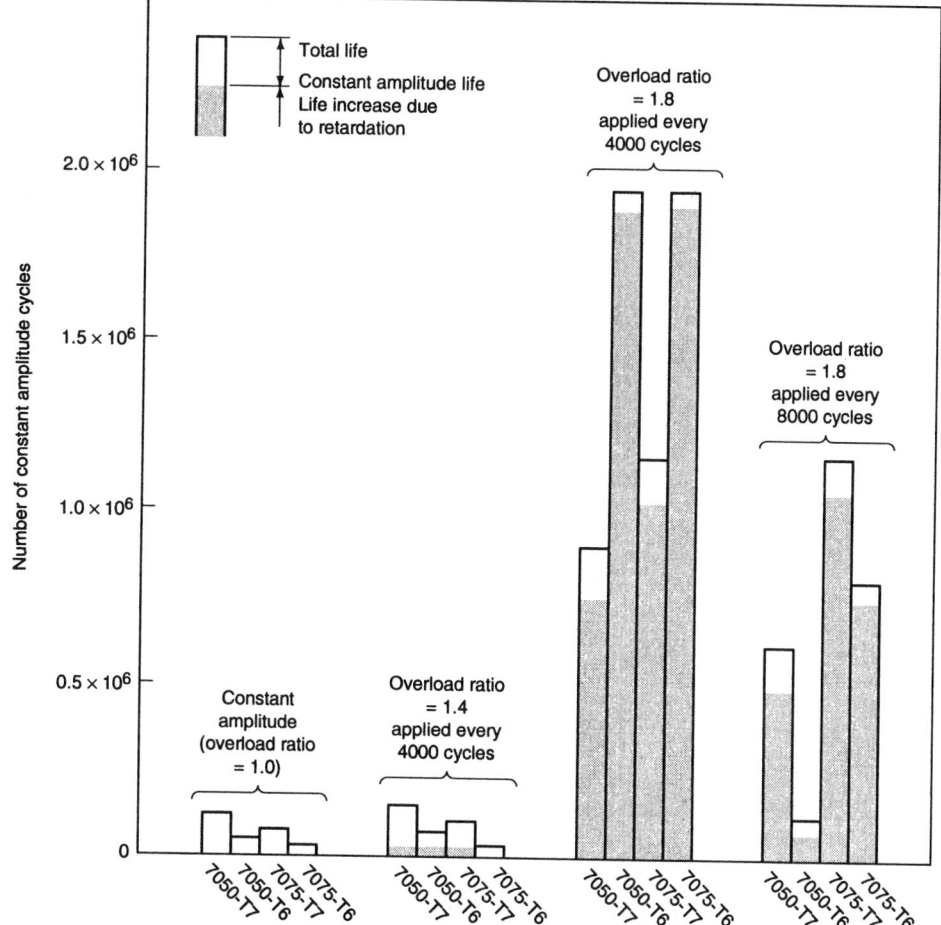

Fig. 64 Relative ranking of fatigue life of 7075 and 7050 aluminum alloys under constant amplitude and periodic single overload conditions. Source: Ref 119

effect, as illustrated by the data for 5083-O in Fig. 61 (Ref 116).

Effects of Exposure Temperature. Although no significant amount of fatigue testing has been done at temperatures above room temperature, there has been a great amount of testing at subzero temperature, particularly at –196 °C (–320 °F). In general, fatigue crack growth rates below room temperature are about the same as, or lower than those, at room temperature (Fig. 62).

Effect of Humidity. The role of humidity and environment is a well-known factor affecting crack growth. As shown in Fig. 62, growth rates for alloy 5083-O are appreciably higher in moist air than in dry air (Ref 116). Growth rates in water solutions of sodium chloride are similar to those in moist air.

Data for Alclad 7075-T6 in Fig. 63 (Ref 117) illustrate that even relatively low levels of moisture can accelerate crack growth rates. Unless relative humidity is below 3 to 5%, it seems best to consider that accelerated fatigue crack growth rates are likely in service.

Effect of Load-Time History. Selection of the appropriate type of load cycle to be used in evaluating fatigue crack propagation rates of aluminum alloys has been found to be particularly critical. Staley (Ref 106) and Bucci et al. (Ref 118) noted that the variable amplitude crack propagation testing offers the following advantages: (a) increased sensitivity to microstructural effects; (b) relevancy to design through consideration of important effects of load history, especially overload/plastic zone interactions; (c) practical interpretation when results are expressed in terms of crack size versus fatigue life; and (d) suitability for automated testing and analysis. Bucci further indicated that there are important interactions between microstructure and crack growth under variable amplitude cyclic loading that are not accounted for in constant amplitude testing. Ratings of alloys based on constant amplitude testing are not very likely to provide realistic indications of fatigue performance under the usual service-type variable amplitude loading. The primary cause of this difference is that variable amplitude loading includes the effects of crack growth retardation on fatigue crack propagation—effects that cannot be demonstrated in constant amplitude testing. Crack growth retardation is caused by tension overloading and consequent plastic deformation. The variable amplitude test is believed to be more sensitive to alloy difference, and it clearly provides more useful information for alloy development investigations.

For example, as illustrated by the data for alloys 7075 and 7050 in Fig. 64 (Ref 119), quite different results are obtained in constant amplitude tests than in tests with single overloads every 4000 or 8000 cycles. Thus, information on the variation in load level during fatigue cycling is required for correct characterization of the fatigue behavior of aluminum alloys. More detailed information is provided in the article "Fatigue Crack Growth under Variable Amplitude Loading" in this Volume.

tankage for liquefied natural gas; growth rates in specimens from four orientations were well within the range for replicate tests in any one orientation (Ref 116). It would be expected that in the high ΔK range, differences in growth rate

would reflect differences in toughness, and therefore would indicate somewhat higher growth rates for stressing normal to the plane of the product—that is, in the S-L and S-T orientations. Product thickness also seems to have a small

Effect of Load Ratio, R. The effect of stress or load ratio on fatigue crack growth rate (load ratio, R, is the ratio of minimum to maximum load in the fatigue cycle) is well known. This effect is shown for alloy 7075-T6 sheet about 2.5 mm (0.1 in.) in Fig. 65 (Ref 120). In general, and in all the data presented here, an increase in R at the same ΔK level causes an increase in growth rate. A modification of the Paris equation [$da/dN = C(\Delta K)^n$] that accounts for the effects of load ratio is the Forman equation, which is as follows:

$$\frac{da}{dN} = \frac{C(\Delta K)^n}{(1-R)K_c - \Delta K} \qquad \text{(Eq 7)}$$

For the data in Fig. 65, the Forman equation gives a good representation of the variation of da/dN with R for a wide range of crack growth rates.

Crack Growth in Alloy Selection and Design

Generally, fatigue crack growth resistance has only modest variations among present high-strength aluminum alloy mill products. Environmental factors, particularly moisture and chlorides, are more significant than material differences in affecting crack growth rates (Fig. 66). Separation of material differences in fatigue behavior is further confounded by the generally accepted practice of plotting results on log coordinates where relatively small shifts in a data trend could have a significant impact on the life of a part. For example, the width of the da/dN band in Fig. 17 represents a factor of ten, affording considerable room for improvement if alloys can be selected or developed that confine their behavior to the crack growth rate lower bound. Moreover, many designers of high-performance structures will concede that a 50% life improvement or 10% weight reduction afforded by design to higher stress without reduction in fatigue strength is significant.

Although material differences generally have only a modest effect on fatigue crack growth rates, research work has established statistically significant effects of alloy microstructure and composition on fatigue crack growth resistance of high-strength aluminum alloys (Ref 43-46). These programs, which consisted of a set of highly controlled experiments on laboratory-fabricated 2XXX (Al-Cu-Mg-Mn) and 7XXX (Al-Zn-Mg-Cu) microstructures,* permitted the following conclusions on fatigue crack growth resistance under constant amplitude loading:

- High-purity and lower-copper versions of alloy 2024 (i.e., 2124 and 2048) provide im-

*Laboratory-fabricated microstructures were designed to simulate commercial material but were developed with greater controls of processing variables in order to systematically investigate their effect. Conclusions from these programs have been verified on commercial alloys.

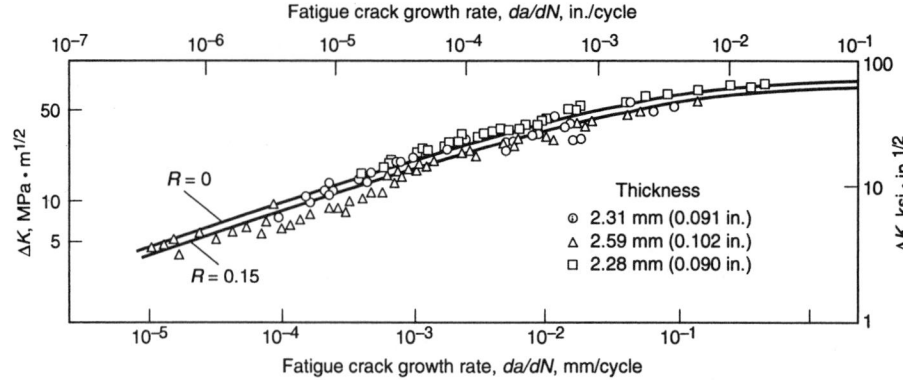

(a)

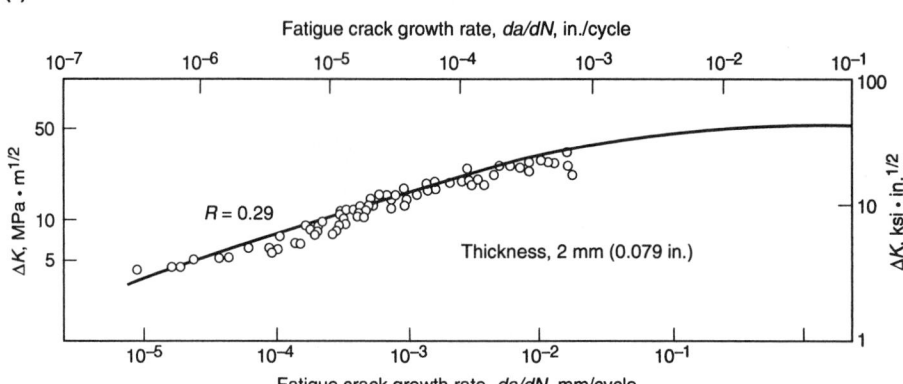

(b)

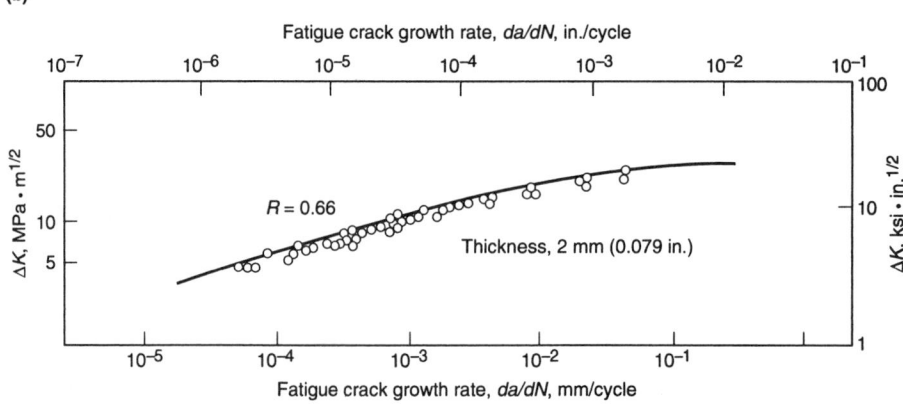

(c)

Fig. 65 Effect of load ratio, R, on fatigue crack growth rates in aluminum alloy 7075-T6. Source: Ref 120

proved resistance to regions 2 and 3 fatigue crack growth over their 2024 counterpart because of improved fracture toughness.
- Improved toughness through increased purity resulted in greatest improvements in 2XXX alloy fatigue crack propagation life in artificially aged T8-type tempers.
- Toughness and resistance to fatigue crack propagation increase as strength decreases in underaged and peak-aged tempers of Al-Cu-Mg-Mn alloys containing low levels of cold work. Overaged 2XXX tempers provide inferior combinations of strength, toughness, and resistance to fatigue crack growth.
- Compared to conventional 7XXX alloys, such as 7075, "improved" 7475 and 7050 alloys provide greater fatigue crack growth resis-

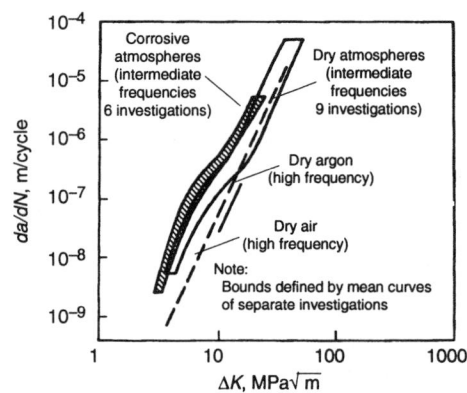

Fig. 66 Influence of environment and cycle frequency on fatigue crack growth ($R = 0$) of aluminum alloy 7075-T6. Source: Ref 121

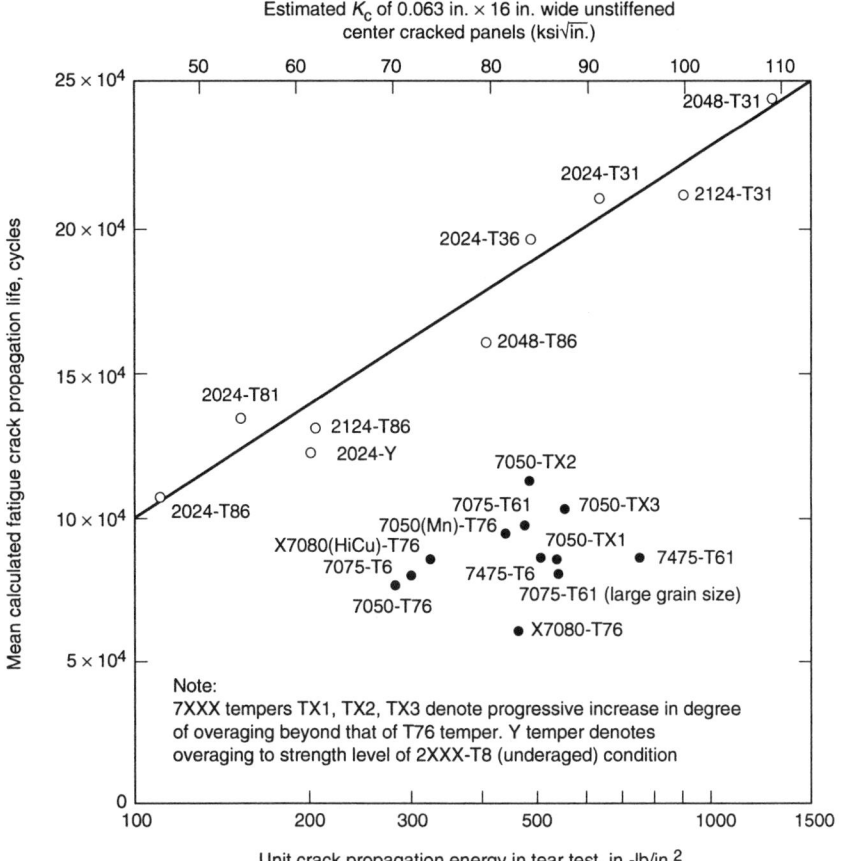

Fig. 67 Relationship between fatigue crack propagation performance and fracture toughness for laboratory-fabricated 2XXX and 7XXX aluminum alloy sheet. Mean crack growth life is that life averaged over four experimental conditions for each alloy. The four conditions were two frequencies, 2 and 20 Hz, and two environments, low (<5%) and high (>90%) relative humidity air. Life was determined from tests of 75 mm (3 in.) wide center crack tension panels where ΔK varied from 6.6 to 16.5 MPa $\sqrt{m}$ (6 to 15 ksi $\sqrt{in.}$) under a constant-stress amplitude of $\Delta\sigma = 36.5$ MPa (5.3 ksi) and $R = 0.33$.

tance, in addition to better combinations of strength, toughness, and SCC resistance.

- Overaging 7XXX alloys to T7-type tempers and increasing the copper content to levels of alloy 7050 appears to increase resistance to fatigue crack growth by increasing resistance to corrosive attack by water vapor. (Note: The nominal copper contents are 1.6 and 2.3% for aluminum alloys 7075 and 7050, respectively.)

Fatigue resistance of 2XXX and 7XXX microstructures have been rated by constant-amplitude crack growth life in 76 mm (3 in.) wide center crack tension specimens (Ref 45). Life was then correlated with toughness (Fig. 67) and yield strength (Fig. 68). Good fatigue crack growth resistance of 2XXX alloys shows high correlation with increasing toughness and/or decreasing strength. Fatigue crack growth resistance of 7XXX alloys shows less correlation to toughness and/or strength. Instead, 7XXX alloy rating is better characterized by microstructure and resistance to environment.

Effects of Load History on Fatigue Resistance of Aluminum Alloys. Constant-amplitude fatigue data provide a basic reference and are a prerequisite for making and improving cumulative damage computations. However, an impor-

tant factor to be considered in addressing fatigue performance of aluminum alloys is the interaction of in-service load history (generally variable amplitude rather than constant amplitude) with material and environmental parameters (Ref 40, 43, 44, 122, 123). It has been amply demonstrated under variable load history that high tensile overloads (at levels of stress intensity that do not promote significant tearing as K approaches K_c) produce significant delays in crack growth during subsequent fatigue cycles at lower amplitudes (see, for example, Fig. 69). The crack growth retardation is generally attributed to a reduction in crack tip stress intensity caused by residual plastic deformation or crack branching resulting from the overload. Compressive loads, hold times, or environmental effects are known to remove some of the benefits produced by the high overloads. However, fatigue crack propagation lives of components subjected to load histories with high tensile overloads are generally greater than linear damage life predictions assuming no load interaction effect (e.g., Fig. 70). Overload-delay phenomena have been well documented in the literature (Ref 124, 125).

Crack growth under variable amplitude loading sometimes shows alloy differences that are not readily apparent from constant-amplitude tests,

the method most commonly used to rate alloy fatigue performance (Ref 123). Factors such as cyclic hardening, crack growth retardation characteristics, alloy strength-toughness combination (as varied by temper), and alloy-microstructure interaction with load history and environment all may have appreciable influence on spectrum fatigue life of a particular alloy. For example, Fig. 70 shows that ratings of fatigue crack growth resistance for several aluminum alloys tested under constant amplitude and flight simulation loading differ (Ref 126). Figure 71 shows that an approximate 12% strength reduction from T76 to T73 temper variation in alloy 7475 resulted in a life improvement which, in effect, made the difference between meeting and not meeting a two lifetime damage tolerant design requirement on a fracture-critical aircraft part (Ref 127). However, the magnitude of the improvement was not predictable from constant amplitude data.

Figure 72 is replotted data from Ref 128 that shows fatigue crack growth lives of aluminum alloy 7050 established from flight simulation testing of precracked center crack tension panels. Strength level was varied by heat treatment. Results show that up to a yield strength of about 500 MPa, monotonic decrease in life is associated with monotonic increase in strength. However, increases in 7050 strength beyond 500 MPa results in increased life until a strength level of about 580 MPa, where resistance to fatigue crack propagation diminishes again. The transition behavior indicated by results of Fig. 72 can be related to competing mechanisms that control the fatigue crack growth process. Related work (Ref 43, 44) showed that the dominant mechanism depends on the interaction of loading conditions, specimen configuration, and alloy microstructure.

The point to be emphasized by these illustrations is that complex competing alloy load-interaction mechanisms may be present in variable amplitude fatigue situations. Those mechanisms that dominate are, in part, application-dependent. Therefore, constant-amplitude loading may not always be sufficient or, for that matter, appropriate for rating alloys for optimum selection and/or design of fatigue-critical parts, and spectrum testing may be necessary. Though appreciable effort has been directed at establishing understanding and predictability of crack growth under variable load history, relatively little has been done to qualitatively rate crack growth retardation characteristics from one alloy to another. This work is confounded by the fact that alloy rating is somewhat spectrum- and environment-dependent. Use of appropriate standardized spectra for certain classes of applications (e.g., fighter, bomber, transport aircraft, automotive spectra) and standard test environments provides a database for basic alloy comparisons and improved understanding.

As illustrated by the preceding discussion, caution should be exercised in design use of K_{Ic} for anything other than calculating critical flaw size, since K_{Ic} may not necessarily be correlated with d*a*/d*N*, retardation characteristics, or alloy-crack growth interactions with environment.

REFERENCES

1. J.G. Kaufman and M. Holt, "Fracture Characteristics of Aluminum Alloys," Technical Paper 18, Alcoa Research Laboratories, 1965
2. J.G. Kaufman, "Aluminum Alloys for Arctic Applications," paper presented at the Conference on Materials Engineering in the Arctic (St. Jovite, Quebec), 1976
3. Aluminum Standards and Data, Aluminum Association, 1976
4. J.G. Kaufman, F.G. Nelson, and R.H. Wygonik, "Large Scale Fracture Toughness Tests of Thick 5083-O Plate and 5183 Welded Panels at Room Temperature, –260 and –320 °F," STP 556, ASTM, 1974
5. R.A. Kelsey, G.E. Nordmark, and J.W. Clark, "Fatigue Crack Growth in Aluminum Alloy 5083-O Thick Plate and Weld for Liquefied Natural Gas Tanks," STP 556, ASTM, 1974
6. "Aircraft Structural Integrity Program, Airplane Requirements," MIL-STD 1530, U.S. Air Force, 1972
7. H.P. Van Leeuwen and L. Schra, "Rate Effects on Residual Strength of Flawed Structures and Materials," NLR-TR76004U, National Aerospace Laboratory, NLR, The Netherlands, 1975
8. G.T. Hahn and A.R. Rosenfield, "Relations between Microstructure and the Fracture Toughness of Metals," Plenery Lecture III-211, Third International Conf. on Fracture (Munich), 1973
9. G.T. Hahn and A.R. Rosenfield, Metallurgical Factors Affecting Toughness of Aluminum Alloys, *Met. Trans.*, Vol 6A, 1975, p 653-668
10. J.T. Staley, "Microstructure and Toughness of Higher Strength Aluminum Alloys," STP 605, ASTM, 1976, p 71-103
11. M.V. Hyatt, New Aluminum Aircraft Alloys for the 1980s, *Met.*, Vol 46 (No. 2), 1977
12. J.T. Staley, "Update on Aluminum Alloy and Process Developments for the Aerospace Industry," paper presented at the Western Metal and Tool Exposition and Conference (WESTEC) (Los Angeles, CA), 1975
13. J.G. Kaufman, "Design of Aluminum Alloys for High Toughness and High Fatigue Strength," paper presented at the Conference on Alloy Design for Fatigue and Fracture Resistance (Brussels, Belgium), 1975
14. R.R. Senz and E.H. Spuhler, Fracture Mechanics Impact on Specifications and Supply, *Metals Progress*, 1975, p 64-66
15. D.O. Sprowls and E.H. Spahler, Avoiding SCC in High Strength Aluminum Alloys, Alcoa Green Letter GL188, Rev 1982-01
16. D.O. Sprowls, "Environmental Cracking—Does It Affect You?," ASTM Standardization News, Vol 24, No. 4, April 1996, p 2-7
17. M.O. Speidel, Hydrogen Embrittlement and Stress Corrosion Cracking of Aluminum Alloys, *Hydrogen Embrittlement and Stress Corrosion Cracking*, American Society for Metals, 1984, p 271-295
18. T.J. Summerson and D.O. Sprowls, "Corrosion Behavior of Aluminum Alloys," Plenary Paper during The International Conference in Celebration of the Centennial of the Hall-Heroult Process, University of Virginia, Charlottesville, VA, 15-20 June 1986, *Vol III of the Conference Proceedings*, Engineering Materials Advisory Services, Ltd., p 1576-1662
19. R.H. Jones and R.E. Ricker, "Mechanisms of Stress-Corrosion Cracking," in *Stress-Corro-*

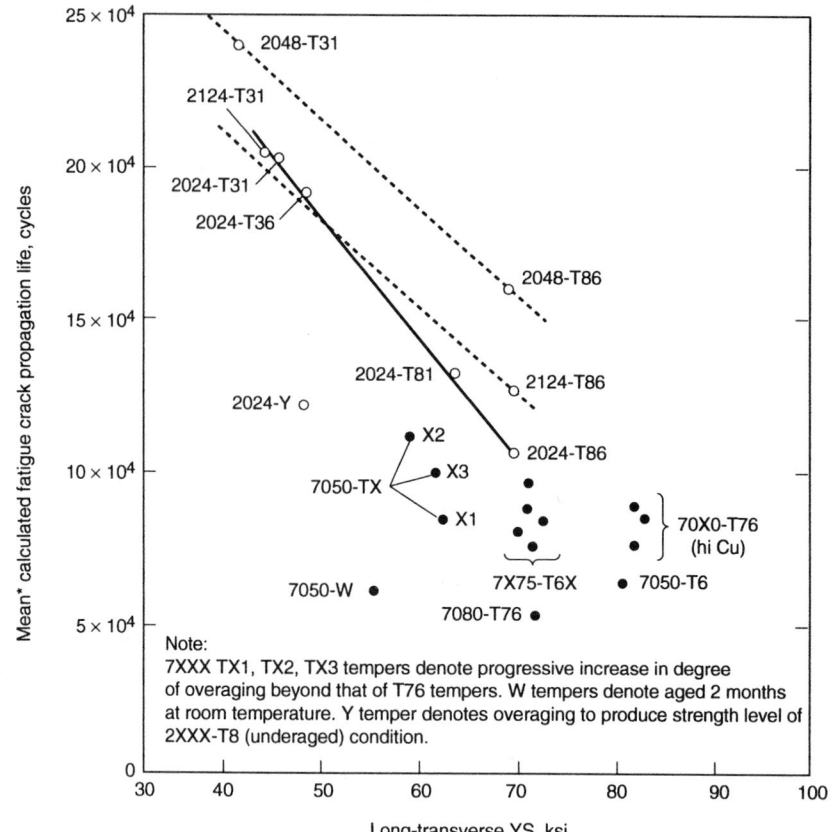

Fig. 68 Relationship between fatigue crack propagation performance and yield strength for laboratory-fabricated 2XXX and 7XXX aluminum alloy sheet. Mean crack growth life is that life averaged over four experimental conditions for each alloy. The four conditions were two frequencies, 2 and 20 Hz, and two environments, low (<5%) and high (>90%) relative humidity air. Life was determined from tests of 75 mm (3 in.) wide center crack tension panels where Δ*K* varied from 6.6 to 16.5 MPa $\sqrt{m}$ (6 to 15 ksi $\sqrt{in.}$) under a constant-stress amplitude of Δσ = 36.5 MPa (5.3 ksi) and *R* = 0.33.

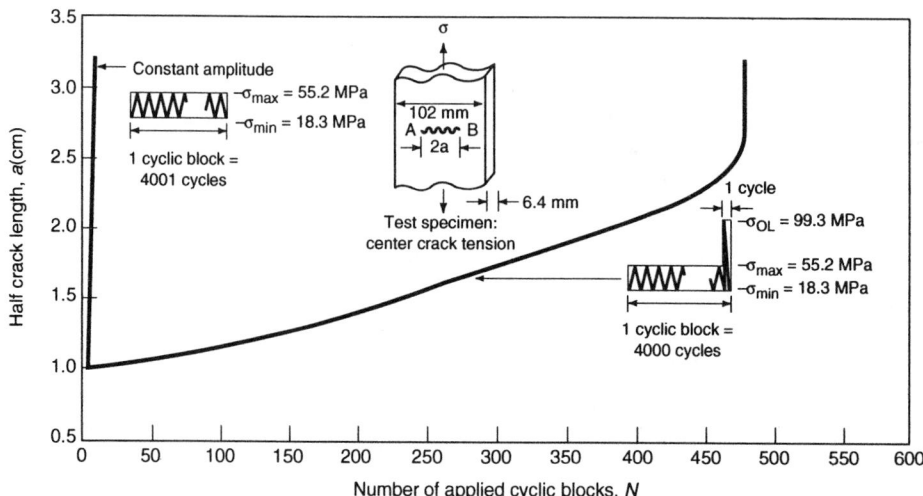

Fig. 69 Crack growth retardation produced by periodic single spike overload in aluminum alloy 7075-T6. Source: Ref 43

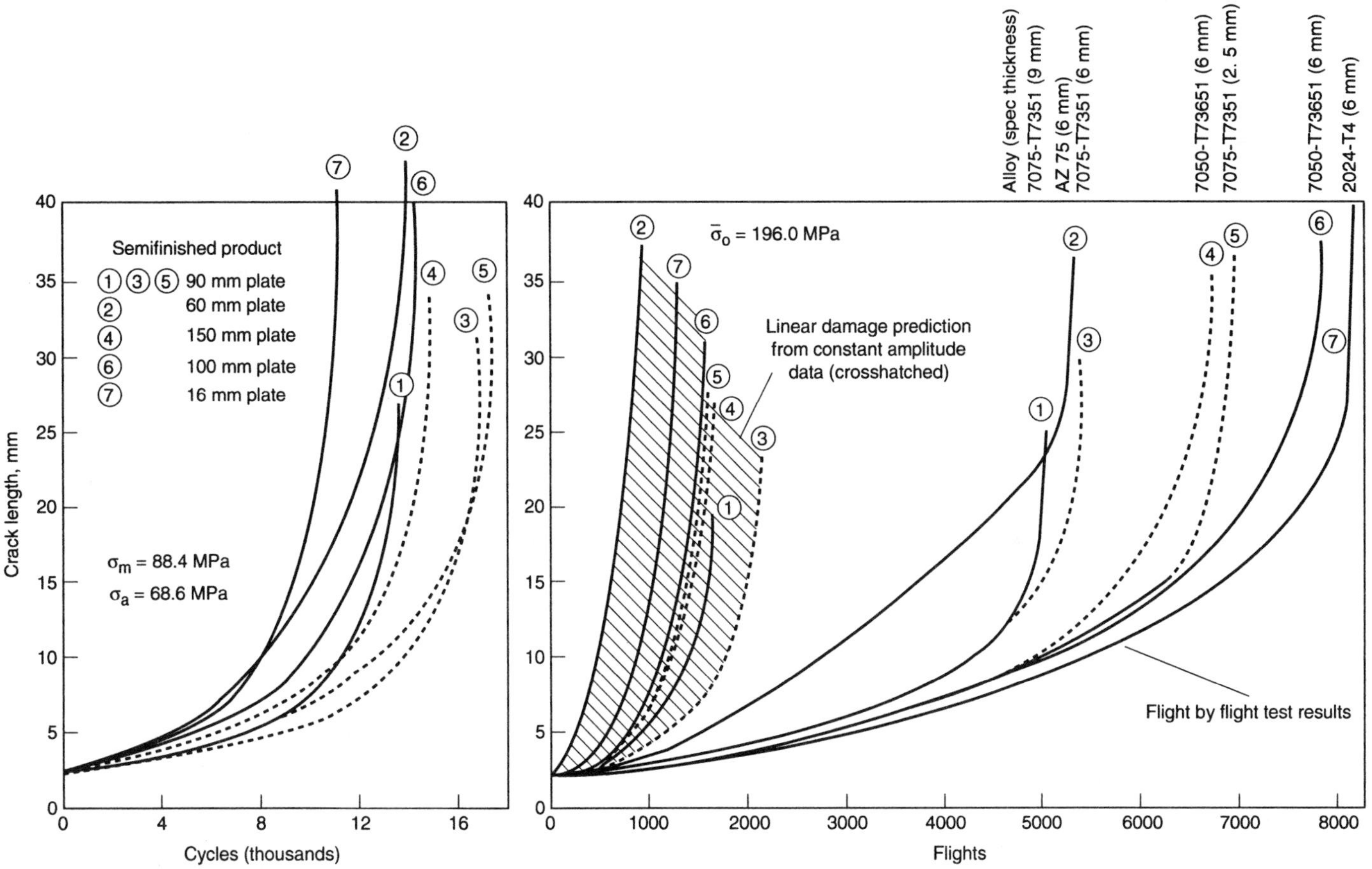

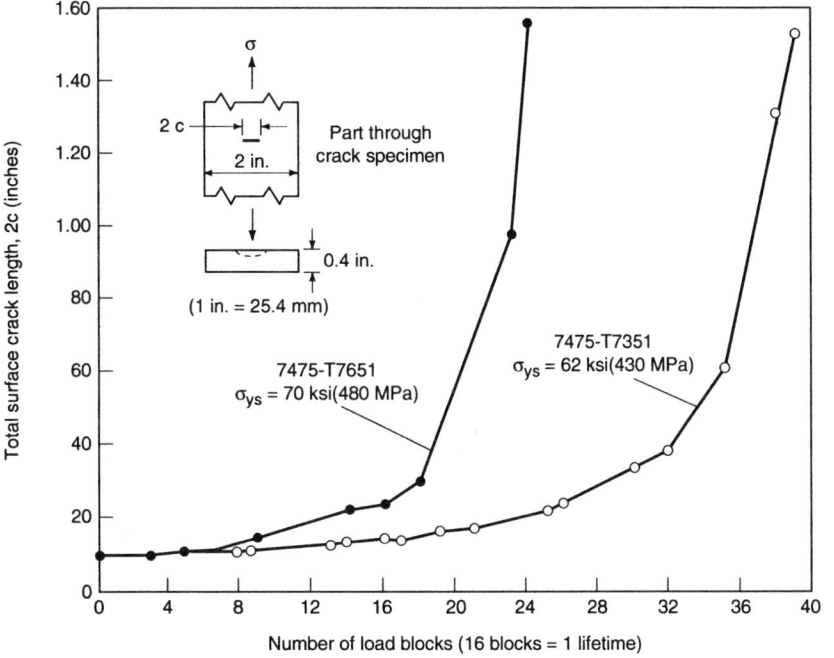

Fig. 70 Crack propagation tests under flight-by-flight and constant-amplitude loading for different aluminum alloys. Source: Ref 126

Fig. 71 Effect of overaging on part through crack growth of aluminum alloy 7475 subject to 500 h block flight-by-flight fighter spectrum loading in sump tank water. Source: Ref 127

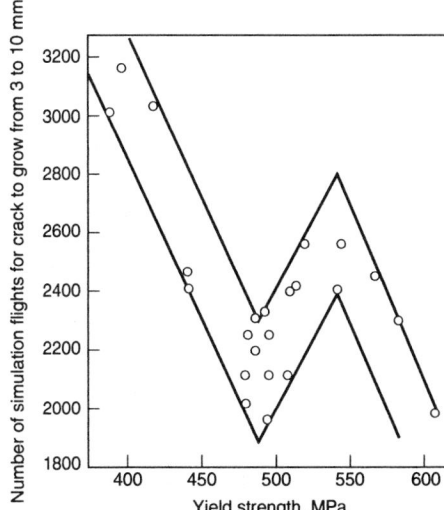

Fig. 72 Effect of yield strength on fatigue crack propagation life of alloy 7050 forging under flight simulation loading

Corrosion Tests and Standards: Application and Interpretation, ASTM Manual 20, Robert Baborian, Ed., 1995, p 447-457

22. ISO 7539-1, International Standard on Corrosion of Metals—Stress Corrosion Testing—Part I: General Guidance on Testing Procedures, International Organization for

sion Cracking—*Materials Performance and Evaluation*, Russell H. Jones, Ed., ASM International, 1992, p 23

20. R.N. Parkins, "Current Understanding of Stress-Corrosion Cracking," *Journal of Metals*, Dec 1992, p 12-19

21. B.W. Lifka, "Aluminum (and Alloys)," Chapter 46 of Section VI on Materials Testing in

Standardization, Geneva, Switzerland, 1987

23. D.O. Sprowls, "Evaluation of Stress-Corrosion Cracking" in *Stress-Corrosion Cracking: Materials Performance and Evaluation*, Russell H. Jones, Ed., ASM International, 1992, p 316-405

24. W.B. Lisagor, "Environmental Cracking—Stress Corrosion," Chapter 25 of Section IV on Testing for Corrosion Types in *Corrosion Tests and Standards: Application and Interpretation, ASTM Manual 20*, Robert Baborian, Ed., 1995, p 240-252

25. ASTM G64, Standard Classification of Resistance to Stress-Corrosion Cracking of Heat-Treatable Aluminum Alloys, *Annual Book of ASTM Standards*, Section 3, Vol 03.02

26. ASTM G47, Standard Test Method for Determining Susceptibility to Stress-Corrosion Cracking of High Strength Aluminum Alloy Products, *Annual Book of ASTM Standards*, Section 3, Vol 03.02

27. D.O. Sprowls, "High Strength Aluminum Alloys with Improved Resistance to Corrosion and Stress-Corrosion Cracking," *Aluminum*, Vol 54 (No. 3), 1978, p 214-217

28. E.H. Hollingsworth and H.Y. Hunsicker, "Corrosion of Aluminum and Aluminum Alloys," in *Metals Handbook, 9th ed., Vol 13, Corrosion*, ASM International, 1987, p 583-609

29. M.O. Speidel, "Stress Corrosion Cracking of Aluminum Alloys," *Met. Trans. A*, Vol 6A, 1975, p 631-651

30. B.W. Lifka, "SCC Resistant Aluminum Alloy 7075-T73 Performance in Various Environments," *Aluminum*, Vol 53 (No. 12), 1977, p 750-752

31. M.V. Hyatt, "Use of Precracked Specimen in Stress Corrosion Testing of High Strength Aluminum Alloys," *Corrosion*, Vol 26 (No. 11), 1970, p 487-503

32. R.E. Davies, G.E. Nordmark, and J.D. Walsh, "Design Mechanical Properties, Fracture Toughness, Fatigue Properties, Exfoliation and Stress Corrosion Resistance of 7050 Sheet, Plate, Extrusions, Hand Forgings and Die Forgings," Final Report Naval Air Systems, Contract N00019-72-C-0512, July 1975

33. B. Sarker, M. Marek, and E.A. Stacke, Jr., *Met. Trans. A*, Vol 12A, 1981, p 1939

34. R.L. Templin, "Fatigue of Aluminum," H.W. Gillette Memorial Lecture, presented at the 57th Annual Meeting of ASTM, 1954

35. G.A. Butz and G.E. Nordmark, "Fatigue Resistance of Aluminum and Its Products," paper presented at the National Farm, Construction, and Industrial Machinery Meeting (Milwaukee, WI), Society of Automotive Engineers, 1964

36. T.W. Crooker, "Crack Propagation in Aluminum Alloys under High Amplitude Cyclic Load," Report 7286, Naval Research Laboratory, 1971

37. W.G. Clark, Jr., How Fatigue Crack Initiation and Growth Properties Affect Material Selection and Design Criteria, *Metals Eng. Quart.*, 1974, p 16-22

38. "Standard Recommended Practice for Constant Amplitude Axial Fatigue Tests of Metallic Materials," Designation E466-76, *Annual Book of ASTM Standards*, Part 10, 1976, p 502-506

39. C.M. Hudson and S.K. Seward, A Literature Review and Inventory of the Effects of Environment on the Fatigue Behavior of Metals, *Eng. Fracture Mech.*, Vol 8 (No. 2), 1976, p 315-329

40. "Corrosion Fatigue of Aircraft Materials," AGARD Report 659, North Atlantic Treaty Organization, 1977

41. C.Q. Bowles, "The Role of Environment, Frequency, and Wave Shape during Fatigue Crack Growth in Aluminum Alloys," Report LR-270, Delft University of Technology, The Netherlands, 1978

42. G.E. Nordmark, B.W. Lifka, M.S. Hunter, and J.G. Kaufman, "Stress Corrosion and Corrosion Fatigue Susceptibility of High Strength Alloys," Technical Report AFML-TR-70-259, Wright-Patterson Air Force Base, 1970

43. T.H. Sanders, R.R. Sawtell, J.T. Staley, R.J. Bucci, and A.B. Thakker, "Effect of Microstructure on Fatigue Crack Growth of 7XXX Aluminum Alloys under Constant Amplitude and Spectrum Loading," Final Report, Contract N00019-76-C-0482, Naval Air Systems Command, 1978

44. J.T. Staley, "How Microstructure Affects Fatigue and Fracture of Aluminum Alloys," paper presented at the International Symposium on Fracture Mechanics (Washington, DC), 1978

45. W.G. Truckner, J.T. Staley, R.J. Bucci, and A.B. Thakker, "Effects of Microstructure on Fatigue Crack Growth of High Strength Aluminum Alloys," Report AFML-TR-76-169, U.S. Air Force Materials Laboratory, 1976

46. J.T. Staley, W.G. Truckner, R.J. Bucci, and A.B. Thakker, Improving Fatigue Resistance of Aluminum Aircraft Alloys, *Aluminum*, Vol 53, 1977, p 667-669

47. M.V. Hyatt, "Program to Improve the Fracture Toughness and Fatigue Resistance of Aluminum Sheet and Plate for Airframe Applications," Technical Report AFML-TR-73-224, Wright-Patterson Air Force Base, 1973

48. C.R. Owen, R.J. Bucci, and R.J. Kegarise, Aluminum Quality Breakthrough for Aircraft Structural Reliability, *Journal of Aircraft*, Vol 26 (No. 2), Feb 1989, p 178-184

49. P.E. Magnusen, A.J. Hinkle, W.T. Kaiser, R.J. Bucci, and R.L. Rolf, Durability Assessment Based on Initial Material Quality, *Journal of Testing and Evaluation*, Vol 18 (No. 6), Nov 1990, p 439-445

50. A.J. Hinkle and M.R. Emptage, Analysis of Fatigue Life Data Using the Box-Cox Transformation, *Fatigue and Fracture of Engineering Materials and Structures*, Vol 14 (No. 5), 1991, p 591-600

51. N.E. Dowling, Fatigue Failure Predictions for Complicated Stress-Strain Histories, *J. Materials*, Vol 7, 1971, p 71

52. R.W. Landgraf and R.M. Wetzel, Cyclic Deformation and Fatigue Damage, *Proc. Int. Conf. Mechanical Behavior of Materials*, Vol 2, 1972

53. R.W. Landgraf, F.D. Richards, and N.R. LaPointe, "Fatigue Life Predictions for a Notched Member under Complex Load Histories," paper presented at the Automotive Engineering Congress (Detroit, MI), Society of Automotive Engineers, 1975

54. T.H. Sanders, J.T. Staley, and D.A. Mauney, "Strain Control Fatigue as a Tool to Interpret Fatigue Initiation of Aluminum Alloys," paper presented at the Tenth Annual International Symposium on Materials Science (Seattle, WA), 1975

55. R.W. Landgraf, The Resistance of Metals to Cyclic Deformation, *Achievement of High Fatigue Resistance in Metals and Alloys*, STP 467, ASTM, 1970, p 3-36

56. S.S. Manson and M. Hirschberg, *Fatigue—An Interdisciplinary Approach*, Syracuse University Press, 1964, p 133

57. R. Smith, M. Hirschberg, and S.S. Manson, "Fatigue Behavior of Materials under Strain Cycling in Low and Intermediate Life Range," Report NASA-TN-D-1574, National Aeronautics and Space Administration, April 1963

58. S.S. Manson, "Fatigue: A Complex Subject—Some Simple Approximations," The William M. Murray Lecture, presented at the Annual Meeting of the Society of Experimental Stress Analysis (Cleveland, OH), 1964

59. H.O. Fuchs and R.I. Stephens, *Metal Fatigue in Engineering*, John Wiley & Sons, 1980

60. Special Publication P-109, in *Proceedings of the SAE Fatigue Conference*, Society of Automotive Engineers, 1982

61. R.C. Rice, Ed., *Fatigue Design Handbook*, 2nd ed., Society of Automotive Engineers, 1988

62. J.B. Conway and L.H. Sjodahl, *Analysis and Representation of Fatigue Data*, ASM International, 1991

63. T.H. Sanders and E.A. Starke, *Metall. Trans. A*, Vol 7A, 1976, p 1407

64. F.S. Lin, Ph.D. thesis, Georgia Institute of Technology, 1978

65. Pelloux and R.E. Stoltz, *Proc. 4th International Conf. on Strength of Metals and Alloys*, 1976, p 1023

66. R. von Mises, *Z. Angew. Math. Mech.*, Vol 8, 1928, p 161

67. E.A. Calnan and C.J.B. Clews, *Philos. Mag.*, Vol 42, 1951, p 616

68. R.E. Sanders and E.A. Starke, *Mater. Sci Eng.*, Vol 28, 1977, p 53

69. G. Lutjering, T. Hamajima, and A. Gysler, *Proc. 4th Int. Conf. Fracture* (Waterloo, Canada), Vol 12, 1977, p 7

70. A.J. McEvily, J.B. Clark, and A.P. Bond, *Trans. ASM*, Vol 60, 1967, p 661

71. A.J. McEvily, R.L. Snyder, and J.B. Clark, *Trans. TMS-AIME*, Vol 227, 1963, p 452

72. F. Ostermann, *Metall. Trans.*, Vol 2, 1971, p 2897

73. W.H. Reimann and A.W. Brisbane, *Eng. Fract. Mech.*, Vol 5, 1973, p 67

74. D. Broek and C.Q. Bowles, *J. Inst. Met.*, Vol 99, 1971, p 255

75. E. DiRusso, M. Conserva, F. Gatto, and H. Markus, *Metall. Trans.*, Vol 4, 1973, p 1133

76. F. Mehrpay et al., *Metall. Trans.*, Vol 7A, 1976, p 761

77. W.G. Truckner, A.B. Thakker, and R.J. Bucci, "Research on the Investigation of Metallurgical Factors on the Crack Growth Rate of High Strength Aluminum Alloys," report to AFML, May 1975

78. R.A. Smith, *Fatigue Crack Growth—30 Years of Progress*, Pergamon Press, 1984, p 35

79. W.G. Truckner, J.T. Staley, R.J. Bucci, and A.B. Thakker, "Effects of microstructure on fatigue crack growth of high strength aluminum alloys," *U.S. Air Force Materials Laboratory Rep.*, AFML-TR-76-169, 1976

80. J.T. Staley, W.G. Truckner, R.J. Bucci, and A.B. Thakker, Improving fatigue resistance of aluminum aircraft alloys, *Aluminum*, Vol 54, 1977, p 667-669

81. M.V. Hyatt, "Program to improve the fracture toughness and fatigue resistance of aluminum sheet and plate for airframe applications," Wright-Patterson AFB, *Tech Rep.*, AFML-TR-73-224, 1973

82. R.J.H. Wanhill and G.F.J.A. Van Gestel, Fatigue fracture of aluminum alloy sheet materials at high growth rates, *Aluminum*, Vol 52, 1976, p 436-443

83. J.T. Staley, How microstructure affects fatigue and fracture of aluminum alloys, *Presented at Int. Symp. Fracture Mech.*, Washington, DC, 1978

84. D.L. Davidson and J. Lankford, Fatigue Crack Growth Mechanisms in Metals and Alloys: Mechanisms and Micromechanics, *International Materials Reviews*, Vol 37 (No. 2), 1992, p 45-76

85. C.M. Hudson and S. Seward, A Compendium of Sources of Fracture Toughness and Fatigue-Crack Growth for Metallic Alloys, Part I, *International Journal of Fracture*, Vol 14, 1978

86. C.M. Hudson and S. Seward, A Compendium of Sources of Fracture Toughness and Fatigue-Crack Growth for Metallic Alloys, Part II, *International Journal of Fracture*, Vol 20, 1982

87. C.M. Hudson and S. Seward, A Compendium of Sources of Fracture Toughness and Fatigue-Crack Growth for Metallic Alloys, Part III, *International Journal of Fracture*, Vol 39, 1989

88. C.M. Hudson and J. Ferrainolo, A Compendium of Sources of Fracture Toughness and Fatigue-Crack Growth for Metallic Alloys, Part IV, *International Journal of Fracture*, Vol 48, 1991

89. Damage Tolerant Design Handbook: A Compilation of Fracture and Crack-Growth Data for High-Strength Alloys, HB-01R, Volumes 1 through 4, MCIC, Dec 1983; with update and revision underway as of this printing

90. H. Boyer, *Atlas of Fatigue Curves*, ASM, 1986, p 322

91. C.T. Hahn and R. Simon, A Review of Fatigue Crack Growth in High Strength Aluminum Alloys and the Relevant Metallurgical Factors, *Engineering Fracture Mechanics*, Vol 5 (No. 3), Sept 1973, p 523-540

92. M.V. Hyatt, "Program to Improve the Fracture Toughness and Fatigue Resistance of Aluminum Sheet and Plate for Airframe Applications," Report AFML-TR-73-224, Wright-Patterson Air Force Base, 1973

93. L.H. Glassman and A.J. McEvily, Jr., "Effects of Constituent Particles on the Notch-Sensitivity and Fatigue Crack Propagation Characteristics of Aluminum-Zinc-Magnesium Alloys," NASA Technical Note, April 1962

94. D. Broek, The Effect of Intermetallic Particles on Fatigue Crack Propagation in Aluminum Alloys, *Fracture 1969*, Proceedings of the 2nd International Conference on Fracture, Chapman and Hall Ltd, 1969, p 754

95. E. DiRusso, M. Conserva, M. Buratti, and F. Gatto, A New Thermomechanical Procedure for Improving the Ductility and Toughness of Al-Zn-Mg-Cu Alloys in the Transverse Directions, *Materials Science and Engineering*, Vol 14 (No. 1), April 1974, p 23-26

96. J. Waldeman, H. Sulinski, and H. Markus, The Effect of Ingot Processing Treatments on the Grain Size and Properties of Aluminum Alloy 7075, *Metallurgical Transactions*, Vol 5, 1974, p 573-584

97. A.R. Rosenfeld and A.J. McEvily, "Some Recent Developments in Fatigue and Fracture," AGARD Report 610, Metallurgical Aspects of Fatigue and Fracture Toughness, NATO Advisory Group for Aerospace Research and Development, Dec 1973, p 23-55

98. C.E. Feltner and P. Beardmore, *Strengthening Mechanisms in Fatigue*, STP 476, American Society for Testing and Materials, 1969, p 77-112

99. J.C. Grosskreutz, Strengthening and Fracture in Fatigue (Approaches for Achieving High Fatigue Strength), *Metallurgical Transactions*, Vol 3 (No. 5), May 1972, p 1255-1262

100. B.K. Park, V. Greenhut, G. Lutjering, and S. Weissman, "Dependence of Fatigue Life and Flow Stress on the Microstructure of Precipitation-Hardened Al-Cu Alloys," Report AFML-TR-70-195, Wright-Patterson Air Force Base, Aug 1970

101. "Mechanism of Fatigue Enhancement in Selected High Strength Aluminum Alloys," Progress Report NADC-MA-7171, Naval Air Development Center, 10 Dec 1971

102. S.M. El-Sondoni and R.M. Peiloux, Influence of Inclusion Content on Fatigue Crack Propagation in Aluminum Alloys, *Metallurgical Transactions*, Vol 14 (No. 2), Feb 1973, p 519-531

103. G. Lutjering, H. Doker, and D. Munz, Microstructure and Fatigue Behavior of Al-Alloys, *The Microstructure and Design of Alloys*, Proceedings of the 3rd International Conference on Strength of Metals and Alloys (Cambridge, England), Vol 1, Aug 1973, p 427-431

104. L.P. Karjalainen, The Effect of Grain Size on the Fatigue of an Al-Mg Alloy, *Scripta Metallurgica*, Vol 7 (No. 1), Jan 1973, p 43-48

105. D. Broek and C.Q. Bowles, The Effect of Precipitate Size on Crack Propagation and Fracture of an Al-Cu-Mg Alloy, *Journal of the Institute for Metals*, Vol 99, Aug 1971, p 255-257

106. J.T. Staley, How Microstructure Affects Fatigue and Fracture of Aluminum Alloys, *Fracture Mechanics*, N. Perrone et al., Ed., University Press of Virginia, 1978, p 671

107. M.V. Hyatt and W.E. Quist, AFML Tech. Rept. TR-67-329, 1967

108. J.T. Staley, Tech. Rept. Naval Air Systems Command Contract M00019-71-C-0131, May 1972

109. J.G. Kaufman, *Design of Aluminum Alloys for High Toughness and High Fatigue Strength*, Presented at 40th Meeting of the Structures and Materials Panel, NATO, Brussells, Belgium, April 1975

110. J.T. Staley, *Microstructure and Toughness of High-Strength Aluminum Alloys*, presented at ASTM Symposium on Properties Related to Toughness, Montreal, Canada, June 1975

111. J.S. Santner, AFML Tech. Rept. TR-76-200, March 1977

112. R.E. Sanders, Jr. and E.A. Starke, Jr., *Met. Trans.*, Vol 9A, 1978, p 1087

113. J.C. Williams and E.A. Starke, Jr., The Role of Thermomechanical Processing in Tailoring the Properties of Aluminum and Titanium Alloys, in *Deformation, Processing, and Structure*, G. Krauss, Ed., ASM, 1984

114. Fu-Shiong Lin and E.A. Starke, Jr., *Mater. Sci. Eng.*, Vol 43, 1980, p 65

115. R.D. Carter, E.W. Lee, E.A. Starke, Jr., and C.J. Beevers, "An Experimental Investigation of the Effects of Microstructure and Environment on Fatigue Crack Closure of 7575," *Met. Trans.*, Vol 15A, 1984, p 558

116. J.G. Kaufman and R.A. Kelsey, Fracture Toughness and Fatigue Properties of 5083-O Plate and 5183 Welds for Liquefied Natural Gas Application, STP 579, American Society for Testing and Materials, 1975, p 138-158

117. R.P. Wei, Some Aspects of Environmentally Enhanced Fatigue Crack Growth, *Engineering Fracture Mechanics*, Vol 1 (No. 4), April 1970, p 633-651

118. R.J. Bucci, A.B. Thakker, T.H. Sanders, R.R. Sawtell, and J.T. Staley, Ranking 7XXX Aluminum Alloy Fatigue Crack Growth Resistance Under Constant Amplitude and Spectrum Loading, STP 714, American Society for Testing and Materials, Oct 1980, p 41-78

119. T.H. Sanders, R.R. Sawtell, J.T. Staley, R.J. Bucci, and A.B. Thakker, "Effect of Microstructure on Fatigue Crack Growth of 7XXX Aluminum Alloys Under Constant Strain and Spectrum Loading," Final Report, Contract N00019-76-C-0482, Naval Air Systems Command, 1978

120. R.G. Forman, R.E. Kearney, and R.M. Engle, Numerical Analysis of Crack Propagation in Cyclic Loaded Structures, *Transactions of ASME, Journal of Basic Engineering*, Vol 89 (No. 3), Sept 1967, p 459-464

121. C.T. Hahn and R. Simon, A Review of Crack Growth in High Strength Aluminum Alloys, *Eng. Fracture Mech.*, Vol 5 (No. 3), 1973, p 523-540

122. J. Schijve, The Accumulation of Fatigue Damage in Aircraft Materials and Structures, *AGARDograph*, No. 157, 1972

123. R.J. Bucci, "Spectrum Loading—A Useful Tool to Screen the Role of Microstructure on Fatigue Crack Growth Resistance of Metal Alloys," STP 631, ASTM, 1977, p 388-401

124. H.A. Wood, The Use of Fracture Mechanics Principles in the Design and Analysis of Damage Tolerant Aircraft Structures, *AGARDograph*, No. 176, 1973

125. Fatigue Crack Growth under Spectrum Loads: Proceedings of Symposium presented at the 78th Annual Meeting of the American Society for Testing and Materials, STP 595, ASTM, 1976

126. K.O. Sippel and D. Weisgerber, "Crack Propagation in Flight by Flight Tests on Different Materials," paper presented at the ICAF Colloquium, 1975

127. W.S. Johnson and J.W. Hagemeyer, Yield Strength Considerations for Selecting Material Subjected to Spectrum Loads, *Int. J. Fracture Mech.*, 1977

128. L. Schra and H.P. Van Leeuwen, "Heat Treatment Studies of Aluminum Alloy Type 7075 Forgings," Interim Report 2, NLR-TR-76008C, National Aerospace Laboratory, NLR, The Netherlands, 1976

Fatigue and Fracture Properties of Aluminum Alloy Castings

Theodore L. Reinhart, Boeing Commercial Airplane Group

CASTING QUALITY has steadily improved over the years such that castings have become a viable alternative in structural applications for military and commercial aircraft. The steady improvement in casting quality is driven by several factors. In some cases, casting is the only viable alternative manufacturing method. In addition, the flexibility of the casting can be used to great advantage in reducing overall manufacturing costs, particularly for complex, three-dimensional structures. Competition and cost pressure also have intensified efforts by aircraft manufacturers to reduce costs through the use of near-net shape manufacturing methods such as casting. However, even with an economic advantage, a casting must meet the same performance criteria as competing design using wrought members and various joining technologies. Such cases are an area of great interest for aircraft structural designers.

With improvements in casting quality and the potential economic advantage of casting technology, cast aluminum alloys are being considered for weight and cost reductions of primary aircraft structures. Despite their acknowledged importance, aluminum castings have until recently been relegated to secondary applications in commercial aircraft structure. Castings have been used extensively in engines and systems, for example, in flight controls, wheels, and so forth, but not in many "primary" structural (carrying significant flight loads) applications. More recently, however, the use of castings in nontraditional applications has been facilitated by improvements in both casting alloys and processes. Casting producers have invested considerable money in applied research and, where possible, borrowed from the lessons learned by the aluminum producers. This has translated into cleaner castings with reduced levels of internal discontinuities, intermetallics and oxides, and better control of microstructures for improved strength, and fatigue and fracture resistance.

In this context, this article reviews the fatigue and fracture properties of aluminum alloy castings from the perspective of both design and manufacturing considerations. Fatigue and fracture resistance is often a limiting factor in the design of primary structures, and thus this article provides an overview of the roles played by microstructure, manufacturing processes, and test conditions in determining the fatigue and fracture properties of aluminum casting alloys. Because it is not possible to cover all the work done on all aluminum casting alloys, the following two alloy systems commonly used in aviation and aerospace are emphasized:

- Alloys A356 and A357/D357 (all-T6), members of the dual-phase Al-Si-Mg system, the most commonly specified
- Alloy A201-T7 from the single-phase Al-Cu system, with higher mechanical properties but more limited producibility

The role of casting design is also briefly discussed with respect to fatigue and fracture performance.

Casting Design and Manufacture

The ability of an engineered component to perform a particular function is governed as much by design as by material properties. Trade studies involving engineered materials (in this case aluminum alloys) sometimes go no further than the properties tables of a reference book. However, attempts to compare diverse design concepts can be misleading. For most design properties, wrought alloys can be found that show advantage over the cast alloy. However, a thorough trade study must take into account the capabilities of the manufacturing processes. For example, the casting process affords the designer a number of unique advantages (Ref 1). In terms of structural shapes with a high degree of complexity, casting allows greater flexibility in design. Improved fatigue design is also possible by virtue of:

- Continuous load paths (intersecting members are continuous; in a built-up design, the engineer is often faced with breaking one member or another into two and then adding splice details to reconnect them during assembly)
- "Soft" section transitions

- Fillet and corner radii can be tailored to reduce stress concentrations
- Fewer rivet/fastener holes (potential crack starters)
- Fewer lap joints (potential corrosion sites)

The engineer is often able to produce a cast design that weighs no more than the traditional design (e.g., Ref 2). In fact, a modest weight reduction is sometimes achieved (Ref 1, 3, 4). Cost reduction is achieved by virtue of reduced part count, assembly labor, and related costs. Typical cost reductions are 35% or more when replacing a complex, three-dimensional built-up assembly.

With respect to casting processes, the great majority of development and production programs have used the so-called "sand-chill composite" mold process (hereafter referred to simply as "sand" process for brevity). Recent work by investment casting suppliers has, however, enabled a few companies to produce aluminum alloy castings (for example, D357) with tensile properties on a par with sand castings. Some European aircraft manufacturers apparently prefer castings made by the investment process. They are, in fact, somewhat ahead of their American counterparts in terms of applications. For example, the first structural casting went into service on an Airbus aircraft circa 1985 (Ref 1).

Weld Rework on Castings and its Effect on Fatigue and Fracture. Foundries often perform modest amounts of welding on castings (always prior to heat treatment) to correct surface defects. This action returns the casting to the "drawing specified" condition and is thus called "rework" (as opposed to "repair," which is discussed in the next section). The static properties of weld-reworked castings have been reasonably well studied, but few references on fatigue and fracture are available. With respect to alloys, only the Al-Si-Mg system has been studied in depth. The Al-Cu system, being inherently hot short, is welded less frequently and is not as well characterized.

The majority of work seems to show that fatigue life is not degraded when welding is properly performed. Apodaca et al. (Ref 5) studied the properties of weld-reworked F-5 pylon castings. In these smooth-bar, axial tests no significant differences were noted between results for welded and unwelded (parent) material. Ozelton et al. (Ref 6) found similar results. Mitchell et al. (Ref 2) reported slightly longer life for weld-reworked A357-T6 specimens tested with $R = +0.02$ and 260 MPa (38 ksi) maximum stress: 48,388 cycles average for welded and 41,880 for parent metal. However, Oswalt et al. (Ref 7) found different results: slightly longer life for A357-T6 parent metal for stresses below about 100 MPa (15 ksi). However, comparisons with other works are difficult at best, due to differences in casting properties, quality, welding procedures, and fatigue testing procedures.

For fracture properties, the only readily available work (Ozelton et al., Ref 6) found that weld rework had no apparent effect on crack growth rates in D357-T6. Fracture toughness, on the other hand, was actually somewhat higher for the welded material.

In general, welding produces a very fine microstructure in the fusion zone of Al-Si-Mg castings. It also tends to produce a distribution of very fine gas pores, sometimes in excess of that found in the adjacent parent metal. The interactions of these features in determining fatigue life has not been studied in depth.

Weld repair differs from rework in that postweld heat treatment is not done; it is sometimes employed by aircraft maintenance crews to render damaged castings usable. Of course, weldment tensile properties are greatly reduced as a result. For example, in alloy A357-T6, yield strength may be reduced by 50% or more. Consequently, weld repairs are performed only in cases where service stresses are low.

A review of the literature turned up two cases in which weld repair was performed on fatigue test articles:

- Hanson et al. (Ref 8) performed fatigue tests on two alloy A201-T7 spoilers with weld repairs. They concluded that weld repair procedures were adequate for field repair functions.
- Ozelton et al. (Ref 6) performed weld repair on a precracked open-hole fatigue specimen to observe crack growth behavior in the weld. Prior to welding, the specimen was subjected to two lifetimes of a modified F-18 lower wing root spectrum. The resulting 0.09 in. crack was welded, the hole was reamed, and the test was continued to failure.

Effect of Hot Isostatic Pressing (HIP). It is well known that the elimination of internal voids by HIPing has a generally favorable effect on the smooth-bar fatigue life of castings. Both Al-Si-Mg and Al-Cu cast alloys show similar results. However, there is some discord in the literature. For example, Kennerknecht et al. (Ref 9) found that HIPing did not have a significant effect on fatigue life in notched specimens ($K_t = 1.5$) and open-hole specimens ($K_t = 3.0$) in alloy D357-T6. They did, however, report slightly lower crack growth rates in HIPed material.

In addition, Rading et al. (Ref 10) compared the fatigue crack growth rates of alloy A206-T7 (Al-4.75% Cu) with varying degrees of porosity. They concluded that HIPing did little to improve da/dn, even though it reduced porosity from 4 to 0.4%. An explanation was offered to the effect that HIPing closes pores but cannot "heal" them due to the presence of oxides on the pore surfaces. However, no metallographic evidence was offered in support of this explanation. If this did in fact occur, then one would expect a reduction in fatigue life due to early crack initiation at the hypothetical collapsed, unhealed pores; however, the available fatigue data indicate the opposite (e.g., Ref 9).

Effect of Casting Discontinuities on Fatigue and Fracture. Casting properties depend greatly on the effect of various discontinuities, but it can be difficult to quantify the effect of a single type of discontinuity. This is so because the processing steps taken to produce castings with fine microstructure also tend to minimize the presence of unwanted actors—dross (oxide inclusions), gas porosity, microshrinkage, and the like. This being the case, some investigators have resorted to using nonstandard practices in an attempt to introduce various quantities of a certain discontinuity type into the test castings. Data produced from such programs are, unfortunately, of questionable relevance from an engineering perspective. However, few other data on this subject are to be found and so are presented here only for the purposes of illustration. The effects of two often-discussed discontinuities, gas porosity and dross, are briefly reviewed.

Gas Porosity. A number of investigators have studied the effects of gas porosity, possibly because it is relatively easy to quantify by metallographic or radiographic inspection (although standards for the latter method are semiquantitative at best). Briefly, the effects of gas porosity on the properties of Al-Si-Mg alloys are as follows:

- Fatigue life is reduced as porosity level is increased, based on a number of investigations (such as Ref 6) and discussed later in the section "Effects on Fatigue Crack Initiation" in this article.
- Fatigue crack growth rates are not greatly affected (Ref 6), although investigators (Ref 9) reported that alloy D357-T6, when subjected to porosity-eliminating HIPing, showed reduced rates.
- Fracture toughness is not greatly affected (Ref 6, 11).

Dross. The role of dross has been studied to a lesser extent than that of gas porosity, possibly because dross is much more difficult to detect and quantify. Also, the presence of dross and gas porosity are related. Aluminum oxide inclusions have been reported to serve as nucleation sites for gas pores (Ref 12).

Therefore, at any given level of dissolved hydrogen in a melt, castings poured with "dirty" metal will contain more gas porosity than those poured with "clean" metal. As a result, it is somewhat difficult to characterize the effects of the two independent of one another. Despite these difficulties, it is clear that dross can adversely affect alloy strength (Ref 13). Ozelton et al. (Ref 6) attempted to introduce dross ("less dense foreign material") into alloy D357-T6 test plates by pouring molten metal down the mold risers rather than the downspure. Castings with foreign material corresponding with ASTM E155 grades A/B, B, and C were more susceptible to crack initiation based on results of stress-life tests, as would be expected. By way of contrast, the investigators found that crack growth rates and fracture toughness levels were not significantly affected (Ref 6).

Al-Si-Mg Cast Alloys

Cast aluminum alloys A356 and A357/D357 (Tables 1 and 2) are part of the Al-Si-Mg alloy family and are the most commonly specified cast aluminum alloys for aircraft applications. This alloy system is a favorite of foundrymen and their customers, as the fluidity provided by silicon

Table 1 Compositions of key aerospace aluminum casting alloys

Alloys	Si	Mg	Cu	Mn	Ag	Fe	Be	Ti	Zn	Others each	Others total
A356(a)											
Min	6.5	0.25	...	...	...	...	...	...	...	...	...
Max	7.5	0.45	0.20	0.10	...	0.20	...	0.20	0.10	0.05	0.15
A357(a)											
Min	6.5	0.40	...	...	...	...	0.04	0.04	...	...	...
Max	7.5	0.70	0.20	0.10	...	0.20	0.07	0.20	0.10	0.05	0.15
D357(b)											
Min	6.5	0.55	...	...	...	...	0.04	0.10	...	...	...
Max	7.5	0.6	...	0.10	...	0.12	0.07	0.20	...	0.05	0.15
A201(a)											
Min	...	0.15	4.0	0.20	0.40	...	...	0.15	...	...	...
Max	0.05	0.35	5.0	0.40	1.0	0.10	...	0.35	...	0.03	0.15
B201(c)											
Min	...	0.20	4.5	0.20	0.40	...	...	0.15	...	...	...
Max	0.05	0.30	5.0	0.50	0.80	0.05	...	0.35	...	0.05	0.15

(a) Per MIL-A-21180. (b) Per AMS 4241. (c) Per AMS 4242

Table 2 Minimum tensile properties of selected aluminum casting alloys

Alloy	Strength class	Tensile properties				Elongation, %
		Ultimate tensile strength		Tensile yield strength		
		MPa	ksi	MPa	ksi	
A356-T6(a)	1	262	38.0	193	28.0	5
	2	275	40.0	206	30.0	3
A357-T6(a)	1	310	45.0	241	35.0	3
	2`	345	50.0	275	40.0	5
D357-T6(b)	Nondesignated	310	45	248	36	2
	Designated	345	50	275	40	3
A201-T7(a)	1	413	60.0	345	50.0	3
	2	413	60.0	345	50.0	5
B201-T7(c)	Nondesignated	385	56	330	48	2
	Designated	413	60	345	50	3

(a) Per MIL-A-21180. (b) Per AMS 4241. (c) Per AMS 4242

additions allows casting of complex, thin-walled components.

High-strength (ultimate tensile strength of 345 MPa, or 50 ksi) alloy A357-T6 castings were in production in the United States by the late 1960s. The alloy was used in the F-5 Northrop fighter wing pylon, which is still remarkable in size, complexity, uniformity of mechanical properties and production quantity (more than 15,000 in a twenty-year span) (Ref 2). The principles of controlled solidification were used to obtain a fine microstructure and a high degree of soundness in an alloy of tightly controlled composition.

The capability to produce high-strength castings with thin walls (less than 2.5 mm, or 0.1 in.) was developed over several decades with some initial work in the mid-1960s. Work by Goehler et al. (Ref 14) led to several new applications, for example, the alloy A357 Krueger flap for the Boeing model 737 aircraft. Several significant programs from the 1970s demonstrated both the technical viability and economic benefits of thin-wall castings in aircraft components:

- Vought/USAF A-7 cast aluminum spoiler (1976, Ref 8)
- Boeing/USAF YC-14 bulkhead ("CAST") program (1976-1979, Ref 3)

In both programs, full-scale cast test articles were subjected to static, fatigue, and damage tolerance tests. Both test articles met aircraft performance specifications. The CAST program in particular is viewed by some investigators (Ref 4) as the technical foundation for aircraft structural castings. The U.S. Air Force/Boeing Air-Launched Cruise Missile AGM-86B took advantage of the results of the CAST program to produce the entire missile body. The cast design enabled the Air Force to realize tremendous savings in both cost and schedule.

Investigations at Boeing (Ref 15) and Northrop (Ref 6) led to the further refinement of alloy A357 and the associated production controls. The product, registered as alloy D357 (AMS 4241), found application in the F-16 air inlet duct. A cast replacement duct design enabled the contractor to widen the throat of the duct while maintaining the existing external profile (made necessary by a switch to a more powerful engine). Destructive test data generated during production of the inlet and other castings were used to formulate statistically based design allowables for the alloy (Ref 16)—a unique feature of alloy D357. The allowables were the product of tighter restrictions on alloy composition, nondestructive inspection criteria, and production practices. Since 1986, the alloy has been specified for other critical castings in both military and civil aircraft.

Microstructure and Properties

The Al-Si-Mg alloys are two-phase systems, where magnesium provides age-hardening, and silicon is present in the microstructure of the age-hardened alloy as both interdendritic eutectic and in the Mg_2Si strengthening precipitate. In general, the microstructural features that most strongly affect mechanical properties are:

- Dendrite-arm spacing
- Size and distribution of second-phase particles and inclusions

Both the primary aluminum dendrite size and the size and morphology of the silicon eutectic depend heavily on processing parameters such as solidification rate, the presence of chemical modifiers, and solution heat treatment. The size and distribution of Mg_2Si precipitate particles depend on quench rate, magnesium content, and aging treatment.

Ideally, these features would be the only considerations in the design of casting processes. However, because of the physical and chemical properties of molten aluminum alloys, the founders and users must also take into account such phenomena as shrinkage, gas porosity, and nonmetallic inclusions usually from dross (i.e., thin aluminum oxide films formed during melting and pouring). In addition, if undesirable trace elements (especially iron) are present, large acicular intermetallic compounds may also be present. Figures 1(a) and (b) illustrate two extremes of microstructure.

All these features influence the static, fatigue, and fracture properties of the alloys, and a tremendous amount of work by producers and users has been done to characterize and improve the product; a few examples are given in Ref 3, 6, and 15. It is beyond the scope of this paper to summarize them here. However, the key role of microstructural fineness is summarized here due to its

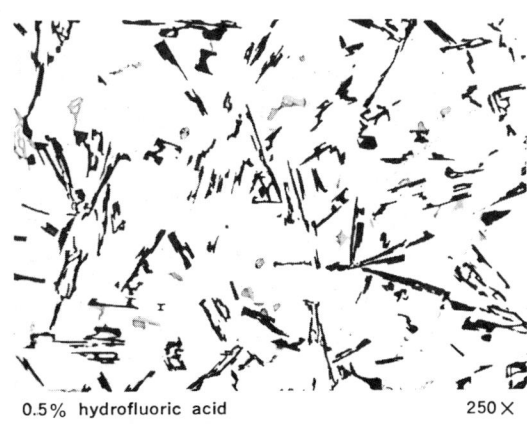

0.5% hydrofluoric acid   250 ×

(a)

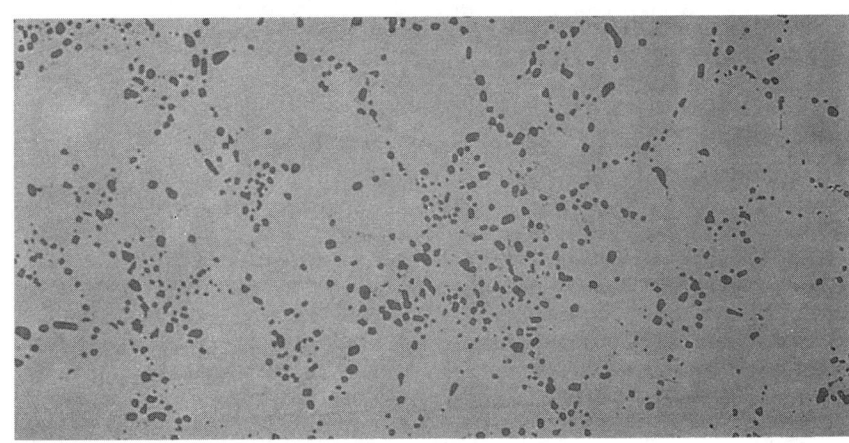

(b)

Fig. 1 Typical microstructure of Al-Si-Mg casting. (a) Photomicrograph showing coarse dendritic structure and acicular silicon eutectic (investment cast 354-F alloy at 250×). (b) Photomicrograph showing fine dendritic structure and heavily modified silicon eutectic (200×)

Table 3 Typical mechanical properties of premium-quality aluminum alloy castings and elevated-temperature aluminum casting alloys

Alloy and temper	Hardness, HB(a)	Ultimate tensile strength MPa	ksi	Tensile yield strength MPa	ksi	Elongation in 50 mm (2 in.), %	Compressive yield strength MPa	ksi	Shear strength MPa	ksi	Fatigue strength(b) MPa	ksi
Premium-quality castings(c)												
A201.0-T7	...	495	72	448	65	6	...	...	...	...	97	14
A206.0-T7	...	445	65	405	59	6	...	...	...	...	90	13
224.0-T7	...	420	61	330	48	4	...	...	...	...	86	12.5
249.0-T7	...	470	68	407	59	6	...	...	...	...	75	11
354.0-T6	...	380	55	283	41	6	...	...	...	...	135(d)	19.5(d)
C355.0-T6	...	317	46	235	34	6	...	...	...	...	97	14
A356.0-T6	...	283	41	207	30	10	...	...	...	...	90	13
A357.0-T6	...	360	52	290	42	8	...	...	...	...	90	13
Piston and elevated-temperature sand cast alloys												
222.0-T2	80	185	27	138	20	1	...	...	...	...	...	...
222.0-T6	115	283	41	275	40	<0.5	...	...	...	...	...	...
242.0-T21	70	185	27	125	18	1	...	...	145	21	55	8
242.0-T571	85	220	32	207	30	0.5	...	...	180	26	75	11
242.0-T77	75	207	30	160	23	2	165	24	165	24	72	10.5
A242.0-T75	...	215	31	...	...	...	...	...	...	...	...	...
243.0	95	207	30	160	23	2	200	29	70	10	70	10
328.0-F	...	220	32	130	19	2.5	...	...	...	...	...	...
328.0-T6	85	290	42	185	27	4.0	180	26	193	28	...	...
Piston and elevated-temperature alloys (permanent mold castings)												
222.0-T55	115	255	37	240	35	...	295	43	207	30	59	8.5
222.0-T65	...	...	...	...	...	...	...	...	...	...	...	...
242.0-T571	105	275	40	235	34	1	...	...	207	30	72	10.5
242.0-T61	110	325	47	290	42	0.5	...	...	240	35	65	9.5
332.0-T551	105	248	36	193	28	0.5	193	28	193	28	90	13
332.0-T5	105	248	36	193	28	1	200	29	193	28	90	13
336.0-T65	125	325	47	295	43	0.5	193	28	248	36	...	...
336.0-T551	105	248	36	193	28	0.5	193	28	193	28	...	...

(a) 10 mm (0.4 in.) ball with 500 kgf (1100 lbf) load. (b) Rotating beam test at 5×10^8 cycles. (c) Typical values of premium-quality castings are not minimums and are not classified by the area from which the specimen is cut. (d) Fatigue strength for 10^6 cycles

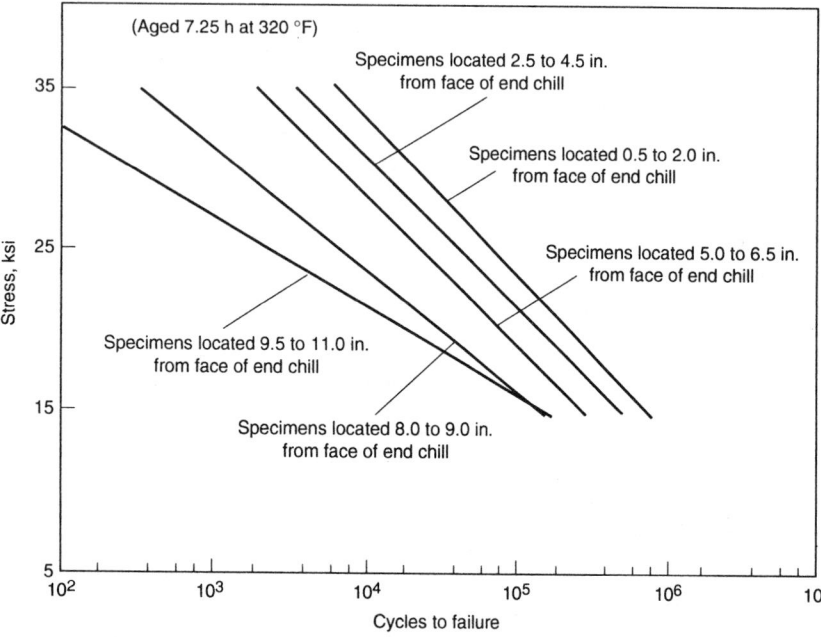

Fig. 2 Stress-life curves for specimens taken at various distances from a cast iron chill embedded in the mold. Alloy 356 with beryllium addition. From Ref 18

well-known beneficial influence on strength, toughness, and overall fatigue and fracture resistance.

Premium (high-strength, high-toughness) castings combine a number of conventional cast-

ing procedures in a selective manner to provide premium strength, soundness, dimensional tolerances, surface finish, or a combination of these characteristics. As a result, premium engineered castings typically have somewhat higher fatigue

resistance than conventional castings. Table 3 gives typical fatigue limits on various premium castings.

The production of premium-quality castings is an example of understanding and using the solidification process to good advantage. Depending on the casting process, various methods are used to engineer the rates and directions of molten metal solidification within the mold. For example, in the sand-chill composite process, metallic chills placed at selected locations within the mold provide greatly enhanced solidification rates. In the shell investment casting process, since the ceramic shell molds are routinely preheated prior to pour, metallic chills are of limited value. Alternative heat extraction methods such as the use of cryogenic or other fluids are therefore used to promote more rapid solidification.

Effects of Microstructural Fineness

Measurement of Microstructural Fineness. Microstructural fineness is related to the size of the primary aluminum dendrites and the interdendritic silicon eutectic phase. Fineness is measured by quantitative systems such as the dendrite arm spacing (DAS) (Ref 17) or the cell count method (Ref 15). Silicon particles are usually characterized by an aspect ratio (Ref 15). When considering effects on properties, it is somewhat difficult to consider the two separately because the steps taken to achieve fine dendrite size (enhanced solidification) also modify the silicon,

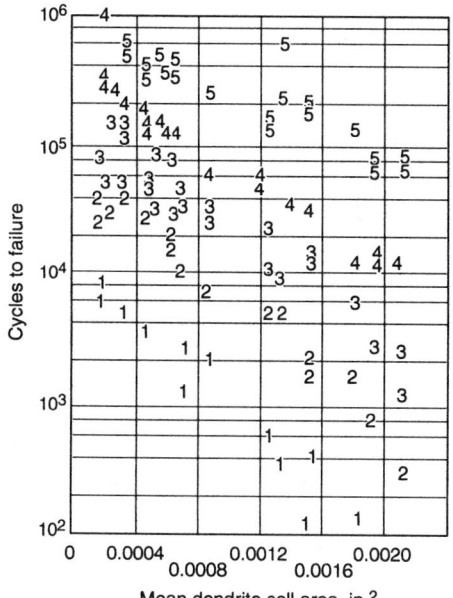

Fig. 3 Fatigue life as a function of stress level and mean dendrite cell size for alloy 356 plus beryllium. 1, 35 ± 2.5 ksi; 2, 30 ± 2.5 ksi; 3, 25 ± 2.5 ksi; 4, 20 ± 2.5 ksi; 5, 15 ± 2.5 ksi. From Ref 18

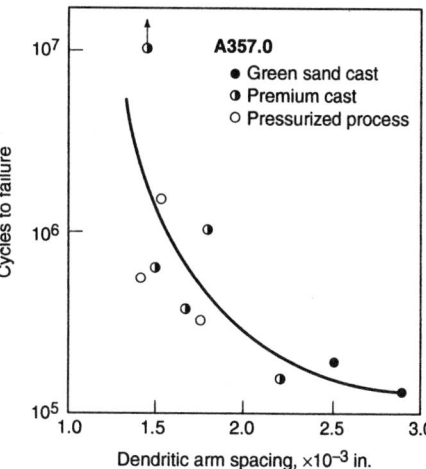

Fig. 4 Fatigue life at 165 MPa (24 ksi) maximum stress for alloy A357-T6 as a function of dendrite arm spacing. From Ref 19

that is, particle size and aspect ratio are reduced. The two will therefore be treated together.

Effect on Fatigue Life. The beneficial effect of refined microstructures for improved fatigue life has been known for decades, and one of the best demonstrations of this is an early work (Ref 18) that shows with good clarity the systematic variations of microstructure, tensile, and fatigue properties with respect to distance from a chill in alloy 356 plus beryllium test casting. The investigator showed that both fatigue strength and ultimate tensile strength decreased with increasing distance from the chill. Figure 2 shows the stress-life relationships for specimens taken at various distances from the chill; the microstructural explanation (or part of it in terms of dendrite cell size) is shown in Fig. 3—at any given maximum stress level, fatigue life is related to microstructu-

ral fineness in terms of the mean dendrite cell size.

Other works show similar trends. For example, Fig. 4 (Ref 19) shows fatigue life as a function of DAS for alloy A357-T6. Ozelton et al. (Ref 6) studied the effects of "nonoptimum" processing on the fatigue life of alloy D357-T6. Smooth specimens with coarser microstructure had shorter fatigue lives, but the effect was less pronounced for notched specimens (Fig. 5).

Effects on Fatigue Crack Initiation. Many investigators (e.g., Ref 6, 19-23) cite the role of various discontinuities—usually gas porosity, microshrinkage, or dross—in fatigue crack initiation. The effect of microstructural refinement on fatigue life may be explained in part by the fact that rapid solidification produces not only fine dendrites and highly modified silicon, but also tends to distribute any available dissolved hydrogen as numerous, minute, rounded pores. Slower solidification, on the other hand, provides time needed for the formation of larger gas pores. A similar trend applies to microshrinkage. Larger pore/cavity size is generally accepted as provid-

ing sites for early crack initiation and, hence, shorter fatigue life. One investigator (Ref 19) concluded that casting discontinuities of size similar to the silicon eutectic particles in alloy A357-T6 will not have a detrimental effect on fatigue strength.

When specimens with stress concentration factor (K_t) greater than 1.0 are considered, initiation is often found to occur at geometric features (holes/notches) as opposed to process-related discontinuities. Comparisons of smooth and open-hole specimens machined from cast alloy D357-T6 test plates ($K_t = 1.0$ and 3.0, respectively) by the author (Ref 24) showed that cracks initiated at small dross (aluminum oxide) inclusions at the polished surfaces of $K_t = 1.0$ specimens. In the latter type, initiation occurred at the hole shoulder or at "rifle" marks in the bore.

Effects on Crack Growth Rates. Results are divided, to some extent, with respect to the effect of microstructural refinement on crack growth in Al-Si-Mg alloys. At least one work (Ref 21) concluded that coarser microstructure gave favorable results; however, others such as Ozelton et al. (Ref 6) report little or no apparent effect. Figure 6 shows da/dn versus ΔK curves for alloy D357-T6 with typical microstructures shown in Fig. 7(a) and (b). Note the relatively large dendrites and the rodlike silicon in the material labeled "nonoptimum." The investigators concluded that this had no effect on da/dn; however, Fig. 6 seems to suggest higher rates for the "nonoptimum" material at higher ΔK levels (above 10 ksi$\sqrt{in.}$).

Lee et al. (Ref 25) found that the fatigue crack growth characteristics of cast alloy A356-T6 were a function of a variety of factors; prominent among them was the effect of silicon particle size and morphology (degree of modification). They found different mechanisms at work in cast alloy with unmodified versus strongly modified silicon:

- Crack growth in unmodified material occurred by cleavage/cracking of silicon particles. The tendency for particle cracking increases with particle size (aspect ratio).

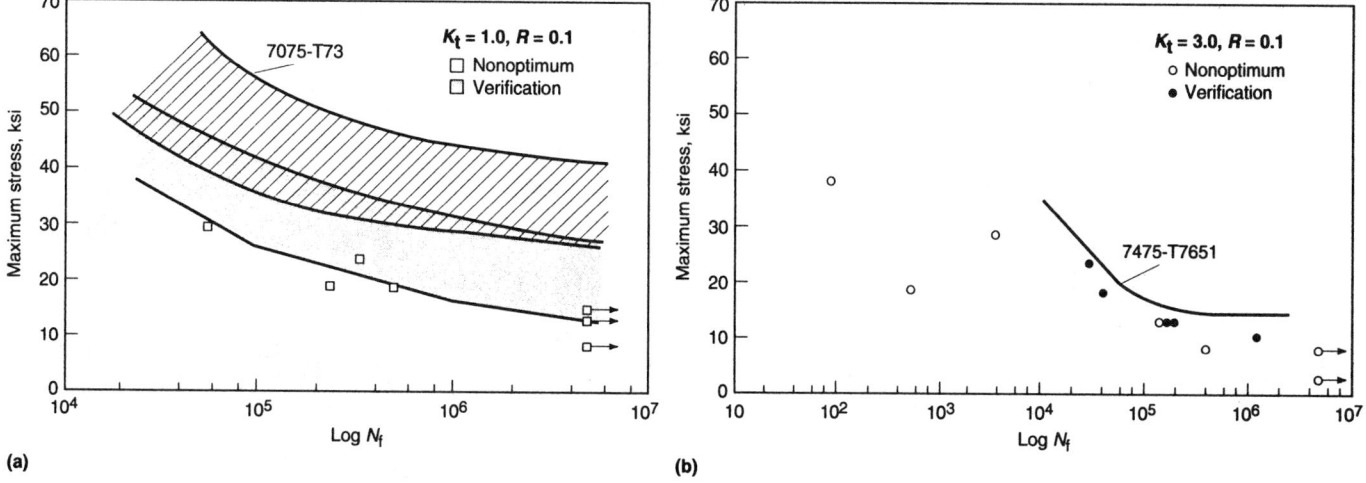

Fig. 5 Stress-life fatigue data for D357-T6 with nonoptimum (relatively coarse) microstructure. (a) Smooth ($K_t = 1.0$) specimens. (b) Notched ($K_t = 3.0$) specimens. From Ref 6

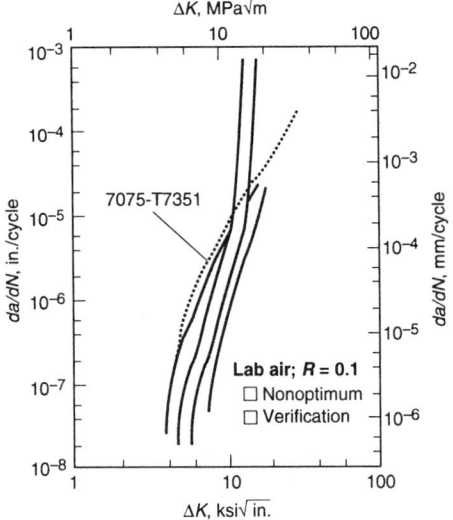

Fig. 6 Fatigue crack growth rate of alloy D357-T6 with nonoptimum (relatively coarse) microstructure. From Ref 6

- In alloy with modified silicon, decohesion of the silicon particles and the matrix occurs. It is believed that this method of creating a new interface requires more energy than does particle cracking (which may explain the slower crack growth rates in the modified material).

Effect on Fracture Toughness. Several examples illustrate the role between refined microstructure and improved fracture toughness. For example, Ozelton et al. (Ref 6) found that alloy D357-T6 with coarse microstructure had significantly lower plane strain fracture toughness (K_{Ic} in the 17 to 18 ksi$\sqrt{\text{in.}}$ range) than material with fine structure (K_Q about 24 ksi$\sqrt{\text{in.}}$). Other examples are demonstrated for dynamic fracture toughness (Ref 11) and Charpy impact energy of

alloy A356-T6 subjected to different solidification rates, chemical modification, and various solution treatment times (Ref 12). As expected, processing conditions that yielded fine dendrites and silicon particles also imparted relatively high impact energy levels.

The significance of fine microstructure is a key factor for improved fatigue life, slower crack growth, and higher fracture toughness. However, other metallurgical factors have important effects as well. Yield strength, for example, is determined primarily by magnesium content and heat treatment (Ref 3, 15, 26). Oswalt and Maloit (Ref 26) characterized the yield strength-toughness relationship for alloy D357-T6. Figure 8 shows the effects of aging at 160 °C (325 °F) up to 12 h; data from Ref 11 are shown superimposed. The investigators also found a correlation between the notch yield ratio (NYR = notch tensile strength divided by tensile yield strength) and fracture toughness; a similar relationship for alloy A357 is reported in Ref 13.

Composition Effects

Residual Iron. Several investigators have studied the effects of residual elements, especially iron, on the properties of Al-Si-Mg casting alloys. Iron forms an intermetallic beta phase with an acicular platelet morphology in many aluminum alloys (both cast and wrought). These platelets act as stress raisers in the alloys and are known to have adverse effects on mechanical properties and fabrication characteristics. And, as it turns out, both fatigue life and fracture toughness can be affected as well. Wickberg et al. (Ref 27) found that an increase in the nominal iron content of alloy A356-T6 from 0.2 to 0.4% resulted in a noticeable reduction in fatigue life—refer to Fig. 9.

Beryllium was originally included in alloy A357-T6 in the composition to mitigate the effects of iron. It does so by modifying the intermetallic morphology from that of a platelet to an indistinct, more rounded shape. Tan et al. (Ref 28) studied the effects of beryllium additions on the fracture toughness of permanent-mold cast alloy A357 with relatively high and low iron levels. They found the following:

- The fracture toughness of beryllium-free alloys was reduced slightly (K_{Ic} from about 25 to 21 MPa$\sqrt{\text{m}}$ by an increase from 0.01 to 0.15% Fe.
- Beryllium is effective in mitigating the effects of iron in alloys with 0.15% Fe.
- Beryllium has no apparent beneficial effect on low-iron (0.01%) alloy.

Al-Cu Cast Alloys

The Al-Cu (and, in some cases, Ag) alloy system offers higher strength than the Al-Si-Mg alloys—at the expense of producibility. Because of the tendency of the Al-Cu system toward hot shortness, only relatively simple casting designs can usually be produced. Nonetheless, alloys such as A201 are used in various applications such as jet engine front frames. Alloy composition and mechanical property minimum values are given in Tables 1 and 2, respectively.

Microstructure and Mechanical Properties. Because Al-Cu alloys lack a pronounced second phase, the effects of microstructure on tensile properties are more subtle. Ozelton et al. (Ref 6) characterized the effects of grain size on the properties of HIPed alloy A201-T7. HIPing reduces internal porosity and microshrinkage to very low levels; when this was done, it was found that grain size, and hence solidification rate, had little

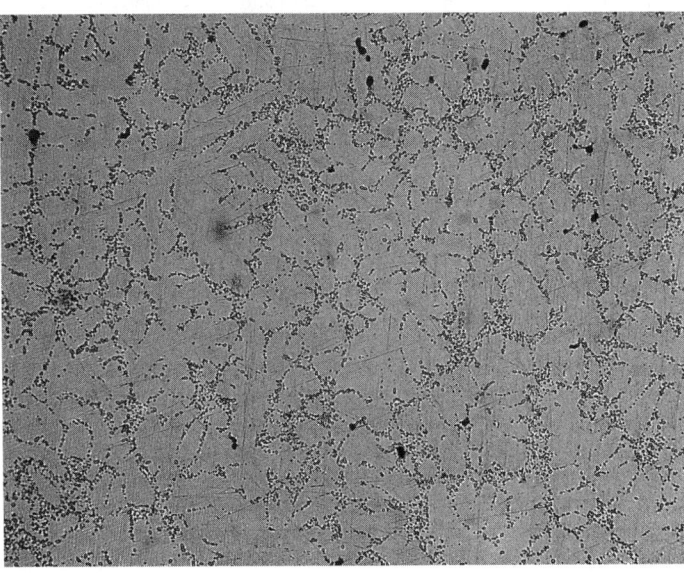

(a)

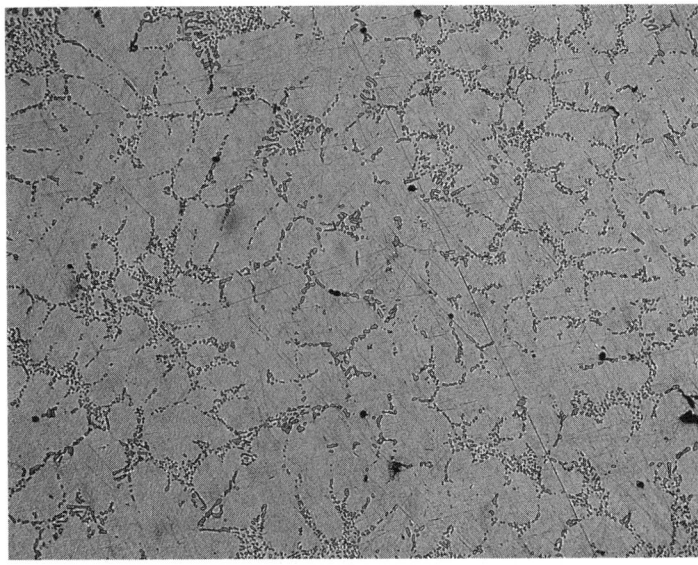

(b)

Fig. 7 Typical microstructures for alloy D357-T6. (a) Relatively fine. As-polished. 50×. (b) Relatively coarse (nonoptimum). 50×. From Ref 6

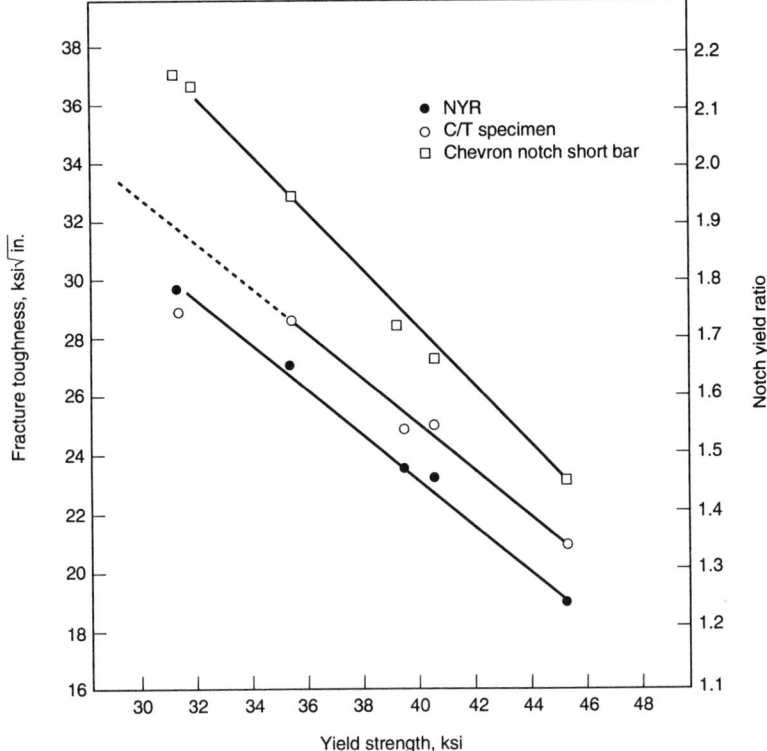

Fig. 8 Yield strength-toughness relationship for alloy D357-T6 aged at 160 °C (325 °F) (0 to 12 h). NYR, notch tensile strength divided by tensile yield strength. From Ref 26

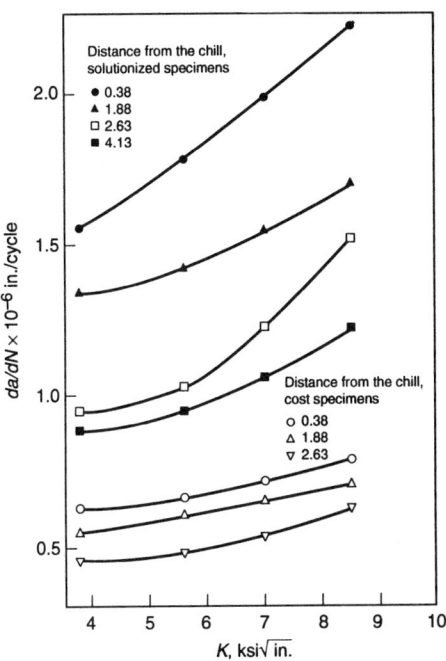

Fig. 11 Crack growth rate curves for alloy KO-1 (now A201) as a function of distance from mold chill. From Ref 29

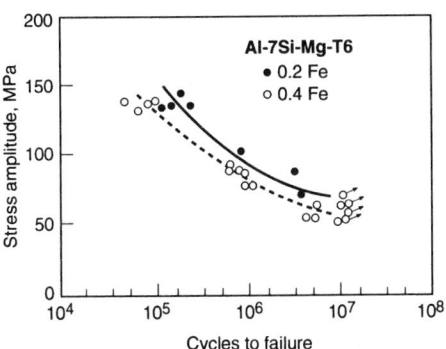

Fig. 9 Influence of alloy A356-T6 iron content on fatigue life. From Ref 27

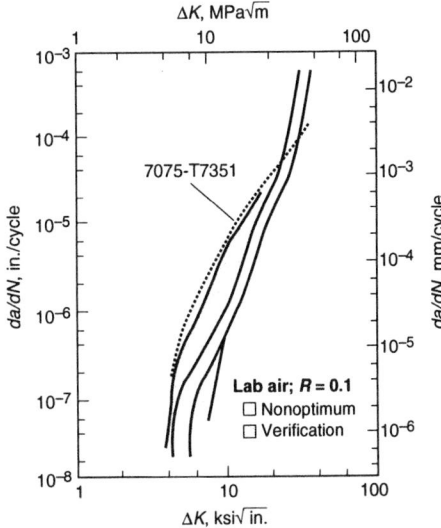

Fig. 10 Crack growth rate curves for coarse- and fine-grained HIPed alloy B201-T7. From Ref 6

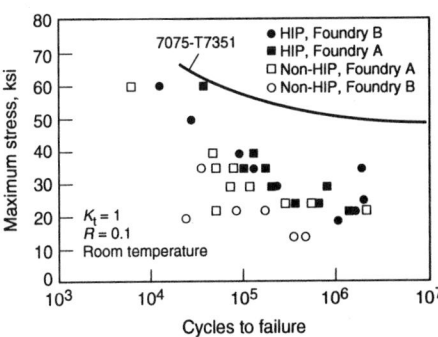

Fig. 12 Stress-life fatigue data for alloy A201-T7 castings with and without HIPing. From Ref 30

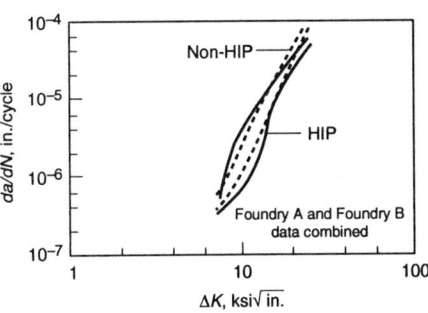

Fig. 13 Fatigue crack growth rates for alloy A201-T7 castings with and without HIPing. From Ref 30

effect on tensile properties. On the other hand, in the absence of HIPing, solidification rate determines the concentration of internal voids in alloy A201 (Ref 29). In this case, tensile elongation is higher under conditions of rapid solidification.

Fatigue life is similarly affected by solidification rates and its consequences on microstructural fineness and the presence of voids. In general, fatigue life drops with distance from the chill, as noted for all conditions of alloy KO-1 (later designated A201) in Ref 29. The decrease is attributed to the increase in microporosity with distance from the chill. Ozelton et al. (Ref 6) found that grain size had no apparent effect on the fatigue life of HIPed alloy B201-T7 (B201 is a modified version of alloy A201).

Likewise, Ozelton et al. (Ref 6) found that grain size had no apparent effect on the crack growth rate (Fig. 10) or fracture toughness in HIPed alloy B201. Chien et al. (Ref 29), on the other hand, found that crack growth rates in alloy KO-1 were slower for coarse-grained alloy (see Fig. 11). Despite the slower *da/dN*, fatigue life in coarse-grained KO-1 was relatively short (Ref 29), indicating that initiation occurred earlier (probably the result of higher microshrinkage

content). Thus, it appears that grain size has only an indirect effect on crack growth in single-phase alloys such as A201.

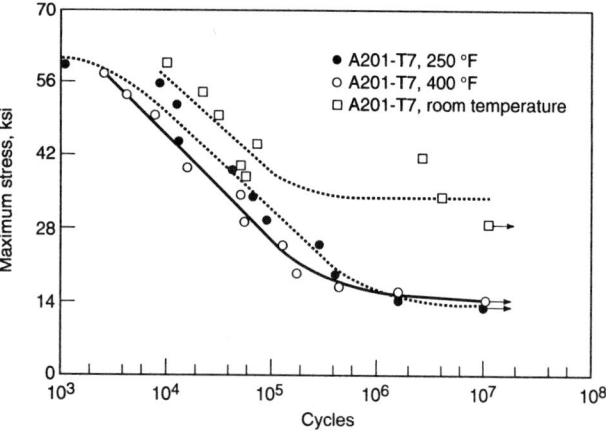

Fig. 14 Stress-life fatigue data at three test temperatures for alloy A201-T7. From Ref 32

Fig. 15 Stress-life fatigue data at three test temperatures for alloy A357-T6. From Ref 32

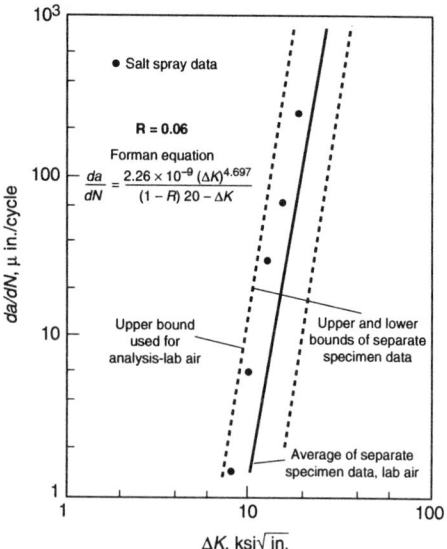

Fig. 16 Fatigue crack growth rates for alloy A357-T6 tested in normal laboratory air and in "high humidity" air. From Ref 33

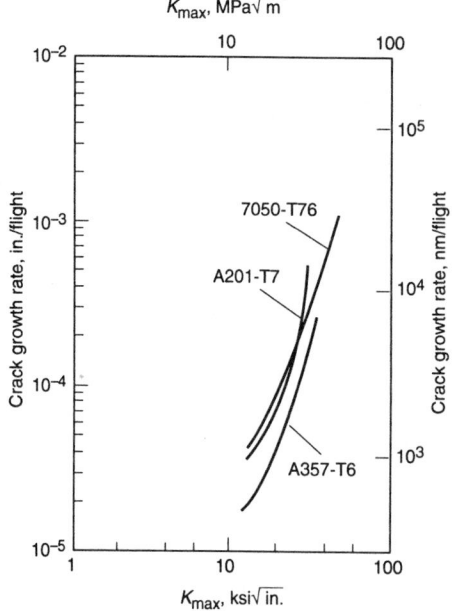

Fig. 17 Fatigue crack growth rates of alloys 7050-T76, A201-T7, and A357-T6 under spectrum loading conditions. From Ref 35

Effect of Discontinuities. The presence of internal discontinuities is a significant factor in determining fatigue and fracture properties. Several studies on the effects of HIPing on alloy A201-T7 (Ref 30 and 31) provide relevant information. For example, both investigations reported improved fatigue life in HIPed alloy, such as shown in Fig. 12. However, Mocarski et al. (Ref 30) reported no effect on crack growth rates (see Fig. 13). They did find a modest increase (about 6%) in plane strain fracture toughness as a result of HIPing.

Environmental Effects. Fatigue and fracture properties tests are expensive, and tests involving special conditions (such as high/low temperature, the presence of corrosives, etc.) add to the costs. This may explain, in part, the limited data on this subject as noted below.

Temperature Effects. The author is aware of two substantive studies in this area. Tirpak (Ref

32) documented fatigue and fracture properties for alloys A357-T6 and A201-T7 from ambient temperature to 200 °C (392 °F); fatigue relationships are shown in Fig. 14 and 15. Misra et al (Ref 33) showed that fracture toughness and crack growth rates for alloy A357-T6 are not affected by temperature over the range from ambient to −196 °C (−320 °F).

Humidity Effects on Fatigue Crack Growth. Doyle et al. (Ref 34) reported that high (but unspecified) humidity level increased the crack growth rate in specimens machined from alloy A357-T6 F-16 vertical stabilizer frames. However, when the data are compared with that from the CAST program (Ref 3), it is seen to be within the scatter band of tests conducted in normal laboratory air (see Fig. 16).

Corrosives (Salt) Effects. Only two references were found regarding corrosion effects in salt environments. Czyryca (Ref 35) studied the corrosion-fatigue behavior in the low-stress, long-life regime for alloy A356-T6. Rotating beam tests ($R = -1$) conducted in air and seawater produced failure stress levels of 10 and 5 ksi, respectively, at 100 million cycles. Metallographic examinations showed fatigue cracks originating at pits caused by selective phase attack (corrosion of the silicon eutectic). Corrosion-fatigue cracks followed a path through the interdendritic region.

Mocarski et al. (Ref 30) performed crack growth tests on alloy A201-T7 specimens in laboratory air in 3.5% saltwater. They found no significant difference between the two sets of results.

Effects of Loading Conditions. Materials screening tests are usually performed using constant-amplitude loading conditions. When evaluating the suitability of metallic materials for military aircraft applications, several investigators have used custom-tailored loading conditions ("spectrum fatigue"). Tirpak and Ruschau (Ref 36) studied the responses of alloys 7050-T76, A201-T7, and A357-T6 to a tension-compression load history chosen to represent that of a lower wing location on a fighter aircraft. They found that A357-T6 had the lowest crack growth rates over the range of stress intensities tested. Figure 17 summarizes the data. Under spectrum fatigue loading, the results differ from those under constant amplitude, tension-only loading. For purposes of comparison, refer back to Fig. 6; note that the curve for the wrought reference material (7075-T7351) crosses over the D357 curve in this case. This crossover was not found by Tirpak et al. for spectrum loading conditions.

Component Tests

A number of programs involving static, fatigue, and damage tolerance testing of full-scale cast test articles are documented in the literature.

A brief review of several prominent programs is offered below.

- *Vought A-7 cast spoiler (Ref 8):* Ten alloy A201-T7 cast components were subjected to static and/or fatigue tests. A baseline built-up spoiler was also tested. The cast components met all performance criteria—static, fatigue, and residual strength.
- *"CAST" Program YC-14 forward bulkhead (Ref 3):* Static, fatigue, and damage tolerance tests were performed on a full-scale bulkhead. The alloy A357-T6 test article met all program requirements.
- *USAF/General Dynamics F-16 vertical stabilizer substructure (Ref 34):* The substructure was assembled into an F-16 vertical stabilizer, complete with graphite-polyimide composite skins, and subjected to static, fatigue, and damage tolerance tests. The assembled stabilizer met all program requirements.
- *Aerospatiale A320 access door substructure:* A conventional built-up access door and a door with cast substructure were subjected to cyclic pressure tests. Crack initiation in the conventional door occurred at 10^5 cycles; initiation occurred at 3×10^5 cycles in the door with cast substructure.

These results were achieved using cast alloys with relatively modest properties. Thus, the influence of design on component performance is evident.

Considerations for Future Work

This article attempts to briefly summarize the general relationships between microstructure, processing, coupon properties, and component performance. In so doing, some points for possible future consideration may include the following:

- *Weld rework of Al-Si-Mg castings:* Knowledge of the interaction between the very fine weld metal microstructure and gas porosity, and how it affects weld fatigue life, would be useful. In addition, the development of optimized welding procedures for alloys such as D357-T6 would be of great value to both casting suppliers and users.
- *Fatigue life modeling:* A study of fatigue life of alloy D357-T6 as a function of distance from a sand mold chill, as was done by Bailey (Ref 18) for A356, would be useful to those engaged in casting design and/or the development of computer modeling tools.
- *Performance analysis of cast components:* Projects to gain a better understanding of casting component performance are needed. For example, the way that loads are redistributed when a crack forms in a cast component is not well understood.
- *Alloy development:* The existing alloys hold a great deal of potential for increased utilization. A new casting alloy would have to provide significant improvements in properties while still providing excellent castability and weldability—at a competitive price, of course.

REFERENCES

1. R. Genoux, "Casting Design, from Theory to Practice," from "Casting Airworthiness," AGARD Report No. 762, NATO, 1989
2. J.W. Mitchell and M.R. Dunkle, "Reliability of Premium Quality Cast Aluminum Structures for the Aerospace/Defense Industry," 33rd International SAMPE Symposium, 7-10 March 1988, Society for the Advancement of Material and Process Engineering
3. J.W. Faber, "Cast Aluminum Structures Technology (CAST)," Final Report, AFWAL-TR-80-3021, April 1980
4. D. Mietrach, Advanced Castings in Aircraft Structures: A Way to Reduce Costs, *Design for Manufacturing Strategies, Principles and Techniques*, J. Corbett et al., Ed., Addison-Wesley, 1991
5. D.R. Apodaca and J.G. Louvier, "Static and Fatigue Properties of Repair Welded Aluminum and Magnesium Premium Quality Castings," presented at the WESTEC Conference (Los Angeles), 7-11 March 1966
6. M.W. Ozelton, S.J. Mocarski, and P.G. Porter, "Durability and Damage Tolerance of Aluminum Castings," WL-TR-91-4111, Oct 1991
7. K.J. Oswalt and Y. Lii, "Manufacturing Methods for Process Effects on Aluminum Casting Allowables," AFWAL-TR-84-4117, March 1985
8. B. Hanson, "Low Cost Cast Spoilers," Final Report, AFFDL-TR-75-81, July 1975
9. S. Kennerknecht, "Fatigue Testing of D357 Alloy Investment Castings," Cercast Division, July 1994
10. G.O. Rading, J. Li, and J.T. Berry, Fatigue Crack Growth in Cast Al-Cu Alloy A206 with Different Levels of Porosity, *AFS Trans.*, Vol 94-045
11. M.F. Hafiz and T. Kobayashi, The Contribution of Microstructure to the Fracture Toughness of Hypoeutectic Al-Si Casting Alloys, *Proc. 4th International Conf. Aluminum Alloys*, Sept 1994, p 107-114
12. S. Shivkumar, L. Wang, and C. Keller, Impact Properties of A356-T6 Alloys, *J. Mater. Eng. Perf.*, Vol 3 (No. 1), Feb 1994, p 83-90
13. N.R. Green and J. Campbell, "The Influence of Oxide Film Filling Defects on the Strength of Al-7Si-Mg Alloy Castings," 1994
14. Technical Proposal, "Cast Aluminum Structures Technology (CAST)," submitted by the Boeing Co. in response to U.S. Air Force RFP No. F33615-76-R-3111, 16 April 1976
15. D.L. McLellan and M.M. Tuttle, "Manufacturing Methodology Improvement for Aluminum Casting Ductility," AFWAL-TR-82-4135, Dec 1982
16. *Metallic Materials and Elements for Aerospace Vehicle Structures*, MIL-HDBK-5
17. "Determination and Acceptance of Dendrite Arm Spacing in Aluminum Castings," Aerospace Recommended Practice ARP 1947
18. W.A. Bailey, How Solidification Affects Fatigue of 356 Aluminum Alloy, *Foundry*, 1965
19. P.C. Inguanti, "Cast Aluminum Fatigue Property/Microstructure Relationship," 17th National SAMPE Technical Conference, Society for the Advancement of Material and Process Engineering, Oct 1985
20. M.J. Couper, A.E. Neeson, and J.R. Griffiths, Casting Defects and the Fatigue Behavior of an Aluminum Casting Alloy, *Fat. Fract. Eng. Mater. Struct.*, Vol 13 (No. 3), 1990, p 213-227
21. S.E. Stanzl-Tschegg, H.R. Mayer, A. Beste, and S. Kroll, Fatigue and Fatigue Crack Propagation in Al7SiMg Cast Alloys under In-Service Loading Conditions, *Int. J. Fatigue*, Vol 17 (No. 2), 1995, p 149-155
22. D.C. Wei, Rim Section Fatigue Results of Various Cast Aluminum Wheels, *AFS Trans.*, Vol 98, 1987, p 681-688
23. B. Skallerud, Fatigue Life Assessment of Aluminum Alloys with Casting Defects, *Eng. Fract. Mech.*, Vol 44 (No. 6), 1993, p 857-874
24. T.L. Reinhart, "Characterization of Alloy D357-T6," presented at Fifth Annual Aerospace Materials and Processes Conference (Anaheim, CA), May 1994
25. F.T. Lee, J.F. Major, and F.H. Samuel, Effect of Silicon Particles on the Fatigue Crack Growth Characteristics of Al-12Wt.Pct.Si-0.35Wt.Pct.Mg-(0 to 0.2) Wt.Pct. Sr Casting Alloys, *Metall. Mater. Trans. A*, Vol 26A, June 1995, p 1553-1570
26. K.J. Oswalt and A. Maloit, A Study of the Relationship of Notch Yield Ratio and Fracture Toughness of Structural Aircraft Quality D357 Casting Alloy, *AFS Trans.*, Vol 90-171, p 865-877
27. A. Wickberg, G. Gustafsson, and L.-E.A. Larsson, "Microstructural Effects on the Fatigue Properties of a Cast Al-7SiMg Alloy," Paper 840121, Society of Automotive Engineers, March 1984
28. Y.-H. Tan, S.-L. Lee, and Y.-L. Lin, Effects of Be and Fe Content of Plane Strain Fracture Toughness in A357 Alloys, *Metall. Mater. Trans. A*, Vol 26A, Nov 1995, p 2937-2945
29. K.-H. Chien, T.Z. Kattamis, and F.R. Mollard, Cast Microstructure and Fatigue Behavior of a High Strength Aluminum Alloy (KO-1), *Metall. Trans.*, Vol 4, April 1973
30. S.J. Mocarski, G.V. Scarich, and K.C. Wu, Effect of Hot Isostatic Pressure on Cast Aluminum Airframe Components, *AFS Trans.*, Vol 91-02, p 77-81
31. S.J. Vonk, G.S. Hoppin, and K.W. Benn, "Hot Isostatic Pressing of Aluminum Alloy Castings," U.S. Navy contract No. N00019-79-C-0649, Final Report, Feb 1981
32. J.D. Tirpak, "Elevated Temperature Properties of Cast Aluminum Alloys A201-T7 and A357-T6," AFWAL-TR-85-4114, Nov 1985
33. M.S. Misra, Evaluation of Tensile and Fracture Properties, Society of Manufacturing Engineers, Paper CM 79-393

34. "Cast Aluminum Primary Aircraft Structure," AFWAL-TR-84-4070, June 1984
35. E.J. Czyryca, "Long-Life Corrosion Fatigue Crack Initiation in Cast A356-T6 Aluminum Alloy," U.S. Naval Surface Warfare Center, 7 April 1992
36. J.D. Tirpak and J.J. Ruschau, "Spectrum Fatigue Crack Growth Rate Characteristics of Cast Aluminum Alloys A201-T7 and A357-T6," AFWAL-TR-86-4047, Dec 1986

SELECTED REFERENCES

- A. Ferran, Y. Barbaux, M. Maurens, and M. Dehaye, "High Quality Aluminum Casting to Produce Structural Civil Aircraft Parts Subjected to Fatigue Loads," *Advanced Aluminum and Magnesium Alloys*, ASM International, 1990, p 459-467
- E. Starke and J. Staley, *Application of Modern Aluminum Alloys to Aircraft*, ALCOA Technical Center, March 1995

Fatigue Strength of Aluminum Alloy Welds

Jeffrey S. Crompton, Edison Welding Institute

THE USE OF ALUMINUM for structural applications has increased significantly in recent years, and more applications are likely to emerge as designers and engineers are required to decrease the weight of structural components. Many of these structures and structural components are used in environments that subject them to dynamic loading. Despite the difficulties in designing against fatigue failure, aluminum is used successfully in many fatigue critical applications in structures such as aircraft, ships, trucks, cars, and bridges. The continued safe operation of these structures requires reliable information on the performance of the parent aluminum alloys, the welded aluminum joints necessary to form a component, and the larger-scale structural details.

In general, the response of a joint is determined by both global and local parameters. Global influences include the level of load that the structure is designed to support, the level of redundancy that is built into the structure, the structural load path, and so forth. Often of more significance in determining fatigue performance is the influence of local effects, such as local stress fields, defect conditions, residual stress state, and material properties. This article briefly reviews the factors that affect the fatigue strength of aluminum alloy weldments.

The progressive failure mechanisms that operate in welded joints are no different from those in the metals which they join. However, welds are usually located at a structural (global) stress concentration, typically introduce a severe notch (i.e., local stress concentration), often contain microstructures that represent an unfavorable alteration of the parent metal microstructure, frequently contain planar or volumetric discontinuities, and typically contain high tensile residual stresses. Most of the data presented for welded details have been obtained from test programs conducted on relatively small specimens (Ref 1). Due to the significance of residual stresses on the fatigue performance of welded construction (Ref 2, 3), direct applicability of these data to welded aluminum components is limited (Ref 4-8). These programs have generally shown that the data obtained from large specimens indicate a lower fatigue performance than smaller-specimen data. As a result, recent revisions of both North American and European specifications (Ref 9, 10) reflect the results of tests on full-scale structures.

Fatigue Performance of Aluminum Welds

A number of factors, both global and local, influence the fatigue performance of welded aluminum joints. Among the most significant are:

- Magnitude, nature, and range of the applied stresses
- Properties of the base alloy
- Stress concentrations associated with the weldment, which are strongly influenced by a number of other parameters, such as joint type, internal and external geometry, and joint size
- Level of residual stress, which can be influenced by postweld heat treatment, weld procedure, and postweld mechanical stress relief

Of these, by far the most significant in determining the fatigue life of aluminum alloy weldments are the stress concentrations associated with the weldment and the residual stresses present in the weld. Base alloy properties lead to secondary influences.

The magnitude, nature, and range of applied stresses influence the fatigue failure of aluminum weldments in much the same manner as for other materials and weldments. As the mean level of stress or the range of applied stress increases, the fatigue life of the specimen, joint, or structure decreases. Similarly, the nature of the loading influences the observed fatigue life through its effect on local stresses. Loading in either tension, bending, or torsion will give rise to differing stress states in the specimen, and since localized stress is the cause of fatigue failure, significant changes in fatigue performance may result.

Table 1 Typical endurance limit values of wrought aluminum alloys

Alloy and temper	Fatigue endurance limit, MPa(a)
2014-T6	124.11
2219-T62	103.43
2219-T81	103.43
5052-H32	117.22
5052-H34	124.11
5056-O	137.9
5083-H321	158.59
5154-O	117.22
5154-H34	131.01
6061-T4	96.53
6061-T6	96.53
6063-T63	68.95
7079-T651	158.59

(a) Unnotched, $R = -1$ at 10^7 cycles

Table 2 Effect of reinforcement on weldment fatigue ($R = 0$)

Alloy	Ultimate strength of parent material, MPa	Average fatigue strength, MPa				No. of tests
		$N = 10^4$	$N = 10^5$	$N = 10^6$	$N = 10^7$	
Reinforcement on, as-welded						
5052-H32	227.54	172.38	103.43	75.85	55.16	12
5052-H34	262.01	199.96	137.90	82.74	62.06	12
5083-H113	317.17	...	124.11	89.64	68.95	45
7004-T41	358.54	...	124.11	96.53	82.74	54
Reinforcement off						
5052-H32	227.54	...	193.06	117.22	103.43	9
5083-H113	317.17	...	165.48	137.90	117.22	20

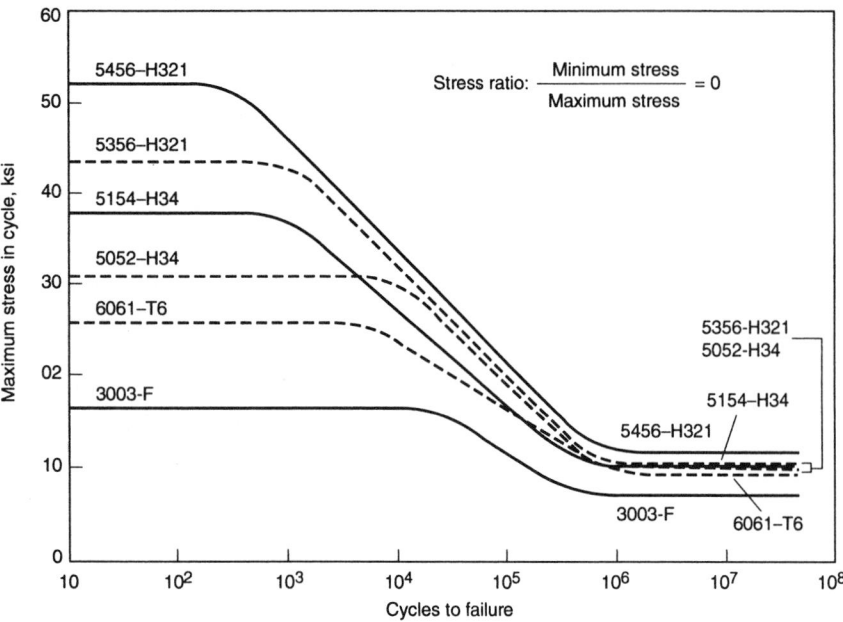

Fig. 1 Results of axial fatigue tests of aluminum alloys as-welded butt joints in 3/8 in. plate. Source: Ref 1

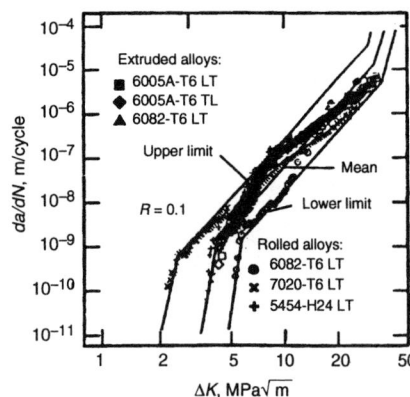

Fig. 2 Fatigue crack propagation rates of wrought aluminum alloys. Source: Ref 7

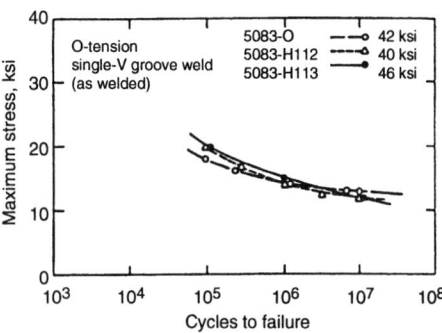

Fig. 3 Effect of temper on axial-tension fatigue behavior of 5083 butt-welded joints. Source: Ref 1

Table 3 Effect of reinforcement on weldment fatigue ($R = 0$)

Alloy parent	Ultimate strength of parent metal, MPa	Average fatigue strength, MPa				No. of tests
		$N = 10^4$	$N = 10^5$	$N = 10^6$	$N = 10^7$	
Reinforcement on, as-welded						
2014-T6 (61)	482.65	220.64	124.11	96.53	96.53	23
2219-31	358.54	220.64	124.11	96.53	96.53	20
2219-T62	413.7	296.49	179.27	151.69	137.9	9
2219-T81	455.07	220.64	124.11	96.53	96.53	21
3003-F	...	110.32	75.88	48.27	48.27	7
5052-H34	262.01	186.17	124.11	82.74	68.95	21
5083-O	289.59	...	137.9	103.43	82.74	15
5083-H112	275.8	...	137.9	96.53	82.74	14
5083-H113	317.17	186.17	131.01	96.53	82.74	99
5086-H32	289.59	...	124.11	82.74	68.95	29
5154-H34	289.59	186.17	117.22	75.85	68.95	34
5356-H321	...	206.85	131.01	75.85	68.95	14
5456-O	310.28	...	...	103.43	82.74	7
5456-H321	351.65	206.85	131.01	89.64	68.95	33
6061-T6	310.28	158.59	110.32	68.95	62.06	26
7039-T61	413.7	220.64	151.69	117.22	89.64	28
7106-T63	434.39	179.27	124.11	82.74	75.85	31
7139-T63	441.28	186.17	117.22	68.95	68.95	13
Reinforcement off						
2219-T31	358.54	255.12	193.06	158.59	137.9	8
2219-T81	455.07	...	193.06	172.38	172.38	10
5052-H34	262.01	186.17	165.48	103.43	89.64	6
5083-O	289.59	...	...	131.01	117.22	9
5083-H112	275.8	...	186.17	144.80	131.01	13
5083-H113	317.17	...	172.38	131.01	117.22	52
5086-H32	289.59	...	186.17	124.11	103.43	40
5154-H34	289.59	...	199.96	117.22	75.85	4
5356-H321	...	199.96	158.59	124.11	82.74	4
5456-H321	351.65	262.01	199.96	137.9	110.32	12
6061-T6	310.28	172.38	131.01	89.64	62.06	3
7039-T61	413.7	...	213.75	144.80	110.32	24
7106-T63	434.39	...	206.85	144.80	131.01	15
7139-T63	441.28	...	193.06	137.9	131.01	6

Parent Alloy Effects

Some aspects of fatigue behavior can be affected by parent alloy selection, although this is generally a secondary factor in terms of the fatigue performance of welded aluminum joints. Typical endurance limit values for smooth-sided specimens of general structural alloys without welds are given in Table 1. It has been observed (Ref 11) that the endurance limit is approximately half of the tensile strength when considering general structural alloys with a tensile strength between 240 and 480 MPa. Alloys whose strength exceeds this upper value tend not to show this behavior, since their increased notch sensitivity limits fatigue life.

In welded aluminum joints, the effect of the mechanical properties of the parent alloy on fatigue strength is not as significant as that reported for welded steel joints. The results of investigations of butt-welded aluminum joints (Ref 12) show that for ultimate tensile strengths up to approximately 345 MPa, there is no significant increase in fatigue strength for either 5xxx- or 7xxx-series alloys. However, 2xxx alloys with ultimate tensile strengths greater than 345 MPa exhibit higher average fatigue strengths. A comparison (Ref 13) of the fatigue performance of as-welded smooth butt joints in a range of structural aluminum alloys is shown in Fig. 1. It is evident that any significant differences in fatigue behavior occur away from the fatigue endurance limit.

At short lives, fatigue life is more closely related to the different strengths of the materials used. At long lives, alloy type exerts little influence, and a common curve is often used for design approaches. At these long lives the predominant mechanism of failure is by crack growth under fatigue loading. Figure 2 shows that there is little influence of alloy type on fatigue crack

Table 4 Effect of reinforcement on weldment fatigue ($R = 0$)

Alloy	Ultimate strength of parent metal, MPa	Average fatigue strength, MPa			No. of tests
		$N = 10^5$	$N = 10^6$	$N = 10^7$	
Reinforcement on, as welded					
5083-O	289.59	103.43	89.64	75.85	13
5083-H113	317.17	131.01	103.43	82.74	65
5456-H321	351.65	124.11	82.74	68.95	19
Reinforcement off					
5083-O	289.59	...	110.32	89.64	6
5083-H113	317.17	172.38	131.01	110.32	33
5456-H321	351.65	...	137.9	110.32	16

Table 5 Fatigue with single-V and double-V reinforcement ($R = 0$)

Alloy	Ultimate strength of parent metal, MPa	Average fatigue strength, MPa			No. of tests
		$N = 10^5$	$N = 10^6$	$N = 10^7$	
Longitudinal single-V, reinforcement on					
5052-H34	262.01	144.8	96.53	68.95	5
5083-O	289.59	...	117.22	82.74	16
5083-H113	317.17	...	110.32	89.64	15
5086-H32	289.59	131.01	96.53	68.95	28
5456-O	310.28	...	103.43	89.64	9
5456-H321	351.65	137.9	82.74	68.95	15
6061-T6	310.28	144.8	89.64	75.85	7
Longitudinal double-V, reinforcement on					
5083-H113	317.17	...	131.01	103.43	24
5456-O	310.28	...	96.53	82.74	9
5456-H321	351.65	...	96.53	82.74	11

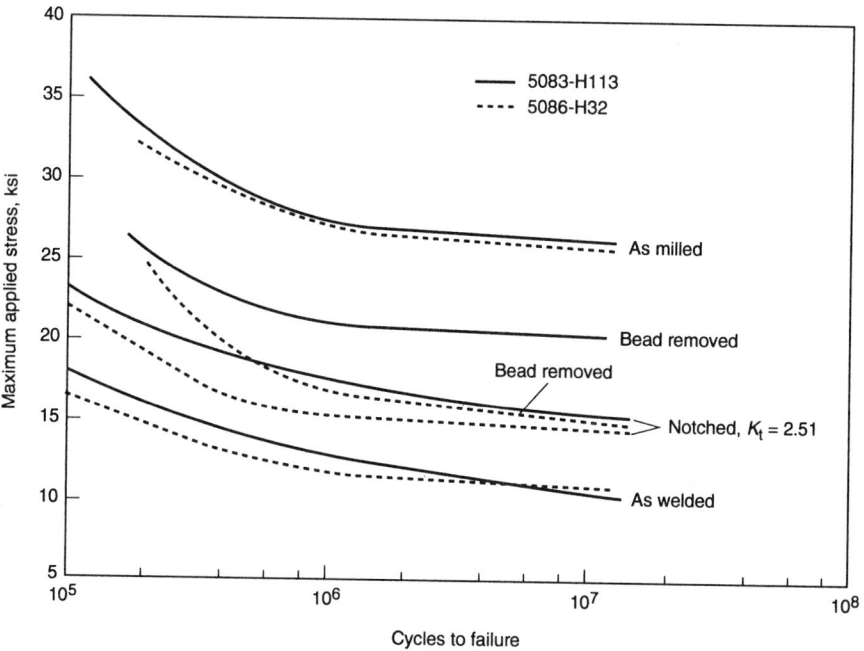

Fig. 4 Comparison of base metal and weldments of 5083-H113 and 5086-H32 under axial fatigue ($R = 0$). Source: Ref 15

growth behavior (Ref 7). Consequently, little difference in fatigue life defined by crack growth approaches would be expected. Similarly, although the heat treatment and temper of an alloy may significantly affect the mechanical properties of the parent alloy, it has been reported (Ref 12) that alloy temper has little effect on the fatigue performance of as-welded smooth butt joints (Fig. 3).

Joint Configuration

The fatigue life of a welded joint or structure is almost always limited by the fatigue life of the weld details. In general, the fatigue life of a component can be broken down into two phases: initiation and propagation. For smooth components, the crack initiation period represents the largest proportion of the total fatigue life. This is particularly noticeable at high fatigue lives, where the fatigue crack initiation period may exceed 95% of the fatigue life. In welded structures, the positioning of welds at structural discontinuities and the localized stress concentrations obtained in a weld detail change this response. The stress concentrations increase the local stresses and act to reduce the initiation period. Alternatively, the discontinuities associated with the weld toe region behave as preexisting cracks, and the fatigue life is then determined by the fatigue crack propagation behavior.

The high stress concentration inherent in the geometry of a fillet weld causes low fatigue strength in such joints. The joint designs with poorest fatigue resistance are the nonsymmetrical types, such as strap, tee, and lap joints in which large secondary stresses are developed. Poor fatigue resistance is also observed in non-load-carrying attachments or reinforcements to a plate (Tables 2-4).

These effects are widely documented and are the basis for classification of structural joints for which statistical minima of fatigue strength have been developed (Ref 14). These classifications are appropriate for alloys in the endurance limit regime since there is little significant difference between alloy performance.

The effect of localized stress concentration on the fatigue performance of welded aluminum butt joints is illustrated in Fig. 4 to 6. As-welded specimens possessed a weld bead, which itself provides a source of localized stress concentration. In specimens where the weld bead was removed, the stress concentration was removed, resulting in longer fatigue lives.

Many stress concentration sites exist in more commonly used joint configurations. These may be simple geometrical discontinuities from weld joint design or discontinuities that arise from the welding procedure. Of these, factors that produce a defective weld (e.g., excessive porosity, lack of fusion, etc.) act to produce severe internal stress concentrations, which may or may not override the effects produced by external factors such as structural discontinuities.

In general, the choice of filler metal and method of edge preparation exert little influence on welded joint fatigue behavior (Fig. 7 and 8). Single-V welds are usually superior (Table 5); lack of fusion and poor root pass problems often occur in double-V or square joints.

The fatigue strength of welded joints is relatively unaffected by the tensile strength of the material. This is because the majority of the fatigue life of a welded joint is spent in the propagation of fatigue cracks to critical sizes. Although tensile strength may be affected by changes in alloys, heat treatment, or temper, crack propagation rates are relatively insensitive to such changes.

Residual Stress Effects

Residual stresses in weldments are produced by the local thermal expansion, plastic deformation, and subsequent shrinkage associated with

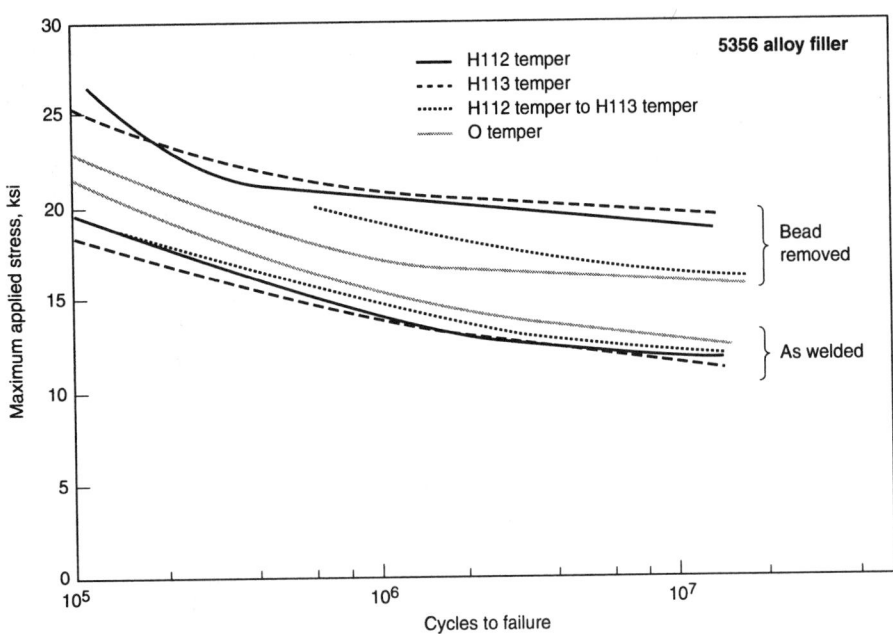

Fig. 5 Effect of weld bead on axial fatigue ($R = 0$) of butt welds in various tempers of 5083 plate with 5356 filler metal. Source: Ref 15

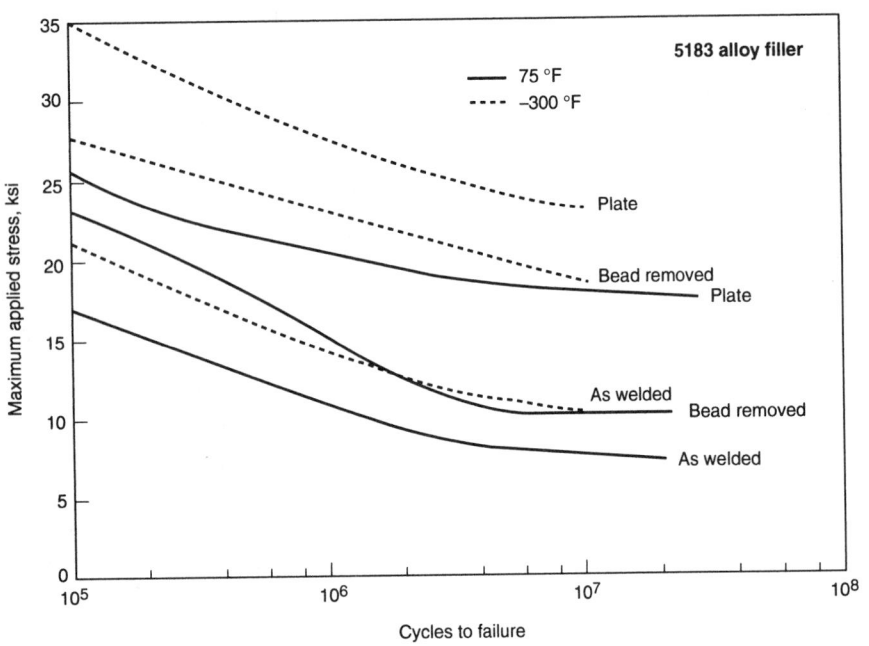

Fig. 6 Effect of cryogenic temperature on reversed bending fatigue ($R = 1$) of transverse double-V butt welds in 5083-H113 0.375 in. plate and filler metal. Source: Ref 15

the thermal cycling that occurs during welding. The precise distribution of residual stresses in a welded joint is complex and depends on a number of factors. Fatigue performance is significantly affected by the level of residual stress present in the weldment. If tensile residual stresses are present, the crack initiation phase is shortened, resulting in a reduction in fatigue life.

Extensive measurements and experimental data on residual stress distribution and its effect on fatigue life have been generated (Ref 16-20).

Typical residual stress values may be as low as 20% of the parent alloy yield value or up to 45% of the heat-affected zone (HAZ) yield value. The influence of residual stresses on fatigue performance has been noted by a number of workers (Ref 21, 22) (Fig. 9). The effect of the residual stress depends on the maximum value and on the stress ratio of the applied stress. For a stress ratio of 0.5, there is little difference in fatigue strength between as-welded and stress-relieved specimens. For a stress ratio of –0.5, fatigue life improves for

the stress-relieved specimens. At a stress ratio of zero, the fatigue strength at 1×10^7 cycles is approximately 40% higher for the stress-relieved material.

The residual stress distribution is also influenced by the welding sequence and welding process (Ref 24), which may explain some of the differences in fatigue response that have been attributed to the influence of weld processing parameters (Ref 12, 25, 26). The buildup of restraint across the weld is a primary influence on the residual stress distribution across the weld. The restraint can be affected by factors such as preheat, weld bead, positioning, and end restraint, and thus determination of the effect of weld processing not attributable to the change in residual stress distribution is difficult.

Improvement of Fatigue Performance

As has been shown, the fatigue life of aluminum weldments is primarily determined by two factors: the stress concentrations associated with the weld detail and the residual stresses associated with the weld. Consequently, there are two primary categories of techniques that can result in improved fatigue life:

Techniques that modify the weld toe geometry

- Machining
- Grinding
- Water jet erosion
- Remelting techniques (e.g., tungsten-inert gas, or TIG, dressing)

These techniques improve fatigues strength because they remove the slag intrusions that form at the weld toes. They also modify the weld toe geometry, thereby reducing the stress concentration effect.

Techniques that impose a compressive residual stress field

- Hammer peening
- Shot peening
- Needle peening
- Spot heating
- Local compression

The introduction of compressive residual stresses in the surface material effectively reduces the tensile component of any applied stress, and thus reduces the crack propagation rate during the early stages of crack growth.

Fatigue performance can also be enhanced by using improved welding procedures. Recent developments in solid-state welding technology (Ref 27) have enabled the production of welds in aluminum that do not possess some of the geometrical and material inhomogeneities commonly found with conventional welding technology. Friction stir welding (FSW) has been found to produce fatigue properties superior to those obtained from specimens produced by metal-inert gas (MIG) and plasma keyhole welding (Ref 27). These data are shown in Fig. 10 and 11 for

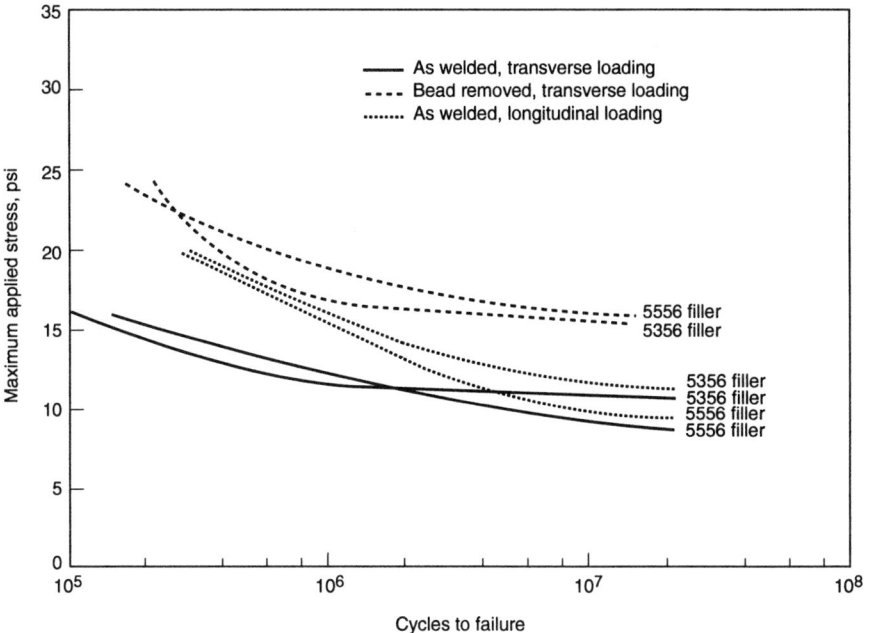

Fig. 7 Effect of loading direction, weld-bead removal, and filler metal on axial fatigue ($R = 0$) of single-V butt welds in 5086-H32. Source: Ref 15

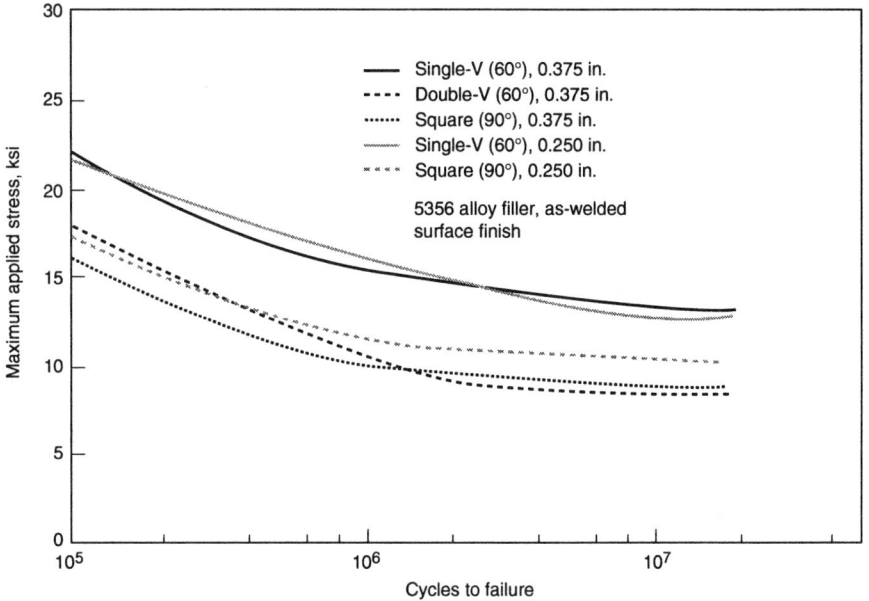

Fig. 8 Influence of edge preparation on axial fatigue ($R = 0$) of transverse butt welds in 5083-H113 plate, as welded, with 5356 filler metal. Source: Ref 15

alloys 6005 and 6082 for a range of temper conditions.

Size Effects on Fatigue Performance

Considering the significant influence of weld geometry and residual stress distribution on the fatigue performance of welded aluminum joints, it is not surprising that the fatigue results obtained from small-scale specimens demonstrate distinctly higher fatigue limits than those obtained in full-scale sections. In a study of the effect of changing attachment plate size from 3 to 24 mm while maintaining a constant local weld toe geometry, Maddox (Ref 28) has shown significant differences in *S-N* data. This has led to the proposal that scale correction factors can be used that allow estimation of the fatigue strength of components of differing thicknesses. This scale correction factor has been proposed to be:

$$\text{Scale factor} = \left(\frac{12}{T_{\text{eff}}}\right)^{0.25} \text{ for plate thicknesses } T \geq 12\,\text{mm}$$

and

$$\text{Scale factor} = \left(\frac{2T}{L}\right)^{0.25}\left(\frac{12}{T}\right)^{0.13} \text{ for plate thicknesses}$$

$$T < 12\,\text{mm}$$

Here T_{eff} is the effective thickness, which is 0.5 L if $L/T \leq 2$, where L is the toe separation of the welds, and T is $L/T > 2$.

In a similar manner, many results (Ref 5, 7, 15, 29, 30) have indicated that larger-scale beam structures have lower fatigue lives than those predicted from small-scale specimens. Obviously, one explanation for this behavior is the potential difference in the residual stress distribution of the welded joints brought about by the different degrees of restraint imposed on the weld by the structural fabrication techniques. In addition, the complex stress distributions associated with the gradients affect both the localized stress, thereby potentially influencing crack initiation, and the driving force for crack growth. By considering the combined effects of residual stress, complex local stress distributions, and steep stress gradients, Jaccard (Ref 7) has been able to evaluate accurately the fatigue life of complex structural welded aluminum components.

REFERENCES

1. W.W. Sanders, Fatigue Behavior of Aluminum Alloy Weldments, *Weld. Res. Counc. Bull.*, Vol 171, 1972
2. R.A. Kelsey and G.E. Nordmark, *Aluminum*, Vol 55, 1979, p 391
3. D. Kosteas, "Influence of Residual Stress on S-N Curve Parameters of Aluminum Weldments," Doc. X111-1239-87, International Institute of Welding, 1987
4. D. Kosteas, "European Recommendations for Fatigue Design of Aluminum Structures," presented at 5th Aluminum Conf. (Munich), Aluminum-Zentrala, 1992
5. C.C. Menzemer, "Fatigue Behavior of Welded Aluminum Structures," Ph.D. dissertation, Lehigh University, 1992
6. F. Soetens, T.J. Van Straalen, and O. Dijkstra, "European Research on Fatigue of Aluminum Structures," presented at 6th Int. Conf. Aluminum Weldments (Cleveland), American Welding Society, 1995
7. R. Jaccard, "Fatigue Life Prediction of Aluminum Components Based on Local Stress Fields, Defect Conditions, and Fatigue Crack Propagation," presented at 3rd Int. Conf. Aluminum Weldments (Munich), Aluminum Verlag, 1985
8. D. Kosteas, K. Polas, and U. Graf, "Results of Welded Beam Program," presented at 3rd Int. Conf. Aluminum Weldments (Munich), Aluminum Verlag, 1985
9. "Specifications for Aluminum Alloy Structures," Aluminum Association, 1986
10. "European Recommendations for Aluminum Alloy Structures Fatigue Design," European Convention for Constructional Steelwork, ECCS Committee, T2, 1992

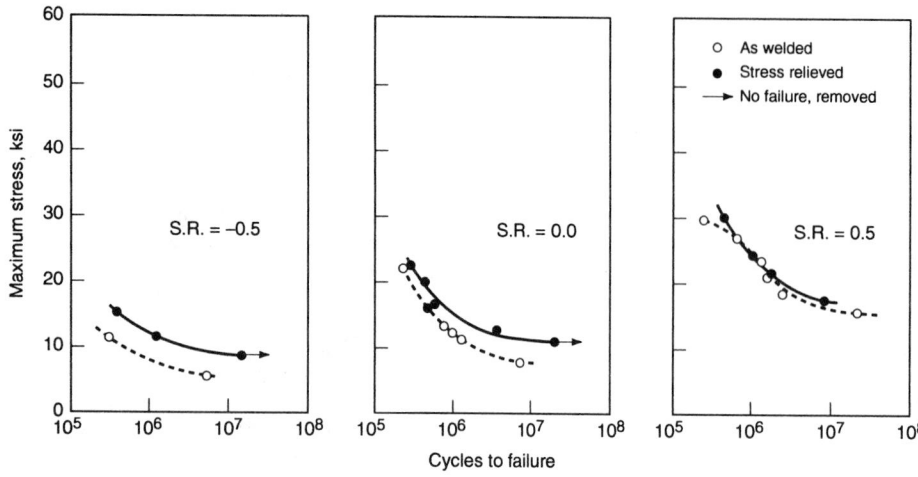

Fig. 9 Effect of stress relief on fatigue strength of 5456-H3221 longitudinal butt welds. Source: Ref 23

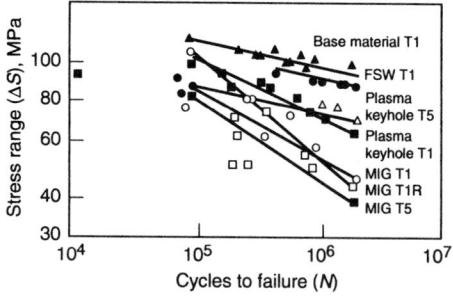

Fig. 10 Mean *S-N* curves and data points for alloy 6005. Source: Ref 21

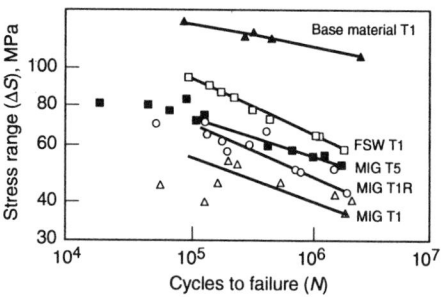

Fig. 11 Mean *S-N* curves and data points for alloy 6082. Source: Ref 21

11. T.R. Gurney, "Fatigue of Welded Structures," Cambridge University Press, 1968

12. N.L. Person, Fatigue of Aluminum Alloy Welded Joints, *Weld. J.*, Vol 50 (No. 2), 1971, p 77

13. I.D. Eaton, "Axial Stress Fatigue Strengths of As-Welded Aluminum Alloy Butt Joints," Alcoa Research Lab Report, 1962

14. "Specifications for Aluminum Structures," *Aluminum Construction Manual*, Section 1, Aluminum Association, 1986

15. H. Mindlin and C. Jaske, Reynolds Aluminum Report, Fatigue Properties of Aluminum Sheet and Plate-Alloys 5083, 5086, 5456, 1968

16. R.A. Kelsey and G.E. Nordmark, *Aluminum*, Vol 55, 1979, p 391

17. F.M. Mazzolani in *Proc. 2nd Int. Conf. Aluminum Weldments*, Aluminum Verlag 1982

18. K. Masubichi, Residual Stresses and Distortion in Welded Aluminum Structures and Their Effects on Service Performance, *Weld. Res. Counc. Bull.*, Vol 174, 1972

19. K. Masubichi, Investigation of Methods of Controlling and Reducing Weld Distortion in Aluminum Structures, *Weld. Res. Counc. Bull.*, Vol 237, 1978

20. S.J. Maddox, The Influence of Residual Stress on the Fatigue Behavior of Al-Zn-Mg Alloy Fillet, Welds, *Proc. SEECO Int. Conf.*, Cambridge University Press, 1976

21. M. Ranes, A.O. Kluken, and O.T. Midling, "Fatigue Properties of As-Welded AA6005 and AA6082 Aluminum Alloys in Ti and TS Temper Condition," presented at *Int. Conf. Advances in Welding Research*, ASM International, 1995

22. T.R. Gurney, "Influence of Artificially-Induced Residual Stresses on the Fatigue Strength of Welded Light Alloy Specimens," BWRA Report, British Welding Research Association, 1961

23. H.N. Hill, Residual Stresses in Aluminum Alloys, *Met. Prog.*, 1961, p 92-96

24. J.C. Downey, D.W. Hood, and D.D. Keiser, Shrinkage in Mechanized Welded Stainless Pipe, *Weld. J.*, Vol 54, 1975, p 170

25. H. Mundler and C.E. Jasper, "Fatigue Properties of 5083, 5088, and 5456 Aluminum Alloy Sheet and Plate," Reynolds Metal Co., 1980

26. P.K. Ghosh, P.C. Gupta, and R. Rathi, Fracture Characteristics of Pulsed MIG-Welded Al-Zn-Mg Alloy, *J. Mater. Sci.*, Vol 26, 1991, p 6161

27. "Improvements Related to Friction Stir Welding: Friction Stir Butt Welding," Patent application PCT/GB92/02203, Welding Institute, Cambridge, U.K., 1992

28. S.J. Maddox, "Scale Effect in Fatigue of Fillet-Welded Aluminum Alloys," presented at 6th Int. Conf. Aluminum Weldments (Cleveland), American Welding Society, 1995

29. R. Ondra and D. Kosteas, Studies of the Fatigue Strength of Welded Joints in Major Aluminum Structural Members, *Weld. Cut.*, Vol 10, 1992, p 11

30. F. Soetens, I.J. Van Straaten, and O.D. Dijktra, "European Research of Fatigue of Welded Aluminum Structures," presented at 6th Int. Conf. Aluminum Weldments (Cleveland), American Welding Society, 1995

Fatigue and Fracture Properties of Titanium Alloys

TITANIUM is used for two primary reasons: structural efficiency, which derives from its combination of high strength and low density; and resistance to corrosion by chlorides and oxidizing media, which derives from its strong passivation tendencies. Titanium, like most structural materials, is supplied in all mill product forms, and several titanium alloys are available to meet specific needs.

In general, titanium and titanium alloys have a well-earned reputation for reliability in service. In no small measure, this is a result of the double and triple vacuum arc melting procedures employed throughout the industry in producing the alloys. That reputation is protected also by the excellent corrosion resistance exhibited by titanium. Titanium does not corrode in salt water. Crack initiation in titanium is almost always mechanically induced; only under very special circumstances will cracks initiate due to a combination of an environment and static stress.

This article summarizes the key variables that affect the fracture toughness, fatigue life, and subcritical crack growth of titanium alloys. Several metallurgical and environmental variables influence the fracture behavior of titanium alloys. Because of the many possible effects of chemistry, microstructure, texture, environment, and loading, it is not possible to quantify the crack growth behavior of titanium alloys unless these factors are closely controlled. Alloys within a given class, such as alpha-beta alloys, show parallel trends in their fracture toughness and crack propagation behaviors. To the extent that they have been studied, the trends for interstitial effects are similar for all alloys, the higher levels of interstitials leading to faster fatigue crack propagation and lower K_{Ic}. A similar trend is observed for variations in microstructure. Those microstructures (Widmanstätten or recrystallization annealed) that give the highest K_{Ic} values generally yield the lowest crack growth rates, whether under fatigue or sustained loads. However, increases in K_{Ic} from microstructural changes do not always lead to lower fatigue crack growth rates. For example, if small surface cracks are considered, a higher toughness can lead to higher fatigue crack growth rates.

Strength and Toughening

Table 1 summarizes some of the common grades of commercially pure titanium and alloy types. These grades have varying amounts of impurities (e.g., carbon, hydrogen, iron, nitrogen, and oxygen). Some "modified" grades also contain small palladium additions (Ti-0.2Pd) and nickel-molybdenum additions (Ti-0.3Mo-0.8Ni). Because small amounts of interstitial impurities greatly affect the mechanical properties of pure titanium, it is not convenient to distinguish be-

Table 1 Commercially pure and modified grades of titanium

	Impurity limits, % max				Tensile properties(a)				Minimum elongation, %
					Ultimate strength		Yield strength		
Designation	C	O	N	Fe	MPa	ksi	MPa	ksi	
JIS class 1	...	0.15	0.05	0.20	275-410	40-60	165(b)	24(b)	27
ASTM grade 1 (UNS R50250)	0.10	0.18	0.03	0.20	240	35	170-310	25-45	24
DIN 3.7025	0.08	0.10	0.05	0.20	295-410	43-60	175	25.5	30
GOST BT1-00	0.05	0.10	0.04	0.20	295	43	...	...	20
BS 19-27t/in.^{2}	...	...	...	0.20	285-410	41-60	195	28	25
JIS class 2	...	0.20	0.05	0.25	343-510	50-74	215(b)	31(b)	23
ASTM grade 2 (UNS R50400)	0.10	0.25	0.03	0.30	343	50	275-410	40-60	20
DIN 3.7035	0.08	0.20	0.06	0.25	372	54	245	35.5	22
GOST BT1-0	0.07	0.20	0.04	0.30	390-540	57-78	...	...	20
BS 25-35t/in.^{2}	...	...	...	0.20	382-530	55-77	285	41	22
JIS class 3	...	0.30	0.07	0.30	480-617	70-90	343(b)	50(b)	18
ASTM grade 3 (UNS R50500)	0.10	0.35	0.05	0.30	440	64	377-520	55-75	18
ASTM grade 4 (UNS R50700)	0.10	0.40	0.05	0.50	550	80	480	70	15
DIN 3.7055	0.10	0.25	0.06	0.30	460-590	67-85	323	47	18
ASTM grade 7 (UNS R52400)	0.10	0.25	0.03	0.30	343	50	275-410	40-60	20
ASTM grade 11 (UNS R52250)	0.10	0.18	0.03	0.20	240	35	170-310	25-45	24
ASTM grade 12 (UNS R53400)	0.10	0.25	0.03	0.30	480	70	380	55	12

(a) Unless a range is specified, all listed values are minimums. (b) Only for sheet, plate, and coil

Table 2 Titanium alloy designations and tensile strengths

Designation	Tensile strength (min)		0.2 % yield strength (min)		Impurity limits, wt %, max				
	MPa	ksi	MPa	ksi	N	C	H	Fe	O
Alpha and near-alpha alloys									
Ti-5Al-2.5Sn	790	115	760	110	0.05	0.08	0.02	0.50	0.20
Ti-5Al-2.5Sn-ELI	690	100	620	90	0.07	0.08	0.0125	0.25	0.12
Ti-8Al-1Mo-1V	900	130	830	120	0.05	0.08	0.015	0.30	0.12
Ti-6Al-2Sn-4Zr-2Mo	900	130	830	120	0.05	0.05	0.0125	0.25	0.15
Ti-6Al-2Nb-1Ta-0.8Mo	790	115	690	100	0.02	0.03	0.0125	0.12	0.10
Ti-2.25Al-11Sn-5Zr-1Mo	1000	145	900	130	0.04	0.04	0.008	0.12	0.17
Ti-5Al-5Sn-2Zr-2Mo(a)	900	130	830	120	0.03	0.05	0.0125	0.15	0.13
Alpha-beta alloys									
Ti-6Al-4V(b)	900	130	830	120	0.05	0.10	0.0125	0.30	0.20
Ti-6Al-4V-ELI(b)	830	120	760	110	0.05	0.08	0.0125	0.25	0.13
Ti-6Al-6V-2Sn(b)	1030	150	970	140	0.04	0.05	0.015	1.0	0.20
Ti-8Mn(b)	860	125	760	110	0.05	0.08	0.015	0.50	0.20
Ti-7Al-4Mo(b)	1030	150	970	140	0.05	0.10	0.013	0.30	0.20
Ti-6Al-2Sn-4Zr-6Mo(c)	1170	170	1100	160	0.04	0.04	0.0125	0.15	0.15
Ti-5Al-2Sn-2Zr-4Mo-4Cr(a)(c)	1125	163	1055	153	0.04	0.05	0.0125	0.30	0.13
Ti-6Al-2Sn-2Zr-2Mo-2Cr(a)(b)	1030	150	970	140	0.03	0.05	0.0125	0.25	0.14
Ti-10V-2Fe-3Al(a)(c)	1170	170	1100	160	0.05	0.05	0.015	2.5	0.16
Ti-3Al-2.5V(d)	620	90	520	75	0.015	0.05	0.015	0.30	0.12
Beta alloys									
Ti-13V-11Cr-3Al(c)	1170	170	1100	160	0.05	0.05	0.025	0.35	0.17
Ti-8Mo-8V-2Fe-3Al(a)(c)	1170	170	1100	160	0.05	0.05	0.015	2.5	0.17
Ti-3Al-8V-6Cr-4Mo-4Zr(a)(b)	900	130	830	120	0.03	0.05	0.020	0.25	0.12
Ti-11.5Mo-6Zr-4.5Sn(b)	690	100	620	90	0.05	0.10	0.020	0.35	0.18

(a) Semicommercial alloy; mechanical properties and composition limits subject to negotiation with suppliers. (b) Mechanical properties given for annealed condition; may be solution treated and aged to increase strength. (c) Mechanical properties given for solution treated and aged condition; alloy not normally applied in annealed condition. Properties may be sensitive to section size and processing. (d) Primarily a tubing alloy; may be cold drawn to increase strength

tween the various grades of unalloyed titanium on the basis of chemical analysis. Titanium mill products are more readily distinguished by mechanical properties, as noted in Table 1 for various specification grades.

Titanium alloys are grouped into classes (Table 2), depending on the room-temperature composition of the alpha crystal (hexagonal close-packed) and/or the beta crystal (body-centered cubic) phase structure. Stable transformation and phase compositions from the high-temperature beta phase are altered by alloying additions, which are typically classified as alpha stabilizers or beta stabilizers. Vanadium, iron, and hydrogen are beta stabilizers, while aluminum, oxygen, and nitrogen are examples of alpha stabilizers. The following sections briefly review the general strengthening and toughening principles for the three groups of titanium alloys. The common Ti-6Al-4V alloy is described separately.

Alpha-Beta Alloys

Ti-6Al-4V. The most important titanium alloy is Ti-6Al-4V, which has found application for a wide variety of aerospace components and fracture-critical parts. With a strength-to-density ratio of 25×10^6 mm (1×10^6 in.), Ti-6Al-4V is an effective lightweight structural material and has strength-toughness combinations between those of steel and aluminum alloys.

The Ti-6Al-4V alloy is most commonly used in the annealed condition. However, the alloy contains a small amount of beta phase (about 10 vol% at room temperature), which imparts some

hardenability by heat treatment in sections less than 25 mm (1 in.) thick. The key to hardening is to quench from high in the alpha-beta field and then age at a lower temperature. A typical strengthening heat treatment consists of heating for 1 h at 955 °C (1750 °F) and water quenching, followed by heating for 4 h at 540 °C (1000 °F) and air cooling. When alloy Ti-6Al-4V is processed improperly after heating into the beta field, alpha phase can form preferentially along the prior beta grains. Extensive hot work is required to break up such structures. Because cracks tend to propagate in or near interfaces, this type of structure can provide loci for crack initiation and propagation and thereby lead to premature failure.

In the annealed condition, the Ti-6Al-4V alloy derives its annealed strength from several sources, the principal source being substitutional and interstitial alloying of elements in solid solution in both alpha and beta phases. Oxygen, nitrogen, hydrogen, and carbon are the interstitial elements, which generally increase strength and decrease ductility. Aluminum is the most important substitutional solid solution strengthener. Its effect on strength is linear. Other, less important sources of strengthening are interstitial solid solution strengthening, grain size effects, second-phase (beta) effects, ordering in alpha, age hardening, and effects of crystallographic texture.

At room temperature, Ti-6Al-4V is about 90 vol% alpha, and thus the alpha phase dominates the physical and mechanical properties of the alloy. However, the overall effects of processing history and heat treatment on microstructure are complex. The microstructure depends on both

processing history and heat treatment, and the microstructure that combines highest strength and ductility is not necessarily the microstructure that provides optimum fracture toughness or resistance to crack growth. Improvements in K_{Ic} can be obtained by providing either a microstructure of transformed beta or an equiaxed structure composed mainly of regrowth alpha that has both low dislocation densities and low concentrations of aluminum and oxygen (the so-called "recrystallization annealed" structure).

Transformed structures enhance toughness, primarily because fractures in such structures must proceed along tortuous, many-faceted crack paths. Transformed, Widmanstätten-like microstructure occurs from $\beta \rightarrow \alpha + \beta$ transformation during cooling from the beta-phase field. The effect on toughness is summarized in Tables 3 and 4.

Another microstructure of special interest in obtaining maximum fracture toughness is one produced by so-called recrystallization annealing. In this process, the alloy is heated to a temperature about 70 °C (125 °F) below the transformation temperature, held for a time, and then very slowly cooled. Most of this structure is continuous primary alpha phase and regrowth alpha (that is, alpha that has formed on the preexisting primary alpha phase). The primary alpha existed at the upper temperature and nucleated the regrowth alpha during cooling. This procedure is well established for the Ti-6Al-4V extra-low interstitial (ELI) alloy, and typical data for the recrystallization annealed alloy are shown in Fig. 1. The slow cooling should be terminated at about 700 °C (1300 °F) to avoid ordering the Ti₃Al formation.

Table 3 Typical fracture toughness of high-strength titanium alloys

Alloy	Alpha morphology	Yield strength MPa	Yield strength ksi	K_{Ic} MPa√m	K_{Ic} ksi√in.
Ti-6Al-4V	Equiaxed	910	130	44-66	40-60
	Transformed	875	125	88-110	80-100
Ti-6Al-6V-2Sn	Equiaxed	1085	155	33-55	30-50
	Transformed	980	140	55-77	50-70
Ti-6Al-2Sn-4Zr-6Mo	Equiaxed	1155	165	22-23	20-30
	Transformed	1120	160	33-55	30-50

Table 4 Relationship between K_{Ic} and fraction of transformed structure in alloy Ti-6Al-4V

Heat treating temperature(a) °C	°F	Fraction of transformed structure, %	K_{Ic} MPa√m(b)	ksi√in.
1050	1922	100	69.0 (69.9)	64
950	1742	70	61.5 (60.4)	55
850	1562	20	46.5 (44.6)	40
750	1382	10	39.5 (41.5)	38

(a) Heated for 1 h at indicated temperature and then air cooled. (b) Values in parentheses calculated from linear least-squares expression relating % transformation to K_{Ic}

Table 5 Effect of hydrogen content on room-temperature K_{Ic} in alloy Ti-6Al-4V after furnace cooling from 927 °C (1700 °F)

Hydrogen content, ppm	K_{Ic} at room temperature(a) MPa√m	ksi√in.
At 0.16 wt % oxygen		
8	145	132
36	118	107
53	104	95
122	100	91
At 0.05 wt % oxygen		
9	133	121
36	125	114
50	96	87
125	101	92

(a) Specimens were tested in accord with ASTM E 399 but were loaded rapidly (total testing time = 10 s). Source: *Met. Trans.*, Vol 5A, 1974, p 2405

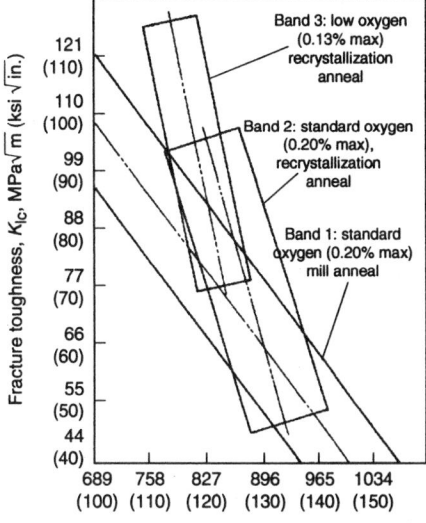

Fig. 1 Fracture toughness and yield strength of Ti-6Al-4V grades. Source: *Metals Progress*, March 1977

This is less important for the ELI grade than for the standard grade. A real virtue of the recrystallization annealed microstructure is that substantial toughness is achieved while maintaining ductility and high strength (Fig. 1). Such a microstructure also tends to reduce scatter in the data, thus permitting higher design criteria.

Within the permissible range of chemistry for a specific titanium alloy and grade, oxygen is the most important impurity insofar as its effect on toughness and strength. As might be expected, hydrogen also has an effect on toughness (Table 5). Very low hydrogen contents (less than about 40 ppm) enhance toughness. This effect is particularly dramatic with hydrogen contents below 10 ppm.

Hydrogen Embrittlement and Stress-Corrosion Cracking. Hydrogen damage of titanium and titanium alloys is manifested as a loss of ductility (embrittlement) and/or a reduction in the stress-intensity threshold for crack propagation. The damage is caused by hydrides, which form as hydrogen diffuses into the material during exposure with either gaseous or cathodic hydrogen. Because the phenomenon depends on both hydrogen diffusion and hydride formation, there may be a peak in hydrogen embrittlement as a function of temperature. The exact level of hydrogen at which a separate hydride phase is formed depends on the composition of the alloy and the previous metallurgical history.

Stress-corrosion cracking (SCC) is related to hydrogen embrittlement as it falls into two general categories: anodic-assisted cracking and cathodic-assisted cracking. Anodic SCC (active path corrosion) involves the dissolution of metal during the initiation and propagation of cracks. Cathodic SCC (embrittlement by corrosion product hydrogen) involves the deposition of hydrogen at cathodic sites on the metal surface or on the

walls of a fissure or crack and its subsequent absorption into the metal lattice. The close interaction of these two mechanisms accounts for the diversity of SCC phenomena. Typical SCC thresholds of Ti-6Al-4V are summarized in various conditions in Fig. 2 and Table 6.

In aerospace applications, there is a natural concern that chemical environmental factors such as water, salt water, or jet fuel will alter toughness in critical components. Data obtained on annealed standard grade Ti-6Al-4V indicate the following environmental effects on the apparent value of fracture toughness:

- Laboratory air, 56 MPa√m (51 ksi√in.)
- JP4 fuel, 47 MPa√m (43 ksi√in.)
- 3.5% salt solution, 34 MPa√m (31 ksi√in.)

Deformation Modes. Crucial to the toughness question is the size of the plastic zone that can form ahead of a propagating crack. That size, simplistically, depends on the yield strength. Having a high yield strength and a relatively low elastic modulus, Ti-6Al-4V can store more energy elastically than most metals before plastic deformation begins. The elastic modulus depends on the direction in which it is measured in a single crystal of titanium.

The plastic zone ahead of an advancing crack is not uniform in cross section because of the anisotropic alpha phase. It varies in a macroscopic sense in response to the microstructure phase and morphology. It also varies from grain to grain in accordance with crystal type, whether alpha or beta, and with crystal orientation. To

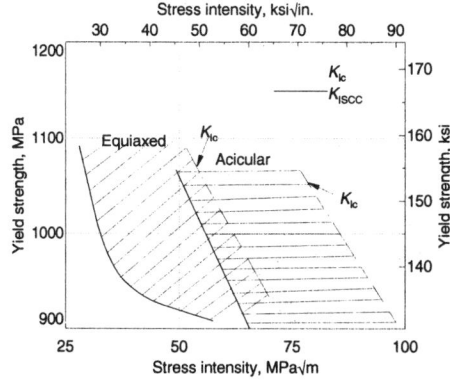

Fig. 2 Relationship between yield stress and stress intensity for either unstable fast fracture (K_{Ic}) or stress-corrosion cracking (K_{Iscc}) of Ti-6Al-4V in salt water. Source: A.W. Thompson and I.M. Bernstein, The Role of Metallurgical Variables in Hydrogen-Assisted Environmental Fracture, *Corrosion Science and Technology*, Vol 7, M.G. Fontana and R.W. Staehle, Ed., Plenum Press, 1980, p 111

complicate matters further, Poisson's ratio and its plastic counterparts necessarily depend on direction on both the macro and micro scales. Events occurring in and around a crack tip and its associated plastic zone in Ti-6Al-4V are, for these reasons, complex indeed. Toughness in the Ti-6Al-4V alloy is not yet quantifiable from first principles. Nevertheless, there is a great deal of empirical information available for Ti-6Al-4V from which some general rules can be developed.

High-strength alpha-beta alloys include Ti-6Al-6V-2Sn and Ti-6Al-2Sn-4Zr-6Mo. Alpha is the dominant phase in these alloys, but to a lesser extent than in Ti-6Al-4V. These alloys are stronger and more readily heat treated than Ti-6Al-4V. These features arise from the increased solid solution strengthening afforded by tin and zirconium, which have relatively small effects on the transformation temperature, and from the in-

Table 6 Stress-intensity thresholds of stress-corrosion cracking (K_{Iscc}) for Ti-6Al-4V in various media

Medium	Alloy condition(a)	K_{Iscc}, ksi√in.	Reference(b)
Freon TF ($C_2Cl_3F_3$)	STA	33	1
Freon TF ($C_2Cl_3F_3$)	Mill annealed	52	2
Freon MF (CCl_3F)	STA	23	3
Freon MF (CCl_3F)	Mill annealed	25	2
Deionized water	Mill annealed	73.4	4
3.5% NaCl, pH = 4.5	Mill annealed	52.7	4
3.5% NaCl, pH = 9.0	Mill annealed	48.0	4
3.5% NaCl, pH = 11.0	Mill annealed	42.2	4
3.5% NaCl + 0.5% FeCl, pH = 2.2	Mill annealed	48.9	4
3.5% NaCl + 1.0% AlCl$_3$, pH = 7.0	Mill annealed	46.1	4
3.5% NaCl + 0.1% Dupanol (wetting agent)	Mill annealed	45.8	4
Kerosene	Mill annealed	70.5	4
Dodecyl alcohol	Mill annealed	66.7	4
Methyl alcohol	Mill annealed	70.3	4
Mercury	Mill annealed	22 (average)	4

(a) STA, solution treated and aged. (b) (1) S.V. Glorioso, *Lunar Module Pressure Vessel Operating Criteria*, Specification SE-V-0024, NASA/MSC, Oct 1968. (2) C.C. Seastrom and R.A. Gorski, The Influence of Fluorocarbon Solvents on Titanium Alloys, Accelerated Crack Propagation of Titanium by Methanol Halogenated Hydrocarbons, *and Other Solutions*, DMIC Memorandum 228, 6 March 1967, p 20. (3) C.F. Tiffany and J.N. Masters, *Investigation of the Flaw Growth Characteristics of Ti-6Al-4V Titanium Used in Apollo Spacecraft Pressure Vessels*, CR-65586, NASA, March 1967. (4) D. Williams, R. Wood, E. White, W. Boyd, and H. Ogden, "Studies of the Mechanism of Crack Propagation in Salt Water Environments of Candidate Supersonic Transport Titanium Alloy Materials," FAA Report SST-66-1, 1966

Table 7 Summary of heat treatments for alpha-beta titanium alloys

Heat treatment designation	Heat treatment cycle	Microstructure
Duplex anneal	Solution treat at 50-75 °C below T_β(a), air cool and age for 2-8 h at 540-675 °C	Primary α, plus Widmanstätten α + β regions
Solution treat and age	Solution treat at ~40 °C below T_β, water quench(b) and age for 2-8 h at 535-675 °C	Primary α, plus tempered α' or a β + α mixture
Beta anneal	Solution treat at ~15 °C above T_β, air cool and stabilize at 650-760 °C for 2 h	Widmanstätten α + β colony microstructure
Beta quench	Solution treat at ~15 °C above T_β, water quench and temper at 650-760 °C for 2 h	Tempered α'
Recrystallization anneal	925 °C for 4 h, cool at 50 °C/h to 760 °C, air cool	Equiaxed α with β at grain-boundary triple points
Mill anneal	α + β hot work, anneal at 705 °C for 30 min to several hours, air cool	Incompletely recrystallized α with a small volume fraction of small β particles

(a) T_β is the β-transus temperature for the particular alloy in question. (b) In more heavily β-stabilized alloys, such as Ti-6Al-2Sn-4Zr-6Mo or Ti-6Al-6V-2Sn, solution treatment is followed by air cooling. Subsequent aging causes precipitation of α phase to form an α + β mixture.

creased amounts of beta phase that result from the larger vanadium and molybdenum additions.

The microstructure of these alloys follows some trends similar to those for Ti-6Al-4V, although the combination of strength and toughness tends to be higher than in Ti-6Al-4V (Fig. 3). These alloys generally follow the trends of Ti-6Al-4V relating microstructure and mechanical properties. As a general rule, the yield stress of alpha-beta alloys is highest when the microstructure consists of a mixture of primary alpha and a fine alpha-beta mixture. In Ti-6Al-4V, high strength is achieved by a martensitic product (α'), which is produced by a solution treatment followed by water quenching and an aging treatment, or often called "solution treat and age" (STA). In more heavily beta-stabilized alpha-beta alloys such as Ti-6Al-2Sn-4Zr-6Mo, solution

treatment is followed by air cooling, which retains much of the beta phase in a metastable state. Both of these treatments (STA in Table 7) are followed by an aging reaction, although subsequent aging of the as-quenched α' or the metastable beta phase leads to different metallurgical reactions. Other common heat treatments for alpha-beta alloys are summarized in Table 7.

In general, higher fracture toughness at comparable strength is obtained in alpha-beta alloys with an acicular (i.e., transformed) structure from beta treatments or STA than from alloys with equiaxed structures. This is characteristic of all alpha-beta alloys but is especially pronounced in higher-strength alloys such as Ti-6Al-2Sn-4Zr-6Mo. The data in Table 8 show that beta-processed forgings of Ti-6Al-2Sn-4Zr-6Mo have decidedly superior strength-toughness combina-

tions. Moreover, in the alpha-beta processed condition this alloy can exhibit very low toughness values. For example, at a yield-strength level of 1150 MPa, a toughness of 26 MPa√m corresponds to a critical flaw size of ~0.5 mm. The high toughness of the beta-forged material correlated well with the change in fracture topography from a flat transgranular fracture in alpha-beta forged material to an irregular intergranular fracture in beta-forged material, as discussed above. This alloy is currently used almost exclusively as a forging alloy, and, as a result, no data are available for commercially produced plate.

The data in Table 8 for Ti-6Al-2Sn-4Zr-2Mo also show that this alloy is generally quite tough. Thus, the improvement in toughness that can be achieved by beta forging is less significant than that attainable in Ti-6Al-2Sn-4Zr-6Mo. The effect of forging on the toughness and yield strength of Ti-6Al-6V-2Sn is shown in Fig. 4.

The effect of texture on toughness has been studied to a limited degree for Ti-6Al-4V and Ti-6Al-2Sn-4Zr-6Mo. In general, the alloys with relatively intense transverse basal textures have greater fracture resistance when the crack propagation direction is perpendicular to the majority of the basal planes (i.e., the transverse direction). The one exception to this appears to be Ti-6Al-4V ELI, for which the opposite appears to be true. Fracture-resistance data obtained from R curves of materials with strong basal transverse textures are shown in Table 9. Note that the R curves for the ELI material differ from the others inasmuch as the incremental crack growth, Δa, that preceded unstable fracture was smaller by a factor of

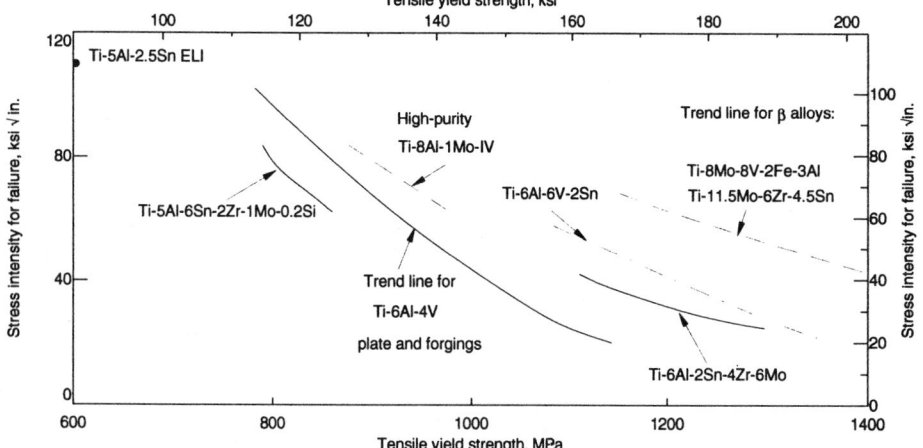

Fig. 3 Fracture toughness vs. yield strength of titanium alloys. Alloys in the annealed condition were Ti-5Al-2.5Sn and Ti-8Al-1Mo-1V, and alloys in the solution heat treated and aged (STA) condition were Ti-6Al-2Sn-4Zr-6Mo, Ti-6Al-6V-2Sn, and β alloys Ti-11.5Mo-6Zr-4.5Sn and Ti-8Mo-8V-2Fe-3Al. The Ti-6Al-4V trend line represents both the annealed and STA conditions. Source: *Titanium and Titanium Alloys*, MIL-HDBK-697A, 1974, p 44

Table 8 Fracture-toughness variations in alpha-beta titanium alloys

Condition	Yield strength, MPa	Tensile strength, MPa	Elongation, %	K_{Ic} (K_Q), MPa√m
Ti-6Al-4V				
Plate				
α + β roll + MA(a)	1096	1171	14	32
α + β roll + RA(a)	1054	1144	13	51
α + β roll + DA (954 °C/1.5 h/AC + 760 °C/1 h/AC)	882	971	14	59
BA (α + β roll + 1038 °C/20 min/AC + 732 °C/2 h/AC)	875	951	11	87
α + β roll + RA(b)	785	882	…	94
Forging				
α + β forge + RA	710	875	12	(121)
α + β forge (15% primary α) + DA	813	889	12	(123)
α + β forge (50% primary α) + DA	854	909	15	(111)
β forge	772	854	11	(119)
BQ	861	930	6	101
α + β forge + STA	875	937	15	(92)
α + β forge + STOA	903	971	16	84
Ti-6Al-2Sn-4Zr-6Mo				
Forging				
α + β forge + STA(c) (10% primary α)	1116	1213	13	34
α + β forge + STA(c) (50% primary α)	1150	1240	14	26
α + β forge + ann(d) (50% primary α)	1061	1130	13	26
β forge + STA(c)	1047	1199	7	57
Ti-6Al-2Sn-4Zr-2Mo				
Forging				
α + β forge-β ST + age	903	…	12.5	81
β forge-α + β ST + age	896	…	11.0	84

MA, mill anneal; RA, recrystallization anneal; DA, duplex anneal; AC, air cool; BQ, β forge + quench; ST, solution treatment; STA, solution treatment and aging; STOA, solution treatment and overaging. (a) Standard-oxygen (~0.20 wt%) material. (b) Lower-oxygen (~0.13 wt%) material. (c) 885 °C/1 h/AC + 593 °C/8 h/AC. (d) 704 °C/1 h/AC. Source: *Deformation Processing and Structure*, American Society for Metals, 1984, p 326

Table 9 Toughness directionality in textured Ti alloys

Alloy (grade)	R-curve toughness(a), MPa√m	
	TL orientation	LT orientation
Ti-6Al-4V (ELI)	123(b)	113(b)
Ti-6Al-4V	113	124
Ti-6Al-2Sn-4Zr-6Mo	106	124
Ti-6Al-4V plate	46	80

(a) Maximum K prior to onset of unstable fracture. (b) Δa corresponding to K_R^{max} was 5 times smaller for the LT orientation than for the TL orientation.

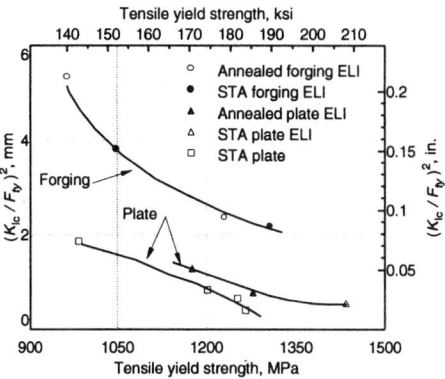

Fig. 4 Influence of yield strength on fracture toughness of Ti-6Al-6V-2Sn. ELI, extra-low interstitial; STA, solution treated and aged. F_{ty}, tensile yield strength

5 in the LT specimens compared with the TL specimens. This may account for the apparent reversal in the effect of texture on K_R in this material. One possible explanation for this apparent reversal is based on the relative amounts of twinning that occur in the crack-tip plastic zone. Both oxygen content and texture affect the extent of twinning. Because twinning occurs in response to *c*-axis loading, the extent of twinning in material of basal transverse texture should be considerably greater in TL specimens than in LT specimens. Extensive twinning would make the material more compliant, which is consistent with the *R* curves for low-oxygen material. Increasing oxygen dramatically reduces the incidence of twinning. Thus, in the low-oxygen material, crack-tip plasticity may be enhanced by twinning, with an attendant increase in toughness.

Alpha Alloys

Alpha alloys such as Ti-6Al-2Sn-4Zr-2Mo and Ti-8Al-1Mo-1V (Table 2) are used primarily in jet engine applications and are useful at temperatures above the normal range for Ti-6Al-4V. These alloys have better creep resistance than Ti-6Al-4V, and creep resistance is enhanced with a fine acicular (Widmanstätten) structure.

Generally speaking, alpha alloys contain less beta phase than Ti-6Al-4V. Age hardening treatments are thus not very effective and are, moreover, deleterious to creep resistance. These alloys therefore are usually employed in the solution annealed and stabilized condition. Solution annealing may be done at a temperature some 35 °C (63 °F) below the transformation temperature, and stabilization is commonly produced by heating for 8 h at about 590 °C (1100 °F). These alloys are more susceptible to the formation of ordered Ti₃Al (α_2) structures, which promotes SCC as described below for three common alloys in this class. However, alloy Ti-3Al-2.5V is also sometimes considered an alpha alloy, as it often contains less beta than Ti-6Al-4V. The alloy Ti-3Al-2.5V is essentially immune to SCC in boiling sea water and simulated sour-gas well brines at room temperature. Like CP titanium, Ti-3Al-2.5V is also immune to hot-salt cracking.

Ti-5Al-2.5Sn

Ti-5Al-2.5Sn (UNS R54520) is a medium-strength, all-alpha titanium alloy with very high fracture toughness. The ELI grade of Ti-5Al-2.5Sn (UNS R54521) is especially well suited for service at cryogenic temperatures and exhibits an excellent combination of strength and toughness at –250 °C (–420 °F). The standard grade is used in applications demanding good weld fabricability, oxidation resistance, and intermediate

strength at service temperatures up to 480 °C (900 °F). Typical combinations of strength and toughness of the two grades are summarized in Table 10.

Ti-5Al-2.5Sn ELI is employed for liquid hydrogen tankage and high-pressure vessels at temperatures below –195 °C (–320 °F), structural members, and turbine parts. However, the ELI grade is quite difficult to hot work into some product forms, particularly when converting from ingot to billet, because of shear cracking, often referred to as strain-induced porosity.

Because tin and aluminum promote the formation of ordered Ti₃Al (α_2) structures, Ti-5Al-2.5Sn is one of the titanium alloys most susceptible to SCC. Like step-cooled Ti-8Al-1Mo-1V, the Ti-5Al-2.5Sn alloy is susceptible to corrosion cracking in distilled water (Fig. 5). The elevated-temperature stress-corrosion resistance of this alloy in the presence of solid salt also is lower than that of other commonly used titanium alloys.

Ti-6242S

Ti-6Al-2Sn-4Zr-2Mo-0.08Si (Ti-6242S or Ti-6242Si) is one of the most creep-resistant titanium alloys and has an outstanding combination of tensile strength, creep strength, toughness, and high-temperature stability for long-term applications at temperatures up to 425 °C (800 °F). Ti-6242S is sometimes described as a near-alpha or superalpha alloy, but in its normal heat-treated condition this alloy has a structure better described as alpha-beta, as noted in Table 7 and 8. An example of thermomechanical processing

Table 10 Fracture toughness of Ti-5Al-2.5Sn

Heat treatment variable(a)	Test temperature		Stress intensity, K_{Ic}		Specimen, orientation(b) and type(c)	Yield strength(d)	
	K	°F	MPa√m	ksi√in.		MPa	ksi
Standard grade							
Air cooled	295	72	71.4	65	LT-CT	876	127
	77	–320	53.8	49	LT-CT	1338	194
	20	–423	51.6	47	LT-B	1482	215
	20	–423	50.5	46	LS-B	...	...
Furnace cooled	295	72	65.9	60	LT-CT	882	128
	77	–320	57.1	52	LT-CT	1379	200
	20	–423	47.2	43	LT-B	1517	220
	20	–423	52.7	48	LS-B	...	...
ELI grade							
Air cooled	295	72	118.7	108(e)	LT-CT	703	102
	77	–320	111.0	101	LT-CT	1179	171
	20	–423	91.2	83	LT-B	1303	189
	20	–423	106.6	97	LS-B	...	...
Furnace cooled	295	72	115.4	105(e)	LT-CT	682	99
	77	–320	82.4	75	LT-CT	1179	171
	20	–423	68.1	62	LT-B	1303	189
	20	–423	80.2	73	LS-B	...	...

(a) Air cooled or furnace cooled from annealing temperature. (b) Orientation notation per ASTM E 399-74. (c) CT, compact tension specimen; B, bend specimen. (d) 0.2% offset. (e) Invalid toughness values (not 100% plane-strain conditions). Source: *Metals Handbook*, 9th ed., Vol 3, American Society for Metals, 1980, p 384

Table 11 Fracture toughness of Ti-6242 forgings from various thermomechanical processing (TMP) routes

Property	TMP route 1(a)	TMP route 2(b)	TMP route 3(c)
Tensile yield strength, MPa (ksi)	937 (136)	903 (131)	917 (133)
Ultimate tensile strength, MPa (ksi)	979 (142)	986 (143)	1006 (146)
Elongation (in 4D), %	16	12	12
Reduction of area, %	34	24	26
Fracture toughness, MPa√m (ksi √in.)	56 (51)	78 (71)	76 (69)
Creep at 510 °C, 241 MPa (950 °F, 35 ksi), h to 0.1%	117	447	430
Fatigue crack growth rate ($da/dN \times 10^{-6}$ in./cycle at ΔK 10 ksi√in.)	1.2	0.7	0.8

(a) TMP route 1 (baseline): α-β hot die forged plus sub-β transus duplex heat treatment. (b) TMP route 2: β hot die forged plus duplex heat treatment. (c) TMP route 3: β hot die forged plus direct age. Source: G. Kuhlman, T. Yu, A. Chakrabarti, and R. Pishko, Mechanical Property Tailoring Titanium Alloys for Jet Engine Applications, *Titanium 1986: Products and Applications*, Titanium Development Association, 1987, p 122

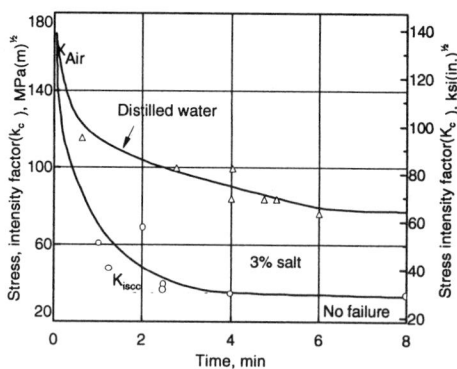

Fig. 5 Ti-5Al-2.5Sn stress-corrosion cracking in distilled water. Source: T.L. Mackay and C.B. Gilpin, "Stress Corrosion Cracking of Titanium Alloys at Ambient Temperature in Aqueous Solutions," Missile & Space Systems Division, Astropower Laboratory, Douglas Aircraft Company, Report SM-49105-F1, June 1967

(TMP) effects or the strength-toughness combination is given in Table 11.

Stress-Corrosion Cracking. Under stress, Ti-6Al-2Sn-4Zr-2Mo has been shown to be subject to SCC at room temperature in the presence of aqueous chloride solution and a preexisting crack (the so-called accelerated crack-growth type of salt-stress corrosion) and at elevated temperatures in the presence of a halogen salt (e.g., NaCl). The SCC susceptibility of Ti-6242 in hot salt appears to be less than that of Ti-8Al-1Mo-1V and Ti-6Al-4V. In ambient salt solution, the SCC threshold of Ti-6242 in the STA condition is comparable to mill-annealed Ti-8Al-1Mo-1V (Table 12).

A substantial amount of laboratory testing on the hot-salt SCC behavior of Ti-6242 has been performed, but like other susceptible alloys (such as Ti-8Al-1Mo-1V), no failure in service has been attributed to hot-salt cracking. The likely reason for this is the critical relationship between stress and environment, which needs to peak simultaneously for extended periods of time for cracking to occur. Actual engine environmental conditions are unique and are considered less conducive to stress corrosion than laboratory exposure conditions. Possible ameliorating engine conditions are high air velocities, high pressures, salt-air conditions, oil contamination, and/or unique operating cycles.

Hot-salt SCC behavior of Ti-6242 in laboratory tests is influenced by several factors. Oxygen is necessary for hot-salt cracking to occur. At least one study has shown that cracking will not occur in more susceptible Ti-5Al-2.5Sn when the environmental pressure is reduced below 10 μm. Although the role of water (moisture) has not been clearly established, it appears that water is also a necessary environmental component in the cracking process.

Effect of Processing. There appears to be little difference between mill annealing and duplex annealing of Ti-6242 bar. Heavy forging can reduce the embrittlement and cracking thresholds. The cracking thresholds reported for sheet at 100 h exposure are generally lower than those reported for bar. Triplex annealing may improve cracking slightly more than a duplex anneal.

Ti-8Al-1Mo-1V

Ti-8Al-1Mo-1V (Ti-811) can be characterized as a near-alpha alloy with several alpha alloy characteristics, such as good creep strength and weldability. However, the alloy does have alpha-beta characteristics, such as a mild degree of hardenability. Ti-811 is generally used in the annealed condition, where lamellar alpha morphology from transformed beta is produced by duplex annealing for toughness (Table 13). An acicular alpha from beta fabrication or heat treatment improves fracture toughness in air but has less effect on fracture toughness in salt water (Table 13). Triplex annealing produces enhanced creep resistance.

Ti-811 is used for airframe and turbine engine applications demanding short-term strength, long-term creep resistance, thermal stability, and stiffness. Ti-811 is predominantly an engine alloy and is available in three grades, including a "premium grade" (triple melted) and a "rotating grade," for use in rotating engine components.

Like the alpha-beta alloys, Ti-811 is susceptible to hydrogen embrittlement in hydrogenating solutions at room temperature, in air or in reducing atmospheres at elevated temperatures, and even in pressurized hydrogen at cryogenic temperatures. Ti-811 is susceptible to SCC in hot salts (especially chlorides) and to accelerated crack propagation in aqueous solutions at ambient temperatures. The environment in which this alloy is to be used should be selected carefully to prevent material degradation.

Stress-Corrosion Cracking. Ti-8Al-1Mo-1V is one of the most susceptible titanium alloys to SCC which stems from the increased tendency to form the highly ordered Ti3Al (α2) phase when aluminum content exceeds 5 wt%. This low-ductility ordered phase forms in the 400 to 700 °C (750 to 1290 °F) temperature range and increases

Table 12 Comparative toughness of titanium alloys in air and stress-corrosion threshold in 3.5% NaCl solution at 25 °C

Alloy	Thickness		Heat treatment(a)	Tensile yield strength		K_{Ic} or K_c		K_{Iscc} or K_{scc}	
	mm	in.		MPa	ksi	MPa√m	ksi√in.	MPa√m	ksi√in.
Ti-6Al-2Sn-4Zr-2Mo	13	0.50	STA	1048	152	58	53	29	27
Ti-6Al-2Sn-4Zr-6Mo	13	0.50	MA	1103	160	60	55	22	20
			DA	1034	150	88	80	49	45
	7	0.30	MA	965	140	57	52	28	26
			β STA	1172	170	89	81	49	45
Ti-8Al-1Mo-1V	1.3	0.05	MA	999	145	82	75	33	30
	1.3	0.05	DA	930	135	176	160	55	50
	13	0.5	MA	999	145	52	48	22	20
	13	0.5	DA	930	135	110	100	35	32
	13	0.5	MA, WQ	841	122	>110	>100	46	42
	13	0.5	βST, WQ	868	126	>110	>100	>110	>100
Ti-6Al-4V (standard grade)	13	0.50	MA	944	137	66	60	38	35
			DA	917	133	77	70	57	52
			STA	1103	160	51	47	27	25
			βSTA	1068	155	77	70	49	45

Note: The data were generated in ambient neutral 3.5% NaCl solution. It should be cautioned that these K_{Iscc} values are highly dependent on alloy composition, metallurgical condition, and product form and thickness. Therefore, they may or may not be representative of alloy product materials commercially available. (a) STA, solution treat and age; MA, mill anneal; DA, duplex anneal; WQ, water quench; ST, solution treat. Source: R. Schutz, Stress-Corrosion Cracking of Titanium Alloys, *Stress-Corrosion Cracking: Materials Performance and Evaluation*, ASM International, 1992

Table 13 Typical toughness of Ti-8Al-1Mo-1V at room temperature

Heat treatment	Yield strength		K_{Ic} or K_c		Aqueous solution K_{Iscc} or K_{scc}	
	MPa	ksi	MPa√m	ksi√in.	MPa√m	ksi√in.
1.3 mm (0.05 in.) plate						
Mill annealed	999	145	82	75	33	30
Duplex annealed	930	135	176	160	55	50
13 mm (0.50 in.) plate						
Mill annealed	999	145	52	48	22	20
Duplex annealed	930	135	110	100	35	32
Mill annealed, WQ	841	122	>110	>100	46	42
βST, WQ	868	126	>110	>100	>110	>100

WQ, water quench; ST, solution treated. Source: R.W. Schutz, Stress Corrosion of Titanium Alloys, *Stress Corrosion*, ASM International, 1992

crack velocity and decreases K_{Iscc} as volume fraction increases. Oxygen levels above 0.20 to 0.25 wt% also promote SCC, again due to a transition from wavy to planar slip. The SCC fracture mode in Ti-811 and other susceptible alpha alloys is transgranular, with the fracture path highly oriented in unidirectionally processed (textured) material and more random in martensitic structures.

Ti-811 generally remains susceptible to SCC whether the structure is equiaxed alpha or martensitic (quenched from the beta phase field). In general, SCC susceptibility of equiaxed alpha alloys diminishes with decreasing grain size, whereas step-cooled Ti-811 has the greatest susceptibility in that smooth specimens will produce cracking in ambient neutral salt solutions. In contrast, mill-annealed Ti-811 has susceptibility in the presence of stress risers, such as a fatigue crack or machined notch. Step-cooled Ti-811 also exhibits SCC in distilled water.

Effect of Halide Solutions. Addition of halide ions such as Cl^-, Br^-, and I^- increases the SCC susceptibility of Ti-811 and can induce susceptibility in other conditions that are immune to SCC in distilled water. Other ionic species in solution can have a neutral or even an inhibitive effect on

Ti-811 SCC if the alloy is not highly susceptible. These species include NO_3^-, SO_4^{-2}, F^-, OH^-, CrO_4^{-2}, and PO_4^{-3}. The inhibitive influence diminishes as halide levels increase. Little influence of nonoxidizing cations such as Na^+, K^+, or Li^+ is noted. However, oxidizing cations such as Cu^{+2} may raise K_{Iscc} values, depending on alloy heat treatment.

K_{Iscc} Behavior. Because titanium alloys exhibit no stage I type crack growth in neutral solutions (i.e., the slowest crack velocities measured are approximately 10^{-3} cm/s), it may be concluded that true K_{Iscc} thresholds exist below which cracks will not propagate. This is not the case in highly acidic solutions, where both stage I and II cracking behavior is observed. Increasing acidity generally reduces K_{Iscc} and increases stage II cracking velocity. Increasing alkalinity appears to have no obvious or significant effect on SCC behavior relative to neutral conditions. As hydroxide concentrations exceed $1M$, increasing inhibition may be expected.

Effect of Potential. In most SCC-susceptible titanium alloys, anodic or cathodic polarization tends to inhibit SCC and increase K_{Iscc} values. However, the anodic and cathodic polarization inhibition phenomenon is not as apparent in

highly susceptible alloys, such as step-cooled Ti-811. In highly acidic solutions, however, stage II crack velocity is independent of applied potential. As a result, inhibition via cathodic polarization is not achievable in highly acidic solutions.

Cracking. Ti-811 is one of the least resistant titanium alloys to hot-salt cracking. Chloride, bromide, and iodide salts have all been shown to produce cracking; fluoride and hydroxide salts have not. The cation associated with the salt has also been reported to affect cracking susceptibility. The severity of attack has been shown to increase as follows:

$$MgCl_2 > SrCl_2 > CsCl > CaCl_2$$
$$> KCl > BaCl_2 > NaCl > LiCl$$

Cracking is normally intergranular in nature, but it depends largely on alloy type. Alpha alloys exhibit both transgranular and intergranular fracture, depending on whether the material was annealed above or below the beta transus, respectively. Alpha-beta alloys exhibit predominantly intergranular fracture.

From a practical standpoint, hot-salt cracking appears to be restricted to the laboratory. No in-service failure has been attributed to hot-salt cracking.

Beta Alloys

An alloy is considered to be a beta alloy if it contains sufficient beta stabilizer alloying element to retain the beta phase without transformation to martensite on quenching to room temperature. A number of titanium alloys (Table 2) contain more than this minimum amount of beta stabilizer alloy addition. In a strict sense there is no truly stable beta alloy because, even the most highly alloyed beta will, on holding at elevated temperatures, begin to precipitate omega, alpha, Ti_3Al, or silicides, depending on temperature, time, and alloy composition. All beta alloys contain a small amount of aluminum, an alpha stabilizer, in order to strengthen alpha that may be present after heat treating. The composition of the precipitating alpha is not constant and will depend on the temperature of heat treatment. The higher the temperature in the alpha-beta phase field, the higher the aluminum content of alpha will be.

The beta- and beta-rich alpha-beta alloys offer the opportunity to tailor the strength-toughness properties combinations to a specific application. That is, moderate strength with high toughness or high strength with moderate toughness can be achieved. This is generally not possible for other types of titanium alloys because they cannot be heat treated over a very wide range. At moderate strength levels, say 965 MPa (140 ksi) and above, the fracture toughness of the beta alloys can be processed to achieve higher values than for the other types (alpha and alpha-beta alloys).

To accomplish these higher toughnesses, however, the processing window is tighter than that normally used for the other alloy types. For the less highly beta-stabilized alloys, such as Ti-10V-2Fe-3Al for example, the thermomechanical process is critical to the properties combinations achieved as this has a strong influence on the final microstructure and the resultant tensile strength and fracture toughnesses that may be achieved.

This is somewhat less important in the more highly beta-stabilized alloys, such as Ti-3Al-8V-6Cr-4Mo-4Zr (Beta C) and Ti-15V-3Cr-3Al-3Sn. In these the final microstructure, precipitated alpha, is so fine that microstructural manipulation through thermomechanical processing is not as effective. In these cases the aging heat treatments, sequence and temperature, are more critical. The key is to obtain a uniform precipitation. This may be obtained by a low-high aging sequence or, with residual cold or warm work, possibly a high-low aging sequence. When highly alloyed beta alloys, such as Beta C, are cold worked prior to aging, high strength can be obtained with good ductility because cold work induces finer and more uniform precipitation.

The thermomechanical processing must, however, be controlled to provide a uniform microstructure throughout the cross section of the material and, in conjunction with the heat treatment, avoid the occurrence of extensive grain boundary alpha or a precipitate-free zone near the grain boundaries.

Solute-Lean Beta Alloys

Solute-lean beta alloys are sometimes classified as beta-rich alpha-beta alloys, and this class includes Ti-10V-2Fe-3Al and proprietary alloys such as Ti-17 (Ti-5Al-2Sn-2Zr-4Mo-4Cr) and Beta CEZ (Ti-5Al-2Sn-4Zr-4Mo-2Cr). As noted above, toughness depends on TMP. For example, TMP of Ti-10V-2Fe-3Al achieves the desired final microstructure through manipulation of alpha phase morphology. Microstructural objectives range from fully transformed, aged beta structures to controlled amounts of elongated primary alpha in an aged beta matrix, characterized by extremely fine secondary (aged) alpha. The latter microstructure is preferred for most aerospace

Table 14 Fracture toughness of Ti-10V-2Fe-3Al forgings

Forgings	Ultimate tensile strength		Tensile yield strength		Elongation, %	Plane-strain fracture toughness	
	MPa	ksi	MPa	ksi		MPa√m	ksi√in.
High strength condition							
Isothermal forgings	1300-1380	188-200	1200-1255	174-182	3-6	29	26
Conventional forgings	1230-1350	178-196	1145-1280	166-186	4-10	44-60	40-54
Pancake forgings	1275-1310	185-190	1150-1160	167-168	5-8	47	43
Extrusions	1240	179	1170	169	4	...	...
Reduced strength condition							
Isothermal forgings	1060-1100	153-159	985-1060	143-153	8-12	70	64
Pancake forgings	965	140	930	135	16	100	91
Extrusions	1110-1170	161-169	1000-1105	145-160	6-7	45-48	41-44
AMS specification (forgings)							
AMS 4984	1190	173	1100	160	4 (in 4D)	44	40
AMS 4986	1100	160	1000	145	6 (in 4D)	60	55
AMS 4987	965	140	895	130	8 (in 4D)	88	80

Source: R. Boyer, D. Eylon, and F. Froes, Comparative Evaluation of Ti-10V-2Fe-3Al Cast, P/M, and Wrought Product Forms, *Titanium Science and Technology*, Vol 2, G. Lütjering, U. Zwicker, and W. Bunk, Ed., Deutsche Gesellschaft für Metallkunde e.V., Germany, 1985, p 1307

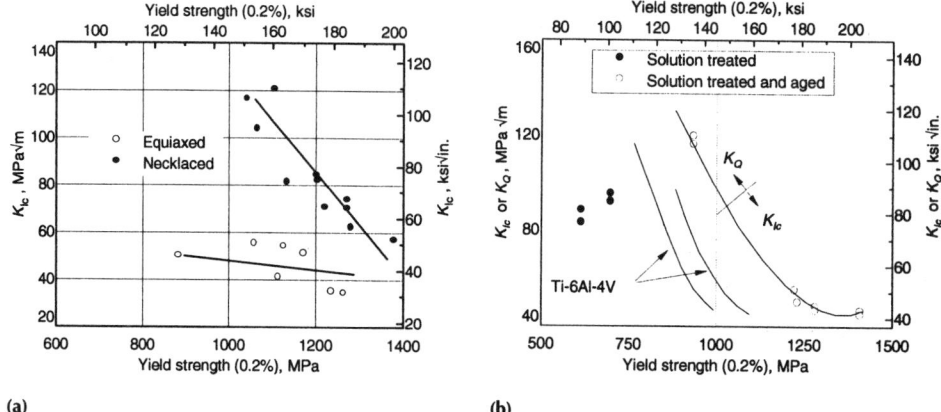

Fig. 6 Yield strength vs. toughness of two solute-lean beta titanium alloys. (a) Beta-CEZ: Fracture toughness vs. yield strength comparison. Specimens were 70 mm (2.7 in.) diam α + β rolled bar (equiaxed structure) and 80 mm (3.1 in.) diam "through the β transus" forged bar (necklaced structure). (b) Ti-10V-2Fe-3Al fracture toughness vs. yield strength. Source: *Materials Properties Handbook: Titanium Alloys*, ASM International, 1993

applications (specifications) and forms the basis for most commercial use of the alloy in forgings. There would also appear to be an optimum amount of primary alpha to achieve maximum toughness (a 10% volume fraction of elongated primary alpha has significantly higher fracture toughness than 30 vol%).

Ti-10V-2Fe-3Al is a deep-hardening, metastable, near-beta alloy that may be thermomechanically processed to a range of strength levels combined with excellent fracture toughness (Table 14). The preferred forging process to meet the above mechanical-property criteria is controlled beta forging followed by controlled alpha-beta forging. This, in combination with final thermal treatment, provides the optimum combination of strength, ductility, toughness, fatigue, and fracture-related properties. Ti-10V-2Fe-3Al may be conventionally alpha-beta forged and thermally treated. With such conventional processes, the alloy achieves high strength and fatigue properties and superior ductility, but

poor toughness and fracture-related properties. The stress-corrosion threshold has been reported to be at least 80% of K_{Ic} except when it is stressed in the short transverse direction, where it is 70% of K_{Ic}.

Ti-5Al-2Sn-4Zr-4Mo-2Cr (Beta CEZ) is another beta-rich alpha-beta alloy with toughness enhanced by TMP. Equiaxed microstructures are characterized by toughness ranging from 45 to 55 MPa√m (40 to 50 ksi√in.). Toughness of the lamellar structure ranges from 60 to 90 MPa√m (54 to 82 ksi√in.), whereas the necklaced microstructure has a toughness ranging from 65 to 95 MPa√m (59 to 86 ksi√in.) (see Fig. 6). Low-temperature toughness usually ranges from 30 to 45 MPa√m (27 to 41 ksi√in.) at –253 °C (–423 °F).

Solute-Rich Beta Alloys

Highly beta-stabilized alloys, such as Ti-3Al-8V-6Cr-4Mo-4Zr (Beta C) and Ti-15V-3Cr-3Al,

Table 15 Fracture toughness of Beta C solution treated and aged billet

Treatment	Test direction	Fracture toughness (K_{Ic})		Ultimate tensile strength		Tensile yield strength (2 % offset)		Elongation, %	Reduction of area, %
		MPa√m	ksi√in.	MPa	ksi	MPa	ksi		
Water quench	L	96.7	88.0	1189	172.5	1125	163.2	9.5	19.6
	T	62	56.4	1188	172.4	1145	166.0	3.0	5.6
Air cool	L	89.9	81.8	1208	175.2	1150	166.8	9.2	17.4
	T	63.3	57.6	1242	180.2	1184	171.7	3.5	6.6

Note: Specimens were 150 mm (6 in.) billet solution treated 815 °C (1500 °F), 15 min, water quenched and air cooled, then aged 12 h at 565 °C (1050 °F) and air cooled. Source: RMI Co., reported in *Industrial Applications of Titanium and Zirconium: Fourth Volume*, C.S. Young and J.C. Durham, Ed., STP 917, ASTM, 1986, p 155

Table 16 Fracture toughness of Ti-15-3 solution treated and aged plate

Heat treatment	Orientation	Tensile yield strength		Ultimate tensile strength		Elongation, %	Fracture toughness (K_{Ic})	
		MPa	ksi	MPa	ksi		MPa√m	ksi√in.
800 °C (1470 °F), 20 min, AC	L-T	1253	182	1376	199	6.2	44.3	40.3
480 °C (895 °F), 14 h, AC	T-L	1304	189	1421	206	6.6	46.8	42.6
800 °C (1470 °F), 20 min, AC	L-T	1213	176	1337	194	7.8	42.1	38.3
510 °C (950 °F), 14 h, AC	T-L	1263	183	1382	200	6.9	43.4	39.5

AC, air cooled. Note: Hot rolled plate had a chemical composition (wt%) of 3.37 Al, 0.004 C, 3.36 Cr, 0.17 Fe, 0.0061 H, 0.0080 N, 0.14 O, 3.04 Sn, and 15.10 V. It was solution treated at 800 °C (1470 °F) for 20 min, air cooled, then aged at 510 °C (950 °F) for 8 or 14 h. Source: C. Ouchi, H. Suenaga, H. Sakuyama, and H. Takatori, Effects of Thermomechanical Processing Variables on Mechanical Properties of Ti-15V-3Cr-3Sn-3Al Alloy Plate, *Designing with Titanium*, 1986, p 130

obtain a good combination of strength and toughness in the aged condition (Tables 15 and 16). Trend lines for some other near-beta alloys are compared in Fig. 3 with other types of titanium alloys.

Hydrogen Effects. Beta alloys can absorb large quantities of hydrogen, on the order of 4000-5000 ppm, but hydrides do not form in beta despite the high degree of supersaturation. However, when alpha is present, hydrides can form at alpha-beta interfaces at relatively low hydrogen contents at room temperature. Hydrogen tolerance depends on the particular alloy and condition, and considerable work remains to be done to understand the behavior of hydrogen in both alpha-beta and beta alloys.

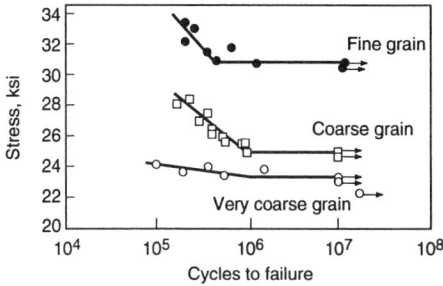

(a)

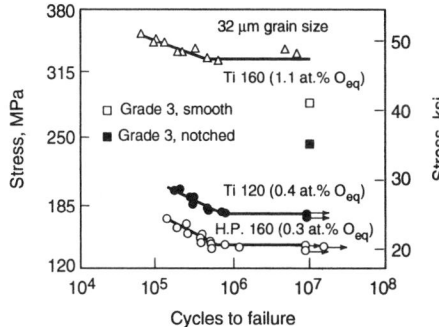

(b)

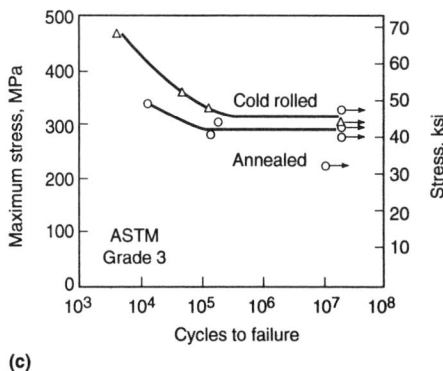

(c)

Fig. 7 S-N curves ($R = -1$) in unalloyed titanium. (a) Effect of grain size. (b) Effect of oxygen content. (c) Effect of cold work. HP-Ti is high-purity titanium with oxygen equivalent (O_{eq}) defined in Ref 4. Source: Ref 4 and *Metals Handbook*, Vol 3, 1980, p 376

Fatigue Life Behavior

L. Wagner, Technical University of Brandenburg at Cottbus, Germany

The fatigue life (N_F) of structural parts is the sum of crack nucleation life (N_I) and crack propagation life to final fracture (N_P): $N_F = N_I + N_P$. Fatigue tests on titanium alloys have shown that at high stress or strain amplitudes, the ratio N_I/N_F can be as small as 0.01 (Ref 1, 2). Therefore, the resistance to crack propagation determines fatigue life in the low-cycle fatigue (LCF) regime. With regard to crack propagation in structural parts, it is useful to distinguish between small surface cracks (microcracks) and long through-cracks (macrocracks), because the dependence of crack growth on microstructural parameters such as grain size or phase dimensions can be contradictory (Ref 3). As a rule of thumb, the fatigue life for small, highly stressed components (small critical flaw size) is mainly controlled by microcrack growth, while for large components operating at low stress levels (large critical flaw size), macrocrack growth behavior is more im-

portant. The purpose of this section is to outline the current understanding of the effects of the main variables (i.e., microstructure and texture) on the fatigue life of titanium and its alloys. As opposed to steels or aluminum alloys, titanium alloys are generally free of defects such as inclusions or pores. However, fatigue behavior in titanium alloys is very sensitive to surface preparation. Thus, this section also deals with the effect of mechanical surface treatments on fatigue life. Macrocrack growth behavior in titanium alloys is treated in the next section, "Fatigue Crack Growth," while this section is limited to fatigue crack nucleation and microcrack propagation.

Effect of Microstructure

Unalloyed Titanium. Fatigue life in unalloyed titanium depends on grain size, interstitial

contents, and degree of cold work, as illustrated in Fig. 7. Decreasing the grain size from 110 to 6 μm improves the fatigue limit (10^7 cycles) in commercially pure titanium from 180 to 240 MPa (26 to 35 ksi) (Fig. 7a) (Ref 4). The effect of oxygen content on the *S-N* curves of high-purity titanium, shown in Fig. 7(b), correlates to the increase in yield stress. Cold work also increases yield stress and thus improves the fatigue performance as shown in Fig. 7(c). Fatigue limits of unalloyed titanium depend on interstitial contents, but the ratio of fatigue limits and yield strength appears relatively constant at room temperature (Ref 5). The ratio of fatigue limit and yield stress does show a temperature dependence (Ref 4). Temperature effects on a low-iron grade of ASTM grade 2 are shown in Fig. 8.

Full Alpha Experimental Alloy. Fatigue life of an alpha alloy is mainly affected by grain size, degree of age hardening, and oxygen content of

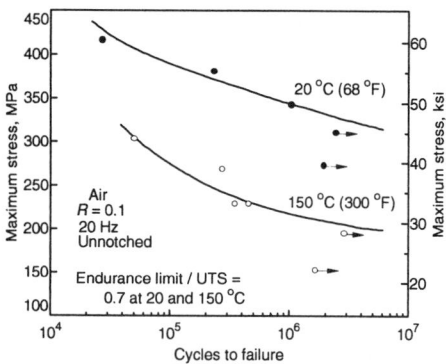

Fig. 8 Fatigue at 150 °C for grade 2 titanium with low iron (0.03 wt% or less). Source: P. Russo, ACHEMA 82, 1982

Table 17 Tensile properties in Ti-6Al-4V (aged 24 h at 500 °C)

Microstructure	Width of α lamellae, μm	0.2% yield strength, MPa	RA, %
Fully lamellar	0.5	1040	16
	1	980	18
	10	930	14
Fully equiaxed	2	1120	46
	6	1070	44
	12	1060	43
Duplex	0.5	1045	39
(10 μm αp size)	1	975	34

α_p, primary alpha. RA, reduction in area. Source: Ref 10-12

the alloy. Fatigue cracks in an experimental Ti-8.5Al binary alloy usually nucleate along planar slip bands within the alpha grains (Ref 6). For a given ΔK, microcracks in the coarse-grained material grow faster than in its fine-grained counterpart. This is related to the higher density of grain boundaries in the fine-grained material, because grain boundaries act as strong obstacles to crack growth in this material. The superior performance of the fine-grained material with respect to microcrack growth is in contrast to macrocrack results (Ref 7) showing better ranking of the coarse-grained material. The latter is mainly caused by crack front geometry effects (Ref 8) and crack closure (Ref 7, 9), which are not significant for microcracks but can be pronounced for macrocracks, particularly in coarse-grained material. Both crack front geometry effects and crack closure slow the growth rate of macrocracks. Age-hardening results in faster microcrack growth in Ti-8.5Al, presumably due to the marked loss in tensile ductility.

Near-Alpha and Alpha-Beta Alloys. In addition to alpha grain size, degree of age hardening, and oxygen content, the fatigue properties of the two-phase near-alpha and alpha-beta alloys are strongly affected by the morphology and arrangement of both alpha and beta phases. Basically, fully lamellar, fully equiaxed, and duplex (primary, alpha phase in a lamellar matrix) microstructures can be developed in near-alpha and alpha-beta alloys. Important microstructural pa-

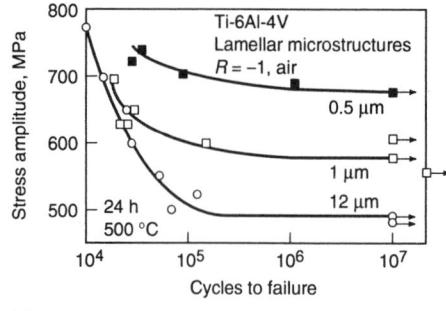

(a)

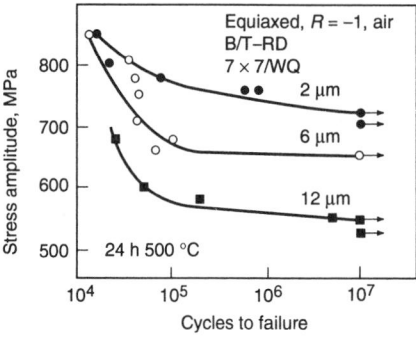

(b)

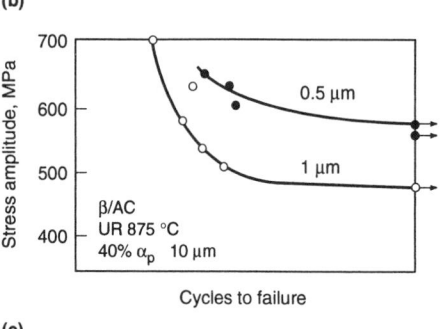

(c)

Fig. 9 S-N curves (R = −1) in Ti-6Al-4V. B/T-RD, basal/transverse texture, rolling direction; WQ, water quench. (a) Fully lamellar microstructure. Effect of width of α lamellae. (b) Fully equiaxed microstructure. Effect of α grain size. (c) Duplex microstructure. Effect of width of a lamellae. Source: Ref 10, 12

rameters are the prior beta grain size or colony size of the alpha and beta lamellae and the width of the alpha lamellae in fully lamellar microstructures. Additional parameters for duplex structures are grain size and volume fraction of the primary alpha (α_p) phase. Tensile properties of fully lamellar, fully equiaxed, and duplex microstructures of Ti-6Al-4V are compared in Table 17.

The fatigue life of these microstructures is shown in Fig. 9. A decrease in the width of the alpha lamellae from 10 to 0.5 μm in fully lamellar microstructures improves the fatigue strength from 480 to 650 MPa (Fig. 9a) (Ref 10). Similarly, a decrease in the alpha grain size from 12 to 2 μm in fully equiaxed structures increases the fatigue strength from 550 to 720 MPa (Fig. 9b) (Ref 10). A decrease in the width of the alpha lamellae from 0.5 to 1 μm in duplex microstructures increases the fatigue strength from 480 to 575 MPa (Fig. 9c) (Ref 12).

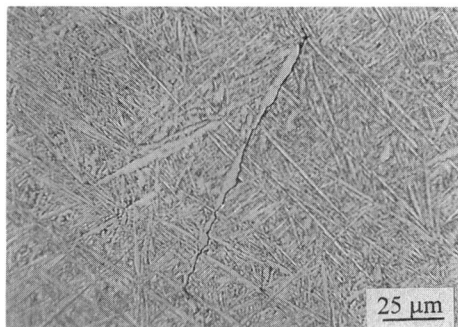

(a)

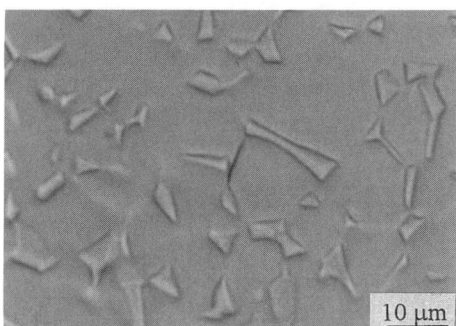

(b)

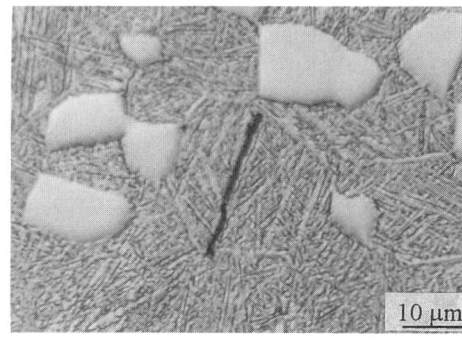

(c)

Fig. 10 Fatigue crack nucleation sites in Ti-6Al-4V. (a) Fully lamellar microstructure. (b) Fully equiaxed microstructure. (c) Duplex microstructure. Source: Ref 13

Typical crack nucleation sites are shown in Fig. 10 (Ref 13). Fatigue cracks in fully lamellar microstructures nucleate at slip bands within the alpha lamellae (Fig. 10) or at alpha zones along prior beta grain boundaries (Ref 14). Because both resistance to dislocation motion and resistance to fatigue crack nucleation depend on the width of the alpha lamellae, there is a direct correlation between fatigue strength and yield stress (Table 17). Fatigue cracks in the fully equiaxed microstructure nucleate at slip bands within the alpha grains (Fig. 10b). Thus, the fatigue strength correlates to the alpha grain size dependence of the yield stress (Table 17). Fatigue cracks in duplex microstructures can nucleate in the lamellar matrix, at the interface between lamellar matrix and the α_p phase, or within the α_p phase. The exact nucleation site depends on cooling rate (Ref 14), volume fraction, and size of the

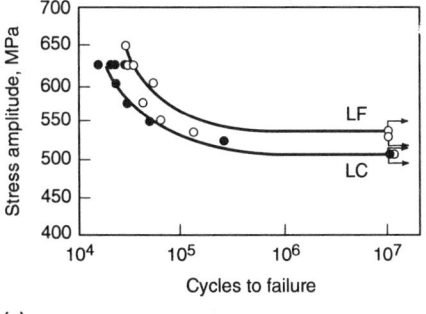

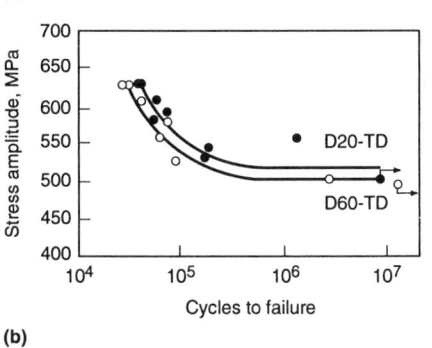

Fig. 11 S-N curves ($R = -1$) in TIMETAL 1100. (a) Fully lamellar microstructures. Effect of prior β grain size. (b) Duplex microstructures. Effect of α_p content. Source: Ref 20, 21

Table 18 Tensile properties of TIMETAL 1100 (aged 8 h at 650 °C)

Prior grain size, μm	0.2 % yield strength, MPa	RA, %
Fully lamellar (0.5 μm width of α lamellae)		
500	955	10
160	940	22
Duplex (15 μm α_p size, 1 μm width of α lamellae)		
20	965	27
60	955	30

α_p, primary alpha phase. Source: Ref 20, 21

α_p phase (Ref 15, 16). An example of crack nucleation within lamellar regions is shown in Fig. 10(c).

Owing to the high silicon content (0.45 wt%) in the near-alpha alloy TIMETAL 1100 and the resulting high silicide solvus temperature, T_{SS}, fine prior beta grain sizes and correspondingly small colony sizes can be realized in TIMETAL 1100 by beta annealing the material below T_{SS} (Ref 17-19). Tensile properties of fine- and coarse-grained fully lamellar microstructures and those of duplex structures with two different α_p volume fractions are compared in Table 18.

Decreasing the prior beta grain size in fully lamellar structures (Table 18) significantly increases tensile ductility. No marked differences in ductility were observed in duplex structures by varying α_p volume fractions between 20 and 60%. Note that lamellar and duplex structures have similar yield stresses due to the same cooling rate used from the solution anneal. S-N curves

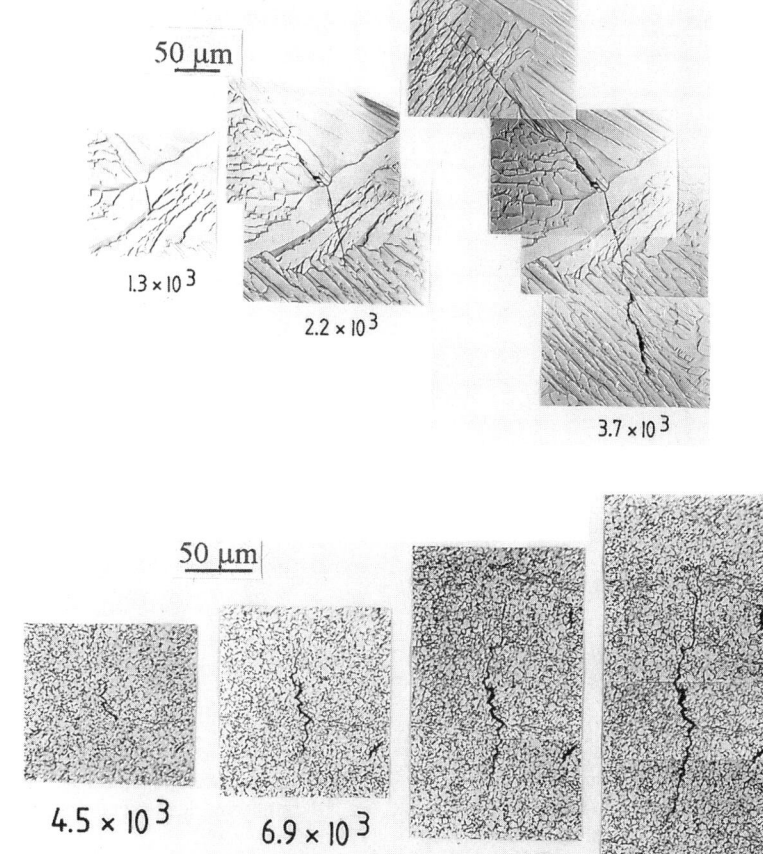

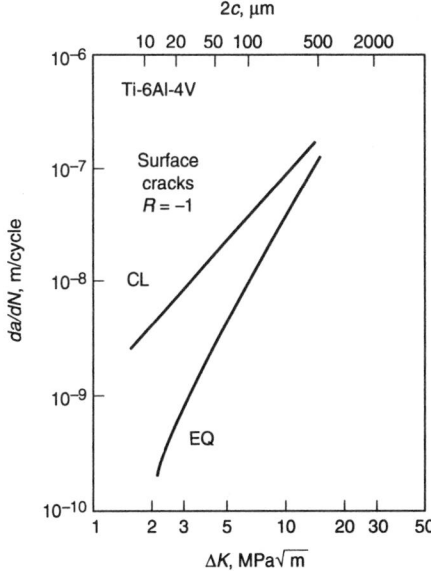

Fig. 12 Microcrack growth in Ti-6Al-4V. (a) Coarse lamellar. Stress amplitude = 775 MPa. (b) Equiaxed. Source: Ref 8

of TIMETAL 1100 are given in Fig. 11. Reducing the prior beta grain size in fully lamellar microstructure (Fig. 11a) as well as decreasing the α_p volume fraction in duplex structures (Fig. 11b) improves fatigue life in both the low-cycle fatigue (LCF) and high-cycle fatigue (HCF) (Ref 20, 21).

Typical growth behavior of microcracks in two extreme microstructures of Ti-6Al-4V, a coarse lamellar and an equiaxed structure is shown in Fig. 12 (Ref 8). Plotting da/dN versus ΔK (Fig. 13), the crack growth resistance in the coarse lamellar microstructure is clearly seen to be inferior to the equiaxed. Similar results are reported on the beta eutectoid alloy Ti-2.5Cu (Ref 22). This is mainly related to the comparably low density of phase boundaries in the coarse lamellar microstructure that is also reflected in the low tensile ductility (Table 17). Crack growth studies on fine lamellar and duplex structures in Ti-6Al-4V have shown that their da/dN-ΔK curves lie in between coarse lamellar and equiaxed structures (Ref 8).

Reducing the prior beta grain size in lamellar microstructures of TIMETAL 1100 (Fig. 14a) and decreasing the α_p volume fraction of duplex

Fig. 13 da/dN-ΔK curves of microcracks in Ti-6Al-4V. CL, coarse lamellar; EQ, equiaxed. Source: Ref 8

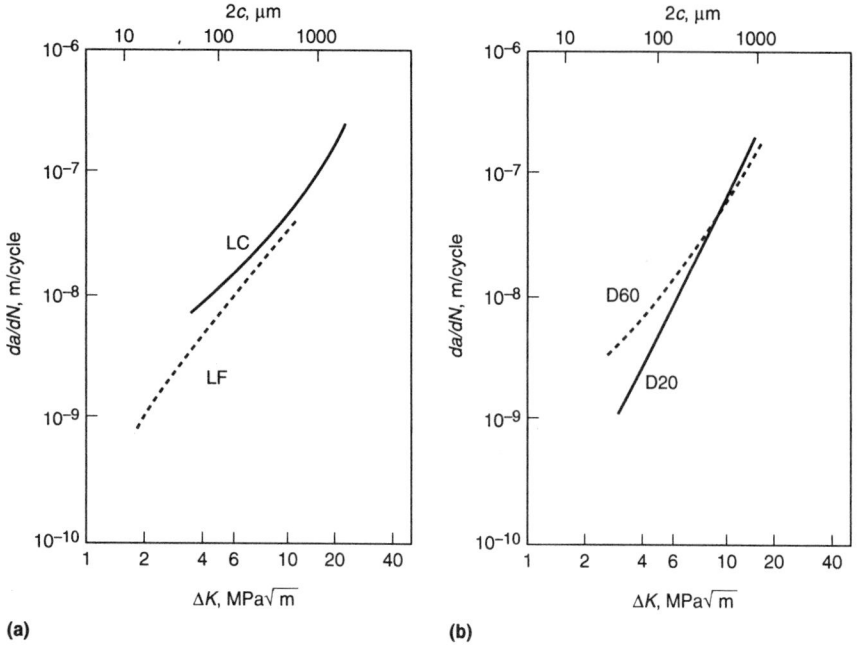

Fig. 14 da/dN-ΔK curves of microcracks in TIMETAL 1100. (a) Fully lamellar microstructure. Effect of prior β grain size. (b) Duplex microstructure. Effect of α_p content. Source: Ref 20, 21

Table 19 Tensile properties in Beta-C

Treatment	0.2% yield strength, MPa	RA, %
As SHT	850	62
16 h at 540 °C	1085	23
4 h at 440 °C, 16 h at 560 °C	1085	24

SHT, solution heat treated. Source: Ref 24

structures (Fig. 14b) improves crack growth resistance and thus fatigue life in the LCF regime (Ref 20, 21). The latter seems to be related to the absence or less frequent occurrence of α_p clusters at low α_p volume fractions. These α_p clusters in duplex structures with high α_p volume fractions can act as large single grains through which microcracks can easily propagate.

Beta Alloys. Depending on the alloy class (solute-rich or solute-lean) the following microstructural parameters are important in determining fatigue life: beta grain size, degree of age hardening, and precipitate-free regions in the solute-rich alloys such as Beta C. In addition, grain boundary α, α_{GB}, α_p size and volume fraction are important in solute-lean alloys such as Ti-10-2-3 (Ref 23).

Typical microstructures in Beta C are shown in Fig. 15 comparing a conventionally aged (Fig. 12a) and a two-step aged (Fig. 12b) condition (Ref 24). Compared to conventional aging, two-step aging results in a more homogeneous precipitation of alpha particles in the beta grain interiors. Tensile properties of these structures are listed in Table 19. The two-step aging of 4 h at 440 °C plus 16 h at 560 °C was chosen to enable a comparison of fatigue behavior with the conventional aging cycle (16 h at 540 °C) on the basis of identical yield stresses.

The *S-N* curves of the various conditions of Beta C are shown in Fig. 16. While aging clearly improves HCF strength due to the increase in yield stress, two-step aging is superior to conventional aging by about 50 MPa. The inferior per-

formance of conventional aging is caused by the presence of precipitate-free regions within the beta grains (Fig. 15a), which were identified as fatigue crack nucleation sites (Fig. 17). Microhardness measurements have shown that these regions have strengths comparable to the solution-heat-treated reference (Ref 24).

Microcrack growth in the various conditions of Beta C is shown in Fig. 18 in terms of da/dN-ΔK curves (Ref 25). No significant differences were observed among the various conditions. This result is in agreement with studies on long through-cracks in C(T)-type specimens (Ref 26). Obviously, the wide variation in fatigue crack growth resistance as observed in the class of alpha-beta alloys is not possible in solute-rich beta alloys.

The alloy Ti-10-2-3 represents the class of solute-lean beta alloys in which the microstructure can be varied to a larger extent through the possible presence of α_p in volume fractions similar to those in alpha-beta alloys (Ref 27-30). Typical microstructures in Ti-10-2-3 with 15 and 30 vol% α_p are shown in Fig. 19 and 20, respectively (Ref 31). In addition, a beta-annealed structure (0% α_p) and a structure with 5 vol% α_p are compared with the tensile properties of these typical conditions in Table 20.

The *S-N* curves of the various microstructures are shown in Fig. 21. Typical fatigue crack nucleation sites can be seen in Fig. 22. Fatigue cracks nucleated at beta grain boundaries in the beta-annealed structure (Fig. 22a), within the beta grains (Fig. 22b), and at thick alpha zones along beta grain boundaries (Fig. 22c) for material with 15 and 30% α_p, respectively. The superior per-

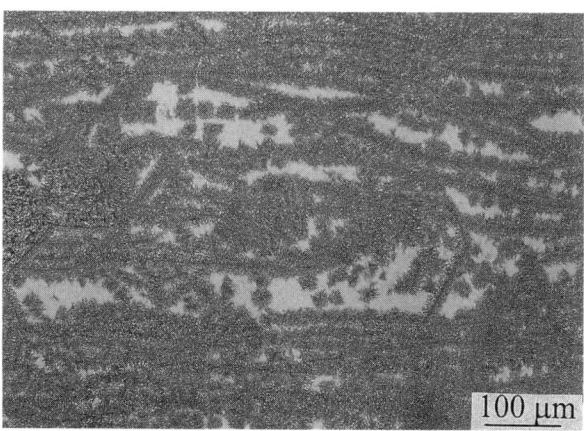

(a)

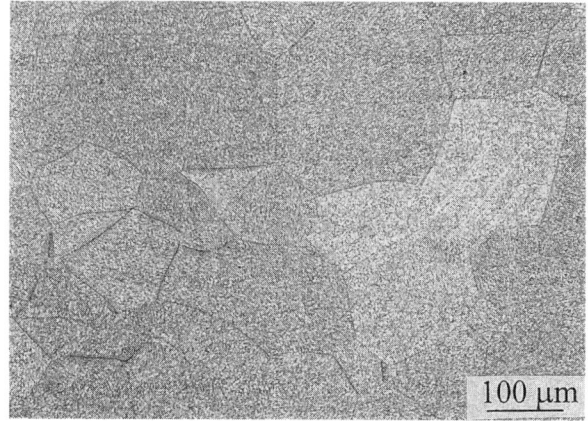

(b)

Fig. 15 Microstructures in Ti-3Al-8V-6Cr-4Mo-4Zr. (a) Conventionally aged. (b) Two-step aged. Source: Ref 24

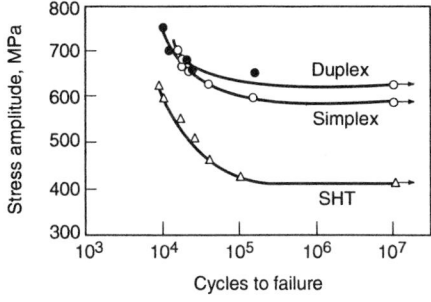

Fig. 16 S-N curves ($R = -1$) in Ti-3Al-8V-6Cr-4Mo-4Zr. Duplex structure, two-step aging; simplex structure, conventional aging. SHT, solution heat treated. Source: Ref 24

Fig. 17 Fatigue crack nucleation in conventionally aged Ti-3Al-8V-6Cr-4Mo-4Zr. Source: Ref 24

Table 20 Tensile properties in Ti-10-2-3 (aged 8 h at 480 °C)

Solution treatment	Primary α phase content, %	0.2% yield strength, MPa	RA, %
0.5 h at 830 °C/AC	0	1555	2.0
0.5 h at 785 °C/AC	5	1370	8.6
0.5 h at 775 °C/AC	15	1330	11.3
0.5 h at 725 °C/AC	30	1195	22.1

AC, air cool. Source: Ref 31

formance of microstructures with low α_p contents of 5 and 15% seems to be related to both the absence of a continuous α_{GB} layer and their comparatively high strengths (Table 20). This is also reflected in the high ratio $\sigma_a(10^7)/\sigma_{0.2}$ amounting to 0.58 and 0.55 for the microstructures with α_p volume fractions of 15 and 5%, respectively. This ratio is only 0.45 for the beta-annealed microstructure, indicating the presence of a soft α_{GB} film that not only dramatically affects ductility (Table 20) but also reduces the resistance to fatigue crack nucleation without affecting the macroscopic yield stress. Increasing the α_p content to 30% increases the ratio $\sigma_a(10^7)/\sigma_{0.2}$ to 0.52. This corresponds to a decrease in strength differential between α_{GB} and the grain interior, because the loss in macroscopic yield stress now reflects the contribution of the high volume fraction of the soft α_p phase in the beta matrix and the concomitant low contribution of precipitation hardening by α_s (Ref 31).

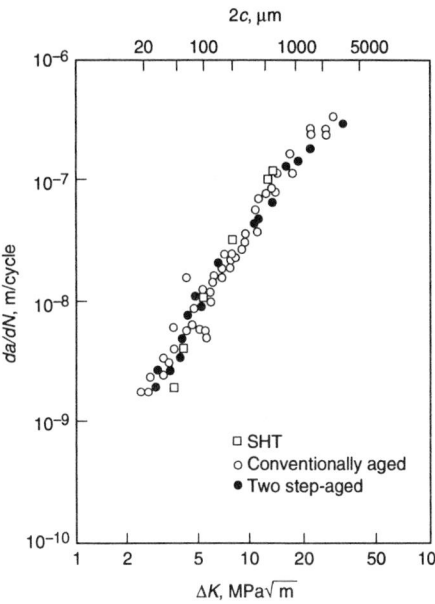

Fig. 18 da/dN-ΔK curves of microcracks in Ti-3Al-8V-6Cr-4Mo-4Zr. SHT, solution heat treated. Source: Ref 25

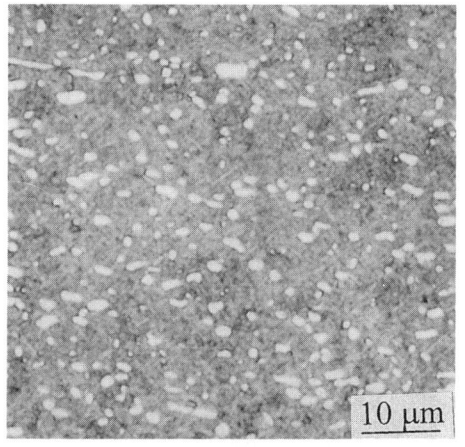

Fig. 19 Microstructure of Ti-10V-2Fe-3Al with 15% α_p content, which is considered about optimal for strength and toughness. Source: Ref 31

Effect of Texture on Fatigue Life. While beta alloys are considered fairly isotropic due to their body-centered cubic structure, the mechanical properties of alpha and alpha-beta alloys can be quite anisotropic. Due to the anisotropy of the hexagonal alpha lattice structure and the resulting directionality of mechanical properties in single crystals, polycrystals in textured material with high volume fractions of the alpha phase can be quite anisotropic with respect to fatigue performance. Because the texture in alpha-beta alloys can be varied to a larger extent than in alpha alloys, the effect of crystallographic texture on fatigue behavior has been mainly studied in Ti-6Al-4V (Ref 32-35). In contrast to beta-annealed (fully lamellar) microstructures, which normally have a nearly random texture, various types of textures

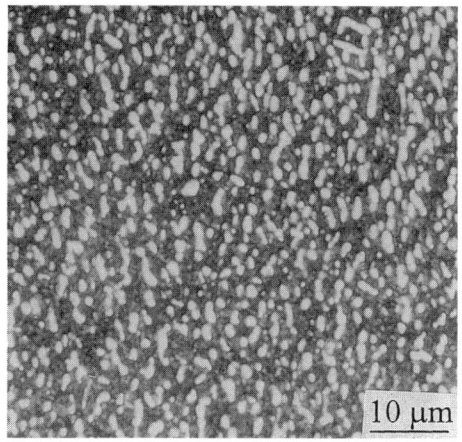

Fig. 20 Ti-10V-2Fe-3Al with 30% α_p Source: Ref 31

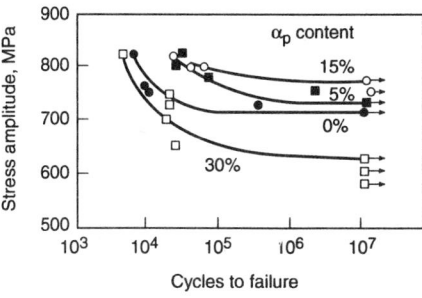

Fig. 21 S-N curves ($R = -1$) in Ti-10V-2Fe-3Al. α_p, primary alpha phase. Source: Ref 31

can be developed in the alpha phase of the fully equiaxed and duplex microstructures.

The four basic types of textures are basal (B), transverse (T), mixed basal/transverse (B/T), and weakly textured (WT). These textures are achieved by appropriate TMP, and the S-N curves of fully equiaxed microstructures of Ti-6Al-4V with various types of sharp textures are shown in Fig. 23 and 24 (Ref 10). The highest HCF strength, about 725 MPa, was obtained by testing the B/T type of texture parallel to the rolling direction (RD). The lowest HCF strength, 580 MPa, was measured for a T-type of texture tested perpendicular to RD in the rolling plane (TD). Comparing these results from fatigue testing in air (Fig. 23) with those in vacuum (Fig. 24), it is seen that laboratory air acts as a corrosive environment for Ti-6Al-4V and that the loss in fatigue strength due to the air environment strongly depends on crystallographic texture and loading condition. For example, the drop in fatigue strength is most pronounced for B/T- and T-textures in loading directions where higher yield stresses are observed due to the stress axis perpendicular to the basal planes (B/T-TD, T-TD). This behavior is thought to be due to the stress-corrosion susceptibility of alpha-beta titanium alloys for loading directions perpendicular to the basal planes (Ref 36). Compared to fully equiaxed and fully lamellar microstructures, the HCF strength of duplex structures is less affected by environment, presumably due to the absence

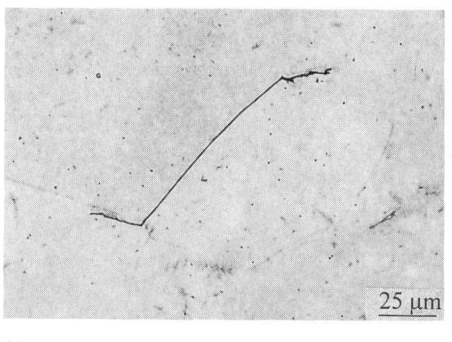

(a)

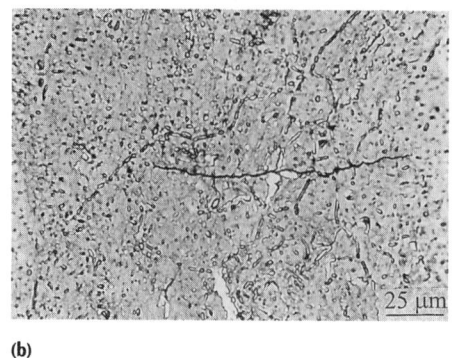

(b)

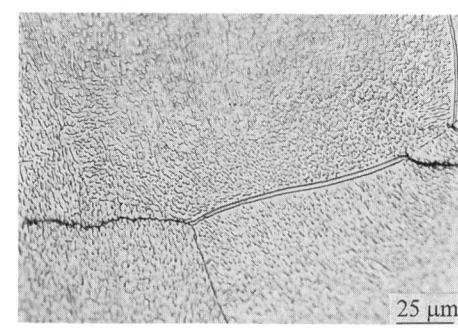

(c)

Fig. 22 Fatigue crack nucleation sites in Ti-10V-2Fe-3Al. α_p, primary alpha phase. Source: Ref 31

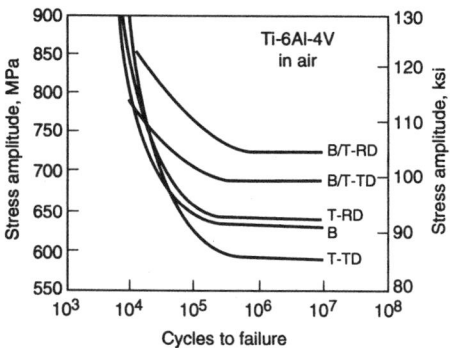

Fig. 23 S-N curves in Ti-6Al-4V (air). Effect of texture and loading direction. B, basal; T, transverse; RD, rolling direction; TD, transverse direction. Source: Ref 10

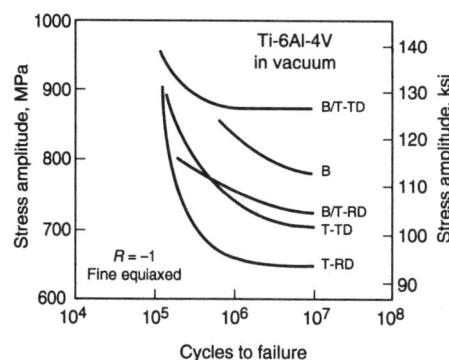

Fig. 24 S-N curves in Ti-6Al-4V (vacuum). Effect of texture and loading direction. B, basal; T, transverse; RD, rolling direction; TD, transverse direction. Source: Ref 10

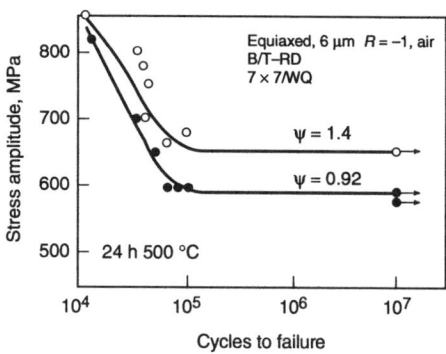

Fig. 25 S-N curves in Ti-6Al-4V. Effect of degree of texture. Source: Ref 37

of alpha/alpha phase boundaries (isolated α_p grains) (Ref 10). The effect of degree of texture can be seen in Fig. 25. A decrease in deformation degree for unidirectional rolling at 800 °C from 75 to 60% results in a slight loss of the HCF strength, from 650 to 590 MPa (Ref 37).

High-Cycle Fatigue Strength

Fatigue limits (or endurance limits) represent the value of stress below which a material can presumably endure an infinite number of cycles. For many conditions, fatigue limits may be observed at 10^7 cycles or more. In other cases, however, fatigue limits are not observed. These cases are generally attributed to periodic over-strains or the absence of hardening, as with very low oxygen levels (Fig. 26). Also, the absence of a fatigue limit in Ti-6Al-4V alloy can be associated with subsurface initiations, especially at cryogenic temperatures (Ref 39).

About 80 to 90% of HCF life involves the nucleation of surface cracks, which can be greatly influenced by stress concentrations (such as corrosion pits or notches) or surface residual stress. These factors are briefly reviewed and discussed below, and the next section describes various surface treatments for improved HCF.

Notch Effects. The HCF strength of wrought Ti-6Al-4V at 10^7 cycles is typically between about 135 to 275 MPa (20 to 40 ksi) ranges for

notched (K_t = 3) specimens (Fig. 27). The static tensile strength has even less influence on fatigue strength for notched specimens than for unnotched specimens (Fig. 28). It should also be noted that fatigue limits of unnotched Ti-6Al-4V and other titanium alloys reveal more scatter than quenched-and-tempered low-alloy steels. For example, extensive tensile and smooth-bar fatigue testing on Ti-6Al-4V and regression analysis has been performed to see if a correlation exists between 10^7 cycle fatigue strength and yield or tensile strength (Ref 42). In both cases the coefficient of correlation was smaller than 0.1, indicating that essentially no correlation exists. This tends to point out an important difference between titanium alloys and steels, that the effects of microstructures on fatigue is a more complex variable, especially for the different variations of microstructures associated with dual-phase materials such as alpha-beta alloys or the solute-lean beta alloys such as Ti-10V-2Fe-3Al.

Effect of Residual Stress. Surface residual stress can be a predominant factor influencing fatigue, and the residual stress effect is most pronounced in the infinite-life stress range of the endurance-limit regime (Fig. 29). Surface residual stress is important, but the effect is not simple because of the combined influences of residual stress, cold-worked structure, and surface roughness on HCF strength. However, when surface

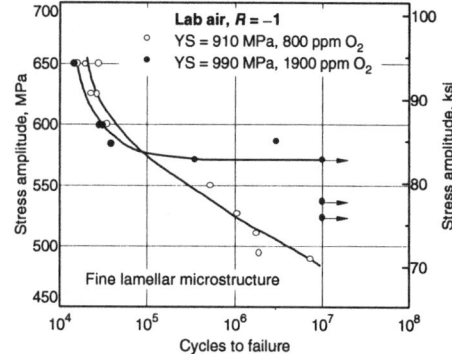

Fig. 26 Effect of low oxygen and yield strength on Ti-6Al-4V fatigue limits. Source: Ref 38

roughness is within a reasonable range of 2.5 to 5 μm (100 to 200 μin., arithmetic average), residual surface stress can be a more potent indicator of fatigue resistance than surface roughness (Ref 43). For many machining and finishing operations, residual stress and surface roughness are closely related, which accounts for the traditional correlation between fatigue strength and surface roughness.

Effect of Mean Stress on Fatigue Life. It has long been recognized that near-alpha and alpha-beta alloys can be highly mean stress sensitive with regard to HCF strength (Ref 44, 45). This so-called anomalous mean stress effect is best

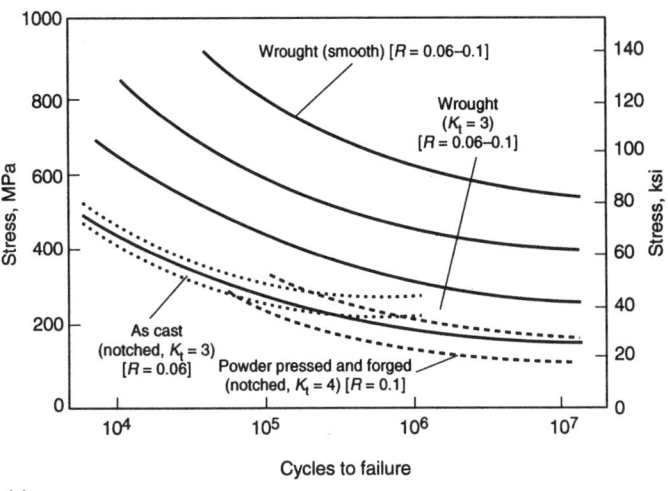

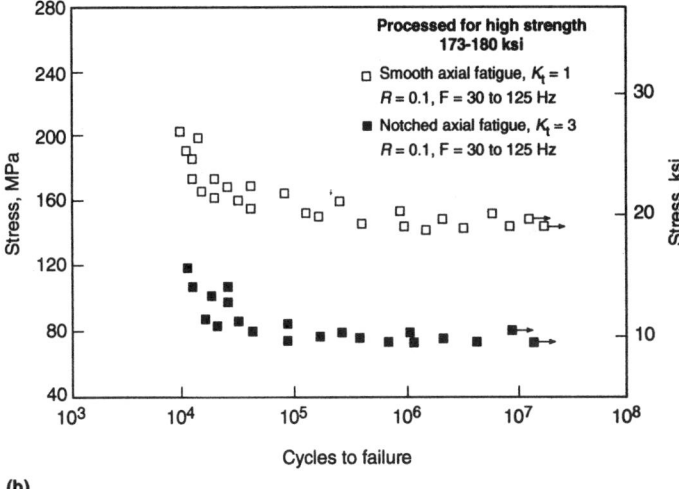

Fig. 27 Notch effects on (a) Ti-6Al-4V and (b) Ti-10V-2Fe-3Al. Source: MIL HDBK-697A, 1974, and Ref 40

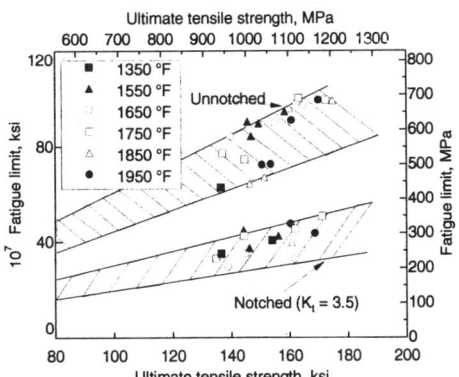

Fig. 28 Notched and unnotched fatigue limit of Ti-6Al-4V. 6.35 mm (0.25 in.) specimens cut from as-rolled bar, solution treated at indicated temperatures, and cooled at various rates (furnace, air, water quench). Rotating-beam fatigue at 8000 rpm. Fatigue limits at 10^7 cycles determined by highest stress amplitude at which three specimens ran 10^7 cycles without failure. Ref 41

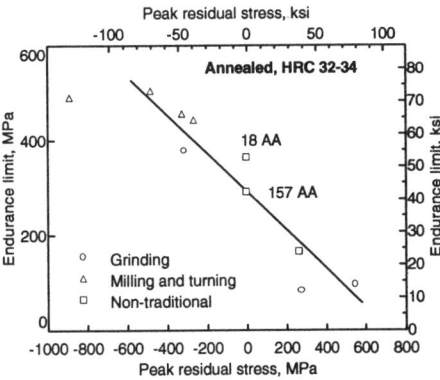

Fig. 29 Endurance limit of Ti-6Al-4V vs. residual stress from various finishing methods. AA, arithmetic average. Source: Ref 43

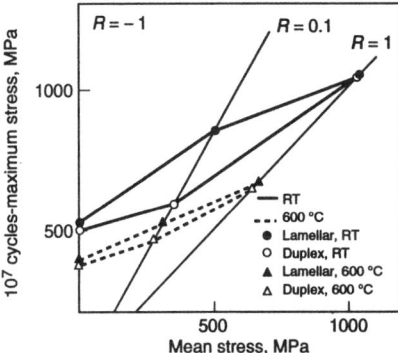

Fig. 30 Smith diagram of TIMETAL 1100 at room and elevated (600 °C) temperature. Source: Ref 46 and 48

seen in a Smith diagram where the maximum allowable stress is plotted versus the mean stress (Fig. 30) (Ref 46). At low tensile mean stresses, the fatigue performance of duplex structures is markedly inferior to fully lamellar structures in TIMETAL 1100. These results confirm earlier work on Ti-6Al-4V (Ref 47). While the exact mechanism for this anomalous mean stress sensitivity is still not fully understood, from Ref 46 and 47 the following conclusions can be drawn:

- The anomalous mean stress effect is crack nucleation controlled.
- The anomalous mean stress effect is also present in inert environments.
- Beta-annealed (fully lamellar) microstructures usually do not show this effect.
- In alpha-beta annealed (fully equiaxed or duplex) microstructures, the degree of the anomalous mean stress sensitivity depends on crystallographic texture and loading direction.

New results on TIMETAL 1100 (Ref 48) indicate that the anomalous mean stress sensitivity is absent at elevated temperatures (Fig. 30).

Effects of Surface Treatments

Mechanical surface treatments such as shot peening, polishing, or surface rolling can be used to improve the fatigue life in titanium alloys (Ref 49, 50). In most cases, three surface properties are altered:

- Surface roughness
- Degree of cold work or dislocation density
- Residual stresses

Because fatigue failure represents the sum of both crack nucleation and crack propagation life, the changes induced by such treatments can have contradictory influences on the fatigue strength. The surface roughness determines whether fatigue strength is primarily crack nucleation controlled (smooth) or crack propagation controlled (rough) (Ref 50). For smooth surfaces, a work-hardened surface layer can delay crack nucleation owing to

the increase in strength. In rough surfaces, the crack nucleation phase can be absent, and a work-hardened surface layer is detrimental to crack propagation owing to the reduced ductility (Ref 51). Near-surface residual compressive stresses are clearly beneficial, as they can significantly retard microcrack growth once cracks are present (Table 21, Ref 49).

It must be kept in mind that the changes induced by mechanical surface treatment are not necessarily stable. In particular:

- Residual stresses can be reduced or eliminated by a stress-relief treatment
- Degree of cold work (dislocation density) can be removed by recrystallizing
- Surface roughness can be reduced by an additional surface treatment (e.g., polishing)

Furthermore, the beneficial residual compressive stresses can be reduced by cyclic plastic deformation (i.e., during fatigue loading in service).

These factors can be taken into account to design surface treatments for titanium alloys that are appropriate to the application. For example, Fig. 31 shows a series of S-N curves for Ti-6Al-4V at room temperature and at elevated temperature (Ref 52). As a reference condition, an electropolished surface is considered that is free of residual stresses and cold work and has a mirror finish. Compared with EP, shot peening (SP) significantly improves the HCF strength at room temperature. At elevated temperature, however, SP lowers HCF strength to values below that of the reference condition. This can be explained by considering the individual contributions of the surface properties to fatigue life. For example, stress relieving (SR) 1 h at 600 °C after SP (SP + SR) decreases the endurance limit of the SP condition at room temperature but does not alter the HCF strength at 500 °C. In effect, SR is redundant when cyclic loading occurs at high temperature. If the compressive residual stresses were the only mechanism operating, one could conclude that SP cannot be used to improve fatigue performance at elevated temperatures. However, an additional surface treatment that reduces the surface roughness, in the present case electropolishing (SP + EP), demonstrates that work hardening the surface layer can also be exploited to improve HCF strength, irrespective of whether a stress relief treatment is applied (SP + SR + EP) or not. The degree of near-surface cold work (dislocation density) is not significantly altered by SR and thus raises the HCF strength.

Microcrack growth rates in the conditions SP and SP + SR are compared to the reference EP in Fig. 32. At room temperature (Fig. 32a), crack growth is drastically retarded in SP, but it is accelerated in SP + SR as compared to EP. The difference in growth rate between curves SP and SP + SR is caused by the residual compressive stresses in SP, while the difference between curves SP + SR and EP is related to the high dislocation density in the SP + SR condition. At elevated temperature (Fig. 32b), no difference in microcrack growth was measured between curves SP and SP + SR owing to residual stress relaxation at that temperature. The inferior performance of SP and SP + SR compared to the reference is due to the negative effect of high dislocation densities (low residual ductility) on crack growth resistance (see Table 21).

Thermomechanical Surface Treatments. Surface microstructure of titanium can be tailored or varied to enhance fatigue crack initiation resistance at the surface, while still maintaining a bulk interior to meet differing requirements (e.g., as in the carburizing of steels). As demonstrated in the following examples, cold working induced by mechanical surface treatments can be used to develop a surface microstructure that is different from that in the bulk, thus combining the optimum features of both, even in cases where conventional TMP may not be practical, as in thick sections. A distinct advantage to be gained by altering the surface microstructure in this way is that such alterations are more stable than those induced by mechanical surface treatments alone.

Surface Treatment of Alpha-Alloys. A mechanical surface treatment in combination with a subsequent recrystallization makes it possible to combine the high strengths and endurance limits associated with fine grains with the superior long

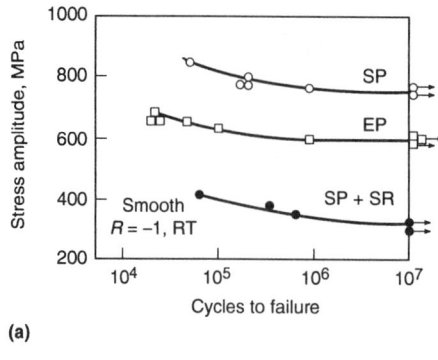

(a)

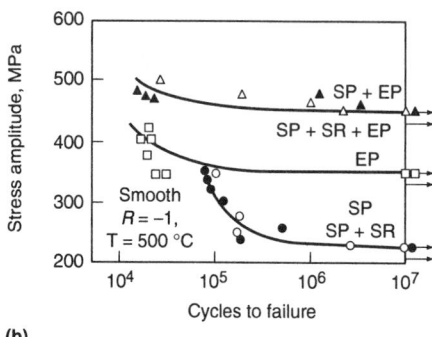

(b)

Fig. 31 S-N curves ($R = -1$) for Ti-6Al-4V with a fine lamellar microstructure. SP, shot peened; EP, electropolished; SR, stress relieved. (a) Room temperature. (b) 500 °C. Source: Ref 52

through-crack fatigue crack growth behavior and fracture toughness of the coarse grains. To maximize the total fatigue life in thicker sections, fine grains are needed on the surface, where good resistance to crack initiation and microcrack growth is critical, and coarse grains are needed in the interior, where they can reduce the driving force to long crack growth. In one study, shot peening followed by a heat treatment of 1 h at 820 °C was performed on coarse-grained Ti-8.5Al to cold work and recrystallize the surface. The improvement in fatigue limit owing to the fine (20 μm) surface grains as opposed to the coarse grains in the bulk (100 μm) was significant, roughly 50 MPa at 350 °C (Fig. 33) (Ref 52).

Near-alpha and alpha-beta alloys are often intended for elevated-temperature service (e.g., in gas turbines), so creep resistance is an important consideration. On this basis, lamellar microstructures would be preferable. However, these microstructures have poor fatigue resistance, particularly in the LCF regime, where surface crack growth determines fatigue life. In such cases, a variation in phase morphology between the surface and the core can be obtained by mechanically working the surface by shot peening and then heat treating. The improvement in *S-N* behavior (at high temperature) gained by this thermomechanical surface treatment is shown in Fig. 34 for Ti-6242 with a creep-resistant lamellar core and a fatigue-resistant fine equiaxed surface layer (Ref 52).

Beta Alloys. Both shot peening and surface rolling in combination with specially developed aging treatments have been applied to Ti-3Al-8V-

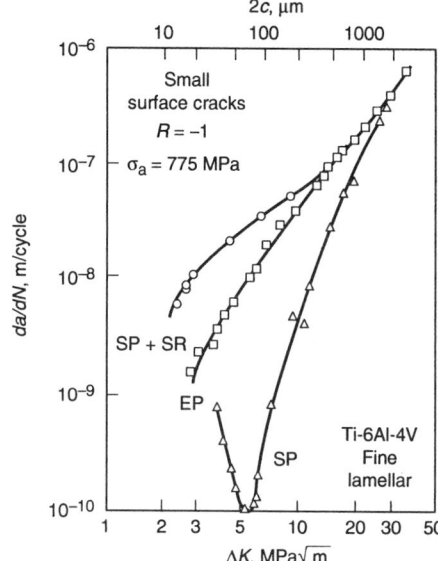

(a)

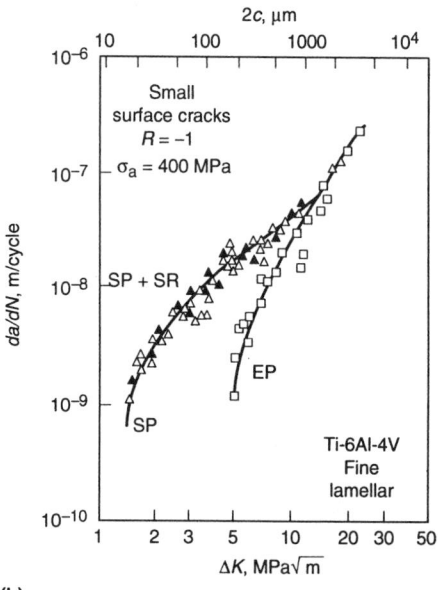

(b)

Fig. 32 Microcrack growth in Ti-6Al-4V. SP, shot peened; EP, electropolished; SR, stress relieved. (a) Room temperature. (b) 500 °C. Source: Ref 52

Table 21 Effects of surface properties on fatigue life of titanium

Surface property	Crack nucleation effect	Microcrack propagation effect
Surface roughness	Accelerates	No effect
Degree of cold work or dislocation density	Retards	Accelerates
Residual compressive stresses	Minor or no effect	Retards

Source: Ref 49

6Cr-4Mo-4Zr (Ref 53, 54) to selectively age harden only the surface. This new thermomechanical surface treatment shows promise for improving properties of high-strength springs and fas-

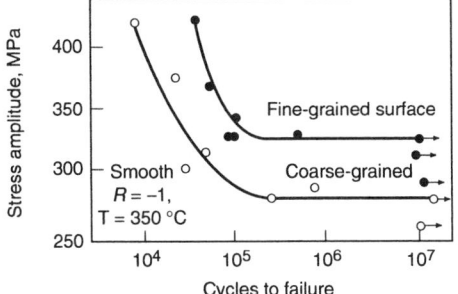

Fig. 33 S-N curves for coarse-grained Ti-8.5Al at 350 °C with and without thermomechanical surface treatment for local grain refinement. Source: Ref 49

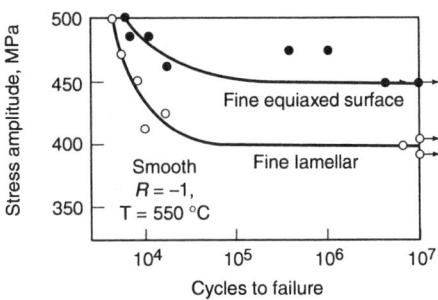

Fig. 34 S-N curves for Ti-6242 at 550 °C with a fine lamellar microstructure and without a thermomechanical surface treatment. Source: Ref 49

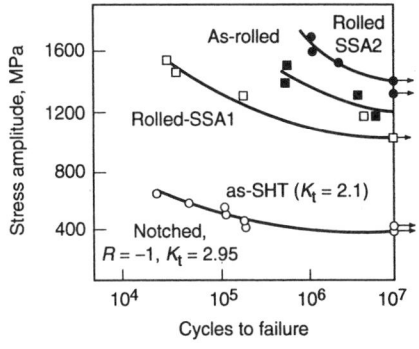

Fig. 35 S-N curves for Ti-3Al-8V-6Cr-4Mo-4Zr. SSA, selective surface aging; SHT, solution heat treated. Source: Ref 49

teners. Figure 35 shows the near-surface region for shot-peened material both without and with a selective surface aging (SSA) treatment. The high strength of the surface increases the fatigue limit to values above those of a conventional bulk aging treatment, while the high ductility of the solution-heat-treated (SHT) condition in the interior provides good notched ductility and toughness (Ref 49). Fully reversed notched fatigue behavior for this alloy after surface rolling with

and without SSA is shown in Fig. 35. Rolling alone increases the notched fatigue limit of the SHT condition (expressed as nominal stress times a concentration factor) from the low value of 400 MPa to 1100 MPa. Depending on the subsequent aging treatment, the *S-N* behavior can deteriorate slightly (SSA1) or be even further improved (SSA2). These results suggest that the residual compressive stresses present in the as-rolled condition are significantly relieved by the SSA1 treatment but not the SSA2 treatment (Ref 49).

Mechanical surface treatments such as shot peening and surface rolling can be applied alone or in combination with heat treatments to obtain optimum properties in mechanically loaded titanium parts. The particular treatment applied should reflect the type of alloy (alpha, alpha-beta, or beta) to make use of its characteristic response to heat treatment and/or TMP.

Fatigue Crack Growth of Titanium Alloys

J.K. Gregory, Martin-Luther University Halle-Wittenberg

Just as K_{Ic} is important for calculating allowable loads in the presence of a flaw, it is also important to estimate remaining life when fatigue is one of the primary failure mechanisms. If inspections are regularly performed to detect cracks, as is the case in aircraft, fracture mechanics methods can be used to predict residual life and specify inspection intervals, providing that fatigue crack growth data are available that correspond to the material and loading conditions appropriate to the application. When plotted as *da/dN* versus ΔK, fatigue crack growth (FCG) rates of titanium alloys generally lie between those of steels and aluminum alloys. This results from the use of ΔK to characterize the stress field at the crack tip, because K, as a linear elastic fracture mechanics parameter, by nature depends itself on the elastic modulus. It should be kept in mind that ΔK is not strictly valid for titanium alloys with a significant volume fraction of alpha grains, because the alpha phase is anisotropic, and the derivation of K assumes isotropic elastic behavior. Nonetheless, K is widely used to describe crack growth behavior in titanium alloys. Typical FCG behavior for titanium as compared to steel and aluminum at a low load ratio is shown in Fig. 36. More so than in steel and aluminum, microstructural variations can significantly influence FCG rates and thresholds, but they are more or less pronounced according to specific loading and environmental conditions.

The sections below first outline the interaction between microstructure, crack geometry, and load ratio in determining FCG and threshold, giving examples for commercial-purity (CP) Ti, near-alpha, alpha-beta, near-beta, and metastable beta alloys. FCG is commonly held to be slower in coarse microstructures than in fine microstructures. However, this is only true when cracks are long and the stress ratio is low (i.e., less than roughly 0.3).

The importance of environment is discussed briefly. In general, FCG is more strongly affected by the environment in the alpha phase than in the beta phase. The influence of temperature is presented for alloys that are recommended for use at high temperatures (near-alpha, alpha-beta, and in some cases metastable beta) and cryogenic temperatures (ELI grades with enhanced toughness). Finally, the available data for FCG in weldments are presented and analyzed in terms of microstructure and residual stresses.

Metallurgical Effects on FCG

The possible influences of microstructure are specific to the alloy type. For example, in CP Ti or alpha alloys, grain size, interstitial (oxygen) content, and cold work can be varied. Figure 37 shows the influence of grain size on FCG at *R* = 0.07 for different grades of CP Ti. In general, the

higher the oxygen content, the greater the influence of grain size, and vice versa (Ref 57). FCG rates in fine-grained material, (i.e., 20-35 μm or ASTM 7-8) are relatively insensitive to oxygen content. When grain size is increased to roughly 220 μm, or ASTM 1.5, FCG rates not only are significantly lower than in fine-grained material, but a higher oxygen content can further lower FCG rates (Ref 57). Cold work would presumably cause slightly increased FCG rates by reducing residual ductility.

In near-alpha and alpha-beta alloys, the phase morphology (lamellar vs. equiaxed) is the most important microstructural feature (Ref 58, 59). Figure 38 (Ref 59-61) shows FCG rates at *R* = 0.1 for Ti-6Al-4V in both mill-annealed and coarse lamellar (beta-annealed) conditions. The fine, mill-annealed microstructure exhibits much higher FCG rates than the lamellar microstructure, particularly at lower ΔK values. Although this is commonly observed for long-crack data measured on standard fracture mechanics specimens, it must be kept in mind that when cracks are small or when specimen dimensions are small compared to the length of the crack front, coarse lamellar microstructures exhibit much higher FCG rates than fine microstructures (Ref 58). This is discussed below in the section "Small/Short Cracks."

When both alpha grains and lamellar regions are present, as in duplex microstructures, FCG

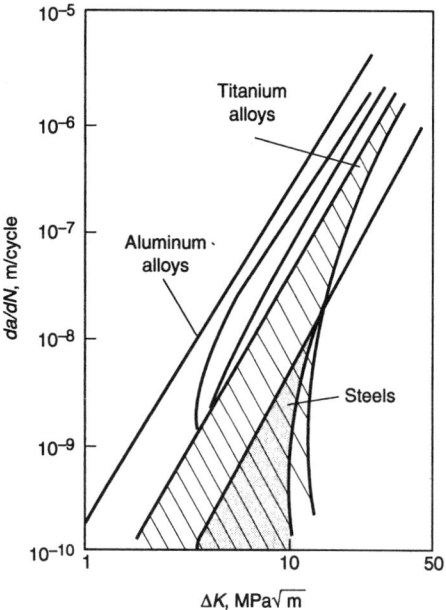

Fig. 36 Range of fatigue crack growth rates in titanium alloys, which lie between those of steels and aluminum alloys. Source: Ref 55, 56

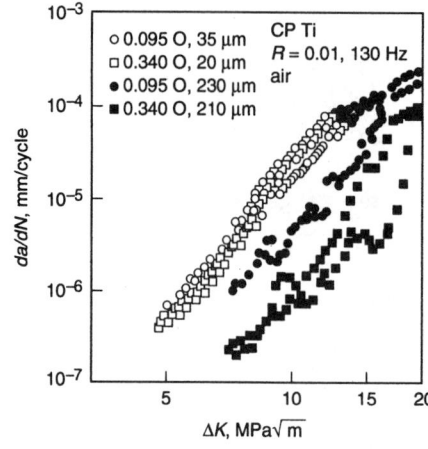

Fig. 37 Influence of grain size on fatigue crack growth at $R = 0.07$ for CP Ti with varying oxygen contents. Under these loading conditions, fatigue crack growth is lowered by coarse grains and increased oxygen content. Source: Ref 57

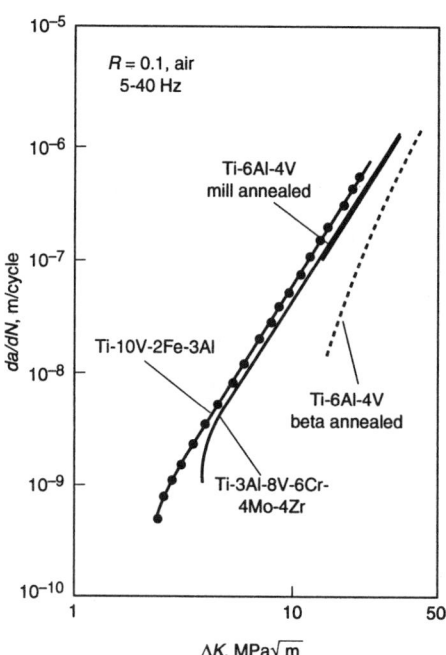

Fig. 38 Typical fatigue crack growth rates for Ti-6Al-4V in a lamellar and fine equiaxed condition and for Ti-10V-2Fe-3Al and Ti-3Al-8V-6Cr-4Mo-4Zr. Fatigue crack growth rates tend to be lower in lamellar microstructures than in fine equiaxed microstructures in Ti-6Al-4V. Fatigue crack growth in near-β and metastable β alloys is usually slightly higher than that of fine equiaxed ($\alpha + \beta$) alloys. Source: Ref 59-61

rates are intermediate between those of fine-grained and fully lamellar microstructures (Ref 62). Oxygen content also influences FCG rates in two-phase alloys in the same manner as in CP Ti; however, the effect is so minor as to be easily masked by other microstructural changes, presumably because the grain sizes in these alloys are always less than 20 μm.

As was shown for CP Ti, FCG in fine-grained material is not very sensitive to interstitial content. The difference between FCG rates in mill-annealed Ti-6Al-4V in both the standard and ELI grades is roughly a factor of 2 (Ref 63).

In near-beta and metastable beta alloys, grain size, the degree of age-hardening, and amount of cold work are possible microstructural parameters to be considered. The few data available suggest that FCG in these alloys is insensitive to grain size and degree of cold work (or residual deformation from hot working). Age hardening has only a minor influence on FCG (Ref 60, 61). Figure 38 shows typical FCG rates for the near-beta and metastable beta alloys Ti-10V-2Fe-3Al and Ti-3Al-8V-6Cr-4Mo-4Zr, respectively. FCG rates are slightly higher than those of fine-grained near-alpha and alpha-beta alloys. This is partially because of the slightly lower elastic modulus of the beta matrix, which tends to shift FCG curves to lower ΔK values. Although significantly lower FCG rates have been reported for Ti-10V-2Fe-3Al which was solution heat treated with no subsequent aging cycle (Ref 60), this condition exhibits virtually nil ductility and very poor fracture toughness.

In CP-Ti and in near-alpha and alpha-beta alloys, a preferred crystallographic orientation, or texture in the alpha phase, can affect FCG. A preferred orientation in the alpha phase can be indirectly assessed by measuring the elastic modulus. In commercial alloys, values of roughly

100 GPa indicate that few basal planes are loaded in tension, while values as high as 130 GPa can be found when many basal planes are loaded in tension. Normally, a higher elastic modulus is expected to lower FCG rates at a given ΔK. However, a high proportion of basal planes in the crack plane leads to more rapid crack growth in CP Ti (Ref 64) and alpha-beta alloys (Ref 65, 66), because cleavage is the preferred mode of crack advance on these planes. Figure 39 shows FCG rates in air for Ti-6Al-4V in a fine equiaxed condition (alpha grain size of 1-2 μm) loaded parallel and perpendicular to the basal planes. It was noted in Ref 66 that this texture effect is absent in vacuum, and in Ref 65 that an influence of texture becomes more pronounced in a salt water environment, suggesting that the acceleration mechanism is related to environment. For near-beta and metastable beta alloys, little data are available regarding the effect of texture on FCG. Where different sample orientations in rolled plate have been investigated, virtually no difference in FCG behavior has been found, suggesting that the influence of texture is negligible (Ref 67).

Fatigue Fracture Modes

A wide variety of characteristic features on fracture surfaces have been observed, and they depend on both the underlying microstructure and loading and environmental conditions. Crack advance can proceed through the alpha phase by a structure-insensitive striation-forming mechanism (i.e., one that involves plastic deformation, or structure-sensitive faceted fracture, which produces cleavage-like facets, Ref 64, or cyclic cleavage, Ref 68). In the latter case, the facets are identical to those produced by cleavage during fast fracture. However, while conventional cleavage during FCG in titanium is associated with high FCG rates and salt water environments, two conditions must generally be met in order to observe cyclic cleavage: the principal tensile stress

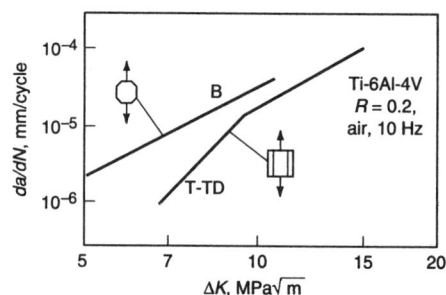

Fig. 39 Influence of texture on fatigue crack growth in Ti-6Al-4V. Fatigue crack growth rates are higher when basal planes are loaded in tension. The elastic modulus in tension for the basal texture (B) is 109 GPa; for the transverse texture (T), 126 GPa. The yield stress is roughly 1150 MPa for both. TD, transverse texture tested perpendicular to the rolling direction. Source: Ref 66

must be within 50° of the normal to the basal plane, and the FCG rates must be sufficiently low (Ref 64). For this purpose, FCG rates are considered low if the calculated cyclic plastic zone size is comparable to the alpha grain size (Ref 69, 70). When the tensile axis lies almost in the basal plane, cleavage and/or cyclic cleavage does not occur, and striations whose spacing correlates reasonably well with measured FCG rates can be found in alpha grains.

Titanium alloys that contain both alpha and beta phases can consist of alpha grains with the beta phase at grain boundary triple points ("equiaxed" or "mill annealed"), consist of alpha grains separated by lamellar regions ("duplex"),

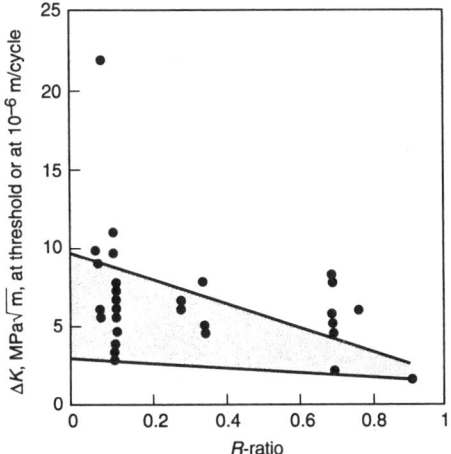

Fig. 40 Threshold ΔK values from Ref 56 together with ΔK at 10^{-9} m/cycle from Ref 57, 59, and 72-78 as a function of R-ratio. Widely differing values at low R-ratio converge to roughly 2 MPa√m as R-ratio increases.

Table 22 ΔK at 10^{-9} m/cycle for fatigue crack growth in titanium in air

Alloy	Condition(a)	R-ratio	Frequency, Hz	ΔK, MPa√m	Ref
Ti115	d_α 35 μm	0.07	130	5.3	57
		0.35		4.3	
		0.7		4.0	
Ti115	d_α 230 μm	0.07	130	9.0	57
		0.7		5.0	
Ti130	d_α 40 μm	0.07	130	6.0	57
		0.35		5.0	
		0.7		4.3	
Ti155	d_α 20 μm	0.07	130	6.0	57
		0.35		4.3	
		0.7		4.0	
Ti155	d_α 210 μm	0.07	130	10.0	57
		0.35		8.0	
		0.7		6.0	
IMI 834	Duplex, 5-25% α, d_α 16-22 μm	0.1	25	6.5-8.0	73
		0.7		3.5-5.0	
Ti 1100	Duplex, 40% α	0.1	50	9.6	74
	Basket weave, prior β > 1 mm	0.1		11.0	
Ti-2.5Cu	Equiaxed, d_α 20 μm	0.1	10	5.0	75
		0.7		2.3	
	Lamellar, LP 300 μm, LW 30 μm	0.1		22.0	
		0.7		8.0	
Ti-6Al-4V	Equiaxed, d_α 12 μm	0.3	30	6.5	76
	Duplex, d_α 7 μm	0.3	30	6.0	
	Fine lamellar, LP 40 μm, LW 2 μm	0.1	20	8.0	
		0.3	30	7.1	
		0.75	20	6.0	
Ti-6Al-4V	Equiaxed, d_α 6 μm	0.1	30	6.0	77
	Duplex, d_α 25 μm			8.0	
	Fine lamellar, LP 100 μm, LW 1 μm			6.0	
	Coarse lamellar, LP 100 μm, LW 1 μm			8.0	
Ti-10V-2Fe-3Al	Almost all	0.1	40	2.3-4	59
Beta-CEZ	Almost all	0.1	...	4-5	78
Ti-3Al-8V-6Cr-4Mo-4Zr	As-solution heat treated	0.1	10	4-5	72
	Solution treated and aged	0.1		3-4	

(a) $d\alpha$, α grain size; LP, lamellar packet size; LW, lamellar width

or be fully lamellar ("transformed beta"). The alpha grains can give rise to fracture surface features, as previously described. Depending on the orientation of the lamellar packets, lamellar regions can either exhibit parallel markings that can potentially be mistaken for striations (Ref 71) or cause "blocky" fracture (Ref 70). The latter results from the tendency for cracks to propagate alternately along the basal planes of the alpha lamellae or between the lamellae. This blocky fracture has an asperity height that is comparable to the lamellar packet size and therefore corresponds to a high degree of out-of-plane cracking. Because duplex microstructures contain both alpha grains and lamellar regions, which themselves can have various orientations relative to the loading axis, they also exhibit complex fatigue fracture surfaces (Ref 70). By comparison, the near-beta and metastable beta alloys have simple fracture modes. If material is in the solution-heat-treated condition, planar slip in the beta phase can cause a high roughness that is comparable to the grain size, as well as slip traces on the fracture surface. If material has been age hardened, fatigue fracture is transgranular with a very low degree of roughness, presumably because no microstructural features are present that can cause the crack to deviate from the macroscopic plane (Ref 72).

Metallurgical Effects on Threshold

The threshold stress-intensity factor, ΔK_{th}, measured at low R-ratios is strongly dependent on microstructure in titanium alloys that consist primarily of the alpha phase. An increase in grain size combined with an increased oxygen content can cause higher thresholds in CP Ti and alpha alloys. The highest thresholds are observed in lamellar microstructures for near-alpha and alpha-beta alloys, as well as the beta-eutectoid alloy Ti-2.5Cu (which consists of alpha plus a mixture of alpha and Ti$_2$Cu rather than alpha plus beta). Low threshold values of roughly 3 MPa√m are found in age-hardened metastable beta alloys. Thresholds are slightly higher for material in the solution-heat-treated condition. Table 22 shows values for threshold ΔK for several alloys in various microstructural conditions, tested in air at different R-ratios. Because most of the older available data were gathered prior to any general agreement regarding the measurement of ΔK_{th}, values of ΔK measured at FCG rates of 10^{-9} m/cycle are given.

Influence of Stress Ratio

When comparing FCG data, in particular thresholds, care should be taken to ensure that the stress ratios are identical. The high-threshold ΔK values measured at low R-ratios are not intrinsic material constants, but rather extrinsic parameters (i.e., they depend on the geometry of the crack relative to the sample size). In titanium, the extrinsic contribution to ΔK_{th} is primarily caused by roughness-induced crack closure (Ref 79). As the R-ratio increases, threshold ΔK values for all alloys and microstructures converge to a value of about 2 MPa√m, independent of composition and/or metallurgical factors. Figure 40 shows the scatterband from Ref 56, showing the influence of R-ratio on measured threshold for various alloys and microstructural conditions. Also shown on this diagram are the ΔK values corresponding to FCG rates of 10^{-9} m/cycle from Table 22.

If the extrinsic contribution of ΔK that is related to closure is subtracted from the applied ΔK, the effective stress intensity that drives FCG, ΔK_{eff}, is obtained. For numerical modeling of FCG rates, this ΔK_{eff} is considerably more useful than the applied ΔK. For example, the FCG rates at R-ratios ranging from 0.7 to –5.0 in a duplex microstructure of Ti-6Al-4V were found in Ref 68 to be brought into coincidence very well when plotted as da/dN versus ΔK_{eff}, where ΔK_{eff} was empirically calculated as: $\Delta K_{eff} = 1.63/(1.73 - R) \Delta K$.

Small/Short Cracks

The behavior of small cracks under cyclic loading is particularly important for parts that experience high-frequency fatigue, such as turbine blades and vanes. Of special significance for service is the fact that the high-threshold ΔK values obtained on standard fracture mechanics specimens are not necessarily applicable to small or short cracks. In fact, any metallurgical changes that decrease FCG rates in standard specimens tend to increase them in small crack specimens. For example, in Ti-6Al-4V, the most rapid FCG rates for small cracks are observed in coarse lamellar microstructures, and the lowest FCG rates are observed in fine-grained or duplex microstructures. Fine lamellar microstructures exhibit intermediate FCG rates (Ref 77). Figure 41

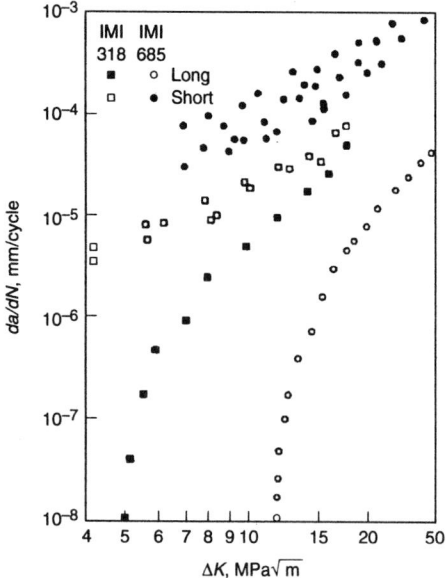

Fig. 41 Fatigue crack growth rates for the near-α alloy IMI 685 having an aligned lamellar microstructure and the fine-grained α-β alloy IMI 318. Lamellar microstructures exhibit lower fatigue crack growth rates than fine microstructures if long through-cracks are considered, but higher fatigue crack growth rates when small/short cracks are considered. Source: Ref 80

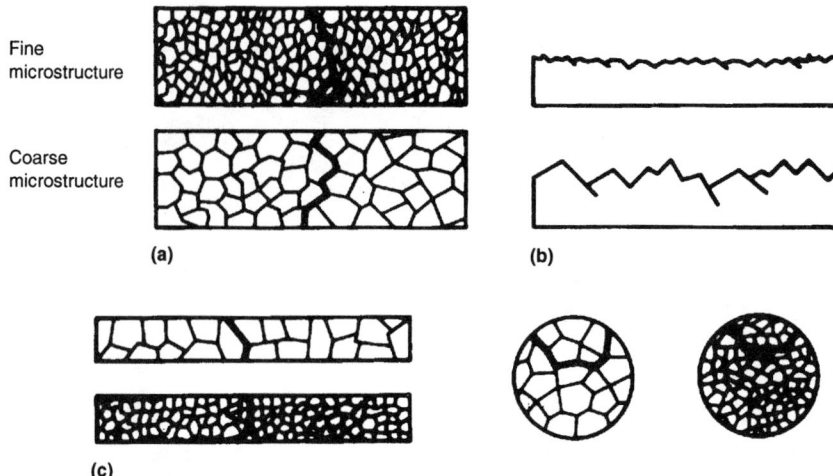

Fig. 42 Schematic relationship between grain size and (a) cylindrical specimens with a small surface crack and (b) thin standard C(T) specimens. Here the high density of grain boundaries hinders crack growth. (c) For thick standard C(T) specimens, this effect is overcompensated by roughness-induced crack closure caused by the large asperity height.

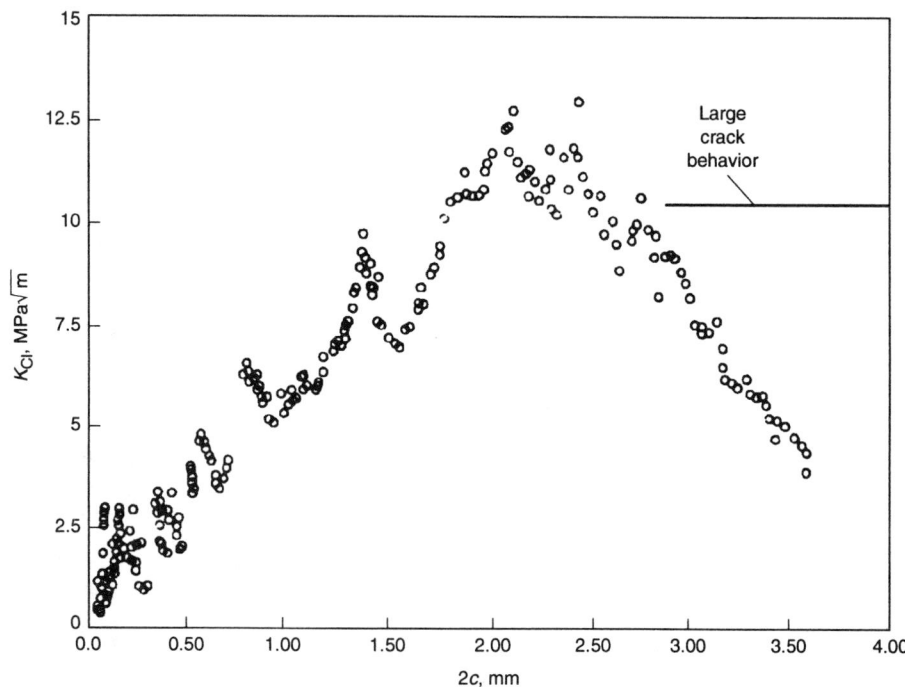

Fig. 43 Contribution of ΔK to closure for small cracks in Ti-8Al as a function of crack length. Closure levels (K_{cl}) comparable to those of long cracks are achieved at a length of approximately 2 mm. Source: Ref 81

shows an example of this effect, comparing the near-alpha alloy IMI 685, which has an aligned lamellar microstructure, with the fine-grained alpha-beta alloy IMI 318. This discrepancy between FCG in long versus short cracks arises because in short cracks, the contribution to the extrinsic threshold owing to closure is small, leaving only the intrinsic threshold, which is determined primarily by the density of microstructural barriers to crack propagation, namely grain or lamellar packet boundaries (Ref 77). This is depicted schematically for various specimen geometries as relationships between grain size and crack front in Fig. 42. As short cracks grow, asperities in the crack wake combine with shear ahead of the crack tip to develop closure effects that prevent the crack from experiencing the complete range of the applied ΔK. In situ measurements using laser interferometry on a model alloy, Ti-8Al, have demonstrated that a minimum of roughly 2 mm in length measured as the surface trace is required for a crack to have developed closure behavior comparable to that of a crack in a standard specimen (Ref 81). These data are shown in Fig. 43 as closure level as a function of crack length. Empirical correlations in IMI 685 having a coarse aligned lamellar microstructure showed that lengths of 3.5 mm are necessary in order for short corner cracks to behave as long cracks (Ref 82). Hence, surface cracks can be expected to exhibit higher FCG rates than long cracks at low R-ratios if the depth is less than several millimeters. It has been suggested that FCG testing on standard specimens can be performed in lieu of the more difficult small crack experiments if data are obtained on long crack specimens at a sufficiently high R-ratio (Ref 83).

However, R-ratios greater than 0.7 are required in order to reasonably reproduce small crack data for lamellar (Ref 75) and duplex two-phase (Ref 73) microstructures. Furthermore, if the crack size is comparable to the microstructural unit size, the validity of ΔK can be doubtful, because the material cannot be considered a continuum.

Environmental Effects

For titanium alloys, air must be considered an aggressive environment. Compared to FCG rates in vacuum or inert gas, FCG is much faster for CP Ti (Ref 57), near-alpha (Ref 84, 85) and alpha-beta alloys (Ref 65, 66, 86, 87) when tested in air. Both oxygen and hydrogen are elements with which titanium can readily react under ambient conditions. The accelerating effect of air on FCG has been attributed to residual moisture (Ref 66, 88), which is thought to oxidize the fresh titanium surface and release hydrogen, resulting in local embrittlement (Ref 84, 88). At a given frequency, FCG rates increase as the environment becomes more aggressive. In particular, liquids that contain halide ions, which can destroy the protective oxide layer of titanium, have been identified as being extremely detrimental to FCG resistance. A variety of aggressive environments were investigated in Ref 86 and demonstrate that FCG rates

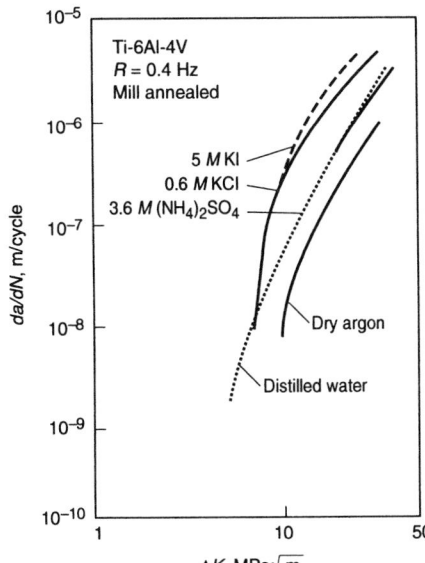

Fig. 44 Fatigue crack growth rates for Ti-6Al-4V in various liquid environments. Water is aggressive compared to air, and a high concentration of halide ions causes rapid fatigue crack growth rates. Source: Ref 86

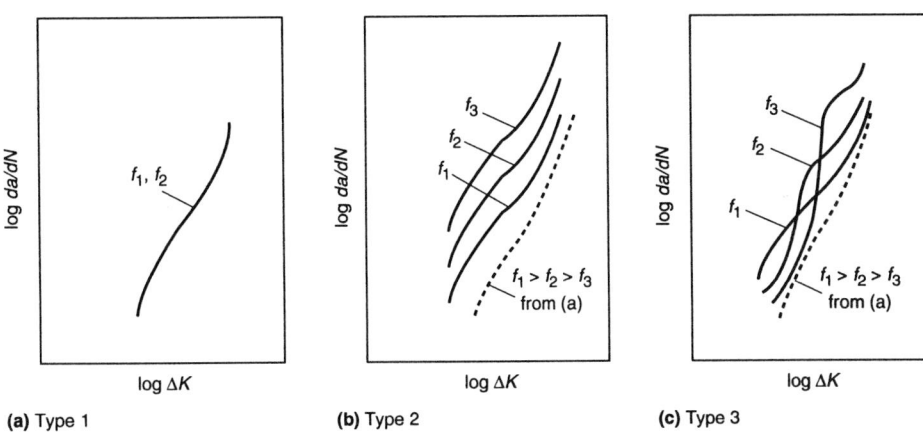

(a) Type 1   (b) Type 2   (c) Type 3

Fig. 45 Influence of loading frequency on fatigue crack growth for three classes of environments (schematic). (a) Little or no effect of frequency, as in vacuum, inert gas, or air. (b) Fatigue crack growth increases with decreasing frequency, as in methanol. (c) Cyclic stress-corrosion cracking effect, as in salt water. Source: Ref 89

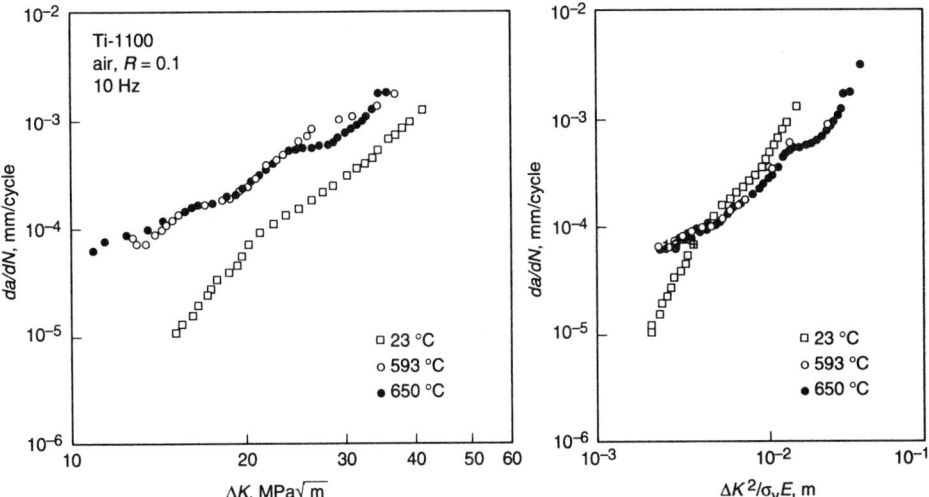

Fig. 46 Fatigue crack growth in beta-processed Ti-1100 at 23, 593, and 650 °C. (a) da/dN-ΔK curves. (b) The same data plotted as da/dN vs. $(\Delta K)^2/E\sigma_y$. The correlation with a CTOD-like parameter does not hold well for the highly anisotropic α-phase, because the dependence of fatigue crack growth on elastic modulus is complex. Source: Ref 95

in solutions that contain Cl^- or I^- ions are up to 10 times faster than in distilled water (Fig. 44).

Although air is an aggressive environment for titanium alloys with a large fraction of the alpha phase, the loading frequency does not have a pronounced effect on FCG rates under ambient conditions. In more aggressive environments, the effect of loading frequency becomes significant. In Ref 89, environments were classified into three groups according to the possible influence of loading frequency. These are shown schematically in Fig. 45. In nominally inert environments such as vacuum, helium, argon, or air, FCG rates in titanium exhibit little or no effect of frequency (Fig. 45a). In liquids such as methanol, a "normal" frequency effect is found, in that higher FCG rates are found at lower frequencies (Fig. 45b). In halide-containing solutions such as salt water, "cyclic SCC" with a characteristic discontinuity in the da/dN-ΔK curve is found (Fig. 45c). The lower the loading frequency, the lower the ΔK value at which the discontinuity is observed, where the limiting value is such that $K_{max} = K_{Iscc}$. This observation is valid for near-alpha and alpha-beta alloys for which K_{Iscc} is significantly lower than K_{Ic}. For near-beta and metastable beta alloys, no significant acceleration in FCG was found in aqueous 3.5% salt solutions as compared to air (Ref 72), and, equivalently, no effect of loading frequency in salt water has been observed (Ref 72, 90).

Temperature Effects

As temperature is increased, the mechanical properties of titanium alloys change in that yield stress and elastic modulus decrease. The tendency to react with the environment, most importantly the degree of oxidation, becomes much more pronounced. Titanium hydrides, which can

form in the alpha phase and cause rapid FCG, are less stable at elevated temperature (Ref 91). Thus, at high temperatures, FCG rates depend on both environment and frequency.

Near-alpha alloys intended for use up to 550 °C (1020 °F) have been the most frequent subjects of investigation. One unusual effect found is that at slightly elevated temperatures (i.e., at 150 °C, or 300 °F), FCG rates are lower than at room temperature (Ref 92). This is presumably related to the fact that needle-like hydrides that can form ahead of the crack tip are less stable at the higher temperature, so that FCG behavior actually improves. At still higher temperatures, FCG rates increase again so that at 260 to 300 °C (500 to 570 °F), behavior similar to that at room temperature is found for both long (Ref 93) and short (Ref 94) cracks. At still higher temperatures, between 400 to 650 °C (750 to 1200 °F), FCG rates are significantly higher than those measured under ambient conditions for ΔK values of up to approximately 25 MPa√m (Ref 94, 95). An example of the magnitude of this influence is shown in Fig.

46(a). The temperature influence in near-alpha alloys cannot be rationalized using only simple mechanics considerations. Attempts to correlate FCG data obtained at different temperatures, using an elastic-plastic parameter similar to a crack-tip opening displacement (CTOD) calculated as $(\Delta K)^2/E\sigma_y$, have not been successful, as demonstrated in Fig. 46(b). This is likely to result from the facts that elastic modulus does not have the effect on FCG in the alpha phase that would normally be expected and that the anisotropy of slip and fracture (cleavage vs. striations) can change with temperature. In general, microstructures that exhibit lower FCG rates at room temperature also exhibit lower FCG rates at elevated temperature (Ref 74). This is shown in Fig. 47 for long cracks in Ti-1100 having both a basket weave and a duplex microstructure. These data also show the substantial decrease in threshold ΔK caused by the high temperature.

Although the metastable beta alloys are not intended for elevated-temperature service as monolithic materials (the upper limit for the service temperature is approximately 350 °C), FCG in laminates of β21S has been investigated owing to interest in this alloy as a potential matrix material for composites with boron-type filaments at up to 760 °C (Ref 96). While FCG rates at room temperature and 482 °C are similar, temperatures greater than 650 °C (1200 °F) cause both substantial increases in FCG rate and decreases in threshold (Fig. 48a). At these temperatures, the ductility is so high that linear elastic fracture mechanics is not applicable, and elastic-plastic fracture mechanics parameters must be used. In contrast to the near-alpha alloys, data obtained at various temperatures could be correlated reasonably well using $(\Delta K)^2/E\sigma_y$ (Ref 92), as shown in Fig. 48.

At low temperatures, many materials become brittle as plastic deformation becomes more difficult. Titanium hydrides are thermodynamically more stable at low temperatures; however, diffusion becomes less rapid (Ref 91). Hence, FCG rates at low temperatures depend on environment as well as on internal hydrogen content. Although titanium alloys that consist primarily of the alpha phase do not exhibit the pronounced ductile-to-brittle transitions common to steels, low-temperature applications usually specify the ELI grades of titanium, because they have higher toughness than their conventional counterparts. This is particularly true for applications at cryogenic temperatures. Hence, FCG data at very low temperatures tend to be available only for these alloys. Both Ti-5Al-2.5Sn (ELI) (Ref 97) and the compositionally similar alloy VT5-1ct (Ref 98) exhibit FCG rates at cryogenic temperatures (20 and 11 K, respectively), very similar to those at room temperatures at ΔK values of up to 40 MPa√m. At higher ΔK values, the significantly reduced fracture toughness at the low temperature causes the transition from the linear regime to the fast fracture regime to shift to lower values of ΔK in the da/dN-ΔK curve. This can be seen in Fig. 49 as scatterbands for Ti-5Al-2.5Sn (ELI). A similar result was obtained at $R = 0.5$ (Ref 97). While Ref 97 tested at room temperature in air and at low temperature in liquid hydrogen, Ref 98 determined the influence of temperature and environment by measuring FCG in both air and vacuum. Taking FCG under ambient conditions as a baseline, da/dN was significantly reduced and the threshold ΔK increased from roughly 7 to 15 MPa√m when tested in vacuum, irrespective of whether the temperature was 93 or 293 K. Environment can thus be said to be far more important than low temperature, at least for these alloys.

In metastable beta alloys, distinct ductile-to-brittle transitions have been observed and have been found to shift to higher temperatures as internal hydrogen content is increased (Ref 99). This is to be expected from the body-centered cubic beta phase and should be kept in mind when these alloys are considered for service at low temperatures. The influence of decreasing temperature on FCG rates in the binary alloy Ti-30Mo depend in a complex manner on internal hydrogen content (Ref 100). For a low (22 wt ppm) hydrogen content, comparable to that of

commercial metastable beta alloys, FCG at 123 K was slightly lower than at 340 K (Fig. 50). At 190 and 233 K, FCG rates were lower than those at 123 K by roughly a factor of 2. At a high (1200 wt ppm) internal hydrogen content, FCG rates were more rapid at 123 K than at 340 K by about a factor of 3. The effects of temperature and/or hydrogen content were more pronounced at FCG rates lower than 10^{-9} m/cycle. The contribution of closure to ΔK in the near-threshold regime was found to be favored by ductility; hence, high hydrogen contents and low temperatures reduce the extrinsic contribution to threshold.

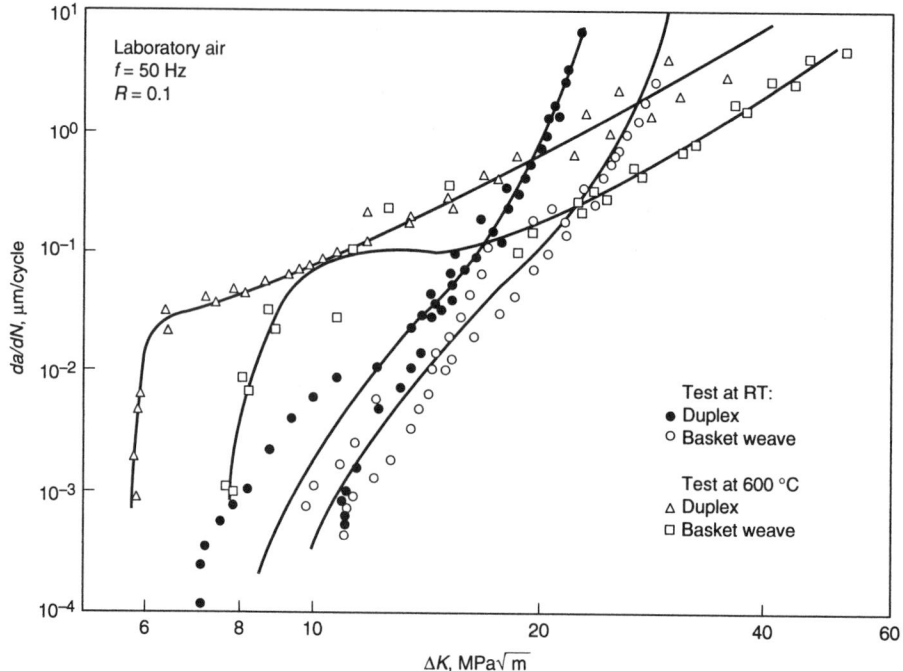

Fig. 47 Fatigue crack growth rates at room temperature and at 600 °C for Ti-1100 in two microstructural conditions. Microstructural dependencies found at room temperature generally hold at elevated temperature. Source: Ref 74

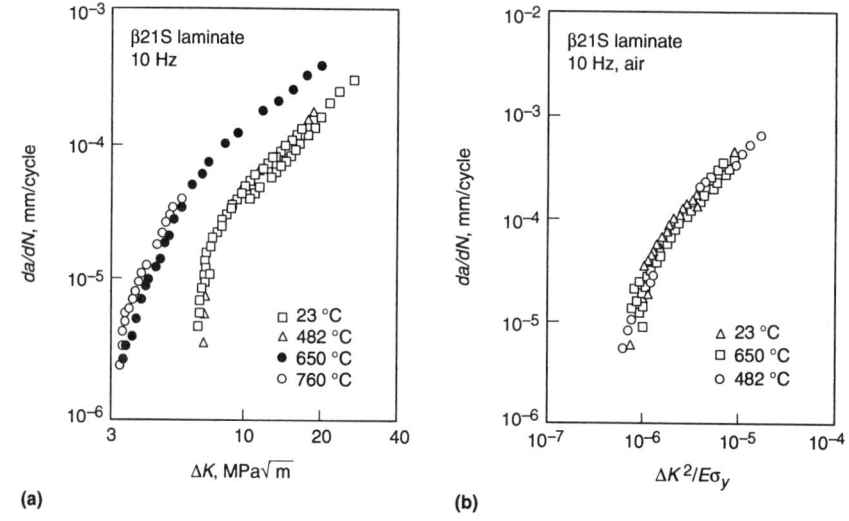

(a) **(b)**

Fig. 48 Fatigue crack growth in laminates of β-21S at various temperatures between 23-760 °C. (a) da/dN-ΔK curves. (b) Data at 23-482 °C plotted as da/dN vs. $(\Delta K)^2/E\sigma_y$. The correlation using the CTOD-like parameter is good for the bcc β phase. Source: Ref 96

Weldments

In all types of titanium alloys, microstructures in the fusion zone and in the heat-affected zone can be completely different from that in the base metal. Grain coarsening can occur in CP Ti, near-alpha, alpha-beta, near-beta, and metastable beta alloys. In near-alpha and alpha-beta alloys, the temperature excursion above the beta transus results in lamellar microstructures. In metastable beta alloys, alpha precipitates can form, coarsen, or be dissolved, depending on the local time-temperature cycle. Owing to the reactivity of titanium

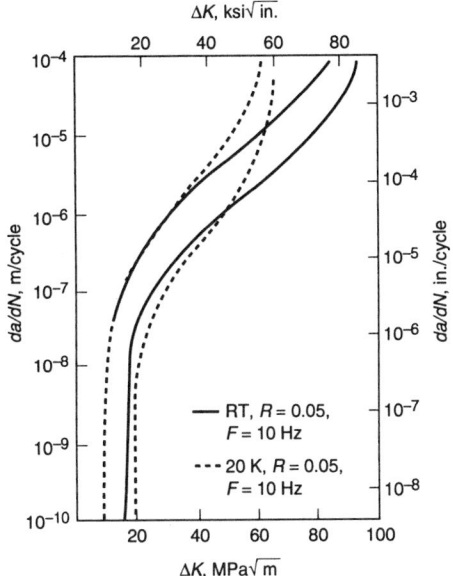

Fig. 49 Comparison between fatigue crack growth at room temperature in air and at 20 K in liquid hydrogen for Ti-5Al-2.5Sn tested at $R = 0.05$, shown as scatterbands. The reduced fracture toughness at 20 K causes the transition to the fast fracture range to shift to lower ΔK values. Source: Ref 97

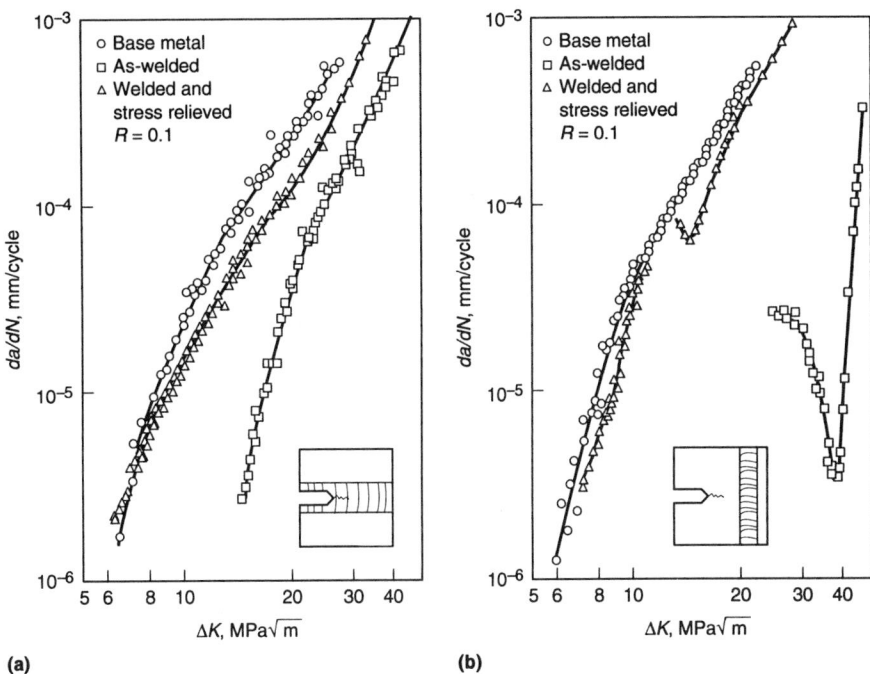

(a)   **(b)**

Fig. 51 Fatigue crack growth in laser beam weldments of Ti-6Al-4V both without and with a postweld stress relief treatment of 4.5 h at 625 °C (1160 °F). (a) Fatigue crack growth parallel to weld. (b) Fatigue crack growth perpendicular to weld. These data suggest that residual stresses in the weldment contribute to low fatigue crack growth rates and/or high thresholds. Source: Ref 103

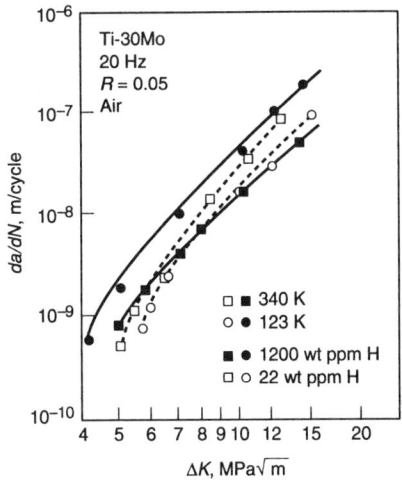

Fig. 50 Comparison between fatigue crack growth at 340 and 123 K for Ti-30Mo with both a low (22 wt ppm) and a high (1200 wt ppm) hydrogen content (after Ref 100). The temperature dependence of fatigue crack growth in the β phase depends on the internal hydrogen content. Source: Ref 100

with oxygen, great care must be taken to prevent oxygen take-up during welding, which can influence FCG rates as well as other mechanical properties. Finally, the rapid cooling rates from elevated temperatures cause residual stresses, which when superposed on applied stresses can have a great influence on FCG rates.

In grade 2 CP Ti, closure levels are significantly higher in weld metal than in the base metal (Ref 101), which is attributed to an increase in grain size. Experimental determination of FCG rates in various weldments of near-alpha and al-

pha-beta alloys (Ref 102-105) using C(T) specimens have shown that FCG in the fusion or heat-affected zone is always much lower than in the untreated base metal, whether welding was done by gas-tungsten arc (Ref 102, 105), electron beam (Ref 104), or laser beam (Ref 103). An example is shown in Fig. 51 for crack orientations parallel and perpendicular to the welding direction. While a change in microstructure from mill-annealed (Ref 102) or duplex (Ref 104) to fine lamellar could qualitatively rationalize the improved FCG resistance when FCG is parallel to the weld, FCG in weldments is in fact more strongly dependent on residual stresses than on microstructure. On the one hand, the threshold ΔK values obtained at the low R-ratios of 0 or 0.1 lie between 15 and 20 MPa$\sqrt{m}$, which is significantly higher than the values of 8 to 12 MPa$\sqrt{m}$ commonly observed in these alloys. On the other hand, the measurement (Ref 105) or elimination of residual stresses by a postweld heat treatment (Ref 103, 104), as well as the testing of FCG in weldments for cracks normal to the weld (Fig. 51b), unambiguously demonstrate the importance of residual stresses. Independent of orientation, a stress relief treatment of 4.5 h at 625 °C (1160 °F) caused FCG rates in weldments to increase almost to those of the base metal (Fig. 51). Furthermore, the retardation in FCG rates found for cracks normal to the weld (Fig. 51b) cannot be explained solely by microstructure.

FCG data for the metastable beta alloy Ti-15V-3Cr-3Al-3Sn can be interpreted similarly, in part because FCG in these alloys is so insensitive to microstructure. Aging treatments of either 8 h at 480 °C or 8 h at 510 °C were performed after gas-tungsten arc welding, resulting in a factor-of-

2 increase in FCG rate at ΔK values greater than 20 MPa$\sqrt{m}$ over the as-welded condition (Ref 106). The FCG behavior of the postweld heat-treated material is almost identical to that of conventionally aged material, suggesting that the lower FCG rates in the as-welded condition could also conceivably be caused by residual stresses.

REFERENCES

1. C.H. Wells and C.P. Sullivan, *Trans. ASM*, Vol 62, 1969, p 263
2. D.K. Benson, J.C. Grosskreutz, and G.G. Shaw, *Met. Trans.*, Vol 3A, 1972, p 1239
3. L. Wagner, J.K. Gregory, A. Gysler, and G. Lütjering, in *Small Fatigue Cracks*, R.O. Ritchie and J. Lankford, Ed., TMS, 1986, p 117
4. N.G. Turner and W.T. Roberts, *Trans. TMS-AIME*, Vol 242, 1968, p 1223
5. Conrad et al., Critical Review: Deformation and Fracture, *Titanium Science and Technology*, Plenum Press, 1973, p 996-1000
6. A. Gysler, J. Lindigkeit, and G. Lütjering, in *Strength of Metals and Alloys*, P. Haasen, V. Gerold, and G. Kostorz, Ed., Pergamon Press, 1979, p 1113
7. G.T. Gray and G. Lütjering, in *Fatigue '84*, C.J. Beevers, Ed., EMAS, 1984, p 707
8. L. Wagner and G. Lütjering, in *Titanium '88: Science and Technology*, Les Editions de Physique, 1988, p 345
9. J.M. Larsen, T. Nicholas, A.W. Thompson, and J.C. Williams, in *Small Fatigue Cracks*, R.O. Ritchie and J. Lankford, Ed., TMS, 1986, p 499

10. M. Peters and G. Lütjering, Report CS-2933, Electric Power Research Institute

11. W. Trojahn, thesis, Ruhr University, Bochum, Germany, 1980

12. L. Wagner, G. Lütjering, and R.I. Jaffee, in *Microstructure/Property Relationships in Titanium Aluminides and Alloys*, TMS-AIME, 1991, p 521

13. L. Wagner and G. Lütjering, *Z. Metallkde*, Vol 87, 1987, p 369

14. G.W. Kuhlman, in *Microstructure/Property Relationships in Titanium Aluminides and Alloys*, TMS-AIME, 1991, p 465

15. R.R. Boyer and J.A. Hall, in *Titanium '92: Science and Technology*, F.H. Froes and I. Caplan, Ed., TMS, 1993, p 77

16. H. Puschnik, J. Fladischer, G. Lütjering, and R.I. Jaffee, *Titanium '92: Science and Technology*, F.H. Froes and I. Caplan, Ed., TMS, 1993, p 131

17. M. Peters, Y.T. Lee, K.J. Grundhoff, H. Schurmann, and G. Welsch, in *Microstructure/Property Relationships in Titanium Aluminides and Alloys*, TMS-AIME, 1991, p 533

18. M. Peters, V. Bachmann, K.H. Trautmann, H. Schurmann, Y.T. Lee, and C.H. Ward, *Titanium '92: Science and Technology*, F.H. Froes and I. Caplan, Ed., TMS, 1993, p 303

19. A. Styczynski, L. Wagner, C. Müller, and H.E. Exner, *Microstructure/Property Relationships of Titanium Alloys*, S. Ankem and J.A. Hall, Ed., TMS-AIME, 1994, p 83

20. A. Berg, J. Kiese, and L. Wagner, in *Light-Weight Alloys for Aerospace Applications III*, E.W. Lee, K.V. Jata, N.J. Kim, and W.E. Frazier, Ed., TMS, 1995, p 407

21. J. Lindemann, A. Styczynski, and L. Wagner, *Light-Weight Alloys for Aerospace Applications III*, E.W. Lee, K.V. Jata, N.J. Kim, and W.E. Frazier, Ed., TMS, 1995, p 391

22. J.K. Gregory and L. Wagner, in *Fatigue '90*, H. Kitagawa and T. Tanaka, Ed., MCEP, 1990, p 191

23. G. Terlinde and G. Fischer, in *Titanium '95*, Science and Technology (in press)

24. J.K. Gregory and L. Wagner, Report 92/E/7, GKSS

25. J.K. Gregory, L. Wagner, and C. Müller, in *Beta Titanium Alloys*, A. Vassel, D. Eylon, and Y. Combres, Ed., Editions de la Revue de Métallurgie, 1994, p 229

26. H.-E. Krugmann and J.K. Gregory, *Microstructure/Property Relationships in Titanium Aluminides and Alloys*, TMS-AIME, 1991, p 549

27. T.W. Duerig, G. Terlinde, and J.C. Williams, *Met. Trans.*, Vol 11A, 1980, p 1987

28. T.W. Duerig, J.E. Allison, and J.C. Williams, *Met. Trans.*, Vol 16A, 1985, p 739

29. R.R. Boyer, *J. Metals*, 1980, p 61

30. C.C. Chen, J.A. Hall, and R.R. Boyer, in *Titanium '80: Science and Technology*, H. Kimura and O. Izumi, Ed., 1980, p 457

31. J. Kiese and L. Wagner, in *Fatigue '96*, G. Lütjering and H. Nowack, Ed., in press

32. F. Larson and A. Zarkades, MCIC Report 74-20, Battelle Columbus Labs, 1974

33. F. Larson, A. Zarkades, and D.H. Avery, *Titanium Sci. Technol.*, Vol 2, 1973, p 1169

34. A.W. Sommer and M. Creager, Report TR-76-222, AFML, 1977

35. M.J. Blackburn, J.A. Feeney, and T.R. Beck, *Adv. Corros. Sci. Technol.*, Vol 3, 1973, p 67

36. D.A. Meyer and G. Sandoz, *Trans. AIME*, Vol 245, 1969, p 1253

37. G. Lütjering and L. Wagner, in *Directional Properties of Materials*, H.J. Bunge, Ed., DGM, 1988, p 177

38. Fatigue and Microstructure, American Society for Metals, 1979, p 237

39. Y. Ito et al., *Sixth World Conf. Titanium*, Les Editions de Physique, 1989, p 87

40. Beta Titanium Alloys in the 1990's, TMS, 1993, p 508-509

41. L. Bartlo, STP 459, ASTM, 1969

42. J.C. Williams and E. Starke, *Deformation, Processing and Structure*, American Society for Metals, 1984, p 326

43. W. Koster, in *Practical Applications of Residual Stress Technology*, ASM International, 1991

44. R.K. Steele and A.J. McEvily, *Eng. Fract. Mech.*, Vol 8, 1976, p 31

45. R.K. Steele and A.J. McEvily, in *Fracture Mechanics and Technology*, G.C. Sih and C.L. Chow, Ed., Sijthoff and Nordhoff, Netherlands, 1977, p 33

46. J. Lindemann, A. Berg, and L. Wagner, in *Fatigue '96*, G. Lütjering and H. Nowack, Ed., in press

47. S. Adachi, L. Wagner, and G. Lütjering, in *Fatigue Eng. Mat. Struct.*, IMechE, 1986, p 67

48. J. Lindemann, A. Styczynski, and L. Wagner, *Intl. Conf. on Strength of Materials (ICSMA 11)*, 1997

49. M. Bauccio, Technical Note 9: Descaling and Special Surface Treatments, *Materials Properties Handbook: Titanium Alloys*, ASM International, 1994, p 1145-1158

50. L. Wagner and G. Lütjering, in *Shot Peening*, Niku Lari, Ed., Pergamon Press, 1982, p 453

51. L. Wagner and G. Lütjering, in *Fatigue '90*, H. Kitagawa and T. Tanaka, Ed., MCEP, 1990, p 323

52. H. Gray, L. Wagner, and G. Lütjering, in *Shot Peening*, H. Wohlfahrt, R. Kopp, and O. Vöhringer, Ed., DGM, 1987, p 467

53. J.K. Gregory, L. Wagner, and C. Müller, in *Surface Engineering*, P. Mayr, Ed., DGM, 1993, p 435

54. J.K. Gregory and L. Wagner, in *Fatigue '93*, J.-P. Bailon and J.I. Dickson, Ed., EMAS, 1993, p 177

55. M.O. Speidel, Stress Corrosion Cracking and Corrosion Fatigue Fracture Mechanics, *Corrosion in Power Generating Equipment*, M.O. Speidel and A. Atrens, Ed., Plenum Press, 1983, p 85-132

56. P.K. Liaw, T.R. Leax, and W.A. Logsdon, Near-Threshold Fatigue Crack Growth Behavior in Metals, *Acta Metall.*, Vol 31, 1983, p 1581-1587

57. J.L. Robinson and C.J. Beevers, The Effects of Load Ratio, Interstitial Content and Grain Size on Low-Stress Fatigue Crack Propagation in α-Titanium, *Met. Sci. J.*, Vol 7, 1973, p 153-159

58. D. Eylon, J.A. Hall, C.M. Pierce, and D.L. Ruckle, Microstructure and Mechanical Properties Relationships in the Ti-11 Alloy at Room and Elevated Temperatures, *Met. Trans.*, Vol 7A, 1976, p 1817-1826

59. G.R. Yoder, L.A. Cooley, and T.W. Crooker, Observations on Microstructurally Sensitive Fatigue Crack Growth in a Widmanstätten Ti-6Al-4V Alloy, *Met. Trans.*, Vol 8A, 1977, p 1737-1743

60. T.W. Duerig, J.E. Allison, and J.C. Williams, Microstructural Influences on Fatigue Crack Propagation in Ti-10V-2Fe-3Al, *Met. Trans.*, Vol 16A, 1985, p 739-751

61. C.G. Rhodes and N.E. Paton, The Influence of Microstructure on Mechanical Properties in Ti-3Al-6Cr-4Mo-4Zr (Beta-C), *Met. Trans.*, Vol 8A, 1977, p 1749-1761

62. M. Peters, A. Gysler, and G. Lütjering, Influence of Microstructure on the Fatigue Behavior of Ti-6Al-4V, *Titanium '80: Science and Technology*, TMS-AIME, 1980, p 1777-1786

63. P. Bania and D. Eylon, Fatigue Crack Propagation of Titanium Alloys under Dwell-Time Conditions, *Met. Trans.*, Vol 9A, 1978, p 847-855

64. C.M. Ward-Close and C.J. Beevers, The Influence of Grain Orientation on the Mode and Rates of Fatigue Crack Growth in α-Titanium, *Met. Trans.*, Vol 11A, 1980, p 1007-1017

65. R.J.H. Wanhill, Environmental Crack Propagation in Ti-6Al-4V Sheet, *Met. Trans.*, Vol 7A, 1976, p 1365-1373

66. M. Peters, A. Gysler, and G. Lütjering, Influence of Texture on Fatigue Properties of Ti-6Al-4V, *Met. Trans.*, Vol 15A, 1984, p 1597-1605

67. H.W. Rosenberg, Ti-15-3: A New Cold-Formable Sheet Titanium Alloy, *J. Metals*, 1983, p 30-34

68. A. Yuen, S.W. Hopkins, G.R. Leverant, and C.A. Rau, Correlations between Fracture Surface Appearance and Fracture Mechanics Parameters for Stage II Fatigue Crack Propagation in Ti-6Al-4V, *Met. Trans.*, Vol 5, 1974, p 1833-1842

69. P.E. Irving and C.J. Beevers, Microstructural Influences on Fatigue Crack Growth in Ti-6Al-4V, *Mater. Sci. Eng.*, Vol 14, 1974, p 229-238

70. R.J.H. Wanhill, R. Galatolo, and C.E.W. Loojie, Fractographic and Microstructural Analysis of FCG in a Ti-6Al-4V Fan Disc Forging, *Int. J. Fatigue*, Vol 11, 1989, p 407-416

71. J.A. Ruppen and A.J. McEvily, Influence of Microstructure and Environment on the Fatigue Crack Growth Fracture Topography of Ti-6Al-2Sn-4Zr-2Mo-0.1Si, *Fractography and Materials Science*, L.N. Gilbertson and R.D. Zipp, Ed., STP 733, ASTM, 1981, p 32-50

72. H.-E. Krugmann and J.K. Gregory, Microstructure and Crack Propagation in Ti-3Al-8V-6Cr-4Mo-4Zr, *Microstructure/Property Relationships in Titanium Alloys and Alu-*

minides, Y.-W. Kim and R.R. Boyer, Ed., TMS-AIME, 1991, p 549-561

73. A.L. Dawson, A.C. Hollis, and C.J. Beevers, The Effect of the Alpha-Phase Volume Fraction and Stress Ratio on the Fatigue Crack Growth Characteristics of the Near-Alpha IMI 834 Ti Alloy, *Int. J. Fatigue*, Vol 14, 1992, p 261-270

74. M. Peters, V. Bachmann, K.-H. Trautmann, H. Schurmann, Y.T. Lee, and C.H. Ward, Room and Elevated Temperature Properties of Ti-1100, *Titanium '92: Science and Technology*, TMS-AIME, p 303-310

75. J.K. Gregory and L. Wagner, Microstructure and Crack Growth in the Titanium Alloy Ti-2.5Cu, *Fatigue '90*, Materials and Component Engineering Publications Ltd., 1990, p 191-196

76. J.C. Chesnutt and J.A. Wert, Effect of Microstructure and Load Ratio on ΔK_{th} in Titanium Alloys, *Fatigue Crack Growth Threshold Concepts*, D. Davidson and S. Suresh, Ed., TMS-AIME, 1984, p 83-97

77. L. Wagner and G. Lütjering, Microstructural Influence on Propagation Behavior of Short Cracks in an ($\alpha + \beta$) Ti-Alloy, *Z. Metallk.*, Vol 78, 1987, p 369-375

78. Y. Combres and B. Champin, β-CEZ Properties, *Beta Titanium Alloys in the 1990's*, D. Eylon, R.R. Boyer, and D.A. Koss, Ed., TMS-AIME, 1993, p 477-483

79. M.D. Halliday and C.J. Beevers, Some Aspects of Fatigue Crack Closure in Two Contrasting Titanium Alloys, *J. Test. Eval.*, Vol 9, 1981, p 195-201

80. M.A. Hicks and C.W. Brown, A Comparison of Short Crack Growth Behaviour in Engineering Alloys, *Fatigue '84*, Engineering Materials Advisory Services, Ltd., UK, Vol III, 1984, p 1337-1347

81. J.M. Larsen, T. Nicholas, A.W. Thompson, and J.C. Williams, Small Crack Growth in Titanium-Aluminum Alloys, *Small Fatigue Cracks*, R.O. Ritchie and J. Lankford, Ed., TMS-AIME, 1986, p 499-512

82. C.W. Brown and M.A. Hicks, A Study of Short Fatigue Crack Growth Behaviour in Titanium Alloy IMI 685, *Fatigue Fract. Eng. Mater. Struct.*, Vol 6, 1983, p 67-76

83. W.A. Herman, R.W. Hertzberg, and R. Jaccard, A Simplified Laboratory Approach for the Prediction of Short Crack Behavior in Engineering Structures, *Fatigue Fract. Eng. Mater. Struct.*, Vol 11, 1988, p 303-320

84. D.A. Meyn, An Analysis of Frequency and Amplitude Effects on Corrosion-Fatigue Crack Propagation in Ti-8Al-1Mo-1V, *Met. Trans.*, Vol 2, 1971, p 853-865

85. H. Döker and D. Munz, Influence of Environment on the Fatigue Crack Propagation of Two Titanium Alloys, *The Influence of Environment on Fatigue*, Mechanical Engineering Publications, Ltd., London, 1977, p 123-130

86. M.O. Speidel, M.J. Blackburn, T.R. Beck, and J.A. Feeney, Corrosion Fatigue and Stress Corrosion Crack Growth in High Strength Aluminum Alloys, Magnesium Alloys and Titanium Alloys, *Corrosion-Fatigue: Chemistry, Mechanics and Microstructure*, O. Devereaux, A.J. McEvily, and R.W. Staehle, Ed., NACE, 1972, p 324-343

87. P.E. Irving and C.J. Beevers, The Effect of Air and Vacuum Environments on Fatigue Crack Growth Rates in Ti-6Al-4V, *Met. Trans.*, Vol 5, 1974, p 391-398

88. S.J. Gao, G.W. Simmons, and R.P. Wei, Fatigue Crack Growth and Surface Reactions for Titanium Alloys Exposed to Water Vapor, *Mater. Sci. Eng.*, Vol 62, 1984, p 65-78

89. D.B. Dawson and R.M. Pelloux, Corrosion Fatigue Crack Growth Rates of Titanium Alloys in Aqueous Environments, *Met. Trans.*, Vol 5, 1974, p 723-731

90. G.R. Yoder, R.R. Boyer, and L.A. Cooley, Corrosion Fatigue Resistance of Ti-10V-2Fe-3Al Alloy in Salt Water, *Sixth World Conference on Titanium*, P. Lacombe, R. Tricot, and G. Béranger, Ed., 1988, p 1741-1746

91. R.R. Boyer and W.F. Spurr, Characteristics of Sustained-Load Cracking and Hydrogen Effects in Ti-6Al-4V, *Met. Trans.*, Vol 9A, 1978, p 23-29

92. W.J. Evans and C.R. Gostelow, The Effect of Hold Time on the Fatigue Properties of a β-Processed Titanium Alloy, *Met. Trans.*, Vol 10A, 1979, p 1837-1846

93. G.C. Salivar, J.E. Heine, and F.K. Haake, The Effect of Stress Ratio on the Near-Threshold Fatigue Crack Growth Behavior of Ti-8Al-1Mo-1V at Elevated Temperature, *Eng. Fract. Mech.*, Vol 32, 1989, p 807-817

94. S.H. Spence, W.J. Evans, and A. Goulding, Small Crack Growth at Elevated Temperatures in a Near Alpha Titanium Alloy, *Titanium '92: Science and Technology*, TMS-AIME, p 1749-1756

95. R. Foerch, A. Madsen, and H. Ghonem, Environmental Interactions in High Temperature Fatigue Crack Growth of Ti-1100, *Met. Trans.*, Vol 24A, 1993, p 1321-1332

96. H. Ghonem, Y. Wen, D. Zheng, M. Thompson, and G. Linsey, Effects of Temperature and Frequency on Fatigue Crack Growth in Ti β21S Monolithic Laminate, *Mater. Sci. Eng.*, Vol A161, 1993, p 45-53

97. J.T. Ryder and W.E. Witzell, Effect of Low Temperature on Fatigue and Fracture Properties of Ti-5Al-2.5Sn (ELI) for Use in Engine Components, *Fatigue at Low Temperatures*, R.I. Stephens, Ed., STP 857, ASTM, 1985, p 210-237

98. N.M. Grinberg, A.R. Smirnov, V.A. Moskalenko, E.N. Aleksenko, L.F. Yakovenko, and V.I. Zmievsky, Dislocation Structure and Fatigue Crack Growth in Titanium Alloy VT5-1ct at Temperatures of 293-11 K, *Mater. Sci. Eng.*, Vol A165, 1993, p 125-131

99. R.J. Lederich, D.S. Schwartz, and S.M.L. Sastry, Effects of Internal Hydrogen on Microstructures and Mechanical Properties of β21S and Ti-15-3, *Beta Titanium Alloys in the 1990's*, D. Eylon, R.R. Boyer, and D.A. Koss, Ed., TMS-AIME, 1993, p 159-169

100. K.V. Jata, W.W. Gerberich, and C.J. Beevers, Low Temperature Fatigue Crack Propagation in a β-Titanium Alloy, *Fatigue at Low Temperatures*, R.I. Stephens, Ed., STP 857, ASTM, 1985, p 102-120

101. L.M. Plaza and A.M. Irisarri, Crack Closure Behaviour of 12 mm Thick Titanium Grade 2 Plates, *Reliability and Structural Integrity of Advanced Materials*, Engineering Materials Advisory Services, Ltd., 1992, p 465-470

102. M. Peters and J.C. Williams, Microstructure and Mechanical Properties of a Welded ($\alpha + \beta$) Ti Alloy, *Met. Trans.*, Vol 15A, 1984, p 1589-1596

103. T.S. Baker, Effect of Stress Relief on Fatigue Crack Propagation and Fracture Toughness of Laser Beam Welded 4 mm Thick Ti-6Al-4V Alloy Sheet in Designing with Titanium, The Institute of Metals, London, 1986, p 267-276

104. M.A. Dauebler and W.H. Stoll, Influence of Welding Parameters and Heat Treatments on Microstructures and Mechanical Properties of Electron Beam Welded Ti-6242, *Sixth World Conference on Titanium*, P. Lacombe, R. Tricot, and G. Béranger, Ed., 1988, p 1409-1414

105. V.T. Troshchenko, V.V. Pokrovskii, V.L. Yarusevich, V.I. Mikhailov, and V.A. Sher, Growth Rate of Fatigue Cracks in Fields of Residual Stresses in Titanium Welded Joints with Different Content of Embrittling Impurities, *Strength Mater.* (USSR), Vol 22, 1990, p 1562-1568

106. A.E. Leach and J.D. McDonough, "Weldability of Formable Sheet Titanium Alloy Ti-15V-3Cr-3Al-3Sn," AFML-TR-78-199, 1978

SELECTED REFERENCES

- R.J.H. Wanhill, Ambient Temperature Crack Growth in Titanium Alloys and its Significance for Aircraft Structures, *Aeronautical J. Royal Aeronautical Soc.*, 1977, p 68-82
- A.R. Rosenfield, An Analysis of Reported Fatigue Crack Growth Rate Data with Special Reference to Ti-6Al-4V, *Eng. Fract. Mech.*, Vol 9, 1977, p 509-520
- Microstructure, Fracture Toughness and Fatigue Crack Growth Rate in Titanium Alloys, A.K. Chakrabarti and J.C. Chesnutt, Ed., The Metallurgical Society, 1987
- J.K. Gregory, Fatigue Crack Propagation in Titanium Alloys, *Handbook of Fatigue Crack Propagation in Metallic Structures*, A. Carpinteri, Ed., Elsevier Science B.V., 1994, p 281-322

Fatigue and Fracture of Nickel-Base Superalloys

Bruce F. Antolovich, Metallurgical Research Consultants, Inc.

NICKEL-BASE SUPERALLOYS have been in continuous use since the late 1930s, when the Nimonic alloys were introduced in the first jet aircraft engines (Ref 1). Some authors consider superalloys to have been developed by Wilhelm Rohn in Germany in the 1920s and 1930s (Ref 2). These nickel-iron-chromium alloys were developed primarily for heat and corrosion resistance and were an early application of vacuum induction melting (Ref 3). Nonetheless, the vast majority of use by tonnage of nickel-base superalloys is found in turbines both for aerospace applications and for land-based power generation. These applications require a material with high strength, good creep and fatigue resistance, good corrosion resistance, and the ability to be operated continuously at elevated temperatures.

Nickel-base superalloys are used primarily in turbine blades (called "buckets" in land-based power turbines), turbine disks, burner cans, and vanes. The operating temperatures of these components range from the relatively mild temperature of 150 °C (300 °F) up to almost 1500 °C (2730 °F). Additionally, several components experience large temperature gradients; for example, turbine disks range from 150 °C (300 °F) at the center to 550 °C (1020 °F) at the rim where the blades are attached. In addition to the high temperatures they must endure, the blades are also subject to an extremely corrosive environment—namely, the products of combustion. The primary loading, which results mainly from centripetal acceleration of the rotating blades and disk, in conjunction with the high temperature, leads to creep deformation. Finally, fatigue cycles result from each engine startup and shutdown as the load changes from zero to maximum and back to zero. For some military engines, thrust settings are varied so greatly that they can also be considered as a fatigue cycle.

Turbine components thus experience thermomechanical loading and fatigue as well as creep-fatigue interactions. The good combination of strength and toughness, as well as an unusual yield behavior (in which the yield strength increases with increased temperature up to about 700 °C, or 1290 °F), continues to make nickel-base superalloys the material of choice for high-performance, high-temperature applications.

Other uses, both actual and proposed, for nickel-base superalloys include:

- Cryogenic applications, such as the compressor section of liquid rocket engines
- Airframe skins for high-speed aircraft and reentry vehicles
- Superconducting applications

Polycrystalline superalloys, although still being developed, are nevertheless quite well understood and will be covered only briefly. The interested reader should consult several other comprehensive reviews (Ref 4-7). This article will cover fracture, fatigue, and creep of nickel-base superalloys, with additional emphasis on directionally solidified and single-crystal applications.

Physical Metallurgy

The microstructure of nickel-base superalloys has a profound effect on their performance. Fortunately, their microstructure is quite simple—consisting of a solid-solution-strengthened austenitic face-centered cubic (fcc) matrix, coherent intermetallic precipitates with an $L1_2$ crystal structure, along with carbides and other phases occurring either in the matrix or at grain boundaries (Ref 8). This microstructure can be significantly influenced by appropriate thermomechanical treatments and composition modifications to create specific microstructures that are resistant to creep, oxidation/corrosion, fatigue crack propagation, and so forth. Compositional modifications are made to affect the microstructure, improve castability of single-crystal and directionally solidified components, enhance various mechanical properties, and decrease susceptibility to environmental attack. These compositional modifications can be quite complex: The dichotomy between an elegant, simple microstructure and a rather complex composition is quite striking.

Although nickel-base superalloys are strengthened by both precipitates and solute atoms, an extensive treatment of these strengthening mechanisms is beyond the scope of this article. Instead, they will be discussed only as they pertain to fracture, fatigue, and creep. Precipitation hardening of nickel-base superalloys has been exhaustively studied and has an elegant fundamental basis. For more detailed information on strengthening mechanisms in nickel-base superalloys, consult Ref 9 to 19.

Phases

The microstructure of nickel-base superalloys consists primarily of the following phases:

- Austenitic matrix, usually called γ
- Coherent intermetallic precipitates, usually called γ'
- Carbides
- Borides
- Topologically close-packed (TCP) phases

Early superalloys had volume fractions of 30 to 40% γ'; more modern superalloys have volume fractions of up to 75%.

Austenitic Matrix. The γ matrix, as stated before, has an fcc crystal structure that can contain solute elements, thereby giving rise to solid-solution strengthening. Typical of solid-solution strengthening, the degree of strengthening corresponds to the difference in atomic size between the base material and the substitutional atoms. Chromium, molybdenum, and tungsten have all been observed to be strong solid-solution strengtheners, whereas others, including cobalt, iron, titanium, and aluminum, have only minor strengthening effects (Ref 20, 21). Both groups of solid-solution elements have been noted to have atomic sizes that vary from nickel by 1 to 13% (Ref 21). Finally, some of these alloying elements serve to increase the flow stress of the alloy as a whole by decreasing the stacking fault energy of the matrix, thereby reducing cross-slip of dislocations (Ref 21). Quantitative estimates of the increase in flow stress are given in Ref 22 and 23.

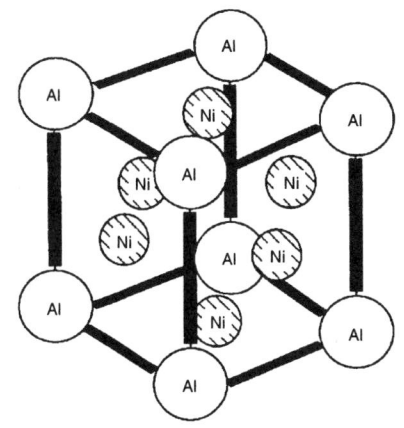

Fig. 1 Unit cell for Ni₃Al. Various elements such as titanium may substitute extensively for the aluminum and significantly affect the mechanical properties of the precipitates.

The γ′ precipitates possess a nominal Ni₃Al composition and have an $L1_2$ crystal structure, often referred to as a derivative fcc structure. The aluminum atoms assume corner positions and the nickel atoms assume the face positions, as shown in Fig. 1. As with the matrix, element substitutions do occur. In fact, γ′ is often referred to as (Ni,Co)₃(Al,Ti) instead of Ni₃Al. The precipitates play a crucial role in strengthening nickel-base superalloys. Additionally, their unusual yield behavior imparts an unusual yield behavior to the entire alloy. (Recall that Ni₃Al has a yield strength that increases with temperature up to about 700 °C, or 1290 °F, and then decreases with increasing temperature.)

Gamma-prime precipitates are spherical or cuboidal, depending on their size and misfit parameter, δ. Mismatch is defined as:

$$\delta = \frac{a_p - a_m}{\bar{a}} \qquad \text{(Eq 1)}$$

where a_p is the lattice parameter of the precipitate, a_m is the lattice parameter of the matrix, and $\bar{a}$ is the average lattice parameter.

The shape of the precipitate is that which will reduce the free energy of the system. The two competing components of free energy are surface energy and strain energy, both of which vary with size and shape. In a general way, the surface energy varies with the square of the characteristic precipitate dimension, whereas the strain energy varies with the cube of this dimension. Thus, at small sizes, where the surface energy dominates, the surface area assumes the smallest possible value by forcing the precipitate to have a spherical geometry. As the size increases, the strain energy dominates and the precipitate assumes a shape to minimize it, even at the expense of the surface energy.

Keeping in mind that fcc structures have cubic elastic symmetry and the [001] directions are "soft," it can be seen that as the size increases, a shape in which strains occur primarily in the [001] directions will dominate since strain along

these directions involves a lower energy. A cube with parallel {001}<100> systems in both the precipitates and the matrix meets these requirements. Increasing γ/γ′ mismatch reduces the required precipitate size to transition to a cuboidal shape, since increasing the mismatch increases the amount of strain that must be accommodated. Of course, the precipitates can change geometry under the effect of mechanical loading provided that there is a nonzero mismatch. This effect has been manifested in the formation of γ′ "rafts," which form perpendicular to the loading direction under creep conditions and may lead to improved creep properties (Ref 24-27). It has been reported that the formation of rafts is usually, though not always, associated with negative mismatch (Ref 27-29).

Carbides occur both in the matrix and at the grain boundaries. Typical carbide compositions include M₂₃C₆, M₆C, and MC. Although carbides were initially thought to be deleterious to the creep behavior of nickel-base superalloys, subsequent experience has shown them to increase the creep resistance of polycrystalline alloys by making grain-boundary sliding more difficult.

The effects that these carbides have on the mechanical behavior of the superalloys depends on their morphology and location. MC carbides have an fcc structure, usually have a coarse cubic morphology, and are distributed randomly throughout the alloy (Ref 21). When they occur in grain interiors, they are often found interdendritically (Ref 21). Common compositions include TiC, TaC, HfC, and CbC (Ref 21). Their effects on mechanical behavior are minimal; rather, their importance is related to their effects on the formation of certain microstructures. M₂₃C₆ carbides, on the other hand, play a very large role in the mechanical behavior of superalloys. They have been observed to form primarily on grain boundaries in alloys with a high chromium content (Ref 21). As stated before, they are generally beneficial and increase creep resistance by preventing grain-boundary sliding. However, they can cause early rupture failures by forming cellular structures at the grain boundary. Finally, M₆C carbides serve to control grain size due to their high-temperature stability in comparison with the other carbides (Ref 21). In essence, they act as zener pinning points for grain boundaries.

Carbides serve other purposes as well, primarily for control and development of microstructure. The interested reader is directed to Ref 21 and 30 to 32.

Borides tend to segregate to the grain boundaries. They generally are of the type M₃B₆ and are quite hard. Their hardness reduces grain-boundary tearing and thus increases the creep resistance of the alloy. Furthermore, boron tends to occupy vacancies on the grain boundaries, thereby reducing the diffusion rates and increasing the creep resistance of the alloy.

Deleterious Phases. In addition to the matrix, γ′ precipitates, carbides, and borides, several additional phases occur that tend to be deleterious to the mechanical properties of the alloy. These

include η, σ, μ, and Laves phases. The η phase has a hexagonal close-packed (hcp) structure and Ni₃X composition and tends to occur at the grain boundaries. Unlike the borides, this is not a desirable phase, for it forms cellular structures at the grain boundary and thus decreases the notch stress-rupture strength. It has also been observed to precipitate intergranularly in a Widmanstätten morphology, reducing strength but not ductility (Ref 4). The σ, μ, and Laves phases all occur in platelike TCP form and generally occur at the grain boundaries (Ref 21). These phases are deleterious at both low and high temperatures. At low temperatures, their hardness and geometry lead to intergranular cracking. At higher temperatures, they reduce the solid-solution strengthening of the γ matrix by depleting alloying elements and also lead to intergranular fracture for the same reasons as listed for low temperatures (Ref 21, 33, 34).

Effects of Grain Boundary and Grain Size

As expected, many of the mechanical properties of superalloys depend on grain size. Furthermore, since it is widely recognized that many failure processes initiate at grain boundaries, the importance of understanding these two parameters is quite evident.

Grain-size effects are in essence a trade-off between good creep resistance and premature failure due to grain sliding and reduced tensile strength (Ref 4). Fine grains are known to reduce creep resistance through increased grain-boundary sliding. On the other hand, excessively large grains can cause the formation of large microcracks. The relative size of the grain to the component has also been found to be important. It has been shown that rupture life and creep resistance increased with increasing component size to grain size ratio (Ref 35). Furthermore, for equivalent component size to grain size ratios, thinner sections exhibited lower rupture strengths. These effects are related to constraint; in thin sections there is little material to impede grain-boundary sliding and creep cracks open up prematurely. Grain size is typically controlled through carbides, γ′ precipitates, and other particles that provide pinning points through a zener mechanism.

As one would expect, grain-boundary chemistry plays an important role. Small additions of zirconium and boron, which segregate to the grain boundaries presumably because of their odd size compared to the other alloying elements, have been found to dramatically increase creep properties (Ref 36-39). Although the reasons for the increased life are not completely clear, it has been postulated that the zirconium and boron atoms segregate to the grain boundaries and fill vacancies, thus reducing the coefficient of diffusion and hence creep (Ref 40). Finally, the grain-boundary chemistry has been modified to allow casting of more intricate shapes through the addition of hafnium. Hafnium additions serve to form stable, randomly arranged, fine MC carbides on the grain boundaries (Ref 34). These stable carbides serve to inhibit the amount of $M_{23}C_6/M_6C$ carbides formed on the grain boundaries. If there were no inhibition, there would be an excess of these carbides, which could interconnect and provide an easy path for cracking. Furthermore, the addition of hafnium raises the oxidation resistance of the superalloy as a whole.

Single-Crystal and Directionally Solidified Alloys

Single-crystal (SX) and directionally solidified (DX) components were recognized quite early as a possible technique to increase the temperature creep resistance of nickel-base superalloys—primarily through the elimination of boundary-initiated cracking mechanisms. Although SX and DX alloys are designed to overcome many of the same problems, DX alloys are also affected by the presence of grain boundaries parallel to the primary loading direction. Complex part geometries and high thermal gradients produce enough loading perpendicular to the grain boundaries that grain-boundary strengthening must be addressed. Even with these complications, the dramatically lower production costs associated with DX blades has proved to be a continuing driving force behind their development. In fact, the second generation of DX alloys now outperforms their first-generation SX counterparts (Ref 41). However, the second generation of SX alloys outperforms their DX counterparts.

In general, most of the comments for polycrystalline alloys apply to SX and DX alloys. Obvious exceptions include grain-boundary effects for single crystals—although even this is not quite straightforward in engineering practice, for it is quite rare that SX components are true monocrystals. Manufacturing techniques have necessarily included compositional modifications to improve the castability of these alloys. It is important to remember that even though the microstructures of polycrystalline, DX, and SX alloys appear similar in the grain interiors, the grain-boundary effects play an important role in determining the mechanical behavior of these alloys. Attention will be focused on the SX alloys.

Single crystals, similar to individual grains within polycrystalline alloys, typically develop dendrites along <100> directions. The concentration of carbides such as $M_{23}C_6$ is much lower than their polycrystalline counterparts. This is because there is no requirement for grain-boundary hardening, and carbide formers can be kept at a low level. Therefore, the carbon content is generally kept below 50 ppm.

Casting Defects and Compositional Modifications. The principal defects found in SX and DX alloys include spurious grain boundaries, equiaxed grains, "freckles," low-angle boundaries, subgrains, splaying, recrystallized grains, and porosity (Ref 42). Compositional additions are used to control these defects and, with one important exception, are very similar for SX and DX alloys.

Table 1 First-generation SX superalloys

Alloy	Nominal composition, wt %										
	Cr	Co	Mo	W	Ta	V	Nb	Al	Ti	Hf	Ni
PWA 1480	10	5	...	4	12	...	...	5	1.5	...	bal
René N4	9	8	2	6	4	...	0.5	3.7	4.2	...	bal
SRR 99	8	5	...	10	3	...	...	5.5	2.2	...	bal
RR 2000	10	15	3	...	...	1	...	5.5	4	...	bal
AM1	8	6	2	6	9	...	...	5.2	1.2	...	bal
AM3	8	6	2	5	4	...	...	6	2	...	bal
CMSX-2	8	5	0.6	8	6	...	...	5.6	1	...	bal
CMSX-3	8	5	0.6	8	6	...	...	5.6	1	0.1	bal
CMSX-6	10	5	3	...	2	...	...	4.8	4.7	0.1	bal
AF 56	12	8	2	4	5	...	...	3.4	4.2	...	bal

Table 2 Second-generation SX superalloys

Alloy	Nominal composition, wt %									
	Cr	Co	Mo	W	Ta	Re	Al	Ti	Hf	Ni
CMSX-4	7	9	0.6	6	7	3	5.6	1	0.1	bal
PWA 1484	5	10	2	6	9	3	5.6	...	0.1	bal
SC 180	5	10	2	5	9	3	5.2	1	0.1	bal
MC2	8	5	2	8	6	...	5	1.5	...	bal

Freckles consist of small chains of equiaxed grains oriented parallel to the <001> solidification direction (Ref 43). They have been attributed to convective instabilities associated with compositional segregation during the solidification process. A good correlation has been established between increased tendency to freckle and high levels of rhenium and tungsten and low levels of tantalum. As rhenium is a desirable compositional addition due to its strengthening effect, many of the SX alloys have elevated tantalum compositions (Ref 44) in an effort to counterbalance the effect of high levels of rhenium. Other elements do not significantly affect freckling (Ref 43). Freckle formation has also been shown to be a function of dendrite arm spacing, which in turn is a function of the cooling rate. Increasing the cooling rate to a critical level has also been shown to suppress freckle formation (Ref 16).

Spurious grain growth affects both SX and DX superalloys. Simply put, spurious grain growth occurs when a grain grows in a completely arbitrary direction as a result of constitutional supercooling and/or nucleating agents in the melt ahead of the liquid/solid interface. Usually, these grains are not equiaxed but rather are long and slender, having aspect ratios of up to 3:1 (Ref 43). As with the case of freckles, compositional modifications (e.g., increased tantalum levels) and appropriate heat treatments (e.g., increased cooling rates) can reduce or eliminate these defects.

Composition Modifications for Improved Performance. Not all compositional modifications are made in order to address castability issues. Several elements have been shown to be very effective in increasing the high-temperature performance of SX and DX nickel-base superalloys. One compositional modification that has been associated with the transition from first- to second-generation SX alloys has been the addition of rhenium. This element acts as a solid-solution strengthener by partitioning to the γ matrix and increasing the γ/γ′ misfit (Ref 45). Furthermore,

small rhenium clusters form and act as additional barriers to dislocation movement (Ref 46).

Tables 1 and 2 show the compositions of first- and second-generation SX alloys, respectively. Comparable alloys from Canon-Muskegon and Pratt & Whitney show this rhenium addition quite clearly, as evidenced by the CMSX-2 to CMSX-4 and PWA 1480 to PWA 1484 developments. A third generation of SX alloys is continuing this trend; for example, CMSX-10 has 6 wt% Re (Ref 47). In fact, rhenium has been such a success as a solid-solution strengthener that its use is spreading to polycrystalline nickel-base superalloys. Unfortunately, rhenium tends to promote freckling and usually requires increased levels of tantalum to counteract this tendency.

Fatigue Crack Propagation

Fatigue crack propagation (FCP) in nickel-base superalloys is very important. Rates of FCP depend on a variety of intrinsic and extrinsic parameters. Intrinsic parameters include the physical metallurgy, mechanical metallurgy, and microstructure of the alloy (as well as others), and extrinsic parameters include factors such as temperature, environment, and loading histories. Considering the number of variables, development of an all-inclusive model to predict rates of FCP has proved to be impossible; instead, empirical or phenomenological approaches are typically used for quantitative rate predictions. Such predictive models usually are based on fracture mechanics. In addition to quantitative FCP rate modeling, the qualitative effects of intrinsic and extrinsic parameters have been the subject of much research. Finally, with the introduction over the last two decades of SX and DX components, the existing FCP rate models have become less useful. The fracture behavior of SX and DX components does not correspond to the tradi-

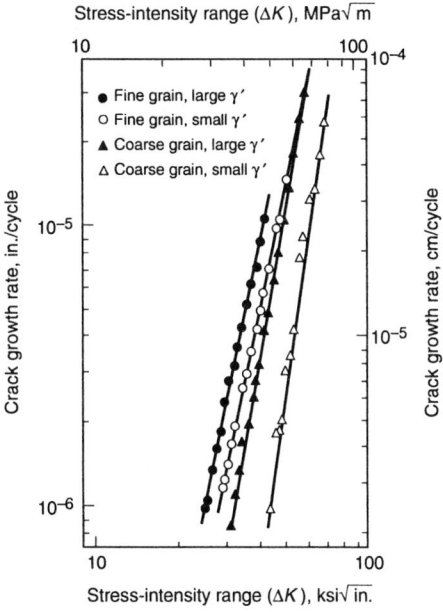

Fig. 2 Dependence of FCP rate on γ′ size and grain size in Waspaloy tested at room temperature. Source: Ref 50

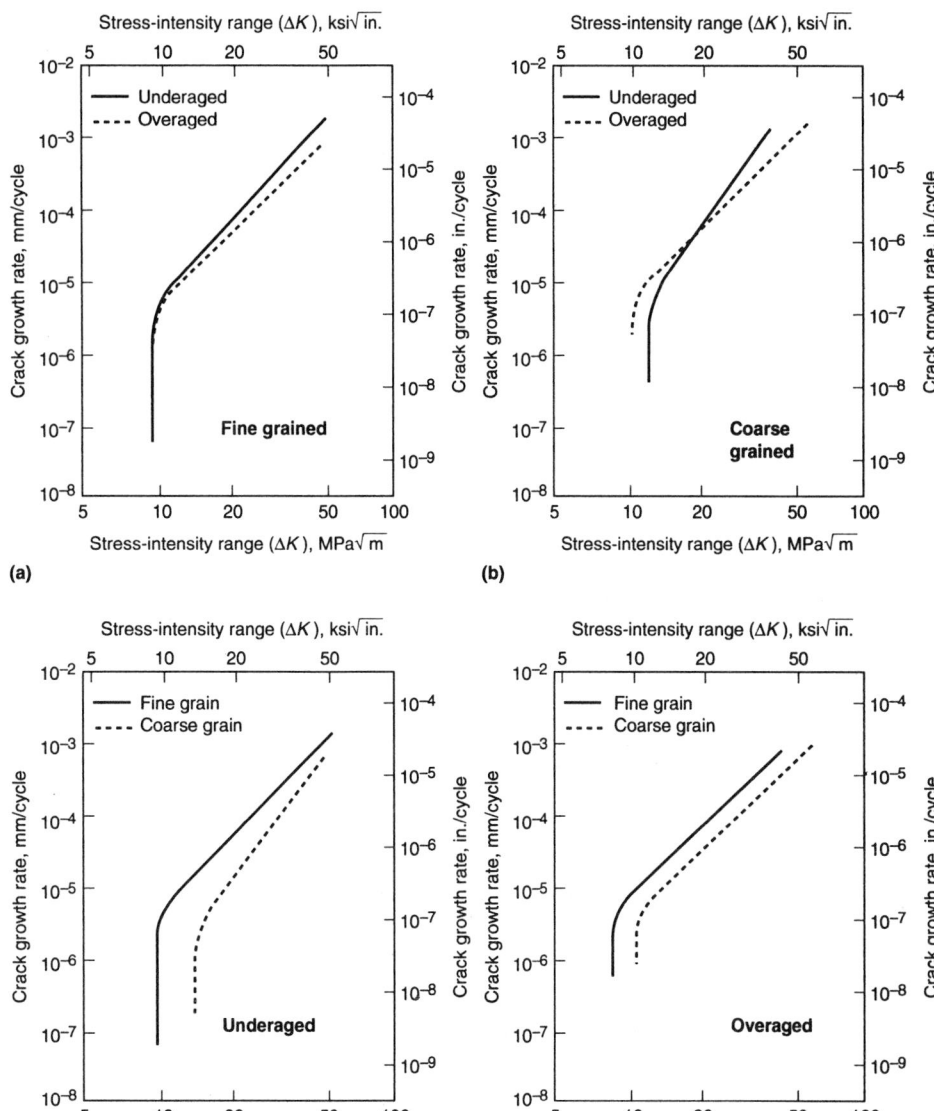

Fig. 3 Effect of precipitate and grain size on FCP rates in Inconel 718 tested in air at 425 °C (800 °F), $R = 0.05$, 0.33 Hz. (a) Fine grain; underaged versus overaged. (b) Coarse grain; underaged versus overaged. (c) Underaged; fine grain versus coarse grain. (d) Overaged; fine grain versus coarse grain. Source: Ref 51

tional fracture mechanics basis of many of these models.

Effects of Microstructure

The effects of microstructure on FCP of nickel-base superalloys have been studied and reported on extensively. Usually, papers with experimental data present the results of a study on the behavior of a single nickel-base superalloy for which the effects of microstructural variables such as γ′ size and morphology and grain size are examined. The experiments are conducted under certain external conditions, such as temperature, environment, waveform loading type, and so on. These conditions are rarely constant from paper to paper; as such, it is somewhat difficult to establish trends that can be applied to all alloys. Instead, the predictions should be viewed as a starting point.

The effects of grain size and γ′ were studied for four different alloys: Inconel 718, Waspaloy, Astroloy, and René (Ref 48-51). The results of these studies universally showed a high dependence of FCP rate on grain size and a possible dependence on γ′ size. Increasing the grain size and decreasing the γ′ size were observed to decrease the FCP rate. This dependence on grain size has been reported in several other studies (Ref 48-50, 52, 53). Additionally, it has been reported that FCP rates increase with increasing strength (Ref 53).

Results from a study on Waspaloy (Ref 50) examined the effects of grain size and γ′ size. Heat treatments were used to provide specimens with grain sizes of either ASTM 3 or 9 with γ′ sizes of either 8 or 90 nm. Regardless of γ′ size, coarse-grain specimens always had lower propagation rates than fine-grain specimens. Specimens with equivalent grain sizes showed lower FCP rates for small γ′ rather than large γ′. These results are shown in Fig. 2. Studies taken from

both low-cycle fatigue (LCF) and FCP studies were used to examine the deformation mechanisms. The large-grain/small-precipitate specimens were found to exhibit particle shearing by the dislocations, whereas the small-grain/large-precipitate specimens deformed by looping. The intermediate rates were found to depend slightly on the degree of deformation inhomogeneity; lower rates of FCP corresponded to greater degrees of inhomogeneity.

In a study of Inconel 718, the effects of γ″ size and grain size were likewise examined (Ref 51). Grain sizes of 250 and 25 μm were examined, with precipitate sizes of 150 and 20 nm diam disks. Rates of FCP were found to exhibit the same dependence on grain size as in the previously cited study. The FCP dependence on γ″ precipitate size, however, was not the same. For

the fine-grain specimens, larger precipitates were observed to have lower FCP rates than small precipitates. Coarse-grain specimens exhibited two regimes: one in which specimens with small precipitates exhibited lower FCP rates and another in which they exhibited higher FCP rates (Fig. 3). These results, when viewed only in terms of precipitate size effect, seem to contradict the results of the previous study. Investigation of the deformation mechanisms suggested that shearing was operative regardless of precipitate size. Furthermore, deformation in the specimens with large precipitates was found to exhibit faulting, with correspondingly more inhomogeneous slip than specimens with small precipitates. When viewed in terms of deformation inhomogeneity, the Inconel 718 results appear to be in accordance with the previously cited Waspaloy results.

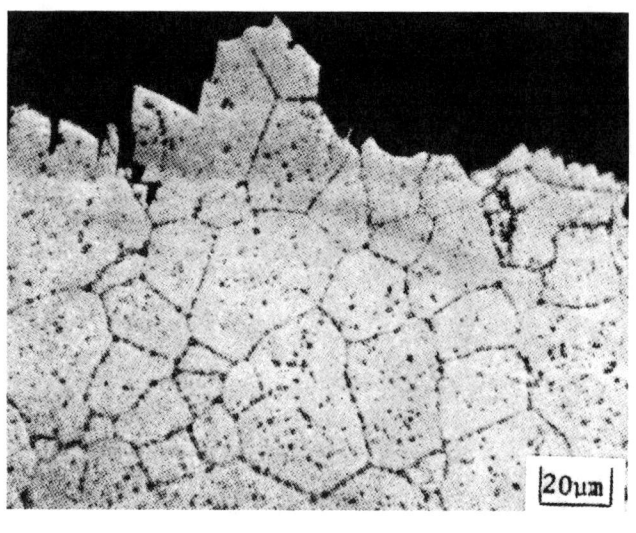

(a)

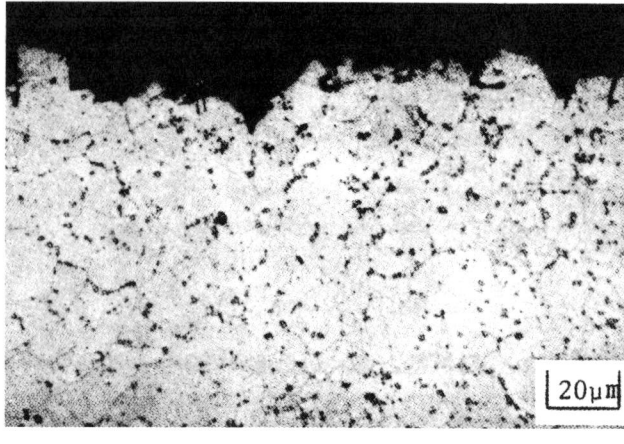

(b)

Fig. 4 Fracture surfaces of René 95 tested at 540 °C (1000 °F) and 0.34 Hz. (a) Material had coarse grain size, fine γ′ precipitates (~0.1 μm), and a crystalline appearance. (b) Material had fine grain size and contained large γ′ precipitates (~0.3 μm), and the fracture surface was flatter than that in (a). Both micrographs correspond to ΔK of 65 MPa√m (60 ksi√in.). Source: Ref 111

Another study examined the effects of microstructural variables such as antiphase boundary (APB) energy, volume fraction of γ′ precipitates, and γ/γ′ mismatch using model alloys of varying composition and heat treatments (Ref 54). The results, when viewed solely in terms of slip planarity and reversibility, show that increases in both factors have the effect of reducing the rate of FCP (Ref 55). Other published works are in agreement with these conclusions (Ref 56).

A possible explanation for the correlation between increased slip planarity and lowered FCP rates may be based on surface roughness effects (Ref 57). Although the effects and mechanisms of roughness have been extensively investigated, essentially surface roughness lowers the effective value of ΔK by contact of the opposing fracture surfaces prior to the complete release of load. If an alloy exhibits high slip planarity, the fracture surfaces usually will take on a faceted appearance, where the facets occur on the octahedral planes of individual grains. Although each facet will present a smooth face, the unoriented polycrystalline nature of the material ensures that these facets will have different orientations, thus creating surface roughness.

Figure 4 shows the surface roughness of René 95 specimens with differing grain sizes. The surface area of this crack can be considerably larger than a projected surface area that would correspond to a "smoother" crack surface, as shown in Fig. 5. Thus, crack growth rates, which are usually measured macroscopically and correspond to projected crack lengths, may in fact be higher than indicated. It has also been proposed that the local stress-intensity factor is lower due to the "slant" nature of the local crack, resulting in lower propagation rates. Such an effect would be expected to be secondary in nature, for numerical results show that small cracks that deviate from the expected path normal to the load by up to 30°

have a stress-intensity factor virtually identical to cracks normal to the loading direction (Ref 58). Nonetheless, the effects of surface roughness on lowering FCP rates are well documented.

Although the effects of intrinsic parameters on FCP rates can be easily seen, the complexity of the interrelationships between various intrinsic parameters prevents a single conclusion based on independent parameters such as γ′ size and morphology or grain size. Rather, the microstructure and physical metallurgy must be controlled to produce a desired effect—namely, the promotion of planar slip through methods such as lowered APB and/or γ/γ′ mismatch or large grains and small γ′ precipitates.

Effects of Extrinsic Parameters

Some common parameters that are considered to be extrinsic in nature include temperature, frequency, environment, specimen geometry, K_{max}, K_{min}, ΔK, R ratio, and overload ratio (Ref 59). The effects of extrinsic parameters are very difficult to address individually; they all, to one extent or another, affect one another. In addition to the normal extrinsic parameters relating to loading, the environmental parameters are particularly important for nickel-base superalloys, which are almost invariably used in an extremely high-temperature, corrosive environment.

Temperature/Environment Interactions. The importance of environmental effects has been discussed and researched extensively (Ref 60). A study of René 95 compared the LCF rates of specimens with subsurface cracks with those of specimens containing surface cracks. Presumably the specimens with subsurface cracks would be "shielded" from the effects of environment. As expected, the lives of the specimens with surface

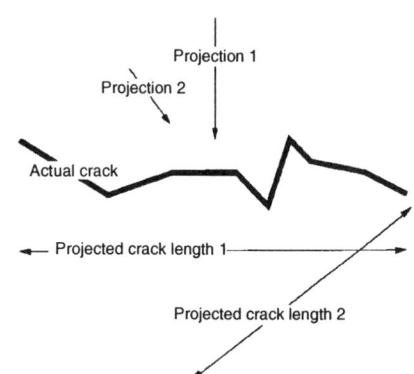

Fig. 5 Various measures of crack length illustrating the difference between projected and surface measurements. Note that the projected lengths are always less than the total length as measured by the "perimeter." Assuming concurrency of loading direction and direction 1, projected crack length 1 would correspond to that measured using a traditional measuring technique, such as traveling microscope or electropotential drop. It also corresponds to crack lengths of "smooth" cracks.

cracks were much shorter than those with subsurface cracks.

These findings were extended by a study of environmental effects on FCP (Ref 61). For a given value of ΔK, da/dN was found to be heavily dependent, although not monotonically, on temperature (Fig. 6). However, in the region most applicable to turbine engines (i.e., at temperatures above 500 °C, or 930 °F), increasing temperature was observed to increase the rate of crack propagation. For an assumed thermally activated singular environmental interaction, an Arrhenius-type equation is expected to describe the effects of temperature on FCP rate:

$$\frac{da}{dN} = A \exp\left\{\frac{-Q\,(\Delta K)}{RT}\right\} \qquad \text{(Eq 2)}$$

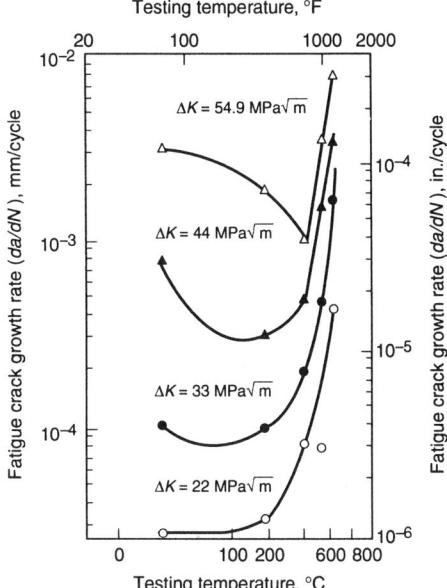

Fig. 6 Effects of temperature on FCP rates of René 95 for constant ΔK. Source: Ref 61

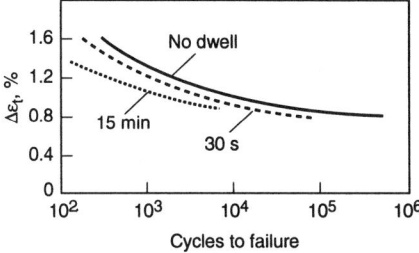

Fig. 7 Effect of dwell time on fatigue life of powder metallurgy Inconel 100 tested at 650 °C (1200 °F). Tests with no dwell were conducted at 0.33 Hz. Source: Ref 69

on the crack surface and at the crack tip accompanied by lower FCP rates. Reduction of FCP rate for the case of oxidation products forming at the crack tip has been reported to be caused by the prevention of crack resharpening during unloading (Ref 66). Alternatively, it has been proposed that microcreep at the crack tip is responsible for crack-tip blunting with consequent reduction in stress intensification (Ref 4). Such an explanation would obviously introduce a hold time and/or frequency effect as well.

Without crack-tip resharpening, the stress intensification is reduced along with FCP rates. Such an explanation would predict decreasing FCP rates with increasing temperatures and/or tensile hold times/lower frequencies. Experimental studies of René 95 revealed these trends (Ref 67). However, there is a clear changeover point on a da/dN versus ΔK curve. Above a certain ΔK level, the FCP rates show a discontinuous increase—possibly associated with the onset of crack-tip oxide cracking. As previously mentioned, oxides have also been observed to form as asperities on the fracture surface. These oxides then may contribute to surface-roughness-induced closure, thus reducing the effective ΔK levels and consequent crack growth rates.

The previous examples show several modes of affecting FCP rates. One type of mode is purely extrinsic, such as directly controlling ΔK. In another, extrinsic parameters such as environment indirectly affect other extrinsic parameters such as oxide-induced closure (as opposed to the purely intrinsic parameter of grain-size effects on surface roughness and consequent closure). Finally, some extrinsic parameters affect intrinsic parameters such as cracking mechanisms. Often there are competing effects. For example, a study on the effect of temperature and environment on FCP rates in model nickel-base superalloys showed that although the intrinsic fatigue resistance as measured by FCP rates was up to 30 times greater in high vacuum than in air, the rates in vacuum were not 30 times lower due to an absence of oxide-induced closure effects (Ref 70). (As an aside, this study pointed out that as FCP rates increase, the behavior becomes more and more like that of specimens tested in vacuum, due to a lack of time for diffusion penetration at the crack tip.)

Although the effects of loading waveform and frequency have been discussed briefly, a few additional remarks are in order. First, a large effect on fatigue life is shown not only by time per cycle (Ref 71) but also by waveform (Ref 4). For example, the loading rate of a triangle waveform will be lower than that of a square waveform with the same frequency. The change in loading rate can affect the deformation mechanisms and hence the FCP rates.

Second, the retardance or acceleration of FCP rates depends strongly on temperature and frequency. Changing deformation and cracking mechanisms and formation of oxides all are temperature dependent and can accelerate or retard crack growth. Extremely high time-based crack growth rates due to high ΔK levels or high load-

ing frequencies tend to "shield" a material from environmental effects.

Third, the loading waveform has a significant effect on FCP lives. Introduction of a dwell time through a trapezoid or square waveform can have differing effects, depending on material properties, length of dwell, and load ratio. The effects of compressive dwell, as shown in Fig. 7, seem to be largest for fine-grain superalloys (Ref 69). It also should be noted that dwell effects cannot be explained entirely in terms of creep or microcreep. Compressive dwell (negative R ratio) has been found to be particularly damaging (Ref 69-72).

Unfortunately, the complexity of environmental interactions makes it difficult to predict quantitatively their effects on FCP rates. In fact, the large number of competing mechanisms even makes it difficult to predict qualitatively the effects on FCP rates. Experimental results can help the engineer understand which competing mechanisms are operating and eventually which will dominate. However, even with this understanding, it is quite difficult, even under laboratory conditions, to quantitatively predict FCP rates for untested environment/temperature combinations. For the present, in order to accurately predict FCP rates, tests generally must be run under conditions that are as close as possible to those met in service by the part. Regardless of the difficulty of quantitative prediction, understanding the mechanisms of crack advance are very useful in making incremental design changes for new alloys.

Modeling FCP Rates

Most, though not all, of the modeling of FCP rates in nickel-base superalloys used for life prediction has been empirical in nature and frequently based on fracture mechanics arguments and parameters. As such, results are usually valid only for a single temperature/environment/loading combination. For example, one empirical model is given by (Ref 73):

$$\log \left(\frac{da}{dN} \right) = C_1 \sinh \left(C_2 \log \Delta K + C_3 \right) + C_4 \qquad \text{(Eq 3)}$$

where C_1, C_2, C_3, and C_4 are empirical constants that are condition specific; and ΔK is the stress-intensity range ($K_{max} - K_{min}$). Aside from the obvious dependence on the temperature/environment combination, the empirical constants C_1 to C_4 are also

where A is a constant and $Q(\Delta K)$ is the apparent activation energy. As implied, the apparent activation energy is a function of ΔK. Possible explanations include a dilatation of the lattice due to the imposed strains ahead of the crack tip (Ref 62). This dilatation would be expected to be linear with the strains and hence with $(\Delta K)^2$. The apparent activation energy would then be expected to decrease linearly with $(\Delta K)^2$. Such behavior has been noted in other materials, such as steel (Ref 63) and titanium (Ref 64). It should be strongly emphasized that this Arrhenius-type equation only describes thermally activated mechanisms. For example, temperature-dependent crack-tip oxide cracking would not be modeled well by the Arrhenius equation over all temperature ranges.

Not all temperature-dependent increases in FCP rates can be modeled with Arrhenius-type equations. Rather, environmental effects can change the cracking mechanisms. For example, fracture surfaces may undergo a transition from intergranular to transgranular cracking with changes in environment, as was shown for Inconel 718 (Ref 65). Tests conducted at 650 °C (1200 °F) at a frequency of 0.1 Hz with sinusoidal loading and $R = 0$ showed transition from transgranular to intergranular with a change from helium to oxygen and sulfur-bearing environments. Such results suggest a weakening of grain boundaries and consequent formation of preferential crack paths along the boundaries. Fatigue crack propagation rates increased dramatically under the presence of the hostile oxygen and sulfur-bearing environments. Interestingly, unstressed prior exposure to the hostile environments did produce significant surface attack, thus precluding exposure tests as an indicator of potential environmental effects.

Environmental effects do not always lead to higher FCP rates (Ref 4). One frequently observed environmental effect is oxide formation

heavily dependent on R, loading frequency, and loading waveform. While it is easy to determine the values of C_1 to C_4 through experimentation and regression analysis, a significant amount of testing must be done to cover appropriate temperature/environment/loading combinations. (This form of equation also suggests symmetry about an inflection point—behavior which has not been observed experimentally or justified physically.) An alternative that also allows the direct introduction of R-ratio effects is given by (Ref 74):

$$\frac{1}{\left(\frac{da}{dN}\right)} = \frac{A_1}{(\Delta K)^{n_1}} + \frac{A_2}{(\Delta K)^{n_2}} + \frac{A_2}{[K_c\,(1-R)]^{n_2}} \qquad \text{(Eq 4)}$$

where A_1 and A_2 are functions of the load ratio, R. Again, a fair amount of experimental work must be carried out in order to develop the constants. Another possible model incorporates not only the load ratio but also the fracture toughness of the material (Ref 75):

$$\frac{da}{dN} = \frac{C\,\Delta K^n}{[(2-R)\,K_{Ic} - \Delta K]} \qquad \text{(Eq 5)}$$

where K_{Ic} is the fracture toughness of the material. Finally, one of the most widely used and recognized models is the classic Paris law equation (Ref 76):

$$\frac{da}{dN} = C\,(\Delta K)^n \qquad \text{(Eq 6)}$$

For superalloys at room temperature in intermediate crack growth, a correlation exists between the FCP rate and Young's modulus of the alloy and can be incorporated into the Paris law (Ref 77):

$$\frac{da}{dN} = 1.7 \times 10^6 \left(\frac{\Delta K}{E}\right)^{35} \text{ m/cycle} \qquad \text{(Eq 7)}$$

As can be seen, there are many available empirical models, ranging from the simple and effective Paris law to more complicated models. It must be noted that these equations are useful solely for life predictions but do not give insight into the mechanisms of crack propagation or techniques to improve life under FCP. A compilation of models of this class can be found in Ref 78.

A great deal of work has attempted to integrate intrinsic parameters into life prediction models. Some are based on dislocation arguments (Ref 79, 80) and are summarized in Ref 59. Other models, originating with McClintock (Ref 81), try to incorporate LCF models with FCP models based on the assumption that the cracking directly at the crack tip is essentially an LCF-type behavior in which damage occurs in a process zone (Ref 82-85). A common element of these two types of models is extraordinarily complicated equation forms. For example, the dislocation-based model presented by Yokobori et al. (Ref 80) can take the form (Ref 59):

Fig. 8 Crystallographic crack morphology from a CT specimen of CMSX-2 tested at room temperature with a secondary orientation of [110]. Source: Ref 103

$$\frac{da}{dN} \propto \left(\frac{\Delta K}{\sigma_{cy}\sqrt{s}}\right)^{[2n'\,(p+1)^2/(1+n')(p+2)] + [1/(1+n')]} \qquad \text{(Eq 8)}$$

where σ_{cy} is the initial cyclic yield stress, n' is the cyclic strain-hardening exponent, p is the exponent on the stress in the equation for dislocation velocity, and s is the characteristic distance near the crack tip over which applied shear stress is averaged.

The integrated LCF/FCP model of Antolovich et al. (Ref 82) takes on the form:

$$\frac{da}{dN} = \frac{C}{(\sigma'_{yc}\,\varepsilon'_f E)^{1/\beta}} \cdot \frac{1}{L^{1/(\beta-1)}} \Delta K^{2/\beta} \qquad \text{(Eq 9)}$$

where σ'_{yc} is cyclic yield, E is Young's modulus, C is a constant, and L is the assumed LCF process zone size. As can be seen, there is direct integration of intrinsic material properties into each of these models, but each retains a ΔK dependence.

Single-Crystal Alloys

Fatigue modeling of SX nickel-base superalloys has been the subject of much research (Ref 70, 88-102). Although numerous alloy systems and testing variables have been examined, many observations have been made that appear to be generally applicable to SX nickel-base superalloys with large volume fractions of γ' precipitates. These generally applicable observations apply both to macroscopic fracture surface morphologies and to the effect of changing testing variables (e.g., environment, temperature, and load ratio). However, on the microstructural level, generalization of observations does not appear to be as applicable.

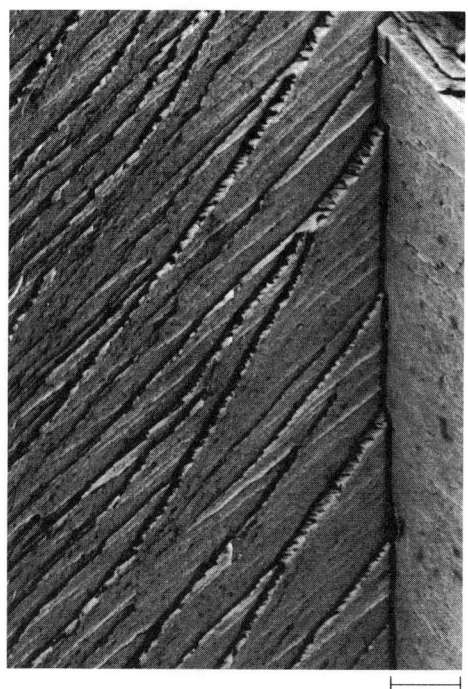

Fig. 9 High-magnification view of Fig. 8 showing crystallographic cracking. Note that the cracking occurs on multiple {111} planes. Source: Ref 103

Fracture Surface Morphologies and Crack Advance Mechanism. Fracture surfaces on most SX nickel-base superalloy specimens, regardless of temperature, environment, crystallographic orientation, or even applied state of stress, appear to have three common aspects (Ref 86-88, 90, 92, 94, 96, 97, 99, 101, 102):

- Regions in which the crack appears to propagate on {111} planes and shows no evidence of being deflected by γ' precipitates. Figures 8 and 9 show an example of this cracking morphology (Ref 103).
- Regions in which the crack propagates macroscopically on a plane normal to the loading direction, with γ' precipitates and the residual dendritic structure clearly visible on the surface. Figure 10 shows an example of this precipitate avoidance morphology (Ref 103).
- Regions in which the crack grows macroscopically normal to the loading direction, with no clear crystallographic plane or γ' precipitates visible

The relative extent of these regions and their locations all change with changing variables and materials. However, all three regions appear on all specimens, with few exceptions. Figure 11 shows fracture surfaces from specimens with two different orientations that exhibit all three morphological aspects (Ref 105).

Crystallographic crack propagation has been observed in both single-crystal and polycrystal nickel-base superalloys (Ref 87, 88, 90, 92, 94, 96, 97, 99, 101, 102). These regions have been referred to as "faceted" in polycrystals, as

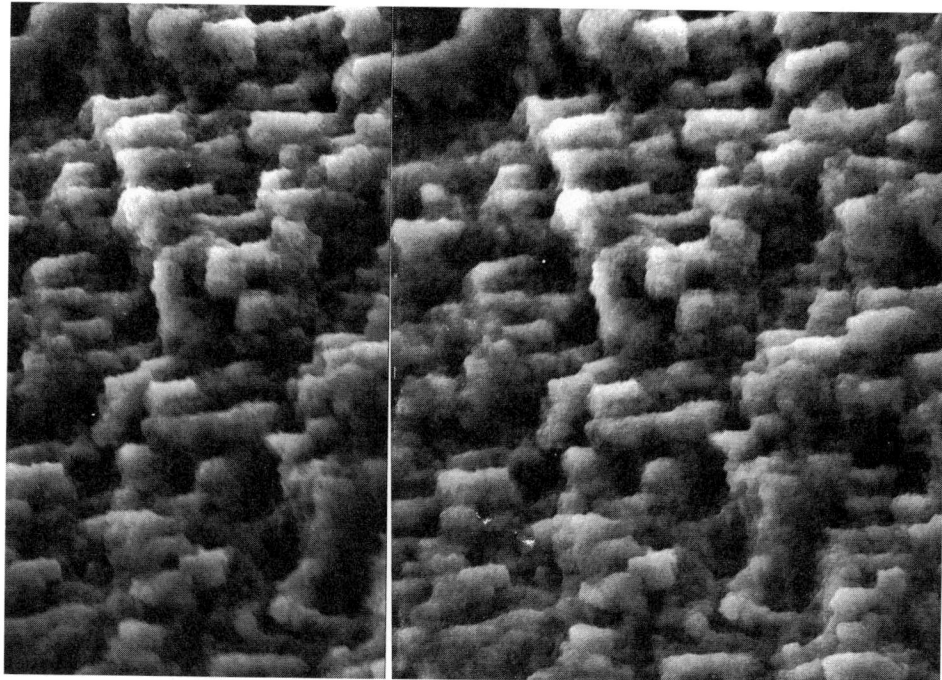

Fig. 10 Stereo view of precipitate avoidance morphology on crack surface. Gamma-prime precipitates are clearly visible. Source: Ref 103

⊢————⊣
2 μm

well as in single crystals where the length of cracking on a particular {111} plane is small. In single crystals, the crystallographic crack growth appears always to occur on {111} planes (i.e., no crystallographic cracking on cube planes). For those specimens tested under nominal single-mode conditions such as those created by compact-tension (CT)-type specimens, the crack transitions between intersecting {111} planes in such a fashion as to attempt to maintain a macroscopic crack plane normal to the loading direction (Ref 90, 101). However, depending on the loading direction relative to the crystallographic orientation, the crack may grow macroscopically on a plane that is not normal to the loading direction (Ref 94). Even so, as the crack grows farther and farther from the plane normal to the loading direction, it will transition to an intersecting {111} plane that returns it closer to the normal plane (Ref 94, 96, 102).

Noncrystallographic Crack Propagation. Two types of noncrystallographic crack propagation are typically observed: one in which the γ' precipitates are clearly visible on the fracture surface and another in which they are not (Ref 90, 96). Gamma-prime precipitates do not appear on the fracture surface near catastrophic failure where a finite amount of crack growth appears with each cycle. Since this region is of limited interest, it will not be discussed further. When the fracture surface does contain γ' precipitates, the relative amount of this noncrystallographic crack growth appears to be affected by crystallographic orientation, temperature, environment, and applied ΔK. The effect of these parameters on the promotion of noncrystallographic crack growth is

inverse to that for crystallographic crack growth. This is to be expected, because the fracture surface is composed only of noncrystallographic regions and crystallographic regions.

Influence of External Variables on Crack Growth Morphologies. The extent of crystallographic crack growth appears to be influenced by temperature, environment, crystallographic orientation, and applied ΔK_I. The effects of changing crystallographic direction do not lend themselves to generalization. Increasing the temperature generally decreases the amount of crystallographic crack growth, regardless of environment. There is not widespread agreement on the mechanisms responsible for this behavior.

The effects of environment can be shown most clearly by comparing the results of tests in laboratory air and in vacuum, where other testing conditions are held constant. At lower temperatures, the effect of changing the environment appears to be inconsequential to the amount of crystallographic crack growth. At elevated temperatures, specimens tested in vacuum have more crystallographic crack growth than those tested in air. Those tested in air sometimes exhibit no observable crystallographic crack growth. It has been proposed that the reduction of crystallographic crack growth is tied to diffusion of oxygen (Ref 99). Although single crystals have no grain boundaries for oxygen diffusion, the residual dendritic structure has been proposed as a diffusion path for oxygen (Ref 97). According to this proposal, the diffusion paths are weakened by the oxidation, resulting in a preferred path for crack propagation. The diffusion arguments are particularly compelling—explaining both the en-

vironmental effects and the temperature effects. A transition from noncrystallographic to crystallographic and back to noncrystallographic crack growth has been reported when going from low to high ΔK values. The physical basis for this correlation has not been established. Furthermore, the applicability of ΔK for this class of material is uncertain. Nonetheless, it has been widely reported that crystallographic cracking occurs only at intermediate values of ΔK_I.

The fact that crystallographic cracking occurs on {111} planes suggests that cracking is intimately tied to glissile dislocation movement on slip planes. This has led to predictive models for planes of crack propagation based on stress fields affecting dislocation movement. One proposed model suggests that dislocation movement on {111} planes damages these planes, thereby weakening them (Ref 90). A normal stress to these planes can then break them apart, advancing the crack. A parameter based on the product of the shear stress resolved in the direction of the Burgers vector and the normal stress to the plane containing the Burgers vector has been proposed as a crack driving force (Ref 90). The slip system that maximizes this parameter will dictate the plane on which crack propagation will occur. This proposed model provides a rational mechanism for the observed non-self-similar crack growth.

Fatigue Crack Growth Rate Modeling. Although many researchers have investigated fatigue behavior in single crystals or in materials with a very large grain size, only a few have proposed new models or modified existing models (Ref 90, 92, 94, 96, 97, 102). These models are based on either micromechanics or macroscale observations. The former approach is usually dislocation based, whereas the latter is usually based on global quantities such as energy release rate. Two factors that further complicate modeling efforts include the elastic anisotropy of the material and the non-self-similar crack growth. An example of the elastic anisotropy of PWA 1480 is shown in Fig. 12.

A significant body of experimental evidence suggests that fatigue crack growth in both SX and polycrystalline materials is intimately tied to dislocation emission from the crack tip (Ref 104-106). Three models have been proposed in which this association with crack advance is the basis for proposed correlating factors for crack growth rates (Ref 90, 96, 102). A common element of all three approaches is the utilization of the resolved shear stress in the direction of the Burgers vector, τ_{rss}. The resolved shear stress depends on all elements of the stress tensor, which in turn depends on all three modes of stress-intensity factor (Ref 107):

$$\sigma_{ij} = \frac{K_i}{\sqrt{2\pi r}} f_{ij}(\theta, \mu) \qquad \text{(Eq 10)}$$

i = Ia, IIa, or IIIa (anisotropic modes)

Note that the anisotropic forms of the field equations are necessary due to the anisotropy of single crystals

CMSX-2 Specimen B7

Environment: Vaccuum

Machine Notch

Precrack

FCP

Fast Fracture

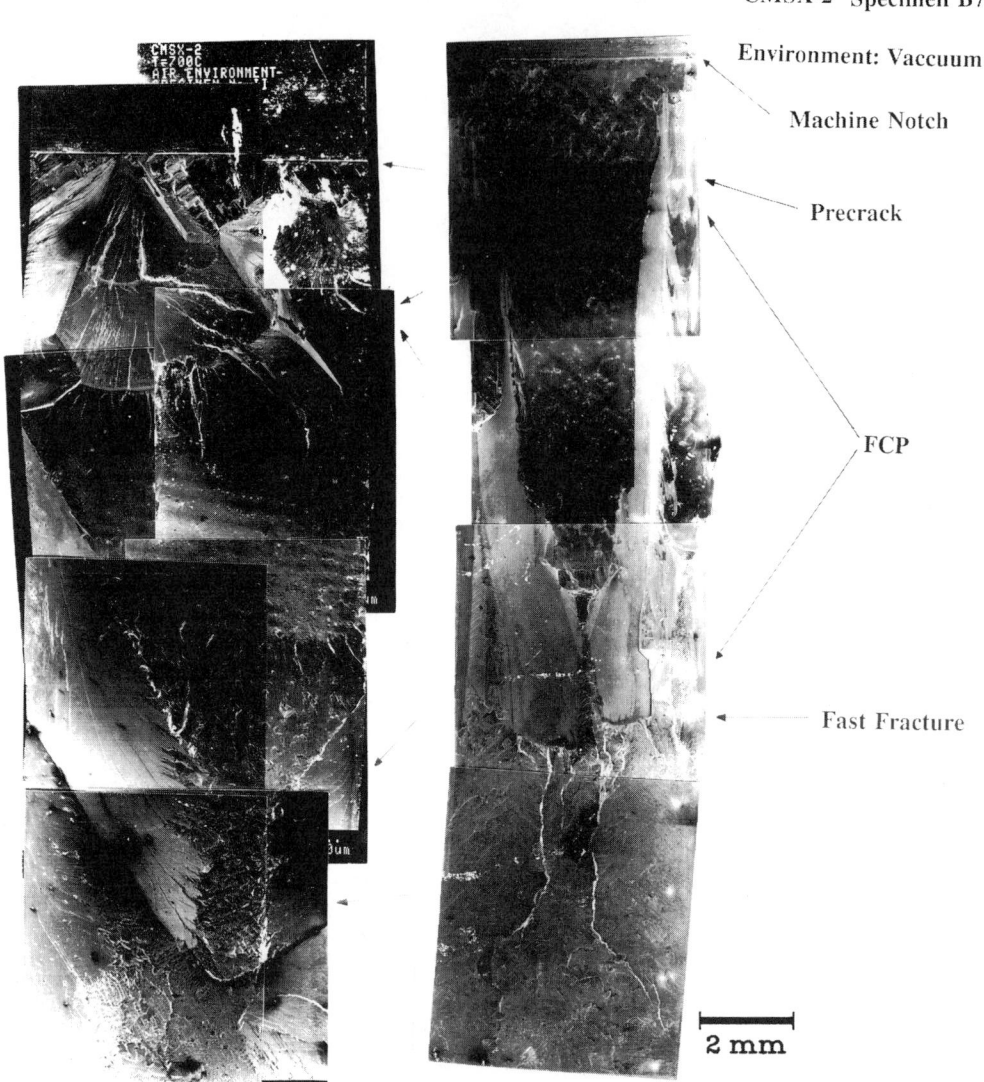

2 mm

Fig. 11 Micrographs showing fracture surfaces of CMSX-2 specimens with [010] and [110] orientations. Each specimen clearly shows evidence of dendritic macrostructure in the noncrystallographic portions of the fracture surface. Crystallographic faces correspond to {111} planes. Vacuum environment. Source: Ref 103

that exhibit cubic symmetry. The values of μ are developed from the characteristic equation:

$$C_{11}\mu^4 - 2C_{16}\mu^3 + (2C_{12} + C_{66})\mu^2 - 2C_{16}\mu + C_{22} = 0 \quad \text{(Eq 11)}$$

where C_{ij} are the elements of the elastic stiffness matrix.

One correlating factor for crack growth rates based on the shear stress resolved in the direction of the Burgers vector is the resolved shear stress-intensity coefficient (RSSIC) (Ref 102), which is defined as:

$$\tau_{rss} = \frac{RSSIC}{\sqrt{2\pi r}} \quad \text{(Eq 12)}$$

Equation 12 implies a number of different values of RSSIC—one for each slip system. This implies that specific values of θ must be used when calculating the stress tensor. This equation can be combined

with the full set of anisotropic field equations for a more direct definition of RSSIC:

$$RSSIC = [b_{i''}] [C_{i''j'}]$$
$$[K'_{Ia}, K'_{IIa}, K'_{IIIa}f(\theta_s)] [C_{i''j'}]^T [n_{j''}] \quad \text{(Eq 13)}$$

where double primes indicate material principal axes, single primes indicate specimen principal axes, $C_{i''j'}$ is the matrix of direction cosines, b_i is the Burgers vector, and n_j is the unit vector normal to the slip plane.

It must be reiterated that this definition of RSSIC is valid only for a particular slip system, leading up to 12 independent values of RSSIC for materials with fcc crystal structures. The choice of which RSSIC value to use as a correlating variable adds subjectivity and limits the usefulness of this approach. Furthermore, the fracture mechanics basis of this approach is two dimen-

sional. As will be shown later, this is a rather significant limitation.

Other researchers have extended the RSSIC-based models (Ref 98). A slightly more rigorous definition of the resolved shear stress-intensity parameter is given by:

$$K_{rss} = \lim_{r \to \infty} \tau_{rss} \sqrt{2\pi r} \quad \text{(Eq 14)}$$

In a direct comparison of crack growth rates for a single specimen using ΔK_{rss} and ΔK_I as correlating functions, no apparent benefit is seen. The curves for each correlating parameter follow each other except for a fixed offset on the ΔK axis. Additionally, ΔK_I was naturally shown to depend on whether a state of plane stress or plane strain was assumed. Holding all other conditions constant, the value of ΔK_I was shown to be approximately twice as high under conditions of plane stress than under plane strain. Although ΔK_I was

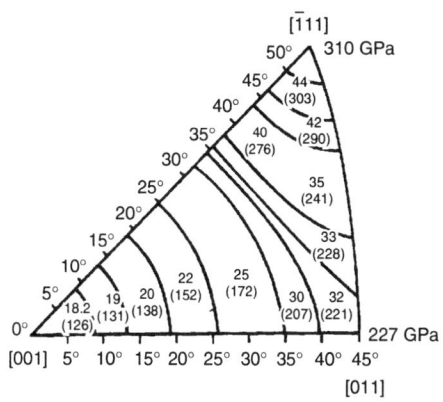

Fig. 12 Elastic modulus of PWA 1480 at room temperature. Source: Ref 42

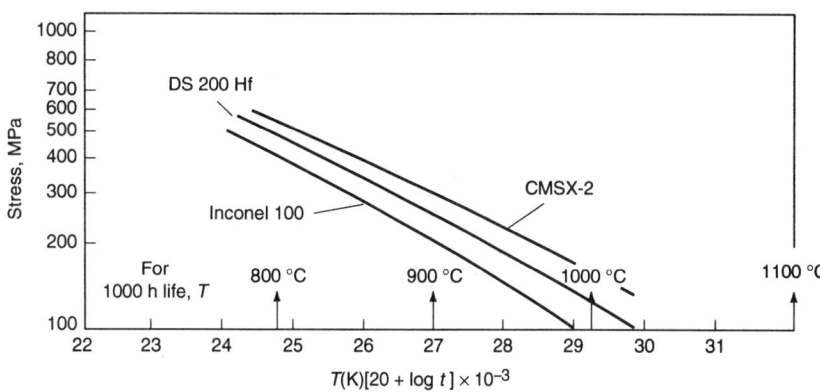

Fig. 14 Stress-rupture lives of single-crystal CMSX-2, directionally solidified DS 200 Hf, and equiaxed Inconel 100 vs Larson-Miller parameter. Source: Ref 25

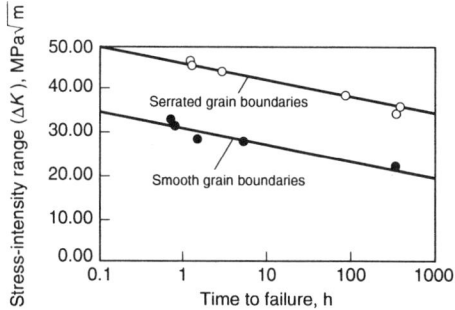

Fig. 13 Effects of serrated grain boundaries upon time to failure for IN-792 at 704 °C (1300 °F). Source: Ref 117

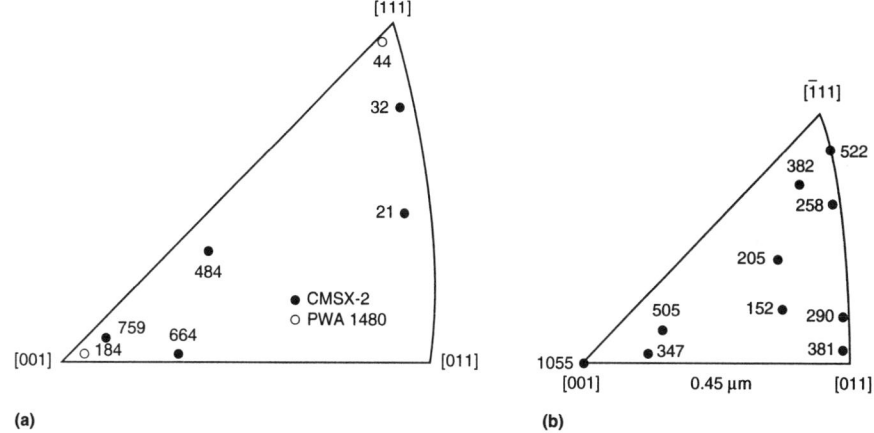

Fig. 15 Stress-rupture lives as a function of orientation. (a) Specimens tested at 760 °C (1400 °F). Source: Ref 25. (b) Specimens tested at 960 °C (1760 °F). Source: Ref 119

not evaluated critically as to its applicability for use with single crystals, the investigators did show that it was an effective aid in qualitatively understanding crack growth mechanisms, as will be discussed later.

Several researchers have used crack growth models that are based on energy arguments (Ref 92, 94, 98). Although the maximum energy dissipation rate criterion has been used for predicting which crack planes will be subject to cracking, this function has not been used extensively as a correlating parameter for predicting rates of crack growth (Ref 98). Nonetheless, this type of criterion does show some promise.

A different approach that has been used more extensively is a modification of the ΔK_{eff} concept (Ref 92, 94). The term ΔK_{eff} is defined following the definitions of total energy release rate (Ref 107):

$$\Delta K_{eff} = \left(\frac{G}{C_1}\right)^{1/2} = \left[\Delta K_I^2 + \frac{C_2}{C_1}\Delta K_{II}^2 + \frac{C_3}{C_1}\Delta K_{III}^2\right] \quad \text{(Eq 15)}$$

Using this definition of ΔK_{eff} as a correlating function, crack growth rates for various orientations from various types of specimens appear to initially correlate well. Testing under multiaxial conditions shows that changing the loading path in stress space results in poor correlation between da/dN and ΔK_{eff}. Critical examination of the theoretical basis and the practical implementation of this model reveals the

sources of some of these deficiencies. The theoretical basis of this model is the linear addition of energy release rates associated with each loading mode. This total energy release rate is then normalized by one of the constants in the characteristic equation (Eq 11). Although the addition of the scalar quantity of energy release rate is clearly on sound mathematical ground, its use as a correlating function makes several material behavior assumptions that are subject to debate. The essential assumption is that each mode of loading is equally damaging to the material (i.e., that equivalent energy release rates, regardless of which mode generates them, would result in equivalent crack growth rates). The physical basis of this assumption is not obvious. Furthermore, the difficulty in accommodating different loading paths in stress space suggests that it is probably incorrect.

Examination of the experimental technique for finding the stress-intensity factors for each mode also reveals possible shortcomings. Mode I and II stress-intensity factors have been calculated using the well-characterized boundary integral equation (BIE) technique (Ref 58, 108-110). This technique has been shown to show close agreement with ASTM results for the CT-type specimen. Stress-intensity factors for mode III are estimated based on an assumed crack geometry.

The assumed crack is self-similar and is inclined relative to the thickness of the specimen. K_{III} was calculated based on the through-thickness shear stresses. These stresses were in turn estimated by taking the state of stress calculated from the two-dimensional approach and applying boundary corrections. The assumption of self-similarity is open to discussion, as pointed out in Ref 94. Furthermore, the non-self-similarity of the crack raises questions as to the applicability of the applied boundary corrections.

Creep

As previously stated, the relatively constant rotational velocities of turbines, in conjunction with their high operating temperatures, produce creep loading. Fortunately, nickel-base superalloys possess quite good creep resistance and creep rupture strengths. Creep crack growth (CCG) must be considered since it can lead to a catastrophic failure while creep elongation is also quite important from a design point of view due to the tight dimensional tolerances typically found in turbine engines. A full treatment of both subjects is beyond the scope of this article, in-

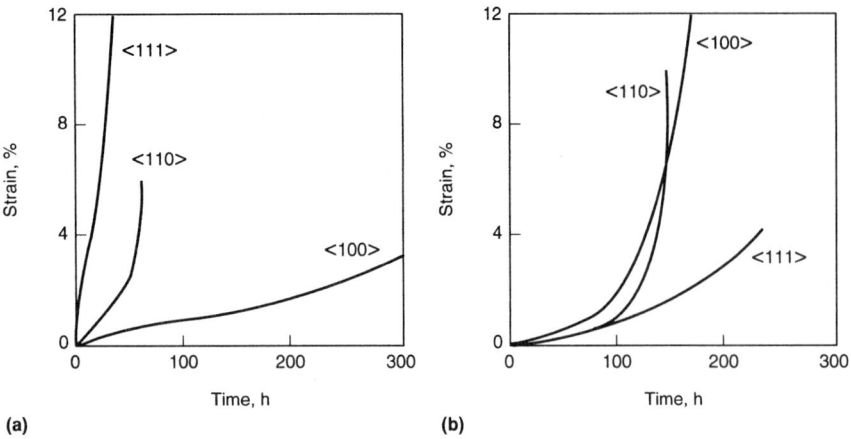

(a)   **(b)**

Fig. 16 Creep elongation of PWA 1480 as a function of temperature and orientation. (a) 760 °C (1400 °F). (b) 980 °C (1800 °F). Source: Ref 42

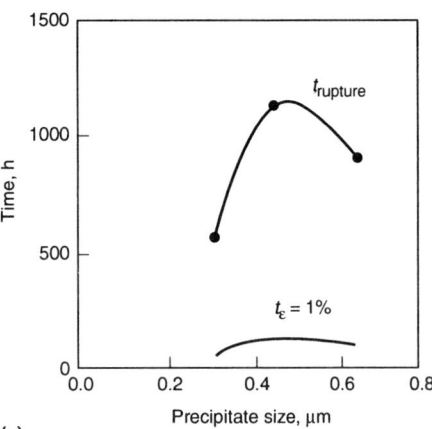

(a)

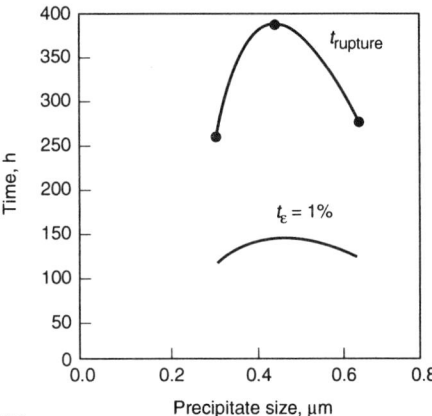

(b)

Fig. 17 Creep-rupture times as a function of γ' size for CMSX-2. (a) 760 °C (1400 °F); 750 MPa (110 ksi). (b) 950 °C (1740 °F); 240 MPa (35 ksi). Source: Ref 24

stead, attention will be focused upon creep crack growth.

Modeling Creep and Creep-Fatigue Crack Growth Rates

Because there are superimposed fatigue cycles and creep cycles, crack growth is usually addressed as a CCG problem. Experimental work usually consists of square wave loading (with low or very low frequencies) or triangle wave loading. In order to facilitate the analysis of overall crack growth rates, much effort has been expended to partition the total crack growth rates into fatigue and creep components. Because of the extensive database of FCP rates in which ΔK is treated as an independent variable for correlation of FCP rates in terms of crack extension per cycle (da/dN), there has been an effort to cast similarly cast CCG rates in terms of ΔK. A study of CCG rates for typical superalloys has resulted in an expression correlating K with 100 h of life prior to failure. One of the assumptions is that there exists a critical strain ahead of the crack tip that can serve as a criterion for CCG (Ref 111). The value of K required to produce failure within 100 h is given by:

$$K = (3\delta_0 \sigma_{ys} E + 0.3 \sigma_{ys} E D \varepsilon^*)^{1/2} \tag{Eq 16}$$

where δ_0 is the initial crack-tip displacement, σ_{ys} is the yield strength, E is Young's modulus, ε^* is the critical strain, and D is the grain diameter. Equation 16 captures the association between resistance to CCG and large grain size, which has been validated experimentally elsewhere (Ref 110). However, this model does not address the problem of crack growth rate (da/dt).

Several parameters based on modifications to the J-integral exist to predict crack growth rates, including C^*, $C(t)$, and C_t (see the article "Elevated-Temperature Crack Growth" in this Volume). For example, C^* is a relatively straightforward modification of the J-integral in which strain components are replaced by strain-rate components:

$$C^* = \int_\Gamma W_s^* \, dy - T_i \left(\frac{\partial \dot{u}_i}{\partial x} \right) ds \tag{Eq 17}$$

where

$$W_s^* = \int_0^{\dot{\varepsilon}_{min}} \sigma_{ij} \, d\dot{\varepsilon}_{ij}$$

and Γ is the path contour from the lower crack surface to the upper crack surface, ds is the incremental arc length along Γ, T_i is the traction vector along path Γ, x and y are Cartesian coordinates with their origin at the crack tip, u_i is the displacement vector, σ_{ij} is the stress tensor, and $\dot{\varepsilon}_{ij}$ is the strain-rate tensor.

It should be noted that Eq 17 is valid only for secondary creep in which the creep strains dominate any existing elastic or plastic strains. Typical constitutive equations describing this type of creep take on a power-law expression:

$$\dot{\varepsilon} = A\sigma^n \tag{Eq 18}$$

where A and n are the creep coefficient and exponent, respectively.

For the case of transient creep, Saxena (Ref 112) has defined the C_t parameter:

$$C_t = -\frac{1}{B} \frac{\partial U_t^*}{\partial a} \tag{Eq 19}$$

where B is specimen thickness and U_t^* is instantaneous stress power. This expression represents the instantaneous stress power dissipation rate, whereas C^* represented the steady-state rate. The expressions for C_t take on a variety of forms, depending on the material and creep regime. A good summary is provided in Ref 113.

As mentioned before, creep and fatigue combine to create creep-fatigue loading. Modeling of crack growth rates under creep-fatigue loading usually is based on a partitioning of crack growth between fatigue and creep components:

$$\frac{da}{dN} = \left(\frac{da}{dN} \right)_{cycle} + \left(\frac{da}{dN} \right)_{time} \tag{Eq 20}$$

Establishment of expressions for $(da/dN)_{time}$ based on C_t and the modified parameter, $(C_t)_{avg}$, have been investigated (Ref 114, 115). For small-scale creep:

$$\left(\frac{da}{dt} \right)_{avg} = \frac{1}{t_h} \left(\frac{da}{dN} \right)_{hold} \tag{Eq 21a}$$

$$(C_t)_{avg} = \frac{1}{t_h} \int_0^{t_h} C_t \, dt \tag{Eq 21b}$$

The following two expressions can then be used:

$$\frac{da}{dt} = b_2 [(C_t)_{avg}]^{p_2} \tag{Eq 22a}$$

$$\left(\frac{da}{dN} \right)_{time} = b'_2 \left(K^4 t_h^{-(n-3)/(n-1)} \frac{F'}{F} \right)^{p_2} \tag{Eq 22b}$$

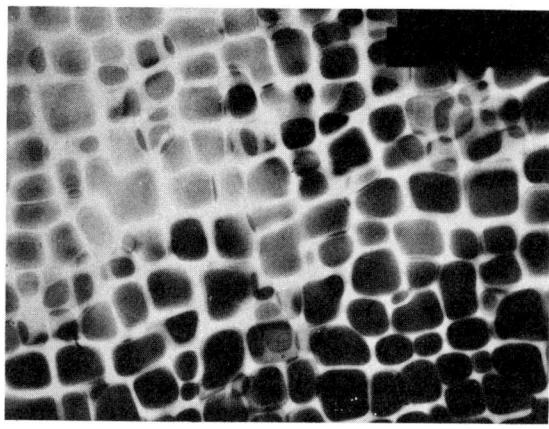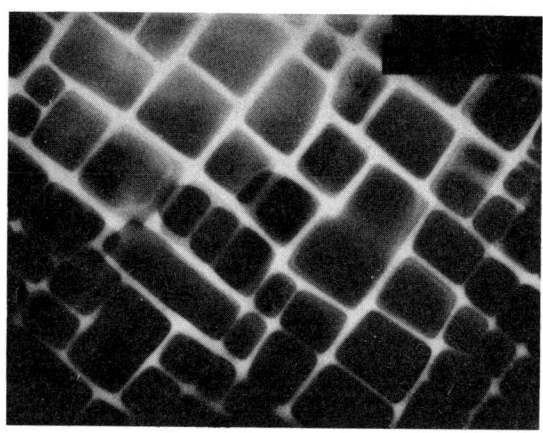

Fig. 18 Differing γ' morphologies as a function of heat treatment. Source: Ref 24

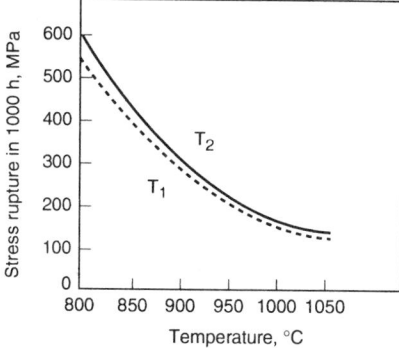

Fig. 19 Effect of precipitate morphology on creep-rupture properties. Curve T_1 corresponds to specimens with a heat treatment that produced irregularly shaped precipitates. Curve T_2 corresponds to regular cuboidal precipitates. Source: Ref 24

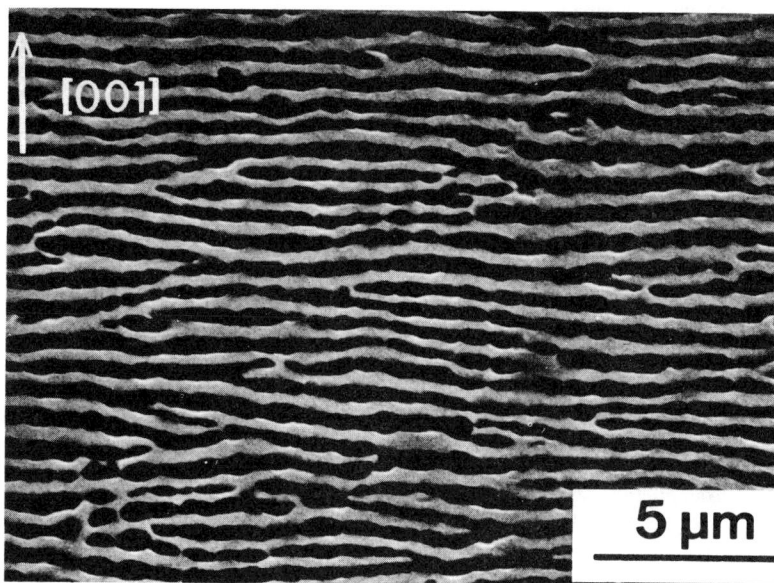

Fig. 20 Coalescence of the γ' phase into "rafts" under creep loading. Source: Ref 24

These equations can then be used in conjunction with expressions for $(da/dN)_{cycle}$ to model total crack growth rates (Ref 113).

Influence of Microstructure and Physical Metallurgy

As mentioned earlier, larger grains tend to lower CCG rates as well as FCP rates. Furthermore, serrated grain boundaries, produced by slow cooling of a warm-worked alloy and associated zener pinning as γ' nucleates at the grain boundaries, have also been shown to lower CCG rates. Lowered creep rates due to serrated boundaries have been reported in Ref 116 and 117. Raj and Ashby (Ref 116) proposed the following model correlating boundary sliding to parameters that quantify the serrated grain boundaries:

$$\dot{U} = \frac{8}{\pi} \frac{\tau_a \Omega}{kT} \frac{\lambda}{h^2} D_v \left(1 + \frac{\pi \delta}{\lambda} \frac{D_B}{D_v}\right) \quad \text{(Eq 23)}$$

where $\dot{U}$ is the grain-boundary sliding rate, τ_a is applied shear stress, λ is the periodicity of the boundary perturbation, h is the double amplitude of boundary perturbation, D_B is the grain-boundary

diffusion coefficient, D_v is the volume diffusion coefficient, Ω is atomic volume, and δ is the thickness of grain-boundary diffusion zone. Clearly, increased serration height decreases the rate of grain-boundary sliding and consequent creep. These general observations are corroborated by Ref 117, as shown in Fig. 13.

Single Crystal Considerations

As was the case with FCP, creep for single crystals merits closer examination of the unique effects that monocrystalline form imposes on creep behavior. In general, SX and DX components exhibit vastly superior creep properties than their polycrystalline counterparts. This is shown in Fig. 14, which compares the creep behavior of polycrystalline equiaxed Inconel 100, directionally solidified DS 200 HF, and single-crystal CMSX-2 (Ref 118). For SX alloys, crystallographic orientation also plays a major role in creep-rupture lives and creep elongation. How-

ever, these effects are highly temperature dependent, as shown in Fig. 15 and 16.

The microstructure of SX nickel-base superalloys has been found to play a significant role in creep resistance. Gamma-prime precipitate size and geometry are particularly important. Analysis of these effects is somewhat complicated by the microstructural changes that affect the precipitates. The influence of γ' size is well illustrated for the case of CMSX-2, where it is varied between approximately 0.25 and 0.65 μm through appropriate heat treatments (Ref 119). Figure 17 shows the effect of γ' size on creep-rupture times as well as time to reach a strain of 1% for two different temperatures. In each case, γ' precipitates of about 0.5 μm seem to produce the best creep resistance.

Precipitate morphology is another important factor. Two different heat treatments were used to produce irregularly shaped precipitates (combinations of spheres and cubes) or regularly shaped cuboidal precipitates. Examination of the deformation mechanisms for the two microstructures showed relatively inhomogeneous and homogeneous deformation for the irregular and regularly shaped precipitates, respectively (Fig. 18). Correspondingly, the creep-rupture behavior for the specimens with a microstructure consisting of regularly shaped γ' precipitates was more favorable than that of the irregularly shaped γ' precipitate microstructure (Fig. 19). As mentioned before, under certain conditions the γ' precipitates coalesce and form "rafts" (Fig. 20). These rafts form perpendicular to the loading direction and can dramatically improve the creep resistance of the alloy.

Conclusions

The FCP and creep behavior of nickel-base superalloys are heavily interdependent upon extrinsic and intrinsic factors. Furthermore, intrinsic microstructural features such as γ' size and morphology, which directly affect FCP and creep performance, are subject to change under the combined effects of loading and temperature. Directionally solidified and single crystals must also include orientation dependence for FCP and creep behavior. Changes in extrinsic parameters such as temperature may be quite beneficial for one environment but detrimental for another. Considerable care must be exercised in order to take into account all extrinsic and intrinsic factors when making quantitative or qualitative FCP or CCG rate predictions. Nonetheless, with appropriate care, an engineer can make meaningful predictions of FCP or CCG behavior.

While polycrystalline superalloys remain quite useful due to lower production costs than DX and SX alloys, their performance is generally lower. The lower production costs for DX components in comparison with SX counterparts ensures their future use and development. Unfortunately, many of the life prediction models developed for polycrystalline alloys are not directly applicable to their SX and DX counterparts. Nonetheless, their superior performance for military and commercial applications ensures that their use will continue to increase. Considerable efforts are being made in the development of appropriate life prediction models and seem likely to continue.

REFERENCES

1. R.M. Brick, A.W. Pense, and R.B. Gordon, *Structure and Properties of Engineering Materials*, McGraw-Hill, 1977, p 381
2. W. Rohn, Jr., The Reduction of Shrinkage Cavities and Vacuum Melting, *J. of the Institute of Metals,* Vol 42, 1929, p 203-219
3. W. Boesch, in *Superalloys, Supercomposites and Superceramics*, J.K. Tien and T. Caulfield, Ed., Academic Press, 1989, p 3
4. S.D. Antolovich and J.E. Campbell, in *Application of Fracture Mechanics for Selection of Metallic Structural Materials*, J.E. Campbell, W.W. Gerberich, and J.H. Underwood, Ed., American Society for Metals, 1983, p 253-310
5. J.K. Tien and T. Caulfield, Ed., *Superalloys, Supercomposites and Superceramics*, Academic Press, 1989
6. M.J. Donachie, Ed., *Superalloys Source Book*, American Society for Metals, 1984
7. E.F. Bradley, Ed., *Source Book on Materials for Elevated Temperature Applications*, American Society for Metals, 1979
8. J.M. Oblak and B.H. Kear, Analysis of Microstructures in Nickel Base Alloys: Implications for Strength and Alloy Design, *Electron Microscopy and the Structure of Materials*, G. Thomas, R.M. Fulrath, and R.M. Fisher, Ed., University of California Press, 1972, p 565-616
9. L.M. Brown and R.K. Ham, *Strengthening Methods in Crystals*, Applied Science Publishers, 1971, p 9-135
10. S.M. Copley and B.H. Kear, A Dynamic Theory of Coherent Precipitation with Application to Nickel-Base Superalloys, *Trans. TMS-AIME*, Vol 239, 1967, p 984-992
11. S. Takeuchi and E. Kuramoto, Temperature and Orientation Dependence of the Yield Stress in Ni_3Ga Single Crystals, *Acta Metall.*, Vol 21, 1973, p 415-425
12. C. Lall, S. Chin, and D.P. Pope, The Orientation and Temperature Dependence of the Yield Stress of $Ni_3(Al,Nb)$ Single Crystals, *Metall. Trans. A*, Vol 10A, 1979, p 1323-1332
13. M.H. Yoo, On the Theory of Anomalous Yield Behavior of Ni_3Al—Effect of Elastic Anisotropy, *Scr. Metall.*, Vol 20, 1986, p 915-920
14. A. deBussac, G. Webb, and S.D. Antolovich, A Model for the Strain-Rate Dependence of Yielding in Ni_3Al Alloys, *Metall. Trans. A*, Vol 22A, 1991, p 125-128
15. W. Milligan and S.D. Antolovich, The Mechanism and Temperature Dependence of Superlattice Stacking Formation in the Single Crystal Superalloy PWA 1480, *Metall. Trans. A*, Vol 22A, 1991, p 2309-2318
16. N.S. Stoloff, Ordered Alloys—Physical Metallurgy and Structural Applications, *Int. Met. Rev. Ordered Alloys*, Vol 29, 1984, p 123-135
17. V. Paider, D.P. Pope, and V. Vitek, A Theory of the Anomalous Yield Behavior in $L1_2$ Ordered Alloys, *Acta Metall.*, Vol 32 (No. 3), 1984, p 435-448
18. M.H. Yoo, J.A. Horton, and C.T. Liu, Micromechanisms of Yield and Flow in Ordered Intermetallic Alloys, *Acta Metall.*, Vol 36 (No. 11), 1988, p 2935-2946
19. S.M. Copley and B.H. Kear, Temperature and Orientation Dependence of the Flow Stress in Off-Stoichiometric Ni_3Al γ' Phase, *Trans. TMS-AIME*, Vol 239, 1967, p 977-984
20. R.F. Decker and C.T. Sims, The Metallurgy of Ni Base Alloys, *The Superalloys*, C.T. Sims and W.C. Hagel, Ed., John Wiley & Sons, 1972, p 33-77
21. E.W. Ross and C.T. Sims, Nickel-Base Alloys, *Superalloys II*, C.T. Sims, N.S. Stoloff, and W.C. Hagel, Ed., John Wiley & Sons, 1987, p 97-133
22. B.E.P. Beeston, I.L. Dillamore, and R.E. Smallman, *Met. Sci. J.*, Vol 2, 1960, p 12
23. B.E.P. Beeston and L. France, *J. Inst. Met.*, Vol 96, 1968, p 105
24. T. Khan, P. Caron, D. Fournier, and K. Harris, Single Crystal Superalloys for Turbine Blades: Characterization and Optimization of CMSX-2 Alloy, *Steels and Special Alloys for Aerospace*, 1985
25. P. Caron and T. Khan, Improved of Creep Strength in a Nickel-Base Single-Crystal Superalloy by Heat Treatment, *Mater. Sci. Eng.*, Vol 61, 1983, p 173
26. R.A. MacKay and L.J. Ebert, Factors Which Influence Directional Coarsening of γ' during Creep in Nickel Base Superalloy Single Crystals, *Superalloys 1984*, The Metallurgical Society of AIME, 1984, p 135-144
27. A. Fredholm and J.L. Strudel, On the Creep Resistance of Some Nickel Base Single Crystals, The Metallurgical Society of AIME, *Superalloys 1984*, 1984, p 211-220
28. T. Miyazaki, K. Nakamura, and H. Mori, *J. Mater. Sci.*, Vol 14, 1979, p 1827
29. C. Carry and J.L. Strudel, *Acta Metall.*, Vol 26, 1978, p 859
30. C.T. Sims, *J. Met.*, Vol 18, Oct 1966, p 1119
31. E.L. Raymond, *Trans. AIME*, Vol 239, 1967, p 1415
32. B.J. Piearcey and R.W. Smashey, *Trans. AIME*, Vol 239, 1967, p 451
33. E.W. Ross, "Recent Research on IN-100," presented at AIME Annual Meeting (Dallas), 1963
34. E.W. Ross, *J. Met.*, Vol 19, Dec 1967, p 12
35. E.G. Richards, *J. Inst. Met.*, Vol 96, 1968, p 365
36. C.G. Bieber, The Melting and Hot Rolling of Nickel and Nickel Alloys, *Metals Handbook*, American Society for Metals, 1948
37. R.W. Koffler, W.J. Pennington, and F.M. Richmond, R&D Report No. 48, Universal-Cyclops Steel Corp., Bridgeville, PA, 1956
38. R.F. Decker, J.P. Rowe, and J.W. Freeman, NACA Technical Note 4049, Washington, DC, June 1957
39. K.E. Volk and A.W. Franklin, *Z. Metallkd.*, Vol 51, 1960, p 172
40. R.F. Decker, "Strengthening Mechanisms in Ni Base Superalloys," presented at Climax Molybdenum Symp. (Zurich), 5-6 May 1969
41. A.D. Cetel and D.N. Duhl, Second Generation Columnar Grain Nickel-Base Superalloy, *Superalloys 1992*, S.D. Antolovich et al., Ed., Minerals, Metals & Materials Society, 1992, p 287-296
42. D.N. Duhl, Single Crystal Superalloys, *Superalloys, Superceramics and Supercomposites*, J.K. Tien and T. Caulfield, Ed., Academic Press, 1989, p 149-182
43. T.M. Pollock, W.H. Murphy, E.H. Goldman, D.L. Uram, and J.S. Tu, Grain Defect Formation during Directional Solidification of Nickel Base Single Crystals, *Superalloys 1992*, S.D. Antolovich et al., Ed., Minerals, Metals & Materials Society, 1992, p 125-134
44. D.J. Frasier, J.R. Whetstone, K. Harris, G.L. Erickson, and R.E. Schwer, Process and Alloy Optimization for CMSX-4® Superalloy Single

Crystal Airfoils, *Conf. Proc. High Temperature Materials for Power Engineering 1990*, 1990

45. A.F. Giamei and D.L. Anton, *Metall. Trans. A*, Vol 16A, 1985, p 1997

46. K. Harris, G.L. Erickson, S.L. Sikkenga, W.D. Brentnall, J.M. Aurrecoechea, and K.G. Kubarych, Development of the Rhenium Containing Superalloys CMSX-4® & CM 186 LC® for Single Crystal Blade and Directionally Solidified Vane Applications in Advanced Turbine Engines, *Superalloys 1992*, S.D. Antolovich et al., Ed., Minerals, Metals & Materials Society, 1992, p 297-306

47. Single-Crystal Engine Alloy Resists Creep, High Heat, *Adv. Mater. Processes*, Vol 149 (No. 3), March 1996, p 7

48. H.F. Merrick and S. Floreen, The Effect of Microstructure on Elevated Temperature Crack Growth in Ni Base Alloys, *Metall. Trans. A*, Vol 9A (No. 2), Feb 1978, p 231-233

49. J. Bartos and S.D. Antolovich, Effect of Grain Size and γ′ Size on FCP in René 95, *Fracture 1977*, Vol 2, D.M.R. Taplin, Ed., University of Waterloo Press, 1977, p 996-1006

50. S.D. Antolovich, C. Bathias, B. Lawless, and B. Boursier, The Effect of Microstructure on the FCP Properties of Waspaloy, *Fracture: Interactions of Microstructure, Mechanisms and Mechanics*, J.M. Wells and D. Landes, Ed., TMS-AIME, 1985, p 285-301

51. D. Krueger, S.D. Antolovich, and R.H. Vanstone, *Metall. Trans. A*, Vol 18A, 1987, p 1431-1449

52. W.J. Mills and L.A. James, Publication 7-WA/PUP-3, American Society of Mechanical Engineers, 1979

53. R. Miner and J. Gayda, *Metall. Trans. A*, Vol 14A, 1983, p 2301-2308

54. R. Bowman, M.S. thesis, Georgia Institute of Technology, 1985

55. B. Lerch and S.D. Antolovich, Cyclic Deformation, Fatigue and Fatigue Crack Propagation in Nickel-Base Superalloys, *Superalloys, Supercomposites and Superceramics*, J.K. Tien and T. Caulfield, Ed., Academic Press, 1989, p 363-411

56. J. Lindigkeit, G. Terlinde, A. Gysler, and G. Lütering, *Acta Metall.*, Vol 27, 1979, p 1717-1726

57. M. Clavel, C. Levaillant, and A. Pineau, Influence of Micromechanisms of Cyclic Deformation at Elevated Temperature on Fatigue Behavior, *Creep-Fatigue-Environment Interactions*, R.M. Pelloux and N.S. Stoloff, Ed., American Institute of Mining, Metallurgical and Petroleum Engineers, 1980, p 24-45

58. T.A. Cruse and K.S. Chan, Stress Intensity Factors for Anisotropic Compact-Tension Specimens with Inclined Cracks, *Eng. Fract. Mech.*, Vol 23 (No. 5), 1986, p 863-874

59. J.P. Baïlon and S.D. Antolovich, Effect of Microstructure on Fatigue Crack Propagation: A Review of Existing Models and Suggestions for Further Research, *STP 811*, ASTM, 1983, p 313-347

60. S. Bashir, Ph. Taupin, and S.D. Antolovich, Low Cycle Fatigue of As-HIP and HIP + Forged René 95, *Metall. Trans. A*, Vol 10A (No. 10), Oct 1979, p 1481-1490

61. P. Domas, Crack Propagation under Thermal Mechanical Cycling: An Interim Progress Report to the Air Force Materials Laboratory, Contract F336115-C-5193, Period 9/1/77-1/15/79 (section written by S.D. Antolovich)

62. L.A. Girifalco and R.O. Welch, *Point Defects and Diffusion in Strained Metals*, Gordon & Breach, 1967, p 124

63. L.C. Jea, "Environment Assisted Fatigue Crack Growth in TRIP Steels," M.S. thesis, University of Cincinnati, 1974

64. P. Bania and S.D. Antolovich, Activation Energy Dependence on Stress Intensity in SCC and Corrosion Fatigue, *STP 610*, ASTM, 1976, p 157-175

65. S. Floreen and R.H. Kane, Effects of Environment on High Temperature Fatigue Crack Growth in a Superalloy, *Metall. Trans. A*, Vol 10A (No. 11), Nov 1979, p 113-153

66. M. Gell and G.R. Leverant, Mechanisms of High Temperature Fatigue, *STP 520*, ASTM, 1973, p 37-67

67. V. Shahani and H.G. Popp, "Evaluation of Cyclic Behavior of Aircraft Turbine Disk Alloys," NASA Report NASA-CR-159433, National Aeronautics and Space Administration, 1978

68. A. deBussac, M.S. thesis, Georgia Institute of Technology, 1990

69. R.V. Miner, Fatigue, *Superalloys II*, C.T. Sims, N.S. Stoloff, and W.C. Hagel, Ed., John Wiley & Sons, 1987, p 263-289

70. C.H. Wells and C.P. Sulivan, *Trans. ASM*, Vol 60, 1967, p 217

71. M.H. Hirschberg and G.R. Halford, NASA TN D-8072, National Aeronautics and Space Administration, Jan 1976

72. W.J. Ostergren, in *1976 ASME-MPC Symposium on Creep-Fatigue Interaction*, American Society of Mechanical Engineers, 1976, p 179

73. M.O. Speidel, Fatigue Crack Growth at High Temperatures, *High Temperature Materials in Gas Turbine Engines*, P.R. Sahm and M.O. Speidel, Ed., Elsevier, 1974, p 208-255

74. R.M. Wallace, C.G. Annis, Jr., and D. Sims, "Application of Fracture Mechanics at Elevated Temperatures," Report AFML-TR-76-176, Part II, 1976

75. R.G. Foreman, V.E. Kearney, and R.M. Engle, Numerical Analysis of Crack Propagation in Cyclic-Loaded Structures, *J. Basic Eng.*, Vol 89, 1967, p 459-464

76. P.C. Paris and F. Erdogan, A Critical Analysis of Crack Propagation Laws, *J. Basic Eng. (Trans. ASME)*, Vol 85, Dec 1963, p 528-534

77. W. Hoffelner, in *Superalloys 1984*, TMS-AIME, 1984, p 771

78. J. Masounave, J.P. Baïlon, and J.I. Dickson, *La Fatigue des Matériaux et Des Structures*, C. Bathias and J.P. Baïlon, Ed., Maloine S.A., Paris, 1980, Chap. 6

79. J. Weertman, Fatigue Crack Propagation Theories, *Fatigue and Microstructure*, American Society for Metals, 1979, p 279-306

80. T. Yokobori, S. Konosu, and A.T. Yokoburi, Jr., in *Fracture 1977*, Vol 1, D.R.M. Taplin, Ed., University of Waterloo Press, 1977, p 665-681

81. F.A. McClintock, *Fracture of Solids*, John Wiley & Sons, p 65-102

82. S.D. Antolovich, A. Saxena, and G.R. Chanani, *Eng. Fract. Mech.*, Vol 7, 1975, p 649-652

83. G.R. Chanani, S.D. Antolovich, and W.W. Gerberich, *Metall. Trans.*, Vol 3, 1972, p 649-652

84. J. Lanteigne and J.P. Baïlon, *Metall. Trans. A*, Vol 12A, 1981, p 459-466

85. S.B. Chakrabortty, *Fatigue Eng. Mater. Struct.*, Vol 2, 1979, p 331-344

86. G.R. Leverant and M. Gell, The Influence of Temperature and Cyclic Frequency on the Fatigue Fracture of Cube Oriented Nickel Base Superalloy Single Crystals, *Metall. Trans. A*, Vol 6A, 1975, p 367-371

87. M. Gell and G.R. Leverant, The Fatigue of the Nickel Base Superalloy Mar-M200 in Single Crystal and Columnar Grains Formed at Room Temperature, *Trans. AIME*, Vol 242, 1968, p 1869-1879

88. D.J. Duquette and M. Gell, The Effects of Environment on the Elevated Temperature Fatigue Behavior of Nickel Base Superalloy Single Crystals, *Metall. Trans. A*, Vol 3A, 1972, p 1899-1905

89. T.P. Gabb, J. Gayda, and R.V. Miner, Orientation and Temperature Dependence of Some Mechanical Properties of the Single Crystal Nickel Base Superalloy René N4: Part II, Low Cycle Fatigue Behavior, *Metall. Trans. A*, Vol 17A, 1986, p 497-505

90. B.A. Lerch and S.D. Antolovich, Fatigue Crack Propagation Behavior of a Single Crystalline Superalloy, *Metall. Trans. A*, Vol 21A, 1990, p 2169-2177

91. P.S. Chen and R.C. Wilcox, Fracture of Single Crystals of the Nickel Base Superalloy PWA 1480E in Helium at 22 °C, *Metall. Trans. A*, Vol 22A, 1991, p 731-737

92. K.S. Chan, J.E. Hack, and G.R. Leverant, Fatigue Crack Propagation in Ni-Base Superalloy Single Crystals under Multi-axial Cyclic Loads, *Metall. Trans. A*, Vol 17A, 1986, p 1739-1750

93. G.K. Bouse, Fatigue Crack Propagation Rate Testing of Single Crystal Superalloys NASAIR 1000 and CMSX-2 at 982°, *Superalloys 1988*, 1988, p 751-759

94. K.S. Chan, J.E. Hack, and G. Leverant, Fatigue Crack Growth in MAR-M200 Single Crystals, *Metall. Trans. A*, Vol 18A, 1987, p 581-591

95. P.K. Wright, Oxidation-Fatigue Interactions in a Single Crystal Superalloy, *STP 942*, 1988, p 558-575

96. J. Telesman and L.J. Ghosn, The Unusual Near-Threshold FCG Behavior of a Single Crystal Superalloy and the Resolved Shear

Stress as the Crack Driving Force, *Eng. Fract. Mech.*, Vol 34 (No. 5/6), 1989, p 1183-1196

97. M. Khobaib, T. Nicholas, and S.V. Ram, Role of Environment in Elevated Temperature Crack Growth Behavior of René N4 Single Crystal, *STP 1049*, ASTM, 1990, p 319-333

98. J.S. Short and D.W. Hoeppner, The Maximal Dissipation Rate Criterion—II: Analysis of Fatigue Crack Propagation in FCC Single Crystals, *Eng. Fract. Mech.*, Vol 34 (No. 1), 1989, p 15-30

99. J.E. King, Fatigue Crack Propagation in Nickel-Base Superalloys—Effects of Microstructure, Load Ratio and Temperature, *Mater. Sci. Technol.*, Vol 3, 1987, p 750-764

100. G.R. Halford, T.G. Meyer, R.S. Nelson, D.M. Nissley, and G.A. Swanson, Fatigue Life Prediction Modeling for Turbine Hot Section Materials, *J. Eng. Gas Turbines Power*, Vol 111, 1989, p 279-285

101. J.S. Cromption and J.W. Martin, Crack Growth in a Single Crystal Superalloy at Elevated Temperature, *Metall. Trans. A*, Vol 15A, 1984, p 1711-1719

102. Q. Chen and H.W. Liu, Resolved Shear Stress Intensity Coefficient and Fatigue Crack Growth in Large Crystals, *Theor. Appl. Fract. Mech.*, Vol 10, 1988, p 111-122

103. B.F. Antolovich, A. Saxena, and S.D. Antolovich, Fatigue Crack Propagation in Single Crystal CMSX-2 at Elevated Temperature, *J. Mater. Eng. Perform.*, Vol 2 (No. 4), Aug 1993,

p 489-495

104. A.J. McEvily and R.C. Boettner, On Fatigue Crack Propagation in FCC Metals, *Acta Metall.*, Vol 11, 1963, p 725-743

105. D.J. Duquette, M. Gell, and J.W. Piteo, A Fractographic Study of Stage I Fatigue Cracking in a Nickel-base Superalloy Single Crystal, *Metall. Trans.*, Vol 1, 1970, p 3107-3115

106. M. Nageswararao and V. Gerold, Fatigue Crack Propagation in Stage I in an Aluminum-Zinc-Magnesium Alloy: General Characteristics, *Metall. Trans. A*, Vol 7A, 1976, p 1847-1855

107. P.C. Paris and G.C. Sih, Fracture Toughness Testing and Its Applications, *STP 381*, ASTM, 1964, p 30-81

108. T.A. Cruse, Boundary Element Analysis in Computational Fracture Mechanics, Kluwer Academic Publishers, 1988

109. T.A. Cruse and M.D. Snyder, Boundary Integral Equation Analysis of Cracked Anisotropic Plates, *International Journal of Fracture,* Vol 11, No. 2, 1975, p 315-328

110. T.A. Cruse, Two-Dimensional (BIE) Fracture Mechanics Analysis, Applied Mathematical Modeling, Vol 2 1978, p 287-293

111. S. Floreen and R.H. Kane, A Critical Strain Model for the Creep Fracture of Nickel Base Superalloys, *Metall. Trans. A*, Vol 7A (No. 8), 1976, p 1157-1160

112. A. Saxena, Creep Crack Growth Under Nonsteady-State Conditions, *Fracture Mechanics*, Vol 17, J.H. Underwood et al., Ed., ASTM,

1986

113. K.G. Yoon, "Characterization of Creep Fatigue Crack Growth Behavior Using the C_t Parameter," Ph.D. thesis, Georgia Institute of Technology, 1990

114. A. Saxena and B. Gieseke, Transients in Elevated Temperature Crack Growth, *Proc. MECAMET Int. Seminar High Temperature Fracture Mechanisms and Mechanics*, Vol 3, Mechanical Engineering Publications, 1987, p 19-36

115. B. Gieseke and A. Saxena, Correlation of Creep-Fatigue Crack Growth Rates Using Crack Tip Parameters, *Proc. 7th Int. Conf. Fracture: ICF-7*, K. Salama et al., Ed., Pergamon Press, 1989, p 189-196

116. R. Raj and M.F. Ashby, On Grain Boundary Sliding and Diffusional Creep, *Metall. Trans.*, Vol 2 (No. 4), p 1113-1127

117. J.M. Larson and S. Floreen, Metallurgical Factors Affecting the Crack Growth Resistance of a Superalloy, *Metall. Trans. A*, Vol 8A (No. 1), Jan 1977, p 51-55

118. K. Harris, G.L. Erickson, and R.E. Schwer, MAR M 247 Derivations—CM 247 LC DS Alloy CMSX Single Crystal Alloys Properties and Performance, *Superalloys 1984*, TMS-AIME, 1984, p 221-230

119. P. Caron, Y. Ohta, Y.G. Nakagawa, and T. Khan, Creep Deformation Anisotropy in Single Crystal Superalloys, *Superalloys 1988*, The Metallurgical Society, 1988, p 215-224

Fatigue Properties of Copper Alloys

Jack Crane, Consultant, John O. Ratka, Brush Wellman, and John F. Breedis, Olin Brass

COMPARED to most structural materials, relatively few applications of copper alloys involve cycling stressing. The most common use of copper alloys under dynamic loading is in rotating electrical machinery used for power generation. This application involves copper and very dilute copper alloys, which are not covered in this article. The reader is referred to Ref 1 for information on fatigue of copper.

One application of copper alloys requiring resistance to fracture after two or three highly stressed reverse bend cycles relates to the use of leadframe in insertion mounted microelectronic devices. The "leadbend fatigue" test involves repeated 90 degree bending and straightening of the leadframe leads. The test was designed to ensure that materials used for leadframes would not be susceptible to fracture if leads were straightened after inadvertent deformation during handling. This subject is covered in Ref 2 and 3, which describe the test method, list many of the alloys used for leadframes, and provide comparative leadbend fatigue data. With the increasing use of surface-mounted devices, the leadbend fatigue test and this property have become less important.

Copper alloys whose fatigue characteristics are covered in this article are used in applications involving repeated flexing: springs used for contacts and connectors, bellows, and Bourdon tubes. Alloys used for these applications include the brasses, bronzes (tin-, silicon-, aluminum-, and combinations thereof), and beryllium coppers. Copper-nickel-tin spinodally hardened alloys are also used in connectors and contacts. Flexural fatigue properties of all these classes of alloys in strip form are presented in this article. Heavier copper alloy sections subjected to cyclic loading are largely confined to the beryllium coppers for applications such as aircraft landing gear bushings, races and rollers for rolling-element bearings, and oil and gas downhole hardware such as antigalling thread-saver subs and instrument housings. Rotating-beam fatigue results are reported for beryllium copper alloy C17200.

The alloy designations and compositions of alloys covered in this article are given in Table 1. Copper alloys are classified by the International Unified Numbering System (UNS) designations, which identify alloy groups by major alloying element.

The brasses, nickel silvers, and bronzes covered in this article are strengthened by cold work, the exception being C70250, the copper-nickel-silicon-magnesium alloy, which is strengthened by combinations of cold work and precipitation hardening. The beryllium coppers are strengthened by cold work and/or precipitation hardening. The spinodally strengthened copper-nickel-tin alloy is cold worked and aged. The temper designations used for the materials tested are listed in Table 2. These designations are used throughout the text and figures. Because of their ability to be precipitation hardened, C70250 beryllium copper alloys can be tailored across a wide range of strength and conductivity combinations.

Alloy Metallurgy and General Mechanical Properties

Brasses. Both the standard cartridge brass (C26000, 70Cu-30Zn) and the higher-strength aluminum brass (C68800) offer excellent combinations of strength and formability and are widely used as spring materials. The microstructure of C26000 is an all-α solid solution. In addition to a copper-zinc-aluminum solid solution, C68800 contains a second-phase cobalt aluminide that acts as a grain refiner.

Nickel-Silvers. The copper-nickel-zinc alloy C76200 is a nickel-modified brass. The nickel modification offers improved strength over conventional brasses. This family of alloys is used in springs of all types.

Bronzes. Tin bronze (C51000) is one of the most widely used alloys for springs that require strength higher than standard brass. It is also used for bellows and Bourdon tubes. It is essentially a single-phase alloy.

The silicon-aluminum-bronze alloy (C63800) and the silicon-tin-bronze alloy (C65400) both are used in a broad range of electronic and electrical springs. The microstructure is a single-phase solid solution, with a coarse second-phase particulate (cobalt silicide in the case of C63800).

The nickel-silicon alloy C70250 has wide application in electrical and electronic springs. It combines moderately high conductivity and formability with high strength. The alloy microstructure consists of an α solid solution and a second-phase nickel silicide that provides precipitation hardening.

Table 1 Alloy designations and compositions

UNS designation	Nominal composition, %
C17200	1.9 Be, 0.25 Co
C17410	0.3 Be, 0.4 Co
C17510	0.4 Be, 1.8 Ni
C2600	30 Zn
C51000	5.0 Sn, 0.2 P
C63800	2.8 Al, 1.8 Si, 0.4 Co
C65400	3.1 Si, 1.6 Sn, 0.05 Cr
C68800	23 Zn, 3.4 Al, 0.4 Co
C72050	3.2 Ni, 0.75 Si, 0.2 Mg
C72900	15 Ni, 8 Sn
C76200	29 Zn, 12 Ni, 0.5 Mn

Table 2 Copper alloy temper designations

ASTM B601 temper designation	Process
O60	Annealed
	Cold work to:
H01	Quarter-hard
H02	Half hard
H03	Three-quarter hard
H04	Hard
H06	Extra hard
H08	Spring
H10	Extra spring
H14	Super spring
TB00	Solution treat (ST)
TF00	ST and age
TD01	ST + cold work (CW)
TD02	ST + CW
TD04	ST + CW
TH01	ST + CW + age
TH02	ST + CW + age
TH04	ST + CW + age
TM00	CW + age (mill hardened)
TM02	CW + age (mill hardened)
TM03	CW + age (mill hardened)
TM04	CW + age (mill hardened)
TM06	CW + age (mill hardened)
TM08	CW + age (mill hardened)

Note: TD, TH, and TM tempers increase in number with increasing cold work.

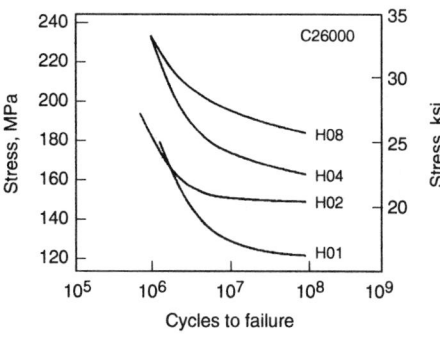

Fig. 1 S-N curves for C26000. Longitudinal loading, R = –1

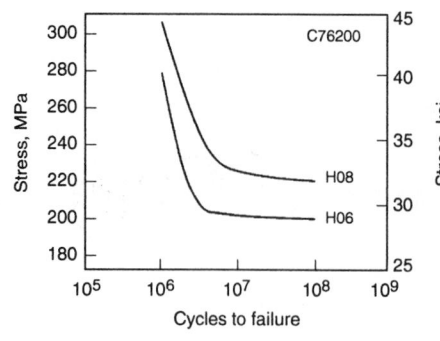

Fig. 3 S-N curves for C76200. Longitudinal loading, R = –1

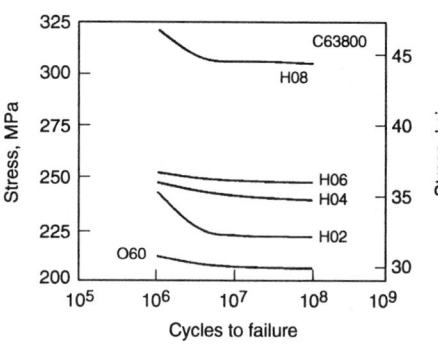

Fig. 5 S-N curves for C63800. Longitudinal loading, R = –1

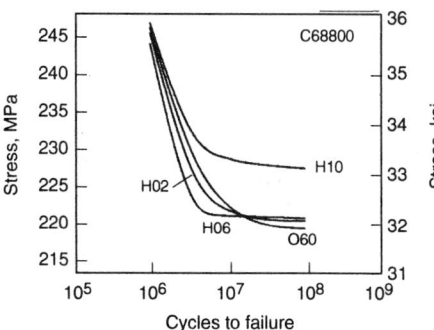

Fig. 2 S-N curves for C68800. Longitudinal loading, R = –1

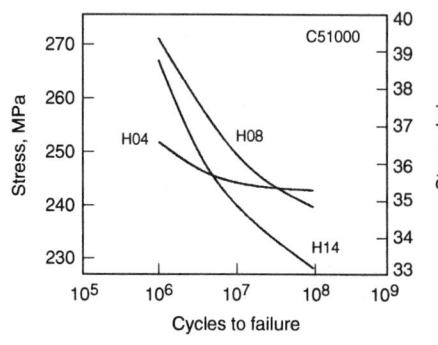

Fig. 4 S-N curves for C51000. Longitudinal loading, R = –1

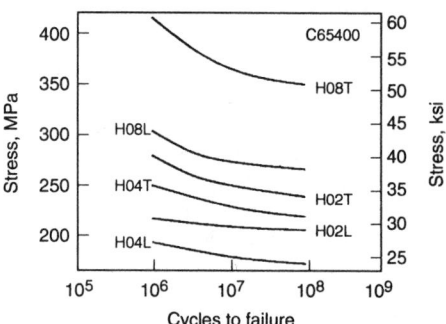

Fig. 6 S-N curves for C65400. Longitudinal and transverse loading, R = –1

Beryllium Coppers. Commercial wrought beryllium copper alloys contain from 0.2 to 2.0 wt% Be and 0.2 to 2.7 wt% Co (or up to 2.2 wt% Ni), with the balance primarily copper. Within this compositional band, two distinct classes of commercial materials have been developed: high-strength alloys and moderate-conductivity alloys.

Wrought high-strength alloys (C17000 and C17200) contain 1.6 to 2.0 wt% Be and nominal 0.25 wt% Co. Wrought moderate-conductivity alloys (C17500 and C17510) contain 0.2 to 0.7 wt% Be and nominal 2.5% Co (or 2 wt% Ni). The leanest alloy is C17410, which contains less than 0.4 wt% Be and 0.6 wt% Co.

Additional detailed information on the composition, physical metallurgy, mechanical properties, and thermal treatments of beryllium copper alloys, including casting alloys and special tempers and alloys, can be found in Ref 4.

Spinodal Alloys. The family of copper-nickel-tin alloys spinodally strengthened during aging is represented here by the highest-strength version, the 15Ni-8Sn alloy C72900 used for connectors. Optimum strength and formability are obtained by a combination of cold working followed by aging.

Fatigue Testing

Strip. Bend fatigue testing of strip was performed in conformance with the ASTM B 593 standard method for copper alloy spring materials. This method employs a fixed-cantilever, constant-deflection machine. The tapered test sample is held as a cantilever beam in a clamp at one end and deflected near the opposite end of the apex of the tapered section. Testing was done in a Krouse machine, with the force applied to the sample by a cam and rod linkage. A wide range of bending load ratios can be applied with this method. Typically, load ratios are chosen between R = –1 and 0 to simulate reverse bending and unidirectional bending, respectively. The test frequency is approximately 20 Hz.

Test samples were made from commercial materials. Gages covered 0.20 to 0.38 mm (0.008 to 0.015 in.) for the beryllium coppers and spinodal alloys and 0.25 to 1.5 mm (0.010 to 0.060 in.) for the remaining alloys. Samples of the as-rolled strip were milled to the required test specimen geometry. The rolled surface was left intact and the milled edges deburred.

The required deflection is determined by using either the cantilever simple beam equation or measured with strain-gaged samples under dynamic conditions. The maximum outer fiber bending stress is calculated by:

$$S = \frac{6PL}{bd^2}$$

where S is the desired bending stress, P is the applied load at the connecting pin (apex of the sample triangle), L is the distance between the connecting pin and the point of stress, b is the specimen width at length L from the point of load application, and d is the specimen thickness.

A load cell at the fixed end of the sample is used to detect change in the sample loading resulting from a macroscopic crack initiation. The load cell information is relayed to a monitoring circuit that determines the test completion based on the failure criteria required.

Rotating Beam. Fatigue tests of materials representing heavy section products made from rod, bar, and plate were conducted by the rotating-beam method following ASTM E 647 guidelines.

Fatigue Data

Strip

Nonaging Alloys. All flexural fatigue data are reported as S-N curves, where S is the maximum stress in flexure and N is the number of cycles to failure. Failure is defined as complete specimen fracture. Fatigue strength is defined here as the maximum stress without failure after 100 million cycles of reversed bending. Figures 1 to 6 present S-N curves for the solid-solution (nonaging) strengthened alloys (with or without a second-phase grain refiner). In general, fatigue strength follows tensile strength monotonically, but there are exceptions where fatigue strength is relatively insensitive to temper and a few cases, most notably C51000 but also C65400, where crossovers occur. This behavior has been observed before in tin-bronze alloys (Ref 5, 6).

For many copper alloy systems, transverse tensile and yield strengths are substantially higher than longitudinal strengths, and this characteristic is reflected normally by higher transverse fatigue strengths, as shown in Fig. 6. Although

Table 3 Tensile and fatigue strengths of the copper alloys in Fig. 1 to 7

Alloy	Temper	0.2% yield strength		Ultimate tensile strength		Elongation, %	10^8 cycles
		MPa	ksi	MPa	ksi		
C26000	H01	310	45	421	61.1	33	18
	H02	427	62	476	69	17	22
	H04	538	78	572	83	5	24
	H08	641	93	676	98	2	26
C51000	H04	581	84	597	86	8	35
	H08	712	103	732	106	3	34
	H14	745	108	788	114	3	33
C63800	O60	290	42	517	75	42	30
	H02	614	89	696	101	13	32
	H04	696	101	772	112	7	35
	H06	765	111	834	121	4	36
	H08	793	115	869	126	3	44
C65400	H02L	568	82	657	95	18	29
	H02T	593	86	701	101	10	34
	H04L	746	108	832	120	4	24
	H04T	731	106	866	125	3	31
	H08L	846	122	934	135	3	37
	H08T	859	124	991	143	2	50
C68800	O60	331	48	538	78	37	32
	H02	621	90	676	98	7	32
	H06	731	106	814	118	2	32
	H10	758	110	862	125	1	33
C70250	TM00	568	82	712	103	15	33
	TM02	643	93	718	104	12	33
C76200	H06	724	105	737	107	4	29
	H08	772	112	786	114	3	32

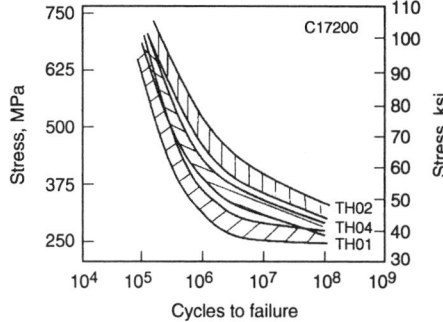

Fig. 9 Bending fatigue curves for heat-treated (peak-aged) C17200 strip. Longitudinal loading, $R = -1$

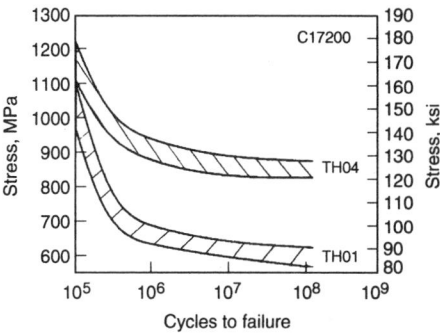

Fig. 10 Bending fatigue curves for heat-treated (peak-aged) C17200 strip. Longitudinal loading, $R = 0$

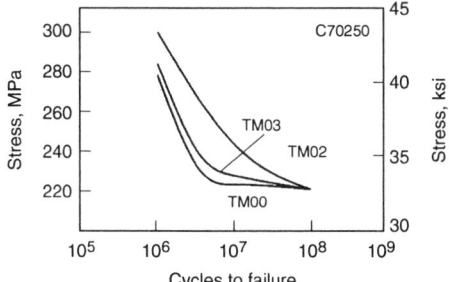

Fig. 7 S-N curves for C70250. Longitudinal loading, $R = -1$

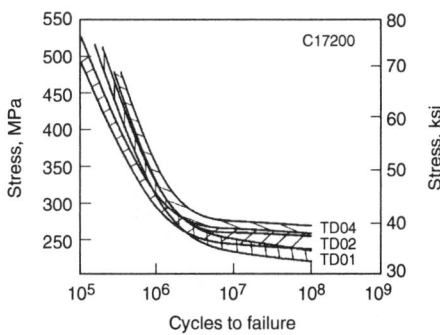

Fig. 8 Bending fatigue curves for beryllium copper C17200 strip in the heat-treatable condition. Longitudinal loading, $R = -1$

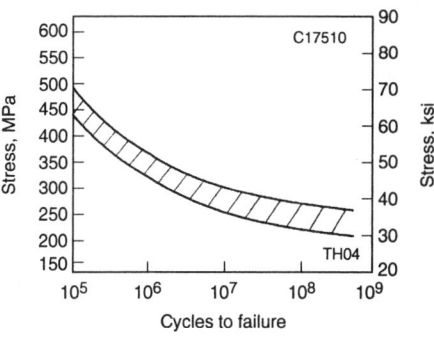

Fig. 11 Bending fatigue curves for C17510 TH04 strip. Longitudinal and transverse loading, $R = -1$

transverse data are not shown for the other alloys, similar behavior is observed. In general, low-stacking-fault-energy (SFE) alloys such as C68800 and C63800, as well as C65400, exhibit greater directionality, more so than the beryllium coppers or C70250.

The C70250 alloys are shown in Fig. 7 for three tempers. At higher cycles, the temper effect is minimal; at lower cycles, there is some temper dependence.

All of the curves shown in Fig. 1 to 7 represent fatigue stresses at which 50% of the samples would be expected to fail. The data should therefore be treated as representative and not be used for design purposes. Table 3 lists tensile properties and fatigue strengths for the alloys shown in Fig. 1 to 7.

Beryllium Coppers: Heat-Treatable and Heat-Treated Alloys. Fatigue data shown here for beryllium copper alloys are represented by a band, the lower bound determined by the lowest stress to cause failure and the upper bound determined by regression analysis of failure data from four or more commercial lots of materials. These

upper bound curves are equivalent to the 50% failure curves shown in Fig. 1-7.

Beryllium copper in the solution-annealed or cold-rolled condition prior to age hardening is referred to as being in the heat-treatable temper. Examples of the bending fatigue behavior of C17200 strip heat-treatable tempers are shown in Fig. 8. The effect of cold reduction, up to 37% for the TD04 temper, has only a small effect on the fatigue response.

Figure 9 demonstrates the effect of age hardening to peak strength after cold work on fatigue behavior—about a 70 MPa (10 ksi) increase in fatigue strength. The differences due to prior cold work become pronounced at $R = 0$ (unidirectional) stressing (Fig. 10). The benefit available for unidirectional stressing is useful for switch designs that operate in unidirectional bending.

The fatigue response for age hardened C17510 (TH04), the higher-conductivity alloy with lower strength (Fig. 11), demonstrates that fatigue strengths achievable at this higher conductivity

level are comparable to the higher-strength beryllium copper C172 in the solution-treated and cold-worked, unaged condition (see Fig. 8).

Beryllium Coppers: Mill-Hardened Strip. Mill hardening consists of age hardening to a specific strength level as part of the manufacturing process. This process can reduce or eliminate the need for age hardening after component forming that is required for the age-hardenable tempers. The data in Fig. 12 represent two mill-hardened tempers of C17200 strip. The TM04 temper is a medium-strength product (760 to 930 MPa, or 110 to 135 ksi, yield strength); the TM08 temper offers the greatest strength available (1035 to 1240 MPa, or 150 to 180 ksi, yield strength). The TM08 temper shows greater stress to failure at high cycles than the TM04 temper.

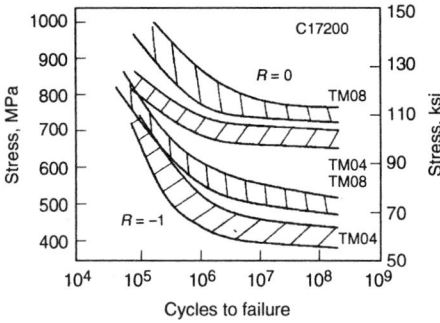

Fig. 12 Bending fatigue curves for C17200 TM04 and TM08 strip tempers. Longitudinal loading, R = 0 and –1

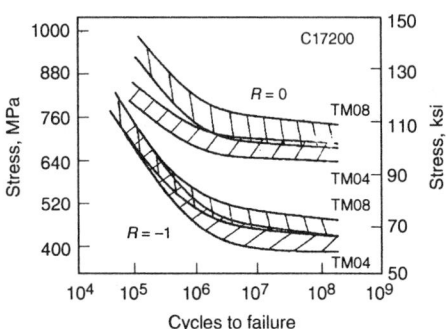

Fig. 13 Bending fatigue curves for C17200 TM04 and TM08 strip tempers. Transverse loading, R = 0 and –1

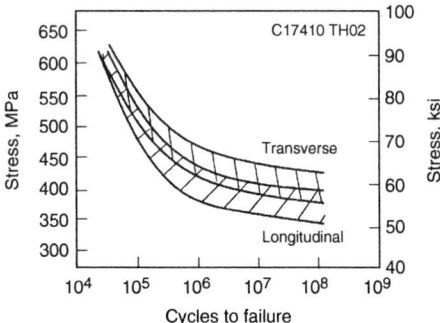

Fig. 14 Bending fatigue curves for C17410 TH02 strip. Longitudinal and transverse loading, R = –1

Table 4 Tensile and fatigue strengths of selected tempers of the beryllium copper alloys in Fig. 8 to 15

Alloy	Temper	0.2 % yield strength		Ultimate tensile strength		Elongation,	10^8 cycles
		MPa	ksi	MPa	ksi	%	(R = –1)
C17200	TD01	415-550	60-80	515-605	75-88	30-45	31-36
	TD02	515-655	75-95	585-690	85-100	12-30	32-38
	TD04	620-795	90-115	690-825	100-120	2-18	35-39
	TH01	1035-1275	150-185	1205-1415	175-205	3-10	40-45
	TH02	1105-1345	160-195	1275-1480	185-215	1-8	42-47
	TH04	1140-1415	165-205	1310-1515	190-220	1-6	45-50
	TM04	760-930	110-135	930-1035	135-150	9-20	45-52
	TM08	1035-1240	150-180	1205-1310	175-190	3-12	50-60
C17510	TH04	655-825	95-120	760-930	110-135	8-20	42-47
C17410	TH02	550-690	80-100	655-790	95-115	10 (min)	45
	TH04	690-825	100-120	760-895	110-130	7 (min)	45

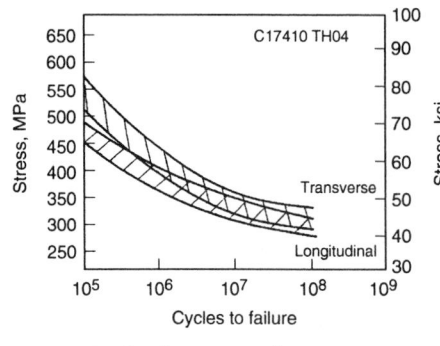

Fig. 15 Bending fatigue curves for C17410 TH04 strip. Longitudinal and transverse loading, R = –1

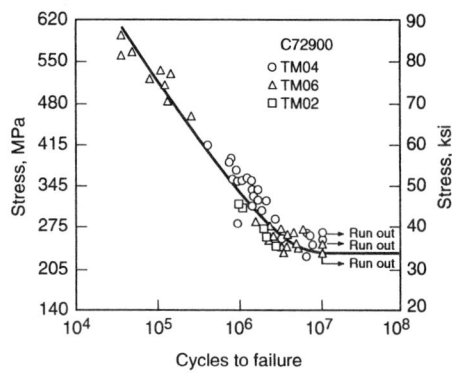

Fig. 16 S-N curve for C72900 in TM02, TM04, and TM06 tempers. Longitudinal loading, R = –1

Both tempers show very little directionality at either load ratio, as illustrated by comparing Fig. 12 and 13. Compared to the heat-treated tempers in Fig. 9, the TM08 temper displays the greater fatigue strength in reverse bending; however, in contrast, the TH04 temper displays the greatest fatigue strength in unidirectional bending (Fig. 10).

C17410 is manufactured in two mill-hardened strip tempers, designated by their manufacturer as TH02 and TH04. Typical mill-hardened tempers are designated TM*x*. These alloys do not require additional aging by the customer. Figures 14 and 15 show the fatigue curves for these alloys. The lower-strength (TH02) temper generally shows greater fatigue strength in reverse bending than the higher-strength temper. This is in contrast to the tensile strength/fatigue trend seen in C17200 alloys. The difference between the two tempers is diminished in unidirectional bending. Fatigue strengths around 550 MPa (80 ksi) are achieved for unidirectional bending. In general, the reverse bending fatigue response of this alloy compares to the fatigue strength of the C17200 mill-hardened strip, but shows slightly reduced response unidirectionally.

Edge condition can severely affect the fatigue response of strip products, particularly high-strength alloys. Electrical and electronic spring contacts are usually manufactured by stamping, slitting, electrodischarge machining, or chemically etching. Each of these operations can impart some degree of damage to the affected edge. The effect of slit edges, simulating stamped conditions, on fatigue response were addressed in Ref 7, which studied stamped versus milled edge samples. The results confirmed that high-cycle fatigue is surface dependent in strip and that careful application of fatigue data to a design is critical.

Table 4 lists tensile and fatigue strengths of the beryllium copper alloys shown in Fig. 8 to 15.

Spinodal Alloy. S-N data for alloy C72900 are shown in Fig. 16 for TM02, 04, and 06 tempers. Fatigue strengths at 100 million cycles in reversed bending are in the range of 220 to 275 MPa (32 to 40 ksi) for all three tempers. These curves represent averaged data comparable to S-N curves in Fig. 1-7.

Heavy-Section Beryllium Copper

Rotating-beam S-N curves are shown in Fig. 17 for C17200 in age-hardened tempers TF00 and TH04. Generally speaking, the smaller the diameter or size, the greater the fatigue strength. This behavior is linked directly to the microstructure, which will be discussed in the following section.

Discussion

As noted earlier, fatigue strength for copper alloys is usually defined as the stress sustainable without failure for 100 million cycles. [For the average (50% failure) curves, this actually means 50% will fail at this stress level.] For this high-cycle condition, most of the life is spent in crack nucleation, assuming the component in question does not have mechanical defects or other notches in the high-stress area.

High SFE alloys exhibit fatigue crack nucleation within persistent slip bands—markings produced by cyclic microplastic deformation. This is not an apparent mode of crack nucleation for low-SFE alloys. In general, cold work has a more beneficial effect on fatigue strength of low-SFE alloys versus copper and high-SFE copper alloys. Reducing grain size generally increases fatigue strength. This effect diminishes with increasing cold work.

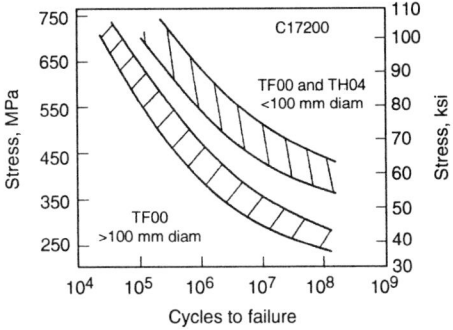

Fig. 17 Rotating-beam fatigue curves for C17200 TF00 and TH04 rod as a function of diameter

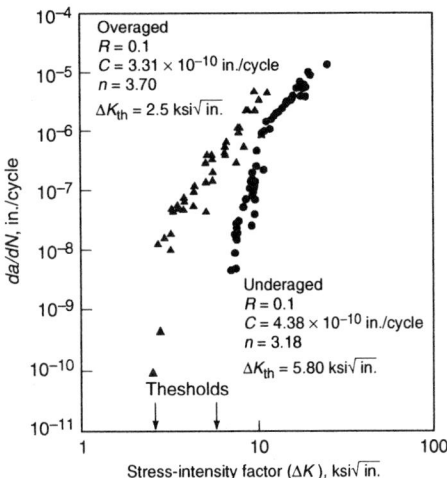

Fig. 18 Fatigue crack propagation rate of C17200 in the overaged and underaged conditions. Material aged to 760 MPa (110 ksi) yield strength

Resistance to microplastic deformation and enhancement of fatigue life can be produced by precipitation hardening if the precipitate phase is stable under cyclic loading. This is the case for beryllium copper and copper-nickel-tin precipitates.

Beryllium Coppers. Microstructure plays an important role in the fatigue performance of beryllium copper alloys. Age-hardened structures contain a mixture of metastable precipitates within a copper alloy matrix. The metastable precipitates dominate the deformation behavior of these alloys because of the small size, high volume fraction, homogeneous distribution, and high elastic strain contribution within the copper matrix. Aging, therefore, significantly affects fatigue behavior.

The high-strength C17200 alloys and C17410 contain a stable cobalt beryllide intermetallic, and C17510 contains a nickel beryllide. These beryllides are considerably larger than the precipitates and spaced sufficiently apart (i.e., long mean free path) so they do not contribute significantly to the fatigue performance.

An overaged structure will generally result in improved bending or rotating fatigue performance. Even though overaging results in somewhat lower strength, the net result is an improvement in fatigue life, because overaging prolongs the crack nucleation stage. Overaging causes the precipitate/matrix orientation relationship to progress to form incoherent γ phase from the metastable coherent Guinier-Preston zones (GPZ).

Conversely, the fatigue crack propagation behavior worsens as the alloys are overaged (Fig. 18). Deformation becomes highly localized at the grain boundaries as a result of the heterogeneous cellular reaction forming the equilibrium γ phase while creating adjacent regions within the grain boundaries of solute-depleted copper. Dislocation pileups at the grain boundaries are not adequately blunted and propagate easily into the boundaries, resulting in a crack path "short circuit." Nonetheless, the increase in crack propagation rate for overaged versus peak-aged material does not alter the overall improved life associated with the overaged condition.

ACKNOWLEDGMENTS

The authors would like to thank W.D. Smith, who performed much of the testing of the nonaging alloys and provided additional information for the test and W. Woodside who performed the testing of the Be-Cu alloys; M. McCowen, who typed and compiled the manuscript; W. Malcolm, who did most of the graphics; and A. Moses, who provided data on the Ametek alloy C72900. The authors also wish to acknowledge the Brass Group of Olin Corporation, Brush Wellman Inc., and Ametek for permission to publish the data in this article.

REFERENCES

1. C. Laird, Mechanisms and Theories of Fatigue, *Fatigue and Microstructure*, American Society for Metals, 1979, p 195
2. D. Mahulikar and T.D. Hann, "Factors Affecting Lead Bend Fatigue in P-Dips," presented at Microelectronic and Processing Engineers Conference (Sunnyvale, CA), 13 Feb 1985
3. J. Crane, J.F. Breedis, and R.M. Fritzche, in *Electronic Materials Handbook*, Vol 1, *Packaging*, ASM International, 1989, p 482
4. J.C. Harkness, W.D. Spiegelberg, and W.R. Cribb, Beryllium Copper and Other Beryllium-Containing Alloys, *Metals Handbook*, 10th ed., Vol 2, *Properties and Selection: Nonferrous Alloys and Special-Purpose Materials*, ASM International, 1993
5. G.R. Gohn, J.P. Guerard, and H.S. Freynik, *The Mechanical Properties of Wrought Phosphor Bronze Alloys*, STP 183, ASTM, 1956
6. A. Fox, Reversed Bending Fatigue Characteristics of Copper Alloy 510 Strip, *J. Mater.*, Vol 5 (No. 2), 1970
7. S.J. Schriver, J.O. Ratka, and W.D. Peregrim, Comparison of Fatigue Performance for Beryllium Copper Alloys by Two Laboratory Techniques, *Proc. ASM 3rd Conf. Electronic Packaging Materials and Processes and Corrosion in Microelectronics*, ASM International, 1987

SELECTED REFERENCES

- J. Crane and J. Winter, Copper: Selection of Wrought Alloys, *Encyclopedia of Materials Science and Engineering*, Pergamon Press, 1986, p 866-871
- J. Crane and J. Winter, Copper: Properties and Alloying, *Encyclopedia of Materials Science and Engineering*, Pergamon Press, 1986, p 848-855
- H.A. Murray, I.J. Zatz, and J.O. Ratka, Fracture Testing and Performance of Beryllium Copper Alloy C17510, *Cyclic Deformation, Fracture and Nondestructive Evaluation of Advanced Materials*, Vol 2, STP 1184, ASTM, 1994, p 109-133

Fatigue and Fracture Resistance of Magnesium Alloys

Magnesium possesses the lowest density of all structural metals, having about 25% the density of iron and approximately 33% that of aluminum. Because of this low density, both cast and wrought magnesium alloys (Tables 1 and 2) have been developed for a wide variety of structural applications in which low weight is important, if not a requirement. In this context, this article briefly summarizes the fatigue and fracture resistance of magnesium alloys.

Fatigue

Most of the fatigue data for magnesium alloys are *S-N* curves dating from the 1930s to 1960s. Strain-life (ε-*N*) curves for magnesium alloys are very rare, and most fatigue crack growth behavior data have originated from work conducted in the former Soviet Union. Data from these sources are compiled in Ref 1.

Like other alloys, fatigue strength of magnesium alloys depends on tensile strength (Fig. 1). However, the ratio of fatigue strength to tensile strength is not as well defined for magnesium

Table 1 Nominal composition, typical tensile properties, and characteristics of selected magnesium casting alloys

ASTM designation	British designation	Al	Zn	Mn	Si	Cu	Zr	RE (MM)	RE (Nd)	Th	Y	Ag	Condition	0.2% yield strength, MPa	Ultimate tensile strength, MPa	Elongation, %	Characteristics
AZ63	...	6	3	0.3	...	...	...	...	...	...	...	...	As-sand cast T6	75	180	4	Good room-temperature strength and ductility
														110	230	3	
AZ81	A8	8	0.5	0.3	...	...	...	...	...	...	...	...	As-sand cast T4	80	140	3	Tough, leak-tight castings
														80	220	5	With 0.0015 Be, used for pressure die casting
AZ91	AZ91	9.5	0.5	0.3								...	As-sand cast	95	135	2	General-purpose alloys used for sand and die castings
													T4	80	230	4	
													T6	120	200	3	
													As-chill cast	100	170	2	
													T4	80	215	5	
													T6	120	215	2	
AM50	...	5	...	0.3	...	...	...	...	...	...	...	...	As-die cast	125	200	7	High-pressure die castings
AM20	...	2	...	0.5	...	...	...	...	...	...	...	...	As-die cast	105	135	10	Good ductility and impact strength
AS41	...	4	...	0.3	1	...	...	...	...	...	...	...	As-die cast	135	225	4.5	Good creep properties to 150 °C
AS21	...	2	...	0.4	1	...	...	...	...	...	...	...	As-die cast	110	170	4	Good creep properties to 150 °C
ZK51	Z5Z	...	4.5	...	...	...	0.7	...	...	...	...	...	T5	140	235	5	Sand castings, good room-temperature strength and ductility
ZK61	...	...	6	...	...	...	0.7	...	...	...	...	...	T5	175	275	5	As for ZK51
ZE41	RZ5	...	4.2	...	...	...	0.7	1.3	...	...	...	...	T5	135	180	2	Sand castings, good room-temperature strength, improved castability
ZC63	ZC63	...	6	0.5	...	3	...	...	...	...	...	...	T6	145	240	5	Pressure-tight castings, good elevated-temperature strength, weldable
EZ33	ZRE1	...	2.7	...	...	...	0.7	3.2	...	...	...	...	Sand cast T5	95	140	3	Good castability, pressure tight, weldable, creep resistant to 250 °C
													Chill cast T5	100	155	3	
HK31	MTZ	...	...	...	...	...	0.7	...	...	3.2	...	...	Sand cast T6	90	185	4	Sand castings, good castability, weldable, creep resistant to 350 °C
HZ32	ZT1	...	2.2	...	...	...	0.7	...	...	3.2	...	...	Sand or chill cast T5	90	185	4	As for HK31
QE22	MSR	...	...	...	...	...	0.7	...	2.5	...	...	2.5	Sand or chill cast T6	185	240	2	Pressure tight and weldable, high yield strength to 250 °C
QH21	QH21	...	...	...	...	...	0.7	...	1	1	...	2.5	As-sand cast T6	185	240	2	Pressure tight, weldable, good creep resistance and yield strength to 300 °C
WE54	WE54	...	...	...	...	...	0.5	...	3.25	...	5.1	...	T6	200	285	4	High strength at room and elevated temperatures. Good corrosion resistance, weldable
WE43	WE43	...	...	...	...	...	0.5	...	3.25	...	4	...	T6	190	250	7	

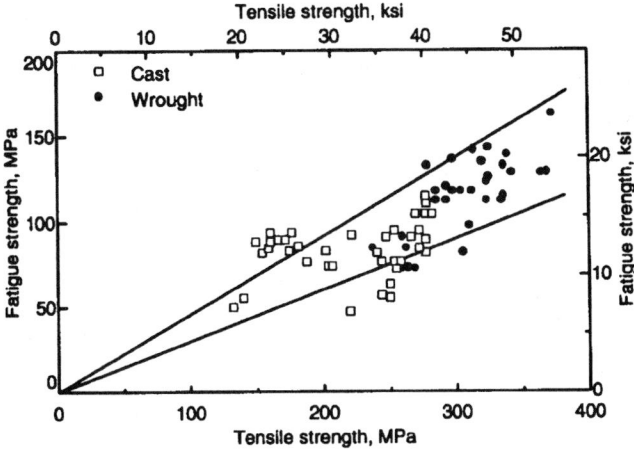

Fig. 1 Rotating bending fatigue strength vs. ultimate tensile strength of magnesium alloys (small smooth specimens). Source: Ref 2

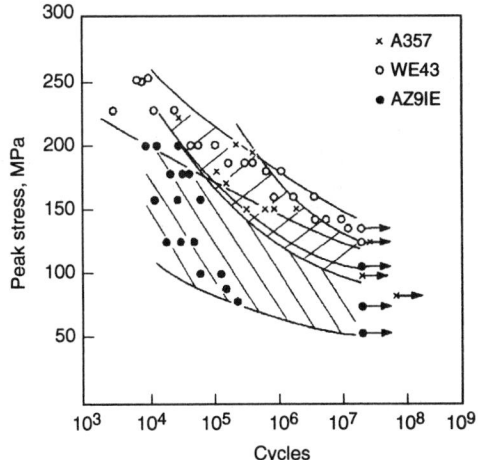

Fig. 2 Fatigue properties of A357, AZ91E, and WE43. $R = 0.1$. Source: Ref 4

Table 2 Nominal composition, typical tensile properties, and characteristics of selected wrought magnesium alloys

ASTM designation	British designation	Nominal composition							Condition	0.2% yield strength, MPa	Ultimate tensile strength, MPa	Elongation, %	Characteristics
		Al	Zn	Mn	Zr	Th	Cu	Li					
M1	AM503	…	…	…	1.5	…	…	…	Sheet, plate F	70	200	4	Low- to medium-strength alloy, weldable, corrosion resistant
									Extrusions F	130	230	4	
									Forgings F	105	200	4	
AZ31	AZ31	3	1	0.3 (0.20 min)	…	…	…	…	Sheet, plate O	120	240	11	Medium-strength alloy, weldable, good formability
									H24	160	250	6	
									Extrusions F	130	230	4	
									Forgings F	105	200	4	
AZ61	AZM	6.5	1	0.3 (0.15 min)	…	…	…	…	Extrusions F	180	260	7	High-strength alloy, weldable
									Forgings F	160	275	7	
AZ80	AZ80	8.5	0.5	0.2 (0.12 min)	…	…	…	…	Forgings T6	200	290	6	High-strength alloy
ZM21	ZM21	…	2	1	…	…	…	…	Sheet, plate O	120	240	11	Medium-strength alloy, good formability, good damping capacity
									H24	165	250	6	
									Extrusions	155	235	8	
									Forgings	125	200	9	
ZMC711	…	…	…	6.5	0.75	…	…	1.25	Extrusions T6	300	325	3	High-strength alloy
LA141	…	1.2	…	0.15 min	…	…	…	14	Sheet, plate T7	95	115	10	Ultralight weight (specific gravity 1.35)
ZK31	ZW3	…	3	…	0.6	…	…	…	Extrusions T5	210	295	8	High-strength alloy, some weldability
									Forgings T5	205	290	7	
ZK61	…	…	6	…	0.8	…	…	…	Extrusions F	210	285	6	High-strength alloy
									T5	240	305	4	
									Forgings T5	?160	275	7	
HK31	…	…	…	…	0.7	3.2	…	…	Sheet, plate H24	170	230	4	High creep resistance to 350 °C, weldable
									Extrusions T5	180	255	4	
HM21	…	…	…	0.8	…	2	…	…	Sheet, plate T8	135	215	6	High creep resistance to 350 °C, short time exposure to 425 °C, weldable
									T81	180	255	4	
									Forgings T5	175	225	3	
HZ11	ZTY	…	0.6	…	0.6	0.8	…	…	Extrusions F	120	215	7	Creep resistance to 350 °C, weldable

alloys as for steels. This is due, in part, to the effect of strengthening mechanisms on fatigue strength. For example, solid-solution strengthening increases the fatigue strength of magnesium alloys, whereas cold working and precipitation strengthening produce little improvement in fatigue strength at longer lives (Ref 3).

Axial fatigue *S-N* curves for AZ91E and WE43 are shown in Fig. 2, along with comparative data for A357 aluminum. The flat curve typical of magnesium alloys contrasts with that of aluminum where there is a marked change in slope between low- and high-cycle regimes. These different shapes of curve indicate that, although A357 performs well at low cycles, the situation changes so that WE43 has the better properties at high cycles. AZ91E has significantly lower properties in the low-cycle regime as a result of lower strength and porosity, but at high cycles the difference is not so marked.

Fatigue Mechanisms. The initiation of fatigue cracks in magnesium alloys is related to slip in preferably oriented grains and is often related to the existence of micropores. For pure magnesium, crack orientation is more strongly influenced by grain boundaries than the slip (Ref 3).

The initial stage of fatigue crack growth usually occurs from quasicleavage, which is common in hexagonal close-packed structures such as magnesium. Further crack growth micromechanisms can be brittle or ductile and trans- or intergranular, depending on metallurgical structure and environmental influence. Some

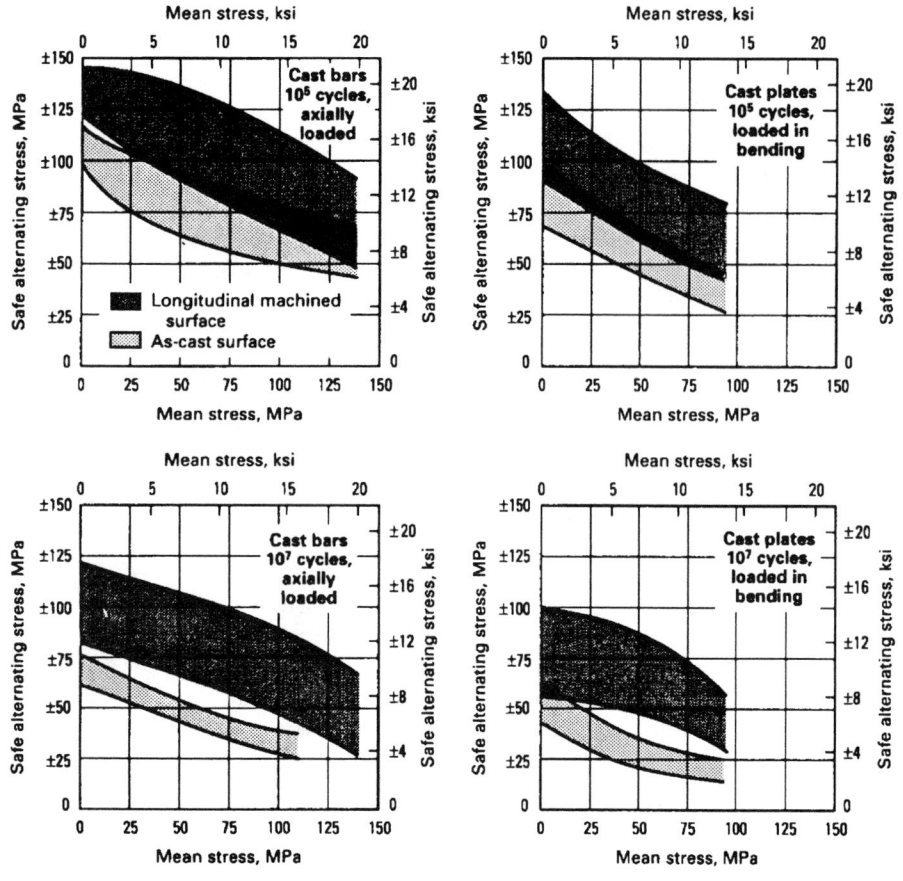

Fig. 3 Effect of surface type on the fatigue properties of cast magnesium-aluminum-zinc alloys. Source: *Metals Handbook*, 9th ed., Vol 2, ASM International, 1979, p 461

severity of stress raisers. Use of generous fillets in re-entrant corners and gradual changes of section greatly increase fatigue life. Conditions in which the effects of one stress raiser overlap those of another should be eliminated. Further improvement in fatigue strength can be obtained by inducing stress patterns conducive to long life. Cold working the surfaces of critical regions by rolling or peening to achieve appreciable plastic deformation produces residual compressive surface stress and increases fatigue life.

Surface rolling of radii is especially beneficial to fatigue resistance because radii generally are the locations of higher-than-normal stresses. In surface rolling, the size and shape of the roller, as well as the feed and pressure, are controlled to obtain definite plastic deformation of the surface layers for an appreciable depth (0.25 to 0.38 mm, or 0.010 to 0.015 in.). In all surface working processes, caution must be exercised to avoid surface cracking, which decreases fatigue life. For example, if shot peening is used, the shot must be smooth and round. The use of broken shot or grit can result in surface cracks.

Test Effects. As with other alloys, several test variables affect the fatigue strength of magnesium alloys. As expected, notched specimens and increasing R ratios decrease fatigue strength (Fig. 4a and 4b). The size of parts also reduces bending fatigue strength.

Generally, thicker portions of castings have greater microporosity and thus reduced fatigue strength, and thick extended bars (>75 mm diameter) and large forgings can experience reduced fatigue strength and increased notch sensitivity. Fatigue strength is also affected by specimen size (Fig. 5), because larger specimens provide greater surface area for crack initiation.

magnesium alloys can have either a hexagonal or body-centered cubic structure, depending on their chemical composition.

Effect of Surface Condition. High-cycle fatigue strength is influenced primarily by surface condition. Sharp notches, small radii, fretting, and corrosion are more likely to reduce fatigue

life than variations in chemical compositions or heat treatment. For example, removing the relatively rough as-cast surfaces of castings by machining improves fatigue properties of the castings (see Fig. 3).

When fatigue is the controlling factor in design, every effort should be made to decrease the

Fatigue Crack Growth

As previously mentioned, availability of fatigue crack growth data on magnesium alloys is rather limited because most fatigue crack growth data were generated in the former Soviet Union.

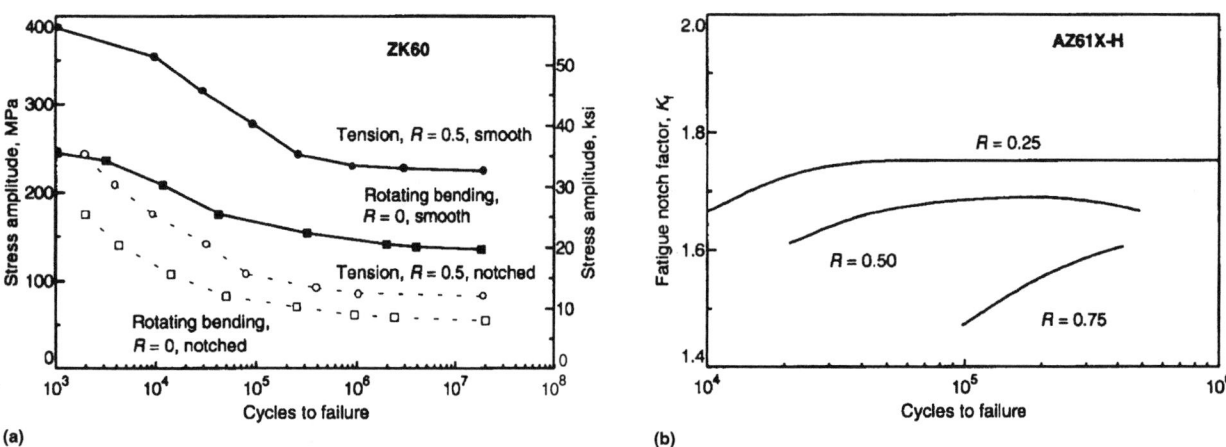

Fig. 4 Effect of stress ratio and notches on fatigue of two magnesium alloys. (a) Rotating bending and tension-compression *S-N* curves of ZK60. (b) Fatigue life of AZ61X-H with different notch factors. Source: Ref 5

However, in examining fatigue crack growth rate curves for many materials, the striking feature is the similarity of curves normalized by modulus. Although from an engineering viewpoint, the differences in crack growth rate between alloys can be important when integrating along da/dN curves to obtain the lifetime of a structure, a large range of metals can be represented by a single curve if the driving force is normalized by modulus, as illustrated in Fig. 6 (Ref 7). The effect of environment is important, particularly water vapor (see Fig. 7 for magnesium alloy ZK60A).

Fracture Toughness

Typical values of magnesium alloy toughness are summarized in Table 3 (Ref 9). The critical stress intensity factor, K_{Ic}, a material constant, is the largest stress intensity the material will support, under conditions of plane strain, without failing catastrophically. If K_{Ic} is known for the material, and the geometry and stress are known for the part, the largest crack that can be tolerated can be calculated. The larger the critical stress intensity factor, the larger the flaw size that can be tolerated.

One of the most difficult problems in fracture mechanics is the prediction of failure when section stresses approach or exceed yield values. Under these conditions, the critical stress intensity (K_c) lies outside the domain of linear elastic fracture mechanics and is not a material constant. In such cases, the *apparent* K_{Ic} depends on specimen geometry and flaw size. An example of variations in apparent K_{Ic} values outside the domain of linear conditions is shown in Fig. 8 for magnesium alloy HM21A-T8 (Ref 10).

The J-integral method has been used as a fracture criterion for nonlinear fracture mechanics. From tests on various alloys including magnesium alloy AZ31B, the J-integral is a valid fracture criterion for monotonic loading of thin section metals by Mode I stress systems. The results indicate that for a wide range of material behavior and specimen size J_c is not a function of crack length or specimen geometry. Additional details of the results are given in Tables 4 and 5. Statistical data for CT (compact tension) specimens are presented in Table 4 and compared with the mean values for data from CC and DEC (double edge cracked) specimens in Table 5. Standard deviations for J_c from CT specimens are seen to be on the order of ±11% of the mean for most of the alloys. These data can be put in the perspective of linear elastic fracture mechanics by the conversion:

$$K_c = \sqrt{(EJ_c)}$$

Stress-Corrosion Cracking

Wrought and cast magnesium alloys (particularly those containing aluminum) are susceptible to stress-corrosion cracking (SCC) when statically loaded below their yield strengths in some

Table 3 Typical toughness of magnesium alloys

Alloy	Temper	Temperature, °C	Tensile strength, ksi Unnotched	Tensile strength, ksi Notched	Ratio	Charpy V-notch, J	K_{Ic}, ksi √(in.)
Sand castings							
AZ81A	T4	25	...	...	...	6.1	...
AZ91C	F	25	...	...	...	0.79	...
	T4	25	...	...	0.90	4.1	...
	T6	25	...	...	0.86(a)	1.4	10.4
AZ92A	F	25	...	...	...	0.7	...
	T4	25	...	...	...	4.1	...
	T6	25	...	...	...	1.4	...
EQ21A	T6	20	...	...	...	...	14.9
EZ32A	T5	20	...	...	...	1.5(b)	...
HZ32A	T5	20	...	...	...	2.2(b)	...
QE22A	T6	25	...	...	1.06(a)	2.0	12.0
WE54A	T6	20	...	...	...	...	10.4
QH21A	T6	25	...	...	...	...	17.0
ZE41A	T5	25	...	...	...	1.4	14.1
ZE63A	T6	25	...	...	...	0.07	19.1
ZH62A	T5	25	...	...	...	3.4(b)	...
ZK51A	T5	20	...	...	...	3.5(b)	...
Extruded alloys							
AZ31B	F	25	...	...	...	3.4	25.5
AZ61A	F	25	...	...	...	4.4	27.3
AZ80A	F	25	46	34(c)	0.75	1.3	26.4
		−195	61	25(c)	0.40	...	...
	T5	25	50	22(c)	0.45	1.4	14.75
		−195	65	14(c)	0.22	...	...
	T6	25	...	...	...	1.4	...
HM31A	T5	25	44	38(c)	0.87	...	...
		−78	51	38(c)	0.75	...	...
		−195	59	39(c)	0.66	...	...
ZK30	F	25	...	...	...	4.0	41.8(d)
ZK60A	T5	25	51	49(c)	0.96	3.4	31.4
		0	...	...	...	2.2	...
		−78	...	...	...	2.2	...
		−195	74	45(c)	0.61	...	...
Sheet							
AZ31B-O	...	24	38	31(e)	0.83	5.9	...
	...	−196	58	33(e)	0.53	...	...
AZ31B-H24	...	24	41	33(e)	0.81	...	26
	...	−196	60	24(e)	0.40	...	...
HK31A-O	...	24	31	27(e)	0.86	4.0	30
	...	−196	52	30(e)	0.57	...	...
HK31A-H24	...	24	39	33(e)	0.85	3.0	23
	...	−196	58	36(e)	0.63	...	...
HM21A-T8	...	24	36	33(e)	0.94	...	23
	...	−78	46	30(e)	0.67	...	...
	...	−196	54	32(e)	0.59	...	...
ZE10-O	...	24	33	29(e)	0.87	6.6	21
	...	−78	44	30(e)	0.69	...	...
	...	−196	53	31(e)	0.59	...	...
ZE10A-H24	...	24	38	38(e)	1.00	...	28
	...	−78	46	38(e)	0.83	...	...
	...	−196	54	30(e)	0.56	...	...
ZH11A-H24	...	20	...	...	...	4.4(b)	...

(a) Notched/unnotched tensile strength ratio with a notch radius of 0.008 mm. (b) Izod specimen. (c) The notched specimen has a reduced section of 0.06 in. × 1 in., a 60° V-notch, a 0.700 in. notched width, and a notch root radius of 0.0003 in. (d) Value is for J_{Ic} since specimen was too small for an accurate K_{Ic} value; true value for K_{Ic} is lower. (e) Specimen dimensions: Total width, 1 in.; notched width, 0.700 in.; thickness, 0.60 in.; 60° V-notch with 0.0003 in. radius. Source: Adapted from Ref 9

Table 4 Fracture toughness of various alloys

Alloy	Mean value, J_c, J/mm^2	Standard deviation, J_c, J/mm^2	Mean value, K_c, MPa√m	Standard deviation, K_c, MPa√m
6061-0	0.125	0.009	93.0	3.4
7075-0	0.075	0.008	71.7	3.5
70/30	0.282	0.029	176.0	9.1
AZ31B	0.052	0.003	48.4	1.4
1018	0.342	0.039	266.0	15.0
4130	0.218	0.021	212.0	10.0
HP9-4-20	0.245	0.023	218.0	11.0

Compact tension (CT) specimens

Table 5 Comparison of mean values of J_c for various specimen geometries and alloys

Alloy	Mean values of J_c, J/mm^2 Compact tension (CT) specimens	Mean values of J_c, J/mm^2 Center cracked (CC) specimens	Mean values of J_c, J/mm^2 Double edge cracked (DEC) specimens
6061-0	0.125	0.115	...
7075-0	0.075	0.078	0.065
70/30	0.282	0.285	...
AZ31B	0.052	0.054	0.048
1018	0.342	0.368	...
4130	0.218	0.248	0.216

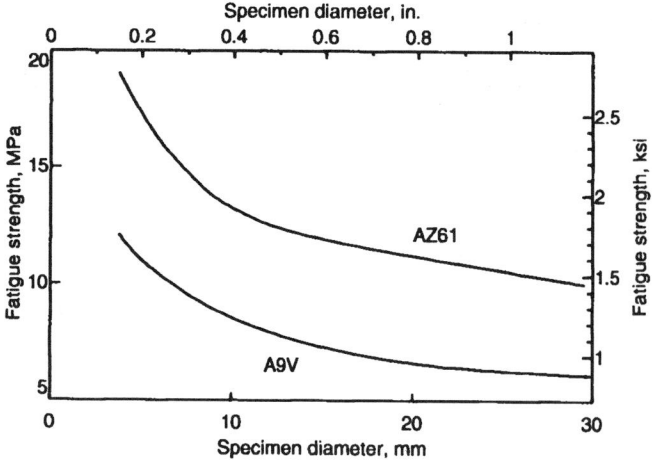

Fig. 5 Effect of specimen size on fatigue strength of magnesium alloys (smooth, rotating bending specimens). Source: Ref 6

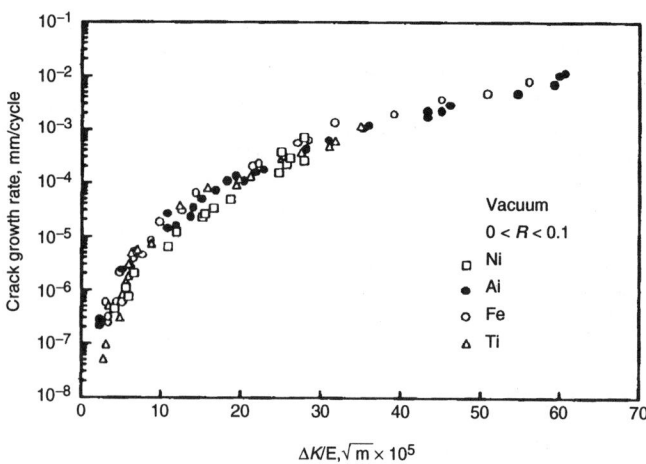

Fig. 6 Crack growth rate curves for several metals compared on the basis of driving force normalized by modulus. Original work includes a much larger range of materials, including polymers. Source: Ref 7

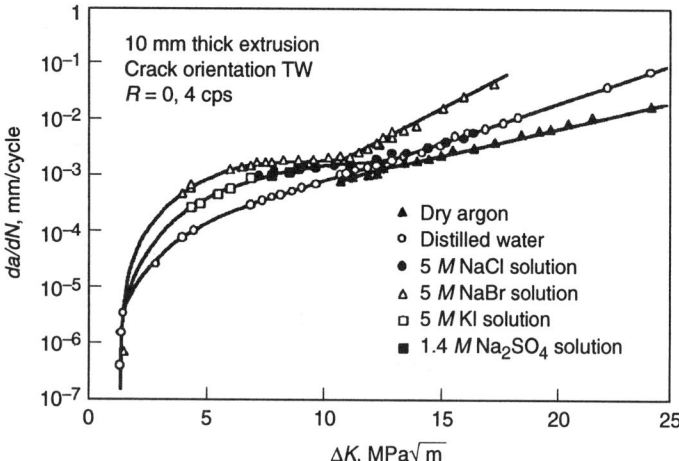

Fig. 7 Corrosion-fatigue crack growth curves for ZK60A-T5 in different environments. Source: Ref 8

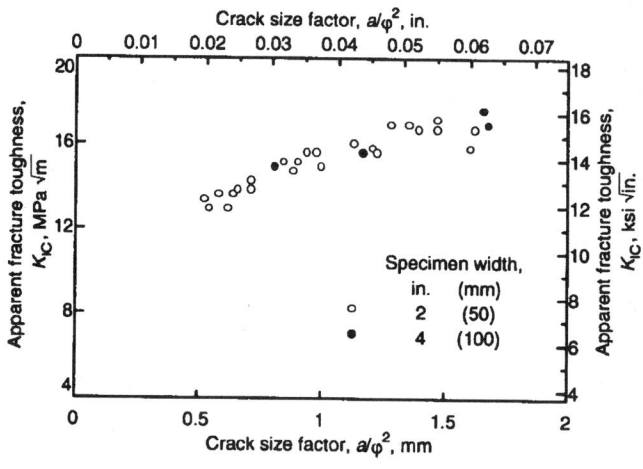

Fig. 8 Variation of apparent fracture toughness (K_{Ic}) with crack size. Source: Ref 10

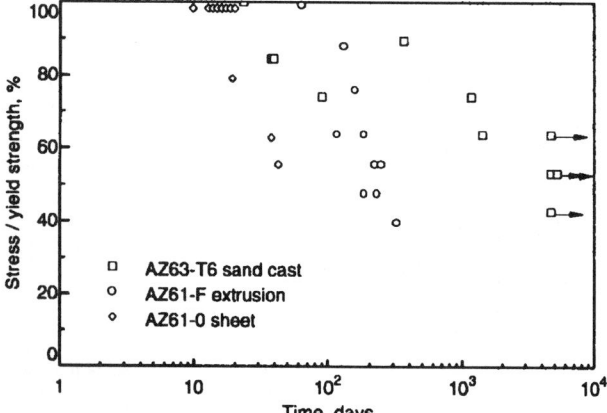

Fig. 9 Data comparing similar cast and wrought magnesium alloys during long-term stress-corrosion cracking (SCC). Long-term rural-atmosphere SCC data compare similar-composition AZ61 sheet, extruded AZ61, and sand-cast AZ63. Although there is a great deal of scatter in these data, all three materials exhibited similar behavior at the higher stress levels. At lower stress levels, the cast alloys became more resistant to SCC. Source: Ref 11

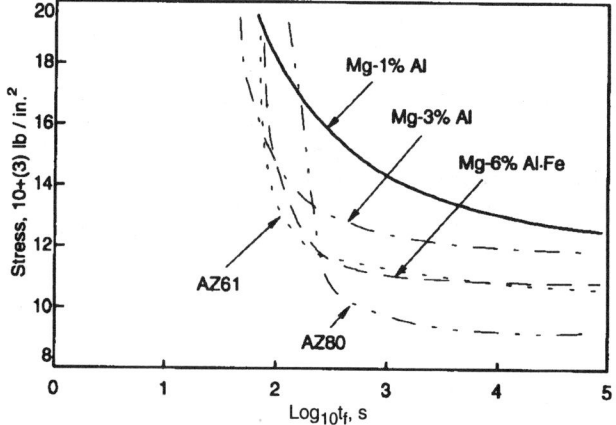

Fig. 10 Stress vs. time-to-failure (t_f) for magnesium-aluminum alloys in aqueous 40 g/L NaCl + 40 g/L Na_2CrO_4. Source: Ref 28

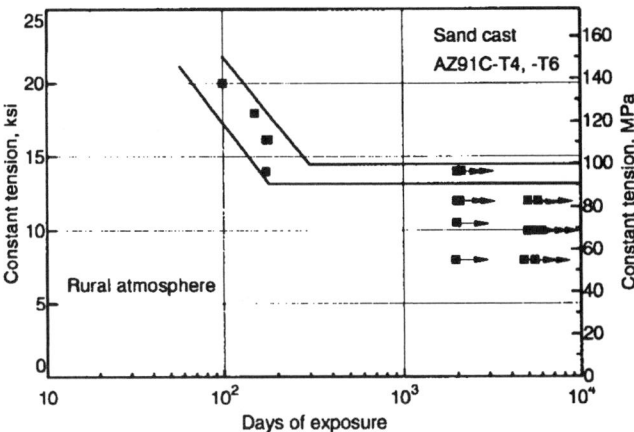

Fig. 11 Stress corrosion of sand-cast AZ91C (T4 and T6) in rural atmosphere. Source: Ref 9

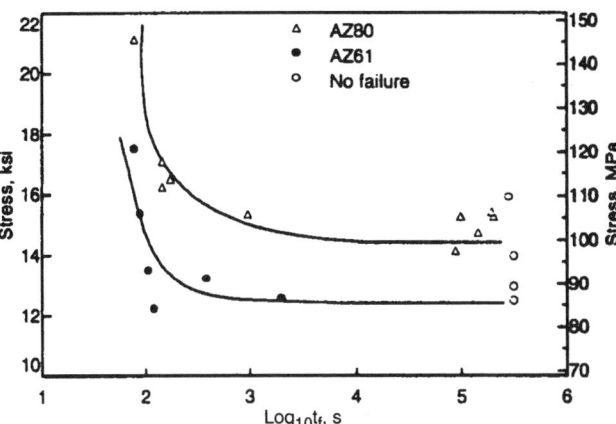

Fig. 12 Stress vs. time-to-failure (t_f) for the two-phase alloys AZ80 (Mg-8.5Al-0.5Zn) and AZ61 (Mg-6Al-1Zn) in aqueous 40 g/L NaCl + 40 g/L Na_2CrO_4

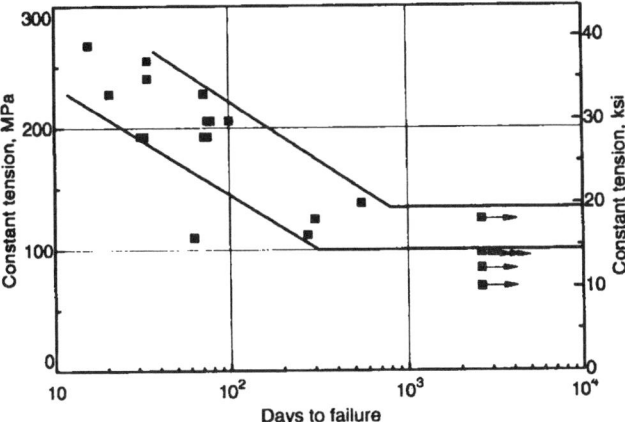

Fig. 13 Stress corrosion of ZK60A-T5 extrusion in rural atmosphere. Source: Ref 9

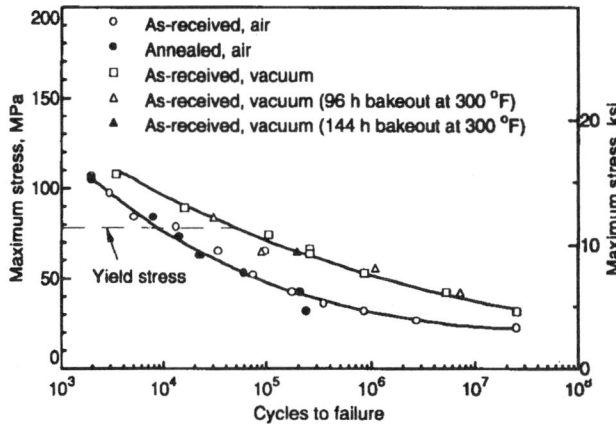

Fig. 14 Fatigue of commercial pure 9980A magnesium (UNS M19980) in air and in vacuum. Conditions: cantilever bending, $R = -1$, 30 Hz, room temperature. Source: *J. Spacecraft Rockets*, Vol 5, 1968, p 700-704

environments. Early studies found susceptibility for wrought forms but little or no susceptibility for cast forms. However, the presence of aluminum as an alloying element is the main factor affecting SCC susceptibility, rather than cast or wrought form. For aluminum-containing magnesium alloys, SCC susceptibility is comparable to that of wrought and cast forms of similar composition (Fig. 9, Ref 11).

While all magnesium alloys will stress corrode to some extent, the most susceptible are those containing aluminum as an alloying element. Magnesium alloys containing aluminum should not be designed for prolonged exposure to stresses near the yield point. Residual stresses from operations such as welding or machining must also be relieved by heat treatment (Ref 9). Service failures normally result from excessive residual stress produced during fabrication (Ref 12-18). Speidel (Ref 19), in a comprehensive review of more than 3000 unclassified failure reports from aerospace companies, government agencies, and research laboratories in the United States and five Western European countries, estimated that approximately 10 to 60 magnesium

aerospace component SCC service failures occurred each year from 1960 to 1970. Of this total, more than 70% involved either cast alloy AZ91-T6 or wrought alloy AZ80-F, both of which contain aluminum. In contrast, magnesium alloys without aluminum do not, in practice, have a stress-corrosion problem (Ref 9). The broad class of alloys containing zirconium are also sufficiently insensitive that SCC is not a problem in practice.

Stress-corrosion cracking of magnesium alloys can occur in many environments. In general, the only solutions that do not induce SCC are either those that are nonactive to magnesium, such as dilute alkalies, concentrated hydrofluoric acid, and chromic acid, or those that are highly active, in which general corrosion predominates. More information on the effect of air, water, and aqueous solutions is contained in Ref 20.

Effects of Alloy Composition. Pure magnesium is not susceptible to SCC when loaded up to its yield strength in atmospheric and most aqueous environments (Ref 13, 21-24). The only reports of SCC of pure magnesium have emanated from laboratory tests in which specimens were

immersed in very severe SCC solutions (Ref 25-27).

As previously mentioned, aluminum-containing magnesium alloys have the highest SCC susceptibility, with the sensitivity increasing with increasing aluminum content, as illustrated in Fig. 10 (Ref 28). An aluminum content above a threshold level of 0.15 to 2.5% is reportedly required to induce SCC behavior (Ref 14, 29, 30), with the effect peaking at approximately 6% Al (Ref 31). The aluminum- and zinc-bearing AZ alloys, which are the most commonly used magnesium alloys, have the greatest susceptibility to SCC. Alloys with higher aluminum content, such as AZ61, AZ80, and AZ91, can be very susceptible to SCC in atmospheric (Fig. 11) and more severe environments (Fig. 12), while lower-aluminum AZ31 is generally more resistant. However, it too can suffer SCC under certain conditions.

Magnesium-zinc alloys that are alloyed with either zirconium or RE elements, but not with aluminum, such as ZK60 and ZE10, have intermediate SCC resistance (Fig. 13), and in some cases SCC has not been a serious problem. However, SCC can still occur in atmospheric environments at stresses as low as 50% of the yield strength, although life is significantly longer than for Mg-Al-Zn alloys.

Magnesium alloys that contain neither aluminum nor zinc are the most SCC resistant. Magnesium-manganese alloys, such as M1, are among the alloys with the highest resistance to SCC, and they are generally considered to be immune when loaded up to the yield strength in normal environments. In fact, SCC of Mg-Mn alloys has been reported only in tests involving stresses higher than the yield strength (Ref 11, 32) and/or exposure to very severe laboratory environments (Ref 33). Alloys QE22, HK31, and HM21 are also resistant to SCC, exhibiting SCC thresholds at approximately 70 to 80% of the yield strength in rural-atmosphere tests (Ref 66).

Magnesium-lithium alloys are of commercial interest because of their higher stiffness and lower density compared with other magnesium alloys. Tests in humid air have resulted in SCC failures of Mg-Li-Al alloys, but SCC did not occur during testing of Mg-Li alloys strengthened with zinc, silicon, and/or silver instead of aluminum (Ref 34).

The earliest investigations of SCC of magnesium and magnesium alloys focused on the influence of alloy chemistry and microstructure, which were just gaining recognition as controlling factors in magnesium corrosion behavior (Ref 35, 36). The overwhelming majority of these studies used a chromate-chloride electrolyte (typically 40 g–1 each) because of its relevance to service conditions, in which chloride ions present in the environment attempt to penetrate a chromate-inhibited magnesium surface. These solutions cause especially severe cracking, but magnesium is also affected similarly by neutral solutions containing only chlorides or even distilled water (Ref 37).

A recent study (Ref 38) sought to compare the stress-corrosion behavior of rapidly solidified alloys to that of cast Mg-Al alloys, paying special attention to the role played by hydrogen and repassivation kinetics. This investigation, which was the first for rapidly solidified Mg-Al alloys, showed that all the alloys, as well as pure magnesium, failed by transgranular SCC in a chromate-chloride electrolyte at displacement rates between 5×10^{-5} and 9×10^{-3} mm s^{-1}. This failure mode was manifested in quasicleavage on the fracture surfaces and in lower maxima in stress intensity and displacement, and it was concluded that transgranular SCC probably occurs in these materials as a result of hydrogen embrittlement. Results from constant displacement rate testing were explained by a hydride formation model using realistic estimates for the diffusivity of hydrogen in magnesium. Based on repassivation results, dissolution appears incapable of achieving the observed crack growth rates. Potential pulse and scratching electrode experiments demonstrated superior repassivation behavior for rapidly solidified Mg-Al alloys compared with their as-cast counterparts, indicating that homogeneity retards pit nucleation and thereby retards the de-

velopment of local environments that impair repassivation. Increasing the aluminum content from 1 to 9% improved the repassivation rate of rapidly solidified alloys. This study also showed that repassivation participates in this SCC mechanism, probably by localizing the corrosion reactions and controlling the amount of hydrogen that enters the unprotected alloy surface when film rupture occurs.

Corrosion Fatigue

Substantial reductions in fatigue strength are shown in laboratory tests using NaCl spray or drops. Such tests are useful for comparing alloys, heat treatments, and protective coatings. Effective coatings, by excluding the corrosive environment, provide the primary defense against corrosion fatigue.

A fundamental study of the corrosion fatigue of magnesium alloys is that of Speidel et al. on high-strength magnesium alloy ZK60A (Ref 8, 39) (Fig. 7). All magnesium alloys behave similarly with respect to environmentally enhanced subcritical crack growth, according to Speidel et al. They found that both stress-corrosion and corrosion-fatigue cracks propagate in a mixed transgranular-intergranular mode. They measured the corrosion-fatigue crack growth for all of the aqueous environments shown in Fig. 7 and compared the corrosion fatigue with stress-corrosion behavior. They found that:

- Corrosion-fatigue crack growth rate is accelerated by the same environments as those that accelerate stress-corrosion crack growth (i.e., sulphate and halide ions).
- The boundary between regions II and III in NaBr solutions of the da/dN versus ΔK curve is higher than the stress-corrosion threshold (K_{Iscc}), which occurs at a much lower stress intensity.
- There is a distinct boundary between regions II and III for all the media given in Fig. 7 (except dry argon). This boundary occurs at about the same stress intensity (~14 MPa $\sqrt{m}$) as K_{Issc} in distilled water.

Surface Protection. It is common practice to protect the surface of magnesium and its alloys, and such protection is essential where contact occurs with other structural metals because this may lead to severe galvanic corrosion. Methods available for magnesium are summarized below. Additional information is contained in Volume 5 of the *ASM Handbook*.

- Fluoride anodizing involves alternating current anodizing at up to 120 V in a bath of 25% ammonium bifluoride, which removes surface impurities and produces a thin, pearly white film of MgF$_2$. This film is normally stripped in boiling chromic acid before further treatment because it gives poor adhesion to organic treatments.
- Chemical treatments involve pickling and conversion of the oxide coating. Components are

dipped in chromate solutions, which clean and passivate the surface to some extent through formation of a film of Mg(OH)$_2$ and a chromium compound. Such films have only slight protective value, but they form a good base for subsequent organic coatings.

- Electrolytic anodizing includes proprietary treatments that deposit a hard ceramic-like coating, which offers some abrasion resistance in addition to corrosion protection (e.g., Dow 17, HEA, and MGZ treatments). Such films are very porous and provide little protection in the unsealed state, but they may be sealed by immersion in a solution of hot dilute sodium dichromate and ammonium bifluoride, followed by draining and drying. A better method is to impregnate with a high-temperature curing epoxy resin (see below). Resin-sealed anodic films offer very high resistance to both corrosion and abrasion, and in some instances they can even be honed to provide a bearing surface. Impregnation is also used to achieve pressure tightness in casting that are susceptible to microporosity.
- Sealing with epoxy resins: The component is heated to 200 to 220 °C to remove moisture, cooled to approximately 60 °C, and dipped in the resin solution. After removal from this solution, draining, and air drying to evaporate solvents, the component is baked at 200 to 220 °C to polymerize the resin. Heat treatment may be repeated once or twice to build up the desired coating thickness, which is commonly 0.025 mm.
- Standard paint finishes: The surface of the component should be prepared as in the methods described above, after which it is preferable to apply a chromate-inhibited primer followed by a good-quality top coat.
- Vitreous enameling can be applied to alloys that do not possess too low a solidus temperature. Surface preparation involves dipping the work in a chromate solution before applying the frit.
- Electroplating: Several stages of surface cleaning and the application of pretreatments, such as a zinc conversion coating, are required before depositing chromium, nickel, or some other metal.

Magnesium alloy components for aerospace applications require maximum protection. Schemes involving chemical cleaning by fluoride anodizing, pretreatment by chromating or anodizing, and sealing with epoxy resin, a chromate primer, and a top coat are sometimes mandatory.

REFERENCES

1. Magnesium Alloys Fatigue and Fracture, in *Fatigue Data Book: Light Structural Alloys*, ASM International, 1995, p 140-179
2. R.B. Heywood, *Designing Against Fatigue of Metals*, Reinhold, 1962
3. R.I. Stechens and V.V. Ogarevic, *Ann. Rev. Mater. Sci.*, Vol 20, 1990, p 141-177

4. B. Geary, Corrosion Resistant Magnesium Casting Alloys, *Advanced Aluminum and Magnesium Alloys*, ASM, 1990
5. Prod. Eng., Vol 22, 1951, p 159-163, and *Proceed. ASTM*, Vol 46, 1946, p 783-798
6. Prevention of the Failure of Metals Under Repeated Stress, John Wiley & Sons, 1941
7. M.O. Speidel, in *High-Temperature Materials in Gas Turbines*, P.R. Sahm and M.O. Speidel, Ed., Elsevier, 1974, p 207-251
8. M.O. Speidel, M.J. Blackburn, T.R. Beck, and J.A. Feeney, *Proc. Corrosion Fatigue: Chem. Mech. Microstructure*, Storrs, CT, 1971, p 324-325
9. R.S. Busk, *Magnesium Products Design*, Marcel Dekker, 1987
10. T.W. Orange, *Engr. Fracture Mechanics*, Vol 3, 1971, p 53-69
11. "Exterior Stress Corrosion Resistance of Commercial Magnesium Alloys," Report Mt 19622, Dow Chemical USA, 8 March 1966
12. W.S. Loose and H.A. Barbian, Stress-Corrosion Testing of Magnesium Alloys, *Symp. Stress-Corrosion Cracking of Metals*, ASTM, 1945, p 273-292
13. W.S. Loose, Magnesium and Magnesium Alloys, *The Corrosion Handbook*, H.H. Uhlig, Ed., John Wiley and Sons, 1948, p 232-250
14. H.L. Logan, Magnesium Alloys, *The Stress Corrosion of Metals*, John Wiley and Sons, 1966, p 217-237
15. Magnesium: Designing Around Corrosion, Dow Chemical Co., 1982, p 16
16. J.D. Hanawalt, Joint Discussion on Aluminum and Magnesium, *Symp. Stress-Corrosion Cracking of Metals*, ASTM, 1945
17. M. Vialatte, Study of the SCC Behavior of the Alloy Mg-8% Al, *Symp. Engineering Practice to Avoid Stress Corrosion Cracking*, NATO, 1970, p 5-1 to 5-10
18. J.J. Lourens, Failure Analysis as a Basis for Design Modification of Military Aircraft, *Fracture and Fracture Mechanics Case Studies*, R.B. Tait and G.G. Garrett, Ed., Pergamon Press, 1985, p 47-56
19. M.O. Speidel, Stress Corrosion Cracking of Aluminum Alloys, *Metall. Trans. A*, Vol 6, 1975, p 631-651
20 SCC of Magnesium Alloys in *Stress Corrosion Cracking*, ASM International, 1992
21. G. Siebel, The Influence of Stress on the Corrosion of Electron Metals, *Jahrbuch der deutschen Luftfahrtforschung*, Part 1, 1937, p 528-531
22. A. Beck, *The Technology of Magnesium and Its Alloys*, 2nd ed., F.A. Hughes and Co., 1940, p 294-297
23. N.D. Tomashov, *Theory of Corrosion and Protection of Metals* (transl.), B.H. Tytell, I. Geld, and H.S. Preiser, Ed., MacMillan, 1966, p 626
24. M.J. Blackburn and M.O. Speidel, The Influence of Microstructure on the Stress Corrosion Cracking of Light Alloys, *Electron Microscopy and Structure of Materials*, G. Thomas, Ed., University of California Press, 1972, p 905-919
25. E.I. Meletis and R.F. Hochman, Crystallography of Stress Corrosion Cracking in Pure Magnesium, *Corrosion*, Vol 40 (No. 1), 1984, p 39-45
26. R.S. Stampella, R.P.M. Procter, and V. Ashworth, Environmentally-Induced Cracking of Magnesium, *Corros. Sci.*, Vol 24 (No. 4), 1984, p 325-341
27. S.P. Lynch and P. Trevena, Stress Corrosion Cracking and Liquid Metal Embrittlement in Pure Magnesium, *Corrosion*, Vol 44 (No. 2), 1988, p 133-124
28. J.A. Beavers, G.H. Koch, and W.E. Berry, "Corrosion of Metals in Marine Environments," Metals and Ceramics Information Center, Battelle Columbus Laboratories, July 1986
29. Metals Handbook, 9th ed., Vol 13, *Corrosion*, ASM International, 1987, p 745
30. R.D. Heidenreich, C.H. Gerould, and R.E. McNulty, Electron Metallographic Methods and Some Results for Magnesium Alloys, *Trans. AIME*, Vol 166, 1946, p 15
31. E.F. Emley, *Principles of Magnesium Technology*, Pergamon Press, 1966
32. H.L. Logan and H. Hessing, Stress Corrosion of Wrought Magnesium Base Alloys, *J. Res. Natl. Bur. Stand.*, Vol 44, 1950, p 233-243
33. E.C.W. Perryman, Stress-Corrosion of Magnesium Alloys, *J. Inst. Met.*, Vol 78, 1951, p 621-642
34. J.C. Kiszka, Stress Corrosion Tests of Some Wrought Magnesium-Lithium Base Alloys, *Mater. Protect.*, Vol 4 (No. 2), 1965, p 28-29
35. D.K. Priest, F.H. Beck, and M.G. Fontana, in *Trans. ASM*, Vol 48, 1955, p 473-492
36. R.D. Heidenreich, C.H. Gerould, and R.E. McNulty, in *Trans. AIME*, Vol 166, 1946, p 15-36
37. W.S. Loose, in *Magnesium*, Proceedings of lecture series presented at National Metal Cong. and Exposition, Cleveland, OH, 1946, American Society for Metals, p 244
38. G.L. Makar, J. Kruger, and K. Sieradzki, *Corros. Sci.*, 1993
39. M.O. Speidel, M.J. Blackburn, T.R. Beck, and J.A. Feeney, in *Corrosion Fatigue: Chemistry, Mechanics and Microstructure*, O. Devereux et al., Ed., National Association of Corrosion Engineers, 1986, p 331

Fatigue of Solders and Electronic Materials

Aleksander Zubelewicz, IBM
Semyon Vaynman, Northwestern University
Srinivas T. Rao, Solectron

AN UNDERSTANDING of the mechanical and fatigue properties of solders used in electronic packaging is a requirement for better design of solder joints and the development of accelerated tests and improved solders. Fatigue failures of a solder joint in an electronic device result from the imposition of strain caused by the joining of materials with different thermal expansivity under conditions of thermal cycling. This process of thermal fatigue in a given device is controlled by the total strain. The strain levels are determined by thermal expansion mismatch of materials used in a package, thermal gradients in a package, temperature excursions during service, the geometry of the solder joint, and compliance of the joint system.

However, even though failure of solder joints is due to thermal fatigue, most of the data available are for isothermal fatigue of solders. Isothermal tests are easier to control, conduct, and interpret, and they are less costly than thermomechanical fatigue tests. Methods to relate isothermal fatigue data to thermal fatigue data are being developed. The long-range objective of solder fatigue research is to use isothermal properties along with short-time thermomechanical behavior to predict solder joint lifetime. It will be shown (based on the experimental data, metallurgical observations, and the micromechanically based theory of the solder behavior) that the thermomechanical fatigue tests that are used to verify the quality of solder joints can be replaced with isothermal tests. In fact, an isothermal fatigue test called the mechanical deflection system (MDS) test was developed and implemented for reliability assessment of solder joints (Ref 1).

Variables Affecting Isothermal Fatigue of Solders. The most important variables during isothermal fatigue of solders are mode of loading, strain range, ramp time, hold times, temperature, and environment (as discussed later in this article). Factors such as solder composition, microstructure, aging conditions, specimen design, and fatigue life definition should be critically reviewed.

Because of processing needs and operational requirements, solders of different compositions are used in electronic packaging. While there are some similarities in fatigue behavior of different solders, all are different materials with specific microstructural, mechanical, and fatigue properties. It is important to stress that because of the very high sensitivity of solder microstructure to impurities and aging conditions, solders of the same composition may have different mechanical properties if aged differently. Therefore, application of solder fatigue and mechanical property data available in the literature to a specific solder and application must consider all of the conditions under which the data were obtained to avoid erroneous assessment of the reliability of a solder joint.

Isothermal fatigue data for solders have been developed for specimens of different design, ranging from a solder joint in a real device (Ref 2-4), two plates joined by solder (Ref 5-9) to a bulk specimen (Ref 10-14) tested under different loading conditions: tension (Ref 11-13), shear (Ref 3-9), bending (Ref 15), and torsion (Ref 2, 10). Different fatigue life criteria have been used at different laboratories: visible cracking proportioned from total fracture (Ref 15), a predetermined drop in load (Ref 2-6, 10, 14, 16, 17), start of the drop of the maximum tensile stress (Ref 11-14) or of the tensile stress-compressive stress ratio (Ref 14), and predetermined increase in resistance (Ref 3, 15). A different definition of failure leads to a different fatigue life. For example, when end of fatigue life in near-eutectic solder was defined as the number of cycles to reduce the initial tensile stress to 75, 50, or 10% of its initial value, the different definitions of fatigue life led to different slopes of Coffin-Manson plots (Ref 18). Thus, it is very often difficult, if not impossible, to compare data developed at different laboratories and correlate these data to the fatigue life of the solder joint in the device.

Microstructural Properties of Solders and Other Ductile Polycrystalline Materials. The behavior of solders operating in a high-tempera-ture environment is very complex and far from understood. Temperature, strain rate, and strain range alter the mechanical properties of the materials, shift the source of material nonlinearity within different levels of the internal structure, and frequently define the fracture regime. Intergranular failure occurs when the inelastic behavior is localized near grain boundaries; hence dislocation creep and grain-boundary slip dominate, both of which are associated with the cavitation process (Ref 19, 20). Dislocation glide is typical at the upper stress extreme and leads to transgranular failure mechanism, while diffusional creep is characteristic at lower stresses.

In general, the total deformation is composed of elastic, recoverable inelastic, and plastic portions. The irreversible (plastic) deformation is responsible for damage processes, while the microstructural recovery allows for relaxation of the residual stresses stored in the internal structure and brings the microstructure into its equilibrium configuration. The damage and recovery processes minimize free energy: The first leads to the energy dissipation; the second minimizes the internal energy inside grains. Recovery dominates within the range of conditions that favor creep mechanisms, and the amount of the recovered inelastic strain is proportional to the magnitude of stress applied during the creep process (Ref 21).

The recovery phenomenon can be driven by several mechanisms, among them: partial recovery of grain shape, relaxation of grain-boundary dislocation pileups leading to reversible grain-boundary slip, or void sintering. In all these cases, grain boundaries are explicitly involved in the recovery process. Frequently, these mechanisms operate together. Extensive studies of these phenomena for tin-lead solders were conducted by Schneibel (Ref 22), Betrabet and Raman (Ref 19), and Tien and Attarwala (Ref 23). All the experimental data suggest that the continuously evolving internal structure has a critical influence on the behavior of solders, especially when fatigue load conditions are applied.

A fatigue model of solders, if predictive, should incorporate the overall creep behavior and the microstructural recovery processes. Such a model was derived by Zubelewicz (Ref 24) and presented at the Lead Free Solders Workshop (Ref 25). It was shown that the solder fatigue life is explicitly dependent on the recoverable and irreversible evolution of the solder microstructure. Particularly, this theory explains the effect of high versus low frequency of cycling, and the differences in fatigue life caused by the isothermal versus thermomechanical test conditions. The experimental results and the understanding of the behavior of solders lead to the conclusion that an isothermal test should and does provide meaningful reliability data for solder joints used in electronic industry.

Isothermal Fatigue of Solder Materials

Effect of Strain Range on Fatigue Life. Increasing the strain range during cycling of materials leads to a decrease in the number of cycles to failure. Low-temperature (<0.3 absolute melting point, T_m), low-cycle fatigue data for many metals may be fit to the empirical Coffin-Manson relationship (Ref 26, 27):

$$N_f^\beta \, \varepsilon_p = C \qquad \text{(Eq 1)}$$

where N_f is the number of cycles to failure, ε_p is the plastic strain range, and β and C are constants. For most metals the constant β is equal to approximately 0.5. Despite the fact that this relationship was developed for fatigue of metals at low homologous temperature, this relationship and its modifications are used widely for prediction of fatigue lives of solders.

Isothermal fatigue data for high-lead, 95Pb-5Sn solder (Ref 15), developed under bending conditions, was reported to obey the Coffin-Manson relationship in the temperature range of 25 to 85 °C. In this temperature range, the exponent β was found to be equal to 0.43 to 0.46 in the strain range of 0.1 to 5.0% and a frequency of 1800 cycles per day. The Coffin-Manson relationship failed to describe the experimental data for this solder at temperatures equal to or higher than 120 °C. At strain ranges higher than 1%, the exponent β became approximately 1 (Ref 15).

95Pb-5Sn solder was found to obey approximately the Coffin-Manson relationship when tested in torsion from 0.05 to 2.0% at frequency of 1 Hz (Fig. 1) (Ref 10). At 1.0% total strain at room temperature, transgranular cracking was observed. At 150 °C, the solder failed at the grain boundaries.

Indalloy 151 (92.5Pb-5Sn-2.5Ag) tested in shear in the 0.5 to 3% plastic shear range at 35 °C was found to obey approximately the Coffin-Manson relationship (Fig. 2) (Ref 8). However, the exponent β was found to be higher than 0.5; the actual value for β varied from 0.68 to 0.95, depending on fatigue life definition. The fractured surfaces were relatively featureless examples of ductile failure. Very shallow holes as well

as occasional perpendicular cracks were observed.

Solders have low melting points, and room temperature is close to or above 0.5 T_m. The Coffin-Manson relationship is not expected to describe the fatigue data for solders even at room temperature because deviation from the Coffin-Manson relationship at low strain ranges (below 1%) is found for many steels fatigued at high temperatures (Ref 28). Indeed, it is evident that the data for 96.5Pb-3.5Sn solder (Fig. 3) (Ref 11) tested at low strain ranges in tension-tension cannot be well fit to a single log-log straight line. The data are much better represented by two straight lines with a breakpoint at approximately 0.3% plastic strain. The break in the lines in the Coffin-Manson plot has been attributed to the change in the fracture mode from intergranular at low strains to mixed transgranular-intergranular at high strains (Ref 11). In part, the deviation from the Coffin-Manson plot for this solder at low strain ranges can be attributed to environmental reactions, namely oxidation (Ref 29, 30).

The Coffin-Manson relationship was found to be valid for 60Sn-40Pb solder (Ref 5, 6, 18) tested in shear in the high 1 to 10% plastic strain range at 35 and 150 °C at a frequency of 0.3 Hz. The slope β was approximately 0.5 at temperatures below 150 °C and decreased to 0.37 at 150 °C. An intergranular mode of failure was observed. Practically no deviation from the Coffin-Manson relationship was detected for 60Sn-40Pb solder tested in 0.5 to 10% shear strain range (Ref 31). However, the choice of failure criterion significantly influenced the slope of the Coffin-Manson plot. No deviation from Coffin-Manson

relation was found for 60Sn-40Pb in tests of lap joints without hold time under strain as well as under load control (Fig. 4) (Ref 32).

The importance of the testing procedure (compliance of the testing system, in particular) on the solder fatigue results was demonstrated during fatigue studies of 63Sn-37Pb solder (Ref 33) (Fig. 5). The solder was tested under strain and stroke controls. For the same total strain range, the fatigue life was an order of magnitude longer in stroke-controlled tests (a more compliant system) than in strain-controlled tests (stiffer system).

62Sn-36Pb-2Ag solder that was tested in tension-tension isothermal fatigue over 0.3 to 3.0% total strain range did not obey the Coffin-Manson relationship at low strain ranges (Ref 34). The data for this solder (bulk specimens) are plotted in Fig. 6 together with the results of tests for 60Sn-40Pb solder (small solder joints) performed under plastic shear strain control (Ref 5). To compare the fatigue data obtained in tensile fatigue test to data obtained in shear fatigue tests, the equivalent von Mises strain (ε_{VM}) is used. The same definition of failure must be used. Tensile strain (ε_{tens}) is equivalent to von Mises strain. Shear strain (γ) relates to von Mises strain as follows:

$$\gamma = \sqrt{3} \, \varepsilon_{VM} \qquad \text{(Eq 2)}$$

or:

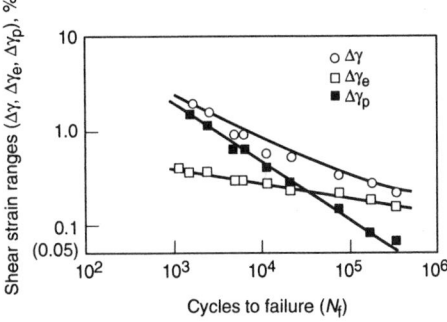

Fig. 1 Fatigue of 95Pb-5Sn solder in torsion at 25 °C. Source: Ref 10

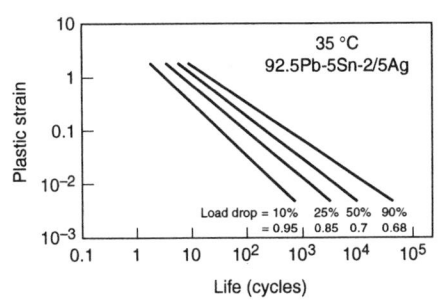
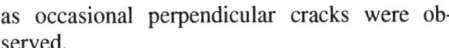

Fig. 2 Fatigue of In151 (92.5Pb-5Sn-2.5Ag) solder in shear at 35 °C. Source: Ref 18

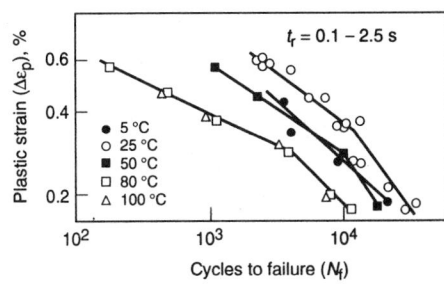

Fig. 3 Fatigue of 96.5Pb-3.5Sn solder in tension at 25 °C. Ramp time (t_r) is 0.1 to 2.5 s. Source: Ref 11

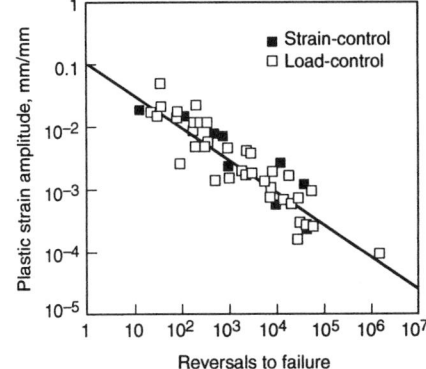

Fig. 4 Fatigue of 60Sn-40Pb solder in shear at 25 °C. Source: Ref 32

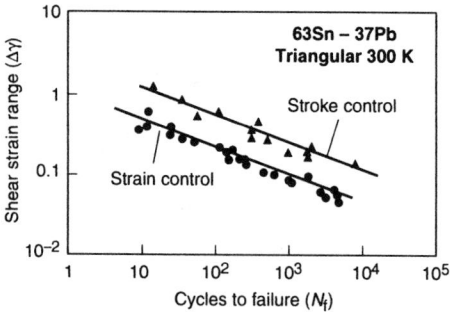

Fig. 5 Fatigue of 63Sn-37Pb solder in shear at 25 °C. Stress ratio (R) is –1; Frequency is 0.1 Hz. Source: Ref 33

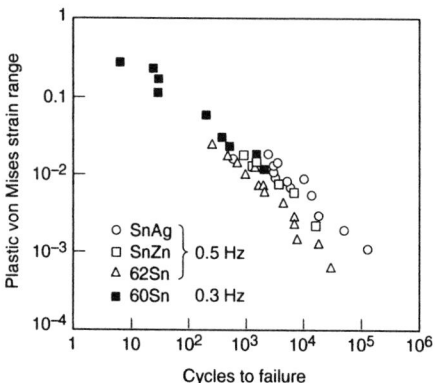

Fig. 6 Fatigue of 62Sn-36Pb-2Ag, 97Sn-3Ag, 91Sn-9Zn (bulk, tension, total strain control), and 60Sn-40Pb (small specimen, shear, plastic strain control) solders. Source: Ref 17

$$\varepsilon_{tens} = \sqrt{1/3}\gamma = 0.577\gamma \qquad \text{(Eq 3)}$$

This figure is very important because it shows that bulk and solder joint specimens of very different geometry and tested in different loading modes can give comparable results.

As already discussed, for low-cycle, high-stress conditions, inelastic deformation is dominant, and a plastic-strain criterion (the Coffin-Manson relationship) is used. However, use of only plastic strain as a failure criterion may not be sufficient for a reliable assessment of solders. A more accurate failure criterion is needed to combine the effects of both stress and strain. A hysteresis-energy criterion of fatigue failure appears to be a very good choice in solder-fatigue research. The hysteresis energy is a scalar quantity and has a cumulative effect.

The fundamental hypothesis in deriving the energy-based failure criterion is that, under cycling of stress and strain, fatigue damage is proportional to the change in the internal strain energy. A change in the internal strain energy is assumed to be reflected in the hysteresis stress-strain curve.

A damage function that includes stress as well as inelastic strain range was proposed as the basic measure of low-cycle fatigue damage at elevated temperature (Ref 35). It is assumed that low-cycle fatigue is essentially a crack propagation process;

therefore, only the tensile part of the cycle is considered. In general, a damage function of the form $\sigma_{max} \varepsilon_p$ or $\tau_{max} \gamma_p$ is considered to be a sufficiently accurate approximation of the proposed measure of fatigue damage. This damage function replaced the plastic strain in the Coffin-Manson and frequency-modified Coffin-Manson relationships. A good correlation between the number of cycles to failure and the damage function was found for steels and nickel-base superalloys tested at elevated temperatures (Ref 35).

Plastic hysteresis energy per cycle is assumed to be proportional to the damage function of the form:

$$D = \sigma_{max}\varepsilon_p \quad \text{or} \quad D = \tau_{max}\gamma_p \qquad \text{(Eq 4)}$$

where σ_{max} and τ_{max} are the maximum tensile and shear stresses, respectively, and ε_p and γ_p are plastic tensile and shear strains, respectively.

The relationship between the fatigue life and the damage per cycle (damage function) for 60Sn-40Pb and 62Sn-36Pb-2Ag solders is depicted in Fig. 7 (Ref 36, 37). It is obvious that the data points for these solders are much closer than in the case where the von Mises plastic-strain criterion is used (Fig. 4). Similarly, the energy-based criteria were shown to describe the fatigue properties of solder joints better than the Coffin-Manson relationship (Ref 38).

Separation of the lead-rich and tin-rich phases and cracking through the tin-rich matrix were found to be the main modes of fracture in near-eutectic solders (Ref 8, 9, 14, 39). Failure of near-eutectic fine-grain solders under high-strain-range conditions was found to proceed through a coarsening mechanism. The solder coarsened first, and then cracks formed and propagated in the coarsened region (Ref 7-9). Practically no coarsening of solder is observed at low strain ranges (Ref 14, 39), and in coarse-grain solder at higher strains (Ref 5, 6).

The fatigue data for 97Sn-3Ag and 91Sn-9Zn bulk solders (tension, total strain control) (Ref 16, 17) are compared with the data for bulk 62Sn-36Pb-2Ag (tension, total strain control) and 60Sn-40Pb simulated solder joint specimens (shear, plastic strain control) (Ref 5, 6). Figure 6 uses the equivalent von Mises strain (ε_{VM}). It is evident that fatigue life of tin-silver and tin-zinc solders is longer than that of near-eutectic solders at the same plastic strain range. In Fig. 7, fatigue data are plotted versus the damage function. It is obvious that the data points for all solders are practically on the same line; that is, damage criterion brings the data points closer than the plastic-strain-range criterion. The reason for this is the fact that the damage function considers not only strain, but also stress. This figure shows that bulk and solder-joint specimens of different compositions, sizes, geometries, and fatigued in different loading modes can give very close results.

Effect of Frequency on the Fatigue Life (No Hold in the Cycle). At temperatures well below one-half of the absolute melting point, frequency has little effect on the fatigue life of most metals. When the temperature is increased over half of

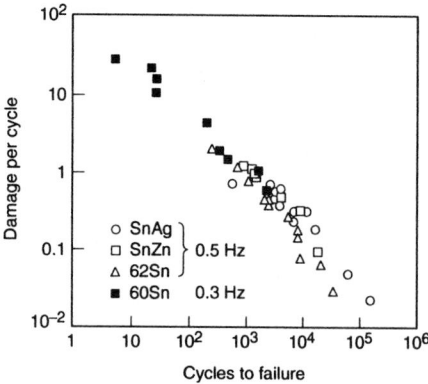

Fig. 7 Fatigue of 62Sn-36Pb-2Ag, 97Sn-3Ag, 91Sn-9Zn (bulk, tension, total strain control), and 60Sn-40Pb (small specimen, shear, plastic strain control) solders at 25 °C. Source: Ref 17

the absolute melting point, a reduction in frequency decreases the number of cycles to failure in isothermal fatigue for many metals including solders (Ref 5, 6, 11-14, 40-58). This behavior arises because at high temperatures, processes such as creep, void formation, and environmental attack, which are time dependent, play important roles in damage accumulation.

J.F. Eckel (Ref 40) found during studies of the fatigue behavior of lead in a rotating bend test that at the strain ranges of 0.27 to 0.65% there was a marked influence of frequency on fatigue life. The time to failure, t_f, and the number of cycles to failure were related to frequency, f, by the following empirical relations:

$$t_f = bf^{-m} \qquad \text{(Eq 5)}$$

$$N_f = bf^{1-m} \qquad \text{(Eq 6)}$$

where b and m are constants. The frequency exponent m for lead in these tests was found to be equal to 0.7 at 43 °C at all strains tested (Ref 40).

A combination of the Coffin-Manson relationship (Eq 1) and Eckel's Eq 6 led to the frequency-modified Coffin-Manson relation of the form:

$$N_f^b \varepsilon_p f^{1-m} = C \qquad \text{(Eq 7)}$$

Equations 5 and 6 were found to be valid for chemical grade lead and lead with 1% Sb (Ref 49) tested at room temperature in reverse bending (zero mean strain) in the strain range from 0.1 to 1.0%. However, the value of m was found to depend on the strain range and to vary from 0.4 to 0.8. The effect of frequency on fatigue life was found to be stronger at high strains. It was also observed that the stiffer the lead alloy the less its fatigue life was affected by the variation in frequency.

The number of cycles to failure for 60Sn-40Pb solder systematically tested in shear at 35 and 150 °C was reduced substantially when the frequency of cycling was decreased below approximately 3 $\times 10^{-4}$ Hz at 35 °C and 3 $\times 10^{-3}$ at 150 °C (Ref 5, 6). The frequency constant was 0.42 for frequen-

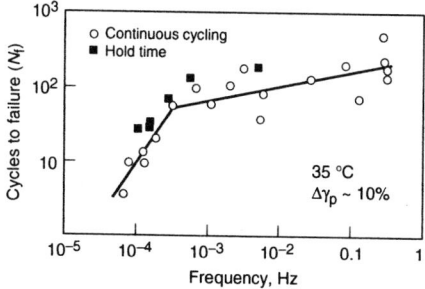

Fig. 8 Effect of cycle frequency on fatigue life of 60Sn-40Pb solder at 35 °C. Plastic shear strain ($\Delta\gamma_p$) ~10%. Source: Ref 6

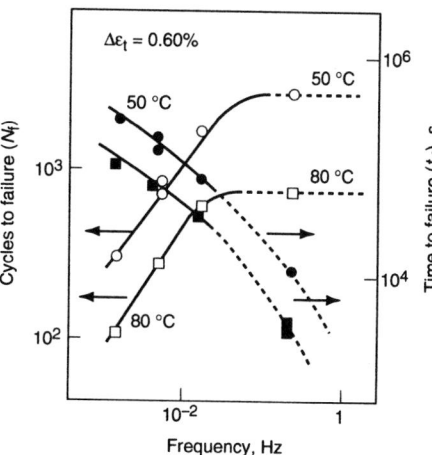

Fig. 9 Effect of cycle frequency on fatigue life of 96.5Pb-3.5Sn solder. Total strain range ($\Delta\varepsilon_t$) is 0.60%. Source: Ref 11

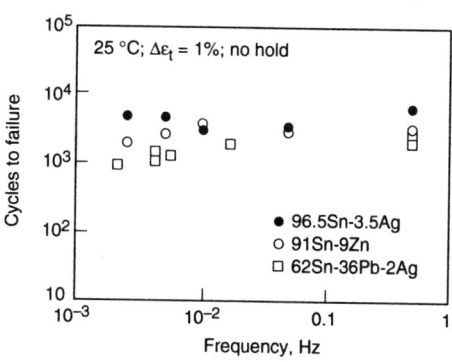

Fig. 10 Effect of cycle frequency on fatigue life of 62Sn-36Pb-2Ag, 97Sn-3Ag and 91Sn-9Zn solders at 25 °C, no hold. Total strain range ($\Delta\varepsilon_t$) is 1%. Source: Ref 17

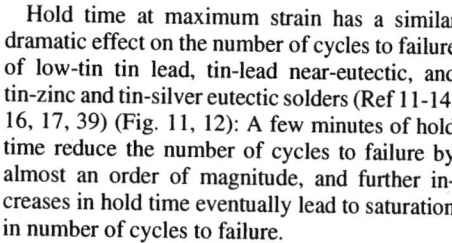

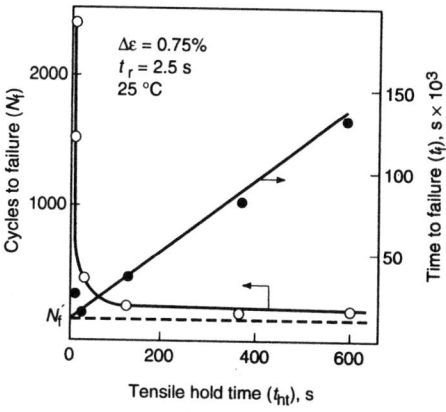

Fig. 11 Effect of tensile hold time on fatigue life of 96.5Pb-3.5Sn solder at 25 °C. Ramp time (t_r) is 2.5 s. Strain range ($\Delta\varepsilon$) is 0.75%. Source: Ref 11

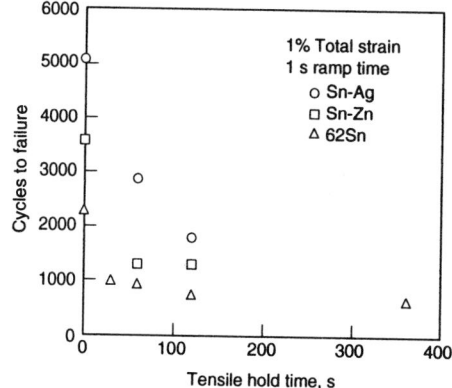

Fig. 12 Effect of tensile hold time on fatigue life of 62Sn-36Pb-2Ag, 97Sn-3Ag and 91Sn-9Zn solders. Ramp time (t_r) is 1 s. Total strain is 1%. Source: Ref 17

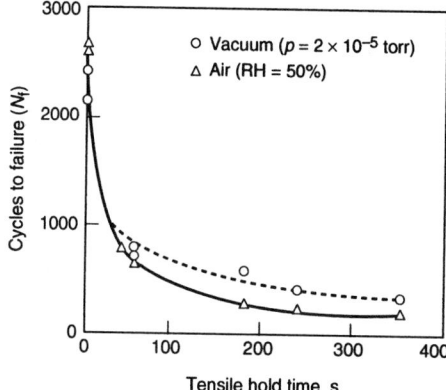

Fig. 13 Effect of tensile hold time on fatigue life of 96.5Pb-3.5Sn solder in air and vacuum at room temperature. Source: Ref 29

cies below 3×10^{-4} Hz in tests with plastic shear strain of 10% at 35 °C (Fig. 8). At frequencies above the 3×10^{-4} Hz, the decrease in the number of cycles to failure with increasing time per cycle was less significant (the frequency exponent was 0.84).

The effect of frequency on fatigue life of 95Pb-5Sn solder was evaluated in bending tests ranging from 0.033 to 192 s/cycle and in strain ranges from 0.5 to 4.0% (Ref 15). Eckel's relations (Eq 6 and 7) were found to be valid with a frequency exponent m equal to 0.75 at all strains tested.

For 96.5Pb-3.5Sn solder, practically no change in the number of cycles to failure was observed on varying frequency in the comparatively high frequency region (Ref 11, 12). A reduction in the number of cycles to failure with decreasing frequency below 10^{-2} Hz was found at 25, 50, and 80 °C (Fig. 9) for all strain ranges tested. The behavior of this solder obeys the general trends found by Eckel.

Lowering the cycle frequency below 0.1 Hz at 35 °C and below 0.01 Hz at 150 °C reduced the number of cycles to failure of 92.5Pb-5Sn-2.5Ag solder (Ref 18). The frequency exponents were approximately 0.75 at 150 °C and 0.61 at 35 °C.

The effect of frequency on fatigue life of tin-zinc and tin-silver appears to be less significant than for tin-lead near-eutectic solders (Fig. 10) (Ref 16, 17).

Effect of Hold Time on Fatigue Life. Hold (dwell) times introduced in the cycle at maximum and/or minimum strains, stresses, or temperatures affect the fatigue life of materials depending on the kind of material and testing conditions.

Tensile hold time is very damaging during high-temperature fatigue of many materials such as steels (Ref 43, 51-55), other metal alloys (Ref 47, 53, 56, 57), and solders (Ref 11-14, 16, 17, 39, 58-60). The number of cycles to failure decreases when tensile hold time is increased. A saturation in the number of cycles to failure with increasing hold time is detected for a number of metals.

Often the reduction in fatigue life of metals due to tensile hold is accompanied by a transition from the transgranular to the intergranular mode of fracture. Cavitation on grain boundaries is observed. Therefore, the effect of tensile hold time on fatigue life is attributed mainly to creep (cavitation) (Ref 46-48, 61, 62).

Hold time at maximum strain has a similar dramatic effect on the number of cycles to failure of low-tin tin lead, tin-lead near-eutectic, and tin-zinc and tin-silver eutectic solders (Ref 11-14, 16, 17, 39) (Fig. 11, 12): A few minutes of hold time reduce the number of cycles to failure by almost an order of magnitude, and further increases in hold time eventually lead to saturation in number of cycles to failure.

The saturation in the number of cycles to failure with increasing tensile hold time per cycle may be explained either by exhaustion of creep processes or saturation of environmental attack during holds.

Tests performed for low-tin lead-base solder in air and vacuum (Ref 29, 30) at 0.80% total strain range indicate that only some of the fatigue life reduction with increasing tensile hold time can be attributed to environmental attack. The number of cycles to failure saturated in vacuum as it did in air. However, the fatigue life in vacuum was found to be only slightly greater than in air (Fig. 13). Therefore, most of the damage during fatigue with hold time at this strain range has to be attributed to creep and probably to the formation and growth of voids on grain boundaries. Voids were found on grain boundaries during fatigue of

98Pb-2Sn (Ref 63) and 97.5Pb-3.5Sn (Ref 29) solders.

When the effects of ramp time with no holds in the cycle and effect of tensile hold time on fatigue life of solders are compared, it becomes obvious that the effect of tensile hold time on fatigue life of solders is much more dramatic than the effect

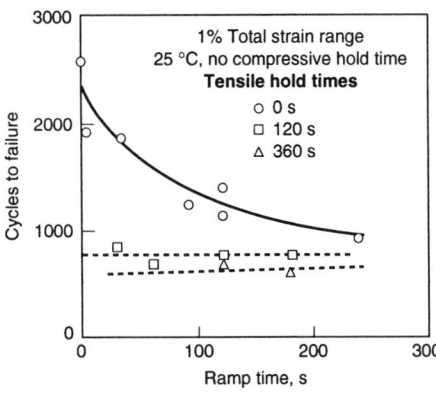

Fig. 14 Effect of ramp time on fatigue life of 62Sn-36Pb-2Ag solder in tests with and without tensile hold time (t_{ht}) (at 25 °C; no compressive hold time). Source: Ref 34

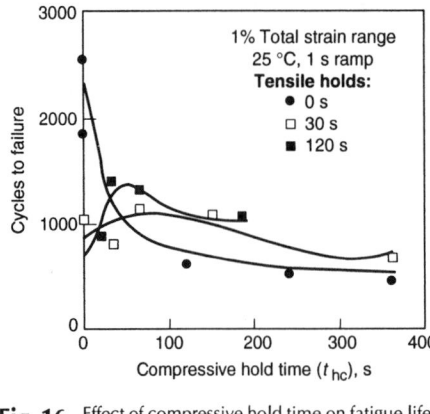

Fig. 16 Effect of compressive hold time on fatigue life of 62Sn-36Pb-2Ag solder in tests with and without tensile hold time (t_{ht}) at 25 °C. Total strain range is 1%. Ramp time (t_r) is 1 s. Source: Ref 34

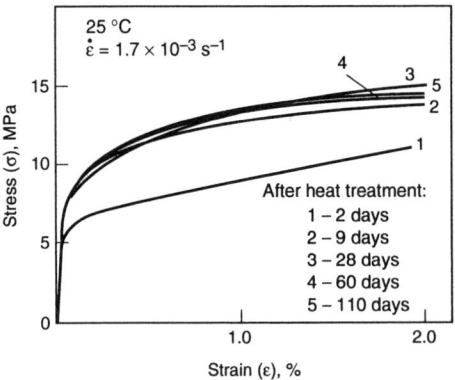

Fig. 18 Effect of aging on tensile properties of 96.5Pb-3.5Sn solder at 25 °C. Strain rate is 1.7×10^{-3} s^{-1}. Source: Ref 11

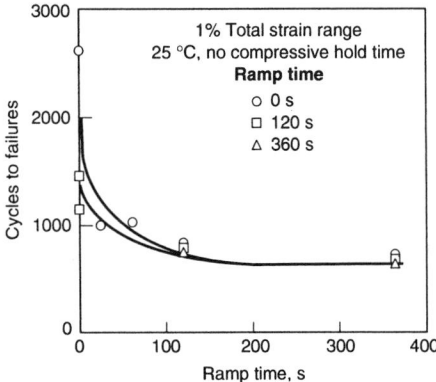

Fig. 15 Effect of tensile hold time on fatigue life of 62Sn-36Pb-2Ag solder in tests with different ramp times (t_r) (at 25 °C; no compressive hold time). Total strain range is 1%. Source: Ref 34

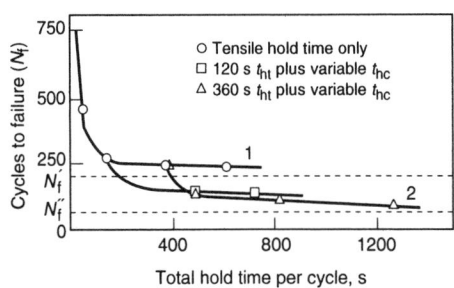

Fig. 17 Effect of tensile hold time (t_{ht}) and compressive hold times (t_{hc}) on fatigue life of 96.5Pb-3.5Sn solder at 25 °C. Ramp time (t_r) is 2.5 s. Total strain range ($\Delta\varepsilon_t$) is 0.75%. Source: Ref 11

of ramp time (Fig. 14, 15) (Ref 34). The difference in the effects of ramp time and hold time on solder fatigue life was attributed to differences in strain rates operating in solder during ramp and hold time. Strain rates during very short hold times are much less (by an order [s] of magnitude) than strain rates during even very long ramp times (Ref 64). Because the fatigue failure micromechanisms for solders are very sensitive to strain rates, the effects of hold time and ramp times on fatigue of solders are dramatically different.

Isothermal fatigue life of 60Sn-40Pb solder in tests with equal tensile and compressive hold times in the cycle was found to be unaffected by such combinations of hold times (Fig. 8) (Ref 6, 65). For 62Sn-36Pb-2Ag solder, addition of compressive hold time to a cycle with tensile hold time increased the number of cycles to failure (Fig. 16) (Ref 34).

Combination of both tensile and compressive hold times during testing of 96.5Sn-3.5Pb solder in air was found to be more damaging than tensile or compressive hold times separately (Fig. 17) (Ref 11).

It is obvious that the effect of hold time is very profound during fatigue of solders. Effects of different hold times on the fatigue life of solders

depend on the type of solders and on testing procedures; thus, this should be considered carefully during estimations of fatigue lives of solders.

Effect of Temperature on Isothermal Fatigue of Solders. As a rule, the isothermal fatigue life of metals decreases as temperature increases. However, the degree of fatigue life change depends on the material and testing conditions. For near-eutectic tin-lead solders, such as 63Sn-37Pb in the 25 to 80 °C range (Ref 14, 39, 66) and 60Sn-40Pb in 35 to 150 °C range (Ref 5, 6) and the –60 to 150 °C (Ref 10) temperature ranges, the effect of temperature was insignificant for all strain ranges and modes of loading investigated.

The effect of temperature on fatigue of low-tin tin-lead solders is much more pronounced than for near-eutectic tin-lead solders. The fatigue life of 95Pb-5Sn solder at strains lower than 1.0% was found to be much shorter at 150 °C than at room temperature or –60 °C (Ref 9).

The fatigue life of 95Pb-5Sn solder was shown to decrease when temperature was increased from room temperature to 120 °C and to be practically the same at 120 and 150 °C (Ref 15).

Fatigue life of 96.5Pb-3.5Sn solder at 80 °C was found to be almost the same as at 100 °C (Ref 11, 12, 66). The fatigue life of this solder was found to be the shortest at these temperatures. The solder fatigue life dependence on tempera-

ture appears to follow an Arrhenius-type equation only between 25 and 80 °C.

The effect of temperature on fatigue life of this low-tin tin-lead solder is a function of frequency: the change in the number of cycles to failure with temperature at a frequency less than 0.01 Hz is less than that at higher frequency. This converts to an apparent activation energy of 24 to 26 kJ/mole at low frequency and 40.6 kJ/mole at high-frequency testing conditions.

The temperature effect on fatigue life of low-tin lead-base solder also is different in tests with hold time and without hold time. The effective activation energies are 24 kJ/mole and 40 kJ/mole, respectively (Ref 11, 66). Thus, fatigue of low-tin tin-lead solder under low-strain-rate conditions (long ramp time and/or hold time in the cycle) is characterized by low activation energy. Different effective activation energies result from different failure mechanisms under different testing conditions.

Aging Effects in Solders. Solders are multiphase low-temperature melting materials, where the solubility in each phase decreases with a decrease in temperature. Precipitation of the second phase out of solid solution and phase coarsening occur at room temperature after the solder joint is solidified. Such precipitation affects the mechanical and fatigue properties of solders. The effect of aging on solder behavior was found to be a function of composition and initial microstructure of the solder.

Room-temperature aging increases the tensile strength of comparatively coarse low-tin tin-lead solder (Fig. 18) (Ref 11) and affects the fatigue life of coarse 63Sn-37Pb solder (Fig. 19, Ref 39), but the effect saturates after approximately one week of aging.

The cooling (fast or slow) during solder solidification affects the initial microstructure of 60Sn-40Pb solder. The fine microstructure of rapidly cooled solders leads to higher initial strength (Fig. 20, Ref 67, 68). However, room-temperature aging leads to coarsened microstructures in both fine and coarse solders, and after 100 days the slow-cooled solder was the less coarsened of the two. The shear strength of both solders was reduced to almost the same level. Fatigue resis-

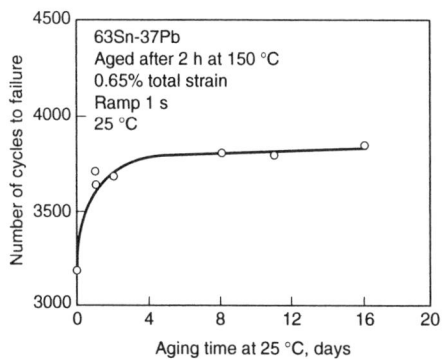

Fig. 19 Effect of aging (after 2 h at 150 °C) on fatigue life of 63Sn-37Pb solder at 25 °C. Total strain is 0.65%. Ramp time is 1 s. Source: Ref 39

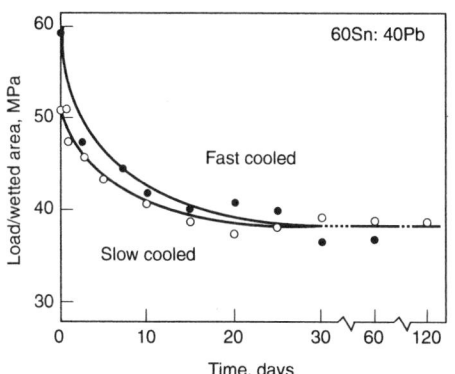

Fig. 20 Effect of cooling rate and aging on shear strength of 60Sn-40Pb solder. Source: Ref 67

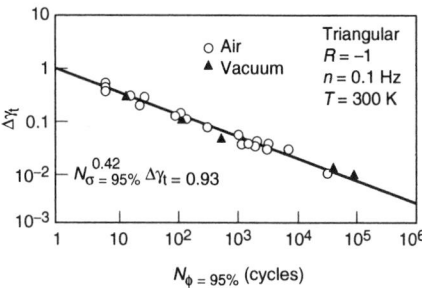

Fig. 21 Fatigue life of 90Pb-10Sn solder in air and vacuum. Stress ratio (R) is –1; frequency (n) is 0.1 Hz. (T) is 300 K. Source: Ref 70

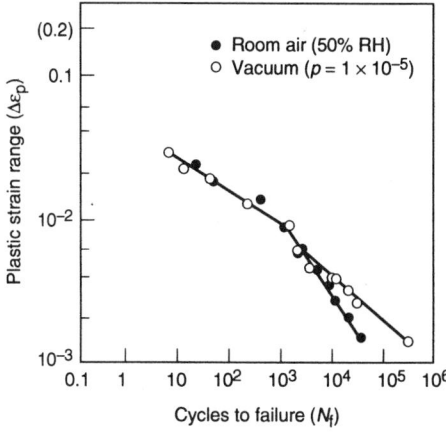

Fig. 22 Fatigue life of 96.5Pb-3.5Sn solder in air and vacuum. Source: Ref 29

tance of this solder was found to gradually decrease with time. The main conclusion from this work (Ref 68) was that aging may neutralize the effects of initial microstructure on mechanical properties of solders.

In order to carry out a systematic study of fatigue properties of solders, the microstructure must be standardized or varied in a controlled manner. Researchers frequently do not give the treatment schedules followed, and this leads to ambiguity when comparing results between studies.

Environmental Effects in Solder Fatigue. The environment affects the fatigue lives of many materials including lead (Ref 51) and solders (Ref 29, 30). A critical partial pressure for oxygen (10^{-2} torr) was found below which the fatigue life of lead was greatly increased (Ref 69). At the same time, the failure mode was changed from intergranular at higher partial pressure to transgranular at lower partial pressure of oxygen.

Fatigue life of metals in vacuum is usually higher than in air. However, the difference between the fatigue lives in air and in vacuum for a given material is a function of strain range and strain rate; the higher the strain range and the higher the strain rate, the smaller the effect of an aggressive environment (Ref 29, 30, 70). For both high-lead (Fig. 21) and tin-lead eutectics, fatigue lives in air and vacuum are practically the same at strain ranges greater than 1%.

At lower strains, fatigue life of high-lead tin-lead solder is different in air and in vacuum (Fig. 22) (Ref 29, 30): The behavior of this solder does not follow the Coffin-Manson relationship in vacuum or in air (Ref 11-13). However, the deviation from the Coffin-Manson relationship is less in vacuum; thus, only part of the deviation in air is due to environmental attack. The increase in fatigue life of solder in vacuum over that in air and the less frequent intergranular failure observed in 96.5Pb-3.5Sn solder in vacuum indicate that the intergranular mode of failure is accelerated by air, presumably due to oxidation at grain boundaries.

Solder Joint Reliability

Solder joints are known as one of the weakest elements of electronic assemblies. Throughout the years, a significant effort has been made to develop specifications and procedures to control solder joint quality. The principal assumption is that the standard reliability test should mimic the electronic device on/off cycles in an accelerated manner. Among the most common stress tests is an accelerated thermal cycling (ATC) test. Test vehicles or functional parts are placed in a thermal chamber and tested long enough to generate meaningful statistical results that can be used for reliability assessment of assemblies. Although the ATC test is referred to as accelerated, it takes several weeks before solder joints start failing, and much more time to make some statistical sense of the data. Therefore, some engineers tend to shorten the evaluation process and even further accelerate it without a complete understanding of the test limitations. This can lead to confusing reliability data, faulty quality assessments, and erroneous business decisions. It is not surprising that the electronic industry is looking for alternative test techniques that both accurately predict product quality and shorten the verification process. Unfortunately, thermomechanical cycling can not be accelerated above the limits that are imposed by thermal mass of the tested product. Therefore, the most encouraging option is to develop an isothermal mechanical test that would replicate the ATC joint failures in a much faster manner.

Solder Joint Reliability Assessment Using Isothermal Fatigue Testing. From experimental data, metallurgical observations, and elements of a micromechanically based theory of the behavior of solder materials developed by the authors, the traditional ATC test used may possibly be replaced with the much faster and more controllable isothermal mechanical deflection system (MDS) test in the future. The MDS test does have some limitations. It is valid only if the MDS fracture mechanisms match the expected field failure mechanisms. The experimental data and the theoretical investigations reviewed in this article show that the ATC and MDS mechanisms can closely approximate each other for testing component durability.

The mechanical deflection system (MDS) test is an isothermal test where cyclic out-of-plane deformation is imposed on an assembled printed wiring board at isothermal conditions with the temperature usually set above the ambient condition (Ref 1). A portion of the applied deformation is transferred to the solder joints and causes the joints to fail. Proper selection of the deformation, test temperature, and cyclic frequency promotes the creep/fatigue failure mechanisms that are typical at field or the ATC test environment. Extensive research has demonstrated that MDS failure mechanisms match quite well the ATC mechanisms while reducing test time up to two orders of magnitude.

The most powerful applications of the MDS test are in the area of an assembly process optimization. It can also be used for design verification and reliability projection or as a near-real-time assembly line monitor.

The MDS test was verified for several technologies, among them J-leaded and gull-wing interconnects, ball and column grid array, and other surface mount interconnects. Usually, the MDS test is capable of failing nearly all the monitored solder joints during the test with a test time of hours instead of weeks or months. The test sensitivity in detecting defective joints is greater than under the ATC test conditions. Mechanical deflection system testing, when compared with thermal cycling, imposes a greater magnitude of axial deformation to the solder joints, and therefore the fully developed cracks in solder joints

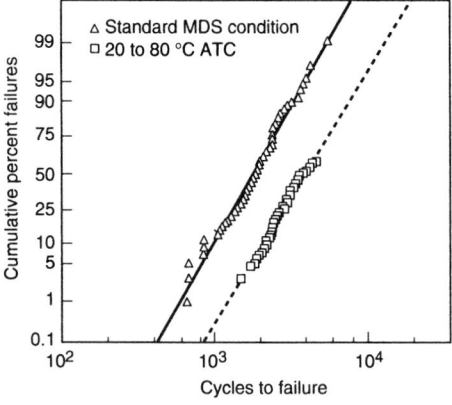

Fig. 23 Log normal probability plots of solder ball failures in MDS and ATC tests. Source: Ref 1

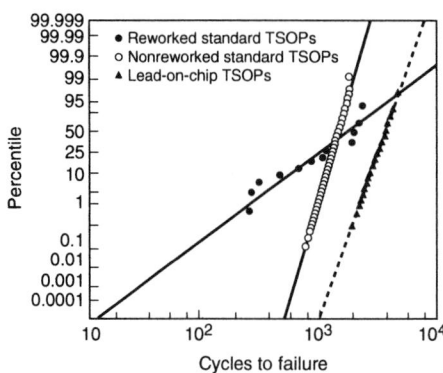

Fig. 24 Log normal probability plots for TSOP joint failures. Test cases: reworked standard TSOPs, nonreworked standard TSOPs, and lead-on-chip TSOPs. Source: Ref 1

produce an electrical open condition in the monitored circuit. Assemblies subjected to thermal stress conditions require very accurate resistance measurements (usually four-point measurements) to detect failed solder joints. Furthermore, the MDS test conditions are precisely controlled. This includes both the out-of-plane deformation and the constant-temperature test environment.

Examples of MDS Testing. Mechanical deflection system testing has proved itself as an excellent technique for optimization of assembly processes. The test sensitivity in detecting defects as well as a very short test time make MDS the best, or perhaps, the only candidate for this application. Four test cases will be presented here to illustrate the MDS test capabilities. First, the ceramic ball grid array (CBGA) technology developed by IBM was tested to correlate MDS and ATC fatigue data. Second, thirty-two-leaded thin small outline package (TSOP) technology was tested to optimize the rework process of removing and reassembling new packages. Next, ceramic quad flat package (CQFP) and IBM-developed fine-pitch technology were tested to optimize assembly process windows. In the last example, MDS testing was used to optimize assembly process parameters to ensure good quality of plastic and ceramic ball grid array packages.

MDS versus ATC Comparison for CBGA Packages. The goal of the evaluation was to verify the MDS-induced joint failures when compared with the ATC failures and to correlate the MDS and ATC failure distributions for CBGA packages. For a true comparison, MDS and ATC tests were conducted concurrently using identical test vehicles and assembly processes. The CBGA solder joints were composed of a high-lead ball (90Pb-10Sn) attached by eutectic tin-lead solder to the substrate and the card pads. In all the MDS and ATC test cases, the fatal cracks propagated near the copper-tin intermetallic layer or along the interface of the high-lead ball and eutectic solders. First, thermal cycling was conducted at near-field low-stress conditions, where the temperature cycling was set between 20 and 55 °C at a frequency of 0.001 Hz. This test was very time consuming, taking more than six months to generate first failure. The number of modules on test was insufficient to draw statistically valid conclusions. Therefore, the next cell of CBGA modules was ATC tested at a higher stress level to accelerate the damage process of the joints while preserving the same failure mechanism. This time, the temperature cycling was set between 20 and 80 °C with a frequency of 0.0005 Hz.

On the other hand, the MDS test was conducted at ambient temperature, two opposite edges of the board were twisted against each other with the magnitude of 0.6 degrees/in. of the board length, and the frequency of cycling was 0.1 Hz. When comparing the MDS and ATC failure mechanisms, both the mechanisms are very similar. Mean fatigue lives are not the same, but can be brought to the same values when needed (Fig. 23).

The MDS and ATC tests generated very similar results, but were accomplished in different time frames. The MDS test needed only 16 h for all the modules to fail, while the ATC test took nine weeks to bring 50% of the modules to failure.

MDS Rework Process Assessment for TSOPs. Forty TSOP modules were tested to evaluate rework process. The rework process consisted of removing previously assembled components and replacing them with new ones. Also, the evaluation was designed to compare two types of TSOPs, standard TSOPs with short leads and TSOPs with longer and more compliant leads.

The MDS- and ATC-induced failure mechanisms were found to be similar, with a more significant grain coarsening in the ATC cell. In both cases, cracks initiated in the joint hill and propagated near the lead/solder interface. Mechanical deflection system testing was conducted at ambient temperature. Out-of-plane deformation was cyclically applied to the test vehicle with the angle of twist equal to 0.9 degrees/in. of the card length at frequency of 0.1 Hz. The statistical distributions of failures are presented in Fig. 24.

The rework process was not optimized at the time the test was conducted. There was a significant population of early failures, but also the very same process produced very good joints (note the upper tail of the distribution). Based on the data, it was quite easy to optimize the process against the best-performing population of joints. Note that the standard ATC test, being very lengthy in time, would expose only the defective joints, and therefore the rework process would be improved but not truly optimized by eliminating the population of the worst-performing joints.

Also, the TSOPs that have the longer (more compliant) leads increased fatigue life by approximately two times when compared with the standard TSOPs.

Assembly Process Optimization for CQFPs. Fine-pitch technologies can exhibit excellent reliability, but require an optimized and tightly controlled assembly process. The CQFPs have very compliant leads that reduce the stress in solder joints. In some cases, the leads may even fail sooner than the joints, depending on the lead material, shape, and also on the test conditions. However, solder joints are still considered the prime reliability concern for these technologies. Because the isothermal MDS cycling is executed through a considerably compliant means (lead), this induces a creep-fatigue failure mechanism assisted with microstructural recovery processes that were discussed earlier. This phenomenon needs special consideration at higher frequency cycling (about 0.1 Hz) and becomes less important when allowing solder to creep by applying a few minutes dwell at the cycle reversals and/or elevating the test temperature. When taking all of the above under consideration, the MDS test conditions were set as follows: angle of twist was set at 0.75 degrees/in. of the board length; temperature was equal to 115 °C; cyclic frequency was

Table 1 Summary of fatigue life results for MDS tests

Pad No.	Nominal pad size, mils	Fatigue life and scatter (a), cycles × 10³, for solder volume			
		1	2	3	4
I	Author to provide	11.6 (first failure) 20.3 (mean) σ = 0.21	14.4 21.4 (mean) σ = 0.18		
II	Author to provide		14.0 22.0 (mean) σ = 0.19	9.0 15.2 (mean) σ = 0.19	
III	Author to provide			10.0 19.9 (mean) σ = 0.24	1.8 14.8 (mean) σ = 0.59

(a) Top value for each pad is the number of cycles (× 10³) to the first failure. The second number is the mean fatigue life with a lognormal fit with standard deviation, σ.

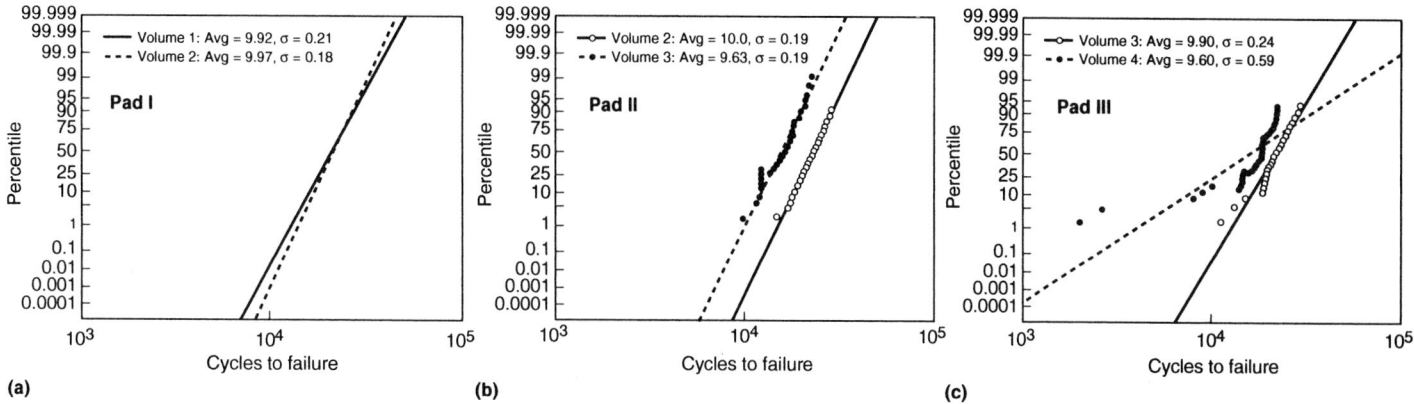

Fig. 25 Multiple-sample log normal probability plots of solder joints in MDS test of ceramic quad flat package. (a) Pad I; solder volumes 1 and 2. (b) Pad II; solder volumes 2 and 3. (c) Pad III, solder volumes 3 and 4. Source: Ref 1

Table 2 Fatigue test results of dog-bone and flag pads of plastic and ceramic BGA packages

Surface metallurgy	Package type	Dog-bone mean cycles	Deviation (σ)	Flag mean cycles	Deviation (σ)
HASL	PBGA	16,806	0.35195	14,338	0.38099
HASL	CBGA	11,949	0.46592	15,491	0.3056
OCC	PBGA	17,904	0.26247	14,057	0.43629
OCC	CBGA	9,318	0.48876	13,898	0.33654

HASL. Hot air solder leveled; OCC, organic coating on copper

- Ball grid array package types

Specifically, the test conducted used pads that were "dog bone" or "flag," with either hot-air solder-leveling or bare copper with an organic coating over the copper pad. The two package types investigated included plastic as well as ceramic BGAs. The results of the study conducted by Ninohira, Yee, and Rao (Ref 71) for the team are summarized in Table 2.

set at 0.1 Hz. In this case, the temperature was set notably higher than in all the other test cases. The objective of the test was to optimize the assembly process for the best combination of the pad sizes and solder volumes. Six test cells were selected as shown in Table 1. The test results are presented using three quantities. In each test case, the first upper number represents the first failure in the test, the second number is the mean fatigue life, and the third is scattering of the fatigue data (sigma of log normal statistical plot). The mean life and sigma were determined from log normal probability plots (Fig. 25).

First, the highest and medium solder volumes (1 and 2) were analyzed for the largest pad (I) (Fig. 25a). The two probability plots look almost identical. This indicates that there is no reliability exposure in either of the two cases. However, the highest volume (No. 1) caused some bridging of the solder between pads; therefore, the combination of the highest solder volume and largest pad is not acceptable (yield problem).

Next, test results for solder volumes 2 and 3 and medium pad (II) were analyzed and presented in Fig. 25(b). Both the plots have similar scattering of failures, but the joints assembled with the lower volume of solder (No. 3) exhibit lower mean fatigue life. The smallest pad (III) in combination with the two lowest volumes of solder (3 and 4) caused significant reliability problems (Fig. 25c). Much earlier failures were observed in these cells. The fatigue data show a bimodal statistical distribution of the failures. The main population of the failures seems to follow the previous results; however, the tail of the distributions indicates that the defective joints in the cell may cause field exposure.

Assembly Process Optimization for Plastic and Ceramic Ball Grid Array Packages. New technology deployment into volume manufacturing is often a serious concern. The cost of new technology development including the development of the infrastructure for timely introduction of a product is an extremely costly process. The costs are even higher when the expenses related to the lengthy assessments of the product reliability are added. However, the risk of not testing for reliability is even greater.

In the case of new technology deployment, it is imperative that one understand and recognize the product exposures. Assessment of pad geometry, pad metallurgy or finish, package selection, and solder metallurgy and therefore reflow process are all factors that play very important roles. When there is a choice, any user would like to know the reliability impact of the selection. Mechanical deformation system testing was deemed the best approach for evaluating some of these factors in the selection of ball grid array (BGA) packaging technology. A study was conducted by a member team with interest in developing the infrastructure for the success of the BGA technology. The team consisted of an electronic systems manufacturer, a possible user of the technology, a supplier of BGA packaging technology, and a printed circuit board fabricator. Boards were fabricated to exercise the following parameters:

- Pad geometry
- Pad metallurgy

REFERENCES

1. A. Zubelewicz et al., Mechanical Deflection System—An Innovative Test Method for SMT Assemblies, *INTERPAC-95* (Lahaina, HI), 26-30 March 1995, Vol 2, ASME, p 1167-1178
2. M.C. Shine and L.R. Fox, Fatigue of Solder Joints in Surface Mount Devices, *Low Cycle Fatigue*, STP 942, H.D. Solomon et al., Ed., 1988, p 588-610
3. H.D. Solomon, Low Cycle Fatigue of Surface Mounted Chip Carrier/Printed Wiring Board Joints, *Proc. 39th Electronic Components Conf.*, Institute of Electrical and Electronics Engineers, 1989, p 277-292
4. H.D. Solomon, Influence of Temperature on the Low Cycle Fatigue of Surface Mounted Chip Carrier/Printed Wiring Board Joints, *J. IES*, Jan-Feb 1990, p 17-25
5. H.D. Solomon, Low-Cycle Fatigue of 60/40 Solder-Plastic Strain Limited versus Displacement Limited Testing, *Electronic Packaging: Materials and Processes*, J.A. Sortell, Ed., American Society for Metals, 1985, p 29-49
6. H.D. Solomon, Low-Frequency, High-Temperature Low-Cycle Fatigue of 60Sn/40Pb Solder, *Low Cycle Fatigue*, STP-942, H.D. Solomon et al., Ed., ASTM, 1988, p 342-369
7. D. Frear, D. Grivas, and J.W. Morris, Jr., Thermal Fatigue Failures in Solder Joints, *J. Met.*, Vol 40, 1988, p 18-22
8. D. Frear, D. Grivas, M. McCormack, D. Tribula, and J.W. Morris, Jr., Fatigue and Thermal Fatigue Testing of Pb-Sn Solder Joints, *Proc. Third Annual Electronic Packaging and*

Corrosion in Microelectronics Conf., M.E. Nicholson, Ed., ASM International, 1987, p 269-274

9. D. Frear, D. Grivas, M. McCormack, D. Tribula, and J.W. Morris, Jr., Fatigue and Thermal Fatigue of Pb-Sn Solder Joints, *Proc. Effect of Load and Thermal Histories on Mechanical Behavior Symposium*, P.K. Liaw and T. Nicholas, Ed., TMS, 1987, p 113-126

10. M. Kitano, T. Shimizu, and T. Kumazava, Statistical Fatigue Life Estimation: The Influence of Temperature and Composition on Low-Cycle Fatigue of Tin-Lead Solders, *Current Japanese Materials Research*, Vol 2, Aug 1987, p 235-250

11. S. Vaynman, M.E. Fine, and D.A. Jeannotte, Isothermal Fatigue of Low Tin Lead Based Solder, *Metall. Trans.*, Vol 19A, 1988, p 1051-1056

12. S. Vaynman, M.E. Fine, and D.A. Jeannotte, Prediction of Fatigue Life of Lead-Base Low Tin Solder, *Proc. 37th Electronic Components Conf.*, Institute of Electrical and Electronics Engineers, 1987, p 598-603

13. S. Vaynman, M.E. Fine, and D.A. Jeannotte, Isothermal Fatigue Failure Mechanisms in Low Tin Lead Based Solder, *Proc. Effect of Load and Thermal Histories on Mechanical Behavior Symp.*, P.K. Liaw and T. Nicholas, Ed., TMS, 1987, p 127-137

14. S. Vaynman and M.E. Fine, Fatigue of Low-Tin Lead-Based and Tin-Lead Eutectic Solders, *Microelectronic Packaging Technology: Materials and Processes*, Proc. Second ASM International Electronic Materials and Processing Congress, W.T. Shieh, Ed., ASM International, 1989, p 255-259

15. H.S. Rathore, R.C. Yih, and A.R. Edenfeld, Fatigue Behavior of Solders Used in Flip-Chip Technology, *J. Test. Eval.*, Vol 1, 1978, p 170-178

16. H. Mavoori, S. Vaynman, J. Chin, B. Moran, L.M. Keer, and M.E. Fine, Mechanical Behavior of Eutectic Sn-Ag and Sn-Zn Solders, *Electronic Packaging Materials Science VIII*, Proc. 1995 MRS Meeting, Materials Research Society, Vol 391, 1995, p 161-176

17. S. Vaynman, H. Mavoori, and M.E. Fine, Comparison of Isothermal Fatigue of Lead-Free Solders with Lead-Tin Solders, *Advances in Electronic Packaging*, Proc. International Intersociety Electronic Packaging Conf.—INTERPAC-95, American Society of Mechanical Engineers, 1995, p 657-662

18. H.D. Solomon, Room Temperature Low Cycle Fatigue of a High Pb Solder (INDALLOY 151), *Microelectronic Packaging Technology: Materials and Processes*, Proc. Second ASM International Electronic Materials and Processing Congress, W.T. Shieh, Ed., ASM International, 1989, p 135-146

19. H.S. Betrabet and V. Raman, Microstructural Observations in Cyclically Deformed Pb-Sn Solid Solution Alloy, *Metall. Trans.*, Vol 19A, 1988, p 1437-1443

20. J.G. Cabanas-Moreno, J.L. Gonzalez-Velazquez, and J.R. Weertman, Interrelationships among Grain Boundaries, Cavitation and Dislocation Structures, *Scr. Metall.*, Vol 25, 1991, p 1093-1097

21. V. Raman and T.C. Reiley, Cyclic Deformation and Fracture in Pb-Sn Solid Solution Alloy, *Metall. Trans.*, Vol 19A, 1988, p 1533-1546

22. J.H. Schneibel and P.M. Hazzledine, Superplasticity in Superplastic Sn-Pb Alloys, *Acta Metall.*, Vol 30, 1982, p 1223-1230

23. J.K. Tien and A.I. Attarwala, "Complications in Life Prediction Estimates at Elevated Temperatures in Lead/Tin Solders during Accelerated Cycling," paper presented at 41st Electronic Components Conf. (Atlanta, GA), IEEE, 1991

24. A. Zubelewicz, Micromechanical Study of Ductile Polycrystalline Materials, *J. Mech. Solids*, Vol 41, 1993, p 1711-1722

25. A. Zubelewicz, "Micromechanical Approach to Model Solder Materials," paper presented at Lead Free Solders Conf. (Evanston, IL), 24-26 July 1995, The Institute of Mechanics and Materials, UC San Diego

26. L.F. Coffin, A Study of the Effect of Cyclic Thermal Stresses on a Ductile Metal, *Trans. ASME*, Vol 76, 1954, p 931-950

27. S.S. Manson, Fatigue: A Complex Subject—Some Simple Approximations, *Exp. Mech.*, Vol 5 (No. 7), 1965, p 193-226

28. R.P. Skelton, *Fatigue at High Temperature*, Applied Science Publishers, 1983

29. R. Berriche, S. Vaynman, M.E. Fine, and D.A. Jeannotte, The Effect of Environment on Fatigue of Low Tin Lead Base Alloys, *Proc. Third Annual Electronic Packaging and Corrosion in Microelectronic Conf.*, M.E. Nicholson, Ed., ASM International, 1987, p 169-174

30. R. Berriche, "Environmental Effects on Time Dependent Fatigue of Low Tin-High Lead Solder," Ph.D. thesis, Northwestern University, 1989

31. G. Engberg, L.E. Larsson, M. Nylen, and H. Steen, Low Cycle Fatigue of Soldered Joints, *Brazing Soldering*, No. 11, Autumn 1986, p 62-65

32. N.F. Enke, T.J. Kilinski, S.A. Schroeder, and J.R. Lesniak, Mechanical Behaviors of 60/40 Tin-Lead Solder Lap Joints, *Proc. 39th Electronic Components Conf.*, Institute of Electrical and Electronics Engineers, 1989, p 264-272

33. K. Bae et al., Effect of Compliance on the Fatigue of Solder Joints in Surface-Mounted Electronic Packages, *Microelectronic Packaging Technology: Materials and Processes*, Proc. Second ASM International Electronic Materials and Processing Congress, W.T. Shieh, Ed., ASM International, 1989, p 109-119

34. S. Vaynman and M.E. Fine, Isothermal Fatigue of 62Sn-36Pb-2Ag Solder, *Mechanical Behavior of Materials and Structures in Microelectronics*, MRS Symposium Proc., Vol 226, Materials Research Society, 1991, p 3-13

35. W.J. Ostergren, A Damage Function and Associated Failure Equations for Predicting Hold Time and Frequency Effects at Elevated Temperature, Low Cycle Fatigue, *J. Test. Eval.*, Vol 4, 1976, p 327-339

36. S. Vaynman and S.A. McKeown, Energy-Based Methodology for Fatigue Life Prediction of Solder Materials, *IEEE Trans. Compon. Hybrids Manuf. Technol*, Vol CHMT-16, 1993, p 317-322

37. S. Vaynman, Energy-Based Methodology for Fatigue Life Prediction of Time-Dependent Process in Solder Materials, *Advances in Electronic Packaging*, Proc. International Intersociety Electronic Packaging Conference—INTERPAC-95, American Society of Mechanical Engineers, 1995, p 663-673

38. V. Sarihan, Energy Based Methodology for Damage and Life Prediction of Solder Joints Under Thermal Cycling, *IEEE Trans. Compon. Hybrids Manuf. Technol.*, Vol CHMT-17B, 1994, p 626-631

39. E.C. Cutiongco, S. Vaynman, M.E. Fine, and D.A. Jeannotte, Isothermal Fatigue of 63Sn-37Pb Solder, *J. Electron. Packag.*, Vol 112, 1990, p 110-114

40. J.F. Eckel, Influence of Frequency on the Repeated Bending Life of Acid Lead, *Proc. ASTM*, Vol 51, 1957, p 745-760

41. H.D. Solomon, Frequency Dependent Low Cycle Fatigue Crack Propagation, *Metall. Trans.*, Vol 3, 1972, p 341-347

42. L.A. James, Effect of Frequency upon the Fatigue Crack Growth of Type 304 Stainless Steel at 1000 °F, *Stress Analysis and Growth of Cracks*, STP 513, ASTM, 1972, p 218-229

43. J.R. Conway, J.T. Berling, and R.H. Stentz, Strain Rate and Hold Time Saturation in Low Cycle Fatigue: Design Parameter Plots, *Fatigue at Elevated Temperatures*, STP 520, ASTM, 1973, p 637-647

44. S.S. Manson, The Challenge to Unify Treatment of High Temperature Fatigue—A Partisan Proposal Based on Strain Range Partitioning, *Fatigue at Elevated Temperatures*, STP 520, ASTM, 1973, p 744-782

45. D. Sidney and L.F. Coffin, Low-Cycle Fatigue Damage Mechanisms at High Temperature, *Fatigue Mechanisms*, STP 675, ASTM, 1979, p 528-568

46. S. Baik and R. Raj, Mechanisms of Creep-Fatigue Interaction, *Metall. Trans.*, Vol 13A, 1982, p 1215-1221

47. J.K. Tien, S.V. Nair, and V.C. Nardone, Creep-Fatigue Interaction in Structural Materials, *Flow and Fracture at Elevated Temperatures*, R. Raj, Ed., ASTM, 1985, p 179-214

48. R. Raj, Mechanisms of Creep-Fatigue Interaction, *Flow and Fracture at Elevated Temperatures*, R. Raj, Ed., ASTM, 1985, p 215-249

49. G.R. Gohn and W.C. Ellis, The Fatigue Tests as Applied to Lead Cable Sheath, *Proc. ASTM*, Vol 51, 1951, p 721-740

50. H.D. Solomon and L.F. Coffin, Effect of Frequency and Environment on Fatigue Crack Growth in A286 at 1100 °F, *Fatigue at Elevated Temperatures*, STP 520, ASTM, 1973, p 112-122

51. P.G. Forrest, *Fatigue of Metals*, Pergamon Press, 1962

52. J. Wareing, Fatigue Crack Growth in a Type 316 Stainless Steel and a 20% Cr/25% Ni/Nb Stain-

less Steel at Elevated Temperature, *Metall. Trans.*, Vol 6A, 1975, p 1367-1377

53. J. Wareing, Creep-Fatigue Interaction in Austenitic Stainless Steel, *Metall. Trans.*, Vol 8A, 1977, p 711-721

54. P.S. Maiya and S. Majumdar, Elevated-Temperature Low-Cycle Fatigue Behavior of Different Heats of Type 304 Stainless Steel, *Metall. Trans.*, Vol 8A, 1977, p 1651-1660

55. S. Majumdar and P.S. Maiya, A Mechanistic Model for Time-Dependent Fatigue, *Eng. Mater. Technol.*, Vol 102, 1980, p 159-167

56. M.F. Day and G.B. Thomas, Microstructural Assessment of Fractional Life Approach to Low Cycle Fatigue at High Temperatures, *Metall. Sci.*, Vol 13, 1979, p 25-33

57. A.H. Meleka, Combined Creep and Fatigue Properties, *Metall. Rev.*, Vol 7, 1962, p 43-93

58. A.R. Ellozy, P.M. Dixon, and R.N. Wild, Combined Low-Cycle Fatigue and Stress Relaxation of Some Pb-In and 96/4 PbSn Alloy at Room Temperature, *Proc. of the Second International Conf. Mechanical Behavior of Materials*, ASME, 1976, p 903-907

59. H.J. Shah and J.H. Kelly, *Effect of Dwell Time on Thermal Cycling of the Flip-Chip Joint*, ISHM, International Society for Hybrid Microelectronics, 1970, p 3.4.1-3.4.6

60. B.N. Agarwala, Thermal Fatigue Damage in Pb-In Solder Interconnections, *23rd Annual Proc., Reliability Physics IEEE/IRPS*, 1985, p 198-205

61. R. Hales, A Quantitative Metallographic Assessment of Structural Degradation of Type 304 Stainless Steel during Creep Fatigue, *Fat. Eng. Mater. Struct.*, Vol 3, 1980, p 339-356

62. D.A. Miller, R.H. Priest, and E.G. Ellison, A Review of Material Response and Life Prediction under Fatigue-Creep Loading Conditions, *High Temp. Mater. Process.*, Vol 6, 1984, p 155-194

63. P.J. Greenwood, T.C. Reiley, V. Raman, and J.K. Tien, Cavitation in a Pb/Low-Sn Solder during Low Cycle Fatigue, *Scr. Metall.*, Vol 22, 1988, p 1465-1468

64. S. Vaynman, Effect of Strain Rate on Fatigue of Low-Tin Lead-Base Solder, *IEEE Trans. Compon. Hybrids Manuf. Technol.*, Vol CHMT-12, 1989, p 469-472

65. H.D. Solomon, The Influence of Hold Time and Fatigue Cycle Wave Shape on the Low Cycle Fatigue of 60/40 Solder, *38th Electronic Components Conf*, IEEE, 1988, p 7-12

66. S. Vaynman, Effect of Temperature on Isothermal Fatigue of Solders, *Proc. InterSociety Conf. Thermal Phenomena in Electronic Systems (I-Therm II)*, Institute for Electrical and Electronics Engineers, 1990, p 16-20

67. A.C. Chilton, M.A. Whitmore, and W. Hampshire, Isothermal Mechanical Fatigue of a Model SMD Joint, *Microelectronic Packaging Technology: Materials and Processes*, Proc. Second ASM International Electronic Materials and Processing Congress, W.T. Shieh, Ed., ASM International, 1989, p 159-176

68. M.A. Whitmore, P.G. Harris, K.S. Chaggar, and A.C. Chilton, "Metallurgical Aspects in the Fatigue of SMD Solder Joints," paper presented at NEPCON Europe, March 1990

69. K.U. Snowden, The Effect of Atmosphere on the Fatigue of Lead, *Acta Metall.*, Vol 12, 1964, p 295-303

70. Z. Guo, A.F. Sprecher, D.Y. Jung, and H. Conrad, Influence of Environment on the Fatigue of Pb-Sn Solder Joints, *IEEE Trans. Compon. Hybrids Manuf. Technol.*, Vol CHMT-14, 1991, p 833-837

71. Y. Ninohira, S. Yee, and S. Rao, to be published

Section 7: Fatigue and Fracture of Composites, Ceramics, and Glasses

Fracture and Fatigue of DRA Composites

John J. Lewandowski and Preet M. Singh, Case Western Reserve University

DISCONTINUOUSLY REINFORCED ALUMINUM (DRA) composites with SiC particulate (SiC_p) or alumina particulate (Al_2O_{3p}) reinforcements are increasingly being considered for structural applications in the aerospace and automotive industries. The reinforcements have higher strength, higher elastic modulus, and lower Poisson ratio than the matrix alloy (Table 1). The use of metal-matrix composites in structural applications is attractive because of their exceptionally good stiffness-to-weight and strength-to-weight ratios, while processing may be conducted using casting, powder metallurgy, or in situ techniques. Secondary processing can significantly improve the strength and ductility of DRA materials (Ref 3-5), as shown in Fig. 1, which illustrates the effects of hot deformation processing (e.g., swaging) on the subsequent uniaxial tension properties of squeeze-cast A-356 DRA (Ref 5). Despite such improvements via hot deformation processing, the composites still possess lower ductility and toughness than unreinforced alloys. It is often these fracture-related properties that limit the further use of metal-matrix composites in fracture-critical applications.

This article reviews the tensile properties and toughness characteristics of DRA composites in terms of particle spacing, particle size, volume fraction, matrix alloy, and matrix microstructure. Both fracture toughness data and impact toughness data are summarized. Where available, information on J_{Ic} and tearing modulus data for Al_2O_3 and SiC-reinforced aluminum alloys are presented. Included in the discussion on tensile deformation are the effects of confining pressure on the ductility of such materials, as this data are relevant to the behavior of these materials under more complex stress states and also correlates with the formability of such materials. A brief summary of DRA fatigue resistance is also included.

Uniaxial Tension Properties

Various authors have studied microstructure/mechanical property relationships for DRA composites (Ref 6-13) and other composite materials (Ref 14-16). A simplistic rule of mixtures is often used to explain the trends in the elastic properties of the composites (Ref 14), whereas other, more precise models account for the size, shape, and distribution of particulates on the elastic properties (Ref 15-24). Elastic modulus values in excess of 100 GPa are possible with reinforcement levels in excess of 20 vol% SiC_p. It has been demonstrated that the use of finer-sized particulates increases the 0.2% offset yield strength (Ref 9, 12, 25, 26) and the work hardening rate, while reducing the level of damage via reinforcement fracture events during tensile straining, as discussed below.

Particle-Associated Damage Accumulation during Tensile Straining. Various investigators (Ref 6, 9, 10, 12, 13) have shown that reinforcement-related damage evolves during straining of DRA materials, and that the extent and type of damage depends on SiC_p size, matrix heat treatment, strength, microstructure, and alloy type. Figures 2 and 3 show the effects of global plastic strain on the percentage and number of cracked SiC_p on the gage surface of duplicate and triplicate MB78 (i.e., Al-7Zn-2Mg-2Cu powder metallurgy alloy) composite specimens in the underaged (UA), overaged (OA), and solution-annealed (SA) conditions, respectively. The percentage of cracked SiC_p shown in Fig. 2 and 3 is the average of results from every selected area on each specimen, which comprises over 10,000 particles for each specimen at each strain. Included in all figures are the *absolute number* of cracked SiC_p at each strain. The results, consistent with previous work (Ref 6, 9, 10, 12, 13, 27-29), clearly show that the number of cracked SiC_p for each condition increases with an increase in applied global plastic strain, e_{gpl}. Comparison of Fig. 2 and 3 shows that the total average numbers of SiC_p cracked at the fracture strain for the 13 μm OA and the 13 μm SA specimens were far less than that obtained in the 13 μm UA specimens. At very low strains (i.e., <1%) all specimens provided similar average values for cracked SiC_p, while higher plastic strains pro-

Table 1 Elastic properties of commonly used reinforcements in DRA materials

Material	Elastic modulus, GPa	Poisson ratio
SiC	400	0.14-0.17
Al_2O_3	379	0.26
B_4C	448	0.17-0.21
Si_3N_4	207	0.27
Si	112	0.42

Source: Ref 1, 2

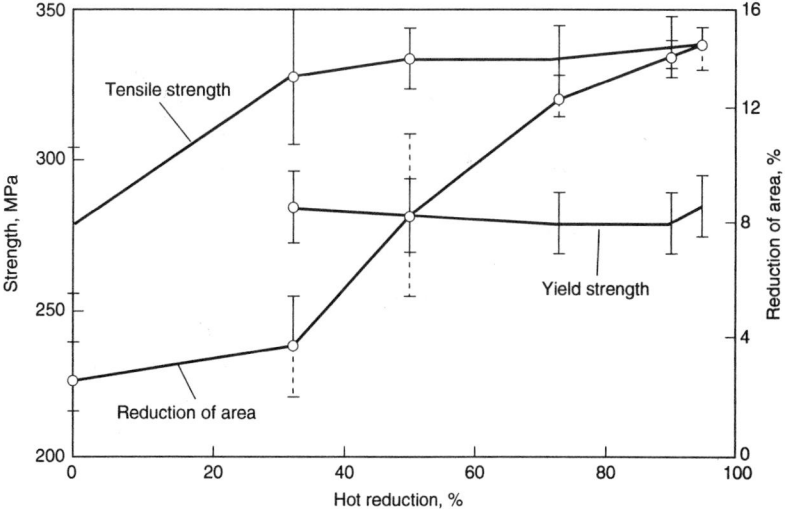

Fig. 1 Effects of the amount of hot work on tensile properties of squeeze-cast composites. Source: Ref 5

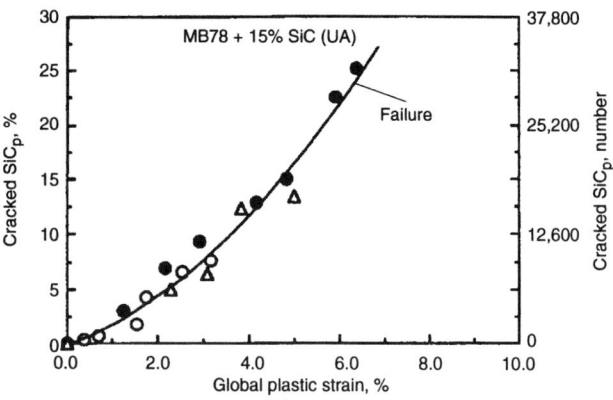

Fig. 2 Percentage and absolute number of cracked SiC$_p$ on grided specimen surface as a function of the applied global plastic strain for 13 mm underaged (UA) specimens. Triplicate specimens tested on identical materials. Source: Ref 12

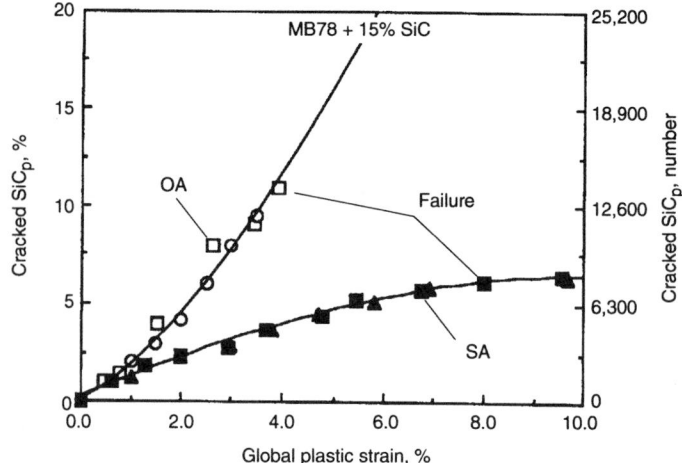

Fig. 3 Percentage and number of cracked SiC$_p$ on grided specimen surface as a function of the applied global plastic strain for 13 mm overaged (OA) and 13 mm solution-annealed (SA) specimens. Duplicate specimens tested for OA and SA material. Source: Ref 12

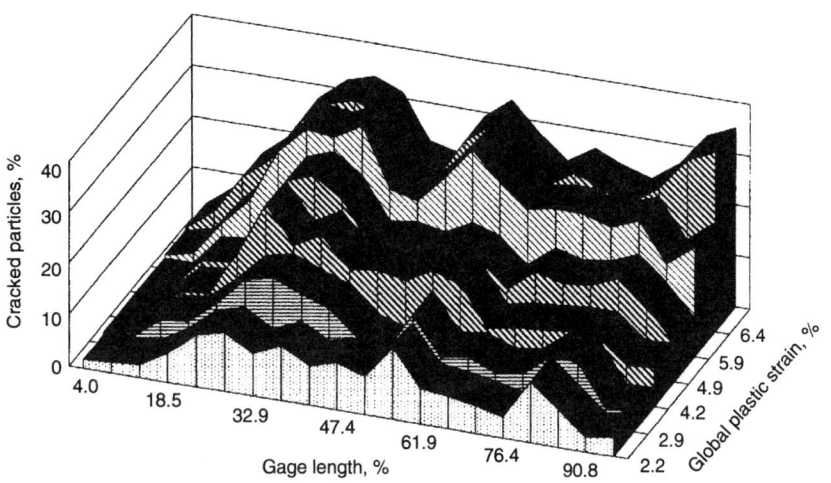

Fig. 4 Distribution of the local percentage of cracked SiC$_p$ along the gage length of an MB78 + 15% SiC$_p$ 13 mm underaged tensile specimen. Source: Ref 12

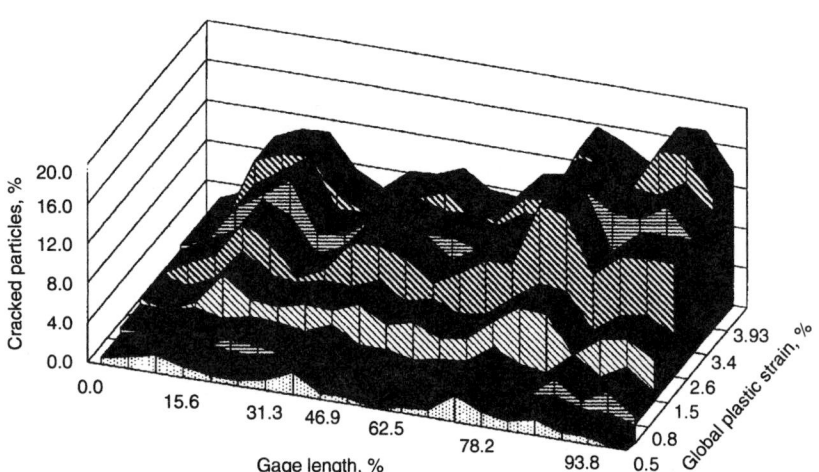

Fig. 5 Distribution of the local percentage of cracked SiC$_p$ along the gage length of an MB78 + 15% SiC$_p$ 13 mm overaged specimen. Source: Ref 12

comparison to the 13 µm OA specimens, in addition to a greater fracture strain.

Figures 4 to 6 show the distribution of cracked SiC$_p$ along the gage length at each strain for each heat treatment for MB78 material containing 13 µm SiC$_p$. Both the percentage of cracked SiC$_p$ and the number of cracked SiC$_p$ at each location are plotted as a function of position along the specimen for each global strain. Integration of the number of cracked SiC$_p$ at each location and strain provided in Fig. 4 to 6 will produce the total number of cracked SiC$_p$ on the specimen gage surface summarized in Fig. 2 and 3. Figures 4 to 6 clearly show that both the percentage and absolute number of the cracked SiC$_p$ at each strain were not constant along the gage length, and that a low background level of cracked SiC$_p$ exists at 0% strain. However, for any given local area, the number of cracked particles increased with an increase in the global plastic strain, e_{gpl}, although the absolute numbers remained different for different areas of the specimens throughout the test.

Metallographic examination showed that SiC$_p$ with higher aspect ratios was preferentially aligned along the extrusion direction. Particles larger than the mean particle size and those with larger aspect ratios preferentially cracked in all heat treatments, consistent with previous reports (Ref 6, 8, 12, 29, 30, 31, 32). Figure 7 shows that large and elongated SiC$_p$ are cracked in the 13 µm UA specimen unloaded at 6.3% e_{gpl} and that there is no evidence of matrix failure.

In addition to SiC$_p$ fracture, the 13 µm OA material exhibited matrix cracks which were readily apparent at around 2.0% e_{gpl}. In contrast, only cracks in the SiC$_p$ were apparent at 2.0% e_{gpl} in the 13 µm UA material. Figure 8 shows some examples of the matrix cracks and cracks near the SiC$_p$/matrix interfaces that were present on the surface of the 13 µm OA material strained to 3.5% e_{gpl}. In some cases, the matrix cracks appeared to be associated with grain boundary regions, as shown in Fig. 8. The 13 µm SA specimen exhibited the minimum number of SiC$_p$ cracked of all of the tested conditions, while Fig.

duced greater differences between the SA specimens and either the UA or OA specimens. The 13 µm UA specimens exhibited a slightly greater average amount of SiC$_p$ fracture at each strain in

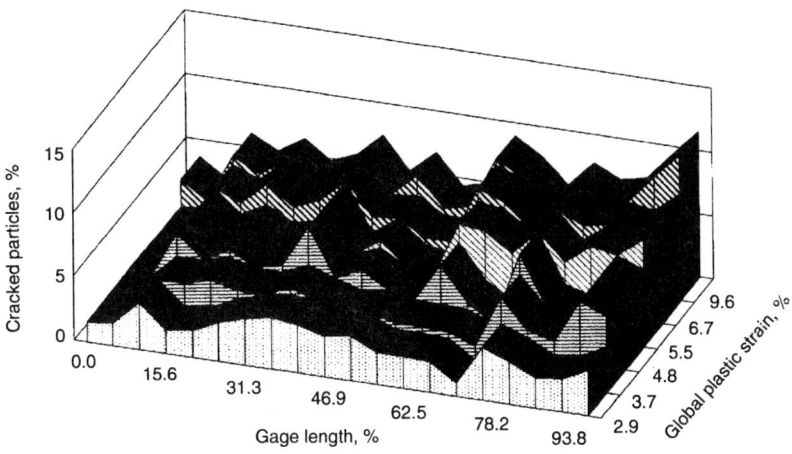

Fig. 6 Distribution of the local percentage of cracked SiC_p along the gage length of an MB78 + 15% SiC_p 13 mm solution-annealed specimen. Source: Ref 12

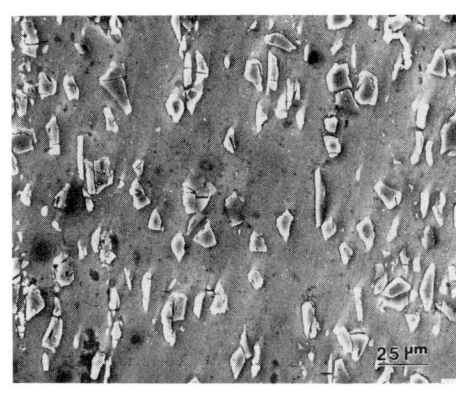

Fig. 7 Micrograph showing cracked SiC_p on the surface of a 13 mm underaged specimen after about 6.3% global plastic strain, e_{gpl}. Source: Ref 12

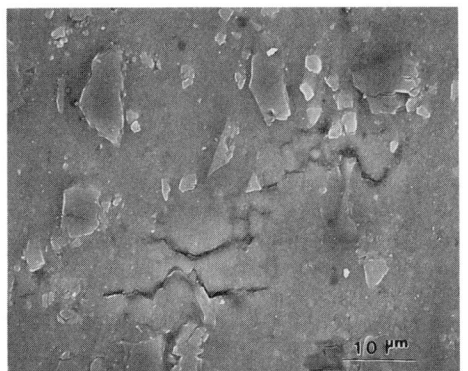

Fig. 8 Micrograph showing cracks in the matrix and near the SiC/matrix interface on the surface of a 13 mm overaged specimen after about 3.5% global plastic strain, e_{gpl}. Straight arrow locates matrix cracks; curved arrows locate voids near SiC_p/matrix interface. Source: Ref 12

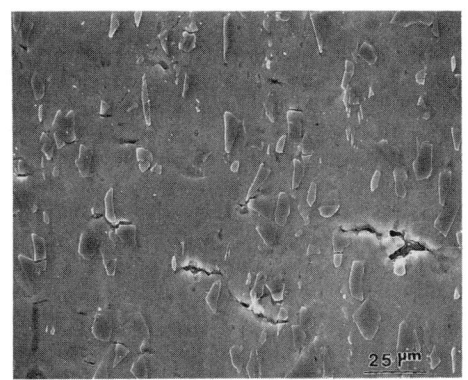

Fig. 9 Micrograph showing cracks in SiC_p and in the matrix on the surface of a 13 mm solution-annealed specimen after about 9.9% global plastic strain, e_{gpl}. Source: Ref 12

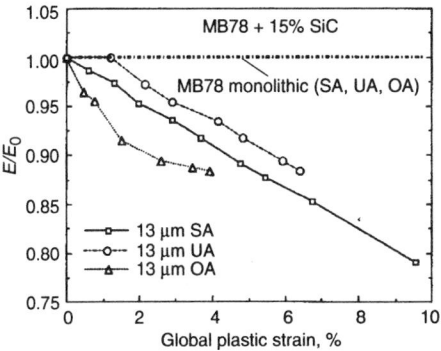

Fig. 10 Effect of global plastic strain on E/E_0 for differently heat-treated MB78 + 15% 13 mm SiC_p DRA materials. SA, solution annealed; UA, underaged; OA, overaged. Source: Ref 12

9 also shows examples of surface matrix cracks exhibited in a 13 μm SA specimen after 9.9% e_{gpl}, which is near the fracture strain of the material. In contrast to the OA material, the matrix cracks in the SA material did not appear to be associated with grain boundary regions. The total fracture strain was similarly affected by heat treatment: the 13 μm SA specimen fractured at 10% e_{gpl}, as compared to 4% e_{gpl} for the 13 μm OA specimen and about 6% e_{gpl} for the 13 μm UA specimen.

The unloading modulus and reloading modulus were measured on the interrupted tensile tests on the MB78 monolithic materials in the UA, OA, and SA conditions. There was no significant change in the modulus during unloading of the MB78 monolithic materials in any condition, strained to e_{gpl} = 10%. In contrast, increases in e_{gpl} produced large decreases in the subsequent unloading and reloading elastic modulus of the MB78 DRA composites. This change in the elastic modulus was exhibited for each of the heat treatments, although to different extents. Figure 10 plots the ratio of the "measured" modulus at each strain, E, to the starting modulus, E_0, for each condition tested. A value of 1.0 for E/E_0

indicates no change in modulus. The number of cracked SiC_p increased with increasing strain in the MB78 DRA materials containing 13 μm average size SiC_p, as exhibited by Fig. 2 to 6. In particular, the area near the location of catastrophic fracture showed a large increase in the number of cracked particles.

Figures 11, 12, and 13 summarize results from the various studies (Ref 10, 12, 13) that have measured the evolution of particle-related damage on different DRA materials. Figure 11 shows the percentage of cracked SiC_p as a function of the fracture strain in the respective specimens. Details of the matrix material, volume fraction and size of reinforcement, and heat treatment are available elsewhere (Ref 13). These results indicate that there is no correlation between fracture strain and the percentage of cracked SiC_p. However, the data collected in Fig. 11 to 13 combine various matrices and types of reinforcement. Figures 12 and 13 provide separate data on DRA materials with Al_2O_3 and SiC reinforcements, respectively. Again there is no direct correlation between the fracture strain and the percentage of cracked particles. These results agree with the results shown in Fig. 2 and 3, which reveal that

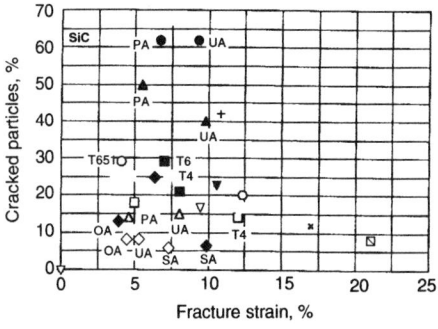

Fig. 11 Data collected from published work of various authors showing the percentage of cracked SiC_p particles as a function of fracture strain for a variety of DRA materials. SA, solution annealed; UA, underaged; OA, overaged; PA, peak aged. **Ref and Material:** □ Ref 10, 6061/10%/6,5 μm; ■ Ref 10, 6061/20%/10 μm; O Ref 30, 2618/15%/12 μm; Δ Ref 13, 201/9%/23 μm; ▲ Ref 13, 201/9%/63 μm; ● Ref 13, 201/9%/142 μm; ▼ Ref 13, 2618/15%; ◆ Ref 33, MB78/15%/13 μm; ◇ Ref 33, MB78/15%/5 μm; ▽ Ref 13, 2080/20%/37 μm; × Ref 13, 2080/20%/12 μm; ⊠ Ref 13, 2080/20%/4 μm; + Ref 13, Al-12Si/17.3%/5 μm

the SA-MB78 DRA specimens had the highest strain to fracture but did not have the highest percentage of cracked particles at fracture. The

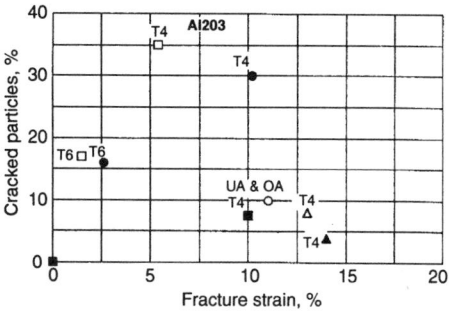

Fig. 12 Data collected from published work of various authors showing the percentage of cracked Al$_2$O$_{3p}$ as a function of fracture strain for a variety of DRA materials. UA, underaged; OA, overaged. **Ref and Material:** □ Ref 13, 2014/14.4%/12 μm; ● Ref 13, 6061/14.9%/12 μm; △ Ref 13, 6061/15%/8 μm; ○ Ref 27, 6061/15%/13 μm

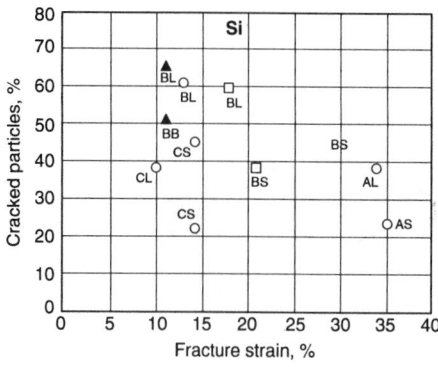

Fig. 13 Data collected from published work of various authors showing the percentage of cracked SiC$_p$ as a function of fracture strain for a variety of DRA materials. SA, solution annealed; UA, underaged; OA, overaged. **Ref and Material:** □ Ref 9, Al-Si-Mg/10%; ○ Ref 9, Al-Si-Mg/15%; ▲ Ref 9, Al-Si-Mg/20%

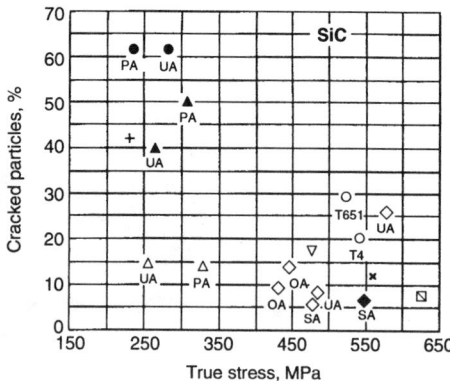

Fig. 14 Data collected from published work of various authors showing the percentage of cracked SiC$_p$ at fracture as a function of true stress for a variety of DRA materials. **Ref and Material:** ○ Ref 30, 2618/15%/12 μm; + Ref 13, Al-12Si/17.3%/5 μm; △ Ref 13, 201/9%/23 μm; ▲ Ref 13, 201/9%/63 μm; ● Ref 13, 201/9%/142 μm; ◆ Ref 33, MB78/15%/13 μm; ◇ Ref 33, MB78/15%/5 μm; ▽ Ref 13, 2080/20%/37 μm; × Ref 13, 2080/20%/12 μm; ◨ Ref 13, 2080/20%/4 μm

results also indicate that the local stress/strain experienced by the reinforcement depends on the microstructural details of the DRA constituents (e.g., matrix composition and microstructure, reinforcement type, size, orientation, and matrix/reinforcement interface characteristics).

Figures 14, 15, and 16 show results taken from the same studies (Ref 10, 12, 13) where the percentage of cracked particulates is plotted versus the true stress at failure. In most cases, this stress was simply the ultimate tensile strength, because the failure strains were typically less than 5%. Figures 14 and 15 fail to reveal any correlation of applied true stress to the percentage of cracked particles, even for DRA materials of the same matrix and particle size. Thus, Fig. 11 to 15 clearly show that both global strain-based models and global stress-based models are unable to explain particle damage or total damage without taking microstructural effects into account. Although global stress and strain are important parameters in damage evolution, as the amount of particle-related damage always increases with an increase in global stress or strain, the mechanism and extent of damage evolution is controlled by the microstructure and conditions that control local stress conditions.

It is clear from a review of Fig. 7 to 9 and previous work (Ref 12) that other modes of damage apart from particle damage contribute to the changes in the elastic modulus or Poisson ratio (Ref 33) with increasing strain. Very little work (Ref 7, 33) has been carried out to monitor Poisson ratios with applied stress/strain, but some researchers have monitored changes in the elastic modulus with applied strain. Figure 17 summarizes data for elastic modulus changes from the work of various authors (Ref 10, 12, 13, 30, 34) and shows that the extent and path of overall damage are different for different DRA materials. The results indicate that different heat treatments can change the fracture mechanisms and extent of overall damage for nominally the same DRA material.

Effects of Confining Pressure on Ductility and Damage Evolution in DRA Materials. The deformation and fracture behavior under complex stress states is often accomplished via testing with superposed hydrostatic pressure. Initial work conducted on the effects of superposed pressure indicated that significant enhancements in ductility can be obtained in both particulate-reinforced (Ref 35-39) and whisker-reinforced (Ref 36) systems, although the magnitude of the pres-

sure-induced ductility increase in both the monolithic and composite systems appears to be dependent on the nature of operative damage mechanisms (Ref 12, 35-38). These early works also demonstrated that the flow stress of such materials can be pressure sensitive (Ref 25, 26). Initial work proposed that such pressure sensitivity is a result of the pressure-induced suppression of the damage (Ref 35, 36), while subsequent work showed that pressure-induced dislocation generation near the reinforcement/matrix interfaces is also partly responsible for local hardening around the particles (Ref 25).

Figures 18 and 19 are cumulative plots summarizing the effects of superposed pressure on the ductility (i.e., reduction in area) for a variety of

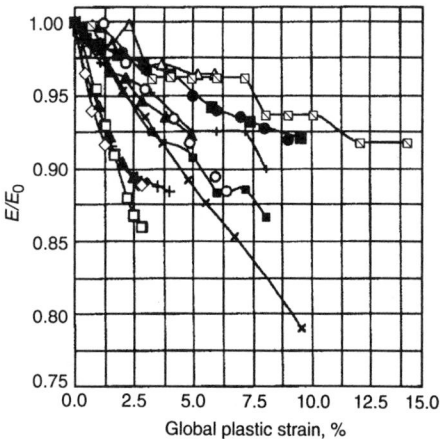

Fig. 17 Data collected from published work of various authors showing the effect of global plastic strain on the reduction in elastic modulus. **Ref and Material:** ◇ Ref 30, 2618-T651/15%/12 μm; □ Ref 13, 2618/15%; ▲ Ref 34, A356-T61/20%; △ Ref 34, A356-OA30/20%; ■ Ref 10, 6061-T4/20%; + Ref 10, 6061-T4/15%; ◨ Ref 10, 6061-T4/10%; + Ref 33, MB78-OA/15%/13 μm; × Ref 33, MB78-SA/15%/13 μm; ○ Ref 33, MB78-UA/15%/13 μm

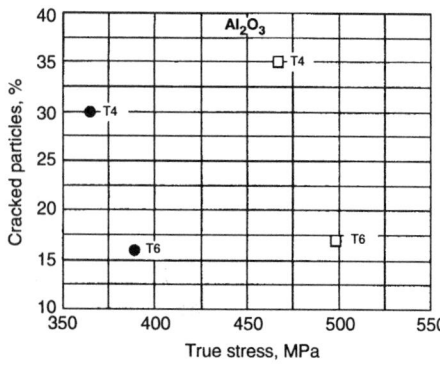

Fig. 15 Data collected from published work of various authors showing the percentage of cracked Al$_2$O$_{3p}$ at fracture as a function of true stress for a variety of DRA materials. **Ref and Material:** □ Ref 13, 2024/14.4%/12 μm; ● Ref 13, 6061/14.9%/12 μm

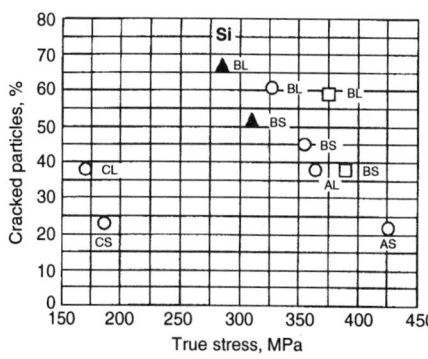

Fig. 16 Data collected from published work of various authors showing the percentage of cracked SiC$_p$ at fracture as a function of true stress for a variety of DRA materials. **Ref and Material:** □ Ref 9, Al-Si-Mg/10%; ○ Ref 9, Al-Si-Mg/15%; ▲ Ref 9, Al-Si-Mg/20%

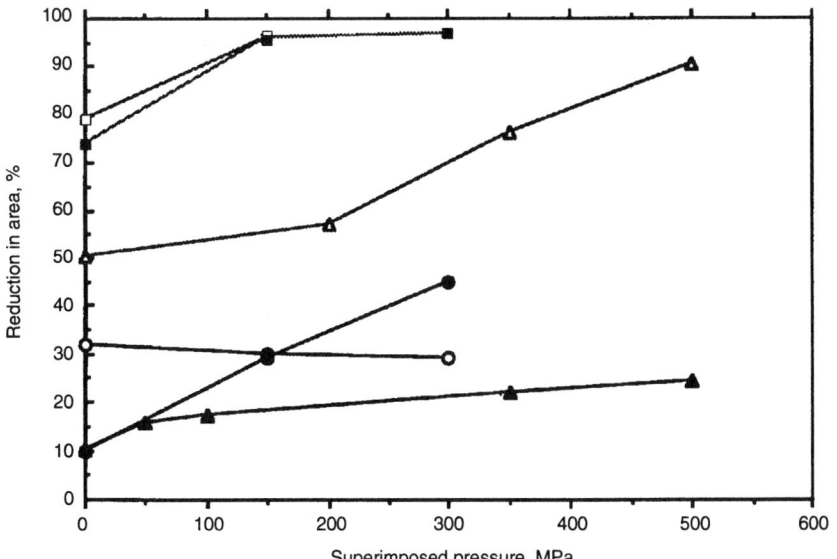

Fig. 18 Effects of superposed pressure on the ductility of unreinforced alloys. **Ref and Material:** ▲ Ref 40, AZ91 Mg-T4; O Ref 28, MB85 Al-UA; ● Ref 28, MB85 Al-OA; □ Ref 27, 6061 Al-UA; ■ Ref 27, 6061 Al-OA; △ Ref 40, A356 Al-T6

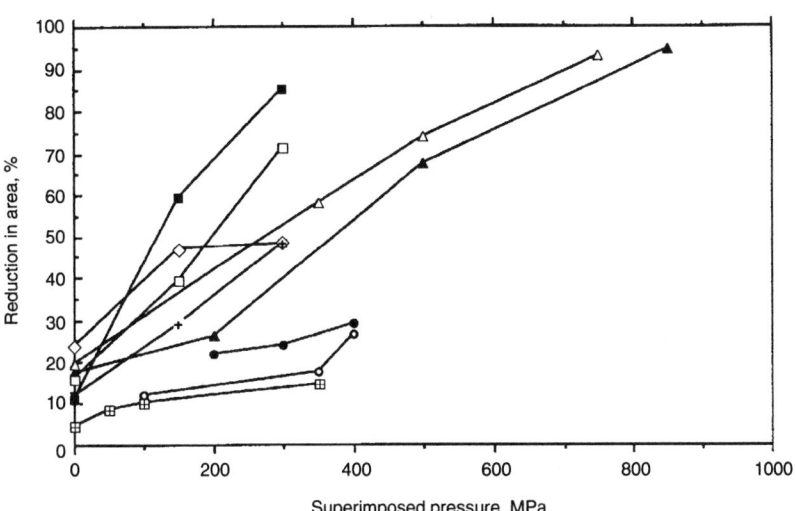

Fig. 19 Effects of superposed pressure on the ductility of DRA. **Ref and Material:** ⊞ Ref 40, AZ91/SiC/20p-T4; ◊ Ref 28, MB85/SiC/15p-UA; + Ref 28, MB85/SiC/15p-OA; □ Ref 27, 6061/SiC/20p-UA; ■ Ref 27, 6061/SiC/20p-OA; △ Ref 40, A356/SiC/10p-T6; ▲ Ref 40, A356/SiC/20p-T6; O Ref 40, 6090/SiC/25p-T6; ● Ref 40, 6090/SiC/25p-SA

metallic alloys and composites, respectively. Embury et al. (Ref 39) assessed the damage in the deformed and undeformed A356/SiC$_P$ system in the T6 condition with detailed metallographic studies and subsequent image analysis. Their examination of polished longitudinal sections revealed that the large plastic deformation accompanying the high ductility promoted by superposed pressure produces a change in SiC$_p$ distribution. Embury and coworkers (Ref 39) found a subsequent enhancement in ductility in both the unreinforced A356 and the A356 composite material under the influence of hydrostatic pressure, as indicated in Fig. 18 and 19. Work by Lewandowski et al. (Ref 12, 25, 35, 37) has similarly investigated the effects of superposed pressure on ductility and damage evolution in

2xxx, 7xxx, and 6061 aluminum alloys and composites in the UA and OA conditions. A significant pressure dependence on ductility was found for all of the composite systems and microstructures, as shown in Fig. 19. The unreinforced alloys also showed a dependence of ductility on pressure, with the exception of the 2xxx (Ref 35, 37) and 7xxx (Ref 41) materials tested in the UA condition. The unreinforced 2xxx and 7xxx materials deformed and failed in shear by the initiation and propagation of localized shear, which is not significantly affected by the levels of pressure used. These results are a clear indication that microstructural changes have a significant impact on the type of damage (i.e., shear, void nucleation, and growth) and that ductility depends on superposed hydrostatic pressure. The differences

in the magnitude of ductility response to pressure of the remaining monolithic materials shown in Fig. 18 are also partly due to the different ratios of superposed pressure to the flow stress of the matrix employed. Damage evolution in the monolithic materials was quantified in terms of the area fraction of voids (Ref 27) and was shown to decrease with increasing levels of superposed hydrostatic pressure, consistent with previous work (Ref 42, 43).

Considerable work on DRA materials has shown that damage may evolve via reinforcement fracture, failure of reinforcement/matrix interfaces, failure in the matrix, or a combination of the above (Ref 6, 12, 36). The superposition of hydrostatic pressure did not change the brittle nature of the reinforcement in such studies, although damage evolution was suppressed with increasing levels of confining pressure (Ref 12, 35–39). Singh and Lewandowski (Ref 12) conducted extensive metallographic and image analyses on longitudinal sections of samples fractured in tension with superposed pressure, 7xxx composites in the UA and OA conditions. Figures 20 and 21 show the percentage of cracked SiC versus the distance from the fracture surface in longitudinal sections taken in multiple specimens, tested at either 0.1 MPa or with superposed pressure. Despite the large increase in strain to failure with superposed pressure, a reduction in the amount of particle cracking was measured. Studies at 0.1 MPa additionally showed that the composite in the OA condition failed with a smaller amount of particle cracking than did the UA specimens, although considerable matrix damage and particle/matrix cavitation was obtained in the former. The effect of pressure was to inhibit all types of damage (i.e., matrix, particle/matrix interface, SiC$_p$ cracking). Recent work on AZ91 magnesium and AZ91 magnesium composites have revealed some similar features, as shown in Fig. 20 and 21, although the details are under investigation. Accompanying this suppression of damage was an elevation in the flow stress of the composites upon superposition of pressure (Ref 25, 36). Transmission electron microscopy of composites simply pressurized revealed that such elevation of initial flow stress was also partly caused by the pressure-induced generation of dislocations in the matrix adjacent to the reinforcement (Ref 25). Recent work on the mechanical behavior of 1100 Al/Al$_2$O$_3$ composites containing either 10 or 15 vol% reinforcement also reveals an increase in the yield strength of this system upon testing under pressure.

Toughness

Toughness of DRA Systems. To the extent possible, this section summarizes published data about the effects of systematic changes in reinforcement volume fraction, reinforcement size, and similar parameters on the resulting toughness of DRA. Figures 22 and 23 summarize the effects of reinforcement volume fraction on the fracture toughness of various DRA systems. The J_{Ic} data presented were collected from works where J_{Ic}

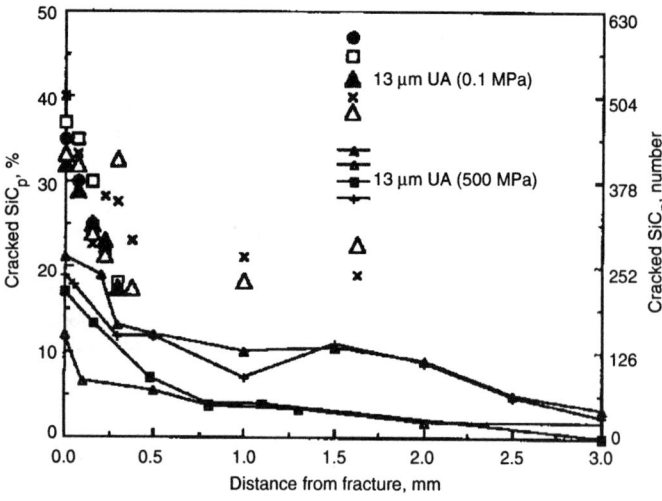

Fig. 20 Effects of superposed pressure on the damage evolution in the MB78/15% SiCp underaged (UA) system.

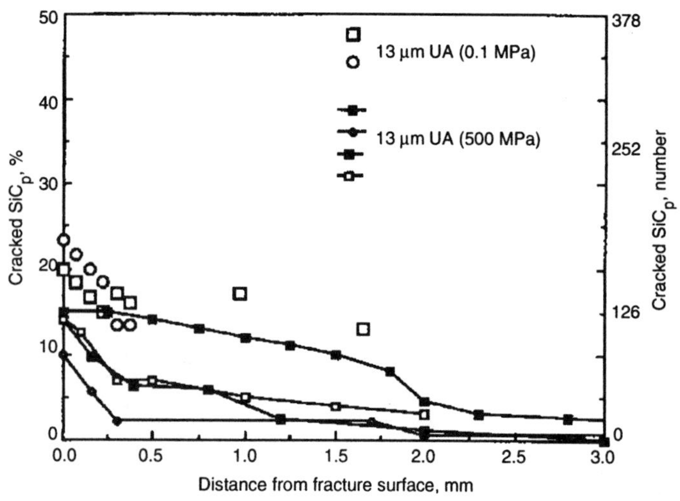

Fig. 21 Effects of superposed pressure on the damage evolution in the MB78/15% SiCp overaged (OA) system.

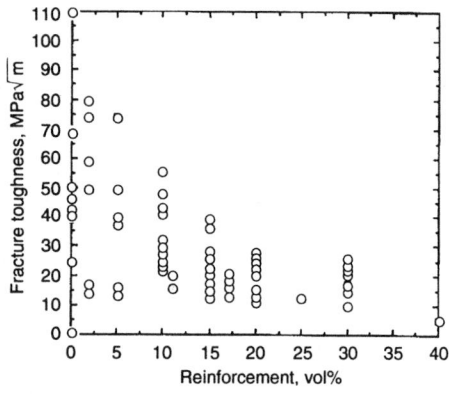

Fig. 22 Compilation of fracture toughness data vs. volume fraction of reinforcement. Reference 44 contains data sources.

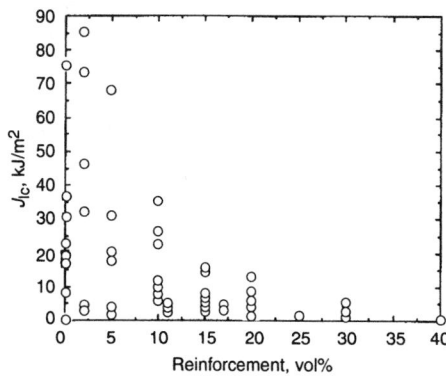

Fig. 23 Compilation of J_{Ic} data vs. volume fraction of reinforcement. Some data were converted from K_{Ic}, K_Q, and K_{EE} data. Reference 44 contains data sources.

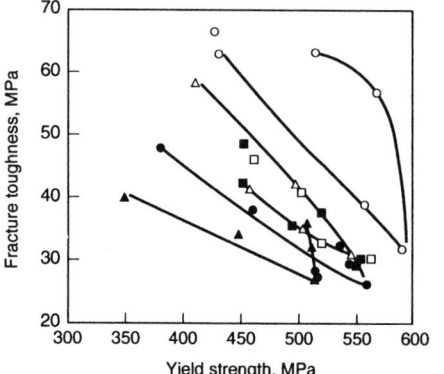

Fig. 24 Compilation of fracture toughness data for monolithic materials vs. yield strength. Reference 44 contains data sources. **Ref and Material:** O Ref 31, MB78-short rod; □ Ref 45, 7178-250F; ■ Ref 45, 7178-290F; △ Ref 45, 7178-340F; ▲ Ref 45, 7075-163C; ● Ref 45, 7075-120C

testing was conducted as well as by converting some of the K_{Ic}, K_Q, and K_{EE} data to J_{Ic} data via the relationship $J = K^2(1 - v^2)/E$. The data demonstrate the general loss in toughness with increasing volume fraction of reinforcement, up to a reinforcement level of 40 vol%, as well as the significant effects of matrix selection and matrix temper. Figures 22 and 23 include data for DRA based on various aluminum alloys, including 1100, 2xxx, 6xxx, 7xxx, 8xxx, and model Al-Si alloys containing brittle Si particles.

The effects of matrix strength and heat treatment on the fracture toughness of both monolithic and DRA materials are more fully shown in Fig. 24 to 26, which plot the fracture toughness versus yield strength and tensile strength. For a given alloy system, the changes in strength were obtained via heat treatment. In the unreinforced alloys, it has generally been observed that the toughness decreases with an increase in strength up to peak strength, and then the toughness recovers upon additional aging (Ref 31, 45, 46), as shown in Fig. 24. This has been demonstrated on a variety of 7xxx monolithic materials (Ref 31, 45, 46), as well as 6xxx and 2xxx (Ref 47) sys-

tems. In contrast, none of the composite systems appear to recover the toughness upon overaging, as shown in Fig. 25 and 26 for DRA materials based on 2xxx (Ref 48, 49), 6xxx (Ref 48, 50), or 7xxx (Ref 6, 8, 31) systems.

This lack of toughness recovery upon overaging has been attributed to a change in the mechanisms of fracture nucleation and growth in such systems (Ref 6, 8, 12, 48). In the 7xxx DRA system, analysis of sequentially strained tensile specimens has indicated that the mechanisms of damage accumulation are significantly affected by aging treatment (Ref 12). Underaged specimens exhibited a predominance of reinforcement fracture at all ranges of strain, while the overaged specimens exhibited matrix failure and failure near the reinforcement/matrix interfaces. The rate of damage accumulation, as monitored by the decay in elastic modulus, was also shown to occur at a faster rate in the overaged material (Ref 12). Analysis of fracture surface features of underaged and overaged DRA specimens has similarly revealed differences in the fracture morphologies between underaged and overaged specimens, also indicative of the effects of micro-

structure on the operative damage mechanisms occurring ahead of a crack tip in such materials (Ref 6, 31, 48). In situ straining/observation of fracture evolution ahead of crack tips has similarly revealed an effect of matrix microstructure on the nature of crack propagation (Ref 6, 8).

Reinforcement size and aspect ratio have been considered. One of the more dramatic differences in behavior exists between whisker- and particle-reinforced materials. Whisker materials show substantially different strength and toughness behavior and exhibit greater strengthening, greater anisotropy, and lower toughness (Ref 51, 52). On a more limited scale, reinforcement particle size effects have been examined in powder metallurgy DRA materials. Recent experiments on x2080/SiC extrusions, which examined reinforcement particles of ~4 μm average size (Federation of European Producers and Abrasives [FEPA] grade F-1000), ~10 μm average size (FEPA grade F-600), and ~36 μm average size (FEPA grade F-280) at particle loadings of 10, 20, and 30 vol%, have proven that the tough-

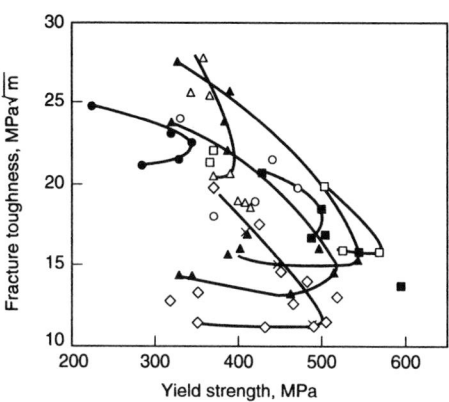

Fig. 25 Compilation of fracture toughness data for DRA materials vs. yield strength. Reference 44 contains data sources. ○ 2014/Al₂O₃/16.2%; ● 6061/Al₂O₃/14.5%; □ 7050/SiC/11%; ■ 7050/SiC/17%; △ 2080/SiC/15%; ▲ MB78/SiC; ◊ 8090/20SiC

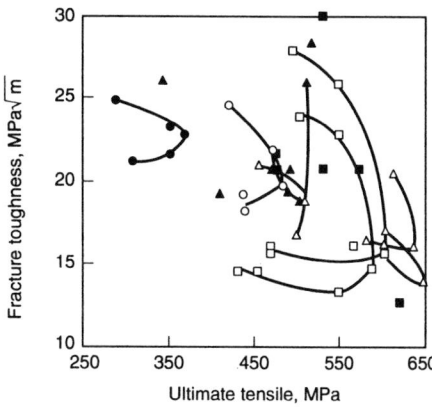

Fig. 26 Compilation of fracture toughness data for DRA materials vs. tensile strength. Reference 44 contains data sources. ○ 2014/Al₂O₃/16.2%; ● 6061/Al₂O₃/14.5%; △ 7050/SiC; ▲ 2080/SiC/ 15%; □ MB78/SiC/20%; ■ 2080-T6/SiC

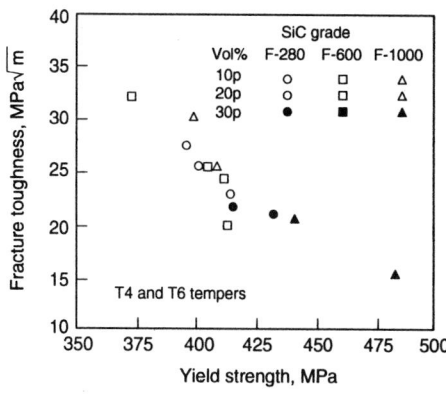

Fig. 27 Toughness/yield strength relationships for x2080/SiC/xxp DRA materials in which both particle volume percent (p) and particle size are varied. Source: Ref 53

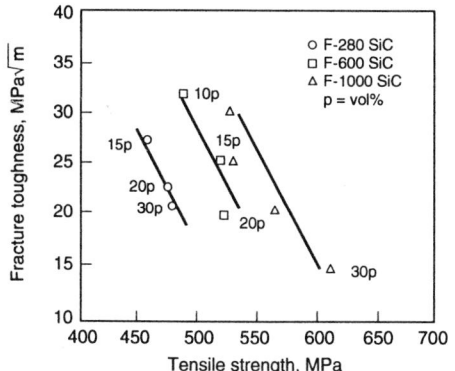

Fig. 28 Toughness/tensile strength relationships for x2080/SiC/xxp DRA materials demonstrating the effect of particle size. Source: Ref 53

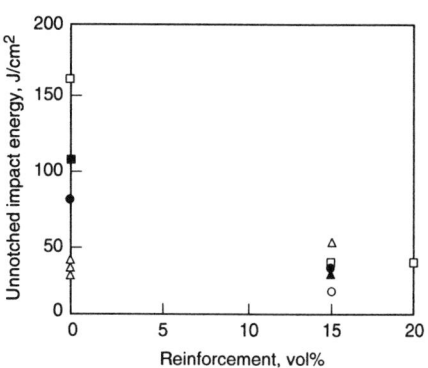

Fig. 29 Effects of reinforcement on unnotched Charpy impact toughness. Reference 44 contains data sources. **Ref and Material:** ■ Ref 60, 3003; □ Ref 60, MB85/UA, MB85/UA/SiC; ● Ref 62, MB78/AR; ○ Ref 62, MB78/ T7E92/SiC; △ Ref 44, 6061/T651, 6061/AR/Al₂O₃; ▲ 2014/ AR/Al₂O₃

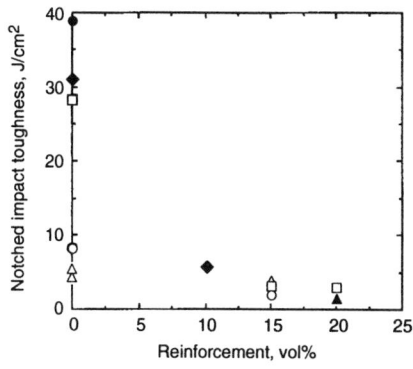

Fig. 30 Effects of reinforcement on notched Charpy impact toughness. Reference 44 contains data sources. **Ref and Material:** ◊ Ref 44, 2014/AR/Al₂O₃; ◆ Ref 44, 2024/AR, 2024/AR/Al₂O₃; □ Ref 60, MB85/UA, MB85/UA/SiC; ○ Ref 66, MB78/ T7E92, MB78/T7E92/SiC; ● Ref 62, MB78/UA; △ Ref 44, 6061/T651, 6061/T6/SiC; ▲ Ref 70, 2080/T6/SiC

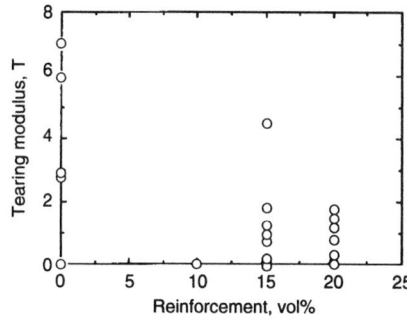

Fig. 31 Tearing modulus data for various DRA systems including 6061/Al₂O₃, MB78/SiC, and Al-Si-Mg/Si. Reference 44 contains data sources.

ness/yield strength relationship is not greatly affected by particle size (Fig. 27), while the toughness/tensile strength relationship (Fig. 28) favors the finer-particle materials (Ref 53). Such behavior is consistent with the higher work hardening and lower damage development rate (e.g., via particle cracking) (Ref 25) in the finer-particle materials. Recent work (Ref 12) has shown higher work hardening and less particle fracture in the finer-particle material.

The Charpy impact toughness of aluminum alloys is similarly reduced via the introduction of discontinuous reinforcement. Figures 29 and 30 summarize recent data obtained on 2xxx, 6xxx, and 7xxx materials in the unnotched and notched conditions, respectively. Although the data are somewhat limited in comparison to the fracture toughness data presented in Fig. 22 to 26, it again appears that there are both matrix alloy and matrix temper effects on the impact energy of the monolithic materials, while the impact toughness of the composites is generally low regardless of heat treatment.

Crack Growth Resistance. It is clear that the flaw tolerance of both monolithic materials and DRA materials is of critical importance in design. While a number of investigators have used the

approach of Rice and Johnson (Ref 44, 54) to characterize the toughness of DRA materials, it has also been observed in fracture toughness tests on such materials that fracture often propagates catastrophically. Thus, the issue of crack growth resistance of such materials may also be relevant. While the above data reviewed for fracture toughness generally characterize the crack initiation resistance of such materials, it is also important to characterize the crack growth resistance of such materials. Of the various methods proposed to evaluate crack growth toughness of materials, the tearing modulus concept introduced by Paris et al. (Ref 55) has found widest acceptance. The tearing modulus parameter is a measure of the resistance of the material to tearing and an indication of the stability of crack growth. The tearing modulus, proportional as it is to dJ/da, is a measure of the strain energy that must be provided to the crack tip to enable it to advance by a unit crack length. Few investigations have measured both the crack initiation and crack growth resistance of both monolithic and DRA materials.

Figure 31 summarizes the available data on 6xxx and 7xxx systems as well as a model Al-Si

alloy containing Si particles. The data again suggest a large effect of matrix alloy composition and heat treatment on the tearing modulus for the materials systems summarized. It is shown that reinforcement additions significantly reduce the crack growth resistance, while matrix aging to an overaged condition in the 7xxx material further reduces the crack growth resistance. In addition,

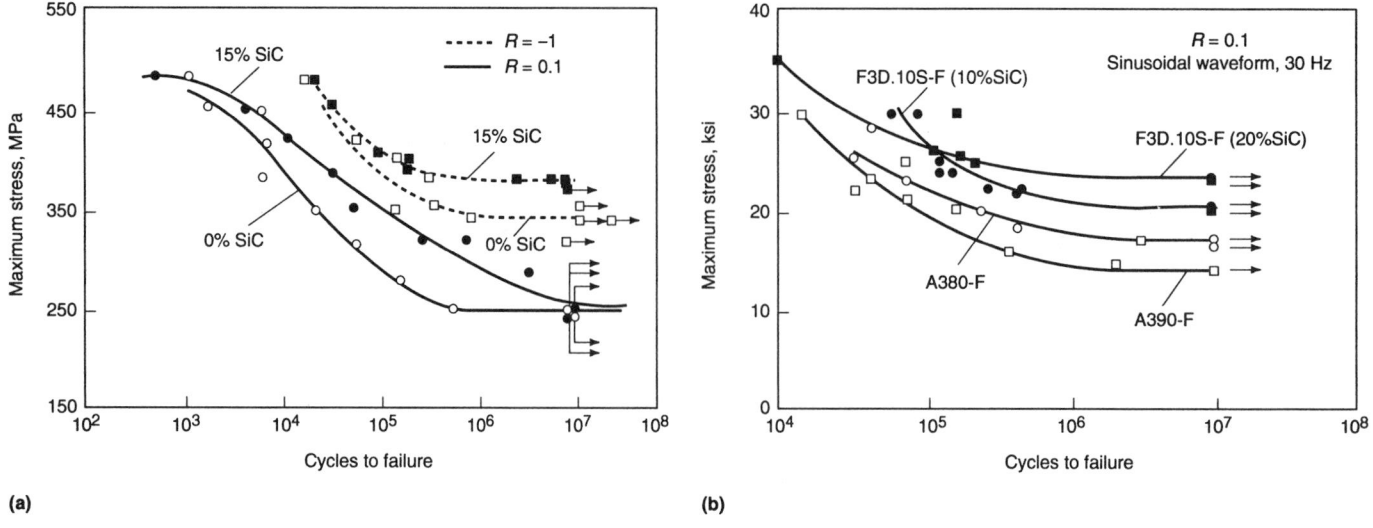

Fig. 32 Effect of reinforcement on stress life. (a) x2080-T4 aluminum alloy at $R = -1$ and $R = 0.1$. (b) Duralcan grade F3D.xxS-F axial fatigue. Source: Ref 74, 75

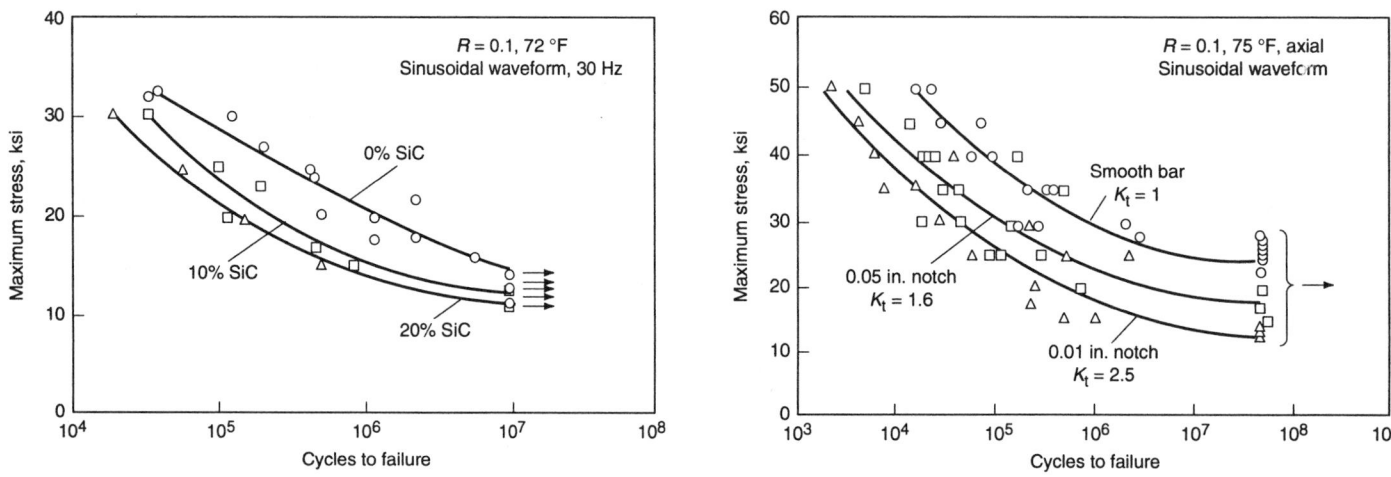

Fig. 33 Axial fatigue of DRA composite F3A.xxT61. Source: Ref 75

Fig. 34 Notched fatigue strength of DRA composite (Duralcan W2A.15-T6). Source: Ref 75

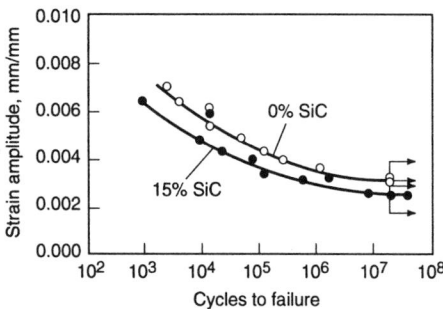

Fig. 35 Strain-life fatigue curve of x2080-T4 composite with and without reinforcement. Stress controlled, $R = -1$, at room temperature. Source: Ref 73

The Next Generation of DRA Composites for Improved Toughness. Additional improvements to the strength/toughness combinations of DRA materials may be possible via improvements in both primary and secondary processing routes as well as judicious selection of matrices, reinforcement, and heat treatments (Ref 53, 56). Recent works (Ref 6, 44, 53, 57-72) have shown that extrinsic toughening approaches such as lamination, microstructural toughening, and ductile-phase toughening may increase toughness without sacrificing the strength/toughness combination or the enhanced stiffness. Recent reviews (Ref 44, 53, 56) have summarized some of the potential benefits and processing routes for such approaches.

Fatigue Behavior

Discontinuously reinforced aluminum composites have complex fatigue behavior (Ref 73),

the effects of matrix strength and ductility on the crack growth resistance are shown in the model Al-SiC materials, where aging to the T6 condition eliminates any tearing resistance compared to the T4 condition.

which can be improved or degraded by the incorporation of reinforcement, depending on the loading mode and the type of fatigue test (*S-N*, *ε-N*, or crack growth). The general fatigue process is also complex. In composite materials, as in many unreinforced metals, factors that improve resistance to the initiation of a crack lead to a decrease in the resistance to the growth of a crack. An example of this is the influence of reinforcement on the fatigue behavior of aluminum alloys. In underaged aluminum alloys, reinforcement leads to an improvement in stress-controlled fatigue behavior but produces an increase in fatigue crack propagation rates. In contrast, in overaged aluminum alloys, reinforcement improves not only the stress-controlled fatigue behavior but also the resistance to fatigue crack propagation. Under stress-controlled high-cycle fatigue conditions, a high volume fraction of fine reinforcement particles is preferred for fatigue crack initiation resistance. For fatigue crack growth resistance, a high volume fraction of coarser particles is preferred.

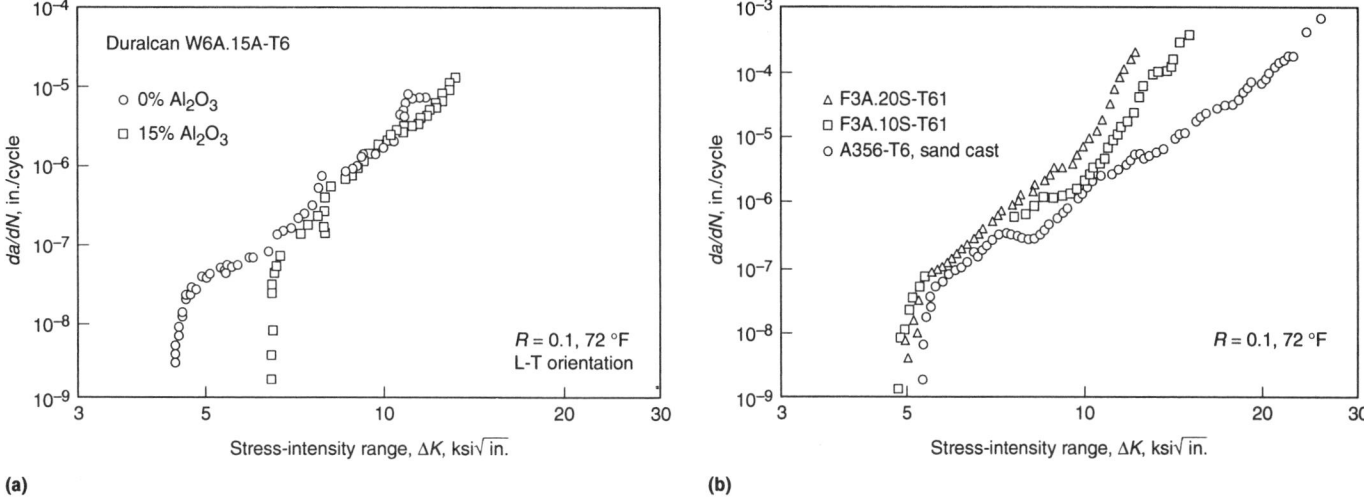

Fig. 36 Effect of reinforcement on fatigue crack growth. (a) Effect on threshold. (b) Effect on crack growth rates. Source: Ref 75

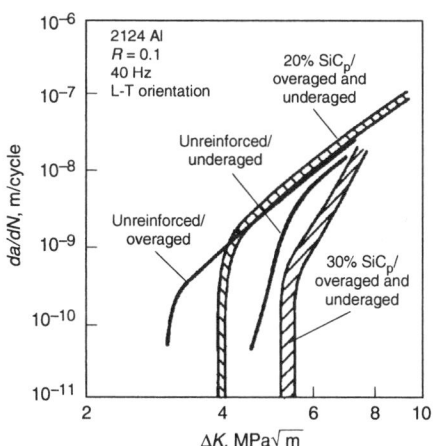

Fig. 37 Effect of volume fraction on fatigue crack growth rates in 2124 DRA composites. Source: Ref 74

Stress-Life Behavior. The high-cycle *S-N* fatigue life of discontinuously reinforced metals is generally higher than that of comparable unreinforced alloys (Fig. 32). This behavior can be understood by the influence of reinforcement on modulus and work hardening. In general, the incorporation of reinforcement leads to an increase in work hardening rate and a decrease in ductility under both monotonic and cyclic loading. Thus, the high-cycle fatigue strength (at lower or intermediate stresses) is generally improved by reinforcement, although there are exceptions (Fig. 33). Notch effects are shown in Fig. 34.

In low-cycle (high-stress) testing, the fatigue strength converges with that of unreinforced counterparts. In this regime, fatigue is ductility controlled, and therefore the lower tensile and cyclic ductilities of reinforced metals become a more dominant factor.

Strain-Life Behavior. In contrast to stress-life, the total strain-life behavior of a discontinuously reinforced matrix is almost without exception lower than that of an unreinforced matrix (Fig. 35) (Ref 73). This behavior is again understood

in terms of the higher modulus of reinforced metal. Under strain control, the elastic strains are lower in the composite than in an unreinforced matrix, and the increased portion of plastic strain accelerates the formation of fatigue damage (Ref 73).

Fatigue Crack Propagation. At low *R* values (under about 0.3), the introduction of a reinforcement phase results in higher thresholds (Fig. 36a), a steeper Paris region (Fig. 36b), and a transition to unstable fracture at lower ΔK levels. The latter behavior is characteristic of all discontinuously reinforced metal-matrix composites and is a direct consequence of the lower fracture toughness in composites compared to unreinforced alloys (Ref 73). However, the influence of reinforcement on the near-threshold and midrange growth behavior can be more complicated than this. In some cases, threshold levels, as well as near-threshold and midrange growth rates in composites, may be higher than, lower than, or roughly equivalent to the values observed in unreinforced alloys (Fig. 37). Although only limited data are available on the effects of volume fraction, it appears that the effect on crack growth is consistent with roughness-induced closure (Ref 73).

ACKNOWLEDGMENTS

The authors acknowledge ARO-DAALO3-89-K-0068, NSF-PYI-DMR-8958326, ONR-N00014-91-J-1370, ALCOA, and ALCAN for partial support of some of the work reviewed presently, as well as supply of materials and interactions on various aspects of the work.

REFERENCES

1. B. Paul, *Trans. Metall. Soc. AIME*, Vol 218, Feb 1960, p 36
2. A.L. Geiger and J.A. Walker, *JOM*, Vol 43, 1991, p 8-15
3. I. Dutta, C.F. Tiedemann, and T.R. McNelley, *Scripta Met.*, Vol 24, 1990, p 1233-1238
4. T.M. Osman, J.J. Lewandowski, and W.H. Hunt, Jr., *Fabrication of Particulate Reinforced Metal Composites*, J. Masenauve, Ed., ASM International, 1990, p 209
5. G.A. Rozak, A.A. Altmisolgu, J.J. Lewandowski, and J.F. Wallace, *J. Comp. Mater.*, Vol 26, 1992, p 2076-2106
6. J.J. Lewandowski, C. Liu, and W.H. Hunt, *Mater. Sci. Eng.*, Vol A107, 1989, p 142-155
7. P.M. Singh and J.J. Lewandowski, *Scripta Metall. Mater.*, Vol 29, 1993, p 199-204
8. M. Manoharan and J.J. Lewandowski, *Acta Metall. Mater.*, Vol 38, 1990, p 489-496
9. W.H. Hunt, Jr., J.R. Brockenbrough, and P.E. Magnusen, *Scripta Metall. Mater.*, Vol 25, 1991, p 15-20
10. D.J. Lloyd, *Acta Metall.*, Vol 39 (No. 1), 1991, p 59-71
11. R.J. Arsenault, N. Shi, C.R. Feng, and L. Wang, *Mater. Sci. Eng.*, Vol A131, 1991, p 59-68
12. P.M. Singh and J.J. Lewandowski, *Met. Trans.*, Vol 24A, 1993, p 2531-2543
13. P.M. Singh and J.J. Lewandowski, *Intrinsic and Extrinsic Fracture Mechanisms in Inorganic Composites*, J.J. Lewandowski and W. Hunt, Jr., Ed., TMS, 1995, p 57-68
14. A. Ashton, J.C. Halpin, and P.H. Petit, *Primer on Composite Materials*, Technomic Publishing, Westport, CT, 1969
15. H.M. Ledbetter and S.K. Dutta, *J. Acoustical Soc. America*, Vol 79, 1986, p 239-248
16. G.P. Tandon and G.J. Weng, *Polymer Comp.*, Vol 5, 1984, p 327-333
17. G.P. Tandon and G.J. Weng, *Comp. Sci. Technol.*, Vol 27, 1986, p 111-132
18. G. Bao, J.W. Hutchinson, and R.M. McMeeking, *Acta Metall.*, Vol 39, 1991, p 1871-1882
19. M. Mori and K. Tanaka, *Acta Metall.*, Vol 21, 1973, p 571-574
20. M. Taya and R. Arsenault, *Metal Matrix Composites: Thermomechanical Behavior*, Pergamon Press, 1989
21. H. Banno, *Ceram. Bull.*, Vol 66 (No. 9), 1987

22. Z. Hashin and S. Shtrikmen, *J. Mech. Phys. Solids*, Vol 11 (No. 2), 1963, p 127
23. L.J. Ebert and P.K. Wright, *Composite Materials—Interfaces in Metal Matrix Composites*, A.G. Metcalfe, Ed., Academic Press, 1974, p 31-64
24. W.S. Johnson, S.J. Lubowinski, and A.L. Highsmith, *Thermal and Mechanical Behavior of Metal Matrix and Ceramic Matrix Composites*, STP 1080, J.M. Kennedy, H.H. Moeller, and W.S. Johnson, Ed., American Society for Testing and Materials, 1990, p 193-218
25. J.J. Lewandowski, D.S. Liu, and C. Liu, *Scripta Metall.*, Vol 25, 1991, p 21-26
26. A.L. Grow and J.J. Lewandowski, Paper 950260, SAE, 1995
27. D.S. Liu and J.J. Lewandowski, *Met. Trans.*, Vol 24A, 1993, p 609-615
28. D.S. Liu, M. Manoharan, and J.J. Lewandowski, *Met. Trans.*, Vol 20A, 1989, p 2409-2417
29. Y. Brechet, J.D. Embury, S. Tao, and L. Luo, *Acta Metall.*, Vol 39, 1991, p 1781-1786
30. J. Llorca, A. Martin, J. Riuz, and M. Elices, *Met. Trans.*, Vol 24A, 1993, p 1575-1588
31. J.J. Lewandowski, C. Liu, and W.H. Hunt, Jr., *Processes and Properties for Powder Metallurgy Composites*, P. Kumar, K. Vedula, and A. Ritter, Ed., The Metallurgical Society, 1988, p 117-131
32. S.I. Hong and G.T. Gray, *Acta Metall.*, Vol 38, 1990, p 1581
33. P.M. Singh and J.J. Lewandowski, *Met. Trans.*, Vol 26A, 1995, p 2911-2921
34. S.F. Corbin and D.S. Wilkinson, *Acta Metall. Mater.*, Vol 42 (No. 4), 1994, p 1329-1335
35. D.S. Liu, M. Manoharan, and J.J. Lewandowski, *Scripta Metall.*, Vol 23, 1989, p 253-256
36. A.K. Vasudevan, O. Richmond, F. Zok, and J.D. Embury, *Mater. Sci. Eng.*, Vol A107, 1989, p 63-69
37. D.S. Liu, M. Manoharan, and J.J. Lewandowski, *Met. Trans.*, Vol 20A, 1989, p 2409-2417
38. D.S. Liu, M. Manoharan, and and J.J. Lewandowski, *J. Mater. Sci. Lett.*, Vol 8, 1989, p 1447-1448
39. J.D. Embury, J. Newell, and S. Tao, *Proc. 12th Riso Symp. on Materials Science: Metal Matrix Composites—Processing, Microstructure and Properties*, Riso National Laboratory, Rosklide, Denmark, 1991
40. A. Vaidya and J.J. Lewandowski, *Intrinsic and Extrinsic Fracture Mechanisms in Inorganic Composites*, J.J. Lewandowski and W. Hunt, Jr., Ed., TMS, 1995, p 57-68
41. A. Korbel, V.S. Raghunathan, D. Teirlinck, W. Spitzig, O. Richmond, and J.D. Embury, *Acta Metall.*, 1984, p 511-519
42. I.E. French and P.F. Weinrich, *Scripta Metall.*, Vol 8, 1974, p 87
43. I.E. French and P.F. Weinrich, *Acta Metall.*, Vol 21, 1973, p 1533-1537
44. J.J. Lewandowski and P.M. Singh, *Intrinsic and Extrinsic Fracture Mechanisms in Inorganic Composite Systems*, J.J. Lewandowski and W.H. Hunt, Jr., Ed., TMS, 1995, p 129-147
45. G.T. Hahn and A.R. Rosenfield, *Met. Trans.*, Vol 6A, 1975, p 653
46. G.G. Garrett and J.F. Knott, *Met. Trans.*, Vol 9, 1978, p 1187
47. R. Develay, *Met. Mater.*, Vol 6, 1972, p 404
48. T.F. Klimowicz and K.S. Vecchio, *Fundamental Relationships between Microstructure and Mechanical Properties in Metal Matrix Composites*, M. Gungor and P.K. Liaw, Ed., TMS, 1990, p 255
49. T.M. Osman, J.J. Lewandowski, and W.H. Hunt, Jr., *Fabrication of Particulates Reinforced Metal Composites*, J. Masenauve and F.G. Hamel, Ed., ASM International, 1990, p 209
50. M. Manoharan and J.J. Lewandowski, unpublished research, Case Western Reserve University, 1988
51. A.L. Geiger and J.A. Walker, *JOM*, Vol 43 (No. 8), 1991, p 8-15
52. S.V. Nair, J.K. Tien, and R.C. Bates, *Int. Met. Rev.*, Vol 30, 1985, p 275
53. W.H. Hunt, T.M. Osman, and J.J. Lewandowski, *JOM*, Vol 45, 1993, p 30-35
54. J.R. Rice and M.A. Johnson, *Inelastic Behavior of Solids*, M.F. Kanninen, H. Lilholt, and O.B. Pederson, Ed., McGraw-Hill, 1969, p 641
55. P.C. Paris, H. Tada, A. Zahoor, and H. Ernst, STP 668, ASTM, 1979, p 5
56. W.H. Hunt, Jr., in *Processing and Fabrication of Advanced Materials III*, V.A. Ravi, T.S. Srivatsan, and J.J. Moore, Ed., TMS, p 663-683
57. J.J. Zhang and J.J. Lewandowski, *J. Mater. Sci.*, 1996, in press
58. M. Manoharan, L.Y. Ellis, and J.J. Lewandowski, *Scripta Metall.*, Vol 24, 1990, p 1515
59. L.Y. Ellis and J.J. Lewandowski, *J. Mater. Sci. Lett.*, Vol 10, 1991, p 461
60. L.Y. Ellis and J.J. Lewandowski, *Mater. Sci. Eng.*, Vol A183, 1994, p 59
61. T.M. Osman and J.J. Lewandowski, *Scripta Metall.*, Vol 31, 1994, p 191
62. L.Y. Ellis, M.S. thesis, Case Western Reserve University, 1992
63. T.M. Osman, M.S. thesis, Case Western Reserve University, 1993
64. C.K. Syn, D.R. Lesuer, and O.D. Sherby, *High Performance Ceramic and Metal Matrix Composites*, TMS, 1994, p 125
65. C.K. Syn, D.R. Lesuer, K.L. Caldwell, O.D. Sherby, and K.R. Brown, *Developments in Ceramic and Metal Matrix Composites*, K. Upadhya, Ed., TMS, 1992, p 311
66. T.M. Osman, J.J. Lewandowski, D.R. Lesuer, C.K. Syn, and W.H. Hunt, Jr., *Aluminum Alloys: Their Physical and Mechanical Properties*, Proc. ICAA-4, T. Sanders and E.A. Starke, Ed., Georgia Institute of Technology, 1994, p 706
67. W.H. Hunt, Jr., T.M. Osman, and J.J. Lewandowski, *JOM*, Vol 45, 1993, p 30
68. V.C. Nardone, J.R. Strife, and K. Prewo, *Met. Trans.*, Vol 22A, 1991, p 171
69. V.C. Nardone and J.R. Strife, *Met. Trans.*, Vol 22A, 1991, p 183
70. V.C. Nardone, *Intrinsic and Extrinsic Fracture Mechanisms in Particulate Reinforced Materials*, W.H. Hunt, Jr. and J.J. Lewandowski, Ed., TMS, 1995
71. T.M. Osman, J.J. Lewandowski, and W.H. Hunt, Jr., *Advances in P/M and Particulate Materials—1994*, C. Lall and A. Neupaver, Ed., MPIF, 1994, p 351
72. T.M. Osman, J.J. Lewandowski, and W.H. Hunt, Jr., *Met. Trans. A*, in press, 1996
73. J. Allison and J. Wayne Jones, Fatigue Behavior of Discontinuously Reinforced Metal-Matrix Composites, *Fundamentals of Metal Matrix Composites*, S. Suresh, A. Mortenson, and A. Needleman, Ed., Butterworth-Heinemann, 1993
74. J.J. Bonnen et al., *Met. Trans.*, Vol 22A, 1991, p 1007-1019
75. J. Klimowicz, "Duralcan Fatigue Tests," Duralcan USA, San Diego

Fatigue of Composite Laminates*

KNOWLEDGE of fatigue behavior at the laminate level is essential for understanding the fatigue life of a laminated composite structure. In fact, most published fatigue data have addressed laminates with or without a notch. Relatively little fatigue data exist on composite structures. Investigations have demonstrated that sufficient fatigue life for a composite structure is achieved if it is designed to satisfy static strength requirements (Ref 1, 2). Fatigue failure mechanisms, S-N relations and characterization, and statistical life prediction models of composite laminates are described in this article. Data on aramid-aluminum (ARALL) and titanium matrix composites are also included. More detailed coverage on polymer composites is contained in Ref 3.

Fatigue Failure

Composite materials exhibit very complex failure mechanisms under static and fatigue loading because of the anisotropic characteristics of their strength and stiffness. Fatigue failure is usually accompanied by extensive damage that multiplies throughout specimen volume, in contrast to the localized formation of a predominant, single crack, as is common in isotropic, brittle materials. The four basic failure mechanisms in composites are layer cracking, delamination, fiber breakage, and fiber-matrix interfacial debonding. Any combination of these can cause fatigue damage that will result in reduced strength and stiffness. Both the type and degree of damage vary widely, depending on material properties, laminations (including stacking sequence), and type of fatigue loading. It has been observed that, in general, damage development under fatigue loading is similar to that under static loading, except that fatigue at a given stress level causes additional damage as fatigue cycles increase.

Layer Cracking. In multidirectional laminates under in-plane loading, failure from layer cracking usually occurs in succession from the weakest layer to the strongest. For example, successive transverse cracks in the off-axis layers of a $(0°/90°/±45°)_s$ graphite-epoxy laminate subjected to uniaxial tension are expected to occur as the load increases. The first crack occurs in the

90° layers; with increasing load, more cracks develop, but they are still confined to the 90° layers. As the load increases further, cracks occur in the adjacent 45° layers, appearing at the tips of the 90° cracks in most cases, and extending to the interface of the +45°/–45° layers (see Fig. 1). Subsequently, the number of cracks increases with the load (Fig. 2) until final laminate failure takes place. However, some laminates reach a crack density limit, after which no new cracks occur before final failure, despite additional loading. The crack density limit for a given layer depends on its thickness and appears to be independent of laminate type (Ref 4-7).

During fatigue loading, more cracks occur in each layer, and reach a crack density limit, than during static loading. For instance, many cracks in the –45° layers of the $(0°/90°/±45°)_s$ laminate occur during fatigue loading, whereas few or none occur during static loading. In addition, many axial cracks initiate at the tips of the transverse cracks and extend along the axial direction as fatigue cycles increase. The multiplication process of transverse cracks in the course of the

fatigue cycle is shown in Fig. 3. The damage ratio is defined here as the ratio of the crack density at n cycles to the crack density at final failure. Cracks in all off-axis plies are grouped together. Most of the crack multiplication occurs during the first 20% of the fatigue life. A significant amount of fatigue life remains after reaching the crack density limit. Cracks in the 90° layers of the $(0°/90°/±45°)_s$ laminate are found at fewer than one million cycles at a fatigue stress level of 170 MPa (25 ksi), which is less than the 280 MPa (40 ksi) stress level at which the first layer fails under static loading.

The 0° layers are also susceptible to cracking in the fiber direction because of the transverse stress resulting from a mismatch in Poisson's ratio between different layers. Because this transverse stress is usually small, axial cracking in 0° layers may not occur under static loading. However, under fatigue loading, axial cracking can occur, for example, in cross-ply laminates such as $(0°/90°)_s$.

Delamination. In composite laminates, free-edge delamination under in-plane axial loading is

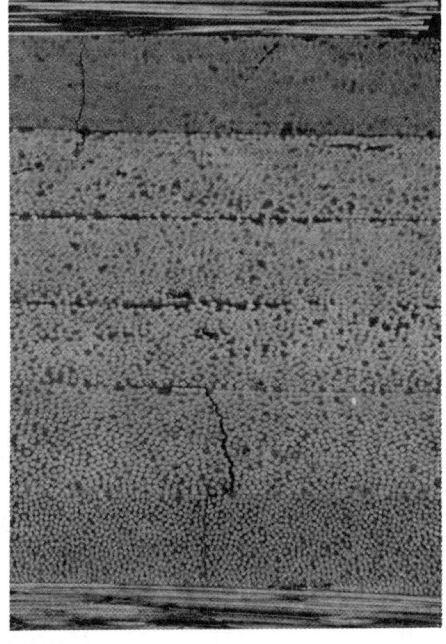 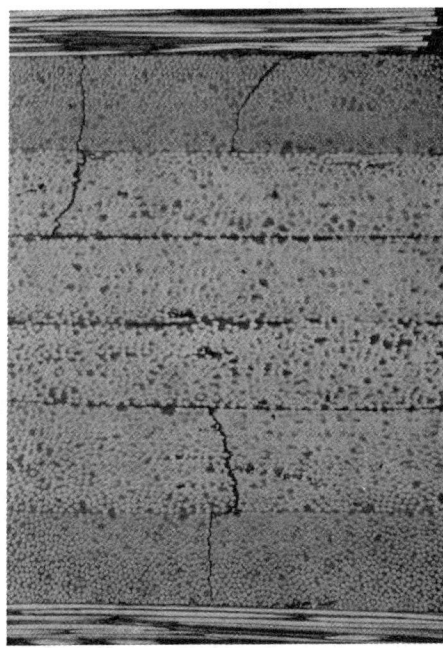

* Portions reprinted from the article by R.Y. Kim, "Fatigue Strength" in *Engineered Materials Handbook*, Volume 1, *Composites*, ASM International, p 436-440

Fig. 1 Photomicrographs showing crack patterns under static loading: $[0°/90°/±45°]_s$. (a) 440 MPa or 65 ksi. (b) 483 MPa or 70 ksi

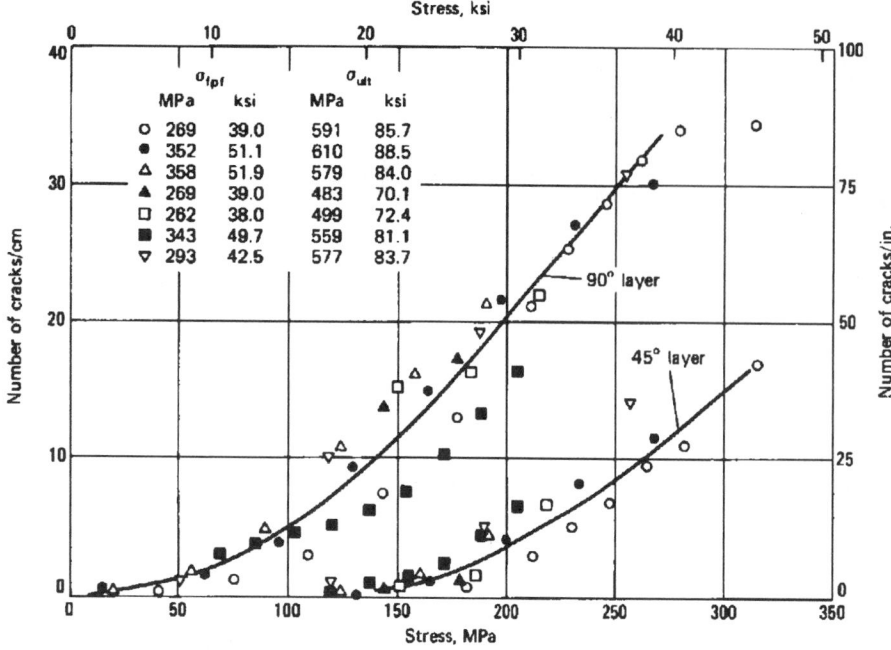

	σ_{fpf}		σ_{ult}	
	MPa	ksi	MPa	ksi
○	269	39.0	591	85.7
●	352	51.1	610	88.5
△	358	51.9	579	84.0
▲	269	39.0	483	70.1
□	262	38.0	499	72.4
■	343	49.7	559	81.1
▽	293	42.5	577	83.7

Fig. 2 Increase of number of cracks for [0°/90°/±45°]ₛ laminate subjected to static loading

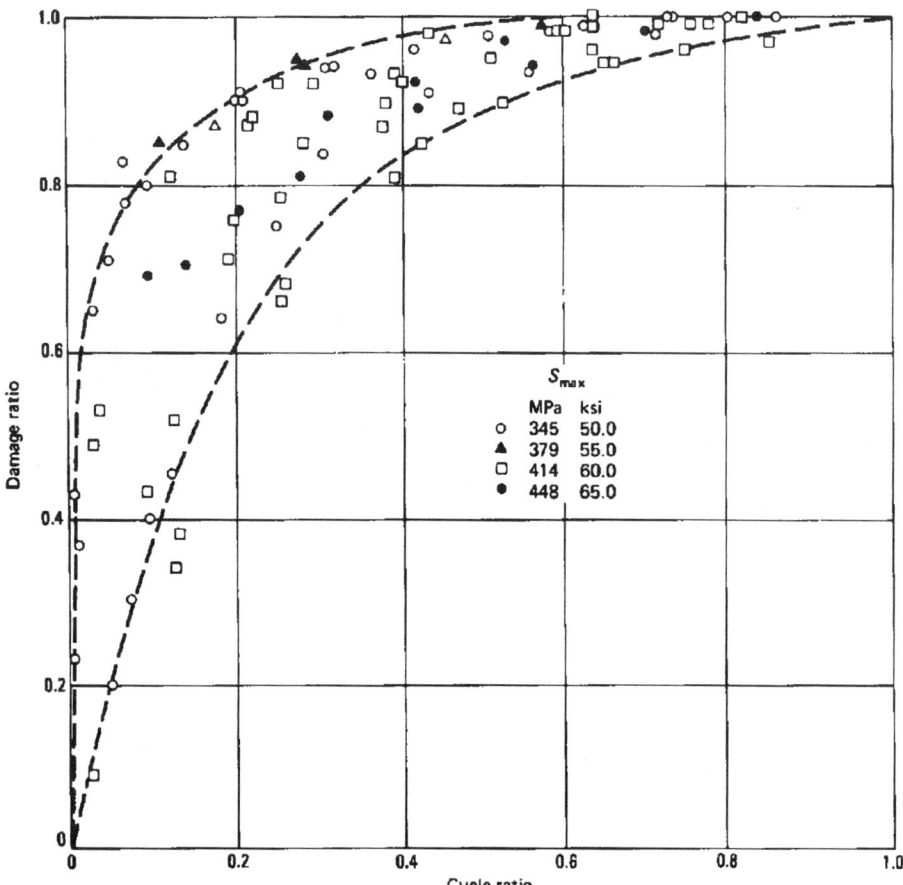

	S_{max}	
	MPa	ksi
○	345	50.0
▲	379	55.0
□	414	60.0
●	448	65.0

Fig. 3 Increase of number of cracks for [0°/90°/±45°]ₛ laminate subjected to fatigue loading

(a)

(b)

Fig. 4 Delamination in [0°/±45°/90°] graphite-epoxy subjected to static loading. (a) Micrograph of a free edge. 35×. (b) An x-ray of the width. 0.2×

caused by interlaminar stresses that are highly localized around the free edge (Ref 8). The nature of interlaminar stresses with regard to their magnitude and sign can be accurately calculated using an analytical model (Ref 9). The magnitude and distribution of interlaminar stress components vary widely depending on the type of laminate, stacking sequence, properties of the constituent materials, and type of loading. For a given laminate, the stacking sequence significantly influences free-edge delamination. As an example, consider a quasi-isotropic laminate with two stacking sequences, (0°/90°/±45°)ₛ and (0°/±45°/90°)ₛ. The (0°/90°/±45°)ₛ laminate does not show any delamination under static tension, whereas the latter laminate shows extensive delamination under static tension. Figure 4 shows a photomicrograph of a free edge and an x-ray radiograph of the specimen width showing delamination in the (0°/±45°/90°)ₛ laminate under static tension. At about 350 MPa (50 ksi) of the applied tension, delamination occurred and extended almost instantly over the entire length of the specimen, taking irregular paths at the interfaces of the 90°/90°, 90°–45°, and 45°/–45° plies. The change in the delamination path happened for the most part at the transverse crack tips. The delamination also grew continuously toward the middle of the specimen from both free edges as the load increased (Fig. 4). In fatigue, the onset of delamination occurred very early (and at fatigue stress levels that were smaller than static stress levels) and propagated rapidly toward the middle of the specimen width as the cycles increased. The lower bound of the delamination-free stress level for a predetermined life cycle is expected to vary, depending upon the type of laminate. In addition to interlaminar tensile stress, other fac-

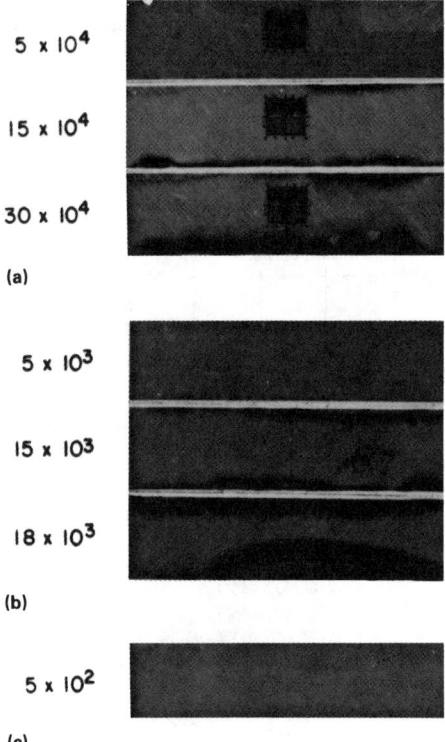

between the ±45° and −45° plies. The interlaminar shear stress at the +45°/−45° interface is not large enough to reach the interlaminar shear strength under static loading, but under fatigue loading, the shear stress becomes significant because of the high fatigue sensitivity of the epoxy matrix. Figure 5 shows delamination growth as a function of fatigue cycle for a (0°/90°/±45°)ₛ graphite-epoxy T300/5208 laminate. However, the growth of delamination in this case is much slower than is the case with interlaminar tension.

Fiber break and interface debonding differ widely, depending on the properties of the constituent materials and fiber defects. In most advanced composites, such as boron and graphite fiber with polymer matrix, the strain-to-failure is greater in the matrix than in the fibers. Therefore, fibers can break, because of defects or weakness, before the matrix fails. The crack created by a fiber break grows into the matrix, as load increases, along a path varying with matrix and interface properties. If bonding is strong, the crack grows into the matrix, resulting in a fairly smooth fracture surface across the section. With a weak bond, the crack is more likely to lead to interfacial debonding and extensive fiber pull-out. An intermediate bond shows irregular failure surfaces with some fiber pull-out. These failure mechanisms occur under static and fatigue loading, although fatigue failure depends on the sensitivity of the matrix, interface, and fiber. Because in most advanced composites the matrix remains elastic until the composite fails, fatigue damage in the interface is negligible, except at the site of fiber breaks. Consequently, the modulus and strength of a unidirectional laminate subjected to fatigue loading along the fiber direction do not decrease until fracture is imminent.

S-N Relation

The fatigue behavior of a material is basically expressed in the form of a relation between applied maximum fatigue stress (S) and fatigue life (number of cycles to failure, N). In composite laminates, the S-N relation primarily depends on the constituent material properties (Ref 11, 12). Most advanced fibers are very insensitive to fatigue, and the resulting composites show good fatigue resistance. Figure 6(a) shows the S-N curves for a variety of laminates of AS 4/3502 graphite-epoxy. The ordinate represents the fatigue strength ratio, which is the ratio of fatigue stress to static strength. The laminates with matrix-dominant failure modes exhibit a smaller fatigue resistance than the laminates with fiber-dominant failure modes. The slope of the S-N curve for the $[0°]_{8T}$ laminate is relatively flat because of the fatigue insensitivity of the graphite fiber. Consequently, the slope of the S-N curve tends to increase as the content of the 0° ply decreases, as shown in Fig. 7 (Ref 13). The fatigue strength of a $(±45°)_s$ laminate is frequently used to determine the longitudinal shear fatigue strength. The fatigue strength in compression-compression fatigue is slightly greater than that in tension-tension fatigue (Fig. 8) for a $(0°/45°/90°/−45°)_{2s}$ graphite-epoxy laminate. The static tensile and compressive strengths for this laminate are practically identical. The extensive damage incurred during tension-tension fatigue is mainly responsible for the reduction in tensile fatigue strength. The fatigue strength of multidirectional laminates is closely related to the 0° fatigue strength, as shown in Fig. 9, unless the laminate undergoes extensive damage during fatigue.

Mean Stress. A constant-amplitude fatigue loading may be regarded as a fully reversed cyclic loading superimposed on a constant load. Consequently, the fatigue strength at a chosen cycle may be expressed in terms of alternating the stress, S_r, and mean stress, S_m. The values of S_m and S_r are given by the following equations:

$$S_m = (S_{max} + S_{min})/2$$

Fig. 5 X-ray radiographs of delamination growth as a function of fatigue cycles for a $[0°/90°/±45°]_s$ carbon-fiber-reinforced plastic T300-5208 laminate. (a) $S_{max} = 345$ MPa (50.0 ksi). $N = 312{,}000$ cycles. (b) $S_{max} = 414$ MPa (60.0 ksi). $N = 19{,}400$ cycles. (c) $S_{max} = 483$ MPa (70.0 ksi). $N = 620$ cycles

tors, such as transverse cracking and interlaminar shear stress, appear to affect the onset and growth of delamination (Ref 10). The $(0°/90°/±45°)_s$ laminate, which has a compressive interlaminar normal stress, does not show any delamination under static tension, but under tensile fatigue, considerable delamination occurs at the interface

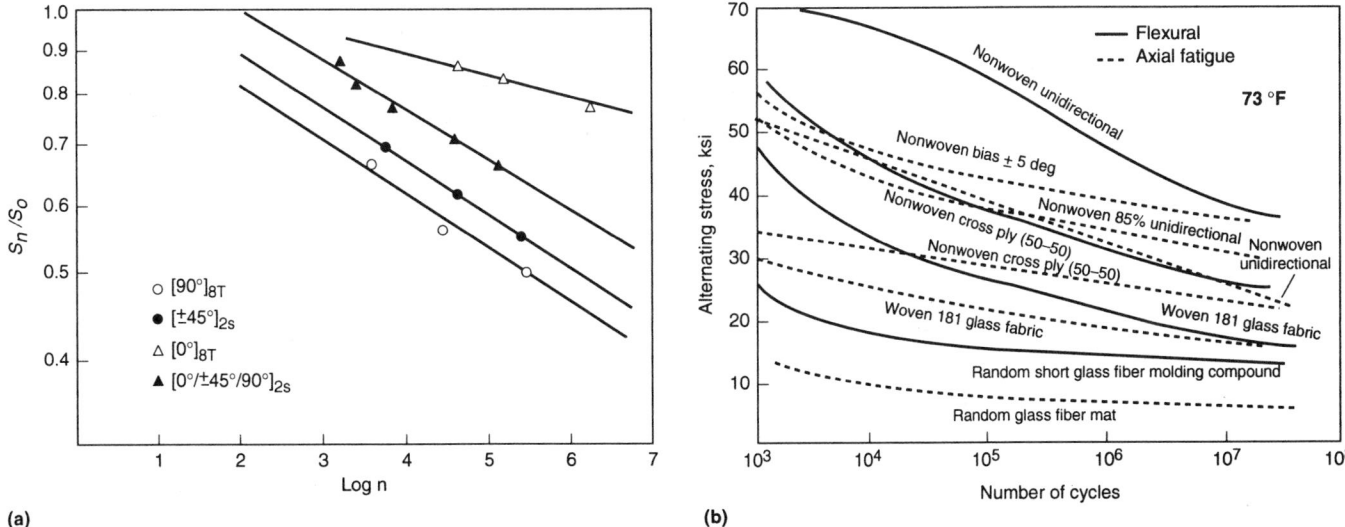

Fig. 6 S-N curves for (a) various AS4-epoxy laminates and (b) glass-fiber polymer laminates at various ply orientations. Source: (a) *Engineered Materials Handbook*, Vol 1, *Composites*, ASM International, 1987, p 438 and (b) C. Osgood, *Fatigue Design*, 2nd ed., 1982, Pergamon Press, p 530

$$S_r = (S_{max} - S_{min})/2$$

where S_{max} and S_{min} are the maximum and minimum fatigue stresses, respectively.

Figure 10 shows the effect of mean stress on fatigue life in the form of a constant-life diagram (Ref 14). This diagram is practically symmetrical with respect to the alternating stress axis, S_r ($R = -1$), although the peaks of constant-life curves tend to lie slightly in the tension-dominant fatigue region. This indicates that the S-N behavior of the composite is independent of the type of fatigue loading (that is, tension-tension, tension-

compression, or compression-compression) as long as the tensile and compressive strengths are equal. In tension-compression fatigue, the fatigue failure in the tension-dominant region (T > C) in Fig. 10 was accompanied by severe matrix damage. This effect on the S-N relation appears to be balanced out by the fact that the compressive stress is much more detrimental to the residual strength of damaged specimens than is the tensile stress. The peak of the constant-life curve for a $(0°/\pm30°)_s$ graphite-epoxy lies in the positive mean stress region because of the lower static compressive strength (Ref 15).

Notch. With metals, the problem of stress concentration resulting from various notches (circular holes and cracks) is complicated by the many

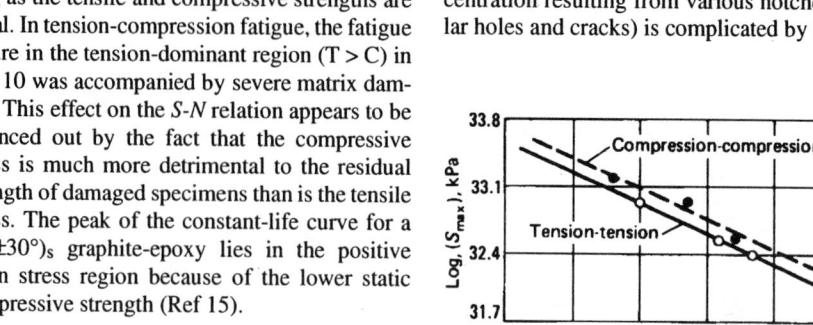

Fig. 8 Tension-tension and compression-compression fatigue tests of $[0°/45°/90°/-45°]_{2s}$ carbon-fiber-reinforced plastic T300/5208

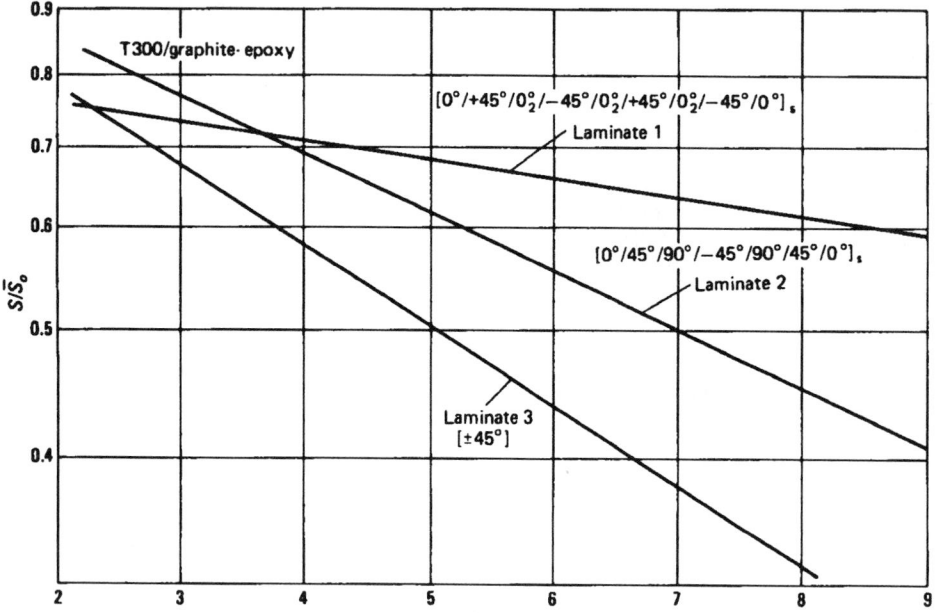

Fig. 7 Effect of 0° layer fraction on S-N relation

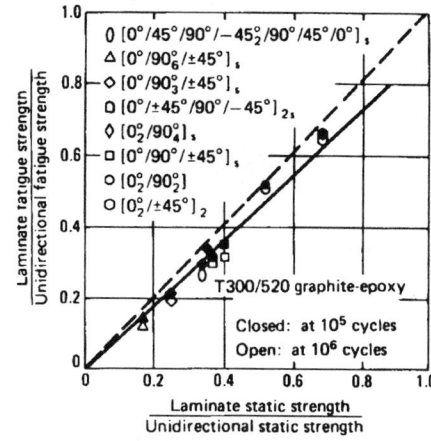

Fig. 9 Fatigue and static strengths normalized with respect to unidirectional tensile strengths

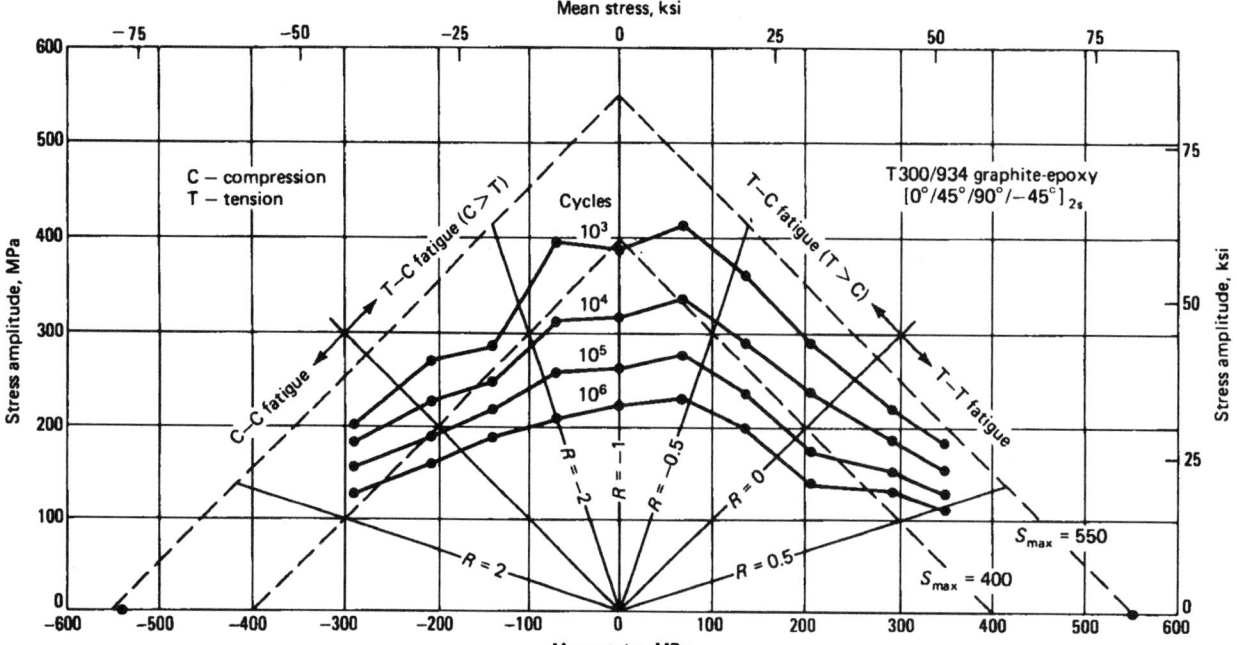

Fig. 10 Effect of mean stress on fatigue life of $[0°/45°/90°/-45°]_{2s}$ carbon-fiber-reinforced plastic T300/5208

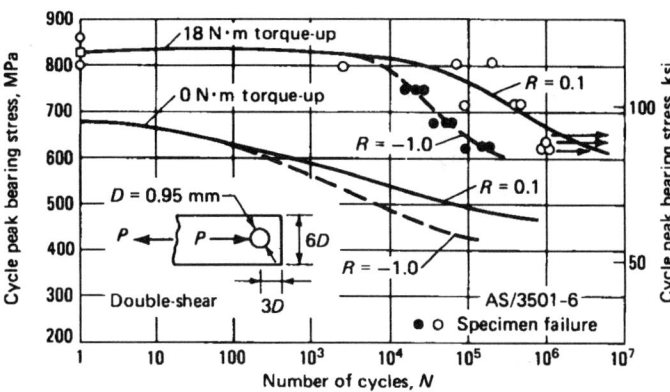

Fig. 11 Effect of torque-up on bolted joint fatigue life

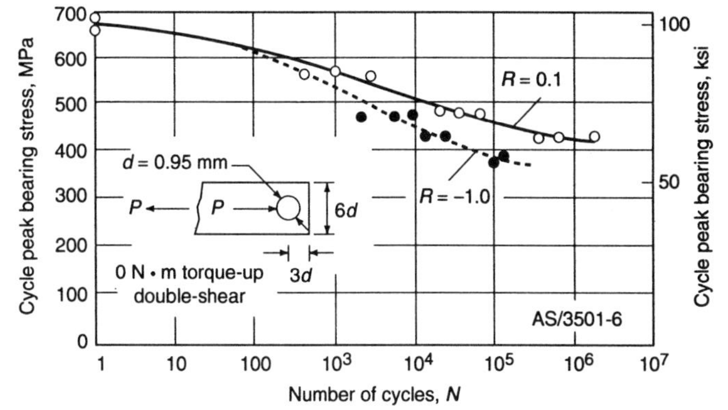

Fig. 12 Effect of R ratio on fatigue life

factors that influence behavior under cyclic stresses. In notched specimens, combined stresses are produced and vary widely in the region around the notch. Furthermore, any damage that occurs during fatigue will change the stress distribution around the notch. The foregoing factors have made it difficult to predict the fatigue strengths of notched specimens from the fatigue strengths of the unnotched specimens. In contrast to metals, the reduction in fatigue strength resulting from the presence of a notch is found to be insignificant in most composite laminates (Ref 14, 16, 17). The value of fatigue notch factor (the ratio of the fatigue strength of an unnotched specimen to the fatigue strength of a notched specimen at N cycles) is much smaller than the static stress-concentration factor, and is close to unity in many cases. The good fatigue resistance of notched specimens is mainly due to the fact that damage relaxes the stress concentration around the notch.

Bolted Joint Fatigue. Fatigue behavior of a bolted joint is similar to the fatigue behavior of notched composites. The residual strength generally increases with the fatigue cycle toward the unnotched net section strength (Ref 18). This strength increase is mainly due to damage that occurs around fasteners, reducing the stress concentration. The residual strength of a bolted joint depends on the degree of fitness of the bolt. When a loose bolt in a hole is stressed back and forth, the hole elongates at an increasing rate, and failure is rapid. This is why it is important that all bolts in composites be fit tightly. A tight fit also promotes uniform sharing of load between multiple fasteners.

Bolt torque pressure has a significant effect on both static strength and fatigue strength (Ref 18, 19). Each strength is improved considerably, compared to the pin-bearing case, which has no torque pressure, as shown in Fig. 11 (Ref 18). High torque virtually eliminates hole elongation that is due to fatigue and also results in a smaller temperature increase. With loose-fit bolts, a large increase in bolt temperature due to friction at high frequency is one of the major problems. This is associated with a decrease in mechanical properties because of temperature elevation.

Fatigue strength of a bolted joint is also influenced by many other factors, such as type of joint

(single or double), geometry of fastener head, interference fit, and so forth. Figure 12 shows the typical S-N relation of a bolted joint of graphite-epoxy laminate of $[0°_5/\pm45°_2/90°]_s$ orientation with two different fatigue loadings. The fatigue life is indicated in terms of fatigue cycles required to produce a 0.50 mm (0.02 in.) elongation of the fastener hole. Tension-compression fatigue ($R = -1$) reduces fatigue strength of the bolted joint for most laminate types. The stacking sequence also influences fatigue strength.

Helicopter Rotor Blade Fatigue. The helicopter rotor blade is the first large-volume use of a composite material as a primary airframe structure. During its service life, the helicopter experiences a severe high-cycle fatigue environment, typically on the order of 10^7 to 10^8 cycles. The blades show excellent performance with unlimited fatigue life when the composite materials are properly used in rotor blade design (Ref 20-22). The results of tests performed on the tail and main rotor blades of the Westland Sea King helicopter are reported in Ref 20. Blade construction is typically of a hollow D-spar composed largely of unidirectional glass-reinforced epoxy tape with internal and external woven wraps. A bolted joint is used at the root end with local reinforcement of the spar. The fatigue testing involved both structural part and complete (full-sized) structure tests. The structural part fatigue tests were performed for the four critical areas of the blade chosen for the test under a programmed series of loading blocks designed to correspond closely to actual flight conditions. The early design of tail rotor blade root end specimens indicated a weakness in the blade spar. Design and manufacturing changes were made at the root end to provide a new standard blade. After the fatigue testing at a fully factored, programmed load of two root end specimens, no failure occurred and unlimited fatigue life was shown.

Fatigue Data Analysis and Life Prediction

Weibull Distribution and Parameters Estimation. A random variable, X, has a Weibull distribution with shape parameter α and scale

parameter β if its cumulative distribution function, $F_X(x)$, is:

$$F_X(x) = P_r\{X \le x\} = 1 - \exp\left[-\left(\frac{X}{\beta}\right)^\alpha\right] \quad \text{(Eq 1)}$$

where $X > 0$ and $F_X(x)$ represents the probability of the random number X being less than or equal to x-experimental. On the other hand, the distribution for the n-ordered data ($x_1 \le x_2, ..., \le X_n$) is given by the medium rank:

$$F_X(x_i) = 1 - \frac{i - 0.3}{n + 0.4} \quad \text{(Eq 2)}$$

The following relation exists:

$$u = \text{expected value of } X = \beta \cdot \Gamma\left(1 + \frac{1}{\alpha}\right)$$

$$\alpha^2 = \text{variance of } X = \beta^2\left\{\Gamma\left(1 + \frac{2}{\alpha}\right) - \Gamma^2\left(1 + \frac{1}{\alpha}\right)\right\} \text{(Eq 3)}$$

where Γ is the gamma function.

A maximum-likelihood method is widely used for estimating Weibull parameters α and β:

$$\frac{\sum\limits_{i=1}^{n} x_i^{\hat{\alpha}} \ln X_i}{\sum\limits_{i=1}^{n} x_i^{\hat{\alpha}}} - \frac{1}{\hat{\alpha}} - \frac{\sum\limits_{i=1}^{n} \ln x_i}{n} = 0 \quad \text{(Eq 4)}$$

and

$$\hat{\beta} = \left[\frac{1}{n}\sum\limits_{i=1}^{n} x_i^{\hat{\alpha}}\right]^{1/\hat{\alpha}} \quad \text{(Eq 5)}$$

where $\hat{\alpha}$ is the Weibull shape parameter estimated from fatigue data.

An efficient iterative technique for obtaining the solution of $f(\hat{\alpha}) = 0$ is to use the Newton-Raphson method, in which the $(j + 1)$st successive approximation, $\hat{\alpha}_{j+1}$ to $\hat{\alpha}_j$, is given by:

Table 1 Percentage points l_γ^* such that P_r $\{\hat{\alpha} \ln (\hat{\beta}/\beta) < l_\gamma^*\} = \gamma$

$n\backslash\gamma$	0.90	0.95	0.98
5	0.772	1.107	1.582
6	0.666	0.939	1.291
7	0.598	0.829	1.120
8	0.547	0.751	1.003
9	0.507	0.691	0.917
10	0.475	0.644	0.851
12	0.425	0.572	0.752
14	0.389	0.520	0.681
16	0.360	0.480	0.627
18	0.338	0.447	0.584
20	0.318	0.421	0.549
22	0.302	0.398	0.519
24	0.288	0.379	0.494
26	0.276	0.362	0.472
28	0.265	0.347	0.453
30	0.256	0.334	0.435
40	0.222	0.285	0.371
50	0.195	0.253	0.328

Source: Ref 18

Table 2 Percentage points, l_γ, such that $P_r(\hat{\alpha} \ln (\hat{\beta}/\beta) < l_\gamma^*) = \gamma$

m	$\gamma\backslash n$	5	6	7	8	9	10
3	0.90	0.655	0.578	0.533	0.488	0.453	0.420
3	0.95	0.875	0.768	0.701	0.639	0.589	0.548
3	0.98	1.116	0.997	0.905	0.816	0.751	0.710
4	0.90	0.656	0.571	0.521	0.480	0.445	0.422
4	0.95	0.849	0.746	0.679	0.628	0.581	0.555
4	0.98	1.094	0.955	0.858	0.781	0.736	0.700
5	0.90	0.648	0.570	0.513	0.466	0.436	0.416
5	0.95	0.836	0.737	0.660	0.606	0.569	0.537
5	0.98	1.063	0.940	0.845	0.764	0.710	0.671

Note: l_γ obtained from Ref 18

$$\hat{\alpha}_{j+1} = \hat{\alpha}_j - f(\hat{\alpha}_j)/f'(\hat{\alpha}_j) \tag{Eq 6}$$

The maximum-likelihood method has the advantage in that the confidence intervals for α and β can be computed and it can be applied to a life-test model in which censoring is progressive, that is, a model in which a portion of the survivors is withdrawn from life test several times during the test.

The 95% confidence interval (one-sided) for β_i can be obtained from Table 1 (Ref 23) as follows:

$$P_r\{\hat{\alpha}\ln(\hat{\beta}_i/\beta_i) \le l_{0.95}^*\} = 0.95 \tag{Eq 7}$$

Let us denote the lower bound of 95% confidence interval for β_i by $\underline{\beta}_i$. Then:

$$\underline{\beta}_i = \hat{\beta}_i \exp\left[-(l_{0.95}^*/\hat{\alpha})\right] \tag{Eq 8}$$

An A-allowable (or B-allowable) material fatigue design value is defined by the probabilistic statement that we can be 95% confident of the assertion that the probability of surviving the A-allowable value is at least 0.99 (or 0.90 for B-allowable). Therefore, A-allowable X_{Ai} under the stress level S_i, where $i = 1, 2, ..., m$, can be obtained by solving the following:

$$R(x_{Ai}) = \exp\left[-\left(\frac{x_{Ai}}{\underline{\beta}_i}\right)^{\hat{\alpha}}\right] = 0.99 \tag{Eq 9}$$

where $\underline{\beta}_i$ is the 95% confidence lower limit for β_i, that is:

$$x_{Ai} = \underline{\beta}_i[-\ln(0.99)]^{1/\hat{\alpha}} \tag{Eq 10}$$

Pooling Technique. To obtain the S-N relationship, m levels of stress are employed with n specimens tested at each level. The shape parameter and scale parameter (characteristic fatigue life) can be obtained from the pooled fatigue data $x_{i1}, x_{i2}, ..., x_{in}$ (under its level

of stress), where $i = 1, 2, ..., m$. By assuming that the shape parameter α for each stress level is equal to that for every other level, the pooled estimation of α can be obtained by normalizing the data, that is, by letting $Y_{ij} = X_{ij}/\beta_i$, for $i = 1, 2, ..., n$. The maximum-likelihood equations for the normalized data are:

$$\frac{\sum\limits_{i=1}^{m}\sum\limits_{j=1}^{n} y_{ij}^{\hat{\alpha}} \ln Y_{ij}}{\sum\limits_{i=1}^{m}\sum\limits_{j=1}^{n} y_{ij}^{\hat{\alpha}}} - \frac{1}{\hat{\alpha}} - \frac{\sum\limits_{i=1}^{m}\sum\limits_{j=1}^{n} \ln Y_{ij}}{n \cdot m} = 0 \tag{Eq 11}$$

and

$$\hat{\beta}_i = \left[\frac{1}{n}\sum\limits_{j=1}^{n} X_{ij}^{\hat{\alpha}}\right]^{1/\hat{\alpha}} \quad \text{where } i = 1, 2, ..., m \tag{Eq 12}$$

This pooled estimate is more efficient than the one introduced in Eq 4 and 5. As before, the solutions of Eq 11 and 12 can be obtained by the iterative technique of the Newton-Raphson method. Table 2 (Ref 24) is useful in computing the confidence interval of β.

S-N Curve Characterization. Traditionally, the fatigue behavior of materials is expressed by the relationship between fatigue stress and fatigue life, that is, cycle to failure. Although no universally accepted rules are available for S-N characterization, the following discussion may be used as a general guide. Constant-amplitude fatigue tests are normally conducted under three to five different stress levels. The values of the extreme stress levels are approximately chosen so that the fatigue cycles under the extreme stress levels range between 10^3 and 10^6 cycles. The values of other intermediate stress levels are arbitrarily chosen. The number of specimens to be tested at each stress level depends on the size of scatter in the data and the availability of specimens and test time, but approximately four to ten specimens are typical for composite laminates. Large variation in fatigue life data is common in composites, and hence an S-N curve obtained from mean lifetimes only may be in error. In view of the foregoing variability in test results, a procedure developed for a more accurate evaluation of fatigue strength is discussed below.

Let us assume that the fatigue data follows a classical power law and a two-parameter Weibull distribution. The S-N curve takes the form:

$$KS^bN = 1 \tag{Eq 13}$$

where K and b are parameters. The K and b can be estimated using the least squares linear regression, because Eq 13 becomes a straight line after logarithmic transformation. The fatigue life, N_i, for each stress level, S_i, can be replaced by Weibull fatigue characteristic life, $\hat{\beta}_i$. Taking natural logarithms, we obtain:

$$\ln \hat{\beta}_i = -b \ln S_i - \ln K \tag{Eq 14}$$

The values of b and K can be easily determined by applying the least squares linear regression analysis.

Fatigue Life Prediction. Because composite laminates exhibit very complex failure processes, no one analytical model has been able to account for all possible processes. Consequently, statistical life prediction methodologies have frequently been adopted. A strength degradation model that is capable of predicting the statistical distribution of both fatigue life and residual strength is discussed in Ref 25 and 26.

ARALL and GLARE Fiber-Metal Laminates

Glenn Nordmark, Alcoa Technical Center (retired)

Fiber-metal laminates are a relatively new material developed for the aerospace industry. They offer a combination of improved strength and fatigue properties together with reduced density, giving airframe manufacturers the opportunity to achieve 20 to 30% weight savings. This section describes how fiber-metal laminates are produced, discusses their various product forms, re-

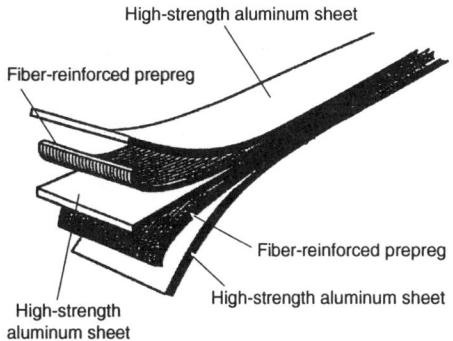

Fig. 13 Fiber-metal laminates structure, a standard 3/2 lay-up: three layers of aluminum, two layers of prepreg. Source: Ref 27

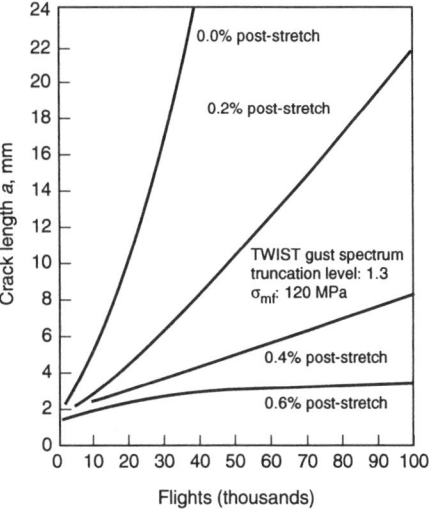

Fig. 14 The influence of post stretching on the fatigue properties of 3/2 ARALL-2 laminates under typical fuselage fatigue loading. Source: Ref 27

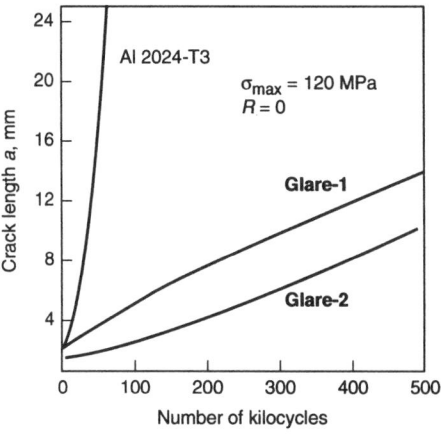

Fig. 15 Fatigue properties of unidirectional and biaxially reinforced GLARE laminates. Source: Ref 27

views the effects of stretching on their fatigue properties, and summarizes their residual strength and durability.

Fiber-metal laminates are a hybrid material composed of alternating layers of high-strength aluminum alloy sheet and fiber-reinforced epoxy adhesive (Fig. 13). The principal benefit of fiber-metal laminates is their ability to impede and self-arrest fatigue crack growth. In a monolithic aluminum sheet, a fatigue crack can grow until the panel fails. In fiber-metal laminates, a fatigue crack develops only in the aluminum layer or layers. The fibers remain intact in the wake of the crack due to their high strength and toughness. As the crack grows, the fibers bridging the crack carry an increasing portion of the load, decreasing the stress intensity at the crack tip and consequently arresting its growth. In addition to the superior fatigue crack growth properties, fiber metal laminates are stronger, lighter, and more damage tolerant than aluminum alloy 2024-T3 sheet.

Fiber-Metal Laminate Types

The properties of fiber-metal laminates can be modified by varying fiber-resin systems, aluminum alloy and sheet gage, stacking sequence, fiber orientation, surface preparation techniques, and the degree of post stretching. A variety of lay-up configurations are available for aramid fiber-metal laminates, ARALL, and glass fibermetal laminates, GLARE.* The lay-up configurations are represented by two numbers: the first number represents the layers of aluminum; the second represents the interspaced adhesive layers. Typical laminate configurations vary from 2/1 (two layers of aluminum alloy sheet with one adhesive layer sandwiched in between), to 5/4 (five layers of aluminum sheet with four adhesive layers interspaced), though laminates with more layers may be produced as required.

---

*ARALL and GLARE are registered trademarks of ALCOA and AKZO.

Before lay-up and cure, the aluminum unclad surfaces are anodized and primed to provide good bond integrity and to inhibit aluminum bondline corrosion in the event of moisture intrusion. Either chromic acid or phosphoric acid anodizing surface treatment for the aluminum layers may be used. For added corrosion protection, or polished sheet applications, cladding can be applied to one or both of the laminate outer surfaces.

During curing, due to the difference in thermal expansion coefficients between the aluminum layers and the fiber-adhesive layers, a small tensile residual stress will occur in the aluminum sheets and a corresponding compressive strength in the fiber layers. This is called the "as-cured" condition. However, the sign of the residual stresses can be reversed favorably by plastically stretching the laminate after curing. This is called "post stretching."

ARALL laminates are aramid fiber-metal laminates, that are unidirectionally reinforced with a 50% fiber volume adhesive prepreg of high-modulus aramid fibers. These laminates were developed for tension-dominated fatigue applications where the stress field is primarily unidirectional. The standard ARALL laminate product types are:

- ARALL-1 laminate
  Alloy 7475-T6
  120 °C (250 °F) cure prepreg
  0.4% permanent stretch
- ARALL-2 laminate
  Alloy 2024-T3
  120 °C (250 °F) cure prepreg
  With or without 0.4% stretch
- ARALL-3 laminate
  Alloy 7475-T76
  120 °C (250 °F) cure prepreg
  0.4% permanent stretch
- ARALL-4 laminate
  Alloy 2024-T8

175 °C (350 °F) cure prepreg
With or without 0.4% stretch

The influence of post stretching on the fatigue properties is shown in Fig. 14.

GLARE laminates are glass fiber-metal laminates and are a more recent development that complement ARALL laminates. GLARE laminates are unidirectional or biaxially reinforced with a 60% fiber volume adhesive prepreg of high-strength glass fibers. The high-strength glass fiber with improved tensile strength, compression properties, and adhesive properties relative to aramid fiber make it attractive as biaxial reinforcement for fiber-metal laminates. The higher tensile strength of the glass fiber allows for the laminate to be layed-up in cross-ply configurations and still offer improved tensile strengths in both the L and LT directions relative to ARALL. The improved compression properties of the glass fiber allow GLARE laminates to be used in the as-cured condition for tension-tension, tension-compression, and $R = 0$ fatigue loading conditions. The standard types of GLARE laminates are:

- GLARE-1
  7475-T6 aluminum alloy
  Unidirectional glass prepreg
  0.5% post-stretched
- GLARE-2
  2024-T3 aluminum alloy
  Unidirectional glass prepreg
- GLARE-3
  2024-T3 aluminum alloy
  Cross-ply glass prepreg
  50% of fibers in L direction
  50% of fibers in LT direction
- GLARE-4
  2024-T3 aluminum alloy
  Cross-ply glass prepreg
  30% of fibers in L direction
  70% of fibers in LT direction

Figure 15 shows the fatigue properties of both unidirectional and biaxially reinforced GLARE

Table 3 Typical mechanical properties of fiber-metal laminates

Mechanical properties	Glare 1		Glare 2		Glare 3		Glare 4		ARALL 2	ARALL 3	
	2/1	3/2	2/1	3/2	2/1	3/2	2/1	3/2	3/2	3/2	2024-T3
Tensile ultimate strength, MPa											
L	1077	1282	992	1214	662	717	843	1027	717	821	455
LT	436	352	331	317	653	716	554	607	317	379	448
Tensile yield strength, MPa											
L	525	545	347	360	315	305	321	352	365	607	359
LT	342	333	244	228	287	283	250	255	228	324	324
Tensile modulus, GPa											
L	66	65	67	65	60	58	60	57	66	68	72
LT	54	50	55	50	60	58	54	50	53	49	72
Ultimate strain, %											
L	4.2	4.2	4.7	4.7	4.7	4.7	4.7	4.7	2.5	2.2	19
LT	7.7	7.7	10.8	10.8	4.7	4.7	4.7	4.7	12.7	8.8	19
Compressive yield strength, MPa											
L	447	424	390	414	319	309	349	365	255	345	303
LT	427	403	253	236	318	306	299	285	234	365	345
Compressive modulus, GPa											
L	63	67	69	67	63	60	62	60	65	66	74
LT	56	51	56	52	62	60	57	54	53	50	74

laminates. Typical mechanical properties of AR-ALL and GLARE are given in Table 3.

Fatigue Properties

The fibers in ARALL and GLARE laminates provide extraordinary resistance to the development of fatigue cracks. In fact, conventional *S/N* fatigue data are rarely reported for unnotched ARALL and GLARE specimens. First, the stress required to develop any sort of fatigue failure is a high percentage of the strength of the aluminum sheet. Second, although microcracks will initiate at stresses of 275 MPa (40 ksi) and higher, the presence of the fibers restrains the crack opening so that the propagation slows and stops at short lengths. As the loading is continued, more and more microcracks develop and the specimen elongates until a saturation point is reached at which the fibers are carrying all of the load. Tests are generally terminated on the basis of decreased stiffness or increased elongation rather than a failure of the test section.

For structural use, the performance in the presence of stress concentrations or cracks is generally critical; accordingly, test programs have concentrated on obtaining data relevant to these structural concerns. Tests at Delft University and Alcoa Technical Center have evaluated fiber-metal laminates under various structural conditions and the influence of temperature, humidity, thermal cycling, outdoor exposure station monitoring, the effect of static and dynamic loading on corrosion behavior, and the influence of pitting corrosion on fatigue crack initiation (Ref 27-29). The durability of ARALL and GLARE laminates appears to be very good. An advantage for AR-ALL in no-load transfer joints has also been demonstrated for practical structures.

Tests at several laboratories showed that mildly notched ARALL-1 specimens had higher fatigue strengths than monolithic aluminum specimens (Fig. 16). The only two results that plot below the

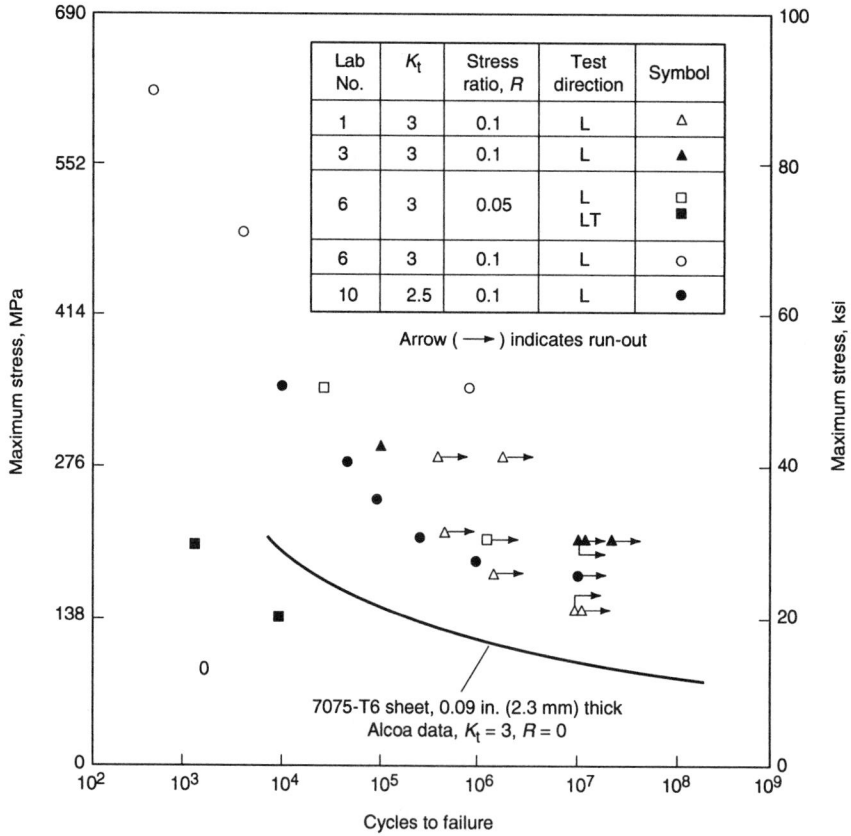

Fig. 16 Notch fatigue behavior of 3/2 ARALL-1 laminate, nominally 0.053 in. (1.3 mm) thick. Source: Ref 27

average curve for 7075-T6 sheet are specimens whose fibers were transverse to the direction of loading. Naturally, the fibers do not resist transverse loadings and only serve to lessen the thickness of aluminum resisting the loads in that direction. Beyond 10^6 cycles, the advantage for the longitudinal specimens was even greater than the plot indicates, because cracks had not developed in the test sections when their tests were terminated.

As demonstrated in Fig. 17, the fatigue crack growth rates obtained for center-cracked ARALL and GLARE specimens are 10 to 100 times slower than those determined for aluminum alloy sheet. The advantage comes primarily from the crack bridging of the fibers in the wake of the crack. The fibers are fully effective only if they are aligned within a few degrees of the loading direction. The drop-off as the angle between the fiber orientation and the direction of loading in-

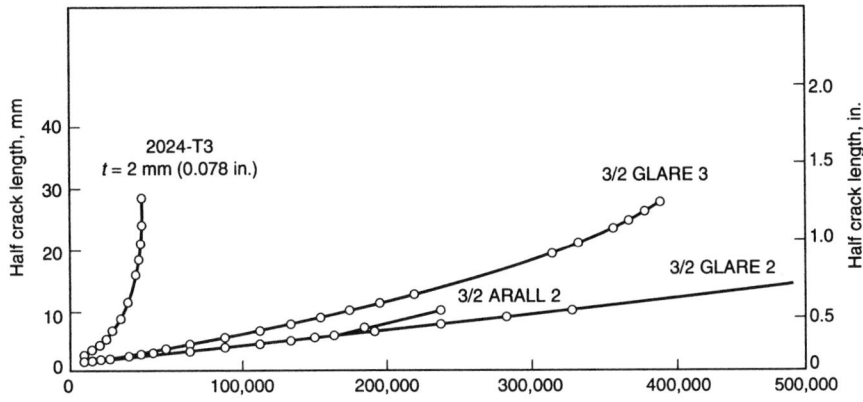

Fig. 17 Crack growth behavior of GLARE fiber-metal laminate and 2024-T3 aluminum alloy. S_{max} = 120 MPa (17.4 ksi). R = 0.05. Frequency = 10 Hz.

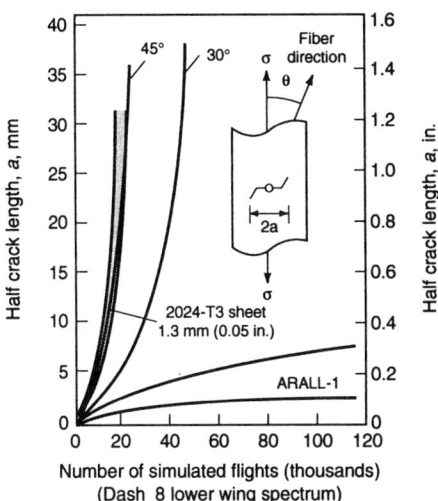

Fig. 18 Effect of fiber orientation on 3/2 ARALL-1 laminate (0.053 in.) fatigue crack growth. Source: Ref 27

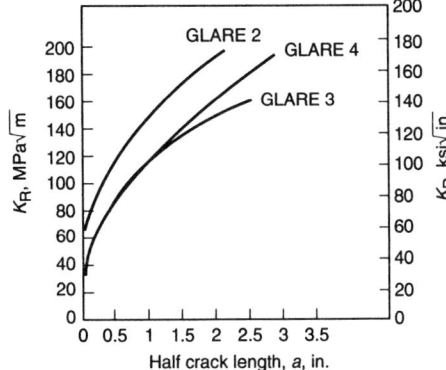

Fig. 19 Fracture toughness of GLARE fiber-metal laminate

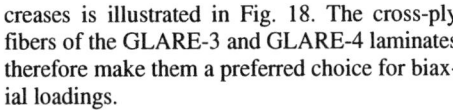

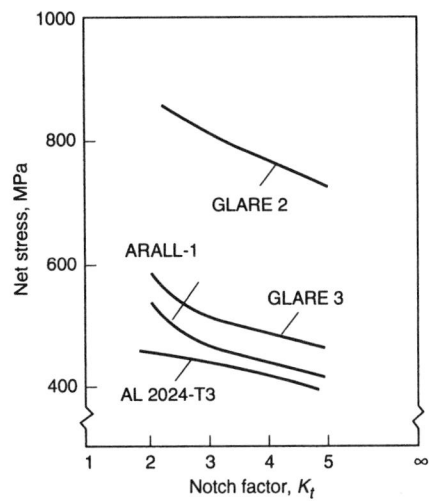

Fig. 20 A comparison of residual strength as a function of K_t for 2024-T3 aluminum and various fiber-metal laminate types

Table 4 Quasistatic Indentation Test

Material	Max force, N	Max displacement, mm	Absorbed energy, J
2024-T3	6.2	6.6	17.3
ARALL-2	3.2	5.6	8.5
GLARE-3	6.4	7.4	18.3
Carbon/PEI	2.5	4.7	4.3
Aramid/PEI	2.8	6.4	6.3

Note: Clamped plate, diameter 80 mm (3.15 in.). Impactor R = 7.5 mm (0.3 in.). Sheet thickness t = 1.4 mm (0.055 in.).

creases is illustrated in Fig. 18. The cross-ply fibers of the GLARE-3 and GLARE-4 laminates therefore make them a preferred choice for biaxial loadings.

Damage Tolerance

Fiber-metal laminates exhibit excellent damage tolerance properties. The residual strength of the material after fatigue, corrosion, or impact damage is superior to that of monolithic aluminum. In all cases, damage will be visible and the detectability and inspectability will be comparable, but less critical, than similar damage in aluminum structures. The fracture toughness of GLARE (Fig. 19) is comparable to that of high-toughness aluminum alloys (e.g., 2024-T3) and outperforms any composite material.

ARALL panels with sawcut flaws have a lower residual strength than monolithic 2024-T3 with sawcut flaws. However, ARALL panels with fatigue precracks have a higher residual strength than fatigue precracks 2024-T3 panels, for two reasons: the fibers in the wake of the crack remain unbroken, and the delamination zone around the crack effectively enlarges the fibers' "strain-length." Using glass fibers instead of aramid fi-

bers further improves the residual strength of fiber-metal laminate panels (Fig. 20).

Impact Toughness. Impact causes the most critical types of damage in composite materials, decreasing the static strength of an unnotched specimen by 25 to 40%. In ARALL laminates, the fibers are the weakest link. The fibers in the outer aramid layer opposite the impact site fail first. This is immediately followed by failure of the aluminum layers, and a crack appears perpendicular to the fiber direction. in GLARE, the aluminum layer opposite the point of impact fails, creating a crack in the fiber direction, while the fibers remain intact. (Only 2/1 and 3/2 lay-ups of GLARE have been tested.) The use of glass fibers improves damage resistance significantly (Table 4). A danger for pure composite materials is that they often show no visible signs of damage. After impact, this is not the case for fiber-metal laminates because the dent in the outer aluminum layer makes impact damage easily detectable (Ref 30).

Strength reduction in fatigue of post-impact specimens is small, about 10%. In practice, crack extension is on the order of 5 mm combined with a decreasing crack growth rate and, in most cases, crack arrest.

Repairability. The damage-tolerant repair of fiber-metal laminates can be accomplished using the same procedures used for monolithic aluminum structures (conventional riveted or bonded patch techniques). Patch materials can be of the same material and thickness as the original structure or of substantially thicker aluminum sheets.

A potential application for GLARE 3 and GLARE 4 is the damage-tolerant repair of cracked aluminum skins. Its high blunt notch strength, moderate stiffness, and excellent fatigue resistance make GLARE an ideal material for repairing aluminum structures.

The primary reason for this fatigue-life improvement is the favorable bearing load distribution that occurs in the repair joint. The lower modulus of GLARE 3, combined with the thinner GLARE patch needed (both as compared to 2024-T3) effectively reduces the critical first rivet row bearing loads. The result is a 25% lighter repair that is two to three times more fatigue resistant than a standard repair, while avoiding the costly and time consuming steps of producing a tapered aluminum patch.

Fatigue of Fiber-Reinforced Metal-Matrix Composites

W.S. Johnson, Georgia Institute of Technology

CONTINUOUS-FIBER-REINFORCED metal-matrix composites (MMCs) offer several attractive properties, such as high strength- and stiffness-to-weight ratios. In general they are more resistant to environments than polymer-matrix composites and can be used at higher temperatures. The major drawback to the use of MMCs is their cost. Therefore, many of their potential applications are limited to where there exists no viable lower-cost alternative.

Titanium MMCs are currently being considered as structural materials for high-temperature applications where weight savings is a premium. Some potential applications of these materials are hypersonic flight vehicles, advanced aircraft engines, missiles, advanced supersonic transports, and advanced fighter aircraft. All of these applications expose the material to repeated mechanical loadings and thermal cycles. In addition, stresses are induced in the composite constituents due to temperature change because of the coefficient of thermal expansion mismatch between the fiber and matrix materials. This, coupled with the different strengths and failure modes of the fiber, matrix, and fiber/matrix interface, contributes to a complex problem in predicting and tracking damage initiation and progression in laminated composites.

The damage initiation and progression process for a number of MMCs has been summarized elsewhere, with the emphasis on aluminum-matrix composites (Ref 31, 32). This section briefly describes examples of fatigue damage and some of the controlling parameters of the fatigue behavior of several titanium-matrix laminates (notched and unnotched) at room temperature and the influence of the fiber/matrix interface strength and the thermal residual stresses on the fatigue damage propagation mode. Coverage is limited to a compilation of previous research on titanium MMCs done primarily by the author and his colleagues at the NASA Langley Research Center using the SCS-6/Ti-15-3 composite system, which is based on titanium alloy Ti-15V-3Cr-3Al-3Sn (Ti-15-3). This metastable beta strip alloy is considered a potential matrix material for high-temperature MMCs because it can be economically cold formed into relatively thin sheets while retaining good mechanical properties (Ref 33). The composite laminates are made by hot pressing Ti-15-3 foils between unidirectional tapes of silicon-carbide fibers. The fibers (designated SCS-6 by Textron Specialty Materials) are 0.142 mm in diameter and are assumed to have isotropic properties with a modulus of 400 GPa. Textron Specialty Materials fabricated panels of each of the following lay-ups: [0]$_8$,

[0$_2$/±45]$_s$, [0/90]$_{2s}$, [0/90/0], and [0/±45/90]$_s$. The fiber volume fractions ranged from 32.5 to 39%, depending on the particular panel. A panel of "fiberless" composite was made identically to the composite but without placing the fiber mats between the foils during fabrication.

Other possible matrix materials include beta alloy TiMetal* 21S and several titanium aluminides, which also have been studied with SCS-6 fibers. The fabrication processes for these materials are very similar to that described for Ti-15-3. The failure mechanisms are also similar, although the resistance to environmental attack at high temperature varies with the matrix material, as does the fiber/matrix interfacial strength. These differences do result in differences in damage initiation sites and crack propagation behavior. The tests were conducted at NASA Langley Research Center as described in Ref 34. Failure mechanisms are covered in Ref 35.

This research has established a good fundamental understanding of fatigue damage initiation and propagation in continuous-fiber-reinforced titanium matrix composites at both room and elevated temperatures. The causes of initial damage on both the global and local levels are becoming well defined. Seemingly insignificant factors, such as thermal residual stresses and interfacial strengths, play profound roles in almost every aspect of the fatigue life, from initiation to fracture, and thus they must not be overlooked. Further understanding is required of environmental effects at very high temperatures and time-at-temperature effects. A more complete collection of papers on this subject, including life prediction approaches for titanium MMCs, can be found in the volume cited in Ref 35.

Fatigue of Unnotched Specimens

In the unnotched laminates, damage starts as fiber/matrix interface failures in the off-axis plies. This results in a stiffness loss early in the fatigue life. These failed interfaces in the off-axis plies result in a bilinear stress-strain response for the laminate. This could present potential design problems.

The fatigue life of the unnotched laminates tested under the same temperatures, cyclic frequencies, and time-at-temperature conditions was shown to be a function of the stress range in the 0° fibers for isothermal, nonisothermal, and thermomechanical fatigue conditions. However, the relationship between the 0° fiber stress

*TiMetal is a registered trademark of Timet.

range/cycles to failure did vary for different temperatures and time-at-temperature conditions, indicating a fiber/matrix reaction effect, matrix oxidation effect, and/or an accumulative strain effect at elevated temperatures.

Unnotched fatigue tests were conducted at 650 °C on the SCS-6/Ti-15-3 laminates described above (Ref 36). The cyclic stress range in the 0° fibers versus number of cycles to failure is plotted with the room-temperature data in Fig. 21. The elevated-temperature data are more scattered than the room-temperature data. At this elevated temperature the matrix could creep during cycling, and creep ratcheting (an accumulation of strain) would result. As the strain accumulates, the maximum stress in the fiber increases. The laminates with off-axis plies would be expected to creep more than the [0]$_8$ laminate, and thus the cyclic R-ratio of the fiber stress will be higher for the [0/±45/90]$_s$ and [0/90]$_{2s}$ laminates. This may, in part, explain why the [0/±45/90]$_s$ and [0/90]$_{2s}$ laminate results fall below the [0]$_8$ laminate results in Fig. 21 when based only on the measured 0° fiber stress range.

Figure 22 compares the maximum strain versus cycles to failure of a unidirectional composite to the maximum strain versus cycles to failure of the matrix loaded in strain control (Ref 36). The fatigue test of the unidirectional specimen was considered to be an in situ fatigue test of the fibers. For high maximum strains (short lives), the initial damage developed in the fibers and the composite had a shorter life than the matrix alone. For low maximum strains (long lives), the initial damage developed in the matrix and the matrix had a shorter life than the composite. Where the two curves intersect, both the matrix and the fibers were equally likely to develop fatigue damage. During high stress (short life) testing, there are many fiber breaks but no significant matrix cracking. Medium-life testing results in both fiber breaks and some short matrix cracks in the specimens.

Similar results were found by Sanders and Mall (Ref 37) under stain-controlled testing at 427 °C. The low-stress, long-life specimen exhibits no fiber breaks but several long matrix cracks. Therefore, it is reasonable to use the fatigue response of the matrix materials and the in situ 0° fiber to predict at which strain levels the fiber will fail before matrix cracking and at which strain levels matrix cracking will precede fiber failure. Majumdar and Newaz (Ref 38) showed that the fatigue life was more a function of strain range than of applied stress range. They also showed a number of fatigue damage mechanisms in the Ti-15-3/SCS-6 system with 15 and 41% fiber volume fraction.

Nicholas and Russ (Ref 39) presented data on SCS-6/Ti-24Al-11Nb that show the effect of stress ratio, frequency, and hold time on isothermal elevated-temperature fatigue. Elevated-temperature testing in a vacuum results in fatigue lives as much as 10× longer than identical tests in air. This has been shown for Ti-15-3 matrix (Ref 40) and for Ti-22Al-23Nb (Ref 41).

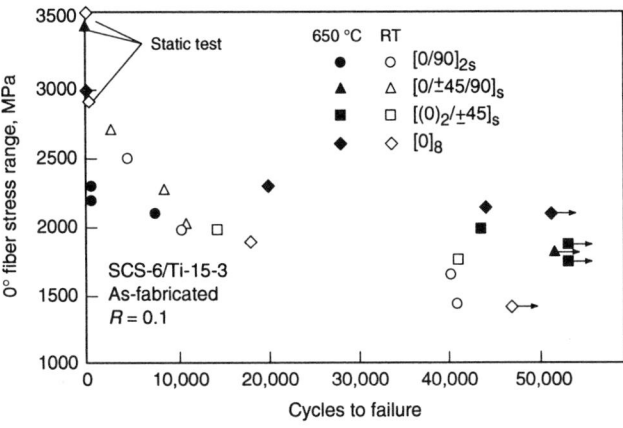

Fig. 21 0° fiber stress range versus number of cycles to failure at 650 °C and room temperature. Source: Ref 36

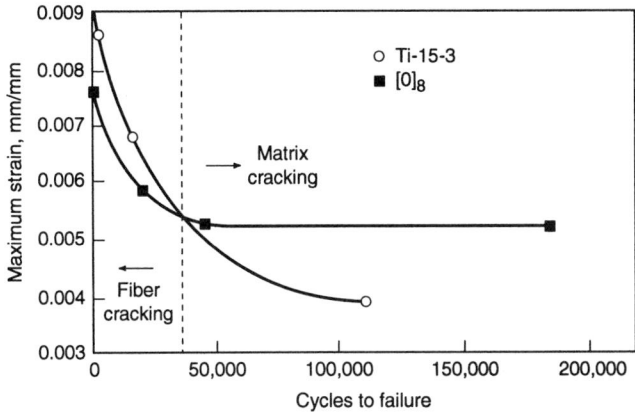

Fig. 22 The life of the unidirectional composite and the matrix as a function of maximum strain at 650 °C. Source: Ref 36

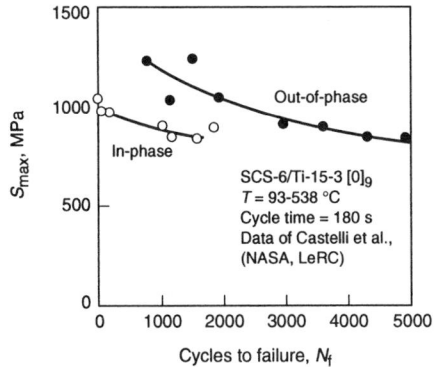

Fig. 23 The maximum applied stress as a function of cycles to failure under thermomechanical fatigue (TMF). Source: Ref 45

Fig. 24 The stress range in the 0° fibers as a function of cycles to failure. Data from both the in-phase and out-of-phase tests form one band. Source: Ref 45

Thermomechanical Fatigue. Both thermal and mechanical stresses are developed in the constituents of a laminate during thermomechanical fatigue (TMF). Constituent stresses are not necessarily related to the measurable laminate strains, so an analysis is needed to calculate the constituent and laminate behavior of arbitrary lay-ups subjected to arbitrary combinations of mechanical and thermal loading. The VISCO-PLY code is a micromechanics analysis (Ref 42) based on constituent properties developed by NASA Langley. The program uses the vanishing fiber diameter (VFD) constitutive model to calculate the orthotropic properties of the plies. These ply properties are then used in a laminated plate analysis to predict overall laminate response. Both the fiber and the matrix can be modeled as a thermo-viscoplastic material. Combinations of thermal and mechanical loads can be modeled. Sequential jobs can be run for various load and temperature profiles. Fiber and matrix average stresses and strains in each ply and the overall composite response under thermomechanical loading conditions are calculated. The VISCO-PLY program, due to the nature of the VFD model, does not account for local stress concentrations around the fiber or for lateral constraint in a ply due to the presence of the fiber. However, based upon the author's experience using the

VFD model for over 12 years on numerous MMCs, the model predicts general laminate response very well, usually within 10% of measured response for laminates containing at least some 0° fibers.

The VISCOPLY program was used to predict the laminate response of unidirectional SCS-6/Ti-15-3 composites (Ref 43) and to determine thermo-viscoplastic material constants (Ref 44). The prediction captures the essence of the loading-unloading behavior of the composite. The fatigue behavior as a function of maximum applied stress for in-phase and out-of-phase TMF is shown in Fig. 23. The in-phase loadings resulted in earlier failures, but the specimens lost very little stiffness prior to failure. On the other hand, the out-of-phase loadings resulted in significant stiffness loss due to matrix cracking prior to failure. The VISCOPLY program was used to predict the fiber and matrix stresses during an in-phase and an out-of-phase TMF cycle between 93 and 538 °C and between 45 and 896 MPa. The fiber stresses were highest for the in-phase test, explaining the earlier laminate fatigue failures. However, the matrix stresses were higher for the out-of-phase loadings, explaining the earlier matrix cracking and the resulting stiffness loss measured during the out-of-phase loadings (Ref 43, 44). When each test shown in Fig. 23 was analyzed and the 0° fiber stress range was plotted

versus the number of cycles to failure, the in-phase and out-of-phase data collapsed into a narrow band, as shown in Fig. 24.

Other TMF and isothermal data were analyzed and are plotted in Ref 43. Within a given test condition (temperature, loading frequency, time at temperature, etc.), the 0° fiber stress range seems to correlate with the number of cycles to failure. Gao and Zhao (Ref 45) also have determined that failure is controlled by the 0° fibers. However, as the test conditions change the fatigue behavior of the fibers appears to change. Because the plot is of stress range in the fiber, the increased loading of the fiber due to matrix stress relaxation is not accounted for. Higher temperatures and slower cycling would both contribute to the shifting of more load to the fiber from the matrix. Additional time at temperature could also cause additional fiber/matrix interface reactions that could affect the mechanical behavior (Ref 46, 47).

Effect of Interface Strength and Residual Stresses on Fatigue. Naik et al. (Ref 47) investigated the effect of a high-temperature cycle (1010 °C) used to simulate a superplastic forming/diffusion bonding (SPF/DB) fabrication cycle on the mechanical properties of a [0/90/0] lay-up of SCS-6/Ti-15-3. The high-temperature SPF/DB cycle increased the strength and stiffness of the matrix material, had little or no effect on the fiber properties, but significantly reduced the static and fatigue strength of the laminate, as shown in Fig. 25. The fracture surfaces showed a change in the failure mode due to the SPF/DB cycle. In the as-fabricated (ASF) specimen, the fiber/matrix interface was weaker and the thermal residual stresses were less compressive than in the SPF/DB specimen; thus, the damage tended to grow around the fibers, debonding the fiber/matrix interface. This produced a tortuous crack path in the matrix and extensive fiber pull-out. In the SPF/DB specimen the residual stresses were high enough and the fiber/matrix interface strong enough to cause the matrix cracks to grow through the fiber, resulting in a planar failure surface with little or no fiber pull-out. A shear-lag analysis indicated that the latter failure mode produced a greater stress concentration in the first

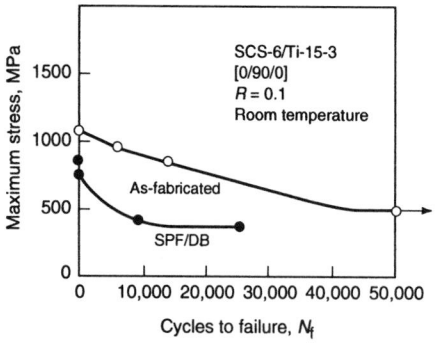

Fig. 25 *S-N curve for the as-fabricated and the super-plastic forming/diffusion bonding (SPF/DB) specimens. The static strength (N = 0) is 25% higher for the as-fabricated specimen.*

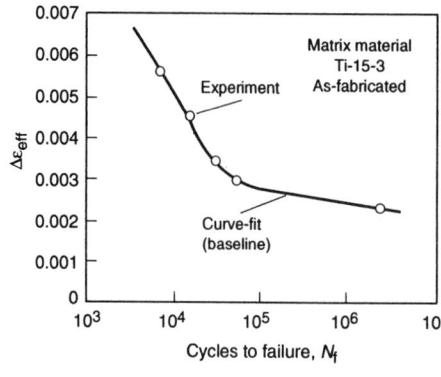

Fig. 26 Strain-controlled matrix fatigue data plotted in terms of the effective strain criterion. Source: Ref 49

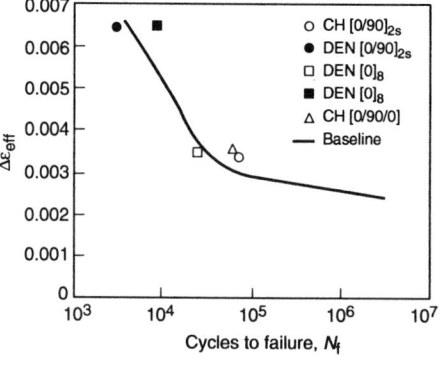

Fig. 27 The solid line is the effective strain versus number of cycles to failure for the matrix material. The symbols are the calculated effective strains in the matrix at the notch tip versus number of cycles to observed crack initiation. Source: Ref 49. Additional data can be seen in Ref 63.

fiber ahead of the crack, thus explaining the reduction in strength of the SPF/DB specimen.

These results (Ref 47) indicate that for a high-strength matrix material, such as a titanium alloy, one can make the fiber/matrix interface too strong, thus sacrificing laminate strength by changing the mode of failure. Conversely, if the interface is too weak, the required load transfer between fiber and matrix necessary for optimum moduli and shear properties will not occur.

Fatigue of Notched Specimens

Fatigue behavior near local stress concentrations can be complex. The initiation process at a notch may be different from the subsequent damage growth process. Depending on the design philosophy adopted, one process may be more important than the other; however, both damage initiation and growth need to be understood.

Fatigue Crack Initiation. Fatigue damage initiation at notches can be predicted using unnotched matrix data and an effective strain parameter that accounts for orthotropic stress concentration and residual thermal stresses in the matrix. Good correlation between the predicted and experimental results has been obtained for several different lay-ups for both center-hole and edge-notched specimens. Matrix fatigue cracks have been found to grow long distances in the laminate without breaking the 0° fibers in the crack path. Fiber/matrix interface debonding early in the fatigue life results in a significant reduction in the stress concentration in 0° fibers at a notch tip. The calculated stress concentrations for virgin notched specimens can be used to calculate damage initiation in a virgin specimen, but they do not apply to damage propagation or fracture in the specimen after damage has initiated.

From fatigue crack initiation tests for several different lay-ups and notch geometries, cracks initiated in the matrix material and grew without breaking fibers (Ref 48). To explain the matrix crack initiation process, Hillberry and Johnson (Ref 49) developed strain-controlled matrix fatigue data by replotting stress data in terms of the Smith-Watson-Topper effective strain parameter,

$\Delta\varepsilon_{eff}$ (Ref 50), as shown in Fig. 26. Hillberry and Johnson (Ref 49) modified the Smith-Watson-Topper effective strain parameter to predict cycles to fatigue damage initiation in the matrix material next to a notch. The modification includes the calculated thermal residual stresses in the matrix (σ^r) and the orthotropic stress concentration factor, K_t, as follows:

$$\Delta\varepsilon_{eff} = [(K_t\varepsilon_{max} + \sigma^r/E_m) K_t(\Delta\varepsilon/2)]^{1/2}$$

where E_m is the matrix modulus, ε_{max} is the maximum applied strain to the specimen, and $\Delta\varepsilon$ is the applied strain range.

For each experimental test, the effective strain parameter was calculated using the above equation and plotted versus the number of cycles to observed crack initiation. The results are shown in Fig. 27. Very good agreement was found between the experiments and predictions based on the matrix fatigue data alone.

The residual stress in the matrix is an important contributor to the total effective strain. At lower values of $\Delta\varepsilon_{eff}$ in Fig. 27 (e.g., 0.0035), σ^r/E is 35% of the ($K_t\varepsilon_{max} + \sigma^r/E$) term. Without the residual stress term, the effective strain parameter would be reduced by 59%. At higher values of $\Delta\varepsilon_{eff}$ in Fig. 27 (e.g., 0.0065), the effective strain parameter would be reduced by 46% because the σ^r/E is 21% of the ($K_t\varepsilon_{max} + \sigma^r/E$) term. Obviously, if the composite data points were shifted down by these percentages, the effective strain parameter would not correlate the data. This illustrates the importance of including the residual stress term in the modified effective strain parameter.

Fatigue Crack Growth from Notches. Fatigue cracks were observed in notched $[0]_8$, $[0/90]_{2s}$, and $[0/90/0]$ laminates at cyclic stress levels above those required for crack initiation but well below the laminate net section static strength (Ref 49). The specimens were periodically radiographed during the tests to monitor fiber breaks. After the tests (but prior to specimen failure), the specimen surfaces were either polished or acid etched to reveal the matrix cracks and fibers in the surface ply. Multiple matrix cracks had initiated at and grown from the notches. Harmon and Saff (Ref 51) reported simi-

lar results for unidirectional SCS-6/Ti-15-3 center-hole specimens. Figure 28 shows a typical matrix crack growing from the edge of a hole in a $[0/90]_{2s}$ specimen. The polished surface clearly shows that the matrix crack grew around and past the fibers. Also, a long debond between the fiber next to the hole and the matrix on the side toward the hole was seen. This type of debonding can significantly reduce the stress concentration at the edge of the notch. One can nominally predict the stress level at which the fibers next to the notch should fail, based on the applied loading and calculated stress concentration for the undamaged state. However, because of damage such as that shown in Fig. 28, the stress concentration may be reduced so that only the matrix would fail and not the fibers.

Bigelow and Naik (Ref 52) developed a macro-micromechanics finite element approach to calculate the local stresses around a continuous fiber at a global stress discontinuity, such as a notch tip. This approach was used by Bigelow and Johnson (Ref 53) to predict the notch-tip interface stresses in the matrix material due to a remote stress of 1 MPa on a double-edge-notched $[0/90]_{2s}$ SCS-6/Ti-15-3 specimen with the same geometry as the specimen described above. This type of analysis can give good insight into the cause of fiber/matrix interface failure such as that shown in Fig. 28. The axial stress is high due to the local stress concentration. Once the notch-tip 0° fiber/matrix interface debonds, the stress concentration will decrease by approximately one-third (Ref 53).

Prediction of Growth Rates in Bridged Cracks. Bakuckas and Johnson (Ref 54) conducted an analytical and experimental investigation of the effect of fiber bridging on fatigue crack growth in $[0]_8$ SCS-6/Ti-15-3. The specimens tested contained center slits with initial notch-length-to-width ratios of 0.30 and 0.35. Under constant-amplitude loading the crack growth rate decreased as the crack length increased. The fiber bridging reduced the crack driving force as the crack grew. Because the crack in the composite grows only in the matrix material, the crack

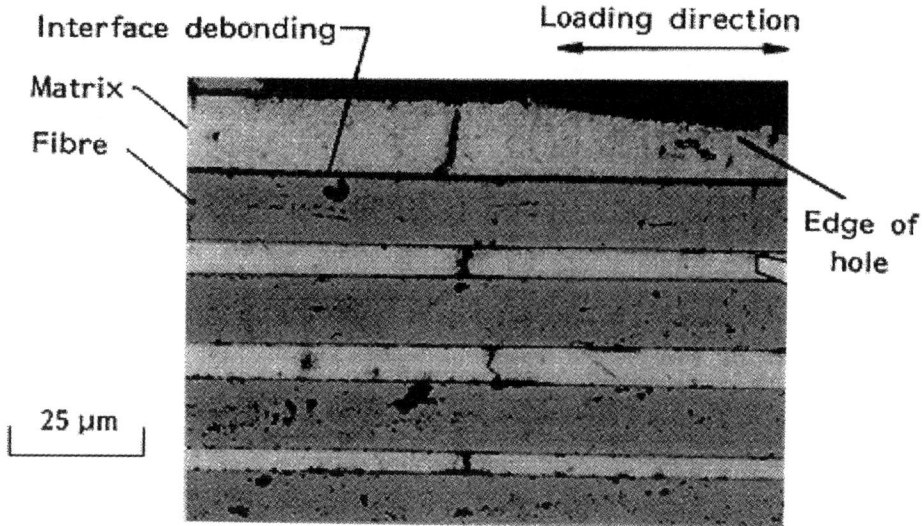

Fig. 28 Matrix crack growing past fibers. Localized debonding of the matrix from the fiber can be seen. Source: Ref 48

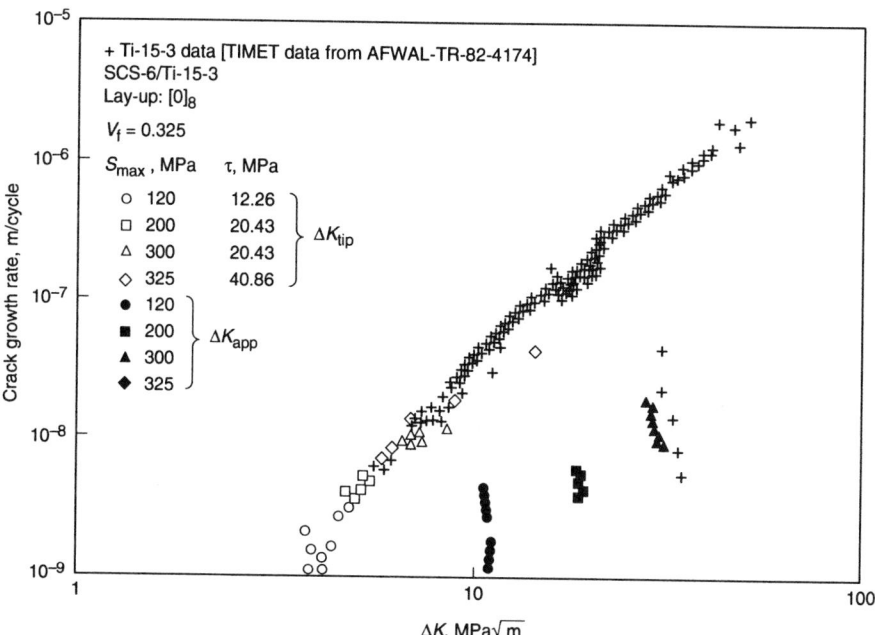

Fig. 29 Crack growth rate data from the Ti-15-3 material compared with the crack growth data from the composite. The solid symbols are the composite crack growth data assuming no fiber bridging. The open symbol data are corrected for bridging, assuming the indicated shear stresses. Source: Ref 54

tive model, because one value of τ was able to correlate all the data for a given fiber and matrix combination.

Larsen et al. (Ref 59) have documented crack growth and the associated mechanisms in SCS-6/Timetal 21S using several lay-ups, R-ratios, and stress levels. They have also developed an effective stress-intensity approach to correlate the crack growth data. Zheng and Ghonem (Ref 60) have conducted similar work on the SM1240/Timetal 21S unidirectional composites. Bowen (Ref 61) has studied the crack growth behavior of unidirectional lay-ups of SCS-6/Ti-15-3, SCS-6/Ti-6Al-4V, SCS-6/Timetal 21S, and SM1240/Ti-6Al-4V. These tests were conducted by cyclic load on a notched bend specimen. Davidson (Ref 62) has performed microstructural observations pertaining to the fiber bridging phenomenon in SCS-6/Ti-15-3. Emphasis was placed on determining the slip that occurs between the matrix and the fiber as a result of debonding during the fiber bridging process.

Effect of Matrix Cracking on Residual Strength. Matrix cracks initiate and grow from notches at cyclic stress levels of less than one-third of the ultimate strength of notched virgin specimen. In cases where only matrix cracks occur with no accompanying fiber failures, one needs to be able to assess the effect of this damage on residual strength. Bakuckas et al. (Ref 63) conducted an experimental and analytical study of [0/90]$_{2s}$ laminate containing a center-hole specimen with a ratio of hole diameter to specimen width of 0.33. The residual strengths were experimentally determined for a virgin specimen and for a specimen that was saturated with matrix cracks after being subjected to cyclic loading between 25 and 250 MPa for 200,000 cycles. The residual strengths were 525 MPa and 325 MPa, respectively. The matrix cracks resulted in a strength reduction of 38%. Thus, the effect of the matrix cracks was significant.

The three-dimensional finite element program PAFAC (Ref 64) was used to determine the effects of the matrix cracks on the stress concentration in the 0° fibers and to predict the residual strengths of both the virgin and post-fatigued specimens. For the virgin condition the composite was initially modeled with the undamaged properties of the matrix and fiber and assuming a perfect fiber/matrix interface. As the virgin specimen was analytically loaded, the fiber/matrix interfaces were assumed to fail in the 90° plies, as previously discussed. For the post-fatigued test the modulus of the matrix was reduced by 69% to account for the saturated state of the matrix cracks. The modulus reduction was determined by calculating the required matrix modulus to account for the stiffness loss in the unnotched crack-saturated portion of the specimen. The fiber strength was determined from unnotched coupons by recording the strain to failure and multiplying it by the fiber modulus. The predicted and measured residual strengths are shown in Fig. 30. The point at which the predicted notch-tip 0° fiber stress crosses the fiber strength line represents the predicted laminate failure. The

growth rate in the composite should correlate with the crack growth rate in the matrix material alone if the correct crack driving force in the matrix, ΔK_{mat}, is defined. Figure 29 shows crack growth rate versus ΔK data for the Ti-15-3 material. The figure also shows the composite data plotted with ΔK_{app}, the crack driving force calculated without accounting for the fiber bridging. The ΔK_{app} does not collapse the composite crack growth data to the Ti-15-3 data. The effect of the fibers bridging the matrix crack must be incorporated into the definition of ΔK_{mat}.

Several fiber bridging models that combine a continuum fracture mechanics analysis and a micromechanics analysis have been investigated

(Ref 55-57). In all of these models, the intact fibers in the wake of the matrix crack are modeled using a continuous closure pressure. Fiber/matrix debonding occurs as the crack progresses past each fiber. An unknown constant shear stress τ is assumed to act on the debonded fiber/matrix interface. The model proposed by McMeeking (Ref 57) provided the most accurate predictions of the measured fiber/matrix debond length and crack opening displacements for test data, as well as collapsing the da/dN data, as shown in Fig. 29. The values of τ used to fit the data are also shown in Fig. 29. Bakuckas and Johnson (Ref 58) modified the fiber bridging models to include thermal residual stresses. This resulted in a more predic-

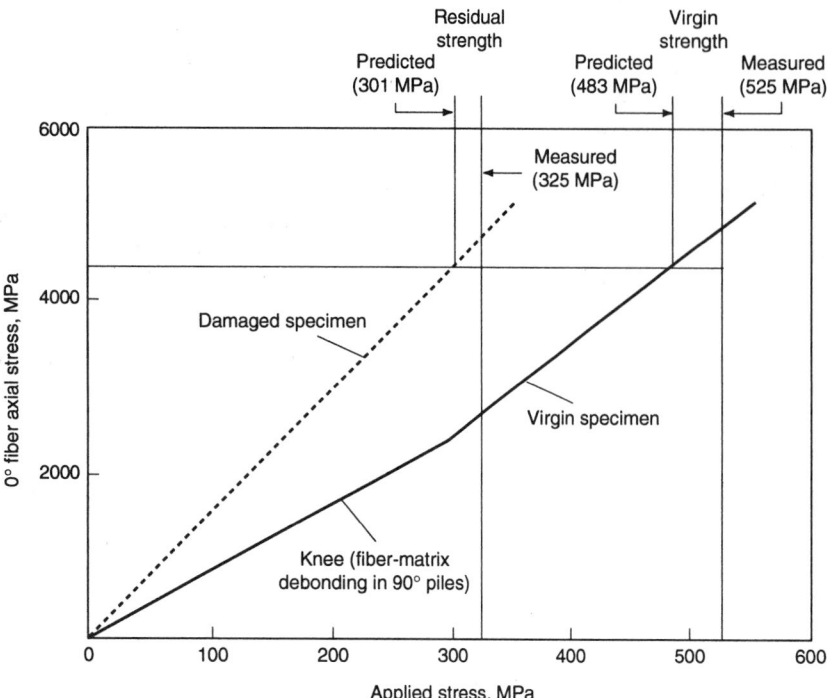

Fig. 30 Axial stress in the 0° fiber next to the hole as a function of applied stress for the virgin and post-fatigued condition in [0/90]$_s$ SCS-6/Ti-15-3 laminates, V_f = 0.355. The applied stress at which the 0° fiber stress reaches the fiber strength corresponds to the notched strength for both conditions. Source: Ref 63

[0/90]$_{2s}$ laminate failure was observed to be catastrophic (i.e., when the first fiber failed, the specimen failed). Using this micromechanics-based strength prediction methodology for the notched laminate resulted in excellent predictions.

Acknowledgment. The author would like to acknowledge the significant contributions of the following colleagues with whom he has had the pleasure of working over the past ten years at the NASA Langley Research Center: Dr. Yahei Bahei-El-Din, Dr. John Bakuckas, Dr. Cathy Bigelow, Dr. Alton Highsmith, Dr. Ben Hillberry, Steve Lubowinski, Dr. Massoud Mirdamadi, Dr. Rajiv Naik, and Bill Pollock. Further, the author would like to acknowledge the help of Jeff Calcaterra, a Ph.D. candidate at Georgia Tech, in assembling this review.

REFERENCES

1. R.A. Weinberger et al., "U.S. Navy Certification of Composite Wings for the F-18 and Advanced Carrier Aircraft," AIAA Paper 77-466, American Institute of Aeronautics and Astronautics, 1977
2. G.C. Grimes, "Investigation of Stress Levels Causing Significant Damage in Composites," AFML-TR-75-33, Air Force Materials Laboratory, 1975
3. B.Z. Jang, *Advanced Polymer Composites: Principles and Applications*, ASM International, 1994
4. K.L. Reifsnider, E.G. Henneke II, and W.W. Stinchcomb, "Defect-Property Relationships in Composite Materials," Technical Report AFML-TR-76-81, Part IV, Air Force Materials Laboratory, 1979
5. A.S.D. Wang and F.W. Crossman, Initiation and Growth of Transverse Cracks and Edge Delamination in Composite Laminates, Part 1: An Energy Method, *J. Compos. Mater.*, Supplement, 1980
6. K.W. Garrett and J.E. Bailey, Multiple Transverse Fracture in 90° Cross-Ply Laminates of a Glass Fiber-Reinforced Polyester, *J. Mater. Sci.*, Vol 12, 1977
7. A. Parvizi and J.E. Baily, On Multiple Transverse Cracking in Glass Fiber Epoxy Cross-Ply Laminates, *J. Mater. Sci.*, Vol 13, 1978
8. N.J. Pagano and R.B. Pipes, Some Observations on the Interlaminar Strength of Composite Laminates, *Int. J. Mech. Sci.*, Vol 15, 1973
9. N.J. Pagano and S.R. Soni, Global-Local Laminate Variational Model, *Int. J. Solids Struct.*, Vol 19 (No. 3), 1983
10. S.R. Soni and R.Y. Kim, Delamination of Composite Laminates Stimulated by Interlaminar Shear, *Composite Materials: Testing and Design (Seventh Conference)*, STP 893, J.M. Whitney, Ed., ASTM, 1986
11. H.T. Hahn, Fatigue Behavior and Life Prediction of Composite Laminates, *Composite Materials: Testing and Design (Fifth Conference)*, STP 674, S. Tsai, Ed., ASTM, 1979
12. M.J. Salkind, *Composite Materials: Testing and Design (Second Conference)*, STP 497, ASTM, 1972, p 143
13. J.M. Whitney, "Fatigue Characterization of Composite Materials," AFWAL-TR-79-4111, Air Force Materials Laboratory, Oct 1979
14. R.Y. Kim, University of Dayton Research Institute, unpublished research, 1986
15. S.V. Ramani and D.P. Williams, Axial Fatigue of [0/±30]$_{3s}$ Graphite/Epoxy, *Failure Modes in Composites*, Vol III, Feb 1976, p 115
16. T.R. Porter, Evaluation of Flawed Composite Structures under Static and Cyclic Loading, *Fatigue of Filamentary Composite Materials*, STP 636, R. Evans, Ed., ASTM, 1977
17. R.W. Walter, R.W. Johnson, R.R. June, and J.E. McCarty, Design for Integrity in Long-Life Composite Aircraft Structures, *Fatigue of Filamentary Composite Materials*, STP 636, R. Evans, Ed., ASTM, Society for Testing and Materials, 1977
18. S.P. Garbo and J.M. Ogonowski, "Effects of Variances and Manufacturing Tolerances on the Design Strength and Life of Mechanically Fastened Composite Structures," AFWAL-TR-81-3041, Vol II, Air Force Wright Aeronautical Laboratories, April 1981
19. J.H. Crews, Jr., Bolt-Bearing Fatigue of a Graphite/Epoxy Laminate, *Joining of Composite Materials*, STP 749, K. Kedward, Ed., ASTM, 1981, p 131-144
20. A.J. Barnard, "Fatigue and Damage Propagation in Composite Rotor Blades," AGARD 288, North Atlantic Treaty Organization, April 1980
21. K. Brunsch, "Service Experience with GRC Helicopter Blades (BO-105)," AGARD 288, North Atlantic Treaty Organization, April 1980
22. E. Jarosch and A. Stepan, *J. Am. Helicopter Soc.*, Vol 15 (No. 1), Jan 1970
23. D.R. Thoman, L.J. Bain, and C.E. Antle, Inferences on the Parameters of the Weibull Distribution, *Technometrics*, Vol 11, 1979, p 257-269
24. W.J. Park, "Pooled Estimate of the Parameters on Weibull Distributions," AFML-TR-29-4112, Air Force Materials Laboratory, 1979
25. H.T. Hahn and R.Y. Kim, Fatigue Behavior of Composite Laminate, *J. Compos. Mater.*, Vol 10, April 1976
26. J.N. Yang, Reliability Prediction of Composites under Proof Tests in Service, *Composite Materials: Testing and Design (Fourth Conference)*, STP 617, J.G. Davis, Jr., Ed., ASTM, March 1977, p 272-295
27. M.A. Gregory and C.H.J.J. Roebroeks, Fiber-Metal Laminates: A Solution to Weight, Strength, and Fatigue Problems, 30th Annual Conference of Metallurgists, Metallurgical Society of CIM, Canada, 1991
28. M.L.C.E. Verbruggen, "ARALL Adhesion Problems and Environmental Effects," doctoral thesis, Delft University of Technology, The Netherlands, Feb 1987
29. R.J. Bucci, L.N. Mueller, M.A. Gregory, and R.M. Bentley, "ARALL Laminates Scale-up from R&D to Flying Articles," paper presented at USAF Structural Integrity Program Conference, San Antonio, TX, Nov 1988
30. A. Vlot, "Impact Tests on Aluminium 2024-T3, Aramid and Glass Reinforced Aluminium Laminates and Thermoplastic Composites," Report LR-534, Delft University of Technology, The Netherlands, Oct 1987

31. W.S. Johnson, Fatigue Testing and Damage Development in Continuous Fiber Reinforced Metal Matrix Composites, *Metal Matrix Composites: Testing, Analysis and Failure Modes*, STP 1032, W.S. Johnson, Ed., ASTM, 1989, p 194-221

32. W.S. Johnson, Fatigue of Metal Matrix Composites, *Fatigue of Composite Materials*, K.L. Reifsnider, Ed., Elsevier, 1991, p 199-229

33. H.W. Rosenberg, *J. Met.*, Vol 35 (No. 11), Nov 1986, p 30-34

34. W.S. Johnson, S.J. Lubowinski, and A.L. Highsmith, Mechanical Characterization of SCS-6/Ti-15-3 Metal Matrix Composites at Room Temperature, *Thermal and Mechanical Behavior of Ceramic and Metal Matrix Composites*, STP 1080, J.M. Kennedy, H.H. Moeller, and W.S. Johnson, Ed., ASTM, 1990, p 193-218

35. T.W. Clyne, P. Feillard, and A.F. Kalton, Interfacial Mechanics and Macroscopic Failure in Titanium-Based Composites, *Life Prediction Methodology for Titanium Matrix Composites*, STP 1253, W.S. Johnson, J.M. Larsen, and B.N. Cox, Ed., ASTM, 1995

36. W.D. Pollock and W.S. Johnson, Characterization of Unnotched SCS-6/Ti-15-3 Metal Matrix Composites at 650 °C, *Composite Materials: Testing and Design (Tenth Volume)*, STP 1120, Glenn Grimes, Ed., ASTM, 1992, p 175-191

37. B.P. Sanders and S. Mall, Longitudinal Fatigue Response of a Metal Matrix Composite under Strain Controlled Mode at Elevated Temperature, *J. Comp. Technol. Res.*, Vol 16 (No. 4), Oct 1994, p 304-313

38. B.S. Majumdar and G.M. Newaz, *Isothermal Fatigue Mechanisms in Ti-Based Metal Matrix Composites*, NASA Contractor Report 191181, Sept 1993, p 62

39. T. Nicholas and S.M. Russ, Elevated Temperature Fatigue Behavior of SCS-6/Ti-24Al-11Nb, *Mater. Sci. Eng.*, Vol A152, 1992, p 514-519

40. J. Gayda, T.P. Gabb, and B.A. Lerch, Fatigue-Environment Interactions in a SiC/Ti-15-3 Composite, *Int. J. Fatigue*, Vol 15 (No. 1), 1993, p 41-45

41. S.M. Russ, A.H. Rosenberger, and D.A. Stubbs, Isothermal Fatigue of a SCS-6/Ti-22Al-23Nb Composite in Air and Vacuum, paper presented at the joint ASME Applied Mechanics and Materials Meeting, Los Angeles, CA, 28-30 June 1995

42. W.S. Johnson, M. Mirdamadi, and Y.A. Bahei-El-Din., Stress-Strain Analysis of a [0/90]₂s Titanium Matrix Laminate Subjected to a Generic Hypersonic Flight Profile, *J. Comp. Technol. Res.*, Vol 15 (No. 4), Winter 1993, p 297-303

43. M. Mirdamadi, W.S. Johnson, Y.A. Bahei-El-Din, and M.G. Castelli, "Analysis of Thermomechanical Fatigue of Unidirectional Titanium Metal Matrix Composites," NASA TM 104105, July 1991

44. M.G. Castelli, P.A. Bartolotta, and J.R. Ellis, Thermomechanical Fatigue Testing of High Temperature Composites: Thermomechanical Fatigue (TMF) Behavior of SiC(SCS-6)/Ti-15-3, *Composite Materials: Testing and Design (Tenth Volume)*, STP 1120, Glenn Grimes, Ed., ASTM, 1992, p 70-86

45. Z. Gao and H. Zhoa, Life Predictions of Metal Matrix Composite Laminates under Isothermal and Nonisothermal Fatigue, *J. Comp. Mater.*, 1995

46. S.M. Jeng, C.J. Yang, P. Alassoeur, and J.-M. Yang, Deformation and Fracture Mechanisms of Fiber-Reinforced Titanium Alloy Matrix Composite, *Composites Design, Manufacture, and Application*, Paper 25-C, ICCM/VIII, S.W. Tsai and G.S. Springer, Ed., 1991

47. R.A. Naik, W.S. Johnson, and W.D. Pollock, Effect of a High Temperature Cycle on the Mechanical Properties of Silicon Carbide/Titanium Metal Matrix Composite, *J. Mater. Sci.*, Vol 26, 1991, p 2913-2920

48. R.A. Naik and W.S. Johnson, Observations of Fatigue Crack Initiation and Damage Growth in Notched Titanium Matrix Composites, *Composite Materials: Fatigue and Fracture (Third Volume)*, STP 1110, T.K. O'Brien, Ed., ASTM, 1991, p 753-771

49. B.M. Hillberry and W.S. Johnson, Prediction of Matrix Fatigue Crack Initiation in Notched SCS-6/Ti-15-3 Metal Matrix Composites, *J. Comp. Technol. Res.*, Fall 1992

50. K.N. Smith, P. Watson, and T.H. Topper, A Stress-Strain Function for Fatigue of Metals," *J. Met.*, Vol 5 (No. 4), Dec 1970, p 767-778

51. D.M. Harmon and C.R. Saff, Damage Initiation and Growth in Fiber Reinforced Metal Matrix Composites, *Metal Matrix Composites: Testing, Analysis and Failure Modes*, STP 1032, W.S. Johnson, Ed., ASTM, 1989, p 237-250

52. C.A. Bigelow and R.A. Naik, A Macro-Micromechanics Analysis of a Notched Metal Matrix Composite, *Composite Materials: Testing and Design (Tenth Volume)*, STP 1120, Glenn Grimes, Ed., ASTM, 1992, p 222-233

53. C.A. Bigelow and W.S. Johnson, Effect of Fiber-Matrix Debonding on Notched Strength of Titanium Metal Matrix Composites, NASA TM-104131, Washington, D.C., Aug 1991

54. J.G. Bakuckas, Jr. and W.S. Johnson, Application of Crack Bridging Models to Fatigue Crack Growth in Titanium Matrix Composites, NASA TM 107588, Washington, D.C., April 1992

55. D.B. Marshall, B.N. Cox, and A.G. Evans, The Mechanics of Matrix Cracking in Brittle-Matrix Fiber Composites, *Acta Metall.*, Vol 33 (No. 11), 1985, p 2013-2021

56. L.N. McCartney, Mechanics of Matrix Cracking in Brittle-Matrix Fibre-Reinforced Composites, *Proc. R. Soc. Lond.*, Vol A409, 1987, p 329-350

57. R.M. McMeeking and A.G. Evans, Matrix Fatigue Cracking in Fiber Composites, *Mech. Mater.*, Vol 9, 1990, p 217-227

58. J.G. Bakuckas and W.S. Johnson, Implementation of Thermal Residual Stresses in the Analysis of Fiber Bridged Matrix Crack Growth in Titanium Matrix Composites, NASA TM 109082, Feb 1994

59. J.M. Larsen, J.R. Jira, R. John, and N.E. Ashbaugh, Crack Bridging Effects in Notch Fatigue of SCS-6/Timel 21S Composite Laminates, *Life Prediction Methodology for Titanium Matrix Composites*, STP 1253, W.S. Johnson, J.M. Larsen, and B.N. Cox, Ed., ASTM, 1995

60. D. Zheng and H. Ghonem, High Temperature/High Frequency Fatigue Crack Growth Damage Mechanisms in Titanium Metal Matrix Composites, *Life Prediction Methodology for Titanium Matrix Composites*, STP 1253, W.S. Johnson, J.M. Larsen, and B.N. Cox, Ed., ASTM, 1995

61. P. Bowen, Characterization of Crack Growth Resistance under Cyclic Loading in the Presence of an Unbridged Defect in Fiber Reinforced Titanium Metal Matrix Composites, *Life Prediction Methodology for Titanium Matrix Composites*, STP 1253, W.S. Johnson, J.M. Larsen, and B.N. Cox, Ed., ASTM, 1995

62. D.L. Davidson, Fiber-Matrix Micromechanics and Microstructural Observations under Tensile and Cyclic Loading, *Life Prediction Methodology for Titanium Matrix Composites*, STP 1253, W.S. Johnson, J.M. Larsen, and B.N. Cox, Ed., ASTM, 1995

63. J.G. Bakuckas, W.S. Johnson, and C.A. Bigelow, Fatigue Damage in Crossply Titanium Metal Matrix Composites Containing Center Holes, *J. Eng. Mater. Technol.*, Vol 115, Oct 1993, p 404-410

64. C.A. Bigelow and Y.A. Bahei-El-Din, Plastic and Failure Analysis of Composites (PAFAC), LAR-13183, COSMIC, University of Georgia, 1983

Residual Strength of Composite Aircraft Structures with Damage

C.C. Poe, Jr., NASA Langley Research Center

OVER THE PAST DECADE or so, secondary and primary structures of numerous commercial and military airplanes have been made of composites. Secondary structures are much more numerous than primary structures. Some of the primary structures being made of carbon-epoxy are vertical tails of the Airbus A300-340 series and the Boeing 777 transports and the wings and fuselages of the B-2 bomber. NASA and its contractors have completed two of three phases of the Advanced Composite Technology (ACT) program to develop composite wings and fuselages for commercial transport airplanes with costs that are competitive with those of current metal airplanes. Structural design criteria for these airplanes include damage tolerance requirements that specify the ability of a structure to operate safely with initial defects or in-service damage. Defects such as porosity, delamination, lack of bond between co-cured parts, and wrinkles in fibers can develop when composite materials are processed. Composite parts can also be scratched and gouged during manufacturing, shipping, and assembly. Low-velocity impacts can cause serious damage to carbon-epoxy even when the damage is not readily visible. Sources of low-velocity impacts are falling tools and equipment, runway debris, hail, birds, and even collisions with other airplanes and ground vehicles. Both metal and composite airplanes can be damaged by high-velocity impacts from parts of rotating machinery that fail in turbofan engines and penetrate the skin and supporting structure (discrete source damage). All of the aforementioned defects and damage are relevant to commercial and military airplanes. In addition, military airplanes can suffer ballistic damage, which is analogous to discrete source damage in civilian airplanes.

For U.S. commercial airplanes, certification criteria regarding defects and damage in composite materials are given in the Federal Airworthiness Regulations (FAR) Part 25 and FAA Advisory Circular (AC) 20-107A. They can be summarized as follows:

- *Nondetectable defects and damage:* Structure must withstand ultimate load 1.5 times limit load
- *Detectable damage:* Structure must withstand limit load (maximum expected load)
- *Discrete source damage:* Structure must carry 70% of limit maneuver loads and 40% of limit gust loads combined with maximum appropriate cabin pressure. (Discrete source damage is self-evident, allowing the crew to fly with minimum loads.)

For military airplanes, similar requirements are contained in the Aircraft Structures, General Specifications for AFGS 87-221.

With regard to detectable damage, AC 20-107A specifies that "The extent of initially detectable damage should be established and be consistent with the inspection techniques employed during manufacture and in service." Simple, inexpensive visual methods are preferred for in-service inspections. The AFGS 87-221 specifies the lower limit for visible damage to be a 2.5 mm deep dent or damage from a 25.4 mm diam impacter with 135 J, whichever is least. However, more recently, arguments are being made for a 1.3 mm-dent threshold for commercial transports.

For metal skins of commercial transport structures, discrete source damage is usually represented by a cut. The length of cut has traditionally been two bays of skin including one severed stiffener or frame (see Fig. 1). Similar configurations are cited in MIL-A-83444 for "fail safe crack arrest structure." For composite laminates, cuts are more likely to give a lower bound to tension strengths. See the results in Fig. 2 for cuts, impact damage, and holes (Ref 1-3). As will be shown subsequently, the strengths for cuts in Fig. 2 would probably have been lower had the laminates been as thick as those with holes. In the NASA ACT program, the two-bay crack has also been used to represent the upper limit for detectable damage. Thus, a strength analysis of the two-bay crack is important for damage tolerance certification of composite structures.

This paper was organized around the damage tolerance criteria listed above. It includes sections on defects, low-velocity impacts, and strength analysis for a two-bay crack. The section on defects is relatively short because impact and discrete source damage tend to be more critical. The section on impact includes subsections on resin toughness, laminate thickness, specimen size and impacter mass, and post-impact fatigue. An understanding of how specimen size and thickness and impacter mass affect impact response is es-

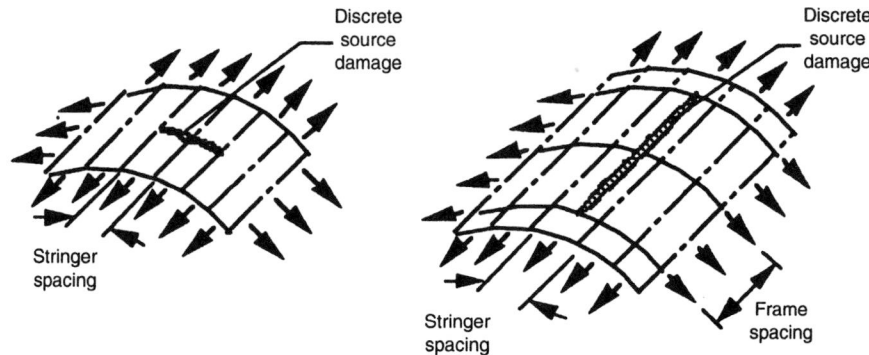

Fig. 1 Sketches of discrete source damage for fuselage

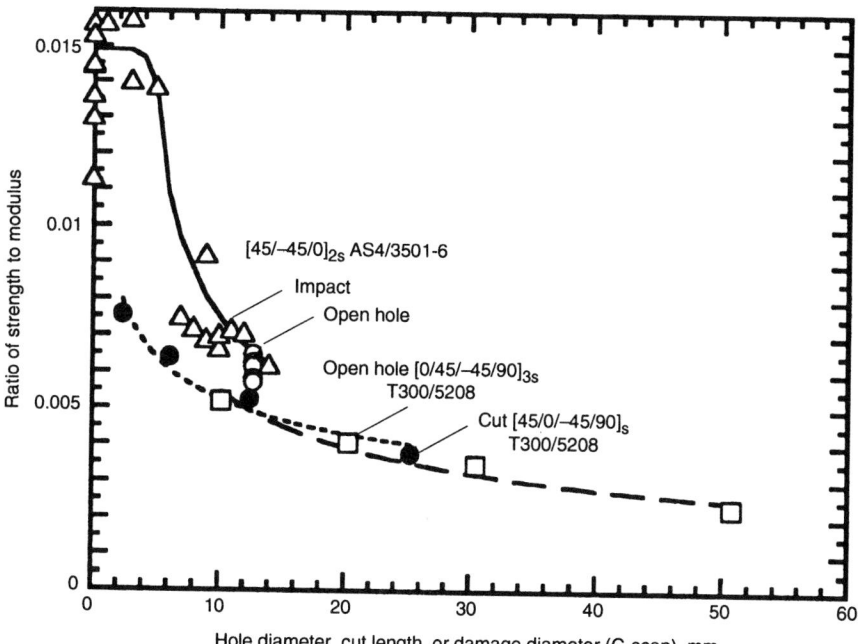

Fig. 2 Tension strengths for impact damage, open holes, and cuts

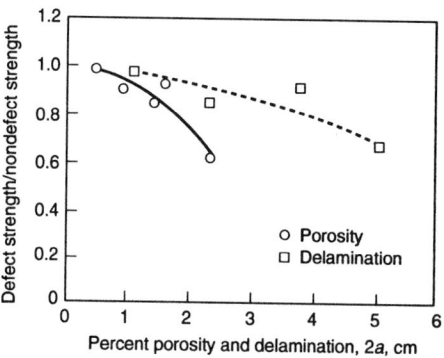

Fig. 3 Compressive strength versus defect size for AS4/3501-6 wing skin laminate

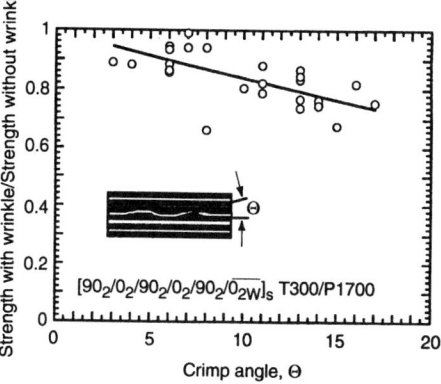

Fig. 4 Effect of manufactured wrinkle on compressive strength

sential to relating coupon and structural test results. The section on strength analysis for a two-bay crack presents both the linear elastic fracture mechanics (LEFM) and the R-curve methods for predicting strength.

Defects

The reduction in compressive strength with increasing size of porosity and delamination (Ref 4) is shown in Fig. 3 for an AS4/3501-6 wing skin laminate. The strengths were normalized by strength without defects or damage. Porosity and delaminations less than 2% and 4 cm in diameter, respectively, caused less than 20% reduction in compressive strength. Defects of this size can be detected initially in the factory using conventional ultrasonic methods. Tensile strengths are largely unaffected by these types of matrix defects. Thus, for both tension and compression, porosity and delaminations tend not to be critical.

The effect of wrinkles is shown in Fig. 4 for $[90_2/0_2/90_2/0_2/90_2/0_{2W}]_s$ T300/P1700 polysulfone laminates (Ref 5). Compressive strength with wrinkles normalized by compressive strength without wrinkles is plotted against crimp angle of the wrinkles. Each ply was 0.20 mm thick. The wrinkles were made in the middle two plies as indicated by the subscript W (see Fig. 4). The curve was fit to the data. The wrinkles reduced compressive strengths as much as 25 to 35% on the average for the largest crimp angle. These wrinkles will also reflect acoustic waves, but to a lesser degree than delaminations, and perhaps can also be detected using conventional ultrasonic methods (or by x-ray if tracer fibers are used). Sometimes wrinkles are visible on the surface and can be detected by visual inspection. The author had an experience where wrinkles in 16-ply quasi-isotropic T300/5208 laminates caused a significant reduction in tensile strength. As will be shown subsequently, the defects in Fig. 3 and 4 are generally less severe than low-velocity impact damage.

Low-Velocity Impacts

Resin Toughness. Low-velocity impacts from hail, runway debris, dropped tools, and so forth, present a special problem for carbon composites. Impacts, especially those made by a blunt object, may cause considerable internal damage without producing visible indications on the surface. For relatively thin laminates, damage to the resin can be widespread without damage to the fibers. A schematic of matrix damage for a $[-45/0/45/90]_{3s}$ T300/934 laminate (Ref 6) is shown in Fig. 5. (The loading direction and 0° fiber direction are synonymous throughout this paper.) In the left-hand figure, the delaminations within each interface resemble a pair of wedges that begin at a ply crack (transverse shear crack) in the lower ply and terminate at a ply crack in the upper ply. In the right-hand figure, an assembly of delaminations is shown for four consecutive interfaces (one repeating group of plies). The wedge-shaped delaminations resemble a spiral staircase. Although not shown in Fig. 5, the diameter of the wedges varies through the thickness much as a parabolic transverse shear stress distribution.

For the more common carbon/epoxy composites, numerous investigators have demonstrated that static indentation tests produce essentially the same damage as low-velocity impact tests for a given impact force. Figure 6 contains a graph of damage diameter d_o versus contact force F for static indentation and falling-weight impact tests (Ref 7). Open and solid symbols indicate data

from static indentation and impact tests, respectively. The 7.0 mm thick $[45/0/-45/90]_{6s}$ AS4/3501-6 and IM7/8551-7 plates were made from prepreg tape. The 3501-6 resin is a "brittle" epoxy, and the 8551-7 resin is a "toughened" epoxy. The plates were clamped over a 12.7 cm diam circular opening for the static indentation tests and a 12.7 cm square opening for the impact tests. A 12.7 mm diam hemisphere was used for the indenter and tup. The mass of the impacter was 4.63 kg. The damage diameters were calculated for a circle with area equal to the damage area in C-scan images. Both composite materials were penetrated in the static indentation tests as indicated by the vertical lines on the right. For both materials, the data for static indentation and impact tests are in agreement. The contact force to initiate damage F_1 (initiate in the sense that damage was visible in a C-scan image) is indicated by the vertical lines on the left. After initiation, damage diameter increased in proportion to contact force, which corresponds to a constant value of transverse shear strength per unit length Q^*.

$$Q^* = \frac{F}{\pi d_o} \qquad \text{(Eq 1)}$$

Both Q^* and F_1 are measures of damage resistance. The values of Q^* and F_1 for the toughened

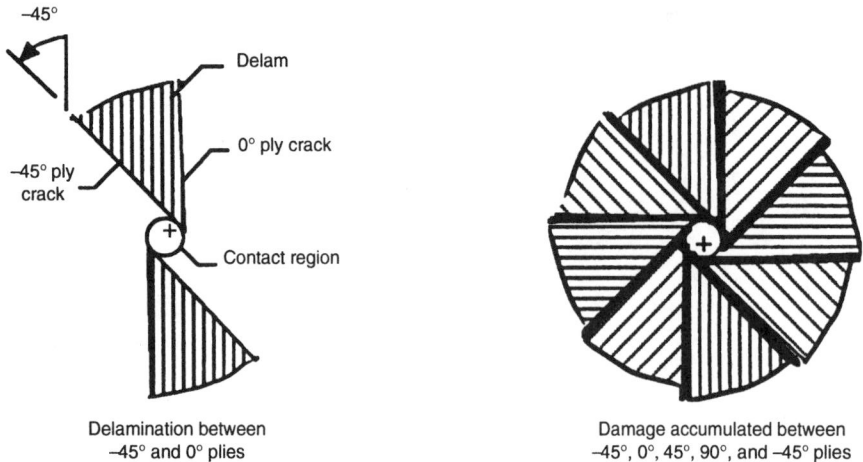

Fig. 5 Schematic showing plan view of impact damage through five consecutive interior plies of a $[-45/0/45/90]_{3s}$ T300/934 laminate

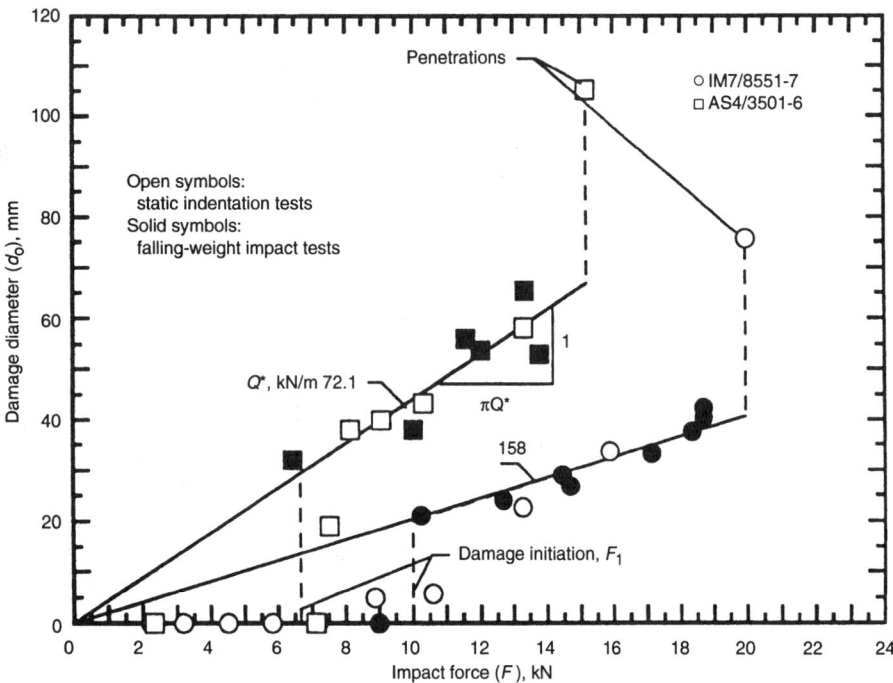

Fig. 6 Damage diameter for $[45/0/-45/90]_{6s}$ IM7/8551-7 and AS4/3501-6 laminates with 12.7 mm diam indenter

when the impacts caused damage large enough to be visible in a C-scan. Impact forces calculated using Eq 2 are also plotted for comparison. A value of $n_o = 4.52$ GPa determined by experiment was used instead of Eq 3. The differences between elastic constants for the two materials were too small to warrant plotting two curves for the energy balance equation. As will be shown subsequently, the relatively large impacter mass should result in a quasistatic response as assumed in Eq 2. Indeed, the measured and calculated impact forces agree for the one test for which damage was not visible in the C-scan. However, for all the impact tests with damage visible in C-scans, the test values of impact force were less than the predicted values. The discrepancies, which increased with increasing damage size, were caused by the reduction in flexural stiffness due to the delamination-type damage. Equation 2 does not account for impact damage. The measured impact forces were probably greater for IM7/8551-7 than for AS4/3501-6 because the damage was smaller (see Fig. 6).

Postimpact compressive strengths for the 7.0 mm thick $[45/0/-45/90]_{6s}$ AS4/3501-6 and IM7/8551-7 plates in Fig. 6 and 7 are plotted in Fig. 8. The static indentation specimens were 12.7 cm square, and the impact specimens were 12.7 cm wide and 25.4 cm long. All edges were simply supported during the strength tests. Undamaged strengths are plotted at $d_o = 0$. The strengths were normalized by modulus to represent far-field failing strains, and the lines were fit to the data. Because the delamination-type damage prevents adjacent plies from fully stabilizing one another in compression, failure initiates prematurely in the damaged region by sublaminate buckling and strengths are significantly below the undamaged strengths. The average undamaged strength ratio for the AS4/3501-6 material was 20% greater than that for the IM7/8551-7 material. However, when d_o exceeded 20 mm, the postimpact strengths for the two materials were about the same for a given d_o. For values of d_o greater than 40 mm (about 30% of the specimen width), the free edges probably exaggerated the strength reductions. That is, very wide specimens would have been stronger for the same d_o.

Laminate Thickness. Figure 9 contains a graph of damage diameter, d_o, versus contact force F for 16-, 24-, 32-, and 48-ply laminates from static indentation tests (Ref 8) and falling-weight impact tests (Ref 9). The $[45/0/-45/90]_{ns}$ AS4/3501-6 uniweave laminates were made using a resin-film-infusion (RFI) process. The average thickness per ply was 0.14 mm. The falling-weight impact tests were conducted with a 5.31 kg weight and a 12.7 mm diam hemispherical tup. The plates were clamped over a 7.62 cm square opening for the static indentation tests and over a 7.62 by 12.7 cm rectangular opening for the impact tests. The damage diameters were calculated for a circle with area equal to the damage area in C-scan images. The results for static indentation and impact tests are in reasonably good agreement. After damage initiated, damage diameter increased in proportion to contact force (in a

IM7/8551-7 material were 216% and 150%, respectively, of those for the AS4/3501-6 material.

Impact force can be calculated from the following equations (Ref 7) using energy balance considerations and assuming that the transverse displacements are given by the quasistatic solution.

$$\frac{1}{2} m_i v_i^2 = \frac{1}{2} \frac{F_{max}^2}{k} + \frac{2}{5} \frac{F_{max}^{5/3}}{n_o^{2/3} R_i^{1/3}} \qquad \text{(Eq 2)}$$

where

$$n_o \approx \frac{4}{3} E_2 \text{ for } E_i \gg E_2 \qquad \text{(Eq 3)}$$

and where F_{max} is the impact force, m_i is the mass of the impacter, v_i is the velocity of the impacter, R_i is the radius of the spherical impacter, E_i is the modulus of the impacter, k is the flexural stiffness of the plate for a quasistatically applied impact force, and E_2 is the modulus of the plate in the thickness direction. The first term on the right-hand side of Eq 2 accounts for flexural displacements, and the second term accounts for Hertzian indentation. For thin plates, the Hertzian term is negligible, and impact force increases in proportion to the square root of the product of kinetic energy and flexural stiffness.

Impact forces for the falling-weight tests in Fig. 6 are plotted against kinetic energy of the impacter in Fig. 7. The solid symbols indicate

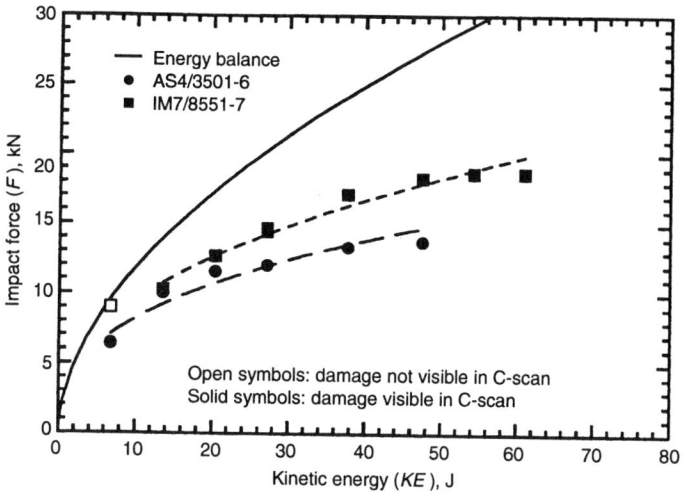

Fig. 7 Impact force versus impacter kinetic energy for $[45/0/-45/90]_{6s}$ IM7/8551-7 and AS4/3501-6 laminates and a 12.7 mm diam tup

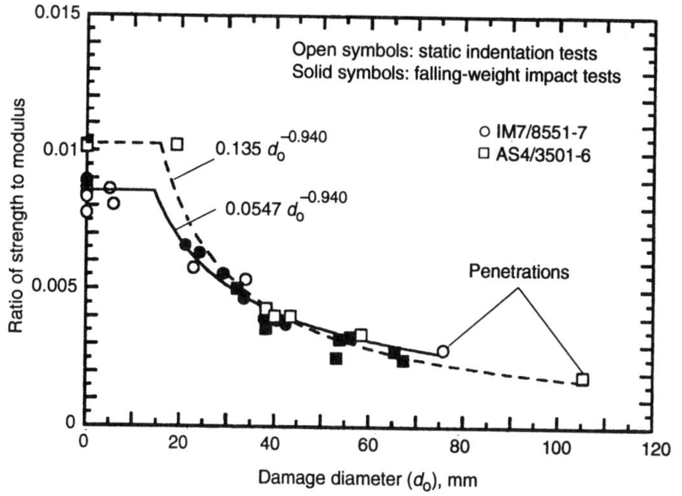

Fig. 8 Postimpact compressive strengths for $[45/0/-45/90]_{6s}$ IM7/8551-7 and AS4/3501-6 laminates with 12.7 mm diam indenter

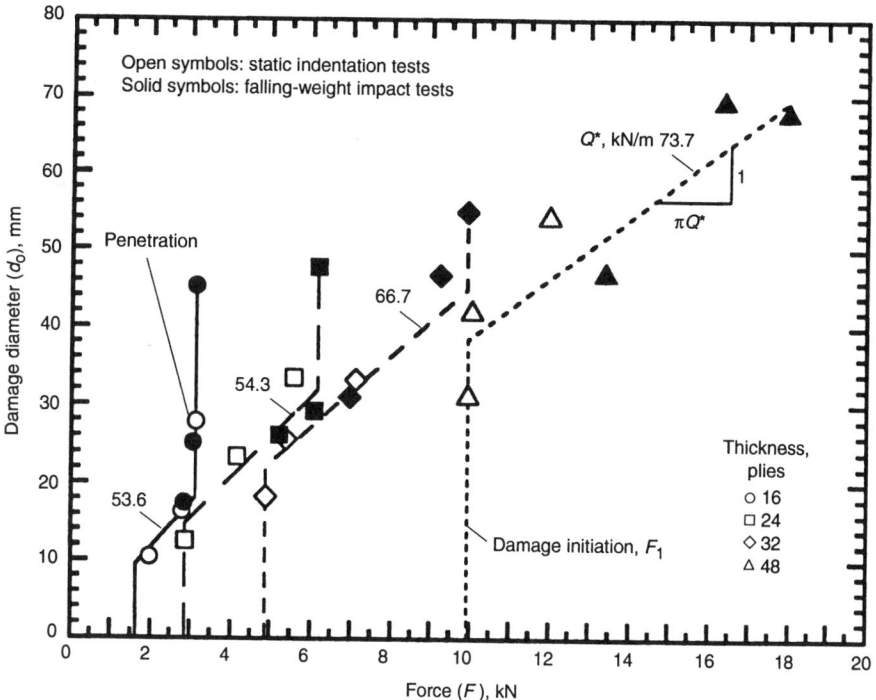

Fig. 9 Damage diameter for $[45/0/-45/90]_{ns}$ AS4/3501-6 uniweave (RFI) with a 12.7 mm diam indenter

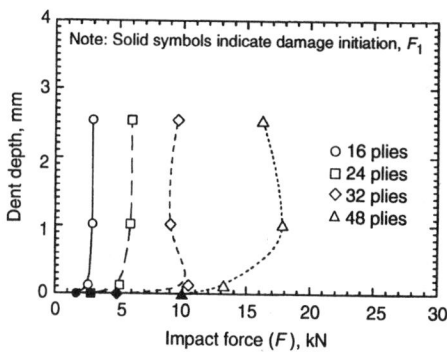

Fig. 10 Impact force to produce dents of various depths for $[45/0/-45/90]_{ns}$ AS4/3501-6 uniweave (RFI) with a 12.7 mm diam indenter

manner similar to that shown in Fig. 6). Also, the values of Q^* for the 48-ply RFI laminate and the AS4/3501-6 tape laminate in Fig. 6 are in good agreement. However, the value of F_1 for the 48-ply RFI laminate is more than 30% greater than that for the AS4/3501-6 tape laminate in Fig. 6. Values of F_1 increased with increasing thickness to approximately the 3/2 power, and the values of Q^* increased with increasing thickness to approximately the 0.3 power. Thus, damage resistance increased with increasing thickness, more so for F_1 than Q^*.

Only the 16-ply composite in Fig. 9 was actually penetrated in the static indentation tests, and no laminates were penetrated in the impact tests. However, for the largest impact forces, the im-

pacter appears to be on the verge of penetrating the 16-, 24-, and 32-ply laminates. The largest impact forces correspond to 2.54 mm dents. Although the forces were essentially equal for penetrating the 16-ply laminate and for making a 2.54 mm deep dent, d_0 was considerably greater for the dent than the penetration. The difference between the sizes of damage in the impact and static indentation tests could be associated with viscoelastic effects.

Values of average impact force for the impact tests in Fig. 9 are plotted against dent depth in Fig. 10. Values of F_1 in Fig. 9 are plotted at zero dent depth. The impact force to cause a dent of a given depth increased with increasing thickness much as F_1. In general, impact force exhibits

asymptotic behavior with increasing dent depth; the impact force to make a 2.54 mm deep dent was not significantly greater than that to make a 0.127 mm deep dent. For the two thickest laminates, the departure from monotonically increasing impact force is probably not statistically significant.

Impact forces for the 16-, 24-, 32-, and 48-ply $[45/0/-45/90]_{ns}$ AS4/3501-6 uniweave RFI plates are plotted against kinetic energy of the impacter in Fig. 11. All of the specimens were dented, and damage was visible in C-scans. Impact forces predicted using Eq 2 with $n_0 = 4.52$ GPa are also plotted for comparison. The predicted impact forces increase approximately with thickness to the 3/2 power. The test values of impact force also increased with increasing thickness, but not as much as predicted. The discrepancy increased with increasing thickness. The delamination-type damage reduced the impact force (in a manner similar to that shown in Fig. 7). The reduction in flexural stiffness is proportionately less for a thin laminate than a thick laminate. Thus, the test and predicted values of impact force for the 16- and 24-ply laminates

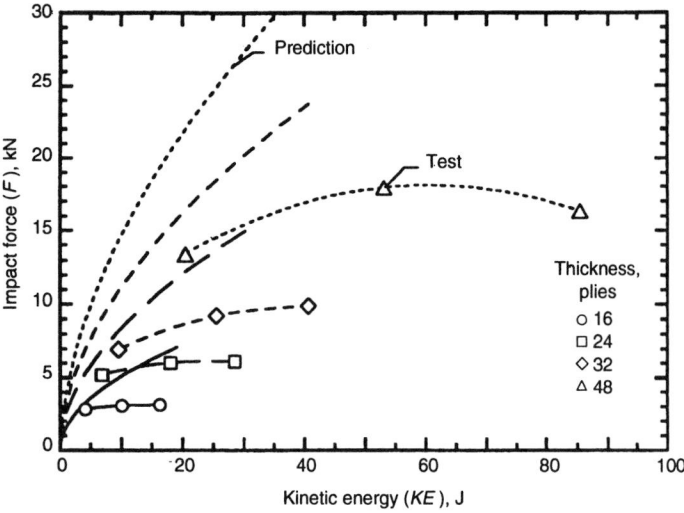

Fig. 11 Impact force versus impacter kinetic energy for $[45/0/–45/90]_{ns}$ AS4/3501-6 uniweave (RFI) and a 12.7 mm diam tup

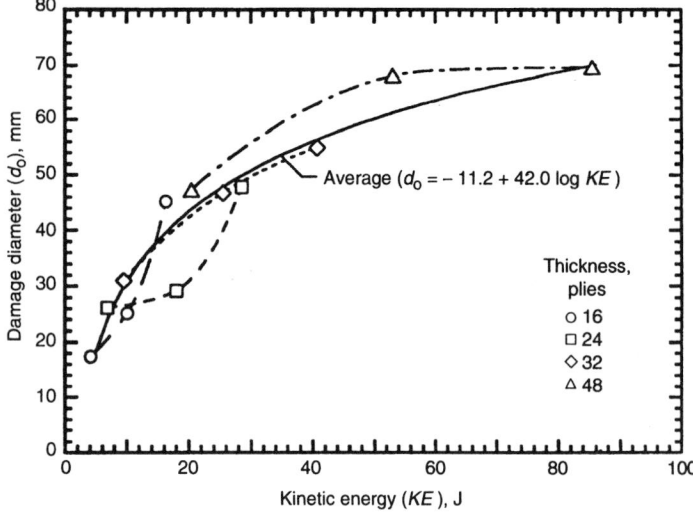

Fig. 12 Damage diameter versus impacter kinetic energy for $[45/0/–45/90]_{ns}$ AS4/3501-6 uniweave (RFI) and a 12.7 mm diam tup

with the least damage were in reasonably good agreement.

As shown previously, both impact force and damage resistance increase with increasing thickness. Thus, the net effect of increasing thickness is not obvious. For this reason, damage diameters in Fig. 9 for the falling-weight tests are plotted against kinetic energy of the impacter in Fig. 12. For the most part, damage size is independent of thickness, and the data are well represented by the average solid curve. The reader should be aware that the results in Fig. 12 should be quite different for a sandwich with face sheets of the same thickness because a sandwich is much stiffer than the face sheets alone and the impact forces would tend to be much larger.

Average postimpact compressive and tensile strengths are plotted against damage diameter in Fig. 13 for the falling-weight tests in Fig. 9 to 12. The compression and tension specimens were 10.2 cm wide by 15.2 cm long and 10.2 cm wide by 25.4 cm long, respectively. For the compression specimens, sides and ends were simply supported. For the tension specimens, sides were unsupported and 10.2 cm of each end were placed in a set of hydraulic grips. The strengths were normalized by modulus to represent far-field failing strains. Undamaged strengths for a similar 48-ply RFI laminate (Ref 10) are plotted as horizontal lines for reference. The strengths were higher for tension than compression, especially for the smallest damage sizes, and generally decreased with increasing damage size. The damage extended from 20 to 80% across the specimen width. Thus, the strength reductions were probably exaggerated by edge effects, much as in Fig. 8. The best-fit curve for compressive strength in Fig. 8 for the 48-ply AS4/3501-6 tape material is in good agreement with the 48-ply AS4/3501-6 RFI material.

The 16- and 24-ply compression specimens in Fig. 13 gave evidence of premature Euler buckling, probably contributing to the relatively low compressive strengths for the two smallest values

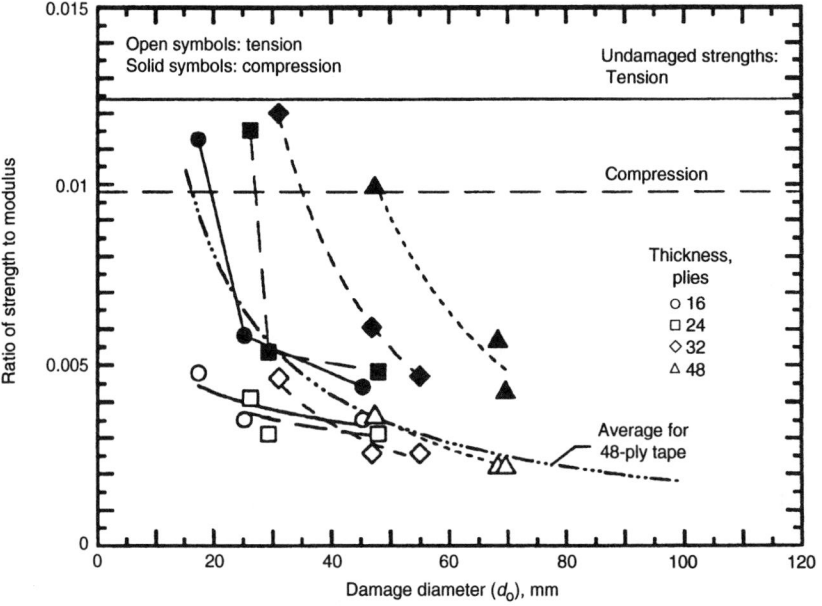

Fig. 13 Postimpact tensile and compressive strengths for $[45/0/–45/90]_{ns}$ AS4/3501-6 uniweave (RFI) and a 12.7 mm diam tup

of d_o. Thus, without buckling, the compressive strengths probably would have varied little with thickness, and C-scan damage size would be a convenient matrix for assessing postimpact compressive strength. On the other hand, tensile strengths increased significantly with increasing thickness. As noted previously, delamination-type damage reduces compressive strength but not tensile strength. Because fiber damage is confined to the contact region, fiber damage is much smaller in size than matrix damage. Thus, C-scan damage size is matrix damage size and not a convenient metric for assessing postimpact tensile strength. Because damage sizes in Fig. 12 were largely independent of thickness for a given impacter kinetic energy, a graph of strength ver-

sus impacter kinetic energy would have an appearance similar to that of Fig. 13 with regard to thickness variations.

Dent depth is an important metric for assessing postimpact strength. Thus, the postimpact tensile and compressive strengths in Fig. 13 are plotted against dent depth in Fig. 14. The strengths exhibit asymptotic behavior with dent depth. For the 0.1 mm deep dents, the tensile strengths were 80 to 97% of undamaged strength, but the compressive strengths were only 37 to 49%. For the 1 mm deep dents, the tensile strengths were 43 to 49% of undamaged strength, and the compressive strengths were 22 to 36%. For the 2.5 mm deep dents, tensile strengths were 34 to 39% of undamaged strength, and the compressive strengths

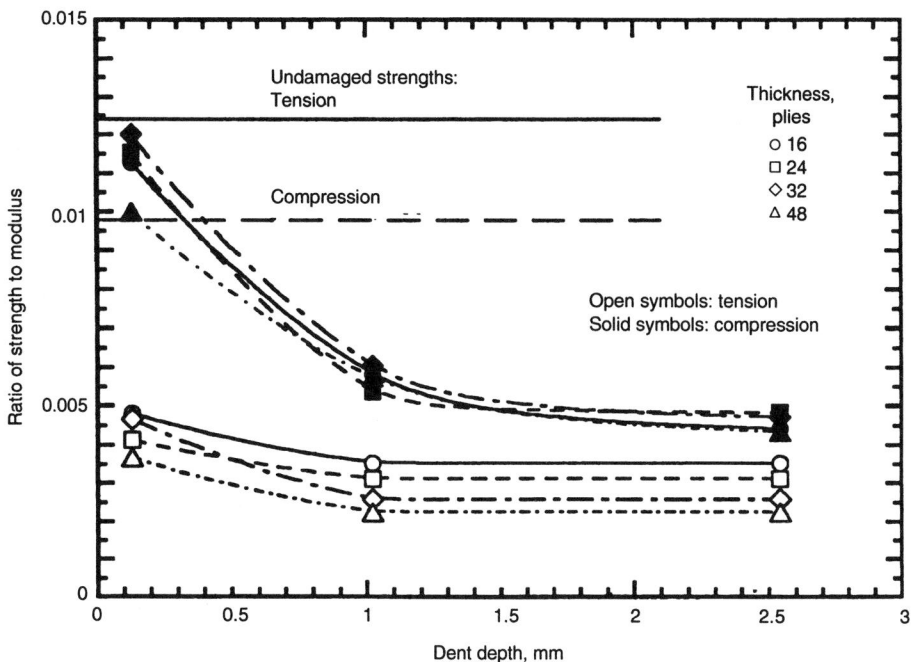

Fig. 14 Postimpact tensile and compressive strengths versus dent depth for [45/0/–45/90]$_{ns}$ AS4/3501-6 uniweave (RFI) and a 12.7 mm diam

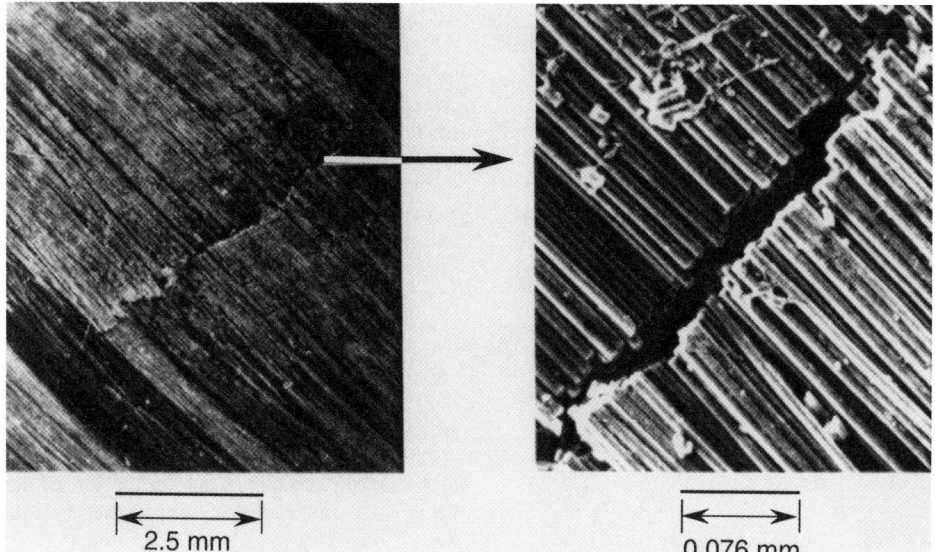

Fig. 15 Fiber damage in seventh layer of 36 mm thick AS4/HBRF-55A filament-wound case

were 22 to 36%. Thus, when dent depth increased from 1 to 2.5 mm, the tensile strengths decreased only 10 to 24% and the compressive strengths were unchanged.

In contrast to tensile strengths plotted against C-scan damage sizes in Fig. 13, the tensile strengths plotted against dent depths in Fig. 14 are largely independent of thickness, indicating that fiber damage is also largely independent of thickness. On the other hand, compressive strengths decrease significantly with increasing thickness when plotted against dent depths. Therefore, dent depth is a more convenient metric than C-scan damage size for assessing postim-

pact tensile strength, but C-scan damage size is a more convenient metric than dent depth for assessing postimpact compressive strength.

In very thick laminates, low-velocity impacts cause mostly fiber damage and very little delamination-type damage (Ref 11). Fiber breaks in the seventh layer of a 36 mm thick filament-wound motor case made of AS4/HBRF-55A are shown in Fig. 15. (Fibers were damaged only in layers 1 through 7 of the 76-layer laminate.) An enlargement of the fiber breaks is shown on the right-hand side of Fig. 15. The layers were separated following pyrolysis of the specimen. The layup was {(±56.5)$_2$/0/[±56.5)$_2$/0]$_3$/(±56.5)$_2$/0]$_7$/

(±56.5/0$_2$)$_4$/(±56.5)$_2$/(0/90)$_c$} where 90° is the axial direction of the cylinder. The underlined layers were about 1.6 times as thick as the other layers, and the innermost (0/90)$_c$ layer is a layer of plane weave cloth. The damage was made by static indentation using a 25.4 mm diam hemispherical indenter with a contact force of 54.3 kN. The locus of fiber breaks has the appearance of a "crack." More than one "crack" was present in damaged layers nearer the surface, and the direction of the cracks in each layer was generally perpendicular to the fiber direction in that layer. Delaminations outside of the contact region were not observed in radiographs.

The maximum depth of fiber damage δ normalized by contact radius, r_c (Ref 11) is plotted against average contact pressure p_c in Fig. 16. The damage was made by static indentation with spherical indenters of three diameters. The contact radius was calculated assuming Hertzian contact. Damage depths were calculated using Love's solution for hemispherical pressure applied to a semi-infinite, isotropic body, and a maximum shear stress criterion with a shear stress of 262 MPa. Damage is predicted to initiate at a critical value of contact pressure below the surface at a depth of δ/r_c = 0.482. The damage does not extend much beyond the contact region for pressures up to 1000 MPa as shown in the figure. The isotropic solution models the maximum damage depth quite well. However, in contradiction to the shear stress criterion, damage was always observed in contiguous layers from the surface down. One should expect the isotropic solution to be in greatest error near the surface where highly anisotropic layers dominate. Also, friction, which was not taken into account, may alter the stresses significantly near the surface.

Several effects of indenter radius are not obvious in Fig. 16. For Hertzian contact (Ref 11), the contact force can be written

$$F = (\pi p_c)^3 \left(\frac{R_i}{n_o}\right)^2 \quad \text{(Eq 4)}$$

Thus, the contact force to initiate damage for the 50.8 mm diam indenter is 16 times that for the 12.7 mm diam indenter. Similarly, the contact radius can be written

$$r_c = \pi p_c R_i/n_o \quad \text{(Eq 5)}$$

Thus, damage size for the 50.8 mm diam indenter is 4 times that for the 12.7 mm diam indenter for a given average contact pressure. Therefore, the sharpness of the indenter has a significant effect on the force to initiate damage and on damage size. The former is intuitive, but not the latter.

Postimpact tensile strengths (Ref 12) are plotted against impact force in Fig. 17 for the 36 mm thick filament-wound motor case. The strengths were normalized by the average of 19 undamaged tensile strengths. The range associated with ±1 standard deviation of undamaged strengths is shown on the ordinate. The damage was made by falling weights with 12.7 and 25.4 mm diam

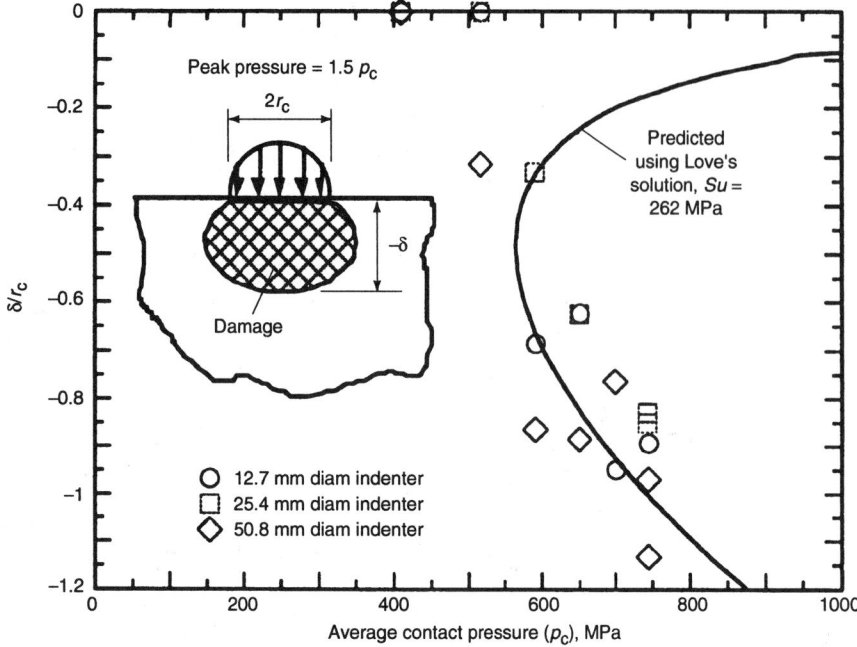

Fig. 16 Depth of fiber damage normalized by contact radius in 36 mm thick AS4/HBRF-55A filament-wound case

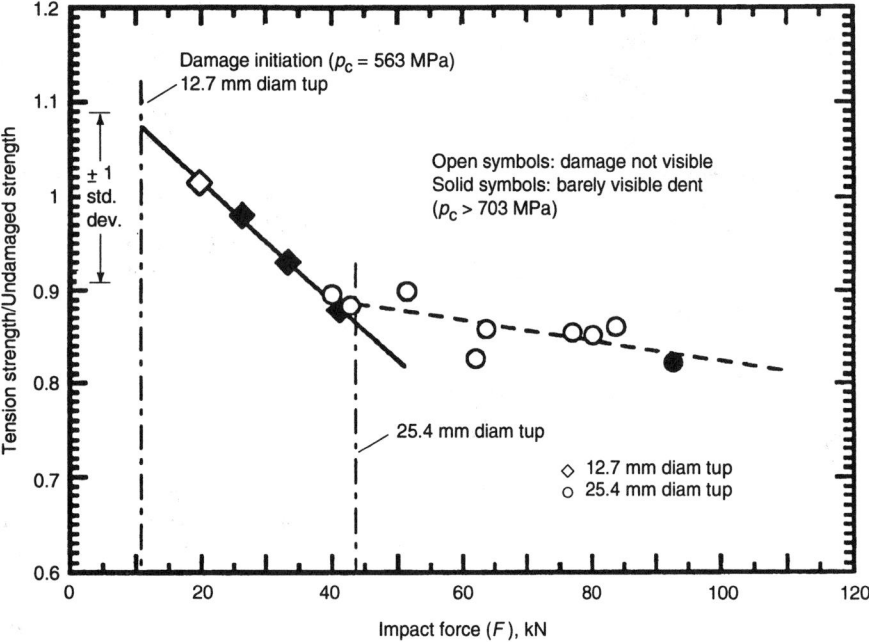

Fig. 17 Postimpact tensile strength for a 36 mm thick AS4/HBRF-55A filament-wound case

diam tup and the shallowest dent, about 80 to 97% of the undamaged strengths.

Specimen Size and Impacter Mass. Previously, test specimens were small and impacters were relatively massive, resulting in quasistatic response. Scaling small coupon results to large structures that do not respond quasistatically can be difficult. In Ref 7, dynamic finite element analyses were made of 7.0 mm thick AS4/3501-6 [45/0/−45/90]$_{6s}$ plates impacted with various masses having a 12.7 mm diam tup. In Fig. 18, the maximum calculated values of impact force are plotted against the frequency ratio $\omega^2/(k/m_i)$ for a constant value of impacter kinetic energy equal to 13.5 J. In the frequency ratio, ω is the natural frequency of the plate and $\sqrt{k/m_i}$ is a pseudonatural frequency of the impacter mass m_i attached to a spring that has a stiffness k equal to the flexural stiffness of the plate for a quasistatically applied impact force. For homogeneous uniformly thick plates, $\omega^2/(k/m_i) = (m_i/m_p)/\alpha^2$, where $\alpha = 0.471$ and 0.371 for simply supported and clamped boundaries, respectively. Results are plotted for three square plates with various sizes and boundary conditions: one was 12.7 cm with clamped boundaries, one was 12.7 cm with simply supported boundaries, and one was 25.4 cm with simply supported boundaries. Although the curves contain cusps and impact force does not vary monotonically with impacter mass, the general trend is for impact force to decrease with increasing impacter mass, particularly for the largest plate. Also, impact force is generally smaller for the larger plate and is less for simply supported than clamped boundary conditions.

The time of contact from the dynamic finite element analyses decreased with decreasing impacter mass. Thus, the number of reflections from the boundaries before the impact force reached a peak decreased with decreasing impacter mass. When impacter mass was to the left of the vertical dashed lines in Fig. 18, no reflections returned before the impact force reached a peak. To the right of the vertical lines, one or more reflections returned before the impact force reached a peak. A cusp occurs in the curves whenever the number of reflections changes. Thus, to the left of the vertical lines, impact force is not affected by plate size or boundary conditions. (Had impact force been plotted against the absolute value of impacter mass, the three curves would have coincided to the first cusp.) The horizontal dashed lines in Fig. 18 represent calculations with Eq 2 and $n_o = 4.52$ GPa, which was derived assuming quasistatic response. For large values of impacter mass, the dynamic analysis curves asymptotically approach Eq 2 as the number of reflections become large.

Experimental values of impact force (Ref 13) are plotted against impacter kinetic energy in Fig. 19 for plates of three sizes (masses): 10.2 cm diam circular plates (0.073 kg), a 24.1 cm square plate (0.50 kg), and a 53.3 cm square plate (2.5 kg). (The plate sizes were actually the sizes of the openings over which the plates were clamped.) Each of the two largest plates was tested multiple times. The plates were made from AS4/3501-6

spherical indenters. The straight lines were fit to the data. The solid symbols indicate that the impacts caused barely visible dents, probably less than 0.5 mm deep. These dents correspond to average contact pressures p_c less than 703 MPa. The vertical lines, which indicate impact forces to initiate damage for the two indenter diameters, correspond to an average contact pressure of 563 MPa (see also Fig. 16). As noted previously, calculations assuming Hertzian contact give a factor of four between impact forces for the two indenters to initiate damage. A similar factor exists between the impact forces to cause barely visible dents. The strengths for the nonvisible dents were lower for the 25.4 mm diam indenter than those for the 12.7 mm diam indenter. Thus, the more blunt indenter resulted in a lower allowable for nonvisible impact damage.

The origin for the strength axis is not shown in Fig. 17 because the strength reductions are less than 20%. Notice that the minimum strength is 88% of the undamaged strengths for the 12.7 mm diam tup. For thin laminates, the tension strengths in Fig. 13 were also high for a 12.7 mm tape with a layup of [45/0/–45/90]$_{6s}$ and a nominal thickness of 6.7 mm. The impacter was a falling weight with a 5.31 kg mass and a 12.7 mm diam hemispherical tup. Predictions were made using Eq 2 with n_0 = 4.52 GPa and plotted for comparison. The corresponding values of $(m_i/m_p)/\alpha^2$ were 15, 77, and 530 in the order of decreasing plate size. Based on the analytical results in Fig. 18, only the impact response for the smallest plates should have been quasistatic. Thus, without impact damage, one would expect the predicted values to be greater than the measured values for the two largest plates, especially for the 53.3 cm square plate that is twice the size of the largest plate in Fig. 18. As noted previously, damage made the measured impact forces fall below the predictions for the smallest plate. However, the test and predicted values for the midsize plate are in agreement, suggesting that the damage offset the transient nature of the response. And for the largest plate, the test values are indeed larger than the predicted values, even with damage. For two of the three impacts to the 53.3 cm square plate, the impacts did not cause visible damage in C-scans as indicated by the open symbols. The results with and without damage follow the same trend line indicating that damage had negligible effect on flexural stiffness for the largest plate, much as in the case of the thinnest plates in Fig. 11. Therefore, it is less important to account for damage when predicting impact force for plates with low flexural stiffness.

Minimum impacter kinetic energy to reduce burst pressure (Ref 14) is plotted in Fig. 20 for filament-wound pressure vessels of two sizes. The cylindrical sections of the 14.6 and 45.7 cm diam vessels were wet-wound with the same numbers of hoop and helical layers so that only size was different. The minimum kinetic energy for the large cylinder is about ten times that for the small cylinder. This size effect is consistent with that in Fig. 19 where the kinetic energy to initiate damage for the largest plate is more than four times that for the smallest plate.

Postimpact compression failing strains (Ref 4) are plotted against impacter kinetic energy in Fig. 21 for coupons and three-spar stiffened panels. The "hard" skins were nominally 6.3 mm thick and were made with [38/50/12] and [42/50/8] layups for coupons and panels, respectively. (The notation [38/50/12] indicates the percentage of 0° plies, ±45° plies, and 90° plies, respectively.) The 17.8 by 30.5 cm coupons were clamped over a 12.7 cm square opening and impacted using a 4.5 kg impacter with a 12.7 mm diam tup. The coupons were then trimmed to a 12.7 by 25.4 cm compression after impact specimen. (More details for this test method are given in Ref 15.) The stiffened panels were 61 by 46 cm. The spacing of the bolted titanium spars (stiffeners) was 14 cm, and the distance between inside edges of the C-section spars was 12 cm. A 25.4 mm diam tup and 11 kg impacter was used for the panel tests. The two panels impacted with 54 and 81 J energies were impacted two times in the transverse centerline (over skin only), once midbay of the center spar and left-hand spar and once midbay of

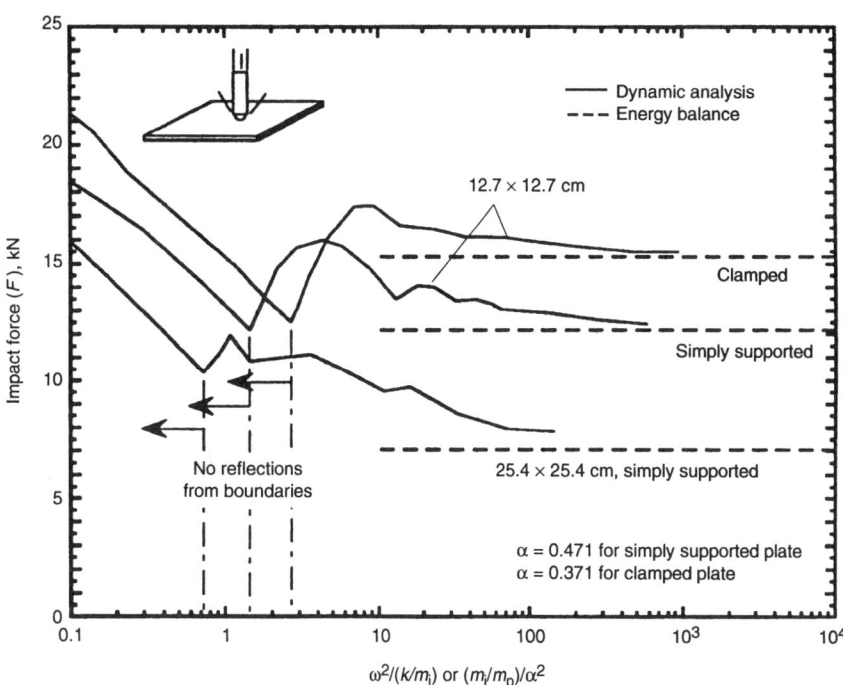

Fig. 18 Calculated impact force versus natural frequency ratio squared for AS4/3501-6 [45/0/–45/90]$_{6s}$ and an impacter kinetic energy = 13.5 J

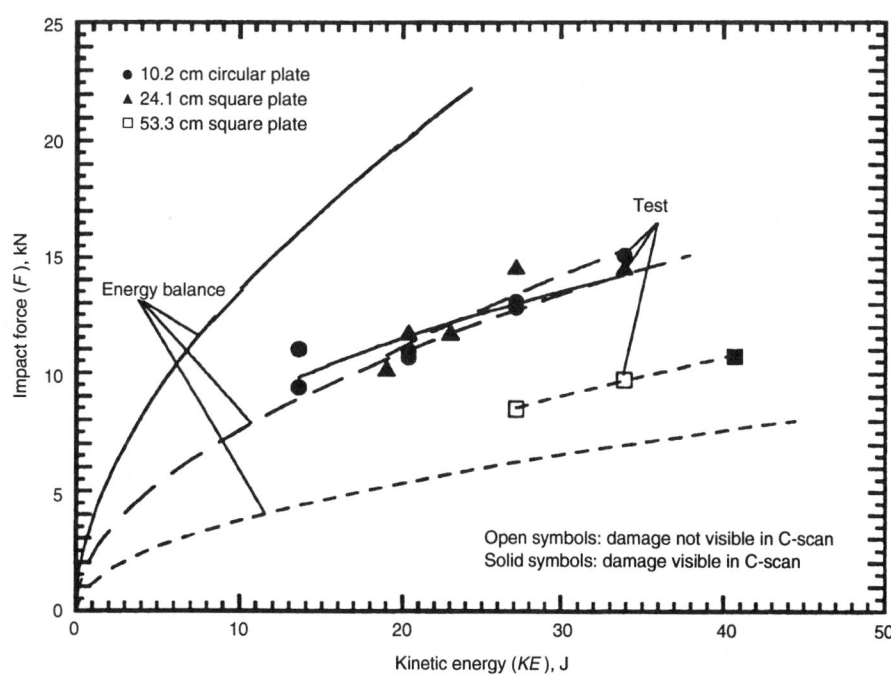

Fig. 19 Impact force versus impacter kinetic energy for [45/0/–45/90]$_{6s}$ AS4/3501-6 laminates of various sizes

the center spar and right-hand spar. The 27 J impact was at only one midbay location. The 135 J impacts were at three locations on each panel: once midbay, once over the skin only but near the edge of a spar, and once over a spar. A curve was fit to the coupon results. Except for panels with multiple 135 J impacts, failures were catastrophic, and failing strains were essentially equal for coupons and for panels. Failures of the

panels with multiple 135 J impacts were not catastrophic. After fracture arrest by the spars, the loads were increased 36 and 61% to cause complete failure. The initial failing strains for the panels with multiple 135 J impacts agreed with an extrapolation of the coupon data.

The agreement between failing strains of the coupons and three-spar panels in Fig. 21 indicates that the damage in the panels at the midbay im-

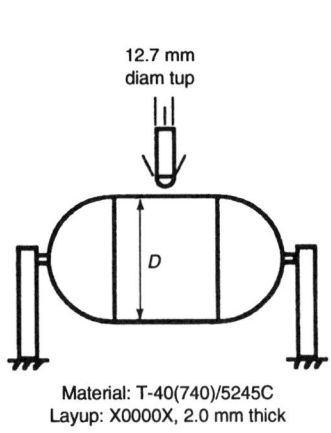

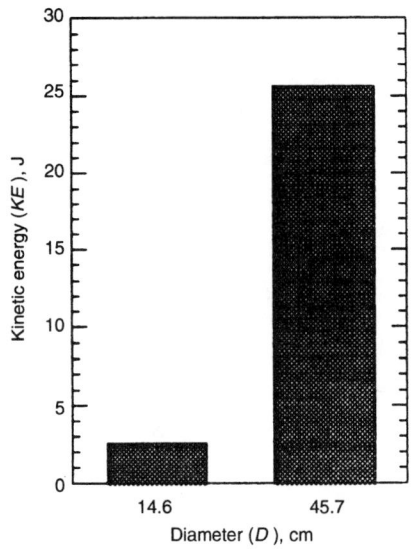

12.7 mm
diam tup

Material: T-40(740)/5245C
Layup: X0000X, 2.0 mm thick

Fig. 20 Minimum impacter kinetic energy to reduce burst pressure of small and large pressure vessels

pact site was critical and was equal in size to the damage in the coupons. Equal damage implies equal impact forces. In order for the impact forces to have been equal, the flexural stiffnesses of the panels and coupons would have to be equal. The flexural stiffness of a spar was estimated to be more than 60 times that of the midbay skin. Assuming that most of the out-of-plane deformation during a midbay impact takes place between spars, the stiffened panels can be treated as semi-infinite clamped plates with a width of 12 cm (distance between spars). Using equations in Ref 16 for a transverse force applied over a small radius to an isotropic clamped plate, the difference between flexural stiffnesses of a 12.7 cm square plate and a 12 cm semi-infinite plate is only 14%. Thus, for midbay impacts, the spars reduced the effective size of the stiffened panel during impact to that of the coupons.

Falling-weight tests are used to simulate tool drops, whereas gas-gun tests are used to simulate hail or runway stones and debris. Stones and hail have very small masses relative to tools. Damage size (Ref 13) is plotted against impacter kinetic energy in Fig. 22 for falling-weight and gas-gun tests conducted on plates of three sizes. Impact forces for the falling-weight data were plotted in Fig. 19. The mass of the falling-weight impacter was 5.31 kg, and a 12.7 mm diam hemispherical tup was attached to the end. The diameter and mass of the gas-gun projectile was 10.2 mm and 0.036 kg, respectively. Thus, the velocities of the gas-gun projectile were 12 times those of the falling weight. For the small and midsize plates, the damage sizes were essentially equal for the falling-weight and gas-gun tests. However, the falling weight did not even damage the largest plate for energies less than 40J; whereas, the gas

gun damaged the large plate as much as the midsize plate.

As noted previously, the mass ratios for the falling-weight tests were $(m_i/m_p)/\alpha^2 = 15, 77$, and 530 in the order of decreasing plate size. The corresponding mass ratios for the gas-gun tests were 0.10, 0.53, and 3.6. The small ratios indicate that the response was transient for all the gas-gun tests (see also Fig. 18). Thus, without damage, the impact forces for the gas-gun tests should have been larger than those for the falling-weight tests. Indeed, for the largest plate, the damage sizes for the gas gun tests were larger than those for the falling-weight tests. However, for the small and midsize plates, damage for the gas-gun tests was not larger than that for the falling-weight tests. As noted previously, damage reduces impact force, and probably ameliorated the differences between impact forces and resulting damage sizes for the gas-gun and falling-weight tests. The mass ratios for the two largest plates are so small that the reflections probably did not return before the peak impact forces were attained. Thus, the damage caused by the gas gun was essentially equal in size for the midsize and largest plates.

Impact forces (Ref 12) are plotted against impacter mass in Fig. 23 for 36 mm thick filament-wound rings. (Strengths for specimens cut from these rings were plotted in Fig. 17.) The impacter kinetic energy was approximately 73 J, and the impacter masses ranged from 2.8 to 18.6 kg. A 25.4 mm tup was attached to the end of the impacters. The 30 cm long rings were cut from a 76 cm diam case. One ring was empty, and one was filled with inert rocket propellant. The masses of the empty and filled rings were 40 and 288 kg, respectively. As in the gas gun tests, impacter masses were less than ring masses. Damage was not visible for any of these tests. Impact force decreased with increasing impacter mass, as also shown by the graph of finite element results in Fig. 18, indicating transient response. Impact forces were larger for the filled ring than the empty ring, indicating that the mass of the

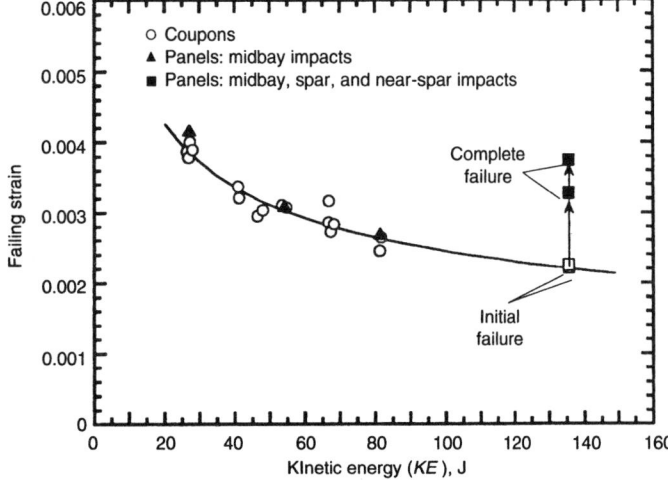

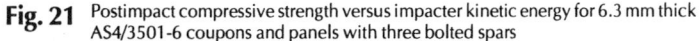

Fig. 21 Postimpact compressive strength versus impacter kinetic energy for 6.3 mm thick AS4/3501-6 coupons and panels with three bolted spars

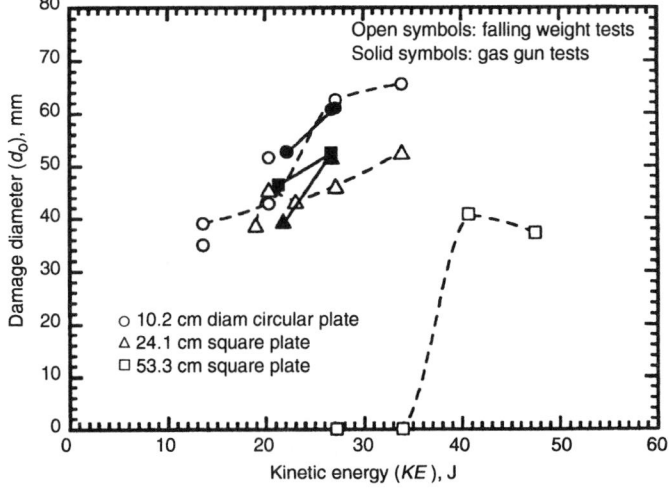

Fig. 22 Damage diameter versus impacter kinetic energy for falling-weight impact tests and gas-gun tests conducted on $[45/0/-45/90]_{6s}$ AS4/3501-6 laminates

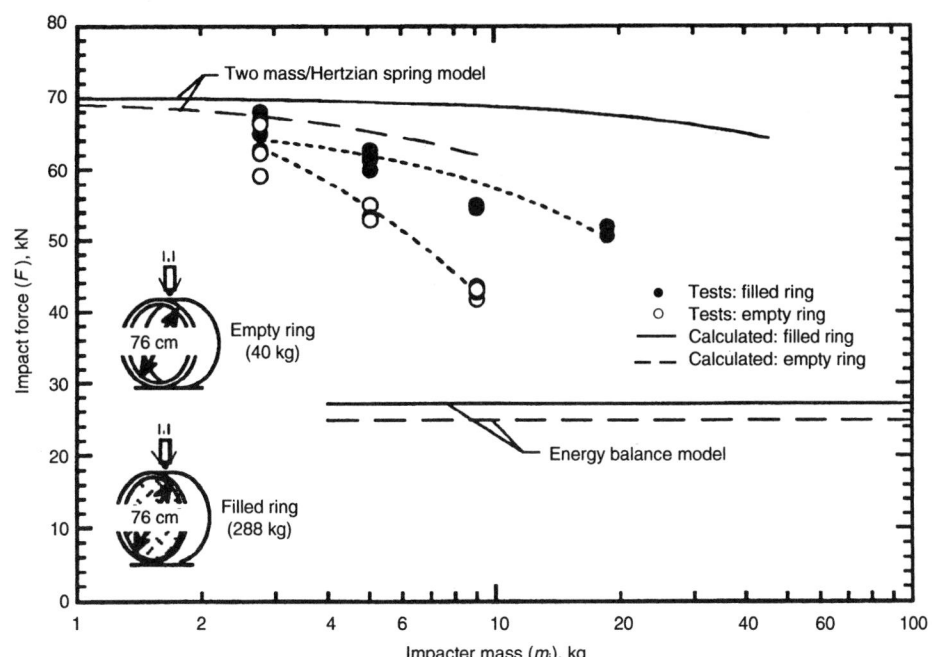

Fig. 23 Impact force versus impacter mass for 73 J impacts with a 25.4 mm diam tup to filled and empty 36 mm thick AS4/HBRF-55A filament-wound cases

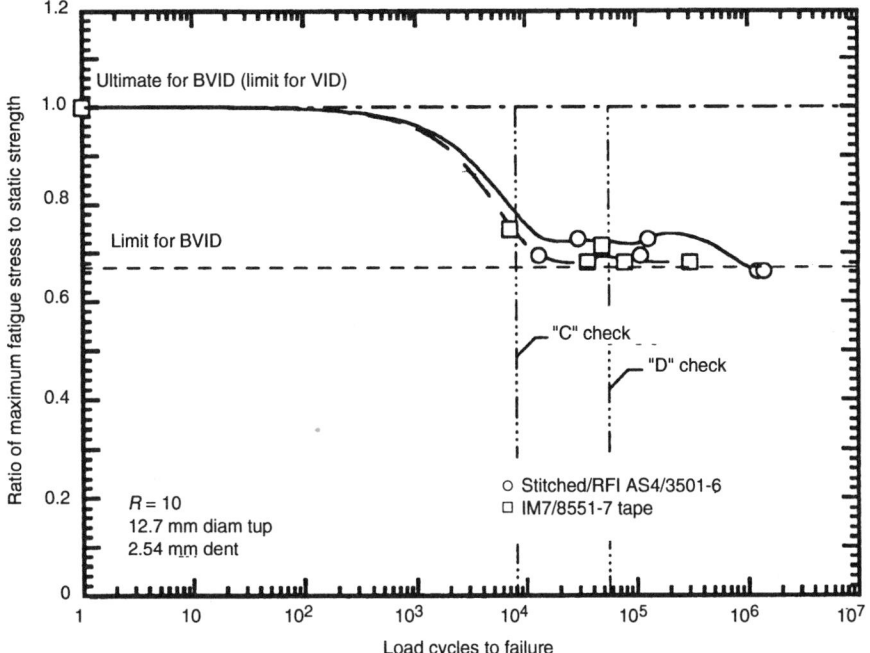

Fig. 24 Constant amplitude compression fatigue for $[45/0/-45/90]_{6s}$ laminates with impact damage resulting in 2.54 mm dent

inert propellant caused the impacts to be even more transient, more to the left in Fig. 18.

Calculations of impact force are plotted in Fig. 23 that bound the test results. The lower bounds were made with the energy balance model (Eq 2 with $n_0 = 4.52$ GPa). The values of flexural stiffness for the empty and filled rings, which were determined experimentally, were 5.08 and 6.34 MN/m, respectively. The upper bounds were calculated with the following "two mass Hertzian spring model" (Ref 12).

$$\frac{1}{2} m_i v_i^2 = \frac{2 (1 + m_i/m_p) F_{max}^{5/3}}{5 \, n_0^{2/3} R_i^{1/3}}$$

(Eq 6)

Equation 6 considers Hertzian contact of two hemispherical bodies, one is soft (plate) and the other is

hard (impacter). The masses of both bodies are present because peak contact force is assumed to occur when the relative velocity of the two bodies is zero. The values of impact force calculated with Eq 2 were less than one-half of those calculated with Eq 6 for the range of masses shown. The test values for the smallest impacter mass were in reasonable agreement with Eq 6. However, the test values diverge from Eq 6 and approach Eq 2 with increasing mass. From Fig. 16, impacter mass would have to exceed 10 times the mass of a flat plate before the test values and Eq 2 would be in agreement.

Postimpact Fatigue. For in-plane loading, carbon/epoxy composites have "superb fatigue properties" compared to metals (Ref 17). However, growth of impact damage during fatigue loading and the resulting reduction in compressive strength between inspections may be critical (Ref 4). The ratio of maximum compressive fatigue stress to static strength for two carbon/epoxy composites with impact damage (Ref 18) is plotted against fatigue cycles to failure in Fig. 24. The average static strengths are represented as 1 cycle. The "smoothed" solid and dashed curves were fit through the test data. The fatigue loading was constant amplitude with a minimum compressive stress of one-tenth the maximum compressive stress. An impacter with a 12.7 mm diam hemispherical tup was used to make 2.54 mm dents on the average. The specimens were 12.7 by 25.4 cm. One composite was made of toughened IM7/8551-7 prepreg tape, and the other was made by laying down layers of dry uniweave fabric, stitching the stack, and introducing resin by the RFI process. The layups for the two composites were $[45/0/-45/90]_{6s}$, and the typical thicknesses were 7.45 and 9.22 mm for the tape and stitched material, respectively. Cycles to failure for the two materials nearly coincide. The fatigue stress ratios asymptotically approached 0.67.

Recall from the introduction that the FAA criteria require that structure carry ultimate load with barely visible impact damage (BVID) and limit load with visible impact damage (VID). Thus, the compressive fatigue stress ratios of 1 and $\frac{2}{3}$ in Fig. 24 correspond to ultimate and limit-load conditions, respectively, with BVID. Taking a 2.54 mm dent as the threshold for VID and assuming that values of normalized compressive fatigue stress would not decrease with increasing dent depth and tup diameter, the stress ratios in Fig. 24 represent an allowable compression stress with BVID. Because the compression fatigue stress ratios are greater than $\frac{2}{3}$ (the maximum expected load) up to 10^6 cycles, compressive fatigue is not critical for these two materials with BVID. It is also not expected that compressive fatigue would be critical for conventional epoxy tape laminates with BVID.

Although structure must carry ultimate load with BVID, structure need only carry limit load with VID. Now the compressive fatigue stress ratio of 1 in Fig. 24 corresponds to limit-load condition. Because the stress ratios are below limit-load condition, factors of safety may be required for growth of VID between inspections.

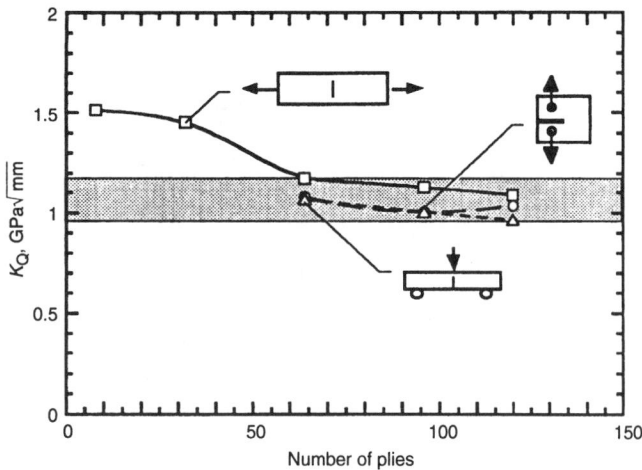

Fig. 25 Fracture toughness for center-crack, compact, and three-point-bend specimens for [0/±45/90]ns T300/5208 laminates

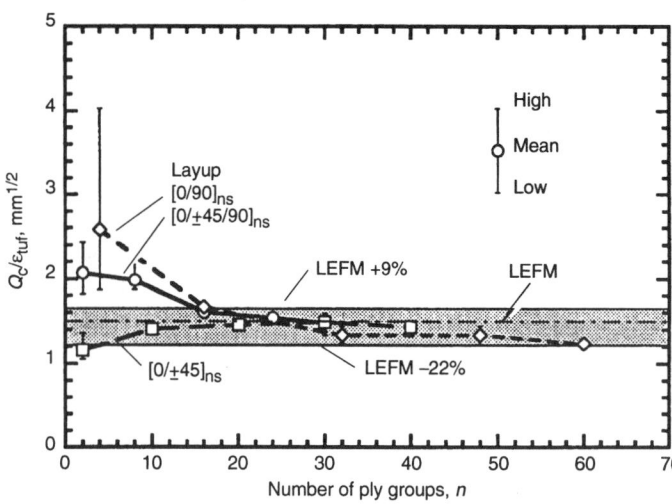

Fig. 26 Fracture toughness versus number of ply groups for several [0/±45/90k]ns T300/5208 laminates

Airlines typically inspect all exposed, exterior surfaces for obvious damage approximately every 2000 flight hours and exterior and interior structure that involve disassembly approximately every 14,000 flight hours. These inspections are sometimes called "C" and "D" checks, respectively. In addition to visual inspection methods, nondestructive inspection methods (ultrasonics, radiography, eddy current, etc.) are used for "D" checks. Assuming 1-h flights and one load cycle per flight, the "C" and "D" checks are shown as vertical lines in Fig. 24 for a factor of safety of four. (Factors of safety greater than four would move the vertical lines even farther to the right.) The vertical lines for "C" and "D" checks intersect the lower curve at 0.74 and 0.69, respectively. These ratios correspond to factors of safety of 1.35 (1/0.74) and 1.45 (1/0.69). Note that a factor of safety of 1.5 would be equivalent to treating VID as BVID. Assuming one load cycle per flight of limit load is probably conservative. Spectrum tests would be necessary to establish accurate factors of safety.

Tension Strength Analysis for Two-Bay Crack

Numerous models and methods have been developed to predict strength of composites with cracklike cuts and tension loads. The following is a partial list given in the order of increasing requirements for computational resources and fracture parameters.

1. Linear elastic fracture mechanics (LEFM) (Ref 3, 19, 20)
2. Point stress (Ref 20)
3. Average stress (Ref 20)
4. Mar-Lin model (Ref 20)
5. Damage zone criterion (Ref 21)
6. R-curve method (Ref 19)
7. Damage zone model (Ref 22)
8. Strain softening method (Ref 23)

All of the above models and methods represent a composite material as an anisotropic continuum amenable to classical lamination theory. Methods 1 to 6 generally require only linear analyses. Method 1 requires no tests to determine fracture parameters, but is generally accurate only when matrix damage at the notch tips is small compared with notch length. Methods 2 to 6 require several fracture parameters per laminate and method 7 requires a fracture curve for each laminate. Methods 7 and 8 require nonlinear finite element analyses as well as fracture parameters for each laminate. Only methods 5 to 8 explicitly account for damage growth at the notch tips (stable tearing). The LEFM and R-curve methods, which require only linear analyses, will be used below to illustrate the bounds for linear and nonlinear behavior.

Linear Elastic Fracture Mechanics (LEFM) Method. Values of fracture toughness (Ref 24) are plotted against thickness (number of plies) in Fig. 25 for middle-cracked, compact, and three-point-bend specimens made from a quasi-isotropic laminate. Compact and three-point-bend specimens with thickness less than 64 plies could not be tested because of large bearing stresses. Fracture toughness values were calculated using stress-intensity factor equations with initial cut length and failure load. Anisotropy generally has little effect on stress-intensity factor solutions for plane problems. Thus, stress-intensity factor solutions for isotropic, homogeneous plates were used throughout this section. For the middle-cracked specimens, fracture toughness decreased with increasing thickness and asymptotically approached a constant value. For laminates 64 plies and thicker, values of fracture toughness were approximately a constant (between 0.97 and 1.18 GPa√mm) for all three types of specimens. For thin laminates, broken fibers in 0° plies and matrix cracks and delaminations were revealed at the ends of the cuts when the laminates were deplied following an application of load to near failure (Ref 25). For the thick laminates, broken fibers in 0° plies at the ends of the cuts were found throughout, but the matrix cracks and delamina-

tions were confined to the outermost four plies; the interior plies were well bonded. Thus, the strengths for thin laminates were dominated by matrix cracks and delaminations at the ends of the cut but not those for the thick laminates.

The fracture toughness of laminates (Ref 26) was expressed in terms of the elastic constants using a point-strain-type failure criterion as follows:

$$K_Q = Q_c E_x \xi^{-1} \qquad \text{(Eq 7)}$$

where

$$\xi = (1 - \sqrt{v_{xy} v_{yx}}) (\cos^2 \alpha + \sqrt{E_x E_y^{-1}} \sin^2 \alpha) \qquad \text{(Eq 8)}$$

$$Q_c = \varepsilon_{tuf} \sqrt{2\pi \, \overline{d_o}} \qquad \text{(Eq 9)}$$

E is Young's modulus, v is Poisson's ratio, x and y are Cartesian coordinates, α is the angle that the principal load-carrying fibers make with a line perpendicular to the cut, ε_{tuf} is the tensile failing strain of the fibers, and d_o is the "characteristic distance." (The angle $\alpha = 0$ for laminates in which 0° plies carry the majority of the load.) A median value of $\sqrt{2\pi \, \overline{d_o}} = 1.5$ mm$^{1/2}$ was determined by analyzing fracture data for numerous layups and materials.

Values of $Q_c/\varepsilon_{tuf} = \sqrt{2\pi \, \overline{d_o}}$ (Ref 27) are plotted against number of ply groups in Fig. 26 for three T300/5208 laminates including the laminate in Fig. 25. For laminates thicker than 16 ply groups, the values of Q_c/ε_{tuf} are between 1.23 and 1.65 mm$^{1/2}$, and the standard deviations are small. For thinner laminates, the mean values range between 1 and 4 mm$^{1/2}$, and the standard deviations are large. As noted previously, the discrepancy between values of fracture toughness for thin and thick laminates is largely associated with matrix cracks and delaminations at the ends of the cut. For thin laminates, the mean fracture toughness of the three layups decreases with decreasing number of 90° plies. Accordingly, the fracture toughness of the thin [0/±45]ns laminates increased with increasing thickness and that of the

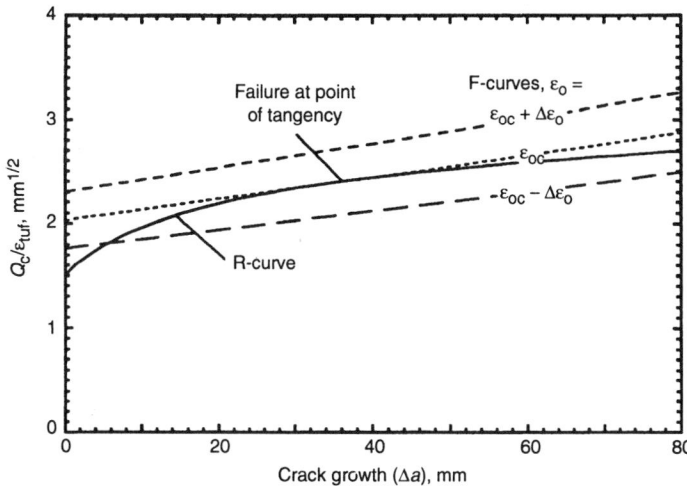

Fig. 27 Graphical solution for R-curve method using F-curves for uniaxially loaded sheet with central crack

Fig. 28 R-curve for $[\overline{+45/0/90/+30/0}]_s$ AS4/938 fuselage crown laminate ($2a_0 = 23$ cm and $W = 91$ cm)

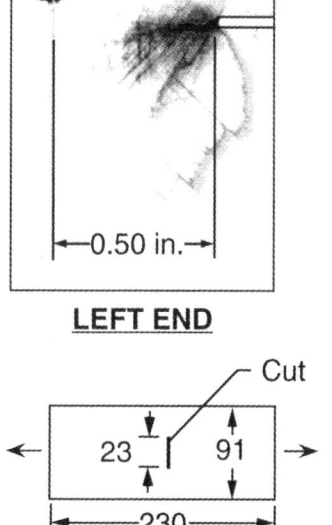

LEFT END

23   91

—230—

All dimensions in centimeters.

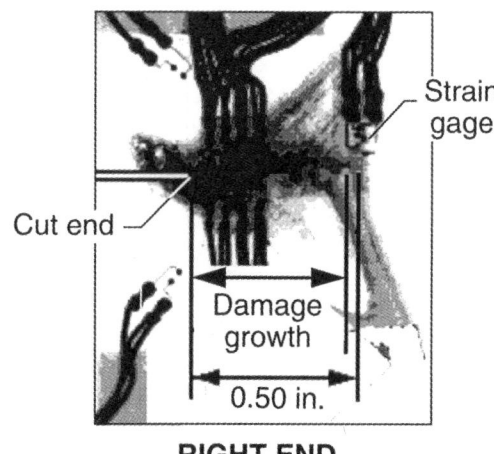

Strain gage

Cut end

Damage growth

0.50 in.

RIGHT END

Far-field strain = 0.00232
(85.5% of failure)

Fig. 29 Radiographs of damage growth at ends of cut immediately before failure for $[\overline{+45/0/90/+30/0}]_s$ AS4/938 fuselage crown laminate

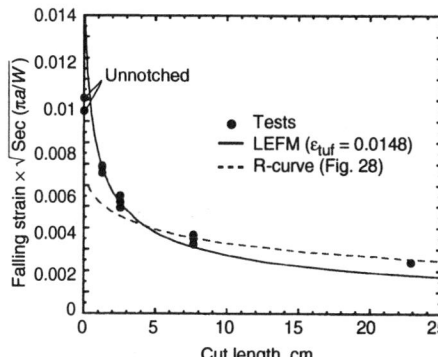

Fig. 30 Predicted failing strains using R-curve and LEFM for $[\overline{+45/0/90/+30/0}]_s$ AS4/938 fuselage crown laminates. $W \geq 8a_0$

$[0/90]_{ns}$ and $[0/\pm45/90]_{ns}$ laminates decreased with increasing thickness. Therefore, LEFM is not generally valid for thin laminates, and the application of LEFM to thin laminates may be conservative or unconservative, depending on layup and resin toughness.

R-Curve Method. The R-curve method can be used for materials that exhibit significant stable damage growth at the ends of cuts. In ASTM E 561 (Ref 28), crack-growth-resistance curves (R-curves) and crack-driving-force curves (F-curves) are expressed in terms of stress-intensity factor. However, for composites, it is convenient to use Q_c/ε_{tuf} instead of stress-intensity factor to normalize for layup and material. (Of course, d_0 could be used as well.) The R- and F-curves can be calculated by:

$$\frac{Q_c}{\varepsilon_{tuf}} = \frac{\xi K}{\varepsilon_{tuf} E_x} \qquad (Eq\ 10)$$

where K is the stress-intensity factor. For a uniaxially loaded sheet of width W containing a central crack of length $2a$,

$$K = \varepsilon_0 \sqrt{\pi a \sec (\pi a/W)} \qquad (Eq\ 11)$$

where ε_0 is the far-field strain, $a = a_0 + \Delta a$, a_0 is cut half-length, and Δa is the damage growth at each end. Failure is assumed to occur when the increase in crack driving force exceeds the increase in crack growth resistance, which can be expressed by the following partial derivative:

$$\frac{\partial}{\partial \Delta a} \left(\frac{Q_c}{\varepsilon_{tuf}} \right)_F \bigg|_{\varepsilon_0 = \varepsilon_{oc}} \geq \frac{\partial}{\partial \Delta a} \left(\frac{Q_c}{\varepsilon_{tuf}} \right)_R \qquad (Eq\ 12)$$

where ε_{oc} is the far-field failing strain and the F and R indicate F- and R-curves, respectively. The R-curve method is illustrated in Fig. 27, where an R-curve and several F-curves for a uniaxially loaded 91 cm wide sheet with a 23 cm long central crack are plotted. The F-curve labeled $\varepsilon_0 = \varepsilon_{oc}$ is tangent to the R-curve satisfying Eq 12. The F-curves labeled $\varepsilon_{oc} + \Delta\varepsilon_0$ and $\varepsilon_{oc} - \Delta\varepsilon_0$ are in the neighborhood of the solution to Eq 12. For all the R-curve predictions in this paper, Eq 12 was solved graphically for far-field failing strain by calculating F-curves for increments of ε_0 until the F-curve was tangent to the R-curve.

An R-curve is shown in Fig. 28 for a 13-ply fuselage crown laminate made of 190 gm/m² prepreg tape using a tow-placement process. Values of Q_c/ε_{tuf} (Ref 19) are plotted against damage growth. Values of Q_c/ε_{tuf} are calculated using Eq 10 and 11. The values of Δa, which were measured in radiographs like those in Fig. 29 and from measurements of crack-opening displacements (COD), agree well. The COD was measured by a

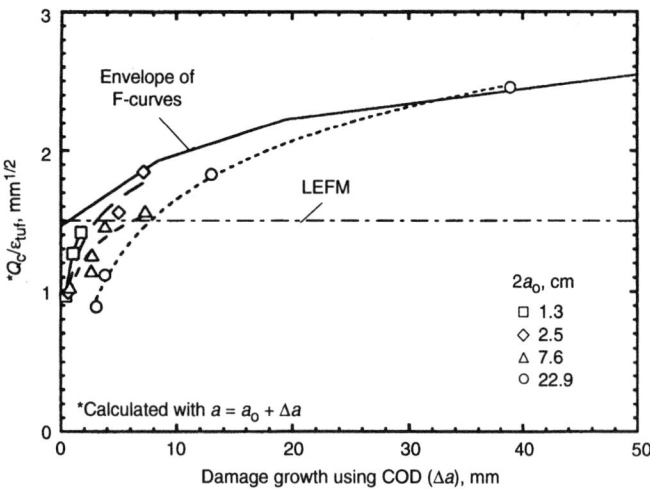

Fig. 31 R-curves for $[\overline{+45}/0/90/\overline{+30}/\overline{0}]_s$ AS4/938 fuselage crown laminates with various cut lengths. $W \geq 8a_0$

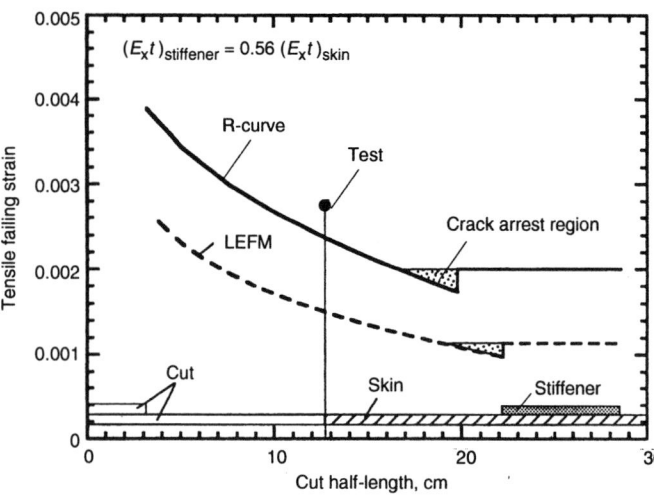

Fig. 32 Test and predicted failing strains for three-stringer $[\overline{+45}/0/90/\overline{+30}/\overline{0}]_s$ AS4/938 fuselage crown panel (76 by 213 cm)

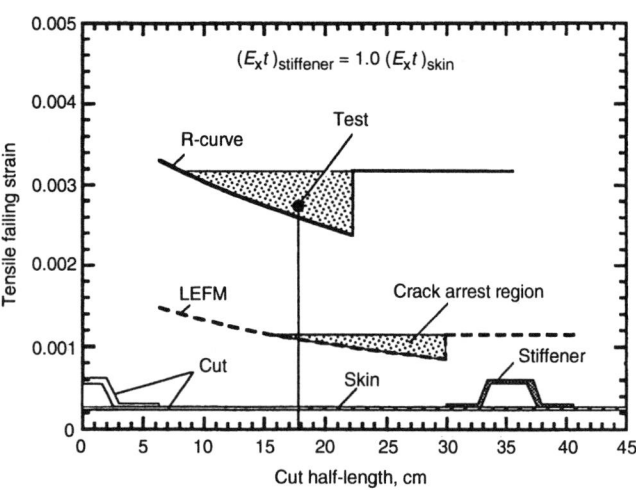

Fig. 33 Test and predicted failing strains for five-stringer $[\overline{+45}/0/90/\overline{+30}/\overline{0}]_s$ AS4/938 fuselage crown panel (160 by 348 cm)

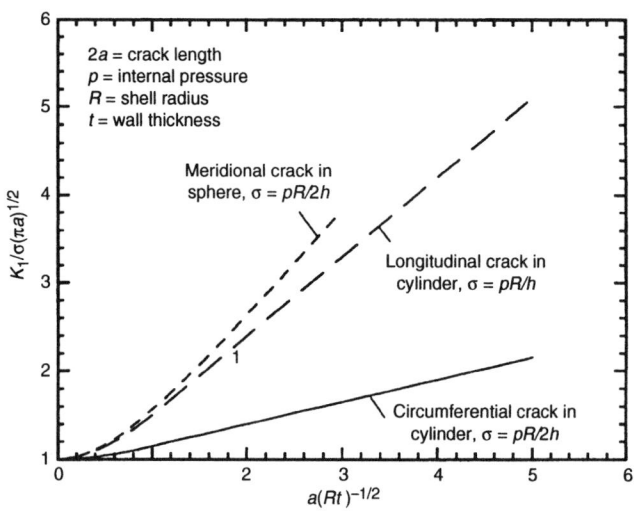

Fig. 34 Stress-intensity correction factors for pressurized shells with cuts

compliance gage located midway between the ends of the cut. The intense dark regions in the radiographs at the ends of the cuts in Fig. 29 indicate broken 0° fibers. The growth of 0° fiber breaks occurred incrementally; each growth resulted in a jump of the COD and an audible sound. The COD is given by

$$COD = 4a\,\varepsilon_o\,\beta^{-1} \qquad \text{(Eq 13)}$$

and the growth was calculated using

$$\Delta a = COD\,\beta\,(4\varepsilon_o)^{-1} - a_o \qquad \text{(Eq 14)}$$

where β is a factor that corrects for anisotropy. The value of β, which is unity for an isotropic material, was determined experimentally.

The maximum value of Q_c/ε_{tuf} in Fig. 28 is about 63% greater than the LEFM value, and the maximum growth was 34% of the cut length. Of

the 63% increase, approximately 48% was due to the increase in strength and only 15% to the increase in cut length. Thus, strength predictions using LEFM would result in a margin of approximately 48%.

Tensile failing strain predictions using the R-curve and LEFM (Ref 19) are plotted against cut length in Fig. 30. The F-curves were calculated using Eq 10 and 11 with $\varepsilon_{tuf} = 0.0148$. Failing strains were multiplied by the finite width correction factor $\sqrt{\sec(\pi a/W)}$ to account for variations in specimen width. The tests and LEFM predictions agree only for cuts ≤ 2.5 cm, and the tests and R-curve predictions agree only for cuts ≥ 7.6 cm. Thus, both fracture toughness and R-curve depend on cut length for this laminate.

R-curves for short cuts along with that for the 23 cm long cut in Fig. 28 are plotted in Fig. 31 for the $[\overline{+45}/0/90/\overline{+30}/\overline{0}]_s$ fuselage crown laminate (Ref 19). All of the R-curves were calculated

using COD measurements. Indeed, the crown laminate does not exhibit a unique R-curve. For a given Q_c/ε_{tuf}, the amount of damage growth increases with increasing cut length. The envelope of F-curves were calculated using test values of ε_{oc} for the various cut lengths in Fig. 30. Failing strains calculated using the "envelope" R-curve will by definition be in agreement with the test results for all cut lengths. R-curves used in the following section were determined using the envelope of the F-curves for tests of specimens with various cut lengths.

Applications to Stiffened Panels. Tensile failing strains for large flat fuselage panels with straps and hat-section stiffeners (Ref 19) are plotted against cut length in Fig. 32 and 33, respectively. The panel with straps in Fig. 32 contained a 25 cm long cut and that with hat-section stiffeners in Fig. 33 contained a 35 cm long cut. The central stiffener in both panels was also cut, and

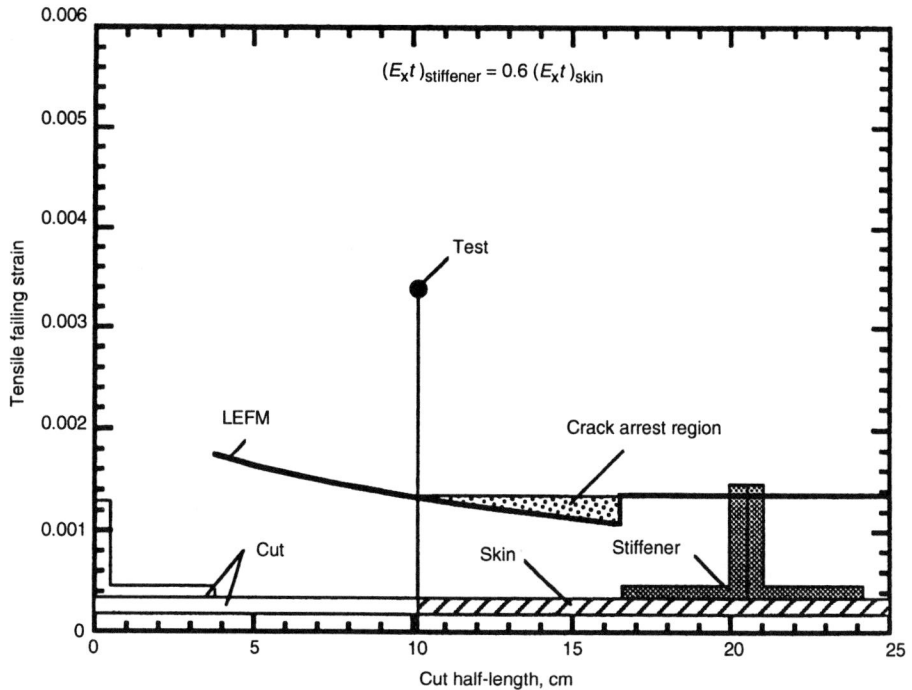

Fig. 35 Test and predicted failing strains for five-stringer, stitched $[0/45/0/-45/0/45/0]_{3s}$ AS4/3501-6 wing panel (102 by 203 cm)

the skins were $[\overline{+45/0/90/+30/0}]_s$ AS4/938 tow-placed fuselage crown laminates. The stiffness ratio $E_x t$ of the straps was 56% of that of the skin and that of the hat-section stiffeners was 100% of that of the skin, where E_x is the modulus in the loading direction and t is the thickness. The panel with straps in Fig. 32 failed catastrophically at an applied strain of 0.00275 with about 25 mm of damage growth at each end of the cut. The damage at the cut ends in the panel with hat-section stiffeners in Fig. 33 grew into the stiffeners (about 18 cm at each end of the cut) before catastrophic failure at an applied strain of 0.00274.

Tension failing strains calculated using LEFM and an R-curve are also plotted against cut length in Fig. 32 and 33. The envelope of F-curves in Fig. 31 was used for the R-curve. Approximate, closed-form equations for the stress-intensity factor in Ref 19 were used with Eq 10 to calculate F-curves for the various cut lengths. The elevation of the stress-intensity factor by the cut stiffener was taken into account. The stiffener was assumed to disbond after the crack tip passed the edge of the intact stiffener, and the failing strain was assumed to be constant thereafter. The jumps in failing strains occur when the end of the cut (LEFM) or the end of the cut plus damage growth (R-curve) coincides with the edge of the intact stiffener. The shaded regions indicate cut lengths for fracture arrest and subsequent increase in failing strain to catastrophic failure. For cut lengths to the left of a shaded region, failure is catastrophic. The failure of the panel in Fig. 32 was catastrophic, but the damage in Fig. 33 grew into the stiffener before failure as predicted. The tests and R-curve calculations were in best agreement, 14% below and 16% above the test values for the

straps and hat-section stiffeners, respectively. Failing strains calculated using LEFM were only 66 and 36% of those calculated using the R-curve for the straps and hat-section stiffeners, respectively.

It should be noted that flat panel results cannot be applied directly to shells with longitudinal cracks and internal pressure because stress-intensity factors for pressurized shells can be much greater than those for flat plates. In Fig. 34, stress-intensity correction factors from Ref 29 are plotted against $a/\sqrt{Rt}$ for isotropic pressurized cylinders and spheres. For a wide body fuselage with a cut equal to two times the frame spacing, $a/\sqrt{Rt}$ can be as large as five. In that case, the stress-intensity factor for an unstiffened cylinder would be more than five times that for a flat unstiffened plate. Analytical results for specially orthotropic cylinders are given in Ref 30 and 31. These results were experimentally verified for 30 cm diam pressurized composite cylinders with longitudinal cuts in Ref 32. Frames and tear straps are able not only to reduce the stress-intensity factor (Ref 33), they can also turn a fracture and limit a failure (Ref 34).

The tensile failing strain for a large flat wing panel with blade stiffeners (Ref 19 and 35) is plotted against cut length in Fig. 35. The five-stringer panel was made using a stitched multi-axial knit fabric and a Resin Film Infusion (RFI) process. The skin was made from 6 layers of multi-axial warp-knit fabric that were stitched together using 6 stitches per square cm of 1600 denier Kevlar 29 thread (3 mm pitch with 5 mm spacing). The skin was 8.4-mm thick, and each layer of fabric was equivalent to a tape layup of $[45_{153}/-45_{153}/0_{320}/90_{173}/0_{320}/-45_{153}/45_{153}/]$,

where the subscripts give the areal weight of the layer in grams per square meter. The T-section stiffeners were made by stitching together 8 layers of the multi-axial warp-knit fabric to form the blades and folding out 4 layers each to form the two flanges. The flanges were stitched to the skin using 6 stitches per square cm. The stitched, stiffened skin preform was infiltrated with 3501-6 resin.

The panel contained a 7-in. long cut that also severed the central stiffener. The fracture initiated at the ends of the cut at a strain of 0.0023, propagated to the edge of the stiffener, and was arrested. With increasing load, the fracture turned and grew parallel to the stiffener. At a strain of 0.0034, failure occurred at the grips. Thus, the stitched stiffeners resulted in a 48% increase in failing strain.

Failing strains calculated using LEFM are also plotted against cut length in Fig. 35. Fracture data were not available to determine an R-curve. The LEFM calculations, which were made similar to those in Fig. 32 and 33, greatly underestimated the failing strain. The F-curve equations (driving force) from Ref 19 were based on failure of the adhesive bond between skin and stiffeners as the fracture in the skin advanced beneath the stiffener. The bond failure ameliorated the effectiveness of the stiffener in reducing the crack driving force. However, the stitches bridged the adhesive bond much as mechanical fasteners do and greatly increased the effectiveness of the stitched stiffeners over that of the bonded stiffeners.

More accurate F-curves can be calculated using numerical techniques such as finite elements. For example, an F-curve was calculated for a curved panel with stiffeners, pressure loading, and a two-bay crack in Ref 36. Large out-of-plane displacements were taken into account. Using an R-curve, the residual strength was predicted successively.

Summary

Defects

- Porosity and delaminations that can reduce compression strengths significantly are likely to be detected during manufacture using conventional ultrasonic methods. Tension strengths are largely unaffected by matrix defects.

- Ply wrinkles can reduce tension and compression strength more than porosity and delaminations and are more difficult to detect using ultrasonics.

Low-Velocity Impacts

Resin Toughness

- Damage resistance was greater for thin laminates made from toughened epoxy resins than those made from conventional epoxy resins. On the other hand, compression after impact strengths were essentially equal for the toughened and conventional epoxy resins when damage size in C-scans was equal.

Laminate Thickness

- Impact damage in thin laminates (2 to 7 mm) initiates as matrix cracking and delaminations. With increasing impact force, fibers also break.
- Damage resistance of thin laminates increased markedly with increasing thickness for a given impact force but tended to be independent of thickness for a given impact energy.
- Tension and compression strengths of thin laminates were reduced 50% or more from impacts that caused dents deeper than 1 mm.
- Impact damage in thick (36 mm) laminates consists of matrix cracks and broken fibers in the outermost laminae directly below the contact region.
- For thick laminates, impact force to initiate damage is proportional to the impacter radius squared, and the depth of damage is proportional to the impacter radius.
- For nondetectable impact damage, tension strengths of thick laminates decreased with increasing impacter radius.

Specimen Size and Impacter Mass

- Impacts are approximately quasistatic when the impacter mass is greater than 100 times the plate mass.
- For thin laminates, impact damage reduced impact force markedly for small plates, but not for large plates.
- Impact energy to reduce burst pressure of large, thin pressure vessels was an order of magnitude times that of small pressure vessels with the same membrane laminate.
- Failing strains of thin coupon type specimens and three-stringer panels with bolted stiffeners were equal for a given impact energy. When energy levels were greater than 135 J, the initial failures were arrested by the stiffeners and final failure strains were 136 to 161% of the initial failure strains.
- For large plates and a given impact energy, impacts by small masses (runway debris, etc.) caused much more damage to thin laminates than impacts by large masses (tools, etc.). For small plates, the damage was of similar size in C-scans.

Post-Impact Fatigue

- Constant amplitude compression fatigue tests indicate that fatigue lives for nondetectable impact damage are probably adequate but those for detectable damage may need to be verified by spectrum tests.

Tension Strength Analysis for Two-Bay Crack

- Linear Elastic Fracture Mechanics (LEFM) can be used to predict the strength of thick laminates, but the use of a nonlinear theory such as the R-curve may be necessary to avoid undue conservatism for thin laminates.

- Special techniques may be required to determine an R-curve that is independent of crack length.
- Approximate equations for stress-intensity factors for the two-bay crack and bonded stiffeners gave reasonably accurate strength predictions.

REFERENCES

1. D.S. Cairns, "Impact and Post-Impact Response of Graphite/Epoxy and Kevlar/Epoxy Structures," Technology Laboratory for Advanced Composites (TELAC), TELAC Report 87-15, Dept. of Aeronautics and Astronautics, Massachusetts Institute of Technology, Cambridge, MA, Aug 1987
2. M.D. Rhodes, M.M. Mikulus, Jr., and P.E. McGowan, Effects of Orthotropy and Width on the Compression Strength of Graphite-Epoxy Panels with Holes, *AIAA J.*, Vol 22 (No. 9), Sept 1984, p 1283-1292
3. C.C. Poe, Jr., "Fracture Toughness of Fibrous Composite Materials," TP 2370, National Aeronautics and Space Administration, Nov 1984
4. R.E. Horton and R.S. Whitehead, *Damage Tolerance of Composites*, Vol I, *Development of Requirements and Compliance Demonstration*, AFWAL-TR-87-3030, July 1988
5. D.O. Adams and M.W. Hyer, Effects of Layer Waviness on the Compression Strength of Thermoplastic Composite Laminates, *J. Reinf. Plast. Compos.*, Vol 12, April 1993, p 414-429
6. J.H. Gosse and P.B.Y. Mori, Impact Damage Characterization of Graphite/Epoxy Laminates, *Proc. American Society for Composites*, Technomic Publishing, Nov 1988, p 344-353
7. W.C. Jackson and C.C. Poe, Jr., The Use of Impact Force as a Scale Parameter for the Impact Response of Composite Laminates, *J. Compos. Technol. Res.*, Vol 15 (No. 4), Winter 1993, p 282-289
8. W.C. Jackson and M. Portanova, "Out-of-Plane Properties," CP 3311, Part 2, National Aeronautics and Space Administration, Oct 1995, p 315-348
9. M. Portanova, "Impact Testing of Textile Composite Materials," CP 3311, Part 2, National Aeronautics and Space Administration, Oct 1995, p 391-424
10. P.J. Minguet and C.K. Gunther, "Test Methods for Textile Composites," CR 4609, National Aeronautics and Space Administration, July 1994
11. C.C. Poe, Jr., Simulated Impact Damage in a Thick Graphite/Epoxy Laminate Using Spherical Indenters, *Proc. American Society for Composites*, Technomic Publishing, Nov 1988, p 334-343
12. C.C. Poe, Jr., Impact Damage and Residual Tension Strength of a Thick Graphite/Epoxy Rocket Motor Case, *J. Spacecraft and Rockets*, Vol 29 (No. 3), May-June 1992, p 394-404
13. W.C. Jackson, M.A. Portanova, and C.C. Poe, Jr., "Effect of Plate Size on Impact Damage,"

presented at Fifth ASTM Symposium on Composite Materials (Atlanta, GA), 4-6 May 1993
14. B.A. Lloyd and G.K. Knight, "Impact Damage Sensitivity of Filament-Wound Composite Pressure Vessels," 1986 JANNAF Propulsion Meeting, CPIA Publication 455 Vol 1, Aug 1986, p 7-15
15. ACEE Composites Project Office, compiler, "NASA/Aircraft Industry Standard Specification for Graphite Fiber/Toughened Thermoset Resin Composite Material," RP 1142, National Aeronautics and Space Administration, June 1985
16. R.J. Roark and W.C. Young, *Formulas for Stress and Strain*, 5th ed., McGraw-Hill, 1975
17. R.S. Whitehead, "Lessons Learned for Composite Aircraft Structures Qualification," *Proc. 1987 Aircraft Engine Structural Integrity Program (ASIPENSIP) Conference*, Air Force Wright Aeronautical Laboratory, p 9-43
18. M.A. Portanova, C.C. Poe, and J.D. Whitcomb, Open Hole and Postimpact Compressive Fatigue of Stitched and Unstitched Carbon-Epoxy Composites, *Composite Materials: Testing and Design*, 10th Volume, STP 1120, G.C. Grimes, Ed., ASTM, 1992, p 37-53
19. C.C. Poe, Jr., C.E. Harris, T.W. Coats, and T.H. Walker, "Tension Strength with Discrete Source Damage," *Fifth NASA/DoD Advanced Composites Technology Conference*, Vol I, Part 1, CP-3294, p 369-437
20. J. Awerbuch and M.S. Madhukar, Notched Strength of Composite Laminates: Predictions and Experiments—A Review, *J. Reinf. Plast. Compos.*, Vol 4 (No. 1), Jan 1985
21. I. Eriksson and C.-G. Aronsson, Strength of Tensile Loaded Graphite/Epoxy Laminates Containing Cracks, Open and Filled Holes, *J. Compos. Mater.*, Vol 24, 1990, p 456-482
22. C.-G. Aronsson, Stacking Sequence Effects on Fracture of Notched Carbon Fibre/Epoxy Composites, *Compos. Sci. Technol.*, Vol 24 (No. 3), 1985, p 179-198
23. T.H. Walker, L.B. Ilcewicz, J.B. Bodine, D.P. Murphy, and E.F. Dost, "Benchmark Panels," CP-194969, Aug 1994
24. C.E. Harris and D.H. Morris, A Comparison of the Fracture Behavior of Thick Laminated Composites Utilizing Compact Tension, Three-Point Bend, and Center-Cracked Tension Specimens, *Fracture Mechanics*, 17th Volume, STP 905, J.H. Underwood, R. Chait, C.W. Smith, D.P. Wilhem, W.A. Andrews, and J.C. Newman, Ed., ASTM, 1986, p 124-135
25. C.E. Harris and D.H. Morris, A Fractographic Investigation of the Influence of Stacking Sequence on the Strength of Notched Laminated Composites, *Fractography of Modern Engineering Materials: Composites and Metals*, STP 948, J.E. Masters and J.J. Au, Ed., ASTM, 1987, p 131-153
26. C.C. Poe, Jr., A Unifying Strain Criterion for Fracture of Fibrous Composite Laminates, *Eng. Fract. Mech.*, Vol 17 (No. 2), 1983, p 153-171
27. C.E. Harris and D.H. Morris, "Fracture Behavior of Thick, Laminated Graphite/Epoxy Com-

posites," CR-3784, National Aeronautics and Space Administration, 1984

28. "Standard Practice for R-Curve Determination," E 561-86, *Annual Book of ASTM Standards*, Vol 03.01, 1988, p 563-574

29. H. Tada, P.C. Paris, and G. Irwin, *The Stress Analysis of Crack Handbook*, 2nd ed., Paris Production (and Del Research), 1985

30. F. Erdogan, M. Ratwani, and U. Yuceoglu, On the Effect of Orthotropy in a Cracked Cylindrical Shell, *Int. J. Fract.*, Vol 10, 1974, p 117-160

31. F. Erdogan, Crack Problems in Cylindrical and Spherical Shells, *Mechanics of Fracture*, Vol 3, Noordhoff International, 1977, p 161-199

32. M.J. Graves and P.A. Lagace, Damage Tolerance of Composite Cylinders, *Compos. Struct.*, Vol 4, 1985, p 75-91

33. O.S. Yahsi and F. Erdogan, "The Crack Problem in a Reinforced Cylindrical Shell," CP-178140, National Aeronautics and Space Administration, June 1986

34. C.U. Ranniger, P.A. Lagace, and M.J. Graves, Damage Tolerance and Arrest Characteristics of Pressurized Graphite/Epoxy Tape Cylinders, *Fifth Symposium on Composite Materials*, STP 1230, ASTM, 1995

35. J. Sutton, Y. Kropp, D. Jegley, and D. Banister-Hendsbee, Design, Analysis, and Tests of Composite Primary Wing Structure Repairs, *Fifth NASA/DoD Advanced Composites Technology Conference*, Vol I, Part 2, CP-3294, National Aeronautics and Space Administration, p 913-934

36. J.T. Wang, D.Y. Xue, D.W. Sleight, and J.M. Housner, "Computation of Energy Release Rates for Cracked Composite Panels with Nonlinear Deformation," Paper No. 95-1463, in proceedings of the 36th AIAA/ASME/ASCE/AHS Structures, Structural Dynamics, and Materials Conference, American Institute of Aeronautics and Astronautics, 10-14 April 1995

Fatigue of Brittle Materials*

THE DEVELOPMENT of toughened ceramics (based on the technologies and understanding of toughening mechanisms such as *in situ* phase transformation, fiber bridging, ductile-particle toughening, and other toughening methods) is the basis for closer consideration of ceramic materials for possible structural applications. As a result, more attention is being given to the fatigue of ceramics and other brittle materials, because of the importance of cyclic loading in many of the potential applications of ceramics, such as high-temperature engine components and biomedical materials. However, because ceramic and other highly brittle materials have been considered free of cyclic fatigue effects, only limited property data on ceramic fatigue are available.

This article summarizes some of the current understanding of the mechanisms and mechanical effects of fatigue processes in highly brittle materials, with particular emphasis on ceramics. Topics include room-temperature fatigue crack growth in monolithic (single-phase) ceramics, transformation-toughened ceramics, and ceramic composites under cyclic compression; cyclic damage zones ahead of tensile fatigue cracks; crack propagation under cyclic tension or tension-compression loads; and elevated-temperature fatigue crack growth in monotonic and composite ceramics.

It is perhaps appropriate at this juncture to clarify the terminology used in the description of cyclic failure. In the metallurgy, polymer science, and mechanical engineering communities, the word *fatigue* is a well-accepted term for describing the deformation and failure of materials under cyclic loading conditions. However, in the ceramics literature, the expression *static fatigue* refers to stable crack propagation under sustained loads in the presence of an embrittling environment (which is commonly known as stress-corrosion crack growth in the metallurgy and engineering literature). The expression *cyclic fatigue* is used in the ceramics community to describe cyclic load deformation and fracture. In keeping with the well-established universal conventions, and in an attempt to avoid confusion, the term *fatigue*

is used in this article to denote deterioration and fracture of both metals and nonmetals due only to cyclic loads.

Crack Growth Behavior

Brittle solids are known to be susceptible to subcritical crack growth in monotonic tension under the influence of an environment. This stress corrosion or static fatigue process has been studied extensively in a wide variety of glasses, oxide ceramics, transformation-toughened ceramics, and other brittle materials. For most inorganic glasses, moisture is responsible for enhanced subcritical flaw growth and for deterioration in strength and other mechanical properties. Even small amounts of water vapor can markedly reduce the lifetime of the component under static loading conditions. Most oxide ceramics also contain glassy silicate phases at grain boundaries and interfaces; these amorphous regions are prone to environmental degradation comparable to that found in glass ceramics. Even if the amount of glass phase present in a ceramic is negligible, the embrittlement from water vapor or impurities can lead to strength deterioration.

The study of sustained load cracking has been more predominant than the study of cyclic fatigue effects, due in part to the very limited crack-tip plasticity of ceramics and other highly brittle materials (Ref 1). Highly brittle materials (such as rocks, concrete, silicate glasses, zincblende structures, diamond structures, aluminum oxides, mica, boron carbides, nitrides, and other ceramics at room temperature) have strong covalent or ionic bonding, which essentially limits any pronounced mobility of point defects and dislocations. In contrast, dislocation motion (slip) has traditionally been considered a necessary condition for fatigue, given the close relationship between cyclic stresses and the motion of dislocations in ductile metals. Therefore, highly brittle solids have long been considered free of true cyclic fatigue effects, based on a typical understanding of cyclic slip as the basis of metal fatigue.

Experimentally, unequivocal demonstrations of a true cyclic-fatigue phenomenon in ceramics are also rare, due in part to the difficulty of performing subcritical crack growth tests on such brittle materials. Another reason for not considering fatigue of brittle solids is the considerable

scatter in stress-life data for brittle materials. The scatter in stress-life data for both monotonic and cyclic loading of brittle materials makes it difficult to decipher intrinsic fatigue effects from the data. Most explanations of fatigue in ceramics and other brittle materials are based on so-called "static fatigue," which is considered to be a noncyclic-load effect based on sustained-load processes or environmentally assisted (stress-corrosion) cracking processes under monotonic loads.

However, persuasive evidence now exists on true cyclic fatigue effects in ceramics and other highly brittle materials. The first unequivocal demonstrations of fatigue effects in brittle solids are based on both tensile fatigue (Fig. 1) (Ref 2, 3) and compressive fatigue effects on crack initiation and fatigue crack growth in a wide variety of notched ceramics and ceramic composites (Ref 4-7), attributable solely to cyclic variations in imposed loads. This section briefly reviews the key mechanisms and mechanical features of cyclic fatigue in ceramics. Fractographic features of ceramic fatigue can resemble fractographs of monotonic overload failures, because fatigue fracture in ceramics rarely shows evidence of fatigue striations. However, cyclic loads do have distinguishable mechanistic and mechanical effects on ceramics.

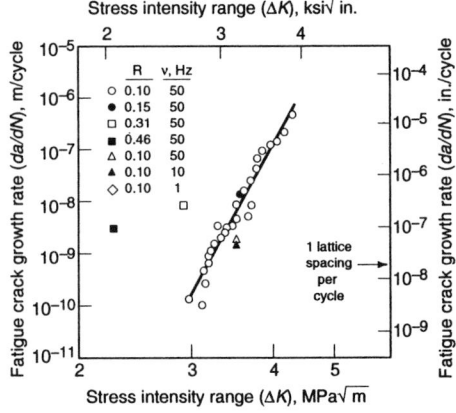

Fig. 1 Variation of fatigue crack propagation rate, *da/dN*, with applied stress intensity range. Δ*K*, for MgO-PSZ ceramic, tested in room-temperature air (45% relative humidity) at a load ratio *R* of 0.10 and a cyclic frequency of 50 Hz. Data for varying load ratio (0.10 to 0.46) and cyclic frequency (1 to 50 Hz) are included for comparison. Source: Ref 2

*Adapted with permission from Chapter 13, "Fatigue of Brittle Solids," in *Fatigue of Materials* (S. Suresh, Cambridge University Press, 1991) and from "Fatigue of Advanced Materials: Part II," in *Advanced Materials and Processes* (R.H. Dauskardt, R.O. Ritchie, and B.N. Cox, August 1993, Vol 144, No. 8, p 30-35).

Table 1 Sources of irreversible microscopic deformation in ductile metals and brittle solids

Materials	Mechanism
Ductile alloys	
Crack-tip region	
	Cross-slip of screw dislocations during a fatigue cycle
	Formation of impediments to dislocation motion such as locks, nodes, or jogs
	Slip irreversibility due to shape changes in body-centered cubic crystals (as a result of slip asymmetry)
	Production of point defects
	Oxidation of freshly formed slip steps or adsorption of an embrittling species on slip steps
Crack-wake region	
	Mechanical contact and frictional sliding between mating asperities on fracture surfaces as a consequence of crack closure involving plastic wake, oxidation, crack-face roughness, or transformed wake
	Periodic deflections in the path of the crack
	Bridging of the crack faces by debris particles, fibers, or discontinuous reinforcements
Brittle solids	
Crack-tip region	
	Frictional sliding along the faces of microcracks
	Wedging of microcracks by debris particles
	Inelastic strains generated by dilatational and shear (phase) transformations
	Microcracking due to the release of thermal residual stresses
	Frictional sliding displacements along interfaces
	Viscous flow of glass phases that are formed *in situ* or left over from the processing of the material
	Creep cavitation
	Fracture of the reinforcement phase
Crack-wake region	
	Bridging of the crack faces by debris particles, unbroken grains, glassy films, or reinforcements
	Transformation-induced or roughness-induced crack closure
	"Squeezing out" of the glass films or debris particles by the pumping action of the crack walls

Source: Ref 8

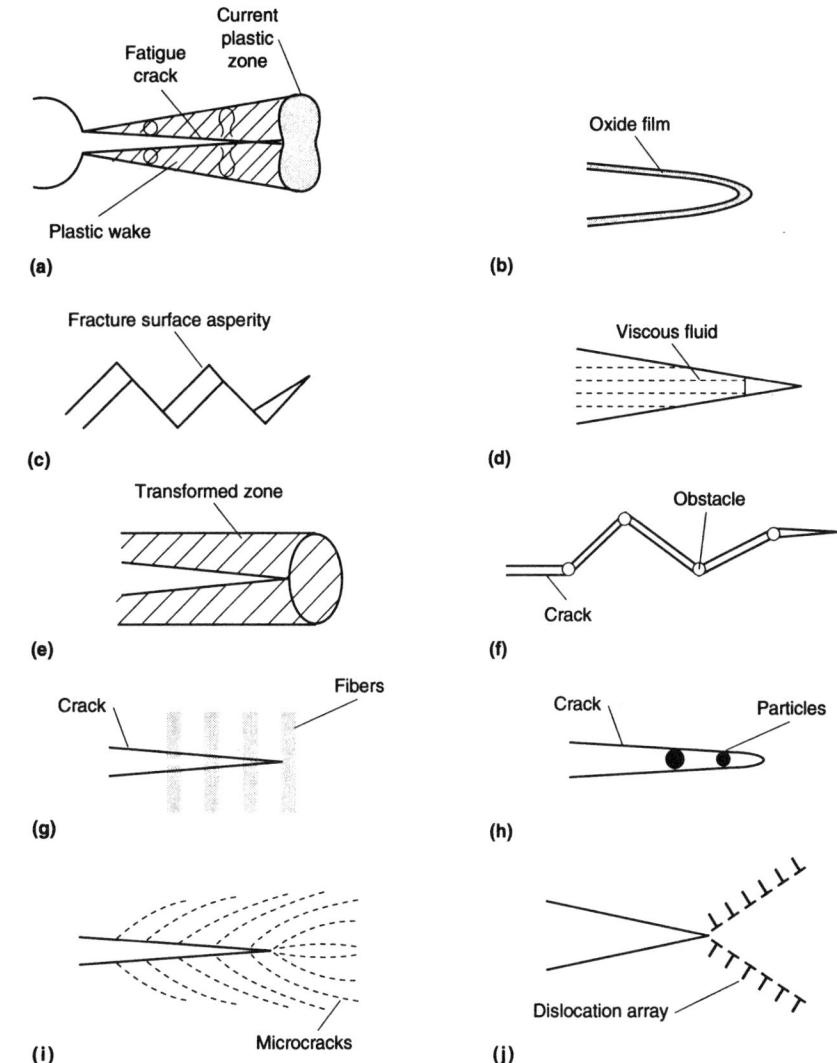

Fig. 2 A schematic illustration of the various mechanisms that modify the effective driving force crack advance in both ductile and brittle solids. (a)-(e) Crack closure due to plasticity, oxides, fracture surface roughness, viscous fluids, and phase transformations, respectively. (f) Crack deflection. (g)-(h) Crack bridging by fibers and particles, respectively. (i)-(j) Shielding of the crack tip by microcracks and dislocations, respectively. Source: Ref 8

Fatigue Mechanisms in Ceramics. Despite the absence of dislocation plasticity, many ceramics exhibit cyclic fatigue degradation at room and elevated temperatures. This is based on a broader understanding of fatigue damage as a consequence of kinematic irreversibility in microscopic deformation (i.e., the irreversibility of the to-and-fro motion of dislocations). The processes that impart kinematic irreversibility to microscopic deformation during the fatigue of brittle solids and ductile metals are summarized in Table 1.

Kinematic irreversibility of cyclic slip in a ductile metal or alloy can be promoted by a variety of mechanistic processes (see Table 1 and Fig. 2). The roughened surface profile created at the surface by the persistant slip bands (PSBs) and the strain incompatibility between the PSBs and the adjoining matrix render the PSB-matrix interface a prime site for the nucleation of fatigue cracks.

The principal mechanism for fatigue crack initiation in both ductile metals and brittle solids is the generation of permanent strains in the vicinity of the notch tip by kinematically irreversible deformation. The mechanisms of this permanent deformation can be as diverse as dislocation plasticity, microcracking, martensitic transformations, creep, interfacial sliding, crazing, or shear flow.

Although the basic mechanisms of cyclic crack growth in ceramics are still preliminary and not clearly defined, two basic classes of mechanisms (Ref 9, 10) are possible where failure is associated with a dominant crack (in many materials, both types may operate in concert):

- *Intrinsic mechanisms*, where, as in metals, crack advance results from damage processes in the crack-tip region that are unique to cyclic loading (just as striation growth under cyclic loading is distinct from microvoid coalescence or cleavage under monotonic loading in metals).
- *Extrinsic mechanisms*, where the crack-advance mechanism is identical to that for monotonic loading, but the unloading cycle promotes accelerated crack growth by reducing the effect of crack-tip shielding.

Schematics of both mechanisms are shown in Fig. 3. A few examples have been tentatively identified in real-world ceramics. For instance, the enhanced microcracking zones ahead of fatigue cracks in SiC_w-Al_2O_3 composites at high temperatures (where SiC_w represents SiC whiskers) may represent an intrinsic mechanism, and the wearing away of interlocking grain bridges in fatigue cracks in coarse-grained Al_2O_3 at ambient temperatures may be an extrinsic mechanism (Ref 10). Much more research is required to elucidate these mechanisms and to model them quantitatively before realistic guidelines can be established for designing ceramic microstructures that have improved fatigue resistance.

Tensile Fatigue Crack Growth. One of the first demonstrations of ceramic fatigue crack growth is based on tension-tension cycling in a magnesia/partially stabilized zirconia (MgO-PSZ) ceramic (Fig. 1). Tests were conducted based on the premise that other inelastic deformation mechanisms may prevail in these materials,

Intrinsic mechanisms

1. Accumulated localized microplasticity/microcracking (damage)

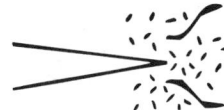

2. Mode II and III crack propagation on unloading

3. Crack tip blunting/resharpening

   a. Continuum

   b. Alternating shear

4. Relaxation of residual stresses

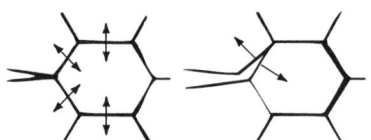

Extrinsic mechanisms

1. Degradation of transformation toughening

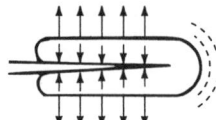

- Degree of reversibility of transformation
- Cyclic accommodation of transformation strain
- Cyclic modification of zone morphology

2. Damage to bridging zone
   a. Friction and wear degradation of...

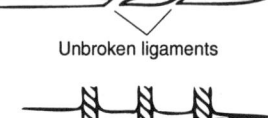

Unbroken ligaments

Whisker/fiber reinforcements

   b. Crushing of asperities and interlocking zones

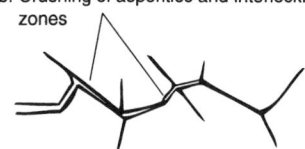

3. Fatigue of ductile reinforcing phase

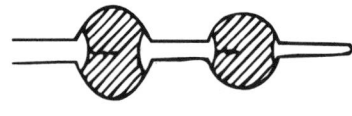

Fig. 3 Schematics of some possible crack-advance micromechanisms that may occur during cyclic fatigue crack growth in ceramics and ceramic-matrix composites. Source: *Advanced Materials & Processes*, Aug 1993

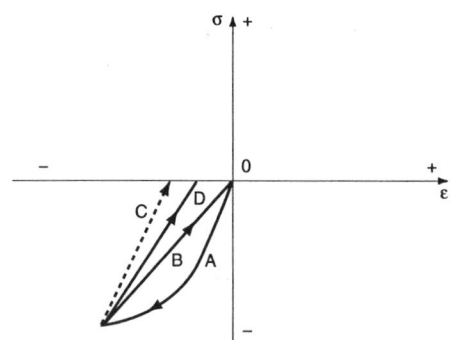

Fig. 4 Idealized schematic of the constitutive behavior of a brittle solid in cyclic compression. Path A: Nonlinear deformation during compression loading. Path B: The idealized situation where microscopic deformation is fully reversible. Paths C and D: Unloading behavior where permanent strains are retained within the deformation zone—Path C for elastic unloading and Path D for linear unloading that is between paths B and C.

- Crack closure plays an important role in determining the rate of crack growth.

Under compressive loading, if the deformation response of the material at the notch tip is characterized by the unloading path "B" in Fig. 4—where no permanent strains are retained upon unloading—no residual stresses are generated. However, if it is characterized by the unloading path "C" or "D" in Fig. 4—where permanent deformation occurs—large residual tensile stresses are generated locally at the notch tip.

Constitutive models for metallic and ceramic materials have shown that the local tensile stresses are of sufficiently high magnitude to induce crack initiation perpendicular to the compression axis (i.e., along the plane of the notch). Because residual stresses are self-equilibrating, a tensile residual stress in the immediate vicinity of the notch tip is accompanied by a compressive residual stress field in regions away from the notch tip. The tensile residual field is fully embedded within a zone of material that is compressively stressed.

As the fatigue crack advances from the notch tip, there is a progressive reduction in the extent of the residual tensile field and an increased level of crack closure due to the contacting fracture surfaces. These two factors lead to a gradual deceleration of the compression fatigue crack, with the crack arresting completely after propagating a certain distance "a" (Fig. 5).

The saturation crack growth distance can be as much as tens of millimeters in thin sheets of metals subjected to cyclic compression, whereas in ceramic materials it is typically a millimeter or less. The rate and total distance of compressive fatigue crack growth depends on variables such as grain size, reinforcement content, specimen geometry, notch geometry, environment, and the number of debris particles formed within the crack due to repeated contact between the crack faces in cyclic compression. Figure 5 shows the effect of debris removal after every 5000 compression cycles.

including microcracking, transforming, or transformation "plasticity" (Ref 2, 3). Data indicate that crack growth rates are a power-law function of the stress-intensity range ΔK, that they are sensitive to frequency and load ratio, and that they show evidence of crack closure analogous to behavior in metals (Ref 11). Cyclic deformation of MgO-PSZ ceramics has also been demonstrated (Ref 12).

Compressive Load Fatigue. As previously mentioned, stable crack growth, attributable solely to cyclic stresses (even in the absence of embrittling environments), was first demonstrated for notched ceramics and ceramic composites under compressive loads (Ref 4, 5). During cyclic compressive loading, cracks display mode I behavior by advancing along the plane of the notch in a direction macroscopically perpendicular to the compression axis. This mode I fatigue crack growth in compression occurs in metals, hard materials (such as cemented carbides), ceramics, polymers, cement mortar, and a variety of organic and inorganic composites (Ref 4, 7, 13-15). Both crystalline and amorphous sol-

ids also have mode I fatigue crack growth behavior under cyclic compression, despite differences in their fatigue constitutive response and failure mechanisms.

This fatigue crack growth phenomenon is regarded as a true mechanical fatigue effect on account of the following experimental observations:

- Mode I fatigue crack growth occurs *perpendicular* to the compression axis in brittle solids subjected only to cyclic loads. Monotonic compressive stresses promote a splitting mode of failure *parallel* to the stress axis.
- There is a gradual increase in crack length with an increase in the number of compression cycles.
- Although an embrittling environment can modify the rate of crack growth, mode I fracture in cyclic compression is not a consequence of environmental effects in that it can take place even in a vacuum.
- The rate of crack growth is strongly influenced by such mechanical fatigue variables as mean stress, stress range, and stress state.

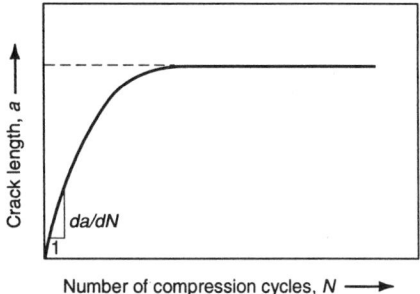

Fig. 5 Schematic representation of fatigue crack length, a, as a function of the number of compression cycles, N

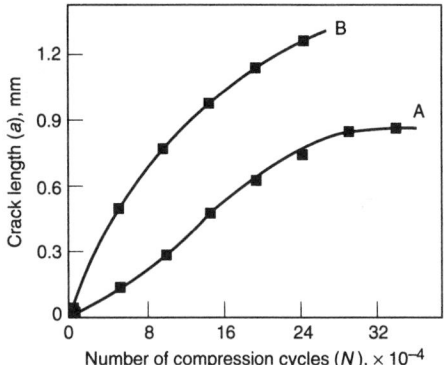

Fig. 6 Variation of fatigue crack length a, measured from the notch tip, as a function of the number of compression cycles in α-alumina of grain size = 18 μm (curve A). Curve B shows the increase in crack growth rates as a consequence of reducing crack closure by the removal of debris particles frome the crack after every 5000 compression cycles using ultrasonic cleaning. Source: Ref 4

Experiments based on crack initiation in cyclic compression in ceramic composites have been useful in studying the effects of fatigue loads on the service life of structural ceramics (see, e.g., Ref 16). Consider, for example, the case of hot-pressed Si_3N_4 that is reinforced with SiC whiskers. It is known that the addition of 20 to 30 vol% SiC to the Si_3N_4 matrix can lead to an increase in fracture toughness by more than a factor of two over that of the unreinforced matrix material. However, when the composite contains stress concentrations, the application of cyclic compressive loads causes fatigue cracks to initiate more easily in the composite than in monolithic Si_3N_4. This effect can be rationalized by noting that the stress-strain curve for the composite is more nonlinear than that for the matrix material. Unloading from a far-field compressive stress can promote a higher degree of permanent deformation which, in turn, may lead to larger residual tensile stresses ahead of the stress concentration.

Mode I flaw growth ahead of stress concentrations also offers a capability to introduce controlled fatigue precracks in brittle materials prior to the determination of fracture toughness, creep crack growth, or fatigue crack growth in tension. An advantage of this technique is that fatigue cracks can be introduced in brittle solids such as ceramics and ceramic composites and in ductile metals, using similar cyclic compression test conditions and specimen geometries. Furthermore, the cyclic compression precracking method is the only known technique for introducing fatigue precracks in circumferentially notched cylindrical rods of brittle solids (which are used for quasistatic and dynamic mode I and mode III fracture tests). A drawback of this method, however, is that the depth of the fatigue crack is typically less than a millimeter in brittle solids.

Cyclic Damage Zone. The phenomenon of crack growth under cyclic compression is a consequence of the development of a cyclic damage zone ahead of the stress concentration. Cyclic damage zones evolve when permanent strains are retained within the deformation zone ahead of a flaw or a notch. The evolution of permanent strains in the damage zone is evident from the existence of open microcracks in the fully unloaded state.

Fatigue damage zones develop ahead of cracks subjected to tensile or compressive loads. Cyclic damage in tensile fatigue of ceramics is qualitatively similar to that in metals. However, the cyclic damage zone from microcracking in ceramics is significantly smaller than the reverse plastic zone for a metal or elastic-plastic solid.

Ceramic Fatigue Data

Fatigue data for ceramics are extremely limited. While stress or strain versus life (S/N) data have been developed for many monolithic ceramics (Ref 17-18), there have been only a few studies of the cyclic constitutive behavior. Similar information for ceramic-matrix composites is also hard to find, although the situation is improving. A few laboratories are generating fatigue crack propagation data for many monolithic and composite ceramic systems. Traditional metal-fatigue variables, such as mean stress (Ref 19), variable-amplitude loading (Ref 17), and environment (Ref 20), are also beginning to be studied. Other current research is documented in the articles listed in "Selected References" at the end of this article.

Test results show that the fatigue effect can be significant in many ceramics. For example, fatigue thresholds are generally as low as 50% of the material's fracture toughness, K_{Ic}, and the velocities of cyclic-fatigue cracks in ceramics can exceed those produced under sustained (static fatigue) loading at the same stress intensity by many orders of magnitude.

Although the fatigue database is rapidly expanding, little work has been done on crack-initiation phenomena. Some limited studies have addressed ceramic fatigue at high temperatures (Ref 21, 22), but little work has addressed the effects of environment on ceramic fatigue at high temperatures. The preliminary work that has been done suggests that creep under sustained loading might predominate over cyclic fatigue for some ceramic materials at very high temperatures.

Fatigue Crack Growth Testing. The general method of fatigue crack growth testing of ceramics can be identical to that given in

ASTM Standard E 647-86A for measuring fatigue crack growth rates in metallic alloys. Crack growth rates are determined between approximately 10^{-11} m/cycle (close to the fatigue threshold ΔK_{TH}) and 10^{-5} m/cycle (approaching instability). They are measured under either manual or computer-controlled K-decreasing and K-increasing conditions, with a normalized K-gradient set typically at about 0.08 mm^{-1}.

The creation of the precrack is perhaps the most critical aspect of the test, because it is extremely difficult to directly initiate a precrack in a machined notch of a ceramic without breaking the sample. Precracking is successfully done using cyclic compression loads, but the practice is not always reliable for all tests. Another technique is to machine a wedge-shaped starter notch into the specimen and apply a low-frequency cyclic load (e.g., a 50 Hz sine wave) at a load ratio of about 0.1 until crack growth of any degree is electronically detected. Loading is then immediately reduced. This precracking process may take as long as several days and is where most inadvertent failures occur. However, once a through-thickness crack of several millimeters in length has been obtained, further crack growth testing is comparatively routine.

Typical fatigue crack growth data are shown in Fig. 7 for MgO-PSZ ceramics with varying levels of toughness by heat treatment (Ref 3). The data indicate a conventional Paris-law relation, da/dN–ΔK, with a log-log slope (m) of 21 to 42, compared to an m value of 2 to 4 for most metals. Table 2 is a summary of tensile fatigue crack growth characteristics of a variety of ceramic materials. A comparison with metal alloys is shown in Fig. 8.

Fractographic features of ceramic fatigue failures resemble monotonic overload failures, because fatigue fracture in ceramics is predominantly transgranular with no evidence of fatigue striations. This is one reason that it was previously held that there are no cyclic fatigue effects in ceramics. However, cyclic loads do have a real effect on the fracture resistance of ceramics. A comparison of the rates of fatigue crack growth measured under cyclic loads with rates of stress-corrosion cracking measured under sustained loads at comparable stress intensities indicated that cyclic crack growth rates in ceramics can be many orders of magnitude faster, and can occur at stress intensities far lower than that required for sustained-load cracking (Ref 2, 3). This is illustrated in Fig. 9 for mid-toughness MgO-PSZ tested in moist air. Here, cyclic crack velocities plotted as a function of time (da/dt) are compared with corresponding stress-corrosion crack velocities measured under sustained loads. At equivalent K-levels, cyclic crack growth rates can be seen to be up to seven orders of magnitude faster under cyclic loads.

High-Temperature Fatigue

High-performance ceramics are being developed as candidate materials for structural components capable of operating at elevated tempera-

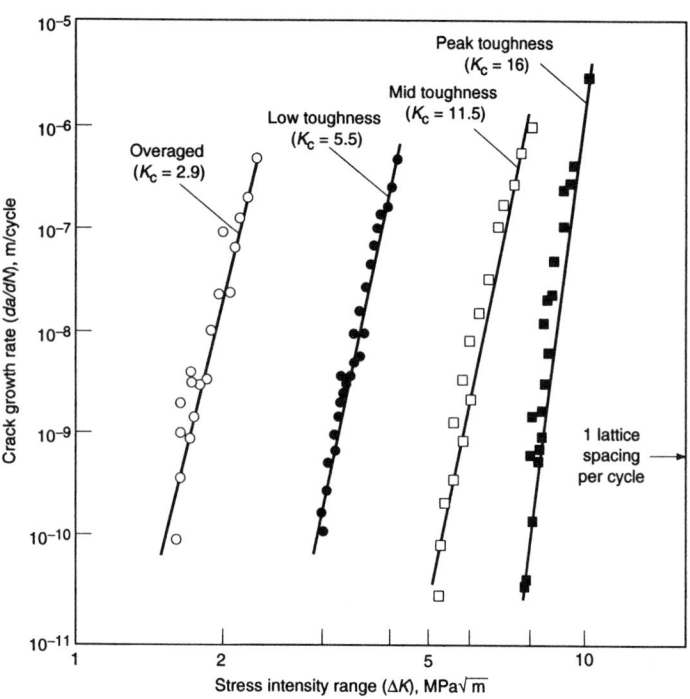

Fig. 7 Cyclic fatigue crack growth behavior, in terms of growth rates per cycle, da/dN, as a function of the stress intensity range, ΔK, for MgO-PSZ, subeutectoid aged to a range of K_c toughnesses from 2.9 to 15.5 MPa√m. Data were obtained on C(T) samples in a room-air environment at 50 Hz frequency with a load ratio ($R = K_{min}/K_{max}$) of 0.1. Source: Ref 3

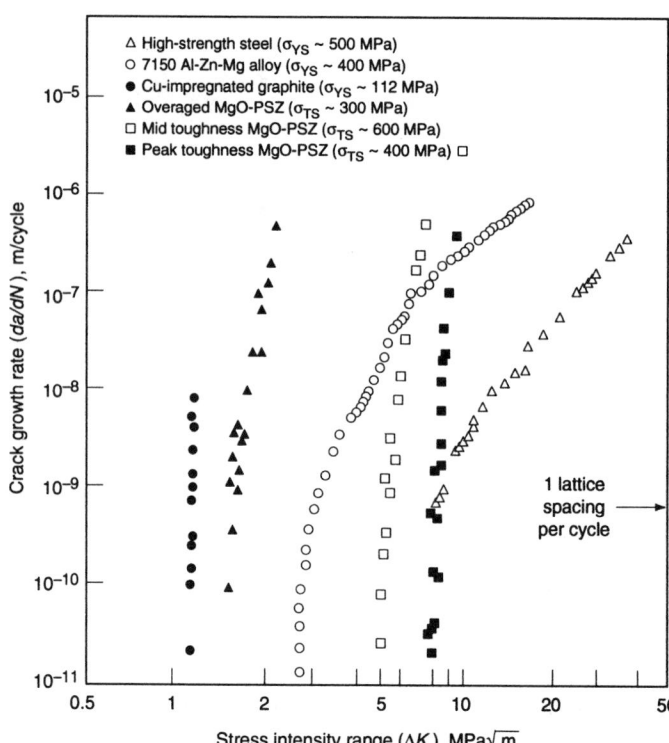

Fig. 8 Fatigue-crack propagation behavior of a number of MgO-PSZ ceramics and a copper-impregnated graphite material compared with that of similar-strength steel and aluminum alloys. Source: Ref 17, 18

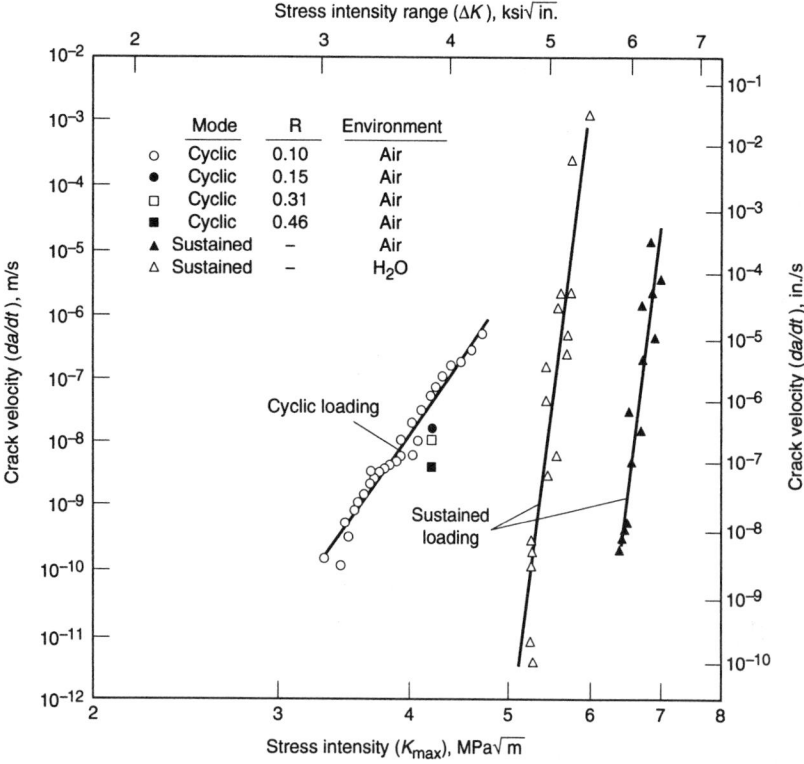

Fig. 9 Subcritical crack growth behavior in MgO-PSZ showing a comparison of cyclic crack velocities, da/dt, from the present study with sustained-load cracking data of Becher for 55% relative humidity air and distilled water. Note how fatigue-induced crack growth is much faster than environmentally assisted crack growth under monotonic loading conditions. Source: Ref 2

tures. However, very little experimental information exists on the resistance of these materials to cyclic loads at elevated temperatures. This lack of data is partly a reflection of the difficulties in conducting crack growth experiments at temperatures typically in excess of 1000 °C. A limited amount of experimental work conducted in the 1970s (Ref 23, 24) indicated that the rates of cyclic crack growth in some ceramic materials at elevated temperatures can be derived on the basis of the time-integration of static crack growth data (with the time to failure in fatigue expressed by an integral number of cycles plus a fraction of a cycle). However, more comprehensive experimental studies on both monolithic ceramics and ceramic composites have shown that, in general, cyclic crack growth rates cannot be predicted on the basis of static crack growth results (Ref 16, 21, 25). In addition, transmission electron microscopy of the crack-tip damage zones indicate differences in microscopic failure mechanisms between static and cyclic crack growth.

Typical fatigue crack growth characteristics (in the Paris regime) of a 90% pure alumina (grain size range = 2-10 μm) in 1050 °C air environment are listed in Table 2. Also presented in this table are similar results for an Al_2O_3/33 vol% SiC_w composite fatigue tested in 1400 °C. Both these materials exhibit subcritical crack growth with Paris exponents that are significantly lower than those listed in Table 2 for the other ceramic materials at room temperature.

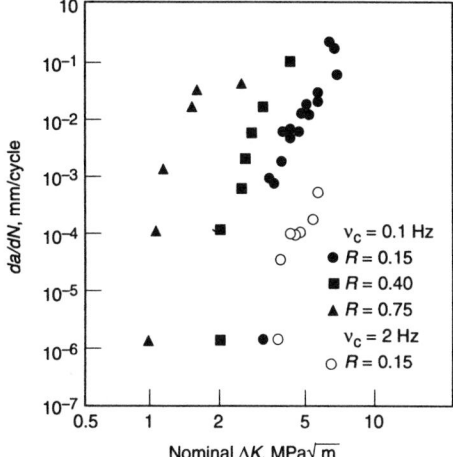

Fig. 10 Crack growth characteristics of Al$_2$O$_3$-33 vol% SiC whisker composite in 1400 °C air. Source: Ref 21

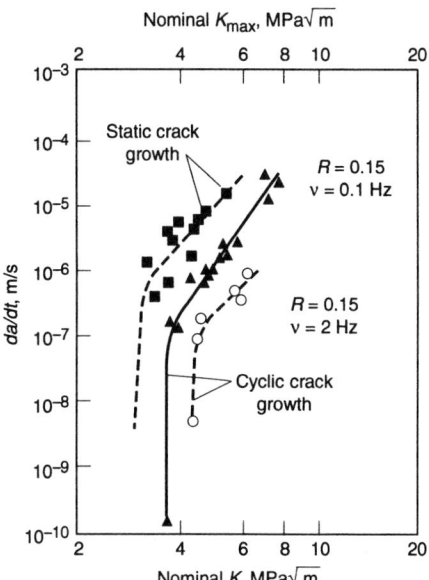

Fig. 11 Crack growth velocity da/dt plotted as a function of the stress intensity factor (K_I or K_{mass}) for static and cyclic load fracture in Al$_2$O$_3$-33 vol% SiC composite in air at 1400 °C. Source: Ref 21

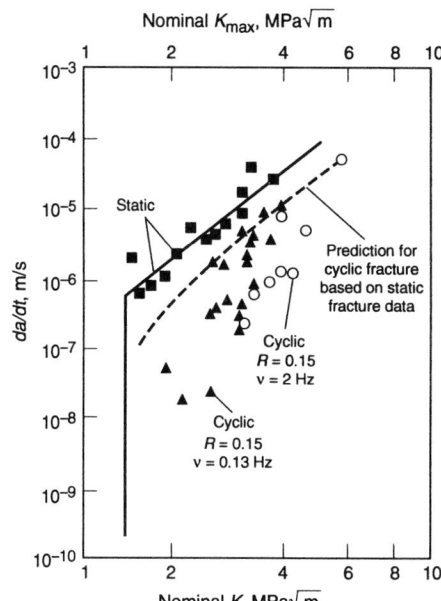

Fig. 12 Experimentally determined sustained-load crack velocity as a function of K_I for 90% pure alumina in 1050 °C air. Source: Ref 8

Figure 10 illustrates the effect of load ratio R and cyclic frequency on the elevated-temperature fatigue crack growth behavior of fiber-reinforced alumina (Ref 21). The influence of R on crack growth in this material is similar to that seen in ductile metals in the near-threshold regime of fatigue. Figure 11 shows the crack velocity da/dt for the composite as a function of the stress intensity factor K_I under monotonic loads in air at high temperature. In Fig. 11, the crack velocities under cyclic loads are *lower* than those under static loads. Furthermore, the cyclic crack growth response is affected by the frequency of the stress cycle. A similar behavior is also found for the 90% pure alumina ceramic (Fig. 12). The differences between static and cyclic crack growth response seen in Fig. 11 and 12 cannot be explained on the basis of the time-dependent crack growth models presented by Evans and Fuller (Ref 23), or by a time-integration model, all of which assume that the static and cyclic crack growth mechanisms are identical. The trend of slower fatigue crack growth under cyclic tensile loads than in monotonic tension (flexure) is also counter to that seen for room-temperature tests where tension-compression fatigue loads accelerate crack growth rates compared to the static load.

In commercially synthesized alumina, such as the 90% pure alumina for which results are shown in Table 2, the grain-boundary regions consist of an amorphous phase that is introduced during fabrication. Upon exposure to elevated temperature, the amorphous phase becomes viscous. The flow of the glass phase (under the influence of an applied stress) along grain boundaries imparts a nonlinear (creep) deformation at the crack tip. The resultant crack-tip damage causes crack growth to occur under both static and cyclic loading conditions. The overall mechanisms of monotonic and cyclic fracture are thus very similar. However, the pumping of the molten glass film within the crack by the cyclic loads (which squeeze the film out of the crack) and the fretting contact between the crack faces due to crack closure are processes specific to

fatigue. Furthermore, the strain-rate sensitivity of the viscous flow of the amorphous film, in conjunction with the time dependence of environmental interactions with SiC$_w$, renders the crack growth process frequency-sensitive under cyclic loads. Recent work (Ref 25) has shown that in fine-grained alumina ceramics of very high purity (i.e., with little pre-existing amorphous phases at grain boundaries), the propensity for stable crack growth is essentially suppressed under both static and cyclic loading conditions. The pre-existing amorphous films along grain boundaries thus appear to play a major role in governing the quasi-static and cyclic crack growth characteristics of monolithic alumina and in determining the differences in apparent crack growth rates between monotonic and fatigue fracture.

In certain materials, such as Al$_2$O$_3$-SiC$_w$ composite, the oxidation of SiC$_w$ at temperatures typically above 1250 °C can lead to the formation of an amorphous glass phase, even if the material retains essentially no amorphous phases after processing. The molten flow of this glass phase

causes cavities to nucleate and grow along the whisker-matrix interface and matrix grain boundaries. The resulting development of microscopic flaws in the composite is manifested as a diffuse microcrack zone ahead of a tensile fatigue crack. The presence of open microcracks in the fully unloaded state is indicative of the development of permanent strains within the crack-tip damage zone. Cyclic loads affect the viscous flow of the glass phase and the resulting interfacial cavitation in different ways than monotonic loads. Furthermore, bridging of the faces of the microcracks by the whiskers (and the associated crack closure effect) or frictional sliding along the debonded matrix-whisker interface in the damage zone ahead of the crack tip may modify the near-tip stress intensity factor under cyclic loads differently than under monotonic loads.

Summary. In summary, a limited amount of experimental work conducted in the 1970s on ceramic materials in the elevated-temperature environment indicated that any apparent cyclic fatigue effect might merely be a manifestation of

Table 2 A survey of some experimental results on tension or fully reversed fatigue of ceramic materials

Material	Test conditions	C	m	ΔK range
Al$_2$O$_3$ (99% pure)(a)	$\nu_c = 5$ Hz, $R = -1.0$, room temperature	1.1×10^{-11}	14 ± 5	2.7-4.0
Al$_2$O$_3$ (90% pure)(b)	$\nu_c = 0.13$ Hz, $R = 0.15$, $T = 1050$ °C	2.8×10^{-10}	10	1.0-3.0
	$\nu_c = 2$ Hz, $R = 0.15$, $T = 1050$ °C	6.3×10^{-11}	8	2.0-3.5
Mg-PSZ (over-aged)(c)	$\nu_c = 50$ Hz, $R = 0.1$, room temperature	2.0×10^{-14}	21	1.5-2.1
Mg-PSZ (peak strength)(c)	$\nu_c = 50$ Hz, $R = 0.1$, room temperature	5.7×10^{-28}	24	5.2-7.2
Al$_2$O$_3$-33vol%SiC whiskers(d)	$\nu_c = 0.1$ Hz, $R = 0.15$, $T = 1400$ °C	4.5×10^{-10}	7	3.5-6.0
	$\nu_c = 2$ Hz, $R = 0.15$, $T = 1400$ °C	4.0×10^{-10}	4	3.5-6.0

Note: C, ΔK, and m are the material constants in the Paris relationship for fatigue crack growth, $da/dN = C(\Delta K)^m$. C and ΔK are listed in units of m cycle, (MPa√m)$^{-m}$ and MPa√m, respectively. (a) Direct push-pull tests on wedge-opening load specimens with $\Delta K = K_{max}$. (b) Four-point flexure specimens. (c) Compact tension specimens. The peak strength and peak toughness of MgO-PSZ were obtained by heat treating at 1100 °C for 3 and 7 h, respectively. (d) Four-point flexure specimens. Source: Ref 8

creep crack growth, and that cyclic crack growth rates could be predicted on the basis of static fracture data because of the overall similarity in the mechanisms of failure under static and cyclic loads. However, in recent years, more controlled experiments covering a wider range of test variables, microstructures, and temperatures have shown that, in general, cyclic crack growth data are distinct from sustained load fracture results (Ref 17-19). The following general trends are observed with elevated-temperature crack growth (Ref 8):

- Static-load cracks propagate much faster than cyclic fatigue cracks in certain ceramics under specific conditions of temperature, frequency, etc., whereas the reverse may be seen under different conditions.
- Crack growth under both static and cyclic tensile loads occurs by a predominantly intergranular failure mode.
- Cyclic crack growth rates are sensitive to both the loading rate (i.e., test frequency) and waveform. An increase in test frequency results in a decrease in crack velocities, because the crack is subjected to a smaller amount of time at the peak tensile stress. As the frequency is reduced or the maximum stress intensity factor is raised, the cyclic fracture data begin to approach the static fracture data. Cyclic load waveforms with increasing hold-times at the peak tensile stress bring the fatigue life closer to the static lifetime at comparable values of maximum applied stress (Ref 21, 22).
- The presence of a glass phase along the grain boundaries has a strong effect on the mechanisms of crack growth. While stable crack growth occurs in the 90% pure alumina over a range of stress intensity factor spanning 1.5 to 5 MPa√m, such subcritical fracture is essentially suppressed in a fine-grained, 99.9% pure alumina, apparently due to the absence of a critical amount of glass phase.
- It is not possible to predict accurately the cyclic crack growth rates on the basis of static data.

Fracture patterns from both static and cyclic crack growth are characterized by microscopically undulating fracture paths along grain boundaries, periodic branching or jumping of the crack front, bridging of the crack faces by ligaments of the viscous amorphous film, and bridging of the crack faces by grain facets. However, under cyclic loading conditions, further changes in the effective driving force seem to take place as a consequence of:

- The formation of debris particles of the alumina ceramic due to repeated contact between the crack faces
- Fretting and increased (roughness-induced) crack closure arising from mismatch between fracture surface asperities
- "Squeezing out" of the amorphous glass film and debris particles from within the crack by the pumping action of the crack walls

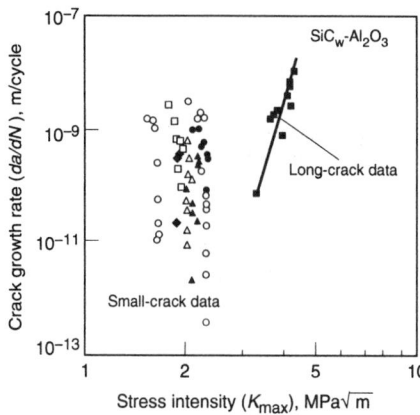

Fig. 13 Small-crack growth rates as a function of applied stress intensity (K_{max} of the loading cycle) for a SiC_w-Al_2O_3 composite. The small cracks propagate at applied stress intensity levels well below the long-crack threshold, ΔK_{TH}, and show an apparent negative dependency on K_{max}. Source: Ref 17, 18

- The dependence of viscous deformation within the glass phase on the loading rate and cyclic frequency

Life Prediction

Life prediction is a major problem for ceramics or ceramic-matrix composites. Because the measured slopes (exponents) of fatigue crack growth rate data (Fig. 8) and the reciprocal slopes of *S/N* curves are so large, the sensitivity of projected life to applied stress can be unacceptably high. Life prediction is further complicated by "small" fatigue cracks that can grow at rates far higher than those for equivalent long cracks and at applied stress intensities far below the long-crack threshold (Fig. 13, 14). "Small" in this context is defined relative to the size of the equilibrium or steady-state shielding zone (i.e., comparable in size to the extent of the crack bridging zone in the wake of the crack tip). The explanation for this seemingly anomalous behavior is that small cracks, by virtue of their limited wake, are unable to develop the same extent of crack-tip shielding. (The effect is identical to the small-crack effect widely reported for metals, where fatigue crack closure provides the dominant shielding mechanism.)

With detailed knowledge and quantitative modeling of the shielding mechanism, it is often possible to normalize long- and small-crack data by characterizing them in terms of the near-tip or net stress intensity (i.e., after subtracting the stress intensity due to shielding) (Ref 48). However, it is questionable whether this technique is practical for routine engineering use with ceramics in actual service.

Fracture Mechanic Lifetime Prediction Compared with That for Metals. In safety-critical applications of metals, damage-tolerant design and life-prediction procedures generally rely on the integration of crack velocity/stress intensity (*da/dN* vs. Δ*K*) curves to estimate the

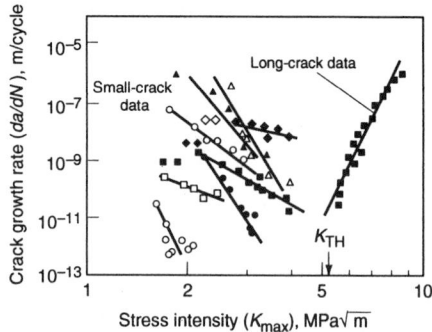

Fig. 14 Small-crack growth of MgO-PSZ ceramic with a mid-toughness (K_c~10 MPa √m) microstructure. Source: Ref 17, 18

time or number of cycles for a presumed initial defect of size a_0 to grow to critical size, a_c (Eq 1). The cracked structure is subjected to an alternating stress, Δσ, and the applied (far-field) stress-intensity factor K can be defined in terms of the applied stress, σ, and a geometry factor, Q, so that:

$$da/dN = C\Delta K^m = C[Q\Delta\sigma\sqrt{(\pi a)}]^m \qquad (Eq\ 1)$$

where $K = Q\sigma\sqrt{(\pi a)}$ (Eq 2)

The projected number of cycles, N_f, to grow a crack to critical size is then given by:

$$N_f = \frac{2}{(m-2)\,C\,Q^m\,\pi^{m/2}\,\Delta\sigma^m}\left[\frac{1}{\frac{m-2}{a_0^2}} - \frac{1}{\frac{m-2}{a_c^2}}\right] \qquad (Eq\ 3)$$

assuming a crack growth relationship of the form of Eq 1, where $m \neq 2$.

Although crack growth data are now available for many advanced materials, the approach may prove difficult to apply in practice because of the large power-law dependence of growth rate on stress intensity, which implies that the projected life will be proportional to the reciprocal of the applied stress raised to a large power. So, for example, for metallic structures ($m \sim 2$ to 4), a factor of two increase in applied stress, σ, reduces the projected life, N_f, by roughly an order of magnitude. For intermetallic structures ($m \sim 10$), the reduction may be three orders of magnitude. And, for ceramic structures, where m generally exceeds 20 (and can be as high as 50 to 100), a factor of two increase in applied stress reduces the projected life by 6 to 30 orders of magnitude.

Thus, conventional damage-tolerant criteria for ceramics imply that marginal differences in assumptions of in-service stresses will lead to significant variations in component life projections. Furthermore, any fluctuation of applied stress or the imposition of overloads (which are frequently encountered in service) may also result in highly nonconservative design lives. Use of this approach also is complicated by two other facts:

- Acceptable projected life for the ceramic may be guaranteed only by restricting the initial defect to an extremely small size.
- It is difficult to include small-crack effects.

Fatigue Threshold. An alternative to the fracture mechanics approach is to redefine the critical crack size, a_c, in terms of the fatigue threshold, ΔK_{TH}, below which crack growth is presumed to be dormant. This is essentially a crack-initiation criterion, where ΔK_{TH} is taken as the effective toughness, rather than the fracture toughness K_c. However, this procedure does not address small-crack effects, which may arise at loads considerably below those required for the growth of long cracks.

Also, as for many high-strength metallic materials, crack-initiation effects in ceramics and intermetallics may involve a very significant portion of lifetime under alternating loads. Therefore, despite its potential limitations, the fracture mechanics approach provides results that are typically highly conservative in the sense that they assume crack growth from the first loading cycle. Although the degree of this conservatism is difficult to quantify, due in part to the paucity of data from crack-initiation studies, the disadvantages of the approach may in fact be considerably outweighed by neglecting the cycles required for initiation. As a result, selection of a damage-tolerant methodology is currently on a case-by-case basis, with the choice often affected by the requirements of specific regulatory agencies.

S/N **Data.** The importance of fatigue crack initiation does, however, suggest an additional design criterion for ceramic components. This more traditional approach is based on *S/N* data. In addition to simple consideration of the fatigue limit, a more sophisticated methodology might include a damage mechanics assessment that uses detailed finite-element stress analyses for components of complex shape, and a statistical evaluation of the pre-existing defect population. While currently limited to large structures where numerous defects may be present, this approach also could be appropriate for small ceramic parts.

Fatigue and Bioceramics. The susceptibility of ceramics to fatigue degradation under cyclic loading has important consequences for bioceramics. While some life predictions involving static (environmentally enhanced) fatigue of ceramics for artificial joint applications have been conducted, very little cyclic fatigue data and few life prediction procedures exist for ceramics operating under physiological loads and environments. Such analyses are crucial for reliable use of bioceramics. For example, artificial prostheses such as pyrolytic-carbon heart-valve devices must be designed for fatigue lives longer than patient life; because the human heart typically beats 38 million times each year, fatigue life must exceed 1 billion cycles. Indeed, several recent structural failures of mechanical heart valves manufactured from pyrolytic carbon have been attributed to progressive fracture by fatigue (Ref 48). Note, however, that the morphology of fracture surfaces formed under cyclic loading conditions in ceramics is almost identical to that produced by fracture itself. This resemblance, which indicates that fracture and fatigue crack-advance mechanisms are similar, greatly complicates failure analysis. References 26-48 provide information on fatigue and fracture of bioceramic materials.

REFERENCES

1. A.G. Evans, Fatigue in Ceramics, *Int. J. Fract.*, Vol 16, 1980, p 485-498
2. R.H. Dauskardt, W. Yu, and R.O. Ritchie, Fatigue Crack Propagation in Transformation-Toughened Zirconia Ceramic, *J. Am. Ceram. Soc.*, Vol 70, 1987, p C-248 to C-252
3. R.H. Dauskardt, D.B. Marshall, and R.O. Ritchie, *J. Am. Ceram. Soc.*, Vol 73 (No. 4), April 1990, p 893-903
4. L. Ewart and S. Suresh, *J. Mater. Sci. Lett.*, Vol 5, 1986, p 774
5. L. Ewart and S. Suresh, *J. Mater. Sci.*, Vol 22, 1987, p 1173
6. J.R. Brockenbrough and S. Suresh, *J. Mech. Phys. Solids*, Vol 35, 1987, p 721
7. S. Suresh and J.R. Brockenbrough, *Acta Metall.*, Vol 36, 1988, p 1455
8. S. Suresh, Fatigue Crack Growth in Brittle Materials, *J. Hard Mater.*, Vol 2 (No. 1-2), 1991, p 29-54
9. R.O. Ritchie et al., Cyclic Fatigue-Crack Propagation, Stress-Corrosion, and Fracture-Toughness Behavior in Pyrolytic Carbon-Coated Graphite for Prosthetic Heart Valve Applications, *J. Biomed. Mater. Res.*, Vol 24, 1990, p 189-206
10. R.H. Dauskardt, A Frictional-Wear Mechanism for Fatigue-Crack Growth in Grain Bridging Ceramics, *Acta Metall. Mater.*, Vol 41, 1993, p 2765-2781
11. R.O. Ritchie, Mechanisms of Fatigue Crack Propagation in Metals, Ceramics and Composites: Role of Crack-Tip Shielding, *Mater. Sci. Eng.*, Vol 103A, 1988, p 15-28
12. K.J. Bowman, P.E. Reyes-Morel, and I-Wei Chen, Reversible Transformation Plasticity in Uniaxial Tension Compression Cycling of Mg-PSZ, *Advanced Structural Ceramics*, Vol 78, *Materials Research Society Symposium Proceedings*, 1987, p 75-88
13. S. Suresh and L.A. Sylva, *Mater. Sci. Eng.*, Vol 83, 1986, p L7
14. S. Suresh, *Int. J. Fract.*, Vol 42, 1990, p 41-56
15. S. Suresh and L. Pruitt, *Deformation, Yield and Fracture of Polymers*, R.J. Young, Ed., The Plastics and Rubber Institute, London, 1991, p 32-1 to 32-4
16. S. Suresh, Mechanics and Micromechanisms of Fatigue Crack Growth in Brittle Solids, *Int. J. Fract.*, Vol 42, 1990, p 41-56
17. R.O. Ritchie and R.H. Dauskardt, Cyclic Fatigue of Ceramics: A Fracture Mechanics Approach to Subcritical Crack Growth and Life Prediction, *J. Ceram. Soc. Jpn.*, Vol 99, 1991, p 1049-1062
18. H. Kishimoto, Cyclic Fatigue in Ceramics, *JSME Int. J.*, Vol 34, 1991, p 393-403
19. R.H. Dauskardt et al., Cyclic Fatigue-Crack Propagation in a Silicon-Carbide Whisker-Reinforced Alumina Composite: Role of Load Ratio, *J. Mater. Sci.*, Vol 28, 1993, p 3258-3266
20. D.S. Jacobson and I-Wei Chen, Mechanical and Environmental Factors in the Cyclic and Static Fatigue of Si_3N_4, *J. Am. Ceram. Soc.*, Vol 76, 1993
21. L.X. Han and S. Suresh, High Temperature Failure on an Al_2O_3-SiC Composite under Cyclic Loads: Mechanisms of Fatigue Crack-Tip Damage, *J. Am. Ceram. Soc.*, Vol 72, 1989, p 1233-1238
22. S. Suresh, *Fatigue of Materials*, Cambridge University Press, 1991, p 451-455
23. A.G. Evans and E.R. Fuller, Crack Propagation in Ceramic Materials under Cyclic Loading Conditions, *Metall. Trans. A*, Vol 5, 1974, p 27-33
24. A.G. Evans, L.R. Russell, and D.W. Richerson, Slow Crack Growth in Ceramic Materials at Elevated Temperatures, *Metall. Trans. A*, Vol 6, 1975, p 707-716
25. S. Suresh, Fatigue Crack Growth in Ceramic Materials at Ambient and Elevated Temperatures, *Fatigue 90*, Vol II, H. Kitagawa and T. Tanaka, Ed., Materials and Components Engineering Publications, 1990, p 759-768
26. S.M. Barinov and Y.V. Baschenko, Application of Ceramic Composites as Implants: Result and Problem, *Conf. Proc. Bioceramics and the Human Body* (Faenza, Italy, 2-5 April 1991), Elsevier Applied Science, Essex, U.K., 1992, p 206-210
27. S.M. Barinov and M.A. Malkov, Dynamic Fatigue of Beta Tricalcium Phosphate Bioceramics, *J. Mater. Sci. Lett.*, Vol 12 (No. 13), 1993, p 1039-1040
28. K.J. Chillag, E. Berg, G. Heimke, and E. Lunceford, Jr., The Lindenhof Acetabular Component: An 11 Year Follow-Up Study, *Bioceramics*, Vol 2, Sept 1989, p 429-435
29. G. Gasser, W. Miller, and R. Mathys, Jr., Preliminary Tests to Determine the Influence of Sterilization and Storage on Compressive Strength of Hydroxyapatite Cylinders, *Conf. Proc. Bioceramics and the Human Body* (Faenza, Italy, 2-5 April 1991), Elsevier Applied Science, Essex, U.K., 1992, p 491-496
30. G. Heimke, Recent Developments in Bioceramics, *Engineering Ceramics*, Vol 3, *Conf. Proc. Euro-Ceramics* (Maastricht, The Netherlands, 18-23 June 1989), Elsevier Science Publishers, Essex, U.K., 1989
31. S.F. Hulbert, M.A. Wack, and D.L. Powers, The Biocompatible and Mechanical Evaluation of a Composite Hip System, *Bioceramics*, Vol 2, Sept 1989, p 330-340
32. T. Kasuga and K. Nakajima, Mechanical Properties of Bioactive Glass-Ceramics/Tetragonal Zirconia Composites, *Bioceramics*, Vol 2, Sept 1989, p 303-310
33. R. Mongiorgi, C. Prati, E. Toschi, and G. Bertocchi, Tensile Bond Strength of Dental Porcelain to Dental Composite Resins, *Conf. Proc. Bioceramics and the Human Body* (Faenza, It-

aly, 2-5 April 1991), Elsevier Applied Science, Essex, U.K., 1992, p 265-269

34. G. Monticelli, L. Romanini, and O. Moreschini, A Review on the Aseptic Total Hip Replacement Failures, *Conf. Proc. Bioceramics and the Human Body* (Faenza, Italy, 2-5 April 1991), Elsevier Applied Science, Essex, U.K., 1992, p 35-45

35. P. Passi, Clinical Results of Intra Mobile Zylinder Dental Implants, *Conf. Proc. Bioceramics and the Human Body* (Faenza, Italy, 2-5 April 1991), Elsevier Applied Science, Essex, U.K., 1992, p 107-112

36. H. Plenk, Jr., M. Bohler, M. Salzer, K. Knahr, and A. Walter, 15 Years' Experience with Alumina-Ceramic Total Hip-Joint Endoprostheses: A Clinical, Histological and Tribological Analysis, *Conf. Proc. Bioceramics and the Human Body* (Faenza, Italy, 2-5 April 1991), Elsevier Applied Science, Essex, U.K., 1992, p 17-25

37. C. Prati, G. Montanari, E. Toschi, and A. Savino, Direct Composite-Ceramic Restorations: a Clinical Study, *Conf. Proc. Bioceramics and the Human Body* (Faenza, Italy, 2-5 April 1991), Elsevier Applied Science, Essex, U.K., 1992, p 113-117

38. E. Ravagli and E. Maggiore, Finite Element Analysis of a Ceramic Hip-Joint Head and Its Failure Mode Due to a Crack in the Material, *Conf. Proc. Bioceramics and the Human Body* (Faenza, Italy, 2-5 April 1991), Elsevier Applied Science, Essex, U.K., 1992, p 477-485

39. A. Ravaglioli and A. Krajewski, Skeletal Implants: From Metals, to Polymers, to Ceramics, *Conf. Proc. Bioceramics and the Human Body* (Faenza, Italy, 2-5 April 1991), Elsevier Applied Science, Essex, U.K., 1992, p 1-16

40. L. Sedel, L. Kerboull, P. Christel, A. Meunnier, and J. Witvoet, Results and Survival Analysis of Alumina-Alumina Total Hip Prosthesis in Active Patients under 50, *Bioceramics*, Vol 2, Sept 1989, p 421-428

41. F. Sernetz, Translucent Alumina for Orthodontic Application, *Bioceramics*, Vol 2, Sept 1989, p 180-192

42. L. Specchia, A. Moroni, L. Ponziani, G. Rollo, S. Pavone, and V. Vendemia, Hip Anatomical Uncemented Ceramic Arthroplasty (ANCA): Results at a 3 Years Follow-Up, *Conf. Proc. Bioceramics and the Human Body* (Faenza, Italy, 2-5 April 1991), Elsevier Applied Science, Essex, U.K., 1992, p 124-129

43. M.V. Swain and V. Zelizko, Comparison of Static and Cyclic Stressing of Alumina in Ringer's Solution, *Bioceramics*, Vol 2, Sept 1989, p 135-144

44. A. Toni, A. Sudanese, P.P. Montina, et al., Bio-Functional Adaptive Behavior to Ceramic Implants, *Conf. Proc. Bioceramics and the Human Body* (Faenza, Italy, 2-5 April 1991), Elsevier Applied Science, Essex, U.K., 1992, p 26-34

45. M. Winter, P. Griss, G. Scheller, and T. Moser, 10-14 Years Results of a Ceramic-Metal-Composite Hip Prosthesis with a Cementless Socket, *Bioceramics*, Vol 2, Sept 1989, p 436-444

46. J.G.C. Wolke, C.P.A.T. Klein, and K. de Groot, Bioceramics for Maxillofacial Applications, *Conf. Proc. Bioceramics and the Human Body* (Faenza, Italy, 2-5 April 1991), Elsevier Applied Science, Essex, U.K., 1992, p 166-180

47. F. Zarotti, Solving of Prosthetic Problems through Bioceramics, *Conf. Proc. Bioceramics and the Human Body* (Faenza, Italy, 2-5 April 1991), Elsevier Applied Science, Essex, U.K., 1992, p 46-48

48. R.H. Dauskardt et al., *Journal Biomedical Materials Research*, Vol 28, 1994, p 791-804

SELECTED REFERENCES

• D. Bolsch and J. Bressers, High Temperature Uniaxial Alternating Fatigue Testing of Engineering Ceramics, *Engineering Ceramics*, Vol 3, *Conf. Proc. Euro-Ceramics* (Maastricht, The Netherlands, 18-23 June 1989), Elsevier Science Publishers, Essex, U.K., 1989, p 3.205 to 3.209

• S.R. Choi, A. Chulya, and J.A. Salem, "Analysis of Precracking Parameters and Fracture Toughness for Ceramic Single-Edge-Precracked-Beam Specimens," Technical Memorandum 105568, National Aeronautics and Space Administration, 1992, p 24

• M.J. Cima and H.K. Bowen, The Future Role of Alumina in Ceramics Technology, *Alumina Chemicals: Science and Technology Handbook*, American Ceramic Society, 1990, p 551-553

• E.L. Courtright, Engineering Property Limitations of Structural Ceramics and Ceramic Composites above 1600 °C, *Ceram. Eng. Sci. Proc.*, Vol 12 (No. 9-10), Sept-Oct 1991, p 1725-1744

• R.T. Cundill, Ceramic Materials for Bearing Components, *Rolling Element Bearings: Towards the 21st Century*, Mechanical Engineering Publications, Suffolk, U.K., 1990, p 31-40

• R. Danzer, Reliability Analysis of Advanced Ceramics, *Conf. Proc. High Temperature Materials for Power Engineering II* (Liege, Belgium, 24-27 Sept 1990), Kluwer Academic Publishers, 1990, p 1575-1598

• L.R. Dharani, Analysis of a Ceramic Matrix Composite Flexure Specimen, *Thermal and Mechanical Behavior of Metal Matrix and Ceramic Matrix Composites*, STP 1080, ASTM, 1990, p 87-97

• A.G. Evans and F.W. Zok, Cracking and Fatigue in Fiber-Reinforced Metal and Ceramic Matrix Composites, *Topics in Fracture and Fatigue*, Springer-Verlag, 1992, p 271-308

• B. Freudenberg, H.-A. Lindner, E. Gugel, and P. Thometzek, Thermomechanical Properties of Aluminium Titanate Ceramic between 20 and 1000 °C, *Properties of Ceramics*, Vol 2, *Conf. Proc. Euro-Ceramics* (Maastricht, The Netherlands, 18-23 June 1989), Elsevier Applied Science, 1989, p 2.64 to 2.71

• Z.D. Guan and W.J. Liu, Study of High Temperature Strength, K_{Ic} and Creep Fracture of Structural Ceramics (Retroactive Coverage), *Conf. Proc. Mechanical Behaviour of Materials V* (Beijing, China, 3-6 June 1987), Vol 2, Pergamon Press, 1988, p 1277-1282

• F. Guiu, Fatigue in Structural Ceramics, *Ceramic Technology International*, Sterling Publications Ltd., London, 1994, p 141-143

• R.H.J. Hannink and M.V. Swain, Progress in Transformation Toughening of Ceramics, *Ann. Rev. Mater. Sci.*, Vol 24, 1994, p 359-408

• T.P. Herbell and A.J. Eckel, "Ceramic Composites for Rocket Engine," Technical Memorandum 103743, National Aeronautics and Space Administration, 1991, p 8

• D.L. Hindman and J.R. Keiser, Performance of Ceramic-Ceramic Composites in an Industrial Waste Incinerator, Paper 191, NACE International, 1994, p 16

• M.G. Jenkins, A.S. Kobayashi, and R.C. Bradt, Crack Length Measurements of Ceramics at Elevated Temperatures Using the Laser Interferometric Displacement Gauge, *Fatigue Crack Measurement: Techniques and Applications*, Engineering Materials Advisory Services Ltd., West Midlands, U.K., 1991, p 335-373

• M. Kato, M. Kajimoto, Y. Inui, T. Kubohori, T. Hayami, and T. Ikuta, Study on Cutting of Ceramics, *Proc. of the 50th Conference of the Thermal Spraying Society of Japan* (Fukuoka, Japan, 16-17 Nov 1989), Thermal Spraying Society of Japan, 1989, p 53-58

• S. Kim, Advanced Ceramics as Cutting Tool Materials, Part I: An Overview, *Can. Ceram. Q.*, Vol 61 (No. 1), 1992, p 51-58

• H. Kishimoto, Cyclic Fatigue in Ceramics, *JSME Int. J.*, Series I, Vol 34 (No. 4), 1991, p 393-403

• A. Kitagawa and A. Matsunawa, Three-Dimensional Shaping of Ceramics by Using CO_2 Laser and Its Optimum Processing Condition, *Laser Materials Processing: Conf. Proc. ICALEO '90* (Boston, MA, 4-9 Nov 1990), Laser Institute of America, 1991, p 294-301

• P. Lamicq, Ceramic Matrix Composites: A New Concept for New Challenges, *Conf. Proc. High Temperature Materials for Power Engineering II* (Liege, Belgium, 24-27 Sept 1990), Kluwer Academic Publishers, 1990, p 1559-1574

• J. Lamon, Structural Reliability of Ceramics, *Engineering Ceramics*, Vol 3, *Conf. Proc. Euro-Ceramics* (Maastricht, The Netherlands, 18-23 June 1989), Elsevier Science Publishers, Essex, U.K., 1989, p 3.100 to 3.105

• G. Lascar, Ceramic-Metal Joints: Joining Techniques, Mechanical and Physicochemical Interactions, *Conf. Proc. Advances in Joining Newer Structural Materials/Progres Pour l'Assemblage des Materiaux de Construction les Plus Recents* (Montreal, Canada, 23-25 July 1990), Pergamon Press, Oxford, U.K., 1990, p 103-108

• X.Y. Li, Y. Xu, and C. Gui, Friction and Wear of Toughened Ceramics, *Proc. of the Japan International Tribology Conference III* (Nagoya, Japan, 29 Oct-1 Nov 1990), Japanese Society of Tribologists, 1990, p 1449-1454

- D. Munz, G. Martin, and H. Riesch-Oppermann, Probabilistic Assessment of Non-Destructively Determined Flaws in Ceramic Materials, *Conf. Proc. ISTFA '90* (Los Angeles, CA, 29 Oct-2 Nov 1990), ASM International, 1990, p 471-475
- D.C. Phillips, Ceramic Composites at Extremes of Performance (Retroactive Coverage), *Conf. Proc. Mechanical and Physical Behaviour of Metallic and Ceramic Composites* (Roskilde, Denmark, 5-9 Sept 1989), Riso National Laboratory, Roskilde, Denmark, 1988, p 183-199
- C.Q. Rousseau, Monotonic and Cyclic Behavior of a Silicon Carbide/Calcium-Aluminosilicate Ceramic Composite, *Thermal and Mechanical Behavior of Metal Matrix and Ceramic Matrix Composites,* STP 1080, ASTM, 1990, p 136-151

- A.S. Shatalin, Some Features of the Fracture Behaviour of Structural Ceramics, *Conf. Proc. New Materials and Their Applications* (University of Warwick, U.K., 10-12 April 1990), IOP Publishing, Bristol, U.K., 1990, p 509-519
- J.J. Swab, Failure Mechanisms in Advanced Structural Ceramics, *Ceramic Technology International,* Sterling Publications Ltd., London, 1993, p 144-146
- H. Takebayashi and K. Ueda, Rolling Contact Fatigue of Ceramics and Effect of Defects in Material, *Proc. of the Japan International Tribology Conference I* (Nagoya, Japan, 29 Oct-1 Nov 1990), Japanese Society of Tribologists, 1990, p 663-666
- K. Tanaka, Y. Akininwa, and H. Tanaka, Fracture Mechanics of Small Cracks in Metals, Ceramics and Composites, *Computational and Experimental Fracture Mechanics: Developments in Japan,* Computational Mechanics Publications, Southampton, U.K., 1994, p 291-315
- S. Tanaka, Y. Hirose, and K. Tanaka, X-Ray Fractographic Study on Alumina and Zirconia Ceramics, *Conf. Proc. Advances in X-Ray Analysis* (Steamboat Springs, CO, 30 July-3 Aug 1990), Plenum Publishing, 1991, p 719-727
- R.E. Tressler, Environmental Effects on Long Term Reliability of SiC and Si_3N_4 Ceramics, *Conf. Proc. Corrosion and Corrosive Degradation of Ceramics* (Anaheim, CA, 1-2 Nov 1989), American Ceramic Society, 1990, p 99-124
- Z. Yajima, Y. Hirose, and K. Tanaka, X-Ray Examination of Fracture Surfaces of Silicon Nitride Ceramics, *Conf. Proc. Advances in X-Ray Analysis* (Denver, CO, 31 July-4 Aug 1989), Plenum Publishing, 1990, p 319-326

Toughening and Strengthening Models for Nominally Brittle Materials

A.G. Evans, Harvard University

IN MATERIALS that do not develop macroscopic inelastic strain prior to failure, stress concentrations arise at strain intensification sites. The design procedure for such materials requires that the ultimate tensile strength (UTS) be compared with these concentrated stresses. This design strategy has deficiencies caused by the weakest link scaling of the UTS and the extreme value nature of the stochastics. It can be alleviated by using higher-toughness materials. A more robust design strategy can be implemented if the material has a capacity to exhibit inelastic strain. Such strain diminishes stress concentrations, leading to a decreased sensitivity to manufacturing flaws, notches, and impacts. The effects are analogous to plasticity in metals. This article addresses these issues and the opportunities for enhancing the "design friendliness" by optimizing toughness and by inducing inelastic strain.

The design of structural components with nominally brittle materials is largely determined by their elastic moduli, density, and tensile strength. The fracture toughness does not enter in an explicit manner. It is understood that a minimum toughness is needed to ensure reliability, but the required toughness is not well defined; it is experience based. Inelastic strain (ductility) is more tangible than toughness because of the associated ability to redistribute stress. Some of the factors involved in design and reliability are addressed in this article, through considerations of the toughness and "ductility" of nominally brittle materials. This article also describes toughening by various bridging mechanisms, as well as process zone effects and their interaction with the bridging rupture zone. Finally, phenomena that give rise to exceptional toughness and notch-insensitive mechanical behavior are described.

Toughness Models

The propagation of cracks in brittle solids is resisted by microstructure (Ref 1-5). The associated toughening phenomena all involve inelastic deformations that occur in reinforcing phases through transformations (Ref 6-9), microcracking (Ref 10, 11), and twinning, as well as plasticity (Ref 13, 14), and internal friction (Ref 4, 5). Toughening is manifest as a fracture resistance, Γ_R, that increases with crack extension, Δa (Fig. 1) (Ref 2, 5-7). Concepts of initiation toughness, Γ_0, steady-state toughness, $\Delta\Gamma_S$, and tearing index, λ, are needed to characterize this behavior.

A simplification of the actual resistance behavior provides perspectives on the role of these quantities in the performance of toughened materials. A simplified relation for transient toughening ($\Delta a \leq L$) having a sound theoretical basis (Ref 15) is given by:

$$\Delta\Gamma_R = \Gamma_R - \Gamma_0 \approx \Delta\Gamma_S \sqrt{(\Delta a/L)} \qquad \text{(Eq 1a)}$$

and for steady-state ($\Delta a \geq L$):

$$\Delta\Gamma_R = \Delta\Gamma_S \qquad \text{(Eq 1b)}$$

where L is a reference length related to the inelastic zone size. The tearing index is a measure of the slope of the resistance curve and is defined as:

$$\lambda = (\Delta\Gamma_S/\Gamma_0)^2/L \qquad \text{(Eq 2)}$$

It has units of length. The UTS of a material, S, relates to the fracture resistance through the size of the flaws, a_0, and the tearing index in accordance with (Ref 16):

$$S/S^* = [\Lambda/2(\sqrt{1+\Lambda} - 1)]^{1/2} \qquad \text{(Eq 3)}$$

where Λ is a non-dimensional strengthening index, related to the tearing index by:

$$\Lambda = \lambda a_0 \qquad \text{(Eq 4)}$$

and S^* is the UTS of the untoughened material. The larger Λ, the more beneficial the effect of the toughening on the strength and reliability. The implication is that the attainment of "beneficial" toughening requires an optimization based on L, as well as $\Delta\Gamma_S$,

through λ and Λ. That is, the toughening models must address both quantities.

The toughening involves inelastic deformation mechanisms. These mechanisms enable dissipation and energy storage as cracks extend (Ref 2, 6, 17). They can be modeled in terms of stress/displacement constitutive laws for representative volume elements (Fig. 2). This is achieved by homogenizing the properties of the material around the crack. This approach rigorously describes high-toughness materials in which the inelastic region around the crack is appreciably larger than the spacing between the relevant microstructural entities.

The known mechanisms involve inelastic zones having two distinct configurations: zones that concentrate in a thin region around the crack plane and rupture as the crack extends, referred to as bridging or Dugdale zones (Fig. 2); and regions that extend normal to the crack plane and remain in a deformed but intact state after the crack has propagated through the material, designated process zones. Process and bridging zone mechanisms may operate simultaneously and synergistically (Ref 18). Toughening by bridging is caused by tractions along the crack surface induced by intact inelastic material ligaments. Usually, the tractions soften as the crack extends. Then, in steady state the toughening $\Delta\Gamma_S$ is (Ref 14, 15):

$$\Delta\Gamma_S = b (t_0 u_c) \qquad \text{(Eq 5)}$$

where u_c is the crack opening at the edge of the bridging zone, t_0 is the peak traction, and $b \approx 1/2$. The inelastic zone length is (Ref 15):

$$L \approx c [Eu_c/t_0] \qquad \text{(Eq 6)}$$

where $c \approx 0.12\pi$. The tearing index is thus:

$$\lambda = (b^2c)^{-1} [t_0^3 u_c/E\Gamma_0^2] \qquad \text{(Eq 7)}$$

Note the particularly strong influence of the traction t_0 on the strengthening. Ductile reinforcements, as

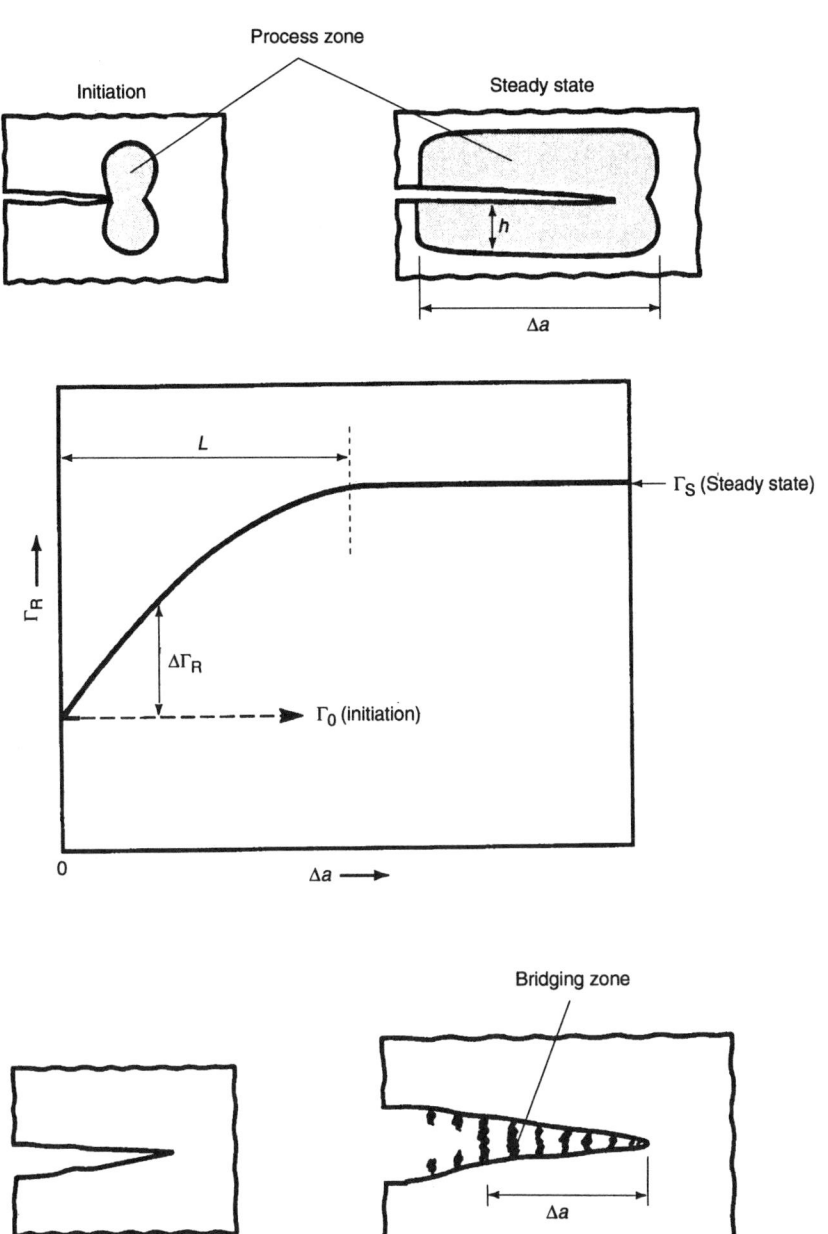

Fig. 1 A schematic of a resistance curve showing the evolution of the inelastic zones and the parameters that characterize the behavior

strengthening flexibility for process zone toughening is appreciably less than that for the bridging mechanisms. Transformation, microcrack, and twin toughening are process zone mechanisms. Plastic deformation also provides process zone toughening, but it is characterized in a different manner.

These strengthening and toughening predictions are often optimistic, because there are important short crack effects (Ref 19). For $a < 5L$, interactions between the inelastic zones at the opposite crack tips reduce the toughening. Modeling of these effects requires numerical procedures with a full constitutive description of the inelastic deformations. This topic is not addressed in this article, though it is crucially important. Some appreciation for the magnitude of the effect is provided by results obtained for transformation toughening (Fig. 3).

Conversely, in some materials, the inelastic zone can become larger than all of the relevant dimensions, such as the crack length and the ligament size, leading to large scale bridging (LSB) or large scale yielding (LSY) (Ref 5, 20). LSB or LSY result in exceptional toughness. In such materials, the strengthening given by Eq 3 is not the most useful engineering quantity. Instead, the notch sensitivity or notch strength, S_n, relative to the strength of the unnotched material, S_0, has greater practical significance (Ref 21). In the limit, LSB causes steady-state multiple cracking (Ref 4, 5), wherein the stress needed to extend cracks becomes independent of the crack length. Then, the toughness is so large that other properties become more critical, particularly embrittlement and creep.

Bridging Mechanisms

Reinforcing elements can be either ductile or brittle. The former comprise dual-phase alloys, polymer blends, and metal-toughened ceramics. Such reinforcement schemes rely on plasticity to create ligaments and dissipate energy (Fig. 4). When the elements are brittle, bridging requires either microstructural residual stresses or weak interfaces. Large local residual stresses caused by thermal expansion mismatch are capable of suppressing local crack propagation and, thereby, may allow intact ligaments to exist behind the crack front (Ref 22). When these ligaments eventually fail in the crack wake, energy is dissipated through acoustic waves and causes toughening. Low-fracture-energy interfaces are more effective. They cause the crack to deflect and debond the interfaces. The debonds acquire mode II (shear) characteristics, leading to friction, stability, and intact ligaments (Ref 23). As the crack extends, further debonding occurs, subject to friction. Eventually, the bridging material fails, either by debonding around the ends or by fracture. Following reinforcement failure, additional friction may occur along the debonded surfaces. The energy dissipation upon crack propagation thus includes terms from the energy of the debonded interfaces, the acoustic energy dissipated upon reinforcement failure, and frictional dissi-

well as brittle fibers and anisotropic grains, toughen by means of bridging tractions.

Process zone toughening may be characterized by the product of the critical stress for activating the inelastic strain mechanism, σ_0, the associated stress-free strain, ε_T, and the zone height, h, in accordance with the stress-strain hysteresis of material elements within the process zone (Fig. 2) (Ref 6):

$$\Delta\Gamma_S = 2h\sigma_0\varepsilon_T \qquad (Eq\ 8)$$

The reference length is governed by the zone height (Ref 6, 7):

$$L = ch \qquad (Eq\ 9)$$

where $c \approx 3$. Moreover, the process zone height is related to the critical stress by (Ref 6, 7):

$$h = d\ [E\Gamma_0/\sigma_0^2] \qquad (Eq\ 10)$$

where d is a mechanism dependent coefficient of order, 0.1. The tearing index is thus:

$$\lambda = (4d/c)\ [\varepsilon_T^2 E/\Gamma_0] \qquad (Eq\ 11)$$

Note that at this level of simplification, the only inelastic material property affecting the strengthening is the transformation strain, ε_T. The critical stress and the zone height are of secondary importance, in contrast to their primary influence on the steady-state toughening. Because of this, the

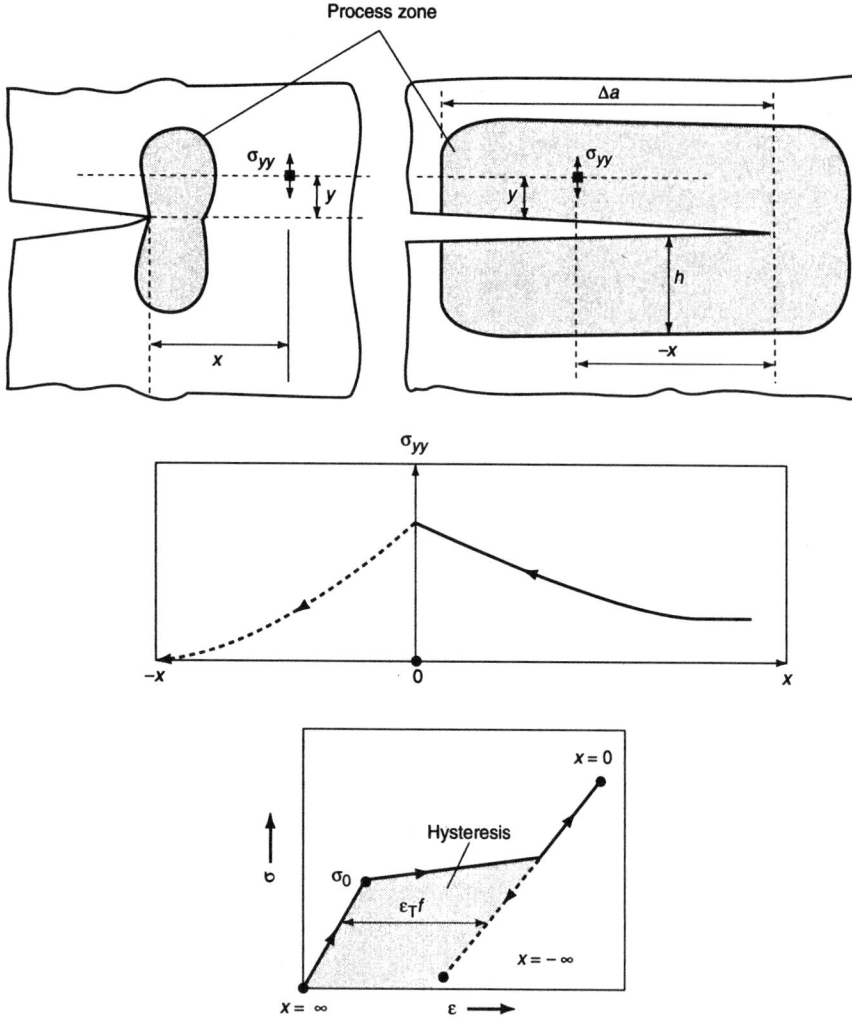

Fig. 2 A schematic of crack extension in the presence of a process zone and the strain/stress history experienced by an element of material within the zone

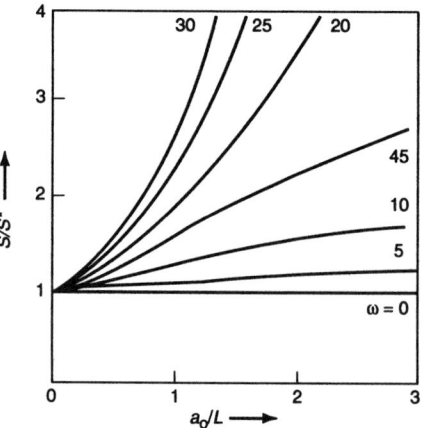

Fig. 3 The effect of crack length on the strengths enabled by transformation toughening

$$L = \beta RE/Y \qquad \text{(Eq 12b)}$$

where f is the volume fraction of the reinforcement. χ and β are "ductility" parameters that also depend on the extent of the interface debonding, d. (Note that Y can increase as R decreases because of strain gradient effects, Ref 27). Values of χ and β have been obtained both by calculation and by experiment (Ref 13, 14, 26, 28, 29). For well-bonded interfaces with ductile phases that fail by necking to a point, $\chi \approx 0.5$ and $\beta \approx 0.1$, both increasing as the strain hardening increases. Less ductile ligaments that rupture prematurely have correspondingly smaller χ and β. In systems subject to debonding, χ and β depend on the reinforcement morphology. For interpenetrating or continuous reinforcing phases, χ and β are increased by debonding, because the dissipation is spread laterally away from the crack plane: χ approaches 8 for large d/R. For discontinuous phases, debonding around the ends of the reinforcement diminishes χ and β.

The tearing index for such materials is obtained from Eq 2 and 12 as:

$$\Lambda = [(\chi f)^2/\beta] \, [RY^3/ET_0^2] \qquad \text{(Eq 13)}$$

The important implication is that the scaling that controls strength and reliability is most strongly affected by the yield strength of the reinforcing material, Y, and its ductility, through χ. Other factors are relatively unimportant.

Brittle Reinforcements

The highest toughnesses are achievable in systems containing "weak" interfaces, which enable debonding and allow dissipation by internal friction. There are major effects on the toughness of the morphology of the debond planes (roughness), the residual stresses, the failure stochastics of the reinforcements, and the stiffnesses of the constituent phases (Ref 30). A prerequisite to such toughening is that the debond criterion be satisfied (Ref 31). Initial interface debonding at the crack front requires that the relative toughness

pation (Fig. 4) (Ref 2, 23). The latter is typically dominant. Moreover, the internal friction can become exceptionally large, resulting in toughness approaching those for ductile metals (Ref 23, 24).

Ductile Phases

The material systems that exhibit plasticity-induced toughening have three distinct microstructures: isolated ductile reinforcements in an elastic matrix, interpenetrating ductile/elastic networks, and a ductile matrix with a dispersed elastic phase. The first two represent most metal-toughened ceramics and intermetallics. The latter includes most metal-matrix composites and rubber-toughened polymers. An important difference between the first two microstructures and the third concerns the potential for macroscopic plastic strain. Plastic strain in the former is limited by the elastic network, such that the only ductile material experiencing extensive strain is that stretching between the crack surfaces in the bridging zone. The latter materials develop an additional plastic zone that enables further, often

substantial, inelastic dissipation. This behavior and its coupling to the rupture zone are elaborated in the following section.

Ductile ligament toughening in an otherwise elastic material is contingent on the ligament failure mechanism, through the stress/stretch relation. Without debonding, the stress attains high levels because of the elastic constraint of the matrix (Ref 13, 25), but then decreases as the crack opens because of necking. Debonding reduces the constraint but increases the plastic stretch to failure (Ref 13, 26). The latter contribution to the dissipation dominates, causing the dissipation to increase as the debond length increases. The toughness attributed to ductile bridging can be reexpressed by noting that the stress scales with the uniaxial yield strength, Y, of the ligaments and that the plastic stretch is proportional to the radius of the cross section of the reinforcements, R. Consequently, the asymptotic toughening is (Ref 26):

$$\Delta\Gamma_S = \chi f R Y \qquad \text{(Eq 12a)}$$

and the zone length is:

of the interface (Γ_i) and the reinforcement (Γ_p) be small enough to lie within the debond zone depicted in Fig. 5 (typically $\Gamma_i/\Gamma_p \leq 1/4$). The extent of initial debonding is small, but further debonding is induced in the crack wake as the crack extends. The extent of further debonding is governed largely by the residual field, the debond surface roughness, and the friction coefficient (Ref 32). Reinforcement failure involves stochastics, subject to a friction stress τ (Ref 33-36). Large τ causes the stress to vary rapidly and induces reinforcement failure close to the crack, leading to a small pullout length, p, and vice versa. The consequent tractions are relatively complex (Ref 37). Insight is gained from solutions for aligned, brittle fibers. At the simplest level, the behavior of short, strong, aligned reinforcements, subject to friction, can be explored. For this case (Ref 16, 37):

$$\Delta\Gamma_S = f\tau\,(p/R)^2\,R \qquad \text{(Eq 14a)}$$

$$L = (c/2)ER/f\tau \qquad \text{(Eq 14b)}$$

where R is the reinforcement radius and p their length, such that p/R is the aspect ratio. The corresponding tearing index is:

$$\lambda = (2/c)\,[(f\tau)^3\,R\,(p/R)^4/E\Gamma_0^2] \qquad \text{(Eq 14c)}$$

Note the major influences of the aspect ratio and the friction stress. However, this toughening and strengthening situation occurs only if the reinforcement strength, S, satisfies (Ref 37):

$$S > \tau\,(p/R) \qquad \text{(Eq 15)}$$

When the reinforcements are long and aligned, but susceptible to fracturing as the crack extends, the tractions are more complex (Ref 37):

$$t/fS_L = \sqrt{w}\,\exp\,[-\alpha_m w^{(m+1)/2}] \qquad \text{(Eq 16)}$$

where m is the Weibull shape parameter associated with the reinforcements, S_L is their average strength at length L_r, and:

$$w = 8\lambda^2\,(u/L_r)$$
$$\alpha_m = [\Gamma(1 + 1/m)]^m/[m + 1]$$
$$\lambda = [E/2E_m\,(1 - f)]\,\sqrt{(E_r/S_L)}$$
$$L_r = S_L R/\tau$$

where the subscripts m and r refer to the matrix and reinforcements, respectively, and Γ is the gamma function. Even in this simplified case, the friction stress and the reinforcement strength have interactive effects on toughening and strengthening that are not evident without detailed analysis. Consequently, for practical implementation, the integrated effect of these variables on the behavior in the presence of strain concentrators, such as notches, is more useful, as manifest in the notch sensitivity (Ref 21, 23). These effects are elaborated in the following section.

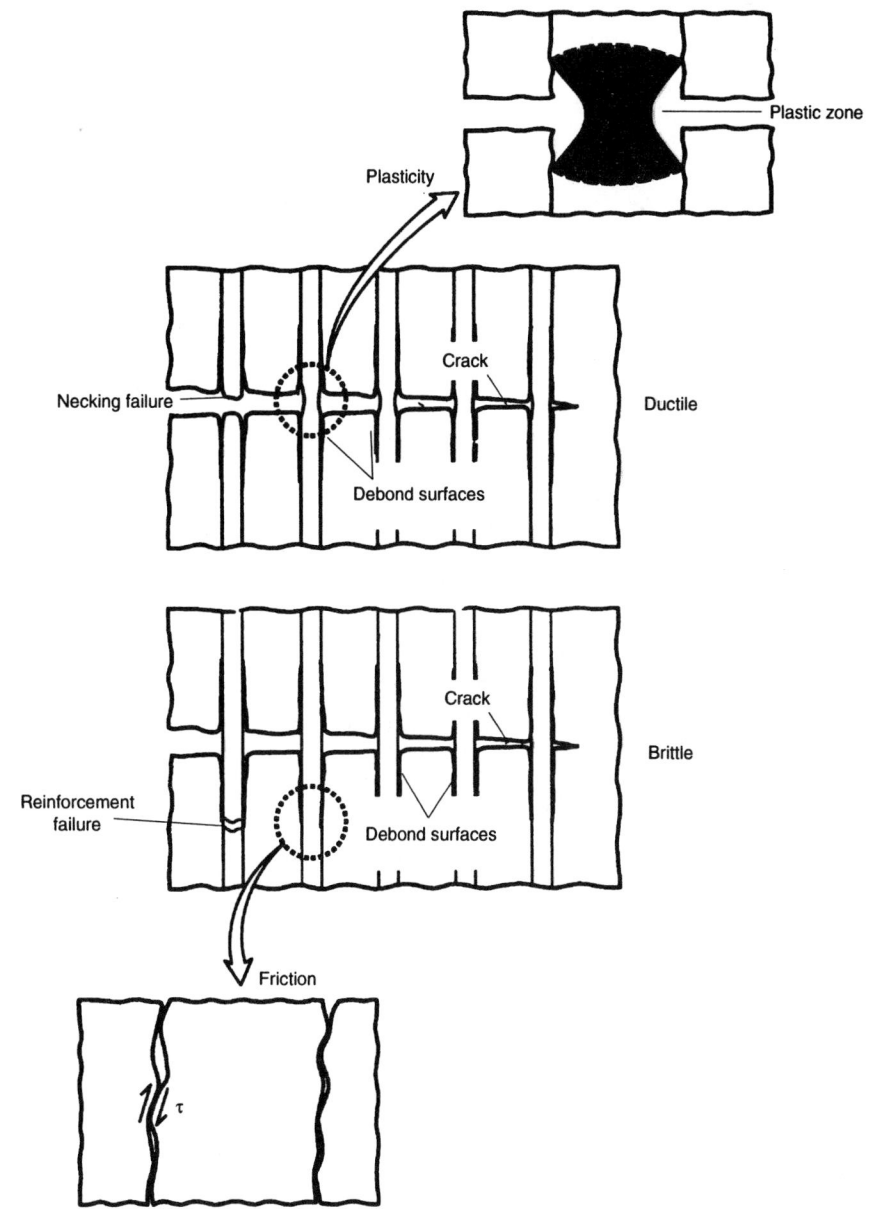

Fig. 4 A schematic indicating the dissipation that occurs with a bridging zone. (a) Ductile ligaments. (b) Brittle reinforcements

Tough materials having the above characteristics are made by creating the appropriate interphase between the reinforcement and matrix. The most common approach comprises a dual coating: the inner coating satisfies the debonding, friction, and compliance requirements, while the outer coating provides protection against the matrix during processing (Ref 30, 38, 39).

Anisotropic Grains

Low-fracture-energy planes or grain boundaries can allow debonding, as in reinforced materials with weak interfaces, such that toughening involves the same considerations. Certain anisotropic ceramics with elongated grains (particularly alumina and silicon nitride) exhibit such

toughening (Ref 40). In these materials, the dominant effect is the friction that operates along the rough, nonaligned, debonded grain boundaries (Ref 16, 41). The trends are thus broadly consistent with the above results for short, strong reinforcements. That is, the toughening and the strengthening increase as the grain radius, R, and their aspect ratio, p/R, increase, provided that the flaws are unchanged (that is, a_0 is fixed).

Intermetallics such as Ti-Al also toughen in this manner because of extreme anisotropy in the cleavage energies (Ref 42, 43). But in this case, the energy of the additional surface created by debonding and ligament formation appears to be more important than the friction. The relevant toughening and strengthening parameters are then (Ref 43):

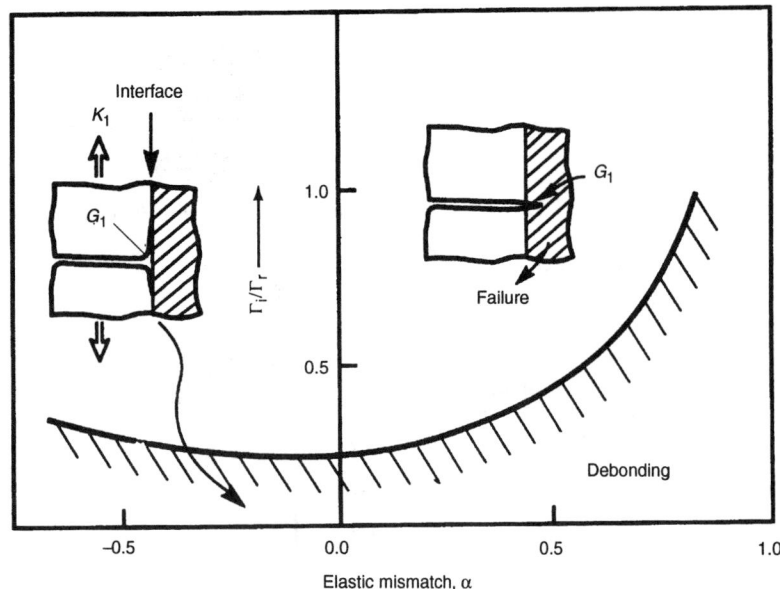

Fig. 5 A debond diagram for brittle reinforcements

$$\Delta\Gamma_S/\Gamma_L = \omega$$
$$L = cR\omega/\chi^2 \qquad \text{(Eq 17)}$$
$$\lambda = (\Gamma_L/\Gamma_0)^2 \,\omega\, \chi^2/R$$

where Γ_L is the cleavage energy on the "tough," transverse planes, and ω and χ are non-dimensional quantities that depend only on the aspect ratio and orientation of the ligaments formed by debonding (Fig. 6).

The salient point is that, when such a mechanism prevails, there can be no effect of the ligament width, R, on the toughening. The only quantities affecting this behavior are the aspect ratio and the orientations of the ligaments that form, as well as the cleavage energies. Conversely, the strengthening is influenced by the ligament size, with the opposite trend to that associated with

frictional toughening. That is, the strengthening increases as the microstructure is refined and R is diminished.

Process Zone Mechanisms

Many materials remain nonlinear once inelastic deformation has commenced (beyond σ_0). In such materials, the tip energy release rate G_{tip} is zero (Ref 44). It is possible to model the process zone dissipation only upon coupling with a failure zone along the crack plane (Ref 45, 46). This situation is elaborated below for plastic zone dissipation. This complexity does not exist when the nonlinear deformation process saturates, such that the stress-strain curve becomes linear at large strains. In this case, G_{tip} is finite, because there is

no dissipation in the elastic saturation zone (Ref 6-9). The crack growth criterion then requires that G_{tip} attain the local toughness of the material immediately ahead of the crack, which may have been modified from that in the pristine material. This saturation approach has been applied to transformation and microcrack toughening.

Saturation Toughening. Transformation toughening is dominated by a dilatational stress-free strain. Then a frontal process zone has no effect on G_{tip}, and initial crack growth must occur without toughening (Fig. 2). Upon crack extension, process-zone elements unload in the wake, hysteresis occurs, and toughening develops. Steady-state toughening is attained at $L/h \approx 3$. The material within the zone behind the crack tip experiences unloading as the crack extends, so the J-integral becomes path dependent and must be superseded by a conservation integral I. At the tip, the familiar result is (Ref 6):

$$I = J = G_{tip} \qquad \text{(Eq 18a)}$$

But, remote from the tip (Ref 6):

$$I = G_\infty - 2\int_0^h U(y)\,dy \qquad \text{(Eq 18b)}$$

where $U(y)$ is the residual energy density in the wake. Equating I for the near-tip and remote paths gives:

$$\Delta\Gamma_R = 2\int_0^h U(y)\,dy \qquad \text{(Eq 18c)}$$

Material elements in the process zone undergo a complete loading/unloading cycle as they translate from the front to the rear of the crack tip during crack advance. Hence, each element is subject to the residual stress work contained by the hysteresis loop (Fig. 2). The residual energy density is (Ref 6):

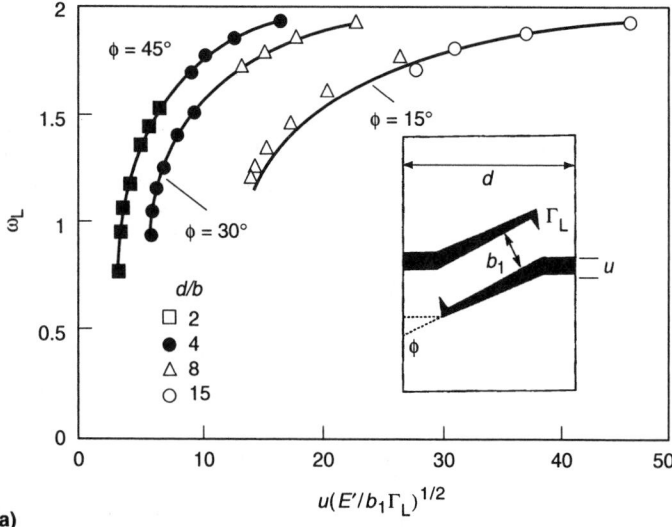

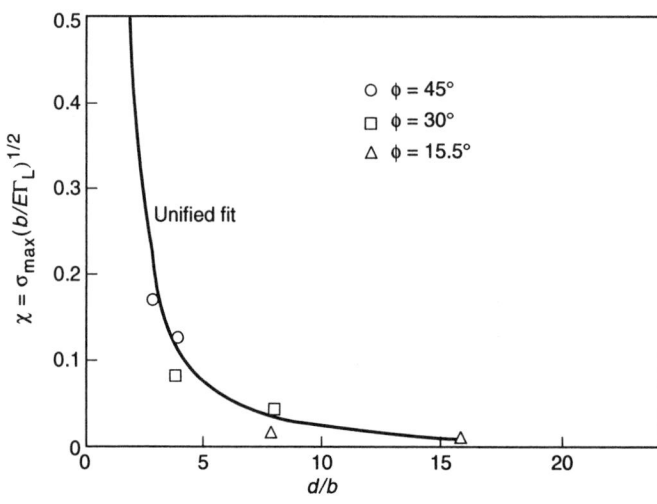

(a)

(b)

Fig. 6 Non-dimensional parameters that affect ligament toughening

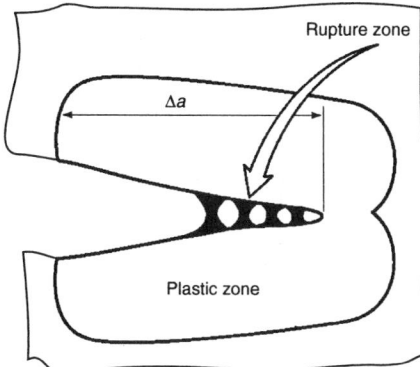

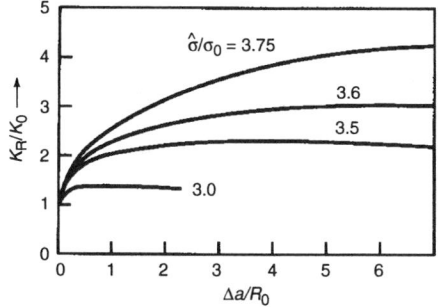

Fig. 7 Resistance curve for plastic zone toughening in the presence of a rupture zone

$$U(y) = \sigma_0\,\varepsilon_T f + B^*\,(f\varepsilon_T)^2/2[1 - B^*/B]$$
$$+ E(\varepsilon_T f)^2/9(1-\nu) \qquad \text{(Eq 19)}$$

where B is the bulk modulus and B^* is the slope of the stress-strain curve of the transforming material. Note that B and f are invariant when the transformation is supercritical, that is, when all of the transformable material undergoes transformation within the process zone ($y < h$). In this case (Ref 6):

$$B^* = 2E/3(1+\nu) \qquad \text{(Eq 20)}$$

and the latter two terms in Eq 19 cancel. The toughening is then given exactly by Eq 8 and the tearing index by Eq 11. Typically, the transformation is subcritical, leading to a gradient in the fraction of transformed material within the process zone. That is, B and f depend on y. The resultant toughening is lower than Eq 8. It must be determined from Eq 19 by imposing the function $f(y)$, obtained by either measurement or calculation (Ref 9, 47).

Similar phenomena occur in materials susceptible to microcracking (Ref 2, 10). Stress-activated microcracks occur within regions of local residual tension, caused for example by thermal expansion mismatch between phases. There is an associated dilatation, governed by the residual opening of the microcracks. The microcracks also reduce the elastic modulus within the process zone. The dilatational contribution to the toughening has precisely the same form as that associated with transformations, but with $f\varepsilon_T$ replaced by θ_T. For penny-shaped microcracks (Ref 10):

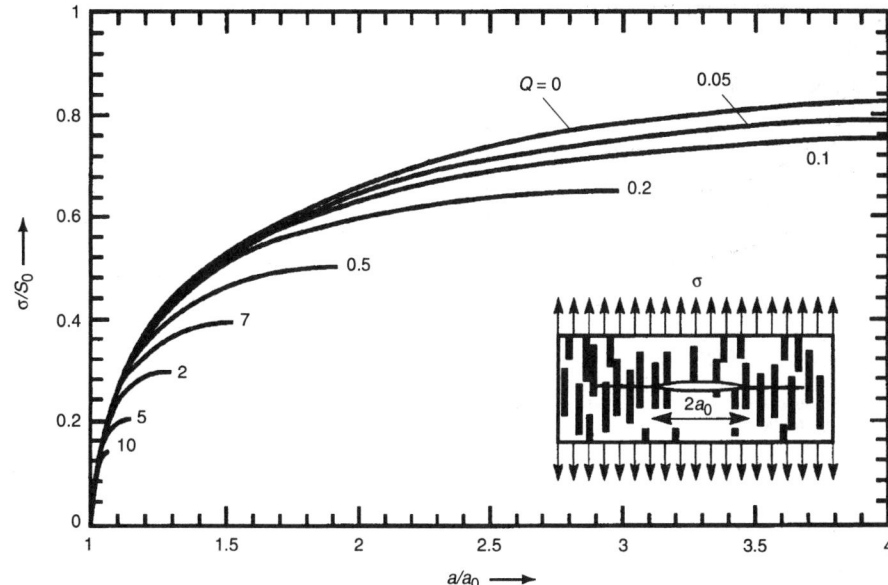

Fig. 8 Crack extension stress for materials reinforced with short, strong, brittle reinforcements

$$\theta_T = (32/27)(1+\nu)n\varepsilon_0 \qquad \text{(Eq 21)}$$

where n is the microcrack density and ε_0 is the misfit strain between the phases. However, other phenomena are also involved in determining the toughening and strengthening. The modulus change reduces the stress ahead of the crack, but it has only a small (and subtle) effect on the energy release rate. The toughening is thus modest and may be zero. This effect depends on the microcrack density but not on the zone height. In general, toughening due to the dilatational and modulus effects are not additive; interaction terms are involved. Moreover, the microcracks have a degrading influence on the toughening that partially counteracts these effects. That is, they reduce the tip toughness through the area fraction of the crack plane they occupy.

Plastic Zone Toughening. The dissipation occurring within a plastic zone can be simulated only by invoking a finite rupture region along the crack plane (Ref 45, 46). This zone couples the stresses and displacements into the plastic zone. All realistic rupture laws have the form depicted in Fig. 7, characterized by a peak stress, σ^*, and a dissipation, Γ^*. For inelastic bridging in an elastic material, described above, σ^* equates with t_0 (Ref 9). When plasticity occurs outside the rupture zone, there is an extra contribution to the toughening. Small scale yielding (SSY) simulations of the resistance for a material subject to power law hardening (uniaxial yield strength, σ_0, and hardening rate, N) are summarized in Fig. 7. When the stresses that operate in the rupture zone are relatively small (σ^*/σ_0 less than approximately 2), a plastic zone is unable to develop before rupture occurs. Then, the toughness is strictly that associated with bridging and rupture (Ref 12). For relatively stronger rupture zones, a plastic zone is induced. The resulting dissipation increases rapidly as σ^*/σ_0 and N increase. When σ^*/σ_0 reaches ~5, the plastic zone extends indefi-

nitely and the crack cannot propagate: that is, the SSY toughness is unbounded. In this case, LSY simulations would be needed to address crack growth. The SSY toughening parameters are:

$$\Delta\Gamma_S/\Gamma_0 = g(\sigma^*/\sigma_0, N) + 1 \qquad \text{(Eq 22a)}$$

$$L = R_0\xi(\sigma^*/\sigma_0, N) \qquad \text{(Eq 22b)}$$

where R_0 is the plastic zone size:

$$R_0 = \phi(E\Gamma_0/\sigma_0^2) \qquad \text{(Eq 22c)}$$

and where g, ξ, and ϕ are non-dimensional functions having the approximate form:

$$g \approx [\sigma^*/\sigma_0 - 2]^{1/\alpha N}$$
$$\xi \approx g^{2\alpha N}$$

where $\alpha \approx 2$ and $\phi \approx 0.1$.

Notch Sensitivity and Inelastic Strain

Notch Strength. For engineering application, the notch sensitivity is most useful (Ref 21). Simulation of the stress σ needed to extend a dominant crack, length a, from an initial unbridged region (or notch), length a_0, is achieved by using the bridging tractions (such as Eq 16). As this occurs, a maximum stress is reached coincident with simultaneous failure of the intact reinforcements. This stress is the notch strength, S_n. The ratio of this strength to that for the unnotched material, S_0, is the relevant design parameter. Such behavior is illustrated for a material with strong, short, aligned reinforcements in Fig. 8. This behavior has the envelope (Ref 37):

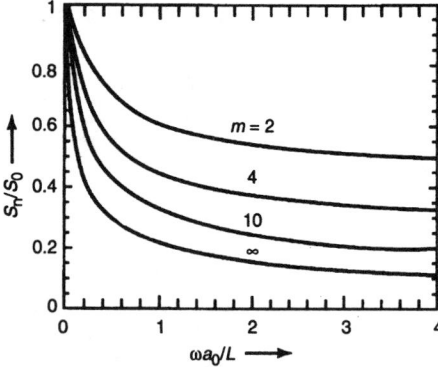

Fig. 9 The calculated notch sensitivity for a material with long, brittle reinforcements that may fracture as the crack extends

$\sigma/S_0 = (2/\pi) \cos^{-1} (a/a_0)$

$S_0 = f\tau p/2R$ (Eq 23)

where this envelope occurs for small Q, representative of small values of the friction stress where

$Q = (2f E_r/EA) (a_0\tau/RE_r)$ (Eq 24)

and where A is an anisotropy parameter of order unity. As Q increases, crack growth is truncated by material failure, at diminishing strength levels. The resulting notch strength is given by:

$S_n/S_0 = \sqrt{(1 + 3\pi Q/2)}$ (Eq 25)

The corresponding notch behavior when the reinforcements are longer and susceptible to failure as the crack grows is summarized in Fig. 9 (Ref 37). Note the minimal notch sensitivity when the Weibull shape parameter, m, is small. When the toughening enabled by internal friction becomes large, a steady-state cracking condition is reached (Ref 5). At this condition, a matrix crack may extend across the component without failing the reinforcements. In this case, multiple matrix cracking is induced. This occurs prior to attainment of the UTS (Ref 23, 24). The consequence is macroscopic inelastic strain, with "ductilities" on the order of 1%. These inelastic strains redistribute stress in an efficient manner, such that the material may become notch insensitive (Fig. 10) (Ref 48, 49). The toughness then has no practical significance. The approach used to characterize these materials, as with metals, involves characterizing the inelastic strains (Ref 50).

Inelastic Strains. A mechanism-based strategy for characterizing and implementing inelastic strains and for calculating stress redistribution is illustrated using results for ceramic-matrix composites (Ref 23). These composites exhibit inelastic deformations when matrix microcracks are stabilized. This is achieved by using either fiber coatings or porous matrices to deviate cracks toward the loading axis (Fig. 5). The resultant composite microstructure and the ensuing mechanical responses resemble those found in various naturally occurring materials, such as

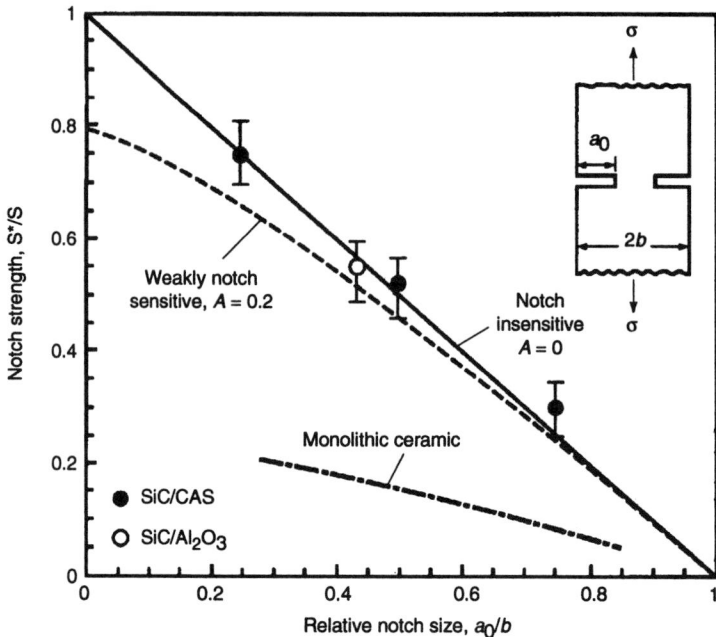

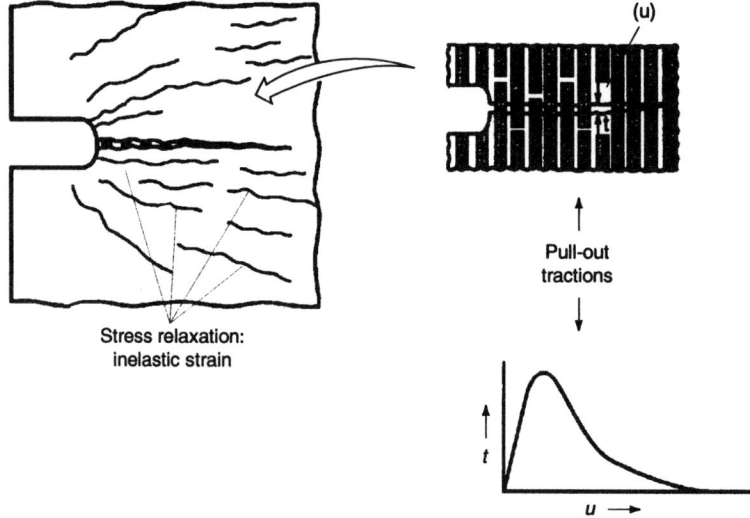

Fig. 10 The notch insensitivity found experimentally for ceramic-matrix composites caused by stress redistribution

wood. The inelastic strain is governed primarily by the number density of cracking sites and the friction stress that operates along the debonded crack surfaces. A cell model represents most of the features (Fig. 11). Two stresses characterize the inelastic strain: a friction stress τ and a debond stress σ_i (Fig. 11). The latter is related to the debond toughness for the coating and the residual stress (Ref 51). The consequent inelastic tensile strain ε has linear and parabolic terms, given by (Ref 52):

$\varepsilon = (1 + \Sigma^T)\sigma/E^* + 2L\sigma^2(1 - \Sigma_i)(1$

$+ \Sigma_i + 2\Sigma^2) - \sigma^T/E$ (Eq 26)

where E^* is the diminished elastic modulus caused by matrix cracking, σ^T is the misfit stress (related to the residual stress), Σ_i is the non-dimensional de-

bond stress, $\Sigma_i = \sigma_i/\sigma$, $\Sigma^T = \sigma^T/\sigma$, and L is an interface friction index, given by (Ref 52):

$L = \dfrac{b(1 - af)^2 R}{4f^2 \tau d E_m}$ (Eq 27)

where d is the crack spacing. The coefficients a and b are of order unity. The parameters E^*, L, and Σ_i (evaluated from hysteresis loops) (Ref 52) provide understanding about the separate influences of debonding, friction, and matrix cracking on the inelastic strain. They also provided the insight needed to develop a constitutive law $\sigma(\varepsilon)$ compatible with finite element codes.

A law derived on this basis that characterizes all plane stress states in the (1,2) directions is (Ref 50):

$$\sigma_1 = \frac{E_0}{1 - v_0^2}(\varepsilon_1 + v_0\varepsilon_2) + \Delta\sigma_1 \cos^2\theta + \Delta\sigma_{II}\sin^2\theta$$

$$\sigma_2 = \frac{E_0}{1 - v_0^2}(\varepsilon_2 + v_0\varepsilon_1) + \Delta\sigma_1 \sin^2\theta$$

$$+ \Delta\sigma_{II}\cos^2\theta \qquad \text{(Eq 28)}$$

$$\tau = \frac{E_{45}}{2(1 - v_{45})}\gamma_{12} - (\Delta\sigma_1 - \Delta\sigma_{II})\sin\theta\cos\theta$$

where τ is now the macroscopic shear stress; $\varepsilon_{1,2}$ are the normal strains; γ_{12} is the shear strain; $\Delta\sigma_{I,II}$ are the stress drops upon matrix cracking at fixed strain, parallel and normal to the fiber directions; θ is the angle between (1) and the fiber direction; $E_{0,45}$ are the Young's modulus in the fiber direction and at 45°, respectively; and $v_{0,45}$ are the corresponding values of the Poisson ratio. Some calculations of stress redistribution in a pin-loaded configuration are summarized in Fig. 12. A comparison with experiment is also shown. These results establish that appreciable stress reduction is enabled by the inelastic strain, resulting in a response insensitive to the strain concentration caused by the hole (Ref 50, 53):

Future Work

Some important phenomena have yet to be adequately explored, particularly short crack effects induced by overlap of the strain fields from the two crack fronts and the free surfaces. These effects diminish the tearing index and adversely affect the strengthening that can be gained by activating the toughening mechanisms. Establishing these characteristics is an important priority for future modeling.

Ductilizing has a much greater influence on reliability than increasing the toughness. A future endeavor would address new methods for improving ductility.

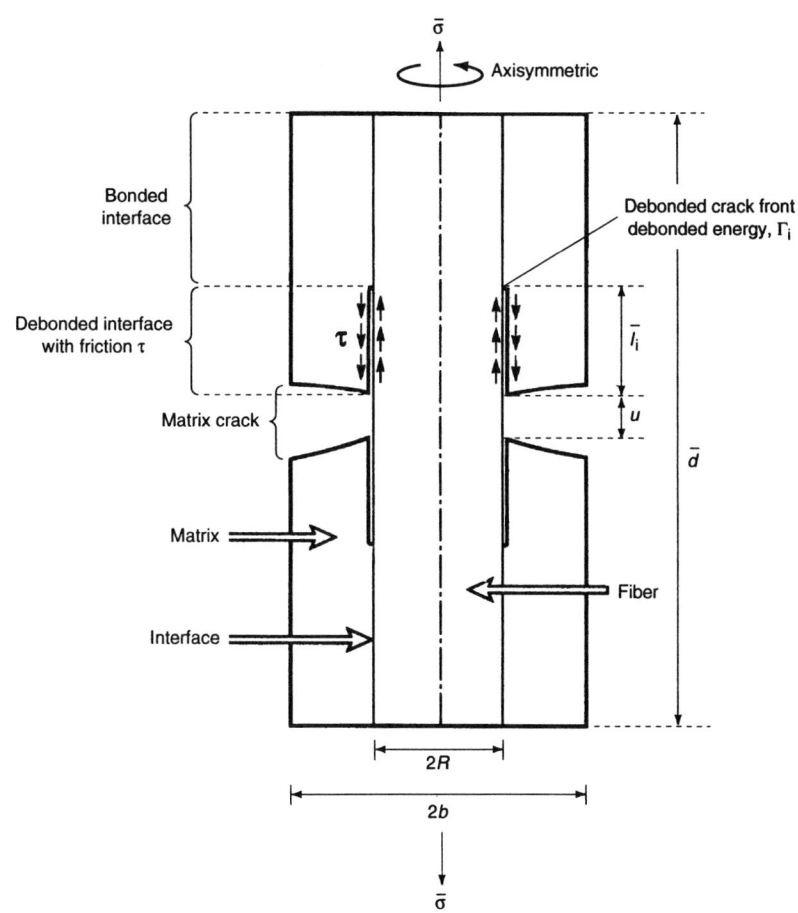

Fig. 11 Basic cell model (hysteresis loops) used to characterize the inelastic strains that occur in ceramic-matrix composites and their dependence on the interface friction. $\varepsilon = u/d$, $f = R/b$.

REFERENCES

1. B.R. Lawn, *Fracture of Brittle Steels*, The Cambridge Press, Cambridge, 1993
2. A.G. Evans, *J. Am. Ceram. Soc.*, Vol 73 (No. 2), 1990, p 187-206
3. P.F. Becher, *J. Am. Ceram. Soc.*, Vol 73, 1991, p 255-269
4. J. Aveston, G.A. Cooper, and A. Kelly, in *NPL Conf. Proc.*, IPC Publishing, 1971, p 15-26
5. D.B. Marshall, B.N. Cox, and A.G. Evans, *Acta Metall.*, Vol 33, 1985, p 2013
6. B. Budiansky, J.W. Hutchinson, and A.G. Evans, *Int. J. Solids Struct.*, Vol 19, 1983, p 337-355
7. R.M. McMeeking and A.G. Evans, *J. Am. Ceram. Soc.*, Vol 65, 1982, p 242-246
8. D.J. Green, R.H.J. Hannink, and M.V. Swain, *Transformation Toughening*, CRC Press, 1989
9. A.G. Evans and R.M. Cannon, *Acta Metall.*, Vol 34, 1986, p 761-800
10. J.W. Hutchinson, *Acta Metall.*, Vol 35, 1987, p 1605-1619
11. A.G. Evans and K.T. Faber, *J. Am. Ceram. Soc.*, Vol 64 (No. 4), 1984, p 255-260
12. H.E. Deve and A.G. Evans, *Acta Metall.*, Vol 39 (No. 6), 1991, p 1171-1176
13. M.F. Ashby, F.J. Blunt, and M. Bannister, *Acta Metall.*, Vol 37, 1989, p 1947-1957
14. A.G. Evans and R.M. McMeeking, *Acta Metall.*, Vol 34, 1986, p 2435-2441
15. B. Bao and C.Y. Hui, *Int. J. Solids Struct.*, Vol 26, 1990, p 631
16. A.G. Evans, *Acta Metall. Mater.*, in press
17. B. Budiansky and J.C. Amazigo, *Int. J. Solids Struct.*, Vol 24 (No. 7), 1988
18. B. Budiansky and J.C. Amazigo, *J. Mech. Phys. Solids*, Vol 36 (No. 5), 1988
19. B. Budiansky and D.M. Stump, *Acta Metall.*, Vol 37 (No. 12), 1989
20. B.N. Cox, *Acta Metall.*, Vol 39, 1991, p 1189-1201
21. G. Bao and Z. Suo (V.C. Li, Ed.), ASME, 1992, p 355-366
22. J.W. Hutchinson and D.K.M. Shum, *Mech. Mater.*, Vol 9, 1990, p 17
23. A.G. Evans and F.W. Zok, *J. Mater. Sci.*, Vol 29, 1995, p 3857
24. A.G. Evans, *Phil. Trans. R. Soc. London*, Vol 315A, 1995, p 511-525
25. J.W. Hutchinson, Y. Huang, and V. Tvergaard, *J. Mech. Phys. Solids*, Vol 39 (No. 2), 1991
26. H. Deve, A.G. Evans, and R. Mehrabian, *Mat. Res. Soc. Symp. Proc.*, Vol 170, 1990, p 33-38
27. J.W. Hutchinson, N.A. Fleck, G.M. Muller, and M.F. Ashby, *Acta Metall.*, Vol 42 (No. 2), 1994, p 475-487
28. P.A. Mataga, *Acta Metall.*, Vol 37, 1989, p 3349-3359
29. B.D. Flinn, C.S. Lo, F.W. Zok, and A.G. Evans, *J. Am. Ceram. Soc.*, Vol 76, 1993, p 369-375
30. R. Kerans, *Scripta Metall. Mater.*, Vol 32, 1994, p 1075
31. M.-Y. He and J.W. Hutchinson, *J. Applied Mech.*, Vol 56, 1989, p 270-278
32. J.W. Hutchinson, M.-Y. He, B.-X. Wu, and A.G. Evans, *Mech. Mater.*, Vol 18, 1994, p 213-229
33. W.A. Curtin, *J. Am. Ceram. Soc.*, Vol 74, 1991, p 2837
34. M.D. Thouless and A.G. Evans, *Acta Metall.*, Vol 36, 1988, p 517
35. M. Sutcu, *Acta Metall.*, Vol 37, 1989, p 651
36. W.A. Curtin and S.J. Zhou, *J. Mech. Phys. Solids*, Vol 43, 1995, p 343
37. B. Budiansky and J.C. Amazigo, paper presented at IUTAM Symposium on Nonlinear Analysis of Fracture, International Union of Theoretical and Applied Mechanics, Sept 1995
38. P.E.D. Morgan and D.B. Marshall, *J. Am. Ceram. Soc.*, Vol 78, 1995, p 1553

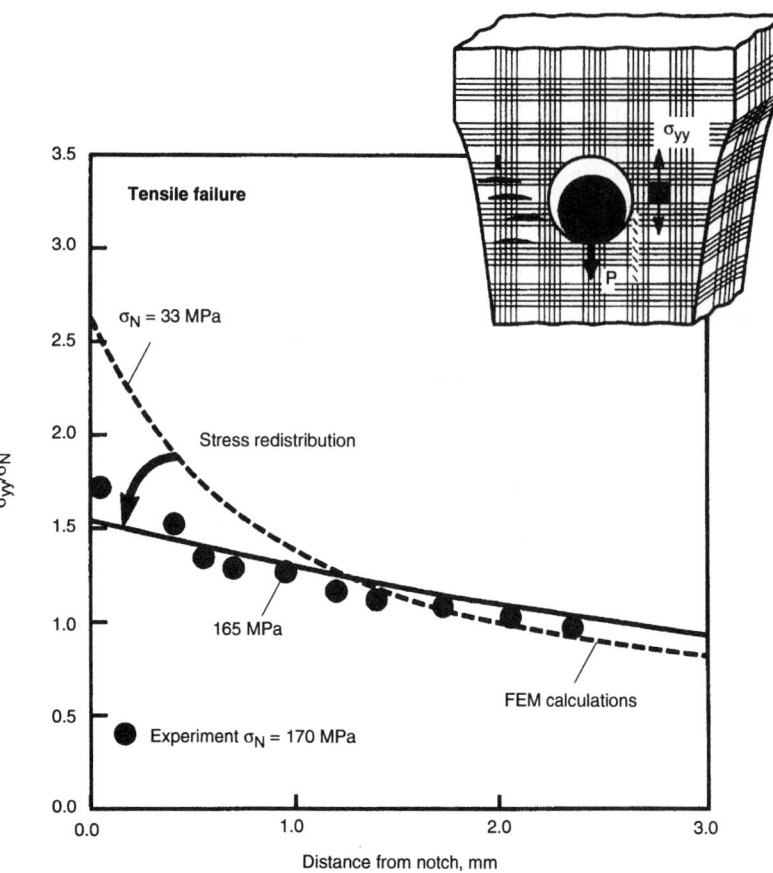

Fig. 12 Stress redistribution around a hole in a pin-loaded ceramic-matrix composite: experiments and finite-element modeling (FEM) calculations

39. J.B. Davis, J.P.A. Löfvander, and A.G. Evans, *J. Am. Ceram. Soc.*, Vol 76 (No. 5), 1993, p 1249-1257
40. C.-W. Li and J. Yamanis, *Ceram. Eng. Sci. Proc.*, Vol 10, 1989, p 632-645
41. P.F. Becher, C.H. Hsueh, P. Angelini, and T.N. Tiegs, *Mat. Sci. Eng.*, Vol A107, 1989, p 257
42. K.S. Chan and Y.W. Kim, *Met. Trans.*, Vol 24A, 1993, p 113
43. D.J. Wissuchek, M.-Y. He, and A.G. Evans, *Acta Metall. Mater.*, in press
44. J.W. Hutchinson, *Advances in Fracture Mechanics*, Vol 6, Pergamon Press, 1982
45. J.W. Hutchinson and V. Tvergaard, *J. Mech. Phys. Solids*, Vol 40 (No. 6), 1992
46. J.W. Hutchinson and V. Tvergaard, *J. Mech. Phys. Solids*, Vol 41, 1993, p 1119-1135
47. M. Rühle, A.G. Evans, R.M. McMeeking, P.G. Charalambides, and J.W. Hutchinson, *Acta Metall.*, Vol 35 (No. 11), 1987, p 2701-2710
48. C.M. Cady, T.J. Mackin, and A.G. Evans, *J. Am. Ceram. Soc.*, Vol 78 (No. 1), 1995, p 77-82
49. F.A. Heredia, A.G. Evans, and C.E. Anderson, *J. Am. Ceram. Soc.*, Vol 78, 1995, p 2790
50. G. Genin and J.W. Hutchinson, *Trans. Am. Ceram. Soc.*, 1995, in press
51. J.W. Hutchinson and H. Jensen, *Mech. Mater.*, Vol 9, 1990, p 139
52. J.-M. Domergue, E. Vagaggini, A.G. Evans, and J. Parenteau, *J. Am. Ceram. Soc.*, Vol 78, 1995, p 2721
53. C. Cady, A.G. Evans, and K.E. Perry, Jr., *Composites*, Vol 26 (No. 10), 1995, p 683

Fatigue and Fracture Behavior of Glasses

James R. Varner, New York State College of Ceramics at Alfred University

SILICATE GLASSES, which comprise nearly all glasses of commercial interest, are linearly elastic, brittle materials. The cyclic fatigue exhibited by metals thus is not a problem in glasses. However, glasses do experience static fatigue, which is time-dependent failure of glass placed under constant load in an environment containing water or water vapor. The mechanism is stress corrosion, crack growth caused by stress-assisted reaction with liquid water or water vapor at the crack tip. Stress corrosion occurs under constant or changing loads. Therefore, glasses subjected to cyclic loading will experience incremental crack growth each time the stress at the crack tip exceeds a critical value, and failure may occur, but the mechanism is stress corrosion, not cyclic fatigue. The development of glass fiber-optic cables for voice and data transmission was accompanied by renewed interest in fatigue of glasses. Glass fibers in fiber-optic systems must be able to withstand sustained loads for many years in environments that may contain significant amounts of water. Engineers must be able to make lifetime predictions of fiber-optic systems based on reliable data and tested theories. Static fatigue and stress-corrosion cracking of glasses have been intensely researched, but these phenomena are far from well understood. Important nuances are still being debated in the literature.

This article discusses only silicate glasses—more specifically, soda-lime-silicate glass (the most widely used glass, used in products such as windows, containers, drinking glasses, etc.), borosilicate glass (used in cookware and laboratory glassware, for example), and vitreous silica (the basis for fiber-optic systems). Fracture behavior will be described first, in order to discuss the brittle nature of glass and provide the background needed for understanding fatigue in glasses. The phenomenon of static fatigue will be documented using examples from the literature, and explanations for it given. Lifetime prediction will be addressed by discussing methods of testing—dynamic fatigue and slow-crack-growth studies—and methods of calculation. Finally, the role of surface damage in strength and fatigue behavior will be discussed.

Fracture Behavior

At ambient temperatures, glass fractures in a completely brittle way. There is no evidence of plastic or viscous flow, either macroscopically (i.e., no necking or other permanent change in dimensions) or microscopically (i.e., no slip bands or flow lines). Permanent deformation of glass at ambient temperatures does occur in complex, concentrated loading situations, such as during indentation with a sharp indenter or during abrasion or machining. Shear bands are observed under indentations in soda-lime-silica glasses, for example. However, brittle fracture in the form of very complex crack formation accompanies this permanent deformation when the applied load exceeds a critical value. All modes of loading of practical importance include a tensile component, and it is this tension that causes cracks to form and glass parts to fail. Fracture of brittle materials is discussed in the previous article in this Volume. For the present discussion of fatigue, it is useful to emphasize some key features about brittle fracture of glasses.

Applied stresses are concentrated at crack tips. In fact, stresses at crack tips are often orders of magnitude larger than the applied stress. The stresses become this high because of (1) the geometry of the crack and (2) the lack of stress-relieving plastic deformation. Both factors are important, since stresses are also concentrated at crack tips in metals, but local plastic deformation may keep these stresses from reaching the value needed for catastrophic failure. Fracture occurs when the stresses at the crack tip exceed the bond strengths in the glass network. The well-known Griffith equation (Ref 1) provides one way to relate crack geometry to fracture strength:

$$\sigma_F = \left(\frac{2E\gamma}{\pi c}\right)^{1/2} \tag{Eq 1}$$

where σ_F is fracture strength, E is modulus of elasticity, γ is surface energy, and c is crack length (if it is a surface crack). For a given glass, therefore, the crack size determines the strength. While it is true that Eq 1 is based on geometrical assumptions that are not exactly valid for all surface cracks, the in-

verse relationship between strength and crack size that is expressed still holds. Using fracture mechanics, a similar equation can be developed, which also shows that the strength is inversely proportional to the size (length) of the crack:

$$\sigma_F = \frac{YK_{Ic}}{c^{1/2}} \tag{Eq 2}$$

where Y is a geometrical factor related to crack and specimen geometry, and K_{Ic} is the critical stress-intensity factor (fracture toughness) of the glass. Equation 2 is valid for any crack, as long as an appropriate value for Y is known.

With rare exceptions, surface cracks are the sources of failure, the fracture origins, in glasses. There are four reasons for this:

1. Glasses usually have few internal flaws, especially flaws that are effective in concentrating stresses.
2. Relatively large cracks are produced on glass surfaces due to contacts between the glass and hard objects. This will be discussed later in more detail.
3. For many loading situations, the maximum tensile stress is on the surface.
4. Surface cracks are exposed to the environment.

As will be discussed, the last three of these reasons also relate to static fatigue of glass.

Static Fatigue

If a glass rod at room temperature in normal air is subjected to a constant load in three- or four-point bending, failure often occurs after some time has passed, rather than at the instant the load was applied. The rod was able to support the load at first; therefore, the maximum tensile stress must have been below the fracture stress, σ_F. Over time, something happened to cause strength degradation. In terms of Eq 2, there must have been either a change in Y (i.e., a change in the geometry, or shape, of the crack) or a change (i.e., an increase) in c, the crack length, since K_{Ic} is a

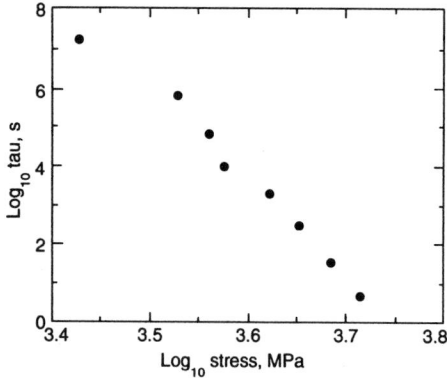

Fig. 1 Log-log static fatigue plot of average time to failure, *t*, versus applied stress. Data are for vitreous silica and are taken from Ref 3. Source: Ref 2

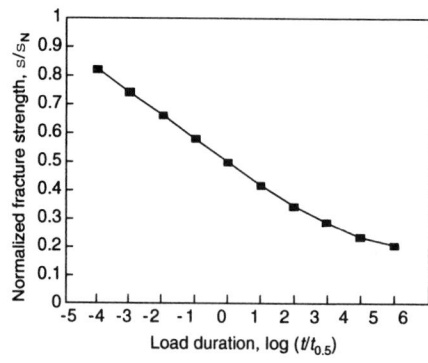

Fig. 2 Universal fatigue plot of normalized stress (stress on specimen divided by fracture stress at –196 °C), σ/σ_N versus log $(t/t_{0.5})$ (load duration divided by the load duration at $\sigma/\sigma_N = 0.5$). Data are for abraded soda-lime-silica specimens and are taken from Ref 6.

Table 1 Time to failure at $\sigma_F/\sigma_N = 0.5$ (designated $t_{0.5}$) for soda-lime-silica specimens tested in flexure

Condition	$t_{0.5}$, s	σ_N, MPa
(a) Different humidities		
Water	8.7	86
43% RH	200	86
0.5% RH	3470	86
(b) Different surface treatments		
Severe grit blast	2.9	86
Mild grit blast	8.8	93
Abraded perpendicular to stress		
600-grit paper	0.0043	134
320-grit paper	0.149	95
150-grit paper	0.56	70
Abraded parallel to stress		
150-grit paper	0.14	165

RH, relative humidity. Data taken from Ref 6

material constant. This rod experiment is one example of the phenomenon of static fatigue, and it is also the classic way of measuring static fatigue—namely, by measuring the time to failure at a given applied load. Static fatigue of fibers may be measured by two-point bending or by applying uniaxial tension. If this experiment is performed using different loads, the time to failure will be seen to increase as the applied load decreases. Figure 1 shows an example of this type of experiment (Ref 2).

The role of water in static fatigue of glass was recognized early on (Ref 4), even though the mechanism was not understood. Consequently, researchers examined different glass compositions and confirmed that those that are known to be more resistant to attack by water exhibit longer times to failure at a given load (Ref 5). Another approach was to try the time-to-failure experiments in air containing different amounts of water vapor, as well as in water itself (Ref 6). As shown in section (a) of Table 1, static fatigue is more pronounced—that is, times to failure are shorter at a given load—when there is more water vapor present, and shorter still in liquid water.

It was discovered that one effective way to eliminate the influence of water was to conduct the experiments in liquid nitrogen (Ref 4), and the results thus obtained were accepted as being free from static fatigue. The practice developed of normalizing strength at failure with respect to the liquid nitrogen strength, σ_N. Static fatigue data can then be presented on a graph of σ/σ_N versus time to failure, *t*. Mould and Southwick (Ref 6), however, showed that using normalized time-to-failure values produced what they termed a "universal fatigue curve." One example of this is shown in Fig. 2, in which σ/σ_N is plotted versus log $(t/t_{0.5})$, where *t* is the time to failure and $t_{0.5}$ is the time to failure at $\sigma/\sigma_N = 0.5$ (Ref 6). The data presented in section (b) of Table 1 fall on the curve shown in Fig. 2. Plotting σ/σ_N versus time to failure (not normalized) results in six separate curves for the six conditions of surface abrasion used in the experiments. Normalizing the time to failure produces the universal fatigue curve. Mould and Southwick (Ref 6) showed that the surface condition of the glass influences the results, as shown by the data presented in section (b) of Table 1. The time to failure at $\sigma/\sigma_N = 0.5$ of abraded rods tested in water is shorter when the abrasion is more severe.

Still other experiments showed that average strengths increase when the specimens are stored in water (for 24 h, for example) prior to testing (Ref 5, 7). This last observation seems to conflict with those that show more rapid fatigue (i.e., strength degradation) in static fatigue tests in water. The difference is that the specimens stored in water are not under load. Once the mechanism of static fatigue has been explained, the reason for the different effects in these two cases will be clear.

To summarize, water plays an important role in static fatigue of glasses. The susceptibility of a given glass to static fatigue increases when more water is present and when the glass exhibits more severe surface damage. Glasses that are inherently more resistant to water attack, such as vitreous silica, are less susceptible to static fatigue. Aging abraded rods in water with no load on the specimens increases the average fracture

strength. On the other hand, aging pristine vitreous silica fibers in water with no load on specimens can decrease strength (Ref 26). Several theories that attempt to explain all these observations will now be examined.

Charles and Hillig (Ref 8) provided concepts of stress corrosion that are still held to be essentially correct, albeit with some modifications. Tensile stress at a crack tip increases the rate of reaction of water with the glass. As a consequence, in the theory proposed by Charles and Hillig, the crack tip becomes sharper, and the crack grows longer; they considered the sharpening to be the more important of the two phenomena. [Today, the prevalent view is that "large" cracks are sharp, and that it is crack growth which is the mechanism responsible for fatigue. More on the issue of blunt and sharp cracks can be found in Ref 28.] As this slow crack growth occurs, the stress concentration at the crack tip increases, since a sharper, longer crack is a more effective concentrator of stress. Eventually, the stress reaches the bond strength of the network-forming structural units in the glass, and the piece breaks. This qualitative description of the Charles-Hillig theory of stress corrosion can be used to explain some of the observations documented earlier.

First of all, static fatigue itself occurs because of slow crack growth due to stress corrosion. Under a constant load, surface cracks exposed to water in the environment grow until they become long enough to cause failure at that load, that is, until they bring the stress concentration at the crack tip to the required level. When larger cracks are present initially, less time is required at a given load before the required amount of slow crack growth takes place—hence the dependence of static fatigue on surface condition. Slow crack growth is faster at higher applied loads, because the stresses at the crack tips are higher; therefore, the time needed for the required amount of slow crack growth is shorter, and times to failure are shorter. Static fatigue is not seen at liquid nitrogen temperature, because the rate of chemical reaction of the water with the glass is extremely low. Slow crack growth is slower in glasses that are more resistant to water attack; therefore, these glasses exhibit lower susceptibility to static fatigue. The positive effects on strength produced by aging abraded glass in water are due to rounding or blunting of crack tips when no applied stress, or very low applied stress, is present. A resharpening must occur when a high enough load is applied before slow crack growth can proceed. Higher strengths are measured. The lower strengths measured after aging pristine vitreous silica fibers in water (Ref 26) may be explained by surface roughening, as shown by the work by Inniss et al (Ref 27). Charles and Hillig (Ref 8) developed a quantitative approach for predicting the velocity of slow crack growth and, therefore, the time needed before failure occurs. As discussed next, more recent models have been developed that are better at lifetime prediction.

Michalske and Freiman (Ref 9) and Michalske and Bunker (Ref 10) proposed a chemical mechanism for reactions at crack tips that predicts slow

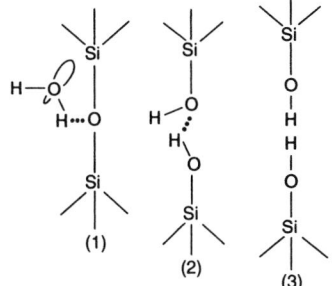

Fig. 3 Schematic of proposed model for the reaction sequence between H_2O and a strained Si–O–Si bond at a crack tip. Source: Ref 11

crack growth (static fatigue) in water and similar liquids. Their model for the reaction between water and vitreous silica involves a three-step process, as shown in Fig. 3. Si–O–Si bonds at crack tips are strained as a result of the stresses concentrated there. In step 1, an H_2O molecule comes close to the oxygen in a strained Si–O–Si bond and orients itself such that one set of the lone electron-pair orbitals on the oxygen of the water molecule aligns toward one Si. At the same time, a hydrogen bond develops between one H and the O. Step 2 really involves two simultaneous transfers at these two sites: (1) electron transfer from the O in the water molecule to the adjacent Si and (2) proton transfer to the O of the Si–O–Si bond. This forms two new bonds. The strain on the Si–O–Si bond enhances the tendency for this to happen. This leaves a hydrogen bond between the O of the water molecule and the transferred H. Step 3 is the breaking of this hydrogen bond, leaving two Si–OH "surfaces." Other liquids or gases may behave similarly if they have structure and bonding similar to water —that is, lone-pair orbitals at one location on the molecule and proton donor sites at another. In fact, ammonia (NH_3), hydrazine (N_2H_4), and formamide (CH_3NO) have been shown to cause slow crack growth in vitreous silica (Ref 11).

At this point, the picture seems clear: The reaction of water with glass is enhanced at crack tips owing to high strains in the network-forming bonds in those regions. Cracks lengthen (and sharpen?), which eventually causes fracture of the piece under load. However, models of water attack are just that—models. The specific attack mechanisms are still not known with certainty. Also, when equations for predicting lifetimes are developed from these theories, and when fatigue susceptibility is tested in other types of experiments, the picture loses its clarity. There is no question that water is involved in fatigue, or that stress-assisted crack growth in some form is involved, but there are questions about what happens in detail. In order to proceed further, these other types of experiments need to be explained and illustrated with some results.

Dynamic Fatigue

The obvious disadvantages of time-to-failure studies are the extremely long times needed for

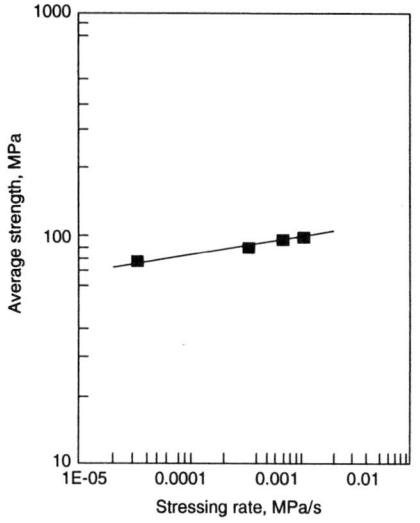

Fig. 4 Example of a log-log dynamic fatigue plot of breaking stress versus stressing rate. Data are for wetted soda-lime-silica rods and are taken from Ref 29. The slope of the line [1/(N+1)] is 14.

failure at low loads and the need for testing large numbers of specimens owing to the statistical nature of glass strength. One way to address the issue of long times is to use the technique called dynamic fatigue. Instead of applying a constant load (stress) to the specimen, a constant *rate* of loading (stress/second) is applied. Sets of specimens tested at lower rates of loading break at lower average strengths than sets tested at higher rates. The principle involved is that there is more time for crack growth to occur when the load is being increased slowly; therefore, the lowering of the strength due to slow crack growth is more pronounced. At very high loading rates, there is virtually no time for significant crack growth to occur before the stress at the crack tip reaches the critical value for fracture. Dynamic fatigue experiments are often done with the specimens immersed in water in order to investigate the maximum fatigue effect.

Average values of stress at failure are plotted versus stressing rate on a log-log graph. The slope of the best-fit straight line through the data points is $1/(N + 1)$, where N is the fatigue resistance parameter. An example is shown in Fig. 4 (Ref 29). The equation upon which this is based is from Charles (Ref 13):

$$\sigma_{af} = C\beta^{1/N + 1} \qquad \text{(Eq 3)}$$

where σ_{af} is the applied stress at which failure occurs, C is a constant, and β is the rate of stress application.

The parameter N provides a simple way to characterize the fatigue resistance of glasses. A high value of N is associated with a high fatigue resistance. According to Ritter et al. (Ref 12), N for soda-lime-silica glass is about 13 to 17; for borosilicate glass, 27 to 40; and for vitreous silica, 32 to 38. These are for measurements done in water.

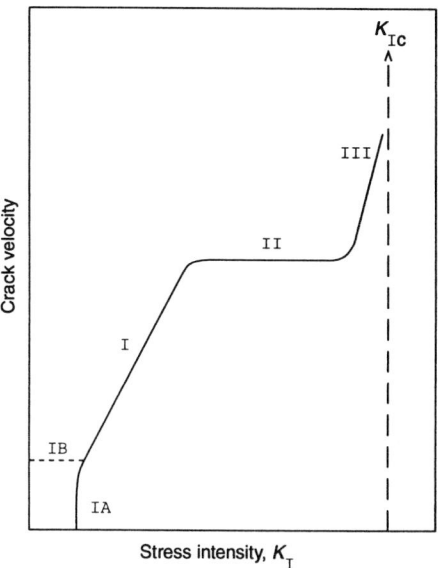

Fig. 5 Schematic V-K curve for a glass showing all possible regions (explained in text); the slope of region I represents the fatigue resistance parameter. Source: Ref 11

Crack Velocity Measurements

Yet another method for studying fatigue susceptibility of glasses is to examine slow crack growth directly. This is done using crack velocity measurements, so-called V-K experiments, in which controlled growth of well-defined macroscopic cracks in fracture mechanics specimens is monitored. Regardless of the particular test geometry, the goal is to obtain the V-K curve, which gives the relationship between crack velocity, V, and the stress-intensity factor, K, for the glass. A schematic of such a curve is shown in Fig. 5 (Ref 11). This is a semilog plot: log(velocity) versus K_I. Although the curve is divided into several regions, region I is of primary interest in characterizing fatigue behavior, since crack growth in this region is controlled by the rate of reaction of water (or water vapor) with the glass. Region II is a transition region in which part of the crack tip is exposed to water and part is running dry (Ref 14). In region III, the crack is moving so fast that water cannot reach the crack tip fast enough to have any effect on crack growth. Cracks in laboratory tests of static or dynamic fatigue and cracks in glass articles in service grow for the greatest length of time in region I. The velocity of crack propagation is exponentially dependent on the stress-intensity factor in region I, so the V-K data in region I should lie on a straight line when plotted on semilog paper. The slope of this line is the fatigue resistance parameter, N.

Wiederhorn (Ref 15), among others, showed that the velocity of crack propagation at a given K_I depends on the relative humidity, with higher velocities for higher humidities or liquid water. However, the slope of region I of the V-K curve does not depend on the relative humidity. It does depend on glass composition and on pH, as shown in Fig. 6 (Ref 16).

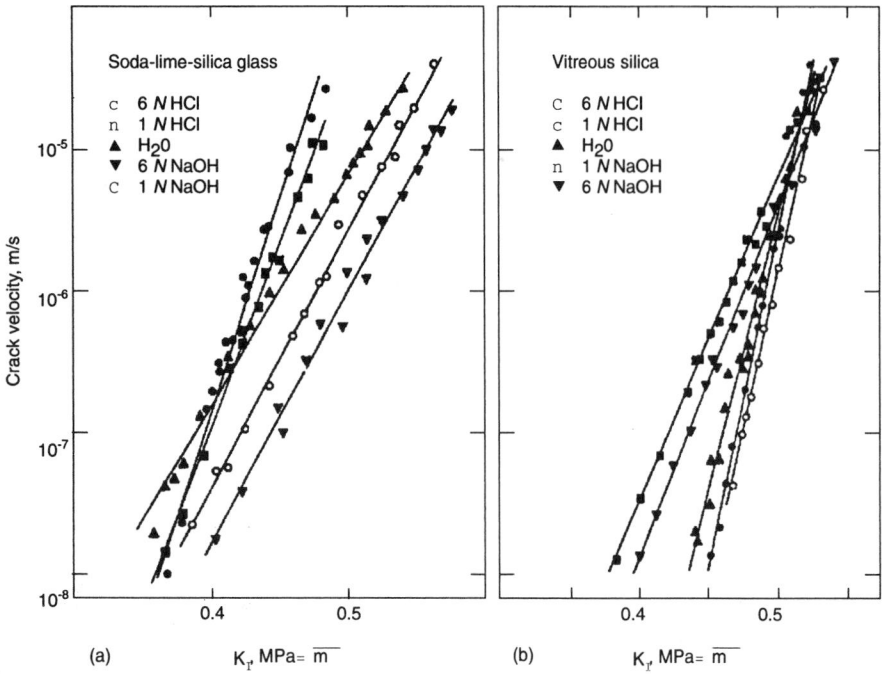

Fig. 6 V-K curves showing the effect of pH of crack growth in soda-lime-silica glass (a) and vitreous silica (b). Data taken from Ref 16. Source: Ref 11

Table 2 Values of N for three glass compositions obtained from the slope of the universal fatigue curve (UFC) or from the slope of the V-K curve

Glass	Curve (surface treatment)	Test condition	Fatigue resistance parameter, N
Soda-lime-silica	UFC (abraded)	Wet	29
	UFC (HF polished)	50 and 100% RH	74
	V-K	Water	35
Borosilicate	UFC (abraded)	100% RH	31
	UFC (HF polished)	100% RH	58
	V-K	Water	62
Vitreous silica	UFC (abraded)	50% RH	31
	UFC (pristine fiber)	100% RH	68
	V-K	Water	72

Source: Ref 17

Comparison of Results

Static fatigue, dynamic fatigue, and crack velocity measurements are three different ways of determining the fatigue resistance parameter. Ideally, results obtained using the three techniques should be the same for tests done on the same glass composition under the same environmental conditions. However, this is not necessarily the case. For example, Ritter (Ref 17) showed that N values obtained using universal fatigue curves and V-K curves for three glass compositions differed. These results are summarized in Table 2. There is no agreement among the N values obtained from the universal fatigue curves and from

the V-K curves for these three glasses. In fact, the N values from the universal fatigue curves show no dependence on composition, whereas the N values from V-K data seem to be obviously different for the three glasses. Notice too that the N values for the abraded specimens are consistently much lower than those for the specimens that were polished in hydrofluoric acid or tested as pristine fibers (i.e., freshly drawn fibers tested without being touched along the length put under maximum stress).

Explanations for these discrepancies can be found by exploring differences in the specimens and methods of testing in more detail. The abraded specimens had many small surface cracks, surface cracks in the HF-polished specimens had been transformed into shallow etch pits, and macroscopic "through" cracks (i.e., cracks extending through the entire thickness of the specimen) were used in the V-K tests. The reason for the very different N values obtained for abraded and for pristine or HF-polished specimens of the same glass is that the flaws were very different in these two cases. Sharp surface cracks were present in the abraded specimens, so crack growth could begin as soon as sufficient stress was applied. Any flaws present on the pristine surfaces were so small that a crack nucleation step probably was involved before crack growth could occur. The time needed to nucleate a crack might be longer than the time needed for that crack to grow to the point where fracture occurs. Unfortunately, there are discrepancies about this in the literature, with reports of lower N values for specimens with smaller flaws (Ref 18). The smaller flaws were produced using diamond indenters, which could produce yet another surface

condition. As for the lack of a composition influence in the data presented by Ritter (Ref 17) for the abraded specimens, residual stresses associated with the surface flaws could be a contributing factor, as pointed out by Lawn et al. (Ref 18, 19).

These conflicting results can cause confusion on the part of someone trying to understand the fundamentals of fatigue of glasses. It is important to realize the implications of different surface conditions, especially when surface cracks or other discontinuities are produced in different ways. A pristine or acid-polished surface has very different flaws than does a surface that has been abraded, or one that has been abraded and annealed, or one that has been indented at low loads, or one that has been indented at high loads. These are just some of the surface treatments that have been used to investigate static and dynamic fatigue of glasses. Crack velocity studies are different again, in that they involve macroscopic cracks that extend through the entire thickness of the glass.

What is one to do when faced with a decision about conducting fatigue tests? Crack velocity tests should provide the best information about the susceptibility of the glass to reaction with water. The slope of the V-K curve in region I (see Fig. 5) is related directly to the rate of reaction of water with the glass, as discussed earlier. Therefore, these tests should be used to provide this kind of fundamental information. On the other hand, real cracks encountered in service conditions are seldom through cracks. Surface cracks or other surface flaws represent the vast majority of cases involving fatigue.

It is most useful to select a surface condition and a test geometry that most nearly duplicate what is expected in service. For example, glass optical fibers are coated immediately after being drawn, and thus have pristine surfaces. The fibers are subjected to uniaxial tension and bending when they are placed in service. Testing of abraded fibers would involve examining a surface condition that is very unlikely to be encountered in service. On the other hand, testing of the coated fibers, or of freshly drawn uncoated fibers, or even of fibers with carefully made low-load indentations would duplicate service conditions. The fibers should be loaded using uniaxial tension or using extreme bending. Simulation of service conditions is a desirable goal in realistic fatigue testing.

Lifetime Prediction

The primary goal of investigating fatigue of glasses is to predict the lifetime of a glass article in service. Given that crack growth occurs when glass is under load in the presence of water, how long will it be before the glass breaks? Alternatively, how significant is the strength degradation when a piece of glass has been loaded under certain conditions? Will glass optical fibers, for example, remain intact for 25 years under the expected service conditions? The key to predicting lifetimes is knowing how cracks grow. Using

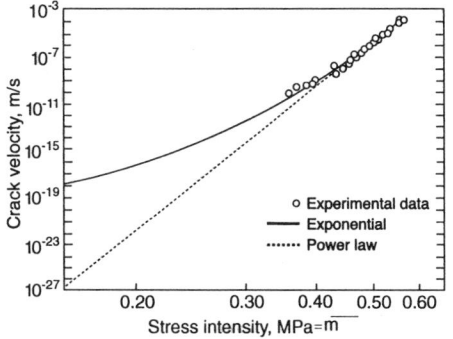

Fig. 7 *V-K* curve for vitreous silica in the low-velocity regime; experiments were done in water. Extrapolated curves for power-law and exponential behavior are shown. Source: Ref 21

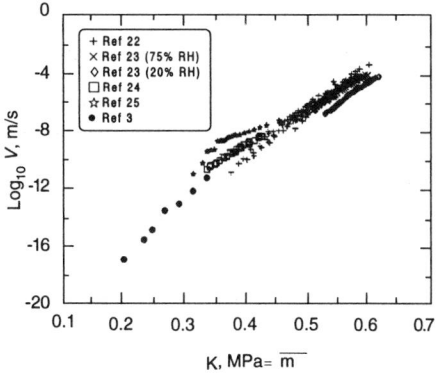

Fig. 8 *V-K* data for vitreous silica in air from a number of sources (Ref 2-5) and low-velocity data calculated from the data in Ref 3 using the method described in Ref 2. Source: Ref 2

fracture mechanics, the issue is knowing how crack velocity, *V*, is related to the stress-intensity factor, *K*.

Earlier, it was stated that *V* depends exponentially on *K* in region I of the *V-K* curve. In fact, that is one commonly used *V(K)* function; another is a power law:

$$V = V_c \left(\frac{K}{K_c}\right)^N \qquad \text{(Eq 4)}$$

where V_c is a constant, K_c is the critical stress-intensity factor (fracture toughness), and *N* is the fatigue parameter. Both equations work equally well in the velocity range where data are taken (Ref 20). However, neither equation accounts for stage IA of Fig. 5, and extrapolations to low velocities differ markedly, as illustrated in Fig. 7 (Ref 21). The velocities predicted by the two approaches differ by orders of magnitude at a stress intensity of 0.2 MPa$\sqrt{m}$. In this case, the exponential law seems to agree well with the data down to the lowest measured velocity, about 10^{-10} m/s. Gupta et al. (Ref 2) recently proposed a way to generate the *V-K* relationship from static fatigue data. Their method involves no assumptions about the form of the *V-K* function, and they are able to calculate *V-K* curves from static fatigue data. Figure 8 shows values of *V* versus *K* obtained using their proposed method (the filled circles labeled Ref 3) and values from other published reports for vitreous silica at room temperature in air. The relative humidities in the other results ranged from 20 to 100%. Note how the calculated *V(K)* data (the Ref 3 results) seem to merge with the other high-*K* results. In their opinion, this confirms the dominance of crack growth in fatigue. Assuming this is the case, all the data in Fig. 8, including the calculated values, can best be represented by a power law:

$$\log V = 1.05 + 10.7 \log K \qquad \text{(Eq 5)}$$

Note that velocities down to 10^{-16} m/s were analyzed using this method, far lower than can be measured in crack velocity experiments, which usually go no lower than about 10^{-10} m/s.

Surface Damage

This final section deals with the surface cracks that are, for the most part, responsible for static fatigue failure of glass articles in service. Few glass products are free from surface cracks. Glass fibers used in optical waveguides may be one such product, since the pristine surfaces are coated immediately after forming, thus preventing direct contact with the glass surface. Static fatigue still occurs, but most probably involves a crack nucleation phase that may be more important than the crack growth phase, as noted earlier. The surfaces of most glass products are exposed to contact with hard objects, however, and this often results in cracks. The literature on contact damage on glass surfaces is extensive; only the essential elements will be presented here. The reader is referred to the list of Selected References at the end of this article for sources of additional information.

Contact damage can be produced by objects that are blunt (spheres, shoulders of bottles) or sharp (grains of an abrasive). Objects can be large (bottles, hammers, stones) or small (sand particles). Contact can be slow (indentation) or fast (impact). The type of damage that results from contact depends primarily on whether the object is blunt or sharp and on whether the glass tends to densify under pressure. The rate of contact generally has little influence on the type of damage produced. (Rate is important at temperatures high enough for the glass to be in the regime where it behaves viscoelastically, such as during some stages of the forming process.) In general, contact with a blunt object will produce shallow, circular cracks on the surface, which may flare out into cone-shaped cracks that extend into the glass. A sharp object tends to produce some cracks that are normal to the glass surface and which extend radially away from the location of contact and some cracks that are nearly parallel to the surface and which may intersect the surface, resulting in the loss of chips of glass next to the contact site. In the case of both blunt and sharp objects, the cracks that extend into the glass are primarily responsible for strength degradation and, there-

fore, are more directly involved in fatigue. Residual stresses are present, even after crack formation occurs, and these can cause crack growth even in the absence of externally applied stresses.

Contact damage can occur anytime during the lifetime of a glass article. Some damage may occur in production, for example, when bottles bump against one another on conveyor belts. Most contact damage occurs once the object is placed in service. Think of the opportunities for contact damage to be produced on an automobile windshield, for example. There is abrasion from wiper blades and from the cloths or towels used in cleaning; there are impacts from particles or even stones kicked up by other vehicles. The nearly daily production of additional contact damage in a case like this provides yet another complication in lifetime prediction, since long-term growth of small flaws could be overshadowed by short-term growth of even one large flaw produced much later. The complex morphologies and residual stresses of contact damage sites must be considered when attempting to simulate these conditions in laboratory experiments, especially when the goal is obtaining data for lifetime prediction.

REFERENCES

1. A.A. Griffith, Fracture Behavior of Brittle Materials, *Philos. Trans. R. Soc.*, Vol A221, 1920, p 163-198
2. P.K. Gupta, D. Inniss, C.R. Kurkjian, and D.I. Brownlow, Determination of Crack Velocity as a Function of Stress Intensity from Static Fatigue Data, *J. Am. Ceram. Soc.*, Vol 77 (No. 9), 1994, p 2445-2449
3. B.A. Proctor, I. Whitney, and J.W. Johnson, The Strength of Fused Silica, *Proc. R. Soc (London)*, Vol A297, 1967, p 534-557
4. T.C. Baker and F.W. Preston, Fatigue of Glass under Static Loads, *J. Appl. Phys.*, Vol 17 (No. 3), 1946, p 170-178
5. R.H. Doremus, Fatigue in Glass, *Strength of Inorganic Glass*, C.R. Kurkjian, Ed., Plenum Press, 1985, p 231-242
6. R.E. Mould and R.D. Southwick, Strength and Static Fatigue of Abraded Glass under Controlled Ambient Conditions: II, Effect of Various Abrasions and the Universal Fatigue Curve, *J. Am. Ceram. Soc.*, Vol 42, 1959, p 582-592
7. W.-T. Han and M. Tomozawa, Crack Initiation and Mechanical Fatigue of Silica Glass, *J. Non-Cryst. Solids*, Vol 122, 1990, p 90-100
8. R.J. Charles and W.B. Hillig, The Kinetics of Glass Failure by Stress Corrosion, *Symposium on Mechanical Strength of Glass and Ways of Improving It*, Union Sci. Cont. du Verre, Charleroi, Belgium, 1962, p 511-527
9. T.A. Michalske and S.W. Freiman, A Molecular Mechanism for Stress Corrosion in Vitreous Silica, *J. Am. Ceram. Soc.*, Vol 66 (No. 4), 1983, p 284-288
10. T.A. Michalske and B.C. Bunker, Chemical Kinetics Model for Glass Fracture, *J. Am.*

Ceram. Soc., Vol 76 (No. 10), 1993, p 2613-2618

11. S.W. Freiman, Environmentally Enhanced Crack Growth in Glasses, *Strength of Inorganic Glass*, C.R. Kurkjian, Ed., Plenum Press, 1985, p 197-215

12. J.E. Ritter, Jr., K. Jakus, K. Buckman, G. Young, and J.S. Haggerty, Strength and Fatigue Behavior of a Borosilicate Glass with an Antireflective Surface, *Glass Technol.*, Vol 23 (No. 2), 1982, p 125-130

13. R.J. Charles, Dynamic Fatigue of Glass, *J. Appl. Physics*, Vol 29, 1958, p 1657-1662

14. J.R. Varner and V.D. Frechette, Mechanisms of Slow Crack Growth, *Amorphous Materials*, R.W. Douglas, Ed., Wiley-Interscience, London, 1972, p 507-512

15. S.M. Wiederhorn, Influence of Water Vapor on Crack Propagation in Soda-Lime Glass, *J. Am. Ceram. Soc.*, Vol 50 (No. 8), 1967, p 407-414

16. S.M. Wiederhorn and H. Johnson, Effect of Electrolyte pH on Crack Propagation in Glass, *J. Am. Ceram. Soc.*, Vol 56 (No. 4), 1973, p 192-197

17. J.E. Ritter, Jr., Strength and Fatigue of Silicate Glasses, *Strength of Inorganic Glass*, C.R. Kurkjian, Ed., Plenum Press, 1985, p 261-270

18. B.R. Lawn, D.B. Marshall, and T.P. Dabbs, Fatigue Strength of Glass: A Controlled Flaw Study, *Strength of Inorganic Glass*, C.R. Kurkjian, Ed., Plenum Press, 1985, p 249-259

19. T.P. Dabbs and B.R. Lawn, Strength and Fatigue Properties of Optical Glass Fibers Containing Microindentation Flaws, *J. Am. Ceram. Soc.*, Vol 68 (No. 11), 1985, p 563-569

20. K. Jakus, J.E. Ritter, and J.M. Sullivan, Dependency of Fatigue Prediction on the Form of the Crack Velocity Equation, *J. Am. Ceram. Soc.*, Vol 64 (No. 6), 1981, p 372-374

21. T.A. Michalske, W.L. Smith, and B.C. Bunker, Fatigue Mechanisms in High-Strength Silica-Glass Fibers, *J. Am. Ceram. Soc.*, Vol 74 (No. 8), 1991, p 1993-1996

22. S.M. Wiederhorn, A.G. Evans, and D.E. Roberts, A Fracture Mechanics Study of the Skylab Windows, *Fracture Mechanics of Ceramics*, Vol 2, R.C. Bradt, D.P.H. Hasselman, and F.F. Lange, Ed., Plenum Press, 1974, p 829

23. S. Sakaguchi, Y. Sawaki, Y. Abe, and T. Kawasaki, Delayed Failure in Silica Glass, *J. Mater. Sci.*, Vol 17, 1982, p 2878-2886

24. M. Muraoka, K. Ebata, and H. Abe, Effect of Humidity on Small-Crack Growth in Silica Optical Fibers, *J. Am. Ceram. Soc.*, Vol 76 (No. 6), 1993, p 1545

25. S.T. Gulati, Mechanical Properties of SiO_2-TiO_2 Bulk Glasses and Fibers, *Mater. Res. Soc. Symp. Proc.*, Vol 144, 1992, p 67-85

26. M.J. Matthewson and C.R. Kurkjian, Environmental Effects on the Static Fatigue of Silica Optical Fiber, *J. Am. Ceram. Soc.*, Vol 71, No. 3, 1988, p 177-183

27. D. Inniss, W. Zhong, and C.R. Kurkjian, Chemically Corroded Pristine Silica Fibers: Blunt or Sharp Flaws, *J. Am. Ceram. Soc.*, Vol 76, No. 12, 1993, p 3173-3177

28. T. Michalske, Fractography of Stress Corrosion Cracking in Glass, in *Fractography of Glass*, R.C. Bradt and R.E. Tressler, Ed., Plenum Press, New York, 1994, p 111-142

29. J.E. Ritter, Jr., Dynamic Fatigue of Soda-Lime-Silica Glass, *J. Appl. Physics*, Vol 40, No. 1, 1969, p 340-344

SELECTED REFERENCES

- R.F. Cook and G.M. Pharr, Direct Observation and Analysis of Indentation Cracking in Glasses and Ceramics, *J. Am. Ceram. Soc.*, Vol 73 (No. 4), 1990, p 787-817

- C.R. Kurkjian, Ed., *Strength of Inorganic Glass*, Plenum Press, 1985

- M. Tomozawa and R.H. Doremus, Ed., *Treatise on Materials Science and Technology*, Vol 22, *Glass III*, Academic Press, 1981

Section 8: Appendices

Parameters for Estimating Fatigue Life

IT HAS LONG BEEN RECOGNIZED that fatigue data, when resolved into elastic and plastic terms, can be represented as linear functions of life on a logarithmic scale. Figure 1 schematically shows this representation of elastic and plastic components, which together define the total fatigue life curve of a material. The general fatigue-life relation, expressed in terms of the strain range ($\Delta\varepsilon$, where $\Delta\varepsilon$ is the strain change from cyclic loading), is as follows:

$$\Delta\varepsilon = \Delta\varepsilon_e + \Delta\varepsilon_p \qquad \text{(Eq 1)}$$

where $\Delta\varepsilon_e$ is the elastic strain range, $\Delta\varepsilon_p$ is the plastic strain range, and where:

$$\frac{\Delta\varepsilon_e}{2} = \frac{\sigma'_f}{E}(2N_f)^b \qquad \text{(Eq 2)}$$

$$\frac{\Delta\varepsilon_p}{2} = \varepsilon'_f (2N_f)^c \qquad \text{(Eq 3)}$$

Therefore, the total strain amplitude (or half the total strain range, $\Delta\varepsilon/2$) can be expressed as the sum of Eq 2 and 3 such that:

$$\frac{\Delta\varepsilon}{2} = \varepsilon'_f (2N_f)^c + \frac{\sigma'_f}{E}(2N_f)^b \qquad \text{(Eq 4)}$$

where ε'_f is the fatigue ductility coefficient, σ'_f is the fatigue strength coefficient, b is the fatigue strength exponent, c is the fatigue ductility exponent, and N_f is the number of cycles to failure.

These four empirical constants (b, c, σ'_f, ε'_f) form the basis of modeling strain-life behavior for many alloys, although it must be noted that some materials (such as some high-strength aluminum alloys and titanium alloys) cannot be represented by Eq 4.

For many steels and other structural alloys, substantial data have been collected for the four parameters in Eq 4. In many cases, the four fatigue constants have been defined by curve fitting of existing fatigue life data (e.g., Ref 3). A collection of this data is tabulated at the end of this article.

The four fatigue constants can also be estimated from monotonic tensile properties. With the availability of extensive data, however, these techniques are not widely used (Ref 4). Nonetheless, this article briefly summarizes the "four-

point method" as a method to estimate fatigue life behavior from tensile properties. This method can be compared with the fatigue and tensile properties tabulated at the end of this article.

In addition, it should also be mentioned that the four fatigue constants are also related to the following parameters:

$$K' = \frac{\sigma'_f}{(\varepsilon'_f)^{n'}} \qquad \text{(Eq 5)}$$

$$n' = \frac{b}{c} \qquad \text{(Eq 6)}$$

where K' is the cyclic strength coefficient and n' is the cyclic strain hardening exponent in the power-law relation for a log-log plot of the completely reversed stabilized cyclic true stress (σ) versus true plastic strain (ε_p) such that $\sigma = K'(\varepsilon_p)^{n'}$. The use of power-law relationship is not based on physical principles, although the relationships in Eq 5 and 6 may be convenient for mathematical purposes. The parameters K' and n' are usually obtained from a curve fit of cyclic stress-strain data (Ref 4).

Four-Point Method

Numerous studies have been devoted to the development of techniques for estimating strain-controlled fatigue characteristics (per Eq 4). For the most part, these studies dealt with data generated under completely reversed strain cycling (i.e., $R = -1$, or $A = \infty$) and usually attempted to relate fatigue properties with tensile properties. Extensions of these studies have carried the estimating procedures a step further, addressing the correlation of fatigue data obtained at various strain ratios (R).

Two common methods for approximating the shape of a fatigue curve are the "method of universal slopes" and the "four-point correlation" method. These two methods have been known for many years and are described in various references (for example, Ref 2 and 5). The method of universal slopes, first proposed by Manson (Ref 2), is based on the relation:

$$\Delta\varepsilon = 3.5\frac{\text{UTS}}{E}(N_f)^{-0.12} + \varepsilon_f^{0.6}(N_f)^{-0.6} \qquad \text{(Eq 7)}$$

where UTS is ultimate strength, E is modulus of elasticity, and ε_f is true fracture ductility, or $\ln[1/(1-\text{RA})]$.

This approximation thus requires only tensile strength, modulus, and reduction in area (RA). However, note that it is based on strain range ($\Delta\varepsilon$) rather than strain amplitude ($\Delta\varepsilon/2$).

The four-point method also allows construction of fatigue life curves from more readily available handbook data (i.e., monotonic tensile data). This method can be compared with the traditional strain-based approach (Fig. 1) or a stress-based approach. In both cases, the four-point method is based on the premise that total fatigue life per Eq 4 can be estimated as the sum of elastic strain (Eq 2) and plastic strain (Eq 3) components. The step-by-step process for locating points on the plastic- and elastic-strain-life lines is described below for both strain-based and stress-based data.

Strain-Based Four-Point Method. The four-point method initially was developed in terms of strain range by Manson (Fig. 1). The four points in Fig. 1 are determined as follows:

1. *Point P_1 on the elastic strain line* is positioned at $N_f = 0.25$ cycles (where a monotonic test is $\frac{1}{4}$ of one fatigue cycle) and at an elastic strain range of $2.5\,\sigma_f/E$ (where σ_f is the fracture stress in a tensile test and E is the elastic modulus).

2. *Point P_2 on the elastic strain line* is positioned at $N_f = 10^5$ cycles and at an elastic strain range of $0.9\,\text{UTS}/E$, where UTS is the conventional ultimate tensile strength.

3. *Point P_3 on the plastic strain line* is positioned at $N_f = 10$ cycles, where the plastic strain range is $0.25D^{3/4}$ and D is the conventional logarithmic ductility (also known as ε_f).

4. *Point P_4 on the plastic strain line* is positioned at $N_f = 10^4$ cycles, where the plastic strain range is given by $\Delta\varepsilon_p$ (at 10^4 cycles) = $0.0069 - 0.525\,\Delta\varepsilon_e$ (at 10^4 cycles), where the elastic strain-range line at $N_f = 10^4$ cycles; ($\Delta\varepsilon_e$ at 10^4) is shown as $\Delta\varepsilon_e^*$ in Fig. 1.

Fracture Stress. Point P_1 depends on fracture stress, which is not readily available in literature. However, fracture stress (which is the load at fracture divided by the area as measured after fracture) can be estimated (Ref 2) by means of the following approximate relationship among frac-

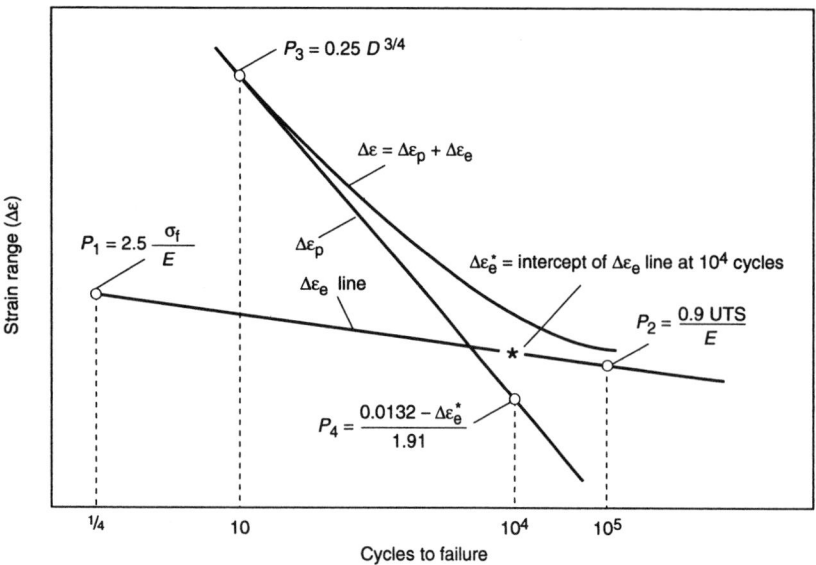

Fig. 1 Schematic of fatigue life curve with the Manson four-point criteria for the elastic and plastic strain lines. D is the tensile ductility, σ_f is the fracture stress (load at fracture divided by cross-sectional area after fracture, and UTS is conventional ultimate tensile strength. Source: Ref 1, 2

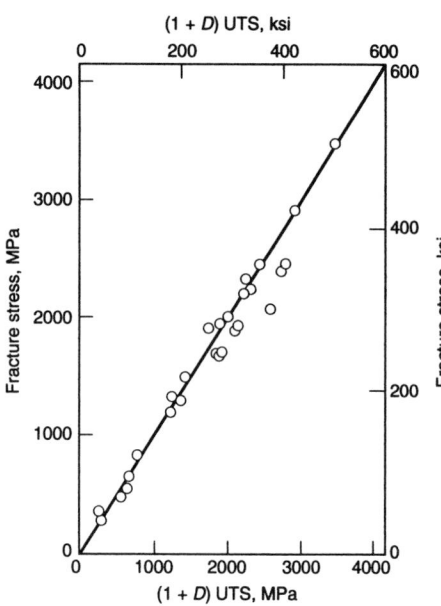

Fig. 2 Fracture stress versus tensile ductility. Source: Ref 5

ture stress, ultimate tensile stress, and fracture ductility; thus,

$$\sigma_f = \text{UTS}(1+D) \qquad \text{(Eq 8)}$$

This relation follows from Fig. 2, where each point is fixed by the data for one material. This calculation and that relating to the four-point method for defining the elastic and plastic strain-range lines were developed by Manson based on the materials listed in Table 1.

Graphical Solution by Manson. On the basis of the approximate equality in Eq 8, Manson noted that when E is known, only two tensile properties—ultimate tensile strength (UTS), and the reduction in area (to give D)—are needed to position the lines in Fig. 1 and thus obtain a prediction of fatigue behavior. Figure 3 shows a convenient graphical solution by Manson for locating the four points in Fig. 1. For example, if UTS/E is 0.01 and the reduction in area is 50% ($D = 0.694$),

the value of P_2 from the right-hand scale is 0.009 and that of P_3 from the top scale is 0.18. Locating the point with the coordinates UTS/$E = 0.01$ and reduction in area equal to 50% gives values for P_1 and P_4 of 0.042 and 0.0009, respectively. These points will locate the two strain-range lines, and the total strain-range curve can then be positioned to relate $\Delta\varepsilon$ and N_f for the material in question.

Stress-Based Four-Point Method. The four-point also applies to the construction of a stress-based *S-N* fatigue curve, as shown in Fig. 4. The four points A, B, C, and D in Fig. 4 can be defined

in terms of either stress or strain. In terms of strain, the points are identical to points P_1, P_2, P_3, and P_4 in Fig. 1. For construction of an *S-N* fatigue curve, the points are determined as described below.

Point A (in terms of stress) is simply the ultimate tensile strength of the metal, plotted on the vertical axis of the graph at $N = \frac{1}{4}$. As in the strain-based approach, this assumes that the simple tensile test represents one-fourth of a single, completely reversed fatigue cycle—the peak positive value of the applied stress.

Table 1 Materials used in low-cycle fatigue study for Fig. 3

4130 soft	Ti-6Al-4V
4130 hard	Ti-5Al-2.5Sn
4130X-hard	Magnesium AZ31B-F
4340 annealed	1100 aluminum
4340 hard	5456-H311 aluminum
304 annealed	2014-T6 aluminum
304 hard	2024-T4 aluminum
52100 hard	7075-T6 aluminum
52100X-hard	Silver (0.99995% pure)
Am-350 annealed	Beryllium
Am-350 hard	Inconel X
310 stainless	A-286 aged
Vascomax 300 CVM	A-286 34% cold reduced and aged
Vascojet MA	D-979
Vascojet 1000	

Source: Ref 2

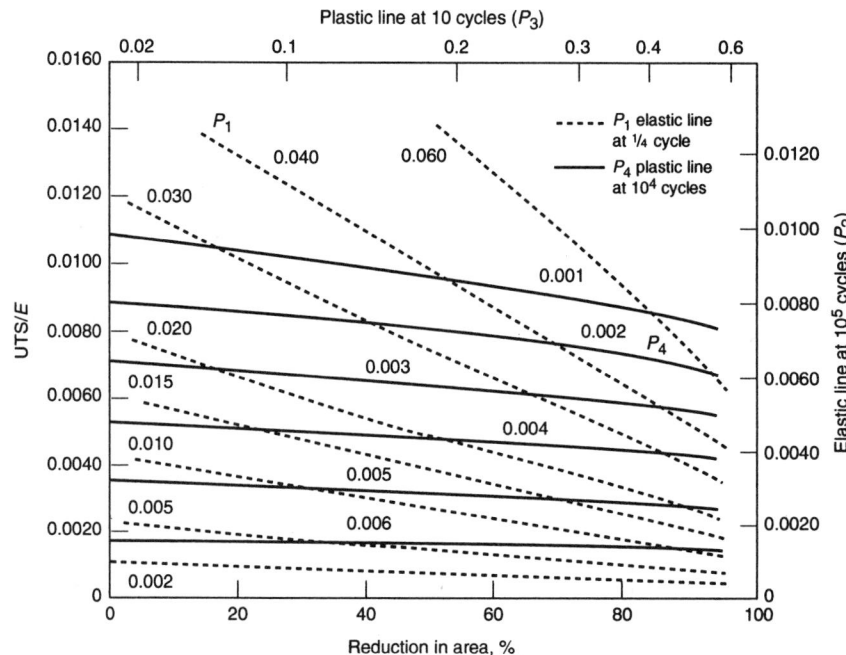

Fig. 3 Graphical solution to obtain the four points (P_1, P_2, P_3, and P_4 in Fig. 1) to position the elastic and plastic strain-range lines. Based on analysis of fatigue data for materials in Table 1. Source: Ref 2

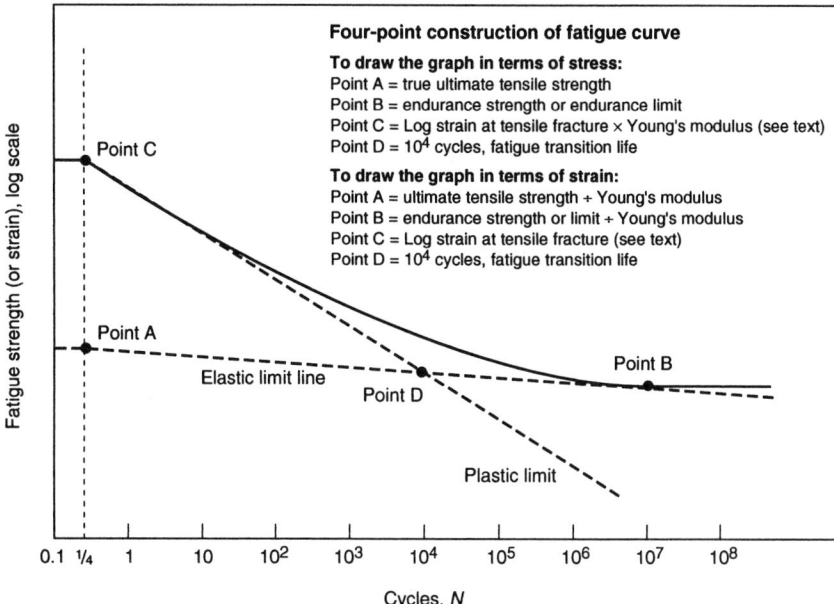

Fig. 4 Schematic summary of four-point method for estimating fatigue strength or strain life. Source: Ref 6

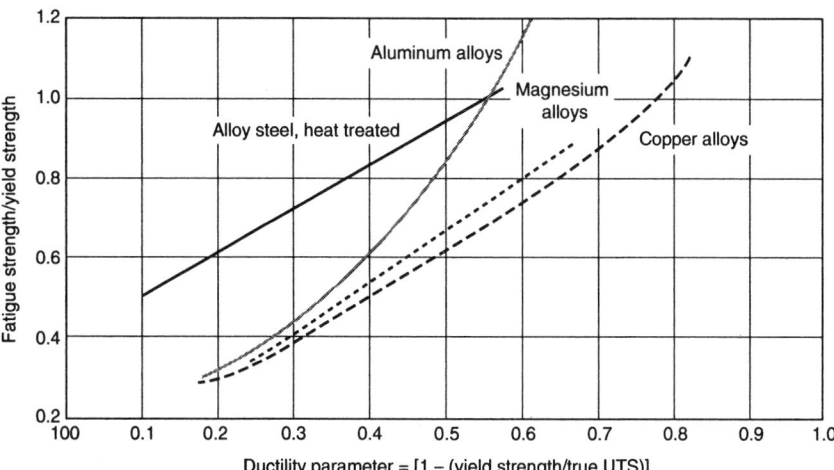

Fig. 5 Plot for estimating fatigue-endurance limits (point B in Fig. 4) for common structural alloy groups. Source: Ref 6

Point B, the right-hand locator of the elastic curve, is defined as the fatigue-endurance limit, if the metal has one; otherwise, point B is the endurance strength. Some ferrous alloys have an endurance limit, that is, a stress level below which fatigue failure will never occur, regardless of number of cycles. This is generally around 10^7 or 10^6 cycles, at which point the fatigue curve approaches zero-slope, or a horizontal line.

Many metals, particularly those that do not work harden, have no detectable endurance limit. Their long-life fatigue curves never become truly horizontal. For these metals, a pseudo-endurance limit, called endurance strength, is reported. Usually, this value is defined as the failure stress at some large number of cycles, for example, 10^7 to 10^{10}.

Point B can also be obtained from tensile-test data—a virtue of this technique, as handbook values for fatigue-endurance strengths or limits are often not available. Use Fig. 5 (Ref 6) to find the fatigue-endurance value from yield strength and true ultimate tensile strength for the material. Simply calculate the "ductility parameter" from handbook tensile data using the equation on the horizontal axis of the graph. Then find the "endurance-to-yield" strength ratio for the appropriate material. Multiply this ratio by "yield strength" to find the endurance value, which is point B.

Beyond point B, the ratio of ultimate tensile strength to yield strength can be used to approximate the slopes of the long-life portion of the fatigue curve. According to many researchers, a ratio greater than 1.2 suggests that the material strain hardens sufficiently to produce a pronounced endurance limit value, and the curve assumes a zero slope. For ratios less than 1.2,

however, the curve will continue to drop beyond point B. The lower the ratio below 1.2, the further the fatigue curve deviates from a horizontal, zero-slope line beyond point B.

Because both endurance strength and endurance limit are reported in terms of stress, this value must be divided by Young's modulus for the metal if the fatigue curve is being constructed in terms of strain.

Point C is a value known as "fracture ductility." If natural, or true, strain at fracture for a simple tension test is known (which would be the distance between gage points at fracture divided by initial gage length), fracture ductility is the natural log of this value.

In most cases, however, reduction of area for a simple tensile test is given in handbooks. As before in the discussion on the universal slopes method, fracture ductility, ε_f, is estimated.

$$\varepsilon_f = \ln \frac{100}{100 - RA} \qquad \text{(Eq 9)}$$

where RA is reduction of area in %.

Because fracture ductility is in units of strain, this value must be multiplied by Young's modulus to obtain point C in terms of stress. In all cases, point C is also plotted at $N = \frac{1}{4}$.

Point D is defined as the intersection of the plastic and elastic curves at 10^4 cycles. (According to the theory of "universal slopes," elastic and plastic strain curves intersect at $N = 10^4$). Thus, locate point D on the elastic curve and draw the plastic curve between points C and D. Now the fatigue curve can be drawn as the arithmetic summation of the elastic and plastic lines.

Comparison with Data for Steel, Aluminum, and Copper Alloys. To demonstrate the validity of the method described here, actual fatigue test results for various steel, aluminum, and copper alloys were compared with curves approximated from handbook data (Ref 6). In addition, a more recent analysis by J.H. Ong (*Int. J. Fatigue*, Vol 15, 1993, p 13-19) on 49 steels demonstrates that the predicted values by the four-point correlation method and the universal slopes method give satisfactory agreement with experimental data. The analysis by Ong shows that the four-point methods gives the best estimates for predicting fatigue properties from uniaxial tension tests.

Of the six comparisons shown (Fig. 6a-6f), fatigue data for steels and aluminum were taken from published sources. The measurements for copper fatigue are original, taken from tests on simulated squirrel-cage rotor, bar-to-end ring joints for induction motors.

Because these parts had been brazed prior to testing, the copper fatigue test data were assumed to represent essentially annealed material. Traceability of the data is not "ideal" in these cases, as handbook tensile data for the approximated curves were selected for truly annealed materials. Nevertheless, correlation between fatigue test data and the curves drawn from annealed tensile data is quite good, indicating that this technique

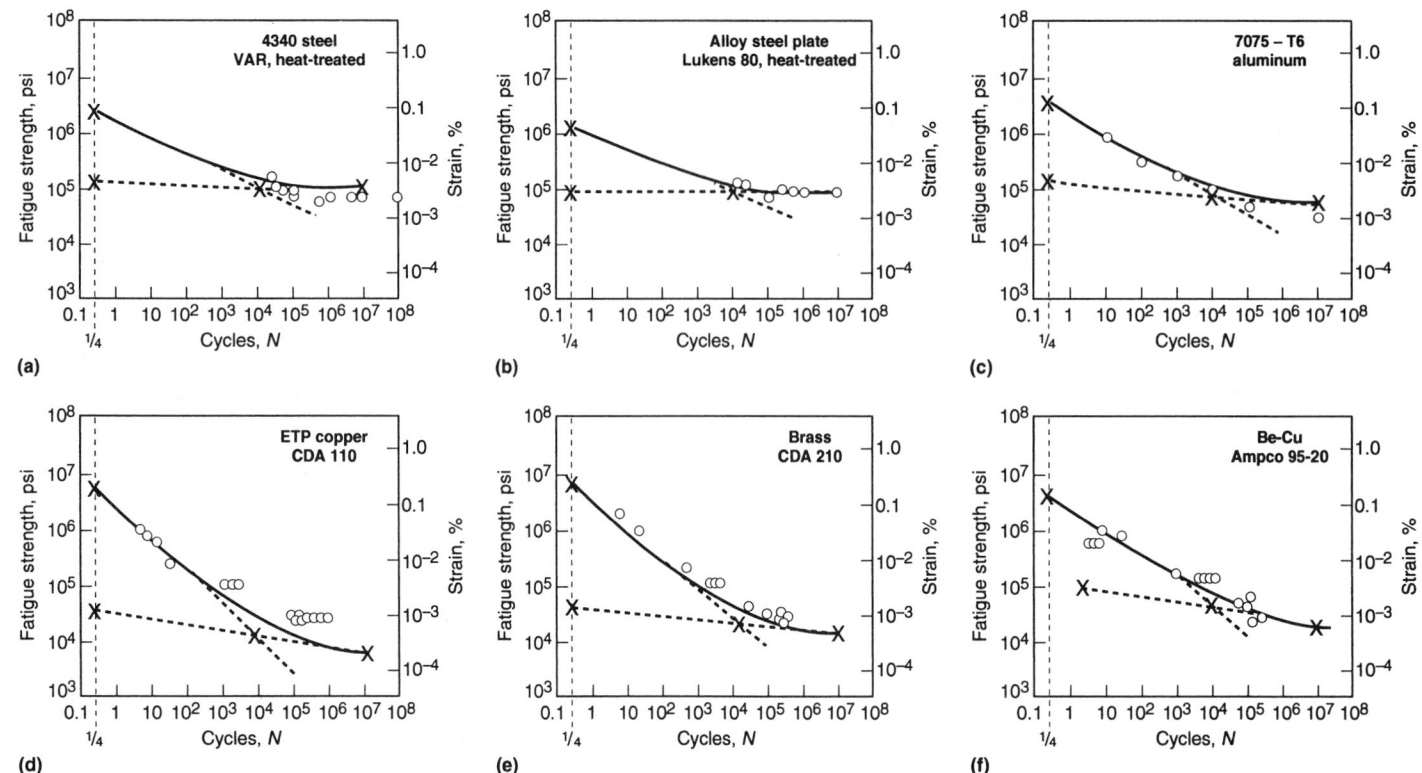

Fig. 6 Comparison of actual fatigue test results (open circles) with fatigue curves constructed by the four-point method from tensile data. Total fatigue life is a solid line and elastic and plastic components are dashed lines constructed from tensile data point (shown by X's). (a) 4340 steel. (b) Alloy steel plate (Lukens 80). (c) 7075-T6 aluminum. (d) Electrolytic-tough-pitch (ETP) copper. (e) Brass. (f) Be-Cu alloy. Source: Ref 6

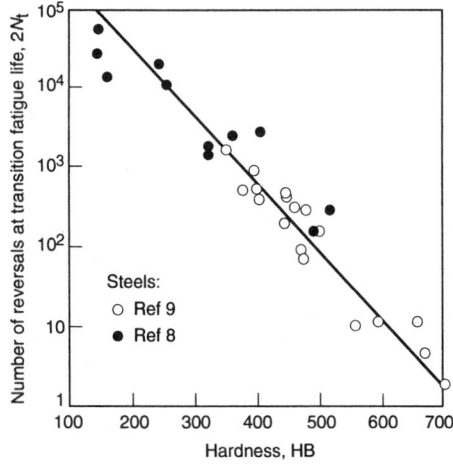

Fig. 7 Transition fatigue life (expressed in terms of load reversal) as a function of hardness for steel. Source: Ref 7

Fig. 8 Schematic of the strain range-life relationship as a function of strain ratio, R_ε. The life is affected by R_ε in regime I only. In regime II, the mean stress is always nearly zero, and R_ε has little or no effect on life. Source: Ref 10

by elastic strains or life dominated by elastic strains) does occur approximately at approximately 10^4 cycles for quenched-and-tempered carbon and low-alloy steels. In general, however, the fatigue transition life is expected to increase as the ductility of a material increases and strength decreases. Figure 7 (Ref 7-9), for example, shows this effect for various steels as a function of hardness. Higher-strength steels resist plastic deformation over a larger range, while softer steels may undergo plastic strain at a higher fatigue life. Transition fatigue life and tensile properties for some nonferrous alloys also are given in Table 2.

Effect of Mean Stress or Strain (*R* ratio) on Transition Life. The fatigue transition life in one study (Ref 10) was determined to be independent of the strain ratio (R_ε) as shown schematically in Fig. 8. Furthermore, in plastic strain region (regime II in Fig. 8), the effects of strain ratio are considered negligible. In regime II, where there are large amounts of plastic strain present, there are no effects of strain ratio, and the fatigue life behavior is essentially as if all testing is done in a completely reversed mode. In this regime, the elastic and plastic slopes for all strain ratios are equivalent to those slopes for the completely reversed data of regime I, and a universal strain-normalized life equation describes the fatigue behavior of the material. These characteristics allow development of a fairly simple basis for defining the effects on strain ratios on fatigue life, as described in more detail in Ref 10.

appears to be perfectly acceptable for copper alloys as well.

Figures 6(a) to (f) were prepared from actual fatigue-test data (open circles) and from handbook tensile data (X's). Fatigue curves constructed according to the techniques outlined in this article are shown in solid curves. Elastic and plastic strain curves used in the construction of the fatigue curves are dashed. While this is no substitute for thorough, conventional fatigue testing, reasonable correlation between the actual fatigue data and the simulated curves indicates that this technique can be a quick shortcut for approximating fatigue-life information.

Transition Fatigue Life

The four-point method and the method of universal slopes are based, in part, on an assumed fatigue transition life at $N_t = 10^4$ cycles. The fatigue transition life (which is considered the point of division between fatigue life dominated

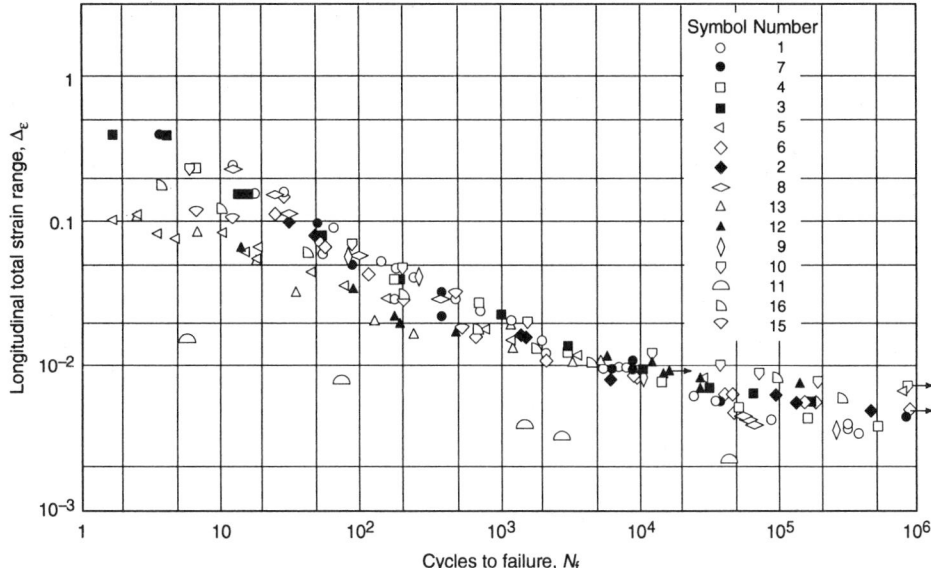

Fig. 9 Total strain versus cyclic life for various alloys. (See Table 3 for material identification.) Source: Ref 11

Summary of Strain-Life Constants

The four fatigue-life parameters in Eq 1 can be determined either by curve fitting actual fatigue life test data or by approximating the constants from tensile properties. Each method has its limitations. Nevertheless, the four constants in Eq 1 do provide a model for fatigue life for many alloys (for example, those in Fig. 9 and Table 3 from the early work by Manson and others in Ref 11).

Curve Fitting from Fatigue Test Data. Tables 4 through 7 at the end of this article summarize fatigue-life constants determined from curve fitting of fatigue test data for several alloys. As previously mentioned, some materials (such as some high-strength aluminum alloys and titanium alloys) cannot be adequately represented by the four fatigue constants in Eq 1. The strain range for the actual data is also an important factor. Errors can occur when extrapolating fatigue life estimates outside the range of the origi-

nal data. Nevertheless, the basis for the four constants in Eq 1 do provide a model for many structural alloys.

Fatigue Constant Approximations. Although the four fatigue constants are often adequately defined from available data, the following approximations can be useful in estimating their values.

Fatigue strength coefficient, σ_f'. A fairly good approximation is $\sigma_f' \approx \sigma_f$ (where σ_f is fracture stress with necking corrections). For steels with hardnesses below 500 HB: $\sigma_f \approx$ UTS + 50 ksi. Other values in estimating σ_f' include (Ref 7):

$$\sigma_f' = \sigma_f$$
$$\sigma_f' = 1.09\,\sigma_f$$
$$\sigma_f' = 0.92\,\sigma_f\ (b = -0.12)$$
$$\sigma_f' = 1.15\,\sigma_f\ (b = -0.12)$$

Fatigue strength exponent, b. The exponent b varies from -0.05 to -0.12 for most metals with an average of -0.085.

Fatigue ductility coefficient, ε_f'. As described earlier for the universal slopes method, a fairly good approximation for this coefficient is $\varepsilon_f' \approx \varepsilon_f$ (monotonic fracture ductility) with $\varepsilon_f = 1$ ln $[1/(1 - $ RA$)]$ and where RA is the reduction in area. Other values in estimating ε_f' include (Ref 7):

$$\varepsilon_f' = 0.35\,\varepsilon_f$$
$$\varepsilon_f' = 0.5\,\varepsilon_f\ (c = -0.6)$$
$$\varepsilon_f' = 0.75\left(\varepsilon_f\right)^{0.75}\ (c = -0.6)$$
$$\varepsilon_f' = 0.76\left(\varepsilon_f\right)^{0.6}\ (c = -0.6)$$
$$\varepsilon_f' = 0.50\,\varepsilon_f$$
$$\varepsilon_f' = 0.71\,\varepsilon_f$$

Fatigue ductility exponent, c. The exponent c is not as well defined as the other parameters. A rule-of-thumb approach must be followed rather than an empirical equation (Ref 4). Estimated values include:

Table 2 Transition fatigue life and tensile data for Inco 706 and two titanium alloys

Material	N_t, cycles	$\Delta\varepsilon_{TR}$, %	Elongation, %	Reduction of area, %	Fracture ductility, ε_f	$b, R = -1.0$	$c, R = -1.0$
Inco 706	194	1.19	14.5	41.5	0.536	−0.0880	−0.44
Ti-5Al-5Sn-2Zr-2Mo	252	1.25	18.5	53.1	0.757	−0.0520	−0.55
Ti-6Al-4V	690	1.45	22.3	61.0	0.942	−0.1212	−1.11

Table 3 Tensile properties for alloys in Fig. 9

Number in Fig. 9	Material	Young's modulus, 10^6 psi	Poisson's ratio	Ductility	Fracture stress, ksi	Ultimate stress, ksi	Endurance stress, ksi
1	4130 soft	32.0	0.290	1.120	245.0	130.0	45.0
2	304 hard	25.0	0.340	1.165	295.0	138.0	40.0
3	4340 hard	29.0	0.300	0.477	278.0	213.0	55.0
4	4340 annealed	28.0	0.320	0.570	174.0	120.0	50.0
5	52100	30.0	0.290	0.119	323.0	292.0	80.0
6	304 annealed	27.0	0.270	1.368	278.0	108.0	40.0
7	4130 hard	29.0	0.280	0.792	302.0	207.0	70.0
8	AISI 310	28.0	0.300	1.006	197.0	93.3	18.0
9	Inconel X (nickel alloy)	31.0	0.310	0.223	219.0	176.0	55.0
10	Ti-6Al-4V	17.0	0.330	0.530	249.0	179.0	70.0
11	Beryllium	42.0	0.024	0.017	47.7	46.9	24.0
12	AM350 hard	26.0	0.300	0.233	328.0	276.0	90.0
13	AM350 annealed	28.0	0.320	0.737	339.0	191.0	55.0
14	1100 aluminum	10.0	0.330	2.090	…	16.2	5.0
15	5456-H311 aluminum	10.0	0.330	0.424	81.7	57.8	20.0
16	2014-T6 aluminum	10.0	0.330	0.288	91.4	73.6	25.0

Source: Ref 11

- Coffin found c to be about –0.5.
- Manson found c to be about –0.6.
- Morrow found that c varied between –0.5 and –0.7.

Fairly ductile metals (where $\varepsilon_f \approx 1$) have average value of $c = -0.6$. For strong metals (where $\varepsilon_f \approx 0.5$) a value of $c = -0.5$ is probably more reasonable (Ref 4).

REFERENCES

1. S.S. Manson, *J. Basic Eng. (Trans. ASME)*, Vol 84 (No. 4), Dec 1962, p 537-541
2. S.S. Manson, "Fatigue: A Complex Subject—Some Simple Approximations," NASA-TM-X-52084, National Aeronautics and Space Administration, 1965
3. *Materials Data for Cyclic Loading*, Vol 42A to 42E, C. Boller and T. Seeger, Ed., Elsevier Science, 1987
4. J.A. Bannantine, J.J. Comer, and J.L. Handrock, *Fundamentals of Metal Fatigue Analysis*, Prentice-Hall, 1990, p 63, 83, problem 2.31 on the method of universal slopes
5. J. Conway and L. Sjodahl, *Analysis and Representation of Fatigue Data*, ASM International, 1991, p 37-68
6. P. Weihsmann, *Mater. Eng.*, March 1980, p 53
7. R. Landgraf, The Resistance of Metals to Cyclic Deformation, in *Achievement of High Fatigue Resistance in Metals and Alloys*, STP 467, ASTM, 1970, p 3-36
8. R.W. Smith, M.H. Hirschberg, and S.S. Manson, "Fatigue Behavior of Materials Under Strain Cycling in Low and Intermediate Life Range," NASA-TN-D-1574, National Aeronautics and Space Administration, April 1963
9. R.W. Landgraf, "Cyclic Deformation and Fatigue Behavior of Hardened Steels," Report No. 320, Department of Theoretical and Applied Mechanics, University of Illinois, Urbana, Nov 1968
10. T. Pollack and M. Doner, Modelling of Strain Ratio Effects on Low Cycle Fatigue Life, *Fatigue Life Analysis and Prediction*, American Society for Metals, 1986, p 341-347
11. R.W. Smith, M.H. Hirschberg, and S.S. Manson, *Behavior of Materials under Strain Cycling in Low and Intermediate Life Range*, TN-D-1574, National Aeronautics and Space Administration, March 1963, and reported in *Fatigue—An Interdisciplinary Approach*, J.J. Burke, N.L. Reed, and V. Weiss, Ed., Syracuse University Press, 1964, p 154
12. C. Boller and T. Seeger, *Materials Data for Cyclic Loading*, Parts A-E, Elsevier, 1987

Table 4(a) Monotonic and fatigue properties for carbon, low-alloy, and HSLA steels ($R = -1$, longitudinal direction)

Material	Hardness(a)	Product condition(b)	UTS, MPa	TYS, MPa	σ_f, MPa	RA, %	EL, %	ε_f	E, GPa	n'	K, MPa	σ_f', MPa	b	ε_f'	c
Material: SAE steel															
950		As-received	556	321	990	67		1.06	206	0.18	1026	799	–0.09	0.22	–0.46
	150 HB	As-received	565	317	999	69		1.19	206	0.20	1151	846	–0.08	0.20	–0.42
	150 HB	As-received	565	317	999	69		1.19	206	0.15	1036	1005	–0.09	0.82	–0.57
	137 HB	As-received	523	432		54			207	0.17	1003	783	–0.08	0.25	–0.48
	137 HB	As-received	523	432		54			207	0.16	957	772	–0.08	0.34	–0.52
	137 HB	As-received	423	432		54			207	0.16	869	1008	–0.11	2.79	–0.79
	154 HB	As-received	565	499		50			207	0.12	744	710	–0.07	0.77	–0.57
		As-received	458	396		63			199	0.12	705	623	–0.07	0.40	–0.62
950X	150 HB	As-received	441	344	751	65		1.06	206	0.20	1024	611	–0.07	0.07	–0.36
	150 HB	As-received	441	344	751	65		1.06	206	0.19	995	474	–0.06	0.02	–0.30
	124 HB	As-received	445	405		59			207	0.17	883	672	–0.08	0.18	–0.44
	120/132 HB	As-received	478	372	1024	75		1.41	206	0.28	1689	1376	–0.15	0.33	–0.52
	157 HB	As-received	497	450		51			207	0.14	812	609	–0.06	0.17	–0.43
		As-received		358					206	0.14	848	869	–0.10	1.28	–0.72
		As-received	428	347		69			194	0.14	826	704	–0.08	0.52	–0.64
		As-received	438	370		64			199	0.14	796	800	–0.10	1.23	–0.72
		As-received	441	361	981	73		1.31	207	0.13	726	732	–0.09	1.38	–0.70
		As-received	441	358	972	68		1.14	207	0.14	802	686	–0.07	0.34	–0.54
		As-received	441	362		68			196	0.13	768	742	–0.09	0.70	–0.65
	120 HB	As-received	444	348	963	78		1.52	193	0.30	1685	1039	–0.13	0.21	–0.45
		As-received	447	382	915	67		1.13	207	0.13	839	923	–0.10	2.32	–0.75
		As-received	451	384		64			224	0.14	829	810	–0.10	0.60	–0.66
		As-received	455	369		69			198	0.14	861	843	–0.10	0.95	–0.69
		As-received	459	396		63			199	0.12	709	664	–0.08	0.88	–0.70
		As-received	460	374	923	64		1.04	207	0.10	661	589	–0.06	0.34	–0.56
		As-received	465	362		77			200	0.15	918	857	–0.10	1.14	–0.72
		As-received	473	427	1234	76		1.38	207	0.18	1176	1383	–0.14	3.09	–0.81
		As-received	474	421		60			204	0.13	810	847	–0.10	0.51	–0.66
		As-received	476	380		69			198	0.14	835	897	–0.10	0.70	–0.62
		As-received	478	402	1108	69		1.19	207	0.17	1121	999	–0.10	0.56	–0.62
		As-received	491	416	1381	74		1.37	207	0.15	1102	1217	–0.12	1.93	–0.77
		As-received	496	439					206	0.08	575	574	–0.06	1.63	–0.77
	146 HB	As-received	510	391	978	74		1.34	193	0.13	939	824	–0.08	0.42	–0.57
	156 HB	As-received	531	335	999	71		1.24	188	0.22	1368	965	–0.10	0.24	–0.46
	183 HB	As-received	565	434	1203	67		1.13	217	0.22	1824	992	–0.08	0.13	–0.43
	167 HB	As-received	584	458	1071	72		1.26	191	0.09	868	876	–0.06	1.09	–0.68
	187 HB	As-received	587	472	1096	73		1.29	187	0.09	900	1083	–0.08	6.12	–0.92
960X		As-received	480	415					206	0.14	969	895	–0.09	0.46	–0.65
		As-received	506	358	1070	60		0.92	182	0.17	1054	1069	–0.11	1.22	–0.64
980X	167 HB	As-received	692	575	1238	68		1.14	202	0.18	1547	1071	–0.09	0.21	–0.51
	225 HB	As-received	695	578	1219	68		1.15	194	0.12	1178	1046	–0.08	6.81	–1.01
	225 HB	As-received	695	578	1219	68		1.15	194	0.25	2658	1184	–0.10	0.09	–0.48
		As-received	585	503					206	0.19	1140	880	–0.10	0.27	–0.52
		As-received	616	557					206	0.13	1229	1355	–0.12	2.22	–0.88
		As-received	627	520	1236	69		1.19	207	0.13	1053	1231	–0.10	4.20	–0.84
		As-received	634	584	1225	60		0.93	207	0.14	1254	1382	–0.11	1.98	–0.82
	157 HB	As-received	644	531	963	64		1.05	190	0.24	1984	991	–0.09	0.06	–0.38

(continued)

(a) C = Core hardness, S = surface hardness. (b) N = nitrided, SR = Stress relief, CR = cold-rolled. Adapted from Ref 12 and other sources

Table 4(a) (continued)

Material	Hardness(a)	Product condition(b)	UTS, MPa	TYS, MPa	σf, MPa	RA, %	EL, %	εf	E, GPa	n'	K, MPa	σf, MPa	b	εf	c
Material: SAE steel (continued)															
980X (continued)	157 HB	As-received	644	580	1173	63		0.44	207	0.12	1175	1452	−0.11	5.29	−0.96
		As-received	652	579	1391	75		1.21	296	0.13	1135	1146	−0.09	1.10	−0.72
		As-received	656	595	1380	68		1.16	207	0.15	1395	1514	−0.12	1.52	−0.77
	207 HB	As-received	664	581	135	64		1.02	207	0.14	1316	898	−0.06	0.10	−0.46
	209 HB	As-received	691	584	1268	72		1.26	197	0.12	1134	1167	−0.09	1.31	−0.72
		As-received	696	568	1268	58		0.88	207	0.13	1428	1426	−0.11	0.84	−0.76
	229/248 HB	As-received	717	553	1267	69		1.18	191	0.08	998	1145	−0.08	2.16	−0.80
SAE 450 XK	183 HB	As-received	566	435	1205	68		1.13	217.51	0.176	1381	875	−0.063	0.594	−0.613
SAE 1015		Fully annealed	392	263	746	55		0.806	196.793	0.193	824	807	−0.117	0.415	−0.528
SAE 1045 (forged)	260 HB	As-received	915	622	1784	59		0.90	199.7	0.226	2407	2350	−0.148	0.447	−0.561
	260 HB	As-received	915	622	1784	59		0.90	199.7	0.191	1762	2022	−0.151	0.517	−0.631
	260 HB	As-received	915	622	1784	59		0.90	199.7	0.132	1186	4911	−0.236	1441.7	−1.442
Material: Carbon steels (AISI)															
1005		Hot-rolled	321	225	784	73		1.32	207	0.27	1240	886	−0.14	0.28	−0.50
		Hot-rolled	355	236	1031	81		1.63	207	0.24	1064	878	−0.13	0.46	−0.54
		Hot-rolled	358	266	951	70		1.21	207	0.20	709	536	−0.09	0.24	−0.46
1006	85 HB	Hot-rolled	318	248		73			207	0.24	1028	629	−0.09	0.15	−0.40
	85 HB	Hot-rolled	318	248		73			207	0.28	1352	802	−0.12	0.48	−0.52
	85 HB	Hot-rolled	318	248		73			207	0.21	813	756	−0.13	1.22	−0.67
1008	82/99 HB	Hot-rolled	363	253	808	77		1.58	209	0.35	1953	1297	−0.18	0.93	−0.59
1015	80 HB	Normalized	415	227	725	68		1.13	206	0.24	1058	976	−0.14	0.76	−0.59
1018	106 HB	QT	354	250					200	0.27	1259	782	−0.11	0.19	−0.41
	118 HB	QT	496	290	678			1.06	207	0.25	640	391	−0.09	0.13	−0.35
	209 HB	Hot-rolled	696	572	741			1.15	207	0.24	862	423	−0.07	0.07	−0.30
1020	108 HB	Annealed	392	254	661	64		1.02	186	0.26	1206	850	−0.12	0.44	−0.51
	105/109 HB	Hot-rolled	441	260	713	61		0.96	203	0.24	1221	895	−0.11	0.29	−0.47
1025		Hot-rolled	547	306	1193	62		0.98	204	0.20	1082	961	−0.10	0.56	−0.51
		Hot-rolled	566	387	880	57		0.87	207	0.19	1178	953	−0.09	0.34	−0.50
1030	128 HB	Hot-rolled	454	289	764	59		0.90	206	0.29	1545	902	−0.12	0.17	−0.42
	128 HB	Hot-rolled	454	289	764	59		0.90	206	0.27	1278	568	−0.09	0.03	−0.29
1035		Hot-rolled	476	250	751	56		0.27	196	0.24	1185	906	−0.11	0.33	−0.47
		QT	1605	1395	1942	42		0.04	200	0.13	2269	3062	−0.12	1.83	−0.78
		QT	2195	1514	2402	9		0.08	200	0.23	5684	3690	−0.13	0.20	−0.63
1045	225 HB	Annealed	751	516	998	44			203	0.17	1178	960	−0.08	0.50	−0.52
		Hot-rolled	671	327	1061	44		0.59	216	0.22	1402	1099	−0.11	0.52	−0.54
	500 HB	QT	1956	1728	2306	38			202	0.20	4634	2888	−0.09	0.23	−0.56
		QT	2067	1825	2135	2		0.02	199	0.10	4264	2416	−0.07	0.002	−0.47
	390 HB	QT	1343	1274	1860	59		0.89	206	0.09	1492	1408	−0.07	1.51	−0.85
	450 HB	QT	1584	1515	2101	55		0.81	206	0.09	1874	1686	−0.06	0.97	−0.83
	500 HB	QT	1825	1688	2273	51		0.71	206	0.12	2636	2165	−0.08	0.22	−0.66
	595 HB	QT	2239	1860	2721	41		0.52	206	0.10	3498	3047	−0.10	0.13	−0.79
1080	421 HB	Austempered	1349	978	1645	32			204	0.21	3177	2364	−0.10	0.51	−0.59
	371/410 HB	QT	1298	1118	1645	29			206	0.13	1820	1787	−0.09	1.00	−0.69
	371/410 HB	QT	1432	1260	1750	34			201	0.16	2284	1916	−0.09	0.36	−0.54
Material: Carbon steels (other)															
10B21		Hot-rolled	503	293	1139	68		1.16	186	0.19	955	767	−0.09	0.51	−0.51
	243 HB	QT	814	742	1230	69		1.18	200	0.08	876	833	−0.05	1.42	−0.74
	242 HB	QT	814	742	1230	69		1.18	205	0.08	877	860	−0.05	1.42	−0.74
	318 HB	QT	1048	999	1499	67		1.13	197	0.06	989	1209	−0.06	4.33	−0.85
	318 HB	QT	1048	999	1499	67		1.13	198	0.06	990	1036	−0.04	4.33	−0.85
	362 HB	QT	1240	1172	2414	64		1.03	205	0.09	1315	1242	−0.05	1.47	−0.70
	363 HB	QT	1241	1172	2415	64		1.03	204	0.08	1281	2924	−0.15	1.88	−0.73
10B22	250/260 HB	QT	833	806	1502			1.09	203	0.08	911	830	−0.04	2.11	−0.75
	250/260 HB	QT	833	806	1502			1.09	203	0.12	1098	970	−0.07	2.59	−0.76
10B30	362 HB	QT	1240	1140	1706	63		1.00	195	0.10	1392	1287	−0.05	1.50	−0.70
	362 HB	QT	1240	1140	1706	63		1.00	195	0.10	1396	1289	−0.05	1.50	−0.70
1522	289 HB	Hot-rolled	1005	902	1281	49		0.68	200	0.20	2118	1253	−0.08	0.07	−0.36
	304 HB	Hot-rolled	1088	1005	1612	61		0.95	200	0.19	2278	1464	−0.08	0.28	−0.51
1541	362 HB	QT	1200	1096	1599	54		0.80	209	0.12	1613	2980	−0.15	0.68	−0.61
1561	234 HB	Hot-rolled	836	447	1052	29		0.34	197	0.19	1448	1278	−0.11	0.53	−0.54
15B27	264 HB	QT	916	854	1426	66		1.10	198	0.05	857	909	−0.05	2.17	−0.82
	319 HB	QT	1054	1005	1598	67		1.10	199	0.06	1030	1172	−0.06	8.40	−1.00
15B35			2073	1454	2768	41		0.57	204	0.23	7795	4541	−0.13	0.41	−0.78
		QT	1679	1431	2085	52		0.74	200	0.08	1962	3304	−0.12	3.53	−0.89
DQSK		Hot-rolled	319	239		69			185	0.33	1572	962	−0.16	0.23	−0.49
		Hot-rolled	326	220		72			186	0.32	1512	726	−0.13	0.10	−0.41
		Hot-rolled	335	213		79			188	0.35	1991	837	−0.14	0.08	−0.39
Material: Low-alloy (AISI)															
P&O	83/89 HB	Hot-rolled	331	234	748	77		1.56	207	0.34	1597	1211	−0.18	0.53	−0.56
4130	253/265 HB	QT	895	778	1419	67		1.12	220	0.13	1359	1273	−0.08	1.51	−0.72

(continued)

(a) C = Core hardness, S = surface hardness. (b) N = nitrided, SR = Stress relief, CR = cold-rolled. Adapted from Ref 12 and other sources

Table 4(a) (continued)

Material	Hardness(a)	Product condition(b)	UTS, MPa	TYS, MPa	σ_f, MPa	RA, %	EL, %	ε_t	E, GPa	n'	K, MPa	σ'_f, MPa	b	ε'_f	c
Material: Low-alloy (AISI) (continued)															
4130 (continued) 362/371 HB		QT	1426	1357	1819	54		0.79	199	0.13	1837	1731	−0.08	0.84	−0.68
4142		QT	1412	1378	1825	48		0.66	206	0.14	2259	1820	−0.08	0.65	−0.76
	400 HB	QT	1550	1446	1894	47		0.63	199	0.07	1556	1796	−0.08	1.42	−0.88
	450 HB	QT	1757	1584	1998	42		0.54	206	0.11	2359	2017	−0.08	0.85	−0.90
	450 HB	QT	1929	1860	2101	37		0.46	199	0.10	2000	1804	−0.07	3.44	−1.01
	475 HB	QT	1929	1722	2170	35		0.43	206	0.11	2713	2209	−0.08	0.68	−0.98
	475 HB	QT	2032	1894	2067	20		0.22	199	0.08	2073	2036	−0.08	2.75	−1.20
	560 HB	QT	2239	1688	2652	27		0.31	206	0.13	4222	3247	−0.12	0.07	−0.81
	670 HB	QT	2446	1619	2583	6		0.06	199	0.07	3484	2727	−0.08	0.06	−1.47
4340	237/247 HB	Hot-rolled	826	634	1088	43		0.57	192	0.17	1384	1232	−0.10	0.53	−0.56
		QT	1171	1102	1632	56		0.83	205	0.13	1603	1649	−0.09	1.39	−0.72
		QT	1171	1102	1632	56		0.83	205	0.12	1471	1244	−0.06	7.37	−0.88
	350 HB	QT	1240	1178	1653	57		0.84	192	0.14	1863	1944	−0.10	1.22	−0.73
	409 HB	QT	1467	1371	1557	38		0.48	199	0.13	1950	1898	−0.09	0.67	−0.64
5160	407/460 HB	QT	1584	1487	1929	39		0.51	200	0.13	2432	2063	−0.08	9.56	−1.05
52100	512/521 HB	Solution treated	2011	1922	2191	11		0.12	206	0.15	3402	2647	−0.09	0.16	−0.58
8630	254 HB	Normalized	785	709	840	16		0.17	199	0.08	961	1049	−0.11	0.20	−0.86
8640	361 HB	QT	1373	1306	1583	52		0.74	223	0.14	1951	1487	−0.06	0.60	−0.61
9262	260 HB	Normalized	923	455	1041	14		0.16	206	0.12	1206	1178	−0.08	0.83	−0.68
	271 HB	QT	999	786	1220	33		0.41	193	0.14	1525	1477	−0.09	0.82	−0.68
Material: Unalloyed steel, bar, tested at 23 °C															
St 37		Stress-relieved	435	295	835	64		1.02	214	0.207	988	895	−0.111	0.7051	−0.569
		Stress-relieved	435	310	835	64		1.02	214	0.239	1170	929	−0.117	0.3908	−0.495
St 37-3		Annealed		297					210	0.201	999	437	−0.053	0.020	−0.287
St 52-3		Annealed	597	400	1083	63		0.98	210	0.185	1228	1193	−0.11	0.6601	−0.553
		Annealed	597	400	1083	63		0.98	210	0.207	1337	1293	−0.115	0.4954	−0.515
St 52-3		Annealed		378					210	0.139	871	733	−0.078	0.305	−0.569
		Annealed		438					210	0.100	679	573	−0.070	0.141	−0.669
		Annealed		385					210	0.100	653	448	−0.065	0.009	−0.496
		Annealed		346					210	0.125	688	389	−0.062	0.008	−0.446
		Annealed		306					210	0.064	400	318	−0.051	0.006	−0.559
Ck01		Annealed	308	190			33		210	0.139	502	682	−0.105	7.594	−0.742
Ck10		Annealed	377	235			28		210	0.098	505	1245	−0.141	14.078	−0.839
	(C)	Annealed	340	185			32		210	0.087	448	599	−0.083	30.752	−0.963
Ck15	108 HV 30 (C)	Normalized	379	240		58	53		210	0.135	607	657	−0.086	1.403	−0.622
	108 HV 30 (C)	Normalized (CR)	410	259		61	44		210	0.048	343	465	−0.054	0.265	−0.475
	287 HV 30(S)	Normalized (N)	404	316		4	6		210	0.161	954	580	−0.057	0.0387	−0.339
Ck22		Annealed	463	260			21		210	0.127	613	700	−0.089	2.489	−0.690
S 25 C	142 HV 30	Annealed	507	346	1027	63	37		210	0.252	1345	959	−0.114	0.265	−0.453
	130 HV 30	Annealed	464	307	982	65	41		210	0.223	1111	965	−0.117	0.53	−0.525
	149 HV 30	Annealed	527	366	997	60	36		210	0.224	1217	925	−0.105	0.298	−0.472
S 35 C	175 HV 30	Annealed	617	414	1150	58	32		210	0.229	1355	1050	−0.107	0.329	−0.469
	174 HV 30	Annealed	593	394	1169	62	34		210	0.238	1460	1226	−0.121	0.503	−0.512
	176 HV 30	Annealed	565	396	1134	63	36		210	0.254	1534	1173	−0.119	0.349	−0.470
	242 HV 30	Normalized	780	587	1514	67	24		210	0.140	1106	1100	−0.087	0.941	−0.619
	222 HV 30	Normalized	656	480	1468	74	28		210	0.156	1033	1019	−0.092	0.947	−0.595
	245 HV 30	Normalized	733	596	1541	71	23		210	0.134	1027	880	−0.068	0.321	−0.513
	222 HV 30	Normalized	730	542	1473	68	25		210	0.149	1087	1004	−0.085	0.584	−0.568
	211 HV 30	Normalized	669	513	1417	70	29		210	0.165	1081	845	−0.075	0.235	−0.460
Material: Unalloyed steel, bar, tested at 23 °C															
Ck45		Annealed	680	415			19		210	0.135	808	2621	−0.210	46593	−0.753
		Annealed	680	415			19		210	0.151	932	1199	−0.117	5.000	−0.770
		Annealed		275					210	0.110	621	1037	−0.124	83.96	−1.102
		Annealed	680	415			19		210	0.125	758	1243	−0.129	45.60	−1.020
		Annealed	680	415			19		210	0.163	945	2581	−0.202	396.6	−1.222
		Annealed		345					210	0.175	933	1122	−0.124	2.937	−0.714
Ck45	207 HV 30	Stress-relieved	678	457	?				210	0.093	599	519	−0.044	0.124	−0.424
	2440 MPa (HB 10)	QT	790	540	1400	60	23		206	0.115	980	987	−0.083	0.994	−0.715
	246 HV 30 (S)	Stress-relieved (CR)	705	396		49			210	0.121	780	695	−0.062	0.132	−0.419
	333 HV 30 (S)	Stress-relieved	700	468		6			210	0.049	514	474	−0.023	0.049	−0.358
Ck70		Annealed	865	465			12		210	0.117	814	1194	−0.110	10.224	−0.852
Ck80		Annealed	948	482			10		210	0.157	1102	1436	−0.128	9.152	−0.870
Ck100		Annealed	550	260			26		210	0.205	1028	802	−0.099	0.210	−0.448
		Annealed	795	293			13		210	0.228	1298	1343	−0.135	0.874	−0.564
		Annealed	960	465			9		210	0.167	1188	1500	−0.123	1.552	−0.647
Material: Unalloyed steel, plate, tested at 23 °C															
09 G 2		Stress-relieved	512	340		70	34		198	0.122	680	596	−0.063	0.306	−0.507
St 44-2		As-received							204.4	0.137	497	499	−0.083	0.750	−0.584
SB 46 (ST46)		As-received	500	310		64	30		210	0.218	1118	1000	−0.118	0.619	−0.546
		As-received	500	310		64	30		210	0.226	1204	1074	−0.122	0.630	−0.544

(continued)

(a) C = Core hardness, S = surface hardness. (b) N = nitrided, SR = Stress relief, CR = cold-rolled. Adapted from Ref 12 and other sources

Table 4(a) (continued)

Material	Hardness(a)	Product condition(b)	UTS, MPa	TYS, MPa	σ_f, MPa	RA, %	EL, %	ε_t	E, GPa	n'	K, MPa	σ_f', MPa	b	ε_f'	c
Material: Unalloyed steel, plate, tested at 23 °C (continued)															
SB 49	143 HV 20	Annealed, SR	524	306	1062	62	32		210	0.206	1094	914	−0.109	0.422	−0.529
SM50B (St50)	156 HV 10	…	541	400			25		206	0.208	1137	690	−0.080	0.056	−0.342
	156 HV 10	…	541	400			25		206	0.176	953	814	−0.099	0.251	−0.507
	154 HV 10	…	534	385			31		206	0.184	998	693	−0.080	0.077	−0.377
	154 HV 10	…	534	385			31		206	0.172	957	829	−0.098	0.415	−0.565
	149 HV 10	…	526	350			35		206	0.204	1116	691	−0.081	0.083	−0.383
	149 HV 10	…	526	385			35		206	0.156	814	770	−0.093	0.282	−0.515
St 52 (1.0841)		Hot-rolled		378					210	0.139	871	733	−0.078	0.305	−0.569
		Hot-rolled		438					210	0.100	679	573	−0.070	0.141	−0.669
		Hot-rolled		385					210	0.100	653	448	−0.065	0.009	−0.496
Material: ASTM A 36															
ASTM A 36	160 HB	Hot-rolled	413	224	799	69		1.19	189	0.25	1075	780	−0.11	0.28	−0.45
	243 HB	Weld HAZ	666	534	820	52		0.74	188	0.17	1192	838	−0.07	0.13	−0.43
	303 HB	Plate	414	224	953	70		1.19	190	0.238	1050	600	−0.090	0.103	−0.384
Material: Unalloyed steel, plate, tested at 200 °C															
SB 49 (St 49)	143 HV 20	Annealed, SR	474	277	924	60	26		196	0.185	974	817	−0.095	0.383	−0.513
	143 HV 20	Annealed, SR	474	277	924	60	26		196	0.163	950	791	−0.083	0.298	−0.497
Material: Unalloyed steel, plate, tested at 300 °C															
SB 46 (ST 46)		As-received	548	213		65	30		197	0.138	949	987	−0.097	1.270	−0.699
		As-received	520	208		57	23		197	0.109	830	809	−0.068	0.753	−0.620
SB 49	143 HV 20	Annealed	510	211	885	54	24		210	0.064	644	632	−0.044	0.700	−0.679
	143 HV 20	Annealed	510	211	885	54	24		210						
Material: Unalloyed steel, plate, tested at 400 °C															
SB 46		As-received	441	197		78	33		193	0.182	872	1136	−0.155	3.647	−0.831
		As-received	457	191		77	31		103	0.182	1015	1143	−0.133	2.018	−0.736
SB 49	143 HV 20	Annealed, SR	462	191	928	69	32		206	0.101	650	535	−0.048	0.172	−0.498
	143 HV 20	Annealed, SR	462	191	928	69	32		206	0.074	540	477	−0.038	0.391	−0.591
	143 HV 20	Annealed, SR	462	191	928	69	32		206	0.116	664	678	−0.088	1.752	−0.811
Material: Unalloyed steel, plate, tested at 500 °C															
SB 46		As-received	256	162		87	40		170	0.103	321	365	−0.087	3.197	−0.831
		As-received	340	173		87	37		170	0.150	585	697	−0.118	3.163	−0.784
SB 49	143 HV 20	Annealed, SR	340	175	742	77	40		192	0.161	630	557	−0.096	0.458	−0.592
	143 HV 20	Annealed, SR	340	175	742	77	40		192	0.140	513	472	−0.091	0.552	−0.653
	143 HV 20	Annealed, SR	340	175	742	77	40		192	0.118	411	406	−0.095	0.978	−0.808
Material: Unalloyed steel, shaft, tested at 23 °C															
Ck15		Cold formed	603	505					210	0.073	677	1978	−0.152	75.359	−1.103
		Cold formed							210	0.129	1150	4937	−0.225	326383	−1.863
		Cold formed							210	0.344	4803	7951	−0.271	3.251	−0.763
									210	0.113	1002	3423	−0.197	73.888	−1.112
Ck45		QT	790	531	1271	60	23	0.777	210.5	0.133	1078	1405	−0.11	0.6065	−0.545
		QT	790	531	1271	60	23	0.777	210.5	0.133	1078	767	−0.064	0.135	−0.426
		QT	790	531	1271	60	23	0.777	210.5	0.133	1078	606	−0.035	0.025	−0.282
Material: Unalloyed steel, rounds, tested at 23 °C															
SAE 1045	416 HB	Cold drawn	1471	1367	2002	57		0.85	186.5	0.118	2235	1661	−0.058	0.062	0.433
	488 HB	Cold drawn	1706	1547	2223	52		0.74	231	0.095	2230	2276	−0.081	1.223	−0.857
	555 HB	Cold drawn	1968	1747	2458	46		0.62	231	0.122	2950	2768	−0.100	0.630	−0.828
	630 HB	Cold drawn	2327	1899	2831	37		0.45	207	0.107	3784	2666	−0.075	0.035	−0.686

(a) C = Core hardness, S = surface hardness. (b) N = nitrided, SR = Stress relief, CR = cold-rolled. Adapted from Ref 12 and other sources

Table 4(b) Monotonic and fatigue properties for low-alloy steels ($R = -1$, longitudinal direction)

Material	Hardness	Product condition(a)	UTS, MPa	TYS, MPa	σ_f, MPa	RA, %	EL, %	ε_t	E, GPa	n'	K, MPa	σ_f', MPa	b	ε_f'	c
Material: Low-alloy steel, tested at 538 °C															
2¼Cr-1Mo		NT							193	0.115	531	419	−0.047	0.116	−0.407
		NT							175	0.092	467	436	−0.052	0.414	−0.550
Material: Low-alloy steel, tested at 23 °C															
HY-80			849	725	1402			1.23	193.3	0.153	1367	1326	−0.093	0.821	−0.608
HY-130			1105	1015	1547			0.92	193.3	0.110	1573	1548	−0.071	0.857	−0.639
42 CrMo 4	230 HV 0.1	Normalized	740	400			19		190.5	0.115	673	1001	−0.111	46.512	−1.001
	230 HV 0.1	Normalized	740	400			19		190.5	0.097	637	894	−0.094	19.095	−0.936
	405 HV 0.1	QT	1375	1315			7		190.5	0.040	1543	3118	−0.155	0.885	−1.244
	350 HV 0.1	QT	1120	1000			12		190.5	0.065	1234	1435	−0.077	0.462	−0.787

(continued)

(a) NT = normalized and tempered, QT = quenched and tempered, ann. = annealed. Source: Adapted from Ref 12 and other sources

Table 4(b) (continued)

Material	Hardness	Product condition(a)	UTS, MPa	TYS, MPa	σ_f, MPa	RA, %	EL, %	ε_t	E, GPa	n'	K, MPa	σ_f', MPa	b	ε_f'	c
Material: Low-alloy steel, tested at 23 °C (continued)															
	208 HV 0.1	QT	735	680			27		190.5	0.087	807	1036	−0.091	2.251	−0.837
	305 HV 0.1	QT	940	875			15		190.5	0.079	1086	1675	−0.115	37.3559	−1.301
	305 HV 0.1	QT	940	875			15		190.5	0.054	789	1481	−0.112	11.431	−1.020
	350 HV 0.1	QT	1120	1000			7		190.5	0.067	1097	1166	−0.057	0.3519	−0.642
IN 787	188 HRB	Hot-rolled	623	454	1229	76		1.450	206	0.127	1111	926	−0.062	0.842	−0.627
SAE 4340		QT	1243	1181	1913	57		0.840	193	0.117	1634	1655	−0.084	1.169	−0.727
		QT	1174	1105	1636	56		0.83	206	0.131	1639	1655	−0.091	1.078	−0.692
		Prestrained	1174	1105	1636	56		0.83	207	0.118	1464	1127	−0.057	7.196	−0.864
Material: Low-alloy steel (nickel alloy), tested at 23 °C															
EN 25		Tempered	1103	841					207	0.162	1654	1752	−0.119	1.388	−0.734
		Tempered	1780	1172					207	0.118	2392	2294	−0.100	0.731	−0.855
Material: Low-alloy steel, rolled beam, tested at 23 °C															
StE 460		As-received	682	510	574	32		0.661	208	0.161	1181	1124	−0.094	0.1925	−0.437
		As-received	682	510	574	32		0.661	208	0.170	1176	849	−0.070	0.152	−0.415
		As-received	682	510	574	32		0.661	208	0.185	1288	1158	−0.101	0.2005	−0.440
		As-received	682	510	574	32		0.661	208	0.150	1011	962	−0.088	0.5508	−0.563
		As-received	682	510	936	32		0.389	208	0.164	1146	1030	−0.103	0.666	−0.663
StE 690		As-received	825	767		75	24		209	0.070	947	928	−0.046	1.681	−0.769
		As-received	825	767		75	24		209	0.092	1094	1064	−0.072	0.234	−0.623
		As-received	825	767		75	24		209	0.123	1240	1153	−0.093	0.9658	−0.881
Material: Low-alloy steel, sheet, tested at 23 °C															
HSB 55 C		Normalized	667	560	1171	61		0.932	210	0.128	1194	1201	−0.085	0.8638	−0.625
StE 690		As-received (rolled)	872	810	1446		18	0.867	214	0.128	1167	1191	−0.09	0.9113	−0.674
N-A-Xtra 70		As-received	929	833	1446	58		0.867	214	0.102	1252	1282	−0.070	1.185	−0.681
		As-received	929	833	1446	58		0.867	214	0.100	1112	1354	−0.092	2.443	−0.762
		As-received	929	833	1446	58		0.867	214	0.125	1310	1465	−0.105	1.320	−0.719
HSB 77 V		QT	852	745	1327	57		0.834	210	0.088	1145	1194	−0.067	1.176	−0.701
Material: Low-alloy steel, plate, tested at −60 °C															
2 MnCr 7 8		Finish-rolled air cooled		449					210	0.107	1047	1082	−0.079	1.392	−0.743
4 MnMo 7		Controlled-rolling, controlled-cooling		520					210	0.107	1004	1023	−0.072	1.114	−0.668
Material: Low-alloy steel, plate, tested at 21 °C															
2 MnCr 7 8		High-temperature finish-rolled, air-cooled		449					210	0.102	995	978	−0.071	0.731	−0.676
4 MnMo 7		Controlled-rolled, spray-quenching		483					210	0.090	806	833	−0.062	1.176	−0.661
Material: Low-alloy steel, plate, tested at 23 °C															
HT 60		QT (rolled)	657	549	1517	59		0.892	210	0.126	985	941	−0.090	0.589	−0.682
A302B		QT (rim part)	608	495	1313	72		1.27	210	0.134	1145	1175	−0.107	1.207	−0.798
		QT (core part)	603	486	1226	69		1.19	210	0.140	1139	1153	−0.107	1.264	−0.797
StE 690		As-received (rolled)	872	810			18		214	0.098	1048	954	−0.054	0.8429	−0.659
HT 80		QT (rolled)	885	743	1675	57		0.851	210	0.139	1294	1222	−0.088	0.685	−0.641
		QT (rolled)	885	743	1675	57		0.851	210	0.154	1340	1207	−0.095	0.508	−0.619
HY 80		As-received	850	778	1606	66		1.071	206	0.188	1528	1378	−0.113	0.577	−0.601
		As-received	949	821	1660	55		0.807	206	0.167	1440	1233	−0.094	0.393	−0.561
		As-received	944	836	1660	63		0.998	206	0.158	1589	1425	−0.092	10.533	−0.599
BHW 25			614	460	1010			0.800	209	0.168	1196	1108	−0.096	0.4646	−0.543
8 Mn 6	64 HRA	Cast	965	862	1579	57	12	0.85	198	0.125	1256	1087	−0.068	0.312	−0.549
	64 HRA	Cast	869	821	1434	53	13	0.75	198	0.101	1258	1073	−0.055	0.203	−0.541
SPV 50	207 HV 10	Hot-rolled	628	579			40		210	0.111	799	784	−0.066	0.745	−0.577
	207 HV 10	Hot-rolled	628	579			40		210	0.100	789	841	−0.066	1.890	−0.669
	207 HV 10	Hot-rolled	628	579			40		210	0.060	580	568	−0.034	1.975	−0.646
	207 HV 10	Hot-rolled	628	579			40		210	0.065	575	571	−0.038	0.706	−0.567
Van-80	225 HB	As-received	697	580	1222	68		1.15	206	0.158	1436	1339	−0.111	1.777	−0.845
	225 HB	As-received	697	580	1222	68		1.15	206	0.242	2376	925	−0.090	0.062	−0.470
RQC 100			960	910	1090	43		0.561	206	0.107	1313	1227	−0.068	0.765	−0.706
AOS 1122 B	304 HB		1091	1008	1616	61		0.95	207	0.145	1616	2177	−0.129	0.840	−0.688
A 516, Gr 70		Normalized		325	993			0.386	205	0.202	1148	873	−0.098	0.295	−0.486
A 516, Gr 70	165 HV	Hot-rolled							193	0.221	1215	1095	−0.122	0.665	−0.561
A 516, Gr 70		Normalized		325	993			0.386	205	0.215	1238	1128	−0.128	0.605	−0.586
		Normalized		340					205	0.231	1288	973	−0.114	0.308	−0.498
Man-Ten			580	330	950	64		1.03	204	0.193	1151	1101	−0.111	0.495	−0.516
			580	330	950	64		1.03	204	0.216	1255	1119	−0.122	0.724	−0.584
SCMV 4	175 HV 10	NT	585	366		79	28		210	0.125	851	803	−0.073	0.626	−0.583
	175 HV 10	NT	585	366		79	28		210	0.075	677	799	−0.075	4.034	−0.900
	193 HV 10	NT	627	457	1580	81	24		210	0.118	842	736	−0.065	0.266	−0.527
SCMV 2	146 HV 10	NT	480	330		75	33		210	0.171	922	802	−0.092	0.455	−0.543
	146 HV 10	NT	480	330		75	33		210	0.176	930	783	−0.092	0.374	−0.519

(continued)

(a) NT = normalized and tempered, QT = quenched and tempered, ann. = annealed. Source: Adapted from Ref 12 and other sources

Table 4(b) (continued)

Material	Hardness	Product condition(a)	UTS, MPa	TYS, MPa	σ_f, MPa	RA, %	EL, %	ε_t	E, GPa	n'	K, MPa	σ'_f, MPa	b	ε'_f	c
Material: Low-alloy steel, plate, tested at 23 °C (continued)															
SCMV 3	155 HV 10	NT	535	331		73	29		210	0.149	937	863	−0.085	0.584	−0.575
	155 HV 10	NT	535	331		73	29		210	0.138	880	826	−0.080	0.628	−0.585
	164 HV 20	NT	556	393		74	30		210	0.143	896	781	−0.078	0.382	−0.546
A 514	303 HB	NT	939	891	1491	63		0.994	209	0.090	1080	1010	−0.054	0.9734	−0.693
TT StE 32			558	375	995			1.00	209	0.159	983	994	−0.095	0.2436	−0.464
USST-1	256 HB	As-received	808	725	1215	66		1.08	208	0.122	1227	1272	−0.085	11.557	−0.907
	256 HB	As-received	808	725	1215	66		1.08	208	0.131	1221	1031	−0.070	0.737	−0.644
Material: Chromium-molybdenum low-alloy steel, plate, tested at 300 °C															
SCMV 4		NT	553	401	1330	78	19		204.1	0.102	691	675	−0.064	0.782	−0.628
		NT	553	401	1330	78	19		204.1	0.136	855	727	−0.071	0.305	−0.522
SCMV 3		NT	532	282	1147	68	23		210	0.105	774	684	−0.058	0.293	−0.546
		NT	532	282	1147	68	23		210	0.070	670	694	−0.053	1.807	−0.771
Material: Chromium-molybdenum low-alloy steel, plate, tested at 400 °C															
SCMV 4 (2.25Cr-1Mo)		NT	481	299		75	23		187.8	0.153	802	683	−0.082	0.363	−0.539
		NT	481	299		75	23		187.8	0.136	760	596	−0.062	0.175	−0.460
		NT	530	383	1170	76	19		187.8	0.102	683	638	−0.066	0.424	−0.617
		NT	530	383	1170	76	19		187.8	0.092	633	596	−0.058	0.546	−0.638
SCMV 2	146 HV	NT	471	224		73	27		188.8	0.082	688	709	−0.062	1.511	−0.763
		NT	471	224		73	27		188.8	0.075	693	716	−0.058	1.574	−0.779
SCMV 3	155 HV 10	NT	453	234		73	26		195	0.128	758	794	−0.093	1.616	−0.741
		NT	453	234		73	26		195	0.071	572	584	−0.052	1.247	−0.724
		NT	522	266	1193	76	28		195	0.048	549	543	−0.033	0.862	−0.701
		NT	522	266	1193	76	28		195	0.082	693	617	−0.045	0.285	−0.570
Material: Chromium-molybdenum low-alloy steel, plate, tested at 500 °C															
SCMV 4		NT	411	272		82	29		184.8	0.130	650	728	−0.105	2.176	−0.793
		NT	411	272		82	29		184.8	0.106	519	550	−0.082	1.470	−0.755
		NT	466	356	1002	83	23		184.8	0.077	497	473	−0.051	0.320	−0.601
		NT	466	356	1002	83	23		184.8	0.070	450	439	−0.051	0.616	−0.715
SCMV 2	146 HV	NT	416	206		78	26		191.7	0.058	517	560	−0.052	3.442	−0.878
		NT	416	206		78	26		191.7	0.100	612	636	−0.081	1.471	−0.808
SCMV 3	155 HV 10	NT	390	225		81	33		190.2	0.090	549	602	−0.074	3.986	−0.873
		NT	390	225		81	33		190.2	0.089	488	505	−0.068	1.602	−0.775
		NT	438	244	1128	82	29		190.2	0.076	550	556	−0.058	0.903	−0.737
		NT	438	244	1128	82	29		190.2	0.126	684	628	−0.085	0.449	−0.657
		NT	438	244	1128	82	29		190.2	0.126	625	636	−0.105	1.143	−0.829
Material: Chromium-molybdenum low-alloy steel, plate, tested at 550 °C															
$2\frac{1}{4}$Cr-1 Mo		Normalized	336	218		82	33		170	0.215	657	808	−0.197	2.471	−0.907
		Normalized	336	218		82	33		170	0.191	700	882	−0.176	3.558	−0.929
		Normalized	336	218		82	33		170	0.180	740	816	−0.147	1.603	−0.810
Material: Chromium-molybdenum low-alloy steel, plate, tested at 600 °C															
SCMV 4		NT	313	225		90	44		162	0.113	481	480	−0.080	0.968	−0.708
		NT	313	225		90	44		162	0.079	309	327	−0.066	2.109	−0.837
		NT	344	264	950	91	30		162	0.057	327	316	−0.038	0.576	−0.681
		NT	344	264	950	91	30		162	0.027	242	249	−0.025	1.272	−0.817
		NT	344	264	950	91	30		162	0.088	279	293	−0.076	21.637	−1.253
SCMV 2	146 HV 10	NT	285	180		88	48		152.5	0.117	481	495	−0.086	1.364	−0.740
		NT	285	180		88	48		152.5	0.224	654	787	−0.190	1.991	−0.831
SCMV 3	155 HV 10	NT	285	191		90	48		172.4	0.127	516	530	−0.091	1.306	−0.723
		NT	285	191		90	48		172.4	0.152	460	537	−0.130	3.041	−0.870
		NT	294	188	1058	90	40		172.4	0.136	486	466	−0.093	0.729	−0.682
		NT	294	188	1058	90	40		172.4	0.121	386	443	−0.109	2.812	−0.884
		NT	294	188	1058	90	40		172.4	0.110	288	263	−0.077	0.522	−0.726
Material: Low-alloy steel, rail, tested at 23 °C															
AISI 1080		As-received	820	359	965			0.251	206	0.167	1143	1116	−0.105	0.911	−0.636
Material: Low-alloy steel, drum, tested at 350 °C															
19 Mn 5		As-received	630	266		49	22		178	0.040	680	743	−0.044	8.097	−1.086
Material: Low alloy steel, blank, tested at 23 °C															
AISI 4130	25-27 HRC	QT	898	780	1692	67		1.12	221	0.138	1366	1211	−0.071	1.008	−0.652
	39-40 HRC	QT	1429	1360	2085	55		0.79	200.25	0.124	1758	1691	−0.080	0.814	−0.674
AISI 52100	52-53 HRC	QT	2016	1927	2230	12		0.12	207	0.146	3328	2620	−0.093	0.145	−0.560
Material: Low-alloy steel, gear blank, tested at 23 °C															
AISI 8620H	82 HRB	Carburized	1586	1200	1669	3		0.026	195	0.135	2759	2313	−0.095	0.0892	−0.576
	47 HRC	Forged, norm.	1510	1200	2034	42		0.549	198	0.112	2149	3046	−0.141	0.542	−0.783
Material: Low-alloy steel, strip, tested at 23 °C															
A36	140 HB	Hot-rolled	540	351	1173	67		1.100	200	0.214	1219	985	−0.106	0.373	−0.493

(continued)

(a) NT = normalized and tempered, QT = quenched and tempered, ann. = annealed. Source: Adapted from Ref 12 and other sources

Table 4(b) (continued)

Material	Hardness	Product condition(a)	UTS, MPa	TYS, MPa	σ_f, MPa	RA, %	EL, %	ε_t	E, GPa	n'	K, MPa	σ'_f, MPa	b	ε'_f	c
Material: Low-alloy steel, truck frame with few service loads, tested at 23 °C															
AOS 1122 A	289 HB		1008	904	1284	49		0.68	207	0.183	1900	1137	−0.069	1.199	−0.740
Material: Low-alloy steel, forged squares, tested at 23 °C															
C30MB	293 HB	QT	950	820	1445	64	19	1.068	206	0.166	1618	1482	−0.106	2.799	−0.824
41CrM4B	293 HB	QT	930	800	1390	62	19	0.96	207.28	0.143	1340	1271	−0.082	1.248	−0.653
40 Cr Mo 4	293 HB	QT	940	840	1440	64	19	1.035	208.78	0.130	1307	1274	−0.078	1.674	−0.677
Material: Low-alloy steel, shaft, tested at 23 °C															
49 MnVS 3			840	566	1152	19	44	0.380	210.2	0.159	1396	1440	−0.105	0.6025	−0.574
			840	566	1152	19	44	0.380	210.2	0.169	1360	790	−0.055	0.043	−0.332
			840	566	1152	19	44	0.380	210.2	0.193	1776	1056	−0.072	0.067	−0.372
28CrMoNiV 4 9		QT	759	616		66	21		197.4	0.050	748	821	−0.049	3.557	−0.917
30 CrNiMo 8 (DIN 1.6580)		QT	910	700	1168	66	20	0.708	206	0.085	972	1087	−0.070	1.312	−0.705
		QT	910	700	1168	66	20	0.708	206	0.029	636	761	−0.036	0.731	−0.632
		QT	910	700	1168	66	20	0.708	206	0.041	656	636	−0.024	0.190	−0.507
42 CrMo 4		QT	910	700	1168	66	20	0.708	206	0.095	995	1013	−0.064	1.184	−0.673
		QT	1111	998	1525	60	23	0.496	211.4	0.104	1367	1454	−0.075	1.508	−0.716
		QT	1111	998	1525	60	23	0.496	211.4	0.206	2400	1234	−0.076	0.045	−0.383
		QT	1111	998	1525	60	23	0.496	211.4	0.084	1146	1034	−0.044	0.271	−0.519
Material: Low-alloy steel, shaft, tested at 525 °C															
28CrMoNiV 4 9		QT	507	449		74	21		161	0.060	552	553	−0.051	1.137	−0.853
Material: Low-alloy steel, rail head, tested at 23 °C															
ASTM A1	25.8 HRC	Hot-rolled	931	502	1060	16	15	0.174	187.5	0.225	1859	1249	−0.101	0.175	−0.451
Material: Low-alloy steel, bar, tested at 20 °C															
55 Cr 3		Rolled, ann., hardened	1300						210	0.114	2105	3248	−0.181	39.630	−1.566
Material: Low-alloy steel bar, tested at 23 °C															
AISI 4340	44 HRC	QT	1471	1374	1920	38		0.48	193.5	0.118	1890	1880	−0.086	0.706	−0.662
	22-24 HRC	Hot-rolled, ann.	829	635	1201	43		0.57	193.5	0.167	1332	1206	−0.095	0.536	−0.568
Material: Low-alloy steel, bar, tested at 525 °C															
10CrMo 9 10		NT	442	332		78	25		163.4	0.137	724	645	−0.090	0.385	−0.646
Material: Low-alloy steel, round bar, tested at 20 °C															
13 CrMo 44		QT	472	350		76	31		208	0.232	1220	811	−0.102	0.161	−0.428
		QT	514	369		77	30		210	0.105	608	453	−0.038	0.075	−0.385
34 CrNiMo 6		QT	1104	1015		58	16		206	0.088	1330	1217	−0.056	0.269	−0.598
Material: Low-alloy steel, round bar, tested at 350 °C															
13 CrMo 44		QT	523	264		65	24		179	0.075	609	643	−0.065	2.493	−0.882
Material: Low-alloy steel, block, tested at 23 °C															
30 CrNiMo 8		Forged	929	775	1629	64	21		208.5	0.045	755	829	−0.041	8.768	−0.926
		QT	1197	1142	1949	56	15		200	0.042	1193	1555	−0.066	2.042	−0.850
28 NiCrMo 74		Forged	1023	908	1777	63	18		200	0.049	1049	1289	−0.064	0.667	−0.707
Material: Low-alloy steel, tube, tested at 20 °C															
15 Mo 3		As-received	481	335		65	33		209	0.133	794	641	−0.064	0.205	−0.481
Material: Low-alloy steel, tube, tested at 23 °C															
30 CrNiMo 8		QT	1002	845	1954	70	18		208.3	0.016	666	694	−0.015	8.937	−0.901
14 MoV 63		QT	570	390		66	25		210	0.153	986	890	−0.091	0.513	−0.593
Material: Low-alloy steel, tube, tested at 350 °C															
15 Mo 3		As-received	513	197		59	30		179.5	0.070	701	693	−0.054	0.748	−0.751
Material: Low-alloy steel tube, tested at 530 °C															
14 MoV 63		QT	387	248		76	59		170	0.097	676	657	−0.076	1.265	−0.855

(a) NT = normalized and tempered, QT = quenched and tempered, ann. = annealed. Source: Adapted from Ref 12 and other sources

Table 5 Monotonic and fatigue properties for high-alloy steels ($R = -1$, longitudinal direction)

Material	Hardness	Product condition(a)	UTS, MPa	TYS, MPa	σf, MPa	RA, %	EL, %	εt	E, GPa	n'	K, MPa	σf, MPa	b	εf	c
Tested at 22 °C															
Incoloy 800 H		ST							210	0.257	1568	1061	-0.108	0.400	-0.494
AISI 304		ST							210	0.546	6693	5813	-0.324	0.194	-0.416
Tested at 23 °C															
AISI 304		...	650	325	1400	80		1.61	183	0.291	1628	986	-0.117	0.170	-0.399
		...							185	0.291	1675	1008	-0.117	0.171	-0.400
X 10 CrNiTi 18 9		As-received		265					200	0.201	1170	1019	-0.110	0.1325	-0.394
Tested at 23 °C, H2, 10 MPa															
Incoloy 800H		STA	550	240			55		210	0.339	2071	1121	-0.142	0.164	-0.420
Tested at 427 °C															
AISI 304		ST							179	0.435	2795	1942	-0.222	0.1352	-0.394
Tested at 538 °C															
Incoloy 800 H		ST							193	0.242	1455	1295	-0.146	0.4768	-0.570
AISI 304		ST							193	0.226	954	1315	-0.186	1.0389	-0.650
Tested at 593 °C															
Incoloy 800H		ST							171	0.177	1052	1133	-0.136	0.6617	-0.658
AISI 304		ST							171	0.223	797	360	-0.063	0.023	-0.261
Tested at 600 °C															
X 6 Cr Ni 18 11		...							143.2	0.319	1074	677	-0.146	0.234	-0.459
Tested at 871 °C															
Hastelloy X		ST							138	0.291	1350	478	-0.073	0.758	-0.758
Material: Bar, tested at 23 °C															
SUH 660-B	337 HV 30	STA	1158	777		52	23		210	0.125	1543	1574	-0.083	1.109	-0.661
	337 HV 30	STA	1158	777		52	23		210	0.132	1617	1537	-0.080	0.662	-0.606
SUS 304-B	142 HV 10	STQ	611	207		75	66		210	0.455	3331	1470	-0.179	0.161	-0.389
	142 HV 10	STQ	611	207		83	79		210	0.434	3001	1268	-0.160	0.134	-0.366
SUS 316-B	138 HV 10	STQ	587	231		78	68		210	0.388	2755	2508	-0.277	0.758	-0.582
	149 HV 10	STQ	655	228		81	67		210	0.376	2674	1595	-0.163	0.252	-0.433
	138 HV 10	STQ	587	230		78	68		210	0.299	1644	1491	-0.568	0.708	-0.568
	149 HV 10	STQ	665	228		81	67		210	0.336	2081	1999	-0.190	0.928	-0.572
SUS 321-B	163 HV 10	STQ	677	211		69	51		210	0.321	3647	1968	-0.138	0.110	-0.393
	163 HV 10	STQ	677	211		67	51		210	0.469	8384	2046	-0.133	0.046	-0.276
	135 HV 10	STQ	668	182		68	56		210	0.435	6179	1998	-0.144	0.081	-0.341
	135 HV 10	STQ	668	182		68	56		210	0.620	17,743	3005	-0.183	0.061	-0.304
		STQ	516	177		74	70		210	0.375	2264	1969	-0.208	0.652	-0.548
	108 HV 10	STQ	516	177		74	70		210	0.292	1535	1806	-0.193	1.720	-0.659
	122 HV 10	STA	535	177		77	70		210	0.424	3080	3079	-0.253	0.976	-0.593
	122 HV 10	STA	535	177		77	70		210	0.348	2097	1868	-0.194	0.803	-0.570
	122 HV 10	STA	529	214		74	63		210	0.357	2086	1858	-0.199	0.713	-0.558
	122 HV 10	STA	529	214		74	63		210	0.306	1682	1740	-0.186	1.136	-0.608
SUS 347-B	150 HV 10	STQ	615	237		72	52		210	0.319	1967	1005	-0.121	0.127	-0.384
	150 HV 10	STQ	615	237		72	52		210	0.289	1667	1060	-0.123	0.230	-0.436
SUS 403-B	238 HV 10	QT	736	598		70	21		210	0.110	987	962	-0.075	0.798	-0.680
	238 HV 10	QT	736	598		70	21		210	0.128	1056	932	-0.074	0.380	-0.579
SUH 310-B	155 HV 10	STQ	630	271		69	45		210	0.334	2302	1512	-0.153	0.301	-0.465
	155 HV 10	STQ	630	271		69	45		210	0.332	2242	1492	-0.152	0.289	-0.454
SUH 616-B	317 HV 10	STA	1013	795		47	15		210	0.093	1301	1216	-0.062	0.490	-0.664
	317 HV 10	STA	1013	795		47	15		210	0.099	1325	1249	-0.066	0.556	-0.664
H11 Mod (X 40 CrMoV 20 5)	60 HRC	Hot-worked	2576	2030		33			213	0.095	4577	4569	-0.094	0.426	-0.844
Material: Rolled bar, tested at 23 °C															
Remanit 1880		As-received	635	245	1908	79	75	1.563	204	0.331	2397	2032	-0.183	0.3249	-0.441
Material: Bar, tested at 400 °C															
SUS 403-B		QT	585	488		69	15		185.2	0.097	691	668	-0.066	0.701	-0.684
		QT	585	488		69	15		185.2	0.079	641	585	-0.046	0.339	-0.594
SUH 616-B		STA	816	647		49	12		191.8	0.100	1096	1193	-0.087	1.891	-0.840
		STA	816	647		49	12		191.8	0.084	1034	1049	-0.066	1.259	-0.797
Material: Bar, tested at 450 °C															
SUH 660-B		STA	1011	711		49	18		174.95	0.117	1574	1477	-0.075	0.570	-0.642
		STA	1011	711		49	18		174.95	0.120	1638	1559	-0.081	0.665	-0.678
SUS 304-B		STQ	435	106		71	47		170.5	0.514	4497	2528	-0.247	0.325	-0.481
		STQ	435	106		71	47		170.5	0.375	2363	1700	-0.202	0.386	-0.529
		STQ	435	106		71	47		170.5	0.340	2193	1890	-0.212	0.653	-0.627
SUS 316-B		STQ	465	128		70	47		170	0.516	5290	1947	-0.207	0.143	-0.399
		STQ	529	145		69	46		135.3	0.349	2292	2131	-0.214	0.825	-0.615
		STQ	465	128		70	47		170	0.380	2294	1774	-0.192	0.250	-0.504

(continued)

(a) ST = solution treated, STA = solution treated and aged or annealed, STQ = solution treated, quenched, QT = quenched and tempered. Source: Adapted from Ref 12

Table 5 (continued)

Material	Hardness	Product condition(a)	UTS, MPa	TYS, MPa	σf, MPa	RA, %	EL, %	εt	E, GPa	n'	K, MPa	σf, MPa	b	εf	c
Material: Bar, tested at 450 °C (continued)															
SUS 316-B (continued)															
		STQ	465	128		70	47		170	0.354	2090	1546	-0.172	0.178	-0.493
		STQ	529	145		69	46		135.3	0.326	2541	1881	-0.198	0.321	-0.579
SUS 321-B		STQ	416	135		62	39		194	0.404	2329	917	-0.143	0.097	-0.351
		STQ	416	135		62	39		194	0.371	2211	1140	-0.170	0.160	-0.452
		STQ	404	117		66	42		201.25	0.523	4290	1713	-0.212	0.172	-0.404
		STQ	404	117		66	42		201.25	0.417	2917	2763	-0.263	0.844	-0.626
	108 HV 10	STQ	379	108		72	46		168.75	0.398	2223	1219	-0.180	0.227	-0.456
		STA	379	108		72	46		168.75	0.284	1419	1151	-0.163	0.431	-0.563
		STA	367	116		72	40		149.7	0.373	1710	946	-0.163	0.210	-0.440
		STA	367	116		72	40		149.7	0.314	1474	1096	-0.173	0.381	-0.549
		STA	375	148		68	37		171	0.352	1636	1054	-0.169	0.293	-0.482
		STA	375	148		68	37		171	0.334	1676	1344	-0.190	0.510	-0.568
SUS 347-B		STQ	449	157		63	37		172	0.442	3151	1189	-0.168	0.110	-0.378
		STQ	449	157		63	37		172	0.347	2155	1158	-0.155	0.159	-0.439
SUH 310-B		STQ	521	178		60	42		165.65	0.299	1754	1235	-0.144	0.272	-0.468
		STQ	521	178		60	42		165.65	0.246	1627	1130	-0.122	0.251	-0.509
Material: Bar, tested at 500 °C															
SUS 403-B		QT	492	404		76	17		197.25	0.131	715	624	-0.077	0.412	-0.605
		QT	492	404		76	17		197.25	0.097	507	466	-0.056	0.418	-0.585
SUH 616-B		STA	700	538		71	22		185.85	0.132	1151	1103	-0.093	0.468	-0.647
		STA	700	538		71	22		185.85	0.108	910	855	-0.071	0.576	-0.657
Material: Bar, tested at 600 °C															
SUH 660-B		STA	912	713		56	22		162.8	0.174	1842	1217	-0.076	0.093	-0.440
		STA	912	713		56	22		162.8	0.270	3221	1120	-0.080	0.019	-0.287
SUS 304-B		STQ	373	92		73	42		158	0.316	1544	1009	-0.156	0.268	-0.499
		STQ	373	92		73	42		158	0.236	1031	728	-0.118	0.224	-0.499
		STQ	373	92		73	42		158	0.074	437	394	-0.041	0.262	-0.564
SUS 316-B		STQ	405	114		71	48		167	0.206	1104	850	-0.113	0.282	-0.548
		STQ	481	135		70	44		170	0.312	1995	1396	-0.172	0.280	-0.535
		STQ	405	114		71	48		167	0.194	1063	755	-0.099	0.179	-0.513
		STQ	481	135		70	44		170	0.202	1147	925	-0.122	0.345	-0.603
		STQ	405	114		71	48		167	0.146	799	601	-0.078	0.139	-0.534
		STQ	481	135		70	44		170	0.137	778	579	-0.063	0.170	-0.513
SUS 321-B		STQ	349	129		66	39		196.2	0.235	891	587	-0.105	0.169	-0.446
		STQ	349	129		66	39		196.2	0.221	804	486	-0.090	0.104	-0.410
		STQ	353	113		66	42		207.85	0.295	1293	893	-0.147	0.309	-0.509
		STQ	353	113		66	42		207.85	0.194	776	471	-0.075	0.081	-0.394
		STQ	333	97		71	43		167.8	0.270	1116	784	-0.139	0.285	-0.520
		STQ	333	97		71	43		167.8	0.140	630	522	-0.083	0.263	-0.598
		STA	312	101		72	40		151.7	0.301	1171	817	-0.157	0.307	-0.525
		STA	312	101		72	40		151.7	0.148	533	384	-0.069	0.122	-0.476
		STA	319	127		73	40		158	0.271	1041	763	-0.143	0.327	-0.532
		STA	319	127		73	40		158	0.166	610	453	-0.084	0.168	-0.509
SUS 347-B		STQ	409	155		63	36		169.4	0.303	1547	863	-0.131	0.155	-0.441
		STQ	409	155		63	36		196.4	0.224	1028	559	-0.076	0.069	-0.347
		STQ	409	155		63	36		196.4	0.184	820	508	-0.073	0.071	-0.392
SUS 403-B		QT	342	232		89	25		173.5	0.116	453	451	-0.079	0.923	-0.676
		QT	342	232		89	25		173.5	0.070	275	255	-0.040	0.356	-0.573
SUH 310-B		STQ	456	161		54	40		163.55	0.246	1404	945	-0.117	0.223	-0.489
		STQ	456	161		54	40		163.55	0.177	1026	792	-0.101	0.236	-0.575
SUH 616-B		STA	492	356		92	36		166.95	0.136	859	844	-0.096	0.823	-0.694
		STA	492	356		92	36		166.95	0.038	355	356	-0.027	0.888	-0.683
Material: Bar, tested at 700 °C															
SUS 660-B		STA	751	595		55	34		164.65	0.208	1691	1049	-0.095	0.086	-0.435
		STA	751	595		55	34		164.65	0.332	2929	891	-0.108	0.025	-0.315
SUS 304-B		STQ	278	81		77	55		152	0.147	473	382	-0.075	0.255	-0.523
		STQ	278	81		77	55		152	0.212	587	389	-0.094	0.138	-0.4392
		STQ	192	81		82	76		152	0.154	372	286	-0.076	0.182	-0.493
SUS 316-B		STQ	313	113		66	57		159	0.214	859	667	-0.122	0.305	-0.569
		STQ	326	114		67	43		168	0.223	936	735	-0.127	0.359	-0.578
		STQ	313	113		66	57		159	0.152	573	467	-0.085	0.262	-0.563
		STQ	326	114		67	43		168	0.240	893	528	-0.099	0.108	-0.409
		STQ	313	113		66	57		159	0.211	728	503	-0.112	0.173	-0.529
		STQ	326	114		67	43		168	0.151	471	389	-0.082	0.297	-0.550
SUS 321-B		STQ	261	120		80	50		198.9	0.163	496	360	-0.072	0.151	-0.448
		STQ	261	120		80	50		198.9	0.122	368	297	-0.062	0.143	-0.483
		STQ	267	115		72	53		202.05	0.155	508	440	-0.088	0.311	-0.538
		STQ	267	115		72	53		202.05	0.168	523	369	-0.083	0.109	-0.477
		STQ	250	94		32	28		170.9	0.184	609	510	-0.110	0.377	-0.597
		STQ	250	94		32	28		170.9	0.133	421	312	-0.072	0.092	-0.523

(continued)

(a) ST = solution treated, STA = solution treated and aged or annealed, STQ = solution treated, quenched, QT = quenched and tempered. Source: Adapted from Ref 12

Table 5 (continued)

Material	Hardness	Product condition(a)	UTS, MPa	TYS, MPa	σ_f, MPa	RA, %	EL, %	ε_t	E, GPa	n'	K, MPa	σ_f', MPa	b	ε_f'	c
Material: Bar, tested at 700 °C (continued)															
SUS 321-B (continued)															
		STA	247	109		59	46		168.5	0.210	665	577	–0.130	0.439	–0.614
		STA	247	109		59	46		168.5	0.148	423	306	–0.079	0.106	–0.528
		STA	255	129		55	48		174	0.202	636	545	–0.122	0.466	–0.606
		STA	255	129		55	48		174	0.123	355	300	–0.074	0.252	–0.596
SUS 347-B		STQ	319	160		61	48		168.75	0.151	593	440	–0.068	0.135	–0.447
		STQ	319	160		61	48		168.75	0.189	704	428	–0.073	0.070	–0.384
		STQ	319	160		61	48		168.75	0.196	676	412	–0.083	0.073	–0.414
SUH 310-B		STQ	357	156		58	44		164	0.171	700	527	–0.081	0.194	–0.478
		STQ	357	156		58	44		164	0.199	617	386	–0.081	0.095	–0.408
Material: Bar, tested at 800 °C															
SUS 304-B		STQ	192	73		82	76		160	0.159	365	283	–0.077	0.210	–0.488
		STQ	192	73		82	76		160	0.063	146	141	–0.043	0.451	–0.653
SUS 316-B		STQ							123.6	0.200	567	540	–0.141	0.800	–0.705
		STQ							123.6	0.056	187	182	–0.040	0.869	–0.756
SUS 347-B		STQ	216	127		53	49		169.35	0.149	423	354	–0.086	0.352	–0.606
		STQ	216	127		61	48		169.35	0.159	290	237	–0.096	0.272	–0.604
Material: Blank, tested at 23 °C															
Inconel X	34-35 HRC	Annealed	1215	704	1512	20		0.22	214.055	0.128	2130	1990	–0.105	0.177	–0.599
AM 350	50-52 HRC	Cold drawn	1906	1864	2265	20		0.23	179.53	0.101	3072	2947	–0.110	0.176	–0.767
		Hot-rolled, pickled, annealed by supplier	1319	442	2341	52		0.74	193.34	0.388	11,825	2777	–0.137	0.014	–0.287
AISI 304	34-36 HRC	Cold drawn	953	746	2037	69		1.16	172.625	0.155	2313	2067	–0.112	0.301	–0.649
AISI ELC	82-84 HRB	Hot rolled and annealed	746	255	1920	74		1.37	186.435	0.309	4634	2377	–0.152	0.068	–0.428
AISI 310	75-81 HRB	Hot-rolled, pickled, annealed by supplier	642	220	1360	63		1.01	193.34	0.271	2267	1646	–0.154	0.302	–0.568
Material: Cast cylinders, tested at 23 °C															
Inconel 713C	37 HRC	ST	932	787			2		208.8	0.121	1899	1264	–0.062	0.032	–0.500
	36 HRC	ST	1001	780			10		208.5	0.099	1576	1140	–0.053	0.035	–0.492
Material: Plate, tested at 21 °C															
Type 316	71-77 HRB	ST							205	0.154	691	660	–0.079	0.341	–0.453
Material: Plate, tested at 23 °C															
AISI 304		...	601	280		46			192	0.419	2807	1936	–0.202	0.412	–0.483
SUS 316-HP	149 HV 10	Hot-rolled	606	257	1830	79	57		210	0.298	2000	1314	–0.132	0.249	–0.445
Material: Plate, tested at 400 °C															
SUS 316-HP		Hot-rolled	513	184	1020	62	40		171.5	0.328	2204	975	–0.115	0.096	–0.368
		Hot-rolled	513	184	990	62	40		171.5	0.271	1789	973	–0.108	0.113	–0.407
		Hot-rolled	513	184	1020	62	40		171.5	0.215	1504	1190	–0.133	0.341	–0.622
Material: Plate, tested at 500 °C															
SUS 316-HP		Hot-rolled	493	178	990	62	41		178	0.224	1406	891	–0.101	0.127	–0.446
		Hot-rolled	493	178	990	62	41		178	0.175	1155	696	–0.065	0.078	–0.413
		Hot-rolled	493	178	990	62	41		178	0.189	1409	811	–0.082	0.062	–0.454
Material: Plate, tested at 600 °C															
AISI 304		...							149	0.272	1022	635	–0.121	0.177	–0.446
		...							149	0.249	836	576	–0.138	0.226	–0.557
		...							149	0.248	861	530	–0.112	0.141	–0.452
		...							149	0.282	1080	625	–0.119	0.145	–0.422
SUS 316-HP		Hot-rolled	449	164	860	62	41		168.5	0.223	1230	681	–0.087	0.072	–0.393
		Hot-rolled	449	164	630	62	41		168.5	0.228	1262	615	–0.079	0.041	–0.343
		Hot-rolled	449	164	860	62	41		168.5	0.067	991	869	–0.042	0.141	–0.624
Material: Plate, tested at 700 °C															
SUS 316-HP		Hot-rolled	319	155	630	68	61		175.5	0.221	818	586	–0.117	0.205	–0.521
		Hot-rolled	319	155	630	68	61		175.5	0.177	550	395	–0.086	0.148	–0.479
		Hot-rolled	319	155	630	68	61		175.5	0.147	395	355	–0.098	0.512	–0.676
Material: Turbine wheel forging, tested at 23 °C															
Waspaloy A	22 HRC	STA		545					210	0.154	1814	1340	–0.075	0.1495	–0.497
	35 HRC	STA	1091	787			7		210	0.128	1931	1510	–0.076	0.147	–0.591
Waspaloy	36 HRC	STA		815					210	0.180	2725	1862	–0.090	0.112	–0.487

(a) ST = solution treated, STA = solution treated and aged or annealed, STQ = solution treated, quenched, QT = quenched and tempered. Source: Adapted from Ref 12

Table 6 Monotonic and fatigue properties for aluminum and titanium alloys ($R = -1$, longitudinal direction)

Material	Hardness	Product condition(a)	UTS, MPa	TYS, MPa	σ_f, MPa	RA, %	EL, %	ε_f	E, GPa	n'	K, MPa	σ_f', MPa	b	ε_f'	c
Material: Aluminum alloy, cold-rolled sheet, tested at 23 °C															
Al 99.5		CR-SR	73	19			43		70	0.265	255	95	–0.088	0.022	–0.328
		CR-SR	73	19			43		70	0.337	453	117	–0.109	0.017	–0.315
Material: Aluminum alloy, bar stocks, tested at 23 °C															
1100 Al	~26 HRB	As-received	110	97		87.6		2.09	69.05	0.159	184	159	–0.092	0.467	–0.613
Material: Aluminum alloy, extruded round bar, tested at 23 °C															
Alloy 6082		STA		290					64	0.051	397	611	–0.099	1.085	–0.857
Material: Aluminum alloy, bar, tested at 23 °C															
Al Mg Si 0.8		Cold age-hard	260	157		58	32		66.7	0.060	392	481	–0.084	1.095	–0.867
		Warm age-hard	329	294		22.5	15		67.2	0.050	451	542	–0.075	0.700	–0.816
Al Mg Si 1		Cold age-hard	348	250		25.5	16		74.5	0.052	454	445	–0.054	0.116	–0.641
		Warm age-hard	383	348		45.5	14		74.55	0.046	478	554	–0.068	5.375	–1.208
Al Cu Mg 2		Cold age-hard	446	275		24	20.5		74.1	0.062	648	687	–0.074	0.514	–0.830
		Soft-annealed	245	88		38	18		74.6	0.201	453	314	–0.091	0.162	–0.452
Material: Aluminum alloy, bar stocks, tested at 23 °C															
5456-H311	56 HRB	As-received	400	235	566	34.6		0.42	69.05	0.084	636	702	–0.102	0.200	–0.655
2014-T6	81-84 HRB	As-received	511	463	628	25.0		0.29	69.05	0.072	704	776	–0.091	0.269	–0.742
Material: Aluminum alloy, extruded profile, tested at 23 °C															
Al Mg Si 0.8		STA							66	0.052	454	444	–0.063	0.005	–0.427
Al Mg 1 Si Cu		STA							75	0.120	364	381	–0.102	0.039	–0.452
		STA							80			2172	–0.224	0.004	–0.249
Material: Aluminum alloy, plate, tested at 23 °C															
Al Mg 4.5 Mn		…	363	298			13		71.5	0.125	693	654	–0.089	0.450	–0.755
		…	363	298			13		71.5	0.07	535	629	–0.086	0.329	–0.684
		…	363	298			13		71.5	0.080	561	723	–0.108	1.613	–0.871
		Prestrained	363	298			13		71.5	0.220	1103	1576	–0.201	1.303	–0.879
		Prestrained	363	298			13		71.5	0.137	777	546	–0.070	0.034	–0.436
7075-T61		…							70	0.074	852	1231	–0.122	0.263	–0.806
7075-T7351		…	462	382			8.4		71	0.094	695	989	–0.140	6.812	–1.198
Material: Aluminum alloy, rod, tested at 23 °C															
7075-T6		STA	580	470	801	33		0.410	71.12	0.088	913	886	–0.076	0.446	–0.759
2024-T4		STA	476	304	684	35		0.430	70.43	0.098	808	764	–0.075	0.334	–0.649
Material: Aluminum alloy, sheet, tested at 23 °C															
Al Mg 4.5 Mn		Rolled	348	226		22.5	17		85	0.056	450	906	–0.148	52.058	–1.441
Al Cu Mg 2		…	490	396		16	16		73.3	0.039	557	782	–0.082	0.197	–0.644
		…	490	476		16	16		73.3	0.074	669	891	–0.103	4.206	–1.056
2024-T3		…	486	378			17.3		74.5	0.040	590	1044	–0.114	1.765	–0.927
			486	378			17.3		74.5	0.040	590	3148	–0.247	0.069	–0.634
			486	378			17.3		82	0.040	590	1044	–0.114	1.765	–0.927
Material: Aluminum alloy, stock, tested at 23 °C															
7075-T65									72	0.032	646	1294	–0.125	10.202	–1.231
									74	0.319	4958	1999	–0.198	0.045	–0.599
Material: Titanium alloy, tested at 23 °C															
Titanium		Annealed	520	275			20		120	0.530	7383	1671	–0.197	0.063	–0.378
Ti-6Al-4V		Annealed	845	805	738	22		0.252	121.5	0.095	1288	1293	–0.088	0.260	–0.721
			1034	1006			14.5		120.35	0.127	1702	2207	–0.126	2.802	–0.860
			1034	1006			14.5		120.35	0.096	1413	1619	–0.080	0.605	–0.670
Material: Titanium alloy, centerless ground, tested at 23 °C															
Ti-6Al-4V	40-41 HRC	ST	1236	1188	1719	41.0		0.53	117.385	0.108	1938	1797	–0.085	0.396	–0.684
Material: Titanium alloy, rod, tested at 23 °C															
Ti-8Al-1Mo-1V		Heat-treated	1022	1008	1650	48		0.66	117.4	0.124	1685	1825	–0.095	1.829	–0.765
Material: Titanium alloy, forged, extruded, swaged to 20.8 mm diam, tested at 23 °C															
Ti-0.4Mn		ST & worked	434	303		69	42		100	0.253	1719	4986	–0.339	312.263	–1.555
Ti-5Mn		ST & worked	814	627		50			100	0.161	2023	1708	–0.101	0.402	–0.648
		ST & worked	855	676		38			100	0.192	2659	1862	–0.114	0.150	–0.590
		ST & worked	834	662		51	29		100	0.303	4151	2848	–0.170	0.283	–0.559
		ST & worked	883	772		50			100	0.204	2675	2189	–0.128	0.380	–0.631
Ti-8Mn		ST & worked	945	883		51	30		100	0.099	1686	3298	–0.182	1939	–1.958
Ti-10Mn		ST & worked	986	979		51			100	0.059	1448	1578	–0.066	3.809	–1.098

(a) CR = cold rolled, SR = stress relieved, STA = solution treated and aged. Source: Adapted from Ref 12

Table 7 Monotonic and fatigue properties for cast and welded metals ($R = -1$, longitudinal direction)

Material	Hardness	Product condition(a)	UTS, MPa	TYS, MPa	σ_f, MPa	RA, %	EL, %	ε_t	E, GPa	n'	K, MPa	σ_f', MPa	b	ε_f'	c
Material: Welds at 23 °C															
A36		Sim. HAZ	667	535	918	52		0.745	189	0.146	1029	887	–0.081	0.288	–0.531
A514		Sim. HAZ	1409	1181	2251	53		0.750	209	0.096	1703	1714	–0.072	0.624	–0.679
A516	225 HV	Sim. HAZ							194	0.244	1667	1356	–0.135	0.277	–0.505
X 2 Cr Ni 18 9		Weld metal	600	340			35		216	0.143	939	914	–0.089	0.734	–0.612
		Weld metal							204	0.151	9120	875	–0.101	0.774	–0.668
HI-Form 60	2276 MPa HV	Sim. HAZ	532	436	877	61		0.93	219.6	0.087	601	611	–0.057	0.549	–0.592
	2472 MPa HV	Sim. HAZ	546	441	856	62		0.91	224.4	0.014	372	1023	–0.095	0.031	–0.335
	1825 MPa HV	Sim. HAZ	463	376	677	54		0.78	206.5	0.021	363	451	–0.032	0.433	–0.561
	1511 MPa HV	Sim. HAZ	474	371	608	49		0.68	209.2	0.109	621	614	–0.066	0.005	–0.151
	1952 MPa HV	Sim. HAZ	501	376	746	54		0.77	214	0.406	5925	1405	–0.125	0.029	–0.307
MS4361		Sim. HAZ	625	414	918	61		0.745	198.2	0.169	1010	775	–0.078	0.207	–0.463
E70T-1		Weld metal	760	463	1464	48		0.928	212.3	0.148	1313	1133	–0.076	0.419	–0.526
Material: Cast irons, tested at 23 °C															
GG 25	137 HB			215	260			0.008	90	0.172	234	123	–0.058	0.041	–0.440
	212 HB			215	260			0.008	90	0.150	468	353	–0.115	0.037	–0.582
	212 HB			215	260			0.008	90	0.139	455	342	–0.105	0.033	–0.581
	174 HB			215	260			0.008	90	0.095	180	241	–0.115	0.008	–0.360
GG 35	195 bulk	Sand cast	438	345				0.015	134	0.237	1549	696	–0.114	0.016	–0.383
GG 40	215 bulk	Sand cast	570	420				0.031	140	0.153	1012	645	–0.078	0.037	–0.457
GGG 40			437	297			24		165	0.068	582	586	–0.058	0.619	–0.816
GGG 60			632	382			7		158	0.076	750	720	–0.058	0.759	–0.789
GTS 55			600	389			5		155	0.128	911	711	–0.073	0.149	–0.575
GGG 60		As-cast, Y-block		553	716			0.041	164	0.070	770	1010	–0.091	0.608	–0.776
Material: Cast plate, tested at 23 °C															
GSMnNi 6 3			501	312			26		203	0.151	925	677	–0.084	0.1151	–0.526
			501	312			26		203	0.150	944	494	–0.066	0.195	–0.698
			501	312			26		203	0.109	695	318	–0.018	0.001	–0.107
GS-22Mo4		Annealed	497	299		60	28		210	0.166	921	739	–0.095	0.304	–0.590
G-X22CrMoV12 1		Annealed	824	648		45	18		199	0.121	1222	1124	–0.086	0.451	–0.688
		Annealed	824	648		45	18		199	0.204	2148	1096	–0.101	0.040	–0.506
Material: Cast plate, tested at 530 °C															
GS-22Mo4		Annealed	322	176		75	36		175	0.113	476	493	–0.093	1.355	–0.824
G-X22CrMoV12 1		Annealed	494	349		57	24		170	0.055	514	471	–0.035	0.513	–0.773
		Annealed	494	349		57	24		170	0.098	650	654	–0.081	0.666	–0.759
Material: Cast plate, tested at 600 °C															
X 2 CrNi 18 9		Weld metal							165	0.115	514	488	–0.072	0.792	–0.657
		Weld metal							165	0.142	600	605	–0.102	1.155	–0.728
		Weld metal							165	0.140	550	566	–0.111	1.405	–0.812
		Weld metal							165	0.096	411	429	–0.084	1.540	–0.874
Material: Cast blocks, tested at 23 °C															
0030	137 HB	Annealed	496	303	751	46		0.62	207	0.169	887	704	–0.091	0.2119	–0.518
		Annealed	544	317	620	30		0.36	209	0.210	1396	1021	–0.112	0.252	–0.550
GS 34 Mn 5	206 HB	Annealed	703	544	751	31		0.37	211	0.170	1551	1245	–0.102	0.2906	–0.614
GS 34 Mn 5		QT	758	558	861	30		0.36	210	0.167	1429	1270	–0.112	0.7028	–0.715
0050A	192 HB	Annealed	785	413	866	19		0.21	209	0.256	2128	1374	–0.128	0.220	–0.524
0050A		Annealed		434	923	16		0.17	209	0.177	1496	1391	–0.117	0.479	–0.629
8630	305 HB	QT	1144	985	1268	29		0.35	207	0.182	2363	2099	–0.131	0.4207	–0.696
8630		QT	1178	999	1254	28		0.33	214	0.119	1654	1763	–0.103	0.687	–0.759
Material: Gear casting, tested at 23 °C															
MPS Cast 8630	254 HB	Annealed	787	711	842	16		0.17	200.25	0.057	876	1178	–0.117	0.348	–0.920
	254 HB	Annealed	787	711	842	16		0.17	200.25	0.100	1065	979	–0.111	3.130	–1.548
Material, Cast sticks, tested at 900 °C															
IN 738 LC		STA							130	0.079	908	1181	–0.113	15.049	–1.344
									130	0.112	994	878	–0.088	0.330	–0.782
Material: Structural part, tested at 23 °C															
GK-AlSi7Mg		Warm age hardened							71.7	0.083	398	388	–0.075	0.114	–0.614
Material: Weld plate, tested at 23 °C															
SPV 50		HAZ							210	0.153	1051	959	–0.081	0.461	–0.516
		HAZ							210	0.109	752	912	–0.079	0.7136	–0.425
SWT32		Weld metal							209	0.094	697	697	–0.060	0.765	–0.612
SPV 50		Weld metal	642						210	0.106	888	949	–0.074	1.836	–0.697
		Weld metal	642						210	0.064	597	711	–0.051	0.004	–0.194
Material: Butt-welded joint of A36, tested at 23 °C															
E60S-3		Weld metal	711	581	987	45		0.59	189	0.122	848	800	–0.067	0.644	–0.552
		Weld metal	580	409	1015	61		0.933	189	0.139	895	841	–0.079	0.6352	–0.567
Material: Butt-welded joint of A514, tested at 23 °C															
E110		Weld metal	1036	836	2210	58		0.857	209	0.113	1406	1628	–0.103	1.509	–0.805
		Weld metal	911	760	1664	59		0.899	209	0.112	1238	1095	–0.057	0.630	–0.589

(a) Sim. HAZ = simulated heat-affected zone, STA = solution treated and aged, QT = quenched and tempered. Source: Adapted from Ref 12

Summary of Stress-Intensity Factors

Alan Liu, Rockwell International (retired)

LINEAR ELASTIC FRACTURE MECHANICS (LEFM), which constitutes the majority of practical fracture mechanics applications, can be based on either energy calculations or stress-intensity calculations (Ref 1). The methods are related and provide identical results in predicting fracture loads of structures containing sharp flaws of known size and location. However, the stress-intensity approach is more generally used because it deals directly with crack-tip stresses and strains, which are more commonly used in engineering.

The stress-intensity concept is based on the parameter K, which quantifies the stresses at a crack tip. Using the conventional theory of elasticity, it is possible to calculate the stress field at the tip of a crack in an arbitrary body with an arbitrary crack under arbitrary loading. Using the coordinate system shown in Fig. 1, the crack-tip stresses for Mode I loading are:

$$\sigma_{ij} = \frac{K_I}{\sqrt{\pi 2r}} f_{ij}(\theta) + C_1 r^0 + C_2 r^{1/2} + \dots \qquad (\text{Eq 1})$$

If r is very small, the first term of the solution is very large (infinite for $r = 0$); therefore, the other terms can be neglected. Because all cracking and fracturing take place at or very near the crack tip (where $r \geq 0$), it is justifiable to use only the first term of the solution to describe the stress field in the area of interest. For the stress in the y-direction along the plane $\theta = 0$, the function $f_{yy}(\theta) = 1$, so that:

$$\sigma_{yy} = \frac{K_I}{\sqrt{2\pi r}} \text{ for } \theta = 0 \qquad (\text{Eq 2})$$

where the subscript I denotes Mode I loading.

Equation 2 shows that the crack-tip stress in the σ_{yy} direction depends on the distance r from the crack tip. A similar relation applies to other directions, or if the entire stress field is taken into account. As a general solution, Eq 2 applies to any geometry under Mode I loading. A specific solution is shown in Fig. 2 for a through-thickness crack in a plate section.

The stress near the crack tip (Eq 2) is also directly proportional to the applied stress (σ), as long as stresses are in the elastic regime. Therefore, under elastic conditions:

$$\sigma(\text{applied}) \propto \sigma_{yy} = K_I / \sqrt{2\pi r} \qquad (\text{Eq 3})$$

which means that a stress intensity, K_I, can be defined in terms of the applied stress and a distance r near the crack tip. The precise relation between the applied stress and K_I depends on geometry, as described in this article for various part configurations and crack geometries. This forms the basis for defining stress-intensity factors (K_I) for fracture mechanics.

The discussions remain valid even if there is some plastic deformation in the region near the crack tip. In reality, a material exhibits some plastic deformation near the crack tip when stresses exceed the yield strength (σ_{YS}) for small values of r (as r approaches zero in Eq 2). However, it can be shown that the size of the so-called plastic zone at the crack tip (Fig. 3) is determined by the value of K (Ref 2). Consequently, for equal values of K, the plastic zones are equal, and the stresses and strains acting on the boundary of the plastic zone are equal. This being the case, equal behavior will take place inside the plastic zones (similitude for equal K). Thus, if the material exhibits a fracture for a certain value of K, it will always exhibit fracture at that value of K.

However, those arguments require that one condition be satisfied: The plastic zone must be so small that its size is determined fully by K and by K only. This will be the case if the plastic zone does not extend beyond a value of r, at which the first term in Eq 1 is still much larger than all other terms; otherwise, the constants C_1, C_2, and so on in Eq 1 will become significant. In general, this will not occur until the value of K_I/σ_{YS} exceeds about 2 (this also depends on geometry). If it does occur, LEFM is no longer valid, and elastic plastic fracture mechanics (EPFM) must be used, although the use of LEFM can be stretched further with simple approximations (Ref 2).

LEFM Geometry Factors

This article summarizes and describes some Mode I stress-intensity factors (K) for various crack geometries commonly found in structural components. A great majority of all practical fracture cases are Mode I problems.

Combined-mode loading is more difficult to deal with, but if the modes are in phase (and remain proportional), the crack in a very early stage of development will turn into a direction in which it experiences only Mode I, unless it is prevented from doing so due to geometrical confinement. (When the modes are independent and/or out of phase, combined-mode loading presents a problem for which no accepted solution is available.) In view of this, fracture mechanics is generally confined to Mode I.

Equation 2 is a general solution for Mode I crack problems, such that the parameter K_I ac-

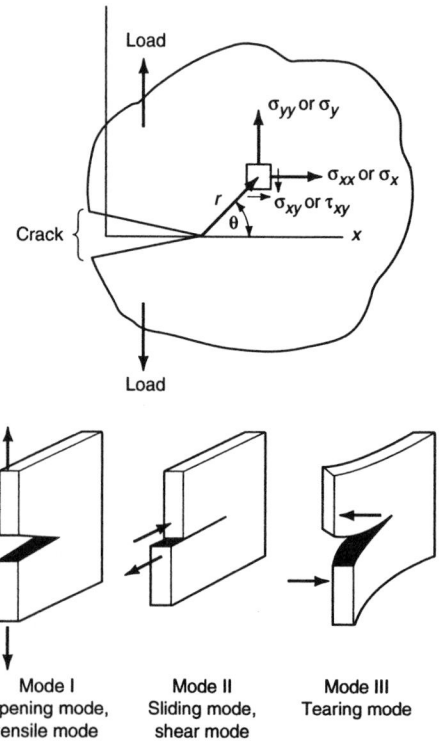

Fig. 1 Arbitrary body and coordinate systems under Mode I loading

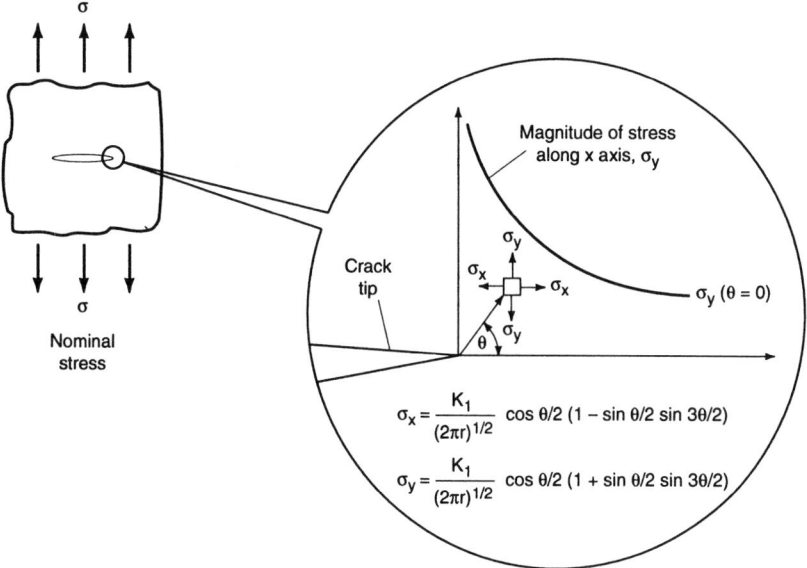

Fig. 2 Distribution of stresses near the tip of a through-thickness crack in a plate. Source: Ref 1

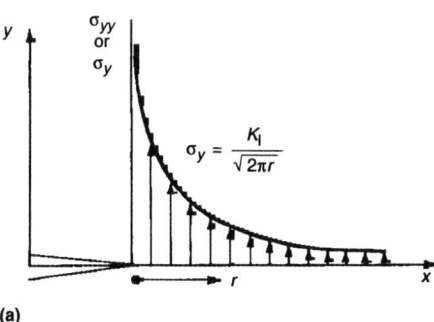

(a)

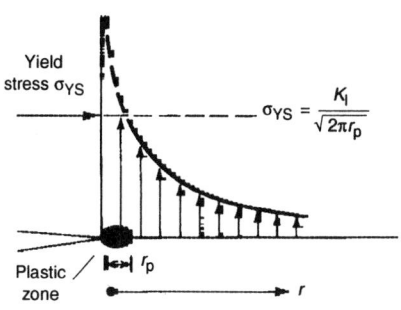

(b)

Fig. 3 Crack-tip stress distribution under (a) elastic conditions and (b) elastic-plastic conditions

counts for all effects of loading and geometry and thus defines the crack-tip stress field in its entirety. However, K_I in Eq 2 is an undefined quantity in terms of a general arbitrary case; it does not define the crack-tip stress field parameter (K) in terms of the applied load and the particular geometry of the crack and part configuration. Therefore, the practical definition of K depends on the relation between the crack-tip stress (σ_{yy}) in terms of the applied stress (Eq 3). This relation between applied stress (σ) and crack-tip stress (σ_{yy}) is governed by the particular geometry of a crack in a part.

For a given structural configuration and applied stress, the definition of K can be determined by a number of analytical methods. For purposes of illustration and a simple introduction, however, first consider a plate of width W containing a through-thickness crack of length $2a$ (Fig. 4a). As long as the stresses are elastic, the stress at any point is proportional to the applied stress ($\sigma_{yy} \propto \sigma$). It must be expected that the crack-tip stress will also depend on the crack size. Because σ_{yy} depends on $1/\sqrt{r}$ according to Eq 2, it is inevitable that it depends on $\sqrt{a}$; otherwise, the dimensions would be wrong. Hence, a simple argument shows that:

$$\sigma_{yy} \propto \frac{\sqrt{a}}{\sqrt{2\pi r}} \tag{Eq 4}$$

This proportional relation can be used to define an explicit relation between the applied stress (σ) and the crack-tip stress (σ_{yy}) such that:

$$\sigma_{yy} = \frac{\beta\sigma\sqrt{\pi a}}{\sqrt{2\pi r}} \tag{Eq 5}$$

where β is a dimensionless constant. The crack-tip stresses will be higher when W is smaller. Thus, β must depend on W. It is known that β must be

dimensionless, yet β cannot be dimensionless a depend on W at the same time, unless β depends W/a or a/W, that is, $\beta = f(a/W)$. Comparison of Eq and 5, then, shows that:

$$K_I = \beta\left(\frac{a}{W}\right)\sigma\sqrt{\pi a} \tag{Eq 6}$$

where the first term β is expressed as a function of a/W. If the crack-tip stress is affected by other geometric parameters—for example, if a crack emanates from a hole, the crack-tip stress will depend on the size of the hole—the only effect on the stress-intensity factor (and the crack-tip stresses) will be in β. Consequently, β will be a function of all geometric factors affecting the crack-tip stress: $\beta = \beta(a/W, a/L, a/D)$. Crack-tip stresses are always given by Eq 2; the value of K in Eq 2 is always given by Eq 6. All effects of geometry are reflected in one geometric parameter, β.

The geometric factor or function (termed β elsewhere in this Volume, as in Eq 6) has been calculated and compiled for many generic geometries in various handbooks (Ref 3-6) and other references listed at the end of this article. The general expression for stress intensity (Eq 6) is re-expressed in some cases as $K = \sigma Y\sqrt{\pi a}$, where Y is the function $\beta(a/W)$ in Eq 6. In this article, the symbol β is used as a term for specific geometric factors as defined in specific cases in this article. In such cases, the term Y represents the complete geometric function of $\beta(a/W)$ in Eq 6. In general, the expression for stress intensity is defined by the geometric factor [Y, or $\beta(a/W)$] that depends on a given structural configuration and crack geometry. The value of K can be determined by a number of analytical methods (e.g., finite element, boundary integral equation, weight function, boundary collocation, etc.). This article will not get into the technicality of these methods. Only the results (mainly extracted from

open literature) are presented here. However, it is significant to point out that each K value can be determined for a specific crack size in an explicit structural (or test specimen) configuration. In this case, the dimensionless factor (β or Y) would be defined when K is normalized by σ and $\sqrt{(\pi a)}$. After a series of K values for various crack sizes across the entire crack plane are determined, a close-form equation can then be obtained by fitting a curve through all the geometric factors.

This article is only intended to be a summary of some Mode I stress-intensity factors for crack geometries commonly found in structural parts. The following sections consider

- A crack in a sheet (or plate)
- A crack originated at a circular hole in a sheet (or plate)
- A pin-loaded lug
- A round bar (which can be a pin or a bolt)
- A hollow cylinder (i.e., a tube)

The crack may be a through-the-thickness crack or a part-through crack (commonly known as a surface crack, or corner crack). The part may be subjected to uniaxial tension, or bending, or it may be loaded by a pair of concentrated forces or the combination of the pin and the reaction loads (e.g., an attachment lug). Many of these stress-intensity solutions have become available only recently and have not been incorporated into any existing handbooks. Although very important in the design and analysis of air transport structures, stress-intensity solutions for a metal sheet (or plate) attached with metal stringers (or manufactured with integral stiffeners) are not included here. Readers are referred to the "Selected

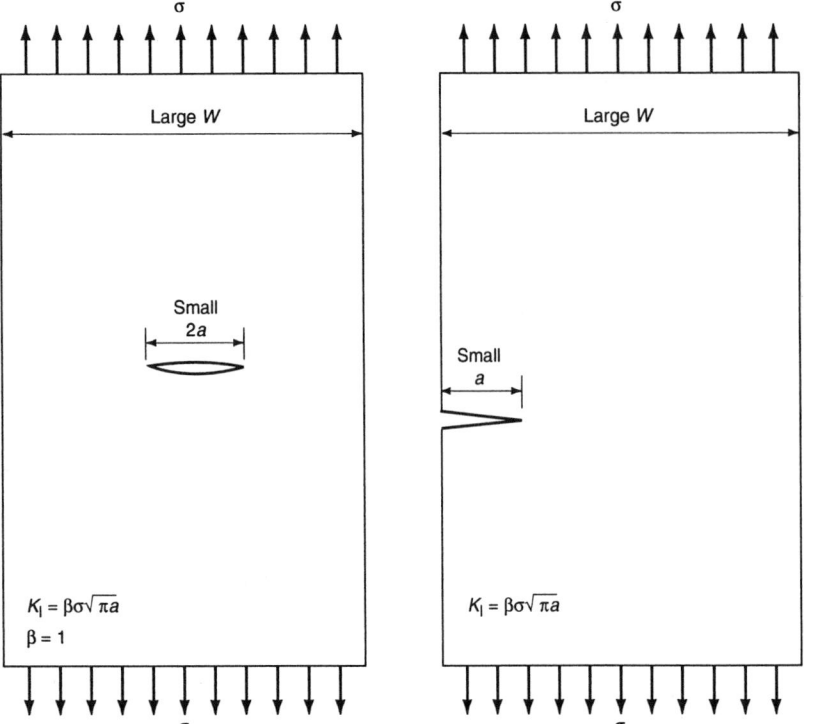

Fig. 4 Crack-tip stress-intensity (K_l) in terms of applied stress (σ) for (a) a center-cracked plate and (b) an edge-cracked plate under uniform tension. The crack shown in (a) is defined as $2a$ and the crack shown in (b) is defined as a, because of the convention that any crack with two tips is defined as $2a$ and any crack with one tip is defined as a. There is no objection to defining the crack in (a) as a, but all quantitative expressions will then differ by a factor of $\sqrt{2}$.

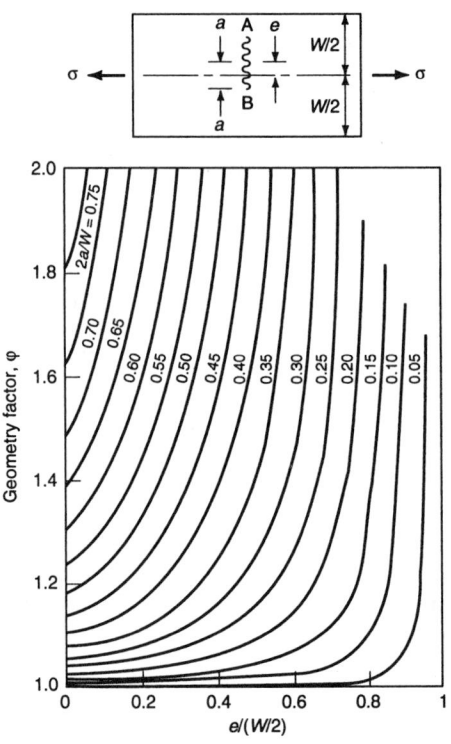

Fig. 6 Stress-intensity factors for an eccentrically cracked plate loaded in tension

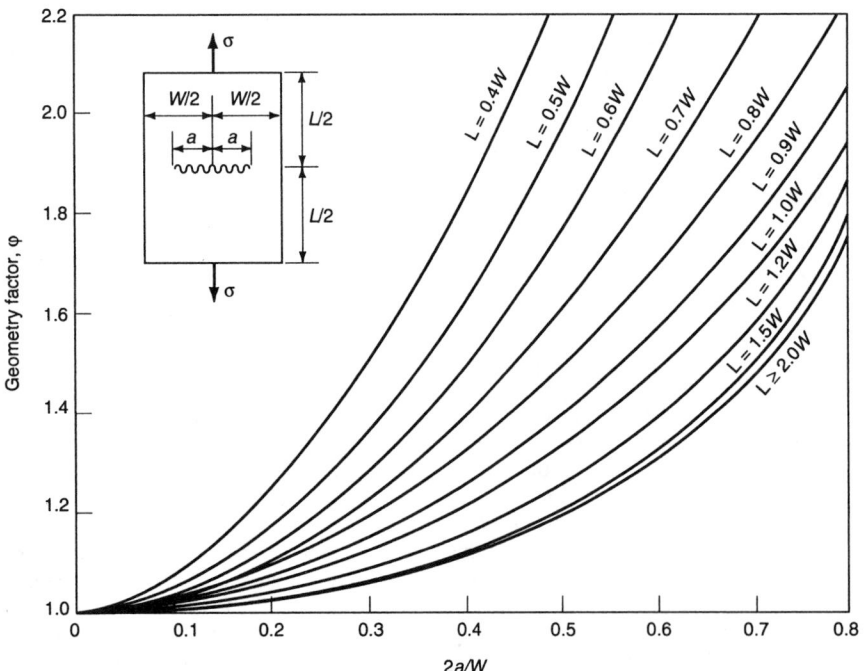

Fig. 5 Stress-intensity factors for a through-the-thickness crack loaded in tension, where $K = \sigma\varphi\sqrt{\pi a}$

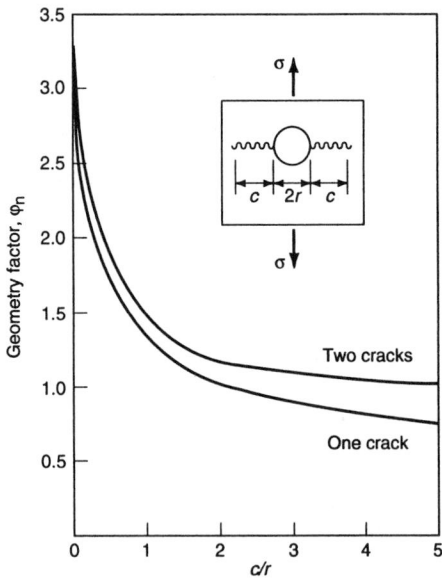

Fig. 7 Stress-intensity factors for crack(s) coming out from a circular hole

References" section at the end of this article. A bibliography of stress intensity factors for mechanical fasteners is also contained in the article "Fatigue of Mechanically Fastened Joints" in this Volume.

Compounding of Geometric Factors. When β is a function of more than one variable, it is desirable to separate the β function into a series of dimensionless parametric functions, where

each segment represents an explicit boundary condition. Therefore Eq 6 can be rewritten as:

$$K = [\sigma\sqrt{(\pi a)}] \, \Pi\alpha \qquad \text{(Eq 7a)}$$

where $\Pi\alpha$ is the product of a series of dimensionless parametric functions, or factors, accounting for the influence of the part and the crack geometries and loading condition. In the absence of any geometric

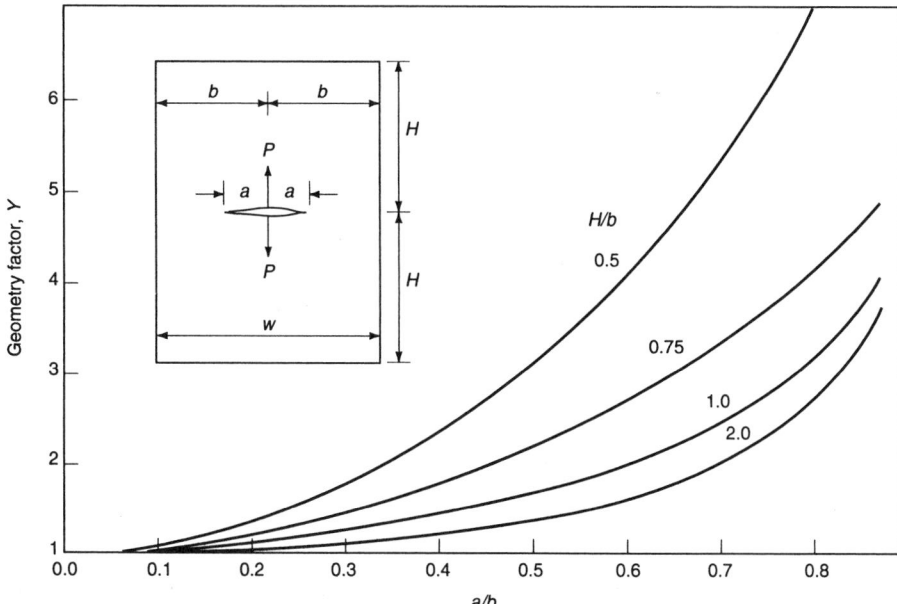

Fig. 8 Stress-intensity factors for a central crack in a rectangular plate with opposing forces at the center of the crack, from the collocation solution by Newman (Ref 15), where $K = PY/\sqrt{\pi a}$, where P is the force (load) per unit thickness

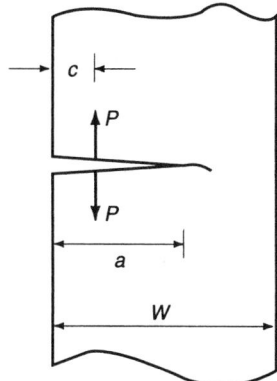

Fig. 9 An edge crack loaded by a pair of point forces

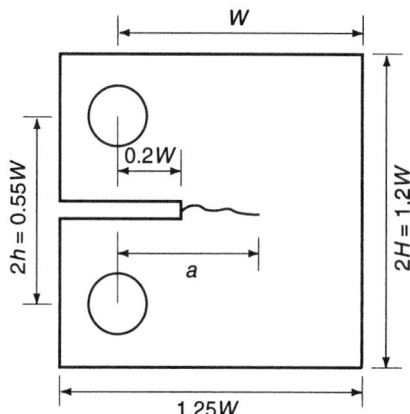

Fig. 10 The compact specimen configuration. Pin hole diameter is $0.25W$.

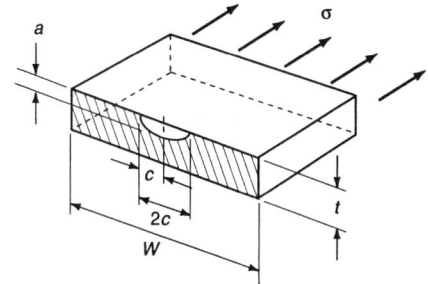

Fig. 11 Configuration of a semi-elliptical surface crack. This figure shows the C-tip is at either end of the major axis (along the surface). The A-tip is at the maximum depth of the minor axis. In this article C-tip(s) are always on the surface, A-tip is always at the maximum depth whether c is greater than a, or vice versa. This definition also is applicable to the corner crack(s) at a circular hole configuration.

influence, for example a through-the-thickness crack in an infinitely wide sheet under uniform farfield tension, $\Pi\alpha$ approaches unity ($\beta = 1$) and Eq 7(a) reduces to the basic stress-intensity expression:

$$K = \sigma\sqrt{(\pi a)} \qquad \text{(Eq 7b)}$$

where σ is the applied stress and a is the half-length of the through-the-thickness crack (Fig. 4). As before, K is a physical quantity (not a factor) with dimensional units of ksi$\sqrt{\text{in.}}$ or MPa$\sqrt{\text{m}}$. In this article the term *stress intensity factor* is loosely defined; it might mean K, or a geometric factor (such as α in Eq 7a or β or Y, as discussed earlier).

The advantage of separating geometric factors into several individual segments is obvious. The fracture mechanics analysis can build around a compounded equation to suit the crack model under consideration by combining several known solutions. The method of compounding is quite simple, but complete coverage is beyond the scope of this article. Several articles that describe the compounding technique are listed in the "Selected References" section at the end of this article.

Through-the-Thickness Crack in a Plate

Through-the-thickness cracks may be located in the middle of a plate (in which case they are called center cracks, Fig. 4a); at the edge of a plate (edge cracks, Fig. 4b); or at the edge of a hole inside a plate.

Uniform Farfield Loading

There are many stress-intensity factor solutions for the center crack subjected to farfield uniform

tension stress. Among them, the Isida solution (Ref 7, 8) is regarded as being the most accurate by the fracture mechanics community. It accounts for the boundary effect (the free edges that define the width and length of the panel). As derived from Eq 7(b), which is for a crack in an infinite sheet, the stress-intensity expression can be written as:

$$K = [\sigma\sqrt{(\pi a)}]\varphi \qquad \text{(Eq 8)}$$

where σ is gross cross-sectional area stress (i.e., ignoring the crack). The geometric factor φ is a dimensionless function of the ratio of crack length to panel width and the ratio of panel length to panel width, as given in Fig. 5. For a panel having its length greater than two times the width, the effect of panel length on K vanishes. According to Feddersen (Ref 9), the finite-width correction factors (i.e., the curve labeled as $L \geq 2W$ in Fig. 5) can be represented by a secant function (sec). That is:

$$\varphi_w = \sqrt{\sec{(\pi a/W)}} \qquad \text{(Eq 9a)}$$

Here the secant function, in radians, is commonly known as the width correction factor for the center crack panel configuration. If the crack is located off the centerline of the plate, Eq 9(a) is replaced by the eccentricity correction factors given in Fig. 6. According to Newman (Ref 10), this set of curves can be approximated by the following equations:

$$\varphi_A = \sqrt{\sec{[(\pi a/2b_1) \cdot (1 - 0.22(e/b)^3]}} \qquad \text{(Eq 9b)}$$

$$\varphi_B = \sqrt{\sec{[(\pi a/2b_1) \cdot (1 - 0.37\,(e/b)^{0.5} + 0.1\,(a/b_1)^{15}]}}$$
$$\text{(Eq 9c)}$$

where the subscripts "A" and "B" refer to the two crack tips indicated in the figure, e is the distance

from the center of the crack to the centerline of the specimen, $b = W/2$, and $b_1 = b - e$ (the distance from the center of the crack to the nearest edge of the specimen). When $e = 0$, either Eq 9(b) or 9(c) is reduced to Eq 9(a).

Edge Crack. For a limiting case in which the crack is right at the edge of the plate (edge crack), the φ factor is given by (Ref 11) as:

$$\varphi_e = \sec{\beta}[(\tan\beta)/\beta]^{1/2} \cdot [0.752 + 2.02\,(a/W) + 0.37\,(1 - \sin\beta)^3] \qquad \text{(Eq 9d)}$$

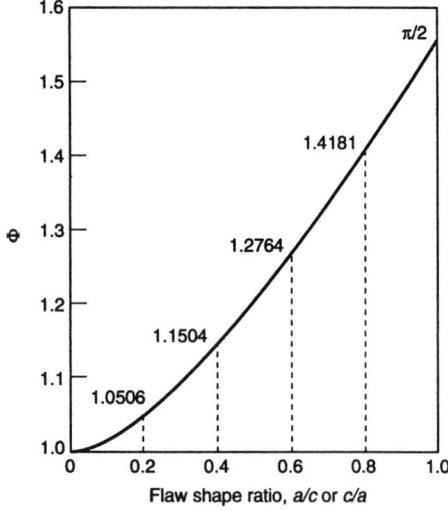

Fig. 12 Elliptical integral Φ

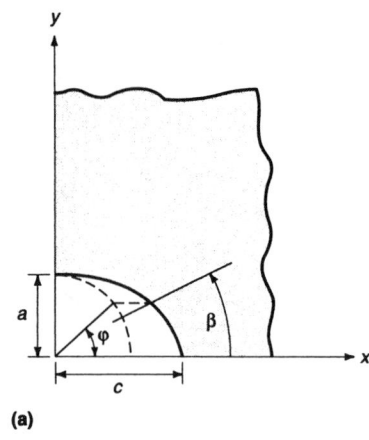

Fig. 13 Definition of φ and β for an elliptical crack. (a) $a/c \leq 1$. (b) $a/c \geq 1$.

where a is the total crack length measured from the edge of the plate across the width W, and β is a specific geometric factor for this case such that $\beta = \pi a/(2W)$. The subscript "e" stands for the edge crack, $\varphi_e = 1.122$ at $a/W = 0$.

Crack Emanating from a Hole. In the case where a crack is originated at the edge of a circular hole inside an infinitely wide sheet,

$$K = [\sigma\sqrt{(\pi c)}]\, \varphi_n \qquad (\text{Eq } 10)$$

where σ is the gross cross-sectional area stress (i.e., again ignoring the hole and the crack), c is the crack length measured from the edge of the hole (not from the center of the hole), and φ_n is the Bowie solution for cracks coming out from a circular hole (Ref 11 and 12). According to Ref 13, the Bowie solution (see Fig. 7) can be fitted by the following equations:

$$\varphi_1 = 0.707 - 0.18 \cdot \lambda + 6.55 \cdot \lambda^2$$
$$- 10.54 \cdot \lambda^3 + 6.85 \cdot \lambda^4 \qquad (\text{Eq } 11a)$$

and

$$\varphi_2 = 1 - 0.15 \cdot \lambda + 3.46 \cdot \lambda^2$$
$$- 4.47 \cdot \lambda^3 + 3.52 \cdot \lambda^4 \qquad (\text{Eq } 11b)$$

where $\lambda = (1 + c/r)^{-1}$ and r is the radius of the hole. The subscripts 1 and 2 stand for a single crack, and two symmetric cracks, respectively. Alternatively, the Bowie solution can be approximated by the following equation (Ref 14):

$$\varphi_n = F_1 \cdot (F_2 + c/r)^{-1} + F_3 \qquad (\text{Eq } 11c)$$

where $F_1 = 0.8733$, $F_2 = 0.3245$, and $F_3 = 0.6762$ for a single crack and $F_1 = 0.6865$, $F_2 = 0.2772$, and $F_3 = 0.9439$ for double symmetrical cracks.

To include the finite width effect:

$$K = [\sigma\sqrt{(\pi c)}] \cdot \varphi_n \cdot f_n \qquad (\text{Eq } 11d)$$

where

$$f_1 = \sqrt{\sec[(\pi/2) \cdot (2r + a)/(W - a)]}$$
$$\times \sqrt{\sec[(\pi/2) \cdot (2r/W)]} \qquad (\text{Eq } 11e)$$

and

$$f_2 = \sqrt{\sec[(\pi/2) \cdot (r + a)/(W/2)]}$$
$$\times \sqrt{\sec[(\pi/2) \cdot (2r/W)]} \qquad (\text{Eq } 11f)$$

where $n = 1$ for a single crack, $n = 2$ for two symmetric cracks, and the hole is located in the center of the plate.

Point Loading of a Center Crack

A crack of length $2a$, subjected to forces per unit thickness P, acting at the center of the crack surfaces, is located centrally in a rectangular plate of height $2H$ and width $2b$ (Fig. 8). The expression for the stress-intensity factor in this case is $K = PY/\sqrt{\pi a}$, where Y is the geometric factor, as shown in Fig. 8 (Ref 15).

Point Loading of Edge Crack

In case the forces are applied at a point on the surface of an edge crack (Fig. 9), the stress-intensity factor is defined as $K = 2P \cdot Y/\sqrt{\pi a}$, where the geometric factor Y is defined by the Tada solution (Ref 3) as:

$$Y = \alpha_1 - \alpha_2 + \alpha_3 \cdot \alpha_4 \qquad (\text{Eq } 12a)$$

where

$$\alpha_1 = 3.52\,(1 - \lambda)/(1 - \beta)^{3/2} \qquad (\text{Eq } 12b)$$

$$\alpha_2 = (4.35 - 5.28\lambda)/(1 - \beta)^{1/2} \qquad (\text{Eq } 12c)$$

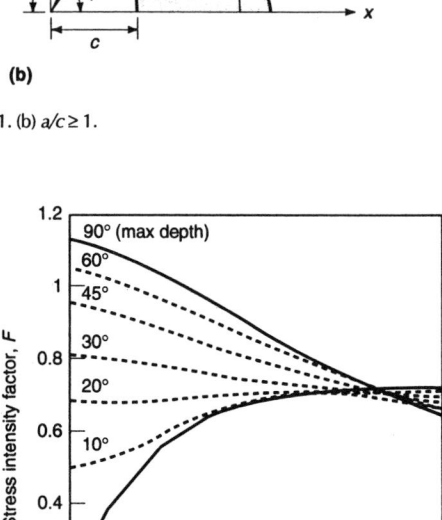

Fig. 14 Variation of stress-intensity factors for a shallow crack in a semi-infinite solid, where $\sigma F = K/\sqrt{\pi a}$ and $F = F_s/\sqrt{Q}$, according to Eq 17(a)

$$\alpha_3 = [(1.3 - 0.3\lambda^{3/2})/(1 - \lambda^2)^{1/2}]$$
$$+ 0.83 - 1.76\lambda \qquad (\text{Eq } 12d)$$

$$\alpha_4 = 1 - (1 - \lambda)\beta \qquad (\text{Eq } 12e)$$

where $\lambda = c/a$ and $\beta = a/b$. When the load is applied right at the edge of the plate (i.e., $\lambda = 0$), Eq 12(a) reduces to:

$$Y = 3.52/(1 - \beta)^{3/2} - 4.35/(1 - \beta)^{1/2} + 2.13(1 - \beta) \qquad (\text{Eq } 12f)$$

The Compact Specimen

Although it is not a common configuration to be considered for structural analysis, the compact specimen (Fig. 10) is specified in ASTM E 399 and E 647 as the standard specimen geometry for generating material K_{Ic} and fatigue crack growth rate data. Therefore, the ASTM-recommended stress-intensity equation for the compact speci-

Table 1 Correction factors ($K/\sigma\sqrt{\pi a/Q}$) for stress intensity at shallow surface cracks under tension

2c/W	a/c	a/t				
		0.0	0.20	0.50	0.80	1.0
At the C-tip: tensile loading						
0.0	0.20	0.5622	0.6110	0.7802	1.1155	1.4436
0.0	0.40	0.6856	0.7817	0.9402	1.1583	1.3383
0.0	1.00	1.1365	1.1595	1.2328	1.3772	1.5145
0.1	0.20	0.5685	0.6133	0.7900	1.1477	1.5014
0.1	0.40	0.6974	0.7824	0.9456	1.2008	0.4256
0.1	1.00	1.1291	1.1544	1.2389	1.3892	1.5273
0.4	0.20	0.5849	0.6265	0.8438	1.3154	1.7999
0.4	0.40	0.7278	0.8029	1.0127	1.4012	1.7739
0.4	1.00	1.1366	1.1969	1.3475	1.5539	1.7238
0.6	0.20	0.5939	0.6415	0.9045	1.5056	2.1422
0.6	0.40	0.7385	0.8351	1.1106	1.6159	2.1036
0.6	1.00	1.1720	1.2855	1.5215	1.8229	2.0621
0.8	0.20	0.6155	0.6739	1.0240	1.8964	2.8650
0.8	0.40	0.7778	0.9036	1.3151	2.1102	2.9068
0.8	1.00	1.2630	1.4957	1.9284	2.4905	2.9440
1.0	0.20	0.6565	0.7237	1.2056	2.6060	4.2705
1.0	0.40	0.8375	1.0093	1.6395	2.9652	4.3596
1.0	1.00	1.3956	1.8446	2.6292	3.6964	4.5865
At the A-tip: tensile loading						
0.0	0.20	1.1120	1.1445	1.4504	1.7620	1.9729
0.0	0.40	1.0900	1.0945	1.2409	1.3672	1.4404
0.0	1.00	1.0400	1.0400	1.0672	1.0883	1.0800
0.1	0.20	1.1120	1.1452	1.4595	1.7744	1.9847
0.1	0.40	1.0900	1.0950	1.2442	1.3699	1.4409
0.1	1.00	1.0400	1.0260	1.0579	1.0846	1.0820
0.4	0.20	1.1120	1.1577	1.5126	1.8662	2.1012
0.4	0.40	1.0900	1.1140	1.2915	1.4254	1.4912
0.4	1.00	1.0400	1.0525	1.1046	1.1093	1.0863
0.6	0.20	1.1120	1.1764	1.5742	1.9849	2.2659
0.6	0.40	1.0900	1.1442	1.3617	1.5117	1.5761
0.6	1.00	1.0400	1.1023	1.1816	1.1623	1.0955
0.8	0.20	1.1120	1.2047	1.6720	2.2010	2.5895
0.8	0.40	1.0900	1.1885	1.4825	1.6849	1.7727
0.8	1.00	1.0400	1.1685	1.3089	1.2767	1.1638
1.0	0.20	1.1120	1.2426	1.8071	2.5259	3.0993
1.0	0.40	1.0900	1.2500	1.6564	1.9534	2.0947
1.0	1.00	1.0400	1.2613	1.4890	1.4558	1.3010

Note: These values are built into the NASA/FLAGRO program (Ref 20). Source: Ref 19

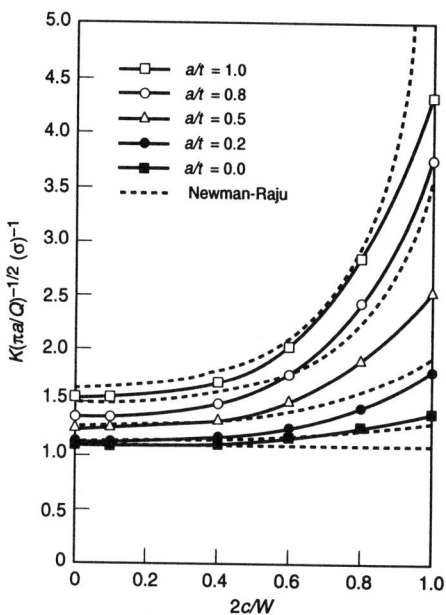

Fig. 17 New and old solutions for C-tip surface crack (Fig. 11) with a/c = 1.0 in tension. Source: Ref 19

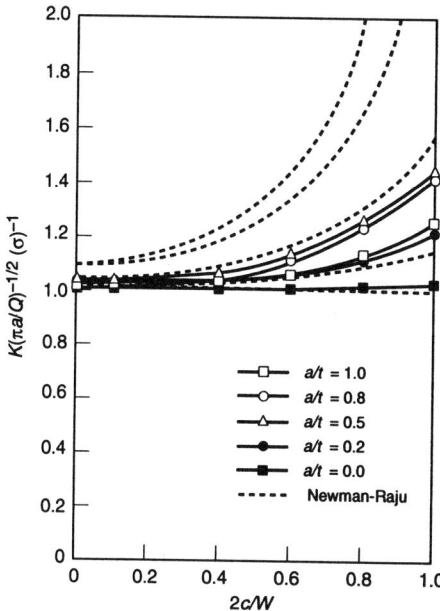

Fig. 18 New and old solutions for A-tip surface crack (Fig. 11) with a/c = 1.0 in tension. Source: Ref 19

men is included here. The equation given in the *1995 Annual Book of ASTM Standards*, Volume 03.01, is:

$$K = (P/(t\sqrt{W})) \cdot F(\alpha_1) \cdot F(\alpha_2) \qquad \text{(Eq 13)}$$

where

$$F(\alpha_1) = 0.886 + 4.64\alpha - 13.32\alpha^2$$
$$+ 14.72\alpha^3 - 5.6\alpha^4 \qquad \text{(Eq 13a)}$$

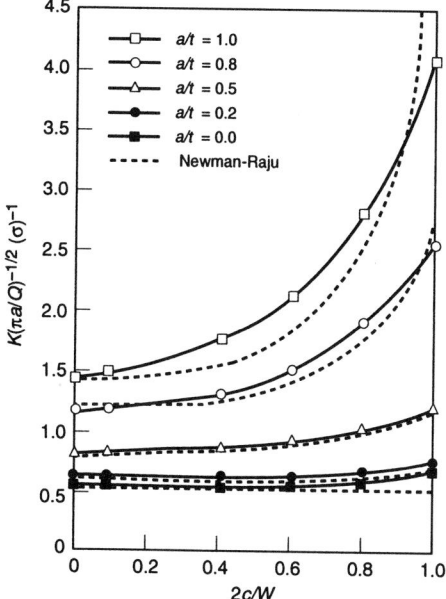

Fig. 15 New and old solutions for C-tip surface crack (Fig. 11) with a/c = 0.2 in tension. Source: Ref 19

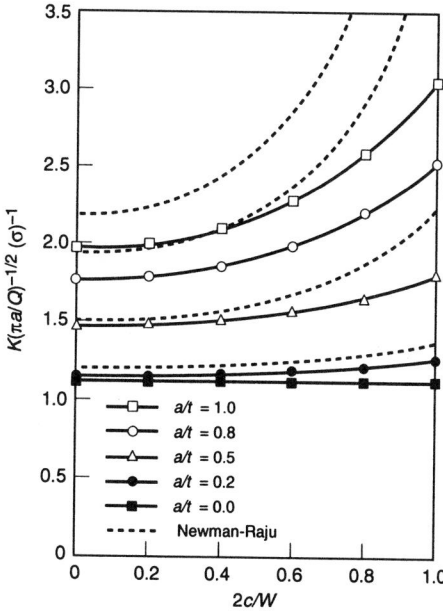

Fig. 16 New and old solutions for A-tip surface crack (Fig. 11) with a/c = 0.2 in tension. Source: Ref 19

Table 2 Correction factors ($K/\sigma\sqrt{\pi a/Q}$) for stress intensity at shallow surface cracks in bending

2c/W	a/c	a/t				
		0.0	0.20	0.50	0.80	1.0
At the C-tip: bending loading						
0.0	0.20	0.5622	0.5772	0.6464	0.7431	0.8230
0.0	0.40	0.6856	0.7301	0.7694	0.7358	0.6729
0.0	1.00	1.1365	1.0778	1.0184	0.9716	0.9474
0.1	0.20	0.5685	0.5809	0.6524	0.7646	0.8624
0.1	0.40	0.6974	0.7315	0.7856	0.8008	0.7895
0.1	1.00	1.1291	1.0740	1.0114	0.9652	0.9435
0.4	0.20	0.5849	0.5981	0.6934	0.8654	1.0249
0.4	0.40	0.7278	0.7519	0.8327	0.9312	1.0068
0.4	1.00	1.1366	1.1079	1.0634	1.0358	1.0268
0.6	0.20	0.5939	0.6158	0.7438	0.9704	1.1802
0.6	0.40	0.7385	0.7816	0.8906	1.0215	1.1211
0.6	1.00	1.1720	1.1769	1.1759	1.1820	1.1900
0.8	0.20	0.6155	0.6446	0.8320	1.1794	1.5113
0.8	0.40	0.7778	0.8386	1.0150	1.2791	1.5073
0.8	1.00	1.2630	1.3633	1.4785	1.5360	1.5431
1.0	0.20	0.6565	0.6848	0.9593	1.5053	2.0518
1.0	0.40	0.8375	0.9232	1.2285	1.7607	2.2637
1.0	1.00	1.3956	1.6821	2.0140	2.1482	2.1446
At the A-tip: bending loading						
0.0	0.20	1.1120	0.8825	0.6793	0.3063	−0.0497
0.0	0.40	1.0900	0.8292	0.5291	0.1070	−0.2489
0.0	1.00	1.0400	0.7411	0.3348	−0.1149	−0.4396
0.1	0.20	1.1120	0.8727	0.6697	0.3071	−0.0348
0.1	0.40	1.0900	0.8243	0.5170	0.1047	−0.2336
0.1	1.00	1.0400	0.7398	0.3322	−0.1172	−0.4408
0.4	0.20	1.1120	0.8683	0.6794	0.3439	0.0291
0.4	0.40	1.0900	0.8330	0.5270	0.1257	−0.1989
0.4	1.00	1.0400	0.7602	0.3572	−0.1080	−0.4543
0.6	0.20	1.1120	0.8904	0.7248	0.4033	0.0915
0.6	0.40	1.0900	0.8625	0.5803	0.1678	−0.1874
0.6	1.00	1.0400	0.7982	0.4072	−0.0856	−0.4750
0.8	0.20	1.1120	0.9191	0.7925	0.5102	0.2254
0.8	0.40	1.0900	0.8987	0.6619	0.2524	−0.1300
0.8	1.00	1.0400	0.8556	0.4981	−0.0329	−0.4960
1.0	0.20	1.1120	0.9545	0.8827	0.6666	0.4351
1.0	0.40	1.0900	0.9417	0.7723	0.3810	−0.0250
1.0	1.00	1.0400	0.9323	0.6312	0.0505	−0.5249

Note: These values are built into the NASA/FLAGRO program (Ref 20). Source: Ref 19

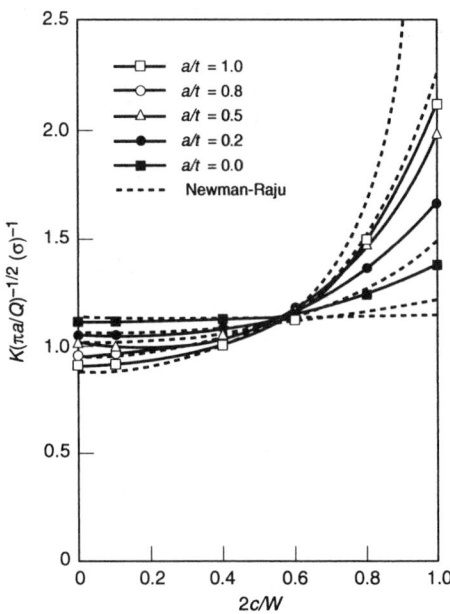

Fig. 21 New and old solutions for C-tip surface crack (Fig. 11) with a/c = 1.0 in bending. Source: Ref 19

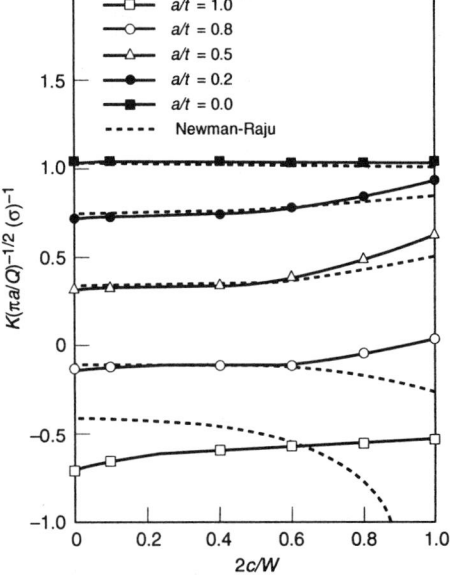

Fig. 22 New and old solutions for A-tip surface crack (Fig. 11) with a/c = 1.0 in bending. Source: Ref 19

and

$$F(\alpha_2) = (2 + \alpha)/(1 - \alpha)^{3/2} \qquad \text{(Eq 13b)}$$

where $\alpha = a/W$ and where again P is the force per unit thickness. Equation 13 is accurate within 0.5% over the range of a/W from 0.2 to 1. As shown in Fig. 10, the height of the specimen (H), the length of the chevron notch, and the loading pin location (h), are functions of the specimen width (W). The ASTM-recommended dimension for the thickness B is as follows:

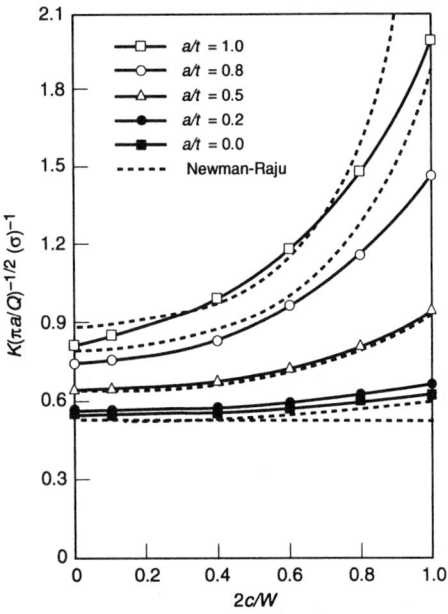

Fig. 19 New and old solutions for C-tip surface crack (Fig. 11) with a/c = 0.2 in bending. Source: Ref 19

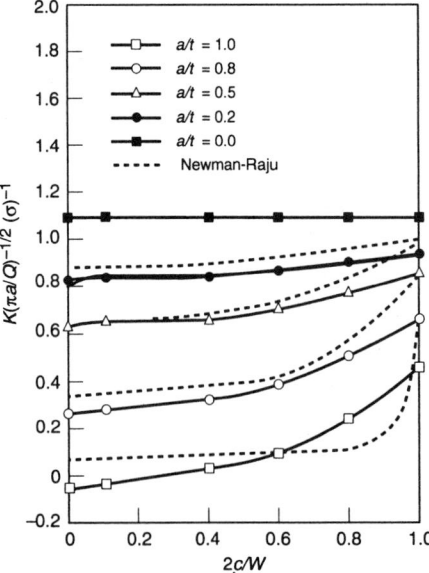

Fig. 20 New and old solutions for A-tip surface crack (Fig. 11) with a/c = 0.2 in bending. Source: Ref 19

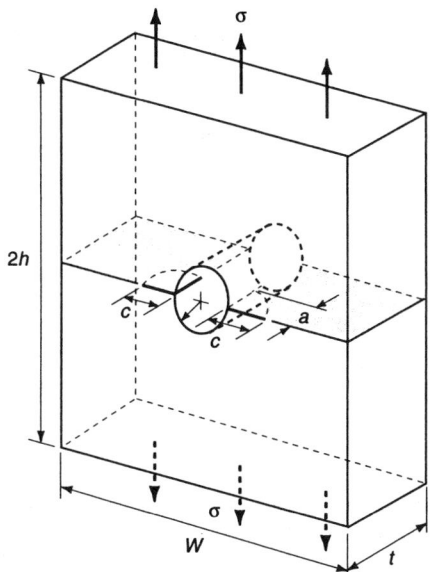

Fig. 23 Corner cracks at an open hole under uniform tension

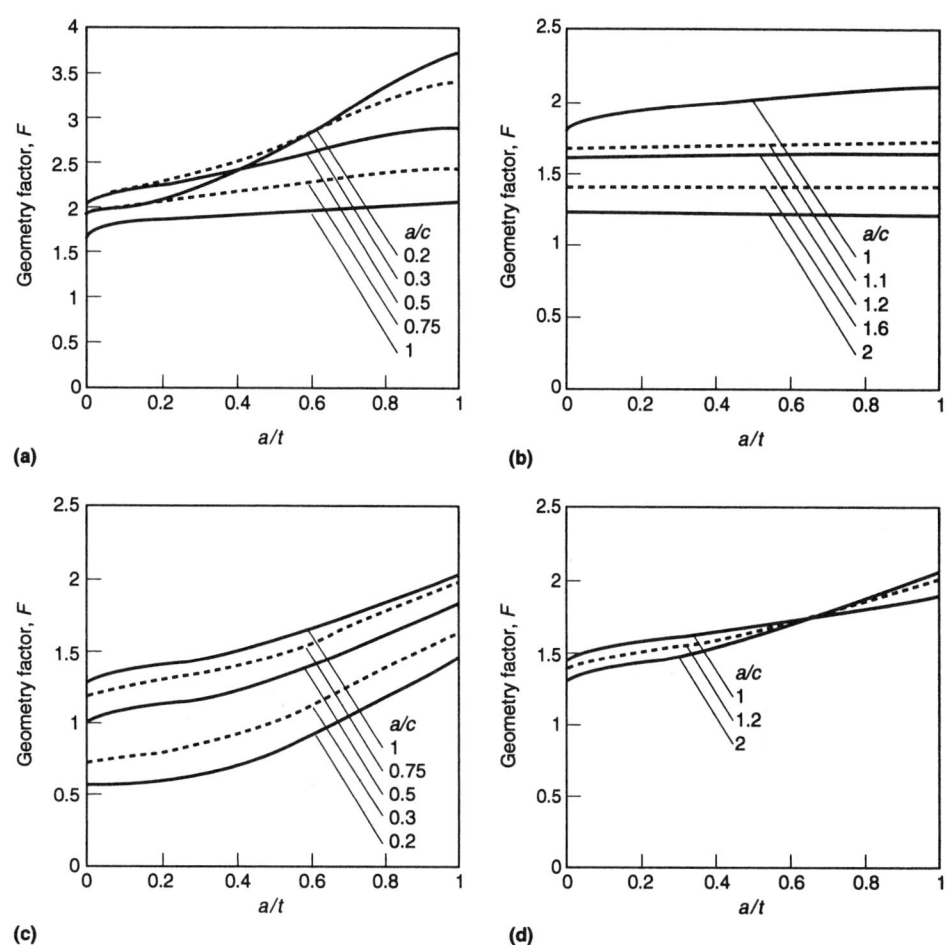

Fig. 24 Stress-intensity factors for a corner crack at an open hole under uniform tension, $r = 3.175$ mm (0.125 in.). (a) $a/c \leq 1$, A-tip surface crack. (b) $a/c \geq 1$, A-tip surface crack. (c) $a/c \leq 1$, C-tip surface crack. (d) $a/c \geq 1$, C-tip surface crack

- For K_{Ic} tests, the preferred thickness is $W/2$. Alternatively, $2 \leq W/B \leq 4$ (with no change in other proportions) is allowed.
- For fatigue crack growth rate tests, B can be in the range of $W/20$ to $W/4$.

For test data reported in the earlier literature, an older version of the K-equation (e.g., one from the 1972 edition of the ASTM standard) might have been used. That is:

$$F(\alpha_1) = 29.6\alpha^{1/2} - 185.5\alpha^{3/2} + 655.7\alpha^{5/2}$$
$$-1017.0\alpha^{7/2} + 754.6\alpha^{9/2} \qquad \text{(Eq 13c)}$$

for specimens where $H/W = 0.6$, or

$$F(\alpha_1) = 30.96\alpha^{1/2} - 195.8\alpha^{3/2} + 730.6\alpha^{5/2}$$
$$-1186.3\alpha^{7/2} + 638.9\alpha^{9/2} \qquad \text{(Eq 13d)}$$

for specimens where $H/W = 0.486$. The second part of the geometric correction factor, $F(\alpha_2)$ did not exist in earlier versions of the equation.

Table 3 Coefficients of closed-form stress-intensity equations for threaded and unthreaded cylinders

Coefficient	Tension		Bending	
	No thread(a)	Threaded(b)	No thread(a)	Threaded(c)
λ	...	2.4371	...	2.295
β	...	−36.5	...	−44.0
A	0.6647	0.5154	0.666	0.654
B	−1.2425	0.4251	−1.2628	−0.9
C	27.998	2.4134	10.737	0.8
D	−162.44	−15.4491	−50.539	10.5
E	472.23	36.157	139.29	−26.2
F	−629.63	...	−183.85	25.9
G	326.05	...	96.347	...

(a) Equivalent to the trigonometric equations of Ref 30. (b) Equation 8(b) of Ref 26 or Eq 9 of Ref 27. (c) Equation 8(d) of Ref 26

Part-Through Crack in a Plate

A part-through crack originated on the surface of a plate is usually modeled as one-half of an ellipse. As shown in Fig. 11, either the major or the minor axis of the ellipse may be placed on the front surface of the plate, depending on the configuration. The length and depth dimensions of the crack are designated as $2c$, and a, respectively. The crack shape is described by an aspect ratio $a/2c$ (or a/c).

Farfield Tensile Loading

For a plate subjected to uniform farfield tension, the expression for K, for some point on the periphery of a semielliptical crack in a semi-infinite solid, is adopted from Ref 16 as:

$$K = [\sigma \sqrt{(\pi a/Q)}] \cdot f_\varphi \cdot \alpha_f \qquad \text{(Eq 14)}$$

where the applied stress σ is equated to the applied load (force) over the full cross section of the plate

(ignoring the crack). Both Q and f_φ are parameters accounting for the shape of the ellipse. Initially, Q was presented in the literature as a function of Φ and the σ/F_{ty} ratio, where Φ is a crack shape parameter (the complete elliptical integral of the second kind presented in Fig. 12) and F_{ty} is the material tensile yield strength. As explained in Ref 2 and 3, the term σ/F_{ty} is used for converting the physical crack length to the effective crack length (to include the effect of crack tip plasticity, Fig. 3). Having this term deleted from Q, the elastic K is solely related to Φ.

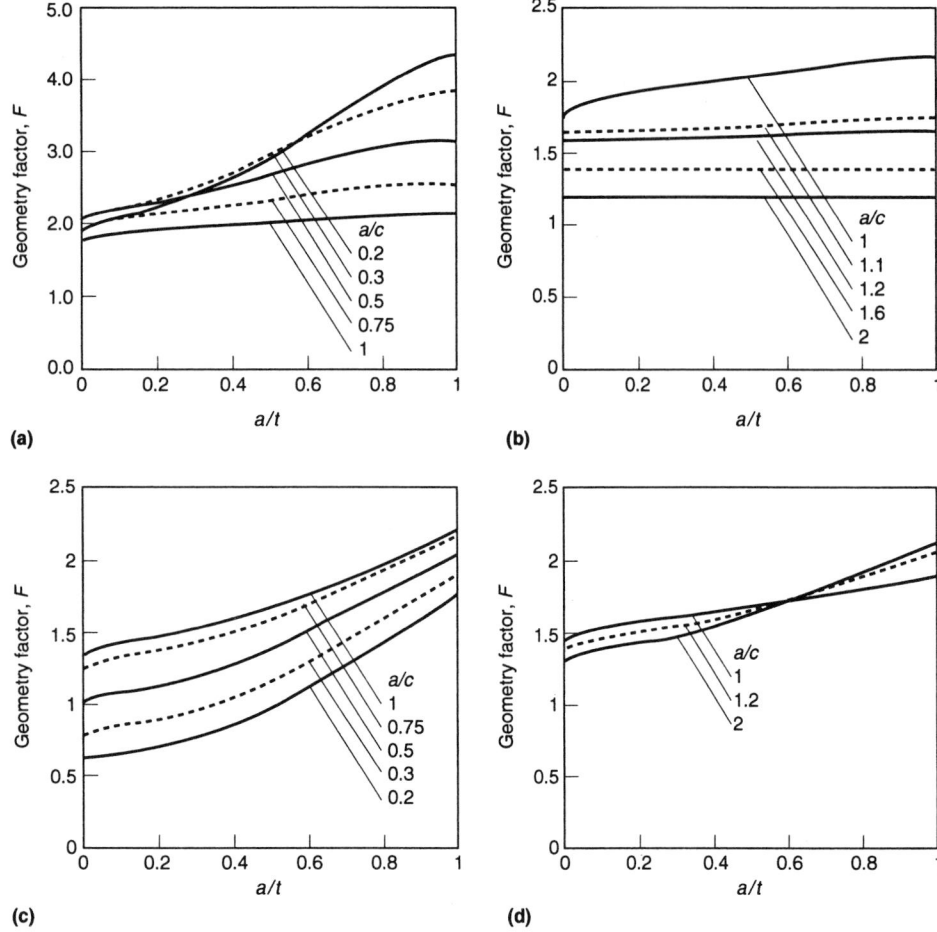

Fig. 25 Stress-intensity factors for a pair of corner cracks at an open hole under uniform tension, $r = 3.175$ mm (0.125 in.). (a) $a/c \le 1$, A-tip surface crack. (b) $a/c \ge 1$, A-tip surface crack. (c) $a/c \le 1$, C-tip surface crack. (d) $a/c \ge 1$, C-tip surface crack

Table 4 Polynomial coefficients for Eq 27(b) in tension

Coefficient	R/t				
	110	55	30	15	7
A	1.0679	1.0775	1.0996	1.0199	1.0203
B	0.0005	−0.0145	−0.0251	−0.015	−0.0145
C	0.0038	0.0037	0.0037	0.0025	0.0019
D	−0.0001	−0.0001	−0.0001	-8×10^{-5}	-6×10^{-5}
E	2×10^{-6}	2×10^{-6}	2×10^{-6}	1×10^{-6}	1×10^{-6}
F	-2×10^{-8}	-1×10^{-8}	-1×10^{-8}	-1×10^{-8}	-7×10^{-9}
G	5×10^{-11}	4×10^{-11}	4×10^{-11}	3×10^{-11}	2×10^{-11}

Note: Values are for circumferential through-the-thickness cracks in hollow cylinders.

Table 5 Polynomial coefficients for Eq 27(b) in bending

Coefficient	R/t				
	110	55	30	15	7
A	0.9524	0.9798	1.0158	1.0083	1.0107
B	0.0297	0.0101	−0.0039	−0.007	−0.008
C	0.0015	0.0019	0.0021	0.0017	0.0012
D	-6×10^{-5}	-6×10^{-5}	-6×10^{-5}	-5×10^{-5}	-4×10^{-5}
E	1×10^{-6}	1×10^{-6}	1×10^{-6}	7×10^{-7}	5×10^{-7}
F	-8×10^{-9}	-8×10^{-9}	-7×10^{-9}	-5×10^{-9}	-4×10^{-9}
G	2×10^{-11}	2×10^{-11}	2×10^{-11}	2×10^{-11}	1×10^{-11}

Note: Values are for circumferential through-the-thickness cracks in hollow cylinders.

In the remainder of this article Φ and $\sqrt{Q}$ are regarded as the same (without plasticity correction).

The angular function f_φ has the following form:

$$f_\varphi = [(a/c)^2 \cdot \cos^2\varphi + \sin^2\varphi]^{1/4} \qquad \text{(Eq 15a)}$$

for $a/c \le 1$, and

$$f_\varphi = [(c/a)^2 \cdot \sin^2\varphi + \cos^2\varphi]^{1/4} \qquad \text{(Eq 15b)}$$

for $a/c > 1$.

In Eq 15(a) and (b), φ is a parametric angle measured from the plate surface toward the center of the crack (i.e., $\varphi = 0°$ is on the plate surface and $\varphi = 90°$ is at the maximum depth of the crack). This terminology is used in all open literature for defining the position of a point on the ellipse. However, it is not an angle that actually connects the center of the ellipse to a specific point on the physical crack periphery. To translate φ to β (the angle between the plate surface and a specific point on the periphery of the ellipse, Fig. 13), the following relationship between φ and the geometric angle β can be used:

$$\beta = \tan^{-1}[(a/c) \cdot \tan\varphi] \qquad \text{(Eq 16)}$$

Finally, the parameter α_f in Eq 15(a) and (b) is called the front face influence factor. It is a function of a/c and φ. For a given a/c ratio, f_φ is a function of φ. Therefore, the combination of α_f, f_φ, and $1/\sqrt{Q}$ is the source of the variance in K values along the crack periphery. Each point along the crack front grows a different amount in different directions. As a result, the crack shape continuously changes as the crack extends. In making structural life prediction, a minimum of two K values (i.e., at the maximum depth, and on the surface) are required for each crack size and its corresponding aspect ratio. A demonstration of the fundamentals of K variation (as a function of a/c and φ) is given in Fig. 14. Further discussion of Fig. 14, along with discussions of finite element solutions, is presented in the following section of this article.

Part-Through Crack in a Finite Plate

For a crack in a rectangular plate of finite thickness and width, the solution for K, (with a given crack size and shape) should account for the influence of the width and the front and back faces of the plate. Finite element solutions (for cracks subjected to tension or bending) have been developed by Newman and Raju (Ref 17, 18) and updated by Raju et al. (Ref 19). The new data (the values of $K/(\sigma\sqrt{\pi a/Q})$ are presented in Tables 1 and 2. These data have been built into the crack library in the NASA/FLAGRO computer program (Ref 20), with which interpolations are accomplished by using a nonlinear table lookup routine to obtain stress-intensity factors that are not available in the tables. Comparisons of the new and old data are shown in Fig. 15 to 22. In some cases the differences are significant.

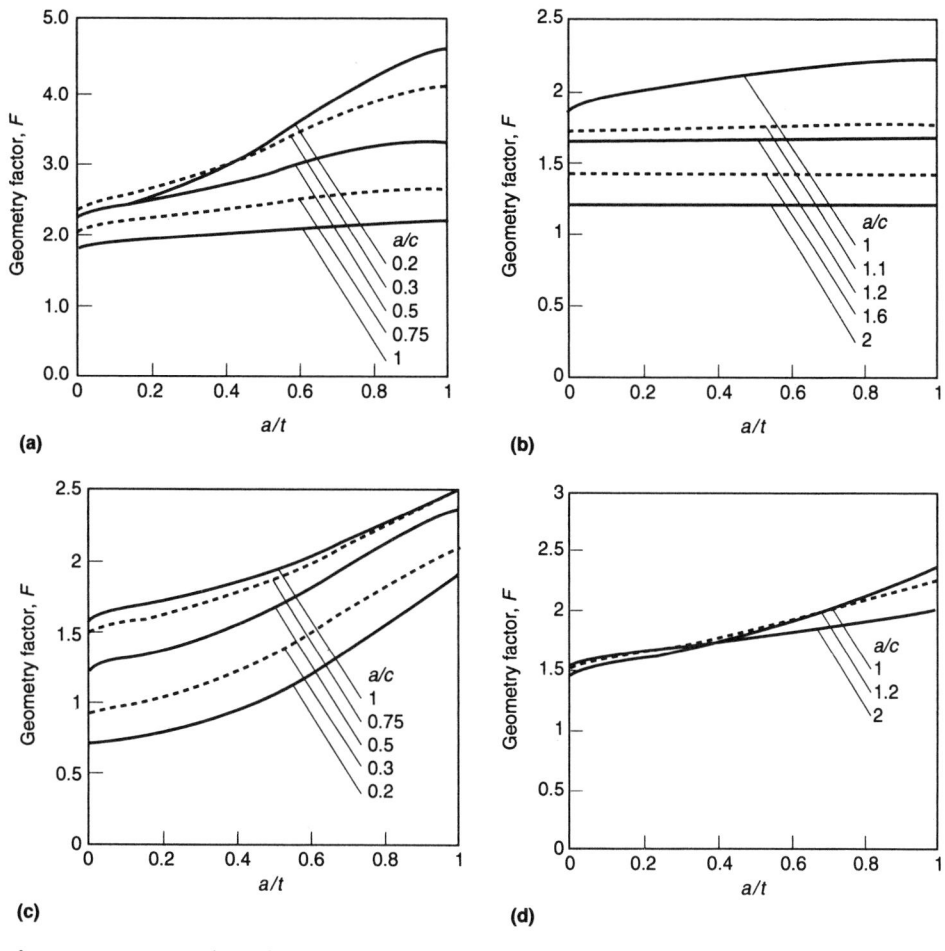

(a)

(b)

(c)

(d)

Fig. 26 Stress-intensity factors for a corner crack at an open hole under uniform tension, $r = 6.35$ mm (0.25 in.). (a) $a/c \leq 1$, A-tip surface crack. (b) $a/c \geq 1$, A-tip surface crack. (c) $a/c \leq 1$, C-tip surface crack. (d) $a/c \geq 1$, C-tip surface crack

The old data were curve fitted, and a general expression for K, which included a group of correction factors for the width and the front and back faces of the plate, was developed by Newman and Raju (Ref 17, 18). These stress-intensity correction factors are included here because the equations are in close form, covering a full range of geometric combinations, and have been used for some time by fracture mechanics analysts in the aircraft/aerospace industry:

$$K = \sigma \cdot F_s \cdot \sqrt{(\pi a / Q)} \qquad \text{(Eq 17a)}$$

where $\sqrt{Q}$ is Φ (Fig. 12), as previously discussed, and F_s is a function of a/c, a/t, c/W, and φ such that:

$$F_s = [M_1 + M_2(a/t)^2 + M_3(a/t)^4] \cdot G_1 \cdot f_\varphi \cdot f_w \qquad \text{(Eq 17b)}$$

For $a/c \leq 1$:

$$M_1 = 1.13 - 0.09(a/c) \qquad \text{(Eq 18a)}$$

$$M_2 = -0.54 + 0.89/[0.2 + (a/c)] \qquad \text{(Eq 18b)}$$

$$M_3 = 0.5 - 1/[0.65 + (a/c)] + 14.0(1 - a/c)^{24} \qquad \text{(Eq 18c)}$$

$$G_1 = 1 + [0.1 + 0.35(a/t)^2] \cdot (1 - \sin \varphi)^2 \qquad \text{(Eq 18d)}$$

$$f_w = \sqrt{\sec[(\pi c/W) \cdot \sqrt{(a/t)}]} \qquad \text{(Eq 18e)}$$

and f_φ is given by Eq 15(a).

By inputting large values of W and t in Eq 17(b), the configuration of a shallow crack ($a/c \leq$

Table 6 F-factors for internal thumbnail cracks on circumferential plane of a hollow cylinder

	a/c = 0.2					a/c = 0.4					a/c = 0.6					a/c = 0.8					a/c = 1.0				
R/t	a/t=0	a/t=0.2	a/t=0.5	a/t=0.8	a/t=1.0	a/t=0	a/t=0.2	a/t=0.5	a/t=0.8	a/t=1.0	a/t=0	a/t=0.2	a/t=0.5	a/t=0.8	a/t=1.0	a/t=0	a/t=0.2	a/t=0.5	a/t=0.8	a/t=1.0	a/t=0	a/t=0.2	a/t=0.5	a/t=0.8	a/t=1.0
C-tip, uniform loading																									
1.0	0.580	0.593	0.610	0.846	1.117	0.630	0.650	0.665	0.841	1.041	0.670	0.688	0.702	0.831	0.976	0.695	0.709	0.722	0.817	0.919	0.700	0.713	0.726	0.796	0.872
2.0	0.600	0.617	0.671	0.824	0.975	0.660	0.669	0.714	0.837	0.956	0.695	0.703	0.741	0.838	0.930	0.715	0.721	0.752	0.828	0.898	0.710	0.722	0.747	0.806	0.860
4.0	0.613	0.633	0.726	0.898	1.049	0.664	0.681	0.746	0.894	1.014	0.698	0.712	0.772	0.880	0.974	0.716	0.727	0.774	0.858	0.930	0.718	0.727	0.762	0.827	0.883
10.0	0.591	0.644	0.785	1.000	1.178	0.651	0.689	0.797	0.967	1.108	0.692	0.718	0.799	0.930	1.041	0.714	0.732	0.791	0.891	0.975	0.717	0.730	0.774	0.849	0.913
300.0	0.538	0.583	0.747	1.075	1.398	0.601	0.679	0.818	1.023	1.199	0.668	0.722	0.829	0.969	1.074	0.700	0.739	0.817	0.919	0.996	0.726	0.736	0.785	0.878	0.960
C-tip, bending loading																									
1.0	0.337	0.265	0.111	0.080	0.050	0.358	0.308	0.216	0.150	0.120	0.370	0.338	0.293	0.253	0.230	0.375	0.355	0.340	0.343	0.354	0.371	0.360	0.359	0.378	0.400
2.0	0.400	0.403	0.410	0.420	0.430	0.430	0.443	0.450	0.465	0.493	0.460	0.470	0.482	0.520	0.559	0.480	0.485	0.503	0.548	0.590	0.482	0.486	0.505	0.547	0.587
4.0	0.498	0.510	0.569	0.678	0.775	0.539	0.550	0.602	0.698	0.782	0.567	0.577	0.622	0.704	0.775	0.581	0.590	0.628	0.696	0.756	0.583	0.590	0.620	0.675	0.723
10.0	0.544	0.595	0.722	0.915	1.072	0.605	0.637	0.732	0.888	1.019	0.646	0.664	0.734	0.858	0.965	0.668	0.677	0.728	0.824	0.909	0.670	0.675	0.713	0.786	0.851
300.0	0.538	0.583	0.747	1.075	1.398	0.601	0.679	0.818	1.023	1.199	0.668	0.722	0.829	0.969	1.074	0.700	0.739	0.817	0.919	0.996	0.726	0.736	0.785	0.878	0.960
A-tip, uniform loading																									
1.0	0.960	0.987	1.064	1.665	2.406	0.875	0.888	0.944	1.360	1.857	0.795	0.799	0.841	1.119	1.437	0.720	0.721	0.754	0.941	1.146	0.650	0.653	0.684	0.823	0.969
2.0	0.990	1.022	1.093	1.380	1.685	0.900	0.911	0.961	1.163	1.377	0.800	0.813	0.847	0.985	1.130	0.710	0.726	0.751	0.846	0.943	0.620	0.652	0.674	0.745	0.815
4.0	1.031	1.045	1.141	1.332	1.504	0.920	0.926	0.991	1.123	1.243	0.819	0.821	0.862	0.951	1.031	0.729	0.729	0.756	0.814	0.868	0.650	0.652	0.672	0.713	0.751
10.0	0.983	1.059	1.189	1.337	1.440	0.888	0.936	1.020	1.120	1.192	0.800	0.827	0.878	0.941	0.989	0.718	0.732	0.761	0.801	0.831	0.642	0.651	0.671	0.697	0.717
300.0	1.059	1.090	1.384	1.682	1.881	0.948	0.951	1.079	1.188	1.251	0.792	0.832	0.888	0.940	0.971	0.720	0.733	0.754	0.777	0.792	0.642	0.656	0.675	0.691	0.700
A-tip, bending loading																									
1.0	0.520	0.545	0.659	1.074	1.523	0.470	0.493	0.597	0.919	1.254	0.430	0.446	0.542	0.792	1.039	0.385	0.405	0.494	0.693	0.879	0.350	0.368	0.454	0.621	0.771
2.0	0.700	0.719	0.821	1.088	1.352	0.630	0.643	0.728	0.935	1.135	0.560	0.575	0.648	0.808	0.957	0.503	0.515	0.579	0.706	0.819	0.448	0.463	0.523	0.629	0.720
4.0	0.839	0.865	0.974	1.173	1.347	0.748	0.767	0.849	0.997	1.126	0.666	0.681	0.743	0.852	0.946	0.592	0.606	0.654	0.735	0.805	0.528	0.542	0.583	0.648	0.702
10.0	0.902	0.985	1.120	1.267	1.366	0.822	0.871	0.959	1.064	1.141	0.744	0.770	0.824	0.897	0.954	0.669	0.682	0.715	0.765	0.806	0.597	0.607	0.631	0.667	0.695
300.0	1.059	1.090	1.384	1.682	1.881	0.948	0.951	1.079	1.188	1.251	0.792	0.832	0.888	0.940	0.971	0.720	0.733	0.754	0.777	0.792	0.642	0.656	0.675	0.691	0.700

Source: Ref 33

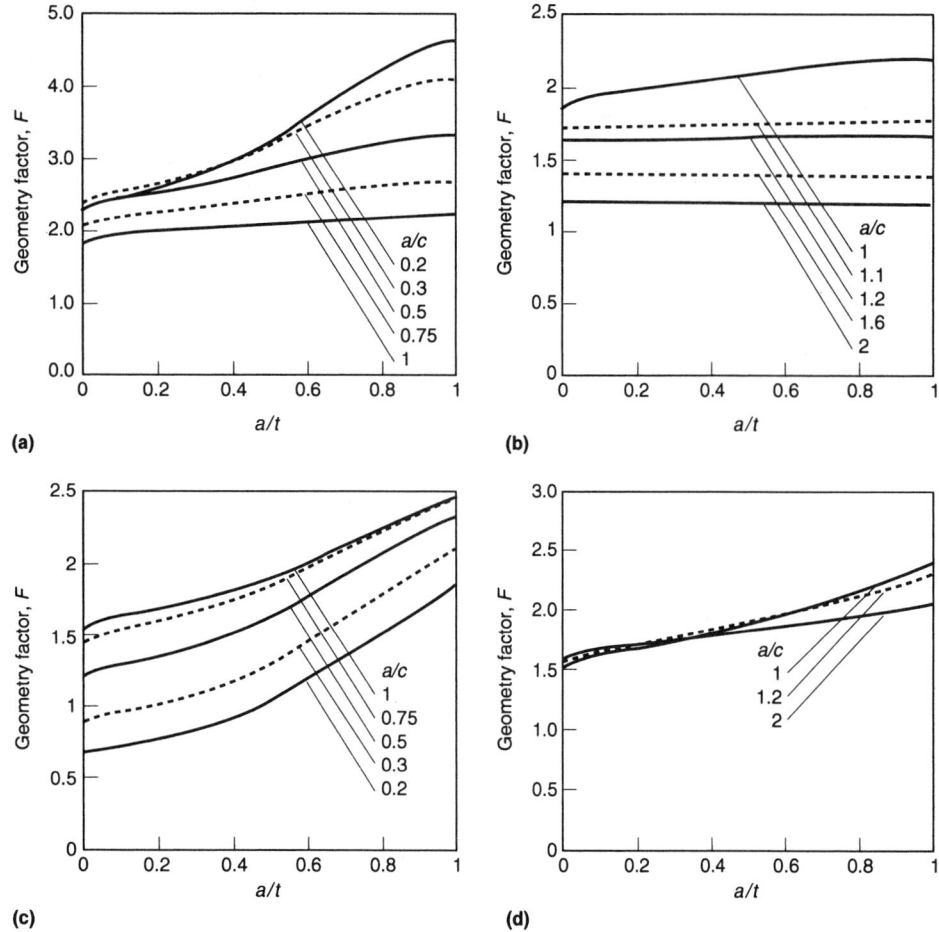

1) in a semi-infinite solid (i.e., without the presence of other boundaries such as width and back face) is obtained. The stress-intensity factor $F = K/\sigma\sqrt{\pi a}$ which is equal to $F_s/\sqrt{Q}$ (Eq 17a) has been computed for several points on the crack periphery (for several a/c ratios) and is plotted in Fig. 14. Conceptually, the location of the highest K value is at the maximum depth (for $a < c$). However, the highest K value is on the surface of the plate when $a \geq c$. As shown in Fig. 14, the switching actually takes place at $a/c \cong 0.8$. For this flaw shape (i.e., $a/c \cong 0.8$), all the F values along the crack boundary are approximately the same. Thus, crack growth rates at each point along the crack boundary are approximately the same. Therefore, whether the crack starts as a scratch (having $a \ll c$) or a deep cavity (having $c \ll a$), given time its flaw shape will eventually stabilize at $a/c \cong 0.8$.

For $a/c > 1$:

$$M_1 = [1.0 + 0.04(c/a)] \cdot (c/a)^{1/2} \qquad \text{(Eq 19a)}$$

$$M_2 = 0.2(c/a)^4 \qquad \text{(Eq 19b)}$$

$$M_3 = -0.11(c/a)^4 \qquad \text{(Eq 19c)}$$

$$G_1 = 1 + [0.1 + 0.35(c/a)(a/t)^2] \cdot (1 - \sin\varphi)^2 \qquad \text{(Eq 19d)}$$

and f_w is given by Eq 18(e) and f_φ is given by Eq 15(b).

In case the plate is subjected to bending load, Eq 17(a) becomes

$$K = \sigma \cdot F_s \cdot H \cdot \sqrt{(\pi a/Q)} \qquad \text{(Eq 20)}$$

Fig. 27 Stress-intensity factors for a pair of corner cracks at an open hole under uniform tension, $r = 6.35$ mm (0.25 in.). (a) $a/c \leq 1$, A-tip surface crack. (b) $a/c \geq 1$, A-tip surface crack. (c) $a/c \leq 1$, C-tip surface crack. (d) $a/c \geq 1$, C-tip surface crack

Table 7   F-factors for external thumbnail crack on circumferential plane of a hollow cylinder

R/t	$a/c = 0.2$ $a/t = 0$	$a/t = 0.2$	$a/t = 0.5$	$a/t = 0.8$	$a/t = 1.0$	$a/c = 0.4$ $a/t = 0$	$a/t = 0.2$	$a/t = 0.5$	$a/t = 0.8$	$a/t = 1.0$	$a/c = 0.6$ $a/t = 0$	$a/t = 0.2$	$a/t = 0.5$	$a/t = 0.8$	$a/t = 1.0$	$a/c = 0.8$ $a/t = 0$	$a/t = 0.2$	$a/t = 0.5$	$a/t = 0.8$	$a/t = 1.0$	$a/c = 1.0$ $a/t = 0$	$a/t = 0.2$	$a/t = 0.5$	$a/t = 0.8$	$a/t = 1.0$
C-tip, uniform loading																									
1.0	0.590	0.672	0.893	1.249	1.552	0.664	0.713	0.871	1.138	1.368	0.712	0.739	0.846	1.039	1.209	0.734	0.747	0.818	0.954	1.075	0.731	0.739	0.788	0.882	0.966
2.0	0.560	0.660	0.876	1.177	1.416	0.643	0.706	0.859	1.086	1.271	0.699	0.734	0.838	1.006	1.148	0.727	0.744	0.814	0.938	1.046	0.728	0.737	0.787	0.881	0.964
4.0	0.540	0.653	0.873	1.162	1.383	0.630	0.701	0.858	1.081	1.257	0.691	0.731	0.839	1.006	1.145	0.722	0.742	0.815	0.940	1.046	0.725	0.735	0.786	0.880	0.962
10.0	0.542	0.646	0.867	1.172	1.414	0.630	0.697	0.855	1.087	1.275	0.689	0.728	0.838	1.010	1.153	0.720	0.741	0.815	0.941	1.049	0.722	0.734	0.785	0.879	0.961
300.0	0.538	0.583	0.747	1.075	1.398	0.601	0.679	0.818	1.023	1.199	0.668	0.722	0.829	0.969	1.074	0.700	0.739	0.817	0.919	0.996	0.726	0.736	0.785	0.878	0.960
C-tip, bending loading																									
1.0	0.592	0.643	0.742	0.870	0.967	0.659	0.690	0.761	0.861	0.940	0.704	0.720	0.768	0.844	0.908	0.727	0.731	0.760	0.819	0.871	0.729	0.724	0.740	0.785	0.829
2.0	0.552	0.645	0.798	0.972	1.092	0.632	0.691	0.801	0.939	1.040	0.687	0.720	0.795	0.902	0.987	0.716	0.731	0.780	0.863	0.932	0.720	0.724	0.757	0.820	0.876
4.0	0.545	0.645	0.835	1.075	1.254	0.624	0.690	0.827	1.014	1.158	0.678	0.717	0.814	0.956	1.069	0.706	0.728	0.794	0.899	0.987	0.710	0.722	0.767	0.845	0.912
10.0	0.524	0.633	0.850	1.136	1.357	0.612	0.684	0.840	1.057	1.229	0.672	0.715	0.823	0.984	1.115	0.703	0.727	0.801	0.918	1.016	0.705	0.721	0.772	0.859	0.932
300.0	0.538	0.583	0.747	1.075	1.398	0.601	0.679	0.818	1.023	1.199	0.668	0.722	0.829	0.969	1.074	0.700	0.739	0.817	0.919	0.996	0.726	0.736	0.785	0.878	0.960
A-tip, uniform loading																									
1.0	1.140	1.189	1.469	2.179	2.898	1.000	1.019	1.188	1.583	1.969	0.860	0.872	0.960	1.140	1.303	0.737	0.748	0.785	0.847	0.899	0.644	0.647	0.660	0.685	0.708
2.0	1.126	1.167	1.370	1.759	2.112	0.975	1.005	1.132	1.362	1.564	0.844	0.865	0.935	1.051	1.149	0.733	0.746	0.780	0.827	0.866	0.640	0.648	0.663	0.683	0.698
4.0	1.099	1.157	1.320	1.576	1.790	0.959	0.999	1.103	1.260	1.388	0.835	0.862	0.923	1.006	1.072	0.728	0.746	0.777	0.815	0.843	0.637	0.649	0.666	0.683	0.693
10.0	1.079	1.146	1.284	1.470	1.615	0.945	0.993	1.083	1.198	1.284	0.827	0.859	0.914	0.977	1.020	0.724	0.745	0.776	0.806	0.825	0.636	0.650	0.668	0.684	0.693
300.0	1.059	1.090	1.384	1.682	1.881	0.948	0.951	1.079	1.188	1.251	0.792	0.832	0.888	0.940	0.971	0.720	0.733	0.754	0.777	0.792	0.642	0.656	0.675	0.691	0.700
A-tip, bending loading																									
1.0	1.110	1.124	1.252	1.676	2.120	0.945	0.958	1.008	1.207	1.419	0.850	0.816	0.810	0.857	0.914	0.729	0.697	0.656	0.624	0.608	0.641	0.601	0.545	0.495	0.466
2.0	1.115	1.121	1.236	1.487	1.721	0.966	0.964	1.018	1.144	1.263	0.836	0.827	0.837	0.876	0.915	0.726	0.711	0.694	0.681	0.675	0.635	0.615	0.585	0.554	0.534
4.0	1.097	1.124	1.242	1.429	1.586	0.949	0.969	1.035	1.138	1.223	0.822	0.834	0.863	0.905	0.937	0.716	0.720	0.725	0.728	0.730	0.630	0.625	0.619	0.605	0.593
10.0	1.042	1.117	1.248	1.403	1.514	0.918	0.969	1.051	1.142	1.205	0.808	0.839	0.885	0.930	0.959	0.709	0.727	0.750	0.767	0.775	0.623	0.634	0.645	0.649	0.649
300.0	1.059	1.090	1.384	1.682	1.881	0.948	0.951	1.079	1.188	1.251	0.792	0.832	0.888	0.940	0.971	0.720	0.733	0.754	0.777	0.792	0.642	0.656	0.675	0.691	0.700

Source: Ref 33

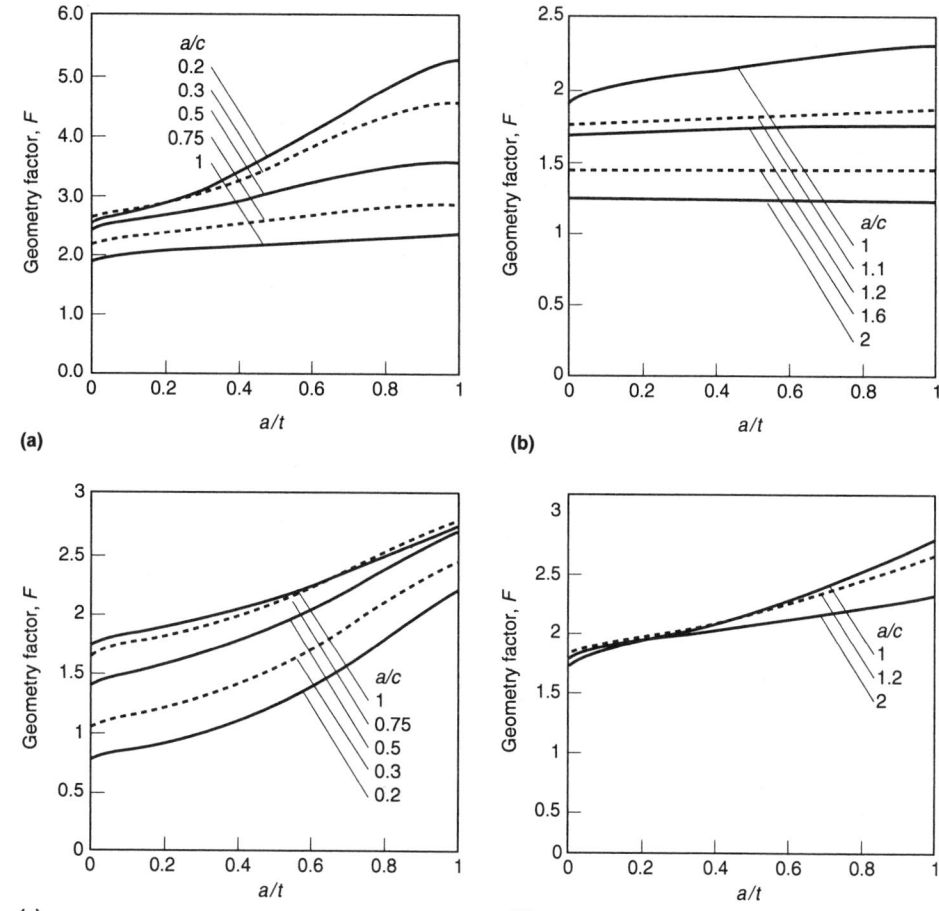

Fig. 28 Stress-intensity factors for a corner crack at an open hole under uniform tension, $r = 9.525$ mm (0.375 in.). (a) $a/c \leq 1$, A-tip surface crack. (b) $a/c \geq 1$, A-tip surface crack. (c) $a/c \leq 1$, C-tip surface crack. (d) $a/c \geq 1$, C-tip surface crack

where $\sqrt{Q}$ is Φ (Fig. 12) and F_s is given in Eq 17(b) for uniform tension. The applied stress (σ) is equal to Mc/I (where M is the bending moment and I is the moment of inertia). For $a/c \leq 1$, the function H has the form

$$H = H_1 + (H_2 - H_1) \cdot \sin^\rho \varphi \qquad (Eq\ 21a)$$

where

$$\rho = 0.2 + a/c + 0.6(a/t) \qquad (Eq\ 21b)$$

$$H_1 = 1 - 0.34(a/t) - 0.11(a/c) \cdot (a/t) \qquad (Eq\ 21c)$$

$$H_2 = 1 + F_1(a/t) + F_2(a/t)^2 \qquad (Eq\ 21d)$$

In Eq 21(d), the factors F_1 and F_2 are as follows:

$$F_1 = -1.22 - 0.12(a/c) \qquad (Eq\ 21e)$$

$$F_2 = 0.55 - 1.05(a/c)^{0.75} + 0.47(a/c)^{1.5} \qquad (Eq\ 21f)$$

A final note on part-through crack growth behavior is directed at the estimate of stress-intensity factors while the crack front is approaching the back face of the plate. Usually, there is a discontinuity where the calculation of K is suddenly switched from the part-through crack solution to the through crack solution. This is commonly known as the *transition phenomenon*. Literature containing discussion of the techniques in handling this problem is listed in the "Selected References" at the end of this article.

Corner Crack(s) at a Circular Hole

The problem of corner crack(s) at a circular hole configuration (Fig. 23) has received considerable attention from the fracture mechanics community because of its effect on the life of aircraft structures. The first set of test data (on fatigue and fracture strength of specimens containing a circular hole) was presented at the Air Force Conference in 1969 (Ref 21). Since then, considerable efforts have been directed at the development of analytical and semiempirical solutions. The latest study was published in 1995 (Ref 22). Prior to the development of a complete set of finite element solutions by Newman and Raju in 1983 (Ref 18), Liu's one-dimensional semiempirical equation (Ref 23) was widely

adopted by fracture mechanics analysts in the aircraft/aerospace industry.

The two-dimensional, curve-fitted equations of Newman and Raju (Ref 18) are presented in the following paragraphs. This group of equations covers a full range of geometric variables and has been accepted by the fracture mechanics community for more than a decade.

Double Crack Configuration

For the double symmetric crack configuration:

$$K = \sigma \cdot F_{ch} \cdot \sqrt{(\pi a/Q)} \qquad (Eq\ 22)$$

where σ again is the applied stress (load over cross section ignoring the hole and the crack) and $\sqrt{Q}$ is Φ (Fig. 12). If the panel length ($2h$) is sufficiently long, F_{ch} is a function of a/c, a/t, r/t, r/W, c/W, and φ. That is:

$$F_{ch} = [M_1 + M_2(a/t)^2 + M_3(a/t)^4] \cdot G_1 \cdot G_2 \cdot G_3 \cdot f_\varphi \cdot F_W \qquad (Eq\ 23a)$$

For $a/c \leq 1$:

$$G_2 = (1 + 0.358\lambda + 1.425\lambda^2 - 1.578\lambda^3 + 2.156\lambda^4)/(1 + 0.13\lambda^2) \qquad (Eq\ 23b)$$

where

$$\lambda = \{1 + (c/r) \cdot \cos(0.85\varphi)\}^{-1} \qquad (Eq\ 23c)$$

The functions G_3 and F_W are as follows:

$$G_3 = [0.1 + 0.04(a/c)] \cdot [1 + 0.1(1 - \cos \varphi)^2] \cdot [0.85 + 0.15(a/t)^{1/4}] \qquad (Eq\ 23d)$$

$$F_W = \frac{}{\sqrt{\sec[(\pi r/W) \cdot \sec(\pi(2r + nc)/2(W - 2c + nc)) \cdot \sqrt{(a/t)}]}} \qquad (Eq\ 23e)$$

where $n = 1$ for a single crack and $n = 2$ for double cracks. The other functions (i.e., M_1, M_2, M_3, G_1, and f_φ) are the same as those previously given for the surface flaw.

For $a/c > 1$:

$$G_3 = [1.13 + 0.09(c/a)] \cdot [1 + 0.1(1 - \cos \varphi)^2] \cdot [0.85 + 0.15(a/t)^{1/4}] \qquad (Eq\ 23f)$$

The functions G_2 and F_W are given by Eq 23(b) and (e), respectively. The other functions (i.e., M_1, M_2, M_3, G_1, and f_φ) are the same as those previously given for the surface flaw.

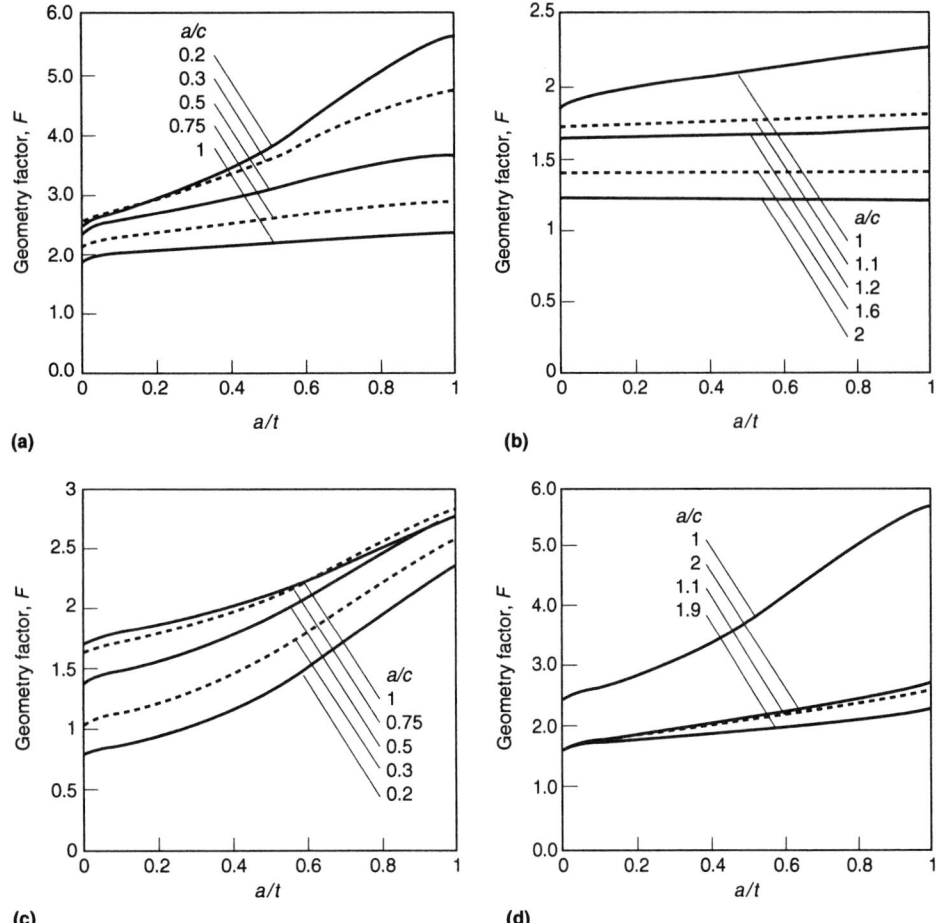

(a)

(b)

(c)

(d)

Fig. 29 Stress-intensity factors for a pair of corner cracks at an open hole under uniform tension, $r = 9.525$ mm (0.375 in.). (a) $a/c \leq 1$, A-tip surface crack. (b) $a/c \geq 1$, A-tip surface crack. (c) $a/c \leq 1$, C-tip surface crack. (d) $a/c \geq 1$, C-tip surface crack

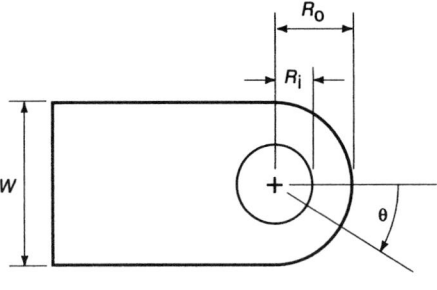

(a)

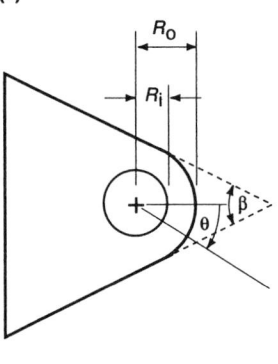

(b)

Fig. 30 Attachment lug configurations. (a) Straight. (b) Tapered

Single Corner Crack

The stress-intensity factors for a single corner crack at a hole can be estimated by using Eq 22 with a conversion factor given by (Ref 24):

$$K_{\text{one crack}} = K_{\text{two cracks}}$$
$$\cdot \sqrt{(4/\pi + ac/2tr)/(4/\pi + ac/tr)} \qquad \text{(Eq 23g)}$$

Implementation of these equations may be cumbersome because of their complexity. To simplify the lengthy calculations in Eq 22 let $F = F_{\text{ch}}/F_{\text{W}}/\sqrt{Q}$ so that Eq 22 can be rewritten as:

$$K = \sigma \cdot F \cdot F_{\text{W}} \sqrt{(\pi a)} \qquad \text{(Eq 24)}$$

Parametric curves (for F versus a/t) for three commonly used hole diameters ($2r = 6.35$, 12.7, and 19.05 mm, or 0.25, 0.5, and 0.75 in.) have been developed and are graphically presented in Fig. 24 to 29. Again, the techniques to handle the transition from part-through crack to through crack are documented in the open literature. Some of the publications are listed in the "Selected References" section at the end of this article.

Crack at Pin Hole in a Lug

Two types of attachment lugs are considered in this article, the straight lug and the tapered lug (Fig. 30). The stress-intensity factors for cracks at the bore of the pin hole are summarized below.

The Straight Lug

Figure 31 presents the finite element solutions of stress-intensity factors for an axially loaded straight attachment lug. The pin load is applied at the direction normal to the base of the lug ($\theta = 0°$). Either the bearing stress, σ_{br}, or the gross area stress, σ_0, can be used to compute stress intensity. That is, for a single through-the-thickness crack on one side of the pin hole (perpendicular to the direction of loading, P) according to Ref 25:

$$K = \sigma_{\text{br}} \cdot F_{\text{RB}} \cdot \sqrt{(\pi c)} \qquad \text{(Eq 25a)}$$

Because $\sigma_{\text{br}} = P/2R_i t$ and $\sigma_0 = P/2R_o t$, Eq 25(a) can be rewritten as:

$$K = \sigma_0 \cdot F_{\text{RB}} \cdot R_o/R_i \cdot \sqrt{(\pi c)} \qquad \text{(Eq 25b)}$$

The values of F_{RB} for five R_o/R_i ratios are given in Fig. 31.

Equations 25(a) and (b) can be modified to analyze a corner crack. Two methods have been proposed (Ref 25). The method that involves analysis at two crack tips (i.e., Points A and C of the corner crack) is rather complicated and incomplete and will not be discussed here. In the one-parameter method, the crack shape (i.e., a/c ratio) is assumed to be constant and equal to 1.33. The stress-intensity factor at the lug surface point (i.e., Point C) is computed using a corner crack correction factor:

$$\Phi_{71} = 1 - 0.2886/[1 + 2(a/c)^2 + (c/t)^2] \qquad \text{(Eq 25c)}$$

and Eq 25(a) or (b) becomes:

$$K^C = K \cdot \Phi_{71} \qquad \text{(Eq 25d)}$$

where the superscript C stands for corner crack and K is the stress-intensity factor for the through-the-thickness crack, given by Eq 25(a) or (b).

The Tapered Lug

All the equations listed above for the straight lug are also applicable to the tapered lug (with a new set of F_{RB} factors). The F_{RB} factors for 0° loading are presented in Fig. 32. Limited finite element data are available for tapered lugs loaded in other directions. The through crack solutions for $\theta = 180°$, $-45°$, and $-90°$ are presented in Fig. 33 to 35, respectively.

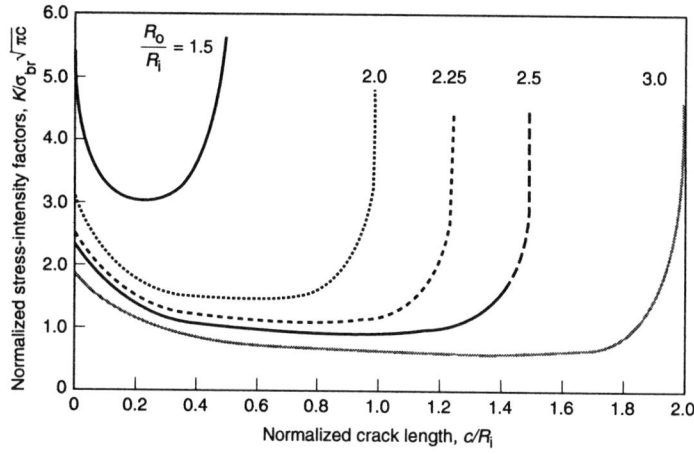

Fig. 31 Normalized stress-intensity factors for single through-the-thickness cracks emanating from a straight lug subjected to a pin loading applied in the 0° loading direction. Source: Ref 25

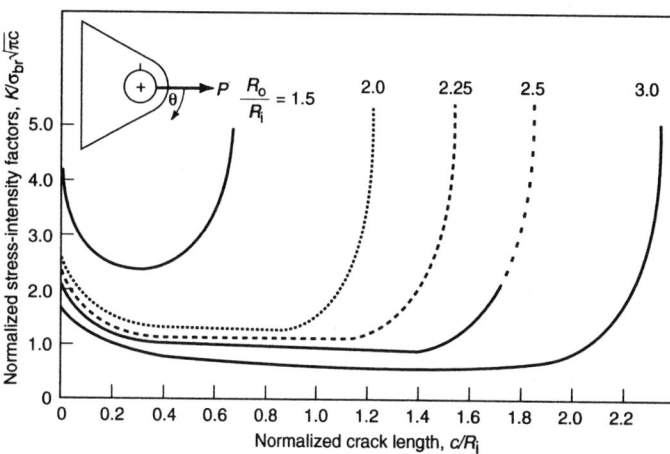

Fig. 32 Normalized stress-intensity factors for single through-the-thickness cracks emanating from a tapered lug subjected to a pin loading applied in the 0° loading direction. Source: Ref 25

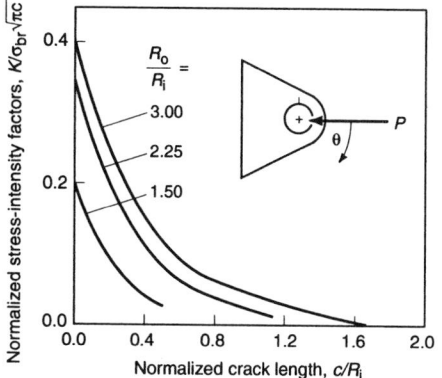

Fig. 33 Normalized stress-intensity factors for single through-the-thickness cracks emanating from a tapered lug subjected to a pin loading applied in the 180° loading direction. Source: Ref 25

Crack in a Solid Cylinder

This type of crack usually has an almond shape, as shown schematically in Fig. 36. In the remainder of this section we will refer it as a crack on the shank of a bolt, or at the thread of a bolt. However, one should keep in mind that this kind of crack can be found in any circumferential plane of a solid round bar. The stress-intensity solutions presented herein are applicable to both.

The nomenclature for this crack type is defined in Fig. 36, where a is the crack depth, the point (Point A) that travels through the diameter of the cylinder; b is the crack length (i.e., one-half of the crack tip-to-tip circumferential arc); d is the minor diameter at the thread of a bolt (or a notched round bar); and D is the diameter of a rod or the

diameter at the unthreaded portion of a bolt, depending on the application.

Due to the geometric difference between a cylinder and a plate, this crack cannot be treated as an edge crack or as a thumbnail crack in a rectangular cross section. It is in a class by itself. Many fracture mechanics analysts believe (Ref 26, 27) that this type of crack has the shape of a circular arc, rather than one-half of an ellipse. The center of the circle floats in between the free surface of the cylinder and a point infinitely far away from the cylindrical surface. As the crack front passes through the center of the rod, the crack front curvature will become flattened, approaching a straight crack front despite the fact that a circular shape (by geometric definition) had been maintained at all times. Experimental observations on

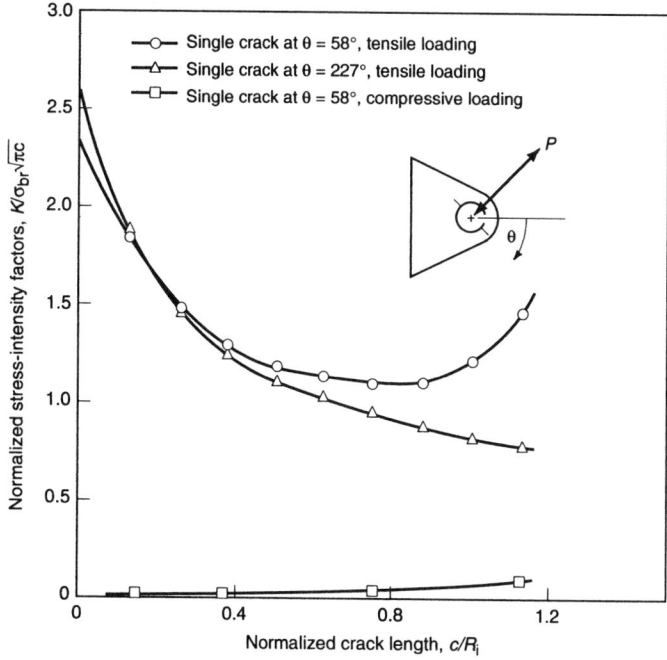

Fig. 34 Normalized stress-intensity factors for single through-the-thickness cracks emanating from a tapered lug subjected to a pin loading applied in the –45° loading direction and its reversed directions, $R_o/R_i = 2.25$. Source: Ref 25

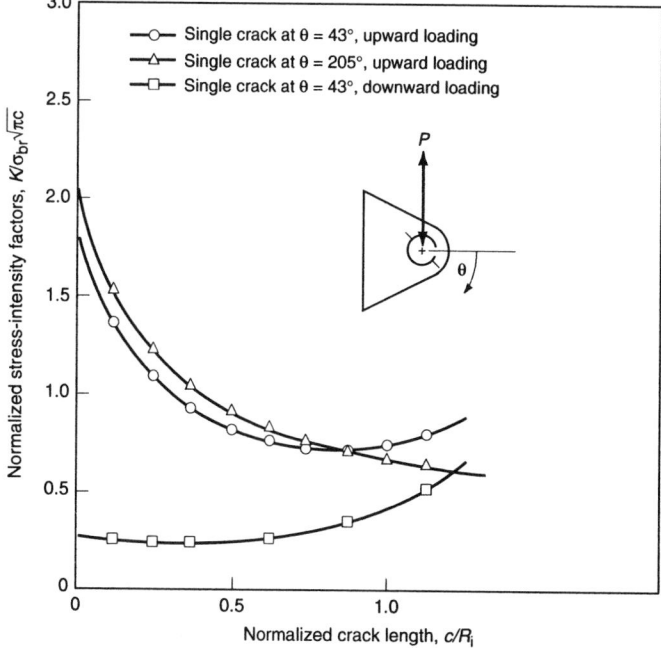

Fig. 35 Normalized stress-intensity factors for single through-the-thickness cracks emanating from a tapered lug subjected to a pin loading applied in the –90° loading direction and its reversed directions, $R_o/R_i = 2.25$. Source: Ref 25

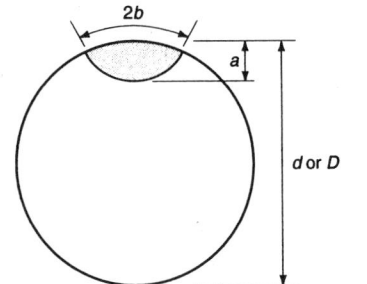

Fig. 36 Definition of crack dimensions for an almond-shaped crack in a solid cylinder

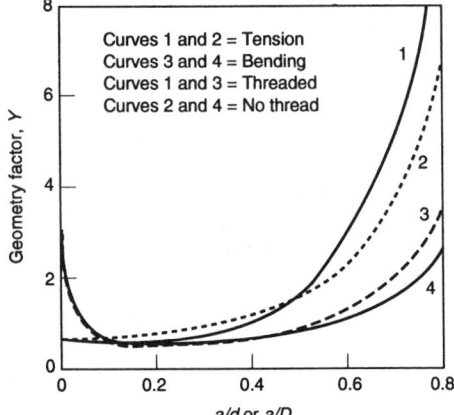

Fig. 37 Stress-intensity factors for threaded and un-threaded cylinders

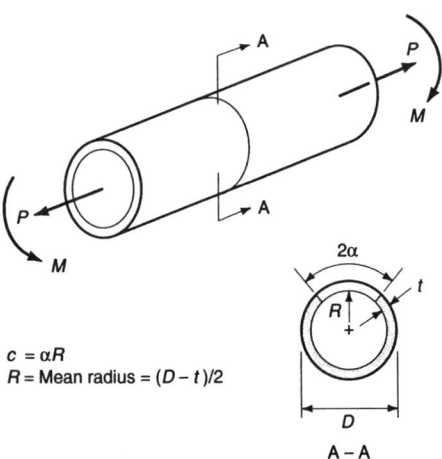

$c = \alpha R$
R = Mean radius = $(D - t)/2$

Fig. 38 A circumferential through-the-thickness crack in a hollow cylinder

Partly-Circular Crack in a Bolt

For an almond-shaped crack in a bolt, a large portion of the crack growth activity will take place in only one-half of the cylindrical crack growth in unthreaded and threaded rods (Ref 28-32) seem to support this claim.

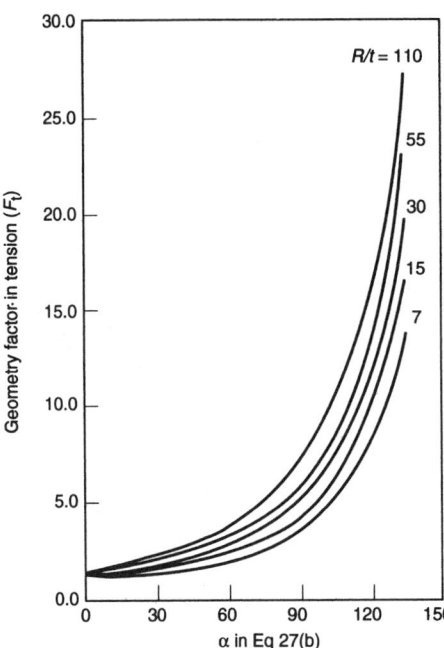

Fig. 39 Stress-intensity factors for circumferential through cracks in hollow cylinders subjected to tension (Eq 27a and b)

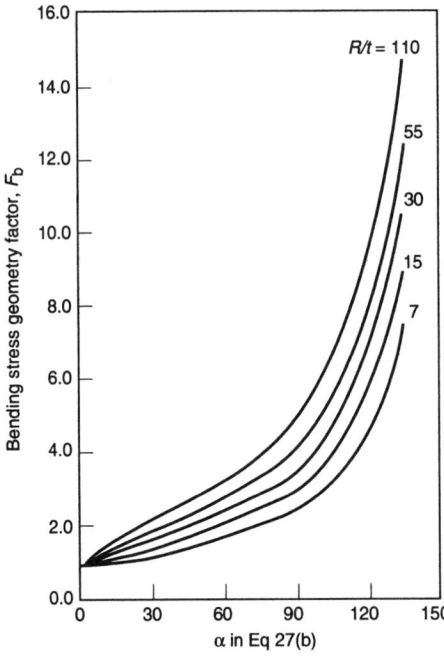

Fig. 40 Stress-intensity factors for circumferential through cracks in hollow cylinders subjected to bending (Eq 27a and b)

plane, that is, where $a \leq D/2$ (or $a \leq d/2$), before it becomes critical. Experimental data has also shown that the crack maintains a circular shape and a constant a/b ratio in this region. Referring back to the semielliptical flaw problem, a constant flaw shape during crack growth implies that the K values along the crack periphery are nearly

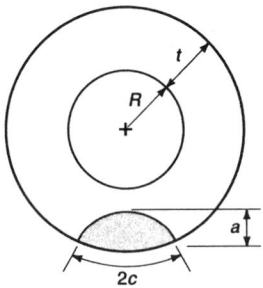

Fig. 41 Part-through cracks on the circumferential plane in a hollow cylinder

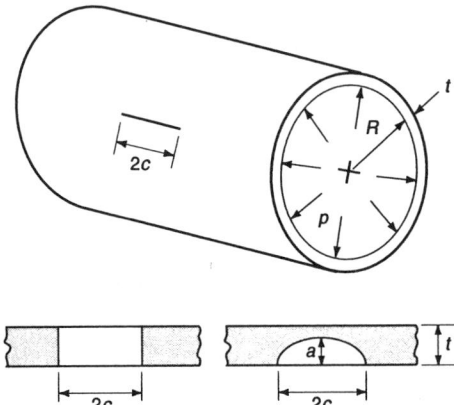

Fig. 42 Axial cracks in a hollow cylinder

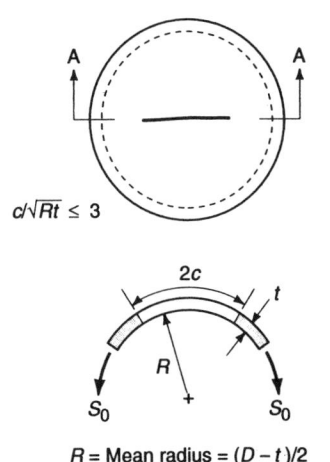

R = Mean radius = $(D - t)/2$

Fig. 43 Through-the-thickness crack in a spherical shell, section of sphere AA

constant. However, this is not the case here because the crack tip on the surface, which travels along the circumference of the cylindrical cross section, has to travel a longer distance than the crack middle point (which propagates in the depth direction), in order to keep up with the crack shape aspect ratios. That is, $db/dN > da/dN$ in each increment of crack extension. This means that the K value at Point B would be higher than that at Point A. Therefore, a two-dimensional crack growth scheme is more suitable for this

Table 8 F-factors for axial cracks in a pressurized cylinder (Eq 30d), $R/t = 1$

a/t	a/c = 0.2		a/c = 0.4		a/c = 0.6		a/c = 0.8		a/c = 1.0	
	F (90°)	F (0°)	F (90°)	F (0°)	F (90°)	F (0°)	F (90°)	F (0°)	F (90°)	F (0°)
Inside crack										
0	2.73	1.542	2.433	1.713	2.15	1.754	1.898	1.769	1.683	1.771
0.08	2.86	1.567	2.533	1.74	2.252	1.79	2.001	1.812	1.785	1.821
0.2	3.099	1.608	2.725	1.786	2.448	1.851	2.198	1.885	1.982	1.902
0.3	3.723	1.88	3.12	1.973	2.782	1.991	2.479	1.984	2.22	1.968
0.4	4.457	2.181	3.598	2.181	3.192	2.147	2.829	2.096	2.519	2.044
0.5	5.325	2.516	4.174	2.416	3.69	2.323	3.26	2.224	2.891	2.13
0.6	7.272	3.337	5.23	2.919	4.528	2.663	3.911	2.427	3.388	2.222
0.7	9.54	4.252	6.489	3.484	5.533	3.047	4.699	2.66	3.995	2.33
0.8	12.188	5.277	7.986	4.123	6.732	3.483	5.645	2.928	4.73	2.46
0.9	16.391	6.885	10.159	5.067	8.421	4.111	6.924	3.293	5.667	2.61
1	21.273	8.687	12.73	6.133	10.425	4.826	8.447	3.715	6.791	2.791
Outside crack										
0	0.789	0.479	0.663	0.514	0.577	0.508	0.501	0.498	0.437	0.487
0.08	0.807	0.441	0.687	0.5	0.598	0.505	0.52	0.503	0.454	0.5
0.2	0.836	0.382	0.726	0.476	0.633	0.499	0.552	0.512	0.482	0.52
0.3	0.957	0.396	0.81	0.506	0.698	0.528	0.6	0.54	0.517	0.547
0.4	1.086	0.411	0.901	0.537	0.768	0.558	0.653	0.57	0.555	0.576
0.5	1.224	0.427	0.997	0.569	0.844	0.59	0.71	0.601	0.598	0.606
0.6	1.593	0.581	1.188	0.695	0.981	0.684	0.804	0.668	0.655	0.652
0.7	1.985	0.742	1.391	0.827	1.129	0.783	0.905	0.739	0.717	0.699
0.8	2.402	0.91	1.608	0.964	1.287	0.886	1.014	0.813	0.785	0.749
0.9	3.091	1.232	1.924	1.2	1.51	1.053	1.159	0.923	0.868	0.812
1	3.823	1.568	2.261	1.446	1.75	1.228	1.316	1.039	0.957	0.879

Table 9 F-factors for axial cracks in a pressurized cylinder (Eq 30d), $R/t = 1.5$

a/t	a/c = 0.2		a/c = 0.4		a/c = 0.6		a/c = 0.8		a/c = 1.0	
	F (90°)	F (0°)	F (90°)	F (0°)	F (90°)	F (0°)	F (90°)	F (0°)	F (90°)	F (0°)
Inside crack										
0	2.097	1.194	1.93	1.326	1.713	1.395	1.52	1.436	1.355	1.462
0.08	2.189	1.212	1.974	1.352	1.757	1.416	1.563	1.453	1.396	1.475
0.2	2.343	1.241	2.055	1.397	1.836	1.452	1.639	1.481	1.47	1.497
0.3	2.692	1.411	2.255	1.522	1.999	1.543	1.771	1.544	1.575	1.536
0.4	3.078	1.592	2.48	1.656	2.185	1.641	1.922	1.612	1.698	1.579
0.5	3.505	1.784	2.736	1.801	2.397	1.747	2.097	1.685	1.842	1.626
0.6	4.402	2.237	3.223	2.078	2.775	1.945	2.383	1.818	2.05	1.704
0.7	5.391	2.719	3.769	2.373	3.201	2.157	2.708	1.96	2.291	1.788
0.8	6.485	3.234	4.381	2.69	3.682	2.385	3.077	2.113	2.568	1.88
0.9	8.152	4.019	5.259	3.135	4.356	2.702	3.578	2.323	2.925	2.002
1	9.995	4.859	6.242	3.614	5.113	3.044	4.144	2.55	3.333	2.134
Outside crack										
0	0.824	0.522	0.716	0.548	0.63	0.554	0.554	0.553	0.488	0.549
0.08	0.856	0.489	0.744	0.54	0.653	0.554	0.573	0.559	0.504	0.56
0.2	0.906	0.437	0.788	0.528	0.69	0.553	0.604	0.568	0.53	0.577
0.3	1.042	0.457	0.872	0.566	0.754	0.587	0.65	0.598	0.561	0.604
0.4	1.184	0.478	0.961	0.605	0.821	0.622	0.698	0.63	0.595	0.633
0.5	1.333	0.5	1.055	0.645	0.891	0.658	0.75	0.662	0.63	0.662
0.6	1.678	0.646	1.229	0.768	1.016	0.752	0.832	0.731	0.679	0.71
0.7	2.039	0.797	1.412	0.894	1.147	0.848	0.92	0.801	0.73	0.759
0.8	2.417	0.953	1.604	1.025	1.285	0.947	1.013	0.874	0.786	0.81
0.9	3.009	1.239	1.876	1.237	1.475	1.102	1.135	0.98	0.853	0.876
1	3.63	1.534	2.161	1.456	1.676	1.261	1.265	1.09	0.924	0.945

type of crack geometry. However, if the relationship between a/D and a/b is clearly established, and the K values (or the geometric factor of β or Y) at locations on the crack propagation path are specifically determined for the expected crack geometries, crack growth can be predicted by monitoring only one point on the crack periphery (i.e., by treating the crack configuration as if it were a one-dimensional crack). In this section the geometry factor Y will be expressed as a function of a/D (or a/d) for stress-intensity factors at Point A. Justifications for using this approach were made by showing good correlations between the stress-intensity equations (presented below) and available test data (Ref 27). A bibliography of stress-intensity factors for mechanical joints is also given in the article "Fatigue of Mechanically Fastened Joints" in this Volume.

Crack on Bolt Shank or Thread

For a crack originated on the circumferential surface of an unthreaded (or threaded) cylinder, the one-dimensional crack-tip stress intensity, K, can be defined as:

$$K = S \cdot Y \cdot \sqrt{(\pi a)} \qquad \text{(Eq 26a)}$$

where S is the applied stress [equal to $4P/(\pi D^2)$, $4P/(\pi d^2)$, $32M/(\pi D^3)$, or $32M/(\pi d^3)$, depending on the application]. The equation below is suitable for cracks on the shank of a bolt, or at the root of a screw thread, subjected to either tension or bending:

$$Y = \lambda \cdot \exp[\beta x] + A + Bx + Cx^2 + Dx^3$$
$$+ Ex^4 + Fx^5 + Gx^6 \qquad \text{(Eq 26b)}$$

Here, x equals a/D or a/d, whichever is appropriate. In the case of an unthreaded cylinder the first two terms in Eq 26(b) do not exist, because these terms cover the local stress concentration caused by the screw thread. The magnitude of the exponential term decreases rapidly as crack length increases. The values of the coefficients are given in Table 3. A graphical presentation of Eq 26(b) is shown in Fig. 37.

It should be noted that λ and β are associated with the stress concentration factor and the local stress distribution near the notch root (or screw thread). It is expected that their values may vary, depending on the depth, the root radius, and the pitch angle of the thread. Therefore, it would be conceptually impossible to have a single curve for Y cover a wide range of geometric details. The values listed in Table 3 for λ and β represent a mixture of geometries that were used in the development of analytical and experimental data. The finite element models had a 45° pitch angle with root radii in a range of 0 to 0.125 mm. The test data were generated from notched specimens that either contained machined grooves that simulated the 12-UNF-3A screw thread (having a root radius of 0.38 mm) or had an unknown geometry. Although these values (for λ and β) certainly will not be universally suitable for all the local geometries encountered in practice, it is conceivable that they are applicable to most thread geometries as long as the crack size is not too small. In any event, Table 3 is a guide to making crack growth life estimates.

Crack on the Circumferential Plane of a Hollow Cylinder

In this category of cracks we consider the through-the-thickness crack and the thumbnail crack originated on the inner or outer wall of a hollow cylinder (a tube; a pipe; or a pressure vessel) that is subjected to axial loads (i.e., uniform tension, bending, or both).

Through-the-Thickness Crack

The nomenclature for this crack type is defined in Fig. 38, where M is the bending moment; P is the tensile load; α is the angle that represents one-half of the crack tip-to-tip circumferential arc; t is the thickness of the cylinder wall; R is the average radius of the inner and outer cylinder wall; c is half the crack length ($R\alpha$); S_t is the tension stress ($P/(2\pi Rt)$); and S_b is the bending stress ($M/(\pi tR^2)$). The stress-intensity equation is:

$$K = S \cdot F \cdot \sqrt{(\pi c)} \qquad \text{(Eq 27a)}$$

where S can be designated as S_t or S_b, depending on the loading condition. Likewise, F can be designated as F_t or F_b. The dimensionless factors (F_t and F_b) are given in Eq 27(a) as a group of close-form solutions. However, application of these equations may be cumbersome because of their complexity. To help readers cut through the lengthy calculations, these factors have been computed for several specific R/t ratios (with an assumed Poisson ratio of 0.3). The results are presented in Fig. 39 and 40. Each of these curves was fitted by a polynomial equation:

$$F = A + B\alpha + C\alpha^2 + D\alpha^3 + E\alpha^4 + F\alpha^5 + G\alpha^6 \qquad \text{(Eq 27b)}$$

The coefficients are listed in Tables 4 and 5. For R/t ratios not included in these tables, interpolated values for F_t or F_b can be used.

It should be noted that the solution of Eq 27(a) was developed on the basis of the thin-wall theory (for $R/t \geq 30$). Figures 39 and 40 and Tables 4 and 5 can be used for tubes having thicker walls or smaller curvatures (i.e., $R/t < 30$). Although $R/t = 7$ is included in these figures and tables, it is recommended that $R/t = 15$ be considered the limit of applicability.

Part-Through Crack

For a thumbnail crack on the inner or outer wall of a hollow cylinder (Fig. 41), the expression for K is given by Ref 33 as:

$$K = S \cdot F \cdot \sqrt{(\pi a)} \qquad \text{(Eq 28)}$$

where S can be S_t or S_b, depending on the loading condition. In either case, F is a function of a/c, R/t, and a/t. The values of F for internal and external cracks subjected to tension or bending are given in Tables 6 and 7.

Pressurized Cylinder and Sphere

Axial Cracks in Cylinder

In this section we consider a long cylinder subjected to internal pressure. A through-the-wall-thickness crack or a part-through crack is placed along the length of the cylinder (Fig. 42). The length of the cylinder is sufficiently long (as compared to the length of the crack), and the end effect on crack-tip stress intensity is not considered.

Through-the-Thickness Crack. In a pressurized thin-wall cylinder (Fig. 42), the elastic stress-intensity factor of Folias/Erdogan is given by Ref 34 as:

$$K = S \cdot Y \cdot \sqrt{(\pi c)} \qquad \text{(Eq 29a)}$$

where

$$Y = (1 + 0.52\lambda + 1.29\lambda^2 - 0.074\lambda^3)^{1/2} \qquad \text{(Eq 29b)}$$

where $\lambda = c/\sqrt{(Rt)}$, S is the pR/t stress, and R is the inner radius of the cylinder. Equation 29(a) accounts for the effects of shell curvature on stress intensity. Poisson's ratio is assumed to be $1/3$. The applicable range is $0 \leq \lambda \leq 10$.

Part-Through Crack. Now consider a pair of thumbnail cracks symmetrically located on the inner or outer wall of a hollow cylinder (Fig. 42). The expression for K can be written as:

$$K = S \cdot \alpha \cdot \sqrt{(\pi a/Q)} \qquad \text{(Eq 30a)}$$

Table 10 F-factors for axial cracks in a pressurized cylinder (Eq 30d), R/t = 2

a/t	a/c = 0.2 F (90°)	F (0°)	a/c = 0.4 F (90°)	F (0°)	a/c = 0.6 F (90°)	F (0°)	a/c = 0.8 F (90°)	F (0°)	a/c = 1.0 F (90°)	F (0°)
Inside crack										
0	1.796	1.027	1.68	1.141	1.495	1.216	1.33	1.264	1.188	1.296
0.08	1.873	1.043	1.705	1.165	1.518	1.231	1.35	1.271	1.207	1.297
0.2	1.996	1.066	1.749	1.208	1.558	1.257	1.387	1.283	1.241	1.298
0.3	2.249	1.197	1.883	1.308	1.665	1.329	1.471	1.332	1.304	1.328
0.4	2.52	1.332	2.03	1.413	1.782	1.405	1.563	1.384	1.376	1.359
0.5	2.81	1.474	2.19	1.524	1.911	1.485	1.665	1.438	1.456	1.392
0.6	3.378	1.798	2.508	1.723	2.156	1.634	1.848	1.545	1.587	1.465
0.7	3.988	2.137	2.853	1.931	2.422	1.79	2.048	1.658	1.733	1.541
0.8	4.644	2.491	3.226	2.148	2.713	1.953	2.268	1.775	1.894	1.621
0.9	5.6	3.019	3.762	2.442	3.122	2.176	2.571	1.938	2.108	1.733
1	6.63	3.572	4.341	2.75	3.567	2.41	2.902	2.108	2.345	1.852
Outside crack										
0	0.846	0.548	0.748	0.567	0.662	0.581	0.585	0.586	0.519	0.586
0.08	0.885	0.517	0.777	0.564	0.686	0.583	0.604	0.592	0.534	0.596
0.2	0.946	0.469	0.824	0.559	0.723	0.586	0.634	0.601	0.558	0.611
0.3	1.09	0.493	0.907	0.601	0.785	0.622	0.677	0.633	0.586	0.638
0.4	1.238	0.517	0.994	0.645	0.849	0.66	0.722	0.665	0.616	0.666
0.5	1.393	0.542	1.085	0.689	0.916	0.698	0.77	0.698	0.647	0.694
0.6	1.722	0.684	1.248	0.81	1.032	0.791	0.845	0.767	0.689	0.743
0.7	2.063	0.829	1.418	0.933	1.152	0.885	0.924	0.837	0.734	0.794
0.8	2.418	0.978	1.595	1.059	1.278	0.982	1.007	0.909	0.781	0.845
0.9	2.953	1.243	1.84	1.257	1.448	1.128	1.115	1.013	0.838	0.913
1	3.507	1.515	2.095	1.459	1.626	1.279	1.229	1.118	0.898	0.982

Table 11 F-factors for axial cracks in a pressurized cylinder (Eq 30d), R/t = 3

a/t	a/c = 0.2 F (90°)	F (0°)	a/c = 0.4 F (90°)	F (0°)	a/c = 0.6 F (90°)	F (0°)	a/c = 0.8 F (90°)	F (0°)	a/c = 1.0 F (90°)	F (0°)
Inside crack										
0	1.476	0.848	1.377	0.972	1.23	1.03	1.098	1.066	0.985	1.089
0.08	1.547	0.864	1.408	0.993	1.255	1.044	1.118	1.075	1.001	1.094
0.2	1.657	0.89	1.456	1.025	1.295	1.067	1.15	1.089	1.026	1.102
0.3	1.848	0.992	1.551	1.103	1.367	1.124	1.204	1.129	1.064	1.127
0.4	2.047	1.096	1.651	1.182	1.444	1.182	1.261	1.179	1.105	1.154
0.5	2.253	1.204	1.756	1.264	1.525	1.243	1.321	1.212	1.149	1.181
0.6	2.625	1.434	1.943	1.415	1.664	1.355	1.421	1.292	1.215	1.233
0.7	3.012	1.671	2.139	1.571	1.811	1.47	1.526	1.373	1.286	1.287
0.8	3.416	1.915	2.346	1.73	1.966	1.589	1.638	1.457	1.362	1.342
0.9	3.977	2.268	2.627	1.946	2.175	1.749	1.786	1.57	1.459	1.417
1	4.564	2.63	2.923	2.168	2.396	1.913	1.943	1.687	1.563	1.493
Outside crack										
0	0.85	0.516	0.763	0.577	0.681	0.6	0.607	0.612	0.544	0.619
0.08	0.904	0.51	0.801	0.584	0.71	0.608	0.629	0.622	0.56	0.629
0.2	0.986	0.502	0.859	0.596	0.755	0.621	0.664	0.636	0.585	0.644
0.3	1.134	0.544	0.939	0.646	0.813	0.663	0.703	0.669	0.609	0.671
0.4	1.287	0.586	1.022	0.698	0.873	0.705	0.744	0.703	0.634	0.698
0.5	1.443	0.629	1.106	0.75	0.935	0.747	0.786	0.737	0.66	0.725
0.6	1.737	0.752	1.238	0.865	1.026	0.835	0.844	0.803	0.691	0.771
0.7	2.04	0.877	1.374	0.982	1.121	0.925	0.904	0.869	0.722	0.818
0.8	2.35	1.004	1.514	1.102	1.218	1.016	0.966	0.936	0.756	0.866
0.9	2.792	1.204	1.693	1.276	1.341	1.146	1.042	1.029	0.792	0.928
1	3.246	1.407	1.878	1.454	1.468	1.279	1.12	1.124	0.831	0.991

Table 12 *F*-factors for axial cracks in a pressurized cylinder (Eq 30d), *R/t* = 4

a/t	a/c = 0.2		a/c = 0.4		a/c = 0.6		a/c = 0.8		a/c = 1.0	
	F (90°)	F (0°)	F (90°)	F (0°)	F (90°)	F (0°)	F (90°)	F (0°)	F (90°)	F (0°)
Inside crack										
0	1.325	0.762	1.235	0.889	1.105	0.939	0.988	0.969	0.888	0.989
0.08	1.393	0.78	1.267	0.907	1.13	0.952	1.008	0.979	0.903	0.995
0.2	1.496	0.806	1.317	0.936	1.17	0.974	1.038	0.994	0.925	1.006
0.3	1.661	0.895	1.397	1.003	1.229	1.024	1.08	1.03	0.953	1.029
0.4	1.83	0.986	1.479	1.072	1.29	1.075	1.124	1.066	0.983	1.053
0.5	2.004	1.078	1.564	1.142	1.354	1.127	1.17	1.103	1.014	1.078
0.6	2.3	1.27	1.702	1.273	1.455	1.224	1.24	1.171	1.058	1.122
0.7	2.604	1.466	1.845	1.406	1.56	1.322	1.313	1.24	1.104	1.167
0.8	2.918	1.666	1.994	1.542	1.669	1.423	1.389	1.311	1.153	1.213
0.9	3.339	1.95	2.187	1.726	1.809	1.558	1.485	1.405	1.212	1.273
1	3.773	2.24	2.387	1.914	1.956	1.696	1.585	1.501	1.275	1.334
Outside crack										
0	0.853	0.498	0.772	0.582	0.692	0.61	0.62	0.627	0.558	0.637
0.08	0.914	0.507	0.814	0.595	0.724	0.623	0.643	0.638	0.574	0.647
0.2	1.007	0.52	0.878	0.616	0.773	0.641	0.679	0.655	0.6	0.662
0.3	1.158	0.572	0.956	0.671	0.828	0.685	0.716	0.689	0.621	0.689
0.4	1.312	0.624	1.035	0.726	0.885	0.729	0.754	0.724	0.644	0.715
0.5	1.468	0.677	1.116	0.783	0.943	0.774	0.793	0.759	0.666	0.742
0.6	1.744	0.789	1.231	0.895	1.022	0.86	0.841	0.822	0.69	0.787
0.7	2.025	0.902	1.348	1.009	1.102	0.946	0.891	0.886	0.715	0.832
0.8	2.311	1.017	1.468	1.124	1.185	1.034	0.942	0.951	0.74	0.878
0.9	2.705	1.182	1.613	1.286	1.282	1.155	1.001	1.038	0.766	0.936
1	3.106	1.348	1.762	1.45	1.383	1.278	1.061	1.126	0.793	0.996

Table 13 *F*-factors for axial cracks in a pressurized cylinder (Eq 30d), *R/t* = 6

a/t	a/c = 0.2		a/c = 0.4		a/c = 0.6		a/c = 0.8		a/c = 1.0	
	F (90°)	F (0°)	F (90°)	F (0°)	F (90°)	F (0°)	F (90°)	F (0°)	F (90°)	F (0°)
Inside crack										
0	1.179	0.691	1.097	0.801	0.986	0.847	0.885	0.874	0.799	0.892
0.08	1.247	0.706	1.133	0.82	1.012	0.861	0.904	0.885	0.811	0.9
0.2	1.351	0.728	1.187	0.849	1.052	0.883	0.932	0.902	0.829	0.912
0.3	1.491	0.806	1.253	0.909	1.1	0.929	0.964	0.935	0.849	0.935
0.4	1.634	0.884	1.32	0.97	1.148	0.974	0.997	0.968	0.869	0.957
0.5	1.779	0.963	1.388	1.032	1.198	1.021	1.031	1.001	0.89	0.98
0.6	2.004	1.129	1.484	1.143	1.267	1.103	1.077	1.059	0.917	1.018
0.7	2.232	1.296	1.582	1.256	1.337	1.187	1.124	1.118	0.945	1.056
0.8	2.464	1.466	1.682	1.371	1.409	1.272	1.172	1.178	0.973	1.095
0.9	2.76	1.707	1.804	1.524	1.495	1.384	1.23	1.256	1.007	1.145
1	3.062	1.951	1.928	1.678	1.584	1.497	1.289	1.335	1.014	1.195
Outside crack										
0	0.872	0.522	0.79	0.602	0.711	0.632	0.64	0.65	0.578	0.661
0.08	0.935	0.529	0.833	0.616	0.743	0.644	0.662	0.66	0.593	0.67
0.2	1.031	0.539	0.899	0.636	0.792	0.662	0.697	0.676	0.616	0.684
0.3	1.176	0.595	0.97	0.691	0.842	0.706	0.73	0.71	0.634	0.709
0.4	1.322	0.652	1.042	0.747	0.893	0.749	0.763	0.743	0.652	0.734
0.5	1.47	0.709	1.115	0.804	0.944	0.793	0.796	0.777	0.671	0.759
0.6	1.72	0.837	1.21	0.916	1.009	0.878	0.835	0.838	0.689	0.801
0.7	1.973	0.966	1.306	1.029	1.074	0.963	0.875	0.9	0.708	0.844
0.8	2.229	1.097	1.404	1.143	1.14	1.049	0.915	0.962	0.726	0.886
0.9	2.572	1.288	1.517	1.303	1.215	1.168	0.959	1.046	0.744	0.941
1	2.919	1.48	1.632	1.465	1.292	1.287	1.003	1.13	0.763	0.996

where $S = pR/t$, and α is a function of R/t, a/c, a/t, φ, and the nonuniform tangential stresses acting on the crack plane. Having the nonuniform stress distribution normalized to the pR/t stress, the α function can be written as (Ref 35):

$$\alpha_i = (t/R) \cdot r^2/(r^2 - R^2) \cdot [2H_0 - 2H_1(a/R)$$
$$+ 3H_2(a/R)^2 - 4H_3(a/R)^3] \qquad \text{(Eq 30b)}$$

for an internal crack, or

$$\alpha_o = (t/R) \cdot R^2/(r^2 - R^2) \cdot [2G_0 + 2G_1(a/r) +$$
$$3G_2(a/r)^2 + 4G_3(a/r)^3] \qquad \text{(Eq 30c)}$$

for an external crack.

Here r and R are the outer and inner radius of the cylinder, respectively. The H and G values are functions of R/t, a/c, a/t, and φ. Each H or G value corresponds to a particular loading distribution. The subscript 0 corresponds to uniform tension; subscript 1 to linear distribution; subscript 2 to quadratic; and subscript 3 to cubic. Using the H

and G values given in Ref 33, α_i and α_o values for the pressurized cylinder were computed for 11 R/t ratios (1, 1.5, 2, 3, 4, 6, 8, 10, 20, 30, and ≥50).

Let $F = \alpha/\sqrt{Q}$ so that Eq 30(a) is reduced to:

$$K = S \cdot F \cdot \sqrt{(\pi a)} \qquad \text{(Eq 30d)}$$

The computed F values are presented in Tables 8 to 18. In Ref 33, the stress-intensity factor tables listed H and G values for only five R/t ratios (1, 2, 4, 10, and ≥50), in combination with three a/c ratios (0.2, 0.4, and 1.0) and five a/t ratios (0, 0.2, 0.5, 0.8, and 1.0). For other R/t ratios listed in Tables 8 to 18, the F values have been obtained by linear interpolation. Likewise, interpolated F values can be used for other parametric combinations not included in these tables.

Although the α factors were originally derived for the double crack configuration, they are also applicable to the single crack configuration. When Eq 30(a) or 30(d) is used for the single crack configuration, the maximum error is 4% (higher than the actual value), depending on the a/c ratio.

Crack in Pressurized Sphere

Currently, the Erdogan solution for a through-the-thickness crack (Ref 20) is the only solution available for the pressurized spherical shell (Fig. 43): $K = S \cdot Y \cdot \sqrt{(\pi c)}$, where

$$Y = (1 + 3\lambda^{1.9})^{0.4}, \qquad \text{(Eq 31)}$$

$\lambda = c/\sqrt{(Rt)}$, S is the $pR/2t$ stress; and R is the average of the inner and outer radii. This equation accounts for the effects of shell curvature on stress intensity. The Poisson ratio is assumed to be 1/3. The applicable range is $0 \le \lambda \le 3$.

REFERENCES

1. G.R. Irwin, *Hanbuch der Physik*, Vol VI, Springer, 1958, p 551
2. D. Broek, *Elementary Engineering Fracture Mechanics*, 3rd ed., Nijhoff, 1981
3. H. Tada, P.C. Paris, and G.R. Irwin, *Stress Analysis of Cracks Handbook*, Del Research Corp., 1973
4. G.C. Sih, *Handbook of Stress-Intensity Factors*, Lehigh University, Bethlehem, PA, 1973
5. D.P. Rooke and D.J. Cartwright, *Compendium of Stress-Intensity Factors*, Her Majesty's Stationery Office, London, 1976
6. V. Kumar et al., "An Engineering Approach for Elastic-Plastic Fracture Analysis," EPRI NP-1931, Electric Power Research Institute, Palo Alto, CA, 1981
7. M. Isida, Stress-Intensity Factor for the Tension of an Eccentrically Cracked Strip, *Journal of Applied Mechanics*, Vol 33, *Trans. ASME*, Series E, 1966, p 674-675
8. M. Isida, Effect of Width and Length on Stress-Intensity Factors of Internally Cracked Plate under Various Boundary Conditions, *International Journal of Fracture Mechanics*, Vol 7, 1971

Table 14 F-factors for axial cracks in a pressurized cylinder (Eq 30d), R/t = 8

a/t	a/c = 0.2 F(90°)	F(0°)	a/c = 0.4 F(90°)	F(0°)	a/c = 0.6 F(90°)	F(0°)	a/c = 0.8 F(90°)	F(0°)	a/c = 1.0 F(90°)	F(0°)
Inside crack										
0	1.108	0.656	1.031	0.759	0.928	0.802	0.835	0.828	0.755	0.845
0.08	1.176	0.669	1.068	0.777	0.955	0.816	0.853	0.839	0.766	0.853
0.2	1.28	0.69	1.124	0.805	0.995	0.838	0.881	0.856	0.783	0.866
0.3	1.409	0.762	1.183	0.862	1.038	0.881	0.909	0.888	0.799	0.888
0.4	1.54	0.834	1.243	0.92	1.08	0.925	0.937	0.919	0.815	0.909
0.5	1.672	0.908	1.304	0.978	1.124	0.968	0.966	0.951	0.832	0.931
0.6	1.866	1.061	1.383	1.081	1.179	1.045	1.002	1.005	0.852	0.967
0.7	2.062	1.215	1.462	1.184	1.235	1.122	1.038	1.059	0.873	1.002
0.8	2.26	1.371	1.543	1.289	1.293	1.199	1.076	1.114	0.894	1.038
0.9	2.505	1.593	1.636	1.428	1.358	1.301	1.118	1.185	0.917	1.083
1	2.753	1.816	1.73	1.568	1.424	1.404	1.161	1.256	0.941	1.129
Outside crack										
0	0.883	0.534	0.8	0.613	0.721	0.644	0.65	0.662	0.588	0.673
0.08	0.946	0.54	0.844	0.627	0.753	0.656	0.672	0.672	0.603	0.382
0.2	1.043	0.55	0.91	0.647	0.802	0.673	0.706	0.688	0.624	0.696
0.3	1.185	0.608	0.977	0.702	0.849	0.716	0.736	0.72	0.64	0.72
0.4	1.327	0.667	1.045	0.758	0.897	0.76	0.767	0.753	0.657	0.744
0.5	1.471	0.726	1.114	0.814	0.945	0.804	0.798	0.787	0.674	0.768
0.6	1.707	0.862	1.198	0.926	1.001	0.887	0.847	0.744	0.688	0.809
0.7	1.946	0.999	1.284	1.039	1.059	0.972	0.907	0.774	0.704	0.85
0.8	2.186	1.138	1.37	1.153	1.117	1.057	0.968	0.805	0.719	0.891
0.9	2.504	1.342	1.467	1.312	1.181	1.174	1.05	0.838	0.733	0.943
1	2.824	1.548	1.565	1.472	1.245	1.292	0.974	1.132	0.747	0.996

Table 15 F-factors for axial cracks in a pressurized cylinder (Eq 30d), R/t = 10

a/t	a/c = 0.2 F(90°)	F(0°)	a/c = 0.4 F(90°)	F(0°)	a/c = 0.6 F(90°)	F(0°)	a/c = 0.8 F(90°)	F(0°)	a/c = 1.0 F(90°)	F(0°)
Inside crack										
0	1.067	0.635	0.992	0.733	0.894	0.775	0.806	0.8	0.729	0.817
0.08	1.135	0.648	1.03	0.752	0.921	0.79	0.823	0.812	0.739	0.826
0.2	1.237	0.667	1.086	0.78	0.961	0.812	0.85	0.829	0.755	0.839
0.3	1.361	0.736	1.142	0.835	1.001	0.853	0.876	0.859	0.769	0.86
0.4	1.485	0.805	1.199	0.89	1.041	0.895	0.902	0.89	0.784	0.881
0.5	1.61	0.875	1.256	0.945	1.081	0.937	0.928	0.921	0.798	0.902
0.6	1.786	1.021	1.324	1.043	1.129	1.01	0.959	0.972	0.815	0.936
0.7	1.964	1.168	1.394	1.142	1.177	1.083	0.99	1.024	0.831	0.97
0.8	2.144	1.316	1.464	1.241	1.227	1.157	1.021	1.076	0.849	1.004
0.9	2.361	1.527	1.541	1.372	1.281	1.253	1.056	1.143	0.867	1.047
1	2.58	1.739	1.62	1.504	1.335	1.35	1.09	1.21	0.885	1.09
Outside crack										
0	0.889	0.542	0.806	0.62	0.728	0.651	0.657	0.669	0.595	0.681
0.08	0.953	0.547	0.85	0.633	0.76	0.663	0.678	0.679	0.609	0.689
0.2	1.051	0.556	0.917	0.653	0.808	0.68	0.712	0.695	0.629	0.703
0.3	1.19	0.616	0.981	0.709	0.853	0.723	0.74	0.727	0.644	0.726
0.4	1.33	0.676	1.047	0.765	0.899	0.766	0.769	0.76	0.659	0.75
0.5	1.471	0.736	1.113	0.821	0.945	0.81	0.798	0.792	0.675	0.774
0.6	1.699	0.878	1.191	0.933	0.997	0.893	0.829	0.852	0.688	0.814
0.7	1.929	1.02	1.27	1.045	1.05	0.977	0.86	0.912	0.701	0.854
0.8	2.16	1.163	1.35	1.159	1.103	1.062	0.891	0.972	0.714	0.894
0.9	2.462	1.375	1.436	1.317	1.16	1.178	0.923	1.053	0.726	0.945
1	2.766	1.589	1.524	1.477	1.217	1.295	0.956	1.134	0.738	0.996

national Journal of Fracture, Vol 11, 1975, p 283-294

15. J.C. Newman, Jr., "An Improved Method of Collocation for the Stress Analysis of Cracked Plates with Various Shaped Boundaries," NASA TN D-6376, National Aeronautics and Space Administration, Aug 1971

16. G.R. Irwin, Crack-Extension Force for a Part-Through Crack in a Plate, *Journal of Applied Mechanics,* Vol 84, *Trans. ASME,* Series E, 1962, p 651-654

17. J.C. Newman, Jr. and I.S. Raju, An Empirical Stress-Intensity Factor Equation for the Surface Crack, *Engineering Fracture Mechanics,* Vol 15 (No. 1-2), 1981, p 185-192

18. J.C. Newman, Jr. and I.S. Raju, Stress-Intensity Factor Equations for Cracks in Three-Dimensional Finite Bodies, *Fracture Mechanics: Fourteenth Symposium—Volume I: Theory and Analysis,* STP 791, American Society for Testing and Materials, 1983, p I-238 to I-265

19. I.S. Raju, S.R. Mettu, and V. Shivakumar, Stress-Intensity Factor Solutions for Surface Cracks in Flat Plates Subjected to Nonuniform Stresses, *Fracture Mechanics: Twenty-Fourth Volume,* STP 1207, American Society for Testing and Materials, 1994, p 560-580

20. R.G. Forman, V. Shivakumar and J.C. Newman, *Fatigue Crack Growth Computer Program NASA/FLAGO,* Version 2.0, JSC-22267A, National Aeronautics and Space Administration, May 1994

21. L.R. Hall and R.W. Finger, Fracture and Fatigue Growth of Partially Embedded Flaws, *Proceeding, Air Force Conference on Fatigue and Fracture of Aircraft Structures and Materials,* AFFDL-TR-70-144, Air Force Flight Dynamics Laboratory, Wright-Patterson Air Force Base, 1970, p 235-262

22. W. Zhao and S.N. Atluri, Stress-Intensity Factors for Surface and Corner-Cracked Fastener Holes by the Weight Function Method, *Structural Integrity of Fasteners,* STP 1236, American Society for Testing and Materials, 1995, p 95-107

23. A.F. Liu, Stress-Intensity Factor for a Corner Flaw, *Engineering Fracture Mechanics,* Vol 4, Pergamon Press, 1972, p 175-179

24. R.C. Shah, Stress-Intensity Factors for Through and Part-Through Cracks Originating at Fastener Holes, *Mechanics of Crack Growth,* STP 590, American Society for Testing and Materials, 1976, p 429-459

25. K. Katherisan, T.M. Hsu, and T.R. Brussat, "Advanced Life Analysis Methods-Crack Growth Analysis Methods for Attachment Lugs," AFWAL TR-84-3080, Vol II, Flight Dynamics Laboratory, Wright-Patterson Air Force Base, Sept 1984

26. A.F. Liu, Evaluation of Current Analytical Methods for Crack Growth in a Bolt, *Durability and Structural Reliability of Airframes—ICAF 17,* Vol 2, EMAS Ltd., West Midlands, U.K., 1993, p 1141-1155

27. A.F. Liu, Behavior of Fatigue Cracks in a Tension Bolt, *Structural Integrity of Fasteners,* STP

9. W.F. Brown, Jr. and J.E. Srawley, *Plane Strain Crack Toughness Testing of High Strength Metallic Materials,* STP 410, American Society for Testing and Materials, 1966, p 77-79

10. E.P. Phillips, The Influence of Crack Closure on Fatigue Crack Growth Thresholds in 2024-T3 Aluminum Alloy, *Mechanics of Fatigue Crack Closure,* STP 982, American Society for Testing and Materials, 1988, p 515

11. P.C. Paris and G. Sih, *Stress Analysis of Cracks,* ASTM STP 381, American Society for Testing and Materials, 1965, p 30-81

12. O.L. Bowie, Analysis of an Infinite Plate Containing Radial Cracks Originating at the Boundaries of an Internal Circular Hole, *Journal of Mathematics and Physics,* Vol 35, 1956, p 60

13. J.C. Newman, Jr., "Predicting Failure of Specimens with Either Surface Cracks or Corner Cracks at Holes," NASA TN D-8244, National Aeronautics and Space Administration, June 1976

14. A.F. Grandt, Jr., Stress-Intensity Factors for Some Through-Cracked Fastener Holes, *Inter-*

Table 16 F-factors for axial cracks in a pressurized cylinder (Eq 30d), R/t = 20

a/t	a/c = 0.2		a/c = 0.4		a/c = 0.6		a/c = 0.8		a/c = 1.0	
	F (90°)	F (0°)	F (90°)	F (0°)	F (90°)	F (0°)	F (90°)	F (0°)	F (90°)	F (0°)
Inside crack										
0	0.972	0.594	0.92	0.677	0.833	0.722	0.753	0.751	0.684	0.771
0.08	1.05	0.607	0.96	0.698	0.86	0.738	0.77	0.762	0.692	0.777
0.2	1.167	0.626	1.021	0.73	0.901	0.761	0.795	0.778	0.704	0.788
0.3	1.278	0.688	1.07	0.781	0.935	0.799	0.816	0.806	0.714	0.807
0.4	1.39	0.75	1.12	0.832	0.969	0.838	0.837	0.835	0.725	0.827
0.5	1.501	0.812	1.169	0.884	1.003	0.878	0.858	0.863	0.735	0.847
0.6	1.631	0.94	1.219	0.968	1.037	0.942	0.879	0.911	0.745	0.88
0.7	1.762	1.068	1.268	1.053	1.071	1.006	0.9	0.958	0.755	0.914
0.8	1.893	1.197	1.318	1.137	1.105	1.071	0.921	1.006	0.766	0.947
0.9	2.033	1.378	1.366	1.246	1.138	1.154	0.941	1.067	0.775	0.991
1	2.175	1.56	1.414	1.354	1.171	1.237	0.961	1.128	0.785	1.034
Outside crack										
0	0.884	0.546	0.829	0.62	0.751	0.659	0.679	0.684	0.617	0.701
0.08	0.958	0.555	0.971	0.683	0.78	0.673	0.697	0.694	0.627	0.708
0.2	1.069	0.569	0.933	0.655	0.823	0.693	0.725	0.709	0.641	0.718
0.3	1.185	0.627	0.985	0.716	0.858	0.732	0.746	0.738	0.651	0.738
0.4	1.3	0.684	1.037	0.766	0.894	0.771	0.768	0.766	0.662	0.759
0.5	1.416	0.742	1.09	0.817	0.93	0.81	0.79	0.795	0.673	0.779
0.6	1.565	0.866	1.143	0.905	0.965	0.877	0.811	0.845	0.682	0.814
0.7	1.715	0.99	1.196	0.994	1.001	0.944	0.832	0.894	0.691	0.849
0.8	1.865	1.114	1.25	1.082	1.037	1.012	0.854	0.944	0.7	0.884
0.9	2.039	1.293	1.302	1.2	1.071	1.101	0.873	1.009	0.708	0.929
1	2.214	1.473	1.354	1.317	1.106	1.19	0.893	1.074	0.715	0.975

Table 17 F-factors for axial cracks in a pressurized cylinder (Eq 30d), R/t = 30

a/t	a/c = 0.2		a/c = 0.4		a/c = 0.6		a/c = 0.8		a/c = 1.0	
	F (90°)	F (0°)	F (90°)	F (0°)	F (90°)	F (0°)	F (90°)	F (0°)	F (90°)	F (0°)
Inside crack										
0	0.941	0.581	0.897	0.659	0.812	0.705	0.735	0.735	0.669	0.755
0.08	1.022	0.593	0.938	0.681	0.84	0.72	0.752	0.745	0.676	0.761
0.2	1.144	0.612	0.999	0.713	0.881	0.744	0.777	0.761	0.688	0.771
0.3	1.251	0.672	1.046	0.763	0.913	0.782	0.796	0.788	0.697	0.79
0.4	1.358	0.732	1.094	0.813	0.946	0.82	0.816	0.816	0.705	0.809
0.5	1.466	0.792	1.141	0.864	0.968	0.858	0.835	0.844	0.714	0.828
0.6	1.582	0.914	1.185	0.944	1.007	0.919	0.853	0.89	0.723	0.862
0.7	1.698	1.036	1.229	1.024	1.037	0.981	0.871	0.936	0.731	0.895
0.8	1.814	1.159	1.272	1.104	1.067	1.043	0.889	0.983	0.739	0.928
0.9	1.932	1.331	1.311	1.205	1.093	1.122	0.905	1.042	0.747	0.972
1	2.05	1.503	1.35	1.306	1.12	1.2	0.921	1.102	0.755	1.015
Outside crack										
0	0.883	0.547	0.837	0.62	0.759	0.662	0.687	0.689	0.625	0.708
0.08	0.96	0.558	0.878	0.64	0.786	0.676	0.704	0.699	0.633	0.714
0.2	1.076	0.574	0.939	0.669	0.828	0.698	0.729	0.714	0.645	0.723
0.3	1.183	0.63	0.986	0.718	0.86	0.735	0.749	0.741	0.654	0.742
0.4	1.29	0.687	1.034	0.767	0.892	0.772	0.768	0.769	0.663	0.761
0.5	1.397	0.744	1.082	0.816	0.925	0.809	0.788	0.796	0.672	0.781
0.6	1.519	0.862	1.126	0.896	0.955	0.871	0.805	0.842	0.679	0.814
0.7	1.642	0.98	1.171	0.976	0.985	0.933	0.823	0.888	0.687	0.847
0.8	1.765	1.098	1.216	1.056	1.015	0.995	0.841	0.935	0.695	0.88
0.9	1.896	1.266	1.256	1.16	1.041	1.075	0.857	0.994	0.702	0.924
1	2.028	1.434	1.296	1.263	1.068	1.155	0.872	1.054	0.708	0.967

1236, American Society for Testing and Materials, 1995, p 124-140

28. T.L. Mackay and B.J. Alperin, Stress-Intensity Factors for Fatigue Cracking in High Strength Bolts, *Engineering Fracture Mechanics,* Vol 21, 1985, p 391-397

29. D. Wilhem, J. FitzGerald, J. Carter, and D. Dittmer, An Empirical Approach to Determining K for Surface Cracks, *Advances in Fracture Research,* Vol 1, Pergamon Press, 1981, p 11-21

30. R.G. Forman and V. Shivakumar, Growth Behavior of Surface Cracks in the Circumferential Plane of Solid and Hollow Cylinders, *Fracture Mechanics: Seventeenth Volume,* STP 905, American Society for Testing and Materials, 1986, p 59-74

31. R.G. Forman and S.R. Mettu, Behavior of Surface and Corner Cracks Subjected to Tensile and Bending Loads in Ti-6Al-4V Alloy, *Fracture Mechanics: Twenty Second Symposium (Volume I),* STP 1131, American Society for Testing and Materials, 1992, p 519-546

32. R.R. Cervay, "Empirical Fatigue Crack Growth Data for a Tension Loaded Threaded Fastener," AFWAL TR-88-4002, Flight Dynamics Laboratory, Wright-Patterson Air Force Base, Feb 1988

33. S.R. Mettu, I.S. Raju, and R.G. Forman, "Stress Intensity Factors for Part-Through Surface Cracks in Hollow Cylinders," Johnson Space Center/Lockheed Engineering Services Company Joint Publication, JSC Report 25685/LESC Report 30124, July 1992

34. J.C. Newman, Jr., "Fracture Analysis of Surface and Through Cracks in Cylindrical Pressure Vessels," NASA TN D-8325, National Aeronautics and Space Administration, Dec 1976

35. I.S. Raju and J.C. Newman, Jr., Stress-Intensity Factors for Internal and External Surface Cracks in Cylindrical Vessels, *Journal of Pressure Vessel Technology,* Vol 104, *Trans. ASME,* 1982, p 293-298

SELECTED REFERENCES

Handbooks of Stress-Intensity Factors

- J.P. Gallagher, F.J. Giessler, A.P. Berens, and R.M. Engle, Jr., *USAF Damage Tolerant Design Handbook: Guidelines for the Analysis and Design of Damage Tolerant Aircraft Structures,* AFWAL-TR-82-3073, Flight Dynamics Laboratory, Wright-Patterson Air Force Base, May 1984

- Y. Murakami et al., Ed., *Stress-Intensity Factors Handbook,* Vol 1 and 2, Pergamon Press, 1987

- P.C. Paris and G. Sih, *Stress Analysis of Cracks,* ASTM STP 381, 1965, with a good historical paper on stress-intensity factors, p 30-81

- D.P. Rooke and D.J. Cartwright, *Compendium of Stress-Intensity Factors,* Her Majesty's Stationery Office, London, U.K., 1976

- G. Sih, *Handbook of Stress-Intensity Factors,* Institute of Fracture and Solid Mechanics, Lehigh University, Bethlehem, PA, 1973 [Note: The Sih handbook defines crack-tip singularity and hence K differently than other handbooks (e.g., Sih, $K = \sigma\sqrt{a}$, where in other handbooks $K = \sigma\sqrt{\pi a}$).

- H. Tada, P.C. Paris, and G.R. Irwin, *Stress Analysis of Cracks Handbook,* Paris Productions, Inc., St. Louis, MO, 1985

Skin-Stringer Structures

- J.M. Bloom, The Effect of Riveted Stringer on the Stress in a Sheet with Circular Cutout, *Journal of Applied Mechanics,* Vol 33, *Trans. ASME,* Series E, 1966, p 198-199

- J.M. Bloom and J.L. Sanders, The Effect of Riveted Stringer on the Stress in Cracked Sheet, *Journal of Applied Mechanics,* Vol 33, *Trans. ASME,* Series E, 1966, p 561-570

- M. Creager and A.F. Liu, The Effect of Reinforcement on the Slow Stable Tear and Catastrophic Failure of Thin Metal Sheet, *Journal of Engineering Materials and Technology,* Vol 96, *Trans. ASME,* Series H, 1974, p 49-55

Table 18 *F*-factors for axial cracks in a pressurized cylinder (Eq 30d), *R/t* = 50

	a/c = 0.2		a/c = 0.4		a/c = 0.6		a/c = 0.8		a/c = 1.0	
a/t	F (90°)	F (0°)	F (90°)	F (0°)	F (90°)	F (0°)	F (90°)	F (0°)	F (90°)	F (0°)
Inside crack										
0	0.917	0.57	0.878	0.645	0.796	0.691	0.721	0.722	0.657	0.743
0.08	1	0.583	0.919	0.667	0.824	0.707	0.738	0.732	0.664	0.748
0.2	1.125	0.601	0.982	0.7	0.866	0.73	0.762	0.747	0.674	0.757
0.3	1.229	0.659	1.027	0.749	0.896	0.767	0.781	0.774	0.682	0.776
0.4	1.333	0.717	1.073	0.798	0.927	0.805	0.799	0.802	0.69	0.795
0.5	1.438	0.776	1.119	0.848	0.958	0.842	0.817	0.829	0.698	0.814
0.6	1.543	0.893	1.158	0.924	0.984	0.902	0.833	0.874	0.705	0.847
0.7	1.648	1.011	1.197	1.001	1.011	0.961	0.849	0.919	0.712	0.88
0.8	1.753	1.129	1.237	1.078	1.037	1.021	0.864	0.964	0.719	0.913
0.9	1.854	1.294	1.269	1.173	1.059	1.096	0.877	1.022	0.725	0.957
1	1.955	1.459	1.301	1.269	1.081	1.172	0.89	1.08	0.731	1
Outside crack										
0	0.882	0.548	0.844	0.62	0.765	0.664	0.693	0.694	0.631	0.714
0.08	0.961	0.56	0.884	0.641	0.792	0.679	0.709	0.703	0.638	0.719
0.2	1.081	0.578	0.943	0.673	0.832	0.702	0.733	0.718	0.648	0.727
0.3	1.181	0.634	0.987	0.72	0.861	0.738	0.75	0.744	0.656	0.746
0.4	1.281	0.689	1.031	0.767	0.891	0.773	0.768	0.77	0.663	0.764
0.5	1.382	0.745	1.075	0.815	0.921	0.809	0.785	0.797	0.671	0.782
0.6	1.482	0.858	1.113	0.888	0.946	0.867	0.8	0.84	0.678	0.814
0.7	1.584	0.972	1.151	0.962	0.971	0.924	0.815	0.883	0.684	0.846
0.8	1.685	1.085	1.188	1.036	0.997	0.981	0.831	0.927	0.691	0.878
0.9	1.781	1.243	1.219	1.128	1.017	1.054	0.843	0.983	0.697	0.919
1	1.878	1.402	1.25	1.22	1.038	1.126	0.856	1.038	0.702	0.961

- C.K. Gunther and J.T. Wozumi, Critical Failure Modes in Cracked Mechanically Fastened Stiffened Panels, *Design of Fatigue and Fracture Resistant Structures*, STP 761, American Society for Testing and Materials, 1982, p 310-327
- A.F. Liu, Fracture Control Methods for Space Vehicles, *Volume 1: Fracture Control Design Methods*, NASA CR-134596, National Aeronautics and Space Administration, Aug 1974
- T. Nishimura, Stress-Intensity Factors of Multiple Cracked Sheet with Riveted Stiffeners, *Journal of Engineering Materials and Technology*, Vol 113, *Trans. ASME*, Series H, 1991, p 280-284
- C.C. Poe, Jr., "The Effect of Riveted and Uniformly Spaced Stringers on the Stress-Intensity Factor of a Cracked Sheet," M.S. thesis, Virginia Polytechnic Institute, 1969
- C.C. Poe, Jr., Fatigue Crack Propagation in Stiffened Panels, *Damage Tolerance in Aircraft Structures*, STP 486, American Society for Testing and Materials, 1971, p 79-97
- C.C. Poe, Jr., "Stress-Intensity Factor for a Cracked Sheet with Riveted and Uniformly Spaced Stringers," NASA TR-R-358, National Aeronautics and Space Administration, Langley Research Center, Hampton, VA, May 1971
- R.C. Shah and F.T. Lin, Stress-Intensity Factors of Stiffened Panels with Partially Cracked Stiffeners, *Fracture Mechanics: Fourteenth Symposium—Volume I: Theory and Analysis*, STP 791, American Society for Testing and Materials, 1983, p I-157 to I-171
- S.V. Shkarayev and E.T. Mayer, Jr., Edge Cracks in Stiffened Plates, *Engineering Fracture Mechanics*, Vol 27, Pergamon Press, 1987, p 127-134
- T. Swift, Fracture Analysis of Stiffened Structure, *Damage Tolerance of Metallic Structures*, STP 842, American Society for Testing and Materials, 1984, p 69-107
- T. Swift, "Widespread Fatigue Monitoring—Issues and Concerns," NASA P-3274, Part 2, National Aeronautics and Space Administration, Langley Research Center, Hampton, VA, Sept 1994, p 829-870
- H. Vlieger, The Residual Strength Characteristics of Stiffened Panels Containing Fatigue Cracks, *Engineering Fracture Mechanics*, Vol 5, Pergamon Press, 1973, p 447-477

Transition of Part-Through Crack or Corner Crack to Through Crack

- J.C. Ekvall, T.R. Brussat, A.F. Liu, and M. Creager, "Engineering Criteria and Analysis Methodology for the Appraisal of Potential Fracture Resistant Primary Aircraft Structure," AFFDL-TR-72-80, Air Force Flight Dynamics Laboratory, Wright-Patterson Air Force Base, 1972
- A.F. Grandt, Jr., J.A. Harter, and B.J. Heath, The Transition of Part-Through Cracks at Holes into Through-the-Thickness Flaws, *Fracture Mechanics: Fifteenth Symposium*, STP 833, American Society for Testing and Materials, 1984, p 7-23
- K. Katherisan, T.M. Hsu, and T.R. Brussat, "Advanced Life Analysis Methods—Crack Growth Analysis Methods for Attachment Lugs," AFWAL-TR-84-3080, Vol II, Flight Dynamics Laboratory, Wright-Patterson Air Force Base, Sept 1984
- J.M. Waraniak and A.F. Liu, "Fatigue and Crack Propagation Analysis of Mechanically Fastened Joints," AIAA Paper 83-0839, presented at the AIAA/ASME/ASCE/AHS 24th Structures, Structural Dynamics and Materials Conference, Lake Tahoe, Nevada, 2-4 May 1983 (synopsis appears in *Journal of Aircraft*, Vol 21, 1984, p 225-226)

The Principle of Compounding

- D.J. Cartwright and D.P. Rooke, Approximate Stress-Intensity Factors for Compound from Known Solutions, *Engineering Fracture Mechanics*, Vol 6, 1974, p 563-571
- J.P. Gallagher, F.J. Giessler, A.P. Berens, and R.M. Engle, Jr., *USAF Damage Tolerant Design Handbook: Guidelines for the Analysis and Design of Damage Tolerant Aircraft Structures*, AFWAL-TR-82-3073, Flight Dynamics Laboratory, Wright-Patterson Air Force Base, May 1984
- K. Katherisan, T.M. Hsu, and T.R. Brussat, "Advanced Life Analysis Methods—Crack Growth Analysis Methods for Attachment Lugs," AFWAL TR-84-3080, Vol II, Flight Dynamics Laboratory, Wright-Patterson Air Force Base, Sept 1984

Metric Conversion Guide

This Section is intended as a guide for expressing weights and measures in the Système International d'Unités (SI). The purpose of SI units, developed and maintained by the General Conference of Weights and Measures, is to provide a basis for worldwide standardization of units and measure. For more information on metric conversions, the reader should consult the following references:

- "Standard for Metric Practice," E 380, *Annual Book of ASTM Standards,* America Society for Testing and Materials, 1916 Race Street, Philadelphia, PA 19103

- "Metric Practice," ANSI/IEEE 268-1982, American National Standards Institute, 1430 Broadway, New York, NY 10018
- *The International System of Units,* SP 330, 1986, National Institute of Standards and Technology. Order from Superintendent of Documents, U.S. Government Printing Office, Washington, DC 20402-9325
- *Metric Editorial Guide,* 4th ed. (revised), 1985, American National Metric Council, 1010 Vermont Avenue NW, Suite 1000, Washington, DC 20005-4960
- *ASME Orientation and Guide for Use of SI (Metric) Units,* ASME Guide SI 1, 9th ed., 1982, The American Society of Mechanical Engineers, 345 East 47th Street, New York, NY 10017

Base, supplementary, and derived SI units

Measure	Unit	Symbol
Base units		
Amount of substance	mole	mol
Electric current	ampere	A
Length	meter	m
Luminous intensity	candela	cd
Mass	kilogram	kg
Thermodynamic temperature	kelvin	K
Time	second	s
Supplementary units		
Plane angle	radian	rad
Solid angle	steradian	sr
Derived units		
Absorbed dose	gray	Gy
Acceleration	meter per second squared	m/s^2
Activity (of radionuclides)	becquerel	Bq
Angular acceleration	radian per second squared	rad/s^2
Angular velocity	radian per second	rad/s
Area	square meter	m^2
Capacitance	farad	F
Concentration (of amount of substance)	mole per cubic meter	mol/m^3
Conductance	siemens	S
Current density	ampere per square meter	A/m^2
Density, mass	kilogram per cubic meter	kg/m^3
Electric charge density	coulomb per cubic meter	C/m^3
Electric field strength	volt per meter	V/m
Electric flux density	coulomb per square meter	C/m^2
Electric potential, potential difference, electromotive force	volt	V
Electric resistance	ohm	Ω
Energy, work, quantity of heat	joule	J
Energy density	joule per cubic meter	J/m^3
Entropy	joule per kelvin	J/K
Force	newton	N

Measure	Unit	Symbol
Frequency	hertz	Hz
Heat capacity	joule per kelvin	J/K
Heat flux density	watt per square meter	W/m^2
Illuminance	lux	lx
Inductance	henry	H
Irradiance	watt per square meter	W/m^2
Luminance	candela per square meter	cd/m^2
Luminous flux	lumen	lm
Magnetic field strength	ampere per meter	A/m
Magnetic flux	weber	Wb
Magnetic flux density	tesla	T
Molar energy	joule per mole	J/mol
Molar entropy	joule per mole kelvin	$J/mol \cdot K$
Molar heat capacity	joule per mole kelvin	$J/mol \cdot K$
Moment of force	newton meter	$N \cdot m$
Permeability	henry per meter	H/m
Permittivity	farad per meter	F/m
Power, radiant flux	watt	W
Pressure, stress	pascal	Pa
Quantity of electricity, electric charge	coulomb	C
Radiance	watt per square meter steradian	$W/m^2 \cdot sr$
Radiant intensity	watt per steradian	W/sr
Specific heat capacity	joule per kilogram kelvin	$J/kg \cdot K$
Specific energy	joule per kilogram	J/kg
Specific entropy	joule per kilogram kelvin	$J/kg \cdot K$
Specific volume	cubic meter per kilogram	m^3/kg
Surface tension	newton per meter	N/m
Thermal conductivity	watt per meter kelvin	$W/m \cdot K$
Velocity	meter per second	m/s
Viscosity, dynamic	pascal second	$Pa \cdot s$
Viscosity, kinematic	square meter per second	m^2/s
Volume	cubic meter	m^3
Wavenumber	1 per meter	1/m

Conversion factors

To convert from	to	multiply by
Angle		
degree	rad	1.745 329 E−02
Area		
in.^{2}	mm^2	6.451 600 E+02
in.^{2}	cm^2	6.451 600 E+00
in.^{2}	m^2	6.451 600 E−04
ft^2	m^2	9.290 304 E−02
Bending moment or torque		
lbf · in.	N · m	1.129 848 E−01
lbf · ft	N · m	1.355 818 E+00
kgf · m	N · m	9.806 650 E+00
ozf · in.	N · m	7.061 552 E−03
Bending moment or torque per unit length		
lbf · in./in.	N · m/m	4.448 222 E+00
lbf · ft/in.	N · m/m	5.337 866 E+01
Current density		
A/in.^{2}	A/cm^2	1.550 003 E−01
A/in.^{2}	A/mm^2	1.550 003 E−03
A/ft^2	A/m^2	1.076 400 E+01
Electricity and magnetism		
gauss	T	1.000 000 E−04
maxwell	µWb	1.000 000 E−02
mho	S	1.000 000 E+00
Oersted	A/m	7.957 700 E+01
Ω · cm	Ω · m	1.000 000 E−02
Ω circular-mil/ft	µΩ · m	1.662 426 E−03
Energy (impact , other)		
ft · lbf	J	1.355 818 E+00
Btu (thermochemical)	J	1.054 350 E+03
cal (thermochemical)	J	4.184 000 E+00
kW · h	J	3.600 000 E+06
W · h	J	3.600 000 E+03
Flow rate		
ft^3/h	L/min	4.719 475 E−01
ft^3/min	L/min	2.831 000 E+01
gal/h	L/min	6.309 020 E−02
gal/min	L/min	3.785 412 E+00
Force		
lbf	N	4.448 222 E+00
kip (1000 lbf)	N	4.448 222 E+00
tonf	kN	8.896 443 E+00
kgf	N	9.806 650 E+00
Force per unit length		
lbf/ft	N/m	1.459 390 E+01
lbf/in.	N/m	1.751 268 E+02
Fracture toughness		
ksi $\sqrt{\text{in.}}$	MPa $\sqrt{\text{m}}$	1.098 800 E+00
Heat content		
Btu/lb	kJ/kg	2.326 000 E+00
cal/g	kJ/kg	4.186 800 E+00
Heat input		
J/in.	J/m	3.937 008 E+01
kJ/in.	kJ/m	3.937 008 E+01

(a) kg × 10^3 = 1 metric ton or 1 megagram (Mg)

To convert from	to	multiply by
Length		
Å	nm	1.000 000 E−01
µin.	µm	2.540 000 E−02
mil	µm	2.540 000 E+01
in.	mm	2.540 000 E+01
in.	cm	2.540 000 E+00
ft	m	3.048 000 E−01
yd	m	9.144 000 E−01
mile	km	1.609 300 E+00
Mass		
oz	kg	2.834 952 E−02
lb	kg	4.535 924 E−01
ton (short, 2000 lb)	kg	9.071 847 E+02
ton (short, 2000 lb)	kg × 10^3(a)	9.071 847 E−01
ton (long, 2240 lb)	kg	1.016 047 E+03
Mass per unit area		
oz/in.^{2}	kg/m^2	4.395 000 E+01
oz/ft^2	kg/m^2	3.051 517 E−01
oz/yd^2	kg/m^2	3.390 575 E−02
lb/ft^2	kg/m^2	4.882 428 E+00
Mass per unit length		
lb/ft	kg/m	1.488 164 E+00
lb/in.	kg/m	1.785 797 E+01
Mass per unit time		
lb/h	kg/s	1.259 979 E−04
lb/min	kg/s	7.559 873 E−03
lb/s	kg/s	4.535 924 E−01
Mass per unit volume (includes density)		
g/cm^3	kg/m^3	1.000 000 E+03
lb/ft^3	g/cm^3	1.601 846 E−02
lb/ft^3	kg/m^3	1.601 846 E+01
lb/in.^{3}	g/cm^3	2.767 990 E+01
lb/in.^{3}	kg/m^3	2.767 990 E+04
Power		
Btu/s	kW	1.055 056 E+00
Btu/min	kW	1.758 426 E−02
Btu/h	W	2.928 751 E−01
erg/s	W	1.000 000 E−07
ft · lbf/s	W	1.355 818 E+00
ft · lbf/min	W	2.259 697 E−02
ft · lbf/h	W	3.766 161 E−04
hp (550 ft · lbf/s)	kW	7.456 999 E−01
hp (electric)	kW	7.460 000 E−01
Power density		
W/in.^{2}	W/m^2	1.550 003 E+03
Press capacity		
See Force		
Pressure (fluid)		
atm (standard)	Pa	1.013 250 E+05
bar	Pa	1.000 000 E+05
in. Hg (32 °F)	Pa	3.386 380 E+03
in. Hg (60 °F)	Pa	3.376 850 E+03
lbf/in.^{2} (psi)	Pa	6.894 757 E+03
torr (mm Hg, 0 °C)	Pa	1.333 220 E+02

To convert from	to	multiply by
Specific heat		
Btu/lb · °F	J/kg · K	4.186 800 E+03
cal/g · °C	J/kg · K	4.186 800 E+03
Stress (force per unit area)		
tonf/in.^{2} (tsi)	MPa	1.378 951 E+01
kgf/mm^2	MPa	9.806 650 E+00
ksi	MPa	6.894 757 E+00
lbf/in.^{2} (psi)	MPa	6.894 757 E−03
MN/m^2	MPa	1.000 000 E+00
Temperature		
°F	°C	5/9 · (°F − 32)
°R	°K	5/9
Temperature interval		
°F	°C	5/9
Thermal conductivity		
Btu · in./s · ft^2 · °F	W/m · K	5.192 204 E+02
Btu/ft · h · °F	W/m · K	1.730 735 E+00
Btu · in./h · ft^2 · °F	W/m · K	1.442 279 E−01
cal/cm · s · °C	W/m · K	4.184 000 E+02
Thermal expansion		
in./in. · °C	m/m · K	1.000 000 E+00
in./in. · °F	m/m · K	1.800 000 E+00
Velocity		
ft/h	m/s	8.466 667 E−05
ft/min	m/s	5.080 000 E−03
ft/s	m/s	3.048 000 E−01
in./s	m/s	2.540 000 E−02
km/h	m/s	2.777 778 E−01
mph	km/h	1.609 344 E+00
Velocity of rotation		
rev/min (rpm)	rad/s	1.047 164 E−01
rev/s	rad/s	6.283 185 E+00
Viscosity		
poise	Pa · s	1.000 000 E−01
stokes	m^2/s	1.000 000 E−04
ft^2/s	m^2/s	9.290 304 E−02
in.^{2}/s	mm^2/s	6.451 600 E+02
Volume		
in.^{3}	m^3	1.638 706 E−05
ft^3	m^3	2.831 685 E−02
fluid oz	m^3	2.957 353 E−05
gal (U.S. liquid)	m^3	3.785 412 E−03
Volume per unit time		
ft^3/min	m^3/s	4.719 474 E−04
ft^3/s	m^3/s	2.831 685 E−02
in.^{3}/min	m^3/s	2.731 177 E−07
Wavelength		
Å	nm	1.000 000 E−01

Abbreviations and Symbols

δ crack tip opening displacement, CTOD

δ_m CTOD at maximum load in measurement of fracture toughness

ε actual or local strain at stress concentration

ε_e elastic strain

ε_{el} elastic strain

ε_{in} inelastic strain

ε_m mean strain

$\dot{\varepsilon}_m, \dot{\varepsilon}_{mech}$ mechanical strain rate

ε_{max} maximum strain

ε_{min} minimum strain

ε_p plastic strain

ε_{pl} plastic strain

$\dot{\varepsilon}_{pl}$ plastic strain rate

$\varepsilon_{pl,cum}$ cumulative plastic strain

$\varepsilon_{pl,max}$ maximum plastic strain

$\varepsilon_{pl,min}$ minimum plastic strain

$\varepsilon_{pl,s}$ plastic strain amplitude of saturation

$\dot{\varepsilon}^{th}$ thermal strain rate

$\Delta\varepsilon$ strain range

$\Delta\varepsilon, \Delta\varepsilon_{net}$ net strain range

$\Delta\varepsilon/2$ strain amplitude

$\Delta\varepsilon_e/2$ elastic strain amplitude

$\Delta\varepsilon_m, \Delta\varepsilon_{mech}$ mechanical strain range

$\Delta\varepsilon_p/2$ plastic strain amplitude

$\Delta\varepsilon_p, \Delta\varepsilon_{pl}$ plastic strain range

$\Delta\varepsilon_{th}$ thermal strain range

η_{pl} coefficient in J_{pl} calculation

μ shear modulus

γ shear strain or stacking fault energy

γ_{ap} amplitude of plastic shear strain

$\gamma_{ap,M}$ shear strain amplitude in matrix

$\gamma_{ap,PSB}$ shear strain amplitude in PSB

γ_{SF} stacking fault energy

ρ dislocation density, mass density or accumulated plastic strain

σ actual or local stress at a stress concentration

σ_0 mean stress in strain life method

σ_a stress amplitude, $\sigma_a = (\sigma_{max} - \sigma_{min})/2$

σ_{ce} critical stress in emergency condition

σ_{cu} critical stress in upset condition

σ_F friction stress amplitude

σ_H hydrostatic energy

σ_m mean stress, $\sigma_m = (\sigma_{max} + \sigma_{min})/2$

σ_{max} maximum local stress

σ_{min} minimum local stress

σ_p minimum permissible residual strength

σ_s saturation stress amplitude

σ_y yield stress

σ_{uts} ultimate tensile strength

σ_{ys} yield strength

σ_Y effective yield strength

σ_{YD} dynamic yield strength

$\Delta\sigma$ stress range, or alternating component of σ

$\Delta\sigma$ stress range

$\Delta\sigma/2$ stress amplitude

τ_a amplitude of resolved shear stress

τ_s shear stress amplitude in saturation

τ_f flow shear stress

τ_G athermal flow stress component

τ^* thermal flow stress component

ν Poisson's ratio or displacement

ν_{pl} plastic compenent of displacement

ϕ phase

a crack length, size, or size effect in fatigue notch factor equation

a_0 initial or starting crack length

a_c critical crack size

a_d detected crack size

a_e effective crack length

a_f final crack length

a_p maximum permissible crack size

Δa crack extension

Δa_e effective crack extension

A the ratio of alternating to mean load values, $A = \sigma_a/\sigma_m$

A(B)   arc bend specimen

AMT   accelerated mission test

A(T)   arc tension specimen

A(T)   arc-shaped specimen (in tension)

ATC   accelerated thermal cycling

a/W crack length to depth ratio

b uncracked ligament length

b_n Burgers vector

B specimen thickness

bcc   body-centered cubic

BFS   back face strain

CA   constant amplitude

CC   center cracked

CG   center of gravity

CT   compact type

CCT   center-cracked tensile (specimen)

C(T)   compact tension specimen

CTE   coefficient of thermal expansion

CTOD   crack tip opening displacement

C(W)   crack-line-wedge-loaded

d distance between corresponding partical dislocations

da/dN fatigue crack growth per cycle

d_c dislocation cell diameter of saturation

DC(T)   disk-shaped compact tension specimen

DCCW   diamond counterclockwise

DCW   diamond clockwise

e nominal strain measured remotely from stress concentration

E elastic (Young's) modulus

EHL   elastohydrodynamic lubrication

EMF   electromagnetic field

ENSIP   Engine Structural Integrity Program

EPFM   elastic-plastic fracture mechanics

f frequency

f_{PSB} volume fraction of persistent slip bands

F_{tu} ultimate tensile strength

F_{ty} yield strength

fcc   face-centered cubic

FCGR   fatigue crack growth rate

FEM   finite-element method

FEMA   failure effects and modes analysis

G shear modulus

H total time of safe operation, or time available for fracture control

ΔH activation energy

hcp   hexagonal close-packed

HR   Hui-Riedel (stress field)

HRR   Hutchinson-Rice-Rosengren (stress field)

I moment of inertia

IF   isothermal fatigue

IP   in-phase

IST   incremental step test

J J integral fracture parameter

J_2 Second Invariant of Deviatoric Stress

J_c J measurement of fracture toughness for unstable fracture

J_{el} elastic component of J

J_I J based fracture initiation fracture toughness

J_{pl} plastic component of J

J_Q provisional value of J_{Ic}

J_u J measurement of fracture toughness for unstable fracture

ΔJ cyclic J-Integral

k yield stress in shear

K crack-tip stress intensity factor

K_c critical stress intensity for static failure, or plane-stress fracture toughness

k_f or K_f fatigue notch factor, k_f = unnotched fatigue strength/notched fatigue strength

K_I stress intensity in Mode I (tension-compression)

K_{Ic} plane strain fracture toughness

K_{Id} dynamic fracture toughness

K_{Jc} critical stress intensity based on J integral

K_{max} maximum value of K in a K-R curve test

K_Q provisional value of K_{Ic} test

k_t or K_t theoretical (elastic) stress concentration factor

ΔK stress intensity range

ΔK_{eff} effective stress intensity range, $\Delta K_{eff} = \Delta K - \Delta K_{closure}$

ΔK_I mode I stress intensity factor range

$\Delta K_{I,th}$ mode I fatigue crack growth threshold

ΔK_{th} threshold stress intensity range

LCF   low cycle fatigue

LEFM   linear elastic fracture mechanics

M bending moment

M(T)   center-cracked tension specimen

M(T)   middle tension specimen geometry

M_d maximum temperature below which deformation-induced martensite formation occurs

MDS   mechanical deflection system

M_s martensite start temperature

MSD   multiple-site damage

N number of cycles

N_f number of cycles to failure

$2N_f$ number of reversals to failure

$2N_t$ transition fatigue life

NDT   nil ductility transition temperature

NDT   nil ductility

OL   overload

OP   out-of-phase

P applied load

P_a load amplitude or S_a

P_f load for fatigue precracking

P_m mean load (also S_m)

P_{max} maximum load in fracture toughness test

P_{min} minimum load

P_p minimum permissible residual strength load

P_Q provisional load

P_s service load

P_u design load

ΔP range of applied load in fatigue

PLB   persistent Lüders band

POD   probability of detection

PSB   persistent slip bands

PSE   principal structural elements

q notch sensitivity index, $q = (K_f - 1)/(K_t - 1)$

R ratio of minimum load to maximum load, $R = \sigma_{min}/\sigma_{max}$, or $R = (1-A)/(1+A)$

r_{cp} cyclic plastic zone radius

r_p rotation factor in CTOD test, plastic zone radius

r_p plastic zone radius

S nominal stress or stress measured remotely from stress concentration

S_0 mean stress

S_a stress amplitude, $S_a = \Delta S/2$

$S_{a,max}$ maximum stress amplitude of a stress spectrum

S_{ar} completely reversed stress amplitude

S_{er} fatigue limit stress for a notched component (with completely reversed stress, S_{ar})

S_{ij} deviatoric stress

S_{ij}^c deviatoric back stress

S_m mean stress (see also S_0)

S_{mf} mean stress in flight

ΔS stress range

SE(B)   single edge-notched bend

SE(B)   single-edge notched bending specimen

SE(T)   single-edge notched specimen in tension

SEN   single-edge notched (specimen)

SiC_p particulate silicon carbide reinforcement

SiC_w wisker silicon carbide reinforcement

SRP   strain range partitioning

SSC   small-scale secondary creep

SSY   small scale yielding

t thickness or time

t_c cycle time (seconds, minutes)

t_f time to failure

T_0 transition temperature

T_m melting temperature

T_{max} maximum temperature in a cycle

T_{min} minimum temperature in a cycle

TEM   transmission electron microscope/microscopy

TF   thermal fatigue, or triaxiality factor

TMF   thermo-mechanical fatigue

TMF DCCW   thermo-mechanical fatigue diamond counter-clockwise

TMF DCW   thermo-mechanical fatigue diamond clockwise

TMF OP   thermo-mechanical fatigue out-of-phase

TMP IP   thermo-mechanical fatigue in-phase

U plastic work per unit area to advance crack

VA   variable-amplitude

W specimen width

WFD   widespread fatigue damage

Y yield strength or generalized geometry factor

Index